THESAURUS
OF SCIENTIFIC,
TECHNICAL, AND
ENGINEERING TERMS

THESAURUS OF SCIENTIFIC, TECHNICAL, AND ENGINEERING TERMS

SCIENCE INFORMATION RESOURCE CENTER

⊙ **HEMISPHERE PUBLISHING CORPORATION**

Cambridge   Philadelphia   San Francisco   Washington
London   Mexico City   São Paulo   Singapore   Sydney

THESAURUS OF SCIENTIFIC, TECHNICAL, AND ENGINEERING TERMS

1 2 3 4 5 6 7 8 9 0   E B E B   8 9 8 7

Library of Congress Cataloging-in-Publication Data

Thesaurus of scientific, technical, and engineering terms.

   Bibliography: p.
   Includes index.
   1. Subject headings—Science.   2. Subject headings—Engineering.   3. Subject headings—Technology.   I. Science Information Resource Center (Philadelphia, Pa.)
Z695.1.S3T52   1988   025.4'95   87-21113
ISBN 0-89116-794-3   Hemisphere Publishing Corporation

CONTENTS

FOREWORD

The term "Thesaurus" has been in general use since 1805 when Peter Mark Roget started work on his *Thesaurus of English Words and Phrases Classified and Arranged so as to Facilitate the Expression of Ideas and Assist in Literary Composition.* The first printed edition of Roget's Thesaurus was published in 1852. Information specialists started using the terms during the 1950s. It has only been in more recent years, with the availability of computerized data bases, automation and the greatly advanced speeds and techniques for communication of information, that the demands for standardized or authoritative vocabularies has increased. This is especially the case in the scientific and technical disciplines.

A thesaurus is not a dictionary, index, or a classification system. A dictionary is used to find the meaning of words, while the thesaurus is used to find the most appropriate term for a given concept. Ambiguity in terminology, created by such factors as synonyms, homographs, acronyms, differences in spelling, word forms, and hierarchical treatment, tend to become barriers to high-speed communication, information storage, and retrieval systems.

One of the main purposes of this *Thesaurus of Scientific, Technical, and Engineering Terms* is to deal with these ambiguities by emphasizing uniqueness and developing single concepts and then grouping these into related areas.

Many terms thought to be unique to a specialized area of technology are essentially synonymous with or clearly related to terms used in other areas of technology. It is quite common for scientists in one area to borrow terms from another. This *Thesaurus* shows these relationships and thereby establishes the selection of terms that will improve communication both within and across boundaries of scientific, technical and engineering specialties. Consequently, the *Thesaurus* will be most useful to authors, editors, indexers, searchers, abstractors, documentalists, librarians, and information specialists engaged in the storage and retrieval of information in these disciplines.

EXPLANATION OF THE HIERARCHICAL LISTING

The *Thesaurus of Scientific, Technical, and Engineering Terms* is based primarily on the indexing vocabulary developed by NASA for their *Thesaurus,* which has undergone extensive revision with the last edition. Other thesauri, notably the *DOD Thesaurus of Scientific and Engineering Terms* have provided additional terms. This *Thesaurus* comprises two volumes:

Volume 1—**Hierarchical Listing** and Volume 2—**Access Vocabulary.** The Hierarchical Listing contains all subject terms and USE cross references currently approved for use. The Listing includes all terms appearing in the NASA Thesaurus and its supplements published through 1986.

DEFINITIONS AND CONVENTIONS

The definitions and conventions employed in the Thesaurus of Scientific, Technical and Engineering Terms follow.

Postable terms. Subject terms that have been approved for use in indexing and thus, can be "posted."

Nonpostable terms. Terms that are included for cross reference information and cannot be used for indexing.

Term Selection. Subject terms have been chosen on the basis of their significance and use in aerospace literature and their effectiveness in incorporating productive retrieval concepts. Particular consideration has been given to frequency of use in earlier indexing and search vocabularies, to relationships with other terms in the vocabulary, and to precise scientific and technical usage.

Noun Usage. In general, subject terms are presented in the noun form.

Singular vs. Plural. The plural form has in general been used for subject terms. The singular form, however, is occasionally employed for specific processes, properties, conditions, and hardware.

Term Length. No more than 42 characters, including spaces, are used for any subject term. Various words in longer terms are often truncated. With this edition scope notes are used to spell out truncated terms.

Term Ambiguity. When subject terms have more than one meaning in scientific or engineering usage, or when distinction between terms must be made, clarification is provided in one of two ways:

    a) Parenthetical qualifying expressions or glosses are added, becoming part of the subject term. For example:

        AGING (BIOLOGY)
        AGING (MATERIALS)
        AGING (METALLURGY)

    b) Parenthetical scope notes are added for explanation or definition; they do not become part of the subject term. For example:

CONTROLLERS
    SN   (DEVICES WHICH EMPLOY AN OUTSIDE SOURCE OF ENERGY AND USUALLY UTILIZE FEED-
            BACK)

Direct Entry. Subject terms that consist of more than one word are listed for direct entry, i.e., in their natural word order rather than in the inverted form. Inverted forms appear in the *Access Vocabulary.* For example:

    HALLEY'S COMET not COMET, HALLEY'S
    RADIATIVE HEAT TRANSFER not HEAT TRANSFER, RADIATIVE

Abbreviations and Acronyms. Some abbreviations and acronyms that are in common use in the technical or aerospace community are employed in this *Thesaurus.* In most cases USE cross references are made from the unabbreviated forms. For example:

    ORBITAL TEST SATELLITE (ESA)
        USE   OTS (ESA)

    METAL INSULATION SEMICONDUCTORS
        USE   MIS (SEMICONDUCTORS)

Synonyms. When candidate subject terms are true synonyms, one is chosen to be the valid, or postable term, and the other is provided with a USE cross reference. For example:

    OXYGEN DEFICIENCY
        USE   HYPOXIA

    VITAMIN C
        USE   ASCORBIC ACID

Array Terms. Subject terms with meaning either too broad or ambiguous for effective indexing or retrieval of information, have been designated array terms and carry the following scope note: (USE OF A MORE SPECIFIC TERM IS RECOMMENDED—CONSULT THE TERMS LISTED BELOW). Relationships with other postable terms are shown by the Related Term (RT) reference only. For example:

    ∞ **ENERGY SOURCES**
      SN   *(USE OF A MORE SPECIFIC TERM IS RECOMMENDED-CONSULT THE TERMS LISTED BELOW)*
      RT   ATMOSPHERIC ENERGY SOURCES
             AUXILIARY POWER SOURCES
             BIOMASS ENERGY PRODUCTION
             ELECTRIC BATTERIES
             ELECTRIC GENERATORS
             ELECTRON SOURCES
             ENERGY CONSUMPTION
             ENERGY REQUIREMENTS
             ENERGY TECHNOLOGY
             GEOTHERMAL RESOURCES
             HEAT SOURCES
             LITHIUM SULFUR BATTERIES
             OCEAN THERMAL ENERGY CONVERSION
             PLASMA POWER SOURCES
             POINT SOURCES
             PROPELLANTS
             RECTIFIERS
             SPACECRAFT POWER SUPPLIES
             TIDEPOWER
             WATERWAVE ENERGY CONVERSION

An infinity symbol ∞ precedes an array term in each of its appearances in Volume 1.

Identifiers. In this *Thesaurus* identifiers, i.e., subject terms that include a numeric or alphabetic designation, or both, for a specific model or item, are treated as regular subject terms and are provided complete cross references. For example:

    IBM 7094 COMPUTER
      GS   DATA PROCESSING EQUIPMENT
            . COMPUTERS
            .. DIGITAL COMPUTERS
            ... IBM 7000 SERIES COMPUTERS
            .... **IBM 7094 COMPUTER**
            .. IBM COMPUTERS

```
. . . IBM 7000 SERIES COMPUTERS
. . . . IBM 7094 COMPUTER
```

CROSS REFERENCE STRUCTURE

Cross reference relationships in the *Hierarchical Listing* are shown as follows:

Cross references	Notation
Broader Term	GS
Narrower Term	GS
Related Term	RT
Use	USE
Used For	UF

These cross references have the following applications:

Broader Term. This reference indicates that the term represents more inclusive concepts. In the Generic Structure (GS), the Broader Terms appear above and to the left of the term referenced. For example:

```
HIMALAYAS
    GS    LANDFORMS
          . MOUNTAINS
          . . HIMALAYAS
```

LANDFORMS and MOUNTAINS are Broader Terms to HIMALAYAS.

Narrower Term. This reference indicates that the term represents more specific concepts. In the Generic Structure (GS), the Narrower Terms appear below and to the right (indented) of the term referenced. For example:

```
GS    POLLUTION
      . ENVIRONMENT POLLUTION
      . . AIR POLLUTION
      . . . GLOBAL AIR POLLUTION
      . . . INDOOR AIR POLLUTION
      . . WATER POLLUTION
      . . . OIL POLLUTION
      . NOISE POLLUTION
      . THERMAL POLLUTION
```

ENVIRONMENTAL POLLUTION, NOISE POLLUTION and THERMAL POLLUTION are Narrower Terms to POLLUTION. INDOOR AIR POLLUTION is narrower to AIR POLLUTION, ENVIRONMENT POLLUTION and POLLUTION.

The number of narrow terms is not limited. For example, MEASURING INSTRUMENTS has over 300 narrower terms.

Related Term (RT). This reference indicates that the two terms are closely related conceptually but are not structured within the broader or narrower "tree," or hierarchy. The reciprocal of the RT reference "a" is the RT reference "b" and vice versa.

```
(a) SUPERCONDUCTORS
        RT    CRYOTRONS

(b) CRYOTRONS
        RT    SUPERCONDUCTORS
```

Use (USE). This reference indicates that the term is not "postable," i.e. not a valid term, and that the following term or terms should be used instead. For example:

```
THINNERS
USE    SOLVENTS

HEAT DISSIPATION
USE    COOLING
```

Used For (UF). This is a reciprocal of the USE cross reference and identifies valid, or "postable," terms. For example:

ALPHABETIZATION

The ordering of subject terms into an alphabetical arrangement can be accomplished in several ways. The most commonly used methods are the letter-by-letter, word-by-word, and the computer sorting order. In the absence of any universal agreement on a standardized approach, a word-oriented modification of the computer sorting technique has been adopted in this *Thesaurus* as the most useful and economic for this purpose.

Nonalphabetic characters are filed either at the beginning of the alphabet, at the end of the alphabet or are ignored altogether. Thus parens are filed before the alphabet in Volume 1. Parens are ignored for filing in Volume 2 due to permuting. Hyphens, slashes and periods follow blank spaces.

RELATIONSHIP TO ACCESS VOCABULARY

The *Thesaurus of Scientific, Technical, and Engineering Terms, Volume 2—Access Vocabulary* is a useful part of this volume and its use is encouraged. It provides thousands of additional "access points" to the terms whose hierarchies are listed in Volume 1.

TYPICAL HIERARCHICAL LISTING ENTRIES

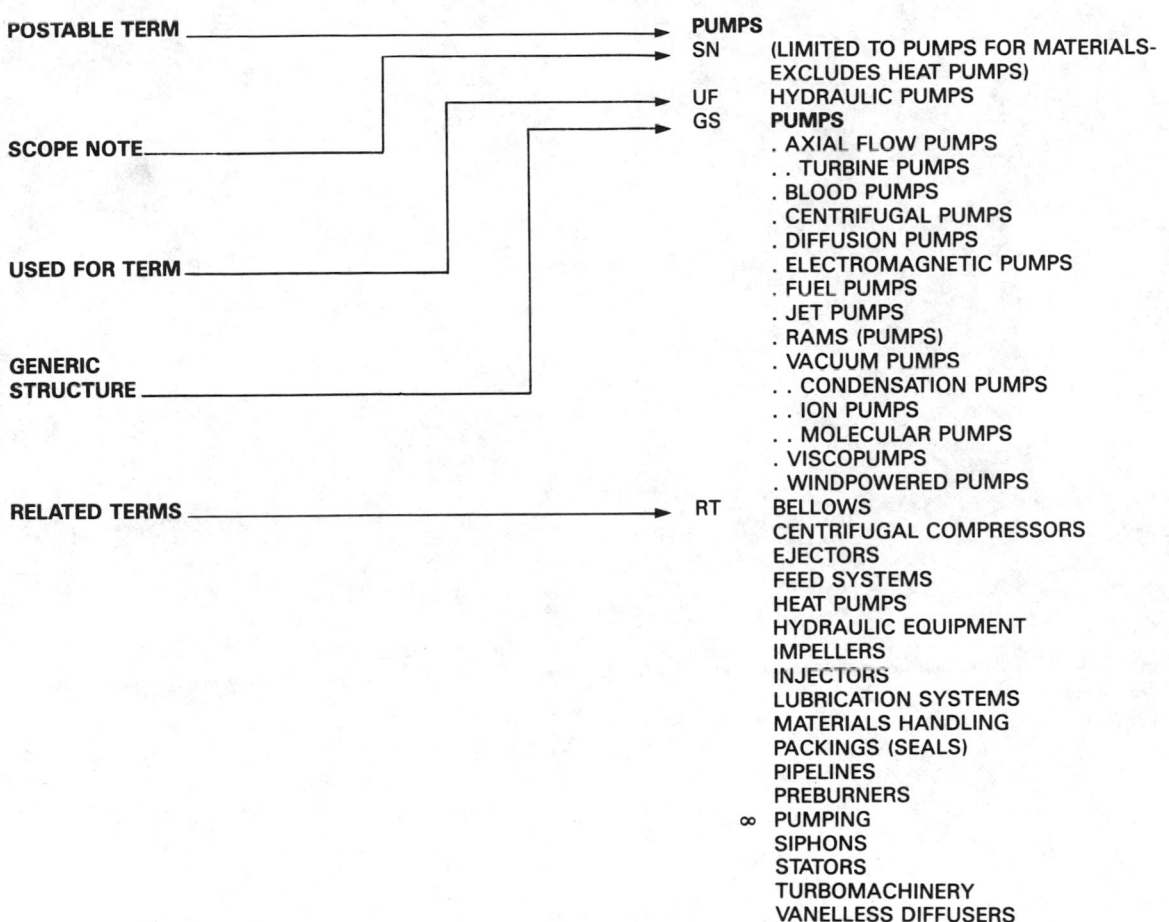

POSTABLE TERM — **PUMPS**

SN (LIMITED TO PUMPS FOR MATERIALS-
EXCLUDES HEAT PUMPS)

SCOPE NOTE

UF HYDRAULIC PUMPS

GS **PUMPS**
. AXIAL FLOW PUMPS
. . TURBINE PUMPS
. BLOOD PUMPS
. CENTRIFUGAL PUMPS
. DIFFUSION PUMPS
. ELECTROMAGNETIC PUMPS
. FUEL PUMPS
. JET PUMPS
. RAMS (PUMPS)
. VACUUM PUMPS
. . CONDENSATION PUMPS
. . ION PUMPS
. . MOLECULAR PUMPS
. VISCOPUMPS
. WINDPOWERED PUMPS

USED FOR TERM

GENERIC
STRUCTURE

RELATED TERMS

RT BELLOWS
CENTRIFUGAL COMPRESSORS
EJECTORS
FEED SYSTEMS
HEAT PUMPS
HYDRAULIC EQUIPMENT
IMPELLERS
INJECTORS
LUBRICATION SYSTEMS
MATERIALS HANDLING
PACKINGS (SEALS)
PIPELINES
PREBURNERS
∞ PUMPING
SIPHONS
STATORS
TURBOMACHINERY
VANELLESS DIFFUSERS

USE CROSS REFERENCE

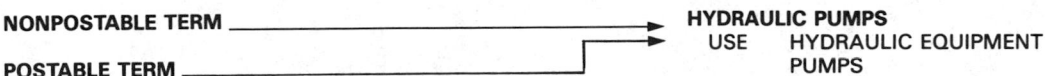

NONPOSTABLE TERM — **HYDRAULIC PUMPS**

USE HYDRAULIC EQUIPMENT
PUMPS

POSTABLE TERM

EXPLANATION OF THE ACCESS VOCABULARY

The *Access Vocabulary* is made available as a ready reference tool to provide better access to Volume 1 of the *Thesaurus of Scientific, Technical, and Engineering Terms,* the *Hierarchical Listing.* It utilizes pseudoterms (permuted terms), embedded terms, other word entries, nonpostable terms (cross references), and postable terms.

PSEUDOTERMS

Pseudoterms are permuted terms where each word in the term is rearranged by the computer to give access to any word in the term. By looking up any word in a term, the user can locate the postable term.

As an example of the potential use of permuted terms, suppose that a user wants to find information on a specific function that he knows is named for a person, but he cannot remember the person's name. By looking up the word function, the user will find 61 types of functions. If the function the user was trying to remember was the Legendre Function, he would find it listed and its presence would probably jog his memory. Without the *Access Vocabulary* this might be difficult if not impossible.

Functions, Analytic
    USE   ANALYTIC FUNCTIONS

Functions, Legendre
    USE   LEGENDRE FUNCTIONS

EMBEDDED TERMS

Embedded terms are rearrangements of parts of a word that contain other words within the term. The feature of permuting such a word is valuable and provides access to information that might otherwise be unavailable. The word waveforms is thus permuted to become forms, wave and can be located under forms in the *Access Vocabulary.* Permutations are also made in terms such as magneto · hydro · dynamics. Access is available through Hydro and Dynamics. These terms are manually selected and segmented for subsequent computer manipulation.

Forms, Wave
    USE   WAVEFORMS

OTHER WORD ENTRIES

These include chemical abbreviations and abbreviations of states.

Ar
    USE   ARGON
NY
    USE   NEW YORK

NONPOSTABLE AND POSTABLE TERMS

These terms without their hierarchies are included for the convenience of the user. Consult the *Hierarchical Listing* for complete information.

NUMBERS

One feature of a permuted index is that numbers are also permuted. You can look up any number that appears in a term. Numbers are found at the end of the alphabet.

239, Plutonium
    USE   PLUTONIUM 239

GLOSSES

A part of a term, usually at the end of a term, that is put in parentheses and qualifies the main term is called a *gloss*. These glosses which are usually terms for broader qualifiers are accessible in the *Access Vocabulary*. For example there are 9 entries under the gloss (Meteorology). Parens are ignored in filing glosses due to permutation factors.

(Meteorology), Fronts
    USE   FRONTS (METEOROLOGY)

(Meteorology), Storms
    USE   STORMS (METEOROLOGY)

(Meteorology), Wind
    USE   WIND (METEOROLOGY)

As another example there are 46 entries under the gloss (Mathematics).

(Mathematics), Combinations
    USE   COMBINATIONS (MATHEMATICS)

(Mathematics), Manifolds
    USE   MANIFOLDS (MATHEMATICS)

(Mathematics), Rings
    USE   RINGS (MATHEMATICS)

RELATIONSHIP TO HIERARCHICAL LISTING

The *Access Vocabulary* is meant to be a complementary tool to the *Hierarchical Listing*. For convenience, the postable terms without their hierarchies and the nonpostable "Use" terms have been repeated. The remainder of the *Access Vocabulary* contains unique "access points" to the hierarchies in Volume 1. Once the desired postable term has been located the complete hierarchical information for the term should be consulted in the *Hierarchical Listing*.

TYPICAL ACCESS VOCABULARY ENTRIES

Nonpostable term in natural language order. Postable term reference.

Pseudoterms (permutations) derived from nonpostable multiword term. Postable term reference follows USE.

Diodes, Barrier Injection Transit Time
    USE   BARRITT DIODES

Barrier Injection Transit Time Diodes
    USE   BARRITT DIODES

Injection Transit Time Diodes, Barrier
    USE   BARRITT DIODES

Transit Time Diodes, Barrier Injection
    USE   BARRITT DIODES

Time Diodes, Barrier Injection
    USE   BARRITT DIODES

Embedded Term.	**BIOINSTRUMENTATION**
Pseudoterms (permutations) derived from embedded terms.	**Instrumentation, Bio** USE BIOINSTRUMENTATION
	PHOTOLUMINESCENCE
	Luminescence, Photo USE PHOTOLUMINESCENCE
Postable multiword term.	**LUNAR EXPLORATION SYSTEM FOR APOLLO**
Pseudoterms derived from multiword term.	**Apollo, Lunar Exploration System For** USE LUNAR EXPLORATION SYSTEM FOR APOLLO
	Exploration System For Apollo, Lunar USE LUNAR EXPLORATION SYSTEM FOR APOLLO
Typical OTHER WORD entry (abbreviation) with postable term reference.	**CA** USE CALIFORNIA
Typical OTHER WORD entry (chemical symbol) with postable term reference.	**Ca** USE CALCIUM

EXPLANATION OF THE SUPPLEMENT

A supplement to Volume I can be found at the end of the volume. The supplement contains entries that were not included in the original volume. These entries are alphabetical, and will be called out in Volume I by daggers next to the entry that alphabetically precedes the entry in the supplement.

VOLUME 1
HIERARCHICAL LISTING

A

A STARS
GS   CELESTIAL BODIES
    . A STARS
RT   BLUE STARS
    PECULIAR STARS
    WOLF-RAYET STARS

A-1 AIRCRAFT
UF   SKYRAIDER AIRCRAFT
GS   ATTACK AIRCRAFT
    . A-1 AIRCRAFT
    MCDONNELL DOUGLAS AIRCRAFT
    . A-1 AIRCRAFT
    MONOPLANES
    . A-1 AIRCRAFT
RT   ∞AIRCRAFT

A-2 AIRCRAFT
UF   SAVAGE AIRCRAFT
GS   ATTACK AIRCRAFT
    . BOMBER AIRCRAFT
    . . A-2 AIRCRAFT
    JET AIRCRAFT
    . A-2 AIRCRAFT
    MONOPLANES
    . A-2 AIRCRAFT
    NORTH AMERICAN AIRCRAFT
    . A-2 AIRCRAFT
    OBSERVATION AIRCRAFT
    . A-2 AIRCRAFT
RT   ∞AIRCRAFT

A-3 AIRCRAFT
UF   A3D AIRCRAFT
    SKYWARRIOR AIRCRAFT
GS   ATTACK AIRCRAFT
    . BOMBER AIRCRAFT
    . . A-3 AIRCRAFT
    JET AIRCRAFT
    . A-3 AIRCRAFT
    MCDONNELL DOUGLAS AIRCRAFT
    . DOUGLAS AIRCRAFT
    . . A-3 AIRCRAFT
    MONOPLANES
    . A-3 AIRCRAFT
RT   ∞AIRCRAFT

A-4 AIRCRAFT
UF   A4D AIRCRAFT
    SKYHAWK AIRCRAFT
GS   ATTACK AIRCRAFT
    . BOMBER AIRCRAFT
    . . A-4 AIRCRAFT
    JET AIRCRAFT
    . A-4 AIRCRAFT
    MCDONNELL DOUGLAS AIRCRAFT
    . DOUGLAS AIRCRAFT
    . . A-4 AIRCRAFT
    MONOPLANES
    . A-4 AIRCRAFT
RT   ∞AIRCRAFT
    J-65 ENGINE

A-5 AIRCRAFT
UF   A3J AIRCRAFT
    VIGILANTE AIRCRAFT
GS   ATTACK AIRCRAFT
    . BOMBER AIRCRAFT
    . . A-5 AIRCRAFT
    JET AIRCRAFT
    . A-5 AIRCRAFT
    MONOPLANES
    . A-5 AIRCRAFT
    NORTH AMERICAN AIRCRAFT
    . A-5 AIRCRAFT
    SUPERSONIC AIRCRAFT
    . A-5 AIRCRAFT

A-5 AIRCRAFT-*(CONT.)*
RT   ∞AIRCRAFT

A-6 AIRCRAFT
UF   A2F AIRCRAFT
    INTRUDER AIRCRAFT
GS   ATTACK AIRCRAFT
    . BOMBER AIRCRAFT
    . . A-6 AIRCRAFT
    GRUMMAN AIRCRAFT
    . A-6 AIRCRAFT
    JET AIRCRAFT
    . A-6 AIRCRAFT
    MONOPLANES
    . A-6 AIRCRAFT
RT   ∞AIRCRAFT

A-7 AIRCRAFT
UF   CORSAIR AIRCRAFT
GS   ATTACK AIRCRAFT
    . A-7 AIRCRAFT
    JET AIRCRAFT
    . TURBOFAN AIRCRAFT
    . . A-7 AIRCRAFT
    LING-TEMCO-VOUGHT AIRCRAFT
    . A-7 AIRCRAFT
    MONOPLANES
    . A-7 AIRCRAFT
RT   ∞AIRCRAFT

A-9 AIRCRAFT
GS   ATTACK AIRCRAFT
    . A-9 AIRCRAFT
    NORTHROP AIRCRAFT
    . A-9 AIRCRAFT
    RECONNAISSANCE AIRCRAFT
    . A-9 AIRCRAFT
RT   ∞AIRCRAFT

A-10 AIRCRAFT
GS   ATTACK AIRCRAFT
    . A-10 AIRCRAFT
    REPUBLIC AIRCRAFT
    . A-10 AIRCRAFT
RT   ∞AIRCRAFT

A-11 SATELLITE
USE   ECHO 1 SATELLITE

A-12 SATELLITE
USE   ECHO 2 SATELLITE

A-37 AIRCRAFT
GS   ATTACK AIRCRAFT
    . A-37 AIRCRAFT
    CESSNA AIRCRAFT
    . A-37 AIRCRAFT
    MONOPLANES
    . A-37 AIRCRAFT
RT   ∞AIRCRAFT
    ∞MILITARY AIRCRAFT
    T-37 AIRCRAFT

A-300 AIRCRAFT
GS   COMMERCIAL AIRCRAFT
    . EUROPEAN AIRBUS
    . . A-300 AIRCRAFT
    JET AIRCRAFT
    . EUROPEAN AIRBUS
    . . A-300 AIRCRAFT
    PASSENGER AIRCRAFT
    . EUROPEAN AIRBUS
    . . A-300 AIRCRAFT
    TRANSPORT AIRCRAFT
    . A-300 AIRCRAFT
RT   ∞AIRCRAFT
    INTERNATIONAL COOPERATION
    SWEPT WINGS

A-310 AIRCRAFT
GS   COMMERCIAL AIRCRAFT
    . EUROPEAN AIRBUS
    . . A-310 AIRCRAFT
    JET AIRCRAFT
    . EUROPEAN AIRBUS
    . . A-310 AIRCRAFT
    PASSENGER AIRCRAFT
    . EUROPEAN AIRBUS
    . . A-310 AIRCRAFT
    TRANSPORT AIRCRAFT
    . SHORT HAUL AIRCRAFT
    . . EUROPEAN AIRBUS
    . . . A-310 AIRCRAFT
RT   INTERNATIONAL COOPERATION
    SWEPT WINGS

A-320 AIRCRAFT
GS   COMMERCIAL AIRCRAFT
    . EUROPEAN AIRBUS
    . . A-320 AIRCRAFT
    JET AIRCRAFT
    . EUROPEAN AIRBUS
    . . A-320 AIRCRAFT
    PASSENGER AIRCRAFT
    . EUROPEAN AIRBUS
    . . A-320 AIRCRAFT
    TRANSPORT AIRCRAFT
    . SHORT HAUL AIRCRAFT
    . . EUROPEAN AIRBUS
    . . . A-320 AIRCRAFT
RT   INTERNATIONAL COOPERATION
    SWEPT WINGS

AAP 1 MISSION
RT   APOLLO APPLICATIONS PROGRAM
    APOLLO PROJECT
    SKYLAB PROGRAM

AAP 2 MISSION
RT   APOLLO APPLICATIONS PROGRAM
    APOLLO PROJECT
    SKYLAB PROGRAM

AAP 3 MISSION
RT   APOLLO APPLICATIONS PROGRAM
    APOLLO PROJECT
    SKYLAB PROGRAM

AAP 4 MISSION
RT   APOLLO APPLICATIONS PROGRAM
    APOLLO PROJECT
    SKYLAB PROGRAM

ABDOMEN
GS   ANATOMY
    . ABDOMEN
    VISCERA
    . ABDOMEN
RT   DIGESTIVE SYSTEM
    GASTROINTESTINAL SYSTEM
    INTESTINES
    PERITONEUM
    STOMACH
    VENTRAL SECTIONS

ABEL FUNCTION
GS   ANALYSIS (MATHEMATICS)
    . REAL VARIABLES
    . . ABEL FUNCTION
    FUNCTIONS (MATHEMATICS)
    . ABEL FUNCTION
RT   SERIES (MATHEMATICS)

ABERRATION
RT   ABNORMALITIES
    ASPHERICITY
    BLURRING
    ∞COMA

ABERRATION-*(CONT.)*
    CRYSTAL OPTICS
    DEVIATION
    DISTORTION
    GRAZING INCIDENCE
    SPATIAL FILTERING

ABILITIES
UF   PROFICIENCY
      SKILLS
GS   ARTS
      . **ABILITIES**
RT   CONSISTENCY
      EFFORT
      HUMAN FACTORS ENGINEERING
      INCOMPATIBILITY
      TRANSFER OF TRAINING

ABIOGENESIS
GS   EVOLUTION (DEVELOPMENT)
      . BIOLOGICAL EVOLUTION
      . . **ABIOGENESIS**
RT   AUTOCATALYSIS
      CHEMICAL EVOLUTION
      LIFE SCIENCES
      PANSPERMIA
      SPERMATOGENESIS

ABLATED NOSETIPS
USE   PANT PROGRAM

ABLATION
RT   ABLATIVE MATERIALS
      ABLATIVE NOSE CONES
      AERODYNAMIC HEAT TRANSFER
      AERODYNAMIC HEATING
      AEROTHERMOCHEMISTRY
      ATMOSPHERIC ENTRY
      BURNTHROUGH (FAILURE)
      CHARRING
      COOLING
      DECOMPOSITION
      EROSION
      GAS-METAL INTERACTIONS
      HEAT SHIELDING
      IMPINGEMENT
      JET IMPINGEMENT
      MASS TRANSFER
      MELTING
      PYROLYSIS
      REENTRY
      REENTRY EFFECTS
      REENTRY PHYSICS
      REENTRY SHIELDING
      SUBLIMATION
      TEMPERATURE EFFECTS
      THERMAL ABSORPTION
      THERMAL DECOMPOSITION
      VAPORIZING

ABLATIVE MATERIALS
RT   ABLATION
      COOLING
      HEAT SHIELDING
      HEAT SINKS
   ∞MATERIALS
      NOSE CONES
      NOZZLE INSERTS
      PYROLYTIC MATERIALS
      REFRACTORY MATERIALS
      TEMPERATURE
      THERMAL CONTROL COATINGS
      THERMAL PROTECTION

ABLATIVE NOSE CONES
GS   CONES
      . NOSE CONES
      . . **ABLATIVE NOSE CONES**
RT   ABLATION
      HEAT SHIELDING
      REENTRY SHIELDING
      REENTRY VEHICLES
      ROCKET NOSE CONES
      SHIELDING

ABLESTAR LAUNCH VEHICLE
GS   LAUNCH VEHICLES
      . **ABLESTAR LAUNCH VEHICLE**
      ROCKET VEHICLES
      . **ABLESTAR LAUNCH VEHICLE**
RT   LIQUID PROPELLANT ROCKET ENGINES

ABM
USE   APOGEE BOOST MOTORS

ABNORMALITIES
RT   ABERRATION
      DEVIATION
      DISTORTION
      ECCENTRICITY
      IRREGULARITIES
      UNIQUENESS

ABORIGINES
RT   ANTHROPOLOGY
      HUMAN BEINGS
      INHABITANTS

ABORT APPARATUS
GS   SAFETY DEVICES
      . **ABORT APPARATUS**
RT   ABORTED MISSIONS
      AIRCRAFT SAFETY
      ARRESTING GEAR
   ∞BARRIERS
      BRAKES (FOR ARRESTING MOTION)
      DRAG DEVICES
      EJECTION SEATS
   ∞EQUIPMENT
      ESCAPE CAPSULES
      ESCAPE ROCKETS
      FLYING EJECTION SEATS

ABORT TRAJECTORIES
GS   TRAJECTORIES
      . **ABORT TRAJECTORIES**
RT   ABORTED MISSIONS
      MATTS (SYSTEMS)

ABORTED MISSIONS
RT   ABORT APPARATUS
      ABORT TRAJECTORIES
      DESTRUCTION
      ENGINE FAILURE
      ESCAPE CAPSULES
      ESCAPE ROCKETS
      FAILURE
      MALFUNCTIONS
   ∞MISSIONS

ABRASION
RT   ABRASIVES
      CHIPPING
      CLEANING
      CUTTING
      DRY FRICTION
      EROSION
      FILES (TOOLS)
      FRICTION
      GRINDING (MATERIAL REMOVAL)
      LESIONS
      METALLOGRAPHY
      POLISHING
      SCORING
      SOIL EROSION
      TRIBOLOGY
      WEAR

ABRASION RESISTANCE
GS   MECHANICAL PROPERTIES
      . **ABRASION RESISTANCE**
RT   HARDNESS
   ∞RESISTANCE
      TOUGHNESS

ABRASIVES
RT   ABRASION
      ALUMINUM OXIDES
      CARBORUNDUM (TRADEMARK)
      CERAMICS
      DIAMONDS
      GRIT
      PUMICE
      QUARTZ
      SILICON CARBIDES

ABRIKOSOV THEORY
RT   CRYSTAL STRUCTURE
      ELECTROMAGNETIC FIELDS
      SUPERCONDUCTIVITY
      SUPERCONDUCTORS
   ∞THEORIES
      VORTICES

ABSOLUTE ZERO
GS   TEMPERATURE
      . **ABSOLUTE ZERO**
RT   CRYOGENICS
      SUBZERO TEMPERATURE

ABSOLUTE ZERO-*(CONT.)*
    TEMPERATURE EFFECTS
    TEMPERATURE SCALES
    ULTRALOW TEMPERATURES
    ZERO POINT ENERGY

ABSORBENTS
UF   MOLECULAR SIEVES
GS   SORBENTS
      . **ABSORBENTS**
RT   ∞ABSORBERS
      ABSORBERS (EQUIPMENT)
      ABSORBERS (MATERIALS)
      ADSORBENTS
      AIR CONDITIONING EQUIPMENT
      DESICCANTS
      LOW DENSITY MATERIALS
      MATERIAL ABSORPTION
   ∞MATERIALS

∞**ABSORBERS**
SN   *(USE OF A MORE SPECIFIC TERM IS*
      RECOMMENDED--CONSULT THE TERMS
      LISTED BELOW)
RT   ABSORBENTS
      ABSORBERS (EQUIPMENT)
      ABSORBERS (MATERIALS)
      ATTENUATORS
      CLEANERS
      OSCILLATION DAMPERS
      SHOCK ABSORBERS
      VIBRATION ISOLATORS

ABSORBERS (EQUIPMENT)
SN   (EXCLUDES EQUIPMENT FOR
      ABSORBING ENERGY)
RT   ABSORBENTS
   ∞ABSORBERS
      ABSORBERS (MATERIALS)
      AIR CONDITIONING EQUIPMENT
      CLEANERS
      COLUMNS (PROCESS ENGINEERING)
      CONDENSERS (LIQUEFIERS)
      COOLING SYSTEMS
      DEGASSING
      DRYING APPARATUS
   ∞EQUIPMENT
      MATERIAL ABSORPTION
      REFRIGERATING MACHINERY
      SHOCK ABSORBERS

ABSORBERS (MATERIALS)
SN   (EXCLUDES ABSORBENTS--LIMITED TO
      MATERIALS FOR ABSORBING RADIATION
      RATHER THAN OTHER MATERIALS)
RT   ABSORBENTS
   ∞ABSORBERS
      ABSORBERS (EQUIPMENT)
      ACOUSTIC RETROFITTING
      ATTENUATORS
      CLEANERS
      ELECTROMAGNETIC ABSORPTION
      ELECTROMAGNETIC WAVE FILTERS
      ENERGY ABSORPTION
   ∞FILTERS
      HEAT SINKS
      INSULATION
      JACKETS
      LOW DENSITY MATERIALS
   ∞MATERIALS
      NEUTRON ABSORBERS
      RADAR ABSORBERS
      RADIATION SHIELDING
      REFRIGERANTS
      SHIELDING
      SINKS
      SOLAR ENERGY ABSORBERS
      STOPPING POWER
      SUPPRESSORS

ABSORPTANCE
GS   ELECTROMAGNETIC PROPERTIES
      . **ABSORPTANCE**
RT   ALBEDO
      CAPTURE EFFECT
      COSMIC RAY ALBEDO
      DENSITY (MASS/VOLUME)
      EARTH ALBEDO
      ELECTROMAGNETIC ABSORPTION
      LIGHT TRANSMISSION
      LUNAR ALBEDO
      OPACITY
      REFLECTANCE
      SURFACE PROPERTIES
      TRANSMISSION

ABSORPTANCE-*(CONT.)*
　　　　TRANSMISSIVITY
　　　　TRANSMITTANCE
　　　　TRANSPARENCE
　　　　TURBIDITY

∞ **ABSORPTION**
　SN　　*(USE OF A MORE SPECIFIC TERM IS*
　　　　RECOMMENDED--CONSULT THE TERMS
　　　　LISTED BELOW)
　RT　　ABSORPTION COOLING
　　　　ABSORPTION CROSS SECTIONS
　　　　ABSORPTION SPECTRA
　　　　ABSORPTION SPECTROSCOPY
　　　　ABSORPTIVITY
　　　　ACTIVATED CARBON
　　　　ADSORPTION
　　　　ATOMIC COLLISIONS
　　　　ATTENUATION
　　　　AURORAL ABSORPTION
　　　　BENEFICIATION
　　　　CAPTURE EFFECT
　　　　COLLISION PARAMETERS
　　　　COSMIC RAY ALBEDO
　　　　DAMPING
　　　　DESORPTION
　　　　DIFFUSION
　　　　DRYING
　　　　ELECTROMAGNETIC ABSORPTION
　　　　ENERGY ABSORPTION
　　　　ENERGY ABSORPTION FILMS
　　　　GAMMA RAY ABSORPTION
　　　　INFRARED ABSORPTION
　　　　INFRARED SPECTRA
　　　　MATERIAL ABSORPTION
　　　　MATERIALS RECOVERY
　　　　MODERATION (ENERGY ABSORPTION)
　　　　MOLECULAR ABSORPTION
　　　　MULTIPHOTON ABSORPTION
　　　　PERMEATING
　　　　PHOTOABSORPTION
　　　　PLANETARY ATMOSPHERES
　　　　POLAR CAP ABSORPTION
　　　　RADIATION ABSORPTION
　　　　SELF ABSORPTION
　　　　SORPTION
　　　　SOUND TRANSMISSION
　　　　THERMAL ABSORPTION
　　　　ULTRAVIOLET ABSORPTION
　　　　VISIBLE SPECTRUM
　　　　X RAY ABSORPTION

ABSORPTION BANDS
　USE　　ABSORPTION SPECTRA

ABSORPTION COEFFICIENT
　USE　　ABSORPTIVITY

ABSORPTION COOLING
　GS　　COOLING
　　　　　ABSORPTION COOLING
　RT　　∞ ABSORPTION
　　　　AMMONIA
　　　　MAGNETIC COOLING
　　　　REFRIGERANTS

ABSORPTION CROSS SECTIONS
　UF　　CAPTURE CROSS SECTIONS
　RT　　∞ ABSORPTION
　　　　∞ CROSS SECTIONS
　　　　IONIZATION CROSS SECTIONS
　　　　NEUTRON CROSS SECTIONS
　　　　RADIATION ABSORPTION
　　　　SCATTERING CROSS SECTIONS
　　　　STOPPING POWER

ABSORPTION SPECTRA
　UF　　ABSORPTION BANDS
　　　　SPECTRAL ABSORPTION
　GS　　SPECTRA
　　　　　ABSORPTION SPECTRA
　　　　. . FRAUNHOFER LINES
　　　　. . HERZBERG BANDS
　　　　. . TELLURIC LINES
　RT　　∞ ABSORPTION
　　　　BALMER SERIES
　　　　∞ BANDS
　　　　CONTINUOUS RADIATION
　　　　D LINES
　　　　ELECTROMAGNETIC ABSORPTION
　　　　ELECTROMAGNETIC SPECTRA
　　　　ELECTRON SPECTROSCOPY
　　　　ELECTRONIC SPECTRA
　　　　EMISSION SPECTRA
　　　　ENERGY SPECTRA

ABSORPTION SPECTRA-*(CONT.)*
　　　　FRAUNHOFER LINE DISCRIMINATORS
　　　　GALACTIC NUCLEI
　　　　GAMMA RAY ABSORPTIOMETRY
　　　　H ALPHA LINE
　　　　H BETA LINE
　　　　H GAMMA LINE
　　　　H LINES
　　　　IONIZING RADIATION
　　　　K LINES
　　　　LASER SPECTROMETERS
　　　　LINE SPECTRA
　　　　MICROWAVE SPECTRA
　　　　MOLECULAR SPECTRA
　　　　MOLECULAR SPECTROSCOPY
　　　　OSCILLATOR STRENGTHS
　　　　PARAMAGNETIC RESONANCE
　　　　PASCHEN SERIES
　　　　PHOTOLUMINESCENT BANDS
　　　　PHOTON ABSORPTIOMETRY
　　　　RAMAN SPECTRA
　　　　RYDBERG SERIES
　　　　SCHUMANN-RUNGE BANDS
　　　　SELF ABSORPTION
　　　　SOLAR SPECTRA
　　　　SOLAR SPECTROMETERS
　　　　SPECTRUM ANALYSIS
　　　　SPIN TEMPERATURE
　　　　STELLAR SPECTRA
　　　　SYMBIOTIC STARS
　　　　ULTRAVIOLET SPECTRA
　　　　VISIBLE SPECTRUM

ABSORPTION SPECTROSCOPY
　GS　　SPECTROSCOPY
　　　　. **ABSORPTION SPECTROSCOPY**
　　　　. . OPTOGALVANIC SPECTROSCOPY
　RT　　∞ ABSORPTION
　　　　FRAUNHOFER LINES
　　　　INFRARED SPECTROSCOPY
　　　　OPTICAL EQUIPMENT
　　　　OPTICAL MEASURING INSTRUMENTS
　　　　ULTRAVIOLET SPECTROSCOPY

ABSORPTIVE INDEX
　USE　　ABSORPTIVITY

ABSORPTIVITY
　UF　　ABSORPTION COEFFICIENT
　　　　ABSORPTIVE INDEX
　GS　　ELECTROMAGNETIC PROPERTIES
　　　　. OPTICAL PROPERTIES
　　　　. . **ABSORPTIVITY**
　RT　　∞ ABSORPTION
　　　　BEER LAW
　　　　BOUGUER LAW
　　　　DENSITY (MASS/VOLUME)
　　　　ELECTROMAGNETIC ABSORPTION
　　　　KIRCHHOFF LAW OF RADIATION
　　　　OPACITY
　　　　OSCILLATOR STRENGTHS
　　　　SCATTERING COEFFICIENTS
　　　　SELF ABSORPTION
　　　　TRANSMISSIVITY
　　　　TRANSPARENCE

ABSTRACTS
　GS　　DOCUMENTS
　　　　. **ABSTRACTS**
　　　　SUMMARIES
　　　　. **ABSTRACTS**
　RT　　ANNOTATIONS
　　　　BIBLIOGRAPHIES
　　　　INDEXES (DOCUMENTATION)
　　　　INFORMATION RETRIEVAL
　　　　TECHNICAL WRITING

ABUNDANCE
　UF　　ELEMENT ABUNDANCE
　RT　　AVAILABILITY
　　　　ENERGY POLICY
　　　　METALLIC STARS
　　　　METALLICITY
　　　　RESERVES
　　　　RESOURCES
　　　　STELLAR COMPOSITION

AC (CURRENT)
　USE　　ALTERNATING CURRENT

AC GENERATORS
　UF　　ALTERNATING CURRENT GENERATORS
　　　　ALTERNATORS (GENERATORS)
　GS　　ELECTRIC GENERATORS
　　　　. **AC GENERATORS**

AC GENERATORS-*(CONT.)*
　　　　. . STATIC ALTERNATORS
　RT　　COMPULSATORS
　　　　∞ GENERATORS
　　　　TURBOGENERATORS

AC-1 AIRCRAFT
　USE　　DHC 4 AIRCRAFT

ACCELERATED LIFE TESTS
　RT　　ACCEPTABILITY
　　　　EVALUATION
　　　　FATIGUE LIFE
　　　　LIFE (DURABILITY)
　　　　PERFORMANCE TESTS
　　　　QUALITY CONTROL
　　　　SERVICE LIFE
　　　　∞ TESTS

ACCELERATING AGENTS
　RT　　∞ ACCELERATORS
　　　　ADMIXTURES
　　　　∞ AGENTS
　　　　CATALYSTS
　　　　RETARDANTS

∞ **ACCELERATION**
　SN　　*(USE OF A MORE SPECIFIC TERM IS*
　　　　RECOMMENDED--CONSULT THE TERMS
　　　　LISTED BELOW)
　RT　　ACCELERATION (PHYSICS)
　　　　ACCELERATION PROTECTION
　　　　ACCELERATION STRESSES
　　　　　(PHYSIOLOGY)
　　　　∞ ACCELERATORS
　　　　ANGULAR ACCELERATION
　　　　CATALYSIS
　　　　ELECTROMAGNETIC ACCELERATION
　　　　ELECTRON ACCELERATION
　　　　GRAVIMETRY
　　　　HIGH ACCELERATION
　　　　HIGH GRAVITY ENVIRONMENTS
　　　　HUMAN TOLERANCES
　　　　IMPACT ACCELERATION
　　　　PARTICLE ACCELERATION
　　　　PHYSIOLOGICAL ACCELERATION
　　　　PLASMA ACCELERATION
　　　　TRANSVERSE ACCELERATION

ACCELERATION (PHYSICS)
　UF　　BOOST
　　　　G FORCE
　GS　　RATES (PER TIME)
　　　　. **ACCELERATION (PHYSICS)**
　　　　. . ANGULAR ACCELERATION
　　　　. . DECELERATION
　　　　. . . SPIN REDUCTION
　　　　. . ELECTRON ACCELERATION
　　　　. . HIGH ACCELERATION
　　　　. . HIGH GRAVITY ENVIRONMENTS
　　　　. . IMPACT ACCELERATION
　　　　. . PARTICLE ACCELERATION
　　　　. . PLASMA ACCELERATION
　　　　. . TRANSVERSE ACCELERATION
　RT　　∞ ACCELERATION
　　　　ACCELERATION STRESSES
　　　　　(PHYSIOLOGY)
　　　　ACCELEROMETERS
　　　　BODY KINEMATICS
　　　　EXPULSION
　　　　FLIGHT STRESS (BIOLOGY)
　　　　∞ FORCE
　　　　KINEMATICS
　　　　KINETICS
　　　　MECHANICAL SHOCK
　　　　∞ MOTION
　　　　PHYSIOLOGICAL ACCELERATION
　　　　STRESS (PHYSIOLOGY)
　　　　THRUST
　　　　THRUST-WEIGHT RATIO
　　　　VELOCITY

ACCELERATION PROTECTION
　GS　　PROTECTION
　　　　. **ACCELERATION PROTECTION**
　RT　　∞ ACCELERATION
　　　　EMBEDDING
　　　　SUPINE POSITION

ACCELERATION STRESSES (PHYSIOLOGY)
　GS　　STRESS (PHYSIOLOGY)
　　　　. **ACCELERATION STRESSES**
　　　　　(PHYSIOLOGY)
　　　　. . CENTRIFUGING STRESS
　RT　　∞ ACCELERATION

ACCELERATION STRESSES-*(CONT.)*
    ACCELERATION (PHYSICS)
    AEROSPACE MEDICINE
    ARTIFICIAL GRAVITY
    BODY KINEMATICS
    GRAVITATIONAL EFFECTS
    GRAVITATIONAL PHYSIOLOGY
    HIGH ACCELERATION
    LOWER BODY NEGATIVE PRESSURE
    MOTION SICKNESS
    PHYSIOLOGICAL ACCELERATION
    TRANSVERSE ACCELERATION

ACCELERATION TOLERANCE
SN   (LIMITED TO ABILITY OF ORGANISMS TO
    WITHSTAND ACCELERATION--FOR
    EFFECTS ON EQUIPMENT, USE SHOCK
    RESISTANCE AND MECHANICAL SHOCK)
GS   TOLERANCES (PHYSIOLOGY)
    . **ACCELERATION TOLERANCE**
RT   BLACKOUT (PHYSIOLOGY)
    BLACKOUT PREVENTION
    CENTRIFUGING STRESS
    GRAVITATIONAL EFFECTS
    HIGH ACCELERATION
    HUMAN CENTRIFUGES
    HUMAN TOLERANCES
    ∞ RESISTANCE

∞ **ACCELERATORS**
SN   (USE OF A MORE SPECIFIC TERM IS
    RECOMMENDED--CONSULT THE TERMS
    LISTED BELOW)
RT   ACCELERATING AGENTS
    ∞ ACCELERATION
    COAXIAL PLASMA ACCELERATORS
    CYCLIC ACCELERATORS
    CYCLOPS PLASMA ACCELERATOR
    ELECTRON ACCELERATORS
    GEOCYCLOTRONS
    HALL ACCELERATORS
    HYPERVELOCITY GUNS
    ION ACCELERATORS
    LINEAR ACCELERATORS
    NIMROD ACCELERATOR
    PARTICLE ACCELERATOR TARGETS
    PARTICLE ACCELERATORS
    PLASMA ACCELERATORS
    RACETRACKS (PARTICLE
      ACCELERATORS)
    RAILGUN ACCELERATORS
    SEPAC (PAYLOAD)
    STORAGE RINGS (PARTICLE
      ACCELERATORS)
    SYNCHROPHASOTRONS
    VAN DE GRAAFF ACCELERATORS

ACCELEROMETERS
GS   MEASURING INSTRUMENTS
    . **ACCELEROMETERS**
    . . STRAIN GAGE ACCELEROMETERS
RT   ACCELERATION (PHYSICS)
    GRAVIMETERS
    GRAVIMETRY
    GYROSCOPIC PENDULUMS
    MECHANICAL MEASUREMENT
    PENDULUMS
    SEISMOGRAPHS
    SHOCK MEASURING INSTRUMENTS
    SPEED INDICATORS
    THRUST MEASUREMENT
    VELOCITY MEASUREMENT
    VIBRATION METERS

ACCEPTABILITY
UF   ACCEPTANCE
RT   ACCELERATED LIFE TESTS
    COMPATIBILITY
    EVALUATION
    EXAMINATION
    FIGURE OF MERIT
    INSPECTION
    PERFORMANCE TESTS
    PROVING
    QUALITY CONTROL
    REJECTION
    RELIABILITY
    RISK
    SAMPLES
    STANDARDS
    SUITABILITY
    ∞ TESTS
    TOLERANCES (MECHANICS)
    VALIDITY

ACCEPTANCE
USE   ACCEPTABILITY

ACCEPTOR MATERIALS
GS   SEMICONDUCTORS (MATERIALS)
    . **ACCEPTOR MATERIALS**
RT   CARRIER DENSITY (SOLID STATE)
    ELECTRONS
    HOLES (ELECTRON DEFICIENCIES)
    ∞ MATERIALS

ACCESS CONTROL
RT   COMMUNICATION NETWORKS
    ∞ CONTROL
    DATA TRANSMISSION
    MULTIPLE ACCESS
    MULTIPLEXING
    RADIO COMMUNICATION
    TELECOMMUNICATION

ACCESS TIME
GS   TIME
    . **ACCESS TIME**
RT   DATA PROCESSING
    RATES (PER TIME)
    TIME CONSTANT
    ∞ TIME RESPONSE

ACCESSORIES
UF   ATTACHMENTS
RT   ∞ COMPONENTS
    EXTENSIONS
    FITTINGS
    INSERTS
    SUBASSEMBLIES

ACCIDENT INVESTIGATION
GS   INVESTIGATION
    . **ACCIDENT INVESTIGATION**
    . . AIRCRAFT ACCIDENT INVESTIGATION
RT   ACCIDENTS
    AUTOMOBILE ACCIDENTS
    WRECKAGE

ACCIDENT PREVENTION
UF   PRECAUTIONS
GS   PREVENTION
    . **ACCIDENT PREVENTION**
RT   ACCIDENTS
    AEROSPACE SAFETY
    AIR BAG RESTRAINT DEVICES
    AUTOMOBILE ACCIDENTS
    AVOIDANCE
    FIRE PREVENTION
    HAZARDS
    PROTECTION
    SAFETY
    SAFETY DEVICES
    SAFETY MANAGEMENT
    WARNING
    WARNING SYSTEMS

ACCIDENT PRONENESS
RT   SAFETY DEVICES
    SAFETY FACTORS

ACCIDENTS
GS   **ACCIDENTS**
    . BIRD-AIRCRAFT COLLISIONS
    . LOSS OF COOLANT
RT   ACCIDENT INVESTIGATION
    ACCIDENT PREVENTION
    AIR BAG RESTRAINT DEVICES
    AIRCRAFT ACCIDENTS
    AUTOMOBILE ACCIDENTS
    CRASH INJURIES
    CRASHES
    DESTRUCTION
    DISASTERS
    EMERGENCIES
    EXPLOSIONS
    FIRES
    FIRST AID
    HAZARDS
    INDUSTRIAL SAFETY
    INJURIES
    SABOTAGE
    SAFETY
    SAFETY DEVICES
    TRAFFIC
    WRECKAGE

ACCLIMATIZATION
UF   DEACCLIMATIZATION

ACCLIMATIZATION-*(CONT.)*
GS   ADAPTATION
    . **ACCLIMATIZATION**
    . . ALTITUDE ACCLIMATIZATION
    . . COLD ACCLIMATIZATION
    . . HEAT ACCLIMATIZATION
RT   HOMEOSTASIS
    LIQUID BREATHING
    STRESS (PHYSIOLOGY)
    TOLERANCES (PHYSIOLOGY)

ACCOMMODATION
RT   ADAPTATION
    CORRECTION
    EYE (ANATOMY)
    FOCUSING
    VISUAL ACCOMMODATION

ACCOMMODATION COEFFICIENT
UF   THERMAL ACCOMMODATION
      COEFFICIENTS
GS   COEFFICIENTS
    . **ACCOMMODATION COEFFICIENT**
RT   HEAT TRANSFER COEFFICIENTS

ACCOUNTING
RT   BUDGETING
    COSTS
    FINANCE

ACCRETION
USE   DEPOSITION

ACCRETION DISKS
RT   ASTROPHYSICS
    BINARY STARS
    BLACK HOLES (ASTRONOMY)
    DISKS (SHAPES)
    ECLIPSING BINARY STARS
    GALACTIC NUCLEI
    ROTATING DISKS
    STELLAR MASS ACCRETION
    X RAY BINARIES

ACCUMULATIONS
RT   ACQUISITION
    AGGLOMERATION
    ASSEMBLIES
    COAGULATION
    COLLECTION
    CONCENTRATING
    DEPOSITION
    FILLING
    GROWTH
    INCREASING
    INPUT
    NUCLEATION
    SETTLING
    STOCKPILING

ACCUMULATORS
UF   COLLECTORS
GS   **ACCUMULATORS**
    . ACCUMULATORS (COMPUTERS)
    . DUST COLLECTORS
    . SOLAR COLLECTORS
    . . SOLAR REFLECTORS
RT   ANODES
    CONCENTRATORS
    ENTRAPMENT
    FUEL SYSTEMS
    PRESSURE VESSELS
    PRESSURIZING

ACCUMULATORS (COMPUTERS)
GS   ACCUMULATORS
    . **ACCUMULATORS (COMPUTERS)**
    PERIPHERAL EQUIPMENT (COMPUTERS)
    . **ACCUMULATORS (COMPUTERS)**
RT   ADDING CIRCUITS
    COUNTERS
    ∞ EQUIPMENT

ACCURACY
UF   ERROR BAND
    FIDELITY
GS   **ACCURACY**
    . GEODETIC ACCURACY
    . GEOMETRIC ACCURACY
RT   ANGULAR RESOLUTION
    CALIBRATING
    CONSISTENCY
    CORRECTION
    ∞ DEFINITION

ACCURACY-*(CONT.)*
  DRIFT (INSTRUMENTATION)
  DYNAMIC CHARACTERISTICS
  ERRORS
  HIGH RESOLUTION
  HYSTERESIS
  LINEARITY
  ∞MEASUREMENT
  MISS DISTANCE
  PRECISION
  QUALITY
  RANGE ERRORS
  RELIABILITY
  RESOLUTION
  SEQUENTIAL CONTROL
  STANDARDS
  SURVEYS
  ∞TESTS
  TOLERANCES (MECHANICS)
  VALIDITY
  VIRTUAL PROPERTIES

ACEE PROGRAM
UF AIRCRAFT ENERGY EFFICIENCY
   PROGRAM
  ENERGY EFFICIENCY TRANSPORT
   PROGRAM
GS PROGRAMS
 . **ACEE PROGRAM**
RT AIRCRAFT ENGINES
  COMBUSTION EFFICIENCY

ACETALDEHYDE
GS ALDEHYDES
 . **ACETALDEHYDE**
  ALIPHATIC COMPOUNDS
 . **ACETALDEHYDE**

ACETALS
GS ALIPHATIC COMPOUNDS
 . **ACETALS**
  ETHERS
 . **ACETALS**

ACETANILIDE
UF PHENACETIN
GS NITROGEN COMPOUNDS
 . **ACETANILIDE**

ACETATES
GS **ACETATES**
 . COBALT ACETATES
 . LEAD ACETATES
 . SODIUM CHLORODIFLUOROACETATES
 . TRIACETIN
RT ACETIC ACID
  ACETYLSALICYLIC ACID
  ALIPHATIC COMPOUNDS
  ESTERS
  ETHYLENEDIAMINETETRAACETIC ACIDS

ACETATION
USE ACETYLATION

ACETAZOLAMIDE
GS HETEROCYCLIC COMPOUNDS
 . AZOLES
  . **ACETAZOLAMIDE**
RT CARBONIC ANHYDRASE
  DIURETICS

ACETIC ACID
GS ACIDS
 . FATTY ACIDS
  . CARBOXYLIC ACIDS
   . **ACETIC ACID**
    . ETHYLENEDIAMINETETRAACETIC
     ACIDS
    . IODOACETIC ACID
  ALIPHATIC COMPOUNDS
 . FATTY ACIDS
  . **ACETIC ACID**
   . ETHYLENEDIAMINETETRAACETIC
    ACIDS
   . IODOACETIC ACID
  ORGANIC COMPOUNDS
 . FATTY ACIDS
  . **ACETIC ACID**
   . ETHYLENEDIAMINETETRAACETIC
    ACIDS
   . IODOACETIC ACID
RT ACETATES
  ACETYL COMPOUNDS
  TRIACETIN

ACETONE
GS ALIPHATIC COMPOUNDS
 . KETONES
  . **ACETONE**
RT ACETYLACETONE
  PENTANONE

ACETYL COMPOUNDS
GS ALIPHATIC COMPOUNDS
 . **ACETYL COMPOUNDS**
RT ACETIC ACID
  ACETYLATION
  ALDEHYDES
  ∞CHEMICAL COMPOUNDS
  ESTERS
  ORGANIC COMPOUNDS

ACETYLACETONE
GS ALIPHATIC COMPOUNDS
 . KETONES
  . **ACETYLACETONE**
RT ACETONE
  PENTANONE

ACETYLATION
UF ACETATION
GS CHEMICAL REACTIONS
 . ACYLATION
  . **ACETYLATION**
RT ACETYL COMPOUNDS

ACETYLENE
GS ALIPHATIC COMPOUNDS
 . ALKYNES
  . **ACETYLENE**
  HYDROCARBONS
 . **ACETYLENE**
RT HYDROCARBON FUELS
  OXYACETYLENE

ACETYLSALICYLIC ACID
UF ASA
GS ACIDS
 . FATTY ACIDS
  . CARBOXYLIC ACIDS
   . **ACETYLSALICYLIC ACID**
RT ACETATES
  SALICYLATES

ACHIEVEMENT
RT COMPLETENESS
  GOALS
  LEARNING

ACHONDRITES
GS CELESTIAL BODIES
 . METEORITES
  . STONY METEORITES
   . **ACHONDRITES**
    . BONDOC METEORITE
    . KAPOETA ACHONDRITE
    . NORTON COUNTY ACHONDRITE
RT CHONDRITES
  IRON METEORITES

ACID BASE EQUILIBRIUM
GS CHEMICAL EQUILIBRIUM
 . **ACID BASE EQUILIBRIUM**
RT ∞EQUILIBRIUM
  HOMEOSTASIS
  PH
  PH FACTOR
  STABILIZATION
  THERMODYNAMIC EQUILIBRIUM

ACID RAIN
GS PRECIPITATION (METEOROLOGY)
 . RAIN
  . **ACID RAIN**
RT AIR POLLUTION
  ATMOSPHERIC CHEMISTRY
  ATMOSPHERIC MOISTURE
  CLOUDS (METEOROLOGY)
  DEW
  METEOROLOGY
  PH
  RAINSTORMS
  SNOW
  SULFUR OXIDES

ACIDITY
GS CHEMICAL PROPERTIES
 . **ACIDITY**
RT HYDROGEN IONS

ACIDITY-*(CONT.)*
  ION CONCENTRATION
  PH
  TITRATION

ACIDOSIS
RT ALKALOSIS
  HYPERVENTILATION
  PH
  PH FACTOR
  TOXICITY

ACIDS
GS **ACIDS**
 . AMINO ACIDS
  . ALANINE
   . PHENYLALANINE
  . ASPARTIC ACID
  . CYSTEINE
  . FOLIC ACID
  . GLUTAMIC ACID
  . GLUTAMINE
  . GLUTATHIONE
  . GLYCINE
  . HIPPURIC ACID
  . HISTIDINE
  . LEUCINE
   . NORLEUCINE
  . LYSINE
  . MELANOIDIN
  . METHIONINE
  . PAPAIN
  . PEPTIDES
   . HYPERTENSIN
   . PROTOPROTEINS
  . PYRIDINE NUCLEOTIDES
  . PYRUVATES
  . THYROXINE
  . TRYPTOPHAN
  . TYROSINE
  . URIDYLIC ACID
 . AMOBARBITAL
 . ASCORBIC ACID
 . BORIC ACIDS
 . BUTYRIC ACID
 . CARBONIC ACID
 . CHROMIC ACID
 . CYANURIC ACID
 . CYTIDYLIC ACID
 . FATTY ACIDS
  . CARBOXYLIC ACIDS
   . ACETIC ACID
    . ETHYLENEDIAMINETETRAACETIC
     ACIDS
    . IODOACETIC ACID
   . ACETYLSALICYLIC ACID
   . BENZILIC ACID
   . BENZOIC ACID
   . DICARBOXYLIC ACIDS
    . TEREPHTHALATE
   . OLEIC ACID
   . PROPIONIC ACID
   . SEBACIC ACID
   . VALERIC ACID
  . LIPOIC ACID
  . PALMITIC ACID
 . HYDRAZOIC ACID
 . HYDROBROMIC ACID
 . HYDROCHLORIC ACID
 . HYDROCYANIC ACID
 . HYDROFLUORIC ACID
 . NITRIC ACID
 . NUCLEIC ACIDS
  . DEOXYRIBONUCLEIC ACID
  . GUANOSINES
  . RIBONUCLEIC ACIDS
  . URIDYLIC ACID
 . OXIDASE
 . PERCHLORIC ACID
 . PHOSPHORIC ACID
 . SULFONIC ACID
 . SULFURIC ACID
 . THYMIDINE
 . THYMINE
 . URIC ACID
 . XANTHIC ACIDS
RT ADRENOCORTICOTROPIN (ACTH)
  ANHYDRIDES
  HYDROGEN COMPOUNDS
  INORGANIC COMPOUNDS
  ORGANIC COMPOUNDS
  ∞OXYGEN COMPOUNDS

ACOUSTIC ATTENUATION
GS ATTENUATION

ACOUSTIC ATTENUATION-*(CONT.)*
   . ACOUSTIC ATTENUATION
   . . SHOCK WAVE ATTENUATION
RT   ACOUSTICS
   ANECHOIC CHAMBERS
   ATMOSPHERIC ATTENUATION
   BIOACOUSTICS
   GRAZING FLOW
   NOISE REDUCTION
   SOUND AMPLIFICATION
   WAVE PROPAGATION
   ZERO SOUND

ACOUSTIC COMBUSTION
USE   COMBUSTION STABILITY

ACOUSTIC DELAY LINES
UF   SONIC WAVEGUIDES
GS   DELAY LINES
   . ACOUSTIC DELAY LINES
RT   COMPUTER STORAGE DEVICES
   DELAY CIRCUITS
   SURFACE ACOUSTIC WAVE DEVICES
   TRANSMISSION LINES

ACOUSTIC DUCTS
GS   DUCTS
   . ACOUSTIC DUCTS
RT   GRAZING FLOW
   NOISE REDUCTION
   SPATIAL MARCHING

ACOUSTIC EMISSION
GS   DECAY
   . ACOUSTIC EMISSION
RT   ACOUSTIC MEASUREMENT
   CRACK PROPAGATION
   FAILURE ANALYSIS
   FATIGUE TESTING MACHINES
   STRESS WAVES

ACOUSTIC EXCITATION
GS   EXCITATION
   . WAVE EXCITATION
   . . ACOUSTIC EXCITATION
RT   ACOUSTICS
   SOUND AMPLIFICATION
   SURFACE NOISE INTERACTIONS

ACOUSTIC FATIGUE
UF   SONIC FATIGUE
GS   FATIGUE (MATERIALS)
   . ACOUSTIC FATIGUE
† RT   ACOUSTICS

ACOUSTIC GENERATORS
USE   SOUND GENERATORS

ACOUSTIC IMPEDANCE
GS   ACOUSTIC PROPERTIES
   . ACOUSTIC IMPEDANCE
   IMPEDANCE
   . ACOUSTIC IMPEDANCE
RT   ACOUSTICS
   GRAZING FLOW

ACOUSTIC INSTABILITY
GS   ACOUSTIC PROPERTIES
   . ACOUSTIC INSTABILITY
   STABILITY
   . ACOUSTIC INSTABILITY
RT   SIGNAL FADING

ACOUSTIC LEVITATION
GS   LEVITATION
   . ACOUSTIC LEVITATION
RT   BUOYANCY
   SPACE PROCESSING

ACOUSTIC MEASUREMENT
SN   (MEASUREMENT OF PROPERTIES,
   QUANTITIES OR CONDITIONS
   ASSOCIATED WITH ELASTIC WAVES)
UF   SOUND MEASUREMENT
GS   ACOUSTIC MEASUREMENT
   . NOISE MEASUREMENT
RT   ACOUSTIC EMISSION
   ANECHOIC CHAMBERS
   AUDIO FREQUENCIES
   AUDIOMETRY
   CEPSTRAL ANALYSIS
   EFFECTIVE PERCEIVED NOISE LEVELS
   FREQUENCY MEASUREMENT
   GRAZING FLOW

ACOUSTIC MEASUREMENT-*(CONT.)*
   ∞ MEASUREMENT
   MECHANICAL MEASUREMENT
   NOISE METERS
   SEISMOGRAPHS
   SOUND PRESSURE
   SOUND WAVES
   ULTRASONIC TESTS

ACOUSTIC MICROSCOPES
UF   SCANNING LASER ACOUSTIC
     MICROSCOPE (SLAM)
GS   MICROSCOPES
   . ACOUSTIC MICROSCOPES
RT   ACOUSTIC PROPAGATION
   IMAGING TECHNIQUES
   MICROWAVE FREQUENCIES
   OPTICAL EQUIPMENT
   WAVE PROPAGATION

ACOUSTIC NOZZLES
RT   ∞ NOZZLES
   SONIC NOZZLES
   SOUND GENERATORS

ACOUSTIC PROPAGATION
GS   TRANSMISSION
   . ACOUSTIC PROPAGATION
RT   ACOUSTIC MICROSCOPES
   ACOUSTICAL HOLOGRAPHY
   ACOUSTICS
   ELASTIC WAVES
   ∞ PROPAGATION
   SOUND PROPAGATION

ACOUSTIC PROPERTIES
GS   ACOUSTIC PROPERTIES
   . ACOUSTIC IMPEDANCE
   . ACOUSTIC INSTABILITY
   . ACOUSTIC SCATTERING
   . . REVERBERATION
   . ACOUSTIC VELOCITY
   . SOUND INTENSITY
   . . ZERO SOUND
RT   FIELD STRENGTH
   GRAZING FLOW
   LAMB WAVES
   MECHANICAL PROPERTIES
   ∞ PHYSICAL PROPERTIES
   ∞ PROPERTIES
   ∞ RESISTANCE
   SOUND WAVES
   WAVE DISPERSION

ACOUSTIC RADIATION
USE   SOUND WAVES

ACOUSTIC RETROFITTING
GS   RETROFITTING
   . ACOUSTIC RETROFITTING
RT   ABSORBERS (MATERIALS)
   AERODYNAMIC NOISE
   AIRCRAFT DESIGN
   AIRCRAFT NOISE
   JET AIRCRAFT NOISE
   MUFFLERS
   NOISE REDUCTION
   VIBRATION ISOLATORS

ACOUSTIC SCATTERING
GS   ACOUSTIC PROPERTIES
   . ACOUSTIC SCATTERING
   . . REVERBERATION
   SCATTERING
   . WAVE SCATTERING
   . . ACOUSTIC SCATTERING
   . . . REVERBERATION
RT   ACOUSTIC SOUNDING
   ACOUSTICS
   DEEP SCATTERING LAYERS
   RECIPROCITY THEOREM
   SODAR
   SOUND DETECTING AND RANGING
   SURFACE NOISE INTERACTIONS
   UNDERWATER ACOUSTICS

ACOUSTIC SIMULATION
GS   SIMULATION
   . ENVIRONMENT SIMULATION
   . . ACOUSTIC SIMULATION
RT   ELASTIC WAVES
   FLIGHT SIMULATION

ACOUSTIC SOUNDING
GS   SOUNDING
   . ACOUSTIC SOUNDING
RT   ACOUSTIC SCATTERING
   ACOUSTICS
   EARTH ATMOSPHERE
   METEOROLOGY
   ROCKET SOUNDING
   ROCKET VEHICLES
   SOUNDING ROCKETS
   ULTRASONIC TESTS
   UNDERGROUND ACOUSTICS
   UPPER ATMOSPHERE

ACOUSTIC STABILITY
USE   FREQUENCY STABILITY

ACOUSTIC STREAMING
RT   FLUID FLOW
   FLUID SWITCHING ELEMENTS
   FRAGMENTATION
   SOUND WAVES
   STREAMLINING

ACOUSTIC VELOCITY
UF   SONIC SPEED
   SOUND BARRIER
   SOUND VELOCITY
GS   ACOUSTIC PROPERTIES
   . ACOUSTIC VELOCITY
   RATES (PER TIME)
   . ACOUSTIC VELOCITY
   VELOCITY
   . ACOUSTIC VELOCITY
RT   ∞ BARRIERS
   EXHAUST VELOCITY
   GUTENBERG ZONE
   MACH CONES
   MACH NUMBER
   SONIC BOOMS
   SOUND PRESSURE
   SUBSONIC SPEED
   SUPERSONIC SPEEDS
   TRANSONIC SPEED

ACOUSTIC VIBRATIONS
USE   SOUND WAVES

ACOUSTICAL HOLOGRAPHY
UF   SONOHOLOGRAPHY
   SOUND HOLOGRAPHY
GS   IMAGERY
   . ACOUSTICAL HOLOGRAPHY
   PHOTOGRAPHY
   . HOLOGRAPHY
   . . ACOUSTICAL HOLOGRAPHY
RT   ACOUSTIC PROPAGATION
   IMAGING TECHNIQUES
   SOUND WAVES
   WAVE FRONT RECONSTRUCTION

ACOUSTICS
UF   SOUND
GS   ACOUSTICS
   . AEROACOUSTICS
   . BIOACOUSTICS
   . GEOMETRICAL ACOUSTICS
   . MAGNETOACOUSTICS
   . MICROSONICS
   . PSYCHOACOUSTICS
   . UNDERWATER ACOUSTICS
RT   ACOUSTIC ATTENUATION
   ACOUSTIC EXCITATION
   ACOUSTIC FATIGUE
   ACOUSTIC IMPEDANCE
   ACOUSTIC PROPAGATION
   ACOUSTIC SCATTERING
   ACOUSTIC SOUNDING
   ANECHOIC CHAMBERS
   ARCHITECTURE
   AUDITORY PERCEPTION
   AUDITORY STIMULI
   AUDITORY TASKS
   COMFORT
   EARPHONES
   ECHOES
   EFFECTIVE PERCEIVED NOISE LEVELS
   ELASTIC WAVES
   HARMONIC EXCITATION
   HARMONIC GENERATIONS
   HARMONIC OSCILLATION
   HARMONICS
   HUM
   INFRASONIC FREQUENCIES
   LAMB WAVES

ACOUSTICS-(CONT.)
    LAME WAVE EQUATIONS
    LOUDNESS
    NOISE (SOUND)
    NOISE POLLUTION
    NOISE PROPAGATION
    NOISE REDUCTION
    OCTAVES
    OPACITY
    PHONETICS
    POWER SPECTRA
    ∞SCIENCE
    SIMPLE HARMONIC MOTION
    SONIC ANEMOMETERS
    SOUND AMPLIFICATION
    SOUND FIELDS
    SOUND PROPAGATION
    SOUND TRANSMISSION
    SOUND WAVES
    SPEECH
    STEREOPHONICS
    ULTRASONIC CLEANING
    ULTRASONIC SCANNERS
    ULTRASONICS
    VERBAL COMMUNICATION
    VIBRATION
    VIBRATION DAMPING
    VOICE COMMUNICATION
    ZERO SOUND

ACOUSTO-OPTICS
RT   CRYSTAL OPTICS
    ELECTRO-OPTICS
    GEOMETRICAL OPTICS
    IMAGERY
    MAGNETO-OPTICS
    OPTICAL PROPERTIES
    ∞OPTICS

ACPL (SPACELAB)
USE ATMOSPHERIC CLOUD PHYSICS LAB
    (SPACELAB)

ACQUISITION
GS   **ACQUISITION**
    . DATA ACQUISITION
    . TARGET ACQUISITION
RT   ACCUMULATIONS
    COLLECTION
    DETECTION
    DOCUMENTATION
    ∞RECEIVING
    RECOGNITION

ACRIFLAVINE
GS   HETEROCYCLIC COMPOUNDS
    . **ACRIFLAVINE**
RT   ANTISEPTICS
    DYES

ACROBATICS
RT   FLIGHT CONTROL
    MANEUVERS

ACROLEINS
GS   ALDEHYDES
    . **ACROLEINS**
    ALIPHATIC COMPOUNDS
    . **ACROLEINS**
RT   TOXICITY AND SAFETY HAZARD

ACRYLATES
GS   ALIPHATIC COMPOUNDS
    . **ACRYLATES**
    ESTERS
    . **ACRYLATES**
RT   RESINS

ACRYLIC ACID
GS   ALIPHATIC COMPOUNDS
    . **ACRYLIC ACID**

ACRYLIC RESINS
UF   METHACRYLATE RESINS
    POLYACRYLATES
GS   PLASTICS
    . SYNTHETIC RESINS
    . . ADDITION RESINS
    . . . **ACRYLIC RESINS**
    RESINS
    . **ACRYLIC RESINS**
RT   LATEX
    THERMOPLASTIC RESINS

ACRYLONITRILES
UF   VINYL CYANIDE
GS   ALIPHATIC COMPOUNDS
    . **ACRYLONITRILES**
    NITRILES
    . **ACRYLONITRILES**
RT   PLASTICS

ACTH
USE ADRENOCORTICOTROPIN (ACTH)

ACTINIDE SERIES
GS   CHEMICAL ELEMENTS
    . **ACTINIDE SERIES**
    . . ACTINIUM
    . . RADIUM
    . . . RADIUM ISOTOPES
    . . . . RADIUM 226
    . . THORIUM
    . . . THORIUM ISOTOPES
    . . TRANSURANIUM ELEMENTS
    . . . AMERICIUM
    . . . . AMERICIUM ISOTOPES
    . . . . . AMERICIUM 241
    . . . BERKELIUM
    . . . CALIFORNIUM
    . . . . CALIFORNIUM ISOTOPES
    . . . CURIUM
    . . . . CURIUM ISOTOPES
    . . . . . CURIUM 242
    . . . . . CURIUM 244
    . . . EINSTEINIUM
    . . . FERMIUM
    . . . LAWRENCIUM
    . . . MENDELEVIUM
    . . . NEPTUNIUM
    . . . . NEPTUNIUM ISOTOPES
    . . . NOBELIUM
    . . . PLUTONIUM
    . . . . PLUTONIUM ISOTOPES
    . . . . . PLUTONIUM 238
    . . . . . PLUTONIUM 239
    . . . . . PLUTONIUM 240
    . . . . . PLUTONIUM 241
    . . . . . PLUTONIUM 244
    . . . SERGENIUM
    . . URANIUM
    . . . URANIUM ISOTOPES
    . . . . URANIUM 232
    . . . . URANIUM 233
    . . . . URANIUM 234
    . . . . URANIUM 235
    . . . . URANIUM 238
    . . . URANIUM PLASMAS
    METALS
    . **ACTINIDE SERIES**
    . . ACTINIUM
    . . RADIUM
    . . . RADIUM ISOTOPES
    . . . . RADIUM 226
    . . THORIUM
    . . . THORIUM ISOTOPES
    . . TRANSURANIUM ELEMENTS
    . . . AMERICIUM
    . . . . AMERICIUM ISOTOPES
    . . . . . AMERICIUM 241
    . . . BERKELIUM
    . . . CALIFORNIUM
    . . . . CALIFORNIUM ISOTOPES
    . . . CURIUM
    . . . . CURIUM ISOTOPES
    . . . . . CURIUM 242
    . . . . . CURIUM 244
    . . . EINSTEINIUM
    . . . FERMIUM
    . . . LAWRENCIUM
    . . . MENDELEVIUM
    . . . NEPTUNIUM
    . . . . NEPTUNIUM ISOTOPES
    . . . NOBELIUM
    . . . PLUTONIUM
    . . . . PLUTONIUM ISOTOPES
    . . . . . PLUTONIUM 238
    . . . . . PLUTONIUM 239
    . . . . . PLUTONIUM 240
    . . . . . PLUTONIUM 241
    . . . . . PLUTONIUM 244
    . . . SERGENIUM
    . . URANIUM
    . . . URANIUM ISOTOPES
    . . . . URANIUM 232
    . . . . URANIUM 233
    . . . . URANIUM 234
    . . . . URANIUM 235
    . . . . URANIUM 238

ACTINIDE SERIES-(CONT.)
    . . . URANIUM PLASMAS
RT   RADIOACTIVE ISOTOPES
    RADIOACTIVE MATERIALS
    TRANSITION METALS

ACTINIDE SERIES COMPOUNDS
GS   **ACTINIDE SERIES COMPOUNDS**
    . CURIUM COMPOUNDS
    . NEPTUNIUM COMPOUNDS
    . PLUTONIUM COMPOUNDS
    . . PLUTONIUM FLUORIDES
    . . PLUTONIUM OXIDES
    . THORIUM COMPOUNDS
    . . THORIUM FLUORIDES
    . . THORIUM OXIDES
    . URANIUM COMPOUNDS
    . . URANIUM CARBIDES
    . . URANIUM FLUORIDES
    . . URANIUM OXIDES
RT   ∞CHEMICAL COMPOUNDS
    ∞GROUP 3B COMPOUNDS

ACTINIUM
GS   CHEMICAL ELEMENTS
    . ACTINIDE SERIES
    . . **ACTINIUM**
    METALS
    . ACTINIDE SERIES
    . . **ACTINIUM**

ACTINOGRAPHS
USE ACTINOMETERS

ACTINOMETERS
UF   ACTINOGRAPHS
    EMISSOGRAPHS
GS   MEASURING INSTRUMENTS
    . RADIATION MEASURING INSTRUMENTS
    . . **ACTINOMETERS**
    . . . INFRARED SPECTROMETERS
    . . . PYRANOMETERS
    . . . RADIOMETERS
    . . . . DICKE RADIOMETERS
    . . . . INFRARED DETECTORS
    . . . . INFRARED SCANNERS
    . . . . MICROWAVE RADIOMETERS
    . . . . PASSIVE L-BAND RADIOMETERS
    . . . . PRESSURE MODULATOR
        RADIOMETERS
    . . . . SPECTRORADIOMETERS
    . . . . SOLAR SPECTROMETERS
    . . . SPECTROHELIOGRAPHS
    . . . SPECTROPHOTOMETERS
    . . . . INFRARED
        SPECTROPHOTOMETERS
    . . . . ULTRAVIOLET
        SPECTROPHOTOMETERS
    . . . ULTRAVIOLET SPECTROMETERS
RT   DOSIMETERS
    FABRY-PEROT SPECTROMETERS
    FIELD INTENSITY METERS
    SPECTROMETERS

ACTINOMYCETES
GS   FUNGI
    . **ACTINOMYCETES**
    MICROORGANISMS
    . BACTERIA
    . . **ACTINOMYCETES**

ACTINOMYCIN
GS   DRUGS
    . **ACTINOMYCIN**
    STEROIDS
    . **ACTINOMYCIN**

ACTIVATED CARBON
GS   CHEMICAL ELEMENTS
    . CARBON
    . . CHARCOAL
    . . . **ACTIVATED CARBON**
    FUELS
    . **ACTIVATED CARBON**
RT   ∞ABSORPTION
    FILTRATION
    HEMOPERFUSION
    WATER TREATMENT

ACTIVATED SLUDGE
GS   SLUDGE
    . **ACTIVATED SLUDGE**
RT   BIODEGRADATION
    HUMAN WASTES
    METABOLIC WASTES

ACTIVATED SLUDGE-*(CONT.)*
    SEWAGE
    WASTES

ACTIVATION
  RT   ACTUATION
        CATALYSIS
        DEACTIVATION
        ELECTROMAGNETIC ABSORPTION
        EXCITATION
        FLOTATION
        INITIATION
        IONIZATION POTENTIALS
        IRRADIATION
        SENSITIZING
        STARTING
        STIMULATION

ACTIVATION (BIOLOGY)
  RT   ACTIVATION ENERGY
        ∞BIOLOGY
        ∞CELLS
        ENZYMES
        STIMULATION

ACTIVATION ANALYSIS
  GS   **ACTIVATION ANALYSIS**
        . NEUTRON ACTIVATION ANALYSIS
  RT   ∞ANALYZING

ACTIVATION ENERGY
  RT   ACTIVATION (BIOLOGY)
        DAMKOHLER NUMBER
        ELECTRON ENERGY
        ∞ENERGY
        HEAT
        NUCLEAR BINDING ENERGY
        NUCLEAR CAPTURE
        PROTON ENERGY
        ROTONS
        SURFACE ENERGY

ACTIVE CONTROL
  GS   AUTOMATIC CONTROL
        . ADAPTIVE CONTROL
        . . **ACTIVE CONTROL**
  RT   AIRCRAFT CONTROL
        ∞CONTROL
        INTERACTIVE CONTROL
        SELF ADAPTIVE CONTROL SYSTEMS
        SELF ALIGNMENT
        SERVOMECHANISMS

ACTIVE GLACIERS
  USE   GLACIERS

ACTIVE MAGNETO PARTICLE TRACER EXPLORERS
  USE   AMPTE (SATELLITES)

ACTIVE SATELLITES
  GS   SATELLITES
        . **ACTIVE SATELLITES**
        . . SYNCOM SATELLITES
        . . . EARLY BIRD SATELLITES
        . . . SYNCOM 1 SATELLITE
        . . . SYNCOM 2 SATELLITE
        . . . SYNCOM 3 SATELLITE
  RT   ADVENT PROJECT
        EXPLORER 29 SATELLITE
        EXPLORER 36 SATELLITE
        GEODETIC SATELLITES
        GEOS 1 SATELLITE
        GEOS 2 SATELLITE
        GEOS 3 SATELLITE
        NAVIGATION SATELLITES
        NAVSTAR SATELLITES
        PASSIVE SATELLITES
        SYNCHRONOUS SATELLITES

ACTIVE VOLCANOES
  USE   VOLCANOES

∞ ACTIVITY
  SN   *(USE OF A MORE SPECIFIC TERM IS*
        RECOMMENDED--CONSULT THE TERMS
        LISTED BELOW)
  RT   ACTIVITY (BIOLOGY)
        EXTRAVEHICULAR ACTIVITY
        FACULAE
        INTRAVEHICULAR ACTIVITY
        RADIOACTIVITY
        SOLAR ACTIVITY

ACTIVITY (BIOLOGY)
  UF   BIOLOGICAL ACTIVITY
  RT   ∞ACTIVITY
        BIOLOGICAL EFFECTS
        ∞BIOLOGY
        CATALYTIC ACTIVITY

ACTIVITY CYCLES (BIOLOGY)
  GS   CYCLES
        . **ACTIVITY CYCLES (BIOLOGY)**
  RT   ∞BIOLOGY
        CIRCADIAN RHYTHMS
        PHENOLOGY
        RHYTHM (BIOLOGY)

ACTUATION
  RT   ACTIVATION
        ACTUATORS
        EXCITATION
        INITIATION
        NUTATION
        SENSITIZING
        STARTING
        STIMULATION

ACTUATOR DISKS
  GS   DISKS (SHAPES)
        . **ACTUATOR DISKS**
  RT   ∞DISKS
        ∞FANS
        PROPELLERS

ACTUATORS
  UF   CARTRIDGE ACTUATED DEVICES
        HYDRAULIC ACTUATORS
        TRIGGERS
  RT   ACTUATION
        AIRCRAFT HYDRAULIC SYSTEMS
        AUTOMATIC CONTROL VALVES
        CAMS
        CONTROL VALVES
        CONTROLLERS
        EXPLOSIVE DEVICES
        ∞INSTRUMENTS
        MISSILE CONTROL
        PROPELLANT ACTUATED INSTRUMENTS
        REGULATORS
        SERVOMECHANISMS
        SERVOMOTORS
        SHAPE CONTROL
        SOLENOIDS
        STARTERS
        STEPPING MOTORS
        TORQUE MOTORS

ACUITY
  GS   **ACUITY**
        . VISUAL ACUITY
        . . HYPEROPIA
  RT   ADAPTATION
        DISCRIMINATION
        ∞FREQUENCY RESPONSE
        PERCEPTION
        SENSITIVITY
        THRESHOLDS (PERCEPTION)

ACYLATION
  GS   CHEMICAL REACTIONS
        . **ACYLATION**
        . . ACETYLATION
  RT   FRIEDEL-CRAFT REACTION

AD-A SATELLITE
  USE   EXPLORER 19 SATELLITE

AD/I B
  USE   EXPLORER 25 SATELLITE

AD/I SATELLITE
  USE   EXPLORER 24 SATELLITE

ADA (PROGRAMMING LANGUAGE)
  GS   LANGUAGES
        . **ADA (PROGRAMMING LANGUAGE)**
  RT   COMPUTER PROGRAMMING
        EMBEDDED COMPUTER SYSTEMS

ADAPTATION
  GS   **ADAPTATION**
        . ACCLIMATIZATION
        . . ALTITUDE ACCLIMATIZATION
        . . COLD ACCLIMATIZATION
        . . HEAT ACCLIMATIZATION
        . DESERT ADAPTATION

ADAPTATION-*(CONT.)*
        . RETINAL ADAPTATION
        . . DARK ADAPTATION
        . . LIGHT ADAPTATION
  RT   ACCOMMODATION
        ACUITY
        CORRECTION
        FITTING
        HIBERNATION
        HOMEOSTASIS
        PERCEPTION
        REACTION TIME
        RETRAINING
        SENSITIVITY
        THRESHOLDS (PERCEPTION)
        VISION

ADAPTERS
  GS   **ADAPTERS**
        . MULTIPLE DOCKING ADAPTERS
  RT   CONNECTORS
        EXTENSIONS
        FITTINGS
        JOINTS (JUNCTIONS)

ADAPTIVE CONTROL
  UF   ADAPTIVE CONTROL SYSTEMS
  GS   AUTOMATIC CONTROL
        . **ADAPTIVE CONTROL**
        . . ACTIVE CONTROL
        . . LEARNING MACHINES
        . . SELF ADAPTIVE CONTROL SYSTEMS
  RT   AUTOMATA THEORY
        AUTONOMY
        ∞CONTROL
        CONTROL THEORY
        CYBERNETICS
        DYNAMIC CONTROL
        FEEDBACK CONTROL
        FEEDFORWARD CONTROL
        OPTIMAL CONTROL
        SELF ALIGNMENT

ADAPTIVE CONTROL SYSTEMS
  USE   ADAPTIVE CONTROL

ADAPTIVE FILTERS
  RT   BANDPASS FILTERS
        BANDSTOP FILTERS
        ELECTRIC FILTERS
        ELECTROMAGNETIC WAVE FILTERS
        ∞FILTERS
        LINEAR FILTERS
        OPTICAL FILTERS
        TRACKING FILTERS

ADAPTIVE OPTICS
  RT   ATMOSPHERIC OPTICS
        COMPUTER TECHNIQUES
        FEEDBACK CONTROL
        IMAGING TECHNIQUES
        INSTRUMENT COMPENSATION
        OPTICAL CORRECTION PROCEDURE
        OPTICAL TRANSFER FUNCTION
        ∞OPTICS
        SELF ADAPTIVE CONTROL SYSTEMS

ADDERS (CIRCUITS)
  USE   ADDING CIRCUITS

ADDING CIRCUITS
  UF   ADDERS (CIRCUITS)
        BINARY SUMMATORS
  GS   CIRCUITS
        . **ADDING CIRCUITS**
  RT   ACCUMULATORS (COMPUTERS)
        BINARY INTEGRATION
        COMPUTER COMPONENTS
        LOGIC CIRCUITS

ADDITION
  RT   ADDITION THEOREM
        AMOUNT
        ARITHMETIC
        COMPUTATION
        NUMBER THEORY
        RESTORATION

ADDITION RESINS
  SN   *(CARBON CHAIN POLYMERS--FOR*
        HETEROATOM CHAIN POLYMERS, USE
        POLYETHER RESINS)
  GS   PLASTICS
        . SYNTHETIC RESINS

ADDITION RESINS-*(CONT.)*
```
        . . ADDITION RESINS
        . . . ACRYLIC RESINS
        . . . . VINYL COPOLYMERS
        RESINS
        . SYNTHETIC RESINS
        . . ADDITION RESINS
        . . . VINYL COPOLYMERS
RT      CROSSLINKING
        POLYBUTADIENE
        POLYETHYLENE TEREPHTHALATE
        POLYETHYLENES
        POLYISOBUTYLENE
        POLYPROPYLENE
        POLYSTYRENE
        POLYVINYL ALCOHOL
        POLYVINYL CHLORIDE
        SYNTHESIS (CHEMISTRY)
        SYNTHETIC FIBERS
        VULCANIZED ELASTOMERS
```

ADDITION THEOREM
```
GS      NUMBER THEORY
        . ADDITION THEOREM
        THEOREMS
        . ADDITION THEOREM
RT      ADDITION
```

ADDITIVES
```
UF      DOPING (ADDITIVES)
GS      ADDITIVES
        . ADMIXTURES
        . ANTIFREEZES
        . ANTIICING ADDITIVES
        . ANTIKNOCK ADDITIVES
        . ANTIOXIDANTS
        . OIL ADDITIVES
        . OPACIFIERS
        . PLASTICIZERS
        . PROPELLANT ADDITIVES
        . . PROPELLANT BINDERS
        . . . SOLID ROCKET BINDERS
RT      ∞AGENTS
        ANTIMISTING FUELS
        ANTIOXIDANTS
        BINDERS (MATERIALS)
        CARRIER INJECTION
        CATALYSTS
        COATINGS
        DILUENTS
        DOPES
        FILLERS
        HIGH ENERGY FUELS
        INHIBITORS
        INTERSTITIALS
        LUBRICANTS
        MAJORITY CARRIERS
        MINORITY CARRIERS
        NEUTRALIZERS
        PIGMENTS
        PRESERVATIVES
        RETARDANTS
        SOLVENTS
        STABILIZERS (AGENTS)
        SUPPRESSORS
        TETRAHYDROFURAN
        THICKENERS (MATERIALS)
        TRAVELING SOLVENT METHOD
        VINYL COPOLYMERS
```

ADDRESSING
```
RT      CODING
        COMPUTER PROGRAMMING
```

ADDUCTS
```
RT      ∞CHEMICAL COMPOUNDS
```

ADEN
```
USE     SOUTHERN YEMEN
```

ADENINES
```
GS      ALIPHATIC COMPOUNDS
        . NUCLEOSIDES
        . . ADENINES
        CARBOHYDRATES
        . ADENINES
        HETEROCYCLIC COMPOUNDS
        . ADENINES
        ORGANIC COMPOUNDS
        . NUCLEOTIDES
        . . ADENINES
        PHOSPHORUS COMPOUNDS
        . ADENINES
        PROTEINS
        . NUCLEOTIDES
```

ADENINES-*(CONT.)*
```
        . . ADENINES
        PURINES
        . ADENINES
```

ADENOSINE DIPHOSPHATE
```
UF      ADP
GS      ALIPHATIC COMPOUNDS
        . NUCLEOSIDES
        . . ADENOSINES
        . . . ADENOSINE DIPHOSPHATE
        CARBOHYDRATES
        . NUCLEOSIDES
        . . ADENOSINES
        . . . ADENOSINE DIPHOSPHATE
        HETEROCYCLIC COMPOUNDS
        . ADENOSINES
        . . ADENOSINE DIPHOSPHATE
        ORGANIC COMPOUNDS
        . NUCLEOTIDES
        . . ADENOSINES
        . . . ADENOSINE DIPHOSPHATE
        PHOSPHORUS COMPOUNDS
        . PHOSPHATES
        . . DIPHOSPHATES
        . . . ADENOSINE DIPHOSPHATE
        PROTEINS
        . NUCLEOTIDES
        . . ADENOSINES
        . . . ADENOSINE DIPHOSPHATE
```

ADENOSINE TRIPHOSPHATE
```
UF      ATP
GS      ALIPHATIC COMPOUNDS
        . NUCLEOSIDES
        . . ADENOSINES
        . . . ADENOSINE TRIPHOSPHATE
        CARBOHYDRATES
        . NUCLEOSIDES
        . . ADENOSINES
        . . . ADENOSINE TRIPHOSPHATE
        HETEROCYCLIC COMPOUNDS
        . ADENOSINES
        . . ADENOSINE TRIPHOSPHATE
        ORGANIC COMPOUNDS
        . NUCLEOTIDES
        . . ADENOSINES
        . . . ADENOSINE TRIPHOSPHATE
        PHOSPHORUS COMPOUNDS
        . PHOSPHATES
        . . ADENOSINE TRIPHOSPHATE
        PROTEINS
        . NUCLEOTIDES
        . . ADENOSINES
        . . . ADENOSINE TRIPHOSPHATE
RT      AMINO ACIDS
```

ADENOSINES
```
GS      ALIPHATIC COMPOUNDS
        . NUCLEOSIDES
        . . ADENOSINES
        . . . ADENOSINE DIPHOSPHATE
        . . . ADENOSINE TRIPHOSPHATE
        CARBOHYDRATES
        . NUCLEOSIDES
        . . ADENOSINES
        . . . ADENOSINE DIPHOSPHATE
        . . . ADENOSINE TRIPHOSPHATE
        HETEROCYCLIC COMPOUNDS
        . ADENOSINES
        . . ADENOSINE DIPHOSPHATE
        . . ADENOSINE TRIPHOSPHATE
        . . CYCLIC AMP
        ORGANIC COMPOUNDS
        . NUCLEOTIDES
        . . ADENOSINES
        . . . ADENOSINE DIPHOSPHATE
        . . . ADENOSINE TRIPHOSPHATE
        . . . CYCLIC AMP
        PROTEINS
        . NUCLEOTIDES
        . . ADENOSINES
        . . . ADENOSINE DIPHOSPHATE
        . . . ADENOSINE TRIPHOSPHATE
        . . . CYCLIC AMP
```

ADENOVIRUSES
```
GS      MICROORGANISMS
        . VIRUSES
        . . ADENOVIRUSES
```

ADEQUACY
```
RT      QUALITY
        VALIDITY
```

ADHEROMETERS
```
USE     ADHESION TESTS
```

ADHESION
```
GS      SURFACE PROPERTIES
        . ADHESION
RT      ADHESION TESTS
        ADHESIVE BONDING
        AGGLUTINATION
        BONDING
        COLD WELDING
        FUSION (MELTING)
        INTERFACIAL ENERGY
        INTERNAL PRESSURE
        ∞JOINING
        METAL BONDING
        PEELING
        SEALING
        SPREADING
        TACKINESS
        TRACTION
        WETTABILITY
```

ADHESION TESTS
```
UF      ADHEROMETERS
        FOKKER BOND TESTERS
RT      ADHESION
        BONDING
        NONDESTRUCTIVE TESTS
        ∞TESTS
        WETTABILITY
```

ADHESIVE BONDING
```
GS      BONDING
        . ADHESIVE BONDING
RT      ADHESION
        AGGLUTINATION
        CEMENTATION
        ∞JOINING
        METAL BONDING
        METAL-METAL BONDING
        RESIN BONDING
        SEALING
```

ADHESIVES
```
UF      BINDERS (ADHESIVES)
GS      ADHESIVES
        . GLUES
        . PASTES
        . TETRAETHYL ORTHOSILICATE
RT      AGGLUTINATION
        ALKYD RESINS
        BINDERS (MATERIALS)
        CEMENTS
        EPOXY RESINS
        FASTENERS
        FURAN RESINS
        JOINTS (JUNCTIONS)
        METAL-METAL BONDING
        PHENOLIC EPOXY RESINS
        PLASTIC TAPES
        SEALERS
        SEAMS (JOINTS)
        ∞TAPES
```

ADIABATIC CONDITIONS
```
GS      CONDITIONS
        . ADIABATIC CONDITIONS
RT      CARNOT CYCLE
        COMPRESSING
        ENTHALPY
        ENVIRONMENTS
        EXPANSION
        ISENTROPE
        ISOENERGETIC PROCESSES
        ISOTHERMAL PROCESSES
        POLYTROPIC PROCESSES
        TEMPERATURE
        THERMAL ENVIRONMENTS
        THERMODYNAMIC CYCLES
        THERMODYNAMIC EQUILIBRIUM
```

ADIABATIC DEMAGNETIZATION COOLING
```
RT      MAGNETIC COOLING
```

ADIABATIC EQUATIONS
```
RT      ∞EQUATIONS
        EQUATIONS OF STATE
        HEAT TRANSMISSION
        NONADIABATIC THEORY
        SHOCK WAVES
```

ADIABATIC FLOW
```
GS      FLUID FLOW
        . ADIABATIC FLOW
```

ADIABATIC FLOW-*(CONT.)*
RT   STAGNATION TEMPERATURE

ADIPOSE TISSUES
GS   TISSUES (BIOLOGY)
    . ADIPOSE TISSUES
RT   CONNECTIVE TISSUE
    FATS

ADIPRENE (TRADEMARK)
GS   RUBBER
    . ADIPRENE (TRADEMARK)

ADIRONDACK MOUNTAINS (NY)
GS   LANDFORMS
    . MOUNTAINS
    . . ADIRONDACK MOUNTAINS (NY)
RT   NEW YORK

ADJOINTS
GS   ALGEBRA
    . VECTOR SPACES
    . . MATRICES (MATHEMATICS)
    . . . ADJOINTS
RT   DATA PROCESSING
    NUMERICAL ANALYSIS

ADJUSTING
UF   ADJUSTMENT
    READJUSTMENT
RT   ALIGNMENT
    CLEARANCES
    COLLIMATION
    CORRECTION
    FITTING
    FOCUSING
    LEVELING
    MATCHING
    OPTICAL CORRECTION PROCEDURE
    POSITIONING
    REVISIONS
    ∞ SETTING
    SMOOTHING

ADJUSTMENT
USE   ADJUSTING

ADMINISTRATION
USE   MANAGEMENT

ADMITTANCE
USE   ELECTRICAL IMPEDANCE

ADMIXTURES
GS   ADDITIVES
    . ADMIXTURES
    MIXTURES
    . ADMIXTURES
RT   ACCELERATING AGENTS
    CATALYSTS
    ∞ COMBINATION
    CONCRETES
    FORMULATIONS
    INGREDIENTS
    MIXERS
    MORTARS (MATERIAL)
    SURFACTANTS

ADOBE FLATS
USE   FLATS (LANDFORMS)

ADP
USE   ADENOSINE DIPHOSPHATE

ADRENAL GLAND
GS   ANATOMY
    . GLANDS (ANATOMY)
    . . ENDOCRINE GLANDS
    . . . ADRENAL GLAND
    VISCERA
    . ENDOCRINE GLANDS
    . . ADRENAL GLAND
RT   ADRENOCORTICOTROPIN (ACTH)
    EPINEPHRINE

ADRENAL METABOLISM
GS   METABOLISM
    . ADRENAL METABOLISM
RT   ALDOSTERONE
    CORTICOSTEROIDS
    CORTISONE
    HYDROXYCORTICOSTEROID

ADRENALINE
USE   EPINEPHRINE

ADRENERGICS
UF   SYMPATHOMIMETICS
GS   DRUGS
    . ADRENERGICS
RT   ANTIADRENERGICS
    ANTICOAGULANTS
    CYCLIC AMP

ADRENOCORTICOTROPIN (ACTH)
UF   ACTH
GS   SECRETIONS
    . ENDOCRINE SECRETIONS
    . . HORMONES
    . . . PITUITARY HORMONES
    . . . . . ADRENOCORTICOTROPIN (ACTH)
RT   ACIDS
    ADRENAL GLAND
    AMINO ACIDS
    PROTEINS

ADRIATIC SEA
GS   SEAS
    . MEDITERRANEAN SEA
    . . ADRIATIC SEA
RT   ITALY
    SEAS
    YUGOSLAVIA

ADSORBENTS
GS   SORBENTS
    . ADSORBENTS
RT   ABSORBENTS
    ADSORPTION
    AIR CONDITIONING EQUIPMENT
    CHARCOAL
    DESICCANTS
    HEMOPERFUSION

ADSORPTION
GS   SORPTION
    . ADSORPTION
    . . CHEMISORPTION
RT   ∞ ABSORPTION
    ADSORBENTS
    ADSORPTIVITY
    BENEFICIATION
    CHROMATOGRAPHY
    CONCENTRATING
    DESORPTION
    DIFFUSION
    ELECTROSTATIC PRECIPITATORS
    ELUTION
    GAS CHROMATOGRAPHY
    GAS-METAL INTERACTIONS
    GIBBS ADSORPTION EQUATION
    ∞ SEPARATION
    WATER TREATMENT

ADSORPTIVITY
GS   SURFACE PROPERTIES
    . ADSORPTIVITY
RT   ADSORPTION
    CHEMICAL PROPERTIES
    CHEMISORPTION
    ∞ PHYSICAL PROPERTIES

ADVANCED AIRBORNE COMMAND POST
USE   E-4A AIRCRAFT

ADVANCED EVA PROTECTION SYSTEMS
USE   AEPS

ADVANCED ORBITING SOLAR OBSERVATORY
USE   AOSO

ADVANCED RANGE INSTRUMENTATION
AIRCRAFT
RT   AIRBORNE EQUIPMENT
    ∞ AIRCRAFT
    APOLLO PROJECT
    C-135 AIRCRAFT
    DATA ACQUISITION
    TELEMETRY

ADVANCED RANGE INSTRUMENTATION SHIP
UF   ARIS INSTRUMENTATION SHIP
GS   WATER VEHICLES
    . SHIPS
    . . ADVANCED RANGE
        INSTRUMENTATION SHIP
RT   ∞ INSTRUMENTS

ADVANCED RANGE INSTRUMENTATION-*(CONT.)*
    MANNED SPACE FLIGHT NETWORK
    SPACECRAFT TRACKING
    TRACKING NETWORKS

ADVANCED RECONN ELECTRIC SPACECRAFT
UF   ARES (SPACECRAFT)
GS   INTERPLANETARY SPACECRAFT
    . MARS PROBES
    . . ADVANCED RECONN ELECTRIC
        SPACECRAFT
    UNMANNED SPACECRAFT
    . SPACE PROBES
    . . MARS PROBES
    . . . ADVANCED RECONN ELECTRIC
        SPACECRAFT
RT   ∞ SPACECRAFT

ADVANCED SODIUM COOLED REACTOR
UF   ASCR REACTOR
GS   NUCLEAR REACTORS
    . LIQUID COOLED REACTORS
    . . LIQUID METAL COOLED REACTORS
    . . . ADVANCED SODIUM COOLED
        REACTOR

ADVANCED TECHNOLOGY LABORATORY
GS   LABORATORIES
    . ADVANCED TECHNOLOGY
        LABORATORY
RT   SPACE SHUTTLES
    SPACELAB

ADVANCED TECHNOLOGY LIGHT TWIN
AIRCRAFT
USE   ATLIT PROJECT

ADVANCED TEST REACTORS
UF   ATR REACTOR
GS   NUCLEAR REACTORS
    . NUCLEAR RESEARCH AND TEST
        REACTORS
    . . ADVANCED TEST REACTORS

ADVANCED VIDICON CAMERA SYSTEM (AVCS)
UF   AVCS
GS   COMMUNICATION EQUIPMENT
    . ADVANCED VIDICON CAMERA SYSTEM
        (AVCS)
    TELEVISION SYSTEMS
    . ADVANCED VIDICON CAMERA SYSTEM
        (AVCS)
RT   ∞ SYSTEMS
    VIDEO EQUIPMENT
    VIDICONS

ADVANCED X RAY ASTROPHYSICAL FACILITY
USE   X RAY ASTROPHYSICS FACILITY

ADVANCED X RAY ASTROPHYSICS FACILITY
USE   X RAY ASTROPHYSICS FACILITY

ADVANCING GLACIERS
USE   GLACIERS

ADVANCING SHORELINES
USE   BEACHES

ADVECTION
RT   ATMOSPHERIC CIRCULATION
    CONVECTION
    HEAT TRANSFER
    PECLET NUMBER

ADVENT PROJECT
GS   PROGRAMS
    . PROJECTS
    . . ADVENT PROJECT
RT   ACTIVE SATELLITES
    COMMUNICATION SATELLITES
    COURIER SATELLITE
    RELAY SATELLITES

AE-A SATELLITE
USE   EXPLORER 17 SATELLITE

AE-B SATELLITE
USE   EXPLORER 32 SATELLITE

AE-C SATELLITE
USE   EXPLORER 51 SATELLITE

AE-D SATELLITE
USE   EXPLORER 54 SATELLITE

AE-E SATELLITE
USE   EXPLORER 55 SATELLITE

AEOLIAN TONES
RT   ELASTIC WAVES
      FREQUENCIES
      KARMAN VORTEX STREET
      NOISE (SOUND)
      SOUND WAVES
      WIND (METEOROLOGY)

AEOLOTROPISM
GS   TROPISM
      . **AEOLOTROPISM**
RT   ANISOTROPY

AEPS
UF   ADVANCED EVA PROTECTION SYSTEMS
GS   SUPPORT SYSTEMS
      . LIFE SUPPORT SYSTEMS
      . . EMERGENCY LIFE SUSTAINING
          SYSTEMS
      . . . **AEPS**
      . . PORTABLE LIFE SUPPORT SYSTEMS
      . . . **AEPS**
RT   EXTRAVEHICULAR ACTIVITY
      LUNAR BASES
      MARS LANDING
      OXYGEN SUPPLY EQUIPMENT
      SPACE SHUTTLES
      SPACE STATIONS
      SURVIVAL EQUIPMENT

AERATION
RT   AGITATION
      BENEFICIATION
      BLOWING
      BUBBLES
      CORROSION PREVENTION
      DEGASSING
      DISSOLVED GASES
      DISSOLVING
      ENTRAINMENT
      MIXERS
      MIXING
      OXYGENATION
      PURIFICATION
      ∞SEPARATION
      SPRAYING
      STIRRING
      SUSPENDING (MIXING)
      WATER TREATMENT

AERIAL EXPLOSIONS
SN   (EXPLOSIONS OCCURRING AT HEIGHTS
      LESS THAN 50 KM)
UF   AIR BLASTS
GS   EXPLOSIONS
      . **AERIAL EXPLOSIONS**
RT   BLAST LOADS
      CHEMICAL EXPLOSIONS
      NUCLEAR EXPLOSIONS
      THERMONUCLEAR EXPLOSIONS

AERIAL IMAGERY
USE   AERIAL PHOTOGRAPHY

AERIAL PHOTOGRAPHY
UF   AERIAL IMAGERY
GS   IMAGERY
      . **AERIAL PHOTOGRAPHY**
      PHOTOGRAPHY
      . **AERIAL PHOTOGRAPHY**
RT   ASTRONOMICAL PHOTOGRAPHY
      CHANGE DETECTION
      CLOUD PHOTOGRAPHS
      CLOUD PHOTOGRAPHY
      COLOR PHOTOGRAPHY
      EARTH OBSERVATIONS (FROM SPACE)
      EARTH RESOURCES SURVEY AIRCRAFT
      FOREST FIRE DETECTION
      GEOGRAPHIC INFORMATION SYSTEMS
      GRAY SCALE
      GROUND TRUTH
      ICE MAPPING
      IMAGE MOTION COMPENSATION
      INFRARED PHOTOGRAPHY
      ORTHOPHOTOGRAPHY
      PHOTOGEOLOGY
      PHOTOGRAMMETRY
      PHOTOINTERPRETATION
      PHOTOMAPPING

AERIAL PHOTOGRAPHY-*(CONT.)*
      PHOTOMAPS
      PLANT STRESS
      ROCKET-BORNE PHOTOGRAPHY
      SATELLITE-BORNE PHOTOGRAPHY
      SEA TRUTH
      SPACEBORNE PHOTOGRAPHY
      STEREOPHOTOGRAPHY
      TIMBER INVENTORY
      ULTRAVIOLET PHOTOGRAPHY

AERIAL RECONNAISSANCE
GS   RECONNAISSANCE
      . **AERIAL RECONNAISSANCE**
      . . AIRBORNE INTEGRATED
          RECONNAISSANCE SYSTEM
RT   AEROMAGNETISM
      CHANGE DETECTION
      EARTH RESOURCES SURVEY AIRCRAFT
      GROUND TRUTH
      HS-801 AIRCRAFT
      INFRARED RADIOMETERS
      METEOROLOGICAL FLIGHT
      PHOTORECONNAISSANCE
      RECONNAISSANCE AIRCRAFT
      RECONNAISSANCE SPACECRAFT
      THERMAL MAPPING

AERIAL RUDDERS
GS   AIRFOILS
      . **AERIAL RUDDERS**
      CONTROL SURFACES
      . **AERIAL RUDDERS**
RT   FINS
      HORIZONTAL TAIL SURFACES
      MARINE RUDDERS
      STABILIZERS (FLUID DYNAMICS)
      TABS (CONTROL SURFACES)
      TAIL ASSEMBLIES

AEROACOUSTICS
GS   ACOUSTICS
      . **AEROACOUSTICS**
RT   AERODYNAMICS
      ∞AERONAUTICS
      AIRCRAFT NOISE
      GRAZING FLOW
      NOISE PREDICTION (AIRCRAFT)
      ∞SCIENCE
      SURFACE NOISE INTERACTIONS

AEROASSIST
RT   AEROBRAKING
      AEROCAPTURE
      AEROMANEUVERING
      ATMOSPHERIC ENTRY
      INTERPLANETARY TRANSFER ORBITS
      TRANSFER ORBITS

AEROBEE ROCKET VEHICLE
GS   ROCKET VEHICLES
      . SOUNDING ROCKETS
      . . **AEROBEE ROCKET VEHICLE**

AEROBES
GS   MICROORGANISMS
      . **AEROBES**
RT   ANAEROBES
      BACTERIA
      SEWAGE TREATMENT

AEROBIOLOGY
RT   AIR POLLUTION
      AIRBORNE INFECTION
      ∞BIOLOGY
      ENVIRONMENT POLLUTION
      POLLEN

AEROBRAKING
RT   AEROASSIST
      AEROCAPTURE
      AEROMANEUVERING
      INTERPLANETARY TRANSFER ORBITS
      TRANSFER ORBITS

AEROCAPTURE
RT   AEROASSIST
      AEROBRAKING
      AEROCAPTURE
      AEROMANEUVERING
      ATMOSPHERIC ENTRY
      INTERPLANETARY TRANSFER ORBITS
      TRANSFER ORBITS

AERODONTALGIA
USE   TOOTH DISEASES

AERODYNAMIC AXIS
USE   AERODYNAMIC BALANCE

AERODYNAMIC BALANCE
UF   AERODYNAMIC AXIS
      AERODYNAMIC CENTER
      DRAG BALANCE
      TRIM (BALANCE)
GS   AERODYNAMIC CHARACTERISTICS
      . **AERODYNAMIC BALANCE**
RT   AIRCRAFT STABILITY
      BALANCE
      DYNAMIC CHARACTERISTICS
      HORIZONTAL FLIGHT
      LIFT DRAG RATIO
      MASS DISTRIBUTION
      SPACECRAFT MOTION
      SPACECRAFT STABILITY
      STATIC AERODYNAMIC
        CHARACTERISTICS
      TURNING FLIGHT

AERODYNAMIC BRAKES
GS   BRAKES (FOR ARRESTING MOTION)
      . **AERODYNAMIC BRAKES**
      . . BALLUTES
      . . DRAG CHUTES
      . . PARAVULCOONS
      . . SPLIT FLAPS
      . . WING FLAPS
      . . . LEADING EDGE SLATS
      . . . TRAILING EDGE FLAPS
      DRAG DEVICES
      . **AERODYNAMIC BRAKES**
      . . BALLUTES
      . . DRAG CHUTES
      . . PARAVULCOONS
      . . SPLIT FLAPS
      . . WING FLAPS
      . . . LEADING EDGE SLATS
      . . . TRAILING EDGE FLAPS
RT   AIRCRAFT BRAKES
      CONTROL SURFACES
      FLAPERONS
      FLAPS (CONTROL SURFACES)
      RETRACTABLE EQUIPMENT
      SPOILERS

AERODYNAMIC BUZZ
USE   FLUTTER

AERODYNAMIC CENTER
USE   AERODYNAMIC BALANCE

AERODYNAMIC CHARACTERISTICS
GS   **AERODYNAMIC CHARACTERISTICS**
      . AERODYNAMIC BALANCE
      . AERODYNAMIC DRAG
      . . SUPERSONIC DRAG
      . AERODYNAMIC STABILITY
      . INTERFERENCE DRAG
      . LIFT
      . . INTERFERENCE LIFT
      . . JET LIFT
      . . ROTOR LIFT
      . . ZERO LIFT
      . STATIC AERODYNAMIC
        CHARACTERISTICS
RT   AERODYNAMIC NOISE
      ANGLE OF ATTACK
      ASPECT RATIO
      ∞CHARACTERISTICS
      CROSS FLOW
      DYNAMIC CHARACTERISTICS
      ENGINE AIRFRAME INTEGRATION
      ROTOR BODY INTERACTIONS
      UNDER SURFACE BLOWING
      UPPER SURFACE BLOWING
      WIND TUNNEL TESTS

AERODYNAMIC CHORDS
USE   AIRFOIL PROFILES
       CHORDS (GEOMETRY)

AERODYNAMIC COEFFICIENTS
UF   LIFT COEFFICIENTS
GS   COEFFICIENTS
      . **AERODYNAMIC COEFFICIENTS**
RT   ∞DRAG COEFFICIENTS
      FLOW COEFFICIENTS
      FLOW DISTORTION
      FORCE DISTRIBUTION

AERODYNAMIC COEFFICIENTS-*(CONT.)*
    LIFT
    LIFT DRAG RATIO
    PITCHING MOMENTS
    PRESSURE DISTRIBUTION
    ROLLING MOMENTS
    YAWING MOMENTS

AERODYNAMIC CONFIGURATIONS
  SN   (AERODYNAMIC VEHICLE SHAPES--FOR
        LIFTING OR THRUSTING SURFACES USE
        AIRFOILS)
  GS   **AERODYNAMIC CONFIGURATIONS**
        . DROOPED AIRFOILS
        . WING NACELLE CONFIGURATIONS
  RT   AIRCRAFT CONFIGURATIONS
        AIRCRAFT DESIGN
        AIRFOILS
        BLUNT BODIES
        BODIES OF REVOLUTION
        BODY-WING AND TAIL CONFIGURATIONS
        BODY-WING CONFIGURATIONS
        CANARD CONFIGURATIONS
        CHANNEL WINGS
        CONES
        ∞CONFIGURATIONS
        CONTROL SURFACES
        ∞DESIGN
        DISKS (SHAPES)
        DRAG
        ENGINE AIRFRAME INTEGRATION
        FAIRINGS
        FINNED BODIES
        HALF CONES
        ∞HEMISPHERES
        INTAKE SYSTEMS
        LAUNCH VEHICLE CONFIGURATIONS
        LIFT
        LIFTING BODIES
        MISSILE CONFIGURATIONS
        MONOPLANES
        NACELLES
        NOSE TIPS
        OBLIQUE WINGS
        PROPULSION SYSTEM CONFIGURATIONS
        PROTUBERANCES
        PYLON MOUNTING
        REYNOLDS EQUATION
        RING STRUCTURES
        ROTOR BODY INTERACTIONS
        SATELLITE CONFIGURATIONS
        SCALE MODELS
        SEMISPAN MODELS
        SLENDER BODIES
        SLENDER CONES
        SPACECRAFT CONFIGURATIONS
        SPHERES
        STRAKES
        STREAMLINED BODIES
        THREE DIMENSIONAL BODIES
        WEDGES
        WIND TUNNEL MODELS
        WING ROOTS

AERODYNAMIC DRAG
  GS   AERODYNAMIC CHARACTERISTICS
        . **AERODYNAMIC DRAG**
        . . SUPERSONIC DRAG
        AERODYNAMIC FORCES
        . **AERODYNAMIC DRAG**
        . . SUPERSONIC DRAG
        DYNAMIC CHARACTERISTICS
        . DRAG
        . . FRICTION DRAG
        . . . **AERODYNAMIC DRAG**
        . . . . SUPERSONIC DRAG
        FRICTION
        . **AERODYNAMIC DRAG**
        . . SUPERSONIC DRAG
  RT   BALLISTICS
        BASE PRESSURE
        ∞DRAG COEFFICIENTS
        DRAG MEASUREMENT
        DRAG REDUCTION
        GROUND EFFECT (AERODYNAMICS)
        HYPERSONIC FORCES
        LIFT
        LIFT DRAG RATIO
        ORBIT DECAY
        PRESSURE DRAG
        ∞RESISTANCE
        SATELLITE DRAG
        TURBULENCE
        VORTEX FLAPS

AERODYNAMIC FORCES
  UF   GLAUERT COEFFICIENT
  GS   **AERODYNAMIC FORCES**
        . AERODYNAMIC DRAG
        . . SUPERSONIC DRAG
        . AERODYNAMIC INTERFERENCE
        . AERODYNAMIC LOADS
        . . BLAST LOADS
        . . GUST LOADS
        . HYPERSONIC FORCES
        . LIFT
        . . INTERFERENCE LIFT
        . . JET LIFT
        . . ROTOR LIFT
        . . ZERO LIFT
        . WING LOADING
  RT   ∞FORCE
        LEADING EDGE THRUST
        THRUST DISTRIBUTION

AERODYNAMIC HEAT TRANSFER
  GS   TRANSMISSION
        . HEAT TRANSMISSION
        . . HEAT TRANSFER
        . . . **AERODYNAMIC HEAT TRANSFER**
        . . . . HYPERSONIC HEAT TRANSFER
        . . . . SUPERSONIC HEAT TRANSFER
  RT   ABLATION
        AEROTHERMODYNAMICS
        TURBULENT HEAT TRANSFER

AERODYNAMIC HEATING
  GS   HEATING
        . **AERODYNAMIC HEATING**
        . . SHOCK HEATING
  RT   ABLATION
        AERODYNAMICS
        AEROTHERMOCHEMISTRY
        AEROTHERMODYNAMICS
        ATMOSPHERIC ENTRY
        BOUNDARY LAYER PLASMAS
        COMPRESSIBLE FLUIDS
        CONVECTIVE HEAT TRANSFER
        HYPERSONIC REENTRY
        REENTRY
        REENTRY EFFECTS
        REENTRY SHIELDING
        SKIN FRICTION
        SKIN TEMPERATURE (NON-BIOLOGICAL)
        TRANSIENT HEATING
        UNCONTROLLED REENTRY
          (SPACECRAFT)

AERODYNAMIC INTERFERENCE
  GS   AERODYNAMIC FORCES
        . **AERODYNAMIC INTERFERENCE**
  RT   AERODYNAMICS
        AIR FLOW
        AIRCRAFT CONFIGURATIONS
        AIRCRAFT STRUCTURES
        AIRFOIL PROFILES
        CONTROL SURFACES
        ∞INTERFERENCE
        PROTUBERANCES
        TURBULENT FLOW
        WING PROFILES

AERODYNAMIC LIFT
  USE   LIFT

AERODYNAMIC LOADS
  GS   AERODYNAMIC FORCES
        . **AERODYNAMIC LOADS**
        . . BLAST LOADS
        . . GUST LOADS
        LOADS (FORCES)
        . DYNAMIC LOADS
        . . **AERODYNAMIC LOADS**
        . . . BLAST LOADS
        . . . GUST LOADS
  RT   AXIAL COMPRESSION LOADS
        AXIAL LOADS
        COMPRESSION LOADS
        CRITICAL LOADING
        EDGE LOADING
        FORCE DISTRIBUTION
        LOADING MOMENTS
        PRESSURE DISTRIBUTION
        SHOCK LOADS
        STATIC LOADS
        STRUCTURAL DESIGN CRITERIA
        THRUST LOADS
        TRANSIENT LOADS
        VIBRATORY LOADS
        WING LOADING

AERODYNAMIC MOMENTS
  USE   STABILITY DERIVATIVES

AERODYNAMIC NOISE
  UF   BOUNDARY LAYER NOISE
  GS   ELASTIC WAVES
        . SOUND WAVES
        . . NOISE (SOUND)
        . . . **AERODYNAMIC NOISE**
  RT   ACOUSTIC RETROFITTING
        AERODYNAMIC CHARACTERISTICS
        AEROELASTICITY
        AIRCRAFT NOISE
        BLADE SLAP NOISE
        FLUTTER
        JET AIRCRAFT NOISE
        NOISE MEASUREMENT
        NOISE REDUCTION
        PANEL FLUTTER
        SHOCK WAVES
        SONIC BOOMS
        SURFACE NOISE INTERACTIONS

AERODYNAMIC STABILITY
  UF   FLYING PLATFORM STABILITY
  GS   AERODYNAMIC CHARACTERISTICS
        . **AERODYNAMIC STABILITY**
        DYNAMIC CHARACTERISTICS
        . DYNAMIC STABILITY
        . . MOTION STABILITY
        . . . **AERODYNAMIC STABILITY**
        STABILITY
        . DYNAMIC STABILITY
        . . MOTION STABILITY
        . . . **AERODYNAMIC STABILITY**
  RT   AEROELASTICITY
        AIRCRAFT STABILITY
        ATTITUDE STABILITY
        BALLAST (MASS)
        BOUNDARY LAYER STABILITY
        BUFFETING
        DIRECTIONAL STABILITY
        DYNAMIC TESTS
        FLIGHT STABILITY TESTS
        FLOW STABILITY
        FLUTTER
        GROUND RESONANCE
        HELICOPTER PERFORMANCE
        HOVERING
        LATERAL STABILITY
        LIQUID SLOSHING
        LONGITUDINAL STABILITY
        LOW SPEED STABILITY
        MASS DISTRIBUTION
        PILOT INDUCED OSCILLATION
        PRESSURE DISTRIBUTION
        REENTRY
        RICHARDSON NUMBER
        SPACECRAFT MOTION
        SPACECRAFT STABILITY
        STABILITY AUGMENTATION
        STATIC AERODYNAMIC
          CHARACTERISTICS
        TURBULENCE EFFECTS
        VORTEX AVOIDANCE
        WIND TUNNEL STABILITY TESTS
        WING OSCILLATIONS
        YAW

AERODYNAMIC STALLING
  RT   AIRCRAFT PERFORMANCE
        AIRCRAFT SPIN
        AIRSPEED
        ANGLE OF ATTACK
        BOUNDARY LAYER SEPARATION
        LIFT DRAG RATIO
        LOW SPEED STABILITY
        ∞STALLING
        SWEEP ANGLE
        ZERO LIFT

AERODYNAMIC VEHICLES
  USE   AIRCRAFT

AERODYNAMICS
  UF   HYDROAEROMECHANICS
  GS   FLUID MECHANICS
        . FLUID DYNAMICS
        . . GAS DYNAMICS
        . . . **AERODYNAMICS**
        . . . . AEROTHERMODYNAMICS
        . . . . . HYPERSONICS
        . . . . ROTOR AERODYNAMICS
        . . . . SUPERSONICS
  RT   AEROACOUSTICS

AERODYNAMICS-(CONT.)
    AERODYNAMIC HEATING
    AERODYNAMIC INTERFERENCE
    AEROELASTICITY
    AERONAUTICAL ENGINEERING
  ∞ AERONAUTICS
  ∞ AEROSPACE SCIENCES
  ∞ AIRCRAFT
    AIRFOILS
    BLUNT BODIES
    BODIES OF REVOLUTION
    BOUNDARY LAYER CONTROL
    COMPRESSIBLE FLOW
    CONTROL SURFACES
    DRAG
  ∞ DYNAMICS
  ∞ FLIGHT
    FLIGHT CHARACTERISTICS
    FLIGHT MECHANICS
  ∞ FLOW
    FLOW THEORY
    FREE WING AIRCRAFT
    GROUND EFFECT (AERODYNAMICS)
    HYPERSONIC FLIGHT
    HYPERSONIC FLOW
    INCOMPRESSIBLE FLOW
    INVISCID FLOW
    LAMINAR FLOW
    LIFT
    MACH NUMBER
    MACH-ZEHNDER INTERFEROMETERS
    REENTRY
  ∞ SCIENCE
    SLENDER BODIES
    SUBSONIC FLOW
    SUPERSONIC FLOW
    THERMODYNAMICS
    TRANSONIC FLOW
    TURBULENT FLOW
    UNIFORM FLOW
    UNSTEADY FLOW
    VISCOUS FLOW
    WIND MEASUREMENT
    WIND TUNNELS

AEROELASTIC RESEARCH WINGS
GS   AIRFOILS
    . WINGS
    . . **AEROELASTIC RESEARCH WINGS**
RT   AIRCRAFT DESIGN
    STRUCTURAL DESIGN

AEROELASTICITY
GS   MECHANICAL PROPERTIES
    . ELASTIC PROPERTIES
    . . **AEROELASTICITY**
    . . . AEROTHERMOELASTICITY
RT   AERODYNAMIC NOISE
    AERODYNAMIC STABILITY
    AERODYNAMICS
    AEROTHERMODYNAMICS
    AIRCRAFT STRUCTURES
    DAST PROGRAM
    FLUTTER
    INFLUENCE COEFFICIENT
    PANEL FLUTTER
    RIGID WINGS
    THERMOELASTICITY
    WING LOADING

AEROEMBOLISM
GS   EMBOLISMS
    . **AEROEMBOLISM**
RT   DECOMPRESSION SICKNESS
    FAT EMBOLISMS
    STRESS (PHYSIOLOGY)

AEROGYRO HELICOPTERS
USE   XH-51 HELICOPTER

AEROLOGY
GS   METEOROLOGY
    . **AEROLOGY**
RT   ATMOSPHERIC & OCEANOGRAPHIC
      INFORM SYS
    GLOBAL ATMOSPHERIC RESEARCH
      PROGRAM
    METEOROLOGICAL PARAMETERS
    POLAR METEOROLOGY
    SEA BREEZE
    WIND (METEOROLOGY)

AEROMAGNETISM
RT   AERIAL RECONNAISSANCE
    GEOMAGNETISM

AEROMAGNETISM-(CONT.)
    MAGNETIC ANOMALIES
    MAGNETIC SURVEYS
    MAGNETIC VARIATIONS
    REMOTE SENSING

AEROMAGNETO FLUTTER
USE   FLUTTER

AEROMANEUVERING
RT   AEROASSIST
    AEROBRAKING
    AEROCAPTURE
    ATMOSPHERIC ENTRY
    INTERPLANETARY TRANSFER ORBITS
    TRANSFER ORBITS

AEROMANEUVERING ORBIT TO ORBIT SHUTTLE
UF   AMOOS
GS   TRANSPORTATION
    . SPACE TRANSPORTATION
    . . SPACE TRANSPORTATION SYSTEM
    . . . SPACE SHUTTLE BOOSTERS
    . . . . **AEROMANEUVERING ORBIT TO**
           ORBIT SHUTTLE
RT   ORBIT MANEUVERING ENGINE (SPACE
      SHUTTLE)
    ORBITAL MECHANICS
    REUSABLE LAUNCH VEHICLES
    SPACE SHUTTLES

AERONAUTICAL ENGINEERING
GS   AEROSPACE ENGINEERING
    . **AERONAUTICAL ENGINEERING**
RT   AERODYNAMICS
  ∞ AERONAUTICS
  ∞ AIRCRAFT
    AIRCRAFT DESIGN
    AIRCRAFT INDUSTRY
    AUXILIARY PROPULSION
    COMPOUND HELICOPTERS
  ∞ ENGINEERING
    FUNCTIONAL DESIGN SPECIFICATIONS
    MECHANICAL ENGINEERING
    PROPULSION
    STRUCTURAL ENGINEERING

AERONAUTICAL SATELLITES
GS   SATELLITES
    . ARTIFICIAL SATELLITES
    . . COMMUNICATION SATELLITES
    . . . **AERONAUTICAL SATELLITES**
    . . . . AEROSAT SATELLITES
RT ∞ AERONAUTICS
    AIR TRAFFIC CONTROL
    AIRCRAFT APPROACH SPACING
    AIRCRAFT COMMUNICATION
    GROUND-AIR-GROUND COMMUNICATION
    RESCUE OPERATIONS
    SATELLITE NETWORKS

∞ **AERONAUTICS**
SN   *(USE OF A MORE SPECIFIC TERM IS*
    RECOMMENDED--CONSULT THE TERMS
    LISTED BELOW)
UF   AVIATION
RT   AEROACOUSTICS
    AERODYNAMICS
    AERONAUTICAL ENGINEERING
    AERONAUTICAL SATELLITES
    AEROSPACE ENGINEERING
  ∞ AEROSPACE SCIENCES
    AIR LAW
  ∞ AIRCRAFT
    AIRPORTS
    AVIONICS
    CIVIL AVIATION
  ∞ FLIGHT
    GENERAL AVIATION AIRCRAFT
    HUMAN FACTORS ENGINEERING
  ∞ MILITARY AVIATION
  ∞ SCIENCE
    TACT PROGRAM

AERONOMY
RT   AIRGLOW
    ALPINE METEOROLOGY
    ATMOSPHERIC COMPOSITION
    ATMOSPHERIC PHYSICS
    AURORAS
    DIAL SATELLITE
    GEOPHYSICS
    MESOMETEOROLOGY
    METEOROLOGY
    POLAR CUSPS

AERONOMY-(CONT.)
    UPPER ATMOSPHERE

AEROPHYSICS
USE   ATMOSPHERIC PHYSICS

AEROQUATIC VEHICLES
RT   AIRCRAFT DESIGN
    ATTACK AIRCRAFT
  ∞ MILITARY VEHICLES
    UNDERWATER PROPULSION
    UNDERWATER VEHICLES

AEROS SATELLITE
GS   SATELLITES
    . ARTIFICIAL SATELLITES
    . . METEOROLOGICAL SATELLITES
    . . . **AEROS SATELLITE**
    . . SYNCHRONOUS SATELLITES
    . . . **AEROS SATELLITE**

AEROSAT SATELLITES
GS   SATELLITES
    . ARTIFICIAL SATELLITES
    . . COMMUNICATION SATELLITES
    . . . AERONAUTICAL SATELLITES
    . . . . **AEROSAT SATELLITES**
    . . ESA SATELLITES
    . . . **AEROSAT SATELLITES**
    . . NAVIGATION SATELLITES
    . . . **AEROSAT SATELLITES**
    . . SYNCHRONOUS SATELLITES
    . . . **AEROSAT SATELLITES**
RT   EUROPEAN SPACE PROGRAMS
    SATELLITE NETWORKS

AEROSINUSITIS
GS   DISEASES
    . RESPIRATORY DISEASES
    . . **AEROSINUSITIS**
RT   AEROSPACE MEDICINE
    ALTITUDE SICKNESS

AEROSOLS
GS   MIXTURES
    . DISPERSIONS
    . . COLLOIDS
    . . . **AEROSOLS**
    . . . . FOG
    . . LIQUID-GAS MIXTURES
    . . . **AEROSOLS**
    . . . . FOG
    PARTICLES
    . **AEROSOLS**
    . . FOG
RT   AIR POLLUTION
    AITKEN NUCLEI
    ATMOSPHERIC EFFECTS
    ATOMIZING
    CONDENSATION NUCLEI
    CROP DUSTING
    DUST
    ENTRAINMENT
    ENVIRONMENT POLLUTION
    ENVIRONMENTAL SURVEYS
    FOG DISPERSAL
    FUMES
    GAS ATOMIZATION
    MIST
    MIXERS
    PHOTOPHORESIS
    POLLUTION TRANSPORT
    SAGE SATELLITE
    SMOKE
    SMOKE ABATEMENT
    SPRAYING

AEROSPACE ENGINEERING
UF   SPACE SYSTEMS ENGINEERING
GS   **AEROSPACE ENGINEERING**
    . AERONAUTICAL ENGINEERING
RT ∞ AERONAUTICS
  ∞ AEROSPACE SCIENCES
  ∞ AIRCRAFT
  ∞ ENGINEERING
    MECHANICAL ENGINEERING
    MISSILE DESIGN
    STRUCTURAL ENGINEERING

AEROSPACE ENVIRONMENTS
UF   SPACE ENVIRONMENT
GS   ENVIRONMENTS
    . **AEROSPACE ENVIRONMENTS**
    . . CISLUNAR SPACE
    . . DEEP SPACE

AEROSPACE ENVIRONMENTS-(CONT.)

```
        . . . INTERPLANETARY SPACE
        . . . INTERSTELLAR SPACE
RT    ∞ AEROSPACE SCIENCES
        ARGON-OXYGEN ATMOSPHERES
      ∞ ASTRONAUTICS
        BIOASTRONAUTICS
        BIOPROCESSING
        BIOSATELLITES
        COSMIC RAYS
        EARTH ATMOSPHERE
        ELECTROMAGNETIC RADIATION
        EXOBIOLOGY
        EXTRATERRESTRIAL ENVIRONMENTS
        EXTRATERRESTRIAL LIFE
        EXTRATERRESTRIAL RADIATION
        EXTRAVEHICULAR ACTIVITY
        GEOPHYSICAL FLUID FLOW CELLS
        HAZARDOUS MATERIAL DISPOSAL (IN
          SPACE)
        HELIUM-OXYGEN ATMOSPHERES
        JUPITER ATMOSPHERE
        LIFE SUPPORT SYSTEMS
        LUNAR ENVIRONMENT
        MANNED SPACE FLIGHT
        MARS ATMOSPHERE
        NEPTUNE ATMOSPHERE
        PANSPERMIA
        PLANETARY ENVIRONMENTS
        RADIATION BELTS
        SOLAR RADIATION
        SPACE EXPLORATION
        SPACE FLIGHT
        SPACE HABITATS
        SPACE MANUFACTURING
        SPACEBORNE EXPERIMENTS
        SPACECRAFT CABIN SIMULATORS
        THERMAL ENVIRONMENTS
        URANUS ATMOSPHERE
        VACUUM
        VENUS ATMOSPHERE
```

AEROSPACE INDUSTRY

```
GS    INDUSTRIES
        . AEROSPACE INDUSTRY
        . . AIRCRAFT INDUSTRY
RT    ∞ AIRCRAFT
        COMMERCIAL SPACECRAFT
        SPACE COMMERCIALIZATION
```

AEROSPACE MEDICINE

```
GS      AEROSPACE MEDICINE
        . AVIATION PSYCHOLOGY
        . SPACE PSYCHOLOGY
RT      ACCELERATION STRESSES
          (PHYSIOLOGY)
        AEROSINUSITIS
      ∞ AEROSPACE SCIENCES
        ALTITUDE SICKNESS
        BIOASTRONAUTICS
        BIOFEEDBACK
      ∞ BIOLOGY
        BIOMEDICAL DATA
        CLOSED ECOLOGICAL SYSTEMS
        FASTING
        GRAVITATIONAL PHYSIOLOGY
        MEDICAL SCIENCE
      ∞ MEDICINE
        MOBILE QUARANTINE FACILITY
        MOTION SICKNESS
        RADIOLOGY
      ∞ SCIENCE
        SPACE ADAPTATION SYNDROME
        SPACECRAFT ENVIRONMENTS
        SPORTS MEDICINE
        WEIGHTLESSNESS
```

AEROSPACE SAFETY

```
GS      SAFETY
        . AEROSPACE SAFETY
RT      ACCIDENT PREVENTION
        AIRCRAFT SAFETY
        FLIGHT SAFETY
        RANGE SAFETY
        SAFETY FACTORS
        SAFETY MANAGEMENT
```

∞ AEROSPACE SCIENCES

```
SN      (USE OF A MORE SPECIFIC TERM IS
          RECOMMENDED--CONSULT THE TERMS
          LISTED BELOW)
UF      SPACE SCIENCES
RT      AERODYNAMICS
      ∞ AERONAUTICS
        AEROSPACE ENGINEERING
```

AEROSPACE SCIENCES-(CONT.)

```
        AEROSPACE ENVIRONMENTS
        AEROSPACE MEDICINE
        AEROSPACE SYSTEMS
        ASTRONOMY
        COMMITTEE ON SPACE RESEARCH
        ENVIRONMENTAL ENGINEERING
        EXTRATERRESTRIAL RADIATION
        SPACE LABORATORIES
```

AEROSPACE SYSTEMS

```
RT    ∞ AEROSPACE SCIENCES
        CONTROL SYSTEMS DESIGN
        MISSILE SYSTEMS
      ∞ SYSTEMS
        SYSTEMS ENGINEERING
```

AEROSPACE TECHNOLOGY TRANSFER

```
GS      TECHNOLOGY TRANSFER
        . AEROSPACE TECHNOLOGY TRANSFER
RT      CANADIAN SPACE PROGRAM
        INFORMATION FLOW
        REPORTS
        TECHNOLOGICAL FORECASTING
        TECHNOLOGY UTILIZATION
```

AEROSPACE VEHICLES

```
GS      AEROSPACE VEHICLES
        . FLEXIBLE SPACECRAFT
RT    ∞ AIRCRAFT
        COMMERCIAL SPACECRAFT
      ∞ SPACECRAFT
        TRANSATMOSPHERIC VEHICLES
```

AEROSPACEPLANES

```
GS      MANEUVERABLE SPACECRAFT
        . AEROSPACEPLANES
        . . ASTROPLANE
        MANNED SPACECRAFT
        . AEROSPACEPLANES
        . . ASTROPLANE
        REENTRY VEHICLES
        . AEROSPACEPLANES
        . . ASTROPLANE
        SOFT LANDING SPACECRAFT
        . AEROSPACEPLANES
        . . ASTROPLANE
RT    ᴍ AIRCRAFT
        ASTRO VEHICLE
        BOOSTGLIDE VEHICLES
        GLIDERS
        HYPERSONIC AIRCRAFT
        HYPERSONIC GLIDERS
        LAUNCH VEHICLES
        LIFTING REENTRY VEHICLES
        LIQUID AIR CYCLE ENGINES
        MILITARY SPACECRAFT
        RESEARCH AIRCRAFT
        ROCKET PLANES
        TRANSATMOSPHERIC VEHICLES
        X-20 AIRCRAFT
```

AEROSTATICS

```
GS      STATICS
        . AEROSTATICS
RT      BUOYANCY
      ∞ DYNAMICS
      ∞ EQUILIBRIUM
        FLUID MECHANICS
        HYDROSTATICS
```

AEROSTATS

```
USE     AIRSHIPS
```

AEROTHERMOCHEMISTRY

```
GS      ENVIRONMENTAL CHEMISTRY
        . AEROTHERMOCHEMISTRY
        THERMOCHEMISTRY
        . AEROTHERMOCHEMISTRY
RT      ABLATION
        AERODYNAMIC HEATING
        AEROTHERMODYNAMICS
        ATMOSPHERIC CHEMISTRY
        CHEMICAL ENGINEERING
      ∞ CHEMISTRY
        NOZZLE FLOW
        PHYSICAL CHEMISTRY
        PYROMETALLURGY
        REENTRY PHYSICS
        REENTRY SHIELDING
        REENTRY VEHICLES
```

AEROTHERMODYNAMICS

```
GS      FLUID MECHANICS
        . FLUID DYNAMICS
```

AEROTHERMODYNAMICS-(CONT.)

```
        . . GAS DYNAMICS
        . . . AERODYNAMICS
        . . . . AEROTHERMODYNAMICS
        THERMODYNAMICS
        . AEROTHERMODYNAMICS
RT      AERODYNAMIC HEAT TRANSFER
        AERODYNAMIC HEATING
        AEROELASTICITY
        AEROTHERMOCHEMISTRY
        ASSET PROJECT
        BOUNDARY LAYER PLASMAS
      ∞ CHEMISTRY
        COMBUSTION PHYSICS
      ∞ DYNAMICS
        HYPERSONIC HEAT TRANSFER
        HYPERSONIC REENTRY
        HYPERSONICS
        RANKINE-HUGONIOT RELATION
        REENTRY
        REENTRY PHYSICS
      ∞ SCIENCE
        SKIN TEMPERATURE (NON-BIOLOGICAL)
        SUPERSONICS
        THERMOELASTICITY
```

AEROTHERMOELASTICITY

```
GS      MECHANICAL PROPERTIES
        . ELASTIC PROPERTIES
        . . AEROELASTICITY
        . . . AEROTHERMOELASTICITY
        . . THERMOELASTICITY
        . . . AEROTHERMOELASTICITY
```

AEROZINE

```
SN      (HYDRAZINE-UDMH MIXTURE)
GS      LIQUIDS
        . LIQUID FUELS
        . . LIQUID ROCKET PROPELLANTS
        . . . MONOPROPELLANTS
        . . . . AEROZINE
        PROPELLANTS
        . ROCKET PROPELLANTS
        . . LIQUID ROCKET PROPELLANTS
        . . . MONOPROPELLANTS
        . . . . AEROZINE
RT      DIMETHYLHYDRAZINES
        HYDRAZINES
```

AFC (CONTROL)

```
USE     AUTOMATIC FREQUENCY CONTROL
```

AFCS (CONTROL SYSTEM)

```
USE     AUTOMATIC FLIGHT CONTROL
```

AFFECTS

```
USE     EFFECTS
```

AFFERENT NERVOUS SYSTEMS

```
GS      NERVOUS SYSTEM
        . AFFERENT NERVOUS SYSTEMS
RT      SENSORIMOTOR PERFORMANCE
      ∞ SYSTEMS
```

AFFINITY

```
RT      ATTRACTION
        COMPATIBILITY
```

AFGHANISTAN

```
GS      NATIONS
        . AFGHANISTAN
RT      ASIA
```

AFRICA

```
GS      CONTINENTS
        . AFRICA
RT      AFRICAN RIFT SYSTEM
        ALGERIA
        ANGOLA
        ARCOMSAT
        BENIN
        BOTSWANA
        BURKINA
        BURUNDI
        CAMEROON
        CAPE VERDE
        CENTRAL AFRICAN REPUBLIC
        CHAD
        CONGO (BRAZZAVILLE)
        EGYPT
        ETHIOPIA
        GABON
        GAMBIA
        GHANA
```

AFRICA-*(CONT.)*
    GUINEA
    IVORY COAST
    KALIHARI BASIN (AFRICA)
    KENYA
    LESOTHO
    LIBERIA
    LIBYA
    LIBYAN DESERT
    MALAGASY REPUBLIC
    MALAWI
    MALI
    MAURITANIA
    MOROCCO
    MOZAMBIQUE
    NAMIBIA
    NATIONS
    NIGER
    NIGERIA
    RED SEA
    REPUBLIC OF SOUTH AFRICA
    RWANDA
    SAHARA DESERT (AFRICA)
    SENEGAL
    SIERRA LEONE
    SOMALIA
    SPANISH SAHARA
    SUDAN
    SWAZILAND
    TANZANIA
    TOGO
    TUNISIA
    UGANDA
    ZAIRE
    ZAMBIA
    ZIMBABWE

AFRICAN RIFT SYSTEM
    GS   GEOLOGICAL FAULTS
        . **AFRICAN RIFT SYSTEM**
    RT   AFRICA
        ∞SYSTEMS

AFTERBODIES
    UF   CYLINDRICAL AFTERBODIES
        STERNS
    GS   AIRCRAFT STRUCTURES
        AFTERBODIES
    RT   BASE HEATING
        BOATTAILS
        ∞BODIES
        CENTERBODIES
        CONICAL BODIES
        CYLINDRICAL BODIES
        FLARED BODIES
        FOREBODIES
        SKIRTS
        SWING TAIL ASSEMBLIES
        TAIL ASSEMBLIES

AFTERBURNERS
    USE   AFTERBURNING

AFTERBURNING
    UF   AFTERBURNERS
    GS   COMBUSTION
        . **AFTERBURNING**
    RT   BURNERS
        EXHAUST SYSTEMS
        INFRARED SUPPRESSION
        INTERNAL COMBUSTION ENGINES
        J-57 ENGINE
        J-57-P-20 ENGINE
        JET ENGINES
        THRUST AUGMENTATION

AFTERGLOWS
    GS   **AFTERGLOWS**
        . HELIUM AFTERGLOW
        . OXYGEN AFTERGLOW
    RT   ATMOSPHERIC IONIZATION
        GAS DISCHARGES
        GAS IONIZATION
        LIGHT SCATTERING
        LUMINESCENCE
        PHOSPHORESCENCE
        PLASMA DECAY

AFTERIMAGES
    GS   IMAGES
        . **AFTERIMAGES**
    RT   CRITICAL FLICKER FUSION
        ILLUSIONS
        PSYCHOLOGICAL EFFECTS
        SENSORY PERCEPTION

AFTERIMAGES-*(CONT.)*
        VISUAL PERCEPTION

AGC (CONTROL)
    USE   AUTOMATIC GAIN CONTROL

AGE DETERMINATION
    USE   CHRONOLOGY

AGE FACTOR
    RT   AGING (BIOLOGY)
        GERONTOLOGY
        LIFE SPAN

AGE HARDENING
    USE   PRECIPITATION HARDENING

AGENA A ROCKET VEHICLE
    GS   ROCKET VEHICLES
        . SINGLE STAGE ROCKET VEHICLES
        . . AGENA ROCKET VEHICLES
        . . . **AGENA A ROCKET VEHICLE**
    RT   DISCOVERER SATELLITES
        THOR AGENA LAUNCH VEHICLE

AGENA B RANGER PROGRAM
    GS   PROGRAMS
        . NASA PROGRAMS
        . . RANGER PROJECT
        . . . **AGENA B RANGER PROGRAM**
        . NASA SPACE PROGRAMS
        . . RANGER PROJECT
        . . . **AGENA B RANGER PROGRAM**
        . PROJECTS
        . . RANGER PROJECT
        . . . **AGENA B RANGER PROGRAM**
    RT   THOR AGENA LAUNCH VEHICLE

AGENA B ROCKET VEHICLE
    GS   ROCKET VEHICLES
        . SINGLE STAGE ROCKET VEHICLES
        . . AGENA ROCKET VEHICLES
        . . . **AGENA B ROCKET VEHICLE**
    RT   DISCOVERER SATELLITES
        ECHO SATELLITES
        EGO
        GEMINI PROJECT
        MARINER PROGRAM
        OAO
        POGO
        RANGER PROJECT

AGENA C ROCKET VEHICLE
    GS   ROCKET VEHICLES
        . SINGLE STAGE ROCKET VEHICLES
        . . AGENA ROCKET VEHICLES
        . . . **AGENA C ROCKET VEHICLE**

AGENA D ROCKET VEHICLE
    GS   ROCKET VEHICLES
        . SINGLE STAGE ROCKET VEHICLES
        . . AGENA ROCKET VEHICLES
        . . . **AGENA D ROCKET VEHICLE**

AGENA ROCKET VEHICLES
    GS   ROCKET VEHICLES
        . SINGLE STAGE ROCKET VEHICLES
        . . **AGENA ROCKET VEHICLES**
        . . . AGENA A ROCKET VEHICLE
        . . . AGENA B ROCKET VEHICLE
        . . . AGENA C ROCKET VEHICLE
        . . . AGENA D ROCKET VEHICLE
    RT   ATLAS AGENA B LAUNCH VEHICLE
        ATLAS AGENA LAUNCH VEHICLES
        DISCOVERER SATELLITES
        ECHO SATELLITES
        GEMINI PROJECT
        MARINER PROGRAM
        RANGER PROJECT
        THOR AGENA LAUNCH VEHICLE

∞ **AGENTS**
    SN   *(USE OF A MORE SPECIFIC TERM IS RECOMMENDED--CONSULT THE TERMS LISTED BELOW)*
    RT   ACCELERATING AGENTS
        ADDITIVES
        ANTICOAGULANTS
        ANTIFOULING
        ANTIOXIDANTS
        DILUENTS
        NEUTRALIZERS
        OPACIFIERS
        OXIDIZERS

AGENTS-*(CONT.)*
        PENETRANTS
        PRESERVATIVES
        STABILIZERS (AGENTS)
        SURFACTANTS

AGGLOMERATION
    RT   ACCUMULATIONS
        CEMENTATION
        CLUMPS
        COAGULATION
        COALESCING
        COMPACTING
        CONCENTRATING
        CRYSTALLIZATION
        DENSIFICATION
        FLOCCULATING
        GALACTIC CLUSTERS
        LUMPING
        PLUGGING
        PRECIPITATION (CHEMISTRY)
        ∞SEPARATION
        SETTLING
        SINTERING
        VIRGO GALACTIC CLUSTER

AGGLUTINATION
    GS   BONDING
        . **AGGLUTINATION**
    RT   ADHESION
        ADHESIVE BONDING
        ADHESIVES
        CEMENTATION
        CHEMICAL BONDS
        COHESION

AGGREGATES
    RT   CONCRETE STRUCTURES
        CONCRETES
        ∞CONSTRUCTION MATERIALS
        DOLOMITE (MINERAL)
        GRAVELS
        LAVA
        LIMESTONE
        ROCKS
        SANDS
        SLAGS

∞ **AGING**
    SN   *(USE OF A MORE SPECIFIC TERM IS RECOMMENDED--CONSULT THE TERMS LISTED BELOW)*
    RT   AGING (BIOLOGY)
        AGING (MATERIALS)
        AGING (METALLURGY)
        RADIOACTIVE AGE DETERMINATION

AGING (BIOLOGY)
    RT   AGE FACTOR
        ∞AGING
        ∞BIOLOGY
        GERIATRICS
        GERONTOLOGY
        LIFE SCIENCES
        LIFE SPAN
        MORTALITY

AGING (MATERIALS)
    GS   **AGING (MATERIALS)**
        . AGING (METALLURGY)
    RT   ∞AGING
        HARDENING (MATERIALS)
        ∞MATERIALS
        MECHANICAL PROPERTIES
        MICROSTRUCTURE
        STRAIN HARDENING

AGING (METALLURGY)
    GS   AGING (MATERIALS)
        . **AGING (METALLURGY)**
    RT   ∞AGING
        HARDENING (MATERIALS)
        HEAT TREATMENT
        MICROSTRUCTURE
        SOLID SOLUTIONS
        STRAIN HARDENING
        SUPERCOOLING
        TIME TEMPERATURE PARAMETER

AGITATION
    GS   **AGITATION**
        . ULTRASONIC AGITATION
    RT   AERATION
        BLOWING
        CHEMICAL REACTION CONTROL

AGITATION-(CONT.)
    COALESCING
    COLLOIDING
    DISPERSING
    DISPOSAL
    HOMOGENIZING
    MIXERS
    MIXING
    ∞SEPARATION
    SETTLING
    SHAKING
    SIZING SCREENS
    SPLASHING
    SUSPENDING (MIXING)
    SWIRLING
    TURBULENT MIXING
    VORTICES
    WATER TREATMENT

AGREEMENTS
RT   CONTRACTS
      CONVENTIONS
      SUBCONTRACTS

AGRICULTURAL AIRCRAFT
GS   GENERAL AVIATION AIRCRAFT
      . AGRICULTURAL AIRCRAFT
RT   AGRICULTURE
      ∞AIRCRAFT
      CROP DUSTING
      LIGHT AIRCRAFT
      SWATH WIDTH

AGRICULTURE
RT   AGRICULTURAL AIRCRAFT
      AGRISTARS PROJECT
      AGROCLIMATOLOGY
      AGROMETEOROLOGY
      AGROPHYSICAL UNITS
      ALFALFA
      BARLEY
      ∞BIOLOGY
      BOTANY
      CITRUS TREES
      CONSERVATION
      CORN
      CROP DUSTING
      CROP GROWTH
      CROP IDENTIFICATION
      CROP INVENTORIES
      CROP VIGOR
      ∞CROPS
      FARM CROPS
      FARMLANDS
      FRUITS
      GRASSLANDS
      GREAT PLAINS CORRIDOR (NORTH
        AMERICA)
      HALOPHILES
      HAY
      HYDROCARBON FUEL PRODUCTION
      HYDROPONICS
      IRRIGATION
      LARGE AREA CROP INVENTORY
        EXPERIMENT
      LEGUMINOUS PLANTS
      OATS
      ORCHARDS
      PLANT STRESS
      PLANTING
      PLANTS (BOTANY)
      PLOWING
      PLOWS
      RURAL AREAS
      RURAL LAND USE
      SILVICULTURE
      SOIL SCIENCE
      SORGHUM
      SUGAR BEETS
      SUGAR CANE
      SUNFLOWERS
      TRACTORS
      VEGETATION GROWTH
      VINEYARDS

AGRISTARS PROJECT
UF   CROP INVENTORIES BY REMOTE
        SENSING
GS   PROGRAMS
      . PROJECTS
      .. AGRISTARS PROJECT
RT   AGRICULTURE
      AGROPHYSICAL UNITS
      CROP INVENTORIES
      FARM CROPS

AGRISTARS PROJECT-(CONT.)
    FRESH WATER
    LAND USE
    LANDSAT SATELLITES
    METEOROLOGICAL SATELLITES
    NASA PROGRAMS
    REMOTE SENSORS
    VEGETATIVE INDEX

AGROCLIMATOLOGY
GS   CLIMATOLOGY
      . AGROCLIMATOLOGY
RT   AGRICULTURE
      AGROMETEOROLOGY
      HYDROCLIMATOLOGY
      METEOROLOGICAL PARAMETERS
      METEOROLOGY
      MICROCLIMATOLOGY

AGROMETEOROLOGY
GS   METEOROLOGY
      . AGROMETEOROLOGY
RT   AGRICULTURE
      AGROCLIMATOLOGY
      HYDROMETEOROLOGY
      MICROMETEOROLOGY
      THERMAL RESOURCES
      TROPICAL METEOROLOGY

AGROPHYSICAL UNITS
RT   AGRICULTURE
      AGRISTARS PROJECT
      FARMLANDS
      LARGE AREA CROP INVENTORY
        EXPERIMENT

AGT
USE   AUTOMATED GUIDEWAY TRANSIT
         VEHICLES

AH-1G HELICOPTER
GS   V/STOL AIRCRAFT
      . ROTARY WING AIRCRAFT
      .. HELICOPTERS
      ... MILITARY HELICOPTERS
      .... AH-1G HELICOPTER
RT   ∞MILITARY AIRCRAFT
      TERRAIN FOLLOWING AIRCRAFT

AH-63 HELICOPTER
GS   ATTACK AIRCRAFT
      . AH-63 HELICOPTER
      BELL AIRCRAFT
      . AH-63 HELICOPTER
RT   ∞MILITARY AIRCRAFT
      TERRAIN FOLLOWING AIRCRAFT

AH-64 HELICOPTER
GS   ATTACK AIRCRAFT
      . AH-64 HELICOPTER
      HUGHES AIRCRAFT
      . AH-64 HELICOPTER
      V/STOL AIRCRAFT
      . ROTARY WING AIRCRAFT
      .. HELICOPTERS
      ... MILITARY HELICOPTERS
      .... AH-64 HELICOPTER
RT   ∞MILITARY AIRCRAFT
      TERRAIN FOLLOWING AIRCRAFT

∞ AIDS
SN   (USE OF A MORE SPECIFIC TERM IS
      RECOMMENDED--CONSULT THE TERMS
      LISTED BELOW)
RT   LANDING AIDS
      NAVIGATION AIDS
      VISUAL AIDS

AILERONS
GS   AIRFOILS
      . AILERONS
      .. FLAPERONS
      .. SPOILER SLOT AILERONS
      CONTROL SURFACES
      . AILERONS
      .. FLAPERONS
      .. SPOILER SLOT AILERONS
RT   ELEVATORS (CONTROL SURFACES)
      ELEVONS
      LATERAL CONTROL
      TABS (CONTROL SURFACES)

AIMP-D
USE   EXPLORER 33 SATELLITE

AIMP-E
USE   EXPLORER 35 SATELLITE

AIMP-1
USE   EXPLORER 33 SATELLITE

AIMP-2
USE   EXPLORER 35 SATELLITE

AIR
GS   GASES
      . AIR
      .. ALVEOLAR AIR
      .. COMPRESSED AIR
      .. EXPIRED AIR
      .. HIGH TEMPERATURE AIR
      .. LIQUID AIR
RT   ∞ATMOSPHERES
      ATMOSPHERIC COMPOSITION
      EARTH ATMOSPHERE
      ENVIRONMENTS
      MIDDLE ATMOSPHERE

AIR BAG RESTRAINT DEVICES
GS   BAGS
      . AIR BAG RESTRAINT DEVICES
      SAFETY DEVICES
      . AIR BAG RESTRAINT DEVICES
RT   ACCIDENT PREVENTION
      ACCIDENTS
      AUTOMOBILES
      COLLISIONS
      CRASHES
      ∞DEVICES
      HIGHWAYS
      PNEUMATIC EQUIPMENT
      SAFETY

AIR BEARINGS
USE   GAS BEARINGS

AIR BLASTS
USE   AERIAL EXPLOSIONS

AIR BREATHING BOOSTERS
RT   AIR BREATHING ENGINES
      BOOSTER ROCKET ENGINES
      ∞BOOSTERS

AIR BREATHING ENGINES
GS   ENGINES
      . AIR BREATHING ENGINES
      .. GAS TURBINE ENGINES
      ... JET ENGINES
      .... RAMJET ENGINES
      ..... INTEGRAL ROCKET RAMJETS
      ..... LOW VOLUME RAMJET ENGINES
      ..... PULSEJET ENGINES
      ..... SUPERSONIC COMBUSTION
              RAMJET ENGINES
      ..... TURBORAMJET ENGINES
      .... TURBOJET ENGINES
      ..... BRISTOL-SIDDELEY OLYMPUS
              593 ENGINE
      ..... BRISTOL-SIDDELEY VIPER
              ENGINE
      ..... DUCTED FAN ENGINES
      ..... J-33 ENGINE
      ..... J-34 ENGINE
      ..... J-47 ENGINE
      ..... J-57 ENGINE
      ..... J-57-P-20 ENGINE
      ..... J-65 ENGINE
      ..... J-69-T-25 ENGINE
      ..... J-71 ENGINE
      ..... J-73 ENGINE
      ..... J-75 ENGINE
      ..... J-79 ENGINE
      ..... J-85 ENGINE
      ..... J-93 ENGINE
      ..... RA-28 ENGINE
      ..... TURBOFAN ENGINES
      ...... BRISTOL-SIDDELEY BS 53
               ENGINE
      ...... CF-700 ENGINE
      ...... J-97 ENGINE
      ...... TF-41 ENGINE
      .... TURBOPROP ENGINES
      ..... T-53 ENGINE
      ..... T-56 ENGINE
      ..... T-64 ENGINE
      ..... T-74 ENGINE
      .... TURBORAMJET ENGINES
      ..... T-58-GE-8B ENGINE
RT   AIR BREATHING BOOSTERS

AIR CARGO
UF AIR FREIGHT
GS CARGO
. **AIR CARGO**
.. AIR MAIL
RT AIRDROPS
AIRFIELD SURFACE MOVEMENTS
AIRLINE OPERATIONS
BAGGAGE
CARGO AIRCRAFT
GROUND HANDLING
HEAVY LIFT HELICOPTERS

AIR CONDITIONING
RT BLOWERS
COMFORT
CONDENSERS (LIQUEFIERS)
CONTROLLED ATMOSPHERES
COOLANTS
COOLERS
COOLING
COOLING SYSTEMS
∞DIFFUSERS
EXHAUST SYSTEMS
FREON
HEAT PUMPS
HEATING
HEATING EQUIPMENT
HUMIDITY
INFILTRATION
LIFE SUPPORT SYSTEMS
MODULAR INTEGRATED UTILITY SYSTEM
REFRIGERANTS
REFRIGERATING
REFRIGERATING MACHINERY
SPACE HEATING (BUILDINGS)
TEMPERATURE
TEMPERATURE CONTROL
TEMPERATURE DISTRIBUTION
THERMAL INSULATION
∞TREATMENT
VENTILATION

AIR CONDITIONING EQUIPMENT
RT ABSORBENTS
ABSORBERS (EQUIPMENT)
ADSORBENTS
BLOWERS
COMPRESSORS
CONDENSERS (LIQUEFIERS)
COOLERS
COOLING SYSTEMS
∞DIFFUSERS
∞EQUIPMENT
EVAPORATORS
∞FANS
HEAT PUMPS
HEATING EQUIPMENT
OXYGEN SUPPLY EQUIPMENT
REFRIGERATING MACHINERY

AIR CONDUCTIVITY
RT ATMOSPHERIC CONDUCTIVITY
ELECTRICAL RESISTIVITY
THERMAL CONDUCTIVITY

AIR COOLING
SN (COOLING WITH AIR)
GS COOLING
. **AIR COOLING**
RT COOLANTS
COOLERS
COOLING SYSTEMS
LIQUID COOLING
REFRIGERATING
VENTILATION

AIR CURRENTS
GS FLUID FLOW
. GAS FLOW
.. AIR FLOW
... **AIR CURRENTS**
.... JET STREAMS (METEOROLOGY)
.... MERIDIONAL FLOW
.... VERTICAL AIR CURRENTS
RT ATMOSPHERIC CIRCULATION
BAROTROPIC FLOW
BOUNDARY LAYER TRANSITION
BRUNT-VAISALA FREQUENCY
CONVECTION CLOUDS
CONVECTION CURRENTS
∞CURRENTS
GROUND WIND
LEE WAVES
SEA BREEZE

AIR CURRENTS-(CONT.)
UPSTREAM
WIND (METEOROLOGY)
WINDPOWER UTILIZATION

AIR CUSHION LANDING SYSTEMS
RT AIRCRAFT LANDING
CUSHIONS
GROUND EFFECT (AERODYNAMICS)
SKID LANDINGS
∞SYSTEMS

AIR CUSHION VEHICLES
USE GROUND EFFECT MACHINES

† **AIR DEFENSE**
GS **AIR DEFENSE**
. ANTIMISSILE DEFENSE
. SAGE AIR DEFENSE SYSTEM
RT ANTIRADIATION MISSILES
BALLISTIC MISSILE EARLY WARNING
SYSTEM
CAMOUFLAGE
CIVIL DEFENSE
DECEPTION
∞DEFENSE
DEFENSE PROGRAM
DMSP SATELLITES
EARLY WARNING SYSTEMS
ELECTRONIC WARFARE
JAMMERS
OPTICAL COUNTERMEASURES
SABOTAGE
SPACE SURVEILLANCE (GROUND
BASED)
SPACE SURVEILLANCE (SPACEBORNE)
WEAPONS DELIVERY

AIR DENSITY EXPLORER A
USE EXPLORER 19 SATELLITE

AIR DENSITY/INJUN EXPLORER B
USE EXPLORER 25 SATELLITE

AIR DROP OPERATIONS
RT BAILOUT
BALLUTES
CARGO
DELIVERY
FREE FALL
∞OPERATIONS
PARACHUTES
PARAVULCOONS
PARAWINGS

AIR DUCTS
GS DUCTS
. **AIR DUCTS**
RT ANNULAR DUCTS
BLOWERS
EXHAUST NOZZLES
∞FANS
GAS FLOW
VENTILATORS

AIR FILTERS
GS CLEANERS
. **AIR FILTERS**
SEPARATORS
. FLUID FILTERS
.. **AIR FILTERS**
RT COOLING SYSTEMS
DUST COLLECTORS
∞FILTERS
PRECIPITATORS
VENTILATION

AIR FLOW
GS FLUID FLOW
. GAS FLOW
.. **AIR FLOW**
... AIR CURRENTS
.... JET STREAMS (METEOROLOGY)
.... MERIDIONAL FLOW
.... VERTICAL AIR CURRENTS
RT AERODYNAMIC INTERFERENCE
ATMOSPHERIC BOUNDARY LAYER
BAROTROPIC FLOW
BRUNT-VAISALA FREQUENCY
COMPRESSIBLE FLOW
∞CURRENTS
DUCT GEOMETRY
DUCTED FLOW
STREAMLINING

AIR FLOW-(CONT.)
STREAMS
VENTILATION

AIR FREIGHT
USE AIR CARGO

AIR INLETS
USE AIR INTAKES

AIR INTAKES
UF AIR INLETS
GS INTAKE SYSTEMS
. **AIR INTAKES**
.. ENGINE INLETS
.. HYPERSONIC INLETS
.. INLET AIRFRAME CONFIGURATIONS
.. SUPERSONIC INLETS
RT BYPASS RATIO
CONICAL INLETS
COWLINGS
INLET NOZZLES
INLET TEMPERATURE
INTERNAL COMPRESSION INLETS
MANIFOLDS
NACELLES
NOSE INLETS
PLENUM CHAMBERS
SCOOPS
SIDE INLETS
SUPERCHARGERS
SUPERSONIC DIFFUSERS
VENTILATION
VENTILATORS
∞WATER INTAKES

AIR JETS
GS FLUID FLOW
. JET FLOW
.. **AIR JETS**
FLUID JETS
. **AIR JETS**
RT GAS FLOW
GAS JETS
JET STREAMS (METEOROLOGY)
∞JETS
VAPOR JETS

AIR LAND INTERACTIONS
RT ATMOSPHERIC BOUNDARY LAYER
ATMOSPHERIC CIRCULATION
GAS-SOLID INTERACTIONS
∞INTERACTIONS
METEOROLOGY

AIR LAUNCHING
GS LAUNCHING
. **AIR LAUNCHING**
RT MULTISTAGE ROCKET VEHICLES
PIGGYBACK SYSTEMS

AIR LAW
GS LAW (JURISPRUDENCE)
. INTERNATIONAL LAW
.. **AIR LAW**
RT ∞AERONAUTICS
AIRSPACE
CIVIL AVIATION
CONVENTIONS
LEGAL LIABILITY
LIABILITIES
∞MILITARY AVIATION
NATIONAL AIRSPACE UTILIZATION
SYSTEM
PENALTIES
POLITICS
PUBLIC LAW
REGULATIONS
SPACE LAW

AIR LOCKS
GS COMPARTMENTS
. **AIR LOCKS**
RT AIRLOCK MODULES
DOORS
EGRESS
ENCLOSURES
HATCHES
INGRESS (SPACECRAFT PASSAGEWAY)
∞LOCKS
PRESSURE CHAMBERS
SEALS (STOPPERS)

AIR MAIL
GS   CARGO
      . AIR CARGO
      . . **AIR MAIL**

AIR MASSES
RT   ANTICYCLONES
      ATMOSPHERIC CIRCULATION
      BRUNT-VAISALA FREQUENCY
      COLD FRONTS
      FRONTS (METEOROLOGY)
      METEOROLOGY
      SYNOPTIC METEOROLOGY
      WARM FRONTS
      WEATHER FORECASTING
      WINDPOWER UTILIZATION

AIR NAVIGATION
GS   NAVIGATION
      . **AIR NAVIGATION**
      . . ALL-WEATHER AIR NAVIGATION
      . . AREA NAVIGATION
      . . NAP-OF-THE-EARTH NAVIGATION
RT   ASTRONAVIGATION
      BORESIGHT ERROR
      CELESTIAL NAVIGATION
      CELESTIAL REFERENCE SYSTEMS
      COLLISION AVOIDANCE
      ∞CONTROL
      DEAD RECKONING
      DIGITAL NAVIGATION
      DOPPLER NAVIGATION
      FLIGHT INSTRUMENTS
      FLIGHT MANAGEMENT SYSTEMS
      FLIGHT PATHS
      FLIGHT PLANS
      FLIGHT RULES
      GUIDANCE (MOTION)
      HYPERBOLIC NAVIGATION
      INERTIAL NAVIGATION
      INSTRUMENT FLIGHT RULES
      LORAN
      LORAN C
      LORAN D
      NATIONAL AIRSPACE UTILIZATION
        SYSTEM
      NAVIGATION AIDS
      OMEGA NAVIGATION SYSTEM
      POLAR NAVIGATION
      RADAR NAVIGATION
      RADIO NAVIGATION
      SHORAN
      SOLAR COMPASSES
      SPACE NAVIGATION
      TACAN
      VHF OMNIRANGE NAVIGATION
      VISUAL FLIGHT

AIR PIRACY
UF   HIJACKING
RT   AIRCRAFT SAFETY
      AIRPORT SECURITY
      CRIME
      FLIGHT HAZARDS
      FLIGHT SAFETY
      OPERATIONAL HAZARDS

AIR POLLUTION
UF   ATMOSPHERIC IMPURITIES
GS   POLLUTION
      . ENVIRONMENT POLLUTION
      . . **AIR POLLUTION**
      . . . GLOBAL AIR POLLUTION
      . . . INDOOR AIR POLLUTION
RT   ACID RAIN
      AEROBIOLOGY
      AEROSOLS
      ASHES
      ATMOSPHERIC CHEMISTRY
      ATMOSPHERIC COMPOSITION
      ATMOSPHERIC DENSITY
      ATMOSPHERIC EFFECTS
      CLEAN ENERGY
      COMBUSTION PRODUCTS
      CONTAMINATION
      DIFFUSION
      DROPS (LIQUIDS)
      DUST
      EARTH ATMOSPHERE
      EARTH ENVIRONMENT
      EFFLUENTS
      ENVIRONMENT EFFECTS
      ENVIRONMENT PROTECTION
      ENVIRONMENTAL CHEMISTRY
      ENVIRONMENTAL QUALITY

AIR POLLUTION-*(CONT.)*
      ENVIRONMENTAL SURVEYS
      ENVIRONMENTS
      EXHAUST GASES
      EXHAUST SYSTEMS
      FALLOUT
      FLUE GASES
      FLY ASH
      FOREST FIRES
      GLOBAL AIR SAMPLING PROGRAM
      HAZE
      HUMAN WASTES
      METABOLIC WASTES
      MIDDLE ATMOSPHERE
      MIXING HEIGHT
      MUTAGENS
      NITROUS ACID
      ODORS
      ORGANIC PEROXIDES
      OXIDIZERS
      PARTICLES
      PHOTOCHEMICAL OXIDANTS
      POLLEN
      POLLUTION MONITORING
      POLLUTION TRANSPORT
      POLYNUCLEAR ORGANIC COMPOUNDS
      SMOG
      SMOKE
      SMOKE ABATEMENT
      SOOT
      TEMPERATURE INVERSIONS
      WASTE DISPOSAL
      WASTES
      WIND (METEOROLOGY)

AIR PURIFICATION
GS   PURIFICATION
      . **AIR PURIFICATION**
RT   CARBON DIOXIDE CONCENTRATION
      CARBON DIOXIDE REMOVAL
      DECONTAMINATION
      ELECTROSTATIC PRECIPITATORS
      HOPCALITE (TRADEMARK)
      REBREATHING
      STERILIZATION
      VENTILATION

AIR QUALITY
GS   QUALITY
      . **AIR QUALITY**
RT   EARTH ATMOSPHERE
      ENVIRONMENTS
      INDOOR AIR POLLUTION
      POLLUTION CONTROL
      POLLUTION MONITORING

AIR SAMPLING
GS   SAMPLING
      . **AIR SAMPLING**
RT   ELECTROSTATIC PRECIPITATORS
      ENVIRONMENT POLLUTION
      GAS ANALYSIS
      GLOBAL AIR SAMPLING PROGRAM
      INDOOR AIR POLLUTION
      SMOG

AIR SEA ICE INTERACTIONS
GS   GAS-LIQUID INTERACTIONS
      . AIR WATER INTERACTIONS
      . . **AIR SEA ICE INTERACTIONS**
RT   ∞INTERACTIONS
      SEA ICE

AIR SEA INTERACTIONS
USE   AIR WATER INTERACTIONS

AIR SICKNESS
USE   MOTION SICKNESS

AIR SLEW MISSILES
GS   MISSILES
      . **AIR SLEW MISSILES**
RT   MANEUVERABILITY
      ∞ROCKETS
      SOLID PROPELLANT ROCKET ENGINES
      THRUST VECTOR CONTROL

AIR START
GS   STARTING
      . **AIR START**
RT   AIRCRAFT CONTROL
      AIRCRAFT ENGINES
      ENGINE CONTROL
      FLIGHT TESTS

AIR TO AIR MISSILES
UF   AIR TO AIR ROCKETS
GS   MISSILES
      . **AIR TO AIR MISSILES**
      . . FALCON MISSILE
      . . MATRA MISSILE
      . . SIDEWINDER MISSILES
      . . SPARROW MISSILES
      . . . SPARROW 2 MISSILE
      . . . SPARROW 3 MISSILE
RT   ANTIAIRCRAFT MISSILES
      RAMJET MISSILES
      SIAM MISSILES
      SPACE WEAPONS
      SURFACE TO AIR MISSILES

AIR TO AIR REFUELING
GS   REFUELING
      . **AIR TO AIR REFUELING**
RT   TANKER AIRCRAFT

AIR TO AIR ROCKETS
USE   AIR TO AIR MISSILES

AIR TO SURFACE MISSILES
GS   MISSILES
      . **AIR TO SURFACE MISSILES**
      . . BULLPUP MISSILES
      . . CONDOR MISSILE
      . . HARPOON MISSILE
      . . HOUND DOG MISSILE
      . . MAVERICK MISSILES
      . . QUAIL MISSILE
      . . SHRIKE MISSILE
RT   MISS DISTANCE
      ORDNANCE
      SURFACE TO AIR MISSILES
      SURFACE TO SURFACE MISSILES
      ∞SURFACES
      WEAPON SYSTEMS

AIR TRAFFIC
GS   TRAFFIC
      . **AIR TRAFFIC**
RT   AIRCRAFT HAZARDS
      AIRLINE OPERATIONS
      AIRSPACE
      COLLISION AVOIDANCE
      FLIGHT HAZARDS
      FLIGHT PATHS
      FLIGHT PLANS
      NATIONAL AIRSPACE UTILIZATION
        SYSTEM
      NATIONAL AVIATION SYSTEM

AIR TRAFFIC CONTROL
GS   GROUND BASED CONTROL
      . **AIR TRAFFIC CONTROL**
      . . AUTOMATED EN ROUTE ATC
      . . RADAR APPROACH CONTROL
      TRAFFIC CONTROL
      . **AIR TRAFFIC CONTROL**
      . . AUTOMATED EN ROUTE ATC
      . . RADAR APPROACH CONTROL
RT   AERONAUTICAL SATELLITES
      AIRBORNE RADAR APPROACH
      AIRCRAFT APPROACH SPACING
      AIRCRAFT COMMUNICATION
      AIRCRAFT GUIDANCE
      AIRCRAFT SAFETY
      AIRPORT SURFACE DETECTION
        EQUIPMENT
      AIRPORT TOWERS
      AIRPORTS
      AIRSPACE
      APPROACH
      APPROACH CONTROL
      APPROACH INDICATORS
      ATTITUDE CONTROL
      AUTOMATED PILOT ADVISORY SYSTEM
      AUTOMATED RADAR TERMINAL SYSTEM
      BEACON COLLISION AVOIDANCE
        SYSTEM
      COLLISION AVOIDANCE
      COLLISIONS
      ∞CONTROL
      DISCRETE ADDRESS BEACON SYSTEM
      FLIGHT ALTITUDE
      FLIGHT CONTROL
      FLIGHT MANAGEMENT SYSTEMS
      FLIGHT PATHS
      FLIGHT PLANS
      FLIGHT RULES
      FLIGHT SAFETY
      FLIGHT TIME

AIR TRAFFIC CONTROL-*(CONT.)*
GROUND SUPPORT EQUIPMENT
GROUND-AIR-GROUND COMMUNICATION
HELIPORTS
INSTRUMENT FLIGHT RULES
INSTRUMENT LANDING SYSTEMS
LANDING
LANDING AIDS
LANDING INSTRUMENTS
LANDING RADAR
LOCATES SYSTEM
MICROWAVE LANDING SYSTEMS
MIDAIR COLLISIONS
MILITARY AIR FACILITIES
NATIONAL AIRSPACE SYSTEM
NATIONAL AIRSPACE UTILIZATION
SYSTEM
NATIONAL AVIATION SYSTEM
NAVIGATION AIDS
∞ OPERATIONS
RADAR NAVIGATION
RADIO NAVIGATION
ROUTES
SOLAR COMPASSES
SURVEILLANCE RADAR
TAKEOFF
TAXIING
TOWERS
TRACKING (POSITION)
TRANSPONDERS
VORTEX ADVISORY SYSTEM
VORTEX AVOIDANCE

AIR TRAFFIC CONTROLLERS (PERSONNEL)
GS   PERSONNEL
. **AIR TRAFFIC CONTROLLERS
(PERSONNEL)**
RT   AIRPORT TOWERS
GROUND BASED CONTROL
LANDING AIDS
TRAFFIC CONTROL

AIR TRANSPORTATION
GS   TRANSPORTATION
. **AIR TRANSPORTATION**
RT   COMPOUND HELICOPTERS
MARINE TRANSPORTATION
NATIONAL AVIATION SYSTEM
RAPID TRANSIT SYSTEMS
SHORT HAUL AIRCRAFT
TRANSPORT AIRCRAFT

AIR WATER INTERACTIONS
UF   AIR SEA INTERACTIONS
GS   GAS-LIQUID INTERACTIONS
. **AIR WATER INTERACTIONS**
. . AIR SEA ICE INTERACTIONS
RT   ATMOSPHERIC & OCEANOGRAPHIC
INFORM SYS
EL NINO
GYRES
∞ INTERACTIONS
LIQUID-GAS MIXTURES
LIQUID-VAPOR INTERFACES
OCEAN DYNAMICS
OCEAN MODELS
SEA SURFACE TEMPERATURE
WATER TUNNEL TESTS

AIRBORNE EQUIPMENT
GS   ONBOARD EQUIPMENT
. **AIRBORNE EQUIPMENT**
. . AIRBORNE/SPACEBORNE
COMPUTERS
. . LIGHT AIRBORNE MULTIPURPOSE
SYSTEM
. . TERCOM
RT   ADVANCED RANGE INSTRUMENTATION
AIRCRAFT
AIRCRAFT COMMUNICATION
AIRCRAFT EQUIPMENT
AUTOMATIC LANDING CONTROL
AVIONICS
BALLOON-BORNE INSTRUMENTS
∞ ELECTRIC EQUIPMENT
∞ EQUIPMENT
FLIGHT INSTRUMENTS
HYDRAULIC EQUIPMENT
MAP MATCHING GUIDANCE
MATTS (SYSTEMS)
RADAR EQUIPMENT
RADIO EQUIPMENT
VACUUM ARC SWITCHES

AIRBORNE INFECTION
GS   DISEASES
. INFECTIOUS DISEASES
. . **AIRBORNE INFECTION**
RT   AEROBIOLOGY
PARASITIC DISEASES

**AIRBORNE INTEGRATED RECONNAISSANCE
SYSTEM**
UF   AIRS (RECONNAISSANCE SYS)
GS   RECONNAISSANCE
. AERIAL RECONNAISSANCE
. . **AIRBORNE INTEGRATED
RECONNAISSANCE SYSTEM**
RT   GROUND TRUTH
PHOTORECONNAISSANCE
∞ SYSTEMS
TARGETS

AIRBORNE LASERS
GS   ONBOARD EQUIPMENT
. **AIRBORNE LASERS**
STIMULATED EMISSION DEVICES
. LASERS
. . **AIRBORNE LASERS**
RT   LASER APPLICATIONS
LASER RANGER/TRACKER
REMOTE SENSORS
SPACEBORNE LASERS

AIRBORNE RADAR APPROACH
GS   APPROACH
. **AIRBORNE RADAR APPROACH**
RT   AIR TRAFFIC CONTROL
AIRCRAFT APPROACH SPACING
HELICOPTER CONTROL
HELICOPTERS
LANDING AIDS
RADAR APPROACH CONTROL

AIRBORNE RANGE AND ORBIT DETERMINATION
UF   AROD (RANGE-ORBIT DETERMINATION)
GS   RANGEFINDING
. **AIRBORNE RANGE AND ORBIT
DETERMINATION**
RT   ∞ MEASUREMENT
ORBITS

AIRBORNE SURVEILLANCE RADAR
GS   RADAR
. SURVEILLANCE RADAR
. . **AIRBORNE SURVEILLANCE RADAR**
RT   AIRCRAFT INSTRUMENTS
DISPLAY DEVICES
ONBOARD EQUIPMENT

AIRBORNE WARNING AND CONTROL SYSTEM
USE   AWACS AIRCRAFT

AIRBORNE/SPACEBORNE COMPUTERS
UF   FLIGHT COMPUTERS
ONBOARD COMPUTERS
GS   DATA PROCESSING EQUIPMENT
. COMPUTERS
. . EMBEDDED COMPUTER SYSTEMS
. . . **AIRBORNE/SPACEBORNE
COMPUTERS**
ONBOARD EQUIPMENT
. AIRBORNE EQUIPMENT
. . **AIRBORNE/SPACEBORNE
COMPUTERS**
RT   DATA PROCESSING
FLIGHT MANAGEMENT SYSTEMS
HIGHLY MANEUVERABLE AIRCRAFT
MINICOMPUTERS
ONBOARD DATA PROCESSING
SPACECRAFT COMPONENTS
SPACECRAFT ELECTRONIC EQUIPMENT
SYSTEMS INTEGRATION

AIRBUS
USE   EUROPEAN AIRBUS

∞ **AIRCRAFT**
SN   *(USE OF A MORE SPECIFIC TERM IS
RECOMMENDED--CONSULT THE TERMS
LISTED BELOW)*
UF   AERODYNAMIC VEHICLES
RT   A-1 AIRCRAFT
A-2 AIRCRAFT
A-3 AIRCRAFT
A-4 AIRCRAFT
A-5 AIRCRAFT
A-6 AIRCRAFT

AIRCRAFT-*(CONT.)*
A-7 AIRCRAFT
A-9 AIRCRAFT
A-10 AIRCRAFT
A-37 AIRCRAFT
A-300 AIRCRAFT
ADVANCED RANGE INSTRUMENTATION
AIRCRAFT
AERODYNAMICS
AERONAUTICAL ENGINEERING
∞ AERONAUTICS
AEROSPACE ENGINEERING
AEROSPACE INDUSTRY
AEROSPACE VEHICLES
AEROSPACEPLANES
AGRICULTURAL AIRCRAFT
AIRCRAFT ACCIDENT INVESTIGATION
AIRCRAFT ACCIDENTS
AIRCRAFT ANTENNAS
AIRCRAFT APPROACH SPACING
AIRCRAFT BRAKES
AIRCRAFT CARRIERS
AIRCRAFT COMMUNICATION
AIRCRAFT COMPARTMENTS
AIRCRAFT CONFIGURATIONS
AIRCRAFT CONSTRUCTION MATERIALS
AIRCRAFT CONTROL
AIRCRAFT DESIGN
AIRCRAFT DETECTION
AIRCRAFT ENGINES
AIRCRAFT EQUIPMENT
AIRCRAFT FUEL SYSTEMS
AIRCRAFT FUELS
AIRCRAFT GUIDANCE
AIRCRAFT HAZARDS
AIRCRAFT HYDRAULIC SYSTEMS
AIRCRAFT INDUSTRY
AIRCRAFT INSTRUMENTS
AIRCRAFT LANDING
AIRCRAFT LIGHTS
AIRCRAFT MAINTENANCE
AIRCRAFT MANEUVERS
AIRCRAFT NOISE
AIRCRAFT PARTS
AIRCRAFT PERFORMANCE
AIRCRAFT PILOTS
AIRCRAFT PRODUCTION
AIRCRAFT PRODUCTION COSTS
AIRCRAFT RELIABILITY
AIRCRAFT RUNUP
AIRCRAFT SAFETY
AIRCRAFT SPECIFICATIONS
AIRCRAFT SPIN
AIRCRAFT STABILITY
AIRCRAFT STRUCTURES
AIRCRAFT SURVIVABILITY
AIRCRAFT TIRES
AIRCRAFT WAKES
AIRSHIPS
ALADIN 2 AIRCRAFT
ALOUETTE HELICOPTERS
ALPHA JET AIRCRAFT
AMPHIBIOUS AIRCRAFT
AN-2 AIRCRAFT
AN-22 AIRCRAFT
AN-24 AIRCRAFT
ANTISUBMARINE WARFARE AIRCRAFT
ANTONOV AIRCRAFT
ARGOSY MK-1 AIRCRAFT
ASSET GLIDERS
ATLIT PROJECT
ATTACK AIRCRAFT
AVRO 707 AIRCRAFT
AWACS AIRCRAFT
B-1 AIRCRAFT
B-26 AIRCRAFT
B-47 AIRCRAFT
B-50 AIRCRAFT
B-52 AIRCRAFT
B-57 AIRCRAFT
B-58 AIRCRAFT
B-66 AIRCRAFT
B-70 AIRCRAFT
BAC AIRCRAFT
BAC 111 AIRCRAFT
BALLOONS
BEAGLE AIRCRAFT
BEECH 99 AIRCRAFT
BEECHCRAFT AIRCRAFT
BEECHCRAFT 18 AIRCRAFT
BELL AIRCRAFT
BELL 214A HELICOPTER
BIPLANES
BIRD-AIRCRAFT COLLISIONS
BOEING AIRCRAFT
BOEING 707 AIRCRAFT

AIRCRAFT-(CONT.)

BOEING 720 AIRCRAFT
BOEING 727 AIRCRAFT
BOEING 733 AIRCRAFT
BOEING 737 AIRCRAFT
BOEING 747 AIRCRAFT
BOEING 757 AIRCRAFT
BOEING 767 AIRCRAFT
BOEING 2707 AIRCRAFT
BOLKOW AIRCRAFT
BOMBER AIRCRAFT
BOOSTGLIDE VEHICLES
BREGUET AIRCRAFT
BREGUET 940 AIRCRAFT
BREGUET 941 AIRCRAFT
BREGUET 1150 AIRCRAFT
BUCCANEER AIRCRAFT
C-1A AIRCRAFT
C-2 AIRCRAFT
C-5 AIRCRAFT
C-8A AUGMENTOR WING AIRCRAFT
C-9 AIRCRAFT
C-15 AIRCRAFT
C-33 AIRCRAFT
C-35 AIRCRAFT
C-46 AIRCRAFT
C-47 AIRCRAFT
C-54 AIRCRAFT
C-118 AIRCRAFT
C-119 AIRCRAFT
C-121 AIRCRAFT
C-123 AIRCRAFT
C-124 AIRCRAFT
C-130 AIRCRAFT
C-131 AIRCRAFT
C-133 AIRCRAFT
C-135 AIRCRAFT
C-140 AIRCRAFT
C-141 AIRCRAFT
C-160 AIRCRAFT
CANADAIR AIRCRAFT
CANBERRA AIRCRAFT
CARGO AIRCRAFT
CEILING (AIRCRAFT CAPABILITY)
CESSNA AIRCRAFT
CESSNA L-19 AIRCRAFT
CESSNA 172 AIRCRAFT
CESSNA 205 AIRCRAFT
CESSNA 402B AIRCRAFT
CHANCE-VOUGHT AIRCRAFT
CHINESE AIRCRAFT
CL-41 AIRCRAFT
CL-44 AIRCRAFT
CL-84 AIRCRAFT
CL-600 CHALLENGER AIRCRAFT
CL-823 AIRCRAFT
COIN AIRCRAFT
COMET 4 AIRCRAFT
COMMERCIAL AIRCRAFT
COMPOUND HELICOPTERS
CONCORDE AIRCRAFT
CURTISS-WRIGHT AIRCRAFT
CV-340 AIRCRAFT
CV-440 AIRCRAFT
CV-880 AIRCRAFT
CV-990 AIRCRAFT
D-558 AIRCRAFT
DASSAULT AIRCRAFT
DC 3 AIRCRAFT
DC 7 AIRCRAFT
DC 8 AIRCRAFT
DC 9 AIRCRAFT
DC 10 AIRCRAFT
DE HAVILLAND AIRCRAFT
DH 112 AIRCRAFT
DH 115 AIRCRAFT
DH 121 AIRCRAFT
DH 125 AIRCRAFT
DHC 2 AIRCRAFT
DHC 4 AIRCRAFT
DHC 5 AIRCRAFT
DO-27 AIRCRAFT
DO-28 AIRCRAFT
DO-31 AIRCRAFT
DORNIER AIRCRAFT
DOUGLAS AIRCRAFT
DRONE AIRCRAFT
E-2 AIRCRAFT
E-3A AIRCRAFT
E-4A AIRCRAFT
EARTH RESOURCES SURVEY AIRCRAFT
EC-121 AIRCRAFT
ELECTRA AIRCRAFT
ELECTRONIC AIRCRAFT
EUROPEAN AIRBUS
F-2 AIRCRAFT

AIRCRAFT-(CONT.)

F-4 AIRCRAFT
F-5 AIRCRAFT
F-8 AIRCRAFT
F-9 AIRCRAFT
F-14 AIRCRAFT
F-15 AIRCRAFT
F-16 AIRCRAFT
F-17 AIRCRAFT
F-18 AIRCRAFT
F-27 AIRCRAFT
F-28 TRANSPORT AIRCRAFT
F-84 AIRCRAFT
F-86 AIRCRAFT
F-89 AIRCRAFT
F-94 AIRCRAFT
F-100 AIRCRAFT
F-101 AIRCRAFT
F-102 AIRCRAFT
F-104 AIRCRAFT
F-105 AIRCRAFT
F-106 AIRCRAFT
F-111 AIRCRAFT
FAIRCHILD-HILLER AIRCRAFT
FAIREY AIRCRAFT
FAN IN WING AIRCRAFT
FD 2 AIRCRAFT
FIAT AIRCRAFT
FIGHTER AIRCRAFT
FIREBEE 2 TARGET DRONE AIRCRAFT
FLIGHT TEST VEHICLES
FLYING PLATFORMS
FOKKER AIRCRAFT
FOLDING FIN AIRCRAFT ROCKET
    VEHICLE
FREE WING AIRCRAFT
FV-12A AIRCRAFT
G-1 AIRCRAFT
G-91 AIRCRAFT
G-95/4 AIRCRAFT
G-222 AIRCRAFT
GA-5 AIRCRAFT
GENERAL AVIATION AIRCRAFT
GENERAL DYNAMICS AIRCRAFT
GETOL AIRCRAFT
GLIDERS
GROUND EFFECT MACHINES
GRUMMAN AIRCRAFT
GYRODYNE AIRCRAFT
H-60 HELICOPTER
H-126 AIRCRAFT
HAMBURGER AIRCRAFT
HANDLEY PAGE AIRCRAFT
HANG GLIDERS
HARRIER AIRCRAFT
HAWKER SIDDELEY AIRCRAFT
HEAVY LIFT HELICOPTERS
HEINKEL AIRCRAFT
HELICOPTERS
HELIO AIRCRAFT
HFB-320 AIRCRAFT
HIGHLY MANEUVERABLE AIRCRAFT
HILLER AIRCRAFT
HOVERCRAFT GROUND EFFECT
    MACHINES
HP-115 AIRCRAFT
HS-748 AIRCRAFT
HS-801 AIRCRAFT
HUGHES AIRCRAFT
HYPERSONIC AIRCRAFT
HYPERSONIC GLIDERS
IL-14 AIRCRAFT
IL-62 AIRCRAFT
ILYUSHIN AIRCRAFT
INFLATABLE GLIDERS
JAGUAR AIRCRAFT
JET AIRCRAFT
JET AIRCRAFT NOISE
JET PROVOST AIRCRAFT
JETSTREAM AIRCRAFT
JINDIVIK TARGET AIRCRAFT
KA-6 SAILPLANES
KAMAN AIRCRAFT
KAWASAKI AIRCRAFT
L-1011 AIRCRAFT
L-2000 AIRCRAFT
LEAR JET AIRCRAFT
LIFTING REENTRY VEHICLES
LIGHT AIRCRAFT
LIGHT INTRATHEATER TRANSPORT
LIGHT TRANSPORT AIRCRAFT
LING-TEMCO-VOUGHT AIRCRAFT
LOCKHEED AIRCRAFT
LOCKHEED MODEL 18 AIRCRAFT
∞ LOW WING AIRCRAFT
MAN POWERED AIRCRAFT

AIRCRAFT-(CONT.)

MARTIN AIRCRAFT
MCDONNELL AIRCRAFT
MCDONNELL DOUGLAS AIRCRAFT
MERCURE AIRCRAFT
METEOROLOGICAL RESEARCH
    AIRCRAFT
MH-262 AIRCRAFT
MIG AIRCRAFT
MIL AIRCRAFT
MILITARY AIR FACILITIES
∞ MILITARY AIRCRAFT
MILITARY HELICOPTERS
MONOPLANES
MRCA AIRCRAFT
MULTIENGINE VEHICLES
NAVION AIRCRAFT
NIGHT FLIGHTS (AIRCRAFT)
NIHON AIRCRAFT
NOISE PREDICTION (AIRCRAFT)
NORD AIRCRAFT
NORD 1500 AIRCRAFT
NORTH AMERICAN AIRCRAFT
NORTHROP AIRCRAFT
NUCLEAR PROPELLED AIRCRAFT
OBSERVATION AIRCRAFT
ONBOARD EQUIPMENT
OV-1 AIRCRAFT
OV-10 AIRCRAFT
P-3 AIRCRAFT
P-51 AIRCRAFT
P-160 AIRCRAFT
P-166 AIRCRAFT
P-308 AIRCRAFT
P-1052 AIRCRAFT
P-1127 AIRCRAFT
P-1154 AIRCRAFT
PA-34 SENECA AIRCRAFT
PANAVIA MILITARY AIRCRAFT
PASSENGER AIRCRAFT
PD-808 AIRCRAFT
PHANTOM AIRCRAFT
PIAGGIO AIRCRAFT
PIASECKI AIRCRAFT
PILOTLESS AIRCRAFT
PIPER AIRCRAFT
POTEZ AIRCRAFT
POWERED LIFT AIRCRAFT
PROPULSION
QUESTOL
RB-50 AIRCRAFT
RECONNAISSANCE AIRCRAFT
REMOTELY PILOTED VEHICLES
REPUBLIC AIRCRAFT
RESEARCH AIRCRAFT
RF-4 AIRCRAFT
RIGID ROTOR HELICOPTERS
ROCKET PLANES
ROTARY WING AIRCRAFT
ROTOR SYSTEMS RESEARCH AIRCRAFT
ROTORCRAFT AIRCRAFT
RYAN AIRCRAFT
S-2 AIRCRAFT
S-3 AIRCRAFT
SAAB AIRCRAFT
SAAB 37 AIRCRAFT
SAAB 105 AIRCRAFT
SC-1 AIRCRAFT
SC-5 AIRCRAFT
SC-7 AIRCRAFT
SCHLEICHER AIRCRAFT
SCIMITAR AIRCRAFT
SE-210 AIRCRAFT
SHORT HAUL AIRCRAFT
SHORT TAKEOFF AIRCRAFT
SIEBEL AIRCRAFT
SIKORSKY AIRCRAFT
SIKORSKY WHIRLWIND HELICOPTER
SNOW AIRCRAFT
SOLAR POWERED AIRCRAFT
SPANLOADER AIRCRAFT
SUBMERSIBLE AIRCRAFT
∞ SUBSONIC AIRCRAFT
SUD AVIATION AIRCRAFT
SUPERSONIC AIRCRAFT
SUPERSONIC CRUISE AIRCRAFT
    RESEARCH
T-2 AIRCRAFT
T-28 AIRCRAFT
T-33 AIRCRAFT
T-37 AIRCRAFT
T-38 AIRCRAFT
T-39 AIRCRAFT
TACT PROGRAM
TAILLESS AIRCRAFT
TANDEM ROTOR HELICOPTERS

AIRCRAFT-*(CONT.)*
- TANDEM WING AIRCRAFT
- TANKER AIRCRAFT
- TARGET DRONE AIRCRAFT
- TERRAIN FOLLOWING AIRCRAFT
- TEST VEHICLES
- TILT ROTOR AIRCRAFT
- TILT ROTOR RESEARCH AIRCRAFT PROGRAM
- TILT WING AIRCRAFT
- TRAINING AIRCRAFT
- TRANSATMOSPHERIC VEHICLES
- TRANSPORT AIRCRAFT
- TS-11 AIRCRAFT
- TSR-2 AIRCRAFT
- TU-104 AIRCRAFT
- TU-124 AIRCRAFT
- TU-134 AIRCRAFT
- TU-144 AIRCRAFT
- TU-154 AIRCRAFT
- TUPOLEV AIRCRAFT
- TURBOFAN AIRCRAFT
- TURBOPROP AIRCRAFT
- U-2 AIRCRAFT
- U-10 AIRCRAFT
- ULTRALIGHT AIRCRAFT
- UNIDENTIFIED FLYING OBJECTS
- UTILITY AIRCRAFT
- V/STOL AIRCRAFT
- VALIANT AIRCRAFT
- VAMPIRE MK 35 AIRCRAFT
- VATOL AIRCRAFT
- VC-10 AIRCRAFT
- VERTICAL TAKEOFF AIRCRAFT
- VICTOR MK-1 AIRCRAFT
- VISCOUNT AIRCRAFT
- VJ-101 AIRCRAFT
- VULCAN AIRCRAFT
- VZ-2 AIRCRAFT
- VZ-8 AIRCRAFT
- WATER TAKEOFF AND LANDING AIRCRAFT
- WEAPONS DELIVERY
- WEATHER RECONNAISSANCE AIRCRAFT
- WESER AIRCRAFT
- WESTLAND AIRCRAFT
- WESTLAND GROUND EFFECT MACHINES
- WESTLAND WHIRLWIND HELICOPTER
- WING NACELLE CONFIGURATIONS
- ∞ WINGED VEHICLES
- X-1 AIRCRAFT
- X-2 AIRCRAFT
- X-3 AIRCRAFT
- X-5 AIRCRAFT
- X-13 AIRCRAFT
- X-14 AIRCRAFT
- X-15 AIRCRAFT
- X-19 AIRCRAFT
- X-20 AIRCRAFT
- X-21 AIRCRAFT
- X-21A AIRCRAFT
- X-22 AIRCRAFT
- X-22A AIRCRAFT
- X-24 AIRCRAFT
- X-29 AIRCRAFT
- XC-142 AIRCRAFT
- XV-3 AIRCRAFT
- XV-4 AIRCRAFT
- XV-5 AIRCRAFT
- XV-8A AIRCRAFT
- XV-9A AIRCRAFT
- XV-11A AIRCRAFT
- XV-15 AIRCRAFT
- YAK 40 AIRCRAFT
- YC-14 AIRCRAFT
- YF-12 AIRCRAFT
- YF-16 AIRCRAFT

AIRCRAFT ACCIDENT INVESTIGATION
- GS   INVESTIGATION
-         . ACCIDENT INVESTIGATION
-         . . **AIRCRAFT ACCIDENT INVESTIGATION**
- RT   ∞ AIRCRAFT

AIRCRAFT ACCIDENTS
- GS   **AIRCRAFT ACCIDENTS**
-         . BIRD-AIRCRAFT COLLISIONS
- RT   ACCIDENTS
-         ∞ AIRCRAFT
-         AIRCRAFT SAFETY
-         COLLISIONS
-         CRASH LANDING
-         CRASHES
-         CRASHWORTHINESS
-         DITCHING (LANDING)

AIRCRAFT ACCIDENTS-*(CONT.)*
- FLIGHT HAZARDS
- FLIGHT SAFETY
- HUMAN FACTORS ENGINEERING
- MALFUNCTIONS
- MIDAIR COLLISIONS
- PILOT ERROR
- WEATHER

AIRCRAFT ANTENNAS
- GS   ANTENNAS
-         . **AIRCRAFT ANTENNAS**
- RT   ∞ AIRCRAFT
-         LOOP ANTENNAS
-         MICROWAVE ANTENNAS
-         MISSILE ANTENNAS
-         PROTUBERANCES
-         RADAR ANTENNAS
-         RADIO ANTENNAS

AIRCRAFT APPROACH SPACING
- GS   SPACING
-         . **AIRCRAFT APPROACH SPACING**
- RT   AERONAUTICAL SATELLITES
-         AIR TRAFFIC CONTROL
-         AIRBORNE RADAR APPROACH
-         ∞ AIRCRAFT
-         AIRCRAFT SAFETY
-         AIRSPACE
-         APPROACH
-         APPROACH CONTROL
-         COLLISION AVOIDANCE
-         FLIGHT SAFETY
-         GLIDE PATHS
-         GROUND BASED CONTROL
-         INSTRUMENT APPROACH
-         NATIONAL AIRSPACE UTILIZATION SYSTEM
-         NATIONAL AVIATION SYSTEM
-         VORTEX ADVISORY SYSTEM
-         VORTEX AVOIDANCE

AIRCRAFT BASES
- USE   MILITARY AIR FACILITIES

AIRCRAFT BRAKES
- GS   BRAKES (FOR ARRESTING MOTION)
-         . **AIRCRAFT BRAKES**
-         . . SPLIT FLAPS
-         . . WING FLAPS
-         . . . LEADING EDGE SLATS
-         . . . TRAILING EDGE FLAPS
- RT   AERODYNAMIC BRAKES
-         ∞ AIRCRAFT
-         ANTISKID DEVICES
-         BALLUTES
-         DRAG CHUTES
-         DRAG DEVICES
-         THRUST REVERSAL
-         TOWED BODIES
-         WHEEL BRAKES

AIRCRAFT CABINS
- USE   AIRCRAFT COMPARTMENTS

AIRCRAFT CARRIERS
- GS   SURFACE VEHICLES
-         . **AIRCRAFT CARRIERS**
-         WATER VEHICLES
-         . SHIPS
-         . . **AIRCRAFT CARRIERS**
- RT   ∞ AIRCRAFT
-         ARRESTING GEAR
-         ∞ CARRIERS
-         MILITARY AIR FACILITIES
-         ∞ MILITARY AIRCRAFT
-         ∞ MILITARY VEHICLES
-         NAVY
-         NUCLEAR POWERED SHIPS

AIRCRAFT COMMUNICATION
- GS   COMMUNICATING
-         . **AIRCRAFT COMMUNICATION**
-         TELECOMMUNICATION
-         . **AIRCRAFT COMMUNICATION**
- RT   AERONAUTICAL SATELLITES
-         AIR TRAFFIC CONTROL
-         AIRBORNE EQUIPMENT
-         ∞ AIRCRAFT
-         APPROACH CONTROL
-         AVIONICS
-         GROUND-AIR-GROUND COMMUNICATION
-         RADAR BEACONS
-         RADIO COMMUNICATION
-         WIRELESS COMMUNICATION

AIRCRAFT COMPARTMENTS
- UF   AIRCRAFT CABINS
- GS   COMPARTMENTS
-         . **AIRCRAFT COMPARTMENTS**
- RT   ∞ AIRCRAFT
-         BAYS (STRUCTURAL UNITS)
-         CABIN ATMOSPHERES
-         ∞ CABINS
-         COCKPITS
-         GONDOLAS
-         PRESSURIZED CABINS
-         WINDSHIELDS

AIRCRAFT CONFIGURATIONS
- UF   FIXED-WING AIRCRAFT
- GS   **AIRCRAFT CONFIGURATIONS**
-         . DROOPED AIRFOILS
- RT   AERODYNAMIC CONFIGURATIONS
-         AERODYNAMIC INTERFERENCE
-         ∞ AIRCRAFT
-         COMPOUND HELICOPTERS
-         CONTROL CONFIGURED VEHICLES
-         FLARED BODIES
-         ∞ FLIGHT VEHICLES
-         ∞ LOW WING AIRCRAFT
-         MISSILE CONFIGURATIONS
-         PROPULSION SYSTEM CONFIGURATIONS
-         SPACECRAFT CONFIGURATIONS
-         UNDER SURFACE BLOWING
-         UPPER SURFACE BLOWING
-         WING ROOTS

AIRCRAFT CONSTRUCTION
- USE   AIRCRAFT STRUCTURES

AIRCRAFT CONSTRUCTION MATERIALS
- GS   **AIRCRAFT CONSTRUCTION MATERIALS**
-         . AIRFRAME MATERIALS
- RT   ∞ AIRCRAFT
-         AIRFRAMES
-         CERAMIC MATRIX COMPOSITES
-         COMPOSITE MATERIALS
-         ∞ CONSTRUCTION MATERIALS
-         FUSELAGES
-         ∞ MATERIALS
-         PLASTIC AIRCRAFT STRUCTURES
-         SKIN (STRUCTURAL MEMBER)
-         STRUCTURAL MEMBERS
-         WINGS

AIRCRAFT CONTROL
- UF   FLAP CONTROL
- GS   **AIRCRAFT CONTROL**
-         . HELICOPTER CONTROL
- RT   ACTIVE CONTROL
-         AIR START
-         ∞ AIRCRAFT
-         ATTITUDE CONTROL
-         AUTOMATIC CONTROL
-         AUTOMATIC FLIGHT CONTROL
-         ∞ CONTROL
-         CONTROL EQUIPMENT
-         CONTROL SIMULATION
-         CONTROL STABILITY
-         CONTROL STICKS
-         CONTROLLABILITY
-         DAST PROGRAM
-         DIRECTIONAL CONTROL
-         ENGINE CONTROL
-         FLIGHT CONTROL
-         FLIGHT INSTRUMENTS
-         FLY BY TUBE CONTROL
-         FLY BY WIRE CONTROL
-         GROUND BASED CONTROL
-         LATERAL CONTROL
-         LONGITUDINAL CONTROL
-         MANEUVERABILITY
-         MANUAL CONTROL
-         MINOR CIRCLE TURNING FLIGHT
-         PILOT INDUCED OSCILLATION
-         RADIO CONTROL
-         REMOTE CONTROL
-         STABILITY AUGMENTATION
-         TURBOJET ENGINE CONTROL
-         VISUAL CONTROL

AIRCRAFT DESIGN
- GS   **AIRCRAFT DESIGN**
-         . HELICOPTER DESIGN
- RT   ACOUSTIC RETROFITTING
-         AERODYNAMIC CONFIGURATIONS
-         AEROELASTIC RESEARCH WINGS
-         AERONAUTICAL ENGINEERING
-         AEROQUATIC VEHICLES
-         ∞ AIRCRAFT

AIRCRAFT DESIGN-*(CONT.)*
    AIRFOILS
    CHANNEL WINGS
    COMPOUND HELICOPTERS
    COMPUTER AIDED DESIGN
    CONTROL CONFIGURED VEHICLES
    DAST PROGRAM
  ∞DESIGN
    ENGINE AIRFRAME INTEGRATION
    ENGINE DESIGN
    FLIGHT TESTS
    FREE WING AIRCRAFT
    LOFTING
    MISSILE DESIGN
    PANAVIA MILITARY AIRCRAFT
    PRODUCT DEVELOPMENT
    ROTOR SYSTEMS RESEARCH AIRCRAFT
    SHORT HAUL AIRCRAFT
    STREAMLINING
    STRUCTURAL DESIGN
    SYSTEMS ENGINEERING
    TERMINAL CONFIGURED VEHICLE
      PROGRAM
    TRANSATMOSPHERIC VEHICLES
    TU-154 AIRCRAFT
    VORTEX SHEETS
    WEIGHT REDUCTION
    YF-12 AIRCRAFT

AIRCRAFT DETECTION
 GS DETECTION
    . AIRCRAFT DETECTION
 RT ∞AIRCRAFT
    ∞DETECTORS
    IFF SYSTEMS (IDENTIFICATION)
    INFRARED SUPPRESSION
    TRACKING (POSITION)

AIRCRAFT ENERGY EFFICIENCY PROGRAM
 USE ACEE PROGRAM

AIRCRAFT ENGINES
 UF AIRCRAFT POWER SOURCES
 GS **AIRCRAFT ENGINES**
    . HELICOPTER ENGINES
    . J-52 ENGINE
    . J-58 ENGINE
    . J-97 ENGINE
    . T-34 ENGINE
    . T-38 ENGINE
    . T-55 ENGINE
    . T-63 ENGINE
    . T-76 ENGINE
    . T-78 ENGINE
    . TF-30 ENGINE
    . TF-34 ENGINE
    . TF-41 ENGINE
    . VARIABLE CYCLE ENGINES
    . VARIABLE STREAM CONTROL ENGINES
 RT ACEE PROGRAM
    AIR START
    ∞AIRCRAFT
    ENGINE AIRFRAME INTEGRATION
    GAS TURBINE ENGINES
    HYDROGEN ENGINES
    INFRARED SUPPRESSION
    INTERNAL COMBUSTION ENGINES
    JET ENGINES
    JET PROPULSION
    LASER PROPULSION
    NUCLEAR PROPULSION
    PISTON ENGINES
    ∞POWER SUPPLIES
    QUIET ENGINE PROGRAM
    ROCKET ENGINES
    ROTARY ENGINES
    T-58 ENGINE
    T-58-GE-8B ENGINE
    TOPPING CYCLE ENGINES
    TURBINE ENGINES
    WANKEL ENGINES

AIRCRAFT EQUIPMENT
 GS ONBOARD EQUIPMENT
    . AIRCRAFT EQUIPMENT
    . . BOMBING EQUIPMENT
    . . FLYING EJECTION SEATS
    . . TERCOM
 RT AIRBORNE EQUIPMENT
    ∞AIRCRAFT
    AIRCRAFT POWER SUPPLIES
    AUTOMATIC LANDING CONTROL
    AUTOMATIC PILOTS
    AVIONICS
    COMMONALITY

AIRCRAFT EQUIPMENT-*(CONT.)*
    DISPLAY DEVICES
  ∞EQUIPMENT
    FLIGHT INSTRUMENTS
    LANDING AIDS
    LANDING INSTRUMENTS
    LIGHT AIRBORNE MULTIPURPOSE
      SYSTEM
    NAVIGATION AIDS
    NAVIGATION INSTRUMENTS
    RADIO DIRECTION FINDERS

AIRCRAFT FUEL SYSTEMS
 GS FUEL SYSTEMS
    . AIRCRAFT FUEL SYSTEMS
 RT ∞AIRCRAFT
    FUEL PUMPS
    FUEL TANK PRESSURIZATION
    FUEL TANKS
    FUEL VALVES
    ∞SYSTEMS

AIRCRAFT FUELS
 GS FUELS
    . CHEMICAL FUELS
    . . LIQUID FUELS
    . . . **AIRCRAFT FUELS**
 RT ∞AIRCRAFT
    ANTIMISTING FUELS
    AUTOMOBILE FUELS
    HYDROCARBON FUELS
    JET ENGINE FUELS
    LIQUID ROCKET PROPELLANTS
    MONOPROPELLANTS
    SLURRY PROPELLANTS
    SOLID PROPELLANTS
    TANKER AIRCRAFT

AIRCRAFT GUIDANCE
 GS GUIDANCE (MOTION)
    . AIRCRAFT GUIDANCE
 RT AIR TRAFFIC CONTROL
    ∞AIRCRAFT
    APPROACH CONTROL
    AUTOMATED EN ROUTE ATC
    COLLISION AVOIDANCE
    ∞INDICATORS
    INSTRUMENT LANDING SYSTEMS
    RADAR APPROACH CONTROL
    RADAR NAVIGATION
    RADARSCOPES
    RADIO NAVIGATION

AIRCRAFT HAZARDS
 GS HAZARDS
    . AIRCRAFT HAZARDS
 RT AIR TRAFFIC
    ∞AIRCRAFT
    BIRD-AIRCRAFT COLLISIONS
    BIRDS
    COLLISIONS
    CRASH LANDING
    CRASHES
    FLIGHT HAZARDS
    FLIGHT SAFETY
    FOREIGN BODIES
    HUMAN FACTORS ENGINEERING
    MALFUNCTIONS
    MIDAIR COLLISIONS
    NOISE (SOUND)
    OPERATIONAL HAZARDS
    REFUELING
    THREAT EVALUATION
    TOXIC HAZARDS
    WEATHER

AIRCRAFT HYDRAULIC SYSTEMS
 GS HYDRAULIC EQUIPMENT
    . AIRCRAFT HYDRAULIC SYSTEMS
 RT ACTUATORS
    ∞AIRCRAFT
    SERVOCONTROL
    SERVOMECHANISMS
    ∞SYSTEMS

AIRCRAFT INDUSTRY
 GS INDUSTRIES
    . AEROSPACE INDUSTRY
    . . **AIRCRAFT INDUSTRY**
 RT AERONAUTICAL ENGINEERING
    ∞AIRCRAFT
    AIRCRAFT PRODUCTION COSTS

AIRCRAFT INSTRUMENTS
 GS **AIRCRAFT INSTRUMENTS**

AIRCRAFT INSTRUMENTS-*(CONT.)*
    . ALTIMETERS
    . . RADIO ALTIMETERS
    . APPROACH INDICATORS
    . ATTITUDE INDICATORS
    . . GYRO HORIZONS
    . AUTOMATIC PILOTS
    . COMPASSES
    . . GYROCOMPASSES
    . . MAGNETIC COMPASSES
    . . SOLAR COMPASSES
    . FLIGHT RECORDERS
    . POSITION INDICATORS
    . . PLAN POSITION INDICATORS
    . . RADIO DIRECTION FINDERS
    . . SPACECRAFT POSITION INDICATORS
    . RATE OF CLIMB INDICATORS
    . SPEED INDICATORS
    . . ANEMOMETERS
    . . . DRAG FORCE ANEMOMETERS
    . . . HOT-WIRE ANEMOMETERS
    . . . SONIC ANEMOMETERS
    . TACHOMETERS
 RT AIRBORNE SURVEILLANCE RADAR
    AUTOMATIC FLIGHT CONTROL
    AVIONICS
    DISPLAY DEVICES
    FLIGHT CONTROL
    FLIGHT INSTRUMENTS
    FLIGHT PATHS
    FLIGHT TEST INSTRUMENTS
    INDICATING INSTRUMENTS
    INSTRUMENT APPROACH
    INSTRUMENT LANDING SYSTEMS
    ∞INSTRUMENTS
    I2S CAMERAS
    LANDING AIDS
    LANDING INSTRUMENTS
    LIGHT EMITTING DIODES
    ∞MEASUREMENT
    MEASURING INSTRUMENTS
    MONITORS
    NAVIGATION AIDS
    NAVIGATION INSTRUMENTS
    RADAR
    RECORDING INSTRUMENTS

AIRCRAFT LANDING
 GS LANDING
    . AIRCRAFT LANDING
    . . CRASH LANDING
    . . . DITCHING (LANDING)
    . . SKID LANDINGS
 RT AIR CUSHION LANDING SYSTEMS
    ∞AIRCRAFT
    ALL-WEATHER LANDING SYSTEMS
    BLIND LANDING
    CEILINGS (METEOROLOGY)
    CONTROLLABILITY
    CRASHWORTHINESS
    GLIDE LANDINGS
    HARD LANDING
    INSTRUMENT LANDING SYSTEMS
    LANDING AIDS
    LANDING MATS
    LANDING RADAR
    LOW VISIBILITY
    MICROWAVE LANDING SYSTEMS
    RUNWAY ALIGNMENT
    SOFT LANDING
    SPACECRAFT LANDING
    TAKEOFF
    TOUCHDOWN
    VERTICAL LANDING
    VORTEX AVOIDANCE
    WATER LANDING

AIRCRAFT LAUNCHING DEVICES
 UF TAKEOFF SYSTEMS
 GS LAUNCHERS
    . AIRCRAFT LAUNCHING DEVICES
    . . JATO ENGINES
 RT CATAPULTS

AIRCRAFT LIGHTS
 GS LIGHTING EQUIPMENT
    . AIRCRAFT LIGHTS
 RT ∞AIRCRAFT
    BEACONS

AIRCRAFT MAINTENANCE
 GS MAINTENANCE
    . AIRCRAFT MAINTENANCE
 RT ∞AIRCRAFT
    CHECKOUT

AIRCRAFT MAINTENANCE-*(CONT.)*
　　　FLIGHT OPERATIONS
　　　GROUND SUPPORT EQUIPMENT
　　　LOGISTICS

AIRCRAFT MANEUVERS
GS　　MANEUVERS
　　　. **AIRCRAFT MANEUVERS**
RT　　∞AIRCRAFT
　　　APPROACH CONTROL
　　　FLIGHT CHARACTERISTICS
　　　FLIGHT PATHS
　　　HIGHLY MANEUVERABLE AIRCRAFT
　　　MANEUVERABILITY
　　　OBSTACLE AVOIDANCE
　　　TRAJECTORY OPTIMIZATION
　　　TURNING FLIGHT

AIRCRAFT MODELS
GS　　MODELS
　　　. **AIRCRAFT MODELS**
RT　　DYNAMIC MODELS
　　　MATHEMATICAL MODELS
　　　POWERED MODELS
　　　SCALE MODELS
　　　SEMISPAN MODELS
　　　SPACECRAFT MODELS
　　　WIND TUNNEL MODELS

AIRCRAFT NOISE
GS　　ELASTIC WAVES
　　　. SOUND WAVES
　　　. . NOISE (SOUND)
　　　. . . **AIRCRAFT NOISE**
　　　. . . . JET AIRCRAFT NOISE
　　　. . . . SONIC BOOMS
RT　　ACOUSTIC RETROFITTING
　　　AEROACOUSTICS
　　　AERODYNAMIC NOISE
　　　∞AIRCRAFT
　　　BLADE SLAP NOISE
　　　COAXIAL NOZZLES
　　　ENGINE NOISE
　　　FOOTPRINTS
　　　JET AIRCRAFT
　　　MUFFLERS
　　　NOISE INTENSITY
　　　NOISE MEASUREMENT
　　　NOISE PREDICTION (AIRCRAFT)
　　　NOISE REDUCTION
　　　SYNCHROPHASING

AIRCRAFT NOISE PREDICTION
USE　　NOISE PREDICTION (AIRCRAFT)

AIRCRAFT PARTS
RT　　∞AIRCRAFT
　　　AIRFOILS
　　　AIRFRAMES
　　　CHANNEL WINGS
　　　CONTROL SURFACES
　　　FUSELAGES
　　　LANDING GEAR
　　　OBLIQUE WINGS
　　　PROTUBERANCES
　　　SWING TAIL ASSEMBLIES
　　　SWING WINGS
　　　TAIL ASSEMBLIES
　　　WINGS

AIRCRAFT PERFORMANCE
GS　　**AIRCRAFT PERFORMANCE**
　　　. HELICOPTER PERFORMANCE
RT　　AERODYNAMIC STALLING
　　　∞AIRCRAFT
　　　AIRCRAFT SPIN
　　　AIRSPEED
　　　CONTROLLABILITY
　　　DISTANCE
　　　FLIGHT CHARACTERISTICS
　　　MANEUVERABILITY
　　　MINIMUM DRAG
　　　PAYLOADS
　　　∞PERFORMANCE
　　　PILOT PERFORMANCE
　　　SPECIFICATIONS
　　　TAKEOFF RUNS

AIRCRAFT PILOTS
UF　　AVIATORS
　　　COPILOTS
　　　JET PILOTS
GS　　PERSONNEL
　　　. FLYING PERSONNEL
　　　. . PILOTS (PERSONNEL)

AIRCRAFT PILOTS-*(CONT.)*
　　　. . . **AIRCRAFT PILOTS**
　　　. . . . TEST PILOTS
　　　. OPERATORS (PERSONNEL)
　　　. PILOTS (PERSONNEL)
　　　. . . **AIRCRAFT PILOTS**
　　　. . . . TEST PILOTS
RT　　∞AIRCRAFT
　　　AVIATION PSYCHOLOGY
　　　FLIGHT CREWS
　　　∞PILOTS

AIRCRAFT POWER SOURCES
USE　　AIRCRAFT ENGINES

AIRCRAFT POWER SUPPLIES
GS　　ELECTRIC POWER SUPPLIES
　　　. **AIRCRAFT POWER SUPPLIES**
RT　　AIRCRAFT EQUIPMENT
　　　AUXILIARY POWER SOURCES
　　　ELECTRIC GENERATORS
　　　∞POWER SUPPLIES

AIRCRAFT PRODUCTION
UF　　FUSELAGE MOUNTING
RT　　∞AIRCRAFT
　　　COSTS
　　　EQUIPMENT SPECIFICATIONS
　　　PRODUCT DEVELOPMENT
　　　∞PRODUCTION
　　　PRODUCTION ENGINEERING

AIRCRAFT PRODUCTION COSTS
GS　　COSTS
　　　. **AIRCRAFT PRODUCTION COSTS**
　　　PRODUCTION COSTS
　　　. **AIRCRAFT PRODUCTION COSTS**
RT　　∞AIRCRAFT
　　　AIRCRAFT INDUSTRY
　　　COST ESTIMATES
　　　EFFICIENCY
　　　∞ENGINEERING
　　　FINANCIAL MANAGEMENT
　　　INDUSTRIES
　　　MANUFACTURING
　　　PRODUCTION ENGINEERING
　　　PRODUCTION MANAGEMENT
　　　PRODUCTIVITY

AIRCRAFT RELIABILITY
UF　　AIRWORTHINESS
　　　AIRWORTHINESS REQUIREMENTS
GS　　RELIABILITY
　　　. **AIRCRAFT RELIABILITY**
RT　　∞AIRCRAFT
　　　CERTIFICATION
　　　CIRCUIT RELIABILITY
　　　COMPONENT RELIABILITY
　　　HELICOPTER PERFORMANCE
　　　QUALITY CONTROL
　　　STRUCTURAL RELIABILITY
　　　VULNERABILITY

AIRCRAFT RUNUP
GS　　PREFLIGHT OPERATIONS
　　　. **AIRCRAFT RUNUP**
RT　　∞AIRCRAFT
　　　ENGINE NOISE
　　　ENGINE TESTS
　　　GROUND TESTS
　　　JET AIRCRAFT NOISE

AIRCRAFT SAFETY
GS　　SAFETY
　　　. **AIRCRAFT SAFETY**
RT　　ABORT APPARATUS
　　　AEROSPACE SAFETY
　　　AIR PIRACY
　　　AIR TRAFFIC CONTROL
　　　∞AIRCRAFT
　　　AIRCRAFT ACCIDENTS
　　　AIRCRAFT APPROACH SPACING
　　　AIRCRAFT SPIN
　　　ALL-WEATHER LANDING SYSTEMS
　　　ARRESTING GEAR
　　　BEACON COLLISION AVOIDANCE
　　　　SYSTEM
　　　COLLISION AVOIDANCE
　　　COLLISIONS
　　　CRASH LANDING
　　　CRASHES
　　　CRASHWORTHINESS
　　　EJECTION SEATS
　　　FLIGHT HAZARDS
　　　FLIGHT SAFETY

AIRCRAFT SAFETY-*(CONT.)*
　　　FLYING EJECTION SEATS
　　　LANDING AIDS
　　　LANDING RADAR
　　　MICROWAVE LANDING SYSTEMS
　　　MIDAIR COLLISIONS
　　　NATIONAL AIRSPACE SYSTEM
　　　NAVIGATION AIDS
　　　SAFETY DEVICES
　　　SOLAR COMPASSES
　　　THREAT EVALUATION
　　　WEATHER
　　　WHEEL BRAKES

AIRCRAFT SPECIFICATIONS
GS　　SPECIFICATIONS
　　　. **AIRCRAFT SPECIFICATIONS**
RT　　∞AIRCRAFT
　　　AIRSPEED
　　　CEILING (AIRCRAFT CAPABILITY)
　　　CONTROLLABILITY
　　　DISTANCE
　　　FLIGHT CHARACTERISTICS
　　　PAYLOADS

AIRCRAFT SPIN
RT　　AERODYNAMIC STALLING
　　　∞AIRCRAFT
　　　AIRCRAFT PERFORMANCE
　　　AIRCRAFT SAFETY
　　　CONTROL STABILITY
　　　CONTROLLABILITY
　　　CRASH LANDING
　　　FLIGHT HAZARDS
　　　FLIGHT SAFETY
　　　HAZARDS
　　　MANEUVERS
　　　SPIN DYNAMICS

AIRCRAFT STABILITY
GS　　DYNAMIC CHARACTERISTICS
　　　. DYNAMIC STABILITY
　　　. . MOTION STABILITY
　　　. . . **AIRCRAFT STABILITY**
　　　. . . . HOVERING STABILITY
　　　STABILITY
　　　. DYNAMIC STABILITY
　　　. . MOTION STABILITY
　　　. . . **AIRCRAFT STABILITY**
　　　. . . . HOVERING STABILITY
RT　　AERODYNAMIC BALANCE
　　　AERODYNAMIC STABILITY
　　　∞AIRCRAFT
　　　ATTITUDE STABILITY
　　　BUFFETING
　　　CONTROL STABILITY
　　　CONTROLLABILITY
　　　COUNTERBALANCES
　　　DIRECTIONAL STABILITY
　　　HORIZONTAL FLIGHT
　　　LATERAL STABILITY
　　　LIQUID SLOSHING
　　　LONGITUDINAL STABILITY
　　　LOW SPEED STABILITY
　　　PILOT INDUCED OSCILLATION
　　　STATIC STABILITY
　　　STRUCTURAL STABILITY
　　　TURNING FLIGHT
　　　UPPER SURFACE BLOWN FLAPS
　　　WIND TUNNEL STABILITY TESTS

AIRCRAFT STRUCTURES
UF　　AIRCRAFT CONSTRUCTION
GS　　**AIRCRAFT STRUCTURES**
　　　. AFTERBODIES
　　　. AIRFRAMES
　　　. CENTERBODIES
　　　. FOREBODIES
　　　. . NOSES (FOREBODIES)
　　　. FUSELAGES
　　　. PLASTIC AIRCRAFT STRUCTURES
RT　　AERODYNAMIC INTERFERENCE
　　　AEROELASTICITY
　　　∞AIRCRAFT
　　　AIRFOILS
　　　BORON-EPOXY COMPOUNDS
　　　CANARD CONFIGURATIONS
　　　CANOPIES
　　　CHANNEL WINGS
　　　CONTROL SURFACES
　　　FAIRINGS
　　　HULLS (STRUCTURES)
　　　LEADING EDGE FLAPS
　　　OBLIQUE WINGS
　　　PYLON MOUNTING

AIRCRAFT STRUCTURES-(CONT.)
    SHELLS (STRUCTURAL FORMS)
    SPACECRAFT STRUCTURES
    STREAMLINING
  ∞ STRUCTURES
    SWING TAIL ASSEMBLIES
    SWING WINGS
    TAIL ASSEMBLIES
    WINGS

AIRCRAFT SURVIVABILITY
  RT ∞ AIRCRAFT
    COMBAT
    ∞ CONSTRUCTION MATERIALS
    DURABILITY
    FLIGHT CONTROL
    HELICOPTERS
    LIFE (DURABILITY)
    ∞ MILITARY AIRCRAFT
    PLASTIC AIRCRAFT STRUCTURES
    REINFORCED PLASTICS
    RELIABILITY
    SPACECRAFT SURVIVABILITY
    SURVIVAL
    SURVIVAL EQUIPMENT
    VULNERABILITY

AIRCRAFT TIRES
  GS TIRES
    . AIRCRAFT TIRES
  RT ∞ AIRCRAFT
    LANDING GEAR
    VEHICLE WHEELS

AIRCRAFT WAKES
  GS WAKES
    . AIRCRAFT WAKES
    . . HELICOPTER WAKES
    . . SLIPSTREAMS
    . . . PROPELLER SLIPSTREAMS
  RT ∞ AIRCRAFT
    HYPERSONIC WAKES
    LAMINAR WAKES
    SUPERSONIC WAKES
    TURBULENT WAKES
    VORTEX ADVISORY SYSTEM
    VORTEX ALLEVIATION

AIRCREWS
  USE FLIGHT CREWS

AIRDROPS
  RT AIR CARGO
    CARGO
    DELIVERY
    DRAG CHUTES
    PARACHUTES

AIRFIELD SURFACE MOVEMENTS
  RT AIR CARGO
    AIRPORTS
    HANGARS
    MATERIALS HANDLING
    MOBILE LOUNGES
    RUNWAYS
    ∞ SURFACES
    TAXIING

AIRFIELDS
  USE AIRPORTS

AIRFOIL CHARACTERISTICS
  USE AIRFOILS

AIRFOIL FENCES
  GS AIRFOILS
    . AIRFOIL FENCES
  RT BOUNDARY LAYER CONTROL
    ∞ FENCES
    VORTEX GENERATORS
    WINGS

AIRFOIL PROFILES
  UF AERODYNAMIC CHORDS
    AIRFOIL SECTIONS
    AIRFOIL THICKNESS
    CLARK Y AIRFOIL
  GS **AIRFOIL PROFILES**
    . WING PROFILES
    . . WING SPAN
  RT AERODYNAMIC INTERFERENCE
    AIRFOILS
    BLADE TIPS
    ∞ CROSS SECTIONS

AIRFOIL PROFILES-(CONT.)
    JOUKOWSKI TRANSFORMATION
    KUTTA-JOUKOWSKI CONDITION
    LIGHTHILL METHOD
    NOSE TIPS
  ∞ PROFILES
    STREAMLINING
    SUPERCRITICAL AIRFOILS
    THEODORSEN TRANSFORMATION
    THICKNESS
    THICKNESS RATIO
    THIN AIRFOILS
    THIN WINGS
    TIPS
    WEDGES
    WING TIPS

AIRFOIL SECTIONS
  USE AIRFOIL PROFILES

AIRFOIL THICKNESS
  USE AIRFOIL PROFILES

AIRFOILS
  UF AIRFOIL CHARACTERISTICS
  GS **AIRFOILS**
    . AERIAL RUDDERS
    . AILERONS
    . . FLAPERONS
    . . SPOILER SLOT AILERONS
    . AIRFOIL FENCES
    . CIRCULATION CONTROL AIRFOILS
    . DROOPED AIRFOILS
    . ELEVATORS (CONTROL SURFACES)
    . ELEVONS
    . FLAPS (CONTROL SURFACES)
    . . EXTERNALLY BLOWN FLAPS
    . . . UPPER SURFACE BLOWN FLAPS
    . . FLAPERONS
    . . JET FLAPS
    . . SPLIT FLAPS
    . . WING FLAPS
    . . . LEADING EDGE SLATS
    . . . TRAILING EDGE FLAPS
    . . . VORTEX FLAPS
    . HORIZONTAL TAIL SURFACES
    . LAMINAR FLOW AIRFOILS
    . PROPELLER BLADES
    . SPOILERS
    . SUPERCRITICAL AIRFOILS
    . . SUPERCRITICAL WINGS
    . SUPERSONIC AIRFOILS
    . TABS (CONTROL SURFACES)
    . THIN AIRFOILS
    . . THIN WINGS
    . . . INFINITE SPAN WINGS
    . WINGS
    . . AEROELASTIC RESEARCH WINGS
    . . CAMBERED WINGS
    . . CARET WINGS
    . . CHANNEL WINGS
    . . CRUCIFORM WINGS
    . . FIXED WINGS
    . . FLEXIBLE WINGS
    . . PARAWINGS
    . . GAW-1 AIRFOIL
    . . GAW-2 AIRFOIL
    . . LOW ASPECT RATIO WINGS
    . . . DELTA WINGS
    . . . TRAPEZOIDAL WINGS
    . . OBLIQUE WINGS
    . . RIGID WINGS
    . . ROTARY WINGS
    . . . LIFTING ROTORS
    . . . . BEARINGLESS ROTORS
    . . . RIGID ROTORS
    . . . TILTING ROTORS
    . . . TIP DRIVEN ROTORS
    . . SLENDER WINGS
    . . . INFINITE SPAN WINGS
    . . SUPERCRITICAL WINGS
    . . SWEPT WINGS
    . . . SWEPT FORWARD WINGS
    . . . . TRAPEZOIDAL WINGS
    . . . SWEPTBACK WINGS
    . . . . ARROW WINGS
    . . . . DELTA WINGS
    . . . . TRAPEZOIDAL WINGS
    . . SWING WINGS
    . . THIN WINGS
    . . . INFINITE SPAN WINGS
    . . TWISTED WINGS
    . . UNCAMBERED WINGS
    . . RING WINGS
    . . UNSWEPT WINGS

AIRFOILS-(CONT.)
    . . . INFINITE SPAN WINGS
    . . . RECTANGULAR WINGS
    . . . RING WINGS
    . . VARIABLE SWEEP WINGS
  RT AERODYNAMIC CONFIGURATIONS
    AERODYNAMICS
    AIRCRAFT DESIGN
    AIRCRAFT PARTS
    AIRCRAFT STRUCTURES
    AIRFOIL PROFILES
    ASPECT RATIO
  ∞ BLADES
    BLUNT LEADING EDGES
    BLUNT TRAILING EDGES
    BODY-WING CONFIGURATIONS
    CAMBER
    CONTROL SURFACES
    DEICERS
    DEICING
    FINS
  ∞ FOILS
    FOILS (MATERIALS)
    GUIDE VANES
    HYDROFOILS
    INTERACTIONAL AERODYNAMICS
    JET VANES
    LEADING EDGE FLAPS
    LEADING EDGE THRUST
    LEADING EDGES
    LIFT
    LIFTING BODIES
    LIGHTHILL METHOD
    MONOPLANES
    ROTOR BLADES (TURBOMACHINERY)
    ROTORS
    RUDDERS
    STABILIZERS (FLUID DYNAMICS)
    STREAMLINED BODIES
    STREAMLINING
    TAIL ASSEMBLIES
    THICKNESS RATIO
    TRAILING EDGES
    TURBOMACHINE BLADES
    VANES
    WEDGES

AIRFRAME MATERIALS
  GS AIRCRAFT CONSTRUCTION MATERIALS
    . AIRFRAME MATERIALS
  RT AIRFRAMES
    COMPOSITE MATERIALS
    ∞ CONSTRUCTION MATERIALS
    GLASS FIBER REINFORCED PLASTICS
    ∞ MATERIALS
    STRUCTURAL DESIGN
    STRUCTURAL MEMBERS

AIRFRAMES
  GS AIRCRAFT STRUCTURES
    . AIRFRAMES
    FRAMES
    . AIRFRAMES
  RT AIRCRAFT CONSTRUCTION MATERIALS
    AIRCRAFT PARTS
    AIRFRAME MATERIALS
    BAYS (STRUCTURAL UNITS)
    CANOPIES
    CONTROL SURFACES
    ENGINE AIRFRAME INTEGRATION
    FINS
    FUSELAGES
    LANDING GEAR
    MISSILE BODIES
    MISSILE STRUCTURES
    NACELLES
    PROTUBERANCES
    TAIL ASSEMBLIES
    WING NACELLE CONFIGURATIONS
    WINGS

AIRGEEP AIRCRAFT
  USE VZ-8 AIRCRAFT

AIRGLOW
  UF ATMOSPHERIC EMISSION
  GS ATMOSPHERIC RADIATION
    . SKY RADIATION
    . . AIRGLOW
    . . . GEOCORONAL EMISSIONS
    . . . NIGHTGLOW
    . . . TWILIGHT GLOW
    ELECTROMAGNETIC RADIATION
    . LIGHT (VISIBLE RADIATION)
    . . SKY RADIATION

AIRGLOW-(CONT.)
```
         . . . AIRGLOW
         . . . . GEOCORONAL EMISSIONS
         . . . . NIGHTGLOW
         . . . . TWILIGHT GLOW
RT       AERONOMY
         ATMOSPHERIC IONIZATION
         AURORAS
         CHEMILUMINESCENCE
         EARTH ATMOSPHERE
         EMISSION
         FABRY-PEROT SPECTROMETERS
         LIGHT EMISSION
         NIGHT SKY
         OXYGEN SPECTRA
         RADIATIVE RECOMBINATION
         RAYLEIGH SCATTERING
         SKY BRIGHTNESS
```

AIRLINE OPERATIONS
```
RT       AIR CARGO
         AIR TRAFFIC
         CIVIL AVIATION
         COMMERCIAL AIRCRAFT
         OPERATING COSTS
       ∞ OPERATIONS
         PASSENGERS
         SHORT HAUL AIRCRAFT
```

AIRLOCK MODULES
```
GS       MODULES
         . AIRLOCK MODULES
RT       AIR LOCKS
         APOLLO APPLICATIONS PROGRAM
         MULTIPLE DOCKING ADAPTERS
         SATURN WORKSHOPS
         SATURN 1 WORKSHOP
         SATURN 5 WORKSHOP
         SKYLAB PROGRAM
         SKYLAB 1
         SKYLAB 2
         SKYLAB 3
         SKYLAB 4
         SPACECRAFT DOCKING MODULES
```

AIRPORT BEACONS
```
GS       LANDING AIDS
         . AIRPORT BEACONS
         . . DISCRETE ADDRESS BEACON
             SYSTEM
         NAVIGATION AIDS
         . BEACONS
         . . AIRPORT BEACONS
         . . . DISCRETE ADDRESS BEACON
             SYSTEM
RT       RADIO BEACONS
         SOLAR COMPASSES
```

AIRPORT LIGHTS
```
GS       LANDING AIDS
         . AIRPORT LIGHTS
         . . RUNWAY LIGHTS
         LIGHTING EQUIPMENT
         . LUMINAIRES
         . . AIRPORT LIGHTS
         . . . RUNWAY LIGHTS
RT       SEARCHLIGHTS
```

AIRPORT PLANNING
```
GS       PLANNING
         . AIRPORT PLANNING
RT       GROUND SUPPORT EQUIPMENT
         HELIPORTS
         LAND USE
         SITES
```

AIRPORT SECURITY
```
GS       SECURITY
         . AIRPORT SECURITY
RT       AIR PIRACY
         AIRPORTS
         PROTECTION
         VULNERABILITY
```

AIRPORT SURFACE DETECTION EQUIPMENT
```
UF       ASDE
RT       AIR TRAFFIC CONTROL
       ∞ EQUIPMENT
         GROUND BASED CONTROL
         RADAR EQUIPMENT
         SEARCH RADAR
       ∞ SURFACES
         SURVEILLANCE RADAR
```

AIRPORT TOWERS
```
GS       TOWERS
         . AIRPORT TOWERS
RT       AIR TRAFFIC CONTROL
         AIR TRAFFIC CONTROLLERS
             (PERSONNEL)
         AIRPORTS
         GROUND BASED CONTROL
         HELIPORTS
         LANDING AIDS
         TRAFFIC CONTROL
```

AIRPORTS
```
UF       AIRFIELDS
GS       AIRPORTS
         . HELIPORTS
RT     ∞ AERONAUTICS
         AIR TRAFFIC CONTROL
         AIRFIELD SURFACE MOVEMENTS
         AIRPORT SECURITY
         AIRPORT TOWERS
       ∞ FACILITIES
         HANGARS
         INSTRUMENT LANDING SYSTEMS
         LANDING AIDS
         LANDING MATS
         MILITARY AIR FACILITIES
         MOBILE LOUNGES
         MOORING
         NATIONAL AIRSPACE SYSTEM
         NAVIGATION AIDS
       ∞ PORTS
         RUNWAYS
         SITE SELECTION
       ∞ STRIP
```

AIRS (RECONNAISSANCE SYS)
```
USE      AIRBORNE INTEGRATED
             RECONNAISSANCE SYSTEM
```

AIRSHIPS
```
UF       AEROSTATS
         DIRIGIBLES
GS       AIRSHIPS
         . HEAVY LIFT AIRSHIPS
RT     ∞ AIRCRAFT
         BALLOONS
         GONDOLAS
         INFLATABLE STRUCTURES
       ∞ MILITARY AIRCRAFT
```

AIRSPACE
```
RT       AIR LAW
         AIR TRAFFIC
         AIR TRAFFIC CONTROL
         AIRCRAFT APPROACH SPACING
         BOUNDARIES
         COLLISION AVOIDANCE
         FLIGHT PATHS
         NATIONAL AIRSPACE SYSTEM
         NATIONAL AIRSPACE UTILIZATION
             SYSTEM
```

AIRSPEED
```
GS       RATES (PER TIME)
         . AIRSPEED
         VELOCITY
         . AIRSPEED
RT       AERODYNAMIC STALLING
         AIRCRAFT PERFORMANCE
         AIRCRAFT SPECIFICATIONS
         BOUNDARY LAYER SEPARATION
         FLIGHT CHARACTERISTICS
         GROUND SPEED
         HIGH SPEED
         LOW SPEED
         MACH NUMBER
         WIND VELOCITY
```

AIRWORTHINESS
```
USE      AIRCRAFT RELIABILITY
```

AIRWORTHINESS REQUIREMENTS
```
USE      AIRCRAFT RELIABILITY
```

AIRY FUNCTION
```
GS       ANALYSIS (MATHEMATICS)
         . COMPLEX VARIABLES
         . . AIRY FUNCTION
         FUNCTIONS (MATHEMATICS)
         . AIRY FUNCTION
RT       CYLINDRICAL BODIES
         DIFFERENTIAL EQUATIONS
         ELASTIC PROPERTIES
         HARMONIC FUNCTIONS
```

AIRY FUNCTION-(CONT.)
```
         POISSON RATIO
         STRESS ANALYSIS
```

AITKEN NUCLEI
```
GS       CONDENSATION NUCLEI
         . AITKEN NUCLEI
RT       AEROSOLS
         ATMOSPHERIC CHEMISTRY
         ATMOSPHERIC COMPOSITION
         CLOUD PHYSICS
         COAGULATION
         CONDENSATES
         CRYSTAL GROWTH
         DUST
         ICE NUCLEI
         NUCLEATION
       ∞ NUCLEI
         SUPERCOOLING
```

AJ-10 ENGINE
```
GS       ENGINES
         . ROCKET ENGINES
         . . BOOSTER ROCKET ENGINES
         . . . AJ-10 ENGINE
         . LIQUID PROPELLANT ROCKET
             ENGINES
         . . . AJ-10 ENGINE
RT       TARTAR MISSILE
```

AJ-1000 ENGINE
```
USE      M-1 ENGINE
```

AKERMANITE
```
GS       CALCIUM COMPOUNDS
         . CALCIUM CARBONATES
         . . AKERMANITE
         . CALCIUM OXIDES
         . . AKERMANITE
         CARBON COMPOUNDS
         . CARBONATES
         . . CALCIUM CARBONATES
         . . . AKERMANITE
         CHALCOGENIDES
         . AKERMANITE
         MAGNESIUM COMPOUNDS
         . AKERMANITE
         MINERALS
         . GEHLENITE
         . . AKERMANITE
RT       SILICATES
         SILICON COMPOUNDS
         SILICON OXIDES
```

ALABAMA
```
GS       NATIONS
         . UNITED STATES
         . . ALABAMA
RT       GULF OF MEXICO
         TENNESSEE VALLEY (AL-KY-TN)
```

ALADIN 2 AIRCRAFT
```
GS       TRANSPORT AIRCRAFT
         . ALADIN 2 AIRCRAFT
RT     ∞ AIRCRAFT
```

ALAIS METEORITE
```
GS       CELESTIAL BODIES
         . METEORITES
         . . STONY METEORITES
         . . . CHONDRITES
         . . . . CARBONACEOUS METEORITES
         . . . . . ALAIS METEORITE
```

ALANINE
```
GS       ACIDS
         . AMINO ACIDS
         . . ALANINE
         . . . PHENYLALANINE
         ORGANIC COMPOUNDS
         . AMINO ACIDS
         . . ALANINE
         . . . PHENYLALANINE
RT       PROTEINS
```

ALARM PROJECT
```
UF       AUTOMATIC LIGHT AIRCRAFT
             READINESS MONITOR
GS       PROGRAMS
         . PROJECTS
         . . ALARM PROJECT
RT       MONITORS
```

ALARMS
USE   WARNING SYSTEMS

ALASKA
GS   NATIONS
          . UNITED STATES
          . . **ALASKA**
RT   ALEUTIAN ISLANDS (US)
          BEAUFORT SEA (NORTH AMERICA)
          CHENA RIVER BASIN (AK)
          COOK INLET (AK)
          GULF OF ALASKA
          PRINCE WILLIAM SOUND (AK)
          WRANGELL MOUNTAINS (AK)

ALBANIA
GS   NATIONS
          . **ALBANIA**
RT   EUROPE

ALBEDO
GS   **ALBEDO**
          . COSMIC RAY ALBEDO
          . EARTH ALBEDO
          . LUNAR ALBEDO
RT   ABSORPTANCE
          COSMIC RAYS
          EARTH RADIATION BUDGET
            EXPERIMENT
          OPTICAL PROPERTIES
          PLANETARY RADIATION
          REFLECTANCE
          SOLAR RADIATION
          SURFACE PROPERTIES

ALBERTA
GS   NATIONS
          . CANADA
          . . **ALBERTA**

ALBINISM
GS   DISEASES
          . **ALBINISM**
RT   PIGMENTS
          SKIN (ANATOMY)

ALBUMINS
GS   PROTEINS
          . **ALBUMINS**
RT   ELASTIN

ALCOHOLS
GS   HYDROXYL COMPOUNDS
          . **ALCOHOLS**
          . . ETHYL ALCOHOL
          . . GLYCOLS
          . . ISOPROPYL ALCOHOL
          . . METHYL ALCOHOLS
          . . PHENOLS
          . . . BISPHENOLS
          . . . CRESOLS
          . . . PHLOROGLUCINOL
          . . . THYMOL
          . . POLYVINYL ALCOHOL
          . . TRIOLS
          . . . CYANURIC ACID
RT   ALIPHATIC COMPOUNDS
          CARBOHYDRATES
          GASOHOL (FUEL)
          GLYCEROLS
          HYDROXYL RADICALS
          METHOXY SYSTEMS
          THIOLS

ALDEHYDES
GS   **ALDEHYDES**
          . ACETALDEHYDE
          . ACROLEINS
          . CHLORAL
          . FORMALDEHYDE
RT   ACETYL COMPOUNDS
          ALIPHATIC COMPOUNDS
          FURFURYL ALCOHOL
          RETINENE

ALDOLASE
GS   ENZYMES
          . **ALDOLASE**
RT   MUSCLES

ALDOSTERONE
GS   STEROIDS
          . CORTICOSTEROIDS
          . . **ALDOSTERONE**

ALDOSTERONE-*(CONT.)*
RT   ADRENAL METABOLISM

ALERTNESS
RT   AROUSAL
          ATTENTION
          WAKEFULNESS

ALEUTIAN ISLANDS (US)
GS   LANDFORMS
          . ISLANDS
          . . **ALEUTIAN ISLANDS (US)**
RT   ALASKA
          ARCHIPELAGOES
          ISLAND ARCS
          UNITED STATES

ALFALFA
GS   FARM CROPS
          . **ALFALFA**
          PLANTS (BOTANY)
          . **ALFALFA**
RT   AGRICULTURE
          BLIGHT
          BOTANY
          CROP GROWTH
          CROP VIGOR
          ∞CROPS
          CURING
          EARTH RESOURCES
          ∞FOOD
          GRASSES
          IRRIGATION
          SEEDS

ALFVEN WAVES
USE   MAGNETOHYDRODYNAMIC WAVES

ALGAAS
USE   ALUMINUM GALLIUM ARSENIDES

ALGAE
UF   ALGAL BLOOM
GS   **ALGAE**
          . ANABAENA
          . BLUE GREEN ALGAE
          . . NOSTOC
          . CHLORELLA
          . DUNALIELLA
          . MICROCYSTIS
          . PORPHYRA
          . SCENEDESMUS
RT   BIOCHEMICAL OXYGEN DEMAND
          BIOCONVERSION
          CHLOROPHYLLS
          EUGLENA
          LICHENS
          MARINE BIOLOGY
          PHOTOSYNTHESIS
          PLANKTON
          THERMOPHILES
          THERMOPHILIC PLANTS
          WATER POLLUTION

ALGAL BLOOM
USE   ALGAE

ALGEBRA
GS   **ALGEBRA**
          . BINOMIAL THEOREM
          . CURRENT ALGEBRA
          . DETERMINANTS
          . GROUP THEORY
          . . HOMOMORPHISMS
          . . . AUTOMORPHISMS
          . . . MONOIDS
          . . . SUBGROUPS
          . LIE GROUPS
          . SPINOR GROUPS
          . LINEAR EQUATIONS
          . . LINEAR EVOLUTION EQUATIONS
          . LINEAR TRANSFORMATIONS
          . NONLINEAR EQUATIONS
          . . CUBIC EQUATIONS
          . . DUFFING DIFFERENTIAL EQUATION
          . . MONGE-AMPERE EQUATION
          . NONLINEAR EVOLUTION EQUATIONS
          . . QUADRATIC EQUATIONS
          . . QUARTIC EQUATIONS
          . POLYNOMIALS
          . . BINOMIALS
          . . DYADICS
          . HERMITIAN POLYNOMIAL
          . TENSORS
          . . STRESS TENSORS

ALGEBRA-*(CONT.)*
          . VECTOR SPACES
          . . HILBERT SPACE
          . . . BANACH SPACE
          . . . . SOBOLEV SPACE
          . . MATRICES (MATHEMATICS)
          . . . ADJOINTS
          . . . CANONICAL FORMS
          . . . EIGENVALUES
          . . . EIGENVECTORS
          . . . JORDAN FORM
          . . . STIFFNESS MATRIX
          . STOKES THEOREM (VECTOR
            CALCULUS)
          . U SPIN SPACE
          . . VECTORS (MATHEMATICS)
          . . . EIGENVECTORS
          . . . STATE VECTORS
          . . . VORTICITY
RT   ANALYSIS (MATHEMATICS)
          ∞ANALYZING
          BOOLEAN ALGEBRA
          COORDINATES
          FUNCTIONS (MATHEMATICS)
          HOMOTROPY
          ∞MATHEMATICS
          SCHWARTZ INEQUALITY
          ∞SCIENCE
          SEMIEMPIRICAL EQUATIONS
          ∞SPACE
          SUMS
          UNIQUENESS THEOREM

ALGERIA
GS   NATIONS
          . **ALGERIA**
RT   AFRICA

ALGOL
GS   LANGUAGES
          . PROGRAMMING LANGUAGES
          . . **ALGOL**
RT   COMPUTER PROGRAMMING
          MACHINE ORIENTED LANGUAGES

ALGOL ENGINE
GS   ENGINES
          . ROCKET ENGINES
          . . BOOSTER ROCKET ENGINES
          . . . **ALGOL ENGINE**
          . . SOLID PROPELLANT ROCKET
            ENGINES
          . . . **ALGOL ENGINE**
RT   BLUE SCOUT ROCKET VEHICLE
          LITTLE JOE 2 LAUNCH VEHICLE
          SCOUT LAUNCH VEHICLE

ALGORITHMS
GS   MATHEMATICAL LOGIC
          . **ALGORITHMS**
          . . PARSING ALGORITHMS
          . . SIMPLEX METHOD
RT   COMPUTER PROGRAMMING
          COMPUTER PROGRAMS
          COMPUTER SYSTEMS PROGRAMS
          COMPUTERIZED SIMULATION
          CONJUGATE GRADIENT METHOD
          DATA CONVERSION ROUTINES
          DIFFERENTIAL ANALYZERS
          FACTORIZATION
          FUZZY SETS
          FUZZY SYSTEMS
          HESSIAN MATRICES
          MEAN SQUARE VALUES
          NUMERICAL ANALYSIS
          NUMERICAL DIFFERENTIATION
          PARAMETERIZATION
          ROBUSTNESS (MATHEMATICS)
          STATE ESTIMATION

ALIGNMENT
GS   **ALIGNMENT**
          . SELF ALIGNMENT
RT   ADJUSTING
          BEARING (DIRECTION)
          CLEARANCES
          COLLIMATION
          CORRECTION
          DIRECTIVITY
          FITTING
          HORIZONTAL ORIENTATION
          INSTRUMENT ORIENTATION
          LOOK ANGLES (ELECTRONICS)
          ∞ORIENTATION
          PLY ORIENTATION

ALIGNMENT-*(CONT.)*
POLARIZATION (SPIN ALIGNMENT)
POSITIONING
VERTICAL ORIENTATION

ALIPHATIC COMPOUNDS
GS   **ALIPHATIC COMPOUNDS**
. ACETALDEHYDE
. ACETALS
. ACETYL COMPOUNDS
. ACROLEINS
. ACRYLATES
. ACRYLIC ACID
. ACRYLONITRILES
. ALKANES
. . BUTANES
. . CETANE
. . ETHANE
. . HEPTANES
. . HEXENES
. . METHANE
. . NITROPROPANE
. . NONANES
. . OCTANES
. . PARAFFINS
. . . CERESIN
. . PENTANES
. . . NEOPENTANE
. . PROPANE
. ALKENES
. . BUTADIENE
. . BUTENES
. . ETHYLENE
. . . VINYLIDENE
. . HEXENES
. . PROPYLENE
. . TRIENES
. ALKYL COMPOUNDS
. . ALKYLIDENE
. . CETYL COMPOUNDS
. . DIBUTYL COMPOUNDS
. . HEXYL COMPOUNDS
. . ISOPROPYL NITRATE
. . METHYL NITRATE
. . PROPYL NITRATE
. . TRIETHYL COMPOUNDS
. . TRIMETHYL COMPOUNDS
. ALKYLATES
. ALKYNES
. . ACETYLENE
. . OXYACETYLENE
. ALLYL COMPOUNDS
. CARBAMATES (TRADENAME)
. . URETHANES
. CARBAMIDES
. CARBON TETRACHLORIDE
. CARBON TETRAFLUORIDE
. CARBOXYLATES
. CASTOR OIL
. CHLORAL
. CHLOROETHYLENE
. CHLOROFORM
. CHOLINE
. CITRIC ACID
. CYANAMIDES
. CYANOGEN
. CYCLIC HYDROCARBONS
. . ANTHRACENE
. . COLCHICINE
. . CYCLOBUTANE
. . CYCLOPROPANE
. . MENTHOL
. . NAPHTHENES
. DDT
. DIALLYL COMPOUNDS
. DIISOCYANATES
. ETHOXY ETHYLENE
. ETHYL ALCOHOL
. ETHYLENEDIAMINE
. FATTY ACIDS
. . ACETIC ACID
. . . ETHYLENEDIAMINETETRAACETIC
. . . . ACIDS
. . . IODOACETIC ACID
. . LIPOIC ACID
. . OLEIC ACID
. . PALMITIC ACID
. . PROPIONIC ACID
. . SEBACIC ACID
. . VALERIC ACID
. FLUOROAMINES
. . NITROFLUORAMINES
. . TRIFLUOROAMINE OXIDE
. FORMHYDROXAMIC ACID
. FORMIC ACID
. GLUCOSIDES

ALIPHATIC COMPOUNDS-*(CONT.)*
. GLUTAMATES
. . GLUTAMIC ACID
. . GLUTAMINE
. . GLUTATHIONE
. GLYCERIDES
. GLYCEROLS
. GLYCOLS
. GUANIDINES
. . GUANETHIDINE
. . TRIAMINOGUANIDINIUM AZIDE
. HEPTADIENE
. HEXADIENE
. HEXAMETHYLENETETRAMINE
. HEXOGENES (TRADEMARK)
. HEXOKINASE
. HIPPURIC ACID
. HYDRAZIDES
. HYDRAZINE NITROFORM
. HYDRAZINES
. . CHLORPROMAZINE
. . DIHYDRAZINE
. . DIMETHYLHYDRAZINES
. . ETHYLENE DIHYDRAZINE
. . HYDRAZINE BORANE
. . HYDRAZINE PERCHLORATES
. . METHYLHYDRAZINE
. . TETRAFLUOROHYDRAZINE
. ISOPROPYL COMPOUNDS
. . ISOPROPYL ALCOHOL
. KETENES
. KETONES
. . ACETONE
. . ACETYLACETONE
. . ANTHRAQUINONES
. . CAMPHOR
. . NEMBUTAL (TRADEMARK)
. . PENTANONE
. . TRIMETHADIONE
. LACTATES
. LACTIC ACID
. MALEATES
. MEPROBAMATE
. METHYL ALCOHOLS
. METHYL COMPOUNDS
. . METHYL CHLOROSILANES
. . METHYL NITRATE
. METHYLENE DIAMINE
. MONOETHANOLAMINE (MEA)
. NITRATE ESTERS
. . ISOPROPYL NITRATE
. . PROPYL NITRATE
. NITROAMINES
. NITROSAMINE
. NUCLEASE
. NUCLEOSIDES
. . ADENINES
. . ADENOSINES
. . . ADENOSINE DIPHOSPHATE
. . . ADENOSINE TRIPHOSPHATE
. OCTOATES
. OXALIC ACID
. OXAMIC ACIDS
. PHOSGENE
. POLYSACCHARIDES
. . CELLULOSE
. . . FORTISAN (TRADEMARK)
. . CHITIN
. . DEXTRANS
. GLYCOGENS
. . STARCHES
. PROPYL COMPOUNDS
. SODIUM CHLORODIFLUOROACETATES
. STEARATES
. SUGARS
. . DEXTRANS
. . GALACTOSE
. . GLUCOSE
. . HEXOSES
. . INOSITOLS
. . LACTOSE
. . MANNITOL
. . MONOSACCHARIDES
. . . RIBOSE
. . . XYLOSE
. . . PENTOSE
. . . RIBOSE
. . . XYLOSE
. . SUCROSE
. TETRABUTYLS
. THIOLS
. . CYSTEINE
. . DIMERCAPROL
. TRINITRAMINE
. TRIOLS
. . CYANURIC ACID

ALIPHATIC COMPOUNDS-*(CONT.)*
RT   ACETATES
ALCOHOLS
ALDEHYDES
AMIDES
AMINES
CARBONYL COMPOUNDS
CARBOXYLIC ACIDS
∞ CHEMICAL COMPOUNDS
CYANIDES
ESTERS
HYDROCARBONS
HYDROXYL COMPOUNDS
NITRITES
NITROSYLS
PHOSPHORUS COMPOUNDS

ALKALI HALIDES
GS   HALOGEN COMPOUNDS
. HALIDES
. . METAL HALIDES
. . . **ALKALI HALIDES**
. . . . CESIUM HALIDES
. . . . . CESIUM BROMIDES
. . . . . CESIUM FLUORIDES
. . . . . CESIUM IODIDES
. . . . POTASSIUM IODIDES
. . . . SODIUM BROMIDES
. . . . SODIUM CHLORIDES
. . . . SODIUM FLUORIDES
. . . . SODIUM IODIDES

∞ **ALKALI METAL COMPOUNDS**
SN *(USE OF A MORE SPECIFIC TERM IS
RECOMMENDED--CONSULT THE TERMS
LISTED BELOW)*
UF   GROUP 1A COMPOUNDS
RT   CESIUM COMPOUNDS
∞ CHEMICAL COMPOUNDS
LITHIUM COMPOUNDS
∞ METAL COMPOUNDS
POTASSIUM COMPOUNDS
RUBIDIUM COMPOUNDS
SODIUM COMPOUNDS

ALKALI METALS
GS   CHEMICAL ELEMENTS
. **ALKALI METALS**
. . CESIUM
. . . CESIUM ISOTOPES
. . . . CESIUM 133
. . . . CESIUM 134
. . . . CESIUM 137
. . . . CESIUM 144
. . . CESIUM VAPOR
. . FRANCIUM
. . LITHIUM
. . . LIQUID LITHIUM
. . . LITHIUM ISOTOPES
. . POTASSIUM
. . . LIQUID POTASSIUM
. . . POTASSIUM ISOTOPES
. . . . POTASSIUM 38
. . . . POTASSIUM 39
. . . . POTASSIUM 40
. . RUBIDIUM
. . . RUBIDIUM ISOTOPES
. . . . RUBIDIUM 86
. . SODIUM
. . . LIQUID SODIUM
. . . SODIUM ISOTOPES
. . . . SODIUM 22
. . . . SODIUM 24
. . . SODIUM VAPOR
METALS
. ALKALI METALS
. . CESIUM
. . . CESIUM ISOTOPES
. . . . CESIUM 133
. . . . CESIUM 134
. . . . CESIUM 137
. . . . CESIUM 144
. . . CESIUM VAPOR
. . FRANCIUM
. . LITHIUM
. . . LIQUID LITHIUM
. . . LITHIUM ISOTOPES
. . POTASSIUM
. . . LIQUID POTASSIUM
. . . POTASSIUM ISOTOPES
. . . . POTASSIUM 38
. . . . POTASSIUM 39
. . . . POTASSIUM 40
. . RUBIDIUM
. . . RUBIDIUM ISOTOPES

ALKALI METALS-(CONT.)
```
        . . . . RUBIDIUM 86
        . . SODIUM
        . . . LIQUID SODIUM
        . . . SODIUM ISOTOPES
        . . . . SODIUM 22
        . . . . SODIUM 24
        . . . SODIUM VAPOR
   RT   CESIUM ALLOYS
        METAL VAPORS
```

ALKALI VAPOR LAMPS
```
   GS   LIGHTING EQUIPMENT
        . LUMINAIRES
        . . FLASH LAMPS
        . . . ALKALI VAPOR LAMPS
   RT   LASERS
        LUMINESCENCE
        METAL VAPORS
        RARE EARTH ELEMENTS
```

ALKALIES
```
   UF   CAUSTICS
   GS   ALKALIES
        . LITHIUM HYDROXIDES
        . POTASSIUM HYDROXIDES
        . SODIUM HYDROXIDES
   RT   ALKALINITY
        BASES (CHEMICAL)
        CARBONATES
        HYDROXIDES
```

ALKALINE BATTERIES
```
   GS   ELECTRIC GENERATORS
        . DIRECT POWER GENERATORS
        . . PRIMARY BATTERIES
        . . . ALKALINE BATTERIES
        ELECTROCHEMICAL CELLS
        . ALKALINE BATTERIES
   RT   STORAGE BATTERIES
        THERMAL BATTERIES
```

∞ ALKALINE EARTH COMPOUNDS
```
   SN   (USE OF A MORE SPECIFIC TERM IS
        RECOMMENDED--CONSULT THE TERMS
        LISTED BELOW)
   UF   GROUP 2A COMPOUNDS
   RT   ALKALINE EARTH METALS
        ALKALINE EARTH OXIDES
        BARIUM COMPOUNDS
        BERYLLIUM COMPOUNDS
        CALCIUM COMPOUNDS
        ∞CHEMICAL COMPOUNDS
        MAGNESIUM COMPOUNDS
        STRONTIUM COMPOUNDS
```

ALKALINE EARTH METALS
```
   GS   CHEMICAL ELEMENTS
        . ALKALINE EARTH METALS
        . . BARIUM ISOTOPES
        METALS
        . ALKALINE EARTH METALS
        . . BARIUM ISOTOPES
   RT   ∞ALKALINE EARTH COMPOUNDS
```

ALKALINE EARTH OXIDES
```
   GS   CHALCOGENIDES
        . OXIDES
        . . METAL OXIDES
        . . . ALKALINE EARTH OXIDES
        . . . . BARIUM OXIDES
        . . . . BERYLLIUM OXIDES
        . . . . CALCIUM OXIDES
        . . . . MAGNESIUM OXIDES
        . . . . . PERICLASE
   RT   ∞ALKALINE EARTH COMPOUNDS
```

ALKALINITY
```
   RT   ALKALIES
        CHEMICAL ANALYSIS
        CHEMICAL COMPOSITION
        PH
        SALINITY
        WATER POLLUTION
        WATER QUALITY
```

ALKALOIDS
```
   GS   HETEROCYCLIC COMPOUNDS
        . ALKALOIDS
        . . ATROPINE
        . . BETAINES
        . . COLCHICINE
        . . ERGOTAMINE
        . . HYOSCINE
        . . LYSERGINE
```

ALKALOIDS-(CONT.)
```
        . . MORPHINE
        . . NICOTINAMIDE
        . . NICOTINE
        . . PILOCARPINE
        . . RESERPINE
        . . TROPYL COMPOUNDS
        NITROGEN COMPOUNDS
        . ALKALOIDS
        . . ATROPINE
        . . BETAINES
        . . COLCHICINE
        . . ERGOTAMINE
        . . HYOSCINE
        . . LYSERGINE
        . . MORPHINE
        . . NICOTINAMIDE
        . . NICOTINE
        . . PILOCARPINE
        . . QUINOLINE
        . . RESERPINE
        . . TROPYL COMPOUNDS
   RT   CURARE
        DRUGS
        MARIJUANA
        STRYCHNINE
```

ALKALOSIS
```
   RT   ACIDOSIS
        HYPERVENTILATION
        PH
        PH FACTOR
        TOXICITY
```

ALKANES
```
   UF   SATURATED HYDROCARBONS
   GS   ALIPHATIC COMPOUNDS
        . ALKANES
        . . BUTANES
        . . CETANE
        . . ETHANE
        . . HEPTANES
        . . HEXENES
        . . METHANE
        . . NITROPROPANE
        . . NONANES
        . . OCTANES
        . . PARAFFINS
        . . . CERESIN
        . . PENTANES
        . . . NEOPENTANE
        . . PROPANE
        HYDROCARBONS
        . ALKANES
        . . BUTANES
        . . CETANE
        . . ETHANE
        . . HEPTANES
        . . HEXENES
        . . METHANE
        . . NITROPROPANE
        . . NONANES
        . . OCTANES
        . . PARAFFINS
        . . . CERESIN
        . . PENTANES
        . . . NEOPENTANE
        . . PROPANE
   RT   HYDROCARBON FUELS
        WAXES
```

ALKENES
```
   UF   OLEFINS
   GS   ALIPHATIC COMPOUNDS
        . ALKENES
        . . BUTADIENE
        . . BUTENES
        . . ETHYLENE
        . . . VINYLIDENE
        . . HEXENES
        . . PROPYLENE
        . . TRIENES
        HYDROCARBONS
        . ALKENES
        . . BUTADIENE
        . . BUTENES
        . . ETHYLENE
        . . . VINYLIDENE
        . . HEXENES
        . . PROPYLENE
        . . TRIENES
   RT   ALKYNES
        TERPENES
```

ALKYD RESINS
```
   GS   RESINS
        . ALKYD RESINS
   RT   ADHESIVES
        PROTECTIVE COATINGS
```

ALKYL COMPOUNDS
```
   GS   ALIPHATIC COMPOUNDS
        . ALKYL COMPOUNDS
        . . ALKYLIDENE
        . . CETYL COMPOUNDS
        . . DIBUTYL COMPOUNDS
        . . HEXYL COMPOUNDS
        . . ISOPROPYL NITRATE
        . . METHYL NITRATE
        . . PROPYL NITRATE
        . . TRIETHYL COMPOUNDS
        . . TRIMETHYL COMPOUNDS
   RT   ∞CHEMICAL COMPOUNDS
```

ALKYLATES
```
   GS   ALIPHATIC COMPOUNDS
        . ALKYLATES
        ESTERS
        . ALKYLATES
   RT   ALKYLATION
```

ALKYLATION
```
   UF   OXYALKYLATION
   GS   CHEMICAL REACTIONS
        . ALKYLATION
   RT   ALKYLATES
        FRIEDEL-CRAFT REACTION
        METHYLATION
        REFINING
```

ALKYLFERROCENE
```
   GS   IRON COMPOUNDS
        . FERROCENES
        . . ALKYLFERROCENE
        ORGANOMETALLIC COMPOUNDS
        . ALKYLFERROCENE
```

ALKYLIDENE
```
   GS   ALIPHATIC COMPOUNDS
        . ALKYL COMPOUNDS
        . . ALKYLIDENE
```

ALKYNES
```
   GS   ALIPHATIC COMPOUNDS
        . ALKYNES
        . . ACETYLENE
        . . OXYACETYLENE
        HYDROCARBONS
        . ALKYNES
        . . OXYACETYLENE
   RT   ALKENES
        CYCLIC AMP
        CYCLIC HYDROCARBONS
```

ALL SKY PHOTOGRAPHY
```
   GS   IMAGERY
        . ALL SKY PHOTOGRAPHY
        PHOTOGRAPHY
        . ALL SKY PHOTOGRAPHY
   RT   BLACK AND WHITE PHOTOGRAPHY
        CLOUD PHOTOGRAPHS
        CLOUD PHOTOGRAPHY
        WIDE ANGLE LENSES
```

ALL-WEATHER AIR NAVIGATION
```
   GS   NAVIGATION
        . AIR NAVIGATION
        . . ALL-WEATHER AIR NAVIGATION
   RT   DOPPLER NAVIGATION
        INERTIAL NAVIGATION
        NAVIGATION AIDS
        RADAR NAVIGATION
        RADIO NAVIGATION
        SOLAR COMPASSES
        TACAN
```

ALL-WEATHER LANDING SYSTEMS
```
   GS   LANDING AIDS
        . INSTRUMENT LANDING SYSTEMS
        . . ALL-WEATHER LANDING SYSTEMS
   RT   AIRCRAFT LANDING
        AIRCRAFT SAFETY
        FLIGHT SAFETY
        LOW VISIBILITY
        ∞SYSTEMS
```

ALLEGHENY PLATEAU (US)
```
   GS   LAND
```

ALLEGHENY PLATEAU (US)-*(CONT.)*
    . ALLEGHENY PLATEAU (US)
    LANDFORMS
    . TERRACES (LANDFORMS)
    . . PLATEAUS
    . . . **ALLEGHENY PLATEAU (US)**
RT   MARYLAND
      PENNSYLVANIA
      VIRGINIA
      WEST VIRGINIA

ALLENDE METEORITE
GS   CELESTIAL BODIES
      . METEORITES
      . . STONY METEORITES
      . . . CHONDRITES
      . . . . CARBONACEOUS CHONDRITES
      . . . . . **ALLENDE METEORITE**

ALLERGIC DISEASES
RT   ANAPHYLAXIS
      CONTACT DERMATITIS
      IMMUNOLOGY

ALLOCATIONS
UF   ASSIGNMENT
GS   **ALLOCATIONS**
      . RESOURCE ALLOCATION
RT   ALLOWANCES
      BUDGETING
      COMMERCIAL ENERGY
      COST EFFECTIVENESS
      DISTRIBUTING
      ∞DISTRIBUTION
      DOMESTIC ENERGY
      ECONOMIC ANALYSIS
      ECONOMIC FACTORS
      ENGINEERING MANAGEMENT
      ESTIMATES
      FEDERAL BUDGETS
      FINANCIAL MANAGEMENT
      INDUSTRIAL ENERGY
      MATRIX MANAGEMENT
      PROCUREMENT MANAGEMENT
      PROJECT PLANNING
      RESEARCH MANAGEMENT
      REVENUE
      TRANSPORTATION ENERGY

ALLOTROPY
RT   AUSTENITE
      CRYSTAL STRUCTURE
      POLYMORPHISM

ALLOWANCES
RT   ALLOCATIONS
      CLEARANCES
      ∞COMPENSATION
      PRECISION
      PRODUCTIVITY
      REGULATIONS
      RELIABILITY
      SAMPLING
      TOLERANCES (MECHANICS)

ALLOXAN
GS   HETEROCYCLIC COMPOUNDS
      . **ALLOXAN**
      PYRIMIDINES
      . **ALLOXAN**
RT   THYMIDINE
      THYMINE
      URACIL
      URIC ACID

ALLOYS
GS   **ALLOYS**
      . ANTIMONY ALLOYS
      . . BABBITT METAL
      . ARSENIC ALLOYS
      . BARIUM ALLOYS
      . BEARING ALLOYS
      . BINARY ALLOYS
      . BISMUTH ALLOYS
      . BORON ALLOYS
      . CADMIUM ALLOYS
      . CAST ALLOYS
      . CESIUM ALLOYS
      . CHROMIUM ALLOYS
      . . ASTROLOY (TRADEMARK)
      . . CHROMIUM STEELS
      . . RENE 41
      . . RENE 63
      . . RENE 77
      . . RENE 95

ALLOYS-*(CONT.)*
    . COBALT ALLOYS
    . . ASTROLOY (TRADEMARK)
    . . RENE 41
    . . RENE 63
    . . RENE 77
    . . RENE 95
    . CONSTANTAN
    . COPPER ALLOYS
    . . BABBITT METAL
    . . BRASSES
    . . BRONZES
    . . MANGANIN (TRADEMARK)
    . EUTECTIC ALLOYS
    . GALLIUM ALLOYS
    . GERMANIUM ALLOYS
    . GOLD ALLOYS
    . HAFNIUM ALLOYS
    . HEAT RESISTANT ALLOYS
    . . NIMONIC ALLOYS
    . . REFRACTORY METAL ALLOYS
    . . . MOLYBDENUM ALLOYS
    . . . . RENE 41
    . . . . RENE 63
    . . . . RENE 77
    . . . NIOBIUM ALLOYS
    . . . OSMIUM ALLOYS
    . . . RHENIUM ALLOYS
    . . . TANTALUM ALLOYS
    . . . TUNGSTEN ALLOYS
    . . UDIMET ALLOYS
    . . WASPALOY
    . HIGH STRENGTH ALLOYS
    . ASTROLOY (TRADEMARK)
    . HIGH STRENGTH STEELS
    . . MARAGING STEELS
    . INDIUM ALLOYS
    . IRON ALLOYS
    . . STEELS
    . . BAINITIC STEEL
    . . . CARBON STEELS
    . . . . LOW CARBON STEELS
    . . CHROMIUM STEELS
    . . CROLOY
    . . HIGH STRENGTH STEELS
    . . . MARAGING STEELS
    . . . NICKEL STEELS
    . . . STAINLESS STEELS
    . . . . AUSTENITIC STAINLESS STEELS
    . . . . FERRITIC STAINLESS STEELS
    . . . . MARTENSITIC STAINLESS STEELS
    . KOVAR (TRADEMARK)
    . LEAD ALLOYS
    . LIGHT ALLOYS
    . . ALUMINUM ALLOYS
    . BERYLLIUM ALLOYS
    . MAGNESIUM ALLOYS
    . LIQUID ALLOYS
    . LITHIUM ALLOYS
    . MANGANESE ALLOYS
    . MANGANIN (TRADEMARK)
    . MERCURY ALLOYS
    . . MERCURY AMALGAMS
    . MONOTECTIC ALLOYS
    . MULBERRY (ALLOY)
    . NICKEL ALLOYS
    . ASTROLOY (TRADEMARK)
    . . HASTELLOY (TRADEMARK)
    . . INCONEL (TRADEMARK)
    . KAMACITE
    . MONEL (TRADEMARK)
    . NICHROME (TRADEMARK)
    . NITINOL ALLOYS
    . RENE 41
    . RENE 63
    . RENE 77
    . RENE 95
    . UDIMET ALLOYS
    . WASPALOY
    . PALLADIUM ALLOYS
    . PERMALLOYS (TRADEMARK)
    . PLATINUM ALLOYS
    . PLUTONIUM ALLOYS
    . POTASSIUM ALLOYS
    . QUATERNARY ALLOYS
    . RARE EARTH ALLOYS
    . . ERBIUM ALLOYS
    . . GADOLINIUM ALLOYS
    . . LANTHANUM ALLOYS
    . . NEODYMIUM ALLOYS
    . RHODIUM ALLOYS
    . RUTHENIUM ALLOYS
    . SELENIUM ALLOYS
    . SHAPE MEMORY ALLOYS
    . . NITINOL ALLOYS
    . SILICON ALLOYS

ALLOYS-*(CONT.)*
    . SILVER ALLOYS
    . SODIUM ALLOYS
    . SOLDERS
    . SYNTECTIC ALLOYS
    . TELLURIUM ALLOYS
    . TERNARY ALLOYS
    . ASTROLOY (TRADEMARK)
    . THALLIUM ALLOYS
    . THORIUM ALLOYS
    . TIN ALLOYS
    . . BABBITT METAL
    . TITANIUM ALLOYS
    . . NITINOL ALLOYS
    . URANIUM ALLOYS
    . VANADIUM ALLOYS
    . WROUGHT ALLOYS
    . YTTRIUM ALLOYS
    . ZINC ALLOYS
    . ZIRCONIUM ALLOYS
    . . ZIRCALOYS (TRADEMARK)
    . . . ZIRCALOY 2 (TRADEMARK)
RT   BIMETALS
      BINARY SYSTEMS (MATERIALS)
      EUTECTIC COMPOSITES
      EUTECTICS
      FERROUS METALS
      HARDENERS
      HEAT TREATMENT
      INTERMETALLICS
      KONDO EFFECT
      LIQUID PHASES
      METALLOGRAPHY
      METALLOIDS
      ∞METALLURGY
      METALS
      MIXTURES
      PHASE DIAGRAMS
      POWDER METALLURGY
      RHEOCASTING
      SOLID SOLUTIONS
      STRESS RELIEVING
      TERNARY SYSTEMS

ALLUVIUM
GS   SOILS
      . **ALLUVIUM**
RT   CLAYS
      DELTAS
      FANS (LANDFORMS)
      FLOODS
      GRAVELS
      HYDROLOGY
      MUD
      RIVERS
      SANDS
      SEDIMENTARY ROCKS
      SEDIMENTS
      STREAMS
      WATER FLOW

ALLYL COMPOUNDS
GS   ALIPHATIC COMPOUNDS
      . **ALLYL COMPOUNDS**
RT   ∞CHEMICAL COMPOUNDS
      DIALLYL COMPOUNDS

ALMUCANTAR
USE ELEVATION ANGLE

ALOHA SYSTEM
GS   RANDOM ACCESS
      . **ALOHA SYSTEM**
      TELECOMMUNICATION
      . MULTIPLE ACCESS
      . . **ALOHA SYSTEM**
RT   CHANNEL CAPACITY
      CHANNEL NOISE
      CODE DIVISION MULTIPLE ACCESS
      FREQUENCY DIVISION MULTIPLE
        ACCESS
      PACKET TRANSMISSION
      PACKETS (COMMUNICATION)
      SATELLITE TRANSMISSION
      ∞SYSTEMS
      TIME DIVISION MULTIPLE ACCESS
      TRANSMISSION
      TRANSMISSION EFFICIENCY

ALOUETTE B SATELLITE
GS   SATELLITES
      . ARTIFICIAL SATELLITES
      . . ALOUETTE SATELLITES
      . . . **ALOUETTE B SATELLITE**
RT   ISIS-X

ALOUETTE HELICOPTERS
GS   SUD AVIATION AIRCRAFT
    . **ALOUETTE HELICOPTERS**
    . . SA-330 HELICOPTER
    . . SE-3160 HELICOPTER
    V/STOL AIRCRAFT
    . ROTARY WING AIRCRAFT
    . . HELICOPTERS
    . . . **ALOUETTE HELICOPTERS**
    . . . . SA-330 HELICOPTER
    . . . . SE-3160 HELICOPTER
RT   ∞AIRCRAFT

ALOUETTE PROJECT
RT   COSMIC NOISE
    DATA ACQUISITION
    IONOSPHERIC SOUNDING
    ISIS-A

ALOUETTE SATELLITES
GS   CANADIAN SPACECRAFT
    . **ALOUETTE SATELLITES**
    SATELLITES
    . ARTIFICIAL SATELLITES
    . . **ALOUETTE SATELLITES**
    . . . ALOUETTE B SATELLITE
    . . . ALOUETTE 1 SATELLITE
    . . . ALOUETTE 2 SATELLITE
RT   ISIS SATELLITES

ALOUETTE 1 SATELLITE
UF   S-27 SATELLITE
GS   SATELLITES
    . ARTIFICIAL SATELLITES
    . . ALOUETTE SATELLITES
    . . . **ALOUETTE 1 SATELLITE**
RT   IONOSPHERIC SOUNDING

ALOUETTE 2 SATELLITE
GS   SATELLITES
    . ARTIFICIAL SATELLITES
    . . ALOUETTE SATELLITES
    . . . **ALOUETTE 2 SATELLITE**
    . . ISIS SATELLITES
    . . . **ALOUETTE 2 SATELLITE**
RT   IONOSPHERIC SOUNDING

ALOUETTE 3 HELICOPTER
USE   SE-3160 HELICOPTER

ALPHA DECAY
GS   DECAY
    . RADIOACTIVE DECAY
    . . **ALPHA DECAY**
    NUCLEAR REACTIONS
    . **ALPHA DECAY**
RT   FINE STRUCTURE
    SELECTION RULES (NUCLEAR PHYSICS)

ALPHA JET AIRCRAFT
GS   ATTACK AIRCRAFT
    . FIGHTER AIRCRAFT
    . . **ALPHA JET AIRCRAFT**
    JET AIRCRAFT
    . **ALPHA JET AIRCRAFT**
    TRAINING AIRCRAFT
    . **ALPHA JET AIRCRAFT**
RT   ∞AIRCRAFT
    ∞MILITARY AIRCRAFT

ALPHA PARTICLES
SN   (EMITTED BY NUCLEI)
UF   ALPHA RADIATION
GS   IONIZING RADIATION
    . **ALPHA PARTICLES**
    PARTICLES
    . ELEMENTARY PARTICLES
    . . BOSONS
    . . . **ALPHA PARTICLES**
    . NUCLEAR PARTICLES
    . . BOSONS
    . . . **ALPHA PARTICLES**
RT   ALPHATRONS
    CORPUSCULAR RADIATION
    COSMIC RAYS
    DEUTERON IRRADIATION
    DEUTERONS
    FLUX DENSITY
    HELIUM
    HELIUM IONS
    IONS
    NUCLEAR RADIATION
    NUCLEONS
    ∞RADIATION

ALPHA PARTICLES-*(CONT.)*
RT   RADIOACTIVITY
    SOLAR WIND VELOCITY
    TRITONS

ALPHA PLASMA DEVICES
GS   PLASMA ACCELERATORS
    . **ALPHA PLASMA DEVICES**
RT   ∞DEVICES
    HALL ACCELERATORS
    MAGNETOHYDRODYNAMICS
    PLASMA PHYSICS
    PLASMAS (PHYSICS)

ALPHA RADIATION
USE   ALPHA PARTICLES

ALPHABETS
RT   ALPHANUMERIC CHARACTERS
    CODING
    LANGUAGES
    SYMBOLS

ALPHANUMERIC CHARACTERS
GS   **ALPHANUMERIC CHARACTERS**
    . DIGITS
    . . BINARY DIGITS
RT   ALPHABETS
    INSTRUCTION SETS (COMPUTERS)
    LIGHT EMITTING DIODES
    ∞NUMBERS
    SYMBOLS

ALPHATRONS
GS   MEASURING INSTRUMENTS
    . PRESSURE GAGES
    . . VACUUM GAGES
    . . . IONIZATION GAGES
    . . . . **ALPHATRONS**
    VACUUM APPARATUS
    . VACUUM GAGES
    . . IONIZATION GAGES
    . . . **ALPHATRONS**
RT   ALPHA PARTICLES

ALPINE METEOROLOGY
GS   METEOROLOGY
    . **ALPINE METEOROLOGY**
RT   AERONOMY
    CLOUDS (METEOROLOGY)
    NEPHANALYSIS
    PRECIPITATION (METEOROLOGY)
    STORMS (METEOROLOGY)
    WEATHER
    WIND (METEOROLOGY)

ALPS MOUNTAINS (EUROPE)
GS   LANDFORMS
    . MOUNTAINS
    . . **ALPS MOUNTAINS (EUROPE)**
RT   AUSTRIA
    EUROPE
    ITALY
    SWITZERLAND
    WEST GERMANY

ALSEP
USE   APOLLO LUNAR SURFACE EXPERIMENTS
        PACKAGE

ALTAIR ENGINE
USE   X-248 ENGINE

ALTERATION
USE   REVISIONS

ALTERNATING CURRENT
UF   AC (CURRENT)
GS   ELECTRIC CURRENT
    . **ALTERNATING CURRENT**
    ELECTRICITY
    . **ALTERNATING CURRENT**
RT   CURRENT CONVERTERS (AC TO DC)
    DIRECT CURRENT
    INDUCTION MOTORS
    INVERTED CONVERTERS (DC TO AC)
    VOLTAGE CONVERTERS (AC TO AC)

ALTERNATING CURRENT GENERATORS
USE   AC GENERATORS

ALTERNATIONS
GS   VARIATIONS
    . **ALTERNATIONS**

ALTERNATIONS-*(CONT.)*
RT   CYCLES
    INTERVALS
    RHYTHM (BIOLOGY)

ALTERNATIVES
RT   OPTIONS
    SUBSTITUTES
    VARIATIONS

ALTERNATORS (GENERATORS)
USE   AC GENERATORS

ALTIMETERS
GS   AIRCRAFT INSTRUMENTS
    . **ALTIMETERS**
    . . RADIO ALTIMETERS
    MEASURING INSTRUMENTS
    . DISTANCE MEASURING EQUIPMENT
    . . **ALTIMETERS**
    . . . RADIO ALTIMETERS
RT   ALTITUDE
    APPROACH INDICATORS
    ASTROLABES
    BAROMETERS
    FLIGHT INSTRUMENTS
    HYPSOMETERS
    LANDING INSTRUMENTS
    NAVIGATION AIDS
    NAVIGATION INSTRUMENTS
    POSITION INDICATORS
    RANGE FINDERS
    RATE OF CLIMB INDICATORS

ALTITUDE
GS   **ALTITUDE**
    . FLIGHT ALTITUDE
    . HIGH ALTITUDE
    . LOW ALTITUDE
    . MIDALTITUDE
    . SEA LEVEL
RT   ALTIMETERS
    APEXES
    AZIMUTH
    DISTANCE
    ELEVATION
    ELEVATION ANGLE
    HEIGHT
    POSITION (LOCATION)

ALTITUDE ACCLIMATIZATION
GS   ADAPTATION
    . ACCLIMATIZATION
    . . **ALTITUDE ACCLIMATIZATION**
RT   MOUNTAIN INHABITANTS

ALTITUDE CONTROL
RT   ∞CONTROL
    LASER ALTIMETERS
    LATERAL CONTROL
    LONGITUDINAL CONTROL
    SPACING

ALTITUDE SICKNESS
GS   SICKNESSES
    . **ALTITUDE SICKNESS**
RT   AEROSINUSITIS
    AEROSPACE MEDICINE
    DECOMPRESSION SICKNESS

ALTITUDE SIMULATION
UF   SIMULATED ALTITUDE
GS   SIMULATION
    . ENVIRONMENT SIMULATION
    . . **ALTITUDE SIMULATION**
RT   COMPUTERIZED SIMULATION
    FLIGHT SIMULATION
    HIGH ALTITUDE ENVIRONMENTS
    HYPOBARIC ATMOSPHERES
    LANDING SIMULATION
    SPACE ENVIRONMENT SIMULATION
    THERMAL SIMULATION
    TRAINING DEVICES
    VACUUM CHAMBERS

ALTITUDE TESTS
GS   **ALTITUDE TESTS**
    . HIGH ALTITUDE TESTS
RT   ENGINE TESTS
    FLIGHT TESTS
    FULL SCALE TESTS
    HIGH ALTITUDE ENVIRONMENTS
    TEST VEHICLES
    ∞TESTS

ALTITUDE TOLERANCE
GS   TOLERANCES (PHYSIOLOGY)
    . **ALTITUDE TOLERANCE**
RT   HIGH ALTITUDE BREATHING
    HIGH ALTITUDE ENVIRONMENTS
    HIGH ALTITUDE PRESSURE
    HYPOBARIC ATMOSPHERES
    LOW PRESSURE

ALU (COMPUTER COMPONENTS)
USE   ARITHMETIC AND LOGIC UNITS

ALUM
GS   ALUMINUM COMPOUNDS
    . **ALUM**
    POTASSIUM COMPOUNDS
    . **ALUM**
    SULFUR COMPOUNDS
    . **ALUM**

ALUMINA
USE   ALUMINUM OXIDES

ALUMINATES
GS   ALUMINUM COMPOUNDS
    . **ALUMINATES**
RT   ALUMINUM OXIDES
    ∞OXYGEN COMPOUNDS
    SPINEL

ALUMINIZING
USE   ALUMINUM COATINGS

ALUMINUM
GS   CHEMICAL ELEMENTS
    . **ALUMINUM**
    . . ALUMINUM ISOTOPES
    . . . ALUMINUM 26
    . . . ALUMINUM 27
    . . POWDERED ALUMINUM
    . . SINTERED ALUMINUM POWDER
    METALS
    . **ALUMINUM**
    . . ALUMINUM ISOTOPES
    . . . ALUMINUM 26
    . . . ALUMINUM 27
    . . POWDERED ALUMINUM
    . . SINTERED ALUMINUM POWDER
RT   ALUMINUM ALLOYS
    BORAL
    BORSIC (TRADENAME)
    CRYOLITE
    DAWSONITE
    REACTION BONDING
    SIALON

ALUMINUM ALLOYS
GS   ALLOYS
    . LIGHT ALLOYS
    . . **ALUMINUM ALLOYS**
RT   ALUMINUM
    BEARING ALLOYS
    LAMELLA (METALLURGY)

ALUMINUM ANTIMONIDES
GS   ALUMINUM COMPOUNDS
    . **ALUMINUM ANTIMONIDES**
    ANTIMONY COMPOUNDS
    . ANTIMONIDES
    . . **ALUMINUM ANTIMONIDES**

ALUMINUM ARSENIDES
GS   ALUMINUM COMPOUNDS
    . **ALUMINUM ARSENIDES**
RT   SEMICONDUCTORS (MATERIALS)

ALUMINUM BOROHYDRIDES
GS   ALUMINUM COMPOUNDS
    . ALUMINUM HYDRIDES
    . . **ALUMINUM BOROHYDRIDES**
    BORON COMPOUNDS
    . **ALUMINUM BOROHYDRIDES**
    HYDROGEN COMPOUNDS
    . HYDRIDES
    . . BOROHYDRIDES
    . . . **ALUMINUM BOROHYDRIDES**
    . . BORON HYDRIDES
    . . . **ALUMINUM BOROHYDRIDES**
    . METAL HYDRIDES
    . . ALUMINUM HYDRIDES
    . . . . **ALUMINUM BOROHYDRIDES**

ALUMINUM BORON COMPOSITES
GS   COMPOSITE MATERIALS

ALUMINUM BORON COMPOSITES-*(CONT.)*
    . **ALUMINUM BORON COMPOSITES**
RT   BORON FIBERS
    FIBER COMPOSITES

ALUMINUM CARBIDES
GS   ALUMINUM COMPOUNDS
    . **ALUMINUM CARBIDES**
    CARBON COMPOUNDS
    . CARBIDES
    . . **ALUMINUM CARBIDES**

ALUMINUM CHLORIDES
GS   ALUMINUM COMPOUNDS
    . **ALUMINUM CHLORIDES**
    HALOGEN COMPOUNDS
    . CHLORINE COMPOUNDS
    . . CHLORIDES
    . . . **ALUMINUM CHLORIDES**
    . HALIDES
    . . CHLORIDES
    . . . **ALUMINUM CHLORIDES**
    . . METAL HALIDES
    . . . **ALUMINUM CHLORIDES**

ALUMINUM COATINGS
UF   ALUMINIZING
GS   COATINGS
    . **ALUMINUM COATINGS**
    METALS
    . METAL COATINGS
    . . **ALUMINUM COATINGS**

ALUMINUM COMPOUNDS
GS   **ALUMINUM COMPOUNDS**
    . ALUM
    . ALUMINATES
    . ALUMINUM ANTIMONIDES
    . ALUMINUM ARSENIDES
    . ALUMINUM CARBIDES
    . ALUMINUM CHLORIDES
    . ALUMINUM FLUORIDES
    . ALUMINUM HYDRIDES
    . . ALUMINUM BOROHYDRIDES
    . ALUMINUM NITRIDES
    . ALUMINUM OXIDES
    . . SAPPHIRE
    . ALUMINUM PERCHLORATES
    . ALUMINUM SILICATES
    . . ANDESITE
    . . KAOLINITE
    . . MONTMORILLONITE
    . . PYROPHYLLITE
    . BERYL
    . CORDIERITE
    . CRYOLITE
    . FELDSPARS
    . GEHLENITE
    . LITHIUM ALUMINUM HYDRIDES
    . MUSCOVITE
    . NEPHELINE
    . NEPHELITE
    . ORGANIC ALUMINUM COMPOUNDS
    . SPODUMENE
    . TOURMALINE
RT   ∞CHEMICAL COMPOUNDS
    ∞GROUP 3A COMPOUNDS
    ∞METAL COMPOUNDS
    METAL FUELS
    METAL PROPELLANTS

ALUMINUM FLUORIDES
GS   ALUMINUM COMPOUNDS
    . **ALUMINUM FLUORIDES**
    HALOGEN COMPOUNDS
    . FLUORINE COMPOUNDS
    . . FLUORIDES
    . . . METAL FLUORIDES
    . . . . **ALUMINUM FLUORIDES**
    . HALIDES
    . . METAL HALIDES
    . . . **ALUMINUM FLUORIDES**

ALUMINUM GALLIUM ARSENIDES
UF   ALGAAS
GS   ARSENIC COMPOUNDS
    . ARSENIDES
    . . GALLIUM ARSENIDES
    . . . **ALUMINUM GALLIUM ARSENIDES**
    GALLIUM COMPOUNDS
    . GALLIUM ARSENIDES
    . . **ALUMINUM GALLIUM ARSENIDES**
RT   NEGATIVE RESISTANCE DEVICES

ALUMINUM GRAPHITE COMPOSITES
GS   COMPOSITE MATERIALS
    . **ALUMINUM GRAPHITE COMPOSITES**
RT   FIBER COMPOSITES
    GRAPHITE

ALUMINUM HYDRIDES
GS   ALUMINUM COMPOUNDS
    . **ALUMINUM HYDRIDES**
    . . ALUMINUM BOROHYDRIDES
    HYDROGEN COMPOUNDS
    . HYDRIDES
    . . METAL HYDRIDES
    . . . **ALUMINUM HYDRIDES**
    . . . . ALUMINUM BOROHYDRIDES

ALUMINUM ISOTOPES
GS   CHEMICAL ELEMENTS
    . ALUMINUM
    . . **ALUMINUM ISOTOPES**
    . . . ALUMINUM 26
    . . . ALUMINUM 27
    . NUCLIDES
    . ISOTOPES
    . . **ALUMINUM ISOTOPES**
    . . . ALUMINUM 26
    . . . ALUMINUM 27
    METALS
    . ALUMINUM
    . . **ALUMINUM ISOTOPES**
    . . . ALUMINUM 26
    . . . ALUMINUM 27

ALUMINUM NITRIDES
GS   ALUMINUM COMPOUNDS
    . **ALUMINUM NITRIDES**
    NITROGEN COMPOUNDS
    . NITRIDES
    . . **ALUMINUM NITRIDES**
RT   METAL NITRIDES

ALUMINUM OXIDES
UF   ALUMINA
    CORUNDUM
GS   ALUMINUM COMPOUNDS
    . **ALUMINUM OXIDES**
    . . SAPPHIRE
    CHALCOGENIDES
    . OXIDES
    . . METAL OXIDES
    . . . **ALUMINUM OXIDES**
    . . . . SAPPHIRE
RT   ABRASIVES
    ALUMINATES
    BAUXITE
    ENERGY ABSORPTION FILMS
    GEHLENITE
    KAOLINITE
    PYROPHYLLITE
    RUBY
    THERMITES

ALUMINUM PERCHLORATES
GS   ALUMINUM COMPOUNDS
    . **ALUMINUM PERCHLORATES**
    HALOGEN COMPOUNDS
    . CHLORINE COMPOUNDS
    . . PERCHLORATES
    . . . **ALUMINUM PERCHLORATES**

ALUMINUM SILICATES
GS   ALUMINUM COMPOUNDS
    . **ALUMINUM SILICATES**
    . . ANDESITE
    . . KAOLINITE
    . . MONTMORILLONITE
    . . PYROPHYLLITE
    SILICON COMPOUNDS
    . **ALUMINUM SILICATES**
    . . ANDESITE
    . . KAOLINITE
    . . MONTMORILLONITE
    . . PYROPHYLLITE
RT   MINERALS
    MULLITES

ALUMINUM 26
GS   CHEMICAL ELEMENTS
    . ALUMINUM
    . . ALUMINUM ISOTOPES
    . . . **ALUMINUM 26**
    . NUCLIDES
    . . ISOTOPES
    . . . ALUMINUM ISOTOPES
    . . . . **ALUMINUM 26**

ALUMINUM 26-*(CONT.)*
   METALS
   . ALUMINUM
   . . ALUMINUM ISOTOPES
   . . . **ALUMINUM 26**

ALUMINUM 27
  GS CHEMICAL ELEMENTS
   . ALUMINUM
   . . ALUMINUM ISOTOPES
   . . . **ALUMINUM 27**
   . NUCLIDES
   . ISOTOPES
   . . ALUMINUM ISOTOPES
   . . . . **ALUMINUM 27**
   METALS
   . ALUMINUM
   . . ALUMINUM ISOTOPES
   . . . **ALUMINUM 27**

ALVEOLAR AIR
  GS GASES
   . AIR
   . . **ALVEOLAR AIR**
   . GAS MIXTURES
   . . **ALVEOLAR AIR**
  RT EXHALATION
   EXPIRED AIR
   LUNGS

ALVEOLI
  RT LUNG MORPHOLOGY
   LUNGS
   PULMONARY CIRCULATION
   PULMONARY FUNCTIONS
   RESPIRATION
   RESPIRATORY SYSTEM

AMALGAMS
  USE MERCURY AMALGAMS

AMALTHEA
  GS SATELLITES
   . NATURAL SATELLITES
   . . JUPITER SATELLITES
   . . . **AMALTHEA**
  RT JUPITER (PLANET)
   SOLAR SYSTEM

AMAZON REGION (SOUTH AMERICA)
  GS REGIONS
   . TROPICAL REGIONS
   . . **AMAZON REGION (SOUTH AMERICA)**
  RT BRAZIL
   FORESTS
   RIVER BASINS
   RIVERS

AMBERLITE (TRADEMARK)
  RT ASBESTOS
   THERMAL INSULATION

AMBIENCE
  RT ENVIRONMENTS

AMBIENT TEMPERATURE
  UF ENVIRONMENTAL TEMPERATURE
  GS TEMPERATURE
   . **AMBIENT TEMPERATURE**
  RT ATMOSPHERIC TEMPERATURE
   OPERATING TEMPERATURE
   ROOM TEMPERATURE
   SATELLITE TEMPERATURE

AMBIGUITY
  RT INTELLIGIBILITY
   POSITIONING

AMBIPOLAR DIFFUSION
  GS DIFFUSION
   . **AMBIPOLAR DIFFUSION**
  RT ELECTRON DIFFUSION
   ELECTRON MOBILITY
   IONIC DIFFUSION
   IONIC MOBILITY
   PLASMA DIFFUSION

AMBIT
  USE FIELD THEORY (PHYSICS)

AMBULANCES
  RT MEDICAL SERVICES
   ∞MILITARY VEHICLES
   SAFETY DEVICES

AMERICAN INDIANS
  RT ANTHROPOLOGY
   CULTURE (SOCIAL SCIENCES)
   ETHNIC FACTORS
   MINORITIES
   RACES

AMERICIUM
  GS CHEMICAL ELEMENTS
   . ACTINIDE SERIES
   . . TRANSURANIUM ELEMENTS
   . . . **AMERICIUM**
   . . . . AMERICIUM ISOTOPES
   . . . . . AMERICIUM 241
   . NUCLIDES
   . . ISOTOPES
   . . . RADIOACTIVE ISOTOPES
   . . . . TRANSURANIUM ELEMENTS
   . . . . . **AMERICIUM**
   . . . . . . AMERICIUM ISOTOPES
   HEAVY ELEMENTS
   . **AMERICIUM**
   . . AMERICIUM ISOTOPES
   . . . AMERICIUM 241
   METALS
   . ACTINIDE SERIES
   . . TRANSURANIUM ELEMENTS
   . . . **AMERICIUM**
   . . . . AMERICIUM ISOTOPES
   . . . . . AMERICIUM 241

AMERICIUM ISOTOPES
  GS CHEMICAL ELEMENTS
   . ACTINIDE SERIES
   . . TRANSURANIUM ELEMENTS
   . . . AMERICIUM
   . . . . **AMERICIUM ISOTOPES**
   . . . . . AMERICIUM 241
   . NUCLIDES
   . . ISOTOPES
   . . . RADIOACTIVE ISOTOPES
   . . . . TRANSURANIUM ELEMENTS
   . . . . . AMERICIUM
   . . . . . . **AMERICIUM ISOTOPES**
   HEAVY ELEMENTS
   . AMERICIUM
   . . **AMERICIUM ISOTOPES**
   . . . AMERICIUM 241
   . TRANSURANIUM ELEMENTS
   . . **AMERICIUM ISOTOPES**
   . . . AMERICIUM 241
   METALS
   . ACTINIDE SERIES
   . . TRANSURANIUM ELEMENTS
   . . . AMERICIUM
   . . . . **AMERICIUM ISOTOPES**
   . . . . . AMERICIUM 241

AMERICIUM 241
  GS CHEMICAL ELEMENTS
   . ACTINIDE SERIES
   . . TRANSURANIUM ELEMENTS
   . . . AMERICIUM
   . . . . AMERICIUM ISOTOPES
   . . . . . **AMERICIUM 241**
   . NUCLIDES
   . . ISOTOPES
   . . . RADIOACTIVE ISOTOPES
   . . . . TRANSURANIUM ELEMENTS
   . . . . . AMERICIUM
   . . . . . . AMERICIUM ISOTOPES
   HEAVY ELEMENTS
   . AMERICIUM
   . . AMERICIUM ISOTOPES
   . . . **AMERICIUM 241**
   . TRANSURANIUM ELEMENTS
   . AMERICIUM ISOTOPES
   . . . **AMERICIUM 241**
   METALS
   . ACTINIDE SERIES
   . . TRANSURANIUM ELEMENTS
   . . . AMERICIUM
   . . . . AMERICIUM ISOTOPES
   . . . . . **AMERICIUM 241**

AMIDASE
  GS ENZYMES
   . **AMIDASE**
   ORGANIC COMPOUNDS
   . **AMIDASE**
   PROTEINS
   . **AMIDASE**
  RT AMINO ACIDS

AMIDES
  GS NITROGEN COMPOUNDS
   . **AMIDES**
   . . CARBAMIDES
   . . CYANAMIDES
   . . FORMHYDROXAMIC ACID
   . . LYSERGAMIDE
   . . NICOTINAMIDE
   . . OXAMIC ACIDS
   . . POLYIMIDES
   . . SUCCINIMIDES
   . . UREAS
   . . . DIFLUOROUREA
   . . . THIOUREAS
   . . . THIURONIUM
  RT ALIPHATIC COMPOUNDS
   IMIDES

AMINES
  GS **AMINES**
   . AMINOPHYLLINE
   . AMPHETAMINES
   . . METHAMPHETAMINE
   . ANILINE
   . CATECHOLAMINE
   . CYSTEAMINE
   . DIAMINES
   . . ETHYLENEDIAMINE
   . . GUANIDINES
   . . . GUANETHIDINE
   . . . TRIAMINOGUANIDINIUM AZIDE
   . DIFLUOROUREA
   . DIMENHYDRINATE
   . DIMETHYLHYDRAZINES
   . DIPHENYL HYDANTOIN
   . ERGOTAMINE
   . FLUOROAMINES
   . . NITROFLUORAMINES
   . . TRIFLUOROAMINE OXIDE
   . GALLAMINE TRIETHIODIDE
   . HEXAMETHYLENETETRAMINE
   . HISTIDINE
   . HYDROXYLAMINE SULFATE
   . HYOSCINE
   . MECAMYLAMINE
   . MELAMINE
   . METHYLENE DIAMINE
   . MONOETHANOLAMINE (MEA)
   . NITROAMINES
   . NITROSAMINE
   . PROMETHAZINE
   . TETRAFLUOROHYDRAZINE
   . TETRYL
   . THIURONIUM
   . TRINITRAMINE
   . TRYPTAMINES
   . . SEROTONIN
  RT ALIPHATIC COMPOUNDS
   HISTAMINES
   HYDRAZINES
   HYDROCARBON FUELS
   IMINES
   LEWIS BASE
   NITROSYLS
   PHENOLIC EPOXY RESINS

AMINO ACIDS
  GS ACIDS
   . **AMINO ACIDS**
   . . ALANINE
   . . . PHENYLALANINE
   . . ASPARTIC ACID
   . . CYSTEINE
   . . FOLIC ACID
   . . GLUTAMIC ACID
   . . GLUTAMINE
   . . GLUTATHIONE
   . . GLYCINE
   . . HIPPURIC ACID
   . . HISTIDINE
   . . LEUCINE
   . . . NORLEUCINE
   . . LYSINE
   . . MELANOIDIN
   . . METHIONINE
   . . PAPAIN
   . . PEPTIDES
   . . . HYPERTENSIN
   . . PROTOPROTEINS
   . . PYRIDINE NUCLEOTIDES
   . . PYRUVATES
   . . THYROXINE
   . . TRYPTOPHAN
   . . TYROSINE
   . . URIDYLIC ACID

AMINO ACIDS-*(CONT.)*
   ORGANIC COMPOUNDS
   . **AMINO ACIDS**
   . . ALANINE
   . . . PHENYLALANINE
   . . ASPARTIC ACID
   . . CYSTEINE
   . . FOLIC ACID
   . . GLUTAMIC ACID
   . . GLUTAMINE
   . . GLUTATHIONE
   . . GLYCINE
   . . HIPPURIC ACID
   . . HISTIDINE
   . . LEUCINE
   . . . NORLEUCINE
   . . LYSINE
   . . MELANOIDIN
   . . METHIONINE
   . . PAPAIN
   . . PEPTIDES
   . . . HYPERTENSIN
   . . . . POLYPEPTIDES
   . . PROTOPROTEINS
   . . PYRIDINE NUCLEOTIDES
   . . THYROXINE
   . . TRYPTOPHAN
   . . TYROSINE
   . . URIDYLIC ACID
RT   ADENOSINE TRIPHOSPHATE
   ADRENOCORTICOTROPIN (ACTH)
   AMIDASE
   ASPARTATES
   CYCLIC AMP
   CYSTEAMINE
   SYNTHETIC FOOD

AMINOPHYLLINE
GS   AMINES
   . **AMINOPHYLLINE**
   DIURETICS
   . **AMINOPHYLLINE**
   DRUGS
   . **AMINOPHYLLINE**
RT   STIMULANTS

AMMETERS
GS   MEASURING INSTRUMENTS
   . **AMMETERS**
   . . MICROMILLIAMMETERS
   . . THERMOELEMENT AMMETERS
RT   COULOMETERS
   ELECTRIC CURRENT
   ELECTRICAL MEASUREMENT
   GALVANOMETERS
   VOLTMETERS

AMMINES
RT   AMMONIA
   ∞CHEMICAL COMPOUNDS
   COPPER
   INTERMETALLICS
   ∞METAL COMPOUNDS

AMMONIA
GS   INORGANIC COMPOUNDS
   . **AMMONIA**
   . . LIQUID AMMONIA
   NITROGEN COMPOUNDS
   . **AMMONIA**
   . . LIQUID AMMONIA
RT   ABSORPTION COOLING
   AMMINES
   AMMONIUM COMPOUNDS
   AMMONOLYSIS
   ATMOSPHERIC ENERGY SOURCES
   CULTIVATION
   FERTILIZERS
   KJELDAHL METHOD
   NITROGEN HYDRIDES
   REFRIGERANTS

AMMONIUM BROMIDES
GS   AMMONIUM COMPOUNDS
   . **AMMONIUM BROMIDES**
   HALOGEN COMPOUNDS
   . BROMINE COMPOUNDS
   . . BROMIDES
   . . . **AMMONIUM BROMIDES**
   . HALIDES
   . BROMIDES
   . . . **AMMONIUM BROMIDES**

AMMONIUM CHLORIDES
GS   AMMONIUM COMPOUNDS

AMMONIUM CHLORIDES-*(CONT.)*
   . **AMMONIUM CHLORIDES**
   HALOGEN COMPOUNDS
   . CHLORINE COMPOUNDS
   . . CHLORIDES
   . . . **AMMONIUM CHLORIDES**
   . HALIDES
   . CHLORIDES
   . . . **AMMONIUM CHLORIDES**

AMMONIUM COMPOUNDS
GS   **AMMONIUM COMPOUNDS**
   . AMMONIUM BROMIDES
   . AMMONIUM CHLORIDES
   . AMMONIUM NITRATES
   . AMMONIUM PERCHLORATES
   . AMMONIUM PHOSPHATES
   . AMMONIUM PICRATES
   . AMMONIUM SULFATES
   . HYDROXYLAMMONIUM PERCHLORATES
RT   AMMONIA
   ∞CHEMICAL COMPOUNDS
   HEXAMETHONIUM

AMMONIUM NITRATES
GS   AMMONIUM COMPOUNDS
   . **AMMONIUM NITRATES**
   NITROGEN COMPOUNDS
   . NITRATES
   . . INORGANIC NITRATES
   . . . **AMMONIUM NITRATES**
RT   CULTIVATION
   FERTILIZERS

AMMONIUM PERCHLORATES
GS   AMMONIUM COMPOUNDS
   . **AMMONIUM PERCHLORATES**
   HALOGEN COMPOUNDS
   . CHLORINE COMPOUNDS
   . . PERCHLORATES
   . . . **AMMONIUM PERCHLORATES**
RT   SOLID ROCKET PROPELLANTS

AMMONIUM PHOSPHATES
GS   AMMONIUM COMPOUNDS
   . **AMMONIUM PHOSPHATES**
   PHOSPHORUS COMPOUNDS
   . PHOSPHATES
   . . **AMMONIUM PHOSPHATES**

AMMONIUM PICRATES
GS   AMMONIUM COMPOUNDS
   . **AMMONIUM PICRATES**
   NITROGEN COMPOUNDS
   . NITRO COMPOUNDS
   . . PICRATES
   . . . **AMMONIUM PICRATES**
RT   EXPLOSIVES

AMMONIUM SULFATES
GS   AMMONIUM COMPOUNDS
   . **AMMONIUM SULFATES**
   SULFUR COMPOUNDS
   . SULFATES
   . . **AMMONIUM SULFATES**

AMMONOLYSIS
GS   CHEMICAL REACTIONS
   . **AMMONOLYSIS**
   DECOMPOSITION
   . **AMMONOLYSIS**
RT   AMMONIA
   CRACKING (CHEMICAL ENGINEERING)
   HYDROLYSIS

AMMUNITION
GS   **AMMUNITION**
   . INCENDIARY AMMUNITION
RT   BLANKS
   BOMBS (ORDNANCE)
   CASE BONDED PROPELLANTS
   EXPLOSIVE DEVICES
   EXPLOSIVES
   FUSES (ORDNANCE)
   GRENADES
   GUNS (ORDNANCE)
   IGNITERS
   MAGAZINES (SUPPLY CHAMBERS)
   MINES (ORDNANCE)
   MISSILES
   ORDNANCE
   PROJECTILES
   PROPELLANTS
   PYROTECHNICS
   ∞ROCKETS

AMMUNITION-*(CONT.)*
   SHAPED CHARGES
   ∞SHOT
   TORPEDOES
   ∞TRACERS
   WARHEADS
   WEAPONS

AMOBARBITAL
GS   ACIDS
   . **AMOBARBITAL**
RT   CENTRAL NERVOUS SYSTEM
      DEPRESSANTS

AMOEBA
GS   ANIMALS
   . **AMOEBA**
   . . PELOMYXA
   MICROORGANISMS
   . PROTOZOA
   . . **AMOEBA**
   . . . PELOMYXA
RT   PARASITIC DISEASES

AMOOS
USE   AEROMANEUVERING ORBIT TO ORBIT
      SHUTTLE

AMOR ASTEROID
UF   MINOR PLANET 1221
GS   CELESTIAL BODIES
   . ASTEROID BELTS
   . . ASTEROIDS
   . . . **AMOR ASTEROID**
RT   ASTRONOMY
   JUPITER (PLANET)
   MARS (PLANET)
   PLANETARY ORBITS
   SOLAR SYSTEM

AMORPHOUS MATERIALS
RT   ASPHALT
   CRYSTALLINITY
   GLASS
   GRAPHOEPITAXY
   GROUT
   ∞MATERIALS
   SPIN GLASS

AMORPHOUS SEMICONDUCTORS
GS   SEMICONDUCTORS (MATERIALS)
   . **AMORPHOUS SEMICONDUCTORS**
RT   SEMICONDUCTING FILMS

AMOUNT
UF   QUANTITY
RT   ADDITION
   SUMS
   VALUE

AMPERAGE
USE   ELECTRIC CURRENT

AMPHETAMINES
GS   AMINES
   . **AMPHETAMINES**
   . . METHAMPHETAMINE
RT   CENTRAL NERVOUS SYSTEM
   STIMULANTS

AMPHIBIA
GS   ANIMALS
   . POIKILOTHERMIA
   . . **AMPHIBIA**
   . . . FROGS
   . VERTEBRATES
   . . **AMPHIBIA**
   . . . FROGS

AMPHIBIOUS AIRCRAFT
GS   AMPHIBIOUS VEHICLES
   . **AMPHIBIOUS AIRCRAFT**
RT   ∞AIRCRAFT
   SEAPLANES
   WATER TAKEOFF AND LANDING
      AIRCRAFT

AMPHIBIOUS VEHICLES
GS   **AMPHIBIOUS VEHICLES**
   . AMPHIBIOUS AIRCRAFT
RT   BOATS
   ∞MILITARY VEHICLES
   SEAPLANES
   SHIPS

AMPHIBIOUS VEHICLES-(CONT.)
      SURFACE VEHICLES
    ∞ VEHICLES
      WATER VEHICLES

AMPHIBOLES
  GS   MINERALS
     . **AMPHIBOLES**
  RT   CALCIUM SILICATES

AMPHITRITE ASTEROID
  GS   CELESTIAL BODIES
     . ASTEROID BELTS
     .. ASTEROIDS
     ... **AMPHITRITE ASTEROID**
  RT   GALILEO PROJECT

AMPLIDYNES
  GS   ELECTRIC GENERATORS
     . ROTATING GENERATORS
     .. **AMPLIDYNES**
  RT   AMPLIFIERS
     ELECTRIC MOTORS
     POWER AMPLIFIERS
     SERVOMOTORS

AMPLIFICATION
  UF   AMPLIFICATION FACTOR
     GAIN (AMPLIFICATION)
     INTENSIFICATION
  GS   **AMPLIFICATION**
     . POWER GAIN
     . SOUND AMPLIFICATION
     . WAVE AMPLIFICATION
  RT   AMPLIFIERS
     AMPLITUDES
     DYNAMIC CHARACTERISTICS
     DYNAMIC RESPONSE
     FLUID AMPLIFIERS
     FLUIDICS
     HIGH GAIN
     MAGNIFICATION
     POSITIVE FEEDBACK
     SENSITIVITY
     STABILITY
     TRANSFER FUNCTIONS
     TRANSIENT RESPONSE

AMPLIFICATION FACTOR
  USE AMPLIFICATION

AMPLIFIER DESIGN
  RT   AMPLIFIERS
     COMPUTER AIDED DESIGN
    ∞ DESIGN
     LOGIC DESIGN
     OPERATIONAL AMPLIFIERS
     PRODUCT DEVELOPMENT

AMPLIFIERS
  UF   ELECTRONIC AMPLIFIERS
  GS   **AMPLIFIERS**
     . BEAM PLASMA AMPLIFIERS
     . BROADBAND AMPLIFIERS
     . CARCINOTRONS
     . CROSSED FIELD AMPLIFIERS
     . CURRENT AMPLIFIERS
     .. PHOTOMULTIPLIER TUBES
     ... FREQUENCY MODULATION
         PHOTOMULTIPLIERS
     . DIFFERENTIAL AMPLIFIERS
     . DISTRIBUTED AMPLIFIERS
     . FEEDBACK AMPLIFIERS
     . FLUID AMPLIFIERS
     .. JET AMPLIFIERS
     . INTERMEDIATE FREQUENCY
         AMPLIFIERS
     . LIGHT AMPLIFIERS
     . LIMITER AMPLIFIERS
     . LINEAR AMPLIFIERS
     . MAGNETIC AMPLIFIERS
     . MAGNETOSTATIC AMPLIFIERS
     . MICROWAVE AMPLIFIERS
     . OPERATIONAL AMPLIFIERS
     . PARAMETRIC AMPLIFIERS
     . POSTAMPLIFIERS
     . POWER AMPLIFIERS
     . PREAMPLIFIERS
     . PUSH-PULL AMPLIFIERS
     . SERVOAMPLIFIERS
     . TRANSISTOR AMPLIFIERS
     . TRAVELING WAVE AMPLIFIERS
     . VOLTAGE AMPLIFIERS
  RT   AMPLIDYNES
     AMPLIFICATION

AMPLIFIERS-(CONT.)
     AMPLIFIER DESIGN
    ∞ BOOSTERS
     CAPACITORS
     CIRCUITS
    ∞ ELECTRIC CELLS
     ELECTRIC CHOPPERS
     IMAGE INTENSIFIERS
     INTENSIFIERS
     KLYSTRONS
     LASER CAVITIES
     LASERS
     LINEAR CIRCUITS
     MASERS
     MODULATORS
     MULTIVIBRATORS
     OSCILLATORS
     RECEIVERS
     REPEATERS
     SOLID STATE DEVICES
     STIMULATED EMISSION DEVICES
     TRANSFORMERS
     TRAVELING WAVE MASERS

AMPLITRONS (TRADEMARK)
  USE PLANOTRONS

AMPLITUDE DISTRIBUTION ANALYSIS
  UF   AMPLITUDE PROBABILITY ANALYSIS
  GS   STATISTICAL ANALYSIS
     . **AMPLITUDE DISTRIBUTION ANALYSIS**
  RT   PHOTOPEAK
     PULSE AMPLITUDE
     SIGNAL TO NOISE RATIOS
     SIGNATURES

AMPLITUDE MODULATION
  GS   MODULATION
     . **AMPLITUDE MODULATION**
  RT   DEMODULATION
     DEMODULATORS
     FREQUENCY MODULATION
     LIGHT MODULATION
     MODULATORS
     P.A.C.M. TELEMETRY
     PHASE MODULATION
     PULSE MODULATION
     SINGLE SIDEBAND TRANSMISSION

AMPLITUDE PROBABILITY ANALYSIS
  USE AMPLITUDE DISTRIBUTION ANALYSIS

AMPLITUDES
  GS   **AMPLITUDES**
     . PULSE AMPLITUDE
     . SCATTERING AMPLITUDE
  RT   AMPLIFICATION
     CYCLES
     DIMENSIONS
     DISPLACEMENT
     FREQUENCIES
    ∞ INTENSITY
     LEVEL (QUANTITY)
     MAGNITUDE
     OSCILLATIONS
     PICOSECOND PULSES
     PULSES
     STANDING WAVE RATIOS
     VIBRATION

AMPOULES
  RT ∞ CONTAINERS
     LABORATORY EQUIPMENT
     VACUUM SYSTEMS

AMPS (SATELLITE PAYLOAD)
  UF   ATMOSPHERIC AND MAGNETOSPHERIC
        PAYLOAD
     PLASMAS-IN-SPACE PAYLOAD
  GS   MEASURING INSTRUMENTS
     . SATELLITE-BORNE INSTRUMENTS
     .. **AMPS (SATELLITE PAYLOAD)**
     PAYLOADS
     . **AMPS (SATELLITE PAYLOAD)**
  RT   INSTRUMENT PACKAGES

AMPTE (SATELLITES)
  SN   (ACTIVE MAGNETOSPHERIC PARTICLE
       TRACER EXPLORERS)
  UF   ACTIVE MAGNETO PARTICLE TRACER
       EXPLORERS
  GS   SATELLITES
     . ARTIFICIAL SATELLITES
     .. SCIENTIFIC SATELLITES
     ... **AMPTE (SATELLITES)**

AMPTE (SATELLITES)-(CONT.)
  RT   EUROPEAN SPACE PROGRAMS
     MAGNETOSPHERE
     SATELLITE-BORNE INSTRUMENTS
     SOLAR WIND
     SPACE PLASMAS
     SPACEBORNE EXPERIMENTS

AMTV
  USE AUTOMATED MIXED TRAFFIC VEHICLES

AN-2 AIRCRAFT
  GS   ANTONOV AIRCRAFT
     . **AN-2 AIRCRAFT**
     JET AIRCRAFT
     . **AN-2 AIRCRAFT**
     MONOPLANES
     . **AN-2 AIRCRAFT**
     TRANSPORT AIRCRAFT
     . **AN-2 AIRCRAFT**
  RT ∞ AIRCRAFT

AN-22 AIRCRAFT
  UF   ANTHEUS AIRCRAFT
     ANTONOV AN-22 AIRCRAFT
     COCK AIRCRAFT
  GS   ANTONOV AIRCRAFT
     . **AN-22 AIRCRAFT**
     JET AIRCRAFT
     . TURBOPROP AIRCRAFT
     .. **AN-22 AIRCRAFT**
     MONOPLANES
     . **AN-22 AIRCRAFT**
     TRANSPORT AIRCRAFT
     . **AN-22 AIRCRAFT**
  RT ∞ AIRCRAFT
     PASSENGER AIRCRAFT

AN-24 AIRCRAFT
  UF   ANTONOV AN-24 AIRCRAFT
     COKE AIRCRAFT
  GS   ANTONOV AIRCRAFT
     . **AN-24 AIRCRAFT**
     JET AIRCRAFT
     . TURBOPROP AIRCRAFT
     .. **AN-24 AIRCRAFT**
     MONOPLANES
     . **AN-24 AIRCRAFT**
     TRANSPORT AIRCRAFT
     . **AN-24 AIRCRAFT**
  RT ∞ AIRCRAFT
     PASSENGER AIRCRAFT

ANABAENA
  GS   ALGAE
     . **ANABAENA**

ANAEROBES
  GS   MICROORGANISMS
     . **ANAEROBES**
  RT   AEROBES
     BACTERIA
     SEWAGE TREATMENT

ANALGESIA
  RT   ANESTHESIA
     DRUGS
     PAIN

ANALOG CIRCUITS
  GS   CIRCUITS
     . **ANALOG CIRCUITS**
  RT   DATA CONVERTERS
     OPERATIONAL AMPLIFIERS
     RHEOELECTRICAL SIMULATION

ANALOG COMPUTERS
  GS   DATA PROCESSING EQUIPMENT
     . COMPUTERS
     .. **ANALOG COMPUTERS**
     ... EAI 680 COMPUTER
     ... HONEYWELL 600/6000 COMPUTER
     ... SIGMA 5 COMPUTER
     ... UNIVAC 1100 SERIES COMPUTERS
  RT   DIFFERENTIAL AMPLIFIERS
     DIFFERENTIAL ANALYZERS
     DIGITAL COMPUTERS
     DISCRIMINATORS
     FUNCTIONAL INTEGRATION
     HYBRID COMPUTERS
     MISSILE CONTROL
     OPERATIONAL AMPLIFIERS
     RESOLVERS
     SIGNAL ANALYZERS

ANALOG COMPUTERS-*(CONT.)*
    SPECTRAL RESOLUTION

ANALOG DATA
  RT   BINARY DATA
      ∞DATA
      DATA CONVERTERS
      DATA PROCESSING
      DIGITAL DATA
      ∞MEASUREMENT
      VIDEO DATA

ANALOG SIMULATION
  GS   MODELS
      . MATHEMATICAL MODELS
      . . **ANALOG SIMULATION**
      SIMULATION
      . COMPUTERIZED SIMULATION
      . . **ANALOG SIMULATION**
  RT   COMPUTER SYSTEMS SIMULATION
      DIGITAL SIMULATION
      FLIGHT SIMULATION
      RHEOELECTRICAL SIMULATION
      SYSTEMS SIMULATION

ANALOG TO DIGITAL CONVERTERS
  UF   DIGITIZERS
  GS   DATA CONVERTERS
      . **ANALOG TO DIGITAL CONVERTERS**
  RT   CODERS
      CODING
      ∞CONVERTERS
      DATA ACQUISITION
      DIGITAL COMPUTERS
      DIGITAL SYSTEMS
      DIGITAL TO ANALOG CONVERTERS
      ILLIAC 3 COMPUTER
      ILLIAC 4 COMPUTER
      PLOTTING

ANALOGIES
  UF   SIMILARITIES
  GS   **ANALOGIES**
      . HYDRAULIC ANALOGIES
  RT   COMPARISON
      HOMOLOGY
      SIMULATION

ANALOGS
  RT   MODELS
      SIMULATORS

ANALYSIS
  USE   ANALYZING

ANALYSIS (MATHEMATICS)
  SN   (THAT PART OF THE FIELD OF
      MATHEMATICS WHICH ARISES FROM
      THE CALCULUS AND WHICH DEALS
      PRIMARILY WITH FUNCTIONS)
  GS   **ANALYSIS (MATHEMATICS)**
      . APERIODIC FUNCTIONS
      . CALCULUS
      . . CONTINUITY (MATHEMATICS)
      . . DIFFERENTIAL CALCULUS
      . . FOURIER-BESSEL TRANSFORMATIONS
      . . GRAEFF CALCULUS
      . . INTEGRAL CALCULUS
      . . LIMITS (MATHEMATICS)
      . . SERIES (MATHEMATICS)
      . . . ASYMPTOTIC SERIES
      . . . CAMPBELL-HAUSDORFF SERIES
      . . . COSINE SERIES
      . . . FOURIER SERIES
      . . . PADE APPROXIMATION
      . . . POWER SERIES
      . . . . TAYLOR SERIES
      . . . . . MACLAURIN SERIES
      . . . PROGRESSIONS
      . . . PRONY SERIES
      . . . SINE SERIES
      . . VECTOR ANALYSIS
      . . . COLLINEARITY
      . . . COPLANARITY
      . . . CURL (VECTORS)
      . . . . VORTICITY
      . COMBINATORIAL ANALYSIS
      . . COMBINATIONS (MATHEMATICS)
      . . FACTORIALS
      . . PARTITIONS (MATHEMATICS)
      . . PERMUTATIONS
      . COMPLEX VARIABLES
      . . AIRY FUNCTION
      . . ANALYTIC FUNCTIONS
      . . . ENTIRE FUNCTIONS

ANALYSIS (MATHEMATICS)-*(CONT.)*
  . . BESSEL FUNCTIONS
  . . . HANKEL FUNCTIONS
  . . CAUCHY INTEGRAL FORMULA
  . . CONFORMAL MAPPING
  . . CONJUGATES
  . . . CONJUGATE POINTS
  . . EXPONENTIAL FUNCTIONS
  . . . LOGARITHMS
  . . GAMMA FUNCTION
  . . HARMONIC FUNCTIONS
  . . HYPERBOLIC FUNCTIONS
  . . HYPERGEOMETRIC FUNCTIONS
  . . LAGUERRE FUNCTIONS
  . . LEGENDRE FUNCTIONS
  . . LIOUVILLE THEOREM
  . . MATHIEU FUNCTION
  . . MEROMORPHIC FUNCTIONS
  . . . ELLIPTIC FUNCTIONS
  . . . RATIONAL FUNCTIONS
  . . NONHOLONOMIC EQUATIONS
  . . ORTHOGONAL FUNCTIONS
  . . SCHWARZ-CHRISTOFFEL
      TRANSFORMATION
  . . SINGULARITY (MATHEMATICS)
  . . . NAKED SINGULARITIES
  . . . SPHERICAL HARMONICS
  . DEPENDENT VARIABLES
  . FOURIER ANALYSIS
  . FOURIER SERIES
  . FUNCTION SPACE
  . HILBERT SPACE
  . . BANACH SPACE
  . . . SOBOLEV SPACE
  . FUNCTIONAL ANALYSIS
  . HARMONIC ANALYSIS
  . . TESSERAL HARMONICS
  . . ZONAL HARMONICS
  . HILBERT SPACE
  . . BANACH SPACE
  . . . SOBOLEV SPACE
  . . INTEGRAL EQUATIONS
  . . FREDHOLM EQUATIONS
  . . J INTEGRAL
  . . SINGULAR INTEGRAL EQUATIONS
  . . VOLTERRA EQUATIONS
  . . WIENER HOPF EQUATIONS
  . . INTEGRAL TRANSFORMATIONS
  . . CONVOLUTION INTEGRALS
  . . FOURIER TRANSFORMATION
  . . HILBERT TRANSFORMATION
  . . . LAPLACE TRANSFORMATION
  . HALF PLANES
  . HALF SPACES
  . HILL DETERMINANT
  . NUMERICAL ANALYSIS
  . . APPROXIMATION
  . . . BORN APPROXIMATION
  . . . BORN-OPPENHEIMER
      APPROXIMATION
  . . . CHEBYSHEV APPROXIMATION
  . . . EDDINGTON APPROXIMATION
  . . . FINITE DIFFERENCE THEORY
  . . . FINITE ELEMENT METHOD
  . . . HARTREE APPROXIMATION
  . . . LEAST SQUARES METHOD
  . . . MEAN SQUARE VALUES
  . . . MILNE METHOD
  . . . NEWTON-RAPHSON METHOD
  . . . NUMERICAL DIFFERENTIATION
  . . . OSEEN APPROXIMATION
  . . . PADE APPROXIMATION
  . . . PARTICLE IN CELL TECHNIQUE
  . . . POHLHAUSEN METHOD
  . . . RAYLEIGH-RITZ METHOD
  . . . RELAXATION METHOD
      (MATHEMATICS)
  . . . RITZ AVERAGING METHOD
  . . . SOMMERFELD APPROXIMATION
  . . COMPUTATIONAL CHEMISTRY
  . . COMPUTATIONAL FLUID DYNAMICS
  . . DIFFERENCE EQUATIONS
  . . ERROR ANALYSIS
  . . GRAEFF CALCULUS
  . . INTERPOLATION
  . . ITERATION
  . . . CONJUGATE GRADIENT METHOD
  . . MONTE CARLO METHOD
  . . NOMOGRAPHS
  . . NUMERICAL INTEGRATION
  . . . RUNGE-KUTTA METHOD
  . PFAFF EQUATION
  . PHASE-SPACE INTEGRAL
  . REAL VARIABLES
  . . ABEL FUNCTION
  . . ASYMPTOTES

ANALYSIS (MATHEMATICS)-*(CONT.)*
  . . BESSEL FUNCTIONS
  . . . HANKEL FUNCTIONS
  . . BETHE-SALPETER EQUATION
  . . CALCULUS OF VARIATIONS
  . . COMPOSITE FUNCTIONS
  . DELTA FUNCTION
  . DIFFERENTIAL EQUATIONS
  . . BLASIUS EQUATION
  . . . CAUCHY-RIEMANN EQUATIONS
  . . CHANDRASEKHAR EQUATION
  . . COSINE SERIES
  . . DUFFING DIFFERENTIAL EQUATION
  . . FALKNER-SKAN EQUATION
  . . HYPERBOLIC DIFFERENTIAL
      EQUATIONS
  . . LAME WAVE EQUATIONS
  . . PARTIAL DIFFERENTIAL EQUATIONS
  . . . BIHARMONIC EQUATIONS
  . . . BURGER EQUATION
  . . . ELLIPTIC DIFFERENTIAL
      EQUATIONS
  . . . . MONGE-AMPERE EQUATION
  . . . EULER-CAUCHY EQUATIONS
  . . . FOKKER-PLANCK EQUATION
  . . . GAUSS EQUATION
  . . . HELMHOLTZ VORTICITY EQUATION
  . . . LIOUVILLE EQUATIONS
  . . . PARABOLIC DIFFERENTIAL
      EQUATIONS
  . . . VLASOV EQUATIONS
  . . POISSON EQUATION
  . EINSTEIN EQUATIONS
  . EXISTENCE THEOREMS
  . EXTREMUM VALUES
  . . LIMITS (MATHEMATICS)
  . . . MAXIMA
  . . . MINIMA
  . FOURIER-BESSEL TRANSFORMATIONS
  . GREEN'S FUNCTIONS
  . HYPERBOLIC FUNCTIONS
  . HYPERPLANES
  . JACOBI INTEGRAL
  . JACOBI MATRIX METHOD
  . KERNEL FUNCTIONS
  . LIAPUNOV FUNCTIONS
  . LINEAR EQUATIONS
  . . LINEAR EVOLUTION EQUATIONS
  . LIPSCHITZ CONDITION
  . MEASURE AND INTEGRATION
  . . BINARY INTEGRATION
  . . BOREL SETS
  . . FUNCTIONAL INTEGRATION
  . . INTEGRAL CALCULUS
  . . J INTEGRAL
  . . LEBESGUE THEOREM
  . . NUMERICAL INTEGRATION
  . . . RUNGE-KUTTA METHOD
  . . STIELTJES INTEGRAL
  . . WEIGHTING FUNCTIONS
  . NEUMANN PROBLEM
  . NONLINEAR EQUATIONS
  . . CUBIC EQUATIONS
  . . DUFFING DIFFERENTIAL EQUATION
  . . MONGE-AMPERE EQUATION
  . . NONLINEAR EVOLUTION
      EQUATIONS
  . . QUADRATIC EQUATIONS
  . . QUARTIC EQUATIONS
  . NUMERICAL DIFFERENTIATION
  . PERIODIC FUNCTIONS
  . . TRIGONOMETRIC FUNCTIONS
  . . . COSINE SERIES
  . . . SINE SERIES
  . . . . TANGENTS
  . SERIES (MATHEMATICS)
  . . ASYMPTOTIC SERIES
  . . CAMPBELL-HAUSDORFF SERIES
  . . COSINE SERIES
  . . FOURIER SERIES
  . . PADE APPROXIMATION
  . . POWER SERIES
  . . . TAYLOR SERIES
  . . . . MACLAURIN SERIES
  . . PROGRESSIONS
  . . PRONY SERIES
  . . SINE SERIES
  . STURM-LIOUVILLE THEORY
  . VECTOR ANALYSIS
  . . COLLINEARITY
  . . COPLANARITY
  . . CURL (VECTORS)
  . . . VORTICITY
  . . WEIERSTRASS FUNCTIONS
  . . WHITTAKER FUNCTIONS
  RT   ALGEBRA

ANALYSIS (MATHEMATICS)-*(CONT.)*
∞ ANALYZING
∞ APPLICATIONS OF MATHEMATICS
  DISCONTINUITY
  EQUILIBRIUM EQUATIONS
  GEOMETRY
∞ MATHEMATICS
  MONOTONE FUNCTIONS
∞ SPACE
  TREES (MATHEMATICS)
  VECTOR SPACES
  VENN DIAGRAMS
  VORTICITY EQUATIONS

ANALYSIS OF VARIANCE
GS   STATISTICAL ANALYSIS
  . VARIANCE (STATISTICS)
  . . **ANALYSIS OF VARIANCE**
RT   ∞ VARIANCE

ANALYTIC FUNCTIONS
UF   HOLOMORPHISM
GS   ANALYSIS (MATHEMATICS)
  . COMPLEX VARIABLES
  . . **ANALYTIC FUNCTIONS**
  . . . ENTIRE FUNCTIONS
  FUNCTIONS (MATHEMATICS)
  . **ANALYTIC FUNCTIONS**
  . . ENTIRE FUNCTIONS
RT   CAUCHY-RIEMANN EQUATIONS
  ISOPERIMETRIC PROBLEM
  NONHOLONOMIC EQUATIONS
  POWER SERIES

ANALYTIC GEOMETRY
GS   GEOMETRY
  . EUCLIDEAN GEOMETRY
  . . **ANALYTIC GEOMETRY**
  . . . CATENARIES
  . . . CIRCUMFERENCES
  . . . CONICS
  . . . . ELLIPSES
  . . . . HYPERBOLAS
  . . . . PARABOLAS
  . . . CYCLOIDS
  . . . EPICYCLOIDS
  . . . LOCI
  . . . MERCATOR PROJECTION
  . . . QUADRANTS
  . . . S CURVES
  . . . . GOMPERTZ CURVES
  . . . SPHEROIDS
  . . . . OBLATE SPHEROIDS
  . . . . PROLATE SPHEROIDS
  . . . TANGENTS
  . . . TORUSES
  . . . TRIGONOMETRY
RT   ANNULI
  ASYMPTOTES
  CALCULUS
  COORDINATES
  CURVES (GEOMETRY)
  ∞ CYLINDERS
  DESCRIPTIVE GEOMETRY
  DIFFERENTIAL GEOMETRY
  POLYTOPES
  PROJECTIVE GEOMETRY

ANALYTICAL CHEMISTRY
RT   CHEMICAL ANALYSIS
  ∞ CHEMISTRY
  INORGANIC CHEMISTRY
  QUALITATIVE ANALYSIS
  QUANTITATIVE ANALYSIS
  VOLUMETRIC ANALYSIS

ANALYZERS
SN   (EXCLUDES DEVICES FOR PERFORMING
  MATHEMATICAL ANALYSIS)
GS   MEASURING INSTRUMENTS
  . **ANALYZERS**
  . . ENGINE ANALYZERS
  . . SIGNAL ANALYZERS
RT   CONTROLLERS
  ∞ DETECTORS
  MONITORS
  SELECTORS
  ∞ TEST EQUIPMENT

∞ ANALYZING
SN   *(USE OF A MORE SPECIFIC TERM IS*
  RECOMMENDED--CONSULT THE TERMS
  LISTED BELOW)TERMS LISTED BELOW)
UF   ANALYSIS
  INSTRUMENTAL ANALYSIS

ANALYZING-*(CONT.)*
RT   ACTIVATION ANALYSIS
  ALGEBRA
  ANALYSIS (MATHEMATICS)
  CHEMICAL ANALYSIS
  COMBINATORIAL ANALYSIS
  COST ANALYSIS
  CREEP ANALYSIS
  DESIGN ANALYSIS
  DIAGNOSIS
  DIFFERENTIAL GEOMETRY
  DUALITY PRINCIPLE
  ERROR ANALYSIS
  EVALUATION
  EXAMINATION
  FAILURE ANALYSIS
  FIGURE OF MERIT
  FORECASTING
  MANAGEMENT ANALYSIS
  MULTIVARIATE STATISTICAL ANALYSIS
  NETWORK ANALYSIS
  NUMERICAL ANALYSIS
  PHOTOINTERPRETATION
  POSTFLIGHT ANALYSIS
  PREDICTION ANALYSIS TECHNIQUES
  PREFLIGHT ANALYSIS
  RELIABILITY ANALYSIS
  SIGNAL ANALYSIS
  SIGNATURE ANALYSIS
  SPECTRUM ANALYSIS
  STATISTICAL ANALYSIS
  STRESS ANALYSIS
  STRUCTURAL ANALYSIS
  SYSTEMS ANALYSIS
  TERRAIN ANALYSIS
  THERMAL ANALYSIS
  TRAINING ANALYSIS
  TRAJECTORY ANALYSIS
  WEIGHT ANALYSIS
  X RAY ANALYSIS

ANAPHYLAXIS
GS   SENSITIVITY
  . **ANAPHYLAXIS**
RT   ALLERGIC DISEASES
  ANTIGENS
  IMMUNOLOGY
  SENSITIZING

ANASTIGMATISM
RT   OPTOMETRY
  VISION

ANATASE
UF   OCTAHEDRITE
GS   CHALCOGENIDES
  . OXIDES
  . . METAL OXIDES
  . . . TITANIUM OXIDES
  . . . . **ANATASE**
  MINERALS
  . **ANATASE**
  TITANIUM COMPOUNDS
  . TITANIUM OXIDES
  . . **ANATASE**
RT   PIGMENTS
  RUTILE

ANATOMY
GS   **ANATOMY**
  . ABDOMEN
  . BRAIN
  . . BRAIN STEM
  . . CEREBELLUM
  . . CEREBRAL CORTEX
  . . CEREBRAL VENTRICLES
  . . CEREBRUM
  . . HIPPOCAMPUS
  . CARDIOVASCULAR SYSTEM
  . . BLOOD VESSELS
  . . . ARTERIES
  . . . . AORTA
  . . . CAPILLARIES (ANATOMY)
  . . . . GLOMERULUS
  . . . VEINS
  . . CORPUSCLES
  . . DIASTOLE
  . . ERYTHROCYTES
  . . RETICULOCYTES
  . . HEART
  . . . CARDIAC AURICLES
  . . . CARDIAC VENTRICLES
  . . . EPICARDIUM
  . . . MYOCARDIUM
  . . HEMATOPOIESIS

ANATOMY-*(CONT.)*
  . . HEMATOPOIETIC SYSTEM
  . . LEUKOCYTES
  . . . EOSINOPHILS
  . . . LYMPHOCYTES
  . . SYSTOLE
  . . THROMBIN
  . . THROMBOPLASTIN
  . CHEST
  . CIRCULATORY SYSTEM
  . . VASCULAR SYSTEM
  . . . BLOOD VESSELS
  . . . . ARTERIES
  . . . . . AORTA
  . . . . CAPILLARIES (ANATOMY)
  . . . . . GLOMERULUS
  . . . . VEINS
  . DIGESTIVE SYSTEM
  . . ESOPHAGUS
  . . GASTROINTESTINAL SYSTEM
  . . . APPENDIX (ANATOMY)
  . . . INTESTINES
  . . . . RECTUM
  . . . STOMACH
  . . MOUTH
  . . PANCREAS
  . . TEETH
  . . TONGUE
  . GENITOURINARY SYSTEM
  . . BLADDER
  . . OVARIES
  . . PROSTATE GLAND
  . . TESTES
  . . UTERUS
  . GLANDS (ANATOMY)
  . . ENDOCRINE GLANDS
  . . . ADRENAL GLAND
  . . . GONADS
  . . . OVARIES
  . . . PANCREAS
  . . . PARATHYROID GLAND
  . . . PINEAL GLAND
  . . . PITUITARY GLAND
  . . . PROSTATE GLAND
  . . . THYMUS GLAND
  . . . THYROID GLAND
  . . MAMMARY GLANDS
  . . SALIVARY GLANDS
  . . SEBACEOUS GLANDS
  . . SEX GLANDS
  . . . GONADS
  . . . OVARIES
  . . . PROSTATE GLAND
  . . . TESTES
  . HEAD (ANATOMY)
  . . OCCIPITAL LOBES
  . . SKULL
  . . . CRANIUM
  . . . . INTRACRANIAL CAVITY
  . HUMAN BODY
  . LIMBS (ANATOMY)
  . . ARM (ANATOMY)
  . . . ELBOW (ANATOMY)
  . . . FOREARM
  . . HAND (ANATOMY)
  . . . FINGERS
  . . LEG (ANATOMY)
  . . . FEET (ANATOMY)
  . . . KNEE (ANATOMY)
  . MUSCULOSKELETAL SYSTEM
  . . BONES
  . . . CARTILAGE
  . . . CEREBRUM
  . . . FEMUR
  . . . MARROW
  . . . MASTOIDS
  . . . PELVIS
  . . . SCAPULA
  . . . SCIATIC REGION
  . . . SKULL
  . . . . CRANIUM
  . . . . . INTRACRANIAL CAVITY
  . . . STERNUM
  . . . TIBIA
  . . . ULNA
  . . . VERTEBRAE
  . . CHIN
  . . CONNECTIVE TISSUE
  . . . CARTILAGE
  . . . COLLAGENS
  . . . CONGENERS
  . . . MARROW
  . . CONSTRICTORS
  . . FLEXORS
  . . JOINTS (ANATOMY)
  . . . ELBOW (ANATOMY)

ANATOMY-*(CONT.)*
   . . . KNEE (ANATOMY)
   . . . WRIST
   . . VERTEBRAL COLUMN
   . NECK (ANATOMY)
   . ORGANS
   . . BLADDER
   . . ESOPHAGUS
   . . KIDNEYS
   . . LIVER
   . . LUNGS
   . . OVARIES
   . . PITUITARY GLAND
   . . SPLEEN
   . . STOMACH
   . . TESTES
   . RESPIRATORY SYSTEM
   . . BRONCHI
   . . BRONCHIAL TUBE
   . . . PHARYNX
   . . . TRACHEA
   . . DIAPHRAGM (ANATOMY)
   . . LUNGS
   . . NOSE (ANATOMY)
   . SENSE ORGANS
   . . BARORECEPTORS
   . . CHEMORECEPTORS
   . . EAR
   . . . CORTI ORGAN
   . . . EARDRUMS
   . . . LABYRINTH
   . . . . COCHLEA
   . . . . VESTIBULES
   . . . MASTOIDS
   . . . MIDDLE EAR
   . . . SEMICIRCULAR CANALS
   . . EUSTACHIAN TUBES
   . . EYE (ANATOMY)
   . . . CHOROID MEMBRANES
   . . . CONJUNCTIVA
   . . . CORNEA
   . . . OCULOMOTOR NERVES
   . . . PUPILS
   . . . RETINA
   . . . . FOVEA
   . . GRAVIRECEPTORS
   . . . OTOLITH ORGANS
   . . MECHANORECEPTORS
   . . PHOTORECEPTORS
   . . PROPRIOCEPTORS
   . . THERMORECEPTORS
   . THIGH
   . TORSO
RT   APPENDAGES
   BIFURCATION (BIOLOGY)
   BIODYNAMICS
   ∞BIOLOGY
   ∞DIFFERENTIATION
   DIFFERENTIATION (BIOLOGY)
   DORSAL SECTIONS
   EPITHELIUM
   EXOSKELETONS
   HEPATITIS
   LUMBAR REGION
   MORPHOLOGY
   POSTERIOR SECTIONS
   VESTIBULAR NYSTAGMUS

ANCHORS (FASTENERS)
GS   FASTENERS
   . ANCHORS (FASTENERS)
RT   ∞BANDS
   BOLTS
   BRACKETS
   CLIPS
   COUPLINGS
   GUY WIRES
   HOLDERS
   MOORING
   NUTS (FASTENERS)
   SCREWS
   STRAPS
   STUDS (STRUCTURAL MEMBERS)
   TETHERLINES

ANDES MOUNTAINS (SOUTH AMERICA)
GS   LANDFORMS
   . MOUNTAINS
   . . ANDES MOUNTAINS (SOUTH
      AMERICA)
RT   SOUTH AMERICA

ANDESITE
GS   ALUMINUM COMPOUNDS
   ALUMINUM SILICATES

ANDESITE-*(CONT.)*
   . . ANDESITE
   ROCKS
   . ANDESITE
   SILICON COMPOUNDS
   . ALUMINUM SILICATES
   . . ANDESITE
   . SILICATES
   . . ANDESITE
RT   FELDSPARS
   IGNEOUS ROCKS
   MINERALS
   SOILS

ANDORRA
GS   NATIONS
   . ANDORRA
RT   EUROPE
   FRANCE
   PYRENEES MOUNTAINS (EUROPE)
   SPAIN

∞ ANDROMEDA
SN   *(USE OF A MORE SPECIFIC TERM IS*
   RECOMMENDED--CONSULT THE TERMS
   LISTED BELOW)
RT   ANDROMEDA CONSTELLATION
   ANDROMEDA GALAXIES

ANDROMEDA CONSTELLATION
GS   CONSTELLATIONS
   . ANDROMEDA CONSTELLATION
RT   ∞ANDROMEDA
   ANDROMEDA GALAXIES

ANDROMEDA GALAXIES
GS   CELESTIAL BODIES
   . GALAXIES
   . . GALACTIC CLUSTERS
   . . . LOCAL GROUP (ASTRONOMY)
   . . . . ANDROMEDA GALAXIES
RT   ∞ANDROMEDA
   ANDROMEDA CONSTELLATION
   DISK GALAXIES
   SPIRAL GALAXIES

ANECHOIC CHAMBERS
GS   COMPARTMENTS
   . TEST CHAMBERS
   . . ANECHOIC CHAMBERS
   TEST FACILITIES
   . ANECHOIC CHAMBERS
RT   ACOUSTIC ATTENUATION
   ACOUSTIC MEASUREMENT
   ACOUSTICS
   ∞CHAMBERS
   ZERO SOUND

ANELASTICITY
GS   MECHANICAL PROPERTIES
   . ELASTIC PROPERTIES
   . . ANELASTICITY
RT   CREEP PROPERTIES
   INTERNAL FRICTION
   MODULUS OF ELASTICITY
   STRESS RELAXATION

ANEMIAS
GS   DISEASES
   . ANEMIAS
RT   BLOOD
   CORPUSCLES
   HEMATOCRIT RATIO
   HEMOGLOBIN

ANEMOMETERS
GS   AIRCRAFT INSTRUMENTS
   . SPEED INDICATORS
   . . ANEMOMETERS
   . . . DRAG FORCE ANEMOMETERS
   . . . HOT-WIRE ANEMOMETERS
   . . . SONIC ANEMOMETERS
   DISPLAY DEVICES
   . SPEED INDICATORS
   . . ANEMOMETERS
   . . . DRAG FORCE ANEMOMETERS
   . . . HOT-FILM ANEMOMETERS
   . . . HOT-WIRE ANEMOMETERS
   . . . SONIC ANEMOMETERS
   MEASURING INSTRUMENTS
   . INDICATING INSTRUMENTS
   . . SPEED INDICATORS
   . . . ANEMOMETERS
   . . . . DRAG FORCE ANEMOMETERS
   . . . . HOT-FILM ANEMOMETERS

ANEMOMETERS-*(CONT.)*
   . . . . HOT-WIRE ANEMOMETERS
   . . . . LASER ANEMOMETERS
   . . . . SONIC ANEMOMETERS
RT   FLOW MEASUREMENT
   METEOROLOGICAL INSTRUMENTS
   VELOCITY MEASUREMENT
   WIND (METEOROLOGY)
   WIND MEASUREMENT
   WIND VANES
   WIND VELOCITY
   WIND VELOCITY MEASUREMENT

ANEMOMETRY
USE   VELOCITY MEASUREMENT

ANESTHESIA
GS   ANESTHESIA
   . ELECTROANESTHESIA
RT   ANALGESIA
   HYPNOSIS
   SENSORY PERCEPTION
   UNCONSCIOUSNESS

ANESTHESIOLOGY
GS   MEDICAL SCIENCE
   . ANESTHESIOLOGY
RT   CHLOROFORM
   CLINICAL MEDICINE
   DEPRESSANTS
   DIAGNOSIS
   DRUGS
   PHARMACOLOGY

ANESTHETICS
GS   DRUGS
   . ANESTHETICS
   . . CHLOROFORM
   . . CYCLOPROPANE
   . . METHYL CHLORIDE
   . . NOVOCAIN
RT   ETHERS

ANGELS
GS   ECHOES
   . ANGELS
RT   GLINT
   RADAR CROSS SECTIONS
   RADIO ECHOES

ANGINA PECTORIS
GS   DISEASES
   . HEART DISEASES
   . . ANGINA PECTORIS
RT   ANOXIA
   ARTERIOSCLEROSIS
   CORONARY ARTERY DISEASE
   EMOTIONAL FACTORS
   HEART FUNCTION
   HEART RATE
   MYOCARDIUM
   PHYSICAL EXERCISE
   STRESS (PHYSIOLOGY)

ANGIOGRAPHY
GS   IMAGERY
   . RADIOGRAPHY
   . . ANGIOGRAPHY
RT   BRAIN
   CARDIOLOGY
   CARDIOVASCULAR SYSTEM

ANGIOSPERMS
RT   GRAINS (FOOD)
   NUTS (FRUITS)
   PLANTS (BOTANY)
   VEGETABLES

ANGLE OF ATTACK
GS   GEOMETRY
   . EUCLIDEAN GEOMETRY
   . . ANGLES (GEOMETRY)
   . . . ANGLE OF ATTACK
   . . . . ZERO ANGLE OF ATTACK
RT   AERODYNAMIC CHARACTERISTICS
   AERODYNAMIC STALLING
   BOUNDARY LAYER SEPARATION
   LIFT
   SWEEP ANGLE

ANGLES (GEOMETRY)
GS   GEOMETRY
   . EUCLIDEAN GEOMETRY
   . . ANGLES (GEOMETRY)

ANGLES (GEOMETRY)-*(CONT.)*	
	. . . ANGLE OF ATTACK
	. . . . ZERO ANGLE OF ATTACK
	. . BRAGG ANGLE
	. . BREWSTER ANGLE
	. . DIHEDRAL ANGLE
	. . ELEVATION ANGLE
	. . LOOK ANGLES (ELECTRONICS)
	. . LOOK ANGLES (TRACKING)
	. . SWEEP ANGLE
	. . . . SWEEPBACK
	. . . . . LEADING EDGE SWEEP
RT	ANGULAR RESOLUTION
	APSIDES
	AZIMUTH
	COMPLEMENTS (MATHEMATICS)
	CORNERS
	ELONGATION
	GONIOMETERS
	∞GRADE
	GRADIENTS
	INCIDENCE
	OBLIQUENESS
	PHASE SHIFT
	PHOTOGONIOMETERS
	PITCH (INCLINATION)
	∞PROFILES
	PROTRACTORS
	RECIPROCAL THEOREMS
	SLOPES
	TRIANGULATION
	TRIGONOMETRY

ANGOLA
GS   NATIONS
           . **ANGOLA**
RT   AFRICA

ANGULAR ACCELERATION
GS   RATES (PER TIME)
           . ACCELERATION (PHYSICS)
           . . **ANGULAR ACCELERATION**
RT   ∞ACCELERATION
           CENTRIFUGAL FORCE
           CENTRIPETAL FORCE
           DECELERATION
           ROTATION
           SPIN REDUCTION
           TRANSVERSE ACCELERATION
           YO-YO DEVICES

ANGULAR CORRELATION
GS   CORRELATION
           . **ANGULAR CORRELATION**
RT   DATA CORRELATION
           MATTS (SYSTEMS)
           VIEW EFFECTS

ANGULAR DISTRIBUTION
GS   DISTRIBUTION (PROPERTY)
           . **ANGULAR DISTRIBUTION**
RT   ELECTRON DENSITY PROFILES
           ELEMENTARY PARTICLE INTERACTIONS
           FLUX DENSITY
           FORCE DISTRIBUTION
           MASS DISTRIBUTION
           MOMENT DISTRIBUTION
           NUCLEAR SCATTERING
           STAR DISTRIBUTION

ANGULAR MOMENTUM
GS   MOMENTUM
           . **ANGULAR MOMENTUM**
RT   CLASSICAL MECHANICS
           CLEBSCH-GORDAN COEFFICIENTS
           ELECTRON SPIN
           KINETICS
           MOMENTS OF INERTIA
           PARTICLE SPIN
           QUANTUM NUMBERS
           QUANTUM THEORY
           QUENCHING (ATOMIC PHYSICS)
           RACAH COEFFICIENT
           REGGE POLES
           SPIN
           SPIN TESTS
           STELLAR ROTATION
           WIGNER COEFFICIENT

ANGULAR MOTION
USE   ANGULAR VELOCITY

ANGULAR RESOLUTION
GS   RESOLUTION
           . **ANGULAR RESOLUTION**

ANGULAR RESOLUTION-*(CONT.)*
RT   ACCURACY
           ANGLES (GEOMETRY)
           HIGH RESOLUTION
           ∞OPTICS
           RADAR RESOLUTION

ANGULAR VELOCITY
UF   ANGULAR MOTION
GS   RATES (PER TIME)
           . **ANGULAR VELOCITY**
           VELOCITY
           . **ANGULAR VELOCITY**
RT   GYRATION
           ORBITAL VELOCITY
           REVOLVING
           ROTATION
           ROTOR SPEED
           SAGNAC EFFECT
           TACHOMETERS
           TIP SPEED

ANHYDRIDES
GS   CHALCOGENIDES
           . OXIDES
           . . **ANHYDRIDES**
           . . . PEROXIDES
           . . . . INORGANIC PEROXIDES
           . . . . ORGANIC PEROXIDES
           . . . . SODIUM PEROXIDES
RT   ACIDS
           BASES (CHEMICAL)

ANIK A
USE   ANIK 1

ANIK B
USE   ANIK 2

ANIK C
USE   ANIK 3

ANIK SATELLITES
GS   CANADIAN SPACECRAFT
           . **ANIK SATELLITES**
           . . ANIK 1
           . . ANIK 2
           . . ANIK 3
           SATELLITES
           . ARTIFICIAL SATELLITES
           . . SYNCHRONOUS SATELLITES
           . . . **ANIK SATELLITES**
           . . . . ANIK 1
           . . . . ANIK 2
           . . . . ANIK 3
RT   CANADIAN SPACE PROGRAM
           DELTA LAUNCH VEHICLE
           INTERNATIONAL COOPERATION

ANIK 1
UF   ANIK A
           TELESAT CANADA A
GS   CANADIAN SPACECRAFT
           . ANIK SATELLITES
           . . **ANIK 1**
           SATELLITES
           . ARTIFICIAL SATELLITES
           . . SYNCHRONOUS SATELLITES
           . . . ANIK SATELLITES
           . . . . **ANIK 1**
RT   CANADA
           CANADIAN SPACE PROGRAM
           DELTA LAUNCH VEHICLE
           INTERNATIONAL COOPERATION

ANIK 2
UF   ANIK B
           TELESAT CANADA B
GS   CANADIAN SPACECRAFT
           . ANIK SATELLITES
           . . **ANIK 2**
           SATELLITES
           . ARTIFICIAL SATELLITES
           . . SYNCHRONOUS SATELLITES
           . . . ANIK SATELLITES
           . . . . **ANIK 2**
RT   CANADA
           CANADIAN SPACE PROGRAM
           DELTA LAUNCH VEHICLE
           INTERNATIONAL COOPERATION

ANIK 3
UF   ANIK C
           TELESAT CANADA C
           TELESAT CANADA 3
GS   CANADIAN SPACECRAFT
           . ANIK SATELLITES
           . . **ANIK 3**
           SATELLITES
           . ARTIFICIAL SATELLITES
           . . SYNCHRONOUS SATELLITES
           . . . ANIK SATELLITES
           . . . . **ANIK 3**
RT   CANADA
           CANADIAN SPACE PROGRAM
           INTERNATIONAL COOPERATION

ANILINE
GS   AMINES
           . **ANILINE**
RT   DYES

ANIMALS
UF   FAUNA
           METAZOA
GS   **ANIMALS**
           . AMOEBA
           . . PELOMYXA
           . HETEROTROPHS
           . INVERTEBRATES
           . . ARTHROPODS
           . . . ARTEMIA
           . . . CEPHALOPODS
           . . . . MOLLUSKS
           . . . . OCTOPUSES
           . . . . SNAILS
           . . . CRABS
           . . . INSECTS
           . . . . BEES
           . . . . BOLL WEEVILS
           . . . . CHIRONOMUS FLIES
           . . . . COCKROACHES
           . . . . COLEOPTERA
           . . . . CRICKETS
           . . . . . BEETLES
           . . . . . . TRIBOLIA
           . . . . DROSOPHILA
           . . . . FIREFLIES
           . . . . GRASSHOPPERS
           . . . . LARVAE
           . . . . . BOLLWORMS
           . . . . . SILKWORMS
           . . . . LOCUSTS
           . . . . MOTHS
           . . . . . SILKWORMS
           . . . PUPA
           . . SPIDERS
           . . PROTOZOA
           . . FLAGELLATA
           . . . TRYPANOSOME
           . . . PARAMECIA
           . . . ROTIFERA
           . . PELOMYXA
           . . SEA URCHINS
           . SPORES
           . . MICROSPORES
           . WORMS
           . . FLATWORMS
           . LIVESTOCK
           . MOLES
           . POIKILOTHERMIA
           . . AMPHIBIA
           . . . FROGS
           . . REPTILES
           . . . LIZARDS
           . . . SNAKES
           . . . TURTLES
           . VERTEBRATES
           . . AMPHIBIA
           . . . FROGS
           . . BIRDS
           . . . CHICKENS
           . . . PIGEONS
           . . . TURKEYS
           . . . WATERFOWL
           . . FISHES
           . . . SCHOOLS (FISH)
           . . . SHARKS
           . . MAMMALS
           . . . BATS
           . . . BEARS
           . . . CATS
           . . . CATTLE
           . . . . CALVES
           . . . DEER
           . . . . CARIBOUS

ANIMALS-*(CONT.)*
```
        . . . DOGS
        . . DOLPHINS
        . . . GOATS
        . . . HORSES
        . . . MANATEES
        . . . MARINE MAMMALS
        . . . PORPOISES
        . . . PRIMATES
        . . . . APES
        . . . . . CHIMPANZEES
        . . . . BABOONS
        . . . HUMAN BEINGS
        . . . . MONKEYS
        . . . RODENTS
        . . . . GUINEA PIGS
        . . . . HAMSTERS
        . . . . MICE
        . . . . . JERBOAS
        . . . . . POCKET MICE
        . . . . RABBITS
        . . . . RATS
        . . . . SQUIRRELS
        . . . . . GROUND SQUIRRELS
        . . . SEALS (ANIMALS)
        . . SHEEP
        . . SHREWS
        . . SWINE
        . . WHALES
        . . WOLVES
        . . REPTILES
        . . . LIZARDS
        . . . SNAKES
        . . . TURTLES
        . WILDLIFE
        . . BABOONS
        . . BATS
        . . BIRDS
        . . CHIMPANZEES
        . . MONKEYS
RT      ∞BIOLOGY
        BIOMASS
        CARBON CYCLE
        CENSUS
        ENDANGERED SPECIES
        FOOD CHAIN
        GRAZING
        HABITATS
        HOMEOTHERMS
        MICROORGANISMS
        ORGANISMS
        PARASITES
        PLANKTON
        PLANTS (BOTANY)
        PREDATORS
        VIABILITY
        WILDLIFE RADIOLOCATION
        ∞ZOOLOGY
```

ANIMATION
USE   MOTION

ANIONS
```
GS     IONS
        . ANIONS
        PARTICLES
        . CHARGED PARTICLES
        . . ANIONS
RT     ANODES
        CATIONS
        CELL ANODES
        ELECTRODE MATERIALS
        IONIC MOBILITY
        NEGATIVE IONS
```

ANISOLE
```
GS     ETHERS
        . ANISOLE
        HETEROCYCLIC COMPOUNDS
        . ANISOLE
```

ANISOTROPIC FLUIDS
```
GS     MEDIA
        . ANISOTROPIC MEDIA
        . . ANISOTROPIC FLUIDS
RT     ANISOTROPY
        ∞FLUIDS
        INVARIANT IMBEDDINGS
        ISOTROPY
        LIQUID CRYSTALS
        NEWTONIAN FLUIDS
```

ANISOTROPIC MEDIA
```
GS     MEDIA
        . ANISOTROPIC MEDIA
```

ANISOTROPIC MEDIA-*(CONT.)*
```
        . . ANISOTROPIC FLUIDS
RT     ANISOTROPY
        BIREFRINGENCE
        BIREFRINGENT COATINGS
        HOMOGENEITY
        ISOTROPIC MEDIA
        ∞MATERIALS
        POLARIZATION (WAVES)
```

ANISOTROPIC PLATES
```
UF     NONISOTROPIC PLATES
GS     STRUCTURAL MEMBERS
        . PLATES (STRUCTURAL MEMBERS)
        . . ANISOTROPIC PLATES
RT     CANTILEVER PLATES
        END PLATES
        PERFORATED PLATES
        REINFORCED PLATES
```

ANISOTROPIC SHELLS
```
GS     SHELLS (STRUCTURAL FORMS)
        . ANISOTROPIC SHELLS
RT     CORRUGATED SHELLS
        ELASTIC SHELLS
        REINFORCED SHELLS
```

ANISOTROPY
```
UF     NONISOTROPY
        ONISOTROPY
        PHOTOTHERMOTROPISM
        THERMOTROPISM
GS     ANISOTROPY
        . PLASTIC ANISOTROPY
        . . ELASTIC ANISOTROPY
RT     AEOLOTROPISM
        ANISOTROPIC FLUIDS
        ANISOTROPIC MEDIA
        BIREFRINGENCE
        CRYSTAL STRUCTURE
        CRYSTALS
        DIRECTIVITY
        ISOTROPY
        MECHANICAL PROPERTIES
        METALLOGRAPHY
        POLARIZATION (SPIN ALIGNMENT)
        POLARIZATION (WAVES)
        SPATIAL DISTRIBUTION
```

ANNA HURRICANE
```
GS     STORMS
        . CYCLONES
        . . HURRICANES
        . . . ANNA HURRICANE
        . STORMS (METEOROLOGY)
        . TROPICAL STORMS
        . . HURRICANES
        . . . . ANNA HURRICANE
```

ANNA SATELLITES
```
GS     SATELLITES
        . ARTIFICIAL SATELLITES
        . . GEODETIC SATELLITES
        . . . ANNA SATELLITES
        UNMANNED SPACECRAFT
        . GEODETIC SATELLITES
        . . ANNA SATELLITES
RT     EXPLORER 29 SATELLITE
        EXPLORER 36 SATELLITE
        GEOS 1 SATELLITE
        GEOS 2 SATELLITE
        GEOS 3 SATELLITE
```

ANNEALING
```
GS     HEAT TREATMENT
        . ANNEALING
        . . LASER ANNEALING
        . . PULSE HEATING
RT     GRAPHITIZATION
        HARDENING (MATERIALS)
        HEATING
        NORMALIZING (HEAT TREATMENT)
        RECRYSTALLIZATION
        SOFTENING
        STRESS RELIEVING
        TEMPERING
```

ANNIHILATION REACTIONS
```
GS     NUCLEAR REACTIONS
        . ANNIHILATION REACTIONS
        . . POSITRON ANNIHILATION
RT     ANTIPARTICLES
        HIGH ENERGY INTERACTIONS
        PHOTONS
        PROTON-PROTON REACTIONS
```

ANNOTATIONS
```
RT     ABSTRACTS
        INFORMATION
        SUMMARIES
```

ANNUAL VARIATIONS
```
UF     SEASONAL VARIATIONS
GS     VARIATIONS
        . PERIODIC VARIATIONS
        . . ANNUAL VARIATIONS
RT     ATMOSPHERIC CIRCULATION
        BROWN WAVE EFFECT
        CYCLES
        GREEN WAVE EFFECT
        MAGNETIC VARIATIONS
        METEOROLOGICAL PARAMETERS
        METEOROLOGY
        MONSOONS
        SEASONS
        TEMPORAL DISTRIBUTION
        WEATHER
        WIND VARIATIONS
```

ANNULAR CORE PULSE REACTORS
```
GS     NUCLEAR REACTORS
        . ANNULAR CORE PULSE REACTORS
RT     ∞NUCLEAR ENERGY
        NUCLEAR FUEL ELEMENTS
        NUCLEAR FUELS
        REACTOR CORES
        REACTOR DESIGN
        REACTOR MATERIALS
        REACTOR PHYSICS
        REACTOR SAFETY
        REACTOR TECHNOLOGY
        ∞REACTORS
```

ANNULAR DUCTS
```
GS     DUCTS
        . ANNULAR DUCTS
RT     AIR DUCTS
        DUCT GEOMETRY
        DUCTED BODIES
        FLUID FLOW
        INTAKE SYSTEMS
        NOSE INLETS
        OPENINGS
        ORIFICES
        VENTS
```

ANNULAR FLOW
```
GS     FLUID FLOW
        . AXISYMMETRIC FLOW
        . . ANNULAR FLOW
RT     ANNULI
        AXIAL FLOW
        CHANNEL FLOW
        COAXIAL FLOW
        COUETTE FLOW
        ∞FLOW
        FLOW GEOMETRY
        HEAT TRANSMISSION
        NOZZLE FLOW
        ONE DIMENSIONAL FLOW
        TURBULENT FLOW
```

ANNULAR NOZZLES
```
RT     ANNULI
        COAXIAL FLOW
        CONICAL NOZZLES
        EXHAUST NOZZLES
        INLET NOZZLES
        ∞NOZZLES
        PLUG NOZZLES
        ROCKET NOZZLES
        SHROUDED NOZZLES
        SPRAY NOZZLES
```

ANNULAR PLATES
```
GS     STRUCTURAL MEMBERS
        . PLATES (STRUCTURAL MEMBERS)
        . . ANNULAR PLATES
RT     ANNULI
        CIRCULAR PLATES
        FLAT PLATES
```

ANNULAR SUSPENSION AND POINTING SYSTEM
```
GS     FLIGHT CONTROL
        . POINTING CONTROL SYSTEMS
        . . ANNULAR SUSPENSION AND
            POINTING SYSTEM
RT     MAGNETIC SUSPENSION
        PAYLOADS
        SPACE SHUTTLES
        SPACE TRANSPORTATION SYSTEM
```

ANNULAR SUSPENSION AND POINTING-*(CONT.)*
    SPACELAB
    SPACELAB PAYLOADS
    ∞ SYSTEMS

ANNULI
RT   ANALYTIC GEOMETRY
    ANNULAR FLOW
    ANNULAR NOZZLES
    ANNULAR PLATES
    FLOW MEASUREMENT
    ∞ RINGS

ANODES
GS   ELECTRODES
    . **ANODES**
    . . CELL ANODES
    . . SHELL ANODES
    . . TUBE ANODES
RT   ACCUMULATORS
    ANIONS
    CATHODES
    ELECTRODE MATERIALS
    MULTI-ANODE MICROCHANNEL ARRAYS

ANODIC COATINGS
GS   COATINGS
    . INORGANIC COATINGS
    . **ANODIC COATINGS**
    . PROTECTIVE COATINGS
    . . **ANODIC COATINGS**
RT   ANODIZING
    CATHODIC COATINGS
    ELECTRODE MATERIALS
    OXIDES

ANODIC STRIPPING
RT   CLADDING
    COATING
    DELAMINATING
    METAL COATINGS
    PLATING
    REMOVAL
    ∞ STRIPPING

ANODIZING
GS   COATING
    . **ANODIZING**
    DEPOSITION
    . **ANODIZING**
RT   ANODIC COATINGS
    PASSIVITY
    PROTECTIVE COATINGS

ANOLYTES
GS   CONDUCTORS
    . ELECTROLYTES
    . . **ANOLYTES**
RT   CATHOLYTES
    CELL ANODES

ANOMALIES
GS   **ANOMALIES**
    . GEOTHERMAL ANOMALIES
    . GRAVITY ANOMALIES
    . MAGNETIC ANOMALIES
    . . GEOMAGNETIC HOLLOW
RT   ANOMALOUS TEMPERATURE ZONES

ANOMALOUS TEMPERATURE ZONES
RT   ANOMALIES
    GEYSERS
    TEMPERATURE MEASUREMENT
    TEMPERATURE MEASURING
      INSTRUMENTS
    TEMPERATURE SCALES
    TEMPERATURE SENSORS

ANORTHOSITE
GS   ROCKS
    . IGNEOUS ROCKS
    . . **ANORTHOSITE**
RT   FELDSPARS
    SOILS

ANOXIA
RT   ANGINA PECTORIS
    ASPHYXIA
    HYPOXIA
    STRESS (PHYSIOLOGY)

ANS
USE   ASTRONOMICAL NETHERLANDS
      SATELLITE

ANTARCTIC ENVIRONMENT
USE   ICE ENVIRONMENTS

ANTARCTIC REGIONS
UF   ANTARCTICA
GS   REGIONS
    . POLAR REGIONS
    . . **ANTARCTIC REGIONS**
    . . . MCMURDO SOUND
    . . . ROSS ICE SHELF
    . REMOTE REGIONS
    . . **ANTARCTIC REGIONS**
    SOUTHERN HEMISPHERE
    . **ANTARCTIC REGIONS**
    . . MCMURDO SOUND
    . . ROSS ICE SHELF
RT   CLIMATOLOGY
    CONTINENTS
    LAND ICE
    POLAR CAPS

ANTARCTICA
USE   ANTARCTIC REGIONS

ANTARES ROCKET VEHICLE
GS   ROCKET VEHICLES
    . MULTISTAGE ROCKET VEHICLES
    . . **ANTARES ROCKET VEHICLE**
    . SOUNDING ROCKETS
    . . **ANTARES ROCKET VEHICLE**
RT   ATMOSPHERIC IONIZATION
    SOLID PROPELLANT ROCKET ENGINES
    X-254 ENGINE

ANTELOPE MISSILE
GS   MISSILES
    . **ANTELOPE MISSILE**

ANTENNA ARRAYS
GS   ARRAYS
    . **ANTENNA ARRAYS**
    . . LINEAR ARRAYS
    . . . ENDFIRE ARRAYS
    . . . . YAGI ANTENNAS
    . . STEERABLE ANTENNAS
    . . . INERTIALESS STEERABLE
      ANTENNAS
    . . TURNSTILE ANTENNAS
RT   ANTENNAS
    COBRA DANE (RADAR)
    DIPOLE ANTENNAS
    DIRECTIONAL ANTENNAS
    LOG PERIODIC ANTENNAS
    MICROWAVE ANTENNAS
    PHASED ARRAYS
    RETROREFLECTION
    SPACE BASED RADAR

ANTENNA COMPONENTS
GS   **ANTENNA COMPONENTS**
    . ANTENNA COUPLERS
    . . DIRECTIONAL COUPLERS
    . ANTENNA FEEDS
RT   COMMUNICATION EQUIPMENT
    ∞ COMPONENTS
    COUPLERS
    ELECTRONIC EQUIPMENT

ANTENNA COUPLERS
GS   ANTENNA COMPONENTS
    . **ANTENNA COUPLERS**
    . . DIRECTIONAL COUPLERS
    COUPLERS
    . **ANTENNA COUPLERS**
    . COUPLING CIRCUITS
    . . DIPLEXERS
    . . DIRECTIONAL COUPLERS
RT   ANTENNAS
    COUPLES
    COUPLING
    ENERGY TRANSFER
    IMPEDANCE MATCHING
    MICROWAVE COUPLING
    TRANSMISSION LINES

ANTENNA DESIGN
RT   ANTENNAS
    BACKLOBES
    CASSEGRAIN ANTENNAS
    DELTA ANTENNAS
    ∞ DESIGN
    DIPOLE ANTENNAS
    GRAVITATIONAL WAVE ANTENNAS
    GREGORIAN ANTENNAS
    HELICAL ANTENNAS

ANTENNA DESIGN-*(CONT.)*
    HORN ANTENNAS
    LENS ANTENNAS
    ∞ LOBES
    LOG PERIODIC ANTENNAS
    MAYPOLE ANTENNAS
    MONOPOLE ANTENNAS
    PARABOLIC ANTENNAS
    PLASMA ANTENNAS
    PRODUCT DEVELOPMENT
    RHOMBIC ANTENNAS
    SIDELOBES
    SLOT ANTENNAS
    SPACE TECHNOLOGY EXPERIMENTS
    SPIRAL ANTENNAS
    YAGI ANTENNAS

ANTENNA FEEDS
GS   ANTENNA COMPONENTS
    . **ANTENNA FEEDS**
RT   GREGORIAN ANTENNAS
    STRIP TRANSMISSION LINES
    TRANSMISSION LINES
    WAVEGUIDES

ANTENNA FIELDS
USE   ANTENNA RADIATION PATTERNS

ANTENNA RADIATION PATTERNS
UF   ANTENNA FIELDS
GS   DISTRIBUTION (PROPERTY)
    . RADIATION DISTRIBUTION
    . . **ANTENNA RADIATION PATTERNS**
    . . . SIDELOBES
RT   ANTENNAS
    BACKLOBES
    CYLINDRICAL ANTENNAS
    DIRECTIONAL ANTENNAS
    ∞ FANS
    FAR FIELDS
    FIELD THEORY (PHYSICS)
    FOOTPRINTS
    FRESNEL REGION
    GREGORIAN ANTENNAS
    ∞ LOBES
    NEAR FIELDS
    PLASMA ANTENNAS
    ∞ RADIATION
    ROSETTE SHAPES
    SCHELKUNOFF PRINCIPLE
    SOMMERFELD APPROXIMATION
    SUPPORT INTERFERENCE
    SYNTHETIC ARRAYS

ANTENNAS
GS   **ANTENNAS**
    . AIRCRAFT ANTENNAS
    . CASSEGRAIN ANTENNAS
    . CYLINDRICAL ANTENNAS
    . DELTA ANTENNAS
    . DIRECTIONAL ANTENNAS
    . . DIPOLE ANTENNAS
    . . HELICAL ANTENNAS
    . . HORN ANTENNAS
    . . LENS ANTENNAS
    . . LOG PERIODIC ANTENNAS
    . . LOOP ANTENNAS
    . . PARABOLIC ANTENNAS
    . . RADAR ANTENNAS
    . . . RADANT
    . . RHOMBIC ANTENNAS
    . . SLOT ANTENNAS
    . . STEERABLE ANTENNAS
    . . . INERTIALESS STEERABLE
      ANTENNAS
    . . TWO REFLECTOR ANTENNAS
    . . YAGI ANTENNAS
    . FURLABLE ANTENNAS
    . GRAVITATIONAL WAVE ANTENNAS
    . HOOP COLUMN ANTENNAS
    . MISSILE ANTENNAS
    . MONOPULSE ANTENNAS
    . MULTIPLE BEAM INTERVAL SCANNERS
    . OMNIDIRECTIONAL ANTENNAS
    . . MONOPOLE ANTENNAS
    . . . WHIP ANTENNAS
    . . TURNSTILE ANTENNAS
    . PLASMA ANTENNAS
    . RADIO ANTENNAS
    . . MICROWAVE ANTENNAS
    . . . SPACETENNAS
    . SATELLITE ANTENNAS
    . SCHWARZSCHILD ANTENNAS
    . SPACECRAFT ANTENNAS
    . SPHERICAL ANTENNAS

ANTENNAS-*(CONT.)*
```
    . SPIRAL ANTENNAS
    . . LOG SPIRAL ANTENNAS
    . WAVEGUIDE ANTENNAS
    . . HORN ANTENNAS
RT  ANTENNA ARRAYS
    ANTENNA COUPLERS
    ANTENNA DESIGN
    ANTENNA RADIATION PATTERNS
    ARRAYS
    CONDUCTORS
    CORNERS
    CURRENT SHEETS
    ELECTROMAGNETIC RADIATION
    FOLDING STRUCTURES
    GREGORIAN ANTENNAS
    NEAR FIELDS
    RADIATION HARDENING
    ∞RADIATORS
    RADIO EQUIPMENT
    RADIO TELESCOPES
    REFLECTORS
    SLEWING
    SPACE TECHNOLOGY EXPERIMENTS
    TELECOMMUNICATION
    TELESCOPES
    TOWERS
    TRANSMITTERS
```

ANTHEUS AIRCRAFT
```
USE  AN-22 AIRCRAFT
```

ANTHRACENE
```
GS  ALIPHATIC COMPOUNDS
    . CYCLIC HYDROCARBONS
    . . ANTHRACENE
    HYDROCARBONS
    . CYCLIC HYDROCARBONS
    . . ANTHRACENE
RT  ANTHRAQUINONES
    PHENANTHRENE
```

ANTHRAQUINONES
```
GS  ALIPHATIC COMPOUNDS
    . KETONES
    . . ANTHRAQUINONES
RT  ANTHRACENE
    DYES
```

ANTHROPOLOGY
```
RT  ABORIGINES
    AMERICAN INDIANS
    ANTHROPOMETRY
    ARCHAEOLOGY
    ARTIFACTS
    CASE HISTORIES
    CITIES
    CULTURE (SOCIAL SCIENCES)
    ESKIMOS
    HUMAN BEINGS
    MINORITIES
    MUSEUMS
    RACE FACTORS
    RACES
    ∞SCIENCE
    SOCIAL FACTORS
    SOCIOLOGY
```

ANTHROPOMETRY
```
GS  BIOENGINEERING
    . BIOMETRICS
    . . BODY MEASUREMENT (BIOLOGY)
    . . . ANTHROPOMETRY
RT  ANTHROPOLOGY
    BODY SIZE (BIOLOGY)
    ∞ENGINEERING
    HUMAN FACTORS ENGINEERING
```

ANTIADRENERGICS
```
GS  DRUGS
    . ANTIADRENERGICS
RT  ADRENERGICS
```

ANTIAIRCRAFT MISSILES
```
GS  MISSILES
    . ANTIAIRCRAFT MISSILES
    . . BOMARC MISSILES
    . . FALCON MISSILE
    . . MAULER MISSILE
    . . NIKE-AJAX MISSILE
    . . NIKE-HERCULES MISSILE
    . . REDEYE MISSILE
    . . SIAM MISSILES
    . . SIDEWINDER MISSILES
    . . TARTAR MISSILE
```

ANTIAIRCRAFT MISSILES-*(CONT.)*
```
    . . TERRIER MISSILE
RT  AIR TO AIR MISSILES
    ANTIMISSILE MISSILES
    NIKE MISSILES
    RAMJET MISSILES
    SURFACE TO AIR MISSILES
```

ANTIBIOTICS
```
GS  DRUGS
    . ANTIBIOTICS
    . . PENICILLIN
    . . PLEUROTIN
    . . STREPTOMYCIN
    . . TETRACYCLINES
RT  ANTIINFECTIVES AND ANTIBACTERIALS
    MICROORGANISMS
    STEROIDS
```

ANTIBODIES
```
GS  ANTIBODIES
    . GAMMA GLOBULIN
RT  ANTISERUMS
    BIOCOMPATIBILITY
    IMMUNOLOGY
    INOCULUM
    PHYSIOLOGICAL DEFENSES
    VACCINES
```

ANTICHOLINERGICS
```
UF  CHOLINERGIC BLOCKING AGENTS
GS  DRUGS
    . CHOLINERGICS
    . . ANTICHOLINERGICS
RT  CURARE
```

ANTICLINES
```
UF  ANTICLINORIA
RT  DOMES (GEOLOGY)
    GEOSYNCLINES
    ∞LAYERS
    STRATA
    STRATIFICATION
    STRATIGRAPHY
    SYNCLINES
```

ANTICLINORIA
```
USE  ANTICLINES
```

ANTICOAGULANTS
```
RT  ADRENERGICS
    ∞AGENTS
    HEPARINS
    PRESERVATIVES
    STABILIZERS (AGENTS)
```

ANTICONVULSANTS
```
GS  DRUGS
    . ANTICONVULSANTS
RT  HEXAMETHONIUM
```

ANTICYCLONES
```
RT  AIR MASSES
    ATMOSPHERIC PRESSURE
    CYCLONES
    HIGH PRESSURE
    METEOROLOGY
    SYNOPTIC METEOROLOGY
```

ANTIDIURETICS
```
GS  DRUGS
    . ANTIDIURETICS
RT  URINE
```

ANTIDOTES
```
GS  DRUGS
    . ANTIDOTES
RT  INHIBITORS
```

ANTIEMETICS AND ANTINAUSEANTS
```
GS  DRUGS
    . ANTIEMETICS AND ANTINAUSEANTS
RT  NAUSEA
```

ANTIFERROELECTRICITY
```
GS  ELECTROMAGNETIC PROPERTIES
    . ELECTRICAL PROPERTIES
    . . ANTIFERROELECTRICITY
RT  DIELECTRIC PROPERTIES
    FERROELECTRICITY
    HYSTERESIS
    ∞POLARIZATION
```

ANTIFERROMAGNETISM
```
GS  MAGNETIC PROPERTIES
    . ANTIFERROMAGNETISM
RT  FERROMAGNETISM
    HYSTERESIS
    MAGNETIC SWITCHING
    MAGNONS
    NEEL TEMPERATURE
    PARAMAGNETISM
```

ANTIFOULING
```
GS  FOULING
    . ANTIFOULING
RT  ∞AGENTS
    CLEANING
    CONTAMINATION
    CORROSION PREVENTION
    INHIBITORS
    STERILIZATION
```

ANTIFREEZES
```
GS  ADDITIVES
    . ANTIFREEZES
RT  ANTIICING ADDITIVES
    FREEZING
```

ANTIFRICTION BEARINGS
```
GS  BEARINGS
    . ANTIFRICTION BEARINGS
    . . BALL BEARINGS
    . . ROLLER BEARINGS
RT  FRICTION REDUCTION
    GAS BEARINGS
    JOURNAL BEARINGS
    NEEDLE BEARINGS
    ROLLING CONTACT LOADS
    THRUST BEARINGS
```

ANTIGENS
```
RT  ANAPHYLAXIS
    BIOCOMPATIBILITY
    IMMUNOASSAY
    IMMUNOLOGY
    INOCULUM
    PHYSIOLOGICAL DEFENSES
    RADIOIMMUNOASSAY
    RHESUS FACTOR
    VACCINES
```

ANTIGRAVITY
```
RT  GRAVITATION
    REDUCED GRAVITY
```

ANTIHISTAMINICS
```
GS  DRUGS
    . ANTIHISTAMINICS
    . . DIMENHYDRINATE
    . . DIPHENYL HYDANTOIN
    . . PROMETHAZINE
RT  DECONGESTANTS
    HISTAMINES
```

ANTIHYPERTENSIVE AGENTS
```
GS  DRUGS
    . ANTIHYPERTENSIVE AGENTS
RT  RESERPINE
```

ANTIICING ADDITIVES
```
GS  ADDITIVES
    . ANTIICING ADDITIVES
RT  ANTIFREEZES
    DEICERS
    DEICING
    FUEL CONTAMINATION
    ICE PREVENTION
    INHIBITORS
    PROPELLANT ADDITIVES
    RETARDANTS
```

ANTIINFECTIVES AND ANTIBACTERIALS
```
GS  DRUGS
    . ANTIINFECTIVES AND
        ANTIBACTERIALS
RT  ANTIBIOTICS
    ANTISEPTICS
    BACTERICIDES
    CONTAMINATION
    FUNGICIDES
```

ANTIKNOCK ADDITIVES
```
GS  ADDITIVES
    . ANTIKNOCK ADDITIVES
RT  AUTOMOBILE FUELS
    GASOLINE
```

ANTIKNOCK ADDITIVES-*(CONT.)*
　　∞ OCTANE
　　OCTANES
　　RETARDANTS

ANTIMATTER
GS　**ANTIMATTER**
　　. ANTIPARTICLES
　　. . ANTINEUTRINOS
　　. . ANTINUCLEONS
　　. . ANTIPROTONS
　　. . POSITRONS
RT　MATTER (PHYSICS)

ANTIMISSILE DEFENSE
SN　(PROTECTION AGAINST MISSILE
　　ATTACK)
GS　AIR DEFENSE
　　. **ANTIMISSILE DEFENSE**
RT　ANTIRADIATION MISSILES
　　CIVIL DEFENSE
　　∞ DEFENSE
　　DEFENSE INDUSTRY
　　DEFENSE PROGRAM
　　MILITARY TECHNOLOGY
　　MISSILE DEFENSE
　　MISSILES
　　OPTICAL COUNTERMEASURES
　　SAFEGUARD SYSTEM
　　SENTINEL SYSTEM
　　SPACE SURVEILLANCE (GROUND
　　BASED)
　　SPACE SURVEILLANCE (SPACEBORNE)

ANTIMISSILE MISSILES
GS　MISSILES
　　. **ANTIMISSILE MISSILES**
　　. . MAULER MISSILE
　　. . NIKE-ZEUS MISSILE
　　. . SPARTAN MISSILE
　　. . SPRINT MISSILE
RT　ANTIAIRCRAFT MISSILES
　　BALLISTIC MISSILES
　　INFRARED TRACKING
　　MISSILE DEFENSE
　　NIKE MISSILES
　　NIKE X SYSTEMS
　　SENTINEL SYSTEM
　　SIAM MISSILES
　　SPACE WEAPONS
　　SURFACE TO AIR MISSILES

ANTIMISTING FUELS
RT　ADDITIVES
　　AIRCRAFT FUELS
　　FLAME RETARDANTS
　　JET ENGINE FUELS
　　KEROSENE

ANTIMONIDES
GS　ANTIMONY COMPOUNDS
　　. **ANTIMONIDES**
　　. . ALUMINUM ANTIMONIDES
　　. . CADMIUM ANTIMONIDES
　　. . CESIUM ANTIMONIDES
　　. . GALLIUM ANTIMONIDES
　　. . GERMANIUM ANTIMONIDES
　　. . INDIUM ANTIMONIDES
　　. . ZINC ANTIMONIDES

ANTIMONY
GS　CHEMICAL ELEMENTS
　　. METALLOIDS
　　. . **ANTIMONY**
RT　METALS

ANTIMONY ALLOYS
GS　ALLOYS
　　. **ANTIMONY ALLOYS**
　　. . BABBITT METAL
RT　MULBERRY (ALLOY)

ANTIMONY COMPOUNDS
GS　**ANTIMONY COMPOUNDS**
　　. ANTIMONIDES
　　. . ALUMINUM ANTIMONIDES
　　. . CADMIUM ANTIMONIDES
　　. . CESIUM ANTIMONIDES
　　. . GALLIUM ANTIMONIDES
　　. . GERMANIUM ANTIMONIDES
　　. . INDIUM ANTIMONIDES
　　. . ZINC ANTIMONIDES
　　. ANTIMONY FLUORIDES
RT　∞ CHEMICAL COMPOUNDS
　　∞ GROUP 5A COMPOUNDS

ANTIMONY COMPOUNDS-*(CONT.)*
　　∞ METAL COMPOUNDS

ANTIMONY FLUORIDES
GS　ANTIMONY COMPOUNDS
　　. **ANTIMONY FLUORIDES**
　　HALOGEN COMPOUNDS
　　. FLUORINE COMPOUNDS
　　. . FLUORIDES
　　. . . **ANTIMONY FLUORIDES**
　　. HALIDES
　　. . FLUORIDES
　　. . . **ANTIMONY FLUORIDES**

ANTIMONY ISOTOPES
GS　CHEMICAL ELEMENTS
　　. METALLOIDS
　　. . **ANTIMONY ISOTOPES**
　　. NUCLIDES
　　. . ISOTOPES
　　. . . **ANTIMONY ISOTOPES**
　　METALS
　　. **ANTIMONY ISOTOPES**

ANTINEUTRINOS
GS　ANTIMATTER
　　. ANTIPARTICLES
　　. . **ANTINEUTRINOS**
　　PARTICLES
　　. ELEMENTARY PARTICLES
　　. . ANTIPARTICLES
　　. . . **ANTINEUTRINOS**
　　. . FERMIONS
　　. . . LEPTONS
　　. . . . **ANTINEUTRINOS**
　　. NUCLEAR PARTICLES
　　. . ANTIPARTICLES
　　. . . **ANTINEUTRINOS**
RT　CHARGED PARTICLES
　　NEUTRINOS

ANTINODES
RT　NODES (STANDING WAVES)
　　RAREFACTION
　　RESONANT FREQUENCIES
　　STANDING WAVES
　　VIBRATION
　　WAVELENGTHS

ANTINUCLEONS
GS　ANTIMATTER
　　. ANTIPARTICLES
　　. . **ANTINUCLEONS**
　　PARTICLES
　　. ELEMENTARY PARTICLES
　　. . ANTIPARTICLES
　　. . . **ANTINUCLEONS**
　　. NUCLEAR PARTICLES
　　. . ANTIPARTICLES
　　. . . **ANTINUCLEONS**
RT　NUCLEONS

ANTIOXIDANTS
GS　ADDITIVES
　　. **ANTIOXIDANTS**
RT　ADDITIVES
　　∞ AGENTS
　　CORROSION PREVENTION
　　CORROSION RESISTANCE
　　INHIBITORS
　　PRESERVATIVES
　　PROPELLANT ADDITIVES
　　RETARDANTS
　　STABILIZERS (AGENTS)

ANTIPARTICLES
GS　ANTIMATTER
　　. **ANTIPARTICLES**
　　. . ANTINEUTRINOS
　　. . ANTINUCLEONS
　　. . ANTIPROTONS
　　. . POSITRONS
　　PARTICLES
　　. ELEMENTARY PARTICLES
　　. . **ANTIPARTICLES**
　　. . . ANTINEUTRINOS
　　. . . ANTINUCLEONS
　　. . . ANTIPROTONS
　　. . . POSITRONS
　　. NUCLEAR PARTICLES
　　. . **ANTIPARTICLES**
　　. . . ANTINEUTRINOS
　　. . . ANTINUCLEONS
　　. . . ANTIPROTONS
　　. . . POSITRONS

ANTIPARTICLES-*(CONT.)*
RT　ANNIHILATION REACTIONS
　　CHARGED PARTICLES
　　HYPERONS
　　POMERANCHUK THEOREM
　　POSITRON ANNIHILATION

ANTIPODES
RT　APSIDES
　　IONOSPHERIC PROPAGATION
　　PROPAGATION MODES
　　RADIO TRANSMISSION
　　ZENITH

ANTIPROTONS
GS　ANTIMATTER
　　. ANTIPARTICLES
　　. . **ANTIPROTONS**
　　IONS
　　. **ANTIPROTONS**
　　PARTICLES
　　. CHARGED PARTICLES
　　. . **ANTIPROTONS**
　　. ELEMENTARY PARTICLES
　　. . ANTIPARTICLES
　　. . . **ANTIPROTONS**
　　. NUCLEAR PARTICLES
　　. . ANTIPARTICLES
　　. . . **ANTIPROTONS**
RT　PROTONS

ANTIQUITIES
RT　ARTIFACTS
　　TOOLS
　　WEAPONS

ANTIRADAR COATINGS
UF　RADAR ABSORBING MATERIALS
GS　COATINGS
　　. **ANTIRADAR COATINGS**
　　COUNTERMEASURES
　　. ELECTRONIC COUNTERMEASURES
　　. . **ANTIRADAR COATINGS**
RT　ELECTRONIC WARFARE
　　INORGANIC COATINGS
　　METAL COATINGS
　　PLASTIC COATINGS
　　RADAR ABSORBERS
　　∞ RAM

ANTIRADIATION DRUGS
UF　RADIOPROTECTIVE AGENTS
GS　DRUGS
　　. **ANTIRADIATION DRUGS**
RT　NUCLEAR MEDICINE
　　PHARMACOLOGY
　　RADIATION PROTECTION
　　RADIATION SICKNESS
　　RADIOBIOLOGY
　　RADIOPATHOLOGY

ANTIRADIATION MISSILES
GS　MISSILES
　　. **ANTIRADIATION MISSILES**
RT　AIR DEFENSE
　　ANTIMISSILE DEFENSE
　　COUNTERMEASURES
　　DIGITAL RADAR SYSTEMS
　　MILITARY TECHNOLOGY
　　MISSILE DEFENSE
　　REMOTE CONTROL

ANTIREFLECTION COATINGS
GS　COATINGS
　　. **ANTIREFLECTION COATINGS**
RT　LENS DESIGN
　　OPTICAL REFLECTION
　　OPTICAL THICKNESS
　　SOLAR CELLS

ANTISEPTICS
UF　DISINFECTANTS
RT　ACRIFLAVINE
　　ANTIINFECTIVES AND ANTIBACTERIALS
　　BACTERICIDES
　　CHEMICAL STERILIZATION
　　CHEMOTHERAPY
　　CLEANING
　　DECONTAMINATION
　　ENVIRONMENTAL CONTROL
　　FUMIGATION
　　INFECTIOUS DISEASES
　　PURIFICATION
　　STERILIZATION

ANTISERUMS
RT　ANTIBODIES
　　IMMUNOLOGY
　　SERUMS
　　VACCINES

ANTISHIP MISSILES
GS　MISSILES
　　. **ANTISHIP MISSILES**
RT　CRUISE MISSILES
　　SEA LAUNCHING
　　SHIPS
　　SUBMARINES
　　WEAPON SYSTEMS

ANTISHIP WARFARE
GS　WARFARE
　　. **ANTISHIP WARFARE**
RT　MISSILES
　　SEA LAUNCHING
　　SHIPS
　　SUBMARINES
　　WARHEADS
　　WEAPONS

ANTISKID DEVICES
RT　AIRCRAFT BRAKES
　　ARRESTING GEAR
　　AUTOMOBILES
　　BRAKES (FOR ARRESTING MOTION)
　　∞DEVICES
　　LANDING AIDS
　　SAFETY DEVICES
　　TRUCKS
　　WHEEL BRAKES

ANTISTATIC DEVICES
USE　STATIC DISCHARGERS

ANTISUBMARINE WARFARE
GS　WARFARE
　　. **ANTISUBMARINE WARFARE**
RT　ASROC ENGINE
　　MILITARY TECHNOLOGY
　　SONOBUOYS
　　SUBMARINES
　　TORPEDOES
　　UNDERWATER EXPLOSIONS
　　UNDERWATER TRAJECTORIES

ANTISUBMARINE WARFARE AIRCRAFT
GS　**ANTISUBMARINE WARFARE AIRCRAFT**
　　. BREGUET 1150 AIRCRAFT
　　. CL-84 AIRCRAFT
　　. P-3 AIRCRAFT
　　. S-3 AIRCRAFT
　　. SH-3 HELICOPTER
　　. SH-4 HELICOPTER
RT　∞AIRCRAFT
　　ATTACK AIRCRAFT
　　BOMBER AIRCRAFT
　　DRONE AIRCRAFT
　　H-25 HELICOPTER
　　∞MILITARY AIRCRAFT
　　OBSERVATION AIRCRAFT
　　P-531 HELICOPTER
　　RECONNAISSANCE AIRCRAFT
　　S-61 HELICOPTER
　　SUBMERSIBLE AIRCRAFT
　　V/STOL AIRCRAFT
　　WATER TAKEOFF AND LANDING
　　　　AIRCRAFT

ANTISYMMETRY
RT　ASYMMETRY
　　SYMMETRY

ANTITANK MISSILES
GS　MISSILES
　　. SURFACE TO SURFACE MISSILES
　　. . **ANTITANK MISSILES**
　　. . . SHILLELAGH MISSILES
　　. . . . TOW MISSILES

ANTONOV AIRCRAFT
GS　**ANTONOV AIRCRAFT**
　　. AN-2 AIRCRAFT
　　. AN-22 AIRCRAFT
　　. AN-24 AIRCRAFT
RT　∞AIRCRAFT

ANTONOV AN-22 AIRCRAFT
USE　AN-22 AIRCRAFT

ANTONOV AN-24 AIRCRAFT
USE　AN-24 AIRCRAFT

ANVIL CLOUDS
GS　CLOUDS
　　. **ANVIL CLOUDS**
RT　ATMOSPHERIC MOISTURE
　　CLIMATOLOGY
　　CLOUD COVER
　　FOG
　　METEOROLOGY
　　NEPHANALYSIS
　　PRECIPITATION (METEOROLOGY)
　　THUNDERSTORMS
　　WEATHER

ANVILS
RT　COMPRESSING
　　TOOLS

ANXIETY
RT　DETACHMENT
　　FEAR
　　PHOBIAS
　　TAYLOR MANIFEST ANXIETY SCALE

AO-1 AIRCRAFT
USE　OV-1 AIRCRAFT

AOIPS
USE　ATMOSPHERIC & OCEANOGRAPHIC
　　　INFORM SYS

AORTA
GS　ANATOMY
　　. CARDIOVASCULAR SYSTEM
　　. . BLOOD VESSELS
　　. . . ARTERIES
　　. . . . **AORTA**
　　. CIRCULATORY SYSTEM
　　. . VASCULAR SYSTEM
　　. . . BLOOD VESSELS
　　. . . . ARTERIES
　　. . . . . **AORTA**
RT　HEART

AOSO
UF　ADVANCED ORBITING SOLAR
　　　OBSERVATORY
GS　OBSERVATORIES
　　. **AOSO**
　　UNMANNED SPACECRAFT
　　. SOLAR OBSERVATORIES
　　. . OSO
　　. . . **AOSO**
RT　SUN

APACHE ROCKET VEHICLE
GS　ROCKET VEHICLES
　　. SOUNDING ROCKETS
　　. . **APACHE ROCKET VEHICLE**
RT　SONDES

APATITES
USE　CALCIUM PHOSPHATES
　　　MINERALS

APERIODIC FUNCTIONS
GS　ANALYSIS (MATHEMATICS)
　　. **APERIODIC FUNCTIONS**
　　FUNCTIONS (MATHEMATICS)
　　. **APERIODIC FUNCTIONS**
RT　COMPLEX VARIABLES
　　REAL VARIABLES

APERTURES
GS　OPENINGS
　　. **APERTURES**
　　. . IRISES (MECHANICAL APERTURES)
　　. . SYNTHETIC APERTURES
RT　BOUNDARY ELEMENT METHOD
　　CAVITIES
　　DOORS
　　GATES (OPENINGS)
　　INFRARED WINDOWS
　　LOUVERS
　　ORIFICES
　　OUTLETS
　　PINHOLE CAMERAS
　　PORTS (OPENINGS)
　　SLITS
　　SYNTHETIC ARRAYS
　　VENTS
　　WINDOWS (APERTURES)

APES
GS　ANIMALS
　　. VERTEBRATES
　　. . MAMMALS
　　. . . PRIMATES
　　. . . . **APES**
　　. . . . . CHIMPANZEES

APEXES
UF　VERTICES
RT　ALTITUDE
　　APHELIONS
　　APOGEES
　　MAXIMA
　　ORBITS
　　∞PEAKS
　　PLATEAUS
　　TRAJECTORIES
　　ZENITH

APHELIONS
SN　(APASTRONS IN THE SOLAR SYSTEM)
GS　APSIDES
　　. **APHELIONS**
　　ORBITS
　　. **APHELIONS**
RT　APEXES
　　PERIHELIONS

APL (PROGRAMMING LANGUAGE)
GS　LANGUAGES
　　. PROGRAMMING LANGUAGES
　　. . **APL (PROGRAMMING LANGUAGE)**
RT　COMPUTER PROGRAMMING

APNEA
USE　RESPIRATION

APOGEE BOOST MOTORS
UF　ABM
GS　ENGINES
　　. ROCKET ENGINES
　　. . BOOSTER ROCKET ENGINES
　　. . . **APOGEE BOOST MOTORS**
　　. . SOLID PROPELLANT ROCKET
　　　　ENGINES
　　. . . **APOGEE BOOST MOTORS**
RT　∞BOOSTERS
　　MOTORS

APOGEES
GS　APSIDES
　　. **APOGEES**
　　ORBITS
　　. EARTH ORBITS
　　. . **APOGEES**
　　. ELLIPTICAL ORBITS
　　. . **APOGEES**
RT　APEXES
　　PERIGEES

APOLLO APPLICATIONS PROGRAM
GS　PROGRAMS
　　. NASA PROGRAMS
　　. . **APOLLO APPLICATIONS PROGRAM**
　　. NASA SPACE PROGRAMS
　　. . **APOLLO APPLICATIONS PROGRAM**
RT　AAP 1 MISSION
　　AAP 2 MISSION
　　AAP 3 MISSION
　　AAP 4 MISSION
　　AIRLOCK MODULES
　　EARTH RESOURCES PROGRAM
　　EARTH RESOURCES SURVEY PROGRAM
　　SATURN PROJECT
　　SATURN WORKSHOPS
　　SATURN 1 WORKSHOP
　　SATURN 5 WORKSHOP
　　SKYLAB PROGRAM

APOLLO ASTEROIDS
GS　CELESTIAL BODIES
　　. ASTEROID BELTS
　　. . ASTEROIDS
　　. . . **APOLLO ASTEROIDS**
RT　ASTRONOMY
　　CHIRON
　　EARTH ORBITS
　　JUPITER (PLANET)
　　MARS (PLANET)
　　PLANETARY ORBITS
　　SOLAR SYSTEM

APOLLO EXTENSION SYSTEM
RT   EXOBIOLOGY
       EXTRAVEHICULAR ACTIVITY
       LUNAR LANDING MODULES
       MANNED SPACE FLIGHT
       NASA PROGRAMS
       ORBITAL WORKSHOPS
      ∞SYSTEMS

APOLLO FLIGHTS
GS   SPACE FLIGHT
     . MANNED SPACE FLIGHT
     .. **APOLLO FLIGHTS**
     ... APOLLO 5 FLIGHT
     ... APOLLO 6 FLIGHT
     ... APOLLO 7 FLIGHT
     ... APOLLO 8 FLIGHT
     ... APOLLO 9 FLIGHT
     ... APOLLO 10 FLIGHT
     ... APOLLO 11 FLIGHT
     ... APOLLO 12 FLIGHT
     ... APOLLO 13 FLIGHT
     ... APOLLO 14 FLIGHT
     ... APOLLO 15 FLIGHT
     ... APOLLO 16 FLIGHT
     ... APOLLO 17 FLIGHT
RT   SKYLAB PROGRAM

APOLLO LUNAR EXPERIMENT MODULE
GS   LUNAR SPACECRAFT
     . **APOLLO LUNAR EXPERIMENT
        MODULE**
     MANEUVERABLE SPACECRAFT
     . APOLLO SPACECRAFT
     .. **APOLLO LUNAR EXPERIMENT
         MODULE**
     MANNED SPACECRAFT
     . APOLLO SPACECRAFT
     .. **APOLLO LUNAR EXPERIMENT
         MODULE**
     . LUNAR MODULE
     .. **APOLLO LUNAR EXPERIMENT
         MODULE**
     REENTRY VEHICLES
     . RECOVERABLE SPACECRAFT
     .. APOLLO SPACECRAFT
     ... **APOLLO LUNAR EXPERIMENT
          MODULE**
     SOFT LANDING SPACECRAFT
     . APOLLO SPACECRAFT
     .. **APOLLO LUNAR EXPERIMENT
         MODULE**
     . LANDING MODULES
     .. LUNAR LANDING MODULES
     ... LUNAR MODULE
     .... **APOLLO LUNAR EXPERIMENT
           MODULE**
     SPACECRAFT COMPONENTS
     . **APOLLO LUNAR EXPERIMENT
        MODULE**
RT   LUNAR EXPLORATION
       LUNAR LANDING

**APOLLO LUNAR SURFACE EXPERIMENTS
PACKAGE**
UF   ALSEP
GS   PACKAGES
     . **APOLLO LUNAR SURFACE
        EXPERIMENTS PACKAGE**
RT   ∞INSTRUMENTS
       LUNAR EXPLORATION
       LUNAR RETROREFLECTORS
       PAYLOADS
      ∞SURFACES

APOLLO PROJECT
GS   PROGRAMS
     . LUNAR PROGRAMS
     .. **APOLLO PROJECT**
     . NASA PROGRAMS
     .. **APOLLO PROJECT**
     . NASA SPACE PROGRAMS
     .. **APOLLO PROJECT**
     . PROJECTS
     .. **APOLLO PROJECT**
RT   AAP 1 MISSION
       AAP 2 MISSION
       AAP 3 MISSION
       AAP 4 MISSION
       ADVANCED RANGE INSTRUMENTATION
        AIRCRAFT
       COMMAND SERVICE MODULES
       LSSM
       LUNAR EXPLORATION

APOLLO PROJECT-(CONT.)
       LUNAR EXPLORATION SYSTEM FOR
        APOLLO
       LUNAR MOBILE LABORATORIES
       LUNAR PROBES
       MANNED SPACECRAFT
       MARQUARDT R4D ENGINE
       MERCURY PROJECT
       SATURN LAUNCH VEHICLES
       SATURN WORKSHOPS
       SATURN 1 WORKSHOP
       SATURN 5 WORKSHOP
       SIM
       SITE DATA PROCESSORS
       SKYLAB PROGRAM
       SOFT LANDING SPACECRAFT

APOLLO SHORT STACK
RT   SPACECRAFT CONFIGURATIONS

APOLLO SOYUZ TEST PROJECT
UF   ASTP
GS   PROGRAMS
     . PROJECTS
     .. **APOLLO SOYUZ TEST PROJECT**
RT   INTERNATIONAL COOPERATION
       INTERNATIONAL RELATIONS
       MANNED SPACECRAFT
       RENDEZVOUS
       SOYUZ SPACECRAFT
       SPACE FLIGHT
       SPACE MISSIONS
       SPACE PROGRAMS
       SPACE RENDEZVOUS
       SPACECREW TRANSFER
       U.S.S.R. SPACE PROGRAM

APOLLO SPACECRAFT
GS   LUNAR SPACECRAFT
     . **APOLLO SPACECRAFT**
     MANEUVERABLE SPACECRAFT
     . **APOLLO SPACECRAFT**
     .. APOLLO LUNAR EXPERIMENT
         MODULE
     MANNED SPACECRAFT
     . **APOLLO SPACECRAFT**
     .. APOLLO LUNAR EXPERIMENT
         MODULE
     REENTRY VEHICLES
     . RECOVERABLE SPACECRAFT
     .. **APOLLO SPACECRAFT**
     ... APOLLO LUNAR EXPERIMENT
          MODULE
     SOFT LANDING SPACECRAFT
     . **APOLLO SPACECRAFT**
     .. APOLLO LUNAR EXPERIMENT
         MODULE
RT   COMMAND MODULES
       LANDING MODULES
       LUNAR MODULE
       LUNAR MODULE 5
       LUNAR MODULE 7
       MANNED ORBITAL LABORATORIES
       MANNED ORBITAL RESEARCH
        LABORATORIES
       SATURN PROJECT
       SERVICE MODULES
       SKYLAB PROGRAM
       UNIFIED S BAND

APOLLO TELESCOPE MOUNT
GS   SPACECRAFT CONFIGURATIONS
     . **APOLLO TELESCOPE MOUNT**
     TELESCOPES
     . **APOLLO TELESCOPE MOUNT**
RT   SKYLAB PROGRAM

APOLLO 5 FLIGHT
GS   SPACE FLIGHT
     . MANNED SPACE FLIGHT
     .. APOLLO FLIGHTS
     ... **APOLLO 5 FLIGHT**
RT   EARTH-MOON TRAJECTORIES
       LUNAR EXPLORATION
       LUNAR EXPLORATION SYSTEM FOR
        APOLLO
       LUNAR FLIGHT
       LUNAR LANDING
       LUNAR LAUNCH
       LUNAR MODULE
       LUNAR SPACECRAFT
       MOON-EARTH TRAJECTORIES

APOLLO 6 FLIGHT
GS   SPACE FLIGHT

APOLLO 6 FLIGHT-(CONT.)
     . MANNED SPACE FLIGHT
     .. APOLLO FLIGHTS
     ... **APOLLO 6 FLIGHT**
RT   EARTH-MOON TRAJECTORIES
       LUNAR EXPLORATION
       LUNAR EXPLORATION SYSTEM FOR
        APOLLO
       LUNAR FLIGHT
       LUNAR LANDING
       LUNAR LAUNCH
       LUNAR MODULE
       LUNAR SPACECRAFT
       MOON-EARTH TRAJECTORIES

APOLLO 7 FLIGHT
GS   SPACE FLIGHT
     . MANNED SPACE FLIGHT
     .. APOLLO FLIGHTS
     ... **APOLLO 7 FLIGHT**
RT   EARTH-MOON TRAJECTORIES
       LUNAR EXPLORATION
       LUNAR EXPLORATION SYSTEM FOR
        APOLLO
       LUNAR FLIGHT
       LUNAR LANDING
       LUNAR LAUNCH
       LUNAR MODULE
       MANNED SPACECRAFT
       MOON-EARTH TRAJECTORIES

APOLLO 8 FLIGHT
GS   SPACE FLIGHT
     . MANNED SPACE FLIGHT
     .. APOLLO FLIGHTS
     ... **APOLLO 8 FLIGHT**
RT   EARTH-MOON TRAJECTORIES
       LUNAR EXPLORATION
       LUNAR EXPLORATION SYSTEM FOR
        APOLLO
       LUNAR FLIGHT
       LUNAR LANDING
       LUNAR LAUNCH
       LUNAR MODULE
       MANNED SPACECRAFT
       MOON-EARTH TRAJECTORIES

APOLLO 9 FLIGHT
GS   SPACE FLIGHT
     . MANNED SPACE FLIGHT
     .. APOLLO FLIGHTS
     ... **APOLLO 9 FLIGHT**
RT   EARTH-MOON TRAJECTORIES
       LUNAR EXPLORATION
       LUNAR EXPLORATION SYSTEM FOR
        APOLLO
       LUNAR FLIGHT
       LUNAR LANDING
       LUNAR LAUNCH
       LUNAR MODULE

APOLLO 10 FLIGHT
GS   SPACE FLIGHT
     . MANNED SPACE FLIGHT
     .. APOLLO FLIGHTS
     ... **APOLLO 10 FLIGHT**
RT   EARTH-MOON TRAJECTORIES
       LUNAR EXPLORATION
       LUNAR EXPLORATION SYSTEM FOR
        APOLLO
       LUNAR FLIGHT
       LUNAR LANDING
       LUNAR LAUNCH
       LUNAR MODULE
       MANNED SPACECRAFT
       MOON-EARTH TRAJECTORIES

APOLLO 11 FLIGHT
GS   SPACE FLIGHT
     . MANNED SPACE FLIGHT
     .. APOLLO FLIGHTS
     ... **APOLLO 11 FLIGHT**
RT   EARTH-MOON TRAJECTORIES
       LUNAR EXPLORATION
       LUNAR EXPLORATION SYSTEM FOR
        APOLLO
       LUNAR FLIGHT
       LUNAR LANDING
       LUNAR LAUNCH
       LUNAR MODULE
       MANNED SPACECRAFT
       MOON-EARTH TRAJECTORIES

APOLLO 12 FLIGHT
GS   SPACE FLIGHT

APOLLO 12 FLIGHT-(CONT.)
. MANNED SPACE FLIGHT
. . APOLLO FLIGHTS
. . . APOLLO 12 FLIGHT
RT   EARTH-MOON TRAJECTORIES
LUNAR EXPLORATION
LUNAR EXPLORATION SYSTEM FOR
APOLLO
LUNAR FLIGHT
LUNAR LANDING
LUNAR LAUNCH
LUNAR MODULE
MANNED SPACECRAFT
MOON-EARTH TRAJECTORIES

APOLLO 13 FLIGHT
GS   SPACE FLIGHT
. MANNED SPACE FLIGHT
. . APOLLO FLIGHTS
. . . APOLLO 13 FLIGHT
RT   EARTH-MOON TRAJECTORIES
LUNAR EXPLORATION
LUNAR EXPLORATION SYSTEM FOR
APOLLO
LUNAR FLIGHT
LUNAR LANDING
LUNAR LAUNCH
LUNAR MODULE
MANNED SPACECRAFT
MOON-EARTH TRAJECTORIES

APOLLO 14 FLIGHT
GS   SPACE FLIGHT
. MANNED SPACE FLIGHT
. . APOLLO FLIGHTS
. . . APOLLO 14 FLIGHT
RT   EARTH-MOON TRAJECTORIES
LUNAR EXPLORATION
LUNAR EXPLORATION SYSTEM FOR
APOLLO
LUNAR FLIGHT
LUNAR LANDING
LUNAR LAUNCH
LUNAR MODULE
MANNED SPACECRAFT
MOON-EARTH TRAJECTORIES

APOLLO 15 FLIGHT
GS   SPACE FLIGHT
. MANNED SPACE FLIGHT
. . APOLLO FLIGHTS
. . . APOLLO 15 FLIGHT
RT   EARTH-MOON TRAJECTORIES
LUNAR EXPLORATION
LUNAR EXPLORATION SYSTEM FOR
APOLLO
LUNAR FLIGHT
LUNAR LANDING
LUNAR LAUNCH
LUNAR MODULE
MANNED SPACECRAFT
MOON-EARTH TRAJECTORIES
SIM

APOLLO 16 FLIGHT
GS   SPACE FLIGHT
. MANNED SPACE FLIGHT
. . APOLLO FLIGHTS
. . . APOLLO 16 FLIGHT
RT   EARTH-MOON TRAJECTORIES
LUNAR EXPLORATION
LUNAR EXPLORATION SYSTEM FOR
APOLLO
LUNAR FLIGHT
LUNAR LANDING
LUNAR LAUNCH
LUNAR MODULE
MANNED SPACECRAFT
MOON-EARTH TRAJECTORIES

APOLLO 17 FLIGHT
GS   SPACE FLIGHT
. MANNED SPACE FLIGHT
. . APOLLO FLIGHTS
. . . APOLLO 17 FLIGHT
RT   EARTH-MOON TRAJECTORIES
LUNAR EXPLORATION
LUNAR EXPLORATION SYSTEM FOR
APOLLO
LUNAR FLIGHT
LUNAR LANDING
LUNAR LAUNCH
LUNAR MODULE
MANNED SPACECRAFT
MOON-EARTH TRAJECTORIES

APPALACHIAN MOUNTAINS (NORTH AMERICA)
GS   LANDFORMS
. MOUNTAINS
. . APPALACHIAN MOUNTAINS (NORTH
AMERICA)
RT   NORTH AMERICA

APPARATUS
USE   EQUIPMENT

APPEARANCE
RT   IMAGERY
QUALITY
VISIBILITY

APPENDAGES
GS   APPENDAGES
. ARM (ANATOMY)
. . ELBOW (ANATOMY)
. . FOREARM
. HAND (ANATOMY)
. LEG (ANATOMY)
. . KNEE (ANATOMY)
RT   ANATOMY
HUMAN BODY
LIMBS (ANATOMY)

APPENDIX (ANATOMY)
GS   ANATOMY
. DIGESTIVE SYSTEM
. . GASTROINTESTINAL SYSTEM
. . . APPENDIX (ANATOMY)
VISCERA
. APPENDIX (ANATOMY)
RT   INTESTINES

APPLICATION
USE   UTILIZATION

APPLICATIONS EXPLORER SATELLITES
GS   SATELLITES
. ARTIFICIAL SATELLITES
. . EXPLORER SATELLITES
. . . APPLICATIONS EXPLORER
SATELLITES
RT   HEAT CAPACITY MAPPING MISSION

∞ APPLICATIONS OF MATHEMATICS
SN   (USE OF A MORE SPECIFIC TERM IS
RECOMMENDED--CONSULT THE TERMS
LISTED BELOW)
UF   MATHEMATICAL ANALYSIS
RT   ANALYSIS (MATHEMATICS)
APPROXIMATION
COMBINATORIAL ANALYSIS
COMPUTATION
DIMENSIONAL ANALYSIS
DYNAMIC PROGRAMMING
ECONOMETRICS
ERROR ANALYSIS
FINITE ELEMENT METHOD
FRACTALS
FUNCTIONS (MATHEMATICS)
INFORMATION THEORY
KALMAN-SCHMIDT FILTERING
LINEAR PROGRAMMING
MATHEMATICAL MODELS
NONLINEAR PROGRAMMING
NUMERICAL ANALYSIS
OPERATIONAL CALCULUS
OPERATIONS RESEARCH
OPTIMIZATION
PARAMETERIZATION
PROBABILITY THEORY
STATISTICAL ANALYSIS
STOCHASTIC PROCESSES
TIME SERIES ANALYSIS

APPLICATIONS PROGRAMS (COMPUTERS)
GS   COMPUTER PROGRAMS
. APPLICATIONS PROGRAMS
(COMPUTERS)
RT   COMPUTER PROGRAMS
COMPUTERS

APPLICATIONS TECHNOLOGY SATELLITES
USE   ATS

APPROACH
GS   APPROACH
. AIRBORNE RADAR APPROACH
. DELAYED FLAP APPROACH
. INSTRUMENT APPROACH
RT   AIR TRAFFIC CONTROL

APPROACH-(CONT.)
AIRCRAFT APPROACH SPACING
ARRIVALS
DESCENT
FLIGHT PATHS
FLIGHT PLANS
GROUND BASED CONTROL
GUIDANCE (MOTION)
LANDING
LANDING AIDS
PASSAGEWAYS
TOUCHDOWN

APPROACH AND LANDING TESTS (STS)
RT   EVALUATION
HORIZONTAL SPACECRAFT LANDING
LANDING
MANNED SPACECRAFT
PROVING
SPACE SHUTTLES
SPACE TRANSPORTATION SYSTEM
SPACECRAFT LANDING
∞TESTS
TOUCHDOWN

APPROACH CONTROL
GS   APPROACH CONTROL
. RADAR APPROACH CONTROL
RT   AIR TRAFFIC CONTROL
AIRCRAFT APPROACH SPACING
AIRCRAFT COMMUNICATION
AIRCRAFT GUIDANCE
AIRCRAFT MANEUVERS
AUTOMATED EN ROUTE ATC
COLLISION AVOIDANCE
∞CONTROL
FLIGHT PATHS
GLIDE PATHS
GROUND BASED CONTROL
INSTRUMENT APPROACH
INSTRUMENT FLIGHT RULES
INSTRUMENT LANDING SYSTEMS
LANDING AIDS
LANDING RADAR
MICROWAVE LANDING SYSTEMS
NIGHT FLIGHTS (AIRCRAFT)
RUNWAY LIGHTS
TRACKING (POSITION)
TRAFFIC CONTROL
VISUAL CONTROL

APPROACH INDICATORS
GS   AIRCRAFT INSTRUMENTS
. APPROACH INDICATORS
DISPLAY DEVICES
. APPROACH INDICATORS
FLIGHT INSTRUMENTS
. APPROACH INDICATORS
LANDING AIDS
. LANDING INSTRUMENTS
. . APPROACH INDICATORS
MEASURING INSTRUMENTS
. INDICATING INSTRUMENTS
. . APPROACH INDICATORS
RT   AIR TRAFFIC CONTROL
ALTIMETERS
BLIND LANDING
GLIDE PATHS
INSTRUMENT APPROACH
INSTRUMENT LANDING SYSTEMS
MICROWAVE SCANNING BEAM LANDING
SYSTEM
NAVIGATION AIDS
RADAR APPROACH CONTROL
SOLAR COMPASSES
SPEED INDICATORS

APPROPRIATIONS
RT   BUDGETING
COST ESTIMATES
FEDERAL BUDGETS
GRANTS

APPROXIMATION
UF   APPROXIMATION METHODS
NOMINAL VALUES
TRUNCATION (MATHEMATICS)
GS   ANALYSIS (MATHEMATICS)
. NUMERICAL ANALYSIS
. . APPROXIMATION
. . . BORN APPROXIMATION
. . . BORN-OPPENHEIMER
APPROXIMATION
. . . CHEBYSHEV APPROXIMATION
. . . EDDINGTON APPROXIMATION

APPROXIMATION-(CONT.)
```
            . . . FINITE DIFFERENCE THEORY
            . . . FINITE ELEMENT METHOD
            . . . HARTREE APPROXIMATION
            . . . LEAST SQUARES METHOD
            . . . MEAN SQUARE VALUES
            . . . MILNE METHOD
            . . . NEWTON-RAPHSON METHOD
            . . . NUMERICAL DIFFERENTIATION
            . . . OSEEN APPROXIMATION
            . . . PADE APPROXIMATION
            . . . PARTICLE IN CELL TECHNIQUE
            . . . POHLHAUSEN METHOD
            . . . RAYLEIGH-RITZ METHOD
            . . . RELAXATION METHOD
                    (MATHEMATICS)
            . . . RITZ AVERAGING METHOD
            . . . SOMMERFELD APPROXIMATION
    RT    ∞ APPLICATIONS OF MATHEMATICS
            CENSORED DATA (MATHEMATICS)
            DIFFERENCE EQUATIONS
          ∞ EQUATIONS
            FORM FACTORS
            GLAUBER THEORY
          ∞ METHODOLOGY
            MINIMAX TECHNIQUE
            NUMERICAL STABILITY
            PROBLEM SOLVING
          ∞ RELATIONSHIPS
            SPLINE FUNCTIONS
            STATIC MODELS
            STATISTICAL ANALYSIS
```

APPROXIMATION METHODS
```
    USE    APPROXIMATION
```

APSIDAL ANGLES
```
    USE    APSIDES
```

APSIDES
```
    UF    APSIDAL ANGLES
    GS    APSIDES
            . APHELIONS
            . APOGEES
            . PERIGEES
            . PERIHELIONS
            . PERILUNES
    RT    ANGLES (GEOMETRY)
            ANTIPODES
            ELLIPTICAL ORBITS
            ORBITAL MECHANICS
```

APT (PICTURE TRANSMISSION)
```
    USE    AUTOMATIC PICTURE TRANSMISSION
```

APTITUDE
```
    RT    LEARNING
            PERSONNEL SELECTION
```

AQUARID METEOROIDS
```
    GS    CELESTIAL BODIES
            . METEOROID SHOWERS
            . . AQUARID METEOROIDS
            . METEOROIDS
            . . AQUARID METEOROIDS
    RT    ORIONID METEOROIDS
```

AQUATIC PLANTS
```
    GS    PLANTS (BOTANY)
            . AQUATIC PLANTS
    RT    AQUICULTURE
            HYDROPONICS
            MARINE BIOLOGY
```

AQUEOUS SOLUTIONS
```
    GS    MIXTURES
            . SOLUTIONS
            . . AQUEOUS SOLUTIONS
    RT    HYDRATES
            SOLVATION
```

AQUICULTURE
```
    RT    AQUATIC PLANTS
            FISHERIES
            FISHES
            HYDROPONICS
            MARINE BIOLOGY
            MARINE ENVIRONMENTS
            MARINE RESOURCES
            MARINE TECHNOLOGY
          ∞ NUTRIENTS
            TIDAL FLATS
```

AQUIFERS
```
    GS    RESOURCES
            . AQUIFERS
    RT    FRESH WATER
            GRAVELS
            GROUND WATER
            HYDROGEOLOGY
            HYDROLOGY
            HYDROTHERMAL SYSTEMS
            LAKES
            LIMNOLOGY
            OASES
            PERMEABILITY
            PONDS
            POROSITY
            RAIN
            SANDS
            SPRINGS (WATER)
            STREAMS
            WATER
            WATER TABLES
            WELLS
```

ARABIAN COMMERCIAL SATELLITE
```
    USE    ARCOMSAT
```

ARABIAN SEA
```
    GS    SEAS
            . ARABIAN SEA
    RT    INDIAN OCEAN
```

ARABSAT
```
    GS    SATELLITES
            . ARTIFICIAL SATELLITES
            . . ARABSAT
    RT    INTERNATIONAL COOPERATION
            SAUDI ARABIAN SPACE PROGRAM
```

ARAGONITE
```
    GS    CALCIUM COMPOUNDS
            . CALCIUM CARBONATES
            . . ARAGONITE
            CARBON COMPOUNDS
            . CARBONATES
            . . CALCIUM CARBONATES
            . . . ARAGONITE
            MINERALS
            . ARAGONITE
            SILICON COMPOUNDS
            . SILICATES
            . . ARAGONITE
    RT    CALCITE
```

ARC CHAMBERS
```
    RT    ∞ CHAMBERS
            ELECTRIC ARCS
            PLASMA GENERATORS
            THRUST CHAMBERS
```

ARC CLOUDS
```
    GS    CLOUDS
            . CLOUDS (METEOROLOGY)
            . . CONVECTION CLOUDS
            . . . CUMULONIMBUS CLOUDS
            . . . . ARC CLOUDS
    RT    METEOROLOGY
            OBSERVATION AIRCRAFT
            SATELLITE OBSERVATION
```

ARC DISCHARGES
```
    GS    ELECTRIC CURRENT
            . ELECTRIC DISCHARGES
            . . ARC DISCHARGES
    RT    ELECTRIC ARCS
```

ARC GENERATORS
```
    RT    ELECTRIC ARCS
            ELECTRIC GENERATORS
            ELECTROSTATIC GENERATORS
          ∞ GENERATORS
          ∞ INDUCTION
            INDUCTORS
            PLASMA GENERATORS
            SPARK GAPS
            SPARK PLUGS
            VOLTAGE GENERATORS
```

ARC HEATING
```
    UF    GERDIEN ARC HEATERS
    GS    HEATING
            . ARC HEATING
    RT    GAS HEATING
            IMAGE FURNACES
            PLASMA HEATING
```

ARC HEATING-(CONT.)
```
            RESISTANCE HEATING
            SAHA EQUATIONS
```

ARC JET ENGINES
```
    GS    ENGINES
            . ROCKET ENGINES
            . . ELECTRIC ROCKET ENGINES
            . . . ELECTROTHERMAL ENGINES
            . . . . ARC JET ENGINES
    RT    ELECTRIC PROPULSION
            ELECTROSTATIC ENGINES
            ION ENGINES
            PLASMA ENGINES
            RESISTOJET ENGINES
```

ARC LAMPS
```
    GS    LIGHTING EQUIPMENT
            . LUMINAIRES
            . . ARC LAMPS
    RT    CARBON ARCS
            LIGHT SOURCES
            MERCURY ARCS
            SEARCHLIGHTS
            XENON LAMPS
```

ARC MELTING
```
    GS    PHASE TRANSFORMATIONS
            . ARC MELTING
    RT    DROP TRANSFER
            ELECTROSLAG REFINING
            VACUUM MELTING
            ZONE MELTING
```

ARC SPRAYING
```
    UF    PLASMA ARC SPRAYING
    GS    SPRAYING
            . ARC SPRAYING
    RT    METAL SPRAYING
```

ARC WELDING
```
    GS    WELDING
            . ELECTRIC WELDING
            . . ARC WELDING
            . . . GAS TUNGSTEN ARC WELDING
            . . . PLASMA ARC WELDING
    RT    ELECTRON BEAM WELDING
            PRESSURE WELDING
            SPOT WELDS
            SPUTTERING
```

ARCAS ROCKET VEHICLES
```
    GS    ROCKET VEHICLES
            . SINGLE STAGE ROCKET VEHICLES
            . . ARCAS ROCKET VEHICLES
            . SOUNDING ROCKETS
            . . ARCAS ROCKET VEHICLES
    RT    RADIOSONDES
            SOLID PROPELLANT ROCKET ENGINES
          ∞ VEHICLES
```

ARCHAEOLOGY
```
    RT    ANTHROPOLOGY
            CULTURAL RESOURCES
            FOSSILS
            PALEOMAGNETISM
```

ARCHES
```
    RT    PERFORATED SHELLS
            RIGID STRUCTURES
            SHELLS (STRUCTURAL FORMS)
            TRUSSES
```

ARCHIPELAGOES
```
    RT    ALEUTIAN ISLANDS (US)
            ISLANDS
            LANDFORMS
            SEAS
            SPITSBERGEN (NORWAY)
            VIRGIN ISLANDS
```

ARCHITECTURE
```
    RT    ACOUSTICS
          ∞ BUILDINGS
            CONSTRUCTION
          ∞ CONSTRUCTION MATERIALS
            HUMAN FACTORS ENGINEERING
            ILLUMINATING
            PLANT DESIGN
            STARSITE PROGRAM
            STRUCTURAL DESIGN
          ∞ STRUCTURES
```

ARCHITECTURE (COMPUTERS)
RT   COMPUTER DESIGN
       CONCURRENT PROCESSING
       DISTRIBUTED PROCESSING
       LOGIC CIRCUITS
       LOGIC DESIGN
       MEMORY (COMPUTERS)
       MODULARITY
       SOFTWARE TOOLS
       VERY LARGE SCALE INTEGRATION

ARCOMSAT
UF   ARABIAN COMMERCIAL SATELLITE
GS   SATELLITES
       . ARTIFICIAL SATELLITES
       . . COMMUNICATION SATELLITES
       . . . **ARCOMSAT**
RT   AFRICA
       INTERNATIONAL COOPERATION
       SAUDI ARABIAN SPACE PROGRAM
       SYMPHONIE SATELLITES
       SYNCHRONOUS SATELLITES

ARCON ROCKET VEHICLE
GS   ROCKET VEHICLES
       . **ARCON ROCKET VEHICLE**
RT   ∞VEHICLES

∞ **ARCS**
SN   *(USE OF A MORE SPECIFIC TERM IS*
       RECOMMENDED--CONSULT THE TERMS
       LISTED BELOW
RT   AURORAL ARCS
       CURVES (GEOMETRY)
       ELECTRIC ARCS
       ISLAND ARCS
       MAGNETIC ANNULAR ARC
       PLASMA JETS
       RED ARCS

ARCTIC ENVIRONMENTS
USE   ICE ENVIRONMENTS

ARCTIC OCEAN
GS   OCEANS
       . **ARCTIC OCEAN**
RT   BARENTS SEA
       BEAUFORT SEA (NORTH AMERICA)
       GREENLAND
       SPITSBERGEN (NORWAY)

ARCTIC REGIONS
GS   NORTHERN HEMISPHERE
       . **ARCTIC REGIONS**
       REGIONS
       . POLAR REGIONS
       . . **ARCTIC REGIONS**
       . REMOTE REGIONS
       . . **ARCTIC REGIONS**
RT   CHUCKCHI SEA
       CLIMATOLOGY
       GEOGRAPHY
       MUSKEGS
       NUNATAKS
       POLAR CAPS
       SIBERIA
       SUBARCTIC REGIONS
       TUNDRA

AREA
RT   ∞CROSS SECTIONS
       GEOMETRY
       INTEGRAL CALCULUS
       LINE OF SIGHT
       ∞SECTORS
       ∞SURFACES
       VOLUME

AREA NAVIGATION
GS   NAVIGATION
       . AIR NAVIGATION
       . . **AREA NAVIGATION**
RT   FLIGHT PATHS
       GROUND TRACKS

AREND-ROLAND COMET
GS   CELESTIAL BODIES
       . COMETS
       . . **AREND-ROLAND COMET**
RT   SOLAR SYSTEM

ARES (SPACECRAFT)
USE   ADVANCED RECONN ELECTRIC
       SPACECRAFT

ARETS
USE   ARIZONA REGIONAL ECOLOGICAL TEST
       SITE

ARGENTINA
GS   NATIONS
       . **ARGENTINA**
RT   SOUTH AMERICA

ARGO ROCKET VEHICLES
GS   ROCKET VEHICLES
       . MULTISTAGE ROCKET VEHICLES
       . . **ARGO ROCKET VEHICLES**
RT   HONEST JOHN ROCKET VEHICLE
       JAVELIN ROCKET VEHICLE
       NIKE-AJAX MISSILE
       SOLID PROPELLANT ROCKET ENGINES
       SOUNDING ROCKETS
       ∞VEHICLES

ARGON
GS   CHEMICAL ELEMENTS
       . RARE GASES
       . . **ARGON**
       . . . ARGON ISOTOPES
       GASES
       . RARE GASES
       . . **ARGON**
       . . . ARGON ISOTOPES
RT   RADIATION TRAPPING

ARGON ISOTOPES
GS   CHEMICAL ELEMENTS
       . NUCLIDES
       . . ISOTOPES
       . . . **ARGON ISOTOPES**
       . RARE GASES
       . . ARGON
       . . . **ARGON ISOTOPES**
       GASES
       . RARE GASES
       . . ARGON
       . . . **ARGON ISOTOPES**

ARGON LASERS
GS   STIMULATED EMISSION DEVICES
       . LASERS
       . . **ARGON LASERS**
RT   CHEMICAL LASERS
       CONTINUOUS WAVE LASERS
       GAS MASERS
       INFRARED LASERS
       MACH-ZEHNDER INTERFEROMETERS
       MOLECULAR OSCILLATIONS
       PULSED LASERS
       Q SWITCHED LASERS
       STIMULATED EMISSION

ARGON PLASMA
GS   GASES
       . IONIZED GASES
       . . CHARGED PARTICLES
       . . . **ARGON PLASMA**
       PARTICLES
       . CHARGED PARTICLES
       . . ENERGETIC PARTICLES
       . . . PLASMAS (PHYSICS)
       . . . . **ARGON PLASMA**
RT   HELIUM PLASMA
       HYDROGEN PLASMA
       OXYGEN PLASMA

ARGON-OXYGEN ATMOSPHERES
GS   CONTROLLED ATMOSPHERES
       . **ARGON-OXYGEN ATMOSPHERES**
RT   AEROSPACE ENVIRONMENTS
       ∞ATMOSPHERES
       ∞BREATHING
       GAS MIXTURES
       PORTABLE LIFE SUPPORT SYSTEMS
       UNDERWATER BREATHING APPARATUS

ARGOSY MK-1 AIRCRAFT
GS   HAWKER SIDDELEY AIRCRAFT
       . **ARGOSY MK-1 AIRCRAFT**
       JET AIRCRAFT
       . TURBOPROP AIRCRAFT
       . . **ARGOSY MK-1 AIRCRAFT**
       MONOPLANES
       . **ARGOSY MK-1 AIRCRAFT**
       TRANSPORT AIRCRAFT
       . **ARGOSY MK-1 AIRCRAFT**
RT   ∞AIRCRAFT

ARGUMENTS (MATHEMATICS)
USE   INDEPENDENT VARIABLES

ARGUS PROJECT
GS   PROGRAMS
       . PROJECTS
       . . **ARGUS PROJECT**
RT   THERMONUCLEAR EXPLOSIONS

ARIANE LAUNCH VEHICLE
GS   LAUNCH VEHICLES
       . **ARIANE LAUNCH VEHICLE**
       ROCKET VEHICLES
       . MULTISTAGE ROCKET VEHICLES
       . . **ARIANE LAUNCH VEHICLE**
RT   ELDO LAUNCH VEHICLE
       EUROPA LAUNCH VEHICLES
       EUROPEAN SPACE AGENCY
       EUROPEAN SPACE PROGRAMS
       GEOSARI PROJECT

ARID LANDS
GS   LAND
       . **ARID LANDS**
RT   BARREN LAND
       DEATH VALLEY (CA)
       DESERTIFICATION
       DESERTLINE
       DESERTS
       DROUGHT
       EARTH ENVIRONMENT
       EARTH RESOURCES
       EQUATORIAL REGIONS
       GOBI DESERT
       MOJAVE DESERT (CA)
       OASES
       SAHARA DESERT (AFRICA)
       STEPPES
       WADIS

†

ARIEL SATELLITES
GS   SATELLITES
       . ARTIFICIAL SATELLITES
       . . **ARIEL SATELLITES**
       . . . ARIEL 1 SATELLITE
       . . . ARIEL 2 SATELLITE
       . . . ARIEL 3 SATELLITE
       . . . ARIEL 4 SATELLITE
       . . . ARIEL 5 SATELLITE
RT   GEOPHYSICAL SATELLITES
       THOR DELTA LAUNCH VEHICLE

ARIEL 1 SATELLITE
UF   S-51 SATELLITE
GS   SATELLITES
       . ARTIFICIAL SATELLITES
       . . ARIEL SATELLITES
       . . . **ARIEL 1 SATELLITE**

ARIEL 2 SATELLITE
UF   S-52 SATELLITE
GS   SATELLITES
       . ARTIFICIAL SATELLITES
       . . ARIEL SATELLITES
       . . . **ARIEL 2 SATELLITE**

ARIEL 3 SATELLITE
GS   SATELLITES
       . ARTIFICIAL SATELLITES
       . . ARIEL SATELLITES
       . . . **ARIEL 3 SATELLITE**

ARIEL 4 SATELLITE
GS   SATELLITES
       . ARTIFICIAL SATELLITES
       . . ARIEL SATELLITES
       . . . **ARIEL 4 SATELLITE**
RT   IONOSPHERIC ELECTRON DENSITY
       IONOSPHERIC SOUNDING

ARIEL 5 SATELLITE
GS   SATELLITES
       . ARTIFICIAL SATELLITES
       . . ARIEL SATELLITES
       . . . **ARIEL 5 SATELLITE**

ARIES CONSTELLATION
GS   CONSTELLATIONS
       . **ARIES CONSTELLATION**
RT   CELESTIAL BODIES
       CELESTIAL SPHERE
       STARS

ARIES SOUNDING ROCKET
GS   ROCKET VEHICLES
. SOUNDING ROCKETS
. . **ARIES SOUNDING ROCKET**

ARIETID METEOROIDS
GS   CELESTIAL BODIES
. METEOROID SHOWERS
. . **ARIETID METEOROIDS**
. METEOROIDS
. . **ARIETID METEOROIDS**

ARIP (IMPACT PREDICTION)
USE   COMPUTERIZED SIMULATION
IMPACT PREDICTION

ARIS INSTRUMENTATION SHIP
USE   ADVANCED RANGE INSTRUMENTATION
SHIP

ARITHMETIC
GS   NUMBER THEORY
. **ARITHMETIC**
. . DOUBLE PRECISION ARITHMETIC
. . FIXED POINT ARITHMETIC
. . FLOATING POINT ARITHMETIC
RT   ADDITION
CALCULATORS
COMPUTATION
DIVIDING (MATHEMATICS)
EXPONENTS
INTEGERS
MULTIPLICATION
SUBTRACTION
SUMS

ARITHMETIC AND LOGIC UNITS
UF   ALU (COMPUTER COMPONENTS)
GS   CENTRAL PROCESSING UNITS
. **ARITHMETIC AND LOGIC UNITS**
RT   COMPUTER COMPONENTS
COMPUTERS
DOUBLE PRECISION ARITHMETIC

ARIZONA
GS   NATIONS
. UNITED STATES
. . **ARIZONA**
RT   ARIZONA REGIONAL ECOLOGICAL TEST
SITE
COLORADO PLATEAU (US)
COLORADO RIVER (NORTH AMERICA)
GRAND CANYON (AZ)
PHOENIX (AZ)
PHOENIX QUADRANGLE (AZ)

ARIZONA REGIONAL ECOLOGICAL TEST SITE
UF   ARETS
GS   SITES
. **ARIZONA REGIONAL ECOLOGICAL
TEST SITE**
RT   ARIZONA
ECOLOGY
TEST FACILITIES

ARKANSAS
GS   NATIONS
. UNITED STATES
. . **ARKANSAS**

ARM (ANATOMY)
GS   ANATOMY
. LIMBS (ANATOMY)
. . **ARM (ANATOMY)**
. . . ELBOW (ANATOMY)
. . . FOREARM
APPENDAGES
. **ARM (ANATOMY)**
. . ELBOW (ANATOMY)
. . FOREARM
RT   HUMERUS
SCAPULA
ULNA
WRIST

ARMATURES
RT   COMMUTATORS
ELECTRIC GENERATORS
ELECTRIC MOTORS
ELECTRIC RELAYS
INDUCTION MOTORS
∞ROTATING ELECTRICAL MACHINES
ROTORS

ARMED FORCES
GS   **ARMED FORCES**
. ARMED FORCES (FOREIGN)
. ARMED FORCES (UNITED STATES)
. NAVY
RT   ∞MILITARY AIRCRAFT
∞MILITARY AVIATION
MILITARY SPACECRAFT
∞MILITARY VEHICLES
TANKS (COMBAT VEHICLES)

ARMED FORCES (FOREIGN)
GS   ARMED FORCES
. **ARMED FORCES (FOREIGN)**
RT   DISARMAMENT
ENEMY PERSONNEL
∞MILITARY AIRCRAFT
MILITARY TECHNOLOGY
∞MILITARY VEHICLES
WEAPONS

ARMED FORCES (UNITED STATES)
GS   ARMED FORCES
. **ARMED FORCES (UNITED STATES)**
RT   DEFENSE PROGRAM
DISARMAMENT
∞MILITARY AIRCRAFT
MILITARY TECHNOLOGY
∞MILITARY VEHICLES
WEAPONS
WEAPONS INDUSTRY

ARMOR
RT   HELMETS
METAL PLATES
ORDNANCE
PROTECTIVE CLOTHING
SHIELDING

ARMY-NAVY INSTRUMENTATION PROGRAM
GS   PROGRAMS
. **ARMY-NAVY INSTRUMENTATION
PROGRAM**
RT   LOGISTICS
MILITARY TECHNOLOGY

AROD (RANGE-ORBIT DETERMINATION)
USE   AIRBORNE RANGE AND ORBIT
DETERMINATION

∞ **AROMATIC COMPOUNDS**
SN   *(USE OF A MORE SPECIFIC TERM IS
RECOMMENDED--CONSULT THE TERMS
LISTED BELOW)*
UF   ARYL COMPOUNDS
RT   ∞CHEMICAL COMPOUNDS
CHLOROAROMATICS
FURFURYL ALCOHOL
HYDROCARBONS
ORGANIC COMPOUNDS

AROOS METEORITE
GS   CELESTIAL BODIES
. METEORITES
. . IRON METEORITES
. . . **AROOS METEORITE**

AROUSAL
RT   ALERTNESS
ELECTROENCEPHALOGRAPHY
∞STIMULI

ARPA COMPUTER NETWORK
GS   COMPUTER NETWORKS
. **ARPA COMPUTER NETWORK**
RT   COMPUTER TECHNIQUES
∞NETWORKS
QUEUEING THEORY
SPACECRAFT COMMUNICATION
SWITCHING CIRCUITS
TELECOMMUNICATION

ARRAYS
GS   **ARRAYS**
. ANTENNA ARRAYS
. . LINEAR ARRAYS
. . . ENDFIRE ARRAYS
. . . . YAGI ANTENNAS
. . STEERABLE ANTENNAS
. . . INERTIALESS STEERABLE
ANTENNAS
. . TURNSTILE ANTENNAS
. LARGE APERTURE SEISMIC ARRAY
. PHASED ARRAYS

ARRAYS-*(CONT.)*
. SOLAR ARRAYS
. . SOLAR BLANKETS
. SYNTHETIC ARRAYS
RT   ANTENNAS
MATRICES (MATHEMATICS)
MULTI-ANODE MICROCHANNEL ARRAYS
PHOTOMASKS
PUSHBROOM SENSOR MODES
RANKING
∞STATISTICS

∞ **ARRESTERS**
SN   *(USE OF A MORE SPECIFIC TERM IS
RECOMMENDED--CONSULT THE TERMS
LISTED BELOW)*
RT   ARRESTING GEAR
BLOCKING
BRAKES (FOR ARRESTING MOTION)
GAPS
LIGHTNING

ARRESTING GEAR
GS   LANDING AIDS
. **ARRESTING GEAR**
SAFETY DEVICES
. **ARRESTING GEAR**
RT   ABORT APPARATUS
AIRCRAFT CARRIERS
AIRCRAFT SAFETY
ANTISKID DEVICES
∞ARRESTERS
∞BARRIERS
BRAKES (FOR ARRESTING MOTION)
CRASH LANDING
∞GEAR

ARRHYTHMIA
GS   RATES (PER TIME)
. HEART RATE
. . **ARRHYTHMIA**

ARRIVALS
RT   APPROACH
LANDING

ARROW WINGS
GS   AIRFOILS
. WINGS
. . SWEPT WINGS
. . . SWEPTBACK WINGS
. . . . **ARROW WINGS**
PLANFORMS
. WING PLANFORMS
. . SWEPTBACK WINGS
. . . **ARROW WINGS**
RT   CARET WINGS
DELTA WINGS
VARIABLE SWEEP WINGS

ARROYOS
GS   LANDFORMS
. **ARROYOS**
RT   CANYONS
DRAINAGE PATTERNS
EROSION
LIMNOLOGY
RAIN IMPACT DAMAGE
WATER
WATER CURRENTS
WATER EROSION

ARSENATES
GS   ARSENIC COMPOUNDS
. **ARSENATES**
RT   ARSENIDES
∞OXYGEN COMPOUNDS

ARSENIC
GS   CHEMICAL ELEMENTS
. METALLOIDS
. . **ARSENIC**
. . . ARSENIC ISOTOPES
RT   METALS

ARSENIC ALLOYS
GS   ALLOYS
. **ARSENIC ALLOYS**
RT   METALLOIDS

ARSENIC COMPOUNDS
GS   **ARSENIC COMPOUNDS**
. ARSENATES
. ARSENIDES

ARSENIC COMPOUNDS-*(CONT.)*
    . . GALLIUM ARSENIDES
    . . . ALUMINUM GALLIUM ARSENIDES
    . . INDIUM ARSENIDES
    . . PROUSTITE
RT   ∞CHEMICAL COMPOUNDS
    ∞GROUP 5A COMPOUNDS

ARSENIC ISOTOPES
GS   CHEMICAL ELEMENTS
    . METALLOIDS
    . . ARSENIC
    . . . **ARSENIC ISOTOPES**
    . NUCLIDES
    . . ISOTOPES
    . . . RADIOACTIVE ISOTOPES
    . . . . **ARSENIC ISOTOPES**
RT   METALS

ARSENIDES
GS   ARSENIC COMPOUNDS
    . **ARSENIDES**
    . . GALLIUM ARSENIDES
    . . . ALUMINUM GALLIUM ARSENIDES
    . . INDIUM ARSENIDES
    . . PROUSTITE
RT   ARSENATES
    INTERMETALLICS

ARTEMIA
GS   ANIMALS
    . INVERTEBRATES
    . . ARTHROPODS
    . . . **ARTEMIA**

ARTERIES
GS   ANATOMY
    . CARDIOVASCULAR SYSTEM
    . . BLOOD VESSELS
    . . . **ARTERIES**
    . . . . AORTA
    . CIRCULATORY SYSTEM
    . . VASCULAR SYSTEM
    . . . BLOOD VESSELS
    . . . . **ARTERIES**
    . . . . . AORTA
RT   ARTERIOSCLEROSIS
    BIFURCATION (BIOLOGY)
    CAROTID SINUS BODY
    CAROTID SINUS REFLEX
    PHONOARTERIOGRAPHY
    SPHYGMOGRAPHY
    VEINS

ARTERIOSCLEROSIS
UF   ATHEROSCLEROSIS
GS   DISEASES
    . **ARTERIOSCLEROSIS**
RT   ANGINA PECTORIS
    ARTERIES
    CHOLESTEROL
    CIRCULATORY SYSTEM
    CORONARY ARTERY DISEASE
    MYOCARDIAL INFARCTION

ARTHRITIS
GS   DISEASES
    . **ARTHRITIS**
RT   BONES
    CALCIFICATION
    JOINTS (ANATOMY)

ARTHROPODS
GS   ANIMALS
    . INVERTEBRATES
    . . **ARTHROPODS**
    . . . ARTEMIA
    . . . CEPHALOPODS
    . . . . MOLLUSKS
    . . . . OCTOPUSES
    . . . . SNAILS
    . . . CRABS
    . . . INSECTS
    . . . . BEES
    . . . . BOLL WEEVILS
    . . . . CHIRONOMUS FLIES
    . . . . COCKROACHES
    . . . . COLEOPTERA
    . . . . CRICKETS
    . . . . . BEETLES
    . . . . . . TRIBOLIA
    . . . DROSOPHILA
    . . . FIREFLIES
    . . . . GRASSHOPPERS
    . . . . LARVAE

ARTHROPODS-*(CONT.)*
    . . . . BOLLWORMS
    . . . . SILKWORMS
    . . . . LOCUSTS
    . . . . MOTHS
    . . . . . SILKWORMS
    . . . PUPA
    . . . SPIDERS
RT   EXOSKELETONS

ARTICULATION
GS   SPEECH
    . **ARTICULATION**
RT   LANGUAGES
    SPEECH DEFECTS

ARTIFACTS
RT   ANTHROPOLOGY
    ANTIQUITIES
    CULTURE (SOCIAL SCIENCES)
    MUSEUMS

ARTIFICIAL CARDIAC PACEMAKER
GS   MEDICAL EQUIPMENT
    . **ARTIFICIAL CARDIAC PACEMAKER**
RT   BIOTECHNOLOGY
    BLOOD CIRCULATION
    CARDIOLOGY
    CIRCULATION
    CIRCULATORY SYSTEM
    HEART
    PULMONARY CIRCULATION

ARTIFICIAL CLOUDS
GS   CLOUDS
    . CLOUDS (METEOROLOGY)
    . . **ARTIFICIAL CLOUDS**
    . . . CHEMICAL CLOUDS
    . . . . BARIUM ION CLOUDS
RT   WEATHER MODIFICATION

ARTIFICIAL EARS
GS   MEDICAL EQUIPMENT
    . PROSTHETIC DEVICES
    . . **ARTIFICIAL EARS**
RT   EAR

ARTIFICIAL GRAVITY
GS   GRAVITATION
    . **ARTIFICIAL GRAVITY**
RT   ACCELERATION STRESSES
     (PHYSIOLOGY)
    ∞ASTRONAUTICS
    ENVIRONMENTAL CONTROL
    GRAVITY GRADIENT SATELLITES
    HUMAN CENTRIFUGES
    LIFE SUPPORT SYSTEMS
    LOWER BODY NEGATIVE PRESSURE
    ROTATING ENVIRONMENTS
    SPIN DYNAMICS
    WEIGHTLESSNESS

ARTIFICIAL HARBORS
GS   WATERWAYS
    . HARBORS
    . . **ARTIFICIAL HARBORS**
RT   CARGO SHIPS
    DEEPWATER TERMINALS
    DREDGING
    MARINE TECHNOLOGY
    OCEANOGRAPHY
    OFFSHORE DOCKING
    OFFSHORE PLATFORMS
    SHIP TERMINALS
    TANKER SHIPS
    TANKER TERMINALS
    ∞TANKERS
    TERMINAL FACILITIES
    TRANSPORTATION

ARTIFICIAL HEART VALVES
GS   MEDICAL EQUIPMENT
    . **ARTIFICIAL HEART VALVES**
    VALVES
    . **ARTIFICIAL HEART VALVES**
RT   BIOTECHNOLOGY
    BLOOD CIRCULATION
    BLOOD PUMPS
    HEART
    HEART IMPLANTATION

ARTIFICIAL INTELLIGENCE
UF   MACHINE RECOGNITION
GS   INTELLIGENCE

ARTIFICIAL INTELLIGENCE-*(CONT.)*
    . **ARTIFICIAL INTELLIGENCE**
    . . EXPERT SYSTEMS
RT   AUTOMATA THEORY
    BIONICS
    CHARACTER RECOGNITION
    COGNITION
    COMPUTER VISION
    COMPUTERS
    DEPERSONALIZATION
    INTELLECT
    LEARNING MACHINES
    ∞LOGIC
    LOGIC PROGRAMMING
    PERCEPTION
    ROBOTICS
    ROBOTS
    SELF ORGANIZING SYSTEMS
    THEOREM PROVING
    VOICE DATA PROCESSING

ARTIFICIAL RADIATION BELTS
GS   EARTH ATMOSPHERE
    . RADIATION BELTS
    . . **ARTIFICIAL RADIATION BELTS**
    PARTICLES
    . CHARGED PARTICLES
    . . MAGNETICALLY TRAPPED PARTICLES
    . . . RADIATION BELTS
    . . . . **ARTIFICIAL RADIATION BELTS**
    . TRAPPED PARTICLES
    . . MAGNETICALLY TRAPPED PARTICLES
    . . . RADIATION BELTS
    . . . . **ARTIFICIAL RADIATION BELTS**
RT   INNER RADIATION BELT
    NUCLEAR EXPLOSIONS
    OUTER RADIATION BELT
    ∞RADIATION

ARTIFICIAL RESPIRATION
USE   RESUSCITATION

ARTIFICIAL SATELLITES
UF   ORBITING SATELLITES
GS   SATELLITES
    . **ARTIFICIAL SATELLITES**
    . . ALOUETTE SATELLITES
    . . . ALOUETTE B SATELLITE
    . . . ALOUETTE 1 SATELLITE
    . . . ALOUETTE 2 SATELLITE
    . . ARABSAT
    . . ARIEL SATELLITES
    . . . ARIEL 1 SATELLITE
    . . . ARIEL 2 SATELLITE
    . . . ARIEL 3 SATELLITE
    . . . ARIEL 4 SATELLITE
    . . . ARIEL 5 SATELLITE
    . . BESS (SATELLITE)
    . . BIOSATELLITES
    . . . BIOSATELLITE 1
    . . . BIOSATELLITE 2
    . . . BIOSATELLITE 3
    . . . ORBITING FROG OTOLITH
    . . SPUTNIK 2 SATELLITE
    . . COMMUNICATION SATELLITES
    . . . AERONAUTICAL SATELLITES
    . . . AEROSAT SATELLITES
    . . . ARCOMSAT
    . . COMMUNICATIONS TECHNOLOGY
      SATELLITE
    . . . . COMSTAR C
    . . . NATO 3B SATELLITE
    . . . COMSTAR SATELLITES
    . . . EUROPEAN COMMUNICATIONS
      SATELLITE
    . . . INTELSAT SATELLITES
    . . L-SAT
    . . LOW FREQUENCY
      TRANSIONOSPHERIC SATELLITES
    . . . MARECS MARITIME SATELLITES
    . . . MAROTS (ESA)
    . . . MOLNIYA SATELLITES
    . . . PALAPA SATELLITES
    . . . PALAPA 2 SATELLITE
    . . . RADUGA SATELLITE
    . . . RCA SATCOM SATELLITES
    . . . RELAY SATELLITES
    . . . . RELAY 1 SATELLITE
    . . . . RELAY 2 SATELLITE
    . . . SYMPHONIE SATELLITES
    . . . SYNCOM SATELLITES
    . . . . EARLY BIRD SATELLITES
    . . . . SYNCOM 1 SATELLITE
    . . . . SYNCOM 2 SATELLITE
    . . . . SYNCOM 3 SATELLITE

ARTIFICIAL SATELLITES-(CONT.)
```
. . . SMS 2
. . TIROS SATELLITES
. . . IMPROVED TIROS OPERATIONAL
          SATELLITES
. . . . ITOS 1
. . . . ITOS 2
. . . . ITOS 3
. . . . ITOS 4
. . . ITOS SATELLITES
. . . . ITOS 1
. . . . ITOS 2
. . . . ITOS 3
. . . . ITOS 4
. . . TIROS M
. . . TIROS N SERIES SATELLITES
. . . TIROS 1 SATELLITE
. . . TIROS 2 SATELLITE
. . . TIROS 3 SATELLITE
. . . TIROS 4 SATELLITE
. . . TIROS 5 SATELLITE
. . . TIROS 6 SATELLITE
. . . TIROS 7 SATELLITE
. . . TIROS 8 SATELLITE
. . . TIROS 9 SATELLITE
. . . TIROS 10 SATELLITE
. . . VANGUARD 2 SATELLITE
. . MIDAS SATELLITES
. . . MIDAS 2 SATELLITE
. . . MIDAS 3 SATELLITE
. . . MIDAS 4 SATELLITE
. . . MIDAS 5 SATELLITE
. . . MIDAS 6 SATELLITE
. . . MIDAS 7 SATELLITE
. . MULTISPECTRAL RESOURCE
          SAMPLER
. . NAVIGATION SATELLITES
. . . AEROSAT SATELLITES
. . . EXPLORER 22 SATELLITE
. . . NAVIGATION TECHNOLOGY
          SATELLITES
. . . NAVSTAR SATELLITES
. . . NOVA SATELLITES
. . . REFSAT
. . . TRANSIT ATTITUDE CONTROL
          SATELLITE
. . . TRANSIT SATELLITES
. . ORBITAL SPACE STATIONS
. . . EOSS
. . . LONG DURATION EXPOSURE
          FACILITY
. . ORBITAL WORKSHOPS
. . PAS
. PASSIVE SATELLITES
. . BEACON SATELLITES
. . . BEACON EXPLORER A
. . . EXPLORER 22 SATELLITE
. . ECHO SATELLITES
. . . ECHO 1 SATELLITE
. . . ECHO 2 SATELLITE
. . LAGEOS (SATELLITE)
. . PAGEOS SATELLITE
. PEGASUS SATELLITES
. POLYOT SATELLITES
. ROSAT MISSION
. SAGE SATELLITE
. SARSAT
. SCATHA SATELLITE
. SCIENTIFIC SATELLITES
. . AMPTE (SATELLITES)
. . ASTRONOMICAL SATELLITES
. . . ATS
. . . . ATS 1
. . . . ATS 2
. . . . ATS 3
. . . . ATS 4
. . . . ATS 5
. . . . ATS 6
. . . . ATS 7
. . . . ATS 8
. . . AZUR SATELLITE
. . . CANNONBALL 2 SATELLITE
. . . DIAL SATELLITE
. . . ENVIRONMENTAL RESEARCH
          SATELLITES
. . . . ERS 17
. . . . ERS 18
. . . . INTASAT SATELLITE
. . . EXOSAT SATELLITE
. . . EXPLORER 45 SATELLITE
. . . GRAVSAT SATELLITE
. . . HAWKEYE SATELLITES
. . . LZEEBE SATELLITE
. . . MAGSAT A SATELLITE
. . . MAGSAT B SATELLITE
. . . MAGSAT SATELLITES
```

ARTIFICIAL SATELLITES-(CONT.)
```
. . . MAGSAT 1 SATELLITE
. . ORBIS
. . . ORBIS CAL SATELLITE
. . OV-1 SATELLITES
. . OV-2 SATELLITES
. . OV-3 SATELLITES
. . OV-4 SATELLITES
. . OV-5 SATELLITES
. . SMALL SCIENTIFIC SATELLITES
. . UK SATELLITES
. . . UK 4 SATELLITE
. . SCORE SATELLITE
. . SEASAT SATELLITES
. . SEASAT 1
. . . SEASAT-B SATELLITE
. . SHUTTLE PALLET SATELLITES
. . SKYNET SATELLITES
. . SNAPSHOT SATELLITE
. . SOLAR POWER SATELLITES
. . SOLAR RADIATION 1 SATELLITE
. . SOLAR RADIATION 3 SATELLITE
. . SOVIET SATELLITES
. . . COSMOS SATELLITES
. . . COSMOS 2 SATELLITE
. . . COSMOS 3 SATELLITE
. . . COSMOS 5 SATELLITE
. . . COSMOS 6 SATELLITE
. . . COSMOS 14 SATELLITE
. . . COSMOS 44 SATELLITE
. . . COSMOS 54 SATELLITE
. . . COSMOS 71 SATELLITE
. . . COSMOS 110 SATELLITE
. . . COSMOS 137 SATELLITE
. . . COSMOS 144 SATELLITE
. . . COSMOS 149 SATELLITE
. . . COSMOS 166 SATELLITE
. . . COSMOS 186 SATELLITE
. . . COSMOS 188 SATELLITE
. . . COSMOS 206 SATELLITE
. . . COSMOS 213 SATELLITE
. . . COSMOS 224 SATELLITE
. . . COSMOS 225 SATELLITE
. . . COSMOS 381 SATELLITE
. . . COSMOS 954 SATELLITE
. . . COSMOS 1129 SATELLITE
. . . . INTERCOSMOS SATELLITES
. . . COSMOS 782 SATELLITE
. . . COSMOS 936 SATELLITE
. . . MOLNIYA SATELLITES
. . . PROGNOZ SATELLITES
. . . PROTON SATELLITES
. . . . PROTON 1 SATELLITE
. . . . PROTON 2 SATELLITE
. . . . PROTON 3 SATELLITE
. . . . PROTON 4 SATELLITE
. . . RADUGA SATELLITE
. . . SPUTNIK SATELLITES
. . . . SPUTNIK 1 SATELLITE
. . . . SPUTNIK 2 SATELLITE
. . . . SPUTNIK 3 SATELLITE
. . . . SPUTNIK 4 SATELLITE
. . . . SPUTNIK 5 SATELLITE
. . . VENERA SATELLITES
. . . . VENERA 9 SATELLITE
. . . . VENERA 10 SATELLITE
. . . . VENERA 11 SATELLITE
. . . . VENERA 12 SATELLITE
. . SYNCHRONOUS SATELLITES
. . . AEROS SATELLITE
. . . AEROSAT SATELLITES
. . . ANIK SATELLITES
. . . . ANIK 1
. . . . ANIK 2
. . . . ANIK 3
. . . GOES SATELLITES
. . . . GOES 1
. . . . GOES 2
. . . . GOES 3
. . . . GOES 4
. . . . GOES 5
. . . MIRANDA SATELLITE
. . . SIRIO SATELLITE
. . . STORMSAT SATELLITE
. . . SYNCHRONOUS EARTH
          OBSERVATORY SATELLITE
. . . . SMS 1
. . . . SMS 2
. . . SYNCHRONOUS METEOROLOGICAL
          SATELLITE
. . . . SMS 1
. . . . SMS 2
. . . SYNCOM SATELLITES
. . . . EARLY BIRD SATELLITES
. . . . SYNCOM 1 SATELLITE
. . . . SYNCOM 2 SATELLITE
```

ARTIFICIAL SATELLITES-(CONT.)
```
. . . . SYNCOM 3 SATELLITE
. . . TD SATELLITES
. . . . TD-1 SATELLITE
. . TELSTAR SATELLITES
. . . TELSTAR 1 SATELLITE
. . . TELSTAR 2 SATELLITE
. . TETHERED SATELLITES
. . UHURU SATELLITE
. VANGUARD SATELLITES
. . . VANGUARD 1 SATELLITE
. . . VANGUARD 2 SATELLITE
. . . VANGUARD 3 SATELLITE
. . VELA SATELLITES
```
```
RT      FLEXIBLE SPACECRAFT
        HALO ORBIT SPACE STATION
        INFLATABLE SPACECRAFT
        INTERPLANETARY SPACECRAFT
        LUNAR ORBITS
        LUNAR SPACECRAFT
        MANEUVERABLE SPACECRAFT
        MANNED SPACECRAFT
        MILITARY SPACECRAFT
        NATIONAL OCEANIC SATELLITE SYSTEM
        NATURAL SATELLITES
        OBSERVATORIES
        ORBITS
        RECONNAISSANCE SPACECRAFT
        SKYLAB 1
        SKYLAB 2
        SKYLAB 3
        SKYLAB 4
        SPACE CAPSULES
        SPACE LABORATORIES
        SPACE STATIONS
        TELSTAR PROJECT
        UNMANNED SPACECRAFT
```

ARTILLERY
```
GS      WEAPONS
        . ARTILLERY
        . . HOWITZERS
        . . PRECISION GUIDED PROJECTILES
RT      GUN LAUNCHERS
        GUNNERY TRAINING
        MISSILES
        RIFLES
        SABOT PROJECTILES
```

ARTILLERY FIRE
```
RT      ∞BARRAGES
        GUNFIRE
```

ARTS
```
GS      ARTS
        . ABILITIES
        . GRAPHIC ARTS
RT      CREATIVITY
        MUSIC
```

ARYABHATA
```
USE     INDIAN SPACECRAFT
```

ARYL COMPOUNDS
```
USE     AROMATIC COMPOUNDS
```

ASA
```
USE     ACETYLSALICYLIC ACID
```

ASBESTOS
```
GS      MINERALS
        . ASBESTOS
RT      AMBERLITE (TRADEMARK)
        ELECTRICAL INSULATION
        INSULATION
        NONFLAMMABLE MATERIALS
        SERPENTINE
        THERMAL INSULATION
```

ASCENT
```
GS      ASCENT
        . CLIMBING FLIGHT
RT      BALLOONS
        DESCENT
        LUNAR MODULE ASCENT STAGE
        SPACE SHUTTLE ASCENT STAGE
        TAKEOFF
```

ASCENT PROPULSION SYSTEMS
```
GS      PROPULSION
        . ASCENT PROPULSION SYSTEMS
        PROPULSION SYSTEM CONFIGURATIONS
        . ASCENT PROPULSION SYSTEMS
RT      LUNAR MODULE
```

ASCENT PROPULSION SYSTEMS-(CONT.)
    MISSILES
    PROPELLANTS
    ROCKET PROPELLANTS
    SPACE FLIGHT
    SPACE SHUTTLE ASCENT STAGE
    ∞SYSTEMS

ASCENT TRAJECTORIES
GS   TRAJECTORIES
    . **ASCENT TRAJECTORIES**
RT   BALLISTIC TRAJECTORIES
    CLIMBING FLIGHT
    COASTING FLIGHT
    DESCENT TRAJECTORIES
    FLIGHT MECHANICS
    GUIDANCE (MOTION)
    INJECTION GUIDANCE
    LOFTING
    LUNAR MODULE ASCENT STAGE
    MIDCOURSE TRAJECTORIES
    MISSILE TRAJECTORIES
    PARABOLIC FLIGHT
    POST BOOST PROPULSION SYSTEM
    RENDEZVOUS TRAJECTORIES
    SPACE SHUTTLE ASCENT STAGE
    SPACECRAFT TRAJECTORIES

ASCORBIC ACID
UF   VITAMIN C
GS   ACIDS
    . **ASCORBIC ACID**
    HETEROCYCLIC COMPOUNDS
    . **ASCORBIC ACID**
    VITAMINS
    . **ASCORBIC ACID**

ASCORBIC ACID METABOLISM
GS   METABOLISM
    . **ASCORBIC ACID METABOLISM**
RT   VITAMINS

ASCR REACTOR
USE   ADVANCED SODIUM COOLED REACTOR

ASDE
USE   AIRPORT SURFACE DETECTION
        EQUIPMENT

ASHES
GS   **ASHES**
    . FLY ASH
RT   AIR POLLUTION
    COAL
    COMBUSTION PRODUCTS
    CULTIVATION
    FERTILIZERS
    FIRE DAMAGE
    FOREST FIRES
    LIGNITE
    REACTION PRODUCTS
    RESIDUES

ASIA
GS   CONTINENTS
    . **ASIA**
RT   AFGHANISTAN
    BANGLADESH
    BRUNEI
    BURMA
    CAMBODIA
    CEYLON
    CHINA
    HIMALAYAS
    HONG KONG
    INDIA
    IRAN
    IRAQ
    ISRAEL
    JAPAN
    KUWAIT
    LAOS
    LEBANON
    MALAYA
    MONGOLIA
    NATIONS
    NEPAL
    NORTH KOREA
    PAKISTAN
    RED SEA
    SAUDI ARABIA
    SEA OF JAPAN
    SIBERIA
    SIKKIM
    SINGAPORE

ASIA-(CONT.)
    SOUTH KOREA
    SOUTHEAST ASIA
    SOUTHERN YEMEN
    SYRIA
    TAIWAN
    THAILAND
    TIBET
    TUNDRA
    U.S.S.R.
    VIETNAM
    YEMEN

ASP ROCKET VEHICLE
UF   NIKE-ASP ROCKET
GS   ROCKET VEHICLES
    . SOUNDING ROCKETS
    . . **ASP ROCKET VEHICLE**
RT   SONDES

ASPARTATES
GS   ESTERS
    . **ASPARTATES**
    ORGANIC COMPOUNDS
    . **ASPARTATES**
    PROTEINS
    . **ASPARTATES**
RT   AMINO ACIDS
    ASPARTIC ACID

ASPARTIC ACID
GS   ACIDS
    . AMINO ACIDS
    . . **ASPARTIC ACID**
    ORGANIC COMPOUNDS
    . AMINO ACIDS
    . . **ASPARTIC ACID**
RT   ASPARTATES
    PEPTIDES

ASPECT RATIO
GS   RATIOS
    . **ASPECT RATIO**
    . . HIGH ASPECT RATIO
    . . LOW ASPECT RATIO
RT   AERODYNAMIC CHARACTERISTICS
    AIRFOILS
    FINENESS RATIO
    LIFT
    ∞SPAN
    WINGS

ASPERGILLUS
GS   FUNGI
    . **ASPERGILLUS**
RT   INFECTIOUS DISEASES
    ∞MOLD

ASPHALT
GS   PRODUCTS
    . PETROLEUM PRODUCTS
    . . **ASPHALT**
RT   AMORPHOUS MATERIALS
    PAVEMENTS
    PITCH (MATERIAL)
    TARS

ASPHALTENES
RT   COAL
    COAL DERIVED LIQUIDS
    COAL LIQUEFACTION
    HYDROGENATION

ASPHERICITY
RT   ABERRATION
    GEOMETRICAL OPTICS
    ∞OPTICS
    REFRACTION
    SPHERES

ASPHYXIA
RT   ANOXIA
    RESPIRATION
    SIGNS AND SYMPTOMS

ASPIRATION
USE   VACUUM

ASROC ENGINE
GS   ENGINES
    . ROCKET ENGINES
    . . SOLID PROPELLANT ROCKET
        ENGINES
    . . . **ASROC ENGINE**

ASROC ENGINE-(CONT.)
RT   ANTISUBMARINE WARFARE
    TORPEDOES

ASSATEAGUE ISLAND (MD-VA)
GS   LANDFORMS
    . ISLANDS
    . . **ASSATEAGUE ISLAND (MD-VA)**
RT   ATLANTIC OCEAN
    MARYLAND
    VIRGINIA

ASSAULTING
USE   ATTACKING (ASSAULTING)

ASSAYING
RT   CHEMICAL ANALYSIS
    IMMUNOASSAY
    MARS SURFACE SAMPLES
    PARTICULATE SAMPLING
    RADIOIMMUNOASSAY
    SAMPLING

ASSEMBLER ROUTINES
GS   COMPUTER PROGRAMS
    . COMPUTER SYSTEMS PROGRAMS
    . . **ASSEMBLER ROUTINES**
    SOFTWARE ENGINEERING
    . COMPUTER PROGRAMMING
    . . **ASSEMBLER ROUTINES**
RT   COMPILERS
    OPERATING SYSTEMS (COMPUTERS)

ASSEMBLIES
GS   **ASSEMBLIES**
    . SUBASSEMBLIES
    . TAIL ASSEMBLIES
    . . SWING TAIL ASSEMBLIES
RT   ACCUMULATIONS
    ASSEMBLING
    ∞ASSEMBLY
    COLLOCATION
    ∞COMPONENTS
    FABRICATION
    MOSAICS
    STRINGS

ASSEMBLING
GS   **ASSEMBLING**
    . ORBITAL ASSEMBLY
RT   ASSEMBLIES
    ∞ASSEMBLY
    ∞ATTACHMENT
    CLEAN ROOMS
    COLLECTION
    CONSTRUCTION
    . FABRICATION
    FITTING
    INSTALLING
    ∞JOINING
    MOUNTING
    PREPARATION
    RIGGING
    SPACE MANUFACTURING

∞ ASSEMBLY
SN   *(USE OF A MORE SPECIFIC TERM IS
    RECOMMENDED--CONSULT THE TERMS
    LISTED BELOW)*
RT   ASSEMBLIES
    ASSEMBLING
    COLLOCATION

ASSEMBLY LANGUAGE
GS   LANGUAGES
    . PROGRAMMING LANGUAGES
    . . **ASSEMBLY LANGUAGE**
    . . . AUTOCODERS
    . . . COMPASS (PROGRAMMING
        LANGUAGE)
    . . . MAP (PROGRAMMING LANGUAGE)
RT   COMPUTER PROGRAMMING
    COMPUTER PROGRAMS
    COMPUTER SYSTEMS PROGRAMS
    MACHINE ORIENTED LANGUAGES

ASSESS PROGRAM
UF   SPACELAB SIMULATION FLIGHTS
GS   PROGRAMS
    . NASA PROGRAMS
    . . **ASSESS PROGRAM**
RT   SPACE SHUTTLES

ASSESSMENTS
GS   **ASSESSMENTS**
     . DAMAGE ASSESSMENT
     . TECHNOLOGY ASSESSMENT
RT   EVALUATION
     RATINGS
     REVENUE
     VALUE

ASSET GLIDERS
GS   GLIDERS
     . **ASSET GLIDERS**
RT   ∞AIRCRAFT
     HYPERSONIC GLIDERS
     LIFTING REENTRY VEHICLES

ASSET PROJECT
GS   PROGRAMS
     . PROJECTS
     . . **ASSET PROJECT**
RT   AEROTHERMODYNAMICS
     ENVIRONMENTAL TESTS

ASSIGNMENT
USE   ALLOCATIONS

ASSIMILATION
RT   DISPERSING
     DISTRIBUTING
     ∞DISTRIBUTION
     MATERIAL ABSORPTION

ASSOCIATION REACTIONS
GS   CHEMICAL REACTIONS
     . **ASSOCIATION REACTIONS**
     GAS-GAS INTERACTIONS
     . **ASSOCIATION REACTIONS**
RT   ASTROPHYSICS
     CHEMICAL EQUILIBRIUM
     CONDENSING
     ENDOTHERMIC REACTIONS
     EXOTHERMIC REACTIONS
     INTERSTELLAR CHEMISTRY
     MOLECULAR GASES
     OXIDATION
     PHOTOCHEMICAL REACTIONS
     PHOTOOXIDATION
     REACTION KINETICS
     VAPOR PHASES

ASSOCIATIONS
USE   ORGANIZATIONS

ASSOCIATIVE PROCESSING (COMPUTERS)
GS   DATA PROCESSING
     . **ASSOCIATIVE PROCESSING
       (COMPUTERS)**
RT   DIGITAL COMPUTERS
     MULTIPROCESSING (COMPUTERS)
     PARALLEL PROCESSING (COMPUTERS)
     PIPELINING (COMPUTERS)
     ∞PROCESSING

ASSUMPTIONS
RT   HYPOTHESES
     INFERENCE
     RISK
     SIMPLIFICATION
     ∞THEORIES

ASSURANCE
RT   QUALITY CONTROL
     REDUNDANCY
     RELIABILITY

ASTATINE
GS   CHEMICAL ELEMENTS
     . HALOGENS
     . . **ASTATINE**
     METALS
     . **ASTATINE**

ASTATINE ISOTOPES
GS   CHEMICAL ELEMENTS
     . NUCLIDES
     . . ISOTOPES
     . . . RADIOACTIVE ISOTOPES
     . . . . **ASTATINE ISOTOPES**
     METALS
     . **ASTATINE ISOTOPES**

ASTEC SOLAR TURBOELECTRIC GENERATOR
GS   AUXILIARY POWER SOURCES

ASTEC SOLAR TURBOELECTRIC-*(CONT.)*
     . ASTEC SOLAR TURBOELECTRIC
       GENERATOR
     ELECTRIC GENERATORS
     . ROTATING GENERATORS
     . . TURBOGENERATORS
     . . . **ASTEC SOLAR TURBOELECTRIC
         GENERATOR**
     . SOLAR GENERATORS
     . . SOLAR AUXILIARY POWER UNITS
     . . . **ASTEC SOLAR TURBOELECTRIC
         GENERATOR**
     TURBOMACHINERY
     . TURBOGENERATORS
     . . **ASTEC SOLAR TURBOELECTRIC
         GENERATOR**
RT   RANKINE CYCLE
     SUN
     THERMOELECTRIC GENERATORS

ASTEROID BELTS
GS   CELESTIAL BODIES
     . **ASTEROID BELTS**
     . . ASTEROIDS
     . . . AMOR ASTEROID
     . . . AMPHITRITE ASTEROID
     . . . APOLLO ASTEROIDS
     . . . CERES ASTEROID
     . . . CHIRON
     . . . ICARUS ASTEROID
     . . . TORO ASTEROID
     . . . VESTA ASTEROID
RT   ∞BELTS
     METEOROIDS
     REGIONS
     SOLAR SYSTEM
     SPACE DEBRIS

ASTEROID CAPTURE
RT   ASTEROIDS
     CELESTIAL BODIES
     CONTAINMENT
     ENCLOSURES
     PAYLOADS
     RETAINING
     SOLAR SYSTEM

ASTEROID MISSIONS
GS   FLYBY MISSIONS
     . **ASTEROID MISSIONS**
     SPACE MISSIONS
     . **ASTEROID MISSIONS**
RT   ASTEROIDS
     ∞MISSIONS
     SPACE EXPLORATION

ASTEROIDS
GS   CELESTIAL BODIES
     . ASTEROID BELTS
     . . **ASTEROIDS**
     . . . AMOR ASTEROID
     . . . AMPHITRITE ASTEROID
     . . . APOLLO ASTEROIDS
     . . . CERES ASTEROID
     . . . CHIRON
     . . . ICARUS ASTEROID
     . . . TORO ASTEROID
     . . . VESTA ASTEROID
RT   ASTEROID CAPTURE
     ASTEROID MISSIONS
     METEOROIDS
     SOLAR SYSTEM
     SPACE DEBRIS

ASTHENOPIA
GS   DISEASES
     . EYE DISEASES
     . . **ASTHENOPIA**
RT   FATIGUE (BIOLOGY)

ASTHMA
GS   DISEASES
     . RESPIRATORY DISEASES
     . . **ASTHMA**

ASTIGMATISM
GS   DISEASES
     . EYE DISEASES
     . . **ASTIGMATISM**
RT   DISTORTION
     FOCUSING
     GEOMETRICAL OPTICS
     HAPLOSCOPES
     LENSES
     ∞OPTICS

ASTIGMATISM-*(CONT.)*
     REFRACTION
     STIGMATISM

ASTP
USE   APOLLO SOYUZ TEST PROJECT

ASTRIONICS
RT   ∞ASTRONAUTICS
     AVIONICS
     ∞CONTROL
     ∞ELECTRONICS
     GUIDANCE (MOTION)
     SINGLE EVENT UPSETS
     SPACECRAFT COMMUNICATION
     SPACECRAFT ELECTRONIC EQUIPMENT
     SPACECRAFT INSTRUMENTS
     ∞TEST EQUIPMENT

ASTRO MISSIONS (STS)

ASTRO VEHICLE
SN   (EXCLUDES STS)
GS   MANEUVERABLE SPACECRAFT
     . **ASTRO VEHICLE**
     MANNED SPACECRAFT
     . **ASTRO VEHICLE**
     REENTRY VEHICLES
     . RECOVERABLE SPACECRAFT
     . . **ASTRO VEHICLE**
     SOFT LANDING SPACECRAFT
     . **ASTRO VEHICLE**
RT   AEROSPACEPLANES
     ASTROPLANE
     BOOSTGLIDE VEHICLES
     FERRY SPACECRAFT
     LIFTING REENTRY VEHICLES
     ∞SPACECRAFT

ASTROBEE ROCKET VEHICLES
GS   ROCKET VEHICLES
     . MULTISTAGE ROCKET VEHICLES
     . . **ASTROBEE ROCKET VEHICLES**
     . . . ASTROBEE 1500 ROCKET VEHICLE
     . SOUNDING ROCKETS
     . . **ASTROBEE ROCKET VEHICLES**
     . . . ASTROBEE 1600 ROCKET VEHICLE
RT   GENIE ROCKET VEHICLE
     SOLID PROPELLANT ROCKET ENGINES
     ∞VEHICLES

ASTROBEE 1500 ROCKET VEHICLE
GS   ROCKET VEHICLES
     . MULTISTAGE ROCKET VEHICLES
     . . ASTROBEE ROCKET VEHICLES
     . . . **ASTROBEE 1500 ROCKET VEHICLE**
     . SOUNDING ROCKETS
     . . ASTROBEE ROCKET VEHICLES
     . . . **ASTROBEE 1500 ROCKET VEHICLE**
RT   SOLID PROPELLANT ROCKET ENGINES

ASTROBIOLOGY
USE   EXOBIOLOGY

ASTRODYNAMICS
GS   CLASSICAL MECHANICS
     . SPACE MECHANICS
     . . **ASTRODYNAMICS**
RT   ∞ASTRONAUTICS
     ASTRONOMICAL OBSERVATORIES
     CELESTIAL BODIES
     CELESTIAL MECHANICS
     ∞DYNAMICS
     INTERPLANETARY FLIGHT
     ORBITAL MECHANICS
     ORBITS
     ∞SCIENCE
     SPACE EXPLORATION
     SPACE FLIGHT
     SPACE NAVIGATION
     ∞SPACECRAFT
     TRAJECTORY ANALYSIS

ASTROGRAPHY
SN   (EXCLUDES ASTRONOMICAL
       PHOTOGRAPHY)
RT   ASTRONOMICAL MAPS
     MAPPING
     PLANETARY MAPPING

ASTROGUIDE NAVIGATION SYSTEM
GS   NAVIGATION
     . CELESTIAL NAVIGATION
     . . **ASTROGUIDE NAVIGATION SYSTEM**

ASTROGUIDE NAVIGATION SYSTEM-*(CONT.)*
. INERTIAL NAVIGATION
.. **ASTROGUIDE NAVIGATION SYSTEM**
RT   INERTIAL COORDINATES
STAR TRACKERS
∞SYSTEMS

ASTROLABES
GS   MEASURING INSTRUMENTS
. INDICATING INSTRUMENTS
.. POSITION INDICATORS
... **ASTROLABES**
RT   ALTIMETERS
ASTROMETRY
ASTRONOMICAL OBSERVATORIES
ASTRONOMY
CELESTIAL BODIES
POSITION (LOCATION)
POSITION ERRORS
SOLAR POSITION
STAR DISTRIBUTION
STAR TRACKERS
STARS

ASTROLOY (TRADEMARK)
GS   ALLOYS
. CHROMIUM ALLOYS
.. **ASTROLOY (TRADEMARK)**
. COBALT ALLOYS
.. **ASTROLOY (TRADEMARK)**
. HIGH STRENGTH ALLOYS
.. **ASTROLOY (TRADEMARK)**
. NICKEL ALLOYS
.. **ASTROLOY (TRADEMARK)**
. TERNARY ALLOYS
.. **ASTROLOY (TRADEMARK)**

ASTROMASTS
USE   LONGERONS

ASTROMETRY
RT   ASTROLABES
ASTRONOMICAL MAPS
ASTRONOMICAL PHOTOGRAPHY
ASTRONOMY
DOUBLE STARS
HIPPARCOS SATELLITE
∞MEASUREMENT
PARALLAX
SOLAR DIAMETER
STELLAR PARALLAX

ASTRON THERMONUCLEAR REACTOR
GS   NUCLEAR REACTORS
. **ASTRON THERMONUCLEAR REACTOR**
RT   RELATIVISTIC PLASMAS
THERMONUCLEAR POWER GENERATION
THERMONUCLEAR REACTIONS

ASTRONAUT LOCOMOTION
GS   EXTRAVEHICULAR MOBILITY UNITS
. **ASTRONAUT LOCOMOTION**
LOCOMOTION
. **ASTRONAUT LOCOMOTION**
RT   EXTRAVEHICULAR ACTIVITY
INTRAVEHICULAR ACTIVITY
LIFE SUPPORT SYSTEMS
MAN OPERATED PROPULSION SYSTEMS
MANNED MANEUVERING UNITS
ORBITAL WORKERS

ASTRONAUT MANEUVERING EQUIPMENT
GS   EXTRAVEHICULAR MOBILITY UNITS
. **ASTRONAUT MANEUVERING
EQUIPMENT**
.. MANNED MANEUVERING UNITS
RT   EXTRAVEHICULAR ACTIVITY
HUMAN FACTORS ENGINEERING
IMLSS
INTRAVEHICULAR ACTIVITY
SELF MANEUVERING UNITS
WALKING MACHINES

ASTRONAUT PERFORMANCE
GS   HUMAN PERFORMANCE
. **ASTRONAUT PERFORMANCE**
.. BLACKOUT PREVENTION
RT   CONFINEMENT
CONFINING
HUMAN FACTORS ENGINEERING
INTRAVEHICULAR ACTIVITY
MAN MACHINE SYSTEMS
OPERATOR PERFORMANCE
∞PERFORMANCE
PHYSIOLOGICAL FACTORS

ASTRONAUT PERFORMANCE-*(CONT.)*
PILOT PERFORMANCE
PSYCHOLOGICAL FACTORS
SPACE PSYCHOLOGY
SPACECRAFT PERFORMANCE
WEIGHTLESSNESS

ASTRONAUT TRAINING
GS   EDUCATION
. **ASTRONAUT TRAINING**
LEARNING
. **ASTRONAUT TRAINING**
RT   EJECTION TRAINING
FLIGHT TRAINING
PILOT TRAINING
SPACE FLIGHT TRAINING
SPACE MAINTENANCE
SPACE PSYCHOLOGY
TRAINING SIMULATORS

∞ ASTRONAUTICS
SN   *(USE OF A MORE SPECIFIC TERM IS
RECOMMENDED--CONSULT THE TERMS
LISTED BELOW)*
RT   AEROSPACE ENVIRONMENTS
ARTIFICIAL GRAVITY
ASTRIONICS
ASTRODYNAMICS
ASTRONAUTS
ASTRONOMY
AUXILIARY PROPULSION
AVIONICS
BIOASTRONAUTICS
BIOSATELLITE 3
COSMONAUTS
EARTH-VENUS TRAJECTORIES
HUMAN FACTORS ENGINEERING
LUNAR BASES
PROPULSION
SATELLITES
SOFT LANDING
SPACE EXPLORATION
SPACE FLIGHT
SPACE MAINTENANCE
SPACE NAVIGATION
SPACECRAFT DOCKING
WEIGHTLESSNESS

ASTRONAUTS
GS   PERSONNEL
. FLYING PERSONNEL
.. **ASTRONAUTS**
... ORBITAL WORKERS
RT   ∞ASTRONAUTICS
AWARDS
COSMONAUTS
CREW EXPERIMENT STATIONS
CREW OBSERVATION STATIONS
CREW STATIONS
CREW WORKSTATIONS
CREWS
PILOTS (PERSONNEL)
SPACECRAFT ENVIRONMENTS
SPACECREWS

ASTRONAVIGATION
GS   NAVIGATION
. CELESTIAL NAVIGATION
.. **ASTRONAVIGATION**
RT   AIR NAVIGATION
INTERPLANETARY NAVIGATION
INTERSTELLAR TRAVEL
RADIO NAVIGATION
SPACE NAVIGATION

ASTRONOMICAL CATALOGS
GS   DOCUMENTS
. CATALOGS (PUBLICATIONS)
.. **ASTRONOMICAL CATALOGS**
RT   CLASSIFICATIONS
EPHEMERIDES
NORTHERN SKY
SOUTHERN SKY
TABLES (DATA)

ASTRONOMICAL COORDINATES
GS   COORDINATES
. **ASTRONOMICAL COORDINATES**
RT   AZIMUTH
CELESTIAL REFERENCE SYSTEMS
GEOCENTRIC COORDINATES
NORTHERN SKY
PLANETOCENTRIC COORDINATES
PLANISPHERES
POLAR COORDINATES

ASTRONOMICAL COORDINATES-*(CONT.)*
REFERENCE STARS
SOLAR LONGITUDE
SPHERICAL COORDINATES

ASTRONOMICAL MAPS
GS   MAPS
. **ASTRONOMICAL MAPS**
.. PLANISPHERES
RT   ASTROGRAPHY
ASTROMETRY
CELESTIAL REFERENCE SYSTEMS
CELESTIAL SPHERE
LUNAR MAPS

ASTRONOMICAL MODELS
UF   ORRERIES
GS   MODELS
. **ASTRONOMICAL MODELS**
.. DENSITY WAVE MODEL
.. STELLAR MODELS
RT   BIG BANG COSMOLOGY
COROTATION
COSMOLOGY
MATHEMATICAL MODELS
MOLECULAR CLOUDS
PLANETARIUMS
REISSNER-NORDSTROM SOLUTION
SOLAR NEUTRINOS
SOLAR OSCILLATIONS
STELLAR OSCILLATIONS

ASTRONOMICAL NETHERLANDS SATELLITE
UF   ANS
GS   OBSERVATORIES
. ASTRONOMICAL OBSERVATORIES
.. ASTRONOMICAL SATELLITES
... **ASTRONOMICAL NETHERLANDS
SATELLITE**
RT   NETHERLANDS

ASTRONOMICAL OBSERVATORIES
GS   OBSERVATORIES
. **ASTRONOMICAL OBSERVATORIES**
.. ASTRONOMICAL SATELLITES
... ASTRONOMICAL NETHERLANDS
SATELLITE
... HEAO
.... HEAO 1
.... HEAO 2
.... HEAO 3
.. HUBBLE SPACE TELESCOPE
.. INFRARED ASTRONOMY SATELLITE
... IUE
.. MAGELLAN MISSION
.. OAO
.... OAO 1
.... OAO 2
.... OAO 3
.. OSO
.... OSO-1
.... OSO-2
.... OSO-3
.... OSO-4
.... OSO-5
.... OSO-6
.... OSO-7
.... OSO-8
.. QUASAT
... SAS
.... SAS-1
.... SAS-2
.... SAS-3
... SPACE INFRARED TELESCOPE
FACILITY
... SPARTAN SATELLITES
.. GAMMA RAY OBSERVATORY
.. ROSAT MISSION
RT   ASTRODYNAMICS
ASTROLABES
ASTRONOMY
CELESTIAL BODIES
GEOPHYSICAL OBSERVATORIES
JODRELL BANK OBSERVATORY
LUNAR OBSERVATORIES
NORTHERN SKY
RADIO ASTRONOMY
SOUTHERN SKY
SPACEBORNE TELESCOPES
TELESCOPES

ASTRONOMICAL PHOTOGRAPHY
GS   IMAGERY
. **ASTRONOMICAL PHOTOGRAPHY**
PHOTOGRAPHY

ASTRONOMICAL PHOTOGRAPHY-*(CONT.)*
. **ASTRONOMICAL PHOTOGRAPHY**
RT AERIAL PHOTOGRAPHY
ASTROMETRY
ASTRONOMY
ATMOSPHERIC WINDOWS
BAKER-NUNN CAMERA
BLACK AND WHITE PHOTOGRAPHY
CORONAGRAPHS
DIFFRACTION LIMITED CAMERAS
ELECTRO-OPTICAL PHOTOGRAPHY
FAINT OBJECT CAMERA
INFRARED ASTRONOMY
INFRARED PHOTOGRAPHY
LALLEMAND CAMERAS
LUNAR PHOTOGRAPHS
LUNAR PHOTOGRAPHY
REFERENCE STARS
ROCKET-BORNE PHOTOGRAPHY
SATELLITE-BORNE PHOTOGRAPHY
SCHMIDT CAMERAS
SOUTHERN SKY
SPACEBORNE PHOTOGRAPHY
SPACEBORNE TELESCOPES

ASTRONOMICAL PHOTOMETRY
GS OPTICAL MEASUREMENT
. **ASTRONOMICAL PHOTOMETRY**
. . STELLAR SPECTROPHOTOMETRY
RT ATMOSPHERIC WINDOWS
BLINKING
COMETARY ATMOSPHERES
DIAL SATELLITE
INFRARED PHOTOMETRY
PHOTOMETRY
SPECTROPHOTOMETRY
TELEPHOTOMETRY

ASTRONOMICAL SATELLITES
GS OBSERVATORIES
. ASTRONOMICAL OBSERVATORIES
. . **ASTRONOMICAL SATELLITES**
. . . ASTRONOMICAL NETHERLANDS
SATELLITE
. . . HEAO
. . . . HEAO 1
. . . . HEAO 2
. . . . HEAO 3
. . . HUBBLE SPACE TELESCOPE
. . . INFRARED ASTRONOMY SATELLITE
. . . IUE
. . . MAGELLAN MISSION
. . . OAO
. . . . OAO 1
. . . . OAO 2
. . . . OAO 3
. . . OSO
. . . . OSO-1
. . . . OSO-2
. . . . OSO-3
. . . . OSO-4
. . . . OSO-5
. . . . OSO-6
. . . . OSO-7
. . . . OSO-8
. . . QUASAT
. . . SAS
. . . . SAS-1
. . . . SAS-2
. . . . SAS-3
. . . SPACE INFRARED TELESCOPE
FACILITY
. . . SPARTAN SATELLITES
SATELLITES
. ARTIFICIAL SATELLITES
. . SCIENTIFIC SATELLITES
. . . **ASTRONOMICAL SATELLITES**
RT ROSAT MISSION
SPACEBORNE ASTRONOMY

ASTRONOMICAL SPECTROSCOPY
GS SPECTROSCOPY
. **ASTRONOMICAL SPECTROSCOPY**
RT ASTRONOMY
CONTINUOUS SPECTRA
ELECTROMAGNETIC SPECTRA
INFRARED SPECTROSCOPY
ORGANIC SOLIDS
RADIAL VELOCITY
RADIATION SPECTRA
RADIO ASTRONOMY
RADIO SPECTROSCOPY
RAMAN SPECTROSCOPY
SOLAR SPECTRA
SOUTHERN SKY

ASTRONOMICAL SPECTROSCOPY-*(CONT.)*
SPECTRA
SPECTROSCOPIC TELESCOPES
STELLAR SPECTRA
ULTRAVIOLET SPECTROSCOPY
VISIBLE SPECTRUM
X RAY SPECTROSCOPY

ASTRONOMICAL TELESCOPES
GS OPTICAL EQUIPMENT
. **ASTRONOMICAL TELESCOPES**
. . HELIOMETERS
. . . PYROHELIOMETERS
. . . SPECTROSCOPIC TELESCOPES
. . . MULTISPECTRAL TRACKING
TELESCOPES
. . . STRATOSCOPE TELESCOPES
. . X RAY TELESCOPES
TELESCOPES
. **ASTRONOMICAL TELESCOPES**
. . HELIOMETERS
. . . PYROHELIOMETERS
. . INFRARED TELESCOPES
. . . SPACE INFRARED TELESCOPE
FACILITY
. . KILOMETER WAVE ORBITING
TELESCOPE
. . SPECTROSCOPIC TELESCOPES
. . . MULTISPECTRAL TRACKING
TELESCOPES
. . . STRATOSCOPE TELESCOPES
. . STARSAT TELESCOPE
. . ULTRAVIOLET TELESCOPES
. . X RAY TELESCOPES
RT BALLOON-BORNE INSTRUMENTS
CELESCOPES
CORONAGRAPHS
MANNED ORBITAL TELESCOPES
MULTI-ANODE MICROCHANNEL ARRAYS
OPTICAL TRANSFER FUNCTION
PARTICLE TELESCOPES
RADIO TELESCOPES
REFLECTING TELESCOPES
REFRACTING TELESCOPES
SCHMIDT CAMERAS
SCHMIDT TELESCOPES
SOLAR OPTICAL TELESCOPE
SPACEBORNE TELESCOPES

ASTRONOMY
UF CELESTIAL OBSERVATION
GS **ASTRONOMY**
. GAMMA RAY ASTRONOMY
. INFRARED ASTRONOMY
. RADAR ASTRONOMY
. RADIO ASTRONOMY
. SPACEBORNE ASTRONOMY
. ULTRAVIOLET ASTRONOMY
. X RAY ASTRONOMY
. . X RAY SOURCES
. . . X RAY BINARIES
RT ∞AEROSPACE SCIENCES
AMOR ASTEROID
APOLLO ASTEROIDS
ASTROLABES
ASTROMETRY
∞ASTRONAUTICS
ASTRONOMICAL OBSERVATORIES
ASTRONOMICAL PHOTOGRAPHY
ASTRONOMICAL SPECTROSCOPY
ASTROPHYSICS
CELESTIAL BODIES
CELESTIAL MECHANICS
EARTH LIMB
GAMMA RAY TELESCOPES
HALOS
INFRARED TELESCOPES
MASS TO LIGHT RATIOS
METEOROID SHOWERS
MISSING MASS (ASTROPHYSICS)
∞PHYSICAL SCIENCES
RELIC RADIATION
∞SCIENCE
SELENOLOGY
SIDEREAL TIME
SOLAR PARALLAX
SOUTHERN SKY
SPACEBORNE TELESCOPES
STELLAR MAGNITUDE
STELLAR MODELS
STELLAR OSCILLATIONS
TELESCOPES

ASTROPHYSICS
UF GEOASTROPHYSICS

ASTROPHYSICS-*(CONT.)*
GS **ASTROPHYSICS**
. COMPUTATIONAL ASTROPHYSICS
. STELLAR PHYSICS
. . SOLAR PHYSICS
RT ACCRETION DISKS
ASSOCIATION REACTIONS
ASTRONOMY
BRIGHTNESS DISTRIBUTION
BRIGHTNESS TEMPERATURE
CELESTIAL BODIES
CELESTIAL MECHANICS
COSMOLOGY
DENSE PLASMAS
DISK GALAXIES
GALACTIC EVOLUTION
GAMMA RAY ASTRONOMY
GRAVITATIONAL COLLAPSE
HELIOSEISMOLOGY
INTERSTELLAR EXTINCTION
MAGNETIC FIELD CONFIGURATIONS
MASS TO LIGHT RATIOS
MICHELSON INTERFEROMETERS
MISSING MASS (ASTROPHYSICS)
NAKED SINGULARITIES
ORION NEBULA
∞PHYSICS
PLANETARY ROTATION
RADIO INTERFEROMETERS
RELIC RADIATION
∞SCIENCE
SOLAR NEUTRINOS
SPARTAN SATELLITES
SPIN TEMPERATURE
STELLAR CORES
STELLAR ENVELOPES
STELLAR EVOLUTION
STELLAR OSCILLATIONS
THEORETICAL PHYSICS
WOLF-RAYET STARS
X RAY ASTROPHYSICS FACILITY
X RAY BINARIES

ASTROPLANE
GS MANEUVERABLE SPACECRAFT
. AEROSPACEPLANES
. . **ASTROPLANE**
MANNED SPACECRAFT
. AEROSPACEPLANES
. . **ASTROPLANE**
REENTRY VEHICLES
. AEROSPACEPLANES
. . **ASTROPLANE**
. RECOVERABLE SPACECRAFT
. . REUSABLE SPACECRAFT
. . . **ASTROPLANE**
SOFT LANDING SPACECRAFT
. AEROSPACEPLANES
. . **ASTROPLANE**
RT ASTRO VEHICLE
BOOSTGLIDE VEHICLES
HYPERSONIC AIRCRAFT
HYPERSONIC GLIDERS
LAUNCH VEHICLES
LIFTING REENTRY VEHICLES
LIQUID AIR CYCLE ENGINES
MILITARY SPACECRAFT
RESEARCH AIRCRAFT
ROCKET PLANES
X-20 AIRCRAFT

ASYMMETRY
UF DISSYMMETRY
RT ANTISYMMETRY
DEVIATION
DISTORTION
ECCENTRICITY
SHAPES
SKEWNESS
SYMMETRY
VARIATIONS

ASYMPTOTES
GS ANALYSIS (MATHEMATICS)
. REAL VARIABLES
. . **ASYMPTOTES**
FUNCTIONS (MATHEMATICS)
. **ASYMPTOTES**
RT ANALYTIC GEOMETRY
ASYMPTOTIC PROPERTIES
ASYMPTOTIC SERIES
CALCULUS
NUMERICAL ANALYSIS

ASYMPTOTIC METHODS
- GS   PROBLEM SOLVING
-      . **ASYMPTOTIC METHODS**
- RT   ASYMPTOTIC PROPERTIES
-      ITERATIVE SOLUTION
-      LEARNING CURVES
-      ∞ METHODOLOGY

ASYMPTOTIC PROPERTIES
- RT   ASYMPTOTES
-      ASYMPTOTIC METHODS
-      ASYMPTOTIC SERIES
-      DIFFERENTIAL EQUATIONS
-      INTEGRAL EQUATIONS
-      MATHEMATICAL MODELS
-      NORMALITY

ASYMPTOTIC SERIES
- GS   ANALYSIS (MATHEMATICS)
-      . CALCULUS
-      . . SERIES (MATHEMATICS)
-      . . . **ASYMPTOTIC SERIES**
-      . REAL VARIABLES
-      . . SERIES (MATHEMATICS)
-      . . . **ASYMPTOTIC SERIES**
- RT   ASYMPTOTES
-      ASYMPTOTIC PROPERTIES
-      SERIES EXPANSION

ASYNCHRONOUS MOTORS
- GS   MOTORS
-      . ELECTRIC MOTORS
-      . . **ASYNCHRONOUS MOTORS**
- RT   INDUCTION MOTORS
-      SYNCHRONOUS MOTORS

ATARS
- USE   AUTOMATIC TRAFFIC ADVISORY AND
-      RESOLUTION

ATAXIA
- GS   DISEASES
-      . **ATAXIA**
- RT   MUSCLES

ATAXITE
- GS   ROCKS
-      . **ATAXITE**
- RT   BRECCIA
-      SOILS

ATCHAFALAYA RIVER BASIN (LA)
- GS   LANDFORMS
-      . STRUCTURAL BASINS
-      . . RIVER BASINS
-      . . . **ATCHAFALAYA RIVER BASIN (LA)**
- RT   LOUISIANA
-      RIVERS

ATELECTASIS
- GS   DISEASES
-      . **ATELECTASIS**
- RT   LUNGS

ATHENA ROCKET VEHICLE
- GS   ROCKET VEHICLES
-      . MULTISTAGE ROCKET VEHICLES
-      . . **ATHENA ROCKET VEHICLE**
- RT   BE-3 ENGINE
-      REENTRY VEHICLES
-      SOLID PROPELLANT ROCKET ENGINES

ATHEROSCLEROSIS
- USE   ARTERIOSCLEROSIS

ATHLETES
- RT   COMPETITION
-      PHYSICAL EXERCISE
-      PHYSICAL FITNESS
-      SPORTS MEDICINE

ATHODYDS
- USE   RAMJET ENGINES

ATLANTA (GA)
- GS   CITIES
-      . **ATLANTA (GA)**
- RT   GEORGIA

ATLANTIC AIRCRAFT
- USE   BREGUET 1150 AIRCRAFT

ATLANTIC OCEAN
- GS   OCEANS

ATLANTIC OCEAN-(CONT.)
-      . **ATLANTIC OCEAN**
- RT   ASSATEAGUE ISLAND (MD-VA)
-      BERMUDA
-      BLOCK ISLAND SOUND (RI)
-      CAPE HATTERAS (NC)
-      CAPE VERDE
-      DELAWARE BAY (US)
-      ENGLISH CHANNEL
-      GARP ATLANTIC TROPICAL EXPERIMENT
-      GULF STREAM
-      LESSER ANTILLES
-      LOMONOSOV CURRENT
-      LONG ISLAND (NY)
-      OUTER BANKS (NC)
-      SARGASSO SEA
-      WALLOPS ISLAND
-      WEST INDIES

ATLANTIS (ORBITER)
- UF   SPACE SHUTTLE ORBITER 104
- GS   REENTRY VEHICLES
-      . RECOVERABLE SPACECRAFT
-      . . REUSABLE SPACECRAFT
-      . . . **ATLANTIS (ORBITER)**
-      TRANSPORTATION
-      . SPACE TRANSPORTATION
-      . . SPACE TRANSPORTATION SYSTEM
-      . . . SPACE SHUTTLE ORBITERS
-      . . . . **ATLANTIS (ORBITER)**
- RT   MANNED SPACE FLIGHT
-      SPACE SHUTTLE MISSION 51-H
-      SPACE SHUTTLE MISSION 51-J
-      SPACE SHUTTLE MISSION 61-B
-      ∞ SPACECRAFT

ATLAS ABLE 5 LAUNCH VEHICLE
- GS   LAUNCH VEHICLES
-      . ATLAS LAUNCH VEHICLES
-      . . **ATLAS ABLE 5 LAUNCH VEHICLE**
-      ROCKET VEHICLES
-      . MULTISTAGE ROCKET VEHICLES
-      . . ATLAS LAUNCH VEHICLES
-      . . . **ATLAS ABLE 5 LAUNCH VEHICLE**
- RT   LUNAR PROBES
-      SPACE PROBES

ATLAS AGENA B LAUNCH VEHICLE
- GS   LAUNCH VEHICLES
-      . ATLAS LAUNCH VEHICLES
-      . . **ATLAS AGENA B LAUNCH VEHICLE**
-      ROCKET VEHICLES
-      . MULTISTAGE ROCKET VEHICLES
-      . . ATLAS LAUNCH VEHICLES
-      . . . **ATLAS AGENA B LAUNCH VEHICLE**
- RT   AGENA ROCKET VEHICLES
-      MARINER 2 SPACE PROBE
-      MIDAS SATELLITES
-      RANGER LUNAR PROBES
-      RANGER 4 LUNAR PROBE

ATLAS AGENA LAUNCH VEHICLES
- GS   LAUNCH VEHICLES
-      . ATLAS LAUNCH VEHICLES
-      . . **ATLAS AGENA LAUNCH VEHICLES**
-      ROCKET VEHICLES
-      . MULTISTAGE ROCKET VEHICLES
-      . . ATLAS LAUNCH VEHICLES
-      . . . **ATLAS AGENA LAUNCH VEHICLES**
- RT   AGENA ROCKET VEHICLES
-      ENVIRONMENTAL RESEARCH
-      SATELLITES
-      ERS 17
-      MARINER PROGRAM
-      MARINER 5 SPACE PROBE
-      MARINER 6 SPACE PROBE
-      OGO-A
-      ∞ VEHICLES

ATLAS CENTAUR LAUNCH VEHICLE
- GS   LAUNCH VEHICLES
-      . ATLAS LAUNCH VEHICLES
-      . . **ATLAS CENTAUR LAUNCH VEHICLE**
-      . CENTAUR LAUNCH VEHICLE
-      . . **ATLAS CENTAUR LAUNCH VEHICLE**
-      ROCKET VEHICLES
-      . CENTAUR LAUNCH VEHICLE
-      . . **ATLAS CENTAUR LAUNCH VEHICLE**
-      . MULTISTAGE ROCKET VEHICLES
-      . . ATLAS LAUNCH VEHICLES
-      . . . **ATLAS CENTAUR LAUNCH VEHICLE**
- RT   CENTAUR PROJECT
-      OAO 1
-      OAO 2
-      OAO 3

ATLAS CENTAUR LAUNCH VEHICLE-(CONT.)
-      RL-10 ENGINES
-      SPACE SHUTTLE UPPER STAGE A
-      SURVEYOR PROJECT
-      SURVEYOR 1 LUNAR PROBE
-      SURVEYOR 2 LUNAR PROBE
-      SURVEYOR 3 LUNAR PROBE
-      SURVEYOR 4 LUNAR PROBE
-      SURVEYOR 5 LUNAR PROBE
-      SURVEYOR 6 LUNAR PROBE
-      SURVEYOR 7 LUNAR PROBE

ATLAS D ICBM
- GS   MISSILES
-      . BALLISTIC MISSILES
-      . . INTERCONTINENTAL BALLISTIC
-      MISSILES
-      . . . ATLAS ICBM
-      . . . . **ATLAS D ICBM**
-      . SURFACE TO SURFACE MISSILES
-      . . INTERCONTINENTAL BALLISTIC
-      MISSILES
-      . . . ATLAS ICBM
-      . . . . **ATLAS D ICBM**
- RT   CENTAUR LAUNCH VEHICLE
-      STANDARD LAUNCH VEHICLES
-      VEGA LAUNCH VEHICLE

ATLAS E ICBM
- GS   MISSILES
-      . BALLISTIC MISSILES
-      . . INTERCONTINENTAL BALLISTIC
-      MISSILES
-      . . . ATLAS ICBM
-      . . . . **ATLAS E ICBM**
-      . SURFACE TO SURFACE MISSILES
-      . . INTERCONTINENTAL BALLISTIC
-      MISSILES
-      . . . ATLAS ICBM
-      . . . . **ATLAS E ICBM**

ATLAS F ICBM
- GS   MISSILES
-      . BALLISTIC MISSILES
-      . . INTERCONTINENTAL BALLISTIC
-      MISSILES
-      . . . ATLAS ICBM
-      . . . . **ATLAS F ICBM**
-      . SURFACE TO SURFACE MISSILES
-      . . INTERCONTINENTAL BALLISTIC
-      MISSILES
-      . . . ATLAS ICBM
-      . . . . **ATLAS F ICBM**

ATLAS ICBM
- GS   MISSILES
-      . BALLISTIC MISSILES
-      . . INTERCONTINENTAL BALLISTIC
-      MISSILES
-      . . . **ATLAS ICBM**
-      . . . . ATLAS D ICBM
-      . . . . ATLAS E ICBM
-      . . . . ATLAS F ICBM
-      . SURFACE TO SURFACE MISSILES
-      . . INTERCONTINENTAL BALLISTIC
-      MISSILES
-      . . . **ATLAS ICBM**
-      . . . . ATLAS D ICBM
-      . . . . ATLAS E ICBM
-      . . . . ATLAS F ICBM
- RT   MA-2 ENGINE
-      MA-3 ENGINE

ATLAS LAUNCH VEHICLES
- UF   SM-65 MISSILE
- GS   LAUNCH VEHICLES
-      . **ATLAS LAUNCH VEHICLES**
-      . . ATLAS ABLE 5 LAUNCH VEHICLE
-      . . ATLAS AGENA B LAUNCH VEHICLE
-      . . ATLAS AGENA LAUNCH VEHICLES
-      . . ATLAS CENTAUR LAUNCH VEHICLE
-      . . ATLAS SLV-3 LAUNCH VEHICLE
-      ROCKET VEHICLES
-      . MULTISTAGE ROCKET VEHICLES
-      . . **ATLAS LAUNCH VEHICLES**
-      . . . ATLAS ABLE 5 LAUNCH VEHICLE
-      . . . ATLAS AGENA B LAUNCH VEHICLE
-      . . . ATLAS AGENA LAUNCH VEHICLES
-      . . . ATLAS CENTAUR LAUNCH VEHICLE
-      . . . ATLAS SLV-3 LAUNCH VEHICLE
- RT   EGO
-      GEMINI PROJECT
-      MA-5 ENGINE
-      MARINER PROGRAM
-      MERCURY FLIGHTS

ATLAS LAUNCH VEHICLES-*(CONT.)*
   MERCURY MA-1 FLIGHT
   MERCURY MA-2 FLIGHT
   MERCURY MA-3 FLIGHT
   MERCURY MA-4 FLIGHT
   MERCURY MA-5 FLIGHT
   MERCURY MA-6 FLIGHT
   MERCURY MA-7 FLIGHT
   MERCURY MA-8 FLIGHT
   MERCURY MA-9 FLIGHT
   MERCURY PROJECT
   NOMAD LAUNCH VEHICLE
   OAO
   ORBITAL RENDEZVOUS
   RANGER PROJECT
   ∞ VEHICLES

ATLAS SLV-3 LAUNCH VEHICLE
 UF   STANDARD LAUNCH VEHICLE 3
 GS   LAUNCH VEHICLES
      . ATLAS LAUNCH VEHICLES
      . . **ATLAS SLV-3 LAUNCH VEHICLE**
      . STANDARD LAUNCH VEHICLES
      . . **ATLAS SLV-3 LAUNCH VEHICLE**
      ROCKET VEHICLES
      . MULTISTAGE ROCKET VEHICLES
      . ATLAS LAUNCH VEHICLES
      . . . **ATLAS SLV-3 LAUNCH VEHICLE**
      . STANDARD LAUNCH VEHICLES
      . . **ATLAS SLV-3 LAUNCH VEHICLE**
 RT   LIQUID PROPELLANT ROCKET ENGINES
      MA-5 ENGINE

ATLIT PROJECT
 UF   ADVANCED TECHNOLOGY LIGHT TWIN
      AIRCRAFT
 GS   PROGRAMS
      . NASA PROGRAMS
      . . **ATLIT PROJECT**
      . PROJECTS
      . . **ATLIT PROJECT**
 RT   ∞ AIRCRAFT
      GAW-1 AIRFOIL
      PA-34 SENECA AIRCRAFT

ATMOSPHERE EXPLORER A
 USE   EXPLORER 17 SATELLITE

ATMOSPHERE EXPLORER B
 USE   EXPLORER 32 SATELLITE

ATMOSPHERE EXPLORER C
 USE   EXPLORER 51 SATELLITE

ATMOSPHERE EXPLORER D
 USE   EXPLORER 54 SATELLITE

ATMOSPHERE EXPLORER E
 USE   EXPLORER 55 SATELLITE

∞ ATMOSPHERES
 SN   *(USE OF A MORE SPECIFIC TERM IS
      RECOMMENDED--CONSULT THE TERMS
      LISTED BELOW)*
 RT   AIR
      ARGON-OXYGEN ATMOSPHERES
      ATMOSPHERIC PRESSURE
      CABIN ATMOSPHERES
      CONTROLLED ATMOSPHERES
      EARTH ATMOSPHERE
      ENVIRONMENTS
      EQUATORIAL ATMOSPHERE
      GAS MIXTURES
      GASES
      HELIUM-OXYGEN ATMOSPHERES
      HYPOBARIC ATMOSPHERES
      LIFE SUPPORT SYSTEMS
      METEOROLOGY
      MIDDLE ATMOSPHERE
      NEPTUNE ATMOSPHERE
      NEUTRAL ATMOSPHERES
      NONGRAY ATMOSPHERES
      NONGRAY GAS
      PLANETARY ATMOSPHERES
      PRIMITIVE EARTH ATMOSPHERE
      SATELLITE ATMOSPHERES
      SOLAR ATMOSPHERE
      STELLAR ATMOSPHERES
      URANUS ATMOSPHERE

ATMOSPHERIC & OCEANOGRAPHIC INFORM SYS
 SN   (ATMOSPHERIC & OCEANOGRAPHIC
      INFORMATION SYSTEMS)
 UF   AOIPS

ATMOSPHERIC & OCEANOGRAPHIC-*(CONT.)*
 GS   DATA PROCESSING
      . **ATMOSPHERIC & OCEANOGRAPHIC
      INFORM SYS**
      INFORMATION SYSTEMS
      . **ATMOSPHERIC & OCEANOGRAPHIC
      INFORM SYS**
 RT   AEROLOGY
      AIR WATER INTERACTIONS
      DATA SYSTEMS
      IMAGERY
      ISOTHERMS
      METEOROLOGICAL PARAMETERS
      METEOROLOGY
      MINICOMPUTERS
      OCEANOGRAPHIC PARAMETERS
      OCEANOGRAPHY
      ∞ SYSTEMS
      WEATHER

ATMOSPHERIC ABSORPTION
 USE   ATMOSPHERIC ATTENUATION

ATMOSPHERIC AND MAGNETOSPHERIC PAYLOAD
 USE   AMPS (SATELLITE PAYLOAD)

ATMOSPHERIC ATTENUATION
 UF   ATMOSPHERIC ABSORPTION
 GS   ATTENUATION
      . **ATMOSPHERIC ATTENUATION**
      . . AURORAL ABSORPTION
      RADIATION ABSORPTION
      . **ATMOSPHERIC ATTENUATION**
      . . AURORAL ABSORPTION
 RT   ACOUSTIC ATTENUATION
      ATMOSPHERIC LASERS
      COSMIC RAY ALBEDO
      ELECTROMAGNETIC ABSORPTION
      ELECTROMAGNETIC SCATTERING
      ELECTROMAGNETIC WAVE
      TRANSMISSION
      INFRARED ABSORPTION
      MOLECULAR ABSORPTION
      PLANETARY ATMOSPHERES
      RADAR ATTENUATION
      RADAR TRANSMISSION
      RADIO ATTENUATION
      RADIO TRANSMISSION
      SHOCK WAVE ATTENUATION
      SHOCK WAVE PROPAGATION
      THERMAL ABSORPTION
      TRANSMISSION
      VEGETATIVE INDEX
      WAVE ATTENUATION
      WAVE PROPAGATION

ATMOSPHERIC BOUNDARY LAYER
 GS   BOUNDARY LAYERS
      . **ATMOSPHERIC BOUNDARY LAYER**
 RT   AIR FLOW
      AIR LAND INTERACTIONS
      BOUNDARY LAYER FLOW
      EKMAN LAYER
      ∞ LAYERS
      PLANETARY BOUNDARY LAYER
      PRIMITIVE EQUATIONS

ATMOSPHERIC CHEMISTRY
 GS   ENVIRONMENTAL CHEMISTRY
      . **ATMOSPHERIC CHEMISTRY**
 RT   ACID RAIN
      AEROTHERMOCHEMISTRY
      AIR POLLUTION
      AITKEN NUCLEI
      ATMOSPHERIC EFFECTS
      ∞ CHEMISTRY
      FORMYL IONS
      MIDDLE ATMOSPHERE
      NITROUS ACID
      PHOTOCHEMICAL OXIDANTS
      PHOTOCHEMICAL REACTIONS
      PHYSICAL CHEMISTRY
      SATELLITE ATMOSPHERES

ATMOSPHERIC CIRCULATION
 UF   WIND CIRCULATION
 GS   CIRCULATION
      . **ATMOSPHERIC CIRCULATION**
 RT   ADVECTION
      AIR CURRENTS
      AIR LAND INTERACTIONS
      AIR MASSES
      ANNUAL VARIATIONS
      ATMOSPHERIC GENERAL CIRCULATION
      EXPERIMENT

ATMOSPHERIC CIRCULATION-*(CONT.)*
      BAROCLINIC INSTABILITY
      BRUNT-VAISALA FREQUENCY
      CIRCULATION DISTRIBUTION
      CIRCUMPOLAR WESTERLIES
      CLIMATOLOGY
      EARTH ATMOSPHERE
      GROUND WIND
      INTERTROPICAL CONVERGENT ZONES
      JET STREAMS (METEOROLOGY)
      MERIDIONAL FLOW
      MIDDLE ATMOSPHERE
      MIXING HEIGHT
      MONSOONS
      PLANETARY WAVES
      POLLUTION TRANSPORT
      SEA BREEZE
      SUPERROTATION
      TORNADOES
      TROPICAL STORMS
      TURBOPAUSE
      TYPHOONS
      UPWELLING WATER
      VERTICAL AIR CURRENTS
      VORTICITY
      WIND (METEOROLOGY)
      WIND DIRECTION
      WIND PROFILES
      WINDPOWER UTILIZATION

ATMOSPHERIC CLOUD PHYSICS LAB (SPACELAB)
 UF   ACPL (SPACELAB)
      ZERO-G ACPL (SPACELAB)
 GS   LABORATORIES
      . SPACE LABORATORIES
      . . **ATMOSPHERIC CLOUD PHYSICS LAB
      (SPACELAB)**
 RT   CLOUD PHYSICS
      METEOROLOGICAL PARAMETERS
      NEPHANALYSIS
      SPACE SHUTTLES
      SPACECRAFT INSTRUMENTS

ATMOSPHERIC COMPOSITION
 GS   COMPOSITION (PROPERTY)
      . **ATMOSPHERIC COMPOSITION**
      . . ATMOSPHERIC MOISTURE
      . . IONOSPHERIC COMPOSITION
 RT   AERONOMY
      AIR
      AIR POLLUTION
      AITKEN NUCLEI
      CARBON DIOXIDE CONCENTRATION
      CHEMICAL COMPOSITION
      EARTH ATMOSPHERE
      ELECTRON DENSITY (CONCENTRATION)
      EQUATORIAL ATMOSPHERE
      GAS COMPOSITION
      LACATE (EXPERIMENT)
      MIDDLE ATMOSPHERE
      MOISTURE CONTENT
      PLANETARY ATMOSPHERES
      PRIMITIVE EARTH ATMOSPHERE
      RADIO OCCULTATION
      RADIOACTIVE CONTAMINANTS
      SATELLITE ATMOSPHERES
      SATURN ATMOSPHERE
      SOLAR MESOSPHERE EXPLORER
      TITAN

ATMOSPHERIC CONDITIONS
 USE   METEOROLOGY

ATMOSPHERIC CONDUCTIVITY
 GS   TRANSPORT PROPERTIES
      . **ATMOSPHERIC CONDUCTIVITY**
      . . IONOSPHERIC CONDUCTIVITY
 RT   AIR CONDUCTIVITY
      ∞ CONDUCTIVITY
      ELECTRICAL RESISTIVITY
      THERMAL CONDUCTIVITY

ATMOSPHERIC CORRECTION
 GS   CORRECTION
      . **ATMOSPHERIC CORRECTION**
 RT   ATMOSPHERIC EFFECTS
      CLOUDS (METEOROLOGY)
      GEOMETRIC RECTIFICATION (IMAGERY)
      IMAGE PROCESSING
      INFRARED RADIOMETERS
      RADIATIVE TRANSFER
      SATELLITE IMAGERY
      SPATIAL RESOLUTION

ATMOSPHERIC DENSITY
GS   DENSITY (MASS/VOLUME)
      . **ATMOSPHERIC DENSITY**
RT   AIR POLLUTION
      BOLTZMANN DISTRIBUTION
      ∞ DENSITY
      DENSITY (NUMBER/VOLUME)
      ELECTRON DENSITY (CONCENTRATION)
      HUMIDITY
      ION DENSITY (CONCENTRATION)
      MAGNETOSPHERIC ELECTRON DENSITY
      MAGNETOSPHERIC ION DENSITY
      MAGNETOSPHERIC PROTON DENSITY
      METEOROLOGY
      PARTICLE DENSITY (CONCENTRATION)
      PLANETARY ATMOSPHERES
      PLASMA DENSITY
      PROTON DENSITY (CONCENTRATION)
      SPACE DENSITY

ATMOSPHERIC DIFFUSION
GS   DIFFUSION
      . **ATMOSPHERIC DIFFUSION**
RT   BOLTZMANN DISTRIBUTION
      MOLECULAR DIFFUSION
      POLLUTION TRANSPORT
      RADIO SCATTERING
      TURBULENT DIFFUSION

ATMOSPHERIC EFFECTS
RT   AEROSOLS
      AIR POLLUTION
      ATMOSPHERIC CHEMISTRY
      ATMOSPHERIC CORRECTION
      ∞ EFFECTS
      EROSION
      EXPOSURE
      RUSTING
      SOIL EROSION
      TURBULENCE
      VEGETATIVE INDEX
      WIND EFFECTS
      WIND EROSION

ATMOSPHERIC ELECTRICITY
GS   ELECTRICITY
      . **ATMOSPHERIC ELECTRICITY**
      . . IONOSPHERIC CURRENTS
      . . . ELECTROJETS
      . . . . AURORAL ELECTROJETS
      . . . . EQUATORIAL ELECTROJET
RT   ATMOSPHERICS
      BALL LIGHTNING
      CLOUD PHYSICS
      DUST STORMS
      EARTH ATMOSPHERE
      ELECTRIC CORONA
      ELECTRON DENSITY PROFILES
      LIGHTNING
      LIGHTNING SUPPRESSION
      PRIMITIVE EARTH ATMOSPHERE
      RING CURRENTS
      STATIC ELECTRICITY
      TELLURIC CURRENTS

ATMOSPHERIC EMISSION
USE   AIRGLOW

ATMOSPHERIC ENERGY SOURCES
GS   HYDROCARBON FUEL PRODUCTION
      . **ATMOSPHERIC ENERGY SOURCES**
RT   AMMONIA
      ∞ ENERGY SOURCES
      ENERGY TECHNOLOGY
      ETHYL ALCOHOL

ATMOSPHERIC ENTRY
UF   PLANETARY ENTRY
GS   **ATMOSPHERIC ENTRY**
      . REENTRY
      . . HYPERBOLIC REENTRY
      . . HYPERSONIC REENTRY
      . . . UNCONTROLLED REENTRY
              (SPACECRAFT)
      . . MANNED REENTRY
      . . SPACECRAFT REENTRY
      . . . UNCONTROLLED REENTRY
              (SPACECRAFT)
RT   ABLATION
      AEROASSIST
      AEROCAPTURE
      AERODYNAMIC HEATING
      AEROMANEUVERING
      BOLIDES
      DESCENT TRAJECTORIES

ATMOSPHERIC ENTRY-(CONT.)
      EARTH ATMOSPHERE
      ∞ ENTRY
      ENTRY GUIDANCE (STS)
      FALLING
      GALILEO PROJECT
      GAS GUNS
      ORBIT DECAY
      SPACE FLIGHT

ATMOSPHERIC ENTRY SIMULATION
GS   SIMULATION
      . EXHAUST FLOW SIMULATION
      . . **ATMOSPHERIC ENTRY SIMULATION**
RT   ENVIRONMENT SIMULATION
      FLIGHT SIMULATORS
      LANDING SIMULATION
      SPACE ENVIRONMENT SIMULATION

ATMOSPHERIC GENERAL CIRCULATION
EXPERIMENT
GS   PAYLOADS
      . SPACE SHUTTLE PAYLOADS
      . . SPACEBORNE EXPERIMENTS
      . . . **ATMOSPHERIC GENERAL
                CIRCULATION EXPERIMENT**
      . SPACELAB PAYLOADS
      . . **ATMOSPHERIC GENERAL
                CIRCULATION EXPERIMENT**
RT   ATMOSPHERIC CIRCULATION
      EARTH ATMOSPHERE
      SPACE TRANSPORTATION SYSTEM

ATMOSPHERIC HEAT BUDGET
GS   ENERGY BUDGETS
      . HEAT BUDGET
      . . **ATMOSPHERIC HEAT BUDGET**
RT   GREENHOUSE EFFECT
      HEAT BALANCE
      HEAT TRANSFER

ATMOSPHERIC HEATING
SN   (EXCLUDES AERODYNAMIC HEATING)
GS   HEATING
      . **ATMOSPHERIC HEATING**
RT   BOLIDES

ATMOSPHERIC IMPURITIES
USE   AIR POLLUTION

ATMOSPHERIC IONIZATION
UF   METEORITIC IONIZATION
GS   IONIZATION
      . GAS IONIZATION
      . . **ATMOSPHERIC IONIZATION**
      . . . AURORAL IONIZATION
RT   AFTERGLOWS
      AIRGLOW
      ANTARES ROCKET VEHICLE
      ELECTRON DENSITY PROFILES
      IONOSPHERE
      PHOTOIONIZATION
      PLASMASPHERE
      RADIO METEORS
      RIOMETERS

ATMOSPHERIC LASERS
GS   STIMULATED EMISSION DEVICES
      . LASERS
      . . **ATMOSPHERIC LASERS**
RT   ATMOSPHERIC ATTENUATION
      ATMOSPHERIC OPTICS
      ATMOSPHERIC SCATTERING
      LASER OUTPUTS
      TEA LASERS

ATMOSPHERIC LOADING
USE   POLLUTION TRANSPORT

ATMOSPHERIC MODELS
GS   MODELS
      . **ATMOSPHERIC MODELS**
      . . REFERENCE ATMOSPHERES
RT   BAROCLINIC INSTABILITY
      CHAPMAN-FERRARO PROBLEM
      ENVIRONMENT MODELS
      ENVIRONMENT SIMULATION
      MATHEMATICAL MODELS
      NUMERICAL WEATHER FORECASTING
      OCEAN MODELS
      PRIMITIVE EARTH ATMOSPHERE
      SOLAR OSCILLATIONS
      STELLAR OSCILLATIONS
      VENUS CLOUDS

ATMOSPHERIC MODELS-(CONT.)
      WEATHER FORECASTING

ATMOSPHERIC MOISTURE
GS   COMPOSITION (PROPERTY)
      . ATMOSPHERIC COMPOSITION
      . . **ATMOSPHERIC MOISTURE**
      . CONCENTRATION (COMPOSITION)
      . . MOISTURE CONTENT
      . . . **ATMOSPHERIC MOISTURE**
RT   ACID RAIN
      ANVIL CLOUDS
      CAP CLOUDS
      CIRROCUMULUS CLOUDS
      CIRROSTRATUS CLOUDS
      CLOUDS (METEOROLOGY)
      DEW POINT
      HUMIDITY
      PRECIPITATION (METEOROLOGY)
      PSYCHROMETERS
      WATER VAPOR

ATMOSPHERIC NOISE
USE   ATMOSPHERICS

ATMOSPHERIC OPTICS
RT   ADAPTIVE OPTICS
      ATMOSPHERIC LASERS
      CLARITY
      HAZE
      INFRARED ABSORPTION
      LIGHT TRANSMISSION
      OPACITY
      ∞ OPTICS
      TRANSPARENCE
      VEGETATIVE INDEX

ATMOSPHERIC PHYSICS
UF   AEROPHYSICS
GS   **ATMOSPHERIC PHYSICS**
      . CLOUD PHYSICS
RT   AERONOMY
      BRUNT-VAISALA FREQUENCY
      DUST STORMS
      INTERNATIONAL MAGNETOSPHERIC
          STUDY
      METEOROLOGY
      NEUTRAL SHEETS
      ∞ PHYSICS
      SATELLITE ATMOSPHERES
      ∞ SCIENCE
      SECULAR VARIATIONS
      TURBOPAUSE

ATMOSPHERIC PRESSURE
UF   BAROMETRIC PRESSURE
GS   PRESSURE
      . **ATMOSPHERIC PRESSURE**
RT   ANTICYCLONES
      ∞ ATMOSPHERES
      CYCLONES
      GAS PRESSURE
      GEOPOTENTIAL HEIGHT
      HIGH ALTITUDE PRESSURE
      ISOBARS (PRESSURE)
      ISOSTATIC PRESSURE
      PRESSURE GRADIENTS
      RADIO OCCULTATION
      WEATHER

ATMOSPHERIC RADIATION
GS   **ATMOSPHERIC RADIATION**
      . AURORAS
      . . AURORAL ARCS
      . . . RED ARCS
      . . RADIO AURORAS
      . DAWN CHORUS
      . IONOSPHERIC NOISE
      . . WHISTLERS
      . SKY RADIATION
      . AIRGLOW
      . . . GEOCORONAL EMISSIONS
      . . . NIGHTGLOW
      . . . TWILIGHT GLOW
      . . DAYGLOW
      . STRATOSPHERE RADIATION
      . TROPOSPHERIC RADIATION
RT   CORPUSCULAR RADIATION
      ELECTROMAGNETIC RADIATION
      EXTRATERRESTRIAL RADIATION
      GREENHOUSE EFFECT
      IONOSPHERIC HEATING
      LIGHT (VISIBLE RADIATION)
      ∞ RADIATION
      ∞ RAYS

ATMOSPHERIC RADIATION-*(CONT.)*
　　SECONDARY COSMIC RAYS
　　TERRESTRIAL RADIATION
　　VLF EMISSION RECORDERS

ATMOSPHERIC REFRACTION
GS　REFRACTION
　　. **ATMOSPHERIC REFRACTION**
　　. . RADIO WAVE REFRACTION
RT　ELECTROMAGNETIC RADIATION
　　LIGHT TRANSMISSION
　　REFRACTIVITY
　　SOLAR RADIATION
　　WAVE DISPERSION

ATMOSPHERIC SCATTERING
GS　SCATTERING
　　. WAVE SCATTERING
　　. . **ATMOSPHERIC SCATTERING**
　　. . . TROPOSPHERIC SCATTERING
RT　ATMOSPHERIC LASERS
　　CIRCUMSOLAR RADIATION
　　DIFFRACTION
　　DIFFUSION
　　ELECTROMAGNETIC SCATTERING
　　HALOS
　　LIGHT SCATTERING
　　MICROWAVE SCATTERING
　　RADIO SCATTERING
　　SIGNAL FADING
　　VEGETATIVE INDEX

ATMOSPHERIC SHELLS
USE　ATMOSPHERIC STRATIFICATION

ATMOSPHERIC SOUNDING
GS　SOUNDING
　　. **ATMOSPHERIC SOUNDING**
RT　BALLOON SOUNDING
　　IONOSPHERIC SOUNDING
　　ROCKET SOUNDING
　　SATELLITE SOUNDING
　　VISIBLE INFRARED SPIN SCAN
　　　RADIOMETER

ATMOSPHERIC STRATIFICATION
UF　ATMOSPHERIC SHELLS
GS　STRATIFICATION
　　. **ATMOSPHERIC STRATIFICATION**
RT　BRUNT-VAISALA FREQUENCY
　　PLASMA LAYERS
　　SURFACE LAYERS

ATMOSPHERIC TEMPERATURE
GS　TEMPERATURE
　　. **ATMOSPHERIC TEMPERATURE**
　　. . AURORAL TEMPERATURE
　　. . IONOSPHERIC TEMPERATURE
RT　AMBIENT TEMPERATURE
　　GAS TEMPERATURE
　　ISOTHERMS
　　LACATE (EXPERIMENT)
　　PLANETARY ATMOSPHERES
　　PLANETARY TEMPERATURE
　　RADIO OCCULTATION
　　SODAR
　　SOUND DETECTING AND RANGING
　　SUBZERO TEMPERATURE
　　TEMPERATURE GRADIENTS
　　TEMPERATURE INVERSIONS
　　THERMAL RESOURCES
　　WEATHER

ATMOSPHERIC TIDES
GS　TIDES
　　. **ATMOSPHERIC TIDES**
RT　EARTH TIDES
　　LUNAR TIDES

ATMOSPHERIC TURBULENCE
GS　TURBULENCE
　　. **ATMOSPHERIC TURBULENCE**
　　. . CLEAR AIR TURBULENCE
　　. . GUSTS
　　. . LOW LEVEL TURBULENCE
RT　DISSIPATION
　　GUST LOADS
　　HOMOGENEOUS TURBULENCE
　　ISOTROPIC TURBULENCE
　　LAMINAR FLOW
　　METEOROLOGICAL PARAMETERS
　　METEOROLOGY
　　TEPHIGRAMS
　　TURBOPAUSE
　　TURBULENT DIFFUSION

ATMOSPHERIC TURBULENCE-*(CONT.)*
　　TURBULENT FLOW
　　WIND VARIATIONS

ATMOSPHERIC WINDOWS
RT　ASTRONOMICAL PHOTOGRAPHY
　　ASTRONOMICAL PHOTOMETRY

ATMOSPHERICS
UF　ATMOSPHERIC NOISE
　　SFERICS
GS　ELECTROMAGNETIC INTERFERENCE
　　. RADIO FREQUENCY INTERFERENCE
　　. . BLACKOUT (PROPAGATION)
　　. . . ELECTROMAGNETIC NOISE
　　. . . . **ATMOSPHERICS**
　　. . . . . IONOSPHERICS
　　. . . . . . DAWN CHORUS
　　. . . . . . HISS
　　. . . . . SUDDEN ENHANCEMENT OF
　　　　　ATMOSPHERICS
　　. . . . . WHISTLERS
RT　ATMOSPHERIC ELECTRICITY
　　ELECTROMAGNETIC COMPATIBILITY
　　RADIO METEOROLOGY
　　RADIO WAVES
　　STATIC ELECTRICITY
　　THUNDERSTORMS
　　VLF EMISSION RECORDERS

ATOLL REEFS
USE　CORAL REEFS

ATOLLS
GS　LANDFORMS
　　. ISLANDS
　　. . **ATOLLS**
RT　CORAL REEFS
　　LAGOONS
　　REEFS

ATOM CONCENTRATION
GS　COMPOSITION (PROPERTY)
　　. CONCENTRATION (COMPOSITION)
　　. . **ATOM CONCENTRATION**
RT　CHEMICAL COMPOSITION
　　∞DENSITY
　　ELECTRON DENSITY (CONCENTRATION)
　　FLUX DENSITY
　　GAS COMPOSITION
　　GAS DENSITY
　　ION DENSITY (CONCENTRATION)
　　IONOSPHERIC COMPOSITION
　　PLASMA COMPOSITION
　　PLASMA DENSITY
　　PROTON DENSITY (CONCENTRATION)

ATOMIC BATTERIES
USE　RADIOISOTOPE BATTERIES

ATOMIC BEAMS
GS　BEAMS (RADIATION)
　　. PARTICLE BEAMS
　　. . **ATOMIC BEAMS**
RT　ION BEAMS
　　MOLECULAR BEAMS
　　NEUTRAL ATOMS
　　NEUTRAL BEAMS
　　NEUTRON BEAMS
　　PARTICLE DIFFUSION
　　RAREFIED GAS DYNAMICS

ATOMIC BOMBS
USE　FISSION WEAPONS

ATOMIC CLOCKS
GS　MEASURING INSTRUMENTS
　　. TIME MEASURING INSTRUMENTS
　　. . CLOCKS
　　. . . **ATOMIC CLOCKS**
RT　AUTONOMOUS SPACECRAFT CLOCKS
　　CHRONOMETERS
　　CLOCK PARADOX
　　FREQUENCY STANDARDS
　　GAS MASERS
　　MASERS
　　MOLECULAR BEAMS
　　TIME MEASUREMENT

ATOMIC COLLISIONS
GS　ATOMIC INTERACTIONS
　　. **ATOMIC COLLISIONS**
　　COLLISIONS
　　. **ATOMIC COLLISIONS**

ATOMIC COLLISIONS-*(CONT.)*
RT　∞ABSORPTION
　　ATOMIZING
　　AUTOIONIZATION
　　∞CROSS SECTIONS
　　ELASTIC SCATTERING
　　ELECTRON SCATTERING
　　∞INTERACTIONS
　　IONIC COLLISIONS
　　IONIZATION
　　MOLECULAR COLLISIONS
　　PARTICLE COLLISIONS
　　RECOIL IONS
　　RECOMBINATION REACTIONS
　　SCATTERING

ATOMIC ENERGY
USE　NUCLEAR ENERGY

ATOMIC ENERGY LEVELS
UF　TRIPLET EXCITATION
　　TRIPLET STATE
GS　LEVEL (QUANTITY)
　　. ENERGY LEVELS
　　. . **ATOMIC ENERGY LEVELS**
RT　ATOMIC INTERACTIONS
　　EXCITATION
　　GROUND STATE
　　LANDAU FACTOR
　　LINE SPECTRA
　　SPONTANEOUS EMISSION

ATOMIC EXCITATIONS
GS　EXCITATION
　　. **ATOMIC EXCITATIONS**
RT　ENERGY LEVELS
　　HEISENBERG THEORY
　　IONIZATION
　　MOLECULAR EXCITATION
　　PARTICLE COLLISIONS
　　RESONANCE FLUORESCENCE

ATOMIC EXPLOSIONS
USE　NUCLEAR EXPLOSIONS

ATOMIC GASES
USE　MONATOMIC GASES

ATOMIC INTERACTIONS
GS　**ATOMIC INTERACTIONS**
　　. ATOMIC COLLISIONS
RT　ATOMIC ENERGY LEVELS
　　∞INTERACTIONS
　　ION ATOM INTERACTIONS
　　MOLECULAR STRUCTURE
　　QUANTUM MECHANICS

ATOMIC MASS
USE　ATOMIC WEIGHTS

ATOMIC MOBILITIES
GS　MOBILITY
　　. **ATOMIC MOBILITIES**
RT　ELECTRON MOBILITY
　　HOLE MOBILITY
　　IONIC MOBILITY
　　SELF DIFFUSION (SOLID STATE)

ATOMIC PHYSICS
RT　HARTREE-FOCK-SLATER METHOD
　　∞PHYSICS
　　RESONANCE FLUORESCENCE
　　∞SCIENCE

ATOMIC RECOMBINATION
GS　CHEMICAL REACTIONS
　　. **ATOMIC RECOMBINATION**
　　. . OXYGEN RECOMBINATION
　　RECOMBINATION REACTIONS
　　. **ATOMIC RECOMBINATION**
　　. . OXYGEN RECOMBINATION
RT　DEIONIZATION
　　DISSOCIATION
　　EMISSION
　　EMISSION SPECTRA
　　ION RECOMBINATION
　　RADIATIVE RECOMBINATION

ATOMIC SPECTRA
GS　SPECTRA
　　. **ATOMIC SPECTRA**
RT　BALMER SERIES
　　LYMAN ALPHA RADIATION
　　LYMAN BETA RADIATION

ATOMIC SPECTRA-*(CONT.)*
    LYMAN SPECTRA
    PASCHEN SERIES
    RYDBERG SERIES

ATOMIC STRUCTURE
  UF   ELECTRONIC STRUCTURE
  RT   ATOMS
        CONSTITUTION
        CRYSTAL LATTICES
        ELEMENTARY PARTICLES
        ENERGY LEVELS
        FINE STRUCTURE
        GRAVITONS
        HARTREE APPROXIMATION
        HYPERFINE STRUCTURE
        INTERATOMIC FORCES
        ISOELECTRONIC SEQUENCE
        MELTS (CRYSTAL GROWTH)
        MOLECULAR STRUCTURE
        NUCLEAR CHEMISTRY
        NUCLEAR MODELS
        NUCLEAR PHYSICS
        OCTETS
        ORDER-DISORDER TRANSFORMATIONS
        PARTICLE PRECIPITATION
        PAULI EXCLUSION PRINCIPLE
        POLYWATER
    ∞STRUCTURES
        THOMAS-FERMI MODEL

ATOMIC THEORY
  GS   **ATOMIC THEORY**
        . HEISENBERG THEORY
  RT   ELECTRON TRANSITIONS
        GROUND STATE
        LANDAU FACTOR
    ∞NUCLEAR ENERGY
        QUANTUM THEORY
    ∞THEORIES

ATOMIC WEIGHTS
  UF   ATOMIC MASS
  GS   WEIGHT (MASS)
        . **ATOMIC WEIGHTS**
  RT   ∞WEIGHT

ATOMIZATION
  USE   ATOMIZING

ATOMIZERS
  RT   ATOMIZING
        EVAPORATORS
        GRINDING MILLS
    ∞NOZZLES
        SPRAYERS

ATOMIZING
  UF   ATOMIZATION
  GS   **ATOMIZING**
        . GAS ATOMIZATION
        . LIQUID ATOMIZATION
  RT   AEROSOLS
        ATOMIC COLLISIONS
        ATOMIZERS
        COLLOIDAL GENERATORS
        COLLOIDING
        COMMINUTION
        DISINTEGRATION
        FLAKING
        GRINDING (COMMINUTION)
        GRINDING MILLS
        METAL POWDER
        SPRAYING

ATOMS
  GS   **ATOMS**
        . HELIUM ATOMS
        . HOT ATOMS
        . HYDROGEN ATOMS
        . METASTABLE ATOMS
        . NEUTRAL ATOMS
        . NITROGEN ATOMS
        . OXYGEN ATOMS
        . RECOIL ATOMS
  RT   ATOMIC STRUCTURE
        CHEMICAL ELEMENTS
    ∞ELEMENTS
        FREE RADICALS
        IONS
        ISOMERS
        ISOTOPE SEPARATION
        ISOTOPES
        MOLECULES
        MONATOMIC MOLECULES

ATOMS-*(CONT.)*
    NUCLEI (NUCLEAR PHYSICS)
    POLYATOMIC MOLECULES
    POSITIVE IONS
    POSITRONIUM

ATP
  USE   ADENOSINE TRIPHOSPHATE

ATR REACTOR
  USE   ADVANCED TEST REACTORS

ATROPHY
  RT   BIOLOGICAL EFFECTS
        DEGENERATION
        DETERIORATION
        NUTRITIONAL REQUIREMENTS
        PHYSICAL EXERCISE
        TISSUES (BIOLOGY)

ATROPINE
  GS   DRUGS
        . STIMULANTS
        . . **ATROPINE**
        HETEROCYCLIC COMPOUNDS
        . ALKALOIDS
        . . **ATROPINE**
        NITROGEN COMPOUNDS
        . ALKALOIDS
        . . **ATROPINE**

ATS
  UF   APPLICATIONS TECHNOLOGY
        SATELLITES
  GS   SATELLITES
        . ARTIFICIAL SATELLITES
        . . GRAVITY GRADIENT SATELLITES
        . . . **ATS**
        . . . . ATS 1
        . . . . ATS 2
        . . . . ATS 3
        . . . . ATS 4
        . . . . ATS 5
        . . . . ATS 6
        . . . . ATS 7
        . . . . ATS 8
        . . SCIENTIFIC SATELLITES
        . . . **ATS**
        . . . . ATS 1
        . . . . ATS 2
        . . . . ATS 3
        . . . . ATS 4
        . . . . ATS 5
        . . . . ATS 6
        . . . . ATS 7
        . . . . ATS 8
  RT   COMMUNICATION SATELLITES
        EARLY BIRD SATELLITES
        METEOROLOGICAL SATELLITES
        NAVIGATION SATELLITES
        NAVSTAR SATELLITES

ATS 1
  GS   SATELLITES
        . ARTIFICIAL SATELLITES
        . . GRAVITY GRADIENT SATELLITES
        . . . ATS
        . . . . **ATS 1**
        . . SCIENTIFIC SATELLITES
        . . . ATS
        . . . . **ATS 1**

ATS 2
  GS   SATELLITES
        . ARTIFICIAL SATELLITES
        . . GRAVITY GRADIENT SATELLITES
        . . . ATS
        . . . . **ATS 2**
        . . SCIENTIFIC SATELLITES
        . . . ATS
        . . . . **ATS 2**

ATS 3
  GS   SATELLITES
        . ARTIFICIAL SATELLITES
        . . GRAVITY GRADIENT SATELLITES
        . . . ATS
        . . . . **ATS 3**
        . . SCIENTIFIC SATELLITES
        . . . ATS
        . . . . **ATS 3**

ATS 4
  GS   SATELLITES

ATS 4-*(CONT.)*
    . ARTIFICIAL SATELLITES
    . . GRAVITY GRADIENT SATELLITES
    . . . ATS
    . . . . **ATS 4**
    . . SCIENTIFIC SATELLITES
    . . . ATS
    . . . . **ATS 4**

ATS 5
  GS   SATELLITES
        . ARTIFICIAL SATELLITES
        . . GRAVITY GRADIENT SATELLITES
        . . . ATS
        . . . . **ATS 5**
        . . SCIENTIFIC SATELLITES
        . . . ATS
        . . . . **ATS 5**

ATS 6
  GS   SATELLITES
        . ARTIFICIAL SATELLITES
        . . GRAVITY GRADIENT SATELLITES
        . . . ATS
        . . . . **ATS 6**
        . . SCIENTIFIC SATELLITES
        . . . ATS
        . . . . **ATS 6**
  RT   HET EXPERIMENT

ATS 7
  GS   SATELLITES
        . ARTIFICIAL SATELLITES
        . . GRAVITY GRADIENT SATELLITES
        . . . ATS
        . . . . **ATS 7**
        . . SCIENTIFIC SATELLITES
        . . . ATS
        . . . . **ATS 7**

ATS 8
  GS   SATELLITES
        . ARTIFICIAL SATELLITES
        . . GRAVITY GRADIENT SATELLITES
        . . . ATS
        . . . . **ATS 8**
        . . SCIENTIFIC SATELLITES
        . . . ATS
        . . . . **ATS 8**

∞ ATTACHMENT
  SN   *(USE OF A MORE SPECIFIC TERM IS*
        RECOMMENDED--CONSULT THE TERMS
        LISTED BELOW)
  UF   REATTACHMENT
  RT   ASSEMBLING
        COANDA EFFECT
        ELECTRON ATTACHMENT
        MOUNTING
        REATTACHED FLOW

ATTACHMENTS
  USE   ACCESSORIES

∞ ATTACK
  SN   *(USE OF A MORE SPECIFIC TERM IS*
        RECOMMENDED--CONSULT THE TERMS
        LISTED BELOW)
  RT   ATTACKING (ASSAULTING)
        CHEMICAL ATTACK

ATTACK AIRCRAFT
  GS   **ATTACK AIRCRAFT**
        . A-1 AIRCRAFT
        . A-7 AIRCRAFT
        . A-9 AIRCRAFT
        . A-10 AIRCRAFT
        . A-37 AIRCRAFT
        . AH-63 HELICOPTER
        . AH-64 HELICOPTER
        . BOMBER AIRCRAFT
        . . A-2 AIRCRAFT
        . . A-3 AIRCRAFT
        . . A-4 AIRCRAFT
        . . A-5 AIRCRAFT
        . . A-6 AIRCRAFT
        . . B-1 AIRCRAFT
        . . B-26 AIRCRAFT
        . . B-47 AIRCRAFT
        . . B-50 AIRCRAFT
        . . B-52 AIRCRAFT
        . . B-57 AIRCRAFT
        . . B-58 AIRCRAFT
        . . B-66 AIRCRAFT
        . . B-70 AIRCRAFT

ATTACK AIRCRAFT-(CONT.)
.. F-100 AIRCRAFT
.. SHACKLETON BOMBER
.. VALIANT AIRCRAFT
.. VICTOR MK-1 AIRCRAFT
.. VULCAN AIRCRAFT
. BREGUET 1150 AIRCRAFT
. BUCCANEER AIRCRAFT
. CL-41 AIRCRAFT
. DH 112 AIRCRAFT
. DH 115 AIRCRAFT
. FIGHTER AIRCRAFT
. ALPHA JET AIRCRAFT
.. F-2 AIRCRAFT
.. F-4 AIRCRAFT
.. F-5 AIRCRAFT
.. F-8 AIRCRAFT
.. F-9 AIRCRAFT
.. F-14 AIRCRAFT
.. F-15 AIRCRAFT
.. F-16 AIRCRAFT
.. F-17 AIRCRAFT
.. F-18 AIRCRAFT
.. F-27 AIRCRAFT
.. F-84 AIRCRAFT
.. F-86 AIRCRAFT
.. F-89 AIRCRAFT
.. F-94 AIRCRAFT
.. F-100 AIRCRAFT
.. F-101 AIRCRAFT
.. F-102 AIRCRAFT
.. F-104 AIRCRAFT
.. F-105 AIRCRAFT
.. F-106 AIRCRAFT
.. F-111 AIRCRAFT
.. FV-12A AIRCRAFT
.. G-91 AIRCRAFT
.. G-95/4 AIRCRAFT
.. GA-5 AIRCRAFT
.. JAGUAR AIRCRAFT
.. JET PROVOST AIRCRAFT
.. MIG AIRCRAFT
.. MIRAGE AIRCRAFT
... MIRAGE 3 AIRCRAFT
.. P-51 AIRCRAFT
.. P-1127 AIRCRAFT
.. P 1154 AIRCRAFT
.. SAAB 37 AIRCRAFT
.. SCIMITAR AIRCRAFT
.. VAMPIRE MK 35 AIRCRAFT
.. VJ-101 AIRCRAFT
.. YF-12 AIRCRAFT
.. YF-16 AIRCRAFT
. OV-10 AIRCRAFT
. P-308 AIRCRAFT
. T-2 AIRCRAFT
. TSR-2 AIRCRAFT
RT   AEROQUATIC VEHICLES
∞AIRCRAFT
ANTISUBMARINE WARFARE AIRCRAFT
JET AIRCRAFT
∞MILITARY AIRCRAFT
MILITARY HELICOPTERS
MRCA AIRCRAFT
SUPERSONIC AIRCRAFT
TERRAIN FOLLOWING AIRCRAFT
V/STOL AIRCRAFT

ATTACKING (ASSAULTING)
UF   ASSAULTING
GS   VIOLENCE
. **ATTACKING (ASSAULTING)**
RT   ∞ATTACK
∞MILITARY AIRCRAFT
TACTICS
WARFARE

ATTENTION
RT   ALERTNESS
CONSCIOUSNESS

ATTENUATION
GS   **ATTENUATION**
. ACOUSTIC ATTENUATION
.. SHOCK WAVE ATTENUATION
. ATMOSPHERIC ATTENUATION
.. AURORAL ABSORPTION
. MICROWAVE ATTENUATION
. SIDELOBE REDUCTION
. WAVE ATTENUATION
.. RADAR ATTENUATION
.. RADIO ATTENUATION
... MANDELSTAM REPRESENTATION
.. SHOCK WAVE ATTENUATION
RT   ∞ABSORPTION

ATTENUATION-(CONT.)
ATTENUATORS
∞CONDUCTION
DAMPING
DIFFRACTION
DILUTION
DISSIPATION
ELECTROMAGNETIC ABSORPTION
ELECTROMAGNETIC WAVE
TRANSMISSION
ELIMINATION
FADING
IMPINGEMENT
∞INHIBITION
INTERNAL FRICTION
LIGHT (VISIBLE RADIATION)
MECHANICAL IMPEDANCE
∞PROPAGATION
∞REDUCTION
RETARDING
SHIELDING
SIGNAL FADING
SIGNAL TO NOISE RATIOS
SOUND PROPAGATION
SOUND TRANSMISSION
SPATIAL FILTERING
TRANSMISSION
TRANSMISSION LOSS
TRANSMITTERS
VIBRATION DAMPING
WAVE DEGRADATION
WAVE DIFFRACTION
WAVE DISPERSION
WAVE PROPAGATION

ATTENUATION COEFFICIENTS
GS   COEFFICIENTS
. **ATTENUATION COEFFICIENTS**
RT   DIFFUSION COEFFICIENT
FLOW COEFFICIENTS
IMPEDANCE
OPACITY
REFLECTANCE
SCATTERING COEFFICIENTS
TRANSMISSION EFFICIENCY
TRANSMITTANCE

ATTENUATORS
GS   **ATTENUATORS**
. RESISTORS
.. POTENTIOMETERS (RESISTORS)
.. PRINTED RESISTORS
.. THERMISTORS
RT   ∞ABSORBERS
ABSORBERS (MATERIALS)
ATTENUATION
BAFFLES
DEFLECTORS
∞DIFFUSERS
ELECTROMAGNETIC WAVE FILTERS
EQUALIZERS (CIRCUITS)
∞FILTERS
INSULATORS
INVERTERS
ISOLATORS
MUFFLERS
POWER LIMITERS
RADIATION SHIELDING
REFLECTORS
SHIELDING
SILENCERS
SUPPRESSORS

ATTITUDE (INCLINATION)
UF   SPATIAL ORIENTATION
TILT
TILTING
GS   **ATTITUDE (INCLINATION)**
. PITCH (INCLINATION)
. ROLL
. SATELLITE ORIENTATION
. YAW
RT   HORIZONTAL ORIENTATION
INSTRUMENT ORIENTATION
MISALIGNMENT
∞MOTION
∞ORIENTATION
∞POSITION
∞SPACE ORIENTATION
STABILITY AUGMENTATION
TILTMETERS
VERTICAL ORIENTATION

ATTITUDE CONTROL
GS   **ATTITUDE CONTROL**

ATTITUDE CONTROL-(CONT.)
. DIRECTIONAL CONTROL
.. THRUST VECTOR CONTROL
. LATERAL CONTROL
. LONGITUDINAL CONTROL
. SATELLITE ATTITUDE CONTROL
RT   AIR TRAFFIC CONTROL
AIRCRAFT CONTROL
AUTOMATIC CONTROL
COLD GAS
∞CONTROL
CONTROL MOMENT GYROSCOPES
FLIGHT CONTROL
GUIDANCE SENSORS
HELICOPTER CONTROL
HORIZON SCANNERS
MAGNETIC CONTROL
MANUAL CONTROL
MIRANDA SATELLITE
MISSILE CONTROL
ORBITAL LIFETIME
REACTION WHEELS
REMOTE CONTROL
ROCKET ENGINE CONTROL
SATELLITE CONTROL
SOLAR SENSORS
SPACECRAFT CONTROL
SPACING
SPIN STABILIZATION
STAR TRACKERS
THRUST CONTROL
TRAJECTORY CONTROL
VISUAL CONTROL

ATTITUDE GYROS
GS   GYROSCOPES
. **ATTITUDE GYROS**
.. GYRO HORIZONS
RT   CONTROL MOMENT GYROSCOPES
SEA KEEPING

ATTITUDE INDICATORS
UF   HELICOPTER ATTITUDE INDICATORS
YAWMETERS
GS   AIRCRAFT INSTRUMENTS
. **ATTITUDE INDICATORS**
.. GYRO HORIZONS
FLIGHT INSTRUMENTS
. **ATTITUDE INDICATORS**
.. GYRO HORIZONS
MEASURING INSTRUMENTS
. INDICATING INSTRUMENTS
.. **ATTITUDE INDICATORS**
... GYRO HORIZONS
RT   CONTROL MOMENT GYROSCOPES
FLIGHT CONTROL
NAVIGATION AIDS

ATTITUDE STABILITY
UF   SATELLITE ATTITUDE DISTURBANCE
GS   DYNAMIC CHARACTERISTICS
. DYNAMIC STABILITY
.. MOTION STABILITY
... **ATTITUDE STABILITY**
.... DIRECTIONAL STABILITY
..... GYROSCOPIC STABILITY
.... LATERAL STABILITY
.... LONGITUDINAL STABILITY
STABILITY
. DYNAMIC STABILITY
.. MOTION STABILITY
... **ATTITUDE STABILITY**
.... DIRECTIONAL STABILITY
..... GYROSCOPIC STABILITY
.... LATERAL STABILITY
.... LONGITUDINAL STABILITY
RT   AERODYNAMIC STABILITY
AIRCRAFT STABILITY
DISCOS (SATELLITE ATTITUDE
CONTROL)
HOVERING STABILITY
LOW SPEED STABILITY
SATELLITE ATTITUDE CONTROL
SPACECRAFT MOTION
SPACECRAFT STABILITY
TUMBLING MOTION

ATTRACTION
RT   AFFINITY
FIELD THEORY (PHYSICS)
∞FORCE
GRAVITATIONAL FIELDS

ATTRIBUTES
USE   PROPERTIES

ATTRITION (MATERIALS)
USE　COMMINUTION

AUDIO DATA
RT　AUDIO FREQUENCIES
　　∞DATA
　　DATA TRANSMISSION

AUDIO EQUIPMENT
GS　**AUDIO EQUIPMENT**
　　. EARPHONES
　　. LOUDSPEAKERS
　　. MICROPHONES
RT　∞EQUIPMENT
　　MONAURAL SIGNALS

AUDIO FREQUENCIES
SN　(APPROXIMATELY 20 TO 20,000 HZ)
GS　FREQUENCIES
　　. **AUDIO FREQUENCIES**
　　. . QUEFRENCIES
RT　ACOUSTIC MEASUREMENT
　　AUDIO DATA
　　AUDIO SIGNALS
　　AUDITORY PERCEPTION
　　CEPSTRAL ANALYSIS
　　EXTREMELY LOW RADIO FREQUENCIES
　　MONAURAL SIGNALS
　　NOISE POLLUTION
　　RADIO FREQUENCIES
　　SOUND GENERATORS
　　SOUND TRANSMISSION
　　SOUND WAVES
　　VERY LOW FREQUENCIES
　　VOICE

AUDIO SIGNALS
RT　AUDIO FREQUENCIES
　　AUDITORY SIGNALS
　　SIGNAL PROCESSING
　　SIGNAL TRANSMISSION
　　∞SIGNALS

AUDIO VISUAL EQUIPMENT
USE　TRAINING DEVICES
　　VISUAL AIDS

AUDIOLOGY
GS　PHYSIOLOGY
　　. **AUDIOLOGY**
RT　AUDIOMETRY
　　AUDITORY FATIGUE
　　AUDITORY PERCEPTION
　　HEARING

AUDIOMETRY
RT　ACOUSTIC MEASUREMENT
　　AUDIOLOGY
　　AUDITORY DEFECTS
　　AUDITORY FATIGUE
　　AUDITORY PERCEPTION
　　AUDITORY STIMULI
　　HEARING
　　MASKING
　　∞MEASUREMENT
　　THRESHOLDS (PERCEPTION)

AUDITORY DEFECTS
UF　DEAFNESS
　　HEARING LOSS
GS　DEFECTS
　　. **AUDITORY DEFECTS**
RT　AUDIOMETRY
　　BIOACOUSTICS
　　LOSSES

AUDITORY FATIGUE
GS　FATIGUE (BIOLOGY)
　　. **AUDITORY FATIGUE**
RT　AUDIOLOGY
　　AUDIOMETRY
　　HEARING
　　NOISE THRESHOLD

AUDITORY PERCEPTION
UF　SOUND PERCEPTION
GS　PERCEPTION
　　. **AUDITORY PERCEPTION**
RT　ACOUSTICS
　　AUDIO FREQUENCIES
　　AUDIOLOGY
　　AUDIOMETRY
　　AUDITORY SENSATION AREAS
　　BINAURAL HEARING

AUDITORY PERCEPTION-(CONT.)
　　EAR
　　EARPHONES
　　MONAURAL SIGNALS
　　NOISE THRESHOLD
　　PSYCHOACOUSTICS
　　SENSITIVITY
　　SOUND LOCALIZATION
　　SOUND WAVES
　　SPEECH
　　THRESHOLDS (PERCEPTION)
　　WEBER TEST

AUDITORY SENSATION AREAS
RT　AUDITORY PERCEPTION
　　AUDITORY STIMULI
　　BIOACOUSTICS
　　THRESHOLDS (PERCEPTION)

AUDITORY SIGNALS
UF　CHIMES
RT　AUDIO SIGNALS
　　BELLS
　　CUES
　　HORNS
　　MONAURAL SIGNALS
　　PSYCHOACOUSTICS
　　SIGNAL MIXING
　　∞SIGNALS
　　WARNING
　　WARNING SYSTEMS

AUDITORY STIMULI
GS　STIMULATION
　　. **AUDITORY STIMULI**
RT　ACOUSTICS
　　AUDIOMETRY
　　AUDITORY SENSATION AREAS
　　NOISE (SOUND)
　　NOISE INTENSITY
　　SOUND GENERATORS
　　SOUND INTENSITY
　　∞STIMULI
　　THRESHOLDS (PERCEPTION)

AUDITORY TASKS
GS　TASKS
　　. **AUDITORY TASKS**
RT　ACOUSTICS
　　HEARING
　　NOISE (SOUND)

AUFEIS (ICE)
RT　ICE
　　MELTING
　　PERMAFROST
　　RIVERS

AUGER EFFECT
RT　COSMIC RAY SHOWERS
　　∞EFFECTS
　　ELECTRON TRANSITIONS

AUGER SPECTROSCOPY
GS　SPECTROSCOPY
　　. **AUGER SPECTROSCOPY**
RT　CHEMICAL ANALYSIS
　　ELECTRON TRANSITIONS
　　SPECTROSCOPIC ANALYSIS
　　THERMITES

AUGMENTATION
UF　ENHANCEMENT
GS　**AUGMENTATION**
　　. STABILITY AUGMENTATION
　　. THRUST AUGMENTATION
RT　INCREASING
　　SPATIAL FILTERING

AURIGA CONSTELLATION
GS　CONSTELLATIONS
　　. **AURIGA CONSTELLATION**
RT　ZETA AURIGAE STAR

AURORA 7
GS　MANNED SPACECRAFT
　　. MERCURY SPACECRAFT
　　. . **AURORA 7**
　　REENTRY VEHICLES
　　. RECOVERABLE SPACECRAFT
　　. . MERCURY SPACECRAFT
　　. . . **AURORA 7**
　　SOFT LANDING SPACECRAFT
　　. MERCURY SPACECRAFT

AURORA 7-(CONT.)
　　. . **AURORA 7**
　　SPACE CAPSULES
　　. MERCURY SPACECRAFT
　　. . **AURORA 7**
RT　MERCURY MA-7 FLIGHT

AURORAL ABSORPTION
GS　ATTENUATION
　　. ATMOSPHERIC ATTENUATION
　　. . **AURORAL ABSORPTION**
　　ENERGY ABSORPTION
　　. ELECTROMAGNETIC ABSORPTION
　　. . **AURORAL ABSORPTION**
　　RADIATION ABSORPTION
　　. ATMOSPHERIC ATTENUATION
　　. . **AURORAL ABSORPTION**
　　. ELECTROMAGNETIC ABSORPTION
　　. . **AURORAL ABSORPTION**
RT　∞ABSORPTION
　　LIGHT EMISSION
　　RIOMETERS

AURORAL ACTIVITY
USE　AURORAS

AURORAL ARCS
GS　ATMOSPHERIC RADIATION
　　. AURORAS
　　. . **AURORAL ARCS**
　　. . . RED ARCS
RT　∞ARCS

AURORAL ECHOES
GS　ECHOES
　　. **AURORAL ECHOES**
RT　RADAR ECHOES
　　RADIO ECHOES

AURORAL ELECTROJETS
GS　ELECTRIC CURRENT
　　. IONOSPHERIC CURRENTS
　　. . ELECTROJETS
　　. . . **AURORAL ELECTROJETS**
　　ELECTRICITY
　　. ATMOSPHERIC ELECTRICITY
　　. . IONOSPHERIC CURRENTS
　　. . . ELECTROJETS
　　. . . . **AURORAL ELECTROJETS**
RT　EQUATORIAL ELECTROJET
　　TELLURIC CURRENTS

AURORAL IONIZATION
GS　IONIZATION
　　. GAS IONIZATION
　　. . ATMOSPHERIC IONIZATION
　　. . . **AURORAL IONIZATION**
RT　AURORAS
　　EXCITATION
　　LIGHT EMISSION
　　PHOTOIONIZATION
　　RED ARCS

AURORAL IRRADIATION
GS　IRRADIATION
　　. **AURORAL IRRADIATION**
RT　AURORAS
　　ELECTRON IRRADIATION
　　EXCITATION
　　ION IRRADIATION
　　PHOTOIONIZATION

AURORAL SPECTROSCOPY
GS　SPECTROSCOPY
　　. **AURORAL SPECTROSCOPY**
RT　CHANNEL MULTIPLIERS
　　FABRY-PEROT SPECTROMETERS
　　LIGHT EMISSION
　　OPTICAL EMISSION SPECTROSCOPY
　　SPECTROSCOPIC ANALYSIS
　　VISIBLE SPECTRUM

AURORAL TEMPERATURE
GS　TEMPERATURE
　　. ATMOSPHERIC TEMPERATURE
　　. . **AURORAL TEMPERATURE**
RT　AURORAS
　　ION TEMPERATURE
　　IONOSPHERIC TEMPERATURE

AURORAL ZONES
GS　REGIONS
　　. **AURORAL ZONES**
RT　AURORAS

AURORAL ZONES-*(CONT.)*
MAGNETIC POLES
POLAR RADIO BLACKOUT
POLAR REGIONS

AURORAS
UF   AURORAL ACTIVITY
POLAR AURORAS
GS   ATMOSPHERIC RADIATION
. **AURORAS**
. . AURORAL ARCS
. . . RED ARCS
. . RADIO AURORAS
RT   AERONOMY
AIRGLOW
AURORAL IONIZATION
AURORAL IRRADIATION
AURORAL TEMPERATURE
AURORAL ZONES
DAWN CHORUS
EARTH ATMOSPHERE
ELECTRON PRECIPITATION
ESRO 4 SATELLITE
LIGHT EMISSION
MAGNETIC DISTURBANCES
NIGHT SKY
PROTON PRECIPITATION
SKY BRIGHTNESS
SOLAR ACTIVITY
X RAYS

AUSFORMING
GS   FORMING TECHNIQUES
. **AUSFORMING**
METAL WORKING
. **AUSFORMING**
RT   FORGING
∞ROLLING

AUSTENITE
RT   ALLOTROPY
FERRITES
IRON ALLOYS
MARTENSITE
MARTENSITIC TRANSFORMATION
MICROSTRUCTURE
STEELS

AUSTENITIC STAINLESS STEELS
GS   ALLOYS
. IRON ALLOYS
. . STEELS
. . . STAINLESS STEELS
. . . . **AUSTENITIC STAINLESS STEELS**
RT   MARTENSITIC STAINLESS STEELS
TIME TEMPERATURE PARAMETER

AUSTRALIA
GS   CONTINENTS
. **AUSTRALIA**
NATIONS
. **AUSTRALIA**
RT   TASMANIA
TORRES STRAIT

AUSTRALITES
GS   CELESTIAL BODIES
. METEORITES
. . STONY METEORITES
. . . TEKTITES
. . . . **AUSTRALITES**
RT   BEDIASITES

AUSTRIA
GS   NATIONS
. **AUSTRIA**
RT   ALPS MOUNTAINS (EUROPE)
CENTRAL EUROPE
EUROPE

AUTOCATALYSIS
GS   CATALYSIS
. **AUTOCATALYSIS**
RT   ABIOGENESIS
CATALYTIC ACTIVITY
REACTION KINETICS

AUTOCLAVES
RT   AUTOCLAVING
CHEMICAL REACTORS
∞CONTAINERS
PRESSURE VESSELS

AUTOCLAVING
RT   AUTOCLAVES
CURING
HEATING
LEACHING
POWDER METALLURGY

AUTOCODERS
GS   LANGUAGES
. PROGRAMMING LANGUAGES
. . ASSEMBLY LANGUAGE
. . . **AUTOCODERS**
RT   COMPILERS
COMPUTER PROGRAMMING
COMPUTER SYSTEMS PROGRAMS
MACHINE ORIENTED LANGUAGES

AUTOCOLLIMATORS
USE   COLLIMATORS

AUTOCORRELATION
GS   CORRELATION
. **AUTOCORRELATION**
RT   CROSS CORRELATION
DATA CORRELATION
FOURIER ANALYSIS
PERIODIC VARIATIONS
TIME SERIES ANALYSIS

AUTODYNES
GS   CIRCUITS
. **AUTODYNES**
OSCILLATORS
. **AUTODYNES**
RT   ∞DETECTORS
FREQUENCY CONTROL
HETERODYNING
SIGNAL ANALYZERS
SIGNAL DETECTION
SIGNAL DETECTORS
VACUUM TUBE OSCILLATORS

AUTOGYROS
GS   V/STOL AIRCRAFT
. ROTARY WING AIRCRAFT
. . **AUTOGYROS**
. . . AVIAN 2/180 AUTOGIRO

AUTOIONIZATION
GS   DISSOCIATION
. **AUTOIONIZATION**
IONIZATION
. **AUTOIONIZATION**
RT   ATOMIC COLLISIONS
MANY ELECTRON EFFECTS

AUTOKINESIS
GS   PERCEPTION
. SENSORY PERCEPTION
. . PROPRIOCEPTION
. . . **AUTOKINESIS**
. VISUAL PERCEPTION
. . . SPACE PERCEPTION
. . . . **AUTOKINESIS**

AUTOMATA THEORY
RT   ADAPTIVE CONTROL
ARTIFICIAL INTELLIGENCE
∞AUTOMATION
BIONICS
COMPUTERS
CYBERNETICS
DEPERSONALIZATION
HEURISTIC METHODS
INFORMATION THEORY
LEARNING MACHINES
ROBOTICS
ROBOTS
SELF ADAPTIVE CONTROL SYSTEMS
∞THEORIES
TURING MACHINES

AUTOMATED EN ROUTE ATC
GS   GROUND BASED CONTROL
. AIR TRAFFIC CONTROL
. . **AUTOMATED EN ROUTE ATC**
TRAFFIC CONTROL
. AIR TRAFFIC CONTROL
. . **AUTOMATED EN ROUTE ATC**
RT   AIRCRAFT GUIDANCE
APPROACH CONTROL
AUTOMATED PILOT ADVISORY SYSTEM
FLIGHT CONTROL
GROUND-AIR-GROUND COMMUNICATION

AUTOMATED EN ROUTE ATC-*(CONT.)*
MICROWAVE LANDING SYSTEMS

AUTOMATED GUIDEWAY TRANSIT VEHICLES
UF   AGT
GS   SURFACE VEHICLES
. AUTOMATED TRANSIT VEHICLES
. . **AUTOMATED GUIDEWAY TRANSIT VEHICLES**
RT   AUTOMATED MIXED TRAFFIC VEHICLES
CONVEYORS
PASSENGERS
RAIL TRANSPORTATION
RAPID TRANSIT SYSTEMS
TRANSPORTATION
URBAN TRANSPORTATION
∞VEHICLES

AUTOMATED MIXED TRAFFIC VEHICLES
UF   AMTV
GS   RESEARCH VEHICLES
. **AUTOMATED MIXED TRAFFIC VEHICLES**
SURFACE VEHICLES
. MOTOR VEHICLES
. . **AUTOMATED MIXED TRAFFIC VEHICLES**
RT   AUTOMATED GUIDEWAY TRANSIT VEHICLES
PASSENGERS
URBAN TRANSPORTATION
∞VEHICLES

AUTOMATED PILOT ADVISORY SYSTEM
RT   AIR TRAFFIC CONTROL
AUTOMATED EN ROUTE ATC
AUTOMATIC TRAFFIC ADVISORY AND RESOLUTION
∞SYSTEMS

AUTOMATED RADAR TERMINAL SYSTEM
RT   AIR TRAFFIC CONTROL
RADAR EQUIPMENT
RADAR TRACKING
∞SYSTEMS

AUTOMATED TRANSIT VEHICLES
GS   SURFACE VEHICLES
. **AUTOMATED TRANSIT VEHICLES**
. . AUTOMATED GUIDEWAY TRANSIT VEHICLES
RT   CONVEYORS
ELECTRIC MOTOR VEHICLES
PASSENGERS
RAIL TRANSPORTATION
RAPID TRANSIT SYSTEMS
TRANSPORTATION
URBAN TRANSPORTATION
∞VEHICLES

AUTOMATIC CONTROL
UF   SELF REGULATING
GS   **AUTOMATIC CONTROL**
. ADAPTIVE CONTROL
. . ACTIVE CONTROL
. . LEARNING MACHINES
. . SELF ADAPTIVE CONTROL SYSTEMS
. AUTOMATIC FLIGHT CONTROL
. . AUTOMATIC LANDING CONTROL
. AUTOMATIC FREQUENCY CONTROL
. AUTOMATIC GAIN CONTROL
. DYNAMIC CONTROL
. FEEDBACK CONTROL
. . CASCADE CONTROL
. FEEDFORWARD CONTROL
. NUMERICAL CONTROL
. OFF-ON CONTROL
. OPTIMAL CONTROL
. . TIME OPTIMAL CONTROL
. PROPORTIONAL CONTROL
. SELF ALIGNMENT
. SEQUENTIAL CONTROL
RT   AIRCRAFT CONTROL
ATTITUDE CONTROL
∞AUTOMATION
COMBUSTION CONTROL
∞CONTROL
CONTROL EQUIPMENT
CONTROL SYSTEMS DESIGN
CONTROLLERS
DEPERSONALIZATION
DIRECTIONAL CONTROL
DYNAMIC CHARACTERISTICS
ELECTRIC CONTROL
ELECTRONIC AIRCRAFT

AUTOMATIC CONTROL-(CONT.)

ELECTRONIC CONTROL
ENGINE CONTROL
ENVIRONMENTAL CONTROL
FLIGHT CONTROL
GROUND BASED CONTROL
GUIDANCE (MOTION)
HELICOPTER CONTROL
HYDRAULIC CONTROL
∞INSTRUMENTS
JET CONTROL
LANDING INSTRUMENTS
LATERAL CONTROL
LONGITUDINAL CONTROL
MANUAL CONTROL
MEASURING INSTRUMENTS
MISSILE CONTROL
NEGATIVE FEEDBACK
PNEUMATIC CONTROL
RADIO CONTROL
REAL TIME OPERATION
RECORDING INSTRUMENTS
REENTRY GUIDANCE
REGULATORS
RELIEF VALVES
REMOTE CONTROL
ROBOTICS
ROCKET ENGINE CONTROL
SATELLITE ATTITUDE CONTROL
SATELLITE CONTROL
SATELLITE GUIDANCE
SELF ABSORPTION
SERVOCONTROL
SERVOMECHANISMS
SERVOMOTORS
SPACECRAFT CONTROL
SPACECRAFT GUIDANCE
SPEED CONTROL
STABILITY AUGMENTATION
TEMPERATURE CONTROL
TERMINAL CONFIGURED VEHICLE
   PROGRAM
THERMOSTATS
THRUST VECTOR CONTROL
TRACKING PROBLEM
TRANSFER FUNCTIONS
TURBOJET ENGINE CONTROL

AUTOMATIC CONTROL VALVES

GS   VALVES
     . **AUTOMATIC CONTROL VALVES**
     . . PRESSURE REGULATORS
     . . RELIEF VALVES
RT   ACTUATORS
     ∞CONTROL
     DAMPERS (VALVES)
     DYNAMIC CHARACTERISTICS
     FLUID AMPLIFIERS
     FLUID SWITCHING ELEMENTS
     GAS VALVES
     HYDRAULIC EQUIPMENT
     PNEUMATIC CONTROL
     REGULATORS
     SERVOMECHANISMS
     SOLENOID VALVES
     TEMPERATURE CONTROL

AUTOMATIC DATA PROCESSING

USE   DATA PROCESSING

AUTOMATIC FLIGHT CONTROL

UF   AFCS (CONTROL SYSTEM)
GS   AUTOMATIC CONTROL
     . **AUTOMATIC FLIGHT CONTROL**
     . . AUTOMATIC LANDING CONTROL
     FLIGHT CONTROL
     . **AUTOMATIC FLIGHT CONTROL**
     . . AUTOMATIC LANDING CONTROL
RT   AIRCRAFT CONTROL
     AIRCRAFT INSTRUMENTS
     AUTONOMOUS NAVIGATION
     ∞CONTROL
     DISTANCE MEASURING EQUIPMENT
     FLIGHT MANAGEMENT SYSTEMS
     HIGHLY MANEUVERABLE AIRCRAFT
     MISSILE CONTROL
     NAVIGATION
     NAVIGATION AIDS
     RADAR NAVIGATION
     RADIO NAVIGATION
     SOLAR COMPASSES
     TERMINAL CONFIGURED VEHICLE
        PROGRAM
     THRUST VECTOR CONTROL

AUTOMATIC FREQUENCY CONTROL

UF   AFC (CONTROL)
GS   AUTOMATIC CONTROL
     . **AUTOMATIC FREQUENCY CONTROL**
     REGULATORS
     . **AUTOMATIC FREQUENCY CONTROL**
RT   ∞CONTROL
     FEEDBACK CONTROL
     FREQUENCY MODULATION
     OSCILLATORS
     TUNING

AUTOMATIC GAIN CONTROL

UF   AGC (CONTROL)
GS   AUTOMATIC CONTROL
     . **AUTOMATIC GAIN CONTROL**
RT   ∞CONTROL
     FEEDBACK CONTROL
     TUNING

AUTOMATIC LANDING CONTROL

GS   AUTOMATIC CONTROL
     . AUTOMATIC FLIGHT CONTROL
     . . **AUTOMATIC LANDING CONTROL**
     FLIGHT CONTROL
     . AUTOMATIC FLIGHT CONTROL
     . . **AUTOMATIC LANDING CONTROL**
     LANDING AIDS
     . INSTRUMENT LANDING SYSTEMS
     . . **AUTOMATIC LANDING CONTROL**
RT   AIRBORNE EQUIPMENT
     AIRCRAFT EQUIPMENT
     BLIND LANDING
     DISTANCE MEASURING EQUIPMENT
     FLIGHT MANAGEMENT SYSTEMS
     MICROWAVE LANDING SYSTEMS
     TERMINAL CONFIGURED VEHICLE
        PROGRAM

AUTOMATIC LIGHT AIRCRAFT READINESS MONITOR

USE   ALARM PROJECT

AUTOMATIC PATTERN RECOGNITION

USE   PATTERN RECOGNITION

AUTOMATIC PICTURE TRANSMISSION

UF   APT (PICTURE TRANSMISSION)
GS   TRANSMISSION
     . SIGNAL TRANSMISSION
     . . DATA TRANSMISSION
     . . . **AUTOMATIC PICTURE
           TRANSMISSION**
RT   TELEVISION TRANSMISSION
     WAVE PROPAGATION

AUTOMATIC PILOTS

UF   AUTOPILOTS
GS   AIRCRAFT INSTRUMENTS
     . **AUTOMATIC PILOTS**
     FLIGHT INSTRUMENTS
     . **AUTOMATIC PILOTS**
RT   AIRCRAFT EQUIPMENT
     FLIGHT CONTROL
     GYROSCOPES
     HIGHLY MANEUVERABLE AIRCRAFT
     HOMING
     LANDING AIDS
     NAVIGATION AIDS
     ∞PILOTS
     RADIO ALTIMETERS
     SOLAR COMPASSES

AUTOMATIC ROCKET IMPACT PREDICTORS

USE   COMPUTERIZED SIMULATION
      IMPACT PREDICTION

AUTOMATIC TEST EQUIPMENT

RT   ∞EQUIPMENT
     MEASURING INSTRUMENTS
     SNEAK CIRCUIT ANALYSIS
     ∞TEST EQUIPMENT

AUTOMATIC TRAFFIC ADVISORY AND RESOLUTION

SN   (AUTOMATIC TRAFFIC ADVISORY AND
      RESOLUTION SERVICE)
UF   ATARS
RT   AUTOMATED PILOT ADVISORY SYSTEM
     COLLISION AVOIDANCE
     GROUND BASED CONTROL
     NAVIGATION AIDS
     RESOLUTION
     ∞SYSTEMS

AUTOMATIC TYPEWRITERS

UF   FLEXOWRITERS (TRADEMARK)
GS   TYPEWRITERS
     . **AUTOMATIC TYPEWRITERS**
RT   CONSOLES
     DISPLAY DEVICES
     PRINTERS (DATA PROCESSING)
     PUNCHED TAPES

AUTOMATIC WEATHER STATIONS

GS   STATIONS
     . **AUTOMATIC WEATHER STATIONS**
RT   DATA ACQUISITION
     DATA COLLECTION PLATFORMS
     INSTRUMENT PACKAGES
     METEOROLOGICAL SERVICES
     OCEAN DATA ACQUISITIONS SYSTEMS
     REMOTE SENSORS
     WEATHER DATA RECORDERS

∞ AUTOMATION

SN   *(USE OF A MORE SPECIFIC TERM IS
      RECOMMENDED--CONSULT THE TERMS
      LISTED BELOW)*
UF   INSTRUMENTAL ANALYSIS
RT   AUTOMATA THEORY
     AUTOMATIC CONTROL
     COMMAND AND CONTROL
     COMPUTER VISION
     COMPUTERS
     CONTROL SYSTEMS DESIGN
     CONTROLLERS
     CYBERNETICS
     DATA PROCESSING
     DEPERSONALIZATION
     FAIL-SAFE SYSTEMS
     FEEDBACK CONTROL
     FEEDFORWARD CONTROL
     INFORMATION THEORY
     MAN MACHINE SYSTEMS
     MATERIALS HANDLING
     MECHANIZATION
     NUMERICAL CONTROL
     REMOTE CONTROL
     ROBOTICS
     SELF ERECTING DEVICES
     SELF REPAIRING DEVICES
     SERVOMECHANISMS
     SYSTEMS ENGINEERING
     TOOLING

AUTOMOBILE ACCIDENTS

RT   ACCIDENT INVESTIGATION
     ACCIDENT PREVENTION
     ACCIDENTS
     SAFETY DEVICES

AUTOMOBILE ENGINES

RT   EXTERNAL COMBUSTION ENGINES
     INTERNAL COMBUSTION ENGINES
     PISTON ENGINES
     ROTARY ENGINES
     TURBINE ENGINES
     WANKEL ENGINES

AUTOMOBILE FUELS

GS   FUELS
     . CHEMICAL FUELS
     . . LIQUID FUELS
     . . . **AUTOMOBILE FUELS**
RT   AIRCRAFT FUELS
     ANTIKNOCK ADDITIVES
     DIESEL FUELS
     GASOLINE
     HYDROCARBON FUELS
     INTERNAL COMBUSTION ENGINES
     SYNTHANE

AUTOMOBILES

UF   JEEPS
GS   SURFACE VEHICLES
     . MOTOR VEHICLES
     . . **AUTOMOBILES**
     . . . ELECTRIC AUTOMOBILES
RT   AIR BAG RESTRAINT DEVICES
     ANTISKID DEVICES
     CHASSIS
     ELECTRIC HYBRID VEHICLES
     ELECTRIC MOTOR VEHICLES
     FUEL SYSTEMS
     HYDROGEN ENGINES
     IGNITION SYSTEMS
     LUBRICATION SYSTEMS
     ∞MILITARY VEHICLES
     TRAILERS

AUTOMOBILES-*(CONT.)*
. TRUCKS
∞ VEHICLES

AUTOMORPHISMS
GS ALGEBRA
. GROUP THEORY
. . HOMOMORPHISMS
. . . **AUTOMORPHISMS**

AUTONOMIC NERVOUS SYSTEM
GS NERVOUS SYSTEM
. **AUTONOMIC NERVOUS SYSTEM**
. . SYMPATHETIC NERVOUS SYSTEM
RT INVOLUNTARY ACTIONS
∞ SYSTEMS

AUTONOMOUS NAVIGATION
RT AUTOMATIC FLIGHT CONTROL
CELESTIAL NAVIGATION
NAVIGATION AIDS
NAVIGATION INSTRUMENTS
SATELLITE NAVIGATION SYSTEMS
SPACE NAVIGATION
SPACECRAFT GUIDANCE

AUTONOMOUS SPACECRAFT CLOCKS
GS MEASURING INSTRUMENTS
. TIME MEASURING INSTRUMENTS
. . CLOCKS
. . . **AUTONOMOUS SPACECRAFT
CLOCKS**
RT ATOMIC CLOCKS
GLOBAL POSITIONING SYSTEM
SPACECRAFT INSTRUMENTS
TDR SATELLITES

AUTONOMY
RT ADAPTIVE CONTROL
COMMAND AND CONTROL
∞ COMMANDS
∞ DIRECTION
EQUATIONS OF MOTION
MANAGEMENT
SELF ADAPTIVE CONTROL SYSTEMS

AUTOPILOTS
USE AUTOMATIC PILOTS

AUTOPSIES
RT DISSECTION
PATHOLOGY

AUTORADIOGRAPHY
GS IMAGERY
. RADIOGRAPHY
. . **AUTORADIOGRAPHY**
PHOTOGRAPHY
. **AUTORADIOGRAPHY**
RT BLACK AND WHITE PHOTOGRAPHY

AUTOREGRESSIVE PROCESSES
RT FACTOR ANALYSIS
∞ PROCESSES
REGRESSION ANALYSIS
STATISTICAL ANALYSIS

AUTOROTATION
UF WINDMILLING
GS GYRATION
. ROTATION
. . **AUTOROTATION**
RT ROTARY WING AIRCRAFT
ROTOCHUTES

AUTOTROPHS
GS **AUTOTROPHS**
. HYDROGENOMONAS
RT HETEROTROPHS

AUTUMN
GS SEASONS
. **AUTUMN**
RT SPRING (SEASON)
SUMMER
WINTER

AUXILIARY EQUIPMENT (COMPUTERS)
SN (EXCLUDES COMPUTER-CONTROLLED
EQUIPMENT)
GS DATA PROCESSING EQUIPMENT
. **AUXILIARY EQUIPMENT (COMPUTERS)**
. . PRINTERS (DATA PROCESSING)
RT COMPUTER SYSTEMS DESIGN

AUXILIARY EQUIPMENT (COMPUTERS)-*(CONT.)*
DATA PROCESSING
∞ EQUIPMENT

AUXILIARY POWER SOURCES
GS **AUXILIARY POWER SOURCES**
. ASTEC SOLAR TURBOELECTRIC
GENERATOR
. CHEMICAL AUXILIARY POWER UNITS
. NUCLEAR AUXILIARY POWER UNITS
. . SNAP
. . . FISSION ELECTRIC CELLS
. . . . SNAP 2
. . . . SNAP 4
. . . . SNAP 8
. . . . SNAP 10A
. . . SNAP 1
. . . SNAP 3
. . . SNAP 7
. . . SNAP 9A
. . . SNAP 11
. . . SNAP 13
. . . SNAP 15
. . . SNAP 17
. . . SNAP 19
. . . SNAP 21
. . . SNAP 23
. . . SNAP 27
. . . SNAP 29
. . . SNAP 50
. . SPACE POWER REACTORS
. . . FISSION ELECTRIC CELLS
. . . . SNAP 2
. . . . SNAP 4
. . . . SNAP 8
. . . . SNAP 10A
. . . SNAP 50
. . . SPACE POWER UNIT REACTORS
. SOLAR AUXILIARY POWER UNITS
. . SUNFLOWER POWER SYSTEM
RT AIRCRAFT POWER SUPPLIES
DIRECT POWER GENERATORS
ELECTRIC BATTERIES
ELECTRIC GENERATORS
∞ ELECTRIC POWER
ELECTRIC POWER SUPPLIES
∞ ENERGY SOURCES
GROUND SUPPORT EQUIPMENT
∞ POWER SUPPLIES
SPACECRAFT POWER SUPPLIES
VOLTAGE CONVERTERS (AC TO AC)
VOLTAGE CONVERTERS (DC TO DC)

AUXILIARY PROPULSION
GS PROPULSION
. **AUXILIARY PROPULSION**
RT AERONAUTICAL ENGINEERING
∞ ASTRONAUTICS
ENGINES
HYDROGEN OXYGEN ENGINES
MARQUARDT R4D ENGINE
MISSILES
PROPELLANTS
PROPULSION SYSTEM CONFIGURATIONS
ROCKET PROPELLANTS
SPACE FLIGHT
SPACE SHUTTLES
∞ SPACECRAFT
THRUST

AV-8A AIRCRAFT
USE HARRIER AIRCRAFT

AV-8B AIRCRAFT
USE HARRIER AIRCRAFT

AVAILABILITY
RT ABUNDANCE
ENERGY POLICY
RESERVES
RESOURCES

AVALANCHE DIODES
UF IMPATT DIODES
TRAPATT DIODES
ZENER DIODES
GS ELECTRONIC EQUIPMENT
. DIODES
. . SEMICONDUCTOR DIODES
. . . **AVALANCHE DIODES**
. SOLID STATE DEVICES
. . SEMICONDUCTOR DEVICES
. . . **AVALANCHE DIODES**
. . . . CRYOSAR
RECTIFIERS

AVALANCHE DIODES-*(CONT.)*
. **AVALANCHE DIODES**
. . CRYOSAR
RT BARRITT DIODES
ION IMPLANTATION
NEGATIVE CONDUCTANCE
TRAPATT DEVICES
VOLTAGE REGULATORS

AVALANCHES
GS **AVALANCHES**
. ELECTRON AVALANCHE
. TOWNSEND AVALANCHE
RT EARTH MOVEMENTS
ELECTRIC DISCHARGES
ION PRODUCTION RATES
IONIZING RADIATION

AVCS
USE ADVANCED VIDICON CAMERA SYSTEM
(AVCS)

AVERAGE
GS **AVERAGE**
. MEAN
RT DISTRIBUTION MOMENTS
MEDIAN (STATISTICS)
MODE (STATISTICS)
NORMALITY
NORMS
QUALITY CONTROL

AVIAN 2/180 AUTOGIRO
GS RESEARCH AIRCRAFT
. **AVIAN 2/180 AUTOGIRO**
V/STOL AIRCRAFT
. ROTARY WING AIRCRAFT
. . AUTOGYROS
. . . **AVIAN 2/180 AUTOGIRO**

AVIATION
USE AERONAUTICS

AVIATION PSYCHOLOGY
GS AEROSPACE MEDICINE
. **AVIATION PSYCHOLOGY**
PSYCHOLOGY
. **AVIATION PSYCHOLOGY**
RT AIRCRAFT PILOTS
MILITARY PSYCHOLOGY
PILOT TRAINING
PSYCHOLOGICAL EFFECTS
PSYCHOLOGICAL FACTORS
SPACE PSYCHOLOGY

AVIATORS
USE AIRCRAFT PILOTS

AVIONICS
RT ∞ AERONAUTICS
AIRBORNE EQUIPMENT
AIRCRAFT COMMUNICATION
AIRCRAFT EQUIPMENT
AIRCRAFT INSTRUMENTS
ASTRIONICS
∞ ASTRONAUTICS
∞ CONTROL
∞ ELECTRONICS
FLIGHT MANAGEMENT SYSTEMS
GUIDANCE (MOTION)
HEAD-UP DISPLAYS
MODULARITY
SINGLE EVENT UPSETS
SYSTEMS INTEGRATION
∞ TEST EQUIPMENT
VIDEO LANDMARK ACQUISITION AND
TRACKING

AVOIDANCE
GS **AVOIDANCE**
. COLLISION AVOIDANCE
. . BEACON COLLISION AVOIDANCE
SYSTEM
. VORTEX AVOIDANCE
RT ACCIDENT PREVENTION
HAZARDS
TRAFFIC
TRAFFIC CONTROL
WARNING SYSTEMS

AVRO WHITWORTH HS-748 AIRCRAFT
USE HS-748 AIRCRAFT

AVRO 698 AIRCRAFT
USE VULCAN AIRCRAFT

AVRO 707 AIRCRAFT
GS HAWKER SIDDELEY AIRCRAFT
. **AVRO 707 AIRCRAFT**
JET AIRCRAFT
. **AVRO 707 AIRCRAFT**
MONOPLANES
. **AVRO 707 AIRCRAFT**
RESEARCH AIRCRAFT
. **AVRO 707 AIRCRAFT**
TAILLESS AIRCRAFT
. **AVRO 707 AIRCRAFT**
RT ∞AIRCRAFT
DELTA WINGS
VULCAN AIRCRAFT

AWACS AIRCRAFT
UF AIRBORNE WARNING AND CONTROL
SYSTEM
GS **AWACS AIRCRAFT**
. E-2 AIRCRAFT
. E-3A AIRCRAFT
. E-4A AIRCRAFT
RT ∞AIRCRAFT
BOEING AIRCRAFT
COMMAND AND CONTROL
EARLY WARNING SYSTEMS
GRUMMAN AIRCRAFT
∞MILITARY AIRCRAFT
MILITARY TECHNOLOGY

AWARDS
SN (EXCLUDES CONTACTS & GRANTS)
RT ASTRONAUTS
BIOGRAPHY
SCIENTISTS

AXAF
USE X RAY ASTROPHYSICS FACILITY

AXES (COORDINATES)
USE COORDINATES

AXES (REFERENCE LINES)
GS **AXES (REFERENCE LINES)**
. AXES OF ROTATION
. . EARTH AXIS
RT COORDINATES

AXES OF ROTATION
GS AXES (REFERENCE LINES)
. **AXES OF ROTATION**
. . EARTH AXIS
RT BODIES OF REVOLUTION
ROTATING BODIES
ROTATION
SHAFTS (MACHINE ELEMENTS)
SYMMETRICAL BODIES

AXIAL COMPRESSION LOADS
GS LOADS (FORCES)
. AXIAL LOADS
. . **AXIAL COMPRESSION LOADS**
. COMPRESSION LOADS
. . **AXIAL COMPRESSION LOADS**
RT AERODYNAMIC LOADS
COMPRESSING
DYNAMIC LOADS
SHOCK LOADS
STATIC LOADS
STRUCTURAL DESIGN CRITERIA
THRUST LOADS

AXIAL COMPRESSORS
USE TURBOCOMPRESSORS

AXIAL FLOW
GS FLUID FLOW
. **AXIAL FLOW**
RT ANNULAR FLOW
AXISYMMETRIC FLOW
COAXIAL FLOW
COAXIAL NOZZLES
COUNTERFLOW
DISCHARGE COEFFICIENT
FLOW GEOMETRY
ONE DIMENSIONAL FLOW
RADIAL FLOW
THREE DIMENSIONAL FLOW
TWO DIMENSIONAL FLOW

AXIAL FLOW COMPRESSORS
USE TURBOCOMPRESSORS

AXIAL FLOW PUMPS
GS PUMPS
. **AXIAL FLOW PUMPS**
. . TURBINE PUMPS
RT CENTRIFUGAL PUMPS
FUEL PUMPS

AXIAL FLOW TURBINES
GS TURBOMACHINERY
. TURBINES
. . **AXIAL FLOW TURBINES**
RT GAS TURBINE ENGINES
GAS TURBINES
STEAM TURBINES

AXIAL LOADS
GS LOADS (FORCES)
. **AXIAL LOADS**
. . AXIAL COMPRESSION LOADS
RT AERODYNAMIC LOADS
COMPRESSION LOADS
DYNAMIC LOADS
STATIC LOADS
STRUCTURAL DESIGN CRITERIA
THRUST LOADS

AXIAL MODES
GS MODES
. **AXIAL MODES**
RT COMBUSTION STABILITY
LASER MODES
PROPELLANT COMBUSTION
ROCKET ENGINES

AXIAL STRAIN
UF AXISYMMETRIC DEFORMATION
UNIAXIAL STRAIN
GS DEFORMATION
. **AXIAL STRAIN**
RT ELASTIC DEFORMATION
STRESS-STRAIN DIAGRAMS
STRUCTURAL STRAIN

AXIAL STRESS
GS STRESSES
. **AXIAL STRESS**
RT TENSILE STRESS

AXIOMS
UF POSTULATES
GS MATHEMATICAL LOGIC
. **AXIOMS**
RT KNOWLEDGE
∞LOGIC
∞MATHEMATICS

AXISYMMETRIC BODIES
GS SYMMETRICAL BODIES
. **AXISYMMETRIC BODIES**
RT BLUNT BODIES
∞BODIES
BODIES OF REVOLUTION
CONICAL BODIES
DUCTED BODIES
LENTICULAR BODIES
MISSILE BODIES
SLENDER BODIES
SLENDER CONES
STREAMLINED BODIES

AXISYMMETRIC DEFORMATION
USE AXIAL STRAIN

AXISYMMETRIC FLOW
GS FLUID FLOW
. **AXISYMMETRIC FLOW**
. . ANNULAR FLOW
. . KARMAN-BODEWADT FLOW
RT AXIAL FLOW
COAXIAL FLOW
CONICAL FLOW
COUETTE FLOW
CROCCO METHOD
CYLINDRICAL WAVES
FLOW GEOMETRY
HELICAL FLOW
THREE DIMENSIONAL BOUNDARY LAYER

AXISYMMETRY
USE SYMMETRY

AXLES
USE SHAFTS (MACHINE ELEMENTS)

AXONS
GS CELLS (BIOLOGY)
. **AXONS**
NERVOUS SYSTEM
. **AXONS**
RT NEUROTRANSMITTERS

AZEOTROPES
RT BINARY MIXTURES
MIXTURES
SOLUTIONS

AZIDES (INORGANIC)
GS NITROGEN COMPOUNDS
. **AZIDES (INORGANIC)**
. . HYDROGEN AZIDES
. . SODIUM AZIDES

AZIDES (ORGANIC)
GS NITROGEN COMPOUNDS
. **AZIDES (ORGANIC)**
. SODIUM AZIDES
. . TRIAMINOGUANIDINIUM AZIDE
RT EXPLOSIVES

AZIMUTH
UF SOLAR AZIMUTH
RT ALTITUDE
ANGLES (GEOMETRY)
ASTRONOMICAL COORDINATES
BEARING (DIRECTION)
CELESTIAL REFERENCE SYSTEMS
∞DIRECTION
ELEVATION ANGLE
LOOK ANGLES (TRACKING)
NAVIGATION
∞ORIENTATION
POSITION (LOCATION)

AZINES
GS HETEROCYCLIC COMPOUNDS
. **AZINES**
. . CYANURATES
. . CYANURIC ACID
. . MECLIZINE
. . METHYLENE BLUE
. . PHENOTHIAZINES
PYRAZINES
. **AZINES**
. . CYANURATES
. . CYANURIC ACID
. . MECLIZINE
. . METHYLENE BLUE
. . PHENOTHIAZINES
RT DYES

AZO COMPOUNDS
GS NITROGEN COMPOUNDS
. **AZO COMPOUNDS**
. . RDX
RT ∞CHEMICAL COMPOUNDS
DYES

AZOLES
GS HETEROCYCLIC COMPOUNDS
. **AZOLES**
. . ACETAZOLAMIDE
. . OXAZOLE
. . PYRROLES
. . . CARBAZOLES
. . . INDOLES
. . . . TRYPTOPHAN

AZOTOBACTER
GS MICROORGANISMS
. BACTERIA
. . **AZOTOBACTER**

AZULENE
GS HETEROCYCLIC COMPOUNDS
. **AZULENE**
TERPENES
. **AZULENE**

AZUR SATELLITE
GS SATELLITES
. ARTIFICIAL SATELLITES
. . SCIENTIFIC SATELLITES
. . . **AZUR SATELLITE**
RT EUROPEAN SPACE PROGRAMS
INTERNATIONAL COOPERATION

AZUR SATELLITE-*(CONT.)*
　　WEST GERMANY

A2F AIRCRAFT
　USE　A-6 AIRCRAFT

A3D AIRCRAFT
　USE　A-3 AIRCRAFT

A3J AIRCRAFT
　USE　A-5 AIRCRAFT

A4D AIRCRAFT
　USE　A-4 AIRCRAFT

B

B STARS
　UF　HELIUM STARS
　GS　CELESTIAL BODIES
　　. STARS
　　. . HOT STARS
　　. . . **B STARS**
　RT　BLUE STARS
　　　HERBIG-HARO OBJECTS
　　　LIMB BRIGHTENING
　　　LIMB DARKENING
　　　PECULIAR STARS
　　　STELLAR COMPOSITION
　　　WOLF-RAYET STARS

B-A-W DEVICES
　USE　BULK ACOUSTIC WAVE DEVICES

B-1 AIRCRAFT
　GS　ATTACK AIRCRAFT
　　. BOMBER AIRCRAFT
　　. . **B-1 AIRCRAFT**
　　NORTH AMERICAN AIRCRAFT
　　. **B-1 AIRCRAFT**
　RT　∞AIRCRAFT
　　　BOMBING EQUIPMENT
　　　BOMBS (ORDNANCE)
　　　COMBAT
　　　JET AIRCRAFT
　　　∞MILITARY AIRCRAFT
　　　MULTIENGINE VEHICLES
　　　WARFARE
　　　∞WINGED VEHICLES

B-26 AIRCRAFT
　UF　INVADER AIRCRAFT
　GS　ATTACK AIRCRAFT
　　. BOMBER AIRCRAFT
　　. . **B-26 AIRCRAFT**
　　MARTIN AIRCRAFT
　　. **B-26 AIRCRAFT**
　　MONOPLANES
　　. **B-26 AIRCRAFT**
　RT　∞AIRCRAFT

B-47 AIRCRAFT
　UF　RB-47 AIRCRAFT
　　　STRATOJET AIRCRAFT
　　　XB-47 AIRCRAFT
　GS　ATTACK AIRCRAFT
　　. BOMBER AIRCRAFT
　　. . **B-47 AIRCRAFT**
　　BOEING AIRCRAFT
　　. **B-47 AIRCRAFT**
　　JET AIRCRAFT
　　. **B-47 AIRCRAFT**
　　MONOPLANES
　　. **B-47 AIRCRAFT**
　RT　∞AIRCRAFT

B-50 AIRCRAFT
　GS　ATTACK AIRCRAFT
　　. BOMBER AIRCRAFT
　　. . **B-50 AIRCRAFT**
　　BOEING AIRCRAFT
　　. **B-50 AIRCRAFT**
　　JET AIRCRAFT
　　. **B-50 AIRCRAFT**
　　MONOPLANES
　　. **B-50 AIRCRAFT**
　RT　∞AIRCRAFT

B-52 AIRCRAFT
　UF　STRATOFORTRESS AIRCRAFT
　GS　ATTACK AIRCRAFT

B-52 AIRCRAFT-*(CONT.)*
　　. BOMBER AIRCRAFT
　　. . **B-52 AIRCRAFT**
　　BOEING AIRCRAFT
　　. **B-52 AIRCRAFT**
　　JET AIRCRAFT
　　. **B-52 AIRCRAFT**
　　MONOPLANES
　　. **B-52 AIRCRAFT**
　RT　∞AIRCRAFT
　　　TURBOFAN ENGINES

B-57 AIRCRAFT
　UF　CANBERRA BOMBER
　　　RB-57 AIRCRAFT
　GS　ATTACK AIRCRAFT
　　. BOMBER AIRCRAFT
　　. . **B-57 AIRCRAFT**
　　JET AIRCRAFT
　　. **B-57 AIRCRAFT**
　　MARTIN AIRCRAFT
　　. **B-57 AIRCRAFT**
　　MONOPLANES
　　. **B-57 AIRCRAFT**
　RT　∞AIRCRAFT
　　　CANBERRA AIRCRAFT

B-58 AIRCRAFT
　UF　HUSTLER AIRCRAFT
　GS　ATTACK AIRCRAFT
　　. BOMBER AIRCRAFT
　　. . **B-58 AIRCRAFT**
　　GENERAL DYNAMICS AIRCRAFT
　　. **B-58 AIRCRAFT**
　　JET AIRCRAFT
　　. **B-58 AIRCRAFT**
　　MONOPLANES
　　. **B-58 AIRCRAFT**
　　SUPERSONIC AIRCRAFT
　　. **B-58 AIRCRAFT**
　　TAILLESS AIRCRAFT
　　. **B-58 AIRCRAFT**
　RT　∞AIRCRAFT

B-66 AIRCRAFT
　UF　DESTROYER AIRCRAFT
　　　RB-66 AIRCRAFT
　GS　ATTACK AIRCRAFT
　　. BOMBER AIRCRAFT
　　. . **B-66 AIRCRAFT**
　　JET AIRCRAFT
　　. **B-66 AIRCRAFT**
　　MCDONNELL DOUGLAS AIRCRAFT
　　. DOUGLAS AIRCRAFT
　　. . **B-66 AIRCRAFT**
　　MONOPLANES
　　. **B-66 AIRCRAFT**
　RT　∞AIRCRAFT

B-70 AIRCRAFT
　UF　VALKYRIE AIRCRAFT
　　　XB-70 AIRCRAFT
　GS　ATTACK AIRCRAFT
　　. BOMBER AIRCRAFT
　　. . **B-70 AIRCRAFT**
　　JET AIRCRAFT
　　. **B-70 AIRCRAFT**
　　MONOPLANES
　　. **B-70 AIRCRAFT**
　　NORTH AMERICAN AIRCRAFT
　　. **B-70 AIRCRAFT**
　　RESEARCH AIRCRAFT
　　. **B-70 AIRCRAFT**
　　SUPERSONIC AIRCRAFT
　　. **B-70 AIRCRAFT**
　RT　∞AIRCRAFT

B-103 AIRCRAFT
　USE　BUCCANEER AIRCRAFT

BABBITT METAL
　GS　ALLOYS
　　. ANTIMONY ALLOYS
　　. . **BABBITT METAL**
　　. COPPER ALLOYS
　　. . **BABBITT METAL**
　　. TIN ALLOYS
　　. . **BABBITT METAL**
　RT　BEARING ALLOYS

BABOONS
　GS　ANIMALS
　　. VERTEBRATES
　　. . MAMMALS
　　. . . PRIMATES

BABOONS-*(CONT.)*
　　. . . . **BABOONS**
　　. WILDLIFE
　　. . **BABOONS**

BAC AIRCRAFT
　UF　BRITISH AIRCRAFT CORP AIRCRAFT
　GS　**BAC AIRCRAFT**
　　. BAC 111 AIRCRAFT
　　. CANBERRA AIRCRAFT
　　. H-126 AIRCRAFT
　　. JET PROVOST AIRCRAFT
　　. SCIMITAR AIRCRAFT
　　. TSR-2 AIRCRAFT
　　. VALIANT AIRCRAFT
　　. VC-10 AIRCRAFT
　　. VISCOUNT AIRCRAFT
　RT　∞AIRCRAFT

BAC TSR 2 AIRCRAFT
　USE　TSR-2 AIRCRAFT

BAC 111 AIRCRAFT
　GS　BAC AIRCRAFT
　　. **BAC 111 AIRCRAFT**
　　JET AIRCRAFT
　　. TURBOFAN AIRCRAFT
　　. . **BAC 111 AIRCRAFT**
　　MONOPLANES
　　. **BAC 111 AIRCRAFT**
　　PASSENGER AIRCRAFT
　　. **BAC 111 AIRCRAFT**
　　TRANSPORT AIRCRAFT
　　. **BAC 111 AIRCRAFT**
　RT　∞AIRCRAFT

BACILLUS
　GS　MICROORGANISMS
　　. BACTERIA
　　. . **BACILLUS**

BACK INJURIES
　GS　INJURIES
　　. **BACK INJURIES**
　RT　WHIPLASH INJURIES

BACKFIRE
　RT　COMBUSTION
　　　DEFLAGRATION
　　　EXPLOSIONS
　　　FIRES
　　　FLAME DEFLECTORS
　　　FLAME PROPAGATION
　　　FLASHBACK

BACKGROUND NOISE
　RT　CHANNEL NOISE
　　　COSMIC NOISE
　　　ELASTIC WAVES
　　　ELECTROMAGNETIC NOISE
　　　IONOSPHERIC NOISE
　　　∞NOISE
　　　NOISE (SOUND)
　　　NOISE MEASUREMENT
　　　NOISE SPECTRA
　　　NOISE THRESHOLD
　　　∞RADIATION
　　　RANDOM NOISE
　　　∞RAYS
　　　SIGNAL TO NOISE RATIOS
　　　SQUELCH CIRCUITS

BACKGROUND RADIATION
　RT　BIG BANG COSMOLOGY
　　　CONTINUOUS RADIATION
　　　CORPUSCULAR RADIATION
　　　COSMIC BACKGROUND EXPLORER
　　　　SATELLITE
　　　COSMIC NOISE
　　　ELECTROMAGNETIC NOISE
　　　EXTRATERRESTRIAL RADIATION
　　　HIGH ALTITUDE TESTS
　　　IONOSPHERIC NOISE
　　　∞RADIATION
　　　RELIC RADIATION
　　　SKY RADIATION

BACKINGS
　USE　BACKUPS

BACKLOBES
　RT　ANTENNA DESIGN
　　　ANTENNA RADIATION PATTERNS
　　　DIRECTIONAL ANTENNAS

BACKLOBES-*(CONT.)*
    ∞LOBES

BACKSCATTERING
  GS   SCATTERING
      . BACKSCATTERING
  RT   FORWARD SCATTERING
      LASER PLASMA INTERACTIONS
      NUCLEAR SCATTERING
      SCATTER PROPAGATION

BACKSHORES
  USE   BEACHES

BACKUPS
  UF   BACKINGS
  RT   REDUNDANT COMPONENTS
      RESERVES
      WELDING

BACKWARD DIFFERENCING
  RT   DIFFERENTIAL EQUATIONS
      NUMERICAL STABILITY
      PROBLEM SOLVING

BACKWARD FACING STEPS
  UF   REARWARD FACING STEPS
  RT   BOUNDARY LAYER FLOW
      FLOW GEOMETRY
      FLUID BOUNDARIES
      REATTACHED FLOW
      RECIRCULATIVE FLUID FLOW
      STAIRSTEPS
      ∞STEPS

BACKWARD WAVE TUBES
  GS   MICROWAVE EQUIPMENT
      . MICROWAVE TUBES
      . . MICROWAVE OSCILLATORS
      . . . BACKWARD WAVE TUBES
  RT   BEAM CURRENTS
      ELECTRON TRANSFER

BACKWARD WAVES
  RT   ELASTIC WAVES
      ELECTROMAGNETIC RADIATION
      SOLITARY WAVES
      TRANSMISSION LINES
      TRAVELING WAVE TUBES
      TRAVELING WAVES

BACKWASH
  SN   (EXCLUDES PROCESSES OF
      BACKWASHING)
  UF   SIDEWASH
  RT   BOUNDARY LAYER STABILITY
      DOWNWASH
      SLIPSTREAMS
      STROUHAL NUMBER
      TURBULENCE
      WAKES

BACTERIA
  GS   MICROORGANISMS
      . BACTERIA
      . . ACTINOMYCETES
      . . AZOTOBACTER
      . . BACILLUS
      . . CLOSTRIDIUM BOTULINUM
      . . ESCHERICHIA
      . . HYDROGENOMONAS
      . . KLEBSIELLA
      . . NITROBACTER
      . . PSEUDOMONAS
      . . SALMONELLA
      . . SARCINA
      . . SERRATIA
      . . STAPHYLOCOCCUS
      . . STEAROTHERMOPHILUS
      . . STREPTOCOCCUS
      . . STREPTOMYCETES
  RT   AEROBES
      ANAEROBES
      BACTERIOLOGY
      BLIGHT
      COLONIES
      GNOTOBIOTICS
      INVERTEBRATES
      PANSPERMIA
      PATHOGENS
      SAPROPHYTES
      WASTE TREATMENT

BACTERICIDES
  UF   GERMICIDES
  RT   ANTIINFECTIVES AND ANTIBACTERIALS
      ANTISEPTICS
      CHEMICAL STERILIZATION
      FUMIGATION
      STERILIZATION

BACTERIOLOGY
  GS   MICROBIOLOGY
      . BACTERIOLOGY
  RT   BACTERIA
      ∞BIOLOGY
      CLOSTRIDIUM BOTULINUM
      COLONIES
      ENDOTOXINS
      GNOTOBIOTICS
      VACCINES

BACTERIOPHAGES
  GS   MICROORGANISMS
      . VIRUSES
      . . BACTERIOPHAGES
  RT   INTERFERON

BADLANDS
  GS   LAND
      . BADLANDS
  RT   BARREN LAND
      TOPOGRAPHY

BAFFLES
  RT   ATTENUATORS
      ∞BARRIERS
      BLAST DEFLECTORS
      CONICAL FLOW
      DAMPING
      DEFLECTORS
      ∞DIFFUSERS
      DIVERTERS
      DIVIDERS
      DUCTS
      FLAME DEFLECTORS
      LIQUID SLOSHING
      LOUVERS
      MIXERS
      MUFFLERS
      PANELS
      REFLECTORS
      SHIELDING
      SUPPRESSORS

BAGGAGE
  GS   CARGO
      . BAGGAGE
  RT   AIR CARGO
      BAGS
      GROUND HANDLING

BAGS
  GS   **BAGS**
      . AIR BAG RESTRAINT DEVICES
      . GAS BAGS
  RT   BAGGAGE
      ∞CONTAINERS
      PACKAGES

BAHAMAS
  GS   LANDFORMS
      . ISLANDS
      . . WEST INDIES
      . . . BAHAMAS
      NATIONS
      . BAHAMAS
  RT   CARIBBEAN REGION

BAHRAIN
  GS   LANDFORMS
      . ISLANDS
      . . BAHRAIN
      NATIONS
      . BAHRAIN

BAILOUT
  RT   AIR DROP OPERATIONS
      EJECTION
      EJECTION INJURIES
      EJECTION SEATS
      EJECTION TRAINING
      ESCAPE (ABANDONMENT)
      ESCAPE SYSTEMS
      FLYING EJECTION SEATS
      JETTISON SYSTEMS
      JETTISONING

BAILOUT-*(CONT.)*
      PARACHUTE DESCENT

BAINITE
  RT   BAINITIC STEEL
      IRON ALLOYS
      MICROSTRUCTURE
      STEELS

BAINITIC STEEL
  GS   ALLOYS
      . IRON ALLOYS
      . . STEELS
      . . . BAINITIC STEEL
  RT   BAINITE

BAJA CALIFORNIA
  USE   LOWER CALIFORNIA (MEXICO)

BAJADAS
  USE   FANS (LANDFORMS)

BAKELITE (TRADEMARK)
  RT   CERAMICS
      RESINS
      THERMOSETTING RESINS

BAKEOUT
  USE   DEGASSING

BAKER-NUNN CAMERA
  GS   OPTICAL EQUIPMENT
      . CAMERAS
      . . BAKER-NUNN CAMERA
      PHOTOGRAPHIC EQUIPMENT
      . CAMERAS
      . . BAKER-NUNN CAMERA
  RT   ASTRONOMICAL PHOTOGRAPHY
      SCHMIDT CAMERAS

BAKING
  SN   (EXCLUDES FOOD PROCESSING)
  GS   HEATING
      . BAKING
  RT   CASTING
      DEGASSING
      DRYING
      HEAT TREATMENT
      OVENS
      ROASTING
      STERILIZATION

BALANCE
  RT   AERODYNAMIC BALANCE
      COMPENSATORS
      ∞EQUILIBRIUM
      HEAT BALANCE
      ∞MASS BALANCE
      MASS DISTRIBUTION
      MATERIAL BALANCE
      WEIGHT INDICATORS

BALANCE EQUATIONS
  USE   EQUATIONS

BALANCED AMPLIFIERS
  USE   PUSH-PULL AMPLIFIERS

BALANCING
  RT   ECCENTRICITY
      ∞EQUILIBRIUM
      FLYWHEELS
      MAN MACHINE SYSTEMS
      STABILIZATION

BALL BEARINGS
  GS   BEARINGS
      . ANTIFRICTION BEARINGS
      . . BALL BEARINGS
  RT   BALLS
      ELASTOHYDRODYNAMICS
      NEEDLE BEARINGS
      ROLLER BEARINGS
      THRUST BEARINGS

BALL LIGHTNING
  GS   ELECTRIC CURRENT
      . ELECTRIC DISCHARGES
      . . LIGHTNING
      . . . BALL LIGHTNING
  RT   ATMOSPHERIC ELECTRICITY

∞ **BALLAST**
SN   (USE OF A MORE SPECIFIC TERM IS
     RECOMMENDED--CONSULT THE TERMS
     LISTED BELOW)
RT   BALLAST (MASS)
     BALLASTS (IMPEDANCES)

BALLAST (MASS)
RT   AERODYNAMIC STABILITY
     ∞BALLAST
     BUOYANCY
     COUNTERBALANCES
     FLOATING
     FLOATS
     HYDRODYNAMICS
     LOADS (FORCES)
     MASS DISTRIBUTION
     STABILITY
     STATIC LOADS

BALLASTS (IMPEDANCES)
RT   ∞BALLAST
     CAPACITORS
     INDUCTORS
     LUMINAIRES
     RESISTORS
     TRANSFORMERS

BALLISTIC CAMERAS
GS   OPTICAL EQUIPMENT
     . CAMERAS
     . . **BALLISTIC CAMERAS**
     PHOTOGRAPHIC EQUIPMENT
     . CAMERAS
     . . **BALLISTIC CAMERAS**
RT   GROUND SUPPORT EQUIPMENT
     HIGH SPEED CAMERAS
     OPTICAL TRACKING
     RANGEFINDING
     STROBOSCOPES
     TRAJECTORY MEASUREMENT

BALLISTIC MISSILE DECOYS
GS   COUNTERMEASURES
     . **BALLISTIC MISSILE DECOYS**
     DECOYS
     . **BALLISTIC MISSILE DECOYS**
RT   MISSILE DEFENSE
     REENTRY DECOYS

BALLISTIC MISSILE EARLY WARNING SYSTEM
UF   BMEWS
GS   WARNING SYSTEMS
     . EARLY WARNING SYSTEMS
     . . **BALLISTIC MISSILE EARLY WARNING
       SYSTEM**
RT   AIR DEFENSE
     MILITARY TECHNOLOGY
     RADAR TRACKING
     ∞SYSTEMS

BALLISTIC MISSILE SUBMARINES
GS   WATER VEHICLES
     . SHIPS
     . . SUBMARINES
     . . . **BALLISTIC MISSILE SUBMARINES**
     UNDERWATER VEHICLES
     . . SUBMARINES
     . . . **BALLISTIC MISSILE SUBMARINES**
RT   FLEET BALLISTIC MISSILES
     MISSILE LAUNCHERS
     MOBILE MISSILE LAUNCHERS
     NAVY
     POSEIDON MISSILES
     SEA LAUNCHING

BALLISTIC MISSILES
SN   (GUIDED ONLY DURING INITIAL
     POWERED PHASE)
GS   MISSILES
     . **BALLISTIC MISSILES**
     . . FIELD ARMY BALLISTIC MISSILES
     . . . SUBROC MISSILE
     . . INTERCONTINENTAL BALLISTIC
       MISSILES
     . . . ATLAS ICBM
     . . . . ATLAS D ICBM
     . . . . ATLAS E ICBM
     . . . . ATLAS F ICBM
     . . . MINUTEMAN ICBM
     . . . TITAN ICBM
     . . . . TITAN 1 ICBM
     . . . . TITAN 2 ICBM
     . . INTERMEDIATE RANGE BALLISTIC
       MISSILES

BALLISTIC MISSILES-(CONT.)
     . . . BLUE STREAK MISSILE
     . . . JUPITER MISSILE
     . . . POLARIS MISSILES
     . . . . POLARIS A1 MISSILE
     . . . . POLARIS A2 MISSILE
     . . . . POLARIS A3 MISSILE
     . . PERSHING MISSILE
     . . POSEIDON MISSILES
     . . SHORT RANGE BALLISTIC MISSILES
     . . SKYBOLT MISSILE
     . . V-2 MISSILE
RT   ANTIMISSILE MISSILES
     SAFEGUARD SYSTEM
     SURFACE TO SURFACE MISSILES

BALLISTIC RANGES
GS   RANGES (FACILITIES)
     . TEST RANGES
     . . **BALLISTIC RANGES**
     TEST FACILITIES
     . TEST RANGES
     . . **BALLISTIC RANGES**
RT   DOWNRANGE
     HYDROBALLISTICS
     MISSILE RANGES

BALLISTIC TRAJECTORIES
GS   TRAJECTORIES
     . **BALLISTIC TRAJECTORIES**
RT   ASCENT TRAJECTORIES
     BALLISTICS
     COASTING FLIGHT
     DESCENT TRAJECTORIES
     DOWNRANGE
     FREE FALL
     IMPACT PREDICTION
     MIDCOURSE TRAJECTORIES
     MISSILE TRAJECTORIES
     PARABOLIC FLIGHT

∞ **BALLISTIC VEHICLES**
SN   (USE OF A MORE SPECIFIC TERM IS
     RECOMMENDED--CONSULT THE TERMS
     LISTED BELOW)
UF   NONLIFTING VEHICLES
RT   REENTRY VEHICLES
     ROCKET VEHICLES
     TEST VEHICLES
     ∞VEHICLES
     WEAPONS

BALLISTICS
GS   **BALLISTICS**
     . HYDROBALLISTICS
     . INTERIOR BALLISTICS
     . TERMINAL BALLISTICS
RT   AERODYNAMIC DRAG
     BALLISTIC TRAJECTORIES
     GAS GUNS
     HOWITZERS
     HYPERVELOCITY GUNS
     ORDNANCE
     PROJECTILES
     PROPELLANTS
     TRAJECTORIES
     TRAJECTORY ANALYSIS
     TRAJECTORY MEASUREMENT

BALLISTOCARDIOGRAPHY
GS   BIOENGINEERING
     . BIOMETRICS
     . . CARDIOGRAPHY
     . . . **BALLISTOCARDIOGRAPHY**
RT   ELECTROCARDIOGRAPHY
     PHONOCARDIOGRAPHY
     SEISMOCARDIOGRAPHY

BALLOON FLIGHT
RT   ∞FLIGHT
     METEOROLOGICAL FLIGHT
     VERTICAL FLIGHT

BALLOON SOUNDING
GS   SOUNDING
     . **BALLOON SOUNDING**
RT   ATMOSPHERIC SOUNDING
     RADIOSONDES
     SUPERPRESSURE BALLOONS

BALLOON-BORNE INSTRUMENTS
GS   MEASURING INSTRUMENTS
     . **BALLOON-BORNE INSTRUMENTS**
RT   AIRBORNE EQUIPMENT
     ASTRONOMICAL TELESCOPES

BALLOON-BORNE INSTRUMENTS-(CONT.)
     BALLOONS
     HIGH ALTITUDE BALLOONS
     METEOROLOGICAL INSTRUMENTS
     RADIOSONDES

BALLOONING MODES
GS   MODES
     . **BALLOONING MODES**
RT   MAGNETOHYDRODYNAMIC STABILITY
     PLASMA CONTROL
     PLASMA EQUILIBRIUM
     TEARING MODES (PLASMAS)

BALLOONS
GS   EXPANDABLE STRUCTURES
     . INFLATABLE STRUCTURES
     . . **BALLOONS**
     . . . HIGH ALTITUDE BALLOONS
     . . . . JIMSPHERE BALLOONS
     . . . . SKYHOOK BALLOONS
     . . . . SUPERPRESSURE BALLOONS
     . . . METEOROLOGICAL BALLOONS
     . . . . JIMSPHERE BALLOONS
     . . . . ROBIN BALLOONS
     . . . MICROBALLOONS
     . . . TETHERED BALLOONS
RT   ∞AIRCRAFT
     AIRSHIPS
     ASCENT
     BALLOON-BORNE INSTRUMENTS
     BALLUTES
     FOLDING STRUCTURES
     GAS BAGS
     GONDOLAS
     OBSERVATION AIRCRAFT
     PARAVULCOONS
     PILOTLESS AIRCRAFT
     STRATOSCOPE TELESCOPES

BALLS
RT   BALL BEARINGS
     FALLING SPHERES
     JOINTS (JUNCTIONS)
     SPHERES
     VALVES

BALLUTES
GS   BRAKES (FOR ARRESTING MOTION)
     . AERODYNAMIC BRAKES
     . . **BALLUTES**
     DRAG DEVICES
     . AERODYNAMIC BRAKES
     . . **BALLUTES**
     EXPANDABLE STRUCTURES
     . INFLATABLE STRUCTURES
     . . **BALLUTES**
RT   AIR DROP OPERATIONS
     AIRCRAFT BRAKES
     BALLOONS
     DRAG CHUTES
     FOLDING STRUCTURES
     PARACHUTES

BALMER SERIES
GS   SPECTRA
     . RADIATION SPECTRA
     . . ELECTROMAGNETIC SPECTRA
     . . . LINE SPECTRA
     . . . . **BALMER SERIES**
RT   ABSORPTION SPECTRA
     ATOMIC SPECTRA
     ELECTRON TRANSITIONS
     EMISSION SPECTRA
     H BETA LINE
     H GAMMA LINE
     H LINES
     HYDROGEN

BALSA
RT   TREES (PLANTS)
     WOOD

BALTIC SEA
GS   SEAS
     . **BALTIC SEA**
RT   ESTONIA
     LATVIA

BALTIC SHIELD (EUROPE)
GS   ROCKS
     . BEDROCK
     . . **BALTIC SHIELD (EUROPE)**
RT   EARTH RESOURCES
     EUROPE

BALTIC SHIELD (EUROPE)-(CONT.)
    PRECAMBRIAN PERIOD

BANACH SPACE
  GS ALGEBRA
     . VECTOR SPACES
     . . HILBERT SPACE
     . . . **BANACH SPACE**
     . . . . SOBOLEV SPACE
     ANALYSIS (MATHEMATICS)
     . FUNCTION SPACE
     . . HILBERT SPACE
     . . . **BANACH SPACE**
     . . . . SOBOLEV SPACE
     . FUNCTIONAL ANALYSIS
     . . HILBERT SPACE
     . . . **BANACH SPACE**
     . . . . SOBOLEV SPACE
  RT HARMONIC ANALYSIS
     METRIC SPACE

BAND RATIOING
  GS IMAGE PROCESSING
     . **BAND RATIOING**
  RT IMAGE ENHANCEMENT
     MULTISPECTRAL BAND SCANNERS
     REMOTE SENSING
     SPECTRAL BANDS

BAND STRUCTURE OF SOLIDS
  RT BRILLOUIN ZONES
     CONDUCTION BANDS
     ELECTRON TRANSITIONS
     ENERGY GAPS (SOLID STATE)
     FORBIDDEN BANDS
     HETEROJUNCTION DEVICES
     QUANTUM WELLS

BANDGAP
  USE ENERGY GAPS (SOLID STATE)

BANDPASS FILTERS
  GS ELECTROMAGNETIC WAVE FILTERS
     . **BANDPASS FILTERS**
     . . CRYSTAL FILTERS
     . . TRACKING FILTERS
  RT ADAPTIVE FILTERS
     BANDSTOP FILTERS
     BANDWIDTH
     ELECTRIC FILTERS
     ∞FILTERS
     FIR FILTERS
     MICROWAVE FILTERS
     OPTICAL FILTERS
     ULTRAVIOLET FILTERS
     VOCODERS

∞ BANDS
  SN *(USE OF A MORE SPECIFIC TERM IS*
     RECOMMENDED--CONSULT THE TERMS
     LISTED BELOW)
  RT ABSORPTION SPECTRA
     ANCHORS (FASTENERS)
     BANDWIDTH
     BLOCH BAND
     BROADBAND
     CLAMPS
     CLIPS
     CONDUCTION BANDS
     EDGE DISLOCATIONS
     ENERGY BANDS
     FASTENERS
     FORBIDDEN BANDS
     FREQUENCIES
     HERZBERG BANDS
     HOLDERS
     LOW FREQUENCY BANDS
     NARROWBAND
     PHOTOLUMINESCENT BANDS
     PLASTIC DEFORMATION
     RING STRUCTURES
     SCHUMANN-RUNGE BANDS
     SIDEBANDS
     SPECTRAL BANDS
     STRAPS
     SWAN BANDS
     VEGARD-KAPLAN BANDS

BANDSTOP FILTERS
  GS ELECTROMAGNETIC WAVE FILTERS
     . ELECTRIC FILTERS
     . . **BANDSTOP FILTERS**
  RT ADAPTIVE FILTERS
     BANDPASS FILTERS
     BANDWIDTH

BANDSTOP FILTERS-(CONT.)
     CRYSTAL FILTERS
     ∞FILTERS
     HIGH PASS FILTERS
     LOW PASS FILTERS
     MICROWAVE FILTERS
     OPTICAL FILTERS
     TRACKING FILTERS
     WAVEGUIDE FILTERS

BANDWIDTH
  GS **BANDWIDTH**
     . BROADBAND
     . NARROWBAND
     . SPECTRAL LINE WIDTH
  RT BANDPASS FILTERS
     ∞BANDS
     BANDSTOP FILTERS
     BROADBAND AMPLIFIERS
     CHANNEL CAPACITY
     DYNAMIC CHARACTERISTICS
     FREQUENCIES
     FREQUENCY RANGES
     IMPEDANCE
     LASER WINDOWS
     RESONANT FREQUENCIES
     SPEECH BASEBAND COMPRESSION
     TRACKING FILTERS
     TRANSFER FUNCTIONS
     WIDTH
     WINDOWS (INTERVALS)

BANG-BANG CONTROL
  USE OFF-ON CONTROL

BANGLADESH
  UF WEST PAKISTAN
  GS NATIONS
     . **BANGLADESH**
  RT ASIA
     INDIA
     PAKISTAN

BANKING FLIGHT
  USE TURNING FLIGHT

BARANY CHAIR
  GS SEATS
     . **BARANY CHAIR**
  RT ROTATING ENVIRONMENTS
     TOLERANCES (PHYSIOLOGY)
     VERTIGO

BARBADOS
  GS LANDFORMS
     . ISLANDS
     . . WEST INDIES
     . . . **BARBADOS**
     NATIONS
     . **BARBADOS**
  RT CARIBBEAN REGION

BARCHANS
  USE DUNES

BARDEEN APPROXIMATION
  USE BARRIER LAYERS
     ELECTRICAL PROPERTIES
     SURFACE PROPERTIES

BARDEEN-COOPER-SCHRIEFFER THEORY
  USE BCS THEORY

BARENTS SEA
  GS SEAS
     . **BARENTS SEA**
  RT ARCTIC OCEAN
     U.S.S.R.

BARITE
  GS MINERALS
     . **BARITE**
     SULFUR COMPOUNDS
     . SULFATES
     . . **BARITE**

BARIUM
  GS CHEMICAL ELEMENTS
     . **BARIUM**
     . . BARIUM ISOTOPES
     METALS
     . **BARIUM**
     . . BARIUM ISOTOPES

BARIUM ALLOYS
  GS ALLOYS
     . **BARIUM ALLOYS**

BARIUM COMPOUNDS
  GS **BARIUM COMPOUNDS**
     . BARIUM FERRATES
     . BARIUM FLUORIDES
     . BARIUM OXIDES
     . BARIUM SULFIDES
     . BARIUM TITANATES
     . BARIUM ZIRCONATES
  RT ∞ALKALINE EARTH COMPOUNDS
     ∞CHEMICAL COMPOUNDS
     ∞METAL COMPOUNDS

BARIUM FERRATES
  GS BARIUM COMPOUNDS
     . **BARIUM FERRATES**
     IRON COMPOUNDS
     . FERRATES
     . . **BARIUM FERRATES**

BARIUM FLUORIDES
  GS BARIUM COMPOUNDS
     . **BARIUM FLUORIDES**
     HALOGEN COMPOUNDS
     . FLUORINE COMPOUNDS
     . . FLUORIDES
     . . . **BARIUM FLUORIDES**
     . HALIDES
     . . FLUORIDES
     . . . **BARIUM FLUORIDES**
     . METAL HALIDES
     . . **BARIUM FLUORIDES**

BARIUM ION CLOUDS
  GS CLOUDS
     . CLOUDS (METEOROLOGY)
     . . ARTIFICIAL CLOUDS
     . . . CHEMICAL CLOUDS
     . . . . **BARIUM ION CLOUDS**
  RT ELECTRIC FIELDS
     GEOMAGNETISM
     LINES OF FORCE
     MAGNETOSPHERE
     METAL IONS
     ROCKET SOUNDING
     THERMITES

BARIUM ISOTOPES
  GS CHEMICAL ELEMENTS
     . ALKALINE EARTH METALS
     . . **BARIUM ISOTOPES**
     . BARIUM
     . . **BARIUM ISOTOPES**
     . NUCLIDES
     . ISOTOPES
     . . . **BARIUM ISOTOPES**
     METALS
     . ALKALINE EARTH METALS
     . . **BARIUM ISOTOPES**
     . BARIUM
     . . **BARIUM ISOTOPES**

BARIUM OXIDES
  GS BARIUM COMPOUNDS
     . **BARIUM OXIDES**
     CHALCOGENIDES
     . OXIDES
     . . METAL OXIDES
     . . . ALKALINE EARTH OXIDES
     . . . . **BARIUM OXIDES**

BARIUM SULFIDES
  GS BARIUM COMPOUNDS
     . **BARIUM SULFIDES**
     CHALCOGENIDES
     . SULFIDES
     . . INORGANIC SULFIDES
     . . . **BARIUM SULFIDES**
     SULFUR COMPOUNDS
     . SULFIDES
     . . INORGANIC SULFIDES
     . . . **BARIUM SULFIDES**

BARIUM TITANATES
  GS BARIUM COMPOUNDS
     . **BARIUM TITANATES**
     TITANIUM COMPOUNDS
     . TITANATES
     . . **BARIUM TITANATES**
  RT DIELECTRICS

BARIUM ZIRCONATES
GS   BARIUM COMPOUNDS
    . BARIUM ZIRCONATES
    ZIRCONIUM COMPOUNDS
    . ZIRCONATES
    . . **BARIUM ZIRCONATES**

BARKHAUSEN EFFECT
RT   ∞EFFECTS
    ELECTROMAGNETIC MEASUREMENT
    ELECTROMAGNETISM
    OSCILLOGRAPHS

BARLEY
GS   FARM CROPS
    . GRAINS (FOOD)
    . . **BARLEY**
    PLANTS (BOTANY)
    . **BARLEY**
RT   AGRICULTURE
    BLIGHT
    BOTANY
    CROP GROWTH
    CROP VIGOR
    ∞CROPS
    ∞FOOD
    IRRIGATION
    SEEDS

BAROCLINIC INSTABILITY
GS   STABILITY
    . **BAROCLINIC INSTABILITY**
RT   ATMOSPHERIC CIRCULATION
    ATMOSPHERIC MODELS
    BAROCLINIC WAVES
    BAROCLINITY
    FLOW STABILITY
    GEOSTROPHIC WIND
    METEOROLOGY

BAROCLINIC WAVES
GS   ELASTIC WAVES
    . CAPILLARY WAVES
    . . GRAVITY WAVES
    . . . **BAROCLINIC WAVES**
    SURFACE WAVES
    . CAPILLARY WAVES
    . . GRAVITY WAVES
    . . . **BAROCLINIC WAVES**
RT   BAROCLINIC INSTABILITY
    BAROCLINITY
    BAROTROPIC FLOW
    CYCLONES
    DENSITY DISTRIBUTION
    GEOSTROPHIC WIND
    RADIATION PRESSURE
    STRATIFIED FLOW
    WAVE AMPLIFICATION
    ∞WAVES

BAROCLINITY
RT   BAROCLINIC INSTABILITY
    BAROCLINIC WAVES
    BAROTROPIC FLOW
    BAROTROPISM
    ∞ISOBARS
    METEOROLOGICAL SOLENOIDS
    STRATIFIED FLOW

BAROMETERS
GS   MEASURING INSTRUMENTS
    . METEOROLOGICAL INSTRUMENTS
    . . **BAROMETERS**
    . PRESSURE GAGES
    . . **BAROMETERS**
RT   ALTIMETERS
    HYPSOMETERS
    MANOMETERS
    PRESSURE MEASUREMENT
    VACUUM GAGES

BAROMETRIC PRESSURE
USE   ATMOSPHERIC PRESSURE

BARORECEPTORS
GS   ANATOMY
    . SENSE ORGANS
    . . **BARORECEPTORS**
    RECEPTORS (PHYSIOLOGY)
    . **BARORECEPTORS**
RT   PRESSURE
    PROPRIOCEPTORS

BAROTRAUMA
GS   INJURIES
    . **BAROTRAUMA**
RT   DECOMPRESSION SICKNESS
    DIVING (UNDERWATER)

BAROTROPIC FLOW
GS   FLUID FLOW
    . **BAROTROPIC FLOW**
RT   AIR CURRENTS
    AIR FLOW
    BAROCLINIC WAVES
    BAROCLINITY
    BAROTROPISM
    FLOW CHARACTERISTICS
    LEE WAVES
    PLANETARY WAVES
    RAYLEIGH WAVES
    ROSSBY REGIMES
    SEA BREEZE
    VISCOUS FLOW
    WIND (METEOROLOGY)
    WIND SHEAR

BAROTROPISM
GS   **BAROTROPISM**
    . PLANETARY WAVES
RT   BAROCLINITY
    BAROTROPIC FLOW
    ∞ISOBARS

∞ **BARRAGES**
SN   *(USE OF A MORE SPECIFIC TERM IS RECOMMENDED--CONSULT THE TERMS LISTED BELOW)*
RT   ARTILLERY FIRE
    DAMS

BARRED GALAXIES
GS   CELESTIAL BODIES
    . GALAXIES
    . . SPIRAL GALAXIES
    . . . **BARRED GALAXIES**
RT   DISK GALAXIES
    GALACTIC STRUCTURE
    HUBBLE DIAGRAM
    LOCAL GROUP (ASTRONOMY)
    STAR CLUSTERS
    STAR DISTRIBUTION
    STARS
    VIRGO GALACTIC CLUSTER

∞ **BARRELS**
SN   *(USE OF A MORE SPECIFIC TERM IS RECOMMENDED--CONSULT THE TERMS LISTED BELOW)*
RT   BARRELS (CONTAINERS)
    ∞DRUMS
    GUN LAUNCHERS

BARRELS (CONTAINERS)
UF   CASKS
RT   ∞BARRELS
    ∞CONTAINERS
    DRUMS (CONTAINERS)

BARREN LAND
UF   BARRENS
GS   LAND
    . **BARREN LAND**
RT   ARID LANDS
    BADLANDS
    DESERTIFICATION
    DESERTS
    LAND USE
    SAHARA DESERT (AFRICA)
    SITES
    SOILS
    TOPOGRAPHY

BARRENS
USE   BARREN LAND

BARRICADES
USE   BARRIERS

BARRIER INJECTION TRANSIT TIME DIODES
USE   BARRITT DIODES

BARRIER LAYERS
UF   BARDEEN APPROXIMATION
RT   ∞BARRIERS
    BARRITT DIODES
    INTERLAYERS

BARRIER LAYERS-*(CONT.)*
    JFET
    JOINTS (JUNCTIONS)
    JUNCTION DIODES
    JUNCTION TRANSISTORS
    ∞LAYERS
    MBM JUNCTIONS
    NONOHMIC EFFECT
    SEALS (STOPPERS)
    SEMICONDUCTOR DEVICES
    SIS (SEMICONDUCTORS)
    SURFACE LAYERS
    WATERPROOFING
    ZENER EFFECT

∞ **BARRIERS**
SN   *(USE OF A MORE SPECIFIC TERM IS RECOMMENDED--CONSULT THE TERMS LISTED BELOW)*
UF   BARRICADES
    OBSTACLES
RT   ABORT APPARATUS
    ACOUSTIC VELOCITY
    ARRESTING GEAR
    BAFFLES
    BARRIER LAYERS
    BARRIERS (LANDFORMS)
    BARRITT DIODES
    BLOOD-BRAIN BARRIER
    BULKHEADS
    CHAINS
    CLOSURES
    CONSTRICTIONS
    CURTAINS
    DAMS
    DIVIDERS
    ELECTRODE FILM BARRIERS
    ENCLOSURES
    FENCES (BARRIERS)
    GATES (OPENINGS)
    GUARDS (SHIELDS)
    MBM JUNCTIONS
    SAFETY DEVICES
    SCHOTTKY DIODES
    SEALS (STOPPERS)
    SHIELDING
    VAPOR BARRIER CLOTHING
    WALLS
    WIND (METEOROLOGY)
    WINDOWS (APERTURES)

BARRIERS (LANDFORMS)
GS   LANDFORMS
    . **BARRIERS (LANDFORMS)**
    . . OUTER BANKS (NC)
    . . REEFS
RT   ∞BARRIERS
    BARS (LANDFORMS)
    ISLAND ARCS

BARRITT DIODES
UF   BARRIER INJECTION TRANSIT TIME DIODES
GS   ELECTRONIC EQUIPMENT
    . DIODES
    . . SEMICONDUCTOR DIODES
    . . . **BARRITT DIODES**
    . SOLID STATE DEVICES
    . . SEMICONDUCTOR DEVICES
    . . . **BARRITT DIODES**
RT   AVALANCHE DIODES
    BARRIER LAYERS
    ∞BARRIERS
    CARRIER INJECTION
    CRYOSAR
    ELECTRIC POTENTIAL
    INJECTION
    JUNCTION DIODES
    MICROWAVE OSCILLATORS
    RECTIFIERS
    SCHOTTKY DIODES
    SEMICONDUCTOR JUNCTIONS
    SHOT NOISE
    TRANSIT TIME

BARS
GS   BARS
    . ELASTIC BARS
    . PRISMATIC BARS
RT   METAL PLATES
    RODS
    STRUCTURAL MEMBERS

BARS (LANDFORMS)
UF   TOMBOLOS

BARS (LANDFORMS)-*(CONT.)*
GS   LANDFORMS
    . **BARS (LANDFORMS)**
RT   BARRIERS (LANDFORMS)
     BEACHES
     COASTAL PLAINS
     LAGOONS
     LITTORAL DRIFT
     REEFS

BARYCENTER
USE   CENTER OF GRAVITY

BARYON RESONANCE
GS   RESONANCE
    . **BARYON RESONANCE**
RT   BARYONS
     HYPERONS
     SCATTERING CROSS SECTIONS

BARYONS
GS   PARTICLES
    . ELEMENTARY PARTICLES
    . . FERMIONS
    . . . **BARYONS**
    . . . . HYPERONS
    . . . . . XI HYPERONS
    . . . . OMEGA-MESONS
    . . . . RHO-MESONS
    . . . . SIGMA-MESONS
    . . HADRONS
    . . . **BARYONS**
    . . . . OMEGA-MESONS
    . . . . RHO-MESONS
    . . . . SIGMA-MESONS
RT   BARYON RESONANCE
     COLD NEUTRONS
     ETA-MESONS
     FAST NEUTRONS
     GRAVITINOS
     KAONS
     MESON RESONANCE
     MESONS
     MUONS
     NEUTRONS
     NUCLEONS
     PHOTONEUTRONS
     PIONS
     PROTONS
     RECOIL PROTONS
     SOLAR PROTONS
     THERMAL NEUTRONS

BASALT
GS   ROCKS
    . IGNEOUS ROCKS
    . . **BASALT**
RT   CONES (VOLCANOES)
     MARS VOLCANOES
     REGOLITH
     SOILS
     VOLCANOES
     VOLCANOLOGY

BASE FLOW
SN   (FLUID FLOW AT THE BASE OR
     EXTREME AFT END OF A BODY)
GS   FLUID FLOW
    . **BASE FLOW**
RT   HEAD FLOW
     WAKES

BASE HEATING
GS   HEATING
    . **BASE HEATING**
RT   AFTERBODIES
     CONVECTION
     EXHAUST NOZZLES
     JET EXHAUST
     JET IMPINGEMENT
    ∞RADIATION
     ROCKET EXHAUST

BASE PRESSURE
GS   PRESSURE
    . **BASE PRESSURE**
RT   AERODYNAMIC DRAG

BASEMENTS
RT   ∞BUILDINGS
     FLOORS
     FOUNDATIONS

∞ **BASES**
SN   *(USE OF A MORE SPECIFIC TERM IS*
     RECOMMENDED--CONSULT THE TERMS
     LISTED BELOW)
RT   BASES (CHEMICAL)
     DATA BASES
     FOUNDATIONS
     INORGANIC COMPOUNDS
     ION CONCENTRATION
     LUNAR BASES
     SPACE BASES
     STATIONS

BASES (CHEMICAL)
RT   ALKALIES
     ANHYDRIDES
   ∞BASES

BASES (FOUNDATIONS)
USE   FOUNDATIONS

BASIC (PROGRAMMING LANGUAGE)
GS   LANGUAGES
    . PROGRAMMING LANGUAGES
    . . **BASIC (PROGRAMMING LANGUAGE)**
RT   COMPUTER PROGRAMMING

BASINS
USE   STRUCTURAL BASINS

BASINS (CONTAINERS)
RT   TANKS (CONTAINERS)

BASKETS
RT   ∞CONTAINERS
     GONDOLAS

BASTNASITE
GS   CARBON COMPOUNDS
    . CARBONATES
    . . **BASTNASITE**
     MINERALS
    . **BASTNASITE**
     RARE EARTH COMPOUNDS
    . CERIUM COMPOUNDS
    . . **BASTNASITE**

BATCH PROCESSING
GS   DATA PROCESSING
    . **BATCH PROCESSING**
RT   COMPUTER PROGRAMMING
     COMPUTER PROGRAMS
     DATA PROCESSING EQUIPMENT
   ∞PROCESSING

BATHING
RT   COOLING
     HYGIENE
     WASHING
     WASTE WATER

BATHOLITHS
GS   ROCK INTRUSIONS
    . **BATHOLITHS**
RT   GRANITE
     IGNEOUS ROCKS
     ROCKS

BATHS
SN   (EXCLUDES BATHING)
GS   **BATHS**
    . SALT BATHS
RT   DIPPING
     ELECTROPLATING
     HEAT TRANSFER
     QUENCHING (COOLING)
   ∞SOAKING
     SUBMERGING
     WATER IMMERSION

BATHYMETERS
UF   BATHYMETRY
GS   MEASURING INSTRUMENTS
    . **BATHYMETERS**
RT   DEPTH MEASUREMENT
     OCEANOGRAPHY
     SOUNDING
     UNDERWATER RESEARCH
      LABORATORIES

BATHYMETRY
USE   BATHYMETERS

BATHYTHERMOGRAPHS
GS   MEASURING INSTRUMENTS
    . TEMPERATURE MEASURING
      INSTRUMENTS
    . . **BATHYTHERMOGRAPHS**
     RECORDING INSTRUMENTS
    . **BATHYTHERMOGRAPHS**
RT   PRESSURE GRADIENTS
     TEMPERATURE GRADIENTS

BATS
GS   ANIMALS
    . VERTEBRATES
    . . MAMMALS
    . . . **BATS**
    . WILDLIFE
    . . **BATS**

BATTERIES
USE   ELECTRIC BATTERIES

BATTERY CHARGERS
RT   CHARGE EFFICIENCY
   ∞CHARGING
     ELECTRIC BATTERIES
     PULSE CHARGING
     STORAGE BATTERIES

BATTERY SEPARATORS
USE   SEPARATORS

BAUSCHINGER EFFECT
RT   ∞EFFECTS
     FATIGUE (MATERIALS)
     MICROSTRUCTURE

BAUXITE
RT   ALUMINUM OXIDES
     MINERALS
     ROCKS

BAY ICE
GS   ICE
    . **BAY ICE**
RT   FREEZING
     FROST
     ICE FORMATION
     ICE MAPPING
     ICE REPORTING
     LAKE ICE
     LOW TEMPERATURE
     NAVIGATION
     OCEANOGRAPHY
     SEA ICE
     SLUSH
     WATER

BAYARD-ALPERT IONIZATION GAGES
GS   MEASURING INSTRUMENTS
    . PRESSURE GAGES
    . . VACUUM GAGES
    . . . IONIZATION GAGES
    . . . . **BAYARD-ALPERT IONIZATION
      GAGES**
     VACUUM APPARATUS
    . VACUUM GAGES
    . . IONIZATION GAGES
    . . . **BAYARD-ALPERT IONIZATION
      GAGES**
RT   HOT CATHODES

BAYES THEOREM
UF   BAYESIAN STATISTICS
GS   THEOREMS
    . **BAYES THEOREM**
RT   QUALITY CONTROL
     SAMPLING

BAYESIAN STATISTICS
USE   BAYES THEOREM

BAYOUS
GS   LANDFORMS
    . INLETS (TOPOGRAPHY)
    . . **BAYOUS**
RT   LAKES
     MARSHLANDS
     RIVERS

∞ **BAYS**
SN   *(USE OF A MORE SPECIFIC TERM IS*
     RECOMMENDED--CONSULT THE TERMS
     LISTED BELOW)
RT   BAYS (STRUCTURAL UNITS)

BAYS-_(CONT.)_
    BAYS (TOPOGRAPHIC FEATURES)

BAYS (STRUCTURAL UNITS)
RT   AIRCRAFT COMPARTMENTS
    AIRFRAMES
    ∞BAYS
    COMPARTMENTS
    FUSELAGES
    HULLS (STRUCTURES)
    SHELLS (STRUCTURAL FORMS)

BAYS (TOPOGRAPHIC FEATURES)
UF   BIGHTS
    COVES
GS   **BAYS (TOPOGRAPHIC FEATURES)**
    . CHESAPEAKE BAY (US)
    . DELAWARE BAY (US)
    . MONTEREY BAY (CA)
    . SAGINAW BAY (MI)
    . SAN FRANCISCO BAY (CA)
    . SAN PABLO BAY (CA)
RT   ∞BAYS
    GULFS
    INLETS (TOPOGRAPHY)

BBGKY HIERARCHY
GS   CLASSIFICATIONS
    . HIERARCHIES
    . . **BBGKY HIERARCHY**
RT   BOGOLIUBOV THEORY
    BOLTZMANN TRANSPORT EQUATION
    EQUATIONS OF STATE
    FOURIER TRANSFORMATION
    KINETIC EQUATIONS
    PLASMA PHYSICS

BCAS
USE   BEACON COLLISION AVOIDANCE
    SYSTEM

BCC LATTICES
USE   BODY CENTERED CUBIC LATTICES

BCH CODES
UF   BOSE-CHAUDHURI-HOCQUENGHEM
    CODES
RT   BINARY CODES
    ∞CODES
    CODING
    COMPUTER PROGRAMMING
    DECODERS
    DECODING
    DIGITAL TECHNIQUES
    ERROR CORRECTING DEVICES
    INFORMATION THEORY
    PARITY
    RANDOM ERRORS

BCS THEORY
UF   BARDEEN-COOPER-SCHRIEFFER THEORY
RT   MANY BODY PROBLEM
    SUPERCONDUCTIVITY
    ∞THEORIES
    THERMODYNAMIC COUPLING

BE A
USE   BEACON EXPLORER A

BE B
USE   EXPLORER 22 SATELLITE

BE C
USE   EXPLORER 27 SATELLITE

BE-3 ENGINE
GS   ENGINES
    . ROCKET ENGINES
    . . RETROROCKET ENGINES
    . . . **BE-3 ENGINE**
RT   ATHENA ROCKET VEHICLE
    RANGER LUNAR LANDING VEHICLES
    SOLID PROPELLANT ROCKET ENGINES

BEACHES
UF   ADVANCING SHORELINES
    BACKSHORES
    INSHORE ZONES
RT   BARS (LANDFORMS)
    COASTAL CURRENTS
    COASTAL PLAINS
    COASTS
    CUSPS (LANDFORMS)
    DUNES

BEACHES-_(CONT.)_
    LAGOONS
    LAKES
    LITTORAL DRIFT
    MARINE ENVIRONMENTS
    SHOALS
    SHORELINES
    TOPOGRAPHY
    WATERFOWL

BEACON COLLISION AVOIDANCE SYSTEM
UF   BCAS
GS   AVOIDANCE
    . COLLISION AVOIDANCE
    . . **BEACON COLLISION AVOIDANCE
      SYSTEM**
RT   AIR TRAFFIC CONTROL
    AIRCRAFT SAFETY
    MIDAIR COLLISIONS
    RADIO BEACONS
    ∞SYSTEMS
    TRANSPONDERS

BEACON EXPLORER A
UF   BE A
    S-66 SATELLITE
GS   EXPANDABLE STRUCTURES
    . INFLATABLE STRUCTURES
    . . INFLATABLE SPACECRAFT
    . . . BEACON SATELLITES
    . . . . **BEACON EXPLORER A**
    SATELLITES
    . ARTIFICIAL SATELLITES
    . . PASSIVE SATELLITES
    . . . BEACON SATELLITES
    . . . . **BEACON EXPLORER A**
    SPACE ERECTABLE STRUCTURES
    . **BEACON EXPLORER A**
    UNMANNED SPACECRAFT
    . PASSIVE SATELLITES
    . . BEACON SATELLITES
    . . . **BEACON EXPLORER A**
RT   DELTA LAUNCH VEHICLE

BEACON EXPLORER B
USE   EXPLORER 22 SATELLITE

BEACON EXPLORER C
USE   EXPLORER 27 SATELLITE

BEACON SATELLITES
UF   POLAR IONOSPHERE BEACON
GS   EXPANDABLE STRUCTURES
    . INFLATABLE STRUCTURES
    . . INFLATABLE SPACECRAFT
    . . . **BEACON SATELLITES**
    . . . . BEACON EXPLORER A
    . . . . EXPLORER 22 SATELLITE
    SATELLITES
    . ARTIFICIAL SATELLITES
    . . PASSIVE SATELLITES
    . . . **BEACON SATELLITES**
    . . . . BEACON EXPLORER A
    . . . . EXPLORER 22 SATELLITE
    SPACE ERECTABLE STRUCTURES
    . INFLATABLE SPACECRAFT
    . . **BEACON SATELLITES**
    . . . EXPLORER 22 SATELLITE
    UNMANNED SPACECRAFT
    . PASSIVE SATELLITES
    . . **BEACON SATELLITES**
    . . . BEACON EXPLORER A
    . . . EXPLORER 22 SATELLITE
RT   LOCATES SYSTEM

BEACONS
GS   NAVIGATION AIDS
    . **BEACONS**
    . . AIRPORT BEACONS
    . . . DISCRETE ADDRESS BEACON
      SYSTEM
    . . RADAR BEACONS
    . . . DISCRETE ADDRESS BEACON
      SYSTEM
    . . RADIO BEACONS
    . . . OMNIDIRECTIONAL RADIO RANGES
    . . . . SELF CALIBRATING OMNIRANGE
    . . RADIO DIRECTION FINDERS
RT   AIRCRAFT LIGHTS
    BUOYS
    COMPASSES
    HOMING
    HOMING DEVICES
    INSTRUMENT FLIGHT RULES
    ∞MARKERS

BEACONS-_(CONT.)_
    POSITION INDICATORS
    PROJECTORS
    SEARCHLIGHTS
    ∞SIGNALS
    SOLAR COMPASSES
    VISUAL SIGNALS

BEADS
RT   SPOT WELDS
    WELDED JOINTS
    WELDING

BEAGLE AIRCRAFT
RT   ∞AIRCRAFT

BEAM CURRENTS
GS   ELECTRIC CURRENT
    . **BEAM CURRENTS**
RT   BACKWARD WAVE TUBES
    BRILLOUIN FLOW
    ∞CURRENTS
    PLASMA CURRENTS

BEAM INJECTION
RT   ELECTRON BEAMS
    ION BEAMS
    NEUTRAL BEAMS
    PLASMA HEATING
    PLASMA-PARTICLE INTERACTIONS
    TOKAMAK DEVICES
    TOROIDAL PLASMAS

BEAM INTERACTIONS
RT   BEAMS (RADIATION)
    COLLISION PARAMETERS
    HIGH ENERGY INTERACTIONS
    ∞INTERACTIONS

BEAM LEADS
GS   JOINTS (JUNCTIONS)
    . **BEAM LEADS**
    WIRING
    . **BEAM LEADS**
RT   BONDING
    ELECTRIC CONNECTORS
    ∞JOINING
    MICROELECTRONICS
    MICROMODULES
    SOLDERED JOINTS

BEAM NEUTRALIZATION
RT   BEAMS (RADIATION)
    ELECTRON BEAMS
    ION BEAMS
    NEUTRAL BEAMS

BEAM PLASMA AMPLIFIERS
GS   AMPLIFIERS
    . **BEAM PLASMA AMPLIFIERS**
RT   ELECTRON BEAMS
    MILLIMETER WAVES
    PLASMA-PARTICLE INTERACTIONS
    PLASMAS (PHYSICS)
    RELATIVISTIC ELECTRON BEAMS

BEAM RIDER GUIDANCE
GS   GUIDANCE (MOTION)
    . **BEAM RIDER GUIDANCE**
RT   MISSILE CONTROL
    MISSILE SYSTEMS

BEAM SPLITTERS
RT   BEAMS (RADIATION)
    PARTICLE ACCELERATORS
    PARTICLE BEAMS
    SCATTER PLATES (OPTICS)

BEAM SWITCHING
GS   SWITCHING
    . **BEAM SWITCHING**
RT   BEAMS (RADIATION)
    ELECTRON OPTICS
    ION ENGINES
    LASERS
    MAGNETIC SWITCHING
    PACKET SWITCHING

BEAM WAVEGUIDES
GS   TRANSMISSION LINES
    . COMMUNICATION CABLES
    . . WAVEGUIDES
    . . . **BEAM WAVEGUIDES**
RT   COLLIMATORS

BEAM WAVEGUIDES-*(CONT.)*
    PHOTON BEAMS
    PLASMAGUIDES
    RECTANGULAR WAVEGUIDES
    WAVE PROPAGATION
    YOKES

∞ **BEAMS**
  SN   *(USE OF A MORE SPECIFIC TERM IS*
       RECOMMENDED--CONSULT THE TERMS
       LISTED BELOW)
  RT   BEAMS (RADIATION)
       BEAMS (SUPPORTS)

BEAMS (RADIATION)
  GS   **BEAMS (RADIATION)**
      . GAMMA RAY BEAMS
      . MICROBEAMS
      . PARTICLE BEAMS
      . . ATOMIC BEAMS
      . . ELECTRON BEAMS
      . . . RELATIVISTIC ELECTRON BEAMS
      . . ION BEAMS
      . . NEUTRAL BEAMS
      . . . MOLECULAR BEAMS
      . . . NEUTRON BEAMS
      . . NEUTRINO BEAMS
      . . PION BEAMS
      . . PROTON BEAMS
      . PHONON BEAMS
      . PHOTON BEAMS
      . . LIGHT BEAMS
      . RADAR BEAMS
  RT   BEAM INTERACTIONS
       BEAM NEUTRALIZATION
       BEAM SPLITTERS
       BEAM SWITCHING
      ∞ BEAMS
       COHERENT ELECTROMAGNETIC
        RADIATION
       COHERENT RADIATION
       CORPUSCULAR RADIATION
       ELECTROMAGNETIC RADIATION
       EXTREME ULTRAVIOLET RADIATION
       INFRARED RADIATION
       IONIZING RADIATION
       IRRADIATION
       LIGHT (VISIBLE RADIATION)
       LONGITUDINAL WAVES
       MONOCHROMATIC RADIATION
       MULTIBEAM ANTENNAS
       PLANE WAVES
      ∞ RADIATION
      ∞ RAYS
       SUBMILLIMETER WAVES
       ULTRAVIOLET RADIATION

BEAMS (SUPPORTS)
  UF   STRUCTURAL BEAMS
  GS   STRUCTURAL MEMBERS
      . **BEAMS (SUPPORTS)**
      . . BOX BEAMS
      . . CANTILEVER BEAMS
      . . CURVED BEAMS
      . . I BEAMS
      . . RECTANGULAR BEAMS
      . . TIMOSHENKO BEAMS
  RT ∞ BEAMS
       COLUMNS (SUPPORTS)
       GIRDERS
      ∞ HEADERS
       T SHAPE
       TRUSSES

BEAMSHAPING
  USE   COLLIMATION

∞ **BEARING**
  SN   *(USE OF A MORE SPECIFIC TERM IS*
       RECOMMENDED--CONSULT THE TERMS
       LISTED BELOW)
  RT   BEARING (DIRECTION)
       BEARINGS
       INTERNAL COMBUSTION ENGINES

BEARING (DIRECTION)
  RT   ALIGNMENT
       AZIMUTH
      ∞ BEARING
      ∞ DIRECTION
       DIRECTION FINDING
       EXPOSURE
       FIELD OF VIEW
       INSTRUMENT ORIENTATION
      ∞ ORIENTATION

BEARING (DIRECTION)-*(CONT.)*
       POSITION (LOCATION)
       SOUND LOCALIZATION
      ∞ SPACE ORIENTATION

BEARING ALLOYS
  GS   ALLOYS
      . **BEARING ALLOYS**
  RT   ALUMINUM ALLOYS
       BABBITT METAL
       BEARINGS
       CADMIUM ALLOYS
       COPPER ALLOYS
       IRON ALLOYS
       LEAD ALLOYS
       METAL POWDER
       SILVER ALLOYS
       TIN ALLOYS
       ZINC ALLOYS

BEARINGLESS ROTORS
  GS   AIRFOILS
      . WINGS
      . . ROTARY WINGS
      . . . LIFTING ROTORS
      . . . . **BEARINGLESS ROTORS**
      ROTATING BODIES
      . ROTORS
      . . ROTARY WINGS
      . . . LIFTING ROTORS
      . . . . **BEARINGLESS ROTORS**
  RT   HINGES
       RIGID ROTORS

BEARINGS
  GS   **BEARINGS**
      . ANTIFRICTION BEARINGS
      . . BALL BEARINGS
      . . ROLLER BEARINGS
      . FOIL BEARINGS
      . GAS BEARINGS
      . JOURNAL BEARINGS
      . LIQUID BEARINGS
      . MAGNETIC BEARINGS
      . NEEDLE BEARINGS
      . THRUST BEARINGS
  RT ∞ BEARING
       BEARING ALLOYS
       BOUNDARY LUBRICATION
       BUSHINGS
       GIMBALS
       IDLERS
       INTERNAL COMBUSTION ENGINES
       LUBRICATION
       PACKINGS (SEALS)
       PIVOTS
       SHAFTS (MACHINE ELEMENTS)
       SUPPORTS
       SUSPENSION SYSTEMS (VEHICLES)
       SWIVELS
       WHEELS

BEARS
  GS   ANIMALS
      . VERTEBRATES
      . . MAMMALS
      . . . **BEARS**

BEAT
  USE   SYNCHRONISM

BEAT FREQUENCIES
  GS   FREQUENCIES
      . **BEAT FREQUENCIES**
  RT   GROUP VELOCITY
       INTERMEDIATE FREQUENCY AMPLIFIERS
       MOIRE EFFECTS
       RESONANT FREQUENCIES
       STANDING WAVES
       SUPERHETERODYNE RECEIVERS

BEAUFORT SEA (NORTH AMERICA)
  GS   SEAS
      . **BEAUFORT SEA (NORTH AMERICA)**
  RT   ALASKA
       ARCTIC OCEAN
       CANADA

BED REST
  GS   REST
      . **BED REST**
  RT   CALCIUM METABOLISM
       CLINICAL MEDICINE
       ORTHOSTATIC TOLERANCE

BEDDING EQUIPMENT
  RT ∞ BLANKETS
      ∞ EQUIPMENT

BEDIASITES
  GS   CELESTIAL BODIES
      . METEORITES
      . . STONY METEORITES
      . . . TEKTITES
      . . . . **BEDIASITES**
  RT   AUSTRALITES

BEDROCK
  UF   SHIELDS (GEOLOGY)
  GS   ROCKS
      . **BEDROCK**
      . . BALTIC SHIELD (EUROPE)
  RT   EARTH RESOURCES
       GEOLOGY
       REGOLITH
      ∞ SHELVES
       SOILS
       STRATA
       STRATIFICATION
       STRATIGRAPHY
       TUNNELING (EXCAVATION)

BEDS
  RT   BEDS (PROCESS ENGINEERING)
       COUCHES

BEDS (GEOLOGY)
  UF   LAKE BEDS
  GS   LANDFORMS
      . **BEDS (GEOLOGY)**
      . . SALT BEDS
  RT   GEOLOGY
       OCEAN BOTTOM
       STRATA
       STRATIGRAPHY

BEDS (PROCESS ENGINEERING)
  RT   BEDS
       CHEMICAL REACTORS
       EXTRACTION
       FILTRATION
       FLUIDIZED BED PROCESSORS
       ION EXCHANGING
       PERCOLATION

BEECH AIRCRAFT
  USE   BEECHCRAFT AIRCRAFT

BEECH C-33 AIRCRAFT
  USE   C-33 AIRCRAFT

BEECH S-35 AIRCRAFT
  USE   C-35 AIRCRAFT

BEECH 99 AIRCRAFT
  GS   BEECHCRAFT AIRCRAFT
      . **BEECH 99 AIRCRAFT**
      . . BEECHCRAFT 18 AIRCRAFT
      . . C-33 AIRCRAFT
      . . C-35 AIRCRAFT
      LIGHT AIRCRAFT
      . **BEECH 99 AIRCRAFT**
      . . BEECHCRAFT 18 AIRCRAFT
      . . C-33 AIRCRAFT
      . . C-35 AIRCRAFT
  RT ∞ AIRCRAFT
      ∞ LOW WING AIRCRAFT

BEECHCRAFT AIRCRAFT
  UF   BEECH AIRCRAFT
  GS   **BEECHCRAFT AIRCRAFT**
      . BEECH 99 AIRCRAFT
      . . BEECHCRAFT 18 AIRCRAFT
      . . C-33 AIRCRAFT
      . . C-35 AIRCRAFT
  RT ∞ AIRCRAFT

BEECHCRAFT 18 AIRCRAFT
  GS   BEECHCRAFT AIRCRAFT
      . BEECH 99 AIRCRAFT
      . . **BEECHCRAFT 18 AIRCRAFT**
      GENERAL AVIATION AIRCRAFT
      . **BEECHCRAFT 18 AIRCRAFT**
      LIGHT AIRCRAFT
      . BEECH 99 AIRCRAFT
      . . **BEECHCRAFT 18 AIRCRAFT**
      MONOPLANES
      . **BEECHCRAFT 18 AIRCRAFT**
  RT ∞ AIRCRAFT

BEER LAW
RT　ABSORPTIVITY
　　BOUGUER LAW
　　ELECTROMAGNETIC ABSORPTION
　　MOLECULAR ABSORPTION

BEES
GS　ANIMALS
　　. INVERTEBRATES
　　. . ARTHROPODS
　　. . . INSECTS
　　. . . . **BEES**
RT　SWARMING

BEETLES
GS　ANIMALS
　　. INVERTEBRATES
　　. . ARTHROPODS
　　. . . INSECTS
　　. . . . CRICKETS
　　. . . . . **BEETLES**
　　. . . . . . TRIBOLIA
RT　INFESTATION

BEHAVIOR
GS　**BEHAVIOR**
　　. DECONDITIONING
　　. HUMAN BEHAVIOR
RT　CONDITIONING (LEARNING)
　　DIAGNOSIS
　　EDUCATION
　　EXTROVERSION
　　LEARNING
　　MIGRATION
　　SKINNER BOXES

BELFAST AIRCRAFT
USE　SC-5 AIRCRAFT

BELGIAN CONGO
USE　ZAIRE

BELGIUM
GS　NATIONS
　　. **BELGIUM**
RT　EUROPE

BELIZE
UF　BRITISH HONDURAS
GS　NATIONS
　　. **BELIZE**
RT　CARIBBEAN REGION
　　CARIBBEAN SEA
　　CENTRAL AMERICA

BELL AIRCRAFT
GS　**BELL AIRCRAFT**
　　. AH-63 HELICOPTER
　　. BELL 214A HELICOPTER
　　. OH-4 HELICOPTER
　　. OH-13 HELICOPTER
　　. UH-1 HELICOPTER
　　. X-1 AIRCRAFT
　　. X-2 AIRCRAFT
　　. X-5 AIRCRAFT
　　. X-14 AIRCRAFT
　　. X-22 AIRCRAFT
　　. XV-3 AIRCRAFT
　　. XV-15 AIRCRAFT
RT　∞AIRCRAFT

BELL 214A HELICOPTER
GS　BELL AIRCRAFT
　　. **BELL 214A HELICOPTER**
　　V/STOL AIRCRAFT
　　. ROTARY WING AIRCRAFT
　　. . HELICOPTERS
　　. . . **BELL 214A HELICOPTER**
RT　∞AIRCRAFT
　　VERTICAL TAKEOFF AIRCRAFT

BELLMAN THEORY
RT　DYNAMIC PROGRAMMING
　　OPTIMIZATION
　　∞THEORIES

BELLOWS
SN　(EXPANDABLE JOINTS--FOR DEVICES TO
　　MOVE GASES, USE BLOWERS)
GS　EXPANDABLE STRUCTURES
　　. **BELLOWS**
RT　EXPULSION BLADDERS
　　JOINTS (JUNCTIONS)
　　PUMPS

BELLS
RT　AUDITORY SIGNALS
　　PRESSURE VESSELS
　　PSYCHOACOUSTICS
　　∞SIGNALS
　　SOUND GENERATORS
　　WARNING
　　WARNING SYSTEMS

BELTRAMI FLOW
GS　FLUID FLOW
　　. **BELTRAMI FLOW**
RT　INCOMPRESSIBLE FLOW
　　STEADY FLOW
　　VORTICITY

∞ **BELTS**
SN　*(USE OF A MORE SPECIFIC TERM IS
　　RECOMMENDED--CONSULT THE TERMS
　　LISTED BELOW)*
RT　ASTEROID BELTS
　　CABLES (ROPES)
　　FASTENERS
　　GIRDLES
　　PROTON BELTS
　　PULLEYS
　　RADIATION BELTS
　　REGIONS
　　ROUSE BELTS
　　SEAT BELTS
　　TERRESTRIAL DUST BELT

BENARD CELLS
GS　CONVECTION
　　. **BENARD CELLS**
　　FLUID FLOW
　　. CONVECTIVE FLOW
　　. . RAYLEIGH-BENARD CONVECTION
　　. . . **BENARD CELLS**
RT　CONVECTION CURRENTS
　　RAYLEIGH NUMBER
　　SOLAR GRANULATION

BENCHES
USE　SEATS

BEND TESTS
RT　BENDING
　　CRACK PROPAGATION
　　DESTRUCTIVE TESTS
　　FRACTURE MECHANICS
　　FRACTURE STRENGTH
　　∞MATERIALS TESTS
　　∞TESTS

BENDING
GS　**BENDING**
　　. ELASTIC BENDING
RT　BEND TESTS
　　∞BOWS
　　BUCKLING
　　CAMBER
　　DEFLECTION
　　DEFORMATION
　　DISPLACEMENT
　　DISTORTION
　　ELASTIC DEFORMATION
　　FATIGUE TESTS
　　FIBER STRENGTH
　　FLEXIBILITY
　　FLEXING
　　FLUTTER
　　FOLDING
　　HEAVING
　　MODULUS OF ELASTICITY
　　PLASTIC DEFORMATION
　　STIFFNESS
　　STRUCTURAL FAILURE
　　STRUCTURAL STRAIN
　　TEMPERATURE INVERSIONS
　　TWISTING
　　WARPAGE

BENDING DIAGRAMS
GS　DIAGRAMS
　　. **BENDING DIAGRAMS**
RT　DEFLECTION

BENDING FATIGUE
GS　FATIGUE (MATERIALS)
　　. **BENDING FATIGUE**
RT　METAL FATIGUE
　　S-N DIAGRAMS

BENDING MOMENTS
GS　MOMENTS
　　. **BENDING MOMENTS**
RT　LOADING MOMENTS
　　NASTRAN
　　STATIC LOADS
　　STRESS ANALYSIS
　　STRUCTURAL DESIGN CRITERIA
　　TORQUE

BENDING THEORY
RT　STRESS ANALYSIS
　　STRESS INTENSITY FACTORS
　　∞THEORIES

BENDING VIBRATION
GS　VIBRATION
　　. STRUCTURAL VIBRATION
　　. . **BENDING VIBRATION**
RT　BREATHING VIBRATION
　　FLUTTER
　　MISSILE VIBRATION
　　PANEL FLUTTER
　　RANDOM VIBRATION
　　SELF INDUCED VIBRATION

BENDS (PHYSIOLOGY)
USE　DECOMPRESSION SICKNESS

BENEFICIATION
RT　∞ABSORPTION
　　ADSORPTION
　　AERATION
　　CLEAN FUELS
　　COMMINUTION
　　CONCENTRATING
　　∞CONDITIONING
　　ENRICHMENT
　　EXPLOITATION
　　EXTRACTION
　　FILTRATION
　　FLOTATION
　　FOAMING
　　ISOTOPIC ENRICHMENT
　　LEACHING
　　∞METALLURGY
　　MINERALS
　　PURIFICATION
　　REFINING
　　∞SEPARATION
　　SETTLING
　　SIZE SEPARATION
　　SUBLIMATION
　　UPGRADING
　　WASHING
　　WASTES

BENIN
UF　DAHOMEY
GS　NATIONS
　　. **BENIN**
RT　AFRICA

BENTONITE
RT　MONTMORILLONITE
　　SOILS
　　WATER TREATMENT

BENZENE
GS　HYDROCARBONS
　　. **BENZENE**
RT　CHLOROBENZENES
　　CYCLOHEXANE
　　SOLVENT REFINED COAL

BENZENE POISONING
RT　HYDROCARBON POISONING
　　INDUSTRIAL SAFETY
　　∞POISONING
　　TOXICITY AND SAFETY HAZARD
　　TOXICOLOGY

BENZILIC ACID
GS　ACIDS
　　. FATTY ACIDS
　　. . CARBOXYLIC ACIDS
　　. . . **BENZILIC ACID**

BENZOIC ACID
GS　ACIDS
　　. FATTY ACIDS
　　. . CARBOXYLIC ACIDS
　　. . . **BENZOIC ACID**

BERENICE ROCKET VEHICLE
GS   ROCKET VEHICLES
    . MULTISTAGE ROCKET VEHICLES
    . . **BERENICE ROCKET VEHICLE**
RT   HYPERSONIC REENTRY
    SOLID PROPELLANT ROCKET ENGINES

BERGMAN OPERATOR
GS   OPERATORS (MATHEMATICS)
    . **BERGMAN OPERATOR**

BERING SEA
GS   SEAS
    . **BERING SEA**
RT   PACIFIC OCEAN

BERKELIUM
GS   CHEMICAL ELEMENTS
    . ACTINIDE SERIES
    . . TRANSURANIUM ELEMENTS
    . . . **BERKELIUM**
    . NUCLIDES
    . . ISOTOPES
    . . . RADIOACTIVE ISOTOPES
    . . . . TRANSURANIUM ELEMENTS
    . . . . . **BERKELIUM**
    HEAVY ELEMENTS
    . TRANSURANIUM ELEMENTS
    . . **BERKELIUM**
    METALS
    . ACTINIDE SERIES
    . . TRANSURANIUM ELEMENTS
    . . . **BERKELIUM**

BERMUDA
GS   LANDFORMS
    . ISLANDS
    . . **BERMUDA**
RT   ATLANTIC OCEAN

BERNOULLI EQUATION
USE   BERNOULLI THEOREM

BERNOULLI THEOREM
UF   BERNOULLI EQUATION
GS   THEOREMS
    . **BERNOULLI THEOREM**
RT   ∞EQUATIONS
    FLUID FLOW
    ISENTROPIC PROCESSES
    LINEARIZATION
    MAGNUS EFFECT
    PANEL METHOD (FLUID DYNAMICS)

BERNSTEIN ENERGY PRINCIPLE
GS   STRUCTURAL ANALYSIS
    . ENERGY METHODS
    . . **BERNSTEIN ENERGY PRINCIPLE**
RT   ∞ENERGY
    MAGNETIC FIELDS

BERYL
UF   EMERALD
GS   ALUMINUM COMPOUNDS
    . **BERYL**
    BERYLLIUM COMPOUNDS
    . **BERYL**
    MINERALS
    . **BERYL**
    SILICON COMPOUNDS
    . SILICATES
    . . **BERYL**
RT   BERYLLIUM

BERYLLIUM
GS   CHEMICAL ELEMENTS
    . **BERYLLIUM**
    . . BERYLLIUM ISOTOPES
    . . . BERYLLIUM 7
    . . . BERYLLIUM 9
    . . . BERYLLIUM 10
    METALS
    . **BERYLLIUM**
    . . BERYLLIUM ISOTOPES
    . . . BERYLLIUM 7
    . . . BERYLLIUM 9
    . . . BERYLLIUM 10
RT   BERYL
    MODERATORS

BERYLLIUM ALLOYS
GS   ALLOYS
    . LIGHT ALLOYS
    . . **BERYLLIUM ALLOYS**

BERYLLIUM BOROHYDRIDES
GS   BERYLLIUM COMPOUNDS
    . **BERYLLIUM BOROHYDRIDES**
    BORON COMPOUNDS
    . BOROHYDRIDES
    . . **BERYLLIUM BOROHYDRIDES**
    . BORON HYDRIDES
    . . **BERYLLIUM BOROHYDRIDES**
    HYDROGEN COMPOUNDS
    . HYDRIDES
    . . BOROHYDRIDES
    . . . **BERYLLIUM BOROHYDRIDES**
    . . BORON HYDRIDES
    . . . **BERYLLIUM BOROHYDRIDES**

BERYLLIUM CHLORIDES
GS   BERYLLIUM COMPOUNDS
    . **BERYLLIUM CHLORIDES**
    HALOGEN COMPOUNDS
    . CHLORINE COMPOUNDS
    . . CHLORIDES
    . . . **BERYLLIUM CHLORIDES**
    . HALIDES
    . . CHLORIDES
    . . . **BERYLLIUM CHLORIDES**
    . . METAL HALIDES
    . . . **BERYLLIUM CHLORIDES**

BERYLLIUM COMPOUNDS
GS   **BERYLLIUM COMPOUNDS**
    . BERYL
    . BERYLLIUM BOROHYDRIDES
    . BERYLLIUM CHLORIDES
    . BERYLLIUM FLUORIDES
    . BERYLLIUM HYDRIDES
    . BERYLLIUM NITRIDES
    . BERYLLIUM OXIDES
RT   ∞ALKALINE EARTH COMPOUNDS
    ∞CHEMICAL COMPOUNDS
    ∞METAL COMPOUNDS
    METAL FUELS
    METAL PROPELLANTS

BERYLLIUM FLUORIDES
GS   BERYLLIUM COMPOUNDS
    . **BERYLLIUM FLUORIDES**
    HALOGEN COMPOUNDS
    . FLUORINE COMPOUNDS
    . . FLUORIDES
    . . . METAL FLUORIDES
    . . . . **BERYLLIUM FLUORIDES**
    . HALIDES
    . . METAL HALIDES
    . . . **BERYLLIUM FLUORIDES**

BERYLLIUM HYDRIDES
GS   BERYLLIUM COMPOUNDS
    . **BERYLLIUM HYDRIDES**
    HYDROGEN COMPOUNDS
    . HYDRIDES
    . . METAL HYDRIDES
    . . . **BERYLLIUM HYDRIDES**

BERYLLIUM ISOTOPES
GS   CHEMICAL ELEMENTS
    . BERYLLIUM
    . . **BERYLLIUM ISOTOPES**
    . . . BERYLLIUM 7
    . . . BERYLLIUM 9
    . . . BERYLLIUM 10
    . NUCLIDES
    . . ISOTOPES
    . . . **BERYLLIUM ISOTOPES**
    . . . . BERYLLIUM 7
    . . . . BERYLLIUM 9
    . . . . BERYLLIUM 10
    METALS
    . BERYLLIUM
    . . **BERYLLIUM ISOTOPES**
    . . . BERYLLIUM 7
    . . . BERYLLIUM 9
    . . . BERYLLIUM 10

BERYLLIUM NITRIDES
GS   BERYLLIUM COMPOUNDS
    . **BERYLLIUM NITRIDES**
    NITROGEN COMPOUNDS
    . NITRIDES
    . . **BERYLLIUM NITRIDES**
RT   METAL NITRIDES

BERYLLIUM OXIDES
GS   BERYLLIUM COMPOUNDS
    . **BERYLLIUM OXIDES**
    CHALCOGENIDES

BERYLLIUM OXIDES-*(CONT.)*
    . OXIDES
    . . METAL OXIDES
    . . . ALKALINE EARTH OXIDES
    . . . . **BERYLLIUM OXIDES**

BERYLLIUM POISONING
RT   INDUSTRIAL SAFETY
    ∞POISONING
    RESPIRATORY DISEASES
    TOXICITY AND SAFETY HAZARD
    TOXICOLOGY

BERYLLIUM 7
GS   CHEMICAL ELEMENTS
    . BERYLLIUM
    . . BERYLLIUM ISOTOPES
    . . . **BERYLLIUM 7**
    . NUCLIDES
    . ISOTOPES
    . . . BERYLLIUM ISOTOPES
    . . . . **BERYLLIUM 7**
    . . . RADIOACTIVE ISOTOPES
    . . . . **BERYLLIUM 7**
    METALS
    . BERYLLIUM
    . . BERYLLIUM ISOTOPES
    . . . **BERYLLIUM 7**

BERYLLIUM 9
GS   CHEMICAL ELEMENTS
    . BERYLLIUM
    . . BERYLLIUM ISOTOPES
    . . . **BERYLLIUM 9**
    . NUCLIDES
    . . ISOTOPES
    . . . BERYLLIUM ISOTOPES
    . . . . **BERYLLIUM 9**
    . . . RADIOACTIVE ISOTOPES
    . . . . **BERYLLIUM 9**
    METALS
    . BERYLLIUM
    . . BERYLLIUM ISOTOPES
    . . . **BERYLLIUM 9**

BERYLLIUM 10
GS   CHEMICAL ELEMENTS
    . BERYLLIUM
    . . BERYLLIUM ISOTOPES
    . . . **BERYLLIUM 10**
    . NUCLIDES
    . . ISOTOPES
    . . . BERYLLIUM ISOTOPES
    . . . . **BERYLLIUM 10**
    . . . RADIOACTIVE ISOTOPES
    . . . . **BERYLLIUM 10**
    METALS
    . BERYLLIUM
    . . BERYLLIUM ISOTOPES
    . . . **BERYLLIUM 10**

BESS (SATELLITE)
UF   BIOMEDICAL EXPERIMENT SCIENTIFIC
    SATELLITE
GS   SATELLITES
    . ARTIFICIAL SATELLITES
    . . **BESS (SATELLITE)**
RT   MULTIMISSION MODULAR SPACECRAFT
    SPACE SHUTTLES

BESSEL FUNCTIONS
GS   ANALYSIS (MATHEMATICS)
    . COMPLEX VARIABLES
    . . **BESSEL FUNCTIONS**
    . . . HANKEL FUNCTIONS
    . REAL VARIABLES
    . . **BESSEL FUNCTIONS**
    . . . HANKEL FUNCTIONS
RT   BOUNDARY VALUE PROBLEMS
    DIFFERENTIAL EQUATIONS
    HYPERGEOMETRIC FUNCTIONS
    ORTHOGONAL FUNCTIONS
    POWER SERIES

BESSEL-BREDICHIN THEORY
RT   COMETS
    KOHOUTEK COMET
    RADIATION PRESSURE
    ∞THEORIES

BETA FACTOR
RT   DENSE PLASMAS
    FLUID PRESSURE
    FUSION REACTORS
    MAGNETIC FIELDS

BETA FACTOR-(CONT.)
    MAGNETIC FLUX
    MAGNETOHYDRODYNAMIC STABILITY
    PLASMA CONTROL
    PLASMA EQUILIBRIUM
    PLASMA HEATING
    PLASMA PHYSICS
    PRESSURE EFFECTS
    REACTOR PHYSICS
    TOKAMAK DEVICES
    TOROIDAL PLASMAS

BETA INTERACTIONS
USE   WEAK INTERACTIONS (FIELD THEORY)

BETA PARTICLES
GS   IONIZING RADIATION
    . **BETA PARTICLES**
    NUCLEAR RADIATION
    . **BETA PARTICLES**
    PARTICLES
    . CHARGED PARTICLES
    . . ENERGETIC PARTICLES
    . . . PLASMAS (PHYSICS)
    . . . . **BETA PARTICLES**
    . CORPUSCULAR RADIATION
    . . ELECTRON RADIATION
    . . . **BETA PARTICLES**
    . ELEMENTARY PARTICLES
    . . **BETA PARTICLES**
    . NUCLEAR PARTICLES
    . . **BETA PARTICLES**
RT   ELECTRON BEAMS
    ELECTRONS
    FLUX (RATE)
    HOT ATOMS
    N ELECTRONS
    RELATIVISTIC ELECTRON BEAMS
    WEAK ENERGY INTERACTIONS

BETAINES
GS   HETEROCYCLIC COMPOUNDS
    . ALKALOIDS
    . . **BETAINES**
    NITROGEN COMPOUNDS
    . ALKALOIDS
    . . **BETAINES**

BETATRONS
GS   PARTICLE ACCELERATORS
    . CYCLIC ACCELERATORS
    . . **BETATRONS**
    . ELECTRON ACCELERATORS
    . . **BETATRONS**
RT   MICROTRONS
    SYNCHROTRONS

BETHE-HEITLER FORMULA
GS   MATHEMATICAL LOGIC
    . FORMULAS (MATHEMATICS)
    . . **BETHE-HEITLER FORMULA**

BETHE-SALPETER EQUATION
GS   ANALYSIS (MATHEMATICS)
    . REAL VARIABLES
    . . **BETHE-SALPETER EQUATION**
RT   DIFFERENTIAL EQUATIONS
    ∞EQUATIONS
    EQUATIONS OF MOTION
    KINETIC EQUATIONS

BEVATRON
GS   PARTICLE ACCELERATORS
    . CYCLIC ACCELERATORS
    . . SYNCHROTRONS
    . . . **BEVATRON**
RT   SYNCHROCYCLOTRONS

BEVERAGES
GS   LIQUIDS
    . POTABLE LIQUIDS
    . . **BEVERAGES**
    . . . WINES
RT   COFFEE
    DRINKING
    ∞FOOD
    MILK

BHUTAN
RT   HIMALAYAS
    INDIA
    SIKKIM
    TIBET

BIAS
GS   **BIAS**
    . RESPONSE BIAS
RT   COMPENSATORS
    DISPLACEMENT
    ELECTRIC POTENTIAL
    ERRORS
    INSTRUMENT ERRORS
    OPEN CIRCUIT VOLTAGE
    TUBE GRIDS

BIBLIOGRAPHIES
GS   DOCUMENTS
    . **BIBLIOGRAPHIES**
RT   ABSTRACTS
    BIOGRAPHY
    DOCUMENTATION
    HANDBOOKS
    INDEXES (DOCUMENTATION)
    INFORMATION DISSEMINATION
    INFORMATION RETRIEVAL
    LIBRARIES
    LITERATURE
    ∞REFERENCE SYSTEMS
    SPACE GLOSSARIES
    SUMMARIES

BICARBONATES
USE   CARBONATES

BICRYSTALS
GS   CRYSTALS
    . **BICRYSTALS**
RT   POLYCRYSTALS
    SINGLE CRYSTALS

∞ BICYCLE
SN   *(USE OF A MORE SPECIFIC TERM IS RECOMMENDED--CONSULT THE TERMS LISTED BELOW)*
RT   LANDING GEAR
    SURFACE VEHICLES

BIFURCATION (BIOLOGY)
RT   ANATOMY
    ARTERIES
    ∞BIOLOGY
    BLOOD VESSELS
    BRANCHING (PHYSICS)
    VEINS

BIFURCATION (MATHEMATICS)
USE   BRANCHING (MATHEMATICS)

BIG BANG COSMOLOGY
GS   COSMOLOGY
    . **BIG BANG COSMOLOGY**
RT   ASTRONOMICAL MODELS
    BACKGROUND RADIATION
    COSMIC RAYS
    GALACTIC EVOLUTION
    GAMMA RAY BURSTS
    GRAVITATIONAL CONSTANT
    RELATIVITY
    RELIC RADIATION
    UNIVERSE

BIG SHOT PROJECT
GS   PROGRAMS
    . PROJECTS
    . . **BIG SHOT PROJECT**

BIGHORN MOUNTAINS (MT-WY)
GS   LANDFORMS
    . MOUNTAINS
    . . **BIGHORN MOUNTAINS (MT-WY)**
RT   MONTANA
    WYOMING

BIGHTS
USE   BAYS (TOPOGRAPHIC FEATURES)

BIHARMONIC EQUATIONS
GS   ANALYSIS (MATHEMATICS)
    . REAL VARIABLES
    . . DIFFERENTIAL EQUATIONS
    . . . PARTIAL DIFFERENTIAL EQUATIONS
    . . . . **BIHARMONIC EQUATIONS**
RT   ELASTIC PROPERTIES
    ∞EQUATIONS

BILLETS
RT   CASTING
    CASTINGS

BILLETS-(CONT.)
    FORGING
    INGOTS
    METAL PLATES
    METAL STRIPS
    RODS
    SLABS
    WIRE

BIMETALS
RT   ALLOYS
    COMPOSITE MATERIALS
    METAL BONDING
    METALS

BIMETRIC THEORIES
RT   GRAVITATION THEORY
    METRIC SPACE
    SCHWARZSCHILD METRIC
    ∞THEORIES

BINARY ALLOYS
GS   ALLOYS
    . **BINARY ALLOYS**
    BINARY SYSTEMS (MATERIALS)
    . **BINARY ALLOYS**

BINARY CODES
RT   BCH CODES
    BIT ERROR RATE
    ∞CODES
    CONCATENATED CODES
    DIGITAL SYSTEMS

BINARY DATA
RT   ANALOG DATA
    BIT ERROR RATE
    BUBBLE MEMORY DEVICES
    ∞DATA
    DATA PROCESSING
    DECIMAL TO BINARY CONVERTERS
    DIGITAL DATA

BINARY DIGITS
GS   ALPHANUMERIC CHARACTERS
    . DIGITS
    . . **BINARY DIGITS**
RT   BIT ERROR RATE
    BITS
    DIGITAL SYSTEMS

BINARY FLUIDS
GS   BINARY SYSTEMS (MATERIALS)
    . BINARY MIXTURES
    . . **BINARY FLUIDS**
    MIXTURES
    . BINARY MIXTURES
    . . **BINARY FLUIDS**
RT   ∞FLUIDS
    GAS MIXTURES
    KINETIC THEORY
    LENNARD-JONES GAS
    TRANSPORT PROPERTIES

BINARY INTEGRATION
GS   ANALYSIS (MATHEMATICS)
    . REAL VARIABLES
    . . MEASURE AND INTEGRATION
    . . . **BINARY INTEGRATION**
RT   ADDING CIRCUITS
    DIGITAL INTEGRATORS

BINARY MIXTURES
GS   BINARY SYSTEMS (MATERIALS)
    . **BINARY MIXTURES**
    . . BINARY FLUIDS
    . . EUTECTICS
    . . . EUTECTIC ALLOYS
    MIXTURES
    . **BINARY MIXTURES**
    . . BINARY FLUIDS
    . . EUTECTICS
    . . . EUTECTIC ALLOYS
RT   AZEOTROPES
    GAS MIXTURES
    LIQUID-GAS MIXTURES

BINARY STARS
GS   CELESTIAL BODIES
    . STARS
    . . DOUBLE STARS
    . . . **BINARY STARS**
    . . . . COMPANION STARS
    . . . . ECLIPSING BINARY STARS

BINARY STARS-*(CONT.)*
```
        . . . . DWARF NOVAE
RT      ACCRETION DISKS
        LIMB DARKENING
        STAR CLUSTERS
        STELLAR PARALLAX
        TWO BODY PROBLEM
        VARIABLE STARS
```

BINARY SUMMATORS
```
USE     ADDING CIRCUITS
```

BINARY SYSTEMS (DIGITAL)
```
USE     DIGITAL SYSTEMS
```

BINARY SYSTEMS (MATERIALS)
```
UF      TWO PHASE SYSTEMS
GS      BINARY SYSTEMS (MATERIALS)
        . BINARY ALLOYS
        . BINARY MIXTURES
        . . BINARY FLUIDS
        . . EUTECTICS
        . . . EUTECTIC ALLOYS
RT      ALLOYS
        ∞MATERIALS
        PHASE DIAGRAMS
        SOLIDUS
        ∞SYSTEMS
        TERNARY SYSTEMS
```

BINARY TO DECIMAL CONVERTERS
```
GS      DATA CONVERTERS
        . BINARY TO DECIMAL CONVERTERS
RT      COMPUTER COMPONENTS
        ∞CONVERTERS
        DATA PROCESSING
        DECIMAL TO BINARY CONVERTERS
```

BINAURAL HEARING
```
GS      HEARING
        . BINAURAL HEARING
        PERCEPTION
        . BINAURAL HEARING
RT      AUDITORY PERCEPTION
        SOUND LOCALIZATION
        WEBER TEST
```

BINDERS (ADHESIVES)
```
USE     ADHESIVES
```

BINDERS (MATERIALS)
```
GS      BINDERS (MATERIALS)
        . PROPELLANT BINDERS
        . . SOLID ROCKET BINDERS
RT      ADDITIVES
        ADHESIVES
        CEMENTS
        ∞MATERIALS
        MOLDING MATERIALS
        SIZING MATERIALS
        SOLID LUBRICANTS
```

BINDING
```
RT      BONDING
        COLLATING
        FOLDING
        ∞JOINING
        PRINTING
        SEALING
        SEWING
```

BINOCULAR VISION
```
GS      VISION
        . BINOCULAR VISION
RT      HAPLOSCOPES
        MOTION PERCEPTION
        SPACE PERCEPTION
        STEREOSCOPIC VISION
```

BINOCULARS
```
GS      OPTICAL EQUIPMENT
        . BINOCULARS
RT      EYEPIECES
        MICROSCOPES
        PERISCOPES
        TELESCOPES
```

BINOMIAL COEFFICIENTS
```
GS      COEFFICIENTS
        . BINOMIAL COEFFICIENTS
RT      FACTORIALS
```

BINOMIAL THEOREM
```
GS      ALGEBRA
```

BINOMIAL THEOREM-*(CONT.)*
```
        . BINOMIAL THEOREM
        THEOREMS
        . BINOMIAL THEOREM
RT      BINOMIALS
        PROBABILITY DENSITY FUNCTIONS
        PROBABILITY THEORY
        STATISTICAL ANALYSIS
        STATISTICAL DISTRIBUTIONS
```

BINOMIALS
```
GS      ALGEBRA
        . POLYNOMIALS
        . . BINOMIALS
RT      BINOMIAL THEOREM
```

BIOACOUSTICS
```
GS      ACOUSTICS
        . BIOACOUSTICS
RT      ACOUSTIC ATTENUATION
        AUDITORY DEFECTS
        AUDITORY SENSATION AREAS
        BIOENGINEERING
        ∞BIOLOGY
        PSYCHOACOUSTICS
        ∞SCIENCE
        SOUND INTENSITY
```

BIOASSAY
```
UF      BIOLOGICAL ANALYSIS
RT      BIOCHEMISTRY
        BIOLOGICAL EFFECTS
        ∞BIOLOGY
        HISTOCHEMICAL ANALYSIS
```

BIOASTRONAUTICAL ORBITAL SPACE SYSTEM
```
GS      PROGRAMS
        . NASA PROGRAMS
        . . BIOASTRONAUTICAL ORBITAL
              SPACE SYSTEM
RT      ∞SYSTEMS
```

BIOASTRONAUTICS
```
SN      (BIOLOGICAL, BEHAVIORAL AND
        MEDICAL ASPECTS EXPECTED TO BE
        FOUND IN SPACE VEHICLES DESIGNED
        TO TRAVEL IN SPACE AND ON
        CELESTIAL BODIES OTHER THAN ON
        EARTH)
RT      AEROSPACE ENVIRONMENTS
        AEROSPACE MEDICINE
        ∞ASTRONAUTICS
        BIOENGINEERING
        ∞BIOLOGY
        BIOSATELLITE 1
        BIOSATELLITE 2
        BIOSATELLITE 3
        CLOSED ECOLOGICAL SYSTEMS
        EARTH ATMOSPHERE
        EXOBIOLOGY
        LUNAR ENVIRONMENT
        PLANETARY ENVIRONMENTS
        ∞SCIENCE
        SPACE ADAPTATION SYNDROME
        SPACE EXPLORATION
        SPACE FLIGHT
        SPACE STATIONS
        SPACECRAFT ENVIRONMENTS
```

BIOCHEMICAL FUEL CELLS
```
GS      ELECTRIC GENERATORS
        . DIRECT POWER GENERATORS
        . . FUEL CELLS
        . . . BIOCHEMICAL FUEL CELLS
        ELECTROCHEMICAL CELLS
        . FUEL CELLS
        . . BIOCHEMICAL FUEL CELLS
RT      ∞BIOLOGY
        PHOSPHORIC ACID FUEL CELLS
        REGENERATIVE FUEL CELLS
```

BIOCHEMICAL OXYGEN DEMAND
```
UF      BOD
RT      ALGAE
        ∞BIOLOGY
        ECOLOGY
        OXIMETRY
        OXYGEN CONSUMPTION
        PLANTS (BOTANY)
        POLLUTION CONTROL
        WATER POLLUTION
        WATER TREATMENT
```

BIOCHEMISTRY
```
GS      ENVIRONMENTAL CHEMISTRY
```

BIOCHEMISTRY-*(CONT.)*
```
        . BIOCHEMISTRY
        . . BIOGEOCHEMISTRY
        . . ENZYMOLOGY
        . . PHYSIOCHEMISTRY
RT      BIOASSAY
        BIODEGRADATION
        BIOENGINEERING
        ∞BIOLOGY
        CHEMICAL WARFARE
        ∞CHEMISTRY
        CYTOLOGY
        GENETIC ENGINEERING
        HISTOCHEMICAL ANALYSIS
        IMMUNOASSAY
        INTERFERON
        MARINE CHEMISTRY
        METABOLITES
        MOLECULAR BIOLOGY
        MUTAGENS
        NITROGEN METABOLISM
        NUTRITION
        OPTICAL ACTIVITY
        ORGANIC CHEMISTRY
        RADIOIMMUNOASSAY
        VEGETATION GROWTH
```

BIOCLIMATOLOGY
```
USE     BIOMETEOROLOGY
```

BIOCOMPATIBILITY
```
GS      COMPATIBILITY
        . BIOCOMPATIBILITY
RT      ANTIBODIES
        ANTIGENS
        ∞BIOLOGY
        BLOOD
        IMMUNOLOGY
        LEUKOCYTES
        PHYSIOLOGICAL DEFENSES
        VACCINES
```

BIOCONTROL SYSTEMS
```
RT      BIOFEEDBACK
        ∞BIOLOGY
        BIONICS
        PSYCHOMOTOR PERFORMANCE
        ∞SYSTEMS
        TOLERANCES (PHYSIOLOGY)
```

BIOCONVERSION
```
RT      ALGAE
        ∞BIOLOGY
        BIOMASS ENERGY PRODUCTION
        BIOPROCESSING
        ∞CONVERSION
        ENZYME ACTIVITY
        FERMENTATION
        FUELS
        HYDROCARBON FUEL PRODUCTION
        METHANE
        SOLAR HEATING
        VEGETATION
```

BIODEGRADABILITY
```
GS      DISSOCIATION
        . BIODEGRADABILITY
RT      ∞BIOLOGY
        DECAY
        DECOMPOSITION
        DETERIORATION
        ORGANIC MATERIALS
        ∞PROPERTIES
```

BIODEGRADATION
```
GS      DEGRADATION
        . BIODEGRADATION
RT      ACTIVATED SLUDGE
        BIOCHEMISTRY
        ∞BIOLOGY
        DECAY
        DECOMPOSITION
        DETERIORATION
```

BIODYNAMICS
```
UF      BIOMECHANICS
RT      ANATOMY
        BIOENGINEERING
        BIOLOGICAL MODELS (MATHEMATICS)
        ∞BIOLOGY
        BIOPHYSICS
        ∞DYNAMICS
        ∞SCIENCE
        STRESS (PHYSIOLOGY)
```

BIOELECTRIC POTENTIAL
GS   POTENTIAL ENERGY
. ELECTRIC POTENTIAL
. . **BIOELECTRIC POTENTIAL**
RT   BIOELECTRICITY
∞BIOLOGY

BIOELECTRICITY
UF   NEURON TRANSMISSION
RT   BIOELECTRIC POTENTIAL
∞BIOLOGY
BIOMAGNETISM
NEUROMUSCULAR TRANSMISSION
SPIKE POTENTIALS

BIOENGINEERING
GS   **BIOENGINEERING**
. BIOINSTRUMENTATION
. . BIOTELEMETRY
. BIOMETRICS
. . BODY MEASUREMENT (BIOLOGY)
. . . ANTHROPOMETRY
. . . ELECTROPLETHYSMOGRAPHY
. . CARDIOGRAPHY
. . . BALLISTOCARDIOGRAPHY
. . . ELECTROCARDIOGRAPHY
. . . MAGNETOCARDIOGRAPHY
. . . PHONOCARDIOGRAPHY
. . . . ECHOCARDIOGRAPHY
. . . SEISMOCARDIOGRAPHY
. . . VECTORCARDIOGRAPHY
. . ECHOENCEPHALOGRAPHY
. . ELECTROENCEPHALOGRAPHY
. . ELECTROMYOGRAPHY
. . ELECTRORETINOGRAPHY
. . PLETHYSMOGRAPHY
. . . ELECTROPLETHYSMOGRAPHY
. . RADIOCARDIOGRAPHY
RT   BIOACOUSTICS
BIOASTRONAUTICS
BIOCHEMISTRY
BIODYNAMICS
∞BIOLOGY
BIONICS
BIOPAKS
BIOPHYSICS
BONE MINERAL CONTENT
∞ENGINEERING
GENETIC ENGINEERING
HUMAN FACTORS ENGINEERING
UNDERWATER BREATHING APPARATUS
VOICE CONTROL

BIOFEEDBACK
GS   FEEDBACK
. **BIOFEEDBACK**
. . SENSORY FEEDBACK
RT   AEROSPACE MEDICINE
BIOCONTROL SYSTEMS
BLOOD PRESSURE
CONDITIONING (LEARNING)
FEEDBACK CONTROL
HEART RATE
HUMAN FACTORS ENGINEERING
PSYCHOLOGY
SENSORY FEEDBACK

BIOFLAVONOIDS
UF   VITAMIN P
GS   HETEROCYCLIC COMPOUNDS
. **BIOFLAVONOIDS**
VITAMINS
. **BIOFLAVONOIDS**
RT   DRUGS

BIOGENESIS
USE   BIOLOGICAL EVOLUTION

BIOGENY
RT   ∞BIOLOGY
∞EVOLUTION
ONTOGENY

BIOGEOCHEMISTRY
GS   ENVIRONMENTAL CHEMISTRY
. BIOCHEMISTRY
. . **BIOGEOCHEMISTRY**
. GEOCHEMISTRY
. . **BIOGEOCHEMISTRY**
RT   ∞BIOLOGY
BOTANY
∞CHEMISTRY
INTERNATIONAL
GEOSPHERE-BIOSPHERE PROGRAM
MINERALS

BIOGEOCHEMISTRY-*(CONT.)*
PLANTS (BOTANY)

BIOGRAPHY
GS   LITERATURE
. **BIOGRAPHY**
RT   AWARDS
BIBLIOGRAPHIES
CASE HISTORIES
DOCUMENTATION

BIOINSTRUMENTATION
UF   BIOSENSORS
GS   BIOENGINEERING
. **BIOINSTRUMENTATION**
. . BIOTELEMETRY
RT   ∞BIOLOGY
BIOMETRICS
BIONICS
ECHOENCEPHALOGRAPHY
∞ENGINEERING
IMBLMS
∞INSTRUMENTS
MAGNETOCARDIOGRAPHY
MEASURING INSTRUMENTS
RESPIROMETERS
∞SENSORS
SPHYGMOGRAPHY
WILDLIFE RADIOLOCATION

BIOLOGICAL ACTIVITY
USE   ACTIVITY (BIOLOGY)

BIOLOGICAL ANALYSIS
USE   BIOASSAY

BIOLOGICAL CELLS
USE   CELLS (BIOLOGY)

BIOLOGICAL CLOCKS
USE   RHYTHM (BIOLOGY)

BIOLOGICAL EFFECTS
GS   **BIOLOGICAL EFFECTS**
. DESYNCHRONIZATION (BIOLOGY)
. JET LAG
. RELATIVE BIOLOGICAL
EFFECTIVENESS (RBE)
RT   ACTIVITY (BIOLOGY)
ATROPHY
BIOASSAY
∞BIOLOGY
BIOMEDICAL DATA
BONE DEMINERALIZATION
BRAGG CURVE
CHEMICAL EFFECTS
DISORIENTATION
DOSAGE
∞EFFECTS
FLIGHT STRESS (BIOLOGY)
HUMAN REACTIONS
ORBITING FROG OTOLITH
PATHOLOGICAL EFFECTS
PHYSIOLOGICAL EFFECTS
PSYCHOLOGICAL EFFECTS
RADIATION DOSAGE
RADIATION EFFECTS
SPACE ADAPTATION SYNDROME
TEMPERATURE
THERMAL POLLUTION

BIOLOGICAL EVOLUTION
UF   BIOGENESIS
GS   EVOLUTION (DEVELOPMENT)
. **BIOLOGICAL EVOLUTION**
. . ABIOGENESIS
RT   ∞BIOLOGY
CHEMICAL EVOLUTION
GENETICS
LIFE SCIENCES
MUTAGENS
MUTATIONS
PANSPERMIA
PROTEIN SYNTHESIS

BIOLOGICAL MODELS
USE   BIONICS

BIOLOGICAL MODELS (MATHEMATICS)
GS   MODELS
. MATHEMATICAL MODELS
. . **BIOLOGICAL MODELS
(MATHEMATICS)**
RT   BIODYNAMICS

BIOLOGICAL MODELS (MATHEMATICS)-*(CONT.)*
∞BIOLOGY
BIONICS
DIGITAL SIMULATION
DYNAMIC MODELS

BIOLOGICAL RHYTHM
USE   RHYTHM (BIOLOGY)

∞ BIOLOGY
SN   *(USE OF A MORE SPECIFIC TERM IS
RECOMMENDED--CONSULT THE TERMS
LISTED BELOW)*
RT   ACTIVATION (BIOLOGY)
ACTIVITY (BIOLOGY)
ACTIVITY CYCLES (BIOLOGY)
AEROBIOLOGY
AEROSPACE MEDICINE
AGING (BIOLOGY)
AGRICULTURE
ANATOMY
ANIMALS
BACTERIOLOGY
BIFURCATION (BIOLOGY)
BIOACOUSTICS
BIOASSAY
BIOASTRONAUTICS
BIOCHEMICAL FUEL CELLS
BIOCHEMICAL OXYGEN DEMAND
BIOCHEMISTRY
BIOCOMPATIBILITY
BIOCONTROL SYSTEMS
BIOCONVERSION
BIODEGRADABILITY
BIODEGRADATION
BIODYNAMICS
BIOELECTRIC POTENTIAL
BIOELECTRICITY
BIOENGINEERING
BIOGENY
BIOGEOCHEMISTRY
BIOINSTRUMENTATION
BIOLOGICAL EFFECTS
BIOLOGICAL EVOLUTION
BIOLOGICAL MODELS (MATHEMATICS)
BIOLUMINESCENCE
BIOMAGNETISM
BIOMASS
BIOMASS ENERGY PRODUCTION
BIOMEDICAL DATA
BIOMETEOROLOGY
BIOMETRICS
BIONICS
BIOPHYSICS
BIOREACTORS
BIOSATELLITES
BIOSPHERE
BIOSYNTHESIS
BIOTECHNOLOGY
BIOTELEMETRY
BODY COMPOSITION (BIOLOGY)
BODY MEASUREMENT (BIOLOGY)
BODY SIZE (BIOLOGY)
BODY VOLUME (BIOLOGY)
BONE DEMINERALIZATION
BONE MINERAL CONTENT
BOTANY
CARBON CYCLE
CELLS (BIOLOGY)
COMPLEMENT (BIOLOGY)
CYTOGENESIS
CYTOLOGY
DIFFERENTIATION (BIOLOGY)
ECOLOGY
EMBRYOLOGY
EVOLUTION (DEVELOPMENT)
EXOBIOLOGY
FATIGUE (BIOLOGY)
FLIGHT STRESS (BIOLOGY)
GENETIC ENGINEERING
GENETICS
HABITATS
IMMUNOLOGY
IMPLANTED ELECTRODES (BIOLOGY)
INTERFERON
LIFE SCIENCES
MARINE BIOLOGY
MEDICAL SCIENCE
MICROBIOLOGY
MOLECULAR BIOLOGY
MORPHOLOGY
NITROGEN METABOLISM
PALEOBIOLOGY
PROTOBIOLOGY
RADIOBIOLOGY

BIOLOGY-(CONT.)
  RELATIVE BIOLOGICAL EFFECTIVENESS
   (RBE)
  REPRODUCTION (BIOLOGY)
  RHYTHM (BIOLOGY)
  ∞SCIENCE
  SKIN TEMPERATURE (BIOLOGY)
  ∞STRESS (BIOLOGY)
  TISSUES (BIOLOGY)
  VETERINARY MEDICINE

BIOLUMINESCENCE
 GS DECAY
  . EMISSION
  . . LIGHT EMISSION
  . . . LUMINESCENCE
  . . . . **BIOLUMINESCENCE**
 RT ∞BIOLOGY
  PHOSPHORESCENCE

BIOMAGNETISM
 GS MAGNETIC FIELDS
  . **BIOMAGNETISM**
  MAGNETIC PROPERTIES
  . **BIOMAGNETISM**
 RT BIOELECTRICITY
  ∞BIOLOGY
  BIOPHYSICS
  ELECTROMAGNETIC FIELDS
  ELECTROMAGNETIC INTERACTIONS
  RADIOBIOLOGY

BIOMASS
 GS WEIGHT (MASS)
  . **BIOMASS**
 RT ANIMALS
  ∞BIOLOGY
  CARBON CYCLE
  ∞DENSITY
  ORGANISMS
  PLANTS (BOTANY)
  POPULATIONS
  SILVICULTURE
  ∞WEIGHT

BIOMASS ENERGY PRODUCTION
 GS ENERGY CONVERSION
  . **BIOMASS ENERGY PRODUCTION**
 RT BIOCONVERSION
  ∞BIOLOGY
  BIOREACTORS
  ∞CROPS
  ∞ENERGY SOURCES
  ENERGY TECHNOLOGY
  HYDROCARBON FUEL PRODUCTION
  MANURES
  METHANATION
  VEGETATION
  WASTE UTILIZATION

BIOMECHANICS
 USE BIODYNAMICS

BIOMEDICAL DATA
 RT AEROSPACE MEDICINE
  BIOLOGICAL EFFECTS
  ∞BIOLOGY
  BIOMETRICS
  BODY MEASUREMENT (BIOLOGY)
  CARDIOGRAMS
  ∞DATA
  HEART RATE
  IMBLMS

**BIOMEDICAL EXPERIMENT SCIENTIFIC
SATELLITE**
 USE BESS (SATELLITE)

BIOMETEOROLOGY
 UF BIOCLIMATOLOGY
 GS METEOROLOGY
  . **BIOMETEOROLOGY**
 RT ∞BIOLOGY
  COASTAL ECOLOGY
  COASTAL PLAINS
  ECOLOGY
  MICROCLIMATOLOGY
  NIGHTGLOW
  PHENOLOGY

BIOMETRICS
 GS BIOENGINEERING
  . **BIOMETRICS**
  . . BODY MEASUREMENT (BIOLOGY)

BIOMETRICS-(CONT.)
  . . . ANTHROPOMETRY
  . . . ELECTROPLETHYSMOGRAPHY
  . . CARDIOGRAPHY
  . . . BALLISTOCARDIOGRAPHY
  . . . ELECTROCARDIOGRAPHY
  . . . MAGNETOCARDIOGRAPHY
  . . . PHONOCARDIOGRAPHY
  . . . . ECHOCARDIOGRAPHY
  . . . SEISMOCARDIOGRAPHY
  . . . VECTORCARDIOGRAPHY
  . . ECHOENCEPHALOGRAPHY
  . . ELECTROENCEPHALOGRAPHY
  . . ELECTROMYOGRAPHY
  . . ELECTRORETINOGRAPHY
  . . PLETHYSMOGRAPHY
  . . . ELECTROPLETHYSMOGRAPHY
  . . RADIOCARDIOGRAPHY
 RT BIOINSTRUMENTATION
  ∞BIOLOGY
  BIOMEDICAL DATA
  BONE MINERAL CONTENT
  ∞ENGINEERING
  ORBITING FROG OTOLITH
  PUPILLOMETRY
  STATISTICAL ANALYSIS
  ∞STATISTICS

BIONICS
 UF BIOLOGICAL MODELS
  BIOSIMULATION
 RT ARTIFICIAL INTELLIGENCE
  AUTOMATA THEORY
  BIOCONTROL SYSTEMS
  BIOENGINEERING
  BIOINSTRUMENTATION
  BIOLOGICAL MODELS (MATHEMATICS)
  ∞BIOLOGY
  CONTROL SYSTEMS DESIGN
  CYBERNETICS
  HUMAN FACTORS ENGINEERING
  MAN MACHINE SYSTEMS
  NEURISTORS
  RHEOELECTRICAL SIMULATION
  ROBOTS
  SIMULATION
  SYNCODERS
  SYSTEMS ENGINEERING

BIOPAKS
 GS SUPPORT SYSTEMS
  . LIFE SUPPORT SYSTEMS
  . . **BIOPAKS**
 RT BIOENGINEERING
  BIOSATELLITES
  ∞CONTAINERS
  ENCLOSURES
  PORTABLE LIFE SUPPORT SYSTEMS
  PRESERVING

BIOPHYSICS
 GS **BIOPHYSICS**
  . HEALTH PHYSICS
  . . PUBLIC HEALTH
 RT BIODYNAMICS
  BIOENGINEERING
  ∞BIOLOGY
  BIOMAGNETISM
  ∞PHYSICS
  ∞SCIENCE

BIOPROCESSING
 RT AEROSPACE ENVIRONMENTS
  BIOCONVERSION
  BIOTECHNOLOGY
  ELECTROPHORESIS
  ∞MICROGRAVITY APPLICATIONS
  PHARMACOLOGY
  REDUCED GRAVITY
  SPACE PROCESSING
  SPACEBORNE EXPERIMENTS
  WEIGHTLESSNESS

BIOREACTORS
 RT ∞BIOLOGY
  BIOMASS ENERGY PRODUCTION
  BIOTECHNOLOGY

BIOREGENERATION
 USE REGENERATION (PHYSIOLOGY)

BIOREGENERATIVE LIFE SUPPORT SYSTEMS
 USE CLOSED ECOLOGICAL SYSTEMS

BIOS PROJECT
 GS PROGRAMS
  . PROJECTS
  . . **BIOS PROJECT**

BIOSATELLITE 1
 GS SATELLITES
  . ARTIFICIAL SATELLITES
  . . BIOSATELLITES
  . . . **BIOSATELLITE 1**
  UNMANNED SPACECRAFT
  . BIOSATELLITES
  . . **BIOSATELLITE 1**
 RT BIOASTRONAUTICS

BIOSATELLITE 2
 GS SATELLITES
  . ARTIFICIAL SATELLITES
  . . BIOSATELLITES
  . . . **BIOSATELLITE 2**
  UNMANNED SPACECRAFT
  . BIOSATELLITES
  . . **BIOSATELLITE 2**
 RT BIOASTRONAUTICS

BIOSATELLITE 3
 GS SATELLITES
  . ARTIFICIAL SATELLITES
  . . BIOSATELLITES
  . . . **BIOSATELLITE 3**
  UNMANNED SPACECRAFT
  . BIOSATELLITES
  . . **BIOSATELLITE 3**
 RT ∞ASTRONAUTICS
  BIOASTRONAUTICS

BIOSATELLITES
 SN (EXCLUDES MANNED SPACECRAFT)
 GS SATELLITES
  . ARTIFICIAL SATELLITES
  . . **BIOSATELLITES**
  . . . BIOSATELLITE 1
  . . . BIOSATELLITE 2
  . . . BIOSATELLITE 3
  . . . ORBITING FROG OTOLITH
  . . . SPUTNIK 2 SATELLITE
  UNMANNED SPACECRAFT
  . **BIOSATELLITES**
  . . BIOSATELLITE 1
  . . BIOSATELLITE 2
  . . BIOSATELLITE 3
  . . SPUTNIK 2 SATELLITE
 RT AEROSPACE ENVIRONMENTS
  ∞BIOLOGY
  BIOPAKS
  ENVIRONMENTAL CONTROL
  EXTRATERRESTRIAL LIFE
  LIFE DETECTORS
  LIFE SUPPORT SYSTEMS
  MANNED SPACECRAFT
  SPACE CAPSULES
  ∞SPACECRAFT

BIOSENSORS
 USE BIOINSTRUMENTATION

BIOSIMULATION
 USE BIONICS

BIOSPHERE
 GS EARTH ATMOSPHERE
  . LOWER ATMOSPHERE
  . . **BIOSPHERE**
 RT ∞BIOLOGY
  CHEMOSPHERE
  EARTH HYDROSPHERE
  FREE ATMOSPHERE
  HOMOSPHERE
  INTERNATIONAL
   GEOSPHERE-BIOSPHERE PROGRAM

BIOSYNTHESIS
 RT ∞BIOLOGY
  CHEMICAL REACTIONS
  GENETIC ENGINEERING
  METABOLITES
  PROSTAGLANDINS
  ∞SYNTHESIS
  SYNTHETIC FOOD

BIOT METHOD
 RT CALCULUS OF VARIATIONS
  ∞METHODOLOGY

BIOT NUMBER
GS   RATIOS
    . DIMENSIONLESS NUMBERS
    .. **BIOT NUMBER**
RT   HEAT TRANSFER
    ∞NUMBERS

BIOTECHNOLOGY
GS   TECHNOLOGIES
    . **BIOTECHNOLOGY**
RT   ARTIFICIAL CARDIAC PACEMAKER
    ARTIFICIAL HEART VALVES
    ∞BIOLOGY
    BIOPROCESSING
    BIOREACTORS
    BLOOD PUMPS
    HEART IMPLANTATION
    MAN MACHINE SYSTEMS

BIOTELEMETRY
UF   PHYSIOLOGICAL TELEMETRY
GS   BIOENGINEERING
    . BIOINSTRUMENTATION
    .. **BIOTELEMETRY**
    TELECOMMUNICATION
    . TELEMETRY
    .. **BIOTELEMETRY**
RT   ∞BIOLOGY
    COMMUNICATION EQUIPMENT
    ∞ENGINEERING
    ORBITING FROG OTOLITH
    PNEUMOGRAPHY
    WILDLIFE RADIOLOCATION

BIOTIN
UF   VITAMIN B COMPLEX
GS   HETEROCYCLIC COMPOUNDS
    . **BIOTIN**
    VITAMINS
    . **BIOTIN**
RT   DRUGS

BIOTITE
UF   KIMBERLITE
GS   MINERALS
    . MICA
    .. **BIOTITE**

BIPLANES
RT   ∞AIRCRAFT
    DUAL WING CONFIGURATIONS
    LIGHT AIRCRAFT
    MONOPLANES
    TANDEM WING AIRCRAFT
    UTILITY AIRCRAFT

BIPOLAR TRANSISTORS
GS   ELECTRONIC EQUIPMENT
    . SOLID STATE DEVICES
    .. SEMICONDUCTOR DEVICES
    ... TRANSISTORS
    .... **BIPOLAR TRANSISTORS**
RT   BIPOLARITY
    CARRIER INJECTION
    EPITAXY
    MAJORITY CARRIERS
    MINORITY CARRIERS
    N-P-N JUNCTIONS
    SEMICONDUCTORS (MATERIALS)

BIPOLARITY
RT   BIPOLAR TRANSISTORS
    ∞POLARIZATION

BIPROPELLANTS
USE   LIQUID ROCKET PROPELLANTS

BIRD-AIRCRAFT COLLISIONS
GS   ACCIDENTS
    . **BIRD-AIRCRAFT COLLISIONS**
    AIRCRAFT ACCIDENTS
    . **BIRD-AIRCRAFT COLLISIONS**
    COLLISIONS
    . MIDAIR COLLISIONS
    .. **BIRD-AIRCRAFT COLLISIONS**
RT   ∞AIRCRAFT
    AIRCRAFT HAZARDS
    BIRDS
    FLIGHT HAZARDS

BIRDS
GS   ANIMALS
    . VERTEBRATES
    .. **BIRDS**

BIRDS-*(CONT.)*
    ... CHICKENS
    ... PIGEONS
    ... TURKEYS
    ... WATERFOWL
    . WILDLIFE
    .. **BIRDS**
RT   AIRCRAFT HAZARDS
    BIRD-AIRCRAFT COLLISIONS
    EARTH RESOURCES
    ENDANGERED SPECIES
    FLIGHT HAZARDS
    HOMEOTHERMS
    PLUMAGE

BIREFRINGENCE
UF   POCKELS EFFECT
GS   ELECTROMAGNETIC PROPERTIES
    . OPTICAL PROPERTIES
    .. **BIREFRINGENCE**
    REFRACTION
    . **BIREFRINGENCE**
RT   ANISOTROPIC MEDIA
    ANISOTROPY
    BIREFRINGENT COATINGS
    BIREFRINGENT FILTERS
    CALCITE
    ELECTRO-OPTICS
    MOIRE EFFECTS
    NONLINEAR OPTICS
    PHOTOELASTICITY
    POLARIZATION (WAVES)
    REFLECTANCE
    REFRACTIVITY
    TEMPERATURE INVERSIONS
    VOIGT EFFECT

BIREFRINGENT COATINGS
GS   COATINGS
    . **BIREFRINGENT COATINGS**
RT   ANISOTROPIC MEDIA
    BIREFRINGENCE
    BIREFRINGENT FILTERS
    REFRACTIVITY

BIREFRINGENT FILTERS
GS   ELECTROMAGNETIC WAVE FILTERS
    . OPTICAL FILTERS
    .. **BIREFRINGENT FILTERS**
RT   BIREFRINGENCE
    BIREFRINGENT COATINGS
    ∞FILTERS
    OPTICAL PROPERTIES
    REFRACTIVITY

BIRTH
RT   FERTILIZATION
    FETUSES
    PREGNANCY
    ∞REPRODUCTION
    REPRODUCTIVE SYSTEMS

BISMUTH
GS   CHEMICAL ELEMENTS
    . **BISMUTH**
    .. BISMUTH ISOTOPES
    METALS
    . **BISMUTH**
    .. BISMUTH ISOTOPES

BISMUTH ALLOYS
GS   ALLOYS
    . **BISMUTH ALLOYS**

BISMUTH COMPOUNDS
GS   **BISMUTH COMPOUNDS**
    . BISMUTH OXIDES
    . BISMUTH SULFIDES
    . BISMUTH TELLURIDES
RT   ∞CHEMICAL COMPOUNDS
    ∞GROUP 5A COMPOUNDS
    ∞METAL COMPOUNDS

BISMUTH ISOTOPES
UF   BISMUTH 205
GS   CHEMICAL ELEMENTS
    . BISMUTH
    .. **BISMUTH ISOTOPES**
    . NUCLIDES
    . ISOTOPES
    ... **BISMUTH ISOTOPES**
    METALS
    . BISMUTH
    .. **BISMUTH ISOTOPES**

BISMUTH OXIDES
GS   BISMUTH COMPOUNDS
    . **BISMUTH OXIDES**
    CHALCOGENIDES
    . OXIDES
    .. METAL OXIDES
    ... **BISMUTH OXIDES**

BISMUTH SULFIDES
GS   BISMUTH COMPOUNDS
    . **BISMUTH SULFIDES**
    CHALCOGENIDES
    . SULFIDES
    .. INORGANIC SULFIDES
    ... **BISMUTH SULFIDES**
    SULFUR COMPOUNDS
    . SULFIDES
    .. INORGANIC SULFIDES
    ... **BISMUTH SULFIDES**

BISMUTH TELLURIDES
GS   BISMUTH COMPOUNDS
    . **BISMUTH TELLURIDES**
    CHALCOGENIDES
    . TELLURIDES
    .. **BISMUTH TELLURIDES**
    TELLURIUM COMPOUNDS
    . TELLURIDES
    .. **BISMUTH TELLURIDES**

BISMUTH 205
USE   BISMUTH ISOTOPES

BISPHENOLS
GS   HYDROXYL COMPOUNDS
    . ALCOHOLS
    .. PHENOLS
    ... **BISPHENOLS**

BISTABLE AMPLIFIERS
USE   FLIP-FLOPS

BISTABLE CIRCUITS
GS   CIRCUITS
    . **BISTABLE CIRCUITS**
RT   DIGITAL TECHNIQUES
    FLIP-FLOPS
    MULTIVIBRATORS
    TRIGGER CIRCUITS

BISTATIC RADAR
USE   MULTISTATIC RADAR

BISTATIC REFLECTIVITY
RT   BRIGHTNESS
    INCIDENT RADIATION
    REFLECTANCE
    SCATTERING

BIT ERROR RATE
GS   RATES (PER TIME)
    . **BIT ERROR RATE**
RT   BINARY CODES
    BINARY DATA
    BINARY DIGITS
    BIT SYNCHRONIZATION
    BITS
    ERROR ANALYSIS
    ERROR CORRECTING CODES
    ERROR DETECTION CODES
    ERROR SIGNALS
    PULSE COMMUNICATION
    SIGNAL TO NOISE RATIOS
    TRANSMISSION EFFICIENCY

BIT SYNCHRONIZATION
GS   SYNCHRONISM
    . **BIT SYNCHRONIZATION**
RT   BIT ERROR RATE
    FREQUENCY SYNCHRONIZATION

BITERNARY CODE
RT   ∞CODES
    DIFFERENTIAL PULSE CODE
      MODULATION
    DIGITAL SYSTEMS
    PULSE CODE MODULATION

BITS
RT   BINARY DIGITS
    BIT ERROR RATE
    DRILL BITS

BITUMENS
RT   CARBON
     COAL
     COKE
     ∞ CONSTRUCTION MATERIALS
     LIGNITE
     ∞ MATERIALS
     SOLVENT REFINED COAL

BIVARIATE ANALYSIS
GS   STATISTICAL ANALYSIS
     . VARIANCE (STATISTICS)
     . . MULTIVARIATE STATISTICAL
        ANALYSIS
     . . . **BIVARIATE ANALYSIS**
RT   CORRELATION

BL LACERTAE OBJECTS
GS   CELESTIAL BODIES
     . **BL LACERTAE OBJECTS**
RT   EXTRAGALACTIC RADIO SOURCES
     GALAXIES
     LUMINOUS INTENSITY
     POLARIZATION (WAVES)
     RADIANT FLUX DENSITY
     RADIO SOURCES (ASTRONOMY)

BLACK AND WHITE PHOTOGRAPHY
GS   IMAGERY
     . **BLACK AND WHITE PHOTOGRAPHY**
     PHOTOGRAPHY
     . **BLACK AND WHITE PHOTOGRAPHY**
RT   ALL SKY PHOTOGRAPHY
     ASTRONOMICAL PHOTOGRAPHY
     AUTORADIOGRAPHY
     CHRONOPHOTOGRAPHY
     CINEMATOGRAPHY
     CLOUD PHOTOGRAPHY
     COLOR PHOTOGRAPHY
     ELECTRO-OPTICAL PHOTOGRAPHY
     ELECTRON PHOTOGRAPHY
     FRAME PHOTOGRAPHY
     INFRARED PHOTOGRAPHY
     LUNAR PHOTOGRAPHY
     PHOTOMICROGRAPHY
     PHOTORECONNAISSANCE
     RADAR PHOTOGRAPHY
     ROCKET-BORNE PHOTOGRAPHY
     SATELLITE-BORNE PHOTOGRAPHY
     SCHLIEREN PHOTOGRAPHY
     SHADOWGRAPH PHOTOGRAPHY
     SPACEBORNE PHOTOGRAPHY
     SPECTROHELIOGRAPHS
     SPECTROPHOTOGRAPHY
     STEREOPHOTOGRAPHY
     ULTRAVIOLET PHOTOMETRY
     UROGRAPHY

BLACK ARROW LAUNCH VEHICLE
USE   BLACK KNIGHT ROCKET VEHICLE

BLACK BODY RADIATION
GS   ELECTROMAGNETIC RADIATION
     . **BLACK BODY RADIATION**
RT   BRIGHTNESS DISTRIBUTION
     BRIGHTNESS TEMPERATURE
     EMISSIVITY
     HEAT RADIATORS
     HOHLRAUMS
     INFRARED RADIATION
     KIRCHHOFF LAW OF RADIATION
     LIGHT (VISIBLE RADIATION)
     NONGRAY ATMOSPHERES
     NONGRAY GAS
     PLANCKS CONSTANT
     RADIANCE
     ∞ RADIATION
     SUNLIGHT
     THERMAL RADIATION
     ULTRAVIOLET RADIATION

BLACK BRANT SOUNDING ROCKETS
GS   ROCKET VEHICLES
     . SINGLE STAGE ROCKET VEHICLES
     . . **BLACK BRANT SOUNDING ROCKETS**
     . . . BLACK BRANT 1 SOUNDING
        ROCKET
     . . . BLACK BRANT 2 SOUNDING
        ROCKET
     . . . BLACK BRANT 3 SOUNDING
        ROCKET
     . . . BLACK BRANT 4 SOUNDING
        ROCKET
     . . . BLACK BRANT 5 SOUNDING
        ROCKET

BLACK BRANT SOUNDING ROCKETS-(CONT.)
     . SOUNDING ROCKETS
     . . **BLACK BRANT SOUNDING ROCKETS**
     . . . BLACK BRANT 1 SOUNDING
        ROCKET
     . . . BLACK BRANT 2 SOUNDING
        ROCKET
     . . . BLACK BRANT 3 SOUNDING
        ROCKET
     . . . BLACK BRANT 4 SOUNDING
        ROCKET
     . . . BLACK BRANT 5 SOUNDING
        ROCKET
RT   SOLID PROPELLANT ROCKET ENGINES

BLACK BRANT 1 SOUNDING ROCKET
GS   ROCKET VEHICLES
     . SINGLE STAGE ROCKET VEHICLES
     . . BLACK BRANT SOUNDING ROCKETS
     . . . **BLACK BRANT 1 SOUNDING
        ROCKET**
     . SOUNDING ROCKETS
     . . BLACK BRANT SOUNDING ROCKETS
     . . . **BLACK BRANT 1 SOUNDING
        ROCKET**
RT   SOLID PROPELLANT ROCKET ENGINES

BLACK BRANT 2 SOUNDING ROCKET
GS   ROCKET VEHICLES
     . SINGLE STAGE ROCKET VEHICLES
     . . BLACK BRANT SOUNDING ROCKETS
     . . . **BLACK BRANT 2 SOUNDING
        ROCKET** .
     . SOUNDING ROCKETS
     . . BLACK BRANT SOUNDING ROCKETS
     . . . **BLACK BRANT 2 SOUNDING
        ROCKET**
RT   SOLID PROPELLANT ROCKET ENGINES

BLACK BRANT 3 SOUNDING ROCKET
GS   ROCKET VEHICLES
     . SINGLE STAGE ROCKET VEHICLES
     . . BLACK BRANT SOUNDING ROCKETS
     . . . **BLACK BRANT 3 SOUNDING
        ROCKET**
     . SOUNDING ROCKETS
     . . BLACK BRANT SOUNDING ROCKETS
     . . . **BLACK BRANT 3 SOUNDING
        ROCKET**
RT   SOLID PROPELLANT ROCKET ENGINES

BLACK BRANT 4 SOUNDING ROCKET
GS   ROCKET VEHICLES
     . SINGLE STAGE ROCKET VEHICLES
     . . BLACK BRANT SOUNDING ROCKETS
     . . . **BLACK BRANT 4 SOUNDING
        ROCKET**
     . SOUNDING ROCKETS
     . . BLACK BRANT SOUNDING ROCKETS
     . . . **BLACK BRANT 4 SOUNDING
        ROCKET**
RT   SOLID PROPELLANT ROCKET ENGINES

BLACK BRANT 5 SOUNDING ROCKET
GS   ROCKET VEHICLES
     . SINGLE STAGE ROCKET VEHICLES
     . . BLACK BRANT SOUNDING ROCKETS
     . . . **BLACK BRANT 5 SOUNDING
        ROCKET**
     . SOUNDING ROCKETS
     . . BLACK BRANT SOUNDING ROCKETS
     . . . **BLACK BRANT 5 SOUNDING
        ROCKET**
RT   SOLID PROPELLANT ROCKET ENGINES

BLACK HAWK ASSAULT HELICOPTER
USE   H-60 HELICOPTER

BLACK HILLS (SD-WY)
GS   LANDFORMS
     . MOUNTAINS
     . . **BLACK HILLS (SD-WY)**
RT   SOUTH DAKOTA
     WYOMING

BLACK HOLES (ASTRONOMY)
GS   GRAVITATIONAL COLLAPSE
     . **BLACK HOLES (ASTRONOMY)**
RT   ACCRETION DISKS
     GRAVITATIONAL LENSES
     NAKED SINGULARITIES
     REISSNER-NORDSTROM SOLUTION
     SUPERNOVA REMNANTS
     WHITE HOLES (ASTRONOMY)
     X RAY BINARIES

BLACK KNIGHT ROCKET VEHICLE
UF   BLACK ARROW LAUNCH VEHICLE
GS   ROCKET VEHICLES
     . MULTISTAGE ROCKET VEHICLES
     . . **BLACK KNIGHT ROCKET VEHICLE**
     . SINGLE STAGE ROCKET VEHICLES
     . . **BLACK KNIGHT ROCKET VEHICLE**
RT   LIQUID PROPELLANT ROCKET ENGINES

BLACK SEA
GS   SEAS
     . **BLACK SEA**
RT   BULGARIA
     ROMANIA
     TURKEY
     U.S.S.R.

BLACKBURN B-103 AIRCRAFT
USE   BUCCANEER AIRCRAFT

∞ **BLACKOUT**
SN   *(USE OF A MORE SPECIFIC TERM IS
     RECOMMENDED--CONSULT THE TERMS
     LISTED BELOW)*
RT   BLACKOUT (PHYSIOLOGY)
     BLACKOUT (PROPAGATION)

BLACKOUT (PHYSIOLOGY)
GS   SYNCOPE
     . **BLACKOUT (PHYSIOLOGY)**
     . . BLACKOUT PREVENTION
     UNCONSCIOUSNESS
     . **BLACKOUT (PHYSIOLOGY)**
     . . BLACKOUT PREVENTION
RT   ACCELERATION TOLERANCE
     ∞ BLACKOUT
     ∞ COMA

BLACKOUT (PROPAGATION)
UF   IONOSPHERIC BLACKOUT
GS   ELECTROMAGNETIC INTERFERENCE
     . RADIO FREQUENCY INTERFERENCE
     . . **BLACKOUT (PROPAGATION)**
     . . . ELECTROMAGNETIC NOISE
     . . . . ATMOSPHERICS
     . . . . . IONOSPHERICS
     . . . . . . DAWN CHORUS
     . . . . . . HISS
     . . . . . . SUDDEN ENHANCEMENT OF
        ATMOSPHERICS
     . . . . . WHISTLERS
     . . . . COSMIC NOISE
     . . . . IONOSPHERIC NOISE
     . . . . . WHISTLERS
     . . . . SHOT NOISE
     . . . . WHITE NOISE
     . . . . . THERMAL NOISE
     . . . . IONOSPHERIC CROSS MODULATION
     . . . POLAR RADIO BLACKOUT
RT   ∞ BLACKOUT
     ELECTROMAGNETIC FIELDS
     IONOSPHERIC DISTURBANCES
     PLASMA SHEATHS
     PLASMAS (PHYSICS)
     RADIATION EFFECTS
     RADIO COMMUNICATION
     REENTRY COMMUNICATION
     REENTRY EFFECTS
     SOLAR ACTIVITY EFFECTS
     X RAYS

BLACKOUT PREVENTION
GS   HUMAN PERFORMANCE
     . ASTRONAUT PERFORMANCE
     . . **BLACKOUT PREVENTION**
     SYNCOPE
     . BLACKOUT (PHYSIOLOGY)
     . . **BLACKOUT PREVENTION**
     UNCONSCIOUSNESS
     . BLACKOUT (PHYSIOLOGY)
     . . **BLACKOUT PREVENTION**
RT   ACCELERATION TOLERANCE
     ∞ COMA
     WEIGHTLESSNESS

BLADDER
GS   ANATOMY
     . GENITOURINARY SYSTEM
     . . **BLADDER**
     . ORGANS
     . . **BLADDER**
     VISCERA
     . ORGANS
     . . **BLADDER**
RT   PROSTATE GLAND

BLADDER-*(CONT.)*
　　UROLOGY

BLADDERS (MECHANICS)
　USE　DIAPHRAGMS (MECHANICS)

BLADE SLAP NOISE
　RT　AERODYNAMIC NOISE
　　　AIRCRAFT NOISE
　　　BLADE TIPS
　　　HELICOPTERS

BLADE TIPS
　GS　TIPS
　　　. BLADE TIPS
　RT　AIRFOIL PROFILES
　　　BLADE SLAP NOISE
　　　PROPELLER BLADES
　　　ROTARY WINGS
　　　ROTOR BLADES (TURBOMACHINERY)
　　　WING TIPS

∞ **BLADES**
　SN　*(USE OF A MORE SPECIFIC TERM IS*
　　　RECOMMENDED--CONSULT THE TERMS
　　　LISTED BELOW)
　RT　AIRFOILS
　　　BLADES (CUTTERS)
　　　COMPRESSOR BLADES
　　　FINS
　　　HYDROFOILS
　　　PROPELLER BLADES
　　　RIMS
　　　ROTARY WINGS
　　　ROTOR BLADES (TURBOMACHINERY)
　　　STATOR BLADES
　　　TURBINE BLADES
　　　TURBOMACHINE BLADES
　　　VANES

BLADES (CUTTERS)
　GS　CUTTERS
　　　. BLADES (CUTTERS)
　　　. . RAZOR BLADES
　RT　∞BLADES

∞ **BLANKETS**
　SN　*(USE OF A MORE SPECIFIC TERM IS*
　　　RECOMMENDED--CONSULT THE TERMS
　　　LISTED BELOW)
　RT　BEDDING EQUIPMENT
　　　BLANKETS (FISSION REACTORS)
　　　BLANKETS (FUSION REACTORS)
　　　CLOUD COVER
　　　CONTROLLED ATMOSPHERES
　　　SOLAR BLANKETS

BLANKETS (FISSION REACTORS)
　RT　∞BLANKETS
　　　∞DAMPERS
　　　FISSION
　　　REACTOR DESIGN
　　　REACTOR MATERIALS

BLANKETS (FUSION REACTORS)
　RT　∞BLANKETS
　　　FUSION REACTORS
　　　LIMITERS (FUSION REACTORS)
　　　MODERATORS
　　　REACTOR DESIGN
　　　REACTOR MATERIALS

∞ **BLANKING**
　SN　*(USE OF A MORE SPECIFIC TERM IS*
　　　RECOMMENDED--CONSULT THE TERMS
　　　LISTED BELOW)
　RT　BLANKING (CUTTING)
　　　FORMING TECHNIQUES
　　　STAMPING

BLANKING (CUTTING)
　GS　CUTTING
　　　. BLANKING (CUTTING)
　　　FORMING TECHNIQUES
　　　. PRESSING (FORMING)
　　　. . BLANKING (CUTTING)
　RT　∞BLANKING
　　　LASER CUTTING
　　　SHEARING
　　　STAMPING

BLANKS
　RT　AMMUNITION
　　　BRIQUETS

BLANKS-*(CONT.)*
　　FORMS (PAPER)
　　PREFORMS

BLASIUS EQUATION
　GS　ANALYSIS (MATHEMATICS)
　　　. REAL VARIABLES
　　　. . DIFFERENTIAL EQUATIONS
　　　. . . BLASIUS EQUATION
　RT　BOUNDARY LAYER FLOW
　　　∞EQUATIONS
　　　FALKNER-SKAN EQUATION
　　　FLAT PLATES
　　　PRANDTL-MEYER EXPANSION

BLASIUS FLOW
　GS　FLUID FLOW
　　　. LAMINAR FLOW
　　　. . BLASIUS FLOW
　　　. UNIFORM FLOW
　　　. . BLASIUS FLOW
　RT　FLAT PLATES
　　　HEAD FLOW
　　　TOLLMEIN-SCHLICHTING WAVES
　　　TURBULENT FLOW
　　　TWO DIMENSIONAL FLOW
　　　WEDGE FLOW

BLAST DEFLECTORS
　GS　DEFLECTORS
　　　. BLAST DEFLECTORS
　RT　BAFFLES
　　　DIVERTERS
　　　FLAME DEFLECTORS
　　　SHIELDING

BLAST LOADS
　GS　AERODYNAMIC FORCES
　　　. AERODYNAMIC LOADS
　　　. . BLAST LOADS
　　　LOADS (FORCES)
　　　. DYNAMIC LOADS
　　　. . AERODYNAMIC LOADS
　　　. . . BLAST LOADS
　　　. . TRANSIENT LOADS
　　　. . . SHOCK LOADS
　　　. . . . BLAST LOADS
　RT　AERIAL EXPLOSIONS
　　　DYNAMIC PRESSURE
　　　EXPLOSIONS
　　　GUST LOADS
　　　IMPACT LOADS
　　　OVERPRESSURE
　　　PRESSURE
　　　PRESSURE PULSES
　　　RIEMANN WAVES
　　　SHOCK WAVES
　　　WAVE RESISTANCE

BLASTOFF
　USE　ROCKET LAUNCHING

∞ **BLASTS**
　SN　*(USE OF A MORE SPECIFIC TERM IS*
　　　RECOMMENDED--CONSULT THE TERMS
　　　LISTED BELOW)
　RT　EXHAUST GASES
　　　EXPLOSIONS
　　　JET BLAST EFFECTS
　　　SHOCK WAVES
　　　SOUND WAVES

BLATTIDAE
　USE　COCKROACHES

BLEACHING
　RT　CHLORINATION
　　　CLEANING
　　　FADING

BLEED-OFF
　USE　PRESSURE REDUCTION

∞ **BLEEDING**
　SN　*(USE OF A MORE SPECIFIC TERM IS*
　　　RECOMMENDED--CONSULT THE TERMS
　　　LISTED BELOW)
　RT　BOUNDARY LAYER CONTROL
　　　FLUID MECHANICS
　　　HEMORRHAGES
　　　PRESSURE REDUCTION

BLENDS
　USE　MIXTURES

BLIGHT
　UF　DISEASED VEGETATION
　GS　DISEASES
　　　. PARASITIC DISEASES
　　　. . BLIGHT
　RT　ALFALFA
　　　BACTERIA
　　　BARLEY
　　　BOTANY
　　　CITRUS TREES
　　　CORN
　　　CROP GROWTH
　　　CROP VIGOR
　　　FUNGI
　　　ORCHARDS
　　　PARASITES
　　　PLANTS (BOTANY)
　　　RHIZOPUS
　　　RUST FUNGI
　　　VINEYARDS

BLIND LANDING
　GS　LANDING
　　　. BLIND LANDING
　RT　AIRCRAFT LANDING
　　　APPROACH INDICATORS
　　　AUTOMATIC LANDING CONTROL
　　　INSTRUMENT APPROACH
　　　INSTRUMENT FLIGHT RULES
　　　INSTRUMENT LANDING SYSTEMS
　　　LANDING INSTRUMENTS
　　　NIGHT FLIGHTS (AIRCRAFT)

BLINDNESS
　GS　BLINDNESS
　　　. FLASH BLINDNESS
　RT　BRAILLE
　　　EYE DISEASES
　　　OPTOMETRY
　　　VISION

BLINDS
　RT　SHIELDING
　　　∞SHUTTERS

BLINKING
　RT　ASTRONOMICAL PHOTOMETRY
　　　DISPLAY DEVICES
　　　EYE MOVEMENTS
　　　VISUAL PERCEPTION

∞ **BLISTERS**
　SN　*(USE OF A MORE SPECIFIC TERM IS*
　　　RECOMMENDED--CONSULT THE TERMS
　　　LISTED BELOW)
　RT　INFECTIOUS DISEASES
　　　INJURIES
　　　MUCOCELES
　　　PROTUBERANCES
　　　RUPTURING
　　　SKIN (ANATOMY)
　　　VIRUSES

BLOCH BAND
　GS　ENERGY BANDS
　　　. BLOCH BAND
　RT　∞BANDS
　　　SUPERCONDUCTIVITY

BLOCK DIAGRAMS
　GS　DIAGRAMS
　　　. BLOCK DIAGRAMS
　RT　CHARTS
　　　COMPUTER PROGRAMMING
　　　COMPUTER PROGRAMS
　　　FLOW CHARTS
　　　RESEARCH MANAGEMENT
　　　SYSTEMS ANALYSIS

BLOCK ISLAND SOUND (RI)
　GS　SOUNDS (TOPOGRAPHIC FEATURES)
　　　. BLOCK ISLAND SOUND (RI)
　RT　ATLANTIC OCEAN
　　　RHODE ISLAND

BLOCKING
　UF　OBSTRUCTING
　RT　∞ARRESTERS
　　　CLOSING
　　　CLOSURES
　　　CONSTRAINTS
　　　CONSTRICTIONS
　　　CONTAINMENT
　　　PLUGGING

BLOCKING-*(CONT.)*
    PLUGS
    PREVENTION
    RETARDERS (DEVICES)
    RETARDING
    SEALING
    SEALS (STOPPERS)
    STOPPING

BLOCKS
  RT   CUBES (MATHEMATICS)
       PULLEYS
       SLABS

BLOEDITE
  GS MINERALS
      . **BLOEDITE**
  RT   MAGNESIUM SULFATES
       SODIUM COMPOUNDS
       SULFUR COMPOUNDS

BLOOD
  GS BODY FLUIDS
      . **BLOOD**
      . . ERYTHROCYTES
      . . . RETICULOCYTES
      . . LEUKOCYTES
      . . . EOSINOPHILS
      . . . LYMPHOCYTES
      . . THROMBIN
      . . THROMBOPLASTIN
      . . WHITE BLOOD CELLS
  RT   ANEMIAS
       BIOCOMPATIBILITY
       BLOOD CIRCULATION
       BLOOD COAGULATION
       BLOOD FLOW
       BLOOD GROUPS
       BLOOD PLASMA
       BLOOD PRESSURE
       BLOOD PUMPS
       BLOOD VESSELS
       BLOOD VOLUME
       BLOOD-BRAIN BARRIER
       CAPILLARIES (ANATOMY)
       CARBOXYHEMOGLOBIN TEST
       CARDIOVASCULAR SYSTEM
       CIRCULATION
       COAGULATION
       FIBRIN
       HEART
       HEMATOCRIT
       HEMATOPOIESIS
       HEMOGLOBIN
       HEMORRHAGES
       HYPERCAPNIA
       HYPOCAPNIA
      · OXIMETRY
       RHESUS FACTOR
       TRANSFUSION

BLOOD CIRCULATION
  GS CIRCULATION
      . **BLOOD CIRCULATION**
      . . BRAIN CIRCULATION
      . . CORONARY CIRCULATION
      . . INTERCRANIAL CIRCULATION
      . . INTRAVASCULAR SYSTEM
      . . ISCHEMIA
      . . OCULAR CIRCULATION
      . . PERIPHERAL CIRCULATION
      . . PULMONARY CIRCULATION
  RT   ARTIFICIAL CARDIAC PACEMAKER
       ARTIFICIAL HEART VALVES
       BLOOD
       BLOOD-BRAIN BARRIER
       CARBOXYHEMOGLOBIN
       CIRCULATORY SYSTEM
       CYANOSIS
       DIASTOLE
       ELECTROPLETHYSMOGRAPHY
       HEART IMPLANTATION
       HEMODYNAMIC RESPONSES
       HEMODYNAMICS
       HYPERVOLEMIA
       HYPOVOLEMIA
       PHONOARTERIOGRAPHY
       RHEOENCEPHALOGRAPHY
       RHEOMETERS
       TOURNIQUETS
       VASCULAR SYSTEM

BLOOD COAGULATION
  GS COAGULATION
       BLOOD COAGULATION

BLOOD COAGULATION-*(CONT.)*
  RT   BLOOD
       CLOTTING
       HEMOSTATICS
       MYOCARDIAL INFARCTION
       PLATELETS
       THROMBOCYTES
       THROMBOSIS

BLOOD FLOW
  GS FLUID FLOW
      . **BLOOD FLOW**
  RT   BLOOD
       CAPILLARY FLOW
       DIASTOLE
       HEMOPERFUSION
       SYSTOLE
       TOURNIQUETS

BLOOD GROUPS
  RT   BLOOD
       PLATELETS

BLOOD PLASMA
  RT   BLOOD
       BODY FLUIDS
       CORPUSCLES

BLOOD PRESSURE
  GS PRESSURE
      . **BLOOD PRESSURE**
      . . DIASTOLIC PRESSURE
      . . HYPERTENSION
      . . HYPOTENSION
      . . SYSTOLIC PRESSURE
  RT   BIOFEEDBACK
       BLOOD
       DIASTOLE
       HEMODYNAMIC RESPONSES
       HEMOPERFUSION
       MANOMETERS
       OPHTHALMODYNAMOMETRY
       ORTHOSTATIC TOLERANCE
       SPHYGMOGRAPHY
       SYSTOLE
      ∞TENSION

BLOOD PUMPS
  GS MEDICAL EQUIPMENT
      . **BLOOD PUMPS**
      PUMPS
      . **BLOOD PUMPS**
  RT   ARTIFICIAL HEART VALVES
       BIOTECHNOLOGY
       BLOOD
       CIRCULATION
       CIRCULATORY SYSTEM
       HEART
       PULMONARY CIRCULATION

BLOOD VESSELS
  GS ANATOMY
      . CARDIOVASCULAR SYSTEM
      . . **BLOOD VESSELS**
      . . . ARTERIES
      . . . . AORTA
      . . . CAPILLARIES (ANATOMY)
      . . . GLOMERULUS
      . . . VEINS
      . CIRCULATORY SYSTEM
      . . VASCULAR SYSTEM
      . . . **BLOOD VESSELS**
      . . . . ARTERIES
      . . . . . AORTA
      . . . . CAPILLARIES (ANATOMY)
      . . . . GLOMERULUS
      . . . . VEINS
  RT   BIFURCATION (BIOLOGY)
       BLOOD
       CAROTID SINUS BODY
       CAROTID SINUS REFLEX
       CATHETERIZATION
       EMBOLISMS
       ENDOTHELIUM
       FAT EMBOLISMS
       VASOCONSTRICTION
       VASODILATION
      ∞VESSELS

BLOOD VOLUME
  RT   BLOOD
       CARDIOVASCULAR SYSTEM
       CHRONIC CONDITIONS
       CLINICAL MEDICINE
       HEMATOPOIETIC SYSTEM

BLOOD VOLUME-*(CONT.)*
       HEMODYNAMICS
       HYPERVOLEMIA
       HYPOVOLEMIA

BLOOD-BRAIN BARRIER
  RT   ∞BARRIERS
       BLOOD
       BLOOD CIRCULATION
       CENTRAL NERVOUS SYSTEM
       NEURONS

BLOWDOWN WIND TUNNELS
  GS TEST FACILITIES
      . WIND TUNNELS
      . . **BLOWDOWN WIND TUNNELS**
  RT   HOTSHOT WIND TUNNELS
       HYPERSONIC WIND TUNNELS
       HYPERVELOCITY WIND TUNNELS
       LOW DENSITY RESEARCH
       LOW SPEED WIND TUNNELS
       SUBSONIC WIND TUNNELS
       SUPERSONIC WIND TUNNELS
       TRANSONIC WIND TUNNELS

BLOWERS
  RT   AIR CONDITIONING
       AIR CONDITIONING EQUIPMENT
       AIR DUCTS
       BLOWING
       CENTRIFUGAL COMPRESSORS
       COMPRESSORS
       COOLING SYSTEMS
       DUCTED FANS
       EXHAUST SYSTEMS
      ∞FANS
       IMPELLERS
       INJECTORS
       MATERIALS HANDLING
       MIXERS
      ∞NOZZLES
       REFRIGERATING MACHINERY
       SEALING
       SPRAYERS
       SUPERCHARGERS
       TURBOMACHINERY
       VENTILATION
       VENTILATION FANS
       VENTILATORS

BLOWING
  GS SPANWISE BLOWING
      . **BLOWING**
  RT   AERATION
       AGITATION
       BLOWERS
       BOUNDARY LAYER CONTROL
       CIRCULATION
       CIRCULATION CONTROL AIRFOILS
       COMPRESSING
       ENTRAINMENT
       EXHAUSTING
       FORCED CONVECTION
       INJECTION
       MIXING
      ∞PUMPING
       SPRAYING
       WIND (METEOROLOGY)

BLOWN FLAPS
  USE EXTERNALLY BLOWN FLAPS

BLOWOUTS
  RT   FATIGUE LIFE
       TIRES

BLUE GOOSE MISSILE
  GS DECOYS
      . **BLUE GOOSE MISSILE**
      MISSILES
      . SURFACE TO AIR MISSILES
      . . **BLUE GOOSE MISSILE**
  RT   BOOSTER ROCKET ENGINES
       COUNTERMEASURES
       J-85 ENGINE
       SOLID PROPELLANT ROCKET ENGINES

BLUE GREEN ALGAE
  UF CYANOPHYTA
  GS ALGAE
      . **BLUE GREEN ALGAE**
      . . NOSTOC

BLUE SCOUT ROCKET VEHICLE
GS   LAUNCH VEHICLES
    . **BLUE SCOUT ROCKET VEHICLE**
    ROCKET VEHICLES
    . MULTISTAGE ROCKET VEHICLES
    . . **BLUE SCOUT ROCKET VEHICLE**
RT   ALGOL ENGINE
    SOLID PROPELLANT ROCKET ENGINES
    X-248 ENGINE
    X-254 ENGINE
    XM-33 ENGINE

BLUE STARS
GS   CELESTIAL BODIES
    . STARS
    . . HOT STARS
    . . . **BLUE STARS**
    . . . . SYMBIOTIC STARS
RT   A STARS
    B STARS
    O STARS

BLUE STEEL MISSILE
GS   MISSILES
    . **BLUE STEEL MISSILE**
RT   LIQUID PROPELLANT ROCKET ENGINES

BLUE STREAK LAUNCH VEHICLE
GS   LAUNCH VEHICLES
    . **BLUE STREAK LAUNCH VEHICLE**
    ROCKET VEHICLES
    . **BLUE STREAK LAUNCH VEHICLE**
RT   ELDO LAUNCH VEHICLE
    LIQUID PROPELLANT ROCKET ENGINES

BLUE STREAK MISSILE
GS   MISSILES
    . BALLISTIC MISSILES
    . . INTERMEDIATE RANGE BALLISTIC
        MISSILES
    . . . **BLUE STREAK MISSILE**
    . SURFACE TO SURFACE MISSILES
    . . INTERMEDIATE RANGE BALLISTIC
        MISSILES
    . . . **BLUE STREAK MISSILE**
    ROCKET VEHICLES
    . **BLUE STREAK MISSILE**
RT   LIQUID PROPELLANT ROCKET ENGINES

BLUEPRINTS
GS   DOCUMENTS
    . ENGINEERING DRAWINGS
    . . **BLUEPRINTS**
    DRAWINGS
    . ENGINEERING DRAWINGS
    . . **BLUEPRINTS**
RT   LAYOUTS
    REPRODUCTION (COPYING)

BLUFF BODIES
RT   BLUNT BODIES
    ∞BODIES
    DUCTED BODIES
    FOREBODIES
    LIFTING BODIES
    REENTRY VEHICLES
    ROSHKO PREDICTION

BLUFFS (LANDFORMS)
USE   CLIFFS

BLUNT BODIES
RT   AERODYNAMIC CONFIGURATIONS
    AERODYNAMICS
    AXISYMMETRIC BODIES
    BLUFF BODIES
    ∞BODIES
    DUCTED BODIES
    FOREBODIES
    MISSILE BODIES
    NOSE CONES
    SYMMETRICAL BODIES

BLUNT LEADING EDGES
GS   EDGES
    . **BLUNT LEADING EDGES**
RT   AIRFOILS
    FOREBODIES
    TRAILING EDGES

BLUNT TRAILING EDGES
GS   EDGES
    . TRAILING EDGES
    . . **BLUNT TRAILING EDGES**

BLUNT TRAILING EDGES-*(CONT.)*
RT   AIRFOILS
    CONTROL SURFACES
    WINGS

BLURRING
RT   ABERRATION
    RESOLUTION
    SPATIAL FILTERING

BMC
USE   BONE MINERAL CONTENT

BMEWS
USE   BALLISTIC MISSILE EARLY WARNING
        SYSTEM

BO-105 HELICOPTER
GS   BOLKOW AIRCRAFT
    . **BO-105 HELICOPTER**
    PASSENGER AIRCRAFT
    . **BO-105 HELICOPTER**
    UTILITY AIRCRAFT
    . **BO-105 HELICOPTER**
    V/STOL AIRCRAFT
    . ROTARY WING AIRCRAFT
    . . HELICOPTERS
    . . . **BO-105 HELICOPTER**

BOARDS (PAPER)
UF   FIBERBOARD
RT   ∞CONSTRUCTION MATERIALS
    PAPER (MATERIAL)
    PAPERS

BOATS
GS   SURFACE VEHICLES
    . **BOATS**
    . . LIFEBOATS
    WATER VEHICLES
    . **BOATS**
    . . LIFEBOATS
RT   AMPHIBIOUS VEHICLES
    HARBORS
    INFLATABLE STRUCTURES
    KEELS
    ∞MILITARY VEHICLES
    RESEARCH VEHICLES
    SHIPS
    UNDERWATER VEHICLES

BOATTAILS
RT   AFTERBODIES
    SKIRTS
    TAIL ASSEMBLIES

BOD
USE   BIOCHEMICAL OXYGEN DEMAND

∞ **BODIES**
SN   *(USE OF A MORE SPECIFIC TERM IS
    RECOMMENDED--CONSULT THE TERMS
    LISTED BELOW)*
RT   AFTERBODIES
    AXISYMMETRIC BODIES
    BLUFF BODIES
    BLUNT BODIES
    BODIES OF REVOLUTION
    CELESTIAL BODIES
    CENTERBODIES
    DUCTED BODIES
    ELASTIC BODIES
    FINNED BODIES
    FLEXIBLE BODIES
    FOREIGN BODIES
    HERBIG-HARO OBJECTS
    HUMAN BODY
    LENTICULAR BODIES
    LIFTING BODIES
    MANEUVERABLE REENTRY BODIES
    MISSILE BODIES
    PLANFORMS
    PYRAMIDAL BODIES
    REENTRY VEHICLES
    ROTATING BODIES
    SLENDER BODIES
    SOLIDS
    STREAMLINED BODIES
    SYMMETRICAL BODIES
    THREE DIMENSIONAL BODIES
    TOWED BODIES
    TWO DIMENSIONAL BODIES

BODIES OF REVOLUTION
GS   SYMMETRICAL BODIES
    . **BODIES OF REVOLUTION**
    . . CONICAL BODIES
    . . . SLENDER CONES
    . . CYLINDRICAL BODIES
    . . . ROTATING CYLINDERS
    . . PARABOLIC BODIES
    . . SPHERES
    . . . CELESTIAL SPHERE
    . . . CONCENTRIC SPHERES
    . . . FALLING SPHERES
    . . . POINCARE SPHERES
    . . . ROTATING SPHERES
    . . TORUSES
RT   AERODYNAMIC CONFIGURATIONS
    AERODYNAMICS
    AXES OF ROTATION
    AXISYMMETRIC BODIES
    ∞BODIES
    CONES
    DISKS (SHAPES)
    ELLIPSOIDS
    FINNED BODIES
    GEOMETRY
    ∞HEMISPHERES
    HEMISPHERICAL SHELLS
    OGIVES
    ∞RINGS
    SPHERICAL SHELLS
    STREAMLINED BODIES

BODY CENTERED CUBIC LATTICES
UF   BCC LATTICES
GS   CRYSTAL LATTICES
    . **BODY CENTERED CUBIC LATTICES**
RT   CLOSE PACKED LATTICES
    CRYSTALS
    FACE CENTERED CUBIC LATTICES

BODY COMPOSITION (BIOLOGY)
GS   COMPOSITION (PROPERTY)
    . **BODY COMPOSITION (BIOLOGY)**
    PHYSIOLOGY
    . **BODY COMPOSITION (BIOLOGY)**
RT   ∞BIOLOGY
    CHEMICAL COMPOSITION
    EXOSKELETONS

BODY FLUIDS
GS   **BODY FLUIDS**
    . BLOOD
    . . ERYTHROCYTES
    . . . RETICULOCYTES
    . . LEUKOCYTES
    . . . EOSINOPHILS
    . . . LYMPHOCYTES
    . . THROMBIN
    . . THROMBOPLASTIN
    . . WHITE BLOOD CELLS
    . CEREBROSPINAL FLUID
    . ENDOLYMPH
    . LYMPH
    . . LYMPHOCYTES
    . MUCUS
    . SALIVA
    . SWEAT
    . URINE
RT   BLOOD PLASMA
    DIURESIS
    EDEMA
    ∞FLUIDS
    ISOTONICITY
    LYSOZYME
    MINERAL METABOLISM
    OBESITY
    PERSPIRATION
    SECRETIONS
    WATER
    WATER BALANCE

BODY KINEMATICS
GS   KINEMATICS
    . **BODY KINEMATICS**
RT   ACCELERATION (PHYSICS)
    ACCELERATION STRESSES
    (PHYSIOLOGY)
    KINETICS
    PARTICLE THEORY
    VELOCITY

BODY MEASUREMENT (BIOLOGY)
SN   (LIMITED TO BIOLOGICAL APPLICATIONS
    --FOR MEASUREMENT OF
    NON-BIOLOGICAL BODIES USE SIZE
    DETERMINATION)
GS   BIOENGINEERING
   . BIOMETRICS
   . . **BODY MEASUREMENT (BIOLOGY)**
   . . . ANTHROPOMETRY
   . . . ELECTROPLETHYSMOGRAPHY
RT   ∞BIOLOGY
    BIOMEDICAL DATA
    ELECTROCARDIOGRAPHY
    ELECTROENCEPHALOGRAPHY
    ELECTROPHYSIOLOGY
    ∞ENGINEERING
    HUMAN BODY
    HUMAN FACTORS ENGINEERING
    OBESITY
    SIZE DETERMINATION
    ∞SIZING

BODY SIZE (BIOLOGY)
RT   ANTHROPOMETRY
    ∞BIOLOGY
    OBESITY

BODY SWAY TEST
GS   PHYSIOLOGICAL TESTS
   . **BODY SWAY TEST**
RT   ∞EQUILIBRIUM
    VERTICAL PERCEPTION
    VESTIBULAR TESTS

BODY TEMPERATURE
SN   (LIMITED TO TEMPERATURE OF
    BIOLOGICAL BODIES)
GS   TEMPERATURE
   . **BODY TEMPERATURE**
RT   COLD TOLERANCE
    FEVER
    HEAT ACCLIMATIZATION
    HEAT STROKE
    HEAT TOLERANCE
    HOMEOSTASIS
    HOMEOTHERMS
    HUMIDITY
    HYPERTHERMIA
    HYPOTHERMIA
    PERSPIRATION
    POIKILOTHERMIA
    SHIVERING
    THERMORECEPTORS
    THERMOREGULATION
    VASOCONSTRICTION
    VASODILATION

BODY TEMPERATURE (NON-BIOLOGICAL)
USE   TEMPERATURE

BODY TEMPERATURE REGULATION
USE   THERMOREGULATION

BODY VOLUME (BIOLOGY)
GS   VOLUME
   . **BODY VOLUME (BIOLOGY)**
RT   ∞BIOLOGY
    OBESITY

BODY WEIGHT
GS   WEIGHT (MASS)
   . **BODY WEIGHT**
RT   OBESITY
    WEIGHTLESSNESS

BODY-WING AND TAIL CONFIGURATIONS
RT   AERODYNAMIC CONFIGURATIONS
    ∞CONFIGURATIONS
    FUSELAGES
    TAIL ASSEMBLIES
    WINGS

BODY-WING CONFIGURATIONS
RT   AERODYNAMIC CONFIGURATIONS
    AIRFOILS
    DROOPED AIRFOILS
    GAW-2 AIRFOIL
    WINGS

BOEING AIRCRAFT
UF   VERTOL MILITARY HELICOPTERS
GS   **BOEING AIRCRAFT**
   . B-47 AIRCRAFT
   . B-50 AIRCRAFT

BOEING AIRCRAFT-*(CONT.)*
   . B-52 AIRCRAFT
   . BOEING 707 AIRCRAFT
   . BOEING 720 AIRCRAFT
   . BOEING 727 AIRCRAFT
   . BOEING 733 AIRCRAFT
   . BOEING 737 AIRCRAFT
   . BOEING 747 AIRCRAFT
   . BOEING 757 AIRCRAFT
   . BOEING 767 AIRCRAFT
   . BOEING 2707 AIRCRAFT
   . C-135 AIRCRAFT
   . CH-21 HELICOPTER
   . CH-46 HELICOPTER
   . CH-47 HELICOPTER
   . CH-62 HELICOPTER
   . E-3A AIRCRAFT
   . E-4A AIRCRAFT
   . H-25 HELICOPTER
   . RB-50 AIRCRAFT
   . VZ-2 AIRCRAFT
   . X-20 AIRCRAFT
RT   ∞AIRCRAFT
    AWACS AIRCRAFT
    YC-14 AIRCRAFT

BOEING MILITARY AIRCRAFT
USE   MILITARY AIRCRAFT

BOEING 707 AIRCRAFT
GS   BOEING AIRCRAFT
   . **BOEING 707 AIRCRAFT**
   COMMERCIAL AIRCRAFT
   . **BOEING 707 AIRCRAFT**
   JET AIRCRAFT
   . TURBOFAN AIRCRAFT
   . . **BOEING 707 AIRCRAFT**
   MONOPLANES
   . **BOEING 707 AIRCRAFT**
   PASSENGER AIRCRAFT
   . **BOEING 707 AIRCRAFT**
   TRANSPORT AIRCRAFT
   . **BOEING 707 AIRCRAFT**
RT   ∞AIRCRAFT

BOEING 720 AIRCRAFT
GS   BOEING AIRCRAFT
   . **BOEING 720 AIRCRAFT**
   COMMERCIAL AIRCRAFT
   . **BOEING 720 AIRCRAFT**
   JET AIRCRAFT
   . TURBOFAN AIRCRAFT
   . . **BOEING 720 AIRCRAFT**
   MONOPLANES
   . **BOEING 720 AIRCRAFT**
   PASSENGER AIRCRAFT
   . **BOEING 720 AIRCRAFT**
   TRANSPORT AIRCRAFT
   . **BOEING 720 AIRCRAFT**
RT   ∞AIRCRAFT

BOEING 727 AIRCRAFT
GS   BOEING AIRCRAFT
   . **BOEING 727 AIRCRAFT**
   COMMERCIAL AIRCRAFT
   . **BOEING 727 AIRCRAFT**
   JET AIRCRAFT
   . TURBOFAN AIRCRAFT
   . . **BOEING 727 AIRCRAFT**
   PASSENGER AIRCRAFT
   . **BOEING 727 AIRCRAFT**
   TRANSPORT AIRCRAFT
   . **BOEING 727 AIRCRAFT**
RT   ∞AIRCRAFT
    CARGO AIRCRAFT

BOEING 733 AIRCRAFT
GS   BOEING AIRCRAFT
   . **BOEING 733 AIRCRAFT**
   COMMERCIAL AIRCRAFT
   . **BOEING 733 AIRCRAFT**
   JET AIRCRAFT
   . TURBOFAN AIRCRAFT
   . . **BOEING 733 AIRCRAFT**
   MONOPLANES
   . **BOEING 733 AIRCRAFT**
   SUPERSONIC AIRCRAFT
   . **BOEING 733 AIRCRAFT**
   TRANSPORT AIRCRAFT
   . **BOEING 733 AIRCRAFT**
RT   ∞AIRCRAFT
    VARIABLE SWEEP WINGS

BOEING 737 AIRCRAFT
GS   BOEING AIRCRAFT

BOEING 737 AIRCRAFT-*(CONT.)*
   . **BOEING 737 AIRCRAFT**
   COMMERCIAL AIRCRAFT
   . **BOEING 737 AIRCRAFT**
   JET AIRCRAFT
   . TURBOFAN AIRCRAFT
   . . **BOEING 737 AIRCRAFT**
   MONOPLANES
   . **BOEING 737 AIRCRAFT**
   PASSENGER AIRCRAFT
   . **BOEING 737 AIRCRAFT**
   TRANSPORT AIRCRAFT
   . **BOEING 737 AIRCRAFT**
RT   ∞AIRCRAFT
    CARGO AIRCRAFT

BOEING 747 AIRCRAFT
GS   BOEING AIRCRAFT
   . **BOEING 747 AIRCRAFT**
   COMMERCIAL AIRCRAFT
   . **BOEING 747 AIRCRAFT**
   JET AIRCRAFT
   . **BOEING 747 AIRCRAFT**
   PASSENGER AIRCRAFT
   . **BOEING 747 AIRCRAFT**
   TRANSPORT AIRCRAFT
   . **BOEING 747 AIRCRAFT**
RT   ∞AIRCRAFT
    TURBOFAN ENGINES

BOEING 747B AIRCRAFT
USE   E-4A AIRCRAFT

BOEING 757 AIRCRAFT
GS   BOEING AIRCRAFT
   . **BOEING 757 AIRCRAFT**
   COMMERCIAL AIRCRAFT
   . **BOEING 757 AIRCRAFT**
   JET AIRCRAFT
   . TURBOFAN AIRCRAFT
   . . **BOEING 757 AIRCRAFT**
   MONOPLANES
   . **BOEING 757 AIRCRAFT**
   PASSENGER AIRCRAFT
   . **BOEING 757 AIRCRAFT**
   TRANSPORT AIRCRAFT
   . **BOEING 757 AIRCRAFT**
RT   ∞AIRCRAFT

BOEING 767 AIRCRAFT
GS   BOEING AIRCRAFT
   . **BOEING 767 AIRCRAFT**
   COMMERCIAL AIRCRAFT
   . **BOEING 767 AIRCRAFT**
   JET AIRCRAFT
   . TURBOFAN AIRCRAFT
   . . **BOEING 767 AIRCRAFT**
   MONOPLANES
   . **BOEING 767 AIRCRAFT**
   PASSENGER AIRCRAFT
   . **BOEING 767 AIRCRAFT**
   TRANSPORT AIRCRAFT
   . **BOEING 767 AIRCRAFT**
RT   ∞AIRCRAFT
    CARGO AIRCRAFT
    TURBOFAN ENGINES

BOEING 2707 AIRCRAFT
GS   BOEING AIRCRAFT
   . **BOEING 2707 AIRCRAFT**
   COMMERCIAL AIRCRAFT
   . SUPERSONIC COMMERCIAL AIR
    TRANSPORT
   . . **BOEING 2707 AIRCRAFT**
   JET AIRCRAFT
   . **BOEING 2707 AIRCRAFT**
   PASSENGER AIRCRAFT
   . **BOEING 2707 AIRCRAFT**
   SUPERSONIC AIRCRAFT
   . SUPERSONIC TRANSPORTS
   . . SUPERSONIC COMMERCIAL AIR
    TRANSPORT
   . . . **BOEING 2707 AIRCRAFT**
   TRANSPORT AIRCRAFT
   . **BOEING 2707 AIRCRAFT**
RT   ∞AIRCRAFT

BOGOLIUBOV THEORY
RT   BBGKY HIERARCHY
    ∞THEORIES

BOGS
USE   MARSHLANDS

BOHR MAGNETON
GS   CONSTANTS
        . **BOHR MAGNETON**
RT   ELECTRONS
        MAGNETIC MOMENTS

BOHR THEORY
GS   THEORETICAL PHYSICS
        . QUANTUM THEORY
        . . **BOHR THEORY**
RT   ELECTRON TRANSITIONS
        LINE SPECTRA
        ∞ THEORIES

BOILER PLATE
GS   STRUCTURAL MEMBERS
        . PLATES (STRUCTURAL MEMBERS)
        . . METAL PLATES
        . . . **BOILER PLATE**
RT   THICK WALLS

BOILERS
UF   STEAM GENERATORS
GS   HEATING EQUIPMENT
        . **BOILERS**
RT   EXTERNAL COMBUSTION ENGINES
        FURNACES
        ∞ GENERATORS
        HEAT BALANCE
        PRESSURE VESSELS
        STEAM
        VAPORIZERS
        WASTE ENERGY UTILIZATION

BOILING
UF   EBULLITION
GS   PHASE TRANSFORMATIONS
        . VAPORIZING
        . . **BOILING**
        . . . FILM BOILING
        . . . NUCLEATE BOILING
        . . . . LEIDENFROST PHENOMENON
RT   EFFERVESCENCE
        EVAPORATION
        EVOLUTION (LIBERATION)
        HEAT TRANSFER
        HEATING

BOILING WATER REACTORS
GS   NUCLEAR REACTORS
        . LIQUID COOLED REACTORS
        . . WATER COOLED REACTORS
        . . . **BOILING WATER REACTORS**
        . . . . EXPERIMENTAL BOILING WATER
                REACTORS
        . . . . HALDEN BOILING WATER
                REACTOR
        . . . . LOS ALAMOS WATER BOILER
                REACTOR
        . . . . PATHFINDER NUCLEAR REACTOR
        . . . . SPERT REACTORS
RT   NUCLEAR POWER REACTORS
        NUCLEAR RESEARCH AND TEST
          REACTORS

BOLIDES
GS   CELESTIAL BODIES
        . METEOROIDS
        . . **BOLIDES**
        . . . CYRILLID METEOROIDS
RT   ATMOSPHERIC ENTRY
        ATMOSPHERIC HEATING
        ∞ FIREBALLS
        METEOR TRAILS
        METEORITES
        METEOROID SHOWERS
        PRIBRAM METEORITE

BOLIVIA
GS   NATIONS
        . **BOLIVIA**
RT   SOUTH AMERICA

BOLKOW AIRCRAFT
GS   **BOLKOW AIRCRAFT**
        . BO-105 HELICOPTER
RT   ∞ AIRCRAFT

BOLL WEEVILS
GS   ANIMALS
        . INVERTEBRATES
        . . ARTHROPODS
        . . . INSECTS
        . . . . **BOLL WEEVILS**

BOLL WEEVILS-(CONT.)
RT   BOLLWORMS
        COTTON
        INFESTATION

BOLLWORMS
GS   ANIMALS
        . INVERTEBRATES
        . . ARTHROPODS
        . . . INSECTS
        . . . . LARVAE
        . . . . . **BOLLWORMS**
RT   BOLL WEEVILS
        CORN
        COTTON
        FRUITS
        INFESTATION
        MOTHS

BOLOGRAMS
USE   BOLOMETERS

BOLOMETERS
UF   BOLOGRAMS
GS   MEASURING INSTRUMENTS
        . RADIATION MEASURING INSTRUMENTS
        . . **BOLOMETERS**
RT   DICKE RADIOMETERS
        ELECTRICAL MEASUREMENT
        HEAT MEASUREMENT
        INFRARED DETECTORS
        PHOTOMETERS
        POTENTIOMETERS (INSTRUMENTS)
        RADIATION PYROMETERS
        RADIOMETERS
        RESISTANCE THERMOMETERS
        TEMPERATURE MEASUREMENT
        TEMPERATURE MEASURING
          INSTRUMENTS

BOLTS
GS   FASTENERS
        . **BOLTS**
        . . ROCK BOLTS
        . . TIEBOLTS
RT   ANCHORS (FASTENERS)
        COUPLINGS
        HOLDERS
        NUTS (FASTENERS)
        SCREWS
        STUDS (STRUCTURAL MEMBERS)
        THREADS

BOLTZMANN DISTRIBUTION
GS   DISTRIBUTION (PROPERTY)
        . **BOLTZMANN DISTRIBUTION**
RT   ATMOSPHERIC DENSITY
        ATMOSPHERIC DIFFUSION
        KINETIC THEORY
        STATISTICAL MECHANICS
        TWO FLUID MODELS

BOLTZMANN TRANSPORT EQUATION
RT   BBGKY HIERARCHY
        CHAPMAN-ENSKOG THEORY
        ∞ EQUATIONS
        FOKKER-PLANCK EQUATION
        HYDRODYNAMIC EQUATIONS
        KINETIC THEORY
        PARTICLE DIFFUSION
        STATISTICAL MECHANICS
        TRANSPORT PROPERTIES
        TRANSPORT THEORY

BOLTZMANN-VLASOV EQUATION
RT   ∞ EQUATIONS
        HIGH TEMPERATURE PLASMAS
        MAXWELL EQUATION
        PARTIAL DIFFERENTIAL EQUATIONS
        WAVE EQUATIONS

BOLZA PROBLEMS
RT   OPTIMIZATION
        ∞ PROBLEMS

BOMARC A MISSILE
GS   MISSILES
        . SURFACE TO AIR MISSILES
        . . BOMARC MISSILES
        . . . **BOMARC A MISSILE**
RT   LIQUID PROPELLANT ROCKET ENGINES
        SOLID PROPELLANT ROCKET ENGINES

BOMARC B MISSILE
GS   MISSILES
        . SURFACE TO AIR MISSILES
        . . BOMARC MISSILES
        . . . **BOMARC B MISSILE**
RT   LIQUID PROPELLANT ROCKET ENGINES
        SOLID PROPELLANT ROCKET ENGINES

BOMARC MISSILES
GS   MISSILES
        . ANTIAIRCRAFT MISSILES
        . . **BOMARC MISSILES**
        . SURFACE TO AIR MISSILES
        . . **BOMARC MISSILES**
        . . . BOMARC A MISSILE
        . . . BOMARC B MISSILE

BOMB CALORIMETERS
GS   MEASURING INSTRUMENTS
        . CALORIMETERS
        . . **BOMB CALORIMETERS**
RT   DROP CALORIMETERS
        FLAME CALORIMETERS
        HEAT MEASUREMENT
        HIGH TEMPERATURE TESTS
        TEMPERATURE MEASURING
          INSTRUMENTS

∞ **BOMBARDMENT**
SN   (USE OF A MORE SPECIFIC TERM IS
        RECOMMENDED--CONSULT THE TERMS
        LISTED BELOW)
RT   HYPERVELOCITY PROJECTILES
        IRRADIATION
        METEORITIC DAMAGE
        SPUTTERING

BOMBER AIRCRAFT
GS   ATTACK AIRCRAFT
        . **BOMBER AIRCRAFT**
        . . A-2 AIRCRAFT
        . . A-3 AIRCRAFT
        . . A-4 AIRCRAFT
        . . A-5 AIRCRAFT
        . . A-6 AIRCRAFT
        . . B-1 AIRCRAFT
        . . B-26 AIRCRAFT
        . . B-47 AIRCRAFT
        . . B-50 AIRCRAFT
        . . B-52 AIRCRAFT
        . . B-57 AIRCRAFT
        . . B-58 AIRCRAFT
        . . B-66 AIRCRAFT
        . . B-70 AIRCRAFT
        . . F-100 AIRCRAFT
        . . SHACKLETON BOMBER
        . . VALIANT AIRCRAFT
        . . VICTOR MK-1 AIRCRAFT
        . . VULCAN AIRCRAFT
RT   ∞ AIRCRAFT
        ANTISUBMARINE WARFARE AIRCRAFT
        BOMBING EQUIPMENT
        JET AIRCRAFT
        ∞ MILITARY AIRCRAFT
        ∞ MILITARY AVIATION
        RB-50 AIRCRAFT
        TANKER AIRCRAFT
        TRAINING AIRCRAFT
        VAMPIRE MK 35 AIRCRAFT

BOMBING EQUIPMENT
GS   ONBOARD EQUIPMENT
        . AIRCRAFT EQUIPMENT
        . . **BOMBING EQUIPMENT**
RT   B-1 AIRCRAFT
        BOMBER AIRCRAFT
        BOMBS (ORDNANCE)
        ∞ EQUIPMENT
        FIRE CONTROL

∞ **BOMBS**
SN   (USE OF A MORE SPECIFIC TERM IS
        RECOMMENDED--CONSULT THE TERMS
        LISTED BELOW)
RT   BOMBS (ORDNANCE)
        PRECISION GUIDED PROJECTILES
        PRESSURE GAGES
        SAMPLERS

BOMBS (ORDNANCE)
GS   EXPLOSIVE DEVICES
        . **BOMBS (ORDNANCE)**
RT   AMMUNITION
        B-1 AIRCRAFT
        BOMBING EQUIPMENT

BOMBS (ORDNANCE)-*(CONT.)*
∞ BOMBS
  EXPLOSIVES
  INCENDIARY AMMUNITION
  MISSILES
  NUCLEAR WEAPONS
  PROJECTILES
  PYROTECHNICS
  SHAPED CHARGES
  TORPEDOES
  WARHEADS

BOMBS (PRESSURE GAGES)
USE   PRESSURE GAGES

BOMBS (SAMPLERS)
USE   SAMPLERS

BONANZA AIRCRAFT
USE   C-35 AIRCRAFT

BOND GRAPHS
GS   CHARTS
      . GRAPHS (CHARTS)
      . . **BOND GRAPHS**
RT   CONTROL SYSTEMS DESIGN
      DIFFERENTIAL EQUATIONS
      DYNAMIC MODELS
      MATHEMATICAL MODELS
      ∞ MATHEMATICS
      ∞ NETWORKS
      SIMULATION
      SYSTEMS ANALYSIS
      SYSTEMS ENGINEERING

BONDING
GS   **BONDING**
      . ADHESIVE BONDING
      . AGGLUTINATION
      . CERAMIC BONDING
      . EXPLOSIVE WELDING
      . INERTIA BONDING
      . METAL BONDING
      . . METAL-METAL BONDING
      . REACTION BONDING
      . RESIN BONDING
RT   ADHESION
      ADHESION TESTS
      BEAM LEADS
      BINDING
      CEMENTATION
      CHEMICAL BONDS
      COHESION
      COLD WELDING
      DIFFUSION WELDING
      ∞ JOINING
      JOINTS (JUNCTIONS)
      LAMINATES
      LASER WELDING
      SEALING
      SOLDERING
      WELDING

BONDOC METEORITE
GS   CELESTIAL BODIES
      . METEORITES
      . . STONY METEORITES
      . . . ACHONDRITES
      . . . . **BONDOC METEORITE**

BONE DEMINERALIZATION
GS   DEMINERALIZING
      . **BONE DEMINERALIZATION**
      DISEASES
      . **BONE DEMINERALIZATION**
RT   BIOLOGICAL EFFECTS
      ∞ BIOLOGY
      BONES
      OSTEOPOROSIS
      PHYSIOLOGICAL EFFECTS
      WEIGHTLESSNESS

BONE MARROW
RT   CANCER
      ERYTHROCYTES
      LEUKEMIAS

BONE MINERAL CONTENT
UF   BMC
GS   CONTENT
      . **BONE MINERAL CONTENT**
RT   BIOENGINEERING
      ∞ BIOLOGY
      BIOMETRICS

BONE MINERAL CONTENT-*(CONT.)*
      BONES
      CALCIUM CARBONATES
      CALCIUM PHOSPHATES
      COLLAGENS
      MINERALS
      OSTEOPOROSIS

BONES
GS   ANATOMY
      . MUSCULOSKELETAL SYSTEM
      . . **BONES**
      . . . CARTILAGE
      . . . CEREBRUM
      . . . FEMUR
      . . . MARROW
      . . . MASTOIDS
      . . . PELVIS
      . . . SCAPULA
      . . . SCIATIC REGION
      . . . SKULL
      . . . . CRANIUM
      . . . . . INTRACRANIAL CAVITY
      . . . STERNUM
      . . . TIBIA
      . . . ULNA
      . . . VERTEBRAE
RT   ARTHRITIS
      BONE DEMINERALIZATION
      BONE MINERAL CONTENT
      CALCIFICATION
      CHIN
      CONNECTIVE TISSUE
      EXOSKELETONS
      JOINTS (ANATOMY)
      LAMELLA
      OSTEOPOROSIS
      SPINAL CORD
      SPLINTS
      VERTEBRAL COLUMN

BONNE PROJECTION
RT   MAPPING
      MAPS
      ∞ PROJECTION

BOOLEAN ALGEBRA
GS   MATHEMATICAL LOGIC
      . LATTICES (MATHEMATICS)
      . . **BOOLEAN ALGEBRA**
      . . . BOOLEAN FUNCTIONS
RT   ALGEBRA
      ∞ CONJUNCTION
      INSTRUCTION SETS (COMPUTERS)
      ∞ LOGIC
      SET THEORY
      SWITCHING THEORY
      TRANSISTOR LOGIC
      ∞ UNIONS

BOOLEAN FUNCTIONS
GS   FUNCTIONS (MATHEMATICS)
      . **BOOLEAN FUNCTIONS**
      MATHEMATICAL LOGIC
      . LATTICES (MATHEMATICS)
      . . BOOLEAN ALGEBRA
      . . . **BOOLEAN FUNCTIONS**

∞ **BOOM**
SN   *(USE OF A MORE SPECIFIC TERM IS*
      RECOMMENDED--CONSULT THE TERMS
      LISTED BELOW)
RT   BOOMS (EQUIPMENT)
      SONIC BOOMS
      TAIL ASSEMBLIES

BOOMS (EQUIPMENT)
GS   POSITIONING DEVICES (MACHINERY)
      . **BOOMS (EQUIPMENT)**
RT   ∞ BOOM
      CRANES

BOOST
USE   ACCELERATION (PHYSICS)

BOOSTER RECOVERY
RT   EXPENDABLE STAGES (SPACECRAFT)
      RECOVERABLE LAUNCH VEHICLES
      ∞ RECOVERY
      RECOVERY PARACHUTES
      SPACECRAFT RECOVERY

BOOSTER ROCKET ENGINES
UF   ROCKET BOOSTERS

BOOSTER ROCKET ENGINES-*(CONT.)*
GS   ENGINES
      . ROCKET ENGINES
      . . **BOOSTER ROCKET ENGINES**
      . . . AJ-10 ENGINE
      . . . ALGOL ENGINE
      . . . APOGEE BOOST MOTORS
      . . . H-1 ENGINE
      . . . LR-87-AJ-5 ENGINE
      . . . M-1 ENGINE
      . . . M-55 ENGINE
      . . . MA-2 ENGINE
      . . . MA-3 ENGINE
      . . . MA-5 ENGINE
      . . . NIKE BOOSTER ROCKET ENGINES
      . . . P-1 ENGINE
      . . . ROCKET ENGINE 9KS-11000
      . . . X-405 ENGINE
RT   AIR BREATHING BOOSTERS
      BLUE GOOSE MISSILE
      ∞ BOOSTERS
      BURNOUT
      DUCTED ROCKET ENGINES
      EXPENDABLE STAGES (SPACECRAFT)
      F-1 ROCKET ENGINE
      HYBRID PROPELLANT ROCKET ENGINES
      INTERNAL COMBUSTION ENGINES
      LAUNCH VEHICLES
      LIQUID PROPELLANT ROCKET ENGINES
      MACE MISSILES
      NUCLEAR ENGINE FOR ROCKET
        VEHICLES
      NUCLEAR ROCKET ENGINES
      RECOVERABLE SPACECRAFT
      SOLID PROPELLANT ROCKET ENGINES
      SPINNING SOLID UPPER STAGE
      STAGE SEPARATION
      SUSTAINER ROCKET ENGINES
      TURBOROCKET ENGINES
      TX-354 ENGINE

∞ **BOOSTER ROCKETS**
SN   *(USE OF A MORE SPECIFIC TERM IS*
      RECOMMENDED--CONSULT THE TERMS
      LISTED BELOW)
RT   LAUNCH VEHICLES

∞ **BOOSTERS**
SN   *(USE OF A MORE SPECIFIC TERM IS*
      RECOMMENDED--CONSULT THE TERMS
      LISTED BELOW)
RT   AIR BREATHING BOOSTERS
      AMPLIFIERS
      APOGEE BOOST MOTORS
      BOOSTER ROCKET ENGINES
      BOOSTERS (EXPLOSIVES)
      LAUNCH VEHICLES
      SCOUT PROJECT
      SPACE SHUTTLE BOOSTERS
      TITAN PROJECT

BOOSTERS (EXPLOSIVES)
GS   EXPLOSIVE DEVICES
      . INITIATORS (EXPLOSIVES)
      . . **BOOSTERS (EXPLOSIVES)**
      IGNITERS
      . INITIATORS (EXPLOSIVES)
      . . **BOOSTERS (EXPLOSIVES)**
RT   ∞ BOOSTERS
      EXPLODING WIRES

BOOSTGLIDE VEHICLES
GS   GLIDERS
      . **BOOSTGLIDE VEHICLES**
      . . X-20 AIRCRAFT
      REENTRY VEHICLES
      . **BOOSTGLIDE VEHICLES**
      . . X-20 AIRCRAFT
RT   AEROSPACEPLANES
      ∞ AIRCRAFT
      ASTRO VEHICLE
      ASTROPLANE
      GLIDING
      HYPERSONIC AIRCRAFT
      HYPERSONIC GLIDERS
      LIFTING REENTRY VEHICLES
      MANNED SPACECRAFT
      RECOVERABLE SPACECRAFT
      ROCKET PLANES
      ∞ VEHICLES

BOOTS (FOOTWEAR)
GS   CLOTHING
      . **BOOTS (FOOTWEAR)**
RT   SHOES

BORAL
GS   COMPOSITE MATERIALS
   . LAMINATES
   . . **BORAL**
   COMPOSITE STRUCTURES
   . LAMINATES
   . . **BORAL**
RT   ALUMINUM
   BORON CARBIDES
   RADIATION SHIELDING

BORANES
GS   BORON COMPOUNDS
   . BORON HYDRIDES
   . . **BORANES**
   . . . CARBORANE
   . . . HYDRAZINE BORANE
   . . . PENTABORANES
   HYDROGEN COMPOUNDS
   . HYDRIDES
   . . BORON HYDRIDES
   . . . **BORANES**
   . . . . CARBORANE
   . . . . HYDRAZINE BORANE
   . . . . PENTABORANES
HT   BOROHYDRIDES

BORATES
GS   BORON COMPOUNDS
   . **BORATES**
   . . LITHIUM BORATES
RT   BORIC ACIDS
   ∞OXYGEN COMPOUNDS

BORAZON (TRADEMARK)
USE   BORON NITRIDES

BORDERS
RT   BOUNDARIES
   MARGINS
   RIMS

BORDONI PEAKS
RT   ELASTIC DEFORMATION
   PLASTIC DEFORMATION
   RESONANT FREQUENCIES
   STRESS RELAXATION

BOREDOM
RT   DETACHMENT
   HUMAN BEHAVIOR
   HUMAN REACTIONS
   LETHARGY
   MONOTONY
   PSYCHOLOGICAL EFFECTS
   PSYCHOLOGY
   SPACE FLIGHT STRESS

BOREHOLES
RT   CAVITIES
   CLAYS
   DRILLING
   EXCAVATION
   EXPLORATION
   GEOLOGY
   GRAVELS
   ∞HOLES
   MINERALS
   PITS (EXCAVATIONS)
   ROCKS
   SHALES
   SOILS

BOREL SETS
GS   ANALYSIS (MATHEMATICS)
   . REAL VARIABLES
   . . MEASURE AND INTEGRATION
   . . . **BOREL SETS**
   MATHEMATICAL LOGIC
   . SET THEORY
   . . **BOREL SETS**
RT   PROBABILITY THEORY

BORES
USE   CAVITIES

BORESCOPES
USE   ENDOSCOPES

BORESIGHT ERROR
GS   ERRORS
   . **BORESIGHT ERROR**
RT   AIR NAVIGATION
   BORESIGHTS

BORESIGHT ERROR-*(CONT.)*
   DIRECTIONAL ANTENNAS
   DISPLACEMENT
   DISPLACEMENT MEASUREMENT
   ERROR ANALYSIS
   INSTRUMENT ERRORS
   LINE OF SIGHT COMMUNICATION
   NAVIGATION INSTRUMENTS
   OPTICAL TRACKING
   RANGE ERRORS

BORESIGHTS
RT   BORESIGHT ERROR
   DIRECTIONAL ANTENNAS
   OPTICAL TRACKING

BORIC ACIDS
GS   ACIDS
   . **BORIC ACIDS**
   BORON COMPOUNDS
   . **BORIC ACIDS**
RT   BORATES

BORIDES
GS   BORON COMPOUNDS
   . **BORIDES**
   . . CHROMIUM BORIDES
   . . TITANIUM BORIDES
RT   INTERMETALLICS

BORING MACHINES
GS   TOOLS
   . **BORING MACHINES**
RT   DRILLS
   ∞MACHINERY

BORN APPROXIMATION
UF   BORN-MAYER EQUATION
GS   ANALYSIS (MATHEMATICS)
   . NUMERICAL ANALYSIS
   . . APPROXIMATION
   . . . **BORN APPROXIMATION**
RT   ∞EQUATIONS
   QUANTUM MECHANICS
   SCATTERING CROSS SECTIONS

BORN-INFELD THEORY
RT   ELECTRODYNAMICS
   ELECTROSTATICS
   MAXWELL EQUATION
   NONLINEAR EQUATIONS
   ∞THEORIES

BORN-MAYER EQUATION
USE   BORN APPROXIMATION

BORN-OPPENHEIMER APPROXIMATION
GS   ANALYSIS (MATHEMATICS)
   . NUMERICAL ANALYSIS
   . . APPROXIMATION
   . . . **BORN-OPPENHEIMER**
          APPROXIMATION
RT   FRANCK-CONDON PRINCIPLE

BOROHYDRIDES
GS   BORON COMPOUNDS
   . **BOROHYDRIDES**
   . . BERYLLIUM BOROHYDRIDES
   HYDROGEN COMPOUNDS
   . HYDRIDES
   . . **BOROHYDRIDES**
   . . . ALUMINUM BOROHYDRIDES
   . . . BERYLLIUM BOROHYDRIDES
RT   BORANES
   BORON HYDRIDES

BORON
GS   CHEMICAL ELEMENTS
   . METALLOIDS
   . . **BORON**
   . . . BORON ISOTOPES
   . . . . BORON 10
RT   BORON REINFORCED MATERIALS
   BORSIC (TRADENAME)

BORON ALLOYS
GS   ALLOYS
   . **BORON ALLOYS**
RT   METALLOIDS

BORON CARBIDES
GS   BORON COMPOUNDS
   . **BORON CARBIDES**
   CARBON COMPOUNDS

BORON CARBIDES-*(CONT.)*
   . CARBIDES
   . . **BORON CARBIDES**
RT   BORAL

BORON CHLORIDES
GS   BORON COMPOUNDS
   . **BORON CHLORIDES**
   HALOGEN COMPOUNDS
   . CHLORINE COMPOUNDS
   . . CHLORIDES
   . . . **BORON CHLORIDES**
   . HALIDES
   . . CHLORIDES
   . . . **BORON CHLORIDES**

BORON COMPOUNDS
GS   **BORON COMPOUNDS**
   . ALUMINUM BOROHYDRIDES
   . BORATES
   . . LITHIUM BORATES
   . BORIC ACIDS
   . BORIDES
   . . CHROMIUM BORIDES
   . . TITANIUM BORIDES
   . BOROHYDRIDES
   . . BERYLLIUM BOROHYDRIDES
   . BORON CARBIDES
   . BORON CHLORIDES
   . BORON FLUORIDES
   . BORON HYDRIDES
   . . BERYLLIUM BOROHYDRIDES
   . . BORANES
   . . . CARBORANE
   . . . HYDRAZINE BORANE
   . . . PENTABORANES
   . BORON NITRIDES
   . BORON OXIDES
   . BORON PHOSPHIDES
   . BORON-EPOXY COMPOUNDS
   . DIBORANE
   . ORGANIC BORON COMPOUNDS
   . TOURMALINE
RT   ∞CHEMICAL COMPOUNDS
   ∞GROUP 3A COMPOUNDS
   HIGH ENERGY FUELS
   METAL FUELS
   METAL PROPELLANTS

BORON FIBERS
GS   FIBERS
   . REINFORCING FIBERS
   . . **BORON FIBERS**
RT   ALUMINUM BORON COMPOSITES
   BORSIC (TRADENAME)
   CARBON FIBERS
   COMPOSITE MATERIALS
   FIBER COMPOSITES
   FIBER ORIENTATION
   FIBER STRENGTH
   ∞FILAMENTS
   GLASS FIBERS
   METAL MATRIX COMPOSITES
   POLYMER MATRIX COMPOSITES
   REINFORCED PLASTICS

BORON FLUORIDES
UF   BORON TRIFLUORIDE
GS   BORON COMPOUNDS
   . **BORON FLUORIDES**
   HALOGEN COMPOUNDS
   . FLUORINE COMPOUNDS
   . . FLUORIDES
   . . . **BORON FLUORIDES**
   . HALIDES
   . . FLUORIDES
   . . . **BORON FLUORIDES**

BORON HYDRIDES
GS   BORON COMPOUNDS
   . **BORON HYDRIDES**
   . . BERYLLIUM BOROHYDRIDES
   . . BORANES
   . . . CARBORANE
   . . . HYDRAZINE BORANE
   . . . PENTABORANES
   HYDROGEN COMPOUNDS
   . HYDRIDES
   . . **BORON HYDRIDES**
   . . . ALUMINUM BOROHYDRIDES
   . . . BERYLLIUM BOROHYDRIDES
   . . . BORANES
   . . . . CARBORANE
   . . . . HYDRAZINE BORANE
   . . . . PENTABORANES

BORON HYDRIDES-*(CONT.)*
RT   BOROHYDRIDES

BORON ISOTOPES
GS   CHEMICAL ELEMENTS
      . METALLOIDS
      . . BORON
      . . . **BORON ISOTOPES**
      . . . . BORON 10
      . NUCLIDES
      . . ISOTOPES
      . . . **BORON ISOTOPES**
      . . . . BORON 10

BORON NITRIDES
UF   BORAZON (TRADEMARK)
GS   BORON COMPOUNDS
      . **BORON NITRIDES**
      NITROGEN COMPOUNDS
      . NITRIDES
      . . **BORON NITRIDES**

BORON OXIDES
GS   BORON COMPOUNDS
      . **BORON OXIDES**
      CHALCOGENIDES
      . OXIDES
      . . **BORON OXIDES**

BORON PHOSPHIDES
GS   BORON COMPOUNDS
      . **BORON PHOSPHIDES**
      PHOSPHORUS COMPOUNDS
      . PHOSPHIDES
      . . **BORON PHOSPHIDES**

BORON REINFORCED MATERIALS
GS   COMPOSITE MATERIALS
      . **BORON REINFORCED MATERIALS**
      . . BORON-EPOXY COMPOUNDS
RT   BORON
      CERAMIC MATRIX COMPOSITES
      EPOXY RESINS
      FIBER COMPOSITES
      FIBERS
      ∞MATERIALS
      PLASTICS
      REINFORCED PLASTICS
      REINFORCING FIBERS

BORON TRIFLUORIDE
USE   BORON FLUORIDES

BORON 10
GS   CHEMICAL ELEMENTS
      . METALLOIDS
      . . BORON
      . . . BORON ISOTOPES
      . . . . **BORON 10**
      . NUCLIDES
      . . ISOTOPES
      . . . BORON ISOTOPES
      . . . . **BORON 10**

BORON-EPOXY COMPOUNDS
GS   BORON COMPOUNDS
      . **BORON-EPOXY COMPOUNDS**
      COMPOSITE MATERIALS
      . BORON REINFORCED MATERIALS
      . . **BORON-EPOXY COMPOUNDS**
      EPOXY COMPOUNDS
      . **BORON-EPOXY COMPOUNDS**
RT   AIRCRAFT STRUCTURES
      ∞CHEMICAL COMPOUNDS
      COMPOSITE STRUCTURES
      EPOXY RESINS
      FIBER COMPOSITES
      LAMINATES
      PLASTIC AIRCRAFT STRUCTURES
      SPACECRAFT COMPONENTS
      SUPERHYBRID MATERIALS

BOROSILICATE GLASS
UF   PYREX (TRADEMARK)
GS   GLASS
      . **BOROSILICATE GLASS**
RT   GLASSWARE
      SILICON DIOXIDE

BORSIC (TRADENAME)
GS   COMPOSITE MATERIALS
      . METAL MATRIX COMPOSITES
      . . **BORSIC (TRADENAME)**
      MIXTURES

BORSIC (TRADENAME)-*(CONT.)*
      . METAL MATRIX COMPOSITES
      . . **BORSIC (TRADENAME)**
RT   ALUMINUM
      BORON
      BORON FIBERS
      FIBER COMPOSITES
      ∞MATERIALS
      METAL FIBERS
      METALS

BOSE GEOMETRY
GS   GEOMETRY
      . **BOSE GEOMETRY**
RT   EQUATIONS OF STATE

BOSE-CHAUDHURI-HOCQUENGHEM CODES
USE   BCH CODES

BOSE-EINSTEIN STATISTICS
USE   QUANTUM STATISTICS

BOSON FIELDS
RT   FIELD THEORY (PHYSICS)
      ∞FIELDS
      MESONS

BOSONS
GS   PARTICLES
      . ELEMENTARY PARTICLES
      . . **BOSONS**
      . . . ALPHA PARTICLES
      . . . MESONS
      . . . . ETA-MESONS
      . . . . KAONS
      . . . . MESON RESONANCE
      . . . . . X MESONS
      . . . . MUONS
      . . . . PIONS
      . . . . VECTOR MESONS
      . . . . . RHO-MESONS
      . . . . . SIGMA-MESONS
      . . . PHOTONS
      . . . LIGHT BEAMS
      . . . XI HYPERONS
      . NUCLEAR PARTICLES
      . . **BOSONS**
      . . . ALPHA PARTICLES
      . . . MESONS
      . . . . ETA-MESONS
      . . . . KAONS
      . . . . MESON RESONANCE
      . . . . . X MESONS
      . . . . MUONS
      . . . . PIONS
      . . . . VECTOR MESONS
      . . . . . RHO-MESONS
      . . . . . SIGMA-MESONS
      . . . PHOTONS
      . . . XI HYPERONS
RT   CHARGED PARTICLES
      FERMI-DIRAC STATISTICS
      QUANTUM STATISTICS

BOTANY
GS   BOTANY
      . GEOBOTANY
RT   AGRICULTURE
      ALFALFA
      BARLEY
      BIOGEOCHEMISTRY
      ∞BIOLOGY
      BLIGHT
      BROWN WAVE EFFECT
      BRUSH (BOTANY)
      CHAPARRAL
      CITRUS TREES
      CORN
      FARM CROPS
      FRUITS
      GREEN WAVE EFFECT
      HABITATS
      HAY
      LEGUMINOUS PLANTS
      NIGELLA
      OATS
      PLANTS (BOTANY)
      ∞SCIENCE
      SILVICULTURE
      SUGAR BEETS
      SUGAR CANE
      VEGETATION GROWTH
      VINEYARDS
      ∞ZOOLOGY

BOTSWANA
GS   NATIONS
      . **BOTSWANA**
RT   AFRICA
      REPUBLIC OF SOUTH AFRICA

BOTTLES
RT   ∞CONTAINERS
      FLASKS
      GLASSWARE
      TANKS (CONTAINERS)

BOUGUER LAW
UF   LAMBERT LAW
RT   ABSORPTIVITY
      BEER LAW
      ELECTROMAGNETIC ABSORPTION
      LAMBERT SURFACE
      THERMOPLASTICITY

BOULES
GS   CRYSTALS
      . **BOULES**
RT   SINGLE CRYSTALS

BOUNDARIES
UF   PERIPHERIES
GS   **BOUNDARIES**
      . FLUID BOUNDARIES
      . . GAS-SOLID INTERFACES
      . . JET BOUNDARIES
      . . LIQUID-LIQUID INTERFACES
      . . LIQUID-SOLID INTERFACES
      . . LIQUID-VAPOR INTERFACES
      . FREE BOUNDARIES
      . GRAIN BOUNDARIES
RT   AIRSPACE
      BORDERS
      CIRCUMFERENCES
      CONTOUR SENSORS
      DELINEATION
      FENCES (BARRIERS)
      INTERFACES
      REGIONS

BOUNDARY ELEMENT METHOD
GS   STRESS ANALYSIS
      . **BOUNDARY ELEMENT METHOD**
RT   APERTURES

BOUNDARY INTEGRAL METHOD
GS   PROCEDURES
      . **BOUNDARY INTEGRAL METHOD**
RT   BOUNDARY VALUE PROBLEMS
      ∞METHODOLOGY

BOUNDARY LAYER COMBUSTION
GS   COMBUSTION
      . **BOUNDARY LAYER COMBUSTION**
RT   BOUNDARY LAYERS
      COMBUSTIBLE FLOW
      CONVECTIVE HEAT TRANSFER
      DIFFUSION FLAMES
      FLAME PROPAGATION
      LAMINAR BOUNDARY LAYER

BOUNDARY LAYER CONTROL
UF   LAMINAR FLOW CONTROL
GS   **BOUNDARY LAYER CONTROL**
      . POROUS BOUNDARY LAYER CONTROL
RT   AERODYNAMICS
      AIRFOIL FENCES
      ∞BLEEDING
      BLOWING
      BOUNDARY LAYERS
      BUFFETING
      CIRCULATION CONTROL AIRFOILS
      ∞CONTROL
      CONTROL SURFACES
      DRAG DEVICES
      FLUID AMPLIFIERS
      FLUTTER
      JET CONTROL
      LEADING EDGE SLATS
      LIFT AUGMENTATION
      LIFT DEVICES
      SPOILERS
      TURBULENCE
      UPPER SURFACE BLOWN FLAPS
      VACUUM
      VORTEX GENERATORS
      WING SLOTS
      X-21 AIRCRAFT

BOUNDARY LAYER EQUATIONS
RT   BOUNDARY LAYERS
        DIFFERENTIAL EQUATIONS
∞EQUATIONS
        FLOW EQUATIONS
        FLOW THEORY

BOUNDARY LAYER FLOW
GS   FLUID FLOW
        . VISCOUS FLOW
        . . **BOUNDARY LAYER FLOW**
        . . . REATTACHED FLOW
        . . . SECONDARY FLOW
        . . . SEPARATED FLOW
        . . . . BOUNDARY LAYER SEPARATION
RT   ATMOSPHERIC BOUNDARY LAYER
        BACKWARD FACING STEPS
        BLASIUS EQUATION
        CONVECTIVE HEAT TRANSFER
        FLOW DISTRIBUTION
        LIGHTHILL GAS MODEL
        MAGNUS EFFECT
        RECIRCULATIVE FLUID FLOW
        REYNOLDS NUMBER
        STAGNATION FLOW
        TOLLMEIN-SCHLICHTING WAVES
        WALL FLOW

BOUNDARY LAYER NOISE
USE   AERODYNAMIC NOISE
        BOUNDARY LAYERS

BOUNDARY LAYER PLASMAS
GS   GASES
        . IONIZED GASES
        . . CHARGED PARTICLES
        . . . **BOUNDARY LAYER PLASMAS**
        PARTICLES
        . CHARGED PARTICLES
        . ENERGETIC PARTICLES
        . . . PLASMAS (PHYSICS)
        . . . . **BOUNDARY LAYER PLASMAS**
RT   AERODYNAMIC HEATING
        AEROTHERMODYNAMICS
        BOUNDARY LAYERS
        HYPERSONIC REENTRY
        PLASMA PHYSICS
        PLASMA SHEATHS

BOUNDARY LAYER SEPARATION
UF   BREAKAWAY
        FLOW SEPARATION
GS   FLUID FLOW
        . VISCOUS FLOW
        . . BOUNDARY LAYER FLOW
        . . . SEPARATED FLOW
        . . . . **BOUNDARY LAYER SEPARATION**
RT   AERODYNAMIC STALLING
        AIRSPEED
        ANGLE OF ATTACK
        BOUNDARY LAYERS
        CROCCO-LEE THEORY
        ∞DIFFUSERS
        FALKNER-SKAN EQUATION
        FLOW DISTRIBUTION
        INJECTION
        KUTTA-JOUKOWSKI CONDITION
        LIFT DRAG RATIO
        REATTACHED FLOW
        RECIRCULATIVE FLUID FLOW
        REVERSED FLOW
        ROTATING STALLS
        ∞SEPARATION
        STAGNATION FLOW
        ∞STALLING
        SWEEP ANGLE
        VORTEX GENERATORS
        ZERO LIFT

BOUNDARY LAYER STABILITY
GS   DYNAMIC CHARACTERISTICS
        . DYNAMIC STABILITY
        . . MOTION STABILITY
        . . . FLOW STABILITY
        . . . . **BOUNDARY LAYER STABILITY**
        . FLOW CHARACTERISTICS
        . . FLOW STABILITY
        . . . **BOUNDARY LAYER STABILITY**
        STABILITY
        . DYNAMIC STABILITY
        . . MOTION STABILITY
        . . . FLOW STABILITY
        . . . . **BOUNDARY LAYER STABILITY**
RT   AERODYNAMIC STABILITY
        BACKWASH

BOUNDARY LAYER STABILITY-(CONT.)
        BOUNDARY LAYERS
        GOERTLER INSTABILITY
        REYNOLDS NUMBER

BOUNDARY LAYER TRANSITION
RT   AIR CURRENTS
        BOUNDARY LAYERS
        EKMAN LAYER
        GOERTLER INSTABILITY
        KNUDSEN FLOW
        LAMINAR BOUNDARY LAYER
        LAMINAR FLOW
        MOLECULAR FLOW
        REYNOLDS NUMBER
        THREE DIMENSIONAL BOUNDARY LAYER
        TOLLMEIN-SCHLICHTING WAVES
        ∞TRANSITION
        TRANSITION FLOW
        ∞TRANSITION LAYERS
        TRANSITION POINTS
        TURBULENCE
        TURBULENT BOUNDARY LAYER
        TURBULENT FLOW

BOUNDARY LAYERS
UF   BOUNDARY LAYER NOISE
GS   **BOUNDARY LAYERS**
        . ATMOSPHERIC BOUNDARY LAYER
        . COMPRESSIBLE BOUNDARY LAYER
        . HYPERSONIC BOUNDARY LAYER
        . INCOMPRESSIBLE BOUNDARY LAYER
        . LAMINAR BOUNDARY LAYER
        . PLANETARY BOUNDARY LAYER
        . SUPERSONIC BOUNDARY LAYERS
        . THERMAL BOUNDARY LAYER
        . THREE DIMENSIONAL BOUNDARY
            LAYER
        . TURBULENT BOUNDARY LAYER
        . TWO DIMENSIONAL BOUNDARY LAYER
RT   BOUNDARY LAYER COMBUSTION
        BOUNDARY LAYER CONTROL
        BOUNDARY LAYER EQUATIONS
        BOUNDARY LAYER PLASMAS
        BOUNDARY LAYER SEPARATION
        BOUNDARY LAYER STABILITY
        BOUNDARY LAYER TRANSITION
        CROCCO METHOD
        ∞DRAFT
        DRAG
        FLUID BOUNDARIES
        FLUID FLOW
        GAS-SOLID INTERFACES
        ∞LAYERS
        LIQUID-LIQUID INTERFACES
        LIQUID-SOLID INTERFACES
        PANEL METHOD (FLUID DYNAMICS)
        SHEAR LAYERS
        SURFACE LAYERS
        WALL PRESSURE

BOUNDARY LUBRICATION
GS   LUBRICATION
        . **BOUNDARY LUBRICATION**
RT   BEARINGS
        LUBRICANTS
        SQUEEZE FILMS

BOUNDARY VALUE PROBLEMS
UF   INITIAL VALUE PROBLEMS
        POINT MATCHING METHOD
            (MATHEMATICS)
GS   **BOUNDARY VALUE PROBLEMS**
        . CAUCHY PROBLEM
        . DIRICHLET PROBLEM
        . NEUMANN PROBLEM
RT   BESSEL FUNCTIONS
        BOUNDARY INTEGRAL METHOD
        COUNTER ROTATION
        CRANK-NICHOLSON METHOD
        DIFFERENTIAL EQUATIONS
        FINITE ELEMENT METHOD
        FINITE VOLUME METHOD
        HALF PLANES
        HALF SPACES
        HANKEL FUNCTIONS
        LAME FUNCTIONS
        MATHIEU FUNCTION
        MINIMAL SURFACES
        MONGE-AMPERE EQUATION
        OBSERVABILITY (SYSTEMS)
        ∞PROBLEMS
        SOBOLEV SPACE
        THREE DIMENSIONAL BODIES

BOURDON TUBES
GS   TRANSDUCERS
        . PRESSURE SENSORS
        . . **BOURDON TUBES**
RT   PRESSURE GAGES
        PRESSURE MEASUREMENT
        ∞TUBES

BOUSSINESQ APPROXIMATION
RT   CONVECTION
        HEAT TRANSFER
        INCOMPRESSIBLE FLUIDS
        PERTURBATION THEORY
        THERMAL EXPANSION

BOW SHOCK WAVES
USE   BOW WAVES
        SHOCK WAVES

BOW WAVES
UF   BOW SHOCK WAVES
RT   HYPERSONIC WAKES
        MACH CONES
        SHOCK WAVES
        SURFACE WAVES

∞ **BOWS**
SN   (USE OF A MORE SPECIFIC TERM IS
        RECOMMENDED--CONSULT THE TERMS
        LISTED BELOW)
RT   BENDING
        CAMBER
        FOREBODIES
        HEAVING

BOX BEAMS
GS   STRUCTURAL MEMBERS
        . BEAMS (SUPPORTS)
        . . **BOX BEAMS**
RT   ∞BOXES
        CANTILEVER BEAMS
        GIRDERS
        RECTANGULAR BEAMS

∞ **BOXES**
SN   (USE OF A MORE SPECIFIC TERM IS
        RECOMMENDED--CONSULT THE TERMS
        LISTED BELOW)
RT   BOX BEAMS
        BOXES (CONTAINERS)

BOXES (CONTAINERS)
RT   ∞BOXES
        ∞BUCKETS
        CASES (CONTAINERS)
        ∞CONTAINERS
        PACKAGES

BRACKETS
RT   ANCHORS (FASTENERS)
        FASTENERS
        FIXTURES
        HOLDERS
        MOUNTING

BRADYCARDIA
GS   RATES (PER TIME)
        . HEART RATE
        . . **BRADYCARDIA**
        SIGNS AND SYMPTOMS
        . **BRADYCARDIA**
RT   HEART DISEASES

BRAGG ANGLE
GS   GEOMETRY
        . EUCLIDEAN GEOMETRY
        . . ANGLES (GEOMETRY)
        . . . **BRAGG ANGLE**
RT   CRYSTALLOGRAPHY
        DIFFRACTION
        DIFFRACTION PATHS
        ELECTRON DIFFRACTION
        ISOTROPY
        ∞ORIENTATION
        ∞PHYSICAL PROPERTIES
        RADIOGRAPHY

BRAGG CURVE
RT   BIOLOGICAL EFFECTS
        NUCLEAR REACTIONS
        PARTICLE INTERACTIONS
        RADIATION EFFECTS

BRAILLE
RT   BLINDNESS
     EMBOSSING

BRAIN
GS   ANATOMY
     . **BRAIN**
     . . BRAIN STEM
     . . CEREBELLUM
     . . CEREBRAL CORTEX
     . . CEREBRAL VENTRICLES
     . . CEREBRUM
     . . HIPPOCAMPUS
     NERVOUS SYSTEM
     . CENTRAL NERVOUS SYSTEM
     . . **BRAIN**
     . . . BRAIN STEM
     . . . CEREBELLUM
     . . . CEREBRAL CORTEX
     . . . CEREBRAL VENTRICLES
     . . . CEREBRUM
     . . . HIPPOCAMPUS
RT   ANGIOGRAPHY
     BRAIN CIRCULATION
     BRAIN DAMAGE
     CEREBROSPINAL FLUID
     DIENCEPHALON
     ECHOENCEPHALOGRAPHY
     ELECTROENCEPHALOGRAPHY
     ENCEPHALITIS
     HEAD (ANATOMY)
     HYPOTHALAMUS
     INTRACRANIAL PRESSURE
     NEUROLOGY
     PINEAL GLAND
     PSYCHIATRY
     PSYCHOLOGY
     RHEOENCEPHALOGRAPHY
     SPINAL CORD
     THALAMUS

BRAIN CIRCULATION
GS   CIRCULATION
     . BLOOD CIRCULATION
     . . **BRAIN CIRCULATION**
RT   BRAIN
     RHEOENCEPHALOGRAPHY

BRAIN DAMAGE
GS   INJURIES
     . **BRAIN DAMAGE**
RT   BRAIN

BRAIN STEM
GS   ANATOMY
     . BRAIN
     . . **BRAIN STEM**
     NERVOUS SYSTEM
     . CENTRAL NERVOUS SYSTEM
     . . BRAIN
     . . . **BRAIN STEM**

∞ **BRAKES**
SN   *(USE OF A MORE SPECIFIC TERM IS*
     RECOMMENDED--CONSULT THE TERMS
     LISTED BELOW)
RT   BRAKES (FOR ARRESTING MOTION)
     BRAKES (FORMING OR BENDING)

BRAKES (FOR ARRESTING MOTION)
UF   DECELERATORS
     DRAGULATORS
GS   **BRAKES (FOR ARRESTING MOTION)**
     . AERODYNAMIC BRAKES
     . . BALLUTES
     . . DRAG CHUTES
     . . PARAVULCOONS
     . . SPLIT FLAPS
     . . WING FLAPS
     . . . LEADING EDGE SLATS
     . . . TRAILING EDGE FLAPS
     . AIRCRAFT BRAKES
     . . SPLIT FLAPS
     . . WING FLAPS
     . . . LEADING EDGE SLATS
     . . . TRAILING EDGE FLAPS
     . WHEEL BRAKES
RT   ABORT APPARATUS
     ANTISKID DEVICES
     ∞ARRESTERS
     ARRESTING GEAR
     ∞BRAKES
     BRAKING
     CYLINDRICAL CHAMBERS
     DRAG DEVICES

BRAKES (FOR ARRESTING MOTION)-*(CONT.)*
     FLAPS (CONTROL SURFACES)
     LANDING GEAR
     NOSE WHEELS
     PARACHUTES
     RETARDERS (DEVICES)
     THRUST REVERSAL
     TOWED BODIES
     VEHICLE WHEELS
     WHEELS

BRAKES (FORMING OR BENDING)
RT   ∞BRAKES
     METAL WORKING

BRAKING
RT   BRAKES (FOR ARRESTING MOTION)
     DECELERATION
     EDDY CURRENTS
     RETARDERS (DEVICES)
     RETARDING
     THRUST REVERSAL

BRANCHING (MATHEMATICS)
UF   BIFURCATION (MATHEMATICS)
RT   CHAOS
     FUNCTIONS (MATHEMATICS)
     ∞LOGIC
     MATHEMATICAL LOGIC
     SET THEORY
     SWITCHING THEORY

BRANCHING (PHYSICS)
RT   BIFURCATION (BIOLOGY)
     ∞PHYSICS

BRASSES
GS   ALLOYS
     . COPPER ALLOYS
     . . **BRASSES**

BRAVAIS CRYSTALS
GS   CRYSTALS
     . **BRAVAIS CRYSTALS**
RT   CRYSTAL GROWTH
     CRYSTAL LATTICES
     CRYSTAL STRUCTURE
     PACKING DENSITY
     SINGLE CRYSTALS

BRAYTON CYCLE
GS   CYCLES
     . THERMODYNAMIC CYCLES
     . . **BRAYTON CYCLE**
RT   GAS TURBINE ENGINES
     GAS TURBINES
     RANKINE CYCLE

BRAZIL
GS   NATIONS
     . **BRAZIL**
RT   AMAZON REGION (SOUTH AMERICA)
     BRAZILIAN SPACE PROGRAM
     SOUTH AMERICA

BRAZILIAN SPACE PROGRAM
GS   PROGRAMS
     . SPACE PROGRAMS
     . . **BRAZILIAN SPACE PROGRAM**
RT   BRAZIL

BRAZING
GS   WELDING
     . LASER WELDING
     . . FUSION WELDING
     . . . GAS WELDING
     . . . . **BRAZING**
     . . . . . LOW TEMPERATURE BRAZING
RT   FLUXES
     ∞JOINING
     METAL BONDING
     SEALING
     SOLDERING

BRAZZAVILLE
USE CONGO (BRAZZAVILLE)

BREADBOARD MODELS
GS   MODELS
     . **BREADBOARD MODELS**
RT   CIRCUITS
     PRINTED CIRCUITS
     PRODUCT DEVELOPMENT
     PROTOTYPES

BREAKAWAY
USE BOUNDARY LAYER SEPARATION

∞ **BREAKDOWN**
SN   *(USE OF A MORE SPECIFIC TERM IS*
     RECOMMENDED--CONSULT TERMS
     LISTED BELOW)
RT   CLASSIFICATIONS
     ELECTRICAL FAULTS
     FAILURE
     GAPS
     METAL WORKING
     SYSTEM FAILURES

BREAKERS (ELECTRIC)
USE CIRCUIT BREAKERS

BREAKING
RT   DESTRUCTION
     FRAGMENTATION
     ∞SEPARATION

BREAKWATERS
UF   JETTIES
     SEA WALLS
RT   CONCRETE STRUCTURES
     HARBORS
     LITTORAL DRIFT
     LITTORAL TRANSPORT
     OCEANOGRAPHY
     STRUCTURAL DESIGN
     ∞STRUCTURES
     UNDERWATER ENGINEERING
     UNDERWATER STRUCTURES
     WATER WAVES
     ∞WAVES

∞ **BREATHING**
SN   *(USE OF A MORE SPECIFIC TERM IS*
     RECOMMENDED--CONSULT THE TERMS
     LISTED BELOW)
RT   ARGON-OXYGEN ATMOSPHERES
     BREATHING APPARATUS
     BREATHING VIBRATION
     EMERGENCY BREATHING TECHNIQUES
     EXPIRATION
     HELIUM-OXYGEN ATMOSPHERES
     HIGH ALTITUDE BREATHING
     HYPERCAPNIA
     HYPERPNEA
     OXYGEN BREATHING
     RESPIRATION
     RESPIRATORY REFLEXES

BREATHING APPARATUS
GS   **BREATHING APPARATUS**
     . OXYGEN MASKS
     . UNDERWATER BREATHING
       APPARATUS
RT   ∞BREATHING
     ∞EQUIPMENT
     LIFE SUPPORT SYSTEMS
     OXYGEN SUPPLY EQUIPMENT
     PORTABLE LIFE SUPPORT SYSTEMS
     RESPIRATORS

BREATHING VIBRATION
GS   VIBRATION
     . STRUCTURAL VIBRATION
     . . **BREATHING VIBRATION**
RT   BENDING VIBRATION
     ∞BREATHING
     EXHAUSTING
     MISSILE VIBRATION
     VENTING

BRECCIA
GS   ROCKS
     . **BRECCIA**
RT   ATAXITE
     IGNEOUS ROCKS
     REGOLITH
     SEDIMENTARY ROCKS
     SOILS

BREEDER REACTORS
GS   NUCLEAR REACTORS
     . **BREEDER REACTORS**
     . . EXPERIMENTAL BREEDER REACTOR
         1
     . . EXPERIMENTAL BREEDER REACTOR
         2
     . . LIGHT WATER BREEDER REACTORS

BREEDER REACTORS-(CONT.)
. . LIQUID METAL FAST BREEDER
REACTORS
RT   ENRICO FERMI ATOMIC POWER PLANT
NUCLEAR POWER REACTORS

BREEDING (REPRODUCTION)
RT   FERTILITY
GENETICS
HEREDITY
∞REPRODUCTION
REPRODUCTION (BIOLOGY)

BREGUET AIRCRAFT
GS   **BREGUET AIRCRAFT**
. **BREGUET 940 AIRCRAFT**
. **BREGUET 941 AIRCRAFT**
. **BREGUET 1150 AIRCRAFT**
RT   ∞AIRCRAFT
JAGUAR AIRCRAFT

BREGUET 940 AIRCRAFT
GS   BREGUET AIRCRAFT
. **BREGUET 940 AIRCRAFT**
MONOPLANES
. **BREGUET 940 AIRCRAFT**
RESEARCH AIRCRAFT
. **BREGUET 940 AIRCRAFT**
V/STOL AIRCRAFT
. SHORT TAKEOFF AIRCRAFT
. . **BREGUET 940 AIRCRAFT**
RT   ∞AIRCRAFT

BREGUET 941 AIRCRAFT
GS   BREGUET AIRCRAFT
. **BREGUET 941 AIRCRAFT**
JET AIRCRAFT
. TURBOPROP AIRCRAFT
. . **BREGUET 941 AIRCRAFT**
MONOPLANES
. **BREGUET 941 AIRCRAFT**
PASSENGER AIRCRAFT
. **BREGUET 941 AIRCRAFT**
TRANSPORT AIRCRAFT
. CARGO AIRCRAFT
. . **BREGUET 941 AIRCRAFT**
V/STOL AIRCRAFT
. SHORT TAKEOFF AIRCRAFT
. . **BREGUET 941 AIRCRAFT**
RT   ∞AIRCRAFT

BREGUET 1150 AIRCRAFT
UF   ATLANTIC AIRCRAFT
GS   ANTISUBMARINE WARFARE AIRCRAFT
. **BREGUET 1150 AIRCRAFT**
ATTACK AIRCRAFT
. **BREGUET 1150 AIRCRAFT**
BREGUET AIRCRAFT
. **BREGUET 1150 AIRCRAFT**
JET AIRCRAFT
. TURBOPROP AIRCRAFT
. . **BREGUET 1150 AIRCRAFT**
MONOPLANES
. **BREGUET 1150 AIRCRAFT**
OBSERVATION AIRCRAFT
. **BREGUET 1150 AIRCRAFT**
RECONNAISSANCE AIRCRAFT
. **BREGUET 1150 AIRCRAFT**
RT   ∞AIRCRAFT

BREMSSTRAHLUNG
GS   ELECTROMAGNETIC RADIATION
. **BREMSSTRAHLUNG**
RT   CERENKOV RADIATION
ELECTRON PHOTON CASCADES
ELECTRON RADIATION
FAR ULTRAVIOLET RADIATION
GAMMA RAY BURSTS
GAMMA RAYS
NUCLEAR RADIATION
RELATIVISTIC PLASMAS
SYNCHROTRON RADIATION
X RAYS

BREWSTER ANGLE
GS   GEOMETRY
. EUCLIDEAN GEOMETRY
. . ANGLES (GEOMETRY)
. . . **BREWSTER ANGLE**
RT   POLARIZATION CHARACTERISTICS
REFLECTION
REFRACTIVITY

BRICKS
GS   MASONRY

BRICKS-(CONT.)
. **BRICKS**
RT   CEMENTS
CERAMICS
CLAYS
∞CONSTRUCTION MATERIALS
MORTARS (MATERIAL)

∞ **BRIDGES**
SN   *(USE OF A MORE SPECIFIC TERM IS
RECOMMENDED--CONSULT THE TERMS
LISTED BELOW)*
RT   BRIDGES (LANDFORMS)
BRIDGES (STRUCTURES)
ELECTRIC BRIDGES

BRIDGES (LANDFORMS)
GS   LANDFORMS
. **BRIDGES (LANDFORMS)**
RT   ∞BRIDGES
GEOLOGY
∞RIDGES

BRIDGES (STRUCTURES)
RT   ∞BRIDGES
CONSTRUCTION
CONSTRUCTION INDUSTRY
CROSSINGS
CROSSOVERS
HIGHWAYS
RAMPS (STRUCTURES)
∞STRUCTURES
TOWERS

BRIDGMAN METHOD
RT   CRYSTAL GROWTH
∞METHODOLOGY
SINGLE CRYSTALS

BRIGHTNESS
GS   ELECTROMAGNETIC PROPERTIES
. OPTICAL PROPERTIES
. . **BRIGHTNESS**
. . . SOLAR GRANULATION
RT   BISTATIC REFLECTIVITY
BRIGHTNESS DISTRIBUTION
COLOR
DIMMING
EMISSIVITY
FLUX (RATE)
GLARE
HUMAN FACTORS ENGINEERING
ILLUMINANCE
ILLUMINATING
INCANDESCENCE
∞INTENSITY
LIGHT (VISIBLE RADIATION)
LIMB BRIGHTENING
LUMINANCE
LUMINESCENCE
LUMINOSITY
LUMINOUS INTENSITY
LUSTER
RADIANCE
RADIANT FLUX DENSITY
REFLECTANCE
SKY BRIGHTNESS
STELLAR LUMINOSITY
VISIBILITY
VISION

BRIGHTNESS DISCRIMINATION
GS   DISCRIMINATION
. **BRIGHTNESS DISCRIMINATION**
RT   ∞ILLUMINATION
VISUAL PERCEPTION

BRIGHTNESS DISTRIBUTION
GS   DISTRIBUTION (PROPERTY)
. **BRIGHTNESS DISTRIBUTION**
ELECTROMAGNETIC PROPERTIES
. OPTICAL PROPERTIES
. . **BRIGHTNESS DISTRIBUTION**
STATISTICAL DISTRIBUTIONS
. **BRIGHTNESS DISTRIBUTION**
RT   ASTROPHYSICS
BLACK BODY RADIATION
BRIGHTNESS
BRIGHTNESS TEMPERATURE
∞DISTRIBUTION
GALACTIC RADIATION
PHOTOGRAPHY
RADIANT FLUX DENSITY
RADIO ASTRONOMY
SOLAR GRANULATION

BRIGHTNESS DISTRIBUTION-(CONT.)
STELLAR LUMINOSITY

BRIGHTNESS TEMPERATURE
GS   TEMPERATURE
. **BRIGHTNESS TEMPERATURE**
RT   ASTROPHYSICS
BLACK BODY RADIATION
BRIGHTNESS DISTRIBUTION
LIMB BRIGHTENING
METEOROLOGY
PHOTOGRAPHY
RADIO ASTRONOMY
TEMPERATURE MEASUREMENT

BRILLOUIN EFFECT
RT   ∞EFFECTS
FREQUENCY SHIFT
LIGHT SCATTERING
MONOCHROMATIC RADIATION

BRILLOUIN FLOW
GS   ELECTRIC CURRENT
. **BRILLOUIN FLOW**
RT   BEAM CURRENTS
ELECTRON BEAMS
ELECTRON OPTICS
∞FLOW
TRAVELING WAVE TUBES

BRILLOUIN ZONES
GS   REGIONS
. **BRILLOUIN ZONES**
RT   BAND STRUCTURE OF SOLIDS
CONDUCTION BANDS
CRYSTAL LATTICES
FERMI SURFACES
FREE ELECTRONS

BRILLOUIN-WIGNER EQUATION
RT   ∞EQUATIONS

BRINES
RT   COOLANTS
REFRIGERANTS
SALINITY
SALT BATHS
SALT BEDS
SEA WATER

BRIQUETS
RT   BLANKS
PELLETS
TABLETS

BRISTOL-SIDDELEY BS 53 ENGINE
UF   PEGASUS ENGINE
GS   ENGINES
. AIR BREATHING ENGINES
. . GAS TURBINE ENGINES
. . . JET ENGINES
. . . . TURBOJET ENGINES
. . . . . TURBOFAN ENGINES
. . . . . . **BRISTOL-SIDDELEY BS 53
ENGINE**
. INTERNAL COMBUSTION ENGINES
. . GAS TURBINE ENGINES
. . . JET ENGINES
. . . . TURBOJET ENGINES
. . . . . TURBOFAN ENGINES
. . . . . . **BRISTOL-SIDDELEY BS 53
ENGINE**
. TURBINE ENGINES
. . GAS TURBINE ENGINES
. . . JET ENGINES
. . . . TURBOJET ENGINES
. . . . . TURBOFAN ENGINES
. . . . . . **BRISTOL-SIDDELEY BS 53
ENGINE**
RT   P-1127 AIRCRAFT

BRISTOL-SIDDELEY OLYMPUS 593 ENGINE
GS   ENGINES
. AIR BREATHING ENGINES
. . GAS TURBINE ENGINES
. . . JET ENGINES
. . . . TURBOJET ENGINES
. . . . . **BRISTOL-SIDDELEY OLYMPUS
593 ENGINE**
. INTERNAL COMBUSTION ENGINES
. . GAS TURBINE ENGINES
. . . JET ENGINES
. . . . TURBOJET ENGINES

BRISTOL-SIDDELEY OLYMPUS 593-(CONT.)
. . . . . **BRISTOL-SIDDELEY OLYMPUS 593 ENGINE**
. TURBINE ENGINES
. . GAS TURBINE ENGINES
. . . JET ENGINES
. . . . TURBOJET ENGINES
. . . . . **BRISTOL-SIDDELEY OLYMPUS 593 ENGINE**

BRISTOL-SIDDELEY VIPER ENGINE
GS　ENGINES
. AIR BREATHING ENGINES
. . GAS TURBINE ENGINES
. . . JET ENGINES
. . . . TURBOJET ENGINES
. . . . . **BRISTOL-SIDDELEY VIPER ENGINE**
. INTERNAL COMBUSTION ENGINES
. . GAS TURBINE ENGINES
. . . JET ENGINES
. . . . TURBOJET ENGINES
. . . . . **BRISTOL-SIDDELEY VIPER ENGINE**
. TURBINE ENGINES
. . GAS TURBINE ENGINES
. . . JET ENGINES
. . . . TURBOJET ENGINES
. . . . . **BRISTOL-SIDDELEY VIPER ENGINE**

BRITISH AIRCRAFT CORP AIRCRAFT
USE　BAC AIRCRAFT

BRITISH COLUMBIA
GS　NATIONS
. CANADA
. . **BRITISH COLUMBIA**

BRITISH GUINEA
USE　GUYANA

BRITISH HONDURAS
USE　BELIZE

BRITTLE MATERIALS
RT　CLEAVAGE
CRACKING (FRACTURING)
EMBRITTLEMENT
FRACTURE STRENGTH
GRANULAR MATERIALS
HARDNESS
IMPACT STRENGTH
∞MATERIALS
POROUS MATERIALS

BRITTLENESS
GS　MECHANICAL PROPERTIES
. **BRITTLENESS**
RT　CHARPY IMPACT TEST
CLEAVAGE
COLD HARDENING
CRACK CLOSURE
CRACK INITIATION
CRACK PROPAGATION
CRACKING (FRACTURING)
DUCTILITY
EMBRITTLEMENT
FRACTOGRAPHY
FRACTURE STRENGTH
FRACTURING
HARDNESS
IMPACT STRENGTH
IMPACT TESTS
NOTCH STRENGTH
NOTCH TESTS
TOUGHNESS
WELDABILITY

BROADBAND
UF　WIDEBAND
GS　BANDWIDTH
. **BROADBAND**
FREQUENCIES
. **BROADBAND**
RT　∞BANDS
∞FREQUENCY RESPONSE
LOG PERIODIC ANTENNAS
NARROWBAND
SPIRAL ANTENNAS

BROADBAND AMPLIFIERS
GS　AMPLIFIERS
. **BROADBAND AMPLIFIERS**

BROADBAND AMPLIFIERS-(CONT.)
RT　BANDWIDTH
FREQUENCIES
WIDEBAND COMMUNICATION

BROADCASTING
UF　RADIO BROADCASTING
GS　TELECOMMUNICATION
. **BROADCASTING**
RT　COMMUNICATION NETWORKS
RADIO COMMUNICATION
RADIO EQUIPMENT
RADIO SIGNALS
RADIO TRANSMISSION
SYMPHONIE SATELLITES
TRANSMISSION
VOICE OF AMERICA

BROKEN SYMMETRY
UF　SYMMETRY BREAKING
GS　SYMMETRY
. **BROKEN SYMMETRY**
RT　MATHEMATICAL MODELS
THEORETICAL PHYSICS

BROMATES
GS　HALOGEN COMPOUNDS
. BROMINE COMPOUNDS
. . **BROMATES**
RT　∞OXYGEN COMPOUNDS

BROMIDES
GS　HALOGEN COMPOUNDS
. BROMINE COMPOUNDS
. . **BROMIDES**
. . . AMMONIUM BROMIDES
. . . CESIUM BROMIDES
. . . CHROMIUM BROMIDES
. . . DIBROMIDES
. . . HYDROBROMIC ACID
. . . HYDROBROMIDES
. . . MAGNESIUM BROMIDES
. . . POTASSIUM BROMIDES
. . . SILVER BROMIDES
. . . SODIUM BROMIDES
. . . STRONTIUM BROMIDES
. HALIDES
. . **BROMIDES**
. . . AMMONIUM BROMIDES
. . . CESIUM BROMIDES
. . . CHROMIUM BROMIDES
. . . DIBROMIDES
. . . HYDROBROMIC ACID
. . . HYDROBROMIDES
. . . MAGNESIUM BROMIDES
. . . POTASSIUM BROMIDES
. . . SILVER BROMIDES
. . . SODIUM BROMIDES
. . . STRONTIUM BROMIDES
RT　SALT BEDS

BROMINATION
GS　CHEMICAL REACTIONS
. HALOGENATION
. . **BROMINATION**

BROMINE
GS　CHEMICAL ELEMENTS
. HALOGENS
. . **BROMINE**
. . . BROMINE ISOTOPES

BROMINE COMPOUNDS
GS　HALOGEN COMPOUNDS
. **BROMINE COMPOUNDS**
. . BROMATES
. . BROMIDES
. . . AMMONIUM BROMIDES
. . . CESIUM BROMIDES
. . . CHROMIUM BROMIDES
. . . DIBROMIDES
. . . HYDROBROMIC ACID
. . . HYDROBROMIDES
. . . MAGNESIUM BROMIDES
. . . POTASSIUM BROMIDES
. . . SILVER BROMIDES
. . . SODIUM BROMIDES
. . . STRONTIUM BROMIDES
RT　∞CHEMICAL COMPOUNDS
HALOCARBONS

BROMINE ISOTOPES
UF　BROMINE 82
BROMINE 87
GS　CHEMICAL ELEMENTS

BROMINE ISOTOPES-(CONT.)
. HALOGENS
. . BROMINE
. . . **BROMINE ISOTOPES**
. NUCLIDES
. . ISOTOPES
. . . **BROMINE ISOTOPES**

BROMINE 82
USE　BROMINE ISOTOPES

BROMINE 87
USE　BROMINE ISOTOPES

BRONCHI
GS　ANATOMY
. RESPIRATORY SYSTEM
. . **BRONCHI**
RT　LUNGS
TRACHEA

BRONCHIAL TUBE
GS　ANATOMY
. RESPIRATORY SYSTEM
. . **BRONCHIAL TUBE**
. . . PHARYNX
. . . TRACHEA
RT　∞TUBES

BRONZES
GS　ALLOYS
. COPPER ALLOYS
. . **BRONZES**

BROTHS
RT　∞FOOD
NUTRITION

BROWN WAVE EFFECT
RT　ANNUAL VARIATIONS
BOTANY
CHLOROPHYLLS
∞EFFECTS
FOLIAGE
LEAVES

BROWNIAN MOVEMENTS
RT　COLLOIDS
DISPERSIONS
EINSTEIN EQUATIONS
EMULSIONS
FOKKER-PLANCK EQUATION
∞MOTION
∞SUSPENSIONS

BRUCETON TEST
USE　STATISTICAL TESTS

BRUCITE
GS　CHALCOGENIDES
. OXIDES
. . **BRUCITE**
MAGNESIUM COMPOUNDS
. **BRUCITE**
MINERALS
. **BRUCITE**

BRUDERHEIM METEORITE
GS　CELESTIAL BODIES
. METEORITES
. . STONY METEORITES
. . . CHONDRITES
. . . . **BRUDERHEIM METEORITE**

BRUNEI
GS　NATIONS
. **BRUNEI**
RT　ASIA

BRUNT-VAISALA FREQUENCY
GS　CONSTRAINTS
. METEOROLOGICAL PARAMETERS
. . **BRUNT-VAISALA FREQUENCY**
FREQUENCIES
. **BRUNT-VAISALA FREQUENCY**
RT　AIR CURRENTS
AIR FLOW
AIR MASSES
ATMOSPHERIC CIRCULATION
ATMOSPHERIC PHYSICS
ATMOSPHERIC STRATIFICATION
OSCILLATIONS

BRUSH (BOTANY)
- UF   SCRUBS (BOTANY)
- GS   PLANTS (BOTANY)
  - .   **BRUSH (BOTANY)**
  - . . CHAPARRAL
- RT   BOTANY
  - DEFOLIATION
  - EARTH RESOURCES
  - GUAYULE
  - HERBICIDES

BRUSHES
- GS   **BRUSHES**
  - . BRUSHES (ELECTRICAL CONTACTS)
- RT   ELECTRIC CONTACTS
  - ELECTRIC GENERATORS
  - ELECTRIC MOTORS

BRUSHES (ELECTRICAL CONTACTS)
- GS   BRUSHES
  - . **BRUSHES (ELECTRICAL CONTACTS)**
- RT   ELECTRIC CONTACTS
  - ELECTRIC GENERATORS
  - ELECTRIC MOTORS

BRYOPHYTES
- GS   PLANTS (BOTANY)
  - . **BRYOPHYTES**

BSX
- GS   EXPLOSIVES
  - . **BSX**
- RT   NITROMETHANE

BUBBLE CHAMBERS
- GS   IONIZATION CHAMBERS
  - . **BUBBLE CHAMBERS**
- RT   ∞CHAMBERS
  - CLOUD CHAMBERS
  - ELEMENTARY PARTICLES
  - PARTICLE TRAJECTORIES
  - RADIATION COUNTERS
  - SPARK CHAMBERS

BUBBLE MEMORY DEVICES
- GS   MAGNETIC STORAGE
  - . **BUBBLE MEMORY DEVICES**
- RT   BINARY DATA
  - COMPUTER COMPONENTS
  - CORE STORAGE
  - DATA PROCESSING
  - DATA RECORDERS
  - DATA RECORDING
  - DATA STORAGE
  - MAGNETIC CORES
  - MAGNETIC DOMAINS
  - MAGNETIC RECORDING
  - MAGNETIC SWITCHING

BUBBLE TECHNIQUE
- GS   TECHNOLOGIES
  - . **BUBBLE TECHNIQUE**
- RT   DATA RECORDERS
  - ELECTRONIC EQUIPMENT
  - FLIGHT INSTRUMENTS
  - ∞INSTRUMENTS
  - MAGNETIC DOMAINS
  - ONBOARD EQUIPMENT
  - RECORDING INSTRUMENTS
  - SEMICONDUCTOR DEVICES
  - SOLID STATE DEVICES
  - SPACECRAFT INSTRUMENTS

BUBBLES
- RT   AERATION
  - CAVITATION FLOW
  - COANDA EFFECT
  - EFFERVESCENCE
  - FOAMS
  - METAL FOAMS
  - WAKES

BUCCANEER AIRCRAFT
- UF   B-103 AIRCRAFT
  - BLACKBURN B-103 AIRCRAFT
- GS   ATTACK AIRCRAFT
  - . **BUCCANEER AIRCRAFT**
  - HAWKER SIDDELEY AIRCRAFT
  - . **BUCCANEER AIRCRAFT**
  - JET AIRCRAFT
  - . **BUCCANEER AIRCRAFT**
  - MONOPLANES
  - . **BUCCANEER AIRCRAFT**
- RT   ∞AIRCRAFT

BUCCANEER AIRCRAFT-(CONT.)
   HARRIER AIRCRAFT

BUCKET BRIGADE DEVICES
- GS   ELECTRONIC EQUIPMENT
  - . SOLID STATE DEVICES
  - . . SEMICONDUCTOR DEVICES
  - . . . METAL OXIDE SEMICONDUCTORS
  - . . . . CHARGE TRANSFER DEVICES
  - . . . . . **BUCKET BRIGADE DEVICES**
- RT   CHARGE COUPLED DEVICES
  - SEMICONDUCTORS (MATERIALS)

∞ BUCKETS
- SN   *(USE OF A MORE SPECIFIC TERM IS RECOMMENDED--CONSULT THE TERMS LISTED BELOW)*
- RT   BOXES (CONTAINERS)
  - ∞CAPSULES
  - DRUMS (CONTAINERS)
  - TRAYS
  - TURBOMACHINE BLADES

BUCKEYE AIRCRAFT
- USE   T-2 AIRCRAFT

BUCKLING
- GS   **BUCKLING**
  - . CREEP BUCKLING
  - . ELASTIC BUCKLING
  - . EULER BUCKLING
  - . THERMAL BUCKLING
- RT   BENDING
  - COLLAPSE
  - COMPRESSION LOADS
  - DEFORMATION
  - DISTORTION
  - DONNELL EQUATIONS
  - FAILURE
  - FAILURE MODES
  - FLANGE WRINKLING
  - HEAVING
  - ∞RIDGES
  - SHELL STABILITY
  - STRESSES
  - STRUCTURAL FAILURE
  - STRUCTURAL STRAIN
  - TEMPERATURE INVERSIONS
  - TORSION
  - TWISTING
  - WARPAGE
  - WRINKLING

BUDGETING
- RT   ACCOUNTING
  - ALLOCATIONS
  - APPROPRIATIONS
  - ∞BUDGETS
  - COST ANALYSIS
  - COST EFFECTIVENESS
  - COST ESTIMATES
  - ECONOMIC FACTORS
  - ESTIMATING
  - FINANCIAL MANAGEMENT
  - FORECASTING
  - GRANTS
  - INCOME
  - MISSION PLANNING
  - PLANNING
  - PROCUREMENT MANAGEMENT
  - PROJECT PLANNING
  - REVENUE

∞ BUDGETS
- SN   *(USE OF A MORE SPECIFIC TERM IS RECOMMENDED--CONSULT THE TERMS LISTED BELOW)*
- RT   BUDGETING
  - ENERGY BUDGETS
  - ENGINEERING MANAGEMENT
  - FEDERAL BUDGETS
  - FOREIGN POLICY
  - HEAT BUDGET
  - PROCUREMENT MANAGEMENT
  - RESEARCH MANAGEMENT

BUFFALO AIRCRAFT
- USE   DHC 5 AIRCRAFT

BUFFER STORAGE
- GS   PERIPHERAL EQUIPMENT (COMPUTERS)
  - . COMPUTER STORAGE DEVICES
  - . . **BUFFER STORAGE**
- RT   ∞BUFFERS
  - CORE STORAGE

BUFFER STORAGE-(CONT.)
- DATA STORAGE
- ∞STORAGE

∞ BUFFERS
- SN   *(USE OF A MORE SPECIFIC TERM IS RECOMMENDED--CONSULT THE TERMS LISTED BELOW)*
- RT   BUFFER STORAGE
  - BUFFERS (CHEMISTRY)

BUFFERS (CHEMISTRY)
- RT   ∞BUFFERS
  - CHEMICAL EQUILIBRIUM
  - NEUTRALIZERS
  - PH

BUFFETING
- RT   AERODYNAMIC STABILITY
  - AIRCRAFT STABILITY
  - BOUNDARY LAYER CONTROL
  - COMPRESSIBILITY EFFECTS
  - FLIGHT CHARACTERISTICS
  - FLUTTER
  - OSCILLATING FLOW
  - SHAKING
  - SPACECRAFT MOTION
  - SPACECRAFT STABILITY
  - STROUHAL NUMBER
  - TURBULENCE EFFECTS
  - VORTEX AVOIDANCE

BUILDING MATERIALS
- USE   CONSTRUCTION MATERIALS

BUILDING STRUCTURES
- USE   BUILDINGS

∞ BUILDINGS
- SN   *(USE OF A MORE SPECIFIC TERM IS RECOMMENDED--CONSULT THE TERMS LISTED BELOW)*
- UF   BUILDING STRUCTURES
- RT   ARCHITECTURE
  - BASEMENTS
  - CEILINGS (ARCHITECTURE)
  - CHIMNEYS
  - CONSTRUCTION
  - CONSTRUCTION INDUSTRY
  - FLOORS
  - GREENHOUSES
  - HANGARS
  - INDOOR AIR POLLUTION
  - INFLATABLE STRUCTURES
  - MISSILE SILOS
  - MUSEUMS
  - ROOFS
  - SHELTERS
  - SOLAR HOUSES
  - STAIRWAYS
  - STARSITE PROGRAM
  - WALLS

BULBS
- RT   LUMINAIRES
  - PLANT ROOTS
  - PRESSURE VESSELS
  - SYRINGES

BULGARIA
- GS   NATIONS
  - . **BULGARIA**
- RT   BLACK SEA
  - EUROPE

BULGING
- GS   METAL WORKING
  - . **BULGING**
- RT   DEEP DRAWING
  - DIMPLING
  - EXPLOSIVE FORMING
  - FORGING
  - HOT WORKING
  - MAGNETIC FORMING
  - METAL DRAWING
  - STRETCH FORMING

BULK ACOUSTIC WAVE DEVICES
- UF   B-A-W DEVICES
- RT   ∞DEVICES
  - SURFACE ACOUSTIC WAVE DEVICES
  - TRANSDUCERS

BULK MODULUS
GS   MECHANICAL PROPERTIES
. **BULK MODULUS**
RT   COMPRESSIBILITY
DENSITY (MASS/VOLUME)

BULKHEADS
GS   WALLS
. **BULKHEADS**
RT   ∞BARRIERS
END PLATES
HULLS (STRUCTURES)
PARTITIONS (STRUCTURES)
REINFORCEMENT (STRUCTURES)
THICK WALLS
THIN WALLS

BULLPUP B MISSILE
RT   LR-62-RM-2 ENGINE

BULLPUP MISSILES
GS   MISSILES
. AIR TO SURFACE MISSILES
. . **BULLPUP MISSILES**
RT   LR-62-RM-2 ENGINE

BUMBLEBEE PROJECT
GS   MISSILES
. **BUMBLEBEE PROJECT**
PROGRAMS
. PROJECTS
. . **BUMBLEBEE PROJECT**
RT   TALOS MISSILE
TARTAR MISSILE
TERRIER MISSILE
TYPHON WEAPON SYSTEM

BUMPERS
RT   CUSHIONS
METEOROID PROTECTION
METEOROIDS
PROTECTORS

BUMPY TORUSES
RT   FUSION REACTORS
PLASMA CONTROL
PLASMA HEATING
TOKAMAK DEVICES
TOROIDAL PLASMAS

BUNA (TRADEMARK)
GS   RUBBER
. SYNTHETIC RUBBERS
. . **BUNA (TRADEMARK)**
RT   BUTADIENE
STYRENES

BUNCHING
GS   **BUNCHING**
. ELECTRON BUNCHING
RT   QUEUEING THEORY
SPACE CHARGE
VELOCITY MODULATION

BUNDLE DRAWING
RT   ∞DRAWING
METAL DRAWING

BUNDLES
RT   ∞CONTAINERS
PACKAGES
UMBILICAL CONNECTORS
WIRING

BUNKERS (FUEL)
GS   TANKS (CONTAINERS)
. **BUNKERS (FUEL)**
RT   FUEL SYSTEMS

BUOYANCY
RT   ACOUSTIC LEVITATION
AEROSTATICS
BALLAST (MASS)
DENSITY (MASS/VOLUME)
FLOATING
GAS DENSITY
LEVITATION
MECHANICAL PROPERTIES
∞PHYSICAL PROPERTIES
POROSITY
RAYLEIGH NUMBER
VOIDS

BUOYS
RT   BEACONS
COMPASSES
FLOATS
∞MARKERS
NAVIGATION AIDS
OCEAN DATA ACQUISITIONS SYSTEMS

BUREAUS (ORGANIZATIONS)
GS   INSTITUTIONS
. **BUREAUS (ORGANIZATIONS)**
ORGANIZATIONS
. FEDERATIONS
. . **BUREAUS (ORGANIZATIONS)**
RT   PROGRAMS
PROJECTS
TEAMS
UNIVERSITY PROGRAM

BURETTES
GS   MEASURING INSTRUMENTS
. **BURETTES**
RT   GLASSWARE
PIPETTES
∞TUBES

BURGER EQUATION
GS   ANALYSIS (MATHEMATICS)
. REAL VARIABLES
. . DIFFERENTIAL EQUATIONS
. . . PARTIAL DIFFERENTIAL EQUATIONS
. . . . **BURGER EQUATION**
RT   CONTINUUM MECHANICS
∞EQUATIONS
NAVIER-STOKES EQUATION
SHOCK WAVE PROPAGATION

BURKINA
UF   UPPER VOLTA
GS   NATIONS
. **BURKINA**
RT   AFRICA

BURMA
GS   NATIONS
. **BURMA**
RT   ASIA

BURN-IN
RT   FAILURE
FAILURE ANALYSIS
INTEGRATED CIRCUITS
QUALITY CONTROL

BURNERS
RT   AFTERBURNING
CHEMICAL REACTORS
COMBUSTION CHAMBERS
DIFFUSION WELDING
FUEL INJECTION
FURNACES
INCINERATORS
WASTE ENERGY UTILIZATION

BURNING
USE   COMBUSTION

BURNING PROCESS
USE   COMBUSTION

BURNING RATE
GS   RATES (PER TIME)
. **BURNING RATE**
RT   BURNOUT
COMBUSTION
COMBUSTION CONTROL
COMBUSTION EFFICIENCY
COMBUSTION STABILITY
EXPLOSIVES
FLAME PROPAGATION
FLAMMABILITY
FUEL CONSUMPTION
FUEL-AIR RATIO
FUELS
PRESSURE DEPENDENCE
PROPELLANT GRAINS
PROPELLANTS
SOLID PROPELLANT COMBUSTION
SOLID PROPELLANT ROCKET ENGINES
SOLID ROCKET PROPELLANTS
VELOCITY COUPLING

BURNING TIME
UF   FIRING TIME

BURNING TIME-(CONT.)
GS   TIME
. **BURNING TIME**
RT   COMBUSTION
COMBUSTION EFFICIENCY
FIRING (IGNITING)
FLIGHT OPTIMIZATION
FLIGHT TIME
ROCKET ENGINES
ROCKET FIRING
TESTING TIME
THRUST
WINDOWS (INTERVALS)

BURNOUT
SN   (TERMINATION OF COMBUSTION IN A
ROCKET ENGINE BECAUSE OF
EXHAUSTION OF THE PROPELLANT)
RT   BOOSTER ROCKET ENGINES
BURNING RATE
COMBUSTION
∞CUT-OFF
EROSIVE BURNING
EXTINGUISHING
SOLID PROPELLANT ROCKET ENGINES
THRUST TERMINATION

BURNS (INJURIES)
GS   INJURIES
. **BURNS (INJURIES)**
RT   CRASH INJURIES
FIRES
LASER DAMAGE
LESIONS
RADIATION INJURIES

BURNTHROUGH (FAILURE)
GS   FAILURE
. **BURNTHROUGH (FAILURE)**
RT   ABLATION
DAMAGE
MELTING
PERFORATING

BURST TESTS
GS   DESTRUCTIVE TESTS
. **BURST TESTS**
RT   CONTAINMENT
FAILURE ANALYSIS
FRACTURE MECHANICS
FRACTURE STRENGTH
∞MATERIALS TESTS
PRESSURE VESSELS

BURSTS
GS   **BURSTS**
. GAMMA RAY BURSTS
. RADIO BURSTS
. . SOLAR RADIO BURSTS
. . . TYPE 2 BURSTS
. . . TYPE 3 BURSTS
. . . TYPE 4 BURSTS
. . . TYPE 5 BURSTS
RT   ∞DISTURBANCES
EMISSION
EXPLOSIONS
FRAGMENTATION
IMPLOSIONS
RUPTURING

BURUNDI
UF   RUANDA-URUNDI
GS   NATIONS
. **BURUNDI**
RT   AFRICA
RWANDA

BUS CONDUCTORS
GS   CONDUCTORS
. **BUS CONDUCTORS**
RT   ELECTRIC WIRE
FLAT CONDUCTORS
POWER LINES
∞POWER TRANSMISSION

BUSHINGS
RT   BEARINGS
INSERTS
LININGS
SHAFTS (MACHINE ELEMENTS)
SPACERS

BUSINESS MANAGEMENT
USE   INDUSTRIAL MANAGEMENT

C

BUTADIENE
UF VINYL ETHYLENE
GS ALIPHATIC COMPOUNDS
 . ALKENES
 . . **BUTADIENE**
 HYDROCARBONS
 . ALKENES
 . . **BUTADIENE**
 . DIENES
 . . **BUTADIENE**
RT BUNA (TRADEMARK)
 HYDROCARBON FUELS
 POLYBUTADIENE

BUTANES
UF ISOBUTANE
GS ALIPHATIC COMPOUNDS
 . ALKANES
 . . **BUTANES**
 HYDROCARBONS
 . ALKANES
 . . **BUTANES**

BUTENES
UF BUTYLENE
 ISOBUTYLENE
GS ALIPHATIC COMPOUNDS
 . ALKENES
 . . **BUTENES**
 HYDROCARBONS
 . ALKENES
 . . **BUTENES**

BUTT JOINTS
GS JOINTS (JUNCTIONS)
 . **BUTT JOINTS**
RT LAP JOINTS
 METAL JOINTS
 RIVETED JOINTS
 SOLDERED JOINTS
 WELDED JOINTS

BUTTERFLY VALVES
GS VALVES
 . **BUTTERFLY VALVES**
 . . DAMPERS (VALVES)

BUTTES
GS LANDFORMS
 . TERRACES (LANDFORMS)
 . . PLATEAUS
 . . . MESAS
 . . . . **BUTTES**

∞ **BUTTONS**
SN *(USE OF A MORE SPECIFIC TERM IS RECOMMENDED--CONSULT THE TERMS LISTED BELOW)*
RT MANUAL CONTROL

BUTYLENE
USE BUTENES

BUTYLENE OXIDES
USE TETRAHYDROFURAN

BUTYRIC ACID
GS ACIDS
 . **BUTYRIC ACID**
RT FERMENTATION

BY-PRODUCTS
RT MATERIALS RECOVERY
 PRODUCTS
 REACTION PRODUCTS
 WASTES

BYPASS RATIO
RT AIR INTAKES
 ENGINE INLETS
 FLOW GEOMETRY
 HYPERSONIC INLETS
 INLET AIRFRAME CONFIGURATIONS
 INLET FLOW
 INLET NOZZLES
 INTAKE SYSTEMS
 NOSE INLETS
 SIDE INLETS
 SUPERSONIC INLETS

BYPASSES
UF SHUNTS
RT DIVERTERS
 RELIEF VALVES

C BAND
GS FREQUENCIES
 . RADIO FREQUENCIES
 . . MICROWAVE FREQUENCIES
 . . . **C BAND**
RT MILLIMETER WAVES
 SUPERHIGH FREQUENCIES

C-M DIAGRAM
USE COLOR-MAGNITUDE DIAGRAM

C-1A AIRCRAFT
UF TRADER AIRCRAFT
GS GRUMMAN AIRCRAFT
 . **C-1A AIRCRAFT**
 TRANSPORT AIRCRAFT
 . CARGO AIRCRAFT
 . . **C-1A AIRCRAFT**
RT ∞ AIRCRAFT
 ∞ MILITARY AIRCRAFT

C-2 AIRCRAFT
UF COD AIRCRAFT
GS GRUMMAN AIRCRAFT
 . **C-2 AIRCRAFT**
 JET AIRCRAFT
 . TURBOPROP AIRCRAFT
 . . **C-2 AIRCRAFT**
 MONOPLANES
 . **C-2 AIRCRAFT**
 TRANSPORT AIRCRAFT
 . CARGO AIRCRAFT
 . . **C-2 AIRCRAFT**
RT ∞ AIRCRAFT

C-5 AIRCRAFT
UF GALAXY AIRCRAFT
 LOCKHEED C-5 AIRCRAFT
GS JET AIRCRAFT
 . **C-5 AIRCRAFT**
 LOCKHEED AIRCRAFT
 . **C-5 AIRCRAFT**
 TRANSPORT AIRCRAFT
 . CARGO AIRCRAFT
 . . **C-5 AIRCRAFT**
RT ∞ AIRCRAFT
 TURBOFAN ENGINES

C-8A AUGMENTOR WING AIRCRAFT
GS JET AIRCRAFT
 . **C-8A AUGMENTOR WING AIRCRAFT**
 RESEARCH AIRCRAFT
 . **C-8A AUGMENTOR WING AIRCRAFT**
RT ∞ AIRCRAFT

C-9 AIRCRAFT
GS JET AIRCRAFT
 . **C-9 AIRCRAFT**
 MCDONNELL DOUGLAS AIRCRAFT
 . DOUGLAS AIRCRAFT
 . . **C-9 AIRCRAFT**
 . MCDONNELL AIRCRAFT
 . . **C-9 AIRCRAFT**
 TRANSPORT AIRCRAFT
 . CARGO AIRCRAFT
 . . **C-9 AIRCRAFT**
RT ∞ AIRCRAFT
 EVACUATING (TRANSPORTATION)

C-15 AIRCRAFT
UF YC-15 AIRCRAFT
GS V/STOL AIRCRAFT
 . SHORT TAKEOFF AIRCRAFT
 . . **C-15 AIRCRAFT**
RT ∞ AIRCRAFT

C-33 AIRCRAFT
UF BEECH C-33 AIRCRAFT
 DEBONAIR AIRCRAFT
GS BEECHCRAFT
 . BEECH 99 AIRCRAFT
 . . **C-33 AIRCRAFT**
 GENERAL AVIATION AIRCRAFT
 . **C-33 AIRCRAFT**
 LIGHT AIRCRAFT
 . BEECH 99 AIRCRAFT
 . . **C-33 AIRCRAFT**
 MONOPLANES
 . **C-33 AIRCRAFT**
 PASSENGER AIRCRAFT
 . **C-33 AIRCRAFT**
RT ∞ AIRCRAFT

C-35 AIRCRAFT
UF BEECH S-35 AIRCRAFT
 BONANZA AIRCRAFT
GS BEECHCRAFT AIRCRAFT
 . BEECH 99 AIRCRAFT
 . . **C-35 AIRCRAFT**
 GENERAL AVIATION AIRCRAFT
 . **C-35 AIRCRAFT**
 LIGHT AIRCRAFT
 . BEECH 99 AIRCRAFT
 . . **C-35 AIRCRAFT**
 MONOPLANES
 . **C-35 AIRCRAFT**
 PASSENGER AIRCRAFT
 . **C-35 AIRCRAFT**
RT ∞ AIRCRAFT

C-46 AIRCRAFT
UF COMMANDO AIRCRAFT
 CURTISS C-46 AIRCRAFT
GS CURTISS-WRIGHT AIRCRAFT
 . **C-46 AIRCRAFT**
 MONOPLANES
 . **C-46 AIRCRAFT**
 PASSENGER AIRCRAFT
 . **C-46 AIRCRAFT**
 TRANSPORT AIRCRAFT
 . CARGO AIRCRAFT
 . . **C-46 AIRCRAFT**
RT ∞ AIRCRAFT

C-47 AIRCRAFT
UF DAKOTA AIRCRAFT
GS MCDONNELL DOUGLAS AIRCRAFT
 . DOUGLAS AIRCRAFT
 . . **C-47 AIRCRAFT**
 MONOPLANES
 . **C-47 AIRCRAFT**
 TRANSPORT AIRCRAFT
 . CARGO AIRCRAFT
 . . **C-47 AIRCRAFT**
RT ∞ AIRCRAFT

C-54 AIRCRAFT
UF R5D AIRCRAFT
 SKYMASTER AIRCRAFT
GS MCDONNELL DOUGLAS AIRCRAFT
 . DOUGLAS AIRCRAFT
 . . **C-54 AIRCRAFT**
 MONOPLANES
 . **C-54 AIRCRAFT**
 TRANSPORT AIRCRAFT
 . CARGO AIRCRAFT
 . . **C-54 AIRCRAFT**
RT ∞ AIRCRAFT

C-118 AIRCRAFT
GS MCDONNELL DOUGLAS AIRCRAFT
 . DOUGLAS AIRCRAFT
 . . **C-118 AIRCRAFT**
 MONOPLANES
 . **C-118 AIRCRAFT**
 TRANSPORT AIRCRAFT
 . CARGO AIRCRAFT
 . . **C-118 AIRCRAFT**
RT ∞ AIRCRAFT

C-119 AIRCRAFT
GS FAIRCHILD-HILLER AIRCRAFT
 . **C-119 AIRCRAFT**
 JET AIRCRAFT
 . **C-119 AIRCRAFT**
 TRANSPORT AIRCRAFT
 . CARGO AIRCRAFT
 . . **C-119 AIRCRAFT**
RT ∞ AIRCRAFT

C-121 AIRCRAFT
UF LOCKHEED CONSTELLATION AIRCRAFT
 R7V AIRCRAFT
GS LOCKHEED AIRCRAFT
 . **C-121 AIRCRAFT**
 MONOPLANES
 . **C-121 AIRCRAFT**
 TRANSPORT AIRCRAFT
 . CARGO AIRCRAFT
 . . **C-121 AIRCRAFT**
RT ∞ AIRCRAFT
 EC-121 AIRCRAFT

C-123 AIRCRAFT
UF PROVIDER AIRCRAFT
 YC-123 AIRCRAFT
GS FAIRCHILD-HILLER AIRCRAFT
 . **C-123 AIRCRAFT**

C-123 AIRCRAFT-(CONT.)
    MONOPLANES
    . **C-123 AIRCRAFT**
    TRANSPORT AIRCRAFT
    . CARGO AIRCRAFT
    . . **C-123 AIRCRAFT**
    V/STOL AIRCRAFT
    . SHORT TAKEOFF AIRCRAFT
    . . **C-123 AIRCRAFT**
  RT   ∞AIRCRAFT

C-124 AIRCRAFT
  GS   MCDONNELL DOUGLAS AIRCRAFT
    . DOUGLAS AIRCRAFT
    . . **C-124 AIRCRAFT**
    MONOPLANES
    . **C-124 AIRCRAFT**
    TRANSPORT AIRCRAFT
    . CARGO AIRCRAFT
    . . **C-124 AIRCRAFT**
  RT   ∞AIRCRAFT

C-130 AIRCRAFT
  UF   GC-130 AIRCRAFT
    HERCULES AIRCRAFT
    JC-130 AIRCRAFT
    KC-130 AIRCRAFT
    NC-130 AIRCRAFT
  GS   LOCKHEED AIRCRAFT
    . **C-130 AIRCRAFT**
    MONOPLANES
    . **C-130 AIRCRAFT**
    TRANSPORT AIRCRAFT
    . CARGO AIRCRAFT
    . . **C-130 AIRCRAFT**
  RT   ∞AIRCRAFT
    T-56 ENGINE

C-131 AIRCRAFT
  UF   SAMARITAN AIRCRAFT
  GS   GENERAL DYNAMICS AIRCRAFT
    . **C-131 AIRCRAFT**
    MONOPLANES
    . **C-131 AIRCRAFT**
    TRANSPORT AIRCRAFT
    . CARGO AIRCRAFT
    . . **C-131 AIRCRAFT**
  RT   ∞AIRCRAFT

C-133 AIRCRAFT
  UF   CARGOMASTER AIRCRAFT
  GS   JET AIRCRAFT
    . TURBOPROP AIRCRAFT
    . . **C-133 AIRCRAFT**
    MCDONNELL DOUGLAS AIRCRAFT
    . DOUGLAS AIRCRAFT
    . . **C-133 AIRCRAFT**
    MONOPLANES
    . **C-133 AIRCRAFT**
    TRANSPORT AIRCRAFT
    . CARGO AIRCRAFT
    . . **C-133 AIRCRAFT**
  RT   ∞AIRCRAFT
    T-34 ENGINE

C-135 AIRCRAFT
  UF   KC-135 AIRCRAFT
    STRATOTANKER AIRCRAFT
  GS   BOEING AIRCRAFT
    . **C-135 AIRCRAFT**
    JET AIRCRAFT
    . **C-135 AIRCRAFT**
    MONOPLANES
    . **C-135 AIRCRAFT**
    TRANSPORT AIRCRAFT
    . CARGO AIRCRAFT
    . . **C-135 AIRCRAFT**
  RT   ADVANCED RANGE INSTRUMENTATION
      AIRCRAFT
    ∞AIRCRAFT
    TURBOFAN AIRCRAFT

C-140 AIRCRAFT
  UF   JET STAR AIRCRAFT
  GS   JET AIRCRAFT
    . **C-140 AIRCRAFT**
    LOCKHEED AIRCRAFT
    . **C-140 AIRCRAFT**
    MONOPLANES
    . **C-140 AIRCRAFT**
    TRANSPORT AIRCRAFT
    . CARGO AIRCRAFT
    . . **C-140 AIRCRAFT**
    UTILITY AIRCRAFT
    . **C-140 AIRCRAFT**

C-140 AIRCRAFT-(CONT.)
  RT   ∞AIRCRAFT

C-141 AIRCRAFT
  UF   KUIPER AIRBORNE OBSERVATORY
    STARLIFTER AIRCRAFT
  GS   JET AIRCRAFT
    . TURBOFAN AIRCRAFT
    . . **C-141 AIRCRAFT**
    LOCKHEED AIRCRAFT
    . **C-141 AIRCRAFT**
    MONOPLANES
    . **C-141 AIRCRAFT**
    TRANSPORT AIRCRAFT
    . CARGO AIRCRAFT
    . . **C-141 AIRCRAFT**
  RT   ∞AIRCRAFT
    TURBOFAN ENGINES

C-142 AIRCRAFT
  USE   XC-142 AIRCRAFT

C-160 AIRCRAFT
  UF   TRANSALL C-160 AIRCRAFT
  GS   HAMBURGER AIRCRAFT
    . **C-160 AIRCRAFT**
    JET AIRCRAFT
    . TURBOPROP AIRCRAFT
    . . **C-160 AIRCRAFT**
    MONOPLANES
    . **C-160 AIRCRAFT**
    NORD AIRCRAFT
    . **C-160 AIRCRAFT**
    TRANSPORT AIRCRAFT
    . CARGO AIRCRAFT
    . . **C-160 AIRCRAFT**
  RT   ∞AIRCRAFT
    TURBOPROP ENGINES

CABIN ATMOSPHERES
  GS   CONTROLLED ATMOSPHERES
    . **CABIN ATMOSPHERES**
    . . SPACECRAFT CABIN ATMOSPHERES
  RT   AIRCRAFT COMPARTMENTS
    ∞ATMOSPHERES
    COCKPITS
    ENVIRONMENTAL CONTROL
    OXYGEN SUPPLY EQUIPMENT
    PRESSURIZED CABINS
    SPACE CAPSULES

∞ CABINS
  SN   *(USE OF A MORE SPECIFIC TERM IS*
      RECOMMENDED--CONSULT THE TERMS
      LISTED BELOW)
  RT   AIRCRAFT COMPARTMENTS
    COCKPITS
    PRESSURIZED CABINS
    SPACECRAFT CABINS

CABLE FORCE RECORDERS
  GS   RECORDING INSTRUMENTS
    . **CABLE FORCE RECORDERS**
  RT   ∞RECORDERS
    STRAIN GAGES
    TENSIOMETERS

∞ CABLES
  SN   *(USE OF A MORE SPECIFIC TERM IS*
      RECOMMENDED--CONSULT THE TERMS
      LISTED BELOW)
  RT   CABLES (ROPES)
    COAXIAL CABLES
    COMMUNICATION CABLES
    POWER LINES
    SUBMARINE CABLES
    TETHERLINES
    TRANSMISSION LINES

CABLES (ROPES)
  RT   ∞BELTS
    ∞CABLES
    CHAINS
    CORDAGE
    FASTENERS
    REELS
    STRANDS
    TOWING
    WIRE

CAD (DESIGN)
  USE   COMPUTER AIDED DESIGN

CADASTRAL MAPPING
  GS   MAPPING
    . **CADASTRAL MAPPING**
  RT   GEOGRAPHY
    MAPS
    THEMATIC MAPPING

CADMIUM
  GS   CHEMICAL ELEMENTS
    . **CADMIUM**
    . . CADMIUM ISOTOPES
    METALS
    . TRANSITION METALS
    . . **CADMIUM**
    . . . CADMIUM ISOTOPES

CADMIUM ALLOYS
  GS   ALLOYS
    . **CADMIUM ALLOYS**
  RT   BEARING ALLOYS

CADMIUM ANTIMONIDES
  GS   ANTIMONY COMPOUNDS
    . ANTIMONIDES
    . . **CADMIUM ANTIMONIDES**
    CADMIUM COMPOUNDS
    . **CADMIUM ANTIMONIDES**

CADMIUM CHLORIDES
  GS   CADMIUM COMPOUNDS
    . **CADMIUM CHLORIDES**
    HALOGEN COMPOUNDS
    . CHLORINE COMPOUNDS
    . . CHLORIDES
    . . . **CADMIUM CHLORIDES**
    . HALIDES
    . . CHLORIDES
    . . . **CADMIUM CHLORIDES**
    . . METAL HALIDES
    . . . **CADMIUM CHLORIDES**

CADMIUM COMPOUNDS
  GS   **CADMIUM COMPOUNDS**
    . CADMIUM ANTIMONIDES
    . CADMIUM CHLORIDES
    . CADMIUM FLUORIDES
    . CADMIUM SELENIDES
    . CADMIUM SULFIDES
    . CADMIUM TELLURIDES
  RT   ∞CHEMICAL COMPOUNDS
    ∞GROUP 2B COMPOUNDS
    ∞METAL COMPOUNDS

CADMIUM FLUORIDES
  GS   CADMIUM COMPOUNDS
    . **CADMIUM FLUORIDES**
    HALOGEN COMPOUNDS
    . FLUORINE COMPOUNDS
    . . FLUORIDES
    . . . METAL FLUORIDES
    . . . . **CADMIUM FLUORIDES**
    . HALIDES
    . . METAL HALIDES
    . . . **CADMIUM FLUORIDES**

CADMIUM ISOTOPES
  UF   CADMIUM 114
  GS   CHEMICAL ELEMENTS
    . CADMIUM
    . . **CADMIUM ISOTOPES**
    . NUCLIDES
    . . ISOTOPES
    . . . **CADMIUM ISOTOPES**
    METALS
    . TRANSITION METALS
    . . CADMIUM
    . . . **CADMIUM ISOTOPES**

CADMIUM MERCURY TELLURIDES
  USE   MERCURY CADMIUM TELLURIDES

CADMIUM NICKEL BATTERIES
  USE   NICKEL CADMIUM BATTERIES

CADMIUM SELENIDES
  GS   CADMIUM COMPOUNDS
    . **CADMIUM SELENIDES**
    CHALCOGENIDES
    . SELENIDES
    . . **CADMIUM SELENIDES**
    SELENIUM COMPOUNDS
    . SELENIDES
    . . **CADMIUM SELENIDES**

CADMIUM SILVER BATTERIES
USE   SILVER CADMIUM BATTERIES

CADMIUM SULFIDES
GS   CADMIUM COMPOUNDS
. **CADMIUM SULFIDES**
CHALCOGENIDES
. SULFIDES
. . INORGANIC SULFIDES
. . . **CADMIUM SULFIDES**
SULFUR COMPOUNDS
. SULFIDES
. . INORGANIC SULFIDES
. . . **CADMIUM SULFIDES**

CADMIUM TELLURIDES
GS   CADMIUM COMPOUNDS
. **CADMIUM TELLURIDES**
CHALCOGENIDES
. TELLURIDES
. . **CADMIUM TELLURIDES**
TELLURIUM COMPOUNDS
. TELLURIDES
. . **CADMIUM TELLURIDES**

CADMIUM 114
USE   CADMIUM ISOTOPES

CAFFEINE
GS   DRUGS
. STIMULANTS
. . **CAFFEINE**
FUNGICIDES
. **CAFFEINE**
HETEROCYCLIC COMPOUNDS
. XANTHINES
. . **CAFFEINE**
NITROGEN COMPOUNDS
. XANTHINES
. . **CAFFEINE**
PURINES
. XANTHINES
. . **CAFFEINE**

CAI
USE   COMPUTER ASSISTED INSTRUCTION

CAISSONS
RT   CONSTRUCTION
FOUNDATIONS

CAJUN ROCKET VEHICLE
GS   ROCKET VEHICLES
. SOUNDING ROCKETS
. . **CAJUN ROCKET VEHICLE**
RT   NIKE-CAJUN ROCKET VEHICLE
SOLID PROPELLANT ROCKET ENGINES
SONDES

CALCIFEROL
UF   VITAMIN D
GS   VITAMINS
. **CALCIFEROL**

CALCIFICATION
RT   ARTHRITIS
BONES

CALCINATION
USE   ROASTING

CALCITE
GS   CALCIUM COMPOUNDS
. CALCIUM CARBONATES
. . **CALCITE**
CARBON COMPOUNDS
. CARBONATES
. . CALCIUM CARBONATES
. . . **CALCITE**
MINERALS
. **CALCITE**
RT   ARAGONITE
BIREFRINGENCE

CALCIUM
GS   CHEMICAL ELEMENTS
. **CALCIUM**
. . CALCIUM ISOTOPES
METALS
. **CALCIUM**
. . CALCIUM ISOTOPES
RT   GYPSUM

CALCIUM CARBONATES
GS   CALCIUM COMPOUNDS
. **CALCIUM CARBONATES**
. . AKERMANITE
. . ARAGONITE
. . CALCITE
. . CHALK
CARBON COMPOUNDS
. CARBONATES
. . **CALCIUM CARBONATES**
. . . AKERMANITE
. . . ARAGONITE
. . . CALCITE
. . . CHALK
RT   BONE MINERAL CONTENT
LIMESTONE

CALCIUM CHLORIDES
GS   CALCIUM COMPOUNDS
. **CALCIUM CHLORIDES**
HALOGEN COMPOUNDS
. CHLORINE COMPOUNDS
. . CHLORIDES
. . . **CALCIUM CHLORIDES**
. HALIDES
. . CHLORIDES
. . . **CALCIUM CHLORIDES**
. . METAL HALIDES
. . . **CALCIUM CHLORIDES**

CALCIUM COMPOUNDS
GS   **CALCIUM COMPOUNDS**
. CALCIUM CARBONATES
. . AKERMANITE
. . ARAGONITE
. . CALCITE
. . CHALK
. CALCIUM CHLORIDES
. CALCIUM FLUORIDES
. . FLUORSPAR
. CALCIUM OXIDES
. . AKERMANITE
. CALCIUM PHOSPHATES
. CALCIUM SILICATES
. . GEHLENITE
. CALCIUM SULFIDES
. CALCIUM TUNGSTATES
. CALCIUM VANADATES
. FLUORITE
. MERWINITE
. MONTICELLITE
. PEROVSKITES
. SCHEELITE
RT   ∞ ALKALINE EARTH COMPOUNDS
∞ CHEMICAL COMPOUNDS
∞ METAL COMPOUNDS

CALCIUM FLUORIDES
GS   CALCIUM COMPOUNDS
. **CALCIUM FLUORIDES**
. . FLUORSPAR
HALOGEN COMPOUNDS
. FLUORINE COMPOUNDS
. . FLUORIDES
. . . DIFLUORIDES
. . . . **CALCIUM FLUORIDES**
. . . . . FLUORSPAR
. . . METAL FLUORIDES
. . . . **CALCIUM FLUORIDES**
. HALIDES
. . FLUORIDES
. . . DIFLUORIDES
. . . . **CALCIUM FLUORIDES**
. . . . . FLUORSPAR
. . METAL HALIDES
. . . **CALCIUM FLUORIDES**
. . . . FLUORSPAR

CALCIUM ISOTOPES
UF   CALCIUM 45
GS   CHEMICAL ELEMENTS
. CALCIUM
. . **CALCIUM ISOTOPES**
. NUCLIDES
. ISOTOPES
. . . **CALCIUM ISOTOPES**
METALS
. CALCIUM
. . **CALCIUM ISOTOPES**

CALCIUM METABOLISM
GS   METABOLISM
. **CALCIUM METABOLISM**
RT   BED REST
OSTEOPOROSIS

CALCIUM METABOLISM-*(CONT.)*
PARATHYROID GLAND
THYROID GLAND

CALCIUM OXIDES
UF   LIME
GS   CALCIUM COMPOUNDS
. **CALCIUM OXIDES**
. . AKERMANITE
CHALCOGENIDES
. OXIDES
. . METAL OXIDES
. . . ALKALINE EARTH OXIDES
. . . . **CALCIUM OXIDES**

CALCIUM PHOSPHATES
UF   APATITES
GS   CALCIUM COMPOUNDS
. **CALCIUM PHOSPHATES**
PHOSPHORUS COMPOUNDS
. PHOSPHATES
. . **CALCIUM PHOSPHATES**
RT   BONE MINERAL CONTENT

CALCIUM SILICATES
GS   CALCIUM COMPOUNDS
. **CALCIUM SILICATES**
. . GEHLENITE
SILICON COMPOUNDS
. SILICATES
. . **CALCIUM SILICATES**
. . . GEHLENITE
RT   AMPHIBOLES
MINERALS

CALCIUM SULFIDES
GS   CALCIUM COMPOUNDS
. **CALCIUM SULFIDES**
CHALCOGENIDES
. SULFIDES
. . INORGANIC SULFIDES
. . . **CALCIUM SULFIDES**
SULFUR COMPOUNDS
. SULFIDES
. . INORGANIC SULFIDES
. . . **CALCIUM SULFIDES**

CALCIUM TUNGSTATES
GS   CALCIUM COMPOUNDS
. **CALCIUM TUNGSTATES**
TUNGSTEN COMPOUNDS
. **CALCIUM TUNGSTATES**

CALCIUM VANADATES
GS   CALCIUM COMPOUNDS
. **CALCIUM VANADATES**
VANADIUM COMPOUNDS
. **CALCIUM VANADATES**

CALCIUM 45
USE   CALCIUM ISOTOPES

CALCULATION
USE   COMPUTATION

CALCULATORS
RT   ARITHMETIC
COMPUTATION
COMPUTERS

CALCULI
UF   RENAL CALCULI
GS   DEPOSITS
. **CALCULI**
. . DENTAL CALCULI
RT   LITHIASIS
UROLITHIASIS

CALCULUS
SN   (MATHEMATICS CONCERNED WITH
LIMITS AND REAL FUNCTIONS)
GS   ANALYSIS (MATHEMATICS)
. **CALCULUS**
. . CONTINUITY (MATHEMATICS)
. . DIFFERENTIAL CALCULUS
. . FOURIER-BESSEL TRANSFORMATIONS
. . GRAEFF CALCULUS
. . INTEGRAL CALCULUS
. . LIMITS (MATHEMATICS)
. . SERIES (MATHEMATICS)
. . . ASYMPTOTIC SERIES
. . . CAMPBELL-HAUSDORFF SERIES
. . . COSINE SERIES
. . . FOURIER SERIES

CALCULUS-(CONT.)
```
      . . . PADE APPROXIMATION
      . . . POWER SERIES
      . . . . TAYLOR SERIES
      . . . . . MACLAURIN SERIES
      . . . PROGRESSIONS
      . . . PRONY SERIES
      . . SINE SERIES
      . . VECTOR ANALYSIS
      . . . COLLINEARITY
      . . . COPLANARITY
      . . . CURL (VECTORS)
      . . . . VORTICITY
RT    ANALYTIC GEOMETRY
      ASYMPTOTES
      DIFFERENTIAL EQUATIONS
      FUNCTIONS (MATHEMATICS)
      ∞MATHEMATICS
      MONOTONE FUNCTIONS
      OPERATIONAL CALCULUS
      REAL VARIABLES
```

CALCULUS OF VARIATIONS
```
UF    VARIATION METHOD
GS    ANALYSIS (MATHEMATICS)
      . REAL VARIABLES
      . . CALCULUS OF VARIATIONS
RT    BIOT METHOD
      CASTIGLIANO VARIATIONAL THEOREM
      DIFFERENTIAL EQUATIONS
      EULER-LAGRANGE EQUATION
      INTEGRAL EQUATIONS
      INVARIANT IMBEDDINGS
      JACOBI MATRIX METHOD
      MAXIMA
      OPERATIONAL CALCULUS
      PONTRYAGIN PRINCIPLE
      STEEPEST DESCENT METHOD
      VARIATIONAL PRINCIPLES
```

CALDERAS
```
GS    LANDFORMS
      . CALDERAS
RT    CONES (VOLCANOES)
      CRATERS
      LAVA
      MARS VOLCANOES
      VOLCANOES
      VOLCANOLOGY
```

CALENDARS
```
GS    CALENDARS
      . CROP CALENDARS
RT    MONTH
      SCHEDULING
      TIME
```

CALIBRATING
```
UF    GRADUATION
GS    CALIBRATING
      . WIND TUNNEL CALIBRATION
RT    ACCURACY
      INSTRUMENT COMPENSATION
      INSTRUMENT ERRORS
      MEASURING INSTRUMENTS
      ∞SCALING
      SOLAR CELL CALIBRATION FACILITY
      STANDARDIZATION
      STANDARDS
      TEMPERATURE SCALES
```

CALIFORNIA
```
GS    NATIONS
      . UNITED STATES
      . . CALIFORNIA
RT    CASCADE RANGE (CA-OR-WA)
      COACHELLA VALLEY (CA)
      COASTAL RANGES (CA)
      DEATH VALLEY (CA)
      FEATHER RIVER BASIN (CA)
      GREAT BASIN (US)
      IMPERIAL VALLEY (CA)
      LAKE TAHOE (CA-NV)
      MOJAVE DESERT (CA)
      MONTEREY BAY (CA)
      PALO VERDE VALLEY (CA)
      PENINSULAR RANGES (CA)
      SACRAMENTO VALLEY (CA)
      SALTON SEA (CA)
      SAN ANDREAS FAULT
      SAN FRANCISCO
      SAN FRANCISCO BAY (CA)
      SAN JOAQUIN VALLEY (CA)
      SAN PABLO BAY (CA)
      SIERRA NEVADA MOUNTAINS (CA)
```

CALIFORNIA-(CONT.)
```
      SOUTHERN CALIFORNIA
```

CALIFORNIUM
```
GS    CHEMICAL ELEMENTS
      . ACTINIDE SERIES
      . . TRANSURANIUM ELEMENTS
      . . . CALIFORNIUM
      . . . . CALIFORNIUM ISOTOPES
      . NUCLIDES
      . . ISOTOPES
      . . . RADIOACTIVE ISOTOPES
      . . . . TRANSURANIUM ELEMENTS
      . . . . . CALIFORNIUM
      . . . . . . CALIFORNIUM ISOTOPES
      HEAVY ELEMENTS
      . TRANSURANIUM ELEMENTS
      . . CALIFORNIUM
      . . . CALIFORNIUM ISOTOPES
      METALS
      . ACTINIDE SERIES
      . . TRANSURANIUM ELEMENTS
      . . . CALIFORNIUM
      . . . . CALIFORNIUM ISOTOPES
```

CALIFORNIUM ISOTOPES
```
UF    CALIFORNIUM 252
GS    CHEMICAL ELEMENTS
      . ACTINIDE SERIES
      . . TRANSURANIUM ELEMENTS
      . . . CALIFORNIUM
      . . . . CALIFORNIUM ISOTOPES
      . NUCLIDES
      . . ISOTOPES
      . . . RADIOACTIVE ISOTOPES
      . . . . TRANSURANIUM ELEMENTS
      . . . . . CALIFORNIUM
      . . . . . . CALIFORNIUM ISOTOPES
      HEAVY ELEMENTS
      . TRANSURANIUM ELEMENTS
      . . CALIFORNIUM
      . . . CALIFORNIUM ISOTOPES
      METALS
      . ACTINIDE SERIES
      . . TRANSURANIUM ELEMENTS
      . . . CALIFORNIUM
      . . . . CALIFORNIUM ISOTOPES
```

CALIFORNIUM 252
```
USE   CALIFORNIUM ISOTOPES
```

CALLISTO
```
GS    CELESTIAL BODIES
      . NATURAL SATELLITES
      . . JUPITER SATELLITES
      . . . GALILEAN SATELLITES
      . . . . CALLISTO
      SATELLITES
      . NATURAL SATELLITES
      . . JUPITER SATELLITES
      . . . GALILEAN SATELLITES
      . . . . CALLISTO
RT    CHARON
      GANYMEDE
      IO
      JUPITER (PLANET)
```

CALORIC REQUIREMENTS
```
GS    NUTRITIONAL REQUIREMENTS
      . CALORIC REQUIREMENTS
RT    DIETS
      ∞FOOD
      METABOLISM
      MINERAL METABOLISM
      ∞NUTRIENTS
      NUTRITION
```

CALORIC STIMULI
```
RT    ∞STIMULI
```

CALORIMETERS
```
UF    MICROCALORIMETERS
GS    MEASURING INSTRUMENTS
      . CALORIMETERS
      . . BOMB CALORIMETERS
      . . DROP CALORIMETERS
      . . FLAME CALORIMETERS
RT    HEAT MEASUREMENT
      HIGH TEMPERATURE TESTS
      TEMPERATURE MEASURING
         INSTRUMENTS
```

CALORIMETRY
```
USE   HEAT MEASUREMENT
```

CALUTRONS
```
USE   CYCLOTRONS
```

CALVES
```
GS    ANIMALS
      . VERTEBRATES
      . . MAMMALS
      . . . CATTLE
      . . . . CALVES
RT    LIVESTOCK
```

CAM (MANUFACTURING)
```
USE   COMPUTER AIDED MANUFACTURING
```

CAMBER
```
GS    CAMBER
      . CONICAL CAMBER
      . WING CAMBER
RT    AIRFOILS
      BENDING
      ∞BOWS
      CAMBERED WINGS
      CURVATURE
      CURVED BEAMS
      DEFLECTION
      DEFORMATION
      DISTORTION
      FLEXING
      FUSELAGES
      LIFT
      WARPAGE
```

CAMBERED WINGS
```
GS    AIRFOILS
      . WINGS
      . . CAMBERED WINGS
RT    CAMBER
      FIXED WINGS
      TWISTED WINGS
      UNCAMBERED WINGS
      WING CAMBER
```

CAMBODIA
```
UF    KAMPUCHEA
GS    NATIONS
      . CAMBODIA
RT    ASIA
```

CAMEL AIRCRAFT
```
USE   TU-104 AIRCRAFT
```

CAMERA SHUTTERS
```
RT    CAMERAS
      IRISES (MECHANICAL APERTURES)
      KERR CELLS
      PANORAMIC CAMERAS
      ∞SHUTTERS
      STREAK CAMERAS
```

CAMERA TUBES
```
GS    ELECTRON TUBES
      . CAMERA TUBES
      . . IMAGE DISSECTOR TUBES
      . . ORTHICONS
      . . . IMAGE ORTHICONS
      . . VIDICONS
      . . . RETURN BEAM VIDICONS
      . . . . THERMICONS
RT    CAMERAS
      DYNODES
      IMAGE CONVERTERS
      IMAGE TRANSDUCERS
      MONOSCOPES
      PLANOTRONS
      TELEVISION CAMERAS
      VIDEO EQUIPMENT
```

CAMERAS
```
GS    OPTICAL EQUIPMENT
      . CAMERAS
      . . BAKER-NUNN CAMERA
      . . BALLISTIC CAMERAS
      . . DELFT CAMERA
      . . DIFFRACTION LIMITED CAMERAS
      . . FAINT OBJECT CAMERA
      . . HIGH SPEED CAMERAS
      . . . FRAMING CAMERAS
      . . I2S CAMERAS
      . . LALLEMAND CAMERAS
      . . MULTISPECTRAL BAND CAMERAS
      . . PANORAMIC CAMERAS
      . . PINHOLE CAMERAS
      . . SCHMIDT CAMERAS
      . . STREAK CAMERAS
```

CAMERAS-(CONT.)
  . . TELEVISION CAMERAS
  . PHOTOGRAPHIC EQUIPMENT
  . **CAMERAS**
  . . BAKER-NUNN CAMERA
  . . BALLISTIC CAMERAS
  . . DELFT CAMERA
  . . DIFFRACTION LIMITED CAMERAS
  . . FAINT OBJECT CAMERA
  . . HIGH SPEED CAMERAS
  . . . FRAMING CAMERAS
  . . I2S CAMERAS
  . . LALLEMAND CAMERAS
  . . MULTISPECTRAL BAND CAMERAS
  . . PANORAMIC CAMERAS
  . . PINHOLE CAMERAS
  . . SCHMIDT CAMERAS
  . . STREAK CAMERAS
  . . TELEVISION CAMERAS
RT   CAMERA SHUTTERS
       CAMERA TUBES
       CINEMATOGRAPHY
       FOCUSING
       LENSES
       PHOTOGRAPHY
       SIM
       STREAK PHOTOGRAPHY
       ULTRAVIOLET PHOTOGRAPHY
       UNDERWATER PHOTOGRAPHY
       WIDE ANGLE LENSES

CAMEROON
GS   NATIONS
     . **CAMEROON**
RT   AFRICA

CAMOUFLAGE
RT   AIR DEFENSE
       COVERINGS

CAMPBELL-HAUSDORFF SERIES
GS   ANALYSIS (MATHEMATICS)
     . CALCULUS
     . . SERIES (MATHEMATICS)
     . . . **CAMPBELL-HAUSDORFF SERIES**
     . REAL VARIABLES
     . . SERIES (MATHEMATICS)
     . . . **CAMPBELL-HAUSDORFF SERIES**

CAMPHOR
GS   ALIPHATIC COMPOUNDS
     . KETONES
     . . **CAMPHOR**
     TERPENES
     . **CAMPHOR**

CAMS
GS   POSITIONING DEVICES (MACHINERY)
     . **CAMS**
RT   ACTUATORS
       ECCENTRICS
       INTERNAL COMBUSTION ENGINES
       LINKAGES
       MECHANICAL DEVICES

CANADA
GS   NATIONS
     . **CANADA**
     . . ALBERTA
     . . BRITISH COLUMBIA
     . . MANITOBA
     . . NEW BRUNSWICK
     . . NEWFOUNDLAND
     . . NORTHWEST TERRITORIES
     . . NOVA SCOTIA
     . . ONTARIO
     . . PRINCE EDWARD ISLAND
     . . QUEBEC
     . . SASKATCHEWAN
     . . YUKON TERRITORY
RT   ANIK 1
       ANIK 2
       ANIK 3
       BEAUFORT SEA (NORTH AMERICA)
       CANADIAN SPACE PROGRAM
       CANADIAN SPACECRAFT
       COMMUNICATIONS TECHNOLOGY
         SATELLITE
       GREAT LAKES (NORTH AMERICA)
       GREAT PLAINS CORRIDOR (NORTH
         AMERICA)
       INTERNATIONAL FIELD YEAR FOR
         GREAT LAKES
       INTERNATIONAL HYDROLOGICAL
         DECADE

CANADA-(CONT.)
       LABRADOR
       LAKE CHAMPLAIN BASIN (NY-VT)
       NORTH AMERICA
       PACIFIC NORTHWEST (US)
       ROCKY MOUNTAINS (NORTH AMERICA)
       ST LAWRENCE VALLEY (NORTH
         AMERICA)
       WILLISTON BASIN (NORTH AMERICA)

CANADAIR AIRCRAFT
UF   CANADAIR CF-104 AIRCRAFT
       CF-104 AIRCRAFT
GS   GENERAL DYNAMICS AIRCRAFT
     . **CANADAIR AIRCRAFT**
     . . CL-41 AIRCRAFT
     . . CL-44 AIRCRAFT
     . . CL-84 AIRCRAFT
RT   ∞AIRCRAFT

CANADAIR CF-104 AIRCRAFT
USE   CANADAIR AIRCRAFT
       F-104 AIRCRAFT

CANADAIR CL-41 AIRCRAFT
USE   CL-41 AIRCRAFT

CANADAIR CL-44 AIRCRAFT
USE   CL-44 AIRCRAFT

CANADAIR CL-84 AIRCRAFT
USE   CL-84 AIRCRAFT

CANADIAN SHIELD
RT   GEOLOGY
       METEORITE CRATERS
       PRECAMBRIAN PERIOD

CANADIAN SPACE PROGRAM
GS   PROGRAMS
     . SPACE PROGRAMS
     . . **CANADIAN SPACE PROGRAM**
RT   AEROSPACE TECHNOLOGY TRANSFER
       ANIK SATELLITES
       ANIK 1
       ANIK 2
       ANIK 3
       CANADA
       CANADIAN SPACECRAFT
       COMMUNICATIONS TECHNOLOGY
         SATELLITE
       NASA PROGRAMS
       RADARSAT
       SCIENTIFIC SATELLITES
       SYNCHRONOUS SATELLITES
       TECHNOLOGY ASSESSMENT
       TECHNOLOGY UTILIZATION

CANADIAN SPACECRAFT
GS   **CANADIAN SPACECRAFT**
     . ALOUETTE SATELLITES
     . ANIK SATELLITES
     . . ANIK 1
     . . ANIK 2
     . . ANIK 3
     . RADARSAT
RT   CANADA
       CANADIAN SPACE PROGRAM
       ∞SPACECRAFT

CANALS
GS   LANDFORMS
     . **CANALS**
     WATERWAYS
     . **CANALS**
RT   DITCHES
       FLOOD CONTROL
       FLUID FLOW
       GATES (OPENINGS)
       GREAT LAKES (NORTH AMERICA)
       IRRIGATION
       MARS SURFACE
       MATERIALS HANDLING
       PANAMA
       SEEPAGE
       STRAITS
       TROUGHS
       WATER FLOW

CANARD CONFIGURATIONS
RT   AERODYNAMIC CONFIGURATIONS
       AIRCRAFT STRUCTURES
       ∞CONFIGURATIONS
       CONTROL SURFACES

CANARD CONFIGURATIONS-(CONT.)
       SAAB 37 AIRCRAFT
       TANDEM WING AIRCRAFT

CANBERRA AIRCRAFT
UF   ENGLISH ELECTRIC CANBERRA
        AIRCRAFT
GS   BAC AIRCRAFT
     . **CANBERRA AIRCRAFT**
     JET AIRCRAFT
     . **CANBERRA AIRCRAFT**
     MONOPLANES
     . **CANBERRA AIRCRAFT**
RT   ∞AIRCRAFT
       B-57 AIRCRAFT

CANBERRA BOMBER
USE   B-57 AIRCRAFT

CANCELLATION
RT   CONTRACTS
       ELIMINATION
       REMOVAL
       STOPPING

CANCELLATION CIRCUITS
GS   CIRCUITS
     . **CANCELLATION CIRCUITS**
RT   DISPLAY DEVICES
       MOVING TARGET INDICATORS
       PULSE DOPPLER RADAR
       RADAR

CANCER
UF   CARCINOMA
       SARCOMA
GS   DISEASES
     . TUMORS
     . . NEOPLASMS
     . . . **CANCER**
     . . . . LEUKEMIAS
RT   BONE MARROW
       CARCINOGENS
       CELLS (BIOLOGY)
       RADIATION THERAPY
       TISSUES (BIOLOGY)
       ULCERS

CANISTERS
USE   CANS

CANNING
GS   FOOD PROCESSING
     . **CANNING**
RT   ENCAPSULATING
       ∞FOOD

CANNONBALL 2 SATELLITE
GS   SATELLITES
     . ARTIFICIAL SATELLITES
     . . SCIENTIFIC SATELLITES
     . . . **CANNONBALL 2 SATELLITE**

CANNONS
USE   GUNS (ORDNANCE)

CANNULAE
RT   ∞TUBES

CANONICAL FORMS
GS   ALGEBRA
     . VECTOR SPACES
     . . MATRICES (MATHEMATICS)
     . . . **CANONICAL FORMS**
RT   FIBERS (MATHEMATICS)

CANOPIES
RT   AIRCRAFT STRUCTURES
       AIRFRAMES
       COCKPITS
       FAIRINGS
       WINDSHIELDS

CANOPIES (VEGETATION)
GS   VEGETATION
     . **CANOPIES (VEGETATION)**
RT   FOLIAGE
       FORESTS
       GRASSES
       LEAVES
       PLANTS (BOTANY)
       RAIN FORESTS
       SOD
       TREES (PLANTS)

CANOPIES (VEGETATION)-*(CONT.)*
  VEGETATIVE INDEX

CANS
  UF   CANISTERS
  RT   ∞CONTAINERS
        DRUMS (CONTAINERS)

CANT
  USE   SLOPES

CANTILEVER BEAMS
  GS   CANTILEVER MEMBERS
        . **CANTILEVER BEAMS**
        STRUCTURAL MEMBERS
        . BEAMS (SUPPORTS)
        . . **CANTILEVER BEAMS**
  RT   BOX BEAMS
        I BEAMS

CANTILEVER MEMBERS
  GS   **CANTILEVER MEMBERS**
        . CANTILEVER BEAMS
        . CANTILEVER PLATES
  RT   LEVERS

CANTILEVER PLATES
  GS   CANTILEVER MEMBERS
        . **CANTILEVER PLATES**
        STRUCTURAL MEMBERS
        . PLATES (STRUCTURAL MEMBERS)
        . . **CANTILEVER PLATES**
  RT   ANISOTROPIC PLATES

CANTILEVER WINGS
  USE   WINGS

CANYONS
  UF   COULEES
        GORGES
  GS   LANDFORMS
        . **CANYONS**
        . . GRAND CANYON (AZ)
  RT   ARROYOS
        CLIFFS
        FANS (LANDFORMS)
        RAVINES
        RIVERS
        VALLEYS
        WATER EROSION

CAP CLOUDS
  UF   OROGRAPHIC CLOUDS
  GS   CLOUDS
        . CLOUDS (METEOROLOGY)
        . . **CAP CLOUDS**
  RT   ATMOSPHERIC MOISTURE
        CLIMATOLOGY
        CLOUD COVER
        METEOROLOGY
        NEPHANALYSIS
        PRECIPITATION (METEOROLOGY)
        WEATHER

CAPACITANCE
  GS   ELECTROMAGNETIC PROPERTIES
        . ELECTRICAL PROPERTIES
        . . **CAPACITANCE**
  RT   CAPACITANCE-VOLTAGE
        CHARACTERISTICS
        CAPACITORS
        ∞CAPACITY
        DIELECTRIC PROPERTIES
        ELECTRIC CHARGE
        ELECTRICAL IMPEDANCE
        ELECTROSTATIC CHARGE
        INDUCTANCE
        OPEN CIRCUIT VOLTAGE
        RC CIRCUITS
        REACTANCE
        RLC CIRCUITS

CAPACITANCE SWITCHES
  GS   SWITCHES
        . **CAPACITANCE SWITCHES**
  RT   CAPACITORS
        DIELECTRICS
        RLC CIRCUITS
        SWITCHING CIRCUITS

CAPACITANCE-VOLTAGE CHARACTERISTICS
  RT   CAPACITANCE
        ∞CHARACTERISTICS
        ELECTRIC POTENTIAL

CAPACITANCE-VOLTAGE-*(CONT.)*
        ELECTRICAL PROPERTIES
        METAL OXIDE SEMICONDUCTORS
        VOLT-AMPERE CHARACTERISTICS

CAPACITIVE FUEL GAGES
  GS   MEASURING INSTRUMENTS
        . FUEL GAGES
        . . **CAPACITIVE FUEL GAGES**
  RT   DIELECTRICS

CAPACITORS
  RT   AMPLIFIERS
        BALLASTS (IMPEDANCES)
        CAPACITANCE
        CAPACITANCE SWITCHES
        CIRCUIT PROTECTION
        CIRCUITS
        ∞CONDENSERS
        DIELECTRICS
        ELECTRETS
        ELECTRIC BRIDGES
        ELECTRIC ENERGY STORAGE
        ELECTRIC FILTERS
        ELECTRIC REACTORS
        ENERGY STORAGE
        GERDIEN CONDENSERS
        PARALLEL PLATES
        SOLID STATE DEVICES

∞ **CAPACITY**
  SN   *(USE OF A MORE SPECIFIC TERM IS
        RECOMMENDED--CONSULT THE TERMS
        LISTED BELOW)*
  RT   CAPACITANCE
        CHANNEL CAPACITY
        OUTPUT
        PRODUCTION ENGINEERING
        RISK
        VOLUME

CAPE HATTERAS (NC)
  GS   LANDFORMS
        . CAPES (LANDFORMS)
        . . **CAPE HATTERAS (NC)**
  RT   ATLANTIC OCEAN
        NORTH CAROLINA

CAPE KENNEDY LAUNCH COMPLEX
  GS   LAUNCHING BASES
        . **CAPE KENNEDY LAUNCH COMPLEX**
  RT   GROUND SUPPORT EQUIPMENT

CAPE VERDE
  GS   NATIONS
        . **CAPE VERDE**
  RT   AFRICA
        ATLANTIC OCEAN
        ISLANDS

CAPES (LANDFORMS)
  GS   LANDFORMS
        . **CAPES (LANDFORMS)**
        . . CAPE HATTERAS (NC)
  RT   LAND

∞ **CAPILLARIES**
  SN   *(USE OF A MORE SPECIFIC TERM IS
        RECOMMENDED--CONSULT THE TERMS
        LISTED BELOW)*
  RT   CAPILLARIES (ANATOMY)
        CAPILLARY TUBES

CAPILLARIES (ANATOMY)
  GS   ANATOMY
        . CARDIOVASCULAR SYSTEM
        . . BLOOD VESSELS
        . . . **CAPILLARIES (ANATOMY)**
        . CIRCULATORY SYSTEM
        . . VASCULAR SYSTEM
        . . . BLOOD VESSELS
        . . . . **CAPILLARIES (ANATOMY)**
  RT   BLOOD
        ∞CAPILLARIES

CAPILLARY CIRCULATION
  USE   CAPILLARY FLOW

CAPILLARY FLOW
  UF   CAPILLARY CIRCULATION
  GS   FLUID FLOW
        . **CAPILLARY FLOW**
  RT   BLOOD FLOW
        LAMINAR FLOW

CAPILLARY TUBES
  RT   ∞CAPILLARIES
        ∞TUBES

CAPILLARY WAVES
  GS   ELASTIC WAVES
        . **CAPILLARY WAVES**
        . . GRAVITY WAVES
        . . . BAROCLINIC WAVES
        . . RIPPLES
        SURFACE WAVES
        . **CAPILLARY WAVES**
        . . GRAVITY WAVES
        . . . BAROCLINIC WAVES
        . . RIPPLES
  RT   INTERFACIAL TENSION
        TWO DIMENSIONAL FLOW
        WATER WAVES

∞ **CAPS**
  SN   *(USE OF A MORE SPECIFIC TERM IS
        RECOMMENDED--CONSULT THE TERMS
        LISTED BELOW)*
  RT   CAPS (EXPLOSIVES)
        COVERINGS
        NOSE CONES
        POLAR CAPS
        SEALS (STOPPERS)
        SPHERICAL CAPS

CAPS (EXPLOSIVES)
  GS   EXPLOSIVE DEVICES
        . INITIATORS (EXPLOSIVES)
        . . **CAPS (EXPLOSIVES)**
        IGNITERS
        . INITIATORS (EXPLOSIVES)
        . . **CAPS (EXPLOSIVES)**
  RT   ∞CAPS
        DETONATORS
        EXPLODING WIRES
        FUSES (ORDNANCE)
        PRIMERS (EXPLOSIVES)

∞ **CAPSULES**
  SN   *(USE OF A MORE SPECIFIC TERM IS
        RECOMMENDED--CONSULT THE TERMS
        LISTED BELOW)*
  RT   ∞BUCKETS
        ∞CONTAINERS
        FUEL CAPSULES
        SHELLS (STRUCTURAL FORMS)
        SPACE CAPSULES
        TABLETS
        TEST CHAMBERS
        ∞TEST EQUIPMENT
        TEST VEHICLES
        ∞VESSELS

CAPSULES (SPACECRAFT)
  USE   SPACE CAPSULES

CAPTIVE TESTS
  GS   **CAPTIVE TESTS**
        . STATIC TESTS
        . . STATIC FIRING
  RT   ENGINE TESTS
        GROUND TESTS
        MISSILE TESTS
        PREFIRING TESTS
        PRELAUNCH TESTS
        ∞TESTS

CAPTURE CROSS SECTIONS
  USE   ABSORPTION CROSS SECTIONS

CAPTURE EFFECT
  RT   ABSORPTANCE
        ∞ABSORPTION
        ∞EFFECTS
        ELECTRON CAPTURE
        FREQUENCY MODULATION
        FREQUENCY SYNCHRONIZATION
        NUCLEAR CAPTURE
        RECOMBINATION REACTIONS
        TRAJECTORY ANALYSIS

CAPTURED AIR BUBBLE VEHICLES
  GS   SURFACE VEHICLES
        . **CAPTURED AIR BUBBLE VEHICLES**
        WATER VEHICLES
        . **CAPTURED AIR BUBBLE VEHICLES**
  RT   HYDROFOIL CRAFT
        SURFACE EFFECT SHIPS
        SWATH (SHIP)

CAPTURED AIR BUBBLE VEHICLES-(CONT.)
∞ VEHICLES

CARAVELLE AIRCRAFT
USE   SE-210 AIRCRAFT

CARBAMATES (TRADENAME)
GS   ALIPHATIC COMPOUNDS
. **CARBAMATES (TRADENAME)**
. . URETHANES
ESTERS
. **CARBAMATES (TRADENAME)**
. . URETHANES
POISONS
. **CARBAMATES (TRADENAME)**
. . URETHANES

CARBAMIDES
GS   ALIPHATIC COMPOUNDS
. **CARBAMIDES**
NITROGEN COMPOUNDS
. AMIDES
. . **CARBAMIDES**

CARBAZOLES
GS   HETEROCYCLIC COMPOUNDS
. AZOLES
. . PYRROLES
. . . **CARBAZOLES**

CARBENES
RT   FREE RADICALS

CARBIDES
GS   CARBON COMPOUNDS
. **CARBIDES**
. . ALUMINUM CARBIDES
. . BORON CARBIDES
. . CEMENTITE
. . CHROMIUM CARBIDES
. . HAFNIUM CARBIDES
. . MOLYBDENUM CARBIDES
. . NIOBIUM CARBIDES
. . SILICON CARBIDES
. . TANTALUM CARBIDES
. . TITANIUM CARBIDES
. . TUNGSTEN CARBIDES
. . URANIUM CARBIDES
. . VANADIUM CARBIDES
. . ZIRCONIUM CARBIDES
RT   CERAMIC NUCLEAR FUELS
REFRACTORY MATERIALS

CARBOHYDRATE METABOLISM
GS   METABOLISM
. **CARBOHYDRATE METABOLISM**
. . HYPERGLYCEMIA
. . HYPOGLYCEMIA
RT   DIABETES MELLITUS
HYDROGEN METABOLISM

CARBOHYDRATES
UF   SACCHARIDES
GS   **CARBOHYDRATES**
. ADENINES
. CITRIC ACID
. FATS
. . CHOLINE
. GLUCOSIDES
. HEXOKINASE
. NUCLEOSIDES
. . ADENOSINES
. . . ADENOSINE DIPHOSPHATE
. . . ADENOSINE TRIPHOSPHATE
. POLYSACCHARIDES
. . CELLULOSE
. . . FORTISAN (TRADEMARK)
. . CHITIN
. . DEXTRANS
. . GLYCOGENS
. . STARCHES
. SUGARS
. . DEXTRANS
. . GALACTOSE
. . GLUCOSE
. . HEXOSES
. . INOSITOLS
. . LACTOSE
. . MANNITOL
. . MONOSACCHARIDES
. . . RIBOSE
. . . XYLOSE
. . PENTOSE
. . . RIBOSE
. . . XYLOSE

CARBOHYDRATES-(CONT.)
. . SUCROSE
RT   ALCOHOLS
ETHYL ALCOHOL
∞ FOOD
GLYCEROLS
∞ NUTRIENTS
OPTICAL ACTIVITY
∞ OXYGEN COMPOUNDS
PHOTOSYNTHESIS
STEREOCHEMISTRY
SYNTHETIC FOOD

CARBON
GS   CHEMICAL ELEMENTS
. **CARBON**
. . CARBON ISOTOPES
. . . CARBON 12
. . . CARBON 13
. . . CARBON 14
. . CHARCOAL
. . . ACTIVATED CARBON
RT   BITUMENS
COKE
DECARBURIZATION
DIAMONDS
GRAPHITE
SOOT

CARBON ARCS
GS   ELECTRIC CURRENT
. ELECTRIC DISCHARGES
. . ELECTRIC ARCS
. . . **CARBON ARCS**
RT   ARC LAMPS
IMAGE FURNACES

CARBON COMPOUNDS
GS   **CARBON COMPOUNDS**
. CARBIDES
. . ALUMINUM CARBIDES
. . BORON CARBIDES
. . CEMENTITE
. . CHROMIUM CARBIDES
. . HAFNIUM CARBIDES
. . MOLYBDENUM CARBIDES
. . NIOBIUM CARBIDES
. . SILICON CARBIDES
. . TANTALUM CARBIDES
. . TITANIUM CARBIDES
. . TUNGSTEN CARBIDES
. . URANIUM CARBIDES
. . VANADIUM CARBIDES
. . ZIRCONIUM CARBIDES
. CARBONATES
. . BASTNASITE
. . CALCIUM CARBONATES
. . . AKERMANITE
. . . ARAGONITE
. . . CALCITE
. . . CHALK
. . DOLOMITE (MINERAL)
. . POLYCARBONATES
. . . LEXAN (TRADEMARK)
. . SIDERITES
. . SODIUM CARBONATES
. . TETRAETHYL ORTHOCARBONATES
. FLUOROPOLYMERS
. HALOCARBONS
. . CHLOROCARBONS
RT   CARBONACEOUS MATERIALS
∞ CHEMICAL COMPOUNDS
∞ GROUP 4A COMPOUNDS
HYDROCARBONS
SWAN BANDS

CARBON CYCLE
GS   CYCLES
. **CARBON CYCLE**
RT   ANIMALS
∞ BIOLOGY
BIOMASS
ECOLOGY
ORGANISMS
PLANTS (BOTANY)
VIABILITY

CARBON DIOXIDE
GS   CHALCOGENIDES
. OXIDES
. . DIOXIDES
. . . **CARBON DIOXIDE**
GASES
. **CARBON DIOXIDE**
RT   METABOLIC WASTES

CARBON DIOXIDE-(CONT.)
SYNTHANE

CARBON DIOXIDE CONCENTRATION
GS   COMPOSITION (PROPERTY)
. CHEMICAL COMPOSITION
. . **CARBON DIOXIDE CONCENTRATION**
. CONCENTRATION (COMPOSITION)
. . **CARBON DIOXIDE CONCENTRATION**
. GAS COMPOSITION
. . **CARBON DIOXIDE CONCENTRATION**
RT   AIR PURIFICATION
ATMOSPHERIC COMPOSITION
DECONTAMINATION
REBREATHING
SPACECRAFT CABIN ATMOSPHERES

CARBON DIOXIDE LASERS
GS   STIMULATED EMISSION DEVICES
. LASERS
. . GAS LASERS
. . . **CARBON DIOXIDE LASERS**
RT   CHEMICAL LASERS
CONTINUOUS WAVE LASERS
GAS MASERS
INFRARED LASERS
MACH-ZEHNDER INTERFEROMETERS
MOLECULAR OSCILLATIONS
ORGANIC LASERS
POLAR GASES
POWER TRANSMISSION (LASERS)
PULSED LASERS
Q SWITCHED LASERS
STIMULATED EMISSION
TEA LASERS
WAVEGUIDE LASERS

CARBON DIOXIDE REMOVAL
RT   AIR PURIFICATION
DECONTAMINATION
REBREATHING
REMOVAL
SMOKE ABATEMENT

CARBON DIOXIDE TENSION
GS   **CARBON DIOXIDE TENSION**
. HYPERCAPNIA
. HYPOCAPNIA

CARBON DISULFIDE
GS   CHALCOGENIDES
. SULFIDES
. . DISULFIDES
. . . **CARBON DISULFIDE**
SULFUR COMPOUNDS
. SULFIDES
. . DISULFIDES
. . . **CARBON DISULFIDE**

CARBON FIBER REINFORCED PLASTICS
UF   CFRP
GS   COMPOSITE MATERIALS
. FIBER COMPOSITES
. . **CARBON FIBER REINFORCED PLASTICS**
PLASTICS
. **CARBON FIBER REINFORCED PLASTICS**
RT   FIBERS
GRAPHITE-EPOXY COMPOSITES
LAY-UP
REINFORCING FIBERS
SUPERHYBRID MATERIALS

CARBON FIBERS
GS   FIBERS
. REINFORCING FIBERS
. . **CARBON FIBERS**
RT   BORON FIBERS
COMPOSITE MATERIALS
FIBER COMPOSITES
FIBER RELEASE
∞ FILAMENTS

CARBON ISOTOPES
GS   CHEMICAL ELEMENTS
. CARBON
. . **CARBON ISOTOPES**
. . . CARBON 12
. . . CARBON 13
. . . CARBON 14
. NUCLIDES
. . ISOTOPES
. . . **CARBON ISOTOPES**
. . . . CARBON 12

CARBON ISOTOPES-(CONT.)
```
          .... CARBON 13
          .... CARBON 14
```

CARBON LASERS
```
GS   STIMULATED EMISSION DEVICES
     . LASERS
     .. CARBON LASERS
RT   CHEMICAL LASERS
     GAS LASERS
     INFRARED LASERS
     LIQUID LASERS
     ORGANIC LASERS
     STIMULATED EMISSION
```

CARBON MONOXIDE
```
GS   CHALCOGENIDES
     . OXIDES
     .. CARBON MONOXIDE
     GASES
     . CARBON MONOXIDE
RT   HOPCALITE (TRADEMARK)
     SMOG
     SYNTHANE
```

CARBON MONOXIDE LASERS
```
GS   STIMULATED EMISSION DEVICES
     . LASERS
     .. GAS LASERS
     ... CARBON MONOXIDE LASERS
RT   CHEMICAL LASERS
     CONTINUOUS WAVE LASERS
     INFRARED LASERS
     MOLECULAR OSCILLATIONS
     POWER TRANSMISSION (LASERS)
     STIMULATED EMISSION
     TEA LASERS
```

CARBON MONOXIDE POISONING
```
GS   DISEASES
     . TOXIC DISEASES
     .. CARBON MONOXIDE POISONING
     TOXICITY
     . CARBON MONOXIDE POISONING
RT   CARBOXYHEMOGLOBIN
     LETHALITY
     PATHOLOGICAL EFFECTS
     ∞POISONING
```

CARBON STARS
```
GS   CELESTIAL BODIES
     . STARS
     .. S STARS
     ... GIANT STARS
     .... RED GIANT STARS
     ..... CARBON STARS
RT   STELLAR COMPOSITION
     SUBGIANT STARS
     WOLF-RAYET STARS
```

CARBON STEELS
```
GS   ALLOYS
     . IRON ALLOYS
     .. STEELS
     ... CARBON STEELS
     .... LOW CARBON STEELS
RT   HIGH STRENGTH STEELS
```

CARBON SUBOXIDES
```
GS   CHALCOGENIDES
     . OXIDES
     .. CARBON SUBOXIDES
     GASES
     . CARBON SUBOXIDES
RT   ∞OXYGEN COMPOUNDS
```

CARBON TETRACHLORIDE
```
UF   TETRACHLOROMETHANE
GS   ALIPHATIC COMPOUNDS
     . CARBON TETRACHLORIDE
     HALOGEN COMPOUNDS
     . CHLORINE COMPOUNDS
     .. CHLORIDES
     ... CARBON TETRACHLORIDE
     . HALIDES
     .. CHLORIDES
     ... CARBON TETRACHLORIDE
```

CARBON TETRACHLORIDE POISONING
```
RT   INDUSTRIAL SAFETY
     ∞POISONING
     TOXICITY AND SAFETY HAZARD
     TOXICOLOGY
```

CARBON TETRAFLUORIDE
```
GS   ALIPHATIC COMPOUNDS
     . CARBON TETRAFLUORIDE
     HALOGEN COMPOUNDS
     . FLUORINE COMPOUNDS
     .. FLUORO COMPOUNDS
     ... FLUORINE ORGANIC COMPOUNDS
     .... FLUOROHYDROCARBONS
     ..... CARBON TETRAFLUORIDE
     ORGANIC COMPOUNDS
     . FLUORINE ORGANIC COMPOUNDS
     .. FLUOROHYDROCARBONS
     ... CARBON TETRAFLUORIDE
```

CARBON 12
```
GS   CHEMICAL ELEMENTS
     . CARBON
     .. CARBON ISOTOPES
     ... CARBON 12
     . NUCLIDES
     .. ISOTOPES
     ... CARBON ISOTOPES
     .... CARBON 12
```

CARBON 13
```
GS   CHEMICAL ELEMENTS
     . CARBON
     .. CARBON ISOTOPES
     ... CARBON 13
     . NUCLIDES
     .. ISOTOPES
     ... CARBON ISOTOPES
     .... CARBON 13
```

CARBON 14
```
GS   CHEMICAL ELEMENTS
     . CARBON
     .. CARBON ISOTOPES
     ... CARBON 14
     . NUCLIDES
     .. ISOTOPES
     ... CARBON ISOTOPES
     .... CARBON 14
     ... RADIOACTIVE ISOTOPES
     .... CARBON 14
```

CARBON-CARBON COMPOSITES
```
GS   COMPOSITE MATERIALS
     . CARBON-CARBON COMPOSITES
RT   FIBER COMPOSITES
     FRACTURE STRENGTH
     REINFORCING FIBERS
     THERMAL PROTECTION
     THERMAL RESISTANCE
```

CARBONACEOUS CHONDRITES
```
GS   CELESTIAL BODIES
     . METEORITES
     .. STONY METEORITES
     ... CHONDRITES
     .... CARBONACEOUS CHONDRITES
     ..... ALLENDE METEORITE
     ..... MURCHISON METEORITE
RT   STONY METEORITES
```

CARBONACEOUS MATERIALS
```
GS   CARBONACEOUS MATERIALS
     . PEAT
RT   CARBON COMPOUNDS
     COAL
     CRUDE OIL
     FOSSIL FUELS
     LIGNITE
     ∞MATERIALS
     ORGANIC MATERIALS
     SOLVENT REFINED COAL
```

CARBONACEOUS METEORITES
```
GS   CELESTIAL BODIES
     . METEORITES
     .. STONY METEORITES
     ... CHONDRITES
     .... CARBONACEOUS METEORITES
     ..... ALAIS METEORITE
     ..... COLD BOKKEVELD METEORITE
     ..... IVUNA METEORITE
     ..... MURRAY METEORITE
     ..... ORGUEIL METEORITE
     ..... TONK METEORITE
RT   EXOBIOLOGY
     METEORITIC COMPOSITION
```

CARBONACEOUS ROCKS
```
GS   SEDIMENTARY ROCKS
     CARBONACEOUS ROCKS
```

CARBONACEOUS ROCKS-(CONT.)
```
     .. COAL
     .. LIGNITE
RT   CARBONATES
     REGOLITH
     ROCKS
     SHATTER CONES
     SOILS
```

CARBONATES
```
UF   BICARBONATES
GS   CARBON COMPOUNDS
     . CARBONATES
     .. BASTNASITE
     .. CALCIUM CARBONATES
     ... AKERMANITE
     ... ARAGONITE
     ... CALCITE
     ... CHALK
     .. DOLOMITE (MINERAL)
     .. POLYCARBONATES
     ... LEXAN (TRADEMARK)
     .. SIDERITES
     .. SODIUM CARBONATES
     .. TETRAETHYL ORTHOCARBONATES
RT   ALKALIES
     CARBONACEOUS ROCKS
     CARBONIC ACID
     ∞OXYGEN COMPOUNDS
```

CARBONIC ACID
```
GS   ACIDS
     . CARBONIC ACID
RT   CARBONATES
```

CARBONIC ANHYDRASE
```
GS   ENZYMES
     . CARBONIC ANHYDRASE
     PROTEINS
     . CARBONIC ANHYDRASE
RT   ACETAZOLAMIDE
```

CARBONIZATION
```
GS   CHEMICAL REACTIONS
     . CARBONIZATION
RT   CHARRING
     DECARBONATION
```

CARBONYL COMPOUNDS
```
RT   ALIPHATIC COMPOUNDS
     ∞CHEMICAL COMPOUNDS
```

CARBORANE
```
GS   BORON COMPOUNDS
     . BORON HYDRIDES
     .. BORANES
     ... CARBORANE
     HYDROGEN COMPOUNDS
     . HYDRIDES
     .. BORON HYDRIDES
     ... BORANES
     .... CARBORANE
```

CARBORUNDUM (TRADEMARK)
```
RT   ABRASIVES
     REFRACTORY MATERIALS
     SILICON CARBIDES
```

CARBOXYHEMOGLOBIN
```
GS   CELLS (BIOLOGY)
     . HEMOGLOBIN
     .. CARBOXYHEMOGLOBIN
     CIRCULATION
     . CARBOXYHEMOGLOBIN
     ORGANOMETALLIC COMPOUNDS
     . HEMOGLOBIN
     .. CARBOXYHEMOGLOBIN
     PROTEINS
     . CARBOXYHEMOGLOBIN
RT   BLOOD CIRCULATION
     CARBON MONOXIDE POISONING
     ERYTHROCYTES
```

CARBOXYHEMOGLOBIN TEST
```
GS   PHYSIOLOGICAL TESTS
     . CARBOXYHEMOGLOBIN TEST
RT   BLOOD
     HEMATOLOGY
```

CARBOXYL GROUP
```
RT   CARBOXYLIC ACIDS
```

CARBOXYLATES
```
GS   ALIPHATIC COMPOUNDS
```

CARBOXYLATES-*(CONT.)*
```
        . CARBOXYLATES
          ESTERS
        . CARBOXYLATES
RT        CARBOXYLIC ACIDS
```

CARBOXYLATION
```
GS      CHEMICAL REACTIONS
        . CARBOXYLATION
RT      DECARBOXYLATION
```

CARBOXYLIC ACIDS
```
GS      ACIDS
        . FATTY ACIDS
        .. CARBOXYLIC ACIDS
        ... ACETIC ACID
        .... ETHYLENEDIAMINETETRAACETIC
             ACIDS
        .... IODOACETIC ACID
        ... ACETYLSALICYLIC ACID
        .. BENZILIC ACID
        .. BENZOIC ACID
        .. DICARBOXYLIC ACIDS
        .... TEREPHTHALATE
        ... OLEIC ACID
        .. PROPIONIC ACID
        ... SEBACIC ACID
        ... VALERIC ACID
RT      ALIPHATIC COMPOUNDS
        CARBOXYL GROUP
        CARBOXYLATES
```

CARBURETORS
```
UF      INJECTION CARBURETORS
RT      CHOKES (FUEL SYSTEMS)
        CONTACTORS
        ENGINE PARTS
        ENGINES
        FUEL INJECTION
        FUEL SYSTEMS
        INJECTORS
        INTERNAL COMBUSTION ENGINES
      ∞JET NOZZLES
        MIXERS
        PREMIXED FLAMES
        THROATS
```

CARBURIZING
```
GS      HARDENING (MATERIALS)
        . CARBURIZING
RT      DECARBURIZATION
```

CARCINOGENS
```
RT      CANCER
        NEOPLASMS
```

CARCINOMA
```
USE     CANCER
```

CARCINOTRONS
```
GS      AMPLIFIERS
        . CARCINOTRONS
        ELECTRON TUBES
        . VACUUM TUBES
        .. VACUUM TUBE OSCILLATORS
        ... MICROWAVE TUBES
        .... PLANOTRONS
        ..... CARCINOTRONS
        .... TRAVELING WAVE TUBES
        ..... CARCINOTRONS
        MICROWAVE EQUIPMENT
        . MICROWAVE TUBES
        .. PLANOTRONS
        ... CARCINOTRONS
        .. TRAVELING WAVE TUBES
        ... CARCINOTRONS
        OSCILLATORS
        . VACUUM TUBE OSCILLATORS
        .. MICROWAVE TUBES
        ... PLANOTRONS
        .... CARCINOTRONS
RT      HELITRONS
```

CARDIAC AURICLES
```
GS      ANATOMY
        . CARDIOVASCULAR SYSTEM
        .. HEART
        ... CARDIAC AURICLES
RT      HIS BUNDLE
```

CARDIAC VENTRICLES
```
GS      ANATOMY
        . CARDIOVASCULAR SYSTEM
        .. HEART
```

CARDIAC VENTRICLES-*(CONT.)*
```
        ... CARDIAC VENTRICLES
RT      DIASTOLIC PRESSURE
        ECHOCARDIOGRAPHY
        HIS BUNDLE
        SYSTOLE
```

CARDIOGRAMS
```
RT      BIOMEDICAL DATA
        CARDIOGRAPHY
        HEART
```

CARDIOGRAPHY
```
GS      BIOENGINEERING
        . BIOMETRICS
        .. CARDIOGRAPHY
        ... BALLISTOCARDIOGRAPHY
        ... ELECTROCARDIOGRAPHY
        ... MAGNETOCARDIOGRAPHY
        .. PHONOCARDIOGRAPHY
        .... ECHOCARDIOGRAPHY
        ... SEISMOCARDIOGRAPHY
        ... VECTORCARDIOGRAPHY
RT      CARDIOGRAMS
        CARDIOVASCULAR SYSTEM
        HEART
        HEART DISEASES
        MEDICAL EQUIPMENT
        PHYSIOLOGICAL TESTS
```

CARDIOLOGY
```
GS      CARDIOLOGY
        . RADIOCARDIOGRAPHY
RT      ANGIOGRAPHY
        ARTIFICIAL CARDIAC PACEMAKER
        HEART
        HEART DISEASES
        HEART RATE
        RADIOCARDIOGRAPHY
```

CARDIOTACHOMETERS
```
GS      MEDICAL EQUIPMENT
        . CARDIOTACHOMETERS
RT      HEART
```

CARDIOVASCULAR SYSTEM
```
GS      ANATOMY
        . CARDIOVASCULAR SYSTEM
        .. BLOOD VESSELS
        ... ARTERIES
        .... AORTA
        ... CAPILLARIES (ANATOMY)
        ... GLOMERULUS
        ... VEINS
        .. CORPUSCLES
        .. DIASTOLE
        .. ERYTHROCYTES
        ... RETICULOCYTES
        .. HEART
        ... CARDIAC AURICLES
        ... CARDIAC VENTRICLES
        ... EPICARDIUM
        .. MYOCARDIUM
        .. HEMATOPOIESIS
        .. HEMATOPOIETIC SYSTEM
        .. LEUKOCYTES
        ... EOSINOPHILS
        ... LYMPHOCYTES
        .. SYSTOLE
        .. THROMBIN
        .. THROMBOPLASTIN
RT      ANGIOGRAPHY
        BLOOD
        BLOOD VOLUME
        CARDIOGRAPHY
        CAROTID SINUS BODY
        CAROTID SINUS REFLEX
        CEREBRAL VASCULAR ACCIDENTS
        CIRCULATORY SYSTEM
        FAT EMBOLISMS
        HEMODYNAMICS
        HEMORRHAGES
        LOWER BODY NEGATIVE PRESSURE
      ∞SYSTEMS
        VASCULAR SYSTEM
```

CARDS
```
GS      CARDS
        . PUNCHED CARDS
RT      COMPUTER STORAGE DEVICES
        DATA STORAGE
```

CARET WINGS
```
GS      AIRFOILS
        . WINGS
```

CARET WINGS-*(CONT.)*
```
        .. CARET WINGS
          PLANFORMS
        . CARET WINGS
RT        ARROW WINGS
          DELTA WINGS
```

CARETS (TEST SITE)
```
USE     CENTRAL ATLANTIC REGIONAL ECOL
        TEST SITE
```

CARGO
```
UF      FREIGHT
GS      CARGO
        . AIR CARGO
        .. AIR MAIL
        . BAGGAGE
RT      AIR DROP OPERATIONS
        AIRDROPS
        DELIVERY
        FREIGHT COSTS
        HARBORS
        HAULING
        MATERIALS HANDLING
        RAILROAD HUMPING TESTS
        RAPID TRANSIT SYSTEMS
        TRANSPORTATION
        TRANSPORTATION ENERGY
        TRUCKS
```

CARGO AIRCRAFT
```
GS      TRANSPORT AIRCRAFT
        . CARGO AIRCRAFT
        .. BREGUET 941 AIRCRAFT
        .. C-1A AIRCRAFT
        .. C-2 AIRCRAFT
        .. C-5 AIRCRAFT
        .. C-9 AIRCRAFT
        .. C-46 AIRCRAFT
        .. C-47 AIRCRAFT
        .. C-54 AIRCRAFT
        .. C-118 AIRCRAFT
        .. C-119 AIRCRAFT
        .. C-121 AIRCRAFT
        .. C-123 AIRCRAFT
        .. C 124 AIRCRAFT
        .. C-130 AIRCRAFT
        .. C-131 AIRCRAFT
        .. C-133 AIRCRAFT
        .. C-135 AIRCRAFT
        .. C-140 AIRCRAFT
        .. C-141 AIRCRAFT
        .. C-160 AIRCRAFT
        .. CL-44 AIRCRAFT
        .. DC 3 AIRCRAFT
        .. DC 7 AIRCRAFT
        .. F-27 AIRCRAFT
        .. P-166 AIRCRAFT
        .. SPANLOADER AIRCRAFT
        .. YC-14 AIRCRAFT
RT      AIR CARGO
      ∞AIRCRAFT
        BOEING 727 AIRCRAFT
        BOEING 737 AIRCRAFT
        BOEING 767 AIRCRAFT
        COMMERCIAL AIRCRAFT
        HEAVY LIFT HELICOPTERS
        JET AIRCRAFT
        MATERIALS HANDLING
        MERCURE AIRCRAFT
        MH-262 AIRCRAFT
      ∞MILITARY AIRCRAFT
        MONOPLANES
        PASSENGER AIRCRAFT
        SC-7 AIRCRAFT
        SUPERSONIC TRANSPORTS
        T-39 AIRCRAFT
        TU-154 AIRCRAFT
        UTILITY AIRCRAFT
        VC-10 AIRCRAFT
```

CARGO SHIPS
```
UF      LOTS CARGO SHIPS
GS      SURFACE VEHICLES
        . CARGO SHIPS
        .. SAVANNAH NUCLEAR SHIP
        .. TANKER SHIPS
        WATER VEHICLES
        . SHIPS
        .. CARGO SHIPS
        ... SAVANNAH NUCLEAR SHIP
        ... TANKER SHIPS
RT      ARTIFICIAL HARBORS
        DEEPWATER TERMINALS
        NUCLEAR POWERED SHIPS
```

CARGO SHIPS-*(CONT.)*
OFFSHORE DOCKING
OFFSHORE PLATFORMS
SHIPYARDS
TANKER TERMINALS
WHARVES

CARGO SPACECRAFT
RT   FERRY SPACECRAFT
∞ SPACECRAFT

CARGOMASTER AIRCRAFT
USE   C-133 AIRCRAFT

CARIBBEAN REGION
RT   BAHAMAS
BARBADOS
BELIZE
CUBA
DEVELOPING NATIONS
DOMINICAN REPUBLIC
FRENCH GUIANA
GUYANA
HAITI
JAMAICA
MARTINIQUE
SURINAM
TRINIDAD AND TOBAGO
VIRGIN ISLANDS
WEST INDIES

CARIBBEAN SEA
GS   SEAS
. **CARIBBEAN SEA**
RT   BELIZE
CUBA
DOMINICAN REPUBLIC
GULF OF MEXICO
GULF STREAM
HAITI
PANAMA CANAL ZONE
VIRGIN ISLANDS

CARIBOU AIRCRAFT
USE   DHC 4 AIRCRAFT

CARIBOUS
GS   ANIMALS
. VERTEBRATES
. . MAMMALS
. . . DEER
. . . . **CARIBOUS**

CARNITINE
GS   HETEROCYCLIC COMPOUNDS
. **CARNITINE**
VITAMINS
. **CARNITINE**

CARNOT CYCLE
GS   CYCLES
. THERMODYNAMIC CYCLES
. . **CARNOT CYCLE**
RT   ADIABATIC CONDITIONS
RANKINE CYCLE
STIRLING CYCLE

CAROTENE
GS   HYDROCARBONS
. **CAROTENE**
PIGMENTS
. **CAROTENE**
RT   RETINENE
SKIN (ANATOMY)

CAROTID SINUS BODY
RT   ARTERIES
BLOOD VESSELS
CARDIOVASCULAR SYSTEM
CAROTID SINUS REFLEX
CHEMORECEPTORS
CIRCULATORY SYSTEM
NERVES
SINUSES
VASCULAR SYSTEM

CAROTID SINUS REFLEX
GS   REFLEXES
. **CAROTID SINUS REFLEX**
RT   ARTERIES
BLOOD VESSELS
CARDIOVASCULAR SYSTEM
CAROTID SINUS BODY
CIRCULATORY SYSTEM

CAROTID SINUS REFLEX-*(CONT.)*
NERVES
SINUSES
VASCULAR SYSTEM

CARPATHIAN MOUNTAINS (EUROPE)
GS   LANDFORMS
. MOUNTAINS
. . **CARPATHIAN MOUNTAINS (EUROPE)**
RT   EUROPE

CARRIAGES
RT   CARTS
CHASSIS
DOLLIES
FRAMES
LANDING GEAR
SUPPORTS
UNDERCARRIAGES

CARRIER DENSITY (SOLID STATE)
GS   DENSITY (NUMBER/VOLUME)
. PARTICLE DENSITY (CONCENTRATION)
. . ELECTRON DENSITY
(CONCENTRATION)
. . . **CARRIER DENSITY (SOLID STATE)**
RT   ACCEPTOR MATERIALS
CARRIER LIFETIME
CARRIER TRANSPORT (SOLID STATE)
∞ CARRIERS
DONOR MATERIALS
ELECTRON-HOLE DROPS
SEMICONDUCTORS (MATERIALS)
ZENER EFFECT

CARRIER FREQUENCIES
GS   FREQUENCIES
. **CARRIER FREQUENCIES**
RT   CARRIER TO NOISE RATIOS
FREQUENCY DIVISION MULTIPLEXING
HARMONIC GENERATIONS
MODULATION
MULTIPLEXING
RADIO FREQUENCIES
SINGLE CHANNEL PER CARRIER
TRANSMISSION
SWEEP FREQUENCY
UNIFIED S BAND

CARRIER INJECTION
GS   INJECTION
. **CARRIER INJECTION**
RT   ADDITIVES
BARRITT DIODES
BIPOLAR TRANSISTORS
CARRIER LIFETIME
CHARGE CARRIERS
CHARGE TRANSFER
INJECTION LOCKING
ION INJECTION
MAJORITY CARRIERS
MINORITY CARRIERS
RADIATIVE RECOMBINATION
SEMICONDUCTORS (MATERIALS)
SUHL EFFECT
TRAVELING SOLVENT METHOD

CARRIER LIFETIME
GS   LIFE (DURABILITY)
. **CARRIER LIFETIME**
RT   CARRIER DENSITY (SOLID STATE)
CARRIER INJECTION
CARRIER MOBILITY
CARRIER TRANSPORT (SOLID STATE)
CHARGE CARRIERS
MINORITY CARRIERS
SOLAR CELLS

CARRIER MOBILITY
GS   ELECTROMAGNETIC PROPERTIES
. ELECTRICAL PROPERTIES
. . **CARRIER MOBILITY**
. . . ELECTRON MOBILITY
. . . . HOLE MOBILITY
MOBILITY
. **CARRIER MOBILITY**
. . ELECTRON MOBILITY
. . HOLE MOBILITY
TRANSPORT PROPERTIES
. **CARRIER MOBILITY**
. . ELECTRON MOBILITY
. . HOLE MOBILITY
RT   CARRIER LIFETIME
ELECTRICAL RESISTIVITY
ELECTROMAGNETIC PROPERTIES

CARRIER MOBILITY-*(CONT.)*
EXCITONS
HALL EFFECT
ION IMPLANTATION
SUPERCONDUCTORS

CARRIER MODULATION
USE   MODULATION

CARRIER ROCKETS
USE   LAUNCH VEHICLES

CARRIER SYSTEMS
USE   WIRELESS COMMUNICATION

CARRIER TO NOISE RATIOS
RT   CARRIER FREQUENCIES
COMMUNICATION SATELLITES
DATA TRANSMISSION
DOWNLINKING
EARTH TERMINALS
FREQUENCY MODULATION
SIGNAL TO NOISE RATIOS
TRANSMISSION EFFICIENCY
UPLINKING

CARRIER TRANSPORT (SOLID STATE)
RT   CARRIER DENSITY (SOLID STATE)
CARRIER LIFETIME
ENERGY CONVERSION EFFICIENCY
SOLAR CELLS

CARRIER WAVES
UF   SUBCARRIER WAVES
RT   MODULATION
RADIO SPECTRA

∞ **CARRIERS**
SN   *(USE OF A MORE SPECIFIC TERM IS
RECOMMENDED--CONSULT THE TERMS
LISTED BELOW)*
RT   AIRCRAFT CARRIERS
CARRIER DENSITY (SOLID STATE)
CHARGE CARRIERS
ZENER EFFECT

CARRINGTON ROTATION
USE   SOLAR ROTATION

CARTAN SPACE
RT   COMPRESSIBLE FLOW
FLUID FLOW
POTENTIAL FLOW
∞ SPACE

CARTESIAN COORDINATES
UF   CYLINDRICAL COORDINATES
RECTANGULAR COORDINATES
GS   COORDINATES
. **CARTESIAN COORDINATES**
GEOMETRY
. EUCLIDEAN GEOMETRY
. . **CARTESIAN COORDINATES**
RT   OBLIQUE COORDINATES

CARTILAGE
GS   ANATOMY
. MUSCULOSKELETAL SYSTEM
. . BONES
. . . **CARTILAGE**
. . CONNECTIVE TISSUE
. . . **CARTILAGE**
RT   LARYNX

CARTOGRAPHY
USE   MAPPING

CARTRIDGE ACTUATED DEVICES
USE   ACTUATORS
EXPLOSIVE DEVICES

CARTRIDGES
RT   CASES (CONTAINERS)
∞ CONTAINERS
PACKAGES
PROJECTILES
PROPELLANTS

CARTS
RT   CARRIAGES
MATERIALS HANDLING
UNDERCARRIAGES

CASCADE CONTROL
UF   MULTILOOP SYSTEMS
GS   AUTOMATIC CONTROL
       . FEEDBACK CONTROL
       . . **CASCADE CONTROL**
RT   ∞CASCADES
       ∞CONTROL
       ELECTRONIC CONTROL
       REMOTE CONTROL

CASCADE FLOW
GS   FLUID FLOW
       . **CASCADE FLOW**
RT   ∞CASCADES
       OUTLET FLOW
       TURBOMACHINE BLADES

CASCADE RANGE (CA-OR-WA)
GS   LAND
       . **CASCADE RANGE (CA-OR-WA)**
       LANDFORMS
       . MOUNTAINS
       . . **CASCADE RANGE (CA-OR-WA)**
RT   CALIFORNIA
       OREGON
       UNITED STATES
       WASHINGTON

CASCADE WIND TUNNELS
GS   TEST FACILITIES
       . WIND TUNNELS
       . . HYPERSONIC WIND TUNNELS
       . . . **CASCADE WIND TUNNELS**
       . . HYPERVELOCITY WIND TUNNELS
       . . . **CASCADE WIND TUNNELS**
RT   HYPERSONIC FLOW
       SHOCK TUNNELS

∞ **CASCADES**
SN   (USE OF A MORE SPECIFIC TERM IS
       RECOMMENDED--CONSULT THE TERMS
       LISTED BELOW)
RT   CASCADE CONTROL
       CASCADE FLOW
       CIRCUITS
       COSMIC RAY SHOWERS
       ELECTRON PHOTON CASCADES

CASCADES (FLUID DYNAMICS)
USE   FLUID DYNAMICS

CASCODE MOSFET
USE   FIELD EFFECT TRANSISTORS

CASE BONDED PROPELLANTS
GS   PROPELLANTS
       . SOLID PROPELLANTS
       . . **CASE BONDED PROPELLANTS**
RT   AMMUNITION
       COMPOSITE PROPELLANTS
       EXPLOSIVES
       HYBRID PROPELLANTS
       INHIBITORS
       PLASTICIZERS
       SOLID ROCKET PROPELLANTS

CASE HISTORIES
GS   HISTORIES
       . **CASE HISTORIES**
RT   ANTHROPOLOGY
       BIOGRAPHY
       CLINICAL MEDICINE
       DOCUMENTATION
       ETIOLOGY
       PHENOMENOLOGY
       RECORDS
       SOCIOLOGY

CASES (CONTAINERS)
GS   **CASES (CONTAINERS)**
       . ROCKET ENGINE CASES
RT   BOXES (CONTAINERS)
       CARTRIDGES
       ∞CONTAINERS
       MISSILE BODIES
       PACKAGES
       ∞SHELVES

∞ **CASING**
SN   (USE OF A MORE SPECIFIC TERM IS
       RECOMMENDED--CONSULT THE TERMS
       LISTED BELOW)
RT   COVERINGS
       ENCLOSURE

CASING-(CONT.)
       JACKETS
       LININGS
       PIPES (TUBES)
       SHEATHS

CASKS
USE   BARRELS (CONTAINERS)

CASPIAN SEA
GS   SEAS
       . **CASPIAN SEA**
RT   COASTS

CASSEGRAIN ANTENNAS
GS   ANTENNAS
       . **CASSEGRAIN ANTENNAS**
RT   ANTENNA DESIGN
       GREGORIAN ANTENNAS
       PARABOLIC ANTENNAS
       SUBREFLECTORS
       TWO REFLECTOR ANTENNAS

CASSEGRAIN OPTICS
RT   FIBER OPTICS
       GEOMETRICAL OPTICS
       MIRRORS
       ∞OPTICS
       REFLECTING TELESCOPES

CASSIOPEIA A
GS   CELESTIAL BODIES
       . NEBULAE
       . . **CASSIOPEIA A**
       . RADIO SOURCES (ASTRONOMY)
       . . **CASSIOPEIA A**
RT   ORION NEBULA

CASSIOPEIA CONSTELLATION
GS   CONSTELLATIONS
       . **CASSIOPEIA CONSTELLATION**

CAST ALLOYS
GS   ALLOYS
       . **CAST ALLOYS**
RT   CASTINGS
       MECHANICAL PROPERTIES
       MICROSTRUCTURE
       RHEOCASTING

CASTIGLIANO VARIATIONAL THEOREM
GS   THEOREMS
       . **CASTIGLIANO VARIATIONAL THEOREM**
RT   CALCULUS OF VARIATIONS
       ENERGY METHODS
       EULER-LAGRANGE EQUATION
       STRESS ANALYSIS
       STRUCTURAL ANALYSIS

CASTING
GS   FORMING TECHNIQUES
       . **CASTING**
       . . CENTRIFUGAL CASTING
       . . INVESTMENT CASTING
       . . PROPELLANT CASTING
       . . RHEOCASTING
       . . SAND CASTING
       . . SLIP CASTING
RT   BAKING
       BILLETS
       DIES
       EXTRUDING
       FORGING
       INCLUSIONS
       INGOTS
       LIQUID METALS
       MELTING
       METAL WORKING
       ∞METALLURGY
       MICROSTRUCTURE
       MOLDING MATERIALS
       MOLDS
       MUSHY ZONES
       PINHOLES
       POLYMERIC FILMS
       POURING
       PULTRUSION
       SHRINKAGE
       SOLIDIFICATION

CASTING SOLVENTS
USE   PLASTICIZERS

CASTINGS
GS   **CASTINGS**
       . INGOTS
       . PROPELLANT CASTING
RT   BILLETS
       CAST ALLOYS
       DEFECTS
       DEGASSING
       FLAT PATTERNS
       INCLUSIONS
       MICROSTRUCTURE
       MOLDS
       PINHOLES
       POURING
       RISERS
       SOLIDIFICATION

CASTOR OIL
GS   ALIPHATIC COMPOUNDS
       . **CASTOR OIL**
       LIPIDS
       . **CASTOR OIL**
       OILS
       . **CASTOR OIL**
       ORGANIC COMPOUNDS
       . **CASTOR OIL**
RT   FATTY ACIDS

CASTOR 2 ENGINE
USE   TX-354 ENGINE

CASTS
RT   DAMAGE ASSESSMENT
       GAUZE
       PLASTERS
       SPLINTS

CASUALTIES
RT   DEATH
       DISASTERS
       EVACUATING (TRANSPORTATION)

CAT SCANNER
USE   COMPUTER AIDED TOMOGRAPHY

CATABOLISM
GS   METABOLISM
       . **CATABOLISM**
RT   PHYSIOLOGY

CATACLYSMIC VARIABLES
GS   CELESTIAL BODIES
       . STARS
       . . VARIABLE STARS
       . . . **CATACLYSMIC VARIABLES**
RT   DWARF STARS
       ECLIPSING BINARY STARS
       FLARE STARS
       HOT STARS
       NOVAE
       PERIODIC VARIATIONS
       SOLAR OSCILLATIONS
       STELLAR FLARES
       STELLAR MASS EJECTION
       STELLAR OSCILLATIONS
       WHITE DWARF STARS

CATALASE
GS   ENZYMES
       . **CATALASE**
RT   CELLS (BIOLOGY)
       PROTEINS

∞ **CATALOGS**
SN   (USE OF A MORE SPECIFIC TERM IS
       RECOMMENDED--CONSULT THE TERMS
       LISTED BELOW)
RT   CATALOGS (PUBLICATIONS)
       HARDWARE UTILIZATION LISTS
       INDEXES (DOCUMENTATION)
       LISTS

CATALOGS (PUBLICATIONS)
GS   DOCUMENTS
       . **CATALOGS (PUBLICATIONS)**
       . . ASTRONOMICAL CATALOGS
RT   ∞CATALOGS
       CATEGORIES
       DOCUMENTATION
       INFORMATION DISSEMINATION
       LIBRARIES

CATALYSIS
GS   **CATALYSIS**

CATALYSIS-*(CONT.)*
. AUTOCATALYSIS
RT ∞ACCELERATION
ACTIVATION
CATALYTIC ACTIVITY
CRACKING (CHEMICAL ENGINEERING)
FISCHER-TROPSCH PROCESS
REACTION KINETICS

CATALYSTS
GS **CATALYSTS**
. ELECTROCATALYSTS
. HOPCALITE (TRADEMARK)
. ZIEGLER CATALYST
RT ACCELERATING AGENTS
ADDITIVES
ADMIXTURES
COAL DERIVED GASES
COAL DERIVED LIQUIDS
ENZYMES
GRIGNARD REACTIONS
HIGH ENERGY FUELS
INHIBITORS
∞INITIATORS
PLATINUM BLACK
PROPELLANT ADDITIVES
REAGENTS
RETARDANTS

CATALYTIC ACTIVITY
RT ACTIVITY (BIOLOGY)
AUTOCATALYSIS
CATALYSIS
CRACKING (CHEMICAL ENGINEERING)
FISCHER-TROPSCH PROCESS

CATAPULTS
GS LAUNCHERS
. **CATAPULTS**
. . ROCKET CATAPULTS
RT AIRCRAFT LAUNCHING DEVICES
MISSILE LAUNCHERS
SEA LAUNCHING

CATARACTS
GS DISEASES
. EYE DISEASES
. . **CATARACTS**
RT LENSES

CATASTROPHE THEORY
RT DISCONTINUITY
DIVERGENCE
PREDICTIONS
∞THEORIES
TOPOLOGY

CATCHERS
RT ELECTRON BUNCHING
KLYSTRONS
OUTPUT

CATCHMENT AREAS
USE WATERSHEDS

CATECHOLAMINE
GS AMINES
. **CATECHOLAMINE**

CATEGORIES
RT CATALOGS (PUBLICATIONS)
CLASSES
∞GROUPS
∞SECTIONS

CATENARIES
GS GEOMETRY
. CURVES (GEOMETRY)
. . **CATENARIES**
. EUCLIDEAN GEOMETRY
. . ANALYTIC GEOMETRY
. . . **CATENARIES**

CATHETERIZATION
RT BLOOD VESSELS
INTRAVENOUS PROCEDURES

CATHETOMETERS
GS MEASURING INSTRUMENTS
. OPTICAL MEASURING INSTRUMENTS
. . **CATHETOMETERS**
OPTICAL EQUIPMENT
. OPTICAL MEASURING INSTRUMENTS
. . **CATHETOMETERS**

CATHODE GLOW
GS DECAY
. EMISSION
. . LIGHT EMISSION
. . . LUMINESCENCE
. . . . **CATHODE GLOW**
RT CATHODOLUMINESCENCE
GLOW DISCHARGES
RAREFIED PLASMAS

CATHODE RAY TUBES
GS ELECTRON TUBES
. VACUUM TUBES
. . VACUUM TUBE OSCILLATORS
. . . **CATHODE RAY TUBES**
. . . . PICTURE TUBES
OSCILLATORS
. VACUUM TUBE OSCILLATORS
. . **CATHODE RAY TUBES**
. . . PICTURE TUBES
RT DISPLAY DEVICES
ELECTRON GUNS
ELECTRON OPTICS
IMAGE TUBES
MAGNETIC LENSES
OSCILLOSCOPES
PRINTERS
TELEVISION EQUIPMENT
VIDEO EQUIPMENT

CATHODES
GS ELECTRODES
. **CATHODES**
. . CELL CATHODES
. . HOLLOW CATHODES
. . TUBE CATHODES
. . . COLD CATHODES
. . . HOT CATHODES
. . . PHOTOCATHODES
. . . THERMIONIC CATHODES
. . . TUNNEL CATHODES
RT ANODES
COLD CATHODE TUBES
ELECTRODE MATERIALS
ELECTRON EMISSION
∞FILAMENTS
FREQUENCY MODULATION
PHOTOMULTIPLIERS
PHOTOMULTIPLIER TUBES
PHOTOTUBES
THERMIONICS
TUBE ANODES

CATHODIC COATINGS
GS COATINGS
. **CATHODIC COATINGS**
RT ANODIC COATINGS
CLADDING
ELECTRODE MATERIALS
ELECTRODEPOSITION
ELECTROPLATING
METAL OXIDES
OXIDE FILMS
OXIDES
PLATING

CATHODOLUMINESCENCE
GS DECAY
. EMISSION
. . LIGHT EMISSION
. . . LUMINESCENCE
. . . . **CATHODOLUMINESCENCE**
RT CATHODE GLOW
EMISSION
LIGHT SOURCES
VISIBLE SPECTRUM

CATHOLYTES
GS CONDUCTORS
. ELECTROLYTES
. . **CATHOLYTES**
RT ANOLYTES
CELL CATHODES
DIAPHRAGMS (MECHANICS)

CATIONS
GS IONS
. **CATIONS**
. . METAL IONS
. . . FERRIC IONS
. . . MANGANESE IONS
. . . VANADYL RADICAL
PARTICLES
. CHARGED PARTICLES
. . **CATIONS**

CATIONS-*(CONT.)*
. . METAL IONS
. . . . FERRIC IONS
. . . . MANGANESE IONS
RT ANIONS
CELL CATHODES
IONIC MOBILITY
POSITIVE IONS

CATS
GS ANIMALS
. VERTEBRATES
. . MAMMALS
. . . **CATS**

CATT DEVICES
UF CONTROLLED AVALANCHE TRANSIT
TIME DEVICES
RT ELECTRON AVALANCHE
POWER GAIN
TRANSIT TIME
TRIODES

CATTLE
GS ANIMALS
. VERTEBRATES
. . MAMMALS
. . . **CATTLE**
. . . . CALVES
RT GRAZING
LIVESTOCK
RANGELANDS

CAUCASUS MOUNTAINS (U.S.S.R.)
GS LANDFORMS
. MOUNTAINS
. . **CAUCASUS MOUNTAINS (U.S.S.R.)**
RT U.S.S.R.

CAUCHY INTEGRAL FORMULA
GS ANALYSIS (MATHEMATICS)
. COMPLEX VARIABLES
. . **CAUCHY INTEGRAL FORMULA**

CAUCHY PROBLEM
UF RIEMANN PROBLEM
GS BOUNDARY VALUE PROBLEMS
. **CAUCHY PROBLEM**
RT DIFFERENTIAL EQUATIONS
∞PROBLEMS

CAUCHY-RIEMANN EQUATIONS
GS ANALYSIS (MATHEMATICS)
. REAL VARIABLES
. . DIFFERENTIAL EQUATIONS
. . . **CAUCHY-RIEMANN EQUATIONS**
RT ANALYTIC FUNCTIONS
∞EQUATIONS
PARTIAL DIFFERENTIAL EQUATIONS

CAULKING
RT MOISTURE RESISTANCE
PLUGGING
SEALING
WATERPROOFING

CAUSES
RT ∞EFFECTS
ETIOLOGY
∞ORIGINS
∞SOURCES

CAUSTIC LINES
RT FLIGHT PATHS
SHOCK WAVES
SONIC BOOMS
SUPERSONIC FLIGHT
WAVE FRONTS

CAUSTICS
USE ALKALIES

CAUSTICS (OPTICS)
RT DIFFRACTION
FRACTURE MECHANICS
∞OPTICS
POLARIZED RADIATION
∞RAYS
SURFACE DEFECTS

CAVES
RT CAVITIES
KARST
KETTLES (GEOLOGY)

CAVES-*(CONT.)*
　　　　UNDERGROUND STRUCTURES

CAVITATION
　USE　CAVITATION FLOW

CAVITATION CORROSION
　GS　CORROSION
　　　. **CAVITATION CORROSION**

CAVITATION FLOW
　UF　CAVITATION
　　　GASEOUS CAVITATION
　GS　FLUID FLOW
　　　. TURBULENT FLOW
　　　.. **CAVITATION FLOW**
　RT　BUBBLES
　　　EROSION
　　　FLOW DISTRIBUTION
　　　IMPINGEMENT
　　　SEPARATED FLOW
　　　SUPERCAVITATING FLOW
　　　ULTRASONIC CLEANING
　　　VORTICES
　　　WAKES
　　　WATER

CAVITIES
　UF　BORES
　RT　APERTURES
　　　BOREHOLES
　　　CAVES
　　　∞CELLS
　　　CRACK GEOMETRY
　　　CRACKS
　　　DEFECTS
　　　DUCTS
　　　GAS POCKETS
　　　HOLE DISTRIBUTION (MECHANICS)
　　　∞HOLES
　　　∞HOLLOW
　　　INTERSTICES
　　　KARST
　　　KETTLES (GEOLOGY)
　　　LEAKAGE
　　　OPENINGS
　　　ORIFICES
　　　OUTLETS
　　　PASSAGEWAYS
　　　PERFORATED PLATES
　　　PERFORATED SHELLS
　　　∞PERFORATION
　　　PORTS (OPENINGS)
　　　RECESSES
　　　TOOTH DISEASES
　　　VENTS
　　　VOIDS

CAVITONS
　RT　ELECTRIC FIELDS
　　　PLASMA DENSITY
　　　PLASMA PHYSICS
　　　PLASMA RESONANCE

CAVITY RESONATORS
　UF　RESONANT CAVITIES
　GS　RESONATORS
　　　. **CAVITY RESONATORS**
　RT　CIRCULATORS (PHASE SHIFT CIRCUITS)
　　　CYCLOTRON RESONANCE DEVICES
　　　ELECTRON TUBES
　　　HELMHOLTZ RESONATORS
　　　KLYSTRONS
　　　MAGNETRONS
　　　MICROWAVE RESONANCE
　　　MULTIMODE RESONATORS
　　　OSCILLATORS
　　　RESONANT FREQUENCIES
　　　TRAVELING WAVE MASERS
　　　VELOCITY MODULATION

CAVITY VAPOR GENERATORS
　RT　∞GENERATORS
　　　VAPORIZERS
　　　VAPORS

CAYS
　USE　KEYS (ISLANDS)

CC-106 AIRCRAFT
　USE　CL-44 AIRCRAFT

CCD
　USE　CHARGE COUPLED DEVICES

CCD STAR TRACKER
　UF　STELLAR (STAR TRACKER)
　GS　TRACKING (POSITION)
　　　. STAR TRACKERS
　　　.. **CCD STAR TRACKER**
　RT　CELESTIAL NAVIGATION
　　　CHARGE COUPLED DEVICES
　　　SPACECRAFT GUIDANCE

CDC COMPUTERS
　GS　DATA PROCESSING EQUIPMENT
　　　. COMPUTERS
　　　.. **CDC COMPUTERS**
　　　... CDC CYBER 74 COMPUTER
　　　... CDC CYBER 170 SERIES
　　　　　COMPUTERS
　　　.... CDC CYBER 175 COMPUTER
　　　... CDC CYBER 174 COMPUTER
　　　... CDC CYBER 203 COMPUTER
　　　... CDC CYBER 205 COMPUTER
　　　... CDC STAR 100 COMPUTER
　　　... CDC 160-A COMPUTER
　　　... CDC 1604 COMPUTER
　　　... CDC 3100 COMPUTER
　　　... CDC 3200 COMPUTER
　　　... CDC 3600 COMPUTER
　　　... CDC 3800 COMPUTER
　　　... CDC 6000 SERIES COMPUTERS
　　　.... CDC 6400 COMPUTER
　　　.... CDC 6600 COMPUTER
　　　.... CDC 6700 COMPUTER
　　　... CDC 7000 SERIES COMPUTERS
　　　.... CDC 7600 COMPUTER
　　　... CDC 8090 COMPUTER
　RT　DIGITAL COMPUTERS

CDC CYBER 74 COMPUTER
　UF　CYBER 74 COMPUTER
　GS　DATA PROCESSING EQUIPMENT
　　　. COMPUTERS
　　　.. CDC COMPUTERS
　　　... **CDC CYBER 74 COMPUTER**
　　　.. DIGITAL COMPUTERS
　　　... **CDC CYBER 74 COMPUTER**

CDC CYBER 170 SERIES COMPUTERS
　GS　DATA PROCESSING EQUIPMENT
　　　. COMPUTERS
　　　.. CDC COMPUTERS
　　　... **CDC CYBER 170 SERIES
　　　　　COMPUTERS**
　　　.. DIGITAL COMPUTERS
　　　... **CDC CYBER 170 SERIES
　　　　　COMPUTERS**

CDC CYBER 174 COMPUTER
　GS　DATA PROCESSING EQUIPMENT
　　　. COMPUTERS
　　　.. CDC COMPUTERS
　　　... **CDC CYBER 174 COMPUTER**
　　　.. DIGITAL COMPUTERS
　　　... **CDC CYBER 174 COMPUTER**

CDC CYBER 175 COMPUTER
　GS　DATA PROCESSING EQUIPMENT
　　　. COMPUTERS
　　　.. CDC COMPUTERS
　　　... CDC CYBER 170 SERIES
　　　　　COMPUTERS
　　　.... **CDC CYBER 175 COMPUTER**

CDC CYBER 203 COMPUTER
　GS　DATA PROCESSING EQUIPMENT
　　　. COMPUTERS
　　　.. CDC COMPUTERS
　　　... **CDC CYBER 203 COMPUTER**
　　　.. DIGITAL COMPUTERS
　　　... **CDC CYBER 203 COMPUTER**

CDC CYBER 205 COMPUTER
　GS　DATA PROCESSING EQUIPMENT
　　　. COMPUTERS
　　　.. CDC COMPUTERS
　　　... **CDC CYBER 205 COMPUTER**
　　　.. DIGITAL COMPUTERS
　　　... **CDC CYBER 205 COMPUTER**

CDC STAR 100 COMPUTER
　GS　DATA PROCESSING EQUIPMENT
　　　. COMPUTERS
　　　.. CDC COMPUTERS
　　　... **CDC STAR 100 COMPUTER**
　　　.. DIGITAL COMPUTERS
　　　... **CDC STAR 100 COMPUTER**

CDC 160-A COMPUTER
　GS　DATA PROCESSING EQUIPMENT
　　　. COMPUTERS
　　　.. CDC COMPUTERS
　　　... **CDC 160-A COMPUTER**
　　　.. DIGITAL COMPUTERS
　　　... **CDC 160-A COMPUTER**

CDC 1604 COMPUTER
　GS　DATA PROCESSING EQUIPMENT
　　　. COMPUTERS
　　　.. CDC COMPUTERS
　　　... **CDC 1604 COMPUTER**
　　　.. DIGITAL COMPUTERS
　　　... **CDC 1604 COMPUTER**

CDC 3100 COMPUTER
　GS　DATA PROCESSING EQUIPMENT
　　　. COMPUTERS
　　　.. CDC COMPUTERS
　　　... **CDC 3100 COMPUTER**
　　　. DIGITAL COMPUTERS
　　　... **CDC 3100 COMPUTER**

CDC 3200 COMPUTER
　GS　DATA PROCESSING EQUIPMENT
　　　. COMPUTERS
　　　.. CDC COMPUTERS
　　　... **CDC 3200 COMPUTER**
　　　. DIGITAL COMPUTERS
　　　... **CDC 3200 COMPUTER**

CDC 3600 COMPUTER
　GS　DATA PROCESSING EQUIPMENT
　　　. COMPUTERS
　　　.. CDC COMPUTERS
　　　... **CDC 3600 COMPUTER**
　　　.. DIGITAL COMPUTERS
　　　... **CDC 3600 COMPUTER**

CDC 3800 COMPUTER
　GS　DATA PROCESSING EQUIPMENT
　　　. COMPUTERS
　　　.. CDC COMPUTERS
　　　... **CDC 3800 COMPUTER**
　　　.. DIGITAL COMPUTERS
　　　... **CDC 3800 COMPUTER**

CDC 6000 SERIES COMPUTERS
　GS　DATA PROCESSING EQUIPMENT
　　　. COMPUTERS
　　　.. CDC COMPUTERS
　　　... **CDC 6000 SERIES COMPUTERS**
　　　.... CDC 6400 COMPUTER
　　　.... CDC 6600 COMPUTER
　　　.... CDC 6700 COMPUTER
　　　.. DIGITAL COMPUTERS
　　　... **CDC 6000 SERIES COMPUTERS**
　　　.... CDC 6400 COMPUTER
　　　.... CDC 6600 COMPUTER
　　　.... CDC 6700 COMPUTER

CDC 6400 COMPUTER
　GS　DATA PROCESSING EQUIPMENT
　　　. COMPUTERS
　　　.. CDC COMPUTERS
　　　... CDC 6000 SERIES COMPUTERS
　　　.... **CDC 6400 COMPUTER**
　　　.. DIGITAL COMPUTERS
　　　... CDC 6000 SERIES COMPUTERS
　　　.... **CDC 6400 COMPUTER**

CDC 6600 COMPUTER
　GS　DATA PROCESSING EQUIPMENT
　　　. COMPUTERS
　　　.. CDC COMPUTERS
　　　... CDC 6000 SERIES COMPUTERS
　　　.... **CDC 6600 COMPUTER**
　　　.. DIGITAL COMPUTERS
　　　... CDC 6000 SERIES COMPUTERS
　　　.... **CDC 6600 COMPUTER**

CDC 6700 COMPUTER
　GS　DATA PROCESSING EQUIPMENT
　　　. COMPUTERS
　　　.. CDC COMPUTERS
　　　... CDC 6000 SERIES COMPUTERS
　　　.... **CDC 6700 COMPUTER**
　　　.. DIGITAL COMPUTERS
　　　... CDC 6000 SERIES COMPUTERS
　　　.... **CDC 6700 COMPUTER**

CDC 7000 SERIES COMPUTERS
　GS　DATA PROCESSING EQUIPMENT

CDC 7000 SERIES COMPUTERS-*(CONT.)*
. COMPUTERS
. . CDC COMPUTERS
. . . **CDC 7000 SERIES COMPUTERS**
. . . . CDC 7600 COMPUTER
. . DIGITAL COMPUTERS
. . . **CDC 7000 SERIES COMPUTERS**
. . . . CDC 7600 COMPUTER

CDC 7600 COMPUTER
GS   DATA PROCESSING EQUIPMENT
. COMPUTERS
. . CDC COMPUTERS
. . . CDC 7000 SERIES COMPUTERS
. . . . **CDC 7600 COMPUTER**
. . DIGITAL COMPUTERS
. . . CDC 7000 SERIES COMPUTERS
. . . . **CDC 7600 COMPUTER**

CDC 8090 COMPUTER
GS   DATA PROCESSING EQUIPMENT
. COMPUTERS
. . CDC COMPUTERS
. . . **CDC 8090 COMPUTER**
. . DIGITAL COMPUTERS
. . . **CDC 8090 COMPUTER**

CDMA
USE   CODE DIVISION MULTIPLE ACCESS

CEDAR RAPIDS (IA)
GS   CITIES
. **CEDAR RAPIDS (IA)**
RT   IOWA

CEFOAM CHECKOUT EQUIPMENT
RT   CHECKOUT
∞ TEST EQUIPMENT

CEILING (AIRCRAFT CAPABILITY)
RT   ∞ AIRCRAFT
AIRCRAFT SPECIFICATIONS
∞ CEILINGS
FLIGHT ALTITUDE
FLIGHT CHARACTERISTICS

∞ **CEILINGS**
SN   *(USE OF A MORE SPECIFIC TERM IS*
RECOMMENDED--CONSULT THE TERMS
LISTED BELOW)
RT   CEILING (AIRCRAFT CAPABILITY)
CEILINGS (ARCHITECTURE)
CEILINGS (METEOROLOGY)

CEILINGS (ARCHITECTURE)
RT   ∞ BUILDINGS
∞ CEILINGS
∞ DIFFUSERS
FLOORS
INSULATION
PANELS
REFLECTORS

CEILINGS (METEOROLOGY)
RT   AIRCRAFT LANDING
∞ CEILINGS
CLOUD HEIGHT INDICATORS
METEOROLOGICAL PARAMETERS
METEOROLOGY
VISIBILITY

CEILOMETERS
USE   CLOUD HEIGHT INDICATORS

CELESCOPES
GS   ELECTRON TUBES
. VACUUM TUBES
. . VACUUM TUBE OSCILLATORS
. . . MICROWAVE TUBES
. . . . PLANOTRONS
. . . . . **CELESCOPES**
MICROWAVE EQUIPMENT
. MICROWAVE TUBES
. . PLANOTRONS
. . . **CELESCOPES**
MIRRORS
. **CELESCOPES**
OPTICAL EQUIPMENT
. IMAGE CONVERTERS
. . **CELESCOPES**
TELESCOPES
. **CELESCOPES**
RT   ASTRONOMICAL TELESCOPES
SOLAR INSTRUMENTS

CELESTIAL BODIES
GS   **CELESTIAL BODIES**
. A STARS
. ASTEROID BELTS
. . ASTEROIDS
. . . AMOR ASTEROID
. . . AMPHITRITE ASTEROID
. . . APOLLO ASTEROIDS
. . . CERES ASTEROID
. . . CHIRON
. . . ICARUS ASTEROID
. . . TORO ASTEROID
. . . VESTA ASTEROID
. BL LACERTAE OBJECTS
. COMETS
. . AREND-ROLAND COMET
. . COMET HEADS
. . COMET NUCLEI
. . COMET TAILS
. . ENCKE COMET
. . GIACOBINI-ZINNER COMET
. . GRIGG-SKJELLERUP COMET
. . HALLEY'S COMET
. . HUMASON COMET
. . IRAS-ARAKI-ALCOCK COMET
. . KOHOUTEK COMET
. . MOREHOUSE COMET
. . MRKOS COMET
. . SCHWASSMANN-WACHMANN COMET
. . TEMPEL 2 COMET
. . WEST COMET
. GALAXIES
. . DISK GALAXIES
. . DWARF GALAXIES
. . ELLIPTICAL GALAXIES
. . GALACTIC CLUSTERS
. . . LOCAL GROUP (ASTRONOMY)
. . . . ANDROMEDA GALAXIES
. . . . VIRGO GALACTIC CLUSTER
. . MAFFEI GALAXIES
. . RADIO GALAXIES
. . SEYFERT GALAXIES
. . SPIRAL GALAXIES
. . . BARRED GALAXIES
. . . MILKY WAY GALAXY
. GLOBULAR CLUSTERS
. HORIZONTAL BRANCH STARS
. METEORITES
. . HARLETON METEORITE
. . IRON METEORITES
. . . AROOS METEORITE
. . . ODESSA METEORITE
. . . SIKHOTE-ALIN METEORITE
. . LAZAREV METEORITE
. . MICROMETEORITES
. . OKHANSK METEORITE
. . STONY METEORITES
. . . ACHONDRITES
. . . . BONDOC METEORITE
. . . . KAPOETA ACHONDRITE
. . . . NORTON COUNTY ACHONDRITE
. . . CHONDRITES
. . . . BRUDERHEIM METEORITE
. . . . CARBONACEOUS CHONDRITES
. . . . . ALLENDE METEORITE
. . . . . MURCHISON METEORITE
. . . . CARBONACEOUS METEORITES
. . . . . ALAIS METEORITE
. . . . . COLD BOKKEVELD METEORITE
. . . . . IVUNA METEORITE
. . . . . MURRAY METEORITE
. . . . . ORGUEIL METEORITE
. . . . . TONK METEORITE
. . . . HVITTIS CHONDRITE
. . . . PANTAR CHONDRITES
. . . . PRIBRAM METEORITE
. . . TEKTITES
. . . . AUSTRALITES
. . . . BEDIASITES
. . . . TUNGUSK METEORITE
. METEOROID SHOWERS
. . AQUARID METEOROIDS
. . ARIETID METEOROIDS
. . CYRILLID METEOROIDS
. . DRACONID METEOROIDS
. . GEMINID METEOROIDS
. . LEONID METEOROIDS
. . ORIONID METEOROIDS
. . PERSEID METEOROIDS
. . QUADRANTID METEOROIDS
. . TAURID METEOROIDS
. METEOROIDS
. . AQUARID METEOROIDS
. . ARIETID METEOROIDS
. . BOLIDES
. . . CYRILLID METEOROIDS

CELESTIAL BODIES-*(CONT.)*
. . DRACONID METEOROIDS
. . GEMINID METEOROIDS
. . LEONID METEOROIDS
. . MICROMETEOROIDS
. . . METEOROID DUST CLOUDS
. . . . ZODIACAL DUST
. . ORIONID METEOROIDS
. . PERSEID METEOROIDS
. . QUADRANTID METEOROIDS
. . RADIO METEORS
. . SPORADIC METEOROIDS
. . TAURID METEOROIDS
. NATURAL SATELLITES
. CHARON
. DEIMOS
. IAPETUS
. JUPITER SATELLITES
. . GALILEAN SATELLITES
. . . CALLISTO
. . . EUROPA
. . . GANYMEDE
. . . IO
. MOON
. RHEA (ASTRONOMY)
. TITAN
. TRITON
. URANUS SATELLITES
. NEBULAE
. CASSIOPEIA A
. CRAB NEBULA
. GUM NEBULA
. ORION NEBULA
. PLANETARY NEBULAE
. REFLECTION NEBULAE
. PHOBOS
. PLANETARY RINGS
. JUPITER RINGS
. SATURN RINGS
. URANUS RINGS
. PLANETS
. EXTRASOLAR PLANETS
. GAS GIANT PLANETS
. . JUPITER (PLANET)
. . NEPTUNE (PLANET)
. . SATURN (PLANET)
. . URANUS (PLANET)
. . . URANUS RINGS
. . . URANUS SATELLITES
. PLUTO (PLANET)
. TERRESTRIAL PLANETS
. . EARTH (PLANET)
. . MARS (PLANET)
. . MERCURY (PLANET)
. . VENUS (PLANET)
. PROTOPLANETS
. RADIO SOURCES (ASTRONOMY)
. CASSIOPEIA A
. EXTRAGALACTIC RADIO SOURCES
. . RADIO GALAXIES
. . QUASARS
. RADIO STARS
. . PULSARS
. SOLAR SYSTEM
. HALLEY'S COMET
. STAR CLUSTERS
. PRAESEPE STAR CLUSTERS
. VIRGO GALACTIC CLUSTER
. STARS
. COOL STARS
. DOUBLE STARS
. . BINARY STARS
. . . COMPANION STARS
. . . ECLIPSING BINARY STARS
. . . . DWARF NOVAE
. . DWARF STARS
. . DWARF NOVAE
. . FLARE STARS
. . RED DWARF STARS
. . SUBDWARF STARS
. . WHITE DWARF STARS
. EARLY STARS
. . PROTOSTARS
. . . T TAURI STARS
. EXTARS
. HERBIG-HARO OBJECTS
. HOT STARS
. . B STARS
. . BLUE STARS
. . . SYMBIOTIC STARS
. . O STARS
. . WHITE DWARF STARS
. . WOLF-RAYET STARS
. INFRARED STARS
. LAMBDA TAURI STARS
. LATE STARS

CELESTIAL BODIES-(CONT.)
```
    . . M STARS
    . . . FLARE STARS
    . . MAGNETIC STARS
    . . MAIN SEQUENCE STARS
    . . . PRE-MAIN SEQUENCE STARS
    . . METALLIC STARS
    . . NEUTRON STARS
    . . . PULSARS
    . . OMICRON CETI STAR
    . . PECULIAR STARS
    . . . SYMBIOTIC STARS
    . . PRAESEPE STAR CLUSTERS
    . . RADIO STARS
    . . PULSARS
    . . REFERENCE STARS
    . . S STARS
    . . GIANT STARS
    . . . . RED GIANT STARS
    . . . . . CARBON STARS
    . . SIGMA ORIONIS
    . . SUBGIANT STARS
    . . SUN
    . . SUPERGIANT STARS
    . . SUPERMASSIVE STARS
    . . VAN BIESBROECK STAR
    . . VARIABLE STARS
    . . . CATACLYSMIC VARIABLES
    . . . CEPHEID VARIABLES
    . . . NOVAE
    . . . . DWARF NOVAE
    . . . . HERCULES NOVA
    . . . SUPERNOVAE
    . . . SYMBIOTIC STARS
    . . . T TAURI STARS
    . . WHITE HOLES (ASTRONOMY)
    . . ZETA AURIGAE STAR
RT  ARIES CONSTELLATION
    ASTEROID CAPTURE
    ASTRODYNAMICS
    ASTROLABES
    ASTRONOMICAL OBSERVATORIES
    ASTRONOMY
    ASTROPHYSICS
  ∞ BODIES
    CENTAURUS CONSTELLATION
    CORONA BOREALIS CONSTELLATION
    CYGNUS CONSTELLATION
    GRAVITATIONAL WAVES
    IMPACT MELTS
    INTERSTELLAR MATTER
    LYRA CONSTELLATION
    ORBITS
    SPACE FLIGHT
    UNIVERSE
```

CELESTIAL GEODESY
```
SN  (DETERMINATION OF THE FORM OF
    THE EARTH, OF THE EARTHS
    GRAVITATIONAL FIELD, AND OF
    RELATIVE POSITIONS OF SATELLITE
    TRAJECTORIES)
GS  GEODESY
    . CELESTIAL GEODESY
RT  EXPLORER 29 SATELLITE
    EXPLORER 36 SATELLITE
    GEODETIC SATELLITES
    GEOS 1 SATELLITE
    GEOS 2 SATELLITE
    GEOS 3 SATELLITE
    INTERNATIONAL SATELLITE GEODESY
        EXPERIMENT
    TIME
```

CELESTIAL MECHANICS
```
GS  CLASSICAL MECHANICS
    . SPACE MECHANICS
    . . CELESTIAL MECHANICS
RT  ASTRODYNAMICS
    ASTRONOMY
    ASTROPHYSICS
    EPHEMERIDES
    EQUATIONS OF MOTION
    FOUR BODY PROBLEM
    GRAVITATIONAL WAVES
    HYPERBOLIC TRAJECTORIES
    LAGRANGIAN EQUILIBRIUM POINTS
    LONG TERM EFFECTS
    MANY BODY PROBLEM
  ∞ MECHANICS (PHYSICS)
    ORBITAL MECHANICS
    ORBITS
    PERTURBATION THEORY
    PLANETS
    ROCHE LIMIT
```

CELESTIAL MECHANICS-(CONT.)
```
    SCHACH EFFECT
    SOLAR SYSTEM
    STARS
    STELLAR ORBITS
    SUN
    TERRESTRIAL PLANETS
    THREE BODY PROBLEM
    TRAJECTORY ANALYSIS
    TROJAN ORBITS
    TWO BODY PROBLEM
    WOLF-RAYET STARS
```

CELESTIAL NAVIGATION
```
GS  NAVIGATION
    . CELESTIAL NAVIGATION
    . . ASTROGUIDE NAVIGATION SYSTEM
    . . ASTRONAVIGATION
RT  AIR NAVIGATION
    AUTONOMOUS NAVIGATION
    CCD STAR TRACKER
    INERTIAL NAVIGATION
    INJECTION GUIDANCE
    INTERPLANETARY NAVIGATION
    POLAR NAVIGATION
    RADAR NAVIGATION
    RADIO NAVIGATION
    REFERENCE STARS
    SOLAR POSITION
    SPACE NAVIGATION
    SPACECRAFT GUIDANCE
    STAR TRACKERS
    SURFACE NAVIGATION
```

CELESTIAL OBSERVATION
```
USE  ASTRONOMY
```

CELESTIAL REFERENCE SYSTEMS
```
RT  AIR NAVIGATION
    ASTRONOMICAL COORDINATES
    ASTRONOMICAL MAPS
    AZIMUTH
    COORDINATES
    GEOCENTRIC COORDINATES
    INERTIAL REFERENCE SYSTEMS
    INTERPLANETARY NAVIGATION
    INTERSTELLAR TRAVEL
    PLANETOCENTRIC COORDINATES
  ∞ REFERENCE SYSTEMS
    SOLAR LONGITUDE
    SPHERICAL COORDINATES
  ∞ SYSTEMS
```

CELESTIAL SPHERE
```
GS  SYMMETRICAL BODIES
    . BODIES OF REVOLUTION
    . . SPHERES
    . . . CELESTIAL SPHERE
RT  ARIES CONSTELLATION
    ASTRONOMICAL MAPS
    CENTAURUS CONSTELLATION
    CONSTELLATIONS
    CORONA BOREALIS CONSTELLATION
    CYGNUS CONSTELLATION
    HORIZON
    LYRA CONSTELLATION
    ORBITAL POSITION ESTIMATION
    PLANISPHERES
    ZENITH
```

CELL ANODES
```
GS  ELECTRODES
    . ANODES
    . . CELL ANODES
RT  ANIONS
    ANOLYTES
    CELL CATHODES
    ELECTRODE MATERIALS
```

CELL CATHODES
```
GS  ELECTRODES
    . CATHODES
    . . CELL CATHODES
RT  CATHOLYTES
    CATIONS
    CELL ANODES
    ELECTRODE MATERIALS
    ELECTRODEPOSITION
```

CELL DIVISION
```
RT  ∞ DIVISION
    ∞ REPRODUCTION
```

CELLOPHANE
```
RT  CELLULOSE
```

CELLOPHANE-(CONT.)
```
    ∞ POLYMERS
```

∞ CELLS
```
SN  (USE OF A MORE SPECIFIC TERM IS
    RECOMMENDED--CONSULT THE TERMS
    LISTED BELOW)
RT  ACTIVATION (BIOLOGY)
    CAVITIES
    CELLS (BIOLOGY)
    COMPARTMENTS
    CORES
    ELECTROCHEMICAL CELLS
    ELECTROLYTIC CELLS
    FILLERS
    FUEL CELLS
    GEOPHYSICAL FLUID FLOW CELLS
    HEXAGONAL CELLS
    HONEYCOMB STRUCTURES
    KERR CELLS
    LITHIUM SULFUR BATTERIES
    PARTICLE IN CELL TECHNIQUE
    PHOTOCONDUCTIVE CELLS
    PHOTOELECTRIC CELLS
    PHOTOVOLTAIC CELLS
    POROUS MATERIALS
    RESOLUTION CELL
    SOLAR CELLS
    TISSUES (BIOLOGY)
    TOPOLOGY
```

CELLS (BIOLOGY)
```
UF  BIOLOGICAL CELLS
GS  CELLS (BIOLOGY)
    . AXONS
    . CHROMOSOMES
    . CORPUSCLES
    . ERYTHROCYTES
    . . RETICULOCYTES
    . FIBROBLASTS
    . . COLLAGENS
    . GAMETOCYTES
    . HEMATOPOIESIS
    . HEMOCYTES
    . HEMOGLOBIN
    . . CARBOXYHEMOGLOBIN
    . . OXYHEMOGLOBIN
    . LEUKOCYTES
    . . EOSINOPHILS
    . . LYMPHOCYTES
    . MACROPHAGES
    . MITOCHONDRIA
    . NEUROBLASTS
    . NEURONS
    . PROTOPLASTS
RT  ∞ BIOLOGY
    CANCER
    CATALASE
  ∞ CELLS
    CHLOROPHYLLS
    CHLOROPLASTS
    CYTOGENESIS
    CYTOLOGY
    ENDOTHELIUM
    GANGLIA
    HISTOCHEMICAL ANALYSIS
    KREBS CYCLE
    MITOSIS
    MUTAGENS
    MUTATIONS
    NEUROGLIA
    NEUROTRANSMITTERS
  ∞ NUCLEI
    PLASMOLYSIS
    TISSUES (BIOLOGY)
```

CELLULAR MATERIALS (NON BIOLOGICAL)
```
USE  FOAMS
```

CELLULOSE
```
GS  ALIPHATIC COMPOUNDS
    . POLYSACCHARIDES
    . . CELLULOSE
    . . . FORTISAN (TRADEMARK)
    CARBOHYDRATES
    . POLYSACCHARIDES
    . . CELLULOSE
    . . . FORTISAN (TRADEMARK)
RT  CELLOPHANE
    LIGNIN
    MASONITE (TRADEMARK)
    SYNTHETIC FOOD
    TENITE
    WOOD
```

CELLULOSE NITRATE
- UF NITROCELLULOSE
-      PYROXYLIN
- GS ESTERS
-      . ORGANIC NITRATES
-      . . **CELLULOSE NITRATE**
-      EXPLOSIVES
-      . **CELLULOSE NITRATE**
-      NITROGEN COMPOUNDS
-      . NITRATES
-      . . ORGANIC NITRATES
-      . . . **CELLULOSE NITRATE**
- RT DOUBLE BASE PROPELLANTS
-      DOUBLE BASE ROCKET PROPELLANTS

CEMENTATION
- RT ADHESIVE BONDING
-      AGGLOMERATION
-      AGGLUTINATION
-      BONDING
-      HEATING
-      PRECIPITATION (CHEMISTRY)

CEMENTITE
- GS CARBON COMPOUNDS
-      . CARBIDES
-      . . **CEMENTITE**
- RT IRON ALLOYS
-      MICROSTRUCTURE
-      PEARLITE
-      STEELS

CEMENTS
- RT ADHESIVES
-      BINDERS (MATERIALS)
-      BRICKS
-      CONCRETES
-      ∞CONSTRUCTION MATERIALS
-      GROUT
-      MASONRY
-      MORTARS (MATERIAL)
-      SEALING

CEMS SYSTEM
- USE CENTRAL ELECTRONIC MANAGEMENT
-       SYSTEM

CENSORED DATA (MATHEMATICS)
- GS DATA PROCESSING
-      . **CENSORED DATA (MATHEMATICS)**
- RT APPROXIMATION
-      ∞DATA
-      ERROR ANALYSIS
-      PROBABILITY DENSITY FUNCTIONS
-      RELIABILITY
-      SAMPLING
-      STATISTICAL ANALYSIS
-      STATISTICAL DISTRIBUTIONS

CENSUS
- RT ANIMALS
-      DEMOGRAPHY
-      HUMAN BEINGS
-      ∞STATISTICS
-      URBAN PLANNING

CENTAUR LAUNCH VEHICLE
- UF CENTAUR VEHICLE
- GS LAUNCH VEHICLES
-      . **CENTAUR LAUNCH VEHICLE**
-      . . ATLAS CENTAUR LAUNCH VEHICLE
-      ROCKET VEHICLES
-      . **CENTAUR LAUNCH VEHICLE**
-      . . ATLAS CENTAUR LAUNCH VEHICLE
- RT ATLAS D ICBM
-      LIQUID PROPELLANT ROCKET ENGINES
-      SATURN PROJECT
-      TITAN CENTAUR LAUNCH VEHICLE

CENTAUR PROJECT
- GS PROGRAMS
-      . NASA PROGRAMS
-      . . **CENTAUR PROJECT**
-      . NASA SPACE PROGRAMS
-      . . **CENTAUR PROJECT**
-      . PROJECTS
-      . . **CENTAUR PROJECT**
- RT ATLAS CENTAUR LAUNCH VEHICLE
-      LAUNCH VEHICLES
-      MARINER PROGRAM
-      RL-10 ENGINES
-      SURVEYOR PROJECT

CENTAUR VEHICLE
- USE CENTAUR LAUNCH VEHICLE

CENTAURUS CONSTELLATION
- GS CONSTELLATIONS
-      . **CENTAURUS CONSTELLATION**
- RT CELESTIAL BODIES
-      CELESTIAL SPHERE
-      STARS

CENTER OF GRAVITY
- UF BARYCENTER
- RT CENTER OF MASS
-      ∞CENTERS
-      CENTROIDS
-      GRAVITATIONAL FIELDS
-      LUNAR ROTATION
-      MASS
-      MOMENTS OF INERTIA

CENTER OF MASS
- GS MASS
-      . **CENTER OF MASS**
- RT CENTER OF GRAVITY
-      MASCONS
-      WEIGHT (MASS)

CENTER OF PRESSURE
- RT ∞CENTERS
-      HYDROSTATIC PRESSURE
-      MOMENTS OF INERTIA
-      PRESSURE
-      PRESSURE DISTRIBUTION
-      PRESSURE HEADS

CENTERBODIES
- GS AIRCRAFT STRUCTURES
-      . **CENTERBODIES**
- RT AFTERBODIES
-      ∞BODIES
-      CYLINDRICAL BODIES
-      FOREBODIES
-      FUSELAGES

∞ CENTERS
- SN *(USE OF A MORE SPECIFIC TERM IS*
-      *RECOMMENDED--CONSULT THE TERMS*
-      *LISTED BELOW)*
- RT CENTER OF GRAVITY
-      CENTER OF PRESSURE
-      COLOR CENTERS
-      CONCENTRICITY
-      FOCI
-      LOCI
-      WORLD DATA CENTERS

CENTIMETER WAVES
- GS ELECTROMAGNETIC RADIATION
-      . RADIO WAVES
-      . . SHORT WAVE RADIATION
-      . . . MICROWAVES
-      . . . . **CENTIMETER WAVES**
- RT COSMIC NOISE
-      EXTRATERRESTRIAL RADIO WAVES
-      MICROWAVE FREQUENCIES
-      SUPERHIGH FREQUENCIES

CENTRAL AFRICAN REPUBLIC
- GS NATIONS
-      . **CENTRAL AFRICAN REPUBLIC**
- RT AFRICA

CENTRAL AMERICA
- RT BELIZE
-      COSTA RICA
-      EL SALVADOR
-      GUATEMALA
-      HONDURAS
-      NICARAGUA
-      NORTH AMERICA
-      PANAMA
-      PANAMA CANAL ZONE
-      REGIONS
-      SOUTH AMERICA

CENTRAL ATLANTIC REGION (US)
- GS REGIONS
-      . **CENTRAL ATLANTIC REGION (US)**
- RT UNITED STATES

CENTRAL ATLANTIC REGIONAL ECOL TEST SITE
- SN (CENTRAL ATLANTIC REGIONAL
-      ECOLOGICAL TEST SITE)
- UF CARETS (TEST SITE)

CENTRAL ATLANTIC REGIONAL ECOL-*(CONT.)*
- GS SITES
-      . **CENTRAL ATLANTIC REGIONAL ECOL**
-        **TEST SITE**
-      TEST FACILITIES
-      . **CENTRAL ATLANTIC REGIONAL ECOL**
-        **TEST SITE**
- RT ECOLOGY

CENTRAL ELECTRONIC MANAGEMENT SYSTEM
- UF CEMS SYSTEM
- GS DATA PROCESSING
-      . **CENTRAL ELECTRONIC MANAGEMENT**
-        **SYSTEM**
- RT MANAGEMENT
-      ∞SYSTEMS

CENTRAL EUROPE
- RT AUSTRIA
-      CONTINENTS
-      CZECHOSLOVAKIA
-      EAST GERMANY
-      EUROPE
-      HUNGARY
-      POLAND
-      ROMANIA
-      WEST GERMANY

CENTRAL NERVOUS SYSTEM
- GS NERVOUS SYSTEM
-      . **CENTRAL NERVOUS SYSTEM**
-      . . BRAIN
-      . . . BRAIN STEM
-      . . . CEREBELLUM
-      . . . CEREBRAL CORTEX
-      . . . CEREBRAL VENTRICLES
-      . . . CEREBRUM
-      . . . HIPPOCAMPUS
-      . . SPINAL CORD
-      . . . SPINE
-      . . THALAMUS
- RT BLOOD-BRAIN BARRIER
-      PSYCHOPHARMACOLOGY
-      PSYCHOTROPIC DRUGS
-      ∞SYSTEMS

CENTRAL NERVOUS SYSTEM DEPRESSANTS
- GS DEPRESSANTS
-      . **CENTRAL NERVOUS SYSTEM**
-        **DEPRESSANTS**
-      DRUGS
-      . **CENTRAL NERVOUS SYSTEM**
-        **DEPRESSANTS**
-      . . MARIJUANA
- RT AMOBARBITAL
-      PSYCHOPHARMACOLOGY
-      ∞SYSTEMS
-      TRANQUILIZERS

CENTRAL NERVOUS SYSTEM STIMULANTS
- GS DRUGS
-      . STIMULANTS
-      . . **CENTRAL NERVOUS SYSTEM**
-          **STIMULANTS**
- RT AMPHETAMINES
-      PSYCHOPHARMACOLOGY
-      ∞SYSTEMS

CENTRAL PIEDMONT (US)
- GS LANDFORMS
-      . TERRACES (LANDFORMS)
-      . . PLATEAUS
-      . . . PIEDMONTS
-      . . . . **CENTRAL PIEDMONT (US)**
- RT MOUNTAINS

CENTRAL PROCESSING UNITS
- UF PROCESSORS (COMPUTERS)
- GS **CENTRAL PROCESSING UNITS**
-      . ARITHMETIC AND LOGIC UNITS
- RT COMPUTER COMPONENTS
-      COMPUTER STORAGE DEVICES
-      COMPUTERS
-      CONTROL UNITS (COMPUTERS)
-      LOGIC CIRCUITS
-      REGISTERS (COMPUTERS)

CENTRIFUGAL CASTING
- GS FORMING TECHNIQUES
-      . CASTING
-      . . **CENTRIFUGAL CASTING**
- RT INVESTMENT CASTING

CENTRIFUGAL COMPRESSORS
GS   COMPRESSORS
. **CENTRIFUGAL COMPRESSORS**
TURBOMACHINERY
. **CENTRIFUGAL COMPRESSORS**
RT   BLOWERS
COMPRESSOR BLADES
COMPRESSOR ROTORS
IMPELLERS
PUMP IMPELLERS
PUMPS
RADIAL FLOW
ROTORS
SUPERCHARGERS
TURBOCOMPRESSORS

CENTRIFUGAL FORCE
RT   ANGULAR ACCELERATION
CENTRIFUGES
CENTRIPETAL FORCE
∞FORCE
GOERTLER INSTABILITY

CENTRIFUGAL PUMPS
GS   PUMPS
. **CENTRIFUGAL PUMPS**
TURBOMACHINERY
. **CENTRIFUGAL PUMPS**
RT   AXIAL FLOW PUMPS
FUEL PUMPS
IMPELLERS
PUMP IMPELLERS
TURBINE PUMPS
TURBOCOMPRESSORS

CENTRIFUGES
UF   CYCLONES (EQUIPMENT)
GS   **CENTRIFUGES**
. HUMAN CENTRIFUGES
RT   CENTRIFUGAL FORCE
CENTRIFUGING
CENTRIPETAL FORCE
CLASSIFIERS
CONCENTRATORS
EXTRACTION
FLIGHT SIMULATORS
FLUID FILTERS
HIGH GRAVITY ENVIRONMENTS
SEPARATORS
SPACE SIMULATORS
∞TEST EQUIPMENT
TRAINING SIMULATORS

CENTRIFUGING
RT   CENTRIFUGES
CONCENTRATING
EXTRACTION
MATERIALS RECOVERY
∞SEPARATION
SWIRLING

CENTRIFUGING STRESS
GS   INVERSIONS
. **CENTRIFUGING STRESS**
STRESS (PHYSIOLOGY)
. ACCELERATION STRESSES
(PHYSIOLOGY)
. . **CENTRIFUGING STRESS**
RT   ACCELERATION TOLERANCE
GRAVITATIONAL PHYSIOLOGY

CENTRIPETAL FORCE
RT   ANGULAR ACCELERATION
CENTRIFUGAL FORCE
CENTRIFUGES
∞FORCE
REVOLVING

CENTROIDS
RT   CENTER OF GRAVITY
MOMENTS OF INERTIA

CENTURION AIRCRAFT
USE   CESSNA 210 AIRCRAFT

CEPHALAGIA
USE   HEADACHE

CEPHALOPODS
GS   ANIMALS
. INVERTEBRATES
. . ARTHROPODS
. . . **CEPHALOPODS**
. . . . MOLLUSKS

CEPHALOPODS-(CONT.)
. . . . OCTOPUSES
. . . . SNAILS

CEPHEID VARIABLES
GS   CELESTIAL BODIES
. STARS
. . VARIABLE STARS
. . . **CEPHEID VARIABLES**
RT   CEPHEUS CONSTELLATION

CEPHEUS CONSTELLATION
GS   CONSTELLATIONS
. **CEPHEUS CONSTELLATION**
RT   CEPHEID VARIABLES

CEPSTRA
GS   SPECTRA
. POWER SPECTRA
. **CEPSTRA**
RT   QUEFRENCIES

CEPSTRAL ANALYSIS
GS   SPECTRUM ANALYSIS
. **CEPSTRAL ANALYSIS**
RT   ACOUSTIC MEASUREMENT
AUDIO FREQUENCIES
ECHOES
MULTIPATH TRANSMISSION
POWER SPECTRA
SIGNAL REFLECTION
SIGNATURE ANALYSIS
SPECTRAL SIGNATURES
SPEECH RECOGNITION
TIME LAG
VIBRATION MEASUREMENT

CERAMAL PROTECTIVE COATINGS
USE   CERMETS
PROTECTIVE COATINGS

CERAMALS
USE   CERMETS

CERAMIC BONDING
GS   BONDING
. **CERAMIC BONDING**
RT   CERAMIC MATRIX COMPOSITES
CERAMICS

CERAMIC COATINGS
SN   (COATINGS CONSISTING OF CERAMIC
MATERIALS)
GS   COATINGS
. INORGANIC COATINGS
. . **CERAMIC COATINGS**
. PROTECTIVE COATINGS
. . **CERAMIC COATINGS**
RT   FINISHES
METAL COATINGS
PORCELAIN
SPRAYED COATINGS
VACUUM DEPOSITION

† **CERAMIC HONEYCOMBS**
RT   CERAMIC MATRIX COMPOSITES
HONEYCOMB CORES
HONEYCOMB STRUCTURES

CERAMIC MATRIX COMPOSITES
GS   COMPOSITE MATERIALS
. **CERAMIC MATRIX COMPOSITES**
RT   AIRCRAFT CONSTRUCTION MATERIALS
BORON REINFORCED MATERIALS
CERAMIC BONDING
CERAMIC HONEYCOMBS
CERAMICS
CERMETS
COMPOSITE STRUCTURES
MATRIX MATERIALS
REINFORCING FIBERS
SILICON NITRIDES
TITANIUM CARBIDES
TITANIUM NITRIDES

CERAMIC NUCLEAR FUELS
GS   CERAMICS
. **CERAMIC NUCLEAR FUELS**
FUELS
. NUCLEAR FUELS
. . **CERAMIC NUCLEAR FUELS**
RT   CARBIDES
CERMETS
NITRIDES

CERAMIC NUCLEAR FUELS-(CONT.)
PLUTONIUM COMPOUNDS
PLUTONIUM OXIDES
SOL-GEL PROCESSES
THORIUM COMPOUNDS
URANIUM CARBIDES
URANIUM COMPOUNDS
URANIUM OXIDES

CERAMICS
GS   **CERAMICS**
. CERAMIC NUCLEAR FUELS
. PORCELAIN
. PYROCERAM (TRADEMARK)
RT   ABRASIVES
BAKELITE (TRADEMARK)
BRICKS
CERAMIC BONDING
CERAMIC MATRIX COMPOSITES
CERMETS
CLAYS
DIELECTRICS
FRIT
GLASS
GLAZES
INJECTION MOLDING
MASONRY
∞MATERIALS SCIENCE
MORTARS (MATERIAL)
PYROLYTIC MATERIALS
REACTION BONDING
REFRACTORIES
REFRACTORY COATINGS
REFRACTORY MATERIALS
SIALON
SILICON DIOXIDE
TILES
VITRIFICATION

CEREBELLUM
GS   ANATOMY
. BRAIN
. . **CEREBELLUM**
NERVOUS SYSTEM
. CENTRAL NERVOUS SYSTEM
. . BRAIN
. . . **CEREBELLUM**

CEREBRAL CORTEX
GS   ANATOMY
. BRAIN
. . **CEREBRAL CORTEX**
NERVOUS SYSTEM
. CENTRAL NERVOUS SYSTEM
. . BRAIN
. . . **CEREBRAL CORTEX**
RT   ∞CORTEXES
HYPOTHALAMUS

CEREBRAL VASCULAR ACCIDENTS
RT   CARDIOVASCULAR SYSTEM
∞STROKES

CEREBRAL VENTRICLES
GS   ANATOMY
. BRAIN
. . **CEREBRAL VENTRICLES**
NERVOUS SYSTEM
. CENTRAL NERVOUS SYSTEM
. . BRAIN
. . . **CEREBRAL VENTRICLES**
RT   CEREBROSPINAL FLUID

CEREBROSPINAL FLUID
GS   BODY FLUIDS
. **CEREBROSPINAL FLUID**
RT   BRAIN
CEREBRAL VENTRICLES
∞FLUIDS

CEREBRUM
GS   ANATOMY
. BRAIN
. . **CEREBRUM**
. MUSCULOSKELETAL SYSTEM
. . BONES
. . . **CEREBRUM**
NERVOUS SYSTEM
. CENTRAL NERVOUS SYSTEM
. . BRAIN
. . . **CEREBRUM**

CERENKOV COUNTERS
GS   MEASURING INSTRUMENTS
. COUNTERS

CERENKOV COUNTERS-*(CONT.)*
```
. . RADIATION COUNTERS
. . . CERENKOV COUNTERS
. RADIATION MEASURING INSTRUMENTS
. . RADIATION COUNTERS
. . . CERENKOV COUNTERS
RT    SCINTILLATION COUNTERS
```

CERENKOV EFFECT
```
USE   CERENKOV RADIATION
```

CERENKOV RADIATION
```
UF    CERENKOV EFFECT
GS    ELECTROMAGNETIC RADIATION
. CERENKOV RADIATION
RT    BREMSSTRAHLUNG
CORPUSCULAR RADIATION
COSMIC RAYS
∞ EFFECTS
GAMMA RAY BURSTS
GAMMA RAYS
LIGHT (VISIBLE RADIATION)
NUCLEAR RADIATION
∞ RADIATION
ULTRAVIOLET RADIATION
```

CERES ASTEROID
```
GS    CELESTIAL BODIES
. ASTEROID BELTS
. . ASTEROIDS
. . . CERES ASTEROID
```

CERESIN
```
GS    ALIPHATIC COMPOUNDS
. ALKANES
. . PARAFFINS
. . . CERESIN
HYDROCARBONS
. ALKANES
. . PARAFFINS
. . . CERESIN
WAXES
. CERESIN
RT    PHASE CHANGE MATERIALS
```

CERIUM
```
GS    CHEMICAL ELEMENTS
. RARE EARTH ELEMENTS
. . CERIUM
. . . CERIUM ISOTOPES
. . . . CERIUM 137
. . . . CERIUM 144
METALS
. RARE EARTH ELEMENTS
. . CERIUM
. . . CERIUM ISOTOPES
. . . . CERIUM 137
. . . . CERIUM 144
```

CERIUM COMPOUNDS
```
GS    RARE EARTH COMPOUNDS
. CERIUM COMPOUNDS
. . BASTNASITE
. . CERIUM OXIDES
RT    ∞ CHEMICAL COMPOUNDS
∞ METAL COMPOUNDS
```

CERIUM ISOTOPES
```
GS    CHEMICAL ELEMENTS
. NUCLIDES
. . ISOTOPES
. . . CERIUM ISOTOPES
. . . . CERIUM 137
. . . . CERIUM 144
. RARE EARTH ELEMENTS
. . CERIUM
. . . CERIUM ISOTOPES
. . . . CERIUM 137
. . . . CERIUM 144
METALS
. RARE EARTH ELEMENTS
. . CERIUM
. . . CERIUM ISOTOPES
. . . . CERIUM 137
. . . . CERIUM 144
```

CERIUM OXIDES
```
GS    RARE EARTH COMPOUNDS
. CERIUM COMPOUNDS
. . CERIUM OXIDES
```

CERIUM 137
```
GS    CHEMICAL ELEMENTS
. NUCLIDES
```

CERIUM 137-*(CONT.)*
```
. . ISOTOPES
. . . CERIUM ISOTOPES
. . . . CERIUM 137
. . . RADIOACTIVE ISOTOPES
. . . . CERIUM 137
. RARE EARTH ELEMENTS
. . CERIUM
. . . CERIUM ISOTOPES
. . . . CERIUM 137
METALS
. RARE EARTH ELEMENTS
. . CERIUM
. . . CERIUM ISOTOPES
. . . . CERIUM 137
```

CERIUM 144
```
GS    CHEMICAL ELEMENTS
. NUCLIDES
. . ISOTOPES
. . . CERIUM ISOTOPES
. . . . CERIUM 144
. . . RADIOACTIVE ISOTOPES
. . . . CERIUM 144
. RARE EARTH ELEMENTS
. . CERIUM
. . . CERIUM ISOTOPES
. . . . CERIUM 144
METALS
. RARE EARTH ELEMENTS
. . CERIUM
. . . CERIUM ISOTOPES
. . . . CERIUM 144
```

CERMETS
```
UF    CERAMAL PROTECTIVE COATINGS
CERAMALS
GS    COMPOSITE MATERIALS
. CERMETS
RT    CERAMIC MATRIX COMPOSITES
CERAMIC NUCLEAR FUELS
CERAMICS
HEAT RESISTANT ALLOYS
POWDER METALLURGY
REFRACTORIES
REFRACTORY MATERIALS
```

CERTIFICATION
```
RT    AIRCRAFT RELIABILITY
CHECKOUT
EVALUATION
FLIGHT TESTS
PERFORMANCE TESTS
PHYSIOLOGICAL TESTS
PSYCHOLOGICAL TESTS
QUALIFICATIONS
QUALITY CONTROL
SELECTION
SITE SELECTION
TRAINING EVALUATION
```

CESIUM
```
GS    CHEMICAL ELEMENTS
. ALKALI METALS
. . CESIUM
. . . CESIUM ISOTOPES
. . . . CESIUM 133
. . . . CESIUM 134
. . . . CESIUM 137
. . . . CESIUM 144
. . . CESIUM VAPOR
METALS
. ALKALI METALS
. . CESIUM
. . . CESIUM ISOTOPES
. . . . CESIUM 133
. . . . CESIUM 134
. . . . CESIUM 137
. . . . CESIUM 144
. . . CESIUM VAPOR
RT    CESIUM ALLOYS
CESIUM ANTIMONIDES
CESIUM BROMIDES
CESIUM DIODES
CESIUM ENGINES
CESIUM FLUORIDES
CESIUM HALIDES
CESIUM HYDRIDES
CESIUM IODIDES
CESIUM IONS
CESIUM PLASMA
CESIUM VAPOR
```

CESIUM ALLOYS
```
GS    ALLOYS
```

CESIUM ALLOYS-*(CONT.)*
```
. CESIUM ALLOYS
RT    ALKALI METALS
CESIUM
```

CESIUM ANTIMONIDES
```
GS    ANTIMONY COMPOUNDS
. ANTIMONIDES
. . CESIUM ANTIMONIDES
CESIUM COMPOUNDS
. CESIUM ANTIMONIDES
RT    CESIUM
```

CESIUM BROMIDES
```
GS    CESIUM COMPOUNDS
. CESIUM HALIDES
. . CESIUM BROMIDES
HALOGEN COMPOUNDS
. BROMINE COMPOUNDS
. . BROMIDES
. . . CESIUM BROMIDES
. HALIDES
. . BROMIDES
. . . CESIUM BROMIDES
. METAL HALIDES
. . ALKALI HALIDES
. . . . CESIUM HALIDES
. . . . . CESIUM BROMIDES
RT    CESIUM
```

CESIUM COMPOUNDS
```
GS    CESIUM COMPOUNDS
. CESIUM ANTIMONIDES
. CESIUM HALIDES
. . CESIUM BROMIDES
. . CESIUM FLUORIDES
. . CESIUM IODIDES
. CESIUM HYDRIDES
. CESIUM OXIDES
RT    ∞ ALKALI METAL COMPOUNDS
∞ CHEMICAL COMPOUNDS
∞ METAL COMPOUNDS
METAL FUELS
```

CESIUM DIODES
```
GS    ELECTRONIC EQUIPMENT
. DIODES
. . CESIUM DIODES
RT    CESIUM
PLASMA DIODES
THERMIONIC CONVERTERS
```

CESIUM ENGINES
```
GS    ENGINES
. ROCKET ENGINES
. . ELECTRIC ROCKET ENGINES
. . . ION ENGINES
. . . . CESIUM ENGINES
RT    CESIUM
ELECTROSTATIC ENGINES
```

CESIUM FLUORIDES
```
GS    CESIUM COMPOUNDS
. CESIUM HALIDES
. . CESIUM FLUORIDES
HALOGEN COMPOUNDS
. FLUORINE COMPOUNDS
. . FLUORIDES
. . . METAL FLUORIDES
. . . . CESIUM FLUORIDES
. HALIDES
. . METAL HALIDES
. . . ALKALI HALIDES
. . . . CESIUM HALIDES
. . . . . CESIUM FLUORIDES
RT    CESIUM
```

CESIUM HALIDES
```
GS    CESIUM COMPOUNDS
. CESIUM HALIDES
. . CESIUM BROMIDES
. . CESIUM FLUORIDES
. . CESIUM IODIDES
HALOGEN COMPOUNDS
. HALIDES
. . METAL HALIDES
. . . ALKALI HALIDES
. . . . CESIUM HALIDES
. . . . . CESIUM BROMIDES
. . . . . CESIUM FLUORIDES
. . . . . CESIUM IODIDES
RT    CESIUM
```

CESIUM HYDRIDES
```
GS    CESIUM COMPOUNDS
```

CESIUM HYDRIDES-*(CONT.)*
```
      . CESIUM HYDRIDES
      HYDROGEN COMPOUNDS
      . HYDRIDES
      . . METAL HYDRIDES
      . . . CESIUM HYDRIDES
RT    CESIUM
```

CESIUM IODIDES
```
GS    CESIUM COMPOUNDS
      . CESIUM HALIDES
      . . CESIUM IODIDES
      HALOGEN COMPOUNDS
      . HALIDES
      . . METAL HALIDES
      . . . ALKALI HALIDES
      . . . . CESIUM HALIDES
      . . . . . CESIUM IODIDES
      . IODINE COMPOUNDS
      . . IODIDES
      . . . CESIUM IODIDES
RT    CESIUM
```

CESIUM IONS
```
GS    IONS
      . CESIUM IONS
RT    CESIUM
```

CESIUM ISOTOPES
```
GS    CHEMICAL ELEMENTS
      . ALKALI METALS
      . . CESIUM
      . . . CESIUM ISOTOPES
      . . . . CESIUM 133
      . . . . CESIUM 134
      . . . . CESIUM 137
      . . . . CESIUM 144
      . NUCLIDES
      . . ISOTOPES
      . . . CESIUM ISOTOPES
      . . . . CESIUM 133
      . . . . CESIUM 134
      . . . . CESIUM 137
      . . . . CESIUM 144
      METALS
      . ALKALI METALS
      . . CESIUM
      . . . CESIUM ISOTOPES
      . . . . CESIUM 133
      . . . . CESIUM 134
      . . . . CESIUM 137
      . . . . CESIUM 144
```

CESIUM OXIDES
```
GS    CESIUM COMPOUNDS
      . CESIUM OXIDES
      CHALCOGENIDES
      . OXIDES
      . . METAL OXIDES
      . . . CESIUM OXIDES
```

CESIUM PLASMA
```
GS    GASES
      . IONIZED GASES
      . . CHARGED PARTICLES
      . . . METALLIC PLASMAS
      . . . . CESIUM PLASMA
      PARTICLES
      . CHARGED PARTICLES
      . . ENERGETIC PARTICLES
      . . . PLASMAS (PHYSICS)
      . . . . METALLIC PLASMAS
      . . . . . CESIUM PLASMA
RT    CESIUM
      THERMIONIC CONVERTERS
```

CESIUM VAPOR
```
GS    CHEMICAL ELEMENTS
      . ALKALI METALS
      . . CESIUM
      . . . CESIUM VAPOR
      NUCLIDES
      . . ISOTOPES
      . . . CESIUM VAPOR
      METALS
      . ALKALI METALS
      . . CESIUM
      . . . CESIUM VAPOR
      VAPORS
      . CESIUM VAPOR
RT    CESIUM
      MERCURY VAPOR
```

CESIUM 133
```
GS    CHEMICAL ELEMENTS
```

CESIUM 133-*(CONT.)*
```
      . ALKALI METALS
      . . CESIUM
      . . . CESIUM ISOTOPES
      . . . . CESIUM 133
      . NUCLIDES
      . . ISOTOPES
      . . . CESIUM ISOTOPES
      . . . . CESIUM 133
      METALS
      . ALKALI METALS
      . . CESIUM
      . . . CESIUM ISOTOPES
      . . . . CESIUM 133
```

CESIUM 134
```
GS    CHEMICAL ELEMENTS
      . ALKALI METALS
      . . CESIUM
      . . . CESIUM ISOTOPES
      . . . . CESIUM 134
      . NUCLIDES
      . . ISOTOPES
      . . . CESIUM ISOTOPES
      . . . . CESIUM 134
      . . . RADIOACTIVE ISOTOPES
      . . . . CESIUM 134
      METALS
      . ALKALI METALS
      . . CESIUM
      . . . CESIUM ISOTOPES
      . . . . CESIUM 134
```

CESIUM 137
```
GS    CHEMICAL ELEMENTS
      . ALKALI METALS
      . . CESIUM
      . . . CESIUM ISOTOPES
      . . . . CESIUM 137
      . NUCLIDES
      . . ISOTOPES
      . . . CESIUM ISOTOPES
      . . . . CESIUM 137
      . . . RADIOACTIVE ISOTOPES
      . . . . CESIUM 137
      METALS
      . ALKALI METALS
      . . CESIUM
      . . . CESIUM ISOTOPES
      . . . . CESIUM 137
```

CESIUM 144
```
GS    CHEMICAL ELEMENTS
      . ALKALI METALS
      . . CESIUM
      . . . CESIUM ISOTOPES
      . . . . CESIUM 144
      . NUCLIDES
      . . ISOTOPES
      . . . CESIUM ISOTOPES
      . . . . CESIUM 144
      . . . RADIOACTIVE ISOTOPES
      . . . . CESIUM 144
      METALS
      . ALKALI METALS
      . . CESIUM
      . . . CESIUM ISOTOPES
      . . . . CESIUM 144
```

CESSNA AIRCRAFT
```
GS    CESSNA AIRCRAFT
      . A-37 AIRCRAFT
      . CESSNA L-19 AIRCRAFT
      . CESSNA 172 AIRCRAFT
      . CESSNA 205 AIRCRAFT
      . CESSNA 210 AIRCRAFT
      . CESSNA 402B AIRCRAFT
      . T-37 AIRCRAFT
RT    ∞ AIRCRAFT
```

CESSNA L-19 AIRCRAFT
```
GS    CESSNA AIRCRAFT
      . CESSNA L-19 AIRCRAFT
      LIGHT AIRCRAFT
      . CESSNA L-19 AIRCRAFT
      MONOPLANES
      . CESSNA L-19 AIRCRAFT
      OBSERVATION AIRCRAFT
      . CESSNA L-19 AIRCRAFT
      RECONNAISSANCE AIRCRAFT
      . CESSNA L-19 AIRCRAFT
RT    ∞ AIRCRAFT
```

CESSNA MILITARY AIRCRAFT
```
USE   MILITARY AIRCRAFT
```

CESSNA 172 AIRCRAFT
```
GS    CESSNA AIRCRAFT
      . CESSNA 172 AIRCRAFT
      GENERAL AVIATION AIRCRAFT
      . CESSNA 172 AIRCRAFT
      LIGHT AIRCRAFT
      . CESSNA 172 AIRCRAFT
      MONOPLANES
      . CESSNA 172 AIRCRAFT
      PASSENGER AIRCRAFT
      . CESSNA 172 AIRCRAFT
RT    ∞ AIRCRAFT
```

CESSNA 205 AIRCRAFT
```
GS    CESSNA AIRCRAFT
      . CESSNA 205 AIRCRAFT
      GENERAL AVIATION AIRCRAFT
      . CESSNA 205 AIRCRAFT
      LIGHT AIRCRAFT
      . CESSNA 205 AIRCRAFT
      MONOPLANES
      . CESSNA 205 AIRCRAFT
      PASSENGER AIRCRAFT
      . CESSNA 205 AIRCRAFT
RT    ∞ AIRCRAFT
```

CESSNA 210 AIRCRAFT
```
UF    CENTURION AIRCRAFT
GS    CESSNA AIRCRAFT
      . CESSNA 210 AIRCRAFT
      GENERAL AVIATION AIRCRAFT
      . CESSNA 210 AIRCRAFT
      LIGHT AIRCRAFT
      . CESSNA 210 AIRCRAFT
      MONOPLANES
      . CESSNA 210 AIRCRAFT
      PASSENGER AIRCRAFT
      . CESSNA 210 AIRCRAFT
```

CESSNA 402B AIRCRAFT
```
GS    CESSNA AIRCRAFT
      . CESSNA 402B AIRCRAFT
      GENERAL AVIATION AIRCRAFT
      . CESSNA 402B AIRCRAFT
      LIGHT AIRCRAFT
      . CESSNA 402B AIRCRAFT
      MONOPLANES
      . CESSNA 402B AIRCRAFT
      PASSENGER AIRCRAFT
      . CESSNA 402B AIRCRAFT
RT    ∞ AIRCRAFT
```

CETANE
```
GS    ALIPHATIC COMPOUNDS
      . ALKANES
      . . CETANE
      HYDROCARBONS
      . ALKANES
      . . CETANE
```

CETYL COMPOUNDS
```
GS    ALIPHATIC COMPOUNDS
      . ALKYL COMPOUNDS
      . . CETYL COMPOUNDS
RT    ∞ CHEMICAL COMPOUNDS
```

CEYLON
```
USE   SRI LANKA
```

CF-104 AIRCRAFT
```
USE   CANADAIR AIRCRAFT
      F-104 AIRCRAFT
```

CF-700 ENGINE
```
GS    ENGINES
      . AIR BREATHING ENGINES
      . . GAS TURBINE ENGINES
      . . . JET ENGINES
      . . . . TURBOJET ENGINES
      . . . . . TURBOFAN ENGINES
      . . . . . . CF-700 ENGINE
      . INTERNAL COMBUSTION ENGINES
      . . GAS TURBINE ENGINES
      . . . JET ENGINES
      . . . . TURBOJET ENGINES
      . . . . . TURBOFAN ENGINES
      . . . . . . CF-700 ENGINE
      . TURBINE ENGINES
      . . GAS TURBINE ENGINES
      . . . JET ENGINES
      . . . . TURBOJET ENGINES
      . . . . . TURBOFAN ENGINES
      . . . . . . CF-700 ENGINE
```

CF-700 ENGINE-(CONT.)
RT   VERTICAL TAKEOFF AIRCRAFT

CFD
USE   CHARGE FLOW DEVICES

CFRP
USE   CARBON FIBER REINFORCED PLASTICS

CH-3 HELICOPTER
GS   PASSENGER AIRCRAFT
. **CH-3 HELICOPTER**
SIKORSKY AIRCRAFT
. **CH-3 HELICOPTER**
TRANSPORT AIRCRAFT
. **CH-3 HELICOPTER**
V/STOL AIRCRAFT
. ROTARY WING AIRCRAFT
. . HELICOPTERS
. . . MILITARY HELICOPTERS
. . . . **CH-3 HELICOPTER**
. . . RIGID ROTOR HELICOPTERS
. . . . **CH-3 HELICOPTER**
RT   S-61 HELICOPTER

CH-21 HELICOPTER
UF   H-21 HELICOPTER
SHAWNEE HELICOPTER
WORKHORSE HELICOPTER
GS   BOEING AIRCRAFT
. **CH-21 HELICOPTER**
V/STOL AIRCRAFT
. ROTARY WING AIRCRAFT
. . HELICOPTERS
. . . **CH-21 HELICOPTER**

CH-34 HELICOPTER
UF   CHOCTAW HELICOPTER
H-34 HELICOPTER
GS   SIKORSKY AIRCRAFT
. **CH-34 HELICOPTER**
TRANSPORT AIRCRAFT
. **CH-34 HELICOPTER**
V/STOL AIRCRAFT
. ROTARY WING AIRCRAFT
. . HELICOPTERS
. . . MILITARY HELICOPTERS
. . . . **CH-34 HELICOPTER**
RT   S-58 HELICOPTER

CH-46 HELICOPTER
UF   CH-113 HELICOPTER
HRB-1 HELICOPTER
SEA KNIGHT HELICOPTER
VOYAGEUR HELICOPTER
GS   BOEING AIRCRAFT
. **CH-46 HELICOPTER**
PASSENGER AIRCRAFT
. **CH-46 HELICOPTER**
TRANSPORT AIRCRAFT
. **CH-46 HELICOPTER**
V/STOL AIRCRAFT
. ROTARY WING AIRCRAFT
. . HELICOPTERS
. . . MILITARY HELICOPTERS
. . . . **CH-46 HELICOPTER**
. . . TANDEM ROTOR HELICOPTERS
. . . . **CH-46 HELICOPTER**

CH-47 HELICOPTER
UF   CHINOOK HELICOPTER
HC-1 HELICOPTER
GS   BOEING AIRCRAFT
. **CH-47 HELICOPTER**
PASSENGER AIRCRAFT
. **CH-47 HELICOPTER**
TRANSPORT AIRCRAFT
. **CH-47 HELICOPTER**
V/STOL AIRCRAFT
. ROTARY WING AIRCRAFT
. . HELICOPTERS
. . . MILITARY HELICOPTERS
. . . . **CH-47 HELICOPTER**
. . . TANDEM ROTOR HELICOPTERS
. . . . **CH-47 HELICOPTER**

CH-53 HELICOPTER
USE   H-53 HELICOPTER

CH-54 HELICOPTER
UF   S-64 HELICOPTER
SIKORSKY S-64 HELICOPTER
SKYCRANE HELICOPTER
GS   PASSENGER AIRCRAFT

CH-54 HELICOPTER-(CONT.)
. **CH-54 HELICOPTER**
SIKORSKY AIRCRAFT
. **CH-54 HELICOPTER**
TRANSPORT AIRCRAFT
. **CH-54 HELICOPTER**
V/STOL AIRCRAFT
. ROTARY WING AIRCRAFT
. . HELICOPTERS
. . . MILITARY HELICOPTERS
. . . . **CH-54 HELICOPTER**

CH-62 HELICOPTER
GS   BOEING AIRCRAFT
. **CH-62 HELICOPTER**
V/STOL AIRCRAFT
. ROTARY WING AIRCRAFT
. . HELICOPTERS
. . . MILITARY HELICOPTERS
. . . . HEAVY LIFT HELICOPTERS
. . . . . **CH-62 HELICOPTER**
RT   ∞MILITARY AIRCRAFT

CH-113 HELICOPTER
USE   CH-46 HELICOPTER

CHAD
GS   NATIONS
. **CHAD**
RT   AFRICA

CHAFF
GS   COUNTERMEASURES
. ELECTRONIC COUNTERMEASURES
. . **CHAFF**
RT   DECEPTION
ELECTRONIC WARFARE
RADAR ECHOES

CHAINS
SN   (EXCLUDES CHEMICAL BONDS AND
NUCLEAR REACTIONS)
RT   ∞BARRIERS
CABLES (ROPES)
FASTENERS
∞LINKS
MOLECULAR CHAINS

CHAIRS
USE   SEATS

CHALCOGENIDES
GS   **CHALCOGENIDES**
. AKERMANITE
. OXIDES
. . ANHYDRIDES
. . . PEROXIDES
. . . . INORGANIC PEROXIDES
. . . . ORGANIC PEROXIDES
. . . . SODIUM PEROXIDES
. . BORON OXIDES
. . BRUCITE
. . CARBON MONOXIDE
. . CARBON SUBOXIDES
. . CHLORINE OXIDES
. . DIOXIDES
. . . CARBON DIOXIDE
. . . FLINT
. . . HYDROGEN PEROXIDE
. . . SILICON DIOXIDE
. . . . QUARTZ
. . . . . COESITE
. . . SULFUR DIOXIDES
. . GERMANIUM OXIDES
. . HEAVY WATER
. . METAL OXIDES
. . . ALKALINE EARTH OXIDES
. . . . BARIUM OXIDES
. . . . BERYLLIUM OXIDES
. . . . CALCIUM OXIDES
. . . . MAGNESIUM OXIDES
. . . . . PERICLASE
. . . ALUMINUM OXIDES
. . . . SAPPHIRE
. . . BISMUTH OXIDES
. . . CESIUM OXIDES
. . . CHROMIUM OXIDES
. . . . CHROMITES
. . . COBALT OXIDES
. . . COPPER OXIDES
. . . HAFNIUM OXIDES
. . . IRON OXIDES
. . . . CHROMITES
. . . . HEMATITE
. . . . ILMENITE

CHALCOGENIDES-(CONT.)
. . . MAGNETITE
. . . LANTHANUM OXIDES
. . . LEAD OXIDES
. . . LITHIUM OXIDES
. . . MANGANESE OXIDES
. . . . HOPCALITE (TRADEMARK)
. . . MERCURY OXIDES
. . . MIXED OXIDES
. . . MOLYBDENUM OXIDES
. . . NICKEL OXIDES
. . . NIOBIUM OXIDES
. . . PLATINUM OXIDES
. . . PLUTONIUM OXIDES
. . . POTASSIUM OXIDES
. . . SCANDIUM OXIDES
. . . SILVER OXIDES
. . . SODIUM PEROXIDES
. . . TANTALUM OXIDES
. . . THORIUM OXIDES
. . . TIN OXIDES
. . . TITANIUM OXIDES
. . . . ANATASE
. . . . ILMENITE
. . . . RUTILE
. . . TUNGSTEN OXIDES
. . . . SCHEELITE
. . . URANIUM OXIDES
. . . VANADIUM OXIDES
. . . YTTRIUM OXIDES
. . . ZINC OXIDES
. . . ZIRCONIUM OXIDES
. . NITROGEN OXIDES
. . . NITRIC OXIDE
. . . NITROGEN DIOXIDE
. . . NITROGEN TETROXIDE
. . . NITROUS OXIDE
. . PHOSPHORUS OXIDES
. . PYROXENES
. . . ENSTATITE
. . SELENIUM OXIDES
. . SILICON OXIDES
. . . MUSCOVITE
. . . NEPHELITE
. . . SILICON DIOXIDE
. . . . QUARTZ
. . . . . COESITE
. . . SPODUMENE
. . SULFUR OXIDES
. . . SULFUR DIOXIDES
. SELENIDES
. . CADMIUM SELENIDES
. . COPPER SELENIDES
. . GALLIUM SELENIDES
. . LEAD SELENIDES
. . ZINC SELENIDES
. SULFIDES
. . DISULFIDES
. . . CARBON DISULFIDE
. . INORGANIC SULFIDES
. . . BARIUM SULFIDES
. . . BISMUTH SULFIDES
. . . CADMIUM SULFIDES
. . . CALCIUM SULFIDES
. . . COPPER SULFIDES
. . . . ENARGITE
. . . HYDROGEN SULFIDE
. . . INDIUM SULFIDES
. . . LEAD SULFIDES
. . . MOLYBDENUM SULFIDES
. . . . MOLYBDENUM DISULFIDES
. . . POLYSULFIDES
. . . STRONTIUM SULFIDES
. . . ZINC SULFIDES
. . . . WURTZITE
. . . . ZINCBLENDE
. . PYRITES
. . PYRRHOTITE
. . . TROILITE
. TELLURIDES
. . BISMUTH TELLURIDES
. . CADMIUM TELLURIDES
. . INDIUM TELLURIDES
. . LANTHANUM TELLURIDES
. . LEAD TELLURIDES
. . MERCURY TELLURIDES
. . TIN TELLURIDES
. . ZINC TELLURIDES
RT   ∞GROUP 6A COMPOUNDS

CHALK
GS   CALCIUM COMPOUNDS
. CALCIUM CARBONATES
. . **CHALK**
CARBON COMPOUNDS
. CARBONATES

CHALK-(CONT.)
```
        . . CALCIUM CARBONATES
        . . . CHALK
RT      GYPSUM
```

CHALLENGER (ORBITER)
```
UF      SPACE SHUTTLE ORBITER 099
GS      REENTRY VEHICLES
        . RECOVERABLE SPACECRAFT
        . . REUSABLE SPACECRAFT
        . . . CHALLENGER (ORBITER)
        TRANSPORTATION
        . SPACE TRANSPORTATION
        . . SPACE TRANSPORTATION SYSTEM
        . . . SPACE SHUTTLE ORBITERS
        . . . . CHALLENGER (ORBITER)
RT      MANNED SPACECRAFT
        SPACE SHUTTLE MISSION 31-B
        SPACE SHUTTLE MISSION 31-C
        SPACE SHUTTLE MISSION 31-D
        SPACE SHUTTLE MISSION 41-B
        SPACE SHUTTLE MISSION 41-C
        SPACE SHUTTLE MISSION 41-G
        SPACE SHUTTLE MISSION 51-B
        SPACE SHUTTLE MISSION 51-E
        SPACE SHUTTLE MISSION 51-F
        SPACE SHUTTLE MISSION 51-L
        SPACE SHUTTLE MISSION 61-A
        ∞ SPACECRAFT
```

∞ CHAMBERS
```
SN      (USE OF A MORE SPECIFIC TERM IS
        RECOMMENDED--CONSULT THE TERMS
        LISTED BELOW)
RT      ANECHOIC CHAMBERS
        ARC CHAMBERS
        BUBBLE CHAMBERS
        CLOUD CHAMBERS
        COMBUSTION CHAMBERS
        CYLINDRICAL CHAMBERS
        FLEXING
        FLOW CHAMBERS
        HYPERBARIC CHAMBERS
        IONIZATION CHAMBERS
        PLENUM CHAMBERS
        PRESSURE CHAMBERS
        SPARK CHAMBERS
        TEST CHAMBERS
        THRUST CHAMBERS
        VACUUM CHAMBERS
```

CHANCE-VOUGHT AIRCRAFT
```
UF      CHANCE-VOUGHT MILITARY AIRCRAFT
RT      ∞ AIRCRAFT
```

CHANCE-VOUGHT MILITARY AIRCRAFT
```
USE     CHANCE-VOUGHT AIRCRAFT
        MILITARY AIRCRAFT
```

CHANDLER MOTION
```
USE     POLAR WANDERING (GEOLOGY)
```

CHANDRASEKHAR EQUATION
```
GS      ANALYSIS (MATHEMATICS)
        . REAL VARIABLES
        . . DIFFERENTIAL EQUATIONS
        . . . CHANDRASEKHAR EQUATION
RT      ELECTROMAGNETIC ABSORPTION
        ∞ EQUATIONS
```

CHANGE DETECTION
```
GS      DETECTION
        . CHANGE DETECTION
RT      AERIAL PHOTOGRAPHY
        AERIAL RECONNAISSANCE
        EARTH RESOURCES PROGRAM
        IMAGE PROCESSING
        IMAGERY
        LAND USE
        MULTISPECTRAL BAND SCANNERS
        MULTISPECTRAL PHOTOGRAPHY
        PATTERN RECOGNITION
        PHOTOINTERPRETATION
        RADAR IMAGERY
        REMOTE SENSING
        SCENE ANALYSIS
        SIDE-LOOKING RADAR
        TERRAIN ANALYSIS
```

CHANNEL CAPACITY
```
RT      ALOHA SYSTEM
        BANDWIDTH
        ∞ CAPACITY
        DEMAND ASSIGNMENT MULTIPLE
            ACCESS
```

CHANNEL CAPACITY-(CONT.)
```
        FREQUENCIES
        PACKET TRANSMISSION
```

CHANNEL FLOW
```
GS      FLUID FLOW
        . CHANNEL FLOW
        . . OPEN CHANNEL FLOW
RT      ANNULAR FLOW
        CORNER FLOW
        DREDGED MATERIALS
        DUCTED FLOW
        FLOW GEOMETRY
        FLUID INJECTION
        INCOMPRESSIBLE FLUIDS
        OUTLET FLOW
        PIPE FLOW
        WALL FLOW
```

CHANNEL MULTIPLIERS
```
UF      CHANNELTRONS
GS      MULTIPLIERS
        . CHANNEL MULTIPLIERS
RT      AURORAL SPECTROSCOPY
        ELECTRON AVALANCHE
        MICROCHANNEL PLATES
        PHOTOMULTIPLIER TUBES
        RADIATION COUNTERS
```

CHANNEL NOISE
```
RT      ALOHA SYSTEM
        BACKGROUND NOISE
        ELECTROMAGNETIC NOISE
        NOISE SPECTRA
        RANDOM NOISE
        SIGNAL TO NOISE RATIOS
        THERMAL NOISE
        TIME DIVISION MULTIPLE ACCESS
```

CHANNEL WINGS
```
GS      AIRFOILS
        . WINGS
        . . CHANNEL WINGS
        PLANFORMS
        . WING PLANFORMS
        . . CHANNEL WINGS
RT      AERODYNAMIC CONFIGURATIONS
        AIRCRAFT DESIGN
        AIRCRAFT PARTS
        AIRCRAFT STRUCTURES
```

∞ CHANNELS
```
SN      (USE OF A MORE SPECIFIC TERM IS
        RECOMMENDED--CONSULT THE TERMS
        LISTED BELOW)
RT      CHANNELS (DATA TRANSMISSION)
        COMPUTER STORAGE DEVICES
        DREDGED MATERIALS
        DUCTS
        FREQUENCIES
        MEDIA
        PARALLEL PLATES
        STRUCTURAL MEMBERS
        TELECOMMUNICATION
        THROATS
```

CHANNELS (DATA TRANSMISSION)
```
UF      DATA BUSSES
RT      ∞ CHANNELS
        ∞ DATA
        DATA LINKS
        DATA PROCESSING
        DATA TRANSMISSION
        PROTOCOL (COMPUTERS)
        SINGLE CHANNEL PER CARRIER
            TRANSMISSION
```

CHANNELTRONS
```
USE     CHANNEL MULTIPLIERS
```

CHAOS
```
SN      (LIMITED TO PHYSICS)
RT      BRANCHING (MATHEMATICS)
        MATHEMATICAL MODELS
        NONLINEAR SYSTEMS
        STOCHASTIC PROCESSES
        STRANGE ATTRACTORS
```

CHAOTIC CLOUD PATTERNS
```
USE     CLOUDS (METEOROLOGY)
```

CHAPARRAL
```
GS      PLANTS (BOTANY)
        . BRUSH (BOTANY)
```

CHAPARRAL-(CONT.)
```
        . . CHAPARRAL
RT      BOTANY
        EARTH RESOURCES
        TREES (PLANTS)
```

CHAPARRAL MISSILE
```
GS      MISSILES
        . SURFACE TO AIR MISSILES
        . . CHAPARRAL MISSILE
RT      SPACE WEAPONS
```

CHAPLYGIN EQUATION
```
RT      ∞ EQUATIONS
        FLOW EQUATIONS
        HODOGRAPHS
        VECTOR SPACES
```

CHAPMAN SHEAR LAYER
```
USE     SHEAR LAYERS
```

CHAPMAN-ENSKOG THEORY
```
UF      ENSKOG-CHAPMAN THEORY
GS      KINETIC THEORY
        . CHAPMAN-ENSKOG THEORY
RT      BOLTZMANN TRANSPORT EQUATION
        DISTRIBUTION FUNCTIONS
        FLOW DISTRIBUTION
        MONATOMIC GASES
        RAREFIED GAS DYNAMICS
        TEMPERATURE GRADIENTS
        ∞ THEORIES
        THERMAL DIFFUSION
```

CHAPMAN-FERRARO PROBLEM
```
RT      ATMOSPHERIC MODELS
        INTERPLANETARY MAGNETIC FIELDS
        MAGNETOPAUSE
        MAGNETOSPHERE
        ∞ PROBLEMS
        SOLAR WIND
```

CHAPMAN-JOUGET FLAME
```
USE     CHEMICAL EQUILIBRIUM
        DETONATION
        FLAME PROPAGATION
```

CHARACTER RECOGNITION
```
GS      RECOGNITION
        . PATTERN RECOGNITION
        . . CHARACTER RECOGNITION
RT      ARTIFICIAL INTELLIGENCE
        CONTRAST
        ∞ DETECTORS
        GRAPHOLOGY
        HANDWRITING
        LEGIBILITY
        OPTICAL DATA PROCESSING
        OPTICAL SCANNERS
        PERCEPTION
        READERS
        READING
        RESOLUTION
        SCENE ANALYSIS
        ∞ SENSORS
        SYMBOLS
        VISIBILITY
```

CHARACTERISTIC EQUATIONS
```
USE     EIGENVALUES
        EIGENVECTORS
```

CHARACTERISTIC FUNCTIONS
```
USE     EIGENVALUES
        EIGENVECTORS
```

CHARACTERISTIC METHOD
```
USE     METHOD OF CHARACTERISTICS
```

∞ CHARACTERISTICS
```
SN      (USE OF A MORE SPECIFIC TERM IS
        RECOMMENDED--CONSULT THE TERMS
        LISTED BELOW)
RT      AERODYNAMIC CHARACTERISTICS
        CAPACITANCE-VOLTAGE
            CHARACTERISTICS
        DYNAMIC CHARACTERISTICS
        FLIGHT CHARACTERISTICS
        FLOW CHARACTERISTICS
        METHOD OF CHARACTERISTICS
        POLARIZATION CHARACTERISTICS
        SEGRE CHARACTERISTIC
        SPRAY CHARACTERISTICS
```

CHARACTERISTICS-(CONT.)
            STATIC AERODYNAMIC
                CHARACTERISTICS
            VOLT-AMPERE CHARACTERISTICS

CHARACTERIZATION
RT   DESCRIPTIONS
        EXAMINATION
        REPRESENTATIONS

CHARACTERS
USE   SYMBOLS

CHARCOAL
GS   CHEMICAL ELEMENTS
        . CARBON
        . . CHARCOAL
        . . . ACTIVATED CARBON
        FUELS
        . CHARCOAL
RT   ADSORBENTS
        COKE

CHARGE CARRIERS
GS   CHARGE CARRIERS
        . FREE ELECTRONS
        . HOLES (ELECTRON DEFICIENCIES)
        . MAJORITY CARRIERS
        . MINORITY CARRIERS
RT   CARRIER INJECTION
        CARRIER LIFETIME
        ∞CARRIERS
        ELECTRON MOBILITY
        HOLE MOBILITY

CHARGE COUPLED DEVICES
UF   CCD
GS   ELECTRONIC EQUIPMENT
        . SOLID STATE DEVICES
        . . SEMICONDUCTOR DEVICES
        . . . METAL OXIDE SEMICONDUCTORS
        . . . . CHARGE TRANSFER DEVICES
        . . . . . CHARGE COUPLED DEVICES
RT   BUCKET BRIGADE DEVICES
        CCD STAR TRACKER
        CHARGE INJECTION DEVICES
        SEMICONDUCTORS (MATERIALS)

CHARGE DISTRIBUTION
GS   DISTRIBUTION (PROPERTY)
        . CHARGE DISTRIBUTION
RT   CURRENT DISTRIBUTION
        ELECTRON DISTRIBUTION
        ELECTROSTATIC CHARGE
        FORCE DISTRIBUTION
        HOLE DISTRIBUTION (ELECTRONICS)
        ION DISTRIBUTION
        MASS DISTRIBUTION
        NEUTRAL ATOMS
        POLARIZATION (CHARGE SEPARATION)

CHARGE EFFICIENCY
GS   EFFICIENCY
        . CHARGE EFFICIENCY
RT   BATTERY CHARGERS
        ∞CHARGING
        ELECTRIC BATTERIES
        PRIMARY BATTERIES
        RECHARGING
        STORAGE BATTERIES

CHARGE EXCHANGE
SN   (COLLISIONAL TRANSFER OF AN
        ELECTRON FROM A NEUTRAL ATOM OR
        MOLECULE TO AN ION--EXCLUDES
        SEMICONDUCTOR AND PHOTOCHEMICAL
        CHARGE TRANSFER)
GS   EXCHANGING
        . CHARGE EXCHANGE
        . . RESONANCE CHARGE EXCHANGE
RT   ELECTRON TRANSFER
        ION ATOM INTERACTIONS
        ION CHARGE
        ION PRODUCTION RATES
        PLASMA-PARTICLE INTERACTIONS
        RECOIL IONS

CHARGE FLOW DEVICES
UF   CFD
GS   ELECTRONIC EQUIPMENT
        . SOLID STATE DEVICES
        . . SEMICONDUCTOR DEVICES
        . . . TRANSISTORS
        . . . . FIELD EFFECT TRANSISTORS

CHARGE FLOW DEVICES-(CONT.)
        . . . . CHARGE FLOW DEVICES
RT   INTEGRATED CIRCUITS
        ∞SENSORS

CHARGE INJECTION DEVICES
UF   CID
GS   ELECTRONIC EQUIPMENT
        . SOLID STATE DEVICES
        . . SEMICONDUCTOR DEVICES
        . . . METAL OXIDE SEMICONDUCTORS
        . . . . CHARGE TRANSFER DEVICES
        . . . . . CHARGE INJECTION DEVICES
RT   CHARGE COUPLED DEVICES
        ELECTRO-OPTICS
        IMAGING TECHNIQUES
        STAR TRACKERS

CHARGE SEPARATION
USE   POLARIZATION (CHARGE SEPARATION)

CHARGE TRANSFER
SN   (EXCLUDES COLLISIONAL CHARGE
        EXCHANGE)
RT   CARRIER INJECTION
        CHARGED PARTICLES
        ELECTRON TRANSFER
        ION EXCHANGING
        IONIC REACTIONS
        MASS TRANSFER
        PHOTOCHEMICAL REACTIONS
        POLARIZATION (CHARGE SEPARATION)
        TRANSFERRING

CHARGE TRANSFER DEVICES
UF   CTD
GS   ELECTRONIC EQUIPMENT
        . SOLID STATE DEVICES
        . . SEMICONDUCTOR DEVICES
        . . . METAL OXIDE SEMICONDUCTORS
        . . . . CHARGE TRANSFER DEVICES
        . . . . . BUCKET BRIGADE DEVICES
        . . . . . CHARGE COUPLED DEVICES
        . . . . . CHARGE INJECTION DEVICES
        SEMICONDUCTORS (MATERIALS)
        . METAL OXIDE SEMICONDUCTORS
        . . CHARGE TRANSFER DEVICES
RT   ORGANIC CHARGE TRANSFER SALTS

CHARGED PARTICLES
GS   GASES
        . IONIZED GASES
        . . CHARGED PARTICLES
        . . . ARGON PLASMA
        . . . BOUNDARY LAYER PLASMAS
        . . . COLD PLASMAS
        . . . COLLISIONAL PLASMAS
        . . . . STRONGLY COUPLED PLASMAS
        . . . COLLISIONLESS PLASMAS
        . . . CONDUCTION ELECTRONS
        . . . DENSE PLASMAS
        . . . STRONGLY COUPLED PLASMAS
        . . . ELECTRON PLASMA
        . . . ELLIPTICAL PLASMAS
        . . . HELIUM PLASMA
        . . . HIGH TEMPERATURE PLASMAS
        . . . LASER PLASMAS
        . . . METALLIC PLASMAS
        . . . . CESIUM PLASMA
        . . . MICROPLASMAS
        . . . NITROGEN PLASMA
        . . . NUCLEI (NUCLEAR PHYSICS)
        . . . . EVEN-EVEN NUCLEI
        . . . . HEAVY NUCLEI
        . . . . HYPERNUCLEI
        . . . . ODD-EVEN NUCLEI
        . . . . ODD-ODD NUCLEI
        . . . RAREFIED PLASMAS
        . . . RELATIVISTIC PLASMAS
        . . . ROTATING PLASMAS
        . . . SOLAR WIND
        . . . STELLAR WINDS
        . . . THERMAL PLASMAS
        PARTICLES
        . CHARGED PARTICLES
        . . ANIONS
        . . ANTIPROTONS
        . . CATIONS
        . . . METAL IONS
        . . . FERRIC IONS
        . . . MANGANESE IONS
        . . ENERGETIC PARTICLES
        . . ELECTRONS
        . . . CONDUCTION ELECTRONS
        . . . HIGH ENERGY ELECTRONS

CHARGED PARTICLES-(CONT.)
        . . . HOT ELECTRONS
        . . . N ELECTRONS
        . . . NEGATRONS
        . . . PI-ELECTRONS
        . . . NUCLEI (NUCLEAR PHYSICS)
        . . . PLASMAS (PHYSICS)
        . . . ARGON PLASMA
        . . . BETA PARTICLES
        . . . BOUNDARY LAYER PLASMAS
        . . . COLD PLASMAS
        . . . COLLISIONAL PLASMAS
        . . . . STRONGLY COUPLED PLASMAS
        . . . COLLISIONLESS PLASMAS
        . . . COSMIC PLASMA
        . . . CYLINDRICAL PLASMAS
        . . . DENSE PLASMAS
        . . . . PLASMA FOCUS
        . . . . STRONGLY COUPLED PLASMAS
        . . . ELECTRON PLASMA
        . . . ELLIPTICAL PLASMAS
        . . . HELIUM PLASMA
        . . . HIGH TEMPERATURE PLASMAS
        . . . LASER PLASMAS
        . . . METALLIC PLASMAS
        . . . . CESIUM PLASMA
        . . . MICROPLASMAS
        . . . NITROGEN PLASMA
        . . . NONEQUILIBRIUM PLASMAS
        . . . NONUNIFORM PLASMAS
        . . . RAREFIED PLASMAS
        . . . RELATIVISTIC PLASMAS
        . . . ROTATING PLASMAS
        . . . SEMICONDUCTOR PLASMAS
        . . . SOLAR WIND
        . . . SPACE PLASMAS
        . . . SPHERICAL PLASMAS
        . . . STELLAR WINDS
        . . . THERMAL PLASMAS
        . . . TOROIDAL PLASMAS
        . . MAGNETICALLY TRAPPED PARTICLES
        . . . RADIATION BELTS
        . . . ARTIFICIAL RADIATION BELTS
        . . . . INNER RADIATION BELT
        . . . . OUTER RADIATION BELT
        . . . . PROTON BELTS
        . . NEGATIVE IONS
        . . PARTONS
        . . PLASMA CLOUDS
        . . PLASMA LAYERS
        . . . PLASMA SHEATHS
        . . PLASMA SLABS
        . . POSITRONS
        . . PROTONS
        . . . RECOIL PROTONS
        . . . SOLAR PROTONS
RT   ANTINEUTRINOS
        ANTIPARTICLES
        BOSONS
        CHARGE TRANSFER
        CORPUSCULAR RADIATION
        COULOMB COLLISIONS
        COULOMB POTENTIAL
        CYCLOTRON FREQUENCY
        CYCLOTRON RADIATION
        CYCLOTRON RESONANCE
        DEUTERON IRRADIATION
        ELEMENTARY PARTICLES
        ETA-MESONS
        GYROFREQUENCY
        HELIOS PROJECT
        HYPERONS
        ION CHARGE
        KAONS
        LEPTONS
        LORENTZ FORCE
        MESON-NUCLEON INTERACTIONS
        MESONS
        MUON SPIN ROTATION
        MUONS
        NEUTRAL SHEETS
        NEUTRONS
        NONADIABATIC THEORY
        ∞NUCLEI
        NUCLEON-NUCLEON INTERACTIONS
        NUCLEONS
        OMEGA-MESONS
        PARTICLE CHARGING
        PARTICLE PRECIPITATION
        PARTICLE TRAJECTORIES
        PIONS
        REISSNER-NORDSTROM SOLUTION
        RHO-MESONS
        SIGMA-MESONS
        SINGLE EVENT UPSETS
        TRAPPED PARTICLES

∞ **CHARGING**
SN   (USE OF A MORE SPECIFIC TERM IS
      RECOMMENDED--CONSULT THE TERMS
      LISTED BELOW)
RT   BATTERY CHARGERS
      CHARGE EFFICIENCY
      ELECTRIC CHARGE
      ELECTROSTATIC CHARGE
      EXPLOSIVE DEVICES
      EXPLOSIVES
      FILLING
      INJECTION
      MAGNETIC CHARGE DENSITY
      SCATHA SATELLITE

CHARM (PARTICLE PHYSICS)
RT   HADRONS
      LEPTONS
      PARTICLE INTERACTIONS
      PARTICLE THEORY
    ∞PHYSICS
      QUANTUM THEORY
      THEORETICAL PHYSICS

CHARON
GS   CELESTIAL BODIES
     . NATURAL SATELLITES
     . . **CHARON**
     SATELLITES
     . NATURAL SATELLITES
     . . **CHARON**
RT   CALLISTO
      DEIMOS
      EARTH-MOON SYSTEM
      EUROPA
      GALILEAN SATELLITES
      GANYMEDE
      IAPETUS
      IO
      PLANETARY ORBITS
      PLUTO (PLANET)
      SOLAR SYSTEM
      TITAN

CHARPY IMPACT TEST
GS   IMPACT TESTS
     . **CHARPY IMPACT TEST**
     NOTCH TESTS
     . **CHARPY IMPACT TEST**
RT   BRITTLENESS
      DROP TESTS
      HARDNESS
    ∞MATERIALS TESTS
      NOTCH SENSITIVITY

CHARRING
RT   ABLATION
      CARBONIZATION
      COMBUSTION
      DECOMPOSITION
      FIRE DAMAGE
      OXIDATION
      THERMAL ABSORPTION

CHARTS
GS   **CHARTS**
     . FLOW CHARTS
     . GRAPHS (CHARTS)
     . . BOND GRAPHS
     . . GOMPERTZ CURVES
     . . MOLLIER DIAGRAM
     . . PATTERSON MAP
     . METEOROLOGICAL CHARTS
     . NAUTICAL CHARTS
RT   BLOCK DIAGRAMS
      DIAGRAMS
      DISPLAY DEVICES
      DRAWINGS
      GRAPHIC ARTS
      MAPS
      NAVIGATION AIDS
      NOMOGRAPHS
    ∞PLOTS
      STATISTICAL ANALYSIS
      STATISTICAL TESTS
      VISUAL AIDS

CHASSIS
GS   FRAMES
     . **CHASSIS**
RT   AUTOMOBILES
      CARRIAGES
    ∞HEADERS
      STRUTS
      SUPPORTS

CHASSIS-(CONT.)
     UNDERCARRIAGES

CHEBYSHEV APPROXIMATION
GS   ANALYSIS (MATHEMATICS)
     . NUMERICAL ANALYSIS
     . . APPROXIMATION
     . . . **CHEBYSHEV APPROXIMATION**
RT   SERIES (MATHEMATICS)
      STATISTICAL ANALYSIS

CHECKOUT
SN   (SEQUENCE OF TESTS TO DETERMINE
      FUNCTIONAL READINESS OF
      EQUIPMENT)
UF   DEBUGGING
RT   AIRCRAFT MAINTENANCE
      CEFOAM CHECKOUT EQUIPMENT
      CERTIFICATION
      COLD FLOW TESTS
      COUNTDOWN
      FILE MAINTENANCE (COMPUTERS)
      INSPECTION
      MAINTENANCE
      PERFORMANCE TESTS
      PREFIRING TESTS
      PROGRAM VERIFICATION (COMPUTERS)
      SPACE VEHICLE CHECKOUT PROGRAM
      SPACECRAFT MAINTENANCE
    ∞TEST EQUIPMENT
    ∞TESTS

CHECKOUT EQUIPMENT
USE   TEST EQUIPMENT

CHELATE COMPOUNDS
USE   CHELATES

CHELATES
UF   CHELATE COMPOUNDS
RT   CHELATION
    ∞CHEMICAL COMPOUNDS
      ORGANOMETALLIC COMPOUNDS

CHELATION
RT   CHELATES
      CHEMICAL REACTIONS

CHEMICAL ANALYSIS
GS   CHEMICAL TESTS
     . **CHEMICAL ANALYSIS**
     . . ELECTROPHOTOMETRY
     . . GAS ANALYSIS
     . . . OZONOMETRY
     . . . VAN SLYKE METHOD
     . . GAS SPECTROSCOPY
     . . IODIMETRY
     . . KARL FISCHER REAGENT
     . . MICROANALYSIS
     . . NEUTRON ACTIVATION ANALYSIS
     . . PAPER CHROMATOGRAPHY
     . . POTENTIOMETRIC ANALYSIS
     . . QUALITATIVE ANALYSIS
     . . QUANTITATIVE ANALYSIS
     . . . KJELDAHL METHOD
     . . . VAN SLYKE METHOD
     . . SPECTROSCOPIC ANALYSIS
     . . URINALYSIS
     . . VOLUMETRIC ANALYSIS
RT   ALKALINITY
      ANALYTICAL CHEMISTRY
    ∞ANALYZING
      ASSAYING
      AUGER SPECTROSCOPY
    ∞CHEMISTRY
      CHROMATOGRAPHY
      COLORIMETRY
      COULOMETERS
      DENSITY MEASUREMENT
      DIFFRACTOMETERS
      ELECTRON PROBES
      FUEL TESTS
      GAS CHROMATOGRAPHY
      HYDROMETERS
      HYGROMETERS
      IDENTIFYING
      INFRARED SPECTROPHOTOMETERS
      INFRARED SPECTROSCOPY
      ION SELECTIVE ELECTRODES
      ISOTOPIC LABELING
      LIQUID CHROMATOGRAPHY
      MARS SURFACE SAMPLES
      MASS SPECTROMETERS
      MASS SPECTROSCOPY
    ∞MATERIALS TESTS

CHEMICAL ANALYSIS-(CONT.)
    ∞MEASUREMENT
      METALLICITY
      METHYLENE BLUE
      MOISTURE METERS
      MUTAGENS
      NEPHANALYSIS
      OPTICAL MEASUREMENT
      PARTICLE TRACKS
      PARTICULATE SAMPLING
      PHOTOMETRY
      PHYSICAL CHEMISTRY
      POLARIMETERS
      POLAROGRAPHY
      PSYCHROMETERS
      RADIOCHEMISTRY
      REAGENTS
      SAMPLING
      SPECTRAL SIGNATURES
      SPECTROMETERS
      SPECTROPHOTOMETERS
      SPECTROSCOPY
    ∞TESTS
      THERMOGRAVIMETRY
      TITRIMETERS
      X RAY ANALYSIS

CHEMICAL ATTACK
GS   **CHEMICAL ATTACK**
     . INTERGRANULAR CORROSION
     . TRANSGRANULAR CORROSION
RT   ∞ATTACK
      CORROSION
      CORROSION PREVENTION
      CORROSION RESISTANCE
      CORROSION TESTS
      DEGRADATION
      DISSOLVING
      IMPREGNATING
      OXIDATION
      PASSIVITY
      PITTING
      RUSTING
      SCALE (CORROSION)

CHEMICAL AUXILIARY POWER UNITS
GS   AUXILIARY POWER SOURCES
     . **CHEMICAL AUXILIARY POWER UNITS**
RT   ELECTRIC BATTERIES
      FUEL CELLS
      LEAD ACID BATTERIES
      MAGNESIUM CELLS

CHEMICAL BONDS
UF   MOLECULAR BONDS
GS   **CHEMICAL BONDS**
     . COVALENT BONDS
     . HYDROGEN BONDS
RT   AGGLUTINATION
      BONDING
      COUPLED MODES
      COVALENCE
      CRYSTAL LATTICES
      IONIC CRYSTALS
      LIGANDS
      MOLECULES
      MONATOMIC MOLECULES
      OCTETS
      POLYATOMIC MOLECULES
      POLYWATER
      SATURATION (CHEMISTRY)
      SWAN BANDS
      UNSATURATION (CHEMISTRY)
      VALENCE

CHEMICAL CLEANING
GS   **CHEMICAL CLEANING**
     . PICKLING (METALLURGY)
RT   CLEANING
      DESCALING
      DISSOLVING

CHEMICAL CLOUDS
GS   CLOUDS
     . CLOUDS (METEOROLOGY)
     . . ARTIFICIAL CLOUDS
     . . . **CHEMICAL CLOUDS**
     . . . . BARIUM ION CLOUDS
RT   CHEMICAL RELEASE MODULES
      PARTICLES
      PLASMA CLOUDS

CHEMICAL COMPOSITION
GS   COMPOSITION (PROPERTY)
     . **CHEMICAL COMPOSITION**

CHEMICAL COMPOSITION-*(CONT.)*
```
        . . CARBON DIOXIDE CONCENTRATION
        . . STELLAR COMPOSITION
  RT    ALKALINITY
        ATMOSPHERIC COMPOSITION
        ATOM CONCENTRATION
        BODY COMPOSITION (BIOLOGY)
        DISTRIBUTION (PROPERTY)
        GAS COMPOSITION
        IONOSPHERIC COMPOSITION
        LIGANDS
        METALLIC STARS
        METALLICITY
        PLANETARY STRUCTURE
        SPECTRAL SIGNATURES
```

∞ CHEMICAL COMPOUNDS
```
  SN    (USE OF A MORE SPECIFIC TERM IS
        RECOMMENDED--CONSULT THE TERMS
        LISTED BELOW)
  RT    ACETYL COMPOUNDS
        ACTINIDE SERIES COMPOUNDS
        ADDUCTS
        ALIPHATIC COMPOUNDS
     ∞  ALKALI METAL COMPOUNDS
     ∞  ALKALINE EARTH COMPOUNDS
        ALKYL COMPOUNDS
        ALLYL COMPOUNDS
        ALUMINUM COMPOUNDS
        AMMINES
        AMMONIUM COMPOUNDS
        ANTIMONY COMPOUNDS
     ∞  AROMATIC COMPOUNDS
        ARSENIC COMPOUNDS
        AZO COMPOUNDS
        BARIUM COMPOUNDS
        BERYLLIUM COMPOUNDS
        BISMUTH COMPOUNDS
        BORON COMPOUNDS
        BORON-EPOXY COMPOUNDS
        BROMINE COMPOUNDS
        CADMIUM COMPOUNDS
        CALCIUM COMPOUNDS
        CARBON COMPOUNDS
        CARBONYL COMPOUNDS
        CERIUM COMPOUNDS
        CESIUM COMPOUNDS
        CETYL COMPOUNDS
        CHELATES
     ∞  CHEMICALS
        CHLORINE COMPOUNDS
        CHROMIUM COMPOUNDS
        CLATHRATES
        COBALT COMPOUNDS
        COMPLEX COMPOUNDS
        COMPOUND A
     ∞  COMPOUNDS
        COPPER COMPOUNDS
        CURIUM COMPOUNDS
        CYANO COMPOUNDS
        CYCLIC COMPOUNDS
        DEUTERIUM COMPOUNDS
        DIALLYL COMPOUNDS
        DIBASIC COMPOUNDS
        DIBUTYL COMPOUNDS
        DIFLUORO COMPOUNDS
        DIPHENYL COMPOUNDS
        DOPA
        DYSPROSIUM COMPOUNDS
        EPOXY COMPOUNDS
        ERBIUM COMPOUNDS
        ETHYL COMPOUNDS
        ETHYLENE COMPOUNDS
        EUROPIUM COMPOUNDS
        FLUORINE COMPOUNDS
        FLUORINE ORGANIC COMPOUNDS
        FLUORO COMPOUNDS
        FURANS
        GALLIUM COMPOUNDS
        GERMANIUM COMPOUNDS
     ∞  GROUP 1B COMPOUNDS
     ∞  GROUP 2B COMPOUNDS
     ∞  GROUP 3A COMPOUNDS
     ∞  GROUP 3B COMPOUNDS
     ∞  GROUP 4A COMPOUNDS
     ∞  GROUP 4B COMPOUNDS
     ∞  GROUP 5A COMPOUNDS
     ∞  GROUP 5B COMPOUNDS
     ∞  GROUP 6A COMPOUNDS
     ∞  GROUP 6B COMPOUNDS
     ∞  GROUP 7B COMPOUNDS
     ∞  GROUP 8 COMPOUNDS
        HAFNIUM COMPOUNDS
        HALOCARBONS
        HALOGEN COMPOUNDS
        HETEROCYCLIC COMPOUNDS
```

CHEMICAL COMPOUNDS-*(CONT.)*
```
        HEXYL COMPOUNDS
        HYDRAZINIUM COMPOUNDS
        HYDRAZONIUM COMPOUNDS
        HYDROGEN COMPOUNDS
        HYDROXYL COMPOUNDS
        INDIUM COMPOUNDS
        INORGANIC COMPOUNDS
        INTERCALATION
        IODINE COMPOUNDS
        IRON COMPOUNDS
        ISOPROPYL COMPOUNDS
        LANTHANUM COMPOUNDS
        LEAD COMPOUNDS
        LEAD ORGANIC COMPOUNDS
        LITHIUM COMPOUNDS
        LUTETIUM COMPOUNDS
        MAGNESIUM COMPOUNDS
        MANGANESE COMPOUNDS
        MERCURY COMPOUNDS
     ∞  METAL COMPOUNDS
        METHOXY SYSTEMS
        METHYL COMPOUNDS
        MOLECULES
        MOLYBDENUM COMPOUNDS
        MONATOMIC MOLECULES
        NEODYMIUM COMPOUNDS
        NEPTUNIUM COMPOUNDS
        NICKEL COMPOUNDS
        NIOBIUM COMPOUNDS
        NITRO COMPOUNDS
        NITROGEN COMPOUNDS
        NITRONIUM COMPOUNDS
        NITROSO COMPOUNDS
        ORGANIC ALUMINUM COMPOUNDS
        ORGANIC BORON COMPOUNDS
        ORGANIC COMPOUNDS
        ORGANIC GERMANIUM COMPOUNDS
        ORGANIC LITHIUM COMPOUNDS
        ORGANIC PHOSPHORUS COMPOUNDS
        ORGANIC SEMICONDUCTORS
        ORGANIC SILICON COMPOUNDS
        ORGANIC SULFUR COMPOUNDS
        ORGANIC TIN COMPOUNDS
        ORGANOMETALLIC COMPOUNDS
        OSMIUM COMPOUNDS
     ∞  OXYGEN COMPOUNDS
        OXYNITRIDES
        PALLADIUM COMPOUNDS
        PHOSGENE
        PHOSPHONIUM COMPOUNDS
        PHOSPHORUS COMPOUNDS
        PLATINUM COMPOUNDS
        PLUTONIUM COMPOUNDS
        POLONIUM COMPOUNDS
        POLYATOMIC MOLECULES
        POLYNUCLEAR ORGANIC COMPOUNDS
        POLYQUINOXALINES
        POTASSIUM COMPOUNDS
        PROPYL COMPOUNDS
        PROTACTINIUM COMPOUNDS
        RARE EARTH COMPOUNDS
     ∞  RARE GAS COMPOUNDS
        REFRACTORY MATERIALS
        RHENIUM COMPOUNDS
        RHODIUM COMPOUNDS
        RUBIDIUM COMPOUNDS
        RUTHENIUM COMPOUNDS
        SAMARIUM COMPOUNDS
        SCANDIUM COMPOUNDS
        SELENIUM COMPOUNDS
        SILICON COMPOUNDS
        SILVER COMPOUNDS
        SODIUM COMPOUNDS
        STRONTIUM COMPOUNDS
        STYPHNATES
        SULFUR COMPOUNDS
        TANTALUM COMPOUNDS
        TECHNETIUM COMPOUNDS
        TELLURIUM COMPOUNDS
        TETRAHYDROFURAN
        THIOLS
        THORIUM COMPOUNDS
        THULIUM COMPOUNDS
        TIN COMPOUNDS
        TITANIUM COMPOUNDS
        TRIETHYL COMPOUNDS
        TRIMETHYL COMPOUNDS
        TRINITRO COMPOUNDS
        TROPYL COMPOUNDS
        TUNGSTEN COMPOUNDS
        URANIUM COMPOUNDS
        VANADIUM COMPOUNDS
        VANADYL COMPOUNDS
        WISWESSER NOTATIONS
        XENON COMPOUNDS
```

CHEMICAL COMPOUNDS-*(CONT.)*
```
        YTTERBIUM COMPOUNDS
        YTTRIUM COMPOUNDS
        ZINC COMPOUNDS
        ZIRCONIUM COMPOUNDS
```

CHEMICAL DEFENSE
```
  RT    CIVIL DEFENSE
        CLOTHING
        DRUGS
        FIRST AID
        INJURIES
        MASKS
        NEUROLOGY
        PHYSIOLOGICAL FACTORS
        PROTECTIVE CLOTHING
        SAFETY DEVICES
        WARFARE
```

CHEMICAL EFFECTS
```
  GS    STERILIZATION EFFECTS
        . CHEMICAL EFFECTS
  RT    BIOLOGICAL EFFECTS
     ∞  EFFECTS
        TEMPERATURE EFFECTS
```

CHEMICAL ELEMENTS
```
  GS    CHEMICAL ELEMENTS
        . ACTINIDE SERIES
        . . ACTINIUM
        . . RADIUM
        . . . RADIUM ISOTOPES
        . . . . RADIUM 226
        . . THORIUM
        . . . THORIUM ISOTOPES
        . . TRANSURANIUM ELEMENTS
        . . . AMERICIUM
        . . . . AMERICIUM ISOTOPES
        . . . . . AMERICIUM 241
        . . . BERKELIUM
        . . . CALIFORNIUM
        . . . . CALIFORNIUM ISOTOPES
        . . . CURIUM
        . . . . CURIUM ISOTOPES
        . . . . . CURIUM 242
        . . . . . CURIUM 244
        . . . EINSTEINIUM
        . . . FERMIUM
        . . . LAWRENCIUM
        . . . MENDELEVIUM
        . . . NEPTUNIUM
        . . . . NEPTUNIUM ISOTOPES
        . . . NOBELIUM
        . . . PLUTONIUM
        . . . . PLUTONIUM ISOTOPES
        . . . . . PLUTONIUM 238
        . . . . . PLUTONIUM 239
        . . . . . PLUTONIUM 240
        . . . . . PLUTONIUM 241
        . . . . . PLUTONIUM 244
        . . . SERGENIUM
        . . URANIUM
        . . . URANIUM ISOTOPES
        . . . . URANIUM 232
        . . . . URANIUM 233
        . . . . URANIUM 234
        . . . . URANIUM 235
        . . . . URANIUM 238
        . . . URANIUM PLASMAS
        . ALKALI METALS
        . . CESIUM
        . . . CESIUM ISOTOPES
        . . . . CESIUM 133
        . . . . CESIUM 134
        . . . . CESIUM 137
        . . . . CESIUM 144
        . . . CESIUM VAPOR
        . . FRANCIUM
        . . LITHIUM
        . . . LIQUID LITHIUM
        . . . LITHIUM ISOTOPES
        . . POTASSIUM
        . . . LIQUID POTASSIUM
        . . . POTASSIUM ISOTOPES
        . . . . POTASSIUM 38
        . . . . POTASSIUM 39
        . . . . POTASSIUM 40
        . . RUBIDIUM
        . . . RUBIDIUM ISOTOPES
        . . . . RUBIDIUM 86
        . . SODIUM
        . . . LIQUID SODIUM
        . . . SODIUM ISOTOPES
        . . . . SODIUM 22
        . . . . SODIUM 24
```

CHEMICAL ELEMENTS-(CONT.)

. . . SODIUM VAPOR
. . ALKALINE EARTH METALS
. . BARIUM ISOTOPES
. . ALUMINUM
. . . ALUMINUM ISOTOPES
. . . . ALUMINUM 26
. . . . ALUMINUM 27
. . POWDERED ALUMINUM
. . SINTERED ALUMINUM POWDER
. BARIUM
. . BARIUM ISOTOPES
. BERYLLIUM
. . BERYLLIUM ISOTOPES
. . . BERYLLIUM 7
. . . BERYLLIUM 9
. . . BERYLLIUM 10
. BISMUTH
. . BISMUTH ISOTOPES
. CADMIUM
. . CADMIUM ISOTOPES
. CALCIUM
. . CALCIUM ISOTOPES
. CARBON
. . CARBON ISOTOPES
. . . CARBON 12
. . . CARBON 13
. . . CARBON 14
. . . CHARCOAL
. . . . ACTIVATED CARBON
. COBALT
. . COBALT ISOTOPES
. . . COBALT 58
. . . COBALT 60
. COPPER
. . COPPER ISOTOPES
. ELEMENT 104
. ELEMENT 105
. GALLIUM
. . GALLIUM ISOTOPES
. GOLD
. . GOLD ISOTOPES
. . . GOLD 198
. HAFNIUM
. . HAFNIUM ISOTOPES
. HALOGENS
. . ASTATINE
. . BROMINE
. . . BROMINE ISOTOPES
. . CHLORINE
. . FLUORINE
. . . FLUORINE ISOTOPES
. . IODINE
. . . IODINE ISOTOPES
. . . . IODINE 125
. . . . IODINE 131
. . . . IODINE 132
. HYDROGEN
. . HYDROGEN ATOMS
. . HYDROGEN IONS
. . HYDROGEN ISOTOPES
. . . DEUTERIUM
. . . HYDROGEN 4
. . . METALLIC HYDROGEN
. . . TRITIUM
. . HYDROGEN PLASMA
. . . DEUTERIUM PLASMA
. . LIQUID HYDROGEN
. . ORTHO HYDROGEN
. . PARA HYDROGEN
. INDIUM
. IRON
. . IRON ISOTOPES
. . . IRON 57
. . . IRON 58
. . . IRON 59
. LEAD (METAL)
. . LEAD ISOTOPES
. LIGHT ELEMENTS
. MAGNESIUM
. . MAGNESIUM ISOTOPES
. MANGANESE
. . MANGANESE ISOTOPES
. MERCURY (METAL)
. . MERCURY ISOTOPES
. . MERCURY VAPOR
. METALLOIDS
. . ANTIMONY
. . . ANTIMONY ISOTOPES
. . ARSENIC
. . . ARSENIC ISOTOPES
. . BORON
. . . BORON ISOTOPES
. . . . BORON 10
. . GERMANIUM
. . . GERMANIUM ISOTOPES

CHEMICAL ELEMENTS-(CONT.)

. . POLONIUM
. . . POLONIUM ISOTOPES
. . . . POLONIUM 208
. . . . POLONIUM 209
. . . . POLONIUM 210
. . SILICON
. . . SILICON ISOTOPES
. . TELLURIUM
. . . TELLURIUM ISOTOPES
. NICKEL
. . NICKEL ISOTOPES
. NITROGEN
. . LIQUID NITROGEN
. . NITROGEN ATOMS
. . NITROGEN IONS
. . NITROGEN ISOTOPES
. . . NITROGEN 15
. . . NITROGEN 16
. . SOLID NITROGEN
. NUCLIDES
. . ISOTOPES
. . . ALUMINUM ISOTOPES
. . . . ALUMINUM 26
. . . . ALUMINUM 27
. . . ANTIMONY ISOTOPES
. . . ARGON ISOTOPES
. . . BARIUM ISOTOPES
. . . BERYLLIUM ISOTOPES
. . . . BERYLLIUM 7
. . . . BERYLLIUM 9
. . . . BERYLLIUM 10
. . . BISMUTH ISOTOPES
. . . BORON ISOTOPES
. . . . BORON 10
. . . BROMINE ISOTOPES
. . . CADMIUM ISOTOPES
. . . CALCIUM ISOTOPES
. . . CARBON ISOTOPES
. . . . CARBON 12
. . . . CARBON 13
. . . . CARBON 14
. . . CERIUM ISOTOPES
. . . . CERIUM 137
. . . . CERIUM 144
. . . CESIUM ISOTOPES
. . . . CESIUM 133
. . . . CESIUM 134
. . . . CESIUM 137
. . . . CESIUM 144
. . . CESIUM VAPOR
. . . CHROMIUM ISOTOPES
. . . COBALT ISOTOPES
. . . . COBALT 58
. . . . COBALT 60
. . . DYSPROSIUM ISOTOPES
. . . ERBIUM ISOTOPES
. . . EUROPIUM ISOTOPES
. . . FLUORINE ISOTOPES
. . . GADOLINIUM ISOTOPES
. . . GALLIUM ISOTOPES
. . . GERMANIUM ISOTOPES
. . . HAFNIUM ISOTOPES
. . . HELIUM ISOTOPES
. . . HOLMIUM ISOTOPES
. . . HYDROGEN ISOTOPES
. . . . DEUTERIUM
. . . . HYDROGEN 4
. . . . TRITIUM
. . . IODINE ISOTOPES
. . . . IODINE 125
. . . . IODINE 131
. . . . IODINE 132
. . . IRIDIUM ISOTOPES
. . . IRON ISOTOPES
. . . . IRON 57
. . . . IRON 58
. . . . IRON 59
. . . KRYPTON ISOTOPES
. . . . KRYPTON 85
. . . LANTHANUM ISOTOPES
. . . LEAD ISOTOPES
. . . LITHIUM ISOTOPES
. . . LUTETIUM
. . . . LUTETIUM ISOTOPES
. . . MANGANESE ISOTOPES
. . . MERCURY ISOTOPES
. . . NEODYMIUM ISOTOPES
. . . NEON ISOTOPES
. . . NICKEL ISOTOPES
. . . NIOBIUM ISOTOPES
. . . . NIOBIUM 95
. . . NITROGEN ISOTOPES
. . . . NITROGEN 15
. . . . NITROGEN 16
. . . OXYGEN ISOTOPES

CHEMICAL ELEMENTS-(CONT.)

. . . . OXYGEN 18
. . . PHOSPHORUS ISOTOPES
. . . . PHOSPHORUS 32
. . . PLATINUM ISOTOPES
. . . POLONIUM ISOTOPES
. . . . POLONIUM 208
. . . . POLONIUM 209
. . . . POLONIUM 210
. . . POTASSIUM ISOTOPES
. . . . POTASSIUM 38
. . . . POTASSIUM 39
. . . . POTASSIUM 40
. . . PRASEODYMIUM ISOTOPES
. . . PROMETHIUM ISOTOPES
. . . PROTACTINIUM ISOTOPES
. . . RADIOACTIVE ISOTOPES
. . . . ARSENIC ISOTOPES
. . . . ASTATINE ISOTOPES
. . . . BERYLLIUM 7
. . . . BERYLLIUM 9
. . . . BERYLLIUM 10
. . . . CARBON 14
. . . . CERIUM 137
. . . . CERIUM 144
. . . . CESIUM 134
. . . . CESIUM 137
. . . . CESIUM 144
. . . . COBALT 58
. . . . COBALT 60
. . . . GOLD ISOTOPES
. . . . . GOLD 198
. . . . INDIUM ISOTOPES
. . . . IODINE 125
. . . . IODINE 131
. . . . IODINE 132
. . . . IRON 59
. . . . KRYPTON 85
. . . . NIOBIUM 95
. . . . NITROGEN 16
. . . . PHOSPHORUS 32
. . . . POLONIUM 208
. . . . POLONIUM 209
. . . . POLONIUM 210
. . . . POTASSIUM 38
. . . . POTASSIUM 40
. . . . RUBIDIUM 86
. . . . SODIUM 22
. . . . SODIUM 24
. . . . STRONTIUM 85
. . . . STRONTIUM 88
. . . . STRONTIUM 89
. . . . STRONTIUM 90
. . . . TRANSURANIUM ELEMENTS
. . . . . AMERICIUM
. . . . . . AMERICIUM ISOTOPES
. . . . . BERKELIUM
. . . . . CALIFORNIUM
. . . . . . CALIFORNIUM ISOTOPES
. . . . . CURIUM
. . . . . . CURIUM ISOTOPES
. . . . . EINSTEINIUM
. . . . . FERMIUM
. . . . . LAWRENCIUM
. . . . . MENDELEVIUM
. . . . . NEPTUNIUM
. . . . . . NEPTUNIUM ISOTOPES
. . . . . NOBELIUM
. . . . . PLUTONIUM
. . . . . . PLUTONIUM ISOTOPES
. . . . . SERGENIUM
. . . . TRITIUM
. . . . URANIUM 232
. . . . URANIUM 233
. . . . URANIUM 238
. . . . XENON 133
. . . . XENON 135
. . . . ZIRCONIUM 95
. . . RADIUM ISOTOPES
. . . . RADIUM 226
. . . RADON ISOTOPES
. . . RHODIUM ISOTOPES
. . . RUBIDIUM ISOTOPES
. . . . RUBIDIUM 86
. . . RUTHENIUM ISOTOPES
. . . SCANDIUM ISOTOPES
. . . SILVER ISOTOPES
. . . SODIUM ISOTOPES
. . . . SODIUM 22
. . . . SODIUM 24
. . . STRONTIUM ISOTOPES
. . . . STRONTIUM 85
. . . . STRONTIUM 87
. . . . STRONTIUM 89
. . . . STRONTIUM 90
. . . TANTALUM ISOTOPES

CHEMICAL ELEMENTS-*(CONT.)*
- - - TELLURIUM
- - - - TELLURIUM ISOTOPES
- - - TERBIUM ISOTOPES
- - - THORIUM ISOTOPES
- - - THULIUM ISOTOPES
- - - TIN ISOTOPES
- - - TITANIUM ISOTOPES
- - - URANIUM ISOTOPES
- - - - URANIUM 232
- - - - URANIUM 233
- - - - URANIUM 234
- - - - URANIUM 235
- - - - URANIUM 238
- - - VANADIUM ISOTOPES
- - - XENON ISOTOPES
- - - - XENON 129
- - - - XENON 133
- - - - XENON 135
- - - YTTRIUM ISOTOPES
- - - ZINC ISOTOPES
- - - ZIRCONIUM ISOTOPES
- - - - ZIRCONIUM 95
- OXYGEN
- - OXYGEN ATOMS
- - OXYGEN IONS
- - OXYGEN ISOTOPES
- - - OXYGEN 17
- - - OXYGEN 18
- - OXYGEN PLASMA
- PALLADIUM
- PHOSPHORUS
- - PHOSPHORUS ISOTOPES
- - - PHOSPHORUS 32
- PLATINUM
- - PLATINUM BLACK
- - PLATINUM ISOTOPES
- PROTACTINIUM
- - PROTACTINIUM ISOTOPES
- RARE EARTH ELEMENTS
- - CERIUM
- - - CERIUM ISOTOPES
- - - - CERIUM 137
- - - - CERIUM 144
- - DYSPROSIUM
- - - DYSPROSIUM ISOTOPES
- - ERBIUM
- - - ERBIUM ISOTOPES
- - EUROPIUM
- - - EUROPIUM ISOTOPES
- - GADOLINIUM
- - - GADOLINIUM ISOTOPES
- - HOLMIUM
- - - HOLMIUM ISOTOPES
- - LANTHANUM
- - - LANTHANUM ISOTOPES
- - LUTETIUM
- - - LUTETIUM ISOTOPES
- - NEODYMIUM
- - - NEODYMIUM ISOTOPES
- - PRASEODYMIUM
- - - PRASEODYMIUM ISOTOPES
- - PROMETHIUM
- - - PROMETHIUM ISOTOPES
- - SAMARIUM
- - - SAMARIUM ISOTOPES
- - SCANDIUM
- - - SCANDIUM ISOTOPES
- - TERBIUM
- - - TERBIUM ISOTOPES
- - THULIUM
- - - THULIUM ISOTOPES
- - YTTERBIUM
- - - YTTERBIUM ISOTOPES
- - YTTRIUM
- - - YTTRIUM ISOTOPES
- RARE GASES
- - ARGON
- - - ARGON ISOTOPES
- - HELIUM
- - - HELIUM ATOMS
- - - HELIUM FILM
- - - HELIUM ISOTOPES
- - - LIQUID HELIUM
- - - - LIQUID HELIUM 2
- - KRYPTON
- - NEON
- - - LIQUID NEON
- - - NEON ISOTOPES
- - RADON
- - - RADON ISOTOPES
- - XENON
- - - XENON ISOTOPES
- - - - XENON 129
- - - - XENON 133
- - - - XENON 135

CHEMICAL ELEMENTS-*(CONT.)*
- REFRACTORY METALS
- - CHROMIUM
- - - CHROMIUM ISOTOPES
- - IRIDIUM
- - - IRIDIUM ISOTOPES
- - MOLYBDENUM
- - NIOBIUM
- - - NIOBIUM ISOTOPES
- - - - NIOBIUM 95
- - OSMIUM
- - - OSMIUM ISOTOPES
- - TANTALUM
- - - TANTALUM ISOTOPES
- - TUNGSTEN
- - TUNGSTEN ISOTOPES
- RHODIUM
- - RHODIUM ISOTOPES
- RUTHENIUM
- - RUTHENIUM ISOTOPES
- SELENIUM
- SILVER
- - SILVER ISOTOPES
- STRONTIUM
- - STRONTIUM ISOTOPES
- - - STRONTIUM 85
- - - STRONTIUM 87
- - - STRONTIUM 89
- - - STRONTIUM 90
- SULFUR
- - SULFUR ISOTOPES
- TECHNETIUM
- THALLIUM
- - THALLIUM ISOTOPES
- TIN
- - TIN ISOTOPES
- TITANIUM
- - TITANIUM ISOTOPES
- TRACE ELEMENTS
- VANADIUM
- - VANADIUM ISOTOPES
- ZINC
- - ZINC ISOTOPES
- - ZIRCONIUM
- - - ZIRCONIUM ISOTOPES
- - - - ZIRCONIUM 95

RT ATOMS
    ∞CHEMICALS
    ∞ELEMENTS
    FERROUS METALS
    HEAVY ELEMENTS
    IONS
    ISOTOPIC ENRICHMENT
    LIGHT IONS
    METALS
    NONFERROUS METALS
    NUCLEAR ISOBARS
    TRACE CONTAMINANTS

CHEMICAL ENERGY
GS **CHEMICAL ENERGY**
    - ENERGY OF FORMATION
RT ∞ENERGY
    FREE ENERGY
    INTERNAL ENERGY
    KINETIC ENERGY
    ∞LEVEL
    MOLECULAR ENERGY LEVELS
    ∞NUCLEAR ENERGY
    POTENTIAL ENERGY

CHEMICAL ENGINEERING
RT AEROTHERMOCHEMISTRY
    ∞CHEMISTRY
    CRACKING (CHEMICAL ENGINEERING)
    DIFFUSION
    ∞ENGINEERING
    FLUID FLOW
    FURNACES
    HEAT TRANSFER
    MATERIALS HANDLING
    ∞OPERATIONS
    THERMOCHEMISTRY

CHEMICAL EQUILIBRIUM
UF CHAPMAN-JOUGET FLAME
    CHEMICAL SHIFT
GS **CHEMICAL EQUILIBRIUM**
    - ACID BASE EQUILIBRIUM
RT ASSOCIATION REACTIONS
    BUFFERS (CHEMISTRY)
    DISSOCIATION
    ∞EQUILIBRIUM
    HEAT OF DISSOCIATION
    PHASE RULE

CHEMICAL EQUILIBRIUM-*(CONT.)*
    REACTION KINETICS
    THERMODYNAMIC EQUILIBRIUM

CHEMICAL EVOLUTION
GS EVOLUTION (DEVELOPMENT)
    - **CHEMICAL EVOLUTION**
RT ABIOGENESIS
    BIOLOGICAL EVOLUTION
    ∞EVOLUTION
    EXOBIOLOGY
    LIFE SCIENCES
    ORGANIC COMPOUNDS
    PROTEIN SYNTHESIS

CHEMICAL EXPLOSIONS
GS EXPLOSIONS
    - **CHEMICAL EXPLOSIONS**
    - - GAS EXPLOSIONS
RT AERIAL EXPLOSIONS
    COMBUSTION
    DETONABLE GAS MIXTURES
    DETONATION
    EXPLOSIVES
    FLAMMABLE GASES
    UNDERGROUND EXPLOSIONS
    UNDERWATER EXPLOSIONS

CHEMICAL EXTINGUISHERS
USE FIRE EXTINGUISHERS

CHEMICAL FRACTIONATION
GS FRACTIONATION
    - **CHEMICAL FRACTIONATION**
RT DISTILLATION
    REFINING
    ∞SEPARATION

CHEMICAL FUELS
GS FUELS
    - **CHEMICAL FUELS**
    - - ENDOTHERMIC FUELS
    - - HIGH ENERGY FUELS
    - - HYDROCARBON FUELS
    - - - DIESEL FUELS
    - - - GASOLINE
    - - - JET ENGINE FUELS
    - - - - JP-4 JET FUEL
    - - - - JP-5 JET FUEL
    - - - - JP-6 JET FUEL
    - - - - JP-8 JET FUEL
    - - - RP-1 ROCKET PROPELLANTS
    - - - SYNTHANE
    - - LIQUID FUELS
    - - - AIRCRAFT FUELS
    - - - AUTOMOBILE FUELS
    - - - DIESEL FUELS
    - - - GASOLINE
    - - - HYDROGEN FUELS
    - - - JET ENGINE FUELS
    - - - - JP-4 JET FUEL
    - - - - JP-5 JET FUEL
    - - - - JP-6 JET FUEL
    - - - - JP-8 JET FUEL
    - - - KEROSENE
    - - METAL FUELS
    - - SYNTHETIC FUELS
    - - - GASOHOL (FUEL)
    - - - SYNTHANE
RT CLEAN FUELS
    EXPLOSIVES
    FUEL PRODUCTION
    GELLED PROPELLANTS
    GELLED ROCKET PROPELLANTS
    HYBRID PROPELLANTS
    MONOPROPELLANTS
    PLASTIC PROPELLANTS
    PYROTECHNICS
    SOLID PROPELLANTS

CHEMICAL INDICATORS
RT ∞INDICATORS
    METHYLENE BLUE
    PHLOROGLUCINOL

CHEMICAL KINETICS
USE REACTION KINETICS

CHEMICAL LASERS
GS STIMULATED EMISSION DEVICES
    - LASERS
    - - **CHEMICAL LASERS**
    - - - HCL LASERS
RT ARGON LASERS
    CARBON DIOXIDE LASERS

CHEMICAL LASERS-(CONT.)
    CARBON LASERS
    CARBON MONOXIDE LASERS
    GAS LASERS
    HCN LASERS
    HF LASERS
    INFRARED LASERS
    LIQUID LASERS
    ORGANIC LASERS
    Q SWITCHED LASERS
    TEA LASERS
    TUBE LASERS

CHEMICAL MACHINING
    UF   CHEMICAL MILLING
    GS   MACHINING
        . CHEMICAL MACHINING
        . . ELECTROCHEMICAL MACHINING
    RT   MILLING (MACHINING)

CHEMICAL MILLING
    USE   CHEMICAL MACHINING

CHEMICAL PROPERTIES
    GS   CHEMICAL PROPERTIES
        . ACIDITY
        . HEAT OF SOLUTION
        . SALINITY
        . THERMOCHEMICAL PROPERTIES
        . . HEAT OF COMBUSTION
        . . HEAT OF FORMATION
        . . HEAT OF FUSION
        . . HEAT OF VAPORIZATION
    RT   ADSORPTIVITY
        ∞ HIGH RESISTANCE
        HYGROSCOPICITY
        ∞ LOW RESISTANCE
        MOISTURE CONTENT
        PASSIVITY
        ∞ PHYSICAL PROPERTIES
        PROPELLANT PROPERTIES
        ∞ PROPERTIES
        ∞ RESISTANCE
        THERMODYNAMIC PROPERTIES
        TOXICITY
        TOXICITY AND SAFETY HAZARD

CHEMICAL PROPULSION
    UF   CHEMONUCLEAR PROPULSION
    GS   PROPULSION
        . CHEMICAL PROPULSION
        . . HYBRID PROPULSION
    RT   JET PROPULSION
        MARINE PROPULSION
        SPACECRAFT PROPULSION
        UNDERWATER PROPULSION

CHEMICAL REACTION CONTROL
    RT   AGITATION
        ∞ CONTROL
        ∞ REACTION CONTROL
        TEMPERATURE CONTROL

CHEMICAL REACTIONS
    UF   FLAME INTERACTION
    GS   CHEMICAL REACTIONS
        . ACYLATION
        . . ACETYLATION
        . ALKYLATION
        . AMMONOLYSIS
        . ASSOCIATION REACTIONS
        . ATOMIC RECOMBINATION
        . . OXYGEN RECOMBINATION
        . CARBONIZATION
        . CARBOXYLATION
        . COPOLYMERIZATION
        . CRACKING (CHEMICAL ENGINEERING)
        . . PYROLYSIS
        . DECARBONATION
        . DECARBOXYLATION
        . DEFLUORINATION
        . DEHYDROGENATION
        . DEIONIZATION
        . DENITROGENATION
        . DEPOLYMERIZATION
        . DESULFURIZING
        . DIELS-ALDER REACTIONS
        . ENDOTHERMIC REACTIONS
        . EPOXIDATION
        . EXOTHERMIC REACTIONS
        . FERMENTATION
        . FRIEDEL-CRAFT REACTION
        . GLYCOLYSIS
        . GRIGNARD REACTIONS
        . HALOGENATION

CHEMICAL REACTIONS-(CONT.)
        . . BROMINATION
        . . CHLORINATION
        . . FLUORINATION
        . HYDROBORATION
        . HYDROGENOLYSIS
        . HYDROLYSIS
        . ION RECOMBINATION
        . METAL-WATER REACTIONS
        . METHANATION
        . METHYLATION
        . MICHAEL REACTION
        . NITRATION
        . NITRIDING
        . NITROGENATION
        . NITROLYSIS
        . OXIDATION
        . . ELECTROCHEMICAL OXIDATION
        . . PHOTOOXIDATION
        . . RUSTING
        . OXIDATION-REDUCTION REACTIONS
        . OXYGENATION
        . PHOSPHORYLATION
        . PHOTOCHEMICAL REACTIONS
        . . PHOTOCHROMISM
        . . PHOTODECOMPOSITION
        . . PHOTOLYSIS
        . . . RADIOLYSIS
        . . PHOTOSYNTHESIS
        . PYROHYDROLYSIS
        . REDUCTION (CHEMISTRY)
        . . DEOXIDIZING
        . . HYDROGENATION
        . SABATIER REACTION
        . SULFATION
        . SULFIDATION
        . THERMAL DECOMPOSITION
        . . PYROLYSIS
        . THERMAL DISSOCIATION
        . TITRATION
    RT   BIOSYNTHESIS
        CHELATION
        CORROSION
        GAS-METAL INTERACTIONS
        HYDRATION
        INTERSTELLAR CHEMISTRY
        ∞ OPERATIONS
        PARTICLE INTERACTIONS
        PLASMA JET SYNTHESIS
        POLYMERIZATION
        RADIOCHEMICAL SEPARATION
        ∞ REACTION
        REACTION BONDING
        REACTION KINETICS
        REACTIVITY
        REAGENTS
        SODALITE
        SOLVATION
        STOICHIOMETRY
        SURFACE REACTIONS
        ∞ SYNTHESIS
        SYNTHESIS (CHEMISTRY)
        SYNTHETIC FUELS
        THERMOCHEMISTRY

CHEMICAL REACTORS
    RT   AUTOCLAVES
        BEDS (PROCESS ENGINEERING)
        BURNERS
        COLUMNS (PROCESS ENGINEERING)
        CONTACTORS
        CONTRACTORS
        FLUIDIZED BED PROCESSORS
        FURNACES
        GAS GENERATORS
        ∞ GAS REACTORS
        REACTOR DESIGN
        REACTOR MATERIALS
        REACTOR SAFETY
        ∞ REACTORS
        SYNTHESIZERS
        TANKS (CONTAINERS)
        WATER COOLED REACTORS

CHEMICAL RELAXATION
    USE   MOLECULAR RELAXATION

CHEMICAL RELEASE MODULES
    GS   MODULES
        . CHEMICAL RELEASE MODULES
    RT   CHEMICAL CLOUDS
        DISPERSING

CHEMICAL SHIFT
    USE   CHEMICAL EQUILIBRIUM

CHEMICAL STERILIZATION
    GS   STERILIZATION
        . CHEMICAL STERILIZATION
    RT   ANTISEPTICS
        BACTERICIDES
        PURIFICATION
        SEWAGE TREATMENT
        SPACECRAFT STERILIZATION

CHEMICAL TESTS
    GS   CHEMICAL TESTS
        . CHEMICAL ANALYSIS
        . . ELECTROPHOTOMETRY
        . . GAS ANALYSIS
        . . . OZONOMETRY
        . . VAN SLYKE METHOD
        . . GAS SPECTROSCOPY
        . . IODIMETRY
        . . KARL FISCHER REAGENT
        . . MICROANALYSIS
        . . NEUTRON ACTIVATION ANALYSIS
        . . PAPER CHROMATOGRAPHY
        . . POTENTIOMETRIC ANALYSIS
        . . QUALITATIVE ANALYSIS
        . . QUANTITATIVE ANALYSIS
        . . . KJELDAHL METHOD
        . . . VAN SLYKE METHOD
        . . SPECTROSCOPIC ANALYSIS
        . . URINALYSIS
        . . VOLUMETRIC ANALYSIS
        . SALT SPRAY TESTS
    RT   CORROSION RESISTANCE
        HIGH TEMPERATURE TESTS
        INSPECTION
        LIQUID CHROMATOGRAPHY
        LOW TEMPERATURE TESTS
        NONDESTRUCTIVE TESTS
        QUALITY CONTROL
        SAMPLING
        STAINING
        ∞ TESTS

CHEMICAL WARFARE
    GS   WARFARE
        . CHEMICAL WARFARE
    RT   BIOCHEMISTRY
        PHYSIOLOGICAL FACTORS

∞ CHEMICALS
    SN   (USE OF A MORE SPECIFIC TERM IS
        RECOMMENDED--CONSULT THE TERMS
        LISTED BELOW)
    RT   ∞ CHEMICAL COMPOUNDS
        CHEMICAL ELEMENTS

CHEMILUMINESCENCE
    GS   DECAY
        . EMISSION
        . . LIGHT EMISSION
        . . . LUMINESCENCE
        . . . . CHEMILUMINESCENCE
    RT   AIRGLOW
        PHOSPHORESCENCE

CHEMISORPTION
    GS   SORPTION
        . ADSORPTION
        . . CHEMISORPTION
    RT   ADSORPTIVITY
        GAS-METAL INTERACTIONS
        HYDROGEN EMBRITTLEMENT
        MASKING

∞ CHEMISTRY
    SN   (USE OF A MORE SPECIFIC TERM IS
        RECOMMENDED--CONSULT THE TERMS
        LISTED BELOW)
    RT   AEROTHERMOCHEMISTRY
        AEROTHERMODYNAMICS
        ANALYTICAL CHEMISTRY
        ATMOSPHERIC CHEMISTRY
        BIOCHEMISTRY
        BIOGEOCHEMISTRY
        CHEMICAL ANALYSIS
        CHEMICAL ENGINEERING
        CHEMOTHERAPY
        COMPUTATIONAL CHEMISTRY
        CRYOCHEMISTRY
        ELECTROCHEMISTRY
        ENVIRONMENTAL CHEMISTRY
        GEOCHEMISTRY
        HYDROXYL RADICALS
        INORGANIC CHEMISTRY
        INTERSTELLAR CHEMISTRY
        MARINE CHEMISTRY

CHEMISTRY-(CONT.)
    NUCLEAR CHEMISTRY
    ORGANIC CHEMISTRY
    PHOTOELECTROCHEMISTRY
    PHYSICAL CHEMISTRY
    ∞PHYSICAL SCIENCES
    PHYSICS AND CHEMISTRY EXPERIMENT
      IN SPACE
    PHYSIOCHEMISTRY
    PLASMA CHEMISTRY
    POLYMER CHEMISTRY
    PRECIPITATION (CHEMISTRY)
    PROPELLANT CHEMISTRY
    QUANTUM CHEMISTRY
    RADIATION CHEMISTRY
    RADIOCHEMISTRY
    REDUCTION (CHEMISTRY)
    SATURATION (CHEMISTRY)
    STEREOCHEMISTRY
    STOICHIOMETRY
    SYNTHESIS (CHEMISTRY)
    THERMOCHEMISTRY
    UNSATURATION (CHEMISTRY)
    WISWESSER NOTATIONS

CHEMONUCLEAR PROPULSION
  USE   CHEMICAL PROPULSION
        NUCLEAR PROPULSION

CHEMORECEPTORS
  GS   ANATOMY
    . SENSE ORGANS
    . . CHEMORECEPTORS
    RECEPTORS (PHYSIOLOGY)
    . CHEMORECEPTORS
  RT   CAROTID SINUS BODY
    OLFACTORY PERCEPTION
    TASTE

CHEMOSPHERE
  GS   EARTH ATMOSPHERE
    . HOMOSPHERE
    . . MIDDLE ATMOSPHERE
    . . . CHEMOSPHERE
  RT   BIOSPHERE
    HETEROSPHERE
    IONOSPHERE
    LOWER ATMOSPHERE
    MESOSPHERE
    OZONOSPHERE
    PLASMASPHERE
    STRATOSPHERE
    THERMOSPHERE
    TROPOSPHERE
    UPPER ATMOSPHERE

CHEMOTHERAPY
  UF   DRUG THERAPY
  GS   THERAPY
    . CHEMOTHERAPY
  RT   ANTISEPTICS
    ∞CHEMISTRY
    DRUGS

CHENA RIVER BASIN (AK)
  GS   LANDFORMS
    . STRUCTURAL BASINS
    . . RIVER BASINS
    . . . CHENA RIVER BASIN (AK)
  RT   ALASKA

CHESAPEAKE BAY (US)
  GS   BAYS (TOPOGRAPHIC FEATURES)
    . CHESAPEAKE BAY (US)
  RT   MARYLAND
    RIVER BASINS
    SOUNDS (TOPOGRAPHIC FEATURES)
    VIRGINIA

CHEST
  GS   ANATOMY
    . CHEST
  RT   THORAX
    TORSO

CHEWING
  USE   MASTICATION

CHIAPAS (MEXICO)
  RT   MEXICO

CHIASMS
  GS   CROSSINGS
    . CHIASMS

CHICKENS
  GS   ANIMALS
    . VERTEBRATES
    . . BIRDS
    . . . CHICKENS

CHILD DEVICE
  RT   LEARNING
    LEARNING THEORY
    TRAINING DEVICES

CHILD-LANGMUIR LAW
  GS   LAWS
    . CHILD-LANGMUIR LAW
  RT   PERVEANCE
    SPACE CHARGE
    THERMIONIC DIODES

CHILDREN
  RT   FEMALES
    HUMAN BEINGS
    MALES
    PARENTS
    PROGENY

CHILE
  GS   NATIONS
    . CHILE
  RT   SOUTH AMERICA

CHILLING
  USE   COOLING

CHIMES
  USE   AUDITORY SIGNALS

CHIMNEYS
  RT   ∞BUILDINGS
    EXHAUST SYSTEMS
    FLUES
    FURNACES
    PLUMES
    STACKS
    VENTS
    WASTE ENERGY UTILIZATION

CHIMPANZEES
  GS   ANIMALS
    . VERTEBRATES
    . . MAMMALS
    . . . PRIMATES
    . . . . APES
    . . . . . CHIMPANZEES
    . WILDLIFE
    . . CHIMPANZEES
  RT   HUMAN BEINGS

CHIN
  GS   ANATOMY
    . MUSCULOSKELETAL SYSTEM
    . . CHIN
    FACE (ANATOMY)
    . CHIN
  RT   BONES

CHINA
  UF   CHINA (COMMUNIST) MAINLAND
  GS   CHINESE PEOPLES REPUBLIC
    NATIONS
    . CHINA
  RT   ASIA
    CHINESE AIRCRAFT
    HONG KONG
    TAIWAN

CHINA (COMMUNIST) MAINLAND
  USE   CHINA

CHINESE AIRCRAFT
  RT   ∞AIRCRAFT
    CHINA

CHINESE PEOPLES REPUBLIC
  USE   CHINA

CHINESE SPACE PROGRAM
  GS   PROGRAMS
    . SPACE PROGRAMS
    . . CHINESE SPACE PROGRAM
  RT   ∞RESEARCH PROJECTS
    SPACE MISSIONS
    TAIWAN

CHINESE SPACECRAFT
  RT   ∞SPACECRAFT
    TAIWAN

CHINOOK HELICOPTER
  USE   CH-47 HELICOPTER

CHIPPING
  RT   ABRASION
    COMMINUTION
    CUTTING
    FLAKING
    FRACTURING
    FRAGMENTATION
    PITTING
    ∞SEPARATION
    SPALLING
    SPLITTING
    WEAR

CHIPS
  RT   CHIPS (ELECTRONICS)
    FRAGMENTS
    SCRAP

CHIPS (ELECTRONICS)
  RT   CHIPS
    INTEGRATED CIRCUITS
    LARGE SCALE INTEGRATION
    VERY LARGE SCALE INTEGRATION
    VHSIC (CIRCUITS)

CHIPS (MEMORY DEVICES)
  RT   ∞DEVICES
    INTEGRATED CIRCUITS
    METAL-NITRIDE-OXIDE-SEMICONDUCTOR
    S
    SEMICONDUCTOR DEVICES

CHIRAL DYNAMICS
  RT   ∞DYNAMICS
    GROUP THEORY
    LAGRANGE MULTIPLIERS
    MATRICES (MATHEMATICS)

CHIRON
  UF   MINOR PLANET 2060
  GS   CELESTIAL BODIES
    . ASTEROID BELTS
    . . ASTEROIDS
    . . . CHIRON
  RT   APOLLO ASTEROIDS
    METEOROIDS
    PLANETS
    SOLAR SYSTEM
    SPACE DEBRIS

CHIRONOMUS FLIES
  GS   ANIMALS
    . INVERTEBRATES
    . . ARTHROPODS
    . . . INSECTS
    . . . . CHIRONOMUS FLIES
  RT   DROSOPHILA
    INFESTATION

CHIRP
  GS   ELECTROMAGNETIC INTERFERENCE
    . RADIO FREQUENCY INTERFERENCE
    . . CHIRP
    . . . CHIRP SIGNALS

CHIRP SIGNALS
  GS   ELECTROMAGNETIC INTERFERENCE
    . RADIO FREQUENCY INTERFERENCE
    . . CHIRP
    . . . CHIRP SIGNALS
  RT   ELECTROMAGNETIC NOISE
    ∞SIGNALS

CHITIN
  SN   (A POLYSACCHARIDE WHICH IS THE
    PRINCIPAL CONSTITUENT OF THE
    SHELLS OF CRABS AND LOBSTERS,
    THE SHARDS OF BEETLES, AND IS
    ALSO FOUND IN CERTAIN FUNGI)
  GS   ALIPHATIC COMPOUNDS
    . POLYSACCHARIDES
    . . CHITIN
    CARBOHYDRATES
    . POLYSACCHARIDES
    . . CHITIN
  RT   GUMS (SUBSTANCES)
    STARCHES

CHLORAL
GS   ALDEHYDES
    . **CHLORAL**
    ALIPHATIC COMPOUNDS
    . **CHLORAL**

CHLORATES
GS   HALOGEN COMPOUNDS
    . CHLORINE COMPOUNDS
    . . **CHLORATES**
RT   ∞OXYGEN COMPOUNDS
    PERCHLORATES

CHLORELLA
GS   ALGAE
    . **CHLORELLA**

CHLORIDES
UF   PENTACHLORIDES
    TRICHLORIDES
GS   HALOGEN COMPOUNDS
    . CHLORINE COMPOUNDS
    . . **CHLORIDES**
    . . . ALUMINUM CHLORIDES
    . . . AMMONIUM CHLORIDES
    . . . BERYLLIUM CHLORIDES
    . . . BORON CHLORIDES
    . . . CADMIUM CHLORIDES
    . . . CALCIUM CHLORIDES
    . . . CARBON TETRACHLORIDE
    . . . COPPER CHLORIDES
    . . . DICHLORIDES
    . . . GERMANIUM CHLORIDES
    . . . HYDROCHLORIDES
    . . . IRON CHLORIDES
    . . . LANTHANUM CHLORIDES
    . . . LEAD CHLORIDES
    . . . LITHIUM CHLORIDES
    . . . MAGNESIUM CHLORIDES
    . . . NITROSYL CHLORIDES
    . . . NITROXYCHLORIDES
    . . . NITRYL CHLORIDES
    . . . PHOSGENE
    . . . POTASSIUM CHLORIDES
    . . . SILICON TETRACHLORIDE
    . . . SILVER CHLORIDES
    . . . SODIUM CHLORIDES
    . . . SULFUR CHLORIDES
    . . . TETRACHLORIDES
    . . . TITANIUM CHLORIDES
    . . . TUNGSTEN CHLORIDES
    . . . ZINC CHLORIDES
    . HALIDES
    . . **CHLORIDES**
    . . . ALUMINUM CHLORIDES
    . . . AMMONIUM CHLORIDES
    . . . BERYLLIUM CHLORIDES
    . . . BORON CHLORIDES
    . . . CADMIUM CHLORIDES
    . . . CALCIUM CHLORIDES
    . . . CARBON TETRACHLORIDE
    . . . COPPER CHLORIDES
    . . . DICHLORIDES
    . . . GERMANIUM CHLORIDES
    . . . HYDROCHLORIDES
    . . . HYDROGEN CHLORIDES
    . . . . HYDROCHLORIC ACID
    . . . IRON CHLORIDES
    . . . LANTHANUM CHLORIDES
    . . . LEAD CHLORIDES
    . . . LITHIUM CHLORIDES
    . . . MAGNESIUM CHLORIDES
    . . . NITROSYL CHLORIDES
    . . . NITROXYCHLORIDES
    . . . NITRYL CHLORIDES
    . . . PHOSGENE
    . . . POTASSIUM CHLORIDES
    . . . SILICON TETRACHLORIDE
    . . . SILVER CHLORIDES
    . . . SODIUM CHLORIDES
    . . . SULFUR CHLORIDES
    . . . TETRACHLORIDES
    . . . TITANIUM CHLORIDES
    . . . TUNGSTEN CHLORIDES
    . . . ZINC CHLORIDES
RT   METHYL CHLORIDE
    POLYVINYL CHLORIDE
    SALT BEDS

CHLORINATION
GS   CHEMICAL REACTIONS
    . HALOGENATION
    . . **CHLORINATION**
RT   BLEACHING
    HYDROMETALLURGY

CHLORINATION-*(CONT.)*
    PYROMETALLURGY
    WATER TREATMENT

CHLORINE
GS   CHEMICAL ELEMENTS
    . HALOGENS
    . . **CHLORINE**

CHLORINE COMPOUNDS
GS   HALOGEN COMPOUNDS
    . **CHLORINE COMPOUNDS**
    . . CHLORATES
    . . CHLORIDES
    . . . ALUMINUM CHLORIDES
    . . . AMMONIUM CHLORIDES
    . . . BERYLLIUM CHLORIDES
    . . . BORON CHLORIDES
    . . . CADMIUM CHLORIDES
    . . . CALCIUM CHLORIDES
    . . . CARBON TETRACHLORIDE
    . . . COPPER CHLORIDES
    . . . DICHLORIDES
    . . . GERMANIUM CHLORIDES
    . . . HYDROCHLORIDES
    . . . IRON CHLORIDES
    . . . LANTHANUM CHLORIDES
    . . . LEAD CHLORIDES
    . . . LITHIUM CHLORIDES
    . . . MAGNESIUM CHLORIDES
    . . . NITROSYL CHLORIDES
    . . . NITROXYCHLORIDES
    . . . NITRYL CHLORIDES
    . . . PHOSGENE
    . . . POTASSIUM CHLORIDES
    . . . SILICON TETRACHLORIDE
    . . . SILVER CHLORIDES
    . . . SODIUM CHLORIDES
    . . . SULFUR CHLORIDES
    . . . TETRACHLORIDES
    . . . TITANIUM CHLORIDES
    . . . TUNGSTEN CHLORIDES
    . . . ZINC CHLORIDES
    . . CHLORINE FLUORIDES
    . . CHLORINE OXIDES
    . . CHLOROSILANES
    . . DDT
    . . MECLIZINE
    . . PERCHLORATES
    . . . ALUMINUM PERCHLORATES
    . . . AMMONIUM PERCHLORATES
    . . . HYDRAZINE PERCHLORATES
    . . . HYDROGEN PERCHLORATE
    . . . HYDROXYLAMMONIUM
        PERCHLORATES
    . . . LITHIUM PERCHLORATES
    . . . MAGNESIUM PERCHLORATES
    . . . NITRONIUM PERCHLORATE
    . . . POTASSIUM PERCHLORATES
RT   ∞CHEMICAL COMPOUNDS
    HALOCARBONS

CHLORINE FLUORIDES
GS   HALOGEN COMPOUNDS
    . CHLORINE COMPOUNDS
    . . **CHLORINE FLUORIDES**
    . FLUORINE COMPOUNDS
    . . FLUORIDES
    . . . **CHLORINE FLUORIDES**
    . HALIDES
    . . FLUORIDES
    . . . **CHLORINE FLUORIDES**
RT   LIQUID ROCKET PROPELLANTS

CHLORINE OXIDES
GS   CHALCOGENIDES
    . OXIDES
    . . **CHLORINE OXIDES**
    HALOGEN COMPOUNDS
    . CHLORINE COMPOUNDS
    . . **CHLORINE OXIDES**

CHLOROAROMATICS
GS   **CHLOROAROMATICS**
    . CHLOROBENZENES
RT   ∞AROMATIC COMPOUNDS

CHLOROBENZENES
GS   CHLOROAROMATICS
    . **CHLOROBENZENES**
    HYDROCARBONS
    . **CHLOROBENZENES**
RT   BENZENE

CHLOROCARBONS
GS   CARBON COMPOUNDS
    . HALOCARBONS
    . . **CHLOROCARBONS**

CHLOROETHYLENE
GS   ALIPHATIC COMPOUNDS
    . **CHLOROETHYLENE**
    ETHYLENE COMPOUNDS
    . **CHLOROETHYLENE**

CHLOROFORM
GS   ALIPHATIC COMPOUNDS
    . **CHLOROFORM**
    DRUGS
    . ANESTHETICS
    . . **CHLOROFORM**
RT   ANESTHESIOLOGY

CHLOROFORMATE
GS   ESTERS
    . **CHLOROFORMATE**
    FORMATES
    . **CHLOROFORMATE**

CHLOROPHYLLS
GS   MAGNESIUM COMPOUNDS
    . **CHLOROPHYLLS**
    ORGANOMETALLIC COMPOUNDS
    . **CHLOROPHYLLS**
    PIGMENTS
    . **CHLOROPHYLLS**
    PORPHYRINS
    . **CHLOROPHYLLS**
RT   ALGAE
    BROWN WAVE EFFECT
    CELLS (BIOLOGY)
    CHLOROPLASTS
    GREEN WAVE EFFECT
    OCEAN COLOR SCANNER
    PHOTOSYNTHESIS
    PLANTS (BOTANY)
    PORPHINES
    SKIN (ANATOMY)

CHLOROPLASTS
RT   CELLS (BIOLOGY)
    CHLOROPHYLLS
    CYTOPLASM
    PHOTOSYNTHESIS

CHLOROPRENE RESINS
UF   NEOPRENES
GS   RUBBER
    . SYNTHETIC RUBBERS
    . . ELASTOMERS
    . . . **CHLOROPRENE RESINS**

CHLOROSILANES
GS   HALOGEN COMPOUNDS
    . CHLORINE COMPOUNDS
    . . **CHLOROSILANES**
    HYDROGEN COMPOUNDS
    . HYDRIDES
    . . SILANES
    . . . **CHLOROSILANES**
    SILICON COMPOUNDS
    . SILANES
    . . **CHLOROSILANES**

CHLORPROMAZINE
GS   ALIPHATIC COMPOUNDS
    . HYDRAZINES
    . . **CHLORPROMAZINE**

CHOCTAW HELICOPTER
USE   CH-34 HELICOPTER

CHOICE
USE   SELECTION

CHOKES
SN   (EXCLUDES FUEL SYSTEM AND
    ELECTRONIC DEVICES)
RT   CHOKES (RESTRICTIONS)
    ∞DIFFUSERS
    ELECTRIC COILS
    MIXING
    NOZZLE INSERTS
    ∞NOZZLES

CHOKES (FUEL SYSTEMS)
RT   CARBURETORS
    FUEL SYSTEMS

CHOKES (FUEL SYSTEMS)-*(CONT.)*
      ORIFICES
      ∞SYSTEMS

CHOKES (RESTRICTIONS)
RT   CHOKES (FUEL SYSTEMS)
      CLOSURES
      CONSTRICTIONS
      IMPEDANCE
      ∞NOZZLES
      ORIFICES
      THROATS
      VALVES

CHOLERA
GS   DISEASES
      . INFECTIOUS DISEASES
      . . **CHOLERA**
RT   DISORDERS
      HUMAN PATHOLOGY
      KIDNEY DISEASES
      PARASITIC DISEASES
      PATHOGENESIS
      PATHOLOGICAL EFFECTS
      PHYSIOLOGICAL EFFECTS

CHOLESKY FACTORIZATION
RT   CONJUGATES
      FINITE ELEMENT METHOD
      ITERATIVE SOLUTION
      REAL VARIABLES

CHOLESTEROL
GS   STEROIDS
      . **CHOLESTEROL**
RT   ARTERIOSCLEROSIS
      LIQUID CRYSTALS

CHOLINE
GS   ALIPHATIC COMPOUNDS
      . **CHOLINE**
      CARBOHYDRATES
      . FATS
      . . **CHOLINE**
      ORGANIC COMPOUNDS
      . FATS
      . . **CHOLINE**

CHOLINERGIC BLOCKING AGENTS
USE   ANTICHOLINERGICS

CHOLINERGICS
GS   DRUGS
      . **CHOLINERGICS**
      . . ANTICHOLINERGICS
RT   CYCLIC AMP

CHOLINESTERASE
GS   ENZYMES
      . **CHOLINESTERASE**
RT   NEUROMUSCULAR TRANSMISSION

CHONDRITES
GS   CELESTIAL BODIES
      . METEORITES
      . . STONY METEORITES
      . . . **CHONDRITES**
      . . . . BRUDERHEIM METEORITE
      . . . . CARBONACEOUS CHONDRITES
      . . . . . ALLENDE METEORITE
      . . . . . MURCHISON METEORITE
      . . . . CARBONACEOUS METEORITES
      . . . . . ALAIS METEORITE
      . . . . . COLD BOKKEVELD METEORITE
      . . . . . IVUNA METEORITE
      . . . . . MURRAY METEORITE
      . . . . . ORGUEIL METEORITE
      . . . . . TONK METEORITE
      . . . . HVITTIS CHONDRITE
      . . . . PANTAR CHONDRITES
      . . . . PRIBRAM METEORITE
RT   ACHONDRITES
      TEKTITES

CHONDRULE
RT   ENSTATITE

CHOPPERS (ELECTRIC)
USE   ELECTRIC CHOPPERS

CHORDS (GEOMETRY)
UF   AERODYNAMIC CHORDS
GS   GEOMETRY
      . EUCLIDEAN GEOMETRY

CHORDS (GEOMETRY)-*(CONT.)*
      . . LINES (GEOMETRY)
      . . . **CHORDS (GEOMETRY)**
RT   CURVES (GEOMETRY)
      GEODESIC LINES
      TANGENTS

CHOROID MEMBRANES
GS   ANATOMY
      . SENSE ORGANS
      . . EYE (ANATOMY)
      . . . **CHOROID MEMBRANES**
      MEMBRANES
      . **CHOROID MEMBRANES**
RT   VISION

CHORUS (DAWN PHENOMENON)
USE   DAWN CHORUS

CHORUS PHENOMENON
USE   DAWN CHORUS

CHROMATES
UF   DICHROMATES
GS   CHROMIUM COMPOUNDS
      . **CHROMATES**
      . . POTASSIUM CHROMATES
RT   ∞OXYGEN COMPOUNDS

CHROMATOGRAPHY
GS   **CHROMATOGRAPHY**
      . GAS CHROMATOGRAPHY
      . LIQUID CHROMATOGRAPHY
      . PAPER CHROMATOGRAPHY
      . THIN LAYER CHROMATOGRAPHY
RT   ADSORPTION
      CHEMICAL ANALYSIS
      COLORIMETRY
      QUANTITATIVE ANALYSIS
      SORPTION

CHROME
USE   CHROMIUM

CHROMIC ACID
GS   ACIDS
      . **CHROMIC ACID**
      CHROMIUM COMPOUNDS
      . **CHROMIC ACID**

CHROMITES
GS   CHALCOGENIDES
      . OXIDES
      . . METAL OXIDES
      . . . CHROMIUM OXIDES
      . . . . **CHROMITES**
      . . . IRON OXIDES
      . . . . **CHROMITES**
      CHROMIUM COMPOUNDS
      . CHROMIUM OXIDES
      . . **CHROMITES**
      IRON COMPOUNDS
      . IRON OXIDES
      . . **CHROMITES**
      MINERALS
      . **CHROMITES**
RT   PERIDOTITE
      SERPENTINE

CHROMIUM
UF   CHROME
GS   CHEMICAL ELEMENTS
      . REFRACTORY METALS
      . . **CHROMIUM**
      . . . CHROMIUM ISOTOPES
      METALS
      . TRANSITION METALS
      . . REFRACTORY METALS
      . . . **CHROMIUM**
      . . . . CHROMIUM ISOTOPES
      REFRACTORY MATERIALS
      . REFRACTORY METALS
      . . **CHROMIUM**
      . . . CHROMIUM ISOTOPES
RT   STRATEGIC MATERIALS

CHROMIUM ALLOYS
GS   ALLOYS
      . **CHROMIUM ALLOYS**
      . . ASTROLOY (TRADEMARK)
      . . CHROMIUM STEELS
      . . RENE 41
      . . RENE 63
      . . RENE 77

CHROMIUM ALLOYS-*(CONT.)*
      . . RENE 95
RT   HEAT RESISTANT ALLOYS
      INCONEL (TRADEMARK)
      STAINLESS STEELS
      STELLITE (TRADEMARK)
      WASPALOY

CHROMIUM BORIDES
GS   BORON COMPOUNDS
      . BORIDES
      . . **CHROMIUM BORIDES**
      CHROMIUM COMPOUNDS
      . **CHROMIUM BORIDES**

CHROMIUM BROMIDES
GS   CHROMIUM COMPOUNDS
      . **CHROMIUM BROMIDES**
      HALOGEN COMPOUNDS
      . BROMINE COMPOUNDS
      . . BROMIDES
      . . . **CHROMIUM BROMIDES**
      . HALIDES
      . . BROMIDES
      . . . **CHROMIUM BROMIDES**
      . . METAL HALIDES
      . . . **CHROMIUM BROMIDES**

CHROMIUM CARBIDES
GS   CARBON COMPOUNDS
      . CARBIDES
      . . **CHROMIUM CARBIDES**
      CHROMIUM COMPOUNDS
      . **CHROMIUM CARBIDES**

CHROMIUM COMPOUNDS
GS   **CHROMIUM COMPOUNDS**
      . CHROMATES
      . . POTASSIUM CHROMATES
      . CHROMIC ACID
      . CHROMIUM BORIDES
      . CHROMIUM BROMIDES
      . CHROMIUM CARBIDES
      . CHROMIUM FLUORIDES
      . CHROMIUM OXIDES
      . . CHROMITES
RT   ∞CHEMICAL COMPOUNDS
      ∞GROUP 6B COMPOUNDS
      ∞METAL COMPOUNDS

CHROMIUM FLUORIDES
GS   CHROMIUM COMPOUNDS
      . **CHROMIUM FLUORIDES**
      HALOGEN COMPOUNDS
      . FLUORINE COMPOUNDS
      . . FLUORIDES
      . . . METAL FLUORIDES
      . . . . **CHROMIUM FLUORIDES**
      . HALIDES
      . . METAL HALIDES
      . . . **CHROMIUM FLUORIDES**

CHROMIUM ISOTOPES
GS   CHEMICAL ELEMENTS
      . NUCLIDES
      . . ISOTOPES
      . . . **CHROMIUM ISOTOPES**
      . REFRACTORY METALS
      . . CHROMIUM
      . . . **CHROMIUM ISOTOPES**
      METALS
      . TRANSITION METALS
      . . REFRACTORY METALS
      . . . CHROMIUM
      . . . . **CHROMIUM ISOTOPES**
      REFRACTORY MATERIALS
      . REFRACTORY METALS
      . . CHROMIUM
      . . . **CHROMIUM ISOTOPES**

CHROMIUM OXIDES
GS   CHALCOGENIDES
      . OXIDES
      . . METAL OXIDES
      . . . **CHROMIUM OXIDES**
      . . . . CHROMITES
      CHROMIUM COMPOUNDS
      . **CHROMIUM OXIDES**
      . . CHROMITES

CHROMIUM STEELS
GS   ALLOYS
      . CHROMIUM ALLOYS
      . . **CHROMIUM STEELS**
      . IRON ALLOYS

CHROMIUM STEELS-(CONT.)
- . . STEELS
- . . . **CHROMIUM STEELS**
- RT   FERRITIC STAINLESS STEELS

CHROMOSOMES
- GS   CELLS (BIOLOGY)
- . **CHROMOSOMES**
- RT   CONGENITAL ANOMALIES
- CYTOLOGY
- GENETIC CODE
- GENETICS
- MITOSIS
- MUTATIONS
- ∞NUCLEI
- REPRODUCTIVE SYSTEMS
- TETRAD THEORY

CHROMOSPHERE
- GS   ENVIRONMENTS
- . EXTRATERRESTRIAL ENVIRONMENTS
- . . STELLAR ATMOSPHERES
- . . . **CHROMOSPHERE**
- RT   CORONAL LOOPS
- FACULAE
- PHOTOSPHERE
- SOLAR ATMOSPHERE
- SOLAR CORONA
- SOLAR PROMINENCES
- SPICULES
- STELLAR STRUCTURE
- STELLAR WINDS

CHRONAXY
- GS   TIME
- . REACTION TIME
- . . **CHRONAXY**
- RT   RESPONSES
- SENSORY STIMULATION
- THRESHOLDS (PERCEPTION)

CHRONIC CONDITIONS
- GS   CONDITIONS
- . **CHRONIC CONDITIONS**
- RT   BLOOD VOLUME
- DISEASES
- DISORDERS
- HEALTH
- PHYSIOLOGY

† **CHRONOGRAPHS**
- USE   CHRONOMETERS

CHRONOLOGY
- UF   AGE DETERMINATION
- DATING
- GS   **CHRONOLOGY**
- . GEOCHRONOLOGY
- RT   TIME

CHRONOMETERS
- UF   CHRONOGRAPHS
- GS   MEASURING INSTRUMENTS
- . TIME MEASURING INSTRUMENTS
- . . CLOCKS
- . . . **CHRONOMETERS**
- RT   ATOMIC CLOCKS
- CLOCK PARADOX
- TIME MEASUREMENT
- TIMING DEVICES

CHRONOPHOTOGRAPHY
- UF   TIME LAPSE PHOTOGRAPHY
- GS   IMAGERY
- . **CHRONOPHOTOGRAPHY**
- PHOTOGRAPHY
- . **CHRONOPHOTOGRAPHY**
- RT   BLACK AND WHITE PHOTOGRAPHY
- MOTION PICTURES

CHRONOTRONS
- USE   PULSE RATE
- TIME LAG

CHUCKCHI SEA
- GS   SEAS
- . **CHUCKCHI SEA**
- RT   ARCTIC REGIONS

CHUGGING
- USE   COMBUSTION STABILITY

CHUTES
- UF   SLIDES

CHUTES-(CONT.)
- RT   CONVEYORS
- MATERIALS HANDLING

CID
- USE   CHARGE INJECTION DEVICES

CINDER CONES
- USE   CONES (VOLCANOES)

CINEFLUOROGRAPHY
- USE   MOTION PICTURES
- RADIOGRAPHY

CINEMATOGRAPHY
- GS   IMAGERY
- . **CINEMATOGRAPHY**
- PHOTOGRAPHY
- . **CINEMATOGRAPHY**
- RT   BLACK AND WHITE PHOTOGRAPHY
- CAMERAS
- COLOR PHOTOGRAPHY
- INFRARED PHOTOGRAPHY
- MOTION PICTURES
- STEREOPHOTOGRAPHY
- STREAK CAMERAS

CINERADIOGRAPHY
- USE   MOTION PICTURES
- RADIOGRAPHY

CINESPECTROGRAPHS
- RT   OPTICAL MEASURING INSTRUMENTS
- SPECTROSCOPY

CINETHEODOLITES
- GS   MEASURING INSTRUMENTS
- . OPTICAL MEASURING INSTRUMENTS
- . . TRANSITS
- . . . THEODOLITES
- . . . . **CINETHEODOLITES**
- OPTICAL EQUIPMENT
- . OPTICAL MEASURING INSTRUMENTS
- . . TRANSITS
- . . . THEODOLITES
- . . . . **CINETHEODOLITES**
- RT   PHOTOGRAPHIC TRACKING
- SATELLITE TRACKING

CIRCADIAN RHYTHMS
- UF   DIURNAL RHYTHMS
- GS   RHYTHM (BIOLOGY)
- . **CIRCADIAN RHYTHMS**
- RT   ACTIVITY CYCLES (BIOLOGY)

CIRCLES (GEOMETRY)
- GS   GEOMETRY
- . EUCLIDEAN GEOMETRY
- . . **CIRCLES (GEOMETRY)**
- . . . GREAT CIRCLES
- RT   CIRCUMFERENCES
- CURVES (GEOMETRY)
- ELLIPSES
- RADII
- ∞RINGS
- ∞SECTORS
- SEGMENTS
- SPHERES

CIRCUIT BOARDS
- RT   ELECTRONIC PACKAGING
- PRINTED CIRCUITS

CIRCUIT BREAKERS
- UF   BREAKERS (ELECTRIC)
- RT   DISCONNECT DEVICES
- DROPOUTS
- ELECTRIC RELAYS
- ∞FUSES
- SWITCHES
- SWITCHING CIRCUITS

CIRCUIT DIAGRAMS
- UF   SCHEMATICS
- GS   DIAGRAMS
- . **CIRCUIT DIAGRAMS**
- RT   ENGINEERING DRAWINGS
- LAYOUTS
- PHOTOMASKS

CIRCUIT PROTECTION
- GS   PROTECTION
- . **CIRCUIT PROTECTION**
- RT   CAPACITORS

CIRCUIT PROTECTION-(CONT.)
- CIRCUITS
- CURRENT REGULATORS
- ELECTRIC FUSES
- ELECTRIC POWER TRANSMISSION
- ELECTRIC REACTORS
- ELECTRICAL FAULTS
- ELECTRICAL GROUNDING
- ELECTRICAL INSULATION
- EXPULSION
- ∞FUSES
- OVERVOLTAGE
- PHASE CONTROL
- PHASE ERROR
- SNEAK CIRCUIT ANALYSIS
- SUPPRESSORS
- SURGES
- TRANSFORMERS
- TRANSMISSION CIRCUITS
- TRANSMISSION LINES
- VOLTAGE REGULATORS

CIRCUIT RELIABILITY
- GS   RELIABILITY
- . **CIRCUIT RELIABILITY**
- RT   AIRCRAFT RELIABILITY
- COMPONENT RELIABILITY
- DRIFT (INSTRUMENTATION)
- QUALITY CONTROL
- SNEAK CIRCUIT ANALYSIS
- SPACECRAFT RELIABILITY

CIRCUITS
- UF   ELECTRIC CIRCUITS
- EXPLODING CONDUCTOR CIRCUITS
- SHUNTS
- SUBCIRCUITS
- GS   **CIRCUITS**
- . ADDING CIRCUITS
- . ANALOG CIRCUITS
- . AUTODYNES
- . BISTABLE CIRCUITS
- . CANCELLATION CIRCUITS
- . CLAMPING CIRCUITS
- . COINCIDENCE CIRCUITS
- . COMPARATOR CIRCUITS
- . CONJUGATED CIRCUITS
- . COUNTING CIRCUITS
- . . SCALERS
- . COUPLING CIRCUITS
- . DELAY CIRCUITS
- . . PHANTASTRONS
- . DIGITAL INTEGRATORS
- . DIPLEXERS
- . DISCRIMINATORS
- . . FRAUNHOFER LINE DISCRIMINATORS
- . . FREQUENCY DISCRIMINATORS
- . ECHO SUPPRESSORS
- . ELECTRIC BRIDGES
- . . WIRE BRIDGE CIRCUITS
- . . . WHEATSTONE BRIDGES
- . EQUIVALENT CIRCUITS
- . FEEDBACK CIRCUITS
- . FIRE CONTROL CIRCUITS
- . FLUIDIC CIRCUITS
- . . FLIP-FLOPS
- . GATES (CIRCUITS)
- . . THRESHOLD GATES
- . HYBRID CIRCUITS
- . INTEGRATED CIRCUITS
- . . DTL INTEGRATED CIRCUITS
- . . ENCAPSULATED MICROCIRCUITS
- . . LARGE SCALE INTEGRATION
- . . LINEAR INTEGRATED CIRCUITS
- . . TTL INTEGRATED CIRCUITS
- . . VERY LARGE SCALE INTEGRATION
- . . VHSIC (CIRCUITS)
- . ITERATIVE NETWORKS
- . LC CIRCUITS
- . LIMITER CIRCUITS
- . . CLIPPER CIRCUITS
- . LINEAR CIRCUITS
- . LOGIC CIRCUITS
- . . THRESHOLD GATES
- . MAGNETIC CIRCUITS
- . MATRICES (CIRCUITS)
- . MICROWAVE CIRCUITS
- . MIXING CIRCUITS
- . MULTIVIBRATORS
- . . FLIP-FLOPS
- . . MONOSTABLE MULTIVIBRATORS
- . NEGATIVE RESISTANCE CIRCUITS
- . OHMS LAW
- . PHASE DETECTORS
- . . SYNCHROSCOPES

CIRCUITS-(CONT.)
- . PHASE SHIFT CIRCUITS
- . . CIRCULATORS (PHASE SHIFT CIRCUITS)
- . PNEUMATIC CIRCUITS
- . POWER SUPPLY CIRCUITS
- . PRINTED CIRCUITS
- . . LARGE SCALE INTEGRATION
- . . MEDIUM SCALE INTEGRATION
- . RC CIRCUITS
- . RL CIRCUITS
- . . RLC CIRCUITS
- . SQUELCH CIRCUITS
- . SWEEP CIRCUITS
- . SWITCHING CIRCUITS
- . . FLUID SWITCHING ELEMENTS
- . TRANSISTOR CIRCUITS
- . TRANSMISSION CIRCUITS
- . TRIGGER CIRCUITS
- . VARACTOR DIODE CIRCUITS

RT AMPLIFIERS
  BREADBOARD MODELS
  CAPACITORS
 ∞CASCADES
  CIRCUIT PROTECTION
  DIFFERENTIATORS
  DUALITY PRINCIPLE
  DUPLEXERS
  ELECTRIC CONNECTORS
  ELECTRIC CURRENT
 ∞ELECTRIC EQUIPMENT
  ELECTRIC FILTERS
  ELECTRIC MOTORS
  ELECTRIC POWER TRANSMISSION
  ELECTRIC WIRE
  ELECTRICAL GROUNDING
  ELECTROMECHANICS
  ELECTRON TUBES
  FLAT CONDUCTORS
  INDUCTORS
  INTEGRATORS
  KIRCHHOFF LAW OF NETWORKS
  LOOPS
  MICROELECTRONICS
  MICROMINIATURIZATION
  MINIATURE ELECTRONIC EQUIPMENT
  MINIATURIZATION
  MODULES
  NETWORK ANALYSIS
 ∞NETWORKS
  OSCILLATORS
  SELECTORS
  SHORT CIRCUITS
  SIGNAL GENERATORS
  SOLID STATE DEVICES
  SOLIONS
 ∞STRIP
  TRANSMISSION LINES
  TREES (MATHEMATICS)
  UNDERGROUND TRANSMISSION LINES
  VOLTAGE CONTROLLED OSCILLATORS
  WIRING

CIRCULAR CONES
GS CONES
  . CIRCULAR CONES
RT HALF CONES
  NOSE CONES

CIRCULAR CYLINDERS
RT ∞CYLINDERS
  CYLINDRICAL BODIES
  CYLINDRICAL SHELLS
  ELLIPTICAL CYLINDERS

CIRCULAR ORBITS
GS ORBITS
  . CIRCULAR ORBITS
  . . STATIONARY ORBITS
RT EARTH ORBITS
  ECCENTRIC ORBITS
  ELLIPTICAL ORBITS
  EQUATORIAL ORBITS
  GEOSYNCHRONOUS ORBITS
  LUNAR ORBITS
  ORBITAL MECHANICS
  PLANETARY ORBITS
  POLAR ORBITS
  QUADRATURES
  SATELLITE ORBITS
  SOLAR ORBITS
  SPACECRAFT ORBITS
  TWENTY-FOUR HOUR ORBITS

CIRCULAR PLATES
GS STRUCTURAL MEMBERS
  . PLATES (STRUCTURAL MEMBERS)
  . . CIRCULAR PLATES
RT ANNULAR PLATES
  DISKS (SHAPES)
  END PLATES
  FLAT PLATES

CIRCULAR POLARIZATION
GS POLARIZATION (WAVES)
  . CIRCULAR POLARIZATION
RT ELLIPTICAL POLARIZATION
  OPTICAL POLARIZATION

CIRCULAR SHELLS
GS SHELLS (STRUCTURAL FORMS)
  . CIRCULAR SHELLS
RT CYLINDRICAL SHELLS
  HEMISPHERICAL SHELLS
  METAL SHELLS
  SPHERICAL SHELLS

CIRCULAR TUBES
RT CYLINDRICAL SHELLS
  DUCT GEOMETRY
  PIPES (TUBES)
 ∞TUBES

CIRCULAR WAVEGUIDES
GS TRANSMISSION LINES
  . COMMUNICATION CABLES
  . . WAVEGUIDES
  . . . CIRCULAR WAVEGUIDES
RT MICROWAVE TRANSMISSION
  PROPAGATION MODES

CIRCULATION
UF RECIRCULATION
GS CIRCULATION
  . ATMOSPHERIC CIRCULATION
  . BLOOD CIRCULATION
  . . BRAIN CIRCULATION
  . . CORONARY CIRCULATION
  . . INTERCRANIAL CIRCULATION
  . . INTRAVASCULAR SYSTEM
  . . ISCHEMIA
  . . OCULAR CIRCULATION
  . . PERIPHERAL CIRCULATION
  . . PULMONARY CIRCULATION
  . CARBOXYHEMOGLOBIN
  . CONGESTION
  . WATER CIRCULATION
  . . WATER CURRENTS
  . . . OCEAN CURRENTS
  . . . . COASTAL CURRENTS
  . . . . EL NINO
  . . . . GULF STREAM
  . . . . LOMONOSOV CURRENT
RT ARTIFICIAL CARDIAC PACEMAKER
  BLOOD
  BLOOD PUMPS
  BLOWING
  CIRCULATION DISTRIBUTION
 ∞CURRENTS
  DELIVERY
  DIFFUSION
  DISPERSING
  HEART IMPLANTATION
  PURGING
  ROTATION

CIRCULATION CONTROL AIRFOILS
GS AIRFOILS
  . CIRCULATION CONTROL AIRFOILS
RT BLOWING
  BOUNDARY LAYER CONTROL
  COANDA EFFECT
 ∞CONTROL
  LIFT AUGMENTATION
  SHORT TAKEOFF AIRCRAFT
  UNDER SURFACE BLOWING
  UPPER SURFACE BLOWING

CIRCULATION CONTROL ROTORS
GS ROTATING BODIES
  . ROTORS
  . . ROTARY WINGS
  . . . CIRCULATION CONTROL ROTORS
RT ∞CONTROL
  VERTICAL TAKEOFF AIRCRAFT
  X WING ROTORS

CIRCULATION DISTRIBUTION
RT ATMOSPHERIC CIRCULATION

CIRCULATION DISTRIBUTION-(CONT.)
  CIRCULATION
 ∞DISTRIBUTION
  VELOCITY DISTRIBUTION

CIRCULATORS (PHASE SHIFT CIRCUITS)
GS CIRCUITS
  . PHASE SHIFT CIRCUITS
  . . CIRCULATORS (PHASE SHIFT CIRCUITS)
RT CAVITY RESONATORS
  DELAY CIRCUITS
  DUPLEXERS
  FARADAY EFFECT
  HALL GENERATORS
  LIMITER CIRCUITS

CIRCULATORY SYSTEM
GS ANATOMY
  . CIRCULATORY SYSTEM
  . . VASCULAR SYSTEM
  . . . BLOOD VESSELS
  . . . . ARTERIES
  . . . . . AORTA
  . . . . CAPILLARIES (ANATOMY)
  . . . . GLOMERULUS
  . . . . VEINS
RT ARTERIOSCLEROSIS
  ARTIFICIAL CARDIAC PACEMAKER
  BLOOD CIRCULATION
  BLOOD PUMPS
  CARDIOVASCULAR SYSTEM
  CAROTID SINUS BODY
  CAROTID SINUS REFLEX
  EXERCISE PHYSIOLOGY
  HEART
  HYPERVOLEMIA

CIRCUMFERENCES
GS GEOMETRY
  . EUCLIDEAN GEOMETRY
  . . ANALYTIC GEOMETRY
  . . . CIRCUMFERENCES
RT BOUNDARIES
  CIRCLES (GEOMETRY)
  DIAMETERS
  RADII

CIRCUMLUNAR COMMUNICATION
GS TELECOMMUNICATION
  . SPACE COMMUNICATION
  . . LUNAR COMMUNICATION
  . . . CIRCUMLUNAR COMMUNICATION
RT INTERPLANETARY COMMUNICATION
  RADAR
  RADIO COMMUNICATION
  SPACECRAFT COMMUNICATION
  UNIFIED S BAND

CIRCUMLUNAR TRAJECTORIES
GS TRAJECTORIES
  . ROUND TRIP TRAJECTORIES
  . . CIRCUMLUNAR TRAJECTORIES
  . SPACECRAFT TRAJECTORIES
  . . LUNAR TRAJECTORIES
  . . . CIRCUMLUNAR TRAJECTORIES
RT EARTH ORBITS
  EARTH-MOON TRAJECTORIES
  LUNAR FLIGHT
  LUNAR ORBITS
  MOON-EARTH TRAJECTORIES
  REENTRY TRAJECTORIES
  RENDEZVOUS TRAJECTORIES
  TRANSFER ORBITS

CIRCUMPOLAR WESTERLIES
GS WIND (METEOROLOGY)
  . CIRCUMPOLAR WESTERLIES
RT ATMOSPHERIC CIRCULATION
  JET STREAMS (METEOROLOGY)
  WINDS ALOFT

CIRCUMSOLAR RADIATION
GS EXTRATERRESTRIAL RADIATION
  . SOLAR RADIATION
  . . CIRCUMSOLAR RADIATION
RT ATMOSPHERIC SCATTERING
  LIGHT SCATTERING
 ∞RADIATION
  SCATTERING
  SUNLIGHT

CIRCUMSOLAR TELESCOPES
GS TELESCOPES
  CIRCUMSOLAR TELESCOPES

CIRCUMSOLAR TELESCOPES-*(CONT.)*
RT   LENSES
       MEASURING INSTRUMENTS
       MIRRORS
       OPTICAL EQUIPMENT
       RADIATION PYROMETERS
       SOLAR ENERGY
       SOLAR RADIATION

CIRCUMSTELLAR MATTER
USE   STELLAR ENVELOPES

CIRQUES (LANDFORMS)
GS   LANDFORMS
      . STRUCTURAL BASINS
      . . **CIRQUES (LANDFORMS)**
RT   GLACIERS
       ICE
       MOUNTAINS
       SNOW

CIRROCUMULUS CLOUDS
GS   CLOUDS
      . CLOUDS (METEOROLOGY)
      . . **CIRROCUMULUS CLOUDS**
RT   ATMOSPHERIC MOISTURE
       CIRROSTRATUS CLOUDS
       CLIMATOLOGY
       CLOUD COVER
       FOG
       METEOROLOGY
       NEPHANALYSIS
       PRECIPITATION (METEOROLOGY)
       THUNDERSTORMS
       WEATHER

CIRROSTRATUS CLOUDS
GS   CLOUDS
      . CLOUDS (METEOROLOGY)
      . . **CIRROSTRATUS CLOUDS**
RT   ATMOSPHERIC MOISTURE
       CIRROCUMULUS CLOUDS
       CLIMATOLOGY
       CLOUD COVER
       FOG
       METEOROLOGY
       NEPHANALYSIS
       PRECIPITATION (METEOROLOGY)
       THUNDERSTORMS
       WEATHER

CIRRUS CLOUDS
GS   CLOUDS
      . CLOUDS (METEOROLOGY)
      . . CONVECTION CLOUDS
      . . . **CIRRUS CLOUDS**

CIRRUS SHIELDS
GS   CLOUDS
      . CLOUDS (METEOROLOGY)
      . . **CIRRUS SHIELDS**
RT   CLIMATOLOGY
       METEOROLOGY
       WEATHER FORECASTING

CISLUNAR SPACE
GS   ENVIRONMENTS
      . AEROSPACE ENVIRONMENTS
      . . **CISLUNAR SPACE**
      . EXTRATERRESTRIAL ENVIRONMENTS
      . . **CISLUNAR SPACE**
RT   DEEP SPACE
       EARTH-MOON TRAJECTORIES
       INTERPLANETARY SPACE
       LUNAR FLIGHT
       LUNAR ORBITS
       ∞SPACE

CITIES
UF   METROPOLITAN AREAS
       URBAN AREAS
GS   **CITIES**
      . ATLANTA (GA)
      . CEDAR RAPIDS (IA)
      . HOUSTON (TX)
      . MANITOU (CO)
      . MOSCOW
      . NEW HAVEN (CT)
      . NEW YORK CITY (NY)
      . PHOENIX (AZ)
      . PONTIAC (MI)
      . SAN FRANCISCO (CA)
      . VATICAN CITY
RT   ANTHROPOLOGY
       COMMUNITIES

CITIES-*(CONT.)*
       HEAT ISLANDS
       INDUSTRIAL AREAS
       INHABITANTS
       MEGALOPOLISES
       NATIONS
       RESIDENTIAL AREAS
       SOCIOLOGY
       SUBURBAN AREAS
       URBAN DEVELOPMENT
       URBAN PLANNING
       URBAN RESEARCH

CITRATES
RT   CITRIC ACID
       ESTERS

CITRIC ACID
GS   ALIPHATIC COMPOUNDS
      . **CITRIC ACID**
      CARBOHYDRATES
      . **CITRIC ACID**
RT   CITRATES

CITRUS TREES
GS   PLANTS (BOTANY)
      . TREES (PLANTS)
      . . **CITRUS TREES**
RT   AGRICULTURE
       BLIGHT
       BOTANY
       CROP GROWTH
       CROP VIGOR
       ∞CROPS
       CURING
       ∞FOOD
       IRRIGATION
       ORCHARDS
       SEEDS

CIVIL AVIATION
UF   COMMERCIAL AVIATION
RT   ∞AERONAUTICS
       AIR LAW
       AIRLINE OPERATIONS
       COMMERCIAL AIRCRAFT
       GENERAL AVIATION AIRCRAFT
       PASSENGER AIRCRAFT

CIVIL DEFENSE
RT   AIR DEFENSE
       ANTIMISSILE DEFENSE
       CHEMICAL DEFENSE
       ∞DEFENSE
       DEFENSE PROGRAM
       EVACUATING (TRANSPORTATION)
       NUCLEAR EXPLOSIONS
       NUCLEAR WARFARE
       PROTECTION
       SENTINEL SYSTEM
       SHELTERS
       SURVIVAL
       WARNING
       WARNING SYSTEMS

CL-41 AIRCRAFT
UF   CANADAIR CL-41 AIRCRAFT
       CT-114 AIRCRAFT
       TUTOR AIRCRAFT
GS   ATTACK AIRCRAFT
      . **CL-41 AIRCRAFT**
      GENERAL DYNAMICS AIRCRAFT
      . CANADAIR AIRCRAFT
      . . **CL-41 AIRCRAFT**
      JET AIRCRAFT
      . **CL-41 AIRCRAFT**
      MONOPLANES
      . **CL-41 AIRCRAFT**
      TRAINING AIRCRAFT
      . **CL-41 AIRCRAFT**
RT   ∞AIRCRAFT

CL-44 AIRCRAFT
UF   CANADAIR CL-44 AIRCRAFT
       CC-106 AIRCRAFT
       YUKON AIRCRAFT
GS   GENERAL DYNAMICS AIRCRAFT
      . CANADAIR AIRCRAFT
      . . **CL-44 AIRCRAFT**
      JET AIRCRAFT
      . TURBOPROP AIRCRAFT
      . . **CL-44 AIRCRAFT**
      MONOPLANES
      . **CL-44 AIRCRAFT**
      TRANSPORT AIRCRAFT

CL-44 AIRCRAFT-*(CONT.)*
      . CARGO AIRCRAFT
      . . **CL-44 AIRCRAFT**
RT   ∞AIRCRAFT

CL-84 AIRCRAFT
UF   CANADAIR CL-84 AIRCRAFT
GS   ANTISUBMARINE WARFARE AIRCRAFT
      . **CL-84 AIRCRAFT**
      GENERAL DYNAMICS AIRCRAFT
      . CANADAIR AIRCRAFT
      . . **CL-84 AIRCRAFT**
      JET AIRCRAFT
      . TURBOPROP AIRCRAFT
      . . **CL-84 AIRCRAFT**
      OBSERVATION AIRCRAFT
      . **CL-84 AIRCRAFT**
      RECONNAISSANCE AIRCRAFT
      . **CL-84 AIRCRAFT**
      TILT WING AIRCRAFT
      . **CL-84 AIRCRAFT**
      TRANSPORT AIRCRAFT
      . **CL-84 AIRCRAFT**
      V/STOL AIRCRAFT
      . **CL-84 AIRCRAFT**
RT   ∞AIRCRAFT

CL-595 HELICOPTER
USE   XH-51 HELICOPTER

CL-600 CHALLENGER AIRCRAFT
GS   GENERAL AVIATION AIRCRAFT
      . **CL-600 CHALLENGER AIRCRAFT**
RT   ∞AIRCRAFT
       ∞MILITARY AIRCRAFT
       SUPERCRITICAL WINGS

CL-823 AIRCRAFT
UF   LOCKHEED CL-823 AIRCRAFT
GS   JET AIRCRAFT
      . **CL-823 AIRCRAFT**
      LOCKHEED AIRCRAFT
      . **CL-823 AIRCRAFT**
      SUPERSONIC AIRCRAFT
      . SUPERSONIC TRANSPORTS
      . . **CL-823 AIRCRAFT**
      TRANSPORT AIRCRAFT
      . **CL-823 AIRCRAFT**
RT   ∞AIRCRAFT

CLADDING
GS   METAL WORKING
      . **CLADDING**
RT   ANODIC STRIPPING
       CATHODIC COATINGS
       COLD WORKING
       COMPOSITE MATERIALS
       EXPLOSIVE WELDING
       EXTRUDING
       LAMINATES
       METAL COATINGS
       METALLIZING
       PLATING
       PROTECTIVE COATINGS

CLAIMING
RT   ∞LAW
       PATENTS
       PREEMPTING

CLAMPING CIRCUITS
GS   CIRCUITS
      . **CLAMPING CIRCUITS**
RT   LIMITER CIRCUITS
       POWER LIMITERS

CLAMPS
RT   ∞BANDS
       CLIPS
       FASTENERS
       HOLDERS
       JIGS
       MECHANICAL DEVICES
       SEALING
       STRAPS

CLARITY
RT   ATMOSPHERIC OPTICS
       ELECTROMAGNETIC PROPERTIES
       HAZE
       OPACITY
       OPTICAL PROPERTIES
       PURITY
       ∞SHARPNESS

CLARITY-*(CONT.)*
    SOLUBILITY
    TRANSPARENCE
    TURBIDITY

CLARK Y AIRFOIL
  USE   AIRFOIL PROFILES

CLASSES
  RT   CATEGORIES
     ∞ GROUPS
     ∞ SECTIONS

CLASSIC AIRCRAFT
  USE   IL-62 AIRCRAFT

CLASSICAL MECHANICS
  GS   **CLASSICAL MECHANICS**
     . SPACE MECHANICS
     . . ASTRODYNAMICS
     . . CELESTIAL MECHANICS
     . . ORBITAL MECHANICS
     . . . KEPLER LAWS
     . . . MINIMUM VARIANCE ORBIT
          DETERMINATION
  RT   ANGULAR MOMENTUM
     CONTINUUM MECHANICS
     EQUATIONS OF MOTION
     EULER-LAGRANGE EQUATION
     HAMILTONIAN FUNCTIONS
     LAGRANGE COORDINATES
     MAXWELL BODIES
     ∞ MECHANICS (PHYSICS)
     MOMENTUM
     PHASE-SPACE INTEGRAL
     POISSON EQUATION
     QUATERNIONS
     STATISTICAL MECHANICS

CLASSIFICATIONS
  GS   **CLASSIFICATIONS**
     . HIERARCHIES
     . . BBGKY HIERARCHY
     . DICHOTOMIES
     . INDEXES (DOCUMENTATION)
     . . KWIC INDEXES
     . SUBJECTS
  RT   ASTRONOMICAL CATALOGS
     ∞ BREAKDOWN
     ∞ CLASSIFYING
     CLUSTER ANALYSIS
     TAXONOMY
     WISWESSER NOTATIONS

CLASSIFIERS
  GS   SEPARATORS
     . **CLASSIFIERS**
     . . SIZING SCREENS
     . . THICKENERS (EQUIPMENT)
  RT   CENTRIFUGES
     CONCENTRATING
     CONCENTRATORS
     FLOTATION
     ∞ SEPARATION
     SHAKERS
     SIZE DETERMINATION
     SIZE SEPARATION
     SPIRALS (CONCENTRATORS)

∞ CLASSIFYING
  SN   *(USE OF A MORE SPECIFIC TERM IS*
     RECOMMENDED--CONSULT THE TERMS
     LISTED BELOW)
  UF   SORTING
  RT   CLASSIFICATIONS
     CONCENTRATORS
     DISCRIMINANT ANALYSIS (STATISTICS)
     EVALUATION
     SECURITY
     SELECTION
     SIZE SEPARATION
     TAXONOMY

CLATHRATES
  RT   ∞ CHEMICAL COMPOUNDS
     CRYSTAL STRUCTURE
     CRYSTALS
     INCLUSIONS

CLAYS
  GS   **CLAYS**
     . ILLITE
     . KAOLINITE
     . MONTMORILLONITE

CLAYS-*(CONT.)*
     . VERMICULITE
  RT   ALLUVIUM
     BOREHOLES
     BRICKS
     CERAMICS
     COLLOIDS
     FANS (LANDFORMS)
     GROUT
     MASONRY
     MINING
     MOLDING MATERIALS
     MUD
     REFRACTORY MATERIALS
     ROCKS
     SEDIMENTARY ROCKS
     SEDIMENTS
     SHALES
     SIZING MATERIALS
     SOILS
     STRIP MINING

CLEAN ENERGY
  RT   AIR POLLUTION
     ENVIRONMENT POLLUTION
     ENVIRONMENTAL ENGINEERING
     GEOTHERMAL ENERGY CONVERSION
     SOLAR ENERGY
     TIDEPOWER
     WATER POLLUTION
     WATERWAVE ENERGY
     WINDPOWER UTILIZATION

CLEAN FUELS
  GS   FUELS
     . **CLEAN FUELS**
     . . FUEL OILS
  RT   BENEFICIATION
     CHEMICAL FUELS
     HYDROCARBON FUELS
     POLLUTION
     REFINING
     SYNTHETIC FUELS

CLEAN ROOMS
  GS   ROOMS
     . **CLEAN ROOMS**
  RT   ASSEMBLING
     CLEANLINESS
     CONTROLLED ATMOSPHERES
     ENVIRONMENTAL CONTROL

CLEANERS
  GS   **CLEANERS**
     . AIR FILTERS
  RT   ∞ ABSORBERS
     ABSORBERS (EQUIPMENT)
     ABSORBERS (MATERIALS)
     CLEANING
     CLEANLINESS
     SEPARATORS
     ULTRASONIC CLEANING
     WASHERS (CLEANERS)

CLEANING
  GS   **CLEANING**
     . HOUSEKEEPING (SPACECRAFT)
     . ULTRASONIC CLEANING
  RT   ABRASION
     ANTIFOULING
     ANTISEPTICS
     BLEACHING
     CHEMICAL CLEANING
     CLEANERS
     CLEANLINESS
     CLEARING
     CORROSION PREVENTION
     DECONTAMINATION
     DESCALING
     DISSOLVING
     DUST
     FLUSHING
     METAL FINISHING
     METAL POLISHING
     POLISHING
     PURIFICATION
     ∞ REDUCTION
     REFINING
     SCARFING
     SCAVENGING
     SCRUBBERS
     ∞ SEPARATION
     STERILIZATION
     SURFACE FINISHING
     WASHING

CLEANING-*(CONT.)*
     WASTE WATER

CLEANLINESS
  GS   **CLEANLINESS**
     . HOUSEKEEPING (SPACECRAFT)
  RT   CLEAN ROOMS
     CLEANERS
     CLEANING
     HYGIENE
     ORAL HYGIENE

CLEAR AIR TURBULENCE
  GS   TURBULENCE
     . ATMOSPHERIC TURBULENCE
     . . **CLEAR AIR TURBULENCE**
  RT   GUSTS
     JET STREAMS (METEOROLOGY)
     THERMAL INSTABILITY
     TURBULENT DIFFUSION
     WIND SHEAR

CLEARANCES
  RT   ADJUSTING
     ALIGNMENT
     ALLOWANCES
     DATUM (ELEVATION)
     SPACING
     TIGHTNESS
     TOLERANCES (MECHANICS)

CLEARING
  RT   CLEANING
     PURGING
     REMOVAL

CLEARINGS (OPENINGS)
  UF   SLASHES
  RT   DEFORESTATION
     FIREBREAKS
     FORESTS
     TREES (PLANTS)

CLEAVAGE
  UF   SCISSION
  RT   BRITTLE MATERIALS
     BRITTLENESS

CLEBSCH-GORDAN COEFFICIENTS
  GS   COEFFICIENTS
     . **CLEBSCH-GORDAN COEFFICIENTS**
  RT   ANGULAR MOMENTUM
     COUPLING

CLIFFS
  UF   BLUFFS (LANDFORMS)
  RT   CANYONS
     ESCARPMENTS
     FIORDS
     LANDSLIDES
     LEDGES
     ∞ SHELVES
     SLOPES
     TOPOGRAPHY

CLIMATE
  UF   MACROCLIMATE
  RT   CLIMATOLOGY
     LONG TERM EFFECTS
     METEOROLOGY
     WEATHER

CLIMATOLOGY
  UF   MILANKOVITCH THEORY
  GS   **CLIMATOLOGY**
     . AGROCLIMATOLOGY
     . MICROCLIMATOLOGY
  RT   ANTARCTIC REGIONS
     ANVIL CLOUDS
     ARCTIC REGIONS
     ATMOSPHERIC CIRCULATION
     CAP CLOUDS
     CIRROCUMULUS CLOUDS
     CIRROSTRATUS CLOUDS
     CIRRUS SHIELDS
     CLIMATE
     CLOUD COVER
     CLOUD DISPERSAL
     CLOUDS (METEOROLOGY)
     DENDROCHRONOLOGY
     DESERTLINE
     DESERTS
     ENVIRONMENTAL CHEMISTRY
     ENVIRONMENTAL ENGINEERING

CLIMATOLOGY-(CONT.)
    FOG DISPERSAL
    GEOGRAPHY
    HAILSTORMS
    HEAT ISLANDS
    HUMIDITY
    HURRICANES
    HYDROCLIMATOLOGY
    HYDROLOGY
    LIGHTNING SUPPRESSION
    METEOROLOGY
    MIDDLE ATMOSPHERE
    PERIODIC VARIATIONS
    PHENOLOGY
    POLAR METEOROLOGY
    POLAR REGIONS
    PRECIPITATION (METEOROLOGY)
    PRIMITIVE EQUATIONS
    RAINMAKING
    SEA BREEZE
    SEASONS
    SNOWSTORMS
    SOLAR RADIATION
    STORM ENHANCEMENT
    STORM SUPPRESSION
    STORMS
    STORMS (METEOROLOGY)
    SUNLIGHT
    TEMPERATE REGIONS
    TEMPERATURE
    TROPICAL REGIONS
    WEATHER
    WIND (METEOROLOGY)

CLIMBING FLIGHT
GS   ASCENT
    . **CLIMBING FLIGHT**
RT   ASCENT TRAJECTORIES
    COASTING FLIGHT
    ∞FLIGHT
    FLIGHT PATHS
    HORIZONTAL FLIGHT
    PARABOLIC FLIGHT
    ROCKET FLIGHT
    SOARING
    TAKEOFF
    TURNING FLIGHT
    VERTICAL FLIGHT

CLINICAL MEDICINE
RT   ANESTHESIOLOGY
    BED REST
    BLOOD VOLUME
    CASE HISTORIES
    DIAGNOSIS
    EXAMINATION
    HEALING
    HEALTH
    HUMAN BEINGS
    MEDICAL SCIENCE
    ∞MEDICINE
    ∞OPERATIONS
    SPORTS MEDICINE
    SURGERY
    TRANSPLANTATION
    ∞TREATMENT

CLIPPER CIRCUITS
GS   CIRCUITS
    . LIMITER CIRCUITS
    . . **CLIPPER CIRCUITS**
RT   COMPARATOR CIRCUITS
    POWER LIMITERS

CLIPS
RT   ANCHORS (FASTENERS)
    ∞BANDS
    CLAMPS
    COUPLINGS
    FASTENERS
    HOLDERS
    MECHANICAL DEVICES

CLOCK PARADOX
GS   TIME MEASUREMENT
    . **CLOCK PARADOX**
RT   ATOMIC CLOCKS
    CHRONOMETERS
    TIME SIGNALS
    TIMING DEVICES

CLOCKS
UF   WATCHES
GS   MEASURING INSTRUMENTS
    . TIME MEASURING INSTRUMENTS

CLOCKS-(CONT.)
    . . CLOCKS
    . . . ATOMIC CLOCKS
    . . . AUTONOMOUS SPACECRAFT
           CLOCKS
    . . . CHRONOMETERS
RT   TIME MEASUREMENT
    TIMING DEVICES

CLOGGING
USE   PLUGGING

CLOSE PACKED LATTICES
GS   CRYSTAL LATTICES
    . **CLOSE PACKED LATTICES**
RT   BODY CENTERED CUBIC LATTICES
    FACE CENTERED CUBIC LATTICES

CLOSED BASINS
USE   STRUCTURAL BASINS

CLOSED CIRCUIT TELEVISION
GS   COMMUNICATION EQUIPMENT
    . **CLOSED CIRCUIT TELEVISION**
    TELECOMMUNICATION
    . **CLOSED CIRCUIT TELEVISION**
    TELEVISION SYSTEMS
    . **CLOSED CIRCUIT TELEVISION**
RT   COLOR TELEVISION
    EDUCATIONAL TELEVISION
    STEREOTELEVISION
    TELEVISION CAMERAS
    TELEVISION RECEIVERS
    TELEVISION TRANSMISSION
    WIRELESS COMMUNICATION

CLOSED CYCLES
SN   (EXCLUDES CLOSED LOOP CONTROL
      SYSTEMS)
RT   CONTROL THEORY
    COOLING SYSTEMS
    ELECTRIC GENERATORS
    GAS TURBINES
    LOOPS
    NUCLEAR REACTORS
    PLASMA GENERATORS
    THERMAL CYCLING TESTS
    THERMODYNAMIC CYCLES

CLOSED ECOLOGICAL SYSTEMS
UF   BIOREGENERATIVE LIFE SUPPORT
      SYSTEMS
GS   SUPPORT SYSTEMS
    . LIFE SUPPORT SYSTEMS
    . . **CLOSED ECOLOGICAL SYSTEMS**
RT   AEROSPACE MEDICINE
    BIOASTRONAUTICS
    ECOLOGY
    ECOSYSTEMS
    GNOTOBIOTICS
    LONG TERM EFFECTS
    OXYGEN PRODUCTION
    SPACE HABITATS
    SPACECRAFT CABIN ATMOSPHERES
    SPACECRAFT ENVIRONMENTS
    SURVIVAL
    ∞SYSTEMS

CLOSED FAULTS
USE   GEOLOGICAL FAULTS

CLOSED LOOP SYSTEMS
USE   FEEDBACK CONTROL

CLOSING
RT   BLOCKING
    PLUGGING
    SEALING
    STOPPING

CLOSTRIDIUM BOTULINUM
GS   MICROORGANISMS
    . BACTERIA
    . . **CLOSTRIDIUM BOTULINUM**
RT   BACTERIOLOGY
    PATHOGENS
    TOXIC DISEASES

CLOSURE LAW
GS   LAWS
    . **CLOSURE LAW**
RT   FIELD THEORY (PHYSICS)
    STATISTICAL MECHANICS
    TURBULENT FLOW

CLOSURES
RT   ∞BARRIERS
    BLOCKING
    CHOKES (RESTRICTIONS)
    CONSTRICTIONS
    COUPLINGS
    COVERINGS
    ENCLOSURES
    END PLATES
    FASTENERS
    FITTINGS
    ∞GATES
    JOINTS (JUNCTIONS)
    PLUGGING
    PLUGS
    SEALS (STOPPERS)
    TIGHTNESS
    VALVES

CLOTH
USE   FABRICS

CLOTHING
GS   **CLOTHING**
    . BOOTS (FOOTWEAR)
    . COTTON FIBERS
    . COVERALLS
    . FLIGHT CLOTHING
    . GARMENTS
    . GLOVES
    . GOGGLES
    . PROTECTIVE CLOTHING
    . . HELMETS
    . . PRESSURE SUITS
    . . . SPACE SUITS
    . . VAPOR BARRIER CLOTHING
    . SHOES
    . SOCKS
    . SUITS
    . . PRESSURE SUITS
    . . . SPACE SUITS
RT   CHEMICAL DEFENSE
    CONSUMABLES (SPACECREW SUPPLIES)
    COTTON
    CUFFS
    FABRICS
    LEATHER
    TEXTILES
    VESTS

CLOTTING
RT   BLOOD COAGULATION
    EMBOLISMS
    THROMBOCYTES
    THROMBOPLASTIN

CLOUD CHAMBERS
GS   IONIZATION CHAMBERS
    . **CLOUD CHAMBERS**
RT   BUBBLE CHAMBERS
    ∞CHAMBERS
    RADIATION COUNTERS
    SPARK CHAMBERS

CLOUD COVER
UF   OVERCAST
RT   ANVIL CLOUDS
    ∞BLANKETS
    CAP CLOUDS
    CIRROCUMULUS CLOUDS
    CIRROSTRATUS CLOUDS
    CLIMATOLOGY
    CLOUDS (METEOROLOGY)
    FLIGHT CONDITIONS
    METEOROLOGICAL PARAMETERS
    METEOROLOGY
    METEOSAT SATELLITE
    NEPHANALYSIS
    SHADOWS
    SKY
    SKY BRIGHTNESS
    SOLAR RADIATION
    SUNLIGHT
    VENUS CLOUDS
    VENUS SURFACE
    WEATHER FORECASTING

CLOUD DISPERSAL
GS   WEATHER MODIFICATION
    . **CLOUD DISPERSAL**
RT   CLIMATOLOGY
    CLOUDS (METEOROLOGY)
    DISPERSING

CLOUD GLACIATION
GS   ICE FORMATION
      . **CLOUD GLACIATION**
RT   FREEZING
      HAIL
      ICE NUCLEI
      SNOW
      SNOW COVER

CLOUD HEIGHT INDICATORS
UF   CEILOMETERS
GS   MEASURING INSTRUMENTS
      . INDICATING INSTRUMENTS
      . **CLOUD HEIGHT INDICATORS**
      . METEOROLOGICAL INSTRUMENTS
      . . **CLOUD HEIGHT INDICATORS**
RT   CEILINGS (METEOROLOGY)

CLOUD PHOTOGRAPHS
GS   PHOTOGRAPHS
      . **CLOUD PHOTOGRAPHS**
RT   AERIAL PHOTOGRAPHY
      ALL SKY PHOTOGRAPHY
      PHOTOGRAPHY
      SPACEBORNE PHOTOGRAPHY
      TIROS PROJECT

CLOUD PHOTOGRAPHY
GS   IMAGERY
      . **CLOUD PHOTOGRAPHY**
      PHOTOGRAPHY
      . **CLOUD PHOTOGRAPHY**
RT   AERIAL PHOTOGRAPHY
      ALL SKY PHOTOGRAPHY
      BLACK AND WHITE PHOTOGRAPHY
      ESSA SATELLITES
      METEOROLOGICAL SATELLITES
      METEOSAT SATELLITE
      NIMBUS PROJECT
      NIMBUS SATELLITES
      NIMBUS 1 SATELLITE
      NIMBUS 2 SATELLITE
      SPACEBORNE PHOTOGRAPHY
      TIROS OPERATIONAL SATELLITE
        SYSTEM
      TIROS PROJECT
      TIROS SATELLITES

CLOUD PHYSICS
GS   ATMOSPHERIC PHYSICS
      . **CLOUD PHYSICS**
RT   AITKEN NUCLEI
      ATMOSPHERIC CLOUD PHYSICS LAB
        (SPACELAB)
      ATMOSPHERIC ELECTRICITY
      CONDENSATION NUCLEI
      CONDENSING
      CONVECTION CLOUDS
      DROP SIZE
      FOG DISPERSAL
      NEPHANALYSIS
      OPHIUCHI CLOUDS
      ∞ PHYSICS
      PRECIPITATION (METEOROLOGY)
      ∞ SCIENCE
      VENUS CLOUDS
      WEATHER MODIFICATION

CLOUD SEEDING
GS   NUCLEATION
      . **CLOUD SEEDING**
      WEATHER MODIFICATION
      . **CLOUD SEEDING**
RT   CLOUDS (METEOROLOGY)
      PRECIPITATION (METEOROLOGY)
      RAIN
      RAINMAKING
      STIMULATION

CLOUDS
GS   **CLOUDS**
      . ANVIL CLOUDS
      . CLOUDS (METEOROLOGY)
      . . ARTIFICIAL CLOUDS
      . . . CHEMICAL CLOUDS
      . . . . BARIUM ION CLOUDS
      . . CAP CLOUDS
      . . CIRROCUMULUS CLOUDS
      . . CIRROSTRATUS CLOUDS
      . . CIRRUS SHIELDS
      . . CONVECTION CLOUDS
      . . . CIRRUS CLOUDS
      . . . CUMULONIMBUS CLOUDS
      . . . . ARC CLOUDS
      . . . NIMBOSTRATUS CLOUDS

CLOUDS *(CONT.)*
      . . . STRATOCUMULUS CLOUDS
      . . . STRATUS CLOUDS
      . . CUMULUS CLOUDS
      . . NOCTILUCENT CLOUDS
      . ELECTRON CLOUDS
      . HYDROGEN CLOUDS
      . MAGELLANIC CLOUDS
      . MOLECULAR CLOUDS
      . PLASMA CLOUDS
      . VENUS CLOUDS
RT   DROP SIZE
      DUST
      METEOROID DUST CLOUDS
      PARTICLES

CLOUDS (METEOROLOGY)
UF   CHAOTIC CLOUD PATTERNS
GS   CLOUDS
      . **CLOUDS (METEOROLOGY)**
      . . ARTIFICIAL CLOUDS
      . . . CHEMICAL CLOUDS
      . . . . BARIUM ION CLOUDS
      . . CAP CLOUDS
      . . CIRROCUMULUS CLOUDS
      . . CIRROSTRATUS CLOUDS
      . . CIRRUS SHIELDS
      . . CONVECTION CLOUDS
      . . . CIRRUS CLOUDS
      . . CUMULONIMBUS CLOUDS
      . . . . ARC CLOUDS
      . . . NIMBOSTRATUS CLOUDS
      . . . STRATOCUMULUS CLOUDS
      . . . STRATUS CLOUDS
      . . CUMULUS CLOUDS
      . . NOCTILUCENT CLOUDS
RT   ACID RAIN
      ALPINE METEOROLOGY
      ATMOSPHERIC CORRECTION
      ATMOSPHERIC MOISTURE
      CLIMATOLOGY
      CLOUD COVER
      CLOUD DISPERSAL
      CLOUD SEEDING
      CONDENSATION NUCLEI
      DROP SIZE
      FOG
      FOG DISPERSAL
      HYDROGEN CLOUDS
      METEOROLOGY
      NEPHANALYSIS
      PLASMA CLOUDS
      PRECIPITATION (METEOROLOGY)
      SHADOWS
      SKY
      THUNDERSTORMS
      WEATHER

CLUMPS
UF   CLUSTERS
RT   AGGLOMERATION
      PATTERN RECOGNITION
      REGRESSION ANALYSIS

CLUSTER ANALYSIS
RT   CLASSIFICATIONS
      IMAGE ANALYSIS
      IMAGE PROCESSING
      PATTERN RECOGNITION
      REMOTE SENSING
      STATISTICAL ANALYSIS

CLUSTERS
USE   CLUMPS

CLUTCHES
RT   ENGINE PARTS
      MECHANICAL DEVICES
      MECHANICAL DRIVES

CLUTTER
GS   ECHOES
      . RADAR ECHOES
      . . **CLUTTER**
RT   JAMMING
      RADIO FREQUENCY INTERFERENCE

CMOS
UF   COMPLEMENTARY METAL OXIDE
        SEMICONDUCTORS
GS   ELECTRONIC EQUIPMENT
      . SOLID STATE DEVICES
      . SEMICONDUCTOR DEVICES
      . . METAL OXIDE SEMICONDUCTORS
      . . . . **CMOS**

CMOS *(CONT.)*
      SEMICONDUCTORS (MATERIALS)
      . METAL OXIDE SEMICONDUCTORS
      . . **CMOS**
RT   LATCH-UP

CN EMISSION
UF   CYANIDE EMISSION
GS   DECAY
      . EMISSION
      . . RADIO EMISSION
      . . . **CN EMISSION**
      ELECTROMAGNETIC RADIATION
      . RADIO WAVES
      . . RADIO EMISSION
      . . . **CN EMISSION**
RT   ELECTROMAGNETIC RADIATION
      HYDROCYANIC ACID
      MILLIMETER WAVES
      RADIO SOURCES (ASTRONOMY)

CNOIDAL WAVES
RT   ELASTIC WAVES
      GRAVITY WAVES
      SHALLOW WATER
      SOLITARY WAVES
      SURFACE WAVES
      WATER DEPTH
      WATER WAVES
      ∞ WAVES

COACHELLA VALLEY (CA)
GS   VALLEYS
      . **COACHELLA VALLEY (CA)**
RT   CALIFORNIA
      DESERTS

COAGULATION
GS   **COAGULATION**
      . BLOOD COAGULATION
RT   ACCUMULATIONS
      AGGLOMERATION
      AITKEN NUCLEI
      BLOOD
      COALESCING
      CONCENTRATING
      DEPOSITION
      EMBOLISMS
      FIBRIN
      FLOCCULATING
      FLOTATION
      GELATION
      HARDENING (MATERIALS)
      HEMORRHAGES
      LUMPING
      PRECIPITATION (CHEMISTRY)
      ∞ SEPARATION
      ∞ SETTING
      SETTLING
      SOLIDIFICATION
      THROMBOPENIA
      WATER TREATMENT

COAL
GS   RESOURCES
      . EARTH RESOURCES
      . . FOSSIL FUELS
      . . . **COAL**
      . . . . LIGNITE
      . . . . SOLVENT REFINED COAL
      ROCKS
      . **COAL**
      . . LIGNITE
      . . SOLVENT REFINED COAL
      SEDIMENTARY ROCKS
      . CARBONACEOUS ROCKS
      . . **COAL**
      . . . LIGNITE
RT   ASHES
      ASPHALTENES
      BITUMENS
      CARBONACEOUS MATERIALS
      COAL DERIVED GASES
      COAL DERIVED LIQUIDS
      COAL GASIFICATION
      COAL LIQUEFACTION
      COAL UTILIZATION
      COKE
      ENERGY POLICY
      FLY ASH
      HYDROPYROLYSIS
      PEAT
      REGOLITH
      SOILS
      STRIP MINING

COAL-*(CONT.)*
    SYNTHANE

COAL DERIVED GASES
  RT   CATALYSTS
        COAL
        COAL DERIVED LIQUIDS
        COAL GASIFICATION
        COAL UTILIZATION
        METHANE

COAL DERIVED LIQUIDS
  RT   ASPHALTENES
        CATALYSTS
        COAL
        COAL DERIVED GASES
        COAL GASIFICATION
        COAL LIQUEFACTION
        COAL UTILIZATION

COAL GASIFICATION
  GS   GASIFICATION
        . **COAL GASIFICATION**
        . . HYDROPYROLYSIS
  RT   COAL
        COAL DERIVED GASES
        COAL DERIVED LIQUIDS
        COAL LIQUEFACTION
        CRACKING (CHEMICAL ENGINEERING)
        ENERGY POLICY
        FUEL CELL POWER PLANTS
        GASES
        HYDROCARBON FUELS
        HYDROCRACKING
        LIGNITE
        METHANATION
        SYNTHANE
        VOLATILITY

COAL LIQUEFACTION
  GS   PHASE TRANSFORMATIONS
        . LIQUEFACTION
        . . **COAL LIQUEFACTION**
  RT   ASPHALTENES
        COAL
        COAL DERIVED LIQUIDS
        COAL GASIFICATION
        ENERGY POLICY
        HYDROCARBON FUELS
        HYDROCRACKING
        HYDROPYROLYSIS
        LIGNITE
        MELTING
        SOLVENT REFINED COAL

COAL UTILIZATION
  GS   UTILIZATION
        . **COAL UTILIZATION**
  RT   COAL
        COAL DERIVED GASES
        COAL DERIVED LIQUIDS
        ENERGY CONSUMPTION
        ENERGY POLICY
        ENERGY TECHNOLOGY
        HYDROCARBON FUELS
        LIGNITE
        SOLVENT REFINED COAL

COALESCENCE
  USE   COALESCING

COALESCING
  UF   COALESCENCE
  RT   AGGLOMERATION
        AGITATION
        COAGULATION
        CONCENTRATING
        FLOCCULATING
        MIXERS
        ∞SEPARATION
        SETTLING
        THICKENERS (EQUIPMENT)

COANDA EFFECT
  RT   ∞ATTACHMENT
        BUBBLES
        CIRCULATION CONTROL AIRFOILS
        ∞EFFECTS
        ENTRAINMENT
        FLUID AMPLIFIERS
        JET AMPLIFIERS
        JET STREAMS (METEOROLOGY)
        REATTACHED FLOW
        ∞SEPARATION
        THRUST AUGMENTATION

COARSENESS
  RT   FINENESS
        REFLECTANCE
        ROUGHNESS
        SURFACE PROPERTIES
        SURFACE ROUGHNESS
        SURFACE STABILITY
        SURFACE TEMPERATURE

COASTAL CURRENTS
  UF   LITTORAL CURRENTS
        LONGSHORE CURRENTS
  GS   CIRCULATION
        . WATER CIRCULATION
        . . WATER CURRENTS
        . . . OCEAN CURRENTS
        . . . . **COASTAL CURRENTS**
  RT   BEACHES
        COASTS
        ∞CURRENTS
        GYRES
        OCEANOGRAPHY
        OCEANS
        SEA TRUTH
        SEAS
        TIDES
        WETLANDS

COASTAL DUNES
  USE   DUNES

COASTAL ECOLOGY
  GS   ECOLOGY
        . **COASTAL ECOLOGY**
  RT   BIOMETEOROLOGY
        COASTS
        EARTH RESOURCES
        ENVIRONMENT EFFECTS
        ENVIRONMENTS
        MARINE ENVIRONMENTS
        MARINE RESOURCES
        OIL POLLUTION
        PHENOLOGY
        THERMAL POLLUTION
        WATERFOWL
        WETLANDS

COASTAL MARSHLANDS
  USE   MARSHLANDS

COASTAL PLAINS
  GS   LAND
        . PLAINS
        . . **COASTAL PLAINS**
  RT   BARS (LANDFORMS)
        BEACHES
        BIOMETEOROLOGY
        COASTS
        EARTH RESOURCES
        ECOLOGY
        ENVIRONMENTS
        PIEDMONTS
        WETLANDS

COASTAL RANGES (CA)
  GS   LANDFORMS
        . MOUNTAINS
        . . **COASTAL RANGES (CA)**
  RT   CALIFORNIA
        PACIFIC OCEAN

COASTAL WATER
  GS   WATER
        . NEARSHORE WATER
        . . **COASTAL WATER**
  RT   ENVIRONMENT EFFECTS
        OCEAN COLOR SCANNER
        OCEANS
        SEA WATER
        SHELLFISH
        SHORELINES
        VADOSE WATER
        WATER DEPTH
        WETLANDS

COASTAL ZONE COLOR SCANNER
  GS   SCANNERS
        . **COASTAL ZONE COLOR SCANNER**

COASTING FLIGHT
  RT   ASCENT TRAJECTORIES
        BALLISTIC TRAJECTORIES
        CLIMBING FLIGHT
        CRUISING FLIGHT

COASTING FLIGHT-*(CONT.)*
        DESCENT TRAJECTORIES
        ∞FLIGHT
        MIDCOURSE TRAJECTORIES
        PARABOLIC FLIGHT
        ROCKET FLIGHT
        SOARING

COASTS
  RT   BEACHES
        CASPIAN SEA
        COASTAL CURRENTS
        COASTAL ECOLOGY
        COASTAL PLAINS
        CORAL REEFS
        CUSPS (LANDFORMS)
        DUNES
        ESTUARIES
        LAGOONS
        LAKES
        LITTORAL DRIFT
        MARINE ENVIRONMENTS
        OCEANS
        SEAS
        SHORELINES
        STORM SURGES
        TIDAL FLATS
        UPWELLING WATER

COATING
  GS   COATING
        . ANODIZING
        . ELECTROPLATING
        . ENCAPSULATING
        . METALLIZING
  RT   ANODIC STRIPPING
        COATINGS
        CORROSION PREVENTION
        DEPOSITION
        FLAME PLATING
        FLAME SPRAYING
        LINING PROCESSES
        METAL FINISHING
        METAL SPRAYING
        ∞METALLURGY
        PLASMA SPRAYING
        ∞PRIMING
        SEALING
        SILICONIZING
        SPRAYING
        SURFACE FINISHING
        SURFACE PROPERTIES
        VAPOR DEPOSITION

COATINGS
  GS   COATINGS
        . ALUMINUM COATINGS
        . ANTIRADAR COATINGS
        . ANTIREFLECTION COATINGS
        . BIREFRINGENT COATINGS
        . CATHODIC COATINGS
        . ELECTROPLATING
        . ENAMELS
        . ENCAPSULATING
        . GLASS COATINGS
        . GLAZES
        . INORGANIC COATINGS
        . . ANODIC COATINGS
        . . CERAMIC COATINGS
        . LACQUERS
        . MAGNETIC FILMS
        . METAL COATINGS
        . . GOLD COATINGS
        . . NICKEL COATINGS
        . . ZINC COATINGS
        . METALLIZING
        . PAINTS
        . PLASTIC COATINGS
        . PROTECTIVE COATINGS
        . . ANODIC COATINGS
        . . CERAMIC COATINGS
        . . PRIMERS (COATINGS)
        . . REFRACTORY COATINGS
        . RUBBER COATINGS
        . SPRAYED COATINGS
        . THERMAL CONTROL COATINGS
  RT   ADDITIVES
        COATING
        COMPOSITE MATERIALS
        CORROSION
        CORROSION PREVENTION
        COVERINGS
        CRYODEPOSITS
        DEPOSITION
        DEPOSITS

COATINGS-(CONT.)
- DIPPING
- ELECTROLESS DEPOSITION
- ENERGY ABSORPTION FILMS
- EPOXY RESINS
- FABRICS
- ∞ FILMS
- FINISHES
- FLAME SPRAYING
- FURAN RESINS
- HOT CORROSION
- IMPREGNATING
- INHIBITORS
- LAMINATES
- ∞ LAYERS
- LINING PROCESSES
- LININGS
- METAL FILMS
- METAL FINISHING
- METAL SPRAYING
- ∞ METALLURGY
- MOISTURE RESISTANCE
- PASSIVITY
- PAVEMENTS
- PLASMA SPRAYING
- PLASTICIZERS
- ∞ PRIMING
- PROTECTION
- RUSTING
- SEALERS
- SEALING
- ∞ SHEETS
- SILICONIZING
- SOLVENTS
- SPRAYING
- SUBSTRATES
- SURFACE FINISHING
- SURFACE PROPERTIES
- THIN FILMS
- VAPOR DEPOSITION
- VENEERS
- WATERPROOFING
- WAXES
- WEATHERPROOFING
- WINGS

COAXIAL CABLES
- UF COAXIAL TRANSMISSION
- GS TRANSMISSION LINES
  - . COMMUNICATION CABLES
    - . . **COAXIAL CABLES**
- RT ∞ CABLES
  - POWER LINES
  - SUBMARINE CABLES
  - WAVEGUIDES

COAXIAL FLOW
- GS FLUID FLOW
  - . **COAXIAL FLOW**
- RT ANNULAR FLOW
  - ANNULAR NOZZLES
  - AXIAL FLOW
  - AXISYMMETRIC FLOW
  - FLOW GEOMETRY
  - HILSCH TUBES
  - SHEAR FLOW
  - STRATIFIED FLOW
  - TWO DIMENSIONAL FLOW

COAXIAL NOZZLES
- RT AIRCRAFT NOISE
  - AXIAL FLOW
  - FLUID FLOW
  - NOISE REDUCTION
  - NOZZLE GEOMETRY
  - ∞ NOZZLES
  - SUPERSONIC NOZZLES
  - VARIABLE CYCLE ENGINES

COAXIAL PLASMA ACCELERATORS
- GS PLASMA ACCELERATORS
  - . **COAXIAL PLASMA ACCELERATORS**
- RT ∞ ACCELERATORS
  - PLASMA ENGINES
  - PLASMA GUNS

COAXIAL TRANSMISSION
- USE COAXIAL CABLES
  - TRANSMISSION

COBALT
- GS CHEMICAL ELEMENTS
  - . **COBALT**
    - . . COBALT ISOTOPES
      - . . . COBALT 58

COBALT-(CONT.)
- . . . COBALT 60
- METALS
- . TRANSITION METALS
- . . **COBALT**
- . . . COBALT ISOTOPES
- . . . . COBALT 58
- . . . . COBALT 60
- RT STRATEGIC MATERIALS

COBALT ACETATES
- GS ACETATES
  - . **COBALT ACETATES**
  - COBALT COMPOUNDS
  - . **COBALT ACETATES**
  - ESTERS
  - . **COBALT ACETATES**

COBALT ALLOYS
- GS ALLOYS
  - . **COBALT ALLOYS**
  - . . ASTROLOY (TRADEMARK)
  - . . RENE 41
  - . . RENE 63
  - . . RENE 77
  - . . RENE 95
- RT HEAT RESISTANT ALLOYS
  - KOVAR (TRADEMARK)
  - STELLITE (TRADEMARK)
  - WASPALOY

COBALT COMPOUNDS
- GS **COBALT COMPOUNDS**
  - . COBALT ACETATES
  - . COBALT FLUORIDES
  - . COBALT OXALATES
  - . COBALT OXIDES
  - . COHENITE
- RT ∞ CHEMICAL COMPOUNDS
  - ∞ GROUP 8 COMPOUNDS
  - ∞ METAL COMPOUNDS

COBALT FLUORIDES
- GS COBALT COMPOUNDS
  - . **COBALT FLUORIDES**
  - HALOGEN COMPOUNDS
  - . FLUORINE COMPOUNDS
  - . . FLUORIDES
  - . . . METAL FLUORIDES
  - . . . . **COBALT FLUORIDES**
  - . HALIDES
  - . . METAL HALIDES
  - . . . **COBALT FLUORIDES**

COBALT ISOTOPES
- GS CHEMICAL ELEMENTS
  - . COBALT
  - . . **COBALT ISOTOPES**
  - . . . COBALT 58
  - . . . COBALT 60
  - . NUCLIDES
  - . . ISOTOPES
  - . . . **COBALT ISOTOPES**
  - . . . . COBALT 58
  - . . . . COBALT 60
  - METALS
  - . TRANSITION METALS
  - . . COBALT
  - . . . **COBALT ISOTOPES**
  - . . . . COBALT 58
  - . . . . COBALT 60

COBALT OXALATES
- GS COBALT COMPOUNDS
  - . **COBALT OXALATES**
  - OXALATES
  - . **COBALT OXALATES**

COBALT OXIDES
- GS CHALCOGENIDES
  - . OXIDES
  - . . METAL OXIDES
  - . . . **COBALT OXIDES**
  - COBALT COMPOUNDS
  - . **COBALT OXIDES**

COBALT 58
- GS CHEMICAL ELEMENTS
  - . COBALT
  - . . COBALT ISOTOPES
  - . . . **COBALT 58**
  - . NUCLIDES
  - . . ISOTOPES
  - . . . COBALT ISOTOPES
  - . . . . **COBALT 58**

COBALT 58-(CONT.)
- . . . RADIOACTIVE ISOTOPES
- . . . . **COBALT 58**
- METALS
- . TRANSITION METALS
- . . COBALT
- . . . COBALT ISOTOPES
- . . . . **COBALT 58**

COBALT 60
- GS CHEMICAL ELEMENTS
  - . COBALT
  - . . COBALT ISOTOPES
  - . . . **COBALT 60**
  - . NUCLIDES
  - . . ISOTOPES
  - . . . COBALT ISOTOPES
  - . . . . **COBALT 60**
  - . . . RADIOACTIVE ISOTOPES
  - . . . . **COBALT 60**
  - METALS
  - . TRANSITION METALS
  - . . COBALT
  - . . . COBALT ISOTOPES
  - . . . . **COBALT 60**

COBE
- USE COSMIC BACKGROUND EXPLORER
  - SATELLITE

COBOL
- GS LANGUAGES
  - . PROGRAMMING LANGUAGES
  - . . **COBOL**
- RT FORTRAN
  - PL/1

COBRA DANE (RADAR)
- GS RADAR
  - . SURVEILLANCE RADAR
  - . . **COBRA DANE (RADAR)**
  - . TRACKING RADAR
  - . . **COBRA DANE (RADAR)**
- RT ANTENNA ARRAYS
  - EARLY WARNING SYSTEMS
  - MISSILE TRAJECTORIES
  - RADAR SIGNATURES

COCCOMYCES
- GS FUNGI
  - . **COCCOMYCES**

COCHLEA
- GS ANATOMY
  - . SENSE ORGANS
  - . . EAR
  - . . . LABYRINTH
  - . . . . **COCHLEA**
- RT CORTI ORGAN

COCK AIRCRAFT
- USE AN-22 AIRCRAFT

COCKPIT SIMULATORS
- GS SIMULATORS
  - . TRAINING SIMULATORS
  - . . FLIGHT SIMULATORS
  - . . . **COCKPIT SIMULATORS**
- RT SPACECRAFT CABIN SIMULATORS
  - TRAINING DEVICES

COCKPITS
- RT AIRCRAFT COMPARTMENTS
  - CABIN ATMOSPHERES
  - ∞ CABINS
  - CANOPIES
  - EJECTION SEATS
  - FLYING EJECTION SEATS
  - FUSELAGES
  - PRESSURIZED CABINS
  - SPACE CAPSULES
  - SPACECRAFT CABIN ATMOSPHERES
  - SPACECRAFT CABINS
  - WINDSHIELDS

COCKROACHES
- UF BLATTIDAE
- GS ANIMALS
  - . INVERTEBRATES
  - . . ARTHROPODS
  - . . . INSECTS
  - . . . . **COCKROACHES**

COCKS
UF STOPCOCKS
GS VALVES
. **COCKS**
RT GAS VALVES
HYDRAULIC EQUIPMENT

COD AIRCRAFT
USE C-2 AIRCRAFT

CODE DIVISION MULTIPLE ACCESS
UF CDMA
GS TELECOMMUNICATION
. MULTIPLE ACCESS
. . **CODE DIVISION MULTIPLE ACCESS**
RADIO COMMUNICATION
. . RADIO RELAY SYSTEMS
. . . **CODE DIVISION MULTIPLE ACCESS**
TRANSMISSION
. SIGNAL TRANSMISSION
. . DATA TRANSMISSION
. . . MULTIPLE ACCESS
. . . . **CODE DIVISION MULTIPLE ACCESS**
RT ALOHA SYSTEM
FREQUENCY DIVISION MULTIPLE ACCESS
MULTICHANNEL COMMUNICATION
MULTIPLEXING
SATELLITE NETWORKS
SWITCHING
WIDEBAND COMMUNICATION

CODE DIVISION MULTIPLEXING
GS TRANSMISSION
. MULTIPLEXING
. . **CODE DIVISION MULTIPLEXING**
RT DATA TRANSMISSION
DEMULTIPLEXING
FREQUENCY DIVISION MULTIPLE ACCESS
RADIO COMMUNICATION
RADIO TRANSMISSION
SATELLITE TRANSMISSION
SIGNAL TRANSMISSION
TELECOMMUNICATION
WAVELENGTH DIVISION MULTIPLEXING

CODERS
UF ENCODERS
RT ANALOG TO DIGITAL CONVERTERS
CODING
DECODERS
PROGRAMMERS

∞ **CODES**
SN *(USE OF A MORE SPECIFIC TERM IS RECOMMENDED--CONSULT THE TERMS LISTED BELOW)*
RT BCH CODES
BINARY CODES
BITERNARY CODE
CODING
COLOR CODING
CONCATENATED CODES
CRYPTOGRAPHY
DIGITS
ERROR CORRECTING CODES
ERROR DETECTION CODES
MORSE CODE
STANDARDS
SYMBOLS

CODING
UF ENCODING
NOTATION
GS **CODING**
. DECODING
. REDUNDANCY ENCODING
. SIGNAL ENCODING
. WISWESSER NOTATIONS
RT ADDRESSING
ALPHABETS
ANALOG TO DIGITAL CONVERTERS
BCH CODES
CODERS
∞ CODES
COLOR CODING
COMPUTER PROGRAMMING
COMPUTER PROGRAMS
CONCATENATED CODES
CRYPTOGRAPHY
DATA TRANSMISSION
DICTIONARIES
DIGITAL TECHNIQUES

CODING-*(CONT.)*
ERROR DETECTION CODES
IDENTIFYING
INFORMATION THEORY
LANGUAGES
PARITY
PULSE COMPRESSION
SYMBOLIC PROGRAMMING
SYMBOLS

COEFFICIENT OF FRICTION
UF FRICTION COEFFICIENT
GS COEFFICIENTS
. **COEFFICIENT OF FRICTION**
SURFACE PROPERTIES
. **COEFFICIENT OF FRICTION**
RT FRICTION
FRICTION FACTOR
FRICTION REDUCTION
KINETIC FRICTION
SLIDING FRICTION
STATIC FRICTION

COEFFICIENTS
GS **COEFFICIENTS**
. ACCOMMODATION COEFFICIENT
. AERODYNAMIC COEFFICIENTS
. ATTENUATION COEFFICIENTS
. BINOMIAL COEFFICIENTS
. CLEBSCH-GORDAN COEFFICIENTS
. COEFFICIENT OF FRICTION
. COHERENCE COEFFICIENT
. CORRELATION COEFFICIENTS
. COUPLING COEFFICIENTS
. DIFFUSION COEFFICIENT
. . SORET COEFFICIENT
. FLOW COEFFICIENTS
. DISCHARGE COEFFICIENT
. HEAT TRANSFER COEFFICIENTS
. INFLUENCE COEFFICIENT
. . STRUCTURAL INFLUENCE COEFFICIENTS
. IONIZATION COEFFICIENTS
. NOZZLE THRUST COEFFICIENTS
. ONSAGER PHENOMENOLOGICAL COEFFICIENT
. RECOMBINATION COEFFICIENT
. REGRESSION COEFFICIENTS
. SCATTERING COEFFICIENTS
. VIRIAL COEFFICIENTS
. WIGNER COEFFICIENT
RT ∞ CONSTANT
CONSTANTS
MECHANICAL PROPERTIES
OPTICAL PROPERTIES
POLYNOMIALS
RACAH COEFFICIENT
SORET COEFFICIENT
STATISTICAL ANALYSIS
∞ WEIGHT

COENZYMES
GS ENZYMES
. **COENZYMES**
. . CYSTEAMINE
PROTEINS
. **COENZYMES**
. . GLUTATHIONE

COERCIVITY
RT MAGNETIC PROPERTIES
MAGNETIZATION

COESITE
GS CHALCOGENIDES
. OXIDES
. . DIOXIDES
. . . SILICON DIOXIDE
. . . . QUARTZ
. . . . . **COESITE**
. . SILICON OXIDES
. . . SILICON DIOXIDE
. . . . QUARTZ
. . . . . **COESITE**
MINERALS
. QUARTZ
. . **COESITE**
SILICON COMPOUNDS
. SILICON OXIDES
. . SILICON DIOXIDE
. . . QUARTZ
. . . . **COESITE**
RT EARTH CRUST
EARTH MANTLE
METEORITES

COESITE-*(CONT.)*
RUTILE
STISHOVITE
STONY METEORITES
TEKTITES

COFFEE
GS FARM CROPS
. **COFFEE**
RT BEVERAGES

COFFIN-MANSON LAW
GS LAWS
. **COFFIN-MANSON LAW**
RT CRACK PROPAGATION
FATIGUE LIFE
FATIGUE TESTS
METAL FATIGUE

COGENERATION
RT ELECTRIC GENERATORS
ELECTRIC POWER PLANTS
ENERGY CONVERSION
∞ GENERATION
HEAT GENERATION
∞ POWER PLANTS
SOLAR ENERGY CONVERSION
THERMAL ENERGY
WASTE ENERGY UTILIZATION

COGNITION
RT ARTIFICIAL INTELLIGENCE
DECISION MAKING
IDENTIFYING
IFF SYSTEMS (IDENTIFICATION)
PERCEPTION

COGNITIVE PSYCHOLOGY
GS PSYCHOLOGY
. **COGNITIVE PSYCHOLOGY**

COGO (PROGRAMMING LANGUAGE)
UF COORDINATE GEOMETRY LANGUAGE
GS LANGUAGES
. PROGRAMMING LANGUAGES
. . **COGO (PROGRAMMING LANGUAGE)**

COHENITE
GS COBALT COMPOUNDS
. **COHENITE**
IRON COMPOUNDS
. **COHENITE**
MINERALS
. **COHENITE**
NICKEL COMPOUNDS
. **COHENITE**

∞ **COHERENCE**
SN *(USE OF A MORE SPECIFIC TERM IS RECOMMENDED--CONSULT THE TERMS LISTED BELOW)*
RT COHERENT RADIATION
COHERENT SCATTERING
COHESION
CONGRUENCES
ELASTIC SCATTERING
INTELLIGIBILITY
LASER OUTPUTS
LASERS
MASER OUTPUTS
PHASE COHERENCE
WAVE DISPERSION
WAVE PROPAGATION

COHERENCE COEFFICIENT
GS COEFFICIENTS
. **COHERENCE COEFFICIENT**
RT COHERENT RADAR
COHERENT RADIATION
∞ INTERFERENCE
NOISE PROPAGATION
PHASE COHERENCE
STOCHASTIC PROCESSES

COHERENT ACOUSTIC RADIATION
GS COHERENT RADIATION
. **COHERENT ACOUSTIC RADIATION**
ELASTIC WAVES
. **COHERENT ACOUSTIC RADIATION**
RT ∞ RADIATION
ULTRASONIC RADIATION
UNDERWATER ACOUSTICS

COHERENT ANTI-STOKES RAMAN
SPECTROSCOPY
USE   RAMAN SPECTROSCOPY

COHERENT ELECTROMAGNETIC RADIATION
GS   COHERENT RADIATION
       . COHERENT ELECTROMAGNETIC
         RADIATION
       . . COHERENT LIGHT
       ELECTROMAGNETIC RADIATION
       . COHERENT ELECTROMAGNETIC
         RADIATION
       . . COHERENT LIGHT
RT   BEAMS (RADIATION)
       HOLOGRAPHY
       INFRARED RADIATION
       INTERSTELLAR MASERS
       IONIZING RADIATION
       KRYPTON FLUORIDE LASERS
       LASERS
       LIGHT (VISIBLE RADIATION)
       MASERS
       MODULATED CONTINUOUS RADIATION
       MONOCHROMATIC RADIATION
       ∞ RADIATION
       RADIO WAVES
       STIMULATED EMISSION
       STIMULATED EMISSION DEVICES
       TRAVELING WAVE MASERS
       ULTRAVIOLET RADIATION

COHERENT LIGHT
GS   COHERENT RADIATION
       . COHERENT ELECTROMAGNETIC
         RADIATION
       . . COHERENT LIGHT
       ELECTROMAGNETIC RADIATION
       . COHERENT ELECTROMAGNETIC
         RADIATION
       . . COHERENT LIGHT
       . LIGHT (VISIBLE RADIATION)
       . . COHERENT LIGHT
RT   GAMMA RAY LASERS
       HCN LASERS
       HOLOGRAPHIC INTERFEROMETRY
       HOLOGRAPHY
       LASER OUTPUTS
       LASERS
       MONOCHROMATIC RADIATION
       NEODYMIUM LASERS
       OPTICAL COMPUTERS
       OPTICAL MEMORY (DATA STORAGE)
       PHASE COHERENCE
       PLASMADYNAMIC LASERS
       RARE GAS-HALIDE LASERS
       SCATTER PLATES (OPTICS)
       SHIVA LASER SYSTEM
       STIMULATED EMISSION
       TWO-WAVELENGTH LASERS
       ULTRAVIOLET LASERS

COHERENT RADAR
GS   RADAR
       . COHERENT RADAR
RT   COHERENCE COEFFICIENT
       CONTINUOUS WAVE RADAR
       DOPPLER RADAR
       MOVING TARGET INDICATORS
       PULSE DOPPLER RADAR
       PULSE RADAR
       RADAR DETECTION
       SEARCH RADAR
       SURVEILLANCE RADAR
       TRACKING RADAR

COHERENT RADIATION
UF   COHERENT SOURCES
       COHERENT TRANSMISSION
GS   COHERENT RADIATION
       . COHERENT ACOUSTIC RADIATION
       . COHERENT ELECTROMAGNETIC
         RADIATION
       . . COHERENT LIGHT
RT   BEAMS (RADIATION)
       ∞ COHERENCE
       COHERENCE COEFFICIENT
       CONTINUOUS RADIATION
       CORPUSCULAR RADIATION
       ELASTIC WAVES
       ELECTROMAGNETIC RADIATION
       LIGHT (VISIBLE RADIATION)
       OPTICAL PROPERTIES
       ∞ RADIATION
       ∞ RAYS
       WAVE PROPAGATION

COHERENT SCATTERING
GS   SCATTERING
       . COHERENT SCATTERING
RT   ∞ COHERENCE
       COMPTON EFFECT
       ELASTIC SCATTERING
       INCOHERENT SCATTERING
       INELASTIC SCATTERING
       NUCLEAR SCATTERING

COHERENT SOURCES
USE   COHERENT RADIATION
       RADIATION SOURCES

COHERENT TRANSMISSION
USE   COHERENT RADIATION

COHESION
RT   AGGLUTINATION
       BONDING
       ∞ COHERENCE
       INTERNAL FRICTION
       INTERNAL PRESSURE
       PLASTIC PROPERTIES
       SPREADING

COHOMOLOGY
USE   HOMOLOGY

∞ COILS
SN   (USE OF A MORE SPECIFIC TERM IS
       RECOMMENDED--CONSULT THE TERMS
       LISTED BELOW)
RT   ELECTRIC COILS
       INDUCTORS
       MAGNET COILS
       MAGNETIC COILS
       SPRINGS (ELASTIC)
       TOROIDS
       WIRE

COIN AIRCRAFT
UF   LARA AIRCRAFT
       LIGHT ARMED RECONNAISSANCE
         AIRCRAFT
GS   COIN AIRCRAFT
       . F-5 AIRCRAFT
       . OV-10 AIRCRAFT
RT   ∞ AIRCRAFT
       LIGHT INTRATHEATER TRANSPORT

COINCIDENCE CIRCUITS
GS   CIRCUITS
       . COINCIDENCE CIRCUITS
RT   GATES (CIRCUITS)
       RADIATION COUNTERS
       SYNCHRONISM

COINING
GS   FORMING TECHNIQUES
       . PRESSING (FORMING)
       . . COINING
       METAL WORKING
       . COINING
RT   COLD PRESSING
       COLD WORKING
       DIES
       FORGING
       HOT PRESSING
       SIZING (SHAPING)
       STAMPING

COKE
GS   FUELS
       . COKE
RT   BITUMENS
       CARBON
       CHARCOAL
       COAL
       LIGNITE

COKE AIRCRAFT
USE   AN-24 AIRCRAFT

COLCHICINE
GS   ALIPHATIC COMPOUNDS
       . CYCLIC HYDROCARBONS
       . . COLCHICINE
       HETEROCYCLIC COMPOUNDS
       . ALKALOIDS
       . . COLCHICINE
       HYDROCARBONS
       . CYCLIC HYDROCARBONS
       . . COLCHICINE

COLCHICINE-(CONT.)
       NITROGEN COMPOUNDS
       . ALKALOIDS
       . . COLCHICINE

COLD ACCLIMATIZATION
GS   ADAPTATION
       . ACCLIMATIZATION
       . . COLD ACCLIMATIZATION
RT   HEAT ACCLIMATIZATION
       SUBZERO TEMPERATURE

COLD BOKKEVELD METEORITE
GS   CELESTIAL BODIES
       . METEORITES
       . . STONY METEORITES
       . . . CHONDRITES
       . . . . CARBONACEOUS METEORITES
       . . . . . COLD BOKKEVELD METEORITE

COLD CATHODE TUBES
GS   MICROWAVE EQUIPMENT
       . MICROWAVE TUBES
       . . COLD CATHODE TUBES
       . . . PHOTOTUBES
       . . . . PHOTOMULTIPLIER TUBES
RT   CATHODES
       ELECTRODES
       GAS DISCHARGES
       ∞ GAS TUBES
       TUBE CATHODES
       TUNNEL CATHODES

COLD CATHODES
GS   ELECTRODES
       . CATHODES
       . . TUBE CATHODES
       . . . COLD CATHODES
RT   GAS DISCHARGES
       TUNNEL CATHODES

COLD DRAWING
RT   DEEP DRAWING
       ∞ DRAWING
       METAL DRAWING

COLD FLOW TESTS
SN   (EXCLUDES MECHANICAL CREEP, TESTS)
GS   ENGINE TESTS
       . COLD FLOW TESTS
       GROUND TESTS
       . COLD FLOW TESTS
RT   CHECKOUT
       FEED SYSTEMS
       PLASTIC PROPERTIES
       PRELAUNCH TESTS
       PROPELLANT TESTS
       PROPULSION SYSTEM PERFORMANCE
       ROCKET ENGINE DESIGN
       STATIC TESTS
       ∞ TESTS

COLD FORMING
USE   COLD WORKING

COLD FRONTS
GS   FRONTS (METEOROLOGY)
       . COLD FRONTS
RT   AIR MASSES
       ∞ FRONTS
       METEOROLOGICAL PARAMETERS
       METEOROLOGY
       STORMS
       SYNOPTIC METEOROLOGY
       THUNDERSTORMS
       TORNADOES
       WARM FRONTS
       WEATHER FORECASTING

COLD GAS
GS   GASES
       . COLD GAS
RT   ATTITUDE CONTROL
       GAS JETS
       JET THRUST

COLD HARDENING
SN   (LIMITED TO HARDENING OF MATERIALS
       BY COOLING TO VERY LOW
       TEMPERATURES-- EXCLUDES
       PRECIPITATION HARDENING AT OR
       NEAR ROOM TEMPERATURE AND
       HARDENING VIA COLD WORKING)
GS   HARDENING (MATERIALS)

COLD HARDENING-*(CONT.)*
. **COLD HARDENING**
RT   BRITTLENESS
HARDNESS
PHASE TRANSFORMATIONS
PRECIPITATION HARDENING
WORK HARDENING

COLD NEUTRONS
GS   PARTICLES
. ELEMENTARY PARTICLES
. . FERMIONS
. . . NEUTRONS
. . . . **COLD NEUTRONS**
. NEUTRAL PARTICLES
. . NEUTRONS
. . . **COLD NEUTRONS**
RT   BARYONS

COLD PLASMAS
UF   LOW TEMPERATURE PLASMAS
GS   GASES
. IONIZED GASES
. . CHARGED PARTICLES
. . **COLD PLASMAS**
PARTICLES
. CHARGED PARTICLES
. . ENERGETIC PARTICLES
. . . PLASMAS (PHYSICS)
. . . . **COLD PLASMAS**
RT   COLLISIONLESS PLASMAS
RAREFIED PLASMAS

COLD PRESSING
RT   COINING
COMPACTING
HOT PRESSING
METAL WORKING
∞PRESSING
PRESSING (FORMING)
UPSETTING

COLD ROLLING
GS   FORMING TECHNIQUES
. COLD WORKING
. . **COLD ROLLING**
RT   METAL WORKING

COLD STRENGTH
GS   MECHANICAL PROPERTIES
. **COLD STRENGTH**
RT   HIGH TEMPERATURE TESTS
LOW TEMPERATURE ENVIRONMENTS
LOW TEMPERATURE TESTS
∞STRENGTH

COLD SURFACES
UF   COLD WALLS
RT   CRYOGENIC FLUID STORAGE
∞SURFACES

COLD TOLERANCE
GS   TOLERANCES (PHYSIOLOGY)
. **COLD TOLERANCE**
RT   BODY TEMPERATURE
EXPOSURE
FROSTBITE
HEAT TOLERANCE
HOMEOSTASIS
SUBZERO TEMPERATURE
THERMOREGULATION
VASOCONSTRICTION

COLD TRAPS
GS   TRAPS
. **COLD TRAPS**
RT   CONDENSERS (LIQUEFIERS)
CRYOGENICS
CRYOTRAPPING
FREEZING
REFRIGERATING
ULTRALOW TEMPERATURES
VACUUM APPARATUS
VAPOR TRAPS

COLD WALLS
USE   COLD SURFACES
WALLS

COLD WATER
GS   WATER
. **COLD WATER**
RT   POTABLE WATER

COLD WEATHER
GS   WEATHER
. **COLD WEATHER**
RT   FROST DAMAGE
LOW TEMPERATURE ENVIRONMENTS
PRESSURE ICE
SNOW COVER
SUBZERO TEMPERATURE
WEATHERPROOFING
WINTER

COLD WEATHER TESTS
GS   ENVIRONMENTAL TESTS
. **COLD WEATHER TESTS**
RT   HIGH TEMPERATURE TESTS
LOW TEMPERATURE TESTS
∞TESTS

COLD WELDING
GS   WELDING
. **COLD WELDING**
RT   ADHESION
BONDING
HIGH VACUUM
VACUUM EFFECTS

COLD WORKING
UF   COLD FORMING
GS   FORMING TECHNIQUES
. **COLD WORKING**
. . COLD ROLLING
. . ELECTROHYDRAULIC FORMING
. . EXPLOSIVE FORMING
RT   CLADDING
COINING
DEEP DRAWING
EXTRUDING
FORGING
∞JOINING
MAGNETIC FORMING
METAL DRAWING
METAL SPINNING
METAL WORKING
PEENING
ROLL FORMING
SHEARING
SHOT PEENING
STAMPING
STRETCH FORMING
STRETCHING
SWAGING
TEMPER (METALLURGY)
UPSETTING
WINDING

COLEOPTERA
GS   ANIMALS
. INVERTEBRATES
. . ARTHROPODS
. . . INSECTS
. . . . **COLEOPTERA**

COLIC
GS   DISEASES
. **COLIC**
RT   GASTROINTESTINAL SYSTEM
INTESTINES

COLLAGENS
GS   ANATOMY
. MUSCULOSKELETAL SYSTEM
. . CONNECTIVE TISSUE
. . . **COLLAGENS**
CELLS (BIOLOGY)
. FIBROBLASTS
. . **COLLAGENS**
RT   BONE MINERAL CONTENT
GELATINS
LEATHER
PROTEINS
SKIN (ANATOMY)

COLLAPSE
RT   BUCKLING
DEFORMATION
FAILURE
STRUCTURAL FAILURE

COLLATING
RT   BINDING
COMPILERS
CORRELATION
INSERTION
POSITION (LOCATION)
POSITIONING

COLLECTION
RT   ACCUMULATIONS
ACQUISITION
ASSEMBLING
INPUT
LUMPING
MUSEUMS
∞RECEIVING
SAMPLING
SELECTION
STOCKPILING

COLLECTORS
USE   ACCUMULATORS

COLLEGES
USE   UNIVERSITIES

COLLIMATION
UF   BEAMSHAPING
RT   ADJUSTING
ALIGNMENT
DIRECTIVITY
MICROBEAMS
∞ORIENTATION
POLARIZATION (WAVES)

COLLIMATORS
UF   AUTOCOLLIMATORS
GS   OPTICAL EQUIPMENT
. **COLLIMATORS**
RT   BEAM WAVEGUIDES
MIRRORS
OPTICAL MEASUREMENT

COLLINEARITY
GS   ANALYSIS (MATHEMATICS)
. CALCULUS
. . VECTOR ANALYSIS
. . . **COLLINEARITY**
. REAL VARIABLES
. . VECTOR ANALYSIS
. . . **COLLINEARITY**
GEOMETRY
. VECTOR ANALYSIS
. . **COLLINEARITY**
LINEARITY
. **COLLINEARITY**

COLLISION AVOIDANCE
UF   COLLISION WARNING DEVICES
GS   AVOIDANCE
. **COLLISION AVOIDANCE**
. . BEACON COLLISION AVOIDANCE
    SYSTEM
RT   AIR NAVIGATION
AIR TRAFFIC
AIR TRAFFIC CONTROL
AIRCRAFT APPROACH SPACING
AIRCRAFT GUIDANCE
AIRCRAFT SAFETY
AIRSPACE
APPROACH CONTROL
AUTOMATIC TRAFFIC ADVISORY AND
    RESOLUTION
COLLISIONS
FLIGHT PATHS
FLIGHT RULES
FLIGHT SAFETY
MIDAIR COLLISIONS
NATIONAL AIRSPACE UTILIZATION
    SYSTEM
RADAR
RADAR NAVIGATION
RADIO NAVIGATION
THREAT EVALUATION
TRAFFIC CONTROL
VISUAL FLIGHT
WARNING
WARNING SYSTEMS

COLLISION PARAMETERS
GS   RATES (PER TIME)
. **COLLISION PARAMETERS**
. . COLLISION RATES
RT   ∞ABSORPTION
BEAM INTERACTIONS
∞CROSS SECTIONS
MEAN FREE PATH
NUCLEAR INTERACTIONS
PARTICLE INTERACTIONS
PARTICLE THEORY
SCATTERING

COLLISION RATES
- GS   RATES (PER TIME)
  - . COLLISION PARAMETERS
  - . . **COLLISION RATES**

COLLISION WARNING DEVICES
- USE   COLLISION AVOIDANCE
  - WARNING SYSTEMS

COLLISIONAL PLASMAS
- GS   GASES
  - . IONIZED GASES
  - . . CHARGED PARTICLES
  - . . . **COLLISIONAL PLASMAS**
  - . . . . STRONGLY COUPLED PLASMAS
  - PARTICLES
  - . CHARGED PARTICLES
  - . . ENERGETIC PARTICLES
  - . . . PLASMAS (PHYSICS)
  - . . . . **COLLISIONAL PLASMAS**
  - . . . . . STRONGLY COUPLED PLASMAS
- RT   ELECTRON RUNAWAY (PLASMA PHYSICS)
  - HIGH TEMPERATURE PLASMAS
  - NUCLEAR FUSION
  - PLASMA CONDUCTIVITY
  - PLASMA DENSITY
  - PLASMA WAVES

COLLISIONLESS PLASMAS
- GS   GASES
  - . IONIZED GASES
  - . . CHARGED PARTICLES
  - . . . **COLLISIONLESS PLASMAS**
  - PARTICLES
  - . CHARGED PARTICLES
  - . . ENERGETIC PARTICLES
  - . . . PLASMAS (PHYSICS)
  - . . . . **COLLISIONLESS PLASMAS**
- RT   COLD PLASMAS
  - IONIC WAVES
  - KELVIN-HELMHOLTZ INSTABILITY
  - LOW DENSITY RESEARCH
  - RAREFIED PLASMAS

COLLISIONS
- GS   **COLLISIONS**
  - . ATOMIC COLLISIONS
  - . COULOMB COLLISIONS
  - . INELASTIC COLLISIONS
  - . IONIC COLLISIONS
  - . METEORITE COLLISIONS
  - . MIDAIR COLLISIONS
  - . . BIRD-AIRCRAFT COLLISIONS
  - . MOLECULAR COLLISIONS
  - . PARTICLE COLLISIONS
- RT   AIR BAG RESTRAINT DEVICES
  - AIR TRAFFIC CONTROL
  - AIRCRAFT ACCIDENTS
  - AIRCRAFT HAZARDS
  - AIRCRAFT SAFETY
  - COLLISION AVOIDANCE
  - CRASHES
  - FLIGHT HAZARDS
  - FLIGHT PATHS
  - GAS ATOMIZATION
  - PILOT ERROR
  - RECOILINGS
  - SCATTERING

COLLOCATION
- RT   ASSEMBLIES
  - ∞ASSEMBLY
  - CONGRUENCES
  - POSITION (LOCATION)
  - POSITIONING

COLLOIDAL GENERATORS
- RT   ATOMIZING
  - DISPERSIONS
  - ∞GENERATORS
  - PLASMA DIFFUSION
  - PLASMA GENERATORS
  - SPRAYERS
  - VAPORIZERS

COLLOIDAL PROPELLANTS
- UF   CORDITE
- GS   MIXTURES
  - . DISPERSIONS
  - . . COLLOIDS
  - . . . **COLLOIDAL PROPELLANTS**
  - PROPELLANTS
  - . **COLLOIDAL PROPELLANTS**
- RT   GELLED PROPELLANTS

COLLOIDAL PROPELLANTS-*(CONT.)*
- SLURRY PROPELLANTS
- SOLID PROPELLANTS
- SOLID SUSPENSIONS

COLLOIDING
- UF   LYOPHILIZATION
- GS   MIXING
  - . **COLLOIDING**
- RT   AGITATION
  - ATOMIZING
  - COLLOIDS
  - COMMINUTION
  - COMPOUNDING
  - DISPERSING
  - FLOCCULATING
  - GELATION
  - HOMOGENIZING
  - PRECIPITATION (CHEMISTRY)
  - SUSPENDING (MIXING)

COLLOIDS
- UF   LYOPHILS
- GS   MIXTURES
  - . DISPERSIONS
  - . . **COLLOIDS**
  - . . . AEROSOLS
  - . . . . FOG
  - . . . COLLOIDAL PROPELLANTS
- RT   BROWNIAN MOVEMENTS
  - CLAYS
  - COLLOIDING
  - ELECTRODIALYSIS
  - ELECTROPHORESIS
  - EMULSIONS
  - FOAMS
  - GELS
  - HOMEOSTASIS
  - NONNEWTONIAN FLUIDS
  - PARTICLES
  - PLASTISOLS
  - ∞SEPARATION

COLOMBIA
- GS   NATIONS
  - . **COLOMBIA**
- RT   LLANOS ORIENTALES (COLOMBIA)
  - MAGDALENA-CAUCA VALLEY (COLOMBIA)
  - SOUTH AMERICA

COLONIES
- RT   BACTERIA
  - BACTERIOLOGY

COLOR
- UF   COLORATION
- GS   ELECTROMAGNETIC PROPERTIES
  - . OPTICAL PROPERTIES
  - . . **COLOR**
  - . . . IRIDESCENCE
  - . . . . WATER COLOR
- RT   BRIGHTNESS
  - COLORIMETRY
  - CONTRAST
  - DARKNESS
  - DICHROISM
  - DISCOLORATION
  - ELECTROCHROMISM
  - FADING
  - HUMAN FACTORS ENGINEERING
  - INCANDESCENCE
  - ISOCHROMATICS
  - LIGHT (VISIBLE RADIATION)
  - PERCEPTION
  - PHOTOTROPISM
  - ∞PHYSICAL PROPERTIES
  - PREWHITENING
  - SPECTRA
  - SURFACE PROPERTIES
  - SYMBOLS
  - THERMOCHROMATIC MATERIALS
  - VEGETATIVE INDEX
  - VISIBILITY
  - VISION
  - WAVE DISPERSION

COLOR (PARTICLE PHYSICS)
- USE   QUANTUM CHROMODYNAMICS

COLOR CENTERS
- UF   F CENTERS
- RT   ∞CENTERS
  - FRANCK-CONDON PRINCIPLE

COLOR CODING
- RT   ∞CODES
  - CODING

COLOR INFRARED PHOTOGRAPHY
- GS   IMAGERY
  - . INFRARED PHOTOGRAPHY
  - . . **COLOR INFRARED PHOTOGRAPHY**
  - PHOTOGRAPHY
  - . MULTISPECTRAL PHOTOGRAPHY
  - . . INFRARED PHOTOGRAPHY
  - . . . **COLOR INFRARED PHOTOGRAPHY**
- RT   COLOR PHOTOGRAPHY
  - INFRARED IMAGERY

COLOR PERCEPTION
- USE   COLOR VISION

COLOR PHOTOGRAPHY
- GS   IMAGERY
  - . **COLOR PHOTOGRAPHY**
  - PHOTOGRAPHY
  - . **COLOR PHOTOGRAPHY**
- RT   AERIAL PHOTOGRAPHY
  - BLACK AND WHITE PHOTOGRAPHY
  - CINEMATOGRAPHY
  - COLOR INFRARED PHOTOGRAPHY
  - ORTHOPHOTOGRAPHY
  - PHOTOCHROMISM
  - PHOTOMAPPING
  - SHADOWGRAPH PHOTOGRAPHY
  - STEREOPHOTOGRAPHY
  - ULTRAVIOLET PHOTOGRAPHY
  - UNDERWATER PHOTOGRAPHY

COLOR TELEVISION
- GS   TELECOMMUNICATION
  - . **COLOR TELEVISION**
  - TELEVISION SYSTEMS
  - . **COLOR TELEVISION**
- RT   CLOSED CIRCUIT TELEVISION
  - COMMUNICATING
  - COMMUNICATION EQUIPMENT
  - EDUCATIONAL TELEVISION
  - SATELLITE TELEVISION
  - SPACECRAFT TELEVISION
  - STEREOTELEVISION
  - TELEVISION RECEPTION
  - TELEVISION TRANSMISSION

COLOR VISION
- UF   COLOR PERCEPTION
- GS   VISION
  - . **COLOR VISION**
- RT   EYE (ANATOMY)
  - YOUNG-HELMHOLTZ THEORY

COLOR-MAGNITUDE DIAGRAM
- UF   C-M DIAGRAM
- GS   DIAGRAMS
  - . **COLOR-MAGNITUDE DIAGRAM**
- RT   GLOBULAR CLUSTERS
  - HERTZSPRUNG-RUSSELL DIAGRAM
  - HORIZONTAL BRANCH STARS
  - MAIN SEQUENCE STARS
  - STAR CLUSTERS
  - STELLAR COLOR
  - STELLAR EVOLUTION
  - STELLAR MAGNITUDE

COLORADO
- GS   NATIONS
  - . UNITED STATES
  - . . **COLORADO**
- RT   COLORADO PLATEAU (US)
  - COLORADO RIVER (NORTH AMERICA)
  - MANITOU (CO)
  - PIKE'S PEAK (CO)
  - SAN JUAN MOUNTAINS (CO)

COLORADO PLATEAU (US)
- GS   LAND
  - . **COLORADO PLATEAU (US)**
  - LANDFORMS
  - . TERRACES (LANDFORMS)
  - . . PLATEAUS
  - . . . **COLORADO PLATEAU (US)**
- RT   ARIZONA
  - COLORADO
  - HIGHLANDS
  - NEW MEXICO
  - UTAH

COLORADO RIVER (NORTH AMERICA)
GS   RIVERS
       . **COLORADO RIVER (NORTH AMERICA)**
RT   ARIZONA
       COLORADO
       MEXICO
       UTAH

COLORATION
USE   COLOR

COLORIMETRY
GS   OPTICAL MEASUREMENT
       . **COLORIMETRY**
RT   CHEMICAL ANALYSIS
       CHROMATOGRAPHY
       COLOR
       ELECTROPHOTOMETRY
       LIQUID CHROMATOGRAPHY
       OCEAN COLOR SCANNER
       OPTICAL MEASURING INSTRUMENTS
       PHOTOMETRY
       SPECTROPHOTOMETRY
       SPECTROSCOPY
       THERMOCHROMATIC MATERIALS

COLS
USE   GAPS (GEOLOGY)

COLUMBIA (ORBITER)
UF   SPACE SHUTTLE ORBITER 102
GS   REENTRY VEHICLES
       . RECOVERABLE SPACECRAFT
       .. REUSABLE SPACECRAFT
       ... **COLUMBIA (ORBITER)**
       TRANSPORTATION
       . SPACE TRANSPORTATION
       .. SPACE TRANSPORTATION SYSTEM
       ... SPACE SHUTTLE ORBITERS
       .... **COLUMBIA (ORBITER)**
RT   MANNED SPACE FLIGHT
       SPACE SHUTTLE MISSION 31-A
       SPACE SHUTTLE MISSION 41-A
       SPACE SHUTTLE MISSION 61-A
       SPACE SHUTTLE MISSION 61-C
       SPACE SHUTTLE MISSION 61-E
       ∞SPACECRAFT

COLUMBIA RIVER BASIN (ID-OR-WA)
GS   LANDFORMS
       . STRUCTURAL BASINS
       .. RIVER BASINS
       ... **COLUMBIA RIVER BASIN**
           (ID-OR-WA)
RT   IDAHO
       OREGON
       RIVERS
       WASHINGTON

COLUMBIUM
USE   NIOBIUM

∞ **COLUMNS**
SN   *(USE OF A MORE SPECIFIC TERM IS*
       RECOMMENDED--CONSULT THE TERMS
       LISTED BELOW)
RT   COLUMNS (PROCESS ENGINEERING)
       COLUMNS (SUPPORTS)

COLUMNS (PROCESS ENGINEERING)
RT   ABSORBERS (EQUIPMENT)
       CHEMICAL REACTORS
       ∞COLUMNS
       CONCENTRATORS
       CONDENSERS (LIQUEFIERS)
       CONTACTORS
       CONTRACTORS
       DEHYDRATION
       DEHYDROGENATION
       DISTILLATION EQUIPMENT
       DRYING APPARATUS
       EXTRACTION
       SCRUBBERS
       SEPARATORS
       VAPORIZERS

COLUMNS (SUPPORTS)
GS   STRUCTURAL MEMBERS
       . **COLUMNS (SUPPORTS)**
       .. TAPERED COLUMNS
RT   BEAMS (SUPPORTS)
       ∞COLUMNS
       PYLON MOUNTING
       PYLONS

COLUMNS (SUPPORTS)-*(CONT.)*
       STRUTS
       STUDS (STRUCTURAL MEMBERS)
       TIMOSHENKO BEAMS
       TOWERS

∞ **COMA**
SN   *(USE OF A MORE SPECIFIC TERM IS*
       RECOMMENDED--CONSULT THE TERMS
       LISTED BELOW)
RT   ABERRATION
       BLACKOUT (PHYSIOLOGY)
       BLACKOUT PREVENTION
       COMET HEADS
       COMET NUCLEI
       COMET TAILS
       COMETARY ATMOSPHERES
       COMETS
       GRIGG-SKJELLERUP COMET
       KOHOUTEK COMET
       SCREEN EFFECT
       TEMPEL 2 COMET
       UNCONSCIOUSNESS

COMBAT
GS   MILITARY OPERATIONS
       . **COMBAT**
       WARFARE
       . **COMBAT**
RT   AIRCRAFT SURVIVABILITY
       B-1 AIRCRAFT
       ELECTRONIC WARFARE

∞ **COMBINATION**
SN   *(USE OF A MORE SPECIFIC TERM IS*
       RECOMMENDED--CONSULT THE TERMS
       LISTED BELOW)
RT   ADMIXTURES
       CONSOLIDATION
       MIXTURES
       PERMUTATIONS

COMBINATIONS (MATHEMATICS)
GS   ANALYSIS (MATHEMATICS)
       . COMBINATORIAL ANALYSIS
       .. **COMBINATIONS (MATHEMATICS)**
RT   PARTITIONS (MATHEMATICS)
       PERMUTATIONS

COMBINATORIAL ANALYSIS
GS   ANALYSIS (MATHEMATICS)
       . **COMBINATORIAL ANALYSIS**
       .. COMBINATIONS (MATHEMATICS)
       .. FACTORIALS
       . PARTITIONS (MATHEMATICS)
       .. PERMUTATIONS
RT   ∞ANALYZING
       ∞APPLICATIONS OF MATHEMATICS
       GRAPH THEORY
       INFORMATION THEORY
       NUMBER THEORY
       PROBABILITY THEORY
       SET THEORY

COMBINED CYCLE POWER GENERATION
RT   ELECTRIC GENERATORS
       ELECTRIC POWER PLANTS
       ENERGY TECHNOLOGY
       GAS TURBINES
       STEAM TURBINES

COMBINED STRESS
GS   STRESSES
       . **COMBINED STRESS**
RT   FATIGUE LIFE
       STRESS ANALYSIS
       STRESS CONCENTRATION
       STRESS INTENSITY FACTORS

COMBUSTIBILITY
USE   FLAMMABILITY

COMBUSTIBLE FLOW
GS   FLUID FLOW
       . **COMBUSTIBLE FLOW**
RT   BOUNDARY LAYER COMBUSTION
       COMBUSTION PHYSICS
       COMBUSTION PRODUCTS
       DETONATION WAVES
       FLAME PROPAGATION
       FUEL FLOW
       TURBULENT FLOW

COMBUSTION
UF   BURNING
       BURNING PROCESS
GS   **COMBUSTION**
       . AFTERBURNING
       . BOUNDARY LAYER COMBUSTION
       . DEFLAGRATION
       . EROSIVE BURNING
       . FUEL COMBUSTION
       .. NUCLEAR FUEL BURNUP
       . HYDROCARBON COMBUSTION
       . HYPERSONIC COMBUSTION
       . METAL COMBUSTION
       . PROPELLANT COMBUSTION
       .. SOLID PROPELLANT COMBUSTION
       ... SOLID PROPELLANT IGNITION
       . SPONTANEOUS COMBUSTION
       . SUPERSONIC COMBUSTION
RT   BACKFIRE
       BURNING RATE
       BURNING TIME
       BURNOUT
       CHARRING
       CHEMICAL EXPLOSIONS
       COMBUSTION CHAMBERS
       COMBUSTION CONTROL
       COMBUSTION EFFICIENCY
       COMBUSTION PHYSICS
       COMBUSTION PRODUCTS
       COMBUSTION STABILITY
       COMBUSTION TEMPERATURE
       COMBUSTION VIBRATION
       DETONATION
       DIFFUSION FLAMES
       EXOTHERMIC REACTIONS
       EXPLOSIONS
       EXTINGUISHING
       FIRE DAMAGE
       FIREBREAKS
       FIRES
       FLAME PROPAGATION
       FLAMEOUT
       FLAMES
       FLAMMABILITY
       FLASHBACK
       FOREST FIRES
       HEAT BALANCE
       HEAT GENERATION
       IGNITION
       IGNITION LIMITS
       INCENDIARY AMMUNITION
       INTERNAL COMBUSTION ENGINES
       OXIDATION
       ∞PHYSICS
       QUENCHING (COOLING)
       SPARK IGNITION
       SUPERSONIC COMBUSTION RAMJET
         ENGINES

COMBUSTION CHAMBERS
UF   COMBUSTORS
RT   BURNERS
       ∞CHAMBERS
       COMBUSTION
       ENGINE PARTS
       ENGINES
       FLAME HOLDERS
       FLAMEOUT
       FURNACES
       INTERNAL COMBUSTION ENGINES
       JET ENGINES
       PISTONS
       REFRACTORIES
       SPARK PLUGS
       THRUST CHAMBERS

† **COMBUSTION CONTROL**
RT   AUTOMATIC CONTROL
       BURNING RATE
       COMBUSTION
       ∞CONTROL
       ENGINE CONTROL
       FUEL CONTROL
       ∞REACTION CONTROL
       TEMPERATURE CONTROL

COMBUSTION EFFICIENCY
GS   EFFICIENCY
       . **COMBUSTION EFFICIENCY**
RT   ACEE PROGRAM
       BURNING RATE
       BURNING TIME
       EXHAUST GASES
       FUEL COMBUSTION
       FUEL CONSUMPTION

COMBUSTION EFFICIENCY-*(CONT.)*
    FUEL-AIR RATIO
    POWER EFFICIENCY
    PROPELLANT COMBUSTION
    PROPULSION SYSTEM PERFORMANCE
    PROPULSIVE EFFICIENCY
    THERMODYNAMIC EFFICIENCY

COMBUSTION HEAT
  USE   HEAT OF COMBUSTION

COMBUSTION INSTABILITY
  USE   COMBUSTION STABILITY

COMBUSTION PHYSICS
  GS   THERMODYNAMICS
      . **COMBUSTION PHYSICS**
  RT   AEROTHERMODYNAMICS
      COMBUSTIBLE FLOW
      COMBUSTION STABILITY
      DAMKOHLER NUMBER
      FLAME PROPAGATION
      HEAT OF COMBUSTION
      IGNITION
      ∞PHYSICS
      PLASMAS (PHYSICS)
      ∞SCIENCE
      THERMOCHEMISTRY

COMBUSTION PRODUCTS
  RT   AIR POLLUTION
      ASHES
      COMBUSTIBLE FLOW
      DILUENTS
      DUST
      EXHAUST GASES
      FIBER RELEASE
      FLUE GASES
      FLY ASH
      HIGH TEMPERATURE GASES
      ODORS
      POLLUTION TRANSPORT
      PRODUCTS
      REACTION PRODUCTS
      ROCKET EXHAUST
      SMOG
      SMOKE
      SOOT
      VAPORS
      WASTES

COMBUSTION STABILITY
  UF   ACOUSTIC COMBUSTION
      CHUGGING
      COMBUSTION INSTABILITY
  GS   DYNAMIC CHARACTERISTICS
      . DYNAMIC STABILITY
      . . **COMBUSTION STABILITY**
      . . . FLAME STABILITY
      STABILITY
      . DYNAMIC STABILITY
      . . **COMBUSTION STABILITY**
      . . . FLAME STABILITY
  RT   AXIAL MODES
      BURNING RATE
      FUEL COMBUSTION
      MOTION STABILITY
      PRESSURE OSCILLATIONS
      PROPELLANT COMBUSTION
      SOLID PROPELLANT COMBUSTION
      THERMAL INSTABILITY
      VELOCITY COUPLING

COMBUSTION TEMPERATURE
  GS   TEMPERATURE
      . **COMBUSTION TEMPERATURE**
  RT   EROSIVE BURNING
      FLAME TEMPERATURE
      FLASH POINT
      IGNITION TEMPERATURE
      OPERATING TEMPERATURE
      SPONTANEOUS COMBUSTION

COMBUSTION VIBRATION
  GS   VIBRATION
      . **COMBUSTION VIBRATION**
  RT   COMBUSTION
      ELASTIC WAVES
      STRUCTURAL STABILITY

COMBUSTION WAVES
  USE   FLAME PROPAGATION

COMBUSTION WIND TUNNELS
  GS   TEST FACILITIES
      . WIND TUNNELS
      . . **COMBUSTION WIND TUNNELS**
  RT   HYPERSONIC WIND TUNNELS
      HYPERVELOCITY WIND TUNNELS

COMBUSTORS
  USE   COMBUSTION CHAMBERS

COMET HEADS
  GS   CELESTIAL BODIES
      . COMETS
      . . **COMET HEADS**
  RT   ∞COMA
      COMETARY ATMOSPHERES
      SOLAR SYSTEM

COMET NUCLEI
  GS   CELESTIAL BODIES
      . COMETS
      . . **COMET NUCLEI**
  RT   ∞COMA
      COMETARY ATMOSPHERES
      SOLAR SYSTEM

COMET TAILS
  GS   CELESTIAL BODIES
      . COMETS
      . . **COMET TAILS**
      ELECTROMAGNETIC RADIATION
      . **COMET TAILS**
  RT   ∞COMA
      COMETARY ATMOSPHERES
      GRIGG-SKJELLERUP COMET
      RADIATION PRESSURE
      SOLAR SYSTEM
      SOLAR WIND

COMET 4 AIRCRAFT
  UF   DE HAVILLAND DH 106 AIRCRAFT
      DH 106 AIRCRAFT
  GS   COMMERCIAL AIRCRAFT
      . **COMET 4 AIRCRAFT**
      DE HAVILLAND AIRCRAFT
      . **COMET 4 AIRCRAFT**
      HAWKER SIDDELEY AIRCRAFT
      . **COMET 4 AIRCRAFT**
      JET AIRCRAFT
      . **COMET 4 AIRCRAFT**
      MONOPLANES
      . **COMET 4 AIRCRAFT**
      PASSENGER AIRCRAFT
      . **COMET 4 AIRCRAFT**
  RT   ∞AIRCRAFT

COMETARY ATMOSPHERES
  RT   ASTRONOMICAL PHOTOMETRY
      ∞COMA
      COMET HEADS
      COMET NUCLEI
      COMET TAILS
      COMETS
      IONOPAUSE

COMETS
  GS   CELESTIAL BODIES
      . **COMETS**
      . . AREND-ROLAND COMET
      . . COMET HEADS
      . . COMET NUCLEI
      . . COMET TAILS
      . . ENCKE COMET
      . . GIACOBINI-ZINNER COMET
      . . GRIGG-SKJELLERUP COMET
      . . HALLEY'S COMET
      . . HUMASON COMET
      . . IRAS-ARAKI-ALCOCK COMET
      . . KOHOUTEK COMET
      . . MOREHOUSE COMET
      . . MRKOS COMET
      . . SCHWASSMANN-WACHMANN COMET
      . . TEMPEL 2 COMET
      . . WEST COMET
  RT   BESSEL-BREDICHIN THEORY
      ∞COMA
      COMETARY ATMOSPHERES
      METEOROID SHOWERS
      METEOROIDS
      SOLAR SYSTEM

COMFORT
  RT   ACOUSTICS
      AIR CONDITIONING
      EFFICIENCY

COMFORT-*(CONT.)*
    ENVIRONMENTAL ENGINEERING
    GLARE
    HUMAN FACTORS ENGINEERING
    HUMIDITY
    ILLUMINATING
    ∞PERFORMANCE
    PHYSIOLOGICAL EFFECTS
    PSYCHOLOGICAL EFFECTS
    REWARD (PSYCHOLOGY)
    RIDING QUALITY
    SEATS
    TEMPERATURE
    VENTILATION

COMMAND AND CONTROL
  UF   COMMAND-CONTROL
  RT   ∞AUTOMATION
      AUTONOMY
      AWACS AIRCRAFT
      ∞COMMANDS
      ∞CONTROL
      DECISION MAKING
      E-2 AIRCRAFT
      E-3A AIRCRAFT
      E-4A AIRCRAFT
      GROUND SUPPORT EQUIPMENT
      LOGISTICS
      MANAGEMENT
      SURVEILLANCE
      TARGETS

COMMAND GUIDANCE
  UF   COMMAND SYSTEMS
  GS   GUIDANCE (MOTION)
      . **COMMAND GUIDANCE**
  RT   ∞COMMANDS
      GROUND SUPPORT EQUIPMENT
      INJECTION GUIDANCE
      MIDCOURSE GUIDANCE
      RENDEZVOUS GUIDANCE
      RENDEZVOUS SPACECRAFT
      SPACECRAFT GUIDANCE
      TERMINAL GUIDANCE

COMMAND LANGUAGES
  GS   LANGUAGES
      . **COMMAND LANGUAGES**
      . . QUERY LANGUAGES
  RT   INFORMATION RETRIEVAL

COMMAND MODULES
  GS   COMPARTMENTS
      . **COMMAND MODULES**
      MODULES
      . SPACECRAFT MODULES
      . . **COMMAND MODULES**
      SPACECRAFT COMPONENTS
      . SPACECRAFT MODULES
      . . **COMMAND MODULES**
  RT   APOLLO SPACECRAFT
      ∞COMMANDS
      MARQUARDT R4D ENGINE
      SERVICE MODULES
      SPACECRAFT DOCKING MODULES
      SPACECRAFT MODULES
      SPACECREW TRANSFER

COMMAND SERVICE MODULES
  UF   CSM
  GS   MODULES
      . SPACECRAFT MODULES
      . . **COMMAND SERVICE MODULES**
  RT   APOLLO PROJECT
      LUNAR ORBITS
      MANNED SPACECRAFT
      SKYLAB 1
      SKYLAB 2
      SKYLAB 3
      SKYLAB 4
      SPACECRAFT DOCKING MODULES

COMMAND SYSTEMS
  USE   COMMAND GUIDANCE

COMMAND-CONTROL
  USE   COMMAND AND CONTROL

COMMANDO AIRCRAFT
  USE   C-46 AIRCRAFT

∞ COMMANDS
SN   *(USE OF A MORE SPECIFIC TERM IS RECOMMENDED--CONSULT THE TERMS LISTED BELOW)*
RT   AUTONOMY
      COMMAND AND CONTROL
      COMMAND GUIDANCE
      COMMAND MODULES
      DECISIONS

COMMERCE
RT   COMMERCIAL SPACECRAFT
      CONSUMPTION
      COSTS
      ECONOMIC DEVELOPMENT
      FINANCE
      GROSS NATIONAL PRODUCT
      INDUSTRIAL AREAS
      INDUSTRIES
    ∞ INVESTMENT
      LIABILITIES
      LOSSES
      MANUFACTURING
      MARKET RESEARCH
      MARKETING
      PERT
      PRODUCT DEVELOPMENT
      PROJECT MANAGEMENT
      RISK
      SUPPLYING

COMMERCE LAB
RT   GOVERNMENT/INDUSTRY RELATIONS
    ∞ MICROGRAVITY APPLICATIONS
      MISSION PLANNING
      SPACE COMMERCIALIZATION
      SPACE SHUTTLE PAYLOADS
      USER REQUIREMENTS

COMMERCIAL AIRCRAFT
UF   COMMERCIAL AVIATION
GS   **COMMERCIAL AIRCRAFT**
    . BOEING 707 AIRCRAFT
    . BOEING 720 AIRCRAFT
    . BOEING 727 AIRCRAFT
    . BOEING 733 AIRCRAFT
    . BOEING 737 AIRCRAFT
    . BOEING 747 AIRCRAFT
    . BOEING 757 AIRCRAFT
    . BOEING 767 AIRCRAFT
    . COMET 4 AIRCRAFT
    . CV-340 AIRCRAFT
    . CV-440 AIRCRAFT
    . CV-880 AIRCRAFT
    . CV-990 AIRCRAFT
    . DC 3 AIRCRAFT
    . DC 7 AIRCRAFT
    . DC 8 AIRCRAFT
    . DC 9 AIRCRAFT
    . DC 10 AIRCRAFT
    . DH 121 AIRCRAFT
    . ELECTRA AIRCRAFT
    . EUROPEAN AIRBUS
    . . A-300 AIRCRAFT
    . . A-310 AIRCRAFT
    . . A-320 AIRCRAFT
    . F-28 TRANSPORT AIRCRAFT
    . IL-62 AIRCRAFT
    . JETSTREAM AIRCRAFT
    . L-1011 AIRCRAFT
    . LEAR JET AIRCRAFT
    . LIGHT TRANSPORT AIRCRAFT
    . P-160 AIRCRAFT
    . SE-210 AIRCRAFT
    . SUPERSONIC COMMERCIAL AIR
      TRANSPORT
    . . BOEING 2707 AIRCRAFT
    . . TU-144 AIRCRAFT
    . TU-104 AIRCRAFT
    . TU-124 AIRCRAFT
    . TU-134 AIRCRAFT
    . TU-154 AIRCRAFT
    . VC-10 AIRCRAFT
RT   ∞ AIRCRAFT
      AIRLINE OPERATIONS
      CARGO AIRCRAFT
      CIVIL AVIATION
      GENERAL AVIATION AIRCRAFT
      GROUND EFFECT MACHINES
      JET AIRCRAFT
      PASSENGER AIRCRAFT
      ROTARY WING AIRCRAFT
      SUPERSONIC TRANSPORTS
      TRANSPORT AIRCRAFT
      UTILITY AIRCRAFT

COMMERCIAL AIRCRAFT-*(CONT.)*
      V/STOL AIRCRAFT
      WATER TAKEOFF AND LANDING
      AIRCRAFT

COMMERCIAL AVIATION
USE   CIVIL AVIATION
      COMMERCIAL AIRCRAFT

COMMERCIAL ENERGY
RT   ALLOCATIONS
      DISTRIBUTING
      DOMESTIC ENERGY
      ECONOMIC FACTORS
    ∞ ENERGY
      ENERGY CONSUMPTION
      ENERGY CONVERSION
      INDUSTRIAL ENERGY
      TRANSPORTATION ENERGY

COMMERCIAL SPACECRAFT
GS   **COMMERCIAL SPACECRAFT**
    . RCA SATCOM SATELLITES
RT   AEROSPACE INDUSTRY
      AEROSPACE VEHICLES
      COMMERCE
      COMMUNICATION SATELLITES
      INDUSTRIES
      SPACE COMMERCIALIZATION
      SPACE INDUSTRIALIZATION
      SPACE MANUFACTURING
      SPACE PROCESSING

COMMINUTION
UF   ATTRITION (MATERIALS)
GS   **COMMINUTION**
    . CRUSHING
    . GRINDING (COMMINUTION)
    . SHREDDING
RT   ATOMIZING
      BENEFICIATION
      CHIPPING
      COLLOIDING
      CRUSHERS
      CUTTING
      DISINTEGRATION
      FLAKING
      FRAGMENTATION
      GAS ATOMIZATION
      GRINDING MILLS
      METAL POWDER
    ∞ MILLING
      PARTICLE PRODUCTION
      POWDER METALLURGY
    ∞ REDUCTION

COMMITTEE ON SPACE RESEARCH
UF   COSPAR (COMMITTEE)
RT   ∞ AEROSPACE SCIENCES
      CONFERENCES
      EUROPEAN SPACE PROGRAMS
      INTERNATIONAL COOPERATION
      NASA PROGRAMS
      PROGRAMS

COMMODITIES
RT   GOVERNMENT PROCUREMENT
      MANUFACTURING
      MARKET RESEARCH
      PROCUREMENT MANAGEMENT
      PRODUCTS

COMMONALITY
GS   STANDARDIZATION
    . **COMMONALITY**
RT   AIRCRAFT EQUIPMENT
      COST REDUCTION
      EFFICIENCY
      EQUIPMENT SPECIFICATIONS
      GROUND SUPPORT SYSTEMS
      SPACECRAFT COMPONENTS
      SPECIFICATIONS

COMMUNICATING
GS   **COMMUNICATING**
    . AIRCRAFT COMMUNICATION
    . CONVERSATION
    . ELECTROCUTANEOUS
      COMMUNICATION
    . GROUND-AIR-GROUND
      COMMUNICATION
    . INFORMATION DISSEMINATION
    . . MESSAGES
    . . SELECTIVE DISSEMINATION OF
      INFORMATION

COMMUNICATING-*(CONT.)*
    . INTERSTELLAR COMMUNICATION
    . LIP READING
    . POINT TO POINT COMMUNICATION
    . . NASCOM NETWORK
    . UNDERGROUND COMMUNICATION
    . VERBAL COMMUNICATION
RT   COLOR TELEVISION
      CROSSTALK
      EDUCATION
      FREQUENCY ASSIGNMENT
      INFORMATION
      INFORMATION FLOW
      INFORMATION MANAGEMENT
      MESSAGE PROCESSING
      MORSE CODE
      STEREOTELEVISION
      SYSTEMS ENGINEERING
      TDR SATELLITES
      TECHNOLOGY TRANSFER
      TELECOMMUNICATION

COMMUNICATION
GS   TELECOMMUNICATION
    . **COMMUNICATION**
    . . FACSIMILE COMMUNICATION
    . . LINE OF SIGHT COMMUNICATION
    . . OPTICAL COMMUNICATION
    . . SHIP TO SHORE COMMUNICATION
    . . UNDERWATER COMMUNICATION
RT   INFORMATION
      INFORMATION FLOW
      INFORMATION MANAGEMENT
      MARISAT SATELLITES
      MESSAGE PROCESSING
      SPREAD SPECTRUM TRANSMISSION
      TECHNOLOGY TRANSFER

COMMUNICATION CABLES
GS   TRANSMISSION LINES
    . **COMMUNICATION CABLES**
    . . COAXIAL CABLES
    . . WAVEGUIDES
    . . . BEAM WAVEGUIDES
    . . . CIRCULAR WAVEGUIDES
      OPTICAL WAVEGUIDES
    . . . PLASMAGUIDES
    . . . RECTANGULAR WAVEGUIDES
RT   ∞ CABLES
      ELECTRIC WIRE
      SUBMARINE CABLES

COMMUNICATION EQUIPMENT
GS   **COMMUNICATION EQUIPMENT**
    . ADVANCED VIDICON CAMERA SYSTEM
      (AVCS)
    . CLOSED CIRCUIT TELEVISION
    . DIPLEXERS
    . INTERPHONES
    . PLAT SYSTEM
    . RADIO RECEIVERS
    . . SUPERHETERODYNE RECEIVERS
    . . TRANSMITTER RECEIVERS
    . . WHISTLER RECORDERS
    . SPACECRAFT TELEVISION
    . . DIGITAL SPACECRAFT TELEVISION
    . . RANGER BLOCK 3 TELEVISION
      SYSTEM
    . SATELLITE TELEVISION
    . STEREOTELEVISION
RT   ANTENNA COMPONENTS
      BIOTELEMETRY
      COLOR TELEVISION
      EARTH TERMINALS
      EDUCATIONAL TELEVISION
    ∞ EQUIPMENT
      FURLABLE ANTENNAS
      INERTIALESS STEERABLE ANTENNAS
      INFORMATION ADAPTIVE SYSTEM
      LOGARITHMIC RECEIVERS
      MATCHED FILTERS
      ORBITING DIPOLES
      P.A.C.M. TELEMETRY
      PULSE FREQUENCY MODULATION
      PULSE FREQUENCY MODULATION
      TELEMETRY
      RADIO COMMUNICATION
      RADIO EQUIPMENT
      RADIO RELAY SYSTEMS
      RADIO TELEGRAPHY
      RADIO TELEMETRY
      SPHERICAL ANTENNAS
      TELEMETRY
      TELEPHONY
      TELEVISION SYSTEMS

COMMUNICATION EQUIPMENT-*(CONT.)*
    UNIFIED S BAND

COMMUNICATION NETWORKS
GS   **COMMUNICATION NETWORKS**
    . NASCOM NETWORK
RT   ACCESS CONTROL
    BROADCASTING
    DATA LINKS
    DEFENSE COMMUNICATIONS SYSTEM
      (DCS)
    DEMAND ASSIGNMENT MULTIPLE
      ACCESS
    ELECTRONIC MAIL
    FREQUENCY DIVISION MULTIPLEXING
    NETWORK CONTROL
    PACKET SWITCHING
    PACKETS (COMMUNICATION)
    PROTOCOL (COMPUTERS)
    PULSE COMMUNICATION
    RADIO COMMUNICATION
    SATELLITE NETWORKS
    TELECOMMUNICATION

COMMUNICATION SATELLITES
GS   SATELLITES
    . ARTIFICIAL SATELLITES
    . . **COMMUNICATION SATELLITES**
    . . . AERONAUTICAL SATELLITES
    . . . . AEROSAT SATELLITES
    . . . ARCOMSAT
    . . . COMMUNICATIONS TECHNOLOGY
      SATELLITE
    . . . . COMSTAR C
    . . . . NATO 3B SATELLITE
    . . . COMSTAR SATELLITES
    . . . EUROPEAN COMMUNICATIONS
      SATELLITE
    . . . INTELSAT SATELLITES
    . . . L-SAT
    . . . LOW FREQUENCY
      TRANSIONOSPHERIC SATELLITES
    . . . MARECS MARITIME SATELLITES
    . . . MAROTS (ESA)
    . . . MOLNIYA SATELLITES
    . . . PALAPA SATELLITES
    . . . . PALAPA 2 SATELLITE
    . . . RADUGA SATELLITE
    . . . RCA SATCOM SATELLITES
    . . . RELAY SATELLITES
    . . . . RELAY 1 SATELLITE
    . . . . RELAY 2 SATELLITE
    . . . SYMPHONIE SATELLITES
    . . . SYNCOM SATELLITES
    . . . . EARLY BIRD SATELLITES
    . . . . SYNCOM 1 SATELLITE
    . . . . SYNCOM 2 SATELLITE
    . . . . SYNCOM 3 SATELLITE
    . . . . SYNCOM 4 SATELLITE
    . . . WESTAR SATELLITES
RT   ADVENT PROJECT
    ATS
    CARRIER TO NOISE RATIOS
    COMMERCIAL SPACECRAFT
    COMSAT PROGRAM
    DEFENSE COMMUNICATIONS SATELLITE
      SYSTEM
    DEMAND ASSIGNMENT MULTIPLE
      ACCESS
    DOMESTIC SATELLITE COMMUNICATIONS
      SYSTEMS
    DOWNLINKING
    EARTH TERMINAL MEASUREMENT
      SYSTEM
    ECHO PROJECT
    ELECTRONIC MAIL
    FLEET SATELLITE COMMUNICATION
      SYSTEM
    GEOPHYSICAL SATELLITES
    GROUND-AIR-GROUND COMMUNICATION
    HET EXPERIMENT
    INDIAN SPACE PROGRAM
    LAND MOBILE SATELLITE SERVICE
    MOBILE COMMUNICATION SYSTEMS
    MSAT
    NETWORK CONTROL
    ORBIT SPECTRUM UTILIZATION
    PASSIVE SATELLITES
    RADIO RELAY SYSTEMS
    SATELLITE NETWORKS
    SKYNET SATELLITES
    SPACE COMMERCIALIZATION
    SPACE COMMUNICATION
    SYNCHRONOUS COMMUNICATIONS
      SATELLITE PROJ

COMMUNICATION SATELLITES-*(CONT.)*
    SYNCHRONOUS METEOROLOGICAL
      SATELLITE
    SYNCHRONOUS PLATFORMS
    SYNCHRONOUS SATELLITES
    TELECOMMUNICATION
    TELECONFERENCING
    TELSTAR PROJECT
    UNMANNED SPACECRAFT
    UPLINKING
    WIRELESS COMMUNICATION

COMMUNICATION SYSTEMS
USE   TELECOMMUNICATION

COMMUNICATION THEORY
UF   STATISTICAL COMMUNICATION THEORY
GS   **COMMUNICATION THEORY**
    . WORDS (LANGUAGE)
    . . SYLLABLES
RT   CROSS COUPLING
    CYBERNETICS
    DATA TRANSMISSION
    HIGH LEVEL LANGUAGES
    INFORMATION THEORY
    INTELLIGIBILITY
    LANGUAGES
    LATTICES (MATHEMATICS)
    MESSAGES
    NETWORK SYNTHESIS
    RANDOM NOISE
    RANDOM PROCESSES
    REDUNDANCY
    SEMANTICS
    SENTENCES
    SIGNAL TO NOISE RATIOS
    SWITCHING THEORY
    ∞ THEORIES

COMMUNICATIONS TECHNOLOGY SATELLITE
UF   HERMES SATELLITE
GS   SATELLITES
    . ARTIFICIAL SATELLITES
    . . COMMUNICATION SATELLITES
    . . . **COMMUNICATIONS TECHNOLOGY**
      SATELLITE
    . . . . COMSTAR C
    . . . . NATO 3B SATELLITE
RT   CANADA
    CANADIAN SPACE PROGRAM
    INTERNATIONAL COOPERATION
    NASA PROGRAMS
    SYNCHRONOUS SATELLITES
    TECHNOLOGY ASSESSMENT
    TECHNOLOGY UTILIZATION

COMMUNITIES
GS   **COMMUNITIES**
    . INHABITANTS
    . . MOUNTAIN INHABITANTS
RT   CITIES
    DEMOGRAPHY
    ETHNIC FACTORS
    INTEGRATED ENERGY SYSTEMS
    MEGALOPOLISES
    MINORITIES
    MODULAR INTEGRATED UTILITY SYSTEM
    NATIONS
    POLICE
    POLITICS
    REGIMES
    SOCIOLOGY
    STARSITE PROGRAM
    UNITED NATIONS
    URBAN DEVELOPMENT
    URBAN PLANNING
    URBAN RESEARCH

COMMUTATION
RT   COMMUTATORS
    INTERPOLATION
    SWITCHING THEORY

COMMUTATORS
GS   **COMMUTATORS**
    . DECOMMUTATORS
RT   ARMATURES
    COMMUTATION
    DISTRIBUTORS
    ELECTRIC CONTACTS
    ELECTRIC MOTORS
    ∞ ROTATING ELECTRICAL MACHINES
    ROTATING GENERATORS

COMPACTING
RT   AGGLOMERATION
    COLD PRESSING
    DENSIFICATION
    HOT PRESSING
    POWDER METALLURGY
    PRESSES
    ∞ PRESSING
    PRESSING (FORMING)
    VIBRATION

COMPACTNESS
USE   VOID RATIO

COMPANDING
RT   FREQUENCY MODULATION
    MODULATION
    RADIO TRANSMISSION
    SIGNAL PROCESSING
    SIGNAL TO NOISE RATIOS

COMPANION STARS
GS   CELESTIAL BODIES
    . STARS
    . . DOUBLE STARS
    . . . BINARY STARS
    . . . . **COMPANION STARS**
RT   PARALLAX
    STELLAR MOTIONS
    VARIABLE STARS
    VISUAL OBSERVATION
    X RAY BINARIES

COMPARATOR CIRCUITS
GS   CIRCUITS
    . **COMPARATOR CIRCUITS**
RT   CLIPPER CIRCUITS
    DELAY CIRCUITS
    DISCRIMINATION
    TIME DISCRIMINATION

COMPARATORS
GS   MEASURING INSTRUMENTS
    . **COMPARATORS**
RT   DISCRIMINATORS
    ERROR SIGNALS
    HARMONIC GENERATORS
    MONOCHROMATORS
    REFLECTOMETERS

COMPARISON
RT   ANALOGIES
    COST ANALYSIS
    ECONOMIC ANALYSIS
    ESTIMATES
    EVALUATION
    EXAMINATION
    MATCHING
    PATTERN REGISTRATION
    RANKING

COMPARTMENTATION
USE   COMPARTMENTS

COMPARTMENTS
UF   COMPARTMENTATION
GS   **COMPARTMENTS**
    . AIR LOCKS
    . AIRCRAFT COMPARTMENTS
    . COMMAND MODULES
    . PRESSURIZED CABINS
    . SPACECRAFT CABINS
    . TEST CHAMBERS
    . . ANECHOIC CHAMBERS
    . . PRESSURE CHAMBERS
    . . . HYPERBARIC CHAMBERS
    . . . VACUUM CHAMBERS
RT   BAYS (STRUCTURAL UNITS)
    ∞ CELLS
    CREW EXPERIMENT STATIONS
    CREW OBSERVATION STATIONS
    CREW STATIONS
    CREW WORKSTATIONS
    ENCLOSURES
    MODULES
    ROOMS
    SPACECRAFT MODULES

COMPASS (PROGRAMMING LANGUAGE)
GS   LANGUAGES
    . PROGRAMMING LANGUAGES
    . . ASSEMBLY LANGUAGE
    . . . **COMPASS (PROGRAMMING**
      LANGUAGE)

COMPASS (PROGRAMMING-(CONT.)
- RT　COMPILERS
- 　　COMPUTER PROGRAMMING

COMPASSES
- GS　AIRCRAFT INSTRUMENTS
- 　. **COMPASSES**
- 　. . GYROCOMPASSES
- 　. . MAGNETIC COMPASSES
- 　. . SOLAR COMPASSES
- 　MEASURING INSTRUMENTS
- 　. **COMPASSES**
- 　. . GYROCOMPASSES
- 　NAVIGATION AIDS
- 　. NAVIGATION INSTRUMENTS
- 　. . **COMPASSES**
- 　. . . GYROCOMPASSES
- 　. . . MAGNETIC COMPASSES
- 　. . . SOLAR COMPASSES
- RT　BEACONS
- 　　BUOYS
- 　　FLIGHT INSTRUMENTS
- 　　RADAR BEACONS
- 　　RADIO DIRECTION FINDERS
- 　　TRANSITS

COMPATIBILITY
- GS　**COMPATIBILITY**
- 　. BIOCOMPATIBILITY
- 　. ELECTROMAGNETIC COMPATIBILITY
- 　. SYSTEMS COMPATIBILITY
- RT　ACCEPTABILITY
- 　　AFFINITY
- 　　CONVENTIONS
- 　　PERMISSIVITY
- 　　STABILITY
- 　　SUITABILITY
- 　　VERSATILITY

∞ **COMPENSATION**
- SN　(USE OF A MORE SPECIFIC TERM IS RECOMMENDED--CONSULT THE TERMS LISTED BELOW)
- RT　ALLOWANCES
- 　　ERRORS
- 　　IMAGE MOTION COMPENSATION
- 　　INSTRUMENT COMPENSATION
- 　　TEMPERATURE COMPENSATION
- 　　TRANSIENT RESPONSE

COMPENSATORS
- RT　BALANCE
- 　　BIAS
- 　　COMPULSATORS
- 　　ERROR SIGNALS
- 　　FEEDBACK
- 　　VIDEO EQUIPMENT

COMPENSATORY TRACKING
- GS　TRACKING (POSITION)
- 　. **COMPENSATORY TRACKING**
- RT　INFRARED TRACKING
- 　　OPTICAL TRACKING
- 　　RADAR TRACKING

COMPETITION
- RT　ATHLETES
- 　　HUMAN PERFORMANCE
- 　　HUMAN REACTIONS
- 　　PHYSICAL FITNESS

COMPILATION (COMPUTERS)
- USE　COMPILERS

COMPILER PROGRAMS
- USE　COMPILERS

COMPILERS
- SN　(PROGRAM-MAKING ROUTINES FOR DIGITAL COMPUTERS)
- UF　COMPILATION (COMPUTERS)
- 　　COMPILER PROGRAMS
- GS　COMPUTER PROGRAMS
- 　. **COMPILERS**
- RT　ASSEMBLER ROUTINES
- 　　AUTOCODERS
- 　　COLLATING
- 　　COMPASS (PROGRAMMING LANGUAGE)
- 　　COMPUTER PROGRAM INTEGRITY
- 　　COMPUTER SYSTEMS PROGRAMS
- 　　DATA CONVERSION ROUTINES
- 　　FORTRAN
- 　　OPERATING SYSTEMS (COMPUTERS)
- 　　PARSING ALGORITHMS

COMPILERS-(CONT.)
- 　　PASCAL (PROGRAMMING LANGUAGE)
- 　　PL/1
- 　　PROGRAMMED INSTRUCTION
- 　　SUBROUTINES

∞ **COMPLEMENT**
- SN　(USE OF A MORE SPECIFIC TERM IS RECOMMENDED--CONSULT THE TERMS LISTED BELOW)
- RT　COMPLEMENT (BIOLOGY)
- 　　COMPLEMENTS (MATHEMATICS)
- 　　PERSONNEL

COMPLEMENT (BIOLOGY)
- RT　∞ BIOLOGY
- 　　∞ COMPLEMENT
- 　　HEMOLYSIS

COMPLEMENTARY METAL OXIDE SEMICONDUCTORS
- USE　CMOS

COMPLEMENTS (MATHEMATICS)
- RT　ANGLES (GEOMETRY)
- 　　∞ COMPLEMENT
- 　　∞ LOGIC

COMPLETENESS
- RT　ACHIEVEMENT
- 　　COMPUTER PROGRAM INTEGRITY
- 　　INTEGRITY

COMPLEX COMPOUNDS
- RT　∞ CHEMICAL COMPOUNDS
- 　　MOLECULAR STRUCTURE
- 　　TRANSITION METALS

COMPLEX NUMBERS
- RT　GEOMETRY
- 　　INTEGERS
- 　　∞ NUMBERS
- 　　REAL NUMBERS

COMPLEX SYSTEMS
- RT　PARAMETER IDENTIFICATION
- 　　RELIABILITY ENGINEERING
- 　　SYSTEM IDENTIFICATION
- 　　∞ SYSTEMS
- 　　SYSTEMS ANALYSIS

COMPLEX VARIABLES
- GS　ANALYSIS (MATHEMATICS)
- 　. **COMPLEX VARIABLES**
- 　. . AIRY FUNCTION
- 　. . ANALYTIC FUNCTIONS
- 　. . . ENTIRE FUNCTIONS
- 　. . BESSEL FUNCTIONS
- 　. . . HANKEL FUNCTIONS
- 　. . CAUCHY INTEGRAL FORMULA
- 　. . CONFORMAL MAPPING
- 　. . CONJUGATES
- 　. . . CONJUGATE POINTS
- 　. . EXPONENTIAL FUNCTIONS
- 　. . . LOGARITHMS
- 　. . GAMMA FUNCTION
- 　. . HARMONIC FUNCTIONS
- 　. . HYPERBOLIC FUNCTIONS
- 　. . HYPERGEOMETRIC FUNCTIONS
- 　. . LAGUERRE FUNCTIONS
- 　. . LEGENDRE FUNCTIONS
- 　. . LIOUVILLE THEOREM
- 　. . MATHIEU FUNCTION
- 　. . MEROMORPHIC FUNCTIONS
- 　. . . ELLIPTIC FUNCTIONS
- 　. . . RATIONAL FUNCTIONS
- 　. . NONHOLONOMIC EQUATIONS
- 　. . ORTHOGONAL FUNCTIONS
- 　. . SCHWARZ-CHRISTOFFEL TRANSFORMATION
- 　. . SINGULARITY (MATHEMATICS)
- 　. . . NAKED SINGULARITIES
- 　. . SPHERICAL HARMONICS
- RT　APERIODIC FUNCTIONS
- 　　COMPLEXITY
- 　　DEPENDENT VARIABLES
- 　　EULER-CAUCHY EQUATIONS
- 　　FUNCTIONAL ANALYSIS
- 　　JOUKOWSKI TRANSFORMATION
- 　　MAXIMUM PRINCIPLE
- 　　REAL VARIABLES
- 　　SCHAUDER FIXPOINT THEOREM
- 　　STABILITY DERIVATIVES
- 　　THEODORSEN TRANSFORMATION

COMPLEX VARIABLES-(CONT.)
- 　　UNIQUENESS THEOREM
- 　　∞ VARIABLE

COMPLEXITY
- UF　COMPLICATION
- GS　**COMPLEXITY**
- 　. TASK COMPLEXITY
- RT　COMPLEX VARIABLES
- 　　FEEDBACK
- 　　∞ PERFORMANCE
- 　　STATISTICAL DISTRIBUTIONS

COMPLIANCE (ELASTICITY)
- USE　MODULUS OF ELASTICITY

COMPLICATION
- USE　COMPLEXITY

COMPONENT RELIABILITY
- GS　RELIABILITY
- 　. **COMPONENT RELIABILITY**
- RT　AIRCRAFT RELIABILITY
- 　　CIRCUIT RELIABILITY
- 　　CUMULATIVE DAMAGE
- 　　PROCESS CONTROL (INDUSTRY)
- 　　QUALITY CONTROL
- 　　RETIREMENT FOR CAUSE
- 　　SPACECRAFT RELIABILITY
- 　　STRUCTURAL RELIABILITY

∞ **COMPONENTS**
- SN　(USE OF A MORE SPECIFIC TERM IS RECOMMENDED--CONSULT THE TERMS LISTED BELOW)
- UF　PARTS
- RT　ACCESSORIES
- 　　ANTENNA COMPONENTS
- 　　ASSEMBLIES
- 　　COMPUTER COMPONENTS
- 　　CONTENT
- 　　ENGINE PARTS
- 　　FRACTIONS
- 　　INGREDIENTS
- 　　MISSILE COMPONENTS
- 　　MODULES
- 　　REDUNDANT COMPONENTS
- 　　SEGMENTS
- 　　SPACECRAFT COMPONENTS
- 　　SPARE PARTS
- 　　STRUCTURAL MEMBERS
- 　　SUBASSEMBLIES

COMPOSITE FUNCTIONS
- GS　ANALYSIS (MATHEMATICS)
- 　. REAL VARIABLES
- 　. . **COMPOSITE FUNCTIONS**
- 　　FUNCTIONS (MATHEMATICS)
- 　. **COMPOSITE FUNCTIONS**

COMPOSITE MATERIALS
- UF　COMPOSITES
- 　　PYROGRAPHALLOY
- 　　REINFORCED MATERIALS
- GS　**COMPOSITE MATERIALS**
- 　. ALUMINUM BORON COMPOSITES
- 　. ALUMINUM GRAPHITE COMPOSITES
- 　. BORON REINFORCED MATERIALS
- 　. . BORON-EPOXY COMPOUNDS
- 　. CARBON-CARBON COMPOSITES
- 　. CERAMIC MATRIX COMPOSITES
- 　. CERMETS
- 　. COMPOSITE PROPELLANTS
- 　. EPOXY MATRIX COMPOSITES
- 　. FIBER COMPOSITES
- 　. . CARBON FIBER REINFORCED PLASTICS
- 　. . GLASS FIBER REINFORCED PLASTICS
- 　. FIBER REINFORCED COMPOSITES
- 　. GLASSY CARBON
- 　. GRAPHITE-POLYIMIDE COMPOSITES
- 　. LAMINATES
- 　. . BORAL
- 　. . PLYWOOD
- 　. METAL MATRIX COMPOSITES
- 　. . BORSIC (TRADENAME)
- 　. . EUTECTIC COMPOSITES
- 　. POLYMER MATRIX COMPOSITES
- 　. REINFORCED PLASTICS
- 　. . GLASS FIBER REINFORCED PLASTICS
- 　. . MICARTA
- 　. RESIN MATRIX COMPOSITES
- 　. . GRAPHITE-EPOXY COMPOSITES
- 　. SUPERHYBRID MATERIALS
- 　. . GRAPHITE-EPOXY COMPOSITES

COMPOSITE MATERIALS-(CONT.)
. THREE DIMENSIONAL COMPOSITES
. WHISKER COMPOSITES
RT AIRCRAFT CONSTRUCTION MATERIALS
AIRFRAME MATERIALS
BIMETALS
BORON FIBERS
CARBON FIBERS
CLADDING
COATINGS
∞CONSTRUCTION MATERIALS
E GLASS
FIBER ORIENTATION
FIBER RELEASE
FIBERS
INSULATION
LAY-UP
LOW DENSITY RESEARCH
∞MATERIALS
∞MATRICES
MATRIX MATERIALS
METALS
MICROMECHANICS
MIXTURES
MODULAR RATIOS
MONOTECTIC ALLOYS
MULTILAYER INSULATION
PLY ORIENTATION
POWDER METALLURGY
PREFORMS
PREPREGS
REINFORCEMENT (STRUCTURES)
REINFORCING FIBERS
REINFORCING MATERIALS
RIGID STRUCTURES
∞ROVINGS
S GLASS
SANDWICH STRUCTURES
SOLID SUSPENSIONS
SPIRAL WRAPPING
THERMOSETTING RESINS

COMPOSITE PROPELLANTS
GS COMPOSITE MATERIALS
. COMPOSITE PROPELLANTS
PROPELLANTS
. SOLID PROPELLANTS
. . COMPOSITE PROPELLANTS
RT CASE BONDED PROPELLANTS
DOUBLE BASE PROPELLANTS
DOUBLE BASE ROCKET PROPELLANTS
EXPLOSIVES
FUEL PRODUCTION
PLASTIC PROPELLANTS
PLASTISOLS
POLYSULFIDES
POLYURETHANE RESINS
PROPELLANT ADDITIVES
PROPELLANT BINDERS
SOLID ROCKET PROPELLANTS

COMPOSITE STRUCTURES
GS COMPOSITE STRUCTURES
. LAMINATES
. . BORAL
. . PLYWOOD
RT BORON-EPOXY COMPOUNDS
CERAMIC MATRIX COMPOSITES
GLASS FIBER REINFORCED PLASTICS
HONEYCOMB CORES
HONEYCOMB STRUCTURES
HYBRID STRUCTURES
LAY-UP
PULTRUSION
STEEL STRUCTURES
∞STRUCTURES

COMPOSITE WRAPPING
RT FIBER COMPOSITES
FILAMENT WINDING
ISOTENSOID STRUCTURES
SPIRAL WRAPPING
∞WRAP

COMPOSITES
USE COMPOSITE MATERIALS

∞ COMPOSITION
SN *(USE OF A MORE SPECIFIC TERM IS RECOMMENDED--CONSULT THE TERMS LISTED BELOW)*
RT COMPOSITION (PROPERTY)
CONTENT
FORMULATIONS
INGREDIENTS

COMPOSITION-(CONT.)
STOICHIOMETRY

COMPOSITION (PROPERTY)
GS COMPOSITION (PROPERTY)
. ATMOSPHERIC COMPOSITION
. . ATMOSPHERIC MOISTURE
. . IONOSPHERIC COMPOSITION
. BODY COMPOSITION (BIOLOGY)
. CHEMICAL COMPOSITION
. . CARBON DIOXIDE CONCENTRATION
. . STELLAR COMPOSITION
. CONCENTRATION (COMPOSITION)
. . ATOM CONCENTRATION
. . CARBON DIOXIDE CONCENTRATION
. . LOW CONCENTRATIONS
. . MASCONS
. . METEOROID CONCENTRATION
. . MOISTURE CONTENT
. . . ATMOSPHERIC MOISTURE
. . GAS COMPOSITION
. . CARBON DIOXIDE CONCENTRATION
. LUNAR COMPOSITION
. METEORITIC COMPOSITION
. PLANETARY COMPOSITION
. PLASMA COMPOSITION
RT ∞COMPOSITION
GRADIENTS
HENRY LAW
LUMPING
MIXTURES
RAOULT LAW
SOLUTIONS
STOICHIOMETRY

COMPOSTING
GS DISPOSAL
. WASTE DISPOSAL
. . COMPOSTING
RT GARBAGE
METABOLIC WASTES
SHREDDING
SOLID WASTES
WASTE TREATMENT
WASTE UTILIZATION

COMPOUND A
GS HALOGEN COMPOUNDS
. FLUORINE COMPOUNDS
. . FLUORIDES
. . . COMPOUND A
RT ∞CHEMICAL COMPOUNDS

COMPOUND HELICOPTERS
GS V/STOL AIRCRAFT
. ROTARY WING AIRCRAFT
. . HELICOPTERS
. . . COMPOUND HELICOPTERS
RT AERONAUTICAL ENGINEERING
AIR TRANSPORTATION
∞AIRCRAFT
AIRCRAFT CONFIGURATIONS
AIRCRAFT DESIGN
HELICOPTER DESIGN
SHORT TAKEOFF AIRCRAFT
VERTICAL TAKEOFF AIRCRAFT

COMPOUNDING
UF MILLING (MIXING)
GS MIXING
. COMPOUNDING
RT COLLOIDING
DISSOLVING
GRINDING (COMMINUTION)
HOMOGENIZING
∞MILLING

∞ COMPOUNDS
SN *(USE OF A MORE SPECIFIC TERM IS RECOMMENDED--CONSULT THE TERMS LISTED BELOW)*
RT ∞CHEMICAL COMPOUNDS
POTTING COMPOUNDS

COMPRESSED AIR
GS GASES
. AIR
. . COMPRESSED AIR
. GAS MIXTURES
. . COMPRESSED AIR
RT COMPRESSORS
DRILLS
ENERGY STORAGE
MAN OPERATED PROPULSION SYSTEMS
OXYGEN SUPPLY EQUIPMENT

COMPRESSED AIR-(CONT.)
PNEUMATIC EQUIPMENT
∞PUMPING

COMPRESSED GAS
GS GASES
. COMPRESSED GAS
. . HIGH PRESSURE OXYGEN
RT COMPRESSORS
GAS PRESSURE
PNEUMATIC CONTROL

COMPRESSIBILITY
GS MECHANICAL PROPERTIES
. COMPRESSIBILITY
RT BULK MODULUS
COMPRESSIBLE FLOW
COMPRESSIVE STRENGTH
DENSITY (MASS/VOLUME)
EQUATIONS OF STATE
GRUNEISEN CONSTANT
HYDROELASTICITY
INCOMPRESSIBILITY
METAL POWDER
POROSITY
POWDER (PARTICLES)

COMPRESSIBILITY EFFECTS
RT BUFFETING
COMPRESSIBLE FLOW
∞EFFECTS
FLUTTER
HEAT TRANSFER
OSCILLATING FLOW
PRESSURE EFFECTS
RELAXATION (PHYSIOLOGY)
SECONDARY FLOW
SUPERSONIC FLOW
TRANSONIC FLOW

COMPRESSIBLE BOUNDARY LAYER
GS BOUNDARY LAYERS
. COMPRESSIBLE BOUNDARY LAYER
RT LAMINAR BOUNDARY LAYER
THREE DIMENSIONAL BOUNDARY LAYER
TURBULENT BOUNDARY LAYER

COMPRESSIBLE FLOW
GS FLUID FLOW
. COMPRESSIBLE FLOW
. . TRANSONIC FLOW
RT AERODYNAMICS
AIR FLOW
CARTAN SPACE
COMPRESSIBILITY
COMPRESSIBILITY EFFECTS
CROCCO METHOD
GAS FLOW
HUGONIOT EQUATION OF STATE
HYPERSONIC FLOW
INCOMPRESSIBLE FLOW
MAGNETOHYDRODYNAMIC FLOW
NEWTON PRESSURE LAW
STAGNATION FLOW
STAGNATION PRESSURE
STAGNATION TEMPERATURE
SUBSONIC FLOW
SUPERSONIC FLOW

COMPRESSIBLE FLUIDS
RT AERODYNAMIC HEATING
CROCCO METHOD
FLUID POWER
∞FLUIDS
HYDROELASTICITY
IDEAL FLUIDS
INCOMPRESSIBLE FLUIDS
MAXWELL FLUIDS
METHOD OF CHARACTERISTICS
P WAVES
SUPERFLUIDITY

COMPRESSING
UF RECOMPRESSION
SQUEEZING
GS COMPRESSING
. PLASMA COMPRESSION
. SPEECH BASEBAND COMPRESSION
RT ADIABATIC CONDITIONS
ANVILS
AXIAL COMPRESSION LOADS
BLOWING
COMPRESSORS
CONCENTRATING
DENSIFICATION

COMPRESSING-*(CONT.)*
      INTERNAL COMPRESSION INLETS
      MAGNETIC COMPRESSION
      MECHANICAL PROPERTIES
      METAL POWDER
      PISTON THEORY
  ∞PRESSING
      PRESSURE
      PRESSURE REDUCTION
      PULSE COMPRESSION
  ∞PUMPING
      RAREFACTION
      SUPERCHARGERS

COMPRESSION LOADS
GS   LOADS (FORCES)
    . **COMPRESSION LOADS**
    . . AXIAL COMPRESSION LOADS
    . . IMPACT LOADS
RT   AERODYNAMIC LOADS
    AXIAL LOADS
    BUCKLING
    COMPRESSIVE STRENGTH
    DYNAMIC LOADS
    EDGE LOADING
    MECHANICAL PROPERTIES
    SHOCK LOADS
    STATIC LOADS
    STRUCTURAL DESIGN CRITERIA
    THRUST LOADS

COMPRESSION RATIO
GS   RATIOS
    . **COMPRESSION RATIO**
RT   EFFICIENCY
    FUEL-AIR RATIO

COMPRESSION TESTERS
USE   COMPRESSION TESTS

COMPRESSION TESTS
UF   COMPRESSION TESTERS
    METEORITE COMPRESSION TESTS
RT   CREEP TESTS
    DESTRUCTIVE TESTS
    HARDNESS TESTS
    IMPACT TESTS
    LOAD TESTS
  ∞MATERIALS TESTS
    STATIC TESTS
  ∞TESTS

COMPRESSION WAVES
GS   ELASTIC WAVES
    . **COMPRESSION WAVES**
RT   P WAVES

COMPRESSIVE STRENGTH
GS   MECHANICAL PROPERTIES
    . **COMPRESSIVE STRENGTH**
RT   COMPRESSIBILITY
    COMPRESSION LOADS
    DUCTILITY
    ELASTIC PROPERTIES
    FIBER STRENGTH
    HIGH STRENGTH
    POISSON RATIO
    RESILIENCE
    SHEAR STRENGTH
  ∞STRENGTH
    TOUGHNESS

COMPRESSOR BLADES
GS   TURBOMACHINE BLADES
    . **COMPRESSOR BLADES**
RT ∞BLADES
    CENTRIFUGAL COMPRESSORS
    FAN BLADES
    ROTATING STALLS
    ROTOR BLADES (TURBOMACHINERY)
    STATOR BLADES
    TURBINE BLADES
    TURBOCOMPRESSORS
    VANES

COMPRESSOR EFFICIENCY
GS   EFFICIENCY
    . **COMPRESSOR EFFICIENCY**
RT   POWER EFFICIENCY
    THERMODYNAMIC EFFICIENCY

COMPRESSOR ROTORS
GS   ROTATING BODIES
    . ROTORS

COMPRESSOR ROTORS-*(CONT.)*
    . . **COMPRESSOR ROTORS**
RT   CENTRIFUGAL COMPRESSORS
    COMPRESSORS
  ∞FANS
    IMPELLERS
    ROTOR BLADES (TURBOMACHINERY)
    TURBINE WHEELS
    TURBOCOMPRESSORS

COMPRESSORS
SN   (EXCLUDES DATA COMPRESSORS)
GS   **COMPRESSORS**
    . CENTRIFUGAL COMPRESSORS
    . SUPERCHARGERS
    . SUPERSONIC COMPRESSORS
    . TRANSONIC COMPRESSORS
    . TURBOCOMPRESSORS
RT   AIR CONDITIONING EQUIPMENT
    BLOWERS
    COMPRESSED AIR
    COMPRESSED GAS
    COMPRESSING
    COMPRESSOR ROTORS
    CONDENSERS (LIQUEFIERS)
    COOLERS
  ∞FANS
    REFRIGERATING MACHINERY
    STATORS
    TURBOMACHINERY
    VACUUM PUMPS
    VANELESS DIFFUSERS

COMPTON EFFECT
GS   SCATTERING
    . **COMPTON EFFECT**
RT   COHERENT SCATTERING
  ∞EFFECTS
    INELASTIC SCATTERING
    NUCLEAR REACTIONS
    PHOTOELECTRICITY

COMPULSATORS
RT   AC GENERATORS
    COMPENSATORS
    ELECTRIC POWER SUPPLIES
    PULSE GENERATORS

COMPUTATION
UF   CALCULATION
GS   **COMPUTATION**
    . ORBIT CALCULATION
    . . MINIMUM VARIANCE ORBIT
         DETERMINATION
RT   ADDITION
  ∞APPLICATIONS OF MATHEMATICS
    ARITHMETIC
    CALCULATORS
    COMPUTATIONAL ASTROPHYSICS
    COMPUTATIONAL FLUID DYNAMICS
    COMPUTERS
    DATA PROCESSING
    DATA REDUCTION
    DIVIDING (MATHEMATICS)
  ∞FORMULAS
    INTERPOLATION
    LINEAR PREDICTION
    MULTIPLICATION
    SUBTRACTION
    SUMS

COMPUTATIONAL ASTROPHYSICS
GS   ASTROPHYSICS
    . **COMPUTATIONAL ASTROPHYSICS**
RT   COMPUTATION
    COMPUTERIZED SIMULATION
    MATHEMATICAL MODELS
  ∞SCIENCE

COMPUTATIONAL CHEMISTRY
GS   ANALYSIS (MATHEMATICS)
    . NUMERICAL ANALYSIS
    . . **COMPUTATIONAL CHEMISTRY**
RT ∞CHEMISTRY
    COMPUTER TECHNIQUES
    COMPUTERIZED SIMULATION
    PHYSICAL CHEMISTRY
  ∞TESTS

COMPUTATIONAL FLUID DYNAMICS
GS   ANALYSIS (MATHEMATICS)
    . NUMERICAL ANALYSIS
    . . **COMPUTATIONAL FLUID DYNAMICS**
    FLUID MECHANICS
    . FLUID DYNAMICS

COMPUTATIONAL FLUID DYNAMICS-*(CONT.)*
    . . **COMPUTATIONAL FLUID DYNAMICS**
RT   COMPUTATION
  ∞DYNAMICS
    EQUATIONS OF MOTION
    FINITE ELEMENT METHOD
    HYDRODYNAMIC COEFFICIENTS
    INTERACTIONAL AERODYNAMICS
    NAVIER-STOKES EQUATION
    PANEL METHOD (FLUID DYNAMICS)
    RELAXATION METHOD (MATHEMATICS)
    SPECTRAL METHODS

COMPUTATIONAL GRIDS
UF   GRIDS (MATHEMATICS)
    MESH (MATHEMATICS)
RT   COORDINATES
    MATHEMATICAL MODELS
    NUMERICAL ANALYSIS
    PROBLEM SOLVING

COMPUTER AIDED DESIGN
UF   CAD (DESIGN)
    COMPUTERIZED DESIGN
GS   COMPUTER TECHNIQUES
    . **COMPUTER AIDED DESIGN**
    . . IPAD
RT   AIRCRAFT DESIGN
    AMPLIFIER DESIGN
    COMPUTER GRAPHICS
    COMPUTER TECHNIQUES
    COMPUTERIZED SIMULATION
  ∞DESIGN
    DRAFTING MACHINES
    ENGINE DESIGN
    HELICOPTER DESIGN
    LENS DESIGN
    LOFTING
    LOGIC DESIGN
    MISSILE DESIGN
    REACTOR DESIGN
    ROBOTICS
    SATELLITE DESIGN
    SPACECRAFT DESIGN
    STRUCTURAL DESIGN

COMPUTER AIDED MANUFACTURING
UF   CAM (MANUFACTURING)
GS   COMPUTER TECHNIQUES
    . **COMPUTER AIDED MANUFACTURING**
    MANUFACTURING
    . **COMPUTER AIDED MANUFACTURING**
RT   COMPUTER GRAPHICS
    COMPUTER TECHNIQUES
    COMPUTERIZED SIMULATION
    ROBOTICS

COMPUTER AIDED MAPPING
GS   COMPUTER TECHNIQUES
    . **COMPUTER AIDED MAPPING**
    MAPPING
    . **COMPUTER AIDED MAPPING**
RT   COMPUTER GRAPHICS
    COMPUTER TECHNIQUES
    COMPUTERIZED SIMULATION
    MAPS
    ROBOTICS

COMPUTER AIDED TOMOGRAPHY
UF   CAT SCANNER
GS   COMPUTER TECHNIQUES
    . **COMPUTER AIDED TOMOGRAPHY**
    IMAGERY
    . RADIOGRAPHY
    . . TOMOGRAPHY
    . . . **COMPUTER AIDED TOMOGRAPHY**
RT   COMPUTER GRAPHICS
    IMAGE PROCESSING

COMPUTER ASSISTED INSTRUCTION
UF   CAI
GS   COMPUTER TECHNIQUES
    . **COMPUTER ASSISTED INSTRUCTION**
    PROGRAMMED INSTRUCTION
    . **COMPUTER ASSISTED INSTRUCTION**
RT   LANGUAGE PROGRAMMING
    SYMBOLIC PROGRAMMING

COMPUTER COMPATIBLE TAPES
GS   PERIPHERAL EQUIPMENT (COMPUTERS)
    . COMPUTER STORAGE DEVICES
    . . MAGNETIC TAPES
    . . . **COMPUTER COMPATIBLE TAPES**
RT   COMPUTERS
    DATA PROCESSING EQUIPMENT

COMPUTER COMPATIBLE TAPES-*(CONT.)*
    DIGITAL COMPUTERS
  ∞ TAPES

COMPUTER COMPONENTS
RT   ADDING CIRCUITS
    ARITHMETIC AND LOGIC UNITS
    BINARY TO DECIMAL CONVERTERS
    BUBBLE MEMORY DEVICES
    CENTRAL PROCESSING UNITS
    ∞ COMPONENTS
    COMPUTERS
    CONSOLES
    CONTROL UNITS (COMPUTERS)
    COUNTERS
    DECIMAL TO BINARY CONVERTERS
    LOGICAL ELEMENTS
    READ-ONLY MEMORY DEVICES
    REMOTE CONSOLES
    SHIFT REGISTERS

COMPUTER DESIGN
SN   (DESIGN OF COMPUTERS--EXCLUDES
    COMPUTERIZED DESIGN AND SYSTEMS
    ENGINEERING)
RT   ARCHITECTURE (COMPUTERS)
    COMPUTERS
    ∞ DESIGN
    FLUID LOGIC
    LOGIC DESIGN
    MEMORY (COMPUTERS)
    MICROPROCESSORS
    OPTICAL COMPUTERS
    PRODUCT DEVELOPMENT
    READ-ONLY MEMORY DEVICES

COMPUTER GRAPHICS
UF   INTERACTIVE GRAPHICS
RT   COMPUTER AIDED DESIGN
    COMPUTER AIDED MANUFACTURING
    COMPUTER AIDED MAPPING
    COMPUTER AIDED TOMOGRAPHY
    COMPUTERS
    DATA PROCESSING TERMINALS
    DISPLAY DEVICES
    PLOTTERS
    REMOTE CONSOLES
    TOMOGRAPHY

COMPUTER INFORMATION SECURITY
GS   SECURITY
    . COMPUTER INFORMATION SECURITY
RT   CRYPTOGRAPHY
    DATA PROCESSING
    OPERATING SYSTEMS (COMPUTERS)
    PRIVACY

COMPUTER METHODS
USE COMPUTER PROGRAMS

COMPUTER NETWORKS
GS   **COMPUTER NETWORKS**
    . ARPA COMPUTER NETWORK
RT   DISTRIBUTED PROCESSING
    ELECTRONIC MAIL
    INTERPROCESSOR COMMUNICATION
    NETWORK CONTROL
    ∞ NETWORKS
    PROTOCOL (COMPUTERS)

COMPUTER PROGRAM INTEGRITY
GS   INTEGRITY
    . COMPUTER PROGRAM INTEGRITY
RT   COMPILERS
    COMPLETENESS
    COMPUTERS
    DIGITAL COMPUTERS
    ERRORS
    PROGRAMS
    REDUNDANCY
    SECURITY

COMPUTER PROGRAMMING
UF   LEGENDRE CODE
GS   SOFTWARE ENGINEERING
    . COMPUTER PROGRAMMING
    . . ASSEMBLER ROUTINES
    . . LANGUAGE PROGRAMMING
    . . LOGIC PROGRAMMING
    . . MICROPROGRAMMING
    . . MULTIPROGRAMMING
    . . ON-LINE PROGRAMMING
    . . PARALLEL PROGRAMMING
    . . SYMBOLIC PROGRAMMING
RT   ADA (PROGRAMMING LANGUAGE)

COMPUTER PROGRAMMING-*(CONT.)*
    ADDRESSING
    ALGOL
    ALGORITHMS
    APL (PROGRAMMING LANGUAGE)
    ASSEMBLY LANGUAGE
    AUTOCODERS
    BASIC (PROGRAMMING LANGUAGE)
    BATCH PROCESSING
    BCH CODES
    BLOCK DIAGRAMS
    CODING
    COMPASS (PROGRAMMING LANGUAGE)
    CONTEXT FREE LANGUAGES
    DATA STRUCTURES
    DIGITAL TECHNIQUES
    EXPERT SYSTEMS
    FILE MAINTENANCE (COMPUTERS)
    FIRMWARE
    FLOW CHARTS
    FORMALISM
    FORMAT
    FORTRAN
    HAL/S (LANGUAGE)
    HEURISTIC METHODS
    KINOFORM
    LINEAR PROGRAMMING
    LISP (PROGRAMMING LANGUAGE)
    LOGIC DESIGN
    MACHINE-INDEPENDENT PROGRAMS
    MAP (PROGRAMMING LANGUAGE)
    MATHEMATICAL PROGRAMMING
    NATURAL LANGUAGE (COMPUTERS)
    NUMERICAL ANALYSIS
    PASCAL (PROGRAMMING LANGUAGE)
    PL/1
    PROGRAM VERIFICATION (COMPUTERS)
    PROGRAMMED INSTRUCTION
    PROGRAMMERS
    ∞ PROGRAMMING
    PROGRAMMING LANGUAGES
    REAL TIME OPERATION
    RESPONSE TIME (COMPUTERS)
    RUN TIME (COMPUTERS)
    SEQUENTIAL CONTROL
    SLEUTH (PROGRAMMING LANGUAGE)
    SOFTWARE TOOLS
    SYSTEMS ANALYSIS
    THEOREM PROVING
    TIME SHARING

COMPUTER PROGRAMS
UF   COMPUTER METHODS
    SOFTWARE (COMPUTERS)
GS   **COMPUTER PROGRAMS**
    . APPLICATIONS PROGRAMS
      (COMPUTERS)
    . COMPILERS
    . COMPUTER SYSTEMS PROGRAMS
    . . ASSEMBLER ROUTINES
    . . INPUT/OUTPUT ROUTINES
    . . OPERATING SYSTEMS (COMPUTERS)
    . . SUBROUTINE LIBRARIES
      (COMPUTERS)
    . EDITING ROUTINES (COMPUTERS)
    . MACHINE-INDEPENDENT PROGRAMS
    . MERGING ROUTINES
    . MULTIPLE OUTPUT PROGRAMS
    . NASTRAN
    . OBJECT PROGRAMS
    . SOURCE PROGRAMS
    . SUBROUTINES
RT   ALGORITHMS
    APPLICATIONS PROGRAMS
      (COMPUTERS)
    ASSEMBLY LANGUAGE
    BATCH PROCESSING
    BLOCK DIAGRAMS
    CODING
    COMPUTERS
    DATA CONVERSION ROUTINES
    DATA FLOW ANALYSIS
    DATA PROCESSING
    DIGITAL COMPUTERS
    ERROR DETECTION CODES
    FIXED POINT ARITHMETIC
    FLOATING POINT ARITHMETIC
    GODDARD TRAJECTORY
      DETERMINATION SYSTEM
    INSTRUCTION SETS (COMPUTERS)
    LASER GUIDANCE
    MACHINE TRANSLATION
    MODULARITY
    NASA INTERACTIVE PLANNING SYSTEM
    NUMERICAL CONTROL
    ON-LINE SYSTEMS

COMPUTER PROGRAMS-*(CONT.)*
    PROGRAMMED INSTRUCTION
    PROGRAMS
    REPORT GENERATORS
    ∞ ROUTINES
    SOFTWARE ENGINEERING
    SOFTWARE TOOLS
    ∞ TRANSLATORS
    USER MANUALS (COMPUTER
      PROGRAMS)

COMPUTER SIMULATION
USE COMPUTERIZED SIMULATION

COMPUTER STORAGE DEVICES
UF   MACHINE STORAGE
GS   PERIPHERAL EQUIPMENT (COMPUTERS)
    . COMPUTER STORAGE DEVICES
    . . BUFFER STORAGE
    . . CRYOGENIC COMPUTER STORAGE
    . . MAGNETIC TAPES
    . . . COMPUTER COMPATIBLE TAPES
    . . OPTICAL DISKS
    . . RANDOM ACCESS MEMORY
    . . READ-ONLY MEMORY DEVICES
    . . REGISTERS (COMPUTERS)
RT   ACOUSTIC DELAY LINES
    CARDS
    CENTRAL PROCESSING UNITS
    ∞ CHANNELS
    CRYOSAR
    ∞ EQUIPMENT
    HOLE BURNING
    MAGNETIC STORAGE
    MEMORY (COMPUTERS)
    MICROPROCESSORS
    OPTICAL MEMORY (DATA STORAGE)
    PARAMETRONS
    PUNCHED CARDS
    PUNCHED TAPES
    RANDOM ACCESS
    SHIFT REGISTERS
    ∞ STORAGE
    THIN FILMS

COMPUTER SYSTEMS DESIGN
GS   SYSTEMS ENGINEERING
    . COMPUTER SYSTEMS DESIGN
RT   AUXILIARY EQUIPMENT (COMPUTERS)
    CONCURRENT PROCESSING
    ∞ DESIGN
    DISTRIBUTED PROCESSING
    INTERPROCESSOR COMMUNICATION
    MAN MACHINE SYSTEMS
    OPERATING SYSTEMS (COMPUTERS)
    READ-ONLY MEMORY DEVICES
    SOFTWARE ENGINEERING
    SOFTWARE TOOLS
    ∞ SYSTEMS
    VIRTUAL MEMORY SYSTEMS

COMPUTER SYSTEMS PERFORMANCE
RT   CONSISTENCY
    DATA SAMPLING
    EFFICIENCY
    EVALUATION
    OPERATOR PERFORMANCE
    OUTPUT
    ∞ PERFORMANCE
    PERFORMANCE TESTS
    QUALITY
    RELIABILITY
    RESPONSE TIME (COMPUTERS)
    ∞ SYSTEMS

COMPUTER SYSTEMS PROGRAMS
UF   SOFTWARE (COMPUTERS)
GS   COMPUTER PROGRAMS
    . COMPUTER SYSTEMS PROGRAMS
    . . ASSEMBLER ROUTINES
    . . INPUT/OUTPUT ROUTINES
    . . OPERATING SYSTEMS (COMPUTERS)
    . . SUBROUTINE LIBRARIES
      (COMPUTERS)
RT   ALGORITHMS
    ASSEMBLY LANGUAGE
    AUTOCODERS
    COMPILERS
    COMPUTERS
    DATA FLOW ANALYSIS
    DATA PROCESSING
    DIGITAL COMPUTERS
    EDITING ROUTINES (COMPUTERS)
    ERROR DETECTION CODES
    REPORT GENERATORS

COMPUTER SYSTEMS PROGRAMS-*(CONT.)*
- ∞ROUTINES
  - SOFTWARE ENGINEERING
  - SOFTWARE TOOLS
- ∞SYSTEMS
  - SYSTEMS ANALYSIS

COMPUTER SYSTEMS SIMULATION
- GS   SIMULATION
  - . **COMPUTER SYSTEMS SIMULATION**
- RT   ANALOG SIMULATION
  - DATA PROCESSING EQUIPMENT
  - DIGITAL SIMULATION
  - MATHEMATICAL MODELS
  - OPERATIONS RESEARCH
  - SIMULATORS
  - ∞SYSTEMS
  - SYSTEMS ANALYSIS

COMPUTER TECHNIQUES
- GS   **COMPUTER TECHNIQUES**
  - . COMPUTER AIDED DESIGN
  - . . IPAD
  - . COMPUTER AIDED MANUFACTURING
  - . COMPUTER AIDED MAPPING
  - . COMPUTER AIDED TOMOGRAPHY
  - . COMPUTER ASSISTED INSTRUCTION
- RT   ADAPTIVE OPTICS
  - ARPA COMPUTER NETWORK
  - COMPUTATIONAL CHEMISTRY
  - COMPUTER AIDED DESIGN
  - COMPUTER AIDED MANUFACTURING
  - COMPUTER AIDED MAPPING
  - COMPUTERS
  - FLIGHT MANAGEMENT SYSTEMS
  - MANAGEMENT INFORMATION SYSTEMS
  - MANAGEMENT METHODS
  - MANAGEMENT SYSTEMS
  - MICROPROCESSORS
  - NASTRAN
  - NUMERICAL DIFFERENTIATION
  - ON-LINE SYSTEMS
  - PARSING ALGORITHMS
  - PERSONAL COMPUTERS
  - WORD PROCESSING

COMPUTER VISION
- RT   ARTIFICIAL INTELLIGENCE
  - ∞AUTOMATION
  - PATTERN RECOGNITION
  - POSITION SENSING
  - ROBOTICS
  - ROBOTS

COMPUTERIZED CONTROL
- USE   NUMERICAL CONTROL

COMPUTERIZED DESIGN
- USE   COMPUTER AIDED DESIGN

COMPUTERIZED SIMULATION
- UF   ARIP (IMPACT PREDICTION)
  - AUTOMATIC ROCKET IMPACT
  -    PREDICTORS
  - COMPUTER SIMULATION
  - IP (IMPACT PREDICTION)
- GS   SIMULATION
  - . **COMPUTERIZED SIMULATION**
  - . . ANALOG SIMULATION
  - . . DIGITAL SIMULATION
- RT   ALGORITHMS
  - ALTITUDE SIMULATION
  - COMPUTATIONAL ASTROPHYSICS
  - COMPUTATIONAL CHEMISTRY
  - COMPUTER AIDED DESIGN
  - COMPUTER AIDED MANUFACTURING
  - COMPUTER AIDED MAPPING
  - COMPUTERS
  - CONTROL SIMULATION
  - DIFFERENTIAL ANALYZERS
  - FLIGHT SIMULATION
  - HIGHLY MANEUVERABLE AIRCRAFT
  - HYDRAULIC ANALOGIES
  - LANDING SIMULATION
  - LENNARD-JONES POTENTIAL
  - MATHEMATICAL MODELS
  - ∞MISSILE SIMULATORS
  - MOTION SIMULATORS
  - NUMERICAL WEATHER FORECASTING
  - OPERATIONS RESEARCH
  - SYSTEMS SIMULATION
  - TARGET SIMULATORS

COMPUTERS
- GS   DATA PROCESSING EQUIPMENT

COMPUTERS-*(CONT.)*
- . COMPUTERS
- . . ANALOG COMPUTERS
- . . . EAI 680 COMPUTER
- . . . HONEYWELL 600/6000 COMPUTER
- . . . SIGMA 5 COMPUTER
- . . . UNIVAC 1100 SERIES COMPUTERS
- . . CDC COMPUTERS
- . . . CDC CYBER 74 COMPUTER
- . . . CDC CYBER 170 SERIES
  -    COMPUTERS
- . . . . CDC CYBER 175 COMPUTER
- . . . CDC CYBER 174 COMPUTER
- . . . CDC CYBER 203 COMPUTER
- . . . CDC CYBER 205 COMPUTER
- . . . CDC STAR 100 COMPUTER
- . . . CDC 160-A COMPUTER
- . . . CDC 1604 COMPUTER
- . . . CDC 3100 COMPUTER
- . . . CDC 3200 COMPUTER
- . . . CDC 3600 COMPUTER
- . . . CDC 3800 COMPUTER
- . . . CDC 6000 SERIES COMPUTERS
- . . . . CDC 6400 COMPUTER
- . . . . CDC 6600 COMPUTER
- . . . . CDC 6700 COMPUTER
- . . . CDC 7000 SERIES COMPUTERS
- . . . . CDC 7600 COMPUTER
- . . . CDC 8090 COMPUTER
- . . COUNTING RATE COMPUTERS
- . . DDP COMPUTERS
- . . . DDP 516 COMPUTER
- . . DIGITAL COMPUTERS
- . . . CDC CYBER 74 COMPUTER
- . . . CDC CYBER 170 SERIES
  -    COMPUTERS
- . . . CDC CYBER 174 COMPUTER
- . . . CDC CYBER 203 COMPUTER
- . . . CDC CYBER 205 COMPUTER
- . . . CDC STAR 100 COMPUTER
- . . . CDC 160-A COMPUTER
- . . . CDC 1604 COMPUTER
- . . . CDC 3100 COMPUTER
- . . . CDC 3200 COMPUTER
- . . . CDC 3600 COMPUTER
- . . . CDC 3800 COMPUTER
- . . . CDC 6000 SERIES COMPUTERS
- . . . . CDC 6400 COMPUTER
- . . . . CDC 6600 COMPUTER
- . . . . CDC 6700 COMPUTER
- . . . CDC 7000 SERIES COMPUTERS
- . . . . CDC 7600 COMPUTER
- . . . CDC 8090 COMPUTER
- . . . EAI 680 COMPUTER
- . . . EAI 8400 COMPUTER
- . . . EAI 8900 COMPUTER
- . . . EMR 6050 COMPUTER
- . . . FERRANTI MERCURY COMPUTER
- . . . GE COMPUTERS
- . . . . GE 625 COMPUTER
- . . . . GE 635 COMPUTER
- . . . HEWLETT-PACKARD COMPUTERS
- . . . HONEYWELL COMPUTERS
- . . . . DDP 516 COMPUTER
- . . . . HONEYWELL ADEPT COMPUTER
- . . . . HONEYWELL DDP 116 COMPUTER
- . . . . HONEYWELL 600/6000 COMPUTER
- . . . IBM 360 COMPUTER
- . . . IBM 370 COMPUTER
- . . . IBM 650 COMPUTER
- . . . IBM 704 COMPUTER
- . . . IBM 709 COMPUTER
- . . . IBM 1130 COMPUTER
- . . . IBM 1401 COMPUTER
- . . . IBM 1410 COMPUTER
- . . . IBM 1620 COMPUTER
- . . . IBM 2250 COMPUTER
- . . . IBM 7000 SERIES COMPUTERS
- . . . . IBM 7030 COMPUTER
- . . . . IBM 7040 COMPUTER
- . . . . IBM 7044 COMPUTER
- . . . . IBM 7070 COMPUTER
- . . . . IBM 7074 COMPUTER
- . . . . IBM 7090 COMPUTER
- . . . . IBM 7094 COMPUTER
- . . . ICL COMPUTERS
- . . . ILLIAC COMPUTERS
- . . . . ILLIAC 3 COMPUTER
- . . . . ILLIAC 4 COMPUTER
- . . . MICROCOMPUTERS
- . . . . PERSONAL COMPUTERS
- . . . MINICOMPUTERS
- . . . . NOVA COMPUTERS
- . . . MODCOMP II COMPUTER
- . . . MODCOMP IV COMPUTER
- . . . PARALLEL COMPUTERS

COMPUTERS-*(CONT.)*
- . . . PDP COMPUTERS
- . . . . PDP 7 COMPUTER
- . . . . PDP 8 COMPUTER
- . . . . PDP 9 COMPUTER
- . . . . PDP 10 COMPUTER
- . . . . PDP 11 COMPUTER
- . . . . PDP 11/20 COMPUTER
- . . . . PDP 11/40 COMPUTER
- . . . . PDP 11/45 COMPUTER
- . . . . PDP 11/50 COMPUTER
- . . . . PDP 11/70 COMPUTER
- . . . . PDP 12 COMPUTER
- . . . PDP 15 COMPUTER
- . . PHILCO 2000 COMPUTER
- . . RAYTHEON COMPUTERS
- . . RCA SPECTRA 70 COMPUTER
- . . SDS 900 SERIES COMPUTERS
- . . . SDS 930 COMPUTER
- . . . SDS 9300 COMPUTER
- . . SEL COMPUTERS
- . . SEQUENTIAL COMPUTERS
- . . SIGMA COMPUTERS
- . . . SIGMA 9 COMPUTER
- . . . SIGMA 5 COMPUTER
- . . SOLOMON COMPUTERS
- . . . UNIVAC LARC COMPUTER
- . . . UNIVAC 80 COMPUTER
- . . . UNIVAC 418 COMPUTER
- . . . UNIVAC 490 COMPUTER
- . . . UNIVAC 494 COMPUTER
- . . . UNIVAC 1100 SERIES COMPUTERS
- . . . . UNIVAC 1105 COMPUTER
- . . . . UNIVAC 1106 COMPUTER
- . . . . UNIVAC 1107 COMPUTER
- . . . . UNIVAC 1108 COMPUTER
- . . . . UNIVAC 1110 COMPUTER
- . . . UNIVAC 1230 COMPUTER
- . . . VAX COMPUTERS
- . . . . VAX-11 SERIES COMPUTERS
- . . . . . VAX-11/780 COMPUTER
- . . EMBEDDED COMPUTER SYSTEMS
- . . . AIRBORNE/SPACEBORNE
  -    COMPUTERS
- . . HYBRID COMPUTERS
- . . IBM COMPUTERS
- . . . IBM 360 COMPUTER
- . . . IBM 370 COMPUTER
- . . . IBM 650 COMPUTER
- . . . IBM 704 COMPUTER
- . . . IBM 709 COMPUTER
- . . . IBM 1130 COMPUTER
- . . . IBM 1401 COMPUTER
- . . . IBM 1410 COMPUTER
- . . . IBM 1620 COMPUTER
- . . . IBM 2250 COMPUTER
- . . . IBM 7000 SERIES COMPUTERS
- . . . . IBM 7030 COMPUTER
- . . . . IBM 7040 COMPUTER
- . . . . IBM 7044 COMPUTER
- . . . . IBM 7070 COMPUTER
- . . . . IBM 7074 COMPUTER
- . . . . IBM 7090 COMPUTER
- . . . . IBM 7094 COMPUTER
- . . MINOS COMPUTER
- . . OPTICAL COMPUTERS
- . . PEGASUS COMPUTER
- . . RCA COMPUTERS
- . . . RCA SPECTRA 70 COMPUTER
- . . . RCA-110 COMPUTERS
- . . SIEMENS 2002 COMPUTER
- . . SITE DATA PROCESSORS
- . . UNIVAC COMPUTERS
- . . . UNIVAC LARC COMPUTER
- . . . UNIVAC 80 COMPUTER
- . . . UNIVAC 418 COMPUTER
- . . . UNIVAC 490 COMPUTER
- . . . UNIVAC 494 COMPUTER
- . . . UNIVAC 1100 SERIES COMPUTERS
- . . . . UNIVAC 1105 COMPUTER
- . . . . UNIVAC 1106 COMPUTER
- . . . . UNIVAC 1107 COMPUTER
- . . . . UNIVAC 1108 COMPUTER
- . . . . UNIVAC 1110 COMPUTER
- . . . UNIVAC 1230 COMPUTER
- RT   APPLICATIONS PROGRAMS
  -    (COMPUTERS)
  - ARITHMETIC AND LOGIC UNITS
  - ARTIFICIAL INTELLIGENCE
  - AUTOMATA THEORY
  - ∞AUTOMATION
  - CALCULATORS
  - CENTRAL PROCESSING UNITS
  - COMPUTATION
  - COMPUTER COMPATIBLE TAPES
  - COMPUTER COMPONENTS

COMPUTERS-(CONT.)
    COMPUTER DESIGN
    COMPUTER GRAPHICS
    COMPUTER PROGRAM INTEGRITY
    COMPUTER PROGRAMS
    COMPUTER SYSTEMS PROGRAMS
    COMPUTER TECHNIQUES
    COMPUTERIZED SIMULATION
    CONTROL DATA (COMPUTERS)
    CONTROL UNITS (COMPUTERS)
    CYBERNETICS
    DATA CONVERTERS
    DATA PROCESSING
    DIGITAL TO VOICE TRANSLATORS
    FILE MAINTENANCE (COMPUTERS)
    FIXED POINT ARITHMETIC
    FLOATING POINT ARITHMETIC
    HAL/S (LANGUAGE)
    HARDWARE
    INFORMATION RETRIEVAL
    INFORMATION THEORY
    INTEL 8080 MICROPROCESSOR
    LOGIC CIRCUITS
    MACHINE-INDEPENDENT PROGRAMS
    ∞ MACHINERY
    MEMORY (COMPUTERS)
    MULTIPROCESSING (COMPUTERS)
    PRINTERS (DATA PROCESSING)
    READ-ONLY MEMORY DEVICES
    REAL TIME OPERATION
    RUN TIME (COMPUTERS)
    TELECOMMUNICATION
    VOCODERS

COMSAT PROGRAM
GS   PROGRAMS
    . **COMSAT PROGRAM**
RT   COMMUNICATION SATELLITES
    EARLY BIRD SATELLITES
    TELSTAR PROJECT
    TELSTAR SATELLITES

COMSTAR C
GS   SATELLITES
    . ARTIFICIAL SATELLITES
    . . COMMUNICATION SATELLITES
    . . . COMMUNICATIONS TECHNOLOGY
            SATELLITE
    . . . . **COMSTAR C**

COMSTAR SATELLITES
GS   SATELLITES
    . ARTIFICIAL SATELLITES
    . . COMMUNICATION SATELLITES
    . . . **COMSTAR SATELLITES**
RT   SATELLITE NETWORKS

CONCATENATED CODES
RT   BINARY CODES
    ∞ CODES
    CODING
    DATA TRANSMISSION
    DECODING
    ERROR CORRECTING CODES
    REDUNDANCY ENCODING
    SIGNAL ENCODING

CONCAVITY
RT   CONTOUR SENSORS
    CONTOURS
    CONVEXITY
    FLATNESS
    SHAPES
    ∞ SURFACE GEOMETRY

CONCENTRATING
RT   ACCUMULATIONS
    ADSORPTION
    AGGLOMERATION
    BENEFICIATION
    CENTRIFUGING
    CLASSIFIERS
    COAGULATION
    COALESCING
    COMPRESSING
    ∞ CONCENTRATION
    CONCENTRATORS
    CONDENSING
    CRYSTALLIZATION
    DISTILLATION
    DRYING
    ENRICHMENT
    EVAPORATION
    EXTRACTION
    FILTRATION

CONCENTRATING-(CONT.)
    FLOCCULATING
    FLOTATION
    PERCOLATION
    PRECIPITATION (CHEMISTRY)
    ∞ SEPARATION
    SEPARATORS
    SETTLING
    SORPTION
    STRESS CONCENTRATION
    UPGRADING
    VAPORIZING

∞ **CONCENTRATION**
SN   *(USE OF A MORE SPECIFIC TERM IS*
    RECOMMENDED--CONSULT THE TERMS
    LISTED BELOW)
RT   CONCENTRATING
    CONCENTRATION (COMPOSITION)
    CROWDING
    FILTRATION
    ISOTOPIC ENRICHMENT

CONCENTRATION (COMPOSITION)
GS   COMPOSITION (PROPERTY)
    . **CONCENTRATION (COMPOSITION)**
    . . ATOM CONCENTRATION
    . . CARBON DIOXIDE CONCENTRATION
    . . LOW CONCENTRATIONS
    . . MASCONS
    . . METEOROID CONCENTRATION
    . . MOISTURE CONTENT
    . . . ATMOSPHERIC MOISTURE
RT   ∞ CONCENTRATION
    DILUTION
    PARTICULATE SAMPLING
    PURITY
    QUALITY
    SAMPLING
    ∞ SATURATION
    SOLUBILITY

CONCENTRATORS
GS   **CONCENTRATORS**
    . SPIRALS (CONCENTRATORS)
RT   ACCUMULATORS
    CENTRIFUGES
    CLASSIFIERS
    ∞ CLASSIFYING
    COLUMNS (PROCESS ENGINEERING)
    CONCENTRATING
    EVAPORATORS
    FILTRATION
    FLUID FILTERS
    PRECIPITATORS
    RADIATIVE HEAT TRANSFER
    SEPARATORS
    SIZE SEPARATION
    SIZING SCREENS
    SOLAR COLLECTORS
    STILLS
    THERMAL RADIATION
    TRAPS
    WASHERS (CLEANERS)

CONCENTRIC CYLINDERS
RT   ∞ CYLINDERS
    CYLINDRICAL SHELLS

CONCENTRIC SPHERES
GS   SYMMETRICAL BODIES
    . BODIES OF REVOLUTION
    . . SPHERES
    . . . **CONCENTRIC SPHERES**
RT   CONCENTRICITY

CONCENTRICITY
RT   ∞ CENTERS
    CONCENTRIC SPHERES
    ECCENTRICITY

CONCORDE AIRCRAFT
GS   JET AIRCRAFT
    . TURBOFAN AIRCRAFT
    . . **CONCORDE AIRCRAFT**
    SUD AVIATION AIRCRAFT
    . **CONCORDE AIRCRAFT**
    SUPERSONIC AIRCRAFT
    . SUPERSONIC TRANSPORTS
    . . **CONCORDE AIRCRAFT**
    TRANSPORT AIRCRAFT
    . **CONCORDE AIRCRAFT**
RT   ∞ AIRCRAFT

CONCRETE STRUCTURES
RT   AGGREGATES
    BREAKWATERS
    CONSTRUCTION
    EARTHQUAKE RESISTANT STRUCTURES
    FOUNDATIONS
    ∞ MATERIALS
    RIGID STRUCTURES
    ∞ STRUCTURES
    TOWERS

CONCRETES
RT   ADMIXTURES
    AGGREGATES
    CEMENTS
    ∞ CONSTRUCTION MATERIALS
    GROUT
    INSULATION
    MASONRY
    MORTARS (MATERIAL)
    PAVEMENTS
    STRUCTURAL MEMBERS

CONCURRENT PROCESSING
GS   DATA PROCESSING
    . **CONCURRENT PROCESSING**
RT   ARCHITECTURE (COMPUTERS)
    COMPUTER SYSTEMS DESIGN
    MULTIPROCESSING (COMPUTERS)
    PARALLEL PROCESSING (COMPUTERS)

CONDENSATES
RT   AITKEN NUCLEI
    ∞ CONDENSATION
    CONDENSERS (LIQUEFIERS)
    CONDENSING
    CONTRAILS
    DROP SIZE
    LIQUEFIED GASES
    PLUMES
    VAPORS

∞ **CONDENSATION**
SN   *(USE OF A MORE SPECIFIC TERM IS*
    RECOMMENDED--CONSULT THE TERMS
    LISTED BELOW)
RT   CONDENSATES
    CONDENSATION NUCLEI
    CONDENSING
    GAS-METAL INTERACTIONS
    LIQUEFACTION
    MAYER PROBLEM
    RECTIFICATION

CONDENSATION NUCLEI
GS   **CONDENSATION NUCLEI**
    . AITKEN NUCLEI
RT   AEROSOLS
    CLOUD PHYSICS
    CLOUDS (METEOROLOGY)
    ∞ CONDENSATION
    CONDENSING
    DROPS (LIQUIDS)
    ICE NUCLEI
    METEOROLOGY
    MICROPARTICLES
    NUCLEATION
    ∞ NUCLEI
    RAIN

CONDENSATION PUMPS
GS   PUMPS
    . VACUUM PUMPS
    . . **CONDENSATION PUMPS**
    VACUUM APPARATUS
    . VACUUM PUMPS
    . . **CONDENSATION PUMPS**

CONDENSATION TRAILS
USE   CONTRAILS

CONDENSER RADIATORS
USE   CONDENSERS (LIQUEFIERS)
    HEAT RADIATORS

∞ **CONDENSERS**
SN   *(USE OF A MORE SPECIFIC TERM IS*
    RECOMMENDED--CONSULT THE TERMS
    LISTED BELOW)
RT   CAPACITORS
    CONDENSERS (LIQUEFIERS)
    JET CONDENSERS
    PHOTOGRAPHIC RECTIFIERS

CONDENSERS (LIQUEFIERS)
UF   CONDENSER RADIATORS
GS   **CONDENSERS (LIQUEFIERS)**
      . JET CONDENSERS
RT   ABSORBERS (EQUIPMENT)
      AIR CONDITIONING
      AIR CONDITIONING EQUIPMENT
      COLD TRAPS
      COLUMNS (PROCESS ENGINEERING)
      COMPRESSORS
      CONDENSATES
      ∞ CONDENSERS
      CONDENSING
      COOLING FINS
      COOLING SYSTEMS
      DISTILLATION EQUIPMENT
      DRYING APPARATUS
      EVAPORATORS
      EXHAUST SYSTEMS
      FILM CONDENSATION
      HEAT EXCHANGERS
      HEAT PUMPS
      LIQUEFIED GASES
      REFRIGERATING MACHINERY
      SEPARATORS
      SPACECRAFT RADIATORS
      VAPORIZERS

CONDENSING
UF   GAS LIQUEFACTION
GS   **CONDENSING**
      . FILM CONDENSATION
RT   ASSOCIATION REACTIONS
      CLOUD PHYSICS
      CONCENTRATING
      CONDENSATES
      ∞ CONDENSATION
      CONDENSATION NUCLEI
      CONDENSERS (LIQUEFIERS)
      COOLING
      DEHUMIDIFICATION
      DEW POINT
      DISTILLATION
      DROP SIZE
      DROPS (LIQUIDS)
      EVAPORATION
      GAS-LIQUID INTERACTIONS
      GAS-METAL INTERACTIONS
      NUCLEATION
      PHASE CHANGE MATERIALS
      PHASE TRANSFORMATIONS
      REFRIGERATING
      ∞ SATURATION
      ∞ SEPARATION
      SUBLIMATION
      SUPERCOOLING
      SUPERSATURATION

CONDITIONED REFLEXES
GS   REFLEXES
      . **CONDITIONED REFLEXES**
RT   CONDITIONING (LEARNING)
      REACTION TIME

CONDITIONED RESPONSES
USE   CONDITIONING (LEARNING)

∞ **CONDITIONING**
SN   (USE OF A MORE SPECIFIC TERM IS
      RECOMMENDED--CONSULT THE TERMS
      LISTED BELOW)
RT   BENEFICIATION
      CONDITIONING (LEARNING)
      POWER CONDITIONING
      PRECONDITIONING

CONDITIONING (LEARNING)
UF   CONDITIONED RESPONSES
GS   LEARNING
      . **CONDITIONING (LEARNING)**
RT   BEHAVIOR
      BIOFEEDBACK
      CONDITIONED REFLEXES
      ∞ CONDITIONING
      HABITUATION (LEARNING)
      INHIBITION (PSYCHOLOGY)

CONDITIONING (TREATING)
USE   TREATMENT

CONDITIONS
GS   **CONDITIONS**
      . ADIABATIC CONDITIONS
      . CHRONIC CONDITIONS
      . FLIGHT CONDITIONS

CONDITIONS-(CONT.)
      . KUTTA-JOUKOWSKI CONDITION
      . LIPSCHITZ CONDITION
      . NONADIABATIC CONDITIONS
      . NONEQUILIBRIUM CONDITIONS
      . RUNWAY CONDITIONS

CONDOR MISSILE
GS   MISSILES
      . AIR TO SURFACE MISSILES
      . . **CONDOR MISSILE**

CONDUCTANCE
USE   RESISTANCE

CONDUCTING
USE   CONDUCTION

CONDUCTING FLUIDS
SN   (EXCLUDES PLASMAS)
RT   CONDUCTORS
      ELECTROLYTES
      MAGNETOHYDRODYNAMICS

CONDUCTING MEDIA
USE   CONDUCTORS

∞ **CONDUCTION**
SN   (USE OF A MORE SPECIFIC TERM IS
      RECOMMENDED--CONSULT THE TERMS
      LISTED BELOW)
UF   CONDUCTING
RT   ATTENUATION
      CONDUCTIVE HEAT TRANSFER
      CONVECTION
      ELECTRIC CONDUCTORS
      ELECTRIC POWER TRANSMISSION
      HEAT TRANSFER
      HEATING
      REFRACTION
      SOUND PROPAGATION
      SOUND TRANSMISSION
      THERMAL CONDUCTORS
      THERMAL DIFFUSION
      TRANSMISSION
      WAVE PROPAGATION

CONDUCTION BANDS
GS   ENERGY BANDS
      . **CONDUCTION BANDS**
RT   BAND STRUCTURE OF SOLIDS
      ∞ BANDS
      BRILLOUIN ZONES
      ELECTRON TRANSITIONS
      FRANCK-CONDON PRINCIPLE
      NDM SEMICONDUCTOR DEVICES
      POLARONS
      QUANTUM WELLS
      SEMICONDUCTORS (MATERIALS)
      TRAPPING

CONDUCTION ELECTRONS
GS   GASES
      . IONIZED GASES
      . . CHARGED PARTICLES
      . . . **CONDUCTION ELECTRONS**
      PARTICLES
      . CHARGED PARTICLES
      . . ENERGETIC PARTICLES
      . . . ELECTRONS
      . . . . **CONDUCTION ELECTRONS**
RT   FREE ELECTRONS
      QUANTUM WELLS
      VALENCE

CONDUCTIVE HEAT TRANSFER
UF   HEAT CONDUCTION
GS   TRANSMISSION
      . HEAT TRANSMISSION
      . . HEAT TRANSFER
      . . . **CONDUCTIVE HEAT TRANSFER**
RT   ∞ CONDUCTION
      CONVECTIVE HEAT TRANSFER
      LAMINAR HEAT TRANSFER
      THERMAL CONDUCTIVITY
      THERMAL CONDUCTORS

∞ **CONDUCTIVITY**
SN   (USE OF A MORE SPECIFIC TERM IS
      RECOMMENDED--CONSULT THE TERMS
      LISTED BELOW)
RT   ATMOSPHERIC CONDUCTIVITY
      ELECTRICAL PROPERTIES
      ELECTRICAL RESISTIVITY

CONDUCTIVITY-(CONT.)
      FLUID FLOW
      IMPEDANCE
      IONOSPHERIC CONDUCTIVITY
      MAGNETORESISTIVITY
      MOBILITY
      OHMS LAW
      PHOTOCONDUCTIVITY
      PLASMA CONDUCTIVITY
      SUPERCONDUCTING POWER
        TRANSMISSION
      SUPERCONDUCTIVITY
      THERMAL CONDUCTIVITY
      TRANSPORT PROPERTIES
      VOID RATIO

CONDUCTIVITY METERS
GS   MEASURING INSTRUMENTS
      . **CONDUCTIVITY METERS**
      . . ELECTRICAL CONDUCTIVITY METERS

CONDUCTORS
UF   CONDUCTING MEDIA
GS   **CONDUCTORS**
      . BUS CONDUCTORS
      . ELECTRIC CONDUCTORS
      . ELECTRIC WIRE
      . ELECTROLYTES
      . . ANOLYTES
      . . CATHOLYTES
      . . ION EXCHANGE MEMBRANE
          ELECTROLYTES
      . . JUMPERS
      . . MOLTEN SALT ELECTROLYTES
      . . NONAQUEOUS ELECTROLYTES
      . . SOLID ELECTROLYTES
      . FLAT CONDUCTORS
      . PHOTOCONDUCTORS
      . SUPERCONDUCTORS
      . THERMAL CONDUCTORS
RT   ANTENNAS
      CONDUCTING FLUIDS
      EXPLODING WIRES
      METALS
      NONFERROUS METALS
      ORGANIC SEMICONDUCTORS
      SEMICONDUCTORS (MATERIALS)
      SUBREFLECTORS

CONES
SN   (LIMITED TO MATERIAL OBJECTS)
UF   CONICAL FLARE
      FUSIFORM SHAPES
GS   **CONES**
      . CIRCULAR CONES
      . CONICAL BODIES
      . . SLENDER CONES
      . NOSE CONES
      . . ABLATIVE NOSE CONES
      . . ROCKET NOSE CONES
      . SHATTER CONES
RT   AERODYNAMIC CONFIGURATIONS
      BODIES OF REVOLUTION
      CONICAL SHELLS
      CONICS
      FRUSTUMS
      HALF CONES
      MACH CONES
      SYMMETRICAL BODIES

CONES (VOLCANOES)
UF   CINDER CONES
GS   GEOLOGY
      . **CONES (VOLCANOES)**
      LANDFORMS
      . **CONES (VOLCANOES)**
RT   BASALT
      CALDERAS
      CRATERS
      EFFUSIVES
      GEOMORPHOLOGY
      LAVA
      MARS VOLCANOES
      MOUNTAINS
      OROGRAPHY
      PALEOMAGNETISM
      PETROLOGY
      ROUSE BELTS
      VOLCANOES
      VOLCANOLOGY

CONFERENCES
UF   MEETINGS
      PROCEEDINGS
RT   COMMITTEE ON SPACE RESEARCH

CONFERENCES-(CONT.)
    CONSULTING
    CONVENTIONS
    ∞DISCUSSION
    DOCUMENTATION
    DOCUMENTS
    PAPERS
    REPORTS
    STARSITE PROGRAM
    TELECONFERENCING

CONFIDENCE
RT   CORRELATION
    ERRORS
    PROBABILITY THEORY
    PSYCHOLOGICAL EFFECTS
    QUALITY CONTROL
    RELIABILITY
    RISK
    STATISTICAL ANALYSIS

CONFIDENCE LIMITS
RT   CONTINGENCY
    ESTIMATES
    FORECASTING
    MAXIMUM LIKELIHOOD ESTIMATES
    ∞MEASUREMENT
    NULL HYPOTHESIS
    PRECISION
    PREDICTIONS
    QUALITY CONTROL
    RANGE (EXTREMES)
    RELIABILITY
    RISK
    SAMPLING
    SIGNIFICANCE
    STANDARD DEVIATION
    STATISTICAL ANALYSIS
    STATISTICAL TESTS
    ∞TESTS
    VARIANCE (STATISTICS)

CONFIGURATION INTERACTION
GS   PARTICLE INTERACTIONS
    . **CONFIGURATION INTERACTION**
    SCATTERING
    . ELECTRON SCATTERING
    . . **CONFIGURATION INTERACTION**
RT   ∞INTERACTIONS
    INTERMOLECULAR FORCES
    MOLECULAR INTERACTIONS
    MOLECULAR STRUCTURE
    ∞STRUCTURES

CONFIGURATION MANAGEMENT
GS   MANAGEMENT
    . **CONFIGURATION MANAGEMENT**
RT   ∞CONFIGURATIONS

∞ CONFIGURATIONS
SN   *(USE OF A MORE SPECIFIC TERM IS*
    RECOMMENDED--CONSULT THE TERMS
    LISTED BELOW)
RT   AERODYNAMIC CONFIGURATIONS
    BODY-WING AND TAIL CONFIGURATIONS
    CANARD CONFIGURATIONS
    CONFIGURATION MANAGEMENT
    LAUNCH VEHICLE CONFIGURATIONS
    MISSILE CONFIGURATIONS
    PROPULSION SYSTEM CONFIGURATIONS
    SPACECRAFT CONFIGURATIONS
    STAGGERING
    TORPEDOES

CONFINEMENT
RT   ASTRONAUT PERFORMANCE
    CONTAINMENT
    ISOLATION
    MAGNETIC COMPRESSION
    NUCLEAR REACTOR CONTROL
    PLASMA CONTROL
    PLASMA EQUILIBRIUM
    SENSORY DEPRIVATION

CONFINING
RT   ASTRONAUT PERFORMANCE
    DEPRIVATION
    ISOLATION
    SENSORY DEPRIVATION

CONFIRMATION
USE   PROVING

CONFLUENCE
USE   CONVERGENCE

CONFORMAL MAPPING
UF   CONFORMAL TRANSFORMATIONS
GS   ANALYSIS (MATHEMATICS)
    . COMPLEX VARIABLES
    . . **CONFORMAL MAPPING**
    FUNCTIONS (MATHEMATICS)
    . **CONFORMAL MAPPING**
RT   COORDINATE TRANSFORMATIONS
    EULER-CAUCHY EQUATIONS
    GRAPHS (CHARTS)
    INVARIANT IMBEDDINGS
    ISOPARAMETRIC FINITE ELEMENTS
    JACOBI INTEGRAL
    LAMBERT SURFACE
    LIGHTHILL METHOD
    MINIMAL SURFACES
    SCHWARZ-CHRISTOFFEL
      TRANSFORMATION
    THEODORSEN TRANSFORMATION

CONFORMAL TRANSFORMATIONS
USE   CONFORMAL MAPPING

CONFUSION
RT   ENTRAPMENT
    TANGLING

CONGENERS
GS   ANATOMY
    . MUSCULOSKELETAL SYSTEM
    . . CONNECTIVE TISSUE
    . . . **CONGENERS**
RT   MUSCLES

CONGENITAL ANOMALIES
UF   CONGENITAL CONDITIONS
RT   CHROMOSOMES
    GENETICS
    HEREDITY
    RHESUS FACTOR

CONGENITAL CONDITIONS
USE   CONGENITAL ANOMALIES

CONGESTION
GS   CIRCULATION
    . **CONGESTION**
RT   ISCHEMIA
    PNEUMONIA
    RESPIRATORY DISEASES
    VASODILATION

CONGO (BRAZZAVILLE)
UF   BRAZZAVILLE
GS   FRENCH EQUATORIAL CONGO
    NATIONS
    . **CONGO (BRAZZAVILLE)**
RT   AFRICA

CONGO (KINSHASA)
USE   ZAIRE

CONGRESSIONAL REPORTS
GS   REPORTS
    . **CONGRESSIONAL REPORTS**
RT   DOCUMENTS
    PRESIDENTIAL REPORTS
    PROCEEDINGS

CONGRUENCES
GS   NUMBER THEORY
    . **CONGRUENCES**
RT   ∞COHERENCE
    COLLOCATION
    DIVIDING (MATHEMATICS)
    GEOMETRY
    IDENTITIES
    INTEGERS
    SYMMETRY

CONICAL BODIES
UF   CONOIDS
GS   CONES
    . **CONICAL BODIES**
    . . SLENDER CONES
    SYMMETRICAL BODIES
    . BODIES OF REVOLUTION
    . . **CONICAL BODIES**
    . . . SLENDER CONES
RT   AFTERBODIES
    AXISYMMETRIC BODIES

CONICAL CAMBER
GS   CAMBER
    . **CONICAL CAMBER**
RT   WING CAMBER

CONICAL FLARE
USE   CONES

CONICAL FLOW
GS   FLUID FLOW
    . **CONICAL FLOW**
RT   AXISYMMETRIC FLOW
    BAFFLES
    ∞DIFFUSERS
    MULTIPHASE FLOW
    SEPARATED FLOW
    THREE DIMENSIONAL FLOW
    WALL FLOW
    WEDGE FLOW

CONICAL INLETS
GS   INTAKE SYSTEMS
    . **CONICAL INLETS**
RT   AIR INTAKES
    CONICAL NOZZLES
    FUNNELS

CONICAL NOZZLES
RT   ANNULAR NOZZLES
    CONICAL INLETS
    CONVERGENT NOZZLES
    CONVERGENT-DIVERGENT NOZZLES
    DIVERGENT NOZZLES
    EXHAUST DIFFUSERS
    EXHAUST NOZZLES
    HYPERSONIC NOZZLES
    INLET NOZZLES
    ∞JET NOZZLES
    NOZZLE GEOMETRY
    NOZZLE INSERTS
    NOZZLE WALLS
    ∞NOZZLES
    PLUG NOZZLES
    ROCKET NOZZLES
    SKIRTS
    SONIC NOZZLES
    SPIKE NOZZLES
    SPRAY NOZZLES
    SUPERSONIC NOZZLES
    TRANSONIC NOZZLES
    TURBINE EXHAUST NOZZLES
    WIND TUNNEL NOZZLES

CONICAL SCANNING
GS   SCANNING
    . **CONICAL SCANNING**
RT   EXAMINATION
    FIELD OF VIEW
    MONITORS
    PANORAMIC SCANNING
    RADAR SCANNING
    READERS
    READING
    SCANNERS
    SEARCHING
    SURVEILLANCE

CONICAL SHELLS
GS   SHELLS (STRUCTURAL FORMS)
    . **CONICAL SHELLS**
RT   CONES

CONICS
GS   GEOMETRY
    . EUCLIDEAN GEOMETRY
    . . ANALYTIC GEOMETRY
    . . . **CONICS**
    . . . . ELLIPSES
    . . . . HYPERBOLAS
    . . . . PARABOLAS
RT   CONES
    HALF CONES
    LOCI

CONIFERS
GS   PLANTS (BOTANY)
    . TREES (PLANTS)
    . . **CONIFERS**
RT   DECIDUOUS TREES
    FORESTS
    TIMBER IDENTIFICATION
    ∞TREES

CONJUGATE GRADIENT METHOD
GS   ANALYSIS (MATHEMATICS)
    . NUMERICAL ANALYSIS
    . . ITERATION
    . . . **CONJUGATE GRADIENT METHOD**
RT   ALGORITHMS
     CONJUGATES
     GRADIENTS
     ITERATIVE SOLUTION

CONJUGATE POINTS
GS   ANALYSIS (MATHEMATICS)
    . COMPLEX VARIABLES
    . . CONJUGATES
    . . . **CONJUGATE POINTS**
RT   LINES OF FORCE
     MAGNETIC FIELDS

CONJUGATED CIRCUITS
GS   CIRCUITS
    . **CONJUGATED CIRCUITS**

CONJUGATES
GS   ANALYSIS (MATHEMATICS)
    . COMPLEX VARIABLES
    . . **CONJUGATES**
    . . . CONJUGATE POINTS
RT   CHOLESKY FACTORIZATION
     CONJUGATE GRADIENT METHOD
     CONJUGATION
     FINITE ELEMENT METHOD

CONJUGATION
GS   **CONJUGATION**
    . PHASE CONJUGATION
RT   CONJUGATES

∞ **CONJUNCTION**
SN   *(USE OF A MORE SPECIFIC TERM IS
     RECOMMENDED--CONSULT THE TERMS
     LISTED BELOW)*
RT   BOOLEAN ALGEBRA
     OCCULTATION
     ORBITS
     PROBABILITY THEORY
     SET THEORY

CONJUNCTIVA
GS   ANATOMY
    . SENSE ORGANS
    . . EYE (ANATOMY)
    . . . **CONJUNCTIVA**
    MEMBRANES
    . **CONJUNCTIVA**
RT   CONJUNCTIVITIS
     KERATITIS
     VISION

CONJUNCTIVITIS
GS   DISEASES
    . EYE DISEASES
    . . **CONJUNCTIVITIS**
    . INFECTIOUS DISEASES
    . . **CONJUNCTIVITIS**
RT   CONJUNCTIVA

CONNECTICUT
GS   NATIONS
    . UNITED STATES
    . . **CONNECTICUT**
RT   NEW HAVEN (CT)

CONNECTIONS
USE   JOINTS (JUNCTIONS)

CONNECTIVE TISSUE
GS   ANATOMY
    . MUSCULOSKELETAL SYSTEM
    . . **CONNECTIVE TISSUE**
    . . . CARTILAGE
    . . . COLLAGENS
    . . . CONGENERS
    . . . MARROW
RT   ADIPOSE TISSUES
     BONES
     EXOSKELETONS
     JOINTS (ANATOMY)
     LIGAMENTS
     TENDONS

CONNECTORS
GS   **CONNECTORS**
    . ELECTRIC CONNECTORS
    . UMBILICAL CONNECTORS

CONNECTORS-*(CONT.)*
    . UNIONS (CONNECTORS)
RT   ADAPTERS
     CORDAGE
     COUPLINGS
     DISCONNECT DEVICES
     FASTENERS
     FITTINGS
     FLANGES
     FLAT CONDUCTORS
     JOINTS (JUNCTIONS)
     JUMPERS
    ∞ JUNCTIONS
     LINKAGES
     SLEEVES
    ∞ TERMINALS
     YOKES

CONNECTORS (ELECTRIC)
USE   ELECTRIC CONNECTORS

CONOIDS
USE   CONICAL BODIES

CONSCIOUSNESS
GS   PERCEPTION
    . SENSORY PERCEPTION
    . . **CONSCIOUSNESS**
RT   ATTENTION
     MENTAL PERFORMANCE
     RECOGNITION
     SLEEP DEPRIVATION

CONSECUTIVE EVENTS
GS   EVENTS
    . **CONSECUTIVE EVENTS**
RT   INTERVALS
     PETRI NETS
     PROBABILITY THEORY
     SCHEDULING
     SEQUENCING
     SEQUENTIAL CONTROL
     TIME MEASUREMENT

CONSERVATION
GS   **CONSERVATION**
    . ENERGY CONSERVATION
RT   AGRICULTURE
     DEFORESTATION
     DROUGHT
     ENERGY POLICY
     ENVIRONMENT MANAGEMENT
     FIREBREAKS
     FOREST MANAGEMENT
     FORESTS
     HABITATS
     LAND USE
     NEWTON SECOND LAW
     NONCONSERVATIVE FORCES
     PARITY
     POTABLE WATER
     REGIONAL PLANNING
     RURAL LAND USE
     SOIL SCIENCE
     SOILS
     WATER MANAGEMENT
     WATER RECLAMATION

CONSERVATION EQUATIONS
RT   CONTINUITY EQUATION
    ∞ EQUATIONS
     NONCONSERVATIVE FORCES
     VORTICITY TRANSPORT HYPOTHESIS

CONSERVATION LAWS
GS   LAWS
    . **CONSERVATION LAWS**
RT   MOMENTUM THEORY
     NEWTON THEORY
     NONCONSERVATIVE FORCES

CONSISTENCY
RT   ABILITIES
     ACCURACY
     COMPUTER SYSTEMS PERFORMANCE
     EFFORT
     ERRORS
     LEVELING
     LINEARITY
    ∞ MEASUREMENT
    ∞ PERFORMANCE
     PRECISION
     QUALITY
     RATINGS
     RELIABILITY

CONSISTENCY-*(CONT.)*
     TOLERANCES (MECHANICS)
     VALIDITY
     VARIABILITY

CONSOLES
GS   PERIPHERAL EQUIPMENT (COMPUTERS)
    . **CONSOLES**
    . . REMOTE CONSOLES
RT   AUTOMATIC TYPEWRITERS
     COMPUTER COMPONENTS
     CONTROL BOARDS
     DATA PROCESSING TERMINALS
     DISPLAY DEVICES
    ∞ EQUIPMENT
     HEAD-UP DISPLAYS
     MAN MACHINE SYSTEMS
     MANUAL CONTROL

CONSOLIDATION
RT   ∞ COMBINATION
     DENSIFICATION
     OVERCONSOLIDATION
     STABILIZATION

CONSONANTS (SPEECH)
RT   SPEECH
     VOWELS
     WORDS (LANGUAGE)

∞ **CONSTANT**
SN   *(USE OF A MORE SPECIFIC TERM IS
     RECOMMENDED--CONSULT THE TERMS
     LISTED BELOW)*
RT   COEFFICIENTS
     CONSTANTS
     INVARIANCE
     TIME CONSTANT

CONSTANT SPEED PROPELLERS
USE   VARIABLE PITCH PROPELLERS

CONSTANT VOLUME BALLOONS
USE   SUPERPRESSURE BALLOONS

CONSTANTAN
GS   ALLOYS
    . **CONSTANTAN**
RT   COPPER
     NICKEL
     THERMOCOUPLES

CONSTANTS
GS   **CONSTANTS**
    . BOHR MAGNETON
    . GRAVITATIONAL CONSTANT
    . GRUNEISEN CONSTANT
    . HUBBLE CONSTANT
    . PLANCKS CONSTANT
    . SOLAR CONSTANT
    . TIME CONSTANT
    . . PERCEPTUAL TIME CONSTANT
RT   COEFFICIENTS
    ∞ CONSTANT

CONSTELLATIONS
GS   **CONSTELLATIONS**
    . ANDROMEDA CONSTELLATION
    . ARIES CONSTELLATION
    . AURIGA CONSTELLATION
    . CASSIOPEIA CONSTELLATION
    . CENTAURUS CONSTELLATION
    . CEPHEUS CONSTELLATION
    . CORONA BOREALIS CONSTELLATION
    . CYGNUS CONSTELLATION
    . LYRA CONSTELLATION
    . ORION CONSTELLATION
    . SAGITTARIUS CONSTELLATION
    . SCORPIUS CONSTELLATION
    . SCUTUM CONSTELLATION
    . TAURUS CONSTELLATION
RT   CELESTIAL SPHERE
     PLANISPHERES
     STARS
     ZODIAC

CONSTITUTION
RT   ATOMIC STRUCTURE
     GOVERNMENTS
     LAW (JURISPRUDENCE)

CONSTITUTIONAL DIAGRAMS
USE   PHASE DIAGRAMS

CONSTITUTIVE EQUATIONS
- RT   ELECTRIC FIELDS
- ∞ EQUATIONS
-   MAGNETIC CHARGE DENSITY
-   MAGNETIC FIELDS
-   MAGNETIC FLUX

CONSTRAINTS
- UF   HINDRANCE
-   LIMITATIONS
-   RESTRAINTS
- GS   **CONSTRAINTS**
-   . METEOROLOGICAL PARAMETERS
-   . . BRUNT-VAISALA FREQUENCY
- RT   BLOCKING
-   CONSTRICTIONS
-   DYNAMIC PROGRAMMING
- ∞ HOLDING
-   LINEAR PROGRAMMING
-   NONLINEAR PROGRAMMING
-   OPERATIONS RESEARCH
-   OPTIMIZATION
-   PENALTY FUNCTION
-   RANGE (EXTREMES)
-   RETAINING

CONSTRICTIONS
- UF   RESTRICTIONS
- RT   ∞ BARRIERS
-   BLOCKING
-   CHOKES (RESTRICTIONS)
-   CLOSURES
-   CONSTRAINTS
-   CONTRACTS
-   IMPEDANCE
-   PLUGGING
- ∞ RESISTANCE
-   RETARDERS (DEVICES)
-   SEALS (STOPPERS)
-   STOPPING

CONSTRICTORS
- GS   ANATOMY
-   . MUSCULOSKELETAL SYSTEM
-   . . **CONSTRICTORS**

CONSTRUCTION
- SN   (EXCLUDES TYPES OF STRUCTURES)
- UF   ERECTION
- RT   ARCHITECTURE
-   ASSEMBLING
-   BRIDGES (STRUCTURES)
- ∞ BUILDINGS
-   CAISSONS
-   CONCRETE STRUCTURES
- ∞ CONSTRUCTION MATERIALS
-   CONTRACTORS
-   CONTRACTS
- ∞ DESIGN
-   EXCAVATION
-   FABRICATION
-   HIGHWAYS
-   INSPECTION
-   INSTALLING
-   LAYOUTS
-   MAINTENANCE
-   MASONRY
-   QUALITY CONTROL
-   RECONSTRUCTION
-   RIGGING
-   SHIPYARDS
-   SPACE MANUFACTURING
-   STARSITE PROGRAM
-   STEEL STRUCTURES
-   STRESS ANALYSIS
-   STRUCTURAL ANALYSIS
-   STRUCTURAL DESIGN
-   STRUCTURAL ENGINEERING
-   STRUCTURAL MEMBERS
-   SURVEYS
-   TUNNELING (EXCAVATION)
-   WELDING

CONSTRUCTION IN SPACE
- USE   ORBITAL ASSEMBLY

CONSTRUCTION INDUSTRY
- GS   INDUSTRIES
-   . **CONSTRUCTION INDUSTRY**
- RT   BRIDGES (STRUCTURES)
- ∞ BUILDINGS
-   CONTRACTORS
-   INDUSTRIAL AREAS
-   INDUSTRIAL PLANTS
-   TOWERS

CONSTRUCTION INDUSTRY-(CONT.)
-   TRUSSES

∞ CONSTRUCTION MATERIALS
- SN   *(USE OF A MORE SPECIFIC TERM IS RECOMMENDED--CONSULT THE TERMS LISTED BELOW)*
- UF   BUILDING MATERIALS
-   STRUCTURAL MATERIALS
- RT   AGGREGATES
-   AIRCRAFT CONSTRUCTION MATERIALS
-   AIRCRAFT SURVIVABILITY
-   AIRFRAME MATERIALS
-   ARCHITECTURE
-   BITUMENS
-   BOARDS (PAPER)
-   BRICKS
-   CEMENTS
-   COMPOSITE MATERIALS
-   CONCRETES
-   CONSTRUCTION
-   GRAPHITE-EPOXY COMPOSITES
-   GROUT
-   INSULATION
-   LATHES
-   MASONITE (TRADEMARK)
-   MASONRY
- ∞ MATERIALS
-   PANELS
-   PLASTICS
-   POLYMER MATRIX COMPOSITES
-   PROTECTIVE COATINGS
-   REACTOR MATERIALS
-   SKIN (STRUCTURAL MEMBER)
-   SPACECRAFT CONSTRUCTION MATERIALS
-   STRUCTURAL MEMBERS

CONSULTING
- RT   CONFERENCES
-   MANAGEMENT PLANNING
-   PERSONNEL
-   RESOURCES

CONSUMABLES (SPACECRAFT)
- GS   **CONSUMABLES (SPACECRAFT)**
-   . CONSUMABLES (SPACECREW SUPPLIES)
-   . . DEHYDRATED FOOD
-   . . SOAPS
-   . . SPACE RATIONS
-   . PROPELLANT STORAGE
-   . SPACE LOGISTICS
-   . STORABLE PROPELLANTS
-   . WORKING FLUIDS
- RT   LOGISTICS

CONSUMABLES (SPACECREW SUPPLIES)
- GS   CONSUMABLES (SPACECRAFT)
-   . **CONSUMABLES (SPACECREW SUPPLIES)**
-   . . DEHYDRATED FOOD
-   . . SOAPS
-   . . SPACE RATIONS
- RT   CLOTHING
- ∞ FOOD
-   HYGIENE
-   PROVISIONING
-   SANITATION
-   SPACE FLIGHT FEEDING
-   SPACE LOGISTICS
-   SURVIVAL EQUIPMENT

CONSUMERS
- RT   CONSUMPTION
-   MARKET RESEARCH
-   MARKETING
-   PRODUCT DEVELOPMENT

CONSUMPTION
- GS   **CONSUMPTION**
-   . ENERGY CONSUMPTION
-   . FUEL CONSUMPTION
-   . OXYGEN CONSUMPTION
-   . WATER CONSUMPTION
- RT   COMMERCE
-   CONSUMERS
-   DEMAND (ECONOMICS)
-   DEPLETION
-   EXHAUSTING
-   EXHAUSTION
-   SUPPLYING
-   UTILIZATION

CONTACT DERMATITIS
- GS   DISEASES
-   . INFECTIOUS DISEASES
-   . . DERMATITIS
-   . . . **CONTACT DERMATITIS**
- RT   ALLERGIC DISEASES
-   DERMATOLOGY
-   EPIDERMIS
-   ITCHING
-   PATCH TESTS
-   SKIN (ANATOMY)

CONTACT LENSES
- GS   LENSES
-   . **CONTACT LENSES**
- RT   EYEPIECES
-   RETICLES

CONTACT POTENTIALS
- GS   POTENTIAL ENERGY
-   . ELECTRIC POTENTIAL
-   . . **CONTACT POTENTIALS**
- RT   ELECTRIC CONTACTS
-   SURFACE PROPERTIES

CONTACT RESISTANCE
- GS   IMPEDANCE
-   . ELECTRICAL IMPEDANCE
-   . . ELECTRICAL RESISTANCE
-   . . . **CONTACT RESISTANCE**
- RT   ELECTRIC CONTACTS
-   NONOHMIC EFFECT
- ∞ RESISTANCE
-   SURFACE PROPERTIES

CONTACTORS
- SN   (EXCLUDES ELECTRIC SWITCHES)
- RT   CARBURETORS
-   CHEMICAL REACTORS
-   COLUMNS (PROCESS ENGINEERING)
-   ELECTRIC SWITCHES
-   MIXERS
-   SPRAYERS

CONTACTS (ELECTRIC)
- USE   ELECTRIC CONTACTS

CONTACTS (GEOLOGY)
- RT   FORMATIONS
-   GEOLOGY
-   METAMORPHISM (GEOLOGY)
-   MINERAL DEPOSITS
-   ROCK INTRUSIONS
-   ROCKS

CONTAINERLESS MELTS
- GS   MELTS (CRYSTAL GROWTH)
-   . **CONTAINERLESS MELTS**
- RT   CRYSTAL GROWTH
-   CRYSTALLIZATION
-   CRYSTALS
-   DIRECTIONAL SOLIDIFICATION (CRYSTALS)
-   LOW GRAVITY MANUFACTURING
-   MANUFACTURING
-   MELTING
-   ORBITAL WORKSHOPS
-   SPACE PROCESSING
-   WEIGHTLESSNESS

∞ CONTAINERS
- SN   *(USE OF A MORE SPECIFIC TERM IS RECOMMENDED--CONSULT THE TERMS LISTED BELOW)*
- UF   RECEPTACLES (CONTAINERS)
- RT   AMPOULES
-   AUTOCLAVES
-   BAGS
-   BARRELS (CONTAINERS)
-   BASKETS
-   BIOPAKS
-   BOTTLES
-   BOXES (CONTAINERS)
-   BUNDLES
-   CANS
- ∞ CAPSULES
-   CARTRIDGES
-   CASES (CONTAINERS)
-   CRUCIBLES
-   DISPOSAL
-   DRUMS (CONTAINERS)
-   ENCLOSURES
-   FUEL TANKS
-   GLASSWARE
-   HOPPERS

CONTAINERS-*(CONT.)*
HOUSINGS
MATERIALS HANDLING
MICROMODULES
PACKAGES
PACKAGING
PRESERVING
PRESSURE VESSELS
PROTECTORS
REELS
SPOOLS
SPRAYERS
TANKS (CONTAINERS)
TRANSPORTER
TRAYS
WING TANKS

CONTAINMENT
RT   ASTEROID CAPTURE
BLOCKING
BURST TESTS
CONFINEMENT
RETAINING
SEALING
STOPPING

CONTAMINANTS
UF   NOXIOUS MATERIALS
POLLUTANTS
GS   **CONTAMINANTS**
. RADIOACTIVE CONTAMINANTS
. TRACE CONTAMINANTS
RT   CONTAMINATION
DECONTAMINATION
DILUENTS
DIRT
DUST
EFFLUENTS
ENVIRONMENT EFFECTS
ENVIRONMENTAL QUALITY
FUEL CONTAMINATION
IMPURITIES
∞MATERIALS
NONPOINT SOURCES
POLLUTION
PURITY
QUALITY
WASTES
WATER TREATMENT

CONTAMINATION
GS   **CONTAMINATION**
. FUEL CONTAMINATION
. SPACECRAFT CONTAMINATION
RT   AIR POLLUTION
ANTIFOULING
ANTIINFECTIVES AND ANTIBACTERIALS
CONTAMINANTS
DECONTAMINATION
ENVIRONMENT EFFECTS
FOULING
INTRUSION
MOLECULAR SHIELDS
NONPOINT SOURCES
POLLUTION
PURITY
RADIOACTIVE WASTES
WATER POLLUTION

CONTENT
GS   **CONTENT**
. BONE MINERAL CONTENT
RT   ∞COMPONENTS
∞COMPOSITION
INGREDIENTS

CONTEXT
RT   NATURAL LANGUAGE (COMPUTERS)
PATTERN RECOGNITION

CONTEXT FREE LANGUAGES
GS   LANGUAGES
. PROGRAMMING LANGUAGES
. . **CONTEXT FREE LANGUAGES**
RT   COMPUTER PROGRAMMING
SYMBOLIC PROGRAMMING

CONTINENTAL DRIFT
RT   CONTINENTS
EARTH CRUST
EARTH PLANETARY STRUCTURE
GEOMAGNETISM
GEOPHYSICS
PALEOMAGNETISM

CONTINENTAL MARGINS
USE   CONTINENTAL SHELVES

CONTINENTAL SHELVES
UF   CONTINENTAL MARGINS
RT   GEOLOGY
OCEAN BOTTOM
SEAMOUNTS
∞SHELVES

CONTINENTS
GS   **CONTINENTS**
. AFRICA
. ASIA
. AUSTRALIA
. EUROPE
. NORTH AMERICA
. SOUTH AMERICA
RT   ANTARCTIC REGIONS
CENTRAL EUROPE
CONTINENTAL DRIFT
CRATONS
GEOGRAPHY
MOUNTAINS
TRANSCONTINENTAL SYSTEMS

CONTINGENCY
RT   CONFIDENCE LIMITS
CORRELATION
ESTIMATES
EXPECTATION
MATERIALS HANDLING
PREDICTIONS
RESERVES
RISK

CONTINUITY
RT   CONTINUITY (MATHEMATICS)
COORDINATION
SCHEDULING
TOPOLOGY
VARIABILITY

CONTINUITY (MATHEMATICS)
GS   ANALYSIS (MATHEMATICS)
. CALCULUS
. . **CONTINUITY (MATHEMATICS)**
RT   CONTINUITY
FUNCTIONS (MATHEMATICS)
ISOPERIMETRIC PROBLEM
NORMAL DENSITY FUNCTIONS
POISSON DENSITY FUNCTIONS
PROBABILITY DENSITY FUNCTIONS
STATISTICAL ANALYSIS
SYMMETRY
TOPOLOGY

CONTINUITY EQUATION
RT   CONSERVATION EQUATIONS
CONTINUUM MECHANICS
CROCCO-LEE THEORY
∞EQUATIONS
EQUATIONS OF MOTION
EQUATIONS OF STATE
FLUID DYNAMICS
NONCONSERVATIVE FORCES
STEADY FLOW

†

CONTINUOUS NOISE
RT   ∞NOISE
NOISE PROPAGATION
SOUND GENERATORS

CONTINUOUS RADIATION
UF   CONTINUOUS WAVES
GS   **CONTINUOUS RADIATION**
. MODULATED CONTINUOUS RADIATION
RT   ABSORPTION SPECTRA
BACKGROUND RADIATION
COHERENT RADIATION
CORPUSCULAR RADIATION
ELASTIC WAVES
ELECTROMAGNETIC RADIATION
EMISSION SPECTRA
PULSED RADIATION
∞RADIATION
∞RAYS

CONTINUOUS SPECTRA
GS   SPECTRA
. **CONTINUOUS SPECTRA**
RT   ASTRONOMICAL SPECTROSCOPY
SOLAR SPECTRA
SPECTRAL EMISSION

CONTINUOUS SPECTRA-*(CONT.)*
STELLAR SPECTRA

CONTINUOUS WAVE LASERS
GS   STIMULATED EMISSION DEVICES
. LASERS
. . **CONTINUOUS WAVE LASERS**
RT   ARGON LASERS
CARBON DIOXIDE LASERS
CARBON MONOXIDE LASERS
LASER STABILITY
SOLID STATE LASERS

CONTINUOUS WAVE RADAR
UF   CW RADAR
GS   RADAR
. **CONTINUOUS WAVE RADAR**
RT   COHERENT RADAR
DOPPLER RADAR
PULSE RADAR
RADAR DETECTION
RADAR RANGE
SEARCH RADAR
SURVEILLANCE RADAR
TRACKING RADAR

CONTINUOUS WAVES
USE   CONTINUOUS RADIATION

CONTINUUM FLOW
GS   FLUID FLOW
. GAS FLOW
. . **CONTINUUM FLOW**
RT   FREE MOLECULAR FLOW
MOLECULAR FLOW
RAREFIED GAS DYNAMICS
SLIP FLOW

CONTINUUM MECHANICS
RT   BURGER EQUATION
CLASSICAL MECHANICS
CONTINUITY EQUATION
CONTINUUM MODELING
∞DYNAMICS
FLOW THEORY
FLUID MECHANICS
MAXWELL BODIES
∞MECHANICS (PHYSICS)
MULTIPOLAR FIELDS
SOLID MECHANICS
STATISTICAL MECHANICS
STRESS TENSORS

CONTINUUM MODELING
RT   CONTINUUM MECHANICS
CONTINUUMS
LARGE SPACE STRUCTURES
MATHEMATICAL MODELS
STRUCTURAL ANALYSIS

CONTINUUMS
RT   CONTINUUM MODELING
PROBABILITY THEORY
REAL VARIABLES
RELATIVITY
TOPOLOGY

CONTOUR SENSORS
RT   BOUNDARIES
CONCAVITY
CONTOURS
CONVEXITY
IMAGERY
IMAGES
∞SENSORS
SHAPES
TOPOGRAPHY

CONTOURS
UF   CURVED SURFACES
RT   CONCAVITY
CONTOUR SENSORS
CONVEXITY
CURVED PANELS
DATUM (ELEVATION)
ELEVATION
FLATNESS
GEOMORPHOLOGY
HYPSOGRAPHY
MAPPING
ROUGHNESS
SHAPES
TOPOGRAPHY

CONTRACT INCENTIVES
- GS   INCENTIVES
- .   **CONTRACT INCENTIVES**
- MOTIVATION
- .   **CONTRACT INCENTIVES**

CONTRACT MANAGEMENT
- GS   MANAGEMENT
- .   **CONTRACT MANAGEMENT**
- RT   CONTRACTS
- DECISION MAKING
- DECISIONS
- PERT
- PROJECT MANAGEMENT
- SCHEDULES
- SUBCONTRACTS
- SYSTEMS ENGINEERING

CONTRACT NEGOTIATION
- RT   CONTRACTORS
- CONTRACTS
- DECISION MAKING
- GOVERNMENT/INDUSTRY RELATIONS
- INDUSTRIES
- MANAGEMENT
- MANUFACTURING
- SUBCONTRACTS

CONTRACTION
- RT   COOLING
- ∞REDUCTION
- SHRINKAGE
- SPASMS

CONTRACTORS
- RT   CHEMICAL REACTORS
- COLUMNS (PROCESS ENGINEERING)
- CONSTRUCTION
- CONSTRUCTION INDUSTRY
- CONTRACT NEGOTIATION
- CONTRACTS
- GOVERNMENT/INDUSTRY RELATIONS
- INDUSTRIES
- QUALIFICATIONS
- SUBCONTRACTS
- TRANSPORTATION

CONTRACTS
- GS   **CONTRACTS**
- .   SUBCONTRACTS
- RT   AGREEMENTS
- CANCELLATION
- CONSTRICTIONS
- CONSTRUCTION
- CONTRACT MANAGEMENT
- CONTRACT NEGOTIATION
- CONTRACTORS
- ESTIMATES
- ESTIMATING
- EXTENSIONS
- FEDERAL BUDGETS
- GOVERNMENT PROCUREMENT
- GOVERNMENT/INDUSTRY RELATIONS
- GRANTS
- LEASING
- LEGAL LIABILITY
- OPTIONS
- PROCUREMENT
- PROJECTS
- REVISIONS
- SUPPLEMENTS

CONTRAILS
- UF   CONDENSATION TRAILS
- VAPOR TRAILS
- RT   CONDENSATES
- WAKES

CONTRALATERAL FUNCTIONS
- RT   ∞FUNCTIONS

CONTRAROTATING PROPELLERS
- GS   PROPELLERS
- .   **CONTRAROTATING PROPELLERS**
- RT   PROPELLER DRIVE
- PROPELLER EFFICIENCY
- TURBOPROP ENGINES

CONTRAST
- GS   **CONTRAST**
- .   IMAGE CONTRAST
- .   PHASE CONTRAST
- RT   CHARACTER RECOGNITION
- COLOR

CONTRAST-(CONT.)
- LEGIBILITY
- PERCEPTION
- PRINTING
- RESOLUTION
- ∞SHARPNESS
- VISIBILITY
- VISION

∞ CONTROL
- SN   (USE OF A MORE SPECIFIC TERM IS
- RECOMMENDED--CONSULT THE TERMS
- LISTED BELOW)
- UF   CONTROL SYSTEMS
- CONTROLLED STABILITY
- REGULATION
- RT   ACCESS CONTROL
- ACTIVE CONTROL
- ADAPTIVE CONTROL
- AIR NAVIGATION
- AIR TRAFFIC CONTROL
- AIRCRAFT CONTROL
- ALTITUDE CONTROL
- APPROACH CONTROL
- ASTRIONICS
- ATTITUDE CONTROL
- AUTOMATIC CONTROL
- AUTOMATIC CONTROL VALVES
- AUTOMATIC FLIGHT CONTROL
- AUTOMATIC FREQUENCY CONTROL
- AUTOMATIC GAIN CONTROL
- AVIONICS
- BOUNDARY LAYER CONTROL
- CASCADE CONTROL
- CHEMICAL REACTION CONTROL
- CIRCULATION CONTROL AIRFOILS
- CIRCULATION CONTROL ROTORS
- COMBUSTION CONTROL
- COMMAND AND CONTROL
- CONTROL BOARDS
- CONTROL CONFIGURED VEHICLES
- CONTROL DATA (COMPUTERS)
- CONTROL EQUIPMENT
- CONTROL MOMENT GYROSCOPES
- CONTROL ROCKETS
- CONTROL RODS
- CONTROL SIMULATION
- CONTROL STABILITY
- CONTROL STICKS
- CONTROL SURFACES
- CONTROL SYSTEMS DESIGN
- CONTROL THEORY
- CONTROL UNITS (COMPUTERS)
- CONTROL VALVES
- CONTROLLABILITY
- CONTROLLED ATMOSPHERES
- CONTROLLED FUSION
- CONTROLLERS
- CRITICAL PATH METHOD
- CYBERNETICS
- DIRECT LIFT CONTROLS
- DIRECTIONAL CONTROL
- DYNAMIC CHARACTERISTICS
- DYNAMIC CONTROL
- ELECTRIC CONTROL
- ELECTRONIC CONTROL
- ELEVATORS (CONTROL SURFACES)
- ENGINE CONTROL
- ENVIRONMENTAL CONTROL
- FEEDBACK CONTROL
- FEEDFORWARD CONTROL
- FIRE CONTROL
- FIRE CONTROL CIRCUITS
- FLAPS (CONTROL SURFACES)
- FLIGHT CONTROL
- FLOOD CONTROL
- FLUIDICS
- FLY BY TUBE CONTROL
- FLY BY WIRE CONTROL
- FREQUENCY CONTROL
- FUEL CONTROL
- GROUND BASED CONTROL
- HARMONIC CONTROL
- HELICOPTER CONTROL
- HYDRAULIC CONTROL
- INTEGRATED MISSION CONTROL
- CENTER
- INTERACTIVE CONTROL
- INVENTORY CONTROLS
- JET CONTROL
- LATERAL CONTROL
- LONGITUDINAL CONTROL
- MAGNETIC CONTROL
- MANUAL CONTROL
- MISSILE CONTROL
- NETWORK CONTROL

CONTROL-(CONT.)
- NUCLEAR REACTOR CONTROL
- NUMERICAL CONTROL
- OFF-ON CONTROL
- OPTIMAL CONTROL
- PAYLOAD CONTROL
- PHASE CONTROL
- PLASMA CONTROL
- PNEUMATIC CONTROL
- POINTING CONTROL SYSTEMS
- POLLUTION CONTROL
- POROUS BOUNDARY LAYER CONTROL
- PROCESS CONTROL (INDUSTRY)
- PROPORTIONAL CONTROL
- QUALITY CONTROL
- RADAR APPROACH CONTROL
- RADIO CONTROL
- ∞REACTION CONTROL
- REGULATIONS
- REGULATORS
- REMOTE CONTROL
- ROCKET ENGINE CONTROL
- SATELLITE ATTITUDE CONTROL
- SATELLITE CONTROL
- SCHEDULING
- SELF ADAPTIVE CONTROL SYSTEMS
- SEQUENTIAL CONTROL
- SERVOCONTROL
- SERVOMECHANISMS
- SHAPE CONTROL
- SHOCK WAVE CONTROL
- SPACECRAFT CONTROL
- SPECTRAL SHIFT CONTROL
- SPECTRAL SHIFT CONTROL REACTOR
- SPEED CONTROL
- STABILIZATION
- STEERING
- SUBMARINE INTEGRATED CONTROL
- PROJECT
- SYSTEMS ENGINEERING
- TABS (CONTROL SURFACES)
- TEMPERATURE CONTROL
- THERMAL CONTROL COATINGS
- THRUST CONTROL
- THRUST VECTOR CONTROL
- TIME OPTIMAL CONTROL
- TRAFFIC CONTROL
- TRAJECTORY CONTROL
- TRANSIT ATTITUDE CONTROL
- SATELLITE
- TRANSPONDER CONTROL GROUP
- TURBOJET ENGINE CONTROL
- VARIABLE STREAM CONTROL ENGINES
- VISUAL CONTROL
- VOICE CONTROL
- WAVE INCIDENCE CONTROL
- WEATHER MODIFICATION

CONTROL BOARDS
- UF   CONTROL PANELS
- RT   CONSOLES
- ∞CONTROL
- DISPLAY DEVICES
- MANUAL CONTROL
- REMOTE CONTROL

CONTROL CONFIGURED VEHICLES
- RT   AIRCRAFT CONFIGURATIONS
- AIRCRAFT DESIGN
- ∞CONTROL
- FLIGHT CONTROL
- TECHNOLOGY UTILIZATION
- ∞VEHICLES

CONTROL DATA (COMPUTERS)
- RT   COMPUTERS
- ∞CONTROL
- ∞DATA
- DATA SYSTEMS

CONTROL DEVICES
- USE   CONTROL EQUIPMENT

CONTROL EQUIPMENT
- UF   CONTROL DEVICES
- EFFECTORS
- GS   **CONTROL EQUIPMENT**
- .   CONTROL STICKS
- .   CRYOSTATS
- .   PRESSURE REGULATORS
- .   PRESSURE SWITCHES
- .   SERVOAMPLIFIERS
- .   SPEED REGULATORS
- .   TELEOPERATORS
- .   THERMOSTATS

CONTROL EQUIPMENT-(CONT.)
RT   AIRCRAFT CONTROL
     AUTOMATIC CONTROL
    ∞CONTROL
     ELECTRIC CONTROL
     ELECTRONIC CONTROL
     FEEDBACK CONTROL
     MANIPULATORS
     MANUAL CONTROL
     NONLINEAR SYSTEMS
     OFF-ON CONTROL
     PNEUMATIC CONTROL
     PROPORTIONAL CONTROL
     RECORDING INSTRUMENTS
     SPEED CONTROL
     TRANSDUCERS

CONTROL MOMENT GYROSCOPES
GS   GYROSCOPES
    . **CONTROL MOMENT GYROSCOPES**
RT   ATTITUDE CONTROL
    ATTITUDE GYROS
    ATTITUDE INDICATORS
    ∞CONTROL
    EQUATIONS OF MOTION
    GIMBALS
    GYRODAMPERS
    INDICATING INSTRUMENTS
    MEASURING INSTRUMENTS
    NUTATION DAMPERS
    SERVOCONTROL
    SERVOMECHANISMS

CONTROL PANELS
USE   CONTROL BOARDS

CONTROL ROCKETS
UF   STEERING ROCKETS
GS   ENGINES
    . TORPEDO ENGINES
    . . VERNIER ENGINES
    . . . **CONTROL ROCKETS**
RT   ∞CONTROL
    RETROROCKET ENGINES
    STEERING
    THRUST CONTROL
    VARIABLE THRUST

CONTROL RODS
GS   RODS
    . **CONTROL RODS**
RT   ∞CONTROL
    NEUTRON ABSORBERS
    NUCLEAR REACTOR CONTROL
    NUCLEAR REACTORS
    POISONING (REACTION INHIBITION)
    REACTOR CORES
    REACTOR SAFETY

CONTROL SIMULATION
GS   SIMULATION
    . **CONTROL SIMULATION**
    SIMULATORS
    . **CONTROL SIMULATION**
RT   AIRCRAFT CONTROL
    COMPUTERIZED SIMULATION
    ∞CONTROL
    FLIGHT SIMULATION
    FLIGHT SIMULATORS
    MOTION SIMULATORS
    SPACECRAFT CONTROL
    SPACECRAFT MANEUVERS
    TRAINING SIMULATORS

CONTROL STABILITY
GS   DYNAMIC CHARACTERISTICS
    . DYNAMIC STABILITY
    . . **CONTROL STABILITY**
    STABILITY
    . DYNAMIC STABILITY
    . . **CONTROL STABILITY**
RT   AIRCRAFT CONTROL
    AIRCRAFT SPIN
    AIRCRAFT STABILITY
    ∞CONTROL
    CONTROLLABILITY
    FLIGHT CONTROL
    MOTION STABILITY
    NYQUIST DIAGRAM
    PILOT INDUCED OSCILLATION
    ROBUSTNESS (MATHEMATICS)
    SPACECRAFT MOTION
    SPACECRAFT STABILITY
    STABILITY AUGMENTATION
    SYSTEMS STABILITY

CONTROL STICKS
GS   CONTROL EQUIPMENT
    . **CONTROL STICKS**
RT   AIRCRAFT CONTROL
    FLIGHT CONTROL
    MANUAL CONTROL

CONTROL SURFACES
GS   **CONTROL SURFACES**
    . AERIAL RUDDERS
    . AILERONS
    . . FLAPERONS
    . . SPOILER SLOT AILERONS
    . ELEVATORS (CONTROL SURFACES)
    . ELEVONS
    . FLAPS (CONTROL SURFACES)
    . . EXTERNALLY BLOWN FLAPS
    . . . UPPER SURFACE BLOWN FLAPS
    . . FLAPERONS
    . . JET FLAPS
    . . SPLIT FLAPS
    . . WING FLAPS
    . . . LEADING EDGE FLAPS
    . . . LEADING EDGE SLATS
    . . . TRAILING EDGE FLAPS
    . GUIDE VANES
    . JET VANES
    . HORIZONTAL TAIL SURFACES
    . RUDDERS
    . . MARINE RUDDERS
    . SPOILERS
    . TABS (CONTROL SURFACES)
RT   AERODYNAMIC BRAKES
    AERODYNAMIC CONFIGURATIONS
    AERODYNAMIC INTERFERENCE
    AERODYNAMICS
    AIRCRAFT PARTS
    AIRCRAFT STRUCTURES
    AIRFOILS
    AIRFRAMES
    BLUNT TRAILING EDGES
    BOUNDARY LAYER CONTROL
    CANARD CONFIGURATIONS
    ∞CONTROL
    DRAG DEVICES
    FINS
    FIRES
    FLIGHT CONTROL
    FREE WING AIRCRAFT
    GUIDANCE (MOTION)
    NOSE FINS
    STABILIZERS (FLUID DYNAMICS)
    ∞SURFACES
    SWEPTBACK TAIL SURFACES
    T TAIL SURFACES
    TAIL ASSEMBLIES
    TAIL SURFACES
    TRAPEZOIDAL TAIL SURFACES
    UPPER SURFACE BLOWN FLAPS
    VANES
    VORTEX FLAPS
    WINGS

CONTROL SYSTEMS
USE   CONTROL

CONTROL SYSTEMS DESIGN
GS   SYSTEMS ENGINEERING
    . **CONTROL SYSTEMS DESIGN**
RT   AEROSPACE SYSTEMS
    AUTOMATIC CONTROL
    ∞AUTOMATION
    BIONICS
    BOND GRAPHS
    ∞CONTROL
    CONTROL THEORY
    CONTROLLERS
    CYBERNETICS
    ∞DESIGN
    DESIGN ANALYSIS
    ELECTRIC CONTROL
    ELECTRONIC CONTROL
    FEEDBACK CONTROL
    MATHEMATICAL MODELS
    NUMERICAL CONTROL
    OPERATIONS RESEARCH
    OPTIMAL CONTROL
    PARAMETER IDENTIFICATION
    SYSTEM IDENTIFICATION
    SYSTEMS ANALYSIS
    SYSTEMS INTEGRATION

CONTROL THEORY
RT   ADAPTIVE CONTROL
    CLOSED CYCLES

CONTROL THEORY-(CONT.)
    CONTROL SIMULATION
    CONTROL SYSTEMS DESIGN
    CONTROLLABILITY
    DISTRIBUTED PARAMETER SYSTEMS
    DYNAMIC CONTROL
    DYNAMICAL SYSTEMS
    FEEDBACK
    FEEDBACK CONTROL
    FEEDFORWARD CONTROL
    INTERACTIVE CONTROL
    OBSERVABILITY (SYSTEMS)
    OFF-ON CONTROL
    OPTIMAL CONTROL
    ROBUSTNESS (MATHEMATICS)
    SERVOCONTROL
    SHAPE CONTROL
    ∞THEORIES
    TRACKING PROBLEM

CONTROL UNITS (COMPUTERS)
RT   CENTRAL PROCESSING UNITS
    COMPUTER COMPONENTS
    COMPUTERS
    ∞CONTROL
    DATA PROCESSING EQUIPMENT

CONTROL VALVES
GS   VALVES
    . **CONTROL VALVES**
RT   ACTUATORS
    ∞CONTROL
    PNEUMATIC CONTROL

CONTROLLABILITY
UF   HANDLING QUALITIES
RT   AIRCRAFT CONTROL
    AIRCRAFT LANDING
    AIRCRAFT PERFORMANCE
    AIRCRAFT SPECIFICATIONS
    AIRCRAFT SPIN
    AIRCRAFT STABILITY
    ∞CONTROL
    CONTROL STABILITY
    CONTROL THEORY
    DIRECTIONAL STABILITY
    FLIGHT CHARACTERISTICS
    HELICOPTER CONTROL
    HELICOPTER PERFORMANCE
    LIQUID SLOSHING
    LOW SPEED STABILITY
    MANEUVERABILITY
    QUALITY
    SPACECRAFT RELIABILITY
    STABILITY
    STEERING
    WHEEL BRAKES

CONTROLLED ATMOSPHERES
GS   **CONTROLLED ATMOSPHERES**
    . ARGON-OXYGEN ATMOSPHERES
    . CABIN ATMOSPHERES
    . . SPACECRAFT CABIN ATMOSPHERES
    . HELIUM-OXYGEN ATMOSPHERES
    . INERT ATMOSPHERE
RT   AIR CONDITIONING
    ∞ATMOSPHERES
    ∞BLANKETS
    CLEAN ROOMS
    ∞CONTROL
    ENVIRONMENTS
    FURNACES
    GAS MIXTURES
    GNOTOBIOTICS
    OXYGEN SUPPLY EQUIPMENT
    SPACECRAFT ENVIRONMENTS

CONTROLLED AVALANCHE TRANSIT TIME DEVICES
USE   CATT DEVICES

CONTROLLED FUSION
SN   (CONTROLLED NUCLEAR FUSION)
GS   NUCLEAR REACTIONS
    . THERMONUCLEAR REACTIONS
    . . NUCLEAR FUSION
    . . . **CONTROLLED FUSION**
RT   ∞CONTROL
    JOINT EUROPEAN TORUS
    LIMITERS (FUSION REACTORS)
    PLASMA COMPRESSION
    PLASMA COOLING
    PLASMA CURRENTS
    PLASMA PHYSICS
    RELATIVISTIC ELECTRON BEAMS

CONTROLLED FUSION-(CONT.)
STRONGLY COUPLED PLASMAS
THERMONUCLEAR POWER GENERATION
ZETA PINCH

CONTROLLED STABILITY
USE   CONTROL

CONTROLLERS
SN   (DEVICES WHICH EMPLOY AN OUTSIDE
     SOURCE OF ENERGY AND USUALLY
     UTILIZE FEEDBACK)
GS   **CONTROLLERS**
    . POWER FACTOR CONTROLLERS
    . SERVOMECHANISMS
    . . SERVOAMPLIFIERS
    . . SERVOMOTORS
RT   ACTUATORS
    ANALYZERS
    AUTOMATIC CONTROL
    ∞AUTOMATION
    ∞CONTROL
    CONTROL SYSTEMS DESIGN
    CRYOSTATS
    CURRENT REGULATORS
    CYBERNETICS
    ELECTRONIC CONTROL
    INSTRUMENT RECEIVERS
    INSTRUMENT TRANSMITTERS
    ∞INSTRUMENTS
    MEASURING INSTRUMENTS
    PNEUMATIC CONTROL
    PRESSURE REGULATORS
    PROPELLANT ACTUATED INSTRUMENTS
    REGULATORS
    REMOTE CONTROL
    ROCKET-BORNE INSTRUMENTS
    SPEED CONTROL
    SPEED REGULATORS
    TEMPERATURE CONTROL
    THERMOSTATS
    VOLTAGE REGULATORS

CONVAIR MILITARY AIRCRAFT
USE   GENERAL DYNAMICS AIRCRAFT
    MILITARY AIRCRAFT

CONVAIR 340 AIRCRAFT
USE   CV-340 AIRCRAFT

CONVAIR 440 AIRCRAFT
USE   CV-440 AIRCRAFT

CONVAIR 880 AIRCRAFT
USE   CV-880 AIRCRAFT

CONVAIR 990 AIRCRAFT
USE   CV-990 AIRCRAFT

CONVECTION
GS   **CONVECTION**
    . BENARD CELLS
    . FORCED CONVECTION
    . FREE CONVECTION
    . . RAYLEIGH-BENARD CONVECTION
    . MARANGONI CONVECTION
RT   ADVECTION
    BASE HEATING
    BOUSSINESQ APPROXIMATION
    ∞CONDUCTION
    FLUID DYNAMICS
    GRASHOF NUMBER
    HEAT TRANSMISSION
    HEATING
    METEOROLOGY
    MIXING HEIGHT

CONVECTION CLOUDS
GS   CLOUDS
    . CLOUDS (METEOROLOGY)
    . . **CONVECTION CLOUDS**
    . . . CIRRUS CLOUDS
    . . . CUMULONIMBUS CLOUDS
    . . . . ARC CLOUDS
    . . . NIMBOSTRATUS CLOUDS
    . . . STRATOCUMULUS CLOUDS
    . . . STRATUS CLOUDS
RT   AIR CURRENTS
    CLOUD PHYSICS
    METEOROLOGY
    NEPHANALYSIS
    SUPERCOOLING
    VERTICAL AIR CURRENTS

CONVECTION CURRENTS
RT   AIR CURRENTS
    BENARD CELLS
    ELECTRON BUNCHING
    FLUID FLOW
    FREE CONVECTION
    MIXING HEIGHT
    RAYLEIGH-BENARD CONVECTION
    SOLAR GRANULATION
    VERTICAL AIR CURRENTS

CONVECTIVE FLOW
UF   THERMAL CURRENTS
GS   FLUID FLOW
    . **CONVECTIVE FLOW**
    . . RAYLEIGH-BENARD CONVECTION
    . . . BENARD CELLS
RT   FREE CONVECTION
    GAS DENSITY
    GEOPHYSICAL FLUID FLOW CELLS
    HEAT TRANSMISSION
    MARANGONI CONVECTION
    MASS FLOW RATE
    MASS TRANSFER
    POROUS BOUNDARY LAYER CONTROL
    TEMPERATURE
    THERMAL DIFFUSION

CONVECTIVE HEAT TRANSFER
GS   TRANSMISSION
    . HEAT TRANSMISSION
    . . HEAT TRANSFER
    . . . **CONVECTIVE HEAT TRANSFER**
RT   AERODYNAMIC HEATING
    BOUNDARY LAYER COMBUSTION
    BOUNDARY LAYER FLOW
    CONDUCTIVE HEAT TRANSFER
    COOLING FINS
    FORCED CONVECTION
    FREE CONVECTION
    LAMINAR HEAT TRANSFER
    MASS TRANSFER
    NUSSELT NUMBER
    RADIATIVE HEAT TRANSFER
    RAYLEIGH-BENARD CONVECTION
    SURFACE COOLING
    TEMPERATURE GRADIENTS
    THERMOHYDRAULICS
    THERMOSIPHONS
    TURBULENT HEAT TRANSFER

CONVENTIONS
RT   AGREEMENTS
    AIR LAW
    COMPATIBILITY
    CONFERENCES
    ∞COOPERATION
    INTERNATIONAL COOPERATION
    INTERNATIONAL LAW
    OUTER SPACE TREATY
    STANDARDS

CONVERGENCE
UF   CONFLUENCE
RT   DIVERGENCE
    TAPERING
    VARIABILITY

CONVERGENT NOZZLES
RT   CONICAL NOZZLES
    FLUID AMPLIFIERS
    NOZZLE GEOMETRY
    NOZZLE WALLS
    ∞NOZZLES
    TURBINE ENGINES
    TURBOJET ENGINES

CONVERGENT-DIVERGENT NOZZLES
UF   DE LAVAL NOZZLES
GS   EXHAUST NOZZLES
    . **CONVERGENT-DIVERGENT NOZZLES**
RT   CONICAL NOZZLES
    NOZZLE GEOMETRY
    NOZZLE INSERTS
    ∞NOZZLES
    ROCKET NOZZLES
    SUPERSONIC NOZZLES
    TRANSONIC NOZZLES
    TURBINE EXHAUST NOZZLES
    WIND TUNNEL NOZZLES

CONVERSATION
GS   COMMUNICATING
    . **CONVERSATION**
    SPEECH

CONVERSATION-(CONT.)
    . **CONVERSATION**
RT   VERBAL COMMUNICATION
    VOICE COMMUNICATION
    WORDS (LANGUAGE)

∞ CONVERSION
SN   (USE OF A MORE SPECIFIC TERM IS
    RECOMMENDED--CONSULT THE TERMS
    LISTED BELOW)
RT   BIOCONVERSION
    CONVERSION TABLES
    DATA CONVERSION ROUTINES
    ELECTRIC GENERATORS
    ENERGY CONVERSION
    ENERGY CONVERSION EFFICIENCY
    EXCHANGING
    FREQUENCY CONVERTERS
    GEOTHERMAL ENERGY CONVERSION
    INTERNAL CONVERSION
    ISOMERIZATION
    LIQUEFACTION
    METRICATION
    OCEAN THERMAL ENERGY CONVERSION
    ORGANIC WASTES (FUEL CONVERSION)
    ORTHO PARA CONVERSION
    PHOTOTHERMAL CONVERSION
    PHOTOVOLTAIC CONVERSION
    REFINING
    SATELLITE SOLAR ENERGY
      CONVERSION
    SOLAR ENERGY CONVERSION
    SOLAR TOTAL ENERGY SYSTEMS
    THERMIONIC POWER GENERATION
    THERMOELECTRIC POWER GENERATION
    TURBOGENERATORS
    WATERWAVE ENERGY CONVERSION

CONVERSION TABLES
GS   TABLES (DATA)
    . **CONVERSION TABLES**
RT   ∞CONVERSION
    DATA CONVERTERS
    INTERNATIONAL SYSTEM OF UNITS
    UNITS OF MEASUREMENT

CONVERTAPLANES
USE   V/STOL AIRCRAFT

∞ CONVERTERS
SN   (USE OF A MORE SPECIFIC TERM IS
    RECOMMENDED--CONSULT THE TERMS
    LISTED BELOW)
RT   ANALOG TO DIGITAL CONVERTERS
    BINARY TO DECIMAL CONVERTERS
    CURRENT CONVERTERS (AC TO DC)
    DATA CONVERTERS
    DIGITAL TO ANALOG CONVERTERS
    DIRECT POWER GENERATORS
    DOWN-CONVERTERS
    ELECTRIC GENERATORS
    FREQUENCY CONVERTERS
    IMAGE CONVERTERS
    INSTRUMENT TRANSFORMERS
    INVERTED CONVERTERS (DC TO AC)
    PARAMETRIC FREQUENCY CONVERTERS
    POWER CONVERTERS
    PULSE WIDTH AMPLITUDE CONVERTERS
    PYROMETALLURGY
    SOLAR BLANKETS
    THERMIONIC CONVERTERS
    TORQUE CONVERTERS
    TRANSDUCERS
    TRANSFORMERS
    UP-CONVERTERS
    VOLTAGE CONVERTERS (AC TO AC)
    VOLTAGE CONVERTERS (DC TO DC)

CONVEXITY
GS   SHAPES
    . **CONVEXITY**
RT   CONCAVITY
    CONTOUR SENSORS
    CONTOURS
    FLATNESS
    LENTICULAR BODIES
    ∞SURFACE GEOMETRY

CONVEYORS
RT   AUTOMATED GUIDEWAY TRANSIT
      VEHICLES
    AUTOMATED TRANSIT VEHICLES
    CHUTES
    CRANES
    ELEVATORS (LIFTS)

CONVEYORS-*(CONT.)*
    FEEDERS
    FORKS
    ∞LIFTS
    MATERIALS HANDLING
    RIBBONS
    ROLLERS
    SCOOPS
    ∞TRACKS
    TRANSPORTATION

CONVOLUTION INTEGRALS
UF   CONVOLUTIONS (MATHEMATICS)
GS   ANALYSIS (MATHEMATICS)
    . FUNCTIONAL ANALYSIS
    . . INTEGRAL TRANSFORMATIONS
    . . . **CONVOLUTION INTEGRALS**

CONVOLUTIONS (MATHEMATICS)
USE   CONVOLUTION INTEGRALS

CONVULSIONS
RT   HUMAN PATHOLOGY
    MUSCLES
    PSYCHOTHERAPY
    SEIZURES
    ∞SHOCK

COOK INLET (AK)
GS   LANDFORMS
    . INLETS (TOPOGRAPHY)
    . . **COOK INLET (AK)**
RT   ALASKA

COOKPOT AIRCRAFT
USE   TU-124 AIRCRAFT

COOL STARS
GS   CELESTIAL BODIES
    . STARS
    . . **COOL STARS**
RT   GIANT STARS
    LATE STARS
    STELLAR ATMOSPHERES
    STELLAR ENVELOPES
    STELLAR SPECTRA
    STELLAR TEMPERATURE

COOLANT LOSS
USE   LOSS OF COOLANT

COOLANTS
GS   **COOLANTS**
    . ENGINE COOLANTS
    . ORGANIC COOLANTS
RT   AIR CONDITIONING
    AIR COOLING
    BRINES
    COOLERS
    COOLING
    COOLING SYSTEMS
    FREON
    GAS COOLING
    HEAT EXCHANGERS
    LIQUID COOLING
    LOSS OF COOLANT
    NUCLEAR REACTORS
    REACTOR MATERIALS
    REFRIGERANTS
    SODIUM COOLING

COOLERS
RT   AIR CONDITIONING
    AIR CONDITIONING EQUIPMENT
    AIR COOLING
    COMPRESSORS
    COOLANTS
    COOLING
    COOLING SYSTEMS
    CRYOGENIC COOLING
    REFRIGERATING
    REFRIGERATING MACHINERY
    REFRIGERATORS

COOLING
UF   CHILLING
    HEAT DISSIPATION
    HEAT DISSIPATION CHILLING
GS   **COOLING**
    . ABSORPTION COOLING
    . AIR COOLING
    . EVAPORATIVE COOLING
    . . FILM COOLING
    . . SWEAT COOLING

COOLING-*(CONT.)*
    . GAS COOLING
    . LIQUID COOLING
    . . FILM COOLING
    . MAGNETIC COOLING
    . PLASMA COOLING
    . PRECOOLING
    . QUENCHING (COOLING)
    . RADIANT COOLING
    . REGENERATIVE COOLING
    . SODIUM COOLING
    . SOLAR COOLING
    . SOLID CRYOGEN COOLING
    . SPACE COOLING (BUILDINGS)
    . SUPERCOOLING
    . . CRYOGENIC COOLING
    . SURFACE COOLING
    . THERMOELECTRIC COOLING
    . THERMOMAGNETIC COOLING
RT   ABLATION
    ABLATIVE MATERIALS
    AIR CONDITIONING
    BATHING
    CONDENSING
    CONTRACTION
    COOLANTS
    COOLERS
    CRYOGENICS
    ENGINE COOLANTS
    FILM CONDENSATION
    FREEZING
    FREON
    GEOTHERMAL ENERGY UTILIZATION
    HEAT EXCHANGERS
    HEAT RADIATORS
    HEAT SHIELDING
    HEAT TRANSFER
    HEATING
    HILSCH TUBES
    JACKETS
    LOW TEMPERATURE
    MELTING
    MUSHY ZONES
    REFRIGERATING
    REUSABLE HEAT SHIELDING
    SPACECRAFT RADIATORS
    TEMPERATURE CONTROL
    TEMPERATURE DISTRIBUTION
    THERMAL CYCLING TESTS
    THERMAL SHOCK
    THERMAL STRESSES
    TRANSPIRATION
    VENTILATION
    VENTILATION FANS
    VENTING
    WETTING

COOLING FINS
GS   FINS
    . **COOLING FINS**
RT   CONDENSERS (LIQUEFIERS)
    CONVECTIVE HEAT TRANSFER
    FINNED BODIES
    HEAT EXCHANGERS
    HEAT RADIATORS
    RADIATIVE HEAT TRANSFER

COOLING SYSTEMS
RT   ABSORBERS (EQUIPMENT)
    AIR CONDITIONING
    AIR CONDITIONING EQUIPMENT
    AIR COOLING
    AIR FILTERS
    BLOWERS
    CLOSED CYCLES
    CONDENSERS (LIQUEFIERS)
    COOLANTS
    COOLERS
    DEHUMIDIFICATION
    ENGINE COOLANTS
    ETTINGSHAUSEN EFFECT
    EVAPORATIVE COOLING
    EVAPORATORS
    EXHAUST SYSTEMS
    FREON
    HEAT EXCHANGERS
    HEAT PUMPS
    HEAT RADIATORS
    HEAT SINKS
    INFRARED SUPPRESSION
    INTAKE SYSTEMS
    LIQUID COOLING
    LUBRICATION SYSTEMS
    REFRIGERANTS
    REFRIGERATING

COOLING SYSTEMS-*(CONT.)*
    REFRIGERATING MACHINERY
    REGISTERS (AIR CIRCULATION)
    SOLAR COOLING
    SOLID CRYOGENS
    SPACE COOLING (BUILDINGS)
    SPACECRAFT RADIATORS
    ∞SYSTEMS
    TEMPERATURE CONTROL
    TEMPERATURE DISTRIBUTION
    TRANSPIRATION
    VENTILATION
    VENTILATION FANS
    VENTS

∞ **COOPERATION**
SN   *(USE OF A MORE SPECIFIC TERM IS*
    RECOMMENDED--CONSULT THE TERMS
    LISTED BELOW)
RT   CONVENTIONS
    EMPLOYEE RELATIONS
    INTERNATIONAL COOPERATION
    PUBLIC RELATIONS
    SEA LAW

COORDINATE GEOMETRY LANGUAGE
USE   COGO (PROGRAMMING LANGUAGE)

COORDINATE SYSTEMS
USE   COORDINATES

COORDINATE TRANSFORMATIONS
GS   FUNCTIONS (MATHEMATICS)
    . **COORDINATE TRANSFORMATIONS**
    TRANSFORMATIONS (MATHEMATICS)
    . **COORDINATE TRANSFORMATIONS**
RT   CONFORMAL MAPPING
    INVARIANT IMBEDDINGS
    ISOPARAMETRIC FINITE ELEMENTS
    ISOTROPIC TURBULENCE
    JOUKOWSKI TRANSFORMATION
    LAMBERT SURFACE
    SCHWARZSCHILD METRIC
    THEODORSEN TRANSFORMATION

COORDINATES
UF   AXES (COORDINATES)
    COORDINATE SYSTEMS
GS   **COORDINATES**
    . ASTRONOMICAL COORDINATES
    . CARTESIAN COORDINATES
    . GEODETIC COORDINATES
    . HYLLERAAS COORDINATES
    . HYPERBOLIC COORDINATES
    . INERTIAL COORDINATES
    . LAGRANGE COORDINATES
    . OBLIQUE COORDINATES
    . PLANETOCENTRIC COORDINATES
    . . GEOCENTRIC COORDINATES
    . POLAR COORDINATES
    . SPHERICAL COORDINATES
RT   ALGEBRA
    ANALYTIC GEOMETRY
    AXES (REFERENCE LINES)
    CELESTIAL REFERENCE SYSTEMS
    COMPUTATIONAL GRIDS
    EARTH AXIS
    EQUATORS
    EUCLIDEAN GEOMETRY
    FRACTALS
    FUJITA METHOD
    GEOMAGNETIC LATITUDE
    GEOMETRY
    ∞GRIDS
    HALF PLANES
    HALF SPACES
    LATITUDE
    LINE OF SIGHT
    LONGITUDE
    MANIFOLDS (MATHEMATICS)
    MAPS
    ∞ORIGINS
    POSITION (LOCATION)
    ∞REFERENCE SYSTEMS

COORDINATION
RT   CONTINUITY
    CORRELATION
    INTERFACES
    SEQUENCING
    TIME SHARING

COORDINATION POLYMERS
RT   ∞POLYMERS

COPERNICUS SPACECRAFT
USE   OAO 3

COPILOTS
USE   AIRCRAFT PILOTS

COPLANARITY
GS   ANALYSIS (MATHEMATICS)
. CALCULUS
. . VECTOR ANALYSIS
. . . **COPLANARITY**
. REAL VARIABLES
. . VECTOR ANALYSIS
. . . **COPLANARITY**
GEOMETRY
. VECTOR ANALYSIS
. . **COPLANARITY**

COPOLYMERIZATION
GS   CHEMICAL REACTIONS
. **COPOLYMERIZATION**
POLYMERIZATION
. **COPOLYMERIZATION**
RT   DIMERIZATION
∞POLYMERS

COPOLYMERS
GS   **COPOLYMERS**
. VITON
RT   KEL-F
∞POLYMERS
VINYL COPOLYMERS

COPPER
GS   CHEMICAL ELEMENTS
. **COPPER**
. . COPPER ISOTOPES
RT   AMMINES
CONSTANTAN
SELENIUM ALLOYS

COPPER ALLOYS
GS   ALLOYS
. **COPPER ALLOYS**
. . BABBITT METAL
. . BRASSES
. . BRONZES
. . MANGANIN (TRADEMARK)
RT   BEARING ALLOYS
LAMELLA (METALLURGY)

COPPER CHLORIDES
GS   COPPER COMPOUNDS
. **COPPER CHLORIDES**
HALOGEN COMPOUNDS
. CHLORINE COMPOUNDS
. . CHLORIDES
. . . **COPPER CHLORIDES**
. HALIDES
. . CHLORIDES
. . . **COPPER CHLORIDES**
. . METAL HALIDES
. . . **COPPER CHLORIDES**

COPPER COMPOUNDS
GS   **COPPER COMPOUNDS**
. COPPER CHLORIDES
. COPPER FLUORIDES
. COPPER OXIDES
. COPPER SELENIDES
. COPPER SULFIDES
. . ENARGITE
RT   ∞CHEMICAL COMPOUNDS
∞GROUP 1B COMPOUNDS
∞METAL COMPOUNDS

COPPER FLUORIDES
GS   COPPER COMPOUNDS
. **COPPER FLUORIDES**
HALOGEN COMPOUNDS
. FLUORINE COMPOUNDS
. . FLUORIDES
. . . METAL FLUORIDES
. . . . **COPPER FLUORIDES**
. HALIDES
. . METAL HALIDES
. . . **COPPER FLUORIDES**

COPPER ISOTOPES
GS   CHEMICAL ELEMENTS
. COPPER
. . **COPPER ISOTOPES**
METALS
. **COPPER ISOTOPES**

COPPER OXIDES
GS   CHALCOGENIDES
. OXIDES
. . METAL OXIDES
. . . **COPPER OXIDES**
COPPER COMPOUNDS
. **COPPER OXIDES**
RT   THERMITES

COPPER SELENIDES
GS   CHALCOGENIDES
. SELENIDES
. . **COPPER SELENIDES**
COPPER COMPOUNDS
. **COPPER SELENIDES**
SELENIUM COMPOUNDS
. SELENIDES
. . **COPPER SELENIDES**

COPPER SULFIDES
GS   CHALCOGENIDES
. SULFIDES
. . INORGANIC SULFIDES
. . . **COPPER SULFIDES**
. . . . ENARGITE
COPPER COMPOUNDS
. **COPPER SULFIDES**
. . ENARGITE
SULFUR COMPOUNDS
. SULFIDES
. . INORGANIC SULFIDES
. . . **COPPER SULFIDES**
. . . . ENARGITE

COPYRIGHTS
RT   DOCUMENTS
LICENSING
PATENT APPLICATIONS
POLICIES
REGULATIONS

CORAL HEADS
USE   CORAL REEFS

CORAL REEFS
UF   ATOLL REEFS
CORAL HEADS
RT   ATOLLS
COASTS
ISLANDS
KEYS (ISLANDS)
REEFS

CORDAGE
RT   CABLES (ROPES)
CONNECTORS
COTTON
FIBERS
∞FILAMENTS
STRANDS
STRINGS
WIRE
YARNS

CORDIERITE
GS   ALUMINUM COMPOUNDS
. **CORDIERITE**
IRON COMPOUNDS
. **CORDIERITE**
MAGNESIUM COMPOUNDS
. **CORDIERITE**
MINERALS
. **CORDIERITE**
SILICON COMPOUNDS
. SILICATES
. . **CORDIERITE**

CORDITE
USE   COLLOIDAL PROPELLANTS
DOUBLE BASE PROPELLANTS

CORE FLOW
GS   FLUID FLOW
. **CORE FLOW**
RT   FLOW GEOMETRY
MAGNETOHYDRODYNAMIC FLOW
ONE DIMENSIONAL FLOW
PLASMAS (PHYSICS)
SHEAR FLOW

CORE SAMPLING
GS   SAMPLING
. **CORE SAMPLING**
RT   CORES

CORE SAMPLING-*(CONT.)*
DEPTH MEASUREMENT
DRILLING
EARTH CRUST
HYDROGEOLOGY
MINES (EXCAVATIONS)
OCEAN BOTTOM
OCEAN CURRENTS
OCEANOGRAPHY
PARTICLE TRACKS
SALINITY
SAMPLERS

CORE STORAGE
UF   MACHINE STORAGE
GS   MAGNETIC STORAGE
. **CORE STORAGE**
RT   BUBBLE MEMORY DEVICES
BUFFER STORAGE
DATA STORAGE
MAGNETIC DISKS
MAGNETIC DRUMS
∞STORAGE

CORES
GS   **CORES**
. HONEYCOMB CORES
. LUNAR CORE
. MAGNETIC CORES
. PLANETARY CORES
. . EARTH CORE
. REACTOR CORES
. STELLAR CORES
RT   ∞CELLS
CORE SAMPLING
MANDRELS
MOLDING MATERIALS

CORIOLIS EFFECT
RT   DISORIENTATION
∞EFFECTS
METEOROLOGY
PLANETARY WAVES
ROTATING ENVIRONMENTS
ROTATION
VESTIBULAR TESTS

CORK (MATERIALS)
GS   WOOD
. **CORK (MATERIALS)**
RT   ∞MATERIALS
ORGANIC MATERIALS
THERMAL INSULATION

CORN
GS   FARM CROPS
. GRAINS (FOOD)
. . **CORN**
PLANTS (BOTANY)
. **CORN**
RT   AGRICULTURE
BLIGHT
BOLLWORMS
BOTANY
CROP GROWTH
∞CROPS
CURING
EARTH RESOURCES
∞FOOD
IRRIGATION
SEEDS

CORNEA
GS   ANATOMY
. SENSE ORGANS
. . EYE (ANATOMY)
. . . **CORNEA**
RT   KERATITIS
VISION

CORNER FLOW
GS   FLUID FLOW
. **CORNER FLOW**
RT   CHANNEL FLOW
DUCTED FLOW
∞FLOW
NOZZLE FLOW
SECONDARY FLOW

CORNERS
RT   ANGLES (GEOMETRY)
ANTENNAS
JOINTS (JUNCTIONS)
SHAPES

CORONA BOREALIS CONSTELLATION
GS   CONSTELLATIONS
.   **CORONA BOREALIS CONSTELLATION**
RT   CELESTIAL BODIES
CELESTIAL SPHERE
STARS

CORONA DISCHARGES
USE ELECTRIC CORONA

CORONAGRAPHS
RT   ASTRONOMICAL PHOTOGRAPHY
ASTRONOMICAL TELESCOPES
SOLAR OBSERVATORIES
SPECTROHELIOGRAPHS
STARSAT TELESCOPE

CORONAL HOLES
GS   CORONAS
.   **CORONAL HOLES**
RT   DECAMETRIC WAVES
∞HOLES
RADIO ASTRONOMY
SOLAR RADIO EMISSION
SOLAR WIND
SOLAR X-RAYS
STELLAR STRUCTURE
ULTRAVIOLET RADIATION

CORONAL LOOPS
GS   CORONAS
.   STELLAR CORONAS
..   SOLAR CORONA
... **CORONAL LOOPS**
RT   CHROMOSPHERE
SOLAR FLARES
SOLAR LIMB

CORONARY ARTERY DISEASE
GS   DISEASES
.   HEART DISEASES
..   **CORONARY ARTERY DISEASE**
RT   ANGINA PECTORIS
ARTERIOSCLEROSIS
MYOCARDIAL INFARCTION

CORONARY CIRCULATION
GS   CIRCULATION
.   BLOOD CIRCULATION
..   **CORONARY CIRCULATION**
RT   HEART
HEART VALVES

CORONAS
GS   **CORONAS**
.   CORONAL HOLES
.   ELECTRIC CORONA
.   STELLAR CORONAS
..   SOLAR CORONA
... CORONAL LOOPS
RT   ELECTRIC ARCS
ELECTRIC DISCHARGES
HALOS
IONIZATION
SOLAR SPECTRA

COROTATION
GS   GYRATION
.   ROTATION
..   **COROTATION**
RT   ASTRONOMICAL MODELS
GALACTIC ROTATION
GALACTIC STRUCTURE
MAGNETOSPHERE
SPIRAL GALAXIES
STELLAR MOTIONS
STELLAR ROTATION

CORPORAL MISSILE
GS   MISSILES
.   SURFACE TO SURFACE MISSILES
..   **CORPORAL MISSILE**
RT   LIQUID PROPELLANT ROCKET ENGINES

CORPUSCLES
GS   ANATOMY
.   CARDIOVASCULAR SYSTEM
..   **CORPUSCLES**
CELLS (BIOLOGY)
.   **CORPUSCLES**
RT   ANEMIAS
BLOOD PLASMA
ERYTHROCYTES
HEMOGLOBIN

CORPUSCLES-(CONT.)
LYMPH
LYMPHOCYTES

CORPUSCULAR RADIATION
SN   (NONELECTROMAGNETIC RADIATION
CONSISTING OF ENERGETIC CHARGED
OR NEUTRAL PARTICLES)
UF   PENETRATING PARTICLES
GS   PARTICLES
.   **CORPUSCULAR RADIATION**
..   CYCLOTRON RADIATION
... ION CYCLOTRON RADIATION
..   ELECTRON PRECIPITATION
..   ELECTRON RADIATION
... BETA PARTICLES
... ELECTRON BEAMS
.... RELATIVISTIC ELECTRON BEAMS
..   PRIMARY COSMIC RAYS
... SOLAR COSMIC RAYS
..   RADIATION BELTS
..   SOLAR CORPUSCULAR RADIATION
... SOLAR ELECTRONS
... SOLAR PROTONS
RT   ALPHA PARTICLES
ATMOSPHERIC RADIATION
BACKGROUND RADIATION
BEAMS (RADIATION)
CERENKOV RADIATION
CHARGED PARTICLES
COHERENT RADIATION
CONTINUOUS RADIATION
COSMIC RAYS
ELECTROMAGNETIC RADIATION
EXTRATERRESTRIAL RADIATION
FLUX (RATE)
GALACTIC RADIATION
INCIDENT RADIATION
INTERSTELLAR RADIATION
IONIZING RADIATION
IONS
MESONS
NEUTRONS
NUCLEAR PARTICLES
NUCLEAR RADIATION
NUCLEI (NUCLEAR PHYSICS)
PARTICLE PRODUCTION
PHONON BEAMS
PULSED RADIATION
∞RADIATION
RADIATION DISTRIBUTION
RADIATION PRESSURE
RADIATION SOURCES
∞RAYS
REFLECTED WAVES
REFRACTED WAVES
SOLAR RADIATION
SOLAR TERRESTRIAL INTERACTIONS
STRATOSPHERE RADIATION

CORRECTION
GS   **CORRECTION**
.   ATMOSPHERIC CORRECTION
.   OPTICAL CORRECTION PROCEDURE
RT   ACCOMMODATION
ACCURACY
ADAPTATION
ADJUSTING
ALIGNMENT
ERROR CORRECTING DEVICES
ERRORS
IMPROVEMENT
INFORMATION THEORY
PARITY
REDUNDANCY
REVISIONS
VEGETATIVE INDEX

CORRELATION
UF   CORRELATION FUNCTIONS
GS   **CORRELATION**
.   ANGULAR CORRELATION
.   AUTOCORRELATION
.   CORRELATION COEFFICIENTS
.   CORRELATION DETECTION
.   CROSS CORRELATION
.   DATA CORRELATION
..   SIGNAL ANALYSIS
.   SPECTRAL CORRELATION
.   STATISTICAL CORRELATION
RT   BIVARIATE ANALYSIS
COLLATING
CONFIDENCE
CONTINGENCY
COORDINATION

CORRELATION-(CONT.)
COVARIANCE
∞ESTIMATORS
EVALUATION
FACTOR ANALYSIS
FORECASTING
INFORMATION THEORY
LEAST SQUARES METHOD
MULTIVARIATE STATISTICAL ANALYSIS
OPTIMIZATION
PROBABILITY THEORY
QUALITY CONTROL
REGRESSION ANALYSIS
REGRESSION COEFFICIENTS
SIGNIFICANCE
STATISTICAL ANALYSIS
TIME SERIES ANALYSIS
VALIDITY
VARIABILITY
VARIANCE (STATISTICS)

CORRELATION COEFFICIENTS
GS   COEFFICIENTS
.   **CORRELATION COEFFICIENTS**
CORRELATION
.   **CORRELATION COEFFICIENTS**
STATISTICAL ANALYSIS
.   **CORRELATION COEFFICIENTS**
RT   QUALITY CONTROL
STATISTICAL CORRELATION

CORRELATION DETECTION
GS   CORRELATION
.   **CORRELATION DETECTION**
DETECTION
.   SIGNAL DETECTION
..   **CORRELATION DETECTION**
RT   ∞DETECTORS
ELECTROMAGNETIC WAVE FILTERS
PHASE LOCK DEMODULATORS
SIGNAL TO NOISE RATIOS

CORRELATION FUNCTIONS
USE CORRELATION

CORRELATORS
SN   (DEVICES THAT DETECT WEAK SIGNALS
IN NOISE BY PERFORMING AN
ELECTRONIC OPERATION)
UF   SYNCHRONOUS DETECTORS
GS   **CORRELATORS**
.   IMAGE CORRELATORS
RT   SYNCHROSCOPES

CORRIDORS
GS   **CORRIDORS**
.   GREAT PLAINS CORRIDOR (NORTH
AMERICA)
.   ST LOUIS-KANSAS CITY CORRIDOR
(MO)
RT   PASSAGEWAYS

CORROSION
UF   METAL CORROSION
GS   **CORROSION**
.   CAVITATION CORROSION
.   ELECTROCHEMICAL CORROSION
.   FRETTING CORROSION
.   FUEL CORROSION
.   HOT CORROSION
.   INTERGRANULAR CORROSION
.   RUSTING
.   SCALE (CORROSION)
.   STRESS CORROSION
.   TRANSGRANULAR CORROSION
RT   CHEMICAL ATTACK
CHEMICAL REACTIONS
COATINGS
DAMAGE
DEGRADATION
DEPOSITS
DETERIORATION
DISSOLVING
DURABILITY
ELECTROCHEMISTRY
ELECTROLYSIS
EROSION
ETCHANTS
ETCHING
FAILURE
FINISHES
FOULING
GAS-METAL INTERACTIONS
HUMIDITY
IMPINGEMENT

CORROSION-(CONT.)
INCOMPATIBILITY
METAL COATINGS
METAL-WATER REACTIONS
∞METALLURGY
OXIDATION
PASSIVITY
PITTING
PROTECTIVE COATINGS
SALT SPRAY TESTS
STERILIZATION EFFECTS
SURFACE PROPERTIES
TRIBOLOGY
WEAR
WEATHERING

CORROSION PREVENTION
GS   PREVENTION
. CORROSION PREVENTION
PROTECTION
. CORROSION PREVENTION
RT   AERATION
ANTIFOULING
ANTIOXIDANTS
CHEMICAL ATTACK
CLEANING
COATING
COATINGS
DESENSITIZING
∞FILMS
FUEL TANKS
∞INHIBITION
INHIBITORS
METAL COATINGS
NICKEL COATINGS
PACKAGING
PASSIVITY
PRESERVING
PROPELLANT ADDITIVES
SENSITIZING
SILICONIZING
SURFACE FINISHING
WATER TREATMENT
WEATHERPROOFING

CORROSION RESISTANCE
GS   CORROSION RESISTANCE
. OXIDATION RESISTANCE
RT   ANTIOXIDANTS
CHEMICAL ATTACK
CHEMICAL TESTS
PASSIVITY
PITTING
∞RESISTANCE
RUSTING
SALT SPRAY TESTS
SILICONIZING
SULFIDATION
SURFACE FINISHING

CORROSION TEST LOOPS
GS   LOOPS
. CORROSION TEST LOOPS
RT   ∞TESTS

CORROSION TESTS
GS   ENVIRONMENTAL TESTS
. CORROSION TESTS
. . SALT SPRAY TESTS
RT   CHEMICAL ATTACK
DESTRUCTIVE TESTS
FUEL TESTS
∞MATERIALS TESTS
PITTING
PROPELLANT TESTS
STABILITY TESTS
STRESS CORROSION CRACKING
∞TESTS
TRANSGRANULAR CORROSION
UNDERWATER TESTS
WEATHERING

CORRUGATED PLATES
GS   STRUCTURAL MEMBERS
. PLATES (STRUCTURAL MEMBERS)
. . CORRUGATED PLATES
RT   CORRUGATING
REINFORCED PLATES

CORRUGATED SHELLS
GS   SHELLS (STRUCTURAL FORMS)
. CORRUGATED SHELLS
RT   ANISOTROPIC SHELLS
CORRUGATING
REINFORCED SHELLS

CORRUGATING
RT   CORRUGATED PLATES
CORRUGATED SHELLS
DEFORMATION
GROOVES
∞PLATES
∞RIDGES
∞WAVES

CORSAIR AIRCRAFT
USE   A-7 AIRCRAFT

∞ CORTEXES
SN   (USE OF A MORE SPECIFIC TERM IS
RECOMMENDED--CONSULT THE TERMS
LISTED BELOW)
RT   CEREBRAL CORTEX
CORTEXES (BOTANY)

CORTEXES (BOTANY)
RT   ∞CORTEXES
PLANTS (BOTANY)

CORTI ORGAN
GS   ANATOMY
. SENSE ORGANS
. . EAR
. . . CORTI ORGAN
RT   COCHLEA

CORTICOSTEROIDS
GS   STEROIDS
. CORTICOSTEROIDS
. . ALDOSTERONE
. . HYDROXYCORTICOSTEROID
. . . CORTISONE
RT   ADRENAL METABOLISM
ENDOCRINE SECRETIONS

CORTISONE
GS   DRUGS
. CORTISONE
STEROIDS
. CORTICOSTEROIDS
. . HYDROXYCORTICOSTEROID
. . . CORTISONE
RT   ADRENAL METABOLISM

CORUNDUM
USE   ALUMINUM OXIDES

CORVUS MISSILE
GS   MISSILES
. CORVUS MISSILE
RT   LIQUID PROPELLANT ROCKET ENGINES

COS-B SATELLITE
GS   ESA SPACECRAFT
. COS-B SATELLITE
SATELLITES
. ARTIFICIAL SATELLITES
. . ESA SATELLITES
. . . COS-B SATELLITE
RT   EUROPA 2 LAUNCH VEHICLE
EUROPEAN SPACE PROGRAMS

COSINE SERIES
GS   ANALYSIS (MATHEMATICS)
. CALCULUS
. . SERIES (MATHEMATICS)
. . . COSINE SERIES
. REAL VARIABLES
. . DIFFERENTIAL EQUATIONS
. . COSINE SERIES
. . PERIODIC FUNCTIONS
. . . TRIGONOMETRIC FUNCTIONS
. . . . COSINE SERIES
. . SERIES (MATHEMATICS)
. . COSINE SERIES
FUNCTIONS (MATHEMATICS)
. TRANSCENDENTAL FUNCTIONS
. . PERIODIC FUNCTIONS
. . . TRIGONOMETRIC FUNCTIONS
. . . . COSINE SERIES

COSMIC BACKGROUND EXPLORER SATELLITE
UF   COBE
GS   SATELLITES
. ARTIFICIAL SATELLITES
. . EXPLORER SATELLITES
. . . COSMIC BACKGROUND EXPLORER
SATELLITE
RT   BACKGROUND RADIATION
RADIATION SPECTRA

COSMIC BACKGROUND EXPLORER-(CONT.)
SPACEBORNE ASTRONOMY

COSMIC DUST
GS   DUST
. COSMIC DUST
. . INTERPLANETARY DUST
. . . METEOROID DUST CLOUDS
. . . . ZODIACAL DUST
RT   INTERGALACTIC MEDIA
INTERSTELLAR MATTER
METEOROIDS
MICROMETEORITES
MICROMETEOROIDS
MOLECULAR CLOUDS
ORGANIC SOLIDS
REFLECTION NEBULAE
SPACE DEBRIS
TERRESTRIAL DUST BELT
VENUS FLY TRAP ROCKET VEHICLE

COSMIC GAMMA RAY BURSTS
USE   GAMMA RAY BURSTS

COSMIC GASES
GS   EXTRATERRESTRIAL MATTER
. COSMIC GASES
. . INTERPLANETARY GAS
. . INTERSTELLAR GAS
. . . NEUTRAL GASES
GASES
. RAREFIED GASES
. COSMIC GASES
. . . INTERPLANETARY GAS
. . . INTERSTELLAR GAS
. . . . NEUTRAL GASES
RT   ELECTRON GAS
INTERGALACTIC MEDIA
IONIZED GASES

COSMIC NOISE
GS   ELECTROMAGNETIC INTERFERENCE
. RADIO FREQUENCY INTERFERENCE
. . BLACKOUT (PROPAGATION)
. . . ELECTROMAGNETIC NOISE
. . . . COSMIC NOISE
RT   ALOUETTE PROJECT
BACKGROUND NOISE
BACKGROUND RADIATION
CENTIMETER WAVES
ELECTROMAGNETIC NOISE
MEASUREMENT
GALACTIC RADIATION
GALACTIC RADIO WAVES
INTERSTELLAR RADIATION
MICROWAVE EMISSION
MICROWAVES
NOISE STORMS
SOLAR RADIATION
SOLAR RADIO EMISSION

COSMIC PLASMA
GS   EXTRATERRESTRIAL MATTER
. COSMIC PLASMA
PARTICLES
. CHARGED PARTICLES
. . ENERGETIC PARTICLES
. . . PLASMAS (PHYSICS)
. . . . COSMIC PLASMA
RT   INTERGALACTIC MEDIA
INTERPLANETARY GAS
PLASMA CLOUDS
PLASMAPAUSE
RELATIVISTIC PLASMAS
SOLAR WIND
STELLAR WINDS
STRONGLY COUPLED PLASMAS

COSMIC RADIATION
USE   COSMIC RAYS

COSMIC RADIO WAVES
USE   EXTRATERRESTRIAL RADIO WAVES

COSMIC RAY ALBEDO
GS   ALBEDO
. COSMIC RAY ALBEDO
RT   ABSORPTANCE
∞ABSORPTION
ATMOSPHERIC ATTENUATION
EARTH ALBEDO
LUNAR ALBEDO
PRIMARY COSMIC RAYS
REFLECTANCE
SECONDARY COSMIC RAYS

COSMIC RAY SHOWERS
UF   MOLIERE FORMULA
GS   IONIZING RADIATION
        . COSMIC RAYS
        . . **COSMIC RAY SHOWERS**
RT   AUGER EFFECT
        ∞ CASCADES
        ELECTRON PHOTON CASCADES
        SECONDARY COSMIC RAYS
        ∞ SHOWERS

COSMIC RAYS
UF   COSMIC RADIATION
GS   IONIZING RADIATION
        . **COSMIC RAYS**
        . . COSMIC RAY SHOWERS
        . . GALACTIC COSMIC RAYS
        . . GAMMA RAY BURSTS
        . . PRIMARY COSMIC RAYS
        . . . SOLAR COSMIC RAYS
        . . SECONDARY COSMIC RAYS
RT   AEROSPACE ENVIRONMENTS
        ALBEDO
        ALPHA PARTICLES
        BIG BANG COSMOLOGY
        CERENKOV RADIATION
        CORPUSCULAR RADIATION
        DEUTERONS
        ELECTROMAGNETIC RADIATION
        ELECTRON ACCELERATION
        ELECTRONS
        EXTRATERRESTRIAL RADIATION
        FORBUSH DECREASES
        GALACTIC RADIATION
        GAMMA RAY TELESCOPES
        GAMMA RAYS
        HELIOSPHERE
        INTERSTELLAR RADIATION
        ION DENSITY (CONCENTRATION)
        MESONS
        NEUTRONS
        NUCLEAR PARTICLES
        NUCLEI (NUCLEAR PHYSICS)
        PARTICLE TRACKS
        PHOTONS
        PROTONS
        ∞ RADIATION
        RADIATION BELTS
        RADIATIVE TRANSFER
        SINGLE EVENT UPSETS
        SOLAR RADIATION
        STELLAR RADIATION
        VLF EMISSION RECORDERS
        X RAYS

COSMIC X RAYS
GS   ELECTROMAGNETIC RADIATION
        . X RAYS
        . . **COSMIC X RAYS**
        IONIZING RADIATION
        . X RAYS
        . . **COSMIC X RAYS**
RT   EXTRATERRESTRIAL RADIATION
        GALACTIC RADIATION
        GAMMA RAY ASTRONOMY
        GAMMA RAY BURSTS
        GAMMA RAYS
        X RAY ASTRONOMY
        X RAY BINARIES

COSMOCHEMISTRY
RT   COSMOLOGY
        EXTRATERRESTRIAL MATTER
        GEOCHEMISTRY
        METEORITIC COMPOSITION

COSMOGONY
USE   COSMOLOGY

COSMOLOGY
UF   COSMOGONY
GS   **COSMOLOGY**
        . BIG BANG COSMOLOGY
        . HUBBLE DIAGRAM
        . MISSING MASS (ASTROPHYSICS)
RT   ASTRONOMICAL MODELS
        ASTROPHYSICS
        COSMOCHEMISTRY
        EXISTENCE
        GALACTIC EVOLUTION
        GRAVITINOS
        HUBBLE CONSTANT
        LOCAL GROUP (ASTRONOMY)
        MASS DISTRIBUTION
        NAKED SINGULARITIES

COSMOLOGY-*(CONT.)*
        PLANETARY EVOLUTION
        PROTOPLANETS
        RED SHIFT
        STAR DISTRIBUTION
        STELLAR EVOLUTION
        STELLAR MASS ACCRETION
        UNIVERSE
        WHITE HOLES (ASTRONOMY)

COSMONAUTS
GS   PERSONNEL
        . FLYING PERSONNEL
        . . **COSMONAUTS**
RT   ∞ ASTRONAUTICS
        ASTRONAUTS
        CREW EXPERIMENT STATIONS
        CREW OBSERVATION STATIONS
        CREW STATIONS
        CREW WORKSTATIONS
        CREWS
        PILOTS (PERSONNEL)
        SPACECRAFT ENVIRONMENTS
        SPACECREWS

∞ **COSMOS**
SN   *(USE OF A MORE SPECIFIC TERM IS*
        RECOMMENDED--CONSULT THE TERMS
        LISTED BELOW)
RT   COSMOS SATELLITES
        UNIVERSE

COSMOS SATELLITES
GS   SATELLITES
        . ARTIFICIAL SATELLITES
        . . GEOPHYSICAL SATELLITES
        . . . **COSMOS SATELLITES**
        . . . . INTERCOSMOS SATELLITES
        . . SOVIET SATELLITES
        . . . **COSMOS SATELLITES**
        . . . . COSMOS 2 SATELLITE
        . . . . COSMOS 3 SATELLITE
        . . . . COSMOS 5 SATELLITE
        . . . . COSMOS 6 SATELLITE
        . . . . COSMOS 14 SATELLITE
        . . . . COSMOS 44 SATELLITE
        . . . . COSMOS 54 SATELLITE
        . . . . COSMOS 71 SATELLITE
        . . . . COSMOS 110 SATELLITE
        . . . . COSMOS 137 SATELLITE
        . . . . COSMOS 144 SATELLITE
        . . . . COSMOS 149 SATELLITE
        . . . . COSMOS 166 SATELLITE
        . . . . COSMOS 186 SATELLITE
        . . . . COSMOS 188 SATELLITE
        . . . . COSMOS 206 SATELLITE
        . . . . COSMOS 213 SATELLITE
        . . . . COSMOS 224 SATELLITE
        . . . . COSMOS 225 SATELLITE
        . . . . COSMOS 381 SATELLITE
        . . . . COSMOS 954 SATELLITE
        . . . . COSMOS 1129 SATELLITE
        . . . . INTERCOSMOS SATELLITES
RT   ∞ COSMOS

COSMOS 2 SATELLITE
GS   SATELLITES
        . ARTIFICIAL SATELLITES
        . . SOVIET SATELLITES
        . . . COSMOS SATELLITES
        . . . . **COSMOS 2 SATELLITE**

COSMOS 3 SATELLITE
GS   SATELLITES
        . ARTIFICIAL SATELLITES
        . . SOVIET SATELLITES
        . . . COSMOS SATELLITES
        . . . . **COSMOS 3 SATELLITE**

COSMOS 5 SATELLITE
GS   SATELLITES
        . ARTIFICIAL SATELLITES
        . . SOVIET SATELLITES
        . . . COSMOS SATELLITES
        . . . . **COSMOS 5 SATELLITE**

COSMOS 6 SATELLITE
GS   SATELLITES
        . ARTIFICIAL SATELLITES
        . . SOVIET SATELLITES
        . . . COSMOS SATELLITES
        . . . . **COSMOS 6 SATELLITE**

COSMOS 14 SATELLITE
GS   SATELLITES

COSMOS 14 SATELLITE-*(CONT.)*
        . ARTIFICIAL SATELLITES
        . . SOVIET SATELLITES
        . . . COSMOS SATELLITES
        . . . . **COSMOS 14 SATELLITE**

COSMOS 44 SATELLITE
GS   SATELLITES
        . ARTIFICIAL SATELLITES
        . . SOVIET SATELLITES
        . . . COSMOS SATELLITES
        . . . . **COSMOS 44 SATELLITE**

COSMOS 54 SATELLITE
GS   SATELLITES
        . ARTIFICIAL SATELLITES
        . . SOVIET SATELLITES
        . . . COSMOS SATELLITES
        . . . . **COSMOS 54 SATELLITE**

COSMOS 71 SATELLITE
GS   SATELLITES
        . ARTIFICIAL SATELLITES
        . . SOVIET SATELLITES
        . . . COSMOS SATELLITES
        . . . . **COSMOS 71 SATELLITE**

COSMOS 110 SATELLITE
GS   SATELLITES
        . ARTIFICIAL SATELLITES
        . . SOVIET SATELLITES
        . . . COSMOS SATELLITES
        . . . . **COSMOS 110 SATELLITE**

COSMOS 137 SATELLITE
GS   SATELLITES
        . ARTIFICIAL SATELLITES
        . . SOVIET SATELLITES
        . . . COSMOS SATELLITES
        . . . . **COSMOS 137 SATELLITE**

COSMOS 144 SATELLITE
GS   SATELLITES
        . ARTIFICIAL SATELLITES
        . . METEOROLOGICAL SATELLITES
        . . . **COSMOS 144 SATELLITE**
        . . SOVIET SATELLITES
        . . . COSMOS SATELLITES
        . . . . **COSMOS 144 SATELLITE**

COSMOS 149 SATELLITE
UF   SPACE ARROW SATELLITE
GS   SATELLITES
        . ARTIFICIAL SATELLITES
        . . SOVIET SATELLITES
        . . . COSMOS SATELLITES
        . . . . **COSMOS 149 SATELLITE**

COSMOS 166 SATELLITE
GS   SATELLITES
        . ARTIFICIAL SATELLITES
        . SOVIET SATELLITES
        . . . COSMOS SATELLITES
        . . . . **COSMOS 166 SATELLITE**

COSMOS 186 SATELLITE
GS   SATELLITES
        . ARTIFICIAL SATELLITES
        . . SOVIET SATELLITES
        . . . COSMOS SATELLITES
        . . . . **COSMOS 186 SATELLITE**

COSMOS 188 SATELLITE
GS   SATELLITES
        . ARTIFICIAL SATELLITES
        . . SOVIET SATELLITES
        . . . COSMOS SATELLITES
        . . . . **COSMOS 188 SATELLITE**

COSMOS 206 SATELLITE
GS   SATELLITES
        . ARTIFICIAL SATELLITES
        . . SOVIET SATELLITES
        . . . COSMOS SATELLITES
        . . . . **COSMOS 206 SATELLITE**

COSMOS 213 SATELLITE
GS   SATELLITES
        . ARTIFICIAL SATELLITES
        . . SOVIET SATELLITES
        . . . COSMOS SATELLITES
        . . . . **COSMOS 213 SATELLITE**

COSMOS 224 SATELLITE
GS   SATELLITES
. ARTIFICIAL SATELLITES
. . SOVIET SATELLITES
. . . COSMOS SATELLITES
. . . . **COSMOS 224 SATELLITE**

COSMOS 225 SATELLITE
GS   SATELLITES
. ARTIFICIAL SATELLITES
. . SOVIET SATELLITES
. . . COSMOS SATELLITES
. . . . **COSMOS 225 SATELLITE**

COSMOS 381 SATELLITE
GS   SATELLITES
. ARTIFICIAL SATELLITES
. . SOVIET SATELLITES
. . . COSMOS SATELLITES
. . . . **COSMOS 381 SATELLITE**

COSMOS 782 SATELLITE
GS   SATELLITES
. ARTIFICIAL SATELLITES
. . SOVIET SATELLITES
. . . **COSMOS 782 SATELLITE**
RT   INTERNATIONAL COOPERATION

COSMOS 936 SATELLITE
GS   SATELLITES
. ARTIFICIAL SATELLITES
. . SOVIET SATELLITES
. . . **COSMOS 936 SATELLITE**
RT   INTERNATIONAL COOPERATION

COSMOS 954 SATELLITE
GS   SATELLITES
. ARTIFICIAL SATELLITES
. . SOVIET SATELLITES
. . . COSMOS SATELLITES
. . . . **COSMOS 954 SATELLITE**
RT   UNCONTROLLED REENTRY
     (SPACECRAFT)

COSMOS 1129 SATELLITE
GS   SATELLITES
. ARTIFICIAL SATELLITES
. . SOVIET SATELLITES
. . . COSMOS SATELLITES
. . . . **COSMOS 1129 SATELLITE**
RT   INTERNATIONAL COOPERATION

COSPAR (COMMITTEE)
USE   COMMITTEE ON SPACE RESEARCH

COSPAS
GS   SATELLITES
. ARTIFICIAL SATELLITES
. . **COSPAS**
RT   RECONNAISSANCE
     RESCUE OPERATIONS
     SARSAT
     SEARCHING

COSSERAT SURFACES
RT   FLAT SURFACES
    ∞SURFACE GEOMETRY
     SURFACE PROPERTIES
    ∞SURFACES

COST ANALYSIS
RT   ∞ANALYZING
     BUDGETING
     COMPARISON
     COSTS
     DESIGN TO COST
     ECONOMIC ANALYSIS
     FEASIBILITY
     FEASIBILITY ANALYSIS
     FINANCIAL MANAGEMENT
     LIFE CYCLE COSTS
     MANAGEMENT
     MANAGEMENT ANALYSIS
     MANAGEMENT PLANNING
     OPTICAL TRANSFER FUNCTION
     PRODUCTION COSTS
     VALUE ENGINEERING
     WAGE SURVEYS

COST EFFECTIVENESS
GS   EFFECTIVENESS
. **COST EFFECTIVENESS**
RT   ALLOCATIONS
     BUDGETING

COST EFFECTIVENESS-(CONT.)
     LIFE CYCLE COSTS

COST ESTIMATES
GS   ESTIMATES
. **COST ESTIMATES**
RT   AIRCRAFT PRODUCTION COSTS
     APPROPRIATIONS
     BUDGETING
     COSTS
     ECONOMIC ANALYSIS
     ECONOMY
    ∞ESTIMATORS
     FEDERAL BUDGETS
     FINANCIAL MANAGEMENT
     MANAGEMENT
     PRODUCTION COSTS
     VALUE ENGINEERING
     WAGE SURVEYS

COST INCENTIVES
RT   EFFICIENCY
     INCENTIVE TECHNIQUES
     MANAGEMENT
     VALUE ENGINEERING

COST REDUCTION
RT   COMMONALITY
     EFFICIENCY
     INCENTIVE TECHNIQUES
     MANAGEMENT
     MANAGEMENT METHODS
     MANAGEMENT PLANNING
     VALUE ENGINEERING
     WAGE SURVEYS

COSTA RICA
GS   NATIONS
. **COSTA RICA**
RT   CENTRAL AMERICA

COSTS
GS   COSTS
. AIRCRAFT PRODUCTION COSTS
. FREIGHT COSTS
. LIFE CYCLE COSTS
. LOW COST
. OPERATING COSTS
RT   ACCOUNTING
     AIRCRAFT PRODUCTION
     COMMERCE
     COST ANALYSIS
     COST ESTIMATES
     DAMAGE ASSESSMENT
     DESIGN TO COST
     ECONOMIC ANALYSIS
     ECONOMIC FACTORS
     ECONOMIC IMPACT
     ECONOMICS
     EFFICIENCY
     ESTIMATING
     EVALUATION
     FEASIBILITY
     FINANCIAL MANAGEMENT
     GROSS NATIONAL PRODUCT
     REVENUE
     TASK COMPLEXITY
     TASKS
     VALUE

COTTON
GS   FARM CROPS
. **COTTON**
RT   BOLL WEEVILS
     BOLLWORMS
     CLOTHING
     CORDAGE
     EARTH RESOURCES
     FABRICS
     FIBERS
     PLANTS (BOTANY)
     TEXTILES
     YARNS

COTTON FIBERS
GS   CLOTHING
. **COTTON FIBERS**
    FIBERS
. **COTTON FIBERS**
    TEXTILES
. **COTTON FIBERS**
RT   CREPE
     ORGANIC MATERIALS

COUCHES
RT   BEDS
     CUSHIONS
     HARNESSES
     PILLOWS
     SEATS
     SPACECRAFT ENVIRONMENTS

COUETTE FLOW
GS   FLUID FLOW
. STEADY FLOW
. . **COUETTE FLOW**
. TWO DIMENSIONAL FLOW
. . **COUETTE FLOW**
. VISCOUS FLOW
. . **COUETTE FLOW**
RT   ANNULAR FLOW
     AXISYMMETRIC FLOW
     HARTMANN FLOW
     ROTATING CYLINDERS

COUGAR AIRCRAFT
USE   F-9 AIRCRAFT

COUGH
GS   REFLEXES
. RESPIRATORY REFLEXES
. . **COUGH**
    SIGNS AND SYMPTOMS
. **COUGH**
RT   EXPELLANTS

COULEES
USE   CANYONS

COULOMB COLLISIONS
GS   COLLISIONS
. **COULOMB COLLISIONS**
RT   CHARGED PARTICLES

COULOMB POTENTIAL
GS   POTENTIAL ENERGY
. ELECTRIC POTENTIAL
. . **COULOMB POTENTIAL**
RT   CHARGED PARTICLES
     COULOMETRY
     ELECTRIC FIELD STRENGTH
     ELECTRIC FIELDS
    ∞POTENTIAL

COULOMETERS
GS   MEASURING INSTRUMENTS
. **COULOMETERS**
RT   AMMETERS
     CHEMICAL ANALYSIS
     COULOMETRY
     ELECTRICAL MEASUREMENT
     ELECTROCHEMISTRY
     ELECTRODEPOSITION
     ELECTROLYSIS
     TITRATION
     VOLTMETERS

COULOMETRY
GS   ELECTRICAL MEASUREMENT
. **COULOMETRY**
    ELECTROCHEMISTRY
. ELECTROLYSIS
. . **COULOMETRY**
RT   COULOMB POTENTIAL
     COULOMETERS

COUNTDOWN
GS   PREFLIGHT OPERATIONS
. **COUNTDOWN**
    SCHEDULES
. **COUNTDOWN**
RT   CHECKOUT
     CREW PROCEDURES (PREFLIGHT)
     LAUNCHING
     PRELAUNCH PROBLEMS
     PRELAUNCH TESTS
     SPACE VEHICLE CHECKOUT PROGRAM
     SPACECRAFT LAUNCHING
     WINDOWS (INTERVALS)

COUNTER ROTATION
GS   GYRATION
. ROTATION
. . **COUNTER ROTATION**
RT   BOUNDARY VALUE PROBLEMS
     COUNTER-ROTATING WHEELS
     COUNTERFLOW
     ROTATING DISKS

COUNTER ROTATION-*(CONT.)*
ROTATING FLUIDS

COUNTER-ROTATING WHEELS
UF   INERTIA WHEELS
GS   WHEELS
    . **COUNTER-ROTATING WHEELS**
RT   COUNTER ROTATION
    FLYWHEELS
    GEARS
    MECHANICAL DRIVES
    REACTION WHEELS

COUNTERBALANCES
RT   AIRCRAFT STABILITY
    BALLAST (MASS)
    DYNAMIC STABILITY
    MASS DISTRIBUTION
    SPACECRAFT STABILITY
    STATIC STABILITY

COUNTERFLOW
GS   FLUID FLOW
    . **COUNTERFLOW**
RT   AXIAL FLOW
    COUNTER ROTATION
    HEAT EXCHANGERS
    HEAT TRANSFER
    TRAPPED VORTEXES
    TURBULENT DIFFUSION
    TURBULENT FLOW
    VORTICES

COUNTERMEASURES
GS   **COUNTERMEASURES**
    . BALLISTIC MISSILE DECOYS
    . ELECTRONIC COUNTERMEASURES
    . . ANTIRADAR COATINGS
    . . CHAFF
    . JAMMING
    . OPTICAL COUNTERMEASURES
    . REENTRY DECOYS
RT   ANTIRADIATION MISSILES
    BLUE GOOSE MISSILE
    DECOYS
    PROTECTION
    QUAIL MISSILE
    RADAR ABSORBERS
    TARGET MASKING
    TORPEDOES

COUNTERS
UF   DEKATRONS
    GAS DISCHARGE COUNTERS
    PULSE RECORDERS
    QUANTIZER
GS   MEASURING INSTRUMENTS
    . **COUNTERS**
    . . RADIATION COUNTERS
    . . . CERENKOV COUNTERS
    . . . ELECTRON COUNTERS
    . . . GEIGER COUNTERS
    . . . NEUTRON COUNTERS
    . . . . NEUTRON SPECTROMETERS
    . . . PARTICLE TELESCOPES
    . . . PROPORTIONAL COUNTERS
    . . . QUANTUM COUNTERS
    . . . SCINTILLATION COUNTERS
    . . . SPARK CHAMBERS
RT   ACCUMULATORS (COMPUTERS)
    COMPUTER COMPONENTS
    COUNTING
    COUNTING CIRCUITS
    DATA RECORDERS
    IONIZATION CHAMBERS
    MONITORS
    RECORDING INSTRUMENTS

COUNTERSINKING
RT   GRINDING (MATERIAL REMOVAL)
    METAL CUTTING

COUNTING
RT   COUNTERS
    DATA ACQUISITION
    ENUMERATION
    ESTIMATING
    ∞ MEASUREMENT
    ∞ NUMBERS
    OBSERVATION
    REPETITION
    SAMPLING

COUNTING CIRCUITS
GS   CIRCUITS

COUNTING CIRCUITS-*(CONT.)*
    . **COUNTING CIRCUITS**
    . . SCALERS
RT   COUNTERS
    LOGIC CIRCUITS

COUNTING RATE COMPUTERS
GS   DATA PROCESSING EQUIPMENT
    . COMPUTERS
    . . **COUNTING RATE COMPUTERS**

COUPLED MODES
UF   MODE COUPLING
GS   MODES
    . **COUPLED MODES**
RT   CHEMICAL BONDS
    COUPLES
    COUPLINGS
    CROSSLINKING
    POLYMERIZATION
    STRONGLY COUPLED PLASMAS
    UNCOUPLED MODES

COUPLERS
SN   (EXCLUDES MECHANICAL DEVICE)
GS   **COUPLERS**
    . ANTENNA COUPLERS
    . . COUPLING CIRCUITS
    . . DIPLEXERS
    . . DIRECTIONAL COUPLERS
RT   ANTENNA COMPONENTS
    COUPLING
    COUPLINGS
    IMPEDANCE MATCHING
    YOKES

COUPLES
GS   COUPLING
    . **COUPLES**
RT   ANTENNA COUPLERS
    COUPLED MODES
    COUPLING CIRCUITS
    CROSS COUPLING
    DIPLEXERS
    OPTICAL COUPLING
    SPIN-SPIN COUPLING
    UNCOUPLED MODES
    YOKES

COUPLING
SN   (FOR MECHANICAL DEVICES, USE
    COUPLINGS)
GS   **COUPLING**
    . COUPLES
    . CROSS COUPLING
    . GYROSCOPIC COUPLING
    . MICROWAVE COUPLING
    . OPTICAL COUPLING
    . SPIN-SPIN COUPLING
    . THERMODYNAMIC COUPLING
RT   ANTENNA COUPLERS
    CLEBSCH-GORDAN COEFFICIENTS
    COUPLERS
    COUPLINGS
    DECOUPLING
    DIRECTIONAL COUPLERS
    LINKAGES
    MECHANICAL DRIVES
    RACAH COEFFICIENT
    VELOCITY COUPLING
    WAVE INTERACTION

COUPLING CIRCUITS
GS   CIRCUITS
    . **COUPLING CIRCUITS**
    COUPLERS
    . ANTENNA COUPLERS
    . . **COUPLING CIRCUITS**
RT   COUPLES
    CROSS COUPLING
    ENERGY TRANSFER
    IMPEDANCE MATCHING
    ∞ NETWORKS
    RC CIRCUITS
    RL CIRCUITS
    TRANSFORMERS

COUPLING COEFFICIENTS
GS   COEFFICIENTS
    . **COUPLING COEFFICIENTS**
RT   FORM FACTORS
    MAGNETIC INDUCTION
    TRANSFER FUNCTIONS

COUPLINGS
RT   ANCHORS (FASTENERS)
    BOLTS
    CLIPS
    CLOSURES
    CONNECTORS
    COUPLED MODES
    COUPLERS
    COUPLING
    DIRECTIONAL COUPLERS
    FASTENERS
    FITTINGS
    ∞ JOINING
    JOINTS (JUNCTIONS)
    LINKAGES
    MECHANICAL DRIVES
    PINS
    RIVETS
    SCREWS
    SLEEVES
    SPLINES
    TRAILERS
    UNIONS (CONNECTORS)

COURIER AIRCRAFT
USE   U-10 AIRCRAFT

COURIER SATELLITE
GS   SATELLITES
    . ARTIFICIAL SATELLITES
    . . **COURIER SATELLITE**
RT   ADVENT PROJECT

COURSES
USE   PATHS

COVALENCE
RT   CHEMICAL BONDS
    COVALENT BONDS

COVALENT BONDS
GS   CHEMICAL BONDS
    . **COVALENT BONDS**
RT   COVALENCE

COVARIANCE
GS   STATISTICAL ANALYSIS
    . VARIANCE (STATISTICS)
    . . MULTIVARIATE STATISTICAL
       ANALYSIS
    . . . **COVARIANCE**
RT   CORRELATION
    EXPERIMENT DESIGN
    FACTOR ANALYSIS
    ORTHOGONALITY
    QUALITY CONTROL
    REGRESSION ANALYSIS
    SIGNIFICANCE
    VARIABILITY

COVERALLS
GS   CLOTHING
    . **COVERALLS**
RT   FLIGHT CLOTHING
    PROTECTIVE CLOTHING

COVERINGS
RT   CAMOUFLAGE
    ∞ CAPS
    ∞ CASING
    CLOSURES
    COATINGS
    ELECTROSTATIC BONDING
    ENCLOSURES
    ∞ ENVELOPES
    GUARDS (SHIELDS)
    HOUSINGS
    JACKETS
    MASKING
    PRESERVING
    SEALING
    SHELLS (STRUCTURAL FORMS)
    SHROUDS
    SPHERICAL CAPS

COVES
USE   BAYS (TOPOGRAPHIC FEATURES)

COWELL METHOD
USE   NUMERICAL INTEGRATION

COWLINGS
GS   HOUSINGS
    . **COWLINGS**

COWLINGS-_(CONT.)_
RT   AIR INTAKES
     FAIRINGS
     NACELLES
     PODS (EXTERNAL STORES)
     PROTUBERANCES
     SHELLS (STRUCTURAL FORMS)

CRAB NEBULA
GS   CELESTIAL BODIES
     . NEBULAE
     . . **CRAB NEBULA**
RT   ORION NEBULA
     SUPERNOVAE
     TAURUS CONSTELLATION

CRABS
GS   ANIMALS
     . INVERTEBRATES
     . . ARTHROPODS
     . . . **CRABS**

CRACK ARREST
RT   CRACK INITIATION
     CRACK PROPAGATION
     CRACK TIPS
     CRACKING (FRACTURING)

CRACK CLOSURE
RT   BRITTLENESS
     CRACKING (FRACTURING)
     CRACKS
     ELBER EQUATION
     FATIGUE (MATERIALS)
     FRACTOGRAPHY
     FRACTURE MECHANICS
     FRACTURE STRENGTH
     FRACTURING
     GRIFFITH CRACK
     METAL FATIGUE
     MICROCRACKS
     STRESS CORROSION CRACKING
     SURFACE CRACKS

CRACK FORMATION
USE   CRACK INITIATION

CRACK GEOMETRY
GS   GEOMETRY
     . **CRACK GEOMETRY**
RT   CAVITIES
     CRACKS
     FATIGUE (MATERIALS)
     FRACTOGRAPHY
     MICROCRACKS
     SURFACE CRACKS
     VOIDS

CRACK INITIATION
UF   CRACK FORMATION
RT   BRITTLENESS
     CRACK ARREST
     CRACK TIPS
     CRACKS
     CRITICAL LOADING
     FRACTURE MECHANICS
     FRACTURE STRENGTH
     J INTEGRAL
     METAL FATIGUE
     METAL SURFACES
     MICROCRACKS
     STRESS CONCENTRATION
     STRESS CORROSION CRACKING
     STRESS INTENSITY FACTORS
     SURFACE CRACKS
     SURFACE DEFECTS
     TOUGHNESS

CRACK PROPAGATION
GS   PROPAGATION (EXTENSION)
     . **CRACK PROPAGATION**
RT   ACOUSTIC EMISSION
     BEND TESTS
     BRITTLENESS
     COFFIN-MANSON LAW
     CRACK ARREST
     CRACK TIPS
     CRACKING (FRACTURING)
     CRACKS
     FATIGUE (MATERIALS)
     FRACTOGRAPHY
     FRACTURE MECHANICS
     FRACTURE STRENGTH
     FRACTURING
     GRIFFITH CRACK

CRACK PROPAGATION-_(CONT.)_
     J INTEGRAL
     METAL FATIGUE
     MICROMECHANICS
     PLANE STRAIN
   ∞ PROPAGATION
     RESIDUAL STRENGTH
   ∞ RESISTANCE
     SEGRE CHARACTERISTIC
     STRESS CORROSION CRACKING
     STRESS INTENSITY FACTORS
     SURFACE CRACKS

CRACK TIPS
GS   CRACKS
     . **CRACK TIPS**
     TIPS
     . **CRACK TIPS**
RT   CRACK ARREST
     CRACK INITIATION
     CRACK PROPAGATION

CRACKING (CHEMICAL ENGINEERING)
GS   CHEMICAL REACTIONS
     . **CRACKING (CHEMICAL ENGINEERING)**
     . . PYROLYSIS
     DECOMPOSITION
     . **CRACKING (CHEMICAL ENGINEERING)**
RT   AMMONOLYSIS
     CATALYSIS
     CATALYTIC ACTIVITY
     CHEMICAL ENGINEERING
     COAL GASIFICATION
     ELECTROLYSIS
     HYDROCARBONS
     HYDROGENOLYSIS
     HYDROLYSIS
     NITROLYSIS
     ORGANIC CHEMISTRY
     PHOTOLYSIS
     THERMAL DISSOCIATION

CRACKING (FRACTURING)
GS   **CRACKING (FRACTURING)**
     . STRESS CORROSION CRACKING
RT   BRITTLE MATERIALS
     BRITTLENESS
     CRACK ARREST
     CRACK CLOSURE
     CRACK PROPAGATION
     CRACKS
     DESTRUCTION
     FAILURE
     FATIGUE (MATERIALS)
     FRACTURING
     J INTEGRAL
     RUPTURING
     STRESS CONCENTRATION
     STRESS CORROSION
     STRESS INTENSITY FACTORS
     STRUCTURAL FAILURE
     STRUCTURAL STRAIN
     TEMPERATURE INVERSIONS

CRACKS
UF   CREVICES
GS   **CRACKS**
     . CRACK TIPS
     . MICROCRACKS
     . SURFACE CRACKS
RT   CAVITIES
     CRACK CLOSURE
     CRACK GEOMETRY
     CRACK INITIATION
     CRACK PROPAGATION
     CRACKING (FRACTURING)
     DEFECTS
     ELBER EQUATION
     FAILURE MODES
     FATIGUE (MATERIALS)
     FRACTURES (MATERIALS)
     INTERSTICES
     LEAKAGE
     OPENINGS
     STRESSES
     TEMPERATURE INVERSIONS
     ULTRASONIC SPECTROSCOPY

CRAFT
USE   VEHICLES

CRAMPS
RT   EPILEPSY
     MUSCULAR FUNCTION
     SEIZURES

CRANES
SN   (EXCLUDES BIRDS)
GS   HANDLING EQUIPMENT
     . **CRANES**
     . . GANTRY CRANES
RT   BOOMS (EQUIPMENT)
     CONVEYORS
   ∞ LIFTS
     LOGISTICS
     MATERIALS HANDLING
     TOWERS
     WINCHES

CRANIUM
GS   ANATOMY
     . HEAD (ANATOMY)
     . . SKULL
     . . . **CRANIUM**
     . . . . INTRACRANIAL CAVITY
     . MUSCULOSKELETAL SYSTEM
     . . BONES
     . . . SKULL
     . . . . **CRANIUM**
     . . . . . INTRACRANIAL CAVITY
RT   INTERCRANIAL CIRCULATION

CRANK-NICHOLSON METHOD
RT   BOUNDARY VALUE PROBLEMS
     DIFFERENTIAL EQUATIONS
     FINITE DIFFERENCE THEORY
     FINITE ELEMENT METHOD
     NUMERICAL ANALYSIS
     PROBLEM SOLVING

CRANKED WINGS
USE   SWEPT WINGS

CRANKS
USE   ECCENTRICS

CRASH INJURIES
GS   INJURIES
     . **CRASH INJURIES**
RT   ACCIDENTS
     BURNS (INJURIES)
     HAZARDS
     WHIPLASH INJURIES

CRASH LANDING
GS   CRASHES
     . **CRASH LANDING**
     . . DITCHING (LANDING)
     LANDING
     . AIRCRAFT LANDING
     . . **CRASH LANDING**
     . . . DITCHING (LANDING)
RT   AIRCRAFT ACCIDENTS
     AIRCRAFT HAZARDS
     AIRCRAFT SAFETY
     AIRCRAFT SPIN
     ARRESTING GEAR
     CRASHWORTHINESS
     FLIGHT HAZARDS
     GLIDE LANDINGS
     HARD LANDING
     HORIZONTAL SPACECRAFT LANDING
     LUNAR LANDING
     PILOT ERROR
     PLANETARY LANDING
     SKID LANDINGS
     SOFT LANDING
     SPACECRAFT LANDING
     WATER LANDING

CRASHES
GS   **CRASHES**
     . CRASH LANDING
     . . DITCHING (LANDING)
RT   ACCIDENTS
     AIR BAG RESTRAINT DEVICES
     AIRCRAFT ACCIDENTS
     AIRCRAFT HAZARDS
     AIRCRAFT SAFETY
     COLLISIONS
     CRASHWORTHINESS
     ENCOUNTERS
     FLIGHT HAZARDS
     FLIGHT SAFETY
     HIGHWAYS
     MIDAIR COLLISIONS
     PILOT ERROR
     SAFETY
     WRECKAGE

CRASHWORTHINESS
RT   AIRCRAFT ACCIDENTS
       AIRCRAFT LANDING
       AIRCRAFT SAFETY
       CRASH LANDING
       CRASHES
       FLIGHT SAFETY
       IMPACT RESISTANCE

CRATERING
GS   **CRATERING**
       . PROJECTILE CRATERING
RT   CRATERS
       EJECTA
       IMPACT DAMAGE
       MARS CRATERS
       METEORITE CRATERS
       METEORITIC DAMAGE
       NUCLEAR EXPLOSIONS

CRATERS
UF   MAARS
       METEOR CRATERS
GS   **CRATERS**
       . LUNAR CRATERS
       . . PTOLEMAEUS CRATER
       . . TYCHO CRATER
       . METEORITE CRATERS
       . PLANETARY CRATERS
       . . MARS CRATERS
RT   CALDERAS
       CONES (VOLCANOES)
       CRATERING
       EJECTA
       IMPACT DAMAGE
       SATELLITE SURFACES

CRATONS
RT   CONTINENTS
       EARTH CRUST
       EARTH SURFACE
       OCEAN BOTTOM

CRAWLER TRACTORS
GS   SURFACE VEHICLES
       . MOTOR VEHICLES
       . . TRACTORS
       . . . **CRAWLER TRACTORS**
RT   ELECTRIC MOTOR VEHICLES
       GROUND SUPPORT EQUIPMENT
       HANDLING EQUIPMENT
       LUNAR SURFACE VEHICLES
       MANNED LUNAR SURFACE VEHICLES
       TRACKED VEHICLES
       ∞ TRANSPORT VEHICLES
       ∞ VEHICLES

CRAY COMPUTERS
GS   SUPERCOMPUTERS
       . **CRAY COMPUTERS**

CRAYONS
RT   ∞ MARKERS
       TEMPERATURE MEASUREMENT

CRAZING
USE   SURFACE CRACKS

CREATINE
GS   CRYSTALS
       . **CREATINE**
RT   JUICES

CREATININE
RT   DISEASES
       URINE

CREATION
USE   CREATIVITY

CREATIVITY
UF   CREATION
RT   ARTS
       EDUCATION
       MORALE

CREEP ANALYSIS
RT   ∞ ANALYZING
       STRESS ANALYSIS
       STRESS RELAXATION
       STRUCTURAL ANALYSIS

CREEP BUCKLING
GS   BUCKLING

CREEP BUCKLING-*(CONT.)*
       . **CREEP BUCKLING**

CREEP DIAGRAMS
GS   DIAGRAMS
       . **CREEP DIAGRAMS**
RT   STRESS RELAXATION
       STRESS-STRAIN-TIME RELATIONS

CREEP PROPERTIES
GS   MECHANICAL PROPERTIES
       . **CREEP PROPERTIES**
       . . SHEAR CREEP
       . . STEADY STATE CREEP
       . . TENSILE CREEP
RT   ANELASTICITY
       DEFORMATION
       DIMENSIONAL STABILITY
       DUCTILITY
       FATIGUE (MATERIALS)
       ∞ FLOW
       PLASTIC DEFORMATION
       PLASTIC FLOW
       ∞ PROPERTIES
       RESIDUAL STRESS
       SHEAR FLOW
       SHEAR PROPERTIES
       STATIC DEFORMATION
       STRESS RELAXATION
       STRESSES
       STRUCTURAL FAILURE
       SUPERPLASTICITY
       TEMPERATURE INVERSIONS

CREEP RESISTANCE
USE   CREEP STRENGTH

CREEP RUPTURE STRENGTH
UF   STRESS RUPTURE STRENGTH
GS   MECHANICAL PROPERTIES
       . **CREEP RUPTURE STRENGTH**
RT   FRACTURE STRENGTH
       J INTEGRAL
       ∞ STRENGTH

CREEP STRENGTH
UF   CREEP RESISTANCE
GS   MECHANICAL PROPERTIES
       . **CREEP STRENGTH**
RT   ∞ RESISTANCE
       ∞ STRENGTH

CREEP TESTS
RT   COMPRESSION TESTS
       FATIGUE TESTS
       LOAD TESTS
       PLASTIC DEFORMATION
       STATIC TESTS
       ∞ TESTS

CREPE
GS   FABRICS
       . **CREPE**
RT   COTTON FIBERS
       SILK

CRESOLS
GS   HYDROXYL COMPOUNDS
       . ALCOHOLS
       . . PHENOLS
       . . . **CRESOLS**

CRESTATRONS
USE   TRAVELING WAVE TUBES

CRESTS
USE   WAVES

CREVASSES
GS   **CREVASSES**
       . GLACIERS
RT   EARTH MOVEMENTS
       GEOLOGICAL FAULTS
       RECESSES
       SEAMOUNTS

CREVICES
USE   CRACKS

CREW EXPERIMENT STATIONS
GS   STATIONS
       . CREW STATIONS
       . . **CREW EXPERIMENT STATIONS**
RT   ASTRONAUTS

CREW EXPERIMENT STATIONS-*(CONT.)*
       COMPARTMENTS
       COSMONAUTS
       CREWS
       PERSONNEL
       SPACECRAFT CABINS
       SPACECREWS

CREW OBSERVATION STATIONS
GS   STATIONS
       . CREW STATIONS
       . . **CREW OBSERVATION STATIONS**
RT   ASTRONAUTS
       COMPARTMENTS
       COSMONAUTS
       CREWS
       PERSONNEL
       SPACECRAFT CABINS
       SPACECREWS

CREW PROCEDURES (INFLIGHT)
GS   FLIGHT OPERATIONS
       . **CREW PROCEDURES (INFLIGHT)**
       PROCEDURES
       . **CREW PROCEDURES (INFLIGHT)**
RT   DISPLAY DEVICES
       FLIGHT CREWS
       IN-FLIGHT MONITORING
       SPACECREWS
       TASKS
       ∞ TESTS

CREW PROCEDURES (PREFLIGHT)
GS   PROCEDURES
       . **CREW PROCEDURES (PREFLIGHT)**
RT   COUNTDOWN
       DISPLAY DEVICES
       FLIGHT CREWS
       FLIGHT OPERATIONS
       GROUND HANDLING
       GROUND TESTS
       IN-FLIGHT MONITORING
       ONBOARD EQUIPMENT
       PREFLIGHT OPERATIONS
       PRELAUNCH TESTS
       SPACECRAFT CONTROL
       SPACECREWS
       TASKS
       ∞ TESTS

CREW SIZE
RT   FLIGHT CREWS

CREW STATIONS
GS   STATIONS
       . **CREW STATIONS**
       . . CREW EXPERIMENT STATIONS
       . . CREW OBSERVATION STATIONS
       . . CREW WORKSTATIONS
RT   ASTRONAUTS
       COMPARTMENTS
       COSMONAUTS
       CREWS
       HELMET MOUNTED DISPLAYS
       PERSONNEL
       SPACECRAFT CABINS
       SPACECREWS

CREW WORKSTATIONS
GS   STATIONS
       . CREW STATIONS
       . . **CREW WORKSTATIONS**
RT   ASTRONAUTS
       COMPARTMENTS
       COSMONAUTS
       CREWS
       PERSONNEL
       SPACECRAFT CABINS
       SPACECREWS

CREWS
GS   PERSONNEL
       . **CREWS**
       . . FLIGHT CREWS
       . . . SPACECREWS
RT   ASTRONAUTS
       COSMONAUTS
       CREW EXPERIMENT STATIONS
       CREW OBSERVATION STATIONS
       CREW STATIONS
       CREW WORKSTATIONS
       FLIGHT NURSES
       PILOTS (PERSONNEL)

CRICKETS
GS ANIMALS
. INVERTEBRATES
. . ARTHROPODS
. . . INSECTS
. . . . **CRICKETS**
. . . . . BEETLES
. . . . . . TRIBOLIA

CRIME
RT AIR PIRACY
LAW (JURISPRUDENCE)
POLICE
REGULATIONS
SECURITY
SOCIAL FACTORS
SURVEILLANCE
VIOLENCE·

CRIMPING
USE FOLDING

CRITERIA
GS **CRITERIA**
. STRUCTURAL DESIGN CRITERIA
RT EVALUATION
FIGURE OF MERIT
∞MEASURES
STANDARDS

CRITICAL EXPERIMENTS
RT EXPERIMENTATION
NUCLEAR FISSION
NUCLEAR REACTIONS

CRITICAL FLICKER FUSION
UF FLICKER FUSION FREQUENCY
GS PERCEPTION
. SENSORY PERCEPTION
. . VISUAL PERCEPTION
. . . **CRITICAL FLICKER FUSION**
RT AFTERIMAGES
FLICKER

CRITICAL FLOW
GS FLUID FLOW
. **CRITICAL FLOW**
RT FLOW CHARACTERISTICS
GAS FLOW
LAMINAR FLOW
LIQUID FLOW
MULTIPHASE FLOW
ORIFICE FLOW
PIPE FLOW
PRESSURE GRADIENTS
SINGLE-PHASE FLOW
STEADY FLOW
STEAM FLOW
SUBCRITICAL FLOW
SUPERCRITICAL FLOW
TURBULENT FLOW
UNSTEADY FLOW

CRITICAL FREQUENCIES
GS FREQUENCIES
. **CRITICAL FREQUENCIES**
RT LIGHT (VISIBLE RADIATION)
RESONANT FREQUENCIES

CRITICAL LOADING
SN (LIMITED TO FORCE LOADS)
UF CRITICAL STRESS
GS LOADS (FORCES)
. **CRITICAL LOADING**
STRESSES
. **CRITICAL LOADING**
RT AERODYNAMIC LOADS
CRACK INITIATION
DYNAMIC LOADS
PROPORTIONAL LIMIT
SHALLOW SHELLS
STATIC LOADS

CRITICAL MACH NUMBER
USE CRITICAL VELOCITY
MACH NUMBER

CRITICAL MASS
GS MASS
. **CRITICAL MASS**
RT NUCLEAR FISSION
NUCLEAR FUEL BURNUP
NUCLEAR REACTIONS
PLASMA CORE REACTORS

CRITICAL MASS-*(CONT.)*
SUBCRITICAL MASS

CRITICAL PATH METHOD
GS NETWORK ANALYSIS
. **CRITICAL PATH METHOD**
RT ∞CONTROL
DYNAMIC PROGRAMMING
ESTIMATING
GERT
MANAGEMENT METHODS
∞METHODOLOGY
MISSION PLANNING
OPERATIONS RESEARCH
∞PATHS
PERT
PLANNING
PROGRAM TREND LINE ANALYSIS
PROGRAMMING (SCHEDULING)
PROJECT MANAGEMENT
RESEARCH
SEQUENCING
SNEAK CIRCUIT ANALYSIS
SYSTEMS ENGINEERING

CRITICAL POINT
GS THERMODYNAMIC PROPERTIES
. THERMOPHYSICAL PROPERTIES
. . **CRITICAL POINT**
RT MAYER PROBLEM

CRITICAL PRESSURE
GS PRESSURE
. **CRITICAL PRESSURE**
THERMODYNAMIC PROPERTIES
. THERMOPHYSICAL PROPERTIES
. . **CRITICAL PRESSURE**
RT HIGH PRESSURE
LIQUID PHASES
SUPERCRITICAL PRESSURES
VAPOR PHASES

CRITICAL REYNOLDS NUMBER
USE CRITICAL VELOCITY
REYNOLDS NUMBER

CRITICAL SPEED
USE CRITICAL VELOCITY

CRITICAL STRESS
USE CRITICAL LOADING

CRITICAL TEMPERATURE
GS TEMPERATURE
. **CRITICAL TEMPERATURE**
THERMODYNAMIC PROPERTIES
. THERMOPHYSICAL PROPERTIES
. . **CRITICAL TEMPERATURE**
RT HEAT TREATMENT
METALLIC HYDROGEN
NONCONDENSABLE GASES
PHASE DIAGRAMS
PHASE TRANSFORMATIONS
ULTRALOW TEMPERATURES

CRITICAL VELOCITY
UF CRITICAL MACH NUMBER
CRITICAL REYNOLDS NUMBER
CRITICAL SPEED
GS RATES (PER TIME)
. **CRITICAL VELOCITY**
VELOCITY
. **CRITICAL VELOCITY**
RT EXHAUST VELOCITY
RESONANT FREQUENCIES
TIP SPEED

CROCCO METHOD
RT AXISYMMETRIC FLOW
BOUNDARY LAYERS
COMPRESSIBLE FLOW
COMPRESSIBLE FLUIDS
ENTROPY
INVISCID FLOW
∞METHODOLOGY
SHOCK WAVE PROPAGATION
STEADY FLOW
VORTICITY

CROCCO-LEE THEORY
RT BOUNDARY LAYER SEPARATION
CONTINUITY EQUATION
GAS FLOW
INVISCID FLOW

CROCCO-LEE THEORY-*(CONT.)*
MASS FLOW
MULTIPHASE FLOW
REATTACHED FLOW
SEPARATED FLOW
∞THEORIES

CROLOY
GS ALLOYS
. IRON ALLOYS
. . STEELS
. . . **CROLOY**

CROP CALENDARS
GS CALENDARS
. **CROP CALENDARS**
RT ∞CROPS
FARM CROPS
GROWTH
SCHEDULING
SEASONS

CROP DUSTING
GS SPRAYING
. **CROP DUSTING**
RT AEROSOLS
AGRICULTURAL AIRCRAFT
AGRICULTURE
DISPERSIONS
FARM CROPS
PESTICIDES
POWDER (PARTICLES)

CROP GROWTH
GS GROWTH
. **CROP GROWTH**
RT AGRICULTURE
ALFALFA
BARLEY
BLIGHT
CITRUS TREES
CORN
∞CROPS
EARTH RESOURCES
FARM CROPS
FARMLANDS
GERMINATION
GRASSLANDS
LARGE AREA CROP INVENTORY
   EXPERIMENT
OATS
ORCHARDS
PHOTOSYNTHESIS
PLANT STRESS
PLANTS (BOTANY)
SUGAR BEETS
SUGAR CANE
THERMAL RESOURCES
VINEYARDS
WHEAT

CROP IDENTIFICATION
GS IDENTIFYING
. **CROP IDENTIFICATION**
RT AGRICULTURE
∞CROPS
EARTH RESOURCES
EVALUATION
FARMLANDS
GROUND TRUTH
IMAGING TECHNIQUES
MULTISPECTRAL PHOTOGRAPHY
RECOGNITION
REMOTE SENSORS
∞SENSORS
SORGHUM
SPECTRAL SIGNATURES
SPOT (FRENCH SATELLITE)
SUNFLOWERS
TIMBER IDENTIFICATION
VEGETATIVE INDEX

CROP INVENTORIES
GS INVENTORIES
. **CROP INVENTORIES**
RT AGRICULTURE
AGRISTARS PROJECT
FARM CROPS
FARMLANDS
LARGE AREA CROP INVENTORY
   EXPERIMENT
REMOTE SENSORS
VEGETATIVE INDEX

CROP INVENTORIES BY REMOTE SENSING
USE   AGRISTARS PROJECT

CROP VIGOR
RT   AGRICULTURE
      ALFALFA
      BARLEY
      BLIGHT
      CITRUS TREES
      FARM CROPS
      FARMLANDS
      IRRIGATION
      OATS
      ORCHARDS
      PHOTOTROPISM
      PLANT STRESS
      PLANTS (BOTANY)
      SUGAR BEETS
      SUGAR CANE
      THERMAL RESOURCES
      VEGETATION GROWTH
      VIABILITY
      VINEYARDS
      WHEAT

CROPLANDS
USE   FARMLANDS

∞ **CROPS**
SN   *(USE OF A MORE SPECIFIC TERM IS*
      RECOMMENDED--CONSULT THE TERMS
      LISTED BELOW)
RT   AGRICULTURE
      ALFALFA
      BARLEY
      BIOMASS ENERGY PRODUCTION
      CITRUS TREES
      CORN
      CROP CALENDARS
      CROP GROWTH
      CROP IDENTIFICATION
      FARM CROPS
      FARMLANDS
      FROST DAMAGE
      LARGE AREA CROP INVENTORY
          EXPERIMENT
      ORCHARDS
      PLANTING
      SORGHUM
      SUNFLOWERS
      VINEYARDS
      WHEAT

CROSS CORRELATION
GS   CORRELATION
      . **CROSS CORRELATION**
RT   AUTOCORRELATION
      DATA CORRELATION

CROSS COUPLING
GS   COUPLING
      . **CROSS COUPLING**
RT   COMMUNICATION THEORY
      COUPLES
      COUPLING CIRCUITS
      MICROWAVE COUPLING
      OPTICAL COUPLING
      RADIO FREQUENCY INTERFERENCE

CROSS FAULTS
USE   GEOLOGICAL FAULTS

CROSS FLOW
GS   FLUID FLOW
      . **CROSS FLOW**
RT   AERODYNAMIC CHARACTERISTICS
      ∞ FLOW
      FLOW CHARACTERISTICS
      FLOW GEOMETRY
      FLUID DYNAMICS
      SPANWISE BLOWING
      WATER TUNNEL TESTS

CROSS POLARIZATION
GS   POLARIZATION (WAVES)
      . **CROSS POLARIZATION**
RT   OPTICAL COUPLING
      OPTICAL PROPERTIES
      POLARIZED ELECTROMAGNETIC
          RADIATION
      POLARONS
      ROTATION

CROSS RELAXATION
RT   MASERS
      ∞ RELAXATION
      RUTILE
      SPIN-SPIN COUPLING

∞ **CROSS SECTIONS**
SN   *(USE OF A MORE SPECIFIC TERM IS*
      RECOMMENDED--CONSULT THE TERMS
      LISTED BELOW)
RT   ABSORPTION CROSS SECTIONS
      AIRFOIL PROFILES
      AREA
      ATOMIC COLLISIONS
      COLLISION PARAMETERS
      DISTRIBUTION (PROPERTY)
      DRAWINGS
      GEOMETRY
      GRADIENTS
      IONIZATION CROSS SECTIONS
      MEAN FREE PATH
      NEUTRON CROSS SECTIONS
      PLANFORMS
      RADAR CROSS SECTIONS
      SCATTERING CROSS SECTIONS
      SHAPES
      STOPPING POWER
      SURVEYS
      TWO DIMENSIONAL BODIES

CROSSBEDDING (GEOLOGY)
GS   GEOLOGY
      . **CROSSBEDDING (GEOLOGY)**
RT   LANDFORMS
      ROCKS
      STRATA
      STRATIFICATION
      STRATIGRAPHY

CROSSED FIELD AMPLIFIERS
GS   AMPLIFIERS
      . **CROSSED FIELD AMPLIFIERS**
RT   ELECTRON TUBES
      MAGNETRONS
      MICROWAVE AMPLIFIERS
      TRAVELING WAVE TUBES

CROSSED FIELD GUNS
RT   ELECTRON GUNS
      ∞ GUNS
      PLASMA CONTROL
      PLASMA GUNS
      PLASMA JETS

CROSSED FIELDS
RT   ELECTRIC FIELDS
      FIELD THEORY (PHYSICS)
      MAGNETIC FIELDS
      MAGNETRONS
      PLASMA CONTROL
      WAVEGUIDES

CROSSINGS
GS   **CROSSINGS**
      . CHIASMS
RT   BRIDGES (STRUCTURES)
      CROSSOVERS
      INTERSECTIONS
      PIPELINES
      RAMPS (STRUCTURES)

CROSSLINKING
GS   **CROSSLINKING**
      . VULCANIZING
RT   ADDITION RESINS
      COUPLED MODES
      CURING
      ∞ JOINING
      PHENOLIC EPOXY RESINS

CROSSOVERS
RT   BRIDGES (STRUCTURES)
      CROSSINGS
      INTERSECTIONS

CROSSTALK
GS   ELECTROMAGNETIC INTERFERENCE
      . **CROSSTALK**
      . . IONOSPHERIC CROSS MODULATION
RT   COMMUNICATING
      ELECTROMAGNETIC COMPATIBILITY
      ∞ INTERFERENCE
      TELEPHONY
      WAVE DIFFRACTION

CROWDING
RT   ∞ CONCENTRATION
      ∞ SATURATION

CRUCIBLES
RT   ∞ CONTAINERS
      HEATING EQUIPMENT

CRUCIFORM WINGS
GS   AIRFOILS
      . WINGS
      . . **CRUCIFORM WINGS**
RT   FIXED WINGS
      LOW ASPECT RATIO WINGS

CRUDE OIL
UF   PETROLEUM
GS   OILS
      . **CRUDE OIL**
      RESOURCES
      . EARTH RESOURCES
      . . FOSSIL FUELS
      . . . **CRUDE OIL**
RT   CARBONACEOUS MATERIALS
      DEPOSITS
      ENERGY POLICY
      FUEL PRODUCTION
      HYDROCARBON FUELS
      OFFSHORE ENERGY SOURCES
      OIL EXPLORATION
      OIL FIELDS
      PETROLEUM PRODUCTS
      RESERVES
      UNDERWATER RESOURCES
      WAXES

CRUISE MISSILES
GS   MISSILES
      . SURFACE TO SURFACE MISSILES
      . . **CRUISE MISSILES**
      . . . NAVAHO MISSILE
      . . . TOMAHAWK MISSILES
RT   ANTISHIP MISSILES

CRUISING FLIGHT
RT   COASTING FLIGHT
      ∞ FLIGHT
      HORIZONTAL FLIGHT

CRUSADER AIRCRAFT
USE   F-8 AIRCRAFT

CRUSHERS
RT   COMMINUTION
      CRUSHING
      DISINTEGRATION
      GRINDING MILLS
      IMPACTORS

CRUSHING
GS   COMMINUTION
      . **CRUSHING**
RT   CRUSHERS
      DISINTEGRATION
      GRINDING (COMMINUTION)

CRUSTAL FRACTURES
GS   FRACTURING
      . **CRUSTAL FRACTURES**
RT   EARTH CRUST
      EARTH MOVEMENTS
      EARTH SURFACE
      EARTHQUAKE RESISTANCE
      EARTHQUAKES
      GEODYNAMICS
      GEOLOGICAL FAULTS
      MICROSEISMS
      P WAVES
      S WAVES
      SAN ANDREAS FAULT
      SEISMIC WAVES
      SEISMOLOGY
      SHATTER CONES
      SHOCK LOADS
      SHOCK WAVES
      SOIL MECHANICS
      SURFACE WAVES

CRUSTS
GS   **CRUSTS**
      . EARTH CRUST
      . LUNAR CRUST
RT   LUNAR MANTLE
      PLANETARY MANTLES

CRYOCHEMISTRY
GS   PHYSICAL CHEMISTRY
      . **CRYOCHEMISTRY**
RT   ∞CHEMISTRY
      CRYOGENIC EQUIPMENT
      CRYOGENICS
      LOW TEMPERATURE PHYSICS

CRYOCYCLE PRINCIPLE
RT   CRYOPUMPING
      SPACECRAFT POWER SUPPLIES

CRYODEPOSITS
GS   DEPOSITS
      . **CRYODEPOSITS**
RT   COATINGS
      ∞CRYOGENIC STORAGE
      CRYOGENICS

CRYOGENIC COMPUTER STORAGE
GS   PERIPHERAL EQUIPMENT (COMPUTERS)
      . COMPUTER STORAGE DEVICES
      . . **CRYOGENIC COMPUTER STORAGE**
RT   ∞CRYOGENIC STORAGE
      CRYOTRONS
      ∞EQUIPMENT
      SUPERCONDUCTORS

CRYOGENIC COOLING
GS   COOLING
      . SUPERCOOLING
      . . **CRYOGENIC COOLING**
RT   COOLERS
      CRYOGENICS
      FREEZING
      HEAT TRANSFER
      REFRIGERATING

CRYOGENIC EQUIPMENT
UF   DEWAR SYSTEMS
RT   CRYOCHEMISTRY
      CRYOGENICS
      ∞EQUIPMENT
      GRAVITATIONAL WAVE ANTENNAS
      REFRIGERATING
      REFRIGERATING MACHINERY
      SOLID CRYOGENS

CRYOGENIC FLUID STORAGE
RT   COLD SURFACES
      ∞CRYOGENIC STORAGE
      EVAPORATIVE COOLING
      FLUID MANAGEMENT
      FUEL TANKS
      MULTILAYER INSULATION
      SPACE STORAGE
      ∞STORAGE
      STORAGE TANKS
      THERMAL INSULATION

CRYOGENIC FLUIDS
GS   LIQUIDS
      . **CRYOGENIC FLUIDS**
      . . FERMI LIQUIDS
      . . FLOX
      . . LIQUID HELIUM
      . . LIQUID HYDROGEN
      . . LIQUID NITROGEN
      . . LIQUID OXYGEN
RT   CRYOGENICS
      CRYOPUMPING
      FLUID MANAGEMENT
      ∞FLUIDS
      ROCKET OXIDIZERS
      SOLID CRYOGEN COOLING
      ULTRALOW TEMPERATURES

CRYOGENIC GYROSCOPES
GS   GYROSCOPES
      . **CRYOGENIC GYROSCOPES**

CRYOGENIC MAGNETS
GS   MAGNETS
      . **CRYOGENIC MAGNETS**
RT   SUPERCONDUCTING MAGNETS

CRYOGENIC ROCKET PROPELLANTS
GS   LIQUIDS
      . LIQUID FUELS
      . . LIQUID ROCKET PROPELLANTS
      . . . **CRYOGENIC ROCKET PROPELLANTS**
      PROPELLANTS
      . ROCKET PROPELLANTS

CRYOGENIC ROCKET PROPELLANTS-*(CONT.)*
      . . LIQUID ROCKET PROPELLANTS
      . . . **CRYOGENIC ROCKET PROPELLANTS**
RT   CRYOGENICS
      ENDOTHERMIC FUELS
      FLUID MANAGEMENT
      GASEOUS ROCKET PROPELLANTS
      GELLED ROCKET PROPELLANTS
      HIGH ENERGY FUELS
      HIGH ENERGY PROPELLANTS
      HYBRID PROPELLANTS
      HYDROGEN FUELS
      HYPERGOLIC ROCKET PROPELLANTS
      LIQUEFIED GASES
      LIQUID HYDROGEN
      LIQUID OXYGEN
      RL-10 ENGINES
      SLUSH
      SPACE STORAGE
      STORABLE PROPELLANTS

∞ **CRYOGENIC STORAGE**
SN   *(USE OF A MORE SPECIFIC TERM IS RECOMMENDED--CONSULT THE TERMS LISTED BELOW)*
RT   CRYODEPOSITS
      CRYOGENIC COMPUTER STORAGE
      CRYOGENIC FLUID STORAGE

CRYOGENIC WIND TUNNELS
GS   TEST FACILITIES
      . WIND TUNNELS
      . . **CRYOGENIC WIND TUNNELS**
RT   FLIGHT SIMULATORS
      TEST CHAMBERS

CRYOGENICS
RT   ABSOLUTE ZERO
      COLD TRAPS
      COOLING
      CRYOCHEMISTRY
      CRYODEPOSITS
      CRYOGENIC COOLING
      CRYOGENIC EQUIPMENT
      CRYOGENIC FLUIDS
      CRYOGENIC ROCKET PROPELLANTS
      CRYOPUMPING
      CRYOSAR
      CRYOSTATS
      CRYOTRONS
      FERMI LIQUIDS
      JOULE-THOMSON EFFECT
      LIQUEFIED GASES
      LOW TEMPERATURE
      LOW TEMPERATURE PHYSICS
      REFRIGERATING
      SOLID CRYOGEN COOLING
      SOLID CRYOGENS
      SOLID NITROGEN
      SOLIDIFIED GASES
      SUPERCONDUCTING POWER TRANSMISSION
      SUPERCONDUCTIVITY
      THERMOELECTRIC COOLING
      THERMOMAGNETIC COOLING
      ULTRALOW TEMPERATURES

CRYOLITE
GS   ALUMINUM COMPOUNDS
      . **CRYOLITE**
      HALOGEN COMPOUNDS
      . FLUORINE COMPOUNDS
      . . FLUORIDES
      . . . **CRYOLITE**
      . . FLUORO COMPOUNDS
      . . . **CRYOLITE**
      MINERALS
      . **CRYOLITE**
      SODIUM COMPOUNDS
      . **CRYOLITE**
RT   ALUMINUM

CRYOPUMPING
RT   CRYOCYCLE PRINCIPLE
      CRYOGENIC FLUIDS
      CRYOGENICS
      ∞PUMPING
      VACUUM PUMPS

CRYOSAR
GS   ELECTRONIC EQUIPMENT
      . SOLID STATE DEVICES
      . . SEMICONDUCTOR DEVICES
      . . . AVALANCHE DIODES

CRYOSAR-*(CONT.)*
      . . . . **CRYOSAR**
      RECTIFIERS
      . AVALANCHE DIODES
      . . **CRYOSAR**
RT   BARRITT DIODES
      COMPUTER STORAGE DEVICES
      CRYOGENICS

CRYOSORPTION
USE   SORPTION

CRYOSTATS
GS   CONTROL EQUIPMENT
      . **CRYOSTATS**
RT   CONTROLLERS
      CRYOGENICS
      ELECTRIC SWITCHES
      HIGH TEMPERATURE TESTS
      LIQUID HELIUM
      LIQUID HELIUM 2
      LOW TEMPERATURE TESTS
      REGULATORS
      TEMPERATURE CONTROL
      THERMOSTATS

CRYOTRAPPING
GS   TRAPPING
      . **CRYOTRAPPING**
RT   COLD TRAPS

CRYOTRONS
GS   ELECTRONIC EQUIPMENT
      . SOLID STATE DEVICES
      . . **CRYOTRONS**
      SWITCHES
      . ELECTRIC SWITCHES
      . . **CRYOTRONS**
RT   CRYOGENIC COMPUTER STORAGE
      CRYOGENICS
      SUPERCONDUCTIVITY
      SUPERCONDUCTORS

CRYPTOGRAPHY
RT   ∞CODES
      CODING
      COMPUTER INFORMATION SECURITY
      DECODING
      INFORMATION THEORY
      MESSAGE PROCESSING

CRYSTAL DEFECTS
UF   LATTICE IMPERFECTIONS
      STACKING FAULTS
GS   DEFECTS
      . **CRYSTAL DEFECTS**
      . . CRYSTAL DISLOCATIONS
      . . . EDGE DISLOCATIONS
      . . . SCREW DISLOCATIONS
      . . POINT DEFECTS
      . . . VACANCIES (CRYSTAL DEFECTS)
      . . . . FRENKEL DEFECTS
RT   CRYSTALLOGRAPHY
      HOLES (ELECTRON DEFICIENCIES)
      IMPURITIES
      INTERSTITIALS
      LATTICE VIBRATIONS
      MECHANICAL TWINNING
      ORDER-DISORDER TRANSFORMATIONS
      PINNING
      POLYGONIZATION
      STACKING FAULT ENERGY
      STACKS
      SURFACE DEFECTS
      TRAPPING
      TWINNING

CRYSTAL DISLOCATIONS
GS   DEFECTS
      . CRYSTAL DEFECTS
      . . **CRYSTAL DISLOCATIONS**
      . . . EDGE DISLOCATIONS
      . . . SCREW DISLOCATIONS
      DISLOCATIONS (MATERIALS)
      . **CRYSTAL DISLOCATIONS**
      . . EDGE DISLOCATIONS
      . . SCREW DISLOCATIONS
RT   FATIGUE (MATERIALS)
      GRAIN BOUNDARIES
      PINNING
      POINT DEFECTS
      SUPERLATTICES
      SUPERPLASTICITY
      SURFACE DEFECTS

CRYSTAL FILTERS
GS   ELECTROMAGNETIC WAVE FILTERS
. BANDPASS FILTERS
. . **CRYSTAL FILTERS**
. ELECTRIC FILTERS
. . **CRYSTAL FILTERS**
RT   BANDSTOP FILTERS
∞FILTERS
INTERMEDIATE FREQUENCY AMPLIFIERS
RADIO EQUIPMENT
RADIO FILTERS

CRYSTAL GROWTH
GS   GROWTH
. **CRYSTAL GROWTH**
. . CZOCHRALSKI METHOD
. . DIRECTIONAL SOLIDIFICATION
    (CRYSTALS)
. . EPITAXY
. . . ELECTROEPITAXY
. . . LIQUID PHASE EPITAXY
. . . MOLECULAR BEAM EPITAXY
. . . VAPOR PHASE EPITAXY
. . HYDROTHERMAL CRYSTAL GROWTH
. . TRAVELING SOLVENT METHOD
. . VERNEUIL PROCESS
RT   AITKEN NUCLEI
BRAVAIS CRYSTALS
BRIDGMAN METHOD
CONTAINERLESS MELTS
CRYSTALLIZATION
CRYSTALLOGRAPHY
CRYSTALS
DOPED CRYSTALS
FLOAT ZONES
INOCULATION
MECHANICAL TWINNING
MELTS (CRYSTAL GROWTH)
NUCLEATION
POLYGONIZATION
RAPID QUENCHING (METALLURGY)
SPACE PROCESSING
TWINNING
VAPOR DEPOSITION

CRYSTAL LATTICES
GS   **CRYSTAL LATTICES**
. BODY CENTERED CUBIC LATTICES
. CLOSE PACKED LATTICES
. CUBIC LATTICES
. . FACE CENTERED CUBIC LATTICES
. SUPERLATTICES
RT   ATOMIC STRUCTURE
BRAVAIS CRYSTALS
BRILLOUIN ZONES
CHEMICAL BONDS
CRYSTALLOGRAPHY
CRYSTALS
EPITAXY
GEOMETRY
GRAPHOEPITAXY
HEXAGONAL CELLS
IONIC CRYSTALS
ISOMORPHISM
KOSSEL PATTERN
LATTICE PARAMETERS
LATTICE VIBRATIONS
∞LATTICES
LAUE METHOD
METAL CRYSTALS
METALLOGRAPHY
MOLECULAR CHAINS
MOLECULAR STRUCTURE
MOSSBAUER EFFECT
ORDER-DISORDER TRANSFORMATIONS
PARTICLE IN CELL TECHNIQUE
PATTERSON MAP
POLYMORPHISM
RAPID QUENCHING (METALLURGY)
SINGLE CRYSTALS
SYNTHETIC METALS
ULTRAPURE METALS

CRYSTAL OPTICS
RT   ABERRATION
ACOUSTO-OPTICS
DIFFRACTION
DOPED CRYSTALS
FIBER OPTICS
GEOMETRICAL OPTICS
∞OPTICS
PHASE MATCHING
PHYSICAL OPTICS

CRYSTAL OSCILLATORS
GS   CRYSTALS
. **CRYSTAL OSCILLATORS**
. . PIEZOELECTRIC CRYSTALS
OSCILLATORS
. **CRYSTAL OSCILLATORS**
RT   ELECTRICAL PROPERTIES
FREQUENCY CONTROL
FREQUENCY STABILITY
OSCILLATIONS
PIEZOELECTRICITY

CRYSTAL RECTIFIERS
UF   SILICON RECTIFIERS
GS   ELECTRONIC EQUIPMENT
. DIODES
. **CRYSTAL RECTIFIERS**
. SOLID STATE DEVICES
. . **CRYSTAL RECTIFIERS**
RECTIFIERS
. **CRYSTAL RECTIFIERS**
RT   CURRENT CONVERTERS (AC TO DC)
SEMICONDUCTOR DEVICES

CRYSTAL STRUCTURE
SN   (AGGLOMERATIONS OF CRYSTALS--
    EXCLUDES CRYSTAL LATTICES)
GS   **CRYSTAL STRUCTURE**
. WIDMANSTATTEN STRUCTURE
RT   ABRIKOSOV THEORY
ALLOTROPY
ANISOTROPY
BRAVAIS CRYSTALS
CLATHRATES
CRYSTALLINITY
CRYSTALLITES
CRYSTALS
DOPED CRYSTALS
EPITAXY
GRAPHOEPITAXY
INTERSTITIALS
ISOMORPHISM
ISOTROPY
LIQUID PHASE EPITAXY
MECHANICAL TWINNING
METAL CRYSTALS
MICROSTRUCTURE
ORDER-DISORDER TRANSFORMATIONS
PACKING DENSITY
PATTERSON MAP
PHONONS
POLYCRYSTALS
POLYMORPHISM
RAPID QUENCHING (METALLURGY)
SPHERULITES
∞STRUCTURES
SUPERLATTICES
TWINNING
VAPOR PHASE EPITAXY

CRYSTAL SURFACES
GS   SOLID SURFACES
. **CRYSTAL SURFACES**
RT   METAL SURFACES
SURFACE LAYERS
∞SURFACES

CRYSTALLINITY
RT   AMORPHOUS MATERIALS
CRYSTAL STRUCTURE

CRYSTALLITES
GS   CRYSTALS
. **CRYSTALLITES**
. . SPHERULITES
RT   CRYSTAL STRUCTURE
MICROCRYSTALS
MINERALS
ROSETTE SHAPES

CRYSTALLIZATION
UF   DEVITRIFICATION
GS   **CRYSTALLIZATION**
. DIRECTIONAL SOLIDIFICATION
    (CRYSTALS)
. MELT SPINNING
. RECRYSTALLIZATION
RT   AGGLOMERATION
CONCENTRATING
CONTAINERLESS MELTS
CRYSTAL GROWTH
DEMINERALIZING
FREEZING
INOCULATION
LIQUIDUS

CRYSTALLIZATION-*(CONT.)*
MATERIALS RECOVERY
MELTS (CRYSTAL GROWTH)
MODULATION
NUCLEATION
PHASE TRANSFORMATIONS
PRECIPITATION (CHEMISTRY)
PURIFICATION
REFINING
∞SEPARATION
SETTLING
SOLID STATE
SOLIDIFICATION
SUBLIMATION
SUPERCOOLING
SUPERSATURATION
ZONE MELTING

CRYSTALLOGRAPHY
RT   BRAGG ANGLE
CRYSTAL DEFECTS
CRYSTAL GROWTH
CRYSTAL LATTICES
CRYSTALS
DEBYE-SCHERRER METHOD
DIRECTIVITY
ISOTROPY
LAMELLA (METALLURGY)
LATTICE PARAMETERS
LAUE METHOD
METALLOGRAPHY
∞METALLURGY
MICROBEAMS
MICROSTRUCTURE
MINERALOGY
NEUTRON DIFFRACTION
ORDER-DISORDER TRANSFORMATIONS
∞ORIENTATION
RADIOGRAPHY
∞SOLID STATE PHYSICS
X RAY ANALYSIS
X RAY DIFFRACTION

CRYSTALS
GS   **CRYSTALS**
BICRYSTALS
. BOULES
. BRAVAIS CRYSTALS
. CREATINE
. CRYSTAL OSCILLATORS
. . PIEZOELECTRIC CRYSTALS
. CRYSTALLITES
. . SPHERULITES
. DENDRITIC CRYSTALS
. DOPED CRYSTALS
. IONIC CRYSTALS
. LIQUID CRYSTALS
. METAL CRYSTALS
. MICROCRYSTALS
. MIXED CRYSTALS
. POLYCRYSTALS
. QUARTZ CRYSTALS
. SINGLE CRYSTALS
. WHISKERS (CRYSTALS)
RT   ANISOTROPY
BODY CENTERED CUBIC LATTICES
CLATHRATES
CONTAINERLESS MELTS
CRYSTAL GROWTH
CRYSTAL LATTICES
CRYSTAL STRUCTURE
CRYSTALLOGRAPHY
ELECTROEPITAXY
FACE CENTERED CUBIC LATTICES
∞GRAINS
ISOTROPY
PACKING DENSITY
PHASE MATCHING
RUBY
SPHERULES

CSM
USE   COMMAND SERVICE MODULES

CT-114 AIRCRAFT
USE   CL-41 AIRCRAFT

CTD
USE   CHARGE TRANSFER DEVICES

CUBA
GS   LANDFORMS
. ISLANDS
. . WEST INDIES
. . . **CUBA**

CUBA-*(CONT.)*
 NATIONS
 . **CUBA**
RT CARIBBEAN REGION
  CARIBBEAN SEA

CUBANE
GS HYDROCARBONS
 . **CUBANE**

CUBES (MATHEMATICS)
GS GEOMETRY
 . EUCLIDEAN GEOMETRY
 . . POLYHEDRONS
 . . . **CUBES (MATHEMATICS)**
RT BLOCKS

CUBIC EQUATIONS
GS ALGEBRA
 . NONLINEAR EQUATIONS
 . . **CUBIC EQUATIONS**
 ANALYSIS (MATHEMATICS)
 . REAL VARIABLES
 . NONLINEAR EQUATIONS
 . . . **CUBIC EQUATIONS**
 FIELD THEORY (ALGEBRA)
 . **CUBIC EQUATIONS**
RT ∞EQUATIONS
  POLYNOMIALS

CUBIC LATTICES
GS CRYSTAL LATTICES
 . **CUBIC LATTICES**
 . . FACE CENTERED CUBIC LATTICES

CUES
RT AUDITORY SIGNALS
  VISUAL SIGNALS

CUESTAS
USE RIDGES

CUFFS
RT CLOTHING
  SEALS (STOPPERS)

CULTIVATION
GS **CULTIVATION**
 . PLOWING
RT AMMONIA
  AMMONIUM NITRATES
  ASHES
  CULTURE TECHNIQUES
  FERTILIZERS
  PLANTING
  SILVICULTURE
  SOILS
  TISSUES (BIOLOGY)

CULTURAL RESOURCES
RT ARCHAEOLOGY
  HUMAN BEINGS

CULTURE (SOCIAL SCIENCES)
RT AMERICAN INDIANS
  ANTHROPOLOGY
  ARTIFACTS
  ESKIMOS
  ETHNIC FACTORS
  GOVERNMENTS
  MINORITIES
  POLITICS
  RACE FACTORS
  RACES
  REGIMES
  SOCIAL FACTORS
  SOCIOLOGY

CULTURE TECHNIQUES
RT CULTIVATION
  MICROBIOLOGY

CUMULATIVE DAMAGE
GS DAMAGE
 . **CUMULATIVE DAMAGE**
RT COMPONENT RELIABILITY
  DEFECTS
  DEGRADATION
  DURABILITY
  FAILURE
  OPERATIONAL HAZARDS
  RELIABILITY
  STRUCTURAL RELIABILITY
  WEAR TESTS

CUMULONIMBUS CLOUDS
GS CLOUDS
 . CLOUDS (METEOROLOGY)
 . . CONVECTION CLOUDS
 . . . **CUMULONIMBUS CLOUDS**
 . . . . ARC CLOUDS
RT CUMULUS CLOUDS
  NIMBOSTRATUS CLOUDS
  PRECIPITATION (METEOROLOGY)
  THUNDERSTORMS
  TORNADOES

CUMULUS CLOUDS
GS CLOUDS
 . CLOUDS (METEOROLOGY)
 . . **CUMULUS CLOUDS**
RT CUMULONIMBUS CLOUDS
  STRATOCUMULUS CLOUDS

∞ **CUPOLAS**
SN *(USE OF A MORE SPECIFIC TERM IS
  RECOMMENDED--CONSULT THE TERMS
  LISTED BELOW)*
RT DOMES (STRUCTURAL FORMS)
  FURNACES
  GUN TURRETS

CURARE
GS POISONS
 . **CURARE**
RT ALKALOIDS
  ANTICHOLINERGICS
  ∞POISONING
  TOXICOLOGY

CURES
RT DISEASES
  DRUGS
  FIRST AID
  HEALING
  THERAPY

CURIE TEMPERATURE
GS MAGNETIC PROPERTIES
 . **CURIE TEMPERATURE**
 TEMPERATURE
 . **CURIE TEMPERATURE**
RT DIAMAGNETISM
  ELECTRETS
  FERROELECTRICITY
  FERROMAGNETISM
  ULTRALOW TEMPERATURES

CURIE-WEISS LAW
RT FERROMAGNETISM
  MAGNETIC PERMEABILITY
  MAGNETIC PROPERTIES
  PARAMAGNETISM

CURING
RT ALFALFA
  AUTOCLAVING
  CITRUS TREES
  CORN
  CROSSLINKING
  DEGRADATION
  DRYING
  FARM CROPS
  OATS
  ORCHARDS
  PRESERVING
 ∞SETTING
  VULCANIZING
  WEATHERING

CURIUM
GS CHEMICAL ELEMENTS
 . ACTINIDE SERIES
 . . TRANSURANIUM ELEMENTS
 . . . **CURIUM**
 . . . . CURIUM ISOTOPES
 . . . . . CURIUM 242
 . . . . . CURIUM 244
 . NUCLIDES
 . . ISOTOPES
 . . . RADIOACTIVE ISOTOPES
 . . . . TRANSURANIUM ELEMENTS
 . . . . . **CURIUM**
 . . . . . . CURIUM ISOTOPES
 HEAVY ELEMENTS
 . TRANSURANIUM ELEMENTS
 . . **CURIUM**
 . . . CURIUM ISOTOPES
 . . . . CURIUM 242
 . . . . CURIUM 244

CURIUM-*(CONT.)*
 METALS
 . ACTINIDE SERIES
 . . TRANSURANIUM ELEMENTS
 . . . **CURIUM**
 . . . . CURIUM ISOTOPES
 . . . . . CURIUM 242
 . . . . . CURIUM 244

CURIUM COMPOUNDS
GS ACTINIDE SERIES COMPOUNDS
 . **CURIUM COMPOUNDS**
RT ∞CHEMICAL COMPOUNDS
  ∞GROUP 3B COMPOUNDS

CURIUM ISOTOPES
GS CHEMICAL ELEMENTS
 . ACTINIDE SERIES
 . . TRANSURANIUM ELEMENTS
 . . . CURIUM
 . . . . **CURIUM ISOTOPES**
 . . . . . CURIUM 242
 . . . . . CURIUM 244
 . NUCLIDES
 . ISOTOPES
 . . RADIOACTIVE ISOTOPES
 . . . TRANSURANIUM ELEMENTS
 . . . . CURIUM
 . . . . . **CURIUM ISOTOPES**
 HEAVY ELEMENTS
 . TRANSURANIUM ELEMENTS
 . . CURIUM
 . . . **CURIUM ISOTOPES**
 . . . . CURIUM 242
 . . . CURIUM 244
 METALS
 . ACTINIDE SERIES
 . . TRANSURANIUM ELEMENTS
 . . . CURIUM
 . . . . **CURIUM ISOTOPES**
 . . . . . CURIUM 242
 . . . . . CURIUM 244

CURIUM 242
GS CHEMICAL ELEMENTS
 . ACTINIDE SERIES
 . . TRANSURANIUM ELEMENTS
 . . . CURIUM
 . . . . CURIUM ISOTOPES
 . . . . . **CURIUM 242**
 . NUCLIDES
 . . ISOTOPES
 . . . RADIOACTIVE ISOTOPES
 . . . . TRANSURANIUM ELEMENTS
 . . . . . CURIUM
 . . . . . . CURIUM ISOTOPES
 HEAVY ELEMENTS
 . TRANSURANIUM ELEMENTS
 . . CURIUM
 . . . CURIUM ISOTOPES
 . . . . **CURIUM 242**
 METALS
 . ACTINIDE SERIES
 . . TRANSURANIUM ELEMENTS
 . . . CURIUM
 . . . . CURIUM ISOTOPES
 . . . . . **CURIUM 242**

CURIUM 244
GS CHEMICAL ELEMENTS
 . ACTINIDE SERIES
 . . TRANSURANIUM ELEMENTS
 . . . CURIUM
 . . . . CURIUM ISOTOPES
 . . . . . **CURIUM 244**
 . NUCLIDES
 . ISOTOPES
 . . . RADIOACTIVE ISCTOPES
 . . . . TRANSURANIUM ELEMENTS
 . . . . . CURIUM
 . . . . . . CURIUM ISOTOPES
 HEAVY ELEMENTS
 . TRANSURANIUM ELEMENTS
 . . CURIUM
 . . . CURIUM ISOTOPES
 . . . . **CURIUM 244**
 METALS
 . ACTINIDE SERIES
 . . TRANSURANIUM ELEMENTS
 . . . CURIUM
 . . . . CURIUM ISOTOPES
 . . . . . **CURIUM 244**

∞ **CURL**
  SN   *(USE OF A MORE SPECIFIC TERM IS*
        RECOMMENDED--CONSULT THE TERMS
        LISTED BELOW)
  RT   CURL (MATERIALS)
        CURL (VECTORS)

CURL (MATERIALS)
  HI   ∞CURL
        DIMENSIONAL STABILITY
        FOLDING
        ∞MATERIALS
        TEXTURES

CURL (VECTORS)
  GS   ANALYSIS (MATHEMATICS)
        . CALCULUS
        . . VECTOR ANALYSIS
        . . . **CURL (VECTORS)**
        . . . . VORTICITY
        . REAL VARIABLES
        . . VECTOR ANALYSIS
        . . . **CURL (VECTORS)**
        . . . . VORTICITY
        GEOMETRY
        . VECTOR ANALYSIS
        . . **CURL (VECTORS)**
        . . . VORTICITY
  RT   ∞CURL

CURRENT ALGEBRA
  GS   ALGEBRA
        . **CURRENT ALGEBRA**
  RT   ∞MATHEMATICS
        NUCLEAR PHYSICS
        VECTOR CURRENTS

CURRENT AMPLIFIERS
  GS   AMPLIFIERS
        . **CURRENT AMPLIFIERS**
        . . PHOTOMULTIPLIER TUBES
        . . . FREQUENCY MODULATION
              PHOTOMULTIPLIERS
  RT   TRANSISTOR AMPLIFIERS
        VOLTAGE AMPLIFIERS

CURRENT CONVERTERS (AC TO DC)
  RT   ALTERNATING CURRENT
        ∞CONVERTERS
        CRYSTAL RECTIFIERS
        DIRECT CURRENT
        ELECTRIC CURRENT
        INVERTED CONVERTERS (DC TO AC)
        RECTIFIERS
        SILICON CONTROLLED RECTIFIERS
        THYRATRONS

CURRENT DENSITY
  GS   RATES (PER TIME)
        . FLUX DENSITY
        . . **CURRENT DENSITY**
  RT   ELECTRIC CURRENT
        ELECTROLYSIS
        ELECTROPLATING
        PINNING

CURRENT DISTRIBUTION
  GS   DISTRIBUTION (PROPERTY)
        . **CURRENT DISTRIBUTION**
  RT   CHARGE DISTRIBUTION
        ELECTRON DISTRIBUTION
        ∞HOLE DISTRIBUTION
        HOLE DISTRIBUTION (ELECTRONICS)
        ION DISTRIBUTION
        MAGNETIC ANNULAR ARC
        NEUTRAL CURRENTS

CURRENT REGULATORS
  UF   CURRENT STABILIZERS
  GS   REGULATORS
        . **CURRENT REGULATORS**
  RT   CIRCUIT PROTECTION
        CONTROLLERS
        ELECTRIC CURRENT
        ∞ELECTRIC EQUIPMENT
        ELECTRIC SWITCHES
        ELECTRONIC CONTROL
        LIMITER CIRCUITS
        POWER FACTOR CONTROLLERS
        POWER SUPPLY CIRCUITS
        SWITCHING CIRCUITS
        TRANSMISSION LOSS
        VOLTAGE REGULATORS

CURRENT SHEETS
  RT   ANTENNAS
        ELECTRIC CURRENT
        MAGNETIC FLUX
        ∞SHEETS

CURRENT STABILIZERS
  USE   CURRENT REGULATORS

∞ **CURRENTS**
  SN   *(USE OF A MORE SPECIFIC TERM IS*
        RECOMMENDED--CONSULT THE TERMS
        LISTED BELOW)
  RT   AIR CURRENTS
        AIR FLOW
        BEAM CURRENTS
        CIRCULATION
        COASTAL CURRENTS
        ELECTRIC CURRENT
        EXTERNAL SURFACE CURRENTS
        FLUID FLOW
        OCEAN CURRENTS
        WATER CURRENTS

CURRENTS (OCEANOGRAPHY)
  USE   WATER CURRENTS

CURTAINS
  RT   ∞BARRIERS
        DIVIDERS
        DOORS
        ENTRANCES
        OPENINGS
        ∞PARTITIONS
        PARTITIONS (STRUCTURES)
        ∞SCREENS
        SEPARATORS
        WALLS
        WINDOWS (APERTURES)

CURTISS C-46 AIRCRAFT
  USE   C-46 AIRCRAFT

CURTISS-WRIGHT AIRCRAFT
  UF   CURTISS-WRIGHT MILITARY AIRCRAFT
  GS   CURTISS-WRIGHT AIRCRAFT
        . C-46 AIRCRAFT
        . X-19 AIRCRAFT
  RT   ∞AIRCRAFT

CURTISS-WRIGHT MILITARY AIRCRAFT
  USE   CURTISS-WRIGHT AIRCRAFT
        MILITARY AIRCRAFT

CURVATURE
  GS   GEOMETRY
        . **CURVATURE**
  RT   CAMBER
        ∞CURVES
        CURVES (GEOMETRY)
        DIFFERENTIAL GEOMETRY
        FLEXING
        ∞PROFILES
        SHAPES
        ZERO FORCE CURVES

CURVE FITTING
  RT   DATA COMPRESSION
        DATA SMOOTHING
        FORECASTING
        LEAST SQUARES METHOD
        MINIMAX TECHNIQUE
        SADDLE POINTS
        STATISTICAL DISTRIBUTIONS
        STATISTICAL TESTS
        TIME SERIES ANALYSIS

CURVED BEAMS
  GS   STRUCTURAL MEMBERS
        . BEAMS (SUPPORTS)
        . . **CURVED BEAMS**
  RT   CAMBER
        I BEAMS

CURVED PANELS
  GS   PANELS
        . **CURVED PANELS**
  RT   CONTOURS
        SHAPES
        WING PANELS

CURVED SURFACES
  USE   CONTOURS
        SHAPES
        SURFACES

∞ **CURVES**
  SN   *(USE OF A MORE SPECIFIC TERM IS*
        RECOMMENDED--CONSULT THE TERMS
        LISTED BELOW)
  RT   CURVATURE
        GRAPHS (CHARTS)
        LEARNING CURVES
        LIGHT CURVE
        TOROIDS
        TRAJECTORIES
        ZERO FORCE CURVES

CURVES (GEOMETRY)
  UF   HELIXES
  GS   GEOMETRY
        . **CURVES (GEOMETRY)**
        . . CATENARIES
        . . CYCLOIDS
        . . EPICYCLOIDS
        . . S CURVES
        . . . GOMPERTZ CURVES
  RT   ANALYTIC GEOMETRY
        ∞ARCS
        CHORDS (GEOMETRY)
        CIRCLES (GEOMETRY)
        CURVATURE
        CUSPS (MATHEMATICS)
        DIFFERENTIAL GEOMETRY
        EUCLIDEAN GEOMETRY
        GEODESIC LINES
        HOMOTOPY THEORY
        INFLECTION POINTS
        LINE SHAPE
        MANIFOLDS (MATHEMATICS)
        MENISCI
        SEGMENTS
        ∞SPIRALS

CURVILINEAR COORDINATES
  USE   SPHERICAL COORDINATES

CUSHIONCRAFT GROUND EFFECT MACHINE
  GS   GROUND EFFECT MACHINES
        . **CUSHIONCRAFT GROUND EFFECT**
          MACHINE
  RT   HOVERING
        VERTICAL TAKEOFF AIRCRAFT

CUSHIONS
  RT   AIR CUSHION LANDING SYSTEMS
        BUMPERS
        COUCHES
        GROUND EFFECT (AERODYNAMICS)
        HYDRAULIC EQUIPMENT
        ∞PAD
        PILLOWS
        PNEUMATIC EQUIPMENT
        SEATS
        SHOCK ABSORBERS
        VIBRATION ISOLATORS

∞ **CUSPS**
  SN   *(USE OF A MORE SPECIFIC TERM IS*
        RECOMMENDED--CONSULT THE TERMS
        LISTED BELOW)
  RT   CUSPS (LANDFORMS)
        CUSPS (MATHEMATICS)
        DOUBLE CUSPS
        POLAR CUSPS

CUSPS (LANDFORMS)
  GS   LANDFORMS
        . **CUSPS (LANDFORMS)**
  RT   BEACHES
        COASTS
        ∞CUSPS
        TOPOGRAPHY

CUSPS (MATHEMATICS)
  GS   GEOMETRY
        . **CUSPS (MATHEMATICS)**
        . . DOUBLE CUSPS
  RT   CURVES (GEOMETRY)
        ∞CUSPS
        EPICYCLOIDS
        MAXIMA
        MINIMA

∞ **CUT-OFF**
   SN   *(USE OF A MORE SPECIFIC TERM IS*
        RECOMMENDED--CONSULT THE TERMS
        LISTED BELOW)
   RT   BURNOUT
        ENGINE FAILURE
        MACHINING

CUT-OUTS
   USE OPENINGS

CUTANEOUS PERCEPTION
   USE TOUCH

CUTTERS
   SN   (EXCLUDES SHIPS)
   GS   **CUTTERS**
        . BLADES (CUTTERS)
        . . RAZOR BLADES
        . DRILL BITS
        . DRILLS
        . SAWS
        . SHEARS
   RT   CUTTING
        DIES
        LASER CUTTING
        MACHINE TOOLS
        SCRAPERS
        TAPS
        TOOLS

CUTTING
   GS   **CUTTING**
        . BLANKING (CUTTING)
        . LASER CUTTING
        . METAL CUTTING
        . MILLING (MACHINING)
        . PLANING
        . SCARFING
        . SHEARING
        . SLICING
        . SPARK MACHINING
   RT   ABRASION
        CHIPPING
        COMMINUTION
        CUTTERS
        DRILLING
        FLAKING
        FORMING TECHNIQUES
        FRACTURING
        GRINDING (MATERIAL REMOVAL)
        GROOVING
        MACHINING
        PEELING
        PERFORATING
        PIERCING
        ∞SEPARATION
        SHREDDING
        SPLITTING
        TORCHES

CV-2 AIRCRAFT
   USE DHC 4 AIRCRAFT

CV-7 AIRCRAFT
   USE DHC 5 AIRCRAFT

CV-340 AIRCRAFT
   UF   CONVAIR 340 AIRCRAFT
   GS   COMMERCIAL AIRCRAFT
        . **CV-340 AIRCRAFT**
        GENERAL DYNAMICS AIRCRAFT
        . **CV-340 AIRCRAFT**
        MONOPLANES
        . **CV-340 AIRCRAFT**
        PASSENGER AIRCRAFT
        . **CV-340 AIRCRAFT**
   RT   ∞AIRCRAFT

CV-440 AIRCRAFT
   UF   CONVAIR 440 AIRCRAFT
        METROPOLITAN AIRCRAFT
   GS   COMMERCIAL AIRCRAFT
        . **CV-440 AIRCRAFT**
        GENERAL DYNAMICS AIRCRAFT
        . **CV-440 AIRCRAFT**
        MONOPLANES
        . **CV-440 AIRCRAFT**
        PASSENGER AIRCRAFT
        . **CV-440 AIRCRAFT**
   RT   ∞AIRCRAFT

CV-880 AIRCRAFT
   UF   CONVAIR 880 AIRCRAFT

CV-880 AIRCRAFT-*(CONT.)*
   GS   COMMERCIAL AIRCRAFT
        . **CV-880 AIRCRAFT**
        GENERAL DYNAMICS AIRCRAFT
        . **CV-880 AIRCRAFT**
        JET AIRCRAFT
        . **CV-880 AIRCRAFT**
        MONOPLANES
        . **CV-880 AIRCRAFT**
        PASSENGER AIRCRAFT
        . **CV-880 AIRCRAFT**
        TRANSPORT AIRCRAFT
        . **CV-880 AIRCRAFT**
   RT   ∞AIRCRAFT

CV-990 AIRCRAFT
   UF   CONVAIR 990 AIRCRAFT
   GS   COMMERCIAL AIRCRAFT
        . **CV-990 AIRCRAFT**
        GENERAL DYNAMICS AIRCRAFT
        . **CV-990 AIRCRAFT**
        JET AIRCRAFT
        . TURBOFAN AIRCRAFT
        . . **CV-990 AIRCRAFT**
        MONOPLANES
        . **CV-990 AIRCRAFT**
        PASSENGER AIRCRAFT
        . **CV-990 AIRCRAFT**
   RT   ∞AIRCRAFT

CW RADAR
   USE CONTINUOUS WAVE RADAR

CYANAMIDES
   GS   ALIPHATIC COMPOUNDS
        . **CYANAMIDES**
        NITROGEN COMPOUNDS
        . AMIDES
        . . **CYANAMIDES**
        . CYANO COMPOUNDS
        . . **CYANAMIDES**

CYANATES
   RT   ESTERS
        URETHANES

CYANIDE EMISSION
   USE CN EMISSION

CYANIDES
   GS   **CYANIDES**
        . CYANOGEN
        . IRON CYANIDES
        . MALONONITRILE
   RT   ALIPHATIC COMPOUNDS
        CYANO COMPOUNDS
        NITROGEN COMPOUNDS

CYANO COMPOUNDS
   GS   NITROGEN COMPOUNDS
        . **CYANO COMPOUNDS**
        . . CYANAMIDES
        . . ISOCYANATES
        . . . DIISOCYANATES
        . . . FULMINATES
   RT   ∞CHEMICAL COMPOUNDS
        CYANIDES
        NITRILES

CYANOCOBALAMIN
   UF   VITAMIN B 12
   GS   HETEROCYCLIC COMPOUNDS
        . **CYANOCOBALAMIN**
        VITAMINS
        . **CYANOCOBALAMIN**

CYANOGEN
   GS   ALIPHATIC COMPOUNDS
        . **CYANOGEN**
        CYANIDES
        . **CYANOGEN**

CYANOPHYTA
   USE BLUE GREEN ALGAE

CYANOSIS
   GS   DISEASES
        . **CYANOSIS**
   RT   BLOOD CIRCULATION
        HEART FUNCTION

CYANURATES
   GS   ESTERS
        . **CYANURATES**

CYANURATES-*(CONT.)*
        HETEROCYCLIC COMPOUNDS
        . AZINES
        . . **CYANURATES**
        PYRAZINES
        . AZINES
        . . **CYANURATES**

CYANURIC ACID
   GS   ACIDS
        . **CYANURIC ACID**
        ALIPHATIC COMPOUNDS
        . TRIOLS
        . . **CYANURIC ACID**
        HETEROCYCLIC COMPOUNDS
        . AZINES
        . . **CYANURIC ACID**
        HYDROXYL COMPOUNDS
        . ALCOHOLS
        . TRIOLS
        . . . **CYANURIC ACID**
        PYRAZINES
        . AZINES
        . . **CYANURIC ACID**

CYBER 74 COMPUTER
   USE CDC CYBER 74 COMPUTER

CYBERNETICS
   RT   ADAPTIVE CONTROL
        AUTOMATA THEORY
        ∞AUTOMATION
        BIONICS
        COMMUNICATION THEORY
        COMPUTERS
        ∞CONTROL
        CONTROL SYSTEMS DESIGN
        CONTROLLERS
        DEPERSONALIZATION
        FEEDBACK
        HUMAN FACTORS ENGINEERING
        INFORMATION THEORY
        LEARNING MACHINES
        MAN MACHINE SYSTEMS
        MANAGEMENT
        NEURAL NETS
        PSYCHOLOGY
        ∞SYSTEMS
        SYSTEMS ENGINEERING

CYCLES
   UF   CYCLING
        PERIODIC PROCESSES
   GS   **CYCLES**
        . ACTIVITY CYCLES (BIOLOGY)
        . CARBON CYCLE
        . SOLAR CYCLES
        . . SUNSPOT CYCLE
        . STRESS CYCLES
        . THERMODYNAMIC CYCLES
        . . BRAYTON CYCLE
        . . CARNOT CYCLE
        . . OTTO CYCLE
        . . RANKINE CYCLE
        . . STIRLING CYCLE
        . WORK-REST CYCLE
   RT   ALTERNATIONS
        AMPLITUDES
        ANNUAL VARIATIONS
        CYCLIC LOADS
        DIURNAL VARIATIONS
        FATIGUE (MATERIALS)
        FREQUENCY DISTRIBUTION
        HARMONICS
        INTERMITTENCY
        LONG TERM EFFECTS
        PERIODIC VARIATIONS
        ∞PHASES
        RECIPROCATION
        RHYTHM (BIOLOGY)
        STARTING
        SUPERHARMONICS

CYCLIC ACCELERATORS
   GS   PARTICLE ACCELERATORS
        . **CYCLIC ACCELERATORS**
        . . BETATRONS
        . . SYNCHROCYCLOTRONS
        . . SYNCHROTRONS
        . . . BEVATRON
        . . . STORAGE RINGS (PARTICLE
              ACCELERATORS)
   RT   ∞ACCELERATORS

CYCLIC ADENOSINE MONOPHOSPHATE
- USE   CYCLIC AMP

CYCLIC AMP
- UF   CYCLIC ADENOSINE MONOPHOSPHATE
- GS   HETEROCYCLIC COMPOUNDS
  - . ADENOSINES
  - . . **CYCLIC AMP**
  - ORGANIC COMPOUNDS
  - . CYCLIC COMPOUNDS
  - . . **CYCLIC AMP**
  - . NUCLEOTIDES
  - . . ADENOSINES
  - . . . **CYCLIC AMP**
  - PHOSPHORUS COMPOUNDS
  - . PHOSPHATES
  - . . **CYCLIC AMP**
  - PROTEINS
  - . NUCLEOTIDES
  - . . ADENOSINES
  - . . . **CYCLIC AMP**
- RT   ADRENERGICS
  - ALKYNES
  - AMINO ACIDS
  - CHOLINERGICS
  - GUANINES
  - PHARMACOLOGY

CYCLIC COMPOUNDS
- GS   ORGANIC COMPOUNDS
  - . **CYCLIC COMPOUNDS**
  - . . CYCLIC AMP
- RT   ∞CHEMICAL COMPOUNDS
  - ORGANIC CHEMISTRY

CYCLIC HYDROCARBONS
- GS   ALIPHATIC COMPOUNDS
  - . **CYCLIC HYDROCARBONS**
  - . . ANTHRACENE
  - . . COLCHICINE
  - . . CYCLOBUTANE
  - . . CYCLOPROPANE
  - . . MENTHOL
  - . . NAPHTHENES
  - HYDROCARBONS
  - . **CYCLIC HYDROCARBONS**
  - . . ANTHRACENE
  - . . COLCHICINE
  - . . CYCLOBUTANE
  - . . CYCLOPROPANE
  - . . MENTHOL
  - . . NAPHTHENES
- RT   ALKYNES

CYCLIC LOADS
- SN   (LIMITED TO FORCE LOADS)
- GS   LOADS (FORCES)
  - . DYNAMIC LOADS
  - . . **CYCLIC LOADS**
- RT   CYCLES
  - ELBER EQUATION
  - INELASTIC STRESS
  - S-N DIAGRAMS
  - STRESS CYCLES
  - STRUCTURAL DESIGN CRITERIA
  - TRANSIENT LOADS
  - VIBRATION
  - VIBRATORY LOADS

CYCLING
- USE   CYCLES

CYCLOBUTANE
- GS   ALIPHATIC COMPOUNDS
  - . CYCLIC HYDROCARBONS
  - . . **CYCLOBUTANE**
  - HYDROCARBONS
  - . CYCLIC HYDROCARBONS
  - . . **CYCLOBUTANE**

CYCLOGENESIS
- GS   STORMS
  - . CYCLONES
  - . . **CYCLOGENESIS**
  - WIND (METEOROLOGY)
  - . **CYCLOGENESIS**

CYCLOHEXANE
- GS   HYDROCARBONS
  - . **CYCLOHEXANE**
- RT   BENZENE
  - HEXENES
  - HYDROGENATION

CYCLOIDS
- GS   GEOMETRY
  - . CURVES (GEOMETRY)
  - . . **CYCLOIDS**
  - . EUCLIDEAN GEOMETRY
  - . . ANALYTIC GEOMETRY
  - . . . **CYCLOIDS**

CYCLONES
- SN   (METEOROLOGICAL--EXCLUDES EQUIPMENT)
- GS   STORMS
  - . **CYCLONES**
  - . . CYCLOGENESIS
  - . . HURRICANES
  - . . . ANNA HURRICANE
  - . . TYPHOONS
- RT   ANTICYCLONES
  - ATMOSPHERIC PRESSURE
  - BAROCLINIC WAVES
  - GROUND WIND
  - LOW PRESSURE
  - METEOROLOGY
  - PRECIPITATION (METEOROLOGY)
  - STORM DAMAGE
  - SYNOPTIC METEOROLOGY
  - TORNADOES
  - TROPICAL STORMS
  - WIND (METEOROLOGY)

CYCLONES (EQUIPMENT)
- USE   CENTRIFUGES

CYCLOPROPANE
- GS   ALIPHATIC COMPOUNDS
  - . CYCLIC HYDROCARBONS
  - . . **CYCLOPROPANE**
  - DRUGS
  - . ANESTHETICS
  - . . **CYCLOPROPANE**
  - HYDROCARBONS
  - . CYCLIC HYDROCARBONS
  - . . **CYCLOPROPANE**
- RT   PROPANE

CYCLOPS PLASMA ACCELERATOR
- GS   PLASMA ACCELERATORS
  - . **CYCLOPS PLASMA ACCELERATOR**
- RT   ∞ACCELERATORS
  - PLASMAS (PHYSICS)

CYCLOTETRAMETHYLENE TETRANITRAMINE
- USE   HMX

CYCLOTRIMETHYLENE TRINITRAMINE
- USE   RDX

CYCLOTRON FREQUENCY
- GS   FREQUENCIES
  - . **CYCLOTRON FREQUENCY**
- RT   CHARGED PARTICLES
  - LARMOR PRECESSION

CYCLOTRON RADIATION
- GS   ELECTROMAGNETIC RADIATION
  - . **CYCLOTRON RADIATION**
  - PARTICLES
  - . CORPUSCULAR RADIATION
  - . . **CYCLOTRON RADIATION**
  - . . . ION CYCLOTRON RADIATION
- RT   CHARGED PARTICLES
  - LARMOR PRECESSION
  - LARMOR RADIUS
  - ∞RADIATION

CYCLOTRON RESONANCE
- GS   RESONANCE
  - . **CYCLOTRON RESONANCE**
- RT   CHARGED PARTICLES
  - DIAMAGNETISM
  - ENERGY TRANSFER
  - FERMI SURFACES
  - ION CYCLOTRON RADIATION
  - PLASMA RESONANCE

CYCLOTRON RESONANCE DEVICES
- UF   GYROTRONS
- GS   MEASURING INSTRUMENTS
  - . RESONANCE PROBES
  - . . **CYCLOTRON RESONANCE DEVICES**
- RT   CAVITY RESONATORS
  - ∞DEVICES
  - KLYSTRONS
  - MILLIMETER WAVES

CYCLOTRON RESONANCE DEVICES-_(CONT.)_
  - POWER AMPLIFIERS
  - TRAVELING WAVE TUBES

CYCLOTRONS
- UF   CALUTRONS
- GS   PARTICLE ACCELERATORS
  - . **CYCLOTRONS**
  - . . GEOCYCLOTRONS
  - . . MICROTRONS
  - . . OAK RIDGE ISOCHRONOUS CYCLOTRON
  - . . OMEGATRONS
  - . . SYNCHROCYCLOTRONS
- RT   SYNCHROTRONS

CYGNUS CONSTELLATION
- GS   CONSTELLATIONS
  - . **CYGNUS CONSTELLATION**
- RT   CELESTIAL BODIES
  - CELESTIAL SPHERE
  - STARS

∞ **CYLINDERS**
- SN   _(USE OF A MORE SPECIFIC TERM IS RECOMMENDED--CONSULT THE TERMS LISTED BELOW)_
- RT   ANALYTIC GEOMETRY
  - CIRCULAR CYLINDERS
  - CONCENTRIC CYLINDERS
  - CYLINDRICAL BODIES
  - CYLINDRICAL CHAMBERS
  - CYLINDRICAL SHELLS
  - ∞DRUMS
  - DRUMS (CONTAINERS)
  - ELASTIC CYLINDERS
  - ELLIPTICAL CYLINDERS
  - HEMISPHERE CYLINDER BODIES
  - MONOCOQUE STRUCTURES
  - ORTHOTROPIC CYLINDERS
  - OSCILLATING CYLINDERS
  - PLASMA CYLINDERS
  - ROTATING CYLINDERS
  - VISCOELASTIC CYLINDERS

CYLINDRICAL AFTERBODIES
- USE   AFTERBODIES
  - CYLINDRICAL BODIES

CYLINDRICAL ANTENNAS
- GS   ANTENNAS
  - . **CYLINDRICAL ANTENNAS**
- RT   ANTENNA RADIATION PATTERNS
  - RADIO EQUIPMENT
  - SATELLITES

CYLINDRICAL BODIES
- UF   CYLINDRICAL AFTERBODIES
  - CYLINDROIDS
- GS   SYMMETRICAL BODIES
  - . BODIES OF REVOLUTION
  - . . **CYLINDRICAL BODIES**
  - . . . ROTATING CYLINDERS
- RT   AFTERBODIES
  - AIRY FUNCTION
  - CENTERBODIES
  - CIRCULAR CYLINDERS
  - ∞CYLINDERS
  - ELASTIC CYLINDERS
  - ELLIPTICAL CYLINDERS
  - FOREBODIES
  - FUSELAGES
  - HEMISPHERE CYLINDER BODIES
  - ORTHOTROPIC CYLINDERS
  - OSCILLATING CYLINDERS
  - PLASMA CYLINDERS
  - ROLLERS
  - VISCOELASTIC CYLINDERS

CYLINDRICAL CHAMBERS
- RT   BRAKES (FOR ARRESTING MOTION)
  - ∞CHAMBERS
  - ∞CYLINDERS

† **CYLINDRICAL COORDINATES**
- USE   CARTESIAN COORDINATES

CYLINDRICAL PLASMAS
- GS   FLUID MECHANICS
  - . FLUID DYNAMICS
  - . . HYDRODYNAMICS
  - . . . MAGNETOHYDRODYNAMICS
  - . . . . **CYLINDRICAL PLASMAS**
  - PARTICLES

CYLINDRICAL PLASMAS-*(CONT.)*
. . . CHARGED PARTICLES
. . . ENERGETIC PARTICLES
. . . . PLASMAS (PHYSICS)
. . . . **CYLINDRICAL PLASMAS**
. . PLASMA CYLINDERS
. **CYLINDRICAL PLASMAS**
RT PINCH EFFECT

CYLINDRICAL SHELLS
GS SHELLS (STRUCTURAL FORMS)
. **CYLINDRICAL SHELLS**
RT CIRCULAR CYLINDERS
CIRCULAR SHELLS
CIRCULAR TUBES
CONCENTRIC CYLINDERS
∞ CYLINDERS
ELASTIC CYLINDERS
ELLIPTICAL CYLINDERS
METAL SHELLS
ORTHOTROPIC CYLINDERS
ORTHOTROPIC SHELLS
OSCILLATING CYLINDERS
PLASMA CYLINDERS
REINFORCED SHELLS
ROTATING CYLINDERS
THIN WALLED SHELLS
VISCOELASTIC CYLINDERS

CYLINDRICAL TANKS
GS TANKS (CONTAINERS)
. **CYLINDRICAL TANKS**
RT FUEL TANKS
PROPELLANT TANKS
STORAGE TANKS

CYLINDRICAL WAVES
RT AXISYMMETRIC FLOW
ELASTIC WAVES
ELECTROMAGNETIC RADIATION
PLANE WAVES
SPHERICAL WAVES
∞ WAVES

CYLINDROIDS
USE CYLINDRICAL BODIES

CYPRUS
GS LANDFORMS
. ISLANDS
. . **CYPRUS**
NATIONS
. **CYPRUS**
RT GREECE

CYRILLID METEOROIDS
GS CELESTIAL BODIES
. METEOROID SHOWERS
. . **CYRILLID METEOROIDS**
. METEOROIDS
. . BOLIDES
. . . **CYRILLID METEOROIDS**
RT NATURAL SATELLITES
TEKTITES

CYSTEAMINE
GS AMINES
. **CYSTEAMINE**
ENZYMES
. COENZYMES
. . **CYSTEAMINE**
RT AMINO ACIDS
PROTEINS

CYSTEINE
GS ACIDS
. AMINO ACIDS
. . **CYSTEINE**
ALIPHATIC COMPOUNDS
. THIOLS
. . **CYSTEINE**
DRUGS
. **CYSTEINE**
ORGANIC COMPOUNDS
. AMINO ACIDS
. . **CYSTEINE**
SULFUR COMPOUNDS
. THIOLS
. . **CYSTEINE**
RT PROTEINS

CYSTIC FIBROSIS
GS DISEASES
. FIBROSIS

CYSTIC FIBROSIS-*(CONT.)*
. . CYSTIC FIBROSIS
RT TISSUES (BIOLOGY)

CYSTS
GS **CYSTS**
. MUCOCELES
RT NEOPLASMS
TISSUES (BIOLOGY)
TUMORS

CYTIDYLIC ACID
GS ACIDS
. **CYTIDYLIC ACID**
HETEROCYCLIC COMPOUNDS
. **CYTIDYLIC ACID**

CYTOCHROMES
GS PIGMENTS
. **CYTOCHROMES**
RT CYTOGENESIS
SKIN (ANATOMY)

CYTOGENESIS
RT ∞ BIOLOGY
CELLS (BIOLOGY)
CYTOCHROMES
CYTOLOGY
CYTOPLASM
DIFFERENTIATION (BIOLOGY)
GENETICS
HEREDITY

CYTOLOGY
RT BIOCHEMISTRY
∞ BIOLOGY
CELLS (BIOLOGY)
CHROMOSOMES
CYTOGENESIS
MITOCHONDRIA
MITOSIS
PLASMOLYSIS

CYTOPLASM
RT CHLOROPLASTS
CYTOGENESIS
EOSINOPHILS
FIBROBLASTS
MITOSIS

CZECHOSLOVAKIA
GS NATIONS
. **CZECHOSLOVAKIA**
RT CENTRAL EUROPE
CZECHOSLOVAKIAN SPACECRAFT
EUROPE

CZECHOSLOVAKIAN SPACECRAFT
RT CZECHOSLOVAKIA
∞ SPACECRAFT

CZOCHRALSKI METHOD
GS GROWTH
. CRYSTAL GROWTH
. . **CZOCHRALSKI METHOD**
RT ∞ METHODOLOGY
VERNEUIL PROCESS

D

D LAYER
USE D REGION

D LINES
GS SPECTRA
. RADIATION SPECTRA
. . ELECTROMAGNETIC SPECTRA
. . . LINE SPECTRA
. . . . **D LINES**
RT ABSORPTION SPECTRA
EMISSION SPECTRA
H LINES
SOLAR SPECTRA

D REGION
UF D LAYER
GS EARTH ATMOSPHERE
. UPPER ATMOSPHERE
. . IONOSPHERE
. . . LOWER IONOSPHERE
. . . . **D REGION**

D REGION-*(CONT.)*
REGIONS
. **D REGION**

D-1 SATELLITE
GS SATELLITES
. ARTIFICIAL SATELLITES
. . FRENCH SATELLITES
. . . **D-1 SATELLITE**

D-2 SATELLITES
UF D-2B SATELLITE
POLAIRE SATELLITE
TOURNESOLE SATELLITE
GS SATELLITES
. ARTIFICIAL SATELLITES
. . FRENCH SATELLITES
. . . **D-2 SATELLITES**
. . METEOROLOGICAL SATELLITES
. . . **D-2 SATELLITES**

D-2B SATELLITE
USE D-2 SATELLITES

D-558 AIRCRAFT
UF DOUGLAS D-558 AIRCRAFT
SKYROCKET AIRCRAFT
SKYSTREAK AIRCRAFT
GS JET AIRCRAFT
. **D-558 AIRCRAFT**
MCDONNELL DOUGLAS AIRCRAFT
. DOUGLAS AIRCRAFT
. . **D-558 AIRCRAFT**
MONOPLANES
. **D-558 AIRCRAFT**
RESEARCH AIRCRAFT
. **D-558 AIRCRAFT**
SUPERSONIC AIRCRAFT
. **D-558 AIRCRAFT**
RT ∞ AIRCRAFT

DACRON (TRADEMARK)
GS FABRICS
. **DACRON (TRADEMARK)**
FIBERS
. SYNTHETIC FIBERS
. . **DACRON (TRADEMARK)**
RT POLYESTER RESINS
REINFORCING FIBERS

DAD EXPLORER
USE DUAL AIR DENSITY EXPLORER

DAEMO (DATA ANALYSIS)
USE DATA PROCESSING
DATA REDUCTION
DATA TRANSMISSION

DAHOMEY
USE BENIN

DAKOTA AIRCRAFT
USE C-47 AIRCRAFT

DALTON LAW
RT GAS COMPOSITION
GAS DYNAMICS
GAS-GAS INTERACTIONS
IDEAL GAS
PARTIAL PRESSURE
VAPOR PRESSURE

DAMA
USE DEMAND ASSIGNMENT MULTIPLE
ACCESS

DAMAGE
GS **DAMAGE**
. CUMULATIVE DAMAGE
. EARTHQUAKE DAMAGE
. FIRE DAMAGE
. FLOOD DAMAGE
. FROST DAMAGE
. IMPACT DAMAGE
. . METEORITIC DAMAGE
. . RAIN IMPACT DAMAGE
. PROTON DAMAGE
. RADIATION DAMAGE
. . LASER DAMAGE
. STORM DAMAGE
RT BURNTHROUGH (FAILURE)
CORROSION
DAMAGE ASSESSMENT
DECAY

DAMAGE-(CONT.)

DECOMPOSITION
DEFECTS
DEFORMATION
DEGRADATION
DESTRUCTION
DETERIORATION
DISCOLORATION
DISINTEGRATION
DURABILITY
FATIGUE (BIOLOGY)
FATIGUE (MATERIALS)
FRACTURES (MATERIALS)
HOT CORROSION
IMMOBILIZATION
IMPAIRMENT
INJURIES
LETHALITY
LOSSES
RADIATION EFFECTS
SABOTAGE
WARPAGE
WEAR
WEATHERING

DAMAGE ASSESSMENT
GS   ASSESSMENTS
     . **DAMAGE ASSESSMENT**
RT   CASTS
     COSTS
     DAMAGE
     ESTIMATES
     MAINTENANCE
     RECOVERABILITY
     REPLACING
     SPARE PARTS
     VALUE

DAMAGE THRESHOLD
USE   YIELD POINT

DAMKOHLER NUMBER
RT   ACTIVATION ENERGY
     COMBUSTION PHYSICS
     DIFFUSION FLAMES
     FLAME PROPAGATION
     ∞ NUMBERS
     REACTION KINETICS

DAMP PROGRAM
USE   DOWNRANGE ANTIMISSILE
      MEASUREMENT PROGRAM

∞ DAMPERS
SN   *(USE OF A MORE SPECIFIC TERM IS*
     RECOMMENDED--CONSULT THE TERMS
     LISTED BELOW)
RT   BLANKETS (FISSION REACTORS)
     DAMPERS (VALVES)
     NUTATION DAMPERS
     OSCILLATION DAMPERS
     VIBRATION ISOLATORS

DAMPERS (VALVES)
GS   VALVES
     . BUTTERFLY VALVES
     .. **DAMPERS (VALVES)**
RT   AUTOMATIC CONTROL VALVES
     ∞ DAMPERS
     GAS VALVES
     VIBRATION ISOLATORS

DAMPING
UF   DAMPING FACTOR
     DAMPING IN PITCH
     DAMPING IN ROLL
     DAMPING IN YAW
     ELASTIC STABILITY
     JET DAMPING
GS   **DAMPING**
     . ELASTIC DAMPING
     .. VISCOELASTIC DAMPING
     . LANDAU DAMPING
     . VIBRATION DAMPING
     . VISCOUS DAMPING
     .. VISCOELASTIC DAMPING
RT   ∞ ABSORPTION
     ATTENUATION
     BAFFLES
     DECELERATION
     DISSIPATION
     DYNAMIC CHARACTERISTICS
     DYNAMIC RESPONSE
     DYNAMIC STABILITY
     ENERGY ABSORPTION

DAMPING-(CONT.)

GYROSCOPE FLUIDS
GYROSCOPIC PENDULUMS
GYROSCOPIC STABILITY
HYSTERESIS
IMPEDANCE
INSULATION
INTERNAL FRICTION
MECHANICAL IMPEDANCE
MUFFLERS
NEGATIVE FEEDBACK
OSCILLATIONS
∞ REDUCTION
∞ RESISTANCE
RESONANT FREQUENCIES
RESONANT VIBRATION
RETARDING
SEA KEEPING
SHOCK ABSORBERS
SILENCERS
STABILITY DERIVATIVES
STOPPING
SUBHARMONIC GENERATORS
SUPPRESSORS
TIME CONSTANT
TRANSFER FUNCTIONS
TRANSIENT OSCILLATIONS
TRANSIENT RESPONSE
VIBRATION ISOLATORS
WAVE INTERACTION

DAMPING FACTOR
USE   DAMPING

DAMPING IN PITCH
USE   DAMPING
     PITCH (INCLINATION)

DAMPING IN ROLL
USE   DAMPING
     ROLL

DAMPING IN YAW
USE   DAMPING
     YAW

DAMPING TESTS
GS   VIBRATION TESTS
     . **DAMPING TESTS**
     .. STROKING TESTS
RT   RESONANCE TESTING
     STABILITY TESTS
     ∞ TESTS
     VIBRATION MEASUREMENT

DAMPNESS
USE   MOISTURE CONTENT

DAMS
RT   ∞ BARRAGES
     ∞ BARRIERS
     FLOOD CONTROL
     HYDROELECTRICITY
     RESERVOIRS
     WHARVES

DANGER
USE   HAZARDS

DARK ADAPTATION
GS   ADAPTATION
     . RETINAL ADAPTATION
     .. **DARK ADAPTATION**
RT   DARKNESS
     NIGHT VISION
     PUPILLOMETRY
     VISION
     VISUAL PIGMENTS

DARKENING
GS   **DARKENING**
     . LIMB DARKENING
RT   DARKNESS
     ∞ ILLUMINATION
     NIGHT
     VISIBILITY

DARKNESS
RT   COLOR
     DARK ADAPTATION
     DARKENING
     DIURNAL VARIATIONS
     ILLUMINATING
     ∞ ILLUMINATION

DARKNESS-(CONT.)

LIGHT (VISIBLE RADIATION)
NIGHT
NIGHT FLIGHTS (AIRCRAFT)
OPTICAL PROPERTIES
SHADOWS

DARKROOMS
GS   ROOMS
     . **DARKROOMS**
RT   PHOTOGRAPHIC PROCESSING
     PHOTOGRAPHIC PROCESSING
      EQUIPMENT
     PHOTOGRAPHY

DART TURBOPROP ENGINES
USE   TURBOPROP ENGINES

DASH HELICOPTER
USE   QH-50 HELICOPTER

DASSAULT AIRCRAFT
GS   **DASSAULT AIRCRAFT**
     . MIRAGE AIRCRAFT
     .. MIRAGE 3 AIRCRAFT
     . MYSTERE 20 AIRCRAFT
RT   ∞ AIRCRAFT
     MYSTERE 50 AIRCRAFT

DASSAULT MIRAGE 3 AIRCRAFT
USE   MIRAGE 3 AIRCRAFT

DASSAULT MYSTERE 20 AIRCRAFT
USE   MYSTERE 20 AIRCRAFT

DASSAULT MYSTERE 50 AIRCRAFT
USE   MYSTERE 50 AIRCRAFT

DAST PROGRAM
SN   (DRONES FOR AERODYNAMIC AND
     STRUCTURAL TESTING)
UF   DRONES FOR AERODYNAMIC AND
     STRUCT TEST
GS   PROGRAMS
     . NASA PROGRAMS
     .. **DAST PROGRAM**
RT   AEROELASTICITY
     AIRCRAFT CONTROL
     AIRCRAFT DESIGN
     DRONE AIRCRAFT
     FLIGHT TESTS
     FLUTTER
     PROGRAMS
     REMOTELY PILOTED VEHICLES
     VIBRATION DAMPING

∞ DATA
SN   *(USE OF A MORE SPECIFIC TERM IS*
     RECOMMENDED--CONSULT THE TERMS
     LISTED BELOW)
RT   ANALOG DATA
     AUDIO DATA
     BINARY DATA
     BIOMEDICAL DATA
     CENSORED DATA (MATHEMATICS)
     CHANNELS (DATA TRANSMISSION)
     CONTROL DATA (COMPUTERS)
     DATA ACQUISITION
     DATA BASE MANAGEMENT SYSTEMS
     DATA BASES
     DATA COLLECTION PLATFORMS
     DATA COMPRESSION
     DATA CONVERSION ROUTINES
     DATA CONVERTERS
     DATA CORRELATION
     DATA LINKS
     DATA MANAGEMENT
     DATA PROCESSING
     DATA PROCESSING EQUIPMENT
     DATA PROCESSING TERMINALS
     DATA RECORDERS
     DATA RECORDING
     DATA REDUCTION
     DATA RETRIEVAL
     DATA SAMPLING
     DATA SMOOTHING
     DATA STORAGE
     DATA SYSTEMS
     DATA TRANSMISSION
     DATUM (ELEVATION)
     DIGITAL DATA
     END-TO-END DATA SYSTEMS
     INFORMATION

DATA-(CONT.)
    INTERSERVICE DATA EXCHANGE
      PROGRAM
   ∞MEASUREMENT
    OCEAN DATA ACQUISITIONS SYSTEMS
    ONBOARD DATA PROCESSING
    OPTICAL DATA PROCESSING
    OPTICAL DATA STORAGE MATERIALS
    OPTICAL MEMORY (DATA STORAGE)
    PRINTERS (DATA PROCESSING)
    RADAR DATA
    RECORDS
    SITE DATA PROCESSORS
    SPACE FLIGHT TRACKING AND DATA
      NETWORK
    STATISTICAL ANALYSIS
    STATISTICAL TESTS
   ∞STATISTICS
    TABLES (DATA)
    VIDEO DATA
    VOICE DATA PROCESSING
    WEATHER DATA RECORDERS
    WORLD DATA CENTERS

DATA ACQUISITION
  GS   ACQUISITION
    . DATA ACQUISITION
  RT   ADVANCED RANGE INSTRUMENTATION
      AIRCRAFT
    ALOUETTE PROJECT
    ANALOG TO DIGITAL CONVERTERS
    AUTOMATIC WEATHER STATIONS
    COUNTING
   ∞DATA
    DEEP SPACE INSTRUMENTATION
      FACILITY
    DETECTION
    EARTH OBSERVATIONS (FROM SPACE)
    END-TO-END DATA SYSTEMS
    FORMS (PAPER)
    GLOBAL TRACKING NETWORK
    GROUND STATIONS
    INFRARED RADIOMETERS
    METEOROLOGICAL RESEARCH
      AIRCRAFT
    NEEDS (DATA SYSTEM)
    NEWS MEDIA
    OBSERVATION
    OCEAN DATA ACQUISITIONS SYSTEMS
    OPTICAL DATA PROCESSING
    OPTICAL SCANNERS
    REMOTE SENSORS
   ∞SENSORS
    SPACE FLIGHT TRACKING AND DATA
      NETWORK
    STDN (NETWORK)
    SURVEYS
    TABLES (DATA)
    TRACKING NETWORKS

DATA ADAPTIVE EVALUATOR/MONITOR
  USE   DATA PROCESSING
      DATA REDUCTION
      DATA TRANSMISSION

DATA ANALYSIS
  USE   DATA PROCESSING
      DATA REDUCTION

DATA BASE MANAGEMENT SYSTEMS
  RT   ∞DATA
    DATA BASES
    DATA MANAGEMENT
    INFORMATION MANAGEMENT
    MANAGEMENT INFORMATION SYSTEMS
    SOFTWARE TOOLS
   ∞SYSTEMS

DATA BASES
  RT   ∞BASES
   ∞DATA
    DATA BASE MANAGEMENT SYSTEMS
    DATA STRUCTURES
    SOFTWARE ENGINEERING

DATA BUSSES
  USE   CHANNELS (DATA TRANSMISSION)

DATA COLLECTION PLATFORMS
  RT   AUTOMATIC WEATHER STATIONS
   ∞DATA
    GROUND STATIONS
    INSTRUMENT PACKAGES
    INTEGRATED GLOBAL OCEAN STATION
      SYSTEMS

DATA COLLECTION PLATFORMS-(CONT.)
    REMOTE SENSORS

DATA COMPACTION
  USE   DATA COMPRESSION

DATA COMPRESSION
  UF   DATA COMPACTION
  RT   CURVE FITTING
   ∞DATA
    DECODING
    FOURIER ANALYSIS
    TELECOMMUNICATION
    TELEMETRY

DATA CONVERSION ROUTINES
  GS   DATA CONVERSION ROUTINES
    . SUBROUTINES
  RT   ALGORITHMS
    COMPILERS
    COMPUTER PROGRAMS
   ∞CONVERSION
   ∞DATA
   ∞ROUTINES

DATA CONVERTERS
  GS   DATA CONVERTERS
    . ANALOG TO DIGITAL CONVERTERS
    . BINARY TO DECIMAL CONVERTERS
    . DECIMAL TO BINARY CONVERTERS
    . DIGITAL TO ANALOG CONVERTERS
  RT   ANALOG CIRCUITS
    ANALOG DATA
    COMPUTERS
    CONVERSION TABLES
   ∞CONVERTERS
   ∞DATA
    DECODERS
    DIGITAL DATA
    TRANSDUCERS
    VIDEO DATA

DATA CORRELATION
  GS   CORRELATION
    . DATA CORRELATION
    . . SIGNAL ANALYSIS
    DATA PROCESSING
    . DATA CORRELATION
    . . SIGNAL ANALYSIS
  RT   ANGULAR CORRELATION
    AUTOCORRELATION
    CROSS CORRELATION
   ∞DATA
    STATISTICAL ANALYSIS
    STATISTICAL CORRELATION
    TEMPERATURE RATIO

DATA FLOW ANALYSIS
  RT   COMPUTER PROGRAMS
    COMPUTER SYSTEMS PROGRAMS
    DATA SIMULATION
    FLOW CHARTS
    NETWORK ANALYSIS
    SEQUENTIAL CONTROL

DATA HANDLING SYSTEMS
  USE   DATA SYSTEMS

DATA INTEGRATION
  RT   DATA MANAGEMENT
    DATA SIMULATION

DATA LINKS
  GS   TELECOMMUNICATION
    . DATA LINKS
  RT   CHANNELS (DATA TRANSMISSION)
    COMMUNICATION NETWORKS
   ∞DATA
    DECOMMUTATORS
    DISCRETE ADDRESS BEACON SYSTEM
    FREQUENCY REUSE
    PROTOCOL (COMPUTERS)
    RADIO RECEIVERS
    RADIO RELAY SYSTEMS
    REMOTE CONSOLES
    SITE DATA PROCESSORS
    TELEMETRY
    WIRELESS COMMUNICATION

DATA MANAGEMENT
  GS   MANAGEMENT
    . DATA MANAGEMENT
  RT   ∞DATA
    DATA BASE MANAGEMENT SYSTEMS

DATA MANAGEMENT-(CONT.)
    DATA INTEGRATION
    DATA SIMULATION
    FRAMES (DATA PROCESSING)
    ON-LINE SYSTEMS
    SURVEYS
    TABLES (DATA)
    VIRTUAL MEMORY SYSTEMS

DATA PROCESSING
  SN   (MECHANICAL OR ELECTRONIC
      MANIPULATION OF DATA)
  UF   AUTOMATIC DATA PROCESSING
    DAEMO (DATA ANALYSIS)
    DATA ADAPTIVE EVALUATOR/MONITOR
    DATA ANALYSIS
  GS   DATA PROCESSING
    . ASSOCIATIVE PROCESSING
      (COMPUTERS)
    . ATMOSPHERIC & OCEANOGRAPHIC
      INFORM SYS
    . BATCH PROCESSING
    . CENSORED DATA (MATHEMATICS)
    . CENTRAL ELECTRONIC MANAGEMENT
      SYSTEM
    . CONCURRENT PROCESSING
    . DATA CORRELATION
    . . SIGNAL ANALYSIS
    . DATA REDUCTION
    . . DATA SMOOTHING
    . DATA RETRIEVAL
    . DATA STORAGE
    . DISTRIBUTED PROCESSING
    . KARHUNEN-LOEVE EXPANSION
    . MULTIPROCESSING (COMPUTERS)
    . . PIPELINING (COMPUTERS)
    . OPTICAL DATA PROCESSING
    . SCENE ANALYSIS
    . PARALLEL PROCESSING (COMPUTERS)
    . SIGNAL PROCESSING
    . VOICE DATA PROCESSING
  RT   ACCESS TIME
    ADJOINTS
    AIRBORNE/SPACEBORNE COMPUTERS
    ANALOG DATA
   ∞AUTOMATION
    AUXILIARY EQUIPMENT (COMPUTERS)
    BINARY DATA
    BINARY TO DECIMAL CONVERTERS
    BUBBLE MEMORY DEVICES
    CHANNELS (DATA TRANSMISSION)
    COMPUTATION
    COMPUTER INFORMATION SECURITY
    COMPUTER PROGRAMS
    COMPUTER SYSTEMS PROGRAMS
    COMPUTERS
    DATA STRUCTURES
    DECIMAL TO BINARY CONVERTERS
    DIGITAL COMPUTERS
    DIGITAL DATA
    EDITING
    EDITING ROUTINES (COMPUTERS)
    END-TO-END DATA SYSTEMS
    FIXED POINT ARITHMETIC
    FLOATING POINT ARITHMETIC
    FRAMES (DATA PROCESSING)
    IMAGE PROCESSING
    INFORMATION RETRIEVAL
    INFORMATION THEORY
    INTERROGATION
    LANGUAGE PROGRAMMING
    MECHANIZATION
    MICROPROCESSORS
    NATURAL LANGUAGE (COMPUTERS)
    NEEDS (DATA SYSTEM)
    ON-LINE SYSTEMS
    PREPROCESSING
   ∞PROCESSING
    PROTOCOL (COMPUTERS)
    RCA COMPUTERS
    RECORDS
    RESPONSE TIME (COMPUTERS)
    SITE DATA PROCESSORS
    SYMBOLS
    SYSTEMS ENGINEERING
    TABLES (DATA)
    TABULATION PROCESSES
    TELECOMMUNICATION
    WORD PROCESSING

DATA PROCESSING EQUIPMENT
  UF   DATA PROCESSORS
  GS   DATA PROCESSING EQUIPMENT
    . AUXILIARY EQUIPMENT (COMPUTERS)
    . . PRINTERS (DATA PROCESSING)

DATA PROCESSING EQUIPMENT-*(CONT.)*
. COMPUTERS
. . ANALOG COMPUTERS
. . . EAI 680 COMPUTER
. . . HONEYWELL 600/6000 COMPUTER
. . . SIGMA 5 COMPUTER
. . . UNIVAC 1100 SERIES COMPUTERS
. . CDC COMPUTERS
. . . CDC CYBER 74 COMPUTER
. . . CDC CYBER 170 SERIES
      COMPUTERS
. . . . CDC CYBER 175 COMPUTER
. . . CDC CYBER 174 COMPUTER
. . . CDC CYBER 203 COMPUTER
. . . CDC CYBER 205 COMPUTER
. . . CDC STAR 100 COMPUTER
. . . CDC 160-A COMPUTER
. . . CDC 1604 COMPUTER
. . . CDC 3100 COMPUTER
. . . CDC 3200 COMPUTER
. . . CDC 3600 COMPUTER
. . . CDC 3800 COMPUTER
. . . CDC 6000 SERIES COMPUTERS
. . . . CDC 6400 COMPUTER
. . . . CDC 6600 COMPUTER
. . . . CDC 6700 COMPUTER
. . . CDC 7000 SERIES COMPUTERS
. . . . CDC 7600 COMPUTER
. . . CDC 8090 COMPUTER
. . COUNTING RATE COMPUTERS
. . DDP COMPUTERS
. . . DDP 516 COMPUTER
. . DIGITAL COMPUTERS
. . . CDC CYBER 74 COMPUTER
. . . CDC CYBER 170 SERIES
      COMPUTERS
. . . CDC CYBER 174 COMPUTER
. . . CDC CYBER 203 COMPUTER
. . . CDC CYBER 205 COMPUTER
. . . CDC STAR 100 COMPUTER
. . . CDC 160-A COMPUTER
. . . CDC 1604 COMPUTER
. . . CDC 3100 COMPUTER
. . . CDC 3200 COMPUTER
. . . CDC 3600 COMPUTER
. . . CDC 3800 COMPUTER
. . . CDC 6000 SERIES COMPUTERS
. . . . CDC 6400 COMPUTER
. . . . CDC 6600 COMPUTER
. . . . CDC 6700 COMPUTER
. . . CDC 7000 SERIES COMPUTERS
. . . . CDC 7600 COMPUTER
. . . CDC 8090 COMPUTER
. . . EAI 680 COMPUTER
. . . EAI 8400 COMPUTER
. . . EAI 8900 COMPUTER
. . . EMR 6050 COMPUTER
. . . FERRANTI MERCURY COMPUTER
. . . GE COMPUTERS
. . . . GE 625 COMPUTER
. . . . GE 635 COMPUTER
. . . HEWLETT-PACKARD COMPUTERS
. . . HONEYWELL COMPUTERS
. . . . DDP 516 COMPUTER
. . . . HONEYWELL ADEPT COMPUTER
. . . . HONEYWELL DDP 116 COMPUTER
. . . HONEYWELL 600/6000 COMPUTER
. . . IBM 360 COMPUTER
. . . IBM 370 COMPUTER
. . . IBM 650 COMPUTER
. . . IBM 704 COMPUTER
. . . IBM 709 COMPUTER
. . . IBM 1130 COMPUTER
. . . IBM 1401 COMPUTER
. . . IBM 1410 COMPUTER
. . . IBM 1620 COMPUTER
. . . IBM 2250 COMPUTER
. . . IBM 7000 SERIES COMPUTERS
. . . . IBM 7030 COMPUTER
. . . . IBM 7040 COMPUTER
. . . . IBM 7044 COMPUTER
. . . . IBM 7070 COMPUTER
. . . . IBM 7074 COMPUTER
. . . . IBM 7090 COMPUTER
. . . . IBM 7094 COMPUTER
. . . ICL COMPUTERS
. . . ILLIAC COMPUTERS
. . . . ILLIAC 3 COMPUTER
. . . . ILLIAC 4 COMPUTER
. . . MICROCOMPUTERS
. . . . PERSONAL COMPUTERS
. . . MINICOMPUTERS
. . . . NOVA COMPUTERS
. . . MODCOMP II COMPUTER
. . . MODCOMP IV COMPUTER
. . . PARALLEL COMPUTERS

DATA PROCESSING EQUIPMENT-*(CONT.)*
. . . PDP COMPUTERS
. . . . PDP 7 COMPUTER
. . . . PDP 8 COMPUTER
. . . . PDP 9 COMPUTER
. . . . PDP 10 COMPUTER
. . . . PDP 11 COMPUTER
. . . . PDP 11/20 COMPUTER
. . . . PDP 11/40 COMPUTER
. . . . PDP 11/45 COMPUTER
. . . . PDP 11/50 COMPUTER
. . . . PDP 11/70 COMPUTER
. . . . PDP 12 COMPUTER
. . . PDP 15 COMPUTER
. . . PHILCO 2000 COMPUTER
. . . RAYTHEON COMPUTERS
. . . RCA SPECTRA 70 COMPUTER
. . . SDS 900 SERIES COMPUTERS
. . . . SDS 930 COMPUTER
. . . SDS 9300 COMPUTER
. . . SEL COMPUTERS
. . . SEQUENTIAL COMPUTERS
. . . SIGMA COMPUTERS
. . . . SIGMA 9 COMPUTER
. . . SIGMA 5 COMPUTER
. . . SOLOMON COMPUTERS
. . . UNIVAC LARC COMPUTER
. . . UNIVAC 80 COMPUTER
. . . UNIVAC 418 COMPUTER
. . . UNIVAC 490 COMPUTER
. . . UNIVAC 494 COMPUTER
. . . UNIVAC 1100 SERIES COMPUTERS
. . . . UNIVAC 1105 COMPUTER
. . . . UNIVAC 1106 COMPUTER
. . . . UNIVAC 1107 COMPUTER
. . . . UNIVAC 1108 COMPUTER
. . . . UNIVAC 1110 COMPUTER
. . . . UNIVAC 1230 COMPUTER
. . . VAX COMPUTERS
. . . VAX-11 SERIES COMPUTERS
. . . . VAX-11/780 COMPUTER
. . EMBEDDED COMPUTER SYSTEMS
. . . AIRBORNE/SPACEBORNE
      COMPUTERS
. . HYBRID COMPUTERS
. . IBM COMPUTERS
. . . IBM 360 COMPUTER
. . . IBM 370 COMPUTER
. . . IBM 650 COMPUTER
. . . IBM 704 COMPUTER
. . . IBM 709 COMPUTER
. . . IBM 1130 COMPUTER
. . . IBM 1401 COMPUTER
. . . IBM 1410 COMPUTER
. . . IBM 1620 COMPUTER
. . . IBM 2250 COMPUTER
. . . IBM 7000 SERIES COMPUTERS
. . . . IBM 7030 COMPUTER
. . . . IBM 7040 COMPUTER
. . . . IBM 7044 COMPUTER
. . . . IBM 7070 COMPUTER
. . . . IBM 7074 COMPUTER
. . . . IBM 7090 COMPUTER
. . . . IBM 7094 COMPUTER
. . MINOS COMPUTER
. . OPTICAL COMPUTERS
. . PEGASUS COMPUTER
. . RCA COMPUTERS
. . . RCA SPECTRA 70 COMPUTER
. . . RCA-110 COMPUTERS
. . SIEMENS 2002 COMPUTER
. . SITE DATA PROCESSORS
. . UNIVAC COMPUTERS
. . . UNIVAC LARC COMPUTER
. . . UNIVAC 80 COMPUTER
. . . UNIVAC 418 COMPUTER
. . . UNIVAC 490 COMPUTER
. . . UNIVAC 494 COMPUTER
. . . UNIVAC 1100 SERIES COMPUTERS
. . . . UNIVAC 1105 COMPUTER
. . . . UNIVAC 1106 COMPUTER
. . . . UNIVAC 1107 COMPUTER
. . . . UNIVAC 1108 COMPUTER
. . . . UNIVAC 1110 COMPUTER
. . . UNIVAC 1230 COMPUTER
. DATA PROCESSING TERMINALS
. MICROPROCESSORS
. . INTEL 8080 MICROPROCESSOR
RT BATCH PROCESSING
    COMPUTER COMPATIBLE TAPES
    COMPUTER SYSTEMS SIMULATION
    CONTROL UNITS (COMPUTERS)
    ∞ DATA
    DIGITAL RADAR SYSTEMS
    ∞ EQUIPMENT
    INTERFACES

DATA PROCESSING EQUIPMENT-*(CONT.)*
    MULTIPROCESSING (COMPUTERS)
    OPTICAL DATA PROCESSING
    PIPELINING (COMPUTERS)
    PRINTERS
    REMOTE CONSOLES
    SIMULATION

DATA PROCESSING TERMINALS
GS DATA PROCESSING EQUIPMENT
    . **DATA PROCESSING TERMINALS**
RT COMPUTER GRAPHICS
    CONSOLES
    ∞ DATA
    MAN MACHINE SYSTEMS
    REMOTE CONSOLES
    ∞ TERMINALS

DATA PROCESSORS
USE DATA PROCESSING EQUIPMENT

DATA READOUT SYSTEMS
USE DATA SYSTEMS
    DISPLAY DEVICES

DATA RECORDERS
RT BUBBLE MEMORY DEVICES
    BUBBLE TECHNIQUE
    COUNTERS
    ∞ DATA
    DISPLAY DEVICES
    MONITORS
    PUNCHED CARDS
    ∞ RECORDERS
    RECORDING INSTRUMENTS
    TAPE RECORDERS
    VIDEO DISKS

DATA RECORDING
GS RECORDING
    . **DATA RECORDING**
RT BUBBLE MEMORY DEVICES
    ∞ DATA
    MAGNETIC RECORDING
    MAGNETIC STORAGE
    OPTICAL DATA STORAGE MATERIALS
    PHOTOGRAPHIC RECORDING
    PUNCHED CARDS
    PUNCHED TAPES
    RECORDING HEADS
    RECORDS
    TABLES (DATA)
    TABULATION PROCESSES
    VIDEO DISKS

DATA REDUCTION
UF DAEMO (DATA ANALYSIS)
    DATA ADAPTIVE EVALUATOR/MONITOR
    DATA ANALYSIS
    TARE (DATA REDUCTION)
GS DATA PROCESSING
    . **DATA REDUCTION**
    . . DATA SMOOTHING
RT COMPUTATION
    ∞ DATA
    EDITING
    PREPROCESSING
    ∞ REDUCTION
    TABLES (DATA)

DATA RETRIEVAL
GS DATA PROCESSING
    . **DATA RETRIEVAL**
    RETRIEVAL
    . **DATA RETRIEVAL**
RT ∞ DATA
    DOCUMENTATION
    INFORMATION MANAGEMENT
    INFORMATION RETRIEVAL
    INTERSERVICE DATA EXCHANGE
      PROGRAM
    LIBRARIES
    MANAGEMENT INFORMATION SYSTEMS
    MICROFILMS
    SEARCH PROFILES
    TABLES (DATA)
    TELEMETRY
    WORLD DATA CENTERS

DATA SAMPLING
UF SAMPLED DATA
    SAMPLED DATA SYSTEMS
GS SAMPLING
    . **DATA SAMPLING**
RT COMPUTER SYSTEMS PERFORMANCE

DATA SAMPLING-(CONT.)
    ∞ DATA
    QUALITY CONTROL
    TELECOMMUNICATION
    TIME SERIES ANALYSIS

DATA SIMULATION
    RT   DATA FLOW ANALYSIS
          DATA INTEGRATION
          DATA MANAGEMENT
          SIMULATION

DATA SMOOTHING
    GS   DATA PROCESSING
          . DATA REDUCTION
          . . DATA SMOOTHING
          RECORDING
          . DATA SMOOTHING
          SMOOTHING
          . DATA SMOOTHING
    RT   CURVE FITTING
          ∞ DATA

DATA STORAGE
    GS   DATA PROCESSING
          . DATA STORAGE
    RT   BUBBLE MEMORY DEVICES
          BUFFER STORAGE
          CARDS
          CORE STORAGE
          ∞ DATA
          DOCUMENT STORAGE
          FLIP-FLOPS
          HOLOGRAPHY
          INFORMATION MANAGEMENT
          INTERSERVICE DATA EXCHANGE
              PROGRAM
          MAGNETIC STORAGE
          MANAGEMENT INFORMATION SYSTEMS
          MICROFILMS
          MICROPHOTOGRAPHS
          OPTICAL DATA STORAGE MATERIALS
          OPTICAL DISKS
          PUNCHED CARDS
          SELECTIVE DISSEMINATION OF
              INFORMATION
          ∞ STORAGE
          VIDEO DISKS
          VIRTUAL MEMORY SYSTEMS
          WHITE LIGHT HOLOGRAPHY
          WORLD DATA CENTERS

DATA STRUCTURES
    RT   COMPUTER PROGRAMMING
          DATA BASES
          DATA PROCESSING

DATA SYSTEMS
    UF   DATA HANDLING SYSTEMS
          DATA READOUT SYSTEMS
    GS   DATA SYSTEMS
          . NEEDS (DATA SYSTEM)
    RT   ATMOSPHERIC & OCEANOGRAPHIC
              INFORM SYS
          CONTROL DATA (COMPUTERS)
          ∞ DATA
          DIGITAL SYSTEMS
          EARTH RESOURCES INFORMATION
              SYSTEM
          END-TO-END DATA SYSTEMS
          GEOGRAPHIC INFORMATION SYSTEMS
          MANAGEMENT INFORMATION SYSTEMS
          ∞ SYSTEMS

DATA TRANSMISSION
    UF   DAEMO (DATA ANALYSIS)
          DATA ADAPTIVE EVALUATOR/MONITOR
          INFORMATION TRANSMISSION
    GS   TRANSMISSION
          . SIGNAL TRANSMISSION
          . . DATA TRANSMISSION
          . . . AUTOMATIC PICTURE
                  TRANSMISSION
          . . . MULTIPLE ACCESS
          . . . . CODE DIVISION MULTIPLE ACCESS
          . . . . FREQUENCY DIVISION MULTIPLE
                    ACCESS
          . . . SINGLE CHANNEL PER CARRIER
                  TRANSMISSION
    RT   ACCESS CONTROL
          AUDIO DATA
          CARRIER TO NOISE RATIOS
          CHANNELS (DATA TRANSMISSION)
          CODE DIVISION MULTIPLEXING
          CODING

DATA TRANSMISSION-(CONT.)
          COMMUNICATION THEORY
          CONCATENATED CODES
          ∞ DATA
          DEEP SPACE INSTRUMENTATION
              FACILITY
          ELECTRONIC MAIL
          FM/PM (MODULATION)
          FREQUENCY DIVISION MULTIPLEXING
          INFORMATION THEORY
          INTERSYMBOLIC INTERFERENCE
          LASER APPLICATIONS
          MODEMS
          MULTIPLEXING
          PACKET SWITCHING
          PACKET TRANSMISSION
          PACKETS (COMMUNICATION)
          PROTOCOL (COMPUTERS)
          PULSE COMMUNICATION
          RADIO TELEMETRY
          RADIO TRANSMISSION
          READING
          REDUNDANCY ENCODING
          SATELLITE TRANSMISSION
          SHIP TO SHORE COMMUNICATION
          TDR SATELLITES
          TELECOMMUNICATION
          TELEMETRY
          TRANSMISSION EFFICIENCY
          VIDEO DATA
          WIRELESS COMMUNICATION

DATING
    USE   CHRONOLOGY
          TIME MEASUREMENT

DATUM (ELEVATION)
    RT   CLEARANCES
          CONTOURS
          ∞ DATA
          ELEVATION ANGLE
          HYPSOGRAPHY
          LEVELING
          MAPS
          SURVEYS

DAWN CHORUS
    UF   CHORUS (DAWN PHENOMENON)
          CHORUS PHENOMENON
    GS   ATMOSPHERIC RADIATION
          . DAWN CHORUS
          ELECTROMAGNETIC INTERFERENCE
          . RADIO FREQUENCY INTERFERENCE
          . . BLACKOUT (PROPAGATION)
          . . . ELECTROMAGNETIC NOISE
          . . . . ATMOSPHERICS
          . . . . . IONOSPHERICS
          . . . . . . DAWN CHORUS
    RT   AURORAS
          MAGNETIC STORMS
          WHISTLERS

DAWSONITE
    GS   MINERALS
          . DAWSONITE
    RT   ALUMINUM
          SODIUM

DAYGLOW
    GS   ATMOSPHERIC RADIATION
          . SKY RADIATION
          . . DAYGLOW
          ELECTROMAGNETIC RADIATION
          . LIGHT (VISIBLE RADIATION)
          . . SKY RADIATION
          . . . DAYGLOW
    RT   GLARE
          LIGHT SOURCES
          SKY
          SOLAR RADIATION
          TWILIGHT GLOW
          ULTRAVIOLET RADIATION

DAYTIME
    RT   DIURNAL VARIATIONS
          EVENING
          MORNING
          NIGHT
          NOON
          SKY BRIGHTNESS

DC (CURRENT)
    USE   DIRECT CURRENT

DC 3 AIRCRAFT
    UF   DOUGLAS DC-3 AIRCRAFT
    GS   COMMERCIAL AIRCRAFT
          . DC 3 AIRCRAFT
          MCDONNELL DOUGLAS AIRCRAFT
          . DOUGLAS AIRCRAFT
          . . DC 3 AIRCRAFT
          MONOPLANES
          . DC 3 AIRCRAFT
          TRANSPORT AIRCRAFT
          . CARGO AIRCRAFT
          . . DC 3 AIRCRAFT
    RT   ∞ AIRCRAFT

DC 7 AIRCRAFT
    UF   DOUGLAS DC-7 AIRCRAFT
    GS   COMMERCIAL AIRCRAFT
          . DC 7 AIRCRAFT
          MCDONNELL DOUGLAS AIRCRAFT
          . DOUGLAS AIRCRAFT
          . . DC 7 AIRCRAFT
          MONOPLANES
          . DC 7 AIRCRAFT
          TRANSPORT AIRCRAFT
          . CARGO AIRCRAFT
          . . DC 7 AIRCRAFT
    RT   ∞ AIRCRAFT
          PASSENGER AIRCRAFT

DC 8 AIRCRAFT
    UF   DOUGLAS DC-8 AIRCRAFT
    GS   COMMERCIAL AIRCRAFT
          . DC 8 AIRCRAFT
          JET AIRCRAFT
          . TURBOFAN AIRCRAFT
          . . DC 8 AIRCRAFT
          MCDONNELL DOUGLAS AIRCRAFT
          . DOUGLAS AIRCRAFT
          . . DC 8 AIRCRAFT
          MONOPLANES
          . DC 8 AIRCRAFT
          PASSENGER AIRCRAFT
          . DC 8 AIRCRAFT
          TRANSPORT AIRCRAFT
          . DC 8 AIRCRAFT
    RT   ∞ AIRCRAFT

DC 9 AIRCRAFT
    UF   DOUGLAS DC-9 AIRCRAFT
    GS   COMMERCIAL AIRCRAFT
          . DC 9 AIRCRAFT
          JET AIRCRAFT
          . DC 9 AIRCRAFT
          MCDONNELL DOUGLAS AIRCRAFT
          . DOUGLAS AIRCRAFT
          . . DC 9 AIRCRAFT
          TRANSPORT AIRCRAFT
          . DC 9 AIRCRAFT
    RT   ∞ AIRCRAFT

DC 10 AIRCRAFT
    GS   COMMERCIAL AIRCRAFT
          . DC 10 AIRCRAFT
          JET AIRCRAFT
          . DC 10 AIRCRAFT
          MCDONNELL DOUGLAS AIRCRAFT
          . DOUGLAS AIRCRAFT
          . . DC 10 AIRCRAFT
          . MCDONNELL AIRCRAFT
          . . DC 10 AIRCRAFT
          PASSENGER AIRCRAFT
          . DC 10 AIRCRAFT
          TRANSPORT AIRCRAFT
          . DC 10 AIRCRAFT
    RT   ∞ AIRCRAFT
          TURBOFAN ENGINES

DDP COMPUTERS
    GS   DATA PROCESSING EQUIPMENT
          . COMPUTERS
          . . DDP COMPUTERS
          . . . DDP 516 COMPUTER
    RT   DIGITAL COMPUTERS

DDP 516 COMPUTER
    GS   DATA PROCESSING EQUIPMENT
          . COMPUTERS
          . . DDP COMPUTERS
          . . . DDP 516 COMPUTER
          . DIGITAL COMPUTERS
          . . HONEYWELL COMPUTERS
          . . . . DDP 516 COMPUTER

DDT
    UF   DICHLORODIPHENYLTRICHLOROETHANE

DDT-(CONT.)
- GS ALIPHATIC COMPOUNDS
  - . DDT
  - HALOGEN COMPOUNDS
  - . CHLORINE COMPOUNDS
  - . . DDT

DE BROGLIE WAVELENGTHS
- GS WAVELENGTHS
  - . DE BROGLIE WAVELENGTHS
- RT ELEMENTARY PARTICLES
  - MASS
  - MOMENTUM
  - PLANCKS CONSTANT
  - QUANTUM THEORY
  - VELOCITY
  - WENTZEL-KRAMER-BRILLOUIN METHOD

DE HAVILLAND AIRCRAFT
- GS DE HAVILLAND AIRCRAFT
  - . COMET 4 AIRCRAFT
  - . DH 112 AIRCRAFT
  - . DH 115 AIRCRAFT
  - . DH 121 AIRCRAFT
  - . DH 125 AIRCRAFT
  - . DHC 2 AIRCRAFT
  - . DHC 4 AIRCRAFT
  - . DHC 5 AIRCRAFT
- RT ∞AIRCRAFT

DE HAVILLAND DH 106 AIRCRAFT
- USE COMET 4 AIRCRAFT

DE HAVILLAND DH 112 AIRCRAFT
- USE DH 112 AIRCRAFT

DE HAVILLAND DH 115 AIRCRAFT
- USE DH 115 AIRCRAFT

DE HAVILLAND DH 121 AIRCRAFT
- USE DH 121 AIRCRAFT

DE HAVILLAND DH 125 AIRCRAFT
- USE DH 125 AIRCRAFT

DE HAVILLAND DHC 4 AIRCRAFT
- USE DHC 4 AIRCRAFT

DE HAVILLAND DHC 5 AIRCRAFT
- USE DHC 5 AIRCRAFT

DE HAVILLAND VENOM AIRCRAFT
- USE DH 112 AIRCRAFT

DE LAVAL NOZZLES
- USE CONVERGENT-DIVERGENT NOZZLES

DEACCLIMATIZATION
- USE ACCLIMATIZATION

DEACTIVATION
- UF INACTIVATION
- RT ACTIVATION
  - PASSIVITY
  - POLARIZATION (CHARGE SEPARATION)
  - POLARIZATION (SPIN ALIGNMENT)
  - SABOTAGE
  - SHUTDOWNS

DEAD RECKONING
- GS NAVIGATION
  - . DEAD RECKONING
- RT AIR NAVIGATION
  - DIGITAL NAVIGATION
  - DOPPLER NAVIGATION
  - INERTIAL NAVIGATION
  - POLAR NAVIGATION
  - RADAR NAVIGATION
  - RADIO NAVIGATION
  - SURFACE NAVIGATION

DEADWEIGHT
- USE STATIC LOADS

DEAFNESS
- USE AUDITORY DEFECTS

DEATH
- RT CASUALTIES
  - EXPIRATION
  - INJURIES
  - LIFE SPAN
  - MORTALITY

DEATH VALLEY (CA)
- GS VALLEYS
  - . DEATH VALLEY (CA)
- RT ARID LANDS
  - CALIFORNIA
  - DESERTIFICATION
  - DESERTS
  - LANDFORMS
  - RIVER BASINS

DEBONAIR AIRCRAFT
- USE C-33 AIRCRAFT

DEBRIS
- GS DEBRIS
  - . SPACE DEBRIS
- RT EJECTA
  - ENVIRONMENT EFFECTS
  - FRAGMENTS
  - GLACIAL DRIFT
  - POLLUTION
  - ∞RADIOACTIVE DEBRIS
  - SCRAP
  - WASTES

DEBUGGING
- USE CHECKOUT

DEBYE LENGTH
- GS DISTANCE
  - . DEBYE LENGTH
- RT PLASMAS (PHYSICS)

DEBYE TEMPERATURE
- USE SPECIFIC HEAT

DEBYE-HUCKEL THEORY
- RT DISSOCIATION
  - ELECTROLYTES
  - PLASMA POTENTIALS
  - ∞THEORIES

DEBYE-SCHERRER METHOD
- RT CRYSTALLOGRAPHY
  - DIFFRACTION
  - ∞METHODOLOGY

DECAMETRIC WAVES
- GS ELECTROMAGNETIC RADIATION
  - . RADIO WAVES
  - . . DECAMETRIC WAVES
- RT CORONAL HOLES
  - HIGH FREQUENCIES
  - VERY HIGH FREQUENCIES

DECARBONATION
- GS CHEMICAL REACTIONS
  - . DECARBONATION
- RT CARBONIZATION

DECARBOXYLATION
- GS CHEMICAL REACTIONS
  - . DECARBOXYLATION
- RT CARBOXYLATION

DECARBURIZATION
- RT CARBON
  - CARBURIZING
  - HEATING
  - METAL WORKING

DECAY
- GS DECAY
  - . ACOUSTIC EMISSION
  - . EMISSION
  - . . EXHAUST EMISSION
  - . . LIGHT EMISSION
  - . . . INCANDESCENCE
  - . . . LUMINESCENCE
  - . . . . BIOLUMINESCENCE
  - . . . . CATHODE GLOW
  - . . . . CATHODOLUMINESCENCE
  - . . . . CHEMILUMINESCENCE
  - . . . . ELECTROLUMINESCENCE
  - . . . . FLUORESCENCE
  - . . . . . PHOSPHORESCENCE
  - . . . . . RESONANCE FLUORESCENCE
  - . . . . . X RAY FLUORESCENCE
  - . . . LUNAR LUMINESCENCE
  - . . . . OPTICAL RESONANCE
  - . . . . PHOTOLUMINESCENCE
  - . . . . . TRIBOLUMINESCENCE
  - . . . . . X RAY FLUORESCENCE
  - . . . . SHOCK WAVE LUMINESCENCE

DECAY-(CONT.)
- . . . . SONOLUMINESCENCE
- . . . . THERMOLUMINESCENCE
- . . MICROWAVE EMISSION
- . . PARTICLE EMISSION
- . . . ELECTRON EMISSION
- . . . . FIELD EMISSION
- . . . . PHOTOELECTRIC EMISSION
- . . . . SECONDARY EMISSION
- . . ION EMISSION
- . . . NEUTRON EMISSION
- . . . THERMIONIC EMISSION
- . . PHOTOELECTRIC EFFECT
- . . PHOTOIONIZATION
- . RADIO EMISSION
- . . CN EMISSION
- . . HYDROXYL EMISSION
- . . SOLAR RADIO EMISSION
- . . . SOLAR RADIO BURSTS
- . . . . TYPE 2 BURSTS
- . . . . TYPE 3 BURSTS
- . . . . TYPE 4 BURSTS
- . . . . TYPE 5 BURSTS
- . . SELF SUSTAINED EMISSION
- . . SPECTRAL EMISSION
- . . SPONTANEOUS EMISSION
- . . STIMULATED EMISSION
- . . . WATER MASERS
- . THERMAL EMISSION
- . . . THERMIONIC EMISSION
- . HALF LIFE
- . NEUTRON DECAY
- . NUCLEAR FISSION
- . PHOTOPRODUCTION
- . PLASMA DECAY
- . RADIOACTIVE DECAY
- . . ALPHA DECAY
- . . NEUTRON EMISSION
- . STRANGENESS
- . WEAK ENERGY INTERACTIONS
- RT BIODEGRADABILITY
  - BIODEGRADATION
  - DAMAGE
  - DEGRADATION
  - DETERIORATION
  - DISINTEGRATION
  - HOT ATOMS
  - RADIATIVE LIFETIME

DECAY RATES
- GS RATES (PER TIME)
  - . DECAY RATES
  - . . ELECTRON DECAY RATE

DECCA NAVIGATION
- GS NAVIGATION
  - . RADIO NAVIGATION
  - . . HYPERBOLIC NAVIGATION
  - . . . DECCA NAVIGATION
- RT DISTANCE MEASURING EQUIPMENT
  - LORAN
  - LORAN C
  - LORAN D
  - NAVIGATION AIDS
  - SHORAN
  - SOLAR COMPASSES
  - SURFACE NAVIGATION

DECELERATION
- UF IMPACT DECELERATION
- GS RATES (PER TIME)
  - . ACCELERATION (PHYSICS)
  - . . DECELERATION
  - . . . SPIN REDUCTION
- RT ANGULAR ACCELERATION
  - BRAKING
  - DAMPING
  - IMPACT
  - IMPACT ACCELERATION
  - LANDING LOADS
  - PHYSIOLOGICAL ACCELERATION
  - ∞REDUCTION
  - RETARDING
  - RETROFIRING
  - RETROTHRUST
  - STOPPING
  - TAPERING
  - THRUST REVERSAL

DECELERATORS
- USE BRAKES (FOR ARRESTING MOTION)

DECEPTION
- RT AIR DEFENSE
  - CHAFF

DECEPTION-*(CONT.)*
       ELECTRONIC COUNTERMEASURES
       ELECTRONIC WARFARE
       OPTICAL COUNTERMEASURES
       SIMULATION

DECIDUOUS TREES
GS   PLANTS (BOTANY)
     . TREES (PLANTS)
     . . **DECIDUOUS TREES**
RT   CONIFERS
     EARTH RESOURCES
     FOLIAGE
     FORESTS
     LEAVES
     TIMBER IDENTIFICATION

DECIMAL TO BINARY CONVERTERS
GS   DATA CONVERTERS
     . **DECIMAL TO BINARY CONVERTERS**
RT   BINARY DATA
     BINARY TO DECIMAL CONVERTERS
     COMPUTER COMPONENTS
     DATA PROCESSING

DECIMALS
RT   DIGITS
     NUMBER THEORY
     ∞NUMBERS

DECIMETER WAVES
GS   ELECTROMAGNETIC RADIATION
     . RADIO WAVES
     . . SHORT WAVE RADIATION
     . . . MICROWAVES
     . . . . **DECIMETER WAVES**
RT   MILLIMETER WAVES
     PLANETARY RADIATION
     SOLAR RADIO EMISSION
     ULTRAHIGH FREQUENCIES

DECISION ELEMENTS
USE   LOGICAL ELEMENTS

DECISION MAKING
RT   COGNITION
     COMMAND AND CONTROL
     CONTRACT MANAGEMENT
     CONTRACT NEGOTIATION
     DECISIONS
     ECONOMY
     JUDGMENTS
     MANAGEMENT
     MANAGEMENT METHODS
     MANAGEMENT PLANNING
     PROBLEM SOLVING
     STARSITE PROGRAM
     SYSTEMS ENGINEERING
     TRADEOFFS

DECISION THEORY
GS   **DECISION THEORY**
     . STATISTICAL DECISION THEORY
RT   DYNAMIC PROGRAMMING
     EXPECTATION
     GAME THEORY
     INFORMATION THEORY
     MARTINGALES
     MATHEMATICAL MODELS
     OPERATIONS RESEARCH
     PROBABILITY THEORY
     RECOMMENDATIONS
     RISK
     SCHEDULING
     STATISTICAL ANALYSIS
     STOCHASTIC PROCESSES
     STRATEGY
     ∞SYNTHESIS
     SYSTEMS ENGINEERING
     ∞THEORIES

DECISIONS
RT   ∞COMMANDS
     CONTRACT MANAGEMENT
     DECISION MAKING
     JUDGMENTS
     LOGIC CIRCUITS
     MANAGEMENT
     PROCUREMENT POLICY
     PROJECT PLANNING
     SELECTION

DECKS (FLOORS)
USE   FLOORS

DECLINATION
RT   MAPPING
     NAVIGATION

DECODERS
RT   BCH CODES
     CODERS
     DATA CONVERTERS
     DECODING
     DEMODULATORS
     ∞TRANSLATORS

DECODING
GS   CODING
     . **DECODING**
RT   BCH CODES
     CONCATENATED CODES
     CRYPTOGRAPHY
     DATA COMPRESSION
     DECODERS
     DEMODULATION
     DICTIONARIES
     ∞INTERPRETATION
     TRANSLATING

DECOMMISSIONING
RT   RADIOACTIVE WASTES
     UNDERGROUND STORAGE

DECOMMUTATORS
GS   COMMUTATORS
     . **DECOMMUTATORS**
RT   DATA LINKS
     DEMODULATORS
     DIFFERENTIAL PULSE CODE
       MODULATION
     ELECTRIC MOTORS
     ∞GENERATORS
     PULSE CODE MODULATION
     TELEMETRY

DECOMPOSITION
GS   **DECOMPOSITION**
     . AMMONOLYSIS
     . CRACKING (CHEMICAL ENGINEERING)
     . GLYCOLYSIS
     . HYDROGENOLYSIS
     . NITROLYSIS
     . PHOTODECOMPOSITION
     . PHOTODISSOCIATION
     . PHOTOLYSIS
     . . RADIOLYSIS
     . PROPELLANT DECOMPOSITION
     . THERMAL DECOMPOSITION
RT   ABLATION
     BIODEGRADABILITY
     BIODEGRADATION
     CHARRING
     DAMAGE
     DEGRADATION
     DETERIORATION
     DISINTEGRATION
     DISSOCIATION
     ELECTROLYSIS
     LATERITES
     OVERVOLTAGE
     STORAGE STABILITY
     THERMAL DISSOCIATION

DECOMPRESSION
USE   PRESSURE REDUCTION

DECOMPRESSION SICKNESS
UF   BENDS (PHYSIOLOGY)
GS   SICKNESSES
     . **DECOMPRESSION SICKNESS**
RT   AEROEMBOLISM
     ALTITUDE SICKNESS
     BAROTRAUMA
     DIVING (UNDERWATER)

DECONDITIONING
GS   BEHAVIOR
     . **DECONDITIONING**
RT   LEARNING
     REFLEXES

DECONGESTANTS
GS   DRUGS
     . **DECONGESTANTS**
RT   ANTIHISTAMINICS

DECONTAMINATION
GS   STERILIZATION EFFECTS

DECONTAMINATION-*(CONT.)*
     . **DECONTAMINATION**
RT   AIR PURIFICATION
     ANTISEPTICS
     CARBON DIOXIDE CONCENTRATION
     CARBON DIOXIDE REMOVAL
     CLEANING
     CONTAMINANTS
     CONTAMINATION
     DEWAXING
     DISPOSAL
     DISSIPATION
     ELIMINATION
     EXHAUSTING
     EXTENSIONS
     ∞FOOD
     POLLUTION
     PURGING
     PURIFICATION
     PURITY
     ∞REDUCTION
     ∞SEPARATION
     SPACECRAFT CONTAMINATION
     STERILIZATION
     WASHING

DECOUPLING
GS   **DECOUPLING**
     . SPIN DECOUPLING
RT   COUPLING
     DISCONNECT DEVICES
     GRAVITINOS
     RELEASING

DECOYS
GS   **DECOYS**
     . BALLISTIC MISSILE DECOYS
     . BLUE GOOSE MISSILE
     . QUAIL MISSILE
     . REENTRY DECOYS
RT   COUNTERMEASURES
     DUMMIES

DECREMENTING
USE   REDUCTION

DEDUCTION
RT   DERIVATION
     INFERENCE

DEEP DRAWING
RT   BULGING
     COLD DRAWING
     COLD WORKING
     EXPLOSIVE FORMING
     MAGNETIC FORMING
     METAL WORKING
     STRETCHING

DEEP SCATTERING LAYERS
RT   ACOUSTIC SCATTERING
     ECHO SOUNDING
     ∞LAYERS
     OCEANOGRAPHY
     ORGANISMS
     SCATTERING
     SOUND WAVES
     UNDERWATER ACOUSTICS

DEEP SPACE
GS   ENVIRONMENTS
     . AEROSPACE ENVIRONMENTS
     . . **DEEP SPACE**
     . . . INTERPLANETARY SPACE
     . . . INTERSTELLAR SPACE
     . EXTRATERRESTRIAL ENVIRONMENTS
     . . **DEEP SPACE**
     . . . INTERPLANETARY SPACE
     . . . INTERSTELLAR SPACE
RT   CISLUNAR SPACE
     FRICTIONLESS ENVIRONMENTS
     LONG DURATION SPACE FLIGHT
     ∞SPACE

DEEP SPACE INSTRUMENTATION FACILITY
UF   DSIF (INSTRUMENTATION FACILITY)
GS   STATIONS
     . GROUND STATIONS
     . . **DEEP SPACE INSTRUMENTATION
        FACILITY**
     . TRACKING STATIONS
     . . **DEEP SPACE INSTRUMENTATION
        FACILITY**
RT   DATA ACQUISITION
     DATA TRANSMISSION

DEEP SPACE INSTRUMENTATION-*(CONT.)*
. RADIO CONTROL

DEEP SPACE NETWORK
GS   TRACKING NETWORKS
. **DEEP SPACE NETWORK**
RT   ∞ NETWORKS
SPACECRAFT TRACKING

DEEP WELL INJECTION (WASTES)
GS   INJECTION
. FLUID INJECTION
. . LIQUID INJECTION
. . . **DEEP WELL INJECTION (WASTES)**
RT   WASTE DISPOSAL

DEEPWATER TERMINALS
RT   ARTIFICIAL HARBORS
CARGO SHIPS
MARINE TECHNOLOGY
MARINE TRANSPORTATION
OCEANOGRAPHY
OFFSHORE DOCKING
OFFSHORE ENERGY SOURCES
OFFSHORE PLATFORMS
SHIP TERMINALS
TANKER SHIPS
TANKER TERMINALS
∞ TANKERS
TERMINAL FACILITIES
TRANSPORTATION

DEER
GS   ANIMALS
. VERTEBRATES
. . MAMMALS
. . . **DEER**
. . . . CARIBOUS
RT   GRAZING
LIVESTOCK

DEFECTS
UF   FLAWS
IMPERFECTIONS
GS   **DEFECTS**
. AUDITORY DEFECTS
. CRYSTAL DEFECTS
. . CRYSTAL DISLOCATIONS
. . . EDGE DISLOCATIONS
. . . SCREW DISLOCATIONS
. . POINT DEFECTS
. . . VACANCIES (CRYSTAL DEFECTS)
. . . . FRENKEL DEFECTS
. INCLUSIONS
. SPEECH DEFECTS
. SURFACE DEFECTS
RT   CASTINGS
CAVITIES
CRACKS
CUMULATIVE DAMAGE
DAMAGE
INHOMOGENEITY
IRREGULARITIES
LEAKAGE
PINHOLES
POROSITY
SCORING
VIGNETTING
VOIDS
X RAY ANALYSIS

DEFENDER PROJECT
GS   PROGRAMS
. PROJECTS
. . **DEFENDER PROJECT**

∞ DEFENSE
SN   *(USE OF A MORE SPECIFIC TERM IS*
RECOMMENDED--CONSULT THE TERMS
LISTED BELOW)
RT   AIR DEFENSE
ANTIMISSILE DEFENSE
CIVIL DEFENSE
DEFENSE COMMUNICATIONS SATELLITE
SYSTEM
DEFENSE COMMUNICATIONS SYSTEM
(DCS)
DEFENSE INDUSTRY
DEFENSE PROGRAM
DMSP SATELLITES
MISSILE DEFENSE
PHYSIOLOGICAL DEFENSES

DEFENSE COMMUNICATIONS SATELLITE SYSTEM
GS   TELECOMMUNICATION

DEFENSE COMMUNICATIONS-*(CONT.)*
. DEFENSE COMMUNICATIONS
SATELLITE SYSTEM
. . FLEET SATELLITE COMMUNICATION
SYSTEM
RT   COMMUNICATION SATELLITES
∞ DEFENSE
RADIO RELAY SYSTEMS
SPACE COMMUNICATION
∞ SYSTEMS

DEFENSE COMMUNICATIONS SYSTEM (DCS)
GS   TELECOMMUNICATION
. **DEFENSE COMMUNICATIONS SYSTEM**
(DCS)
RT   COMMUNICATION NETWORKS
∞ DEFENSE
MILITARY TECHNOLOGY
∞ SYSTEMS

DEFENSE INDUSTRY
GS   INDUSTRIES
. **DEFENSE INDUSTRY**
. . WEAPONS INDUSTRY
RT   ANTIMISSILE DEFENSE
∞ DEFENSE
MILITARY TECHNOLOGY
MISSILE DEFENSE

DEFENSE METEOROLOGICAL SATELLITE
PROGRAM
USE   DMSP SATELLITES

DEFENSE PROGRAM
GS   PROGRAMS
. **DEFENSE PROGRAM**
RT   AIR DEFENSE
ANTIMISSILE DEFENSE
ARMED FORCES (UNITED STATES)
CIVIL DEFENSE
∞ DEFENSE
DMSP SATELLITES
MILITARY TECHNOLOGY
MISSILE DEFENSE
SPACE TRANSPORTATION SYSTEM
WEAPONS DELIVERY

∞ DEFINITION
SN   *(USE OF A MORE SPECIFIC TERM IS*
RECOMMENDED--CONSULT THE TERMS
LISTED BELOW)
RT   ACCURACY
DELINEATION
DESCRIPTIONS
DICTIONARIES
∞ MEASUREMENT
NOMENCLATURES
PRECISION
RESOLUTION

DEFLAGRATION
GS   COMBUSTION
. **DEFLAGRATION**
RT   BACKFIRE
FIRES
FLASHBACK

DEFLATING
USE   INFLATABLE STRUCTURES
PRESSURE REDUCTION

DEFLECTION
RT   BENDING
BENDING DIAGRAMS
CAMBER
DEFORMATION
DIFFRACTION
DISPERSING
DISPLACEMENT
DISTORTION
ELASTIC DEFORMATION
FLEXING
MAXWELL-MOHR METHOD
REFLECTION
REFRACTION
SCATTERING
STRUCTURAL STRAIN
TEMPERATURE INVERSIONS
TORSION
VARIATIONS
WAVE DISPERSION
YOKES

DEFLECTORS
GS   **DEFLECTORS**
. BLAST DEFLECTORS
. FLAME DEFLECTORS
RT   ATTENUATORS
BAFFLES
∞ DIFFUSERS
DIVERTERS
FLOW DEFLECTION
GUST ALLEVIATORS
REFLECTORS
SAFETY DEVICES
SHIELDING
SPOILERS

DEFLUORINATION
GS   CHEMICAL REACTIONS
. **DEFLUORINATION**
RT   FLUORINATION
HALOGENATION

DEFOCUSING
GS   FOCUSING
. **DEFOCUSING**
RT   ∞ OPTICS

DEFOLIANTS
RT   DEFOLIATION
FOLIAGE
FORESTS
HERBICIDES
LEAVES
PLANTS (BOTANY)
TREES (PLANTS)

DEFOLIATION
RT   BRUSH (BOTANY)
DEFOLIANTS
DEFORESTATION
FORESTS
GRASSES
LEAVES
PLANTS (BOTANY)
TREES (PLANTS)

DEFORESTATION
RT   CLEARINGS (OPENINGS)
CONSERVATION
DEFOLIATION
ENVIRONMENT EFFECTS
FORESTS

DEFORMATION
GS   **DEFORMATION**
. AXIAL STRAIN
. ELASTIC DEFORMATION
. . ELASTIC BENDING
. . ELASTIC BUCKLING
. NUCLEAR DEFORMATION
. PLASTIC DEFORMATION
. STATIC DEFORMATION
. TENSILE DEFORMATION
. WAVE FRONT DEFORMATION
RT   BENDING
BUCKLING
CAMBER
COLLAPSE
CORRUGATING
CREEP PROPERTIES
DAMAGE
DEFLECTION
DEFORMETERS
DISPLACEMENT
DISTORTION
ELONGATION
FAILURE
FLEXING
FRACTURES (MATERIALS)
INDENTATION
MECHANICAL PROPERTIES
SET
SKEWNESS
STIFFNESS
STRUCTURAL FAILURE
STRUCTURAL STRAIN
TEMPERATURE INVERSIONS
TOPOLOGY
TORSION
TWISTING
VOLUMETRIC STRAIN
WARPAGE
WRINKLING

DEFORMETERS
GS   MEASURING INSTRUMENTS

DEFORMETERS-*(CONT.)*
. **DEFORMETERS**
RT DEFORMATION
    DIMENSIONAL MEASUREMENT
    EXTENSOMETERS
    MECHANICAL MEASUREMENT
    STRAIN GAGES
    STRESS MEASUREMENT
    TENSOMETERS

DEFROSTING
RT DEICING
    HEATING
    ICE PREVENTION
    MELTING
    REFRIGERATING
    REFRIGERATORS

DEGASSING
UF BAKEOUT
GS **DEGASSING**
    . DEOXYGENATION
RT ABSORBERS (EQUIPMENT)
    AERATION
    BAKING
    CASTINGS
    DEOXIDIZING
    DESORPTION
    GAS EVOLUTION
    OCCLUSION
    OFFGASSING
    OUTGASSING
    PURGING
    SCAVENGING
    ∞ SEPARATION

DEGENERATION
RT ATROPHY
    DETERIORATION
    NEGATIVE FEEDBACK

DEGENERATIVE FEEDBACK
USE NEGATIVE FEEDBACK

DEGRADATION
GS **DEGRADATION**
    . BIODEGRADATION
    . THERMAL DEGRADATION
    . WAVE DEGRADATION
RT CHEMICAL ATTACK
    CORROSION
    CUMULATIVE DAMAGE
    CURING
    DAMAGE
    DECAY
    DECOMPOSITION
    DEPOLYMERIZATION
    DETERIORATION
    DISCOLORATION
    DURABILITY
    EMBRITTLEMENT
    EROSION
    HOT CORROSION
    OXIDATION
    PITTING
    PRESERVING
    RUSTING
    SCALE (CORROSION)
    STERILIZATION EFFECTS
    THERMAL DISSOCIATION
    WEATHERING

DEGREES OF FREEDOM
RT EQUIPARTITION THEOREM
    EXPERIMENT DESIGN
    FACTOR ANALYSIS
    NULL HYPOTHESIS
    PHASE RULE
    QUALITY CONTROL
    SIGNIFICANCE
    THREE DIMENSIONAL MOTION
    TORQUERS
    ∞ VARIANCE

DEHP
USE DIETHYL HYDROGEN PHOSPHITE (DEHP)

DEHUMIDIFICATION
GS DRYING
    . **DEHUMIDIFICATION**
RT CONDENSING
    COOLING SYSTEMS
    DEHYDRATION
    DIFFUSION
    HUMIDITY

DEHUMIDIFICATION-*(CONT.)*
    REFRIGERATING
    ∞ SEPARATION
    SILICA GEL

DEHYDRATED FOOD
GS CONSUMABLES (SPACECRAFT)
    . CONSUMABLES (SPACECREW
      SUPPLIES)
    . . **DEHYDRATED FOOD**
RT DEHYDRATION
    DRYING APPARATUS
    ∞ FOOD
    FOOD PROCESSING
    FREEZE DRYING
    PRESERVING
    SPACE FLIGHT FEEDING

DEHYDRATION
GS DRYING
    . **DEHYDRATION**
RT COLUMNS (PROCESS ENGINEERING)
    DEHUMIDIFICATION
    DEHYDRATED FOOD
    DEWATERING
    EVAPORATION
    FREEZE DRYING
    HYDRATION
    PLASMOLYSIS
    ∞ SEPARATION
    SILICA GEL
    THERMOGRAVIMETRY
    WATER LOSS

DEHYDROGENATION
GS CHEMICAL REACTIONS
    . **DEHYDROGENATION**
RT COLUMNS (PROCESS ENGINEERING)
    HYDROFORMING
    HYDROGENATION
    HYDROGENOLYSIS
    OXIDATION
    REDUCTION (CHEMISTRY)

DEICERS
UF DEICING SYSTEMS
RT AIRFOILS
    ANTIICING ADDITIVES
    DEICING
    ∞ HEATERS
    HEATING EQUIPMENT
    ICE PREVENTION

DEICING
RT AIRFOILS
    ANTIICING ADDITIVES
    DEFROSTING
    DEICERS
    ∞ HEATERS
    HEATING EQUIPMENT
    ICE PREVENTION
    MELTING

DEICING SYSTEMS
USE DEICERS

DEIMOS
GS CELESTIAL BODIES
    . NATURAL SATELLITES
    . . **DEIMOS**
    SATELLITES
    . NATURAL SATELLITES
    . . **DEIMOS**
RT CHARON
    MARS (PLANET)
    PHOBOS

DEIONIZATION
GS CHEMICAL REACTIONS
    . **DEIONIZATION**
RT ATOMIC RECOMBINATION
    DEMINERALIZING
    EXCHANGING
    ION RECOMBINATION
    RADIATIVE RECOMBINATION
    ∞ SEPARATION
    SOFTENING

DEKATRONS
USE COUNTERS

DELAMINATING
RT ANODIC STRIPPING
    PEELING

DELAMINATING-*(CONT.)*
    ∞ SEPARATION

DELAWARE
GS NATIONS
    . UNITED STATES
    . . **DELAWARE**
RT DELAWARE RIVER BASIN (US)
    DELMARVA PENINSULA (DE-MD-VA)

DELAWARE BAY (US)
GS BAYS (TOPOGRAPHIC FEATURES)
    . **DELAWARE BAY (US)**
RT ATLANTIC OCEAN
    GULFS
    INLETS (TOPOGRAPHY)
    NEW JERSEY
    PENNSYLVANIA

DELAWARE RIVER BASIN (US)
GS LANDFORMS
    . STRUCTURAL BASINS
    . . RIVER BASINS
    . . . **DELAWARE RIVER BASIN (US)**
RT DELAWARE
    NEW JERSEY
    NEW YORK
    PENNSYLVANIA
    RIVERS
    STREAMS
    VALLEYS

DELAY
RT DWELL
    ∞ HOLDING
    LATENESS
    STOPPING
    TIME LAG
    ∞ TIME RESPONSE

DELAY CIRCUITS
GS CIRCUITS
    . **DELAY CIRCUITS**
    . . PHANTASTRONS
RT ACOUSTIC DELAY LINES
    CIRCULATORS (PHASE SHIFT CIRCUITS)
    COMPARATOR CIRCUITS
    PHASE SHIFT CIRCUITS

DELAY LINES
GS **DELAY LINES**
    . ACOUSTIC DELAY LINES
    . DELAY LINES (COMPUTER STORAGE)
RT ∞ LINES
    TIME LAG

DELAY LINES (COMPUTER STORAGE)
GS DELAY LINES
    . **DELAY LINES (COMPUTER STORAGE)**
RT SHIFT REGISTERS

DELAYED FLAP APPROACH
UF DFA
GS APPROACH
    . **DELAYED FLAP APPROACH**
RT FLAPS (CONTROL SURFACES)
    FLIGHT PATHS
    LANDING AIDS
    NASA PROGRAMS
    NOISE REDUCTION

DELETION
GS ELIMINATION
    . **DELETION**
RT DISPOSAL
    REMOVAL

DELFIN AIRCRAFT
USE L-29 JET TRAINER

DELFT CAMERA
GS OPTICAL EQUIPMENT
    . CAMERAS
    . . **DELFT CAMERA**
    PHOTOGRAPHIC EQUIPMENT
    . CAMERAS
    . . **DELFT CAMERA**

DELINEATION
RT BOUNDARIES
    ∞ DEFINITION
    ∞ PROFILES

DELIVERY
GS   DELIVERY
      . PAYLOAD DELIVERY (STS)
      . WEAPONS DELIVERY
RT   AIR DROP OPERATIONS
      AIRDROPS
      CARGO
      CIRCULATION
      HAULING
      MATERIALS HANDLING
      OUTPUT
      ∞RECEIVING
      TRANSPORTATION
      TRUCKS

DELMARVA PENINSULA (DE-MD-VA)
GS   LANDFORMS
      . PENINSULAS
      . . DELMARVA PENINSULA (DE-MD-VA)
RT   DELAWARE
      MARYLAND
      VIRGINIA

DELPHI METHOD (FORECASTING)
GS   FORECASTING
      . DELPHI METHOD (FORECASTING)
      MANAGEMENT METHODS
      . DELPHI METHOD (FORECASTING)
RT   ESTIMATING
      ∞METHODOLOGY
      OPERATIONS RESEARCH
      PATTERN METHOD (FORECASTING)
      PLANNING
      PREDICTIONS
      PROBE METHOD (FORECASTING)
      PROFILE METHOD (FORECASTING)
      TECHNOLOGY ASSESSMENT

DELRIN (TRADEMARK)
GS   PLASTICS
      . DELRIN (TRADEMARK)
RT   RESINS

DELTA ANTENNAS
GS   ANTENNAS
      . DELTA ANTENNAS
RT   ANTENNA DESIGN
      RESONATORS
      TRANSMISSION LINES

DELTA DAGGER AIRCRAFT
USE   F-102 AIRCRAFT

DELTA DART AIRCRAFT
USE   F-106 AIRCRAFT

DELTA FUNCTION
GS   ANALYSIS (MATHEMATICS)
      . REAL VARIABLES
      . . DELTA FUNCTION
      FUNCTIONS (MATHEMATICS)
      . DELTA FUNCTION

DELTA LAUNCH VEHICLE
GS   LAUNCH VEHICLES
      . DELTA LAUNCH VEHICLE
RT   ANIK SATELLITES
      ANIK 1
      ANIK 2
      BEACON EXPLORER A
      ESSA 1 SATELLITE
      ESSA 2 SATELLITE
      ESSA 3 SATELLITE
      ESSA 4 SATELLITE
      ESSA 5 SATELLITE
      ESSA 6 SATELLITE
      ESSA 7 SATELLITE
      ESSA 8 SATELLITE
      ESSA 9 SATELLITE
      EXPLORER 10 SATELLITE
      EXPLORER 12 SATELLITE
      EXPLORER 14 SATELLITE
      EXPLORER 15 SATELLITE
      EXPLORER 17 SATELLITE
      EXPLORER 18 SATELLITE
      EXPLORER 21 SATELLITE
      EXPLORER 26 SATELLITE
      EXPLORER 28 SATELLITE
      EXPLORER 29 SATELLITE
      EXPLORER 32 SATELLITE
      EXPLORER 33 SATELLITE
      EXPLORER 38 SATELLITE
      EXPLORER 43 SATELLITE
      EXPLORER 49 SATELLITE
      EXPLORER 55 SATELLITE

DELTA LAUNCH VEHICLE-(CONT.)
      INTERNATIONAL MAGNETOSPHERIC
        EXPLORER
      OSO-C
      OSO-1
      OSO-2
      OSO-4
      OUTER PLANETS EXPLORERS
      PIONEER 6 SPACE PROBE
      PIONEER 7 SPACE PROBE
      RCA SATCOM SATELLITES
      SPACE SHUTTLE UPPER STAGE D
      SYNCOM 1 SATELLITE
      SYNCOM 2 SATELLITE
      SYNCOM 3 SATELLITE
      TIROS 2 SATELLITE
      TIROS 3 SATELLITE
      TIROS 4 SATELLITE
      TIROS 5 SATELLITE
      TIROS 6 SATELLITE
      TIROS 7 SATELLITE
      TIROS 8 SATELLITE
      TIROS 9 SATELLITE
      TIROS 10 SATELLITE

DELTA MODULATION
GS   MODULATION
      . PULSE MODULATION
      . . PULSE CODE MODULATION
      . . . DELTA MODULATION
RT   PULSE COMMUNICATION

DELTA WINGS
UF   TRIANGULAR WINGS
GS   AIRFOILS
      . WINGS
      . . LOW ASPECT RATIO WINGS
      . . . DELTA WINGS
      . . SWEPT WINGS
      . . . SWEPTBACK WINGS
      . . . . DELTA WINGS
      PLANFORMS
      . WING PLANFORMS
      . . SWEPTBACK WINGS
      . . . DELTA WINGS
RT   ARROW WINGS
      AVRO 707 AIRCRAFT
      CARET WINGS
      FD 2 AIRCRAFT
      GA-5 AIRCRAFT
      VARIABLE SWEEP WINGS
      VATOL AIRCRAFT

DELTAS
GS   LANDFORMS
      . DELTAS
      . . MISSISSIPPI DELTA (LA)
      . . RHONE DELTA (FRANCE)
RT   ALLUVIUM
      FANS (LANDFORMS)
      RIVERS
      SANDS
      SOILS

DEMAGNETIZATION
RT   MAGNETIC FIELDS
      ∞REDUCTION

DEMAND (ECONOMICS)
GS   ECONOMICS
      . DEMAND (ECONOMICS)
RT   CONSUMPTION
      SUPPLYING

DEMAND ASSIGNMENT MULTIPLE ACCESS
UF   DAMA
GS   TELECOMMUNICATION
      . MULTIPLE ACCESS
      . . DEMAND ASSIGNMENT MULTIPLE
          ACCESS
RT   CHANNEL CAPACITY
      COMMUNICATION NETWORKS
      COMMUNICATION SATELLITES
      SATELLITE NETWORKS

DEMINERALIZING
GS   DEMINERALIZING
      . BONE DEMINERALIZATION
RT   CRYSTALLIZATION
      DEIONIZATION
      DESALINIZATION
      DISTILLATION
      ION EXCHANGING
      OSMOSIS
      PURIFICATION

DEMINERALIZING-(CONT.)
      REVERSE OSMOSIS
      ∞SEPARATION
      SOFTENING
      WATER TREATMENT

DEMOCRATIC PEOPLES REPUBLIC OF KOREA
USE   NORTH KOREA

DEMODULATION
RT   AMPLITUDE MODULATION
      DECODING
      DEMODULATORS
      ∞DETECTORS
      FREQUENCY MODULATION
      HETERODYNING
      INTERMODULATION
      MODULATION
      PHASE MODULATION
      PULSE MODULATION
      REMODULATION
      TELECOMMUNICATION

DEMODULATORS
GS   DEMODULATORS
      . FREQUENCY COMPRESSION
          DEMODULATORS
      . MODEMS
      . PHASE DEMODULATORS
      . PHASE LOCK DEMODULATORS
RT   AMPLITUDE MODULATION
      DECODERS
      DECOMMUTATORS
      DEMODULATION
      FREQUENCY MODULATION
      MATCHED FILTERS
      MODULATION
      MODULATORS
      PHASE MODULATION
      PULSE MODULATION

DEMOGRAPHY
RT   CENSUS
      COMMUNITIES
      ∞DENSITY
      ∞DISTRIBUTION
      HUMAN BEINGS
      INHABITANTS
      MEGALOPOLISES
      NATIONS
      SOCIOLOGY
      ∞STATISTICS

DEMONSTRATION
USE   PROVING

DEMULTIPLEXING
GS   TRANSMISSION
      . DEMULTIPLEXING
RT   CODE DIVISION MULTIPLEXING
      FREQUENCY DIVISION MULTIPLEXING
      MULTIPLEXING
      TIME DIVISION MULTIPLEXING
      WAVELENGTH DIVISION MULTIPLEXING

DENDRITIC CRYSTALS
GS   CRYSTALS
      . DENDRITIC CRYSTALS
RT   ISOTROPY
      NEEDLES
      WHISKERS (CRYSTALS)

DENDRITIC DRAINAGE
USE   DRAINAGE PATTERNS

DENDROCHRONOLOGY
UF   TREE RING DATING
RT   CLIMATOLOGY
      GEOCHRONOLOGY
      PERIODIC VARIATIONS
      TIMBERLINE
      TREES (PLANTS)

DENITROGENATION
GS   CHEMICAL REACTIONS
      . DENITROGENATION
RT   NITRATION

DENMARK
GS   NATIONS
      . DENMARK
RT   EUROPE
      GREENLAND
      SCANDINAVIA

DENSE PLASMAS
GS   GASES
        . IONIZED GASES
        . . CHARGED PARTICLES
        . . . **DENSE PLASMAS**
        . . . . STRONGLY COUPLED PLASMAS
        PARTICLES
        . CHARGED PARTICLES
        . . ENERGETIC PARTICLES
        . . . PLASMAS (PHYSICS)
        . . . . **DENSE PLASMAS**
        . . . . . PLASMA FOCUS
        . . . . . STRONGLY COUPLED PLASMAS
RT   ASTROPHYSICS
        BETA FACTOR
        ELECTRON SCATTERING
        HIGH TEMPERATURE PLASMAS
        NUCLEAR FUSION
        PARTICLE COLLISIONS
        PLASMA COMPRESSION
        SPHEROMAKS
        STELLAR STRUCTURE

DENSIFICATION
GS   PRESSURE
        . **DENSIFICATION**
RT   AGGLOMERATION
        COMPACTING
        COMPRESSING
        CONSOLIDATION
        LUDOX (TRADEMARK)
        PRESSURIZING

DENSIMETERS
GS   MEASURING INSTRUMENTS
        . **DENSIMETERS**
        . . ULTRASONIC DENSIMETERS
RT   DENSITY (MASS/VOLUME)
        DENSITY MEASUREMENT
        ∞ INSTRUMENTS
        ∞ MEASUREMENT

DENSITOMETERS
GS   MEASURING INSTRUMENTS
        . **DENSITOMETERS**
        . . MICRODENSITOMETERS
RT   GAMMA RAY ABSORPTIOMETRY
        GRAVIMETERS
        OPTICAL EQUIPMENT
        OPTICAL MEASUREMENT
        OPTICAL MEASURING INSTRUMENTS
        PHOTOMETERS
        PHOTON ABSORPTIOMETRY
        TRANSMISSOMETERS

∞ **DENSITY**
SN   (USE OF A MORE SPECIFIC TERM IS
        RECOMMENDED--CONSULT THE TERMS
        LISTED BELOW)
RT   ATMOSPHERIC DENSITY
        ATOM CONCENTRATION
        BIOMASS
        DEMOGRAPHY
        DENSITY (MASS/VOLUME)
        DENSITY (NUMBER/VOLUME)
        FLUX DENSITY
        OPTICAL DENSITY
        POROSITY
        RANKINE-HUGONIOT RELATION

DENSITY (MASS/VOLUME)
UF   SPECIFIC GRAVITY
GS   **DENSITY (MASS/VOLUME)**
        . ATMOSPHERIC DENSITY
        . GAS DENSITY
        . SPACE DENSITY
RT   ABSORPTANCE
        ABSORPTIVITY
        BULK MODULUS
        BUOYANCY
        COMPRESSIBILITY
        DENSIMETERS
        ∞ DENSITY
        DENSITY MEASUREMENT
        HYDROMETERS
        INTERNAL FRICTION
        ISOPYCNIC PROCESSES
        LEWIS NUMBERS
        OPACITY
        PERMEABILITY
        ∞ PHYSICAL PROPERTIES
        POROSITY
        PYCNOMETERS
        STOPPING POWER
        TRANSMISSIVITY

DENSITY (MASS/VOLUME)-(CONT.)
        TRANSMITTANCE
        TRANSPARENCE
        ULTRASONIC DENSIMETERS
        VISCOSITY
        VOID RATIO
        WEIGHT MEASUREMENT

DENSITY (NUMBER/VOLUME)
GS   **DENSITY (NUMBER/VOLUME)**
        . METEOROID CONCENTRATION
        . PACKING DENSITY
        . PARTICLE DENSITY (CONCENTRATION)
        . . ELECTRON DENSITY
                (CONCENTRATION)
        . . . CARRIER DENSITY (SOLID STATE)
        . . . ELECTRON DENSITY PROFILES
        . . . IONOSPHERIC ELECTRON DENSITY
        . . . MAGNETOSPHERIC ELECTRON
                DENSITY
        . . ELECTRON DISTRIBUTION
        . . . ELECTRON DENSITY PROFILES
        . . ION DENSITY (CONCENTRATION)
        . . . IONOSPHERIC ION DENSITY
        . . . MAGNETOSPHERIC ION DENSITY
        . . . . MAGNETOSPHERIC PROTON
                DENSITY
        . . . PROTON DENSITY
                (CONCENTRATION)
        . . . . MAGNETOSPHERIC PROTON
                DENSITY
        . . PLASMA DENSITY
        . SPACE DENSITY
RT   ATMOSPHERIC DENSITY
        ∞ DENSITY

DENSITY (RATE/AREA)
USE   FLUX DENSITY

DENSITY DISTRIBUTION
RT   BAROCLINIC WAVES
        FOKKER-PLANCK EQUATION
        MAXWELL-BOLTZMANN DENSITY
                FUNCTION
        SHOCK DISCONTINUITY
        TAYLOR INSTABILITY

DENSITY MEASUREMENT
GS   **DENSITY MEASUREMENT**
        . GAMMA RAY ABSORPTIOMETRY
        . PHOTON ABSORPTIOMETRY
        . X RAY DENSITY MEASUREMENT
RT   CHEMICAL ANALYSIS
        DENSIMETERS
        DENSITY (MASS/VOLUME)
        HYDROMETERS
        ∞ MEASUREMENT
        MECHANICAL MEASUREMENT
        ULTRASONIC DENSIMETERS
        WIND TUNNEL TESTS

DENSITY WAVE MODEL
GS   MODELS
        . ASTRONOMICAL MODELS
        . . **DENSITY WAVE MODEL**
RT   GALACTIC STRUCTURE
        MASS DISTRIBUTION
        SPIRAL GALAXIES
        WAVE EQUATIONS

DENTAL CALCULI
GS   DEPOSITS
        . CALCULI
        . . **DENTAL CALCULI**
RT   LITHIASIS
        TEETH
        TOOTH DISEASES

DENTISTRY
GS   MEDICAL SCIENCE
        . **DENTISTRY**
RT   MEDICAL EQUIPMENT
        ORAL HYGIENE
        TEETH
        TOOTH DISEASES

DEOXIDIZING
GS   CHEMICAL REACTIONS
        . REDUCTION (CHEMISTRY)
        . . **DEOXIDIZING**
RT   DEGASSING
        ∞ DEOXIFICATION
        DEOXYGENATION
        SCAVENGING

∞ **DEOXIFICATION**
SN   (USE OF A MORE SPECIFIC TERM IS
        RECOMMENDED--CONSULT THE TERMS
        LISTED BELOW)
RT   DEOXIDIZING
        DEOXYGENATION
        STERILIZATION EFFECTS

DEOXYGENATION
GS   DEGASSING
        . **DEOXYGENATION**
RT   DEOXIDIZING
        ∞ DEOXIFICATION
        ∞ REDUCTION
        ∞ SEPARATION

DEOXYRIBONUCLEIC ACID
UF   DNA
GS   ACIDS
        . NUCLEIC ACIDS
        . . **DEOXYRIBONUCLEIC ACID**
RT   THYMIDINE
        THYMINE

DEPENDENCE
UF   DEPENDENCY
GS   **DEPENDENCE**
        . SPATIAL DEPENDENCIES
        . TEMPERATURE DEPENDENCE
        . TIME DEPENDENCE
RT   GROUP DYNAMICS
        SOCIOLOGY

DEPENDENCY
USE   DEPENDENCE

DEPENDENT VARIABLES
GS   ANALYSIS (MATHEMATICS)
        . **DEPENDENT VARIABLES**
RT   COMPLEX VARIABLES
        INDEPENDENT VARIABLES
        OBSERVABILITY (SYSTEMS)
        PARAMETERIZATION
        REAL VARIABLES
        ∞ VARIABLE

DEPERSONALIZATION
RT   ARTIFICIAL INTELLIGENCE
        AUTOMATA THEORY
        AUTOMATIC CONTROL
        ∞ AUTOMATION
        CYBERNETICS
        DETACHMENT
        DISORDERS
        MAN MACHINE SYSTEMS
        MECHANIZATION
        PERSONALITY
        PERSONNEL

DEPLETION
RT   CONSUMPTION
        DEPRECIATION
        DISSIPATION
        ELIMINATION
        ENERGY POLICY
        EXHAUSTION
        EXPLOITATION
        LIFE (DURABILITY)
        LOSSES
        ∞ REDUCTION
        REMOVAL
        RESOURCES
        UTILIZATION

DEPLOYMENT
RT   GAME THEORY
        LOGISTICS
        MILITARY OPERATIONS
        MILITARY TECHNOLOGY
        ∞ OPERATIONS
        PERSONNEL
        STRATEGY
        TACTICS

DEPOLARIZATION
SN   (EXCLUDES CONSIDERATION OF
        OPTICAL DEPOLARIZATION AND
        PARTICLE SPIN DISALIGNMENT)
UF   DEPOLARIZERS
RT   ELECTROLYTIC POLARIZATION
        ELECTROPHYSIOLOGY
        POLARIZATION (CHARGE SEPARATION)
        ∞ REDUCTION
        SPIKE POTENTIALS

DEPOLARIZERS
USE   DEPOLARIZATION

DEPOLYMERIZATION
GS   CHEMICAL REACTIONS
. **DEPOLYMERIZATION**
RT   DEGRADATION
DETERIORATION
POLYMERIZATION

DEPOSITION
UF   ACCRETION
GS   **DEPOSITION**
. ANODIZING
. ELECTRODEPOSITION
. . ELECTROPLATING
. ELECTROLESS DEPOSITION
. VAPOR DEPOSITION
. . VACUUM DEPOSITION
RT   ACCUMULATIONS
COAGULATION
COATING
COATINGS
DEPOSITS
ELECTROFORMING
ELECTRON BOMBARDMENT
FORMING TECHNIQUES
FOULING
MAGNETRON SPUTTERING
METAL COATINGS
PLATING
PRECIPITATION (CHEMISTRY)
SEDIMENTS
∞SEPARATION
SETTLING
SPUTTERING

DEPOSITS
SN   (EXCLUDES BANK MINERAL AND
GEOLOGICAL DEPOSITS)
GS   **DEPOSITS**
. CALCULI
. . DENTAL CALCULI
. CRYODEPOSITS
RT   COATINGS
CORROSION
CRUDE OIL
DEPOSITION
PLATING
SEDIMENTS
SLUDGE

DEPRECIATION
RT   DEPLETION
DETERIORATION
INVESTMENTS
LIFE (DURABILITY)
WEAR

DEPRESSANTS
GS   **DEPRESSANTS**
. CENTRAL NERVOUS SYSTEM
DEPRESSANTS
RT   ANESTHESIOLOGY

∞ **DEPRESSION**
SN   (USE OF A MORE SPECIFIC TERM IS
RECOMMENDED--CONSULT THE TERMS
LISTED BELOW)
RT   DETACHMENT
DISORDERS
EMOTIONS
∞HOLLOW
INTROVERSION
LETHARGY
LOW PRESSURE
NEUROTIC DEPRESSION
PSYCHOTIC DEPRESSION
RECESSION
SCHIZOPHRENIA
TECTONICS
TOPOGRAPHY

DEPRESSIONS (TOPOGRAPHY)
USE   STRUCTURAL BASINS

DEPRESSURIZATION
USE   PRESSURE REDUCTION

DEPRIVATION
GS   **DEPRIVATION**
. SENSORY DEPRIVATION
. SLEEP DEPRIVATION
. WATER DEPRIVATION

DEPRIVATION-(CONT.)
RT   CONFINING
ISOLATION
STRESS (PHYSIOLOGY)

DEPTH
GS   DIMENSIONS
. **DEPTH**
RT   DISTANCE
HEIGHT
THICKNESS

DEPTH MEASUREMENT
RT   BATHYMETERS
CORE SAMPLING
DISTANCE MEASURING EQUIPMENT
ECHO SOUNDING
∞MEASUREMENT
MECHANICAL MEASUREMENT
SOUNDING

DEPTH PERCEPTION
USE   SPACE PERCEPTION

DERIVATION
RT   DEDUCTION
∞INDUCTION
∞ORIGINS
PARAMETERIZATION
∞SOURCES

DERIVATION CALCULUS
USE   DIFFERENTIAL CALCULUS

DERMATITIS
GS   DISEASES
. INFECTIOUS DISEASES
. . DERMATITIS
. . . CONTACT DERMATITIS
RT   DERMATOLOGY
ITCHING
RADIATION HAZARDS
RADIATION SICKNESS
SKIN (ANATOMY)

DERMATOLOGY
GS   MEDICAL SCIENCE
. **DERMATOLOGY**
RT   CONTACT DERMATITIS
DERMATITIS
SKIN (ANATOMY)

DESALINIZATION
RT   DEMINERALIZING
DISTILLATION
OSMOSIS
PURIFICATION
REVERSE OSMOSIS
SALINITY
VAPORIZING
WATER TREATMENT

DESATURATION
RT   DRYING
∞SATURATION

DESCALING
RT   CHEMICAL CLEANING
CLEANING
METAL FINISHING
PICKLING (METALLURGY)
SCALE (CORROSION)
∞SEPARATION
SHOT PEENING

DESCENT
GS   **DESCENT**
. PARACHUTE DESCENT
RT   APPROACH
ASCENT
FLIGHT PATHS
GLIDING
REENTRY
UNCONTROLLED REENTRY
(SPACECRAFT)

DESCENT PROPULSION SYSTEMS
GS   PROPULSION
. **DESCENT PROPULSION SYSTEMS**
PROPULSION SYSTEM CONFIGURATIONS
. **DESCENT PROPULSION SYSTEMS**
RT   SPACECRAFT PROPULSION
∞SYSTEMS

DESCENT TRAJECTORIES
GS   TRAJECTORIES
. **DESCENT TRAJECTORIES**
. . REENTRY TRAJECTORIES
RT   ASCENT TRAJECTORIES
ATMOSPHERIC ENTRY
BALLISTIC TRAJECTORIES
COASTING FLIGHT
FALLING
FLIGHT MECHANICS
MANNED REENTRY
MIDCOURSE TRAJECTORIES
MISSILE TRAJECTORIES
PARABOLIC FLIGHT
REENTRY
REENTRY GUIDANCE
SPACECRAFT TRAJECTORIES
TERMINAL GUIDANCE

DESCRIPTIONS
RT   CHARACTERIZATION
∞DEFINITION
NOMENCLATURES
REPRESENTATIONS

DESCRIPTIVE GEOMETRY
GS   GEOMETRY
. EUCLIDEAN GEOMETRY
. . **DESCRIPTIVE GEOMETRY**
RT   ANALYTIC GEOMETRY
ENGINEERING DRAWINGS
LAYOUTS
∞PROJECTION
PROJECTIVE GEOMETRY
TORUSES

DESENSITIZING
RT   CORROSION PREVENTION
PROTECTIVE COATINGS
RUSTING

DESERT ADAPTATION
GS   ADAPTATION
. **DESERT ADAPTATION**
RT   SURVIVAL

DESERTIFICATION
RT   ARID LANDS
BARREN LAND
DEATH VALLEY (CA)
DESERTLINE
DESERTS
DROUGHT
EARTH ENVIRONMENT
GOBI DESERT
LAND
LAND USE
MAN ENVIRONMENT INTERACTIONS
MOJAVE DESERT (CA)
OASES
REMOTE SENSING
SAHARA DESERT (AFRICA)
STEPPES
WADIS

DESERTLINE
RT   ARID LANDS
CLIMATOLOGY
DESERTIFICATION
LAND
TOPOGRAPHY

DESERTS
GS   LAND
. **DESERTS**
. . GOBI DESERT
. . LIBYAN DESERT
. . MOJAVE DESERT (CA)
. . SAHARA DESERT (AFRICA)
RT   ARID LANDS
BARREN LAND
CLIMATOLOGY
COACHELLA VALLEY (CA)
DEATH VALLEY (CA)
DESERTIFICATION
DUNES
EARTH RESOURCES
IMPERIAL VALLEY (CA)
KALIHARI BASIN (AFRICA)
OASES
PALO VERDE VALLEY (CA)
PLAYAS
REMOTE REGIONS
TOPOGRAPHY
WILDERNESS

DESICCANTS
RT   ABSORBENTS
      ADSORBENTS

DESICCATION
USE   DRYING

DESICCATORS
GS   SEPARATORS
      . DRYING APPARATUS
      . . **DESICCATORS**

∞ **DESIGN**
SN   (USE OF A MORE SPECIFIC TERM IS
      RECOMMENDED--CONSULT THE TERMS
      LISTED BELOW)
UF   TAILORING
RT   AERODYNAMIC CONFIGURATIONS
      AIRCRAFT DESIGN
      AMPLIFIER DESIGN
      ANTENNA DESIGN
      COMPUTER AIDED DESIGN
      COMPUTER DESIGN
      COMPUTER SYSTEMS DESIGN
      CONSTRUCTION
      CONTROL SYSTEMS DESIGN
      DESIGN ANALYSIS
      DESIGN TO COST
      DIMENSIONS
      DRAFTING MACHINES
      ENGINE DESIGN
      ENGINEERING DRAWINGS
      EQUIPMENT SPECIFICATIONS
      ESTIMATING
      EXPERIMENT DESIGN
      FACTORIAL DESIGN
      FUNCTIONAL DESIGN SPECIFICATIONS
      HELICOPTER DESIGN
      IPAD
      LAYOUTS
      LENS DESIGN
      LOGIC DESIGN
      MISSILE DESIGN
      NOZZLE DESIGN
      OPTIMIZATION
      PLANNING
      PLANT DESIGN
      PRESSURE VESSEL DESIGN
      PRODUCT DEVELOPMENT
      REACTOR DESIGN
      RELIABILITY
      RESEARCH
      RESEARCH AND DEVELOPMENT
      ROCKET ENGINE DESIGN
      SATELLITE DESIGN
      SPACECRAFT DESIGN
      STRUCTURAL DESIGN
      STRUCTURAL DESIGN CRITERIA
      ∞ SYNTHESIS
      SYSTEMS ENGINEERING

DESIGN ANALYSIS
RT   ∞ ANALYZING
      CONTROL SYSTEMS DESIGN
      ∞ DESIGN
      LOGIC DESIGN
      MAINTAINABILITY
      OPTIMIZATION
      RELIABILITY
      RELIABILITY ANALYSIS
      SAFETY FACTORS
      VALUE ENGINEERING

DESIGN OF EXPERIMENTS
USE   EXPERIMENT DESIGN

DESIGN TO COST
RT   COST ANALYSIS
      COSTS
      ∞ DESIGN
      LIFE CYCLE COSTS
      PRODUCTION COSTS

DESORPTION
RT   ∞ ABSORPTION
      ADSORPTION
      DEGASSING
      EVOLUTION (LIBERATION)
      OUTGASSING
      PERMEATING
      ∞ SEPARATION
      SUBLIMATION

DESPINNING
USE   SPIN REDUCTION

DESTABILIZATION
RT   SPIN REDUCTION
      TUMBLING MOTION

DESTROYER AIRCRAFT
USE   B-66 AIRCRAFT

DESTRUCTION
RT   ABORTED MISSIONS
      ACCIDENTS
      BREAKING
      CRACKING (FRACTURING)
      DAMAGE
      DESTRUCTIVE TESTS
      FAILURE
      FATIGUE (MATERIALS)
      FLIGHT HAZARDS
      FLIGHT SAFETY
      LETHALITY
      STRESSES

DESTRUCTIVE TESTS
GS   **DESTRUCTIVE TESTS**
      . BURST TESTS
RT   BEND TESTS
      COMPRESSION TESTS
      CORROSION TESTS
      DESTRUCTION
      DROP TESTS
      FATIGUE TESTS
      IMPACT TESTS
      LOAD TESTS
      ∞ MATERIALS TESTS
      NONDESTRUCTIVE TESTS
      TENSILE TESTS
      ∞ TESTS
      VIBRATION TESTS
      WEAR TESTS

DESULFURIZING
GS   CHEMICAL REACTIONS
      . **DESULFURIZING**
RT   FLUE GASES
      REFINING
      ROASTING

DESYNCHRONIZATION (BIOLOGY)
GS   BIOLOGICAL EFFECTS
      . **DESYNCHRONIZATION (BIOLOGY)**
      DISORIENTATION
      . **DESYNCHRONIZATION (BIOLOGY)**
      PSYCHOLOGICAL EFFECTS
      . **DESYNCHRONIZATION (BIOLOGY)**
RT   JET LAG
      PHYSIOLOGICAL RESPONSES
      RHYTHM (BIOLOGY)

DESYNCHRONIZED SLEEP
USE   RAPID EYE MOVEMENT STATE

DETACHMENT
RT   ANXIETY
      BOREDOM
      DEPERSONALIZATION
      ∞ DEPRESSION
      DISORDERS
      DISORIENTATION
      EMOTIONAL FACTORS
      HUMAN BEHAVIOR
      ∞ INHIBITION
      INTROVERSION
      LETHARGY
      PSYCHOLOGY
      PSYCHOSES

DETECTION
UF   SENSING
GS   **DETECTION**
      . AIRCRAFT DETECTION
      . CHANGE DETECTION
      . FOREST FIRE DETECTION
      . HAZE DETECTION
      . HIGH ALTITUDE NUCLEAR DETECTION
      . MISSILE DETECTION
      . . RADAR DETECTION
      . REMOTE SENSING
      . SIGNAL DETECTION
      . . CORRELATION DETECTION
      . TARGET RECOGNITION
      . ULTRASONIC FLAW DETECTION
RT   ACQUISITION
      DATA ACQUISITION
      ∞ DETECTORS
      EARLY WARNING SYSTEMS
      EXAMINATION

DETECTION-(CONT.)
      EXPLORATION
      GAS DETECTORS
      IDENTIFYING
      INSPECTION
      INTELLIGENCE
      MARKING
      ∞ MEASUREMENT
      MISSILE SIGNATURES
      OBSERVATION
      POSITION (LOCATION)
      RADAR SIGNATURES
      SIGNATURE ANALYSIS
      SIGNATURES
      SOUND LOCALIZATION
      SOUND RANGING
      SPACE OBSERVATIONS (FROM EARTH)
      SURVEILLANCE
      TARGET ACQUISITION
      TARGETS
      TRACKING (POSITION)
      WARNING
      WARNING SYSTEMS

∞ **DETECTORS**
SN   (USE OF A MORE SPECIFIC TERM IS
      RECOMMENDED--CONSULT THE TERMS
      LISTED BELOW)
RT   AIRCRAFT DETECTION
      ANALYZERS
      AUTODYNES
      CHARACTER RECOGNITION
      CORRELATION DETECTION
      DEMODULATION
      DETECTION
      DISPLAY DEVICES
      FLIR DETECTORS
      FOREST FIRE DETECTION
      GAS DETECTORS
      HAZARDS
      HELMET MOUNTED DISPLAYS
      INDICATING INSTRUMENTS
      INFRARED DETECTORS
      INSTRUMENT RECEIVERS
      LASER TARGET DESIGNATORS
      LIFE DETECTORS
      MEASURING INSTRUMENTS
      MINE DETECTORS
      MONITORS
      MULTISPECTRAL LINEAR ARRAYS
      PHASE DETECTORS
      RADIATION DETECTORS
      RADIATION MEASURING INSTRUMENTS
      READERS
      RECEIVERS
      REMOTE SENSORS
      SAFETY
      SIGNAL DETECTION
      SIGNAL DETECTORS
      SQUID (DETECTORS)
      TELECOMMUNICATION
      TRANSDUCERS
      ULTRASONIC FLAW DETECTION
      VENTURI TUBES
      WARNING
      WARNING SYSTEMS

DETERGENTS
RT   ETHYLENEDIAMINETETRAACETIC ACIDS
      LUBRICATING OILS
      SOAPS
      SURFACTANTS

DETERIORATION
RT   ATROPHY
      BIODEGRADABILITY
      BIODEGRADATION
      CORROSION
      DAMAGE
      DECAY
      DECOMPOSITION
      DEGENERATION
      DEGRADATION
      DEPOLYMERIZATION
      DEPRECIATION
      DISINTEGRATION
      DURABILITY
      EROSION
      EROSIVE BURNING
      FAILURE
      HOT CORROSION
      RUSTING
      SOIL EROSION
      SYSTEM FAILURES
      WEAR

DETERIORATION-(CONT.)
WEATHERING

DETERMINANTS
GS   ALGEBRA
. **DETERMINANTS**
RT   LINEAR EQUATIONS
MATRICES (MATHEMATICS)

DETERMINATION
USE MEASUREMENT

DETONABLE GAS MIXTURES
GS   GASES
. GAS MIXTURES
. . **DETONABLE GAS MIXTURES**
MIXTURES
. SOLUTIONS
. . GAS MIXTURES
. . . **DETONABLE GAS MIXTURES**
RT   CHEMICAL EXPLOSIONS
FIRING (IGNITING)
FLAMMABILITY
FLAMMABLE GASES
GAS EXPLOSIONS
GAS-GAS INTERACTIONS
OXYACETYLENE

DETONATION
UF   CHAPMAN-JOUGET FLAME
RT   CHEMICAL EXPLOSIONS
COMBUSTION
∞DISCHARGE
EXPLOSIONS
FIRING (IGNITING)
FLAME PROPAGATION
INITIATION
PERCUSSION
PRIMERS (EXPLOSIVES)
PROPELLANT EXPLOSIONS
ROCKET FIRING
SHOCK WAVES

DETONATION WAVES
GS   ELASTIC WAVES
. SHOCK WAVES
. . **DETONATION WAVES**
RT   COMBUSTIBLE FLOW
FLAME PROPAGATION
GAS EXPLOSIONS
SEISMIC WAVES
SOUND WAVES
∞WAVES

DETONATORS
GS   EXPLOSIVE DEVICES
. INITIATORS (EXPLOSIVES)
. . **DETONATORS**
IGNITERS
. INITIATORS (EXPLOSIVES)
. . **DETONATORS**
RT   CAPS (EXPLOSIVES)
EXPLODING WIRES
EXPLOSIVES
FULMINATES
FUSES (ORDNANCE)
PRIMERS (EXPLOSIVES)
SODIUM AZIDES

DEUTERIDES
GS   HYDROGEN COMPOUNDS
. DEUTERIUM COMPOUNDS
. . **DEUTERIDES**
RT   HYDRIDES

DEUTERIUM
UF   HYDROGEN 2
GS   CHEMICAL ELEMENTS
. HYDROGEN
. . HYDROGEN ISOTOPES
. . . **DEUTERIUM**
. NUCLIDES
. . ISOTOPES
. . . HYDROGEN ISOTOPES
. . . . **DEUTERIUM**
GASES
. HYDROGEN
. . HYDROGEN ISOTOPES
. . . **DEUTERIUM**
RT   HEAVY WATER
HYDROGEN FUELS
HYDROGEN PLASMA
NUCLEAR FUELS

DEUTERIUM COMPOUNDS
GS   HYDROGEN COMPOUNDS
. **DEUTERIUM COMPOUNDS**
. . DEUTERIDES
. . DEUTERIUM FLUORIDES
. . HEAVY WATER
RT   ∞CHEMICAL COMPOUNDS

DEUTERIUM FLUORIDE LASERS
USE DF LASERS

DEUTERIUM FLUORIDES
UF   DF
GS   HYDROGEN COMPOUNDS
. DEUTERIUM COMPOUNDS
. . **DEUTERIUM FLUORIDES**
RT   DF LASERS

DEUTERIUM OXIDES
USE HEAVY WATER

DEUTERIUM PLASMA
GS   CHEMICAL ELEMENTS
. HYDROGEN
. . HYDROGEN PLASMA
. . . **DEUTERIUM PLASMA**
GASES
. HYDROGEN
. . HYDROGEN PLASMA
. . . **DEUTERIUM PLASMA**
RT   DEUTERONS

DEUTERON IRRADIATION
GS   IRRADIATION
. ION IRRADIATION
. . **DEUTERON IRRADIATION**
RT   ALPHA PARTICLES
CHARGED PARTICLES
NUCLEAR FUSION
PARTICLES
PLASMAS (PHYSICS)
PROTON IRRADIATION

DEUTERONS
GS   IONS
. **DEUTERONS**
PARTICLES
. ELEMENTARY PARTICLES
. . **DEUTERONS**
RT   ALPHA PARTICLES
COSMIC RAYS
DEUTERIUM PLASMA
PHOTOMAGNETIC EFFECTS
PLASMAS (PHYSICS)
POMERANCHUK THEOREM
PROTONS

DEVELOPERS (PHOTOGRAPHY)
USE PHOTOGRAPHIC DEVELOPERS

DEVELOPING NATIONS
RT   CARIBBEAN REGION
ECONOMIC DEVELOPMENT
ECONOMIC FACTORS
NATIONS
UNITED NATIONS

∞ DEVELOPMENT
SN   (USE OF A MORE SPECIFIC TERM IS
RECOMMENDED--CONSULT THE TERMS
LISTED BELOW)
RT   ENERGY POLICY
EVOLUTION (DEVELOPMENT)
EXPLOITATION
GROWTH
LAND USE
MANAGEMENT PLANNING
MISSILE DESIGN
PERSONNEL DEVELOPMENT
PHOTOGRAPHIC DEVELOPERS
PRODUCT DEVELOPMENT
RURAL LAND USE
STARSITE PROGRAM
TRAINING ANALYSIS
URBAN DEVELOPMENT

DEVIATION
RT   ABERRATION
ABNORMALITIES
ASYMMETRY
∞DISPERSION
DISTORTION
DIVERGENCE
∞DRIFT

DEVIATION-(CONT.)
ECCENTRICITY
HETEROGENEITY
IRREGULARITIES
NONSYNCHRONIZATION
VARIATIONS

∞ DEVICES
SN   (USE OF A MORE SPECIFIC TERM IS
RECOMMENDED--CONSULT THE TERMS
LISTED BELOW)
RT   AIR BAG RESTRAINT DEVICES
ALPHA PLASMA DEVICES
ANTISKID DEVICES
BULK ACOUSTIC WAVE DEVICES
CHIPS (MEMORY DEVICES)
CYCLOTRON RESONANCE DEVICES
ERROR CORRECTING DEVICES
EXPLOSIVE DEVICES
LIFT DEVICES
LIFTING BODIES
MECHANICAL DEVICES
NDM SEMICONDUCTOR DEVICES
NUCLEAR DEVICES
PHOTOELECTROCHEMICAL DEVICES
PLASMA DISPLAY DEVICES
POSITIONING DEVICES (MACHINERY)
SAFETY DEVICES
SELF ERECTING DEVICES
SELF REPAIRING DEVICES
SOLID STATE DEVICES
SURFACE ACOUSTIC WAVE DEVICES
TRAINING DEVICES
TRAPATT DEVICES

DEVITRIFICATION
USE CRYSTALLIZATION

DEW
GS   PRECIPITATION (METEOROLOGY)
. **DEW**
RT   ACID RAIN
FROST
WATER VAPOR

DEW POINT
RT   ATMOSPHERIC MOISTURE
CONDENSING
HYGROMETERS

DEWAR SYSTEMS
USE CRYOGENIC EQUIPMENT

DEWATERING
RT   DEHYDRATION
DRYING
POLLUTION CONTROL
WASTE DISPOSAL
WATER RECLAMATION

DEWAXING
RT   DECONTAMINATION
REFINING

DEWETTING
USE DRYING

DEXTRANS
GS   ALIPHATIC COMPOUNDS
. POLYSACCHARIDES
. . **DEXTRANS**
. SUGARS
. . **DEXTRANS**
CARBOHYDRATES
. POLYSACCHARIDES
. . **DEXTRANS**
. SUGARS
. . **DEXTRANS**

DF
USE DEUTERIUM FLUORIDES

DF LASERS
UF   DEUTERIUM FLUORIDE LASERS
GS   STIMULATED EMISSION DEVICES
. LASERS
. . GAS LASERS
. . . **DF LASERS**
RT   DEUTERIUM FLUORIDES

DFA
USE DELAYED FLAP APPROACH

DH 106 AIRCRAFT
USE   COMET 4 AIRCRAFT

DH 112 AIRCRAFT
UF   DE HAVILLAND DH 112 AIRCRAFT
     DE HAVILLAND VENOM AIRCRAFT
     VENOM AIRCRAFT
GS   ATTACK AIRCRAFT
    . **DH 112 AIRCRAFT**
    DE HAVILLAND AIRCRAFT
    . **DH 112 AIRCRAFT**
    HAWKER SIDDELEY AIRCRAFT
    . **DH 112 AIRCRAFT**
    JET AIRCRAFT
    . **DH 112 AIRCRAFT**
    MONOPLANES
    . **DH 112 AIRCRAFT**
RT   ∞AIRCRAFT

DH 115 AIRCRAFT
UF   DE HAVILLAND DH 115 AIRCRAFT
     VAMPIRE AIRCRAFT
GS   ATTACK AIRCRAFT
    . **DH 115 AIRCRAFT**
    DE HAVILLAND AIRCRAFT
    . **DH 115 AIRCRAFT**
    HAWKER SIDDELEY AIRCRAFT
    . **DH 115 AIRCRAFT**
    JET AIRCRAFT
    . **DH 115 AIRCRAFT**
    MONOPLANES
    . **DH 115 AIRCRAFT**
    TRAINING AIRCRAFT
    . **DH 115 AIRCRAFT**
RT   ∞AIRCRAFT

DH 121 AIRCRAFT
UF   DE HAVILLAND DH 121 AIRCRAFT
     TRIDENT AIRCRAFT
GS   COMMERCIAL AIRCRAFT
    . **DH 121 AIRCRAFT**
    DE HAVILLAND AIRCRAFT
    . **DH 121 AIRCRAFT**
    HAWKER SIDDELEY AIRCRAFT
    . **DH 121 AIRCRAFT**
    JET AIRCRAFT
    . TURBOFAN AIRCRAFT
    . . **DH 121 AIRCRAFT**
    MONOPLANES
    . **DH 121 AIRCRAFT**
    PASSENGER AIRCRAFT
    . **DH 121 AIRCRAFT**
    TRANSPORT AIRCRAFT
    . **DH 121 AIRCRAFT**
RT   ∞AIRCRAFT

DH 125 AIRCRAFT
UF   DE HAVILLAND DH 125 AIRCRAFT
     HS-125 AIRCRAFT
     JET DRAGON AIRCRAFT
GS   DE HAVILLAND AIRCRAFT
    . **DH 125 AIRCRAFT**
    GENERAL AVIATION AIRCRAFT
    . **DH 125 AIRCRAFT**
    HAWKER SIDDELEY AIRCRAFT
    . **DH 125 AIRCRAFT**
    JET AIRCRAFT
    . **DH 125 AIRCRAFT**
    LIGHT AIRCRAFT
    . **DH 125 AIRCRAFT**
    MONOPLANES
    . **DH 125 AIRCRAFT**
    PASSENGER AIRCRAFT
    . **DH 125 AIRCRAFT**
    TRANSPORT AIRCRAFT
    . **DH 125 AIRCRAFT**
RT   ∞AIRCRAFT

DHC BEAVER AIRCRAFT
USE   DHC 2 AIRCRAFT

DHC 2 AIRCRAFT
UF   DHC BEAVER AIRCRAFT
GS   DE HAVILLAND AIRCRAFT
    . **DHC 2 AIRCRAFT**
    GENERAL AVIATION AIRCRAFT
    . **DHC 2 AIRCRAFT**
    JET AIRCRAFT
    . **DHC 2 AIRCRAFT**
    MONOPLANES
    . **DHC 2 AIRCRAFT**
    TRANSPORT AIRCRAFT
    . **DHC 2 AIRCRAFT**
RT   ∞AIRCRAFT

DHC 4 AIRCRAFT
UF   AC-1 AIRCRAFT
     CARIBOU AIRCRAFT
     CV-2 AIRCRAFT
     DE HAVILLAND DHC 4 AIRCRAFT
GS   DE HAVILLAND AIRCRAFT
    . **DHC 4 AIRCRAFT**
    MONOPLANES
    . **DHC 4 AIRCRAFT**
    TRANSPORT AIRCRAFT
    . **DHC 4 AIRCRAFT**
    UTILITY AIRCRAFT
    . **DHC 4 AIRCRAFT**
    V/STOL AIRCRAFT
    . SHORT TAKEOFF AIRCRAFT
    . . **DHC 4 AIRCRAFT**
RT   ∞AIRCRAFT

DHC 5 AIRCRAFT
UF   BUFFALO AIRCRAFT
     CV-7 AIRCRAFT
     DE HAVILLAND DHC 5 AIRCRAFT
GS   DE HAVILLAND AIRCRAFT
    . **DHC 5 AIRCRAFT**
    JET AIRCRAFT
    . TURBOPROP AIRCRAFT
    . . **DHC 5 AIRCRAFT**
    MONOPLANES
    . **DHC 5 AIRCRAFT**
    TRANSPORT AIRCRAFT
    . **DHC 5 AIRCRAFT**
    UTILITY AIRCRAFT
    . **DHC 5 AIRCRAFT**
    V/STOL AIRCRAFT
    . SHORT TAKEOFF AIRCRAFT
    . . **DHC 5 AIRCRAFT**
RT   ∞AIRCRAFT

DIABETES MELLITUS
GS   DISEASES
    . **DIABETES MELLITUS**
RT   CARBOHYDRATE METABOLISM
    ENZYME ACTIVITY
    INSULIN
    PANCREAS
    URINALYSIS

DIADEME SATELLITES
GS   SATELLITES
    . ARTIFICIAL SATELLITES
    . . **DIADEME SATELLITES**

DIAGNOSIS
RT   ∞ANALYZING
    ANESTHESIOLOGY
    BEHAVIOR
    CLINICAL MEDICINE
    DISEASES
    EXAMINATION
    INJURIES
    MEDICAL EQUIPMENT
    MEDICAL SCIENCE
    PATHOLOGY
    PROGNOSIS
    PSYCHOLOGY
    PSYCHOMETRICS
    VETERINARY MEDICINE

DIAGRAMS
GS   **DIAGRAMS**
    . BENDING DIAGRAMS
    . BLOCK DIAGRAMS
    . CIRCUIT DIAGRAMS
    . COLOR-MAGNITUDE DIAGRAM
    . CREEP DIAGRAMS
    . FEYNMAN DIAGRAMS
    . HERTZSPRUNG-RUSSELL DIAGRAM
    . MOLLIER DIAGRAM
    . NYQUIST DIAGRAM
    . PHASE DIAGRAMS
    . S-N DIAGRAMS
    . STRESS-STRAIN DIAGRAMS
    . TEPHIGRAMS
    . VENN DIAGRAMS
RT   CHARTS
    DRAWINGS
    GEOMETRY
    GRAPHIC ARTS
    VISUAL AIDS

DIAL SATELLITE
GS   SATELLITES
    . ARTIFICIAL SATELLITES
    . . SCIENTIFIC SATELLITES
    . . . **DIAL SATELLITE**

DIAL SATELLITE-_(CONT.)_
RT   AERONOMY
    ASTRONOMICAL PHOTOMETRY
    EUROPEAN SPACE PROGRAMS
    SATELLITE-BORNE INSTRUMENTS

DIALLYL COMPOUNDS
GS   ALIPHATIC COMPOUNDS
    . **DIALLYL COMPOUNDS**
RT   ALLYL COMPOUNDS
    ∞CHEMICAL COMPOUNDS

DIALS
UF   POINTERS
RT   DISPLAY DEVICES
    INDICATING INSTRUMENTS

DIALYSIS
GS   **DIALYSIS**
    . ELECTRODIALYSIS
RT   DIAPHRAGMS (MECHANICS)
    DIFFUSION
    EXTRACTION
    PERMEATING
    ∞SEPARATION

DIAMAGNETISM
UF   MEISSNER EFFECT
GS   MAGNETIC PROPERTIES
    . **DIAMAGNETISM**
RT   CURIE TEMPERATURE
    CYCLOTRON RESONANCE
    ELECTRICAL PROPERTIES
    FERROMAGNETISM
    PARAMAGNETISM

DIAMANT LAUNCH VEHICLE
GS   LAUNCH VEHICLES
    . **DIAMANT LAUNCH VEHICLE**
    ROCKET VEHICLES
    . MULTISTAGE ROCKET VEHICLES
    . . **DIAMANT LAUNCH VEHICLE**
RT   LIQUID PROPELLANT ROCKET ENGINES
    SOLID PROPELLANT ROCKET ENGINES

DIAMETERS
GS   DIMENSIONS
    . **DIAMETERS**
RT   CIRCUMFERENCES
    GEOMETRY
    RADII
    THICKNESS

DIAMINES
GS   AMINES
    . **DIAMINES**
    . . ETHYLENEDIAMINE
    . . GUANIDINES
    . . . GUANETHIDINE
    . . . . TRIAMINOGUANIDINIUM AZIDE

DIAMOND WINGS
USE   LOW ASPECT RATIO WINGS
     SWEPT WINGS

DIAMONDS
GS   **DIAMONDS**
    . METEORITIC DIAMONDS
RT   ABRASIVES
    CARBON
    SINGLE CRYSTALS

DIAPHRAGM (ANATOMY)
GS   ANATOMY
    . RESPIRATORY SYSTEM
    . . **DIAPHRAGM (ANATOMY)**
RT   ∞DIAPHRAGMS
    MUSCLES
    THORAX

∞ **DIAPHRAGMS**
SN   _(USE OF A MORE SPECIFIC TERM IS_
    RECOMMENDED--CONSULT THE TERMS
    LISTED BELOW)
RT   DIAPHRAGM (ANATOMY)
    DIAPHRAGMS (MECHANICS)
    ELECTROLYTIC CELLS
    MEMBRANES

DIAPHRAGMS (MECHANICS)
SN   (NON-ANATOMICAL)
UF   BLADDERS (MECHANICS)
GS   **DIAPHRAGMS (MECHANICS)**
    . EXPULSION BLADDERS

DIAPHRAGMS (MECHANICS)-(CONT.)
- RT   CATHOLYTES
  - DIALYSIS
  - ∞DIAPHRAGMS
  - ELECTROLYTIC CELLS
  - MEMBRANE STRUCTURES
  - MEMBRANES
  - OPTICAL FILTERS
  - OSMOSIS
  - THIN PLATES
  - THIN WALLS
  - WEBS (SHEETS)
  - WEBS (SUPPORTS)

DIASTOLE
- GS   ANATOMY
  - . CARDIOVASCULAR SYSTEM
  - . . **DIASTOLE**
- RT   BLOOD CIRCULATION
  - BLOOD FLOW
  - BLOOD PRESSURE
  - DIASTOLIC PRESSURE
  - HEART
  - HEART RATE
  - SYSTOLE

DIASTOLIC PRESSURE
- GS   PRESSURE
  - . BLOOD PRESSURE
  - . . **DIASTOLIC PRESSURE**
- RT   CARDIAC VENTRICLES
  - DIASTOLE

DIATOMIC GASES
- GS   GASES
  - . MOLECULAR GASES
  - . . POLYATOMIC GASES
  - . . . **DIATOMIC GASES**

DIATOMIC MOLECULES
- GS   MOLECULES
  - . **DIATOMIC MOLECULES**
- RT   LOW MOLECULAR WEIGHTS
  - MORSE POTENTIAL
  - TRIATOMIC MOLECULES

DIBASIC COMPOUNDS
- RT   ∞CHEMICAL COMPOUNDS
  - MONOMERS

DIBORANE
- GS   BORON COMPOUNDS
  - . **DIBORANE**
  - HYDROGEN COMPOUNDS
  - . HYDRIDES
  - . . **DIBORANE**

DIBROMIDES
- GS   HALOGEN COMPOUNDS
  - . BROMINE COMPOUNDS
  - . . BROMIDES
  - . . . **DIBROMIDES**
  - . HALIDES
  - . . BROMIDES
  - . . . **DIBROMIDES**

DIBUTYL COMPOUNDS
- GS   ALIPHATIC COMPOUNDS
  - . ALKYL COMPOUNDS
  - . . **DIBUTYL COMPOUNDS**
- RT   ∞CHEMICAL COMPOUNDS

DICARBOXYLIC ACIDS
- GS   ACIDS
  - . FATTY ACIDS
  - . . CARBOXYLIC ACIDS
  - . . . **DICARBOXYLIC ACIDS**
  - . . . . TEREPHTHALATE

DICHLORIDES
- GS   HALOGEN COMPOUNDS
  - . CHLORINE COMPOUNDS
  - . . CHLORIDES
  - . . . **DICHLORIDES**
  - . HALIDES
  - . . CHLORIDES
  - . . . **DICHLORIDES**

DICHLORODIPHENYLTRICHLOROETHANE
- USE   DDT

DICHOTOMIES
- GS   CLASSIFICATIONS
  - . HIERARCHIES

DICHOTOMIES-(CONT.)
  - . . **DICHOTOMIES**

DICHROISM
- GS   ELECTROMAGNETIC PROPERTIES
  - . OPTICAL PROPERTIES
  - . . **DICHROISM**
- RT   COLOR
  - ISOCHROMATICS
  - LIGHT (VISIBLE RADIATION)
  - PHOTOELASTICITY

DICHROMATES
- USE   CHROMATES

DICKE RADIOMETERS
- UF   DICKE TYPE RADIOMETERS
- GS   MEASURING INSTRUMENTS
  - . RADIATION MEASURING INSTRUMENTS
  - . . ACTINOMETERS
  - . . . RADIOMETERS
  - . . . . **DICKE RADIOMETERS**
- RT   BOLOMETERS
  - THERMOPILES

DICKE TYPE RADIOMETERS
- USE   DICKE RADIOMETERS

DICTIONARIES
- UF   GLOSSARIES
- RT   CODING
  - DECODING
  - ∞DEFINITION
  - DOCUMENTS
  - NOMENCLATURES
  - SPACE GLOSSARIES
  - TERMINOLOGY

DIDYMIUM
- RT   LANTHANUM
  - NEODYMIUM
  - OPTICAL FILTERS
  - PRASEODYMIUM

DIELDRIN
- GS   POISONS
  - . PESTICIDES
  - . . INSECTICIDES
  - . . . **DIELDRIN**

DIELECTRIC CONSTANT
- USE   PERMITTIVITY

DIELECTRIC MATERIALS
- USE   DIELECTRICS

DIELECTRIC PERMEABILITY
- GS   PERMEABILITY
  - . **DIELECTRIC PERMEABILITY**
- RT   MAGNETIC PERMEABILITY

DIELECTRIC POLARIZATION
- GS   POLARIZATION (CHARGE SEPARATION)
  - . **DIELECTRIC POLARIZATION**
- RT   DIELECTRICS
  - ELECTRETS
  - ELECTRIC FIELDS

DIELECTRIC PROPERTIES
- GS   ELECTROMAGNETIC PROPERTIES
  - . ELECTRICAL PROPERTIES
  - . . **DIELECTRIC PROPERTIES**
  - . . . PERMITTIVITY
- RT   ANTIFERROELECTRICITY
  - CAPACITANCE
  - FERROELECTRICITY
  - ∞PROPERTIES
  - SOMMERFELD WAVES

DIELECTRICS
- UF   DIELECTRIC MATERIALS
- GS   **DIELECTRICS**
  - . LOSSLESS MATERIALS
  - . RADOME MATERIALS
- RT   BARIUM TITANATES
  - CAPACITANCE SWITCHES
  - CAPACITIVE FUEL GAGES
  - CAPACITORS
  - CERAMICS
  - DIELECTRIC POLARIZATION
  - ELECTRETS
  - ELECTRIC CONDUCTORS
  - ELECTRICAL INSULATION
  - ELECTROMAGNETIC SURFACE WAVES

DIELECTRICS-(CONT.)
  - ∞INSULATED STRUCTURES
  - INSULATORS
  - MAGNETOELECTRIC MEDIA
  - SCREEN EFFECT
  - SPARK GAPS

DIELECTRONIC SATELLITE LINES
- USE   RESONANCE LINES

DIELS-ALDER REACTIONS
- GS   CHEMICAL REACTIONS
  - . **DIELS-ALDER REACTIONS**
- RT   ORGANIC CHEMISTRY

DIENCEPHALON
- GS   NERVOUS SYSTEM
  - . **DIENCEPHALON**
- RT   BRAIN

DIENES
- GS   HYDROCARBONS
  - . **DIENES**
  - . . BUTADIENE
  - . . HEPTADIENE
  - . . HEXADIENE

DIES
- RT   CASTING
  - COINING
  - CUTTERS
  - EXTRUDING
  - INJECTION MOLDING
  - MACHINE TOOLS
  - MOLDS
  - PULTRUSION
  - PUNCHES
  - RHEOCASTING
  - STAMPING

DIESEL ENGINES
- GS   ENGINES
  - . INTERNAL COMBUSTION ENGINES
  - . . **DIESEL ENGINES**
  - . PISTON ENGINES
  - . . **DIESEL ENGINES**
- RT   LOCOMOTIVES

DIESEL FUELS
- GS   FUELS
  - . CHEMICAL FUELS
  - . . HYDROCARBON FUELS
  - . . . **DIESEL FUELS**
  - . LIQUID FUELS
  - . . . **DIESEL FUELS**
- RT   AUTOMOBILE FUELS
  - GASOLINE
  - INTERNAL COMBUSTION ENGINES
  - KEROSENE

DIETHYL ETHER
- GS   ETHERS
  - . **DIETHYL ETHER**

DIETHYL HYDROGEN PHOSPHITE (DEHP)
- UF   DEHP
- GS   PHOSPHORUS COMPOUNDS
  - . **DIETHYL HYDROGEN PHOSPHITE (DEHP)**
- RT   ETHYL COMPOUNDS

DIETS
- RT   CALORIC REQUIREMENTS
  - FASTING
  - ∞FOOD
  - NUTRITION
  - SPACE FLIGHT FEEDING

DIFFERENCE EQUATIONS
- GS   ANALYSIS (MATHEMATICS)
  - . NUMERICAL ANALYSIS
  - . . **DIFFERENCE EQUATIONS**
- RT   APPROXIMATION
  - DIFFERENCES
  - DIFFERENTIAL EQUATIONS
  - ∞EQUATIONS
  - FINITE DIFFERENCE THEORY
  - LINEAR EVOLUTION EQUATIONS
  - NONLINEAR EVOLUTION EQUATIONS
  - NUMERICAL STABILITY

DIFFERENCES
- RT   DIFFERENCE EQUATIONS
  - DIVERGENCE

DIFFERENCES-(CONT.)
  FINITE DIFFERENCE THEORY
  GRADIENTS
  VARIATIONS

DIFFERENTIAL ALGEBRA
 USE DIFFERENTIAL CALCULUS
   MATRICES (MATHEMATICS)

DIFFERENTIAL AMPLIFIERS
 GS AMPLIFIERS
   . **DIFFERENTIAL AMPLIFIERS**
 RT ANALOG COMPUTERS
   ERROR SIGNALS
   OPERATIONAL AMPLIFIERS
   TRANSISTOR AMPLIFIERS

DIFFERENTIAL ANALYZERS
 RT ALGORITHMS
   ANALOG COMPUTERS
   COMPUTERIZED SIMULATION
   DIFFERENTIAL EQUATIONS
   DIGITAL COMPUTERS
   DIGITAL INTEGRATORS

DIFFERENTIAL CALCULUS
 UF DERIVATION CALCULUS
   DIFFERENTIAL ALGEBRA
 GS ANALYSIS (MATHEMATICS)
   . CALCULUS
   . . **DIFFERENTIAL CALCULUS**
 RT ∞ DIFFERENTIATION
   DIFFERENTIATORS
   INTEGRAL CALCULUS
   LIMITS (MATHEMATICS)
   MINIMA
   NUMERICAL DIFFERENTIATION
   OPTIMIZATION
   REAL VARIABLES

DIFFERENTIAL EQUATIONS
 UF DIFFERENTIAL OPERATORS
   INTEGRODIFFERENTIAL EQUATIONS
 GS ANALYSIS (MATHEMATICS)
   . REAL VARIABLES
   . . **DIFFERENTIAL EQUATIONS**
   . . . BLASIUS EQUATION
   . . . CAUCHY-RIEMANN EQUATIONS
   . . . CHANDRASEKHAR EQUATION
   . . . COSINE SERIES
   . . . DUFFING DIFFERENTIAL EQUATION
   . . . FALKNER-SKAN EQUATION
   . . . HYPERBOLIC DIFFERENTIAL
     EQUATIONS
   . . . LAME WAVE EQUATIONS
   . . . PARTIAL DIFFERENTIAL EQUATIONS
   . . . . BIHARMONIC EQUATIONS
   . . . . BURGER EQUATION
   . . . . ELLIPTIC DIFFERENTIAL
     EQUATIONS
   . . . . . MONGE-AMPERE EQUATION
   . . . . EULER-CAUCHY EQUATIONS
   . . . . FOKKER-PLANCK EQUATION
   . . . . GAUSS EQUATION
   . . . . HELMHOLTZ VORTICITY EQUATION
   . . . . LIOUVILLE EQUATIONS
   . . . . PARABOLIC DIFFERENTIAL
     EQUATIONS
   . . . . VLASOV EQUATIONS
   . . . POISSON EQUATION
 RT AIRY FUNCTION
   ASYMPTOTIC PROPERTIES
   BACKWARD DIFFERENCING
   BESSEL FUNCTIONS
   BETHE-SALPETER EQUATION
   BOND GRAPHS
   BOUNDARY LAYER EQUATIONS
   BOUNDARY VALUE PROBLEMS
   CALCULUS
   CALCULUS OF VARIATIONS
   CAUCHY PROBLEM
   CRANK-NICHOLSON METHOD
   DIFFERENCE EQUATIONS
   DIFFERENTIAL ANALYZERS
   DIRICHLET PROBLEM
   DISTRIBUTED PARAMETER SYSTEMS
   ∞ EQUATIONS
   FLOQUET THEOREM
   FOURIER ANALYSIS
   FOURIER-BESSEL TRANSFORMATIONS
   FUNCTIONAL INTEGRATION
   GREEN'S FUNCTIONS
   HALF PLANES
   HANKEL FUNCTIONS
   HILL DETERMINANT

DIFFERENTIAL EQUATIONS-(CONT.)
  INTEGRAL EQUATIONS
  INTEGRALS
  LAGRANGE MULTIPLIERS
  LAME FUNCTIONS
  LAPLACE TRANSFORMATION
  LIAPUNOV FUNCTIONS
  LINEARITY
  LIPSCHITZ CONDITION
  MATHIEU FUNCTION
  MAXIMUM PRINCIPLE
  MILNE METHOD
  NEUMANN PROBLEM
  NONLINEAR EQUATIONS
  NONLINEARITY
  NUMERICAL ANALYSIS
  NUMERICAL DIFFERENTIATION
  NUMERICAL INTEGRATION
  NUMERICAL STABILITY
  OPERATIONAL CALCULUS
  PFAFF EQUATION
  POTENTIAL THEORY
  RICCATI EQUATION
  RIEMANN WAVES
  RIESZ THEOREM
  SCHAUDER FIXPOINT THEOREM
  SCHMIDT METHOD
  SPECTRAL METHODS
  STABILITY DERIVATIVES
  STURM-LIOUVILLE THEORY
  VECTOR ANALYSIS
  WHITTAKER FUNCTIONS

DIFFERENTIAL GEOMETRY
 UF NONEUCLIDIAN GEOMETRY
 GS GEOMETRY
   . **DIFFERENTIAL GEOMETRY**
   . . LIE GROUPS
   . . . SPINOR GROUPS
   . . RIEMANN MANIFOLD
   . . TENSOR ANALYSIS
 RT ANALYTIC GEOMETRY
   ∞ ANALYZING
   CURVATURE
   CURVES (GEOMETRY)
   INVARIANT IMBEDDINGS
   LOFTING
   RELATIVITY

DIFFERENTIAL INTERFEROMETRY
 GS INTERFEROMETRY
   . **DIFFERENTIAL INTERFEROMETRY**

DIFFERENTIAL OPERATORS
 USE DIFFERENTIAL EQUATIONS
   OPERATORS (MATHEMATICS)

DIFFERENTIAL PRESSURE
 GS PRESSURE
   . **DIFFERENTIAL PRESSURE**
 RT PRESSURE DISTRIBUTION
   PRESSURE GRADIENTS
   PRESSURE MEASUREMENT

DIFFERENTIAL PULSE CODE MODULATION
 UF DPCM (MODULATION)
 GS MODULATION
   . PULSE MODULATION
   . . PULSE CODE MODULATION
   . . . **DIFFERENTIAL PULSE CODE
     MODULATION**
 RT BITERNARY CODE
   DECOMMUTATORS
   LINEAR PREDICTION
   P.A.C.M. TELEMETRY
   PCM TELEMETRY
   PULSE COMMUNICATION
   PULSE FREQUENCY MODULATION
   TELEMETRY
   UNIFIED S BAND

DIFFERENTIAL THERMAL ANALYSIS
 USE THERMAL ANALYSIS

∞ DIFFERENTIATION
 SN *(USE OF A MORE SPECIFIC TERM IS
   RECOMMENDED--CONSULT TERMS
   LISTED BELOW)*
 RT ANATOMY
   DIFFERENTIAL CALCULUS
   DIFFERENTIATION (BIOLOGY)
   DISCRIMINATION

DIFFERENTIATION (BIOLOGY)
 RT ANATOMY

DIFFERENTIATION (BIOLOGY)-(CONT.)
  ∞ BIOLOGY
  CYTOGENESIS
  ∞ DIFFERENTIATION
  EMBRYOLOGY
  MORPHOLOGY
  PHYSIOLOGY

DIFFERENTIATORS
 RT CIRCUITS
   DIFFERENTIAL CALCULUS
   INTEGRATORS
   ∞ NETWORKS

DIFFRACTION
 UF INTERFERENCE MONOCHROMATIZATION
   KIRCHHOFF-HUYGENS PRINCIPLE
 GS **DIFFRACTION**
   . ELECTRON DIFFRACTION
   . FRESNEL DIFFRACTION
   . NEUTRON DIFFRACTION
   . PULSE DIFFRACTION
   . WAVE DIFFRACTION
   . X RAY DIFFRACTION
 RT ATMOSPHERIC SCATTERING
   ATTENUATION
   BRAGG ANGLE
   CAUSTICS (OPTICS)
   CRYSTAL OPTICS
   DEBYE-SCHERRER METHOD
   DEFLECTION
   DIFFRACTOMETERS
   ECHELETTE GRATINGS
   ELECTROMAGNETIC RADIATION
   GEOMETRICAL THEORY OF
     DIFFRACTION
   HUYGENS PRINCIPLE
   ISOCHROMATICS
   LAUE METHOD
   MOIRE EFFECTS
   MOSAICS
   OPTICAL PROPERTIES
   RAY TRACING
   REFRACTION
   TRANSMISSION
   WAVE DISPERSION
   WAVE PROPAGATION

DIFFRACTION GRATINGS
 USE GRATINGS (SPECTRA)

DIFFRACTION LIMITED CAMERAS
 GS OPTICAL EQUIPMENT
   . CAMERAS
   . . **DIFFRACTION LIMITED CAMERAS**
   PHOTOGRAPHIC EQUIPMENT
   . CAMERAS
   . . **DIFFRACTION LIMITED CAMERAS**
 RT ASTRONOMICAL PHOTOGRAPHY
   SPACEBORNE PHOTOGRAPHY
   SPACEBORNE TELESCOPES

DIFFRACTION PATHS
 RT BRAGG ANGLE
   ELECTRON TRAJECTORIES
   MULTIPATH TRANSMISSION
   OPTICAL PATHS
   ∞ PATHS
   SPHERICAL WAVES

DIFFRACTION PATTERNS
 UF FRINGE PATTERNS
 GS DISTRIBUTION (PROPERTY)
   . RADIATION DISTRIBUTION
   . . **DIFFRACTION PATTERNS**
   . . . KOSSEL PATTERN
   . . . RAINBOWS
 RT DIFFRACTOMETERS
   FRESNEL INTEGRALS
   FRESNEL REGION
   FRINGE MULTIPLICATION
   GEOMETRICAL THEORY OF
     DIFFRACTION
   HOLOGRAPHIC INTERFEROMETRY
   INTERFEROMETRY
   MOIRE FRINGES
   MOIRE INTERFEROMETRY
   NULL ZONES
   ∞ OPTICS
   ∞ PATTERNS
   PHASE CONTRAST
   POMERANCHUK THEOREM
   SIGNAL FADING
   SPECKLE PATTERNS
   UNDERWATER OPTICS

DIFFRACTION PATTERNS-*(CONT.)*
　　　VERY LONG BASE INTERFEROMETRY

DIFFRACTION PROPAGATION
GS　　TRANSMISSION
　　　. WAVE PROPAGATION
　　　. . **DIFFRACTION PROPAGATION**
RT　　GEOMETRICAL OPTICS
　　　∞OPTICS
　　　∞PROPAGATION
　　　SPHERICAL WAVES
　　　UNDERWATER OPTICS

DIFFRACTION TELESCOPES
USE　　SPECTROSCOPIC TELESCOPES

DIFFRACTOMETERS
GS　　MEASURING INSTRUMENTS
　　　. OPTICAL MEASURING INSTRUMENTS
　　　. . **DIFFRACTOMETERS**
　　　OPTICAL EQUIPMENT
　　　. OPTICAL MEASURING INSTRUMENTS
　　　. . **DIFFRACTOMETERS**
RT　　CHEMICAL ANALYSIS
　　　DIFFRACTION
　　　DIFFRACTION PATTERNS
　　　GONIOMETERS
　　　INTERFEROMETERS
　　　MACH-ZEHNDER INTERFEROMETERS
　　　OPTICAL MEASUREMENT
　　　PHOTOGONIOMETERS
　　　SPECTROMETERS
　　　WAVE FRONT RECONSTRUCTION

DIFFUSE RADIATION
UF　　LUNAR SCATTERING
RT　　HEAT TRANSFER
　　　LIGHT SCATTERING
　　　POINT SOURCES
　　　∞RADIATION
　　　SPECULAR REFLECTION

∞ **DIFFUSERS**
SN　　*(USE OF A MORE SPECIFIC TERM IS*
　　　RECOMMENDED--CONSULT THE TERMS
　　　LISTED BELOW)
UF　　SHOCK DIFFUSERS
RT　　AIR CONDITIONING
　　　AIR CONDITIONING EQUIPMENT
　　　ATTENUATORS
　　　BAFFLES
　　　BOUNDARY LAYER SEPARATION
　　　CEILINGS (ARCHITECTURE)
　　　CHOKES
　　　CONICAL FLOW
　　　DEFLECTORS
　　　DIFFUSION
　　　DIVERTERS
　　　ENGINE INLETS
　　　EXHAUST DIFFUSERS
　　　HYPERSONIC INLETS
　　　∞ILLUMINATION
　　　INLET FLOW
　　　INLET NOZZLES
　　　LOUVERS
　　　MIXERS
　　　MUFFLERS
　　　∞NOZZLES
　　　POROUS WALLS
　　　PRESSURE RECOVERY
　　　SEPARATORS
　　　SPRAYERS
　　　SUPERSONIC DIFFUSERS
　　　VANELESS DIFFUSERS
　　　VENTILATORS

DIFFUSION
UF　　DIFFUSION EFFECT
　　　PERFUSION
GS　　**DIFFUSION**
　　　. AMBIPOLAR DIFFUSION
　　　. ATMOSPHERIC DIFFUSION
　　　. GASEOUS DIFFUSION
　　　. . GASEOUS SELF-DIFFUSION
　　　. MAGNETIC DIFFUSION
　　　. MOLECULAR DIFFUSION
　　　. PARTICLE DIFFUSION
　　　. . ELECTRON DIFFUSION
　　　. . IONIC DIFFUSION
　　　. PLASMA DIFFUSION
　　　. SELF DIFFUSION (SOLID STATE)
　　　. SELF PROPAGATION
　　　. SPECIES DIFFUSION
　　　. SURFACE DIFFUSION
　　　. THERMAL DIFFUSION

DIFFUSION-*(CONT.)*
　　　. TURBULENT DIFFUSION
RT　　∞ABSORPTION
　　　ADSORPTION
　　　AIR POLLUTION
　　　ATMOSPHERIC SCATTERING
　　　CHEMICAL ENGINEERING
　　　CIRCULATION
　　　DEHUMIDIFICATION
　　　DIALYSIS
　　　∞DIFFUSERS
　　　DIFFUSIVITY
　　　DILUTION
　　　DISPERSING
　　　∞DISPERSION
　　　DISSIPATION
　　　DISSOLVING
　　　DISTILLATION
　　　DRYING
　　　∞EFFECTS
　　　∞EQUILIBRIUM
　　　EVAPORATION
　　　EXTRACTION
　　　FICKS EQUATION
　　　GAS-METAL INTERACTIONS
　　　KINETIC THEORY
　　　MIXING
　　　NONPOINT SOURCES
　　　OSMOSIS
　　　PENETRATION
　　　PERCOLATION
　　　PERMEABILITY
　　　PERMEATING
　　　∞PROPAGATION
　　　RADIAL FLOW
　　　REFLECTION
　　　SCATTERING
　　　SELF ABSORPTION
　　　∞SEPARATION
　　　SOUND PROPAGATION
　　　SOUND WAVES
　　　SPRAYING
　　　SPREADING
　　　SUBLIMATION
　　　SURFACE PROPERTIES
　　　TRANSPORT PROPERTIES

DIFFUSION BONDING
USE　　DIFFUSION WELDING

DIFFUSION COEFFICIENT
GS　　COEFFICIENTS
　　　. **DIFFUSION COEFFICIENT**
　　　. . SORET COEFFICIENT
　　　TRANSPORT PROPERTIES
　　　. **DIFFUSION COEFFICIENT**
　　　. . SORET COEFFICIENT
RT　　ATTENUATION COEFFICIENTS
　　　∞EQUILIBRIUM
　　　FICKS EQUATION
　　　GASEOUS DIFFUSION
　　　LEWIS NUMBERS
　　　MASS FLOW RATE
　　　MOLECULAR DIFFUSION
　　　PARTICLE DIFFUSION

DIFFUSION EFFECT
USE　　DIFFUSION

DIFFUSION ELECTRODES
GS　　ELECTRODES
　　　. **DIFFUSION ELECTRODES**
RT　　ELECTROLYTIC CELLS
　　　SEMICONDUCTOR DEVICES

DIFFUSION FLAMES
GS　　FLAMES
　　　. **DIFFUSION FLAMES**
RT　　BOUNDARY LAYER COMBUSTION
　　　COMBUSTION
　　　DAMKOHLER NUMBER

DIFFUSION PUMPS
GS　　PUMPS
　　　. **DIFFUSION PUMPS**
RT　　VACUUM APPARATUS
　　　VACUUM PUMPS

DIFFUSION THEORY
RT　　FOKKER-PLANCK EQUATION
　　　JACOBI INTEGRAL
　　　KINETIC THEORY
　　　KIRKENDALL EFFECT
　　　MONTE CARLO METHOD
　　　∞THEORIES

DIFFUSION THEORY-*(CONT.)*
　　　TRANSPORT THEORY

DIFFUSION WAVES
RT　　ELASTIC WAVES
　　　ELECTRON DIFFUSION
　　　ELECTROSTATIC WAVES
　　　IONIC DIFFUSION
　　　KINETIC THEORY
　　　MOLECULAR DIFFUSION
　　　PLASMA DIFFUSION
　　　PLASMA WAVES

DIFFUSION WELDING
UF　　DIFFUSION BONDING
GS　　WELDING
　　　. PRESSURE WELDING
　　　. . **DIFFUSION WELDING**
RT　　BONDING
　　　BURNERS
　　　KIRKENDALL EFFECT
　　　METAL BONDING
　　　METAL-METAL BONDING

DIFFUSIVITY
RT　　DIFFUSION
　　　FLUID MECHANICS
　　　IMPEDANCE
　　　KIRKENDALL EFFECT
　　　MOBILITY
　　　NDM SEMICONDUCTOR DEVICES
　　　PERMEABILITY
　　　∞PHYSICAL PROPERTIES
　　　∞RESISTANCE
　　　SOLUBILITY
　　　THERMODYNAMIC PROPERTIES

DIFLUORIDES
GS　　HALOGEN COMPOUNDS
　　　. FLUORINE COMPOUNDS
　　　. . FLUORIDES
　　　. . . **DIFLUORIDES**
　　　. . . . CALCIUM FLUORIDES
　　　. . . . . FLUORSPAR
　　　. HALIDES
　　　. . FLUORIDES
　　　. . . **DIFLUORIDES**
　　　. . . . CALCIUM FLUORIDES
　　　. . . . . FLUORSPAR

DIFLUORO COMPOUNDS
GS　　HALOGEN COMPOUNDS
　　　. FLUORINE COMPOUNDS
　　　. . FLUORO COMPOUNDS
　　　. . . **DIFLUORO COMPOUNDS**
　　　. . . . PERFLUOROALKANE
　　　. . . . POLYTETRAFLUOROETHYLENE
RT　　∞CHEMICAL COMPOUNDS

DIFLUOROUREA
GS　　AMINES
　　　. **DIFLUOROUREA**
　　　NITROGEN COMPOUNDS
　　　. AMIDES
　　　. . UREAS
　　　. . . **DIFLUOROUREA**

DIGESTING
RT　　EATING
　　　ENZYMOLOGY
　　　∞FOOD
　　　LYSINE
　　　MASTICATION
　　　SOFTENING

DIGESTIVE SYSTEM
GS　　ANATOMY
　　　. **DIGESTIVE SYSTEM**
　　　. . ESOPHAGUS
　　　. . GASTROINTESTINAL SYSTEM
　　　. . . APPENDIX (ANATOMY)
　　　. . . INTESTINES
　　　. . . . RECTUM
　　　. . . STOMACH
　　　. . MOUTH
　　　. . PANCREAS
　　　. . TEETH
　　　. . TONGUE
RT　　ABDOMEN
　　　ENZYME ACTIVITY
　　　ENZYMOLOGY
　　　GALL
　　　ORGANS
　　　SALIVA
　　　∞SYSTEMS

DIGITAL COMMAND SYSTEMS
RT   NUMERICAL CONTROL
     REMOTE CONTROL
     SERVOCONTROL
    ∞SYSTEMS

DIGITAL COMMUNICATION
USE PULSE COMMUNICATION

DIGITAL COMPUTERS
GS   DATA PROCESSING EQUIPMENT
    . COMPUTERS
    . . **DIGITAL COMPUTERS**
    . . . CDC CYBER 74 COMPUTER
    . . . CDC CYBER 170 SERIES
        COMPUTERS
    . . . CDC CYBER 174 COMPUTER
    . . . CDC CYBER 203 COMPUTER
    . . . CDC CYBER 205 COMPUTER
    . . . CDC STAR 100 COMPUTER
    . . . CDC 160-A COMPUTER
    . . . CDC 1604 COMPUTER
    . . . CDC 3100 COMPUTER
    . . . CDC 3200 COMPUTER
    . . . CDC 3600 COMPUTER
    . . . CDC 3800 COMPUTER
    . . . CDC 6000 SERIES COMPUTERS
    . . . . CDC 6400 COMPUTER
    . . . . CDC 6600 COMPUTER
    . . . . CDC 6700 COMPUTER
    . . . CDC 7000 SERIES COMPUTERS
    . . . . CDC 7600 COMPUTER
    . . . CDC 8090 COMPUTER
    . . . EAI 680 COMPUTER
    . . . EAI 8400 COMPUTER
    . . . EAI 8900 COMPUTER
    . . . EMR 6050 COMPUTER
    . . . FERRANTI MERCURY COMPUTER
    . . . GE COMPUTERS
    . . . . GE 625 COMPUTER
    . . . . GE 635 COMPUTER
    . . . HEWLETT-PACKARD COMPUTERS
    . . . HONEYWELL COMPUTERS
    . . . . DDP 516 COMPUTER
    . . . . HONEYWELL ADEPT COMPUTER
    . . . . HONEYWELL DDP 116 COMPUTER
    . . . . HONEYWELL 600/6000 COMPUTER
    . . . IBM 360 COMPUTER
    . . . IBM 370 COMPUTER
    . . . IBM 650 COMPUTER
    . . . IBM 704 COMPUTER
    . . . IBM 709 COMPUTER
    . . . IBM 1130 COMPUTER
    . . . IBM 1401 COMPUTER
    . . . IBM 1410 COMPUTER
    . . . IBM 1620 COMPUTER
    . . . IBM 2250 COMPUTER
    . . . IBM 7000 SERIES COMPUTERS
    . . . . IBM 7030 COMPUTER
    . . . . IBM 7040 COMPUTER
    . . . . IBM 7044 COMPUTER
    . . . . IBM 7070 COMPUTER
    . . . . IBM 7074 COMPUTER
    . . . . IBM 7090 COMPUTER
    . . . . IBM 7094 COMPUTER
    . . . ICL COMPUTERS
    . . . ILLIAC COMPUTERS
    . . . . ILLIAC 3 COMPUTER
    . . . . ILLIAC 4 COMPUTER
    . . . MICROCOMPUTERS
    . . . . PERSONAL COMPUTERS
    . . . MINICOMPUTERS
    . . . . NOVA COMPUTERS
    . . . MODCOMP II COMPUTER
    . . . MODCOMP IV COMPUTER
    . . . PARALLEL COMPUTERS
    . . . PDP COMPUTERS
    . . . . PDP 7 COMPUTER
    . . . . PDP 8 COMPUTER
    . . . . PDP 9 COMPUTER
    . . . . PDP 10 COMPUTER
    . . . . PDP 11 COMPUTER
    . . . . PDP 11/20 COMPUTER
    . . . . PDP 11/40 COMPUTER
    . . . . PDP 11/45 COMPUTER
    . . . . PDP 11/50 COMPUTER
    . . . . PDP 11/70 COMPUTER
    . . . . PDP 12 COMPUTER
    . . . . PDP 15 COMPUTER
    . . . PHILCO 2000 COMPUTER
    . . . RAYTHEON COMPUTERS
    . . . RCA SPECTRA 70 COMPUTER
    . . . SDS 900 SERIES COMPUTERS
    . . . . SDS 930 COMPUTER
    . . . SDS 9300 COMPUTER

DIGITAL COMPUTERS-*(CONT.)*
    . . . SEL COMPUTERS
    . . . SEQUENTIAL COMPUTERS
    . . . SIGMA COMPUTERS
    . . . . SIGMA 9 COMPUTER
    . . . . SIGMA 5 COMPUTER
    . . . SOLOMON COMPUTERS
    . . . UNIVAC LARC COMPUTER
    . . . UNIVAC 80 COMPUTER
    . . . UNIVAC 418 COMPUTER
    . . . UNIVAC 490 COMPUTER
    . . . UNIVAC 494 COMPUTER
    . . . UNIVAC 1100 SERIES COMPUTERS
    . . . . UNIVAC 1105 COMPUTER
    . . . . UNIVAC 1106 COMPUTER
    . . . . UNIVAC 1107 COMPUTER
    . . . . UNIVAC 1108 COMPUTER
    . . . . UNIVAC 1110 COMPUTER
    . . . UNIVAC 1230 COMPUTER
    . . . VAX COMPUTERS
    . . . . VAX-11 SERIES COMPUTERS
    . . . . . VAX-11/780 COMPUTER
RT   ANALOG COMPUTERS
    ANALOG TO DIGITAL CONVERTERS
    ASSOCIATIVE PROCESSING
     (COMPUTERS)
    CDC COMPUTERS
    COMPUTER COMPATIBLE TAPES
    COMPUTER PROGRAM INTEGRITY
    COMPUTER PROGRAMS
    COMPUTER SYSTEMS PROGRAMS
    DATA PROCESSING
    DDP COMPUTERS
    DIFFERENTIAL ANALYZERS
    HYBRID COMPUTERS
    IBM COMPUTERS
    LOGIC CIRCUITS
    TURING MACHINES
    UNIVAC COMPUTERS

DIGITAL DATA
RT   ANALOG DATA
    BINARY DATA
    ∞DATA
    DATA CONVERTERS
    DATA PROCESSING
    VIDEO DATA

DIGITAL FILTERS
GS   ELECTROMAGNETIC WAVE FILTERS
    . ELECTRIC FILTERS
    . . **DIGITAL FILTERS**
    . . . FIR FILTERS
RT   ∞FILTERS
    MICROWAVE FILTERS

DIGITAL INTEGRATORS
GS   CIRCUITS
    . **DIGITAL INTEGRATORS**
    INTEGRATORS
    . **DIGITAL INTEGRATORS**
RT   BINARY INTEGRATION
    DIFFERENTIAL ANALYZERS
    FUNCTIONAL INTEGRATION
    NUMERICAL INTEGRATION

DIGITAL NAVIGATION
GS   DIGITAL SYSTEMS
    . **DIGITAL NAVIGATION**
    NAVIGATION
    . **DIGITAL NAVIGATION**
RT   AIR NAVIGATION
    DEAD RECKONING
    INERTIAL NAVIGATION
    POLAR NAVIGATION
    SPACE NAVIGATION
    SURFACE NAVIGATION

DIGITAL RADAR SYSTEMS
GS   DIGITAL SYSTEMS
    . **DIGITAL RADAR SYSTEMS**
RT   ANTIRADIATION MISSILES
    DATA PROCESSING EQUIPMENT
    RADAR DETECTION
    RADAR EQUIPMENT
    RADAR RECEIVERS
    RADAR SCANNING
    RADAR TARGETS
    RADAR TRACKING
    RADAR TRANSMISSION
    SIGNAL ANALYSIS
    SURVEILLANCE RADAR
    TRACKING RADAR

DIGITAL SIMULATION
GS   MODELS
    . MATHEMATICAL MODELS
    . . **DIGITAL SIMULATION**
    SIMULATION
    . COMPUTERIZED SIMULATION
    . . **DIGITAL SIMULATION**
RT   ANALOG SIMULATION
    BIOLOGICAL MODELS (MATHEMATICS)
    COMPUTER SYSTEMS SIMULATION
    WAR GAMES

DIGITAL SPACECRAFT TELEVISION
GS   COMMUNICATION EQUIPMENT
    . SPACECRAFT TELEVISION
    . . **DIGITAL SPACECRAFT TELEVISION**
    DIGITAL TELEVISION
    . **DIGITAL SPACECRAFT TELEVISION**
    TELECOMMUNICATION
    . PULSE COMMUNICATION
    . . **DIGITAL SPACECRAFT TELEVISION**
    . SPACECRAFT TELEVISION
    . . **DIGITAL SPACECRAFT TELEVISION**
    TELEVISION SYSTEMS
    . SPACECRAFT TELEVISION
    . . **DIGITAL SPACECRAFT TELEVISION**
RT   WIRELESS COMMUNICATION

DIGITAL SYSTEMS
UF   BINARY SYSTEMS (DIGITAL)
    TERNARY SYSTEMS (DIGITAL)
GS   **DIGITAL SYSTEMS**
    . DIGITAL NAVIGATION
    . DIGITAL RADAR SYSTEMS
RT   ANALOG TO DIGITAL CONVERTERS
    BINARY CODES
    BINARY DIGITS
    BITERNARY CODE
    DATA SYSTEMS
    ∞SYSTEMS
    SYSTEMS INTEGRATION
    TELECOMMUNICATION

DIGITAL TECHNIQUES
RT   BCH CODES
    BISTABLE CIRCUITS
    CODING
    COMPUTER PROGRAMMING
    ERROR CORRECTING CODES
    ERROR DETECTION CODES
    ∞METHODOLOGY
    NUMERICAL CONTROL
    SHIFT REGISTERS
    TERMINAL AREA ENERGY MANAGEMENT

DIGITAL TELEVISION
GS   **DIGITAL TELEVISION**
    . DIGITAL SPACECRAFT TELEVISION
RT   PULSE COMMUNICATION
    TELEVISION SYSTEMS
    TELEVISION TRANSMISSION

DIGITAL TO ANALOG CONVERTERS
GS   DATA CONVERTERS
    . **DIGITAL TO ANALOG CONVERTERS**
RT   ANALOG TO DIGITAL CONVERTERS
    ∞CONVERTERS
    PLOTTERS
    SIGNAL ENCODING
    X-Y PLOTTERS

DIGITAL TO VOICE TRANSLATORS
UF   DIVOT (VOICE TRANSLATORS)
RT   COMPUTERS
    ∞TRANSLATORS
    VOCODERS
    VOICE DATA PROCESSING

DIGITAL TRANSDUCERS
GS   TRANSDUCERS
    . **DIGITAL TRANSDUCERS**
RT   INTERDIGITAL TRANSDUCERS

DIGITALIS
GS   DRUGS
    . **DIGITALIS**

DIGITIZERS
USE ANALOG TO DIGITAL CONVERTERS

DIGITS
SN   (EXCLUDES FINGERS AND TOES)
GS   ALPHANUMERIC CHARACTERS
    . **DIGITS**

DIGITS-*(CONT.)*
```
        . . BINARY DIGITS
RT    ∞CODES
        DECIMALS
        INTEGERS
        NUMBER THEORY
        ∞NUMBERS
        SYMBOLS
```

DIHEDRAL ANGLE
```
GS    GEOMETRY
        . EUCLIDEAN GEOMETRY
        . . ANGLES (GEOMETRY)
        . . . DIHEDRAL ANGLE
RT    LATERAL STABILITY
```

DIHEDRAL EFFECT
```
USE   LATERAL STABILITY
```

DIHYDRAZINE
```
GS    ALIPHATIC COMPOUNDS
        . HYDRAZINES
        . . DIHYDRAZINE
```

DIHYDRIDES
```
GS    HYDROGEN COMPOUNDS
        . HYDRIDES
        . . DIHYDRIDES
```

DIHYDROXYPHENYLALANINE
```
USE   DOPA
```

DIISOCYANATES
```
GS    ALIPHATIC COMPOUNDS
        . DIISOCYANATES
        ESTERS
        . ISOCYANATES
        . . DIISOCYANATES
        NITROGEN COMPOUNDS
        . CYANO COMPOUNDS
        . . ISOCYANATES
        . . . DIISOCYANATES
```

DIKES (GEOLOGY)
```
USE   ROCK INTRUSIONS
```

DILATATION
```
USE   STRETCHING
```

DILATATIONAL WAVES
```
GS    ELASTIC WAVES
        . DILATATIONAL WAVES
RT    LONGITUDINAL WAVES
        P WAVES
        S WAVES
        SEISMIC WAVES
        ∞SHEAR
        STRETCHING
        ∞WAVES
```

DILATOMETERS
```
USE   EXTENSOMETERS
```

DILATOMETRY
```
RT    EXTENSOMETERS
        ∞MEASUREMENT
        THERMAL EXPANSION
```

DILUENTS
```
RT    ADDITIVES
        ∞AGENTS
        COMBUSTION PRODUCTS
        CONTAMINANTS
        DISPERSIONS
        EXHAUST GASES
        SOLVENTS
```

DILUTION
```
GS    DILUTION
        . GEOMETRIC DILUTION OF PRECISION
RT    ATTENUATION
        CONCENTRATION (COMPOSITION)
        DIFFUSION
        DISPERSING
        DISPOSAL
        DISSIPATION
        DISSOLVING
        LOW CONCENTRATIONS
        MIXING
        PURITY
        ∞REDUCTION
        WASTE DISPOSAL
```

DIMENHYDRINATE
```
GS    AMINES
        . DIMENHYDRINATE
        DRUGS
        . ANTIHISTAMINICS
        . . DIMENHYDRINATE
        HETEROCYCLIC COMPOUNDS
        . DIMENHYDRINATE
```

DIMENSIONAL ANALYSIS
```
RT    ∞APPLICATIONS OF MATHEMATICS
        DIMENSIONLESS NUMBERS
        DIMENSIONS
        FIBERS (MATHEMATICS)
        FLUID FLOW
        PARAMETERIZATION
        SCALING LAWS
        SIMILARITY NUMBERS
        UNITS OF MEASUREMENT
```

DIMENSIONAL MEASUREMENT
```
RT    DEFORMETERS
        DISTANCE MEASURING EQUIPMENT
        ∞MEASUREMENT
        MICROMETERS
        SIZE DETERMINATION
```

DIMENSIONAL STABILITY
```
GS    MECHANICAL PROPERTIES
        . DIMENSIONAL STABILITY
        . . STRUCTURAL STABILITY
        . . . SHELL STABILITY
        STABILITY
        . STATIC STABILITY
        . . DIMENSIONAL STABILITY
        . . . STRUCTURAL STABILITY
        . . . . SHELL STABILITY
RT    CREEP PROPERTIES
        CURL (MATERIALS)
        DYNAMIC STABILITY
        ROCHE LIMIT
        THERMAL STABILITY
        TOLERANCES (MECHANICS)
```

DIMENSIONLESS NUMBERS
```
GS    RATIOS
        . DIMENSIONLESS NUMBERS
        . . BIOT NUMBER
        . . FROUDE NUMBER
        . . GRASHOF NUMBER
        . . HARTMAN NUMBER
        . . LAVAL NUMBER
        . . LEWIS NUMBER
        . . MACH NUMBER
        . . NUSSELT NUMBER
        . . PECLET NUMBER
        . . PRANDTL NUMBER
        . . RAYLEIGH NUMBER
        . . REYNOLDS NUMBER
        . . . HIGH REYNOLDS NUMBER
        . . . LOW REYNOLDS NUMBER
        . . RICHARDSON NUMBER
        . . SCHMIDT NUMBER
        . . SIMILARITY NUMBER
        . . STANTON NUMBER
        . . STROUHAL NUMBER
RT    DIMENSIONAL ANALYSIS
        FLUID FLOW
        HEAT TRANSFER
        NUMBERS
        SCALING LAWS
```

DIMENSIONS
```
GS    DIMENSIONS
        . DEPTH
        . DIAMETERS
        . FILM THICKNESS
        . FRACTALS
        . HEIGHT
        . . SCALE HEIGHT
        . LENGTH
        . RADII
        . . LARMOR RADIUS
        . TARGET THICKNESS
        . WIDTH
RT    AMPLITUDES
        ∞DESIGN
        DIMENSIONAL ANALYSIS
        DISTANCE
        DRAWINGS
        ENGINEERING DRAWINGS
        FINENESS RATIO
        GEOMETRY
        MAGNITUDE
        PARTICLE SIZE DISTRIBUTION
```

DIMENSIONS-*(CONT.)*
```
        RELATIVISTIC EFFECTS
        ∞SPAN
        THICKNESS
        TOPOLOGY
        UNITS OF MEASUREMENT
        VOLUME
```

DIMERCAPROL
```
GS    ALIPHATIC COMPOUNDS
        . THIOLS
        . . DIMERCAPROL
        SULFUR COMPOUNDS
        . THIOLS
        . . DIMERCAPROL
```

DIMERIZATION
```
GS    POLYMERIZATION
        . DIMERIZATION
RT    COPOLYMERIZATION
```

DIMERS
```
GS    PREPOLYMERS
        . DIMERS
RT    MONOMERS
        TRIMERS
```

DIMETHYLHYDRAZINES
```
GS    ALIPHATIC COMPOUNDS
        . HYDRAZINES
        . . DIMETHYLHYDRAZINES
        AMINES
        . DIMETHYLHYDRAZINES
RT    AEROZINE
        METHYLHYDRAZINE
```

DIMINUTION
```
USE   REDUCTION
```

DIMMING
```
RT    BRIGHTNESS
        LIGHT EMISSION
        ∞REDUCTION
```

DIMPLING
```
RT    BULGING
        METAL WORKING
        STAMPING
```

DINING PHILOSOPHERS PROBLEM
```
RT    DISTRIBUTED PROCESSING
        INTERPROCESSOR COMMUNICATION
        PROBLEM SOLVING
        SYNCHRONISM
```

DINITRATES
```
GS    NITROGEN COMPOUNDS
        . NITRATES
        . . DINITRATES
```

DIODE-TRANSISTOR-LOGIC INTEG CIRCUITS
```
USE   DTL INTEGRATED CIRCUITS
```

DIODES
```
UF    P-I-N DIODES
GS    ELECTRONIC EQUIPMENT
        . DIODES
        . . CESIUM DIODES
        . . CRYSTAL RECTIFIERS
        . . PLASMA DIODES
        . . SEMICONDUCTOR DIODES
        . . . AVALANCHE DIODES
        . . . BARRITT DIODES
        . . . GERMANIUM DIODES
        . . . GUNN DIODES
        . . . JUNCTION DIODES
        . . . LIGHT EMITTING DIODES
        . . . PARAMETRIC DIODES
        . . . PHOTODIODES
        . . . SCHOTTKY DIODES
        . . . TUNNEL DIODES
        . . . VARACTOR DIODES
RT    ELECTRON TUBES
        ION IMPLANTATION
        RECTIFIERS
        SEMICONDUCTOR DEVICES
        SOLIONS
        TRAPATT DEVICES
        TRIODES
        VARACTOR DIODE CIRCUITS
```

DIONE
```
GS    SATELLITES
        . NATURAL SATELLITES
```

DIONE-(CONT.)
```
       . . SATURN SATELLITES
       . . . DIONE
RT        SATURN (PLANET)
```

DIOPHANTINE EQUATION
```
GS     NUMBER THEORY
       . DIOPHANTINE EQUATION
RT     ∞EQUATIONS
```

DIORITE
```
GS     ROCKS
       . IGNEOUS ROCKS
       . . DIORITE
RT     MINERALS
       SOILS
```

DIOXIDES
```
GS     CHALCOGENIDES
       . OXIDES
       . . DIOXIDES
       . . . CARBON DIOXIDE
       . . . FLINT
       . . . HYDROGEN PEROXIDE
       . . . SILICON DIOXIDE
       . . . . QUARTZ
       . . . . . COESITE
       . . . SULFUR DIOXIDES
RT     KARL FISCHER REAGENT
       PEROXIDES
       SULFUR OXIDES
       THORIUM OXIDES
       TITANIUM OXIDES
```

DIPHENYL COMPOUNDS
```
GS     HYDROCARBONS
       . DIPHENYL COMPOUNDS
       . . DIPHENYL HYDANTOIN
RT     ∞CHEMICAL COMPOUNDS
```

DIPHENYL HYDANTOIN
```
GS     AMINES
       . DIPHENYL HYDANTOIN
       DRUGS
       . ANTIHISTAMINICS
       . . DIPHENYL HYDANTOIN
       HYDROCARBONS
       . DIPHENYL COMPOUNDS
       . . DIPHENYL HYDANTOIN
```

DIPHOSPHATES
```
GS     PHOSPHORUS COMPOUNDS
       . PHOSPHATES
       . . DIPHOSPHATES
       . . . ADENOSINE DIPHOSPHATE
```

DIPHTHERIA
```
GS     DISEASES
       . INFECTIOUS DISEASES
       . . DIPHTHERIA
RT     TOXIC DISEASES
```

DIPLEXERS
```
GS     CIRCUITS
       . DIPLEXERS
       COMMUNICATION EQUIPMENT
       . DIPLEXERS
       COUPLERS
       . ANTENNA COUPLERS
       . . DIPLEXERS
RT     COUPLES
       RADAR ANTENNAS
       RADAR EQUIPMENT
       TELEVISION EQUIPMENT
       TRANSFORMERS
```

DIPOLE ANTENNAS
```
SN     (SINGLE DIPOLE ANTENNAS)
GS     ANTENNAS
       . DIRECTIONAL ANTENNAS
       . . DIPOLE ANTENNAS
RT     ANTENNA ARRAYS
       ANTENNA DESIGN
       ∞DIPOLES
       DIRECTORS (ANTENNA ELEMENTS)
       LENS ANTENNAS
       LINEAR ARRAYS
       LOG PERIODIC ANTENNAS
       LOG SPIRAL ANTENNAS
       MONOPOLE ANTENNAS
       OMNIDIRECTIONAL ANTENNAS
       RADAR ANTENNAS
       TURNSTILE ANTENNAS
       YAGI ANTENNAS
```

DIPOLE MOMENTS
```
GS     MOMENTS
       . DIPOLE MOMENTS
       . . ELECTRIC MOMENTS
       . . MAGNETIC MOMENTS
RT     DOMAINS
       ELECTRICAL PROPERTIES
       MAGNETIC DOMAINS
       MAGNETIC PROPERTIES
       VAN DER WAAL FORCES
```

∞ DIPOLES
```
SN     (USE OF A MORE SPECIFIC TERM IS
       RECOMMENDED--CONSULT THE TERMS
       LISTED BELOW)
RT     DIPOLE ANTENNAS
       ELECTRIC CHARGE
       ELECTRIC DIPOLES
       MAGNETIC DIPOLES
       MAGNETIC POLES
       MONOPOLES
       ORBITING DIPOLES
       POLARITY
       ∞POLES
       QUADRUPOLES
```

DIPPING
```
RT     BATHS
       COATINGS
       QUENCHING (COOLING)
       SUBMERGING
       WETTING
```

DIRAC EQUATION
```
GS     WAVE EQUATIONS
       . DIRAC EQUATION
RT     ∞EQUATIONS
       FIELD THEORY (PHYSICS)
       KLEIN-GORDON EQUATION
       LORENTZ TRANSFORMATIONS
       QUANTUM THEORY
```

DIRECT CURRENT
```
UF     DC (CURRENT)
GS     ELECTRIC CURRENT
       . DIRECT CURRENT
RT     ALTERNATING CURRENT
       CURRENT CONVERTERS (AC TO DC)
       HOMOPOLAR GENERATORS
       INVERTED CONVERTERS (DC TO AC)
       VOLTAGE CONVERTERS (DC TO DC)
```

DIRECT LIFT CONTROLS
```
RT     ∞CONTROL
       LIFT DEVICES
```

DIRECT POWER GENERATORS
```
UF     ENERGY CONVERTERS
GS     ELECTRIC GENERATORS
       . DIRECT POWER GENERATORS
       . . ELECTROSTATIC GENERATORS
       . . FUEL CELLS
       . . . BIOCHEMICAL FUEL CELLS
       . . . HYDROGEN OXYGEN FUEL CELLS
       . . . PHOSPHORIC ACID FUEL CELLS
       . . . REGENERATIVE FUEL CELLS
       . . MAGNETOHYDRODYNAMIC
           GENERATORS
       . . PHOTOELECTRIC GENERATORS
       . . PRIMARY BATTERIES
       . . . ALKALINE BATTERIES
       . . . DRY CELLS
       . . . . MAGNESIUM CELLS
       . . . . NICKEL ZINC BATTERIES
       . . . METAL AIR BATTERIES
       . . . . ZINC-OXYGEN BATTERIES
       . . . THERMAL BATTERIES
       . . RADIOISOTOPE BATTERIES
       . . . SNAP 7
       . . . SNAP 9A
       . . . SNAP 11
       . . . SNAP 13
       . . . SNAP 15
       . . . SNAP 17
       . . . SNAP 19
       . . . SNAP 21
       . . . SNAP 23
       . . . SNAP 27
       . . . SNAP 29
       . . SOLAR CELLS
       . . . HOMOJUNCTIONS
       . . . VERTICAL JUNCTION SOLAR CELLS
       . . THERMIONIC CONVERTERS
       . . . SNAP 13
       . . . SOLAR BLANKETS
```

DIRECT POWER GENERATORS-(CONT.)
```
       . . THERMOELECTRIC GENERATORS
       . . . SNAP 3
       . . . SNAP 7
       . . . SNAP 9A
       . . . SNAP 10A
       . . . SNAP 11
       . . . SNAP 15
       . . . SNAP 17
       . . . SNAP 19
       . . . SNAP 21
       . . . SNAP 23
       . . . SNAP 27
       . . . SNAP 29
       . . . SOLAR SEA POWER PLANTS
RT     AUXILIARY POWER SOURCES
       ∞CONVERTERS
       ELECTRIC BATTERIES
       ∞ELECTRIC CELLS
       ELECTRIC ENERGY STORAGE
       ENERGY ABSORPTION FILMS
       ENERGY CONVERSION
       ENERGY CONVERSION EFFICIENCY
       ∞GENERATORS
       HEAT GENERATION
       PHOTOELECTRIC CELLS
       SOLAR GENERATORS
       SOLAR TOTAL ENERGY SYSTEMS
       SPACECRAFT POWER SUPPLIES
```

∞ DIRECTION
```
SN     (USE OF A MORE SPECIFIC TERM IS
       RECOMMENDED--CONSULT THE TERMS
       LISTED BELOW)
RT     AUTONOMY
       AZIMUTH
       BEARING (DIRECTION)
       DIRECTIVITY
       LINE OF SIGHT
       MANAGEMENT
       REVERSING
```

DIRECTION FINDERS (RADIO)
```
USE    RADIO DIRECTION FINDERS
```

DIRECTION FINDING
```
RT     BEARING (DIRECTION)
       RADIO DIRECTION FINDERS
       SIGNAL PROCESSING
```

DIRECTIONAL ANTENNAS
```
UF     TRACKING ANTENNAS
GS     ANTENNAS
       . DIRECTIONAL ANTENNAS
       . . DIPOLE ANTENNAS
       . . HELICAL ANTENNAS
       . . HORN ANTENNAS
       . . LENS ANTENNAS
       . . LOG PERIODIC ANTENNAS
       . . LOOP ANTENNAS
       . . PARABOLIC ANTENNAS
       . . RADAR ANTENNAS
       . . . RADANT
       . . RHOMBIC ANTENNAS
       . . SLOT ANTENNAS
       . . STEERABLE ANTENNAS
       . . . INERTIALESS STEERABLE
             ANTENNAS
       . . TWO REFLECTOR ANTENNAS
       . . YAGI ANTENNAS
RT     ANTENNA ARRAYS
       ANTENNA RADIATION PATTERNS
       BACKLOBES
       BORESIGHT ERROR
       BORESIGHTS
       ENDFIRE ARRAYS
       MICROWAVE ANTENNAS
       MICROWAVE COUPLING
       MISSILE ANTENNAS
       MONOPULSE ANTENNAS
       OMNIDIRECTIONAL ANTENNAS
       RADIO ANTENNAS
       SOMMERFELD APPROXIMATION
       SPINNERS
```

DIRECTIONAL CONTROL
```
UF     VECTOR CONTROL
GS     ATTITUDE CONTROL
       . DIRECTIONAL CONTROL
       . . THRUST VECTOR CONTROL
RT     AIRCRAFT CONTROL
       AUTOMATIC CONTROL
       ∞CONTROL
       HELICOPTER CONTROL
       JET CONTROL
```

DIRECTIONAL CONTROL-*(CONT.)*
    LATERAL CONTROL
    LONGITUDINAL CONTROL
    MANUAL CONTROL
    MISSILE CONTROL
  ∞REACTION CONTROL
    ROCKET ENGINE CONTROL
    SATELLITE ATTITUDE CONTROL
    SATELLITE CONTROL
    YAW

DIRECTIONAL COUPLERS
  GS   ANTENNA COMPONENTS
    . ANTENNA COUPLERS
    . . **DIRECTIONAL COUPLERS**
    COUPLERS
    . ANTENNA COUPLERS
    . . **DIRECTIONAL COUPLERS**
  RT   COUPLING
    COUPLINGS
    IMPEDANCE MATCHING
    MICROSTRIP TRANSMISSION LINES
    MICROWAVE COUPLING
    TRANSMISSION LINES
    YOKES

DIRECTIONAL SOLIDIFICATION (CRYSTALS)
  GS   CRYSTALLIZATION
    . **DIRECTIONAL SOLIDIFICATION
      (CRYSTALS)**
    GROWTH
    . CRYSTAL GROWTH
    . . **DIRECTIONAL SOLIDIFICATION
      (CRYSTALS)**
  RT   CONTAINERLESS MELTS
    EUTECTIC COMPOSITES
    PHASE TRANSFORMATIONS

DIRECTIONAL STABILITY
  GS   DYNAMIC CHARACTERISTICS
    . DYNAMIC STABILITY
    . . MOTION STABILITY
    . . . ATTITUDE STABILITY
    . . . . **DIRECTIONAL STABILITY**
    . . . . . GYROSCOPIC STABILITY
    STABILITY
    . DYNAMIC STABILITY
    . . MOTION STABILITY
    . . . ATTITUDE STABILITY
    . . . . **DIRECTIONAL STABILITY**
    . . . . . GYROSCOPIC STABILITY
  RT   AERODYNAMIC STABILITY
    AIRCRAFT STABILITY
    CONTROLLABILITY
    FLOW STABILITY
    HORIZONTAL ORIENTATION
    HOVERING STABILITY
    LATERAL OSCILLATION
    LATERAL STABILITY
    LONGITUDINAL STABILITY
    ROTARY STABILITY
    SPACECRAFT STABILITY
    STABILITY AUGMENTATION
    VERTICAL ORIENTATION
    YAW

DIRECTIVITY
  RT   ALIGNMENT
    ANISOTROPY
    COLLIMATION
    CRYSTALLOGRAPHY
  ∞DIRECTION
    FIELD STRENGTH
    INSTRUMENT ORIENTATION
    ISOTROPY
    LOOK ANGLES (ELECTRONICS)
  ∞ORIENTATION

DIRECTORIES
  RT   HANDBOOKS
    MANUALS

DIRECTORS (ANTENNA ELEMENTS)
  RT   DIPOLE ANTENNAS
    RADIO RECEIVERS
    REFLECTOMETERS
    REFLECTORS
    RODS
    YAGI ANTENNAS

DIRICHLET PROBLEM
  GS   BOUNDARY VALUE PROBLEMS
    . **DIRICHLET PROBLEM**
  RT   DIFFERENTIAL EQUATIONS
    HYPERBOLIC DIFFERENTIAL EQUATIONS

DIRICHLET PROBLEM-*(CONT.)*
  ∞PROBLEMS

DIRIGIBLES
  USE AIRSHIPS

DIRT
  GS   SOILS
    . **DIRT**
  RT   CONTAMINANTS
    DUST
    IMPURITIES
    PARTICLES
    ROCKS

DISARMAMENT
  RT   ARMED FORCES (FOREIGN)
    ARMED FORCES (UNITED STATES)
    INTERNATIONAL COOPERATION
    WEAPONS

DISASTERS
  RT   ACCIDENTS
    CASUALTIES
    EMERGENCIES
    FIRST AID
    SABOTAGE

∞ **DISCHARGE**
  SN   *(USE OF A MORE SPECIFIC TERM IS
      RECOMMENDED--CONSULT THE TERMS
      LISTED BELOW)*
  RT   DETONATION
    DISPERSING
    DISPOSAL
    DRAINAGE
    EFFLUENTS
    EJECTION
    ELECTRIC DISCHARGES
    ELECTRODELESS DISCHARGES
    ELIMINATION
    EMISSION
    EXHAUSTING
    EXPELLANTS
    EXPLOSIONS
    OUTLETS
    RELEASING
    RELIEVING
    RING DISCHARGE
    UNLOADING
    VENTING

DISCHARGE COEFFICIENT
  GS   COEFFICIENTS
    . FLOW COEFFICIENTS
    . . **DISCHARGE COEFFICIENT**
  RT   AXIAL FLOW
    FLOW VELOCITY
    INFLUENCE COEFFICIENT
    MASS FLOW FACTORS
    NOZZLE FLOW
    NOZZLE GEOMETRY
    NOZZLE THRUST COEFFICIENTS
    WALL FLOW

DISCHARGE TUBES
  USE GAS DISCHARGE TUBES

DISCHARGERS
  GS   **DISCHARGERS**
    . STATIC DISCHARGERS
  RT   DISSIPATION
    NEUTRALIZERS

DISCIPLINING
  RT   LIABILITIES
    MORALE
    PENALTIES

DISCOLORATION
  RT   COLOR
    DAMAGE
    DEGRADATION
    FADING
    STAINING

DISCONNECT DEVICES
  UF   DISCONNECTORS
  RT   CIRCUIT BREAKERS
    CONNECTORS
    DECOUPLING
    DUMPING
    EJECTION
    ELECTRIC CONNECTORS

DISCONNECT DEVICES-*(CONT.)*
    ELECTRIC FUSES
    ELECTRIC RELAYS
  ∞RELAY
    RELEASING

DISCONNECTORS
  USE DISCONNECT DEVICES

DISCONTINUITY
  GS   **DISCONTINUITY**
    . SHOCK DISCONTINUITY
  RT   ANALYSIS (MATHEMATICS)
    CATASTROPHE THEORY
    GIBBS PHENOMENON
    INCOHERENCE
    VORTEX STREETS

DISCOS (SATELLITE ATTITUDE CONTROL)
  RT   ATTITUDE STABILITY
    NOVA SATELLITES
    SATELLITE PERTURBATION
    SPACECRAFT STABILITY
    TRANSIT SATELLITES

DISCOVERER RECOVERY CAPSULES
  UF   DRC (CAPSULE)
  GS   SPACE CAPSULES
    . **DISCOVERER RECOVERY CAPSULES**
  RT   RECOVERY PARACHUTES
    SPACECRAFT RECOVERY

DISCOVERER SATELLITES
  GS   SATELLITES
    . ARTIFICIAL SATELLITES
    . . **DISCOVERER SATELLITES**
  RT   AGENA A ROCKET VEHICLE
    AGENA B ROCKET VEHICLE
    AGENA ROCKET VEHICLES
    THOR AGENA LAUNCH VEHICLE

DISCOVERING
  USE EXPLORATION

DISCOVERY (ORBITER)
  UF   SPACE SHUTTLE ORBITER 103
  GS   REENTRY VEHICLES
    . RECOVERABLE SPACECRAFT
    . . REUSABLE SPACECRAFT
    . . . **DISCOVERY (ORBITER)**
    TRANSPORTATION
    . SPACE TRANSPORTATION
    . . SPACE TRANSPORTATION SYSTEM
    . . . SPACE SHUTTLE ORBITERS
    . . . . **DISCOVERY (ORBITER)**
  RT   MANNED SPACE FLIGHT
    SPACE SHUTTLE MISSION 41-D
    SPACE SHUTTLE MISSION 51-A
    SPACE SHUTTLE MISSION 51-C
    SPACE SHUTTLE MISSION 51-D
    SPACE SHUTTLE MISSION 51-G
    SPACE SHUTTLE MISSION 51-I
  ∞SPACECRAFT

DISCRETE ADDRESS BEACON SYSTEM
  GS   LANDING AIDS
    . AIRPORT BEACONS
    . . **DISCRETE ADDRESS BEACON
      SYSTEM**
    NAVIGATION AIDS
    . BEACONS
    . . AIRPORT BEACONS
    . . . **DISCRETE ADDRESS BEACON
      SYSTEM**
    . . RADAR BEACONS
    . . . **DISCRETE ADDRESS BEACON
      SYSTEM**
    RADAR EQUIPMENT
    . **DISCRETE ADDRESS BEACON SYSTEM**
  RT   AIR TRAFFIC CONTROL
    DATA LINKS
    GROUND-AIR-GROUND COMMUNICATION
    SECONDARY RADAR
  ∞SYSTEMS

DISCRETE FUNCTIONS
  GS   FUNCTIONS (MATHEMATICS)
    . **DISCRETE FUNCTIONS**
  RT   DISTRIBUTION FUNCTIONS
    HISTOGRAMS
    NORMAL DENSITY FUNCTIONS
    POISSON DENSITY FUNCTIONS
    PROBABILITY DENSITY FUNCTIONS
    PROBABILITY DISTRIBUTION FUNCTIONS

DISCRETE FUNCTIONS-*(CONT.)*
    STATISTICAL ANALYSIS

DISCRIMINANT ANALYSIS (STATISTICS)
UF   DISCRIMINANT FUNCTIONS
GS   FUNCTIONS (MATHEMATICS)
    . **DISCRIMINANT ANALYSIS
      (STATISTICS)**
    STATISTICAL ANALYSIS
    . **DISCRIMINANT ANALYSIS
      (STATISTICS)**
RT   ∞CLASSIFYING
    MULTIVARIATE STATISTICAL ANALYSIS
    POPULATIONS

DISCRIMINANT FUNCTIONS
USE   DISCRIMINANT ANALYSIS (STATISTICS)

DISCRIMINATION
GS   **DISCRIMINATION**
    . BRIGHTNESS DISCRIMINATION
    . SENSORY DISCRIMINATION
    . . TACTILE DISCRIMINATION
    . . VISUAL DISCRIMINATION
RT   ACUITY
    COMPARATOR CIRCUITS
    ∞DIFFERENTIATION
    SELECTIVITY
    SIGNAL DETECTION
    SIGNAL DETECTORS
    TARGET RECOGNITION

DISCRIMINATORS
GS   CIRCUITS
    . **DISCRIMINATORS**
    . . FRAUNHOFER LINE DISCRIMINATORS
    . . FREQUENCY DISCRIMINATORS
RT   ANALOG COMPUTERS
    COMPARATORS
    ERROR SIGNALS
    INTERMODULATION
    RC CIRCUITS

∞ DISCUSSION
SN   *(USE OF A MORE SPECIFIC TERM IS
    RECOMMENDED--CONSULT THE TERMS
    LISTED BELOW)*
RT   CONFERENCES
    EVALUATION
    EXAMINATION
    REPORTS
    REVIEWING

DISEASED VEGETATION
USE   BLIGHT

DISEASES
GS   **DISEASES**
    . ALBINISM
    . ANEMIAS
    . ARTERIOSCLEROSIS
    . ARTHRITIS
    . ATAXIA
    . ATELECTASIS
    . BONE DEMINERALIZATION
    . COLIC
    . CYANOSIS
    . DIABETES MELLITUS
    . EDEMA
    . ENCEPHALITIS
    . EPILEPSY
    . EYE DISEASES
    . . ASTHENOPIA
    . . ASTIGMATISM
    . . CATARACTS
    . . CONJUNCTIVITIS
    . . GLAUCOMA
    . . KERATITIS
    . . PHORIA
    . FAT EMBOLISMS
    . FIBROSIS
    . . CYSTIC FIBROSIS
    . HEADACHE
    . HEART DISEASES
    . . ANGINA PECTORIS
    . . CORONARY ARTERY DISEASE
    . INFARCTION
    . . MYOCARDIAL INFARCTION
    . INFECTIOUS DISEASES
    . . AIRBORNE INFECTION
    . . CHOLERA
    . . CONJUNCTIVITIS
    . . DERMATITIS
    . . . CONTACT DERMATITIS
    . . DIPHTHERIA

DISEASES-*(CONT.)*
    . . HEPATITIS
    . . INFLUENZA
    . . MENINGITIS
    . . POLIOMYELITIS
    . . SMALLPOX
    . . SYPHILIS
    . . TUBERCULOSIS
    . . TYPHOID
    . . TYPHUS
    . KIDNEY DISEASES
    . NEPHRITIS
    . LITHIASIS
    . METABOLIC DISEASES
    . MILIARIA
    . NARCOLEPSY
    . NEURASTHENIA
    . NEURITIS
    . OSTEOPOROSIS
    . PARALYSIS
    . PARASITIC DISEASES
    . . BLIGHT
    . PARKINSON DISEASE
    . PULMONARY LESIONS
    . RADIATION SICKNESS
    . RESPIRATORY DISEASES
    . . AEROSINUSITIS
    . . ASTHMA
    . . EMPHYSEMA
    . . INFLUENZA
    . . PNEUMONIA
    . . TUBERCULOSIS
    . RHEUMATIC DISEASES
    . TACHYCARDIA
    . THROMBOPENIA
    . THROMBOSIS
    . TOOTH DISEASES
    . TOXIC DISEASES
    . . CARBON MONOXIDE POISONING
    . . LEAD POISONING
    . TUMORS
    . . NEOPLASMS
    . . . CANCER
    . . . . LEUKEMIAS
    . ULCERS
    . UROLITHIASIS
RT   CHRONIC CONDITIONS
    CREATININE
    CURES
    DIAGNOSIS
    ETIOLOGY
    MEDICAL SCIENCE
    PATHOGENESIS
    PATHOLOGICAL EFFECTS
    PNEUMOTHORAX
    PROPHYLAXIS
    SIGNS AND SYMPTOMS
    SYMPTOMOLOGY
    THERAPY
    VACCINES
    VETERINARY MEDICINE

DISHES
USE   PARABOLIC REFLECTORS

DISILICIDES
GS   SILICON COMPOUNDS
    . SILICIDES
    . . **DISILICIDES**
RT   SILANES
    SILICATES

DISINFECTANTS
USE   ANTISEPTICS

DISINTEGRATION
RT   ATOMIZING
    COMMINUTION
    CRUSHERS
    CRUSHING
    DAMAGE
    DECAY
    DECOMPOSITION
    DETERIORATION
    FLAKING
    GRINDING (COMMINUTION)
    IONIZATION
    LYSOGENESIS

DISK GALAXIES
GS   CELESTIAL BODIES
    . GALAXIES
    . . **DISK GALAXIES**
RT   ANDROMEDA GALAXIES
    ASTROPHYSICS

DISK GALAXIES-*(CONT.)*
    BARRED GALAXIES
    ELLIPTICAL GALAXIES
    GALACTIC CLUSTERS
    GALACTIC EVOLUTION
    GALACTIC NUCLEI
    GALACTIC ROTATION
    GALACTIC STRUCTURE
    LOCAL GROUP (ASTRONOMY)
    RADIO GALAXIES
    SPIRAL GALAXIES
    STAR CLUSTERS
    VIRGO GALACTIC CLUSTER

∞ DISKS
SN   *(USE OF A MORE SPECIFIC TERM IS
    RECOMMENDED--CONSULT THE TERMS
    LISTED BELOW)*
RT   ACTUATOR DISKS
    DISKS (SHAPES)
    INTERVERTEBRAL DISKS
    MAGNETIC DISKS

DISKS (SHAPES)
GS   **DISKS (SHAPES)**
    . ACTUATOR DISKS
    . INTERVERTEBRAL DISKS
    . ROTATING DISKS
RT   ACCRETION DISKS
    AERODYNAMIC CONFIGURATIONS
    BODIES OF REVOLUTION
    CIRCULAR PLATES
    ∞DISKS
    ∞PLATES
    VIDEO DISKS

DISLOCATIONS (MATERIALS)
GS   **DISLOCATIONS (MATERIALS)**
    . CRYSTAL DISLOCATIONS
    . . EDGE DISLOCATIONS
    . . SCREW DISLOCATIONS
RT   DISPLACEMENT
    FLOW THEORY
    ∞MATERIALS

DISORDERS
GS   DISORIENTATION
    . **DISORDERS**
RT   CHOLERA
    CHRONIC CONDITIONS
    DEPERSONALIZATION
    ∞DEPRESSION
    DETACHMENT
    ∞DISTURBANCES
    DITHERS
    EMOTIONAL FACTORS
    HUMAN BEHAVIOR
    JET LAG
    PSYCHOLOGY
    PSYCHOSES
    VIOLENCE

DISORIENTATION
SN   (EXCLUDES PHYSICAL OR
    MATHEMATICAL MISALIGNMENT)
GS   **DISORIENTATION**
    . DESYNCHRONIZATION (BIOLOGY)
    . DISORDERS
    . JET LAG
RT   BIOLOGICAL EFFECTS
    CORIOLIS EFFECT
    DETACHMENT
    DITHERS
    IRRATIONALITY
    MISALIGNMENT
    PSYCHOLOGICAL EFFECTS
    PSYCHOLOGY
    STAGGERING
    WEIGHTLESSNESS

DISPATCHING
USE   DISTRIBUTING

DISPENSERS
RT   DISTRIBUTORS
    EJECTORS
    FEEDERS
    MATERIALS HANDLING
    ROLLERS
    SPRAYERS

DISPERSING
SN   (OF MATERIALS OR PARTICLES)
RT   AGITATION
    ASSIMILATION

DISPERSING-(CONT.)
　　　CHEMICAL RELEASE MODULES
　　　CIRCULATION
　　　CLOUD DISPERSAL
　　　COLLOIDING
　　　DEFLECTION
　　　DIFFUSION
　　　DILUTION
　∞ DISCHARGE
　∞ DISPERSION
　　　DISPERSIONS
　　　DISPOSAL
　　　DISSIPATION
　　　DISTRIBUTING
　∞ DISTRIBUTION
　　　ENTRAINMENT
　　　EXHAUSTING
　　　FOG DISPERSAL
　　　HOMOGENIZING
　　　LANGEVIN FORMULA
　　　PERMEATING
　　　POLLUTION TRANSPORT
　∞ REDUCTION
　　　RELEASING
　　　SCATTERING
　∞ SEPARATION
　　　SHAKING
　　　SPRAYING
　　　SPREADING
　　　STIRRING
　　　SUSPENDING (MIXING)
　　　SWIRLING

∞ **DISPERSION**
　SN　　(USE OF A MORE SPECIFIC TERM IS
　　　　RECOMMENDED--CONSULT THE TERMS
　　　　LISTED BELOW)
　RT　　DEVIATION
　　　　DIFFUSION
　　　　DISPERSING
　　　　DISPERSIONS
　　　　DUST
　　　　KRAMERS-KRONIG FORMULA
　　　　MAGNETIC DISPERSION
　　　　MIXERS
　　　　RANDOM ERRORS
　　　　STATISTICAL ANALYSIS
　　　　VARIABILITY
　　　　WAVE DISPERSION

DISPERSION PRECIPITATION HARDENING
　USE　　PRECIPITATION HARDENING

DISPERSIONS
　GS　　MIXTURES
　　　. **DISPERSIONS**
　　　. . COLLOIDS
　　　. . . AEROSOLS
　　　. . . . FOG
　　　. . . COLLOIDAL PROPELLANTS
　　　. . EMULSIONS
　　　. . . PHOTOGRAPHIC EMULSIONS
　　　. . . . NUCLEAR EMULSIONS
　　　. . LIQUID-GAS MIXTURES
　　　. . . AEROSOLS
　　　. . . . FOG
　　　. . PLASTISOLS
　　　. . . SMOKE
　RT　　BROWNIAN MOVEMENTS
　　　　COLLOIDAL GENERATORS
　　　　CROP DUSTING
　　　　DILUENTS
　　　　DISPERSING
　∞ DISPERSION
　　　　DUST
　　　　FERROFLUIDS
　　　　FOG DISPERSAL
　　　　FUMES
　　　　MIST
　　　　PARTICLES
　　　　SLURRIES
　　　　SLURRY PROPELLANTS
　　　　SUSPENDING (MIXING)
　∞ SUSPENSIONS

DISPLACEMENT
　RT　　AMPLITUDES
　　　　BENDING
　　　　BIAS
　　　　BORESIGHT ERROR
　　　　DEFLECTION
　　　　DEFORMATION
　　　　DISLOCATIONS (MATERIALS)
　　　　DISTORTION
　　　　DIVERGENCE

DISPLACEMENT-(CONT.)
　　　　ENGINES
　　　　FLEXIBLE SPACECRAFT
　　　　HEAVING
　　　　LEVEL (QUANTITY)
　　　　MAGNITUDE
　∞ MOTION
　　　　NUTATION
　　　　POSITIONING
　　　　SKEWNESS
　　　　TEMPERATURE INVERSIONS
　　　　VARIATIONS
　　　　VIBRATION

DISPLACEMENT MEASUREMENT
　SN　　(MEASUREMENT IN CHANGE OF
　　　　POSITION)
　GS　　MECHANICAL MEASUREMENT
　　　. **DISPLACEMENT MEASUREMENT**
　RT　　BORESIGHT ERROR

DISPLAY DEVICES
　UF　　DATA READOUT SYSTEMS
　　　　DISPLAY SYSTEMS
　　　　VISUAL DISPLAYS
　GS　　**DISPLAY DEVICES**
　　　. APPROACH INDICATORS
　　　. FLOW DIRECTION INDICATORS
　　　. . WIND VANES
　　　. GYRO HORIZONS
　　　. HEAD-UP DISPLAYS
　　　. HELMET MOUNTED DISPLAYS
　　　. KINOFORM
　　　. MICROVISION LANDING AID
　　　. PLASMA DISPLAY DEVICES
　　　. POSITION INDICATORS
　　　. . PLAN POSITION INDICATORS
　　　. . RADIO DIRECTION FINDERS
　　　. . SPACECRAFT POSITION INDICATORS
　　　. RADARSCOPES
　　　. . PLAN POSITION INDICATORS
　　　. SPEED INDICATORS
　　　. . ANEMOMETERS
　　　. . . DRAG FORCE ANEMOMETERS
　　　. . . HOT-FILM ANEMOMETERS
　　　. . . HOT-WIRE ANEMOMETERS
　　　. . . SONIC ANEMOMETERS
　　　. . TACHOMETERS
　RT　　AIRBORNE SURVEILLANCE RADAR
　　　　AIRCRAFT EQUIPMENT
　　　　AIRCRAFT INSTRUMENTS
　　　　AUTOMATIC TYPEWRITERS
　　　　BLINKING
　　　　CANCELLATION CIRCUITS
　　　　CATHODE RAY TUBES
　　　　CHARTS
　　　　COMPUTER GRAPHICS
　　　　CONSOLES
　　　　CONTROL BOARDS
　　　　CREW PROCEDURES (INFLIGHT)
　　　　CREW PROCEDURES (PREFLIGHT)
　　　　DATA RECORDERS
　∞ DETECTORS
　　　　DIALS
　　　　ELECTROCHROMISM
　　　　FLIGHT CONTROL
　　　　FLIGHT INSTRUMENTS
　　　　FLYING SPOT SCANNERS
　　　　IMAGE RECONSTRUCTION
　　　　IMAGE TUBES
　　　　IMAGERY
　　　　IMAGES
　　　　INDICATING INSTRUMENTS
　　　　INSTRUMENT LANDING SYSTEMS
　　　　INSTRUMENT RECEIVERS
　∞ INSTRUMENTS
　　　　LIGHT EMITTING DIODES
　　　　LISTS
　　　　MAN MACHINE SYSTEMS
　　　　MAP MATCHING GUIDANCE
　　　　MONITORS
　　　　NAVIGATION AIDS
　　　　PERCEPTUAL ERRORS
　　　　PHOTOGRAPHS
　　　　PICTURE TUBES
　　　　PLANETARIUMS
　∞ PLOTS
　　　　PLOTTERS
　　　　PLOTTING
　　　　PRINTERS (DATA PROCESSING)
　　　　PROMOTION
　　　　RADAR
　　　　RADAR RESOLUTION
　　　　RAPID BALLISTICS IDENTIFICATION
　　　　READING

DISPLAY DEVICES-(CONT.)
　　　　READOUT
　　　　REAL TIME OPERATION
　　　　RECEIVERS
　　　　REMOTE CONSOLES
　∞ SCREENS
　　　　SOLAR COMPASSES
　∞ STRIP
　∞ SYSTEMS
　　　　TARGET SIMULATORS
　　　　TERCOM
　　　　VIDEO DATA
　　　　VIDEO EQUIPMENT
　　　　VIEWING
　　　　VISUAL AIDS
　　　　VISUAL CONTROL
　　　　WARNING SYSTEMS

DISPLAY SYSTEMS
　USE　　DISPLAY DEVICES

DISPOSAL
　GS　　**DISPOSAL**
　　　. WASTE DISPOSAL
　　　. . COMPOSTING
　　　. . HAZARDOUS MATERIAL DISPOSAL (IN
　　　　　SPACE)
　RT　　AGITATION
　∞ CONTAINERS
　　　　DECONTAMINATION
　　　　DELETION
　　　　DILUTION
　∞ DISCHARGE
　　　　DISPERSING
　　　　DISSIPATION
　　　　DISTRIBUTING
　∞ DISTRIBUTION
　　　　DUMPING
　　　　EJECTION
　　　　ELIMINATION
　　　　EMPTYING
　　　　EXHAUSTING
　　　　EXPULSION
　　　　ISOLATION
　　　　JETTISONING
　　　　MATERIALS HANDLING
　　　　MATERIALS RECOVERY
　　　　REMOVAL
　　　　SINKS
　　　　SPREADING
　∞ STORAGE
　　　　UNLOADING

DISRUPTING
　RT　　∞ INTERFERENCE
　　　　RUPTURING

DISSECTION
　RT　　AUTOPSIES
　　　　PATHOLOGY

DISSIPATION
　UF　　DISSIPATORS
　GS　　**DISSIPATION**
　　　. ENERGY DISSIPATION
　　　. OHMIC DISSIPATION
　RT　　ATMOSPHERIC TURBULENCE
　　　　ATTENUATION
　　　　DAMPING
　　　　DECONTAMINATION
　　　　DEPLETION
　　　　DIFFUSION
　　　　DILUTION
　　　　DISCHARGERS
　　　　DISPERSING
　　　　DISPOSAL
　　　　EXHAUSTING
　　　　POLLUTION
　　　　PURIFICATION
　∞ REDUCTION
　　　　REMOVAL
　　　　WASTE DISPOSAL

DISSIPATORS
　USE　　DISSIPATION

DISSOCIATION
　UF　　MOLECULAR DISSOCIATION
　GS　　**DISSOCIATION**
　　　. AUTOIONIZATION
　　　. BIODEGRADABILITY
　　　. GAS DISSOCIATION
　　　. PHOTODISSOCIATION
　　　. THERMAL DISSOCIATION
　RT　　ATOMIC RECOMBINATION

DISSOCIATION-*(CONT.)*
    CHEMICAL EQUILIBRIUM
    DEBYE-HUCKEL THEORY
    DECOMPOSITION
    ELECTRODISSOLUTION
    HEAT OF DISSOCIATION
    IONIZATION
    MOLECULAR DIFFUSION
    MOLECULAR INTERACTIONS

DISSOLUTION
USE   DISSOLVING

DISSOLVED GASES
GS   GASES
    . **DISSOLVED GASES**
RT   AERATION
    DISSOLVING
    MIXTURES
    OXYGENATION
    SOLUBILITY
    SOLUTIONS

DISSOLVING
UF   DISSOLUTION
GS   MIXING
    . **DISSOLVING**
RT   AERATION
    CHEMICAL ATTACK
    CHEMICAL CLEANING
    CLEANING
    COMPOUNDING
    CORROSION
    DIFFUSION
    DILUTION
    DISSOLVED GASES
    EXTRACTION
    HOMOGENIZING
    LEACHING
    PRECIPITATION (CHEMISTRY)
    ∞SEPARATION
    SOFTENING
    SOLUBILITY
    SOLUTES
    ∞SOLUTION
    SOLVENT RETENTION
    SOLVENTS
    WASHING

DISSYMMETRY
USE   ASYMMETRY

DISTANCE
GS   **DISTANCE**
    . DEBYE LENGTH
    . MISS DISTANCE
    . OPTICAL SLANT RANGE
    . RADAR RANGE
    . RADIO RANGE
    . RANGE AND RANGE RATE TRACKING
    . REENTRY RANGE
RT   AIRCRAFT PERFORMANCE
    AIRCRAFT SPECIFICATIONS
    ALTITUDE
    DEPTH
    DIMENSIONS
    FOCUSING
    GEOMETRY
    HEIGHT
    LENGTH
    POSITION (LOCATION)
    PROXIMITY
    RADAR NAVIGATION
    ∞RANGE
    RANGE (EXTREMES)
    TAKEOFF RUNS
    ∞TRAVEL

DISTANCE MEASURING EQUIPMENT
GS   MEASURING INSTRUMENTS
    . **DISTANCE MEASURING EQUIPMENT**
    . . ALTIMETERS
    . . . RADIO ALTIMETERS
    . . GEODIMETERS
    . . RANGE FINDERS
    . . . OPTICAL RANGE FINDERS
    . . . . LASER RANGE FINDERS
    . . STADIMETERS
    . . TELLUROMETERS
RT   AUTOMATIC FLIGHT CONTROL
    AUTOMATIC LANDING CONTROL
    DECCA NAVIGATION
    DEPTH MEASUREMENT
    DIMENSIONAL MEASUREMENT
    LORAC NAVIGATION SYSTEM

DISTANCE MEASURING EQUIPMENT-*(CONT.)*
    LORAN
    LUNAR RANGEFINDING
    MICROMETERS
    NAVIGATION
    NAVIGATION AIDS
    OMNIDIRECTIONAL RADIO RANGES
    POSITION INDICATORS
    RADAR
    RADAR EQUIPMENT
    RADAR MEASUREMENT
    RADAR NAVIGATION
    RADIO NAVIGATION
    RANGE ERRORS
    SHORAN
    SOLAR COMPASSES
    SONAR
    SOUND RANGING

DISTANCE PERCEPTION
USE   SPACE PERCEPTION

DISTILLATION
GS   **DISTILLATION**
    . STRIPPING (DISTILLATION)
RT   CHEMICAL FRACTIONATION
    CONCENTRATING
    CONDENSING
    DEMINERALIZING
    DESALINIZATION
    DIFFUSION
    EVAPORATION
    FLASHING (VAPORIZING)
    MATERIALS RECOVERY
    PURGING
    PURIFICATION
    RECTIFICATION
    REFINING
    ∞SEPARATION
    TAR SANDS
    VAPORIZING
    WASHING

DISTILLATION EQUIPMENT
RT   COLUMNS (PROCESS ENGINEERING)
    CONDENSERS (LIQUEFIERS)
    ∞EQUIPMENT
    STILLS

DISTORTION
GS   **DISTORTION**
    . FLOW DISTORTION
    . SIGNAL DISTORTION
    . SURFACE DISTORTION
RT   ABERRATION
    ABNORMALITIES
    ASTIGMATISM
    ASYMMETRY
    BENDING
    BUCKLING
    CAMBER
    DEFLECTION
    DEFORMATION
    DEVIATION
    DISPLACEMENT
    EXPANSION
    FAILURE
    FLEXING
    FOLDING
    GEOMETRIC ACCURACY
    GHOSTS
    HEAVING
    REFRACTION
    SKEWNESS
    STRETCHING
    SWELLING
    TEMPERATURE INVERSIONS
    TWISTING
    VARIATIONS
    WARPAGE
    WRINKLING

DISTRIBUTED AMPLIFIERS
GS   AMPLIFIERS
    . **DISTRIBUTED AMPLIFIERS**
RT   ∞FREQUENCY RESPONSE
    TRANSMISSION LINES

DISTRIBUTED FEEDBACK LASERS
GS   STIMULATED EMISSION DEVICES
    . LASERS
    . . **DISTRIBUTED FEEDBACK LASERS**
RT   FEEDBACK AMPLIFIERS
    FEEDBACK CONTROL
    HETEROJUNCTION DEVICES

DISTRIBUTED FEEDBACK LASERS-*(CONT.)*
    LASER OUTPUTS
    LASING
    SEMICONDUCTOR LASERS
    SOLID STATE LASERS

DISTRIBUTED PARAMETER SYSTEMS
RT   CONTROL THEORY
    DIFFERENTIAL EQUATIONS
    INDEPENDENT VARIABLES
    INTEGRAL EQUATIONS
    LINEAR CIRCUITS
    LINEAR SYSTEMS
    NETWORK ANALYSIS
    NONLINEAR SYSTEMS
    ∞SYSTEMS

DISTRIBUTED PROCESSING
GS   DATA PROCESSING
    . **DISTRIBUTED PROCESSING**
RT   ARCHITECTURE (COMPUTERS)
    COMPUTER NETWORKS
    COMPUTER SYSTEMS DESIGN
    DINING PHILOSOPHERS PROBLEM
    MICROPROCESSORS

DISTRIBUTING
UF   DISPATCHING
RT   ALLOCATIONS
    ASSIMILATION
    COMMERCIAL ENERGY
    DISPERSING
    DISPOSAL
    ∞DISTRIBUTION
    DOMESTIC ENERGY
    ∞FOOD
    INDUSTRIAL ENERGY
    INVENTORY CONTROLS
    MATERIALS HANDLING
    POSITIONING
    PROPORTION
    RESOURCE ALLOCATION
    TRANSPORTATION
    TRANSPORTATION ENERGY

∞ **DISTRIBUTION**
SN   *(USE OF A MORE SPECIFIC TERM IS*
    RECOMMENDED--CONSULT THE TERMS
    LISTED BELOW)
RT   ALLOCATIONS
    ASSIMILATION
    BRIGHTNESS DISTRIBUTION
    CIRCULATION DISTRIBUTION
    DEMOGRAPHY
    DISPERSING
    DISPOSAL
    DISTRIBUTING
    DISTRIBUTION (PROPERTY)
    KURTOSIS
    LOAD DISTRIBUTION (FORCES)
    MASS DISTRIBUTION
    MATERIALS HANDLING
    POSITIONING
    PRESSURE DISTRIBUTION
    SPECTRAL ENERGY DISTRIBUTION
    STATISTICAL DISTRIBUTIONS
    THRUST DISTRIBUTION
    TRANSPORTATION

DISTRIBUTION (PROPERTY)
UF   PATTERN DISTRIBUTION
GS   **DISTRIBUTION (PROPERTY)**
    . ANGULAR DISTRIBUTION
    . BOLTZMANN DISTRIBUTION
    . BRIGHTNESS DISTRIBUTION
    . CHARGE DISTRIBUTION
    . CURRENT DISTRIBUTION
    . ELECTRON DISTRIBUTION
    . . ELECTRON DENSITY PROFILES
    . ENERGY DISTRIBUTION
    . . SPECTRAL ENERGY DISTRIBUTION
    . FLOW DISTRIBUTION
    . FORCE DISTRIBUTION
    . FREQUENCY DISTRIBUTION
    . KURTOSIS
    . HOLE DISTRIBUTION (ELECTRONICS)
    . HOLE DISTRIBUTION (MECHANICS)
    . INTERFERENCE LIFT
    . ION DISTRIBUTION
    . LOAD DISTRIBUTION (FORCES)
    . MASS DISTRIBUTION
    . MOMENT DISTRIBUTION
    . NEUTRON DISTRIBUTION
    . PRESSURE DISTRIBUTION
    . RADIAL DISTRIBUTION

DISTRIBUTION (PROPERTY)-*(CONT.)*
. RADIATION DISTRIBUTION
. . ANTENNA RADIATION PATTERNS
. . . SIDELOBES
. . DIFFRACTION PATTERNS
. . . KOSSEL PATTERN
. . . RAINBOWS
. SPATIAL DISTRIBUTION
. . STAR DISTRIBUTION
. STRESS CONCENTRATION
. TEMPERATURE DISTRIBUTION
. VELOCITY DISTRIBUTION
. VERTICAL DISTRIBUTION
. . STAR DISTRIBUTION
RT   CHEMICAL COMPOSITION
∞ CROSS SECTIONS
∞ DISTRIBUTION
DYNAMIC CHARACTERISTICS
FIELD THEORY (PHYSICS)
GRADIENTS
JET LIFT
LIFT
∞ PATTERNS
∞ PROFILES
ROTOR LIFT
STATISTICAL DISTRIBUTIONS
SYNTHETIC ARRAYS
ZERO LIFT

DISTRIBUTION FUNCTIONS
GS   FUNCTIONS (MATHEMATICS)
. **DISTRIBUTION FUNCTIONS**
RT   CHAPMAN-ENSKOG THEORY
DISCRETE FUNCTIONS
MAXIMUM ENTROPY METHOD
PROBABILITY THEORY
STATISTICAL DISTRIBUTIONS

DISTRIBUTION MOMENTS
UF   STATISTICAL MOMENTS
GS   MOMENTS
. **DISTRIBUTION MOMENTS**
. . MEAN
. . ORTHOGONALITY
. . STANDARD DEVIATION
RT   AVERAGE
MEDIAN (STATISTICS)
METHOD OF MOMENTS
MODE (STATISTICS)
SKEWNESS
STATISTICAL DISTRIBUTIONS
VARIANCE (STATISTICS)

DISTRIBUTORS
RT   COMMUTATORS
DISPENSERS
FEEDERS
IGNITION SYSTEMS
INTERNAL COMBUSTION ENGINES
MATERIALS HANDLING
ROLLERS
SPRAYERS

DISTRICT OF COLUMBIA
RT   POTOMAC RIVER VALLEY (MD-VA-WV)
UNITED STATES

DISTURBANCE THEORY
USE   PERTURBATION THEORY

∞ **DISTURBANCES**
SN   *(USE OF A MORE SPECIFIC TERM IS
RECOMMENDED--CONSULT THE TERMS
LISTED BELOW)*
RT   BURSTS
DISORDERS
ELECTROMAGNETIC INTERFERENCE
IONOSPHERIC DISTURBANCES
IONOSPHERIC STORMS
MAGNETIC DISTURBANCES
PERTURBATION
RADIO AURORAS
RADIO BURSTS
SOLAR ACTIVITY
STORMS
SUDDEN IONOSPHERIC DISTURBANCES
VORTICES

DISTURBING FUNCTIONS
GS   FUNCTIONS (MATHEMATICS)
. **DISTURBING FUNCTIONS**
RT   PERTURBATION THEORY

DISULFIDES
GS   CHALCOGENIDES

DISULFIDES-*(CONT.)*
. SULFIDES
. . **DISULFIDES**
. . . CARBON DISULFIDE
SULFUR COMPOUNDS
. SULFIDES
. . **DISULFIDES**
. . . CARBON DISULFIDE

DITCHES
RT   CANALS
IRRIGATION
LANDFORMS
TROUGHS

∞ **DITCHING**
SN   *(USE OF A MORE SPECIFIC TERM IS
RECOMMENDED--CONSULT THE TERMS
LISTED BELOW)*
RT   DITCHING (LANDING)
EXCAVATION

DITCHING (EXCAVATION)
USE   EXCAVATION

DITCHING (LANDING)
GS   CRASHES
. CRASH LANDING
. . **DITCHING (LANDING)**
LANDING
. AIRCRAFT LANDING
. . CRASH LANDING
. . . **DITCHING (LANDING)**
. WATER LANDING
. . **DITCHING (LANDING)**
RT   AIRCRAFT ACCIDENTS
∞ DITCHING
GLIDE LANDINGS

DITHERS
GS   SHAKING
. **DITHERS**
SHIVERING
. **DITHERS**
RT   DISORDERS
DISORIENTATION
EMOTIONAL FACTORS
HUMAN BEHAVIOR
∞ INHIBITION
IRRATIONALITY
VACILLATION

DITHIOLS
USE   THIOLS

DIURESIS
RT   BODY FLUIDS
EDEMA
URINATION

DIURETICS
GS   **DIURETICS**
. AMINOPHYLLINE
RT   ACETAZOLAMIDE
UREAS

DIURNAL RHYTHMS
USE   CIRCADIAN RHYTHMS

DIURNAL VARIATIONS
GS   VARIATIONS
. PERIODIC VARIATIONS
. . **DIURNAL VARIATIONS**
RT   CYCLES
DARKNESS
DAYTIME
MAGNETIC VARIATIONS
NIGHT
NOCTURNAL VARIATIONS
TROPOPAUSE
WIND VARIATIONS

DIVERGENCE
GS   **DIVERGENCE**
. MAGNETIC CHARGE DENSITY
RT   CATASTROPHE THEORY
CONVERGENCE
DEVIATION
DIFFERENCES
DISPLACEMENT
∞ DRIFT
FOURIER ANALYSIS
FUNCTIONS (MATHEMATICS)
GEOSTROPHIC WIND

DIVERGENCE-*(CONT.)*
REFRACTION
SERIES (MATHEMATICS)
SERIES EXPANSION
VARIATIONS
VORTICES

DIVERGENT NOZZLES
RT   CONICAL NOZZLES
EXHAUST NOZZLES
NOZZLE GEOMETRY
NOZZLE WALLS
∞ NOZZLES
ROCKET NOZZLES
THRUST CHAMBERS
WIND TUNNEL NOZZLES

DIVERTERS
RT   BAFFLES
BLAST DEFLECTORS
BYPASSES
DEFLECTORS
∞ DIFFUSERS
DIVIDERS
FLAME DEFLECTORS
∞ SEPARATION
SEPARATORS
SHIELDING
VALVES

DIVIDERS
SN   *(EXCLUDES VOLTAGE AND FREQUENCY
DIVIDERS)*
GS   SEPARATORS
. **DIVIDERS**
RT   BAFFLES
∞ BARRIERS
CURTAINS
DIVERTERS
PANELS
SPACERS

DIVIDES (LANDFORMS)
GS   LANDFORMS
. **DIVIDES (LANDFORMS)**
RT   DRAINAGE PATTERNS
MOUNTAINS
WATERSHEDS

DIVIDING (MATHEMATICS)
GS   NUMBER THEORY
. **DIVIDING (MATHEMATICS)**
RT   ARITHMETIC
COMPUTATION
CONGRUENCES
∞ DIVISION
QUOTIENTS

DIVING (UNDERWATER)
GS   SUBMERGED BODIES
. **DIVING (UNDERWATER)**
RT   BAROTRAUMA
DECOMPRESSION SICKNESS
HUMAN TOLERANCES
MEDICAL PHENOMENA
PHYSIOLOGICAL EFFECTS
UNDERWATER PHYSIOLOGY
UNDERWATER TESTS

∞ **DIVISION**
SN   *(USE OF A MORE SPECIFIC TERM IS
RECOMMENDED--CONSULT THE TERMS
LISTED BELOW)*
RT   CELL DIVISION
DIVIDING (MATHEMATICS)
NUMBER THEORY
∞ SEPARATION
SUBDIVISIONS
SUBSIDIARIES

DIVOT (VOICE TRANSLATORS)
USE   DIGITAL TO VOICE TRANSLATORS

DME-A SATELLITE
USE   EXPLORER 31 SATELLITE

DMSP SATELLITES
UF   DEFENSE METEOROLOGICAL SATELLITE
PROGRAM
GS   MILITARY SPACECRAFT
. **DMSP SATELLITES**
RT   AIR DEFENSE
∞ DEFENSE
DEFENSE PROGRAM

DMSP SATELLITES-(CONT.)
METEOROLOGY
PHOTOMAPPING
PHOTORECONNAISSANCE
REMOTE SENSING
SATELLITE-BORNE PHOTOGRAPHY

DNA
USE   DEOXYRIBONUCLEIC ACID

DO-27 AIRCRAFT
UF   DORNIER DO-27 AIRCRAFT
GS   DORNIER AIRCRAFT
    . **DO-27 AIRCRAFT**
    GENERAL AVIATION AIRCRAFT
    . **DO-27 AIRCRAFT**
    LIGHT AIRCRAFT
    . **DO-27 AIRCRAFT**
    MONOPLANES
    . **DO-27 AIRCRAFT**
    PASSENGER AIRCRAFT
    . **DO-27 AIRCRAFT**
    UTILITY AIRCRAFT
    . **DO-27 AIRCRAFT**
RT   ∞AIRCRAFT

DO-28 AIRCRAFT
UF   DORNIER DO-28 AIRCRAFT
GS   DORNIER AIRCRAFT
    . **DO-28 AIRCRAFT**
    GENERAL AVIATION AIRCRAFT
    . **DO-28 AIRCRAFT**
    LIGHT AIRCRAFT
    . **DO-28 AIRCRAFT**
    MONOPLANES
    . **DO-28 AIRCRAFT**
    PASSENGER AIRCRAFT
    . **DO-28 AIRCRAFT**
    UTILITY AIRCRAFT
    . **DO-28 AIRCRAFT**
RT   ∞AIRCRAFT

DO-31 AIRCRAFT
UF   DORNIER DO-31 AIRCRAFT
GS   DORNIER AIRCRAFT
    . **DO-31 AIRCRAFT**
    JET AIRCRAFT
    . TURBOFAN AIRCRAFT
    . . **DO-31 AIRCRAFT**
    MONOPLANES
    . **DO-31 AIRCRAFT**
    TRANSPORT AIRCRAFT
    . **DO-31 AIRCRAFT**
    V/STOL AIRCRAFT
    . **DO-31 AIRCRAFT**
RT   ∞AIRCRAFT

DOCKING
USE   SPACECRAFT DOCKING

DOCUMENT STORAGE
RT   DATA STORAGE
    DOCUMENTATION
    ∞FILES
    REPRODUCTION (COPYING)
    ∞STORAGE

DOCUMENTATION
SN   (THE ASSEMBLING, CODING,
    KNOWLEDGE FOR GIVING
    DOCUMENTARY INFORMATION MAXIMUM
    ACCESSIBILITY AND USABILITY)
GS   LITERATURE
    . **DOCUMENTATION**
RT   ACQUISITION
    BIBLIOGRAPHIES
    BIOGRAPHY
    CASE HISTORIES
    CATALOGS (PUBLICATIONS)
    CONFERENCES
    DATA RETRIEVAL
    DOCUMENT STORAGE
    DOCUMENTS
    HISTORIES
    INDEXES (DOCUMENTATION)
    INFORMATION
    INFORMATION DISSEMINATION
    INFORMATION RETRIEVAL
    KNOWLEDGE
    LIBRARIES
    NEWS
    RECORDS
    ∞REFERENCE SYSTEMS
    REPORTS

DOCUMENTATION-(CONT.)
SELECTIVE DISSEMINATION OF
    INFORMATION
SPACE GLOSSARIES
SUMMARIES
TECHNICAL WRITING
TECHNOLOGY TRANSFER
TRANSLATING

DOCUMENTS
UF   PUBLICATIONS
GS   **DOCUMENTS**
    . ABSTRACTS
    . BIBLIOGRAPHIES
    . CATALOGS (PUBLICATIONS)
    . . ASTRONOMICAL CATALOGS
    . ENGINEERING DRAWINGS
    . . BLUEPRINTS
    . HANDBOOKS
    . . USER MANUALS (COMPUTER
        PROGRAMS)
    . MANUALS
    . . INSTALLATION MANUALS
    . . USER MANUALS (COMPUTER
        PROGRAMS)
    . PAPERS
    . PERIODICALS
    . POSTLAUNCH REPORTS
    . PRESIDENTIAL REPORTS
    . RECORDS
    . . VIDEO DISKS
    . SUPPLEMENTS
    . TEXTBOOKS
    . TEXTS
    . THESES
RT   CONFERENCES
    CONGRESSIONAL REPORTS
    COPYRIGHTS
    DICTIONARIES
    DOCUMENTATION
    DRAWINGS
    FORMAT
    HARDWARE UTILIZATION LISTS
    INDEXES (DOCUMENTATION)
    INFORMATION RETRIEVAL
    LIBRARIES
    LITERATURE
    REPORTS
    TECHNOLOGY TRANSFER

DODGE SATELLITE
GS   SATELLITES
    . ARTIFICIAL SATELLITES
    . . **DODGE SATELLITE**

DOGHOUSES (ELECTRONICS)
GS   HOUSINGS
    . **DOGHOUSES (ELECTRONICS)**
    RADAR EQUIPMENT
    . **DOGHOUSES (ELECTRONICS)**
RT   ENCLOSURES
    RADAR ANTENNAS

DOGS
GS   ANIMALS
    . VERTEBRATES
    . . MAMMALS
    . . . **DOGS**

DOLLIES
GS   SURFACE VEHICLES
    . **DOLLIES**
RT   CARRIAGES
    MATERIALS HANDLING
    SLEDS
    TRUCKS
    UNDERCARRIAGES

DOLOMITE (MINERAL)
GS   CARBON COMPOUNDS
    . CARBONATES
    . . **DOLOMITE (MINERAL)**
    MAGNESIUM COMPOUNDS
    . **DOLOMITE (MINERAL)**
    MINERALS
    . **DOLOMITE (MINERAL)**
RT   AGGREGATES
    LIMESTONE
    ROCKS
    SEDIMENTARY ROCKS

DOLPHINS
GS   ANIMALS
    . VERTEBRATES
    . . MAMMALS

DOLPHINS-(CONT.)
    . . . **DOLPHINS**
RT   MARINE MAMMALS

DOMAIN WALL
RT   DOMAINS
    MAGNETIC DOMAINS
    ∞MOTION

DOMAINS
GS   **DOMAINS**
    . MAGNETIC DOMAINS
RT   DIPOLE MOMENTS
    DOMAIN WALL
    ELECTRICAL PROPERTIES
    RANGE (EXTREMES)

∞ **DOMES**
SN   (USE OF A MORE SPECIFIC TERM IS
    RECOMMENDED--CONSULT THE TERMS
    LISTED BELOW)
RT   DOMES (GEOLOGY)
    DOMES (STRUCTURAL FORMS)

DOMES (GEOLOGY)
RT   ANTICLINES
    ∞DOMES
    GEOLOGY
    GEOSYNCLINES
    SYNCLINES

DOMES (STRUCTURAL FORMS)
GS   SHELLS (STRUCTURAL FORMS)
    . **DOMES (STRUCTURAL FORMS)**
    . . RADOMES
RT   ∞CUPOLAS
    ∞DOMES
    HEMISPHERICAL SHELLS
    HOUSINGS
    PRESSURE VESSELS
    PROTUBERANCES

DOMESTIC ENERGY
RT   ALLOCATIONS
    COMMERCIAL ENERGY
    DISTRIBUTING
    ECONOMIC FACTORS
    ∞ENERGY
    ENERGY CONSUMPTION
    ENERGY CONVERSION
    INDUSTRIAL ENERGY
    SOLAR COOLING
    SOLAR HOUSES
    TRANSPORTATION ENERGY
    WATER HEATING

DOMESTIC SATELLITE COMMUNICATIONS SYSTEMS
RT   COMMUNICATION SATELLITES
    MICROWAVE TRANSMISSION
    RCA SATCOM SATELLITES
    SATELLITE NETWORKS
    SATELLITE TRANSMISSION
    SATELLITES
    ∞SYSTEMS

DOMINANCE
GS   **DOMINANCE**
    . EYE DOMINANCE
RT   GENETICS

DOMINICA
GS   LANDFORMS
    . ISLANDS
    . . WEST INDIES
    . . . **DOMINICA**
    NATIONS
    . **DOMINICA**

DOMINICAN REPUBLIC
GS   NATIONS
    . **DOMINICAN REPUBLIC**
RT   CARIBBEAN REGION
    CARIBBEAN SEA

DOMINO PROPELLANTS
GS   PROPELLANTS
    . HIGH ENERGY PROPELLANTS
    . . **DOMINO PROPELLANTS**
RT   PLASTICIZERS
    ROCKET OXIDIZERS
    SOLID ROCKET PROPELLANTS

DONNELL EQUATIONS
RT   BUCKLING
      ∞EQUATIONS
      STRESS ANALYSIS

DONOR MATERIALS
GS   SEMICONDUCTORS (MATERIALS)
      . **DONOR MATERIALS**
RT   CARRIER DENSITY (SOLID STATE)
      ELECTRONS
      HOLES (ELECTRON DEFICIENCIES)
      ∞MATERIALS

DOORS
UF   EXITS (DOORS)
RT   AIR LOCKS
      APERTURES
      CURTAINS
      EGRESS
      ENTRANCES
      FLOORS
      GATES (OPENINGS)
      HATCHES
      INGRESS (SPACECRAFT PASSAGEWAY)
      OPENINGS
      OUTLETS
      ∞THRESHOLDS
      WINDOWS (APERTURES)

DOPA
UF   DIHYDROXYPHENYLALANINE
GS   ORGANIC COMPOUNDS
      . **DOPA**
RT   ∞CHEMICAL COMPOUNDS
      MELANIN
      OXIDATION
      PIGMENTS

DOPED CRYSTALS
GS   CRYSTALS
      . **DOPED CRYSTALS**
RT   CRYSTAL GROWTH
      CRYSTAL OPTICS
      CRYSTAL STRUCTURE

DOPES
RT   ADDITIVES
      FILLERS
      FINISHES
      GELS
      PRIMERS (COATINGS)
      SEALERS

DOPING (ADDITIVES)
USE   ADDITIVES

DOPPLER EFFECT
UF   DOVAP
      STELLAR DOPPLER SHIFT
GS   **DOPPLER EFFECT**
      . DOPPLER-FIZEAU EFFECT
RT   ∞EFFECTS
      ELASTIC WAVES
      ELECTROMAGNETIC RADIATION
      FIZEAU EFFECT
      FREQUENCY SHIFT
      OPTICAL HETERODYNING
      RADIAL VELOCITY
      RED SHIFT
      SATELLITE DOPPLER POSITIONING
      STELLAR MOTIONS

DOPPLER NAVIGATION
GS   NAVIGATION
      . **DOPPLER NAVIGATION**
RT   AIR NAVIGATION
      ALL-WEATHER AIR NAVIGATION
      DEAD RECKONING
      RADAR NAVIGATION
      RADIO NAVIGATION
      SATELLITE DOPPLER POSITIONING

DOPPLER RADAR
GS   RADAR
      . **DOPPLER RADAR**
      . . MULTISTATIC RADAR
RT   COHERENT RADAR
      CONTINUOUS WAVE RADAR
      MONOPULSE RADAR
      MOVING TARGET INDICATORS
      POLYSTATION DOPPLER TRACKING
         SYSTEM
      PULSE RADAR
      RADAR DETECTION

DOPPLER RADAR-*(CONT.)*
GS   RADAR EQUIPMENT
      RADAR NAVIGATION
      RADAR NETWORKS
      RADAR TRACKING
      SATELLITE DOPPLER POSITIONING
      SURVEILLANCE RADAR

DOPPLER-FIZEAU EFFECT
GS   DOPPLER EFFECT
      . **DOPPLER-FIZEAU EFFECT**
RT   ∞EFFECTS
      FIZEAU EFFECT
      FREQUENCY SHIFT
      RADAR NAVIGATION
      RED SHIFT
      STELLAR MOTIONS

DORNIER AIRCRAFT
GS   **DORNIER AIRCRAFT**
      . DO-27 AIRCRAFT
      . DO-28 AIRCRAFT
      . DO-31 AIRCRAFT
RT   ∞AIRCRAFT

DORNIER DO-27 AIRCRAFT
USE   DO-27 AIRCRAFT

DORNIER DO-28 AIRCRAFT
USE   DO-28 AIRCRAFT

DORNIER DO-31 AIRCRAFT
USE   DO-31 AIRCRAFT

DORNIER PARAGLIDER ROCKET VEHICLE
GS   ROCKET VEHICLES
      . SINGLE STAGE ROCKET VEHICLES
      . . **DORNIER PARAGLIDER ROCKET
            VEHICLE**
      . SOUNDING ROCKETS
      . . **DORNIER PARAGLIDER ROCKET
            VEHICLE**
RT   LIQUID PROPELLANT ROCKET ENGINES

DORSAL SECTIONS
RT   ANATOMY
      POSTERIOR SECTIONS

DOSAGE
UF   DOSE
GS   **DOSAGE**
      . RADIATION DOSAGE
      . SUBLETHAL DOSAGE
RT   BIOLOGICAL EFFECTS
      DOSIMETERS
      DRUGS
      RADIATION MEASUREMENT

DOSE
USE   DOSAGE

DOSIMETERS
UF   DOSIMETRY
GS   MEASURING INSTRUMENTS
      . RADIATION MEASURING INSTRUMENTS
      . . RADIATION DETECTORS
      . . . **DOSIMETERS**
      . . . . THRESHOLD DETECTORS
               (DOSIMETERS)
RT   ACTINOMETERS
      DOSAGE
      EXPOSURE
      FLUX (RATE)
      FLUX DENSITY
      GEIGER COUNTERS
      IONIZATION CHAMBERS
      IRRADIATION
      NEUTRON COUNTERS
      NUCLEAR EMULSIONS
      PHOTOGRAPHIC MEASUREMENT
      PROPORTIONAL COUNTERS
      RADIANT FLUX DENSITY
      RADIATION COUNTERS
      RADIATION DOSAGE
      RADIATION EFFECTS
      RADIATION HAZARDS
      RADIATION MEASUREMENT
      RADIOBIOLOGY

DOSIMETRY
USE   DOSIMETERS

DOUBLE BASE PROPELLANTS
UF   CORDITE

DOUBLE BASE PROPELLANTS-*(CONT.)*
GS   PROPELLANTS
      . **DOUBLE BASE PROPELLANTS**
      . . DOUBLE BASE ROCKET
            PROPELLANTS
RT   CELLULOSE NITRATE
      COMPOSITE PROPELLANTS
      ENDOTHERMIC FUELS
      EXPLOSIVES
      FUELS
      NITROGLYCERIN
      PLASTISOLS
      PYROTECHNICS

DOUBLE BASE ROCKET PROPELLANTS
GS   FUELS
      . SOLID ROCKET PROPELLANTS
      . . **DOUBLE BASE ROCKET
            PROPELLANTS**
      GELS
      . **DOUBLE BASE ROCKET
            PROPELLANTS**
      PROPELLANTS
      . DOUBLE BASE PROPELLANTS
      . . **DOUBLE BASE ROCKET
            PROPELLANTS**
      . ROCKET PROPELLANTS
      . . SOLID ROCKET PROPELLANTS
      . . . **DOUBLE BASE ROCKET
               PROPELLANTS**
      . SOLID PROPELLANTS
      . . SOLID ROCKET PROPELLANTS
      . . . **DOUBLE BASE ROCKET
               PROPELLANTS**
RT   CELLULOSE NITRATE
      COMPOSITE PROPELLANTS
      EXPLOSIVES
      NITROGLYCERIN

DOUBLE CUSPS
UF   OSCULATIONS
GS   GEOMETRY
      . CUSPS (MATHEMATICS)
      . . **DOUBLE CUSPS**
RT   ∞CUSPS

DOUBLE PRECISION ARITHMETIC
GS   NUMBER THEORY
      . ARITHMETIC
      . . **DOUBLE PRECISION ARITHMETIC**
RT   ARITHMETIC AND LOGIC UNITS
      ∞NUMBERS

DOUBLE SIDEBAND TRANSMISSION
GS   TRANSMISSION
      . ELECTROMAGNETIC WAVE
         TRANSMISSION
      . . RADIO TRANSMISSION
      . . . **DOUBLE SIDEBAND TRANSMISSION**
      . SIGNAL TRANSMISSION
      . . RADIO TRANSMISSION
      . . . **DOUBLE SIDEBAND TRANSMISSION**
RT   MODULATION
      SIDEBANDS
      SINGLE SIDEBAND TRANSMISSION
      TELEVISION TRANSMISSION
      WAVE PROPAGATION

DOUBLE STARS
GS   CELESTIAL BODIES
      . STARS
      . . **DOUBLE STARS**
      . . . BINARY STARS
      . . . . COMPANION STARS
      . . . . ECLIPSING BINARY STARS
      . . . . . DWARF NOVAE
      STELLAR MOTIONS
      . **DOUBLE STARS**
RT   ASTROMETRY

DOUGHNUT SHAPE WHEELS
USE   TOROIDAL WHEELS

DOUGLAS AIRCRAFT
UF   DOUGLAS MILITARY AIRCRAFT
GS   MCDONNELL DOUGLAS AIRCRAFT
      . **DOUGLAS AIRCRAFT**
      . . A-3 AIRCRAFT
      . . A-4 AIRCRAFT
      . . B-66 AIRCRAFT
      . . C-9 AIRCRAFT
      . . C-47 AIRCRAFT
      . . C-54 AIRCRAFT
      . . C-118 AIRCRAFT
      . . C-124 AIRCRAFT

DOUGLAS AIRCRAFT-*(CONT.)*
  . . C-133 AIRCRAFT
  . . D-558 AIRCRAFT
  . . DC 3 AIRCRAFT
  . . DC 7 AIRCRAFT
  . . DC 8 AIRCRAFT
  . . DC 9 AIRCRAFT
  . . DC 10 AIRCRAFT
  . . PD-808 AIRCRAFT
  . . X-3 AIRCRAFT
  RT ∞AIRCRAFT

DOUGLAS D-558 AIRCRAFT
  USE D-558 AIRCRAFT

DOUGLAS DC-3 AIRCRAFT
  USE DC 3 AIRCRAFT

DOUGLAS DC-7 AIRCRAFT
  USE DC 7 AIRCRAFT

DOUGLAS DC-8 AIRCRAFT
  USE DC 8 AIRCRAFT

DOUGLAS DC-9 AIRCRAFT
  USE DC 9 AIRCRAFT

DOUGLAS MILITARY AIRCRAFT
  USE DOUGLAS AIRCRAFT
        MILITARY AIRCRAFT

DOUGLAS PD-808 AIRCRAFT
  USE PD-808 AIRCRAFT

DOVAP
  USE DOPPLER EFFECT

DOWN-CONVERTERS
  GS FREQUENCY CONVERTERS
     . **DOWN-CONVERTERS**
  RT ∞CONVERTERS
       FREQUENCY DIVIDERS

DOWNLINKING
  RT CARRIER TO NOISE RATIOS
       COMMUNICATION SATELLITES
       FREQUENCY REUSE
       GROUND STATIONS
       MICROWAVE TRANSMISSION
       SATELLITE TRANSMISSION
       TRANSMISSION EFFICIENCY
       UPLINKING

DOWNRANGE
  RT BALLISTIC RANGES
       BALLISTIC TRAJECTORIES
       FLIGHT TESTS
       IMPACT PREDICTION
       MISSILE RANGES
       RECOVERY ZONES
       TEST RANGES
       TOUCHDOWN
       TRAJECTORIES

DOWNRANGE ANTIMISSILE MEASUREMENT PROGRAM
  UF DAMP PROGRAM
  GS PROGRAMS
     . **DOWNRANGE ANTIMISSILE**
        MEASUREMENT PROGRAM
  RT ∞MEASUREMENT

DOWNRANGE MEASUREMENT
  RT ∞MEASUREMENT
       TEST RANGES

DOWNTIME
  GS TIME
     . **DOWNTIME**
  RT FAILURE
       INVENTORY MANAGEMENT
       LOGISTICS
       MAINTENANCE
       MALFUNCTIONS
       MTBF
       RELIABILITY
       SPARE PARTS
       SYSTEM FAILURES
       TURNAROUND (STS)

DOWNWASH
  RT BACKWASH
      ∞DRAFT
       GROUND EFFECT (AERODYNAMICS)

DOWNWASH-*(CONT.)*
       HELICOPTER WAKES
       LIFT AUGMENTATION
       PERIPHERAL JET FLOW
       UPWASH
       WAKES

DPCM (MODULATION)
  USE DIFFERENTIAL PULSE CODE
        MODULATION

DRACONID METEOROIDS
  GS CELESTIAL BODIES
     . METEOROID SHOWERS
     . . **DRACONID METEOROIDS**
     . METEOROIDS
     . . **DRACONID METEOROIDS**
  RT GIACOBINI-ZINNER COMET

∞ DRAFT
  SN *(USE OF A MORE SPECIFIC TERM IS*
      RECOMMENDED--CONSULT THE TERMS
      LISTED BELOW)
  RT BOUNDARY LAYERS
       DOWNWASH
       DRAFT (GAS FLOW)
       DRAFTING (DRAWING)
       UPWASH
       WAKES

DRAFT (GAS FLOW)
  RT ∞DRAFT
       FLUES
       VENTILATION

DRAFTING (DRAWING)
  RT ∞DRAFT
      ∞DRAWING
       DRAWINGS
       GRAPHIC ARTS

DRAFTING MACHINES
  RT COMPUTER AIDED DESIGN
      ∞DESIGN
      ∞MACHINERY

DRAG
  UF DRAG EFFECT
  GS DYNAMIC CHARACTERISTICS
     . **DRAG**
     . . ELECTROSTATIC DRAG
     . . FRICTION DRAG
     . . . AERODYNAMIC DRAG
     . . . . SUPERSONIC DRAG
     . . . VISCOUS DRAG
     . . MINIMUM DRAG
     . . PRESSURE DRAG
     . . . SUPERSONIC DRAG
     . . . WAVE DRAG
     . . . . INTERFERENCE DRAG
     . . SATELLITE DRAG
  RT AERODYNAMIC CONFIGURATIONS
       AERODYNAMICS
       BOUNDARY LAYERS
      ∞DRAG COEFFICIENTS
       DRAG MEASUREMENT
       FRICTION
       GRAVITATION
       GROUND EFFECT (AERODYNAMICS)
       LIFT
       SKIN FRICTION
       WAKES

DRAG BALANCE
  USE AERODYNAMIC BALANCE
        LIFT DRAG RATIO

DRAG CHUTES
  UF DROGUE PARACHUTES
  GS BRAKES (FOR ARRESTING MOTION)
     . AERODYNAMIC BRAKES
     . . **DRAG CHUTES**
     DRAG DEVICES
     . AERODYNAMIC BRAKES
     . . **DRAG CHUTES**
     PARACHUTES
     . **DRAG CHUTES**
  RT AIRCRAFT BRAKES
       AIRDROPS
       BALLUTES
       RIBBON PARACHUTES
       TOWED BODIES

∞ DRAG COEFFICIENTS
  SN *(USE OF A MORE SPECIFIC TERM IS*
      RECOMMENDED--CONSULT THE TERMS
      LISTED BELOW)
  RT AERODYNAMIC COEFFICIENTS
       AERODYNAMIC DRAG
       DRAG
       HYDRODYNAMIC COEFFICIENTS

DRAG DEVICES
  UF DRAGULATORS
  GS **DRAG DEVICES**
     . AERODYNAMIC BRAKES
     . . BALLUTES
     . . DRAG CHUTES
     . PARAVULCOONS
     . . SPLIT FLAPS
     . . WING FLAPS
     . . . LEADING EDGE SLATS
     . . . TRAILING EDGE FLAPS
     . SPOILERS
  RT ABORT APPARATUS
       AIRCRAFT BRAKES
       BOUNDARY LAYER CONTROL
       BRAKES (FOR ARRESTING MOTION)
       CONTROL SURFACES
       FLAPS (CONTROL SURFACES)
       LIFT DEVICES
       SKIN FRICTION
       VORTEX ALLEVIATION

DRAG EFFECT
  USE DRAG

DRAG FORCE ANEMOMETERS
  GS AIRCRAFT INSTRUMENTS
     . SPEED INDICATORS
     . . ANEMOMETERS
     . . . **DRAG FORCE ANEMOMETERS**
     DISPLAY DEVICES
     . SPEED INDICATORS
     . . ANEMOMETERS
     . . . **DRAG FORCE ANEMOMETERS**
     MEASURING INSTRUMENTS
     . INDICATING INSTRUMENTS
     . . SPEED INDICATORS
     . . . ANEMOMETERS
     . . . . **DRAG FORCE ANEMOMETERS**
  RT FLOW MEASUREMENT
      ∞INSTRUMENTS
       VELOCITY MEASUREMENT

DRAG MEASUREMENT
  GS MECHANICAL MEASUREMENT
     . **DRAG MEASUREMENT**
  RT AERODYNAMIC DRAG
       DRAG
       ELECTROSTATIC DRAG
       FLOW MEASUREMENT
      ∞MEASUREMENT
       MEASURING INSTRUMENTS

DRAG REDUCTION
  RT AERODYNAMIC DRAG
       FLUID FLOW
       FRICTION
       LIFT DRAG RATIO
      ∞REDUCTION
       WINGLETS

DRAGULATORS
  USE BRAKES (FOR ARRESTING MOTION)
        DRAG DEVICES

DRAINAGE
  UF DRAINING
      RUNOFFS
  RT ∞DISCHARGE
       EVACUATING (VACUUM)
       EXCAVATION
       FLOOD CONTROL
       HYDROLOGY
       HYDROLOGY MODELS
       IRRIGATION
       LIQUID WASTES
       MINES (EXCAVATIONS)
       PERMEABILITY
      ∞PUMPING
       SEEPAGE
       SEWERS
       SUMPS
       TUNNELING (EXCAVATION)
       WASTE DISPOSAL
       WATER FLOW
       WATER RUNOFF

DRAINAGE-(CONT.)
    WATER TABLES

DRAINAGE PATTERNS
UF   DENDRITIC DRAINAGE
     INTERLACING DRAINAGE
     RADIAL DRAINAGE PATTERNS
     RECTANGULAR DRAINAGE
RT   ARROYOS
     DIVIDES (LANDFORMS)
     FLOOD DAMAGE
     HYDROLOGY
     IRRIGATION
     MISSISSIPPI RIVER (US)
    ∞PATTERNS
     PRECIPITATION (METEOROLOGY)
     TRIBUTARIES
     WATER EROSION
     WATER FLOW
     WATERSHEDS

DRAINING
USE   DRAINAGE

∞ DRAWING
SN   (USE OF A MORE SPECIFIC TERM IS
     RECOMMENDED--CONSULT THE TERMS
     LISTED BELOW)
RT   BUNDLE DRAWING
     COLD DRAWING
     DRAFTING (DRAWING)
     DRAWINGS
     EXTRUDING
     LAYOUTS
     METAL DRAWING
     PULLING
     RECORDS
     STRETCH FORMING
     STRETCHING
     TEMPERING

DRAWINGS
UF   ELEVATIONS (DRAWINGS)
GS   DRAWINGS
     . ENGINEERING DRAWINGS
     . . BLUEPRINTS
RT   CHARTS
    ∞CROSS SECTIONS
     DIAGRAMS
     DIMENSIONS
     DOCUMENTS
     DRAFTING (DRAWING)
    ∞DRAWING
     GRAPHIC ARTS
     INKS
     LAYOUTS
    ∞PLANS
    ∞PROJECTION
     REPRESENTATIONS
     REPRODUCTION (COPYING)
     SPECIFICATIONS
    ∞TRACING
     VISUAL AIDS

DRC (CAPSULE)
USE   DISCOVERER RECOVERY CAPSULES

DREAMS
RT   RAPID EYE MOVEMENT STATE
     SLEEP

DREDGED MATERIALS
RT   CHANNEL FLOW
    ∞CHANNELS
    ∞MATERIALS
     SEDIMENTS

DREDGING
RT   ARTIFICIAL HARBORS
     HARBORS
     MINERAL DEPOSITS
     MINING
     UNDERWATER RESOURCES

∞ DRIFT
SN   (USE OF A MORE SPECIFIC TERM IS
     RECOMMENDED--CONSULT THE TERMS
     LISTED BELOW)
RT   DEVIATION
     DIVERGENCE
     DRIFT (INSTRUMENTATION)
     DRIFT RATE
     FLIGHT PATHS
     IONOSPHERIC DRIFT

DRIFT-(CONT.)
    STABILITY

DRIFT (INSTRUMENTATION)
UF   INSTRUMENT DRIFT
RT   ACCURACY
     CIRCUIT RELIABILITY
    ∞DRIFT
     DRIFT RATE
     DYNAMIC STABILITY
     ERRORS
     INSTRUMENT ERRORS
     STATIC STABILITY
     TOLERANCES (MECHANICS)

DRIFT RATE
GS   RATES (PER TIME)
     . DRIFT RATE
RT   ∞DRIFT
     DRIFT (INSTRUMENTATION)
     IONOSPHERIC DRIFT
     MOBILITY
     ORBIT PERTURBATION
     ORBITAL MECHANICS
     ROTATING PLASMAS
     STABILITY
     TRAJECTORY CONTROL

DRILL BITS
GS   CUTTERS
     . DRILL BITS
     TOOLS
     . DRILL BITS
RT   BITS
     DRILLING
     DRILLS

DRILLING
GS   DRILLING
     . LASER DRILLING
RT   BOREHOLES
     CORE SAMPLING
     CUTTING
     DRILL BITS
     DRILLS
     EXPLORATION
     MACHINING
     NATURAL GAS EXPLORATION
     OFFSHORE ENERGY SOURCES
     OIL EXPLORATION
     OIL FIELDS
     PENETRATION
     PERFORATING
     PIERCING
     TUNNELING (EXCAVATION)
     WELLS

DRILLS
GS   CUTTERS
     . DRILLS
RT   BORING MACHINES
     COMPRESSED AIR
     DRILL BITS
     DRILLING
     MACHINE TOOLS
     TAPS
     TOOLS

DRINKING
GS   INGESTION (BIOLOGY)
     . DRINKING
RT   BEVERAGES
     SWALLOWING

∞ DRIVES
SN   (USE OF A MORE SPECIFIC TERM IS
     RECOMMENDED--CONSULT THE TERMS
     LISTED BELOW)
RT   MECHANICAL DRIVES
     MOTIVATION
     PROPULSION
     SEX
     SLEEP
     WIND TUNNEL DRIVES

DROGUE PARACHUTES
USE   DRAG CHUTES

DROGUES
USE   TOWED BODIES

DRONE AIRCRAFT
UF   DRONE HELICOPTERS
GS   DRONE VEHICLES

DRONE AIRCRAFT-(CONT.)
    . DRONE AIRCRAFT
    . . TARGET DRONE AIRCRAFT
    . . . FIREBEE 2 TARGET DRONE
       AIRCRAFT
    . . . JINDIVIK TARGET AIRCRAFT
     PILOTLESS AIRCRAFT
    . DRONE AIRCRAFT
    . . TARGET DRONE AIRCRAFT
    . . . FIREBEE 2 TARGET DRONE
       AIRCRAFT
    . . . JINDIVIK TARGET AIRCRAFT
RT   ∞AIRCRAFT
     ANTISUBMARINE WARFARE AIRCRAFT
     DAST PROGRAM
     LIGHT AIRCRAFT
    ∞MILITARY AIRCRAFT
     OBLIQUE WINGS
     REMOTELY PILOTED VEHICLES
     RESEARCH AIRCRAFT
     V/STOL AIRCRAFT

DRONE HELICOPTERS
USE   DRONE AIRCRAFT
     HELICOPTERS

DRONE VEHICLES
GS   DRONE VEHICLES
    . DRONE AIRCRAFT
    . . TARGET DRONE AIRCRAFT
    . . . FIREBEE 2 TARGET DRONE
       AIRCRAFT
    . . . JINDIVIK TARGET AIRCRAFT
RT   ∞MILITARY AIRCRAFT
     PILOTLESS AIRCRAFT
     SANDPIPER TARGET MISSILE
    ∞VEHICLES
    ∞WINGED VEHICLES

DRONES FOR AERODYNAMIC AND STRUCT TEST
USE   DAST PROGRAM

DROOPED AIRFOILS
GS   AERODYNAMIC CONFIGURATIONS
     . DROOPED AIRFOILS
     AIRCRAFT CONFIGURATIONS
     . DROOPED AIRFOILS
     AIRFOILS
     . DROOPED AIRFOILS
RT   BODY-WING CONFIGURATIONS
     WING ROOTS
     WINGS

∞ DROP
SN   (USE OF A MORE SPECIFIC TERM IS
     RECOMMENDED--CONSULT THE TERMS
     LISTED BELOW)
RT   DROPS (LIQUIDS)
     GRADIENTS

DROP CALORIMETERS
GS   MEASURING INSTRUMENTS
     . CALORIMETERS
     . . DROP CALORIMETERS
RT   BOMB CALORIMETERS
     FLAME CALORIMETERS
     HEAT MEASUREMENT
     HIGH TEMPERATURE TESTS
     TEMPERATURE MEASURING
       INSTRUMENTS

DROP SIZE
RT   CLOUD PHYSICS
     CLOUDS
     CLOUDS (METEOROLOGY)
     CONDENSATES
     CONDENSING
     FOG
     HUMIDITY
     HYDROGEN CLOUDS
     NUCLEATION
     PARTICLE DIFFUSION
     PARTICLE SIZE DISTRIBUTION
    ∞PRECIPITATION
     PRECIPITATION PARTICLE
       MEASUREMENT
     RAINDROPS
     SIZE DISTRIBUTION

DROP TESTS
UF   DROP WEIGHT TESTS
RT   CHARPY IMPACT TEST
     DESTRUCTIVE TESTS
     IMPACT TESTING MACHINES
     IMPACT TESTS

DROP TESTS-(CONT.)
- NOTCH TESTS
- SHOCK TESTS
- ∞ TESTS

DROP TOWERS
- UF DROP TUBES
- RT FALLING SPHERES
  - GRAVITATIONAL EFFECTS
  - LOW GRAVITY MANUFACTURING
  - REDUCED GRAVITY
  - WEIGHTLESSNESS

DROP TRANSFER
- GS TRANSFERRING
  - . **DROP TRANSFER**
- RT ARC MELTING
  - MELTING
  - PLASMA JETS
  - REFINING

DROP TUBES
- USE DROP TOWERS

DROP WEIGHT TESTS
- USE DROP TESTS

DROPOUTS
- RT CIRCUIT BREAKERS
  - ELECTRIC CONTACTS
  - ELECTRIC SWITCHES
  - SWITCHES

DROPS (LIQUIDS)
- UF LIQUID DROPS
- GS PARTICLES
  - . **DROPS (LIQUIDS)**
  - . . RAINDROPS
- RT AIR POLLUTION
  - CONDENSATION NUCLEI
  - CONDENSING
  - ∞ DROP
  - SPRAYERS

DROPSONDES
- GS MEASURING INSTRUMENTS
  - . METEOROLOGICAL INSTRUMENTS
  - . . **DROPSONDES**
  - . SONDES
  - . . **DROPSONDES**
- RT METEOROLOGICAL BALLOONS
  - RADIOSONDES
  - RAWINSONDES

DROSOPHILA
- GS ANIMALS
  - . INVERTEBRATES
  - . . ARTHROPODS
  - . . . INSECTS
  - . . . . **DROSOPHILA**
- RT CHIRONOMUS FLIES

DROUGHT
- UF DROUGHT CONDITIONS
- RT ARID LANDS
  - CONSERVATION
  - DESERTIFICATION
  - FLOODS
  - HYDROLOGY
  - POTABLE WATER
  - PRECIPITATION (METEOROLOGY)
  - WATER CONSUMPTION
  - WATER MANAGEMENT
  - WATER POLLUTION
  - WATER RECLAMATION

DROUGHT CONDITIONS
- USE DROUGHT

DROWSINESS
- USE SLEEP

DRUG THERAPY
- USE CHEMOTHERAPY

DRUGS
- GS **DRUGS**
  - . ACTINOMYCIN
  - . ADRENERGICS
  - . AMINOPHYLLINE
  - . ANESTHETICS
  - . . CHLOROFORM
  - . . CYCLOPROPANE
  - . . METHYL CHLORIDE

DRUGS-(CONT.)
- . . NOVOCAIN
- . ANTIADRENERGICS
- . ANTIBIOTICS
- . . PENICILLIN
- . . PLEUROTIN
- . . STREPTOMYCIN
- . . TETRACYCLINES
- . ANTICONVULSANTS
- . ANTIDIURETICS
- . ANTIDOTES
- . ANTIEMETICS AND ANTINAUSEANTS
- . ANTIHISTAMINICS
- . . DIMENHYDRINATE
- . . DIPHENYL HYDANTOIN
- . . PROMETHAZINE
- . ANTIHYPERTENSIVE AGENTS
- . ANTIINFECTIVES AND ANTIBACTERIALS
- . ANTIRADIATION DRUGS
- . CENTRAL NERVOUS SYSTEM
  - DEPRESSANTS
- . . MARIJUANA
- . CHOLINERGICS
- . . ANTICHOLINERGICS
- . CORTISONE
- . CYSTEINE
- . DECONGESTANTS
- . DIGITALIS
- . EPINEPHRINE
- . . NOREPINEPHRINE
- . ERGOTAMINE
- . HEMOSTATICS
- . HISTAMINES
- . INSULIN
- . METHAMPHETAMINE
- . METRAZOL
- . MOTION SICKNESS DRUGS
- . MUSCLE RELAXANTS
- . NARCOTICS
- . . MORPHINE
- . NEMBUTAL (TRADEMARK)
- . PENTOBARBITAL SODIUM
- . . RESERPINE
- . PSYCHOTROPIC DRUGS
- . SEDATIVES
- . STIMULANTS
- . . ATROPINE
- . . CAFFEINE
- . . CENTRAL NERVOUS SYSTEM
  - STIMULANTS
- . . NORADRENALINE
- . . NOREPINEPHRINE
- . . TRANQUILIZERS
- . . TRIMETHADIONE
- . VASOCONSTRICTOR DRUGS
- . . HYPERTENSIN
- RT ALKALOIDS
  - ANALGESIA
  - ANESTHESIOLOGY
  - BIOFLAVONOIDS
  - BIOTIN
  - CHEMICAL DEFENSE
  - CHEMOTHERAPY
  - CURES
  - DOSAGE
  - ETHERS
  - LYSERGAMIDE
  - PENTOBARBITAL
  - PHARMACOLOGY
  - PHENOBARBITAL
  - QUINOLINE
  - SALICYLATES
  - SUBLETHAL DOSAGE
  - VACCINES
  - VITAMINS

DRUMLINS
- USE GLACIAL DRIFT

∞ DRUMS
- SN *(USE OF A MORE SPECIFIC TERM IS RECOMMENDED--CONSULT THE TERMS LISTED BELOW)*
- RT ∞ BARRELS
  - ∞ CYLINDERS
  - DRUMS (CONTAINERS)
  - MAGNETIC DRUMS
  - MAGNETIC STORAGE

DRUMS (CONTAINERS)
- SN (EXCLUDE MAGNETIC COMPUTER MEMORIES)
- RT BARRELS (CONTAINERS)
  - ∞ BUCKETS
  - CANS

DRUMS (CONTAINERS)-(CONT.)
- ∞ CONTAINERS
- ∞ CYLINDERS
- ∞ DRUMS
- TANKS (CONTAINERS)

DRY CELLS
- GS ELECTRIC GENERATORS
  - . DIRECT POWER GENERATORS
  - . . PRIMARY BATTERIES
  - . . . **DRY CELLS**
  - . . . . MAGNESIUM CELLS
  - . . . . NICKEL ZINC BATTERIES
  - ELECTROCHEMICAL CELLS
  - . ELECTRIC BATTERIES
  - . . PRIMARY BATTERIES
  - . . . **DRY CELLS**
  - . . . . MAGNESIUM CELLS
  - . . . . NICKEL ZINC BATTERIES
- RT METAL AIR BATTERIES
  - NICKEL CADMIUM BATTERIES
  - STORAGE BATTERIES
  - THERMAL BATTERIES

DRY FRICTION
- GS FRICTION
  - . **DRY FRICTION**
- RT ABRASION
  - KINETIC FRICTION
  - SLIDING FRICTION
  - STATIC FRICTION

DRY HEAT
- GS HEAT
  - . **DRY HEAT**
- RT GEOTHERMAL RESOURCES
  - GEOTHERMAL TECHNOLOGY
  - HIGH TEMPERATURE ENVIRONMENTS
  - HUMIDITY
  - OVENS

DRYDOCKS
- RT ∞ PORTS
  - SEA LAUNCHING

DRYERS (EQUIPMENT)
- USE DRYING APPARATUS

DRYING
- UF DESICCATION
  - DEWETTING
- GS **DRYING**
  - . DEHUMIDIFICATION
  - . DEHYDRATION
  - . . FREEZE DRYING
- RT ∞ ABSORPTION
  - BAKING
  - CONCENTRATING
  - CURING
  - DESATURATION
  - DEWATERING
  - DIFFUSION
  - ENTHALPY
  - EVAPORATION
  - FIRING (IGNITING)
  - ROASTING
  - ∞ SEPARATION
  - SILICA GEL
  - WATER LOSS

DRYING APPARATUS
- UF DRYERS (EQUIPMENT)
- GS SEPARATORS
  - . **DRYING APPARATUS**
  - . . DESICCATORS
- RT ABSORBERS (EQUIPMENT)
  - COLUMNS (PROCESS ENGINEERING)
  - CONDENSERS (LIQUEFIERS)
  - DEHYDRATED FOOD
  - EVAPORATORS
  - FURNACES

DSIF (INSTRUMENTATION FACILITY)
- USE DEEP SPACE INSTRUMENTATION FACILITY

DSN HELICOPTER
- USE QH-50 HELICOPTER

DTA (ANALYSIS)
- USE THERMAL ANALYSIS

DTL INTEGRATED CIRCUITS
- SN ITS)

DTL INTEGRATED CIRCUITS-_(CONT.)_
UF   DIODE-TRANSISTOR-LOGIC INTEG
       CIRCUITS
GS   CIRCUITS
     . INTEGRATED CIRCUITS
     . . **DTL INTEGRATED CIRCUITS**
RT   ELECTRONIC PACKAGING
       LARGE SCALE INTEGRATION
       MICROMINIATURIZATION
       MOLECULAR ELECTRONICS
       TRANSISTOR CIRCUITS

DTMB-111 GROUND EFFECT MACHINE
USE   GROUND EFFECT MACHINES

DTMB-430 GROUND EFFECT MACHINE
USE   GROUND EFFECT MACHINES

DUAL AIR DENSITY EXPLORER
UF   DAD EXPLORER
GS   SATELLITES
     . ARTIFICIAL SATELLITES
     . . EXPLORER SATELLITES
     . . . **DUAL AIR DENSITY EXPLORER**

DUAL MODE PROPULSION
USE   HYBRID PROPULSION

DUAL SPIN SPACECRAFT
RT   OSO-7
       ∞SPACECRAFT
       SPACECRAFT STABILITY
       SPIN STABILIZATION

DUAL THRUST NOZZLES
GS   ROCKET NOZZLES
     . **DUAL THRUST NOZZLES**
RT   ∞NOZZLES
       THRUST

DUAL WING CONFIGURATIONS
RT   BIPLANES
       TANDEM WING AIRCRAFT
       WINGS

DUALITY PRINCIPLE
RT   ∞ANALYZING
       CIRCUITS
       EQUIVALENT CIRCUITS
       NETWORK ANALYSIS
       ∞PATHS
       ∞PRINCIPLES
       SIGNAL FLOW GRAPHS

DUALITY THEOREM
GS   THEOREMS
     . **DUALITY THEOREM**
RT   HOMOLOGY
       ISOMORPHISM
       ∞MATHEMATICS
       ∞RELATIONSHIPS

DUCT GEOMETRY
GS   GEOMETRY
     . **DUCT GEOMETRY**
RT   AIR FLOW
       ANNULAR DUCTS
       CIRCULAR TUBES
       FLUID FLOW
       INTAKE SYSTEMS
       OPENINGS
       SPATIAL MARCHING

DUCTED BODIES
RT   ANNULAR DUCTS
       AXISYMMETRIC BODIES
       BLUFF BODIES
       BLUNT BODIES
       ∞BODIES
       DUCTS
       INTAKE SYSTEMS
       NACELLES
       NOSE INLETS
       SHROUDS
       SLENDER BODIES
       TWO DIMENSIONAL BODIES

DUCTED FAN ENGINES
GS   ENGINES
     . AIR BREATHING ENGINES
     . . GAS TURBINE ENGINES
     . . . JET ENGINES
     . . . . TURBOJET ENGINES
     . . . . . **DUCTED FAN ENGINES**

DUCTED FAN ENGINES-_(CONT.)_
     . INTERNAL COMBUSTION ENGINES
     . . GAS TURBINE ENGINES
     . . . JET ENGINES
     . . . . TURBOJET ENGINES
     . . . . . **DUCTED FAN ENGINES**
     . TURBINE ENGINES
     . . GAS TURBINE ENGINES
     . . . JET ENGINES
     . . . . TURBOJET ENGINES
     . . . . . **DUCTED FAN ENGINES**
RT   TURBOFAN ENGINES

DUCTED FANS
RT   BLOWERS
       FAN BLADES
       ∞FANS
       LIFT FANS
       PROPELLER FANS
       RING WINGS
       SHROUDED PROPELLERS
       TURBOFANS
       VENTILATION FANS

DUCTED FLOW
GS   FLUID FLOW
     . **DUCTED FLOW**
     . . KNUDSEN FLOW
RT   AIR FLOW
       CHANNEL FLOW
       CORNER FLOW
       FLOW GEOMETRY
       FUEL FLOW
       HEAT TRANSMISSION
       WALL FLOW

DUCTED PROPELLERS
USE   SHROUDED PROPELLERS

DUCTED ROCKET ENGINES
GS   ENGINES
     . ROCKET ENGINES
     . . **DUCTED ROCKET ENGINES**
RT   BOOSTER ROCKET ENGINES
       ∞HYBRID ROCKET ENGINES
       INTERNAL COMBUSTION ENGINES
       LIQUID PROPELLANT ROCKET ENGINES
       RESTARTABLE ROCKET ENGINES
       SOLID PROPELLANT ROCKET ENGINES
       SUSTAINER ROCKET ENGINES

DUCTILITY
GS   MECHANICAL PROPERTIES
     . **DUCTILITY**
RT   BRITTLENESS
       COMPRESSIVE STRENGTH
       CREEP PROPERTIES
       ELONGATION
       FATIGUE (MATERIALS)
       FLATTENING
       FRACTOGRAPHY
       FRACTURE STRENGTH
       HARDNESS
       IMPACT STRENGTH
       MALLEABILITY
       METAL DRAWING
       NOTCH STRENGTH
       PLASTIC PROPERTIES
       SHEAR PROPERTIES
       SOFTNESS
       STRESS RELAXATION
       STRETCHING
       TEMPER (METALLURGY)
       TENSILE STRENGTH
       TOUGHNESS
       TRESCA FLOW
       WELDABILITY

DUCTS
GS   DUCTS
     . ACOUSTIC DUCTS
     . AIR DUCTS
     . ANNULAR DUCTS
RT   BAFFLES
       CAVITIES
       ∞CHANNELS
       DUCTED BODIES
       EXHAUST SYSTEMS
       FLUES
       INTAKE SYSTEMS
       NOSE INLETS
       OPENINGS
       ORIFICES
       OUTLETS
       PIPES (TUBES)

DUCTS-_(CONT.)_
       PLENUM CHAMBERS
       PORTS (OPENINGS)
       SCOOPS
       THROATS
       ∞TUBES
       VENTILATION
       VENTS
       WINDOWS (APERTURES)

DUFFING DIFFERENTIAL EQUATION
GS   ALGEBRA
     . NONLINEAR EQUATIONS
     . . **DUFFING DIFFERENTIAL EQUATION**
       ANALYSIS (MATHEMATICS)
     . REAL VARIABLES
     . . DIFFERENTIAL EQUATIONS
     . . . **DUFFING DIFFERENTIAL EQUATION**
     . . NONLINEAR EQUATIONS
     . . . **DUFFING DIFFERENTIAL EQUATION**
RT   ∞EQUATIONS
       PROBABILITY THEORY

DULLNESS
USE   LUSTER

DUMMIES
RT   DECOYS
       MODELS
       SIMULATORS

DUMMY LOADS
USE   IMPEDANCE
       LOADING
       OUTPUT

DUMPING
RT   DISCONNECT DEVICES
       DISPOSAL
       EJECTION
       EMPTYING
       EXPULSION
       JETTISONING
       MATERIALS HANDLING
       OIL SLICKS
       RELEASING
       SPILLING
       SPREADING
       UNLOADING

DUNALIELLA
GS   ALGAE
     . **DUNALIELLA**

DUNES
UF   BARCHANS
       COASTAL DUNES
       SAND DUNES
GS   LANDFORMS
     . **DUNES**
RT   BEACHES
       COASTS
       DESERTS
       LAGOONS
       SAHARA DESERT (AFRICA)
       SANDS
       TOPOGRAPHY
       WIND EFFECTS

DUNGEYS WIND SHEAR MECHANISM
USE   WIND SHEAR

DUNITE
GS   ROCKS
     . IGNEOUS ROCKS
     . . **DUNITE**
RT   MINERALS
       OLIVINE
       PERIDOTITE
       SOILS

DUOCHROMATORS
RT   ELECTROMAGNETIC RADIATION
       ∞GENERATORS
       LIGHT SOURCES
       MEASURING INSTRUMENTS
       MONOCHROMATORS
       RADIATION SOURCES
       SPECTROPHOTOMETERS

DUOPLASMATRONS
GS   ION SOURCES
     . PLASMATRONS
     . . **DUOPLASMATRONS**

DUOPLASMATRONS-(CONT.)
```
        PLASMA GENERATORS
        . PLASMATRONS
        . . DUOPLASMATRONS
RT   ELECTRIC DISCHARGES
     ION PROPULSION
     PLASMA PROPULSION
     PLASMAS (PHYSICS)
     SPUTTERING
```

DUPLEX OPERATION
```
RT   ∞ METALLURGY
     PHASE SHIFT CIRCUITS
     SWITCHING CIRCUITS
```

DUPLEXERS
```
RT   CIRCUITS
     CIRCULATORS (PHASE SHIFT CIRCUITS)
     MAGIC TEES
     MONOPULSE RADAR
     RECEIVERS
     SWITCHING CIRCUITS
     TRANSMITTERS
```

DUPLICATING
```
USE   REPRODUCTION (COPYING)
```

DURABILITY
```
RT   AIRCRAFT SURVIVABILITY
     CORROSION
     CUMULATIVE DAMAGE
     DAMAGE
     DEGRADATION
     DETERIORATION
     ∞ ENDURANCE
     LIFE (DURABILITY)
     LONG TERM EFFECTS
     MECHANICAL PROPERTIES
     ∞ PHYSICAL PROPERTIES
     QUALITY
     RELIABILITY
     ∞ RESISTANCE
     RUGGEDNESS
     STABILITY
     VULNERABILITY
     WEAR
```

DURATION
```
USE   TIME
```

DURENE
```
GS   HYDROCARBONS
     . DURENE
```

DUST
```
GS   DUST
     . COSMIC DUST
     . . INTERPLANETARY DUST
     . . . METEOROID DUST CLOUDS
     . . . . ZODIACAL DUST
     . LUNAR DUST
     . TERRESTRIAL DUST BELT
RT   AEROSOLS
     AIR POLLUTION
     AITKEN NUCLEI
     CLEANING
     CLOUDS
     COMBUSTION PRODUCTS
     CONTAMINANTS
     DIRT
     ∞ DISPERSION
     DISPERSIONS
     FUMES
     PARTICLES
     POLLEN
     POWDER (PARTICLES)
     SANDS
     SMOKE
     SPACE DEBRIS
```

DUST COLLECTORS
```
GS   ACCUMULATORS
     . DUST COLLECTORS
     SEPARATORS
     . DUST COLLECTORS
RT   AIR FILTERS
     ELECTROSTATIC PRECIPITATORS
     EXHAUST SYSTEMS
     PRECIPITATORS
```

DUST STORMS
```
GS   STORMS
     . STORMS (METEOROLOGY)
     . . DUST STORMS
```

DUST STORMS-(CONT.)
```
RT   ATMOSPHERIC ELECTRICITY
     ATMOSPHERIC PHYSICS
     MARS (PLANET)
     MARS ENVIRONMENT
     MARS SURFACE
     WIND EFFECTS
```

DWARF GALAXIES
```
GS   CELESTIAL BODIES
     . GALAXIES
     . . DWARF GALAXIES
RT   LOCAL GROUP (ASTRONOMY)
```

DWARF NOVAE
```
GS   CELESTIAL BODIES
     . STARS
     . . DOUBLE STARS
     . . . BINARY STARS
     . . . . ECLIPSING BINARY STARS
     . . . . . DWARF NOVAE
     . . DWARF STARS
     . . . DWARF NOVAE
     . . VARIABLE STARS
     . . . NOVAE
     . . . . DWARF NOVAE
RT   HERCULES NOVA
     STELLAR MASS ACCRETION
     STELLAR MASS EJECTION
     WHITE DWARF STARS
```

DWARF STARS
```
GS   CELESTIAL BODIES
     . STARS
     . . DWARF STARS
     . . . DWARF NOVAE
     . . . FLARE STARS
     . . . RED DWARF STARS
     . . . SUBDWARF STARS
     . . . WHITE DWARF STARS
RT   CATACLYSMIC VARIABLES
     LATE STARS
     MAIN SEQUENCE STARS
     SUBGIANT STARS
```

DWELL
```
RT   DELAY
     IGNITION SYSTEMS
     TIMING DEVICES
```

DYADICS
```
GS   ALGEBRA
     . POLYNOMIALS
     . . DYADICS
RT   VECTORS (MATHEMATICS)
```

DYE LASERS
```
GS   STIMULATED EMISSION DEVICES
     . LASERS
     . . ORGANIC LASERS
     . . . DYE LASERS
RT   DYES
     INFRARED LASERS
     LASER OUTPUTS
     LIQUID LASERS
     OPTICAL COMMUNICATION
     TUNING
     TWO-WAVELENGTH LASERS
```

DYES
```
GS   DYES
     . METHYLENE BLUE
     . THIAZINE (TRADEMARK)
RT   ACRIFLAVINE
     ANILINE
     ANTHRAQUINONES
     AZINES
     AZO COMPOUNDS
     DYE LASERS
     ∞ MARKERS
     METHYLENE
     PHENANTHRENE
     STILBENE
```

DYNA-SOAR SPACE GLIDER
```
USE   X-20 AIRCRAFT
```

DYNAMIC CHARACTERISTICS
```
UF   DYNAMIC PROPERTIES
GS   DYNAMIC CHARACTERISTICS
     . DRAG
     . . ELECTROSTATIC DRAG
     . . FRICTION DRAG
     . . . AERODYNAMIC DRAG
```

DYNAMIC CHARACTERISTICS-(CONT.)
```
     . . . . SUPERSONIC DRAG
     . . . VISCOUS DRAG
     . . . MINIMUM DRAG
     . . PRESSURE DRAG
     . . . SUPERSONIC DRAG
     . . WAVE DRAG
     . . . INTERFERENCE DRAG
     . . SATELLITE DRAG
     . DYNAMIC PRESSURE
     . DYNAMIC STABILITY
     . . COMBUSTION STABILITY
     . . . FLAME STABILITY
     . . CONTROL STABILITY
     . . FREQUENCY STABILITY
     . . MOTION STABILITY
     . . . AERODYNAMIC STABILITY
     . . . AIRCRAFT STABILITY
     . . . . HOVERING STABILITY
     . . . ATTITUDE STABILITY
     . . . . DIRECTIONAL STABILITY
     . . . . . GYROSCOPIC STABILITY
     . . . . LATERAL STABILITY
     . . . . LONGITUDINAL STABILITY
     . . . FLOW STABILITY
     . . . . BOUNDARY LAYER STABILITY
     . . . . FLAME STABILITY
     . . . . MAGNETOHYDRODYNAMIC
             STABILITY
     . . . . . WEIBEL INSTABILITY
     . . . LOW SPEED STABILITY
     . . ROTARY STABILITY
     . . . GYROSCOPIC STABILITY
     . . SPACECRAFT STABILITY
     . FLOW CHARACTERISTICS
     . . FLOW DISTRIBUTION
     . . FLOW STABILITY
     . . . BOUNDARY LAYER STABILITY
     . . . FLAME STABILITY
     . . . MAGNETOHYDRODYNAMIC
             STABILITY
     . . . . WEIBEL INSTABILITY
     . . FLOW VELOCITY
     . LIFT
     . . INTERFERENCE LIFT
     . . JET LIFT
     . . ROTOR LIFT
     . . ZERO LIFT
     . TRANSIENT RESPONSE
RT   ACCURACY
     AERODYNAMIC BALANCE
     AERODYNAMIC CHARACTERISTICS
     AMPLIFICATION
     AUTOMATIC CONTROL
     AUTOMATIC CONTROL VALVES
     BANDWIDTH
     ∞ CHARACTERISTICS
     ∞ CONTROL
     DAMPING
     DISTRIBUTION (PROPERTY)
     ∞ DYNAMICS
     ∞ EQUILIBRIUM
     ERRORS
     ∞ FREQUENCY RESPONSE
     HYSTERESIS
     IMPEDANCE
     LINEARITY
     OCEAN DYNAMICS
     PRECISION
     ∞ PROPERTIES
     RANGE (EXTREMES)
     REACTION TIME
     RELIABILITY
     REMOTE CONTROL
     RESOLUTION
     RESONANT FREQUENCIES
     SENSITIVITY
     STABILITY
     TIME CONSTANT
     TRANSFER FUNCTIONS
```

DYNAMIC CONTROL
```
GS   AUTOMATIC CONTROL
     . DYNAMIC CONTROL
RT   ADAPTIVE CONTROL
     ∞ CONTROL
     CONTROL THEORY
     ∞ DYNAMICS
     FEEDBACK CONTROL
```

DYNAMIC LOADS
```
GS   LOADS (FORCES)
     . DYNAMIC LOADS
     . . AERODYNAMIC LOADS
     . . . BLAST LOADS
```

DYNAMIC LOADS-*(CONT.)*
- . . . GUST LOADS
- . . CYCLIC LOADS
- . . ROLLING CONTACT LOADS
- . . THRUST LOADS
- . . TRANSIENT LOADS
- . . . GUST LOADS
- . . . IMPACT LOADS
- . . . LANDING LOADS
- . . . SHOCK LOADS
- . . . . BLAST LOADS
- . . VIBRATORY LOADS
- . . WING LOADING
- RT AXIAL COMPRESSION LOADS
-     AXIAL LOADS
-     COMPRESSION LOADS
-     CRITICAL LOADING
-     ∞DYNAMICS
-     EDGE LOADING
-     GAS-SOLID INTERACTIONS
-     NASTRAN
-     RANDOM LOADS
-     STATIC LOADS
-     STRUCTURAL DESIGN CRITERIA
-     WIND PRESSURE

DYNAMIC MODELS
- GS MODELS
-     **DYNAMIC MODELS**
- RT AIRCRAFT MODELS
-     BIOLOGICAL MODELS (MATHEMATICS)
-     BOND GRAPHS
-     ∞DYNAMICS
-     MATHEMATICAL MODELS
-     OCEAN MODELS
-     PETRI NETS
-     POWERED MODELS
-     SIMILARITY THEOREM
-     SPACECRAFT MODELS
-     STATIC MODELS
-     SYSTEMS SIMULATION
-     WIND TUNNEL MODELS

DYNAMIC MODULUS OF ELASTICITY
- GS MECHANICAL PROPERTIES
- . ELASTIC PROPERTIES
- . . MODULUS OF ELASTICITY
- . . . **DYNAMIC MODULUS OF**
-        **ELASTICITY**
- RT ∞DYNAMICS
-     ULTRASONIC TESTS

DYNAMIC PRESSURE
- GS DYNAMIC CHARACTERISTICS
- . **DYNAMIC PRESSURE**
-     PRESSURE
- . **DYNAMIC PRESSURE**
- RT BLAST LOADS
-     ∞DYNAMICS
-     IMPACT LOADS
-     KINETIC THEORY
-     OVERPRESSURE
-     RIEMANN WAVES

DYNAMIC PROGRAMMING
- GS OPERATIONS RESEARCH
- . **DYNAMIC PROGRAMMING**
-     RESEARCH
- . MATHEMATICAL PROGRAMMING
- . . **DYNAMIC PROGRAMMING**
- RT ∞APPLICATIONS OF MATHEMATICS
-     BELLMAN THEORY
-     CONSTRAINTS
-     CRITICAL PATH METHOD
-     DECISION THEORY
-     ∞DYNAMICS
-     FORMALISM
-     LINEAR PROGRAMMING
-     MATHEMATICAL MODELS
-     NONLINEAR SYSTEMS
-     OPERATIONS RESEARCH
-     ∞PROGRAMMING
-     STEEPEST DESCENT METHOD

DYNAMIC PROPERTIES
- USE DYNAMIC CHARACTERISTICS

DYNAMIC RESPONSE
- GS RESPONSES
- . **DYNAMIC RESPONSE**
- . . TRANSIENT RESPONSE
- RT AMPLIFICATION
-     DAMPING
-     ∞DYNAMICS
-     FIBER ORIENTATION

DYNAMIC RESPONSE-*(CONT.)*
- ∞FREQUENCY RESPONSE
-   IMPEDANCE
-   MODAL RESPONSE
-   PARAMETER IDENTIFICATION
-   RAMP FUNCTIONS
-   REACTION TIME
-   RESPONSE BIAS
-   SENSITIVITY
-   STEP FUNCTIONS
-   STROKING TESTS
-   SYSTEM IDENTIFICATION
-   TIME CONSTANT
-   TRANSFER FUNCTIONS

DYNAMIC STABILITY
- GS DYNAMIC CHARACTERISTICS
- . **DYNAMIC STABILITY**
- . . COMBUSTION STABILITY
- . . . FLAME STABILITY
- . . CONTROL STABILITY
- . . FREQUENCY STABILITY
- . . MOTION STABILITY
- . . . AERODYNAMIC STABILITY
- . . . AIRCRAFT STABILITY
- . . . . HOVERING STABILITY
- . . . . ATTITUDE STABILITY
- . . . . DIRECTIONAL STABILITY
- . . . . . GYROSCOPIC STABILITY
- . . . . . LATERAL STABILITY
- . . . . LONGITUDINAL STABILITY
- . . . FLOW STABILITY
- . . . . BOUNDARY LAYER STABILITY
- . . . . FLAME STABILITY
- . . . . MAGNETOHYDRODYNAMIC
-          STABILITY
- . . . . WEIBEL INSTABILITY
- . . . LOW SPEED STABILITY
- . . . ROTARY STABILITY
- . . . . GYROSCOPIC STABILITY
- . . . SPACECRAFT STABILITY
- STABILITY
- . **DYNAMIC STABILITY**
- . . COMBUSTION STABILITY
- . . . FLAME STABILITY
- . . CONTROL STABILITY
- . . FREQUENCY STABILITY
- . . MOTION STABILITY
- . . . AERODYNAMIC STABILITY
- . . . AIRCRAFT STABILITY
- . . . . HOVERING STABILITY
- . . . . ATTITUDE STABILITY
- . . . . DIRECTIONAL STABILITY
- . . . . . GYROSCOPIC STABILITY
- . . . . LATERAL STABILITY
- . . . LONGITUDINAL STABILITY
- . . . FLOW STABILITY
- . . . . BOUNDARY LAYER STABILITY
- . . . . FLAME STABILITY
- . . . . MAGNETOHYDRODYNAMIC
-          STABILITY
- . . . LOW SPEED STABILITY
- . . . ROTARY STABILITY
- . . . . GYROSCOPIC STABILITY
- . . . SPACECRAFT STABILITY
- RT COUNTERBALANCES
-     DAMPING
-     DIMENSIONAL STABILITY
-     DRIFT (INSTRUMENTATION)
-     ∞DYNAMICS
-     HORIZONTAL ORIENTATION
-     MISSING MASS (ASTROPHYSICS)
-     RESONANT VIBRATION
-     SPACECRAFT MOTION
-     STABLE OSCILLATIONS
-     STATIC STABILITY
-     SURFACE STABILITY
-     SYSTEMS STABILITY
-     TRANSIENT RESPONSE
-     VERTICAL ORIENTATION

DYNAMIC STRUCTURAL ANALYSIS
- UF STRUCTURAL DYNAMICS
- GS STRUCTURAL ANALYSIS
- . **DYNAMIC STRUCTURAL ANALYSIS**
- RT ∞DYNAMICS
-     FLAT PLATES
-     SHOCK SPECTRA

DYNAMIC TESTS
- RT AERODYNAMIC STABILITY
-     ∞DYNAMICS
-     FLIGHT TESTS
-     LOW SPEED STABILITY
-     MOTION STABILITY

DYNAMIC TESTS-*(CONT.)*
- SPIN DYNAMICS
- SPIN TESTS
- STATIC TESTS
- ∞TESTS
- VIBRATION TESTS

DYNAMICAL SYSTEMS
- RT CONTROL THEORY
-     MATHEMATICAL MODELS
-     NONLINEAR SYSTEMS
-     SYSTEMS SIMULATION

∞ DYNAMICS
- SN *(USE OF A MORE SPECIFIC TERM IS*
-     *RECOMMENDED--CONSULT THE TERMS*
-     *LISTED BELOW)*
- RT AERODYNAMICS
-     AEROSTATICS
-     AEROTHERMODYNAMICS
-     ASTRODYNAMICS
-     BIODYNAMICS
-     CHIRAL DYNAMICS
-     COMPUTATIONAL FLUID DYNAMICS
-     CONTINUUM MECHANICS
-     DYNAMIC CHARACTERISTICS
-     DYNAMIC CONTROL
-     DYNAMIC LOADS
-     DYNAMIC MODELS
-     DYNAMIC MODULUS OF ELASTICITY
-     DYNAMIC PRESSURE
-     DYNAMIC PROGRAMMING
-     DYNAMIC RESPONSE
-     DYNAMIC STABILITY
-     DYNAMIC STRUCTURAL ANALYSIS
-     DYNAMIC TESTS
-     ELASTODYNAMICS
-     ELECTRODYNAMICS
-     EQUATIONS OF MOTION
-     FIELD THEORY (PHYSICS)
-     FLUID DYNAMICS
-     FLUID MECHANICS
-     GAS DYNAMICS
-     GEODYNAMICS
-     GROUP DYNAMICS
-     HAMILTONIAN FUNCTIONS
-     HEMODYNAMICS
-     HYDRODYNAMICS
-     KINEMATICS
-     KINETICS
-     MAGNETOHYDRODYNAMICS
-     ∞MECHANICS (PHYSICS)
-     MOMENTUM
-     MOMENTUM TRANSFER
-     NUTATION
-     OCEAN DYNAMICS
-     PLASMA DYNAMICS
-     QUANTUM CHROMODYNAMICS
-     RAREFIED GAS DYNAMICS
-     RESONANT FREQUENCIES
-     RESONANT VIBRATION
-     SPIN DYNAMICS
-     STABILIZERS (FLUID DYNAMICS)
-     STATICS
-     TERRADYNAMICS
-     THERMODYNAMICS
-     VARIATIONAL PRINCIPLES
-     VELOCITY
-     VIBRATION

DYNAMICS EXPLORER SATELLITES
- GS SATELLITES
- . ARTIFICIAL SATELLITES
- . . EXPLORER SATELLITES
- . . . **DYNAMICS EXPLORER SATELLITES**
- . . . . DYNAMICS EXPLORER 1
-          SATELLITE
- . . . . DYNAMICS EXPLORER 2
-          SATELLITE

DYNAMICS EXPLORER 1 SATELLITE
- GS SATELLITES
- . ARTIFICIAL SATELLITES
- . . EXPLORER SATELLITES
- . . . DYNAMICS EXPLORER SATELLITES
- . . . . **DYNAMICS EXPLORER 1**
-          **SATELLITE**

DYNAMICS EXPLORER 2 SATELLITE
- GS SATELLITES
- . ARTIFICIAL SATELLITES
- . . EXPLORER SATELLITES
- . . . DYNAMICS EXPLORER SATELLITES
- . . . . **DYNAMICS EXPLORER 2**
-          **SATELLITE**

DYNAMITE
GS   EXPLOSIVES
       . **DYNAMITE**
RT   NITROGLYCERIN

DYNAMO THEORY
RT   EARTH CORE
       GEOMAGNETISM
       TELLURIC CURRENTS
       ∞ THEORIES

DYNAMOMETERS
UF   ELECTRODYNAMOMETERS
GS   ELECTRIC GENERATORS
       . ROTATING GENERATORS
       . . **DYNAMOMETERS**
       MEASURING INSTRUMENTS
       . **DYNAMOMETERS**
RT   ERGOMETERS
       MECHANICAL MEASUREMENT
       ∞ TEST EQUIPMENT
       THRUST MEASUREMENT
       TORQUEMETERS

DYNAMOS
USE   ROTATING GENERATORS

DYNODES
GS   ELECTRODES
       . **DYNODES**
RT   CAMERA TUBES
       PHOTOMULTIPLIER TUBES
       SECONDARY EMISSION

DYSON THEORY
RT   HEISENBERG THEORY
       QUANTUM MECHANICS
       ∞ THEORIES

DYSPNEA
GS   RATES (PER TIME)
       . RESPIRATORY RATE
       . . **DYSPNEA**
       SIGNS AND SYMPTOMS
       . **DYSPNEA**

DYSPROSIUM
GS   CHEMICAL ELEMENTS
       . RARE EARTH ELEMENTS
       . . **DYSPROSIUM**
       . . . DYSPROSIUM ISOTOPES
       METALS
       . RARE EARTH ELEMENTS
       . . **DYSPROSIUM**
       . . . DYSPROSIUM ISOTOPES

DYSPROSIUM COMPOUNDS
GS   RARE EARTH COMPOUNDS
       . **DYSPROSIUM COMPOUNDS**
RT   ∞ CHEMICAL COMPOUNDS
       ∞ METAL COMPOUNDS

DYSPROSIUM ISOTOPES
UF   DYSPROSIUM 161
GS   CHEMICAL ELEMENTS
       . NUCLIDES
       . . ISOTOPES
       . . . **DYSPROSIUM ISOTOPES**
       . RARE EARTH ELEMENTS
       . . DYSPROSIUM
       . . . **DYSPROSIUM ISOTOPES**
       METALS
       . RARE EARTH ELEMENTS
       . . DYSPROSIUM
       . . . **DYSPROSIUM ISOTOPES**

DYSPROSIUM 161
USE   DYSPROSIUM ISOTOPES

E

E GLASS
GS   GLASS
       . **E GLASS**
       . . S GLASS
RT   COMPOSITE MATERIALS
       GLASS FIBER REINFORCED PLASTICS
       GLASS FIBERS
       SILICON DIOXIDE

E LAYERS
USE   E REGION

E REGION
UF   E LAYERS
       NIGHT E LAYER
GS   EARTH ATMOSPHERE
       . UPPER ATMOSPHERE
       . . IONOSPHERE
       . . . **E REGION**
       . . . . E-1 LAYER
       . . . . E-2 LAYER
       . . . . SPORADIC E LAYER
       REGIONS
       . **E REGION**
RT   LOWER IONOSPHERE
       UPPER IONOSPHERE

E-1 LAYER
GS   EARTH ATMOSPHERE
       . UPPER ATMOSPHERE
       . . IONOSPHERE
       . . . E REGION
       . . . . **E-1 LAYER**
RT   SPORADIC E LAYER

E-2 AIRCRAFT
UF   HAWKEYE AIRCRAFT
       W2F AIRCRAFT
GS   AWACS AIRCRAFT
       . **E-2 AIRCRAFT**
       GRUMMAN AIRCRAFT
       . **E-2 AIRCRAFT**
       OBSERVATION AIRCRAFT
       . **E-2 AIRCRAFT**
RT   ∞ AIRCRAFT
       COMMAND AND CONTROL
       EARLY WARNING SYSTEMS
       ∞ MILITARY AIRCRAFT
       MILITARY TECHNOLOGY
       PASSENGER AIRCRAFT
       TURBOPROP ENGINES

E-2 LAYER
GS   EARTH ATMOSPHERE
       . UPPER ATMOSPHERE
       . . IONOSPHERE
       . . . E REGION
       . . . . **E-2 LAYER**
RT   SPORADIC E LAYER

E-3A AIRCRAFT
GS   AWACS AIRCRAFT
       . **E-3A AIRCRAFT**
       BOEING AIRCRAFT
       . **E-3A AIRCRAFT**
RT   ∞ AIRCRAFT
       COMMAND AND CONTROL
       EARLY WARNING SYSTEMS
       ∞ MILITARY AIRCRAFT
       MILITARY TECHNOLOGY

E-4A AIRCRAFT
UF   ADVANCED AIRBORNE COMMAND POST
       BOEING 747B AIRCRAFT
GS   AWACS AIRCRAFT
       . **E-4A AIRCRAFT**
       BOEING AIRCRAFT
       . **E-4A AIRCRAFT**
RT   ∞ AIRCRAFT
       COMMAND AND CONTROL
       EARLY WARNING SYSTEMS
       ∞ MILITARY AIRCRAFT
       MILITARY TECHNOLOGY

EAI 680 COMPUTER
GS   DATA PROCESSING EQUIPMENT
       . COMPUTERS
       . . ANALOG COMPUTERS
       . . . **EAI 680 COMPUTER**
       . DIGITAL COMPUTERS
       . . **EAI 680 COMPUTER**

EAI 8400 COMPUTER
GS   DATA PROCESSING EQUIPMENT
       . COMPUTERS
       . . DIGITAL COMPUTERS
       . . . **EAI 8400 COMPUTER**

EAI 8900 COMPUTER
GS   DATA PROCESSING EQUIPMENT
       . COMPUTERS
       . . DIGITAL COMPUTERS
       . . . **EAI 8900 COMPUTER**

EAR
GS   ANATOMY
       . SENSE ORGANS
       . . **EAR**
       . . . CORTI ORGAN
       . . . EARDRUMS
       . . . LABYRINTH
       . . . . COCHLEA
       . . . . . VESTIBULES
       . . . MASTOIDS
       . . . MIDDLE EAR
       . . . SEMICIRCULAR CANALS
RT   ARTIFICIAL EARS
       AUDITORY PERCEPTION
       ENDOLYMPH
       EUSTACHIAN TUBES
       HEARING
       LABYRINTHECTOMY
       OTOLARYNGOLOGY
       OTOLOGY

EAR PRESSURE TEST
GS   PHYSIOLOGICAL TESTS
       . **EAR PRESSURE TEST**
RT   MIDDLE EAR PRESSURE
       PRESSURE
       VERTIGO
       VESTIBULAR TESTS

EAR PROTECTORS
GS   PROTECTORS
       . **EAR PROTECTORS**
RT   NOISE INJURIES
       NOISE REDUCTION

EARDRUMS
GS   ANATOMY
       . SENSE ORGANS
       . . EAR
       . . . **EARDRUMS**
RT   EUSTACHIAN TUBES
       MIDDLE EAR PRESSURE
       SEMICIRCULAR CANALS

EARLY APOLLO SURFACE EXPERIMENTS PACKAGE
USE   EASEP

EARLY BIRD SATELLITES
GS   SATELLITES
       . ACTIVE SATELLITES
       . . SYNCOM SATELLITES
       . . . **EARLY BIRD SATELLITES**
       . ARTIFICIAL SATELLITES
       . . COMMUNICATION SATELLITES
       . . . SYNCOM SATELLITES
       . . . . **EARLY BIRD SATELLITES**
       . . SYNCHRONOUS SATELLITES
       . . . SYNCOM SATELLITES
       . . . . **EARLY BIRD SATELLITES**
RT   ATS
       COMSAT PROGRAM

EARLY STARS
GS   CELESTIAL BODIES
       . STARS
       . . **EARLY STARS**
       . . . PROTOSTARS
       . . . . T TAURI STARS
RT   LATE STARS
       MAIN SEQUENCE STARS

EARLY WARNING SYSTEMS
GS   WARNING SYSTEMS
       . **EARLY WARNING SYSTEMS**
       . . BALLISTIC MISSILE EARLY WARNING
           SYSTEM
RT   AIR DEFENSE
       AWACS AIRCRAFT
       COBRA DANE (RADAR)
       DETECTION
       E-2 AIRCRAFT
       E-3A AIRCRAFT
       E-4A AIRCRAFT
       MISSILE DETECTION
       OVER-THE-HORIZON RADAR
       RADAR TARGETS
       RADAR TRACKING
       SYNCHRONOUS EARTH OBSERVATORY
           SATELLITE
       ∞ SYSTEMS
       WARNING

EARPHONES
UF   HEADSETS

EARPHONES-(CONT.)
```
GS   AUDIO EQUIPMENT
     . EARPHONES
RT   ACOUSTICS
     AUDITORY PERCEPTION
     INTERPHONES
     SOUND TRANSMISSION
     TELEPHONES
```

EARTH (PLANET)
```
UF   WORLD
GS   CELESTIAL BODIES
     . PLANETS
     . . TERRESTRIAL PLANETS
     . . . EARTH (PLANET)
RT   EASTERN HEMISPHERE
     GEODESY
     GEOELECTRICITY
     GEOGRAPHY
     GEOLOGY
     GEOMAGNETISM
     GEOPHYSICS
     ∞ GLOBES
     PLANETARY CRATERS
     POLAR CAPS
     TERRESTRIAL RADIATION
     WESTERN HEMISPHERE
```

EARTH & OCEAN PHYSICS APPLICATIONS PROGRAM
```
UF   EOPAP
GS   PROGRAMS
     . NASA PROGRAMS
     . . EARTH & OCEAN PHYSICS
           APPLICATIONS PROGRAM
     . NASA SPACE PROGRAMS
     . . EARTH & OCEAN PHYSICS
           APPLICATIONS PROGRAM
     . PROJECTS
     . . EARTH & OCEAN PHYSICS
           APPLICATIONS PROGRAM
RT   OCEANOGRAPHY
     ∞ RESEARCH PROJECTS
```

EARTH ALBEDO
```
GS   ALBEDO
     . EARTH ALBEDO
RT   ABSORPTANCE
     COSMIC RAY ALBEDO
     EARTH RADIATION BUDGET
        EXPERIMENT
     EBERT SPECTROMETERS
     LUNAR ALBEDO
     REFLECTANCE
     TERRESTRIAL RADIATION
```

EARTH ATMOSPHERE
```
GS   EARTH ATMOSPHERE
     . FREE ATMOSPHERE
     . HETEROSPHERE
     . HOMOSPHERE
     . . MIDDLE ATMOSPHERE
     . . . CHEMOSPHERE
     . . . STRATOSPHERE
     . LOWER ATMOSPHERE
     . . BIOSPHERE
     . . OZONOSPHERE
     . . TROPOSPHERE
     . MIDLATITUDE ATMOSPHERE
     . PRIMITIVE EARTH ATMOSPHERE
     . RADIATION BELTS
     . . ARTIFICIAL RADIATION BELTS
     . . INNER RADIATION BELT
     . . OUTER RADIATION BELT
     . . PROTON BELTS
     . STRATOPAUSE
     . TROPOPAUSE
     . UPPER ATMOSPHERE
     . . EXOSPHERE
     . . IONOSPHERE
     . . . E REGION
     . . . . E-1 LAYER
     . . . . E-2 LAYER
     . . . . SPORADIC E LAYER
     . . . LOWER IONOSPHERE
     . . . . D REGION
     . . . UPPER IONOSPHERE
     . . . . F REGION
     . . . . . F 1 REGION
     . . . . . F 2 REGION
     . . MAGNETOSPHERE
     . . . GEOMAGNETIC TAIL
     . . . MAGNETOPAUSE
     . . MESOPAUSE
     . . MESOSPHERE
```

EARTH ATMOSPHERE-(CONT.)
```
     . . THERMOSPHERE
     . . . TURBOPAUSE
RT   ACOUSTIC SOUNDING
     AEROSPACE ENVIRONMENTS
     AIR
     AIR POLLUTION
     AIR QUALITY
     AIRGLOW
     ∞ ATMOSPHERES
     ATMOSPHERIC CIRCULATION
     ATMOSPHERIC COMPOSITION
     ATMOSPHERIC ELECTRICITY
     ATMOSPHERIC ENTRY
     ATMOSPHERIC GENERAL CIRCULATION
        EXPERIMENT
     AURORAS
     BIOASTRONAUTICS
     GEOPOTENTIAL HEIGHT
     GLOBAL AIR POLLUTION
     GREENHOUSE EFFECT
     METEOR TRAILS
     OPEN PROJECT
     PLANETARY ATMOSPHERES
     PLASMASPHERE
     SATELLITE ATMOSPHERES
     SCALE HEIGHT
     SUPERROTATION
     WEATHERING
```

EARTH AXIS
```
GS   AXES (REFERENCE LINES)
     . AXES OF ROTATION
     . . EARTH AXIS
RT   COORDINATES
     GEODESY
     POLAR WANDERING (GEOLOGY)
```

EARTH CORE
```
GS   CORES
     . PLANETARY CORES
     . . EARTH CORE
     LITHOSPHERE
     . EARTH CORE
RT   DYNAMO THEORY
     GEOPHYSICAL FLUIDS
     STRUCTURAL PROPERTIES (GEOLOGY)
```

EARTH CRUST
```
GS   CRUSTS
     . EARTH CRUST
     LITHOSPHERE
     . EARTH CRUST
RT   COESITE
     CONTINENTAL DRIFT
     CORE SAMPLING
     CRATONS
     CRUSTAL FRACTURES
     EARTHQUAKE DAMAGE
     FOLDS (GEOLOGY)
     LUNAR CRUST
     MASSIFS
     PLATES (TECTONICS)
     SAN ANDREAS FAULT
     STISHOVITE
     STRUCTURAL PROPERTIES (GEOLOGY)
```

EARTH CURRENTS
```
USE   TELLURIC CURRENTS
```

EARTH ENERGY BUDGET EXPERIMENT
```
USE   LZEEBE SATELLITE
```

EARTH ENVIRONMENT
```
GS   ENVIRONMENTS
     . EARTH ENVIRONMENT
RT   AIR POLLUTION
     ARID LANDS
     DESERTIFICATION
```

EARTH FIGURE
```
USE   GEODESY
```

EARTH HYDROSPHERE
```
UF   HYDROSPHERE (EARTH)
RT   BIOSPHERE
     HYDROLOGY
     LAKES
     LIMNOLOGY
     OCEANS
     SEAS
```

EARTH LIMB
```
RT   ASTRONOMY
```

EARTH LIMB-(CONT.)
```
     LIBRATION
     ∞ LIMBS
     PLANETARY LIMB
```

EARTH MANTLE
```
UF   MANTLE (EARTH STRUCTURE)
GS   LITHOSPHERE
     . EARTH MANTLE
     PLANETARY MANTLES
     . EARTH MANTLE
RT   COESITE
     LUNAR MANTLE
     PLATES (TECTONICS)
     REGOLITH
     STISHOVITE
     STRUCTURAL PROPERTIES (GEOLOGY)
     SUBDUCTION (GEOLOGY)
```

∞ EARTH MOTION
```
SN   (USE OF A MORE SPECIFIC TERM IS
      RECOMMENDED--CONSULT THE TERMS
      LISTED BELOW)
RT   EARTH MOVEMENTS
     EARTH ROTATION
     POLAR WANDERING (GEOLOGY)
     SOLAR ORBITS
```

EARTH MOVEMENTS
```
GS   EARTH MOVEMENTS
     . EARTHQUAKES
     . LANDSLIDES
RT   AVALANCHES
     CREVASSES
     CRUSTAL FRACTURES
     ∞ EARTH MOTION
     EARTHQUAKE DAMAGE
     GEODYNAMICS
     LARGE APERTURE SEISMIC ARRAY
     SEISMIC WAVES
     SEISMOLOGY
     TECTONICS
     TSUNAMI WAVES
```

EARTH OBSERVATIONS (FROM SPACE)
```
GS   OBSERVATION
     . EARTH OBSERVATIONS (FROM
        SPACE)
     . . SATELLITE OBSERVATION
     . . SPOT (FRENCH SATELLITE)
RT   AERIAL PHOTOGRAPHY
     DATA ACQUISITION
     EARTHNET
     FEATURE IDENTIFICATION AND
        LOCATION EXPER
     INTERNATIONAL
        GEOSPHERE-BIOSPHERE PROGRAM
     LANDSAT SATELLITES
     MULTISPECTRAL BAND SCANNERS
     MULTISPECTRAL PHOTOGRAPHY
     PHOTOGRAPHY
```

EARTH ORBITAL RENDEZVOUS
```
UF   EOR (RENDEZVOUS)
GS   MANEUVERS
     . ORBITAL RENDEZVOUS
     . . EARTH ORBITAL RENDEZVOUS
     RENDEZVOUS
     . EARTH ORBITAL RENDEZVOUS
RT   LUNAR ORBITAL RENDEZVOUS
     ORBITAL MECHANICS
     RENDEZVOUS TRAJECTORIES
     SPACECRAFT TRAJECTORIES
     TRANSFER ORBITS
```

EARTH ORBITING SPACE STATIONS
```
USE   EOSS
```

EARTH ORBITS
```
SN   (ORBITS AROUND THE EARTH)
GS   ORBITS
     . EARTH ORBITS
     . . APOGEES
     . . PERIGEES
RT   APOLLO ASTEROIDS
     CIRCULAR ORBITS
     CIRCUMLUNAR TRAJECTORIES
     ELLIPTICAL ORBITS
     EQUATORIAL ORBITS
     HANSEN LUNAR THEORY
     HILL LUNAR THEORY
     HILL METHOD
     LUNAR ORBITS
     ORBITAL LIFETIME
     ORBITAL MECHANICS
```

EARTH ORBITS-*(CONT.)*
    PARKING ORBITS
    PLANETARY ORBITS
    POLAR ORBITS
    SATELLITE ORBITS
    SPACECRAFT ORBITS
    STATIONARY ORBITS
    TRANSFER ORBITS
    TWENTY-FOUR HOUR ORBITS

EARTH PLANETARY STRUCTURE
RT   CONTINENTAL DRIFT
    GEOLOGY
    GEOPHYSICS
    HYDROLOGY
    LITHOSPHERE
    OCEANOGRAPHY
    PLANETARY COMPOSITION
    PLANETARY STRUCTURE
    PLATES (TECTONICS)
    PRIMITIVE EARTH ATMOSPHERE
    STRUCTURAL PROPERTIES (GEOLOGY)
  ∞STRUCTURES
    TECTONICS

EARTH RADIATION
USE TERRESTRIAL RADIATION

EARTH RADIATION BUDGET EXPERIMENT
UF   ERBE
GS   PAYLOADS
  . SPACE SHUTTLE PAYLOADS
  .. SPACEBORNE EXPERIMENTS
  ... **EARTH RADIATION BUDGET
      EXPERIMENT**
RT   ALBEDO
    EARTH ALBEDO
  ∞RADIATION
    RADIATION MEASURING INSTRUMENTS
    TERRESTRIAL RADIATION

EARTH RESOURCES
GS   RESOURCES
  . **EARTH RESOURCES**
  .. FORESTS
  ... RAIN FORESTS
  .. FOSSIL FUELS
  ... COAL
  .... LIGNITE
  ..... SOLVENT REFINED COAL
  ... CRUDE OIL
  .. GLACIERS
  .. ICEBERGS
  .. LAND ICE
  .. MARINE RESOURCES
  .. OIL FIELDS
  .. RANGE RESOURCES
  .. SPRINGS (WATER)
  .. TAR SANDS
  .. THERMAL RESOURCES
  ... GEOTHERMAL RESOURCES
  .... GEYSERS
  .. UNDERWATER RESOURCES
  .. WATER RESOURCES
RT   ALFALFA
    ARID LANDS
    BALTIC SHIELD (EUROPE)
    BEDROCK
    BIRDS
    BRUSH (BOTANY)
    CHAPARRAL
    COASTAL ECOLOGY
    COASTAL PLAINS
    CORN
    COTTON
    CROP GROWTH
    CROP IDENTIFICATION
    DECIDUOUS TREES
    DESERTS
    EARTHNET
    ENERGY POLICY
    ENERGY TECHNOLOGY
    ENVIRONMENT MANAGEMENT
    ENVIRONMENT POLLUTION
    ENVIRONMENTAL SURVEYS
    EROS (SATELLITES)
    FARM CROPS
    FARMLANDS
    FISHES
    FLATS (LANDFORMS)
  ∞FOOD
    FOREST MANAGEMENT
    GEOGRAPHIC APPLICATIONS PROGRAM
    GEOTHERMAL ENERGY CONVERSION
    GRAINS (FOOD)

EARTH RESOURCES-*(CONT.)*
    GRANITE
    GRASSLANDS
    GREAT BASIN (US)
    GREAT LAKES (NORTH AMERICA)
    GREAT SALT LAKE (UT)
    GROUND WATER
    HABITATS
    HAY
    ICE MAPPING
    IMAGERY
    KEROGEN
    KETTLES (GEOLOGY)
    KEYS (ISLANDS)
    LAND USE
    LARGE AREA CROP INVENTORY
      EXPERIMENT
    LAVA
    LEGUMINOUS PLANTS
    LIMESTONE
    MAMMALS
    MARSHLANDS
    MILLET
    MINERAL DEPOSITS
    MINERALS
    MISSISSIPPI RIVER (US)
    NASA INTERACTIVE PLANNING SYSTEM
    OATS
    OCEANOGRAPHY
    PHOTOGRAPHY
    PHOTOMAPPING
    PLANTS (BOTANY)
    RECONNAISSANCE
    REGOLITH
    REMOTE SENSING
    REMOTE SENSORS
    RESOURCES MANAGEMENT
    RIVERS
    ROCKS
    RURAL LAND USE
    SANDS
    SANDSTONES
    SCANNING
    SHALES
    SOILS
    SORGHUM
    SPACEBORNE PHOTOGRAPHY
    SPECTRAL RECONNAISSANCE
    SPOT (FRENCH SATELLITE)
    STRIP MINING
    SUGAR BEETS
    SUGAR CANE
    SUNFLOWERS
    SURFACE WATER
    SURVEILLANCE
    TERRAIN ANALYSIS
    THERMAL MAPPING
    TIDEPOWER
    TIMBER IDENTIFICATION
    TIMBER INVENTORY
    TRIBUTARIES
    VEGETATION
    VINEYARDS
    WATERWAVE ENERGY
    WATERWAVE ENERGY CONVERSION
    WHARVES
    WINDPOWER UTILIZATION

EARTH RESOURCES EXPERIMENT PACKAGE
USE EREP

EARTH RESOURCES INFORMATION SYSTEM
GS   INFORMATION SYSTEMS
  . **EARTH RESOURCES INFORMATION
     SYSTEM**
RT   DATA SYSTEMS
    NASA PROGRAMS
    PROGRAMS
    SKYLAB PROGRAM
  ∞SYSTEMS

EARTH RESOURCES OBSERVATION SATELLITES
USE EROS (SATELLITES)

EARTH RESOURCES PROGRAM
GS   PROGRAMS
  . NASA PROGRAMS
  .. **EARTH RESOURCES PROGRAM**
  ... EARTH RESOURCES SURVEY
      PROGRAM
  .... SEASAT PROGRAM
  . NASA SPACE PROGRAMS
  .. **EARTH RESOURCES PROGRAM**
  ... EARTH RESOURCES SURVEY
      PROGRAM

EARTH RESOURCES PROGRAM-*(CONT.)*
  .... SEASAT PROGRAM
RT   APOLLO APPLICATIONS PROGRAM
    CHANGE DETECTION
    GEOGRAPHIC APPLICATIONS PROGRAM
    INFRARED RADIOMETERS
    LARGE AREA CROP INVENTORY
      EXPERIMENT
    PLANT STRESS
    SATELLITE OBSERVATION
    SKYLAB PROGRAM

EARTH RESOURCES SHUTTLE IMAGING RADAR
GS   RADAR
  . PULSE RADAR
  .. PULSE DOPPLER RADAR
  ... **EARTH RESOURCES SHUTTLE
      IMAGING RADAR**
RT   RADAR IMAGERY
    SYNTHETIC ARRAYS

EARTH RESOURCES SURVEY AIRCRAFT
GS   RECONNAISSANCE AIRCRAFT
  . **EARTH RESOURCES SURVEY
     AIRCRAFT**
RT   AERIAL PHOTOGRAPHY
    AERIAL RECONNAISSANCE
  ∞AIRCRAFT
    PHOTOGEOLOGY
    PHOTORECONNAISSANCE

EARTH RESOURCES SURVEY PROGRAM
GS   PROGRAMS
  . NASA PROGRAMS
  .. EARTH RESOURCES PROGRAM
  ... **EARTH RESOURCES SURVEY
      PROGRAM**
  .... SEASAT PROGRAM
  . NASA SPACE PROGRAMS
  .. EARTH RESOURCES PROGRAM
  ... **EARTH RESOURCES SURVEY
      PROGRAM**
  .... SEASAT PROGRAM
RT   APOLLO APPLICATIONS PROGRAM
    SKYLAB PROGRAM

EARTH RESOURCES TECHNOLOGY SATELLITE B
USE LANDSAT 2

EARTH RESOURCES TECHNOLOGY SATELLITE C
USE LANDSAT 3

EARTH RESOURCES TECHNOLOGY SATELLITE D
USE LANDSAT 4

EARTH RESOURCES TECHNOLOGY SATELLITE E
USE LANDSAT E

EARTH RESOURCES TECHNOLOGY SATELLITE F
USE LANDSAT F

EARTH RESOURCES TECHNOLOGY SATELLITE 1
USE LANDSAT 1

EARTH RESOURCES TECHNOLOGY SATELLITES
USE LANDSAT SATELLITES

EARTH ROTATION
GS   GYRATION
  . ROTATION
  .. **EARTH ROTATION**
RT ∞EARTH MOTION
    SIDEREAL TIME
    SUPERROTATION

EARTH SHAPE
USE GEODESY

EARTH SURFACE
GS   LITHOSPHERE
  . **EARTH SURFACE**
RT   CRATONS
    CRUSTAL FRACTURES
    EQUATORIAL REGIONS
    GEODETIC ACCURACY
    MARSHLANDS
    OCEAN SURFACE
    PLANETARY SURFACES
    STRUCTURAL PROPERTIES (GEOLOGY)
  ∞SURFACES
    TERRADYNAMICS
    TOPOGRAPHY

EARTH TERMINAL MEASUREMENT SYSTEM
- RT   COMMUNICATION SATELLITES
-     ELECTROMAGNETIC MEASUREMENT
-     ELECTRONIC EQUIPMENT TESTS
-     GROUND SUPPORT EQUIPMENT
-     ∞ MEASUREMENT
-     RADIO RELAY SYSTEMS
-     ∞ SYSTEMS
-     ∞ TEST EQUIPMENT

EARTH TERMINALS
- GS   STATIONS
-     . GROUND STATIONS
-     . . **EARTH TERMINALS**
- RT   CARRIER TO NOISE RATIOS
-     COMMUNICATION EQUIPMENT
-     RADIO RELAY SYSTEMS
-     SATELLITE TRANSMISSION
-     SPACECRAFT COMMUNICATION
-     TELEVISION SYSTEMS

EARTH TIDES
- GS   TIDES
-     . **EARTH TIDES**
- RT   ATMOSPHERIC TIDES
-     LUNAR TIDES

EARTH VIEWING APPLICATIONS LABORATORY
- UF   EVAL
- GS   LABORATORIES
-     . SPACE LABORATORIES
-     . . **EARTH VIEWING APPLICATIONS LABORATORY**
-     PAYLOADS
-     . **EARTH VIEWING APPLICATIONS LABORATORY**
- RT   SAIL PROJECT
-     SPACE SHUTTLES

EARTH-MARS TRAJECTORIES
- GS   TRAJECTORIES
-     . SPACECRAFT TRAJECTORIES
-     . . INTERPLANETARY TRAJECTORIES
-     . . . **EARTH-MARS TRAJECTORIES**
- RT   ELLIPTICAL ORBITS
-     ORBITAL MECHANICS
-     TRANSFER ORBITS

EARTH-MERCURY TRAJECTORIES
- GS   TRAJECTORIES
-     . SPACECRAFT TRAJECTORIES
-     . . INTERPLANETARY TRAJECTORIES
-     . . . **EARTH-MERCURY TRAJECTORIES**
- RT   ELLIPTICAL ORBITS
-     ORBITAL MECHANICS
-     TRANSFER ORBITS

EARTH-MOON SYSTEM
- RT   CHARON
-     GRAVITATIONAL FIELDS
-     GRAVITATIONAL WAVES
-     LUNAR RETROREFLECTORS
-     MOON
-     NATURAL SATELLITES
-     ORBITAL MECHANICS
-     SOLAR SYSTEM
-     ∞ SYSTEMS
-     TWO BODY PROBLEM

EARTH-MOON TRAJECTORIES
- GS   TRAJECTORIES
-     . SPACECRAFT TRAJECTORIES
-     . . LUNAR TRAJECTORIES
-     . . . **EARTH-MOON TRAJECTORIES**
- RT   APOLLO 5 FLIGHT
-     APOLLO 6 FLIGHT
-     APOLLO 7 FLIGHT
-     APOLLO 8 FLIGHT
-     APOLLO 9 FLIGHT
-     APOLLO 10 FLIGHT
-     APOLLO 11 FLIGHT
-     APOLLO 12 FLIGHT
-     APOLLO 13 FLIGHT
-     APOLLO 14 FLIGHT
-     APOLLO 15 FLIGHT
-     APOLLO 16 FLIGHT
-     APOLLO 17 FLIGHT
-     CIRCUMLUNAR TRAJECTORIES
-     CISLUNAR SPACE
-     INTERPLANETARY TRAJECTORIES
-     LUNAR FLIGHT
-     LUNAR ORBITS
-     MOON-EARTH TRAJECTORIES
-     PARKING ORBITS
-     RENDEZVOUS TRAJECTORIES

EARTH-MOON TRAJECTORIES-(CONT.)
-     ROUND TRIP TRAJECTORIES
-     TRANSFER ORBITS

EARTH-VENUS TRAJECTORIES
- GS   TRAJECTORIES
-     . SPACECRAFT TRAJECTORIES
-     . . **EARTH-VENUS TRAJECTORIES**
- RT   ∞ ASTRONAUTICS
-     FLIGHT OPTIMIZATION
-     INTERPLANETARY FLIGHT
-     INTERPLANETARY TRAJECTORIES
-     ∞ MISSIONS
-     ORBITS
-     SPACE MISSIONS
-     SPACE NAVIGATION
-     SPACECRAFT REENTRY
-     TRANSFER ORBITS

EARTHNET
- RT   EARTH OBSERVATIONS (FROM SPACE)
-     EARTH RESOURCES
-     ESA SATELLITES
-     EUROPEAN SPACE PROGRAMS
-     LANDSAT SATELLITES
-     REMOTE SENSORS
-     SYNTHETIC APERTURE RADAR

EARTHQUAKE DAMAGE
- GS   DAMAGE
-     . **EARTHQUAKE DAMAGE**
- RT   EARTH CRUST
-     EARTH MOVEMENTS
-     GEOLOGICAL FAULTS
-     MICROSEISMS
-     SEISMIC ENERGY
-     SEISMIC WAVES
-     SEISMOLOGY
-     SHOCK WAVES
-     TSUNAMI WAVES

EARTHQUAKE RESISTANCE
- GS   MECHANICAL PROPERTIES
-     . **EARTHQUAKE RESISTANCE**
- RT   CRUSTAL FRACTURES
-     EARTHQUAKES
-     FRACTURE STRENGTH
-     IMPACT STRENGTH
-     LANDFORMS
-     ∞ RESISTANCE
-     SEISMIC WAVES
-     SHOCK RESISTANCE
-     SHOCK WAVES
-     TREMORS

EARTHQUAKE RESISTANT STRUCTURES
- RT   CONCRETE STRUCTURES
-     ELASTIC BENDING
-     ∞ ELASTIC SYSTEMS
-     SEISMIC WAVES
-     SHOCK WAVES
-     STRUCTURAL VIBRATION
-     ∞ STRUCTURES

EARTHQUAKES
- GS   EARTH MOVEMENTS
-     . **EARTHQUAKES**
- RT   CRUSTAL FRACTURES
-     EARTHQUAKE RESISTANCE
-     GEOLOGICAL FAULTS
-     LARGE APERTURE SEISMIC ARRAY
-     MICROSEISMS
-     PLANETARY QUAKES
-     PLATES (TECTONICS)
-     ROUSE BELTS
-     SAN ANDREAS FAULT
-     SAN ANDREAS FAULT EXPERIMENT
-     SEISMIC WAVES
-     SEISMOLOGY
-     SHOCK WAVES
-     SUBDUCTION (GEOLOGY)
-     TREMORS
-     TSUNAMI WAVES

EASEP
- UF   EARLY APOLLO SURFACE EXPERIMENTS PACKAGE
- GS   PACKAGES
-     . INSTRUMENT PACKAGES
-     . . **EASEP**
- RT   ∞ INSTRUMENTS
-     LUNAR EXPLORATION
-     PAYLOADS
-     ∞ SURFACES

EAST GERMANY
- UF   GERMAN DEMOCRATIC REPUBLIC
-     PEOPLES DEMOCRATIC REPUBLIC OF GERMANY
- GS   NATIONS
-     . **EAST GERMANY**
- RT   CENTRAL EUROPE
-     EUROPE
-     ∞ GERMANY
-     WEST GERMANY

EASTERN HEMISPHERE
- RT   EARTH (PLANET)
-     GEOGRAPHY
-     ∞ HEMISPHERES
-     WESTERN HEMISPHERE

EATING
- GS   INGESTION (BIOLOGY)
-     . **EATING**
- RT   DIGESTING
-     ∞ FOOD
-     MASTICATION
-     SPACE FLIGHT FEEDING
-     SWALLOWING
-     SYNTHETIC FOOD

EBERT SPECTROMETERS
- GS   MEASURING INSTRUMENTS
-     . OPTICAL MEASURING INSTRUMENTS
-     . . **EBERT SPECTROMETERS**
-     . RADIATION MEASURING INSTRUMENTS
-     . . **EBERT SPECTROMETERS**
-     . SPECTROMETERS
-     . . **EBERT SPECTROMETERS**
-     OPTICAL EQUIPMENT
-     . OPTICAL MEASURING INSTRUMENTS
-     . . **EBERT SPECTROMETERS**
- RT   EARTH ALBEDO
-     FILTER WHEEL INFRARED SPECTROMETERS
-     INFRARED SPECTROMETERS
-     ULTRAVIOLET SPECTROMETERS

EBF
- USE   EXTERNALLY BLOWN FLAPS

EBR-1 REACTOR
- USE   EXPERIMENTAL BREEDER REACTOR 1

EBR-2 REACTOR
- USE   EXPERIMENTAL BREEDER REACTOR 2

EBULLITION
- USE   BOILING

EBWR (REACTOR)
- USE   EXPERIMENTAL BOILING WATER REACTORS

EC-121 AIRCRAFT
- UF   R7V AIRCRAFT
-     WARNING STAR AIRCRAFT
- GS   LOCKHEED AIRCRAFT
-     . **EC-121 AIRCRAFT**
-     MONOPLANES
-     . **EC-121 AIRCRAFT**
-     TRANSPORT AIRCRAFT
-     . **EC-121 AIRCRAFT**
- RT   ∞ AIRCRAFT
-     C-121 AIRCRAFT

ECCENTRIC GEOPHYSICAL OBSERVATORY
- USE   EGO

ECCENTRIC ORBIT GEOPHYSICAL OBSERVATORY
- USE   EGO

ECCENTRIC ORBITS
- GS   ORBITS
-     . **ECCENTRIC ORBITS**
- RT   CIRCULAR ORBITS
-     ELLIPTICAL ORBITS
-     EXOSAT SATELLITE
-     LISSAJOUS FIGURES

ECCENTRICITY
- RT   ABNORMALITIES
-     ASYMMETRY
-     BALANCING
-     CONCENTRICITY
-     DEVIATION
-     ELLIPTICITY
-     ELONGATION

ECCENTRICITY-*(CONT.)*
    SKEWNESS
    SYMMETRY
    VARIABILITY
    VARIATIONS

ECCENTRICS
  UF   CRANKS
  RT   CAMS
        LINKAGES

ECHELETTE GRATINGS
  GS   GRATINGS (SPECTRA)
        . ECHELETTE GRATINGS
  RT   DIFFRACTION
        REFLECTION

ECHELON FAULTS
  USE   GEOLOGICAL FAULTS

ECHO PROJECT
  GS   PROGRAMS
        . NASA PROGRAMS
        . . ECHO PROJECT
        . NASA SPACE PROGRAMS
        . . ECHO PROJECT
        . PROJECTS
        . . ECHO PROJECT
  RT   COMMUNICATION SATELLITES
        PASSIVE SATELLITES

ECHO SATELLITES
  GS   SATELLITES
        . ARTIFICIAL SATELLITES
        . . PASSIVE SATELLITES
        . . . ECHO SATELLITES
        . . . . ECHO 1 SATELLITE
        . . . . ECHO 2 SATELLITE
        UNMANNED SPACECRAFT
        . PASSIVE SATELLITES
        . . ECHO SATELLITES
        . . . ECHO 1 SATELLITE
        . . . ECHO 2 SATELLITE
  RT   AGENA B ROCKET VEHICLE
        AGENA ROCKET VEHICLES

ECHO SOUNDING
  GS   SOUNDING
        . ECHO SOUNDING
  RT   DEEP SCATTERING LAYERS
        DEPTH MEASUREMENT
        ECHOES
        NAVIGATION AIDS
        SONAR
        SOUND LOCALIZATION
        SOUND RANGING
        UNDERWATER ACOUSTICS

ECHO SUPPRESSORS
  GS   CIRCUITS
        . ECHO SUPPRESSORS
        SUPPRESSORS
        . ECHO SUPPRESSORS
  RT   NOISE REDUCTION
        PULSE RADAR
        RADIOTELEPHONES
        SONAR
        SWITCHES
        TELEPHONY
        VOICE COMMUNICATION

ECHO 1 CARRIER ROCKET
  USE   THOR DELTA LAUNCH VEHICLE

ECHO 1 SATELLITE
  UF   A-11 SATELLITE
  GS   SATELLITES
        . ARTIFICIAL SATELLITES
        . . PASSIVE SATELLITES
        . . . ECHO SATELLITES
        . . . . ECHO 1 SATELLITE
        UNMANNED SPACECRAFT
        . PASSIVE SATELLITES
        . . ECHO SATELLITES
        . . . ECHO 1 SATELLITE
  RT   THOR DELTA LAUNCH VEHICLE

ECHO 2 SATELLITE
  UF   A-12 SATELLITE
  GS   SATELLITES
        . ARTIFICIAL SATELLITES
        . . PASSIVE SATELLITES
        . . . ECHO SATELLITES
        . . . . ECHO 2 SATELLITE

ECHO 2 SATELLITE-*(CONT.)*
        UNMANNED SPACECRAFT
        . PASSIVE SATELLITES
        . . ECHO SATELLITES
        . . . ECHO 2 SATELLITE

ECHOCARDIOGRAPHY
  GS   BIOENGINEERING
        . BIOMETRICS
        . . CARDIOGRAPHY
        . . . PHONOCARDIOGRAPHY
        . . . . ECHOCARDIOGRAPHY
  RT   CARDIAC VENTRICLES
        HEART DISEASES
        HEART FUNCTION

ECHOENCEPHALOGRAPHY
  GS   BIOENGINEERING
        . BIOMETRICS
        . . ECHOENCEPHALOGRAPHY
  RT   BIOINSTRUMENTATION
        BRAIN
        ELECTROPHYSIOLOGY
        MEDICAL ELECTRONICS
        MEDICAL EQUIPMENT

ECHOES
  GS   **ECHOES**
        . ANGELS
        . AURORAL ECHOES
        . LUNAR ECHOES
        . . LUNAR RADAR ECHOES
        . RADAR ECHOES
        . . CLUTTER
        . . LUNAR RADAR ECHOES
        . . SOLAR RADAR ECHOES
        . . VENUS RADAR ECHOES
        . RADIO ECHOES
        . SIGNAL REFLECTION
  RT   ACOUSTICS
        CEPSTRAL ANALYSIS
        ECHO SOUNDING
        GROUND EFFECT (COMMUNICATIONS)
        NOISE (SOUND)
        REVERBERATION

ECLIPSE PROJECT
  GS   PROGRAMS
        . PROJECTS
        . . ECLIPSE PROJECT

ECLIPSES
  GS   **ECLIPSES**
        . LUNAR ECLIPSES
        . SOLAR ECLIPSES
  RT   ECLIPSING BINARY STARS
        LUNAR SHADOW
        OCCULTATION
        PENUMBRAS
        UMBRAS

ECLIPSING BINARY STARS
  GS   CELESTIAL BODIES
        . STARS
        . . DOUBLE STARS
        . . . BINARY STARS
        . . . . ECLIPSING BINARY STARS
        . . . . . DWARF NOVAE
  RT   ACCRETION DISKS
        CATACLYSMIC VARIABLES
        ECLIPSES
        STELLAR OCCULTATION
        SYMBIOTIC STARS
        VARIABLE STARS
        X RAY BINARIES

ECLIPTIC
  RT   PLANETS
        SOLAR ORBITS
        ZODIAC

ECLOGITE
  GS   ROCKS
        . IGNEOUS ROCKS
        . . ECLOGITE
  RT   GARNETS
        PYROXENES
        SOILS

ECOLOGICAL SYSTEMS
  USE   ECOLOGY

ECOLOGY
  UF   ECOLOGICAL SYSTEMS

ECOLOGY-*(CONT.)*
  GS   **ECOLOGY**
        . COASTAL ECOLOGY
  RT   ARIZONA REGIONAL ECOLOGICAL TEST
          SITE
        BIOCHEMICAL OXYGEN DEMAND
    ∞BIOLOGY
        BIOMETEOROLOGY
        CARBON CYCLE
        CENTRAL ATLANTIC REGIONAL ECOL
          TEST SITE
        CLOSED ECOLOGICAL SYSTEMS
        COASTAL PLAINS
        ECOSYSTEMS
        ENDANGERED SPECIES
        ENERGY POLICY
        ENVIRONMENTS
        HABITABILITY
        HABITATS
        PHENOLOGY
        PREDATORS
        SYMBIOSIS
        VEGETATION GROWTH

ECONOMETRICS
  RT   ∞APPLICATIONS OF MATHEMATICS
        ECONOMICS
        GROSS NATIONAL PRODUCT
        STATISTICAL CORRELATION

ECONOMIC ANALYSIS
  RT   ALLOCATIONS
        COMPARISON
        COST ANALYSIS
        COST ESTIMATES
        COSTS
        ECONOMY
        EFFICIENCY
        MANAGEMENT
        OPERATING COSTS
        VALUE ENGINEERING

ECONOMIC DEVELOPMENT
  RT   COMMERCE
        DEVELOPING NATIONS
        ECONOMY
        GEOGRAPHY
        INDUSTRIES
        LAND USE
        MANUFACTURING
        RESOURCES
        SPACE INDUSTRIALIZATION
        URBAN DEVELOPMENT

ECONOMIC FACTORS
  RT   ALLOCATIONS
        BUDGETING
        COMMERCIAL ENERGY
        COSTS
        DEVELOPING NATIONS
        DOMESTIC ENERGY
        ECONOMY
        EFFICIENCY
        ENERGY POLICY
        FEASIBILITY ANALYSIS
        INDUSTRIAL ENERGY
        MANAGEMENT
        RESERVES
        RESOURCES
        TRANSPORTATION ENERGY

ECONOMIC IMPACT
  GS   IMPACT
        . ECONOMIC IMPACT
  RT   COSTS
        ENVIRONMENTS
        INDUSTRIES
        INVESTMENTS
        RESOURCES

ECONOMICS
  GS   **ECONOMICS**
        . DEMAND (ECONOMICS)
  RT   COSTS
        ECONOMETRICS
        EVALUATION
        FIDUCIARIES
        FOREIGN TRADE
        INCOME
        INTERNATIONAL TRADE
        INVESTMENTS
        PREJUDICES
        PROGRESS
        RECESSION
        RESOURCES

ECONOMICS-*(CONT.)*
    STATISTICAL ANALYSIS

ECONOMY
RT   COST ESTIMATES
        DECISION MAKING
        ECONOMIC ANALYSIS
        ECONOMIC DEVELOPMENT
        ECONOMIC FACTORS
        FINANCIAL MANAGEMENT
        LOW COST
        MANAGEMENT PLANNING
        RECYCLING

ECOSYSTEMS
RT   CLOSED ECOLOGICAL SYSTEMS
        ECOLOGY
        ENDANGERED SPECIES
        FOOD CHAIN
        PREDATORS
        ∞ SYSTEMS

ECS
USE   EUROPEAN COMMUNICATIONS
            SATELLITE

ECUADOR
GS   NATIONS
        . ECUADOR
RT   SOUTH AMERICA

EDDIES
USE   VORTICES

EDDINGTON APPROXIMATION
GS   ANALYSIS (MATHEMATICS)
        . NUMERICAL ANALYSIS
        . . APPROXIMATION
        . . . EDDINGTON APPROXIMATION

EDDY CURRENTS
SN   (LIMITED TO ELECTRIC CURRENTS)
GS   ELECTRIC CURRENT
        . EDDY CURRENTS
RT   BRAKING
        ELECTRIC CONDUCTORS
        ELECTRICAL PROPERTIES
        HYSTERESIS
        LOSSES
        MAGNETIC PROPERTIES
        ∞ PHYSICAL PROPERTIES
        PLASMA CURRENTS
        VORTICITY TRANSPORT HYPOTHESIS

EDDY DIFFUSION
USE   TURBULENT DIFFUSION

EDDY VISCOSITY
GS   TRANSPORT PROPERTIES
        . VISCOSITY
        . . EDDY VISCOSITY
RT   FLOW CHARACTERISTICS
        FLOW RESISTANCE
        INTERNAL FRICTION
        TURBULENT FLOW
        VISCOUS DRAG
        VISCOUS FLOW

EDEMA
GS   DISEASES
        . EDEMA
RT   BODY FLUIDS
        DIURESIS
        WATER BALANCE

EDGE DISLOCATIONS
UF   SLIP BANDS
GS   DEFECTS
        . CRYSTAL DEFECTS
        . . CRYSTAL DISLOCATIONS
        . . . EDGE DISLOCATIONS
        DISLOCATIONS (MATERIALS)
        . CRYSTAL DISLOCATIONS
        . . EDGE DISLOCATIONS
RT   ∞ BANDS
        SCREW DISLOCATIONS

EDGE LOADING
GS   LOADS (FORCES)
        . EDGE LOADING
RT   AERODYNAMIC LOADS
        COMPRESSION LOADS
        DYNAMIC LOADS
        STATIC LOADS

EDGE LOADING-*(CONT.)*
    WING LOADING

EDGES
GS   EDGES
        . BLUNT LEADING EDGES
        . LEADING EDGES
        . . SHARP LEADING EDGES
        . TRAILING EDGES
        . . BLUNT TRAILING EDGES
RT   MARGINS
        RIMS
        SCALLOPING
        SIDES
        TIPS

EDITING
RT   DATA PROCESSING
        DATA REDUCTION
        EDITING ROUTINES (COMPUTERS)
        FORMAT
        TECHNICAL WRITING

EDITING ROUTINES (COMPUTERS)
GS   COMPUTER PROGRAMS
        . EDITING ROUTINES (COMPUTERS)
RT   COMPUTER SYSTEMS PROGRAMS
        DATA PROCESSING
        EDITING

EDTA
USE   ETHYLENEDIAMINETETRAACETIC ACIDS

EDUCATION
UF   INSTRUCTIONS
        TEACHING
        TRAINING
GS   EDUCATION
        . ASTRONAUT TRAINING
        . EJECTION TRAINING
        . FLIGHT TRAINING
        . . SPACE FLIGHT TRAINING
        . GUNNERY TRAINING
        . MAINTENANCE TRAINING
        . PILOT TRAINING
RT   BEHAVIOR
        COMMUNICATING
        CREATIVITY
        EXPERIENCE
        HUMAN FACTORS ENGINEERING
        INSTRUCTORS
        KNOWLEDGE
        LEARNING
        LEARNING THEORY
        LECTURES
        MEMORY
        ∞ ORIENTATION
        PSYCHOMETRICS
        QUALIFICATIONS
        RETRAINING
        SAFETY MANAGEMENT
        SCHOOLS
        STUDENTS
        ∞ TESTS
        TEXTBOOKS
        TRAINING ANALYSIS
        TRAINING DEVICES
        TRANSFER OF TRAINING
        UNIVERSITIES

EDUCATIONAL TELEVISION
GS   TELECOMMUNICATION
        . EDUCATIONAL TELEVISION
        TELEVISION SYSTEMS
        . EDUCATIONAL TELEVISION
RT   CLOSED CIRCUIT TELEVISION
        COLOR TELEVISION
        COMMUNICATION EQUIPMENT
        LEARNING
        STEREOTELEVISION
        TRAINING DEVICES

EEG (ELECTROENCEPHALOGRAMS)
USE   ELECTROENCEPHALOGRAPHY

EFFECTIVE PERCEIVED NOISE LEVELS
UF   EPNL
GS   LEVEL (QUANTITY)
        . EFFECTIVE PERCEIVED NOISE LEVELS
RT   ACOUSTIC MEASUREMENT
        ACOUSTICS
        LOUDNESS
        ∞ NOISE
        NOISE (SOUND)
        NOISE INTENSITY

EFFECTIVE PERCEIVED NOISE LEVELS-*(CONT.)*
    NOISE REDUCTION
    SOUND INTENSITY

EFFECTIVENESS
GS   EFFECTIVENESS
        . COST EFFECTIVENESS
        . SYSTEM EFFECTIVENESS
RT   EFFICIENCY

EFFECTORS
USE   CONTROL EQUIPMENT

∞ **EFFECTS**
SN   (USE OF A MORE SPECIFIC TERM IS
        RECOMMENDED--CONSULT THE TERMS
        LISTED BELOW)
UF   AFFECTS
RT   ATMOSPHERIC EFFECTS
        AUGER EFFECT
        BARKHAUSEN EFFECT
        BAUSCHINGER EFFECT
        BIOLOGICAL EFFECTS
        BRILLOUIN EFFECT
        BROWN WAVE EFFECT
        CAPTURE EFFECT
        CAUSES
        CERENKOV RADIATION
        CHEMICAL EFFECTS
        COANDA EFFECT
        COMPRESSIBILITY EFFECTS
        COMPTON EFFECT
        CORIOLIS EFFECT
        DIFFUSION
        DOPPLER EFFECT
        DOPPLER-FIZEAU EFFECT
        ELECTRO-OPTICAL EFFECT
        ENVIRONMENT EFFECTS
        ETTINGSHAUSEN EFFECT
        FARADAY EFFECT
        FIELD EFFECT TRANSISTORS
        FIZEAU EFFECT
        FORBUSH DECREASES
        GALVANOMAGNETIC EFFECTS
        GREEN WAVE EFFECT
        GREENHOUSE EFFECT
        GROUND EFFECT (AERODYNAMICS)
        GROUND EFFECT (COMMUNICATIONS)
        GROUND EFFECT MACHINES
        GUNN EFFECT
        HALL EFFECT
        HYDRODYNAMIC RAM EFFECT
        ISOTOPE EFFECT
        JAHN-TELLER EFFECT
        JET BLAST EFFECTS
        JOULE-THOMSON EFFECT
        ∞ KERR EFFECTS
        KERR ELECTROOPTICAL EFFECT
        KERR MAGNETOOPTICAL EFFECT
        KIRKENDALL EFFECT
        KONDO EFFECT
        LONG TERM EFFECTS
        LUNAR EFFECTS
        LUNAR GRAVITATIONAL EFFECTS
        LUXEMBOURG EFFECT
        MAGNETIC EFFECTS
        MAGNUS EFFECT
        MOIRE EFFECTS
        MOSSBAUER EFFECT
        NERNST-ETTINGSHAUSEN EFFECT
        NONOHMIC EFFECT
        NUCLEAR EXPLOSION EFFECT
        OVERHAUSER EFFECT
        PATHOLOGICAL EFFECTS
        PELTIER EFFECTS
        PENNING EFFECT
        PHOTOELECTRIC EFFECT
        PHOTOELECTROMAGNETIC EFFECTS
        PHOTOMAGNETIC EFFECTS
        PHOTOMECHANICAL EFFECT
        PHOTOVOLTAIC EFFECT
        PHYSIOLOGICAL EFFECTS
        PINCH EFFECT
        POGO EFFECTS
        POYNTING-ROBERTSON EFFECT
        PRESSURE EFFECTS
        PROXIMITY EFFECT (ELECTRICITY)
        PSYCHOLOGICAL EFFECTS
        RADIATION EFFECTS
        RAMSAUER EFFECT
        REENTRY EFFECTS
        RELATIVISTIC EFFECTS
        SCALE EFFECT
        SCHACH EFFECT

EFFECTS-(CONT.)
- SCREEN EFFECT
- SEEBECK EFFECT
- SOLAR ACTIVITY EFFECTS
- STARK EFFECT
- STERILIZATION EFFECTS
- SUHL EFFECT
- SURFACE EFFECT SHIPS
- SURFACE ROUGHNESS EFFECTS
- SWEEP EFFECT
- TEMPERATURE EFFECTS
- THERMOMAGNETIC EFFECTS
- TURBULENCE EFFECTS
- UMKEHR EFFECT
- VACUUM EFFECTS
- VIBRATION EFFECTS
- VIEW EFFECTS
- VOIGT EFFECT
- WIND EFFECTS
- ZEEMAN EFFECT
- ZENER EFFECT

EFFERENT NERVOUS SYSTEMS
- UF   MOTOR SYSTEMS (BIOLOGY)
- GS   NERVOUS SYSTEM
-    . EFFERENT NERVOUS SYSTEMS
- RT   SENSORIMOTOR PERFORMANCE
-    ∞ SYSTEMS

EFFERVESCENCE
- RT   BOILING
-    BUBBLES
-    SURFACE PROPERTIES

EFFICIENCY
- GS   EFFICIENCY
-    . CHARGE EFFICIENCY
-    . COMBUSTION EFFICIENCY
-    . COMPRESSOR EFFICIENCY
-    . ENERGY CONVERSION EFFICIENCY
-    . NOZZLE EFFICIENCY
-    . POWER EFFICIENCY
-    . PROPULSIVE EFFICIENCY
-    . . PROPELLER EFFICIENCY
-    . THERMODYNAMIC EFFICIENCY
-    . TRANSMISSION EFFICIENCY
- RT   AIRCRAFT PRODUCTION COSTS
-    COMFORT
-    COMMONALITY
-    COMPRESSION RATIO
-    COMPUTER SYSTEMS PERFORMANCE
-    COST INCENTIVES
-    COST REDUCTION
-    COSTS
-    ECONOMIC ANALYSIS
-    ECONOMIC FACTORS
-    EFFECTIVENESS
-    FEASIBILITY
-    FIGURE OF MERIT
-    HUMAN FACTORS ENGINEERING
-    INCENTIVE TECHNIQUES
-    INDEXES (RATIOS)
-    OPTIMIZATION
-    ∞ PERFORMANCE
-    PRODUCTIVITY
-    RATIOS
-    UTILIZATION

EFFLUENTS
- RT   AIR POLLUTION
-    CONTAMINANTS
-    ∞ DISCHARGE
-    ENVIRONMENT PROTECTION
-    EXHAUST GASES
-    FILTRATION
-    LIQUID WASTES
-    REACTION PRODUCTS
-    SETTLING
-    SEWAGE
-    SEWERS
-    WASTE DISPOSAL
-    WASTES

EFFLUX
- RT   EMISSION
-    OUTPUT

EFFORT
- RT   ABILITIES
-    CONSISTENCY
-    FATIGUE (BIOLOGY)
-    ∞ PERFORMANCE
-    PHYSICAL WORK

EFFUSIVES
- GS   EFFUSIVES
-    . LAVA
- RT   CONES (VOLCANOES)
-    IGNEOUS ROCKS
-    MARS VOLCANOES
-    ROCKS
-    VOLCANOES
-    VOLCANOLOGY

EGCR (REACTOR)
- USE   EXPERIMENTAL GAS COOLED
-    REACTORS

EGGS
- RT   EMBRYOS
-    FETUSES
-    ∞ FOOD
-    OVARIES

EGO
- UF   ECCENTRIC GEOPHYSICAL
-    OBSERVATORY
-    ECCENTRIC ORBIT GEOPHYSICAL
-    OBSERVATORY
-    EOGO
- GS   OBSERVATORIES
-    . GEOPHYSICAL OBSERVATORIES
-    . . OGO
-    . . . EGO
-    SATELLITES
-    . ARTIFICIAL SATELLITES
-    . . GEOPHYSICAL SATELLITES
-    . . . OGO
-    . . . . EGO
- RT   AGENA B ROCKET VEHICLE
-    ATLAS LAUNCH VEHICLES
-    POGO

EGRESS
- RT   AIR LOCKS
-    DOORS
-    HATCHES
-    INGRESS (SPACECRAFT PASSAGEWAY)
-    OPENINGS
-    OUTLETS

EGYPT
- GS   NATIONS
-    . EGYPT
- RT   AFRICA

EIGENFUNCTIONS
- USE   EIGENVECTORS

EIGENSTATES
- USE   EIGENVECTORS

EIGENVALUES
- UF   CHARACTERISTIC EQUATIONS
-    CHARACTERISTIC FUNCTIONS
- GS   ALGEBRA
-    . VECTOR SPACES
-    . . MATRICES (MATHEMATICS)
-    . . . EIGENVALUES
- RT   EIGENVECTORS
-    HILL DETERMINANT
-    JACOBI MATRIX METHOD
-    JORDAN FORM
-    POLYNOMIALS
-    ROOTS OF EQUATIONS

EIGENVECTORS
- UF   CHARACTERISTIC EQUATIONS
-    CHARACTERISTIC FUNCTIONS
-    EIGENFUNCTIONS
-    EIGENSTATES
- GS   ALGEBRA
-    . VECTOR SPACES
-    . . MATRICES (MATHEMATICS)
-    . . . EIGENVECTORS
-    . . VECTORS (MATHEMATICS)
-    . . . EIGENVECTORS
- RT   EIGENVALUES
-    JACOBI MATRIX METHOD
-    MATHIEU FUNCTION
-    POLYNOMIALS

EIKONAL EQUATION
- GS   WAVE EQUATIONS
-    . EIKONAL EQUATION
- RT   ∞ EQUATIONS
-    POMERANCHUK THEOREM
-    REFRACTED WAVES

EIKONAL EQUATION-(CONT.)
-    WAVE FRONTS
-    ∞ WAVES

EINSTEIN EQUATIONS
- GS   ANALYSIS (MATHEMATICS)
-    . REAL VARIABLES
-    . . EINSTEIN EQUATIONS
- RT   BROWNIAN MOVEMENTS
-    ∞ EQUATIONS
-    EQUATIONS OF MOTION
-    KINETIC EQUATIONS
-    PROBABILITY THEORY
-    UNIFIED FIELD THEORY

EINSTEIN OBSERVATORY
- USE   HEAO 2

EINSTEINIUM
- GS   CHEMICAL ELEMENTS
-    . ACTINIDE SERIES
-    . . TRANSURANIUM ELEMENTS
-    . . . EINSTEINIUM
-    . NUCLIDES
-    . . ISOTOPES
-    . . . RADIOACTIVE ISOTOPES
-    . . . . TRANSURANIUM ELEMENTS
-    . . . . . EINSTEINIUM
-    HEAVY ELEMENTS
-    . TRANSURANIUM ELEMENTS
-    . . EINSTEINIUM
-    METALS
-    . ACTINIDE SERIES
-    . . TRANSURANIUM ELEMENTS
-    . . . EINSTEINIUM

EISCAT RADAR SYSTEM (EUROPE)
- UF   EUROPEAN INCOHERENT SCATTER
-    RADAR
- GS   RADAR
-    . INCOHERENT SCATTER RADAR
-    . . EISCAT RADAR SYSTEM (EUROPE)
- RT   INCOHERENT SCATTERING
-    INTERNATIONAL COOPERATION
-    IONOSPHERIC PROPAGATION
-    RADAR SCATTERING
-    RADAR TRANSMISSION
-    ∞ SYSTEMS
-    ULTRAHIGH FREQUENCIES

EJECTA
- RT   CRATERING
-    CRATERS
-    DEBRIS
-    EJECTION
-    FRAGMENTS
-    IMPACT DAMAGE
-    MARS CRATERS
-    METEORITE CRATERS
-    METEORITIC DAMAGE
-    PROJECTILE CRATERING
-    WOLF-RAYET STARS

EJECTION
- GS   EJECTION
-    . STELLAR MASS EJECTION
- RT   BAILOUT
-    ∞ DISCHARGE
-    DISCONNECT DEVICES
-    DISPOSAL
-    DUMPING
-    EJECTA
-    EJECTORS
-    EMISSION
-    EMPTYING
-    ESCAPE (ABANDONMENT)
-    ESCAPE SYSTEMS
-    EVACUATING (TRANSPORTATION)
-    EVACUATING (VACUUM)
-    EXHAUSTING
-    EXPULSION
-    EXPULSION BLADDERS
-    FLUSHING
-    JETTISON SYSTEMS
-    JETTISONING
-    MATERIALS HANDLING
-    PARACHUTE DESCENT
-    RELEASING
-    REMOVAL
-    SHEDDING
-    THROWING
-    UNLOADING

EJECTION INJURIES
- GS   INJURIES

EJECTION INJURIES-*(CONT.)*
. **EJECTION INJURIES**
RT   BAILOUT
       PILOT TRAINING

EJECTION SEATS
GS   SAFETY DEVICES
     . **EJECTION SEATS**
     . . FLYING EJECTION SEATS
     SEATS
     . **EJECTION SEATS**
     . . FLYING EJECTION SEATS
RT   ABORT APPARATUS
       AIRCRAFT SAFETY
       BAILOUT
       COCKPITS
       EJECTORS
       ESCAPE CAPSULES
       ESCAPE SYSTEMS
       JETTISON SYSTEMS
     ∞PROPELLANT ACTUATED DEVICES

EJECTION TRAINING
GS   EDUCATION
     . **EJECTION TRAINING**
RT   ASTRONAUT TRAINING
       BAILOUT
       ESCAPE (ABANDONMENT)
       FLIGHT TRAINING
       PARACHUTE DESCENT
       PILOT TRAINING

EJECTORS
RT   DISPENSERS
       EJECTION
       EJECTION SEATS
       EXHAUST DIFFUSERS
       EXHAUST NOZZLES
       EXHAUST SYSTEMS
       FLYING EJECTION SEATS
       INJECTORS
       JET ENGINES
       JET PUMPS
       MATERIALS HANDLING
       PUMPS
       ROCKET ENGINES
       SPRAYERS
       VACUUM PUMPS

EKMAN LAYER
RT   ATMOSPHERIC BOUNDARY LAYER
       BOUNDARY LAYER TRANSITION
     ∞LAYERS
       POROUS BOUNDARY LAYER CONTROL
       TURBULENT BOUNDARY LAYER

EL NINO
GS   CIRCULATION
     . WATER CIRCULATION
     . . WATER CURRENTS
     . . . OCEAN CURRENTS
     . . . . **EL NINO**
RT   AIR WATER INTERACTIONS
       OCEAN TEMPERATURE
       PACIFIC OCEAN
       PERIODIC VARIATIONS
       TROPICAL METEOROLOGY

EL SALVADOR
GS   NATIONS
     . **EL SALVADOR**
RT   CENTRAL AMERICA

ELASTIC ANISOTROPY
GS   ANISOTROPY
     . PLASTIC ANISOTROPY
     . . **ELASTIC ANISOTROPY**

ELASTIC BARS
GS   BARS
     . **ELASTIC BARS**

ELASTIC BENDING
GS   BENDING
     . **ELASTIC BENDING**
     DEFORMATION
     . ELASTIC DEFORMATION
     . . **ELASTIC BENDING**
RT   EARTHQUAKE RESISTANT STRUCTURES

ELASTIC BODIES
RT   ∞BODIES
     ∞ELASTIC SYSTEMS
     ELASTODYNAMICS

ELASTIC BODIES-*(CONT.)*
     ELASTOSTATICS

ELASTIC BUCKLING
GS   BUCKLING
     . **ELASTIC BUCKLING**
     DEFORMATION
     . ELASTIC DEFORMATION
     . . **ELASTIC BUCKLING**
RT   FAILURE MODES

ELASTIC COLLISIONS
USE   ELASTIC SCATTERING

ELASTIC CONSTANTS
USE   ELASTIC PROPERTIES

ELASTIC CYLINDERS
RT   ∞CYLINDERS
      CYLINDRICAL BODIES
      CYLINDRICAL SHELLS

ELASTIC DAMPING
GS   DAMPING
     . **ELASTIC DAMPING**
     . . VISCOELASTIC DAMPING
     ELASTODYNAMICS
     . **ELASTIC DAMPING**
RT   RESONANCE TESTING
      VIBRATION DAMPING
      VISCOUS DAMPING

ELASTIC DEFORMATION
GS   DEFORMATION
     . **ELASTIC DEFORMATION**
     . . ELASTIC BENDING
     . . ELASTIC BUCKLING
RT   AXIAL STRAIN
      BENDING
      BORDONI PEAKS
      DEFLECTION
      ELASTODYNAMICS
      ELASTOSTATICS
      FLEXIBLE SPACECRAFT
      PLANE STRAIN
      PLASTIC DEFORMATION
      PRESTRESSING
      STRESS-STRAIN RELATIONSHIPS
      STRETCHING
      STRUCTURAL STRAIN
      TENSILE DEFORMATION

ELASTIC MEDIA
GS   MEDIA
     . **ELASTIC MEDIA**
RT   ELASTODYNAMICS
      ELASTOSTATICS

ELASTIC MODULUS
USE   MODULUS OF ELASTICITY

ELASTIC PLATES
GS   STRUCTURAL MEMBERS
     . PLATES (STRUCTURAL MEMBERS)
     . . **ELASTIC PLATES**

ELASTIC PROPERTIES
UF   ELASTIC CONSTANTS
      ELASTICITY
GS   MECHANICAL PROPERTIES
     . **ELASTIC PROPERTIES**
     . . AEROELASTICITY
     . . . AEROTHERMOELASTICITY
     . . ANELASTICITY
     . . ELASTOPLASTICITY
     . . HYDROELASTICITY
     . . HYPOELASTICITY
     . . MAGNETOSTRICTION
     . . MODULUS OF ELASTICITY
     . . . DYNAMIC MODULUS OF ELASTICITY
     . . PHOTOELASTICITY
     . . . PHOTOVISCOELASTICITY
     . . PROPORTIONAL LIMIT
     . . THERMOELASTICITY
     . . . AEROTHERMOELASTICITY
     . . VISCOELASTICITY
     . . . PHOTOVISCOELASTICITY
     . . . THERMOVISCOELASTICITY
RT   AIRY FUNCTION
      BIHARMONIC EQUATIONS
      COMPRESSIVE STRENGTH
      ELASTODYNAMICS
      ELASTOMETERS
      ELASTOSTATICS

ELASTIC PROPERTIES-*(CONT.)*
     FLEXIBILITY
     HOOKES LAW
     HYBRID STRUCTURES
     INFLUENCE COEFFICIENT
     MECHANICAL PROPERTIES
     MICROSONICS
     MICROYIELD STRENGTH
    ∞PHYSICAL PROPERTIES
     PIEZOELECTRICITY
     PLASTIC PROPERTIES
     POISSON RATIO
     PROPELLANT PROPERTIES
     RESILIENCE
     SOFTNESS
     STRESS TENSORS
     TENSILE PROPERTIES
     TENSILE STRENGTH
     YIELD STRENGTH

ELASTIC SCATTERING
UF   ELASTIC COLLISIONS
GS   SCATTERING
     . **ELASTIC SCATTERING**
RT   ATOMIC COLLISIONS
     ∞COHERENCE
      COHERENT SCATTERING
      ELECTRON SCATTERING
      GLAUBER THEORY
      INELASTIC SCATTERING
      NUCLEAR SCATTERING
      PHOTON-ELECTRON INTERACTION
      POMERANCHUK THEOREM

ELASTIC SHEETS
RT   GIRDER WEBS
     ∞SHEETS
      WEBS (SHEETS)
      WEBS (SUPPORTS)

ELASTIC SHELLS
GS   SHELLS (STRUCTURAL FORMS)
     . **ELASTIC SHELLS**
RT   ANISOTROPIC SHELLS

ELASTIC STABILITY
USE   DAMPING

ELASTIC STRENGTH
USE   PROPORTIONAL LIMIT

∞ **ELASTIC SYSTEMS**
SN   *(USE OF A MORE SPECIFIC TERM IS*
     RECOMMENDED--CONSULT THE TERMS
     LISTED BELOW)
RT   EARTHQUAKE RESISTANT STRUCTURES
      ELASTIC BODIES
     ∞SYSTEMS

ELASTIC WAVES
UF   EXPANSION WAVES
      LOADING WAVES
      PRESSURE WAVES
      RAREFACTION WAVES
GS   **ELASTIC WAVES**
     . CAPILLARY WAVES
     . . GRAVITY WAVES
     . . . BAROCLINIC WAVES
     . . RIPPLES
     . COHERENT ACOUSTIC RADIATION
     . COMPRESSION WAVES
     . DILATATIONAL WAVES
     . IONIC WAVES
     . MAGNETOELASTIC WAVES
     . MAGNETOACOUSTIC WAVES
     . MAGNETOHYDRODYNAMIC WAVES
     . PLASMA WAVES
     . . . ELECTROSTATIC WAVES
     . P WAVES
     . PHONONS
     . . PHONON BEAMS
     . POLARIZED ELASTIC WAVES
     . S WAVES
     . SEISMIC WAVES
     . . LOVE WAVES
     . . MICROSEISMS
     . . RAYLEIGH WAVES
     . SHOCK WAVES
     . . DETONATION WAVES
     . . MACH CONES
     . . NORMAL SHOCK WAVES
     . . OBLIQUE SHOCK WAVES
     . . RIEMANN WAVES
     . . SONIC BOOMS
     . SOUND WAVES

ELASTIC WAVES-(CONT.)
```
         . . ELECTROACOUSTIC WAVES
         . . ION ACOUSTIC WAVES
         . . LAMB WAVES
         . . NOISE (SOUND)
         . . . AERODYNAMIC NOISE
         . . . AIRCRAFT NOISE
         . . . . JET AIRCRAFT NOISE
         . . . . SONIC BOOMS
         . . . ENGINE NOISE
         . . . . ROCKET ENGINE NOISE
         . . . THERMAL NOISE
         . STRESS WAVES
         . TOLLMEIN-SCHLICHTING WAVES
         . ULTRASONIC RADIATION
         . UNLOADING WAVES
    RT   ACOUSTIC PROPAGATION
         ACOUSTIC SIMULATION
         ACOUSTICS
         AEOLIAN TONES
         BACKGROUND NOISE
         BACKWARD WAVES
         CNOIDAL WAVES
         COHERENT RADIATION
         COMBUSTION VIBRATION
         CONTINUOUS RADIATION
         CYLINDRICAL WAVES
         DIFFUSION WAVES
         DOPPLER EFFECT
         ELASTODYNAMICS
         ELASTOHYDRODYNAMICS
         LAME WAVE EQUATIONS
         LONGITUDINAL WAVES
         MAGNETOHYDRODYNAMIC STABILITY
         PLANE WAVES
         POLARIZED RADIATION
         PRESSURE
         PULSED RADIATION
        ∞RADIATION
         RADIATION DISTRIBUTION
         RADIATION PRESSURE
         RAREFACTION
         REFLECTED WAVES
         REFRACTED WAVES
         SINE WAVES
         SOLITARY WAVES
         SOUND TRANSMISSION
         SPHERICAL WAVES
         STRESS PROPAGATION
         SURFACE WAVES
         TRANSVERSE WAVES
         TRAVELING WAVES
         TROPOSPHERIC WAVES
         UNDERWATER ACOUSTICS
         VIBRATION
         WAVE DISPERSION
        ∞WAVES
```

ELASTICITY
```
    USE   ELASTIC PROPERTIES
```

ELASTICIZERS
```
    USE   PLASTICIZERS
```

ELASTIN
```
    GS    PROTEINS
          . ELASTIN
    RT    ALBUMINS
```

ELASTODYNAMICS
```
    GS    ELASTODYNAMICS
          . ELASTIC DAMPING
          . ELASTOHYDRODYNAMICS
    RT   ∞DYNAMICS
          ELASTIC BODIES
          ELASTIC DEFORMATION
          ELASTIC MEDIA
          ELASTIC PROPERTIES
          ELASTIC WAVES
          ELASTOSTATICS
```

ELASTOHYDRODYNAMICS
```
    GS    ELASTODYNAMICS
          . ELASTOHYDRODYNAMICS
          FLUID MECHANICS
          . FLUID DYNAMICS
          . . HYDRODYNAMICS
          . . . ELASTOHYDRODYNAMICS
          . HYDROMECHANICS
          . . HYDRODYNAMICS
          . . . ELASTOHYDRODYNAMICS
    RT    BALL BEARINGS
          ELASTIC WAVES
          FRICTION MEASUREMENT
          LUBRICATION
```

ELASTOHYDRODYNAMICS-(CONT.)
```
          ROTATING CYLINDERS
          SQUEEZE FILMS
          WATER WAVES
```

ELASTOMERS
```
    GS    RUBBER
          . SYNTHETIC RUBBERS
          . . ELASTOMERS
          . . . CHLOROPRENE RESINS
          . . . THIOPLASTICS
          . . . VITON RUBBER (TRADEMARK)
          . . . . VULCANIZED ELASTOMERS
    RT    LATEX
          ORGANIC MATERIALS
          PLASTICS
         ∞POLYMERS
          SILICONE RUBBER
          SOLITHANES
          SPONGES (MATERIALS)
```

ELASTOMETERS
```
    GS    MEASURING INSTRUMENTS
          . ELASTOMETERS
    RT    ELASTIC PROPERTIES
          EXTENSOMETERS
          STRAIN GAGES
```

ELASTOPLASTICITY
```
    GS    MECHANICAL PROPERTIES
          . ELASTIC PROPERTIES
          . . ELASTOPLASTICITY
          . PLASTIC PROPERTIES
          . . ELASTOPLASTICITY
    RT    J INTEGRAL
```

ELASTOSTATICS
```
    RT    ELASTIC BODIES
          ELASTIC DEFORMATION
          ELASTIC MEDIA
          ELASTIC PROPERTIES
          ELASTODYNAMICS
          STATICS
```

ELBER EQUATION
```
    RT    CRACK CLOSURE
          CRACKS
          CYCLIC LOADS
         ∞EQUATIONS
          FRACTOGRAPHY
          FRACTURE MECHANICS
          MICROCRACKS
          STRESS CONCENTRATION
          STRESS CYCLES
```

ELBOW (ANATOMY)
```
    GS    ANATOMY
          . LIMBS (ANATOMY)
          . . ARM (ANATOMY)
          . . . ELBOW (ANATOMY)
          . MUSCULOSKELETAL SYSTEM
          . . JOINTS (ANATOMY)
          . . . ELBOW (ANATOMY)
          APPENDAGES
          . ARM (ANATOMY)
          . . ELBOW (ANATOMY)
    RT    HUMERUS
          ULNA
```

ELDO LAUNCH VEHICLE
```
    GS    LAUNCH VEHICLES
          . ELDO LAUNCH VEHICLE
          ROCKET VEHICLES
          . MULTISTAGE ROCKET VEHICLES
          . . ELDO LAUNCH VEHICLE
    RT    ARIANE LAUNCH VEHICLE
          BLUE STREAK LAUNCH VEHICLE
          EUROPA LAUNCH VEHICLES
          EUROPEAN SPACE AGENCY
          EUROPEAN 1 SPACECRAFT
```

ELECTRA AIRCRAFT
```
    GS    COMMERCIAL AIRCRAFT
          . ELECTRA AIRCRAFT
          JET AIRCRAFT
          . TURBOPROP AIRCRAFT
          . . ELECTRA AIRCRAFT
          LOCKHEED AIRCRAFT
          . ELECTRA AIRCRAFT
          MONOPLANES
          . ELECTRA AIRCRAFT
          PASSENGER AIRCRAFT
          . ELECTRA AIRCRAFT
          TRANSPORT AIRCRAFT
          . ELECTRA AIRCRAFT
```

ELECTRA AIRCRAFT-(CONT.)
```
    RT   ∞AIRCRAFT
```

ELECTRETS
```
    RT    CAPACITORS
          CURIE TEMPERATURE
          DIELECTRIC POLARIZATION
          DIELECTRICS
          ELECTRIC ENERGY STORAGE
          ELECTRIC FIELDS
          ENERGY STORAGE
          MAGNETS
          POLARIZATION (CHARGE SEPARATION)
```

ELECTRIC AIRCRAFT
```
    USE   FLY BY WIRE CONTROL
```

ELECTRIC APPLIANCES
```
    USE   ELECTRIC EQUIPMENT
```

ELECTRIC ARCS
```
    GS    ELECTRIC CURRENT
          . ELECTRIC DISCHARGES
          . . ELECTRIC ARCS
          . . . CARBON ARCS
          . . . MERCURY ARCS
    RT    ARC CHAMBERS
          ARC DISCHARGES
          ARC GENERATORS
         ∞ARCS
          CORONAS
          ELECTRICAL FAULTS
          FLASHOVER
          GAS DISCHARGES
          GLOW DISCHARGES
          IONIZATION
          LIGHT SOURCES
          LIGHTNING
          MAGNETOHYDRODYNAMICS
          PLANOTRONS
          PLASMA GENERATORS
          PLASMAS (PHYSICS)
          SAHA EQUATIONS
          SHORT CIRCUITS
```

ELECTRIC AUTOMOBILES
```
    GS    SURFACE VEHICLES
          . MOTOR VEHICLES
          . . AUTOMOBILES
          . . . ELECTRIC AUTOMOBILES
    RT    TRANSPORTATION
```

ELECTRIC BATTERIES
```
    SN    (INCLUDES BOTH RECHARGEABLE OR
          STORAGE BATTERIES AND
          NON-RECHARGEABLE BATTERIES FOR
          GENERATING CURRENT FROM A
          STORED CHEMICAL ENERGY SOURCE)
    UF    BATTERIES
    GS    ELECTROCHEMICAL CELLS
          . ELECTRIC BATTERIES
          . . NICKEL IRON BATTERIES
          . . PRIMARY BATTERIES
          . . . DRY CELLS
          . . . . MAGNESIUM CELLS
          . . . . NICKEL ZINC BATTERIES
          . . . METAL AIR BATTERIES
          . . . . ZINC-OXYGEN BATTERIES
          . . . SODIUM SULFUR BATTERIES
          . . . THERMAL BATTERIES
          . . REDOX CELLS
          . . STORAGE BATTERIES
          . . . LEAD ACID BATTERIES
          . . . NICKEL CADMIUM BATTERIES
          . . . NICKEL HYDROGEN BATTERIES
          . . . NICKEL ZINC BATTERIES
          . . . SILVER CADMIUM BATTERIES
          . . . SILVER HYDROGEN BATTERIES
          . . . SILVER ZINC BATTERIES
          . . . ZINC-BROMIDE BATTERIES
          . . . ZINC-CHLORINE BATTERIES
          . . WET CELLS
    RT    AUXILIARY POWER SOURCES
          BATTERY CHARGERS
          CHARGE EFFICIENCY
          CHEMICAL AUXILIARY POWER UNITS
          DIRECT POWER GENERATORS
         ∞ELECTRIC CELLS
          ELECTRODES
          ELECTROLYTES
          ELECTROLYTIC CELLS
          ELECTROMOTIVE FORCES
         ∞ENERGY SOURCES
          ENERGY STORAGE
          NONAQUEOUS ELECTROLYTES
```

ELECTRIC BATTERIES-*(CONT.)*
- ∞POWER SUPPLIES
  PULSE CHARGING
  RADIOISOTOPE BATTERIES
  ROADWAY POWERED VEHICLES
  SPACECRAFT POWER SUPPLIES
  VOLTAGE CONVERTERS (DC TO DC)

ELECTRIC BRIDGES
- GS   CIRCUITS
  . **ELECTRIC BRIDGES**
  . . WIRE BRIDGE CIRCUITS
  . . . WHEATSTONE BRIDGES
- RT   ∞BRIDGES
  CAPACITORS
  ELECTRICAL MEASUREMENT
  MEASURING INSTRUMENTS
  SOLID STATE DEVICES

∞ ELECTRIC CELLS
- SN   *(USE OF A MORE SPECIFIC TERM IS*
  RECOMMENDED--CONSULT THE TERMS
  LISTED BELOW)
- RT   AMPLIFIERS
  DIRECT POWER GENERATORS
  ELECTRIC BATTERIES
  ELECTRIC GENERATORS
  ELECTROCHEMICAL CELLS
  ELECTROLYTIC CELLS
  FISSION ELECTRIC CELLS
  FUEL CELLS
  KERR CELLS
  LEAD ACID BATTERIES
  LITHIUM SULFUR BATTERIES
  NONAQUEOUS ELECTROLYTES
  PHOTOELECTRIC CELLS
  SODIUM SULFUR BATTERIES
  SOLAR CELLS
  WET CELLS

ELECTRIC CHARGE
- GS   **ELECTRIC CHARGE**
  . ELECTRIC DIPOLES
  . . ORBITING DIPOLES
  . ELECTROSTATIC CHARGE
  . ION CHARGE
  . SPACE CHARGE
  . TRAVELING CHARGE
- RT   CAPACITANCE
  ∞CHARGING
  ∞DIPOLES
  ELECTRICAL PROPERTIES
  ELECTROMETERS
  POLARITY
  POLARIZATION (CHARGE SEPARATION)
  PULSE CHARGING
  SCATHA SATELLITE

ELECTRIC CHOPPERS
- SN   (DEVICES FOR CONVERTING DC TO AC)
- UF   CHOPPERS (ELECTRIC)
- RT   AMPLIFIERS
  MECHANICAL OSCILLATORS

ELECTRIC CIRCUITS
- USE   CIRCUITS

ELECTRIC COILS
- GS   **ELECTRIC COILS**
  . MAGNETIC COILS
- RT   CHOKES
  ∞COILS
  IGNITION SYSTEMS
  IMPEDANCE
  MAGNETIC CORES
  TRANSFORMERS

ELECTRIC CONDUCTORS
- UF   ELECTRICAL LEADS
- GS   CONDUCTORS
  . **ELECTRIC CONDUCTORS**
- RT   ∞CONDUCTION
  DIELECTRICS
  EDDY CURRENTS
  ELECTRICAL INSULATION
  ELECTRICAL RESISTIVITY
  ELECTROSTATIC SHIELDING
  INSULATORS
  RESISTORS
  SEMICONDUCTORS (MATERIALS)
  SOMMERFELD WAVES
  THERMAL CONDUCTORS
  TRANSMISSION LINES

ELECTRIC CONNECTORS
- UF   CONNECTORS (ELECTRIC)
  JACKS (ELECTRICAL)
- GS   CONNECTORS
  . **ELECTRIC CONNECTORS**
- RT   BEAM LEADS
  CIRCUITS
  DISCONNECT DEVICES
  FLAT CONDUCTORS
  ∞JACKS
  SWITCHES

ELECTRIC CONTACTS
- UF   CONTACTS (ELECTRIC)
- RT   BRUSHES
  BRUSHES (ELECTRICAL CONTACTS)
  COMMUTATORS
  CONTACT POTENTIALS
  CONTACT RESISTANCE
  DROPOUTS
  FLAT CONDUCTORS
  ∞RELAY
  ∞SLIDING CONTACT
  SLIDING FRICTION
  SWITCHES

ELECTRIC CONTROL
- UF   ELECTROHYDRAULIC CONTROL
- RT   AUTOMATIC CONTROL
  ∞CONTROL
  CONTROL EQUIPMENT
  CONTROL SYSTEMS DESIGN
  ELECTRONIC CONTROL
  ENGINE CONTROL
  NUMERICAL CONTROL
  REMOTE CONTROL
  SOLENOID VALVES
  VOLTAGE CONTROLLED OSCILLATORS

ELECTRIC CORONA
- UF   CORONA DISCHARGES
- GS   CORONAS
  . **ELECTRIC CORONA**
  ELECTRIC CURRENT
  . ELECTRIC DISCHARGES
  . . **ELECTRIC CORONA**
- RT   ATMOSPHERIC ELECTRICITY
  ELECTROHYDRODYNAMICS
  GLOW DISCHARGES
  IONIZATION
  SOLAR CORONA
  STATIC ELECTRICITY

ELECTRIC CURRENT
- UF   AMPERAGE
  ELECTROSEISMIC EFFECT
  .   HALL CURRENTS
  PHOTOCURRENTS
- GS   **ELECTRIC CURRENT**
  . ALTERNATING CURRENT
  . BEAM CURRENTS
  . BRILLOUIN FLOW
  . DIRECT CURRENT
  . EDDY CURRENTS
  . ELECTRIC DISCHARGES
  . . ARC DISCHARGES
  . . ELECTRIC ARCS
  . . . CARBON ARCS
  . . . MERCURY ARCS
  . . ELECTRIC CORONA
  . . ELECTRIC SPARKS
  . . ELECTRODELESS DISCHARGES
  . . FLASHOVER
  . . GLOW DISCHARGES
  . . LIGHTNING
  . . . BALL LIGHTNING
  . . MULTIPACTOR DISCHARGES
  . . PENNING DISCHARGE
  . . RADIO FREQUENCY DISCHARGE
  . . SAINT ELMO FIRE
  . . TOWNSEND DISCHARGE
  . . . GAS DISCHARGES
  . . . . TOROIDAL DISCHARGE
  . . . . . RING DISCHARGE
  . HIGH CURRENT
  . IONOSPHERIC CURRENTS
  . . ELECTROJETS
  . . . AURORAL ELECTROJETS
  . . . EQUATORIAL ELECTROJET
  . LINE CURRENT
  . LOW CURRENTS
  . PLASMA CURRENTS
  . RING CURRENTS
  . SHORT CIRCUIT CURRENTS
  . TELLURIC CURRENTS

ELECTRIC CURRENT-*(CONT.)*
  . THRESHOLD CURRENTS
- RT   AMMETERS
  CIRCUITS
  CURRENT CONVERTERS (AC TO DC)
  CURRENT DENSITY
  CURRENT REGULATORS
  CURRENT SHEETS
  ∞CURRENTS
  ELECTRICAL RESISTIVITY
  ELECTRICITY
  HIGH VOLTAGES
  HYDROELECTRICITY
  INVERTED CONVERTERS (DC TO AC)
  KIRCHHOFF LAW OF NETWORKS
  LEVITATION MELTING
  LIENARD POTENTIAL
  LOW CONDUCTIVITY
  MICROMILLIAMMETERS
  OHMS LAW
  POWER CONDITIONING
  SYSTEM GENERATED
    ELECTROMAGNETIC PULSES
  TRANSMISSION LINES
  VOLT-AMPERE CHARACTERISTICS

ELECTRIC DIPOLES
- GS   ELECTRIC CHARGE
  . **ELECTRIC DIPOLES**
  . . ORBITING DIPOLES
- RT   ∞DIPOLES
  MAGNETIC DIPOLES

ELECTRIC DISCHARGES
- GS   ELECTRIC CURRENT
  . **ELECTRIC DISCHARGES**
  . . ARC DISCHARGES
  . . ELECTRIC ARCS
  . . . CARBON ARCS
  . . . MERCURY ARCS
  . . ELECTRIC CORONA
  . . ELECTRIC SPARKS
  . . ELECTRODELESS DISCHARGES
  . . FLASHOVER
  . . GLOW DISCHARGES
  . . LIGHTNING
  . . . BALL LIGHTNING
  . . MULTIPACTOR DISCHARGES
  . . PENNING DISCHARGE
  . . RADIO FREQUENCY DISCHARGE
  . . SAINT ELMO FIRE
  . . TOWNSEND DISCHARGE
  . . . GAS DISCHARGES
  . . . . TOROIDAL DISCHARGE
  . . . . . RING DISCHARGE
- RT   AVALANCHES
  CORONAS
  ∞DISCHARGE
  DUOPLASMATRONS
  ELECTRON EMISSION
  ELECTROSTATIC CHARGE
  ∞FLASH
  IONIZATION
  LIGHTNING SUPPRESSION
  MOLNIYA SATELLITES
  PLASMA CURRENTS
  SPACE CHARGE
  ZENER EFFECT

ELECTRIC ENERGY STORAGE
- GS   ENERGY STORAGE
  . **ELECTRIC ENERGY STORAGE**
- RT   CAPACITORS
  DIRECT POWER GENERATORS
  ELECTRETS
  INDUCTORS
  POTENTIAL ENERGY

∞ ELECTRIC EQUIPMENT
- SN   *(USE OF A MORE SPECIFIC TERM IS*
  RECOMMENDED--CONSULT THE TERMS
  LISTED BELOW)
- UF   ELECTRIC APPLIANCES
- RT   AIRBORNE EQUIPMENT
  CIRCUITS
  CURRENT REGULATORS
  ELECTRIC GENERATORS
  ELECTRIC POWER TRANSMISSION
  ELECTRICITY
  ELECTROMECHANICAL DEVICES
  ELECTRONIC EQUIPMENT
  HEATING EQUIPMENT
  HOMOPOLAR GENERATORS
  LIGHTING EQUIPMENT
  LOGISTICS

ELECTRIC EQUIPMENT-*(CONT.)*
    MINIATURE ELECTRONIC EQUIPMENT
    MOTORS
    SOLENOID VALVES
    UTILITIES
    VOLTAGE CONVERTERS (AC TO AC)
    VOLTAGE CONVERTERS (DC TO DC)
    WELDING MACHINES

ELECTRIC EQUIPMENT TESTS
SN   (CHECKOUT OF ELECTRICAL
        EQUIPMENT)
RT   ELECTRICAL MEASUREMENT
       ELECTRONIC EQUIPMENT TESTS
       GROUND TESTS
       ∞TESTS

ELECTRIC FIELD STRENGTH
GS   FIELD STRENGTH
      . **ELECTRIC FIELD STRENGTH**
RT   COULOMB POTENTIAL
       ∞FORCE
       H WAVES
       ∞STRENGTH

ELECTRIC FIELDS
UF   ELECTROSTATIC FIELDS
GS   **ELECTRIC FIELDS**
      . EXTERNAL SURFACE CURRENTS
RT   BARIUM ION CLOUDS
       CAVITONS
       CONSTITUTIVE EQUATIONS
       COULOMB POTENTIAL
       CROSSED FIELDS
       DIELECTRIC POLARIZATION
       ELECTRETS
       ELECTRODYNAMICS
       ELECTROMAGNETISM
       ELECTROMECHANICS
       ELECTROSTATIC CHARGE
       ELECTROSTATICS
       EXTERNAL SURFACE CURRENTS
       FIELD EMISSION
       FIELD STRENGTH
       ∞FIELDS
       LIENARD POTENTIAL
       MAGNETIC FIELDS
       PERMITTIVITY
       POLARITY
       SPACECRAFT CHARGING
       SPARK GAPS
       STARK EFFECT
       STATIC ELECTRICITY

ELECTRIC FILTERS
GS   ELECTROMAGNETIC WAVE FILTERS
      . **ELECTRIC FILTERS**
      . . BANDSTOP FILTERS
      . . CRYSTAL FILTERS
      . . DIGITAL FILTERS
      . . . FIR FILTERS
      . . MICROWAVE FILTERS
      . . RADAR FILTERS
      . . RADIO FILTERS
      . . TRACKING FILTERS
      . . WAVEGUIDE FILTERS
RT   ADAPTIVE FILTERS
       BANDPASS FILTERS
       CAPACITORS
       CIRCUITS
       ELECTRONIC FILTERS
       ∞FILTERS
       HIGH PASS FILTERS
       INFRARED FILTERS
       KALMAN FILTERS
       LINEAR FILTERS
       LOW PASS FILTERS
       NONLINEAR FILTERS
       OPTICAL FILTERS
       RC CIRCUITS
       RECEIVERS
       REDUCED ORDER FILTERS
       RESISTORS
       TRANSFORMERS
       ULTRAVIOLET FILTERS
       WIENER FILTERING

ELECTRIC FURNACES
GS   HEATING EQUIPMENT
      . FURNACES
      . . **ELECTRIC FURNACES**
RT   ∞MATERIALS
       SPACE PROCESSING

ELECTRIC FUSES
RT   CIRCUIT PROTECTION
       DISCONNECT DEVICES
       ∞FUSES

ELECTRIC GENERATORS
UF   ELECTRIC POWER CONVERSION
       ELECTROGENERATORS
       POWER GENERATORS
GS   **ELECTRIC GENERATORS**
      . AC GENERATORS
      . . STATIC ALTERNATORS
      . DIRECT POWER GENERATORS
      . . ELECTROSTATIC GENERATORS
      . . FUEL CELLS
      . . . BIOCHEMICAL FUEL CELLS
      . . . HYDROGEN OXYGEN FUEL CELLS
      . . . PHOSPHORIC ACID FUEL CELLS
      . . . REGENERATIVE FUEL CELLS
      . . MAGNETOHYDRODYNAMIC
         GENERATORS
      . . PHOTOELECTRIC GENERATORS
      . . PRIMARY BATTERIES
      . . . ALKALINE BATTERIES
      . . . DRY CELLS
      . . . . MAGNESIUM CELLS
      . . . NICKEL ZINC BATTERIES
      . . . METAL AIR BATTERIES
      . . . . ZINC-OXYGEN BATTERIES
      . . THERMAL BATTERIES
      . . RADIOISOTOPE BATTERIES
      . . . SNAP 7
      . . . SNAP 9A
      . . . SNAP 11
      . . . SNAP 13
      . . . SNAP 15
      . . . SNAP 17
      . . . SNAP 19
      . . . SNAP 21
      . . . SNAP 23
      . . . SNAP 27
      . . . SNAP 29
      . . SOLAR CELLS
      . . . HOMOJUNCTIONS
      . . . VERTICAL JUNCTION SOLAR CELLS
      . . THERMIONIC CONVERTERS
      . . . SNAP 13
      . . . SOLAR BLANKETS
      . . THERMOELECTRIC GENERATORS
      . . . SNAP 3
      . . . SNAP 7
      . . . SNAP 9A
      . . . SNAP 10A
      . . . SNAP 11
      . . . SNAP 15
      . . . SNAP 17
      . . . SNAP 19
      . . . SNAP 21
      . . . SNAP 23
      . . . SNAP 27
      . . . SNAP 29
      . . . SOLAR SEA POWER PLANTS
      . ROTATING GENERATORS
      . . AMPLIDYNES
      . . DYNAMOMETERS
      . . HOMOPOLAR GENERATORS
      . . STATIC ALTERNATORS
      . . TURBOGENERATORS
      . . . ASTEC SOLAR TURBOELECTRIC
         GENERATOR
      . SOLAR GENERATORS
      . . SOLAR AUXILIARY POWER UNITS
      . . . ASTEC SOLAR TURBOELECTRIC
         GENERATOR
      . . . SUNFLOWER POWER SYSTEM
      . . SOLAR CELLS
      . . . HOMOJUNCTIONS
      . . . VERTICAL JUNCTION SOLAR CELLS
RT   AIRCRAFT POWER SUPPLIES
       ARC GENERATORS
       ARMATURES
       AUXILIARY POWER SOURCES
       BRUSHES
       BRUSHES (ELECTRICAL CONTACTS)
       CLOSED CYCLES
       COGENERATION
       COMBINED CYCLE POWER GENERATION
       ∞CONVERSION
       ∞CONVERTERS
       ∞ELECTRIC CELLS
       ∞ELECTRIC EQUIPMENT
       ELECTRICAL ENGINEERING
       ELECTROMOTIVE FORCES
       ∞ENERGY SOURCES
       ∞GENERATORS

ELECTRIC GENERATORS-*(CONT.)*
    NUCLEAR ELECTRIC POWER
      GENERATION
    ∞POWER
    POWER CONDITIONING
    ∞POWER SUPPLIES
    SNAP
    SOLAR PONDS (HEAT STORAGE)
    STATIC INVERTERS
    THERMONUCLEAR POWER GENERATION
    TIDE POWERED GENERATORS
    WINDMILLS (WINDPOWERED MACHINES)
    WINDPOWERED GENERATORS

ELECTRIC HYBRID VEHICLES
GS   SURFACE VEHICLES
      . **ELECTRIC HYBRID VEHICLES**
RT   AUTOMOBILES
       INTERNAL COMBUSTION ENGINES
       ∞ROTATING ELECTRICAL MACHINES
       ∞VEHICLES

ELECTRIC IGNITION
GS   IGNITION
      . **ELECTRIC IGNITION**
RT   IGNITERS
       IGNITION SYSTEMS
       SPARK IGNITION
       SQUIBS
       STARTING

ELECTRIC IMPULSES
USE   ELECTRIC PULSES

ELECTRIC MOMENTS
GS   MOMENTS
      . DIPOLE MOMENTS
      . . **ELECTRIC MOMENTS**
RT   MAGNETIC MOMENTS
       POLARIZATION (CHARGE SEPARATION)

ELECTRIC MOTOR VEHICLES
GS   SURFACE VEHICLES
      . MOTOR VEHICLES
      . . **ELECTRIC MOTOR VEHICLES**
RT   AUTOMATED TRANSIT VEHICLES
       AUTOMOBILES
       CRAWLER TRACTORS
       RESEARCH VEHICLES
       ROADWAY POWERED VEHICLES
       TEST VEHICLES
       TRACTORS
       TRUCKS
       ∞VEHICLES

ELECTRIC MOTORS
GS   MOTORS
      . **ELECTRIC MOTORS**
      . . ASYNCHRONOUS MOTORS
      . . INDUCTION MOTORS
      . . MICROMOTORS
      . . STEPPING MOTORS
      . . SYNCHRONOUS MOTORS
      . . TORQUE MOTORS
RT   AMPLIDYNES
       ARMATURES
       BRUSHES
       BRUSHES (ELECTRICAL CONTACTS)
       CIRCUITS
       COMMUTATORS
       DECOMMUTATORS
       POWER FACTOR CONTROLLERS
       ∞ROTATING ELECTRICAL MACHINES
       SERVOMECHANISMS
       SERVOMOTORS
       STATORS
       TRANSFORMERS

ELECTRIC NETWORKS
RT   IMPEDANCE MATCHING
       ∞NETWORKS
       SNEAK CIRCUIT ANALYSIS
       VOLTAGE CONTROLLED OSCILLATORS

ELECTRIC OUTLETS
RT   ∞POWER TRANSMISSION

ELECTRIC POTENTIAL
UF   VOLTAGE
GS   POTENTIAL ENERGY
      . **ELECTRIC POTENTIAL**
      . . BIOELECTRIC POTENTIAL
      . . CONTACT POTENTIALS
      . . COULOMB POTENTIAL

ELECTRIC POTENTIAL-*(CONT.)*
　　. . LIENARD POTENTIAL
　　. . LOW VOLTAGE
　　. . OPEN CIRCUIT VOLTAGE
　　. . PHOTOVOLTAGES
　　. . QUANTUM WELLS
　　. . SPIKE POTENTIALS
RT　　BARRITT DIODES
　　　BIAS
　　　CAPACITANCE-VOLTAGE
　　　　CHARACTERISTICS
　　　ELECTROMOTIVE FORCES
　　　GIBBS-HELMHOLTZ EQUATIONS
　　　HIGH VOLTAGES
　　　IONIZATION POTENTIALS
　　　KIRCHHOFF LAW OF NETWORKS
　　　OVERVOLTAGE
　　∞POTENTIAL
　　　POTENTIOMETERS (INSTRUMENTS)
　　　POWER CONDITIONING
　　　STATIC ELECTRICITY
　　　VOLT-AMPERE CHARACTERISTICS

∞ **ELECTRIC POWER**
SN　　*(USE OF A MORE SPECIFIC TERM IS*
　　　RECOMMENDED--CONSULT THE TERMS
　　　LISTED BELOW)
UF　　ELECTRICAL ENERGY
RT　　AUXILIARY POWER SOURCES
　　　ELECTRIC BATTERIES
　　　ELECTRIC CURRENT
　　　ELECTRIC POWER PLANTS
　　　ELECTRIC PROPULSION
　　　ELECTRICAL PROPERTIES
　　　ELECTRICITY
　　　ELECTRIFICATION
　　　GEOTHERMAL ENERGY UTILIZATION
　　　HYDROELECTRICITY
　　　INDUCTION MOTORS
　　　POYNTING THEOREM
　　　TOKAMAK DEVICES
　　　TURBOGENERATORS
　　　UTILITIES
　　　VOLTAGE CONVERTERS (AC TO AC)
　　　VOLTAGE CONVERTERS (DC TO DC)

ELECTRIC POWER CONVERSION
USE　　ELECTRIC GENERATORS

ELECTRIC POWER PLANTS
GS　　**ELECTRIC POWER PLANTS**
　　. FUEL CELL POWER PLANTS
　　. NUCLEAR POWER PLANTS
　　. . ENRICO FERMI ATOMIC POWER
　　　　PLANT
　　. . HALLAM NUCLEAR POWER FACILITY
　　. . ML-1 NUCLEAR POWER PLANT
　　. SOLAR THERMAL ELECTRIC POWER
　　　　PLANTS
RT　　COGENERATION
　　　COMBINED CYCLE POWER GENERATION
　　∞ELECTRIC POWER
　　　ELECTRICAL ENGINEERING
　　∞FACILITIES
　　　FLUE GASES
　　　FLY ASH
　　　HYDROELECTRIC POWER STATIONS
　　　INTEGRATED ENERGY SYSTEMS
　　　MODULAR INTEGRATED UTILITY SYSTEM
　　∞POWER PLANTS
　　　SOLAR SEA POWER PLANTS

ELECTRIC POWER SUPPLIES
GS　　**ELECTRIC POWER SUPPLIES**
　　. AIRCRAFT POWER SUPPLIES
　　. SPACECRAFT POWER SUPPLIES
RT　　AUXILIARY POWER SOURCES
　　　COMPULSATORS
　　　INDUCTION MOTORS
　　　LINE CURRENT
　　　PAYLOAD DELIVERY (STS)
　　∞POWER SUPPLIES

ELECTRIC POWER TRANSMISSION
GS　　TRANSMISSION
　　. **ELECTRIC POWER TRANSMISSION**
RT　　CIRCUIT PROTECTION
　　　CIRCUITS
　　∞CONDUCTION
　　∞ELECTRIC EQUIPMENT
　　　ELECTRICAL ENGINEERING
　　　ELECTRIFICATION
　　　HYDROELECTRIC POWER STATIONS
　　∞NETWORKS
　　　POLES (SUPPORTS)

ELECTRIC POWER TRANSMISSION-*(CONT.)*
　　　POWER LINES
　　∞POWER TRANSMISSION
　　　POWER TRANSMISSION (LASERS)
　　　SUPERCONDUCTING POWER
　　　　TRANSMISSION
　　　TRANSMISSION CIRCUITS
　　　TRANSMISSION LINES
　　　TRANSMISSION LOSS
　　　UNDERGROUND TRANSMISSION LINES

ELECTRIC PROPULSION
SN　　(EXCLUDES PROPULSION USING
　　　ELECTRIC MOTORS AS PRIME MOVERS)
GS　　PROPULSION
　　. **ELECTRIC PROPULSION**
　　. . ELECTROMAGNETIC PROPULSION
　　. . ELECTROSTATIC PROPULSION
　　. . . ION PROPULSION
　　. . LASER PROPULSION
　　. . PLASMA PROPULSION
　　. . SOLAR ELECTRIC PROPULSION
RT　　ARC JET ENGINES
　　∞ELECTRIC POWER
　　　LOW THRUST PROPULSION
　　　MARINE PROPULSION
　　　NUCLEAR ELECTRIC PROPULSION
　　　PLASMA POWER SOURCES
　　　RIFT (REACTOR IN FLIGHT TEST)
　　　SERT 1 SPACECRAFT
　　　SERT 2 SPACECRAFT
　　　SPACECRAFT PROPULSION
　　　TWO STAGE PLASMA ENGINES
　　　UNDERWATER PROPULSION

ELECTRIC PULSES
UF　　ELECTRIC IMPULSES
GS　　PULSES
　　. **ELECTRIC PULSES**
RT　　ELECTROMAGNETIC PULSES
　　　PULSE AMPLITUDE
　　　PULSE DURATION
　　　PULSE GENERATORS
　　　PULSE MODULATION
　　　PULSE RATE
　　∞SIGNALS
　　　SYSTEM GENERATED
　　　　ELECTROMAGNETIC PULSES

ELECTRIC REACTORS
GS　　**ELECTRIC REACTORS**
　　. SATURABLE REACTORS
RT　　CAPACITORS
　　　CIRCUIT PROTECTION
　　∞REACTORS
　　　RESISTORS
　　　TRANSFORMERS

ELECTRIC RELAYS
SN　　(EXCLUDES COMMUNICATION SYSTEM
　　　REPEATERS)
GS　　SWITCHES
　　. **ELECTRIC RELAYS**
RT　　ARMATURES
　　　CIRCUIT BREAKERS
　　　DISCONNECT DEVICES
　　　INTERRUPTION
　　∞RELAY
　　　SELECTORS
　　　SOLENOID VALVES
　　　SOLENOIDS
　　　SWITCHING CIRCUITS
　　　TIME LAG

ELECTRIC ROCKET ENGINES
GS　　ENGINES
　　. ROCKET ENGINES
　　. . **ELECTRIC ROCKET ENGINES**
　　. . . ELECTROSTATIC ENGINES
　　. . . ELECTROTHERMAL ENGINES
　　. . . . ARC JET ENGINES
　　. . . . RESISTOJET ENGINES
　　. . . ION ENGINES
　　. . . . CESIUM ENGINES
　　. . . . MERCURY ION ENGINES
　　. . . . RIT ENGINES
RT　　MICROROCKET ENGINES
　　　PULSED JET ENGINES
　　　RESTARTABLE ROCKET ENGINES
　　　SERT 1 SPACECRAFT
　　　SERT 2 SPACECRAFT
　　　SPACE ELECTRIC ROCKET TESTS
　　　SUSTAINER ROCKET ENGINES
　　　VERNIER ENGINES

ELECTRIC SPARKS
UF　　SPARK DISCHARGES
GS　　ELECTRIC CURRENT
　　. ELECTRIC DISCHARGES
　　. . **ELECTRIC SPARKS**
　　　SPARKS
　　. **ELECTRIC SPARKS**
RT　　FLASHOVER
　　　GAS DISCHARGES
　　　IONIZATION
　　　LIGHTNING
　　　SPARK CHAMBERS
　　　SPARK GAPS
　　　SPARK IGNITION
　　　SPARK PLUGS
　　　STATIC ELECTRICITY

ELECTRIC STIMULI
RT　　PHYSIOLOGY
　　∞STIMULI

ELECTRIC SWITCHES
GS　　SWITCHES
　　. **ELECTRIC SWITCHES**
　　. . CRYOTRONS
　　. . STEPPING SWITCHES
　　. . THERMOSTATS
　　. . VACUUM ARC SWITCHES
RT　　CONTACTORS
　　　CRYOSTATS
　　　CURRENT REGULATORS
　　　DROPOUTS
　　　ELECTRONIC CONTROL
　　　PRESSURE SWITCHES
　　　SOLENOID VALVES
　　　SWITCHING CIRCUITS
　　　VOLTAGE REGULATORS

ELECTRIC TERMINALS
RT　　∞TERMINALS

ELECTRIC WELDING
GS　　WELDING
　　. **ELECTRIC WELDING**
　　. . ARC WELDING
　　. . . GAS TUNGSTEN ARC WELDING
　　. . . PLASMA ARC WELDING
　　. . ELECTRON BEAM WELDING
　　. . ELECTROSLAG WELDING
RT　　FLASH WELDING
　　　FUSION WELDING
　　　PRESSURE WELDING
　　　SPOT WELDS
　　　WELDING MACHINES

ELECTRIC WIRE
UF　　ELECTRIC WIRING
GS　　CONDUCTORS
　　. **ELECTRIC WIRE**
　　　WIRE
　　. **ELECTRIC WIRE**
RT　　BUS CONDUCTORS
　　　CIRCUITS
　　　COMMUNICATION CABLES
　　　ELECTRICAL INSULATION
　　　EXPLODING WIRES
　　　FLAT CONDUCTORS
　　　POWER LINES
　　　TRANSMISSION LINES
　　　WIRE BRIDGE CIRCUITS

ELECTRIC WIRING
USE　　ELECTRIC WIRE
　　　WIRING

ELECTRICAL BREAKDOWN
USE　　ELECTRICAL FAULTS

ELECTRICAL CONDUCTIVITY
USE　　ELECTRICAL RESISTIVITY

ELECTRICAL CONDUCTIVITY METERS
GS　　MEASURING INSTRUMENTS
　　. CONDUCTIVITY METERS
　　. . **ELECTRICAL CONDUCTIVITY METERS**
RT　　OHMMETERS

ELECTRICAL ENERGY
USE　　ELECTRIC POWER

ELECTRICAL ENGINEERING
RT　　ELECTRIC GENERATORS
　　　ELECTRIC POWER PLANTS
　　　ELECTRIC POWER TRANSMISSION

ELECTRICAL ENGINEERING-(CONT.)
- ∞ ELECTRONICS
- ∞ ENGINEERING
- ∞ POWER TRANSMISSION
- SYSTEMS ENGINEERING
- TRANSMISSION LINES
- TURBOGENERATORS

ELECTRICAL FAULTS
- UF   ELECTRICAL BREAKDOWN
-       VOLTAGE BREAKDOWN
- GS   **ELECTRICAL FAULTS**
-       . SHORT CIRCUITS
- RT   ∞ BREAKDOWN
-       CIRCUIT PROTECTION
-       ELECTRIC ARCS
-       FAILURE
-       ∞ FAULTS
-       FLASHOVER
-       SNEAK CIRCUIT ANALYSIS
-       SPARK GAPS

ELECTRICAL GROUNDING
- RT   CIRCUIT PROTECTION
-       CIRCUITS
-       NOISE REDUCTION
-       TRANSFORMERS

ELECTRICAL IMPEDANCE
- UF   ADMITTANCE
-       IMMITTANCE
- GS   IMPEDANCE
-       . **ELECTRICAL IMPEDANCE**
-       . . ELECTRICAL RESISTANCE
-       . . . CONTACT RESISTANCE
-       . . . SKIN RESISTANCE
-       . . REACTANCE
- RT   CAPACITANCE
-       IMPEDANCE MATCHING
-       IMPEDANCE MEASUREMENT
-       INDUCTANCE
-       LATCH-UP
-       OHMMETERS
-       SMITH CHART

ELECTRICAL INSULATION
- UF   NONCONDUCTORS
- GS   INSULATION
-       . **ELECTRICAL INSULATION**
- RT   ASBESTOS
-       CIRCUIT PROTECTION
-       DIELECTRICS
-       ELECTRIC CONDUCTORS
-       ELECTRIC WIRE
-       EXCITONS
-       ∞ INSULATED STRUCTURES
-       INSULATORS
-       WIRING

ELECTRICAL LEADS
- USE   ELECTRIC CONDUCTORS

ELECTRICAL MEASUREMENT
- SN   (MEASUREMENT OF ELECTRICAL
-       PROPERTIES, QUANTITIES, OR
-       CONDITIONS)
- UF   VOLTAGE MEASUREMENT
- GS   **ELECTRICAL MEASUREMENT**
-       . COULOMETRY
-       . POLAROGRAPHY
- RT   AMMETERS
-       BOLOMETERS
-       COULOMETERS
-       ELECTRIC BRIDGES
-       ELECTRIC EQUIPMENT TESTS
-       ELECTROMAGNETIC MEASUREMENT
-       ELECTROMETERS
-       ELECTRONIC EQUIPMENT TESTS
-       FLOWMETERS
-       IMPEDANCE MEASUREMENT
-       MAGNETOMETERS
-       ∞ MEASUREMENT
-       MEASURING INSTRUMENTS
-       MICROMILLIAMMETERS
-       MISMATCH (ELECTRICAL)
-       OHMMETERS
-       OSCILLOGRAPHS
-       POTENTIOMETERS (INSTRUMENTS)
-       WATTMETERS

ELECTRICAL PROPERTIES
- UF   BARDEEN APPROXIMATION
- GS   ELECTROMAGNETIC PROPERTIES
-       . **ELECTRICAL PROPERTIES**
-       . . ANTIFERROELECTRICITY

ELECTRICAL PROPERTIES-(CONT.)
- . . CAPACITANCE
- . . CARRIER MOBILITY
- . . . ELECTRON MOBILITY
- . . . HOLE MOBILITY
- . . DIELECTRIC PROPERTIES
- . . . PERMITTIVITY
- . . FERROELECTRICITY
- . . INDUCTANCE
- . . . PROXIMITY EFFECT (ELECTRICITY)
- . . PHOTOCONDUCTIVITY
- . . PHOTOVOLTAIC EFFECT
- RT   CAPACITANCE-VOLTAGE
-       CHARACTERISTICS
-       ∞ CONDUCTIVITY
-       CRYSTAL OSCILLATORS
-       DIAMAGNETISM
-       DIPOLE MOMENTS
-       DOMAINS
-       EDDY CURRENTS
-       ELECTRIC CHARGE
-       ∞ ELECTRIC POWER
-       ELECTRICITY
-       FIELD STRENGTH
-       HYSTERESIS
-       IMPEDANCE
-       MAGNETIC PROPERTIES
-       OPEN CIRCUIT VOLTAGE
-       OPTICAL PROPERTIES
-       PHOTOELECTRIC EMISSION
-       ∞ PHYSICAL PROPERTIES
-       ∞ PROPERTIES
-       QUALITY CONTROL
-       ∞ RESISTANCE
-       ∞ SOLID STATE PHYSICS
-       STANDING WAVE RATIOS

ELECTRICAL RESISTANCE
- GS   IMPEDANCE
-       . ELECTRICAL IMPEDANCE
-       . . **ELECTRICAL RESISTANCE**
-       . . . CONTACT RESISTANCE
-       . . . SKIN RESISTANCE
- RT   GALVANIC SKIN RESPONSE
-       ∞ HIGH RESISTANCE
-       LINEAR CIRCUITS
-       ∞ LOW RESISTANCE
-       MANGANIN (TRADEMARK)
-       OHMMETERS
-       OHMS LAW
-       RC CIRCUITS
-       REACTANCE
-       ∞ RESISTANCE
-       RL CIRCUITS
-       RLC CIRCUITS

ELECTRICAL RESISTIVITY
- UF   ELECTRICAL CONDUCTIVITY
-       ELECTROCONDUCTIVITY
-       RESISTIVITY
- GS   TRANSPORT PROPERTIES
-       . **ELECTRICAL RESISTIVITY**
-       . . IONOSPHERIC CONDUCTIVITY
-       . . MAGNETORESISTIVITY
-       . . PHOTOCONDUCTIVITY
-       . . PLASMA CONDUCTIVITY
-       . . SUPERCONDUCTIVITY
-       . . . KONDO EFFECT
- RT   AIR CONDUCTIVITY
-       ATMOSPHERIC CONDUCTIVITY
-       CARRIER MOBILITY
-       ∞ CONDUCTIVITY
-       ELECTRIC CONDUCTORS
-       ELECTRIC CURRENT
-       ELECTRICAL RESISTIVITY
-       ELECTRICAL RESISTIVITY
-       ELECTROMIGRATION
-       LOW CONDUCTIVITY
-       OPEN CIRCUIT VOLTAGE
-       PLASMA CURRENTS
-       ∞ RESISTANCE

ELECTRICALLY SUSPENDED GYROSCOPES
- USE   ELECTROSTATIC GYROSCOPES

ELECTRICITY
- GS   **ELECTRICITY**
-       . ALTERNATING CURRENT
-       . ATMOSPHERIC ELECTRICITY
-       . . IONOSPHERIC CURRENTS
-       . . . ELECTROJETS
-       . . . . AURORAL ELECTROJETS
-       . . . . EQUATORIAL ELECTROJET
-       . GEOELECTRICITY
-       . . TELLURIC CURRENTS

ELECTRICITY-(CONT.)
- . HYDROELECTRICITY
- . STATIC ELECTRICITY
- RT   ELECTRIC CURRENT
-       ∞ ELECTRIC EQUIPMENT
-       ∞ ELECTRIC POWER
-       ELECTRICAL PROPERTIES
-       ELECTROMAGNETISM
-       ∞ ELECTRONICS
-       LIGHTNING
-       MAXWELL EQUATION
-       OHMS LAW
-       PHOTOELECTRICITY
-       PIEZOELECTRICITY
-       PROXIMITY EFFECT (ELECTRICITY)

ELECTRIFICATION
- RT   ∞ ELECTRIC POWER
-       ELECTRIC POWER TRANSMISSION
-       ∞ POWER TRANSMISSION
-       TRANSMISSION LINES

ELECTRO-OPTICAL EFFECT
- RT   ∞ EFFECTS
-       ∞ KERR EFFECTS
-       LIGHT MODULATION
-       NONLINEAR OPTICS
-       ∞ OPTICS

ELECTRO-OPTICAL PHOTOGRAPHY
- UF   ELECTRONIC PHOTOGRAPHY
- GS   IMAGERY
-       . **ELECTRO-OPTICAL PHOTOGRAPHY**
-       PHOTOGRAPHY
-       . **ELECTRO-OPTICAL PHOTOGRAPHY**
- RT   ASTRONOMICAL PHOTOGRAPHY
-       BLACK AND WHITE PHOTOGRAPHY
-       IMAGE RESOLUTION
-       LALLEMAND CAMERAS
-       OPTICAL MEASUREMENT
-       ∞ OPTICS
-       STREAK PHOTOGRAPHY

ELECTRO-OPTICS
- GS   ELECTROPHYSICS
-       . **ELECTRO-OPTICS**
- RT   ACOUSTO-OPTICS
-       BIREFRINGENCE
-       CHARGE INJECTION DEVICES
-       ELECTROCHROMISM
-       ELECTROLUMINESCENCE
-       ELECTRON OPTICS
-       INTEGRATED OPTICS
-       KERR ELECTROOPTICAL EFFECT
-       LASER MICROSCOPY
-       LIGHT VALVES
-       MAGNETO-OPTICS
-       OPTICAL COMPUTERS
-       OPTICAL RELAY SYSTEMS
-       ∞ OPTICS
-       PHOTONICS
-       POSITION SENSING
-       PUSHBROOM SENSOR MODES
-       STARK EFFECT

ELECTROACOUSTIC TRANSDUCERS
- GS   TRANSDUCERS
-       . SOUND TRANSDUCERS
-       . . **ELECTROACOUSTIC TRANSDUCERS**
-       . . . HYDROPHONES
-       . . . LOUDSPEAKERS
-       . . . MICROPHONES
- RT   INTERDIGITAL TRANSDUCERS
-       SURFACE ACOUSTIC WAVE DEVICES

ELECTROACOUSTIC WAVES
- GS   ELASTIC WAVES
-       . SOUND WAVES
-       . . **ELECTROACOUSTIC WAVES**
- RT   ELECTROMAGNETIC RADIATION
-       PLASMA WAVES
-       PRESSURE SENSORS
-       WAVE INTERACTION
-       ∞ WAVES

ELECTROANESTHESIA
- GS   ANESTHESIA
-       . **ELECTROANESTHESIA**
- RT   ELECTRONARCOSIS

ELECTROCARDIOGRAMS
- USE   ELECTROCARDIOGRAPHY

†

ELECTROCARDIOGRAPHY
UF   ELECTROCARDIOGRAMS
GS   BIOENGINEERING
. BIOMETRICS
. . CARDIOGRAPHY
. . . **ELECTROCARDIOGRAPHY**
RT   BALLISTOCARDIOGRAPHY
BODY MEASUREMENT (BIOLOGY)
ELECTROPHYSIOLOGY
HEART DISEASES
MEDICAL ELECTRONICS
MUSCLES
PHONOCARDIOGRAPHY
VECTORCARDIOGRAPHY

ELECTROCATALYSTS
UF   FUEL CELL CATALYSTS
GS   CATALYSTS
. **ELECTROCATALYSTS**
RT   FUEL CELLS

ELECTROCHEMICAL CELLS
GS   **ELECTROCHEMICAL CELLS**
. ALKALINE BATTERIES
. ELECTRIC BATTERIES
. . NICKEL IRON BATTERIES
. . PRIMARY BATTERIES
. . . DRY CELLS
. . . . MAGNESIUM CELLS
. . . . NICKEL ZINC BATTERIES
. . . METAL AIR BATTERIES
. . . . ZINC-OXYGEN BATTERIES
. . . SODIUM SULFUR BATTERIES
. . . THERMAL BATTERIES
. . REDOX CELLS
. . STORAGE BATTERIES
. . . LEAD ACID BATTERIES
. . . NICKEL CADMIUM BATTERIES
. . . NICKEL HYDROGEN BATTERIES
. . . NICKEL ZINC BATTERIES
. . . SILVER CADMIUM BATTERIES
. . . SILVER HYDROGEN BATTERIES
. . . SILVER ZINC BATTERIES
. . . ZINC-BROMIDE BATTERIES
. . . ZINC-CHLORINE BATTERIES
. . WET CELLS
. FUEL CELLS
. . BIOCHEMICAL FUEL CELLS
. . HYDROGEN OXYGEN FUEL CELLS
. . PHOSPHORIC ACID FUEL CELLS
. . REGENERATIVE FUEL CELLS
. . LITHIUM SULFUR BATTERIES
RT   ∞CELLS
∞ELECTRIC CELLS
ELECTROCHEMISTRY
PHOTOELECTRIC CELLS
PHOTOELECTROCHEMICAL DEVICES
PHOTOVOLTAIC CELLS

ELECTROCHEMICAL CORROSION
GS   CORROSION
. **ELECTROCHEMICAL CORROSION**
RT   ELECTRODISSOLUTION
ELECTROLYSIS
METAL-WATER REACTIONS

ELECTROCHEMICAL MACHINING
UF   ELECTROLYTIC GRINDING
GS   MACHINING
. CHEMICAL MACHINING
. . **ELECTROCHEMICAL MACHINING**
RT   ELECTROPOLISHING

ELECTROCHEMICAL OXIDATION
GS   CHEMICAL REACTIONS
. OXIDATION
. . **ELECTROCHEMICAL OXIDATION**

ELECTROCHEMISTRY
GS   **ELECTROCHEMISTRY**
. ELECTROLYSIS
. . COULOMETRY
. PHOTOELECTROCHEMISTRY
RT   ∞CHEMISTRY
CORROSION
COULOMETERS
ELECTROCHEMICAL CELLS
ELECTROCHROMISM
ELECTRODEPOSITION
ELECTRODES
ELECTRODISSOLUTION
ELECTROLYTES
ELECTROLYTIC CELLS
ELECTROPHYSICS
FUEL CELLS

ELECTROCHEMISTRY-(CONT.)
GLASS ELECTRODES
NONAQUEOUS ELECTROLYTES
OXIDATION-REDUCTION REACTIONS
REDOX CELLS

ELECTROCHROMISM
RT   COLOR
DISPLAY DEVICES
ELECTRO-OPTICS
ELECTROCHEMISTRY
THIN FILMS

ELECTROCONDUCTIVITY
USE   ELECTRICAL RESISTIVITY

ELECTROCUTANEOUS COMMUNICATION
GS   COMMUNICATING
. **ELECTROCUTANEOUS
   COMMUNICATION**
RT   PERCEPTION
SENSORY PERCEPTION
TOUCH

ELECTRODE FILM BARRIERS
RT   ∞BARRIERS
ELECTRODES
∞FILMS
POLARIZATION (CHARGE SEPARATION)
THIN FILMS

ELECTRODE MATERIALS
RT   ANIONS
ANODES
ANODIC COATINGS
CATHODES
CATHODIC COATINGS
CELL ANODES
CELL CATHODES
ELECTRODES
PHOTOCATHODES
PHOTOELECTRIC CELLS
PHOTOELECTRIC MATERIALS
PHOTOELECTROCHEMICAL DEVICES
TUBE ANODES

ELECTRODELESS DISCHARGES
GS   ELECTRIC CURRENT
. ELECTRIC DISCHARGES
. . **ELECTRODELESS DISCHARGES**
RT   ∞DISCHARGE
GAS DISCHARGES
GLOW DISCHARGES
LIGHTNING
PENNING DISCHARGE
RADIO FREQUENCY DISCHARGE
RING DISCHARGE
TOROIDAL DISCHARGE
TOWNSEND DISCHARGE

ELECTRODEPOSITION
GS   DEPOSITION
. **ELECTRODEPOSITION**
. . ELECTROPLATING
RT   CATHODIC COATINGS
CELL CATHODES
COULOMETERS
ELECTROCHEMISTRY
ELECTROFORMING
ELECTROLESS DEPOSITION
ELECTROLYSIS
ELECTROLYTES
ELECTROLYTIC CELLS
ELECTROPHORESIS
ELECTROWINNING
METAL MATRIX COMPOSITES
METAL POWDER
PLATING
POWDER METALLURGY
REDUCTION (CHEMISTRY)

ELECTRODERMAL RESPONSE
USE   GALVANIC SKIN RESPONSE

ELECTRODES
GS   **ELECTRODES**
. ANODES
. . CELL ANODES
. . SHELL ANODES
. . TUBE ANODES
. CATHODES
. . CELL CATHODES
. . HOLLOW CATHODES
. . TUBE CATHODES

ELECTRODES-(CONT.)
. . . COLD CATHODES
. . . HOT CATHODES
. . . PHOTOCATHODES
. . . THERMIONIC CATHODES
. . . TUNNEL CATHODES
. DIFFUSION ELECTRODES
. DYNODES
. GLASS ELECTRODES
. IMPLANTED ELECTRODES (BIOLOGY)
. ION SELECTIVE ELECTRODES
. PLASMA ELECTRODES
. SOLID ELECTRODES
. TUBE GRIDS
RT   COLD CATHODE TUBES
ELECTRIC BATTERIES
ELECTROCHEMISTRY
ELECTRODE FILM BARRIERS
ELECTRODE MATERIALS
ELECTROLYSIS
ELECTROPLATING
ELECTROREFINING
ELECTROWINNING
GRAPHITE
PHOTOMULTIPLIER TUBES
PHOTOTUBES
TAFEL LAW

ELECTRODIALYSIS
GS   DIALYSIS
. **ELECTRODIALYSIS**
RT   COLLOIDS
HYDROMETALLURGY
∞SEPARATION

ELECTRODISSOLUTION
RT   DISSOCIATION
ELECTROCHEMICAL CORROSION
ELECTROCHEMISTRY
ELECTROLYSIS

ELECTRODYNAMICS
GS   **ELECTRODYNAMICS**
. ELECTROHYDRODYNAMICS
. ELECTROMECHANICS
. QUANTUM ELECTRODYNAMICS
. . LIGHT-CONE EXPANSION
RT   BORN-INFELD THEORY
∞DYNAMICS
ELECTRIC FIELDS
ELECTROMAGNETIC INTERACTIONS
ELECTROMECHANICAL DEVICES
LINE CURRENT
MAXWELL EQUATION
PONDEROMOTIVE FORCES
TRAVELING CHARGE

ELECTRODYNAMOMETERS
USE   DYNAMOMETERS

ELECTROENCEPHALOGRAM
USE   ELECTROENCEPHALOGRAPHY

ELECTROENCEPHALOGRAPHY
UF   EEG (ELECTROENCEPHALOGRAMS)
ELECTROENCEPHALOGRAM
GS   BIOENGINEERING
. BIOMETRICS
. . **ELECTROENCEPHALOGRAPHY**
RT   AROUSAL
BODY MEASUREMENT (BIOLOGY)
BRAIN
ELECTROPHYSIOLOGY
MEDICAL ELECTRONICS
MEDICAL EQUIPMENT

ELECTROEPITAXY
GS   GROWTH
. CRYSTAL GROWTH
. . EPITAXY
. . . **ELECTROEPITAXY**
RT   CRYSTALS
HYDROTHERMAL CRYSTAL GROWTH
LIQUID PHASES
TRAVELING SOLVENT METHOD

ELECTROEROSION
USE   SPARK MACHINING

ELECTROEXPLOSIVE DEVICES
USE   INITIATORS (EXPLOSIVES)

ELECTROFORMING
GS   FORMING TECHNIQUES

ELECTROFORMING-*(CONT.)*
. **ELECTROFORMING**
RT   DEPOSITION
ELECTRODEPOSITION
ELECTROLESS DEPOSITION
ELECTROPLATING
SPARK MACHINING

ELECTROGENERATORS
USE   ELECTRIC GENERATORS

ELECTROHYDRAULIC CONTROL
USE   ELECTRIC CONTROL
HYDRAULIC CONTROL

ELECTROHYDRAULIC FORMING
GS   FORMING TECHNIQUES
. COLD WORKING
. . **ELECTROHYDRAULIC FORMING**
RT   EXPLOSIVE FORMING
METAL WORKING

ELECTROHYDRODYNAMICS
GS   ELECTRODYNAMICS
. **ELECTROHYDRODYNAMICS**
FLUID MECHANICS
. FLUID DYNAMICS
. . HYDRODYNAMICS
. . . **ELECTROHYDRODYNAMICS**
. HYDROMECHANICS
. . HYDRODYNAMICS
. . . **ELECTROHYDRODYNAMICS**
RT   ELECTRIC CORONA
ELECTROKINETICS
ELECTRON GAS
ELECTRON MOBILITY
ION DISTRIBUTION
IONIC MOBILITY
MAGNETOHYDRODYNAMICS

ELECTROJETS
GS   ELECTRIC CURRENT
. IONOSPHERIC CURRENTS
. . -**ELECTROJETS**
. . . AURORAL ELECTROJETS
. . . EQUATORIAL ELECTROJET
ELECTRICITY
. ATMOSPHERIC ELECTRICITY
. . IONOSPHERIC CURRENTS
. . . **ELECTROJETS**
. . . . AURORAL ELECTROJETS
. . . . EQUATORIAL ELECTROJET
RT   GEOMAGNETISM
IONOSPHERE
IONOSPHERIC CONDUCTIVITY
IONOSPHERIC DRIFT
RING CURRENTS

ELECTROKINETICS
GS   KINETICS
. **ELECTROKINETICS**
RT   ELECTROHYDRODYNAMICS
ELECTROMAGNETIC FIELDS
ELECTROMECHANICS
ELECTROPHYSICS

ELECTROLESS DEPOSITION
GS   DEPOSITION
. **ELECTROLESS DEPOSITION**
RT   COATINGS
ELECTRODEPOSITION
ELECTROFORMING
METAL COATINGS
PLATING
VACUUM DEPOSITION
VAPOR DEPOSITION

ELECTROLUMINESCENCE
UF   ELECTROLUMINESCENT LAMPS
GS   DECAY
. EMISSION
. . LIGHT EMISSION
. . . LUMINESCENCE
. . . . **ELECTROLUMINESCENCE**
RT   ELECTRO-OPTICS
LIGHT EMITTING DIODES
LIGHT SOURCES

ELECTROLUMINESCENT LAMPS
USE   ELECTROLUMINESCENCE
LUMINAIRES

ELECTROLYSIS
GS   ELECTROCHEMISTRY

ELECTROLYSIS-*(CONT.)*
. **ELECTROLYSIS**
. . COULOMETRY
RT   CORROSION
COULOMETERS
CRACKING (CHEMICAL ENGINEERING)
CURRENT DENSITY
DECOMPOSITION
ELECTROCHEMICAL CORROSION
ELECTRODEPOSITION
ELECTRODES
ELECTRODISSOLUTION
ELECTROLYTES
ELECTROLYTIC CELLS
ELECTROPLATING
HYDROGEN PRODUCTION
IONIC MOBILITY
METATHESIS
PASSIVITY
PHOTOLYSIS
REDUCTION (CHEMISTRY)
TAFEL LAW

ELECTROLYTE METABOLISM
GS   METABOLISM
. **ELECTROLYTE METABOLISM**

ELECTROLYTES
GS   CONDUCTORS
. **ELECTROLYTES**
. . ANOLYTES
. . CATHOLYTES
. . ION EXCHANGE MEMBRANE
ELECTROLYTES
. . JUMPERS
. . MOLTEN SALT ELECTROLYTES
. . NONAQUEOUS ELECTROLYTES
. . SOLID ELECTROLYTES
RT   CONDUCTING FLUIDS
DEBYE-HUCKEL THEORY
ELECTRIC BATTERIES
ELECTROCHEMISTRY
ELECTRODEPOSITION
ELECTROLYSIS
ELECTROLYTIC CELLS
ELECTROPLATING
ELECTROREFINING
ELECTROWINNING
FUEL CELLS
IONS
NONELECTROLYTES
PRIMARY BATTERIES
REDOX CELLS
STORAGE BATTERIES
WET CELLS

ELECTROLYTIC CELLS
UF   GALVANIC CELLS
RT   ∞CELLS
∞DIAPHRAGMS
DIAPHRAGMS (MECHANICS)
DIFFUSION ELECTRODES
ELECTRIC BATTERIES
∞ELECTRIC CELLS
ELECTROCHEMISTRY
ELECTRODEPOSITION
ELECTROLYSIS
ELECTROLYTES
ELECTROPLATING
ELECTROREFINING
ELECTROWINNING
IONIC MOBILITY
LEAD ACID BATTERIES
NONAQUEOUS ELECTROLYTES
PHOSPHORIC ACID FUEL CELLS

ELECTROLYTIC GRINDING
USE   ELECTROCHEMICAL MACHINING

ELECTROLYTIC POLARIZATION
GS   POLARIZATION (CHARGE SEPARATION)
. **ELECTROLYTIC POLARIZATION**
RT   DEPOLARIZATION
MAGNESIUM CELLS

ELECTROLYTIC POLISHING
USE   ELECTROPOLISHING

ELECTROMAGNETIC ABSORPTION
UF   IONOSPHERIC ABSORPTION
LIGHT ABSORPTION
MAGNETIC ABSORPTION
OPTICAL ABSORPTION
GS   ELECTROMAGNETIC PROPERTIES
. **ELECTROMAGNETIC ABSORPTION**

ELECTROMAGNETIC ABSORPTION-*(CONT.)*
. . INFRARED ABSORPTION
ENERGY ABSORPTION
. **ELECTROMAGNETIC ABSORPTION**
. . AURORAL ABSORPTION
. . GAMMA RAY ABSORPTION
. . INFRARED ABSORPTION
. . MULTIPHOTON ABSORPTION
. . PHOTOABSORPTION
. . POLAR CAP ABSORPTION
. . ULTRAVIOLET ABSORPTION
. . X RAY ABSORPTION
RADIATION ABSORPTION
. **ELECTROMAGNETIC ABSORPTION**
. . AURORAL ABSORPTION
. . GAMMA RAY ABSORPTION
. . INFRARED ABSORPTION
. . MULTIPHOTON ABSORPTION
. . PHOTOABSORPTION
. . POLAR CAP ABSORPTION
. . ULTRAVIOLET ABSORPTION
. . X RAY ABSORPTION
RT   ABSORBERS (MATERIALS)
ABSORPTANCE
∞ABSORPTION
ABSORPTION SPECTRA
ABSORPTIVITY
ACTIVATION
ATMOSPHERIC ATTENUATION
ATTENUATION
BEER LAW
BOUGUER LAW
CHANDRASEKHAR EQUATION
EXCITATION
FADING
FLUORESCENCE
GAMMA RAY ABSORPTIOMETRY
IRRADIATION
LIGHT SCATTERING
MANDELSTAM REPRESENTATION
MOLECULAR ABSORPTION
MOSSBAUER EFFECT
NUCLEAR PHYSICS
OPACITY
OPTICAL PROPERTIES
OPTICAL REFLECTION
PAIR PRODUCTION
PHOTODECOMPOSITION
PHOTODISSOCIATION
PHOTON ABSORPTIOMETRY
PHOTOPRODUCTION
RADAR ABSORBERS
RADAR ATTENUATION
RADIATION SHIELDING
RADIO ATTENUATION
REFLECTION
RESONANT FREQUENCIES
SIGNAL FADING
SOLAR ENERGY ABSORBERS
TOWNSEND AVALANCHE
TRANSMISSION
TRANSMITTANCE
TRANSPARENCE
WAVE ATTENUATION
WAVE PROPAGATION

ELECTROMAGNETIC ACCELERATION
RT   ∞ACCELERATION
ELECTROMAGNETIC INTERACTIONS
MAGNETIC FIELDS
PARTICLE ACCELERATION
PLASMA ACCELERATORS

ELECTROMAGNETIC COMPATIBILITY
GS   COMPATIBILITY
. **ELECTROMAGNETIC COMPATIBILITY**
RT   ATMOSPHERICS
CROSSTALK
ELECTRONIC COUNTERMEASURES
ELECTRONIC WARFARE
∞INTERFERENCE
NOISE SPECTRA
RADIO FREQUENCY INTERFERENCE

ELECTROMAGNETIC CONTROL
USE   ELECTROMAGNETS
REMOTE CONTROL

ELECTROMAGNETIC DEDUCTION
USE   MAGNETIC INDUCTION

ELECTROMAGNETIC ENVIRONMENT EXPERIMENT
GS   PAYLOADS
. SPACE SHUTTLE PAYLOADS

ELECTROMAGNETIC ENVIRONMENT-*(CONT.)*
　. . ELECTROMAGNETIC ENVIRONMENT
　　　EXPERIMENT

ELECTROMAGNETIC FIELDS
GS　**ELECTROMAGNETIC FIELDS**
　. FAR FIELDS
　. NEAR FIELDS
　. SYSTEM GENERATED
　　　ELECTROMAGNETIC PULSES
RT　ABRIKOSOV THEORY
　　BIOMAGNETISM
　　BLACKOUT (PROPAGATION)
　　ELECTROKINETICS
　　ELECTROMAGNETISM
　　ELECTROMECHANICS
　　EXTERNAL SURFACE CURRENTS
　　FIELD STRENGTH
　　FIELD THEORY (PHYSICS)
　　MAGNETIC FIELD CONFIGURATIONS
　　MAGNETIC FIELD INVERSIONS
　　MAGNETIC FIELDS
　　QUANTUM ELECTRODYNAMICS
　　RECIPROCITY THEOREM
　　SOLAR MAGNETIC FIELD
　　SOMMERFELD APPROXIMATION
　　STELLAR MAGNETIC FIELDS
　　UNIFIED FIELD THEORY
　　WHISTLERS
　　YANG-MILLS FIELDS

ELECTROMAGNETIC HAMMERS
GS　HAMMERS
　. **ELECTROMAGNETIC HAMMERS**
RT　FORMING TECHNIQUES
　　MAGNETIC COILS
　　MAGNETIC FORMING
　　METAL WORKING

ELECTROMAGNETIC INTERACTIONS
GS　**ELECTROMAGNETIC INTERACTIONS**
　. PLASMA-ELECTROMAGNETIC
　　　INTERACTION
　. . LASER PLASMA INTERACTIONS
RT　BIOMAGNETISM
　　ELECTRODYNAMICS
　　ELECTROMAGNETIC ACCELERATION
　　ELECTROSTATICS
　　ELEMENTARY PARTICLE INTERACTIONS
　　FEYNMAN DIAGRAMS
　　∞ INTERACTIONS
　　MESON-MESON INTERACTIONS
　　PLASMA INTERACTIONS
　　PLASMA RESONANCE
　　PLASMA-PARTICLE INTERACTIONS
　　QUANTUM MECHANICS
　　UNIFIED FIELD THEORY
　　WAVE INTERACTION

ELECTROMAGNETIC INTERFERENCE
GS　**ELECTROMAGNETIC INTERFERENCE**
　. CROSSTALK
　. . IONOSPHERIC CROSS MODULATION
　. JAMMING
　. RADIO FREQUENCY INTERFERENCE
　. . BLACKOUT (PROPAGATION)
　. . . ELECTROMAGNETIC NOISE
　. . . . ATMOSPHERICS
　. . . . . IONOSPHERICS
　. . . . . . DAWN CHORUS
　. . . . . . HISS
　. . . . . SUDDEN ENHANCEMENT OF
　　　　　ATMOSPHERICS
　. . . . . WHISTLERS
　. . . . COSMIC NOISE
　. . . . IONOSPHERIC NOISE
　. . . . . WHISTLERS
　. . . . SHOT NOISE
　. . . . WHITE NOISE
　. . . . . THERMAL NOISE
　. . . IONOSPHERIC CROSS MODULATION
　. . . POLAR RADIO BLACKOUT
　. . CHIRP
　. . . CHIRP SIGNALS
RT　∞ DISTURBANCES
　　ELECTRONIC COUNTERMEASURES
　　ELECTRONIC WARFARE
　　ENVIRONMENTS
　　FEEDBACK
　　GROUND EFFECT (COMMUNICATIONS)
　　∞ INTERFERENCE
　　INTERFERENCE IMMUNITY
　　NOISE REDUCTION
　　SCATHA SATELLITE
　　SIGNAL TO NOISE RATIOS

ELECTROMAGNETIC INTERFERENCE-*(CONT.)*
　　SYSTEM GENERATED
　　ELECTROMAGNETIC PULSES

ELECTROMAGNETIC MEASUREMENT
SN　(MEASUREMENT OF ELECTROMAGNETIC
　　　PROPERTIES, QUANTITIES OR
　　　CONDITIONS)
GS　**ELECTROMAGNETIC MEASUREMENT**
　. ELECTROMAGNETIC NOISE
　　　MEASUREMENT
RT　BARKHAUSEN EFFECT
　　EARTH TERMINAL MEASUREMENT
　　　SYSTEM
　　ELECTRICAL MEASUREMENT
　　ELECTROMAGNETISM
　　INFRARED DETECTORS
　　MAGNETIC MEASUREMENT
　　MAGNETIC TRANSDUCERS
　　∞ MEASUREMENT

ELECTROMAGNETIC NOISE
UF　RADIATION NOISE
　　RADIO FREQUENCY NOISE
GS　ELECTROMAGNETIC INTERFERENCE
　. RADIO FREQUENCY INTERFERENCE
　. . BLACKOUT (PROPAGATION)
　. . . **ELECTROMAGNETIC NOISE**
　. . . . ATMOSPHERICS
　. . . . . IONOSPHERICS
　. . . . . . DAWN CHORUS
　. . . . . . HISS
　. . . . SUDDEN ENHANCEMENT OF
　　　　　ATMOSPHERICS
　. . . . WHISTLERS
　. . . COSMIC NOISE
　. . . IONOSPHERIC NOISE
　. . . . WHISTLERS
　. . . SHOT NOISE
　. . . WHITE NOISE
　. . . . THERMAL NOISE
RT　BACKGROUND NOISE
　　BACKGROUND RADIATION
　　CHANNEL NOISE
　　CHIRP SIGNALS
　　EXTRATERRESTRIAL RADIATION
　　GROUND EFFECT (COMMUNICATIONS)
　　MICROWAVES
　　MILLIMETER WAVES
　　∞ NOISE
　　NOISE GENERATORS
　　NOISE INTENSITY
　　NOISE REDUCTION
　　NOISE SPECTRA
　　NOISE STORMS
　　NOISE TEMPERATURE
　　RADAR RECEIVERS
　　∞ RADIATION
　　RADIO RECEIVERS
　　RADIO SPECTRA
　　RADIO WAVES
　　RANDOM NOISE
　　SIGNAL TO NOISE RATIOS
　　SOLAR RADIO EMISSION
　　SQUELCH CIRCUITS
　　SUBMILLIMETER WAVES

ELECTROMAGNETIC NOISE MEASUREMENT
GS　ELECTROMAGNETIC MEASUREMENT
　. **ELECTROMAGNETIC NOISE**
　　　MEASUREMENT
RT　COSMIC NOISE
　　∞ MEASUREMENT
　　THERMAL NOISE
　　WHITE NOISE

ELECTROMAGNETIC PROPAGATION
USE　ELECTROMAGNETIC WAVE
　　　TRANSMISSION

ELECTROMAGNETIC PROPERTIES
GS　**ELECTROMAGNETIC PROPERTIES**
　. ABSORPTANCE
　. ELECTRICAL PROPERTIES
　. . ANTIFERROELECTRICITY
　. . CAPACITANCE
　. . CARRIER MOBILITY
　. . . ELECTRON MOBILITY
　. . . HOLE MOBILITY
　. . DIELECTRIC PROPERTIES
　. . . PERMITTIVITY
　. . FERROELECTRICITY
　. . INDUCTANCE
　. . . PROXIMITY EFFECT (ELECTRICITY)
　. . PHOTOCONDUCTIVITY

ELECTROMAGNETIC PROPERTIES-*(CONT.)*
　. . PHOTOVOLTAIC EFFECT
　. ELECTROMAGNETIC ABSORPTION
　. . INFRARED ABSORPTION
　. FARADAY EFFECT
　. KERR MAGNETOOPTICAL EFFECT
　. OPTICAL PROPERTIES
　. . ABSORPTIVITY
　. . BIREFRINGENCE
　. . BRIGHTNESS
　. . . SOLAR GRANULATION
　. . BRIGHTNESS DISTRIBUTION
　. . COLOR
　. . . IRIDESCENCE
　. . . WATER COLOR
　. . DICHROISM
　. . LUMINOSITY
　. . . STELLAR LUMINOSITY
　. . OPACITY
　. . OPTICAL BISTABILITY
　. . OPTICAL REFLECTION
　. . PHOSPHORESCENCE
　. . PHOTOCONDUCTIVITY
　. . PHOTOELECTRIC EFFECT
　. . . PHOTOIONIZATION
　. . PHOTOELECTRIC EMISSION
　. . PHOTOVISCOELASTICITY
　. . PHOTOVOLTAIC EFFECT
　. . RADIANCE
　. . REFLECTANCE
　. . REFRACTIVITY
　. . SKY BRIGHTNESS
　. . STIGMATISM
　. . TRANSLUCENCE
　. . TRANSMISSIVITY
　. . TRANSMITTANCE
　. . TRANSPARENCE
　. . TURBIDITY
　. PHOTOELASTICITY
　. PHOTOVISCOELASTICITY
RT　CARRIER MOBILITY
　　CLARITY
　　GAMMA RAY ABSORPTION
　　MAGNETIC PROPERTIES
　　∞ PHYSICAL PROPERTIES
　　∞ PROPERTIES
　　SPECTRAL REFLECTANCE

ELECTROMAGNETIC PROPULSION
GS　PROPULSION
　. ELECTRIC PROPULSION
　. . **ELECTROMAGNETIC PROPULSION**
　. LOW THRUST PROPULSION
　. . **ELECTROMAGNETIC PROPULSION**
　. SPACECRAFT PROPULSION
　. . **ELECTROMAGNETIC PROPULSION**
　. . . MASS DRIVERS (PAYLOAD
　　　　DELIVERY)
RT　ELECTROSTATIC PROPULSION
　　ION PROPULSION
　　PHOTONIC PROPULSION
　　PLASMA PROPULSION

ELECTROMAGNETIC PULSES
GS　ELECTROMAGNETIC RADIATION
　. **ELECTROMAGNETIC PULSES**
　. . SYSTEM GENERATED
　　　ELECTROMAGNETIC PULSES
　　PULSED RADIATION
　. **ELECTROMAGNETIC PULSES**
　. . SYSTEM GENERATED
　　　ELECTROMAGNETIC PULSES
　　PULSES
　. **ELECTROMAGNETIC PULSES**
　. . SYSTEM GENERATED
　　　ELECTROMAGNETIC PULSES
RT　ELECTRIC PULSES
　　EXTERNAL SURFACE CURRENTS
　　PICOSECOND PULSES
　　PULSE COMMUNICATION
　　PULSE MODULATION
　　PULSE RADAR
　　RADAR TRANSMISSION

ELECTROMAGNETIC PUMPS
SN　(ENCOMPASSES DEVICES FOR
　　　MATERIALS HANDLING ONLY--EXCLUDES
　　　OPTICAL AND PARTICLE ENERGIZING
　　　DEVICES)
GS　PUMPS
　. **ELECTROMAGNETIC PUMPS**
RT　FUEL PUMPS

ELECTROMAGNETIC RADIATION
UF　ELECTROMAGNETIC WAVES
　　WAVE RADIATION
GS　**ELECTROMAGNETIC RADIATION**
　. BLACK BODY RADIATION
　. BREMSSTRAHLUNG
　. CERENKOV RADIATION
　. COHERENT ELECTROMAGNETIC
　　　RADIATION
　. COHERENT LIGHT
　. COMET TAILS
　. CYCLOTRON RADIATION
　. ELECTROMAGNETIC PULSES
　. . SYSTEM GENERATED
　　　ELECTROMAGNETIC PULSES
　. ELECTROMAGNETIC SURFACE WAVES
　. GAMMA RAY BEAMS
　. GAMMA RAYS
　. . GAMMA RAY BURSTS
　. H WAVES
　. INFRARED RADIATION
　. . FAR INFRARED RADIATION
　. . NEAR INFRARED RADIATION
　. KILOMETRIC WAVES
　. LIGHT (VISIBLE RADIATION)
　. . COHERENT LIGHT
　. . GEGENSCHEIN
　. . LIGHT BEAMS
　. . POLARIZED LIGHT
　. . SKY RADIATION
　. . . AIRGLOW
　. . . . GEOCORONAL EMISSIONS
　. . . . NIGHTGLOW
　. . . . TWILIGHT GLOW
　. . . DAYGLOW
　. . SUNLIGHT
　. . ZODIACAL LIGHT
　. MODULATED CONTINUOUS RADIATION
　. MONOCHROMATIC RADIATION
　. NONEQUILIBRIUM RADIATION
　. PHOTON BEAMS
　. LIGHT BEAMS
　. PLANETARY RADIATION
　. PLASMONS
　. POLARIZED ELECTROMAGNETIC
　　　RADIATION
　. . POLARIZED LIGHT
　. . SYNCHROTRON RADIATION
　. RADIO WAVES
　. . DECAMETRIC WAVES
　. . EXTRATERRESTRIAL RADIO WAVES
　. . . GALACTIC RADIO WAVES
　. . . RADIO BURSTS
　. . . . SOLAR RADIO BURSTS
　. . . . . TYPE 2 BURSTS
　. . . . . TYPE 3 BURSTS
　. . . . . TYPE 4 BURSTS
　. . . . . TYPE 5 BURSTS
　. . . SOLAR RADIO EMISSION
　. . . . SOLAR RADIO BURSTS
　. . . . . TYPE 2 BURSTS
　. . . . . TYPE 3 BURSTS
　. . . . . TYPE 4 BURSTS
　. . . . . TYPE 5 BURSTS
　. . LONG WAVE RADIATION
　. . RADIO EMISSION
　. . . CN EMISSION
　. . . HYDROXYL EMISSION
　. . . RADIO BURSTS
　. . . . SOLAR RADIO BURSTS
　. . . . . TYPE 2 BURSTS
　. . . . . TYPE 3 BURSTS
　. . . . . TYPE 4 BURSTS
　. . . . . TYPE 5 BURSTS
　. . . SOLAR RADIO EMISSION
　. . . . SOLAR RADIO BURSTS
　. . . . . TYPE 2 BURSTS
　. . . . . TYPE 3 BURSTS
　. . . . . TYPE 4 BURSTS
　. . . . . TYPE 5 BURSTS
　. . SHORT WAVE RADIATION
　. . . MICROWAVES
　. . . . CENTIMETER WAVES
　. . . . DECIMETER WAVES
　. . . . MICROWAVE EMISSION
　. . . . MILLIMETER WAVES
　. . . SUBMILLIMETER WAVES
　. . SKY WAVES
　. . SOMMERFELD WAVES
　. TERRESTRIAL RADIATION
　. THERMAL RADIATION
　. . PHONON BEAMS
　. TROPOSPHERIC RADIATION
　. ULTRAVIOLET RADIATION
　. . EXTREME ULTRAVIOLET RADIATION
　. . FAR ULTRAVIOLET RADIATION

ELECTROMAGNETIC RADIATION-*(CONT.)*
　. . . LYMAN ALPHA RADIATION
　. . . LYMAN BETA RADIATION
　. . NEAR ULTRAVIOLET RADIATION
　. X RAYS
　. . COSMIC X RAYS
　. . SOLAR X-RAYS
RT　AEROSPACE ENVIRONMENTS
　　ANTENNAS
　　ATMOSPHERIC RADIATION
　　ATMOSPHERIC REFRACTION
　　BACKWARD WAVES
　　BEAMS (RADIATION)
　　CN EMISSION
　　COHERENT RADIATION
　　CONTINUOUS RADIATION
　　CORPUSCULAR RADIATION
　　COSMIC RAYS
　　CYLINDRICAL WAVES
　　DIFFRACTION
　　DOPPLER EFFECT
　　DUOCHROMATORS
　　ELECTROACOUSTIC WAVES
　　ELECTROMAGNETISM
　　EXTRATERRESTRIAL RADIATION
　　FAR FIELDS
　　FLUX (RATE)
　　FLUX DENSITY
　　GALACTIC RADIATION
　　GAMMA RAY ABSORPTION
　　GAMMA RAY TELESCOPES
　　GAUGE INVARIANCE
　　GLARE
　　HARMONIC RADIATION
　　INCIDENT RADIATION
　　INCOHERENT SCATTERING
　　INTERSTELLAR RADIATION
　　IONIZING RADIATION
　　KERR ELECTROOPTICAL EFFECT
　　LIGHT EMISSION
　　MAGNETO-OPTICS
　　NEAR FIELDS
　　NONLINEAR OPTICS
　　NUCLEAR RADIATION
　　PHASE VELOCITY
　　PHOTONS
　　PLANCKS CONSTANT
　　POLARIZED RADIATION
　　POYNTING THEOREM
　∞PROPAGATION
　　PROPAGATION VELOCITY
　　PULSED RADIATION
　　RADAR
　∞RADIATION
　　RADIATION CHEMISTRY
　　RADIATION DISTRIBUTION
　　RADIATION HAZARDS
　　RADIATION LAWS
　　RADIATION PRESSURE
　　RADIATION SOURCES
　　RADIATIVE TRANSFER
　∞RAYS
　　REFLECTED WAVES
　　REFLECTION
　　REFRACTED WAVES
　　RONCHI TEST
　　SCATTERING
　　SINE WAVES
　　SOLAR RADIATION
　　SOLITARY WAVES
　　SPECTRAL EMISSION
　　SPECTRAL ENERGY DISTRIBUTION
　　SPHERICAL WAVES
　　SPONTANEOUS EMISSION
　　STEFAN-BOLTZMANN LAW
　　STELLAR RADIATION
　　STRATOSPHERE RADIATION
　　TELECOMMUNICATION
　　THOMSON SCATTERING
　　TRANSMISSION
　　TRANSVERSE WAVES
　　TRAVELING WAVES
　　ULTRAVIOLET ASTRONOMY
　　VLF EMISSION RECORDERS
　　WAVE AMPLIFICATION
　　WAVE DISPERSION
　　WAVE GENERATION
　∞WAVES
　　WHITE HOLES (ASTRONOMY)

ELECTROMAGNETIC SCATTERING
GS　SCATTERING
　. WAVE SCATTERING
　. . **ELECTROMAGNETIC SCATTERING**
　. . . IONOSPHERIC F-SCATTER
　　　　PROPAGATION

ELECTROMAGNETIC SCATTERING-*(CONT.)*
　. . . LIGHT SCATTERING
　. . . . HALOS
　. . . MICROWAVE SCATTERING
　. . . MIE SCATTERING
　. . . . RAYLEIGH SCATTERING
　. . . RAMAN SPECTRA
　. . . THOMSON SCATTERING
　. . . X RAY SCATTERING
RT　ATMOSPHERIC ATTENUATION
　　ATMOSPHERIC SCATTERING
　　MAGNETIC DISPERSION
　　RECIPROCITY THEOREM

ELECTROMAGNETIC SHIELDING
GS　SHIELDING
　. **ELECTROMAGNETIC SHIELDING**
　. . RADIO FREQUENCY SHIELDING
RT　MAGNETIC SHIELDING
　　RADIATION SHIELDING

ELECTROMAGNETIC SPECTRA
GS　SPECTRA
　. RADIATION SPECTRA
　. . **ELECTROMAGNETIC SPECTRA**
　. . . GAMMA RAY SPECTRA
　. . . INFRARED SPECTRA
　. . . LINE SPECTRA
　. . . . BALMER SERIES
　. . . . D LINES
　. . . . ELECTRONIC SPECTRA
　. . . . FRAUNHOFER LINES
　. . . . H LINES
　. . . . . H ALPHA LINE
　. . . . . H BETA LINE
　. . . . . H GAMMA LINE
　. . . . K LINES
　. . . LYMAN SPECTRA
　. . . . PASCHEN SERIES
　. . . . RYDBERG SERIES
　. . . . TELLURIC LINES
　. . . RADIO SPECTRA
　. . . . MICROWAVE SPECTRA
　. . . RAMAN SPECTRA
　. . . STELLAR SPECTRA
　. . . . SOLAR SPECTRA
　. . . UBV SPECTRA
　. . . ULTRAVIOLET SPECTRA
　. . . VIBRATIONAL SPECTRA
　. . . VISIBLE SPECTRUM
　. . . X RAY SPECTRA
RT　ABSORPTION SPECTRA
　　ASTRONOMICAL SPECTROSCOPY
　　ELECTRONIC WARFARE
　　EMISSION SPECTRA
　　ENERGY SPECTRA
　　LIGHT (VISIBLE RADIATION)
　　MOLECULAR SPECTRA
　　NOISE SPECTRA
　　SPECTRAL CORRELATION
　　SPECTRAL RECONNAISSANCE

ELECTROMAGNETIC SURFACE WAVES
GS　ELECTROMAGNETIC RADIATION
　. **ELECTROMAGNETIC SURFACE WAVES**
　　SURFACE WAVES
　. **ELECTROMAGNETIC SURFACE WAVES**
RT　DIELECTRICS
　　PROPAGATION MODES
　　RADIO WAVES
　∞SURFACES
　　WAVEGUIDES
　∞WAVES

ELECTROMAGNETIC WAVE FILTERS
GS　**ELECTROMAGNETIC WAVE FILTERS**
　. BANDPASS FILTERS
　. . CRYSTAL FILTERS
　. . TRACKING FILTERS
　. ELECTRIC FILTERS
　. . BANDSTOP FILTERS
　. . CRYSTAL FILTERS
　. . DIGITAL FILTERS
　. . . FIR FILTERS
　. . MICROWAVE FILTERS
　. . RADAR FILTERS
　. . RADIO FILTERS
　. . TRACKING FILTERS
　. . WAVEGUIDE FILTERS
　. MATCHED FILTERS
　. OPTICAL FILTERS
　. . BIREFRINGENT FILTERS
　. . INFRARED FILTERS
　. . ULTRAVIOLET FILTERS
RT　ABSORBERS (MATERIALS)

ELECTROMAGNETIC WAVE FILTERS-*(CONT.)*
ADAPTIVE FILTERS
ATTENUATORS
CORRELATION DETECTION
∞FILTERS
HIGH PASS FILTERS
LINEAR FILTERS
LOW PASS FILTERS
NONLINEAR FILTERS
SCREEN EFFECT

ELECTROMAGNETIC WAVE TRANSMISSION
UF   ELECTROMAGNETIC PROPAGATION
GS   TRANSMISSION
. **ELECTROMAGNETIC WAVE TRANSMISSION**
. . LIGHT TRANSMISSION
. . . LIGHT SCATTERING
. . . . HALOS
. . RADAR TRANSMISSION
. . RADIO TRANSMISSION
. . . DOUBLE SIDEBAND TRANSMISSION
. . . IONOSPHERIC PROPAGATION
. . . . IONOSPHERIC F-SCATTER PROPAGATION
. . . MICROWAVE ATTENUATION
. . . MICROWAVE TRANSMISSION
. . . MULTIPATH TRANSMISSION
. . . SHORT WAVE RADIO TRANSMISSION
. . . SINGLE SIDEBAND TRANSMISSION
. . . SPREAD SPECTRUM TRANSMISSION
. . . TRANSEQUATORIAL PROPAGATION
. . . TRANSHORIZON RADIO PROPAGATION
. . SCATTER PROPAGATION
. . . IONOSPHERIC F-SCATTER PROPAGATION
. . TELEVISION TRANSMISSION
RT   ATMOSPHERIC ATTENUATION
ATTENUATION
FERMAT PRINCIPLE
INCOHERENT SCATTERING
LOSSY MEDIA
MAGNETOIONICS
PLASMA DECAY
PLASMAGUIDES
RADAR ATTENUATION
RADIO ATTENUATION
RADOME MATERIALS
SCREEN EFFECT
TRANSMISSION EFFICIENCY
WAVE PROPAGATION

ELECTROMAGNETIC WAVES
USE   ELECTROMAGNETIC RADIATION

ELECTROMAGNETICS
USE   ELECTROMAGNETISM

ELECTROMAGNETISM
UF   ELECTROMAGNETICS
GS   **ELECTROMAGNETISM**
. MAGNETOSTATICS
RT   BARKHAUSEN EFFECT
ELECTRIC FIELDS
ELECTRICITY
ELECTROMAGNETIC FIELDS
ELECTROMAGNETIC MEASUREMENT
ELECTROMAGNETIC RADIATION
ELECTROMAGNETS
ELECTROPHYSICS
MAGNET COILS
MAGNETIC COILS
MAGNETIC FIELD INVERSIONS
MAGNETIC FIELDS
MAGNETIC PROPERTIES
MAGNETORESISTIVITY
UNIFIED FIELD THEORY

ELECTROMAGNETS
UF   ELECTROMAGNETIC CONTROL
GS   MAGNETS
. **ELECTROMAGNETS**
. . HIGH FIELD MAGNETS
. . SUPERCONDUCTING MAGNETS
RT   ELECTROMAGNETISM
FIELD COILS
MAGNET COILS
RACETRACKS (PARTICLE ACCELERATORS)
SOLENOIDS

ELECTROMECHANICAL DEVICES
RT   ∞ELECTRIC EQUIPMENT

ELECTROMECHANICAL DEVICES-*(CONT.)*
ELECTRODYNAMICS
ELECTROMECHANICS
∞EQUIPMENT
HOMOPOLAR GENERATORS

ELECTROMECHANICS
GS   ELECTRODYNAMICS
. **ELECTROMECHANICS**
RT   CIRCUITS
ELECTRIC FIELDS
ELECTROKINETICS
ELECTROMAGNETIC FIELDS
ELECTROMECHANICAL DEVICES
ELECTROSTATICS
MAGNETIC FIELD INVERSIONS
MAGNETIC FIELDS
MAXWELL EQUATION
∞MECHANICS (PHYSICS)

ELECTROMETERS
GS   MEASURING INSTRUMENTS
. **ELECTROMETERS**
RT   ELECTRIC CHARGE
ELECTRICAL MEASUREMENT
ELECTRON COUNTERS
GALVANOMETERS
POTENTIOMETERS (INSTRUMENTS)
VOLTMETERS
WATTMETERS

ELECTROMIGRATION
RT   ELECTRICAL RESISTIVITY
ELECTRON MOBILITY
HOLE MOBILITY
IONIC MOBILITY
POLARIZATION (CHARGE SEPARATION)
THERMOMIGRATION

ELECTROMOTIVE FORCES
GS   **ELECTROMOTIVE FORCES**
. PONDEROMOTIVE FORCES
RT   ELECTRIC BATTERIES
ELECTRIC GENERATORS
ELECTRIC POTENTIAL
OHMS LAW
OPEN CIRCUIT VOLTAGE

ELECTROMYOGRAMS
USE   ELECTROMYOGRAPHY

ELECTROMYOGRAPHS
USE   ELECTROMYOGRAPHY

ELECTROMYOGRAPHY
UF   ELECTROMYOGRAMS
ELECTROMYOGRAPHS
GS   BIOENGINEERING
. BIOMETRICS
. . **ELECTROMYOGRAPHY**
RT   ELECTROPHYSIOLOGY
MEDICAL ELECTRONICS
MYOELECTRICITY

ELECTRON ACCELERATION
GS   RATES (PER TIME)
. ACCELERATION (PHYSICS)
. . **ELECTRON ACCELERATION**
RT   ∞ACCELERATION
COSMIC RAYS
ELECTRONS
EXTRATERRESTRIAL RADIATION
PARTICLE BEAMS
SOLAR COSMIC RAYS

ELECTRON ACCELERATORS
GS   PARTICLE ACCELERATORS
. **ELECTRON ACCELERATORS**
. . BETATRONS
RT   ∞ACCELERATORS
LINEAR ACCELERATORS
SYNCHROTRONS
VAN DE GRAAFF ACCELERATORS

ELECTRON ATTACHMENT
GS   **ELECTRON ATTACHMENT**
. NUCLEOPHILES
RT   ∞ATTACHMENT
GAS IONIZATION
IONIZATION

ELECTRON AVALANCHE
GS   AVALANCHES
. **ELECTRON AVALANCHE**

ELECTRON AVALANCHE-*(CONT.)*
RT   CATT DEVICES
CHANNEL MULTIPLIERS
FREE ELECTRONS
GAS DISCHARGES
TOWNSEND AVALANCHE

ELECTRON BEAM WELDING
GS   WELDING
. ELECTRIC WELDING
. . **ELECTRON BEAM WELDING**
RT   ARC WELDING
SPIKING

ELECTRON BEAMS
GS   BEAMS (RADIATION)
. PARTICLE BEAMS
. . **ELECTRON BEAMS**
. . . RELATIVISTIC ELECTRON BEAMS
PARTICLES
. CORPUSCULAR RADIATION
. . ELECTRON RADIATION
. . . **ELECTRON BEAMS**
. . . . RELATIVISTIC ELECTRON BEAMS
RT   BEAM INJECTION
BEAM NEUTRALIZATION
BEAM PLASMA AMPLIFIERS
BETA PARTICLES
BRILLOUIN FLOW
ELECTRON CYCLOTRON HEATING
IONIZING RADIATION
MAGNETIC LENSES
MONOSCOPES
PLASMA JETS
SCALLOPING

ELECTRON BOMBARDMENT
RT   DEPOSITION
PARTICLE BEAMS
PLASMA JETS
RELATIVISTIC ELECTRON BEAMS
SPUTTERING

ELECTRON BUNCHING
GS   BUNCHING
. **ELECTRON BUNCHING**
RT   CATCHERS
CONVECTION CURRENTS
KLYSTRONS
TRAVELING WAVE TUBES
VELOCITY MODULATION

ELECTRON CAPTURE
GS   NUCLEAR REACTIONS
. NUCLEAR INTERACTIONS
. . NUCLEAR CAPTURE
. . . **ELECTRON CAPTURE**
. . SPIN-ORBIT INTERACTIONS
. . . **ELECTRON CAPTURE**
PARTICLE INTERACTIONS
. ELEMENTARY PARTICLE INTERACTIONS
. . NUCLEAR CAPTURE
. . . **ELECTRON CAPTURE**
. NUCLEAR INTERACTIONS
. . NUCLEAR CAPTURE
. . . **ELECTRON CAPTURE**
. . SPIN-ORBIT INTERACTIONS
. . . **ELECTRON CAPTURE**
SPIN
. SPIN-ORBIT INTERACTIONS
. . **ELECTRON CAPTURE**
RT   CAPTURE EFFECT
MANY ELECTRON EFFECTS

ELECTRON CLOUDS
GS   CLOUDS
. **ELECTRON CLOUDS**
RT   ORBITRONS
SPACE CHARGE

ELECTRON COLLISIONS
USE   ELECTRON SCATTERING

ELECTRON COMPOUNDS
USE   INTERMETALLICS

ELECTRON COUNTERS
UF   ELECTRON DETECTORS
GS   MEASURING INSTRUMENTS
. COUNTERS
. . RADIATION COUNTERS
. . . **ELECTRON COUNTERS**
. RADIATION MEASURING INSTRUMENTS

ELECTRON COUNTERS-(CONT.)
```
      . . RADIATION COUNTERS
      . . . ELECTRON COUNTERS
RT    ELECTROMETERS
      IONIZATION CHAMBERS
```

ELECTRON CYCLOTRON HEATING
```
GS    HEATING
      . PLASMA HEATING
      . . ELECTRON CYCLOTRON HEATING
RT    ELECTRON BEAMS
      ELECTRON GUNS
      KLYSTRONS
      MAGNETIC PUMPING
```

ELECTRON DECAY RATE
```
GS    RATES (PER TIME)
      . DECAY RATES
      . . ELECTRON DECAY RATE
RT    MUONS
      SECONDARY COSMIC RAYS
```

ELECTRON DENSITY (CONCENTRATION)
```
GS    DENSITY (NUMBER/VOLUME)
      . PARTICLE DENSITY (CONCENTRATION)
      . . ELECTRON DENSITY
            (CONCENTRATION)
      . . . CARRIER DENSITY (SOLID STATE)
      . . . ELECTRON DENSITY PROFILES
      . . . IONOSPHERIC ELECTRON DENSITY
      . . . MAGNETOSPHERIC ELECTRON
            DENSITY
RT    ATMOSPHERIC COMPOSITION
      ATMOSPHERIC DENSITY
      ATOM CONCENTRATION
      FREE ELECTRONS
      ION DENSITY (CONCENTRATION)
      PLASMA DENSITY
      PLASMA FREQUENCIES
      RADIATION BELTS
      SEMICONDUCTORS (MATERIALS)
      SPACE DENSITY
```

ELECTRON DENSITY PROFILES
```
GS    DENSITY (NUMBER/VOLUME)
      . PARTICLE DENSITY (CONCENTRATION)
      . . ELECTRON DENSITY
            (CONCENTRATION)
      . . . ELECTRON DENSITY PROFILES
      . . ELECTRON DISTRIBUTION
      . . . ELECTRON DENSITY PROFILES
      DISTRIBUTION (PROPERTY)
      . ELECTRON DISTRIBUTION
      . . ELECTRON DENSITY PROFILES
      GRADIENTS
      . ELECTRON DENSITY PROFILES
RT    ANGULAR DISTRIBUTION
      ATMOSPHERIC ELECTRICITY
      ATMOSPHERIC IONIZATION
```

ELECTRON DETECTORS
```
USE   ELECTRON COUNTERS
```

ELECTRON DIFFRACTION
```
GS    DIFFRACTION
      . ELECTRON DIFFRACTION
RT    BRAGG ANGLE
      X RAY DIFFRACTION
```

ELECTRON DIFFUSION
```
GS    DIFFUSION
      . PARTICLE DIFFUSION
      . . ELECTRON DIFFUSION
RT    AMBIPOLAR DIFFUSION
      DIFFUSION WAVES
      GASEOUS SELF-DIFFUSION
      IONIC DIFFUSION
      PLASMA DIFFUSION
      THERMAL DIFFUSION
```

ELECTRON DISTRIBUTION
```
GS    DENSITY (NUMBER/VOLUME)
      . PARTICLE DENSITY (CONCENTRATION)
      . . ELECTRON DISTRIBUTION
      . . . ELECTRON DENSITY PROFILES
      DISTRIBUTION (PROPERTY)
      . ELECTRON DISTRIBUTION
      . . ELECTRON DENSITY PROFILES
RT    CHARGE DISTRIBUTION
      CURRENT DISTRIBUTION
      THOMAS-FERMI MODEL
      VERTICAL DISTRIBUTION
```

ELECTRON EMISSION
```
GS    DECAY
      . EMISSION
      . . PARTICLE EMISSION
      . . . ELECTRON EMISSION
      . . . . FIELD EMISSION
      . . . . PHOTOELECTRIC EMISSION
      . . . . SECONDARY EMISSION
RT    CATHODES
      ELECTRIC DISCHARGES
      EMITTERS
      PAIR PRODUCTION
      PHOTOELECTRIC MATERIALS
      PHOTOELECTRON SPECTROSCOPY
      PHOTOELECTRONS
      PHOTOIONIZATION
      PHOTOVOLTAIC EFFECT
      RADIO FREQUENCY DISCHARGE
      SELF SUSTAINED EMISSION
      STIMULATED EMISSION
      THERMAL EMISSION
      THERMIONIC EMISSION
      THERMIONICS
      WORK FUNCTIONS
```

ELECTRON ENERGY
```
UF    ELECTRON TEMPERATURE
      ELECTRONIC LEVELS
GS    PARTICLE ENERGY
      . ELECTRON ENERGY
      . . ELECTRON STATES
RT    ACTIVATION ENERGY
      ELECTROSTATIC PROBES
      ∞ENERGY
      FORBIDDEN BANDS
      HARTREE-FOCK-SLATER METHOD
      INTERFACIAL ENERGY
      IONOSPHERIC TEMPERATURE
      KINETIC ENERGY
      NOISE TEMPERATURE
      PLASMAS (PHYSICS)
      PROTON ENERGY
      SPACE TEMPERATURE
      SURFACE ENERGY
      TEMPERATURE
```

ELECTRON FLUX
```
USE   ELECTRONS
      FLUX (RATE)
```

ELECTRON FLUX DENSITY
```
SN    (ELECTRON EMISSION OR DETECTION
      RATE PER UNIT AREA)
UF    ELECTRON INTENSITY
GS    RATES (PER TIME)
      . FLUX DENSITY
      . . RADIANT FLUX DENSITY
      . . . PARTICLE FLUX DENSITY
      . . . . ELECTRON FLUX DENSITY
RT    IRRADIANCE
      RADIANCY
      SOLAR FLUX DENSITY
```

ELECTRON GAS
```
RT    COSMIC GASES
      ELECTROHYDRODYNAMICS
      FREE ELECTRONS
      IONIZED GASES
      PLASMONS
      RAREFIED GASES
      SCREEN EFFECT
      SUPERCONDUCTORS
```

ELECTRON GUNS
```
RT    CATHODE RAY TUBES
      CROSSED FIELD GUNS
      ELECTRON CYCLOTRON HEATING
      FLYING SPOT SCANNERS
      ∞GUNS
      MAGNETIC LENSES
      PARTICLE ACCELERATORS
      PLASMA GUNS
      TUBE ANODES
      TUBE CATHODES
      TUBE GRIDS
```

ELECTRON IMPACT
```
GS    IMPACT
      . ELECTRON IMPACT
RT    ION IMPACT
      POINT IMPACT
      PROTON IMPACT
```

ELECTRON INTENSITY
```
USE   ELECTRON FLUX DENSITY
```

ELECTRON INTERACTIONS
```
USE   ELECTRON SCATTERING
```

ELECTRON IONIZATION
```
USE   IONIZATION
```

ELECTRON IRRADIATION
```
GS    IRRADIATION
      . ELECTRON IRRADIATION
RT    AURORAL IRRADIATION
      ION IRRADIATION
      SECONDARY EMISSION
```

ELECTRON MASS
```
GS    MASS
      . PARTICLE MASS
      . . ELECTRON MASS
RT    ELECTRONS
```

ELECTRON MICROSCOPES
```
GS    MICROSCOPES
      . ELECTRON MICROSCOPES
RT    FIELD EMISSION
      ION MICROSCOPES
      MAGNETIC LENSES
      MICROANALYSIS
      OPTICAL MICROSCOPES
      PHOTOMICROGRAPHY
      REPLICAS
```

ELECTRON MICROSCOPY
```
GS    MICROSCOPY
      . ELECTRON MICROSCOPY
RT    FIELD EMISSION
      ION MICROSCOPES
      MAGNETIC LENSES
      MICROANALYSIS
      PHASE CONTRAST
```

ELECTRON MOBILITY
```
GS    ELECTROMAGNETIC PROPERTIES
      . ELECTRICAL PROPERTIES
      . . CARRIER MOBILITY
      . . . ELECTRON MOBILITY
      MOBILITY
      . CARRIER MOBILITY
      . . ELECTRON MOBILITY
      TRANSPORT PROPERTIES
      . CARRIER MOBILITY
      . . ELECTRON MOBILITY
RT    AMBIPOLAR DIFFUSION
      ATOMIC MOBILITIES
      CHARGE CARRIERS
      ELECTROHYDRODYNAMICS
      ELECTROMIGRATION
      HIGH ELECTRON MOBILITY
        TRANSISTORS
      HOLE MOBILITY
      MAJORITY CARRIERS
      MINORITY CARRIERS
      NDM SEMICONDUCTOR DEVICES
      SEMICONDUCTOR PLASMAS
      ∞SOLID STATE PHYSICS
      SQUARE WELLS
```

ELECTRON MULTIPLIERS
```
USE   PHOTOMULTIPLIER TUBES
```

ELECTRON OPTICS
```
RT    BEAM SWITCHING
      BRILLOUIN FLOW
      CATHODE RAY TUBES
      ELECTRO-OPTICS
      FLYING SPOT SCANNERS
      ∞OPTICS
      PARTICLE TRAJECTORIES
      STEERING
```

ELECTRON ORBITALS
```
GS    ORBITALS
      . ELECTRON ORBITALS
RT    EXCIMERS
```

ELECTRON OSCILLATIONS
```
GS    OSCILLATIONS
      . ELECTRON OSCILLATIONS
RT    OSCILLATOR STRENGTHS
      PLASMA OSCILLATIONS
      TRANSIENT OSCILLATIONS
```

ELECTRON PARAMAGNETIC RESONANCE
```
UF    ELECTRON SPIN RESONANCE
GS    RESONANCE
      . MAGNETIC RESONANCE
```

ELECTRON PARAMAGNETIC-*(CONT.)*
    . . PARAMAGNETIC RESONANCE
    . . . **ELECTRON PARAMAGNETIC
        RESONANCE**
RT   JAHN-TELLER EFFECT

ELECTRON PATHS
USE   ELECTRON TRAJECTORIES

ELECTRON PHONON INTERACTIONS
RT   ∞INTERACTIONS
    PARTICLE INTERACTIONS
    PLASMA-PARTICLE INTERACTIONS
    POLARONS
    SUPERCONDUCTIVITY
    THERMODYNAMIC COUPLING

ELECTRON PHOTOGRAPHY
GS   IMAGERY
    . **ELECTRON PHOTOGRAPHY**
    PHOTOGRAPHY
    . **ELECTRON PHOTOGRAPHY**
RT   BLACK AND WHITE PHOTOGRAPHY

ELECTRON PHOTON CASCADES
RT   BREMSSTRAHLUNG
    ∞CASCADES
    COSMIC RAY SHOWERS
    PAIR PRODUCTION
    SECONDARY COSMIC RAYS

ELECTRON PLASMA
GS   GASES
    . IONIZED GASES
    . . CHARGED PARTICLES
    . . . **ELECTRON PLASMA**
    PARTICLES
    . CHARGED PARTICLES
    . . ENERGETIC PARTICLES
    . . . PLASMAS (PHYSICS)
    . . . . **ELECTRON PLASMA**
RT   HELIUM PLASMA
    HIGH TEMPERATURE PLASMAS
    LANDAU DAMPING
    METALLIC PLASMAS
    PLASMA WAVES
    PLASMA-PARTICLE INTERACTIONS
    RAREFIED PLASMAS
    RELATIVISTIC PLASMAS
    THERMAL PLASMAS

ELECTRON PRECIPITATION
GS   PARTICLE PRECIPITATION
    . **ELECTRON PRECIPITATION**
    PARTICLES
    . CORPUSCULAR RADIATION
    . . **ELECTRON PRECIPITATION**
RT   AURORAS
    ∞PRECIPITATION
    PROTON PRECIPITATION
    RADIATION BELTS
    SECONDARY COSMIC RAYS
    TRAPPED PARTICLES

ELECTRON PRESSURE
GS   PRESSURE
    . RADIATION PRESSURE
    . . **ELECTRON PRESSURE**

ELECTRON PROBES
GS   MEASURING INSTRUMENTS
    . **ELECTRON PROBES**
RT   CHEMICAL ANALYSIS
    IRRADIATION
    MICROWAVE PLASMA PROBES
    SPECTROMETERS

ELECTRON PUMPING
RT   ENERGY TRANSFER
    EXCIMER LASERS
    GAS LASERS
    LASERS
    NUCLEAR PUMPING
    NUCLEAR RADIATION
    OPTICAL PUMPING
    POPULATION INVERSION
    ∞PUMPING
    STIMULATED EMISSION
    STIMULATED EMISSION DEVICES

ELECTRON RADIATION
SN   (RADIATION CONSISTING OF
    ELECTRONS-- EXCLUDES
    ELECTROMAGNETIC RADIATION)

ELECTRON RADIATION-*(CONT.)*
GS   PARTICLES
    . CORPUSCULAR RADIATION
    . . **ELECTRON RADIATION**
    . . . BETA PARTICLES
    . . . ELECTRON BEAMS
    . . . . RELATIVISTIC ELECTRON BEAMS
RT   BREMSSTRAHLUNG
    NUCLEAR RADIATION
    PLASMA RADIATION
    PROTON IRRADIATION
    ∞RADIATION
    RADIATION EFFECTS

ELECTRON RECOMBINATION
GS   RECOMBINATION REACTIONS
    . **ELECTRON RECOMBINATION**
    . . RADIATIVE RECOMBINATION
RT   ION RECOMBINATION
    NEUTRAL PARTICLES

ELECTRON RING ACCELERATORS
USE   STORAGE RINGS (PARTICLE
    ACCELERATORS)

ELECTRON RUNAWAY (PLASMA PHYSICS)
GS   SCATTERING
    . ELECTRON SCATTERING
    . . **ELECTRON RUNAWAY (PLASMA
        PHYSICS)**
RT   COLLISIONAL PLASMAS
    HIGH ACCELERATION
    PLASMA PHYSICS
    SCATTERING CROSS SECTIONS

ELECTRON SCATTERING
UF   ELECTRON COLLISIONS
    ELECTRON INTERACTIONS
GS   NUCLEAR REACTIONS
    . **ELECTRON SCATTERING**
    SCATTERING
    . **ELECTRON SCATTERING**
    . . CONFIGURATION INTERACTION
    . . ELECTRON RUNAWAY (PLASMA
        PHYSICS)
RT   ATOMIC COLLISIONS
    DENSE PLASMAS
    ELASTIC SCATTERING
    INELASTIC SCATTERING
    ION SCATTERING
    MANY ELECTRON EFFECTS
    NUCLEAR SCATTERING
    PARTICLE INTERACTIONS
    PHOTON-ELECTRON INTERACTION
    RAMSAUER EFFECT
    RECOIL IONS
    RELATIVISTIC ELECTRON BEAMS
    UMKLAPP PROCESS

ELECTRON SOURCES
RT   ∞ENERGY SOURCES
    ION SOURCES
    ∞POWER SUPPLIES
    RADIATION SOURCES
    ∞SOURCES

ELECTRON SPECTROSCOPY
GS   SPECTROSCOPY
    . **ELECTRON SPECTROSCOPY**
RT   ABSORPTION SPECTRA
    EMISSION SPECTRA
    INFRARED SPECTROSCOPY
    MOLECULAR SPECTROSCOPY
    OPTICAL EMISSION SPECTROSCOPY
    X RAY ABSORPTION

ELECTRON SPIN
GS   SPIN
    . PARTICLE SPIN
    . . **ELECTRON SPIN**
RT   ANGULAR MOMENTUM
    NUCLEAR SPIN
    SPIN DYNAMICS

ELECTRON SPIN RESONANCE
USE   ELECTRON PARAMAGNETIC RESONANCE

ELECTRON STATES
GS   LEVEL (QUANTITY)
    . ENERGY LEVELS
    . . **ELECTRON STATES**
    PARTICLE ENERGY
    . ELECTRON ENERGY
    . . **ELECTRON STATES**

ELECTRON STATES-*(CONT.)*
RT   EXCIMERS
    EXCITATION
    GROUND STATE
    MANY ELECTRON EFFECTS
    NOISE TEMPERATURE

ELECTRON SWEEPING
USE   SWEEP FREQUENCY

ELECTRON TELESCOPES
USE   PARTICLE TELESCOPES

ELECTRON TEMPERATURE
USE   ELECTRON ENERGY

ELECTRON TRAJECTORIES
UF   ELECTRON PATHS
GS   TRAJECTORIES
    . PARTICLE TRAJECTORIES
    . . **ELECTRON TRAJECTORIES**
RT   DIFFRACTION PATHS
    MAGNETIC RIGIDITY
    RADIATION BELTS

ELECTRON TRANSFER
RT   BACKWARD WAVE TUBES
    CHARGE EXCHANGE
    CHARGE TRANSFER
    OXIDATION
    TRANSFERRED ELECTRON DEVICES
    TRANSFERRING

ELECTRON TRANSITIONS
RT   ATOMIC THEORY
    AUGER EFFECT
    AUGER SPECTROSCOPY
    BALMER SERIES
    BAND STRUCTURE OF SOLIDS
    BOHR THEORY
    CONDUCTION BANDS
    EXCIMERS
    EXCITATION
    FORBIDDEN TRANSITIONS
    FRANCK-CONDON PRINCIPLE
    JAHN-TELLER EFFECT
    LASING
    MANY ELECTRON EFFECTS
    NUCLEAR CAPTURE
    OPTICAL TRANSITION
    OSCILLATOR STRENGTHS
    PASCHEN SERIES
    RYDBERG SERIES
    ∞TRANSITION
    TRANSITION PROBABILITIES
    X RAY LASERS
    XENON CHLORIDE LASERS
    XENON FLUORIDE LASERS

ELECTRON TUBES
GS   **ELECTRON TUBES**
    . CAMERA TUBES
    . . IMAGE DISSECTOR TUBES
    . . ORTHICONS
    . . . IMAGE ORTHICONS
    . . VIDICONS
    . . . RETURN BEAM VIDICONS
    . . . . THERMICONS
    . VACUUM TUBES
    . . VACUUM TUBE OSCILLATORS
    . . . CATHODE RAY TUBES
    . . . . PICTURE TUBES
    . . . GAS DISCHARGE TUBES
    . . . . IGNITRONS
    . . . . THYRATRONS
    . . . MICROWAVE TUBES
    . . . . KLYSTRONS
    . . . . MAGNETRONS
    . . . . . NIGOTRONS
    . . . . MICROWAVE OSCILLATORS
    . . . . . PLANOTRONS
    . . . . . CARCINOTRONS
    . . . . . CELESCOPES
    . . . . . HELITRONS
    . . . . . IGNITRONS
    . . . . . THERMICONS
    . . . . . THERMIONIC DIODES
    . . . . . THYRATRONS
    . . . . TRAVELING WAVE TUBES
    . . . . . CARCINOTRONS
    . . . . . HELITRONS
RT   CAVITY RESONATORS
    CIRCUITS
    CROSSED FIELD AMPLIFIERS
    DIODES

ELECTRON TUBES-(CONT.)
```
        FIBER OPTICS
    ∞ HEATERS
        MODULATORS
        ORBITRONS
        OSCILLATORS
        PENTODES
        RECTIFIERS
        RESONATORS
        TETRODES
        TRIODES
        TUBE GRIDS
    ∞ TUBES
        TUNNEL CATHODES
        VELOCITY MODULATION
        X RAY TUBES
```

ELECTRON TUNNELING
```
UF    TUNNEL RESISTORS
RT    ENERGY LEVELS
      MIM DIODES
      SEMICONDUCTORS (MATERIALS)
      SUPERCONDUCTIVITY
      TUNNEL DIODES
    ∞ TUNNELING
```

ELECTRON-HOLE DROPS
```
RT    CARRIER DENSITY (SOLID STATE)
      LUMINESCENCE
      MAGNETIC FIELDS
      OPTICAL PUMPING
      PHASE TRANSFORMATIONS
      PLASMA DENSITY
      PLASMA EQUILIBRIUM
      SEMICONDUCTOR PLASMAS
      SINGLE EVENT UPSETS
```

ELECTRON-ION RECOMBINATION
```
GS    RECOMBINATION REACTIONS
      . ELECTRON-ION RECOMBINATION
      . . RADIATIVE RECOMBINATION
RT    ION RECOMBINATION
      PLASMA CONTROL
```

ELECTRONARCOSIS
```
RT    ELECTROANESTHESIA
      ELECTROPHYSIOLOGY
```

ELECTRONIC AIRCRAFT
```
RT  ∞ AIRCRAFT
      AUTOMATIC CONTROL
      ELECTRONIC COUNTERMEASURES
```

ELECTRONIC AMPLIFIERS
```
USE   AMPLIFIERS
```

ELECTRONIC CONTROL
```
RT    AUTOMATIC CONTROL
      CASCADE CONTROL
    ∞ CONTROL
      CONTROL EQUIPMENT
      CONTROL SYSTEMS DESIGN
      CONTROLLERS
      CURRENT REGULATORS
      ELECTRIC CONTROL
      ELECTRIC SWITCHES
      FEEDBACK CONTROL
      HYDRAULIC CONTROL
      PNEUMATIC CONTROL
      REMOTE CONTROL
      TERMINAL CONFIGURED VEHICLE
        PROGRAM
      VOLTAGE REGULATORS
```

ELECTRONIC COUNTERMEASURES
```
GS    COUNTERMEASURES
      . ELECTRONIC COUNTERMEASURES
      . . ANTIRADAR COATINGS
      . . CHAFF
RT    DECEPTION
      ELECTROMAGNETIC COMPATIBILITY
      ELECTROMAGNETIC INTERFERENCE
      ELECTRONIC AIRCRAFT
      ELECTRONIC WARFARE
      JAMMING
      OPTICAL COUNTERMEASURES
      RADAR DETECTION
      RADIO FREQUENCY INTERFERENCE
```

ELECTRONIC EQUIPMENT
```
GS    ELECTRONIC EQUIPMENT
      . DIODES
      . . CESIUM DIODES
      . . CRYSTAL RECTIFIERS
```

ELECTRONIC EQUIPMENT-(CONT.)
```
      . . PLASMA DIODES
      . . SEMICONDUCTOR DIODES
      . . . AVALANCHE DIODES
      . . . BARRITT DIODES
      . . . GERMANIUM DIODES
      . . . GUNN DIODES
      . . . JUNCTION DIODES
      . . . LIGHT EMITTING DIODES
      . . . PARAMETRIC DIODES
      . . . PHOTODIODES
      . . . SCHOTTKY DIODES
      . . . TUNNEL DIODES
      . . . VARACTOR DIODES
      . ELECTRONIC FILTERS
      . ELECTRONIC MODULES
      . . MICROMODULES
      . ELECTRONIC PACKAGING
      . ELECTRONIC RECORDING SYSTEMS
      . ELECTRONIC TRANSDUCERS
      . MINIATURE ELECTRONIC EQUIPMENT
      . SOLID STATE DEVICES
      . . CRYOTRONS
      . . CRYSTAL RECTIFIERS
      . . METAL-NITRIDE-OXIDE-SEMICONDUCT
        ORS
      . . MULTISPECTRAL LINEAR ARRAYS
      . . SEMICONDUCTOR DEVICES
      . . . AVALANCHE DIODES
      . . . . CRYOSAR
      . . . BARRITT DIODES
      . . . GERMANIUM DIODES
      . . . HETEROJUNCTION DEVICES
      . . . HIGH ELECTRON MOBILITY
            TRANSISTORS
      . . . JUNCTION DIODES
      . . . . MIM DIODES
      . . . . STEP RECOVERY DIODES
      . . . LIGHT EMITTING DIODES
      . . . METAL OXIDE SEMICONDUCTORS
      . . . . CHARGE TRANSFER DEVICES
      . . . . . BUCKET BRIGADE DEVICES
      . . . . . CHARGE COUPLED DEVICES
      . . . . . CHARGE INJECTION DEVICES
      . . . . CMOS
      . . . . SOS (SEMICONDUCTORS)
      . . . MIM (SEMICONDUCTORS)
      . . . MIS (SEMICONDUCTORS)
      . . . NDM SEMICONDUCTOR DEVICES
      . . . NEURISTORS
      . . . PARAMETRIC DIODES
      . . . PHOTODIODES
      . . . PHOTOVOLTAIC CELLS
      . . . SCHOTTKY DIODES
      . . . SEMICONDUCTOR LASERS
      . . . . GALLIUM ARSENIDE LASERS
      . . . THERMISTORS
      . . . THYRISTORS
      . . . . SILICON CONTROLLED
              RECTIFIERS
      . . . TRANSFERRED ELECTRON DEVICES
      . . . TRANSISTOR AMPLIFIERS
      . . . TRANSISTORS
      . . . . BIPOLAR TRANSISTORS
      . . . . FIELD EFFECT TRANSISTORS
      . . . . . CHARGE FLOW DEVICES
      . . . . . JFET
      . . . . HIGH ELECTRON MOBILITY
              TRANSISTORS
      . . . . JUNCTION TRANSISTORS
      . . . . . JFET
      . . . . PHOTOTRANSISTORS
      . . . . SILICON TRANSISTORS
      . . . . . SOS (SEMICONDUCTORS)
      . . . TRAPATT DEVICES
      . . . VARACTOR DIODES
      . . . VARISTORS
      . . SIS (SEMICONDUCTORS)
      . . SOLID STATE LASERS
      . . . GALLIUM ARSENIDE LASERS
      . . . RUBY LASERS
      . . . YAG LASERS
      . SPACECRAFT ELECTRONIC EQUIPMENT
RT    ANTENNA COMPONENTS
      BUBBLE TECHNIQUE
    ∞ ELECTRIC EQUIPMENT
    ∞ ELECTRONICS
    ∞ EQUIPMENT
      RADIATION HARDENING
      SPHERICAL ANTENNAS
      SYSTEM GENERATED
        ELECTROMAGNETIC PULSES
```

ELECTRONIC EQUIPMENT TESTS
```
SN    (CHECKOUT OF ELECTRONIC
      EQUIPMENT)
```

ELECTRONIC EQUIPMENT TESTS-(CONT.)
```
RT    EARTH TERMINAL MEASUREMENT
        SYSTEM
      ELECTRIC EQUIPMENT TESTS
      ELECTRICAL MEASUREMENT
      ENVIRONMENTAL TESTS
      NONDESTRUCTIVE TESTS
      OSCILLOSCOPES
      QUALITY CONTROL
      RESONANCE TESTING
      STABILITY TESTS
    ∞ TEST EQUIPMENT
    ∞ TESTS
      VIBRATION TESTS
```

ELECTRONIC FILTERS
```
GS    ELECTRONIC EQUIPMENT
      . ELECTRONIC FILTERS
RT    ELECTRIC FILTERS
    ∞ FILTERS
      FIR FILTERS
```

ELECTRONIC LEVELS
```
USE   ELECTRON ENERGY
      ENERGY LEVELS
```

ELECTRONIC MAIL
```
GS    TELECOMMUNICATION
      . ELECTRONIC MAIL
RT    COMMUNICATION NETWORKS
      COMMUNICATION SATELLITES
      COMPUTER NETWORKS
      DATA TRANSMISSION
```

ELECTRONIC MODULES
```
GS    ELECTRONIC EQUIPMENT
      . ELECTRONIC MODULES
      . . MICROMODULES
      MODULES
      . ELECTRONIC MODULES
      . . MICROMODULES
RT    HARDWARE
      MINIATURE ELECTRONIC EQUIPMENT
      MODULARITY
      SUBMINIATURIZATION
```

ELECTRONIC PACKAGING
```
GS    ELECTRONIC EQUIPMENT
      . ELECTRONIC PACKAGING
      PACKAGING
      . ELECTRONIC PACKAGING
RT    CIRCUIT BOARDS
      DTL INTEGRATED CIRCUITS
      ENCAPSULATING
      HYBRID CIRCUITS
      INTEGRATED CIRCUITS
      LARGE SCALE INTEGRATION
      LINEAR INTEGRATED CIRCUITS
      MEDIUM SCALE INTEGRATION
      MICROMODULES
      PRINTED CIRCUITS
      THICK FILMS
      TTL INTEGRATED CIRCUITS
```

ELECTRONIC PHOTOGRAPHY
```
USE   ELECTRO-OPTICAL PHOTOGRAPHY
```

ELECTRONIC RECORDING SYSTEMS
```
GS    ELECTRONIC EQUIPMENT
      . ELECTRONIC RECORDING SYSTEMS
RT    RECORDING INSTRUMENTS
    ∞ SYSTEMS
      TAPE RECORDERS
```

ELECTRONIC SIGNAL MEASUREMENT
```
USE   SIGNAL MEASUREMENT
```

ELECTRONIC SPECTRA
```
SN    (EMISSION OR ABSORPTION
      MOLECULAR SPECTRA OF AN
      ELECTRON TRANSITION)
GS    SPECTRA
      . ENERGY SPECTRA
      . . ELECTRONIC SPECTRA
      . MOLECULAR SPECTRA
      . . ELECTRONIC SPECTRA
      . RADIATION SPECTRA
      . . ELECTROMAGNETIC SPECTRA
      . . . LINE SPECTRA
      . . . . ELECTRONIC SPECTRA
RT    ABSORPTION SPECTRA
      EMISSION SPECTRA
      LYMAN SPECTRA
      SPECTRAL BANDS
```

ELECTRONIC SPECTRA-*(CONT.)*
      VIBRATIONAL SPECTRA

ELECTRONIC STRUCTURE
   USE   ATOMIC STRUCTURE

ELECTRONIC SWITCHES
   USE   SWITCHING CIRCUITS

ELECTRONIC TRANSDUCERS
   GS   ELECTRONIC EQUIPMENT
         . **ELECTRONIC TRANSDUCERS**
         TRANSDUCERS
         . **ELECTRONIC TRANSDUCERS**
   RT   MAGNETIC TRANSDUCERS
         ∞ SENSORS
         ULTRASONIC WAVE TRANSDUCERS

ELECTRONIC WARFARE
   GS   MILITARY OPERATIONS
         . **ELECTRONIC WARFARE**
         WARFARE
         . **ELECTRONIC WARFARE**
   RT   AIR DEFENSE
         ANTIRADAR COATINGS
         CHAFF
         COMBAT
         DECEPTION
         ELECTROMAGNETIC COMPATIBILITY
         ELECTROMAGNETIC INTERFERENCE
         ELECTROMAGNETIC SPECTRA
         ELECTRONIC COUNTERMEASURES
         EVASIVE ACTIONS
         JAMMING
         MISSILE DETECTION
         PEACETIME
         RADAR DETECTION
         RADIO FREQUENCY INTERFERENCE
         STRATEGY

∞ **ELECTRONICS**
   SN   *(USE OF A MORE SPECIFIC TERM IS*
         RECOMMENDED--CONSULT THE TERMS
         LISTED BELOW)
   UF   PHOTOELECTRONICS
   RT   ASTRIONICS
         AVIONICS
         ELECTRICAL ENGINEERING
         ELECTRICITY
         ELECTRONIC EQUIPMENT
         ELECTROPHYSICS
         MEDICAL ELECTRONICS
         MICROELECTRONICS
         MOLECULAR ELECTRONICS
         NUCLEONICS
         QUANTUM ELECTRONICS
         RADIO ELECTRONICS
         THERMIONICS
         TRANSISTOR CIRCUITS
         VOLT-AMPERE CHARACTERISTICS

ELECTRONOGRAPHY
   RT   ELECTROPHYSIOLOGY
         PRINTING

ELECTRONS
   UF   ELECTRON FLUX
         NONRELATIVISTIC ELECTRONS
   GS   PARTICLES
         . CHARGED PARTICLES
         . . ENERGETIC PARTICLES
         . . . **ELECTRONS**
         . . . . CONDUCTION ELECTRONS
         . . . . HIGH ENERGY ELECTRONS
         . . . . HOT ELECTRONS
         . . . . N ELECTRONS
         . . . . NEGATRONS
         . . . . PI-ELECTRONS
   RT   ACCEPTOR MATERIALS
         BETA PARTICLES
         BOHR MAGNETON
         COSMIC RAYS
         DONOR MATERIALS
         ELECTRON ACCELERATION
         ELECTRON MASS
         EXCITONS
         HOLES (ELECTRON DEFICIENCIES)
         LEWIS BASE
         MAJORITY CARRIERS
         ∞ MATERIALS
         MINORITY CARRIERS
         MUONIUM
         N-TYPE SEMICONDUCTORS
         NUCLEAR RADIATION
         POMERANCHUK THEOREM

ELECTRONS-*(CONT.)*
         QUANTUM NUMBERS
         RADIATION BELTS
         SEMICONDUCTORS (MATERIALS)
         SUHL EFFECT

ELECTRONYSTAGMOGRAPHY
   GS   EYE MOVEMENTS
         . **ELECTRONYSTAGMOGRAPHY**
         PHYSIOLOGICAL TESTS
         . **ELECTRONYSTAGMOGRAPHY**
   RT   NYSTAGMUS
         OPHTHALMOLOGY

ELECTROPHORESIS
   RT   BIOPROCESSING
         COLLOIDS
         ELECTRODEPOSITION
         ELECTROPLATING
         ∞ MICROGRAVITY APPLICATIONS
         PARTICLE MOTION

ELECTROPHOTOMETERS
   UF   PHOTOELECTRIC PHOTOMETERS
   GS   MEASURING INSTRUMENTS
         . OPTICAL MEASURING INSTRUMENTS
         . . PHOTOMETERS
         . . . **ELECTROPHOTOMETERS**
         . RADIATION MEASURING INSTRUMENTS
         . . PHOTOMETERS
         . . . **ELECTROPHOTOMETERS**
         OPTICAL EQUIPMENT
         . OPTICAL MEASURING INSTRUMENTS
         . . PHOTOMETERS
         . . . **ELECTROPHOTOMETERS**
   RT   ELECTROPHOTOMETRY

ELECTROPHOTOMETRY
   GS   CHEMICAL TESTS
         . CHEMICAL ANALYSIS
         . . **ELECTROPHOTOMETRY**
         OPTICAL MEASUREMENT
         . PHOTOMETRY
         . . **ELECTROPHOTOMETRY**
   RT   COLORIMETRY
         ELECTROPHOTOMETERS
         ELECTROPHYSICS
         ∞ MATERIALS TESTS
         MICROANALYSIS
         PHOTOMETERS
         QUALITATIVE ANALYSIS
         QUANTITATIVE ANALYSIS
         SPECTROSCOPIC ANALYSIS
         SPECTROSCOPY

ELECTROPHYSICS
   GS   **ELECTROPHYSICS**
         . ELECTRO-OPTICS
         . MOLECULAR ELECTRONICS
   RT   ELECTROCHEMISTRY
         ELECTROKINETICS
         ELECTROMAGNETISM
         ∞ ELECTRONICS
         ELECTROPHOTOMETRY
         ∞ PHYSICS
         ∞ SCIENCE
         THEORETICAL PHYSICS

ELECTROPHYSIOLOGY
   GS   PHYSIOLOGY
         . **ELECTROPHYSIOLOGY**
   RT   BODY MEASUREMENT (BIOLOGY)
         DEPOLARIZATION
         ECHOENCEPHALOGRAPHY
         ELECTROCARDIOGRAPHY
         ELECTROENCEPHALOGRAPHY
         ELECTROMYOGRAPHY
         ELECTRONARCOSIS
         ELECTRONOGRAPHY
         ELECTROPLETHYSMOGRAPHY
         ELECTRORETINOGRAPHY
         HIS BUNDLE
         NERVOUS SYSTEM
         NEUROLOGY
         ∞ SCIENCE

ELECTROPLATING
   GS   COATING
         . **ELECTROPLATING**
         COATINGS
         . **ELECTROPLATING**
         DEPOSITION
         . ELECTRODEPOSITION
         . . **ELECTROPLATING**
         PLATING

ELECTROPLATING-*(CONT.)*
         . **ELECTROPLATING**
   RT   BATHS
         CATHODIC COATINGS
         CURRENT DENSITY
         ELECTRODES
         ELECTROFORMING
         ELECTROLYSIS
         ELECTROLYTES
         ELECTROLYTIC CELLS
         ELECTROPHORESIS
         METALLIZING
         NICKEL PLATE
         PROTECTIVE COATINGS
         SURFACE FINISHING

ELECTROPLETHYSMOGRAPHY
   GS   BIOENGINEERING
         . BIOMETRICS
         . . BODY MEASUREMENT (BIOLOGY)
         . . . **ELECTROPLETHYSMOGRAPHY**
         . . PLETHYSMOGRAPHY
         . . . **ELECTROPLETHYSMOGRAPHY**
   RT   BLOOD CIRCULATION
         ELECTROPHYSIOLOGY
         MEDICAL ELECTRONICS

ELECTROPOLISHING
   UF   ELECTROLYTIC POLISHING
   GS   METAL FINISHING
         . **ELECTROPOLISHING**
         POLISHING
         . METAL POLISHING
         . . **ELECTROPOLISHING**
   RT   ELECTROCHEMICAL MACHINING
         METALLOGRAPHY
         SURFACE FINISHING

ELECTROREFINING
   GS   REFINING
         . **ELECTROREFINING**
   RT   ELECTRODES
         ELECTROLYTES
         ELECTROLYTIC CELLS
         ELECTROWINNING

ELECTRORETINOGRAPHY
   GS   BIOENGINEERING
         . BIOMETRICS
         . . **ELECTRORETINOGRAPHY**
   RT   ELECTROPHYSIOLOGY
         MEDICAL ELECTRONICS
         RETINA

ELECTROSEISMIC EFFECT
   USE   ELECTRIC CURRENT
         SEISMIC WAVES

∞ **ELECTROSLAG PROCESS**
   SN   *(USE OF A MORE SPECIFIC TERM IS*
         RECOMMENDED--CONSULT THE TERMS
         LISTED BELOW)
   RT   ELECTROSLAG REFINING
         ELECTROSLAG WELDING

ELECTROSLAG REFINING
   GS   REFINING
         . **ELECTROSLAG REFINING**
   RT   ARC MELTING
         ∞ ELECTROSLAG PROCESS
         RESISTANCE HEATING

ELECTROSLAG WELDING
   GS   WELDING
         . ELECTRIC WELDING
         . . **ELECTROSLAG WELDING**
   RT   ∞ ELECTROSLAG PROCESS

ELECTROSTATIC BONDING
   RT   COVERINGS
         ENCAPSULATING
         ENERGY TECHNOLOGY
         GLASS
         SOLAR ARRAYS
         SOLAR CELLS

ELECTROSTATIC CHARGE
   GS   ELECTRIC CHARGE
         . **ELECTROSTATIC CHARGE**
   RT   CAPACITANCE
         CHARGE DISTRIBUTION
         ∞ CHARGING
         ELECTRIC DISCHARGES
         ELECTRIC FIELDS

ELECTROSTATIC CHARGE-(CONT.)
```
        ELECTROSTATICS
        SCATHA SATELLITE
        STATIC ELECTRICITY
        XEROGRAPHY
```

ELECTROSTATIC DRAG
```
GS      DYNAMIC CHARACTERISTICS
        . DRAG
        . . ELECTROSTATIC DRAG
RT      DRAG MEASUREMENT
        SATELLITE DRAG
```

ELECTROSTATIC ENGINES
```
GS      ENGINES
        . ROCKET ENGINES
        . . ELECTRIC ROCKET ENGINES
        . . . ELECTROSTATIC ENGINES
RT      ARC JET ENGINES
        CESIUM ENGINES
        ION ENGINES
        MERCURY ION ENGINES
        MICROROCKET ENGINES
        RESTARTABLE ROCKET ENGINES
        RIT ENGINES
        SUSTAINER ROCKET ENGINES
        VERNIER ENGINES
```

ELECTROSTATIC EROSION
```
USE     SPARK MACHINING
```

ELECTROSTATIC FIELDS
```
USE     ELECTRIC FIELDS
```

ELECTROSTATIC GENERATORS
```
GS      ELECTRIC GENERATORS
        . DIRECT POWER GENERATORS
        . . ELECTROSTATIC GENERATORS
RT      ARC GENERATORS
        ∞GENERATORS
        KLYSTRONS
        MAGNETRONS
        ROTATING GENERATORS
        VOLTAGE GENERATORS
```

ELECTROSTATIC GYROSCOPES
```
UF      ELECTRICALLY SUSPENDED
          GYROSCOPES
        ESG (GYROSCOPES)
GS      GYROSCOPES
        . ELECTROSTATIC GYROSCOPES
RT      LEVITATION
```

ELECTROSTATIC PLASMA
```
USE     PLASMAS (PHYSICS)
```

ELECTROSTATIC PRECIPITATORS
```
GS      SEPARATORS
        . PRECIPITATORS
        . . ELECTROSTATIC PRECIPITATORS
RT      ADSORPTION
        AIR PURIFICATION
        AIR SAMPLING
        DUST COLLECTORS
        FLY ASH
        ∞SEPARATION
```

ELECTROSTATIC PROBES
```
UF      LANGMUIR PROBES
GS      MEASURING INSTRUMENTS
        . PLASMA PROBES
        . . ELECTROSTATIC PROBES
        . RADIATION MEASURING INSTRUMENTS
        . . ELECTROSTATIC PROBES
RT      ELECTRON ENERGY
        PLASMA FREQUENCIES
        RADIATION COUNTERS
        SCATHA SATELLITE
```

ELECTROSTATIC PROPULSION
```
GS      PROPULSION
        . ELECTRIC PROPULSION
        . . ELECTROSTATIC PROPULSION
        . . . ION PROPULSION
        . LOW THRUST PROPULSION
        . . ELECTROSTATIC PROPULSION
        . . . ION PROPULSION
        . SPACECRAFT PROPULSION
        . . ELECTROSTATIC PROPULSION
        . . . ION PROPULSION
RT      ELECTROMAGNETIC PROPULSION
        PLASMA PROPULSION
```

ELECTROSTATIC SHIELDING
```
GS      SHIELDING
        . ELECTROSTATIC SHIELDING
RT      ELECTRIC CONDUCTORS
        SCATHA SATELLITE
```

ELECTROSTATIC WAVES
```
GS      ELASTIC WAVES
        . MAGNETOHYDRODYNAMIC WAVES
        . . PLASMA WAVES
        . . . ELECTROSTATIC WAVES
RT      DIFFUSION WAVES
        IONIC WAVES
        LONGITUDINAL WAVES
        MAGNETOELASTIC WAVES
        SHOCK WAVES
```

ELECTROSTATICS
```
GS      STATICS
        . ELECTROSTATICS
RT      BORN-INFELD THEORY
        ELECTRIC FIELDS
        ELECTROMAGNETIC INTERACTIONS
        ELECTROMECHANICS
        ELECTROSTATIC CHARGE
        MAGNETOSTATICS
        POISSON EQUATION
        STATIC ELECTRICITY
```

ELECTROSTRICTION
```
GS      MECHANICAL PROPERTIES
        . ELECTROSTRICTION
RT      MAGNETOSTRICTION
        PIEZOELECTRICITY
```

ELECTROTHERMAL ENGINES
```
GS      ENGINES
        . ROCKET ENGINES
        . . ELECTRIC ROCKET ENGINES
        . . . ELECTROTHERMAL ENGINES
        . . . . ARC JET ENGINES
        . . . . RESISTOJET ENGINES
RT      HIGH TEMPERATURE PROPELLANTS
        ION ENGINES
        NUCLEAR ELECTRIC PROPULSION
        PULSED JET ENGINES
        RESTARTABLE ROCKET ENGINES
        SUSTAINER ROCKET ENGINES
```

ELECTROWINNING
```
RT      ELECTRODEPOSITION
        ELECTRODES
        ELECTROLYTES
        ELECTROLYTIC CELLS
        ELECTROREFINING
```

ELEKTRON SATELLITES
```
GS      SATELLITES
        . ARTIFICIAL SATELLITES
        . . METEOROLOGICAL SATELLITES
        . . . ELEKTRON SATELLITES
        . . . . ELEKTRON 1 SATELLITE
        . . . . ELEKTRON 2 SATELLITE
        . . . . ELEKTRON 4 SATELLITE
```

ELEKTRON 1 SATELLITE
```
GS      SATELLITES
        . ARTIFICIAL SATELLITES
        . . METEOROLOGICAL SATELLITES
        . . . ELEKTRON SATELLITES
        . . . . ELEKTRON 1 SATELLITE
```

ELEKTRON 2 SATELLITE
```
GS      SATELLITES
        . ARTIFICIAL SATELLITES
        . . METEOROLOGICAL SATELLITES
        . . . ELEKTRON SATELLITES
        . . . . ELEKTRON 2 SATELLITE
```

ELEKTRON 4 SATELLITE
```
GS      SATELLITES
        . ARTIFICIAL SATELLITES
        . . METEOROLOGICAL SATELLITES
        . . . ELEKTRON SATELLITES
        . . . . ELEKTRON 4 SATELLITE
```

ELEMENT ABUNDANCE
```
USE     ABUNDANCE
```

ELEMENT 104
```
GS      CHEMICAL ELEMENTS
        . ELEMENT 104
RT      ∞ELEMENTS
```

ELEMENT 105
```
GS      CHEMICAL ELEMENTS
        . ELEMENT 105
RT      ∞ELEMENTS
```

ELEMENTARY EXCITATIONS
```
UF      QUASI-PARTICLES
GS      ELEMENTARY EXCITATIONS
        . EXCITONS
        . MAGNONS
        . PHONONS
        . . PHONON BEAMS
        . PLASMONS
        . POLARONS
RT      MANY BODY PROBLEM
```

ELEMENTARY PARTICLE INTERACTIONS
```
GS      PARTICLE INTERACTIONS
        . ELEMENTARY PARTICLE
            INTERACTIONS
        . . MESON-MESON INTERACTIONS
        . . MESON-NUCLEON INTERACTIONS
        . . NUCLEAR CAPTURE
        . . . ELECTRON CAPTURE
        . . NUCLEON-NUCLEON INTERACTIONS
RT      ANGULAR DISTRIBUTION
        ELECTROMAGNETIC INTERACTIONS
        ∞INTERACTIONS
        ION ATOM INTERACTIONS
        PHOTON-ELECTRON INTERACTION
        VENEZIANO MODEL
```

ELEMENTARY PARTICLES
```
GS      PARTICLES
        . ELEMENTARY PARTICLES
        . . ANTIPARTICLES
        . . . ANTINEUTRINOS
        . . . ANTINUCLEONS
        . . . ANTIPROTONS
        . . . POSITRONS
        . . BETA PARTICLES
        . . BOSONS
        . . . ALPHA PARTICLES
        . . . MESONS
        . . . . ETA-MESONS
        . . . . KAONS
        . . . . MESON RESONANCE
        . . . . . X MESONS
        . . . . MUONS
        . . . . PIONS
        . . . . VECTOR MESONS
        . . . . . RHO-MESONS
        . . . . . SIGMA-MESONS
        . . . PHOTONS
        . . . . LIGHT BEAMS
        . . XI HYPERONS
        . DEUTERONS
        . FERMIONS
        . . BARYONS
        . . . HYPERONS
        . . . . XI HYPERONS
        . . . . OMEGA-MESONS
        . . . . RHO-MESONS
        . . . . SIGMA-MESONS
        . . . ETA-MESONS
        . . LEPTONS
        . . . ANTINEUTRINOS
        . . . MUONS
        . . . NEUTRINOS
        . . . . SOLAR NEUTRINOS
        . . . MESON RESONANCE
        . . NEUTRONS
        . . . COLD NEUTRONS
        . . . FAST NEUTRONS
        . . . PHOTONEUTRONS
        . . . THERMAL NEUTRONS
        . . PROTONS
        . . . RECOIL PROTONS
        . . . SOLAR PROTONS
        . GLUONS
        . GRAVITINOS
        . GRAVITONS
        . HADRONS
        . . BARYONS
        . . . OMEGA-MESONS
        . . . RHO-MESONS
        . . . SIGMA-MESONS
        . . MESONS
        . . . KAONS
        . . . MUONS
        . . . OMEGA-MESONS
        . . . VECTOR MESONS
        . . . . RHO-MESONS
        . . . . SIGMA-MESONS
        . . MAGNETIC MONOPOLES
```

ELEMENTARY PARTICLES-(CONT.)
```
       . . NUCLEONS
       . . PARTONS
       . . QUARKS
       . . TACHYONS
   RT    ATOMIC STRUCTURE
          BUBBLE CHAMBERS
          CHARGED PARTICLES
          DE BROGLIE WAVELENGTHS
          GEOCYCLOTRONS
          HYPERNUCLEI
          INSTANTONS
          IONIZING RADIATION
          NEUTRON SCATTERING
          NUCLEAR INTERACTIONS
          NUCLEAR PARTICLES
          NUCLEAR RADIATION
          NUCLEI (NUCLEAR PHYSICS)
          PARTICLE ACCELERATORS
          POMERANCHUK THEOREM
          POSITRON ANNIHILATION
          QUANTUM THEORY
          RADIATION BELTS
```

∞ ELEMENTS
```
   SN    (USE OF A MORE SPECIFIC TERM IS
          RECOMMENDED--CONSULT THE TERMS
          LISTED BELOW)
   RT    ATOMS
          CHEMICAL ELEMENTS
          ELEMENT 104
          ELEMENT 105
          HEAVY ELEMENTS
          ISOPARAMETRIC FINITE ELEMENTS
          LIGHT ELEMENTS
          LOGICAL ELEMENTS
          NEUTRAL ATOMS
          NUCLEAR FUEL ELEMENTS
          ORBITAL ELEMENTS
          TASKS
```

ELEVATION
```
   RT    ALTITUDE
          CONTOURS
          HEAD (FLUID MECHANICS)
          HYDROSTATIC PRESSURE
          HYDROSTATICS
          HYPSOGRAPHY
          LOW ALTITUDE
          PRESSURE HEADS
          TOPOGRAPHY
```

ELEVATION ANGLE
```
   UF    ALMUCANTAR
   GS    GEOMETRY
       . EUCLIDEAN GEOMETRY
       . . ANGLES (GEOMETRY)
       . . . ELEVATION ANGLE
   RT    ALTITUDE
          AZIMUTH
          DATUM (ELEVATION)
          FIELD OF VIEW
          LOOK ANGLES (TRACKING)
          TOPOGRAPHY
```

ELEVATIONS (DRAWINGS)
```
   USE    DRAWINGS
```

ELEVATOR ILLUSION
```
   GS    PSYCHOLOGICAL EFFECTS
       . ILLUSIONS
       . . OPTICAL ILLUSION
       . . . ELEVATOR ILLUSION
   RT    VISUAL PERCEPTION
```

ELEVATORS (CONTROL SURFACES)
```
   GS    AIRFOILS
       . ELEVATORS (CONTROL SURFACES)
          CONTROL SURFACES
       . ELEVATORS (CONTROL SURFACES)
   RT    AILERONS
       ∞ CONTROL
          ELEVONS
          HORIZONTAL TAIL SURFACES
          HYDROFOILS
          STABILIZERS (FLUID DYNAMICS)
       ∞ SURFACES
          TABS (CONTROL SURFACES)
          TAIL ASSEMBLIES
          TAIL SURFACES
```

ELEVATORS (LIFTS)
```
   RT    CONVEYORS
          ESCALATORS
       ∞ JACKS
```

ELEVATORS (LIFTS)-(CONT.)
```
       ∞ LIFTS
          WINCHES
```

ELEVONS
```
   GS    AIRFOILS
       . ELEVONS
          CONTROL SURFACES
       . ELEVONS
   RT    AILERONS
          ELEVATORS (CONTROL SURFACES)
          LATERAL CONTROL
          TABS (CONTROL SURFACES)
```

ELIMINATION
```
   GS    ELIMINATION
       . DELETION
   RT    ATTENUATION
          CANCELLATION
          DECONTAMINATION
          DEPLETION
       ∞ DISCHARGE
          DISPOSAL
          EVACUATING (TRANSPORTATION)
          EVACUATING (VACUUM)
          EXCLUSION
          EXHAUST SYSTEMS
          EXHAUSTING
          POLLUTION
          PURIFICATION
       ∞ REDUCTION
          REJECTION
       ∞ SEPARATION
          STOPPING
          WASTE DISPOSAL
```

ELLIPSES
```
   GS    GEOMETRY
       . EUCLIDEAN GEOMETRY
       . . ANALYTIC GEOMETRY
       . . . CONICS
       . . . . ELLIPSES
   RT    CIRCLES (GEOMETRY)
```

ELLIPSOIDS
```
   UF    IZSAK ELLIPSOID
   GS    SYMMETRICAL BODIES
       . ELLIPSOIDS
   RT    BODIES OF REVOLUTION
          ELLIPTICITY
          OGIVES
```

ELLIPSOMETERS
```
   GS    MEASURING INSTRUMENTS
       . OPTICAL MEASURING INSTRUMENTS
       . . ELLIPSOMETERS
          OPTICAL EQUIPMENT
       . OPTICAL MEASURING INSTRUMENTS
       . . ELLIPSOMETERS
   RT    PHOTOMETERS
          POLARIMETERS
```

ELLIPTIC DIFFERENTIAL EQUATIONS
```
   GS    ANALYSIS (MATHEMATICS)
       . REAL VARIABLES
       . . DIFFERENTIAL EQUATIONS
       . . . PARTIAL DIFFERENTIAL EQUATIONS
       . . . . ELLIPTIC DIFFERENTIAL
                 EQUATIONS
       . . . . . MONGE-AMPERE EQUATION
   RT    ∞ EQUATIONS
          HALF SPACES
          MAXIMUM PRINCIPLE
```

ELLIPTIC FUNCTIONS
```
   UF    ELLIPTIC INTEGRALS
   GS    ANALYSIS (MATHEMATICS)
       . COMPLEX VARIABLES
       . . MEROMORPHIC FUNCTIONS
       . . . ELLIPTIC FUNCTIONS
          FUNCTIONS (MATHEMATICS)
       . MEROMORPHIC FUNCTIONS
       . . ELLIPTIC FUNCTIONS
   RT    JACOBI INTEGRAL
          WEIERSTRASS FUNCTIONS
```

ELLIPTIC INTEGRALS
```
   USE    ELLIPTIC FUNCTIONS
```

ELLIPTICAL CYLINDERS
```
   RT    CIRCULAR CYLINDERS
       ∞ CYLINDERS
          CYLINDRICAL BODIES
          CYLINDRICAL SHELLS
```

ELLIPTICAL GALAXIES
```
   GS    CELESTIAL BODIES
       . GALAXIES
       . . ELLIPTICAL GALAXIES
   RT    DISK GALAXIES
          GALACTIC CLUSTERS
          LOCAL GROUP (ASTRONOMY)
          SPIRAL GALAXIES
          STAR CLUSTERS
          VIRGO GALACTIC CLUSTER
```

ELLIPTICAL ORBITS
```
   UF    HOHMANN TRAJECTORIES
          HOHMANN TRANSFER ORBITS
   GS    ORBITS
       . ELLIPTICAL ORBITS
       . . APOGEES
       . . PERIGEES
       . . PERIHELIONS
       . . TRANSFER ORBITS
       . . . INTERPLANETARY TRANSFER
                ORBITS
   RT    APSIDES
          CIRCULAR ORBITS
          EARTH ORBITS
          EARTH-MARS TRAJECTORIES
          EARTH-MERCURY TRAJECTORIES
          ECCENTRIC ORBITS
          ELLIPTICITY
          EQUATORIAL ORBITS
          EULER-LAMBERT EQUATION
          LUNAR ORBITS
          ORBITAL MECHANICS
          PAS
          PLANETARY ORBITS
          POLAR ORBITS
          SATELLITE ORBITS
          SOLAR ORBITS
          SPACECRAFT ORBITS
```

ELLIPTICAL PLASMAS
```
   GS    GASES
       . IONIZED GASES
       . . CHARGED PARTICLES
       . . . ELLIPTICAL PLASMAS
          PARTICLES
       . CHARGED PARTICLES
       . . ENERGETIC PARTICLES
       . . . PLASMAS (PHYSICS)
       . . . . ELLIPTICAL PLASMAS
   RT    MAGNETOHYDRODYNAMIC STABILITY
          PLASMA CONTROL
          TOROIDAL PLASMAS
```

ELLIPTICAL POLARIZATION
```
   GS    POLARIZATION (WAVES)
       . ELLIPTICAL POLARIZATION
   RT    CIRCULAR POLARIZATION
          MAGNETOIONICS
```

ELLIPTICITY
```
   GS    SHAPES
       . ELLIPTICITY
   RT    ECCENTRICITY
          ELLIPSOIDS
          ELLIPTICAL ORBITS
          FLATTENING
          OBLATE SPHEROIDS
```

ELONGATION
```
   RT    ANGLES (GEOMETRY)
          DEFORMATION
          DUCTILITY
          ECCENTRICITY
          EXPANSION
          MECHANICAL PROPERTIES
          PLASTIC DEFORMATION
          STRETCHING
          SUPERPLASTICITY
          TENSILE DEFORMATION
          TENSILE STRENGTH
```

ELUTION
```
   UF    ELUTRIATION
   RT    ADSORPTION
          EXTRACTION
          FLUSHING
          LEACHING
          PURIFICATION
       ∞ SEPARATION
          WASHING
```

ELUTRIATION
```
   USE    ELUTION
```

EMANATION
USE   EMISSION

EMBEDDED COMPUTER SYSTEMS
GS   DATA PROCESSING EQUIPMENT
. COMPUTERS
.. **EMBEDDED COMPUTER SYSTEMS**
... AIRBORNE/SPACEBORNE
COMPUTERS
RT   ADA (PROGRAMMING LANGUAGE)

EMBEDDING
RT   ACCELERATION PROTECTION
ENCAPSULATING
INSERTION

EMBOLISMS
GS   **EMBOLISMS**
. AEROEMBOLISM
. FAT EMBOLISMS
RT   BLOOD VESSELS
CLOTTING
COAGULATION
INFARCTION

EMBOSSING
RT   BRAILLE

EMBRITTLEMENT
RT   BRITTLE MATERIALS
BRITTLENESS
DEGRADATION
TIME TEMPERATURE PARAMETER

EMBRYOLOGY
RT   ∞BIOLOGY
DIFFERENTIATION (BIOLOGY)
EMBRYOS
FETUSES
REPRODUCTION (BIOLOGY)

EMBRYOS
RT   EGGS
EMBRYOLOGY
FETUSES
SEEDS

EMERALD
USE   BERYL

EMERGENCIES
RT   ACCIDENTS
DISASTERS
FAIL-SAFE SYSTEMS

EMERGENCY BREATHING TECHNIQUES
RT   ∞BREATHING
∞METHODOLOGY
PRESSURE BREATHING
RESPIRATORS
RESUSCITATION

EMERGENCY LIFE SUSTAINING SYSTEMS
GS   SUPPORT SYSTEMS
. LIFE SUPPORT SYSTEMS
.. **EMERGENCY LIFE SUSTAINING
SYSTEMS**
... AEPS
RT   ENVIRONMENTAL CONTROL
ESCAPE CAPSULES
FLOATS
HIGH ALTITUDE BREATHING
MEDICAL EQUIPMENT
OXYGEN SUPPLY EQUIPMENT
PORTABLE LIFE SUPPORT SYSTEMS
PRESSURIZED CABINS
PROTECTIVE CLOTHING
SAFETY
SAFETY DEVICES
SURVIVAL EQUIPMENT
∞SYSTEMS

EMERGENCY LOCATOR TRANSMITTERS
GS   TRANSMITTERS
. **EMERGENCY LOCATOR
TRANSMITTERS**

EMERGING
RT   EMISSION
EMISSIVITY
EMITTANCE

EMISSION
UF   EMANATION

EMISSION-*(CONT.)*
GS   DECAY
. **EMISSION**
.. EXHAUST EMISSION
.. LIGHT EMISSION
... INCANDESCENCE
... LUMINESCENCE
.... BIOLUMINESCENCE
.... CATHODE GLOW
.... CATHODOLUMINESCENCE
.... CHEMILUMINESCENCE
.... ELECTROLUMINESCENCE
.... FLUORESCENCE
..... PHOSPHORESCENCE
..... RESONANCE FLUORESCENCE
.... X RAY FLUORESCENCE
.... LUNAR LUMINESCENCE
.... OPTICAL RESONANCE
.... PHOTOLUMINESCENCE
.... TRIBOLUMINESCENCE
..... X RAY FLUORESCENCE
.... SHOCK WAVE LUMINESCENCE
.... SONOLUMINESCENCE
.... THERMOLUMINESCENCE
.. MICROWAVE EMISSION
.. PARTICLE EMISSION
... ELECTRON EMISSION
.... FIELD EMISSION
... PHOTOELECTRIC EMISSION
.... SECONDARY EMISSION
... ION EMISSION
... NEUTRON EMISSION
... THERMIONIC EMISSION
.. PHOTOELECTRIC EFFECT
... PHOTOIONIZATION
.. RADIO EMISSION
... CN EMISSION
... HYDROXYL EMISSION
... SOLAR RADIO EMISSION
.... SOLAR RADIO BURSTS
..... TYPE 2 BURSTS
..... TYPE 3 BURSTS
..... TYPE 4 BURSTS
..... TYPE 5 BURSTS
. SELF SUSTAINED EMISSION
. SPECTRAL EMISSION
.. SPONTANEOUS EMISSION
.. STIMULATED EMISSION
... WATER MASERS
.. THERMAL EMISSION
... THERMIONIC EMISSION
RT   AIRGLOW
ATOMIC RECOMBINATION
BURSTS
CATHODOLUMINESCENCE
∞DISCHARGE
EFFLUX
EJECTION
EMERGING
EMITTERS
EXCITATION
IONIZING RADIATION
IRRADIATION
NUCLEAR REACTIONS
PAIR PRODUCTION
QUANTUM THEORY
∞RADIATION
RADIOACTIVE DECAY
RADIOACTIVITY
RELEASING
SELECTION RULES (NUCLEAR PHYSICS)
SPUTTERING

EMISSION SPECTRA
SN   (ELECTROMAGNETIC RADIATION OF ANY
WAVELENGTH EMITTED FROM EXCITED
MATTER --EXCLUDES PARTICLE
SPECTRA)
GS   SPECTRA
. RADIATION SPECTRA
.. **EMISSION SPECTRA**
RT   ABSORPTION SPECTRA
ATOMIC RECOMBINATION
BALMER SERIES
CONTINUOUS RADIATION
D LINES
ELECTROMAGNETIC SPECTRA
ELECTRON SPECTROSCOPY
ELECTRONIC SPECTRA
EXTARS
FLAME SPECTROSCOPY
GAMMA RAY SPECTRA
GAMMA RAYS
H ALPHA LINE
H BETA LINE
H GAMMA LINE

EMISSION SPECTRA-*(CONT.)*
H LINES
HYDROXYL EMISSION
INFRARED SPECTRA
K LINES
LINE SPECTRA
LYMAN SPECTRA
MOLECULAR SPECTRA
MOLECULAR SPECTROSCOPY
NUCLEAR RADIATION
OPTICAL EMISSION SPECTROSCOPY
OPTICAL TRANSITION
PASCHEN SERIES
PHOTOLUMINESCENT BANDS
PLASMA SPECTRA
RAMAN SPECTRA
RYDBERG SERIES
SCHUMANN-RUNGE BANDS
SOLAR SPECTRA
SOLAR SPECTROMETERS
SPECTRAL SIGNATURES
SPECTRUM ANALYSIS
SPONTANEOUS EMISSION
STELLAR SPECTRA
SWAN BANDS
SYMBIOTIC STARS
ULTRAVIOLET SPECTRA
VEGARD-KAPLAN BANDS
VISIBLE SPECTRUM
X RAYS

EMISSIVITY
UF   PHOTOEMISSIVITY
GS   THERMODYNAMIC PROPERTIES
. THERMOPHYSICAL PROPERTIES
.. **EMISSIVITY**
RT   BLACK BODY RADIATION
BRIGHTNESS
EMERGING
EMITTANCE
HOHLRAUMS
INCANDESCENCE
LUMINOSITY
NONGRAY ATMOSPHERES
NONGRAY GAS
OPTICAL MEASUREMENT
RADIANCE
RADIANT FLUX DENSITY
STEFAN-BOLTZMANN LAW
SURFACE PROPERTIES
TEMPERATURE
THERMAL EMISSION

EMISSOGRAPHS
USE   ACTINOMETERS
RECORDING INSTRUMENTS

EMITTANCE
RT   EMERGING
EMISSIVITY
FLUX (RATE)
LUMINOSITY
LUMINOUS INTENSITY
OPTICAL PROPERTIES
RADIANCE
RADIANT FLUX DENSITY
SPECTRAL EMISSION
THERMODYNAMIC PROPERTIES

EMITTERS
GS   **EMITTERS**
. THERMIONIC CATHODES
. THERMIONIC EMITTERS
RT   ELECTRON EMISSION
EMISSION
SEMICONDUCTORS (MATERIALS)

EMOTIONAL FACTORS
RT   ANGINA PECTORIS
DETACHMENT
DISORDERS
DITHERS
FEEDBACK
FRUSTRATION
HUMAN REACTIONS
MOODS
PANIC
PHOBIAS
PSYCHOLOGICAL FACTORS
PSYCHOLOGY
SENSORY FEEDBACK
SENSORY STIMULATION

EMOTIONS
RT   ∞DEPRESSION

EMOTIONS-(CONT.)
    FEAR
    FEAR OF FLYING
    FRUSTRATION
    HUMAN BEHAVIOR
    LAUGHING
    MOODS
    PANIC
    PSYCHOLOGICAL EFFECTS
    PSYCHOLOGY
    SENSORY FEEDBACK

EMPENNAGE
  USE   TAIL ASSEMBLIES

EMPHYSEMA
  GS   DISEASES
    . RESPIRATORY DISEASES
    . . **EMPHYSEMA**

EMPLOYEE RELATIONS
  RT   ∞COOPERATION
    HUMAN RELATIONS·
    PERSONNEL
    PERSONNEL DEVELOPMENT
    PERSONNEL MANAGEMENT
    POSITION (TITLE)
    PRODUCTION MANAGEMENT
    RETIREMENT
    WAGE SURVEYS

EMPLOYMENT
  RT   PERSONNEL SELECTION
    ∞TESTS

EMPTYING
  RT   DISPOSAL
    DUMPING
    EJECTION
    EXPULSION
    EXPULSION BLADDERS
    JETTISONING
    MATERIALS HANDLING
    RELEASING
    REMOVAL
    SPILLING
    SPREADING
    UNLOADING

EMR 6050 COMPUTER
  GS   DATA PROCESSING EQUIPMENT
    . COMPUTERS
    . . DIGITAL COMPUTERS
    . . . **EMR 6050 COMPUTER**

EMULSIONS
  GS   MIXTURES
    . DISPERSIONS
    . . **EMULSIONS**
    . . . PHOTOGRAPHIC EMULSIONS
    . . . . NUCLEAR EMULSIONS
  RT   BROWNIAN MOVEMENTS
    COLLOIDS
    SLURRIES
    SOLUTIONS

ENAMELS
  GS   COATINGS
    . **ENAMELS**
    FINISHES
    . **ENAMELS**
  RT   PORCELAIN

ENARGITE
  GS   CHALCOGENIDES
    . SULFIDES
    . . INORGANIC SULFIDES
    . . . COPPER SULFIDES
    . . . . **ENARGITE**
    COPPER COMPOUNDS
    . COPPER SULFIDES
    . . **ENARGITE**
    SULFUR COMPOUNDS
    . SULFIDES
    . . INORGANIC SULFIDES
    . . . COPPER SULFIDES
    . . . . **ENARGITE**

ENCAPSULATED MICROCIRCUITS
  GS   CIRCUITS
    . INTEGRATED CIRCUITS
    . . **ENCAPSULATED MICROCIRCUITS**
  RT   MICROELECTRONICS

ENCAPSULATING
  GS   COATING
    . **ENCAPSULATING**
    COATINGS
    . **ENCAPSULATING**
  RT   CANNING
    ELECTRONIC PACKAGING
    ELECTROSTATIC BONDING
    EMBEDDING
    ∞IMBEDDINGS
    MATERIALS HANDLING
    PACKAGING
    PLASTIC COATINGS
    POTTING COMPOUNDS
    PROTECTIVE COATINGS
    SEALING
    SHEATHS

ENCELADUS
  GS   SATELLITES
    . NATURAL SATELLITES
    . . SATURN SATELLITES
    . . . **ENCELADUS**
  RT   SATURN (PLANET)

ENCEPHALITIS
  GS   DISEASES
    . **ENCEPHALITIS**
  RT   BRAIN

ENCKE COMET
  GS   CELESTIAL BODIES
    . COMETS
    . . **ENCKE COMET**

ENCKE METHOD
  RT   ∞METHODOLOGY

ENCLOSURE
  RT   ∞CASING
    HOUSINGS
    PACKAGING

ENCLOSURES
  RT   AIR LOCKS
    ASTEROID CAPTURE
    ∞BARRIERS
    BIOPAKS
    CLOSURES
    COMPARTMENTS
    ∞CONTAINERS
    COVERINGS
    DOGHOUSES (ELECTRONICS)
    ∞ENVELOPES
    HOUSINGS
    ∞PENS
    PERFORATED SHELLS
    PRESSURE CHAMBERS
    PROTECTORS
    ROOMS
    SAFETY DEVICES
    SHELLS (STRUCTURAL FORMS)
    SHIELDING
    SHIPYARDS
    WALLS

ENCODERS
  USE   CODERS

ENCODING
  USE   CODING

ENCOUNTERS
  RT   CRASHES
    SCATTERING

 †

END MORAINES
  USE   GLACIAL DRIFT

END PLATES
  GS   STRUCTURAL MEMBERS
    . PLATES (STRUCTURAL MEMBERS)
    . . **END PLATES**
  RT   ANISOTROPIC PLATES
    BULKHEADS
    CIRCULAR PLATES
    CLOSURES
    FLAT PLATES
    SHALLOW SHELL EQUATIONS

END-TO-END DATA SYSTEMS
  GS   **END-TO-END DATA SYSTEMS**
    . NEEDS (DATA SYSTEM)
  RT   ∞DATA

END-TO-END DATA SYSTEMS-(CONT.)
    DATA ACQUISITION
    DATA PROCESSING
    DATA SYSTEMS
    ∞SYSTEMS

ENDANGERED SPECIES
  RT   ANIMALS
    BIRDS
    ECOLOGY
    ECOSYSTEMS
    HABITATS
    POLLUTION
    TOXICITY
    WILDLIFE

ENDFIRE ARRAYS
  GS   ARRAYS
    . ANTENNA ARRAYS
    . . LINEAR ARRAYS
    . . . **ENDFIRE ARRAYS**
    . . . . YAGI ANTENNAS
  RT   DIRECTIONAL ANTENNAS

ENDOCRINE GLANDS
  GS   ANATOMY
    . GLANDS (ANATOMY)
    . . **ENDOCRINE GLANDS**
    . . . ADRENAL GLAND
    . . . GONADS
    . . . OVARIES
    . . . PANCREAS
    . . . PARATHYROID GLAND
    . . . PINEAL GLAND
    . . . PITUITARY GLAND
    . . . PROSTATE GLAND
    . . . . THYMUS GLAND
    . . . THYROID GLAND
    VISCERA
    . **ENDOCRINE GLANDS**
    . . ADRENAL GLAND
    . . GONADS
    . . OVARIES
    . . PANCREAS
    . . PARATHYROID GLAND
    . . PINEAL GLAND
    . . PITUITARY GLAND
    . . PROSTATE GLAND
    . . THYMUS GLAND
    . . THYROID GLAND
  RT   ENDOCRINOLOGY
    ESTROGENS

ENDOCRINE SECRETIONS
  GS   SECRETIONS
    . **ENDOCRINE SECRETIONS**
    . . HORMONES
    . . . ESTROGENS
    . . . HYDROXYCORTICOSTEROID
    . . . ·PITUITARY HORMONES
    . . . . ADRENOCORTICOTROPIN (ACTH)
    . . . PROSTAGLANDINS
    . . INSULIN
  RT   CORTICOSTEROIDS

ENDOCRINE SYSTEMS
  RT   ENDOCRINOLOGY
    GLANDS (ANATOMY)
    HORMONES
    MINERAL METABOLISM
    ∞SYSTEMS

ENDOCRINOLOGY
  GS   MEDICAL SCIENCE
    . **ENDOCRINOLOGY**
  RT   ENDOCRINE GLANDS
    ENDOCRINE SYSTEMS

ENDOLYMPH
  GS   BODY FLUIDS
    . **ENDOLYMPH**
  RT   EAR

ENDORADIOSONDES
  GS   MEASURING INSTRUMENTS
    . METEOROLOGICAL INSTRUMENTS
    . . RADIOSONDES
    . . . **ENDORADIOSONDES**
    SONDES
    . . RADIOSONDES
    . . . **ENDORADIOSONDES**
    RADIO EQUIPMENT
    . RADIO TRANSMITTERS
    . . RADIOSONDES
    . . . **ENDORADIOSONDES**

ENDORADIOSONDES-*(CONT.)*
TRANSMITTERS
. RADIO TRANSMITTERS
.. RADIOSONDES
... **ENDORADIOSONDES**

ENDOSCOPES
UF   BORESCOPES
GS   MEDICAL EQUIPMENT
    . **ENDOSCOPES**
    OPTICAL EQUIPMENT
    . **ENDOSCOPES**
RT   INSPECTION

ENDOTHELIUM
GS   TISSUES (BIOLOGY)
    . **ENDOTHELIUM**
RT   BLOOD VESSELS
    CELLS (BIOLOGY)

ENDOTHERMIC FUELS
GS   FUELS
    . CHEMICAL FUELS
    .. **ENDOTHERMIC FUELS**
RT   CRYOGENIC ROCKET PROPELLANTS
    DOUBLE BASE PROPELLANTS
    GASEOUS ROCKET PROPELLANTS
    HYDROCARBON FUELS
    PROPELLANT DECOMPOSITION

ENDOTHERMIC REACTIONS
GS   CHEMICAL REACTIONS
    . **ENDOTHERMIC REACTIONS**
RT   ASSOCIATION REACTIONS
    EXOTHERMIC REACTIONS
    HEAT SINKS
    PYROLYSIS
    THERMAL DECOMPOSITION

ENDOTOXINS
GS   POISONS
    . **ENDOTOXINS**
    TOXINS AND ANTITOXINS
    . **ENDOTOXINS**
RT   BACTERIOLOGY
    TOXICOLOGY

ENDRIN
GS   EPOXY COMPOUNDS
    . **ENDRIN**
    HETEROCYCLIC COMPOUNDS
    . **ENDRIN**
RT   INSECTICIDES

∞ **ENDURANCE**
SN   *(USE OF A MORE SPECIFIC TERM IS*
    RECOMMENDED--CONSULT THE TERMS
    LISTED BELOW)
RT   DURABILITY
    FATIGUE (MATERIALS)

ENEMY PERSONNEL
GS   PERSONNEL
    . **ENEMY PERSONNEL**
RT   ARMED FORCES (FOREIGN)

ENERGETIC PARTICLE EXPLORER A
USE   EXPLORER 12 SATELLITE

ENERGETIC PARTICLE EXPLORER B
USE   EXPLORER 14 SATELLITE

ENERGETIC PARTICLE EXPLORER C
USE   EXPLORER 15 SATELLITE

ENERGETIC PARTICLE EXPLORER D
USE   EXPLORER 26 SATELLITE

ENERGETIC PARTICLES
GS   PARTICLES
    . CHARGED PARTICLES
    .. **ENERGETIC PARTICLES**
    ... ELECTRONS
    .... CONDUCTION ELECTRONS
    .... HIGH ENERGY ELECTRONS
    .... HOT ELECTRONS
    .... N ELECTRONS
    .... NEGATRONS
    .... PI-ELECTRONS
    ... NUCLEI (NUCLEAR PHYSICS)
    ... PLASMAS (PHYSICS)
    .... ARGON PLASMA
    .... BETA PARTICLES
    .... BOUNDARY LAYER PLASMAS

ENERGETIC PARTICLES-*(CONT.)*
    .... COLD PLASMAS
    .... COLLISIONAL PLASMAS
    ..... STRONGLY COUPLED PLASMAS
    .... COLLISIONLESS PLASMAS
    .... COSMIC PLASMA
    .... CYLINDRICAL PLASMAS
    .... DENSE PLASMAS
    ..... PLASMA FOCUS
    .... STRONGLY COUPLED PLASMAS
    .... ELECTRON PLASMA
    .... ELLIPTICAL PLASMAS
    .... HELIUM PLASMA
    .... HIGH TEMPERATURE PLASMAS
    .... LASER PLASMAS
    .... METALLIC PLASMAS
    ..... CESIUM PLASMA
    .... MICROPLASMAS
    .... NITROGEN PLASMA
    .... NONEQUILIBRIUM PLASMAS
    .... NONUNIFORM PLASMAS
    .... RAREFIED PLASMAS
    .... RELATIVISTIC PLASMAS
    .... ROTATING PLASMAS
    .... SEMICONDUCTOR PLASMAS
    .... SOLAR WIND
    .... SPACE PLASMAS
    .... SPHERICAL PLASMAS
    .... STELLAR WINDS
    .... THERMAL PLASMAS
    .... TOROIDAL PLASMAS
RT   GALACTIC COSMIC RAYS
    SOLAR COSMIC RAYS

∞ **ENERGY**
SN   *(USE OF A MORE SPECIFIC TERM IS*
    RECOMMENDED--CONSULT THE TERMS
    LISTED BELOW)
RT   ACTIVATION ENERGY
    BERNSTEIN ENERGY PRINCIPLE
    CHEMICAL ENERGY
    COMMERCIAL ENERGY
    DOMESTIC ENERGY
    ELECTRON ENERGY
    ENERGY CONSERVATION
    ENERGY CONVERSION EFFICIENCY
    ENERGY OF FORMATION
    ENTHALPY
    ENTROPY
    FLUX (RATE)
    FLUX DENSITY
    FREE ENERGY
    HEAT
    HYDROGEN-BASED ENERGY
    INDUSTRIAL ENERGY
    INTERFACIAL ENERGY
    INTERNAL ENERGY
    KINETIC ENERGY
    MOLECULAR ENERGY LEVELS
    NUCLEAR BINDING ENERGY
    PARTICLE ENERGY
    POTENTIAL ENERGY
    PROTON ENERGY
    RADIANT HEATING
    SEISMIC ENERGY
    SOLAR ENERGY
    SOLAR TOTAL ENERGY SYSTEMS
    STACKING FAULT ENERGY
    STRAIN ENERGY METHODS
    SURFACE ENERGY
    THERMAL ENERGY
    THERMONUCLEAR POWER GENERATION
    TRANSPORTATION ENERGY
    WATERWAVE ENERGY
    WORK

ENERGY ABSORPTION
UF   NONREFLECTION
GS   **ENERGY ABSORPTION**
    . ELECTROMAGNETIC ABSORPTION
    .. AURORAL ABSORPTION
    .. GAMMA RAY ABSORPTION
    .. INFRARED ABSORPTION
    .. MULTIPHOTON ABSORPTION
    .. PHOTOABSORPTION
    .. POLAR CAP ABSORPTION
    .. ULTRAVIOLET ABSORPTION
    .. X RAY ABSORPTION
    . MODERATION (ENERGY ABSORPTION)
    .. THERMALIZATION (ENERGY
      ABSORPTION)
    ... NEUTRON THERMALIZATION
    . MOLECULAR ABSORPTION
    . SELF ABSORPTION
    . THERMAL ABSORPTION

ENERGY ABSORPTION-*(CONT.)*
    . POLAR CAP ABSORPTION
RT   ABSORBERS (MATERIALS)
    ∞ABSORPTION
    DAMPING
    GAMMA RAY ABSORPTIOMETRY
    HEAT SINKS
    INFRARED RADIATION
    LIGHT (VISIBLE RADIATION)
    PHOTON ABSORPTIOMETRY
    SHOCK ABSORBERS
    SOUND TRANSMISSION
    VIBRATION ISOLATORS

ENERGY ABSORPTION FILMS
GS   THIN FILMS
    . **ENERGY ABSORPTION FILMS**
RT   ∞ABSORPTION
    ALUMINUM OXIDES
    COATINGS
    DIRECT POWER GENERATORS
    GOLAY DETECTOR CELLS
    MONOMOLECULAR FILMS
    PHOTOELECTRIC CELLS
    PHOTOTHERMAL CONVERSION
    PHOTOVOLTAIC CONVERSION
    SELECTIVE SURFACES
    SEMICONDUCTING FILMS
    SOLAR ENERGY

ENERGY BANDS
GS   **ENERGY BANDS**
    . BLOCH BAND
    . CONDUCTION BANDS
    . FORBIDDEN BANDS
RT   ∞BANDS
    EXCITONS
    LASER WINDOWS
    QUANTUM WELLS
    SPECTRAL BANDS
    WINDOWS (INTERVALS)

ENERGY BUDGETS
GS   **ENERGY BUDGETS**
    . HEAT BUDGET
    .. ATMOSPHERIC HEAT BUDGET
RT   ∞BUDGETS

ENERGY CONSERVATION
GS   CONSERVATION
    . **ENERGY CONSERVATION**
RT   ∞ENERGY
    ENERGY POLICY
    POWER FACTOR CONTROLLERS
    RESIDENTIAL ENERGY
    RESOURCE ALLOCATION
    RESOURCES

ENERGY CONSUMPTION
GS   CONSUMPTION
    . **ENERGY CONSUMPTION**
RT   COAL UTILIZATION
    COMMERCIAL ENERGY
    DOMESTIC ENERGY
    ∞ENERGY SOURCES
    FUEL CONSUMPTION
    INDUSTRIAL ENERGY

ENERGY CONVERSION
GS   **ENERGY CONVERSION**
    . BIOMASS ENERGY PRODUCTION
    . GEOTHERMAL ENERGY CONVERSION
    . OCEAN THERMAL ENERGY
      CONVERSION
    . PHOTOTHERMAL CONVERSION
    . SATELLITE SOLAR ENERGY
      CONVERSION
    . SOLAR ENERGY CONVERSION
    .. SOLAR TOTAL ENERGY SYSTEMS
    . WATERWAVE ENERGY CONVERSION
RT   COGENERATION
    COMMERCIAL ENERGY
    ∞CONVERSION
    DIRECT POWER GENERATORS
    DOMESTIC ENERGY
    GEOTHERMAL ENERGY EXTRACTION
    HYDROGEN PRODUCTION
    HYDROTHERMAL SYSTEMS
    INDUSTRIAL ENERGY
    INTEGRATED ENERGY SYSTEMS
    LIGNITE
    ORGANIC WASTES (FUEL CONVERSION)
    POWER CONDITIONING
    SATELLITE SOLAR POWER STATIONS
    SOLAR PONDS (HEAT STORAGE)

ENERGY CONVERSION-*(CONT.)*
    SOLAR SEA POWER PLANTS
    SPACE INDUSTRIALIZATION
    TRANSPORTATION ENERGY
    WASTE ENERGY UTILIZATION

ENERGY CONVERSION EFFICIENCY
GS   EFFICIENCY
    . ENERGY CONVERSION EFFICIENCY
RT   CARRIER TRANSPORT (SOLID STATE)
    ∞CONVERSION
    DIRECT POWER GENERATORS
    ENERGY ABSORPTION
    ENGINES
    FUEL CELLS
    ∞GENERATORS
    MOTORS
    OPEN CIRCUIT VOLTAGE
    PHOTOTHERMAL CONVERSION
    POWER CONDITIONING
    POWER FACTOR CONTROLLERS
    QUANTUM EFFICIENCY
    REDOX CELLS
    SPECTROPHOTOVOLTAICS
    TIDE POWERED GENERATORS
    TRANSDUCERS
    VOLUMETRIC EFFICIENCY
    WATERWAVE ENERGY CONVERSION

ENERGY CONVERTERS
USE   DIRECT POWER GENERATORS

ENERGY DENSITY
USE   FLUX DENSITY

ENERGY DISSIPATION
UF   ENERGY LOSSES
GS   DISSIPATION
    . ENERGY DISSIPATION
RT   FRICTION
    INSERTION LOSS
    LAGRANGE SIMILARITY HYPOTHESIS
    LOSSES
    NONADIABATIC THEORY
    ∞POWER LOSS
    TRAVELING CHARGE

ENERGY DISTRIBUTION
GS   DISTRIBUTION (PROPERTY)
    . ENERGY DISTRIBUTION
    .. SPECTRAL ENERGY DISTRIBUTION
RT   EQUIPARTITION THEOREM
    FLUX DENSITY
    FORCE DISTRIBUTION
    INTEGRATED ENERGY SYSTEMS
    QUANTUM MECHANICS
    STATISTICAL MECHANICS

ENERGY EFFICIENCY TRANSPORT PROGRAM
USE   ACEE PROGRAM

ENERGY EQUIPARTITION
USE   EQUIPARTITION THEOREM

ENERGY EXCHANGE
USE   ENERGY TRANSFER

ENERGY GAPS (SOLID STATE)
UF   BANDGAP
GS   GAPS
    . ENERGY GAPS (SOLID STATE)
RT   BAND STRUCTURE OF SOLIDS
    QUANTUM WELLS
    SOLID STATE
    ∞SOLID STATE PHYSICS

ENERGY LEVELS
UF   ELECTRONIC LEVELS
GS   LEVEL (QUANTITY)
    . ENERGY LEVELS
    .. ATOMIC ENERGY LEVELS
    .. ELECTRON STATES
    .. GROUND STATE
    .. MOLECULAR ENERGY LEVELS
    ... INTERMOLECULAR FORCES
RT   ATOMIC EXCITATIONS
    ATOMIC STRUCTURE
    ELECTRON TUNNELING
    EXCITATION
    FERMI SURFACES
    MOLECULAR EXCITATION
    NUCLEAR CAPTURE
    NUCLEAR MODELS
    NUCLEAR QUADRUPOLE RESONANCE

ENERGY LEVELS-*(CONT.)*
    NUCLEAR SPIN
    NUCLEAR STRUCTURE
    POPULATION INVERSION
    QUANTUM NUMBERS
    QUANTUM THEORY

ENERGY LOSSES
USE   ENERGY DISSIPATION

ENERGY METHODS
GS   STRUCTURAL ANALYSIS
    . ENERGY METHODS
    .. BERNSTEIN ENERGY PRINCIPLE
    .. STRAIN ENERGY METHODS
RT   CASTIGLIANO VARIATIONAL THEOREM
    MATRICES (MATHEMATICS)
    ∞METHODOLOGY
    STRESS ANALYSIS

ENERGY OF FORMATION
GS   CHEMICAL ENERGY
    . ENERGY OF FORMATION
RT   ∞ENERGY
    FREE ENERGY
    MOLECULAR ENERGY LEVELS

ENERGY POLICY
GS   POLICIES
    . ENERGY POLICY
RT   ABUNDANCE
    AVAILABILITY
    COAL
    COAL GASIFICATION
    COAL LIQUEFACTION
    COAL UTILIZATION
    CONSERVATION
    CRUDE OIL
    DEPLETION
    ∞DEVELOPMENT
    EARTH RESOURCES
    ECOLOGY
    ECONOMIC FACTORS
    ENERGY CONSERVATION
    FUEL OILS
    FUELS
    HYDROCARBON FUELS
    LAND USE
    LIGNITE
    LOGISTICS
    MINING
    ∞NUCLEAR ENERGY
    NUCLEAR FUELS
    OIL EXPLORATION
    OILS
    OPERATING COSTS
    POLLUTION
    REFINING
    RESERVES
    RESOURCE ALLOCATION
    RESOURCES
    SAFETY

ENERGY REQUIREMENTS
RT   ∞ENERGY SOURCES
    FUEL CONSUMPTION
    ∞POWER SUPPLIES

∞ **ENERGY SOURCES**
SN   *(USE OF A MORE SPECIFIC TERM IS*
    RECOMMENDED--CONSULT THE TERMS
    LISTED BELOW)
RT   ATMOSPHERIC ENERGY SOURCES
    AUXILIARY POWER SOURCES
    BIOMASS ENERGY PRODUCTION
    ELECTRIC BATTERIES
    ELECTRIC GENERATORS
    ELECTRON SOURCES
    ENERGY CONSUMPTION
    ENERGY REQUIREMENTS
    ENERGY TECHNOLOGY
    GEOTHERMAL RESOURCES
    HEAT SOURCES
    LITHIUM SULFUR BATTERIES
    OCEAN THERMAL ENERGY CONVERSION
    PLASMA POWER SOURCES
    POINT SOURCES
    PROPELLANTS
    RECTIFIERS
    SPACECRAFT POWER SUPPLIES
    TIDEPOWER
    WATERWAVE ENERGY CONVERSION

ENERGY SPECTRA
GS   SPECTRA

ENERGY SPECTRA-*(CONT.)*
    . ENERGY SPECTRA
    .. ELECTRONIC SPECTRA
    .. NEUTRON SPECTRA
RT   ABSORPTION SPECTRA
    ELECTROMAGNETIC SPECTRA
    GAMMA RAY ASTRONOMY
    GRIST (TELESCOPE)
    MASS SPECTRA
    MOLECULAR SPECTRA
    PLASMA SPECTRA
    POWER SPECTRA
    RADIATION SPECTRA
    SHOCK SPECTRA
    SPECTRAL ENERGY DISTRIBUTION
    SPECTROPHOTOVOLTAICS
    VIBRATIONAL SPECTRA

ENERGY STORAGE
UF   ENERGY STORAGE DEVICES
GS   **ENERGY STORAGE**
    . ELECTRIC ENERGY STORAGE
    . HEAT STORAGE
    . MAGNETIC ENERGY STORAGE
RT   CAPACITORS
    COMPRESSED AIR
    ELECTRETS
    ELECTRIC BATTERIES
    FLYWHEELS
    FUEL CELLS
    FUELS
    GEOTHERMAL ENERGY UTILIZATION
    HEAT SOURCES
    INDUCTORS
    LEAD ACID BATTERIES
    NICKEL HYDROGEN BATTERIES
    ∞NUCLEAR ENERGY
    POTENTIAL ENERGY
    REDOX CELLS
    REGENERATORS
    ROADWAY POWERED VEHICLES
    SPRINGS (ELASTIC)
    ∞STORAGE
    SUPERCONDUCTORS

ENERGY STORAGE DEVICES
USE   ENERGY STORAGE

ENERGY TECHNOLOGY
GS   TECHNOLOGIES
    . ENERGY TECHNOLOGY
    .. GEOTHERMAL TECHNOLOGY
RT   ATMOSPHERIC ENERGY SOURCES
    BIOMASS ENERGY PRODUCTION
    COAL UTILIZATION
    COMBINED CYCLE POWER GENERATION
    EARTH RESOURCES
    ELECTROSTATIC BONDING
    ∞ENERGY SOURCES
    FUEL CELL POWER PLANTS
    GAS RECOVERY
    GEOTHERMAL ENERGY EXTRACTION
    HYDROCARBON FUEL PRODUCTION
    HYDROGEN-BASED ENERGY
    LIGNITE
    MAGNETIC ENERGY STORAGE
    OFFSHORE ENERGY SOURCES
    OIL RECOVERY
    PHOSPHORIC ACID FUEL CELLS
    PHOTOELECTROCHEMICAL DEVICES
    PHOTOTHERMAL CONVERSION
    QUANTUM EFFICIENCY
    RESIDENTIAL ENERGY
    SOLAR COOLING
    SOLAR ENERGY CONVERSION
    SOLAR HOUSES
    SPACE COOLING (BUILDINGS)
    TROMBE WALLS
    WASTE HEAT

ENERGY TRANSFER
UF   ENERGY EXCHANGE
GS   **ENERGY TRANSFER**
    . LINEAR ENERGY TRANSFER (LET)
RT   ANTENNA COUPLERS
    COUPLING CIRCUITS
    CYCLOTRON RESONANCE
    ELECTRON PUMPING
    GAS TRANSPORT
    GAS-LIQUID INTERACTIONS
    HEAT TRANSFER
    HEISENBERG THEORY
    LAGRANGE SIMILARITY HYPOTHESIS
    MASS TRANSFER
    MOMENTUM TRANSFER

ENERGY TRANSFER-*(CONT.)*
    NONADIABATIC CONDITIONS
    NONISOTHERMAL PROCESSES
    NUCLEAR PUMPING
    PLASMA HEATING
    POYNTING THEOREM
    RADIATIVE TRANSFER
    TERMINAL BALLISTICS
    TRANSFERRING

ENGINE AIRFRAME INTEGRATION
RT   AERODYNAMIC CHARACTERISTICS
    AERODYNAMIC CONFIGURATIONS
    AIRCRAFT DESIGN
    AIRCRAFT ENGINES
    AIRFRAMES

ENGINE ANALYZERS
GS   MEASURING INSTRUMENTS
   . ANALYZERS
   . . **ENGINE ANALYZERS**

ENGINE CONTROL
GS   **ENGINE CONTROL**
   . ROCKET ENGINE CONTROL
   . TURBOJET ENGINE CONTROL
RT   AIR START
   AIRCRAFT CONTROL
   AUTOMATIC CONTROL
   COMBUSTION CONTROL
 ∞CONTROL
   ELECTRIC CONTROL
   FLIGHT INSTRUMENTS
   FUEL CONTROL
   HYDRAULIC CONTROL
   MANUAL CONTROL
   PNEUMATIC CONTROL
   REMOTE CONTROL
   SPACECRAFT CONTROL
   SPEED CONTROL
   TEMPERATURE CONTROL
   THRUST CONTROL
   . VARIABLE STREAM CONTROL ENGINES

ENGINE COOLANTS
GS   COOLANTS
   . **ENGINE COOLANTS**
RT   COOLING
   COOLING SYSTEMS

ENGINE DESIGN
GS   **ENGINE DESIGN**
   . ROCKET ENGINE DESIGN
RT   AIRCRAFT DESIGN
   COMPUTER AIDED DESIGN
 ∞DESIGN
   HELICOPTER DESIGN
   MISSILE DESIGN
   NOZZLE DESIGN
   PRODUCT DEVELOPMENT
   REACTOR DESIGN
   SPACECRAFT DESIGN
   VOLUMETRIC EFFICIENCY

ENGINE FAILURE
GS   FAILURE
   . **ENGINE FAILURE**
RT   ABORTED MISSIONS
 ∞CUT-OFF
   INGESTION (ENGINES)
 ∞STALLING

ENGINE INLETS
GS   INTAKE SYSTEMS
   . AIR INTAKES
   . . **ENGINE INLETS**
RT   BYPASS RATIO
 ∞DIFFUSERS
   HYPERSONIC INLETS
   INLET AIRFRAME CONFIGURATIONS
   INLET NOZZLES
   INLET TEMPERATURE
   INTERNAL COMPRESSION INLETS
   NACELLES

ENGINE MONITORING INSTRUMENTS
GS   MEASURING INSTRUMENTS
   . **ENGINE MONITORING INSTRUMENTS**
RT   FLIGHT INSTRUMENTS

ENGINE NOISE
GS   ELASTIC WAVES
   . SOUND WAVES
   . . NOISE (SOUND)

ENGINE NOISE-*(CONT.)*
   . . . ENGINE NOISE
   . . . . ROCKET ENGINE NOISE
RT   AIRCRAFT NOISE
   AIRCRAFT RUNUP
   JET AIRCRAFT NOISE
   QUIET ENGINE PROGRAM

ENGINE PARTS
RT   CARBURETORS
   CLUTCHES
   COMBUSTION CHAMBERS
 ∞COMPONENTS
   FLYWHEELS
   INTERNAL COMBUSTION ENGINES
   PISTONS
   RETIREMENT FOR CAUSE
   ROCKET LININGS
   SPARE PARTS
   TURBINE BLADES
   TURBINE WHEELS
   VALVES

ENGINE PRIMERS
RT   INTERNAL COMBUSTION ENGINES
 ∞PRIMERS
   STARTING

ENGINE STARTERS
GS   STARTERS
   . **ENGINE STARTERS**
RT   ENGINES
   INTERNAL COMBUSTION ENGINES
   JET ENGINES

ENGINE TESTING LABORATORIES
GS   LABORATORIES
   . **ENGINE TESTING LABORATORIES**
   TEST FACILITIES
   . **ENGINE TESTING LABORATORIES**
RT   ENGINES

ENGINE TESTS
GS   **ENGINE TESTS**
   . COLD FLOW TESTS
   . PREFIRING TESTS
   . SPACE ELECTRIC ROCKET TESTS
   . STATIC FIRING
RT   AIRCRAFT RUNUP
   ALTITUDE TESTS
   CAPTIVE TESTS
   FLIGHT TESTS
   FUEL TESTS
   FULL SCALE TESTS
   GROUND TESTS
   LUBRICANT TESTS
   MISSILE TESTS
   NONDESTRUCTIVE TESTS
   PRELAUNCH TESTS
   PROPELLANT TESTS
   PROPULSIVE EFFICIENCY
   ROCKET ENGINE DESIGN
   ROCKET TEST FACILITIES
   SERT 1 SPACECRAFT
   SERT 2 SPACECRAFT
   STATIC TESTS
   TEST FIRING
   TEST STANDS
   TESTING TIME
 ∞TESTS
   VIBRATION TESTS

∞ **ENGINEERING**
SN   *(USE OF A MORE SPECIFIC TERM IS*
   RECOMMENDED--CONSULT THE TERMS
   LISTED BELOW)
RT   AERONAUTICAL ENGINEERING
   AEROSPACE ENGINEERING
   AIRCRAFT PRODUCTION COSTS
   ANTHROPOMETRY
   BIOENGINEERING
   BIOINSTRUMENTATION
   BIOMETRICS
   BIOTELEMETRY
   BODY MEASUREMENT (BIOLOGY)
   CHEMICAL ENGINEERING
   ELECTRICAL ENGINEERING
   ENVIRONMENTAL ENGINEERING
   MAN MACHINE SYSTEMS
   MECHANICAL ENGINEERING
   PRODUCTION ENGINEERING
   REACTOR TECHNOLOGY
   RELIABILITY ENGINEERING
   STRUCTURAL ENGINEERING
   SYSTEMS ENGINEERING

ENGINEERING-*(CONT.)*
   UNDERWATER ENGINEERING
   VALUE ENGINEERING

ENGINEERING DEVELOPMENT
USE   PRODUCT DEVELOPMENT

ENGINEERING DRAWINGS
UF   MECHANICAL DRAWINGS
GS   DOCUMENTS
   . **ENGINEERING DRAWINGS**
   . . BLUEPRINTS
   DRAWINGS
   . **ENGINEERING DRAWINGS**
   . . BLUEPRINTS
RT   CIRCUIT DIAGRAMS
   DESCRIPTIVE GEOMETRY
 ∞DESIGN
   DIMENSIONS
   GRAPHIC ARTS
   LAYOUTS
   LOFTING
   REPRODUCTION (COPYING)

ENGINEERING MANAGEMENT
GS   MANAGEMENT
   . INDUSTRIAL MANAGEMENT
   . . **ENGINEERING MANAGEMENT**
RT   ALLOCATIONS
 ∞BUDGETS
   GOALS
   MANPOWER
   PRIORITIES
   RESEARCH MANAGEMENT
   RESOURCE ALLOCATION
   RESOURCES

ENGINEERING TEST REACTORS
UF   ETR (REACTORS)
GS   NUCLEAR REACTORS
   . **ENGINEERING TEST REACTORS**
RT   REACTOR DESIGN
   REACTOR TECHNOLOGY

ENGINES
SN   (MACHINES WITH SELF-CONTAINED
   POWER SOURCES FOR CONTINUOUS
   OPERATION-- SEE MOTORS FOR
   MACHINES UTILIZING EXTERNAL POWER
   SOURCES FOR NORMAL OPERATION)
UF   GAS GENERATOR ENGINES
GS   **ENGINES**
   . AIR BREATHING ENGINES
   . . GAS TURBINE ENGINES
   . . . JET ENGINES
   . . . . RAMJET ENGINES
   . . . . . INTEGRAL ROCKET RAMJETS
   . . . . . LOW VOLUME RAMJET ENGINES
   . . . . . PULSEJET ENGINES
   . . . . . SUPERSONIC COMBUSTION
        RAMJET ENGINES
   . . . . TURBORAMJET ENGINES
   . . . TURBOJET ENGINES
   . . . . BRISTOL-SIDDELEY OLYMPUS
        593 ENGINE
   . . . . BRISTOL-SIDDELEY VIPER
        ENGINE
   . . . . DUCTED FAN ENGINES
   . . . . J-33 ENGINE
   . . . . J-34 ENGINE
   . . . . J-47 ENGINE
   . . . . J-57 ENGINE
   . . . . J-57-P-20 ENGINE
   . . . . J-65 ENGINE
   . . . . J-69-T-25 ENGINE
   . . . . J-71 ENGINE
   . . . . J-73 ENGINE
   . . . . J-75 ENGINE
   . . . . J-79 ENGINE
   . . . . J-85 ENGINE
   . . . . J-93 ENGINE
   . . . . RA-28 ENGINE
   . . . . TURBOFAN ENGINES
   . . . . . BRISTOL-SIDDELEY BS 53
        ENGINE
   . . . . . CF-700 ENGINE
   . . . . . J-97 ENGINE
   . . . . . TF-41 ENGINE
   . . . . TURBOPROP ENGINES
   . . . . . T-53 ENGINE
   . . . . . T-56 ENGINE
   . . . . . T-64 ENGINE
   . . . . . T-74 ENGINE
   . . . . TURBORAMJET ENGINES
   . . . T-58-GE-8B ENGINE

ENRICHMENT-*(CONT.)*
- RT　BENEFICIATION
- 　　CONCENTRATING
- 　　PURIFICATION
- 　　REFINING
- 　　UPGRADING

ENRICO FERMI ATOMIC POWER PLANT
- GS　ELECTRIC POWER PLANTS
- 　. NUCLEAR POWER PLANTS
- 　. . **ENRICO FERMI ATOMIC POWER PLANT**
- 　　NUCLEAR ELECTRIC POWER GENERATION
- 　. NUCLEAR POWER PLANTS
- 　. . **ENRICO FERMI ATOMIC POWER PLANT**
- RT　BREEDER REACTORS
- 　　FAST NUCLEAR REACTORS
- 　　LIQUID METAL COOLED REACTORS
- ∞POWER PLANTS

ENSKOG-CHAPMAN THEORY
- USE　CHAPMAN-ENSKOG THEORY

ENSTATITE
- GS　CHALCOGENIDES
- 　. OXIDES
- 　. . PYROXENES
- 　. . . **ENSTATITE**
- 　　MAGNESIUM COMPOUNDS
- 　. **ENSTATITE**
- 　　MINERALS
- 　. PYROXENES
- 　. . **ENSTATITE**
- 　　SILICON COMPOUNDS
- 　. SILICATES
- 　. . PYROXENES
- 　. . . **ENSTATITE**
- RT　CHONDRULE
- 　　IGNEOUS ROCKS
- 　　REGOLITH
- 　　ROCKS
- 　　SOILS

ENSTROPHY
- USE　VORTICITY

ENTERPRISE (ORBITER)
- UF　SPACE SHUTTLE ORBITER 101
- GS　REENTRY VEHICLES
- 　. RECOVERABLE SPACECRAFT
- 　. . REUSABLE SPACECRAFT
- 　. . . **ENTERPRISE (ORBITER)**
- 　　TRANSPORTATION
- 　. SPACE TRANSPORTATION
- 　. . SPACE TRANSPORTATION SYSTEM
- 　. . . SPACE SHUTTLE ORBITERS
- 　. . . . **ENTERPRISE (ORBITER)**
- RT　MANNED SPACE FLIGHT
- ∞SPACECRAFT

ENTHALPY
- UF　HEAT CONTENT
- GS　THERMODYNAMIC PROPERTIES
- 　. **ENTHALPY**
- RT　ADIABATIC CONDITIONS
- 　　DRYING
- ∞ENERGY
- 　　ENTROPY
- 　　FREE ENERGY
- 　　GIBBS FREE ENERGY
- 　　GIBBS-HELMHOLTZ EQUATIONS
- 　　HEAT
- 　　HEAT MEASUREMENT
- 　　HEAT OF FUSION
- 　　JOULE-THOMSON EFFECT
- 　　MOLLIER DIAGRAM
- 　　SPECIFIC HEAT
- 　　THERMOCHEMISTRY
- 　　THERMODYNAMICS

ENTHALPY-ENTROPY DIAGRAMS
- USE　MOLLIER DIAGRAM

ENTIRE FUNCTIONS
- UF　INTEGRAL FUNCTIONS
- GS　ANALYSIS (MATHEMATICS)
- 　. COMPLEX VARIABLES
- 　. . ANALYTIC FUNCTIONS
- 　. . . **ENTIRE FUNCTIONS**
- 　　FUNCTIONS (MATHEMATICS)
- 　. ANALYTIC FUNCTIONS
- 　. . **ENTIRE FUNCTIONS**

ENTOMOLOGY
- RT　INSECTICIDES
- 　　INSECTS
- ∞SCIENCE
- ∞ZOOLOGY

ENTRAINMENT
- RT　AERATION
- 　　AEROSOLS
- 　　BLOWING
- 　　COANDA EFFECT
- 　　DISPERSING
- 　　SPRAYING
- 　　SUSPENDING (MIXING)

ENTRANCES
- RT　CURTAINS
- 　　DOORS
- 　　INTAKE SYSTEMS
- ∞THRESHOLDS
- 　　TRANSFER TUNNELS

ENTRAPMENT
- RT　ACCUMULATORS
- 　　CONFUSION
- 　　ESCAPE (ABANDONMENT)
- 　　RADIATION BELTS
- 　　TANGLING
- 　　TRAPS

ENTROPY
- GS　THERMODYNAMIC PROPERTIES
- 　. **ENTROPY**
- RT　CROCCO METHOD
- ∞ENERGY
- 　　ENTHALPY
- 　　HEAT
- 　　MAXIMUM ENTROPY METHOD
- 　　MOLLIER DIAGRAM
- 　　NONISENTROPICITY
- 　　SHANNON-WIENER MEASURE
- 　　TEPHIGRAMS
- 　　THERMOCHEMISTRY
- 　　THERMODYNAMICS

ENTROPY (STATISTICS)
- GS　**ENTROPY (STATISTICS)**
- 　. MAXIMUM ENTROPY METHOD
- 　. MINIMUM ENTROPY METHOD
- RT　∞STATISTICS

∞ ENTRY
- SN　*(USE OF A MORE SPECIFIC TERM IS RECOMMENDED--CONSULT THE TERMS LISTED BELOW)*
- RT　ATMOSPHERIC ENTRY
- 　　REENTRY
- 　　UNCONTROLLED REENTRY (SPACECRAFT)

ENTRY GUIDANCE (STS)
- GS　GUIDANCE (MOTION)
- 　. **ENTRY GUIDANCE (STS)**
- RT　ATMOSPHERIC ENTRY
- 　　FLIGHT CONTROL
- 　　HYPERSONIC REENTRY
- 　　POINTING CONTROL SYSTEMS
- 　　SPACE SHUTTLES
- 　　SPACE TRANSPORTATION SYSTEM FLIGHTS
- 　　SPACECRAFT REENTRY
- 　　TERMINAL GUIDANCE

ENUMERATION
- RT　COUNTING
- 　　LISTS
- 　　NUMBER THEORY

∞ ENVELOPES
- SN　*(USE OF A MORE SPECIFIC TERM IS RECOMMENDED--CONSULT THE TERMS LISTED BELOW)*
- RT　COVERINGS
- 　　ENCLOSURES
- 　　LIMITS (MATHEMATICS)

ENVIRONMENT EFFECTS
- SN　(EFFECTS ON ENVIRONMENT)
- RT　AIR POLLUTION
- 　　COASTAL ECOLOGY
- 　　COASTAL WATER
- 　　CONTAMINANTS
- 　　CONTAMINATION
- 　　DEBRIS

ENVIRONMENT EFFECTS-*(CONT.)*
- 　　DEFORESTATION
- ∞EFFECTS
- 　　ENVIRONMENTS
- 　　EUTROPHICATION
- 　　EXHAUST GASES
- 　　GREENHOUSE EFFECT
- 　　HABITATS
- 　　ICE ENVIRONMENTS
- 　　MAN ENVIRONMENT INTERACTIONS
- 　　MARINE BIOLOGY
- 　　MARINE ENVIRONMENTS
- 　　METABOLIC WASTES
- 　　NOISE POLLUTION
- 　　POISONS
- 　　POLLUTION
- 　　SEWAGE
- 　　SOIL EROSION
- 　　THERMAL POLLUTION
- 　　WASTE DISPOSAL
- 　　WASTES
- 　　WATER POLLUTION
- 　　WATER QUALITY
- 　　WATER RESOURCES
- 　　WETLANDS
- 　　WILDLIFE

ENVIRONMENT MANAGEMENT
- GS　MANAGEMENT
- 　. **ENVIRONMENT MANAGEMENT**
- RT　CONSERVATION
- 　　EARTH RESOURCES
- 　　ENVIRONMENTAL MONITORING
- 　　LAND MANAGEMENT
- 　　LAND USE
- 　　MAN ENVIRONMENT INTERACTIONS
- 　　RESOURCES MANAGEMENT
- 　　WATER MANAGEMENT
- 　　WATER RESOURCES

ENVIRONMENT MODELS
- GS　MODELS
- 　. **ENVIRONMENT MODELS**
- RT　ATMOSPHERIC MODELS
- 　　EXOBIOLOGY
- 　　TEST CHAMBERS

ENVIRONMENT POLLUTION
- GS　POLLUTION
- 　. **ENVIRONMENT POLLUTION**
- 　. . AIR POLLUTION
- 　. . . GLOBAL AIR POLLUTION
- 　. . . INDOOR AIR POLLUTION
- 　. . WATER POLLUTION
- 　. . . OIL POLLUTION
- RT　AEROBIOLOGY
- 　　AEROSOLS
- 　　AIR SAMPLING
- 　　CLEAN ENERGY
- 　　EARTH RESOURCES
- 　　ENVIRONMENTAL MONITORING
- 　　ENVIRONMENTAL SURVEYS
- 　　HUMAN WASTES
- 　　METABOLIC WASTES
- 　　NOISE POLLUTION
- 　　OIL SLICKS
- 　　POISONS
- 　　POLLUTION MONITORING
- 　　POLLUTION TRANSPORT
- 　　RADIOACTIVE WASTES
- 　　THERMAL POLLUTION
- 　　WASTE DISPOSAL

ENVIRONMENT PROTECTION
- GS　PROTECTION
- 　. **ENVIRONMENT PROTECTION**
- RT　AIR POLLUTION
- 　　EFFLUENTS
- 　　ENVIRONMENTAL MONITORING
- 　　POLLUTION
- 　　RADIOACTIVE WASTES
- 　　WASTE DISPOSAL
- 　　WATER POLLUTION

ENVIRONMENT SIMULATION
- GS　SIMULATION
- 　. **ENVIRONMENT SIMULATION**
- 　. . ACOUSTIC SIMULATION
- 　. . ALTITUDE SIMULATION
- 　. . SPACE ENVIRONMENT SIMULATION
- 　. . . WEIGHTLESSNESS SIMULATION
- 　. . THERMAL SIMULATION
- RT　ATMOSPHERIC ENTRY SIMULATION
- 　　ATMOSPHERIC MODELS
- 　　ENVIRONMENTAL TESTS

ENVIRONMENT SIMULATION-(CONT.)
  FLIGHT SIMULATION

ENVIRONMENT SIMULATORS
  GS   SIMULATORS
       . **ENVIRONMENT SIMULATORS**
       . . LUNAR GRAVITY SIMULATOR
       . . SOLAR SIMULATORS
       . . SPACE SIMULATORS
       . . . HIGH VACUUM ORBITAL SIMULATOR
       . . . LANGLEY COMPLEX COORDINATOR
  RT   TEST CHAMBERS

ENVIRONMENTAL CHAMBERS
  USE   TEST CHAMBERS

ENVIRONMENTAL CHEMISTRY
  GS   **ENVIRONMENTAL CHEMISTRY**
       . AEROTHERMOCHEMISTRY
       . ATMOSPHERIC CHEMISTRY
       . BIOCHEMISTRY
       . . BIOGEOCHEMISTRY
       . . ENZYMOLOGY
       . . PHYSIOCHEMISTRY
       . GEOCHEMISTRY
       . . BIOGEOCHEMISTRY
  RT   AIR POLLUTION
       ∞CHEMISTRY
       CLIMATOLOGY
       HYDROCARBON FUELS
       PESTICIDES
       SMOG
       WASTE DISPOSAL
       WATER POLLUTION

ENVIRONMENTAL CONTROL
  RT   ANTISEPTICS
       ARTIFICIAL GRAVITY
       AUTOMATIC CONTROL
       BIOSATELLITES
       CABIN ATMOSPHERES
       CLEAN ROOMS
       ∞CONTROL
       EMERGENCY LIFE SUSTAINING
         SYSTEMS
       ENVIRONMENTS
       HABITABILITY
       MANNED REENTRY
       MANNED SPACECRAFT
       PRESSURIZED CABINS
       RESOURCES MANAGEMENT
       SPACECRAFT CABIN ATMOSPHERES
       SPACECRAFT ENVIRONMENTS
       TEMPERATURE CONTROL
       TEST CHAMBERS
       WEATHER MODIFICATION
       WINDSHIELDS

ENVIRONMENTAL ENGINEERING
  RT   ∞AEROSPACE SCIENCES
       CLEAN ENERGY
       CLIMATOLOGY
       COMFORT
       ∞ENGINEERING
       ENVIRONMENTS
       HEATING
       HUMAN FACTORS ENGINEERING
       ILLUMINATING
       LIFE SCIENCES
       LIFE SUPPORT SYSTEMS
       METEOROLOGY
       PHYSIOLOGICAL EFFECTS
       PSYCHOLOGICAL EFFECTS
       SHELTERS
       SPACE HEATING (BUILDINGS)
       STARSITE PROGRAM
       TEMPERATURE CONTROL
       TEMPERATURE DISTRIBUTION
       VENTILATION
       WASTE DISPOSAL

ENVIRONMENTAL INDEX
  RT   PHYSIOLOGICAL TESTS

ENVIRONMENTAL LABORATORIES
  GS   LABORATORIES
       . **ENVIRONMENTAL LABORATORIES**
       TEST FACILITIES
       . **ENVIRONMENTAL LABORATORIES**
  RT   HUMAN FACTORS LABORATORIES
       TEST CHAMBERS

ENVIRONMENTAL MONITORING
  RT   ENVIRONMENT MANAGEMENT
       ENVIRONMENT POLLUTION

ENVIRONMENTAL MONITORING-(CONT.)
       ENVIRONMENT PROTECTION
       INFRARED RADIOMETERS
       METEOROLOGY
       MONITORS
       OCEANOGRAPHY
       WEATHER FORECASTING

ENVIRONMENTAL QUALITY
  GS   QUALITY
       . **ENVIRONMENTAL QUALITY**
       . . WATER QUALITY
  RT   AIR POLLUTION
       CONTAMINANTS
       ENVIRONMENTS
       GLOBAL AIR SAMPLING PROGRAM
       MARINE BIOLOGY
       NOISE POLLUTION
       POLLUTION
       THERMAL POLLUTION
       WATER POLLUTION

ENVIRONMENTAL RESEARCH SATELLITES
  UF   OCTAHEDRAL RESEARCH SATELLITES
  GS   SATELLITES
       . ARTIFICIAL SATELLITES
       . . SCIENTIFIC SATELLITES
       . . . **ENVIRONMENTAL RESEARCH
             SATELLITES**
       . . . . ERS 17
       . . . . ERS 18
       . . . . INTASAT SATELLITE
  RT   ATLAS AGENA LAUNCH VEHICLES

ENVIRONMENTAL SURVEYS
  SN   (LIMITED TO INDEXING ENVIRONMENTAL
       IMPACT STATEMENTS)
  RT   AEROSOLS
       AIR POLLUTION
       EARTH RESOURCES
       ENVIRONMENT POLLUTION
       HUMAN WASTES
       METABOLIC WASTES
       POISONS
       POLLUTION
       POLLUTION CONTROL
       RADIOACTIVE WASTES
       THERMAL POLLUTION
       WASTE DISPOSAL
       WATER POLLUTION

ENVIRONMENTAL TEMPERATURE
  USE   AMBIENT TEMPERATURE

ENVIRONMENTAL TESTS
  GS   **ENVIRONMENTAL TESTS**
       . COLD WEATHER TESTS
       . CORROSION TESTS
       . . SALT SPRAY TESTS
       . HIGH TEMPERATURE TESTS
       . LOW TEMPERATURE TESTS
       . UNDERWATER TESTS
  RT   ASSET PROJECT
       ELECTRONIC EQUIPMENT TESTS
       ENVIRONMENT SIMULATION
       ∞MATERIALS TESTS
       PHYSIOLOGICAL TESTS
       PSYCHOLOGICAL TESTS
       SPIN TESTS
       TEST CHAMBERS
       ∞TESTS
       THERMAL VACUUM TESTS
       VIBRATION TESTS

ENVIRONMENTS
  GS   **ENVIRONMENTS**
       . AEROSPACE ENVIRONMENTS
       . . CISLUNAR SPACE
       . . DEEP SPACE
       . . . INTERPLANETARY SPACE
       . . . . INTERSTELLAR SPACE
       . EARTH ENVIRONMENT
       . EXTRATERRESTRIAL ENVIRONMENTS
       . . CISLUNAR SPACE
       . . DEEP SPACE
       . . . INTERPLANETARY SPACE
       . . . . INTERSTELLAR SPACE
       . . LUNAR ENVIRONMENT
       . . . LUNAR ATMOSPHERE
       . . PLANETARY ENVIRONMENTS
       . . . MARS ENVIRONMENT
       . . . . MARS ATMOSPHERE
       . . . PLANETARY ATMOSPHERES
       . . . . HELIUM HYDROGEN
             ATMOSPHERES

ENVIRONMENTS-(CONT.)
       . . . . JUPITER ATMOSPHERE
       . . . . MARS ATMOSPHERE
       . . . . NEPTUNE ATMOSPHERE
       . . . . SATURN ATMOSPHERE
       . . . . URANUS ATMOSPHERE
       . . . . VENUS ATMOSPHERE
       . . . SATELLITE ATMOSPHERES
       . . . LUNAR ATMOSPHERE
       . . STELLAR ATMOSPHERES
       . . . CHROMOSPHERE
       . . . SOLAR ATMOSPHERE
       . FRICTIONLESS ENVIRONMENTS
       . HETEROSPHERE
       . HIGH ALTITUDE ENVIRONMENTS
       . HIGH GRAVITY ENVIRONMENTS
       . HIGH TEMPERATURE ENVIRONMENTS
       . ICE ENVIRONMENTS
       . INNER RADIATION BELT
       . IONOSPHERE
       . LOWER IONOSPHERE
       . LOW TEMPERATURE ENVIRONMENTS
       . MAGNETOSPHERE
       . . GEOMAGNETIC TAIL
       . . MAGNETOPAUSE
       . MARINE ENVIRONMENTS
       . MESOPAUSE
       . MESOSPHERE
       . MIDLATITUDE ATMOSPHERE
       . ROTATING ENVIRONMENTS
       . SPACECRAFT ENVIRONMENTS
       . THERMAL ENVIRONMENTS
  RT   ADIABATIC CONDITIONS
       AIR
       AIR POLLUTION
       AIR QUALITY
       AMBIENCE
       ∞ATMOSPHERES
       COASTAL ECOLOGY
       COASTAL PLAINS
       CONTROLLED ATMOSPHERES
       ECOLOGY
       ECONOMIC IMPACT
       ELECTROMAGNETIC INTERFERENCE
       ENVIRONMENT EFFECTS
       ENVIRONMENTAL CONTROL
       ENVIRONMENTAL ENGINEERING
       ENVIRONMENTAL QUALITY
       GLOBAL AIR POLLUTION
       GRAVITATION
       HABITABILITY
       HABITATS
       HUMAN FACTORS ENGINEERING
       HUMIDITY
       LIFE SUPPORT SYSTEMS
       NONPOINT SOURCES
       ∞PERFORMANCE
       PHYSIOLOGICAL EFFECTS
       PLANTS (BOTANY)
       PRESSURE
       PSYCHOLOGICAL EFFECTS
       REGIMES
       TEMPERATURE
       THERMAL POLLUTION
       VACUUM EFFECTS
       WEIGHTLESSNESS

ENZYME ACTIVITY
  GS   METABOLISM
       . **ENZYME ACTIVITY**
       . . FERMENTATION
  RT   BIOCONVERSION
       DIABETES MELLITUS
       DIGESTIVE SYSTEM
       TYROSINE

ENZYMES
  GS   **ENZYMES**
       . ALDOLASE
       . AMIDASE
       . CARBONIC ANHYDRASE
       . CATALASE
       . CHOLINESTERASE
       . COENZYMES
       . . CYSTEAMINE
       . HEXOKINASE
       . LYSOZYME
       . NUCLEASE
       . OXIDASE
       . PAPAIN
       . PEPSIN
       . PROTEASE
       . PROTHROMBIN
       . THROMBIN
       . TRYPSIN

ENZYMES-(CONT.)
- RT   ACTIVATION (BIOLOGY)
-      CATALYSTS
-      ENZYMOLOGY

ENZYMOLOGY
- GS   ENVIRONMENTAL CHEMISTRY
-     . BIOCHEMISTRY
-     . . **ENZYMOLOGY**
- RT   DIGESTING
-      DIGESTIVE SYSTEM
-      ENZYMES
-      METABOLISM
-      NITROGEN METABOLISM

EOCR (REACTOR)
- USE   EXPERIMENTAL ORGANIC COOLED
-       REACTORS

EOGO
- USE   EGO

EOLE SATELLITES
- GS   SATELLITES
-     . ARTIFICIAL SATELLITES
-     . . FRENCH SATELLITES
-     . . . **EOLE SATELLITES**
-     . . METEOROLOGICAL SATELLITES
-     . . . **EOLE SATELLITES**
- RT   FRENCH SPACE PROGRAMS
-      GEOLE SATELLITES
-      GEOPHYSICAL SATELLITES

EOPAP
- USE   EARTH & OCEAN PHYSICS
-       APPLICATIONS PROGRAM

EOR (RENDEZVOUS)
- USE   EARTH ORBITAL RENDEZVOUS

EOS
- USE   LANDSAT SATELLITES

EOS-A
- USE   LANDSAT E

EOS-B
- USE   LANDSAT F

EOSINOPHILS
- GS   ANATOMY
-     . CARDIOVASCULAR SYSTEM
-     . . LEUKOCYTES
-     . . . **EOSINOPHILS**
-     BODY FLUIDS
-     . BLOOD
-     . . LEUKOCYTES
-     . . . **EOSINOPHILS**
-     CELLS (BIOLOGY)
-     . LEUKOCYTES
-     . . **EOSINOPHILS**
- RT   CYTOPLASM

EOSS
- UF   EARTH ORBITING SPACE STATIONS
- GS   SATELLITES
-     . ARTIFICIAL SATELLITES
-     . . ORBITAL SPACE STATIONS
-     . . . **EOSS**
- RT   ORBITING LUNAR STATIONS

EPE-A
- USE   EXPLORER 12 SATELLITE

EPE-B
- USE   EXPLORER 14 SATELLITE

EPE-C
- USE   EXPLORER 15 SATELLITE

EPE-D
- USE   EXPLORER 26 SATELLITE

EPHEMERIDES
- GS   **EPHEMERIDES**
-     . PLANET EPHEMERIDES
- RT   ASTRONOMICAL CATALOGS
-      CELESTIAL MECHANICS
-      EPHEMERIS TIME
-      ORBITS
-      POSITION (LOCATION)

EPHEMERIS TIME
- GS   TIME
-     . **EPHEMERIS TIME**
- RT   EPHEMERIDES
-      UNIVERSAL TIME

EPICARDIUM
- GS   ANATOMY
-     . CARDIOVASCULAR SYSTEM
-     . . HEART
-     . . . **EPICARDIUM**
-     TISSUES (BIOLOGY)
-     . **EPICARDIUM**

EPICYCLOIDS
- GS   GEOMETRY
-     . CURVES (GEOMETRY)
-     . . **EPICYCLOIDS**
-     . EUCLIDEAN GEOMETRY
-     . . ANALYTIC GEOMETRY
-     . . . **EPICYCLOIDS**
- RT   CUSPS (MATHEMATICS)

EPIDEMIOLOGY
- GS   MEDICAL SCIENCE
-     . **EPIDEMIOLOGY**
- RT   INFECTIOUS DISEASES
-      VACCINES
-      VETERINARY MEDICINE

EPIDERMIS
- GS   SKIN (ANATOMY)
-     . **EPIDERMIS**
- RT   CONTACT DERMATITIS

EPILEPSY
- GS   DISEASES
-     . **EPILEPSY**
- RT   CRAMPS
-      HUMAN PATHOLOGY
-      SHAKING

EPINEPHRINE
- UF   ADRENALINE
- GS   DRUGS
-     . **EPINEPHRINE**
-     . . NOREPINEPHRINE
- RT   ADRENAL GLAND
-      HEART RATE
-      STIMULANTS

EPITAXY
- GS   GROWTH
-     . CRYSTAL GROWTH
-     . . **EPITAXY**
-     . . . ELECTROEPITAXY
-     . . . LIQUID PHASE EPITAXY
-     . . . MOLECULAR BEAM EPITAXY
-     . . . VAPOR PHASE EPITAXY
- RT   BIPOLAR TRANSISTORS
-      CRYSTAL LATTICES
-      CRYSTAL STRUCTURE
-      JUNCTION TRANSISTORS

EPITHELIUM
- GS   SKIN (ANATOMY)
-     . **EPITHELIUM**
-     TISSUES (BIOLOGY)
-     . **EPITHELIUM**
- RT   ANATOMY
-      HISTOLOGY

EPNL
- USE   EFFECTIVE PERCEIVED NOISE LEVELS

EPOCHS
- USE   TIME MEASUREMENT

EPOXIDATION
- GS   CHEMICAL REACTIONS
-     . **EPOXIDATION**
- RT   OXIDATION

EPOXIDES
- USE   EPOXY COMPOUNDS

EPOXY COMPOUNDS
- UF   EPOXIDES
- GS   **EPOXY COMPOUNDS**
-     . BORON-EPOXY COMPOUNDS
-     . ENDRIN
-     . ETHYLENE OXIDE
-     . HYOSCINE
-     . PROPYLENE OXIDE

EPOXY COMPOUNDS-(CONT.)
- RT   ∞CHEMICAL COMPOUNDS
-      ETHERS

EPOXY MATRIX COMPOSITES
- GS   COMPOSITE MATERIALS
-     . **EPOXY MATRIX COMPOSITES**
- RT   FIBER ORIENTATION
-      LOW DENSITY RESEARCH
-     ∞MATERIALS
-     ∞MATRICES
-      MATRIX MATERIALS
-      PULTRUSION
-      SANDWICH STRUCTURES

EPOXY RESINS
- GS   PLASTICS
-     . SYNTHETIC RESINS
-     . . THERMOSETTING RESINS
-     . . . **EPOXY RESINS**
-     RESINS
-     . SYNTHETIC RESINS
-     . . THERMOSETTING RESINS
-     . . . **EPOXY RESINS**
- RT   ADHESIVES
-      BORON REINFORCED MATERIALS
-      BORON-EPOXY COMPOUNDS
-      COATINGS
-      GRAPHITE-EPOXY COMPOSITES
-      LAY-UP
-      PREPREGS
-      RESIN MATRIX COMPOSITES

EQUALIZERS (CIRCUITS)
- RT   ATTENUATORS
-     ∞FREQUENCY RESPONSE
-      PHASE SHIFT
-      SIGNAL PROCESSING

∞ EQUATIONS
- SN   *(USE OF A MORE SPECIFIC TERM IS*
-      *RECOMMENDED--CONSULT THE TERMS*
-      *LISTED BELOW)*
- UF   BALANCE EQUATIONS
-      FORCED VIBRATORY MOTION
-       EQUATIONS
- RT   ADIABATIC EQUATIONS
-      APPROXIMATION
-      BERNOULLI THEOREM
-      BETHE-SALPETER EQUATION
-      BIHARMONIC EQUATIONS
-      BLASIUS EQUATION
-      BOLTZMANN TRANSPORT EQUATION
-      BOLTZMANN-VLASOV EQUATION
-      BORN APPROXIMATION
-      BOUNDARY LAYER EQUATIONS
-      BRILLOUIN-WIGNER EQUATION
-      BURGER EQUATION
-      CAUCHY-RIEMANN EQUATIONS
-      CHANDRASEKHAR EQUATION
-      CHAPLYGIN EQUATION
-      CONSERVATION EQUATIONS
-      CONSTITUTIVE EQUATIONS
-      CONTINUITY EQUATION
-      CUBIC EQUATIONS
-      DIFFERENCE EQUATIONS
-      DIFFERENTIAL EQUATIONS
-      DIOPHANTINE EQUATION
-      DIRAC EQUATION
-      DONNELL EQUATIONS
-      DUFFING DIFFERENTIAL EQUATION
-      EIKONAL EQUATION
-      EINSTEIN EQUATIONS
-      ELBER EQUATION
-      ELLIPTIC DIFFERENTIAL EQUATIONS
-      EQUATIONS OF MOTION
-      EQUATIONS OF STATE
-      EQUILIBRIUM EQUATIONS
-      EULER EQUATIONS OF MOTION
-      EULER-CAUCHY EQUATIONS
-      EULER-LAGRANGE EQUATION
-      EULER-LAMBERT EQUATION
-      FADDEEV EQUATIONS
-      FALKNER-SKAN EQUATION
-      FICKS EQUATION
-      FLOW EQUATIONS
-      FOKKER-PLANCK EQUATION
-      FREDHOLM EQUATIONS
-      GAUSS EQUATION
-      GIBBS ADSORPTION EQUATION
-     ∞GIBBS EQUATIONS
-      GIBBS-HELMHOLTZ EQUATIONS
-      GLIMM METHOD
-      HAMILTON-JACOBI EQUATION
-     ∞HELMHOLTZ EQUATIONS

EQUATIONS-*(CONT.)*
　　　HELMHOLTZ VORTICITY EQUATION
　　　HUGONIOT EQUATION OF STATE
　　　HYDRODYNAMIC EQUATIONS
　　　HYPERBOLIC DIFFERENTIAL EQUATIONS
　　　IDENTITIES
　　　INHOUR EQUATION
　　　INTEGRAL EQUATIONS
　　　KINEMATIC EQUATIONS
　　　KINETIC EQUATIONS
　　　KLEIN-GORDON EQUATION
　　　KORTEWEG-DEVRIES EQUATION
　　　KROOK EQUATION
　　　LAME WAVE EQUATIONS
　　　LANDAU-GINZBURG EQUATIONS
　　　LAPLACE EQUATION
　　　LINEAR EQUATIONS
　　　LINEAR EVOLUTION EQUATIONS
　　　LINEARIZATION
　　　LIOUVILLE EQUATIONS
　　　MACROSCOPIC EQUATIONS
　　　MATHIEU FUNCTION
　　　MAXWELL EQUATION
　　　MONGE-AMPERE EQUATION
　　　NAVIER-STOKES EQUATION
　　　NONHOLONOMIC EQUATIONS
　　　NONLINEAR EQUATIONS
　　　NONLINEAR EVOLUTION EQUATIONS
　　　ORR-SOMMERFELD EQUATIONS
　　　PARABOLIC DIFFERENTIAL EQUATIONS
　　　PARTIAL DIFFERENTIAL EQUATIONS
　　　PFAFF EQUATION
　　　POISSON EQUATION
　　　POLYNOMIALS
　　　PRIMITIVE EQUATIONS
　　　QUADRATIC EQUATIONS
　　　QUARTIC EQUATIONS
　　　RAYLEIGH EQUATIONS
　　　REYNOLDS EQUATION
　　　ROOTS OF EQUATIONS
　　　SAHA EQUATIONS
　　　SCHROEDINGER EQUATION
　　　SEMIEMPIRICAL EQUATIONS
　　　SHALLOW SHELL EQUATIONS
　　　SIMULTANEOUS EQUATIONS
　　　SINGULAR INTEGRAL EQUATIONS
　　　STOKES-BELTRAMI EQUATION
　　　THERMODYNAMICS
　　　VLASOV EQUATIONS
　　　VOLTERRA EQUATIONS
　　　VON KARMAN EQUATION
　　　VORTICITY EQUATIONS
　　　WAVE EQUATIONS
　　　WIENER HOPF EQUATIONS

EQUATIONS OF MOTION
　UF　MOTION EQUATIONS
　GS　**EQUATIONS OF MOTION**
　　　. EULER EQUATIONS OF MOTION
　　　. KINETIC EQUATIONS
　　　. . HYDRODYNAMIC EQUATIONS
　　　. . . HELMHOLTZ VORTICITY EQUATION
　　　. . KINEMATIC EQUATIONS
　　　. NAVIER-STOKES EQUATION
　　　. REYNOLDS EQUATION
　RT　AUTONOMY
　　　BETHE-SALPETER EQUATION
　　　CELESTIAL MECHANICS
　　　CLASSICAL MECHANICS
　　　COMPUTATIONAL FLUID DYNAMICS
　　　CONTINUITY EQUATION
　　　CONTROL MOMENT GYROSCOPES
　　　∞DYNAMICS
　　　EINSTEIN EQUATIONS
　　　∞EQUATIONS
　　　EQUILIBRIUM EQUATIONS
　　　HAMILTON-JACOBI EQUATION
　　　INERTIA PRINCIPLE
　　　KINEMATICS
　　　LISSAJOUS FIGURES
　　　MACH INERTIA PRINCIPLE
　　　MOMENTS OF INERTIA
　　　MOTION AFTEREFFECTS
　　　SPINNING UNGUIDED ROCKET
　　　　TRAJECTORY
　　　STABILITY
　　　SYSTEMS STABILITY
　　　TRAJECTORIES
　　　TRAJECTORY ANALYSIS
　　　VARIABLE MASS SYSTEMS
　　　VON ZEIPEL METHOD

EQUATIONS OF STATE
　UF　STATE EQUATIONS
　GS　**EQUATIONS OF STATE**

EQUATIONS OF STATE-*(CONT.)*
　　　. HUGONIOT EQUATION OF STATE
　RT　ADIABATIC EQUATIONS
　　　BBGKY HIERARCHY
　　　BOSE GEOMETRY
　　　COMPRESSIBILITY
　　　CONTINUITY EQUATION
　　　∞EQUATIONS
　　　EQUILIBRIUM EQUATIONS
　　　IDEAL FLUIDS
　　　IDEAL GAS
　　　KINETIC THEORY
　　　MOLLIER DIAGRAM
　　　REAL GASES
　　　THERMODYNAMICS
　　　VIRIAL COEFFICIENTS

EQUATORIAL ATMOSPHERE
　RT　∞ATMOSPHERES
　　　ATMOSPHERIC COMPOSITION
　　　METEOROLOGICAL PARAMETERS
　　　MIDDLE ATMOSPHERE
　　　TROPICAL METEOROLOGY
　　　TROPICAL REGIONS

EQUATORIAL ELECTROJET
　GS　ELECTRIC CURRENT
　　　. IONOSPHERIC CURRENTS
　　　. . ELECTROJETS
　　　. . . **EQUATORIAL ELECTROJET**
　　　ELECTRICITY
　　　. ATMOSPHERIC ELECTRICITY
　　　. IONOSPHERIC CURRENTS
　　　. . ELECTROJETS
　　　. . . . **EQUATORIAL ELECTROJET**
　RT　AURORAL ELECTROJETS

EQUATORIAL ORBITS
　GS　ORBITS
　　　. **EQUATORIAL ORBITS**
　　　. . STATIONARY ORBITS
　RT　CIRCULAR ORBITS
　　　EARTH ORBITS
　　　ELLIPTICAL ORBITS
　　　GEOSYNCHRONOUS ORBITS
　　　LUNAR ORBITS
　　　ORBITAL MECHANICS
　　　PLANETARY ORBITS
　　　POLAR ORBITS
　　　SATELLITE ORBITS
　　　SPACECRAFT ORBITS
　　　TWENTY-FOUR HOUR ORBITS

EQUATORIAL REGIONS
　GS　REGIONS
　　　. **EQUATORIAL REGIONS**
　RT　ARID LANDS
　　　EARTH SURFACE
　　　TROPICAL REGIONS

EQUATORS
　GS　**EQUATORS**
　　　. LUNAR EQUATOR
　　　. MAGNETIC EQUATOR
　RT　COORDINATES
　　　ROTATING SPHERES
　　　TRANSEQUATORIAL PROPAGATION

∞ EQUILIBRIUM
　SN　*(USE OF A MORE SPECIFIC TERM IS*
　　　RECOMMENDED--CONSULT THE TERMS
　　　LISTED BELOW)
　RT　ACID BASE EQUILIBRIUM
　　　AEROSTATICS
　　　BALANCE
　　　BALANCING
　　　BODY SWAY TEST
　　　CHEMICAL EQUILIBRIUM
　　　DIFFUSION
　　　DIFFUSION COEFFICIENT
　　　DYNAMIC CHARACTERISTICS
　　　EQUILIBRIUM EQUATIONS
　　　HEAT OF DISSOCIATION
　　　HOMEOSTASIS
　　　ISOSTASY
　　　LIQUID-VAPOR EQUILIBRIUM
　　　LOADS (FORCES)
　　　MAXWELL-MOHR METHOD
　　　NONEQUILIBRIUM CONDITIONS
　　　ONSAGER RELATIONSHIP
　　　PLASMA EQUILIBRIUM
　　　RELAXATION (MECHANICS)
　　　RELAXATION TIME
　　　STABILITY
　　　STABILIZATION

EQUILIBRIUM-*(CONT.)*
　　　STATICS
　　　STEADY STATE
　　　SYSTEMS STABILITY
　　　THERMODYNAMIC EQUILIBRIUM
　　　THERMODYNAMIC PROPERTIES
　　　THERMODYNAMICS
　　　TRANSITION POINTS
　　　UNSTEADY STATE
　　　VARIABILITY
　　　WATER BALANCE

EQUILIBRIUM DIAGRAMS
　USE　PHASE DIAGRAMS

EQUILIBRIUM EQUATIONS
　RT　ANALYSIS (MATHEMATICS)
　　　∞EQUATIONS
　　　EQUATIONS OF MOTION
　　　EQUATIONS OF STATE
　　　∞EQUILIBRIUM

EQUILIBRIUM FLOW
　UF　STEADY STATE FLOW
　GS　FLUID FLOW
　　　. GAS FLOW
　　　. . **EQUILIBRIUM FLOW**
　　　. . . FROZEN EQUILIBRIUM FLOW
　　　. . . . SHIFTING EQUILIBRIUM FLOW
　RT　EYRING THEORY
　　　HEAT TRANSMISSION
　　　NONEQUILIBRIUM FLOW
　　　PLASMA EQUILIBRIUM
　　　QUASI-STEADY STATES
　　　STEADY FLOW

EQUILIBRIUM METHODS
　SN　(LIMITED TO STRUCTURAL ANALYSIS)
　GS　STRUCTURAL ANALYSIS
　　　. **EQUILIBRIUM METHODS**
　RT　MATRIX METHODS
　　　∞METHODOLOGY
　　　VARIATIONAL PRINCIPLES

EQUINOXES
　RT　SEASONS
　　　SOLAR POSITION
　　　SOLSTICES
　　　WINTER

EQUIPARTITION THEOREM
　UF　ENERGY EQUIPARTITION
　GS　THEOREMS
　　　. **EQUIPARTITION THEOREM**
　RT　DEGREES OF FREEDOM
　　　ENERGY DISTRIBUTION
　　　KINETIC ENERGY
　　　SPECIFIC HEAT

∞ EQUIPMENT
　SN　*(USE OF A MORE SPECIFIC TERM IS*
　　　RECOMMENDED--CONSULT THE TERMS
　　　LISTED BELOW)
　UF　APPARATUS
　RT　ABORT APPARATUS
　　　ABSORBERS (EQUIPMENT)
　　　ACCUMULATORS (COMPUTERS)
　　　AIR CONDITIONING EQUIPMENT
　　　AIRBORNE EQUIPMENT
　　　AIRCRAFT EQUIPMENT
　　　AIRPORT SURFACE DETECTION
　　　　EQUIPMENT
　　　AUDIO EQUIPMENT
　　　AUTOMATIC TEST EQUIPMENT
　　　AUXILIARY EQUIPMENT (COMPUTERS)
　　　BEDDING EQUIPMENT
　　　BOMBING EQUIPMENT
　　　BREATHING APPARATUS
　　　COMMUNICATION EQUIPMENT
　　　COMPUTER STORAGE DEVICES
　　　CONSOLES
　　　CRYOGENIC COMPUTER STORAGE
　　　CRYOGENIC EQUIPMENT
　　　DATA PROCESSING EQUIPMENT
　　　DISTILLATION EQUIPMENT
　　　ELECTROMECHANICAL DEVICES
　　　ELECTRONIC EQUIPMENT
　　　GROUND SUPPORT EQUIPMENT
　　　HANDLING EQUIPMENT
　　　HARDWARE
　　　HEATING EQUIPMENT
　　　HYDRAULIC EQUIPMENT
　　　LABORATORY EQUIPMENT
　　　LIGHTING EQUIPMENT
　　　MECHANICAL DEVICES

EQUIPMENT-(CONT.)
- .MEDICAL EQUIPMENT
- MINIATURE ELECTRONIC EQUIPMENT
- ONBOARD EQUIPMENT
- PERIPHERAL EQUIPMENT (COMPUTERS)
- PHOTOGRAPHIC EQUIPMENT
- PNEUMATIC EQUIPMENT
- PORTABLE EQUIPMENT
- RADAR EQUIPMENT
- RIGGING
- SAFETY DEVICES
- SELF ERECTING DEVICES
- SERVICE LIFE
- SPACECRAFT EQUIPMENT
- SURVIVAL EQUIPMENT
- SYRINGES
- TELEVISION EQUIPMENT
- ∞ TEST EQUIPMENT
- WIND TUNNEL APPARATUS
- X RAY APPARATUS

EQUIPMENT SPECIFICATIONS
GS   SPECIFICATIONS
. **EQUIPMENT SPECIFICATIONS**
RT   AIRCRAFT PRODUCTION
COMMONALITY
∞ DESIGN
FUNCTIONAL DESIGN SPECIFICATIONS
MAINTENANCE
PROCUREMENT

EQUIPOTENTIALS
GS   FLUID FLOW
. POTENTIAL FLOW
. . **EQUIPOTENTIALS**
RT   ∞ FLOW GRAPHS
FLOW NETS

EQUIVALENCE
GS   MATHEMATICAL LOGIC
. SET THEORY
. . **EQUIVALENCE**
RT   PARITY
PARTITIONS (MATHEMATICS)

EQUIVALENT CIRCUITS
GS   CIRCUITS
. **EQUIVALENT CIRCUITS**
RT   DUALITY PRINCIPLE
NETWORK ANALYSIS
NETWORK SYNTHESIS
SUPERPOSITION (MATHEMATICS)

ERBE
USE   EARTH RADIATION BUDGET
EXPERIMENT

ERBIUM
GS   CHEMICAL ELEMENTS
. RARE EARTH ELEMENTS
. . **ERBIUM**
. . . ERBIUM ISOTOPES
METALS
. RARE EARTH ELEMENTS
. . **ERBIUM**
. . . ERBIUM ISOTOPES

ERBIUM ALLOYS
GS   ALLOYS
. RARE EARTH ALLOYS
. . **ERBIUM ALLOYS**

ERBIUM COMPOUNDS
GS   RARE EARTH COMPOUNDS
. **ERBIUM COMPOUNDS**
RT   ∞ CHEMICAL COMPOUNDS
∞ METAL COMPOUNDS

ERBIUM ISOTOPES
UF   ERBIUM 169
ERBIUM 171
GS   CHEMICAL ELEMENTS
. NUCLIDES
. . ISOTOPES
. . . **ERBIUM ISOTOPES**
. RARE EARTH ELEMENTS
. . ERBIUM
. . . **ERBIUM ISOTOPES**
METALS
. RARE EARTH ELEMENTS
. . ERBIUM
. . . **ERBIUM ISOTOPES**

ERBIUM 169
USE   ERBIUM ISOTOPES

ERBIUM 171
USE   ERBIUM ISOTOPES

ERECTION
USE   CONSTRUCTION

EREP
UF   EARTH RESOURCES EXPERIMENT
PACKAGE
GS   PACKAGES
. INSTRUMENT PACKAGES
. . **EREP**
RT   ∞ INSTRUMENTS
SKYLAB 1
SKYLAB 2
SKYLAB 3
SKYLAB 4

ERGODIC PROCESS
RT   INFORMATION THEORY
PROBABILITY THEORY
STOCHASTIC PROCESSES
THERMODYNAMICS

ERGOMETERS
GS   MEASURING INSTRUMENTS
. **ERGOMETERS**
RT   DYNAMOMETERS

ERGONOMICS
USE   HUMAN FACTORS ENGINEERING

ERGOTAMINE
GS   AMINES
. **ERGOTAMINE**
DRUGS
. **ERGOTAMINE**
HETEROCYCLIC COMPOUNDS
. ALKALOIDS
. . **ERGOTAMINE**
NITROGEN COMPOUNDS
. ALKALOIDS
. . **ERGOTAMINE**

EROS (SATELLITES)
UF   EARTH RESOURCES OBSERVATION
SATELLITES
GS   SATELLITES
. ARTIFICIAL SATELLITES
. . **EROS (SATELLITES)**
RT   EARTH RESOURCES
OCEANOGRAPHY
REMOTE SENSORS
SATELLITE OBSERVATION
SCANNING
TERRAIN ANALYSIS

EROS PROJECT
USE   EXPERIMENTAL REFLECTOR ORBITAL
SHOT PROJ

EROSION
UF   SCARS (GEOLOGY)
GS   **EROSION**
. RAIN EROSION
. SOIL EROSION
. WATER EROSION
. WIND EROSION
RT   ABLATION
ABRASION
ARROYOS
ATMOSPHERIC EFFECTS
CAVITATION FLOW
CORROSION
DEGRADATION
DETERIORATION
EROSIVE BURNING
ETCHING
FRETTING
HOT CORROSION
HYDROGEOLOGY
IMPINGEMENT
INLIERS (LANDFORMS)
METAL SURFACES
METAL-WATER REACTIONS
PENEPLAINS
PITTING
PLATEAUS
RAIN IMPACT DAMAGE
RAVINES
RIVERS

EROSION-(CONT.)
- SOIL SCIENCE
- SPARK MACHINING
- SURFACE REACTIONS
- TRIBOLOGY
- VALLEYS
- WAVE RESISTANCE
- WEAR
- WEAR TESTS
- WEATHERING
- WIND EFFECTS

EROSIVE BURNING
GS   COMBUSTION
. **EROSIVE BURNING**
RT   BURNOUT
COMBUSTION TEMPERATURE
DETERIORATION
EROSION
EXHAUST GASES
FUEL COMBUSTION
HYPERSONIC COMBUSTION
OXIDATION
PITTING
PROPELLANT COMBUSTION
SOLID PROPELLANT COMBUSTION
TRIBOLOGY

ERROR ANALYSIS
GS   ANALYSIS (MATHEMATICS)
. NUMERICAL ANALYSIS
. . **ERROR ANALYSIS**
RT   ∞ ANALYZING
∞ APPLICATIONS OF MATHEMATICS
BIT ERROR RATE
BORESIGHT ERROR
CENSORED DATA (MATHEMATICS)
FAULT TOLERANCE
MEAN SQUARE VALUES
PROBABILITY THEORY
RANGE ERRORS
RAYLEIGH DISTRIBUTION
ROOT-MEAN-SQUARE ERRORS

ERROR BAND
USE   ACCURACY

ERROR CORRECTING CODES
RT   BIT ERROR RATE
∞ CODES
CONCATENATED CODES
DIGITAL TECHNIQUES
REDUNDANCY ENCODING

ERROR CORRECTING DEVICES
RT   BCH CODES
CORRECTION
∞ DEVICES
INSTRUMENT COMPENSATION
REDUNDANCY ENCODING

ERROR DETECTION CODES
RT   BIT ERROR RATE
∞ CODES
CODING
COMPUTER PROGRAMS
COMPUTER SYSTEMS PROGRAMS
DIGITAL TECHNIQUES
FAULT TOLERANCE
INFORMATION THEORY
PARITY
PROVING
QUALITY CONTROL
REDUNDANCY
REDUNDANCY ENCODING

ERROR FUNCTIONS
GS   FUNCTIONS (MATHEMATICS)
. **ERROR FUNCTIONS**
RT   STATISTICAL DISTRIBUTIONS

ERROR SIGNALS
RT   BIT ERROR RATE
COMPARATORS
COMPENSATORS
DIFFERENTIAL AMPLIFIERS
DISCRIMINATORS
ERRORS
PHASE ERROR
POSITION ERRORS
RANGE ERRORS
SIGNAL MIXING
∞ SIGNALS
SLEWING

ERRORS
UF   INVALIDITY
GS   **ERRORS**
.   BORESIGHT ERROR
.   INSTRUMENT ERRORS
.   PHASE ERROR
.   PILOT ERROR
.   POSITION ERRORS
.   RANDOM ERRORS
.   RANGE ERRORS
.   ROOT-MEAN-SQUARE ERRORS
.   TRUNCATION ERRORS
.   VELOCITY ERRORS
RT   ACCURACY
     BIAS
∞COMPENSATION
     COMPUTER PROGRAM INTEGRITY
     CONFIDENCE
     CONSISTENCY
     CORRECTION
     DRIFT (INSTRUMENTATION)
     DYNAMIC CHARACTERISTICS
     ERROR SIGNALS
     HYSTERESIS
     LINEARITY
     MALFUNCTIONS
     MEDIAN (STATISTICS)
     OPTICAL CORRECTION PROCEDURE
     PRECISION
     QUALITY CONTROL
     RANGE (EXTREMES)
     RELIABILITY
     RESOLUTION
     RESPONSE BIAS
∞SCALING
∞TESTS
     TOLERANCES (MECHANICS)

ERS 17
GS   SATELLITES
.   ARTIFICIAL SATELLITES
. . SCIENTIFIC SATELLITES
. . . ENVIRONMENTAL RESEARCH
       SATELLITES
. . . . **ERS 17**
RT   ATLAS AGENA LAUNCH VEHICLES

ERS 18
GS   SATELLITES
.   ARTIFICIAL SATELLITES
. . SCIENTIFIC SATELLITES
. . . ENVIRONMENTAL RESEARCH
       SATELLITES
. . . . **ERS 18**

ERS-1 (ESA SATELLITE)
GS   SATELLITES
.   ARTIFICIAL SATELLITES
. . ESA SATELLITES
. . . **ERS-1 (ESA SATELLITE)**
. . MARITIME SATELLITES
. . . **ERS-1 (ESA SATELLITE)**
RT   EUROPEAN SPACE AGENCY

ERTS
USE LANDSAT SATELLITES

ERTS-A
USE LANDSAT 1

ERTS-B
USE LANDSAT 2

ERTS-C
USE LANDSAT 3

ERTS-D
USE LANDSAT 4

ERTS-E
USE LANDSAT E

ERTS-F
USE LANDSAT F

ERYTHROCYTES
UF   RED BLOOD CELLS
GS   ANATOMY
.   CARDIOVASCULAR SYSTEM
. . **ERYTHROCYTES**
. . . RETICULOCYTES
     BODY FLUIDS
.   BLOOD
. . **ERYTHROCYTES**

ERYTHROCYTES-(CONT.)
. . . RETICULOCYTES
     CELLS (BIOLOGY)
.   **ERYTHROCYTES**
. . RETICULOCYTES
RT   BONE MARROW
     CARBOXYHEMOGLOBIN
     CORPUSCLES
     HEMATOCRIT RATIO
     HEMOGLOBIN
     HEMOLYSIS
     LEUKOCYTES
     OXYHEMOGLOBIN

ESA
USE EUROPEAN SPACE AGENCY

ESA SATELLITES
SN   (EUROPEAN SPACE RESEARCH
       ORGANIZATION SATELLITES)
UF   ESRO SATELLITES
     EUROPEAN SPACE RESEARCH
       ORGANIZATION SAT
GS   ESA SPACECRAFT
.   **ESA SATELLITES**
. . ESRO 1 SATELLITE
. . L-SAT
. . MARECS MARITIME SATELLITES
. . MAROTS (ESA)
. . METEOSAT SATELLITE
. . OTS (ESA)
. . TD SATELLITES
. . . TD-1 SATELLITE
     SATELLITES
.   ARTIFICIAL SATELLITES
. . **ESA SATELLITES**
. . . AEROSAT SATELLITES
. . . COS-B SATELLITE
. . . ERS-1 (ESA SATELLITE)
. . . ESRO 1 SATELLITE
. . . ESRO 2 SATELLITE
. . . ESRO 4 SATELLITE
. . . EUROPEAN COMMUNICATIONS
         SATELLITE
. . . EXOSAT SATELLITE
. . . GEOS SATELLITES (ESA)
. . . HEOS SATELLITES
. . . . HEOS A SATELLITE
. . . . HEOS B SATELLITE
. . . HIPPARCOS SATELLITE
. . . L-SAT
. . . MAGELLAN MISSION
. . . MARECS MARITIME SATELLITES
. . . MAROTS (ESA)
. . . METEOSAT SATELLITE
. . . OTS (ESA)
. . . TD SATELLITES
. . . . TD-1 SATELLITE
RT   EARTHNET
     EUROPEAN SPACE AGENCY
     EUROPEAN SPACE PROGRAMS
     INTERNATIONAL COOPERATION

ESA SPACECRAFT
SN   EUROPEAN SPACE RESEARCH
       ORGANIZATION SATELLITES)
GS   **ESA SPACECRAFT**
.   COS-B SATELLITE
.   ESA SATELLITES
. . ESRO 1 SATELLITE
. . L-SAT
. . MARECS MARITIME SATELLITES
. . MAROTS (ESA)
. . METEOSAT SATELLITE
. . OTS (ESA)
. . TD SATELLITES
. . . TD-1 SATELLITE
.   GIOTTO MISSION
RT   ∞SPACECRAFT

ESAKI DIODES
USE TUNNEL DIODES

ESCALATORS
RT   ELEVATORS (LIFTS)
     LADDERS
∞LIFTS
     STAIRWAYS

∞ **ESCAPE**
SN   (USE OF A MORE SPECIFIC TERM IS
       RECOMMENDED--CONSULT THE TERMS
       LISTED BELOW)
RT   ESCAPE (ABANDONMENT)
     ESCAPE CAPSULES

ESCAPE-(CONT.)
     ESCAPE ROCKETS
     ESCAPE SYSTEMS
     ESCAPE VELOCITY
     LEAKAGE

ESCAPE (ABANDONMENT)
RT   BAILOUT
     EJECTION
     EJECTION TRAINING
     ENTRAPMENT
∞ESCAPE
     ESCAPE ROCKETS
     ESCAPE SYSTEMS
     JETTISON SYSTEMS
     JETTISONING
     PARACHUTE DESCENT

ESCAPE CAPSULES
GS   SAFETY DEVICES
.   **ESCAPE CAPSULES**
     SPACE CAPSULES
.   **ESCAPE CAPSULES**
RT   ABORT APPARATUS
     ABORTED MISSIONS
     EJECTION SEATS
     EMERGENCY LIFE SUSTAINING
       SYSTEMS
∞ESCAPE
     FLYING EJECTION SEATS
     HIGH ALTITUDE ENVIRONMENTS
     LAUNCH ESCAPE SYSTEMS
     LUNAR ESCAPE DEVICES
     PARACONE
     PRESSURIZED CABINS

ESCAPE ROCKETS
GS   SAFETY DEVICES
.   **ESCAPE ROCKETS**
-
RT   ABORT APPARATUS
     ABORTED MISSIONS
∞ESCAPE
     ESCAPE (ABANDONMENT)
     LAUNCH ESCAPE SYSTEMS
     LUNAR ESCAPE DEVICES
∞ROCKETS
∞SPACECRAFT

ESCAPE SYSTEMS
GS   **ESCAPE SYSTEMS**
.   LAUNCH ESCAPE SYSTEMS
RT   BAILOUT
     EJECTION
     EJECTION SEATS
∞ESCAPE
     ESCAPE (ABANDONMENT)
     JETTISON SYSTEMS
     PARACONE
     SAFETY FACTORS
∞SYSTEMS

ESCAPE VELOCITY
UF   PARABOLIC VELOCITY
GS   RATES (PER TIME)
.   **ESCAPE VELOCITY**
     VELOCITY
.   **ESCAPE VELOCITY**
RT   ∞ESCAPE
     HIGH SPEED
     HYPERBOLIC TRAJECTORIES
∞HYPERVELOCITY
     ORBITAL VELOCITY
     PLANETARY GRAVITATION
     SCHWARZSCHILD METRIC
     VELOCITY ERRORS

ESCARPMENTS
UF   SCARPS
GS   LANDFORMS
.   **ESCARPMENTS**
RT   CLIFFS
     SLOPES
     TOPOGRAPHY

ESCHERICHIA
GS   MICROORGANISMS
.   BACTERIA
. . **ESCHERICHIA**

ESG (GYROSCOPES)
USE ELECTROSTATIC GYROSCOPES

ESKERS
USE GLACIAL DRIFT

ESKIMOS
RT   ANTHROPOLOGY
      CULTURE (SOCIAL SCIENCES)

ESOPHAGUS
GS   ANATOMY
    . DIGESTIVE SYSTEM
    . . **ESOPHAGUS**
    . ORGANS
    . . **ESOPHAGUS**
    VISCERA
    . ORGANS
    . . **ESOPHAGUS**

ESRO
USE   EUROPEAN SPACE AGENCY

ESRO SATELLITES
USE   ESA SATELLITES

ESRO 1 SATELLITE
GS   ESA SPACECRAFT
    . ESA SATELLITES
    . . **ESRO 1 SATELLITE**
    SATELLITES
    . ARTIFICIAL SATELLITES
    . . ESA SATELLITES
    . . . **ESRO 1 SATELLITE**
RT   EUROPEAN SPACE AGENCY
    EUROPEAN SPACE PROGRAMS

ESRO 2 SATELLITE
GS   SATELLITES
    . ARTIFICIAL SATELLITES
    . . ESA SATELLITES
    . . . **ESRO 2 SATELLITE**
RT   EUROPEAN SPACE AGENCY
    EUROPEAN SPACE PROGRAMS

ESRO 4 SATELLITE
GS   SATELLITES
    . ARTIFICIAL SATELLITES
    . . ESA SATELLITES
    . . . **ESRO 4 SATELLITE**
RT   AURORAS
    EUROPEAN SPACE AGENCY
    EUROPEAN SPACE PROGRAMS
    PARTICLE DENSITY (CONCENTRATION)
    SCIENTIFIC SATELLITES

ESSA SATELLITES
GS   SATELLITES
    . ARTIFICIAL SATELLITES
    . . METEOROLOGICAL SATELLITES
    . . . **ESSA SATELLITES**
    . . . . ESSA 1 SATELLITE
    . . . . ESSA 2 SATELLITE
    . . . . ESSA 3 SATELLITE
    . . . . ESSA 4 SATELLITE
    . . . . ESSA 5 SATELLITE
    . . . . ESSA 6 SATELLITE
    . . . . ESSA 7 SATELLITE
    . . . . ESSA 8 SATELLITE
    . . . . ESSA 9 SATELLITE
RT   CLOUD PHOTOGRAPHY
    NIMBUS SATELLITES
    SATELLITE OBSERVATION
    TIROS SATELLITES

ESSA 1 SATELLITE
UF   OT-3
GS   SATELLITES
    . ARTIFICIAL SATELLITES
    . . METEOROLOGICAL SATELLITES
    . . . ESSA SATELLITES
    . . . . **ESSA 1 SATELLITE**
RT   DELTA LAUNCH VEHICLE

ESSA 2 SATELLITE
UF   OT-2
GS   SATELLITES
    . ARTIFICIAL SATELLITES
    . . METEOROLOGICAL SATELLITES
    . . . ESSA SATELLITES
    . . . . **ESSA 2 SATELLITE**
RT   DELTA LAUNCH VEHICLE

ESSA 3 SATELLITE
UF   TOS-A
GS   SATELLITES
    . ARTIFICIAL SATELLITES
    . . METEOROLOGICAL SATELLITES
    . . . ESSA SATELLITES
    . . . . **ESSA 3 SATELLITE**

ESSA 3 SATELLITE-*(CONT.)*
RT   DELTA LAUNCH VEHICLE

ESSA 4 SATELLITE
GS   SATELLITES
    . ARTIFICIAL SATELLITES
    . . METEOROLOGICAL SATELLITES
    . . . ESSA SATELLITES
    . . . . **ESSA 4 SATELLITE**
RT   DELTA LAUNCH VEHICLE

ESSA 5 SATELLITE
GS   SATELLITES
    . ARTIFICIAL SATELLITES
    . . METEOROLOGICAL SATELLITES
    . . . ESSA SATELLITES
    . . . . **ESSA 5 SATELLITE**
RT   DELTA LAUNCH VEHICLE

ESSA 6 SATELLITE
GS   SATELLITES
    . ARTIFICIAL SATELLITES
    . . METEOROLOGICAL SATELLITES
    . . . ESSA SATELLITES
    . . . . **ESSA 6 SATELLITE**
RT   DELTA LAUNCH VEHICLE

ESSA 7 SATELLITE
GS   SATELLITES
    . ARTIFICIAL SATELLITES
    . . METEOROLOGICAL SATELLITES
    . . . ESSA SATELLITES
    . . . . **ESSA 7 SATELLITE**
RT   DELTA LAUNCH VEHICLE

ESSA 8 SATELLITE
GS   SATELLITES
    . ARTIFICIAL SATELLITES
    . . METEOROLOGICAL SATELLITES
    . . . ESSA SATELLITES
    . . . . **ESSA 8 SATELLITE**
RT   DELTA LAUNCH VEHICLE

ESSA 9 SATELLITE
GS   SATELLITES
    . ARTIFICIAL SATELLITES
    . . METEOROLOGICAL SATELLITES
    . . . ESSA SATELLITES
    . . . . **ESSA 9 SATELLITE**
RT   DELTA LAUNCH VEHICLE

ESTERS
GS   **ESTERS**
    . ACRYLATES
    . ALKYLATES
    . ASPARTATES
    . CARBAMATES (TRADENAME)
    . . URETHANES
    . CARBOXYLATES
    . CHLOROFORMATE
    . COBALT ACETATES
    . CYANURATES
    . GLUTAMATES
    . GLYCERIDES
    . ISOCYANATES
    . . DIISOCYANATES
    . . FULMINATES
    . LACTATES
    . LEAD ACETATES
    . MALEATES
    . MEPROBAMATE
    . NITRATE ESTERS
    . . ISOPROPYL NITRATE
    . . PROPYL NITRATE
    . OCTOATES
    . ORGANIC NITRATES
    . . CELLULOSE NITRATE
    . . NITROFORMS
    . . . HYDRAZINE NITROFORM
    . . NITROGLYCERIN
    . . PETN
    . PHTHALATES
    . POLYCARBONATES
    . . LEXAN (TRADEMARK)
    . POLYESTERS
    . POLYETHYLENE TEREPHTHALATE
    . SODIUM CHLORODIFLUOROACETATES
    . SODIUM SALICYLATES
    . STEARATES
    . SULFONATES
    . TRIACETIN
RT   ACETATES
    ACETYL COMPOUNDS
    ALIPHATIC COMPOUNDS
    CITRATES

ESTERS-*(CONT.)*
    CYANATES
    LIPIDS
    NITROSYLS
    PLASTICIZERS
    SALICYLATES
    SKYDROL (TRADEMARK)

ESTIMATES
GS   **ESTIMATES**
    . COST ESTIMATES
RT   ALLOCATIONS
    COMPARISON
    CONFIDENCE LIMITS
    CONTINGENCY
    CONTRACTS
    DAMAGE ASSESSMENT
    ESTIMATING
    EVALUATION
    FORECASTING
    LIKELIHOOD RATIO
    MANAGEMENT METHODS
    MANAGEMENT PLANNING
    NOISE PREDICTION (AIRCRAFT)
    PARAMETER IDENTIFICATION
    PREDICTIONS
    PRODUCTION MANAGEMENT
    PROJECT PLANNING
    QUALITY CONTROL
    RELIABILITY
    RESERVES
    RISK
    STATISTICAL ANALYSIS
    STATISTICAL TESTS
    ∞ STATISTICS
    SUBCONTRACTS
    SYSTEM IDENTIFICATION
    VALUE

ESTIMATING
GS   **ESTIMATING**
    . ORBITAL POSITION ESTIMATION
    . PARAMETER IDENTIFICATION
    . SYSTEM IDENTIFICATION
RT   BUDGETING
    CONTRACTS
    COSTS
    COUNTING
    CRITICAL PATH METHOD
    DELPHI METHOD (FORECASTING)
    ∞ DESIGN
    ESTIMATES
    ∞ ESTIMATORS
    EVALUATION
    FEASIBILITY
    FORECASTING
    ∞ MEASUREMENT
    MISSION PLANNING
    NUMERICAL DIFFERENTIATION
    PATTERN METHOD (FORECASTING)
    PROBE METHOD (FORECASTING)
    PROFILE METHOD (FORECASTING)
    PROJECTS
    QUALITY CONTROL
    RESERVES
    RISK
    SAMPLING
    STANDARD DEVIATION
    STATISTICAL ANALYSIS
    STATISTICAL TESTS
    ∞ STATISTICS
    TECHNOLOGICAL FORECASTING
    VALUE

∞ **ESTIMATORS**
SN   *(USE OF A MORE SPECIFIC TERM IS
    RECOMMENDED--CONSULT THE TERMS
    LISTED BELOW)*
RT   CORRELATION
    COST ESTIMATES
    ESTIMATING
    PARAMETERIZATION
    PERSONNEL

ESTONIA
RT   BALTIC SEA
    EUROPE
    NATIONS

ESTROGENS
GS   SECRETIONS
    . ENDOCRINE SECRETIONS
    . . HORMONES
    . . . **ESTROGENS**
RT   ENDOCRINE GLANDS

ESTROGENS-*(CONT.)*
    SEX GLANDS

ESTUARIES
UF   OUTLETS (GEOLOGY)
RT   COASTS
        FISHERIES
        GEOGRAPHY
        HARBORS
        OCEANOGRAPHY
        RIVERS
        TIDAL FLATS
        TIDES
        TRIBUTARIES

ETA-MESONS
GS   PARTICLES
    . ELEMENTARY PARTICLES
    .. BOSONS
    ... MESONS
    .... **ETA-MESONS**
    . FERMIONS
    ... **ETA-MESONS**
    . NUCLEAR PARTICLES
    .. BOSONS
    ... MESONS
    .... **ETA-MESONS**
RT   BARYONS
        CHARGED PARTICLES
        OMEGA-MESONS
        RHO-MESONS
        SIGMA-MESONS

ETCHANTS
RT   CORROSION
        ETCHING

ETCHING
RT   CORROSION
        ENGRAVING
        EROSION
        ETCHANTS
        METALLOGRAPHY
        PITTING
        ULTRASONIC CLEANING

ETHANE
GS   ALIPHATIC COMPOUNDS
    . ALKANES
    .. **ETHANE**
    HYDROCARBONS
    . ALKANES
    .. **ETHANE**
RT   HYDROCARBON FUELS

ETHERS
GS   **ETHERS**
    . ACETALS
    . ANISOLE
    . DIETHYL ETHER
    . GALLAMINE TRIETHIODIDE
    . POLYPHENYL ETHER
RT   ANESTHETICS
        DRUGS
        EPOXY COMPOUNDS
        PROPARGYL GROUPS

ETHICS
RT   ∞METHODOLOGY
        NORMS
        RESEARCH

ETHIOPIA
GS   NATIONS
    . **ETHIOPIA**
RT   AFRICA

ETHNIC FACTORS
GS   SOCIOLOGY
    . SOCIAL FACTORS
    .. **ETHNIC FACTORS**
RT   AMERICAN INDIANS
        COMMUNITIES
        CULTURE (SOCIAL SCIENCES)
        GROUP DYNAMICS
        SOCIOLOGY

ETHOXY ETHYLENE
GS   ALIPHATIC COMPOUNDS
    . **ETHOXY ETHYLENE**

ETHYL ALCOHOL
GS   ALIPHATIC COMPOUNDS
    . **ETHYL ALCOHOL**

ETHYL ALCOHOL-*(CONT.)*
    HYDROXYL COMPOUNDS
    . ALCOHOLS
    .. **ETHYL ALCOHOL**
RT   ATMOSPHERIC ENERGY SOURCES
        CARBOHYDRATES

ETHYL COMPOUNDS
RT   ∞CHEMICAL COMPOUNDS
        DIETHYL HYDROGEN PHOSPHITE (DEHP)
        TETRAETHYL ORTHOSILICATE
        TRIETHYL COMPOUNDS

ETHYLENE
GS   ALIPHATIC COMPOUNDS
    . ALKENES
    .. **ETHYLENE**
    ... VINYLIDENE
    HYDROCARBONS
    . ALKENES
    .. **ETHYLENE**
    ... VINYLIDENE
RT   HYDROCARBON FUELS
        POLYETHYLENES

ETHYLENE COMPOUNDS
GS   **ETHYLENE COMPOUNDS**
    . CHLOROETHYLENE
    . ETHYLENE DIHYDRAZINE
    . ETHYLENEDIAMINE
    . ETHYLENEDIAMINETETRAACETIC ACIDS
RT   ∞CHEMICAL COMPOUNDS

ETHYLENE DIHYDRAZINE
GS   ALIPHATIC COMPOUNDS
    . HYDRAZINES
    .. **ETHYLENE DIHYDRAZINE**
    ETHYLENE COMPOUNDS
    . **ETHYLENE DIHYDRAZINE**

ETHYLENE OXIDE
GS   EPOXY COMPOUNDS
    . **ETHYLENE OXIDE**

ETHYLENEDIAMINE
GS   ALIPHATIC COMPOUNDS
    . **ETHYLENEDIAMINE**
    AMINES
    . DIAMINES
    .. **ETHYLENEDIAMINE**
    ETHYLENE COMPOUNDS
    . **ETHYLENEDIAMINE**

ETHYLENEDIAMINETETRAACETIC ACIDS
UF   EDTA
GS   ACIDS
    . FATTY ACIDS
    .. CARBOXYLIC ACIDS
    ... ACETIC ACID
    .... **ETHYLENEDIAMINETETRAACETIC ACIDS**
    ALIPHATIC COMPOUNDS
    . FATTY ACIDS
    .. ACETIC ACID
    ... **ETHYLENEDIAMINETETRAACETIC ACIDS**
    ETHYLENE COMPOUNDS
    . **ETHYLENEDIAMINETETRAACETIC ACIDS**
    ORGANIC COMPOUNDS
    . FATTY ACIDS
    .. ACETIC ACID
    ... **ETHYLENEDIAMINETETRAACETIC ACIDS**
RT   ACETATES
        DETERGENTS

ETIOLOGY
RT   CASE HISTORIES
        CAUSES
        DISEASES
        PREVENTION

ETR (REACTORS)
USE   ENGINEERING TEST REACTORS

ETTINGSHAUSEN COOLERS
USE   ETTINGSHAUSEN EFFECT
         THERMOELECTRIC COOLING

ETTINGSHAUSEN EFFECT
UF   ETTINGSHAUSEN COOLERS
RT   COOLING SYSTEMS
        ∞EFFECTS

ETTINGSHAUSEN EFFECT-*(CONT.)*
    TEMPERATURE EFFECTS
    THERMOELECTRIC COOLING
    THERMOELECTRICITY
    THERMOMAGNETIC COOLING
    THERMOMAGNETIC EFFECTS

EUCLIDEAN GEOMETRY
UF   EUCLIDEAN SPACE
GS   GEOMETRY
    . **EUCLIDEAN GEOMETRY**
    .. ANALYTIC GEOMETRY
    ... CATENARIES
    ... CIRCUMFERENCES
    ... CONICS
    .... ELLIPSES
    .... HYPERBOLAS
    .... PARABOLAS
    ... CYCLOIDS
    ... EPICYCLOIDS
    ... LOCI
    ... MERCATOR PROJECTION
    ... QUADRANTS
    ... S CURVES
    .... GOMPERTZ CURVES
    ... SPHEROIDS
    .... OBLATE SPHEROIDS
    .... PROLATE SPHEROIDS
    ... TANGENTS
    ... TORUSES
    ... TRIGONOMETRY
    .. ANGLES (GEOMETRY)
    ... ANGLE OF ATTACK
    .... ZERO ANGLE OF ATTACK
    ... BRAGG ANGLE
    ... BREWSTER ANGLE
    ... DIHEDRAL ANGLE
    ... ELEVATION ANGLE
    ... LOOK ANGLES (ELECTRONICS)
    ... LOOK ANGLES (TRACKING)
    ... SWEEP ANGLE
    .... SWEEPBACK
    ..... LEADING EDGE SWEEP
    .. CARTESIAN COORDINATES
    .. CIRCLES (GEOMETRY)
    ... GREAT CIRCLES
    .. DESCRIPTIVE GEOMETRY
    .. LINES (GEOMETRY)
    ... CHORDS (GEOMETRY)
    ... GEODESIC LINES
    .. POINTS (MATHEMATICS)
    ... FIXED POINTS (MATHEMATICS)
    ... INFLECTION POINTS
    .. POLYGONS
    ... HEXAGONS
    ... TETRAGONS
    ... PARALLELOGRAMS
    ..... RHOMBOIDS
    .... RECTANGLES
    .... SQUARES (MATHEMATICS)
    .... TRAPEZOIDS
    ... TRIANGLES
    .. POLYHEDRONS
    ... CUBES (MATHEMATICS)
    ... ICOSAHEDRONS
    ... OCTAHEDRONS
    ... PARALLELEPIPEDS
    ... PYRAMIDS
    ... RHOMBOHEDRONS
    ... TETRAHEDRONS
    .. PROJECTIVE GEOMETRY
    ... MERCATOR PROJECTION
    .. RADII
RT   COORDINATES
        CURVES (GEOMETRY)
        PHASE-SPACE INTEGRAL
        POLYTOPES
        RIEMANN MANIFOLD
        SOBOLEV SPACE
        SPHERES

EUCLIDEAN SPACE
USE   EUCLIDEAN GEOMETRY

EUDIOMETERS
GS   MEASURING INSTRUMENTS
    . **EUDIOMETERS**
RT   GAS MIXTURES
        SPARK IGNITION

EUGLENA
RT   ALGAE

EULER BUCKLING
GS   BUCKLING

EULER BUCKLING-*(CONT.)*
- . **EULER BUCKLING**
- RT STRESS ANALYSIS

EULER EQUATIONS OF MOTION
- GS EQUATIONS OF MOTION
- . **EULER EQUATIONS OF MOTION**
- RT ∞EQUATIONS
- HYDRODYNAMICS
- MOMENTS OF INERTIA
- PRIMITIVE EQUATIONS
- RIGID STRUCTURES

EULER-CAUCHY EQUATIONS
- GS ANALYSIS (MATHEMATICS)
- . REAL VARIABLES
- . . DIFFERENTIAL EQUATIONS
- . . . PARTIAL DIFFERENTIAL EQUATIONS
- . . . . **EULER-CAUCHY EQUATIONS**
- RT COMPLEX VARIABLES
- CONFORMAL MAPPING
- ∞EQUATIONS
- VECTOR ANALYSIS

EULER-LAGRANGE EQUATION
- UF LAGRANGE EQUATIONS OF MOTION
- RT CALCULUS OF VARIATIONS
- CASTIGLIANO VARIATIONAL THEOREM
- CLASSICAL MECHANICS
- ∞EQUATIONS
- EXTREMUM VALUES

EULER-LAMBERT EQUATION
- RT ELLIPTICAL ORBITS
- ∞EQUATIONS

EURECA (ESA)
- UF EUROPEAN RETRIEVABLE CARRIER
- GS SPACE PLATFORMS
- . **EURECA (ESA)**
- RT SPACE SHUTTLES

EUROPA
- GS CELESTIAL BODIES
- . NATURAL SATELLITES
- . . JUPITER SATELLITES
- . . . GALILEAN SATELLITES
- . . . . **EUROPA**
- SATELLITES
- . NATURAL SATELLITES
- . . JUPITER SATELLITES
- . . . GALILEAN SATELLITES
- . . . . **EUROPA**
- RT CHARON
- JUPITER (PLANET)

EUROPA LAUNCH VEHICLES
- GS LAUNCH VEHICLES
- . **EUROPA LAUNCH VEHICLES**
- . . EUROPA 1 LAUNCH VEHICLE
- . . EUROPA 2 LAUNCH VEHICLE
- . . EUROPA 3 LAUNCH VEHICLE
- . . EUROPA 4 LAUNCH VEHICLE
- RT ARIANE LAUNCH VEHICLE
- ELDO LAUNCH VEHICLE
- EUROPEAN SPACE AGENCY
- EUROPEAN SPACE PROGRAMS
- ∞VEHICLES

EUROPA 1 LAUNCH VEHICLE
- GS LAUNCH VEHICLES
- . EUROPA LAUNCH VEHICLES
- . . **EUROPA 1 LAUNCH VEHICLE**

EUROPA 2 LAUNCH VEHICLE
- GS LAUNCH VEHICLES
- . EUROPA LAUNCH VEHICLES
- . . **EUROPA 2 LAUNCH VEHICLE**
- RT COS-B SATELLITE

EUROPA 3 LAUNCH VEHICLE
- GS LAUNCH VEHICLES
- . EUROPA LAUNCH VEHICLES
- . . **EUROPA 3 LAUNCH VEHICLE**

EUROPA 4 LAUNCH VEHICLE
- GS LAUNCH VEHICLES
- . EUROPA LAUNCH VEHICLES
- . . **EUROPA 4 LAUNCH VEHICLE**

EUROPE
- GS CONTINENTS
- . **EUROPE**
- RT ALBANIA

EUROPE-*(CONT.)*
- ALPS MOUNTAINS (EUROPE)
- ANDORRA
- AUSTRIA
- BALTIC SHIELD (EUROPE)
- BELGIUM
- BULGARIA
- CARPATHIAN MOUNTAINS (EUROPE)
- CENTRAL EUROPE
- CZECHOSLOVAKIA
- DENMARK
- EAST GERMANY
- ESTONIA
- FINLAND
- FRANCE
- GREECE
- HUNGARY
- ITALY
- LATVIA
- LIECHTENSTEIN
- LITHUANIA
- LUXEMBOURG
- MONACO
- NATIONS
- NETHERLANDS
- NORWAY
- POLAND
- PORTUGAL
- ROMANIA
- SAN MARINO
- SPAIN
- SWEDEN
- SWITZERLAND
- TURKEY
- U.S.S.R.
- UNITED KINGDOM
- VATICAN CITY
- WEST GERMANY
- YUGOSLAVIA

EUROPEAN AIRBUS
- UF AIRBUS
- GS COMMERCIAL AIRCRAFT
- . **EUROPEAN AIRBUS**
- . . A-300 AIRCRAFT
- . . A-310 AIRCRAFT
- . . A-320 AIRCRAFT
- JET AIRCRAFT
- . **EUROPEAN AIRBUS**
- . . A-300 AIRCRAFT
- . . A-310 AIRCRAFT
- . . A-320 AIRCRAFT
- PASSENGER AIRCRAFT
- . **EUROPEAN AIRBUS**
- . . A-300 AIRCRAFT
- . . A-310 AIRCRAFT
- . . A-320 AIRCRAFT
- TRANSPORT AIRCRAFT
- . SHORT HAUL AIRCRAFT
- . . **EUROPEAN AIRBUS**
- . . . A-310 AIRCRAFT
- . . . A-320 AIRCRAFT
- RT ∞AIRCRAFT
- INTERNATIONAL COOPERATION

EUROPEAN COMMUNICATIONS SATELLITE
- UF ECS
- GS SATELLITES
- . ARTIFICIAL SATELLITES
- . . COMMUNICATION SATELLITES
- . . . **EUROPEAN COMMUNICATIONS SATELLITE**
- . . ESA SATELLITES
- . . . **EUROPEAN COMMUNICATIONS SATELLITE**
- RT EUROPEAN SPACE PROGRAMS
- OTS (ESA)

EUROPEAN INCOHERENT SCATTER RADAR
- USE EISCAT RADAR SYSTEM (EUROPE)

EUROPEAN LARGE TELECOMM SATELLITE
- USE L-SAT

EUROPEAN RETRIEVABLE CARRIER
- USE EURECA (ESA)

EUROPEAN SPACE AGENCY
- UF ESA
- ESRO
- EUROPEAN SPACE RESEARCH ORGANIZATION
- GS ORGANIZATIONS
- . **EUROPEAN SPACE AGENCY**
- RT ARIANE LAUNCH VEHICLE

EUROPEAN SPACE AGENCY-*(CONT.)*
- ELDO LAUNCH VEHICLE
- ERS-1 (ESA SATELLITE)
- ESA SATELLITES
- ESRO 1 SATELLITE
- ESRO 2 SATELLITE
- ESRO 4 SATELLITE
- EUROPA LAUNCH VEHICLES
- EUROPEAN SPACE PROGRAMS
- EXPOS (SPACELAB PAYLOAD)
- GEOSARI PROJECT
- ICL COMPUTERS
- LIRTS (TELESCOPE)
- MAROTS (ESA)
- METEOSAT SATELLITE
- SPACE PROGRAMS

EUROPEAN SPACE PROGRAMS
- GS PROGRAMS
- . SPACE PROGRAMS
- . . **EUROPEAN SPACE PROGRAMS**
- RT AEROSAT SATELLITES
- AMPTE (SATELLITES)
- ARIANE LAUNCH VEHICLE
- AZUR SATELLITE
- COMMITTEE ON SPACE RESEARCH
- COS-B SATELLITE
- DIAL SATELLITE
- EARTHNET
- ESA SATELLITES
- ESRO 1 SATELLITE
- ESRO 2 SATELLITE
- ESRO 4 SATELLITE
- EUROPA LAUNCH VEHICLES
- EUROPEAN COMMUNICATIONS SATELLITE
- EUROPEAN SPACE AGENCY
- EXOSAT SATELLITE
- FOREIGN POLICY
- FRENCH SATELLITES
- FRENCH SPACE PROGRAMS
- GEOS SATELLITES (ESA)
- HEOS SATELLITES
- HIPPARCOS SATELLITE
- INTERNATIONAL MAGNETOSPHERIC STUDY
- INTERNATIONAL SATELLITE GEODESY EXPERIMENT
- IRIS SATELLITES
- ITALIAN SPACE PROGRAM
- MARECS MARITIME SATELLITES
- METEOSAT SATELLITE
- OTS (ESA)
- QUASAT
- SWEDISH SPACE PROGRAM
- SWISS SPACE PROGRAM
- SYMPHONIE SATELLITES
- U.S.S.R. SPACE PROGRAM

EUROPEAN SPACE RESEARCH ORGANIZATION
- USE EUROPEAN SPACE AGENCY

EUROPEAN SPACE RESEARCH ORGANIZATION SAT
- USE ESA SATELLITES

EUROPEAN 1 SPACECRAFT
- GS SATELLITES
- . ARTIFICIAL SATELLITES
- . . **EUROPEAN 1 SPACECRAFT**
- RT ELDO LAUNCH VEHICLE

EUROPIUM
- GS CHEMICAL ELEMENTS
- . RARE EARTH ELEMENTS
- . . **EUROPIUM**
- . . . EUROPIUM ISOTOPES
- METALS
- . RARE EARTH ELEMENTS
- . . **EUROPIUM**
- . . . EUROPIUM ISOTOPES

EUROPIUM COMPOUNDS
- GS RARE EARTH COMPOUNDS
- . **EUROPIUM COMPOUNDS**
- RT ∞CHEMICAL COMPOUNDS
- ∞METAL COMPOUNDS

EUROPIUM ISOTOPES
- GS CHEMICAL ELEMENTS
- . NUCLIDES
- . . ISOTOPES
- . . . **EUROPIUM ISOTOPES**
- . RARE EARTH ELEMENTS
- . . EUROPIUM

EUROPIUM ISOTOPES-*(CONT.)*
```
        . . . EUROPIUM ISOTOPES
        METALS
        . RARE EARTH ELEMENTS
        . . EUROPIUM
        . . . EUROPIUM ISOTOPES
```

EUSTACHIAN TUBES
```
GS      ANATOMY
        . SENSE ORGANS
        . . EUSTACHIAN TUBES
RT      EAR
        EARDRUMS
        ∞ TUBES
```

EUTECTIC ALLOYS
```
GS      ALLOYS
        . EUTECTIC ALLOYS
        BINARY SYSTEMS (MATERIALS)
        . BINARY MIXTURES
        . . EUTECTICS
        . . . EUTECTIC ALLOYS
        MIXTURES
        . BINARY MIXTURES
        . . EUTECTICS
        . . . EUTECTIC ALLOYS
RT      LAMELLA (METALLURGY)
        SUPERPLASTICITY
        WHISKER COMPOSITES
```

EUTECTIC COMPOSITES
```
GS      COMPOSITE MATERIALS
        . METAL MATRIX COMPOSITES
        . . EUTECTIC COMPOSITES
RT      ALLOYS
        DIRECTIONAL SOLIDIFICATION
            (CRYSTALS)
        EUTECTICS
        FRACTURE STRENGTH
        ∞ MATRICES
        METALS
        MIXTURES
        PRECIPITATION HARDENING
```

EUTECTIC DIAGRAMS
```
USE     PHASE DIAGRAMS
```

EUTECTICS
```
GS      BINARY SYSTEMS (MATERIALS)
        . BINARY MIXTURES
        . . EUTECTICS
        . . . EUTECTIC ALLOYS
        MIXTURES
        . BINARY MIXTURES
        . . EUTECTICS
        . . . EUTECTIC ALLOYS
RT      ALLOYS
        EUTECTIC COMPOSITES
        LIQUID PHASES
        PHASE DIAGRAMS
        SOLID PHASES
        SOLUTIONS
        SYNTECTIC ALLOYS
```

EUTROPHICATION
```
RT      ENVIRONMENT EFFECTS
        LAKES
        ∞ NUTRIENTS
```

EUVE
```
USE     EXTREME ULTRAVIOLET EXPLORER
            SATELLITE
```

EUXENITE
```
GS      MINERALS
        . EUXENITE
RT      NIOBATES
        OXIDES
        TITANATES
```

∞ EVACUATING
```
SN      (USE OF A MORE SPECIFIC TERM IS
        RECOMMENDED--CONSULT THE TERMS
        LISTED BELOW)
RT      EVACUATING (TRANSPORTATION)
        EVACUATING (VACUUM)
```

EVACUATING (TRANSPORTATION)
```
SN      (CLEARANCE OF PERSONNEL, ANIMALS,
        OR MATERIAL FROM A GIVEN
        LOCALITY)
RT      C-9 AIRCRAFT
        CASUALTIES
        CIVIL DEFENSE
```

EVACUATING (TRANSPORTATION)-*(CONT.)*
```
        EJECTION
        ELIMINATION
        ∞ EVACUATING
        HOSPITALS
        MOBILE QUARANTINE FACILITY
        REMOVAL
        TRANSPORTATION
        UNLOADING
```

EVACUATING (VACUUM)
```
UF      GAS EVACUATING
RT      DRAINAGE
        EJECTION
        ELIMINATION
        ∞ EVACUATING
        EXHAUSTING
        GAS POCKETS
        PURGING
        REMOVAL
        SUCTION
        VACUUM
        VACUUM PUMPS
        VENTING
        VENTS
```

EVAL
```
USE     EARTH VIEWING APPLICATIONS
            LABORATORY
```

EVALUATION
```
GS      EVALUATION
        . TRAINING EVALUATION
RT      ACCELERATED LIFE TESTS
        ACCEPTABILITY
        ∞ ANALYZING
        APPROACH AND LANDING TESTS (STS)
        ASSESSMENTS
        CERTIFICATION
        ∞ CLASSIFYING
        COMPARISON
        COMPUTER SYSTEMS PERFORMANCE
        CORRELATION
        COSTS
        CRITERIA
        CROP IDENTIFICATION
        ∞ DISCUSSION
        ECONOMICS
        ESTIMATES
        ESTIMATING
        EXAMINATION
        FEASIBILITY
        FIGURE OF MERIT
        FORECASTING
        ∞ INDICATION
        INSPECTION
        MANAGEMENT
        ∞ MEASUREMENT
        NORMALIZING (STATISTICS)
        OBSERVATION
        ∞ PERFORMANCE
        PERFORMANCE PREDICTION
        POSITION (TITLE)
        PROVING
        QUALITY
        RANKING
        RATINGS
        REJECTION
        RESERVES
        REVIEWING
        SELECTION
        STATISTICAL CORRELATION
        TECHNOLOGY ASSESSMENT
        ∞ TESTS
        TIMBER IDENTIFICATION
        VALUE
```

EVANESCENCE
```
RT      EVAPORATION
        SURFACE PROPERTIES
        TRANSPIRATION
```

EVAPORATION
```
GS      PHASE TRANSFORMATIONS
        . VAPORIZING
        . . EVAPORATION
        . . . EVAPOTRANSPIRATION
        . . . PROPELLANT EVAPORATION
        . . . TRANSPIRATION
RT      BOILING
        CONCENTRATING
        CONDENSING
        DEHYDRATION
        DIFFUSION
        DISTILLATION
```

EVAPORATION-*(CONT.)*
```
        DRYING
        EVANESCENCE
        EVAPORATIVE COOLING
        EVAPOROGRAPHY
        FLASHING (VAPORIZING)
        GAS-LIQUID INTERACTIONS
        GAS-METAL INTERACTIONS
        LIQUID-VAPOR INTERFACES
        PERSPIRATION
        RESERVOIRS
        RESPIRATORY SYSTEM
        ∞ SEPARATION
        SKIN (ANATOMY)
        SUBLIMATION
        VOLATILITY
        WATER LOSS
```

EVAPORATION RATE
```
GS      RATES (PER TIME)
        . EVAPORATION RATE
RT      HEAT TRANSFER COEFFICIENTS
```

EVAPORATIVE COOLING
```
GS      COOLING
        . EVAPORATIVE COOLING
        . . FILM COOLING
        . . SWEAT COOLING
RT      COOLING SYSTEMS
        CRYOGENIC FLUID STORAGE
        EVAPORATION
        PROPELLANT EVAPORATION
        SURFACE COOLING
```

EVAPORATORS
```
GS      HEATING EQUIPMENT
        . VAPORIZERS
        . . EVAPORATORS
        SEPARATORS
        . EVAPORATORS
RT      AIR CONDITIONING EQUIPMENT
        ATOMIZERS
        CONCENTRATORS
        CONDENSERS (LIQUEFIERS)
        COOLING SYSTEMS
        DRYING APPARATUS
        HEAT EXCHANGERS
        REFRIGERATING MACHINERY
```

EVAPOROGRAPHY
```
RT      EVAPORATION
        IMAGES
        INFRARED RADIATION
        PHOTOGRAPHY
```

EVAPOTRANSPIRATION
```
GS      PHASE TRANSFORMATIONS
        . VAPORIZING
        . . EVAPORATION
        . . . EVAPOTRANSPIRATION
RT      TRANSPIRATION
        VADOSE WATER
```

EVASIVE ACTIONS
```
GS      MANEUVERS
        . EVASIVE ACTIONS
RT      ELECTRONIC WARFARE
        OBSTACLE AVOIDANCE
        TACTICS
        WARFARE
```

EVASIVE SATELLITES
```
GS      MANEUVERABLE SPACECRAFT
        . EVASIVE SATELLITES
        SATELLITES
        . ARTIFICIAL SATELLITES
        . . EVASIVE SATELLITES
RT      MILITARY SPACECRAFT
```

EVECTION
```
USE     LUNAR ORBITS
        ORBIT PERTURBATION
        SOLAR GRAVITATION
```

EVEN-EVEN NUCLEI
```
GS      GASES
        . IONIZED GASES
        . . CHARGED PARTICLES
        . . . NUCLEI (NUCLEAR PHYSICS)
        . . . . EVEN-EVEN NUCLEI
RT      NUCLEAR STRUCTURE
        ODD-EVEN NUCLEI
        ODD-ODD NUCLEI
```

EVENING
RT   DAYTIME
        NIGHT
        SUNSET

EVENTS
GS   **EVENTS**
        . CONSECUTIVE EVENTS
RT   OCCURRENCES
        PROBABILITY DENSITY FUNCTIONS
        PROBABILITY THEORY
        STATISTICAL ANALYSIS
        STATISTICAL DISTRIBUTIONS
        STOCHASTIC PROCESSES

EVERGLADES (FL)
RT   FLORIDA

EVOKED RESPONSE (PSYCHOPHYSIOLOGY)
RT   PHYSIOLOGICAL RESPONSES
        PSYCHOPHYSIOLOGY

∞ **EVOLUTION**
SN   *(USE OF A MORE SPECIFIC TERM IS
        RECOMMENDED--CONSULT THE TERMS
        LISTED BELOW)*
RT   BIOGENY
        CHEMICAL EVOLUTION
        EVOLUTION (DEVELOPMENT)
        EVOLUTION (LIBERATION)
        EXISTENCE

EVOLUTION (DEVELOPMENT)
GS   **EVOLUTION (DEVELOPMENT)**
        . BIOLOGICAL EVOLUTION
        . . ABIOGENESIS
        . CHEMICAL EVOLUTION
        . GALACTIC EVOLUTION
        . LUNAR EVOLUTION
        . PLANETARY EVOLUTION
        . STELLAR EVOLUTION
        . . STELLAR MASS ACCRETION
RT   ∞ BIOLOGY
        ∞ DEVELOPMENT
        ∞ EVOLUTION
        EXTINCTION
        GENETICS
        GROWTH
        HEREDITY
        INTERSTELLAR EXTINCTION
        ONTOGENY
        SPECIES DIFFUSION

EVOLUTION (LIBERATION)
GS   **EVOLUTION (LIBERATION)**
        . GAS EVOLUTION
RT   BOILING
        DESORPTION
        ∞ EVOLUTION
        OUTGASSING
        TRANSPIRATION
        VAPORIZING

EXACTNESS
USE   PRECISION

EXAMINATION
GS   **EXAMINATION**
        . EYE EXAMINATIONS
RT   ACCEPTABILITY
        ∞ ANALYZING
        CHARACTERIZATION
        CLINICAL MEDICINE
        COMPARISON
        CONICAL SCANNING
        DETECTION
        DIAGNOSIS
        ∞ DISCUSSION
        EVALUATION
        EXPLORATION
        INSPECTION
        INVESTIGATION
        ∞ MEASUREMENT
        OBSERVATION
        ∞ PERFORMANCE
        PROVING
        REVIEWING
        SCANNING
        ∞ TESTS
        TRAINING EVALUATION
        ULTRASONIC FLAW DETECTION

EXCAVATION
UF   DITCHING (EXCAVATION)

EXCAVATION-*(CONT.)*
GS   **EXCAVATION**
        . TUNNELING (EXCAVATION)
RT   BOREHOLES
        CONSTRUCTION
        ∞ DITCHING
        DRAINAGE
        EXPLORATION
        FOUNDATIONS
        MATERIALS HANDLING
        MINERAL DEPOSITS
        MINERAL EXPLORATION
        MINING
        PITS (EXCAVATIONS)
        STRIP MINING
        UNDERGROUND STRUCTURES

∞ **EXCHANGERS**
SN   *(USE OF A MORE SPECIFIC TERM IS
        RECOMMENDED--CONSULT THE TERMS
        LISTED BELOW)*
RT   HEAT EXCHANGERS

EXCHANGING
GS   **EXCHANGING**
        . CHARGE EXCHANGE
        . . RESONANCE CHARGE EXCHANGE
        . GAS EXCHANGE
        . ION EXCHANGING
        . SPIN EXCHANGE
RT   ∞ CONVERSION
        DEIONIZATION
        ∞ SEPARATION
        ∞ SHIFT
        TRANSFERRING

EXCIMER LASERS
GS   STIMULATED EMISSION DEVICES
        . LASERS
        . . GAS LASERS
        . . . **EXCIMER LASERS**
RT   ELECTRON PUMPING
        FLUORIDES
        HALOGENS
        LASER OUTPUTS
        LASING
        OPTICAL PUMPING
        ∞ RARE GAS COMPOUNDS
        XENON CHLORIDE LASERS
        XENON FLUORIDE LASERS

EXCIMERS
RT   ELECTRON ORBITALS
        ELECTRON STATES
        ELECTRON TRANSITIONS
        INTERMOLECULAR FORCES
        MOLECULAR ENERGY LEVELS
        ∞ RARE GAS COMPOUNDS

EXCITATION
UF   EXCITED STATES
GS   **EXCITATION**
        . ATOMIC EXCITATIONS
        . MOLECULAR EXCITATION
        . SELF EXCITATION
        . WAVE EXCITATION
        . . ACOUSTIC EXCITATION
        . . HARMONIC EXCITATION
RT   ACTIVATION
        ACTUATION
        ATOMIC ENERGY LEVELS
        AURORAL IONIZATION
        AURORAL IRRADIATION
        ELECTROMAGNETIC ABSORPTION
        ELECTRON STATES
        ELECTRON TRANSITIONS
        EMISSION
        ENERGY LEVELS
        IONIZATION
        IRRADIATION
        METASTABLE STATE
        RADIATION TRAPPING
        RELAXATION TIME
        ROTONS
        STARTING
        TRANSITION PROBABILITIES

EXCITED STATES
USE   EXCITATION

EXCITONS
GS   ELEMENTARY EXCITATIONS
        . **EXCITONS**
RT   CARRIER MOBILITY
        ELECTRICAL INSULATION

EXCITONS-*(CONT.)*
        ELECTRONS
        ENERGY BANDS
        HOLES (ELECTRON DEFICIENCIES)
        IONIC CRYSTALS
        LIGHT (VISIBLE RADIATION)
        OPTICAL PROPERTIES
        PHOTOELECTROMAGNETIC EFFECTS
        PLASMONS
        POSITRONIUM
        SEMICONDUCTORS (MATERIALS)
        SPECTRA
        SUHL EFFECT

EXCLUSION
RT   ELIMINATION
        ISOLATION
        PAULI EXCLUSION PRINCIPLE
        REJECTION
        ∞ SEPARATION

EXCRETION
RT   EXPULSION
        FECES
        HUMAN WASTES
        PERSPIRATION
        URINE

EXECUTIVE AIRCRAFT
USE   GENERAL AVIATION AIRCRAFT
        PASSENGER AIRCRAFT

EXERCISE
USE   PHYSICAL EXERCISE

EXERCISE PHYSIOLOGY
GS   PHYSIOLOGY
        . **EXERCISE PHYSIOLOGY**
RT   CIRCULATORY SYSTEM
        HUMAN BODY
        LOCOMOTION
        MUSCULAR TONUS
        PHYSICAL FITNESS
        PHYSIOCHEMISTRY
        PHYSIOLOGICAL EFFECTS
        RESPIRATORY PHYSIOLOGY
        SPORTS MEDICINE
        STRESS (PHYSIOLOGY)

EXERTION
USE   PHYSICAL WORK

EXHALATION
RT   ALVEOLAR AIR
        EXPIRED AIR
        RESPIROMETERS

EXHAUST DIFFUSERS
RT   CONICAL NOZZLES
        ∞ DIFFUSERS
        EJECTORS
        ∞ JET NOZZLES
        SUPERSONIC DIFFUSERS
        VANELESS DIFFUSERS

EXHAUST EMISSION
GS   DECAY
        . EMISSION
        . . **EXHAUST EMISSION**
RT   GAS-GAS INTERACTIONS
        GAS-METAL INTERACTIONS
        HIGH TEMPERATURE GASES
        INFRARED RADIATION
        JET EXHAUST
        PARTICLE EMISSION
        POLLUTION TRANSPORT
        RELEASING
        THERMAL EMISSION

EXHAUST FLOW SIMULATION
GS   SIMULATION
        . **EXHAUST FLOW SIMULATION**
        . . ATMOSPHERIC ENTRY SIMULATION
RT   FLOW DISTRIBUTION
        MATHEMATICAL MODELS
        WIND TUNNELS

EXHAUST GASES
UF   EXHAUST JETS
GS   GASES
        . **EXHAUST GASES**
        . . FLUE GASES
RT   AIR POLLUTION
        ∞ BLASTS

EXHAUST GASES-(CONT.)
     COMBUSTION EFFICIENCY
     COMBUSTION PRODUCTS
     DILUENTS
     EFFLUENTS
     ENVIRONMENT EFFECTS
     EROSIVE BURNING
     EXHAUSTING
     FUMES
     GAS MIXTURES
     GAS RECOVERY
     INFRARED SUPPRESSION
     JET BLAST EFFECTS
     JET EXHAUST
     NOZZLE FLOW
     ODORS
     POLLUTION TRANSPORT
     PROPULSION
     REACTION PRODUCTS
     ROCKET EXHAUST
     SMOG
     SMOKE
     SMOKE ABATEMENT
     VAPORS
     WASTE DISPOSAL
     WASTE ENERGY UTILIZATION
     WASTES

EXHAUST JETS
USE   EXHAUST GASES

EXHAUST NOZZLES
GS   **EXHAUST NOZZLES**
     . CONVERGENT-DIVERGENT NOZZLES
     . PLUG NOZZLES
     . SPIKE NOZZLES
     . TURBINE EXHAUST NOZZLES
RT   AIR DUCTS
     ANNULAR NOZZLES
     BASE HEATING
     CONICAL NOZZLES
     DIVERGENT NOZZLES
     EJECTORS
     ∞ FLOW
     INFRARED SUPPRESSION
     INLET NOZZLES
     JET ENGINES
     ∞ JET NOZZLES
     NOZZLE FLOW
     NOZZLE INSERTS
     ∞ NOZZLES
     OPENINGS
     OUTLETS
     ROCKET ENGINES
     SKIRTS

EXHAUST SYSTEMS
RT   AFTERBURNING
     AIR CONDITIONING
     AIR POLLUTION
     BLOWERS
     CHIMNEYS
     CONDENSERS (LIQUEFIERS)
     COOLING SYSTEMS
     DUCTS
     DUST COLLECTORS
     EJECTORS
     ELIMINATION
     ENGINES
     EXHAUSTING
     FLUES
     FUEL TANK PRESSURIZATION
     INTAKE SYSTEMS
     INTERNAL COMBUSTION ENGINES
     MANIFOLDS
     MUFFLERS
     OPENINGS
     OUTLETS
     PIPE NOZZLES
     PLENUM CHAMBERS
     PORTS (OPENINGS)
     ROCKET EXHAUST
     ∞ SYSTEMS
     TEMPERATURE CONTROL
     VENTILATION
     VENTILATORS
     VENTS
     WASTE DISPOSAL

EXHAUST VELOCITY
GS   RATES (PER TIME)
     . **EXHAUST VELOCITY**
     VELOCITY
     . **EXHAUST VELOCITY**
RT   ACOUSTIC VELOCITY

EXHAUST VELOCITY-(CONT.)
     CRITICAL VELOCITY
     EXPULSION
     FLOW VELOCITY

EXHAUSTING
RT   BLOWING
     BREATHING VIBRATION
     CONSUMPTION
     DECONTAMINATION
     ∞ DISCHARGE
     DISPERSING
     DISPOSAL
     DISSIPATION
     EJECTION
     ELIMINATION
     EVACUATING (VACUUM)
     EXHAUST GASES
     EXHAUST SYSTEMS
     RELIEVING
     REMOVAL
     VENTILATION
     VENTING

EXHAUSTION
RT   CONSUMPTION
     DEPLETION
     FATIGUE (BIOLOGY)
     HYPERKINESIA

EXISTENCE
RT   COSMOLOGY
     ∞ EVOLUTION
     LIFE SPAN
     VALIDITY

EXISTENCE THEOREMS
GS   ANALYSIS (MATHEMATICS)
     . REAL VARIABLES
     . . **EXISTENCE THEOREMS**
     THEOREMS
     . **EXISTENCE THEOREMS**
RT   PROBLEM SOLVING
     ROOTS OF EQUATIONS

EXITS (DOORS)
USE   DOORS

EXOBIOLOGY
UF   ASTROBIOLOGY
     SPACE BIOLOGY
RT   AEROSPACE ENVIRONMENTS
     APOLLO EXTENSION SYSTEM
     BIOASTRONAUTICS
     ∞ BIOLOGY
     CARBONACEOUS METEORITES
     CHEMICAL EVOLUTION
     ENVIRONMENT MODELS
     EXTRATERRESTRIAL LIFE
     LIFE SUPPORT SYSTEMS
     LUNAR ENVIRONMENT
     PANSPERMIA
     PLANETARY ENVIRONMENTS
     SPACECRAFT CONTAMINATION
     SPACECRAFT ENVIRONMENTS
     SPACECRAFT STERILIZATION

EXOPHORIA
USE   HETEROPHORIA

EXOS SOUNDING ROCKET
GS   ROCKET VEHICLES
     . MULTISTAGE ROCKET VEHICLES
     . . **EXOS SOUNDING ROCKET**
     . SOUNDING ROCKETS
     . . **EXOS SOUNDING ROCKET**
RT   HONEST JOHN ROCKET VEHICLE
     NIKE-AJAX MISSILE
     SOLID PROPELLANT ROCKET ENGINES
     XM-33 ENGINE

EXOSAT SATELLITE
UF   HELOS (SATELLITE)
     HIGH ECCENTRIC LUNAR OCCULTATION
        SATELLITE
GS   SATELLITES
     . ARTIFICIAL SATELLITES
     . . ESA SATELLITES
     . . . **EXOSAT SATELLITE**
     . SCIENTIFIC SATELLITES
     . . . **EXOSAT SATELLITE**
     UNMANNED SPACECRAFT
     . . **EXOSAT SATELLITE**
RT   ECCENTRIC ORBITS

EXOSAT SATELLITE-(CONT.)
     EUROPEAN SPACE PROGRAMS
     LUNAR OCCULTATION
     X RAY ASTRONOMY
     X RAY SOURCES

EXOSKELETONS
RT   ANATOMY
     ARTHROPODS
     BODY COMPOSITION (BIOLOGY)
     BONES
     CONNECTIVE TISSUE
     MUSCULOSKELETAL SYSTEM
     PHYSIOLOGY

EXOSPHERE
GS   EARTH ATMOSPHERE
     . UPPER ATMOSPHERE
     . . **EXOSPHERE**
RT   HETEROSPHERE
     IONOSPHERE
     MAGNETOSPHERE
     RADIATION BELTS
     THERMOSPHERE

EXOTHERMIC REACTIONS
GS   CHEMICAL REACTIONS
     . **EXOTHERMIC REACTIONS**
RT   ASSOCIATION REACTIONS
     COMBUSTION
     ENDOTHERMIC REACTIONS
     INCENDIARY AMMUNITION
     PYROLYSIS
     THERMAL DECOMPOSITION

EXPANDABLE STRUCTURES
GS   **EXPANDABLE STRUCTURES**
     . BELLOWS
     . INFLATABLE STRUCTURES
     . . BALLOONS
     . . . HIGH ALTITUDE BALLOONS
     . . . . JIMSPHERE BALLOONS
     . . . . SKYHOOK BALLOONS
     . . . . SUPERPRESSURE BALLOONS
     . . . . METEOROLOGICAL BALLOONS
     . . . . JIMSPHERE BALLOONS
     . . . ROBIN BALLOONS
     . . . MICROBALLOONS
     . . . TETHERED BALLOONS
     . . BALLUTES
     . . GAS BAGS
     . . INFLATABLE GLIDERS
     . . INFLATABLE SPACECRAFT
     . . . BEACON SATELLITES
     . . . . BEACON EXPLORER A
     . . . . EXPLORER 22 SATELLITE
     . . PARAVULCOONS
RT   EXPULSION BLADDERS
     FOLDING STRUCTURES
     LARGE SPACE STRUCTURES
     ORBITAL ASSEMBLY
     SPACE ERECTABLE STRUCTURES
     ∞ SPACECRAFT
     ∞ STRUCTURES
     VARIABLE GEOMETRY STRUCTURES

EXPANSION
UF   ENLARGING
GS   **EXPANSION**
     . GAS EXPANSION
     . KARHUNEN-LOEVE EXPANSION
     . PRANDTL-MEYER EXPANSION
     . SERIES EXPANSION
     . THERMAL EXPANSION
RT   ADIABATIC CONDITIONS
     DISTORTION
     ELONGATION
     EXTENSIONS
     INFLATING
     RAREFACTION
     RELAXATION (MECHANICS)
     SWELLING
     THERMAL BUCKLING

EXPANSION WAVES
USE   ELASTIC WAVES

EXPECTANCY HYPOTHESIS
GS   HYPOTHESES
     . **EXPECTANCY HYPOTHESIS**
RT   MONTE CARLO METHOD
     PROBABILITY DENSITY FUNCTIONS
     STATISTICAL ANALYSIS
     STATISTICAL DISTRIBUTIONS

EXPECTATION
RT   CONTINGENCY
      DECISION THEORY
      FORECASTING
      RELIABILITY

EXPEDITIONS
RT   EXPLORATION
      ∞MISSIONS
      SPACE FLIGHT

EXPELLANTS
RT   COUGH
      ∞DISCHARGE
      EXPULSION
      FLUSHING

EXPENDABLE STAGES (SPACECRAFT)
RT   BOOSTER RECOVERY
      BOOSTER ROCKET ENGINES
      ENGINES
      MULTISTAGE ROCKET VEHICLES
      RECOVERABLE SPACECRAFT
      REUSABLE SPACECRAFT
      ROCKET ENGINES
      SPACE SHUTTLES
      STAGE SEPARATION

EXPERIENCE
RT   EDUCATION
      QUALIFICATIONS
      UPGRADING

EXPERIMENT DESIGN
SN   (DESIGN OF EXPERIMENTS EXCLUDES
      PROTOTYPES)
UF   DESIGN OF EXPERIMENTS
GS   **EXPERIMENT DESIGN**
      . FACTORIAL DESIGN
RT   COVARIANCE
      DEGREES OF FREEDOM
      ∞DESIGN
      FACTOR ANALYSIS
      LABORATORIES
      MATHEMATICAL MODELS
      OPERATIONS RESEARCH
      ORTHOGONALITY
      QUALITY CONTROL
      REGRESSION ANALYSIS
      STATISTICAL ANALYSIS
      SYSTEMS ENGINEERING
      VARIANCE (STATISTICS)

EXPERIMENTAL BOILING WATER REACTORS
UF   EBWR (REACTOR)
GS   NUCLEAR REACTORS
      . LIQUID COOLED REACTORS
      . . WATER COOLED REACTORS
      . . . BOILING WATER REACTORS
      . . . . **EXPERIMENTAL BOILING WATER
              REACTORS**
      . NUCLEAR RESEARCH AND TEST
        REACTORS
      . . **EXPERIMENTAL BOILING WATER
            REACTORS**
      . WATER MODERATED REACTORS
      . . **EXPERIMENTAL BOILING WATER
            REACTORS**

EXPERIMENTAL BREEDER REACTOR 1
UF   EBR-1 REACTOR
GS   NUCLEAR REACTORS
      . BREEDER REACTORS
      . . **EXPERIMENTAL BREEDER REACTOR
            1**
      . FAST NUCLEAR REACTORS
      . . **EXPERIMENTAL BREEDER REACTOR
            1**
      . LIQUID COOLED REACTORS
      . . LIQUID METAL COOLED REACTORS
      . . . **EXPERIMENTAL BREEDER
              REACTOR 1**
      . NUCLEAR RESEARCH AND TEST
        REACTORS
      . . **EXPERIMENTAL BREEDER REACTOR
            1**

EXPERIMENTAL BREEDER REACTOR 2
UF   EBR-2 REACTOR
GS   NUCLEAR REACTORS
      . BREEDER REACTORS
      . . **EXPERIMENTAL BREEDER REACTOR
            2**
      . FAST NUCLEAR REACTORS

EXPERIMENTAL BREEDER REACTOR 2-*(CONT.)*
      . . **EXPERIMENTAL BREEDER REACTOR
            2**
      . LIQUID COOLED REACTORS
      . . LIQUID METAL COOLED REACTORS
      . . . **EXPERIMENTAL BREEDER
              REACTOR 2**
      . NUCLEAR RESEARCH AND TEST
        REACTORS
      . . **EXPERIMENTAL BREEDER REACTOR
            2**

EXPERIMENTAL GAS COOLED REACTORS
UF   EGCR (REACTOR)
GS   NUCLEAR REACTORS
      . GAS COOLED REACTORS
      . . **EXPERIMENTAL GAS COOLED
            REACTORS**
      . NUCLEAR RESEARCH AND TEST
        REACTORS
      . . **EXPERIMENTAL GAS COOLED
            REACTORS**

EXPERIMENTAL ORGANIC COOLED REACTORS
UF   EOCR (REACTOR)
GS   NUCLEAR REACTORS
      . LIQUID COOLED REACTORS
      . . ORGANIC COOLED REACTORS
      . . . **EXPERIMENTAL ORGANIC COOLED
              REACTORS**
      . NUCLEAR RESEARCH AND TEST
        REACTORS
      . . **EXPERIMENTAL ORGANIC COOLED
            REACTORS**
      . ORGANIC MODERATED REACTORS
      . . **EXPERIMENTAL ORGANIC COOLED
            REACTORS**

**EXPERIMENTAL REFLECTOR ORBITAL SHOT
PROJ**
UF   EROS PROJECT
GS   PROGRAMS
      . PROJECTS
      . . **EXPERIMENTAL REFLECTOR
            ORBITAL SHOT PROJ**

**EXPERIMENTAL STOL TRANSPORT RSCH
AIRPLANE**
USE   QUESTOL

EXPERIMENTATION
GS   **EXPERIMENTATION**
      . PHYSICS AND CHEMISTRY
        EXPERIMENT IN SPACE
      . SPHINX
RT   CRITICAL EXPERIMENTS
      EXPLORATION
      INVESTIGATION
      LABORATORIES
      SPACEBORNE EXPERIMENTS

EXPERT SYSTEMS
UF   KNOWLEDGE ENGINEERING
GS   INTELLIGENCE
      . ARTIFICIAL INTELLIGENCE
      . . **EXPERT SYSTEMS**
RT   COMPUTER PROGRAMMING
      ∞LOGIC
      LOGIC PROGRAMMING

EXPIRATION
RT   ∞BREATHING
      DEATH
      EXPIRED AIR
      EXPULSION
      MORTALITY
      RESPIRATION

EXPIRED AIR
GS   GASES
      . AIR
      . . **EXPIRED AIR**
      . GAS MIXTURES
      . . **EXPIRED AIR**
RT   ALVEOLAR AIR
      EXHALATION
      EXPIRATION
      GAS COMPOSITION
      METABOLIC WASTES
      REBREATHING
      RESPIRATION

EXPLODING CONDUCTOR CIRCUITS
USE   CIRCUITS
      EXPLODING WIRES

EXPLODING CONDUCTORS
USE   EXPLODING WIRES

EXPLODING WIRES
UF   EXPLODING CONDUCTOR CIRCUITS
      EXPLODING CONDUCTORS
GS   EXPLOSIVE DEVICES
      . INITIATORS (EXPLOSIVES)
      . . **EXPLODING WIRES**
      WIRE
      . **EXPLODING WIRES**
RT   BOOSTERS (EXPLOSIVES)
      CAPS (EXPLOSIVES)
      CONDUCTORS
      DETONATORS
      ELECTRIC WIRE
      PLASMA GENERATORS
      PRIMERS (EXPLOSIVES)
      RADIATION TRANSPORT
      SHOCK WAVES
      WIRE BRIDGE CIRCUITS

EXPLOITATION
RT   BENEFICIATION
      DEPLETION
      ∞DEVELOPMENT
      EXPLORATION
      GEOLOGY
      LAND USE
      MINES (EXCAVATIONS)
      MINING
      RESERVES
      STRIP MINING

EXPLORATION
UF   DISCOVERING
      PROSPECTING
GS   **EXPLORATION**
      . LUNAR EXPLORATION
      . MINERAL EXPLORATION
      . OIL EXPLORATION
      . SPACE EXPLORATION
RT   BOREHOLES
      DETECTION
      DRILLING
      EXAMINATION
      EXCAVATION
      EXPEDITIONS
      EXPERIMENTATION
      EXPLOITATION
      GEOLOGICAL SURVEYS
      GEOLOGY
      GEOTHERMAL TECHNOLOGY
      INVESTIGATION
      MINES (EXCAVATIONS)
      OSS-1 PAYLOAD
      RESEARCH
      RESERVES
      SAMPLING
      SPACE FLIGHT
      SURVEYS
      UNDERGROUND ACOUSTICS

EXPLORER SATELLITES
GS   SATELLITES
      . ARTIFICIAL SATELLITES
      . . **EXPLORER SATELLITES**
      . . . APPLICATIONS EXPLORER
              SATELLITES
      . . . COSMIC BACKGROUND EXPLORER
              SATELLITE
      . . . DUAL AIR DENSITY EXPLORER
      . . . DYNAMICS EXPLORER SATELLITES
      . . . . DYNAMICS EXPLORER 1
                SATELLITE
      . . . . DYNAMICS EXPLORER 2
                SATELLITE
      . . . EXPLORER 1 SATELLITE
      . . . EXPLORER 2 SATELLITE
      . . . EXPLORER 3 SATELLITE
      . . . EXPLORER 4 SATELLITE
      . . . EXPLORER 5 SATELLITE
      . . . EXPLORER 6 SATELLITE
      . . . EXPLORER 7 SATELLITE
      . . . EXPLORER 8 SATELLITE
      . . . EXPLORER 9 SATELLITE
      . . . EXPLORER 10 SATELLITE
      . . . EXPLORER 11 SATELLITE
      . . . EXPLORER 12 SATELLITE
      . . . EXPLORER 14 SATELLITE
      . . . EXPLORER 15 SATELLITE

EXPLORER SATELLITES-*(CONT.)*
```
. . . EXPLORER 16 SATELLITE
. . . EXPLORER 17 SATELLITE
. . . EXPLORER 18 SATELLITE
. . . EXPLORER 19 SATELLITE
. . . EXPLORER 20 SATELLITE
. . . EXPLORER 21 SATELLITE
. . . EXPLORER 22 SATELLITE
. . . EXPLORER 23 SATELLITE
. . . EXPLORER 24 SATELLITE
. . . EXPLORER 25 SATELLITE
. . . EXPLORER 26 SATELLITE
. . . EXPLORER 27 SATELLITE
. . . EXPLORER 28 SATELLITE
. . . EXPLORER 29 SATELLITE
. . . EXPLORER 30 SATELLITE
. . . EXPLORER 31 SATELLITE
. . . EXPLORER 32 SATELLITE
. . . EXPLORER 33 SATELLITE
. . . EXPLORER 34 SATELLITE
. . . EXPLORER 35 SATELLITE
. . . EXPLORER 36 SATELLITE
. . . EXPLORER 37 SATELLITE
. . . EXPLORER 38 SATELLITE
. . . EXPLORER 39 SATELLITE
. . . EXPLORER 40 SATELLITE
. . . EXPLORER 41 SATELLITE
. . . EXPLORER 43 SATELLITE
. . . EXPLORER 44 SATELLITE
. . . EXPLORER 46 SATELLITE
. . . EXPLORER 47 SATELLITE
. . . EXPLORER 48 SATELLITE
. . . EXPLORER 49 SATELLITE
. . . EXPLORER 50 SATELLITE
. . . EXPLORER 51 SATELLITE
. . . EXPLORER 52 SATELLITE
. . . EXPLORER 53 SATELLITE
. . . EXPLORER 54 SATELLITE
. . . EXPLORER 55 SATELLITE
. . . EXTREME ULTRAVIOLET EXPLORER
        SATELLITE
. . . FAR UV SPECTROSCOPIC
        EXPLORER
. . . IMP
. . . INTERNATIONAL MAGNETOSPHERIC
        EXPLORER
. . . INTERNATIONAL SUN EARTH
        EXPLORERS
. . . . INTERNATIONAL SUN EARTH
        EXPLORER 1
. . . . INTERNATIONAL SUN EARTH
        EXPLORER 2
. . . . INTERNATIONAL SUN EARTH
        EXPLORER 3
. . . MICROMETEOROID EXPLORER
        SATELLITES
. . . RADIO ASTRONOMY EXPLORER
        SATELLITE
. . . SOLAR MESOSPHERE EXPLORER
. . . X RAY TIMING EXPLORER
RT   IUE
     JUNO 1 LAUNCH VEHICLE
     JUPITER C ROCKET VEHICLE
     METEOROID DUST CLOUDS
     MICROMETEOROIDS
     OUTER PLANETS EXPLORERS
     SCOUT PROJECT
     THOR DELTA LAUNCH VEHICLE
     ZODIACAL DUST
```

EXPLORER 1 SATELLITE
```
GS   SATELLITES
     . ARTIFICIAL SATELLITES
     . . EXPLORER SATELLITES
     . . . EXPLORER 1 SATELLITE
```

EXPLORER 2 SATELLITE
```
GS   SATELLITES
     . ARTIFICIAL SATELLITES
     . . EXPLORER SATELLITES
     . . . EXPLORER 2 SATELLITE
```

EXPLORER 3 SATELLITE
```
GS   SATELLITES
     . ARTIFICIAL SATELLITES
     . . EXPLORER SATELLITES
     . . . EXPLORER 3 SATELLITE
```

EXPLORER 4 SATELLITE
```
GS   SATELLITES
     . ARTIFICIAL SATELLITES
     . . EXPLORER SATELLITES
     . . . EXPLORER 4 SATELLITE
```

EXPLORER 5 SATELLITE
```
GS   SATELLITES
     . ARTIFICIAL SATELLITES
     . . EXPLORER SATELLITES
     . . . EXPLORER 5 SATELLITE
```

EXPLORER 6 SATELLITE
```
GS   SATELLITES
     . ARTIFICIAL SATELLITES
     . . EXPLORER SATELLITES
     . . . EXPLORER 6 SATELLITE
     . . GEOPHYSICAL SATELLITES
     . . . EXPLORER 6 SATELLITE
RT   THOR ABLE ROCKET VEHICLE
```

EXPLORER 7 SATELLITE
```
GS   SATELLITES
     . ARTIFICIAL SATELLITES
     . . EXPLORER SATELLITES
     . . . EXPLORER 7 SATELLITE
```

EXPLORER 8 SATELLITE
```
GS   SATELLITES
     . ARTIFICIAL SATELLITES
     . . EXPLORER SATELLITES
     . . . EXPLORER 8 SATELLITE
```

EXPLORER 9 SATELLITE
```
GS   SATELLITES
     . ARTIFICIAL SATELLITES
     . . EXPLORER SATELLITES
     . . . EXPLORER 9 SATELLITE
     . . METEOROLOGICAL SATELLITES
     . . . EXPLORER 9 SATELLITE
RT   SCOUT LAUNCH VEHICLE
```

EXPLORER 10 SATELLITE
```
GS   SATELLITES
     . ARTIFICIAL SATELLITES
     . . EXPLORER SATELLITES
     . . . EXPLORER 10 SATELLITE
     . . GEOPHYSICAL SATELLITES
     . . . EXPLORER 10 SATELLITE
RT   DELTA LAUNCH VEHICLE
```

EXPLORER 11 SATELLITE
```
UF   GAMMA RAY ASTRONOMY EXPLORER
GS   SATELLITES
     . ARTIFICIAL SATELLITES
     . . EXPLORER SATELLITES
     . . . EXPLORER 11 SATELLITE
RT   JUNO 2 LAUNCH VEHICLE
```

EXPLORER 12 SATELLITE
```
UF   ENERGETIC PARTICLE EXPLORER A
     EPE-A
     S-3 SATELLITE
GS   SATELLITES
     . ARTIFICIAL SATELLITES
     . . EXPLORER SATELLITES
     . . . EXPLORER 12 SATELLITE
     . . GEOPHYSICAL SATELLITES
     . . . EXPLORER 12 SATELLITE
RT   DELTA LAUNCH VEHICLE
```

EXPLORER 14 SATELLITE
```
UF   ENERGETIC PARTICLE EXPLORER B
     EPE-B
GS   SATELLITES
     . ARTIFICIAL SATELLITES
     . . EXPLORER SATELLITES
     . . . EXPLORER 14 SATELLITE
RT   DELTA LAUNCH VEHICLE
```

EXPLORER 15 SATELLITE
```
UF   ENERGETIC PARTICLE EXPLORER C
     EPE-C
GS   SATELLITES
     . ARTIFICIAL SATELLITES
     . . EXPLORER SATELLITES
     . . . EXPLORER 15 SATELLITE
RT   DELTA LAUNCH VEHICLE
```

EXPLORER 16 SATELLITE
```
GS   SATELLITES
     . ARTIFICIAL SATELLITES
     . . EXPLORER SATELLITES
     . . . EXPLORER 16 SATELLITE
RT   SCOUT LAUNCH VEHICLE
```

EXPLORER 17 SATELLITE
```
UF   AE-A SATELLITE
     ATMOSPHERE EXPLORER A
     S-6 SATELLITE
```

EXPLORER 17 SATELLITE-*(CONT.)*
```
GS   SATELLITES
     . ARTIFICIAL SATELLITES
     . . EXPLORER SATELLITES
     . . . EXPLORER 17 SATELLITE
     . . METEOROLOGICAL SATELLITES
     . . . EXPLORER 17 SATELLITE
RT   DELTA LAUNCH VEHICLE
```

EXPLORER 18 SATELLITE
```
UF   IMP-A
     IMP-1
     INTERPLANETARY EXPLORER
     S-74 SATELLITE
GS   INTERPLANETARY SPACECRAFT
     . EXPLORER 18 SATELLITE
     SATELLITES
     . ARTIFICIAL SATELLITES
     . . EXPLORER SATELLITES
     . . . EXPLORER 18 SATELLITE
     . . LUNAR SATELLITES
     . . . EXPLORER 18 SATELLITE
     UNMANNED SPACECRAFT
     . SPACE PROBES
     . . EXPLORER 18 SATELLITE
RT   DELTA LAUNCH VEHICLE
```

EXPLORER 19 SATELLITE
```
UF   AD-A SATELLITE
     AIR DENSITY EXPLORER A
GS   SATELLITES
     . ARTIFICIAL SATELLITES
     . . EXPLORER SATELLITES
     . . . EXPLORER 19 SATELLITE
     . . METEOROLOGICAL SATELLITES
     . . . EXPLORER 19 SATELLITE
RT   SCOUT LAUNCH VEHICLE
```

EXPLORER 20 SATELLITE
```
UF   IONOSPHERE EXPLORER A
GS   SATELLITES
     . ARTIFICIAL SATELLITES
     . . EXPLORER SATELLITES
     . . . EXPLORER 20 SATELLITE
RT   SCOUT LAUNCH VEHICLE
```

EXPLORER 21 SATELLITE
```
UF   IMP-B
     IMP-2
GS   SATELLITES
     . ARTIFICIAL SATELLITES
     . . EXPLORER SATELLITES
     . . . EXPLORER 21 SATELLITE
RT   DELTA LAUNCH VEHICLE
```

EXPLORER 22 SATELLITE
```
UF   BE B
     BEACON EXPLORER B
GS   EXPANDABLE STRUCTURES
     . INFLATABLE STRUCTURES
     . . INFLATABLE SPACECRAFT
     . . . BEACON SATELLITES
     . . . . EXPLORER 22 SATELLITE
     SATELLITES
     . ARTIFICIAL SATELLITES
     . . EXPLORER SATELLITES
     . . . EXPLORER 22 SATELLITE
     . . NAVIGATION SATELLITES
     . . . EXPLORER 22 SATELLITE
     . . PASSIVE SATELLITES
     . . . BEACON SATELLITES
     . . . . EXPLORER 22 SATELLITE
     SPACE ERECTABLE STRUCTURES
     . INFLATABLE SPACECRAFT
     . . BEACON SATELLITES
     . . . EXPLORER 22 SATELLITE
     UNMANNED SPACECRAFT
     . NAVIGATION SATELLITES
     . . EXPLORER 22 SATELLITE
     . PASSIVE SATELLITES
     . . BEACON SATELLITES
     . . . EXPLORER 22 SATELLITE
RT   SCOUT LAUNCH VEHICLE
```

EXPLORER 23 SATELLITE
```
GS   SATELLITES
     . ARTIFICIAL SATELLITES
     . . EXPLORER SATELLITES
     . . . EXPLORER 23 SATELLITE
RT   SCOUT LAUNCH VEHICLE
```

EXPLORER 24 SATELLITE
```
UF   AD/I SATELLITE
GS   SATELLITES
     . ARTIFICIAL SATELLITES
```

EXPLORER 24 SATELLITE-*(CONT.)*
```
        . . EXPLORER SATELLITES
        . . . EXPLORER 24 SATELLITE
RT      SCOUT LAUNCH VEHICLE
```

EXPLORER 25 SATELLITE
```
UF      AD/I B
        AIR DENSITY/INJUN EXPLORER B
        INJUN EXPLORER
GS      SATELLITES
        . ARTIFICIAL SATELLITES
        . . EXPLORER SATELLITES
        . . . EXPLORER 25 SATELLITE
        . . INJUN SATELLITES
        . . . EXPLORER 25 SATELLITE
RT      SCOUT LAUNCH VEHICLE
```

EXPLORER 26 SATELLITE
```
UF      ENERGETIC PARTICLE EXPLORER D
        EPE-D
GS      SATELLITES
        . ARTIFICIAL SATELLITES
        . . EXPLORER SATELLITES
        . . . EXPLORER 26 SATELLITE
RT      DELTA LAUNCH VEHICLE
```

EXPLORER 27 SATELLITE
```
UF      BE C
        BEACON EXPLORER C
GS      SATELLITES
        . ARTIFICIAL SATELLITES
        . . EXPLORER SATELLITES
        . . . EXPLORER 27 SATELLITE
RT      SCOUT LAUNCH VEHICLE
```

EXPLORER 28 SATELLITE
```
UF      IMP-C
        IMP-3
GS      SATELLITES
        . ARTIFICIAL SATELLITES
        . . EXPLORER SATELLITES
        . . . EXPLORER 28 SATELLITE
        . . LUNAR SATELLITES
        . . . EXPLORER 28 SATELLITE
RT      DELTA LAUNCH VEHICLE
```

EXPLORER 29 SATELLITE
```
GS      SATELLITES
        . ARTIFICIAL SATELLITES
        . . EXPLORER SATELLITES
        . . . EXPLORER 29 SATELLITE
        . . GEODETIC SATELLITES
        . . . EXPLORER 29 SATELLITE
        UNMANNED SPACECRAFT
        . GEODETIC SATELLITES
        . . EXPLORER 29 SATELLITE
RT      ACTIVE SATELLITES
        ANNA SATELLITES
        CELESTIAL GEODESY
        DELTA LAUNCH VEHICLE
        GEOS 1 SATELLITE
        LARGOS SATELLITE
        PAGEOS SATELLITE
```

EXPLORER 30 SATELLITE
```
UF      SE-A
GS      SATELLITES
        . ARTIFICIAL SATELLITES
        . . EXPLORER SATELLITES
        . . . EXPLORER 30 SATELLITE
RT      SCOUT LAUNCH VEHICLE
```

EXPLORER 31 SATELLITE
```
UF      DME-A SATELLITE
GS      SATELLITES
        . ARTIFICIAL SATELLITES
        . . EXPLORER SATELLITES
        . . . EXPLORER 31 SATELLITE
RT      THOR AGENA LAUNCH VEHICLE
```

EXPLORER 32 SATELLITE
```
UF      AE-B SATELLITE
        ATMOSPHERE EXPLORER B
GS      SATELLITES
        . ARTIFICIAL SATELLITES
        . . EXPLORER SATELLITES
        . . . EXPLORER 32 SATELLITE
RT      DELTA LAUNCH VEHICLE
```

EXPLORER 33 SATELLITE
```
UF      AIMP-D
        AIMP-1
        IMP-D
GS      SATELLITES
```

EXPLORER 33 SATELLITE-*(CONT.)*
```
        . ARTIFICIAL SATELLITES
        . . EXPLORER SATELLITES
        . . . EXPLORER 33 SATELLITE
RT      DELTA LAUNCH VEHICLE
```

EXPLORER 34 SATELLITE
```
UF      IMP-F
        IMP-4
GS      SATELLITES
        . ARTIFICIAL SATELLITES
        . . EXPLORER SATELLITES
        . . . EXPLORER 34 SATELLITE
RT      THOR AGENA LAUNCH VEHICLE
```

EXPLORER 35 SATELLITE
```
UF      AIMP-E
        AIMP-2
        IMP-E
GS      SATELLITES
        . ARTIFICIAL SATELLITES
        . . EXPLORER SATELLITES
        . . . EXPLORER 35 SATELLITE
RT      THOR AGENA LAUNCH VEHICLE
```

EXPLORER 36 SATELLITE
```
GS      SATELLITES
        . ARTIFICIAL SATELLITES
        . . EXPLORER SATELLITES
        . . . EXPLORER 36 SATELLITE
        . . GEODETIC SATELLITES
        . . . EXPLORER 36 SATELLITE
        UNMANNED SPACECRAFT
        . GEODETIC SATELLITES
        . . EXPLORER 36 SATELLITE
RT      ACTIVE SATELLITES
        ANNA SATELLITES
        CELESTIAL GEODESY
        GEOS 2 SATELLITE
        LARGOS SATELLITE
        PAGEOS SATELLITE
        THOR AGENA LAUNCH VEHICLE
```

EXPLORER 37 SATELLITE
```
GS      SATELLITES
        . ARTIFICIAL SATELLITES
        . . EXPLORER SATELLITES
        . . . EXPLORER 37 SATELLITE
RT      SCOUT LAUNCH VEHICLE
```

EXPLORER 38 SATELLITE
```
UF      RAE-1
GS      SATELLITES
        . ARTIFICIAL SATELLITES
        . . EXPLORER SATELLITES
        . . . EXPLORER 38 SATELLITE
RT      DELTA LAUNCH VEHICLE
```

EXPLORER 39 SATELLITE
```
GS      SATELLITES
        . ARTIFICIAL SATELLITES
        . . EXPLORER SATELLITES
        . . . EXPLORER 39 SATELLITE
RT      SCOUT LAUNCH VEHICLE
```

EXPLORER 40 SATELLITE
```
UF      INJUN 5 SATELLITE
GS      SATELLITES
        . ARTIFICIAL SATELLITES
        . . EXPLORER SATELLITES
        . . . EXPLORER 40 SATELLITE
RT      SCOUT LAUNCH VEHICLE
```

EXPLORER 41 SATELLITE
```
UF      IMP-G
        IMP-5
GS      SATELLITES
        . ARTIFICIAL SATELLITES
        . . EXPLORER SATELLITES
        . . . EXPLORER 41 SATELLITE
```

EXPLORER 42 SATELLITE
```
USE     UHURU SATELLITE
```

EXPLORER 43 SATELLITE
```
UF      IMP-I
        IMP-6
GS      SATELLITES
        . ARTIFICIAL SATELLITES
        . . EXPLORER SATELLITES
        . . . EXPLORER 43 SATELLITE
RT      DELTA LAUNCH VEHICLE
```

EXPLORER 44 SATELLITE
```
UF      SOLRAD 10 SATELLITE
GS      SATELLITES
        . ARTIFICIAL SATELLITES
        . . EXPLORER SATELLITES
        . . . EXPLORER 44 SATELLITE
```

EXPLORER 45 SATELLITE
```
GS      SATELLITES
        . ARTIFICIAL SATELLITES
        . . GEOPHYSICAL SATELLITES
        . . . EXPLORER 45 SATELLITE
        . SCIENTIFIC SATELLITES
        . . . EXPLORER 45 SATELLITE
```

EXPLORER 46 SATELLITE
```
UF      METEOROID TECHNOLOGY SATELLITE
GS      SATELLITES
        . ARTIFICIAL SATELLITES
        . . EXPLORER SATELLITES
        . . . EXPLORER 46 SATELLITE
```

EXPLORER 47 SATELLITE
```
UF      IMP-H
        IMP-7
GS      SATELLITES
        . ARTIFICIAL SATELLITES
        . . EXPLORER SATELLITES
        . . . EXPLORER 47 SATELLITE
```

EXPLORER 48 SATELLITE
```
GS      SATELLITES
        . ARTIFICIAL SATELLITES
        . . EXPLORER SATELLITES
        . . . EXPLORER 48 SATELLITE
RT      SAS
        SAS-2
```

EXPLORER 49 SATELLITE
```
UF      RADIO ASTRONOMY EXPLORER B
        RADIO ASTRONOMY EXPLORER 2
        RAE B
        RAE 1
        RAE 2
GS      SATELLITES
        . ARTIFICIAL SATELLITES
        . . EXPLORER SATELLITES
        . . . EXPLORER 49 SATELLITE
RT      DELTA LAUNCH VEHICLE
```

EXPLORER 50 SATELLITE
```
UF      IMP-J
        IMP-8
GS      SATELLITES
        . ARTIFICIAL SATELLITES
        . . EXPLORER SATELLITES
        . . . EXPLORER 50 SATELLITE
```

EXPLORER 51 SATELLITE
```
UF      AE-C SATELLITE
        ATMOSPHERE EXPLORER C
GS      SATELLITES
        . ARTIFICIAL SATELLITES
        . . EXPLORER SATELLITES
        . . . EXPLORER 51 SATELLITE
```

EXPLORER 52 SATELLITE
```
UF      HAWKEYE 1 SATELLITE
GS      SATELLITES
        . ARTIFICIAL SATELLITES
        . . EXPLORER SATELLITES
        . . . EXPLORER 52 SATELLITE
```

EXPLORER 53 SATELLITE
```
GS      SATELLITES
        . ARTIFICIAL SATELLITES
        . . EXPLORER SATELLITES
        . . . EXPLORER 53 SATELLITE
RT      SAS-3
```

EXPLORER 54 SATELLITE
```
UF      AE-D SATELLITE
        ATMOSPHERE EXPLORER D
GS      SATELLITES
        . ARTIFICIAL SATELLITES
        . . EXPLORER SATELLITES
        . . . EXPLORER 54 SATELLITE
```

EXPLORER 55 SATELLITE
```
UF      AE-E SATELLITE
        ATMOSPHERE EXPLORER E
GS      SATELLITES
        . ARTIFICIAL SATELLITES
        . . EXPLORER SATELLITES
```

EXPLORER 55 SATELLITE-*(CONT.)*
        ... **EXPLORER 55 SATELLITE**
        UNMANNED SPACECRAFT
       . **EXPLORER 55 SATELLITE**
RT   DELTA LAUNCH VEHICLE
      SCIENTIFIC SATELLITES

EXPLOSION SUPPRESSION
RT   FIRE PREVENTION
      FOAMS
      RETARDANTS

EXPLOSIONS
GS   **EXPLOSIONS**
     . AERIAL EXPLOSIONS
     . CHEMICAL EXPLOSIONS
     .. GAS EXPLOSIONS
     . NUCLEAR EXPLOSIONS
     .. THERMONUCLEAR EXPLOSIONS
     . UNDERGROUND EXPLOSIONS
     . UNDERWATER EXPLOSIONS
RT   ACCIDENTS
     BACKFIRE
     BLAST LOADS
   ∞BLASTS
     BURSTS
     COMBUSTION
     DETONATION
   ∞DISCHARGE
     EXPLOSIVE DECOMPRESSION
     EXPLOSIVES
     FIRES
     FLAME PROPAGATION
   ∞FLASH
     FLASHBACK
     HAZARDS
     HYDROCARBON COMBUSTION
     IMPLOSIONS
     REACTOR SAFETY
     RIEMANN WAVES
     SAFETY
     SHOCK WAVES
     SOUND PRESSURE
     SPONTANEOUS COMBUSTION
     WARNING SYSTEMS

EXPLOSIVE DECOMPRESSION
GS   PRESSURE REDUCTION
     . **EXPLOSIVE DECOMPRESSION**
RT   EXPLOSIONS
     IMPLOSIONS
     PRESSURE RECOVERY
     PRESSURIZED CABINS

EXPLOSIVE DEVICES
UF   CARTRIDGE ACTUATED DEVICES
GS   **EXPLOSIVE DEVICES**
     . BOMBS (ORDNANCE)
     . INITIATORS (EXPLOSIVES)
     .. BOOSTERS (EXPLOSIVES)
     .. CAPS (EXPLOSIVES)
     .. DETONATORS
     .. EXPLODING WIRES
     .. PRIMERS (EXPLOSIVES)
     . NUCLEAR DEVICES
     . SHAPED CHARGES
     . TORPEDOES
RT   ACTUATORS
     AMMUNITION
   ∞CHARGING
   ∞DEVICES
     IGNITERS
   ∞PROPELLANT ACTUATED DEVICES
     WARHEADS

EXPLOSIVE FORMING
GS   FORMING TECHNIQUES
     . COLD WORKING
     .. **EXPLOSIVE FORMING**
     METAL WORKING
     . **EXPLOSIVE FORMING**
RT   BULGING
     DEEP DRAWING
     ELECTROHYDRAULIC FORMING
     EXTRUDING
     SHAPED CHARGES

EXPLOSIVE GASES
USE FLAMMABLE GASES

EXPLOSIVE WELDING
GS   BONDING
     . **EXPLOSIVE WELDING**
     WELDING
     . PRESSURE WELDING

EXPLOSIVE WELDING-*(CONT.)*
      .. **EXPLOSIVE WELDING**
RT   CLADDING
     METAL BONDING
     METAL JOINTS
     METAL WORKING
     METAL-METAL BONDING

EXPLOSIVES
GS   **EXPLOSIVES**
     . BSX
     . CELLULOSE NITRATE
     . DYNAMITE
     . HYDRAZINE NITROFORM
     . HYDROGEN AZIDES
     . NITRASOL EXPLOSIVES
     . OCTOL (EXPLOSIVE)
     . PENTOLITE
     . RDX
     . STYPHNATES
     . TATB
     . TETRYL
     . TRINITROTOLUENE
RT   AMMONIUM PICRATES
     AMMUNITION
     AZIDES (ORGANIC)
     BOMBS (ORDNANCE)
     BURNING RATE
     CASE BONDED PROPELLANTS
   ∞CHARGING
     CHEMICAL EXPLOSIONS
     CHEMICAL FUELS
     COMPOSITE PROPELLANTS
     DETONATORS
     DOUBLE BASE PROPELLANTS
     DOUBLE BASE ROCKET PROPELLANTS
     EXPLOSIONS
     FIRES
     FLAMMABLE GASES
     FULMINATES
     GUN PROPELLANTS
     GUNS (ORDNANCE)
     HMX
     NITROGLYCERIN
     NITROGUANIDINE
     NITROMETHANE
     NUCLEAR WEAPONS
     ORDNANCE
     PETN
     PLASTIC PROPELLANTS
     POTASSIUM PERCHLORATES
     POWDER (PARTICLES)
     PROPELLANTS
     PYROPHORIC MATERIALS
     PYROTECHNICS
     SHAPED CHARGES
     SODIUM AZIDES
     TAGN
     TORPEDOES
     WARHEADS

EXPONENTIAL FUNCTIONS
GS   ANALYSIS (MATHEMATICS)
     . COMPLEX VARIABLES
     .. **EXPONENTIAL FUNCTIONS**
     ... LOGARITHMS
     FUNCTIONS (MATHEMATICS)
     . TRANSCENDENTAL FUNCTIONS
     .. **EXPONENTIAL FUNCTIONS**
     ... LOGARITHMS
RT   FOURIER ANALYSIS
     HYPERBOLIC FUNCTIONS
     ORTHOGONAL FUNCTIONS
     POISSON DENSITY FUNCTIONS
     PROBABILITY DENSITY FUNCTIONS
     STATISTICAL ANALYSIS
     WEIBULL DENSITY FUNCTIONS

EXPONENTS
GS   NUMBER THEORY
     . **EXPONENTS**
RT   ARITHMETIC
     FRACTALS
     LOGARITHMS

EXPORTS
USE INTERNATIONAL TRADE

EXPOS (SPACELAB PAYLOAD)
UF   X RAY SPECTROPOLARIMETRY
      PAYLOAD
GS   PAYLOADS
     . **EXPOS (SPACELAB PAYLOAD)**
RT   EUROPEAN SPACE AGENCY
     SPACELAB

EXPOSURE
RT   ATMOSPHERIC EFFECTS
     BEARING (DIRECTION)
     COLD TOLERANCE
     DOSIMETERS
     IRRADIATION
     PHOTOGRAPHY
     POSITION (LOCATION)
     POSITIONING
     RADIATION DOSAGE
     TIME
     TRINITROTOLUENE
     WEATHERING

EXPRESSIONS (MATHEMATICS)
USE FORMULAS (MATHEMATICS)

EXPULSION
RT   ACCELERATION (PHYSICS)
     CIRCUIT PROTECTION
     DISPOSAL
     DUMPING
     EJECTION
     EMPTYING
     EXCRETION
     EXHAUST VELOCITY
     EXPELLANTS
     EXPIRATION
     FLUID FLOW
     GRAVITY GRADIENT SATELLITES
     JETTISONING
     PARTICLE EMISSION
     PRESSURIZING
     REMOVAL
     UNLOADING

EXPULSION BLADDERS
GS   DIAPHRAGMS (MECHANICS)
     . **EXPULSION BLADDERS**
RT   BELLOWS
     EJECTION
     EMPTYING
     EXPANDABLE STRUCTURES
     FUEL TANK PRESSURIZATION
     FUEL TANKS
     PRESSURIZING
     PROPELLANT STORAGE
     PROPELLANT TANKS
     STORAGE TANKS

EXTARS
GS   CELESTIAL BODIES
     . STARS
     .. **EXTARS**
RT   EMISSION SPECTRA
     NEUTRON STARS
     RADIATION SOURCES
     STELLAR RADIATION
     UHURU SATELLITE
     X RAY ASTRONOMY
     X RAY BINARIES
     X RAY SOURCES
     X RAY TELESCOPES
     X RAYS

EXTENDED DURATION SPACE FLIGHT
USE LONG DURATION SPACE FLIGHT

EXTENSIONS
GS   **EXTENSIONS**
     . PROLONGATION
RT   ACCESSORIES
     ADAPTERS
     CONTRACTS
     DECONTAMINATION
     EXPANSION
     FILLING
     FITTINGS
     REVISIONS
     SUPPLEMENTS

EXTENSOMETERS
UF   DILATOMETERS
GS   MEASURING INSTRUMENTS
     . **EXTENSOMETERS**
RT   DEFORMETERS
     DILATOMETRY
     ELASTOMETERS
     MECHANICAL MEASUREMENT
     STRAIN GAGES
     STRESS MEASUREMENT
     TENSOMETERS
     THERMAL EXPANSION
     TRANSDUCERS

EXTERNAL COMBUSTION ENGINES
GS   ENGINES
    . **EXTERNAL COMBUSTION ENGINES**
RT   AUTOMOBILE ENGINES
    BOILERS
    GAS TURBINE ENGINES
    INTERNAL COMBUSTION ENGINES
    PISTON ENGINES

EXTERNAL STORE SEPARATION
UF   STORE RELEASE
RT   NACELLES
    PODS (EXTERNAL STORES)
    PROTUBERANCES
    ∞SEPARATION
    ∞STORAGE
    WING TANKS
    WING-FUSELAGE STORES

EXTERNAL STORES
GS   **EXTERNAL STORES**
    . PODS (EXTERNAL STORES)
RT   NACELLES
    PROTUBERANCES
    ∞STORAGE
    WING TANKS
    WING-FUSELAGE STORES

EXTERNAL SURFACE CURRENTS
GS   ELECTRIC FIELDS
    . **EXTERNAL SURFACE CURRENTS**
RT   ∞CURRENTS
    ELECTRIC FIELDS
    ELECTROMAGNETIC FIELDS
    ELECTROMAGNETIC PULSES
    LEVITATION MELTING
    PHOTOELECTRIC EMISSION
    SPACECRAFT CHARGING
    ∞SURFACES
    SYSTEM GENERATED
       ELECTROMAGNETIC PULSES

EXTERNAL TANKS
GS   TANKS (CONTAINERS)
    . **EXTERNAL TANKS**
RT   FUEL TANKS
    NACELLES
    PROPELLANT TANKS
    STORAGE TANKS
    WING TANKS

EXTERNALLY BLOWN FLAPS
UF   BLOWN FLAPS
    EBF
GS   AIRFOILS
    . FLAPS (CONTROL SURFACES)
    .. **EXTERNALLY BLOWN FLAPS**
    ... UPPER SURFACE BLOWN FLAPS
    CONTROL SURFACES
    . FLAPS (CONTROL SURFACES)
    .. **EXTERNALLY BLOWN FLAPS**
    ... UPPER SURFACE BLOWN FLAPS
RT   JET FLAPS
    LIFT
    LIFT DEVICES
    POWERED LIFT AIRCRAFT
    SHORT TAKEOFF AIRCRAFT
    SPANWISE BLOWING
    WING FLAPS
    WING NACELLE CONFIGURATIONS

EXTINCTION
GS   **EXTINCTION**
    . INTERSTELLAR EXTINCTION
RT   EVOLUTION (DEVELOPMENT)
    EXTINGUISHING
    FADING
    FLUORESCENCE

EXTINGUISHERS
USE   FIRE EXTINGUISHERS

EXTINGUISHING
UF   FLAME QUENCHING
RT   BURNOUT
    COMBUSTION
    EXTINCTION
    FLAMEOUT
    OCCULTATION
    ∞QUENCHING
    QUENCHING (COOLING)

EXTRACTION
GS   **EXTRACTION**

EXTRACTION-*(CONT.)*
    . GEOTHERMAL ENERGY EXTRACTION
    . ION EXTRACTION
    . SOLVENT EXTRACTION
RT   BEDS (PROCESS ENGINEERING)
    BENEFICIATION
    CENTRIFUGES
    CENTRIFUGING
    COLUMNS (PROCESS ENGINEERING)
    CONCENTRATING
    DIALYSIS
    DIFFUSION
    DISSOLVING
    ELUTION
    FILTRATION
    FURNACES
    HYDROLYSIS
    LEACHING
    MATERIAL ABSORPTION
    MATERIALS RECOVERY
    MELTING
    OSMOSIS
    PERCOLATION
    RECYCLING
    REFINING
    REMOVAL
    ∞SEPARATION
    SOLVENTS
    SORPTION
    WASHERS (CLEANERS)

EXTRAGALACTIC LIGHT
USE   EXTRATERRESTRIAL RADIATION
    LIGHT (VISIBLE RADIATION)

EXTRAGALACTIC MEDIA
USE   INTERGALACTIC MEDIA

EXTRAGALACTIC RADIO SOURCES
GS   CELESTIAL BODIES
    . RADIO SOURCES (ASTRONOMY)
    .. **EXTRAGALACTIC RADIO SOURCES**
    ... RADIO GALAXIES
RT   BL LACERTAE OBJECTS
    EXTRATERRESTRIAL RADIATION
    EXTRATERRESTRIAL RADIO WAVES
    QUASARS
    RADIATION SOURCES
    RADIO ASTRONOMY
    RADIO EMISSION
    ∞SOURCES

EXTRAPOLATION
RT   FINITE DIFFERENCE THEORY
    FORECASTING
    INTERPOLATION
    PERIODIC VARIATIONS
    QUALITY CONTROL
    STATISTICAL ANALYSIS
    ∞TESTS
    TIME SERIES ANALYSIS
    TRENDS

EXTRASENSORY PERCEPTION
UF   PARAPSYCHOLOGY
GS   PERCEPTION
    . SENSORY PERCEPTION
    .. **EXTRASENSORY PERCEPTION**

EXTRASOLAR PLANETS
GS   CELESTIAL BODIES
    . PLANETS
    .. **EXTRASOLAR PLANETS**
RT   GAS GIANT PLANETS

EXTRATERRESTRIAL COMMUNICATION
GS   TELECOMMUNICATION
    . SPACE COMMUNICATION
    .. **EXTRATERRESTRIAL
       COMMUNICATION**
RT   INFORMATION DISSEMINATION
    INTERPLANETARY COMMUNICATION
    RADIO TELEMETRY

EXTRATERRESTRIAL ENVIRONMENTS
GS   ENVIRONMENTS
    . **EXTRATERRESTRIAL ENVIRONMENTS**
    .. CISLUNAR SPACE
    . DEEP SPACE
    ... INTERPLANETARY SPACE
    ... INTERSTELLAR SPACE
    .. LUNAR ENVIRONMENT
    ... LUNAR ATMOSPHERE
    .. PLANETARY ENVIRONMENTS
    ... MARS ENVIRONMENT

EXTRATERRESTRIAL ENVIRONMENTS-*(CONT.)*
    .... MARS ATMOSPHERE
    ... PLANETARY ATMOSPHERES
    .... HELIUM HYDROGEN
        ATMOSPHERES
    .... JUPITER ATMOSPHERE
    .... MARS ATMOSPHERE
    .... NEPTUNE ATMOSPHERE
    .... SATURN ATMOSPHERE
    .... URANUS ATMOSPHERE
    .... VENUS ATMOSPHERE
    .. SATELLITE ATMOSPHERES
    ... LUNAR ATMOSPHERE
    .. STELLAR ATMOSPHERES
    ... CHROMOSPHERE
    .. SOLAR ATMOSPHERE
RT   AEROSPACE ENVIRONMENTS
    HIGH GRAVITY ENVIRONMENTS
    LONG DURATION SPACE FLIGHT
    SPACE EXPLORATION
    SPACECRAFT ENVIRONMENTS
    VENUS SURFACE

EXTRATERRESTRIAL INTELLIGENCE
GS   INTELLIGENCE
    . **EXTRATERRESTRIAL INTELLIGENCE**
RT   INTERSTELLAR COMMUNICATION
    INTERSTELLAR TRAVEL
    PROJECT SETI
    SPACE COMMUNICATION
    UNIDENTIFIED FLYING OBJECTS

EXTRATERRESTRIAL LIFE
GS   LIFE SCIENCES
    . **EXTRATERRESTRIAL LIFE**
RT   AEROSPACE ENVIRONMENTS
    BIOSATELLITES
    EXOBIOLOGY
    GULLIVER PROGRAM
    LIFE DETECTORS
    PANSPERMIA

EXTRATERRESTRIAL MATTER
GS   **EXTRATERRESTRIAL MATTER**
    . COSMIC GASES
    .. INTERPLANETARY GAS
    .. INTERSTELLAR GAS
    ... NEUTRAL GASES
    . COSMIC PLASMA
RT   COSMOCHEMISTRY
    MATTER (PHYSICS)
    VENUS FLY TRAP ROCKET VEHICLE

EXTRATERRESTRIAL RADIATION
UF   EXTRAGALACTIC LIGHT
    SPACE RADIATION
    STELLAR DOPPLER SHIFT
GS   **EXTRATERRESTRIAL RADIATION**
    . EXTRATERRESTRIAL RADIO WAVES
    .. GALACTIC RADIO WAVES
    ... NORTH POLAR SPUR (ASTRONOMY)
    . RADIO BURSTS
    ... SOLAR RADIO BURSTS
    .... TYPE 2 BURSTS
    .... TYPE 3 BURSTS
    .... TYPE 4 BURSTS
    .... TYPE 5 BURSTS
    .. SOLAR RADIO EMISSION
    ... SOLAR RADIO BURSTS
    .... TYPE 2 BURSTS
    .... TYPE 3 BURSTS
    .... TYPE 4 BURSTS
    .... TYPE 5 BURSTS
    . GALACTIC RADIATION
    .. GALACTIC COSMIC RAYS
    .. GALACTIC RADIO WAVES
    ... NORTH POLAR SPUR (ASTRONOMY)
    . GAMMA RAY BURSTS
    . GEGENSCHEIN
    . INTERSTELLAR RADIATION
    . LUNAR RADIATION
    . PLANETARY RADIATION
    . PRIMARY COSMIC RAYS
    . SOLAR COSMIC RAYS
    . SOLAR RADIATION
    .. CIRCUMSOLAR RADIATION
    .. SOLAR CORPUSCULAR RADIATION
    ... SOLAR ELECTRONS
    ... SOLAR NEUTRINOS
    ... SOLAR PROTONS
    .. SOLAR COSMIC RAYS
    .. SOLAR RADIO EMISSION
    ... SOLAR RADIO BURSTS
    .... TYPE 2 BURSTS
    .... TYPE 3 BURSTS

EXTRATERRESTRIAL RADIATION-(CONT.)

```
         . . . . TYPE 4 BURSTS
         . . . . TYPE 5 BURSTS
         . . SOLAR WIND
         . . SOLAR X-RAYS
         . . SUNLIGHT
         . STELLAR RADIATION
         . . STELLAR WINDS
         . ZODIACAL LIGHT
RT       AEROSPACE ENVIRONMENTS
∞        AEROSPACE SCIENCES
         ATMOSPHERIC RADIATION
         BACKGROUND RADIATION
         CORPUSCULAR RADIATION
         COSMIC RAYS
         COSMIC X RAYS
         ELECTROMAGNETIC NOISE
         ELECTROMAGNETIC RADIATION
         ELECTRON ACCELERATION
         EXTRAGALACTIC RADIO SOURCES
         LIGHT (VISIBLE RADIATION)
         LYMAN ALPHA RADIATION
         LYMAN BETA RADIATION
         MICROWAVE EMISSION
         POLARIZED ELECTROMAGNETIC
            RADIATION
         POLARIZED RADIATION
∞        RADIATION
         RADIATION BELTS
         RADIATIVE TRANSFER
         RADIO WAVES
∞        RAYS
         RELIC RADIATION
         SYNCHROTRON RADIATION
         SYSTEM GENERATED
            ELECTROMAGNETIC PULSES
         TERRESTRIAL RADIATION
         X RAYS
```

EXTRATERRESTRIAL RADIO WAVES

```
UF       COSMIC RADIO WAVES
GS       ELECTROMAGNETIC RADIATION
         . RADIO WAVES
         . . EXTRATERRESTRIAL RADIO WAVES
         . . . GALACTIC RADIO WAVES
         . . . RADIO BURSTS
         . . . . SOLAR RADIO BURSTS
         . . . . . TYPE 2 BURSTS
         . . . . . TYPE 3 BURSTS
         . . . . . TYPE 4 BURSTS
         . . . . . TYPE 5 BURSTS
         . . . SOLAR RADIO EMISSION
         . . . . SOLAR RADIO BURSTS
         . . . . . TYPE 2 BURSTS
         . . . . . TYPE 3 BURSTS
         . . . . . TYPE 4 BURSTS
         . . . . . TYPE 5 BURSTS
         EXTRATERRESTRIAL RADIATION
         . EXTRATERRESTRIAL RADIO WAVES
         . . GALACTIC RADIO WAVES
         . . NORTH POLAR SPUR (ASTRONOMY)
         . RADIO BURSTS
         . . SOLAR RADIO BURSTS
         . . . . TYPE 2 BURSTS
         . . . . TYPE 3 BURSTS
         . . . . TYPE 4 BURSTS
         . . . . TYPE 5 BURSTS
         . . SOLAR RADIO EMISSION
         . . . SOLAR RADIO BURSTS
         . . . . TYPE 2 BURSTS
         . . . . TYPE 3 BURSTS
         . . . . TYPE 4 BURSTS
         . . . . TYPE 5 BURSTS
RT       CENTIMETER WAVES
         EXTRAGALACTIC RADIO SOURCES
         MICROWAVE EMISSION
         MICROWAVES
         MILLIMETER WAVES
         RADIO ASTRONOMY
         RADIO EMISSION
         RADIO FREQUENCY INTERFERENCE
         RADIO SOURCES (ASTRONOMY)
```

EXTRATERRESTRIAL RESOURCES

```
GS       RESOURCES
         . EXTRATERRESTRIAL RESOURCES
RT       LUNAR EXPLORATION
         PLANETARY BASES
         SPACE EXPLORATION
         SPACE LOGISTICS
```

EXTRATERRESTRIAL ROVING VEHICLES

```
USE      ROVING VEHICLES
```

EXTRAVEHICULAR ACTIVITY

```
SN       (ACTIVITY OUTSIDE THE SPACECRAFT)
RT  ∞    ACTIVITY
         AEPS
         AEROSPACE ENVIRONMENTS
         APOLLO EXTENSION SYSTEM
         ASTRONAUT LOCOMOTION
         ASTRONAUT MANEUVERING EQUIPMENT
         IMLSS
         INTRAVEHICULAR ACTIVITY
         MAN OPERATED PROPULSION SYSTEMS
         MANNED MANEUVERING UNITS
         MANNED SPACE FLIGHT
         ORBITAL WORKERS
         SELF MANEUVERING UNITS
         SPACE FLIGHT
         SPACE MAINTENANCE
         SPACE SHUTTLE PAYLOADS
         UMBILICAL CONNECTORS
         WEIGHTLESSNESS
```

EXTRAVEHICULAR MOBILITY UNITS

```
GS       EXTRAVEHICULAR MOBILITY UNITS
         . ASTRONAUT LOCOMOTION
         . ASTRONAUT MANEUVERING
            EQUIPMENT
         . . MANNED MANEUVERING UNITS
RT       IMLSS
         SELF MANEUVERING UNITS
```

EXTREMA

```
USE      RANGE (EXTREMES)
```

EXTREME ULTRAVIOLET EXPLORER SATELLITE

```
UF       EUVE
GS       SATELLITES
         . ARTIFICIAL SATELLITES
         . . EXPLORER SATELLITES
         . . . EXTREME ULTRAVIOLET EXPLORER
            SATELLITE
RT       IUE
         ULTRAVIOLET ASTRONOMY
```

EXTREME ULTRAVIOLET RADIATION

```
GS       ELECTROMAGNETIC RADIATION
         . ULTRAVIOLET RADIATION
         . . EXTREME ULTRAVIOLET RADIATION
         IONIZING RADIATION
         . ULTRAVIOLET RADIATION
         . . EXTREME ULTRAVIOLET RADIATION
RT       BEAMS (RADIATION)
         MAGELLAN MISSION
∞        RADIATION
         SOLAR RADIATION
```

EXTREMELY HIGH FREQUENCIES

```
SN       (30 TO 300 GHZ)
UF       K BAND
         KA BAND
         V BAND
GS       FREQUENCIES
         . RADIO FREQUENCIES
         . . MICROWAVE FREQUENCIES
         . . . EXTREMELY HIGH FREQUENCIES
RT       MILLIMETER WAVES
```

EXTREMELY LOW FREQUENCIES

```
GS       FREQUENCIES
         . EXTREMELY LOW FREQUENCIES
RT       LOW FREQUENCIES
         RADIO FREQUENCIES
         SEAFARER PROJECT
```

EXTREMELY LOW RADIO FREQUENCIES

```
SN       (BELOW 300 H)
UF       ULTRALOW FREQUENCIES
GS       FREQUENCIES
         . RADIO FREQUENCIES
         . . EXTREMELY LOW RADIO
            FREQUENCIES
RT       AUDIO FREQUENCIES
```

EXTREMUM VALUES

```
GS       ANALYSIS (MATHEMATICS)
         . REAL VARIABLES
         . . EXTREMUM VALUES
         . . . LIMITS (MATHEMATICS)
         . . . MAXIMA
         . . . MINIMA
RT       EULER-LAGRANGE EQUATION
         FUNCTIONS (MATHEMATICS)
         OPTIMIZATION
∞        PEAKS
         PROBABILITY THEORY
```

EXTROVERSION

```
RT       BEHAVIOR
         HUMAN BEHAVIOR
         PSYCHOLOGY
```

EXTRUDING

```
UF       HOT EXTRUDING
GS       FORMING TECHNIQUES
         . EXTRUDING
         . . PULTRUSION
RT       CASTING
         CLADDING
         COLD WORKING
         DIES
∞        DRAWING
         EXPLOSIVE FORMING
         INJECTION MOLDING
         INTRUSION
         METAL SPINNING
         METAL WORKING
         PIERCING
         PRESSING (FORMING)
         WET SPINNING
```

EYE (ANATOMY)

```
GS       ANATOMY
         . SENSE ORGANS
         . . EYE (ANATOMY)
         . . . CHOROID MEMBRANES
         . . . CONJUNCTIVA
         . . . CORNEA
         . . . OCULOMOTOR NERVES
         . . . PUPILS
         . . . RETINA
         . . . . FOVEA
RT       ACCOMMODATION
         COLOR VISION
         FACE (ANATOMY)
         HEAD (ANATOMY)
         LENSES
         MIOSIS
         OPHTHALMODYNAMOMETRY
         OPHTHALMOLOGY
         OPTOMETRY
         VESTIBULAR NYSTAGMUS
         VISION
```

EYE DISEASES

```
GS       DISEASES
         . EYE DISEASES
         . . ASTHENOPIA
         . . ASTIGMATISM
         . . CATARACTS
         . . CONJUNCTIVITIS
         . . GLAUCOMA
         . . KERATITIS
         . . PHORIA
RT       BLINDNESS
         OPHTHALMOLOGY
```

EYE DOMINANCE

```
GS       DOMINANCE
         . EYE DOMINANCE
RT       VISION
```

EYE EXAMINATIONS

```
GS       EXAMINATION
         . EYE EXAMINATIONS
RT       HAPLOSCOPES
```

EYE MOVEMENTS

```
GS       EYE MOVEMENTS
         . ELECTRONYSTAGMOGRAPHY
         . NYSTAGMUS
         . SACCADIC EYE MOVEMENTS
RT       BLINKING
         OCULOMETERS
         RAPID EYE MOVEMENT STATE
```

EYE PROTECTION

```
GS       PROTECTION
         . EYE PROTECTION
RT       FLASH BLINDNESS
         GOGGLES
         SUNGLASSES
         VISORS
```

EYEPIECES

```
GS       OPTICAL EQUIPMENT
         . EYEPIECES
RT       BINOCULARS
         CONTACT LENSES
         LENSES
         MICROSCOPES
         PERISCOPES
```

EYEPIECES-(CONT.)
   RETICLES
   SUNGLASSES
   TELESCOPES

EYRING THEORY
GS   KINETIC THEORY
   . TRANSPORT THEORY
   . . EYRING THEORY
RT   EQUILIBRIUM FLOW
   FLUID DYNAMICS
   ∞THEORIES

F

F CENTERS
USE   COLOR CENTERS

F DISPLAYS
USE   F REGION

F LAYER
USE   F REGION

F REGION
UF   F DISPLAYS
   F LAYER
   NIGHT F LAYER
GS   EARTH ATMOSPHERE
   . UPPER ATMOSPHERE
   . . IONOSPHERE
   . . . UPPER IONOSPHERE
   . . . . F REGION
   . . . . . F 1 REGION
   . . . . . F 2 REGION
   REGIONS
   . F REGION
RT   PLASMA BUBBLES

F 1 REGION
GS   EARTH ATMOSPHERE
   . UPPER ATMOSPHERE
   . . IONOSPHERE
   . . . UPPER IONOSPHERE
   . . . . F REGION
   . . . . . F 1 REGION

F 2 REGION
GS   EARTH ATMOSPHERE
   . UPPER ATMOSPHERE
   . . IONOSPHERE
   . . . UPPER IONOSPHERE
   . . . . F REGION
   . . . . . F 2 REGION
RT   SPREAD F
   TRANSEQUATORIAL PROPAGATION

F-1 ROCKET ENGINE
GS   ENGINES
   . ROCKET ENGINES
   . . LIQUID PROPELLANT ROCKET
     ENGINES
   . . . F-1 ROCKET ENGINE
RT   BOOSTER ROCKET ENGINES
   NOVA LAUNCH VEHICLES
   SATURN LAUNCH VEHICLES

F-2 AIRCRAFT
UF   HAWKER HUNTER AIRCRAFT
   HUNTER F-2 AIRCRAFT
GS   ATTACK AIRCRAFT
   . FIGHTER AIRCRAFT
   . . F-2 AIRCRAFT
   HAWKER SIDDELEY AIRCRAFT
   . F-2 AIRCRAFT
   JET AIRCRAFT
   . F-2 AIRCRAFT
   MONOPLANES
   . F-2 AIRCRAFT
RT   ∞AIRCRAFT

F-4 AIRCRAFT
UF   F-110 AIRCRAFT
   F4H AIRCRAFT
GS   ATTACK AIRCRAFT
   . FIGHTER AIRCRAFT
   . . F-4 AIRCRAFT
   JET AIRCRAFT
   . PHANTOM AIRCRAFT
   . . F-4 AIRCRAFT
   MCDONNELL DOUGLAS AIRCRAFT

F-4 AIRCRAFT-(CONT.)
   . MCDONNELL AIRCRAFT
   . . PHANTOM AIRCRAFT
   . . . F-4 AIRCRAFT
   SUPERSONIC AIRCRAFT
   . PHANTOM AIRCRAFT
   . . F-4 AIRCRAFT
RT   ∞AIRCRAFT
   J-79 ENGINE
   RF-4 AIRCRAFT

F-5 AIRCRAFT
UF   FREEDOM FIGHTER AIRCRAFT
   N-156 AIRCRAFT
GS   ATTACK AIRCRAFT
   . FIGHTER AIRCRAFT
   . . F-5 AIRCRAFT
   COIN AIRCRAFT
   . F-5 AIRCRAFT
   JET AIRCRAFT
   . F-5 AIRCRAFT
   MONOPLANES
   . F-5 AIRCRAFT
   NORTHROP AIRCRAFT
   . F-5 AIRCRAFT
   OBSERVATION AIRCRAFT
   . F-5 AIRCRAFT
   RECONNAISSANCE AIRCRAFT
   . F-5 AIRCRAFT
   SUPERSONIC AIRCRAFT
   . F-5 AIRCRAFT
RT   ∞AIRCRAFT

F-8 AIRCRAFT
UF   CRUSADER AIRCRAFT
   F8U AIRCRAFT
   RF-8 AIRCRAFT
GS   ATTACK AIRCRAFT
   . FIGHTER AIRCRAFT
   . . F-8 AIRCRAFT
   JET AIRCRAFT
   . F-8 AIRCRAFT
   LING-TEMCO-VOUGHT AIRCRAFT
   . F-8 AIRCRAFT
   MONOPLANES
   . F-8 AIRCRAFT
   SUPERSONIC AIRCRAFT
   . F-8 AIRCRAFT
RT   ∞AIRCRAFT

F-9 AIRCRAFT
UF   COUGAR AIRCRAFT
   F9F AIRCRAFT
   PANTHER AIRCRAFT
GS   ATTACK AIRCRAFT
   . FIGHTER AIRCRAFT
   . . F-9 AIRCRAFT
   GRUMMAN AIRCRAFT
   . F-9 AIRCRAFT
   JET AIRCRAFT
   . F-9 AIRCRAFT
   MONOPLANES
   . F-9 AIRCRAFT
RT   ∞AIRCRAFT

F-14 AIRCRAFT
GS   ATTACK AIRCRAFT
   . FIGHTER AIRCRAFT
   . . F-14 AIRCRAFT
   GRUMMAN AIRCRAFT
   . F-14 AIRCRAFT
   JET AIRCRAFT
   . F-14 AIRCRAFT
   SUPERSONIC AIRCRAFT
   . F-14 AIRCRAFT
RT   ∞AIRCRAFT

F-15 AIRCRAFT
GS   ATTACK AIRCRAFT
   . FIGHTER AIRCRAFT
   . . F-15 AIRCRAFT
   JET AIRCRAFT
   . F-15 AIRCRAFT
   SUPERSONIC AIRCRAFT
   . F-15 AIRCRAFT
RT   ∞AIRCRAFT

F-16 AIRCRAFT
GS   ATTACK AIRCRAFT
   . FIGHTER AIRCRAFT
   . . F-16 AIRCRAFT
   JET AIRCRAFT
   . F-16 AIRCRAFT
   SUPERSONIC AIRCRAFT
   . F-16 AIRCRAFT

F-16 AIRCRAFT-(CONT.)
RT   ∞AIRCRAFT

F-17 AIRCRAFT
UF   YF-17 AIRCRAFT
GS   ATTACK AIRCRAFT
   . FIGHTER AIRCRAFT
   . . F-17 AIRCRAFT
   JET AIRCRAFT
   . F-17 AIRCRAFT
   MONOPLANES
   . F-17 AIRCRAFT
   SUPERSONIC AIRCRAFT
   . F-17 AIRCRAFT
RT   ∞AIRCRAFT
   F-18 AIRCRAFT

F-18 AIRCRAFT
GS   ATTACK AIRCRAFT
   . FIGHTER AIRCRAFT
   . . F-18 AIRCRAFT
   JET AIRCRAFT
   . F-18 AIRCRAFT
   MCDONNELL DOUGLAS AIRCRAFT
   . F-18 AIRCRAFT
   NORTHROP AIRCRAFT
   . F-18 AIRCRAFT
RT   ∞AIRCRAFT
   F-17 AIRCRAFT

F-20 AIRCRAFT

F-27 AIRCRAFT
UF   FOKKER F 27 AIRCRAFT
   FOKKER FRIENDSHIP AIRCRAFT
GS   ATTACK AIRCRAFT
   . FIGHTER AIRCRAFT
   . . F-27 AIRCRAFT
   FOKKER AIRCRAFT
   . F-27 AIRCRAFT
   JET AIRCRAFT
   . TURBOPROP AIRCRAFT
   . . F-27 AIRCRAFT
   MONOPLANES
   . F-27 AIRCRAFT
   PASSENGER AIRCRAFT
   . F-27 AIRCRAFT
   TRANSPORT AIRCRAFT
   . CARGO AIRCRAFT
   . . F-27 AIRCRAFT
RT   ∞AIRCRAFT

F-28 HELICOPTER
GS   LIGHT AIRCRAFT
   . F-28 HELICOPTER
   PASSENGER AIRCRAFT
   . F-28 HELICOPTER
   V/STOL AIRCRAFT
   . ROTARY WING AIRCRAFT
   . . HELICOPTERS
   . . . RIGID ROTOR HELICOPTERS
   . . . . F-28 HELICOPTER

F-28 TRANSPORT AIRCRAFT
UF   FELLOWSHIP AIRCRAFT
   FOKKER F 28 AIRCRAFT
GS   COMMERCIAL AIRCRAFT
   . F-28 TRANSPORT AIRCRAFT
   FOKKER AIRCRAFT
   . F-28 TRANSPORT AIRCRAFT
   JET AIRCRAFT
   . TURBOFAN AIRCRAFT
   . . F-28 TRANSPORT AIRCRAFT
   MONOPLANES
   . F-28 TRANSPORT AIRCRAFT
   PASSENGER AIRCRAFT
   . F-28 TRANSPORT AIRCRAFT
   TRANSPORT AIRCRAFT
   . F-28 TRANSPORT AIRCRAFT
RT   ∞AIRCRAFT

F-80 AIRCRAFT
USE   T-33 AIRCRAFT

F-84 AIRCRAFT
GS   ATTACK AIRCRAFT
   . FIGHTER AIRCRAFT
   . . F-84 AIRCRAFT
   JET AIRCRAFT
   . F-84 AIRCRAFT
   MONOPLANES
   . F-84 AIRCRAFT
   REPUBLIC AIRCRAFT
   . F-84 AIRCRAFT
RT   ∞AIRCRAFT

F-86 AIRCRAFT
UF   SABRE AIRCRAFT
GS   ATTACK AIRCRAFT
    . FIGHTER AIRCRAFT
    . . **F-86 AIRCRAFT**
    JET AIRCRAFT
    . **F-86 AIRCRAFT**
    MONOPLANES
    . **F-86 AIRCRAFT**
    NORTH AMERICAN AIRCRAFT
    . **F-86 AIRCRAFT**
RT   ∞AIRCRAFT

F-89 AIRCRAFT
GS   ATTACK AIRCRAFT
    . FIGHTER AIRCRAFT
    . . **F-89 AIRCRAFT**
    JET AIRCRAFT
    . **F-89 AIRCRAFT**
    MONOPLANES
    . **F-89 AIRCRAFT**
    NORTHROP AIRCRAFT
    . **F-89 AIRCRAFT**
RT   ∞AIRCRAFT

F-94 AIRCRAFT
GS   ATTACK AIRCRAFT
    . FIGHTER AIRCRAFT
    . . **F-94 AIRCRAFT**
    JET AIRCRAFT
    . **F-94 AIRCRAFT**
    LOCKHEED AIRCRAFT
    . **F-94 AIRCRAFT**
    MONOPLANES
    . **F-94 AIRCRAFT**
RT   ∞AIRCRAFT

F-100 AIRCRAFT
UF   SUPER SABRE AIRCRAFT
GS   ATTACK AIRCRAFT
    . BOMBER AIRCRAFT
    . . **F-100 AIRCRAFT**
    . FIGHTER AIRCRAFT
    . . **F-100 AIRCRAFT**
    JET AIRCRAFT
    . **F-100 AIRCRAFT**
    MONOPLANES
    . **F-100 AIRCRAFT**
    NORTH AMERICAN AIRCRAFT
    . **F-100 AIRCRAFT**
    SUPERSONIC AIRCRAFT
    . **F-100 AIRCRAFT**
RT   ∞AIRCRAFT

F-101 AIRCRAFT
UF   JF 101 AIRCRAFT
    VOODOO AIRCRAFT
GS   ATTACK AIRCRAFT
    . FIGHTER AIRCRAFT
    . . **F-101 AIRCRAFT**
    JET AIRCRAFT
    . **F-101 AIRCRAFT**
    MCDONNELL DOUGLAS AIRCRAFT
    . MCDONNELL AIRCRAFT
    . . **F-101 AIRCRAFT**
    SUPERSONIC AIRCRAFT
    . **F-101 AIRCRAFT**
RT   ∞AIRCRAFT

F-102 AIRCRAFT
UF   DELTA DAGGER AIRCRAFT
    YF-102 AIRCRAFT
GS   ATTACK AIRCRAFT
    . FIGHTER AIRCRAFT
    . . **F-102 AIRCRAFT**
    GENERAL DYNAMICS AIRCRAFT
    . **F-102 AIRCRAFT**
    JET AIRCRAFT
    . **F-102 AIRCRAFT**
    MONOPLANES
    . **F-102 AIRCRAFT**
    SUPERSONIC AIRCRAFT
    . **F-102 AIRCRAFT**
    TAILLESS AIRCRAFT
    . **F-102 AIRCRAFT**
RT   ∞AIRCRAFT

F-104 AIRCRAFT
UF   CANADAIR CF-104 AIRCRAFT
    CF-104 AIRCRAFT
    STARFIGHTER AIRCRAFT
GS   ATTACK AIRCRAFT
    . FIGHTER AIRCRAFT
    . . **F-104 AIRCRAFT**
    JET AIRCRAFT

F-104 AIRCRAFT-*(CONT.)*
    . **F-104 AIRCRAFT**
    LOCKHEED AIRCRAFT
    . **F-104 AIRCRAFT**
    MONOPLANES
    . **F-104 AIRCRAFT**
    SUPERSONIC AIRCRAFT
    . **F-104 AIRCRAFT**
RT   ∞AIRCRAFT

F-105 AIRCRAFT
UF   THUNDERCHIEF AIRCRAFT
GS   ATTACK AIRCRAFT
    . FIGHTER AIRCRAFT
    . . **F-105 AIRCRAFT**
    JET AIRCRAFT
    . **F-105 AIRCRAFT**
    MONOPLANES
    . **F-105 AIRCRAFT**
    REPUBLIC AIRCRAFT
    . **F-105 AIRCRAFT**
RT   ∞AIRCRAFT

F-106 AIRCRAFT
UF   DELTA DART AIRCRAFT
GS   ATTACK AIRCRAFT
    . FIGHTER AIRCRAFT
    . . **F-106 AIRCRAFT**
    GENERAL DYNAMICS AIRCRAFT
    . **F-106 AIRCRAFT**
    JET AIRCRAFT
    . **F-106 AIRCRAFT**
    MONOPLANES
    . **F-106 AIRCRAFT**
    SUPERSONIC AIRCRAFT
    . **F-106 AIRCRAFT**
    TAILLESS AIRCRAFT
    . **F-106 AIRCRAFT**
RT   ∞AIRCRAFT

F-110 AIRCRAFT
USE   F-4 AIRCRAFT

F-111 AIRCRAFT
UF   LASV
    TFX AIRCRAFT
GS   ATTACK AIRCRAFT
    . FIGHTER AIRCRAFT
    . . **F-111 AIRCRAFT**
    GENERAL DYNAMICS AIRCRAFT
    . **F-111 AIRCRAFT**
    GRUMMAN AIRCRAFT
    . **F-111 AIRCRAFT**
    JET AIRCRAFT
    . TURBOFAN AIRCRAFT
    . . **F-111 AIRCRAFT**
    SUPERSONIC AIRCRAFT
    . **F-111 AIRCRAFT**
RT   ∞AIRCRAFT
    VARIABLE SWEEP WINGS

FAB (PROGRAMMING LANGUAGE)
USE   FORTRAN

FABRICATION
GS   **FABRICATION**
    . SPACE MANUFACTURING
RT   ASSEMBLIES
    ASSEMBLING
    CONSTRUCTION
    LOW GRAVITY MANUFACTURING
    MANUFACTURING
    PRODUCTION MANAGEMENT

FABRICS
UF   CLOTH
GS   **FABRICS**
    . CREPE
    . DACRON (TRADEMARK)
    . FELTS
    . FORTISAN (TRADEMARK)
    . GAUZE
    . LINEN
    . PARACHUTE FABRICS
    . SILK
    . WOOL
RT   CLOTHING
    COATINGS
    COTTON
    FIBERS
    ∞FILMS
    FLAME RETARDANTS
    GEOTECHNICAL FABRICS
    GORES
    INTERLAYERS

FABRICS-*(CONT.)*
    LAMINATES
    MESH
    MICARTA
    MULTILAYER INSULATION
    REINFORCING MATERIALS
    RIBBONS
    ∞SHEETS
    SOCKS
    TEXTILES
    WEAVING
    WEBBING
    WEBS (SHEETS)
    WIRE CLOTH

FABRY-PEROT INTERFEROMETERS
GS   MEASURING INSTRUMENTS
    . INTERFEROMETERS
    . . **FABRY-PEROT INTERFEROMETERS**
RT   MICROWAVE INTERFEROMETERS
    PLASMA DIAGNOSTICS

FABRY-PEROT LASERS
USE   LASERS

FABRY-PEROT SPECTROMETERS
GS   MEASURING INSTRUMENTS
    . RADIATION MEASURING INSTRUMENTS
    . . **FABRY-PEROT SPECTROMETERS**
    . SPECTROMETERS
    . . **FABRY-PEROT SPECTROMETERS**
RT   ACTINOMETERS
    AIRGLOW
    AURORAL SPECTROSCOPY
    OPTICAL EQUIPMENT
    OPTICAL MEASURING INSTRUMENTS

FACE (ANATOMY)
GS   **FACE (ANATOMY)**
    . CHIN
    . FOREHEAD
    . NOSE (ANATOMY)
RT   EYE (ANATOMY)
    HEAD (ANATOMY)
    LIPS (ANATOMY)

FACE CENTERED CUBIC LATTICES
UF   FCC LATTICES
GS   CRYSTAL LATTICES
    . CUBIC LATTICES
    . . **FACE CENTERED CUBIC LATTICES**
RT   BODY CENTERED CUBIC LATTICES
    CLOSE PACKED LATTICES
    CRYSTALS

FACETS
USE   FLAT SURFACES

∞ **FACILITIES**
SN   *(USE OF A MORE SPECIFIC TERM IS RECOMMENDED--CONSULT THE TERMS LISTED BELOW)*
RT   AIRPORTS
    ELECTRIC POWER PLANTS
    GROUND HANDLING
    INDUSTRIAL AREAS
    INDUSTRIAL PLANTS
    LAND USE
    LAUNCHING BASES
    LOGISTICS
    LOGISTICS MANAGEMENT
    MILITARY AIR FACILITIES
    MOBILE QUARANTINE FACILITY
    RESEARCH FACILITIES
    ROADS
    SITE SELECTION
    SITES
    SOLAR CELL CALIBRATION FACILITY
    STATIONS
    TERMINAL FACILITIES
    TEST FACILITIES
    X RAY ASTROPHYSICS FACILITY

FACSIMILE COMMUNICATION
UF   FACSIMILE TRANSMISSION
GS   TELECOMMUNICATION
    . COMMUNICATION
    . . **FACSIMILE COMMUNICATION**
RT   INTERPLANETARY COMMUNICATION
    LUNAR COMMUNICATION
    SPACECRAFT COMMUNICATION
    TELETYPEWRITER SYSTEMS
    TELEVISION SYSTEMS
    TRANSOCEANIC COMMUNICATION
    WIRELESS COMMUNICATION

FACSIMILE TRANSMISSION
USE   FACSIMILE COMMUNICATION

FACTOR ANALYSIS
GS   STATISTICAL ANALYSIS
.   **FACTOR ANALYSIS**
RT   AUTOREGRESSIVE PROCESSES
CORRELATION
COVARIANCE
DEGREES OF FREEDOM
EXPERIMENT DESIGN
FACTORIZATION
MATRICES (MATHEMATICS)
ORTHOGONALITY
REGRESSION ANALYSIS
STATISTICAL TESTS
VARIABILITY
VARIANCE (STATISTICS)

FACTORIAL DESIGN
GS   EXPERIMENT DESIGN
.   **FACTORIAL DESIGN**
RT   ∞DESIGN
MATHEMATICAL MODELS
STATISTICAL ANALYSIS

FACTORIALS
GS   ANALYSIS (MATHEMATICS)
.   COMBINATORIAL ANALYSIS
. .   **FACTORIALS**
RT   BINOMIAL COEFFICIENTS
GAMMA FUNCTION

FACTORIES
USE   INDUSTRIAL PLANTS

FACTORIZATION
RT   ALGORITHMS
FACTOR ANALYSIS
FINITE ELEMENT METHOD
REAL VARIABLES

FACTORS
USE   VARIABLE

FACULAE
UF   PLAGES (FACULAE)
SOLAR FACULAE
GS   STELLAR ACTIVITY
.   SOLAR ACTIVITY
. .   **FACULAE**
RT   ∞ACTIVITY
CHROMOSPHERE
PHOTOSPHERE
STARSPOTS
SUNSPOTS

FADDEEV EQUATIONS
RT   ∞EQUATIONS
PARTICLE COLLISIONS
SCATTERING AMPLITUDE
WAVE SCATTERING

FADING
GS   **FADING**
.   SIGNAL FADING
. .   SELECTIVE FADING
RT   ATTENUATION
BLEACHING
COLOR
DISCOLORATION
ELECTROMAGNETIC ABSORPTION
EXTINCTION
RECEPTION DIVERSITY
SIGNAL FADING RATE
WAVE DISPERSION

FAHRENHEIT TEMPERATURE SCALE
USE   TEMPERATURE SCALES

FAIL-SAFE SYSTEMS
RT   ∞AUTOMATION
EMERGENCIES
FAULT TOLERANCE
SAFETY DEVICES
SAFETY MANAGEMENT
∞SYSTEMS

FAILURE
GS   **FAILURE**
.   BURNTHROUGH (FAILURE)
.   ENGINE FAILURE
.   STRUCTURAL FAILURE
.   SYSTEM FAILURES

FAILURE-(CONT.)
RT   ABORTED MISSIONS
∞BREAKDOWN
BUCKLING
BURN-IN
COLLAPSE
CORROSION
CRACKING (FRACTURING)
CUMULATIVE DAMAGE
DEFORMATION
DESTRUCTION
DETERIORATION
DISTORTION
DOWNTIME
ELECTRICAL FAULTS
FAILURE ANALYSIS
FATIGUE (MATERIALS)
FLASHOVER
FRACTURES (MATERIALS)
MALFUNCTIONS
MILLS RATIO
RUPTURING
SHEARING
SHORT CIRCUITS
STRUCTURAL STRAIN
TEMPERATURE INVERSIONS
WEAR

FAILURE ANALYSIS
RT   ACOUSTIC EMISSION
∞ANALYZING
BURN-IN
BURST TESTS
FATIGUE LIFE
FAULT TOLERANCE
LIFE (DURABILITY)
MILLS RATIO
MTBF
PROBABILITY DENSITY FUNCTIONS
RELIABILITY
STATISTICAL ANALYSIS

FAILURE MODES
GS   MODES
.   **FAILURE MODES**
RT   BUCKLING
CRACKS
ELASTIC BUCKLING
∞MODE
MODE (STATISTICS)
MTBF
SHEARING

FAINT OBJECT CAMERA
GS   OPTICAL EQUIPMENT
.   CAMERAS
. .   **FAINT OBJECT CAMERA**
PHOTOGRAPHIC EQUIPMENT
.   CAMERAS
. .   **FAINT OBJECT CAMERA**
RT   ASTRONOMICAL PHOTOGRAPHY
HUBBLE SPACE TELESCOPE
INFRARED PHOTOGRAPHY
OPTICAL MEASURING INSTRUMENTS
SPACEBORNE ASTRONOMY
SPACEBORNE TELESCOPES
ULTRAVIOLET PHOTOGRAPHY

FAINTING
USE   SYNCOPE

FAIRCHILD MILITARY AIRCRAFT
USE   FAIRCHILD-HILLER AIRCRAFT
MILITARY AIRCRAFT

FAIRCHILD-HILLER AIRCRAFT
UF   FAIRCHILD MILITARY AIRCRAFT
GS   **FAIRCHILD-HILLER AIRCRAFT**
.   C-119 AIRCRAFT
.   C-123 AIRCRAFT
.   OH-5 HELICOPTER
.   OH-23 HELICOPTER
.   XC-142 AIRCRAFT
RT   ∞AIRCRAFT

FAIREY AIRCRAFT
GS   **FAIREY AIRCRAFT**
.   FD 2 AIRCRAFT
RT   ∞AIRCRAFT

FAIREY DELTA 2 AIRCRAFT
USE   FD 2 AIRCRAFT

FAIRINGS
GS   SYMMETRICAL BODIES
.   STREAMLINED BODIES
. .   **FAIRINGS**
RT   AERODYNAMIC CONFIGURATIONS
AIRCRAFT STRUCTURES
CANOPIES
COWLINGS
FILLETS
HOUSINGS
LANDING GEAR
NACELLES
OGIVES
PERFORATED SHELLS
PROTECTORS
PROTUBERANCES
SHEATHS
SHELLS (STRUCTURAL FORMS)
STREAMLINING
WING ROOTS

FAITH 7
GS   MANNED SPACECRAFT
.   MERCURY SPACECRAFT
. .   **FAITH 7**
REENTRY VEHICLES
.   RECOVERABLE SPACECRAFT
. .   MERCURY SPACECRAFT
. . .   **FAITH 7**
SOFT LANDING SPACECRAFT
.   MERCURY SPACECRAFT
. .   **FAITH 7**
SPACE CAPSULES
.   MERCURY SPACECRAFT
. .   **FAITH 7**
RT   MERCURY MA-2 FLIGHT
MERCURY MA-9 FLIGHT

FALCON MISSILE
GS   MISSILES
.   AIR TO AIR MISSILES
. .   **FALCON MISSILE**
.   ANTIAIRCRAFT MISSILES
. .   **FALCON MISSILE**
RT   M-46 ENGINE
SOLID PROPELLANT ROCKET ENGINES

FALKNER-SKAN EQUATION
GS   ANALYSIS (MATHEMATICS)
.   REAL VARIABLES
. .   DIFFERENTIAL EQUATIONS
. . .   **FALKNER-SKAN EQUATION**
RT   BLASIUS EQUATION
BOUNDARY LAYER SEPARATION
∞EQUATIONS
LAMINAR FLOW
PRANDTL-MEYER EXPANSION
WEDGE FLOW

FALLING
RT   ATMOSPHERIC ENTRY
DESCENT TRAJECTORIES
PARTICLE MOTION
∞PRECIPITATION
SINKING
VERTICAL MOTION

FALLING SPHERES
GS   SYMMETRICAL BODIES
.   BODIES OF REVOLUTION
. .   SPHERES
. . .   **FALLING SPHERES**
RT   BALLS
DROP TOWERS
FREE FALL
GLOBULES
RAINDROPS
SPHEROIDS

FALLOUT
UF   WASHOUT (RADIOACTIVITY)
RT   AIR POLLUTION
FISSION PRODUCTS
FISSION WEAPONS
NUCLEAR EXPLOSION EFFECT
NUCLEAR EXPLOSIONS
NUCLEAR METEOROLOGY
POST-BLAST NUCLEAR RADIATION
∞RADIATION
RADIATION EFFECTS
RADIATION HAZARDS
RADIOACTIVE CONTAMINANTS
∞RADIOACTIVE DEBRIS
RADIOACTIVITY

FAN BLADES
RT   COMPRESSOR BLADES
       DUCTED FANS
       ∞FANS
       PROPELLER BLADES
       ROTARY WINGS
       TURBINE BLADES
       TURBOMACHINE BLADES
       VENTILATION FANS

FAN IN WING AIRCRAFT
GS   **FAN IN WING AIRCRAFT**
       . XV-5 AIRCRAFT
RT   ∞AIRCRAFT
       LIFT FANS
       RESEARCH AIRCRAFT
       SHORT TAKEOFF AIRCRAFT
       TILT WING AIRCRAFT
       V/STOL AIRCRAFT
       VERTICAL TAKEOFF AIRCRAFT

FANLIFT DEVICES
USE   LIFT FANS

∞ FANS
SN   (USE OF A MORE SPECIFIC TERM IS
       RECOMMENDED--CONSULT THE TERMS
       LISTED BELOW)
RT   ACTUATOR DISKS
       AIR CONDITIONING EQUIPMENT
       AIR DUCTS
       ANTENNA RADIATION PATTERNS
       BLOWERS
       COMPRESSOR ROTORS
       COMPRESSORS
       DUCTED FANS
       FAN BLADES
       PROPELLER FANS
       TURBOFANS
       VENTILATION FANS
       WIND TUNNEL DRIVES

FANS (LANDFORMS)
UF   BAJADAS
GS   LANDFORMS
       FANS (LANDFORMS)
RT   ALLUVIUM
       CANYONS
       CLAYS
       DELTAS
       GRAVELS
       MUD
       SANDS
       SEDIMENTS

FAR FIELDS
UF   FRAUNHOFER REGION
GS   ELECTROMAGNETIC FIELDS
       . **FAR FIELDS**
RT   ANTENNA RADIATION PATTERNS
       ELECTROMAGNETIC RADIATION
       FIELD THEORY (PHYSICS)
       FRESNEL REGION
       NEAR FIELDS
       NOISE PROPAGATION
       RADIANT FLUX DENSITY

FAR INFRARED RADIATION
SN   (30 MICRONS TO ABOUT 1000 MICRONS)
GS   ELECTROMAGNETIC RADIATION
       . INFRARED RADIATION
       . . **FAR INFRARED RADIATION**
RT   LONG WAVE RADIATION
       NEAR INFRARED RADIATION
       ∞RADIATION
       RADIO WAVES
       SHORT WAVE RADIATION
       SUBMILLIMETER WAVES
       TERRESTRIAL RADIATION

FAR ULTRAVIOLET RADIATION
SN   (200 TO 2000 ANGSTROMS)
UF   VACUUM ULTRAVIOLET RADIATION
GS   ELECTROMAGNETIC RADIATION
       . ULTRAVIOLET RADIATION
       . . **FAR ULTRAVIOLET RADIATION**
       . . . LYMAN ALPHA RADIATION
       . . . LYMAN BETA RADIATION
       IONIZING RADIATION
       . ULTRAVIOLET RADIATION
       . . **FAR ULTRAVIOLET RADIATION**
       . . . LYMAN ALPHA RADIATION
       . . . LYMAN BETA RADIATION
RT   BREMSSTRAHLUNG
       MAGELLAN MISSION

FAR ULTRAVIOLET RADIATION-(CONT.)
       NEAR ULTRAVIOLET RADIATION
       ∞RADIATION
       ULTRAVIOLET TELESCOPES
       X RAYS

FAR UV SPECTROSCOPIC EXPLORER
GS   SATELLITES
       . ARTIFICIAL SATELLITES
       . . EXPLORER SATELLITES
       . . . **FAR UV SPECTROSCOPIC
       EXPLORER**

FARADAY DARK SPACE
RT   GAS DISCHARGE TUBES
       GLOW DISCHARGES

FARADAY EFFECT
UF   FARADAY ROTATION
GS   ELECTROMAGNETIC PROPERTIES
       . **FARADAY EFFECT**
RT   CIRCULATORS (PHASE SHIFT CIRCUITS)
       ∞EFFECTS
       HALL GENERATORS
       KERR MAGNETOOPTICAL EFFECT
       MAGNETO-OPTICS
       OPTICAL MEASUREMENT
       OPTICAL PROPERTIES
       POLARIZATION (WAVES)
       POLARIZED ELECTROMAGNETIC
       RADIATION
       ROTATION

FARADAY ROTATION
USE   FARADAY EFFECT

FARM CROPS
GS   **FARM CROPS**
       . ALFALFA
       . COFFEE
       . COTTON
       . FRUITS
       . GRAINS (FOOD)
       . . BARLEY
       . . CORN
       . . OATS
       . . RICE
       . . SORGHUM
       . . WHEAT
       . HAY
       . LEGUMINOUS PLANTS
       . MILLET
       . POTATOES
       . SPINACH
       . SUGAR BEETS
       . SUGAR CANE
       . SUNFLOWERS
RT   AGRICULTURE
       AGRISTARS PROJECT
       BOTANY
       CROP CALENDARS
       CROP DUSTING
       CROP GROWTH
       CROP INVENTORIES
       CROP VIGOR
       ∞CROPS
       CURING
       EARTH RESOURCES
       FARMLANDS
       FROST DAMAGE
       GRASSES
       GRASSLANDS
       IRRIGATION
       LARGE AREA CROP INVENTORY
       EXPERIMENT
       LOCUSTS
       ORCHARDS
       PLANTING
       PLOWING
       SEEDS
       VINEYARDS

FARMLANDS
UF   CROPLANDS
       PLOWED FIELDS
GS   LAND
       . **FARMLANDS**
RT   AGRICULTURE
       AGROPHYSICAL UNITS
       CROP GROWTH
       CROP IDENTIFICATION
       CROP INVENTORIES
       CROP VIGOR
       ∞CROPS
       EARTH RESOURCES

FARMLANDS-(CONT.)
       FARM CROPS
       GRASSES
       GRASSLANDS
       HAY
       IRRIGATION
       LAND USE
       PLAINS
       PLANTING
       PLOWING
       REGIONAL PLANNING
       RURAL AREAS
       RURAL LAND USE
       SOD
       SUGAR BEETS
       SUGAR CANE

FAST FOURIER TRANSFORMATIONS
UF   FFT
GS   TRANSFORMATIONS (MATHEMATICS)
       . INTEGRAL TRANSFORMATIONS
       . . **FAST FOURIER TRANSFORMATIONS**
RT   FUNCTIONS (MATHEMATICS)
       WALSH FUNCTION

FAST NEUTRONS
GS   NUCLEAR RADIATION
       . **FAST NEUTRONS**
       PARTICLES
       . ELEMENTARY PARTICLES
       . . FERMIONS
       . . . NEUTRONS
       . . . . **FAST NEUTRONS**
       . NEUTRAL PARTICLES
       . . NEUTRONS
       . . . **FAST NEUTRONS**
RT   BARYONS
       NUCLEONS
       THERMAL NEUTRONS

FAST NUCLEAR REACTORS
GS   NUCLEAR REACTORS
       . **FAST NUCLEAR REACTORS**
       . . EXPERIMENTAL BREEDER REACTOR
       1
       . . EXPERIMENTAL BREEDER REACTOR
       2
       . . FAST OXIDE REACTORS
       . . FAST TEST REACTORS
       . . GAS COOLED FAST REACTORS
       . . LIQUID METAL FAST BREEDER
       REACTORS
RT   ENRICO FERMI ATOMIC POWER PLANT
       NUCLEAR POWER REACTORS

FAST OXIDE REACTORS
GS   NUCLEAR REACTORS
       . FAST NUCLEAR REACTORS
       . . **FAST OXIDE REACTORS**
RT   NUCLEAR POWER REACTORS

FAST TEST REACTORS
GS   NUCLEAR REACTORS
       . FAST NUCLEAR REACTORS
       . . **FAST TEST REACTORS**
RT   ∞REACTORS

FASTENERS
GS   **FASTENERS**
       . ANCHORS (FASTENERS)
       . BOLTS
       . . ROCK BOLTS
       . . TIEBOLTS
       . LOCKS (FASTENERS)
       . NUTS (FASTENERS)
       . PINS
       . RIVETS
       . SCREWS
       . WASHERS (SPACERS)
       . ZIPPERS
RT   ADHESIVES
       ∞BANDS
       ∞BELTS
       BRACKETS
       CABLES (ROPES)
       CHAINS
       CLAMPS
       CLIPS
       CLOSURES
       CONNECTORS
       COUPLINGS
       FITTINGS
       HOLDERS
       HOOKS
       INSERTS

FASTENERS-*(CONT.)*
    JOINTS (JUNCTIONS)
    LATCHES
    LINKAGES
    LOCKING
    LUGS
    MOORING
    RIBBONS
    SLEEVES
    SPACERS
  ∞SPIKES
    SPLICING
    SPLINES
    STRAPS
    STRUCTURAL MEMBERS
    STUDS (STRUCTURAL MEMBERS)
  ∞TAPES
    UNIONS (CONNECTORS)
    WIRE

FASTING
RT   AEROSPACE MEDICINE
    DIETS
    FOOD INTAKE
    HYPOXIA

FAT EMBOLISMS
GS   DISEASES
   . **FAT EMBOLISMS**
   EMBOLISMS
   . **FAT EMBOLISMS**
RT   AEROEMBOLISM
    BLOOD VESSELS
    CARDIOVASCULAR SYSTEM
    HEART DISEASES

FATIGUE (BIOLOGY)
GS   **FATIGUE (BIOLOGY)**
   . AUDITORY FATIGUE
   . FLIGHT FATIGUE
   . MUSCULAR FATIGUE
RT   ASTHENOPIA
  ∞BIOLOGY
    DAMAGE
    EFFORT
    EXHAUSTION
    HUMAN FACTORS ENGINEERING
    HYPERKINESIA
    MASSAGING
  ∞PERFORMANCE
    PHYSICAL EXERCISE
    STRESS (PHYSIOLOGY)
    STRESS (PSYCHOLOGY)
    WORK-REST CYCLE
    WORKLOADS (PSYCHOPHYSIOLOGY)

FATIGUE (MATERIALS)
UF   STRAIN FATIGUE
    STRUCTURAL FATIGUE
GS   **FATIGUE (MATERIALS)**
   . ACOUSTIC FATIGUE
   . BENDING FATIGUE
   . METAL FATIGUE
   . STRUCTURAL STRAIN
   . THERMAL FATIGUE
   . VOLUMETRIC STRAIN
RT   BAUSCHINGER EFFECT
    CRACK CLOSURE
    CRACK GEOMETRY
    CRACK PROPAGATION
    CRACKING (FRACTURING)
    CRACKS
    CREEP PROPERTIES
    CRYSTAL DISLOCATIONS
    CYCLES
    DAMAGE
    DESTRUCTION
    DUCTILITY
  ∞ENDURANCE
    FAILURE
    FRACTOGRAPHY
    FRETTING
    FRETTING CORROSION
    HARDNESS
  ∞MATERIALS
    MECHANICAL PROPERTIES
    NOTCH SENSITIVITY
    PLASTIC PROPERTIES
    RESIDUAL STRENGTH
    S-N DIAGRAMS
    SHEAR PROPERTIES
    SHOT PEENING
    STRESS CONCENTRATION
    STRESS CYCLES
    STRESS RATIO

FATIGUE (MATERIALS)-*(CONT.)*
    STRESS RELAXATION
    STRESS RELIEVING
    STRESSES
    STRUCTURAL FAILURE
    SURFACE DEFECTS
    SYSTEM FAILURES
    TEMPERATURE INVERSIONS
    THERMAL STRESSES
    VIBRATION

FATIGUE DIAGRAMS
USE S-N DIAGRAMS

FATIGUE LIFE
GS   LIFE (DURABILITY)
   . **FATIGUE LIFE**
   MECHANICAL PROPERTIES
   . **FATIGUE LIFE**
RT   ACCELERATED LIFE TESTS
    BLOWOUTS
    COFFIN-MANSON LAW
    COMBINED STRESS
    FAILURE ANALYSIS
    PALMGREN-MINER RULE
    RETIREMENT FOR CAUSE
    S-N DIAGRAMS
    SERVICE LIFE
    STRESS CYCLES

FATIGUE TESTING MACHINES
RT   ACOUSTIC EMISSION
  ∞MACHINERY
  ∞TEST EQUIPMENT

FATIGUE TESTS
RT   BENDING
    COFFIN-MANSON LAW
    CREEP TESTS
    DESTRUCTIVE TESTS
    FERROGRAPHY
    IMPACT TESTING MACHINES
    IMPACT TESTS
    LOAD TESTS
  ∞MATERIALS TESTS
    NOTCH STRENGTH
    NOTCH TESTS
    RESONANCE TESTING
    S-N DIAGRAMS
    SPECIMEN GEOMETRY
    STATIC TESTS
    STRESS CONCENTRATION
    STRESS CYCLES
    STRESS RATIO
    TENSILE TESTS
    TESTING TIME
  ∞TESTS
    THERMAL CYCLING TESTS
    WEIBULL DENSITY FUNCTIONS
    WELD TESTS

FATS
GS   CARBOHYDRATES
   . **FATS**
   . . CHOLINE
   ORGANIC COMPOUNDS
   . **FATS**
   . . CHOLINE
RT   ADIPOSE TISSUES
  ∞FOOD
    GREASES
    MYELIN
  ∞NUTRIENTS
    OILS
    PALMITIC ACID
    SYNTHETIC FOOD

FATTY ACIDS
GS   ACIDS
   . **FATTY ACIDS**
   . . CARBOXYLIC ACIDS
   . . . ACETIC ACID
   . . . . ETHYLENEDIAMINETETRAACETIC
       ACIDS
   . . . . IODOACETIC ACID
   . . . ACETYLSALICYLIC ACID
   . . . BENZILIC ACID
   . . . BENZOIC ACID
   . . . DICARBOXYLIC ACIDS
   . . . . TEREPHTHALATE
   . . . OLEIC ACID
   . . . PROPIONIC ACID
   . . . SEBACIC ACID
   . . . VALERIC ACID
   . . LIPOIC ACID

FATTY ACIDS-*(CONT.)*
   . . PALMITIC ACID
   ALIPHATIC COMPOUNDS
   . **FATTY ACIDS**
   . . ACETIC ACID
   . . . ETHYLENEDIAMINETETRAACETIC
      ACIDS
   . . . IODOACETIC ACID
   . . LIPOIC ACID
   . . OLEIC ACID
   . . PALMITIC ACID
   . . PROPIONIC ACID
   . . SEBACIC ACID
   . . VALERIC ACID
   ORGANIC COMPOUNDS
   . **FATTY ACIDS**
   . . ACETIC ACID
   . . . ETHYLENEDIAMINETETRAACETIC
      ACIDS
   . . . IODOACETIC ACID
   . . LIPOIC ACID
   . . OLEIC ACID
   . . PROPIONIC ACID
   . . SEBACIC ACID
   . . VALERIC ACID
RT   CASTOR OIL
  ∞NUTRIENTS

FAULT MECHANICS
USE FRACTURE MECHANICS

FAULT TOLERANCE
RT   ERROR ANALYSIS
    ERROR DETECTION CODES
    FAIL-SAFE SYSTEMS
    FAILURE ANALYSIS
    RELIABILITY ENGINEERING

FAULT TREES
GS   TREES (MATHEMATICS)
   . **FAULT TREES**
RT   GRAPHS (CHARTS)
    TOPOLOGY

∞ **FAULTS**
SN   *(USE OF A MORE SPECIFIC TERM IS*
    RECOMMENDED--CONSULT THE TERMS
    LISTED BELOW)
RT   ELECTRICAL FAULTS
    GEOLOGICAL FAULTS
    LANDFORMS
    MASSIFS
    SEAMOUNTS
    TEST PATTERN GENERATORS

FAUNA
USE ANIMALS

FAYALITE
GS   IRON COMPOUNDS
   . **FAYALITE**
   MINERALS
   . **FAYALITE**
   SILICON COMPOUNDS
   . SILICATES
   . . **FAYALITE**

FBFM (MODULATION)
USE FEEDBACK FREQUENCY MODULATION

FBM (MISSILES)
USE FLEET BALLISTIC MISSILES

FCC LATTICES
USE FACE CENTERED CUBIC LATTICES

FD 2 AIRCRAFT
UF   FAIREY DELTA 2 AIRCRAFT
GS   FAIREY AIRCRAFT
   . **FD 2 AIRCRAFT**
   JET AIRCRAFT
   . **FD 2 AIRCRAFT**
   MONOPLANES
   . **FD 2 AIRCRAFT**
   RESEARCH AIRCRAFT
   . **FD 2 AIRCRAFT**
   TAILLESS AIRCRAFT
   . **FD 2 AIRCRAFT**
RT ∞AIRCRAFT
    DELTA WINGS

FDL-5 REENTRY VEHICLE
GS   REENTRY VEHICLES
   . LIFTING REENTRY VEHICLES

FDL-5 REENTRY VEHICLE-*(CONT.)*
. . FDL-5 REENTRY VEHICLE

FDMA
USE   FREQUENCY DIVISION MULTIPLE
           ACCESS

FEAR
GS   PHOBIAS
. FEAR
. . FEAR OF FLYING
RT   ANXIETY
         EMOTIONS
         NEUROSES
         PANIC
         PSYCHOSES

FEAR OF FLYING
GS   PHOBIAS
. FEAR
. . FEAR OF FLYING
RT   ANXIETY
         EMOTIONS
         NEUROSES

FEASIBILITY
RT   COST ANALYSIS
         COSTS
         EFFICIENCY
         ESTIMATING
         EVALUATION

FEASIBILITY ANALYSIS
RT   COST ANALYSIS
         ECONOMIC FACTORS
         MANAGEMENT PLANNING
         RESEARCH MANAGEMENT
         SYSTEMS ANALYSIS
         TECHNOLOGY ASSESSMENT

FEATHER RIVER BASIN (CA)
GS   LANDFORMS
. STRUCTURAL BASINS
. . RIVER BASINS
. . . FEATHER RIVER BASIN (CA)
RT   CALIFORNIA
         RIVERS

FEATHERING
RT   PROPELLER BLADES
         PROPELLERS

FEATURE EXTRACTION
USE   PATTERN RECOGNITION

FEATURE IDENTIFICATION AND LOCATION EXPER
SN   (FEATURE LOCATION AND
           IDENTIFICATION AND LOCATION
           EXPERIMENT)
RT   EARTH OBSERVATIONS (FROM SPACE)
         IMAGE PROCESSING
         PATTERN RECOGNITION
         REMOTE SENSING
         REMOTE SENSORS
         SCENE ANALYSIS
         SPACE SHUTTLE PAYLOADS

FECES
GS   WASTES
. FECES
RT   EXCRETION
         PERSPIRATION
         URINE

FEDERAL BUDGETS
RT   ALLOCATIONS
         APPROPRIATIONS
         ∞BUDGETS
         CONTRACTS
         COST ESTIMATES
         FINANCIAL MANAGEMENT
         GOVERNMENT PROCUREMENT
         PROCUREMENT MANAGEMENT

FEDERAL REPUBLIC OF GERMANY
USE   WEST GERMANY

FEDERATIONS
GS   ORGANIZATIONS
. FEDERATIONS
. . BUREAUS (ORGANIZATIONS)
RT   INSTITUTIONS
         INTERNATIONAL COOPERATION

FEDERATIONS-*(CONT.)*
         NATIONS
         TEAMS
         UNIONIZATION
         UNITED NATIONS

FEED SYSTEMS
RT   COLD FLOW TESTS
         FEEDING (SUPPLYING)
         FUEL TANKS
         INTAKE SYSTEMS
         PUMPS
         ∞SYSTEMS

FEEDBACK
GS   FEEDBACK
. BIOFEEDBACK
. . SENSORY FEEDBACK
. NEGATIVE FEEDBACK
. NONLINEAR FEEDBACK
. POSITIVE FEEDBACK
RT   COMPENSATORS
         COMPLEXITY
         CONTROL THEORY
         CYBERNETICS
         ELECTROMAGNETIC INTERFERENCE
         EMOTIONAL FACTORS
         OSCILLATIONS
         OSCILLATORS
         ∞SYSTEMS
         TRANSFER FUNCTIONS

FEEDBACK AMPLIFIERS
GS   AMPLIFIERS
. FEEDBACK AMPLIFIERS
RT   DISTRIBUTED FEEDBACK LASERS
         NONLINEAR FEEDBACK
         OSCILLATORS
         PHANTASTRONS
         POSITIVE FEEDBACK
         POWER AMPLIFIERS
         SELF OSCILLATION
         SERVOAMPLIFIERS
         TRANSISTOR AMPLIFIERS
         VOLTAGE AMPLIFIERS

FEEDBACK CIRCUITS
GS   CIRCUITS
. FEEDBACK CIRCUITS
RT   FEEDFORWARD CONTROL
         TRANSFER FUNCTIONS

FEEDBACK CONTROL
UF   CLOSED LOOP SYSTEMS
GS   AUTOMATIC CONTROL
. FEEDBACK CONTROL
. . CASCADE CONTROL
RT   ADAPTIVE CONTROL
         ADAPTIVE OPTICS
         AUTOMATIC FREQUENCY CONTROL
         AUTOMATIC GAIN CONTROL
         ∞AUTOMATION
         BIOFEEDBACK
         ∞CONTROL
         CONTROL EQUIPMENT
         CONTROL SYSTEMS DESIGN
         CONTROL THEORY
         DISTRIBUTED FEEDBACK LASERS
         DYNAMIC CONTROL
         ELECTRONIC CONTROL
         KALMAN-SCHMIDT FILTERING
         LEARNING MACHINES
         NEGATIVE FEEDBACK
         NONLINEAR FEEDBACK
         OBSERVABILITY (SYSTEMS)
         OPTIMAL CONTROL
         PROPORTIONAL CONTROL
         ROBUSTNESS (MATHEMATICS)
         SERVOCONTROL
         SERVOMECHANISMS
         STABILITY AUGMENTATION
         TERMINAL CONFIGURED VEHICLE
             PROGRAM
         TRACKING PROBLEM

FEEDBACK FREQUENCY MODULATION
UF   FBFM (MODULATION)
GS   MODULATION
. FREQUENCY MODULATION
. . FEEDBACK FREQUENCY
         MODULATION
RT   PHASE LOCKED SYSTEMS

FEEDERS
SN   (FOR FLUID AND PARTICULATE
           MATERIALS)
RT   CONVEYORS
         DISPENSERS
         DISTRIBUTORS
         FEEDING (SUPPLYING)
         FUEL SYSTEMS
         INJECTORS
         INTAKE SYSTEMS
         LOADING OPERATIONS
         MATERIALS HANDLING
         MIXERS

FEEDFORWARD CONTROL
GS   AUTOMATIC CONTROL
. FEEDFORWARD CONTROL
RT   ADAPTIVE CONTROL
         ∞AUTOMATION
         ∞CONTROL
         CONTROL THEORY
         FEEDBACK CIRCUITS
         OPTIMAL CONTROL

FEEDING (SUPPLYING)
RT   FEED SYSTEMS
         FEEDERS
         INJECTION
         INPUT
         ∞LOADING
         MATERIALS HANDLING

FEELINGS
USE   SENSORY FEEDBACK

FEET (ANATOMY)
GS   ANATOMY
. LIMBS (ANATOMY)
. . LEG (ANATOMY)
. . . FEET (ANATOMY)
RT   LEG (ANATOMY)
         LIMBS (ANATOMY)
         PHYSIOLOGY

FELDSPARS
GS   ALUMINUM COMPOUNDS
. FELDSPARS
         MINERALS
. FELDSPARS
         SILICON COMPOUNDS
. SILICATES
. . FELDSPARS
RT   ANDESITE
         ANORTHOSITE
         FELSITE
         IGNEOUS ROCKS

FELLOWSHIP AIRCRAFT
USE   F-28 TRANSPORT AIRCRAFT

FELSITE
GS   ROCKS
. IGNEOUS ROCKS
. . FELSITE
RT   FELDSPARS
         MINERALS
         QUARTZ

FELTS
GS   FABRICS
. FELTS
RT   WOOL

FEMALES
UF   WOMEN
RT   CHILDREN
         GYNECOLOGY
         HUMAN BEINGS
         MALES
         MENSTRUATION
         SEX
         SEX FACTOR

FEMUR
GS   ANATOMY
. MUSCULOSKELETAL SYSTEM
. . BONES
. . . FEMUR
RT   KNEE (ANATOMY)
         LEG (ANATOMY)

∞ FENCES
SN (USE OF A MORE SPECIFIC TERM IS
    RECOMMENDED--CONSULT THE TERMS
    LISTED BELOW)
RT AIRFOIL FENCES
    FENCES (BARRIERS)
    TRACKING STATIONS

FENCES (BARRIERS)
RT ∞ BARRIERS
    BOUNDARIES
    ∞ FENCES
    GATES (OPENINGS)

FERMAT PRINCIPLE
RT ELECTROMAGNETIC WAVE
       TRANSMISSION
    LIGHT TRANSMISSION
    MULTIPATH TRANSMISSION
    OPTICAL THICKNESS
    VELOCITY

FERMENTATION
GS CHEMICAL REACTIONS
    . FERMENTATION
    METABOLISM
    . ENZYME ACTIVITY
    . . FERMENTATION
RT BIOCONVERSION
    BUTYRIC ACID

FERMI LIQUIDS
GS LIQUIDS
    . CRYOGENIC FLUIDS
    . . FERMI LIQUIDS
RT CRYOGENICS

FERMI SURFACES
RT BRILLOUIN ZONES
    CYCLOTRON RESONANCE
    ENERGY LEVELS
    MAGNETORESISTIVITY
    ∞ SURFACES
    TRANSITION PROBABILITIES

FERMI-DIRAC STATISTICS
RT BOSONS
    FERMIONS
    QUANTUM MECHANICS
    QUANTUM STATISTICS
    ∞ STATISTICS

FERMIONS
GS PARTICLES
    . ELEMENTARY PARTICLES
    . . FERMIONS
    . . . BARYONS
    . . . . HYPERONS
    . . . . . XI HYPERONS
    . . . . OMEGA-MESONS
    . . . RHO-MESONS
    . . . . SIGMA-MESONS
    . . . ETA-MESONS
    . . . LEPTONS
    . . . . ANTINEUTRINOS
    . . . . MUONS
    . . . . NEUTRINOS
    . . . . . SOLAR NEUTRINOS
    . . MESON RESONANCE
    . . . NEUTRONS
    . . . . COLD NEUTRONS
    . . . . FAST NEUTRONS
    . . . . PHOTONEUTRONS
    . . . . THERMAL NEUTRONS
    . . . PROTONS
    . . . . RECOIL PROTONS
    . . . . SOLAR PROTONS
RT FERMI-DIRAC STATISTICS
    PAULI EXCLUSION PRINCIPLE
    QUANTUM STATISTICS

FERMIUM
GS CHEMICAL ELEMENTS
    . ACTINIDE SERIES
    . . TRANSURANIUM ELEMENTS
    . . . FERMIUM
    . NUCLIDES
    . ISOTOPES
    . . . RADIOACTIVE ISOTOPES
    . . . . TRANSURANIUM ELEMENTS
    . . . . . . FERMIUM
    HEAVY ELEMENTS
    . TRANSURANIUM ELEMENTS
    . . FERMIUM
    METALS

FERMIUM-(CONT.)
    . ACTINIDE SERIES
    . . TRANSURANIUM ELEMENTS
    . . . FERMIUM

FERRANTI MERCURY COMPUTER
GS DATA PROCESSING EQUIPMENT
    . COMPUTERS
    . . DIGITAL COMPUTERS
    . . . FERRANTI MERCURY COMPUTER

FERRATES
GS IRON COMPOUNDS
    . FERRATES
    . . BARIUM FERRATES

FERRIC IONS
GS IONS
    . CATIONS
    . . METAL IONS
    . . . FERRIC IONS
    PARTICLES
    . CHARGED PARTICLES
    . . CATIONS
    . . . METAL IONS
    . . . . FERRIC IONS

FERRIMAGNETIC MATERIALS
GS MAGNETIC MATERIALS
    . FERRIMAGNETIC MATERIALS
RT FERROMAGNETIC MATERIALS
    ∞ MATERIALS

FERRIMAGNETISM
GS MAGNETIC PROPERTIES
    . FERRIMAGNETISM
RT FERRIMAGNETS
    MAGNONS

FERRIMAGNETS
GS MAGNETS
    . FERRIMAGNETS
RT FERRIMAGNETISM

FERRITES
GS IRON COMPOUNDS
    . FERRITES
RT AUSTENITE
    FERRITIC STAINLESS STEELS
    GYRATORS
    IRON ALLOYS
    MAGNETIC CORES
    MICROSTRUCTURE
    PEARLITE
    SPINEL
    STEELS
    YTTRIUM-ALUMINUM GARNET
    YTTRIUM-IRON GARNET

FERRITIC STAINLESS STEELS
GS ALLOYS
    . IRON ALLOYS
    . . STEELS
    . . . STAINLESS STEELS
    . . . . FERRITIC STAINLESS STEELS
RT CHROMIUM STEELS
    FERRITES
    HEAT TREATMENT
    MAGNETIC PROPERTIES
    MECHANICAL PROPERTIES

FERROALLOYS
USE IRON ALLOYS

FERROCENES
GS IRON COMPOUNDS
    . FERROCENES
    . . ALKYLFERROCENE
    ORGANOMETALLIC COMPOUNDS
    . FERROCENES

FERROELECTRICITY
GS ELECTROMAGNETIC PROPERTIES
    . ELECTRICAL PROPERTIES
    . . FERROELECTRICITY
RT ANTIFERROELECTRICITY
    CURIE TEMPERATURE
    DIELECTRIC PROPERTIES
    MICROWAVE SWITCHING

FERROFLUIDS
GS LIQUIDS
    . FERROFLUIDS
    MAGNETIC MATERIALS

FERROFLUIDS-(CONT.)
    . FERROMAGNETIC MATERIALS
    . . FERROFLUIDS
RT DISPERSIONS
    ∞ FLUIDS
    MICROPARTICLES
    SUSPENDING (MIXING)
    ∞ SUSPENSIONS

FERROGRAPHY
RT FATIGUE TESTS
    METALLOGRAPHY
    WEAR TESTS

FERROMAGNETIC FILMS
GS MAGNETIC MATERIALS
    . FERROMAGNETIC MATERIALS
    . . FERROMAGNETIC FILMS
    THIN FILMS
    . FERROMAGNETIC FILMS

FERROMAGNETIC MATERIALS
GS MAGNETIC MATERIALS
    . FERROMAGNETIC MATERIALS
    . . FERROFLUIDS
    . . FERROMAGNETIC FILMS
    . . MAGNETITE
    . . PERMALLOYS (TRADEMARK)
RT FERRIMAGNETIC MATERIALS
    MAGNETS
    ∞ MATERIALS
    YOKES

FERROMAGNETIC RESONANCE
GS RESONANCE
    . MAGNETIC RESONANCE
    . . FERROMAGNETIC RESONANCE
RT MAGNETIC FIELDS
    PARAMAGNETIC RESONANCE

FERROMAGNETISM
UF ISING MODEL
GS MAGNETIC PROPERTIES
    . FERROMAGNETISM
RT ANTIFERROMAGNETISM
    CURIE TEMPERATURE
    CURIE-WEISS LAW
    DIAMAGNETISM
    LANGEVIN FORMULA
    MAGNETIC CORES
    MAGNETIC DISPERSION
    MAGNETS
    MAGNONS

FERROUS METALS
GS METALS
    . FERROUS METALS
RT ALLOYS
    CHEMICAL ELEMENTS
    IRON
    IRON ISOTOPES
    ∞ METALLURGY

FERRY SPACECRAFT
UF SPACE BUSES
GS MANEUVERABLE SPACECRAFT
    . FERRY SPACECRAFT
    MANNED SPACECRAFT
    . FERRY SPACECRAFT
RT ASTRO VEHICLE
    CARGO SPACECRAFT
    MARS (MANNED REUSABLE
       SPACECRAFT)
    REENTRY VEHICLES
    RENDEZVOUS SPACECRAFT
    REUSABLE SPACECRAFT
    SOFT LANDING SPACECRAFT
    SPACE STATIONS

FERTILITY
RT BREEDING (REPRODUCTION)
    FERTILIZATION
    ∞ REPRODUCTION
    REPRODUCTION (BIOLOGY)
    REPRODUCTIVE SYSTEMS

FERTILIZATION
RT BIRTH
    FERTILITY
    RECOMBINATION REACTIONS
    ∞ REPRODUCTION
    SPERMATOZOA

FERTILIZERS
RT   AMMONIA
      AMMONIUM NITRATES
      ASHES
      CULTIVATION
      PLANTING
      UREAS
      VEGETATION GROWTH

FET (TRANSISTORS)
USE   FIELD EFFECT TRANSISTORS

FETUSES
UF   FOETUSES
RT   BIRTH
      EGGS
      EMBRYOLOGY
      EMBRYOS
      REPRODUCTION (BIOLOGY)
      REPRODUCTIVE SYSTEMS

FEVER
RT   BODY TEMPERATURE
      HYPERTHERMIA
      SKIN TEMPERATURE (BIOLOGY)

FEYNMAN DIAGRAMS
GS   DIAGRAMS
    . **FEYNMAN DIAGRAMS**
RT   ELECTROMAGNETIC INTERACTIONS
      MINKOWSKI SPACE
      PARTICLE INTERACTIONS
      QUANTUM ELECTRODYNAMICS

FFAR ROCKET VEHICLE
USE   FOLDING FIN AIRCRAFT ROCKET
        VEHICLE

FFT
USE   FAST FOURIER TRANSFORMATIONS

FH-1100 HELICOPTER
USE   OH-5 HELICOPTER

FIAT AIRCRAFT
GS   **FIAT AIRCRAFT**
    . G-91 AIRCRAFT
    . G-95/4 AIRCRAFT
    . G-222 AIRCRAFT
RT   ∞AIRCRAFT

FIAT G-91 AIRCRAFT
USE   G-91 AIRCRAFT

FIAT G-95/4 AIRCRAFT
USE   G-95/4 AIRCRAFT

FIAT G-222 AIRCRAFT
USE   G-222 AIRCRAFT

FIBER COMPOSITES
GS   COMPOSITE MATERIALS
    . **FIBER COMPOSITES**
    . . CARBON FIBER REINFORCED
        PLASTICS
    . . GLASS FIBER REINFORCED PLASTICS
RT   ALUMINUM BORON COMPOSITES
      ALUMINUM GRAPHITE COMPOSITES
      BORON FIBERS
      BORON REINFORCED MATERIALS
      BORON-EPOXY COMPOUNDS
      BORSIC (TRADENAME)
      CARBON FIBERS
      CARBON-CARBON COMPOSITES
      COMPOSITE WRAPPING
      FIBER REINFORCED COMPOSITES
      FILAMENT WINDING
      GRAPHITE
      GRAPHITE-EPOXY COMPOSITES
      LAMINATES
      METAL FIBERS
      METAL MATRIX COMPOSITES
      MICARTA
      POLYMER MATRIX COMPOSITES
      PULTRUSION
      REINFORCED PLASTICS
      REINFORCING FIBERS
      SUPERHYBRID MATERIALS
      THREE DIMENSIONAL COMPOSITES

FIBER OPTICS
RT   CASSEGRAIN OPTICS
      CRYSTAL OPTICS
      ELECTRON TUBES

FIBER OPTICS-(CONT.)
      GEOMETRICAL OPTICS
      GRADIENT INDEX OPTICS
      LIGHT TRANSMISSION
    ∞OPTICS
      PHOTONICS
      PHYSICAL OPTICS
      SAGNAC EFFECT
      VIDICONS

FIBER ORIENTATION
RT   BORON FIBERS
      COMPOSITE MATERIALS
      DYNAMIC RESPONSE
      EPOXY MATRIX COMPOSITES
      GLASS FIBER REINFORCED PLASTICS
      IMPACT LOADS
      LAY-UP
      MECHANICAL PROPERTIES
    ∞ORIENTATION
      REINFORCING FIBERS

FIBER REINFORCED COMPOSITES
GS   COMPOSITE MATERIALS
    . **FIBER REINFORCED COMPOSITES**
RT   FIBER COMPOSITES
      GLASS FIBER REINFORCED PLASTICS
      MATRIX MATERIALS
      REINFORCING MATERIALS

FIBER RELEASE
GS   RELEASING
    . **FIBER RELEASE**
RT   CARBON FIBERS
      COMBUSTION PRODUCTS
      COMPOSITE MATERIALS
      FIBERS
      FIRES
      GRAPHITE

FIBER STRENGTH
GS   MECHANICAL PROPERTIES
    . **FIBER STRENGTH**
RT   BENDING
      BORON FIBERS
      COMPRESSIVE STRENGTH
      HOOKES LAW
      POISSON RATIO
      SHEAR STRENGTH
    ∞STRENGTH
      TENSILE STRENGTH

FIBERBOARD
USE   BOARDS (PAPER)

FIBERGLASS
USE   GLASS FIBERS

FIBERS
UF   FIBROUS MATERIALS
      REFRASIL (TRADEMARK)
GS   **FIBERS**
    . COTTON FIBERS
    . HAIR
    . LINEN
    . METAL FIBERS
    . MICROFIBERS
    . REINFORCING FIBERS
    . . BORON FIBERS
    . . CARBON FIBERS
    . SILK
    . SYNTHETIC FIBERS
    . . DACRON (TRADEMARK)
    . . FORTISAN (TRADEMARK)
    . . GLASS FIBERS
    . . NYLON (TRADEMARK)
    . . RAYON
    . . VYCOR
    . WOOL
RT   BORON REINFORCED MATERIALS
      CARBON FIBER REINFORCED PLASTICS
      COMPOSITE MATERIALS
      CORDAGE
      COTTON
      FABRICS
      FIBER RELEASE
    ∞FILAMENTS
      GLASS FIBER REINFORCED PLASTICS
      METAL MATRIX COMPOSITES
      PAPERS
      POLYMERIC FILMS
      REINFORCING MATERIALS
      SLIVERS
      STRANDS
      TEXTILES

FIBERS-(CONT.)
      WET SPINNING
      WHISKERS (CRYSTALS)
      YARNS

FIBERS (MATHEMATICS)
RT   CANONICAL FORMS
      DIMENSIONAL ANALYSIS
      FUNCTION SPACE
      GROUP THEORY
      HOMOTOPY THEORY
      MANIFOLDS (MATHEMATICS)
      TOPOLOGY

FIBONACCI NUMBERS
RT   NUMBER THEORY
    ∞NUMBERS
      SET THEORY

FIBRILLATION
RT   HEART DISEASES
      MUSCLES
      SEISMOCARDIOGRAPHY

FIBRIN
GS   PROTEINS
    . **FIBRIN**
RT   BLOOD
      COAGULATION

FIBRINOGEN
GS   PROTEINS
    . **FIBRINOGEN**
RT   HEMOSTATICS
      HOMEOSTASIS
      THROMBIN

FIBROBLASTS
GS   CELLS (BIOLOGY)
    . **FIBROBLASTS**
    . . COLLAGENS
RT   CYTOPLASM
      TENDONS
      TISSUES (BIOLOGY)

FIBROSIS
GS   DISEASES
    . **FIBROSIS**
    . . CYSTIC FIBROSIS
RT   TISSUES (BIOLOGY)

FIBROUS MATERIALS
USE   FIBERS

FICKS EQUATION
RT   DIFFUSION
      DIFFUSION COEFFICIENT
    ∞EQUATIONS
      TAFEL LAW

FIDELITY
USE   ACCURACY

FIDUCIARIES
RT   ECONOMICS
      FINANCE
      MANAGEMENT

FIELD ARMY BALLISTIC MISSILES
GS   MISSILES
    . BALLISTIC MISSILES
    . . **FIELD ARMY BALLISTIC MISSILES**
    . . . SUBROC MISSILE
RT   INTERMEDIATE RANGE BALLISTIC
        MISSILES
      SHORT RANGE BALLISTIC MISSILES

FIELD COILS
GS   MAGNET COILS
    . **FIELD COILS**
RT   ELECTROMAGNETS
      HELICAL INDUCERS

FIELD EFFECT TRANSISTORS
UF   CASCODE MOSFET
      FET (TRANSISTORS)
      IGFET
      MESFETS
      MISFETS
      MOSFET
      UNIPOLAR TRANSISTORS
GS   ELECTRONIC EQUIPMENT
    . SOLID STATE DEVICES
    . . SEMICONDUCTOR DEVICES

FIELD EFFECT TRANSISTORS-*(CONT.)*
   . . . TRANSISTORS
   . . . . **FIELD EFFECT TRANSISTORS**
   . . . . . CHARGE FLOW DEVICES
   . . . . . JFET
RT   ∞EFFECTS
   HIGH ELECTRON MOBILITY
     TRANSISTORS
   ION IMPLANTATION

FIELD EMISSION
GS   DECAY
   . EMISSION
   . . PARTICLE EMISSION
   . . . ELECTRON EMISSION
   . . . . **FIELD EMISSION**
RT   ELECTRIC FIELDS
   ELECTRON MICROSCOPES
   ELECTRON MICROSCOPY
   MAGNETIC FIELDS
   SECONDARY EMISSION
   ZENER EFFECT

FIELD INTENSITY METERS
SN   (EMPLOY THIS TERM WHEN TYPE OF
   FIELD INVOLVED IS NOT
   SPECIFIED--OTHERWISE USE A MORE
   SPECIFIC TERM)
GS   MEASURING INSTRUMENTS
   . **FIELD INTENSITY METERS**
RT   ACTINOMETERS
   FLUX DENSITY
   MAGNETOMETERS
   NOISE METERS

FIELD MODE THEORY
RT   ∞THEORIES

FIELD OF VIEW
GS   VIEWING
   . **FIELD OF VIEW**
RT   BEARING (DIRECTION)
   CONICAL SCANNING
   ELEVATION ANGLE
   ∞FIELDS
   LOOK ANGLES (TRACKING)
   VISUAL FIELDS

FIELD STRENGTH
GS   **FIELD STRENGTH**
   . ELECTRIC FIELD STRENGTH
   . MAGNETIC FLUX
RT   ACOUSTIC PROPERTIES
   DIRECTIVITY
   ELECTRIC FIELDS
   ELECTRICAL PROPERTIES
   ELECTROMAGNETIC FIELDS
   FLUX DENSITY
   GRAVITATIONAL FIELDS
   ISOTROPY
   MAGNETIC DIFFUSION
   MAGNETIC FIELDS
   MAGNETIC PROPERTIES
   ∞ORIENTATION
   PERMITTIVITY
   ∞STRENGTH

FIELD THEORY (ALGEBRA)
GS   **FIELD THEORY (ALGEBRA)**
   . CUBIC EQUATIONS
   . QUADRATIC EQUATIONS
RT   ∞FIELDS
   GREEN'S FUNCTIONS
   HOMOMORPHISMS
   NONLINEAR EQUATIONS
   ∞THEORIES

FIELD THEORY (PHYSICS)
UF   AMBIT
   FORCE FIELDS
   WIGHTMAN THEORY
GS   **FIELD THEORY (PHYSICS)**
   . QUANTUM CHROMODYNAMICS
   . . INSTANTONS
   . STRONG INTERACTIONS (FIELD
     THEORY)
   . UNIFIED FIELD THEORY
   . WEAK INTERACTIONS (FIELD THEORY)
RT   ANTENNA RADIATION PATTERNS
   ATTRACTION
   BOSON FIELDS
   CLOSURE LAW
   CROSSED FIELDS
   DIRAC EQUATION
   DISTRIBUTION (PROPERTY)

FIELD THEORY (PHYSICS)-*(CONT.)*
   ∞DYNAMICS
   ELECTROMAGNETIC FIELDS
   FAR FIELDS
   ∞FIELDS
   FLOW DISTRIBUTION
   FLUX (RATE)
   FLUX DENSITY
   FUNCTION SPACE
   GEOMAGNETISM
   GRAVITATIONAL FIELDS
   GREEN'S FUNCTIONS
   MAGNETIC FIELD INVERSIONS
   MAGNETIC FIELDS
   MAGNETOSTATIC FIELDS
   MANY BODY PROBLEM
   MULTIPOLAR FIELDS
   NUCLEAR PHYSICS
   NULL ZONES
   ∞PHYSICS
   POMERANCHUK THEOREM
   POTENTIAL FIELDS
   PRESSURE DISTRIBUTION
   QUANTUM ELECTRODYNAMICS
   QUANTUM THEORY
   RADIATION DISTRIBUTION
   RELATIVITY
   SELF CONSISTENT FIELDS
   SOUND FIELDS
   TEMPERATURE DISTRIBUTION
   TENSORS
   ∞THEORIES
   TRAVELING CHARGE
   YANG-MILLS FIELDS
   YANG-MILLS THEORY
   ZERO POINT ENERGY

∞ **FIELDS**
SN   *(USE OF A MORE SPECIFIC TERM IS
   RECOMMENDED--CONSULT THE TERMS
   LISTED BELOW)*
RT   BOSON FIELDS
   ELECTRIC FIELDS
   FIELD OF VIEW
   FIELD THEORY (ALGEBRA)
   FIELD THEORY (PHYSICS)
   GRAVITATIONAL FIELDS
   MAGNETIC FIELDS
   MILITARY AIR FACILITIES
   SELF CONSISTENT FIELDS
   VISUAL FIELDS

FIGHTER AIRCRAFT
UF   INTERCEPTOR AIRCRAFT
GS   ATTACK AIRCRAFT
   . **FIGHTER AIRCRAFT**
   . . ALPHA JET AIRCRAFT
   . . F-2 AIRCRAFT
   . . F-4 AIRCRAFT
   . . F-5 AIRCRAFT
   . . F-8 AIRCRAFT
   . . F-9 AIRCRAFT
   . . F-14 AIRCRAFT
   . . F-15 AIRCRAFT
   . . F-16 AIRCRAFT
   . . F-17 AIRCRAFT
   . . F-18 AIRCRAFT
   . . F-27 AIRCRAFT
   . . F-84 AIRCRAFT
   . . F-86 AIRCRAFT
   . . F-89 AIRCRAFT
   . . F-94 AIRCRAFT
   . . F-100 AIRCRAFT
   . . F-101 AIRCRAFT
   . . F-102 AIRCRAFT
   . . F-104 AIRCRAFT
   . . F-105 AIRCRAFT
   . . F-106 AIRCRAFT
   . . F-111 AIRCRAFT
   . . FV-12A AIRCRAFT
   . . G-91 AIRCRAFT
   . . G-95/4 AIRCRAFT
   . . GA-5 AIRCRAFT
   . . JAGUAR AIRCRAFT
   . . JET PROVOST AIRCRAFT
   . . MIG AIRCRAFT
   . . MIRAGE AIRCRAFT
   . . . MIRAGE 3 AIRCRAFT
   . . P-51 AIRCRAFT
   . . P-1127 AIRCRAFT
   . . P-1154 AIRCRAFT
   . . SAAB 37 AIRCRAFT
   . . SCIMITAR AIRCRAFT
   . . VAMPIRE MK 35 AIRCRAFT
   . . VJ-101 AIRCRAFT

FIGHTER AIRCRAFT-*(CONT.)*
   . . YF-12 AIRCRAFT
   . . YF-16 AIRCRAFT
RT   ∞AIRCRAFT
   HIGHLY MANEUVERABLE AIRCRAFT
   ∞INTERCEPTORS
   JET AIRCRAFT
   ∞MILITARY AIRCRAFT
   ∞MILITARY AVIATION
   MRCA AIRCRAFT
   PANAVIA MILITARY AIRCRAFT
   SUPERSONIC AIRCRAFT
   TRAINING AIRCRAFT
   V/STOL AIRCRAFT

FIGURE OF MERIT
RT   ACCEPTABILITY
   ∞ANALYZING
   CRITERIA
   EFFICIENCY
   EVALUATION
   MODULATION TRANSFER FUNCTION
   OPTICAL TRANSFER FUNCTION
   ∞PERFORMANCE
   Q FACTORS
   QUALITY
   SELECTION
   VALUE

FILAMENT WINDING
UF   FILAMENT WOUND CONSTRUCTION
GS   WINDING
   . **FILAMENT WINDING**
RT   COMPOSITE WRAPPING
   FIBER COMPOSITES
   ISOTENSOID STRUCTURES
   LAMINATES
   METAL FIBERS
   PREIMPREGNATION

FILAMENT WOUND CONSTRUCTION
USE   FILAMENT WINDING

∞ **FILAMENTS**
SN   *(USE OF A MORE SPECIFIC TERM IS
   RECOMMENDED--CONSULT THE TERMS
   LISTED BELOW)*
RT   BORON FIBERS
   CARBON FIBERS
   CATHODES
   CORDAGE
   FIBERS
   IONIZERS
   REINFORCING MATERIALS
   RESISTORS
   STRANDS
   VORTEX FILAMENTS
   WET SPINNING
   WHISKERS (CRYSTALS)
   WIRE

FILAMENTS (SOLAR PHYSICS)
USE   SOLAR PROMINENCES

FILE MAINTENANCE (COMPUTERS)
GS   MAINTENANCE
   . **FILE MAINTENANCE (COMPUTERS)**
RT   CHECKOUT
   COMPUTER PROGRAMMING
   COMPUTERS
   PROGRAM VERIFICATION (COMPUTERS)
   PROGRAMMERS
   ∞PROGRAMMING

∞ **FILES**
SN   *(USE OF A MORE SPECIFIC TERM IS
   RECOMMENDED--CONSULT THE TERMS
   LISTED BELOW)*
RT   DOCUMENT STORAGE
   FILES (TOOLS)

FILES (TOOLS)
GS   TOOLS
   . **FILES (TOOLS)**
RT   ABRASION
   ∞FILES
   SCRAPERS

FILLERS
RT   ADDITIVES
   ∞CELLS
   DOPES
   OPACIFIERS
   PAINTS

FILLERS-*(CONT.)*
　　PIGMENTS
　　PRIMERS (COATINGS)
　　REINFORCEMENT (STRUCTURES)
　　RESINS
　　SEALERS
　　SIZING MATERIALS
　　VARNISHES

FILLETS
　RT　FAIRINGS
　　　JOINTS (JUNCTIONS)
　　　SEAMS (JOINTS)
　　　WELDING

FILLING
　GS　**FILLING**
　　　. REFILLING
　RT　ACCUMULATIONS
　　　∞CHARGING
　　　EXTENSIONS
　　　INJECTION
　　　INPUT
　　　∞LOADING
　　　REPLENISHMENT
　　　SUPPLYING

FILM BOILING
　GS　PHASE TRANSFORMATIONS
　　　. VAPORIZING
　　　. . BOILING
　　　. . . **FILM BOILING**
　RT　HEAT TRANSFER
　　　LEIDENFROST PHENOMENON
　　　NUCLEATE BOILING

FILM CONDENSATION
　GS　CONDENSING
　　　. **FILM CONDENSATION**
　RT　CONDENSERS (LIQUEFIERS)
　　　COOLING
　　　HEAT TRANSFER

FILM COOLING
　GS　COOLING
　　　. EVAPORATIVE COOLING
　　　. . **FILM COOLING**
　　　. LIQUID COOLING
　　　. . **FILM COOLING**
　RT　LIQUID INJECTION
　　　SURFACE COOLING
　　　SWEAT COOLING

FILM THICKNESS
　GS　DIMENSIONS
　　　. **FILM THICKNESS**
　RT　THICKNESS

∞ FILMS
　SN　*(USE OF A MORE SPECIFIC TERM IS*
　　　RECOMMENDED--CONSULT THE TERMS
　　　LISTED BELOW)
　RT　COATINGS
　　　CORROSION PREVENTION
　　　ELECTRODE FILM BARRIERS
　　　FABRICS
　　　FLUID FILMS
　　　HELIUM FILM
　　　KAPTON (TRADEMARK)
　　　LAMINATES
　　　MAGNETIC FILMS
　　　MEMBRANES
　　　METAL FILMS
　　　MONOMOLECULAR FILMS
　　　OXIDE FILMS
　　　PAPERS
　　　PHOTOGRAPHIC FILM
　　　POLYMERIC FILMS
　　　SEMICONDUCTING FILMS
　　　SILICON FILMS
　　　SQUEEZE FILMS
　　　THERMOPLASTIC FILMS
　　　THICK FILMS
　　　THIN FILMS
　　　WEBS (SHEETS)

FILTER WHEEL INFRARED SPECTROMETERS
　GS　MEASURING INSTRUMENTS
　　　. OPTICAL MEASURING INSTRUMENTS
　　　. . INFRARED SPECTROMETERS
　　　. . . **FILTER WHEEL INFRARED**
　　　　　SPECTROMETERS
　　　. SPECTROMETERS
　　　. . INFRARED SPECTROMETERS

FILTER WHEEL INFRARED-*(CONT.)*
　. . . **FILTER WHEEL INFRARED**
　　　SPECTROMETERS
　　　OPTICAL EQUIPMENT
　　　. OPTICAL MEASURING INSTRUMENTS
　　　. . INFRARED SPECTROMETERS
　　　. . . **FILTER WHEEL INFRARED**
　　　　　SPECTROMETERS
　RT　EBERT SPECTROMETERS
　　　∞FILTERS
　　　INFRARED SPECTROPHOTOMETERS
　　　SOLAR SPECTROMETERS

FILTERING
　USE　FILTRATION

∞ FILTERS
　SN　*(USE OF A MORE SPECIFIC TERM IS*
　　　RECOMMENDED--CONSULT THE TERMS
　　　LISTED BELOW)
　RT　ABSORBERS (MATERIALS)
　　　ADAPTIVE FILTERS
　　　AIR FILTERS
　　　ATTENUATORS
　　　BANDPASS FILTERS
　　　BANDSTOP FILTERS
　　　BIREFRINGENT FILTERS
　　　CRYSTAL FILTERS
　　　DIGITAL FILTERS
　　　ELECTRIC FILTERS
　　　ELECTROMAGNETIC WAVE FILTERS
　　　ELECTRONIC FILTERS
　　　FILTER WHEEL INFRARED
　　　　SPECTROMETERS
　　　FIR FILTERS
　　　FLUID FILTERS
　　　HIGH PASS FILTERS
　　　IMAGE FILTERS
　　　KALMAN FILTERS
　　　LINEAR FILTERS
　　　LOW PASS FILTERS
　　　MATCHED FILTERS
　　　MONOCHROMATIC RADIATION
　　　NONLINEAR FILTERS
　　　OPTICAL FILTERS
　　　RADIO FILTERS
　　　REDUCED ORDER FILTERS
　　　SEPARATORS
　　　SPATIAL FILTERING

FILTRATION
　UF　FILTERING
　GS　**FILTRATION**
　　　. SPATIAL FILTERING
　RT　ACTIVATED CARBON
　　　BEDS (PROCESS ENGINEERING)
　　　BENEFICIATION
　　　CONCENTRATING
　　　∞CONCENTRATION
　　　CONCENTRATORS
　　　EFFLUENTS
　　　EXTRACTION
　　　FLUID FILTERS
　　　HYDROMETALLURGY
　　　MATERIALS RECOVERY
　　　PERCOLATION
　　　PRECIPITATION (CHEMISTRY)
　　　∞SCREENING
　　　∞SEPARATION
　　　SEWAGE TREATMENT
　　　SIZE SEPARATION
　　　WATER TREATMENT

FINANCE
　RT　ACCOUNTING
　　　COMMERCE
　　　FIDUCIARIES
　　　GROSS NATIONAL PRODUCT
　　　INVESTMENTS
　　　MANAGEMENT PLANNING
　　　MARKETING
　　　RISK
　　　WAGE SURVEYS

FINANCIAL MANAGEMENT
　GS　MANAGEMENT
　　　. **FINANCIAL MANAGEMENT**
　RT　AIRCRAFT PRODUCTION COSTS
　　　ALLOCATIONS
　　　BUDGETING
　　　COST ANALYSIS
　　　COST ESTIMATES
　　　COSTS
　　　ECONOMY
　　　FEDERAL BUDGETS

FINANCIAL MANAGEMENT-*(CONT.)*
　　　LIFE CYCLE COSTS
　　　PROCUREMENT MANAGEMENT

∞ FINE
　SN　*(USE OF A MORE SPECIFIC TERM IS*
　　　RECOMMENDED--CONSULT THE TERMS
　　　LISTED BELOW)
　RT　FINE STRUCTURE
　　　FINENESS
　　　FINENESS RATIO
　　　FINES

FINE STRUCTURE
　UF　MULTIPLETS
　RT　ALPHA DECAY
　　　ATOMIC STRUCTURE
　　　∞FINE
　　　HYPERFINE STRUCTURE
　　　LINE SPECTRA
　　　SPECTRAL ENERGY DISTRIBUTION
　　　∞STRUCTURES

FINENESS
　RT　COARSENESS
　　　FINENESS RATIO
　　　PARTICLE SIZE DISTRIBUTION
　　　PURITY
　　　QUALITY
　　　SIZE (DIMENSIONS)
　　　TEXTURES

FINENESS RATIO
　GS　RATIOS
　　　. **FINENESS RATIO**
　RT　ASPECT RATIO
　　　DIMENSIONS
　　　FINENESS
　　　OBLATE SPHEROIDS
　　　SLENDER BODIES
　　　THICKNESS RATIO

FINES
　GS　PARTICLES
　　　. POWDER (PARTICLES)
　　　. . **FINES**
　RT　∞FINE
　　　∞FLOUR
　　　FRACTIONS
　　　PARTICLE SIZE DISTRIBUTION
　　　∞SCREENING

FINGERS
　GS　ANATOMY
　　　. LIMBS (ANATOMY)
　　　. . HAND (ANATOMY)
　　　. . . **FINGERS**
　RT　HAND (ANATOMY)
　　　SENSE ORGANS

FINISHES
　GS　**FINISHES**
　　　. ENAMELS
　　　. GLAZES
　　　. LACQUERS
　RT　CERAMIC COATINGS
　　　COATINGS
　　　CORROSION
　　　DOPES
　　　IMPREGNATING
　　　LUSTER
　　　MACHINING
　　　METALLIZING
　　　PAINTS
　　　PLATING
　　　POLISHING
　　　PRIMERS (COATINGS)
　　　PROTECTIVE COATINGS
　　　SIZING (SURFACE TREATMENT)
　　　SPRAYED COATINGS
　　　SURFACE FINISHING
　　　SURFACE PROPERTIES
　　　VARNISHES
　　　VENEERS
　　　WAXES

FINITE DIFFERENCE THEORY
　GS　ANALYSIS (MATHEMATICS)
　　　. NUMERICAL ANALYSIS
　　　. . APPROXIMATION
　　　. . . **FINITE DIFFERENCE THEORY**
　RT　CRANK-NICHOLSON METHOD
　　　DIFFERENCE EQUATIONS
　　　DIFFERENCES
　　　EXTRAPOLATION

FINITE DIFFERENCE THEORY-*(CONT.)*
    INTERPOLATION
    SIGNIFICANCE
    ∞ THEORIES
    TIME MARCHING

FINITE ELEMENT METHOD
GS   ANALYSIS (MATHEMATICS)
    . NUMERICAL ANALYSIS
    . . APPROXIMATION
    . . . **FINITE ELEMENT METHOD**
    PROCEDURES
    . **FINITE ELEMENT METHOD**
RT   ∞ APPLICATIONS OF MATHEMATICS
    BOUNDARY VALUE PROBLEMS
    CHOLESKY FACTORIZATION
    COMPUTATIONAL FLUID DYNAMICS
    CONJUGATES
    CRANK-NICHOLSON METHOD
    FACTORIZATION
    FRACTURE MECHANICS
    ISOPARAMETRIC FINITE ELEMENTS
    ITERATIVE SOLUTION
    MATRICES (MATHEMATICS)
    ∞ METHODOLOGY
    MINIMAL SURFACES
    NASTRAN
    PANEL METHOD (FLUID DYNAMICS)
    SOLID MECHANICS

FINITE IMPULSE RESPONSE FILTERS
USE   FIR FILTERS

FINITE VOLUME METHOD
GS   PROCEDURES
    . **FINITE VOLUME METHOD**
RT   BOUNDARY VALUE PROBLEMS
    ∞ METHODOLOGY

FINITE-STATE MACHINES
USE   TURING MACHINES

FINLAND
GS   NATIONS
    . **FINLAND**
RT   EUROPE
    SCANDINAVIA

FINNED BODIES
RT   AERODYNAMIC CONFIGURATIONS
    ∞ BODIES
    BODIES OF REVOLUTION
    COOLING FINS
    FINS
    HEAT EXCHANGERS
    MISSILE BODIES
    NOSE FINS
    PROJECTILES
    SYMMETRICAL BODIES

FINS
UF   VERTICAL FINS
GS   **FINS**
    . COOLING FINS
    . NOSE FINS
RT   AERIAL RUDDERS
    AIRFOILS
    AIRFRAMES
    ∞ BLADES
    CONTROL SURFACES
    FINNED BODIES
    HYDROFOILS
    MISSILE COMPONENTS
    RUDDERS
    SAILS
    STABILIZERS (FLUID DYNAMICS)
    TAIL ASSEMBLIES
    VANES
    WINGLETS

FIORDS
GS   LANDFORMS
    . **FIORDS**
RT   CLIFFS
    GEOLOGY
    INLETS (TOPOGRAPHY)
    NORWAY
    OCEANOGRAPHY
    WATER

FIR FILTERS
UF   FINITE IMPULSE RESPONSE FILTERS
GS   ELECTROMAGNETIC WAVE FILTERS
    . ELECTRIC FILTERS

FIR FILTERS-*(CONT.)*
    . . DIGITAL FILTERS
    . . . **FIR FILTERS**
RT   BANDPASS FILTERS
    ELECTRONIC FILTERS
    ∞ FILTERS
    MICROWAVE FILTERS
    RADAR FILTERS
    RECURSIVE FUNCTIONS

FIRE CONTROL
SN   (LIMITED TO CONTROL OF THE FIRING
    OF WEAPONS--EXCLUDES FIRE
    PREVENTION AND FIRE FIGHTING)
RT   BOMBING EQUIPMENT
    ∞ CONTROL
    FIRING (IGNITING)
    GUNFIRE
    GUNNERY TRAINING
    RANGE FINDERS
    WEAPON SYSTEMS
    WEAPONS

FIRE CONTROL CIRCUITS
GS   CIRCUITS
    . **FIRE CONTROL CIRCUITS**
RT   ∞ CONTROL

FIRE DAMAGE
GS   DAMAGE
    . **FIRE DAMAGE**
RT   ASHES
    CHARRING
    COMBUSTION
    FIRES
    FLAMES
    FUMES
    SMOKE
    SOOT

FIRE EXTINGUISHERS
UF   CHEMICAL EXTINGUISHERS
    EXTINGUISHERS
RT   FIREBREAKS

FIRE FIGHTING
SN   (EXCLUDES FIRE CONTROL--CONTROL
    OF THE FIRING OF WEAPONS)
RT   FIREBREAKS

FIRE POINT
RT   FLAMMABILITY
    FLASH POINT
    SPONTANEOUS COMBUSTION

FIRE PREVENTION
SN   (EXCLUDES FIRE CONTROL--CONTROL
    OF THE FIRING OF WEAPONS)
GS   PREVENTION
    . **FIRE PREVENTION**
RT   ACCIDENT PREVENTION
    EXPLOSION SUPPRESSION
    FIREBREAKS
    FIREPROOFING
    FIRES
    FLAME RETARDANTS
    FOREST FIRES
    HIGH PRESSURE OXYGEN
    SAFETY
    SAFETY DEVICES
    SAFETY MANAGEMENT
    SMOKE DETECTORS
    SPONTANEOUS COMBUSTION
    WARNING
    WARNING SYSTEMS

FIRE RESISTANCE
USE   FLAMMABILITY

FIRE RETARDANTS
USE   FLAME RETARDANTS

∞ **FIREBALLS**
SN   *(USE OF A MORE SPECIFIC TERM IS*
    RECOMMENDED--CONSULT THE TERMS
    LISTED BELOW)
RT   BOLIDES
    NUCLEAR EXPLOSIONS

FIREBEE 2 TARGET DRONE AIRCRAFT
GS   DRONE VEHICLES
    . DRONE AIRCRAFT
    . . TARGET DRONE AIRCRAFT

FIREBEE 2 TARGET DRONE AIRCRAFT-*(CONT.)*
    . . . **FIREBEE 2 TARGET DRONE**
        AIRCRAFT
    LIGHT AIRCRAFT
    . **FIREBEE 2 TARGET DRONE AIRCRAFT**
    PILOTLESS AIRCRAFT
    . DRONE AIRCRAFT
    . . TARGET DRONE AIRCRAFT
    . . . **FIREBEE 2 TARGET DRONE**
        AIRCRAFT
    RESEARCH AIRCRAFT
    . **FIREBEE 2 TARGET DRONE AIRCRAFT**
    RYAN AIRCRAFT
    . **FIREBEE 2 TARGET DRONE AIRCRAFT**
    SUPERSONIC AIRCRAFT
    . **FIREBEE 2 TARGET DRONE AIRCRAFT**
RT   ∞ AIRCRAFT
    ∞ MILITARY AIRCRAFT
    TARGETS
    ∞ WINGED VEHICLES

FIREBREAKS
RT   CLEARINGS (OPENINGS)
    COMBUSTION
    CONSERVATION
    FIRE EXTINGUISHERS
    FIRE FIGHTING
    FIRE PREVENTION
    FIRES
    FLAMES
    FOREST FIRES
    FORESTS

FIREFLIES
GS   ANIMALS
    . INVERTEBRATES
    . . ARTHROPODS
    . . . INSECTS
    . . . . **FIREFLIES**

FIREPROOFING
RT   FIRE PREVENTION
    NONFLAMMABLE MATERIALS
    SAFETY

FIRES
GS   **FIRES**
    . FOREST FIRES
RT   ACCIDENTS
    BACKFIRE
    BURNS (INJURIES)
    COMBUSTION
    CONTROL SURFACES
    DEFLAGRATION
    EXPLOSIONS
    EXPLOSIVES
    FIBER RELEASE
    FIRE DAMAGE
    FIRE PREVENTION
    FIREBREAKS
    FIRING (IGNITING)
    FLAMES
    FLASHBACK
    HAZARDS
    SAFETY
    SAINT ELMO FIRE
    WARNING SYSTEMS

FIREWORKS
USE   PYROTECHNICS

FIRING (IGNITING)
GS   **FIRING (IGNITING)**
    . ROCKET FIRING
    . . RETROFIRING
    . TEST FIRING
    . . STATIC FIRING
RT   BURNING TIME
    DETONABLE GAS MIXTURES
    DETONATION
    DRYING
    FIRE CONTROL
    FIRES
    FLAMMABLE GASES
    GUNFIRE
    IGNITION
    STARTING

FIRING TIME
USE   BURNING TIME

FIRMWARE
RT   COMPUTER PROGRAMMING
    HARDWARE
    MICROPROCESSORS

FIRMWARE-*(CONT.)*
    MICROPROGRAMMING

FIRST AID
RT   ACCIDENTS
    CHEMICAL DEFENSE
    CURES
    DISASTERS
    KITS
    MEDICAL EQUIPMENT
    MEDICAL SCIENCE
    MEDICAL SERVICES
    RESUSCITATION
    SPLINTS
    STRETCHERS
    TOURNIQUETS
    TRANSFUSION

FISCHER-TROPSCH PROCESS
RT   CATALYSIS
    CATALYTIC ACTIVITY
    REACTION KINETICS
    SYNTHESIS (CHEMISTRY)
    SYNTHETIC FUELS

FISH
USE   FISHES

FISHBOWL OPERATION
RT   HIGH ALTITUDE TESTS
    NUCLEAR EXPLOSIONS
    ∞OPERATIONS
    VELA SATELLITES

FISHERIES
RT   AQUICULTURE
    ESTUARIES
    FISHES
    MARINE BIOLOGY
    MARINE RESOURCES
    SEA WATER
    SHALLOW WATER
    TIDAL FLATS
    WETLANDS

FISHES
UF   FISH
GS   ANIMALS
    . VERTEBRATES
    . . **FISHES**
    . . . SCHOOLS (FISH)
    . . . SHARKS
RT   AQUICULTURE
    EARTH RESOURCES
    FISHERIES
    ∞FOOD
    ICHTHYOLOGY
    ∞NUTRIENTS
    RED TIDE
    SQUAMA

FISHTAILING
USE   YAW

FISSILE FUELS
GS   FUELS
    . **FISSILE FUELS**
RT   GASEOUS FISSION REACTORS
    NUCLEAR FUELS
    NUCLEAR REACTORS
    RADIOACTIVE MATERIALS

FISSILE MATERIALS
USE   FISSIONABLE MATERIALS

FISSION
RT   BLANKETS (FISSION REACTORS)
    FUEL PRODUCTION
    NUCLEAR FUELS
    SPLITTING

FISSION ELECTRIC CELLS
GS   AUXILIARY POWER SOURCES
    . NUCLEAR AUXILIARY POWER UNITS
    . . SNAP
    . . . **FISSION ELECTRIC CELLS**
    . . . . SNAP 2
    . . . . SNAP 4
    . . . . SNAP 8
    . . . . SNAP 10A
    . SPACE POWER REACTORS
    . . . **FISSION ELECTRIC CELLS**
    . . . . SNAP 2
    . . . . SNAP 4

FISSION ELECTRIC CELLS-*(CONT.)*
    . . . . SNAP 8
    . . . . SNAP 10A
    NUCLEAR ELECTRIC POWER
      GENERATION
    . NUCLEAR AUXILIARY POWER UNITS
    . . SNAP
    . . . **FISSION ELECTRIC CELLS**
    . . . . SNAP 2
    . . . . SNAP 4
    . . . . SNAP 8
    . . . . SNAP 10A
    . . SPACE POWER REACTORS
    . . . **FISSION ELECTRIC CELLS**
    . . . . SNAP 2
    . . . . SNAP 4
    . . . . SNAP 8
    . . . . SNAP 10A
    . NUCLEAR POWER REACTORS
    . . SPACE POWER REACTORS
    . . . **FISSION ELECTRIC CELLS**
    . . . . SNAP 2
    . . . . SNAP 4
    . . . . SNAP 8
    . . . . SNAP 10A
    NUCLEAR REACTORS
    . NUCLEAR POWER REACTORS
    . . SPACE POWER REACTORS
    . . . **FISSION ELECTRIC CELLS**
    . . . . SNAP 2
    . . . . SNAP 4
    . . . . SNAP 8
    . . . . SNAP 10A
RT   ∞ELECTRIC CELLS
    RADIOISOTOPE BATTERIES
    SPACE POWER UNIT REACTORS

FISSION PRODUCTS
RT   FALLOUT
    HIGH ENERGY INTERACTIONS
    NUCLEAR FISSION
    NUCLEAR PARTICLES
    NUCLEAR PUMPING
    NUCLEAR RADIATION
    PRODUCTS
    RADIOACTIVE MATERIALS
    RADIOACTIVE WASTES
    RADIOACTIVITY

FISSION WEAPONS
UF   ATOMIC BOMBS
GS   WEAPONS
    . NUCLEAR WEAPONS
    . . **FISSION WEAPONS**
RT   FALLOUT
    NUCLEAR DEVICES
    THERMONUCLEAR EXPLOSIONS

FISSIONABLE MATERIALS
UF   FISSILE MATERIALS
RT   GASEOUS FISSION REACTORS
    ∞MATERIALS
    NOZZLE FLOW
    NUCLEAR FUELS
    PLUTONIUM
    RADIOACTIVE MATERIALS
    URANIUM

FISSIUM
GS   FUELS
    . . NUCLEAR FUELS
    . . **FISSIUM**

FISSURES (GEOLOGY)
RT   FOLDS (GEOLOGY)
    GEOLOGICAL FAULTS
    STRUCTURAL PROPERTIES (GEOLOGY)
    TECTONICS

FITNESS
GS   **FITNESS**
    . FLIGHT FITNESS
    . PHYSICAL FITNESS
RT   QUALIFICATIONS

FITTING
RT   ADAPTATION
    ADJUSTING
    ALIGNMENT
    ASSEMBLING
    FITTINGS
    GOODNESS OF FIT
    ∞JOINING
    MATCHING
    POSITIONING

FITTINGS
RT   ACCESSORIES
    ADAPTERS
    CLOSURES
    CONNECTORS
    COUPLINGS
    EXTENSIONS
    FASTENERS
    FITTING
    INSERTS
    JOINTS (JUNCTIONS)
    LINKAGES
    SLEEVES
    U BENDS
    UNIONS (CONNECTORS)

FITZGERALD-LORENTZ CONTRACTION
USE   LORENTZ CONTRACTION

FIX
USE   FIXING

FIXED POINT ARITHMETIC
GS   NUMBER THEORY
    . ARITHMETIC
    . . **FIXED POINT ARITHMETIC**
RT   COMPUTER PROGRAMS
    COMPUTERS
    DATA PROCESSING

FIXED POINTS (MATHEMATICS)
GS   GEOMETRY
    . EUCLIDEAN GEOMETRY
    . . POINTS (MATHEMATICS)
    . . . **FIXED POINTS (MATHEMATICS)**
    . TOPOLOGY
    . . **FIXED POINTS (MATHEMATICS)**
RT   MANIFOLDS (MATHEMATICS)
    MAPPING

FIXED WINGS
UF   FIXED-WING AIRCRAFT
GS   AIRFOILS
    . WINGS
    . . **FIXED WINGS**
RT   CAMBERED WINGS
    CRUCIFORM WINGS
    FLEXIBLE WINGS
    LOW ASPECT RATIO WINGS
    RIGID WINGS
    SLENDER WINGS
    SWEPT WINGS
    THIN WINGS
    TWISTED WINGS
    UNCAMBERED WINGS
    UNSWEPT WINGS

FIXED-WING AIRCRAFT
USE   AIRCRAFT CONFIGURATIONS
    FIXED WINGS

∞ FIXING
SN   *(USE OF A MORE SPECIFIC TERM IS*
    RECOMMENDED--CONSULT THE TERMS
    LISTED BELOW)
UF   FIX
RT   MAINTENANCE
    NAVIGATION
    POSITION (LOCATION)
    POSITIONING

FIXTURES
RT   BRACKETS
    HARDWARE
    JIGS
    LUMINAIRES
    TOOLS

FIZEAU EFFECT
RT   DOPPLER EFFECT
    DOPPLER-FIZEAU EFFECT
    ∞EFFECTS

FLAGELLATA
GS   ANIMALS
    . INVERTEBRATES
    . PROTOZOA
    . . . **FLAGELLATA**
    . . . . TRYPANOSOME
    MICROORGANISMS
    . PROTOZOA
    . . **FLAGELLATA**
    . . . TRYPANOSOME

FLAKES
GS   PARTICLES
      . **FLAKES**
RT   FLAKING
      METAL POWDER
      POWDER (PARTICLES)

FLAKING
RT   ATOMIZING
      CHIPPING
      COMMINUTION
      CUTTING
      DISINTEGRATION
      FLAKES
      FRACTURING
      PEELING
      ∞ SEPARATION
      SPALLING
      SPLITTING
      WEAR

FLAME CALORIMETERS
GS   MEASURING INSTRUMENTS
      . CALORIMETERS
      . . **FLAME CALORIMETERS**
RT   BOMB CALORIMETERS
      DROP CALORIMETERS
      HEAT MEASUREMENT
      HIGH TEMPERATURE TESTS
      TEMPERATURE MEASURING
          INSTRUMENTS

FLAME DEFLECTORS
GS   DEFLECTORS
      . **FLAME DEFLECTORS**
RT   BACKFIRE
      BAFFLES
      BLAST DEFLECTORS
      DIVERTERS
      FLASHBACK
      LAUNCHING PADS
      SAFETY DEVICES
      SHIELDING
      TEST STANDS

FLAME FRONTS
USE   FLAME PROPAGATION

FLAME HOLDERS
GS   HOLDERS
      . **FLAME HOLDERS**
RT   COMBUSTION CHAMBERS
      FLAMEOUT
      FLAMES

FLAME INTERACTION
USE   CHEMICAL REACTIONS
      FLAME PROPAGATION

FLAME IONIZATION
GS   IONIZATION
      . GAS IONIZATION
      . . **FLAME IONIZATION**

FLAME PLATING
GS   PLATING
      . **FLAME PLATING**
RT   COATING
      WELDING

FLAME PROBES
GS   MEASURING INSTRUMENTS
      . **FLAME PROBES**
RT   GAS ANALYSIS
      MANOMETERS
      TEMPERATURE MEASURING
          INSTRUMENTS

FLAME PROPAGATION
UF   CHAPMAN-JOUGET FLAME
      COMBUSTION WAVES
      FLAME FRONTS
      FLAME INTERACTION
GS   PROPAGATION (EXTENSION)
      . **FLAME PROPAGATION**
RT   BACKFIRE
      BOUNDARY LAYER COMBUSTION
      BURNING RATE
      COMBUSTIBLE FLOW
      COMBUSTION
      COMBUSTION PHYSICS
      DAMKOHLER NUMBER
      DETONATION
      DETONATION WAVES

FLAME PROPAGATION-(CONT.)
      EXPLOSIONS
      FLAMES
      FLAMMABILITY
      FLASHBACK
      GAS EXPLOSIONS
      GAS-METAL INTERACTIONS
      IGNITION
      PREMIXED FLAMES
      PRESSURE OSCILLATIONS
      PRESSURE PULSES
      ∞ PROPAGATION

FLAME QUENCHING
USE   EXTINGUISHING
      QUENCHING (COOLING)

FLAME RETARDANTS
UF   FIRE RETARDANTS
GS   RETARDANTS
      . **FLAME RETARDANTS**
RT   ANTIMISTING FUELS
      FABRICS
      FIRE PREVENTION
      FLAMMABILITY
      IGNITION LIMITS
      INORGANIC COMPOUNDS
      POLYBROMINATED BIPHENYLS
      SYNTHETIC FIBERS

FLAME SPECTROSCOPY
GS   SPECTROSCOPY
      . GAS SPECTROSCOPY
      . . **FLAME SPECTROSCOPY**
      . SPECTROSCOPIC ANALYSIS
      . . **FLAME SPECTROSCOPY**
      SPECTRUM ANALYSIS
      . **FLAME SPECTROSCOPY**
RT   EMISSION SPECTRA
      LINE SPECTRA
      OPTOGALVANIC SPECTROSCOPY
      QUALITATIVE ANALYSIS

FLAME SPRAYING
GS   SPRAYING
      . **FLAME SPRAYING**
RT   COATING
      COATINGS
      METAL SPRAYING
      METALLIZING
      PLASMA SPRAYING

FLAME STABILITY
GS   DYNAMIC CHARACTERISTICS
      . DYNAMIC STABILITY
      . . COMBUSTION STABILITY
      . . . **FLAME STABILITY**
      . MOTION STABILITY
      . . . FLOW STABILITY
      . . . . **FLAME STABILITY**
      . FLOW CHARACTERISTICS
      . . FLOW STABILITY
      . . **FLAME STABILITY**
      STABILITY
      . DYNAMIC STABILITY
      . . COMBUSTION STABILITY
      . . . **FLAME STABILITY**
      . . MOTION STABILITY
      . . . FLOW STABILITY
      . . . . **FLAME STABILITY**
RT   FLAMEOUT

FLAME TEMPERATURE
GS   TEMPERATURE
      . **FLAME TEMPERATURE**
RT   COMBUSTION TEMPERATURE

FLAMEOUT
RT   COMBUSTION
      COMBUSTION CHAMBERS
      EXTINGUISHING
      FLAME HOLDERS
      FLAME STABILITY
      GAS TURBINE ENGINES
      JET ENGINES

FLAMES
UF   JET FLAMES
      LAMINAR FLAMES
GS   **FLAMES**
      . DIFFUSION FLAMES
      . PREMIXED FLAMES
RT   COMBUSTION
      FIRE DAMAGE
      FIREBREAKS

FLAMES-(CONT.)
      FIRES
      FLAME HOLDERS
      FLAME PROPAGATION
      ∞ FLARES
      FOREST FIRES
      FUELS
      SMOG

FLAMMABILITY
UF   COMBUSTIBILITY
      FIRE RESISTANCE
RT   BURNING RATE
      COMBUSTION
      DETONABLE GAS MIXTURES
      FIRE POINT
      FLAME PROPAGATION
      FLAME RETARDANTS
      FLAMMABLE GASES
      FLASH POINT
      IGNITION
      IGNITION LIMITS
      IGNITION TEMPERATURE
      PYROPHORIC MATERIALS
      ∞ RESISTANCE
      SPONTANEOUS COMBUSTION

FLAMMABLE GASES
UF   EXPLOSIVE GASES
GS   GASES
      . **FLAMMABLE GASES**
      . . LIQUEFIED NATURAL GAS
      . . PYROGEN
RT   CHEMICAL EXPLOSIONS
      DETONABLE GAS MIXTURES
      EXPLOSIVES
      FIRING (IGNITING)
      FLAMMABILITY
      GAS EXPLOSIONS
      HAZARDS

FLANGE WRINKLING
GS   WRINKLING
      . **FLANGE WRINKLING**
RT   BUCKLING

FLANGES
RT   CONNECTORS
      METAL PLATES

FLAP CONTROL
USE   AIRCRAFT CONTROL
      FLAPS (CONTROL SURFACES)

FLAPERONS
GS   AIRFOILS
      . AILERONS
      . . **FLAPERONS**
      . FLAPS (CONTROL SURFACES)
      . . **FLAPERONS**
      CONTROL SURFACES
      . AILERONS
      . . **FLAPERONS**
      . FLAPS (CONTROL SURFACES)
      . . **FLAPERONS**
RT   AERODYNAMIC BRAKES

FLAPPING
RT   FLUTTER
      RESONANT VIBRATION
      ROTOR AERODYNAMICS
      SHAKING
      UNDAMPED OSCILLATIONS
      VIBRATION
      WING OSCILLATIONS

FLAPPING HINGES
GS   HINGES
      . **FLAPPING HINGES**
RT   ROTARY WINGS
      ROTOR AERODYNAMICS

FLAPS (CONTROL SURFACES)
UF   FLAP CONTROL
GS   AIRFOILS
      . **FLAPS (CONTROL SURFACES)**
      . . EXTERNALLY BLOWN FLAPS
      . . . UPPER SURFACE BLOWN FLAPS
      . . FLAPERONS
      . . JET FLAPS
      . . SPLIT FLAPS
      . . WING FLAPS
      . . . LEADING EDGE SLATS
      . . . TRAILING EDGE FLAPS

FLAPS (CONTROL SURFACES)-*(CONT.)*
```
      . . . VORTEX FLAPS
      CONTROL SURFACES
    . FLAPS (CONTROL SURFACES)
    . . EXTERNALLY BLOWN FLAPS
    . . . UPPER SURFACE BLOWN FLAPS
    . . FLAPERONS
    . . JET FLAPS
    . . SPLIT FLAPS
    . . WING FLAPS
    . . . LEADING EDGE FLAPS
    . . . LEADING EDGE SLATS
    . . . TRAILING EDGE FLAPS
RT    AERODYNAMIC BRAKES
      BRAKES (FOR ARRESTING MOTION)
    ∞CONTROL
      DELAYED FLAP APPROACH
      DRAG DEVICES
      GAW-2 AIRFOIL
      LIFT DEVICES
      SPOILERS
    ∞SURFACES
```

FLARE STARS
```
UF    UV CETI STARS
GS    CELESTIAL BODIES
    . STARS
    . . DWARF STARS
    . . . FLARE STARS
    . . M STARS
    . . . FLARE STARS
RT    CATACLYSMIC VARIABLES
      SOLAR FLARES
      STELLAR ACTIVITY
      STELLAR FLARES
      SYMBIOTIC STARS
      VARIABLE STARS
```

FLARED BODIES
```
RT    AFTERBODIES
      AIRCRAFT CONFIGURATIONS
    ∞FLARES
      SPACECRAFT CONFIGURATIONS
      SYMMETRICAL BODIES
```

∞ FLARES
```
SN    (USE OF A MORE SPECIFIC TERM IS
      RECOMMENDED--CONSULT THE TERMS
      LISTED BELOW)
RT    FLAMES
      FLARED BODIES
      ILLUMINATING
      LIGHTING EQUIPMENT
      LUMINAIRES
      PYROTECHNICS
      RUNWAY LIGHTS
      SOLAR FLARES
      SOLAR MAXIMUM MISSION
      SOLAR TERRESTRIAL INTERACTIONS
      STELLAR ACTIVITY
      STELLAR FLARES
```

∞ FLASH
```
SN    (USE OF A MORE SPECIFIC TERM IS
      RECOMMENDED--CONSULT THE TERMS
      LISTED BELOW)
UF    LIGHT DURATION
RT    ELECTRIC DISCHARGES
      EXPLOSIONS
      FLASH WELDING
      FLASHING (VAPORIZING)
      LIGHT (VISIBLE RADIATION)
      RADIOGRAPHY
      SOLAR FLARES
```

FLASH BLINDNESS
```
GS    BLINDNESS
    . FLASH BLINDNESS
RT    EYE PROTECTION
      LIGHT ADAPTATION
      VISION
```

FLASH LAMPS
```
UF    FLASH TUBES
GS    LIGHTING EQUIPMENT
    . LUMINAIRES
    . . FLASH LAMPS
    . . . ALKALI VAPOR LAMPS
RT    LIGHT SOURCES
      XENON LAMPS
```

FLASH POINT
```
GS    TEMPERATURE
    . IGNITION TEMPERATURE
    . . FLASH POINT
```

FLASH POINT-*(CONT.)*
```
RT    COMBUSTION TEMPERATURE
      FIRE POINT
      FLAMMABILITY
      IGNITION
      SPONTANEOUS COMBUSTION
      VAPOR PRESSURE
      VOLATILITY
```

FLASH TUBES
```
USE   FLASH LAMPS
```

FLASH WELDING
```
GS    WELDING
    . FLASH WELDING
RT    ELECTRIC WELDING
    ∞FLASH
      FUSION WELDING
      PRESSURE WELDING
```

FLASHBACK
```
RT    BACKFIRE
      COMBUSTION
      DEFLAGRATION
      EXPLOSIONS
      FIRES
      FLAME DEFLECTORS
      FLAME PROPAGATION
```

FLASHING (VAPORIZING)
```
GS    PHASE TRANSFORMATIONS
    . VAPORIZING
    . . FLASHING (VAPORIZING)
RT    DISTILLATION
      EVAPORATION
    ∞FLASH
      PREVAPORIZATION
    ∞SEPARATION
```

FLASHOVER
```
GS    ELECTRIC CURRENT
    . ELECTRIC DISCHARGES
    . . FLASHOVER
RT    ELECTRIC ARCS
      ELECTRIC SPARKS
      ELECTRICAL FAULTS
      FAILURE
```

FLASKS
```
RT    BOTTLES
      GLASSWARE
```

FLAT COAXIAL TRANSMISSION LINES
```
USE   MICROSTRIP TRANSMISSION LINES
```

FLAT CONDUCTORS
```
GS    CONDUCTORS
    . FLAT CONDUCTORS
RT    BUS CONDUCTORS
      CIRCUITS
      CONNECTORS
      ELECTRIC CONNECTORS
      ELECTRIC CONTACTS
      ELECTRIC WIRE
      WIRE
      WIRING
```

FLAT LAYERS
```
RT    FLATNESS
    ∞LAYERS
      PLANAR STRUCTURES
      STRATA
      STRATIFICATION
```

FLAT PATTERNS
```
RT    CASTINGS
      MOLDS
```

FLAT PLATES
```
GS    STRUCTURAL MEMBERS
    . FLAT PLATES
RT    ANNULAR PLATES
      BLASIUS EQUATION
      BLASIUS FLOW
      CIRCULAR PLATES
      DYNAMIC STRUCTURAL ANALYSIS
      END PLATES
      FLATNESS
      FLUID MECHANICS
      HEAT TRANSFER
      METAL PLATES
      PANELS
      PARALLEL PLATES
      PLANAR STRUCTURES
```

FLAT PLATES-*(CONT.)*
```
      PLATE THEORY
    ∞PLATES
      PLATES (STRUCTURAL MEMBERS)
      RECTANGULAR PLATES
    ∞SHEETS
      SLABS
      THICK PLATES
      THIN PLATES
```

FLAT SURFACES
```
UF    FACETS
RT    COSSERAT SURFACES
      FLATNESS
      PLANAR STRUCTURES
    ∞SURFACE GEOMETRY
      SURFACE PROPERTIES
    ∞SURFACES
```

FLATNESS
```
GS    SHAPES
    . FLATNESS
RT    CONCAVITY
      CONTOURS
      CONVEXITY
      FLAT LAYERS
      FLAT PLATES
      FLAT SURFACES
      FLATTENING
      INTERFEROMETERS
      MECHANICAL PROPERTIES
      PLANAR STRUCTURES
      ROUGHNESS
    ∞SURFACE GEOMETRY
```

FLATS (LANDFORMS)
```
UF    ADOBE FLATS
      SALT FLATS
GS    LANDFORMS
    . FLATS (LANDFORMS)
    . . TIDAL FLATS
RT    EARTH RESOURCES
      MARSHLANDS
      MESAS
      PLAINS
      SALT BEDS
```

FLATTENING
```
RT    DUCTILITY
      ELLIPTICITY
      FLATNESS
      LEVELING
      METAL WORKING
      OBLATE SPHEROIDS
    ∞ROLLING
      SMOOTHING
```

FLATWORMS
```
GS    ANIMALS
    . INVERTEBRATES
    . . WORMS
    . . . FLATWORMS
RT    INFESTATION
```

FLAVOR (PARTICLE PHYSICS)
```
RT    HADRONS
      PARTICLE INTERACTIONS
      PARTICLE THEORY
      QUANTUM THEORY
      QUARKS
      THEORETICAL PHYSICS
```

FLAW DETECTION
```
USE   NONDESTRUCTIVE TESTS
```

FLAWS
```
USE   DEFECTS
```

FLEET BALLISTIC MISSILES
```
UF    FBM (MISSILES)
GS    MISSILES
    . SURFACE TO SURFACE MISSILES
    . . FLEET BALLISTIC MISSILES
    . . . POLARIS A1 MISSILE
    . . . POLARIS A2 MISSILE
    . . . POLARIS A3 MISSILE
    . . . POSEIDON MISSILES
    . . . SUBROC MISSILE
RT    BALLISTIC MISSILE SUBMARINES
      GUIDED MISSILE SUBMARINES
      INTERCONTINENTAL BALLISTIC MISSILES
      INTERMEDIATE RANGE BALLISTIC
        MISSILES
      SEA LAUNCHING
```

FLEET SATELLITE COMMUNICATION SYSTEM
UF   FLEETSATCOM
        FLTSATCOM
GS   TELECOMMUNICATION
     . DEFENSE COMMUNICATIONS
       SATELLITE SYSTEM
     .. **FLEET SATELLITE COMMUNICATION
       SYSTEM**
RT   COMMUNICATION SATELLITES
      MARISAT SATELLITES
      MICROWAVE TRANSMISSION
      MILITARY TECHNOLOGY
      NASCOM NETWORK
      NAVY
      RADIO COMMUNICATION
      SATELLITES
     ∞SYSTEMS
      TELECOMMUNICATION
      ULTRAHIGH FREQUENCIES

FLEETSATCOM
USE   FLEET SATELLITE COMMUNICATION
      SYSTEM

FLEXIBILITY
UF   NONRIGIDITY
GS   MECHANICAL PROPERTIES
     . **FLEXIBILITY**
RT   BENDING
      ELASTIC PROPERTIES
      FLEXING
      NONUNIFORMITY
      PLASTIC PROPERTIES
     ∞RIGIDITY
      SOFTNESS
      STIFFNESS
      VERSATILITY

FLEXIBLE BODIES
GS   **FLEXIBLE BODIES**
     . FLEXIBLE SPACECRAFT
RT   ∞BODIES
      HYBRID STRUCTURES
      INFLATABLE STRUCTURES

FLEXIBLE SPACECRAFT
GS   AEROSPACE VEHICLES
     . **FLEXIBLE SPACECRAFT**
      FLEXIBLE BODIES
     . **FLEXIBLE SPACECRAFT**
RT   ARTIFICIAL SATELLITES
      DISPLACEMENT
      ELASTIC DEFORMATION
      FLEXING
      LARGE SPACE STRUCTURES
      SATELLITE CONTROL
      SATELLITE ORIENTATION
      SATELLITE ROTATION
      SATELLITES
      SHAPE CONTROL
     ∞SPACECRAFT
      SPACECRAFT CONTROL
      SPACECRAFT MOTION
      STRUCTURAL VIBRATION
      VIBRATION DAMPING

FLEXIBLE WINGS
UF   ROGALLO WINGS
GS   AIRFOILS
     . WINGS
     .. **FLEXIBLE WINGS**
     ... PARAWINGS
RT   FIXED WINGS
      GLIDERS
      HANG GLIDERS
      INFINITE SPAN WINGS
      INFLATABLE STRUCTURES
      RIGID WINGS
      THIN WINGS
      TWISTED WINGS
      XV-8A AIRCRAFT

FLEXING
UF   FLEXURE
RT   BENDING
      CAMBER
     ∞CHAMBERS
      CURVATURE
      DEFLECTION
      DEFORMATION
      DISTORTION
      FLEXIBILITY
      FLEXIBLE SPACECRAFT
      FOLDING
      HEAVING

FLEXING-(CONT.)
      LOADING MOMENTS

FLEXORS
GS   ANATOMY
     . MUSCULOSKELETAL SYSTEM
     .. **FLEXORS**
RT   JOINTS (ANATOMY)

FLEXOWRITERS (TRADEMARK)
USE   AUTOMATIC TYPEWRITERS

FLEXURE
USE   FLEXING

FLICKER
RT   CRITICAL FLICKER FUSION
      LIGHT TRANSMISSION

FLICKER FUSION FREQUENCY
USE   CRITICAL FLICKER FUSION

∞ **FLIGHT**
SN   *(USE OF A MORE SPECIFIC TERM IS
     RECOMMENDED--CONSULT THE TERMS
     LISTED BELOW)*
UF   FLYING
      HIGH ALTITUDE FLIGHT
      HIGH SPEED FLIGHT
RT   AERODYNAMICS
     ∞AERONAUTICS
      BALLOON FLIGHT
      CLIMBING FLIGHT
      COASTING FLIGHT
      CRUISING FLIGHT
      FLIGHT ALTITUDE
      FLIGHT CONTROL
      FLIGHT MECHANICS
      FLIGHT OPTIMIZATION
      FLIGHT PATHS
      FLIGHT SAFETY
      FLIGHT TESTS
      FLIGHT TIME
      FREE FLIGHT
      GLIDING
      HORIZONTAL FLIGHT
      HYPERSONIC FLIGHT
      LONG DURATION SPACE FLIGHT
      LUNAR FLIGHT
      METEOROLOGICAL FLIGHT
      PARABOLIC FLIGHT
      ROCKET FLIGHT
      SOARING
      SPACE FLIGHT
      STEERING
      SUBORBITAL FLIGHT
      SUPERSONIC FLIGHT
      TRAJECTORIES
      TRANSOCEANIC FLIGHT
      TRANSONIC FLIGHT
      TURNING FLIGHT
      VERTICAL FLIGHT
      VISUAL FLIGHT

FLIGHT ALTITUDE
GS   ALTITUDE
     . **FLIGHT ALTITUDE**
RT   AIR TRAFFIC CONTROL
      CEILING (AIRCRAFT CAPABILITY)
      MIDALTITUDE

FLIGHT CHARACTERISTICS
UF   FLIGHT PERFORMANCE
      FLYING QUALITIES
RT   AERODYNAMICS
      AIRCRAFT MANEUVERS
      AIRCRAFT PERFORMANCE
      AIRCRAFT SPECIFICATIONS
      AIRSPEED
      BUFFETING
      CEILING (AIRCRAFT CAPABILITY)
     ∞CHARACTERISTICS
      CONTROLLABILITY
      FLUTTER
      HELICOPTER PERFORMANCE
      HIGHLY MANEUVERABLE AIRCRAFT
      LOW SPEED STABILITY
      MANEUVERABILITY
     ∞PERFORMANCE
      QUALITY

FLIGHT CLOTHING
GS   CLOTHING
     . **FLIGHT CLOTHING**

FLIGHT CLOTHING-(CONT.)
RT   COVERALLS
      GARMENTS
      GOGGLES
      HELMETS
      PRESSURE SUITS
      PROTECTIVE CLOTHING

FLIGHT COMPUTERS
USE   AIRBORNE/SPACEBORNE COMPUTERS

FLIGHT CONDITIONS
GS   CONDITIONS
     . **FLIGHT CONDITIONS**
RT   CLOUD COVER
      INSTRUMENT FLIGHT RULES
      METEOROLOGICAL SERVICES
      STORMS (METEOROLOGY)
      VISUAL FLIGHT
      WEATHER FORECASTING

FLIGHT CONTROL
GS   **FLIGHT CONTROL**
     . AUTOMATIC FLIGHT CONTROL
     .. AUTOMATIC LANDING CONTROL
     . FLY BY TUBE CONTROL
     . FLY BY WIRE CONTROL
     . POINTING CONTROL SYSTEMS
     .. ANNULAR SUSPENSION AND
       POINTING SYSTEM
     . THRUST VECTOR CONTROL
RT   ACROBATICS
      AIR TRAFFIC CONTROL
      AIRCRAFT CONTROL
      AIRCRAFT INSTRUMENTS
      AIRCRAFT SURVIVABILITY
      ATTITUDE CONTROL
      ATTITUDE INDICATORS
      AUTOMATED EN ROUTE ATC
      AUTOMATIC CONTROL
      AUTOMATIC PILOTS
     ∞CONTROL
      CONTROL CONFIGURED VEHICLES
      CONTROL STABILITY
      CONTROL STICKS
      CONTROL SURFACES
      DISPLAY DEVICES
      ENTRY GUIDANCE (STS)
      FLIGHT CHARACTERISTICS
      FLIGHT MANAGEMENT SYSTEMS
      GROUND BASED CONTROL
      GROUND SUPPORT EQUIPMENT
      GUIDANCE (MOTION)
      HELICOPTER CONTROL
      IN-FLIGHT MONITORING
      INSTRUMENT APPROACH
      INSTRUMENT LANDING SYSTEMS
      MANEUVERABILITY
      MANEUVERS
      MISSILE CONTROL
      NAVIGATION
      NAVIGATION AIDS
      NAVIGATION INSTRUMENTS
      RADIO NAVIGATION
      REMOTE CONTROL
      ROCKET ENGINE CONTROL
      SOLAR COMPASSES
      SPACECRAFT CONTROL
      STABILITY AUGMENTATION
      TURBOJET ENGINE CONTROL

FLIGHT CREWS
UF   AIRCREWS
GS   PERSONNEL
     . CREWS
     .. **FLIGHT CREWS**
     ... SPACECREWS
     . FLYING PERSONNEL
     .. **FLIGHT CREWS**
     ... SPACECREWS
RT   AIRCRAFT PILOTS
      CREW PROCEDURES (INFLIGHT)
      CREW PROCEDURES (PREFLIGHT)
      CREW SIZE
      NAVIGATORS
      PILOTS (PERSONNEL)

FLIGHT FATIGUE
GS   FATIGUE (BIOLOGY)
     . **FLIGHT FATIGUE**

FLIGHT FITNESS
GS   FITNESS
     . **FLIGHT FITNESS**
RT   ∞FLIGHT STRESS

FLIGHT FITNESS-*(CONT.)*
    FLYING PERSONNEL
    PHYSICAL EXAMINATIONS
    PHYSICAL FITNESS

FLIGHT HAZARDS
GS HAZARDS
    . **FLIGHT HAZARDS**
    . . METEOROID HAZARDS
RT AIR PIRACY
    AIR TRAFFIC
    AIRCRAFT ACCIDENTS
    AIRCRAFT HAZARDS
    AIRCRAFT SAFETY
    AIRCRAFT SPIN
    BIRD-AIRCRAFT COLLISIONS
    BIRDS
    COLLISIONS
    CRASH LANDING
    CRASHES
    DESTRUCTION
    MIDAIR COLLISIONS
    NOISE (SOUND)
    OPERATIONAL HAZARDS
    TOXIC HAZARDS
    WEATHER

FLIGHT INSTRUMENTS
GS **FLIGHT INSTRUMENTS**
    . APPROACH INDICATORS
    . ATTITUDE INDICATORS
    . . GYRO HORIZONS
    . AUTOMATIC PILOTS
    . FLIGHT TEST INSTRUMENTS
    . HORIZON SCANNERS
    . RADIO ALTIMETERS
RT AIR NAVIGATION
    AIRBORNE EQUIPMENT
    AIRCRAFT CONTROL
    AIRCRAFT EQUIPMENT
    AIRCRAFT INSTRUMENTS
    ALTIMETERS
    BUBBLE TECHNIQUE
    COMPASSES
    DISPLAY DEVICES
    ENGINE CONTROL
    ENGINE MONITORING INSTRUMENTS
    HEAD-UP DISPLAYS
    INSTRUMENT APPROACH
    INSTRUMENT FLIGHT RULES
    INSTRUMENT LANDING SYSTEMS
    ∞INSTRUMENTS
    LANDING INSTRUMENTS
    LIGHT AIRBORNE MULTIPURPOSE
      SYSTEM
    MEASURING INSTRUMENTS
    NAVIGATION INSTRUMENTS
    NIGHT FLIGHTS (AIRCRAFT)
    ONBOARD EQUIPMENT
    POSITION INDICATORS
    RADAR
    RADIO DIRECTION FINDERS
    RATE OF CLIMB INDICATORS
    RECORDING INSTRUMENTS
    SATELLITE INSTRUMENTS
    SOLAR COMPASSES
    SPACECRAFT INSTRUMENTS
    SPACECRAFT POSITION INDICATORS
    SPEED INDICATORS
    STAR TRACKERS
    TERCOM

FLIGHT LOAD RECORDERS
GS MEASURING INSTRUMENTS
    . **FLIGHT LOAD RECORDERS**
    RECORDING INSTRUMENTS
    . **FLIGHT LOAD RECORDERS**
RT STRAIN GAGES

FLIGHT MANAGEMENT SYSTEMS
GS MANAGEMENT SYSTEMS
    . **FLIGHT MANAGEMENT SYSTEMS**
RT AIR NAVIGATION
    AIR TRAFFIC CONTROL
    AIRBORNE/SPACEBORNE COMPUTERS
    AUTOMATIC FLIGHT CONTROL
    AUTOMATIC LANDING CONTROL
    AVIONICS
    COMPUTER TECHNIQUES
    FLIGHT CONTROL
    GROUND BASED CONTROL
    NAVIGATION AIDS
    ONBOARD DATA PROCESSING
    SYSTEMS ENGINEERING

FLIGHT MECHANICS
RT AERODYNAMICS
    ASCENT TRAJECTORIES
    DESCENT TRAJECTORIES
    ∞MECHANICS (PHYSICS)
    MISSILE TRAJECTORIES
    ORBIT CALCULATION
    ORBIT DECAY
    ORBITAL MECHANICS
    ∞PLATFORMS
    REENTRY TRAJECTORIES
    RENDEZVOUS
    RENDEZVOUS TRAJECTORIES
    SPACE FLIGHT
    SPACE MECHANICS
    SPACECRAFT REENTRY
    SPACECRAFT TRAJECTORIES
    THRUST PROGRAMMING
    TRAJECTORIES
    TRAJECTORY MEASUREMENT
    TRAJECTORY OPTIMIZATION

FLIGHT NURSES
GS PERSONNEL
    . MEDICAL PERSONNEL
    . . **FLIGHT NURSES**
RT CREWS

FLIGHT OPERATIONS
GS **FLIGHT OPERATIONS**
    . CREW PROCEDURES (INFLIGHT)
RT AIRCRAFT MAINTENANCE
    CREW PROCEDURES (PREFLIGHT)
    GROUND HANDLING
    ONBOARD EQUIPMENT
    REFUELING

FLIGHT OPTIMIZATION
GS OPTIMIZATION
    . **FLIGHT OPTIMIZATION**
RT BURNING TIME
    EARTH-VENUS TRAJECTORIES
    GREAT CIRCLES
    ORBITAL MECHANICS
    ORBITS
    PARKING ORBITS
    SPACE FLIGHT
    THRUST PROGRAMMING
    TRAJECTORIES
    TRAJECTORY OPTIMIZATION

FLIGHT PATHS
GS **FLIGHT PATHS**
    . GLIDE PATHS
RT AIR NAVIGATION
    AIR TRAFFIC
    AIR TRAFFIC CONTROL
    AIRCRAFT INSTRUMENTS
    AIRCRAFT MANEUVERS
    AIRSPACE
    APPROACH
    APPROACH CONTROL
    AREA NAVIGATION
    CAUSTIC LINES
    CLIMBING FLIGHT
    COLLISION AVOIDANCE
    COLLISIONS
    DELAYED FLAP APPROACH
    DESCENT
    ∞DRIFT
    GLIDING
    GLOBAL POSITIONING SYSTEM
    GREAT CIRCLES
    GROUND TRACKS
    GUIDANCE (MOTION)
    HORIZONTAL FLIGHT
    MISSILE TRAJECTORIES
    NATIONAL AIRSPACE UTILIZATION
      SYSTEM
    NAVIGATION
    NAVIGATION AIDS
    ORBITS
    ∞PATHS
    REENTRY
    ROCKET FLIGHT
    SATELLITE GROUND TRACKS
    SOLAR COMPASSES
    SWATH WIDTH
    TACAN
    TRAJECTORIES
    TURNING FLIGHT
    UNCONTROLLED REENTRY
      (SPACECRAFT)
    VERTICAL FLIGHT
    VISUAL FLIGHT

FLIGHT PERFORMANCE
USE FLIGHT CHARACTERISTICS

FLIGHT PLANS
RT AIR NAVIGATION
    AIR TRAFFIC
    AIR TRAFFIC CONTROL
    APPROACH
    INSTRUMENT FLIGHT RULES
    NATIONAL AIRSPACE UTILIZATION
      SYSTEM
    ∞PLANS
    ROUTES
    THRUST PROGRAMMING
    WEATHER

FLIGHT RECORDERS
GS AIRCRAFT INSTRUMENTS
    . **FLIGHT RECORDERS**
    MEASURING INSTRUMENTS
    . **FLIGHT RECORDERS**
    RECORDING INSTRUMENTS
    . **FLIGHT RECORDERS**

FLIGHT RULES
GS RULES
    . **FLIGHT RULES**
    . . INSTRUMENT FLIGHT RULES
    . . VISUAL FLIGHT RULES
RT AIR NAVIGATION
    AIR TRAFFIC CONTROL
    COLLISION AVOIDANCE
    NATIONAL AIRSPACE UTILIZATION
      SYSTEM
    NATIONAL AVIATION SYSTEM
    NOISE REDUCTION

FLIGHT SAFETY
GS SAFETY
    . **FLIGHT SAFETY**
RT AEROSPACE SAFETY
    AIR PIRACY
    AIR TRAFFIC CONTROL
    AIRCRAFT ACCIDENTS
    AIRCRAFT APPROACH SPACING
    AIRCRAFT HAZARDS
    AIRCRAFT SAFETY
    AIRCRAFT SPIN
    ALL-WEATHER LANDING SYSTEMS
    COLLISION AVOIDANCE
    CRASHES
    CRASHWORTHINESS
    DESTRUCTION
    ∞FLIGHT
    ∞FLIGHT STRESS
    FLYING EJECTION SEATS
    MIDAIR COLLISIONS
    ONBOARD EQUIPMENT
    SAFETY DEVICES
    SELF SEALING
    VISUAL FLIGHT

FLIGHT SIMULATION
GS SIMULATION
    . **FLIGHT SIMULATION**
RT ACOUSTIC SIMULATION
    ALTITUDE SIMULATION
    ANALOG SIMULATION
    COMPUTERIZED SIMULATION
    CONTROL SIMULATION
    ENVIRONMENT SIMULATION
    LANDING SIMULATION
    MOTION SIMULATION
    SPACE ENVIRONMENT SIMULATION
    SPACE FLIGHT
    SYSTEMS SIMULATION
    TRAINING SIMULATORS
    WEIGHTLESSNESS SIMULATION

FLIGHT SIMULATORS
GS SIMULATORS
    . TRAINING SIMULATORS
    . . **FLIGHT SIMULATORS**
    . . . COCKPIT SIMULATORS
RT ATMOSPHERIC ENTRY SIMULATION
    CENTRIFUGES
    CONTROL SIMULATION
    CRYOGENIC WIND TUNNELS
    LANGLEY COMPLEX COORDINATOR
    LUNAR ORBIT AND LANDING
      SIMULATORS
    ∞MISSILE SIMULATORS
    MOTION SIMULATION
    MOTION SIMULATORS
    PILOT TRAINING

FLIGHT SIMULATORS-*(CONT.)*
　　SPACE ENVIRONMENT SIMULATION
　　SPACE SIMULATORS
　　TEST FACILITIES
　　TRAINING DEVICES
　　WIND TUNNELS

FLIGHT STABILITY TESTS
　GS　FLIGHT TESTS
　　. **FLIGHT STABILITY TESTS**
　　STABILITY TESTS
　　. **FLIGHT STABILITY TESTS**
　RT　AERODYNAMIC STABILITY
　　∞ TESTS

∞ **FLIGHT STRESS**
　SN　*(USE OF A MORE SPECIFIC TERM IS*
　　RECOMMENDED--CONSULT THE TERMS
　　LISTED BELOW)
　RT　FLIGHT FATIGUE
　　FLIGHT FITNESS
　　FLIGHT SAFETY
　　HUMAN FACTORS ENGINEERING
　　SPACE FLIGHT STRESS
　　STRESS ANALYSIS
　　STRESSES

FLIGHT STRESS (BIOLOGY)
　SN　(EXCLUDES MECHANICAL STRESS AND
　　AND STRAIN)
　GS　**FLIGHT STRESS (BIOLOGY)**
　　. SPACE FLIGHT STRESS
　RT　ACCELERATION (PHYSICS)
　　BIOLOGICAL EFFECTS
　　∞ BIOLOGY
　　JET LAG
　　PHYSIOLOGICAL FACTORS
　　PSYCHOLOGICAL FACTORS
　　STRESS (PHYSIOLOGY)
　　STRESS (PSYCHOLOGY)
　　WEIGHTLESSNESS

FLIGHT SURGEONS
　GS　PERSONNEL
　　. MEDICAL PERSONNEL
　　. . SURGEONS
　　. . . **FLIGHT SURGEONS**

FLIGHT TECHNICAL ERROR
　USE　PILOT ERROR

FLIGHT TEST INSTRUMENTS
　GS　FLIGHT INSTRUMENTS
　　. **FLIGHT TEST INSTRUMENTS**
　RT　AIRCRAFT INSTRUMENTS
　　ROCKET-BORNE INSTRUMENTS
　　SPACECRAFT INSTRUMENTS

FLIGHT TEST VEHICLES
　GS　TEST VEHICLES
　　. **FLIGHT TEST VEHICLES**
　RT　∞ AIRCRAFT
　　LAUNCH VEHICLES
　　MISSILES
　　RESEARCH AIRCRAFT
　　∞ SPACECRAFT
　　∞ VEHICLES

FLIGHT TESTS
　GS　**FLIGHT TESTS**
　　. FLIGHT STABILITY TESTS
　　. SPACE TRANSPORTATION SYSTEM
　　　FLIGHTS
　　. . SPACE TRANSPORTATION SYSTEM 1
　　　FLIGHT
　　. . SPACE TRANSPORTATION SYSTEM 2
　　　FLIGHT
　　. . SPACE TRANSPORTATION SYSTEM 3
　　　FLIGHT
　　. . SPACE TRANSPORTATION SYSTEM 4
　　　FLIGHT
　RT　AIR START
　　AIRCRAFT DESIGN
　　ALTITUDE TESTS
　　CERTIFICATION
　　DAST PROGRAM
　　DOWNRANGE
　　DYNAMIC TESTS
　　ENGINE TESTS
　　FREE FLIGHT TEST APPARATUS
　　FULL SCALE TESTS
　　GROUND TESTS
　　HIGH ALTITUDE TESTS
　　HIGHLY MANEUVERABLE AIRCRAFT
　　IN-FLIGHT MONITORING

FLIGHT TESTS-*(CONT.)*
　　MISSILE DESIGN
　　MISSILE TESTS
　　POSTMISSION ANALYSIS (SPACECRAFT)
　　SPACE ELECTRIC ROCKET TESTS
　　STABILITY TESTS
　　∞ TESTS
　　VIBRATION TESTS
　　WING FLOW METHOD TESTS

FLIGHT TIME
　GS　TIME
　　. **FLIGHT TIME**
　RT　AIR TRAFFIC CONTROL
　　BURNING TIME
　　TESTING TIME
　　TRAJECTORIES
　　TRANSIT TIME
　　TURNAROUND (STS)
　　WINDOWS (INTERVALS)

FLIGHT TRAINING
　GS　EDUCATION
　　. **FLIGHT TRAINING**
　　. . SPACE FLIGHT TRAINING
　RT　ASTRONAUT TRAINING
　　EJECTION TRAINING
　　FLYING PERSONNEL
　　PILOT TRAINING
　　TRAINING SIMULATORS

∞ **FLIGHT VEHICLES**
　SN　*(USE OF A MORE SPECIFIC TERM IS*
　　RECOMMENDED--CONSULT THE TERMS
　　LISTED BELOW)
　RT　AIRCRAFT CONFIGURATIONS
　　GROUND EFFECT MACHINES
　　HYPERSONIC VEHICLES
　　LUNAR FLYING VEHICLES
　　MISSILES
　　REENTRY VEHICLES
　　RESEARCH VEHICLES
　　ROCKET VEHICLES
　　∞ VEHICLES

FLINT
　GS　CHALCOGENIDES
　　. OXIDES
　　. . DIOXIDES
　　. . . **FLINT**
　　SILICON COMPOUNDS
　　. **FLINT**
　RT　QUARTZ

FLIP-FLOPS
　UF　BISTABLE AMPLIFIERS
　GS　CIRCUITS
　　. FLUIDIC CIRCUITS
　　. . **FLIP-FLOPS**
　　. MULTIVIBRATORS
　　. . **FLIP-FLOPS**
　RT　BISTABLE CIRCUITS
　　DATA STORAGE
　　FLUID SWITCHING ELEMENTS
　　OSCILLATORS

FLIR DETECTORS
　UF　FORWARD LOOKING INFRARED
　　　DETECTORS
　GS　MEASURING INSTRUMENTS
　　. RADIATION MEASURING INSTRUMENTS
　　. . INFRARED INSTRUMENTS
　　. . . INFRARED DETECTORS
　　. . . . **FLIR DETECTORS**
　RT　∞ DETECTORS
　　INFRARED RADAR
　　∞ SENSORS

FLOAT ZONES
　RT　CRYSTAL GROWTH
　　MELTS (CRYSTAL GROWTH)
　　SILICON
　　SOLAR CELLS
　　SPACE PROCESSING
　　ZONE MELTING

FLOATING
　RT　BALLAST (MASS)
　　BUOYANCY
　　FLOATS

FLOATING POINT ARITHMETIC
　GS　NUMBER THEORY
　　. ARITHMETIC

FLOATING POINT ARITHMETIC-*(CONT.)*
　　. . **FLOATING POINT ARITHMETIC**
　RT　COMPUTER PROGRAMS
　　COMPUTERS
　　DATA PROCESSING

FLOATS
　UF　FLOTATION SYSTEMS
　RT　BALLAST (MASS)
　　BUOYS
　　EMERGENCY LIFE SUSTAINING
　　　SYSTEMS
　　FLOATING
　　INFLATABLE STRUCTURES
　　LANDING GEAR
　　LIFE RAFTS
　　RAFTS
　　SEPARATORS

FLOCCULATING
　RT　AGGLOMERATION
　　COAGULATION
　　COALESCING
　　COLLOIDING
　　CONCENTRATING
　　FLOTATION
　　PRECIPITATION (CHEMISTRY)
　　SETTLING
　　WATER TREATMENT

FLOOD CONTROL
　RT　CANALS
　　∞ CONTROL
　　DAMS
　　DRAINAGE
　　HYDROLOGY
　　RAINSTORMS
　　STORM DAMAGE
　　STORMS (METEOROLOGY)
　　WATERSHEDS

FLOOD DAMAGE
　GS　DAMAGE
　　. **FLOOD DAMAGE**
　RT　DRAINAGE PATTERNS
　　HYDROLOGY
　　LANDSLIDES
　　PRECIPITATION (METEOROLOGY)
　　SEEPAGE
　　STORMS
　　STORMS (METEOROLOGY)
　　TIDES
　　WATER EROSION
　　WATER FLOW

FLOOD PLAINS
　GS　LAND
　　. PLAINS
　　. . **FLOOD PLAINS**
　RT　FLOODS
　　HYDROGEOLOGY
　　HYDROLOGY

FLOOD PREDICTIONS
　GS　PREDICTIONS
　　. **FLOOD PREDICTIONS**
　RT　FLOODS
　　HYDROGEOLOGY
　　HYDROLOGY
　　PRECIPITATION (METEOROLOGY)
　　RAIN
　　RAINSTORMS
　　∞ SHOWERS
　　STORMS (METEOROLOGY)
　　WEATHER FORECASTING

FLOODS
　RT　ALLUVIUM
　　DROUGHT
　　FLOOD PLAINS
　　FLOOD PREDICTIONS
　　HYDROLOGY
　　HYDROLOGY MODELS
　　MISSISSIPPI RIVER (US)
　　PRECIPITATION (METEOROLOGY)
　　STORM DAMAGE
　　STORMS
　　STORMS (METEOROLOGY)
　　TIDES
　　WATER FLOW
　　WATER MANAGEMENT
　　WATERSHEDS

FLOORS
　UF　DECKS (FLOORS)

FLOORS-(CONT.)
RT　　BASEMENTS
　　　∞BUILDINGS
　　　CEILINGS (ARCHITECTURE)
　　　DOORS
　　　∞PLATFORMS
　　　SUBSTRUCTURES
　　　TILES
　　　WALLS

FLOQUET THEOREM
GS　　THEOREMS
　　　. FLOQUET THEOREM
RT　　DIFFERENTIAL EQUATIONS
　　　LINEAR EQUATIONS
　　　PERIODIC FUNCTIONS

FLORA
USE　　PLANTS (BOTANY)

FLORIDA
GS　　NATIONS
　　　. UNITED STATES
　　　.. FLORIDA
RT　　EVERGLADES (FL)
　　　GULF OF MEXICO
　　　MERRITT ISLAND (FL)

FLOTATION
RT　　ACTIVATION
　　　BENEFICIATION
　　　CLASSIFIERS
　　　COAGULATION
　　　CONCENTRATING
　　　FLOCCULATING
　　　FLUID ROTOR GYROSCOPES
　　　FOAMING
　　　LEVITATION
　　　∞SEPARATION
　　　SETTLING
　　　SIZE SEPARATION
　　　SUSPENSION SYSTEMS (VEHICLES)
　　　WATER TREATMENT

FLOTATION SYSTEMS
USE　　FLOATS

∞ FLOUR
SN　　*(USE OF A MORE SPECIFIC TERM IS
　　　RECOMMENDED--CONSULT THE TERMS
　　　LISTED BELOW)*
RT　　FINES
　　　FLOUR (FOOD)

FLOUR (FOOD)
RT　　∞FLOUR
　　　∞FOOD
　　　MILLET
　　　POWDER (PARTICLES)

∞ FLOW
SN　　*(USE OF A MORE SPECIFIC TERM IS
　　　RECOMMENDED--CONSULT THE TERMS
　　　LISTED BELOW)*
RT　　AERODYNAMICS
　　　ANNULAR FLOW
　　　BRILLOUIN FLOW
　　　CORNER FLOW
　　　CREEP PROPERTIES
　　　CROSS FLOW
　　　EXHAUST NOZZLES
　　　FLOW EQUATIONS
　　　FLOW THEORY
　　　FLOW VELOCITY
　　　FLUID FLOW
　　　GRAZING FLOW
　　　HEAT TRANSMISSION
　　　INFORMATION FLOW
　　　INTERACTIONAL AERODYNAMICS
　　　INVISCID FLOW
　　　LOW DENSITY FLOW
　　　MASS FLOW
　　　ORIFICE FLOW
　　　OUTLET FLOW
　　　PANEL METHOD (FLUID DYNAMICS)
　　　PLASTIC FLOW
　　　SHEAR FLOW
　　　SOLIDS FLOW
　　　STEADY FLOW
　　　TRANSONIC FLOW
　　　UNSTEADY FLOW
　　　VISCOUS FLOW

FLOW CHAMBERS
RT　　∞CHAMBERS

FLOW CHARACTERISTICS
GS　　DYNAMIC CHARACTERISTICS
　　　. FLOW CHARACTERISTICS
　　　.. FLOW DISTRIBUTION
　　　.. FLOW STABILITY
　　　... BOUNDARY LAYER STABILITY
　　　... FLAME STABILITY
　　　... MAGNETOHYDRODYNAMIC
　　　　　　STABILITY
　　　.... WEIBEL INSTABILITY
　　　.. FLOW VELOCITY
RT　　BAROTROPIC FLOW
　　　∞CHARACTERISTICS
　　　CRITICAL FLOW
　　　CROSS FLOW
　　　EDDY VISCOSITY
　　　INVISCID FLOW
　　　LAMINAR FLOW
　　　NONUNIFORM FLOW
　　　OUTLET FLOW
　　　REATTACHED FLOW
　　　SEPARATED FLOW
　　　STEADY FLOW
　　　STROUHAL NUMBER
　　　SUBCRITICAL FLOW
　　　SUPERCRITICAL FLOW
　　　TURBULENCE
　　　TURBULENT FLOW
　　　VISCOSITY
　　　VISCOUS FLOW

FLOW CHARTS
GS　　CHARTS
　　　. FLOW CHARTS
RT　　BLOCK DIAGRAMS
　　　COMPUTER PROGRAMMING
　　　DATA FLOW ANALYSIS
　　　∞FLOW GRAPHS
　　　MATHEMATICAL MODELS

FLOW COEFFICIENTS
GS　　COEFFICIENTS
　　　. FLOW COEFFICIENTS
　　　.. DISCHARGE COEFFICIENT
RT　　AERODYNAMIC COEFFICIENTS
　　　ATTENUATION COEFFICIENTS
　　　MASS FLOW FACTORS
　　　NOZZLE THRUST COEFFICIENTS
　　　REFLECTANCE
　　　TRANSPORT PROPERTIES

FLOW DEFLECTION
RT　　DEFLECTORS

FLOW DIRECTION INDICATORS
GS　　DISPLAY DEVICES
　　　. FLOW DIRECTION INDICATORS
　　　.. WIND VANES
　　　MEASURING INSTRUMENTS
　　　. INDICATING INSTRUMENTS
　　　.. FLOW DIRECTION INDICATORS
　　　... WIND VANES

FLOW DISTORTION
GS　　DISTORTION
　　　. FLOW DISTORTION
RT　　AERODYNAMIC COEFFICIENTS
　　　FLUID FLOW
　　　MULTIPHASE FLOW
　　　ORR-SOMMERFELD EQUATIONS
　　　OSCILLATING FLOW
　　　SMALL PERTURBATION FLOW
　　　VORTICES
　　　WING TIP VORTICES

FLOW DISTRIBUTION
UF　　FLOW FIELDS
　　　FLOW PATTERNS
GS　　DISTRIBUTION (PROPERTY)
　　　. FLOW DISTRIBUTION
　　　DYNAMIC CHARACTERISTICS
　　　. FLOW CHARACTERISTICS
　　　.. FLOW DISTRIBUTION
RT　　BOUNDARY LAYER FLOW
　　　BOUNDARY LAYER SEPARATION
　　　CAVITATION FLOW
　　　CHAPMAN-ENSKOG THEORY
　　　EXHAUST FLOW SIMULATION
　　　FIELD THEORY (PHYSICS)
　　　∞FLOW GRAPHS
　　　HYDRODYNAMIC COEFFICIENTS
　　　ISOTHERMAL FLOW

FLOW DISTRIBUTION-(CONT.)
　　　METHOD OF CHARACTERISTICS
　　　NUMERICAL FLOW VISUALIZATION
　　　REATTACHED FLOW
　　　RHEOELECTRICAL SIMULATION
　　　SEPARATED FLOW
　　　STROUHAL NUMBER
　　　THREE DIMENSIONAL BODIES
　　　TRAPPED VORTEXES
　　　VELOCITY DISTRIBUTION
　　　VORTEX SHEETS
　　　WATER TUNNEL TESTS
　　　WIND TUNNEL TESTS

FLOW EQUATIONS
GS　　FLOW EQUATIONS
　　　. HELMHOLTZ VORTICITY EQUATION
　　　. VON KARMAN EQUATION
　　　. VORTICITY EQUATIONS
RT　　BOUNDARY LAYER EQUATIONS
　　　CHAPLYGIN EQUATION
　　　∞EQUATIONS
　　　FLUID FLOW
　　　PARTICLE IN CELL TECHNIQUE
　　　PERCUS METHOD

FLOW FIELDS
USE　　FLOW DISTRIBUTION

FLOW GEOMETRY
GS　　GEOMETRY
　　　. FLOW GEOMETRY
RT　　ANNULAR FLOW
　　　AXIAL FLOW
　　　AXISYMMETRIC FLOW
　　　BACKWARD FACING STEPS
　　　BYPASS RATIO
　　　CHANNEL FLOW
　　　COAXIAL FLOW
　　　CORE FLOW
　　　CROSS FLOW
　　　DUCTED FLOW
　　　HELICAL FLOW
　　　INLET AIRFRAME CONFIGURATIONS
　　　INLET FLOW
　　　LAMINAR FLOW
　　　MERIDIONAL FLOW
　　　NOZZLE FLOW
　　　ONE DIMENSIONAL FLOW
　　　PARALLEL FLOW
　　　RADIAL FLOW
　　　STEADY FLOW
　　　STRATIFIED FLOW
　　　THREE DIMENSIONAL FLOW
　　　TWO DIMENSIONAL FLOW
　　　WEDGE FLOW

∞ FLOW GRAPHS
SN　　*(USE OF A MORE SPECIFIC TERM IS
　　　RECOMMENDED--CONSULT THE TERMS
　　　LISTED BELOW)*
RT　　EQUIPOTENTIALS
　　　FLOW CHARTS
　　　FLOW DISTRIBUTION
　　　SIGNAL FLOW GRAPHS

FLOW MEASUREMENT
GS　　MECHANICAL MEASUREMENT
　　　. FLOW MEASUREMENT
RT　　ANEMOMETERS
　　　ANNULI
　　　DRAG FORCE ANEMOMETERS
　　　DRAG MEASUREMENT
　　　FLOWMETERS
　　　FLUID FLOW
　　　GAS METERS
　　　HOT-FILM ANEMOMETERS
　　　HOT-WIRE ANEMOMETERS
　　　LASER DOPPLER VELOCIMETERS
　　　∞MEASUREMENT
　　　MULTIPHASE FLOW
　　　∞NOZZLES
　　　ORIFICES
　　　PITOT TUBES
　　　PNEUMATIC PROBES
　　　PRESSURE MEASUREMENT
　　　RHEOLOGY
　　　SOLIDS FLOW
　　　VELOCITY MEASUREMENT
　　　VENTURI TUBES
　　　WATER FLOW
　　　WIND VELOCITY

FLOW NETS
RT　　EQUIPOTENTIALS

FLOW NETS-*(CONT.)*
  SEEPAGE

FLOW PATTERNS
  USE   FLOW DISTRIBUTION

FLOW RATE
  USE   FLOW VELOCITY

FLOW REGULATORS
  GS   REGULATORS
      . **FLOW REGULATORS**
      . . FUEL FLOW REGULATORS
  RT   FLOWMETERS
      OXYGEN REGULATORS
      PRESSURE REGULATORS

FLOW RESISTANCE
  GS   FRICTION
      . **FLOW RESISTANCE**
      . . FRICTION DRAG
      . . . SUPERSONIC DRAG
      . . . VISCOUS DRAG
  RT   EDDY VISCOSITY
      ∞HIGH RESISTANCE
      ∞LOW RESISTANCE
      ∞RESISTANCE
      SKIN FRICTION
      VISCOSITY

FLOW SEPARATION
  USE   BOUNDARY LAYER SEPARATION
      SEPARATED FLOW

FLOW STABILITY
  UF   HYDRODYNAMIC STABILITY
  GS   DYNAMIC CHARACTERISTICS
      . DYNAMIC STABILITY
      . . MOTION STABILITY
      . . . **FLOW STABILITY**
      . . . . BOUNDARY LAYER STABILITY
      . . . . FLAME STABILITY
      . . . . MAGNETOHYDRODYNAMIC
            STABILITY
      . . . . . WEIBEL INSTABILITY
      . FLOW CHARACTERISTICS
      . **FLOW STABILITY**
      . . . BOUNDARY LAYER STABILITY
      . . . FLAME STABILITY
      . . . MAGNETOHYDRODYNAMIC
            STABILITY
      . . . . WEIBEL INSTABILITY
      STABILITY
      . DYNAMIC STABILITY
      . MOTION STABILITY
      . . **FLOW STABILITY**
      . . . . BOUNDARY LAYER STABILITY
      . . . . FLAME STABILITY
      . . . . MAGNETOHYDRODYNAMIC
            STABILITY
  RT   AERODYNAMIC STABILITY
      BAROCLINIC INSTABILITY
      DIRECTIONAL STABILITY
      FLUID FLOW
      GOERTLER INSTABILITY
      HYDRODYNAMIC EQUATIONS
      HYDROFOIL OSCILLATIONS
      KELVIN-HELMHOLTZ INSTABILITY
      LAMINAR FLOW
      LATERAL STABILITY
      LONGITUDINAL STABILITY
      LOW SPEED STABILITY
      ORR-SOMMERFELD EQUATIONS
      ROTARY STABILITY
      STEADY FLOW
      STROUHAL NUMBER
      SUPERSONIC DIFFUSERS
      SYSTEMS STABILITY
      TURBULENT FLOW
      UNSTEADY FLOW
      VISCOUS FLUIDS
      VON KARMAN EQUATION
      VORTEX BREAKDOWN
      VORTEX FILAMENTS
      VORTICES
      VORTICITY

FLOW THEORY
  GS   **FLOW THEORY**
      . MIXING LENGTH FLOW THEORY
  RT   AERODYNAMICS
      BOUNDARY LAYER EQUATIONS
      CONTINUUM MECHANICS
      DISLOCATIONS (MATERIALS)
      FLUID FLOW

FLOW THEORY-*(CONT.)*
      FLUID MECHANICS
      HYDRODYNAMIC EQUATIONS
      HYDRODYNAMICS
      LIGHTHILL METHOD
      MASS FLOW
      NAVIER-STOKES EQUATION
      ORR-SOMMERFELD EQUATIONS
      PANEL METHOD (FLUID DYNAMICS)
      PNEUMATICS
      RHEOLOGY
      SOLIDS FLOW
      ∞THEORIES

FLOW VELOCITY
  UF   FLOW RATE
  GS   DYNAMIC CHARACTERISTICS
      . FLOW CHARACTERISTICS
      . . **FLOW VELOCITY**
      RATES (PER TIME)
      . **FLOW VELOCITY**
      . . SOLAR WIND VELOCITY
      VELOCITY
      . **FLOW VELOCITY**
      . . SOLAR WIND VELOCITY
  RT   DISCHARGE COEFFICIENT
      EXHAUST VELOCITY
      ∞FLOW
      HYDRODYNAMIC COEFFICIENTS
      HYPERSONIC FLOW
      HYPERVELOCITY FLOW
      LASER ANEMOMETERS
      LOW SPEED
      MASS FLOW RATE
      PARALLEL FLOW
      SUBSONIC FLOW
      SUPERSONIC FLOW
      TRANSONIC FLOW
      UNSTEADY FLOW
      VELOCITY DISTRIBUTION
      VELOCITY MEASUREMENT
      VORTEX PRECESSION

FLOW VISUALIZATION
  UF   VISUALIZATION OF FLOW
  GS   **FLOW VISUALIZATION**
      . NUMERICAL FLOW VISUALIZATION
  RT   FLUID FLOW
      HYDRAULIC ANALOGIES
      SCHLIEREN PHOTOGRAPHY
      SHADOWGRAPH PHOTOGRAPHY
      WATER TUNNEL TESTS
      WIND TUNNEL MODELS

FLOWMETERS
  GS   MEASURING INSTRUMENTS
      . **FLOWMETERS**
      . . GAS METERS
      . . HOT-WIRE FLOWMETERS
      . . RHEOMETERS
  RT   ELECTRICAL MEASUREMENT
      FLOW MEASUREMENT
      FLOW REGULATORS
      FLUID FLOW
      FUEL GAGES
      HOT-WIRE ANEMOMETERS
      MECHANICAL MEASUREMENT
      ORIFICES
      PITOT TUBES
      PRESSURE GAGES
      PRESSURE MEASUREMENT
      SONIC ANEMOMETERS
      SPEED INDICATORS
      TURBINE INSTRUMENTS
      VELOCITY MEASUREMENT
      VENTURI TUBES
      VORTEX PRECESSION

FLOX
  UF   FLUORINE-LIQUID OXYGEN
  GS   LIQUIDS
      . CRYOGENIC FLUIDS
      . . **FLOX**
      OXIDIZERS
      . ROCKET OXIDIZERS
      . . **FLOX**
  RT   FLUORINE
      LIQUID OXYGEN

FLTSATCOM
  USE   FLEET SATELLITE COMMUNICATION
      SYSTEM

FLUCTUATION
  USE   VARIATIONS

FLUCTUATION THEORY
  RT   HOMOGENEOUS TURBULENCE
      STATISTICAL MECHANICS
      ∞THEORIES

FLUE GASES
  GS   GASES
      . EXHAUST GASES
      . . **FLUE GASES**
  RT   AIR POLLUTION
      COMBUSTION PRODUCTS
      DESULFURIZING
      ELECTRIC POWER PLANTS
      FLUES
      POLLUTION CONTROL
      SCRUBBERS

FLUENCE
  RT   HEALTH PHYSICS
      IONIZING RADIATION
      RADIATION COUNTERS

FLUERICS
  GS   FLUIDICS
      . **FLUERICS**
  RT   FLUID AMPLIFIERS
      FLUID SWITCHING ELEMENTS
      FLUIDIC CIRCUITS
      HYDRAULIC ANALOGIES

FLUES
  RT   CHIMNEYS
      DRAFT (GAS FLOW)
      DUCTS
      EXHAUST SYSTEMS
      FLUE GASES
      VENTS

FLUID AMPLIFICATION
  USE   FLUID AMPLIFIERS

FLUID AMPLIFIERS
  UF   FLUID AMPLIFICATION
      FLUID JET AMPLIFIERS
  GS   AMPLIFIERS
      . **FLUID AMPLIFIERS**
      . . JET AMPLIFIERS
  RT   AMPLIFICATION
      AUTOMATIC CONTROL VALVES
      BOUNDARY LAYER CONTROL
      COANDA EFFECT
      CONVERGENT NOZZLES
      FLUERICS
      FLUIDIC CIRCUITS
      FLUIDICS
      HYDRAULIC EQUIPMENT
      PNEUMATIC EQUIPMENT
      PRESSURE RECOVERY
      TURBULENT FLOW
      TURBULENT JETS
      WALL JETS

FLUID BOUNDARIES
  GS   BOUNDARIES
      . **FLUID BOUNDARIES**
      . . GAS-SOLID INTERFACES
      . . JET BOUNDARIES
      . . LIQUID-LIQUID INTERFACES
      . . LIQUID-SOLID INTERFACES
      . . LIQUID-VAPOR INTERFACES
      INTERFACES
      . **FLUID BOUNDARIES**
      . . GAS-SOLID INTERFACES
      . . JET BOUNDARIES
      . . LIQUID-LIQUID INTERFACES
      . . LIQUID-SOLID INTERFACES
      . . LIQUID-VAPOR INTERFACES
  RT   BACKWARD FACING STEPS
      BOUNDARY LAYERS
      FREE BOUNDARIES
      HEAT TRANSFER
      INTERFACE STABILITY
      LIQUID LEVELS
      LIQUID SURFACES
      PRESSURE GRADIENTS

FLUID DYNAMICS
  UF   CASCADES (FLUID DYNAMICS)
  GS   FLUID MECHANICS
      . **FLUID DYNAMICS**
      . . COMPUTATIONAL FLUID DYNAMICS
      . . GAS DYNAMICS
      . . . AERODYNAMICS
      . . . . AEROTHERMODYNAMICS
      . . . . . HYPERSONICS

FLUID DYNAMICS-(CONT.)
```
          . . . . ROTOR AERODYNAMICS
          . . . . SUPERSONICS
          . . . INTERACTIONAL AERODYNAMICS
          . . . RAREFIED GAS DYNAMICS
          . . HYDRODYNAMICS
          . . . ELASTOHYDRODYNAMICS
          . . . ELECTROHYDRODYNAMICS
          . . . MAGNETOHYDRODYNAMICS
          . . . . CYLINDRICAL PLASMAS
          . . ROTONS
          . . VORTEX SHEDDING
RT        CONTINUITY EQUATION
          CONVECTION
          CROSS FLOW
        ∞DYNAMICS
          EYRING THEORY
          FLUID MANAGEMENT
          GAS-SOLID INTERACTIONS
          GEOPHYSICAL FLUIDS
          GLIMM METHOD
        ∞HYDRAULICS
          HYDROMECHANICS
          KINETICS
          LOW DENSITY FLOW
          MAGNUS EFFECT
        ∞MECHANICS (PHYSICS)
          OCEAN DYNAMICS
          PANEL METHOD (FLUID DYNAMICS)
          PARALLEL FLOW
          PISTON THEORY
          PRIMITIVE EQUATIONS
          QUASI-STEADY STATES
        ∞SCIENCE
          SLAMMING
          STAGNATION POINT
          STEADY STATE
          STREAMLINING
          THERMOHYDRAULICS
          TURBULENCE
          TURBULENT FLOW
          UNIFORM FLOW
          UNSTEADY FLOW
          UNSTEADY STATE
          VORTEX FILAMENTS
```

FLUID FILLED SHELLS
```
GS        SHELLS (STRUCTURAL FORMS)
          . FLUID FILLED SHELLS
          . . LIQUID FILLED SHELLS
RT        HYDRODYNAMIC RAM EFFECT
          PROPELLANT TANKS
          REINFORCED SHELLS
          SHELL STABILITY
        ∞STORAGE
          TANKS (CONTAINERS)
        ∞VESSELS
```

FLUID FILMS
```
GS        FLUID FILMS
          . SQUEEZE FILMS
RT      ∞FILMS
          GAS BEARINGS
          LIQUID-SOLID INTERFACES
```

FLUID FILTERS
```
UF        MASS FILTERS
          PARTICULATE FILTERS
GS        SEPARATORS
          . FLUID FILTERS
          . . AIR FILTERS
RT        CENTRIFUGES
          CONCENTRATORS
        ∞FILTERS
          FILTRATION
          FLUIDIZED BED PROCESSORS
          SIEVES
          SIZING SCREENS
```

FLUID FLOW
```
UF        INDUCED FLUID FLOW
          ROTATIONAL FLOW
GS        FLUID FLOW
          . ADIABATIC FLOW
          . AXIAL FLOW
          . AXISYMMETRIC FLOW
          . . ANNULAR FLOW
          . . KARMAN-BODEWADT FLOW
          . BAROTROPIC FLOW
          . BASE FLOW
          . BELTRAMI FLOW
          . BLOOD FLOW
          . CAPILLARY FLOW
          . CASCADE FLOW
          . CHANNEL FLOW
```

FLUID FLOW-(CONT.)
```
          . OPEN CHANNEL FLOW
          . COAXIAL FLOW
          . COMBUSTIBLE FLOW
          . COMPRESSIBLE FLOW
          . . TRANSONIC FLOW
          . CONICAL FLOW
          . CONVECTIVE FLOW
          . . RAYLEIGH-BENARD CONVECTION
            . . BENARD CELLS
          . CORE FLOW
          . CORNER FLOW
          . COUNTERFLOW
          . CRITICAL FLOW
          . CROSS FLOW
          . DUCTED FLOW
          . . KNUDSEN FLOW
          . FREE FLOW
          . FUEL FLOW
          . . PROPELLANT TRANSFER
          . GAS FLOW
          . . AIR FLOW
          . . . AIR CURRENTS
          . . . . JET STREAMS (METEOROLOGY)
          . . . . MERIDIONAL FLOW
          . . . . VERTICAL AIR CURRENTS
          . . CONTINUUM FLOW
          . . EQUILIBRIUM FLOW
          . . . FROZEN EQUILIBRIUM FLOW
          . . . SHIFTING EQUILIBRIUM FLOW
          . . FREE MOLECULAR FLOW
          . . KNUDSEN FLOW
          . . MOLECULAR FLOW
          . . SLIP FLOW
          . . TRANSITION FLOW
          . . NONEQUILIBRIUM FLOW
          . . PIPE FLOW
          . HEAD (FLUID MECHANICS)
          . . HEAD FLOW
          . . PRESSURE HEADS
          . HELICAL FLOW
          . HYPERSONIC FLOW
          . HYPERVELOCITY FLOW
          . INCOMPRESSIBLE FLOW
          . . STOKES FLOW
          . INLET FLOW
          . INVISCID FLOW
          . . STAGNATION FLOW
          . ISOTHERMAL FLOW
          . JET FLOW
          . . AIR JETS
          . . PERIPHERAL JET FLOW
          . . SUPERSONIC JET FLOW
          . JET MIXING FLOW
          . LAMINAR FLOW
          . . BLASIUS FLOW
          . . HARTMANN FLOW
          . . STRATIFIED FLOW
          . LIQUID FLOW
          . . OPEN CHANNEL FLOW
          . . WATER FLOW
          . MAGNETOHYDRODYNAMIC FLOW
          . MASS FLOW
          . MULTIPHASE FLOW
          . . TWO PHASE FLOW
          . NONNEWTONIAN FLOW
          . NONUNIFORM FLOW
          . NOZZLE FLOW
          . ONE DIMENSIONAL FLOW
          . ORIFICE FLOW
          . OUTLET FLOW
          . PARALLEL FLOW
          . . PIPE FLOW
          . . THREE DIMENSIONAL FLOW
          . PLASTIC FLOW
          . . TRESCA FLOW
          . POTENTIAL FLOW
          . . EQUIPOTENTIALS
          . RADIAL FLOW
          . RECIRCULATIVE FLUID FLOW
          . REVERSED FLOW
          . SHEAR FLOW
          . SINGLE-PHASE FLOW
          . SMALL PERTURBATION FLOW
          . SOLIDS FLOW
          . STEADY FLOW
          . . COUETTE FLOW
          . . HARTMANN FLOW
          . STEAM FLOW
          . SUBCRITICAL FLOW
          . SUBSONIC FLOW
          . SUPERCRITICAL FLOW
          . SUPERSONIC FLOW
          . TURBULENT FLOW
          . . CAVITATION FLOW
          . . SUPERCAVITATING FLOW
```

FLUID FLOW-(CONT.)
```
          . TWO DIMENSIONAL FLOW
          . . COUETTE FLOW
          . UNIFORM FLOW
          . . BLASIUS FLOW
          . UNSTEADY FLOW
          . . OSCILLATING FLOW
          . VISCOUS FLOW
          . . BOUNDARY LAYER FLOW
          . . . REATTACHED FLOW
          . . . SECONDARY FLOW
          . . . SEPARATED FLOW
          . . . . BOUNDARY LAYER SEPARATION
          . . COUETTE FLOW
          . . KARMAN-BODEWADT FLOW
          . . STOKES FLOW
          . WALL FLOW
          . WEDGE FLOW
RT        ACOUSTIC STREAMING
          ANNULAR DUCTS
          BERNOULLI THEOREM
          BOUNDARY LAYERS
          CANALS
          CARTAN SPACE
          CHEMICAL ENGINEERING
          COAXIAL NOZZLES
        ∞CONDUCTIVITY
          CONVECTION CURRENTS
        ∞CURRENTS
          DIMENSIONAL ANALYSIS
          DIMENSIONLESS NUMBERS
          DRAG REDUCTION
          DUCT GEOMETRY
          EXPULSION
        ∞FLOW
          FLOW DISTORTION
          FLOW EQUATIONS
          FLOW MEASUREMENT
          FLOW STABILITY
          FLOW THEORY
          FLOW VISUALIZATION
          FLOWMETERS
          FLUIDICS
        ∞FLUIDS
          FRICTION
          FROUDE NUMBER
          GEOPHYSICAL FLUID FLOW CELLS
          HEAD (FLUID MECHANICS)
          HEAT TRANSMISSION
        ∞HYDRAULICS
          HYDRODYNAMICS
          HYDROMECHANICS
          INJECTION
          LABYRINTH SEALS
          LEAKAGE
          LEWIS NUMBERS
          MAGNUS EFFECT
          MANNING THEORY
          MATERIALS HANDLING
          MECHANICAL ENGINEERING
          OCEAN CURRENTS
          OCEAN SURFACE
          PIPES (TUBES)
          PLANETARY WAVES
          PRESSURE GRADIENTS
          PRESSURE HEADS
          RAYLEIGH WAVES
          REYNOLDS NUMBER
          SKIN FRICTION
          STREAMS
          SUPERSONIC BOUNDARY LAYERS
          SURGES
          SYRINGES
          THERMOHYDRAULICS
          ULTRASONIC CLEANING
          VORTICES
          WING FLOW METHOD TESTS
```

FLUID INJECTION
```
GS        INJECTION
          . FLUID INJECTION
          . . GAS INJECTION
          . . LIQUID INJECTION
          . . . DEEP WELL INJECTION (WASTES)
          . . . WATER INJECTION
RT        CHANNEL FLOW
          FUEL INJECTION
          INLET FLOW
          LAMINAR MIXING
          NOZZLE FLOW
          SECONDARY INJECTION
```

FLUID JET AMPLIFIERS
```
USE       FLUID AMPLIFIERS
          JET AMPLIFIERS
```

FLUID JETS
GS   **FLUID JETS**
    . AIR JETS
    . FREE JETS
    . GAS JETS
    . HYDRAULIC JETS
    . VAPOR JETS
RT   JET AMPLIFIERS
    JET FLOW
    JET MIXING FLOW
    JET STREAMS (METEOROLOGY)
    ∞JETS
    PLASMA JETS

FLUID LOGIC
RT   COMPUTER DESIGN
    FLUIDIC CIRCUITS
    FLUIDICS
    ∞LOGIC
    LOGIC CIRCUITS

FLUID MANAGEMENT
RT   CRYOGENIC FLUID STORAGE
    CRYOGENIC FLUIDS
    CRYOGENIC ROCKET PROPELLANTS
    FLUID DYNAMICS
    FUEL CONTROL
    REDUCED GRAVITY

FLUID MECHANICS
GS   **FLUID MECHANICS**
    . FLUID DYNAMICS
    . . COMPUTATIONAL FLUID DYNAMICS
    . . GAS DYNAMICS
    . . . AERODYNAMICS
    . . . . AEROTHERMODYNAMICS
    . . . . HYPERSONICS
    . . . . ROTOR AERODYNAMICS
    . . . . SUPERSONICS
    . . . INTERACTIONAL AERODYNAMICS
    . . . RAREFIED GAS DYNAMICS
    . . HYDRODYNAMICS
    . . . ELASTOHYDRODYNAMICS
    . . . ELECTROHYDRODYNAMICS
    . . . MAGNETOHYDRODYNAMICS
    . . . . CYLINDRICAL PLASMAS
    . . ROTONS
    . VORTEX SHEDDING
    . HYDROMECHANICS
    . . HYDRODYNAMICS
    . . ELASTOHYDRODYNAMICS
    . . . ELECTROHYDRODYNAMICS
    . . . MAGNETOHYDRODYNAMICS
    . . HYDROSTATICS
    . . . MAGNETOHYDROSTATICS
    . PNEUMATICS
RT   AEROSTATICS
    ∞BLEEDING
    CONTINUUM MECHANICS
    DIFFUSIVITY
    ∞DYNAMICS
    FLAT PLATES
    FLOW THEORY
    FLUIDICS
    ∞HYDRAULICS
    HYDRODYNAMIC EQUATIONS
    INCOMPRESSIBILITY
    KINETICS
    MAXWELL FLUIDS
    ∞MECHANICS (PHYSICS)
    MICROPOLAR FLUIDS
    ∞SCIENCE
    STATICS
    SUPERCRITICAL FLUIDS
    THERMODYNAMICS

FLUID POWER
RT   COMPRESSIBLE FLUIDS
    FLUID PRESSURE
    FLUIDIC CIRCUITS
    FLUIDICS
    HYDRAULIC CONTROL
    HYDRAULIC EQUIPMENT
    ∞HYDRAULICS
    HYDRODYNAMICS
    INCOMPRESSIBLE FLUIDS
    PNEUMATIC CONTROL
    PNEUMATIC EQUIPMENT
    PNEUMATICS
    ∞POWER
    ∞PRESSURE DROP
    WORKING FLUIDS

FLUID PRESSURE
GS   PRESSURE

FLUID PRESSURE-*(CONT.)*
    . **FLUID PRESSURE**
    . . WATER PRESSURE
RT   BETA FACTOR
    FLUID POWER
    FLUIDICS
    ∞FLUIDS
    HYDRAULIC FLUIDS

FLUID ROTOR GYROSCOPES
GS   GYROSCOPES
    . ROTARY GYROSCOPES
    . . **FLUID ROTOR GYROSCOPES**
RT   FLOTATION
    GIMBALS

FLUID SWITCHING ELEMENTS
GS   CIRCUITS
    . SWITCHING CIRCUITS
    . . **FLUID SWITCHING ELEMENTS**
    SWITCHES
    . SWITCHING CIRCUITS
    . . **FLUID SWITCHING ELEMENTS**
RT   ACOUSTIC STREAMING
    AUTOMATIC CONTROL VALVES
    FLIP-FLOPS
    FLUERICS
    FLUIDIC CIRCUITS
    FLUIDICS
    HYDRAULIC EQUIPMENT
    PNEUMATIC EQUIPMENT

FLUID TRANSMISSION LINES
GS   TRANSMISSION LINES
    . **FLUID TRANSMISSION LINES**
RT   HYDRAULIC FLUIDS
    TRANSMISSION FLUIDS
    WORKING FLUIDS

FLUID TRANSPIRATION
USE   TRANSPIRATION

FLUID-SOLID INTERACTIONS
GS   **FLUID-SOLID INTERACTIONS**
    . GAS-SOLID INTERACTIONS
    . . GAS-METAL INTERACTIONS
RT   GAS-SOLID INTERFACES
    ∞INTERACTIONS
    LIQUID-SOLID INTERFACES
    SURFACE REACTIONS

FLUIDIC CIRCUITS
GS   CIRCUITS
    . **FLUIDIC CIRCUITS**
    . . FLIP-FLOPS
RT   FLUERICS
    FLUID AMPLIFIERS
    FLUID LOGIC
    FLUID POWER
    FLUID SWITCHING ELEMENTS
    FLUIDICS
    FLY BY TUBE CONTROL

FLUIDICS
GS   **FLUIDICS**
    . FLUERICS
RT   AMPLIFICATION
    ∞CONTROL
    FLUID AMPLIFIERS
    FLUID FLOW
    FLUID LOGIC
    FLUID MECHANICS
    FLUID POWER
    FLUID PRESSURE
    FLUID SWITCHING ELEMENTS
    FLUIDIC CIRCUITS
    HYDRAULIC ANALOGIES
    HYDRAULIC CONTROL
    ∞LOGIC
    PNEUMATIC CIRCUITS
    PNEUMATIC CONTROL
    PNEUMATIC EQUIPMENT
    PNEUMATICS

FLUIDIZED BED PROCESSORS
RT   BEDS (PROCESS ENGINEERING)
    CHEMICAL REACTORS
    FLUID FILTERS
    FURNACES
    SEPARATORS

∞ **FLUIDS**
SN   *(USE OF A MORE SPECIFIC TERM IS*
    RECOMMENDED--CONSULT THE TERMS
    LISTED BELOW)
RT   ANISOTROPIC FLUIDS
    BINARY FLUIDS
    BODY FLUIDS
    CEREBROSPINAL FLUID
    COMPRESSIBLE FLUIDS
    CRYOGENIC FLUIDS
    FERROFLUIDS
    FLUID FLOW
    FLUID PRESSURE
    GASES
    GYROSCOPE FLUIDS
    HIGH TEMPERATURE FLUIDS
    HYDRAULIC FLUIDS
    IDEAL FLUIDS
    INCOMPRESSIBLE FLUIDS
    LIQUIDS
    MAXWELL FLUIDS
    MICROPOLAR FLUIDS
    NEWTONIAN FLUIDS
    NONEQUILIBRIUM FLOW
    NONNEWTONIAN FLUIDS
    RHEOLOGY
    ROTATING FLUIDS
    SERUMS
    SIPHONING
    SOLIDS
    SUPERCRITICAL FLUIDS
    SUPERFLUIDITY
    TRANSMISSION FLUIDS
    VISCOUS FLUIDS
    WEIGHTLESS FLUIDS
    WORKING FLUIDS

FLUORESCENCE
UF   FLUORESCENT EMISSION
GS   DECAY
    . EMISSION
    . . LIGHT EMISSION
    . . . LUMINESCENCE
    . . . . **FLUORESCENCE**
    . . . . . PHOSPHORESCENCE
    . . . . . RESONANCE FLUORESCENCE
    . . . . . X RAY FLUORESCENCE
RT   ELECTROMAGNETIC ABSORPTION
    EXTINCTION
    MOSSBAUER EFFECT
    PHOSPHORS
    PHOTOLUMINESCENCE
    PLASMA RADIATION
    TRIBOLUMINESCENCE

FLUORESCENT EMISSION
USE   FLUORESCENCE

FLUORIDES
GS   HALOGEN COMPOUNDS
    . FLUORINE COMPOUNDS
    . . **FLUORIDES**
    . . . ANTIMONY FLUORIDES
    . . . BARIUM FLUORIDES
    . . . BORON FLUORIDES
    . . . CHLORINE FLUORIDES
    . . . COMPOUND A
    . . . CRYOLITE
    . . . DIFLUORIDES
    . . . . CALCIUM FLUORIDES
    . . . . . FLUORSPAR
    . . . HYDROFLUORIC ACID
    . . . METAL FLUORIDES
    . . . . ALUMINUM FLUORIDES
    . . . . BERYLLIUM FLUORIDES
    . . . . CADMIUM FLUORIDES
    . . . . CALCIUM FLUORIDES
    . . . . CESIUM FLUORIDES
    . . . . CHROMIUM FLUORIDES
    . . . . COBALT FLUORIDES
    . . . . COPPER FLUORIDES
    . . . . LANTHANUM FLUORIDES
    . . . . LITHIUM FLUORIDES
    . . . . MAGNESIUM FLUORIDES
    . . . . NICKEL FLUORIDES
    . . . . PLUTONIUM FLUORIDES
    . . . . PROTACTINIUM FLUORIDES
    . . . . SODIUM FLUORIDES
    . . . . STRONTIUM FLUORIDES
    . . . . THORIUM FLUORIDES
    . . . . TUNGSTEN FLUORIDES
    . . . . URANIUM FLUORIDES
    . . . . ZINC FLUORIDES
    . . . NITROGEN FLUORIDES
    . . . OXYFLUORIDES

FLUORIDES-(CONT.)
```
        . . . OXYGEN FLUORIDES
        . . . OZONE FLUORIDE
        . . . PERCHLORYL FLUORIDES
        . . . SULFUR FLUORIDES
        . . . TECHNETIUM FLUORIDES
        . . HALIDES
        . . FLUORIDES
        . . . ANTIMONY FLUORIDES
        . . BARIUM FLUORIDES
        . . BORON FLUORIDES
        . . . CHLORINE FLUORIDES
        . . . DIFLUORIDES
        . . . . CALCIUM FLUORIDES
        . . . . . FLUORSPAR
        . . . HYDROFLUORIC ACID
        . . . NITROGEN FLUORIDES
        . . . OXYFLUORIDES
        . . . OXYGEN FLUORIDES
        . . . OZONE FLUORIDE
        . . . PERCHLORYL FLUORIDES
        . . . SULFUR FLUORIDES
        . . . TECHNETIUM FLUORIDES
RT      EXCIMER LASERS
```

FLUORINATION
```
GS      CHEMICAL REACTIONS
        . HALOGENATION
        . . FLUORINATION
RT      DEFLUORINATION
```

FLUORINE
```
GS      CHEMICAL ELEMENTS
        . HALOGENS
        . . FLUORINE
        . . . FLUORINE ISOTOPES
RT      FLOX
        OXIDIZERS
```

FLUORINE COMPOUNDS
```
GS      HALOGEN COMPOUNDS
        . FLUORINE COMPOUNDS
        . . FLUORIDES
        . . . ANTIMONY FLUORIDES
        . . . BARIUM FLUORIDES
        . . . BORON FLUORIDES
        . . . CHLORINE FLUORIDES
        . . COMPOUND A
        . . . CRYOLITE
        . . . DIFLUORIDES
        . . . . CALCIUM FLUORIDES
        . . . . . FLUORSPAR
        . . . HYDROFLUORIC ACID
        . . . METAL FLUORIDES
        . . . . ALUMINUM FLUORIDES
        . . . . BERYLLIUM FLUORIDES
        . . . . CADMIUM FLUORIDES
        . . . . CALCIUM FLUORIDES
        . . . . CESIUM FLUORIDES
        . . . . CHROMIUM FLUORIDES
        . . . . COBALT FLUORIDES
        . . . . COPPER FLUORIDES
        . . . . LANTHANUM FLUORIDES
        . . . . LITHIUM FLUORIDES
        . . . . MAGNESIUM FLUORIDES
        . . . . NICKEL FLUORIDES
        . . . . PLUTONIUM FLUORIDES
        . . . . PROTACTINIUM FLUORIDES
        . . . . SODIUM FLUORIDES
        . . . . STRONTIUM FLUORIDES
        . . . . THORIUM FLUORIDES
        . . . . TUNGSTEN FLUORIDES
        . . . URANIUM FLUORIDES
        . . . . ZINC FLUORIDES
        . . . NITROGEN FLUORIDES
        . . OXYFLUORIDES
        . . . OXYGEN FLUORIDES
        . . OZONE FLUORIDE
        . . PERCHLORYL FLUORIDES
        . . SULFUR FLUORIDES
        . . . TECHNETIUM FLUORIDES
        . . FLUORITE
        . . FLUORO COMPOUNDS
        . . . CRYOLITE
        . . DIFLUORO COMPOUNDS
        . . . PERFLUOROALKANE
        . . . . POLYTETRAFLUOROETHYLENE
        . . FLUORINE ORGANIC COMPOUNDS
        . . . . FLUOROAMINES
        . . . . . NITROFLUORAMINES
        . . . . . . TRIFLUOROAMINE OXIDE
        . . . . . FLUOROCARBONS
        . . . . . FLUOROHYDROCARBONS
        . . . . . . CARBON TETRAFLUORIDE
        . . . . FLUOROPOLYMERS
```

FLUORINE COMPOUNDS-(CONT.)
```
        . . . . KEL-F
        . . . . PERFLUOROALKANE
        . . . . PERFLUOROGUANIDINE
RT      ∞CHEMICAL COMPOUNDS
        HALOCARBONS
```

FLUORINE ISOTOPES
```
GS      CHEMICAL ELEMENTS
        . HALOGENS
        . . FLUORINE
        . . . FLUORINE ISOTOPES
        . NUCLIDES
        . . ISOTOPES
        . . . FLUORINE ISOTOPES
```

FLUORINE ORGANIC COMPOUNDS
```
UF      ORGANIC FLUORINE COMPOUNDS
GS      HALOGEN COMPOUNDS
        . FLUORINE COMPOUNDS
        . FLUORO COMPOUNDS
        . . FLUORINE ORGANIC COMPOUNDS
        . . . FLUOROAMINES
        . . . . NITROFLUORAMINES
        . . . . . TRIFLUOROAMINE OXIDE
        . . . FLUOROCARBONS
        . . . FLUOROHYDROCARBONS
        . . . . CARBON TETRAFLUORIDE
        . . . FLUOROPOLYMERS
        . . . KEL-F
        . . . PERFLUOROALKANE
        . . . PERFLUOROGUANIDINE
        ORGANIC COMPOUNDS
        . FLUORINE ORGANIC COMPOUNDS
        . . FLUOROAMINES
        . . . NITROFLUORAMINES
        . . . TRIFLUOROAMINE OXIDE
        . . FLUOROCARBONS
        . . FLUOROHYDROCARBONS
        . . . CARBON TETRAFLUORIDE
        . . KEL-F
        . . PERFLUOROALKANE
        . . PERFLUOROGUANIDINE
RT      ∞CHEMICAL COMPOUNDS
```

FLUORINE-LIQUID OXYGEN
```
USE     FLOX
```

FLUORITE
```
GS      CALCIUM COMPOUNDS
        . FLUORITE
        HALOGEN COMPOUNDS
        . FLUORINE COMPOUNDS
        . . FLUORITE
        MINERALS
        . FLUORITE
```

FLUORO COMPOUNDS
```
GS      HALOGEN COMPOUNDS
        . FLUORINE COMPOUNDS
        . . FLUORO COMPOUNDS
        . . . CRYOLITE
        . . DIFLUORO COMPOUNDS
        . . . . PERFLUOROALKANE
        . . . . POLYTETRAFLUOROETHYLENE
        . . . FLUORINE ORGANIC COMPOUNDS
        . . . . FLUOROAMINES
        . . . . . NITROFLUORAMINES
        . . . . . . TRIFLUOROAMINE OXIDE
        . . . . FLUOROCARBONS
        . . . . FLUOROHYDROCARBONS
        . . . . . CARBON TETRAFLUORIDE
        . . . . FLUOROPOLYMERS
        . . . KEL-F
        . . . PERFLUOROALKANE
        . . . PERFLUOROGUANIDINE
RT      ∞CHEMICAL COMPOUNDS
        HALOCARBONS
```

FLUOROAMINES
```
GS      ALIPHATIC COMPOUNDS
        . FLUOROAMINES
        . . NITROFLUORAMINES
        . . TRIFLUOROAMINE OXIDE
        AMINES
        . FLUOROAMINES
        . . NITROFLUORAMINES
        . . TRIFLUOROAMINE OXIDE
        HALOGEN COMPOUNDS
        . FLUORINE COMPOUNDS
        . FLUORO COMPOUNDS
        . . FLUORINE ORGANIC COMPOUNDS
        . . . FLUOROAMINES
        . . . . NITROFLUORAMINES
        . . . . . TRIFLUOROAMINE OXIDE
```

FLUOROAMINES-(CONT.)
```
        ORGANIC COMPOUNDS
        . FLUORINE ORGANIC COMPOUNDS
        . . FLUOROAMINES
        . . . NITROFLUORAMINES
        . . . TRIFLUOROAMINE OXIDE
```

FLUOROCARBONS
```
GS      HALOGEN COMPOUNDS
        . FLUORINE COMPOUNDS
        . . FLUORO COMPOUNDS
        . . . FLUORINE ORGANIC COMPOUNDS
        . . . . FLUOROCARBONS
        . . HALOCARBONS
        . . FLUOROCARBONS
        ORGANIC COMPOUNDS
        . FLUORINE ORGANIC COMPOUNDS
        . . FLUOROCARBONS
RT      FLUOROHYDROCARBONS
        FLUOROPOLYMERS
```

FLUOROHYDROCARBONS
```
GS      HALOGEN COMPOUNDS
        . FLUORINE COMPOUNDS
        . . FLUORO COMPOUNDS
        . . . FLUORINE ORGANIC COMPOUNDS
        . . . . FLUOROHYDROCARBONS
        . . . . . CARBON TETRAFLUORIDE
        ORGANIC COMPOUNDS
        . FLUORINE ORGANIC COMPOUNDS
        . . FLUOROHYDROCARBONS
        . . . CARBON TETRAFLUORIDE
RT      FLUOROCARBONS
        FREON
        REFRIGERANTS
        VITON
```

FLUOROMICA
```
USE     FLUOROSILICATES
        MICA
```

FLUOROPHLOGOPITE
```
GS      MINERALS
        . MICA
        . . FLUOROPHLOGOPITE
```

FLUOROPLASTICS
```
USE     FLUOROPOLYMERS
```

FLUOROPOLYMERS
```
UF      FLUOROPLASTICS
GS      CARBON COMPOUNDS
        . FLUOROPOLYMERS
        HALOGEN COMPOUNDS
        . FLUORINE COMPOUNDS
        . . FLUORO COMPOUNDS
        . . . FLUORINE ORGANIC COMPOUNDS
        . . . . FLUOROPOLYMERS
RT      FLUOROCARBONS
        PLASTICS
        ∞POLYMERS
```

FLUOROSCOPY
```
RT      MEDICAL EQUIPMENT
        X RAY ANALYSIS
```

FLUOROSILICATES
```
UF      FLUOROMICA
GS      SILICON COMPOUNDS
        . SILICATES
        . . FLUOROSILICATES
RT      MINERALS
```

FLUORSPAR
```
GS      CALCIUM COMPOUNDS
        . CALCIUM FLUORIDES
        . . FLUORSPAR
        HALOGEN COMPOUNDS
        . FLUORINE COMPOUNDS
        . . FLUORIDES
        . . . DIFLUORIDES
        . . . . CALCIUM FLUORIDES
        . . . . . FLUORSPAR
        . HALIDES
        . . FLUORIDES
        . . . DIFLUORIDES
        . . . . CALCIUM FLUORIDES
        . . . . . FLUORSPAR
        . . METAL HALIDES
        . . . CALCIUM FLUORIDES
        . . . . FLUORSPAR
        MINERALS
        . FLUORSPAR
```

FLUSHING
- RT CLEANING
-   EJECTION
-   ELUTION
-   EXPELLANTS
-   LEACHING
-   PURGING
-   PURIFICATION
-   ∞SEPARATION
-   VENTING
-   WASHING
-   WASTE WATER

FLUTING
- USE GROOVING

FLUTTER
- UF AERODYNAMIC BUZZ
-   AEROMAGNETO FLUTTER
- GS VIBRATION
-   . STRUCTURAL VIBRATION
-   . . **FLUTTER**
-   . . . PANEL FLUTTER
-   . . . SUBSONIC FLUTTER
-   . . . SUPERSONIC FLUTTER
-   . . . TRANSONIC FLUTTER
- RT AERODYNAMIC NOISE
-   AERODYNAMIC STABILITY
-   AEROELASTICITY
-   BENDING
-   BENDING VIBRATION
-   BOUNDARY LAYER CONTROL
-   BUFFETING
-   COMPRESSIBILITY EFFECTS
-   DAST PROGRAM
-   FLAPPING
-   FLIGHT CHARACTERISTICS
-   FORCED VIBRATION
-   HOVERING
-   HYDROFOIL OSCILLATIONS
-   INFLUENCE COEFFICIENT
-   MISSILE VIBRATION
-   RANDOM VIBRATION
-   RESONANT VIBRATION
-   SELF INDUCED VIBRATION
-   SHAKING
-   SPACECRAFT MOTION
-   TURBULENCE EFFECTS
-   UNDAMPED OSCILLATIONS
-   VIBRATION SIMULATORS
-   VIBRATION TESTS
-   VIBRATIONAL STRESS
-   WING OSCILLATIONS

FLUTTER ANALYSIS
- GS STRUCTURAL ANALYSIS
-   . **FLUTTER ANALYSIS**
- RT STRUCTURAL VIBRATION

∞ FLUX
- SN *(USE OF A MORE SPECIFIC TERM IS RECOMMENDED--CONSULT THE TERMS LISTED BELOW)*
- RT FLUX (RATE)
-   FLUX DENSITY
-   FLUX QUANTIZATION
-   FLUXES
-   LEVEL (QUANTITY)

FLUX (RATE PER UNIT AREA)
- USE FLUX DENSITY

FLUX (RATE)
- SN *(THE TOTAL EMANATION OF ENERGY, MATERIAL OR PARTICLES FROM A SINGLE SOURCE PER UNIT TIME--OFTEN USED ERRONEOUSLY IN LIEU OF FLUX DENSITY, WHICH IS ENERGY, MATERIAL OR PARTICLE RATE PER UNIT AREA)*
- UF ELECTRON FLUX
-   NEUTRON FLUX
-   PARTICLE FLUX
- GS RATES (PER TIME)
-   . **FLUX (RATE)**
-   . . HEAT FLUX
-   . . MAGNETIC FLUX
-   . . SOLAR FLUX
- RT BETA PARTICLES
-   BRIGHTNESS
-   CORPUSCULAR RADIATION
-   DOSIMETERS
-   ELECTROMAGNETIC RADIATION
-   EMITTANCE
-   ∞ENERGY
-   FIELD THEORY (PHYSICS)

FLUX (RATE)-*(CONT.)*
-   FLUX DENSITY
-   GAMMA RAYS
-   ∞INTENSITY
-   LEVEL (QUANTITY)
-   LUMINOUS INTENSITY
-   MAGNETIC CIRCUITS
-   MAGNETIC INDUCTION
-   MAGNETOSTATICS
-   PARTICLE BEAMS
-   PARTICLE DIFFUSION
-   ∞POWER
-   RADIANT FLUX DENSITY
-   ∞RADIATION
-   STEFAN-BOLTZMANN LAW

FLUX DENSITY
- SN *(ENERGY, MATERIAL OR PARTICLE RATE PER UNIT AREA--THE QUANTITY USUALLY MEASURED EXCEPT WHEN THE TOTAL EMANATION RATE FROM A SINGLE SOURCE CAN BE DETERMINED)*
- UF DENSITY (RATE/AREA)
-   ENERGY DENSITY
-   FLUX (RATE PER UNIT AREA)
-   FLUX MAPPING
- GS RATES (PER TIME)
-   . **FLUX DENSITY**
-   . . CURRENT DENSITY
-   . . PHOTON DENSITY
-   . . RADIANT FLUX DENSITY
-   . . . IRRADIANCE
-   . . . . ILLUMINANCE
-   . . . . SOLAR CONSTANT
-   . . . LUMENS
-   . . . LUMINOUS INTENSITY
-   . . . . ILLUMINANCE
-   . . . . LUMINANCE
-   . . . PARTICLE FLUX DENSITY
-   . . . . ELECTRON FLUX DENSITY
-   . . . . NEUTRON FLUX DENSITY
-   . . . . PROTON FLUX DENSITY
-   . . . RADIANCE
-   . . . RADIANCY
-   . . . SOLAR FLUX DENSITY
-   . . . . SOLAR CONSTANT
-   . . SOUND INTENSITY
-   . . . ZERO SOUND
- RT ALPHA PARTICLES
-   ANGULAR DISTRIBUTION
-   ATOM CONCENTRATION
-   ∞DENSITY
-   DOSIMETERS
-   ELECTROMAGNETIC RADIATION
-   ∞ENERGY
-   ENERGY DISTRIBUTION
-   FIELD INTENSITY METERS
-   FIELD STRENGTH
-   FIELD THEORY (PHYSICS)
-   FLUX (RATE)
-   GAMMA RAYS
-   HEAT FLUX
-   ∞INTENSITY
-   IRRADIATION
-   LEVEL (QUANTITY)
-   LOUDNESS
-   MASS DISTRIBUTION
-   METEOROID CONCENTRATION
-   ONSAGER PHENOMENOLOGICAL COEFFICIENT
-   ∞POWER
-   POWER SPECTRA
-   PROTONS
-   ∞RADIATION
-   RADIATION DISTRIBUTION
-   RADIATION HAZARDS
-   REMANENCE
-   SCATTERING FUNCTIONS
-   SOLAR MAXIMUM MISSION
-   SOUND PRESSURE
-   SPECTRA
-   X RAY DENSITY MEASUREMENT

FLUX MAPPING
- USE FLUX DENSITY
-   MAPPING

FLUX PINNING
- GS PINNING
-   . **FLUX PINNING**
- RT LINES OF FORCE
-   MAGNETIC FLUX
-   SUPERCONDUCTIVITY
-   TRAPPED MAGNETIC FIELDS
-   TRAPPING

FLUX PUMPS
- RT MAGNETIC COILS
-   MAGNETIC FIELDS
-   SUPERCONDUCTING MAGNETS
-   SUPERCONDUCTIVITY

FLUX QUANTIZATION
- RT ∞FLUX

FLUXES
- RT BRAZING
-   ∞FLUX
-   LIMESTONE
-   SOLDERING
-   WELDING

FLUXMETERS
- USE MAGNETIC MEASUREMENT
-   MEASURING INSTRUMENTS

FLY ASH
- GS ASHES
-   . **FLY ASH**
- RT AIR POLLUTION
-   COAL
-   COMBUSTION PRODUCTS
-   ELECTRIC POWER PLANTS
-   ELECTROSTATIC PRECIPITATORS
-   POLLUTION CONTROL

FLY BY TUBE CONTROL
- GS FLIGHT CONTROL
-   . **FLY BY TUBE CONTROL**
- RT AIRCRAFT CONTROL
-   ∞CONTROL
-   FLUIDIC CIRCUITS
-   HYDRAULIC EQUIPMENT
-   SERVOAMPLIFIERS

FLY BY WIRE CONTROL
- UF ELECTRIC AIRCRAFT
- GS FLIGHT CONTROL
-   . **FLY BY WIRE CONTROL**
- RT AIRCRAFT CONTROL
-   ∞CONTROL
-   GROUND BASED CONTROL
-   SPACECRAFT CONTROL

FLYBY MISSIONS
- GS **FLYBY MISSIONS**
-   . ASTEROID MISSIONS
-   . GIOTTO MISSION
-   . GRAND TOURS
-   . . MARINER JUPITER-SATURN FLYBY
-   . . MARINER JUPITER-URANUS FLYBY
-   . . VOYAGER 1977 MISSION
-   . MARINER VENUS-MERCURY 1973
-   . MARINER-MERCURY 1973
- RT GALILEO PROJECT
-   GALILEO SPACECRAFT
-   INTERPLANETARY FLIGHT
-   LONG DURATION SPACE FLIGHT
-   LUNAR FLIGHT
-   MARINER MARK 2 SPACECRAFT
-   MARINER PROGRAM
-   ∞MISSIONS
-   OUTER PLANETS EXPLORERS
-   SPACE FLIGHT
-   SPACE MISSIONS
-   SWINGBY TECHNIQUE
-   TOPS (SPACECRAFT)
-   VEGA PROJECT
-   VOYAGER 1 SPACECRAFT
-   VOYAGER 2 SPACECRAFT

FLYING
- USE FLIGHT

FLYING BEDSTEAD AIRCRAFT
- USE FLYING PLATFORMS

FLYING CRANE HELICOPTER
- USE H-17 HELICOPTER

FLYING EJECTION SEATS
- GS ONBOARD EQUIPMENT
-   . AIRCRAFT EQUIPMENT
-   . . **FLYING EJECTION SEATS**
-   SAFETY DEVICES
-   . EJECTION SEATS
-   . . **FLYING EJECTION SEATS**
-   SEATS
-   . EJECTION SEATS
-   . . **FLYING EJECTION SEATS**

FLYING EJECTION SEATS-*(CONT.)*
RT　ABORT APPARATUS
　　AIRCRAFT SAFETY
　　BAILOUT
　　COCKPITS
　　EJECTORS
　　ESCAPE CAPSULES
　　FLIGHT SAFETY
　　JET ENGINES
　　PROTECTION

FLYING PERSONNEL
GS　PERSONNEL
　　. **FLYING PERSONNEL**
　　. . ASTRONAUTS
　　. . . ORBITAL WORKERS
　　. . COSMONAUTS
　　. . FLIGHT CREWS
　　. . . SPACECREWS
　　. . PILOTS (PERSONNEL)
　　. . . AIRCRAFT PILOTS
　　. . . . TEST PILOTS
RT　FLIGHT FITNESS
　　FLIGHT TRAINING
　　NAVIGATORS

FLYING PLATFORM STABILITY
USE　AERODYNAMIC STABILITY
　　FLYING PLATFORMS

FLYING PLATFORMS
UF　FLYING BEDSTEAD AIRCRAFT
　　FLYING PLATFORM STABILITY
GS　V/STOL AIRCRAFT
　　. VERTICAL TAKEOFF AIRCRAFT
　　. . **FLYING PLATFORMS**
RT　∞AIRCRAFT
　　GROUND EFFECT MACHINES
　　JET AIRCRAFT
　　OBSERVATION AIRCRAFT
　　∞PLATFORMS
　　RECONNAISSANCE AIRCRAFT
　　RESEARCH AIRCRAFT
　　∞SUBSONIC AIRCRAFT
　　VZ-8 AIRCRAFT

FLYING QUALITIES
USE　FLIGHT CHARACTERISTICS

FLYING SPOT SCANNERS
GS　SCANNERS
　　. OPTICAL SCANNERS
　　. . **FLYING SPOT SCANNERS**
RT　DISPLAY DEVICES
　　ELECTRON GUNS
　　ELECTRON OPTICS
　　IMAGE TUBES
　　OSCILLOSCOPES
　　PHOTOTUBES
　　PICTURE TUBES
　　TELEVISION EQUIPMENT
　　VIDEO EQUIPMENT

FLYING WING AIRCRAFT
USE　TAILLESS AIRCRAFT

FLYWHEELS
GS　ROTATING BODIES
　　. ROTORS
　　. . **FLYWHEELS**
　　WHEELS
　　. **FLYWHEELS**
RT　BALANCING
　　COUNTER-ROTATING WHEELS
　　ENERGY STORAGE
　　ENGINE PARTS
　　MECHANICAL ENGINEERING
　　REACTION WHEELS

FM/PM (MODULATION)
GS　MODULATION
　　. FREQUENCY MODULATION
　　. . **FM/PM (MODULATION)**
　　. PHASE MODULATION
　　. . **FM/PM (MODULATION)**
RT　DATA TRANSMISSION

FOAMING
RT　BENEFICIATION
　　FLOTATION
　　FOAMS
　　METAL FOAMS
　　∞SEPARATION
　　SURFACE PROPERTIES

FOAMING-*(CONT.)*
　　SWIRLING
　　WETTING

FOAMS
UF　CELLULAR MATERIALS (NON
　　　BIOLOGICAL)
GS　**FOAMS**
　　. METAL FOAMS
RT　BUBBLES
　　COLLOIDS
　　EXPLOSION SUPPRESSION
　　FOAMING
　　LOW DENSITY MATERIALS
　　∞MATERIALS
　　POLYURETHANE FOAM
　　STYROFOAM (TRADEMARK)

FOCI
GS　**FOCI**
　　. PLASMA FOCUS
RT　∞CENTERS
　　FOCUSING
　　GEOMETRY
　　LOCI
　　∞OPTICS
　　POINTS (MATHEMATICS)
　　RESOLUTION

FOCUSING
GS　**FOCUSING**
　　. DEFOCUSING
　　. PREFOCUSING
　　. SELF FOCUSING
RT　ACCOMMODATION
　　ADJUSTING
　　ASTIGMATISM
　　CAMERAS
　　DISTANCE
　　FOCI
　　FRESNEL LENSES
　　GEOMETRICAL OPTICS
　　GRAVITATIONAL LENSES
　　IMAGE CONTRAST
　　IMAGE ENHANCEMENT
　　LASER CUTTING
　　LASER DRILLING
　　LENSES
　　PANORAMIC CAMERAS
　　SOLAR REFLECTORS
　　STEERING
　　STIGMATISM
　　VIGNETTING

FOETUSES
USE　FETUSES

FOG
GS　MIXTURES
　　. DISPERSIONS
　　. . COLLOIDS
　　. . . AEROSOLS
　　. . . . **FOG**
　　. . LIQUID-GAS MIXTURES
　　. . . AEROSOLS
　　. . . . **FOG**
　　PARTICLES
　　. AEROSOLS
　　. . **FOG**
RT　ANVIL CLOUDS
　　CIRROCUMULUS CLOUDS
　　CIRROSTRATUS CLOUDS
　　CLOUDS (METEOROLOGY)
　　DROP SIZE
　　HAZE
　　HAZE DETECTION
　　MIST
　　PRECIPITATION (METEOROLOGY)
　　SMOG
　　SMOKE
　　STEAM
　　STRATUS CLOUDS
　　VISIBILITY

FOG DISPERSAL
GS　WEATHER MODIFICATION
　　. **FOG DISPERSAL**
RT　AEROSOLS
　　CLIMATOLOGY
　　CLOUD PHYSICS
　　CLOUDS (METEOROLOGY)
　　DISPERSING
　　DISPERSIONS
　　MIST
　　PRECIPITATION (METEOROLOGY)

FOIL BEARINGS
GS　BEARINGS
　　. **FOIL BEARINGS**

∞ **FOILS**
SN　*(USE OF A MORE SPECIFIC TERM IS
　　RECOMMENDED-CONSULT THE TERMS
　　LISTED BELOW)*
RT　AIRFOILS
　　FOILS (MATERIALS)
　　HYDROFOILS
　　MULTILAYER INSULATION

FOILS (MATERIALS)
GS　**FOILS (MATERIALS)**
　　. METAL FOILS
RT　AIRFOILS
　　∞FOILS
　　HYDROFOILS
　　∞MATERIALS
　　MULTILAYER INSULATION
　　THIN PLATES

FOKKER AIRCRAFT
GS　**FOKKER AIRCRAFT**
　　. F-27 AIRCRAFT
　　. F-28 TRANSPORT AIRCRAFT
RT　∞AIRCRAFT

FOKKER BOND TESTERS
USE　ADHESION TESTS

FOKKER F 27 AIRCRAFT
USE　F-27 AIRCRAFT

FOKKER F 28 AIRCRAFT
USE　F-28 TRANSPORT AIRCRAFT

FOKKER FRIENDSHIP AIRCRAFT
USE　F-27 AIRCRAFT

FOKKER-PLANCK EQUATION
GS　ANALYSIS (MATHEMATICS)
　　. REAL VARIABLES
　　. . DIFFERENTIAL EQUATIONS
　　. . . PARTIAL DIFFERENTIAL EQUATIONS
　　. . . . **FOKKER-PLANCK EQUATION**
RT　BOLTZMANN TRANSPORT EQUATION
　　BROWNIAN MOVEMENTS
　　DENSITY DISTRIBUTION
　　DIFFUSION THEORY
　　∞EQUATIONS
　　IONIZED GASES
　　STOCHASTIC PROCESSES

FOLDING
UF　CRIMPING
RT　BENDING
　　BINDING
　　CURL (MATERIALS)
　　DISTORTION
　　FLEXING

FOLDING FIN AIRCRAFT ROCKET VEHICLE
UF　FFAR ROCKET VEHICLE
GS　ROCKET VEHICLES
　　. **FOLDING FIN AIRCRAFT ROCKET
　　　VEHICLE**
RT　∞AIRCRAFT
　　SOLID PROPELLANT ROCKET ENGINES

FOLDING STRUCTURES
UF　ROGALLO WINGS
　　TELESCOPING STRUCTURES
GS　**FOLDING STRUCTURES**
　　. SAILWINGS
RT　ANTENNAS
　　BALLOONS
　　BALLUTES
　　EXPANDABLE STRUCTURES
　　FURLABLE ANTENNAS
　　INFLATABLE STRUCTURES
　　PADDLES
　　PARACHUTES
　　PARAGLIDERS
　　PARAVULCOONS
　　PARAWINGS
　　ROTARY WINGS
　　SPACE ERECTABLE STRUCTURES
　　SPACECRAFT STRUCTURES
　　∞STRUCTURES
　　VARIABLE GEOMETRY STRUCTURES
　　VARIABLE SWEEP WINGS

FOLDS (GEOLOGY)
```
UF    NAPPES
RT    EARTH CRUST
      FISSURES (GEOLOGY)
      GEOLOGICAL FAULTS
      GREAT BASIN (US)
   ∞ LAYERS
      OUTCROPS
      ROCKS
      SEAMOUNTS
      STRATA
      STRATIFICATION
```

FOLIAGE
```
GS    PLANTS (BOTANY)
      . FOLIAGE
RT    BROWN WAVE EFFECT
      CANOPIES (VEGETATION)
      DECIDUOUS TREES
      DEFOLIANTS
      GREEN WAVE EFFECT
      HERBICIDES
      LEAVES
      LOCUSTS
      TIMBER VIGOR
```

FOLIC ACID
```
UF    VITAMIN M
GS    ACIDS
      . AMINO ACIDS
      . . FOLIC ACID
      HETEROCYCLIC COMPOUNDS
      . FOLIC ACID
      NITROGEN COMPOUNDS
      . FOLIC ACID
      ORGANIC COMPOUNDS
      . AMINO ACIDS
      . . FOLIC ACID
      VITAMINS
      . FOLIC ACID
```

∞ FOOD
```
SN    (USE OF A MORE SPECIFIC TERM IS
      RECOMMENDED--CONSULT THE TERMS
      LISTED BELOW)
RT    ALFALFA
      BARLEY
      BEVERAGES
      BROTHS
      CALORIC REQUIREMENTS
      CANNING
      CARBOHYDRATES
      CITRUS TREES
      CONSUMABLES (SPACECREW SUPPLIES)
      CORN
      DECONTAMINATION
      DEHYDRATED FOOD
      DIETS
      DIGESTING
      DISTRIBUTING
      EARTH RESOURCES
      EATING
      EGGS
      FATS
      FISHES
      FLOUR (FOOD)
      FOOD CHAIN
      FROZEN FOODS
      FRUITS
      GELATINS
      HAY
      LEGUMINOUS PLANTS
      MILK
      MILLET
      NUTRITION
      OATS
      ORCHARDS
      PEPPERS
      POTATOES
      PRESERVING
      PROTEINS
      PROVISIONING
      RATIONS
      SERVICES
      SOYBEANS
      SPACE FLIGHT FEEDING
      SPACE RATIONS
      SPINACH
      STARCHES
      SUGAR CANE
      SUGARS
      SYNTHETIC FOOD
      VEGETABLES
      VINEYARDS
      VITAMINS
```

FOOD-(CONT.)
```
      YEAST
```

FOOD CHAIN
```
RT    ANIMALS
      ECOSYSTEMS
   ∞ FOOD
      PLANTS (BOTANY)
```

FOOD INTAKE
```
RT    FASTING
      SPACE FLIGHT FEEDING
      SYNTHETIC FOOD
```

FOOD PROCESSING
```
GS    FOOD PROCESSING
      . CANNING
      . PRESERVING
RT    DEHYDRATED FOOD
      FROZEN FOODS
   ∞ PROCESSING
```

† FOOTPRINTS
```
RT    AIRCRAFT NOISE
      ANTENNA RADIATION PATTERNS
      MATHEMATICAL MODELS
```

FORBIDDEN BANDS
```
GS    ENERGY BANDS
      . FORBIDDEN BANDS
RT    BAND STRUCTURE OF SOLIDS
   ∞ BANDS
      ELECTRON ENERGY
      FREE ELECTRONS
      LATTICE VIBRATIONS
      WAVE EQUATIONS
```

FORBIDDEN TRANSITIONS
```
RT    ELECTRON TRANSITIONS
      FRANCK-CONDON PRINCIPLE
      QUANTUM THEORY
      SELECTION RULES (NUCLEAR PHYSICS)
   ∞ SOLID STATE PHYSICS
   ∞ TRANSITION
      WAVE FUNCTIONS
```

FORBUSH DECREASES
```
UF    FORBUSH EFFECT
RT    COSMIC RAYS
   ∞ EFFECTS
      MAGNETIC STORMS
      SOLAR FLARES
      SOLAR FURNACES
      SOLAR STORMS
```

FORBUSH EFFECT
```
USE   FORBUSH DECREASES
```

∞ FORCE
```
SN    (USE OF A MORE SPECIFIC TERM IS
      RECOMMENDED--CONSULT THE TERMS
      LISTED BELOW)
UF    REPULSION
RT    ACCELERATION (PHYSICS)
      AERODYNAMIC FORCES
      ATTRACTION
      CENTRIFUGAL FORCE
      CENTRIPETAL FORCE
      ELECTRIC FIELD STRENGTH
      HIGH IMPULSE
      INERTIA
      KINETICS
      LINES OF FORCE
      LOADS (FORCES)
      LORENTZ FORCE
      NEWTON
      NONCONSERVATIVE FORCES
      NULL ZONES
      PONDEROMOTIVE FORCES
      PRESSURE
      PULLING
      PUSHING
      THRUST
      THRUST MEASUREMENT
      TORQUE
      TORSION
      VAN DER WAAL FORCES
      WEIGHT (MASS)
      ZERO FORCE CURVES
```

FORCE DISTRIBUTION
```
UF    LIFT DISTRIBUTION
      NORMAL FORCE DISTRIBUTION
GS    DISTRIBUTION (PROPERTY)
```

FORCE DISTRIBUTION-(CONT.)
```
      . FORCE DISTRIBUTION
RT    AERODYNAMIC COEFFICIENTS
      AERODYNAMIC LOADS
      ANGULAR DISTRIBUTION
      CHARGE DISTRIBUTION
      ENERGY DISTRIBUTION
      INFLUENCE COEFFICIENT
      LIFT DRAG RATIO
      LOADS (FORCES)
      MASS DISTRIBUTION
      MOMENT DISTRIBUTION
      SCALE EFFECT
      STRESS CONCENTRATION
      STRESS INTENSITY FACTORS
      SWEEP EFFECT
      THRUST DISTRIBUTION
      WING LOADING
```

FORCE FIELDS
```
USE   FIELD THEORY (PHYSICS)
```

FORCE VECTOR RECORDERS
```
GS    MEASURING INSTRUMENTS
      . FORCE VECTOR RECORDERS
      RECORDING INSTRUMENTS
      . FORCE VECTOR RECORDERS
RT    ∞ INSTRUMENTS
```

FORCE-FREE MAGNETIC FIELDS
```
GS    MAGNETIC FIELDS
      . FORCE-FREE MAGNETIC FIELDS
```

FORCED CONVECTION
```
GS    CONVECTION
      . FORCED CONVECTION
RT    BLOWING
      CONVECTIVE HEAT TRANSFER
      FREE CONVECTION
      HEAT TRANSFER
      LAMINAR FLOW
      PRANDTL NUMBER
      RAYLEIGH-BENARD CONVECTION
      STANTON NUMBER
```

FORCED OSCILLATION
```
USE   FORCED VIBRATION
```

FORCED VIBRATION
```
UF    FORCED OSCILLATION
      FORCED VIBRATORY MOTION
      EQUATIONS
GS    VIBRATION
      . FORCED VIBRATION
RT    FLUTTER
      FREE VIBRATION
      RANDOM VIBRATION
      SELF EXCITATION
      SELF INDUCED VIBRATION
```

FORCED VIBRATORY MOTION EQUATIONS
```
USE   EQUATIONS
      FORCED VIBRATION
```

FOREARM
```
GS    ANATOMY
      . LIMBS (ANATOMY)
      . . ARM (ANATOMY)
      . . . FOREARM
      APPENDAGES
      . ARM (ANATOMY)
      . . FOREARM
```

FOREBODIES
```
GS    AIRCRAFT STRUCTURES
      . FOREBODIES
      . . NOSES (FOREBODIES)
RT    AFTERBODIES
      BLUFF BODIES
      BLUNT BODIES
      BLUNT LEADING EDGES
   ∞ BOWS
      CENTERBODIES
      CYLINDRICAL BODIES
      HAMMERHEAD CONFIGURATION
      LEADING EDGES
```

FORECASTING
```
UF    FORECASTS
GS    FORECASTING
      . DELPHI METHOD (FORECASTING)
      . PERFORMANCE PREDICTION
      . . PREDICTION ANALYSIS TECHNIQUES
      . TECHNOLOGICAL FORECASTING
```

FORECASTING-*(CONT.)*
```
    . . PATTERN METHOD (FORECASTING)
    . . PROBE METHOD (FORECASTING)
    . . PROFILE METHOD (FORECASTING)
    . WEATHER FORECASTING
    . . LONG RANGE WEATHER
          FORECASTING
    . . NOWCASTING
    . . NUMERICAL WEATHER FORECASTING
    . . STATISTICAL WEATHER
          FORECASTING
RT  ∞ANALYZING
    BUDGETING
    CONFIDENCE LIMITS
    CORRELATION
    CURVE FITTING
    ESTIMATES
    ESTIMATING
    EVALUATION
    EXPECTATION
    EXTRAPOLATION
    MANAGEMENT
    MANAGEMENT METHODS
    MANAGEMENT PLANNING
    MATHEMATICAL MODELS
    MAXIMUM LIKELIHOOD ESTIMATES
    MISSION PLANNING
    NOISE PREDICTION (AIRCRAFT)
    OPERATIONS RESEARCH
    PLANNING
    PREDICTIONS
    PROBABILITY THEORY
    PROJECT PLANNING
    ∞PROJECTION
    REGRESSION ANALYSIS
    REGRESSION COEFFICIENTS
    RELIABILITY
    RESERVES
    RISK
    SCHEDULING
    STATISTICAL ANALYSIS
    STATISTICAL DISTRIBUTIONS
    SYSTEMS ENGINEERING
    TIME SERIES ANALYSIS
    TRENDS
```

FORECASTS
```
USE  FORECASTING
```

FOREHEAD
```
GS   FACE (ANATOMY)
     . FOREHEAD
RT   HEAD (ANATOMY)
     SKULL
```

FOREIGN BODIES
```
RT   AIRCRAFT HAZARDS
     ∞BODIES
     INJURIES
     METEORITES
```

FOREIGN POLICY
```
GS   FOREIGN POLICY
     . INTERNATIONAL RELATIONS
     . . INTERNATIONAL COOPERATION
     . . . OUTER SPACE TREATY
RT   ∞BUDGETS
     EUROPEAN SPACE PROGRAMS
     INTERNATIONAL HYDROLOGICAL
          DECADE
```

FOREIGN TRADE
```
RT   ECONOMICS
```

FORENSIC SCIENCES
```
USE  LAW (JURISPRUDENCE)
```

FOREST FIRE DETECTION
```
GS   DETECTION
     . FOREST FIRE DETECTION
RT   AERIAL PHOTOGRAPHY
     ∞DETECTORS
     HAZE DETECTION
     INFRARED DETECTORS
     INFRARED INSTRUMENTS
     INFRARED PHOTOGRAPHY
     INFRARED RADIOMETERS
     INFRARED SCANNERS
     MEASURING INSTRUMENTS
     OBSERVATION
     RADIOMETERS
     SATELLITE-BORNE PHOTOGRAPHY
     SURVEILLANCE
```

FOREST FIRES
```
GS   FIRES
     . FOREST FIRES
RT   AIR POLLUTION
     ASHES
     COMBUSTION
     FIRE PREVENTION
     FIREBREAKS
     FLAMES
     FORESTS
     SMOKE
     WASTES
```

FOREST MANAGEMENT
```
GS   MANAGEMENT
     . RESOURCES MANAGEMENT
     . . FOREST MANAGEMENT
     . . . REFORESTATION
RT   CONSERVATION
     EARTH RESOURCES
     FORESTS
     LAND USE
     REGIONAL PLANNING
     TIMBER INVENTORY
```

FORESTS
```
UF   LUMBERING AREAS
GS   RESOURCES
     . EARTH RESOURCES
     . . FORESTS
     . . . RAIN FORESTS
RT   AMAZON REGION (SOUTH AMERICA)
     CANOPIES (VEGETATION)
     CLEARINGS (OPENINGS)
     CONIFERS
     CONSERVATION
     DECIDUOUS TREES
     DEFOLIANTS
     DEFOLIATION
     DEFORESTATION
     FIREBREAKS
     FOREST FIRES
     FOREST MANAGEMENT
     HERBICIDES
     LOGGING (INDUSTRY)
     PLANTS (BOTANY)
     REFORESTATION
     REGIONAL PLANNING
     SILVICULTURE
     TIMBER IDENTIFICATION
     TIMBER INVENTORY
     TIMBER VIGOR
     TIMBERLINE
     TREES (PLANTS)
     WILDERNESS
```

FORGING
```
UF   METAL FORGING
GS   FORMING TECHNIQUES
     . FORGING
     METAL WORKING
     . FORGING
RT   AUSFORMING
     BILLETS
     BULGING
     CASTING
     COINING
     COLD WORKING
     HEAT TREATMENT
     HOT PRESSING
     HOT WORKING
     PIERCING
     PRESSING (FORMING)
     RHEOCASTING
     ∞ROLLING
     STAMPING
```

FORKS
```
RT   CONVEYORS
     HOOKS
```

FORM
```
USE  SHAPES
```

FORM FACTORS
```
RT   APPROXIMATION
     COUPLING COEFFICIENTS
     FUNCTIONAL ANALYSIS
     HARMONIC ANALYSIS
     RECTIFIERS
     SCATTERING COEFFICIENTS
     SERIES (MATHEMATICS)
     SQUARE WAVES
     TRANSDUCERS
     ∞VARIABLE
```

FORM FACTORS-*(CONT.)*
```
     WAVEFORMS
     X RAY SCATTERING
```

FORM PERCEPTION
```
USE  SPACE PERCEPTION
```

FORMALDEHYDE
```
GS   ALDEHYDES
     . FORMALDEHYDE
RT   PHENOL FORMALDEHYDE
```

FORMALISM
```
RT   COMPUTER PROGRAMMING
     DYNAMIC PROGRAMMING
     LINEAR PROGRAMMING
     ∞LOGIC
     NONLINEAR PROGRAMMING
     PARAMETERIZATION
     SCHEDULING
```

FORMAT
```
RT   COMPUTER PROGRAMMING
     DOCUMENTS
     EDITING
     FRAMES (DATA PROCESSING)
     PRINTOUTS
     RECORDS
     SYNTAX
     TEXTS
```

FORMATES
```
GS   FORMATES
     . CHLOROFORMATE
     . NITROFORMATES
RT   FORMIC ACID
     FORMYL IONS
```

∞ FORMATION
```
SN   (USE OF A MORE SPECIFIC TERM IS
     RECOMMENDED -- CONSULT THE
     TERMS LISTED BELOW)
RT   FORMATIONS
     GROWTH
     NUCLEATION
     STRATIGRAPHY
```

FORMATION HEAT
```
USE  HEAT OF FORMATION
```

FORMATIONS
```
RT   CONTACTS (GEOLOGY)
     ∞FORMATION
     FRACTURING
     GAS INJECTION
     GEOLOGICAL FAULTS
     GEOLOGY
     GEOPHYSICS
     GREAT BASIN (US)
     ∞LAYERS
     MOUNTAINS
     OUTCROPS
     OUTLIERS (LANDFORMS)
     PALEONTOLOGY
     PERFORATING
     PERMEABILITY
     PETROLOGY
     POROSITY
     ROCKS
     SHATTER CONES
     SOILS
     STAIRSTEPS
     STRATIGRAPHY
     TERRACES (LANDFORMS)
     WETTABILITY
```

FORMHYDROXAMIC ACID
```
GS   ALIPHATIC COMPOUNDS
     . FORMHYDROXAMIC ACID
     NITROGEN COMPOUNDS
     . AMIDES
     . . FORMHYDROXAMIC ACID
```

FORMIC ACID
```
GS   ALIPHATIC COMPOUNDS
     . FORMIC ACID
RT   FORMATES
     FORMYL IONS
```

FORMICA
```
RT   LAMINATES
     ∞POLYMERS
     THERMOSETTING RESINS
```

FORMING TECHNIQUES
SN (TECHNIQUES OF SHAPING ITEM)
UF METAL FORMING
GS **FORMING TECHNIQUES**
. AUSFORMING
. CASTING
. . CENTRIFUGAL CASTING
. . INVESTMENT CASTING
. . PROPELLANT CASTING
. . RHEOCASTING
. . SAND CASTING
. . SLIP CASTING
. COLD WORKING
. . COLD ROLLING
. . ELECTROHYDRAULIC FORMING
. . EXPLOSIVE FORMING
. ELECTROFORMING
. EXTRUDING
. . PULTRUSION
. FORGING
. HOT WORKING
. INJECTION MOLDING
. MAGNETIC FORMING
. METAL DRAWING
. METAL SPINNING
. . HYDROSPINNING
. PRESSING (FORMING)
. . BLANKING (CUTTING)
. . COINING
. . STAMPING
. ROLL FORMING
RT ∞BLANKING
CUTTING
DEPOSITION
ELECTROMAGNETIC HAMMERS
HOT MACHINING
LASER CUTTING
MACHINING
METAL GRINDING
METAL WORKING
SPRAYING
UPSETTING

FORMS (PAPER)
RT BLANKS
DATA ACQUISITION

∞ **FORMULAS**
SN *(USE OF A MORE SPECIFIC TERM IS*
RECOMMENDED--CONSULT THE TERMS
LISTED BELOW)
RT COMPUTATION
FORMULAS (MATHEMATICS)
FORMULATIONS
KRAMERS-KRONIG FORMULA

FORMULAS (MATHEMATICS)
UF EXPRESSIONS (MATHEMATICS)
GS MATHEMATICAL LOGIC
. **FORMULAS (MATHEMATICS)**
. . BETHE-HEITLER FORMULA
RT ∞FORMULAS
∞MATHEMATICS

FORMULATIONS
RT ADMIXTURES
∞COMPOSITION
∞FORMULAS
INGREDIENTS
MIXTURES
PARAMETERIZATION
STOICHIOMETRY

FORMYL IONS
GS IONS
. **FORMYL IONS**
RADICALS
. **FORMYL IONS**
RT ATMOSPHERIC CHEMISTRY
FORMATES
FORMIC ACID
HYDROXYL RADICALS
INTERSTELLAR CHEMISTRY
INTERSTELLAR MATTER
MOLECULAR IONS
NEGATIVE IONS
POSITIVE IONS

FORSTERITE
GS MAGNESIUM COMPOUNDS
. **FORSTERITE**
MINERALS
. OLIVINE
. . **FORSTERITE**
SILICON COMPOUNDS

FORSTERITE-*(CONT.)*
. SILICATES
. . **FORSTERITE**
RT REFRACTORIES

FORTISAN (TRADEMARK)
GS ALIPHATIC COMPOUNDS
. POLYSACCHARIDES
. . CELLULOSE
. . . **FORTISAN (TRADEMARK)**
CARBOHYDRATES
. POLYSACCHARIDES
. . CELLULOSE
. . . **FORTISAN (TRADEMARK)**
FABRICS
. **FORTISAN (TRADEMARK)**
FIBERS
. SYNTHETIC FIBERS
. . **FORTISAN (TRADEMARK)**
RT PARACHUTE FABRICS

FORTRAN
UF FAB (PROGRAMMING LANGUAGE)
GS LANGUAGES
. PROGRAMMING LANGUAGES
. . **FORTRAN**
RT COBOL
COMPILERS
COMPUTER PROGRAMMING
PL/1

FORWARD LOOKING INFRARED DETECTORS
USE FLIR DETECTORS

FORWARD SCATTERING
GS SCATTERING
. **FORWARD SCATTERING**
RT BACKSCATTERING
INVERSE SCATTERING
LIGHT SCATTERING
NUCLEAR SCATTERING
SCATTER PROPAGATION

FOSSIL FUELS
GS RESOURCES
. EARTH RESOURCES
. . **FOSSIL FUELS**
. . . COAL
. . . . LIGNITE
. . . . . SOLVENT REFINED COAL
. . . CRUDE OIL
RT CARBONACEOUS MATERIALS
UNDERWATER RESOURCES

FOSSIL METEORITE CRATERS
USE FOSSILS
METEORITE CRATERS

FOSSILS
UF FOSSIL METEORITE CRATERS
RT ARCHAEOLOGY
PALEOBIOLOGY
PALEONTOLOGY
PARTICLE TRACKS
RADIOACTIVE AGE DETERMINATION

FOSTER THEORY
RT NETWORK ANALYSIS
REACTANCE
RESONANCE
∞THEORIES

FOULING
GS FOULING
. ANTIFOULING
RT CONTAMINATION
CORROSION
DEPOSITION
ICE FORMATION
PLUGGING
RETARDING

FOUNDATIONS
UF BASES (FOUNDATIONS)
STRUCTURAL FOUNDATIONS
GS **FOUNDATIONS**
. PILE FOUNDATIONS
RT BASEMENTS
∞BASES
CAISSONS
CONCRETE STRUCTURES
EXCAVATION
GEOTECHNICAL ENGINEERING
OVERCONSOLIDATION

FOUNDATIONS-*(CONT.)*
∞PAD
PAVEMENTS
SKIRTS
STRUCTURAL MEMBERS
∞STRUCTURES
SUBSTRUCTURES
SUPPORTS
UNDERGROUND STRUCTURES

FOUNDRIES
GS INDUSTRIAL PLANTS
. **FOUNDRIES**
RT FURNACES
∞METALLURGY
MOLDS

FOUR BODY PROBLEM
RT CELESTIAL MECHANICS
MANY BODY PROBLEM
ORBITS
PERTURBATION
∞PROBLEMS
THREE BODY PROBLEM

FOURIER ANALYSIS
GS ANALYSIS (MATHEMATICS)
. **FOURIER ANALYSIS**
. . FOURIER SERIES
RT AUTOCORRELATION
DATA COMPRESSION
DIFFERENTIAL EQUATIONS
DIVERGENCE
EXPONENTIAL FUNCTIONS
FREQUENCY DISTRIBUTION
HARMONIC ANALYSIS
HARMONIC EXCITATION
HARMONIC FUNCTIONS
HARMONIC GENERATIONS
HARMONIC OSCILLATION
HARMONICS
INFORMATION THEORY
KURTOSIS
LINEAR TRANSFORMATIONS
MEASURE AND INTEGRATION
OPERATIONAL CALCULUS
PERIODIC FUNCTIONS
PERIODIC VARIATIONS
REAL VARIABLES
SIMPLE HARMONIC MOTION
TIME SERIES ANALYSIS

FOURIER LAW
GS LAWS
. **FOURIER LAW**
RT THERMAL CONDUCTIVITY

FOURIER SERIES
GS ANALYSIS (MATHEMATICS)
. CALCULUS
. . SERIES (MATHEMATICS)
. . . **FOURIER SERIES**
. FOURIER ANALYSIS
. . **FOURIER SERIES**
. REAL VARIABLES
. . SERIES (MATHEMATICS)
. . . **FOURIER SERIES**
RT GIBBS PHENOMENON

FOURIER TRANSFORMATION
GS ANALYSIS (MATHEMATICS)
. FUNCTIONAL ANALYSIS
. . INTEGRAL TRANSFORMATIONS
. . . **FOURIER TRANSFORMATION**
FUNCTIONS (MATHEMATICS)
. **FOURIER TRANSFORMATION**
RT BBGKY HIERARCHY
HOLOGRAPHIC SPECTROSCOPY
MAXIMUM ENTROPY METHOD
WALSH FUNCTION

FOURIER-BESSEL TRANSFORMATIONS
GS ANALYSIS (MATHEMATICS)
. CALCULUS
. . **FOURIER-BESSEL**
TRANSFORMATIONS
. REAL VARIABLES
. . **FOURIER-BESSEL**
TRANSFORMATIONS
FUNCTIONS (MATHEMATICS)
. **FOURIER-BESSEL TRANSFORMATIONS**
TRANSFORMATIONS (MATHEMATICS)
. **FOURIER-BESSEL TRANSFORMATIONS**
RT DIFFERENTIAL EQUATIONS
SERIES (MATHEMATICS)

FOVEA
GS   ANATOMY
    . SENSE ORGANS
    . . EYE (ANATOMY)
    . . . RETINA
    . . . . **FOVEA**
RT   SACCADIC EYE MOVEMENTS

FR-1 SATELLITE
GS   SATELLITES
    . ARTIFICIAL SATELLITES
    . . FRENCH SATELLITES
    . . . **FR-1 SATELLITE**

FRACTALS
GS   DIMENSIONS
    . **FRACTALS**
    GEOMETRY
    . **FRACTALS**
RT   ∞APPLICATIONS OF MATHEMATICS
    COORDINATES
    EXPONENTS
    HALF SPACES
    ∞MATHEMATICS
    RATIOS
    SET THEORY
    ∞SPACE
    STRANGE ATTRACTORS

FRACTIONATION
GS   **FRACTIONATION**
    . CHEMICAL FRACTIONATION
    . HYDROCRACKING
RT   REFINING
    RETORT PROCESSING
    ∞SEPARATION
    SOLVENT REFINED COAL

FRACTIONS
SN   (EXCLUDES MATHEMATICAL CONCEPTS)
RT   ∞COMPONENTS
    FINES
    PARTICLE SIZE DISTRIBUTION
    RATIOS

FRACTOGRAPHY
GS   PHOTOGRAPHY
    . **FRACTOGRAPHY**
RT   BRITTLENESS
    CRACK CLOSURE
    CRACK GEOMETRY
    CRACK PROPAGATION
    DUCTILITY
    ELBER EQUATION
    FATIGUE (MATERIALS)
    FRACTURES (MATERIALS)
    ∞METALLURGY

FRACTURE MECHANICS
UF   FAULT MECHANICS
    MOHR CIRCLES
RT   BEND TESTS
    BURST TESTS
    CAUSTICS (OPTICS)
    CRACK CLOSURE
    CRACK INITIATION
    CRACK PROPAGATION
    ELBER EQUATION
    FINITE ELEMENT METHOD
    FRACTURING
    GRIFFITH CRACK
    HOLE GEOMETRY (MECHANICS)
    ISOPARAMETRIC FINITE ELEMENTS
    J INTEGRAL
    ∞MECHANICS (PHYSICS)
    MICROMECHANICS
    PLANE STRAIN
    RESIDUAL STRENGTH
    ROCK MECHANICS
    RUPTURING
    SOIL MECHANICS
    STRESS INTENSITY FACTORS
    STRESS TENSORS
    TIME TEMPERATURE PARAMETER

FRACTURE RESISTANCE
USE   FRACTURE STRENGTH

FRACTURE STRENGTH
UF   FRACTURE RESISTANCE
    FRACTURE TOUGHNESS
GS   MECHANICAL PROPERTIES
    . **FRACTURE STRENGTH**
RT   BEND TESTS
    BRITTLE MATERIALS

FRACTURE STRENGTH-*(CONT.)*
    BRITTLENESS
    BURST TESTS
    CARBON-CARBON COMPOSITES
    CRACK CLOSURE
    CRACK INITIATION
    CRACK PROPAGATION
    CREEP RUPTURE STRENGTH
    DUCTILITY
    EARTHQUAKE RESISTANCE
    EUTECTIC COMPOSITES
    HARDNESS
    J INTEGRAL
    RESIDUAL STRENGTH
    ∞RESISTANCE
    RETIREMENT FOR CAUSE
    ∞STRENGTH
    TOUGHNESS
    YIELD STRENGTH

FRACTURE TOUGHNESS
USE   FRACTURE STRENGTH

FRACTURES (MATERIALS)
RT   CRACKS
    DAMAGE
    DEFORMATION
    FAILURE
    FRACTOGRAPHY
    ∞MATERIALS

FRACTURING
GS   **FRACTURING**
    . CRUSTAL FRACTURES
RT   BRITTLENESS
    CHIPPING
    CRACK CLOSURE
    CRACK PROPAGATION
    CRACKING (FRACTURING)
    CUTTING
    FLAKING
    FORMATIONS
    FRACTURE MECHANICS
    FRAGMENTATION
    FRAGMENTS
    METAL FATIGUE
    PERFORATING
    ∞SEPARATION
    SPALLING
    SPLITTING
    STRESS FUNCTIONS
    STRUCTURAL FAILURE

FRAGMENTATION
UF   SHATTERING
RT   ACOUSTIC STREAMING
    BREAKING
    BURSTS
    CHIPPING
    COMMINUTION
    FRACTURING
    FRAGMENTS
    PENETRATION
    SABOT PROJECTILES
    SHRAPNEL
    SPALLING
    TERMINAL BALLISTICS

FRAGMENTS
RT   CHIPS
    DEBRIS
    EJECTA
    FRACTURING
    FRAGMENTATION
    SHRAPNEL

FRAME PHOTOGRAPHY
GS   PHOTOGRAPHY
    . **FRAME PHOTOGRAPHY**
RT   BLACK AND WHITE PHOTOGRAPHY
    FRAMING CAMERAS
    HIGH SPEED CAMERAS

FRAMES
GS   **FRAMES**
    . AIRFRAMES
    . CHASSIS
    . UNDERCARRIAGES
RT   CARRIAGES
    SPRINGS (ELASTIC)
    ∞STRUCTURES
    STRUTS
    SUPPORTS
    TRUSSES

FRAMES (DATA PROCESSING)
RT   DATA MANAGEMENT
    DATA PROCESSING
    FORMAT
    IMAGE PROCESSING

FRAMING CAMERAS
GS   OPTICAL EQUIPMENT
    . CAMERAS
    . . HIGH SPEED CAMERAS
    . . . **FRAMING CAMERAS**
    PHOTOGRAPHIC EQUIPMENT
    . CAMERAS
    . . HIGH SPEED CAMERAS
    . . . **FRAMING CAMERAS**
RT   FRAME PHOTOGRAPHY
    ROTATING MIRRORS

FRANCE
GS   NATIONS
    . **FRANCE**
    . . FRENCH GUIANA
    . . GUADELOUPE
    . . MARTINIQUE
RT   ANDORRA
    ENGLISH CHANNEL
    EUROPE
    FRENCH SPACE PROGRAMS
    GUADELOUPE
    PYRENEES MOUNTAINS (EUROPE)
    RHONE DELTA (FRANCE)

FRANCIUM
GS   CHEMICAL ELEMENTS
    . ALKALI METALS
    . . **FRANCIUM**
    METALS
    . ALKALI METALS
    . . **FRANCIUM**

FRANCK-CONDON PRINCIPLE
RT   BORN-OPPENHEIMER APPROXIMATION
    COLOR CENTERS
    CONDUCTION BANDS
    ELECTRON TRANSITIONS
    FORBIDDEN TRANSITIONS
    OPTICAL TRANSITION

FRAUNHOFER LINE DISCRIMINATORS
GS   CIRCUITS
    . DISCRIMINATORS
    . . **FRAUNHOFER LINE DISCRIMINATORS**
RT   ABSORPTION SPECTRA
    LUMINESCENCE
    MEASURING INSTRUMENTS
    SPECTROSCOPIC ANALYSIS
    SPECTROSCOPY

FRAUNHOFER LINES
GS   SPECTRA
    . ABSORPTION SPECTRA
    . . **FRAUNHOFER LINES**
    . RADIATION SPECTRA
    . . ELECTROMAGNETIC SPECTRA
    . . . LINE SPECTRA
    . . . . **FRAUNHOFER LINES**
    . SPECTRAL BANDS
    . . **FRAUNHOFER LINES**
RT   ABSORPTION SPECTROSCOPY
    OPTOGALVANIC SPECTROSCOPY
    SOLAR SPECTRA

FRAUNHOFER REGION
USE   FAR FIELDS

FREDHOLM EQUATIONS
UF   FREDHOLM OPERATORS
GS   ANALYSIS (MATHEMATICS)
    . FUNCTIONAL ANALYSIS
    . . INTEGRAL EQUATIONS
    . . . **FREDHOLM EQUATIONS**
RT   ∞EQUATIONS
    POMERANCHUK THEOREM

FREDHOLM OPERATORS
USE   FREDHOLM EQUATIONS
    OPERATORS (MATHEMATICS)

FREE ATMOSPHERE
GS   EARTH ATMOSPHERE
    . **FREE ATMOSPHERE**
RT   BIOSPHERE
    MIDDLE ATMOSPHERE
    PRIMITIVE EARTH ATMOSPHERE

FREE BOUNDARIES
GS   BOUNDARIES
.   **FREE BOUNDARIES**
RT   FLUID BOUNDARIES
INTERFACES
JET FLOW
JET MIXING FLOW
JET STREAMS (METEOROLOGY)
LIQUID SURFACES
LIQUID-LIQUID INTERFACES
LIQUID-VAPOR INTERFACES

FREE CONVECTION
UF   THERMAL CONVECTION
GS   CONVECTION
.   **FREE CONVECTION**
. .   RAYLEIGH-BENARD CONVECTION
RT   CONVECTION CURRENTS
CONVECTIVE FLOW
CONVECTIVE HEAT TRANSFER
FORCED CONVECTION
LAMINAR FLOW
MARANGONI CONVECTION
POROUS BOUNDARY LAYER CONTROL
TEMPERATURE
THERMOSIPHONS
TURBULENT FLOW

FREE ELECTRON LASERS
GS   STIMULATED EMISSION DEVICES
.   LASERS
. .   **FREE ELECTRON LASERS**
RT   WIGGLER MAGNETS

FREE ELECTRONS
GS   CHARGE CARRIERS
.   **FREE ELECTRONS**
RT   BRILLOUIN ZONES
CONDUCTION ELECTRONS
ELECTRON AVALANCHE
ELECTRON DENSITY (CONCENTRATION)
ELECTRON GAS
FORBIDDEN BANDS
PLASMA FREQUENCIES
RECOMBINATION COEFFICIENT

FREE ENERGY
GS   THERMODYNAMIC PROPERTIES
.   **FREE ENERGY**
. .   GIBBS FREE ENERGY
RT   CHEMICAL ENERGY
∞ENERGY
ENERGY OF FORMATION
ENTHALPY
GIBBS-HELMHOLTZ EQUATIONS
INTERNAL ENERGY
∞LEVEL
MOLECULAR ENERGY LEVELS
THERMAL ENERGY
THERMODYNAMICS

FREE FALL
RT   AIR DROP OPERATIONS
BALLISTIC TRAJECTORIES
FALLING SPHERES
PARACHUTE DESCENT
WEIGHTLESSNESS

FREE FLIGHT
RT   ∞FLIGHT
GLIDERS
GLIDING
HANG GLIDERS

FREE FLIGHT TEST APPARATUS
RT   FLIGHT TESTS
∞TEST EQUIPMENT

FREE FLOW
UF   FREE STREAM EFFECTS
FREE STREAMS
GS   FLUID FLOW
.   **FREE FLOW**
RT   VOID RATIO

FREE JETS
GS   FLUID JETS
.   **FREE JETS**
RT   JET BOUNDARIES
JET FLOW
∞JETS

FREE MOLECULAR FLOW
GS   FLUID FLOW

FREE MOLECULAR FLOW-*(CONT.)*
.   GAS FLOW
. .   **FREE MOLECULAR FLOW**
RT   CONTINUUM FLOW
KINETIC THEORY
KNUDSEN FLOW
MOLECULAR BEAMS
RAREFIED GAS DYNAMICS
RAREFIED GASES
SLIP FLOW
TRANSITION FLOW

FREE OSCILLATIONS
USE   FREE VIBRATION

FREE RADICALS
GS   RADICALS
.   **FREE RADICALS**
. .   HYDROXYL RADICALS
RT   ATOMS
CARBENES
IONS
NEGATIVE IONS
OXYGEN IONS
TRIVALENT IONS
VINYL RADICAL

FREE STREAM EFFECTS
USE   FREE FLOW

FREE STREAMS
USE   FREE FLOW

FREE VIBRATION
UF   FREE OSCILLATIONS
GS   VIBRATION
.   **FREE VIBRATION**
RT   FORCED VIBRATION
LINEAR VIBRATION
PROTON PRECESSION
SELF EXCITATION
SELF INDUCED VIBRATION
VIBRATION MODE

FREE WING AIRCRAFT
RT   AERODYNAMICS
∞AIRCRAFT
AIRCRAFT DESIGN
CONTROL SURFACES

FREEDOM FIGHTER AIRCRAFT
USE   F-5 AIRCRAFT

FREEZE DRYING
GS   DRYING
.   **FREEZE DRYING**
RT   DEHYDRATED FOOD
DEHYDRATION
FREEZING
FROZEN FOODS
PRESERVING

FREEZING
GS   PHASE TRANSFORMATIONS
.   **FREEZING**
. .   VIBRATIONAL FREEZING
. .   ZONE MELTING
RT   ANTIFREEZES
BAY ICE
CLOUD GLACIATION
COLD TRAPS
COOLING
CRYOGENIC COOLING
CRYSTALLIZATION
FREEZE DRYING
FROST
ICE FORMATION
ICE NUCLEI
LOW TEMPERATURE
MELTING
PRESERVING
PRESSURE ICE
REFRIGERATING
SEA ICE
SOLIDIFICATION
SOLIDIFIED GASES

FREEZING POINTS
USE   MELTING POINTS

FREIGHT
USE   CARGO

FREIGHT COSTS
GS   COSTS
.   **FREIGHT COSTS**
RT   CARGO
FREIGHTERS
TRANSPORTATION

FREIGHTERS
RT   FREIGHT COSTS
HARBORS
TRANSPORTATION
WHARVES

FRENCH EQUATORIAL CONGO
USE   CONGO (BRAZZAVILLE)

FRENCH GUIANA
GS   NATIONS
.   FRANCE
. .   **FRENCH GUIANA**
RT   CARIBBEAN REGION
SOUTH AMERICA

FRENCH SATELLITES
GS   SATELLITES
.   ARTIFICIAL SATELLITES
. .   **FRENCH SATELLITES**
. . .   D-1 SATELLITE
. . .   D-2 SATELLITES
. . .   EOLE SATELLITES
. . .   FR-1 SATELLITE
. . .   GEOLE SATELLITES
. . .   PEOLE SATELLITES
. . .   POSEIDON SATELLITE
. . .   SPOT (FRENCH SATELLITE)
. . .   SRET SATELLITES
. . . .   SRET 1 SATELLITE
. . . .   SRET 2 SATELLITE
RT   EUROPEAN SPACE PROGRAMS
METEOSAT SATELLITE
SYMPHONIE SATELLITES

FRENCH SPACE PROGRAMS
GS   PROGRAMS
.   SPACE PROGRAMS
. .   **FRENCH SPACE PROGRAMS**
RT   EOLE SATELLITES
EUROPEAN SPACE PROGRAMS
FRANCE
GEOLE SATELLITES
HERMES MANNED SPACEPLANE
INTERNATIONAL COOPERATION
METEOSAT SATELLITE
∞RESEARCH PROJECTS
SPACE EXPLORATION
SPACE MISSIONS
∞SPACECRAFT
SRET SATELLITES
SRET 1 SATELLITE

FRENKEL DEFECTS
GS   DEFECTS
.   CRYSTAL DEFECTS
. .   POINT DEFECTS
. . .   VACANCIES (CRYSTAL DEFECTS)
. . . .   **FRENKEL DEFECTS**

FREON
RT   AIR CONDITIONING
COOLANTS
COOLING
COOLING SYSTEMS
FLUOROHYDROCARBONS
GAS COOLING
REFRIGERANTS
REFRIGERATING

FREQUENCIES
UF   FREQUENCY BANDS
GS   **FREQUENCIES**
.   AUDIO FREQUENCIES
. .   QUEFRENCIES
.   BEAT FREQUENCIES
.   BROADBAND
.   BRUNT-VAISALA FREQUENCY
.   CARRIER FREQUENCIES
.   CRITICAL FREQUENCIES
.   CYCLOTRON FREQUENCY
.   EXTREMELY LOW FREQUENCIES
.   INFRASONIC FREQUENCIES
.   INTERMEDIATE FREQUENCIES
.   IONIZATION FREQUENCIES
.   MAXIMUM USABLE FREQUENCY
.   NYQUIST FREQUENCIES
.   PLASMA FREQUENCIES

FREQUENCIES-*(CONT.)*
- . RADIO FREQUENCIES
- . . EXTREMELY LOW RADIO
    FREQUENCIES
- . . HIGH FREQUENCIES
- . LOW FREQUENCIES
- . . . SUBAUDIBLE FREQUENCIES
- . . . VERY LOW FREQUENCIES
- . . . LOW FREQUENCY BANDS
- . . . VERY LOW FREQUENCIES
- . MICROWAVE FREQUENCIES
- . . . C BAND
- . . . EXTREMELY HIGH FREQUENCIES
- . . . P BAND
- . . . SUPERHIGH FREQUENCIES
- . . ULTRAHIGH FREQUENCIES
- . . . P BAND
- . . VERY HIGH FREQUENCIES
- . . . P BAND
- . RESONANT FREQUENCIES
- . SWEEP FREQUENCY
- RT   AEOLIAN TONES
-     AMPLITUDES
- ∞BANDS
-     BANDWIDTH
-     BROADBAND AMPLIFIERS
-     CHANNEL CAPACITY
- ∞CHANNELS
-     FREQUENCY DISTRIBUTION
-     FREQUENCY RANGES
-     FREQUENCY REUSE
-     HARMONICS
-     LINE SPECTRA
-     LONGITUDINAL WAVES
-     MICROCHANNELS
-     MILLIMETER WAVES
-     NARROWBAND
- ∞PITCH
-     RADIO WAVES
-     SPECTRAL BANDS
-     STANDING WAVES
-     SUBMILLIMETER WAVES
-     SUPERHARMONICS

FREQUENCY ANALYZERS
- RT   HARMONIC ANALYSIS
-     INTERMODULATION
-     OSCILLOSCOPES
-     SELECTIVE FADING
-     SIGNAL ANALYSIS
-     SPECTRUM ANALYSIS
-     SWEEP FREQUENCY
- ∞TEST EQUIPMENT
-     VIBRATION MEASUREMENT

FREQUENCY ASSIGNMENT
- RT   COMMUNICATING
-     FREQUENCY REUSE
-     MAXIMUM USABLE FREQUENCY
-     ORBIT SPECTRUM UTILIZATION

FREQUENCY BANDS
- USE   FREQUENCIES

FREQUENCY COMPRESSION DEMODULATORS
- GS   DEMODULATORS
-     . **FREQUENCY COMPRESSION
    DEMODULATORS**

FREQUENCY CONTROL
- UF   FREQUENCY REGULATION
- GS   REGULATORS
-     . **FREQUENCY CONTROL**
- RT   AUTODYNES
- ∞CONTROL
-     CRYSTAL OSCILLATORS
-     QUARTZ CRYSTALS
-     SIGNAL STABILIZATION

FREQUENCY CONVERSION
- USE   FREQUENCY CONVERTERS

FREQUENCY CONVERTERS
- UF   FREQUENCY CONVERSION
-     FREQUENCY TRANSLATION
- GS   **FREQUENCY CONVERTERS**
-     . DOWN-CONVERTERS
-     . FREQUENCY DIVIDERS
-     . FREQUENCY MULTIPLIERS
-     . FREQUENCY SYNTHESIZERS
-     . PARAMETRIC FREQUENCY
    CONVERTERS
-     . UP-CONVERTERS
- RT   ∞CONVERSION
-     ∞CONVERTERS

FREQUENCY CONVERTERS-*(CONT.)*
-     HARMONIC GENERATORS
-     MIXING CIRCUITS
-     PARAMETRIC AMPLIFIERS
-     PULSE WIDTH AMPLITUDE CONVERTERS

FREQUENCY DISCRIMINATORS
- GS   CIRCUITS
-     . DISCRIMINATORS
-     . . **FREQUENCY DISCRIMINATORS**

FREQUENCY DISTRIBUTION
- SN   (OF CYCLIC VARIATIONS)
- GS   DISTRIBUTION (PROPERTY)
-     . **FREQUENCY DISTRIBUTION**
-     . . KURTOSIS
- RT   CYCLES
-     FOURIER ANALYSIS
-     FREQUENCIES
-     SUBAUDIBLE FREQUENCIES

FREQUENCY DIVIDERS
- GS   FREQUENCY CONVERTERS
-     . **FREQUENCY DIVIDERS**
- RT   DOWN-CONVERTERS

FREQUENCY DIVISION MULTIPLE ACCESS
- UF   FDMA
- GS   TELECOMMUNICATION
-     . MULTIPLE ACCESS
-     . . **FREQUENCY DIVISION MULTIPLE
    ACCESS**
-     TRANSMISSION
-     . SIGNAL TRANSMISSION
-     . . DATA TRANSMISSION
-     . . . MULTIPLE ACCESS
-     . . . . **FREQUENCY DIVISION MULTIPLE
    ACCESS**
- RT   ALOHA SYSTEM
-     CODE DIVISION MULTIPLE ACCESS
-     CODE DIVISION MULTIPLEXING
-     MULTIPLEXING
-     RADIO COMMUNICATION
-     TELECOMMUNICATION
-     TIME DIVISION MULTIPLE ACCESS

FREQUENCY DIVISION MULTIPLEXING
- GS   TRANSMISSION
-     . MULTIPLEXING
-     . . **FREQUENCY DIVISION MULTIPLEXING**
- RT   CARRIER FREQUENCIES
-     COMMUNICATION NETWORKS
-     DATA TRANSMISSION
-     DEMULTIPLEXING
-     MULTIPLE ACCESS
-     PULSE COMMUNICATION
-     RADIO COMMUNICATION
-     SATELLITE TRANSMISSION
-     TELECOMMUNICATION
-     TIME DIVISION MULTIPLEXING
-     WAVELENGTH DIVISION MULTIPLEXING

FREQUENCY HOPPING
- RT   FREQUENCY REUSE
-     FREQUENCY SHIFT KEYING
-     JAMMING
-     SPREAD SPECTRUM TRANSMISSION
-     TRANSMISSION EFFICIENCY

FREQUENCY MEASUREMENT
- RT   ACOUSTIC MEASUREMENT
-     ∞MEASUREMENT
-     TIME MEASUREMENT
-     VIBRATION MEASUREMENT

FREQUENCY MODULATION
- GS   MODULATION
-     . **FREQUENCY MODULATION**
-     . . FEEDBACK FREQUENCY MODULATION
-     . . FM/PM (MODULATION)
-     . . FREQUENCY SHIFT KEYING
-     . . PULSE FREQUENCY MODULATION
-     . . PULSE FREQUENCY MODULATION
    TELEMETRY
- RT   AMPLITUDE MODULATION
-     AUTOMATIC FREQUENCY CONTROL
-     CAPTURE EFFECT
-     CARRIER TO NOISE RATIOS
-     COMPANDING
-     DEMODULATION
-     DEMODULATORS
-     INTERMODULATION
-     LIGHT MODULATION
-     LINE OF SIGHT COMMUNICATION
-     MODULATORS

FREQUENCY MODULATION-*(CONT.)*
-     PHASE MODULATION
-     PULSE MODULATION
-     VOCODERS
-     VOLTAGE CONTROLLED OSCILLATORS

FREQUENCY MODULATION PHOTOMULTIPLIERS
- GS   AMPLIFIERS
-     . CURRENT AMPLIFIERS
-     . . PHOTOMULTIPLIER TUBES
-     . . . **FREQUENCY MODULATION
    PHOTOMULTIPLIERS**
- RT   CATHODES

FREQUENCY MULTIPLIERS
- GS   FREQUENCY CONVERTERS
-     . **FREQUENCY MULTIPLIERS**
- RT   PHASE MATCHING

FREQUENCY RANGES
- GS   RANGE (EXTREMES)
-     . **FREQUENCY RANGES**
-     . . OCTAVES
-     . . RADIO RANGE
-     . . SUBAUDIBLE FREQUENCIES
- RT   BANDWIDTH
-     FREQUENCIES
-     ∞FREQUENCY RESPONSE
-     FREQUENCY REUSE

FREQUENCY REGULATION
- USE   FREQUENCY CONTROL

∞ **FREQUENCY RESPONSE**
- SN   *(USE OF A MORE SPECIFIC TERM IS
    RECOMMENDED--CONSULT THE TERMS
    LISTED BELOW)*
- RT   ACUITY
-     BROADBAND
-     DISTRIBUTED AMPLIFIERS
-     DYNAMIC CHARACTERISTICS
-     DYNAMIC RESPONSE
-     EQUALIZERS (CIRCUITS)
-     FREQUENCY RANGES
-     LINEAR FILTERS
-     LINEAR RECEIVERS
-     LOG PERIODIC ANTENNAS
-     LOGARITHMIC RECEIVERS
-     PERCEPTION
-     PULSE REPETITION RATE
-     RAMP FUNCTIONS
-     RESPONSES
-     SENSITIVITY
-     SMEAR
-     SPECTRAL SENSITIVITY
-     STEP FUNCTIONS
-     STROKING TESTS
-     THRESHOLDS (PERCEPTION)

FREQUENCY REUSE
- RT   DATA LINKS
-     DOWNLINKING
-     FREQUENCIES
-     FREQUENCY ASSIGNMENT
-     FREQUENCY HOPPING
-     FREQUENCY RANGES
-     MAXIMUM USABLE FREQUENCY
-     MICROWAVE TRANSMISSION
-     RADIO TRANSMISSION
-     SATELLITE TRANSMISSION
-     UPLINKING

FREQUENCY SCANNING
- GS   SCANNING
-     . **FREQUENCY SCANNING**
- RT   PANORAMIC SCANNING
-     RADAR SCANNING
-     SPECTRUM ANALYSIS
-     SWEEP CIRCUITS
-     SWEEP FREQUENCY

FREQUENCY SHIFT
- RT   BRILLOUIN EFFECT
-     DOPPLER EFFECT
-     DOPPLER-FIZEAU EFFECT
-     GYROTROPISM
-     ∞SHIFT

FREQUENCY SHIFT KEYING
- GS   KEYING
-     . **FREQUENCY SHIFT KEYING**
-     MODULATION
-     . FREQUENCY MODULATION
-     . . **FREQUENCY SHIFT KEYING**

FREQUENCY SHIFT KEYING-*(CONT.)*
- RT   FREQUENCY HOPPING
         RADIO TRANSMISSION

FREQUENCY STABILITY
- UF   ACOUSTIC STABILITY
- GS   DYNAMIC CHARACTERISTICS
- . DYNAMIC STABILITY
- . . **FREQUENCY STABILITY**
         STABILITY
- . DYNAMIC STABILITY
- . . **FREQUENCY STABILITY**
- RT   CRYSTAL OSCILLATORS
         LASER STABILITY
         OSCILLATORS
         QUARTZ CRYSTALS
         STABLE OSCILLATIONS
         VOLTAGE CONTROLLED OSCILLATORS

FREQUENCY STANDARDS
- GS   STANDARDS
- . **FREQUENCY STANDARDS**
- RT   ATOMIC CLOCKS
         GAS MASERS
         ION STORAGE
         MASERS
         RESONATORS
         TIME SIGNALS

FREQUENCY SYNCHRONIZATION
- GS   SYNCHRONISM
- . **FREQUENCY SYNCHRONIZATION**
- RT   BIT SYNCHRONIZATION
         CAPTURE EFFECT
         HOMODYNE RECEPTION
         SWEEP FREQUENCY
         SYNCHRONIZED OSCILLATORS
         SYNTONY

FREQUENCY SYNTHESIZERS
- GS   FREQUENCY CONVERTERS
- . **FREQUENCY SYNTHESIZERS**
         SIGNAL GENERATORS
- . **FREQUENCY SYNTHESIZERS**
- RT   MIXING CIRCUITS
         OSCILLATORS
         SYNTHESIZERS

FREQUENCY TRANSLATION
- USE   FREQUENCY CONVERTERS

FRESH WATER
- GS   WATER
- . **FRESH WATER**
- RT   AGRISTARS PROJECT
         AQUIFERS
         GROUND WATER
         LIMNOLOGY
         POTABLE WATER
         RESERVOIRS
         SPRINGS (WATER)

FRESNEL DIFFRACTION
- GS   DIFFRACTION
- . **FRESNEL DIFFRACTION**
- RT   GRATINGS (SPECTRA)
         INTERFEROMETRY

FRESNEL INTEGRALS
- UF   FRESNEL-KIRCHHOFF INTEGRALS
- GS   FUNCTIONS (MATHEMATICS)
- . **FRESNEL INTEGRALS**
- RT   DIFFRACTION PATTERNS
         TRIGONOMETRIC FUNCTIONS
         WAVE DIFFRACTION

FRESNEL LENSES
- GS   LENSES
- . **FRESNEL LENSES**
- RT   FOCUSING
- ∞OPTICS

FRESNEL REFLECTORS
- GS   MIRRORS
- . **FRESNEL REFLECTORS**
         REFLECTORS
- . **FRESNEL REFLECTORS**
- RT   INTERFEROMETRY
         SLITS

FRESNEL REGION
- GS   REGIONS
- . **FRESNEL REGION**
- RT   ANTENNA RADIATION PATTERNS

FRESNEL REGION-*(CONT.)*
         DIFFRACTION PATTERNS
         FAR FIELDS

FRESNEL-KIRCHHOFF INTEGRALS
- USE   FRESNEL INTEGRALS

FRETTING
- RT   EROSION
         FATIGUE (MATERIALS)
         TRIBOLOGY
         WEAR TESTS

FRETTING CORROSION
- GS   CORROSION
- . **FRETTING CORROSION**
- RT   FATIGUE (MATERIALS)
         STRESS CORROSION
         WEAR

FRICTION
- GS   **FRICTION**
- . AERODYNAMIC DRAG
- . . SUPERSONIC DRAG
- . DRY FRICTION
- . FLOW RESISTANCE
- . . FRICTION DRAG
- . . . SUPERSONIC DRAG
- . . . VISCOUS DRAG
- . INTERNAL FRICTION
- . KINETIC FRICTION
- . . SLIDING FRICTION
- . SKIN FRICTION
- . FRICTION DRAG
- . . . SUPERSONIC DRAG
- . . . VISCOUS DRAG
- . STATIC FRICTION
- RT   ABRASION
         COEFFICIENT OF FRICTION
         DRAG
         DRAG REDUCTION
         ENERGY DISSIPATION
         FLUID FLOW
         MECHANICAL IMPEDANCE
- ∞PRESSURE DROP
         SCORING
         SURFACE PROPERTIES
         SURFACE ROUGHNESS
         TRACTION
         TRIBOLOGY
         TRIBOLUMINESCENCE
         WEAR
         WEAR TESTS
         WHEEL BRAKES

FRICTION COEFFICIENT
- USE   COEFFICIENT OF FRICTION

FRICTION DRAG
- UF   NONEQUILIBRIUM DRAG
- GS   DYNAMIC CHARACTERISTICS
- . DRAG
- . . **FRICTION DRAG**
- . . . AERODYNAMIC DRAG
- . . . . SUPERSONIC DRAG
- . . . VISCOUS DRAG
         FRICTION
- . FLOW RESISTANCE
- . . **FRICTION DRAG**
- . . . SUPERSONIC DRAG
- . . . VISCOUS DRAG
- . SKIN FRICTION
- . . **FRICTION DRAG**
- . . . SUPERSONIC DRAG
- . . . VISCOUS DRAG
- RT   MINIMUM DRAG
         PRESSURE DRAG
         SATELLITE DRAG
         SURFACE ROUGHNESS EFFECTS
         WAVE DRAG

FRICTION FACTOR
- UF   FRICTION LOSS COEFFICIENT
- RT   COEFFICIENT OF FRICTION
         PRESSURE GRADIENTS
         SKIN FRICTION

FRICTION LOSS COEFFICIENT
- USE   FRICTION FACTOR

FRICTION MEASUREMENT
- GS   MECHANICAL MEASUREMENT
- . **FRICTION MEASUREMENT**
- RT   ELASTOHYDRODYNAMICS

FRICTION MEASUREMENT-*(CONT.)*
         KINETIC FRICTION
- ∞MEASUREMENT
         STATIC FRICTION

FRICTION PRESSURE DROP
- USE   SKIN FRICTION

FRICTION REDUCTION
- RT   ANTIFRICTION BEARINGS
         COEFFICIENT OF FRICTION
         LUBRICATION
- ∞REDUCTION
         STREAMLINING

FRICTION WELDING
- GS   WELDING
- . **FRICTION WELDING**
- RT   PRESSURE WELDING

FRICTIONLESS ENVIRONMENTS
- GS   ENVIRONMENTS
- . **FRICTIONLESS ENVIRONMENTS**
- RT   DEEP SPACE
         LEVITATION

FRIEDEL-CRAFT REACTION
- GS   CHEMICAL REACTIONS
- . **FRIEDEL-CRAFT REACTION**
- RT   ACYLATION
         ALKYLATION

FRIENDSHIP 7
- GS   MANNED SPACECRAFT
- . MERCURY SPACECRAFT
- . . **FRIENDSHIP 7**
         REENTRY VEHICLES
- . RECOVERABLE SPACECRAFT
- . MERCURY SPACECRAFT
- . . . **FRIENDSHIP 7**
         SOFT LANDING SPACECRAFT
- . MERCURY SPACECRAFT
- . . **FRIENDSHIP 7**
         SPACE CAPSULES
- . MERCURY SPACECRAFT
- . . **FRIENDSHIP 7**
- RT   MERCURY MA-6 FLIGHT

FRINGE MULTIPLICATION
- RT   DIFFRACTION PATTERNS
         INTERFERENCE GRATING
         MOIRE EFFECTS
         MOIRE FRINGES
         MULTIPLICATION
         PHOTOELASTIC ANALYSIS
         STRESS ANALYSIS
         STRESS CONCENTRATION

FRINGE PATTERNS
- USE   DIFFRACTION PATTERNS

FRIT
- RT   CERAMICS
         FUSION (MELTING)
         GLAZES
         VITREOUS MATERIALS

FROGS
- GS   ANIMALS
- . POIKILOTHERMIA
- . . AMPHIBIA
- . . . **FROGS**
- . VERTEBRATES
- . . AMPHIBIA
- . . . **FROGS**

FRONTAL AREAS (METEOROLOGY)
- USE   FRONTS (METEOROLOGY)

FRONTAL WAVES
- RT   OCEAN CURRENTS
         OCEANOGRAPHY
         TSUNAMI WAVES
         WATER WAVES
- ∞WAVES

∞ **FRONTS**
- SN   *(USE OF A MORE SPECIFIC TERM IS
         RECOMMENDED--CONSULT THE TERMS
         LISTED BELOW)*
- RT   COLD FRONTS
         FRONTS (METEOROLOGY)
         SHOCK FRONTS
         WARM FRONTS

FRONTS-*(CONT.)*
    WAVE FRONTS

FRONTS (METEOROLOGY)
UF   FRONTAL AREAS (METEOROLOGY)
        WEATHER FRONTS
GS   **FRONTS (METEOROLOGY)**
    . COLD FRONTS
    . WARM FRONTS
RT   AIR MASSES
    ∞FRONTS
        INTERTROPICAL CONVERGENT ZONES
        MARINE METEOROLOGY
        METEOROLOGICAL PARAMETERS
        METEOROLOGY
        STORMS
        SYNOPTIC METEOROLOGY
        THUNDERSTORMS
        TORNADOES

FROST
RT   BAY ICE
        DEW
        FREEZING
        ICE
        LOW TEMPERATURE

FROST DAMAGE
GS   DAMAGE
    . **FROST DAMAGE**
RT   COLD WEATHER
    ∞CROPS
        FARM CROPS
        FRUITS
        LOW TEMPERATURE
        ORCHARDS
        PLANTS (BOTANY)

FROSTBITE
GS   INJURIES
    . **FROSTBITE**
RT   COLD TOLERANCE

FROUDE NUMBER
GS   RATIOS
    . DIMENSIONLESS NUMBERS
    . . **FROUDE NUMBER**
RT   FLUID FLOW
        INERTIA
        KINETIC ENERGY
        POTENTIAL ENERGY
        REYNOLDS NUMBER
        STROUHAL NUMBER

FROZEN EQUILIBRIUM FLOW
GS   FLUID FLOW
    . GAS FLOW
    . . EQUILIBRIUM FLOW
    . . . **FROZEN EQUILIBRIUM FLOW**
RT   SHIFTING EQUILIBRIUM FLOW

FROZEN FOODS
RT   ∞FOOD
        FOOD PROCESSING
        FREEZE DRYING
        PRESERVING
        REFRIGERATING

FROZEN SOILS
USE   PERMAFROST

FRUITS
GS   FARM CROPS
    . **FRUITS**
RT   AGRICULTURE
        BOLLWORMS
        BOTANY
    ∞FOOD
        FROST DAMAGE
        ORCHARDS

FRUSTRATION
RT   EMOTIONAL FACTORS
        EMOTIONS
        HANDICAPS
    ∞INHIBITION
        LETHARGY
        PSYCHOLOGICAL EFFECTS
        PSYCHOLOGY

FRUSTUMS
RT   CONES
        GEOMETRY
        PYRAMIDS

FRUSTUMS-*(CONT.)*
    VOLUME

FUEL CAPSULES
RT   ∞CAPSULES
        NUCLEAR FUELS
        PELLETS
        SPENT FUELS

FUEL CELL CATALYSTS
USE   ELECTROCATALYSTS

FUEL CELL POWER PLANTS
GS   ELECTRIC POWER PLANTS
    . **FUEL CELL POWER PLANTS**
RT   COAL GASIFICATION
        ENERGY TECHNOLOGY
        FUEL CELLS

FUEL CELLS
SN   (EXCLUDES BATTERIES)
GS   ELECTRIC GENERATORS
    . DIRECT POWER GENERATORS
    . . **FUEL CELLS**
    . . . BIOCHEMICAL FUEL CELLS
    . . . HYDROGEN OXYGEN FUEL CELLS
    . . . PHOSPHORIC ACID FUEL CELLS
    . . . REGENERATIVE FUEL CELLS
    ELECTROCHEMICAL CELLS
    . **FUEL CELLS**
    . . BIOCHEMICAL FUEL CELLS
    . . HYDROGEN OXYGEN FUEL CELLS
    . . PHOSPHORIC ACID FUEL CELLS
    . . REGENERATIVE FUEL CELLS
RT   ∞CELLS
        CHEMICAL AUXILIARY POWER UNITS
    ∞ELECTRIC CELLS
        ELECTROCATALYSTS
        ELECTROCHEMISTRY
        ELECTROLYTES
        ENERGY CONVERSION EFFICIENCY
        ENERGY STORAGE
        FUEL CELL POWER PLANTS
        HYDROGEN FUELS
        HYDROGEN-BASED ENERGY
        ION EXCHANGE MEMBRANE
          ELECTROLYTES
        MAGNETOHYDRODYNAMIC GENERATORS
        SOLAR CELLS
        SOLAR GENERATORS
        THERMIONIC CONVERTERS
        THERMOELECTRIC GENERATORS
        WET CELLS

FUEL COMBUSTION
GS   COMBUSTION
    . **FUEL COMBUSTION**
    . . NUCLEAR FUEL BURNUP
RT   COMBUSTION EFFICIENCY
        COMBUSTION STABILITY
        EROSIVE BURNING
        HYDROCARBON COMBUSTION
        HYPERSONIC COMBUSTION
        IGNITION
        METAL COMBUSTION
        OXIDATION
        PROPELLANT COMBUSTION
        SOLID PROPELLANT COMBUSTION
        SPONTANEOUS COMBUSTION
        SUPERSONIC COMBUSTION

FUEL CONSUMPTION
GS   CONSUMPTION
    . **FUEL CONSUMPTION**
RT   BURNING RATE
        COMBUSTION EFFICIENCY
        ENERGY CONSUMPTION
        ENERGY REQUIREMENTS
        ENGINES
        INTERNAL COMBUSTION ENGINES
        REFUELING

FUEL CONTAMINATION
GS   CONTAMINATION
    . **FUEL CONTAMINATION**
RT   ANTIICING ADDITIVES
        CONTAMINANTS
        REFUELING

FUEL CONTROL
RT   COMBUSTION CONTROL
    ∞CONTROL
        ENGINE CONTROL
        FLUID MANAGEMENT
        LIQUID SLOSHING

FUEL CONTROL-*(CONT.)*
    PROPELLANT TRANSFER
    REFUELING
    ROCKET ENGINE CONTROL
    TURBOJET ENGINE CONTROL

FUEL CORROSION
GS   CORROSION
    . **FUEL CORROSION**
RT   PROPELLANT DECOMPOSITION
        PROPELLANT STORABILITY

FUEL ELEMENTS (NUCLEAR REACTORS)
USE   NUCLEAR FUEL ELEMENTS

FUEL FLOW
GS   FLUID FLOW
    . **FUEL FLOW**
    . . PROPELLANT TRANSFER
RT   COMBUSTIBLE FLOW
        DUCTED FLOW
        PARTICLE LADEN JETS

FUEL FLOW REGULATORS
GS   REGULATORS
    . FLOW REGULATORS
    . . **FUEL FLOW REGULATORS**

FUEL GAGES
GS   MEASURING INSTRUMENTS
    . **FUEL GAGES**
    . . CAPACITIVE FUEL GAGES
RT   FLOWMETERS

FUEL INJECTION
UF   INJECTION CARBURETORS
GS   INJECTION
    . **FUEL INJECTION**
RT   BURNERS
        CARBURETORS
        FLUID INJECTION
        GAS INJECTION
        INJECTORS
        INTERNAL COMBUSTION ENGINES
        JET ENGINES
        JET MIXING FLOW
    ∞JET NOZZLES
        LIQUID INJECTION
        PISTON ENGINES
        PROPELLANT SPRAYS
        SPRAY NOZZLES

FUEL OILS
GS   FUELS
    . CLEAN FUELS
    . . **FUEL OILS**
    OILS
    . **FUEL OILS**
RT   ENERGY POLICY
        KEROGEN
        KEROSENE
        SHALE OIL
        SOLVENT REFINED COAL

FUEL PRODUCTION
RT   CHEMICAL FUELS
        COMPOSITE PROPELLANTS
        CRUDE OIL
        FISSION
    ∞FUSION
        HYDROCARBON FUELS
        HYDROGEN FUELS
        LIQUID FUELS
        NUCLEAR FUELS
    ∞PRODUCTION

FUEL PUMPS
GS   PUMPS
    . **FUEL PUMPS**
RT   AIRCRAFT FUEL SYSTEMS
        AXIAL FLOW PUMPS
        CENTRIFUGAL PUMPS
        ELECTROMAGNETIC PUMPS
        INTERNAL COMBUSTION ENGINES
        JET ENGINES
        JET PUMPS
        MATERIALS HANDLING
        TURBINE PUMPS

FUEL SPRAYS
RT   LIQUID INJECTION
        PROPELLANT SPRAYS
        SPRAYERS

FUEL SYSTEMS
GS   **FUEL SYSTEMS**
     . AIRCRAFT FUEL SYSTEMS
RT   ACCUMULATORS
     AUTOMOBILES
     BUNKERS (FUEL)
     CARBURETORS
     CHOKES (FUEL SYSTEMS)
     ENGINES
     FEEDERS
     FUELS
     INJECTORS
     INLET TEMPERATURE
     INTAKE SYSTEMS
     INTERNAL COMBUSTION ENGINES
     MANIFOLDS
     PLENUM CHAMBERS
     PROPELLANT TRANSFER
     REFUELING
     SELF SEALING
     SPRAY NOZZLES
     ∞SYSTEMS

FUEL TANK PRESSURIZATION
GS   PRESSURIZING
     . **FUEL TANK PRESSURIZATION**
RT   AIRCRAFT FUEL SYSTEMS
     EXHAUST SYSTEMS
     EXPULSION BLADDERS
     LIQUID ROCKET PROPELLANTS
     PRESSURE
     PRESSURE REGULATORS
     PRESSURE VESSELS
     PROPELLANT STORAGE
     PROPELLANT TANKS
     PROPULSION
     RELIEF VALVES
     ULLAGE
     VAPOR PRESSURE

FUEL TANKS
GS   TANKS (CONTAINERS)
     . **FUEL TANKS**
     .. WING TANKS
RT   AIRCRAFT FUEL SYSTEMS
     ∞CONTAINERS
     CORROSION PREVENTION
     CRYOGENIC FLUID STORAGE
     CYLINDRICAL TANKS
     EXPULSION BLADDERS
     EXTERNAL TANKS
     FEED SYSTEMS
     FUELS
     HEATING EQUIPMENT
     LIQUID SLOSHING
     PODS (EXTERNAL STORES)
     PRESSURE VESSELS
     PROPELLANT STORAGE
     PROPELLANT TANKS
     PROPELLANTS
     PROPULSION
     PROTUBERANCES
     SPACECRAFT STRUCTURES
     SPHERICAL TANKS
     STORAGE TANKS
     TANKER AIRCRAFT
     ULLAGE

FUEL TESTS
GS   **FUEL TESTS**
     . REACTOR STARTUP TESTS
RT   CHEMICAL ANALYSIS
     CORROSION TESTS
     ENGINE TESTS
     ∞MATERIALS TESTS
     MISSILE TESTS
     PROPELLANT TESTS
     STABILITY TESTS
     TEST FIRING
     ∞TESTS

FUEL VALVES
GS   VALVES
     . **FUEL VALVES**
RT   AIRCRAFT FUEL SYSTEMS
     GAS VALVES
     RELIEF VALVES

FUEL-AIR RATIO
GS   RATIOS
     . **FUEL-AIR RATIO**
RT   BURNING RATE
     COMBUSTION EFFICIENCY
     COMPRESSION RATIO
     GAS MIXTURES

FUEL-AIR RATIO-*(CONT.)*
     IGNITION LIMITS
     PREMIXING
     PRESSURE RATIO
     VOLUMETRIC EFFICIENCY

FUELING
USE   REFUELING

FUELS
GS   **FUELS**
     . ACTIVATED CARBON
     . CHARCOAL
     . CHEMICAL FUELS
     .. ENDOTHERMIC FUELS
     .. HIGH ENERGY FUELS
     .. HYDROCARBON FUELS
     ... DIESEL FUELS
     ... GASOLINE
     ... JET ENGINE FUELS
     .... JP-4 JET FUEL
     .... JP-5 JET FUEL
     .... JP-6 JET FUEL
     .... JP-8 JET FUEL
     ... RP-1 ROCKET PROPELLANTS
     ... SYNTHANE
     .. LIQUID FUELS
     ... AIRCRAFT FUELS
     ... AUTOMOBILE FUELS
     ... DIESEL FUELS
     ... GASOLINE
     ... HYDROGEN FUELS
     ... JET ENGINE FUELS
     .... JP-4 JET FUEL
     .... JP-5 JET FUEL
     .... JP-6 JET FUEL
     .... JP-8 JET FUEL
     ... KEROSENE
     .. METAL FUELS
     .. SYNTHETIC FUELS
     ... GASOHOL (FUEL)
     ... SYNTHANE
     . CLEAN FUELS
     .. FUEL OILS
     . COKE
     . FISSILE FUELS
     . GASEOUS FUELS
     . LIQUEFIED NATURAL GAS
     . NUCLEAR FUELS
     .. CERAMIC NUCLEAR FUELS
     .. FISSIUM
     .. SPENT FUELS
     . SOLID ROCKET PROPELLANTS
     . DOUBLE BASE ROCKET
       PROPELLANTS
     .. METAL PROPELLANTS
RT   BIOCONVERSION
     BURNING RATE
     DOUBLE BASE PROPELLANTS
     ENERGY POLICY
     ENERGY STORAGE
     FLAMES
     FUEL SYSTEMS
     FUEL TANKS
     GASES
     HYDROGEN
     HYDROGEN PRODUCTION
     KEROGEN
     LIQUID AMMONIA
     LIQUID HYDROGEN
     NUCLEAR FUEL ELEMENTS
     OILS
     OPERATING COSTS
     OXIDIZERS
     PREMIXING
     PROPELLANTS
     ROCKET PROPELLANTS
     SHALE OIL
     TRANSPORTATION ENERGY

FUJITA METHOD
RT   COORDINATES
     ∞METHODOLOGY
     ∞TRANSFORMATIONS
     TRANSFORMATIONS (MATHEMATICS)

FULL SCALE TESTS
RT   ALTITUDE TESTS
     ENGINE TESTS
     FLIGHT TESTS
     GROUND TESTS
     HIGH ALTITUDE TESTS
     ∞TESTS

FULMINATES
GS   ESTERS
     . ISOCYANATES
     .. **FULMINATES**
     NITROGEN COMPOUNDS
     . CYANO COMPOUNDS
     .. ISOCYANATES
     ... **FULMINATES**
RT   DETONATORS
     EXPLOSIVES
     PROPELLANTS

FUMES
RT   AEROSOLS
     DISPERSIONS
     DUST
     EXHAUST GASES
     FIRE DAMAGE
     GAS MIXTURES
     GASES
     HAZE DETECTION
     REACTION PRODUCTS
     SMOKE
     SMOKE DETECTORS
     VAPORS
     WASTES

FUMIGATION
RT   ANTISEPTICS
     BACTERICIDES
     SPRAYING
     STERILIZATION

FUNCTION GENERATORS
GS   SIGNAL GENERATORS
     . **FUNCTION GENERATORS**
RT   ∞GENERATORS
     PULSE GENERATORS
     VOLTAGE GENERATORS
     WAVE GENERATION

FUNCTION SPACE
GS   ANALYSIS (MATHEMATICS)
     . **FUNCTION SPACE**
     .. HILBERT SPACE
     ... BANACH SPACE
     .... SOBOLEV SPACE
RT   FIBERS (MATHEMATICS)
     FIELD THEORY (PHYSICS)
     FUNCTIONS (MATHEMATICS)
     ORTHOGONAL FUNCTIONS
     QUANTUM MECHANICS
     SERIES (MATHEMATICS)
     ∞SPACE
     STATISTICAL MECHANICS
     VECTORS (MATHEMATICS)

FUNCTIONAL ANALYSIS
GS   ANALYSIS (MATHEMATICS)
     . **FUNCTIONAL ANALYSIS**
     .. HARMONIC ANALYSIS
     ... TESSERAL HARMONICS
     ... ZONAL HARMONICS
     .. HILBERT SPACE
     ... BANACH SPACE
     .... SOBOLEV SPACE
     .. INTEGRAL EQUATIONS
     ... FREDHOLM EQUATIONS
     ... J INTEGRAL
     ... SINGULAR INTEGRAL EQUATIONS
     ... VOLTERRA EQUATIONS
     ... WIENER HOPF EQUATIONS
     .. INTEGRAL TRANSFORMATIONS
     ... CONVOLUTION INTEGRALS
     ... FOURIER TRANSFORMATION
     ... HILBERT TRANSFORMATION
     ... LAPLACE TRANSFORMATION
RT   COMPLEX VARIABLES
     FORM FACTORS
     FUNCTIONS (MATHEMATICS)
     SERIES (MATHEMATICS)
     WALSH FUNCTION

FUNCTIONAL DESIGN SPECIFICATIONS
GS   SPECIFICATIONS
     . **FUNCTIONAL DESIGN SPECIFICATIONS**
RT   AERONAUTICAL ENGINEERING
     ∞DESIGN
     EQUIPMENT SPECIFICATIONS
     MISSILE DESIGN
     PRODUCT DEVELOPMENT
     SYSTEMS ENGINEERING

FUNCTIONAL INTEGRATION
GS   ANALYSIS (MATHEMATICS)

FUNCTIONAL INTEGRATION-*(CONT.)*
　　　. REAL VARIABLES
　　　. . MEASURE AND INTEGRATION
　　　. . . **FUNCTIONAL INTEGRATION**
RT　ANALOG COMPUTERS
　　　DIFFERENTIAL EQUATIONS
　　　DIGITAL INTEGRATORS
　　　PARTIAL DIFFERENTIAL EQUATIONS

FUNCTIONALS
RT　∞FUNCTIONS
　　　FUNCTIONS (MATHEMATICS)
　　　INTEGRALS

∞ **FUNCTIONS**
SN　*(USE OF A MORE SPECIFIC TERM IS*
　　　RECOMMENDED--CONSULT THE TERMS
　　　LISTED BELOW)
RT　CONTRALATERAL FUNCTIONS
　　　FUNCTIONALS
　　　FUNCTIONS (MATHEMATICS)
　　　MUSCULAR FUNCTION
　　　PARENTERAL FUNCTIONS
　　　PENALTY FUNCTION
　　　PULMONARY FUNCTIONS
　　　RENAL FUNCTION
　　　SCATTERING FUNCTIONS
　　　WORK FUNCTIONS

FUNCTIONS (MATHEMATICS)
GS　**FUNCTIONS (MATHEMATICS)**
　　　. ABEL FUNCTION
　　　. AIRY FUNCTION
　　　. ANALYTIC FUNCTIONS
　　　. . ENTIRE FUNCTIONS
　　　. APERIODIC FUNCTIONS
　　　. ASYMPTOTES
　　　. BOOLEAN FUNCTIONS
　　　. COMPOSITE FUNCTIONS
　　　. CONFORMAL MAPPING
　　　. COORDINATE TRANSFORMATIONS
　　　. DELTA FUNCTION
　　　. DISCRETE FUNCTIONS
　　　. DISCRIMINANT ANALYSIS (STATISTICS)
　　　. DISTRIBUTION FUNCTIONS
　　　. DISTURBING FUNCTIONS
　　　. ERROR FUNCTIONS
　　　. FOURIER TRANSFORMATION
　　　. FOURIER-BESSEL TRANSFORMATIONS
　　　. FRESNEL INTEGRALS
　　　. GAMMA FUNCTION
　　　. GREEN'S FUNCTIONS
　　　. HAMILTONIAN FUNCTIONS
　　　. HANKEL FUNCTIONS
　　　. HARMONIC FUNCTIONS
　　　. HYPERBOLIC FUNCTIONS
　　　. HYPERGEOMETRIC FUNCTIONS
　　　. KERNEL FUNCTIONS
　　　. LAGUERRE FUNCTIONS
　　　. LAME FUNCTIONS
　　　. LAPLACE TRANSFORMATION
　　　. LEGENDRE FUNCTIONS
　　　. LIAPUNOV FUNCTIONS
　　　. LINEAR TRANSFORMATIONS
　　　. LORENTZ TRANSFORMATIONS
　　　. MATHIEU FUNCTION
　　　. MAXWELL-BOLTZMANN DENSITY
　　　　　FUNCTION
　　　. MELLIN TRANSFORMS
　　　. MEROMORPHIC FUNCTIONS
　　　. . ELLIPTIC FUNCTIONS
　　　. . RATIONAL FUNCTIONS
　　　. MONOTONE FUNCTIONS
　　　. ORTHOGONAL FUNCTIONS
　　　. . WALSH FUNCTION
　　　. ORTHONORMAL FUNCTIONS
　　　. PENALTY FUNCTION
　　　. POINT SPREAD FUNCTIONS
　　　. POISSON DENSITY FUNCTIONS
　　　. PROBABILITY DENSITY FUNCTIONS
　　　. . NORMAL DENSITY FUNCTIONS
　　　. . PEARSON DISTRIBUTIONS
　　　. . RAYLEIGH DISTRIBUTION
　　　. . WEIBULL DENSITY FUNCTIONS
　　　. PROBABILITY DISTRIBUTION
　　　　　FUNCTIONS
　　　. RAMP FUNCTIONS
　　　. RECURSIVE FUNCTIONS
　　　. SCHWARZ-CHRISTOFFEL
　　　　　TRANSFORMATION
　　　. SPACE-TIME FUNCTIONS
　　　. SPHERICAL HARMONICS
　　　. SPLINE FUNCTIONS
　　　. STEP FUNCTIONS
　　　. STRESS FUNCTIONS

FUNCTIONS (MATHEMATICS)-*(CONT.)*
　　　. TIME FUNCTIONS
　　　. TRANSCENDENTAL FUNCTIONS
　　　. . EXPONENTIAL FUNCTIONS
　　　. . . LOGARITHMS
　　　. . PERIODIC FUNCTIONS
　　　. . . TRIGONOMETRIC FUNCTIONS
　　　. . . . COSINE SERIES
　　　. . . . SINE SERIES
　　　. . . . TANGENTS
　　　. TRANSFER FUNCTIONS
　　　. . MODULATION TRANSFER FUNCTION
　　　. . OPTICAL TRANSFER FUNCTION
　　　. WEIGHTING FUNCTIONS
　　　. WHITTAKER FUNCTIONS
RT　ALGEBRA
　　　∞APPLICATIONS OF MATHEMATICS
　　　BRANCHING (MATHEMATICS)
　　　CALCULUS
　　　CONTINUITY (MATHEMATICS)
　　　DIVERGENCE
　　　EXTREMUM VALUES
　　　FAST FOURIER TRANSFORMATIONS
　　　FUNCTION SPACE
　　　FUNCTIONAL ANALYSIS
　　　FUNCTIONALS
　　　∞FUNCTIONS
　　　INFINITY
　　　INFLECTION POINTS
　　　LINEARITY
　　　MAPPING
　　　MATHEMATICAL LOGIC
　　　MATHEMATICAL MODELS
　　　∞MATHEMATICS
　　　NONLINEARITY
　　　NUMBER THEORY
　　　NUMERICAL DIFFERENTIATION
　　　OPERATIONS RESEARCH
　　　OPERATORS (MATHEMATICS)
　　　RANDOM VARIABLES
　　　RANGE (EXTREMES)
　　　∞TRANSFORMATIONS
　　　TRANSFORMATIONS (MATHEMATICS)

FUNGI
GS　**FUNGI**
　　　. ACTINOMYCETES
　　　. ASPERGILLUS
　　　. COCCOMYCES
　　　. GIBBERELLINS
　　　. NEUROSPORA
　　　. RHIZOPUS
　　　. RUST FUNGI
　　　. SACCHAROMYCES
　　　. SPORES
　　　. . MICROSPORES
　　　. STREPTOMYCETES
　　　. YEAST
RT　BLIGHT
　　　LICHENS
　　　MITRA
　　　∞MOLD
　　　PANSPERMIA
　　　PARASITIC DISEASES
　　　THERMOPHILES

FUNGICIDES
GS　**FUNGICIDES**
　　　. CAFFEINE
　　　. XANTHINES
　　　. . GUANINES
　　　. . URIC ACID
RT　ANTIINFECTIVES AND ANTIBACTERIALS
　　　TOXICOLOGY

FUNNELS
RT　CONICAL INLETS
　　　∞NOZZLES

FURAN RESINS
GS　PLASTICS
　　　. SYNTHETIC RESINS
　　　. . THERMOSETTING RESINS
　　　. . . **FURAN RESINS**
　　　. . . . POLYAMIDE RESINS
　　　. . . . . KEVLAR (TRADEMARK)
　　　RESINS
　　　. SYNTHETIC RESINS
　　　. . THERMOSETTING RESINS
　　　. . . **FURAN RESINS**
　　　. . . . POLYAMIDE RESINS
　　　. . . . . KEVLAR (TRADEMARK)
RT　ADHESIVES
　　　COATINGS

FURANS
GS　HETEROCYCLIC COMPOUNDS
　　　. **FURANS**
　　　. . TETRAHYDROFURAN
　　　ORGANIC COMPOUNDS
　　　. **FURANS**
　　　. . TETRAHYDROFURAN
RT　∞CHEMICAL COMPOUNDS
　　　PLASTICS
　　　SOLVENTS

FURFURYL ALCOHOL
RT　ALDEHYDES
　　　∞AROMATIC COMPOUNDS

FURLABLE ANTENNAS
GS　ANTENNAS
　　　. **FURLABLE ANTENNAS**
RT　COMMUNICATION EQUIPMENT
　　　FOLDING STRUCTURES
　　　SATELLITE ANTENNAS
　　　SPACE COMMUNICATION
　　　SPACECRAFT ANTENNAS

FURNACES
SN　(EXCLUDES DOMESTIC HEATING
　　　EQUIPMENT)
GS　HEATING EQUIPMENT
　　　. **FURNACES**
　　　. . ELECTRIC FURNACES
　　　. . IMAGE FURNACES
　　　. . SOLAR FURNACES
　　　. . VACUUM FURNACES
RT　BOILERS
　　　BURNERS
　　　CHEMICAL ENGINEERING
　　　CHEMICAL REACTORS
　　　CHIMNEYS
　　　COMBUSTION CHAMBERS
　　　CONTROLLED ATMOSPHERES
　　　∞CUPOLAS
　　　DRYING APPARATUS
　　　EXTRACTION
　　　FLUIDIZED BED PROCESSORS
　　　FOUNDRIES
　　　HEARTHS
　　　HEAT TREATMENT
　　　INCINERATORS
　　　INDUCTION HEATING
　　　MECHANICAL ENGINEERING
　　　MELTING
　　　∞METALLURGY
　　　MUFFLERS
　　　OVENS
　　　REFRACTORIES
　　　SEPARATORS
　　　SINTERING
　　　WASTE ENERGY UTILIZATION

FUSELAGE MOUNTING
USE　AIRCRAFT PRODUCTION

FUSELAGES
GS　AIRCRAFT STRUCTURES
　　　. **FUSELAGES**
RT　AIRCRAFT CONSTRUCTION MATERIALS
　　　AIRCRAFT PARTS
　　　AIRFRAMES
　　　BAYS (STRUCTURAL UNITS)
　　　BODY-WING AND TAIL CONFIGURATIONS
　　　CAMBER
　　　CENTERBODIES
　　　COCKPITS
　　　CYLINDRICAL BODIES
　　　HULLS (STRUCTURES)
　　　WING-FUSELAGE STORES

∞ **FUSES**
SN　*(USE OF A MORE SPECIFIC TERM IS*
　　　RECOMMENDED--CONSULT THE TERMS
　　　LISTED BELOW)
RT　CIRCUIT BREAKERS
　　　CIRCUIT PROTECTION
　　　ELECTRIC FUSES
　　　FUSES (ORDNANCE)
　　　WARHEADS

FUSES (ORDNANCE)
RT　AMMUNITION
　　　CAPS (EXPLOSIVES)
　　　DETONATORS
　　　∞FUSES
　　　INITIATORS (EXPLOSIVES)
　　　WARHEADS
　　　WICKS

FUSIBILITY
GS   THERMODYNAMIC PROPERTIES
        . THERMOPHYSICAL PROPERTIES
        . . **FUSIBILITY**
RT   ∞PHYSICAL PROPERTIES
        ∞RESISTANCE
        WELDING

FUSIFORM SHAPES
USE   CONES

∞ **FUSION**
SN   (USE OF A MORE SPECIFIC TERM IS
        RECOMMENDED--CONSULT THE TERMS
        LISTED BELOW)
RT   FUEL PRODUCTION
        FUSION (MELTING)
        INERTIAL FUSION (REACTOR)
        LASER FUSION
        LIQUID-SOLID INTERFACES
        NUCLEAR FUSION

FUSION (MELTING)
GS   PHASE TRANSFORMATIONS
        . MELTING
        . . **FUSION (MELTING)**
RT   ADHESION
        FRIT
        FUSION WELDING
        HEAT OF FUSION
        ∞JOINING
        PHASE CHANGE MATERIALS

FUSION REACTORS
GS   NUCLEAR REACTORS
        . **FUSION REACTORS**
        . . SPHEROMAKS
RT   BETA FACTOR
        BLANKETS (FUSION REACTORS)
        BUMPY TORUSES
        FUSION-FISSION HYBRID REACTORS
        IMPACT FUSION
        INERTIAL FUSION (REACTOR)
        LIMITERS (FUSION REACTORS)
        MIRROR FUSION
        NUCLEAR FISSION
        NUCLEAR FUSION
        ∞REACTORS

FUSION WEAPONS
UF   HYDROGEN BOMBS
GS   WEAPONS
        . **FUSION WEAPONS**
RT   LASER WEAPONS
        NUCLEAR FUSION

FUSION WELDING
GS   WELDING
        . LASER WELDING
        . . **FUSION WELDING**
        . . . GAS WELDING
        . . . . BRAZING
        . . . . . LOW TEMPERATURE BRAZING
RT   ELECTRIC WELDING
        FLASH WELDING
        PRESSURE WELDING
        SPOT WELDS

FUSION-FISSION HYBRID REACTORS
GS   NUCLEAR REACTORS
        . **FUSION-FISSION HYBRID REACTORS**
RT   FUSION REACTORS
        NUCLEAR FISSION
        NUCLEAR FUSION
        ∞REACTORS

FUZZY SETS
RT   ALGORITHMS
        FUZZY SYSTEMS
        SET THEORY

FUZZY SYSTEMS
RT   ALGORITHMS
        FUZZY SETS
        PROBABILITY THEORY
        SET THEORY
        SYSTEM IDENTIFICATION
        ∞SYSTEMS
        SYSTEMS ANALYSIS

FV-12A AIRCRAFT
GS   ATTACK AIRCRAFT
        . FIGHTER AIRCRAFT
        . . **FV-12A AIRCRAFT**

FV-12A AIRCRAFT-*(CONT.)*
        V/STOL AIRCRAFT
        . **FV-12A AIRCRAFT**
RT   ∞AIRCRAFT
        ∞MILITARY AIRCRAFT

F4H AIRCRAFT
USE   F-4 AIRCRAFT

F8U AIRCRAFT
USE   F-8 AIRCRAFT

F9F AIRCRAFT
USE   F-9 AIRCRAFT

G

G FORCE
USE   ACCELERATION (PHYSICS)

G-1 AIRCRAFT
UF   NAVION G-1 AIRCRAFT
        NAVION RANGEMASTER AIRCRAFT
        RANGEMASTER AIRCRAFT
GS   GENERAL AVIATION AIRCRAFT
        . **G-1 AIRCRAFT**
        LIGHT AIRCRAFT
        . **G-1 AIRCRAFT**
        MONOPLANES
        . **G-1 AIRCRAFT**
        NAVION AIRCRAFT
        . **G-1 AIRCRAFT**
        PASSENGER AIRCRAFT
        . **G-1 AIRCRAFT**
        TRANSPORT AIRCRAFT
        . **G-1 AIRCRAFT**
RT   ∞AIRCRAFT

G-91 AIRCRAFT
UF   FIAT G-91 AIRCRAFT
GS   ATTACK AIRCRAFT
        . FIGHTER AIRCRAFT
        . . **G-91 AIRCRAFT**
        FIAT AIRCRAFT
        . **G-91 AIRCRAFT**
        JET AIRCRAFT
        . **G-91 AIRCRAFT**
        LIGHT AIRCRAFT
        . **G-91 AIRCRAFT**
        MONOPLANES
        . **G-91 AIRCRAFT**
        OBSERVATION AIRCRAFT
        . **G-91 AIRCRAFT**
        RECONNAISSANCE AIRCRAFT
        . **G-91 AIRCRAFT**
        TRAINING AIRCRAFT
        . **G-91 AIRCRAFT**
RT   ∞AIRCRAFT

G-95/4 AIRCRAFT
UF   FIAT G-95/4 AIRCRAFT
GS   ATTACK AIRCRAFT
        . FIGHTER AIRCRAFT
        . . **G-95/4 AIRCRAFT**
        FIAT AIRCRAFT
        . **G-95/4 AIRCRAFT**
        JET AIRCRAFT
        . **G-95/4 AIRCRAFT**
        MONOPLANES
        . **G-95/4 AIRCRAFT**
        OBSERVATION AIRCRAFT
        . **G-95/4 AIRCRAFT**
        RECONNAISSANCE AIRCRAFT
        . **G-95/4 AIRCRAFT**
        SUPERSONIC AIRCRAFT
        . **G-95/4 AIRCRAFT**
        V/STOL AIRCRAFT
        . **G-95/4 AIRCRAFT**
RT   ∞AIRCRAFT

G-222 AIRCRAFT
UF   FIAT G-222 AIRCRAFT
GS   FIAT AIRCRAFT
        . **G-222 AIRCRAFT**
        JET AIRCRAFT
        . TURBOPROP AIRCRAFT
        . . **G-222 AIRCRAFT**
        MONOPLANES
        . **G-222 AIRCRAFT**
        PASSENGER AIRCRAFT
        . **G-222 AIRCRAFT**

G-222 AIRCRAFT-*(CONT.)*
        TRANSPORT AIRCRAFT
        . **G-222 AIRCRAFT**
        V/STOL AIRCRAFT
        . **G-222 AIRCRAFT**
RT   ∞AIRCRAFT

GA-5 AIRCRAFT
UF   GLOSTER GA-5 AIRCRAFT
        JAVELIN AIRCRAFT
GS   ATTACK AIRCRAFT
        . FIGHTER AIRCRAFT
        . . **GA-5 AIRCRAFT**
        HAWKER SIDDELEY AIRCRAFT
        . **GA-5 AIRCRAFT**
        JET AIRCRAFT
        . **GA-5 AIRCRAFT**
        MONOPLANES
        . **GA-5 AIRCRAFT**
RT   ∞AIRCRAFT
        DELTA WINGS

GABON
GS   NATIONS
        . **GABON**
RT   AFRICA

GADOLINIUM
GS   CHEMICAL ELEMENTS
        . RARE EARTH ELEMENTS
        . . **GADOLINIUM**
        . . . GADOLINIUM ISOTOPES
        METALS
        . RARE EARTH ELEMENTS
        . . **GADOLINIUM**
        . . . GADOLINIUM ISOTOPES

GADOLINIUM ALLOYS
GS   ALLOYS
        . RARE EARTH ALLOYS
        . . **GADOLINIUM ALLOYS**
        MIXTURES
        . **GADOLINIUM ALLOYS**
RT   METALS
        MIXTURES

GADOLINIUM ISOTOPES
GS   CHEMICAL ELEMENTS
        . NUCLIDES
        . . ISOTOPES
        . . . **GADOLINIUM ISOTOPES**
        . RARE EARTH ELEMENTS
        . . GADOLINIUM
        . . . **GADOLINIUM ISOTOPES**
        METALS
        . RARE EARTH ELEMENTS
        . . GADOLINIUM
        . . . **GADOLINIUM ISOTOPES**

GAGES
USE   MEASURING INSTRUMENTS

GAIN (AMPLIFICATION)
USE   AMPLIFICATION

GALACTIC CLUSTERS
GS   CELESTIAL BODIES
        . GALAXIES
        . . **GALACTIC CLUSTERS**
        . . . LOCAL GROUP (ASTRONOMY)
        . . . . ANDROMEDA GALAXIES
        . . . VIRGO GALACTIC CLUSTER
RT   AGGLOMERATION
        DISK GALAXIES
        ELLIPTICAL GALAXIES
        METALLICITY
        MISSING MASS (ASTROPHYSICS)
        STAR CLUSTERS
        STAR DISTRIBUTION

GALACTIC COSMIC RAYS
GS   EXTRATERRESTRIAL RADIATION
        . GALACTIC RADIATION
        . . **GALACTIC COSMIC RAYS**
        IONIZING RADIATION
        . COSMIC RAYS
        . . **GALACTIC COSMIC RAYS**
RT   ENERGETIC PARTICLES
        SOLAR ACTIVITY EFFECTS
        SOLAR WIND

GALACTIC EVOLUTION
GS   EVOLUTION (DEVELOPMENT)
        . **GALACTIC EVOLUTION**

GALACTIC EVOLUTION-(CONT.)
  RT   ASTROPHYSICS
        BIG BANG COSMOLOGY
        COSMOLOGY
        DISK GALAXIES
        STAR DISTRIBUTION
        STELLAR EVOLUTION
        STELLAR MASS ACCRETION

GALACTIC MAGNETIC FIELDS
  USE   INTERSTELLAR MAGNETIC FIELDS

GALACTIC NUCLEI
  RT   ABSORPTION SPECTRA
        ACCRETION DISKS
        DISK GALAXIES
        RADIO SOURCES (ASTRONOMY)
        SEYFERT GALAXIES

GALACTIC RADIATION
  GS   EXTRATERRESTRIAL RADIATION
        . GALACTIC RADIATION
        . . GALACTIC COSMIC RAYS
        . . GALACTIC RADIO WAVES
        . . . NORTH POLAR SPUR (ASTRONOMY)
  RT   BRIGHTNESS DISTRIBUTION
        CORPUSCULAR RADIATION
        COSMIC NOISE
        COSMIC RAYS
        COSMIC X RAYS
        ELECTROMAGNETIC RADIATION
        GAMMA RAY ASTRONOMY
        GAMMA RAY BURSTS
        HUBBLE DIAGRAM
        INTERSTELLAR RADIATION
        MASS TO LIGHT RATIOS
        ∞RADIATION
        RADIATIVE TRANSFER
        SOLAR RADIATION 1 SATELLITE
        SOLAR RADIATION 3 SATELLITE
        STELLAR RADIATION
        UHURU SATELLITE

GALACTIC RADIATION EXP BACKGROUND SATS
  USE   GREB SATELLITES

GALACTIC RADIO WAVES
  GS   ELECTROMAGNETIC RADIATION
        . RADIO WAVES
        . . EXTRATERRESTRIAL RADIO WAVES
        . . . GALACTIC RADIO WAVES
        EXTRATERRESTRIAL RADIATION
        . EXTRATERRESTRIAL RADIO WAVES
        . . GALACTIC RADIO WAVES
        . . . NORTH POLAR SPUR (ASTRONOMY)
        . GALACTIC RADIATION
        . . GALACTIC RADIO WAVES
        . . . NORTH POLAR SPUR (ASTRONOMY)
  RT   COSMIC NOISE

GALACTIC ROTATION
  RT   COROTATION
        DISK GALAXIES
        HYDROGEN CLOUDS
        STELLAR MOTIONS
        VELOCITY DISTRIBUTION

GALACTIC STRUCTURE
  RT   BARRED GALAXIES
        COROTATION
        DENSITY WAVE MODEL
        DISK GALAXIES
        GALAXIES
        MISSING MASS (ASTROPHYSICS)
        ∞STRUCTURES

GALACTOSE
  GS   ALIPHATIC COMPOUNDS
        . SUGARS
        . . GALACTOSE
        CARBOHYDRATES
        . SUGARS
        . . GALACTOSE

GALAXIES
  GS   CELESTIAL BODIES
        . GALAXIES
        . . DISK GALAXIES
        . . DWARF GALAXIES
        . . ELLIPTICAL GALAXIES
        . . GALACTIC CLUSTERS
        . . . LOCAL GROUP (ASTRONOMY)
        . . . . ANDROMEDA GALAXIES
        . . . VIRGO GALACTIC CLUSTER

GALAXIES-(CONT.)
        . . MAFFEI GALAXIES
        . . RADIO GALAXIES
        . . SEYFERT GALAXIES
        . . SPIRAL GALAXIES
        . . . BARRED GALAXIES
        . . . MILKY WAY GALAXY
  RT   BL LACERTAE OBJECTS
        GALACTIC STRUCTURE
        GUM NEBULA
        HUBBLE CONSTANT
        HUBBLE DIAGRAM
        METALLICITY
        NEBULAE
        ORION NEBULA
        QUASARS
        RADIO SOURCES (ASTRONOMY)
        RED SHIFT
        STAR CLUSTERS
        STARS

GALAXY AIRCRAFT
  USE   C-5 AIRCRAFT

GALERKIN METHOD
  RT   LINEARIZATION
        ∞METHODOLOGY

GALILEAN SATELLITES
  GS   CELESTIAL BODIES
        . NATURAL SATELLITES
        . . JUPITER SATELLITES
        . . . GALILEAN SATELLITES
        . . . . CALLISTO
        . . . . EUROPA
        . . . . GANYMEDE
        . . . . IO
        SATELLITES
        . NATURAL SATELLITES
        . . JUPITER SATELLITES
        . . . GALILEAN SATELLITES
        . . . . CALLISTO
        . . . . EUROPA
        . . . . GANYMEDE
        . . . . IO
  RT   CHARON
        GALILEO PROJECT
        GALILEO SPACECRAFT
        JUPITER (PLANET)
        TRITON

GALILEO MISSION
  USE   GALILEO PROJECT

GALILEO PROBE
  GS   INTERPLANETARY SPACECRAFT
        . JUPITER PROBES
        . . GALILEO PROBE
        UNMANNED SPACECRAFT
        . SPACE PROBES
        . . JUPITER PROBES
        . . . GALILEO PROBE
  RT   JUPITER (PLANET)
        ∞PROBES
        ∞SPACECRAFT

GALILEO PROJECT
  UF   GALILEO MISSION
  GS   PROGRAMS
        . NASA PROGRAMS
        . . GALILEO PROJECT
        . NASA SPACE PROGRAMS
        . . GALILEO PROJECT
        . PROJECTS
        . . GALILEO PROJECT
  RT   AMPHITRITE ASTEROID
        ATMOSPHERIC ENTRY
        FLYBY MISSIONS
        GALILEAN SATELLITES
        JUPITER ATMOSPHERE
        JUPITER PROBES

GALILEO SPACECRAFT
  GS   INTERPLANETARY SPACECRAFT
        . JUPITER PROBES
        . . GALILEO SPACECRAFT
        UNMANNED SPACECRAFT
        . SPACE PROBES
        . . JUPITER PROBES
        . . . GALILEO SPACECRAFT
  RT   FLYBY MISSIONS
        GALILEAN SATELLITES
        JUPITER (PLANET)
        ∞MISSIONS
        ∞SPACECRAFT

GALL
  RT   DIGESTIVE SYSTEM
        GASTROINTESTINAL SYSTEM
        SECRETIONS

GALLAMINE TRIETHIODIDE
  GS   AMINES
        . GALLAMINE TRIETHIODIDE
        ETHERS
        . GALLAMINE TRIETHIODIDE
        HALOGEN COMPOUNDS
        . IODINE COMPOUNDS
        . . IODIDES
        . . . GALLAMINE TRIETHIODIDE

GALLATES
  GS   GALLIUM COMPOUNDS
        . GALLATES
        . . SODIUM GALLATES

GALLIUM
  GS   CHEMICAL ELEMENTS
        . GALLIUM
        . . GALLIUM ISOTOPES
        METALS
        . GALLIUM
        . . GALLIUM ISOTOPES

GALLIUM ALLOYS
  GS   ALLOYS
        . GALLIUM ALLOYS

GALLIUM ANTIMONIDES
  GS   ANTIMONY COMPOUNDS
        . ANTIMONIDES
        . . GALLIUM ANTIMONIDES
        GALLIUM COMPOUNDS
        . GALLIUM ANTIMONIDES

GALLIUM ARSENIDE LASERS
  GS   ELECTRONIC EQUIPMENT
        . SOLID STATE DEVICES
        . . SEMICONDUCTOR DEVICES
        . . . SEMICONDUCTOR LASERS
        . . . . GALLIUM ARSENIDE LASERS
        . . SOLID STATE LASERS
        . . . GALLIUM ARSENIDE LASERS
        STIMULATED EMISSION DEVICES
        . LASERS
        . . SEMICONDUCTOR LASERS
        . . . GALLIUM ARSENIDE LASERS
        . . SOLID STATE LASERS
        . . . GALLIUM ARSENIDE LASERS
  RT   INJECTION LASERS
        STIMULATED EMISSION
        WAVEGUIDE LASERS

GALLIUM ARSENIDES
  GS   ARSENIC COMPOUNDS
        . ARSENIDES
        . . GALLIUM ARSENIDES
        . . . ALUMINUM GALLIUM ARSENIDES
        GALLIUM COMPOUNDS
        . GALLIUM ARSENIDES
        . . ALUMINUM GALLIUM ARSENIDES
  RT   GUNN DIODES
        HETEROJUNCTION DEVICES
        INJECTION LASERS
        NEGATIVE CONDUCTANCE
        NEGATIVE RESISTANCE DEVICES
        SCHOTTKY DIODES
        SEMICONDUCTOR LASERS
        TRANSFERRED ELECTRON DEVICES

GALLIUM COMPOUNDS
  GS   GALLIUM COMPOUNDS
        . GALLATES
        . . SODIUM GALLATES
        . GALLIUM ANTIMONIDES
        . GALLIUM ARSENIDES
        . . ALUMINUM GALLIUM ARSENIDES
        . GALLIUM NITRIDES
        . GALLIUM OXIDES
        . GALLIUM PHOSPHIDES
        . GALLIUM SELENIDES
  RT   ∞CHEMICAL COMPOUNDS
        ∞GROUP 3A COMPOUNDS
        ∞METAL COMPOUNDS

GALLIUM ISOTOPES
  GS   CHEMICAL ELEMENTS
        . GALLIUM
        . . GALLIUM ISOTOPES
        . NUCLIDES

GALLIUM ISOTOPES-*(CONT.)*
```
. . ISOTOPES
. . . GALLIUM ISOTOPES
METALS
. GALLIUM
. . GALLIUM ISOTOPES
```

GALLIUM NITRIDES
```
GS    GALLIUM COMPOUNDS
      . GALLIUM NITRIDES
RT    SEMICONDUCTORS (MATERIALS)
```

GALLIUM OXIDES
```
GS    GALLIUM COMPOUNDS
      . GALLIUM OXIDES
```

GALLIUM PHOSPHIDES
```
GS    GALLIUM COMPOUNDS
      . GALLIUM PHOSPHIDES
      PHOSPHORUS COMPOUNDS
      . PHOSPHIDES
      . . GALLIUM PHOSPHIDES
```

GALLIUM SELENIDES
```
GS    CHALCOGENIDES
      . SELENIDES
      . . GALLIUM SELENIDES
      GALLIUM COMPOUNDS
      . GALLIUM SELENIDES
      SELENIUM COMPOUNDS
      . SELENIDES
      . . GALLIUM SELENIDES
```

GALVANIC CELLS
```
USE   ELECTROLYTIC CELLS
```

GALVANIC SKIN RESPONSE
```
UF    ELECTRODERMAL RESPONSE
GS    RESPONSES
      . GALVANIC SKIN RESPONSE
RT    ELECTRICAL RESISTANCE
      LIES
```

GALVANIZING
```
USE   ZINC COATINGS
```

GALVANOMAGNETIC EFFECTS
```
UF    GALVANOMAGNETISM
GS    GALVANOMAGNETIC EFFECTS
      . NERNST-ETTINGSHAUSEN EFFECT
RT    ∞EFFECTS
      HALL EFFECT
```

GALVANOMAGNETISM
```
USE   GALVANOMAGNETIC EFFECTS
```

GALVANOMETERS
```
GS    MEASURING INSTRUMENTS
      . GALVANOMETERS
RT    AMMETERS
      ELECTROMETERS
      MICROMILLIAMMETERS
      MILLIVOLTMETERS
      THERMOCOUPLE PYROMETERS
```

GAMBIA
```
GS    NATIONS
      . GAMBIA
RT    AFRICA
```

GAME THEORY
```
GS    GAME THEORY
      . SADDLE POINTS (GAME THEORY)
RT    DECISION THEORY
      DEPLOYMENT
      INFORMATION THEORY
      LINEAR PROGRAMMING
      MARTINGALES
      MATHEMATICAL MODELS
      MATHEMATICAL PROGRAMMING
      MINIMAX TECHNIQUE
      MONTE CARLO METHOD
      OPERATIONS RESEARCH
      PROBABILITY THEORY
      RISK
      SADDLE POINTS
      SIMULATION
      STATISTICAL ANALYSIS
      STATISTICAL DECISION THEORY
      STOCHASTIC PROCESSES
      STRATEGY
      ∞THEORIES
      WAR GAMES
```

GAMETOCYTES
```
UF    OOCYTES
      SPERMATOCYTES
GS    CELLS (BIOLOGY)
      . GAMETOCYTES
RT    SPERMATOGENESIS
```

GAMMA FUNCTION
```
GS    ANALYSIS (MATHEMATICS)
      . COMPLEX VARIABLES
      . . GAMMA FUNCTION
      FUNCTIONS (MATHEMATICS)
      . GAMMA FUNCTION
RT    FACTORIALS
      STATISTICAL DISTRIBUTIONS
```

GAMMA GLOBULIN
```
GS    ANTIBODIES
      . GAMMA GLOBULIN
      PROTEINS
      . GLOBULINS
      . . GAMMA GLOBULIN
```

GAMMA RADIATION
```
USE   GAMMA RAYS
```

GAMMA RAY ABSORPTIOMETRY
```
GS    DENSITY MEASUREMENT
      . GAMMA RAY ABSORPTIOMETRY
RT    ABSORPTION SPECTRA
      DENSITOMETERS
      ELECTROMAGNETIC ABSORPTION
      ENERGY ABSORPTION
      ∞MEASUREMENT
      PHOTON ABSORPTIOMETRY
      RADIATION ABSORPTION
```

GAMMA RAY ABSORPTION
```
GS    ENERGY ABSORPTION
      . ELECTROMAGNETIC ABSORPTION
      . . GAMMA RAY ABSORPTION
      RADIATION ABSORPTION
      . ELECTROMAGNETIC ABSORPTION
      . . GAMMA RAY ABSORPTION
RT    ∞ABSORPTION
      ELECTROMAGNETIC PROPERTIES
      ELECTROMAGNETIC RADIATION
      IONIZING RADIATION
      NUCLEAR RADIATION
      PHOTON ABSORPTIOMETRY
```

GAMMA RAY ASTRONOMY
```
GS    ASTRONOMY
      . GAMMA RAY ASTRONOMY
RT    ASTROPHYSICS
      COSMIC X RAYS
      ENERGY SPECTRA
      GALACTIC RADIATION
      GAMMA RAY BURSTS
      RADIO ASTRONOMY
      X RAY ASTRONOMY
```

GAMMA RAY ASTRONOMY EXPLORER
```
USE   EXPLORER 11 SATELLITE
```

GAMMA RAY BEAMS
```
GS    BEAMS (RADIATION)
      . GAMMA RAY BEAMS
      ELECTROMAGNETIC RADIATION
      . GAMMA RAY BEAMS
      IONIZING RADIATION
      . GAMMA RAY BEAMS
      NUCLEAR RADIATION
      . GAMMA RAY BEAMS
RT    PHOTON BEAMS
      RADIOACTIVE DECAY
```

GAMMA RAY BURSTS
```
UF    COSMIC GAMMA RAY BURSTS
GS    BURSTS
      . GAMMA RAY BURSTS
      ELECTROMAGNETIC RADIATION
      . GAMMA RAYS
      . . GAMMA RAY BURSTS
      EXTRATERRESTRIAL RADIATION
      . GAMMA RAY BURSTS
      IONIZING RADIATION
      . COSMIC RAYS
      . . GAMMA RAY BURSTS
      . GAMMA RAYS
      . . GAMMA RAY BURSTS
      NUCLEAR RADIATION
      . GAMMA RAYS
      . . GAMMA RAY BURSTS
```

GAMMA RAY BURSTS-*(CONT.)*
```
RT    BIG BANG COSMOLOGY
      BREMSSTRAHLUNG
      CERENKOV RADIATION
      COSMIC X RAYS
      GALACTIC RADIATION
      GAMMA RAY ASTRONOMY
      INTERSTELLAR RADIATION
      NUCLEAR PARTICLES
      RADIANT FLUX DENSITY
      STELLAR RADIATION
      X RAY ASTRONOMY
```

GAMMA RAY LASERS
```
GS    STIMULATED EMISSION DEVICES
      . LASERS
      . . GAMMA RAY LASERS
RT    COHERENT LIGHT
      LIGHT TRANSMISSION
      OPTICAL PUMPING
      PULSED RADIATION
```

GAMMA RAY OBSERVATORY
```
GS    OBSERVATORIES
      . ASTRONOMICAL OBSERVATORIES
      . . GAMMA RAY OBSERVATORY
RT    OGO
      TELESCOPES
```

GAMMA RAY SPECTRA
```
GS    SPECTRA
      . RADIATION SPECTRA
      . . ELECTROMAGNETIC SPECTRA
      . . . GAMMA RAY SPECTRA
RT    EMISSION SPECTRA
      IONIZING RADIATION
      RADIATION SPECTRA
      SPECTRA
```

GAMMA RAY SPECTROMETERS
```
GS    MEASURING INSTRUMENTS
      . SPECTROMETERS
      . . GAMMA RAY SPECTROMETERS
RT    OPTICAL MEASUREMENT
      SOLAR MAXIMUM MISSION
      SPECTRA
      SPECTRUM ANALYSIS
```

GAMMA RAY TELESCOPES
```
GS    TELESCOPES
      . GAMMA RAY TELESCOPES
RT    ASTRONOMY
      COSMIC RAYS
      ELECTROMAGNETIC RADIATION
```

GAMMA RAYS
```
SN    (EMITTED BY NUCLEI)
UF    GAMMA RADIATION
GS    ELECTROMAGNETIC RADIATION
      . GAMMA RAYS
      . . GAMMA RAY BURSTS
      IONIZING RADIATION
      . GAMMA RAYS
      . . GAMMA RAY BURSTS
      NUCLEAR RADIATION
      . GAMMA RAYS
      . . GAMMA RAY BURSTS
RT    BREMSSTRAHLUNG
      CERENKOV RADIATION
      COSMIC RAYS
      COSMIC X RAYS
      EMISSION SPECTRA
      FLUX (RATE)
      FLUX DENSITY
      MONOCHROMATIC RADIATION
      MOSSBAUER EFFECT
      PHOTOMAGNETIC EFFECTS
      PHOTONS
      ∞RADIATION
      RADIATION EFFECTS
      RADIATION SHIELDING
      RADIOACTIVE DECAY
      RADIOACTIVITY
      ∞RAYS
      TRANSVERSE OSCILLATION
      TRANSVERSE WAVES
      X RAYS
```

GANGLIA
```
GS    NERVOUS SYSTEM
      . GANGLIA
      . . NERVES
      . . NEURONS
RT    CELLS (BIOLOGY)
      NEUROPHYSIOLOGY
```

GANTRIES
USE   GANTRY CRANES

GANTRY CRANES
UF   GANTRIES
GS   HANDLING EQUIPMENT
       . CRANES
       . . **GANTRY CRANES**
RT   GROUND SUPPORT EQUIPMENT
       LAUNCHING PADS
       LAUNCHING SITES
       UMBILICAL TOWERS

GANYMEDE
GS   CELESTIAL BODIES
       . NATURAL SATELLITES
       . . JUPITER SATELLITES
       . . . GALILEAN SATELLITES
       . . . . **GANYMEDE**
       SATELLITES
       . NATURAL SATELLITES
       . . JUPITER SATELLITES
       . . . GALILEAN SATELLITES
       . . . . **GANYMEDE**
RT   CALLISTO
       CHARON
       IO
       JUPITER (PLANET)

GAPS
GS   **GAPS**
       . ENERGY GAPS (SOLID STATE)
       . SPARK GAPS
RT   ∞ARRESTERS
       ∞BREAKDOWN
       ∞HOLES
       OPENINGS
       ORIFICES
       PASSAGEWAYS
       QUANTUM WELLS
       ∞TUNNELS

GAPS (GEOLOGY)
UF   COLS
       PASSES
GS   LANDFORMS
       . **GAPS (GEOLOGY)**
RT   GEOLOGY
       MOUNTAINS
       ∞RIDGES

GARBAGE
GS   WASTES
       . **GARBAGE**
RT   COMPOSTING
       ORGANIC WASTES (FUEL CONVERSION)
       SEWERS
       SOLID WASTES
       UTILITIES
       WASTE DISPOSAL
       WASTE TREATMENT

GARMENTS
GS   CLOTHING
       . **GARMENTS**
RT   FLIGHT CLOTHING
       SUITS
       VESTS

GARNETS
GS   MINERALS
       . **GARNETS**
       SILICON COMPOUNDS
       . SILICATES
       . . **GARNETS**
       . . . YTTRIUM-ALUMINUM GARNET
       . . . YTTRIUM-IRON GARNET
RT   ECLOGITE
       LASERS

GARP
USE   GLOBAL ATMOSPHERIC RESEARCH
           PROGRAM

GARP ATLANTIC TROPICAL EXPERIMENT
UF   GATE (EXPERIMENT)
GS   PROGRAMS
       . GLOBAL ATMOSPHERIC RESEARCH
           PROGRAM
       . . **GARP ATLANTIC TROPICAL
           EXPERIMENT**
RT   ATLANTIC OCEAN
       INTERTROPICAL CONVERGENT ZONES
       METEOROLOGY

GARP ATLANTIC TROPICAL-(CONT.)
       NASA PROGRAMS
       OCEANOGRAPHY
       TROPICAL METEOROLOGY
       TROPICAL REGIONS
       WEATHER FORECASTING

GAS ANALYSIS
GS   CHEMICAL TESTS
       . CHEMICAL ANALYSIS
       . . **GAS ANALYSIS**
       . . . OZONOMETRY
       . . . VAN SLYKE METHOD
RT   AIR SAMPLING
       FLAME PROBES
       HOPCALITE (TRADEMARK)
       MASS SPECTROMETERS
       ∞MATERIALS TESTS
       OXYGEN ANALYZERS
       QUALITATIVE ANALYSIS
       QUANTITATIVE ANALYSIS
       VOLUMETRIC ANALYSIS

GAS ATOMIZATION
GS   ATOMIZING
       . **GAS ATOMIZATION**
RT   AEROSOLS
       COLLISIONS
       COMMINUTION
       LIQUID ATOMIZATION
       PARTICLES

GAS BAGS
GS   BAGS
       . **GAS BAGS**
       EXPANDABLE STRUCTURES
       . INFLATABLE STRUCTURES
       . . **GAS BAGS**
RT   BALLOONS
       HIGH ALTITUDE BALLOONS

GAS BEARINGS
UF   AIR BEARINGS
       GAS LUBRICATED BEARINGS
GS   BEARINGS
       . **GAS BEARINGS**
RT   ANTIFRICTION BEARINGS
       FLUID FILMS
       HIGH TEMPERATURE LUBRICANTS
       SQUEEZE FILMS
       THRUST BEARINGS
       TURBINE ENGINES

GAS CHROMATOGRAPHY
GS   CHROMATOGRAPHY
       . **GAS CHROMATOGRAPHY**
RT   ADSORPTION
       CHEMICAL ANALYSIS
       PAPER CHROMATOGRAPHY
       SORPTION
       THIN LAYER CHROMATOGRAPHY

GAS COMPOSITION
GS   COMPOSITION (PROPERTY)
       . **GAS COMPOSITION**
       . . CARBON DIOXIDE CONCENTRATION
RT   ATMOSPHERIC COMPOSITION
       ATOM CONCENTRATION
       CHEMICAL COMPOSITION
       DALTON LAW
       EXPIRED AIR
       IONOSPHERIC COMPOSITION
       PLASMA COMPOSITION
       POLAR GASES

GAS COOLED FAST REACTORS
GS   NUCLEAR REACTORS
       . FAST NUCLEAR REACTORS
       . . **GAS COOLED FAST REACTORS**
       . GAS COOLED REACTORS
       . . **GAS COOLED FAST REACTORS**

GAS COOLED REACTORS
UF   GCR (REACTORS)
GS   NUCLEAR REACTORS
       . **GAS COOLED REACTORS**
       . . EXPERIMENTAL GAS COOLED
           REACTORS
       . . GAS COOLED FAST REACTORS
       . . HIGH TEMPERATURE NUCLEAR
           REACTORS
       . . . HIGH TEMPERATURE GAS COOLED
           REACTORS
       . . KIWI REACTORS
       . . . KIWI B REACTORS

GAS COOLED REACTORS-(CONT.)
       . . . . KIWI B-1 REACTOR
       . . . . KIWI B-4 REACTOR
       . . TORY 2 REACTOR
       . . TORY 2-A REACTOR
       . . TORY 2-C REACTOR
RT   ∞GAS REACTORS

GAS COOLING
SN   (COOLING WITH GAS)
GS   COOLING
       . **GAS COOLING**
RT   COOLANTS
       FREON
       HEAT EXCHANGERS

GAS DENSITY
GS   DENSITY (MASS/VOLUME)
       . **GAS DENSITY**
RT   ATOM CONCENTRATION
       BUOYANCY
       CONVECTIVE FLOW
       GASEOUS DIFFUSION
       IDEAL GAS
       PROBABILITY DENSITY FUNCTIONS
       RAREFIED GASES
       REAL GASES

GAS DETECTORS
RT   DETECTION
       ∞DETECTORS
       HAZE DETECTION
       IDENTIFYING
       INDICATING INSTRUMENTS
       MONITORS
       ∞PROBES
       ∞SENSORS
       SMOKE DETECTORS
       WARNING SYSTEMS

GAS DIFFUSION
USE   GASEOUS DIFFUSION

GAS DISCHARGE COUNTERS
USE   COUNTERS
       GAS DISCHARGE TUBES

GAS DISCHARGE TUBES
UF   DISCHARGE TUBES
       GAS DISCHARGE COUNTERS
GS   ELECTRON TUBES
       . VACUUM TUBES
       . . VACUUM TUBE OSCILLATORS
       . . . **GAS DISCHARGE TUBES**
       . . . . IGNITRONS
       . . . . THYRATRONS
       OSCILLATORS
       . VACUUM TUBE OSCILLATORS
       . . **GAS DISCHARGE TUBES**
RT   FARADAY DARK SPACE
       ∞GAS TUBES
       MICROWAVE EQUIPMENT
       MICROWAVE OSCILLATORS
       MICROWAVE TUBES
       PHOTOTUBES
       RADIATION COUNTERS

GAS DISCHARGES
GS   ELECTRIC CURRENT
       . ELECTRIC DISCHARGES
       . . TOWNSEND DISCHARGE
       . . . **GAS DISCHARGES**
       . . . . TOROIDAL DISCHARGE
       . . . . . RING DISCHARGE
RT   AFTERGLOWS
       COLD CATHODE TUBES
       COLD CATHODES
       ELECTRIC ARCS
       ELECTRIC SPARKS
       ELECTRODELESS DISCHARGES
       ELECTRON AVALANCHE
       GLOW DISCHARGES
       LIGHTNING
       POLAR GASES

GAS DISSOCIATION
GS   DISSOCIATION
       . **GAS DISSOCIATION**
RT   THERMAL DISSOCIATION

GAS DYNAMICS
GS   FLUID MECHANICS
       . FLUID DYNAMICS
       . . **GAS DYNAMICS**

GAS DYNAMICS-(CONT.)
```
        . . . AERODYNAMICS
        . . . . AEROTHERMODYNAMICS
        . . . . HYPERSONICS
        . . . . ROTOR AERODYNAMICS
        . . . . SUPERSONICS
        . . . INTERACTIONAL AERODYNAMICS
        . . . RAREFIED GAS DYNAMICS
RT      DALTON LAW
        ∞ DYNAMICS
        GAS PATH ANALYSIS
        GASEOUS DIFFUSION
        GASEOUS SELF-DIFFUSION
        GASES
        HYDRODYNAMIC EQUATIONS
        HYDRODYNAMICS
        JET MEMBRANE PROCESS
        KINETICS
        LORENTZ GAS
        MAGNETOHYDRODYNAMICS
        ∞ MECHANICS (PHYSICS)
        MOLECULAR GASES
        POLAR GASES
        THERMODYNAMICS
```

GAS EVACUATING
```
USE     EVACUATING (VACUUM)
```

GAS EVOLUTION
```
GS      EVOLUTION (LIBERATION)
        . GAS EVOLUTION
RT      DEGASSING
        OUTGASSING
        TRANSPIRATION
```

GAS EXCHANGE
```
GS      EXCHANGING
        . GAS EXCHANGE
RT      OXYGEN PRODUCTION
```

GAS EXPANSION
```
GS      EXPANSION
        . GAS EXPANSION
RT      JOULE-THOMSON EFFECT
        PRESSURE REDUCTION
```

GAS EXPLOSIONS
```
GS      EXPLOSIONS
        . CHEMICAL EXPLOSIONS
        . . GAS EXPLOSIONS
RT      DETONABLE GAS MIXTURES
        DETONATION WAVES
        FLAME PROPAGATION
        FLAMMABLE GASES
        UNDERGROUND EXPLOSIONS
```

GAS FLOW
```
UF      GASEOUS CAVITATION
GS      FLUID FLOW
        . GAS FLOW
        . . AIR FLOW
        . . . AIR CURRENTS
        . . . . JET STREAMS (METEOROLOGY)
        . . . . MERIDIONAL FLOW
        . . . . VERTICAL AIR CURRENTS
        . . CONTINUUM FLOW
        . . EQUILIBRIUM FLOW
        . . . FROZEN EQUILIBRIUM FLOW
        . . . SHIFTING EQUILIBRIUM FLOW
        . . FREE MOLECULAR FLOW
        . . KNUDSEN FLOW
        . . MOLECULAR FLOW
        . . . SLIP FLOW
        . . . TRANSITION FLOW
        . . NONEQUILIBRIUM FLOW
        . . PIPE FLOW
RT      AIR DUCTS
        AIR JETS
        COMPRESSIBLE FLOW
        CRITICAL FLOW
        CROCCO-LEE THEORY
        GAS PATH ANALYSIS
        GASDYNAMIC LASERS
        GASEOUS DIFFUSION
        GASES
        GEOPHYSICAL FLUID FLOW CELLS
        HYDRAULIC ANALOGIES
        HYPERSONIC FLOW
        INCOMPRESSIBLE FLOW
        INVISCID FLOW
        JOULE-THOMSON EFFECT
        LAMINAR FLOW
        LIQUID FLOW
        MAGNETOHYDRODYNAMIC FLOW
        MASS FLOW
```

GAS FLOW-(CONT.)
```
        MOLECULAR RELAXATION
        MOLECULAR TRAJECTORIES
        MULTIPHASE FLOW
        NONUNIFORM FLOW
        ORIFICE FLOW
        ∞ PRESSURE DROP
        RADIAL FLOW
        SINGLE-PHASE FLOW
        STEADY FLOW
        STEAM FLOW
        STREAMS
        SUBCRITICAL FLOW
        SUBSONIC FLOW
        SUPERCRITICAL FLOW
        SUPERSONIC FLOW
        SUPERSONIC JET FLOW
        TRANSONIC FLOW
        TURBULENT FLOW
        TWO PHASE FLOW
        UNIFORM FLOW
        UNSTEADY FLOW
        VAPOR JETS
        VISCOUS FLOW
```

GAS GENERATOR ENGINES
```
USE     ENGINES
        GAS GENERATORS
```

GAS GENERATORS
```
UF      GAS GENERATOR ENGINES
RT      CHEMICAL REACTORS
        ∞ GENERATORS
        PNEUMATIC EQUIPMENT
        PRESSURIZING
        VAPORIZERS
```

GAS GIANT PLANETS
```
GS      CELESTIAL BODIES
        . PLANETS
        . . GAS GIANT PLANETS
        . . . JUPITER (PLANET)
        . . . NEPTUNE (PLANET)
        . . . SATURN (PLANET)
        . . . URANUS (PLANET)
        . . . . URANUS RINGS
        . . . . URANUS SATELLITES
RT      EXTRASOLAR PLANETS
        JUPITER RED SPOT
        NEPTUNE ATMOSPHERE
        PLANETARY COMPOSITION
        SATURN RINGS
        SOLAR SYSTEM
        URANUS ATMOSPHERE
```

GAS GUNS
```
GS      GAS GUNS
        . LIGHT GAS GUNS
RT      ATMOSPHERIC ENTRY
        BALLISTICS
        ∞ GUNS
        HYPERVELOCITY GUNS
        WIND TUNNELS
```

GAS HEATING
```
GS      HEATING
        . GAS HEATING
RT      ARC HEATING
        KINETIC HEATING
        PLASMA HEATING
        RADIANT HEATING
        RESISTANCE HEATING
        THERMAL DIFFUSION
```

GAS INJECTION
```
GS      INJECTION
        . FLUID INJECTION
        . . GAS INJECTION
RT      FORMATIONS
        FUEL INJECTION
        INFLATING
        PERFORATING
        PLASMA PUMPING
        POROSITY
        PRESSURIZING
        STIMULATION
        WATER INJECTION
```

GAS IONIZATION
```
GS      IONIZATION
        . GAS IONIZATION
        . . ATMOSPHERIC IONIZATION
        . . . AURORAL IONIZATION
        . . FLAME IONIZATION
RT      AFTERGLOWS
```

GAS IONIZATION-(CONT.)
```
        ELECTRON ATTACHMENT
        HELIUM AFTERGLOW
        IONIZED GASES
        IONIZERS
        PENNING DISCHARGE
        PENNING EFFECT
        PHOTOIONIZATION
        PLASMA DISPLAY DEVICES
        RING DISCHARGE
```

GAS JETS
```
GS      FLUID JETS
        . GAS JETS
RT      AIR JETS
        COLD GAS
        ∞ JETS
```

GAS LASERS
```
GS      STIMULATED EMISSION DEVICES
        . LASERS
        . . GAS LASERS
        . . . CARBON DIOXIDE LASERS
        . . . CARBON MONOXIDE LASERS
        . . . DF LASERS
        . . . EXCIMER LASERS
        . . . HCL LASERS
        . . . . HCL ARGON LASERS
        . . . HCN LASERS
        . . . HELIUM-NEON LASERS
        . . . HF LASERS
        . . . KRYPTON FLUORIDE LASERS
        . . . NITROGEN LASERS
        . . . TEA LASERS
        . . . ULTRAVIOLET LASERS
        . . . XENON CHLORIDE LASERS
        . . . XENON FLUORIDE LASERS
RT      CARBON LASERS
        CHEMICAL LASERS
        ELECTRON PUMPING
        GASDYNAMIC LASERS
        INFRARED LASERS
        MACH-ZEHNDER INTERFEROMETERS
        MOLECULAR OSCILLATIONS
        NUCLEAR PUMPING
        ORGANIC LASERS
        POLAR GASES
        POWER TRANSMISSION (LASERS)
        PULSED LASERS
        Q SWITCHED LASERS
        RARE GAS-HALIDE LASERS
        STIMULATED EMISSION
        WATER MASERS
```

GAS LIQUEFACTION
```
USE     CONDENSING
```

GAS LUBRICANTS
```
GS      LUBRICANTS
        . GAS LUBRICANTS
RT      GASEOUS DIFFUSION
        HIGH TEMPERATURE LUBRICANTS
        METAL-GAS SYSTEMS
        SOLID LUBRICANTS
        SQUEEZE FILMS
```

GAS LUBRICATED BEARINGS
```
USE     GAS BEARINGS
```

GAS MASERS
```
GS      STIMULATED EMISSION DEVICES
        . MASERS
        . . GAS MASERS
        . . . HYDROGEN MASERS
RT      ARGON LASERS
        ATOMIC CLOCKS
        CARBON DIOXIDE LASERS
        FREQUENCY STANDARDS
        INTERSTELLAR MASERS
        POLAR GASES
        STIMULATED EMISSION
        TEA LASERS
        WATER MASERS
```

GAS METERS
```
GS      MEASURING INSTRUMENTS
        . FLOWMETERS
        . . GAS METERS
RT      FLOW MEASUREMENT
        VENTURI TUBES
```

GAS MIXTURES
```
GS      GASES
        . GAS MIXTURES
        . . ALVEOLAR AIR
```

GAS MIXTURES-*(CONT.)*
```
     . . COMPRESSED AIR
     . . DETONABLE GAS MIXTURES
     . . EXPIRED AIR
     . . LIQUID AIR
     . MIXTURES
     . SOLUTIONS
     . . GAS MIXTURES
         DETONABLE GAS MIXTURES
RT   ARGON-OXYGEN ATMOSPHERES
     ∞ATMOSPHERES
     BINARY FLUIDS
     BINARY MIXTURES
     CONTROLLED ATMOSPHERES
     EUDIOMETERS
     EXHAUST GASES
     FUEL-AIR RATIO
     FUMES
     GASEOUS ROCKET PROPELLANTS
     HELIUM-OXYGEN ATMOSPHERES
     HYDROGEN-BASED ENERGY
     IGNITION LIMITS
     LAMINAR MIXING
     LIGHTHILL GAS MODEL
     LIQUEFIED GASES
     LIQUID-GAS MIXTURES
     PREMIXED FLAMES
     PREMIXING
```

GAS PATH ANALYSIS
```
RT   GAS DYNAMICS
     GAS FLOW
```

GAS PHASES
```
USE  VAPOR PHASES
```

GAS PIPES
```
GS   PIPES (TUBES)
     . GAS PIPES
RT   ∞GAS TUBES
```

GAS POCKETS
```
RT   CAVITIES
     EVACUATING (VACUUM)
```

GAS PRESSURE
```
GS   PRESSURE
     . GAS PRESSURE
RT   ATMOSPHERIC PRESSURE
     COMPRESSED GAS
     INTERNAL PRESSURE
     PARTIAL PRESSURE
```

∞ GAS REACTORS
```
SN   (USE OF A MORE SPECIFIC TERM IS
     RECOMMENDED--CONSULT THE TERMS
     LISTED BELOW)
RT   CHEMICAL REACTORS
     GAS COOLED REACTORS
     GASEOUS FISSION REACTORS
```

GAS RECOVERY
```
GS   RECLAMATION
     . GAS RECOVERY
RT   ENERGY TECHNOLOGY
     EXHAUST GASES
     GASEOUS DIFFUSION
     GASES
     ∞RECOVERY
     WASTES
```

GAS SPECTROSCOPY
```
GS   CHEMICAL TESTS
     . CHEMICAL ANALYSIS
     . . GAS SPECTROSCOPY
     SPECTROSCOPY
     . GAS SPECTROSCOPY
     . . FLAME SPECTROSCOPY
RT   MAGNETIC SPECTROSCOPY
     MASS SPECTROSCOPY
     OPTOGALVANIC SPECTROSCOPY
     SPECTROSCOPIC ANALYSIS
     VACUUM SPECTROSCOPY
     VISIBLE SPECTRUM
```

GAS STREAMS
```
GS   GASES
     . GAS STREAMS
     STREAMS
     . GAS STREAMS
RT   JET FLOW
     LAMINAR FLOW
     TURBULENCE
     WIND TUNNELS
```

GAS TEMPERATURE
```
GS   TEMPERATURE
     . GAS TEMPERATURE
RT   ATMOSPHERIC TEMPERATURE
     INLET TEMPERATURE
     IONIZED GASES
     RAREFIED GASES
     SHOCK TUBES
     SHOCK WAVES
     TEMPERATURE MEASUREMENT
```

GAS TRANSPORT
```
SN   (ENCOMPASSES GAS DYNAMICS--
     EXCLUDES MATERIALS HANDLING)
RT   ENERGY TRANSFER
     GASEOUS DIFFUSION
     HEAT TRANSFER
     KINETIC THEORY
     LIGHTHILL GAS MODEL
     MAGNETOHYDRODYNAMICS
     MASS TRANSFER
     POLLUTION TRANSPORT
     TRANSPORT THEORY
```

∞ GAS TUBES
```
SN   (USE OF A MORE SPECIFIC TERM IS
     RECOMMENDED--CONSULT THE TERMS
     LISTED BELOW)
RT   COLD CATHODE TUBES
     GAS DISCHARGE TUBES
     GAS PIPES
     TRIGATRONS
```

GAS TUNGSTEN ARC WELDING
```
UF   TIG WELDING
     TUNGSTEN INERT GAS WELDING
GS   WELDING
     . ELECTRIC WELDING
     . . ARC WELDING
     . . . GAS TUNGSTEN ARC WELDING
```

GAS TURBINE ENGINES
```
GS   ENGINES
     . AIR BREATHING ENGINES
     . . GAS TURBINE ENGINES
     . . . JET ENGINES
     . . . . RAMJET ENGINES
     . . . . . INTEGRAL ROCKET RAMJETS
     . . . . . LOW VOLUME RAMJET ENGINES
     . . . . . PULSEJET ENGINES
     . . . . . SUPERSONIC COMBUSTION
               RAMJET ENGINES
     . . . . . TURBORAMJET ENGINES
     . . . . TURBOJET ENGINES
     . . . . . BRISTOL-SIDDELEY OLYMPUS
               593 ENGINE
     . . . . . BRISTOL-SIDDELEY VIPER
               ENGINE
     . . . . . DUCTED FAN ENGINES
     . . . . . J-33 ENGINE
     . . . . . J-34 ENGINE
     . . . . . J-47 ENGINE
     . . . . . J-57 ENGINE
     . . . . . J-57-P-20 ENGINE
     . . . . . J-65 ENGINE
     . . . . . J-69-T-25 ENGINE
     . . . . . J-71 ENGINE
     . . . . . J-73 ENGINE
     . . . . . J-75 ENGINE
     . . . . . J-79 ENGINE
     . . . . . J-85 ENGINE
     . . . . . J-93 ENGINE
     . . . . . RA-28 ENGINE
     . . . . . TURBOFAN ENGINES
     . . . . . . BRISTOL-SIDDELEY BS 53
                 ENGINE
     . . . . . . CF-700 ENGINE
     . . . . . . J-97 ENGINE
     . . . . . . TF-41 ENGINE
     . . . . TURBOPROP ENGINES
     . . . . . T-53 ENGINE
     . . . . . T-56 ENGINE
     . . . . . T-64 ENGINE
     . . . . . T-74 ENGINE
     . . . . TURBORAMJET ENGINES
     . . T-58-GE-8B ENGINE
     . INTERNAL COMBUSTION ENGINES
     . . GAS TURBINE ENGINES
     . . . HYDROGEN ENGINES
     . . . JET ENGINES
     . . . . RAMJET ENGINES
     . . . . . INTEGRAL ROCKET RAMJETS
     . . . . . LOW VOLUME RAMJET ENGINES
     . . . . . PULSEJET ENGINES
```

GAS TURBINE ENGINES-*(CONT.)*
```
     . . . . . SUPERSONIC COMBUSTION
               RAMJET ENGINES
     . . . . TURBORAMJET ENGINES
     . . . T-63 ENGINE
     . . . T-76 ENGINE
     . . . TURBOJET ENGINES
     . . . . BRISTOL-SIDDELEY OLYMPUS
             593 ENGINE
     . . . . BRISTOL-SIDDELEY VIPER
             ENGINE
     . . . . DUCTED FAN ENGINES
     . . . . J-33 ENGINE
     . . . . J-34 ENGINE
     . . . . J-47 ENGINE
     . . . . J-52 ENGINE
     . . . . J-57 ENGINE
     . . . . J-57-P-20 ENGINE
     . . . . J-65 ENGINE
     . . . . J-69-T-25 ENGINE
     . . . . J-71 ENGINE
     . . . . J-73 ENGINE
     . . . . J-75 ENGINE
     . . . . J-79 ENGINE
     . . . . J-85 ENGINE
     . . . . J-93 ENGINE
     . . . . RA-28 ENGINE
     . . . . TURBOFAN ENGINES
     . . . . . BRISTOL-SIDDELEY BS 53
               ENGINE
     . . . . CF-700 ENGINE
     . . . . J-97 ENGINE
     . . . . TF-30 ENGINE
     . . . . TF-41 ENGINE
     . . . TURBOPROP ENGINES
     . . . . T-34 ENGINE
     . . . . T-38 ENGINE
     . . . . T-53 ENGINE
     . . . . T-56 ENGINE
     . . . . T-64 ENGINE
     . . . . T-74 ENGINE
     . . . . T-78 ENGINE
     . . . TURBORAMJET ENGINES
     . . T-58 ENGINE
     . T-58-GE-8B ENGINE
     . TURBINE ENGINES
     . . GAS TURBINE ENGINES
     . . . JET ENGINES
     . . . . RAMJET ENGINES
     . . . . LOW VOLUME RAMJET ENGINES
     . . . . PULSEJET ENGINES
     . . . . SUPERSONIC COMBUSTION
             RAMJET ENGINES
     . . . . TURBORAMJET ENGINES
     . . . T-63 ENGINE
     . . . T-76 ENGINE
     . . . TURBOJET ENGINES
     . . . . BRISTOL-SIDDELEY OLYMPUS
             593 ENGINE
     . . . . BRISTOL-SIDDELEY VIPER
             ENGINE
     . . . . DUCTED FAN ENGINES
     . . . . J-33 ENGINE
     . . . . J-34 ENGINE
     . . . . J-47 ENGINE
     . . . . J-52 ENGINE
     . . . . J-57 ENGINE
     . . . . J-57-P-20 ENGINE
     . . . . J-65 ENGINE
     . . . . J-69-T-25 ENGINE
     . . . . J-71 ENGINE
     . . . . J-73 ENGINE
     . . . . J-75 ENGINE
     . . . . J-79 ENGINE
     . . . . J-85 ENGINE
     . . . . J-93 ENGINE
     . . . . RA-28 ENGINE
     . . . . TURBOFAN ENGINES
     . . . . . BRISTOL-SIDDELEY BS 53
               ENGINE
     . . . . CF-700 ENGINE
     . . . . J-97 ENGINE
     . . . . TF-30 ENGINE
     . . . . TF-41 ENGINE
     . . . TURBOPROP ENGINES
     . . . . T-34 ENGINE
     . . . . T-38 ENGINE
     . . . . T-53 ENGINE
     . . . . T-56 ENGINE
     . . . . T-64 ENGINE
     . . . . T-74 ENGINE
     . . . . T-78 ENGINE
     . . . TURBORAMJET ENGINES
     . . T-58 ENGINE
     . . T-58-GE-8B ENGINE
RT   AIRCRAFT ENGINES
```

GAS TURBINE ENGINES-(CONT.)
    AXIAL FLOW TURBINES
    BRAYTON CYCLE
    EXTERNAL COMBUSTION ENGINES
    FLAMEOUT
    STEAM TURBINES
    SUPERSONIC TURBINES
    TURBOGENERATORS
    TWO STAGE TURBINES

GAS TURBINES
GS   TURBOMACHINERY
    . TURBINES
    . . **GAS TURBINES**
RT   AXIAL FLOW TURBINES
    BRAYTON CYCLE
    CLOSED CYCLES
    COMBINED CYCLE POWER GENERATION
    INTERNAL COMBUSTION ENGINES
    SPRAY INGESTION
    STEAM TURBINES
    SUPERSONIC TURBINES
    TURBOGENERATORS
    TWO STAGE TURBINES

GAS VALVES
GS   PNEUMATIC EQUIPMENT
    . **GAS VALVES**
    VALVES
    . **GAS VALVES**
RT   AUTOMATIC CONTROL VALVES
    COCKS
    DAMPERS (VALVES)
    FUEL VALVES
    RELIEF VALVES

GAS VISCOSITY
GS   TRANSPORT PROPERTIES
    . VISCOSITY
    . . **GAS VISCOSITY**
RT   GASEOUS DIFFUSION
    LENNARD-JONES GAS

GAS WELDING
SN   (EXCLUDES ELECTRIC WELDING IN THE
    PRESENCE OF A CONTROLLED
    GASEOUS ATMOSPHERE)
GS   WELDING
    . LASER WELDING
    . . FUSION WELDING
    . . . **GAS WELDING**
    . . . . BRAZING
    . . . . . LOW TEMPERATURE BRAZING
RT   GAS-METAL INTERACTIONS
    PRESSURE WELDING

GAS-GAS INTERACTIONS
GS   **GAS-GAS INTERACTIONS**
    . ASSOCIATION REACTIONS
RT   DALTON LAW
    DETONABLE GAS MIXTURES
    EXHAUST EMISSION
    GASEOUS DIFFUSION
    ∞INTERACTIONS

GAS-ION INTERACTIONS
UF   ION-GAS INTERACTIONS
RT   GASEOUS DIFFUSION
    ∞INTERACTIONS

GAS-LIQUID INTERACTIONS
GS   **GAS-LIQUID INTERACTIONS**
    . AIR WATER INTERACTIONS
    . . AIR SEA ICE INTERACTIONS
RT   CONDENSING
    ENERGY TRANSFER
    EVAPORATION
    GASEOUS DIFFUSION
    HEAT TRANSFER
    ∞INTERACTIONS
    INTERFACIAL TENSION
    MASS TRANSFER
    MOMENTUM TRANSFER
    NONCONDENSABLE GASES
    SURFACE REACTIONS

GAS-METAL INTERACTIONS
GS   FLUID-SOLID INTERACTIONS
    . GAS-SOLID INTERACTIONS
    . . **GAS-METAL INTERACTIONS**
RT   ABLATION
    ADSORPTION
    CHEMICAL REACTIONS
    CHEMISORPTION
    ∞CONDENSATION

GAS-METAL INTERACTIONS-(CONT.)
    CONDENSING
    CORROSION
    DIFFUSION
    EVAPORATION
    EXHAUST EMISSION
    FLAME PROPAGATION
    GAS WELDING
    GASEOUS DIFFUSION
    HOT CORROSION
    HYDROGEN EMBRITTLEMENT
    ∞INTERACTIONS
    METAL COMBUSTION
    METAL VAPORS
    METAL-GAS SYSTEMS
    OCCLUSION
    SOLID PHASES
    SUBLIMATION
    SULFIDATION
    VAPOR PHASES

GAS-SOLID INTERACTIONS
GS   FLUID-SOLID INTERACTIONS
    . **GAS-SOLID INTERACTIONS**
    . . GAS-METAL INTERACTIONS
RT   AIR LAND INTERACTIONS
    DYNAMIC LOADS
    FLUID DYNAMICS
    IMPINGEMENT
    ∞INTERACTIONS
    PANEL METHOD (FLUID DYNAMICS)

GAS-SOLID INTERFACES
GS   BOUNDARIES
    . FLUID BOUNDARIES
    . . **GAS-SOLID INTERFACES**
    INTERFACES
    . FLUID BOUNDARIES
    . . **GAS-SOLID INTERFACES**
RT   BOUNDARY LAYERS
    FLUID-SOLID INTERACTIONS
    HEAT TRANSFER
    INTERFACE STABILITY
    METAL SURFACES
    OCCLUSION
    SOLID PHASES
    SOLID-SOLID INTERFACES
    SOLUBILITY
    SUBLIMATION
    VAPOR PHASES

GASDYNAMIC LASERS
GS   STIMULATED EMISSION DEVICES
    . LASERS
    . . **GASDYNAMIC LASERS**
RT   GAS FLOW
    GAS LASERS
    LASER OUTPUTS
    PLASMADYNAMIC LASERS
    TUBE LASERS

GASEOUS CAVITATION
USE   CAVITATION FLOW
    GAS FLOW

GASEOUS DIFFUSION
UF   GAS DIFFUSION
GS   DIFFUSION
    . **GASEOUS DIFFUSION**
    . . GASEOUS SELF-DIFFUSION
    TRANSPORT PROPERTIES
    . **GASEOUS DIFFUSION**
    . . GASEOUS SELF-DIFFUSION
RT   DIFFUSION COEFFICIENT
    GAS DENSITY
    GAS DYNAMICS
    GAS FLOW
    GAS LUBRICANTS
    GAS RECOVERY
    GAS TRANSPORT
    GAS VISCOSITY
    GAS-GAS INTERACTIONS
    GAS-ION INTERACTIONS
    GAS-LIQUID INTERACTIONS
    GAS-METAL INTERACTIONS
    MOLECULAR DIFFUSION
    POLLUTION TRANSPORT
    THERMAL DIFFUSION

GASEOUS FISSION REACTORS
GS   NUCLEAR REACTORS
    . **GASEOUS FISSION REACTORS**
RT   FISSILE FUELS
    FISSIONABLE MATERIALS
    ∞GAS REACTORS

GASEOUS FISSION REACTORS-(CONT.)
    NUCLEAR LIGHTBULB ENGINES
    NUCLEAR PROPULSION
    PLASMA PROPULSION

GASEOUS FUELS
GS   FUELS
    . **GASEOUS FUELS**
RT   LIGNITE
    LIQUID FUELS

GASEOUS ROCKET PROPELLANTS
GS   PROPELLANTS
    . ROCKET PROPELLANTS
    . . **GASEOUS ROCKET PROPELLANTS**
RT   CRYOGENIC ROCKET PROPELLANTS
    ENDOTHERMIC FUELS
    GAS MIXTURES
    HIGH ENERGY PROPELLANTS
    HYBRID PROPELLANTS
    HYDROGEN FUELS
    LIQUID ROCKET PROPELLANTS
    MAN OPERATED PROPULSION SYSTEMS
    MONOPROPELLANTS
    STORABLE PROPELLANTS

GASEOUS SELF-DIFFUSION
GS   DIFFUSION
    . GASEOUS DIFFUSION
    . . **GASEOUS SELF-DIFFUSION**
    TRANSPORT PROPERTIES
    . GASEOUS DIFFUSION
    . . **GASEOUS SELF-DIFFUSION**
RT   ELECTRON DIFFUSION
    GAS DYNAMICS
    KINETIC THEORY
    MOLECULAR DIFFUSION
    PARTICLE DIFFUSION
    PLASMA DIFFUSION

GASES
GS   **GASES**
    . AIR
    . . ALVEOLAR AIR
    . . COMPRESSED AIR
    . . EXPIRED AIR
    . . HIGH TEMPERATURE AIR
    . . LIQUID AIR
    . CARBON DIOXIDE
    . CARBON MONOXIDE
    . CARBON SUBOXIDES
    . COLD GAS
    . COMPRESSED GAS
    . . HIGH PRESSURE OXYGEN
    . DISSOLVED GASES
    . EXHAUST GASES
    . . FLUE GASES
    . FLAMMABLE GASES
    . . LIQUEFIED NATURAL GAS
    . . PYROGEN
    . GAS MIXTURES
    . . ALVEOLAR AIR
    . . COMPRESSED AIR
    . . DETONABLE GAS MIXTURES
    . . EXPIRED AIR
    . . LIQUID AIR
    . GAS STREAMS
    . GRAY GAS
    . HIGH TEMPERATURE GASES
    . . HIGH TEMPERATURE AIR
    . HYDROGEN
    . . HYDROGEN ATOMS
    . . HYDROGEN IONS
    . . HYDROGEN ISOTOPES
    . . . DEUTERIUM
    . . . HYDROGEN 4
    . . . TRITIUM
    . . HYDROGEN PLASMA
    . . . DEUTERIUM PLASMA
    . . LIQUID HYDROGEN
    . . ORTHO HYDROGEN
    . . PARA HYDROGEN
    . IDEAL GAS
    . IONIZED GASES
    . . CHARGED PARTICLES
    . . . ARGON PLASMA
    . . . BOUNDARY LAYER PLASMAS
    . . . COLD PLASMAS
    . . . COLLISIONAL PLASMAS
    . . . . STRONGLY COUPLED PLASMAS
    . . . COLLISIONLESS PLASMAS
    . . . CONDUCTION ELECTRONS
    . . . DENSE PLASMAS
    . . . . STRONGLY COUPLED PLASMAS
    . . . ELECTRON PLASMA

GASES-*(CONT.)*
. . . ELLIPTICAL PLASMAS
. . . HELIUM PLASMA
. . . HIGH TEMPERATURE PLASMAS
. . . LASER PLASMAS
. . . METALLIC PLASMAS
. . . . CESIUM PLASMA
. . . MICROPLASMAS
. . . NITROGEN PLASMA
. . . NUCLEI (NUCLEAR PHYSICS)
. . . . EVEN-EVEN NUCLEI
. . . . HEAVY NUCLEI
. . . . HYPERNUCLEI
. . . . ODD-EVEN NUCLEI
. . . . ODD-ODD NUCLEI
. . . RAREFIED PLASMAS
. . . RELATIVISTIC PLASMAS
. . . ROTATING PLASMAS
. . . SOLAR WIND
. . . STELLAR WINDS
. . . THERMAL PLASMAS
. LIQUEFIED GASES
. . LIQUEFIED NATURAL GAS
. . LIQUID AIR
. . LIQUID AMMONIA
. . LIQUID FLUORINE
. . LIQUID HELIUM
. . . LIQUID HELIUM 2
. . LIQUID HYDROGEN
. . LIQUID NEON
. . LIQUID NITROGEN
. . LIQUID OXYGEN
. LORENTZ GAS
. MOLECULAR GASES
. . POLAR GASES
. . POLYATOMIC GASES
. DIATOMIC GASES
. MONATOMIC GASES
. NATURAL GAS
. LIQUEFIED NATURAL GAS
. NITROGEN
. . LIQUID NITROGEN
. . NITROGEN IONS
. NONCONDENSABLE GASES
. NONGRAY GAS
. NONPOLAR GASES
. OXYGEN
. . HIGH PRESSURE OXYGEN
. . LIQUID OXYGEN
. . OXYGEN ATOMS
. . OXYGEN IONS
. . OXYGEN ISOTOPES
. . . OXYGEN 18
. . OXYGEN PLASMA
. . OZONE
. . . OZONATES
. . . OZONIDES
. PHOSGENE
. RARE GASES
. . ARGON
. . . ARGON ISOTOPES
. . HELIUM
. . . HELIUM ATOMS
. . . HELIUM FILM
. . . HELIUM ISOTOPES
. . . LIQUID HELIUM
. . . . LIQUID HELIUM 2
. . KRYPTON
. . NEON
. . . LIQUID NEON
. . . NEON ISOTOPES
. . RADON
. . . RADON ISOTOPES
. . XENON
. . . XENON ISOTOPES
. . . . XENON 129
. . . . XENON 133
. . . . XENON 135
. RAREFIED GASES
. . COSMIC GASES
. . . INTERPLANETARY GAS
. . . INTERSTELLAR GAS
. . . . NEUTRAL GASES
. REAL GASES
. RESIDUAL GAS
. SOLIDIFIED GASES
. . SOLID CRYOGENS
RT   ∞ATMOSPHERES
COAL GASIFICATION
∞FLUIDS
FUELS
FUMES
GAS DYNAMICS
GAS FLOW
GAS RECOVERY
HYDROGEN CLOUDS

GASES-*(CONT.)*
METAL-GAS SYSTEMS
NONPOINT SOURCES
ODORS
PLASMAS (PHYSICS)
PNEUMATICS
PREVAPORIZATION
REACTION PRODUCTS
VAPOR PHASES
VAPORS

GASIFICATION
GS   **GASIFICATION**
. COAL GASIFICATION
. . HYDROPYROLYSIS
RT   SYNTHANE

GASKETS
GS   SEALS (STOPPERS)
. **GASKETS**
RT   LABYRINTH SEALS
O RING SEALS
PUMP SEALS

GASOHOL (FUEL)
GS   FUELS
. CHEMICAL FUELS
. . SYNTHETIC FUELS
. . . **GASOHOL (FUEL)**
RT   ALCOHOLS
GASOLINE

GASOLINE
GS   FUELS
. CHEMICAL FUELS
. . HYDROCARBON FUELS
. . . **GASOLINE**
. . LIQUID FUELS
. . . **GASOLINE**
RT   ANTIKNOCK ADDITIVES
AUTOMOBILE FUELS
DIESEL FUELS
GASOHOL (FUEL)
JET ENGINE FUELS
KEROGEN
KEROSENE
OCTANE NUMBER
SHALE OIL
SOLVENT REFINED COAL

GASP
USE   GLOBAL AIR SAMPLING PROGRAM

GASTROINTESTINAL SYSTEM
GS   ANATOMY
. DIGESTIVE SYSTEM
. . **GASTROINTESTINAL SYSTEM**
. . . APPENDIX (ANATOMY)
. . . INTESTINES
. . . . RECTUM
. . . STOMACH
RT   ABDOMEN
COLIC
GALL
GLANDS (ANATOMY)
LIVER
ORGANS
PANCREAS
∞SYSTEMS

GATE (EXPERIMENT)
USE   GARP ATLANTIC TROPICAL EXPERIMENT

∞ GATES
SN   *(USE OF A MORE SPECIFIC TERM IS*
RECOMMENDED--CONSULT THE TERMS
LISTED BELOW)
RT   CLOSURES
GATES (CIRCUITS)
GATES (OPENINGS)

GATES (CIRCUITS)
UF   OR-GATES
GS   CIRCUITS
. **GATES (CIRCUITS)**
. . THRESHOLD GATES
RT   COINCIDENCE CIRCUITS
∞GATES
LOGIC CIRCUITS
LOGICAL ELEMENTS
SWITCHING CIRCUITS
THRESHOLD LOGIC
TRIGGER CIRCUITS

GATES (OPENINGS)
RT   APERTURES
∞BARRIERS
CANALS
DOORS
FENCES (BARRIERS)
∞GATES
HATCHES
HYDRAULIC EQUIPMENT
OPENINGS
OUTLETS
SAFETY DEVICES
VENTS
WALLS

GAUGE INVARIANCE
GS   INVARIANCE
. **GAUGE INVARIANCE**
RT   ELECTROMAGNETIC RADIATION
TRANSFORMATIONS (MATHEMATICS)

GAUGE THEORY
RT   GRAVITATION THEORY
∞THEORIES
YANG-MILLS FIELDS
YANG-MILLS THEORY

GAUSS EQUATION
UF   GAUSS FUNCTION
GS   ANALYSIS (MATHEMATICS)
. REAL VARIABLES
. . DIFFERENTIAL EQUATIONS
. . . PARTIAL DIFFERENTIAL EQUATIONS
. . . . **GAUSS EQUATION**
RT   ∞EQUATIONS
MAXWELL EQUATION

GAUSS FUNCTION
USE   GAUSS EQUATION

GAUSS-MARKOV THEOREM
GS   THEOREMS
. **GAUSS-MARKOV THEOREM**
RT   LEAST SQUARES METHOD
STATISTICAL ANALYSIS
VARIANCE (STATISTICS)

GAUSSIAN DISTRIBUTIONS
USE   NORMAL DENSITY FUNCTIONS

GAUSSIAN NOISE
USE   RANDOM NOISE

GAUSSMETERS
USE   MAGNETOMETERS

GAUZE
GS   FABRICS
. **GAUZE**
RT   CASTS

GAW-1 AIRFOIL
UF   GENERAL AVIATION WHITCOMB AIRFOIL
GS   AIRFOILS
. WINGS
. . **GAW-1 AIRFOIL**
RT   ATLIT PROJECT
PA-34 SENECA AIRCRAFT
WING PROFILES

GAW-2 AIRFOIL
UF   GENERAL AVIATION WHITCOMB AIRFOIL
GS   AIRFOILS
. WINGS
. . **GAW-2 AIRFOIL**
RT   BODY-WING CONFIGURATIONS
FLAPS (CONTROL SURFACES)
GENERAL AVIATION AIRCRAFT
WING PROFILES

GC-130 AIRCRAFT
USE   C-130 AIRCRAFT

GCR (REACTORS)
USE   GAS COOLED REACTORS

GDOP
USE   GEOMETRIC DILUTION OF PRECISION

GE COMPUTERS
UF   GENERAL ELECTRIC COMPUTERS
GS   DATA PROCESSING EQUIPMENT
. COMPUTERS
. . DIGITAL COMPUTERS

GE COMPUTERS-*(CONT.)*
. . . **GE COMPUTERS**
. . . . GE 625 COMPUTER
. . . . GE 635 COMPUTER

GE 625 COMPUTER
GS   DATA PROCESSING EQUIPMENT
. COMPUTERS
. . DIGITAL COMPUTERS
. . . GE COMPUTERS
. . . . **GE 625 COMPUTER**

GE 635 COMPUTER
GS   DATA PROCESSING EQUIPMENT
. COMPUTERS
. . DIGITAL COMPUTERS
. . . GE COMPUTERS
. . . . **GE 635 COMPUTER**

∞ **GEAR**
SN   *(USE OF A MORE SPECIFIC TERM IS RECOMMENDED--CONSULT THE TERMS LISTED BELOW)*
RT   ARRESTING GEAR
GEARS
LANDING GEAR
MECHANICAL DRIVES

GEAR TEETH
RT   GEARS
MECHANICAL DRIVES

GEARS
GS   **GEARS**
. RACKS (GEARS)
RT   COUNTER-ROTATING WHEELS
∞ GEAR
GEAR TEETH
IDLERS
LUBRICATION
MECHANICAL DRIVES
TRANSMISSIONS (MACHINE ELEMENTS)
WHEELS
WINDMILLS (WINDPOWERED MACHINES)

GEGENSCHEIN
GS   ELECTROMAGNETIC RADIATION
. LIGHT (VISIBLE RADIATION)
. . **GEGENSCHEIN**
EXTRATERRESTRIAL RADIATION
. **GEGENSCHEIN**
RT   NIGHT SKY
POLARIZED LIGHT
SKY BRIGHTNESS
SOLAR RADIATION
TERRESTRIAL DUST BELT
ZODIACAL LIGHT

GEHLENITE
GS   ALUMINUM COMPOUNDS
. **GEHLENITE**
CALCIUM COMPOUNDS
. CALCIUM SILICATES
. . **GEHLENITE**
MAGNESIUM COMPOUNDS
. **GEHLENITE**
MINERALS
. **GEHLENITE**
. . AKERMANITE
SILICON COMPOUNDS
. SILICATES
. . CALCIUM SILICATES
. . . **GEHLENITE**
RT   ALUMINUM OXIDES

GEIGER COUNTERS
UF   GEIGER-MUELLER TUBES
GS   IONIZATION CHAMBERS
. **GEIGER COUNTERS**
MEASURING INSTRUMENTS
. COUNTERS
. . RADIATION COUNTERS
. . . **GEIGER COUNTERS**
. RADIATION MEASURING INSTRUMENTS
. . RADIATION COUNTERS
. . . **GEIGER COUNTERS**
RT   DOSIMETERS
NEUTRON COUNTERS
OVERVOLTAGE
PARTICLE TELESCOPES
PROPORTIONAL COUNTERS
RADIATION DETECTORS

GEIGER-MUELLER TUBES
USE   GEIGER COUNTERS

GEL PERMEATION CHROMATOGRAPHY
USE   LIQUID CHROMATOGRAPHY

GELATINS
RT   COLLAGENS
∞ FOOD
GELS
NONNEWTONIAN FLUIDS

GELATION
RT   COAGULATION
COLLOIDING
GELS
SOLIDIFICATION
THIXOTROPY

GELLED PROPELLANTS
GS   PROPELLANTS
. **GELLED PROPELLANTS**
. . GELLED ROCKET PROPELLANTS
RT   CHEMICAL FUELS
COLLOIDAL PROPELLANTS
HIGH TEMPERATURE PROPELLANTS
HYDROGEN FUELS
METAL FUELS
METAL PROPELLANTS
PLASTIC PROPELLANTS
PROPELLANT ADDITIVES
SOLID PROPELLANTS

GELLED ROCKET PROPELLANTS
UF   THIXOTROPIC PROPELLANTS
GS   PROPELLANTS
. GELLED PROPELLANTS
. . **GELLED ROCKET PROPELLANTS**
. ROCKET PROPELLANTS
. . LIQUID ROCKET PROPELLANTS
. . . **GELLED ROCKET PROPELLANTS**
RT   CHEMICAL FUELS
CRYOGENIC ROCKET PROPELLANTS
GELS
HIGH TEMPERATURE PROPELLANTS
HYBRID PROPELLANTS
HYPERGOLIC ROCKET PROPELLANTS
LIQUID FUELS
METAL PROPELLANTS
MONOPROPELLANTS
SLURRY PROPELLANTS
SOLID ROCKET PROPELLANTS
STORABLE PROPELLANTS

GELS
GS   **GELS**
. DOUBLE BASE ROCKET PROPELLANTS
. SILICA GEL
RT   COLLOIDS
DOPES
GELATINS
GELATION
GELLED ROCKET PROPELLANTS
NONNEWTONIAN FLUIDS
SLURRIES
THICKENERS (MATERIALS)
THIXOTROPY

GEMINI (GT-1) SPACECRAFT
GS   MANNED SPACECRAFT
. GEMINI SPACECRAFT
. . **GEMINI (GT-1) SPACECRAFT**
REENTRY VEHICLES
. RECOVERABLE SPACECRAFT
. . GEMINI SPACECRAFT
. . . **GEMINI (GT-1) SPACECRAFT**
SOFT LANDING SPACECRAFT
. GEMINI SPACECRAFT
. . **GEMINI (GT-1) SPACECRAFT**
RT   MANNED SPACE FLIGHT

GEMINI B SPACECRAFT
GS   MANNED SPACECRAFT
. . **GEMINI B SPACECRAFT**
REENTRY VEHICLES
. RECOVERABLE SPACECRAFT
. . **GEMINI B SPACECRAFT**
SOFT LANDING SPACECRAFT
. **GEMINI B SPACECRAFT**
RT   MANNED SPACE FLIGHT

GEMINI FLIGHTS
GS   SPACE FLIGHT
. MANNED SPACE FLIGHT

GEMINI FLIGHTS-*(CONT.)*
. . **GEMINI FLIGHTS**
. . . GEMINI 3 FLIGHT
. . . GEMINI 4 FLIGHT
. . . GEMINI 5 FLIGHT
. . . GEMINI 6 FLIGHT
. . . GEMINI 7 FLIGHT
. . . GEMINI 8 FLIGHT
. . . GEMINI 9 FLIGHT
. . . GEMINI 10 FLIGHT
. . . GEMINI 11 FLIGHT
. . . GEMINI 12 FLIGHT

GEMINI PROJECT
GS   PROGRAMS
. NASA PROGRAMS
. . **GEMINI PROJECT**
. NASA SPACE PROGRAMS
. . **GEMINI PROJECT**
. PROJECTS
. . **GEMINI PROJECT**
RT   AGENA B ROCKET VEHICLE
AGENA ROCKET VEHICLES
ATLAS LAUNCH VEHICLES
INTEGRATED MISSION CONTROL CENTER
MERCURY PROJECT
TITAN PROJECT

GEMINI SPACECRAFT
GS   MANNED SPACECRAFT
. **GEMINI SPACECRAFT**
. . GEMINI (GT-1) SPACECRAFT
. . GEMINI 2 SPACECRAFT
REENTRY VEHICLES
. RECOVERABLE SPACECRAFT
. . **GEMINI SPACECRAFT**
. . . GEMINI (GT-1) SPACECRAFT
. . . GEMINI 2 SPACECRAFT
SOFT LANDING SPACECRAFT
. **GEMINI SPACECRAFT**
. . GEMINI (GT-1) SPACECRAFT
. . GEMINI 2 SPACECRAFT
RT   MANNED SPACE FLIGHT
SPACE CAPSULES
TITAN PROJECT

GEMINI 2 SPACECRAFT
GS   MANNED SPACECRAFT
. GEMINI SPACECRAFT
. . **GEMINI 2 SPACECRAFT**
REENTRY VEHICLES
. RECOVERABLE SPACECRAFT
. . GEMINI SPACECRAFT
. . . **GEMINI 2 SPACECRAFT**
SOFT LANDING SPACECRAFT
. GEMINI SPACECRAFT
. . **GEMINI 2 SPACECRAFT**
RT   MANNED SPACE FLIGHT

GEMINI 3 FLIGHT
GS   SPACE FLIGHT
. MANNED SPACE FLIGHT
. . GEMINI FLIGHTS
. . . **GEMINI 3 FLIGHT**
RT   TITAN LAUNCH VEHICLES

GEMINI 4 FLIGHT
GS   SPACE FLIGHT
. MANNED SPACE FLIGHT
. . GEMINI FLIGHTS
. . . **GEMINI 4 FLIGHT**

GEMINI 5 FLIGHT
GS   SPACE FLIGHT
. MANNED SPACE FLIGHT
. . GEMINI FLIGHTS
. . . **GEMINI 5 FLIGHT**

GEMINI 6 FLIGHT
GS   SPACE FLIGHT
. MANNED SPACE FLIGHT
. . GEMINI FLIGHTS
. . . **GEMINI 6 FLIGHT**

GEMINI 7 FLIGHT
GS   SPACE FLIGHT
. MANNED SPACE FLIGHT
. . GEMINI FLIGHTS
. . . **GEMINI 7 FLIGHT**
RT   TITAN LAUNCH VEHICLES

GEMINI 8 FLIGHT
GS   SPACE FLIGHT

GEMINI 8 FLIGHT-(CONT.)
. MANNED SPACE FLIGHT
. . GEMINI FLIGHTS
. . . **GEMINI 8 FLIGHT**
RT   TITAN LAUNCH VEHICLES

GEMINI 9 FLIGHT
GS   SPACE FLIGHT
. MANNED SPACE FLIGHT
. . GEMINI FLIGHTS
. . . **GEMINI 9 FLIGHT**
RT   TITAN LAUNCH VEHICLES

GEMINI 10 FLIGHT
GS   SPACE FLIGHT
. MANNED SPACE FLIGHT
. . GEMINI FLIGHTS
. . . **GEMINI 10 FLIGHT**
RT   TITAN LAUNCH VEHICLES

GEMINI 11 FLIGHT
GS   SPACE FLIGHT
. MANNED SPACE FLIGHT
. . GEMINI FLIGHTS
. . . **GEMINI 11 FLIGHT**
RT   TITAN LAUNCH VEHICLES

GEMINI 12 FLIGHT
GS   SPACE FLIGHT
. MANNED SPACE FLIGHT
. . GEMINI FLIGHTS
. . . **GEMINI 12 FLIGHT**
RT   TITAN LAUNCH VEHICLES

GEMINID METEOROIDS
GS   CELESTIAL BODIES
. METEOROID SHOWERS
. . **GEMINID METEOROIDS**
. METEOROIDS
. . **GEMINID METEOROIDS**

GENERAL AVIATION AIRCRAFT
UF   EXECUTIVE AIRCRAFT
PRIVATE AIRCRAFT
GS   **GENERAL AVIATION AIRCRAFT**
. AGRICULTURAL AIRCRAFT
. BEECHCRAFT 18 AIRCRAFT
. C-33 AIRCRAFT
. C-35 AIRCRAFT
. CESSNA 172 AIRCRAFT
. CESSNA 205 AIRCRAFT
. CESSNA 210 AIRCRAFT
. CESSNA 402B AIRCRAFT
. CL-600 CHALLENGER AIRCRAFT
. DH 125 AIRCRAFT
. DHC 2 AIRCRAFT
. DO-27 AIRCRAFT
. DO-28 AIRCRAFT
. G-1 AIRCRAFT
. HC-3 HELICOPTER
. YAK 40 AIRCRAFT
RT   ∞AERONAUTICS
∞AIRCRAFT
CIVIL AVIATION
COMMERCIAL AIRCRAFT
GAW-2 AIRFOIL
HELICOPTERS
JET AIRCRAFT
LIGHT AIRCRAFT
∞LOW WING AIRCRAFT
PASSENGER AIRCRAFT
PIPER AIRCRAFT
∞SUBSONIC AIRCRAFT
TRAINING AIRCRAFT
TRANSPORT AIRCRAFT
TURBOPROP AIRCRAFT
UTILITY AIRCRAFT

GENERAL AVIATION WHITCOMB AIRFOIL
USE   GAW-1 AIRFOIL
GAW-2 AIRFOIL

GENERAL DYNAMICS AIRCRAFT
UF   CONVAIR MILITARY AIRCRAFT
GENERAL DYNAMICS MILITARY
AIRCRAFT
GS   **GENERAL DYNAMICS AIRCRAFT**
. B-58 AIRCRAFT
. C-131 AIRCRAFT
. CANADAIR AIRCRAFT
. . CL-41 AIRCRAFT
. . CL-44 AIRCRAFT
. . CL-84 AIRCRAFT
. CV-340 AIRCRAFT
. CV-440 AIRCRAFT

GENERAL DYNAMICS AIRCRAFT-(CONT.)
. CV-880 AIRCRAFT
. CV-990 AIRCRAFT
. F-102 AIRCRAFT
. F-106 AIRCRAFT
. F-111 AIRCRAFT
RT   ∞AIRCRAFT
PA-34 SENECA AIRCRAFT

GENERAL DYNAMICS MILITARY AIRCRAFT
USE   GENERAL DYNAMICS AIRCRAFT
MILITARY AIRCRAFT

GENERAL ELECTRIC COMPUTERS
USE   GE COMPUTERS

GENERALIZATION (PSYCHOLOGY)
RT   TRANSFER OF TRAINING

∞ **GENERATION**
SN   (USE OF A MORE SPECIFIC TERM IS
RECOMMENDED--CONSULT THE TERMS
LISTED BELOW)
RT   COGENERATION
HEAT GENERATION
INITIATION
REGENERATION (ENGINEERING)

∞ **GENERATORS**
SN   (USE OF A MORE SPECIFIC TERM IS
RECOMMENDED--CONSULT THE TERMS
LISTED BELOW)
RT   AC GENERATORS
ARC GENERATORS
BOILERS
CAVITY VAPOR GENERATORS
COLLOIDAL GENERATORS
DECOMMUTATORS
DIRECT POWER GENERATORS
DUOCHROMATORS
ELECTRIC GENERATORS
ELECTROSTATIC GENERATORS
ENERGY CONVERSION EFFICIENCY
FUNCTION GENERATORS
GAS GENERATORS
HALL GENERATORS
HARMONIC GENERATORS
HOMOPOLAR GENERATORS
IMPULSE GENERATORS
MAGNETOHYDRODYNAMIC GENERATORS
MOTORS
NOISE GENERATORS
PHOTOELECTRIC GENERATORS
PLASMA GENERATORS
PULSE GENERATORS
RADIATION SOURCES
REPORT GENERATORS
ROTATING GENERATORS
SHOCK WAVE GENERATORS
SIGNAL GENERATORS
SOLAR SEA POWER PLANTS
SOUND GENERATORS
STATORS
STIMULATED EMISSION DEVICES
SUBHARMONIC GENERATORS
TEST PATTERN GENERATORS
THERMOELECTRIC GENERATORS
TIDE POWERED GENERATORS
TURBOGENERATORS
VAPORIZERS
VOLTAGE GENERATORS
VORTEX GENERATORS
WAVE GENERATION
WINDPOWERED GENERATORS

GENETIC CODE
GS   GENETICS
. **GENETIC CODE**
RT   CHROMOSOMES

GENETIC ENGINEERING
UF   HYBRIDS (BIOLOGY)
RT   BIOCHEMISTRY
BIOENGINEERING
∞BIOLOGY
BIOSYNTHESIS
GENETICS

GENETICS
GS   **GENETICS**
. GENETIC CODE
. MUTATIONS
RT   BIOLOGICAL EVOLUTION
∞BIOLOGY
BREEDING (REPRODUCTION)

GENETICS-(CONT.)
CHROMOSOMES
CONGENITAL ANOMALIES
CYTOGENESIS
DOMINANCE
EVOLUTION (DEVELOPMENT)
GENETIC ENGINEERING
MUTAGENS
NEUROSPORA
NUCLEOGENESIS
SPECIES DIFFUSION

GENIE ROCKET VEHICLE
UF   MB-1 ROCKET VEHICLE
GS   ROCKET VEHICLES
. SINGLE STAGE ROCKET VEHICLES
. . **GENIE ROCKET VEHICLE**
RT   ASTROBEE ROCKET VEHICLES
SOLID PROPELLANT ROCKET ENGINES

GENITOURINARY SYSTEM
GS   ANATOMY
. **GENITOURINARY SYSTEM**
. . BLADDER
. . OVARIES
. . PROSTATE GLAND
. . TESTES
. . UTERUS
RT   GYNECOLOGY
∞SYSTEMS
UROLOGY

†
GEOASTROPHYSICS
USE   ASTROPHYSICS
GEOPHYSICS

GEOBOTANY
GS   BOTANY
. **GEOBOTANY**
RT   PLANTS (BOTANY)
RAIN FORESTS
TREES (PLANTS)

GEOCENTRIC COORDINATES
GS   COORDINATES
. PLANETOCENTRIC COORDINATES
. . **GEOCENTRIC COORDINATES**
RT   ASTRONOMICAL COORDINATES
CELESTIAL REFERENCE SYSTEMS
INERTIAL COORDINATES
PLANET EPHEMERIDES
SPHERICAL COORDINATES

GEOCHEMISTRY
GS   ENVIRONMENTAL CHEMISTRY
. **GEOCHEMISTRY**
. . BIOGEOCHEMISTRY
RT   ∞CHEMISTRY
COSMOCHEMISTRY
GEOCHRONOLOGY
GEOLOGY
GEOPHYSICS
HYDROLOGY
LIMNOLOGY
MARINE CHEMISTRY
MINERALOGY
PALEOBIOLOGY
PALEONTOLOGY
PETROLOGY
RADIOACTIVITY

GEOCHRONOLOGY
GS   CHRONOLOGY
. **GEOCHRONOLOGY**
GEOLOGY
. **GEOCHRONOLOGY**
RT   DENDROCHRONOLOGY
GEOCHEMISTRY
GEOPHYSICS
PALEOBIOLOGY
PALEONTOLOGY
PARTICLE TRACKS
RADIOACTIVE AGE DETERMINATION
STRATIGRAPHY

GEOCORONAL EMISSIONS
GS   ATMOSPHERIC RADIATION
. SKY RADIATION
. . AIRGLOW
. . . **GEOCORONAL EMISSIONS**
ELECTROMAGNETIC RADIATION
. LIGHT (VISIBLE RADIATION)
. . SKY RADIATION
. . . AIRGLOW
. . . . **GEOCORONAL EMISSIONS**

GEOCYCLOTRONS
GS   PARTICLE ACCELERATORS
    . CYCLOTRONS
    . . **GEOCYCLOTRONS**
RT   ∞ACCELERATORS
    ELEMENTARY PARTICLES

GEODESIC LINES
GS   GEOMETRY
    . EUCLIDEAN GEOMETRY
    . . LINES (GEOMETRY)
    . . . **GEODESIC LINES**
RT   CHORDS (GEOMETRY)
    CURVES (GEOMETRY)

GEODESY
UF   EARTH FIGURE
    EARTH SHAPE
    IZSAK ELLIPSOID
GS   **GEODESY**
    . CELESTIAL GEODESY
RT   EARTH (PLANET)
    EARTH AXIS
    GEODETIC ACCURACY
    GEODETIC SURVEYS
    GEOIDS '
    GEOLOGY
    GEOPHYSICS
    GRAVIMETERS
    LUNAR RETROREFLECTORS
    OBLATE SPHEROIDS
    OGO-4
    OGO-5
    PERTURBATION
    PHOTOMAPPING
    POLAR WANDERING (GEOLOGY)
    SATELLITE DOPPLER POSITIONING
    TOPOGRAPHY
    VINTI THEORY

GEODETIC ACCURACY
GS   ACCURACY
    . **GEODETIC ACCURACY**
RT   EARTH SURFACE
    GEODESY
    GEOIDS
    GEOPOTENTIAL HEIGHT
    SATELLITE DOPPLER POSITIONING

GEODETIC COORDINATES
GS   COORDINATES
    . **GEODETIC COORDINATES**
RT   INTERNATIONAL SATELLITE GEODESY
     EXPERIMENT
    LATITUDE
    LONGITUDE
    SATELLITE DOPPLER POSITIONING

GEODETIC SATELLITES
GS   SATELLITES
    . ARTIFICIAL SATELLITES
    . . **GEODETIC SATELLITES**
    . . . ANNA SATELLITES
    . . . EXPLORER 29 SATELLITE
    . . . EXPLORER 36 SATELLITE
    . . . GEOLE SATELLITES
    . . . GEOS 1 SATELLITE
    . . . GEOS 2 SATELLITE
    . . . GEOS 3 SATELLITE
    . . . LARGOS SATELLITE
    . . . PAGEOS SATELLITE
    . . . VANGUARD 1 SATELLITE
    UNMANNED SPACECRAFT
    . **GEODETIC SATELLITES**
    . . ANNA SATELLITES
    . . EXPLORER 29 SATELLITE
    . . EXPLORER 36 SATELLITE
    . . GEOS 1 SATELLITE
    . . GEOS 2 SATELLITE
    . . GEOS 3 SATELLITE
    . . LARGOS SATELLITE
    . . PAGEOS SATELLITE
    . . VANGUARD 1 SATELLITE
RT   ACTIVE SATELLITES
    CELESTIAL GEODESY
    NAVIGATION SATELLITES
    NAVSTAR SATELLITES
    PASSIVE SATELLITES
    SATELLITE DOPPLER POSITIONING
    VANGUARD SATELLITES

GEODETIC SURVEYS
GS   SURVEYS
    . **GEODETIC SURVEYS**
RT   GEODESY

GEODETIC SURVEYS-(CONT.)
    GEOLOGICAL SURVEYS
    PHOENIX QUADRANGLE (AZ)
    SATELLITE DOPPLER POSITIONING
    TOPOGRAPHY

GEODIMETERS
GS   MEASURING INSTRUMENTS
    . DISTANCE MEASURING EQUIPMENT
    . . **GEODIMETERS**
    . OPTICAL MEASURING INSTRUMENTS
    . . **GEODIMETERS**
    OPTICAL EQUIPMENT
    . OPTICAL MEASURING INSTRUMENTS
    . . **GEODIMETERS**
RT   OPTICAL MEASUREMENT
    RANGE FINDERS
    TELLUROMETERS

GEODYNAMIC EXPERIMENTAL OCEAN SATELLITE
USE   GEOS-D SATELLITE

GEODYNAMICS
RT   CRUSTAL FRACTURES
    ∞DYNAMICS
    EARTH MOVEMENTS
    GEOMORPHOLOGY
    GEOPHYSICS
    PLANETARY QUAKES
    SHOCK WAVES
    TERRADYNAMICS

GEOELECTRICITY
GS   ELECTRICITY
    . **GEOELECTRICITY**
    . . TELLURIC CURRENTS
RT   EARTH (PLANET)
    GEOPHYSICS
    GEOPOTENTIAL

GEOFABRICS
USE   GEOTECHNICAL FABRICS

GEOFRACTURES
USE   GEOLOGICAL FAULTS

GEOGRAPHIC APPLICATIONS PROGRAM
GS   PROGRAMS
    . SPACE PROGRAMS
    . . **GEOGRAPHIC APPLICATIONS
     PROGRAM**
RT   EARTH RESOURCES
    EARTH RESOURCES PROGRAM
    GEOGRAPHY
    MAPPING
    NASA PROGRAMS
    REMOTE SENSORS
    SATELLITE-BORNE PHOTOGRAPHY
    SOIL MAPPING
    TERRAIN ANALYSIS

GEOGRAPHIC INFORMATION SYSTEMS
GS   INFORMATION SYSTEMS
    . **GEOGRAPHIC INFORMATION SYSTEMS**
RT   AERIAL PHOTOGRAPHY
    DATA SYSTEMS
    GEOGRAPHY
    IMAGERY
    INFRARED PHOTOGRAPHY
    REMOTE SENSING

GEOGRAPHY
GS   **GEOGRAPHY**
    . HYPSOGRAPHY
    . OROGRAPHY
RT   ARCTIC REGIONS
    CADASTRAL MAPPING
    CLIMATOLOGY
    CONTINENTS
    EARTH (PLANET)
    EASTERN HEMISPHERE
    ECONOMIC DEVELOPMENT
    ESTUARIES
    GEOGRAPHIC APPLICATIONS PROGRAM
    GEOGRAPHIC INFORMATION SYSTEMS
    GEOMORPHOLOGY
    HEAT CAPACITY MAPPING MISSION
    MAPPING
    MAPS
    OCEANOGRAPHY
    OCEANS
    PLAINS
    POLAR REGIONS
    SELENOGRAPHY

GEOGRAPHY-(CONT.)
    TEMPERATE REGIONS
    TROPICAL REGIONS
    TUNDRA
    WESTERN HEMISPHERE

GEOIDS
RT   GEODESY
    GEODETIC ACCURACY
    GEOMETRY
    GEOPHYSICS
    OBLATE SPHEROIDS
    SHAPES
    SPHEROIDS
    SYMMETRICAL BODIES

GEOLE SATELLITES
GS   SATELLITES
    . ARTIFICIAL SATELLITES
    . . FRENCH SATELLITES
    . . . **GEOLE SATELLITES**
    . . GEODETIC SATELLITES
    . . . **GEOLE SATELLITES**
    . . METEOROLOGICAL SATELLITES
    . . . **GEOLE SATELLITES**
RT   EOLE SATELLITES
    FRENCH SPACE PROGRAMS

GEOLOGICAL FAULTS
UF   CLOSED FAULTS
    CROSS FAULTS
    ECHELON FAULTS
    GEOFRACTURES
    GRABENS
    RIFTS
    SPLITS (GEOLOGY)
    STEP FAULTS
    THRUST FAULTS
GS   **GEOLOGICAL FAULTS**
    . AFRICAN RIFT SYSTEM
    . SAN ANDREAS FAULT
RT   CREVASSES
    CRUSTAL FRACTURES
    EARTHQUAKE DAMAGE
    EARTHQUAKES
    ∞FAULTS
    FISSURES (GEOLOGY)
    FOLDS (GEOLOGY)
    FORMATIONS
    INLIERS (LANDFORMS)
    LANDFORMS
    MASSIFS
    PLATES (TECTONICS)
    ROUSE BELTS
    SAN ANDREAS FAULT EXPERIMENT
    SYNCLINES

GEOLOGICAL SURVEYS
GS   SURVEYS
    . **GEOLOGICAL SURVEYS**
RT   EXPLORATION
    GEODETIC SURVEYS
    GEOLOGY
    GEOPHYSICS
    PALEONTOLOGY
    PETROLOGY
    PHOTOGEOLOGY
    RADAR GEOLOGY

GEOLOGY
GS   **GEOLOGY**
    . CONES (VOLCANOES)
    . CROSSBEDDING (GEOLOGY)
    . GEOCHRONOLOGY
    . GEOMORPHOLOGY
    . GLACIOLOGY
    . HYDROGEOLOGY
    . KETTLES (GEOLOGY)
    . LITHOLOGY
    . LUNAR GEOLOGY
    . OROGRAPHY
    . PETROLOGY
    . . PETROGRAPHY
    . PHOTOGEOLOGY
    . RADAR GEOLOGY
    . STRUCTURAL PROPERTIES (GEOLOGY)
    . SUBDUCTION (GEOLOGY)
    . TECTONICS
    . VOLCANOES
    . . MARS VOLCANOES
    . VOLCANOLOGY
RT   BEDROCK
    BEDS (GEOLOGY)
    BOREHOLES
    BRIDGES (LANDFORMS)

GEOLOGY-*(CONT.)*
  CANADIAN SHIELD
  CONTACTS (GEOLOGY)
  CONTINENTAL SHELVES
  DOMES (GEOLOGY)
  EARTH (PLANET)
  EARTH PLANETARY STRUCTURE
  EXPLOITATION
  EXPLORATION
  FIORDS
  FORMATIONS
  GAPS (GEOLOGY)
  GEOCHEMISTRY
  GEODESY
  GEOLOGICAL SURVEYS
  GEOPHYSICAL OBSERVATORIES
  GEOPHYSICS
  GEOPRESSURE
  GEOTEMPERATURE
  GRAVIMETRY
  GREAT BASIN (US)
  INLIERS (LANDFORMS)
  ISTHMUSES
  KREEP
  MASSIFS
  METEOROLOGY
  MINERAL DEPOSITS
  MINERALOGY
  MINERALS
  MORPHOLOGY
  OCEAN BOTTOM
  OCEANOGRAPHY
  OIL EXPLORATION
  OUTCROPS
  PALEOMAGNETISM
  PALEONTOLOGY
  PENEPLAINS
  PHOTOMAPPING
  ∞PHYSICAL SCIENCES
  PRECAMBRIAN PERIOD
  REGOLITH
  ROCK MECHANICS
  ROCKS
  ∞SCIENCE
  SEISMOLOGY
  SHATTER CONES
  SOILS
  STRATIGRAPHY
  STRUCTURAL BASINS

GEOMAGNETIC ANOMALIES
  USE   MAGNETIC ANOMALIES

GEOMAGNETIC CROTCHETS
  USE   SUDDEN IONOSPHERIC DISTURBANCES

GEOMAGNETIC EFFECTS
  USE   MAGNETIC EFFECTS

GEOMAGNETIC EQUATOR
  USE   MAGNETIC EQUATOR

GEOMAGNETIC FIELD
  USE   GEOMAGNETISM

GEOMAGNETIC HOLLOW
  GS   ANOMALIES
         . MAGNETIC ANOMALIES
         . . **GEOMAGNETIC HOLLOW**
  RT   MAGNETOHYDRODYNAMIC FLOW
         MAGNETOSPHERE
         PLASMA CLOUDS

GEOMAGNETIC LATITUDE
  GS   LATITUDE
         . **GEOMAGNETIC LATITUDE**
  RT   COORDINATES
         GEOMAGNETISM
         POLAR CUSPS

GEOMAGNETIC MICROPULSATIONS
  GS   PULSES
         . GEOMAGNETIC PULSATIONS
         . . **GEOMAGNETIC MICROPULSATIONS**
         . MICROPULSATIONS
         . . **GEOMAGNETIC MICROPULSATIONS**
         VARIATIONS
         . MAGNETIC VARIATIONS
         . . GEOMAGNETIC PULSATIONS
         . . . **GEOMAGNETIC MICROPULSATIONS**
  RT   NOCTURNAL VARIATIONS
         TELLURIC CURRENTS

GEOMAGNETIC PULSATIONS
  GS   PULSES
         . **GEOMAGNETIC PULSATIONS**
         . . GEOMAGNETIC MICROPULSATIONS
         VARIATIONS
         . MAGNETIC VARIATIONS
         . . **GEOMAGNETIC PULSATIONS**
         . . . GEOMAGNETIC MICROPULSATIONS
  RT   GEOMAGNETISM
         KP INDEX
         MAGNETOSPHERIC INSTABILITY
         NOCTURNAL VARIATIONS

GEOMAGNETIC STORMS
  USE   MAGNETIC STORMS

GEOMAGNETIC TAIL
  GS   EARTH ATMOSPHERE
         . UPPER ATMOSPHERE
         . . MAGNETOSPHERE
         . . . **GEOMAGNETIC TAIL**
         ENVIRONMENTS
         . MAGNETOSPHERE
         . . **GEOMAGNETIC TAIL**
  RT   GEOMAGNETISM
         MAGNETIC FIELDS
         PLANETARY MAGNETIC FIELDS
         POLAR CUSPS

GEOMAGNETICALLY TRAPPED PARTICLES
  USE   RADIATION BELTS

GEOMAGNETISM
  UF   GEOMAGNETIC FIELD
         TERRESTRIAL MAGNETISM
  GS   MAGNETIC FIELDS
         . **GEOMAGNETISM**
         MAGNETIC PROPERTIES
         . **GEOMAGNETISM**
  RT   AEROMAGNETISM
         BARIUM ION CLOUDS
         CONTINENTAL DRIFT
         DYNAMO THEORY
         EARTH (PLANET)
         ELECTROJETS
         FIELD THEORY (PHYSICS)
         GEOMAGNETIC LATITUDE
         GEOMAGNETIC PULSATIONS
         GEOMAGNETIC TAIL
         GEOPHYSICS
         ∞INCLINATION
         INTERNATIONAL MAGNETOSPHERIC
           STUDY
         KP INDEX
         M REGION
         MAGNETIC ANOMALIES
         MAGNETIC DISTURBANCES
         MAGNETIC EFFECTS
         MAGNETIC EQUATOR
         MAGNETIC POLES
         MAGNETIC SURVEYS
         MAGNETOIONICS
         MAGNETOMETERS
         MAGNETOSPHERE
         MAGSAT A SATELLITE
         MAGSAT B SATELLITE
         MAGSAT SATELLITES
         MAGSAT 1 SATELLITE
         PALEOMAGNETISM
         PLANETARY MAGNETIC FIELDS
         POLAR CUSPS
         SPACE PLASMAS
         VARIOMETERS

GEOMETRIC ACCURACY
  GS   ACCURACY
         . **GEOMETRIC ACCURACY**
  RT   DISTORTION
         GEOMETRIC RECTIFICATION (IMAGERY)
         IMAGE PROCESSING
         IMAGE RESOLUTION

GEOMETRIC DILUTION OF PRECISION
  UF   GDOP
  GS   DILUTION
         . **GEOMETRIC DILUTION OF PRECISION**
  RT   PRECISION

GEOMETRIC RECTIFICATION (IMAGERY)
  GS   IMAGE PROCESSING
         . **GEOMETRIC RECTIFICATION
             (IMAGERY)**
         RECTIFICATION
         . **GEOMETRIC RECTIFICATION
             (IMAGERY)**

GEOMETRIC RECTIFICATION-*(CONT.)*
  RT   ATMOSPHERIC CORRECTION
         GEOMETRIC ACCURACY
         IMAGE ENHANCEMENT
         IMAGERY

GEOMETRICAL ACOUSTICS
  UF   RAY ACOUSTICS
  GS   ACOUSTICS
         . **GEOMETRICAL ACOUSTICS**
  RT   GEOMETRICAL THEORY OF
           DIFFRACTION
         GEOMETRY
         WAVE PROPAGATION

GEOMETRICAL HYDROMAGNETICS
  USE   MAGNETOHYDRODYNAMICS

GEOMETRICAL OPTICS
  UF   RAY OPTICS
  RT   ACOUSTO-OPTICS
         ASPHERICITY
         ASTIGMATISM
         CASSEGRAIN OPTICS
         CRYSTAL OPTICS
         DIFFRACTION PROPAGATION
         FIBER OPTICS
         FOCUSING
         GEOMETRICAL THEORY OF
           DIFFRACTION
         GRADIENT INDEX OPTICS
         LIGHT (VISIBLE RADIATION)
         LIGHT TRANSMISSION
         NONLINEAR OPTICS
         OPTICAL EQUIPMENT
         OPTICAL MEASUREMENT
         OPTICAL PATHS
         OPTICAL PROPERTIES
         OPTICAL REFLECTION
         ∞OPTICS
         PHYSICAL OPTICS
         RAY TRACING
         SNELLS LAW
         UNDERWATER OPTICS

GEOMETRICAL THEORY OF DIFFRACTION
  RT   DIFFRACTION
         DIFFRACTION PATTERNS
         GEOMETRICAL ACOUSTICS
         GEOMETRICAL OPTICS
         RAY TRACING
         REFLECTANCE
         ∞THEORIES
         WAVE DIFFRACTION

GEOMETRODYNAMICS
  USE   RELATIVITY

GEOMETRY
  GS   **GEOMETRY**
         . BOSE GEOMETRY
         . CRACK GEOMETRY
         . CURVATURE
         . CURVES (GEOMETRY)
         . . CATENARIES
         . . CYCLOIDS
         . . EPICYCLOIDS
         . . S CURVES
         . . . GOMPERTZ CURVES
         . CUSPS (MATHEMATICS)
         . . DOUBLE CUSPS
         . DIFFERENTIAL GEOMETRY
         . . LIE GROUPS
         . . . SPINOR GROUPS
         . . RIEMANN MANIFOLD
         . . TENSOR ANALYSIS
         . DUCT GEOMETRY
         . EUCLIDEAN GEOMETRY
         . . ANALYTIC GEOMETRY
         . . . CATENARIES
         . . . CIRCUMFERENCES
         . . . CONICS
         . . . . ELLIPSES
         . . . . HYPERBOLAS
         . . . . PARABOLAS
         . . . CYCLOIDS
         . . . EPICYCLOIDS
         . . . LOCI
         . . . MERCATOR PROJECTION
         . . . QUADRANTS
         . . . S CURVES
         . . . . GOMPERTZ CURVES
         . . . SPHEROIDS
         . . . . OBLATE SPHEROIDS
         . . . . PROLATE SPHEROIDS

GEOMETRY-(CONT.)
```
... TANGENTS
.. TORUSES
.. TRIGONOMETRY
. ANGLES (GEOMETRY)
... ANGLE OF ATTACK
.... ZERO ANGLE OF ATTACK
... BRAGG ANGLE
... BREWSTER ANGLE
.. DIHEDRAL ANGLE
.. ELEVATION ANGLE
.. LOOK ANGLES (ELECTRONICS)
.. LOOK ANGLES (TRACKING)
... SWEEP ANGLE
.... SWEEPBACK
..... LEADING EDGE SWEEP
.. CARTESIAN COORDINATES
. CIRCLES (GEOMETRY)
... GREAT CIRCLES
. DESCRIPTIVE GEOMETRY
. LINES (GEOMETRY)
... CHORDS (GEOMETRY)
.. GEODESIC LINES
. POINTS (MATHEMATICS)
... FIXED POINTS (MATHEMATICS)
.. INFLECTION POINTS
. POLYGONS
... HEXAGONS
... TETRAGONS
.... PARALLELOGRAMS
..... RHOMBOIDS
.... RECTANGLES
.... SQUARES (MATHEMATICS)
.... TRAPEZOIDS
... TRIANGLES
.. POLYHEDRONS
.. CUBES (MATHEMATICS)
... ICOSAHEDRONS
... OCTAHEDRONS
... PARALLELEPIPEDS
... PYRAMIDS
... RHOMBOHEDRONS
... TETRAHEDRONS
.. PROJECTIVE GEOMETRY
. MERCATOR PROJECTION
.. RADII
. FLOW GEOMETRY
. FRACTALS
. NOZZLE GEOMETRY
. SPECIMEN GEOMETRY
. TANK GEOMETRY
. TOPOLOGY
.. FIXED POINTS (MATHEMATICS)
.. HOMOTOPY THEORY
.. IMBEDDINGS (MATHEMATICS)
... INVARIANT IMBEDDINGS
.. LINKS (MATHEMATICS)
.. METRIC SPACE
. VECTOR ANALYSIS
.. COLLINEARITY
.. COPLANARITY
.. CURL (VECTORS)
... VORTICITY
```
```
RT    ANALYSIS (MATHEMATICS)
      AREA
      BODIES OF REVOLUTION
      COMPLEX NUMBERS
      CONGRUENCES
      COORDINATES
   ∞ CROSS SECTIONS
      CRYSTAL LATTICES
      DIAGRAMS
      DIAMETERS
      DIMENSIONS
      DISTANCE
      FOCI
      FRUSTUMS
      GEOIDS
      GEOMETRICAL ACOUSTICS
      HYPERGEOMETRIC FUNCTIONS
      HYPERSPHERES
      INFINITY
   ∞ MATHEMATICS
   ∞ MEASUREMENT
      PLANFORMS
      POINCARE SPHERES
      POSITION (LOCATION)
   ∞ PROFILES
      RECIPROCAL THEOREMS
      RHOMBOIDS
   ∞ SCIENCE
      SHAPES
      SIDES
      SPHERES
   ∞ SURFACE GEOMETRY
      SURVEYS
```

GEOMETRY-(CONT.)
```
      SYMMETRY
      TOROIDS
      UNIQUENESS THEOREM
      VENN DIAGRAMS
      VOLUME
```

GEOMORPHOLOGY
```
UF    PHYSIOGRAPHY
GS    GEOLOGY
      . GEOMORPHOLOGY
      MORPHOLOGY
      . GEOMORPHOLOGY
RT    CONES (VOLCANOES)
      CONTOURS
      GEODYNAMICS
      GEOGRAPHY
      GLACIOLOGY
      ISOSTASY
      LUNAR GEOLOGY
      MOUNTAINS
      OROGRAPHY
      PHOTOGEOLOGY
      SHATTER CONES
      SLUMPING
      TERRAIN
      TOPOGRAPHY
      VOLCANOES
      VOLCANOLOGY
```

GEON (TRADEMARK)
```
USE   POLYVINYL CHLORIDE
```

GEOPHYSICAL FLUID FLOW CELLS
```
GS    PAYLOADS
      . SPACELAB PAYLOADS
      .. GEOPHYSICAL FLUID FLOW CELLS
RT    AEROSPACE ENVIRONMENTS
   ∞ CELLS
      CONVECTIVE FLOW
      FLUID FLOW
      GAS FLOW
      INVESTIGATION
      JUPITER ATMOSPHERE
      SPACE TRANSPORTATION SYSTEM
         FLIGHTS
      SPACEBORNE EXPERIMENTS
      SPACELAB
   ∞ TEST EQUIPMENT
```

GEOPHYSICAL FLUIDS
```
RT    EARTH CORE
      FLUID DYNAMICS
      GEOTHERMAL RESOURCES
      GEOTHERMAL TECHNOLOGY
```

GEOPHYSICAL OBSERVATORIES
```
GS    OBSERVATORIES
      . GEOPHYSICAL OBSERVATORIES
      .. OGO
      ... EGO
      ... OGO-A
      ... OGO-3
      ... OGO-5
      ... POGO
      .... OGO-C
      .... OGO-4
      .... OGO-6
      .. OSO
      ... OSO-C
      ... OSO-1
      ... OSO-2
      ... OSO-3
      ... OSO-4
      ... OSO-5
      ... OSO-6
      ... OSO-7
      ... OSO-8
RT    ASTRONOMICAL OBSERVATORIES
      GEOLOGY
      GEOPHYSICS
```

GEOPHYSICAL SATELLITES
```
GS    SATELLITES
      . ARTIFICIAL SATELLITES
      .. GEOPHYSICAL SATELLITES
      ... COSMOS SATELLITES
      .... INTERCOSMOS SATELLITES
      ... EXPLORER 6 SATELLITE
      ... EXPLORER 10 SATELLITE
      ... EXPLORER 12 SATELLITE
      ... EXPLORER 45 SATELLITE
      ... OGO
      .... EGO
      .... OGO-A
```

GEOPHYSICAL SATELLITES-(CONT.)
```
      .... OGO-3
      .... OGO-5
      .... POGO
      ..... OGO-C
      ..... OGO-4
      ..... OGO-6
      ... OSO
      ... OSO-C
      ... OSO-1
      ... OSO-2
      ... OSO-3
      ... OSO-4
      ... OSO-5
      ... OSO-6
      ... OSO-7
      ... OSO-8
      ... RADIATION AND METEOROID
            SATELLITE
      ... SPUTNIK 3 SATELLITE
      ... VANGUARD 3 SATELLITE
RT    ARIEL SATELLITES
      COMMUNICATION SATELLITES
      EOLE SATELLITES
      METEOROLOGICAL SATELLITES
      PEOLE SATELLITES
      SPACE LABORATORIES
      UNMANNED SPACECRAFT
      VANGUARD SATELLITES
```

GEOPHYSICS
```
UF    GEOASTROPHYSICS
RT    AERONOMY
      CONTINENTAL DRIFT
      EARTH (PLANET)
      EARTH PLANETARY STRUCTURE
      FORMATIONS
      GEOCHEMISTRY
      GEOCHRONOLOGY
      GEODESY
      GEODYNAMICS
      GEOELECTRICITY
      GEOIDS
      GEOLOGICAL SURVEYS
      GEOLOGY
      GEOMAGNETISM
      GEOPHYSICAL OBSERVATORIES
      GRAVIMETERS
      GRAVIMETRY
      HEAT TRANSMISSION
      HYDROGRAPHY
      HYDROLOGY
      INTERNATIONAL GEOPHYSICAL YEAR
      INTERNATIONAL
         GEOSPHERE-BIOSPHERE PROGRAM
      ISOSTASY
      LIMNOLOGY
      METEOROLOGY
      OCEANOGRAPHY
      PALEOMAGNETISM
      PETROLOGY
   ∞ PHYSICS
      PLATES (TECTONICS)
      POLAR CUSPS
   ∞ RADIATION
      RADIOACTIVITY
   ∞ SCIENCE
      SEISMOLOGY
      STRATIGRAPHY
      STRUCTURAL PROPERTIES (GEOLOGY)
      TECTONICS
      THEORETICAL PHYSICS
      TILTMETERS
      TOPOGRAPHY
```

GEOPOTENTIAL
```
GS    GEOPOTENTIAL
      . GEOPOTENTIAL HEIGHT
RT    GEOELECTRICITY
      GRAVITATIONAL FIELDS
      HEIGHT
   ∞ POTENTIAL
      POTENTIAL ENERGY
```

GEOPOTENTIAL HEIGHT
```
GS    GEOPOTENTIAL
      . GEOPOTENTIAL HEIGHT
      POTENTIAL ENERGY
      . GEOPOTENTIAL HEIGHT
RT    ATMOSPHERIC PRESSURE
      EARTH ATMOSPHERE
      GEODETIC ACCURACY
      GRAVITATIONAL FIELDS
      HEAD (FLUID MECHANICS)
      SCALE HEIGHT
```

GEOPRESSURE
- GS   PRESSURE
-     . **GEOPRESSURE**
- RT   GEOLOGY
-      GEOTHERMAL RESOURCES
-      GEOTHERMAL TECHNOLOGY
-      PRESSURE GRADIENTS

GEORGIA
- GS   NATIONS
-     . UNITED STATES
-     . . **GEORGIA**
- RT   ATLANTA (GA)
-      SAND HILLS REGION (GA-NC-SC)

GEOS SATELLITES (ESA)
- UF   GEOS SATELLITES (ESRO)
- GS   SATELLITES
-     . ARTIFICIAL SATELLITES
-     . . ESA SATELLITES
-     . . . **GEOS SATELLITES (ESA)**
- RT   EUROPEAN SPACE PROGRAMS
-      GEOSARI PROJECT
-      MAGNETOSPHERE

GEOS SATELLITES (ESRO)
- USE   GEOS SATELLITES (ESA)

GEOS 1 SATELLITE
- GS   SATELLITES
-     . ARTIFICIAL SATELLITES
-     . . GEODETIC SATELLITES
-     . . . **GEOS 1 SATELLITE**
-     UNMANNED SPACECRAFT
-     . GEODETIC SATELLITES
-     . . **GEOS 1 SATELLITE**
- RT   ACTIVE SATELLITES
-      ANNA SATELLITES
-      CELESTIAL GEODESY
-      EXPLORER 29 SATELLITE
-      LARGOS SATELLITE
-      PAGEOS SATELLITE

GEOS 2 SATELLITE
- UF   GEOS-B SATELLITE
- GS   SATELLITES
-     . ARTIFICIAL SATELLITES
-     . . GEODETIC SATELLITES
-     . . . **GEOS 2 SATELLITE**
-     UNMANNED SPACECRAFT
-     . GEODETIC SATELLITES
-     . . **GEOS 2 SATELLITE**
- RT   ACTIVE SATELLITES
-      ANNA SATELLITES
-      CELESTIAL GEODESY
-      EXPLORER 36 SATELLITE
-      LARGOS SATELLITE
-      PAGEOS SATELLITE

GEOS 3 SATELLITE
- UF   GEOS-C SATELLITE
- GS   SATELLITES
-     . ARTIFICIAL SATELLITES
-     . . GEODETIC SATELLITES
-     . . . **GEOS 3 SATELLITE**
-     UNMANNED SPACECRAFT
-     . GEODETIC SATELLITES
-     . . **GEOS 3 SATELLITE**
- RT   ACTIVE SATELLITES
-      ANNA SATELLITES
-      CELESTIAL GEODESY
-      LARGOS SATELLITE
-      PAGEOS SATELLITE

GEOS-B SATELLITE
- USE   GEOS 2 SATELLITE

GEOS-C SATELLITE
- USE   GEOS 3 SATELLITE

GEOS-D SATELLITE
- UF   GEODYNAMIC EXPERIMENTAL OCEAN
-         SATELLITE
- GS   SATELLITES
-     . ARTIFICIAL SATELLITES
-     . . **GEOS-D SATELLITE**

GEOSARI PROJECT
- GS   PROGRAMS
-     . PROJECTS
-     . . **GEOSARI PROJECT**
- RT   ARIANE LAUNCH VEHICLE
-      EUROPEAN SPACE AGENCY
-      GEOS SATELLITES (ESA)

GEOSTATIONARY OPERATIONAL ENVIRON SATS
- USE   GOES SATELLITES

GEOSTATIONARY OPERATL ENVIRON SATELLITE B
- USE   GOES 2

GEOSTATIONARY PLATFORMS
- USE   SYNCHRONOUS PLATFORMS

GEOSTATIONARY SATELLITES
- USE   SYNCHRONOUS SATELLITES

GEOSTROPHIC WIND
- GS   WIND (METEOROLOGY)
-     . WINDS ALOFT
-     . . **GEOSTROPHIC WIND**
- RT   BAROCLINIC INSTABILITY
-      BAROCLINIC WAVES
-      DIVERGENCE
-      ISOBARS (PRESSURE)
-      SEA BREEZE
-      WIND SHEAR

†

GEOSYNCHRONOUS ORBITS
- GS   ORBITS
-     . SPACECRAFT ORBITS
-     . . SATELLITE ORBITS
-     . . . **GEOSYNCHRONOUS ORBITS**
- RT   CIRCULAR ORBITS
-      EQUATORIAL ORBITS
-      INFRARED ASTRONOMY SATELLITE
-      STATIONARY ORBITS
-      SYNCHRONOUS PLATFORMS
-      TWENTY-FOUR HOUR ORBITS

GEOSYNCLINES
- RT   ANTICLINES
-      DOMES (GEOLOGY)
-     ∞ LAYERS
-      STRATA
-      STRATIFICATION
-      STRATIGRAPHY
-      SYNCLINES

GEOTECHNICAL ENGINEERING
- RT   FOUNDATIONS
-      SOIL MECHANICS
-      STRUCTURAL DESIGN CRITERIA
-      STRUCTURAL ENGINEERING

GEOTECHNICAL FABRICS
- UF   GEOFABRICS
-      GEOTEXTILES
- RT   FABRICS
-      SOIL MECHANICS

GEOTEMPERATURE
- UF   GEOTHERMOMETRY
- RT   GEOLOGY
-      GEOTHERMAL ANOMALIES
-      TEMPERATURE

GEOTEXTILES
- USE   GEOTECHNICAL FABRICS

GEOTHERMAL ANOMALIES
- GS   ANOMALIES
-     . **GEOTHERMAL ANOMALIES**
- RT   GEOTEMPERATURE
-      GEOTHERMAL RESOURCES
-      SURFACE TEMPERATURE
-      THERMAL MAPPING

GEOTHERMAL ENERGY CONVERSION
- GS   ENERGY CONVERSION
-     . **GEOTHERMAL ENERGY CONVERSION**
- RT   CLEAN ENERGY
-     ∞ CONVERSION
-      EARTH RESOURCES
-      ENGINES
-      HEAT TRANSFER
-      HEAT TRANSMISSION
-      OCEAN THERMAL ENERGY CONVERSION
-      THERMAL ENERGY
-      TURBINES
-      TURBOGENERATORS

GEOTHERMAL ENERGY EXTRACTION
- GS   EXTRACTION
-     . **GEOTHERMAL ENERGY EXTRACTION**
- RT   ENERGY CONVERSION
-      ENERGY TECHNOLOGY
-      HEAT EXCHANGERS

GEOTHERMAL ENERGY EXTRACTION-_(CONT.)_
-      HEAT PUMPS
-      HEAT TRANSMISSION
-      HEATING
-      TURBINES
-      TURBOGENERATORS
-      WATER HEATING

GEOTHERMAL ENERGY UTILIZATION
- GS   UTILIZATION
-     . **GEOTHERMAL ENERGY UTILIZATION**
- RT   COOLING
-     ∞ ELECTRIC POWER
-      ENERGY STORAGE
-      HEAT PIPES
-      HEAT STORAGE
-      HEATING
-     ∞ POWER PLANTS
-      TURBOGENERATORS

GEOTHERMAL RESOURCES
- GS   HEAT SOURCES
-     . THERMAL RESOURCES
-     . . **GEOTHERMAL RESOURCES**
-     . . . GEYSERS
-     RESOURCES
-     . EARTH RESOURCES
-     . THERMAL RESOURCES
-     . . . **GEOTHERMAL RESOURCES**
-     . . . . GEYSERS
- RT   DRY HEAT
-     ∞ ENERGY SOURCES
-      GEOPHYSICAL FLUIDS
-      GEOPRESSURE
-      GEOTHERMAL ANOMALIES
-      HEAT TRANSMISSION
-      HYDROTHERMAL SYSTEMS
-      THERMAL ENERGY
-      THERMAL MAPPING
-      UNDERWATER RESOURCES

GEOTHERMAL TECHNOLOGY
- GS   TECHNOLOGIES
-     . ENERGY TECHNOLOGY
-     . . **GEOTHERMAL TECHNOLOGY**
- RT   DRY HEAT
-      EXPLORATION
-      GEOPHYSICAL FLUIDS
-      GEOPRESSURE
-      GEYSERS
-      HEAT SOURCES
-      HEAT TRANSFER
-      OCEAN THERMAL ENERGY CONVERSION
-      RESOURCES
-      THERMAL RESOURCES

GEOTHERMOMETRY
- USE   GEOTEMPERATURE

GEOTROPISM
- GS   TROPISM
-     . **GEOTROPISM**
- RT   GRAVITATIONAL EFFECTS
-      PHYSIOLOGICAL EFFECTS
-      PLANTS (BOTANY)

GEP TELESCOPES
- USE   PARTICLE TELESCOPES

GERDIEN ARC HEATERS
- USE   ARC HEATING
-      HEATING EQUIPMENT

GERDIEN CONDENSERS
- GS   MEASURING INSTRUMENTS
-     . **GERDIEN CONDENSERS**
- RT   CAPACITORS
-      ION DENSITY (CONCENTRATION)

GERIATRICS
- GS   MEDICAL SCIENCE
-     . **GERIATRICS**
- RT   AGING (BIOLOGY)
-      GERONTOLOGY

GERMAN DEMOCRATIC REPUBLIC
- USE   EAST GERMANY

GERMANATES
- GS   GERMANIUM COMPOUNDS
-     . **GERMANATES**
-     . . MAGNESIUM GERMANATES

GERMANIDES
GS GERMANIUM COMPOUNDS
    . **GERMANIDES**
    . . MAGNESIUM GERMANIDES
RT GERMANIUM ALLOYS

GERMANIUM
GS CHEMICAL ELEMENTS
    . METALLOIDS
    . . **GERMANIUM**
    . . . GERMANIUM ISOTOPES

GERMANIUM ALLOYS
GS ALLOYS
    . **GERMANIUM ALLOYS**
RT GERMANIDES

GERMANIUM ANTIMONIDES
GS ANTIMONY COMPOUNDS
    . ANTIMONIDES
    . . **GERMANIUM ANTIMONIDES**
    GERMANIUM COMPOUNDS
    . **GERMANIUM ANTIMONIDES**

GERMANIUM CHLORIDES
GS GERMANIUM COMPOUNDS
    . **GERMANIUM CHLORIDES**
    HALOGEN COMPOUNDS
    . CHLORINE COMPOUNDS
    . . CHLORIDES
    . . . **GERMANIUM CHLORIDES**
    . HALIDES
    . . CHLORIDES
    . . . **GERMANIUM CHLORIDES**

GERMANIUM COMPOUNDS
GS **GERMANIUM COMPOUNDS**
    . GERMANATES
    . . MAGNESIUM GERMANATES
    . GERMANIDES
    . . MAGNESIUM GERMANIDES
    . GERMANIUM ANTIMONIDES
    . GERMANIUM CHLORIDES
    . GERMANIUM OXIDES
    . ORGANIC GERMANIUM COMPOUNDS
RT ∞CHEMICAL COMPOUNDS
    ∞GROUP 4A COMPOUNDS
    ∞METAL COMPOUNDS

GERMANIUM DIODES
UF GERMANIUM RECTIFIERS
GS ELECTRONIC EQUIPMENT
    . DIODES
    . . SEMICONDUCTOR DIODES
    . . . **GERMANIUM DIODES**
    . SOLID STATE DEVICES
    . . SEMICONDUCTOR DEVICES
    . . . **GERMANIUM DIODES**
    RECTIFIERS
    . **GERMANIUM DIODES**
RT JUNCTION DIODES
    TRANSISTORS

GERMANIUM ISOTOPES
GS CHEMICAL ELEMENTS
    . METALLOIDS
    . . GERMANIUM
    . . . **GERMANIUM ISOTOPES**
    . NUCLIDES
    . . ISOTOPES
    . . . **GERMANIUM ISOTOPES**

GERMANIUM OXIDES
GS CHALCOGENIDES
    . OXIDES
    . . **GERMANIUM OXIDES**
    GERMANIUM COMPOUNDS
    . **GERMANIUM OXIDES**
RT ∞OXYGEN COMPOUNDS

GERMANIUM RECTIFIERS
USE GERMANIUM DIODES

∞ **GERMANY**
SN *(USE OF A MORE SPECIFIC TERM IS*
    RECOMMENDED--CONSULT THE TERMS
    LISTED BELOW)
RT EAST GERMANY
    WEST GERMANY

GERMICIDES
USE BACTERICIDES

GERMINATION
RT CROP GROWTH
    GROWTH
    PHYTOTRONS
    VIABILITY

GERMINATORS
USE PHYTOTRONS

GERONTOLOGY
RT AGE FACTOR
    AGING (BIOLOGY)
    GERIATRICS
    LIFE SPAN

GERT
UF GRAPHIC EVALUATION AND REVIEW
      TECHNIQUES
RT CRITICAL PATH METHOD
    MANAGEMENT
    MANAGEMENT ANALYSIS
    MANAGEMENT METHODS
    MANAGEMENT PLANNING
    ∞METHODOLOGY
    PERT
    PROJECT MANAGEMENT

GESTALT THEORY
RT PSYCHOTHERAPY
    ∞THEORIES

GETOL AIRCRAFT
GS GROUND EFFECT MACHINES
    . **GETOL AIRCRAFT**
RT ∞AIRCRAFT
    ∞SUBSONIC AIRCRAFT
    VERTICAL TAKEOFF AIRCRAFT

GETTERS
RT ION PUMPS
    PROPARGYL GROUPS
    PURIFICATION
    RESIDUAL GAS
    VACUUM
    VAPOR TRAPS

GEYSERS
GS HEAT SOURCES
    . THERMAL RESOURCES
    . . GEOTHERMAL RESOURCES
    . . . **GEYSERS**
    RESOURCES
    . EARTH RESOURCES
    . . THERMAL RESOURCES
    . . . GEOTHERMAL RESOURCES
    . . . . **GEYSERS**
RT ANOMALOUS TEMPERATURE ZONES
    GEOTHERMAL TECHNOLOGY
    HYDROGEOLOGY
    HYDROTHERMAL SYSTEMS

GHANA
GS NATIONS
    . **GHANA**
RT AFRICA

GHOSTS
RT DISTORTION
    RADAR ECHOES
    RADIO ECHOES

GIACOBINI-ZINNER COMET
GS CELESTIAL BODIES
    . COMETS
    . . **GIACOBINI-ZINNER COMET**
RT DRACONID METEOROIDS

GIANT STARS
GS CELESTIAL BODIES
    . STARS
    . . S STARS
    . . . **GIANT STARS**
    . . . . RED GIANT STARS
    . . . . . CARBON STARS
RT COOL STARS
    LATE STARS
    M STARS
    MAIN SEQUENCE STARS
    SUBGIANT STARS
    SUPERGIANT STARS

GIBBERELLINS
GS FUNGI
    . **GIBBERELLINS**

GIBBERELLINS-*(CONT.)*
    REGULATORS
    . **GIBBERELLINS**
RT HEMOSTATICS

GIBBS ADSORPTION EQUATION
RT ADSORPTION
    ∞EQUATIONS
    ∞GIBBS EQUATIONS
    INTERFACIAL TENSION

∞ **GIBBS EQUATIONS**
SN *(USE OF A MORE SPECIFIC TERM IS*
    RECOMMENDED--CONSULT THE TERMS
    LISTED BELOW)
RT ∞EQUATIONS
    GIBBS ADSORPTION EQUATION
    GIBBS FREE ENERGY
    GIBBS-HELMHOLTZ EQUATIONS
    PHASE RULE

GIBBS FREE ENERGY
GS THERMODYNAMIC PROPERTIES
    . FREE ENERGY
    . . **GIBBS FREE ENERGY**
RT ENTHALPY
    ∞GIBBS EQUATIONS
    GIBBS-HELMHOLTZ EQUATIONS
    MAYER PROBLEM

GIBBS PHENOMENON
RT DISCONTINUITY
    FOURIER SERIES
    SERIES (MATHEMATICS)

GIBBS-HELMHOLTZ EQUATIONS
RT ELECTRIC POTENTIAL
    ENTHALPY
    ∞EQUATIONS
    FREE ENERGY
    ∞GIBBS EQUATIONS
    GIBBS FREE ENERGY
    INTERNAL ENERGY
    PRESSURE
    TEMPERATURE

GIMBALLESS INERTIAL NAVIGATION
GS NAVIGATION
    . INERTIAL NAVIGATION
    . . **GIMBALLESS INERTIAL NAVIGATION**
RT GYROSCOPES
    INERTIAL PLATFORMS
    NAVIGATION INSTRUMENTS

GIMBALS
RT BEARINGS
    CONTROL MOMENT GYROSCOPES
    FLUID ROTOR GYROSCOPES
    GYRODAMPERS
    GYROSCOPES
    PIVOTS
    STABILIZED PLATFORMS
    SUPPORTS
    SWIVELS

GIOTTO MISSION
GS ESA SPACECRAFT
    . **GIOTTO MISSION**
    FLYBY MISSIONS
    . **GIOTTO MISSION**
    SPACE MISSIONS
    . **GIOTTO MISSION**
    UNMANNED SPACECRAFT
    . SPACE PROBES
    . . **GIOTTO MISSION**
RT HALLEY'S COMET

GIRDER WEBS
GS STRUCTURAL MEMBERS
    . PLATES (STRUCTURAL MEMBERS)
    . . **GIRDER WEBS**
    WEBS (SUPPORTS)
    . **GIRDER WEBS**
RT ELASTIC SHEETS
    GIRDERS
    METAL PLATES

GIRDERS
GS STRUCTURAL MEMBERS
    . **GIRDERS**
RT BEAMS (SUPPORTS)
    BOX BEAMS
    GIRDER WEBS
    PLATES (STRUCTURAL MEMBERS)

GIRDERS-*(CONT.)*
TRUSSES

GIRDLES
RT   ∞BELTS
PELVIS

GLACIAL DRIFT
UF   DRUMLINS
END MORAINES
ESKERS
GLACIOFLUVIAL DEPOSITS
MORAINES
STOSS-AND-LEE TOPOGRAPHY
RT   DEBRIS
GLACIERS
KETTLES (GEOLOGY)
LAND ICE
LANDFORMS
SEA ICE
SEDIMENTS

GLACIERS
UF   ACTIVE GLACIERS
ADVANCING GLACIERS
GS   CREVASSES
. **GLACIERS**
ICE
. **GLACIERS**
RESOURCES
. EARTH RESOURCES
. . **GLACIERS**
RT   CIRQUES (LANDFORMS)
GLACIAL DRIFT
GLACIOLOGY
LAND ICE
SEA ICE

GLACIOFLUVIAL DEPOSITS
USE   GLACIAL DRIFT

GLACIOLOGY
GS   GEOLOGY
. **GLACIOLOGY**
RT   GEOMORPHOLOGY
GLACIERS
HYDROGEOLOGY
ISOSTASY

∞ **GLANDS**
SN   *(USE OF A MORE SPECIFIC TERM IS*
RECOMMENDED--CONSULT THE TERMS
LISTED BELOW)
RT   GLANDS (ANATOMY)
GLANDS (SEALS)
PUMP SEALS

GLANDS (ANATOMY)
GS   ANATOMY
. **GLANDS (ANATOMY)**
. . ENDOCRINE GLANDS
. . . ADRENAL GLAND
. . . GONADS
. . . OVARIES
. . . PANCREAS
. . . PARATHYROID GLAND
. . . PINEAL GLAND
. . . PITUITARY GLAND
. . . PROSTATE GLAND
. . . THYMUS GLAND
. . . THYROID GLAND
. . MAMMARY GLANDS
. . SALIVARY GLANDS
. . SEBACEOUS GLANDS
. . SEX GLANDS
. . . GONADS
. . . OVARIES
. . . PROSTATE GLAND
. . . TESTES
RT   ENDOCRINE SYSTEMS
GASTROINTESTINAL SYSTEM
∞GLANDS
LIVER
ORGANS
SECRETIONS

GLANDS (SEALS)
GS   SEALS (STOPPERS)
. **GLANDS (SEALS)**
RT   ∞GLANDS
LABYRINTH SEALS
O RING SEALS
PACKINGS (SEALS)
PUMP SEALS
SEALING

GLARE
RT   BRIGHTNESS
COMFORT
DAYGLOW
ELECTROMAGNETIC RADIATION
HUMAN FACTORS ENGINEERING
ILLUMINATING
LIGHT (VISIBLE RADIATION)
LUMINANCE
LUSTER
OPTICAL PROPERTIES
RADIANCE
SKY BRIGHTNESS
SPECULAR REFLECTION
SPREAD REFLECTION
VISIBILITY
VISION

GLASS
GS   **GLASS**
. BOROSILICATE GLASS
. E GLASS
. . S GLASS
. GLASS FIBERS
. METALLIC GLASSES
. OBSIDIAN GLASS
. PYROCERAM (TRADEMARK)
. SILICA GLASS
. SPIN GLASS
. VYCOR
RT   AMORPHOUS MATERIALS
CERAMICS
ELECTROSTATIC BONDING
GLASSWARE
GLASSY CARBON
GRIFFITH CRACK
∞MATERIALS
MOLDAVITE
OBSIDIAN
OPTICAL PROPERTIES
PHOTOGRAPHIC PLATES
PORCELAIN
SILICON DIOXIDE
VITREOUS MATERIALS
VITRIFICATION

GLASS COATINGS
SN   (COATINGS CONSISTING OF GLASS)
GS   COATINGS
. **GLASS COATINGS**
RT   GLAZES
METALLIC GLASSES
PROTECTIVE COATINGS
SILICA GLASS

GLASS ELECTRODES
GS   ELECTRODES
. **GLASS ELECTRODES**
RT   ELECTROCHEMISTRY
ION EXCHANGING
SILICA GLASS

GLASS FIBER REINFORCED PLASTICS
GS   COMPOSITE MATERIALS
. FIBER COMPOSITES
. . **GLASS FIBER REINFORCED**
PLASTICS
. REINFORCED PLASTICS
. . **GLASS FIBER REINFORCED**
PLASTICS
RT   AIRFRAME MATERIALS
COMPOSITE STRUCTURES
E GLASS
FIBER ORIENTATION
FIBER REINFORCED COMPOSITES
FIBERS
LAMINATES
PLASTIC AIRCRAFT STRUCTURES
PULTRUSION
REINFORCING FIBERS
S GLASS
THERMOPLASTIC RESINS
THERMOSETTING RESINS

GLASS FIBERS
UF   FIBERGLASS
GS   FIBERS
. SYNTHETIC FIBERS
. . **GLASS FIBERS**
GLASS
. **GLASS FIBERS**
RT   BORON FIBERS
E GLASS
GRADIENT INDEX OPTICS
METALLIC GLASSES

GLASS FIBERS-*(CONT.)*
REINFORCING FIBERS
S GLASS
SILICA GLASS
VYCOR

GLASS LASERS
GS   STIMULATED EMISSION DEVICES
. LASERS
. . **GLASS LASERS**
RT   HIGH POWER LASERS
LASER FUSION
LASER OUTPUTS
LASER PLASMA INTERACTIONS
LASER TARGETS
NEODYMIUM LASERS
OPTICAL PUMPING
PULSED LASERS
ULTRASHORT PULSED LASERS

GLASSWARE
RT   BOROSILICATE GLASS
BOTTLES
BURETTES
∞CONTAINERS
FLASKS
GLASS
LABORATORY EQUIPMENT
PIPETTES
SILICA GLASS

GLASSY CARBON
GS   COMPOSITE MATERIALS
. **GLASSY CARBON**
RT   GLASS
∞MATERIALS

GLAUBER THEORY
RT   APPROXIMATION
ELASTIC SCATTERING
POMERANCHUK THEOREM
∞THEORIES

GLAUCOMA
GS   DISEASES
. EYE DISEASES
. . **GLAUCOMA**
RT   INTRAOCULAR PRESSURE

GLAUERT COEFFICIENT
USE   AERODYNAMIC FORCES
MACH NUMBER

GLAZES
GS   COATINGS
. **GLAZES**
FINISHES
. **GLAZES**
RT   CERAMICS
FRIT
GLASS COATINGS
PORCELAIN
PROTECTIVE COATINGS

GLIDE ANGLES
USE   GLIDE PATHS

GLIDE LANDINGS
GS   LANDING
. **GLIDE LANDINGS**
. . HORIZONTAL SPACECRAFT LANDING
RT   AIRCRAFT LANDING
CRASH LANDING
DITCHING (LANDING)
PLANETARY LANDING
SOFT LANDING
SPACECRAFT LANDING
WATER LANDING

GLIDE PATHS
UF   GLIDE ANGLES
GLIDE SLOPES
GS   FLIGHT PATHS
. **GLIDE PATHS**
SLOPES
. **GLIDE PATHS**
RT   AIRCRAFT APPROACH SPACING
APPROACH CONTROL
APPROACH INDICATORS
GLIDING
INSTRUMENT APPROACH
INSTRUMENT LANDING SYSTEMS
TERMINAL GUIDANCE

GLIDE SLOPES
USE   GLIDE PATHS

GLIDERS
UF   SAILPLANES
GS   **GLIDERS**
. ASSET GLIDERS
. BOOSTGLIDE VEHICLES
. . X-20 AIRCRAFT
. HANG GLIDERS
. HL-10 REENTRY VEHICLE
. HYPERSONIC GLIDERS
. X-20 AIRCRAFT
. JANUS SPACECRAFT
. KA-6 SAILPLANES
. PARAGLIDERS
. . INFLATABLE GLIDERS
. . PARAWINGS
RT   AEROSPACEPLANES
∞AIRCRAFT
FLEXIBLE WINGS
FREE FLIGHT
GLIDING
LIFTING REENTRY VEHICLES
∞MILITARY AIRCRAFT
MONOPLANES
OBSERVATION AIRCRAFT
SAILS
SAILWINGS
SOARING
∞SUBSONIC AIRCRAFT
TOWED BODIES
∞WINGED VEHICLES

GLIDING
RT   BOOSTGLIDE VEHICLES
DESCENT
∞FLIGHT
FLIGHT PATHS
FREE FLIGHT
GLIDE PATHS
GLIDERS
KA-6 SAILPLANES
LIFT
∞MOTION
SOARING

GLIMM METHOD
GS   PROCEDURES
. **GLIMM METHOD**
RT   ∞EQUATIONS
FLUID DYNAMICS
∞METHODOLOGY

GLINT
RT   ANGELS
RADAR ECHOES
SCINTILLATION

GLOBAL AIR POLLUTION
GS   POLLUTION
. ENVIRONMENT POLLUTION
. . AIR POLLUTION
. . . **GLOBAL AIR POLLUTION**
RT   EARTH ATMOSPHERE
ENVIRONMENTS
POLLUTION MONITORING
POLLUTION TRANSPORT

GLOBAL AIR SAMPLING PROGRAM
UF   GASP
RT   AIR POLLUTION
AIR SAMPLING
ENVIRONMENTAL QUALITY
SAMPLING

GLOBAL ATMOSPHERIC RESEARCH PROGRAM
UF   GARP
GS   PROGRAMS
. **GLOBAL ATMOSPHERIC RESEARCH
PROGRAM**
. . GARP ATLANTIC TROPICAL
EXPERIMENT
RT   AEROLOGY
INTEGRATED GLOBAL OCEAN STATION
SYSTEMS
METEOROLOGY
NASA PROGRAMS
WEATHER
WEATHER RECONNAISSANCE AIRCRAFT

GLOBAL COMMUNICATIONS ANTENNA GRID (NAVY)
USE   SEAFARER PROJECT

GLOBAL POSITIONING SYSTEM
RT   AUTONOMOUS SPACECRAFT CLOCKS
FLIGHT PATHS
NAVIGATION
POSITION INDICATORS
SATELLITE NAVIGATION SYSTEMS
SPACE NAVIGATION
∞SYSTEMS

GLOBAL TRACKING NETWORK
UF   GLOTRAC (TRACKING NETWORK)
GS   STATIONS
. TRACKING STATIONS
. . **GLOBAL TRACKING NETWORK**
TRACKING NETWORKS
. **GLOBAL TRACKING NETWORK**
RT   DATA ACQUISITION
GROUND STATIONS
MINITRACK SYSTEM
NASCOM NETWORK
OPTICAL TRACKING
RADIO RELAY SYSTEMS
RANGE AND RANGE RATE TRACKING
SATELLITE TRACKING
SPACE FLIGHT TRACKING AND DATA
NETWORK
STDN (NETWORK)

∞ GLOBES
SN   (USE OF A MORE SPECIFIC TERM IS
RECOMMENDED--CONSULT THE TERMS
LISTED BELOW)
RT   EARTH (PLANET)
LUMINAIRES
MAPS
SPHERES

GLOBULAR CLUSTERS
GS   CELESTIAL BODIES
. **GLOBULAR CLUSTERS**
. . HORIZONTAL BRANCH STARS
RT   COLOR-MAGNITUDE DIAGRAM
METALLICITY
STAR DISTRIBUTION

GLOBULES
RT   FALLING SPHERES
INTERFACIAL TENSION
LIQUIDS
MICROBALLOONS
SPHERES

GLOBULINS
GS   PROTEINS
. **GLOBULINS**
. . GAMMA GLOBULIN

GLOMERULUS
GS   ANATOMY
. CARDIOVASCULAR SYSTEM
. . BLOOD VESSELS
. . . **GLOMERULUS**
. CIRCULATORY SYSTEM
. . VASCULAR SYSTEM
. . . BLOOD VESSELS
. . . . **GLOMERULUS**

GLOSSARIES
USE   DICTIONARIES

GLOSTER GA-5 AIRCRAFT
USE   GA-5 AIRCRAFT

GLOTRAC (TRACKING NETWORK)
USE   GLOBAL TRACKING NETWORK

GLOTTIS
GS   LARYNX
. **GLOTTIS**
RT   VOCAL CORDS

GLOVES
GS   CLOTHING
. **GLOVES**
RT   PROTECTIVE CLOTHING

GLOW
USE   LUMINESCENCE

GLOW DISCHARGES
GS   ELECTRIC CURRENT
. ELECTRIC DISCHARGES
. . **GLOW DISCHARGES**
RT   CATHODE GLOW

GLOW DISCHARGES-(CONT.)
ELECTRIC ARCS
ELECTRIC CORONA
ELECTRODELESS DISCHARGES
FARADAY DARK SPACE
GAS DISCHARGES
LIGHT SOURCES
PLASMA DISPLAY DEVICES
PLASMA RADIATION

GLUCOSE
GS   ALIPHATIC COMPOUNDS
. SUGARS
. . **GLUCOSE**
CARBOHYDRATES
. SUGARS
. . **GLUCOSE**

GLUCOSIDES
UF   GLYCOSIDES
GS   ALIPHATIC COMPOUNDS
. **GLUCOSIDES**
CARBOHYDRATES
. **GLUCOSIDES**
RT   NUCLEOSIDES

GLUES
GS   ADHESIVES
. **GLUES**
RT   PASTES
SIZING MATERIALS
TETRAETHYL ORTHOSILICATE

GLUONS
GS   PARTICLES
. ELEMENTARY PARTICLES
. . **GLUONS**
RT   LEPTONS
MESONS
QUANTUM CHROMODYNAMICS
QUARKS

GLUTAMATES
GS   ALIPHATIC COMPOUNDS
. **GLUTAMATES**
ESTERS
. **GLUTAMATES**
RT   GLUTAMIC ACID

GLUTAMIC ACID
GS   ACIDS
. AMINO ACIDS
. . **GLUTAMIC ACID**
ALIPHATIC COMPOUNDS
. **GLUTAMIC ACID**
ORGANIC COMPOUNDS
. AMINO ACIDS
. . **GLUTAMIC ACID**
RT   GLUTAMATES

GLUTAMINE
GS   ACIDS
. AMINO ACIDS
. . **GLUTAMINE**
ALIPHATIC COMPOUNDS
. **GLUTAMINE**
ORGANIC COMPOUNDS
. AMINO ACIDS
. . **GLUTAMINE**

GLUTATHIONE
GS   ACIDS
. AMINO ACIDS
. . **GLUTATHIONE**
ALIPHATIC COMPOUNDS
. **GLUTATHIONE**
ORGANIC COMPOUNDS
. AMINO ACIDS
. . **GLUTATHIONE**
PROTEINS
. COENZYMES
. . **GLUTATHIONE**

GLYCERIDES
GS   ALIPHATIC COMPOUNDS
. **GLYCERIDES**
ESTERS
. **GLYCERIDES**
RT   GLYCEROLS
NITROGLYCERIN

GLYCERINS
USE   GLYCEROLS

GLYCEROLS
UF   GLYCERINS
GS   ALIPHATIC COMPOUNDS
      . **GLYCEROLS**
RT   ALCOHOLS
      CARBOHYDRATES
      GLYCERIDES
      LIPIDS
      LIQUIDS
      NITROGLYCERIN
      TRIACETIN

GLYCINE
GS   ACIDS
      . AMINO ACIDS
      . . **GLYCINE**
      ORGANIC COMPOUNDS
      . AMINO ACIDS
      . . **GLYCINE**

GLYCOGENS
GS   ALIPHATIC COMPOUNDS
      . POLYSACCHARIDES
      . . **GLYCOGENS**
      CARBOHYDRATES
      . POLYSACCHARIDES
      . . **GLYCOGENS**

GLYCOLS
GS   ALIPHATIC COMPOUNDS
      . **GLYCOLS**
      HYDROXYL COMPOUNDS
      . ALCOHOLS
      . . **GLYCOLS**
RT   HYDROXYL RADICALS

GLYCOLYSIS
GS   CHEMICAL REACTIONS
      . **GLYCOLYSIS**
      DECOMPOSITION
      . **GLYCOLYSIS**

GLYCOSIDES
USE   GLUCOSIDES

GNEISS
GS   ROCKS
      . **GNEISS**
RT   SOILS

GNOMONIC PROJECTION
RT   PHOTOMAPPING
      ∞ PROJECTION
      PROJECTIVE GEOMETRY

GNOTOBIOTICS
RT   BACTERIA
      BACTERIOLOGY
      CLOSED ECOLOGICAL SYSTEMS
      CONTROLLED ATMOSPHERES
      ISOLATION
      MICROBIOLOGY
      MICROORGANISMS
      STERILIZATION

GNP
USE   GROSS NATIONAL PRODUCT

GOAL THEORY
RT   GOALS
      ∞ THEORIES

GOALS
RT   ACHIEVEMENT
      ENGINEERING MANAGEMENT
      GOAL THEORY
      PROJECT PLANNING
      PURPOSES
      RESEARCH MANAGEMENT

GOATS
GS   ANIMALS
      . VERTEBRATES
      . . MAMMALS
      . . . **GOATS**
RT   GRAZING
      LIVESTOCK

GOBI DESERT
GS   LAND
      . DESERTS
      . . **GOBI DESERT**
RT   ARID LANDS
      DESERTIFICATION

GODDARD EXPERIMENT PACKAGE TELESCOPE
USE   PARTICLE TELESCOPES

GODDARD TRAJECTORY DETERMINATION SYSTEM
UF   GTDS
RT   COMPUTER PROGRAMS
      INTERPLANETARY TRAJECTORIES
      MOON-EARTH TRAJECTORIES
      ORBIT CALCULATION
      ORBITAL MECHANICS
      ORBITAL POSITION ESTIMATION
      SPACECRAFT TRAJECTORIES
      ∞ SYSTEMS
      TRAJECTORY ANALYSIS
      TRAJECTORY OPTIMIZATION

GOERTLER INSTABILITY
UF   TAYLOR-GOERTLER INSTABILITY
GS   STABILITY
      . **GOERTLER INSTABILITY**
RT   BOUNDARY LAYER STABILITY
      BOUNDARY LAYER TRANSITION
      CENTRIFUGAL FORCE
      FLOW STABILITY
      LAMINAR BOUNDARY LAYER
      ROTATING FLUIDS
      ROTATING LIQUIDS
      TAYLOR INSTABILITY
      VORTICES
      WALL FLOW

GOES SATELLITES
SN   GEOSTATIONARY OPERATIONAL
      ENVIRONMENTAL SATELLITES
GS   SATELLITES
      . ARTIFICIAL SATELLITES
      . . SYNCHRONOUS SATELLITES
      . . . **GOES SATELLITES**
      . . . . GOES 1
      . . . . GOES 2
      . . . . GOES 3
      . . . . GOES 4
      . . . . GOES 5

GOES 1
GS   SATELLITES
      . ARTIFICIAL SATELLITES
      . . SYNCHRONOUS SATELLITES
      . . . GOES SATELLITES
      . . . . **GOES 1**
RT   METEOROLOGICAL SATELLITES

GOES 2
SN   GEOSTATIONARY OPERATIONAL
      ENVIRONMENTAL SATELLITE B
GS   SATELLITES
      . ARTIFICIAL SATELLITES
      . . SYNCHRONOUS SATELLITES
      . . . GOES SATELLITES
      . . . . **GOES 2**
RT   METEOROLOGICAL SATELLITES
      SMS 1
      SMS 2

GOES 3
GS   SATELLITES
      . ARTIFICIAL SATELLITES
      . . SYNCHRONOUS SATELLITES
      . . . GOES SATELLITES
      . . . . **GOES 3**
RT   METEOROLOGICAL SATELLITES

GOES 4
GS   SATELLITES
      . ARTIFICIAL SATELLITES
      . . SYNCHRONOUS SATELLITES
      . . . GOES SATELLITES
      . . . . **GOES 4**
RT   METEOROLOGICAL SATELLITES

GOES 5
GS   SATELLITES
      . ARTIFICIAL SATELLITES
      . . SYNCHRONOUS SATELLITES
      . . . GOES SATELLITES
      . . . . **GOES 5**
RT   METEOROLOGICAL SATELLITES

GOGGLES
GS   CLOTHING
      . **GOGGLES**
RT   EYE PROTECTION
      FLIGHT CLOTHING

GOGGLES-(CONT.)
      HELMETS
      PROTECTIVE CLOTHING
      SUNGLASSES

GOLAY DETECTOR CELLS
GS   MEASURING INSTRUMENTS
      . RADIATION MEASURING INSTRUMENTS
      . . RADIATION DETECTORS
      . . . **GOLAY DETECTOR CELLS**
RT   ENERGY ABSORPTION FILMS
      PNEUMATIC EQUIPMENT
      RADIATION ABSORPTION

GOLD
GS   CHEMICAL ELEMENTS
      . **GOLD**
      . . GOLD ISOTOPES
      . . . GOLD 198
      METALS
      . NOBLE METALS
      . . **GOLD**
      . . . GOLD ISOTOPES
      . . . . GOLD 198
      . TRANSITION METALS
      . . **GOLD**
      . . . GOLD ISOTOPES
      . . . . GOLD 198

GOLD ALLOYS
GS   ALLOYS
      . **GOLD ALLOYS**

GOLD COATINGS
UF   GOLD PLATE
GS   COATINGS
      . METAL COATINGS
      . . **GOLD COATINGS**
      METALS
      . METAL COATINGS
      . . **GOLD COATINGS**
RT   NICKEL PLATE
      PROTECTIVE COATINGS

GOLD ISOTOPES
GS   CHEMICAL ELEMENTS
      . GOLD
      . . **GOLD ISOTOPES**
      . . . GOLD 198
      . NUCLIDES
      . . ISOTOPES
      . . . RADIOACTIVE ISOTOPES
      . . . . **GOLD ISOTOPES**
      . . . . . GOLD 198
      METALS
      . NOBLE METALS
      . . GOLD
      . . . **GOLD ISOTOPES**
      . . . . GOLD 198
      . TRANSITION METALS
      . . GOLD
      . . . **GOLD ISOTOPES**
      . . . . GOLD 198

GOLD PLATE
USE   GOLD COATINGS

GOLD 198
GS   CHEMICAL ELEMENTS
      . GOLD
      . . GOLD ISOTOPES
      . . . **GOLD 198**
      . NUCLIDES
      . . ISOTOPES
      . . . RADIOACTIVE ISOTOPES
      . . . . GOLD ISOTOPES
      . . . . . **GOLD 198**
      METALS
      . NOBLE METALS
      . . GOLD
      . . . GOLD ISOTOPES
      . . . . **GOLD 198**
      . TRANSITION METALS
      . . GOLD
      . . . GOLD ISOTOPES
      . . . . **GOLD 198**

GOMPERTZ CURVES
GS   CHARTS
      . GRAPHS (CHARTS)
      . . **GOMPERTZ CURVES**
      GEOMETRY
      . CURVES (GEOMETRY)
      . . S CURVES
      . . . **GOMPERTZ CURVES**

GOMPERTZ CURVES-*(CONT.)*
. . EUCLIDEAN GEOMETRY
. . . ANALYTIC GEOMETRY
. . . . S CURVES
. . . . . **GOMPERTZ CURVES**

GONADS
GS   ANATOMY
. GLANDS (ANATOMY)
. . ENDOCRINE GLANDS
. . . **GONADS**
. . SEX GLANDS
. . . **GONADS**
VISCERA
. ENDOCRINE GLANDS
. . **GONADS**
. SEX GLANDS
. . **GONADS**
RT   PHYSIOLOGICAL EFFECTS
∞REPRODUCTION

GONDOLAS
RT   AIRCRAFT COMPARTMENTS
AIRSHIPS
BALLOONS
BASKETS

GONIOMETERS
GS   MEASURING INSTRUMENTS
. **GONIOMETERS**
. . PHOTOGONIOMETERS
. . RADIOGONIOMETERS
RT   ANGLES (GEOMETRY)
DIFFRACTOMETERS
INTERFEROMETERS
MACH-ZEHNDER INTERFEROMETERS
MONOCHROMATORS
OPTICAL MEASURING INSTRUMENTS
REFRACTOMETERS
SPECTROMETERS

GOODNESS OF FIT
GS   STATISTICAL ANALYSIS
. **GOODNESS OF FIT**
RT   FITTING
MATHEMATICAL MODELS
MAXIMUM LIKELIHOOD ESTIMATES
PROBABILITY DISTRIBUTION FUNCTIONS
PROBABILITY THEORY
STATISTICAL DISTRIBUTIONS
STATISTICAL TESTS
VARIANCE (STATISTICS)

GORES
RT   FABRICS
PARACHUTE FABRICS

GORGES
USE   CANYONS

GOSS (SUPPORT SYSTEM)
USE   GROUND OPERATIONAL SUPPORT
SYSTEM

GOVERNMENT PROCUREMENT
GS   PROCUREMENT
. **GOVERNMENT PROCUREMENT**
RT   COMMODITIES
CONTRACTS
FEDERAL BUDGETS
SERVICES

GOVERNMENT/INDUSTRY RELATIONS
RT   COMMERCE LAB
CONTRACT NEGOTIATION
CONTRACTORS
CONTRACTS
PROCUREMENT

GOVERNMENTS
RT   CONSTITUTION
CULTURE (SOCIAL SCIENCES)
POLICIES
POLITICS
REGIMES
VOTING

GOVERNORS
USE   SPEED REGULATORS

GRABENS
USE   GEOLOGICAL FAULTS

∞ GRADE
SN   *(USE OF A MORE SPECIFIC TERM IS*
RECOMMENDED--CONSULT THE TERMS
LISTED BELOW)
RT   ANGLES (GEOMETRY)
GRADIENTS
LEVEL (HORIZONTAL)
POSITION (TITLE)
QUALITY
SLOPES

GRADIENT INDEX OPTICS
RT   FIBER OPTICS
GEOMETRICAL OPTICS
GLASS FIBERS
LENS DESIGN
LENSES
NONLINEAR OPTICS
OPTICAL PROPERTIES
∞OPTICS
PHYSICAL OPTICS
RAY TRACING
REFRACTIVITY

GRADIENTS
GS   **GRADIENTS**
. ELECTRON DENSITY PROFILES
. POTENTIAL GRADIENTS
. PRESSURE GRADIENTS
. TEMPERATURE GRADIENTS
. . THERMOCLINES
RT   ANGLES (GEOMETRY)
COMPOSITION (PROPERTY)
CONJUGATE GRADIENT METHOD
∞CROSS SECTIONS
DIFFERENCES
DISTRIBUTION (PROPERTY)
∞DROP
∞GRADE
GRAVITY GRADIOMETERS
ISOBARS (PRESSURE)
ISOTHERMS
LEVEL (HORIZONTAL)
OPTIMIZATION
∞PROFILES
SLOPES
VARIATIONS
VECTOR ANALYSIS

GRADIOMETERS
USE   MAGNETOMETERS

GRADUATION
USE   CALIBRATING

GRAEFF CALCULUS
GS   ANALYSIS (MATHEMATICS)
. CALCULUS
. . **GRAEFF CALCULUS**
. NUMERICAL ANALYSIS
. . **GRAEFF CALCULUS**

GRAFTING
RT   IMPLANTATION
INSERTION

GRAIN BOUNDARIES
GS   BOUNDARIES
. **GRAIN BOUNDARIES**
RT   CRYSTAL DISLOCATIONS
GRAIN SIZE
INTERGRANULAR CORROSION
INTERSTICES
INTERSTITIALS
TWINNING

GRAIN SIZE
GS   SIZE (DIMENSIONS)
. **GRAIN SIZE**
RT   GRAIN BOUNDARIES
METAL FATIGUE
MICROSTRUCTURE
PARTICLE SIZE DISTRIBUTION

∞ GRAINS
SN   *(USE OF A MORE SPECIFIC TERM IS*
RECOMMENDED--CONSULT THE TERMS
LISTED BELOW)
RT   CRYSTALS
GRAINS (FOOD)
GRANULAR MATERIALS
PROPELLANT GRAINS

GRAINS (FOOD)
GS   FARM CROPS
. **GRAINS (FOOD)**
. . BARLEY
. . CORN
. . OATS
. . RICE
. . SORGHUM
. . WHEAT
RT   ANGIOSPERMS
EARTH RESOURCES
∞GRAINS
GRASSES
MILLET
SEEDS

GRAMMARS
RT   LANGUAGES
PARSING ALGORITHMS
SEMANTICS
SYNTAX
VOWELS
WORDS (LANGUAGE)

GRAND CANYON (AZ)
GS   LANDFORMS
. CANYONS
. . **GRAND CANYON (AZ)**
RT   ARIZONA

GRAND TOURS
UF   OUTER PLANET MISSIONS
GS   FLYBY MISSIONS
. **GRAND TOURS**
. . MARINER JUPITER-SATURN FLYBY
. . MARINER JUPITER-URANUS FLYBY
. . VOYAGER 1977 MISSION
RT   ∞MISSIONS
OUTER PLANETS EXPLORERS
SPACE FLIGHT
SPACE MISSIONS
VOYAGER 1 SPACECRAFT
VOYAGER 2 SPACECRAFT

GRANITE
GS   ROCKS
. IGNEOUS ROCKS
. . **GRANITE**
RT   BATHOLITHS
EARTH RESOURCES
SOILS

GRANTS
RT   APPROPRIATIONS
BUDGETING
CONTRACTS
NASA PROGRAMS
PATENTS
SUBCONTRACTS

GRANULAR MATERIALS
RT   BRITTLE MATERIALS
∞GRAINS
LOW DENSITY MATERIALS
∞MATERIALS
PARTICLES
PELLETS
POWDER (PARTICLES)

GRAPH THEORY
RT   COMBINATORIAL ANALYSIS
GRAPHS (CHARTS)
MATHEMATICAL MODELS
SET THEORY
∞THEORIES
TOPOLOGY
TREES (MATHEMATICS)

GRAPHIC ARTS
GS   ARTS
. **GRAPHIC ARTS**
RT   CHARTS
DIAGRAMS
DRAFTING (DRAWING)
DRAWINGS
ENGINEERING DRAWINGS
IMAGERY
INKS
MOTION PICTURES
PHOTOGRAPHY
∞PROJECTION

**GRAPHIC EVALUATION AND REVIEW
TECHNIQUES**
  USE   GERT

GRAPHITE
  GS   MINERALS
         . **GRAPHITE**
  RT   ALUMINUM GRAPHITE COMPOSITES
         CARBON
         ELECTRODES
         FIBER COMPOSITES
         FIBER RELEASE
         GRAPHITE-EPOXY COMPOSITES
         INTERCALATION
         LUBRICANTS
         MODERATORS
         SINGLE CRYSTALS
         SOLID LUBRICANTS
         SYNTHETIC METALS

GRAPHITE-EPOXY COMPOSITES
  GS   COMPOSITE MATERIALS
         . RESIN MATRIX COMPOSITES
         . . **GRAPHITE-EPOXY COMPOSITES**
         . SUPERHYBRID MATERIALS
         . . **GRAPHITE-EPOXY COMPOSITES**
  RT   CARBON FIBER REINFORCED PLASTICS
         ∞CONSTRUCTION MATERIALS
         EPOXY RESINS
         FIBER COMPOSITES
         GRAPHITE
         GRAPHITE-POLYIMIDE COMPOSITES
         ∞MATERIALS
         REINFORCED PLASTICS
         REINFORCING FIBERS

GRAPHITE-POLYIMIDE COMPOSITES
  GS   COMPOSITE MATERIALS
         . **GRAPHITE-POLYIMIDE COMPOSITES**
  RT   GRAPHITE-EPOXY COMPOSITES

GRAPHITIZATION
  RT   ANNEALING
         HEAT TREATMENT

GRAPHOEPITAXY
  RT   AMORPHOUS MATERIALS
         CRYSTAL LATTICES
         CRYSTAL STRUCTURE

GRAPHOLOGY
  GS   HANDWRITING
         . **GRAPHOLOGY**
         RECOGNITION
         . PATTERN RECOGNITION
         . . **GRAPHOLOGY**
  RT   CHARACTER RECOGNITION

GRAPHS (CHARTS)
  UF   POLARIZATION CHARTS
  GS   CHARTS
         . **GRAPHS (CHARTS)**
         . . BOND GRAPHS
         . . GOMPERTZ CURVES
         . . MOLLIER DIAGRAM
         . . PATTERSON MAP
  RT   CONFORMAL MAPPING
         ∞CURVES
         FAULT TREES
         GRAPH THEORY
         HISTOGRAMS
         ∞NETWORKS
         NOMOGRAPHS
         ∞ORIGINS
         PETRI NETS
         RECORDING INSTRUMENTS
         REPRESENTATIONS
         STATISTICAL ANALYSIS
         TREES (MATHEMATICS)

GRASHOF NUMBER
  GS   RATIOS
         . DIMENSIONLESS NUMBERS
         . . **GRASHOF NUMBER**
  RT   CONVECTION
         REYNOLDS NUMBER

GRASSES
  GS   PLANTS (BOTANY)
         . **GRASSES**
         . . HAY
         . . REEDS (PLANTS)
         . . SEA GRASSES
         . . SORGHUM

GRASSES-*(CONT.)*
  RT   ALFALFA
         CANOPIES (VEGETATION)
         DEFOLIATION
         FARM CROPS
         FARMLANDS
         GRAINS (FOOD)
         GRASSLANDS
         MILLET
         OATS
         SOD

GRASSHOPPERS
  GS   ANIMALS
         . INVERTEBRATES
         . . ARTHROPODS
         . . . INSECTS
         . . . . **GRASSHOPPERS**

GRASSLANDS
  UF   GRAZING LANDS
         MEADOWLANDS
         PRAIRIES
         SAVANNAHS
  GS   LAND
         . **GRASSLANDS**
         . . LLANOS ORIENTALES (COLOMBIA)
  RT   AGRICULTURE
         CROP GROWTH
         EARTH RESOURCES
         FARM CROPS
         FARMLANDS
         GRASSES
         HAY
         LAND USE
         PLAINS
         PLOWING
         RANGELANDS
         RURAL AREAS
         RURAL LAND USE
         SOD
         STEPPES

GRASSMANN ALGEBRA
  USE   VECTOR SPACES

∞ **GRATINGS**
  SN   *(USE OF A MORE SPECIFIC TERM IS
         RECOMMENDED--CONSULT THE TERMS
         LISTED BELOW)*
  RT   GRATINGS (SPECTRA)
         INTERFERENCE GRATING
         OPTICAL FILTERS

GRATINGS (SPECTRA)
  UF   DIFFRACTION GRATINGS
  GS   **GRATINGS (SPECTRA)**
         . ECHELETTE GRATINGS
  RT   FRESNEL DIFFRACTION
         ∞GRATINGS
         OPTICAL FILTERS
         RONCHI TEST
         ROWLAND CIRCLES

GRAVEL DEPOSITS
  USE   GRAVELS

GRAVELS
  UF   GRAVEL DEPOSITS
  GS   SEDIMENTS
         . **GRAVELS**
         SOILS
         . **GRAVELS**
  RT   AGGREGATES
         ALLUVIUM
         AQUIFERS
         BOREHOLES
         FANS (LANDFORMS)
         GRIT
         SANDS

GRAVIMETERS
  GS   MEASURING INSTRUMENTS
         . **GRAVIMETERS**
  RT   ACCELEROMETERS
         DENSITOMETERS
         GEODESY
         GEOPHYSICS
         GRAVIMETRY
         GRAVITATION
         MICRODENSITOMETERS

GRAVIMETRY
  RT   ∞ACCELERATION

GRAVIMETRY-*(CONT.)*
         ACCELEROMETERS
         GEOLOGY
         GEOPHYSICS
         GRAVIMETERS
         GRAVITATIONAL FIELDS
         ∞MEASUREMENT
         QUANTITATIVE ANALYSIS

GRAVIRECEPTORS
  GS   ANATOMY
         . SENSE ORGANS
         . . **GRAVIRECEPTORS**
         . . . OTOLITH ORGANS
         RECEPTORS (PHYSIOLOGY)
         . **GRAVIRECEPTORS**
  RT   OCULOGRAVIC ILLUSIONS
         SENSITOMETRY
         VERTICAL PERCEPTION

GRAVITATION
  UF   GRAVITY
  GS   **GRAVITATION**
         . ARTIFICIAL GRAVITY
         . GRAVITY ANOMALIES
         . LUNAR GRAVITATION
         . LUNAR GRAVITATIONAL EFFECTS
         . PLANETARY GRAVITATION
         . REDUCED GRAVITY
         . SOLAR GRAVITATION
  RT   ANTIGRAVITY
         DRAG
         ENVIRONMENTS
         GRAVIMETERS
         GRAVITATIONAL CONSTANT
         GRAVITATIONAL EFFECTS
         GRAVITATIONAL FIELDS
         GRAVITATIONAL WAVES
         GRAVITY GRADIOMETERS
         GRAVSAT SATELLITE
         HIGH GRAVITY ENVIRONMENTS
         ISOSTASY
         LOW WEIGHT
         LUNAR GRAVITY SIMULATOR
         PENDULUMS
         ROCHE LIMIT
         SIMILITUDE LAW
         SPACECRAFT ENVIRONMENTS
         TERMINAL VELOCITY
         WEIGHT (MASS)
         WEIGHTLESSNESS

GRAVITATION THEORY
  RT   BIMETRIC THEORIES
         GAUGE THEORY
         GRAVITATIONAL FIELDS
         GRAVITATIONAL WAVE ANTENNAS
         GRAVITINOS
         GRAVITONS
         ∞THEORIES
         UNIFIED FIELD THEORY

GRAVITATIONAL COLLAPSE
  GS   **GRAVITATIONAL COLLAPSE**
         . BLACK HOLES (ASTRONOMY)
         . WHITE HOLES (ASTRONOMY)
  RT   ASTROPHYSICS
         NAKED SINGULARITIES
         NEUTRAL CURRENTS
         QUASARS
         RELATIVISTIC PLASMAS
         STELLAR CORES
         SUPERNOVAE

GRAVITATIONAL CONSTANT
  GS   CONSTANTS
         . **GRAVITATIONAL CONSTANT**
  RT   BIG BANG COSMOLOGY

GRAVITATIONAL EFFECTS
  GS   **GRAVITATIONAL EFFECTS**
         . GRAVITATIONAL LENSES
         . LAGRANGIAN EQUILIBRIUM POINTS
         . LUNAR GRAVITATIONAL EFFECTS
         . STELLAR GRAVITATION
  RT   ACCELERATION STRESSES
           (PHYSIOLOGY)
         ACCELERATION TOLERANCE
         DROP TOWERS
         ∞EFFECTS
         GEOTROPISM
         GRAVITATIONAL PHYSIOLOGY
         GRAVITROPISM
         GRAVITY PROBE B
         LANGLEY COMPLEX COORDINATOR

GRAVITATIONAL EFFECTS-(CONT.)
    LOWER BODY NEGATIVE PRESSURE
    REISSNER-NORDSTROM SOLUTION
    STELLAR MASS ACCRETION
    SWINGBY TECHNIQUE
    WEIGHTLESSNESS

GRAVITATIONAL FIELDS
UF   GRAVITATIONAL POTENTIAL
GS   **GRAVITATIONAL FIELDS**
    . STELLAR GRAVITATION
RT   ATTRACTION
    CENTER OF GRAVITY
    EARTH-MOON SYSTEM
    FIELD STRENGTH
    FIELD THEORY (PHYSICS)
    ∞ FIELDS
    GEOPOTENTIAL
    GEOPOTENTIAL HEIGHT
    GRAVIMETRY
    GRAVITATION THEORY
    GRAVITATIONAL LENSES
    GRAVITY ANOMALIES
    GRAVSAT SATELLITE
    LAGRANGIAN EQUILIBRIUM POINTS
    MULTIPOLAR FIELDS
    SATELLITE PERTURBATION
    SCHWARZSCHILD METRIC
    UNIFIED FIELD THEORY
    YANG-MILLS FIELDS

GRAVITATIONAL LENSES
GS   GRAVITATIONAL EFFECTS
    . **GRAVITATIONAL LENSES**
    LENSES
    . **GRAVITATIONAL LENSES**
RT   BLACK HOLES (ASTRONOMY)
    FOCUSING
    GRAVITATIONAL FIELDS
    LIGHT SCATTERING
    NEUTRON STARS
    RELATIVISTIC EFFECTS
    RELATIVITY
    STELLAR GRAVITATION
    WHITE HOLES (ASTRONOMY)

GRAVITATIONAL PHYSIOLOGY
GS   PHYSIOLOGY
    . **GRAVITATIONAL PHYSIOLOGY**
RT   ACCELERATION STRESSES
      (PHYSIOLOGY)
    AEROSPACE MEDICINE
    CENTRIFUGING STRESS
    GRAVITATIONAL EFFECTS
    PHYSIOLOGICAL ACCELERATION
    PHYSIOLOGICAL EFFECTS
    PHYSIOLOGICAL RESPONSES
    SPACE FLIGHT STRESS
    STRESS (PHYSIOLOGY)

GRAVITATIONAL POTENTIAL
USE   GRAVITATIONAL FIELDS

GRAVITATIONAL RADIATION
USE   GRAVITATIONAL WAVES

GRAVITATIONAL WAVE ANTENNAS
GS   ANTENNAS
    . **GRAVITATIONAL WAVE ANTENNAS**
RT   ANTENNA DESIGN
    CRYOGENIC EQUIPMENT
    GRAVITATION THEORY

GRAVITATIONAL WAVES
UF   GRAVITATIONAL RADIATION
RT   CELESTIAL BODIES
    CELESTIAL MECHANICS
    EARTH-MOON SYSTEM
    GRAVITY WAVES
    ∞ RADIATION
    ∞ WAVES

GRAVITINOS
GS   PARTICLES
    . ELEMENTARY PARTICLES
    . . **GRAVITINOS**
    . NEUTRAL PARTICLES
    . . **GRAVITINOS**
RT   BARYONS
    COSMOLOGY
    DECOUPLING
    GRAVITATION THEORY
    GRAVITONS
    NEUTRINOS
    PARTICLE MASS

GRAVITINOS-(CONT.)
    WEAK ENERGY INTERACTIONS

GRAVITONS
GS   PARTICLES
    . ELEMENTARY PARTICLES
    . . **GRAVITONS**
RT   ATOMIC STRUCTURE
    GRAVITATION THEORY
    GRAVITINOS

GRAVITROPISM
GS   TROPISM
    . **GRAVITROPISM**
RT   GRAVITATIONAL EFFECTS
    PLANTS (BOTANY)
    VEGETATION GROWTH

GRAVITY
USE   GRAVITATION

GRAVITY ANOMALIES
GS   ANOMALIES
    . **GRAVITY ANOMALIES**
    GRAVITATION
    . **GRAVITY ANOMALIES**
RT   GRAVITATIONAL FIELDS
    MASCONS

GRAVITY GRADIENT SATELLITES
GS   SATELLITES
    . ARTIFICIAL SATELLITES
    . . **GRAVITY GRADIENT SATELLITES**
    . . . ATS
    . . . . ATS 1
    . . . . ATS 2
    . . . . ATS 3
    . . . . ATS 4
    . . . . ATS 5
    . . . . ATS 6
    . . . . ATS 7
    . . . . ATS 8
    . . . ORBIS CAL SATELLITE
RT   ARTIFICIAL GRAVITY
    EXPULSION
    MANNED SPACECRAFT
    OV-1 SATELLITES
    OV-2 SATELLITES
    OV-3 SATELLITES
    OV-4 SATELLITES
    OV-5 SATELLITES
    SATELLITE ATTITUDE CONTROL
    SATELLITE CONTROL
    SPIN REDUCTION
    UNMANNED SPACECRAFT

GRAVITY GRADIOMETERS
GS   MEASURING INSTRUMENTS
    . **GRAVITY GRADIOMETERS**
RT   GRADIENTS
    GRAVITATION

GRAVITY PROBE B
RT   GRAVITATIONAL EFFECTS
    GYROSCOPES
    NASA PROGRAMS
    RELATIVITY

GRAVITY WAVES
GS   ELASTIC WAVES
    . CAPILLARY WAVES
    . . **GRAVITY WAVES**
    . . . BAROCLINIC WAVES
    SURFACE WAVES
    . CAPILLARY WAVES
    . . **GRAVITY WAVES**
    . . . BAROCLINIC WAVES
RT   CNOIDAL WAVES
    GRAVITATIONAL WAVES
    PLANETARY WAVES
    RIPPLES
    WATER WAVES
    WIND (METEOROLOGY)

GRAVSAT SATELLITE
GS   SATELLITES
    . ARTIFICIAL SATELLITES
    . . SCIENTIFIC SATELLITES
    . . . **GRAVSAT SATELLITE**
RT   GRAVITATION
    GRAVITATIONAL FIELDS

GRAY GAS
GS   GASES

GRAY GAS-(CONT.)
    . **GRAY GAS**
RT   NONGRAY ATMOSPHERES
    RADIATION ABSORPTION
    RAYLEIGH SCATTERING
    THERMAL ABSORPTION

GRAY SCALE
RT   AERIAL PHOTOGRAPHY
    IMAGE CONTRAST
    IMAGE ENHANCEMENT
    IMAGE PROCESSING
    IMAGING TECHNIQUES
    OPTICAL DATA PROCESSING
    PATTERN RECOGNITION

GRAZING
GS   INGESTION (BIOLOGY)
    . **GRAZING**
RT   ANIMALS
    CATTLE
    DEER
    GOATS
    HORSES
    RANGELANDS
    RURAL LAND USE
    SWINE

GRAZING FLOW
RT   ACOUSTIC ATTENUATION
    ACOUSTIC DUCTS
    ACOUSTIC IMPEDANCE
    ACOUSTIC MEASUREMENT
    ACOUSTIC PROPERTIES
    AEROACOUSTICS
    ∞ FLOW
    NOISE REDUCTION
    ORIFICE FLOW
    RESONATORS
    SHEAR FLOW

GRAZING INCIDENCE
GS   INCIDENCE
    . **GRAZING INCIDENCE**
RT   ABERRATION
    OPTICAL MEASUREMENT
    RAY TRACING

GRAZING INCIDENCE SOLAR TELESCOPE
USE   GRIST (TELESCOPE)

GRAZING LANDS
USE   GRASSLANDS

GREASES
GS   PRODUCTS
    . PETROLEUM PRODUCTS
    . . **GREASES**
RT   FATS
    KEROGEN
    LUBRICANTS
    OILS
    THICKENERS (MATERIALS)

GREAT BASIN (US)
GS   LANDFORMS
    . STRUCTURAL BASINS
    . . **GREAT BASIN (US)**
RT   CALIFORNIA
    EARTH RESOURCES
    FOLDS (GEOLOGY)
    FORMATIONS
    GEOLOGY
    NEVADA
    STRUCTURAL PROPERTIES (GEOLOGY)
    UTAH

GREAT BRITAIN
USE   UNITED KINGDOM

GREAT CIRCLES
GS   GEOMETRY
    . EUCLIDEAN GEOMETRY
    . . CIRCLES (GEOMETRY)
    . . . **GREAT CIRCLES**
RT   FLIGHT OPTIMIZATION
    FLIGHT PATHS
    GROUND TRACKS
    TRAJECTORIES

GREAT LAKES (NORTH AMERICA)
GS   LAKES
    . **GREAT LAKES (NORTH AMERICA)**
    . . LAKE ERIE

GREAT LAKES (NORTH AMERICA)-*(CONT.)*
.. LAKE HURON
.. LAKE MICHIGAN
.. LAKE ONTARIO
.. LAKE SUPERIOR
RT   CANADA
CANALS
EARTH RESOURCES
INLAND WATERS
INTERNATIONAL FIELD YEAR FOR
GREAT LAKES
RESOURCES
UNITED STATES
WATER FLOW
WATER RESOURCES

GREAT PLAINS CORRIDOR (NORTH AMERICA)
GS   CORRIDORS
. **GREAT PLAINS CORRIDOR (NORTH
AMERICA)**
RT   AGRICULTURE
CANADA
PLAINS
RURAL LAND USE
UNITED STATES

GREAT SALT LAKE (UT)
GS   LAKES
. **GREAT SALT LAKE (UT)**
WATER
. **GREAT SALT LAKE (UT)**
RT   EARTH RESOURCES
HYDROLOGY
INLAND WATERS
PONDS
UTAH

GREAT SMOKY MOUNTAINS (NC-TN)
GS   LANDFORMS
. MOUNTAINS
.. **GREAT SMOKY MOUNTAINS (NC-TN)**
RT   NORTH CAROLINA
TENNESSEE

GREB SATELLITES
SN   (GALACTIC RADIATION EXPERIMENTAL
BACKGROUND SATELLITES)
UF   GALACTIC RADIATION EXP
BACKGROUND SATS
GS   SATELLITES
. ARTIFICIAL SATELLITES
.. **GREB SATELLITES**

GREECE
GS   NATIONS
. **GREECE**
RT   CYPRUS
EUROPE

GREEN THEOREM
USE   GREEN'S FUNCTIONS

GREEN WAVE EFFECT
RT   ANNUAL VARIATIONS
BOTANY
CHLOROPHYLLS
∞EFFECTS
FOLIAGE
LEAVES

GREEN'S FUNCTIONS
UF   GREEN THEOREM
GS   ANALYSIS (MATHEMATICS)
. REAL VARIABLES
.. **GREEN'S FUNCTIONS**
FUNCTIONS (MATHEMATICS)
. **GREEN'S FUNCTIONS**
RT   DIFFERENTIAL EQUATIONS
FIELD THEORY (ALGEBRA)
FIELD THEORY (PHYSICS)
HALF PLANES
HALF SPACES
JACOBI INTEGRAL
MANY BODY PROBLEM

GREENHOUSE EFFECT
RT   ATMOSPHERIC HEAT BUDGET
ATMOSPHERIC RADIATION
EARTH ATMOSPHERE
∞EFFECTS
ENVIRONMENT EFFECTS
INSOLATION
TERRESTRIAL RADIATION
THERMAL RADIATION

GREENHOUSE EFFECT-*(CONT.)*
VENUS CLOUDS

GREENHOUSES
RT   ∞BUILDINGS
PHYTOTRONS
PLANTS (BOTANY)

GREENLAND
GS   LANDFORMS
. ISLANDS
.. **GREENLAND**
RT   ARCTIC OCEAN
DENMARK

GREGORIAN ANTENNAS
RT   ANTENNA DESIGN
ANTENNA FEEDS
ANTENNA RADIATION PATTERNS
ANTENNAS
CASSEGRAIN ANTENNAS
MICROWAVE ANTENNAS

GRENADES
RT   AMMUNITION
INCENDIARY AMMUNITION
PYROTECHNICS

∞ **GRIDS**
SN   *(USE OF A MORE SPECIFIC TERM IS
RECOMMENDED--CONSULT THE TERMS
LISTED BELOW)*
RT   COORDINATES
IONIZERS
∞MATRICES
MESH
RETICLES
TUBE GRIDS
TURNSTILE ANTENNAS
WIRE GRID LENSES

GRIDS (MATHEMATICS)
USE   COMPUTATIONAL GRIDS

GRIFFITH CRACK
RT   CRACK CLOSURE
CRACK PROPAGATION
FRACTURE MECHANICS
GLASS
∞THEORIES

GRIFFON AIRCRAFT
USE   NORD 1500 AIRCRAFT

GRIGG-SKJELLERUP COMET
GS   CELESTIAL BODIES
. COMETS
.. **GRIGG-SKJELLERUP COMET**
RT   ∞COMA
COMET TAILS
SOLAR SYSTEM
SOLAR WIND

GRIGNARD REACTIONS
GS   CHEMICAL REACTIONS
. **GRIGNARD REACTIONS**
RT   CATALYSTS

∞ **GRINDING**
SN   *(USE OF A MORE SPECIFIC TERM IS
RECOMMENDED--CONSULT THE TERMS
LISTED BELOW)*
RT   GRINDING (COMMINUTION)
GRINDING (MATERIAL REMOVAL)

GRINDING (COMMINUTION)
UF   PULVERIZING
GS   COMMINUTION
. **GRINDING (COMMINUTION)**
RT   ATOMIZING
COMPOUNDING
CRUSHING
DISINTEGRATION
GRINDING (MATERIAL REMOVAL)
MIXING

GRINDING (MATERIAL REMOVAL)
GS   **GRINDING (MATERIAL REMOVAL)**
. METAL GRINDING
RT   ABRASION
COUNTERSINKING
CUTTING
GRINDING (COMMINUTION)
GROOVING

GRINDING (MATERIAL REMOVAL)-*(CONT.)*
MACHINING
METAL CUTTING
PLANING
POLISHING
SCARFING
WEAR

GRINDING MACHINES
GS   TOOLS
. MACHINE TOOLS
.. **GRINDING MACHINES**
RT   LATHES
∞MACHINERY
METAL GRINDING
MILLING MACHINES
SHAPERS
ULTRASONIC CLEANING

GRINDING MILLS
RT   ATOMIZERS
ATOMIZING
COMMINUTION
CRUSHERS
IMPACTORS
MIXERS

GRIST (TELESCOPE)
UF   GRAZING INCIDENCE SOLAR
TELESCOPE
GS   TELESCOPES
. **GRIST (TELESCOPE)**
RT   ENERGY SPECTRA
SOLAR COSMIC RAYS
SPACELAB
SUN

GRIT
RT   ABRASIVES
GRAVELS
PARTICLES
SANDS
SEDIMENTS

GROOVES
GS   **GROOVES**
. V GROOVES
RT   CORRUGATING
GROOVING

GROOVING
UF   FLUTING
RT   CUTTING
GRINDING (MATERIAL REMOVAL)
GROOVES
KNURLING
MACHINING
MILLING (MACHINING)
STRIATION

GROSS NATIONAL PRODUCT
UF   GNP
GS   PRODUCTS
. **GROSS NATIONAL PRODUCT**
RT   COMMERCE
COSTS
ECONOMETRICS
FINANCE
INDUSTRIES

GROUND BASED CONTROL
GS   **GROUND BASED CONTROL**
. AIR TRAFFIC CONTROL
.. AUTOMATED EN ROUTE ATC
.. RADAR APPROACH CONTROL
RT   AIR TRAFFIC CONTROLLERS
(PERSONNEL)
AIRCRAFT APPROACH SPACING
AIRCRAFT CONTROL
AIRPORT SURFACE DETECTION
EQUIPMENT
AIRPORT TOWERS
APPROACH
APPROACH CONTROL
AUTOMATIC CONTROL
AUTOMATIC TRAFFIC ADVISORY AND
RESOLUTION
∞CONTROL
FLIGHT CONTROL
FLIGHT MANAGEMENT SYSTEMS
FLY BY WIRE CONTROL
GUIDANCE (MOTION)
INSTRUMENT LANDING SYSTEMS
INTEGRATED MISSION CONTROL
CENTER

GROUND BASED CONTROL-*(CONT.)*
　　　　LANDING AIDS
　　　　MISSILE CONTROL
　　　　RADAR NAVIGATION
　　　　RADIO CONTROL
　　　　REMOTE CONTROL
　　　　SPACECRAFT CONTROL
　　　　SPACECRAFT GUIDANCE
　　　　TRAFFIC CONTROL

GROUND CREWS
　GS　　PERSONNEL
　　　　. **GROUND CREWS**
　RT　　MAINTENANCE

GROUND EFFECT (AERODYNAMICS)
　RT　　AERODYNAMIC DRAG
　　　　AERODYNAMICS
　　　　AIR CUSHION LANDING SYSTEMS
　　　　CUSHIONS
　　　　DOWNWASH
　　　　DRAG
　　　　∞EFFECTS
　　　　GROUND RESONANCE
　　　　JET BLAST EFFECTS
　　　　LIFT
　　　　PERIPHERAL JET FLOW
　　　　WAKES

GROUND EFFECT (COMMUNICATIONS)
　RT　　ECHOES
　　　　∞EFFECTS
　　　　ELECTROMAGNETIC INTERFERENCE
　　　　ELECTROMAGNETIC NOISE
　　　　RADIO ATTENUATION
　　　　SIGNAL FADING
　　　　WAVE REFLECTION

GROUND EFFECT MACHINES
　UF　　AIR CUSHION VEHICLES
　　　　DTMB-111 GROUND EFFECT MACHINE
　　　　DTMB-430 GROUND EFFECT MACHINE
　　　　HOVERCRAFT
　GS　　**GROUND EFFECT MACHINES**
　　　　. CUSHIONCRAFT GROUND EFFECT
　　　　　MACHINE
　　　　. GETOL AIRCRAFT
　　　　. HOVERCRAFT GROUND EFFECT
　　　　　MACHINES
　　　　. WESTLAND GROUND EFFECT
　　　　　MACHINES
　RT　　∞AIRCRAFT
　　　　COMMERCIAL AIRCRAFT
　　　　∞EFFECTS
　　　　∞FLIGHT VEHICLES
　　　　FLYING PLATFORMS
　　　　HOVERING
　　　　LIFTING ROTORS
　　　　∞MACHINERY
　　　　∞MILITARY AIRCRAFT
　　　　PASSENGER AIRCRAFT
　　　　PERIPHERAL JET FLOW
　　　　RAPID TRANSIT SYSTEMS
　　　　RESEARCH AIRCRAFT
　　　　∞SUBSONIC AIRCRAFT
　　　　SURFACE VEHICLES
　　　　∞TRANSPORT VEHICLES
　　　　V/STOL AIRCRAFT
　　　　∞VEHICLES
　　　　WATER TAKEOFF AND LANDING
　　　　　AIRCRAFT

GROUND HANDLING
　GS　　MATERIALS HANDLING
　　　　. **GROUND HANDLING**
　RT　　AIR CARGO
　　　　BAGGAGE
　　　　CREW PROCEDURES (PREFLIGHT)
　　　　∞FACILITIES
　　　　FLIGHT OPERATIONS
　　　　HANGARS
　　　　MOBILE LOUNGES
　　　　TRACTORS
　　　　TRUCKS

GROUND OPERATIONAL SUPPORT SYSTEM
　UF　　GOSS (SUPPORT SYSTEM)
　GS　　GROUND SUPPORT EQUIPMENT
　　　　. **GROUND OPERATIONAL SUPPORT
　　　　　SYSTEM**
　　　　SUPPORT SYSTEMS
　　　　. **GROUND OPERATIONAL SUPPORT
　　　　　SYSTEM**
　　　　WEAPON SYSTEMS

GROUND OPERATIONAL SUPPORT-*(CONT.)*
　. **GROUND OPERATIONAL SUPPORT
　　SYSTEM**
　RT　　∞SYSTEMS

GROUND RESONANCE
　RT　　AERODYNAMIC STABILITY
　　　　GROUND EFFECT (AERODYNAMICS)
　　　　HELICOPTERS
　　　　ROTARY WINGS
　　　　ROTOR AERODYNAMICS

GROUND SPEED
　GS　　RATES (PER TIME)
　　　　. **GROUND SPEED**
　　　　VELOCITY
　　　　. **GROUND SPEED**
　RT　　AIRSPEED
　　　　HIGH SPEED
　　　　LOW SPEED

GROUND SQUIRRELS
　GS　　ANIMALS
　　　　. VERTEBRATES
　　　　. . MAMMALS
　　　　. . . RODENTS
　　　　. . . . SQUIRRELS
　　　　. . . . . **GROUND SQUIRRELS**

GROUND STATE
　GS　　LEVEL (QUANTITY)
　　　　. ENERGY LEVELS
　　　　. . **GROUND STATE**
　RT　　ATOMIC ENERGY LEVELS
　　　　ATOMIC THEORY
　　　　ELECTRON STATES
　　　　QUANTUM THEORY

GROUND STATIONS
　GS　　STATIONS
　　　　. **GROUND STATIONS**
　　　　. . DEEP SPACE INSTRUMENTATION
　　　　　FACILITY
　　　　. . EARTH TERMINALS
　　　　. . INTEGRATED MISSION CONTROL
　　　　　CENTER
　　　　. . POLYSTATION DOPPLER TRACKING
　　　　　SYSTEM
　　　　. . SPACE DETECTION AND TRACKING
　　　　　SYSTEM
　　　　. . STDN (NETWORK)
　RT　　DATA ACQUISITION
　　　　DATA COLLECTION PLATFORMS
　　　　DOWNLINKING
　　　　GLOBAL TRACKING NETWORK
　　　　HANGARS
　　　　INTEGRATED GLOBAL OCEAN STATION
　　　　　SYSTEMS
　　　　JODRELL BANK OBSERVATORY
　　　　LAND MOBILE SATELLITE SERVICE
　　　　MSAT
　　　　OCEAN DATA ACQUISITIONS SYSTEMS
　　　　POLLUTION MONITORING
　　　　SPACE FLIGHT TRACKING AND DATA
　　　　　NETWORK
　　　　TRACKING STATIONS
　　　　WEATHER STATIONS

GROUND SUPPORT EQUIPMENT
　GS　　**GROUND SUPPORT EQUIPMENT**
　　　　. GROUND OPERATIONAL SUPPORT
　　　　　SYSTEM
　RT　　AIR TRAFFIC CONTROL
　　　　AIRCRAFT MAINTENANCE
　　　　AIRPORT PLANNING
　　　　AUXILIARY POWER SOURCES
　　　　BALLISTIC CAMERAS
　　　　CAPE KENNEDY LAUNCH COMPLEX
　　　　COMMAND AND CONTROL
　　　　COMMAND GUIDANCE
　　　　CRAWLER TRACTORS
　　　　EARTH TERMINAL MEASUREMENT
　　　　　SYSTEM
　　　　∞EQUIPMENT
　　　　FLIGHT CONTROL
　　　　GANTRY CRANES
　　　　HANDLING EQUIPMENT
　　　　LANDING AIDS
　　　　LAUNCHING BASES
　　　　LAUNCHING PADS
　　　　LAUNCHING SITES
　　　　MAINTENANCE
　　　　MISSILE LAUNCHERS
　　　　MISSILE STORAGE
　　　　MISSILES

GROUND SUPPORT EQUIPMENT-*(CONT.)*
　　　　ORDNANCE
　　　　PROPELLANT STORAGE
　　　　RADIO TELEMETRY
　　　　REFUELING
　　　　ROCKET LAUNCHERS
　　　　SATELLITE GROUND SUPPORT
　　　　∞SPACECRAFT
　　　　STORABLE PROPELLANTS
　　　　∞TEST EQUIPMENT
　　　　TRACKING NETWORKS
　　　　TRACKING STATIONS

GROUND SUPPORT SYSTEMS
　GS　　SUPPORT SYSTEMS
　　　　. **GROUND SUPPORT SYSTEMS**
　RT　　COMMONALITY
　　　　∞SYSTEMS

GROUND TESTS
　GS　　**GROUND TESTS**
　　　　. COLD FLOW TESTS
　　　　. PRELAUNCH TESTS
　　　　. . STATIC FIRING
　RT　　AIRCRAFT RUNUP
　　　　CAPTIVE TESTS
　　　　CREW PROCEDURES (PREFLIGHT)
　　　　ELECTRIC EQUIPMENT TESTS
　　　　ENGINE TESTS
　　　　FLIGHT TESTS
　　　　FULL SCALE TESTS
　　　　MISSILE TESTS
　　　　PREFIRING TESTS
　　　　PREFLIGHT OPERATIONS
　　　　SPACE ELECTRIC ROCKET TESTS
　　　　STABILITY TESTS
　　　　STATIC TESTS
　　　　TEST FIRING
　　　　∞TESTS
　　　　WING FLOW METHOD TESTS

GROUND TRACKS
　GS　　**GROUND TRACKS**
　　　　. SATELLITE GROUND TRACKS
　RT　　AREA NAVIGATION
　　　　FLIGHT PATHS
　　　　GREAT CIRCLES
　　　　ORBITS
　　　　∞PATHS
　　　　∞TRACKS

GROUND TRUTH
　RT　　AERIAL PHOTOGRAPHY
　　　　AERIAL RECONNAISSANCE
　　　　AIRBORNE INTEGRATED
　　　　　RECONNAISSANCE SYSTEM
　　　　CROP IDENTIFICATION
　　　　IMAGERY
　　　　PHOTOINTERPRETATION
　　　　PHOTORECONNAISSANCE
　　　　SPECTROPHOTOGRAPHY

GROUND WATER
　GS　　WATER
　　　　. INLAND WATERS
　　　　. . **GROUND WATER**
　RT　　AQUIFERS
　　　　EARTH RESOURCES
　　　　FRESH WATER
　　　　LIMNOLOGY
　　　　LYSIMETERS
　　　　POTABLE WATER
　　　　SPRINGS (WATER)
　　　　SURFACE WATER
　　　　WATER FLOW
　　　　WATER RESOURCES
　　　　WATER RUNOFF
　　　　WATER TABLES
　　　　WELLS

GROUND WAVE PROPAGATION
　GS　　TRANSMISSION
　　　　. WAVE PROPAGATION
　　　　. . **GROUND WAVE PROPAGATION**
　RT　　RADIO WAVES
　　　　SELECTIVE FADING
　　　　SKY WAVES

GROUND WIND
　GS　　WIND (METEOROLOGY)
　　　　. **GROUND WIND**
　RT　　AIR CURRENTS
　　　　ATMOSPHERIC CIRCULATION
　　　　CYCLONES
　　　　GUST LOADS

GROUND WIND-(CONT.)
　　　GUSTS
　　　MONSOONS
　　　SQUALLS
　　　STORMS (METEOROLOGY)
　　　TORNADOES
　　　WIND DIRECTION
　　　WIND EFFECTS
　　　WIND EROSION
　　　WIND PRESSURE
　　　WIND PROFILES
　　　WIND SHEAR
　　　WIND VELOCITY
　　　WINDMILLS (WINDPOWERED MACHINES)
　　　WINDPOWER UTILIZATION
　　　WINDPOWERED GENERATORS

GROUND-AIR-GROUND COMMUNICATION
　GS　　COMMUNICATING
　　　. **GROUND-AIR-GROUND**
　　　　COMMUNICATION
　　　　TELECOMMUNICATION
　　　. **GROUND-AIR-GROUND**
　　　　COMMUNICATION
　RT　　AERONAUTICAL SATELLITES
　　　　AIR TRAFFIC CONTROL
　　　　AIRCRAFT COMMUNICATION
　　　　AUTOMATED EN ROUTE ATC
　　　　COMMUNICATION SATELLITES
　　　　DISCRETE ADDRESS BEACON SYSTEM
　　　　OPTICAL COMMUNICATION
　　　　RADIO COMMUNICATION
　　　　SPACECRAFT COMMUNICATION
　　　　VOICE COMMUNICATION

GROUND-TO-AIR MISSILES
　USE　　SURFACE TO AIR MISSILES

GROUP BEHAVIOR
　USE　　GROUP DYNAMICS

GROUP DYNAMICS
　UF　　GROUP BEHAVIOR
　RT　　DEPENDENCE
　　　∞DYNAMICS
　　　　ETHNIC FACTORS
　　　　PROBLEM SOLVING
　　　　SOCIOLOGY

GROUP THEORY
　GS　　ALGEBRA
　　　. **GROUP THEORY**
　　　. . HOMOMORPHISMS
　　　. . . AUTOMORPHISMS
　　　. . . MONOIDS
　　　. . . SUBGROUPS
　RT　　CHIRAL DYNAMICS
　　　　FIBERS (MATHEMATICS)
　　　　LIE GROUPS
　　　∞THEORIES

GROUP VELOCITY
　GS　　RATES (PER TIME)
　　　. **GROUP VELOCITY**
　　　　VELOCITY
　　　. **GROUP VELOCITY**
　RT　　BEAT FREQUENCIES
　　　　HARMONIC MOTION
　　　　PHASE VELOCITY
　　　　PROPAGATION VELOCITY
　　　　QUANTUM MECHANICS
　　　　WAVE PROPAGATION

GROUP 1A COMPOUNDS
　USE　　ALKALI METAL COMPOUNDS

∞ **GROUP 1B COMPOUNDS**
　SN　　*(USE OF A MORE SPECIFIC TERM IS*
　　　　RECOMMENDED--CONSULT THE TERMS
　　　　LISTED BELOW)
　RT　　∞CHEMICAL COMPOUNDS
　　　　COPPER COMPOUNDS
　　　　NOBLE METALS
　　　　SILVER COMPOUNDS

GROUP 2A COMPOUNDS
　USE　　ALKALINE EARTH COMPOUNDS

∞ **GROUP 2B COMPOUNDS**
　SN　　*(USE OF A MORE SPECIFIC TERM IS*
　　　　RECOMMENDED--CONSULT THE TERMS
　　　　LISTED BELOW)
　RT　　CADMIUM COMPOUNDS
　　　　∞CHEMICAL COMPOUNDS

GROUP 2B COMPOUNDS-*(CONT.)*
　RT　　MERCURY COMPOUNDS
　　　　ZINC COMPOUNDS

∞ **GROUP 3A COMPOUNDS**
　SN　　*(USE OF A MORE SPECIFIC TERM IS*
　　　　RECOMMENDED--CONSULT THE TERMS
　　　　LISTED BELOW)
　RT　　ALUMINUM COMPOUNDS
　　　　BORON COMPOUNDS
　　　　∞CHEMICAL COMPOUNDS
　　　　GALLIUM COMPOUNDS
　　　　INDIUM COMPOUNDS

∞ **GROUP 3B COMPOUNDS**
　SN　　*(USE OF A MORE SPECIFIC TERM IS*
　　　　RECOMMENDED--CONSULT THE TERMS
　　　　LISTED BELOW)
　RT　　ACTINIDE SERIES COMPOUNDS
　　　　∞CHEMICAL COMPOUNDS
　　　　CURIUM COMPOUNDS
　　　　RARE EARTH COMPOUNDS
　　　　SCANDIUM COMPOUNDS
　　　　YTTRIUM COMPOUNDS

∞ **GROUP 4A COMPOUNDS**
　SN　　*(USE OF A MORE SPECIFIC TERM IS*
　　　　RECOMMENDED--CONSULT THE TERMS
　　　　LISTED BELOW)
　RT　　CARBON COMPOUNDS
　　　　∞CHEMICAL COMPOUNDS
　　　　GERMANIUM COMPOUNDS
　　　　LEAD COMPOUNDS
　　　　SILICON COMPOUNDS
　　　　TIN COMPOUNDS

∞ **GROUP 4B COMPOUNDS**
　SN　　*(USE OF A MORE SPECIFIC TERM IS*
　　　　RECOMMENDED--CONSULT THE TERMS
　　　　LISTED BELOW)
　RT　　∞CHEMICAL COMPOUNDS
　　　　HAFNIUM COMPOUNDS
　　　　TITANIUM COMPOUNDS
　　　　ZIRCONIUM COMPOUNDS

∞ **GROUP 5A COMPOUNDS**
　SN　　*(USE OF A MORE SPECIFIC TERM IS*
　　　　RECOMMENDED--CONSULT THE TERMS
　　　　LISTED BELOW)
　UF　　PNICTIDES
　RT　　ANTIMONY COMPOUNDS
　　　　ARSENIC COMPOUNDS
　　　　BISMUTH COMPOUNDS
　　　　∞CHEMICAL COMPOUNDS
　　　　NITROGEN COMPOUNDS
　　　　OXYNITRIDES
　　　　PHOSPHORUS COMPOUNDS

∞ **GROUP 5B COMPOUNDS**
　SN　　*(USE OF A MORE SPECIFIC TERM IS*
　　　　RECOMMENDED--CONSULT THE TERMS
　　　　LISTED BELOW)
　RT　　∞CHEMICAL COMPOUNDS
　　　　NIOBIUM COMPOUNDS
　　　　TANTALUM COMPOUNDS
　　　　VANADIUM COMPOUNDS

∞ **GROUP 6A COMPOUNDS**
　SN　　*(USE OF A MORE SPECIFIC TERM IS*
　　　　RECOMMENDED--CONSULT THE TERMS
　　　　LISTED BELOW)
　RT　　CHALCOGENIDES
　　　　∞CHEMICAL COMPOUNDS
　　　　POLONIUM COMPOUNDS
　　　　SELENIUM COMPOUNDS
　　　　SULFUR COMPOUNDS
　　　　TELLURIUM COMPOUNDS

∞ **GROUP 6B COMPOUNDS**
　SN　　*(USE OF A MORE SPECIFIC TERM IS*
　　　　RECOMMENDED--CONSULT THE TERMS
　　　　LISTED BELOW)
　RT　　∞CHEMICAL COMPOUNDS
　　　　CHROMIUM COMPOUNDS
　　　　MOLYBDENUM COMPOUNDS
　　　　TUNGSTEN COMPOUNDS

GROUP 7A COMPOUNDS
　USE　　HALOGEN COMPOUNDS

∞ **GROUP 7B COMPOUNDS**
　SN　　*(USE OF A MORE SPECIFIC TERM IS*
　　　　RECOMMENDED--CONSULT THE TERMS
　　　　LISTED BELOW)

GROUP 7B COMPOUNDS-*(CONT.)*
　RT　　∞CHEMICAL COMPOUNDS
　　　　MANGANESE COMPOUNDS
　　　　RHENIUM COMPOUNDS
　　　　TECHNETIUM COMPOUNDS

∞ **GROUP 8 COMPOUNDS**
　SN　　*(USE OF A MORE SPECIFIC TERM IS*
　　　　RECOMMENDED--CONSULT THE
　　　　TERMS LISTED BELOW)
　RT　　∞CHEMICAL COMPOUNDS
　　　　COBALT COMPOUNDS
　　　　IRON COMPOUNDS
　　　　NICKEL COMPOUNDS
　　　　OSMIUM COMPOUNDS
　　　　PLATINUM COMPOUNDS
　　　　RHODIUM COMPOUNDS

∞ **GROUPS**
　SN　　*(USE OF A MORE SPECIFIC TERM IS*
　　　　RECOMMENDED--CONSULT THE TERMS
　　　　LISTED BELOW)
　RT　　CATEGORIES
　　　　CLASSES
　　　　SUBDIVISIONS

GROUT
　RT　　AMORPHOUS MATERIALS
　　　　CEMENTS
　　　　CLAYS
　　　　CONCRETES
　　　　∞CONSTRUCTION MATERIALS
　　　　MORTARS (MATERIAL)
　　　　MUD
　　　　PLASTERS
　　　　TILES

GROWTH
　UF　　HYPERTROPHY
　　　　MATURING
　GS　　**GROWTH**
　　　. CROP GROWTH
　　　. CRYSTAL GROWTH
　　　. . CZOCHRALSKI METHOD
　　　. . DIRECTIONAL SOLIDIFICATION
　　　　　(CRYSTALS)
　　　. . EPITAXY
　　　. . . ELECTROEPITAXY
　　　. . . LIQUID PHASE EPITAXY
　　　. . . MOLECULAR BEAM EPITAXY
　　　. . . VAPOR PHASE EPITAXY
　　　. . HYDROTHERMAL CRYSTAL GROWTH
　　　. . TRAVELING SOLVENT METHOD
　　　. . VERNEUIL PROCESS
　　　. VEGETATION GROWTH
　RT　　ACCUMULATIONS
　　　　CROP CALENDARS
　　　　∞DEVELOPMENT
　　　　EVOLUTION (DEVELOPMENT)
　　　　∞FORMATION
　　　　GERMINATION
　　　　INCREASING
　　　　INFLATING
　　　　ONTOGENY
　　　　PHYTOTRONS
　　　　SHRINKAGE
　　　　SINTERING
　　　　SWELLING
　　　　TIMBER VIGOR
　　　　TIMBERLINE
　　　　TRENDS
　　　　VIABILITY
　　　　WARPAGE
　　　　YOUTH

GROWTH CHAMBERS
　USE　　PHYTOTRONS

GRUMMAN AIRCRAFT
　GS　　**GRUMMAN AIRCRAFT**
　　　. A-6 AIRCRAFT
　　　. C-1A AIRCRAFT
　　　. C-2 AIRCRAFT
　　　. E-2 AIRCRAFT
　　　. F-9 AIRCRAFT
　　　. F-14 AIRCRAFT
　　　. F-111 AIRCRAFT
　　　. JETSTREAM AIRCRAFT
　　　. OV-1 AIRCRAFT
　RT　　∞AIRCRAFT
　　　　AWACS AIRCRAFT

GRUMMAN OV-1C AIRCRAFT
　USE　　OV-1 AIRCRAFT

GRUNEISEN CONSTANT
GS   CONSTANTS
    . **GRUNEISEN CONSTANT**
RT   COMPRESSIBILITY
    SPECIFIC HEAT
    THERMAL EXPANSION

GTDS
USE   GODDARD TRAJECTORY
        DETERMINATION SYSTEM

GUADELOUPE
GS   LANDFORMS
    . ISLANDS
    . . WEST INDIES
    . . . **GUADELOUPE**
    NATIONS
    . FRANCE
    . . **GUADELOUPE**
RT   FRANCE

GUAM
GS   LANDFORMS
    . ISLANDS
    . . PACIFIC ISLANDS
    . . . **GUAM**
RT   UNITED STATES

GUANETHIDINE
GS   ALIPHATIC COMPOUNDS
    . GUANIDINES
    . . **GUANETHIDINE**
    AMINES
    . DIAMINES
    . . GUANIDINES
    . . . **GUANETHIDINE**
    HETEROCYCLIC COMPOUNDS
    . **GUANETHIDINE**

GUANIDINES
GS   ALIPHATIC COMPOUNDS
    . **GUANIDINES**
    . . GUANETHIDINE
    . . TRIAMINOGUANIDINIUM AZIDE
    AMINES
    . DIAMINES
    . . **GUANIDINES**
    . . . GUANETHIDINE
    . . . TRIAMINOGUANIDINIUM AZIDE
RT   PERFLUOROGUANIDINE

GUANINES
GS   FUNGICIDES
    . XANTHINES
    . . **GUANINES**
    HETEROCYCLIC COMPOUNDS
    . XANTHINES
    . . **GUANINES**
    NITROGEN COMPOUNDS
    . XANTHINES
    . . **GUANINES**
    PROTEINS
    . **GUANINES**
    PURINES
    . XANTHINES
    . . **GUANINES**
RT   CYCLIC AMP

GUANOSINES
UF   VERNINE
GS   ACIDS
    . NUCLEIC ACIDS
    . . **GUANOSINES**
RT   RIBONUCLEIC ACIDS

GUARDS (SHIELDS)
RT   ∞BARRIERS
    COVERINGS
    HOUSINGS
    SAFETY DEVICES
    SAFETY MANAGEMENT
    SHIELDING

GUATEMALA
GS   NATIONS
    . **GUATEMALA**
RT   CENTRAL AMERICA

GUAYULE
GS   PLANTS (BOTANY)
    . **GUAYULE**
RT   BRUSH (BOTANY)
    RUBBER

GUIDANCE (MOTION)
GS   **GUIDANCE (MOTION)**
    . AIRCRAFT GUIDANCE
    . BEAM RIDER GUIDANCE
    . COMMAND GUIDANCE
    . ENTRY GUIDANCE (STS)
    . INERTIAL GUIDANCE
    . . STRAPDOWN INERTIAL GUIDANCE
    . INJECTION GUIDANCE
    . MAP MATCHING GUIDANCE
    . MIDCOURSE GUIDANCE
    . REENTRY GUIDANCE
    . RENDEZVOUS GUIDANCE
    . SPACECRAFT GUIDANCE
    . . SATELLITE GUIDANCE
    . STANDARDIZED SPACE GUIDANCE
    . TERMINAL GUIDANCE
    . . LASER GUIDANCE
RT   AIR NAVIGATION
    APPROACH
    ASCENT TRAJECTORIES
    ASTRIONICS
    AUTOMATIC CONTROL
    AVIONICS
    CONTROL SURFACES
    FLIGHT CONTROL
    FLIGHT PATHS
    GROUND BASED CONTROL
    HOMING
    HOMING DEVICES
    IMPACT PREDICTION
    LANDING
    MANUAL CONTROL
    MISSILES
    NAVIGATION
    ∞PLATFORMS
    POINTING CONTROL SYSTEMS
    RADIO NAVIGATION
    REMOTE CONTROL
    STATIONKEEPING
    ∞SYSTEMS
    TRAJECTORY CONTROL
    VISUAL CONTROL

GUIDANCE SENSORS
RT   ATTITUDE CONTROL
    IMAGE DISSECTOR TUBES
    OPTICAL MEASURING INSTRUMENTS
    ∞SENSORS
    SOLAR SENSORS
    SPACECRAFT INSTRUMENTS
    STAR TRACKERS

GUIDE VANES
UF   JETAVATORS
GS   CONTROL SURFACES
    . **GUIDE VANES**
    . . JET VANES
    VANES
    . **GUIDE VANES**
    . . JET VANES
RT   AIRFOILS
    HYDROFOILS
    THRUST VECTOR CONTROL

GUIDED MISSILE SUBMARINES
UF   POLARIS SUBMARINES
GS   WATER VEHICLES
    . SHIPS
    . . SUBMARINES
    . . . **GUIDED MISSILE SUBMARINES**
    . UNDERWATER VEHICLES
    . . SUBMARINES
    . . . **GUIDED MISSILE SUBMARINES**
RT   FLEET BALLISTIC MISSILES
    POSEIDON MISSILES

GUINEA
GS   NATIONS
    . **GUINEA**
RT   AFRICA

GUINEA PIGS
GS   ANIMALS
    . VERTEBRATES
    . . MAMMALS
    . . . RODENTS
    . . . . **GUINEA PIGS**

GULF OF ALASKA
GS   GULFS
    . **GULF OF ALASKA**
RT   ALASKA
    PACIFIC OCEAN

GULF OF CALIFORNIA (MEXICO)
GS   GULFS
    . **GULF OF CALIFORNIA (MEXICO)**
RT   MEXICO
    PACIFIC OCEAN

GULF OF MEXICO
GS   GULFS
    . **GULF OF MEXICO**
RT   ALABAMA
    CARIBBEAN SEA
    FLORIDA
    LOUISIANA
    MEXICO
    MISSISSIPPI
    RIO GRANDE (NORTH AMERICA)
    TEXAS

GULF STREAM
GS   CIRCULATION
    . WATER CIRCULATION
    . . WATER CURRENTS
    . . . OCEAN CURRENTS
    . . . . **GULF STREAM**
RT   ATLANTIC OCEAN
    CARIBBEAN SEA
    LOMONOSOV CURRENT
    SARGASSO SEA
    TOPEX

GULFS
GS   **GULFS**
    . GULF OF ALASKA
    . GULF OF CALIFORNIA (MEXICO)
    . GULF OF MEXICO
    . PERSIAN GULF
RT   BAYS (TOPOGRAPHIC FEATURES)
    DELAWARE BAY (US)
    INLETS (TOPOGRAPHY)
    TOPOGRAPHY

GULLIVER PROGRAM
GS   PROGRAMS
    . **GULLIVER PROGRAM**
RT   EXTRATERRESTRIAL LIFE
    SPACE EXPLORATION

GUM NEBULA
GS   CELESTIAL BODIES
    . NEBULAE
    . . **GUM NEBULA**
RT   GALAXIES
    ORION NEBULA

GUM VULCANIZATES
USE   VULCANIZED ELASTOMERS

GUMBEL THEORY
USE   RANGE (EXTREMES)

GUMS (SUBSTANCES)
GS   **GUMS (SUBSTANCES)**
    . ROSIN
RT   CHITIN
    POLYSACCHARIDES
    RUBBER
    TARS

GUN LAUNCHERS
SN   (ORDNANCE DEVICES FOR FIRING
    MISSILES AND ROCKETS WITH INITIAL
    ATTITUDE CONTROL)
GS   LAUNCHERS
    . **GUN LAUNCHERS**
RT   ARTILLERY
    ∞BARRELS
    GUNFIRE
    HOWITZERS
    HYPERVELOCITY LAUNCHERS
    MISSILE LAUNCHERS
    ROCKET CATAPULTS
    ROCKET LAUNCHERS
    SABOT PROJECTILES

GUN PROPELLANTS
UF   GUNPOWDER
GS   PROPELLANTS
    . **GUN PROPELLANTS**
RT   EXPLOSIVES
    GUNS (ORDNANCE)

GUN TURRETS
RT   ∞CUPOLAS
    GUNS (ORDNANCE)

GUN TURRETS-*(CONT.)*
∞ TURRET

GUNFIRE
RT   ARTILLERY FIRE
      FIRE CONTROL
      FIRING (IGNITING)
      GUN LAUNCHERS
      GUNS (ORDNANCE)
      PROJECTILES

GUNN DIODES
GS   ELECTRONIC EQUIPMENT
      . DIODES
      . . SEMICONDUCTOR DIODES
      . . . **GUNN DIODES**
RT   GALLIUM ARSENIDES
      NEGATIVE RESISTANCE DEVICES
      SEMICONDUCTOR DEVICES

GUNN EFFECT
RT   ∞ EFFECTS
      NEGATIVE CONDUCTANCE
      NEGATIVE RESISTANCE DEVICES
      SEMICONDUCTOR DEVICES
      SEMICONDUCTOR LASERS

GUNNERY TRAINING
GS   EDUCATION
      . **GUNNERY TRAINING**
RT   ARTILLERY
      FIRE CONTROL
      GUNS (ORDNANCE)
      HOWITZERS
      WEAPONS

GUNPOWDER
USE   GUN PROPELLANTS

∞ **GUNS**
SN   *(USE OF A MORE SPECIFIC TERM IS*
      RECOMMENDED--CONSULT THE TERMS
      LISTED BELOW)
RT   CROSSED FIELD GUNS
      ELECTRON GUNS
      GAS GUNS
      GUNS (ORDNANCE)
      HYPERVELOCITY GUNS
      PLASMA GUNS

GUNS (ORDNANCE)
UF   CANNONS
GS   WEAPONS
      . **GUNS (ORDNANCE)**
      . . HOWITZERS
      . . PRECISION GUIDED PROJECTILES
      . . RIFLES
RT   AMMUNITION
      EXPLOSIVES
      GUN PROPELLANTS
      GUN TURRETS
      GUNFIRE
      GUNNERY TRAINING
      ∞ GUNS
      HEAT OF COMBUSTION
      HYPERVELOCITY GUNS
      INCENDIARY AMMUNITION
      PROJECTILES
      PROPELLANTS
      SABOT PROJECTILES

GUST ALLEVIATORS
RT   DEFLECTORS
      GUSTS
      SPOILERS
      TURBULENT FLOW
      VORTEX ALLEVIATION

GUST LOADS
GS   AERODYNAMIC FORCES
      . AERODYNAMIC LOADS
      . . **GUST LOADS**
      LOADS (FORCES)
      . DYNAMIC LOADS
      . . AERODYNAMIC LOADS
      . . . **GUST LOADS**
      . . TRANSIENT LOADS
      . . . **GUST LOADS**
      . RANDOM LOADS
      . . **GUST LOADS**
RT   ATMOSPHERIC TURBULENCE
      BLAST LOADS
      GROUND WIND
      GUSTS

GUST LOADS-*(CONT.)*
      STRUCTURAL DESIGN CRITERIA
      WIND PRESSURE
      WING LOADING

GUSTATORY PERCEPTION
USE   TASTE

GUSTS
GS   TURBULENCE
      . ATMOSPHERIC TURBULENCE
      . . **GUSTS**
      WIND (METEOROLOGY)
      . **GUSTS**
RT   CLEAR AIR TURBULENCE
      GROUND WIND
      GUST ALLEVIATORS
      GUST LOADS
      SEA BREEZE
      STORM DAMAGE
      STORMS
      STORMS (METEOROLOGY)
      VORTEX AVOIDANCE

†

GUTENBERG ZONE
GS   MODELS
      . **GUTENBERG ZONE**
      REGIONS
      . **GUTENBERG ZONE**
RT   ACOUSTIC VELOCITY
      SEISMIC WAVES

GUY WIRES
UF   STAYS
GS   WIRE
      . **GUY WIRES**
RT   ANCHORS (FASTENERS)
      STRUCTURAL MEMBERS

GUYANA
UF   BRITISH GUINEA
GS   NATIONS
      . **GUYANA**
RT   CARIBBEAN REGION
      SOUTH AMERICA

GYMNASTICS
USE   PHYSICAL EXERCISE

GYNECOLOGY
GS   MEDICAL SCIENCE
      . **GYNECOLOGY**
RT   FEMALES
      GENITOURINARY SYSTEM

GYPSUM
GS   MINERALS
      . **GYPSUM**
      PLASTERS
      . **GYPSUM**
RT   CALCIUM
      CHALK
      ROCKS
      SEDIMENTARY ROCKS
      SULFATES

GYRATION
GS   **GYRATION**
      . PRECESSION
      . . LARMOR PRECESSION
      . . PROTON PRECESSION
      . . QUENCHING (ATOMIC PHYSICS)
      . REVOLVING
      . ROTATION
      . . AUTOROTATION
      . . COROTATION
      . . COUNTER ROTATION
      . . EARTH ROTATION
      . . MOLECULAR ROTATION
      . . MUON SPIN ROTATION
      . . PLANETARY ROTATION
      . . SATELLITE ROTATION
      . . STELLAR ROTATION
      . . . SOLAR ROTATION
RT   ANGULAR VELOCITY
      ∞ MOTION
      SPIN DYNAMICS

GYRATORS
UF   TELLEGEN THEORY
GS   MICROWAVE EQUIPMENT
      . **GYRATORS**
      . . MICROWAVE FILTERS
RT   FERRITES

GYRATORS-*(CONT.)*
      MICROWAVE SWITCHING
      NETWORK ANALYSIS
      ∞ NETWORKS
      PHASE SHIFT CIRCUITS
      WAVEGUIDES

GYRES
RT   AIR WATER INTERACTIONS
      COASTAL CURRENTS
      OCEAN CURRENTS
      OCEANOGRAPHY

GYRO HORIZONS
GS   AIRCRAFT INSTRUMENTS
      . ATTITUDE INDICATORS
      . . **GYRO HORIZONS**
      DISPLAY DEVICES
      . **GYRO HORIZONS**
      FLIGHT INSTRUMENTS
      . ATTITUDE INDICATORS
      . . **GYRO HORIZONS**
      GYROSCOPES
      . ATTITUDE GYROS
      . . **GYRO HORIZONS**
      MEASURING INSTRUMENTS
      . INDICATING INSTRUMENTS
      . . ATTITUDE INDICATORS
      . . . **GYRO HORIZONS**
RT   HORIZON

GYROCOMPASSES
GS   AIRCRAFT INSTRUMENTS
      . COMPASSES
      . . **GYROCOMPASSES**
      GYROSCOPES
      . **GYROCOMPASSES**
      MEASURING INSTRUMENTS
      . COMPASSES
      . . **GYROCOMPASSES**
      . INDICATING INSTRUMENTS
      . . **GYROCOMPASSES**
      NAVIGATION AIDS
      . NAVIGATION INSTRUMENTS
      . . COMPASSES
      . . . **GYROCOMPASSES**
RT   MAGNETIC COMPASSES
      RADIO DIRECTION FINDERS
      SOLAR COMPASSES

GYRODAMPERS
RT   CONTROL MOMENT GYROSCOPES
      GIMBALS
      GYROSCOPIC STABILITY
      STRUCTURAL VIBRATION
      VIBRATION DAMPING

GYRODYNE AIRCRAFT
GS   **GYRODYNE AIRCRAFT**
      . QH-50 HELICOPTER
RT   ∞ AIRCRAFT

GYRODYNE DSN-3 HELICOPTER
USE   QH-50 HELICOPTER

GYRODYNE MILITARY AIRCRAFT
USE   QH-50 HELICOPTER

GYROFREQUENCY
GS   MAGNETIC PROPERTIES
      . GYROMAGNETISM
      . . **GYROFREQUENCY**
RT   CHARGED PARTICLES
      MAGNETOIONICS

GYROINTERACTION
USE   MAGNETIC RIGIDITY

GYROMAGNETISM
GS   MAGNETIC PROPERTIES
      . **GYROMAGNETISM**
      . . GYROFREQUENCY
RT   LARMOR RADIUS

GYROPLANES
USE   HELICOPTERS

GYROS
USE   GYROSCOPES

GYROSCOPE FLUIDS
RT   DAMPING
      ∞ FLUIDS
      ROTARY GYROSCOPES

GYROSCOPES

GYROSCOPE FLUIDS-*(CONT.)*
    SUSPENDING (HANGING)

GYROSCOPES
UF   GYROS
     GYROSCOPIC DRIFT
     GYROSTATS
GS   **GYROSCOPES**
     . ATTITUDE GYROS
     . . GYRO HORIZONS
     . CONTROL MOMENT GYROSCOPES
     . CRYOGENIC GYROSCOPES
     . ELECTROSTATIC GYROSCOPES
     . GYROCOMPASSES
     . GYROSCOPIC PENDULUMS
     . GYROSTABILIZERS
     . LASER GYROSCOPES
     . NUCLEAR GYROSCOPES
     . OPTICAL GYROSCOPES
     . ROTARY GYROSCOPES
     . . FLUID ROTOR GYROSCOPES
     . TUNING FORK GYROSCOPES
RT   AUTOMATIC PILOTS
     GIMBALLESS INERTIAL NAVIGATION
     GIMBALS
     GRAVITY PROBE B
     GYROSCOPIC STABILITY
     PRECESSION
     ∞ STABILIZERS
     TORQUERS

GYROSCOPIC COUPLING
GS   COUPLING
     . **GYROSCOPIC COUPLING**
RT   NAVIGATION

GYROSCOPIC DRIFT
USE   GYROSCOPES
     GYROSCOPIC STABILITY

GYROSCOPIC PENDULUMS
UF   PENDULOUS GYROSCOPES
GS   GYROSCOPES
     . **GYROSCOPIC PENDULUMS**
     OSCILLATORS
     . MECHANICAL OSCILLATORS
     . . PENDULUMS
     . . . **GYROSCOPIC PENDULUMS**
RT   ACCELEROMETERS
     DAMPING
     SCHULER TUNING

GYROSCOPIC STABILITY
UF   GYROSCOPIC DRIFT
GS   DYNAMIC CHARACTERISTICS
     . DYNAMIC STABILITY
     . . MOTION STABILITY
     . . . ATTITUDE STABILITY
     . . . . DIRECTIONAL STABILITY
     . . . . . **GYROSCOPIC STABILITY**
     . . . ROTARY STABILITY
     . . . . **GYROSCOPIC STABILITY**
     STABILITY
     . DYNAMIC STABILITY
     . . MOTION STABILITY
     . . . ATTITUDE STABILITY
     . . . . DIRECTIONAL STABILITY
     . . . . . **GYROSCOPIC STABILITY**
     . . . ROTARY STABILITY
     . . . . **GYROSCOPIC STABILITY**
RT   DAMPING
     GYRODAMPERS
     GYROSCOPES
     HOVERING STABILITY
     INERTIAL PLATFORMS
     PRECESSION
     ROTARY GYROSCOPES
     SCHULER TUNING
     SEA KEEPING
     STABILIZED PLATFORMS
     STABLE OSCILLATIONS
     YO-YO DEVICES

GYROSTABILIZERS
GS   GYROSCOPES
     . **GYROSTABILIZERS**
RT   NAVIGATION AIDS
     SEA KEEPING
     STABILIZED PLATFORMS
     THRUST VECTOR CONTROL

GYROSTATS
USE   GYROSCOPES

GYROTRONS
USE   CYCLOTRON RESONANCE DEVICES

GYROTROPISM
GS   TROPISM
     . **GYROTROPISM**
RT   FREQUENCY SHIFT

H

H ALPHA LINE
GS   SPECTRA
     . RADIATION SPECTRA
     . . ELECTROMAGNETIC SPECTRA
     . . . LINE SPECTRA
     . . . . H LINES
     . . . . . **H ALPHA LINE**
RT   ABSORPTION SPECTRA
     EMISSION SPECTRA
     SOLAR SPECTRA

H BETA LINE
GS   SPECTRA
     . RADIATION SPECTRA
     . . ELECTROMAGNETIC SPECTRA
     . . . LINE SPECTRA
     . . . . H LINES
     . . . . . **H BETA LINE**
RT   ABSORPTION SPECTRA
     BALMER SERIES
     EMISSION SPECTRA
     SOLAR SPECTRA

H GAMMA LINE
GS   SPECTRA
     . RADIATION SPECTRA
     . . ELECTROMAGNETIC SPECTRA
     . . . LINE SPECTRA
     . . . . H LINES
     . . . . . **H GAMMA LINE**
RT   ABSORPTION SPECTRA
     BALMER SERIES
     EMISSION SPECTRA
     SOLAR SPECTRA

H LINES
SN   (EXCLUDES SURFACES OF CONSTANT
     MAGNETIC FIELD STRENGTH)
GS   SPECTRA
     . RADIATION SPECTRA
     . . ELECTROMAGNETIC SPECTRA
     . . . LINE SPECTRA
     . . . . **H LINES**
     . . . . . H ALPHA LINE
     . . . . . H BETA LINE
     . . . . . H GAMMA LINE
RT   ABSORPTION SPECTRA
     BALMER SERIES
     D LINES
     EMISSION SPECTRA
     K LINES
     LYMAN SPECTRA
     PASCHEN SERIES
     RYDBERG SERIES
     SOLAR SPECTRA
     TELLURIC LINES

H WAVES
GS   ELECTROMAGNETIC RADIATION
     . **H WAVES**
     OSCILLATIONS
     . TRANSVERSE OSCILLATION
     . . **H WAVES**
     TRANSVERSE WAVES
     . **H WAVES**
RT   ELECTRIC FIELD STRENGTH

H-1 ENGINE
GS   ENGINES
     . ROCKET ENGINES
     . . BOOSTER ROCKET ENGINES
     . . . **H-1 ENGINE**
     . . LIQUID PROPELLANT ROCKET
       ENGINES
     . . . **H-1 ENGINE**
RT   SATURN 1 LAUNCH VEHICLES
     SATURN 1B LAUNCH VEHICLES

H-13 HELICOPTER
USE   OH-13 HELICOPTER

H-17 HELICOPTER
UF   FLYING CRANE HELICOPTER
GS   HUGHES AIRCRAFT
     . **H-17 HELICOPTER**
     JET AIRCRAFT
     . **H-17 HELICOPTER**
     RESEARCH AIRCRAFT
     . **H-17 HELICOPTER**
     V/STOL AIRCRAFT
     . ROTARY WING AIRCRAFT
     . . HELICOPTERS
     . . . **H-17 HELICOPTER**

H-19 HELICOPTER
GS   PASSENGER AIRCRAFT
     . **H-19 HELICOPTER**
     SIKORSKY AIRCRAFT
     . **H-19 HELICOPTER**
     TRANSPORT AIRCRAFT
     . **H-19 HELICOPTER**
RT   V/STOL AIRCRAFT

H-21 HELICOPTER
USE   CH-21 HELICOPTER

H-23 HELICOPTER
USE   OH-23 HELICOPTER

H-25 HELICOPTER
GS   BOEING AIRCRAFT
     . **H-25 HELICOPTER**
     V/STOL AIRCRAFT
     . ROTARY WING AIRCRAFT
     . . HELICOPTERS
     . . . TANDEM ROTOR HELICOPTERS
     . . . . **H-25 HELICOPTER**
RT   ANTISUBMARINE WARFARE AIRCRAFT

H-34 HELICOPTER
USE   CH-34 HELICOPTER

H-43 HELICOPTER
GS   KAMAN AIRCRAFT
     . **H-43 HELICOPTER**
     V/STOL AIRCRAFT
     . ROTARY WING AIRCRAFT
     . . HELICOPTERS
     . . . MILITARY HELICOPTERS
     . . . . **H-43 HELICOPTER**

H-51 HELICOPTER
USE   XH-51 HELICOPTER

H-53 HELICOPTER
UF   CH-53 HELICOPTER
     HHX HELICOPTER
     SIKORSKY S-65 HELICOPTER
GS   PASSENGER AIRCRAFT
     . **H-53 HELICOPTER**
     SIKORSKY AIRCRAFT
     . **H-53 HELICOPTER**
     TRANSPORT AIRCRAFT
     . **H-53 HELICOPTER**
     V/STOL AIRCRAFT
     . ROTARY WING AIRCRAFT
     . . HELICOPTERS
     . . . MILITARY HELICOPTERS
     . . . . **H-53 HELICOPTER**

H-54 HELICOPTER
GS   V/STOL AIRCRAFT
     . ROTARY WING AIRCRAFT
     . . HELICOPTERS
     . . . **H-54 HELICOPTER**

H-56 HELICOPTER
GS   PASSENGER AIRCRAFT
     . **H-56 HELICOPTER**
     SIKORSKY AIRCRAFT
     . **H-56 HELICOPTER**
     TRANSPORT AIRCRAFT
     . **H-56 HELICOPTER**
     V/STOL AIRCRAFT
     . ROTARY WING AIRCRAFT
     . . HELICOPTERS
     . . . MILITARY HELICOPTERS
     . . . . **H-56 HELICOPTER**

H-60 HELICOPTER
UF   BLACK HAWK ASSAULT HELICOPTER
GS   SIKORSKY AIRCRAFT
     . **H-60 HELICOPTER**
     V/STOL AIRCRAFT
     . ROTARY WING AIRCRAFT

H-60 HELICOPTER-(CONT.)
```
      . . HELICOPTERS
      . . . MILITARY HELICOPTERS
      . . . . H-60 HELICOPTER
RT    ∞AIRCRAFT
      ∞MILITARY AIRCRAFT
```

H-126 AIRCRAFT
```
UF    HUNTING H-126 AIRCRAFT
GS    BAC AIRCRAFT
      . H-126 AIRCRAFT
      JET AIRCRAFT
      . H-126 AIRCRAFT
      MONOPLANES
      . H-126 AIRCRAFT
      RESEARCH AIRCRAFT
      . H-126 AIRCRAFT
RT    ∞AIRCRAFT
      JET FLAPS
```

HABITABILITY
```
RT    ECOLOGY
      ENVIRONMENTAL CONTROL
      ENVIRONMENTS
      SHELTERS
```

HABITATS
```
SN    (LIMITED TO PLANTS AND ANIMALS)
GS    REGIONS
      . HABITATS
RT    ANIMALS
      ∞BIOLOGY
      BOTANY
      CONSERVATION
      EARTH RESOURCES
      ECOLOGY
      ENDANGERED SPECIES
      ENVIRONMENT EFFECTS
      ENVIRONMENTS
      WILDLIFE
```

HABITS
```
RT    LEARNING
      PSYCHOLOGICAL FACTORS
```

HABITUATION (LEARNING)
```
GS    LEARNING
      . HABITUATION (LEARNING)
RT    CONDITIONING (LEARNING)
```

HADRONS
```
GS    PARTICLES
      . ELEMENTARY PARTICLES
      . . HADRONS
      . . . BARYONS
      . . . . OMEGA-MESONS
      . . . . RHO-MESONS
      . . . . SIGMA-MESONS
      . . . MESONS
      . . . . KAONS
      . . . . MUONS
      . . . . OMEGA-MESONS
      . . . . VECTOR MESONS
      . . . . . RHO-MESONS
      . . . . . SIGMA-MESONS
RT    CHARM (PARTICLE PHYSICS)
      FLAVOR (PARTICLE PHYSICS)
      PARTONS
      QUARK PARTON MODEL
      VECTOR DOMINANCE MODEL
```

HAFNIUM
```
GS    CHEMICAL ELEMENTS
      . HAFNIUM
      . . HAFNIUM ISOTOPES
      METALS
      . TRANSITION METALS
      . HAFNIUM
      . . . HAFNIUM ISOTOPES
```

HAFNIUM ALLOYS
```
GS    ALLOYS
      . HAFNIUM ALLOYS
```

HAFNIUM CARBIDES
```
GS    CARBON COMPOUNDS
      . CARBIDES
      . . HAFNIUM CARBIDES
      HAFNIUM COMPOUNDS
      . HAFNIUM CARBIDES
```

HAFNIUM COMPOUNDS
```
GS    HAFNIUM COMPOUNDS
      . HAFNIUM CARBIDES
```

HAFNIUM COMPOUNDS-(CONT.)
```
      . HAFNIUM IODIDES
      . HAFNIUM OXIDES
RT    ∞CHEMICAL COMPOUNDS
      ∞GROUP 4B COMPOUNDS
      ∞METAL COMPOUNDS
```

HAFNIUM IODIDES
```
GS    HAFNIUM COMPOUNDS
      . HAFNIUM IODIDES
      HALOGEN COMPOUNDS
      . HALIDES
      . . METAL HALIDES
      . . . HAFNIUM IODIDES
      . IODINE COMPOUNDS
      . . IODIDES
      . . . HAFNIUM IODIDES
```

HAFNIUM ISOTOPES
```
GS    CHEMICAL ELEMENTS
      . HAFNIUM
      . . HAFNIUM ISOTOPES
      . NUCLIDES
      . . ISOTOPES
      . . . HAFNIUM ISOTOPES
      METALS
      . TRANSITION METALS
      . . HAFNIUM
      . . . HAFNIUM ISOTOPES
```

HAFNIUM OXIDES
```
GS    CHALCOGENIDES
      . OXIDES
      . . METAL OXIDES
      . . . HAFNIUM OXIDES
      HAFNIUM COMPOUNDS
      . HAFNIUM OXIDES
```

HAIL
```
UF    HAILSTONES
GS    PRECIPITATION (METEOROLOGY)
      . HAIL
RT    CLOUD GLACIATION
      HAILSTORMS
      ICE FORMATION
      STORMS (METEOROLOGY)
      THUNDERSTORMS
```

HAILSTONES
```
USE   HAIL
```

HAILSTORMS
```
GS    STORMS
      . STORMS (METEOROLOGY)
      . . HAILSTORMS
RT    CLIMATOLOGY
      HAIL
      METEOROLOGY
      PRECIPITATION (METEOROLOGY)
      RAINSTORMS
      STORM DAMAGE
      STORM ENHANCEMENT
      STORM SUPPRESSION
      THUNDERSTORMS
```

HAIR
```
GS    FIBERS
      . HAIR
RT    KERATINS
      WOOL
```

HAITI
```
GS    LANDFORMS
      . ISLANDS
      . . WEST INDIES
      . . . HAITI
      NATIONS
      . HAITI
RT    CARIBBEAN REGION
      CARIBBEAN SEA
```

HAL/S (LANGUAGE)
```
GS    LANGUAGES
      . PROGRAMMING LANGUAGES
      . . HAL/S (LANGUAGE)
RT    COMPUTER PROGRAMMING
      COMPUTERS
```

HALDEN BOILING WATER REACTOR
```
UF    HALDEN REACTOR
      HBWR REACTOR
GS    NUCLEAR REACTORS
      . LIQUID COOLED REACTORS
      . . WATER COOLED REACTORS
```

HALDEN BOILING WATER REACTOR-(CONT.)
```
      . . . BOILING WATER REACTORS
      . . . . HALDEN BOILING WATER
              REACTOR
```

HALDEN REACTOR
```
USE   HALDEN BOILING WATER REACTOR
```

HALF CONES
```
RT    AERODYNAMIC CONFIGURATIONS
      CIRCULAR CONES
      CONES
      CONICS
      NOSE CONES
```

HALF LIFE
```
GS    DECAY
      . HALF LIFE
      LIFE (DURABILITY)
      . HALF LIFE
RT    NUCLEAR REACTIONS
      POST-BLAST NUCLEAR RADIATION
      RADIATIVE LIFETIME
      RADIOACTIVE AGE DETERMINATION
      RADIOACTIVE DECAY
      RADIOACTIVITY
      REACTION KINETICS
```

HALF PLANES
```
GS    ANALYSIS (MATHEMATICS)
      . HALF PLANES
RT    BOUNDARY VALUE PROBLEMS
      COORDINATES
      DIFFERENTIAL EQUATIONS
      GREEN'S FUNCTIONS
```

HALF SPACES
```
GS    ANALYSIS (MATHEMATICS)
      . HALF SPACES
RT    BOUNDARY VALUE PROBLEMS
      COORDINATES
      ELLIPTIC DIFFERENTIAL EQUATIONS
      FRACTALS
      GREEN'S FUNCTIONS
```

HALIDES
```
GS    HALOGEN COMPOUNDS
      . HALIDES
      . . BROMIDES
      . . . AMMONIUM BROMIDES
      . . . CESIUM BROMIDES
      . . . CHROMIUM BROMIDES
      . . . DIBROMIDES
      . . . HYDROBROMIC ACID
      . . . HYDROBROMIDES
      . . . MAGNESIUM BROMIDES
      . . . POTASSIUM BROMIDES
      . . . SILVER BROMIDES
      . . . SODIUM BROMIDES
      . . . STRONTIUM BROMIDES
      . . CHLORIDES
      . . . ALUMINUM CHLORIDES
      . . . AMMONIUM CHLORIDES
      . . . BERYLLIUM CHLORIDES
      . . . BORON CHLORIDES
      . . . CADMIUM CHLORIDES
      . . . CALCIUM CHLORIDES
      . . . CARBON TETRACHLORIDE
      . . . COPPER CHLORIDES
      . . . DICHLORIDES
      . . . GERMANIUM CHLORIDES
      . . . HYDROCHLORIDES
      . . . HYDROGEN CHLORIDES
      . . . . HYDROCHLORIC ACID
      . . . IRON CHLORIDES
      . . . LANTHANUM CHLORIDES
      . . . LEAD CHLORIDES
      . . . LITHIUM CHLORIDES
      . . . MAGNESIUM CHLORIDES
      . . . NITROSYL CHLORIDES
      . . . NITROXYCHLORIDES
      . . . NITRYL CHLORIDES
      . . . PHOSGENE
      . . . POTASSIUM CHLORIDES
      . . . SILICON TETRACHLORIDE
      . . . SILVER CHLORIDES
      . . . SODIUM CHLORIDES
      . . . SULFUR CHLORIDES
      . . . TETRACHLORIDES
      . . . TITANIUM CHLORIDES
      . . . TUNGSTEN CHLORIDES
      . . . ZINC CHLORIDES
      . . FLUORIDES
      . . . ANTIMONY FLUORIDES
      . . . BARIUM FLUORIDES
```

HALIDES-*(CONT.)*
    . . . BORON FLUORIDES
    . . . CHLORINE FLUORIDES
    . . . DIFLUORIDES
    . . . . CALCIUM FLUORIDES
    . . . . . FLUORSPAR
    . . . HYDROFLUORIC ACID
    . . . NITROGEN FLUORIDES
    . . . OXYFLUORIDES
    . . . OXYGEN FLUORIDES
    . . . OZONE FLUORIDE
    . . . PERCHLORYL FLUORIDES
    . . . SULFUR FLUORIDES
    . . . TECHNETIUM FLUORIDES
    . . METAL HALIDES
    . . . ALKALI HALIDES
    . . . . CESIUM HALIDES
    . . . . . CESIUM BROMIDES
    . . . . . CESIUM FLUORIDES
    . . . . . CESIUM IODIDES
    . . . . POTASSIUM IODIDES
    . . . . SODIUM BROMIDES
    . . . . SODIUM CHLORIDES
    . . . . SODIUM FLUORIDES
    . . . . SODIUM IODIDES
    . . . ALUMINUM CHLORIDES
    . . . ALUMINUM FLUORIDES
    . . . BARIUM FLUORIDES
    . . . BERYLLIUM CHLORIDES
    . . . BERYLLIUM FLUORIDES
    . . . CADMIUM CHLORIDES
    . . . CADMIUM FLUORIDES
    . . . CALCIUM CHLORIDES
    . . . CALCIUM FLUORIDES
    . . . . FLUORSPAR
    . . . CHROMIUM BROMIDES
    . . . CHROMIUM FLUORIDES
    . . . COBALT FLUORIDES
    . . . COPPER CHLORIDES
    . . . COPPER FLUORIDES
    . . . HAFNIUM IODIDES
    . . . IRON CHLORIDES
    . . . LANTHANUM CHLORIDES
    . . . LANTHANUM FLUORIDES
    . . . LEAD CHLORIDES
    . . . LITHIUM CHLORIDES
    . . . LITHIUM FLUORIDES
    . . . MAGNESIUM BROMIDES
    . . . MAGNESIUM FLUORIDES
    . . . NICKEL FLUORIDES
    . . . NIOBIUM IODIDES
    . . . PLUTONIUM FLUORIDES
    . . . POTASSIUM BROMIDES
    . . . POTASSIUM CHLORIDES
    . . . PROTACTINIUM FLUORIDES
    . . . SILVER HALIDES
    . . . . SILVER BROMIDES
    . . . . SILVER CHLORIDES
    . . . . SILVER IODIDES
    . . . STRONTIUM BROMIDES
    . . . STRONTIUM FLUORIDES
    . . . TECHNETIUM FLUORIDES
    . . . THORIUM FLUORIDES
    . . . TITANIUM CHLORIDES
    . . . TUNGSTEN HALIDES
    . . . . TUNGSTEN CHLORIDES
    . . . . TUNGSTEN FLUORIDES
    . . . URANIUM FLUORIDES
    . . . ZINC CHLORIDES
    . . . ZINC FLUORIDES
    . . . ZIRCONIUM IODIDES
    . . OXYHALIDES
RT HALOGENS
    MOLTEN SALTS
    NITROSYLS

HALITES
UF ROCK SALT
RT MOLTEN SALTS
    ∞ SALTS

HALL ACCELERATORS
RT ∞ ACCELERATORS
    ALPHA PLASMA DEVICES
    MAGNETOHYDRODYNAMICS
    PLASMA PHYSICS

HALL COEFFICIENT
USE HALL EFFECT

HALL CURRENTS
USE ELECTRIC CURRENT
     HALL EFFECT

HALL EFFECT
UF HALL COEFFICIENT
    HALL CURRENTS
RT CARRIER MOBILITY
    ∞ EFFECTS
    GALVANOMAGNETIC EFFECTS
    MAGNETOHYDRODYNAMICS
    MOBILITY
    POLARIZATION (CHARGE SEPARATION)
    SEMICONDUCTOR DEVICES
    TRANSPORT PROPERTIES

HALL GENERATORS
RT CIRCULATORS (PHASE SHIFT CIRCUITS)
    FARADAY EFFECT
    ∞ GENERATORS
    PLASMA GENERATORS
    SIGNAL GENERATORS

HALLAM NUCLEAR POWER FACILITY
UF HNPF (HALLAM NUCLEAR POWER
       FACILITY)
GS ELECTRIC POWER PLANTS
    . NUCLEAR POWER PLANTS
    . . **HALLAM NUCLEAR POWER FACILITY**
    NUCLEAR ELECTRIC POWER
       GENERATION
    . NUCLEAR POWER PLANTS
    . . **HALLAM NUCLEAR POWER FACILITY**
RT ∞ POWER PLANTS
    SODIUM GRAPHITE REACTORS

HALLEY'S COMET
GS CELESTIAL BODIES
    . COMETS
    . . **HALLEY'S COMET**
    . SOLAR SYSTEM
    . . **HALLEY'S COMET**
RT GIOTTO MISSION
    VEGA PROJECT

HALLUCINATIONS
GS PSYCHOLOGICAL EFFECTS
    . ILLUSIONS
    . . **HALLUCINATIONS**
RT SIGNS AND SYMPTOMS

HALO ORBIT SPACE STATION
GS STATIONS
    . SPACE STATIONS
    . . ORBITAL SPACE STATIONS
    . . . **HALO ORBIT SPACE STATION**
RT ARTIFICIAL SATELLITES
    LUNAR SPACECRAFT
    SATELLITES

HALOCARBONS
GS CARBON COMPOUNDS
    . **HALOCARBONS**
    . . CHLOROCARBONS
    HALOGEN COMPOUNDS
    . **HALOCARBONS**
    . . FLUOROCARBONS
RT BROMINE COMPOUNDS
    ∞ CHEMICAL COMPOUNDS
    CHLORINE COMPOUNDS
    FLUORINE COMPOUNDS
    FLUORO COMPOUNDS
    IODINE COMPOUNDS

HALOE
USE HALOGEN OCCULTATION EXPERIMENT

HALOGEN COMPOUNDS
UF GROUP 7A COMPOUNDS
GS **HALOGEN COMPOUNDS**
    . BROMINE COMPOUNDS
    . . BROMATES
    . . BROMIDES
    . . . AMMONIUM BROMIDES
    . . . CESIUM BROMIDES
    . . . CHROMIUM BROMIDES
    . . DIBROMIDES
    . . . HYDROBROMIC ACID
    . . . HYDROBROMIDES
    . . . MAGNESIUM BROMIDES
    . . . POTASSIUM BROMIDES
    . . . SILVER BROMIDES
    . . . SODIUM BROMIDES
    . . . STRONTIUM BROMIDES
    . CHLORINE COMPOUNDS
    . . CHLORATES
    . . CHLORIDES
    . . . ALUMINUM CHLORIDES
    . . . AMMONIUM CHLORIDES

HALOGEN COMPOUNDS-*(CONT.)*
    . . . BERYLLIUM CHLORIDES
    . . . BORON CHLORIDES
    . . . CADMIUM CHLORIDES
    . . . CALCIUM CHLORIDES
    . . . CARBON TETRACHLORIDE
    . . . COPPER CHLORIDES
    . . . DICHLORIDES
    . . . GERMANIUM CHLORIDES
    . . . HYDROCHLORIDES
    . . . IRON CHLORIDES
    . . . LANTHANUM CHLORIDES
    . . . LEAD CHLORIDES
    . . . LITHIUM CHLORIDES
    . . . MAGNESIUM CHLORIDES
    . . . NITROSYL CHLORIDES
    . . . NITROXYCHLORIDES
    . . . NITRYL CHLORIDES
    . . . PHOSGENE
    . . . POTASSIUM CHLORIDES
    . . . SILICON TETRACHLORIDE
    . . . SILVER CHLORIDES
    . . . SODIUM CHLORIDES
    . . . SULFUR CHLORIDES
    . . . TETRACHLORIDES
    . . . TITANIUM CHLORIDES
    . . . TUNGSTEN CHLORIDES
    . . ZINC CHLORIDES
    . CHLORINE FLUORIDES
    . CHLORINE OXIDES
    . CHLOROSILANES
    . DDT
    . MECLIZINE
    . PERCHLORATES
    . . ALUMINUM PERCHLORATES
    . . AMMONIUM PERCHLORATES
    . . HYDRAZINE PERCHLORATES
    . . HYDROGEN PERCHLORATE
    . . HYDROXYLAMMONIUM
         PERCHLORATES
    . . LITHIUM PERCHLORATES
    . . MAGNESIUM PERCHLORATES
    . . NITRONIUM PERCHLORATE
    . . POTASSIUM PERCHLORATES
    . FLUORINE COMPOUNDS
    . . FLUORIDES
    . . . ANTIMONY FLUORIDES
    . . . BARIUM FLUORIDES
    . . . BORON FLUORIDES
    . . . CHLORINE FLUORIDES
    . . . COMPOUND A
    . . . CRYOLITE
    . . . DIFLUORIDES
    . . . . CALCIUM FLUORIDES
    . . . . . FLUORSPAR
    . . . HYDROFLUORIC ACID
    . . . METAL FLUORIDES
    . . . . ALUMINUM FLUORIDES
    . . . . BERYLLIUM FLUORIDES
    . . . . CADMIUM FLUORIDES
    . . . . CALCIUM FLUORIDES
    . . . . CESIUM FLUORIDES
    . . . . CHROMIUM FLUORIDES
    . . . . COBALT FLUORIDES
    . . . . COPPER FLUORIDES
    . . . . LANTHANUM FLUORIDES
    . . . . LITHIUM FLUORIDES
    . . . . MAGNESIUM FLUORIDES
    . . . . NICKEL FLUORIDES
    . . . . PLUTONIUM FLUORIDES
    . . . . PROTACTINIUM FLUORIDES
    . . . . SODIUM FLUORIDES
    . . . . STRONTIUM FLUORIDES
    . . . . THORIUM FLUORIDES
    . . . . TUNGSTEN FLUORIDES
    . . . . URANIUM FLUORIDES
    . . . . ZINC FLUORIDES
    . . . NITROGEN FLUORIDES
    . . . OXYFLUORIDES
    . . . OXYGEN FLUORIDES
    . . . OZONE FLUORIDE
    . . . PERCHLORYL FLUORIDES
    . . . SULFUR FLUORIDES
    . . . TECHNETIUM FLUORIDES
    . . FLUORITE
    . FLUORO COMPOUNDS
    . . CRYOLITE
    . . DIFLUORO COMPOUNDS
    . . . PERFLUOROALKANE
    . . . . POLYTETRAFLUOROETHYLENE
    . . FLUORINE ORGANIC COMPOUNDS
    . . . FLUOROAMINES
    . . . . NITROFLUORAMINES
    . . . . . TRIFLUOROAMINE OXIDE
    . . . . FLUOROCARBONS
    . . . . FLUOROHYDROCARBONS

HALOGEN COMPOUNDS-(CONT.)
```
. . . . CARBON TETRAFLUORIDE
. . . . FLUOROPOLYMERS
. . . . KEL-F
. . . . PERFLUOROALKANE
. . . . PERFLUOROGUANIDINE
. HALIDES
. . BROMIDES
. . . AMMONIUM BROMIDES
. . . CESIUM BROMIDES
. . . CHROMIUM BROMIDES
. . . DIBROMIDES
. . . HYDROBROMIC ACID
. . . HYDROBROMIDES
. . . MAGNESIUM BROMIDES
. . . POTASSIUM BROMIDES
. . . SILVER BROMIDES
. . . SODIUM BROMIDES
. . . STRONTIUM BROMIDES
. . CHLORIDES
. . . ALUMINUM CHLORIDES
. . . AMMONIUM CHLORIDES
. . . BERYLLIUM CHLORIDES
. . . BORON CHLORIDES
. . . CADMIUM CHLORIDES
. . . CALCIUM CHLORIDES
. . . CARBON TETRACHLORIDE
. . . COPPER CHLORIDES
. . . DICHLORIDES
. . . GERMANIUM CHLORIDES
. . . HYDROCHLORIDES
. . . HYDROGEN CHLORIDES
. . . . HYDROCHLORIC ACID
. . . IRON CHLORIDES
. . . LANTHANUM CHLORIDES
. . . LEAD CHLORIDES
. . . LITHIUM CHLORIDES
. . . MAGNESIUM CHLORIDES
. . . NITROSYL CHLORIDES
. . . NITROXYCHLORIDES
. . . NITRYL CHLORIDES
. . . PHOSGENE
. . . POTASSIUM CHLORIDES
. . . SILICON TETRACHLORIDE
. . . SILVER CHLORIDES
. . . SODIUM CHLORIDES
. . . SULFUR CHLORIDES
. . . TETRACHLORIDES
. . . TITANIUM CHLORIDES
. . . TUNGSTEN CHLORIDES
. . . ZINC CHLORIDES
. . FLUORIDES
. . . ANTIMONY FLUORIDES
. . . BARIUM FLUORIDES
. . . BORON FLUORIDES
. . . CHLORINE FLUORIDES
. . . DIFLUORIDES
. . . . CALCIUM FLUORIDES
. . . . FLUORSPAR
. . . HYDROFLUORIC ACID
. . . NITROGEN FLUORIDES
. . . OXYFLUORIDES
. . . OXYGEN FLUORIDES
. . . OZONE FLUORIDE
. . . PERCHLORYL FLUORIDES
. . . SULFUR FLUORIDES
. . . TECHNETIUM FLUORIDES
. . METAL HALIDES
. . . ALKALI HALIDES
. . . CESIUM HALIDES
. . . . CESIUM BROMIDES
. . . . CESIUM FLUORIDES
. . . . CESIUM IODIDES
. . . . POTASSIUM IODIDES
. . . . SODIUM BROMIDES
. . . . SODIUM CHLORIDES
. . . . SODIUM FLUORIDES
. . . . SODIUM IODIDES
. . . ALUMINUM CHLORIDES
. . . ALUMINUM FLUORIDES
. . . BARIUM FLUORIDES
. . . BERYLLIUM CHLORIDES
. . . BERYLLIUM FLUORIDES
. . . CADMIUM CHLORIDES
. . . CADMIUM FLUORIDES
. . . CALCIUM CHLORIDES
. . . CALCIUM FLUORIDES
. . . . FLUORSPAR
. . . CHROMIUM BROMIDES
. . . CHROMIUM FLUORIDES
. . . COBALT FLUORIDES
. . . COPPER CHLORIDES
. . . COPPER FLUORIDES
. . . HAFNIUM IODIDES
. . . IRON CHLORIDES
. . . LANTHANUM CHLORIDES
```

HALOGEN COMPOUNDS-(CONT.)
```
. . . LANTHANUM FLUORIDES
. . . LEAD CHLORIDES
. . . LITHIUM CHLORIDES
. . . LITHIUM FLUORIDES
. . . MAGNESIUM BROMIDES
. . . MAGNESIUM FLUORIDES
. . . NICKEL FLUORIDES
. . . NIOBIUM IODIDES
. . . PLUTONIUM FLUORIDES
. . . POTASSIUM BROMIDES
. . . POTASSIUM CHLORIDES
. . . PROTACTINIUM FLUORIDES
. . . SILVER HALIDES
. . . . SILVER BROMIDES
. . . . SILVER CHLORIDES
. . . . SILVER IODIDES
. . . STRONTIUM BROMIDES
. . . STRONTIUM FLUORIDES
. . . TECHNETIUM FLUORIDES
. . . THORIUM FLUORIDES
. . . TITANIUM CHLORIDES
. . . TUNGSTEN HALIDES
. . . . TUNGSTEN CHLORIDES
. . . . TUNGSTEN FLUORIDES
. . . URANIUM FLUORIDES
. . . ZINC CHLORIDES
. . . ZINC FLUORIDES
. . . ZIRCONIUM IODIDES
. . OXYHALIDES
. HALOCARBONS
. . FLUOROCARBONS
. IODINE COMPOUNDS
. . IODATES
. . . LITHIUM IODATES
. . IODIDES
. . . CESIUM IODIDES
. . . GALLAMINE TRIETHIODIDE
. . . HAFNIUM IODIDES
. . . NIOBIUM IODIDES
. . . POTASSIUM IODIDES
. . . SILVER IODIDES
. . . SODIUM IODIDES
. . . ZIRCONIUM IODIDES
. . IODOACETIC ACID
. . NITROSYLS
. . NITROSYL CHLORIDES
RT    ∞CHEMICAL COMPOUNDS
```

HALOGEN OCCULTATION EXPERIMENT
```
UF    HALOE
GS    PAYLOADS
. SPACE SHUTTLE PAYLOADS
. . HALOGEN OCCULTATION
    EXPERIMENT
RT    OZONE
```

HALOGENATION
```
GS    CHEMICAL REACTIONS
. HALOGENATION
. . BROMINATION
. . CHLORINATION
. . FLUORINATION
RT    DEFLUORINATION
HALOGENS
```

HALOGENS
```
GS    CHEMICAL ELEMENTS
. HALOGENS
. . ASTATINE
. . BROMINE
. . . BROMINE ISOTOPES
. . CHLORINE
. . FLUORINE
. . . FLUORINE ISOTOPES
. . IODINE
. . . IODINE ISOTOPES
. . . . IODINE 125
. . . . IODINE 131
. . . . IODINE 132
RT    EXCIMER LASERS
HALIDES
HALOGENATION
```

HALOPHILES
```
RT    AGRICULTURE
PLANTS (BOTANY)
```

HALOS
```
GS    SCATTERING
. WAVE SCATTERING
. . ELECTROMAGNETIC SCATTERING
. . . LIGHT SCATTERING
. . . . HALOS
TRANSMISSION
```

HALOS-(CONT.)
```
. ELECTROMAGNETIC WAVE
    TRANSMISSION
. . LIGHT TRANSMISSION
. . . LIGHT SCATTERING
. . . . HALOS
. WAVE PROPAGATION
. . LIGHT SCATTERING
. . . HALOS
RT    ASTRONOMY
ATMOSPHERIC SCATTERING
CORONAS
HAZE
IMAGES
RAINBOWS
```

HALPHEN METHOD
```
RT    ∞METHODOLOGY
```

HAMBURGER AIRCRAFT
```
GS    HAMBURGER AIRCRAFT
. C-160 AIRCRAFT
. HFB-320 AIRCRAFT
RT    ∞AIRCRAFT
```

HAMBURGER HFB-320 AIRCRAFT
```
USE   HFB-320 AIRCRAFT
```

HAMILTON-JACOBI EQUATION
```
RT    ∞EQUATIONS
EQUATIONS OF MOTION
HAMILTONIAN FUNCTIONS
RELATIVISTIC PARTICLES
```

HAMILTONIAN FUNCTIONS
```
GS    FUNCTIONS (MATHEMATICS)
. HAMILTONIAN FUNCTIONS
RT    CLASSICAL MECHANICS
∞DYNAMICS
HAMILTON-JACOBI EQUATION
QUANTUM THEORY
VON ZEIPEL METHOD
```

HAMMERHEAD CONFIGURATION
```
RT    FOREBODIES
MISSILE CONFIGURATIONS
```

HAMMERS
```
GS    HAMMERS
. ELECTROMAGNETIC HAMMERS
RT    IMPACTORS
PRESSES
RAMS (PRESSES)
TOOLS
```

HAMSTERS
```
GS    ANIMALS
. VERTEBRATES
. . MAMMALS
. . . RODENTS
. . . . HAMSTERS
```

HAND (ANATOMY)
```
GS    ANATOMY
. LIMBS (ANATOMY)
. . HAND (ANATOMY)
. . . FINGERS
APPENDAGES
. HAND (ANATOMY)
RT    FINGERS
WRIST
```

HANDBOOKS
```
GS    DOCUMENTS
. HANDBOOKS
. . USER MANUALS (COMPUTER
        PROGRAMS)
RT    BIBLIOGRAPHIES
DIRECTORIES
INDEXES (DOCUMENTATION)
MANUALS
SUBJECTS
TEXTBOOKS
TRAINING ANALYSIS
```

HANDEDNESS
```
RT    LATERAL STABILITY
```

HANDICAPS
```
RT    FRUSTRATION
∞INHIBITION
WHEELCHAIRS
```

HANDLES
RT   KNOBS
      LEVERS
      MANUAL CONTROL

HANDLEY PAGE AIRCRAFT
GS   **HANDLEY PAGE AIRCRAFT**
      . HP-115 AIRCRAFT
      . VICTOR MK-1 AIRCRAFT
RT   ∞AIRCRAFT

HANDLEY PAGE HP-115 AIRCRAFT
USE   HP-115 AIRCRAFT

HANDLING EQUIPMENT
GS   **HANDLING EQUIPMENT**
      . CRANES
      . . GANTRY CRANES
RT   CRAWLER TRACTORS
      ∞EQUIPMENT
      GROUND SUPPORT EQUIPMENT
      HARBORS
      LOCOMOTIVES
      PROPELLANT STORAGE
      ∞STORAGE
      TRACTORS
      TRANSPORTATION

HANDLING QUALITIES
USE   CONTROLLABILITY

HANDWRITING
GS   **HANDWRITING**
      . GRAPHOLOGY
RT   CHARACTER RECOGNITION
      ORTHOGRAPHY

HANFORD REACTORS
GS   NUCLEAR REACTORS
      . **HANFORD REACTORS**
RT   REACTOR DESIGN
      REACTOR PHYSICS
      REACTOR TECHNOLOGY

HANG GLIDERS
GS   GLIDERS
      . **HANG GLIDERS**
RT   ∞AIRCRAFT
      FLEXIBLE WINGS
      FREE FLIGHT
      MAN POWERED AIRCRAFT
      PARAWINGS
      SAILWINGS
      SOARING
      ULTRALIGHT AIRCRAFT
      ∞WINGED VEHICLES

HANGARS
RT   AIRFIELD SURFACE MOVEMENTS
      AIRPORTS
      ∞BUILDINGS
      GROUND HANDLING
      GROUND STATIONS
      HELIPORTS
      MILITARY AIR FACILITIES

HANKEL FUNCTIONS
GS   ANALYSIS (MATHEMATICS)
      . COMPLEX VARIABLES
      . . BESSEL FUNCTIONS
      . . . **HANKEL FUNCTIONS**
      . REAL VARIABLES
      . . BESSEL FUNCTIONS
      . . . **HANKEL FUNCTIONS**
      FUNCTIONS (MATHEMATICS)
      . **HANKEL FUNCTIONS**
RT   BOUNDARY VALUE PROBLEMS
      DIFFERENTIAL EQUATIONS
      ORTHOGONAL FUNCTIONS

HANSEN LUNAR THEORY
RT   EARTH ORBITS
      ORBITAL MECHANICS
      PERTURBATION THEORY
      ∞THEORIES

HAPLOSCOPES
GS   MEASURING INSTRUMENTS
      . OPTICAL MEASURING INSTRUMENTS
      . . **HAPLOSCOPES**
RT   ASTIGMATISM
      BINOCULAR VISION
      EYE EXAMINATIONS
      ∞INSTRUMENTS

HAPLOSCOPES-(CONT.)
      OPTOMETRY

HARBORS
GS   WATERWAYS
      . **HARBORS**
      . . ARTIFICIAL HARBORS
RT   BOATS
      BREAKWATERS
      CARGO
      DREDGING
      ESTUARIES
      FREIGHTERS
      HANDLING EQUIPMENT
      MARINE TRANSPORTATION
      OCEANOGRAPHY
      ∞PORTS
      REGIONAL PLANNING
      SHIP TERMINALS
      SHIPS
      TANKER SHIPS
      TERMINAL FACILITIES
      TRAFFIC
      ∞TRAVEL
      WATER VEHICLES
      WHARVES

HARD LANDING
GS   LANDING
      . **HARD LANDING**
RT   AIRCRAFT LANDING
      CRASH LANDING
      LUNAR LANDING
      PLANETARY LANDING
      SOFT LANDING
      SPACECRAFT LANDING
      WATER LANDING

HARDENERS
RT   ALLOYS
      HARDENING (MATERIALS)
      HEAT TREATMENT

∞ **HARDENING**
SN   *(USE OF A MORE SPECIFIC TERM IS*
      RECOMMENDED--CONSULT THE TERMS
      LISTED BELOW)
RT   HARDENING (MATERIALS)
      HARDENING (SYSTEMS)

HARDENING (MATERIALS)
UF   METAL HARDENING
GS   **HARDENING (MATERIALS)**
      . CARBURIZING
      . COLD HARDENING
      . HOT PRESSING
      . NITRIDING
      . PRECIPITATION HARDENING
      . . MARAGING
      . PULSE HEATING
      . SHOT PEENING
      . SILICONIZING
      . WORK HARDENING
      . . STRAIN HARDENING
RT   AGING (MATERIALS)
      AGING (METALLURGY)
      ANNEALING
      COAGULATION
      HARDENERS
      ∞HARDENING
      HEAT TREATMENT
      MARTENSITE
      METAL WORKING
      ∞METALLURGY
      MICROSTRUCTURE
      NORMALIZING (HEAT TREATMENT)
      PEENING
      QUENCHING (COOLING)
      ∞SETTING
      SOFTENING
      TEMPERING

HARDENING (SYSTEMS)
SN   (TECHNIQUES FOR DECREASING THE
      SUSCEPTIBILITY OR VULNERABILITY OF
      WEAPON SYSTEMS AND COMPONENTS)
RT   ∞HARDENING
      MISSILE DEFENSE
      NUCLEAR WARFARE
      ∞SYSTEMS

HARDNESS
GS   MECHANICAL PROPERTIES
      . **HARDNESS**
      . . KNOOP HARDNESS

HARDNESS-(CONT.)
      . . MICROHARDNESS
      . . ROCKWELL HARDNESS
RT   ABRASION RESISTANCE
      BRITTLE MATERIALS
      BRITTLENESS
      CHARPY IMPACT TEST
      COLD HARDENING
      DUCTILITY
      FATIGUE (MATERIALS)
      FRACTURE STRENGTH
      IMPACT STRENGTH
      INDENTATION
      NOTCH TESTS
      PLASTIC PROPERTIES
      SOFTNESS
      SURFACE PROPERTIES
      TEMPER (METALLURGY)
      TOUGHNESS
      WEAR

HARDNESS TESTS
RT   COMPRESSION TESTS
      HIGH TEMPERATURE TESTS
      IMPACT TESTS
      KNOOP HARDNESS
      LOW TEMPERATURE TESTS
      ∞MATERIALS TESTS
      NONDESTRUCTIVE TESTS
      STATIC TESTS
      ∞TESTS
      WEAR TESTS

HARDWARE
RT   COMPUTERS
      ELECTRONIC MODULES
      ∞EQUIPMENT
      FIRMWARE
      FIXTURES
      TOOLS

HARDWARE UTILIZATION LISTS
UF   HUL
GS   LISTS
      . **HARDWARE UTILIZATION LISTS**
RT   ∞CATALOGS
      DOCUMENTS

HARLETON METEORITE
GS   CELESTIAL BODIES
      . METEORITES
      . . **HARLETON METEORITE**
RT   IRON METEORITES
      STONY METEORITES

HARMONIC ANALYSIS
GS   ANALYSIS (MATHEMATICS)
      . FUNCTIONAL ANALYSIS
      . . **HARMONIC ANALYSIS**
      . . . TESSERAL HARMONICS
      . . . ZONAL HARMONICS
RT   BANACH SPACE
      FORM FACTORS
      FOURIER ANALYSIS
      FREQUENCY ANALYZERS
      MICROWAVE RESONANCE

HARMONIC CONTROL
RT   ∞CONTROL
      HARMONIC OSCILLATION
      HARMONICS
      HELICOPTER CONTROL
      ROTARY WINGS
      VIBRATION DAMPING

HARMONIC EXCITATION
GS   EXCITATION
      . WAVE EXCITATION
      . . **HARMONIC EXCITATION**
      HARMONICS
      . **HARMONIC EXCITATION**
RT   ACOUSTICS
      FOURIER ANALYSIS
      SIMPLE HARMONIC MOTION

HARMONIC FUNCTIONS
GS   ANALYSIS (MATHEMATICS)
      . COMPLEX VARIABLES
      . . **HARMONIC FUNCTIONS**
      FUNCTIONS (MATHEMATICS)
      . **HARMONIC FUNCTIONS**
RT   AIRY FUNCTION
      FOURIER ANALYSIS
      LAPLACE EQUATION
      MAXIMUM PRINCIPLE

HARMONIC GENERATIONS
GS   HARMONICS
     . **HARMONIC GENERATIONS**
RT   ACOUSTICS
     CARRIER FREQUENCIES
     FOURIER ANALYSIS
     PHASE MATCHING
     WAVE GENERATION

HARMONIC GENERATORS
RT   COMPARATORS
     FREQUENCY CONVERTERS
     ∞ GENERATORS
     HARMONICS
     OSCILLATORS
     SUBHARMONIC GENERATORS

HARMONIC MOTION
GS   **HARMONIC MOTION**
     . SIMPLE HARMONIC MOTION
RT   GROUP VELOCITY
     ∞ MOTION

HARMONIC OSCILLATION
GS   HARMONICS
     . **HARMONIC OSCILLATION**
     OSCILLATIONS
     . **HARMONIC OSCILLATION**
RT   ACOUSTICS
     FOURIER ANALYSIS
     HARMONIC CONTROL
     TRANSVERSE OSCILLATION

HARMONIC OSCILLATORS
GS   OSCILLATORS
     . **HARMONIC OSCILLATORS**
RT   HARMONICS
     MECHANICAL OSCILLATORS
     SUBHARMONIC GENERATORS

HARMONIC RADIATION
RT   ELECTROMAGNETIC RADIATION
     ∞ RADIATION

HARMONICS
UF   OVERTONES
GS   **HARMONICS**
     . HARMONIC EXCITATION
     . HARMONIC GENERATIONS
     . HARMONIC OSCILLATION
     . SIMPLE HARMONIC MOTION
     . SPHERICAL HARMONICS
     . SUPERHARMONICS
     . TESSERAL HARMONICS
     . ZONAL HARMONICS
RT   ACOUSTICS
     CYCLES
     FOURIER ANALYSIS
     FREQUENCIES
     HARMONIC CONTROL
     HARMONIC GENERATORS
     HARMONIC OSCILLATORS
     NODES (STANDING WAVES)
     RESONANT FREQUENCIES
     SOUND-SOUND INTERACTIONS
     STANDING WAVES
     SUBAUDIBLE FREQUENCIES
     SUBHARMONIC GENERATORS
     VIBRATION
     WAVELENGTHS

HARNESSES
RT   COUCHES
     SAFETY DEVICES
     SEAT BELTS
     SEATS
     TRANSMISSION LINES

HARPOON MISSILE
GS   MISSILES
     . AIR TO SURFACE MISSILES
     . . **HARPOON MISSILE**
RT   SURFACE TO SURFACE MISSILES
     WEAPON SYSTEMS

HARRIER AIRCRAFT
UF   AV-8A AIRCRAFT
     AV-8B AIRCRAFT
GS   HAWKER SIDDELEY AIRCRAFT
     . **HARRIER AIRCRAFT**
RT   ∞ AIRCRAFT
     BUCCANEER AIRCRAFT
     ∞ MILITARY AIRCRAFT
     P-1127 AIRCRAFT

HARRIER AIRCRAFT-*(CONT.)*
     SAAB 37 AIRCRAFT
     VAMPIRE MK 35 AIRCRAFT
     VULCAN AIRCRAFT

HARTMANN FLOW
GS   FLUID FLOW
     . LAMINAR FLOW
     . . **HARTMANN FLOW**
     . STEADY FLOW
     . . **HARTMANN FLOW**
RT   COUETTE FLOW
     MAGNETOHYDRODYNAMIC FLOW
     MAGNETOHYDRODYNAMICS
     TWO DIMENSIONAL FLOW

HARTMANN NUMBER
GS   RATIOS
     . DIMENSIONLESS NUMBERS
     . . **HARTMANN NUMBER**
RT   MAGNETOHYDRODYNAMICS
     VISCOUS DRAG

HARTREE APPROXIMATION
UF   HARTREE-APPLETON APPROXIMATION
     HARTREE-FOCK APPROXIMATION
GS   ANALYSIS (MATHEMATICS)
     . NUMERICAL ANALYSIS
     . . APPROXIMATION
     . . . **HARTREE APPROXIMATION**
RT   ATOMIC STRUCTURE
     MANY BODY PROBLEM
     PERTURBATION THEORY
     SELF CONSISTENT FIELDS
     WAVE FUNCTIONS

HARTREE-APPLETON APPROXIMATION
USE   HARTREE APPROXIMATION

HARTREE-FOCK APPROXIMATION
USE   HARTREE APPROXIMATION

HARTREE-FOCK-SLATER METHOD
RT   ATOMIC PHYSICS
     ELECTRON ENERGY
     ∞ METHODOLOGY
     SLATER ORBITALS

HARVARD RADIO METEOR PROJECT
GS   PROGRAMS
     . PROJECTS
     . . **HARVARD RADIO METEOR PROJECT**
RT   RADIO ECHOES

HASTELLOY (TRADEMARK)
GS   ALLOYS
     . NICKEL ALLOYS
     . . **HASTELLOY (TRADEMARK)**
RT   IRON ALLOYS
     MOLYBDENUM ALLOYS

HATCHES
RT   AIR LOCKS
     DOORS
     EGRESS
     GATES (OPENINGS)
     INGRESS (SPACECRAFT PASSAGEWAY)

HAULING
RT   CARGO
     DELIVERY
     MATERIALS HANDLING
     PACKAGING
     TRANSPORTATION
     TRANSPORTATION ENERGY
     TRUCKS

HAWAII
GS   LANDFORMS
     . ISLANDS
     . . **HAWAII**
     NATIONS
     . UNITED STATES
     . . **HAWAII**

HAWK MISSILE
GS   MISSILES
     . SURFACE TO AIR MISSILES
     . . **HAWK MISSILE**
RT   SOLID PROPELLANT ROCKET ENGINES

HAWKER HUNTER AIRCRAFT
USE   F-2 AIRCRAFT

HAWKER P-1052 AIRCRAFT
USE   P-1052 AIRCRAFT

HAWKER P-1127 AIRCRAFT
USE   P-1127 AIRCRAFT

HAWKER P-1154 AIRCRAFT
USE   P-1154 AIRCRAFT

HAWKER SIDDELEY AIRCRAFT
GS   **HAWKER SIDDELEY AIRCRAFT**
     . ARGOSY MK-1 AIRCRAFT
     . AVRO 707 AIRCRAFT
     . BUCCANEER AIRCRAFT
     . COMET 4 AIRCRAFT
     . DH 112 AIRCRAFT
     . DH 115 AIRCRAFT
     . DH 121 AIRCRAFT
     . DH 125 AIRCRAFT
     . F-2 AIRCRAFT
     . GA-5 AIRCRAFT
     . HARRIER AIRCRAFT
     . HS-748 AIRCRAFT
     . HS-801 AIRCRAFT
     . P-1052 AIRCRAFT
     . P-1127 AIRCRAFT
     . P-1154 AIRCRAFT
     . SHACKLETON BOMBER
     . VAMPIRE MK 35 AIRCRAFT
     . VULCAN AIRCRAFT
RT   ∞ AIRCRAFT

HAWKEYE AIRCRAFT
USE   E-2 AIRCRAFT

HAWKEYE SATELLITES
GS   SATELLITES
     . ARTIFICIAL SATELLITES
     . . SCIENTIFIC SATELLITES
     . . . **HAWKEYE SATELLITES**

HAWKEYE 1 SATELLITE
USE   EXPLORER 52 SATELLITE

HAY
GS   FARM CROPS
     . **HAY**
     PLANTS (BOTANY)
     . GRASSES
     . . **HAY**
RT   AGRICULTURE
     BOTANY
     EARTH RESOURCES
     FARMLANDS
     ∞ FOOD
     GRASSLANDS
     LEGUMINOUS PLANTS

HAYNES STELLITE
USE   STELLITE (TRADEMARK)

HAZARDOUS MATERIAL DISPOSAL (IN SPACE)
GS   DISPOSAL
     . WASTE DISPOSAL
     . . **HAZARDOUS MATERIAL DISPOSAL (IN SPACE)**
RT   AEROSPACE ENVIRONMENTS
     PUBLIC HEALTH
     RADIOACTIVE WASTES
     TOXICITY AND SAFETY HAZARD

HAZARDS
UF   DANGER
     NOISE HAZARDS
GS   **HAZARDS**
     . AIRCRAFT HAZARDS
     . FLIGHT HAZARDS
     . . METEOROID HAZARDS
     . OPERATIONAL HAZARDS
     . RADIATION HAZARDS
     . TOXIC HAZARDS
RT   ACCIDENT PREVENTION
     ACCIDENTS
     AIRCRAFT SPIN
     AVOIDANCE
     CRASH INJURIES
     ∞ DETECTORS
     EXPLOSIONS
     FIRES
     FLAMMABLE GASES
     INCOMPATIBILITY
     INJURIES
     LOW VISIBILITY
     NOISE TOLERANCE

HAZARDS-*(CONT.)*
- PROTECTION
- RISK
- SABOTAGE
- SAFETY
- SAFETY DEVICES
- SAFETY FACTORS
- SAFETY MANAGEMENT
- SPONTANEOUS COMBUSTION
- TOXICOLOGY
- WARNING SYSTEMS

HAZE
- RT　AIR POLLUTION
- ATMOSPHERIC OPTICS
- CLARITY
- FOG
- HALOS
- LIGHT TRANSMISSION
- LOW VISIBILITY
- MIST
- OPACITY
- OPTICAL PROPERTIES
- TRANSPARENCE
- TURBIDITY
- VISIBILITY

HAZE DETECTION
- GS　DETECTION
- . **HAZE DETECTION**
- RT　FOG
- FOREST FIRE DETECTION
- FUMES
- GAS DETECTORS
- MIST
- REMOTE SENSORS
- SMOKE
- VAPORS

HBNQ
- USE　NITROGUANIDINE

HBWR REACTOR
- USE　HALDEN BOILING WATER REACTOR

HC-1 HELICOPTER
- USE　CH-47 HELICOPTER

HC-3 HELICOPTER
- UF　OMNIPOL HC-3 HELICOPTER
- GS　GENERAL AVIATION AIRCRAFT
- . **HC-3 HELICOPTER**
- TRANSPORT AIRCRAFT
- . **HC-3 HELICOPTER**
- UTILITY AIRCRAFT
- . **HC-3 HELICOPTER**
- V/STOL AIRCRAFT
- . ROTARY WING AIRCRAFT
- . . HELICOPTERS
- . . . **HC-3 HELICOPTER**
- RT　PASSENGER AIRCRAFT

HCL ARGON LASERS
- GS　STIMULATED EMISSION DEVICES
- . LASERS
- . . GAS LASERS
- . . . HCL LASERS
- . . . . **HCL ARGON LASERS**

HCL LASERS
- GS　STIMULATED EMISSION DEVICES
- . LASERS
- . . CHEMICAL LASERS
- . . . **HCL LASERS**
- . . GAS LASERS
- . . . **HCL LASERS**
- . . . . HCL ARGON LASERS

HCMM
- USE　HEAT CAPACITY MAPPING MISSION

HCN LASERS
- GS　STIMULATED EMISSION DEVICES
- . LASERS
- . . GAS LASERS
- . . . **HCN LASERS**
- RT　CHEMICAL LASERS
- COHERENT LIGHT
- HYDROCYANIC ACID
- LIGHT AMPLIFIERS
- LIGHT SOURCES
- OPTICAL PUMPING
- STIMULATED EMISSION

HD-1 GROUND EFFECT MACHINES
- USE　HOVERCRAFT GROUND EFFECT
- MACHINES

HEAD (ANATOMY)
- GS　ANATOMY
- . **HEAD (ANATOMY)**
- . . OCCIPITAL LOBES
- . . SKULL
- . . . CRANIUM
- . . . . INTRACRANIAL CAVITY
- RT　BRAIN
- EYE (ANATOMY)
- FACE (ANATOMY)
- FOREHEAD
- LIPS (ANATOMY)
- NOSE (ANATOMY)
- SENSE ORGANS

HEAD (FLUID MECHANICS)
- GS　FLUID FLOW
- . **HEAD (FLUID MECHANICS)**
- . . HEAD FLOW
- . . PRESSURE HEADS
- RT　ELEVATION
- FLUID FLOW
- GEOPOTENTIAL HEIGHT
- HYDROSTATIC PRESSURE
- HYDROSTATICS
- LIQUID FLOW
- PRESSURE
- SCALE HEIGHT

HEAD (PRESSURE)
- USE　PRESSURE HEADS

HEAD FLOW
- GS　FLUID FLOW
- . HEAD (FLUID MECHANICS)
- . . **HEAD FLOW**
- RT　BASE FLOW
- BLASIUS FLOW
- INLET FLOW
- LIQUID FLOW
- ∞PRESSURE DROP

HEAD MOVEMENT
- RT　∞MOTION

HEAD-UP DISPLAYS
- GS　DISPLAY DEVICES
- . **HEAD-UP DISPLAYS**
- RT　AVIONICS
- CONSOLES
- FLIGHT INSTRUMENTS
- IMAGE TUBES
- INDICATING INSTRUMENTS
- LANDING AIDS
- NAVIGATION AIDS
- POSITION INDICATORS
- SPACECRAFT POSITION INDICATORS
- WARNING SYSTEMS

HEADACHE
- UF　CEPHALAGIA
- GS　DISEASES
- . **HEADACHE**
- SIGNS AND SYMPTOMS
- . **HEADACHE**

∞ HEADERS
- SN　*(USE OF A MORE SPECIFIC TERM IS*
- *RECOMMENDED--CONSULT THE TERMS*
- *LISTED BELOW)*
- RT　BEAMS (SUPPORTS)
- CHASSIS
- HERMETIC SEALS
- PIPES (TUBES)
- SUPPORTS
- ∞TERMINALS

HEADSETS
- USE　EARPHONES

HEALING
- GS　HEALING
- . WOUND HEALING
- RT　CLINICAL MEDICINE
- CURES
- THERAPY

HEALTH
- GS　**HEALTH**
- . HEALTH PHYSICS

HEALTH-*(CONT.)*
- . . PUBLIC HEALTH
- . MENTAL HEALTH
- RT　CHRONIC CONDITIONS
- CLINICAL MEDICINE
- HYGIENE
- ORAL HYGIENE
- PSYCHOTHERAPY
- SANITATION

HEALTH PHYSICS
- GS　BIOPHYSICS
- . **HEALTH PHYSICS**
- . . PUBLIC HEALTH
- HEALTH
- . **HEALTH PHYSICS**
- . . PUBLIC HEALTH
- RT　FLUENCE
- INDUSTRIAL SAFETY
- NUCLEAR MEDICINE
- NUCLEAR PHYSICS
- NUCLEAR RADIATION
- ∞PHYSICS
- RADIATION DETECTORS
- RADIATION DOSAGE
- RADIATION EFFECTS
- RADIATION HAZARDS
- RADIATION INJURIES
- RADIATION MEASURING INSTRUMENTS
- RADIATION PROTECTION
- RADIATION SICKNESS
- RADIOBIOLOGY
- SAFETY FACTORS
- ∞SCIENCE

HEALTH PHYSICS RESEARCH REACTOR
- UF　HPRR
- GS　NUCLEAR REACTORS
- . NUCLEAR RESEARCH AND TEST
- REACTORS
- . . **HEALTH PHYSICS RESEARCH**
- **REACTOR**
- RT　∞PHYSICS

HEALTH-EDUCATION TELECOMMUNICATIONS EXP
- USE　HET EXPERIMENT

HEAO
- UF　HIGH ENERGY ASTRONOMY
- OBSERVATORIES
- GS　OBSERVATORIES
- . ASTRONOMICAL OBSERVATORIES
- . . ASTRONOMICAL SATELLITES
- . . . **HEAO**
- . . . . HEAO 1
- . . . . HEAO 2
- . . . . HEAO 3
- RT　OAO

HEAO A
- USE　HEAO 1

HEAO B
- USE　HEAO 2

HEAO C
- USE　HEAO 3

HEAO 1
- UF　HEAO A
- HIGH ENERGY ASTRONOMY
- OBSERVATORY A
- HIGH ENERGY ASTRONOMY
- OBSERVATORY 1
- GS　OBSERVATORIES
- . ASTRONOMICAL OBSERVATORIES
- . . ASTRONOMICAL SATELLITES
- . . . HEAO
- . . . . **HEAO 1**
- UNMANNED SPACECRAFT
- . **HEAO 1**
- RT　OAO

HEAO 2
- UF　EINSTEIN OBSERVATORY
- HEAO B
- HIGH ENERGY ASTRONOMY
- OBSERVATORY B
- HIGH ENERGY ASTRONOMY
- OBSERVATORY 2
- GS　OBSERVATORIES
- . ASTRONOMICAL OBSERVATORIES
- . . ASTRONOMICAL SATELLITES
- . . . HEAO

HEAO 2-(CONT.)
```
      . . . . HEAO 2
      UNMANNED SPACECRAFT
      . HEAO 2
RT    OAO
```

HEAO 3
```
UF    HEAO C
      HIGH ENERGY ASTRONOMY
         OBSERVATORY C
      HIGH ENERGY ASTRONOMY
         OBSERVATORY 3
GS    OBSERVATORIES
      . ASTRONOMICAL OBSERVATORIES
      . . ASTRONOMICAL SATELLITES
      . . . HEAO
      . . . . HEAO 3
      UNMANNED SPACECRAFT
      . HEAO 3
RT    OAO
```

HEARING
```
GS    HEARING
      . BINAURAL HEARING
RT    AUDIOLOGY
      AUDIOMETRY
      AUDITORY FATIGUE
      AUDITORY TASKS
      EAR
      LOUDNESS
      STEREOPHONICS
      THRESHOLDS (PERCEPTION)
```

HEARING LOSS
```
USE   AUDITORY DEFECTS
```

HEART
```
GS    ANATOMY
      . CARDIOVASCULAR SYSTEM
      . . HEART
      . . . CARDIAC AURICLES
      . . . CARDIAC VENTRICLES
      . . . EPICARDIUM
      . . . MYOCARDIUM
RT    AORTA
      ARTIFICIAL CARDIAC PACEMAKER
      ARTIFICIAL HEART VALVES
      BLOOD
      BLOOD PUMPS
      CARDIOGRAMS
      CARDIOGRAPHY
      CARDIOLOGY
      CARDIOTACHOMETERS
      CIRCULATORY SYSTEM
      CORONARY CIRCULATION
      DIASTOLE
      PHONOCARDIOGRAPHY
```

HEART DISEASES
```
GS    DISEASES
      . HEART DISEASES
      . . ANGINA PECTORIS
      . . CORONARY ARTERY DISEASE
RT    BRADYCARDIA
      CARDIOGRAPHY
      CARDIOLOGY
      ECHOCARDIOGRAPHY
      ELECTROCARDIOGRAPHY
      FAT EMBOLISMS
      FIBRILLATION
      HIS BUNDLE
      MYOCARDIAL INFARCTION
      PHONOCARDIOGRAPHY
```

HEART FUNCTION
```
GS    HEART FUNCTION
      . HEART MINUTE VOLUME
RT    ANGINA PECTORIS
      CYANOSIS
      ECHOCARDIOGRAPHY
      HIS BUNDLE
```

HEART IMPLANTATION
```
GS    IMPLANTATION
      . HEART IMPLANTATION
RT    ARTIFICIAL HEART VALVES
      BIOTECHNOLOGY
      BLOOD CIRCULATION
      CIRCULATION
      PULMONARY CIRCULATION
      SURGERY
      TRANSPLANTATION
```

HEART MINUTE VOLUME
```
GS    HEART FUNCTION
```

HEART MINUTE VOLUME-(CONT.)
```
      . HEART MINUTE VOLUME
RT    SPIROMETERS
```

HEART RATE
```
GS    RATES (PER TIME)
      . HEART RATE
      . . ARRHYTHMIA
      . . BRADYCARDIA
      . . SYSTOLE
      . . TACHYCARDIA
RT    ANGINA PECTORIS
      BIOFEEDBACK
      BIOMEDICAL DATA
      CARDIOLOGY
      DIASTOLE
      EPINEPHRINE
      HERING-BREVER REFLEX
      SPHYGMOGRAPHY
```

HEART VALVES
```
GS    VALVES
      . HEART VALVES
RT    CORONARY CIRCULATION
```

HEARTHS
```
RT    FURNACES
      REFRACTORIES
```

HEAT
```
GS    HEAT
      . DRY HEAT
      . HEAT OF DISSOCIATION
      . NUCLEAR HEAT
      . PROCESS HEAT
RT    ACTIVATION ENERGY
      ∞ENERGY
      ENTHALPY
      ENTROPY
      HEAT OF FUSION
      HEATING
      INFRARED RADIATION
      TEMPERATURE
      THERMAL ENERGY
      THERMAL INSULATION
      THERMAL RADIATION
      THERMOCHEMISTRY
      THERMODYNAMIC PROPERTIES
      THERMODYNAMICS
      WORK
```

HEAT ACCLIMATIZATION
```
GS    ADAPTATION
      . ACCLIMATIZATION
      . . HEAT ACCLIMATIZATION
RT    BODY TEMPERATURE
      COLD ACCLIMATIZATION
      HIGH TEMPERATURE ENVIRONMENTS
      HUMAN TOLERANCES
      PERSPIRATION
      PHYSIOLOGICAL EFFECTS
```

HEAT BALANCE
```
RT    ATMOSPHERIC HEAT BUDGET
      BALANCE
      BOILERS
      COMBUSTION
      MATERIAL BALANCE
      PYROMETALLURGY
      THERMOCHEMICAL PROPERTIES
      THERMOCHEMISTRY
      THERMODYNAMIC PROPERTIES
```

HEAT BUDGET
```
GS    ENERGY BUDGETS
      . HEAT BUDGET
      . . ATMOSPHERIC HEAT BUDGET
RT    ∞BUDGETS
      SPECIFIC HEAT
```

HEAT CAPACITY
```
USE   SPECIFIC HEAT
```

HEAT CAPACITY MAPPING MISSION
```
UF    HCMM
RT    APPLICATIONS EXPLORER SATELLITES
      GEOGRAPHY
      MAPPING
      ∞MISSIONS
      PLANETARY MAPPING
      THERMAL MAPPING
```

HEAT CONDUCTION
```
USE   CONDUCTIVE HEAT TRANSFER
```

HEAT CONTENT
```
USE   ENTHALPY
```

HEAT DISSIPATION
```
USE   COOLING
```

HEAT DISSIPATION CHILLING
```
USE   COOLING
```

HEAT EFFECTS
```
USE   TEMPERATURE EFFECTS
```

HEAT EQUATIONS
```
USE   THERMODYNAMICS
```

HEAT EXCHANGERS
```
GS    HEAT EXCHANGERS
      . TUBE HEAT EXCHANGERS
RT    CONDENSERS (LIQUEFIERS)
      COOLANTS
      COOLING
      COOLING FINS
      COOLING SYSTEMS
      COUNTERFLOW
      EVAPORATORS
      ∞EXCHANGERS
      FINNED BODIES
      GAS COOLING
      GEOTHERMAL ENERGY EXTRACTION
      HEATING
      HEATING EQUIPMENT
      REGENERATIVE COOLING
      REGENERATORS
      SNAP
      SNAP 1
      SNAP 2
      SNAP 8
      SNAP 10A
      SPACE COOLING (BUILDINGS)
      SPACE POWER REACTORS
      SPACE POWER UNIT REACTORS
      WASTE HEAT
      WATER HEATING
```

HEAT FLOW
```
USE   HEAT TRANSMISSION
```

HEAT FLUX
```
SN    (HEAT ENERGY TRANSMISSION RATE)
GS    RATES (PER TIME)
      . FLUX (RATE)
      . . HEAT FLUX
RT    FLUX DENSITY
      SOLAR FLUX
```

HEAT GAIN
```
USE   HEATING
```

HEAT GENERATION
```
SN    (EXCLUDES BIOLOGICAL PRODUCTION
         OF HEAT)
RT    COGENERATION
      COMBUSTION
      DIRECT POWER GENERATORS
      ∞GENERATION
      HEATING
      HEATING EQUIPMENT
      PROCESS HEAT
      SOLID PROPELLANT COMBUSTION
```

HEAT ISLANDS
```
RT    CITIES
      CLIMATOLOGY
      URBAN PLANNING
      WEATHER MODIFICATION
```

HEAT MEASUREMENT
```
UF    CALORIMETRY
RT    BOLOMETERS
      BOMB CALORIMETERS
      CALORIMETERS
      DROP CALORIMETERS
      ENTHALPY
      FLAME CALORIMETERS
      ∞MEASUREMENT
      SHELL ANODES
```

HEAT OF COMBUSTION
```
UF    COMBUSTION HEAT
GS    CHEMICAL PROPERTIES
      . THERMOCHEMICAL PROPERTIES
```

HEAT OF COMBUSTION-*(CONT.)*
    . . **HEAT OF COMBUSTION**
    THERMODYNAMIC PROPERTIES
    . THERMOCHEMICAL PROPERTIES
    . . **HEAT OF COMBUSTION**
RT   COMBUSTION PHYSICS
    GUNS (ORDNANCE)

HEAT OF DISSOCIATION
GS   HEAT
    . **HEAT OF DISSOCIATION**
RT   CHEMICAL EQUILIBRIUM
    DISSOCIATION
    ∞EQUILIBRIUM
    REACTION KINETICS
    THERMAL DISSOCIATION
    THERMOCHEMISTRY
    THERMODYNAMIC EQUILIBRIUM

HEAT OF FORMATION
UF   FORMATION HEAT
GS   CHEMICAL PROPERTIES
    . THERMOCHEMICAL PROPERTIES
    . . **HEAT OF FORMATION**
    THERMODYNAMIC PROPERTIES
    . THERMOCHEMICAL PROPERTIES
    . . **HEAT OF FORMATION**

HEAT OF FUSION
UF   LATENT HEAT OF FUSION
GS   CHEMICAL PROPERTIES
    . THERMOCHEMICAL PROPERTIES
    . . **HEAT OF FUSION**
    THERMODYNAMIC PROPERTIES
    . THERMOCHEMICAL PROPERTIES
    . . **HEAT OF FUSION**
    . THERMOPHYSICAL PROPERTIES
    . . **HEAT OF FUSION**
RT   ENTHALPY
    FUSION (MELTING)
    HEAT
    MELTING
    PHASE CHANGE MATERIALS
    PHASE DIAGRAMS
    PHASE TRANSFORMATIONS
    SPECIFIC HEAT
    THERMAL ENERGY
    THERMOCHEMISTRY
    THERMODYNAMICS
    TRANSITION TEMPERATURE

HEAT OF SOLUTION
GS   CHEMICAL PROPERTIES
    . **HEAT OF SOLUTION**
    THERMODYNAMIC PROPERTIES
    . THERMOPHYSICAL PROPERTIES
    . . SPECIFIC HEAT
    . . . **HEAT OF SOLUTION**
RT   MOLECULAR ENERGY LEVELS
    THERMAL ENERGY
    THERMOCHEMISTRY
    THERMODYNAMICS

HEAT OF VAPORIZATION
UF   VAPORIZATION HEAT
GS   CHEMICAL PROPERTIES
    . THERMOCHEMICAL PROPERTIES
    . . **HEAT OF VAPORIZATION**
    THERMODYNAMIC PROPERTIES
    . THERMOCHEMICAL PROPERTIES
    . . **HEAT OF VAPORIZATION**
RT   VAPORIZING

HEAT PIPES
SN   (EXCLUDES PIPES AND TUBES USED
    FOR THE TRANSMISSION OF HEATED
    LIQUIDS OR GASES)
RT   GEOTHERMAL ENERGY UTILIZATION
    SPACECRAFT TEMPERATURE

HEAT PUMPS
RT   AIR CONDITIONING
    AIR CONDITIONING EQUIPMENT
    CONDENSERS (LIQUEFIERS)
    COOLING SYSTEMS
    GEOTHERMAL ENERGY EXTRACTION
    HEATING EQUIPMENT
    PUMPS
    REFRIGERATING MACHINERY
    RESIDENTIAL ENERGY
    SPACE COOLING (BUILDINGS)
    THERMOELECTRIC COOLING
    WASTE HEAT

HEAT RADIATORS
UF   CONDENSER RADIATORS
    HEAT REJECTION DEVICES
GS   **HEAT RADIATORS**
    . SPACECRAFT RADIATORS
RT   BLACK BODY RADIATION
    COOLING
    COOLING FINS
    COOLING SYSTEMS
    HEATING EQUIPMENT
    ∞INSULATED STRUCTURES
    RADIATIVE HEAT TRANSFER
    ∞RADIATORS
    STEFAN-BOLTZMANN LAW

HEAT REGULATION
USE   TEMPERATURE CONTROL

HEAT REJECTION DEVICES
USE   HEAT RADIATORS

HEAT RESISTANCE
USE   THERMAL RESISTANCE

HEAT RESISTANT ALLOYS
UF   HIGH TEMPERATURE ALLOYS
    SUPERALLOYS
GS   ALLOYS
    . **HEAT RESISTANT ALLOYS**
    . . NIMONIC ALLOYS
    . . REFRACTORY METAL ALLOYS
    . . . MOLYBDENUM ALLOYS
    . . . . RENE 41
    . . . . RENE 63
    . . . . RENE 77
    . . . NIOBIUM ALLOYS
    . . . OSMIUM ALLOYS
    . . . RHENIUM ALLOYS
    . . . TANTALUM ALLOYS
    . . . TUNGSTEN ALLOYS
    . . UDIMET ALLOYS
    . . WASPALOY
RT   CERMETS
    CHROMIUM ALLOYS
    COBALT ALLOYS
    NICKEL ALLOYS
    REFRACTORY METALS
    SULFIDATION
    SUPERPLASTICITY

HEAT SHIELDING
UF   THERMAL SHIELDING
GS   SHIELDING
    . **HEAT SHIELDING**
    . . REENTRY SHIELDING
    . . REUSABLE HEAT SHIELDING
RT   ABLATION
    ABLATIVE MATERIALS
    ABLATIVE NOSE CONES
    COOLING
    INFRARED SUPPRESSION
    ∞INSULATED STRUCTURES
    LUDOX (TRADEMARK)
    PYROLYTIC GRAPHITE
    SOLAR REFLECTORS
    SPACECRAFT SHIELDING
    TEMPERATURE
    TEMPERATURE CONTROL
    THERMAL CONTROL COATINGS
    THERMAL INSULATION
    THERMAL PROTECTION

HEAT SINKS
GS   SINKS
    . **HEAT SINKS**
RT   ABLATIVE MATERIALS
    ABSORBERS (MATERIALS)
    COOLING SYSTEMS
    ENDOTHERMIC REACTIONS
    ENERGY ABSORPTION
    REENTRY SHIELDING
    REGENERATORS
    THERMAL ABSORPTION
    THERMAL INSULATION

HEAT SOURCES
UF   HYDRAULIC HEATING SOURCES
GS   **HEAT SOURCES**
    . THERMAL RESOURCES
    . . GEOTHERMAL RESOURCES
    . . . GEYSERS
RT   ∞ENERGY SOURCES
    ENERGY STORAGE
    ENGINES
    GEOTHERMAL TECHNOLOGY

HEAT SOURCES-*(CONT.)*
    LASER HEATING
    LIGHT SOURCES
    ∞POWER SUPPLIES
    RADIATION SOURCES
    THERMODYNAMIC EFFICIENCY

HEAT STORAGE
UF   THERMAL ENERGY STORAGE
GS   ENERGY STORAGE
    . **HEAT STORAGE**
RT   GEOTHERMAL ENERGY UTILIZATION
    HEAT TAPES
    PHASE CHANGE MATERIALS
    SOLAR HOUSES
    TEMPERATURE
    TROMBE WALLS

HEAT STROKE
RT   BODY TEMPERATURE
    HEAT TOLERANCE
    HOT WEATHER
    HYPERTHERMIA
    PHYSIOLOGICAL EFFECTS
    THERMAL COMFORT
    THERMAL ENVIRONMENTS

HEAT TAPES
RT   HEAT STORAGE
    ICE PREVENTION
    ∞TAPES

HEAT TESTS
USE   HIGH TEMPERATURE TESTS

HEAT TOLERANCE
GS   TOLERANCES (PHYSIOLOGY)
    . **HEAT TOLERANCE**
RT   BODY TEMPERATURE
    COLD TOLERANCE
    HEAT STROKE
    HUMAN TOLERANCES

HEAT TRANSFER
SN   (TRANSMISSION ACROSS AN
    INTERFACE)
UF   NONADIABATIC PROCESSES
GS   TRANSMISSION
    . HEAT TRANSMISSION
    . . **HEAT TRANSFER**
    . . . AERODYNAMIC HEAT TRANSFER
    . . . . HYPERSONIC HEAT TRANSFER
    . . . . SUPERSONIC HEAT TRANSFER
    . . . CONDUCTIVE HEAT TRANSFER
    . . . CONVECTIVE HEAT TRANSFER
    . . . LAMINAR HEAT TRANSFER
    . . . RADIATIVE HEAT TRANSFER
    . . . TURBULENT HEAT TRANSFER
RT   ADVECTION
    ATMOSPHERIC HEAT BUDGET
    BATHS
    BIOT NUMBER
    BOILING
    BOUSSINESQ APPROXIMATION
    CHEMICAL ENGINEERING
    COMPRESSIBILITY EFFECTS
    ∞CONDUCTION
    COOLING
    COUNTERFLOW
    CRYOGENIC COOLING
    DIFFUSE RADIATION
    DIMENSIONLESS NUMBERS
    ENERGY TRANSFER
    FILM BOILING
    FILM CONDENSATION
    FLAT PLATES
    FLUID BOUNDARIES
    FORCED CONVECTION
    GAS TRANSPORT
    GAS-LIQUID INTERACTIONS
    GAS-SOLID INTERFACES
    GEOTHERMAL ENERGY CONVERSION
    GEOTHERMAL TECHNOLOGY
    HEATING
    HOT SURFACES
    LEIDENFROST PHENOMENON
    LEWIS NUMBERS
    LIQUID-LIQUID INTERFACES
    LIQUID-SOLID INTERFACES
    LIQUID-VAPOR INTERFACES
    MASS TRANSFER
    MECHANICAL ENGINEERING
    METAL VAPORS
    NONADIABATIC CONDITIONS
    NONGRAY GAS

HEAT TRANSFER-*(CONT.)*
- NONISOTHERMAL PROCESSES
- NUCLEATE BOILING
- NUSSELT NUMBER
- PECLET NUMBER
- PHASE CHANGE MATERIALS
- PRANDTL NUMBER
- RADIATIVE TRANSFER
- RAYLEIGH EQUATIONS
- REUSABLE HEAT SHIELDING
- STANTON NUMBER
- TEMPERATURE PROFILES
- TEMPERATURE RATIO
- THERMAL DIFFUSION
- THERMAL EXPANSION
- THERMAL INSULATION
- THERMAL POLLUTION
- THERMODYNAMICS
- THERMOMIGRATION
- TRANSFERRING
- TRANSPORT PROPERTIES
- WASTE ENERGY UTILIZATION

HEAT TRANSFER COEFFICIENTS
- SN   (HEAT FLUX PER UNIT AREA PER UNIT TEMPERATURE DIFFERENCE)
- GS   COEFFICIENTS
- . **HEAT TRANSFER COEFFICIENTS**
- RT   ACCOMMODATION COEFFICIENT
- EVAPORATION RATE
- HEATING
- MASS FLOW FACTORS
- NUCLEATE BOILING

HEAT TRANSMISSION
- UF   HEAT FLOW
- GS   TRANSMISSION
- . **HEAT TRANSMISSION**
- . . HEAT TRANSFER
- . . . AERODYNAMIC HEAT TRANSFER
- . . . . HYPERSONIC HEAT TRANSFER
- . . . . SUPERSONIC HEAT TRANSFER
- . . . CONDUCTIVE HEAT TRANSFER
- . . . CONVECTIVE HEAT TRANSFER
- . . . LAMINAR HEAT TRANSFER
- . . . RADIATIVE HEAT TRANSFER
- . . . TURBULENT HEAT TRANSFER
- RT   ADIABATIC EQUATIONS
- ANNULAR FLOW
- CONVECTION
- CONVECTIVE FLOW
- DUCTED FLOW
- EQUILIBRIUM FLOW
- ∞FLOW
- FLUID FLOW
- GEOPHYSICS
- GEOTHERMAL ENERGY CONVERSION
- GEOTHERMAL ENERGY EXTRACTION
- GEOTHERMAL RESOURCES
- MASS FLOW FACTORS
- NONEQUILIBRIUM FLOW
- POTENTIAL FLOW
- RADIAL FLOW
- RADIATIVE TRANSFER
- STEADY FLOW
- THERMAL ANALYSIS
- THERMAL INSULATION
- THERMOHYDRAULICS
- UNIFORM FLOW
- UNSTEADY FLOW
- WALL FLOW

HEAT TREATMENT
- GS   **HEAT TREATMENT**
- . ANNEALING
- . . LASER ANNEALING
- . . PULSE HEATING
- . MARAGING
- . NITRIDING
- . NORMALIZING (HEAT TREATMENT)
- . STRESS RELIEVING
- . TEMPERING
- RT   AGING (METALLURGY)
- ALLOYS
- BAKING
- CRITICAL TEMPERATURE
- FERRITIC STAINLESS STEELS
- FORGING
- FURNACES
- GRAPHITIZATION
- HARDENERS
- HARDENING (MATERIALS)
- HEATING
- MARTENSITE
- ∞METALLURGY

HEAT TREATMENT-*(CONT.)*
- MICROSTRUCTURE
- NUCLEATION
- PHASE DIAGRAMS
- PRECIPITATION HARDENING
- QUENCHING (COOLING)
- RECRYSTALLIZATION
- SALT BATHS
- ∞SOAKING
- STABILIZATION
- SUPERCOOLING
- SUPERSATURATION
- TEMPER (METALLURGY)
- TEMPERATURE DISTRIBUTION
- THERMOCHEMISTRY
- THERMOMECHANICAL TREATMENT
- ∞TREATMENT

∞ **HEATERS**
- SN   *(USE OF A MORE SPECIFIC TERM IS RECOMMENDED——CONSULT THE TERMS LISTED BELOW)*
- RT   DEICERS
- DEICING
- ELECTRON TUBES
- HEATING EQUIPMENT
- VAPORIZERS
- WATER HEATING

HEATING
- UF   HEAT GAIN
- PREHEATING
- REHEATING
- WARMING
- GS   **HEATING**
- . AERODYNAMIC HEATING
- . . SHOCK HEATING
- . ARC HEATING
- . ATMOSPHERIC HEATING
- . BAKING
- . BASE HEATING
- . GAS HEATING
- . INDUCTION HEATING
- . IONOSPHERIC HEATING
- . KINETIC HEATING
- . . SHOCK HEATING
- . LASER HEATING
- . MAGNETOHYDRODYNAMIC SHEAR HEATING
- . PASTEURIZING
- . PLASMA HEATING
- . . ELECTRON CYCLOTRON HEATING
- . RADIANT HEATING
- . RADIO FREQUENCY HEATING
- . RESISTANCE HEATING
- . SOLAR HEATING
- . SPACE HEATING (BUILDINGS)
- . SUPERHEATING
- . TRANSIENT HEATING
- . . PULSE HEATING
- . . SHOCK HEATING
- . WATER HEATING
- RT   AIR CONDITIONING
- ANNEALING
- AUTOCLAVING
- BOILING
- CEMENTATION
- ∞CONDUCTION
- CONVECTION
- COOLING
- DECARBURIZATION
- DEFROSTING
- ENVIRONMENTAL ENGINEERING
- GEOTHERMAL ENERGY EXTRACTION
- GEOTHERMAL ENERGY UTILIZATION
- HEAT EXCHANGERS
- HEAT GENERATION
- HEAT TRANSFER
- HEAT TRANSFER COEFFICIENTS
- HEAT TREATMENT
- HEATING EQUIPMENT
- HILSCH TUBES
- HYDROTHERMAL SYSTEMS
- ICE PREVENTION
- INTEGRATED ENERGY SYSTEMS
- JACKETS
- LASER ANNEALING
- LASER WELDING
- MELTING
- MODULAR INTEGRATED UTILITY SYSTEM
- ∞RADIATION
- ROASTING
- SINTERING
- ∞SOAKING
- TEMPERATURE

HEATING-*(CONT.)*
- TEMPERATURE CONTROL
- TEMPERATURE DISTRIBUTION
- THERMAL CYCLING TESTS
- THERMAL SHOCK
- THERMAL STRESSES
- VAPORIZING
- WASTE ENERGY UTILIZATION

HEATING EQUIPMENT
- UF   GERDIEN ARC HEATERS
- PREHEATERS
- GS   **HEATING EQUIPMENT**
- . BOILERS
- . FURNACES
- . . ELECTRIC FURNACES
- . . . IMAGE FURNACES
- . . SOLAR FURNACES
- . . VACUUM FURNACES
- . OVENS
- . VAPORIZERS
- . . EVAPORATORS
- RT   AIR CONDITIONING
- AIR CONDITIONING EQUIPMENT
- CRUCIBLES
- DEICERS
- DEICING
- ∞ELECTRIC EQUIPMENT
- ∞EQUIPMENT
- FUEL TANKS
- HEAT EXCHANGERS
- HEAT GENERATION
- HEAT PUMPS
- HEAT RADIATORS
- ∞HEATERS
- HEATING
- ONBOARD EQUIPMENT
- SPACE HEATING (BUILDINGS)
- TEMPERATURE CONTROL
- THERMAL INSULATION
- WATER HEATING

HEAVING
- RT   BENDING
- ∞BOWS
- BUCKLING
- DISPLACEMENT
- DISTORTION
- FLEXING
- ∞MOTION
- PITCH (INCLINATION)
- WARPAGE

HEAVY COSMIC RAY PRIMARIES
- USE   HEAVY NUCLEI
- PRIMARY COSMIC RAYS

HEAVY ELEMENTS
- GS   **HEAVY ELEMENTS**
- . AMERICIUM
- . . AMERICIUM ISOTOPES
- . . . AMERICIUM 241
- . TRANSURANIUM ELEMENTS
- . . AMERICIUM ISOTOPES
- . . . AMERICIUM 241
- . . BERKELIUM
- . . CALIFORNIUM
- . . . CALIFORNIUM ISOTOPES
- . . CURIUM
- . . . CURIUM ISOTOPES
- . . . . CURIUM 242
- . . . . CURIUM 244
- . . EINSTEINIUM
- . . FERMIUM
- . . LAWRENCIUM
- . . MENDELEVIUM
- . . NEPTUNIUM
- . . . NEPTUNIUM ISOTOPES
- . . NOBELIUM
- . . PLUTONIUM
- . . . PLUTONIUM ISOTOPES
- . . . . PLUTONIUM 238
- . . . . PLUTONIUM 239
- . . . . PLUTONIUM 240
- . . . . PLUTONIUM 241
- . . . . PLUTONIUM 244
- . . SERGENIUM
- RT   CHEMICAL ELEMENTS
- ∞ELEMENTS

HEAVY IONS
- GS   IONS
- . **HEAVY IONS**
- RT   ION STRIPPING
- ISOTOPE SEPARATION

HEAVY IONS-*(CONT.)*
    ISOTOPES
    LIGHT IONS

HEAVY LIFT AIRSHIPS
GS   AIRSHIPS
    . **HEAVY LIFT AIRSHIPS**
RT   MATERIALS HANDLING
    ROTORS

HEAVY LIFT HELICOPTERS
GS   V/STOL AIRCRAFT
    . ROTARY WING AIRCRAFT
    .. HELICOPTERS
    ... MILITARY HELICOPTERS
    .... **HEAVY LIFT HELICOPTERS**
    ..... CH-62 HELICOPTER
RT   AIR CARGO
    ∞AIRCRAFT
    CARGO AIRCRAFT
    HELICOPTER DESIGN

HEAVY LIFT LAUNCH VEHICLES
UF   HLLV
GS   LAUNCH VEHICLES
    . **HEAVY LIFT LAUNCH VEHICLES**
RT   ROCKET ENGINES
    ∞ROCKETS
    SPACECRAFT LAUNCHING
    ∞VEHICLES

HEAVY NUCLEI
UF   HEAVY COSMIC RAY PRIMARIES
GS   GASES
    . IONIZED GASES
    .. CHARGED PARTICLES
    ... NUCLEI (NUCLEAR PHYSICS)
    .... **HEAVY NUCLEI**
RT   PRIMARY COSMIC RAYS

HEAVY WATER
UF   DEUTERIUM OXIDES
    HYDROGEN DEUTERIUM OXIDE
GS   CHALCOGENIDES
    . OXIDES
    .. **HEAVY WATER**
    HYDROGEN COMPOUNDS
    . DEUTERIUM COMPOUNDS
    .. **HEAVY WATER**
    WATER
    . **HEAVY WATER**
RT   DEUTERIUM
    MODERATORS
    TRITIUM

HEAVY WATER COMPONENTS TEST REACTORS
GS   NUCLEAR REACTORS
    . LIQUID COOLED REACTORS
    .. WATER COOLED REACTORS
    ... HEAVY WATER REACTORS
    .... **HEAVY WATER COMPONENTS TEST REACTORS**
    . NUCLEAR RESEARCH AND TEST REACTORS
    .. **HEAVY WATER COMPONENTS TEST REACTORS**
    . WATER MODERATED REACTORS
    .. **HEAVY WATER COMPONENTS TEST REACTORS**

HEAVY WATER REACTORS
GS   NUCLEAR REACTORS
    . LIQUID COOLED REACTORS
    .. WATER COOLED REACTORS
    ... **HEAVY WATER REACTORS**
    .... HEAVY WATER COMPONENTS TEST REACTORS
    .... PLUTONIUM RECYCLE TEST REACTOR
    .... ZERO POWER REACTOR 2
RT   LIGHT WATER BREEDER REACTORS

HEF (HIGH ENERGY FUELS)
USE   HIGH ENERGY FUELS

HEIGHT
GS   DIMENSIONS
    . **HEIGHT**
    .. SCALE HEIGHT
RT   ALTITUDE
    DEPTH
    DISTANCE
    GEOPOTENTIAL
    ∞LEVEL

HEIGHT-*(CONT.)*
    SLOPES

HEINKEL AIRCRAFT
RT   ∞AIRCRAFT

HEISENBERG THEORY
GS   ATOMIC THEORY
    . **HEISENBERG THEORY**
RT   ATOMIC EXCITATIONS
    DYSON THEORY
    ENERGY TRANSFER
    ∞THEORIES

HELICAL ANTENNAS
GS   ANTENNAS
    . DIRECTIONAL ANTENNAS
    .. **HELICAL ANTENNAS**
RT   ANTENNA DESIGN
    MICROWAVE ANTENNAS

HELICAL FLOW
GS   FLUID FLOW
    . **HELICAL FLOW**
RT   AXISYMMETRIC FLOW
    FLOW GEOMETRY
    MAGNETOHYDRODYNAMIC STABILITY
    THREE DIMENSIONAL FLOW

HELICAL INDUCERS
GS   INTAKE SYSTEMS
    . **HELICAL INDUCERS**
RT   FIELD COILS
    PLASMA CONTROL

HELICAL WINDINGS
GS   WINDING
    . **HELICAL WINDINGS**

HELICOPTER ATTITUDE INDICATORS
USE   ATTITUDE INDICATORS
    HELICOPTERS

HELICOPTER CONTROL
GS   AIRCRAFT CONTROL
    . **HELICOPTER CONTROL**
RT   AIRBORNE RADAR APPROACH
    ATTITUDE CONTROL
    AUTOMATIC CONTROL
    ∞CONTROL
    CONTROLLABILITY
    DIRECTIONAL CONTROL
    FLIGHT CONTROL
    HARMONIC CONTROL
    LATERAL CONTROL
    LONGITUDINAL CONTROL
    MANUAL CONTROL
    SPEED CONTROL
    TAIL ROTORS

HELICOPTER DESIGN
GS   AIRCRAFT DESIGN
    . **HELICOPTER DESIGN**
RT   COMPOUND HELICOPTERS
    COMPUTER AIDED DESIGN
    ∞DESIGN
    ENGINE DESIGN
    HEAVY LIFT HELICOPTERS
    PRODUCT DEVELOPMENT
    ROTOR BODY INTERACTIONS
    STREAMLINING
    STRUCTURAL DESIGN
    UH-60A HELICOPTER
    UH-61A HELICOPTER
    WHIRL TOWERS

HELICOPTER ENGINES
GS   AIRCRAFT ENGINES
    . **HELICOPTER ENGINES**
    ENGINES
    . INTERNAL COMBUSTION ENGINES
    .. **HELICOPTER ENGINES**
RT   JET ENGINES
    T-53 ENGINE
    T-55 ENGINE
    T-58 ENGINE
    T-58-GE-8B ENGINE
    T-63 ENGINE
    T-64 ENGINE
    T-74 ENGINE
    T-76 ENGINE

HELICOPTER PERFORMANCE
GS   AIRCRAFT PERFORMANCE

HELICOPTER PERFORMANCE-*(CONT.)*
    . **HELICOPTER PERFORMANCE**
RT   AERODYNAMIC STABILITY
    AIRCRAFT RELIABILITY
    CONTROLLABILITY
    FLIGHT CHARACTERISTICS
    MANEUVERABILITY

HELICOPTER PROPELLER DRIVE
GS   MECHANICAL DRIVES
    . **HELICOPTER PROPELLER DRIVE**
RT   JET PROPULSION
    ROTARY WINGS
    TILTED PROPELLERS
    VARIABLE PITCH PROPELLERS

HELICOPTER ROTORS
USE   ROTARY WINGS

HELICOPTER TAIL ROTORS
GS   ROTATING BODIES
    . ROTORS
    .. TAIL ROTORS
    ... **HELICOPTER TAIL ROTORS**
RT   ROTARY WINGS
    ∞ROTOR BLADES

HELICOPTER WAKES
GS   WAKES
    . AIRCRAFT WAKES
    .. **HELICOPTER WAKES**
RT   DOWNWASH

HELICOPTERS
UF   DRONE HELICOPTERS
    GYROPLANES
    HELICOPTER ATTITUDE INDICATORS
GS   V/STOL AIRCRAFT
    . ROTARY WING AIRCRAFT
    .. **HELICOPTERS**
    ... ALOUETTE HELICOPTERS
    .... SA-330 HELICOPTER
    .... SE-3160 HELICOPTER
    ... BELL 214A HELICOPTER
    ... BO-105 HELICOPTER
    ... CH-21 HELICOPTER
    ... COMPOUND HELICOPTERS
    ... H-17 HELICOPTER
    ... H-54 HELICOPTER
    ... HC-3 HELICOPTER
    ... MILITARY HELICOPTERS
    .... AH-1G HELICOPTER
    .... AH-64 HELICOPTER
    .... CH-3 HELICOPTER
    .... CH-34 HELICOPTER
    .... CH-46 HELICOPTER
    .... CH-47 HELICOPTER
    .... CH-54 HELICOPTER
    .... H-43 HELICOPTER
    .... H-53 HELICOPTER
    .... H-56 HELICOPTER
    .... H-60 HELICOPTER
    .... HEAVY LIFT HELICOPTERS
    ..... CH-62 HELICOPTER
    .... HH-43 HELICOPTER
    .... OH-4 HELICOPTER
    .... OH-5 HELICOPTER
    .... OH-6 HELICOPTER
    .... OH-13 HELICOPTER
    .... OH-23 HELICOPTER
    .... OH-58 HELICOPTER
    .... QH-50 HELICOPTER
    .... S-58 HELICOPTER
    .... S-61 HELICOPTER
    .... SA-330 HELICOPTER
    .... UH-1 HELICOPTER
    .... UH-2 HELICOPTER
    .... UH-34 HELICOPTER
    .... UH-60A HELICOPTER
    .... UH-61A HELICOPTER
    ... P-531 HELICOPTER
    ... RIGID ROTOR HELICOPTERS
    .... CH-3 HELICOPTER
    .... F-28 HELICOPTER
    .... XH-51 HELICOPTER
    ... SH-3 HELICOPTER
    ... SH-4 HELICOPTER
    ... SIKORSKY WHIRLWIND HELICOPTER
    ... TANDEM ROTOR HELICOPTERS
    .... CH-46 HELICOPTER
    .... CH-47 HELICOPTER
    .... H-25 HELICOPTER
    .... TH-55 HELICOPTER
    ... WESTLAND WHIRLWIND HELICOPTER

HELICOPTERS-*(CONT.)*
```
      . . . XV-9A AIRCRAFT
RT    AIRBORNE RADAR APPROACH
      ∞AIRCRAFT
      AIRCRAFT SURVIVABILITY
      BLADE SLAP NOISE
      GENERAL AVIATION AIRCRAFT
      GROUND RESONANCE
      HELIPORTS
      ∞MILITARY AIRCRAFT
      NAP-OF-THE-EARTH NAVIGATION
      RECOVERY VEHICLES
      ROTOR SYSTEMS RESEARCH AIRCRAFT
      SHORT TAKEOFF AIRCRAFT
      ∞SUBSONIC AIRCRAFT
      TILT ROTOR AIRCRAFT
      TILT ROTOR RESEARCH AIRCRAFT
         PROGRAM
      UTILITY AIRCRAFT
      VERTICAL TAKEOFF AIRCRAFT
      WESER AIRCRAFT
      WESTLAND AIRCRAFT
      XV-15 AIRCRAFT
```

HELIO AIRCRAFT
```
UF    HELIO MILITARY AIRCRAFT
GS    HELIO AIRCRAFT
      . U-10 AIRCRAFT
RT    ∞AIRCRAFT
```

HELIO MILITARY AIRCRAFT
```
USE   HELIO AIRCRAFT
```

HELIOCENTRIC ORBITS
```
USE   SOLAR ORBITS
```

HELIOGRAPHS
```
USE   SPECTROHELIOGRAPHS
```

HELIOGRAPHY
```
USE   SPECTROHELIOGRAPHS
```

HELIOMAGNETISM
```
USE   SOLAR MAGNETIC FIELD
```

HELIOMETERS
```
UF    HELIOMETRY
GS    MEASURING INSTRUMENTS
      . HELIOMETERS
      . . PYROHELIOMETERS
      OPTICAL EQUIPMENT
      . ASTRONOMICAL TELESCOPES
      . . HELIOMETERS
      . . . PYROHELIOMETERS
      TELESCOPES
      . ASTRONOMICAL TELESCOPES
      . . HELIOMETERS
      . . . PYROHELIOMETERS
```

HELIOMETRY
```
USE   HELIOMETERS
      PYROHELIOMETERS
```

HELIOS A
```
GS    SATELLITES
      . ARTIFICIAL SATELLITES
      . . HELIOS SATELLITES
      . . . HELIOS A
      UNMANNED SPACECRAFT
      . SPACE PROBES
      . . SOLAR PROBES
      . . . HELIOS A
```

HELIOS B
```
GS    SATELLITES
      . ARTIFICIAL SATELLITES
      . . HELIOS SATELLITES
      . . . HELIOS B
      UNMANNED SPACECRAFT
      . SPACE PROBES
      . . SOLAR PROBES
      . . . HELIOS B
```

HELIOS PROJECT
```
GS    PROGRAMS
      . NASA PROGRAMS
      . . HELIOS PROJECT
      . NASA SPACE PROGRAMS
      . . HELIOS PROJECT
      . PROJECTS
      . . HELIOS PROJECT
RT    CHARGED PARTICLES
      HIGH TEMPERATURE PLASMAS
      SOLAR PROBES
```

HELIOS PROJECT-*(CONT.)*
```
      ZODIACAL LIGHT
```

HELIOS SATELLITES
```
GS    SATELLITES
      . ARTIFICIAL SATELLITES
      . . HELIOS SATELLITES
      . . . HELIOS A
      . . . HELIOS B
      . . . HELIOS 1
      . . . HELIOS 2
RT    MAGNETIC FIELDS
      PARTICLE FLUX DENSITY
      SOLAR FLUX DENSITY
```

HELIOS 1
```
GS    SATELLITES
      . ARTIFICIAL SATELLITES
      . . HELIOS SATELLITES
      . . . HELIOS 1
      UNMANNED SPACECRAFT
      . SPACE PROBES
      . . SOLAR PROBES
      . . . HELIOS 1
```

HELIOS 2
```
GS    SATELLITES
      . ARTIFICIAL SATELLITES
      . . HELIOS SATELLITES
      . . . HELIOS 2
      UNMANNED SPACECRAFT
      . SPACE PROBES
      . . SOLAR PROBES
      . . . HELIOS 2
```

HELIOSEISMOLOGY
```
UF    SOLAR DYNAMICS
      SOLAR SEISMOLOGY
GS    SEISMOLOGY
      . HELIOSEISMOLOGY
RT    ASTROPHYSICS
      ∞SCIENCE
      SOLAR PHYSICS
```

HELIOSPHERE
```
RT    COSMIC RAYS
      INTERPLANETARY SPACE
      INTERSTELLAR GAS
      SOLAR ACTIVITY EFFECTS
      SOLAR WIND
```

HELIOSTATS
```
RT    ∞INSTRUMENTS
      MIRRORS
      REFLECTORS
      SERVOMOTORS
      SOLAR REFLECTORS
      SYNCHRONIZERS
```

HELIPORTS
```
GS    AIRPORTS
      . HELIPORTS
RT    AIR TRAFFIC CONTROL
      AIRPORT PLANNING
      AIRPORT TOWERS
      HANGARS
      HELICOPTERS
      LANDING AIDS
      LANDING SITES
      MILITARY AIR FACILITIES
      NAVIGATION AIDS
      SOLAR COMPASSES
      V/STOL AIRCRAFT
```

HELITRONS
```
GS    ELECTRON TUBES
      . VACUUM TUBES
      . . VACUUM TUBE OSCILLATORS
      . . . MICROWAVE TUBES
      . . . . PLANOTRONS
      . . . . . HELITRONS
      . . . . TRAVELING WAVE TUBES
      . . . . . HELITRONS
      MICROWAVE EQUIPMENT
      . MICROWAVE TUBES
      . . PLANOTRONS
      . . . HELITRONS
      . . TRAVELING WAVE TUBES
      . . . HELITRONS
      OSCILLATORS
      . VACUUM TUBE OSCILLATORS
      . . MICROWAVE TUBES
      . . . PLANOTRONS
      . . . . HELITRONS
RT    CARCINOTRONS
```

HELIUM
```
GS    CHEMICAL ELEMENTS
      . RARE GASES
      . . HELIUM
      . . . HELIUM ATOMS
      . . . HELIUM FILM
      . . . HELIUM ISOTOPES
      . . . LIQUID HELIUM
      . . . . LIQUID HELIUM 2
      GASES
      . RARE GASES
      . . HELIUM
      . . . HELIUM ATOMS
      . . . HELIUM FILM
      . . . HELIUM ISOTOPES
      . . . LIQUID HELIUM
      . . . . LIQUID HELIUM 2
RT    ALPHA PARTICLES
      HELIUM AFTERGLOW
      WOLF-RAYET STARS
```

HELIUM AFTERGLOW
```
GS    AFTERGLOWS
      . HELIUM AFTERGLOW
RT    GAS IONIZATION
      PLASMA DECAY
```

HELIUM ATOMS
```
GS    ATOMS
      . HELIUM ATOMS
      CHEMICAL ELEMENTS
      . RARE GASES
      . . HELIUM
      . . . HELIUM ATOMS
      GASES
      . RARE GASES
      . . HELIUM
      . . . HELIUM ATOMS
```

HELIUM COMPOUNDS
```
RT    ∞RARE GAS COMPOUNDS
```

HELIUM FILM
```
GS    CHEMICAL ELEMENTS
      . RARE GASES
      . . HELIUM
      . . . HELIUM FILM
      GASES
      . RARE GASES
      . . HELIUM
      . . . HELIUM FILM
RT    ∞FILMS
```

HELIUM HYDROGEN ATMOSPHERES
```
GS    ENVIRONMENTS
      . EXTRATERRESTRIAL ENVIRONMENTS
      . . PLANETARY ENVIRONMENTS
      . . . PLANETARY ATMOSPHERES
      . . . . HELIUM HYDROGEN
              ATMOSPHERES
```

HELIUM IONS
```
GS    IONS
      . HELIUM IONS
RT    ALPHA PARTICLES
```

HELIUM ISOTOPES
```
UF    HELIUM 2
      HELIUM 3
      HELIUM 4
GS    CHEMICAL ELEMENTS
      . NUCLIDES
      . . ISOTOPES
      . . . HELIUM ISOTOPES
      . RARE GASES
      . . HELIUM
      . . . HELIUM ISOTOPES
      GASES
      . RARE GASES
      . . HELIUM
      . . . HELIUM ISOTOPES
```

HELIUM PLASMA
```
GS    GASES
      . IONIZED GASES
      . . CHARGED PARTICLES
      . . . HELIUM PLASMA
      PARTICLES
      . CHARGED PARTICLES
      . . ENERGETIC PARTICLES
      . . . PLASMAS (PHYSICS)
      . . . . HELIUM PLASMA
RT    ARGON PLASMA
      ELECTRON PLASMA
      HYDROGEN PLASMA
```

HELIUM PLASMA-*(CONT.)*
  OXYGEN PLASMA

HELIUM STARS
  USE   B STARS

HELIUM 2
  USE   HELIUM ISOTOPES
         LIQUID HELIUM

HELIUM 3
  USE   HELIUM ISOTOPES

HELIUM 4
  USE   HELIUM ISOTOPES

HELIUM-NEON LASERS
  GS   STIMULATED EMISSION DEVICES
         . LASERS
         . . GAS LASERS
         . . . **HELIUM-NEON LASERS**
  RT   LASER MODES
         LASER OUTPUTS

HELIUM-OXYGEN ATMOSPHERES
  GS   CONTROLLED ATMOSPHERES
         . **HELIUM-OXYGEN ATMOSPHERES**
  RT   AEROSPACE ENVIRONMENTS
         ∞ATMOSPHERES
         ∞BREATHING
         GAS MIXTURES
         PORTABLE LIFE SUPPORT SYSTEMS
         UNDERWATER BREATHING APPARATUS

HELIX TUBES
  USE   TRAVELING WAVE TUBES

HELIXES
  USE   CURVES (GEOMETRY)

HELLMANN-FEYNMAN THEOREM
  GS   THEOREMS
         . **HELLMANN-FEYNMAN THEOREM**

HELMET MOUNTED DISPLAYS
  GS   DISPLAY DEVICES
         . **HELMET MOUNTED DISPLAYS**
  RT   CREW STATIONS
         ∞DETECTORS
         IMAGES
         INDICATING INSTRUMENTS
         ∞INSTRUMENTS
         MONITORS
         PERSONNEL

HELMETS
  GS   CLOTHING
         . PROTECTIVE CLOTHING
         . . **HELMETS**
         SAFETY DEVICES
         . **HELMETS**
  RT   ARMOR
         FLIGHT CLOTHING
         GOGGLES
         PRESSURE SUITS

∞ **HELMHOLTZ EQUATIONS**
  SN   *(USE OF A MORE SPECIFIC TERM IS*
         RECOMMENDED--CONSULT THE TERMS
         LISTED BELOW)
  RT   ∞EQUATIONS
         HELMHOLTZ VORTICITY EQUATION
         KELVIN-HELMHOLTZ INSTABILITY
         TIME DEPENDENCE
         WAVE EQUATIONS

HELMHOLTZ RESONATORS
  GS   RESONATORS
         . **HELMHOLTZ RESONATORS**
  RT   CAVITY RESONATORS
         NOISE REDUCTION

HELMHOLTZ VORTICITY EQUATION
  GS   ANALYSIS (MATHEMATICS)
         . REAL VARIABLES
         . . DIFFERENTIAL EQUATIONS
         . . . PARTIAL DIFFERENTIAL EQUATIONS
         . . . . **HELMHOLTZ VORTICITY**
                 EQUATION
         EQUATIONS OF MOTION
         . KINETIC EQUATIONS
         . . HYDRODYNAMIC EQUATIONS
         . . . **HELMHOLTZ VORTICITY EQUATION**
         FLOW EQUATIONS

HELMHOLTZ VORTICITY EQUATION-*(CONT.)*
         . **HELMHOLTZ VORTICITY EQUATION**
  RT   ∞EQUATIONS
         ∞HELMHOLTZ EQUATIONS
         VORTICITY

HELOS (SATELLITE)
  USE   EXOSAT SATELLITE

HEMATITE
  GS   CHALCOGENIDES
         . OXIDES
         . . METAL OXIDES
         . . . IRON OXIDES
         . . . . **HEMATITE**
         IRON COMPOUNDS
         . IRON OXIDES
         . . **HEMATITE**
         MINERALS
         . IRON ORES
         . . **HEMATITE**

HEMATOCRIT
  RT   BLOOD

HEMATOCRIT RATIO
  RT   ANEMIAS
         ERYTHROCYTES
         HEMATOLOGY

HEMATOLOGY
  RT   CARBOXYHEMOGLOBIN TEST
         HEMATOCRIT RATIO
         RETICULOCYTES

HEMATOPOIESIS
  GS   . ANATOMY
         . CARDIOVASCULAR SYSTEM
         . . **HEMATOPOIESIS**
         CELLS (BIOLOGY)
         . **HEMATOPOIESIS**
  RT   BLOOD
         PHYSIOLOGICAL EFFECTS
         RADIATION EFFECTS

HEMATOPOIETIC SYSTEM
  GS   ANATOMY
         . CARDIOVASCULAR SYSTEM
         . . **HEMATOPOIETIC SYSTEM**
         PHYSIOLOGY
         . **HEMATOPOIETIC SYSTEM**
  RT   BLOOD VOLUME
         PHYSIOLOGICAL EFFECTS
         ∞SYSTEMS

HEMATURIA
  GS   SIGNS AND SYMPTOMS
         . **HEMATURIA**
  RT   URINE

HEMISPHERE CYLINDER BODIES
  RT   ∞CYLINDERS
         CYLINDRICAL BODIES
         HEMISPHERICAL SHELLS
         PRESSURE VESSELS

∞ **HEMISPHERES**
  SN   *(USE OF A MORE SPECIFIC TERM IS*
         RECOMMENDED--CONSULT THE TERMS
         LISTED BELOW)
  RT   AERODYNAMIC CONFIGURATIONS
         BODIES OF REVOLUTION
         EASTERN HEMISPHERE
         HEMISPHERICAL SHELLS
         NORTHERN HEMISPHERE
         SOUTHERN HEMISPHERE
         SPHERES

HEMISPHERICAL SHELLS
  GS   SHELLS (STRUCTURAL FORMS)
         . **HEMISPHERICAL SHELLS**
  RT   BODIES OF REVOLUTION
         CIRCULAR SHELLS
         DOMES (STRUCTURAL FORMS)
         HEMISPHERE CYLINDER BODIES
         ∞HEMISPHERES
         METAL SHELLS
         REINFORCED SHELLS
         SPHERES
         SPHERICAL SHELLS

HEMOCYTES
  GS   CELLS (BIOLOGY)
         . **HEMOCYTES**

HEMOCYTES-*(CONT.)*
  RT   INVERTEBRATES

HEMODYNAMIC RESPONSES
  GS   PHYSIOLOGICAL EFFECTS
         . PHYSIOLOGICAL RESPONSES
         . . **HEMODYNAMIC RESPONSES**
         RESPONSES
         . PHYSIOLOGICAL RESPONSES
         . . **HEMODYNAMIC RESPONSES**
  RT   BLOOD CIRCULATION
         BLOOD PRESSURE

HEMODYNAMICS
  GS   **HEMODYNAMICS**
         . LOWER BODY NEGATIVE PRESSURE
  RT   BLOOD CIRCULATION
         BLOOD VOLUME
         CARDIOVASCULAR SYSTEM
         ∞DYNAMICS
         HEMOPERFUSION

HEMOGLOBIN
  GS   CELLS (BIOLOGY)
         . **HEMOGLOBIN**
         . . CARBOXYHEMOGLOBIN
         . . OXYHEMOGLOBIN
         ORGANOMETALLIC COMPOUNDS
         . **HEMOGLOBIN**
         . . CARBOXYHEMOGLOBIN
         . . OXYHEMOGLOBIN
  RT   ANEMIAS
         BLOOD
         CORPUSCLES
         ERYTHROCYTES
         HEMOLYSIS
         POLYCYTHEMIA
         PORPHINES
         PORPHYRINS
         RETICULOCYTES

HEMOLYSIS
  RT   COMPLEMENT (BIOLOGY)
         ERYTHROCYTES
         HEMOGLOBIN
         POLYCYTHEMIA
         RETICULOCYTES

HEMOPERFUSION
  RT   ACTIVATED CARBON
         ADSORBENTS
         BLOOD FLOW
         BLOOD PRESSURE
         HEMODYNAMICS
         TOXICOLOGY

HEMORRHAGES
  GS   **HEMORRHAGES**
         . PETECHIA
  RT   ∞BLEEDING
         BLOOD
         CARDIOVASCULAR SYSTEM
         COAGULATION
         HEMOSTATICS
         HYPOTENSION
         INJURIES
         PATHOLOGY
         POLYCYTHEMIA
         RETICULOCYTES

HEMOSTASIS
  USE   HEMOSTATICS

HEMOSTATICS
  UF   HEMOSTASIS
  GS   DRUGS
         . **HEMOSTATICS**
  RT   BLOOD COAGULATION
         FIBRINOGEN
         GIBBERELLINS
         HEMORRHAGES
         THROMBIN
         THROMBOPLASTIN

HENRY LAW
  RT   COMPOSITION (PROPERTY)
         PARTIAL PRESSURE
         RAOULT LAW
         SOLUBILITY
         SOLUTIONS
         VAPOR PRESSURE

HEOS A SATELLITE
  GS   SATELLITES

HEOS A SATELLITE-*(CONT.)*
. ARTIFICIAL SATELLITES
. . ESA SATELLITES
. . . HEOS SATELLITES
. . . . **HEOS A SATELLITE**

HEOS B SATELLITE
GS   SATELLITES
. ARTIFICIAL SATELLITES
. . ESA SATELLITES
. . . HEOS SATELLITES
. . . . **HEOS B SATELLITE**

HEOS SATELLITES
UF   HIGHLY ECCENTRIC ORBIT SATELLITES
GS   SATELLITES
. ARTIFICIAL SATELLITES
. . ESA SATELLITES
. . . **HEOS SATELLITES**
. . . . HEOS A SATELLITE
. . . . HEOS B SATELLITE
RT   EUROPEAN SPACE PROGRAMS
SOLAR ORBITS

HEPARINS
RT   ANTICOAGULANTS

HEPATITIS
GS   DISEASES
. INFECTIOUS DISEASES
. . **HEPATITIS**
RT   ANATOMY
LIVER

HEPTADIENE
GS   ALIPHATIC COMPOUNDS
. **HEPTADIENE**
HYDROCARBONS
. DIENES
. . **HEPTADIENE**

HEPTANES
GS   ALIPHATIC COMPOUNDS
. ALKANES
. . **HEPTANES**
HYDROCARBONS
. ALKANES
. . **HEPTANES**
RT   HYDROCARBON FUELS

HERBICIDES
RT   BRUSH (BOTANY)
DEFOLIANTS
FOLIAGE
FORESTS
LEAVES
PLANTS (BOTANY)
TOXICITY
TREES (PLANTS)

HERBIG-HARO OBJECTS
GS   CELESTIAL BODIES
. STARS
. . **HERBIG-HARO OBJECTS**
RT   B STARS
∞BODIES
INFRARED STARS
STELLAR RADIATION
STELLAR SPECTRA
T TAURI STARS

HERCULES AIRCRAFT
USE   C-130 AIRCRAFT

HERCULES ENGINE
GS   ENGINES
. ROCKET ENGINES
. . SOLID PROPELLANT ROCKET
    ENGINES
. . . **HERCULES ENGINE**
RT   HONEST JOHN ROCKET VEHICLE
LITTLE JOHN ROCKET VEHICLE

HERCULES NOVA
GS   CELESTIAL BODIES
. STARS
. . VARIABLE STARS
. . . NOVAE
. . . . **HERCULES NOVA**
RT   DWARF NOVAE

HEREDITY
RT   BREEDING (REPRODUCTION)
CONGENITAL ANOMALIES

HEREDITY-*(CONT.)*
CYTOGENESIS
EVOLUTION (DEVELOPMENT)

HERING-BREVER REFLEX
GS   REFLEXES
. RESPIRATORY REFLEXES
. . **HERING-BREVER REFLEX**
RT   HEART RATE

HERMES MANNED SPACEPLANE
GS   MANNED SPACECRAFT
. SPACE SHUTTLES
. . **HERMES MANNED SPACEPLANE**
REENTRY VEHICLES
. RECOVERABLE SPACECRAFT
. . REUSABLE SPACECRAFT
. . . **HERMES MANNED SPACEPLANE**
RT   FRENCH SPACE PROGRAMS
SPACE TRANSPORTATION SYSTEM
SPACECRAFT DESIGN

HERMES SATELLITE
USE   COMMUNICATIONS TECHNOLOGY
    SATELLITE

HERMETIC SEALS
GS   SEALS (STOPPERS)
. **HERMETIC SEALS**
RT   ∞HEADERS
LABYRINTH SEALS
O RING SEALS
PUMP SEALS

HERMITIAN POLYNOMIAL
GS   ALGEBRA
. POLYNOMIALS
. . **HERMITIAN POLYNOMIAL**
RT   JACOBI MATRIX METHOD
MATRICES (MATHEMATICS)
REAL VARIABLES
VECTOR SPACES

HERO REACTOR
GS   NUCLEAR REACTORS
. NUCLEAR RESEARCH AND TEST
    REACTORS
. . **HERO REACTOR**

HERTZSPRUNG-RUSSELL DIAGRAM
UF   HR DIAGRAM
GS   DIAGRAMS
. **HERTZSPRUNG-RUSSELL DIAGRAM**
RT   COLOR-MAGNITUDE DIAGRAM
HORIZONTAL BRANCH STARS
STELLAR EVOLUTION
STELLAR LUMINOSITY
STELLAR SPECTRA

HERZBERG BANDS
GS   SPECTRA
. ABSORPTION SPECTRA
. . **HERZBERG BANDS**
. RADIATION SPECTRA
. . **HERZBERG BANDS**
. SPECTRAL BANDS
. . **HERZBERG BANDS**
RT   ∞BANDS
OXYGEN SPECTRA
SCHUMANN-RUNGE BANDS
ULTRAVIOLET SPECTRA

HESSIAN MATRICES
RT   ALGORITHMS
MATRICES (MATHEMATICS)
OPTIMIZATION

HET EXPERIMENT
UF   HEALTH-EDUCATION
    TELECOMMUNICATIONS EXP
GS   TELECOMMUNICATION
. **HET EXPERIMENT**
TELECONFERENCING
. **HET EXPERIMENT**
RT   ATS 6
COMMUNICATION SATELLITES
SATELLITE NETWORKS
SATELLITES

HETEROCYCLIC COMPOUNDS
GS   **HETEROCYCLIC COMPOUNDS**
. ACRIFLAVINE
. ADENINES
. ADENOSINES

HETEROCYCLIC COMPOUNDS-*(CONT.)*
. . ADENOSINE DIPHOSPHATE
. . ADENOSINE TRIPHOSPHATE
. . CYCLIC AMP
. ALKALOIDS
. . ATROPINE
. . BETAINES
. . COLCHICINE
. . ERGOTAMINE
. . HYOSCINE
. . LYSERGINE
. . MORPHINE
. . NICOTINAMIDE
. . NICOTINE
. . PILOCARPINE
. . RESERPINE
. . TROPYL COMPOUNDS
. ALLOXAN
. ANISOLE
. ASCORBIC ACID
. AZINES
. . CYANURATES
. . CYANURIC ACID
. . MECLIZINE
. . METHYLENE BLUE
. . PHENOTHIAZINES
. AZOLES
. . ACETAZOLAMIDE
. . OXAZOLE
. . PYRROLES
. . . CARBAZOLES
. . . INDOLES
. . . . TRYPTOPHAN
. AZULENE
. BIOFLAVONOIDS
. BIOTIN
. CARNITINE
. CYANOCOBALAMIN
. CYTIDYLIC ACID
. DIMENHYDRINATE
. ENDRIN
. FOLIC ACID
. FURANS
. . TETRAHYDROFURAN
. GUANETHIDINE
. NICOTINIC ACID
. PHTHALOCYANIN
. PHYLLOQUINONE
. PIPERIDINE
. PROMETHAZINE
. PYRIDOXINE
. RDX
. RETINENE
. RIBOFLAVIN
. TETRACYCLINES
. TETRAZOLES
. THIAMINE
. THIAZINE (TRADEMARK)
. THYMIDINE
. THYMINE
. TOCOPHEROL
. TRIMETHADIONE
. URACIL
. XANTHINES
. . CAFFEINE
. . GUANINES
. . URIC ACID
RT   ∞CHEMICAL COMPOUNDS

HETERODYNING
GS   **HETERODYNING**
. OPTICAL HETERODYNING
RT   AUTODYNES
DEMODULATION
INTERMEDIATE FREQUENCY AMPLIFIERS
MIXING CIRCUITS
SUPERHETERODYNE RECEIVERS

HETEROGENEITY
RT   DEVIATION
IMPURITIES
INCLUSIONS
INHOMOGENEITY
RANGE (EXTREMES)
SAMPLING
STANDARD DEVIATION
STATISTICAL TESTS
VARIABILITY
VARIANCE (STATISTICS)

HETEROJUNCTION DEVICES
GS   ELECTRONIC EQUIPMENT
. SOLID STATE DEVICES
. . SEMICONDUCTOR DEVICES
. . . **HETEROJUNCTION DEVICES**

HETEROJUNCTION DEVICES-(CONT.)
```
       . . . . HIGH ELECTRON MOBILITY
                  TRANSISTORS
RT     BAND STRUCTURE OF SOLIDS
       DISTRIBUTED FEEDBACK LASERS
       GALLIUM ARSENIDES
       HOMOJUNCTIONS
       JUNCTION DIODES
       QUANTUM EFFICIENCY
       QUANTUM WELLS
       SEMICONDUCTOR JUNCTIONS
       SOLAR ENERGY CONVERSION
       WAVEGUIDE LASERS
```

HETEROJUNCTIONS
```
RT     HOMOJUNCTIONS
       JUNCTION DIODES
       QUANTUM WELLS
       SEMICONDUCTOR JUNCTIONS
       SILICON JUNCTIONS
       SOLAR CELLS
       THIN FILMS
```

HETEROPHORIA
```
UF     EXOPHORIA
RT     VISION
```

HETEROSPHERE
```
GS     EARTH ATMOSPHERE
       . HETEROSPHERE
       ENVIRONMENTS
       . HETEROSPHERE
RT     CHEMOSPHERE
       EXOSPHERE
       IONOSPHERE
       LOWER ATMOSPHERE
       MAGNETOSPHERE
       MIDDLE ATMOSPHERE
       THERMOSPHERE
       UPPER ATMOSPHERE
```

HETEROTROPHS
```
GS     ANIMALS
       . HETEROTROPHS
RT     AUTOTROPHS
       METABOLISM
       PLANTS (BOTANY)
```

HEURISTIC METHODS
```
RT     AUTOMATA THEORY
       COMPUTER PROGRAMMING
       ∞METHODOLOGY
       SIMULATION
```

HEUS ROCKET ENGINES
```
GS     ENGINES
       . ROCKET ENGINES
       . . HEUS ROCKET ENGINES
RT     ROCKET ENGINE CONTROL
       ROCKET VEHICLES
```

HEWLETT-PACKARD COMPUTERS
```
GS     DATA PROCESSING EQUIPMENT
       . COMPUTERS
       . . DIGITAL COMPUTERS
       . . . HEWLETT-PACKARD COMPUTERS
```

HEXADIENE
```
GS     ALIPHATIC COMPOUNDS
       . HEXADIENE
       HYDROCARBONS
       . DIENES
       . . HEXADIENE
```

HEXAGONAL CELLS
```
RT     ∞CELLS
       CRYSTAL LATTICES
       HONEYCOMB STRUCTURES
```

HEXAGONS
```
GS     GEOMETRY
       . EUCLIDEAN GEOMETRY
       . . POLYGONS
       . . . HEXAGONS
```

HEXAHEDRITE
```
GS     MAGNESIUM COMPOUNDS
       . MAGNESIUM SULFATES
       . . HEXAHEDRITE
       MINERALS
       . HEXAHEDRITE
       SULFUR COMPOUNDS
       . SULFATES
       . . MAGNESIUM SULFATES
```

HEXAHEDRITE-(CONT.)
```
       . . . HEXAHEDRITE
```

HEXAMETHONIUM
```
RT     AMMONIUM COMPOUNDS
       ANTICONVULSANTS
```

HEXAMETHYLENETETRAMINE
```
GS     ALIPHATIC COMPOUNDS
       . HEXAMETHYLENETETRAMINE
       AMINES
       . HEXAMETHYLENETETRAMINE
```

HEXANITROSTILBENE
```
UF     HNST
RT     STILBENE
```

HEXENES
```
GS     ALIPHATIC COMPOUNDS
       . ALKANES
       . . HEXENES
       . ALKENES
       . . HEXENES
       HYDROCARBONS
       . ALKANES
       . . HEXENES
       . ALKENES
       . . HEXENES
RT     CYCLOHEXANE
       HYDROCARBON FUELS
```

HEXOGENES (TRADEMARK)
```
GS     ALIPHATIC COMPOUNDS
       . HEXOGENES (TRADEMARK)
```

HEXOKINASE
```
GS     ALIPHATIC COMPOUNDS
       . HEXOKINASE
       CARBOHYDRATES
       . HEXOKINASE
       ENZYMES
       . HEXOKINASE
```

HEXOSES
```
GS     ALIPHATIC COMPOUNDS
       . SUGARS
       . . HEXOSES
       CARBOHYDRATES
       . SUGARS
       . . HEXOSES
```

HEXYL COMPOUNDS
```
GS     ALIPHATIC COMPOUNDS
       . ALKYL COMPOUNDS
       . . HEXYL COMPOUNDS
RT     ∞CHEMICAL COMPOUNDS
```

HF LASERS
```
GS     STIMULATED EMISSION DEVICES
       . LASERS
       . . GAS LASERS
       . . . HF LASERS
RT     CHEMICAL LASERS
       INFRARED LASERS
       TEA LASERS
```

HFB-320 AIRCRAFT
```
UF     HAMBURGER HFB-320 AIRCRAFT
GS     HAMBURGER AIRCRAFT
       . HFB-320 AIRCRAFT
       JET AIRCRAFT
       . HFB-320 AIRCRAFT
       MONOPLANES
       . HFB-320 AIRCRAFT
       PASSENGER AIRCRAFT
       . HFB-320 AIRCRAFT
       TRANSPORT AIRCRAFT
       . HFB-320 AIRCRAFT
RT     ∞AIRCRAFT
```

HFIR
```
USE    HIGH FLUX ISOTOPE REACTORS
```

HFIR (REACTOR)
```
USE    HIGH FLUX ISOTOPE REACTORS
```

HH-43 HELICOPTER
```
UF     HH-43B HELICOPTER
       HUSKIE HELICOPTER
GS     KAMAN AIRCRAFT
       . HH-43 HELICOPTER
       UTILITY AIRCRAFT
       . HH-43 HELICOPTER
       V/STOL AIRCRAFT
```

HH-43 HELICOPTER-(CONT.)
```
       . ROTARY WING AIRCRAFT
       . . HELICOPTERS
       . . . MILITARY HELICOPTERS
       . . . . HH-43 HELICOPTER
```

HH-43B HELICOPTER
```
USE    HH-43 HELICOPTER
```

HHX HELICOPTER
```
USE    H-53 HELICOPTER
```

HIBERNATION
```
RT     ADAPTATION
       THERMOREGULATION
```

HICAT (RADAR TECHNIQUE)
```
USE    HIGH RESOLUTION COVERAGE
          ANTENNAS
```

HICAT PROJECT
```
USE    HIGH RESOLUTION COVERAGE
          ANTENNAS
```

HIERARCHIES
```
GS     CLASSIFICATIONS
       . HIERARCHIES
       . . BBGKY HIERARCHY
       . . DICHOTOMIES
```

HIGH ACCELERATION
```
GS     RATES (PER TIME)
       . ACCELERATION (PHYSICS)
       . . HIGH ACCELERATION
RT     ∞ACCELERATION
       ACCELERATION STRESSES
          (PHYSIOLOGY)
       ACCELERATION TOLERANCE
       ELECTRON RUNAWAY (PLASMA
          PHYSICS)
       MECHANICAL SHOCK
       ∞MOTION
       SHOCK RESISTANCE
```

**HIGH ALT TARGET AND BACKGROUND
MEASUREMENT**
```
UF     HITAB PROGRAM
RT     ∞MEASUREMENT
       TARGET ACQUISITION
```

HIGH ALTITUDE
```
UF     HIGH ALTITUDE FLIGHT
GS     ALTITUDE
       . HIGH ALTITUDE
RT     MIDALTITUDE
       SKYHOOK BALLOONS
       UPPER ATMOSPHERE
```

HIGH ALTITUDE BALLOONS
```
GS     EXPANDABLE STRUCTURES
       . INFLATABLE STRUCTURES
       . . BALLOONS
       . . . HIGH ALTITUDE BALLOONS
       . . . . JIMSPHERE BALLOONS
       . . . . SKYHOOK BALLOONS
       . . . . SUPERPRESSURE BALLOONS
RT     BALLOON-BORNE INSTRUMENTS
       GAS BAGS
       METEOROLOGICAL BALLOONS
       ROBIN BALLOONS
       ROCKOONS
```

HIGH ALTITUDE BREATHING
```
GS     RESPIRATION
       . HIGH ALTITUDE BREATHING
RT     ALTITUDE TOLERANCE
       ∞BREATHING
       EMERGENCY LIFE SUSTAINING
          SYSTEMS
       HYPOBARIC ATMOSPHERES
       OXYGEN MASKS
```

HIGH ALTITUDE ENVIRONMENTS
```
GS     ENVIRONMENTS
       . HIGH ALTITUDE ENVIRONMENTS
RT     ALTITUDE SIMULATION
       ALTITUDE TESTS
       ALTITUDE TOLERANCE
       ESCAPE CAPSULES
       HYPOBARIC ATMOSPHERES
       LOW PRESSURE
       LOW TEMPERATURE ENVIRONMENTS
       MOUNTAIN INHABITANTS
       THERMAL VACUUM TESTS
```

HIGH ALTITUDE ENVIRONMENTS-*(CONT.)*
    TIMBERLINE
    VACUUM CHAMBERS

HIGH ALTITUDE FLIGHT
USE   FLIGHT
    HIGH ALTITUDE

HIGH ALTITUDE NUCLEAR DETECTION
GS   DETECTION
    . **HIGH ALTITUDE NUCLEAR DETECTION**
RT   SPACE SURVEILLANCE (SPACEBORNE)
    VELA SATELLITES

HIGH ALTITUDE PRESSURE
GS   PRESSURE
    . LOW PRESSURE
    . . **HIGH ALTITUDE PRESSURE**
RT   ALTITUDE TOLERANCE
    ATMOSPHERIC PRESSURE
    HYPOBARIC ATMOSPHERES
    VACUUM CHAMBERS

HIGH ALTITUDE SOUNDING PROJECTILE
USE   WASP SOUNDING ROCKET

HIGH ALTITUDE TESTS
GS   ALTITUDE TESTS
    . **HIGH ALTITUDE TESTS**
RT   BACKGROUND RADIATION
    FISHBOWL OPERATION
    FLIGHT TESTS
    FULL SCALE TESTS
    TEST VEHICLES
    ∞TESTS
    VELA SATELLITES

HIGH ASPECT RATIO
GS   RATIOS
    . ASPECT RATIO
    . . **HIGH ASPECT RATIO**

HIGH ASPECT RATIO WINGS
USE   SLENDER WINGS

HIGH CURRENT
GS   ELECTRIC CURRENT
    . **HIGH CURRENT**
RT   HIGH VOLTAGES
    PLASMA CURRENTS

HIGH DISPERSION SPECTROGRAPHS
GS   MEASURING INSTRUMENTS
    . SPECTROMETERS
    . . ULTRAVIOLET SPECTROMETERS
    . . . **HIGH DISPERSION SPECTROGRAPHS**
    SPECTROGRAPHS
    . **HIGH DISPERSION SPECTROGRAPHS**

HIGH ECCENTRIC LUNAR OCCULTATION
SATELLITE
USE   EXOSAT SATELLITE

HIGH ELECTRON MOBILITY TRANSISTORS
GS   ELECTRONIC EQUIPMENT
    . SOLID STATE DEVICES
    . . SEMICONDUCTOR DEVICES
    . . . HETEROJUNCTION DEVICES
    . . . . **HIGH ELECTRON MOBILITY TRANSISTORS**
    . . . TRANSISTORS
    . . . . **HIGH ELECTRON MOBILITY TRANSISTORS**
RT   ELECTRON MOBILITY
    FIELD EFFECT TRANSISTORS

HIGH ENERGY ASTRONOMY OBSERVATORIES
USE   HEAO

HIGH ENERGY ASTRONOMY OBSERVATORY A
USE   HEAO 1

HIGH ENERGY ASTRONOMY OBSERVATORY B
USE   HEAO 2

HIGH ENERGY ASTRONOMY OBSERVATORY C
USE   HEAO 3

HIGH ENERGY ASTRONOMY OBSERVATORY 1
USE   HEAO 1

HIGH ENERGY ASTRONOMY OBSERVATORY 2
USE   HEAO 2

HIGH ENERGY ASTRONOMY OBSERVATORY 3
USE   HEAO 3

HIGH ENERGY ELECTRONS
GS   PARTICLES
    . CHARGED PARTICLES
    . . ENERGETIC PARTICLES
    . . . ELECTRONS
    . . . . **HIGH ENERGY ELECTRONS**
RT   SCATHA SATELLITE

HIGH ENERGY FUELS
SN   (HEAT CONTENT GREATER THAN OR EQUAL TO APPROXIMATELY 25,000 BTU/LB)
UF   HEF (HIGH ENERGY FUELS)
GS   FUELS
    . CHEMICAL FUELS
    . . **HIGH ENERGY FUELS**
RT   ADDITIVES
    BORON COMPOUNDS
    CATALYSTS
    CRYOGENIC ROCKET PROPELLANTS
    HYBRID PROPELLANTS
    HYDROCARBON FUELS

HIGH ENERGY INTERACTIONS
GS   NUCLEAR REACTIONS
    . **HIGH ENERGY INTERACTIONS**
    . . LIGHT-CONE EXPANSION
RT   ANNIHILATION REACTIONS
    BEAM INTERACTIONS
    FISSION PRODUCTS
    ∞INTERACTIONS
    NUCLEAR EXPLOSIONS
    NUCLEAR FISSION
    NUCLEAR FUSION
    NUCLEAR RADIATION
    NUCLEAR RESEARCH
    PAIR PRODUCTION
    PARTICLE INTERACTIONS
    PARTICLE PRODUCTION
    POMERANCHUK THEOREM
    SPHINX
    THERMONUCLEAR REACTIONS
    VECTOR DOMINANCE MODEL

HIGH ENERGY OXIDIZERS
GS   OXIDIZERS
    . **HIGH ENERGY OXIDIZERS**
RT   ROCKET OXIDIZERS

HIGH ENERGY PROPELLANTS
GS   PROPELLANTS
    . **HIGH ENERGY PROPELLANTS**
    . . DOMINO PROPELLANTS
RT   CRYOGENIC ROCKET PROPELLANTS
    GASEOUS ROCKET PROPELLANTS
    HYBRID PROPELLANTS
    LIQUID ROCKET PROPELLANTS

HIGH FIELD MAGNETS
UF   SUPERMAGNETS
GS   MAGNETS
    . ELECTROMAGNETS
    . . **HIGH FIELD MAGNETS**
RT   SUPERCONDUCTING MAGNETS

HIGH FLUX BEAM REACTORS
RT   NUCLEAR REACTORS

HIGH FLUX ISOTOPE REACTORS
UF   HFIR
    HFIR (REACTOR)
GS   NUCLEAR REACTORS
    . **HIGH FLUX ISOTOPE REACTORS**
RT   NEUTRON FLUX DENSITY

HIGH FREQUENCIES
GS   FREQUENCIES
    . RADIO FREQUENCIES
    . . **HIGH FREQUENCIES**
RT   DECAMETRIC WAVES
    INTERMEDIATE FREQUENCIES
    LOW FREQUENCY BANDS
    MAXIMUM USABLE FREQUENCY
    RING DISCHARGE
    SHORT WAVE RADIATION
    SHORT WAVE RADIO TRANSMISSION
    TOROIDAL DISCHARGE

HIGH GAIN
RT   AMPLIFICATION
    PILOT INDUCED OSCILLATION
    POWER GAIN
    TRANSFER FUNCTIONS

HIGH GRAVITY (ACCELERATION)
USE   HIGH GRAVITY ENVIRONMENTS

HIGH GRAVITY ENVIRONMENTS
UF   HIGH GRAVITY (ACCELERATION)
GS   ENVIRONMENTS
    . **HIGH GRAVITY ENVIRONMENTS**
    RATES (PER TIME)
    . ACCELERATION (PHYSICS)
    . . **HIGH GRAVITY ENVIRONMENTS**
RT   ∞ACCELERATION
    CENTRIFUGES
    EXTRATERRESTRIAL ENVIRONMENTS
    GRAVITATION
    HUMAN CENTRIFUGES
    REDUCED GRAVITY
    ROTATING ENVIRONMENTS

HIGH IMPULSE
RT   ∞FORCE
    IMPULSES
    PROPULSION

HIGH INTENSITY LASERS
USE   HIGH POWER LASERS

HIGH LATITUDES
USE   POLAR REGIONS

HIGH LEVEL LANGUAGES
GS   LANGUAGES
    . PROGRAMMING LANGUAGES
    . . **HIGH LEVEL LANGUAGES**
RT   COMMUNICATION THEORY
    LANGUAGE PROGRAMMING
    SYMBOLS

HIGH MELTING COMPOUNDS
USE   REFRACTORY MATERIALS

HIGH PASS FILTERS
RT   BANDSTOP FILTERS
    ELECTRIC FILTERS
    ELECTROMAGNETIC WAVE FILTERS
    ∞FILTERS
    MICROWAVE FILTERS
    OPTICAL FILTERS

HIGH POLYMERS
RT   ∞POLYMERS

HIGH POWER LASERS
UF   HIGH INTENSITY LASERS
GS   STIMULATED EMISSION DEVICES
    . LASERS
    . . **HIGH POWER LASERS**
    . . . NOVA LASER SYSTEM
    . . . SHIVA LASER SYSTEM
RT   GLASS LASERS
    LASER FUSION
    LASER OUTPUTS
    OPTICAL COMMUNICATION

HIGH PRESSURE
GS   PRESSURE
    . **HIGH PRESSURE**
RT   ANTICYCLONES
    CRITICAL PRESSURE
    HYPERBARIC CHAMBERS
    LOW PRESSURE
    SUPERCRITICAL PRESSURES
    TRANSITION PRESSURE
    VACUUM

HIGH PRESSURE OXYGEN
GS   GASES
    . COMPRESSED GAS
    . . **HIGH PRESSURE OXYGEN**
    . OXYGEN
    . . **HIGH PRESSURE OXYGEN**
RT   FIRE PREVENTION
    PRESSURE
    SPACECRAFT CABIN ATMOSPHERES

HIGH Q
USE   Q FACTORS

∞ **HIGH RESISTANCE**
- SN (USE OF A MORE SPECIFIC TERM IS
  RECOMMENDED--CONSULT THE TERMS
  LISTED BELOW)
- RT CHEMICAL PROPERTIES
  ELECTRICAL RESISTANCE
  FLOW RESISTANCE
  MECHANICAL PROPERTIES
  ∞ PHYSICAL PROPERTIES
  ∞ RESISTANCE
  RUGGEDNESS
  THERMAL RESISTANCE

HIGH RESOLUTION
- GS RESOLUTION
  . **HIGH RESOLUTION**
- RT ACCURACY
  ANGULAR RESOLUTION
  PRECISION
  SPATIAL RESOLUTION

HIGH RESOLUTION COVERAGE ANTENNAS
- UF HICAT (RADAR TECHNIQUE)
  HICAT PROJECT
- RT RADAR ANTENNAS
  RADAR RESOLUTION
  RESOLUTION

HIGH REYNOLDS NUMBER
- SN (RN ABOVE 3,000)
- GS RATIOS
  . DIMENSIONLESS NUMBERS
  . . REYNOLDS NUMBER
  . . . **HIGH REYNOLDS NUMBER**
- RT LOW REYNOLDS NUMBER

HIGH SPEED
- UF HIGH SPEED FLIGHT
- GS RATES (PER TIME)
  . **HIGH SPEED**
  VELOCITY
  . **HIGH SPEED**
- RT AIRSPEED
  ESCAPE VELOCITY
  GROUND SPEED
  HYPERSONIC SPEED
  LANDING SPEED
  LIGHT SPEED
  RELATIVISTIC VELOCITY
  ROTOR SPEED
  SUPERSONIC SPEEDS

HIGH SPEED CAMERAS
- GS OPTICAL EQUIPMENT
  . CAMERAS
  . . **HIGH SPEED CAMERAS**
  . . . FRAMING CAMERAS
  PHOTOGRAPHIC EQUIPMENT
  . CAMERAS
  . . **HIGH SPEED CAMERAS**
  . . . FRAMING CAMERAS
- RT BALLISTIC CAMERAS
  FRAME PHOTOGRAPHY
  HIGH SPEED PHOTOGRAPHY
  ROTATING MIRRORS
  STREAK PHOTOGRAPHY
  STROBOSCOPES

HIGH SPEED FLIGHT
- USE FLIGHT
  HIGH SPEED

HIGH SPEED PHOTOGRAPHY
- GS PHOTOGRAPHY
  . **HIGH SPEED PHOTOGRAPHY**
- RT HIGH SPEED CAMERAS
  PHOTOGRAPHIC RECORDING

HIGH SPEED TRANSPORTATION
- USE RAPID TRANSIT SYSTEMS

HIGH STRENGTH
- GS MECHANICAL PROPERTIES
  . **HIGH STRENGTH**
- RT COMPRESSIVE STRENGTH
  SHEAR STRENGTH
  ∞ STRENGTH
  TENSILE STRENGTH
  TENSILE STRESS
  YIELD STRENGTH

HIGH STRENGTH ALLOYS
- GS ALLOYS
  . **HIGH STRENGTH ALLOYS**

HIGH STRENGTH ALLOYS-(CONT.)
  . . ASTROLOY (TRADEMARK)
  . . HIGH STRENGTH STEELS
  . . . MARAGING STEELS
- RT TENSILE PROPERTIES

HIGH STRENGTH STEELS
- UF LOW ALLOY STEELS
- GS ALLOYS
  . HIGH STRENGTH ALLOYS
  . . **HIGH STRENGTH STEELS**
  . . . MARAGING STEELS
  . IRON ALLOYS
  . . STEELS
  . . . **HIGH STRENGTH STEELS**
  . . . . MARAGING STEELS
- RT CARBON STEELS

HIGH TEMPERATURE
- GS TEMPERATURE
  . **HIGH TEMPERATURE**
- RT SIALON
  TEMPERATURE MEASUREMENT

HIGH TEMPERATURE AIR
- UF HOT AIR
- GS GASES
  . AIR
  . . **HIGH TEMPERATURE AIR**
  . HIGH TEMPERATURE GASES
  . . **HIGH TEMPERATURE AIR**
  HIGH TEMPERATURE FLUIDS
  . HIGH TEMPERATURE GASES
  . . **HIGH TEMPERATURE AIR**

HIGH TEMPERATURE ALLOYS
- USE HEAT RESISTANT ALLOYS

HIGH TEMPERATURE ENVIRONMENTS
- GS ENVIRONMENTS
  . **HIGH TEMPERATURE ENVIRONMENTS**
- RT DRY HEAT
  HEAT ACCLIMATIZATION
  LUNAR TEMPERATURE
  THERMAL ENVIRONMENTS
  THERMAL FATIGUE

HIGH TEMPERATURE FLUIDS
- GS **HIGH TEMPERATURE FLUIDS**
  . HIGH TEMPERATURE GASES
  . . HIGH TEMPERATURE AIR
- RT ∞ FLUIDS
  HYDRAULIC FLUIDS
  PLASMAS (PHYSICS)
  WORKING FLUIDS

HIGH TEMPERATURE GAS COOLED REACTORS
- UF HTGR
- GS NUCLEAR REACTORS
  . GAS COOLED REACTORS
  . . HIGH TEMPERATURE NUCLEAR
  REACTORS
  . . . **HIGH TEMPERATURE GAS COOLED
  REACTORS**
- RT NUCLEAR POWER REACTORS

HIGH TEMPERATURE GASES
- UF HOT GAS SYSTEMS
  HOT GASES
  HOT JET EXHAUST
- GS GASES
  . **HIGH TEMPERATURE GASES**
  . . HIGH TEMPERATURE AIR
  HIGH TEMPERATURE FLUIDS
  . **HIGH TEMPERATURE GASES**
  . . HIGH TEMPERATURE AIR
- RT COMBUSTION PRODUCTS
  EXHAUST EMISSION
  IONIZED GASES
  PNEUMATIC PROBES
  RAREFIED GASES
  SHOCK WAVE PROPAGATION

HIGH TEMPERATURE LUBRICANTS
- GS LUBRICANTS
  . **HIGH TEMPERATURE LUBRICANTS**
- RT GAS BEARINGS
  GAS LUBRICANTS
  THERMAL RESISTANCE

HIGH TEMPERATURE MATERIALS
- USE REFRACTORY MATERIALS

HIGH TEMPERATURE NUCLEAR REACTORS
- UF LOS ALAMOS TURRET REACTOR
  UHTREX (NUCLEAR REACTORS)
- GS NUCLEAR REACTORS
  . GAS COOLED REACTORS
  . . **HIGH TEMPERATURE NUCLEAR
  REACTORS**
  . . . HIGH TEMPERATURE GAS COOLED
  REACTORS
  . NUCLEAR RESEARCH AND TEST
  REACTORS
  . . **HIGH TEMPERATURE NUCLEAR
  REACTORS**
- RT NUCLEAR PROPULSION
  REACTOR DESIGN
  REACTOR TECHNOLOGY
  ∞ REACTORS

HIGH TEMPERATURE PLASMAS
- UF HOT PLASMAS
- GS GASES
  . IONIZED GASES
  . . CHARGED PARTICLES
  . . . **HIGH TEMPERATURE PLASMAS**
  PARTICLES
  . CHARGED PARTICLES
  . . ENERGETIC PARTICLES
  . . . PLASMAS (PHYSICS)
  . . . . **HIGH TEMPERATURE PLASMAS**
- RT BOLTZMANN-VLASOV EQUATION
  COLLISIONAL PLASMAS
  DENSE PLASMAS
  ELECTRON PLASMA
  HELIOS PROJECT
  RELATIVISTIC PLASMAS
  STRONGLY COUPLED PLASMAS
  THERMAL PLASMAS

HIGH TEMPERATURE PROPELLANTS
- GS PROPELLANTS
  . **HIGH TEMPERATURE PROPELLANTS**
- RT ELECTROTHERMAL ENGINES
  GELLED PROPELLANTS
  GELLED ROCKET PROPELLANTS
  ION PROPULSION
  NUCLEAR PROPULSION
  PLASMA ENGINES
  SOLID PROPELLANTS
  STORABLE PROPELLANTS

HIGH TEMPERATURE RESEARCH
- GS RESEARCH
  . **HIGH TEMPERATURE RESEARCH**
- RT PLASMA GENERATORS
  REFRACTORY MATERIALS

HIGH TEMPERATURE TESTS
- UF HEAT TESTS
- GS ENVIRONMENTAL TESTS
  . **HIGH TEMPERATURE TESTS**
- RT BOMB CALORIMETERS
  CALORIMETERS
  CHEMICAL TESTS
  COLD STRENGTH
  COLD WEATHER TESTS
  CRYOSTATS
  DROP CALORIMETERS
  FLAME CALORIMETERS
  HARDNESS TESTS
  LUBRICANT TESTS
  ∞ MATERIALS TESTS
  MELTING POINTS
  NONDESTRUCTIVE TESTS
  TEMPERATURE CONTROL
  ∞ TESTS
  THERMAL EXPANSION
  THERMAL RESISTANCE
  THERMAL SHOCK
  THERMAL STABILITY
  THERMODYNAMIC PROPERTIES
  TRANSPORT PROPERTIES

HIGH THRUST
- GS THRUST
  . **HIGH THRUST**
- RT JET THRUST
  LOW THRUST
  ROCKET THRUST
  THRUST AUGMENTATION
  VARIABLE THRUST

HIGH VACUUM
- GS PRESSURE
  . VACUUM
  . . **HIGH VACUUM**

HIGH VACUUM-*(CONT.)*
RT   COLD WELDING
        LOW VACUUM
        MOLECULAR SHIELDS
        RESIDUAL GAS
        SPACE MANUFACTURING
        ULTRAHIGH VACUUM
        VACUUM APPARATUS
        VACUUM TESTS

HIGH VACUUM ORBITAL SIMULATOR
UF   HIVOS (SIMULATOR)
GS   SIMULATORS
       . ENVIRONMENT SIMULATORS
       . . SPACE SIMULATORS
       . . . **HIGH VACUUM ORBITAL**
            SIMULATOR
RT   SPACE ENVIRONMENT SIMULATION

HIGH VOLTAGES
RT   ELECTRIC CURRENT
        ELECTRIC POTENTIAL
        HIGH CURRENT

HIGHLANDS
RT   COLORADO PLATEAU (US)
        MESAS
        MOUNTAINS
        PLATEAUS
        TOPOGRAPHY

HIGHLY ECCENTRIC ORBIT SATELLITES
USE   HEOS SATELLITES

HIGHLY MANEUVERABLE AIRCRAFT
UF   HIMAT
RT   AIRBORNE/SPACEBORNE COMPUTERS
      ∞AIRCRAFT
        AIRCRAFT MANEUVERS
        AUTOMATIC FLIGHT CONTROL
        AUTOMATIC PILOTS
        COMPUTERIZED SIMULATION
        FIGHTER AIRCRAFT
        FLIGHT CHARACTERISTICS
        FLIGHT TESTS
        REMOTELY PILOTED VEHICLES

HIGHWAYS
GS   ROADS
       . **HIGHWAYS**
RT   AIR BAG RESTRAINT DEVICES
        BRIDGES (STRUCTURES)
        CONSTRUCTION
        CRASHES
        INTERSECTIONS
        PAVEMENTS
        RAMPS (STRUCTURES)
        RAPID TRANSIT SYSTEMS
        REGIONAL PLANNING
        STREETS
        TRANSPORTATION
        TRANSPORTATION NETWORKS
        URBAN PLANNING

HIJACKING
USE   AIR PIRACY

HILBERT SPACE
GS   ALGEBRA
       . VECTOR SPACES
       . . **HILBERT SPACE**
       . . . BANACH SPACE
       . . . . SOBOLEV SPACE
       ANALYSIS (MATHEMATICS)
       . FUNCTION SPACE
       . . **HILBERT SPACE**
       . . . BANACH SPACE
       . . . . SOBOLEV SPACE
       . FUNCTIONAL ANALYSIS
       . . **HILBERT SPACE**
       . . . BANACH SPACE
       . . . . SOBOLEV SPACE

HILBERT TRANSFORMATION
GS   ANALYSIS (MATHEMATICS)
       . FUNCTIONAL ANALYSIS
       . . INTEGRAL TRANSFORMATIONS
       . . . **HILBERT TRANSFORMATION**

HILL CURVES
USE   HILL METHOD

HILL DETERMINANT
GS   ANALYSIS (MATHEMATICS)

HILL DETERMINANT-*(CONT.)*
       . **HILL DETERMINANT**
RT   DIFFERENTIAL EQUATIONS
        EIGENVALUES
        MATHIEU FUNCTION

HILL LUNAR THEORY
RT   EARTH ORBITS
        ORBITAL MECHANICS
        PERTURBATION THEORY
      ∞THEORIES

HILL METHOD
UF   HILL CURVES
RT   EARTH ORBITS
      ∞METHODOLOGY
        ORBITAL MECHANICS
        PERTURBATION THEORY

HILLER AIRCRAFT
UF   HILLER MILITARY AIRCRAFT
GS   **HILLER AIRCRAFT**
       . OH-5 HELICOPTER
RT   ∞AIRCRAFT

HILLER MILITARY AIRCRAFT
USE   HILLER AIRCRAFT
        MILITARY AIRCRAFT

HILSCH TUBES
UF   VORTEX TUBES
RT   COAXIAL FLOW
        COOLING
        HEATING
      ∞TUBES
        VORTEX GENERATORS
        VORTICES

HIMALAYAS
GS   LANDFORMS
       . MOUNTAINS
       . . **HIMALAYAS**
RT   ASIA
        BHUTAN
        INDIA
        PAKISTAN
        SIKKIM
        TIBET

HIMAT
USE   HIGHLY MANEUVERABLE AIRCRAFT

HINDRANCE
USE   CONSTRAINTS

HINGE MOMENTS
USE   TORQUE

HINGED ROTOR BLADES
USE   HINGES
        ROTARY WINGS

HINGELESS ROTORS
USE   RIGID ROTORS

HINGES
UF   HINGED ROTOR BLADES
GS   **HINGES**
       . FLAPPING HINGES
RT   BEARINGLESS ROTORS
        PIVOTS
        SWIVELS

HIPPARCOS SATELLITE
GS   SATELLITES
       . ARTIFICIAL SATELLITES
       . . ESA SATELLITES
       . . . **HIPPARCOS SATELLITE**
RT   ASTROMETRY
        EUROPEAN SPACE PROGRAMS
        SPACEBORNE ASTRONOMY
        STELLAR MOTIONS
        STELLAR PARALLAX

HIPPOCAMPUS
GS   ANATOMY
       . BRAIN
       . . **HIPPOCAMPUS**
       NERVOUS SYSTEM
       . CENTRAL NERVOUS SYSTEM
       . . BRAIN
       . . . **HIPPOCAMPUS**

HIPPURIC ACID
GS   ACIDS
       . AMINO ACIDS
       . . **HIPPURIC ACID**
       ALIPHATIC COMPOUNDS
       . **HIPPURIC ACID**
       ORGANIC COMPOUNDS
       . AMINO ACIDS
       . . **HIPPURIC ACID**

HIS BUNDLE
RT   CARDIAC AURICLES
        CARDIAC VENTRICLES
        ELECTROPHYSIOLOGY
        HEART DISEASES
        HEART FUNCTION
        NERVES

HISS
GS   ELECTROMAGNETIC INTERFERENCE
       . RADIO FREQUENCY INTERFERENCE
       . . BLACKOUT (PROPAGATION)
       . . . ELECTROMAGNETIC NOISE
       . . . . ATMOSPHERICS
       . . . . . IONOSPHERICS
       . . . . . . **HISS**

HISTAMINES
GS   DRUGS
       . **HISTAMINES**
RT   AMINES
        ANTIHISTAMINICS
        ITCHING

HISTIDINE
GS   ACIDS
       . AMINO ACIDS
       . . **HISTIDINE**
       AMINES
       . **HISTIDINE**
       ORGANIC COMPOUNDS
       . AMINO ACIDS
       . . **HISTIDINE**

HISTOCHEMICAL ANALYSIS
RT   BIOASSAY
        BIOCHEMISTRY
        CELLS (BIOLOGY)
        ORGANIC CHEMISTRY
        TISSUES (BIOLOGY)

HISTOGRAMS
RT   DISCRETE FUNCTIONS
        GRAPHS (CHARTS)
        NORMAL DENSITY FUNCTIONS

HISTOLOGY
GS   MEDICAL SCIENCE
       . **HISTOLOGY**
RT   EPITHELIUM
        MORPHOLOGY
        PLATELETS

HISTORIES
GS   **HISTORIES**
       . CASE HISTORIES
RT   DOCUMENTATION
        MUSEUMS
        PALEONTOLOGY
        PEACETIME
        RECORDS

HITAB PROGRAM
USE   HIGH ALT TARGET AND BACKGROUND
        MEASUREMENT

HIVOS (SIMULATOR)
USE   HIGH VACUUM ORBITAL SIMULATOR

HL-10 REENTRY VEHICLE
GS   GLIDERS
       . **HL-10 REENTRY VEHICLE**
       LIFTING BODIES
       . LIFTING REENTRY VEHICLES
       . . **HL-10 REENTRY VEHICLE**
       MANEUVERABLE SPACECRAFT
       . LIFTING REENTRY VEHICLES
       . . **HL-10 REENTRY VEHICLE**
       REENTRY VEHICLES
       . LIFTING REENTRY VEHICLES
       . . **HL-10 REENTRY VEHICLE**
RT   HYPERSONIC GLIDERS

HLD-35 REENTRY VEHICLE
GS   HYPERSONIC VEHICLES
      . LIFTING REENTRY VEHICLES
      . . **HLD-35 REENTRY VEHICLE**
      LIFTING BODIES
      . LIFTING REENTRY VEHICLES
      . . **HLD-35 REENTRY VEHICLE**
      MANEUVERABLE SPACECRAFT
      . LIFTING REENTRY VEHICLES
      . . **HLD-35 REENTRY VEHICLE**
      REENTRY VEHICLES
      . LIFTING REENTRY VEHICLES
      . . **HLD-35 REENTRY VEHICLE**
RT   HYPERSONIC GLIDERS

HLLV
USE   HEAVY LIFT LAUNCH VEHICLES

HMX
UF   CYCLOTETRAMETHYLENE
         TETRANITRAMINE
      TETRANITROTETRAZACYCLOOCTANE
RT   EXPLOSIVES

HNPF (HALLAM NUCLEAR POWER FACILITY)
USE   HALLAM NUCLEAR POWER FACILITY

HNST
USE   HEXANITROSTILBENE

HO-4 HELICOPTER
USE   OH-4 HELICOPTER

HO-5 HELICOPTER
USE   OH-5 HELICOPTER

HO-6 HELICOPTER
USE   OH-6 HELICOPTER

HODOGRAPHS
RT   CHAPLYGIN EQUATION
      KINEMATICS
      VECTOR SPACES

HODOSCOPES
GS   MEASURING INSTRUMENTS
      . RADIATION MEASURING INSTRUMENTS
      . . **HODOSCOPES**
RT   RADIATION COUNTERS

HOGBACKS
USE   RIDGES

HOHLRAUMS
RT   BLACK BODY RADIATION
      EMISSIVITY

HOHMANN TRAJECTORIES
USE   ELLIPTICAL ORBITS
      TRANSFER ORBITS

HOHMANN TRANSFER ORBITS
USE   ELLIPTICAL ORBITS
      TRANSFER ORBITS

HOLDERS
GS   **HOLDERS**
      . FLAME HOLDERS
RT   ANCHORS (FASTENERS)
      ∞ BANDS
      BOLTS
      BRACKETS
      CLAMPS
      CLIPS
      FASTENERS
      JIGS
      LATCHES
      LUGS
      MECHANICAL DEVICES
      NUTS (FASTENERS)
      PINS
      POSITIONING DEVICES (MACHINERY)
      RIVETS
      SCREWS
      ∞ SPIKES
      SPLINES
      STRAPS
      STUDS (STRUCTURAL MEMBERS)
      ZIPPERS

∞ **HOLDING**
SN   *(USE OF A MORE SPECIFIC TERM IS*
      RECOMMENDED--CONSULT THE TERMS
      LISTED BELOW)

HOLDING-*(CONT.)*
RT   CONSTRAINTS
      DELAY
      RETAINING
      STOPPING

HOLE BURNING
RT   COMPUTER STORAGE DEVICES
      HOLOGRAPHY
      LASER APPLICATIONS
      LASERS
      LASING
      MEMORY (COMPUTERS)

∞ **HOLE DISTRIBUTION**
SN   *(USE OF A MORE SPECIFIC TERM IS*
      RECOMMENDED--CONSULT THE TERMS
      LISTED BELOW)
RT   CURRENT DISTRIBUTION
      HOLE DISTRIBUTION (ELECTRONICS)
      HOLE DISTRIBUTION (MECHANICS)

HOLE DISTRIBUTION (ELECTRONICS)
GS   DISTRIBUTION (PROPERTY)
      . **HOLE DISTRIBUTION (ELECTRONICS)**
RT   CHARGE DISTRIBUTION
      CURRENT DISTRIBUTION
      ∞ HOLE DISTRIBUTION
      HOLES (ELECTRON DEFICIENCIES)
      SEMICONDUCTORS (MATERIALS)

HOLE DISTRIBUTION (MECHANICS)
GS   DISTRIBUTION (PROPERTY)
      . **HOLE DISTRIBUTION (MECHANICS)**
RT   CAVITIES
      ∞ HOLE DISTRIBUTION
      PERFORATED SHELLS
      POROSITY
      STRESS CONCENTRATION
      VOID RATIO

HOLE GEOMETRY (MECHANICS)
RT   FRACTURE MECHANICS
      PERFORATED PLATES
      PERFORATED SHELLS
      STRESS CONCENTRATION
      STRESS INTENSITY FACTORS
      STRUCTURAL ANALYSIS

HOLE MOBILITY
GS   ELECTROMAGNETIC PROPERTIES
      . ELECTRICAL PROPERTIES
      . . CARRIER MOBILITY
      . . . **HOLE MOBILITY**
      MOBILITY
      . CARRIER MOBILITY
      . . **HOLE MOBILITY**
      TRANSPORT PROPERTIES
      . CARRIER MOBILITY
      . . **HOLE MOBILITY**
RT   ATOMIC MOBILITIES
      CHARGE CARRIERS
      ELECTROMIGRATION
      ELECTRON MOBILITY
      HOLES (ELECTRON DEFICIENCIES)
      ∞ SOLID STATE PHYSICS

∞ **HOLES**
SN   *(USE OF A MORE SPECIFIC TERM IS*
      RECOMMENDED--CONSULT THE TERMS
      LISTED BELOW)
RT   BOREHOLES
      CAVITIES
      CORONAL HOLES
      GAPS
      HOLES (ELECTRON DEFICIENCIES)

HOLES (ELECTRON DEFICIENCIES)
GS   CHARGE CARRIERS
      . **HOLES (ELECTRON DEFICIENCIES)**
RT   ACCEPTOR MATERIALS
      CRYSTAL DEFECTS
      DONOR MATERIALS
      ELECTRONS
      EXCITONS
      HOLE DISTRIBUTION (ELECTRONICS)
      HOLE MOBILITY
      ∞ HOLES
      MAJORITY CARRIERS
      ∞ MATERIALS
      MINORITY CARRIERS
      ORDER-DISORDER TRANSFORMATIONS
      P-TYPE SEMICONDUCTORS
      SEMICONDUCTOR PLASMAS
      SEMICONDUCTORS (MATERIALS)

HOLES (ELECTRON DEFICIENCIES)-*(CONT.)*
      SUHL EFFECT
      VACANCIES (CRYSTAL DEFECTS)

HOLLAND
USE   NETHERLANDS

∞ **HOLLOW**
SN   *(USE OF A MORE SPECIFIC TERM IS*
      RECOMMENDED--CONSULT THE TERMS
      LISTED BELOW)
RT   CAVITIES
      ∞ DEPRESSION
      RECESSES

HOLLOW CATHODES
GS   ELECTRODES
      . CATHODES
      . . **HOLLOW CATHODES**
RT   TUBE CATHODES
      TUNNEL CATHODES

HOLMIUM
GS   CHEMICAL ELEMENTS
      . RARE EARTH ELEMENTS
      . . **HOLMIUM**
      . . . HOLMIUM ISOTOPES
      METALS
      . RARE EARTH ELEMENTS
      . . **HOLMIUM**
      . . . HOLMIUM ISOTOPES

HOLMIUM ISOTOPES
GS   CHEMICAL ELEMENTS
      . NUCLIDES
      . . ISOTOPES
      . . . **HOLMIUM ISOTOPES**
      . RARE EARTH ELEMENTS
      . . HOLMIUM
      . . . **HOLMIUM ISOTOPES**
      METALS
      . RARE EARTH ELEMENTS
      . . HOLMIUM
      . . . **HOLMIUM ISOTOPES**

HOLOGRAMMETRY
RT   HOLOGRAPHY
      IMAGING TECHNIQUES
      PHOTOGRAPHIC RECORDING
      PHOTOMAPPING
      TERRAIN ANALYSIS

HOLOGRAPHIC INTERFEROMETRY
GS   INTERFEROMETRY
      . **HOLOGRAPHIC INTERFEROMETRY**
RT   COHERENT LIGHT
      DIFFRACTION PATTERNS
      HOLOGRAPHY
      LASER OUTPUTS
      MOIRE INTERFEROMETRY
      SCATTER PLATES (OPTICS)
      WAVE FRONT RECONSTRUCTION

HOLOGRAPHIC SPECTROSCOPY
GS   SPECTROSCOPY
      . **HOLOGRAPHIC SPECTROSCOPY**
RT   FOURIER TRANSFORMATION
      HOLOGRAPHY
      SPECTRUM ANALYSIS
      WAVE FRONT RECONSTRUCTION

HOLOGRAPHIC SUBTRACTION
UF   SELF SUBTRACTION HOLOGRAPHY
RT   HOLOGRAPHY

HOLOGRAPHY
GS   IMAGERY
      . **HOLOGRAPHY**
      . . MICROWAVE HOLOGRAPHY
      . . WHITE LIGHT HOLOGRAPHY
      PHOTOGRAPHY
      . **HOLOGRAPHY**
      . . ACOUSTICAL HOLOGRAPHY
      . . MICROWAVE HOLOGRAPHY
      . . WHITE LIGHT HOLOGRAPHY
RT   COHERENT ELECTROMAGNETIC
         RADIATION
      COHERENT LIGHT
      DATA STORAGE
      HOLE BURNING
      HOLOGRAMMETRY
      HOLOGRAPHIC INTERFEROMETRY
      HOLOGRAPHIC SPECTROSCOPY
      HOLOGRAPHIC SUBTRACTION

HOLOGRAPHY-*(CONT.)*
 IMAGE CORRELATORS
 IMAGE RECONSTRUCTION
 KINOFORM
 LASERS
 OPTICAL MEMORY (DATA STORAGE)
 SCATTER PLATES (OPTICS)
 SPATIAL FILTERING
 SPECKLE PATTERNS
 WAVE FRONT RECONSTRUCTION

HOLOMORPHISM
USE ANALYTIC FUNCTIONS

HOMEOSTASIS
RT ACCLIMATIZATION
 ACID BASE EQUILIBRIUM
 ADAPTATION
 BODY TEMPERATURE
 COLD TOLERANCE
 COLLOIDS
 ∞EQUILIBRIUM
 FIBRINOGEN
 HORMONES
 METABOLISM
 NERVOUS SYSTEM
 OSMOSIS
 PHYSIOLOGY
 RESPIRATORY SYSTEM
 SKIN (ANATOMY)
 STRESS (PHYSIOLOGY)
 THERMOREGULATION
 THROMBOPLASTIN
 WATER BALANCE

HOMEOTHERMS
RT ANIMALS
 BIRDS
 BODY TEMPERATURE
 VERTEBRATES

HOMING
RT AUTOMATIC PILOTS
 BEACONS
 GUIDANCE (MOTION)
 MISSILE CONTROL
 RADIO DIRECTION FINDERS
 TERMINAL GUIDANCE

HOMING DEVICES
UF SEEKERS
RT BEACONS
 GUIDANCE (MOTION)
 INFRARED TRACKING
 LASER GUIDANCE
 MISSILES
 NAVIGATION
 NAVIGATION AIDS
 RADIO BEACONS
 RADIO DIRECTION FINDERS
 RADIO NAVIGATION
 RENDEZVOUS GUIDANCE
 SOLAR COMPASSES
 TRAJECTORY CONTROL

HOMODYNE RECEPTION
RT FREQUENCY SYNCHRONIZATION
 RADIO RECEPTION
 SIGNAL RECEPTION

HOMOGENEITY
RT ANISOTROPIC MEDIA
 HOMOGENIZING
 SAMPLING
 STATISTICAL TESTS
 UNITY
 VARIANCE (STATISTICS)

HOMOGENEOUS TURBULENCE
GS TURBULENCE
 . **HOMOGENEOUS TURBULENCE**
RT ATMOSPHERIC TURBULENCE
 FLUCTUATION THEORY
 ISOTROPIC TURBULENCE
 LOW LEVEL TURBULENCE
 MAGNETOHYDRODYNAMIC TURBULENCE

HOMOGENIZATION
USE HOMOGENIZING

HOMOGENIZING
UF HOMOGENIZATION
GS MIXING
 . **HOMOGENIZING**

HOMOGENIZING-*(CONT.)*
RT AGITATION
 COLLOIDING
 COMPOUNDING
 DISPERSING
 DISSOLVING
 HOMOGENEITY
 PREMIXING
 ∞SEPARATION
 SUSPENDING (MIXING)

HOMOJUNCTIONS
GS ELECTRIC GENERATORS
 . DIRECT POWER GENERATORS
 . . SOLAR CELLS
 . . . **HOMOJUNCTIONS**
 . SOLAR GENERATORS
 . . SOLAR CELLS
 . . . **HOMOJUNCTIONS**
RT HETEROJUNCTION DEVICES
 HETEROJUNCTIONS
 SEMICONDUCTOR JUNCTIONS
 SILICON JUNCTIONS

HOMOLOGY
UF COHOMOLOGY
RT ANALOGIES
 DUALITY THEOREM
 MATCHING
 ∞RELATIONSHIPS
 TOPOLOGY

HOMOMORPHISMS
GS ALGEBRA
 . GROUP THEORY
 . . **HOMOMORPHISMS**
 . . . AUTOMORPHISMS
 . . . MONOIDS
 . . . SUBGROUPS
RT FIELD THEORY (ALGEBRA)
 ISOMORPHISM

HOMOPOLAR GENERATORS
GS ELECTRIC GENERATORS
 . ROTATING GENERATORS
 . . **HOMOPOLAR GENERATORS**
RT DIRECT CURRENT
 ∞ELECTRIC EQUIPMENT
 ELECTROMECHANICAL DEVICES
 ∞GENERATORS

HOMOSPHERE
GS EARTH ATMOSPHERE
 . **HOMOSPHERE**
 . . MIDDLE ATMOSPHERE
 . . . CHEMOSPHERE
 . . . STRATOSPHERE
RT BIOSPHERE
 IONOSPHERE
 LOWER ATMOSPHERE
 MESOSPHERE
 OZONOSPHERE
 STRATOSPHERE
 THERMOSPHERE
 TROPOSPHERE
 UPPER ATMOSPHERE

HOMOTOPY THEORY
GS GEOMETRY
 . TOPOLOGY
 . . **HOMOTOPY THEORY**
RT CURVES (GEOMETRY)
 FIBERS (MATHEMATICS)
 ∞THEORIES

HOMOTROPY
RT ALGEBRA
 PROBLEM SOLVING
 SET THEORY
 TOPOLOGY

HONDURAS
GS NATIONS
 . **HONDURAS**
RT CENTRAL AMERICA

HONEST JOHN ROCKET VEHICLE
GS ROCKET VEHICLES
 . SINGLE STAGE ROCKET VEHICLES
 . . **HONEST JOHN ROCKET VEHICLE**
 . SURFACE TO SURFACE ROCKETS
 . . **HONEST JOHN ROCKET VEHICLE**
RT ARGO ROCKET VEHICLES
 EXOS SOUNDING ROCKET

HONEST JOHN ROCKET VEHICLE-*(CONT.)*
 HERCULES ENGINE
 SOLID PROPELLANT ROCKET ENGINES
 TRAILBLAZER 1 REENTRY VEHICLE

HONEYCOMB CORES
GS CORES
 . **HONEYCOMB CORES**
 HONEYCOMB STRUCTURES
 . **HONEYCOMB CORES**
RT CERAMIC HONEYCOMBS
 COMPOSITE STRUCTURES
 LOW DENSITY MATERIALS
 SANDWICH STRUCTURES

HONEYCOMB STRUCTURES
GS **HONEYCOMB STRUCTURES**
 . HONEYCOMB CORES
RT ∞CELLS
 CERAMIC HONEYCOMBS
 COMPOSITE STRUCTURES
 HEXAGONAL CELLS
 INSULATION
 LAMINATES
 LOW DENSITY MATERIALS
 METAL FOILS
 POROUS MATERIALS
 SANDWICH STRUCTURES
 ∞STRUCTURES

HONEYWELL ADEPT COMPUTER
GS DATA PROCESSING EQUIPMENT
 . COMPUTERS
 . . DIGITAL COMPUTERS
 . . . HONEYWELL COMPUTERS
 . . . . **HONEYWELL ADEPT COMPUTER**

HONEYWELL COMPUTERS
GS DATA PROCESSING EQUIPMENT
 . COMPUTERS
 . . DIGITAL COMPUTERS
 . . . **HONEYWELL COMPUTERS**
 . . . . DDP 516 COMPUTER
 . . . . HONEYWELL ADEPT COMPUTER
 . . . . HONEYWELL DDP 116 COMPUTER
 . . . . HONEYWELL 600/6000 COMPUTER

HONEYWELL DDP 116 COMPUTER
GS DATA PROCESSING EQUIPMENT
 . COMPUTERS
 . . DIGITAL COMPUTERS
 . . . HONEYWELL COMPUTERS
 . . . . **HONEYWELL DDP 116 COMPUTER**

HONEYWELL 600/6000 COMPUTER
GS DATA PROCESSING EQUIPMENT
 . COMPUTERS
 . . ANALOG COMPUTERS
 . . . **HONEYWELL 600/6000 COMPUTER**
 . . DIGITAL COMPUTERS
 . . . HONEYWELL COMPUTERS
 . . . . **HONEYWELL 600/6000 COMPUTER**

HONG KONG
RT ASIA
 CHINA
 NATIONS
 TAIWAN

HONING
RT SCRAPERS
 SMOOTHING

HOOKES LAW
GS LAWS
 . **HOOKES LAW**
RT ELASTIC PROPERTIES
 FIBER STRENGTH
 MAXWELL BODIES
 MODULUS OF ELASTICITY
 SHEAR PROPERTIES
 STRESS-STRAIN DIAGRAMS

HOOKS
RT FASTENERS
 FORKS
 SWIVELS

HOOP COLUMN ANTENNAS
GS ANTENNAS
 . **HOOP COLUMN ANTENNAS**
RT LARGE SPACE STRUCTURES
 SPACECRAFT COMMUNICATION

HOOPS
RT　RING STRUCTURES
　　TENSILE STRESS

HOPCALITE (TRADEMARK)
GS　CATALYSTS
　　. **HOPCALITE (TRADEMARK)**
　　CHALCOGENIDES
　　. OXIDES
　　. . METAL OXIDES
　　. . . MANGANESE OXIDES
　　. . . . **HOPCALITE (TRADEMARK)**
　　MANGANESE COMPOUNDS
　　. **HOPCALITE (TRADEMARK)**
RT　AIR PURIFICATION
　　CARBON MONOXIDE
　　GAS ANALYSIS

HOPPERS
RT　∞CONTAINERS
　　MATERIALS HANDLING
　　PACKAGING

HORIZON
GS　**HORIZON**
　　. RADIO HORIZONS
RT　CELESTIAL SPHERE
　　GYRO HORIZONS
　　RANGE (EXTREMES)

HORIZON SCANNERS
UF　HORIZON SENSING
　　INFRARED HORIZON SCANNERS
GS　FLIGHT INSTRUMENTS
　　. **HORIZON SCANNERS**
　　SCANNERS
　　. **HORIZON SCANNERS**
RT　ATTITUDE CONTROL
　　INFRARED SCANNERS
　　NAVIGATION INSTRUMENTS
　　OPTICAL EQUIPMENT
　　PHOTOMETERS
　　RADIO HORIZONS
　　RADIOMETERS
　　SCANNER PROJECT

HORIZON SENSING
USE　HORIZON SCANNERS

HORIZONTAL BRANCH STARS
GS　CELESTIAL BODIES
　　. GLOBULAR CLUSTERS
　　. . **HORIZONTAL BRANCH STARS**
RT　COLOR-MAGNITUDE DIAGRAM
　　HERTZSPRUNG-RUSSELL DIAGRAM
　　STELLAR EVOLUTION
　　STELLAR LUMINOSITY
　　STELLAR SPECTRA
　　STELLAR SPECTROPHOTOMETRY

HORIZONTAL FLIGHT
RT　AERODYNAMIC BALANCE
　　AIRCRAFT STABILITY
　　CLIMBING FLIGHT
　　CRUISING FLIGHT
　　∞FLIGHT
　　FLIGHT PATHS
　　ROCKET FLIGHT
　　SOARING
　　TURNING FLIGHT

HORIZONTAL ORIENTATION
RT　ALIGNMENT
　　ATTITUDE (INCLINATION)
　　DIRECTIONAL STABILITY
　　DYNAMIC STABILITY
　　∞ORIENTATION
　　STABILIZATION
　　VERTICAL ORIENTATION

HORIZONTAL SPACECRAFT LANDING
GS　LANDING
　　. GLIDE LANDINGS
　　. . **HORIZONTAL SPACECRAFT LANDING**
　　. SPACECRAFT LANDING
　　. . **HORIZONTAL SPACECRAFT LANDING**
RT　APPROACH AND LANDING TESTS (STS)
　　CRASH LANDING
　　PLANETARY LANDING
　　SOFT LANDING
　　WATER LANDING

HORIZONTAL STABILIZERS
USE　STABILIZERS (FLUID DYNAMICS)

HORIZONTAL TAIL SURFACES
UF　TAIL PLANES
GS　AIRFOILS
　　. **HORIZONTAL TAIL SURFACES**
　　CONTROL SURFACES
　　. **HORIZONTAL TAIL SURFACES**
　　STABILIZERS (FLUID DYNAMICS)
　　. **HORIZONTAL TAIL SURFACES**
　　TAIL SURFACES
　　. **HORIZONTAL TAIL SURFACES**
RT　AERIAL RUDDERS
　　ELEVATORS (CONTROL SURFACES)
　　∞SURFACES
　　TAIL ASSEMBLIES
　　TRAPEZOIDAL TAIL SURFACES

HORMONE METABOLISMS
GS　METABOLISM
　　. **HORMONE METABOLISMS**
RT　HORMONES

HORMONES
GS　SECRETIONS
　　. ENDOCRINE SECRETIONS
　　. . **HORMONES**
　　. . . ESTROGENS
　　. . . HYDROXYCORTICOSTEROID
　　. . . PITUITARY HORMONES
　　. . . . ADRENOCORTICOTROPIN (ACTH)
　　. . . PROSTAGLANDINS
RT　ENDOCRINE SYSTEMS
　　HOMEOSTASIS
　　HORMONE METABOLISMS

HORN ANTENNAS
GS　ANTENNAS
　　. DIRECTIONAL ANTENNAS
　　. . **HORN ANTENNAS**
　　. WAVEGUIDE ANTENNAS
　　. . **HORN ANTENNAS**
　　MICROWAVE EQUIPMENT
　　. MICROWAVE ANTENNAS
　　. . **HORN ANTENNAS**
RT　ANTENNA DESIGN
　　LENS ANTENNAS
　　PARABOLIC ANTENNAS
　　RADAR ANTENNAS
　　SCHELKUNOFF PRINCIPLE
　　SIDELOBE REDUCTION
　　SLOT ANTENNAS

HORNS
RT　AUDITORY SIGNALS
　　SCHWARZSCHILD ANTENNAS
　　∞SIGNALS
　　SIRENS
　　SOUND GENERATORS
　　WARNING
　　WARNING SYSTEMS

HORSEPOWER
RT　PHYSICAL WORK
　　∞POWER
　　POWER EFFICIENCY
　　WORK

HORSES
GS　ANIMALS
　　. VERTEBRATES
　　. . MAMMALS
　　. . . **HORSES**
RT　GRAZING
　　LIVESTOCK

HOSES
RT　PIPES (TUBES)
　　∞TUBES

HOSPITALS
RT　EVACUATING (TRANSPORTATION)
　　MEDICAL EQUIPMENT

HOT AIR
USE　HIGH TEMPERATURE AIR

HOT ATOMS
GS　ATOMS
　　. **HOT ATOMS**
RT　BETA PARTICLES
　　DECAY
　　NEUTRON DECAY

HOT CATHODES
GS　ELECTRODES

HOT CATHODES-*(CONT.)*
　　. CATHODES
　　. . TUBE CATHODES
　　. . **HOT CATHODES**
RT　BAYARD-ALPERT IONIZATION GAGES
　　IONIZATION GAGES
　　THERMIONIC CATHODES

HOT CORROSION
GS　CORROSION
　　. **HOT CORROSION**
RT　COATINGS
　　DAMAGE
　　DEGRADATION
　　DETERIORATION
　　EROSION
　　GAS-METAL INTERACTIONS
　　METAL COATINGS
　　OXIDATION
　　PITTING
　　RUSTING
　　SCALE (CORROSION)
　　SURFACE PROPERTIES

HOT CYCLE PROPULSION SYSTEM
USE　TIP DRIVEN ROTORS

HOT ELECTRONS
GS　PARTICLES
　　. CHARGED PARTICLES
　　. . ENERGETIC PARTICLES
　　. . . ELECTRONS
　　. . . . **HOT ELECTRONS**

HOT EXTRUDING
USE　EXTRUDING

HOT FORMING
USE　HOT WORKING

HOT GAS SYSTEMS
USE　HIGH TEMPERATURE GASES

HOT GASES
USE　HIGH TEMPERATURE GASES

HOT JET EXHAUST
USE　HIGH TEMPERATURE GASES
　　JET EXHAUST

HOT JETS
USE　JET FLOW

HOT MACHINING
GS　MACHINING
　　. **HOT MACHINING**
RT　FORMING TECHNIQUES

HOT PLASMAS
USE　HIGH TEMPERATURE PLASMAS

HOT PRESSING
GS　HARDENING (MATERIALS)
　　. **HOT PRESSING**
RT　COINING
　　COLD PRESSING
　　COMPACTING
　　FORGING
　　METAL WORKING
　　∞PRESSING
　　PRESSING (FORMING)
　　SINTERING
　　STAMPING
　　UPSETTING

HOT STARS
GS　CELESTIAL BODIES
　　. STARS
　　. . **HOT STARS**
　　. . . B STARS
　　. . . BLUE STARS
　　. . . . SYMBIOTIC STARS
　　. . . O STARS
　　. . . WHITE DWARF STARS
　　. . . WOLF-RAYET STARS
RT　CATACLYSMIC VARIABLES
　　PECULIAR STARS
　　RED DWARF STARS

HOT SURFACES
RT　HEAT TRANSFER
　　RAYLEIGH-BENARD CONVECTION
　　∞SURFACES

HOT WATER ROCKET ENGINES
GS   ENGINES
    . ROCKET ENGINES
    . . **HOT WATER ROCKET ENGINES**

HOT WEATHER
GS   WEATHER
    . **HOT WEATHER**
RT   HEAT STROKE
    SUMMER
    TROPICAL REGIONS

HOT WORKING
UF   HOT FORMING
GS   FORMING TECHNIQUES
    . **HOT WORKING**
RT   BULGING
    FORGING
    METAL DRAWING
    METAL SPINNING
    METAL WORKING
    PULTRUSION
    SHEARING
    UPSETTING

HOT-FILM ANEMOMETERS
GS   DISPLAY DEVICES
    . SPEED INDICATORS
    . . ANEMOMETERS
    . . . **HOT-FILM ANEMOMETERS**
    MEASURING INSTRUMENTS
    . INDICATING INSTRUMENTS
    . . SPEED INDICATORS
    . . . ANEMOMETERS
    . . . . **HOT-FILM ANEMOMETERS**
RT   FLOW MEASUREMENT
    METEOROLOGICAL INSTRUMENTS
    SONIC ANEMOMETERS
    VELOCITY MEASUREMENT
    WIND (METEOROLOGY)
    WIND MEASUREMENT
    WIND VANES
    WIND VELOCITY
    WIND VELOCITY MEASUREMENT

HOT-WIRE ANEMOMETERS
GS   AIRCRAFT INSTRUMENTS
    . SPEED INDICATORS
    . . ANEMOMETERS
    . . . **HOT-WIRE ANEMOMETERS**
    DISPLAY DEVICES
    . SPEED INDICATORS
    . . ANEMOMETERS
    . . . **HOT-WIRE ANEMOMETERS**
    MEASURING INSTRUMENTS
    . INDICATING INSTRUMENTS
    . . SPEED INDICATORS
    . . . ANEMOMETERS
    . . . . **HOT-WIRE ANEMOMETERS**
RT   FLOW MEASUREMENT
    FLOWMETERS
    METEOROLOGICAL INSTRUMENTS
    VELOCITY MEASUREMENT

HOT-WIRE FLOWMETERS
UF   HOT-WIRE TURBULENCE METERS
GS   MEASURING INSTRUMENTS
    . FLOWMETERS
    . . **HOT-WIRE FLOWMETERS**
RT   PIRANI GAGES
    PLASMA ELECTRODES
    THERMAL CONDUCTIVITY
    TURBULENCE METERS

HOT-WIRE TURBULENCE METERS
USE   HOT-WIRE FLOWMETERS
    TURBULENCE METERS

HOTSHOT WIND TUNNELS
GS   TEST FACILITIES
    . WIND TUNNELS
    . . HYPERSONIC WIND TUNNELS
    . . . **HOTSHOT WIND TUNNELS**
    . . HYPERVELOCITY WIND TUNNELS
    . . . **HOTSHOT WIND TUNNELS**
RT   BLOWDOWN WIND TUNNELS
    SHOCK TUBES
    SHOCK TUNNELS

HOUND DOG MISSILE
GS   MISSILES
    . AIR TO SURFACE MISSILES
    . . **HOUND DOG MISSILE**
RT   TURBOJET ENGINES

HOUSEHOLDER TRANSFORMATIONS
GS   TRANSFORMATIONS (MATHEMATICS)
    . **HOUSEHOLDER TRANSFORMATIONS**
RT   PROBLEM SOLVING

HOUSEKEEPING (SPACECRAFT)
GS   CLEANING
    . **HOUSEKEEPING (SPACECRAFT)**
    CLEANLINESS
    . **HOUSEKEEPING (SPACECRAFT)**
    STERILIZATION
    . **HOUSEKEEPING (SPACECRAFT)**
    WASHING
    . **HOUSEKEEPING (SPACECRAFT)**
RT   HYGIENE
    SANITATION

HOUSINGS
GS   **HOUSINGS**
    . COWLINGS
    . DOGHOUSES (ELECTRONICS)
    . RADOMES
RT   ∞CONTAINERS
    COVERINGS
    DOMES (STRUCTURAL FORMS)
    ENCLOSURE
    ENCLOSURES
    FAIRINGS
    GUARDS (SHIELDS)
    NACELLES
    PERFORATED SHELLS
    PROTECTION
    PROTECTORS
    PROTUBERANCES
    SHELLS (STRUCTURAL FORMS)
    SHIELDING
    WALLS

HOUSTON (TX)
GS   CITIES
    . **HOUSTON (TX)**
RT   TEXAS

HOVERCRAFT
USE   GROUND EFFECT MACHINES

HOVERCRAFT GROUND EFFECT MACHINES
UF   HD-1 GROUND EFFECT MACHINES
GS   GROUND EFFECT MACHINES
    . **HOVERCRAFT GROUND EFFECT**
      MACHINES
RT   ∞AIRCRAFT
    RESEARCH AIRCRAFT
    WATER TAKEOFF AND LANDING
      AIRCRAFT

HOVERING
GS   MANEUVERS
    . **HOVERING**
RT   AERODYNAMIC STABILITY
    CUSHIONCRAFT GROUND EFFECT
      MACHINE
    FLUTTER
    GROUND EFFECT MACHINES
    HOVERING ROCKET VEHICLES
    V/STOL AIRCRAFT
    VERTICAL FLIGHT
    WHIRL TOWERS

HOVERING ROCKET VEHICLES
GS   ROCKET VEHICLES
    . **HOVERING ROCKET VEHICLES**
RT   HOVERING
    SOFT LANDING SPACECRAFT
    ∞VEHICLES

HOVERING STABILITY
GS   DYNAMIC CHARACTERISTICS
    . DYNAMIC STABILITY
    . . MOTION STABILITY
    . . . AIRCRAFT STABILITY
    . . . . **HOVERING STABILITY**
    STABILITY
    . DYNAMIC STABILITY
    . . MOTION STABILITY
    . . . AIRCRAFT STABILITY
    . . . . **HOVERING STABILITY**
RT   ATTITUDE STABILITY
    DIRECTIONAL STABILITY
    GYROSCOPIC STABILITY
    LATERAL STABILITY
    LONGITUDINAL STABILITY
    LOW SPEED STABILITY
    WHIRL TOWERS

HOWITZERS
GS   WEAPONS
    . ARTILLERY
    . . **HOWITZERS**
    . GUNS (ORDNANCE)
    . . **HOWITZERS**
RT   BALLISTICS
    GUN LAUNCHERS
    GUNNERY TRAINING

HP-115 AIRCRAFT
UF   HANDLEY PAGE HP-115 AIRCRAFT
GS   HANDLEY PAGE AIRCRAFT
    . **HP-115 AIRCRAFT**
    JET AIRCRAFT
    . **HP-115 AIRCRAFT**
    MONOPLANES
    . **HP-115 AIRCRAFT**
    RESEARCH AIRCRAFT
    . **HP-115 AIRCRAFT**
    TAILLESS AIRCRAFT
    . **HP-115 AIRCRAFT**
RT   ∞AIRCRAFT
    WING PLANFORMS

HPRR
USE   HEALTH PHYSICS RESEARCH REACTOR

HR DIAGRAM
USE   HERTZSPRUNG-RUSSELL DIAGRAM

HRB-1 HELICOPTER
USE   CH-46 HELICOPTER

HS-125 AIRCRAFT
USE   DH 125 AIRCRAFT

HS-748 AIRCRAFT
UF   AVRO WHITWORTH HS-748 AIRCRAFT
GS   HAWKER SIDDELEY AIRCRAFT
    . **HS-748 AIRCRAFT**
    JET AIRCRAFT
    . TURBOPROP AIRCRAFT
    . . **HS-748 AIRCRAFT**
    MONOPLANES
    . **HS-748 AIRCRAFT**
    PASSENGER AIRCRAFT
    . **HS-748 AIRCRAFT**
RT   ∞AIRCRAFT

HS-801 AIRCRAFT
GS   HAWKER SIDDELEY AIRCRAFT
    . **HS-801 AIRCRAFT**
    JET AIRCRAFT
    . **HS-801 AIRCRAFT**
    RECONNAISSANCE AIRCRAFT
    . **HS-801 AIRCRAFT**
RT   AERIAL RECONNAISSANCE
    ∞AIRCRAFT
    OBSERVATION AIRCRAFT
    PHOTOGRAPHY
    PHOTORECONNAISSANCE

HSS-2 HELICOPTER
USE   SH-3 HELICOPTER

HTGR
USE   HIGH TEMPERATURE GAS COOLED
      REACTORS

HTPB PROPELLANTS
GS   PROPELLANTS
    . SOLID PROPELLANTS
    . . SOLID ROCKET PROPELLANTS
    . . . **HTPB PROPELLANTS**
RT   PLASTIC PROPELLANTS
    POLYBUTADIENE

HU-1 HELICOPTER
USE   UH-1 HELICOPTER

HUBBLE CONSTANT
GS   CONSTANTS
    . **HUBBLE CONSTANT**
RT   COSMOLOGY
    GALAXIES
    RED SHIFT
    VELOCITY MEASUREMENT

HUBBLE DIAGRAM
GS   COSMOLOGY
    . **HUBBLE DIAGRAM**
RT   BARRED GALAXIES
    GALACTIC RADIATION

HUBBLE DIAGRAM-*(CONT.)*
    GALAXIES
    RED SHIFT
    VELOCITY MEASUREMENT

HUBBLE SPACE TELESCOPE
  UF   LARGE SPACE TELESCOPE
       LST
       SPACE TELESCOPE
  GS   OBSERVATORIES
       . ASTRONOMICAL OBSERVATORIES
       . . ASTRONOMICAL SATELLITES
       . . . **HUBBLE SPACE TELESCOPE**
       TELESCOPES
       . SPACEBORNE TELESCOPES
       . . **HUBBLE SPACE TELESCOPE**
  RT   FAINT OBJECT CAMERA
       SPACE SHUTTLES
       SPACE STATIONS
       SPACEBORNE ASTRONOMY

HUBS
  UF   ROTOR HUBS
  RT   SPOKES
       WHEELS

HUDSON RIVER (NY-NJ)
  GS   RIVERS
       . **HUDSON RIVER (NY-NJ)**
  RT   NEW JERSEY
       NEW YORK

HUECKEL THEORY
  RT   ∞THEORIES

HUGHES AIRCRAFT
  UF   HUGHES MILITARY AIRCRAFT
  GS   **HUGHES AIRCRAFT**
       . AH-64 HELICOPTER
       . H-17 HELICOPTER
       . OH-6 HELICOPTER
       . TH-55 HELICOPTER
       . XV-9A AIRCRAFT
  RT   ∞AIRCRAFT

HUGHES MILITARY AIRCRAFT
  USE HUGHES AIRCRAFT
       MILITARY AIRCRAFT

HUGONIOT ADIABAT
  USE HUGONIOT EQUATION OF STATE

HUGONIOT EQUATION OF STATE
  UF   HUGONIOT ADIABAT
  GS   EQUATIONS OF STATE
       . **HUGONIOT EQUATION OF STATE**
  RT   COMPRESSIBLE FLOW
       ∞EQUATIONS
       LOADS (FORCES)
       ONE DIMENSIONAL FLOW
       SHOCK WAVES

HUL
  USE HARDWARE UTILIZATION LISTS

HULLS (STRUCTURES)
  GS   **HULLS (STRUCTURES)**
       . SHIP HULLS
  RT   AIRCRAFT STRUCTURES
       BAYS (STRUCTURAL UNITS)
       BULKHEADS
       FUSELAGES
       HYDROFOILS
       KEELS
       METAL SHELLS
       PERFORATED SHELLS
       SEAPLANES
       SHELLS (STRUCTURAL FORMS)
       SKIN (STRUCTURAL MEMBER)
       STRAKES
       SWATH (SHIP)

HUM
  RT   ACOUSTICS
       ∞INTERFERENCE
       ∞NOISE

HUMAN BEHAVIOR
  GS   BEHAVIOR
       . **HUMAN BEHAVIOR**
  RT   BOREDOM
       DETACHMENT
       DISORDERS
       DITHERS

HUMAN BEHAVIOR-*(CONT.)*
       EMOTIONS
       EXTROVERSION
       INTROVERSION
       LETHARGY
       NEUROPSYCHIATRY
       PANIC

HUMAN BEINGS
  UF   MAN
  GS   ANIMALS
       . VERTEBRATES
       . . MAMMALS
       . . . PRIMATES
       . . . . **HUMAN BEINGS**
  RT   ABORIGINES
       ANTHROPOLOGY
       CENSUS
       CHILDREN
       CHIMPANZEES
       CLINICAL MEDICINE
       CULTURAL RESOURCES
       DEMOGRAPHY
       FEMALES
       MAN ENVIRONMENT INTERACTIONS
       PARENTS
       PATIENTS
       RACE FACTORS
       RACES
       YOUTH

HUMAN BODY
  GS   ANATOMY
       . **HUMAN BODY**
  RT   APPENDAGES
       ∞BODIES
       BODY MEASUREMENT (BIOLOGY)
       EXERCISE PHYSIOLOGY
       LIMBS (ANATOMY)
       LUMBAR REGION
       PHYSIOLOGY
       POSTURE

HUMAN CENTRIFUGES
  UF   PILOTED CENTRIFUGES
  GS   CENTRIFUGES
       . **HUMAN CENTRIFUGES**
  RT   ACCELERATION TOLERANCE
       ARTIFICIAL GRAVITY
       HIGH GRAVITY ENVIRONMENTS

HUMAN ENGINEERING
  USE HUMAN FACTORS ENGINEERING

HUMAN FACTORS ENGINEERING
  UF   ERGONOMICS
       HUMAN ENGINEERING
  GS   **HUMAN FACTORS ENGINEERING**
       . MOTION SIMULATION
  RT   ABILITIES
       ∞AERONAUTICS
       AIRCRAFT ACCIDENTS
       AIRCRAFT HAZARDS
       ANTHROPOMETRY
       ARCHITECTURE
       ASTRONAUT MANEUVERING EQUIPMENT
       ASTRONAUT PERFORMANCE
       ∞ASTRONAUTICS
       BIOENGINEERING
       BIOFEEDBACK
       BIONICS
       BODY MEASUREMENT (BIOLOGY)
       BRIGHTNESS
       COLOR
       COMFORT
       CYBERNETICS
       EDUCATION
       EFFICIENCY
       ENVIRONMENTAL ENGINEERING
       ENVIRONMENTS
       FATIGUE (BIOLOGY)
       ∞FLIGHT STRESS
       GLARE
       ILLUMINATING
       LIFE SUPPORT SYSTEMS
       MAN MACHINE SYSTEMS
       MANNED SPACE FLIGHT
       MANUAL CONTROL
       MONOCULAR VISION
       NOISE (SOUND)
       ∞PERFORMANCE
       PILOT ERROR
       PRODUCTION ENGINEERING
       PSYCHOLOGICAL EFFECTS
       SAFETY DEVICES

HUMAN FACTORS ENGINEERING-*(CONT.)*
       SAFETY MANAGEMENT
       SYSTEMS ENGINEERING
       TELEOPERATORS
       VISIBILITY
       VISION
       WHEELCHAIRS
       WORKSTATIONS

HUMAN FACTORS LABORATORIES
  GS   LABORATORIES
       . **HUMAN FACTORS LABORATORIES**
  RT   ENVIRONMENTAL LABORATORIES

HUMAN PATHOLOGY
  GS   MEDICAL SCIENCE
       . PATHOLOGY
       . . **HUMAN PATHOLOGY**
  RT   CHOLERA
       CONVULSIONS
       EPILEPSY
       PATIENTS

HUMAN PERFORMANCE
  GS   **HUMAN PERFORMANCE**
       . ASTRONAUT PERFORMANCE
       . . BLACKOUT PREVENTION
       . OPERATOR PERFORMANCE
       . PILOT PERFORMANCE
  RT   COMPETITION
       INTRAVEHICULAR ACTIVITY
       MENTAL HEALTH
       MENTAL PERFORMANCE
       ∞PERFORMANCE
       PILOT ERROR
       PSYCHOMOTOR PERFORMANCE
       RACE FACTORS
       SENSORIMOTOR PERFORMANCE
       WORKLOADS (PSYCHOPHYSIOLOGY)

HUMAN REACTIONS
  RT   BIOLOGICAL EFFECTS
       BOREDOM
       COMPETITION
       EMOTIONAL FACTORS
       LAUGHING
       NOISE POLLUTION
       PHYSIOLOGICAL EFFECTS
       PSYCHOLOGICAL EFFECTS
       PSYCHOMOTOR PERFORMANCE
       ∞REACTION
       REACTION TIME
       REWARD (PSYCHOLOGY)
       SENSORIMOTOR PERFORMANCE
       SHOCK (PHYSIOLOGY)
       VACILLATION

HUMAN RELATIONS
  UF   INTERPERSONAL RELATIONS
  RT   EMPLOYEE RELATIONS
       PERSONNEL MANAGEMENT
       SOCIOLOGY

HUMAN RESOURCES
  RT   MANPOWER

HUMAN TOLERANCES
  GS   TOLERANCES (PHYSIOLOGY)
       . **HUMAN TOLERANCES**
  RT   ∞ACCELERATION
       ACCELERATION TOLERANCE
       DIVING (UNDERWATER)
       HEAT ACCLIMATIZATION
       HEAT TOLERANCE
       NOISE POLLUTION
       NOISE TOLERANCE
       ORTHOSTATIC TOLERANCE
       RADIATION TOLERANCE
       SHOCK (PHYSIOLOGY)

HUMAN WASTES
  GS   WASTES
       . METABOLIC WASTES
       . . **HUMAN WASTES**
       . . . URINE
  RT   ACTIVATED SLUDGE
       AIR POLLUTION
       ENVIRONMENT POLLUTION
       ENVIRONMENTAL SURVEYS
       EXCRETION
       LIQUID WASTES
       ORGANIC WASTES (FUEL CONVERSION)
       POLLUTION
       SEWAGE
       SEWERS

HUMAN WASTES-*(CONT.)*
    SOLID WASTES
    WASTE DISPOSAL

HUMASON COMET
GS   CELESTIAL BODIES
   . COMETS
   . . **HUMASON COMET**

HUMERUS
RT   ARM (ANATOMY)
    ELBOW (ANATOMY)

HUMIDITY
RT   AIR CONDITIONING
    ATMOSPHERIC DENSITY
    ATMOSPHERIC MOISTURE
    BODY TEMPERATURE
    CLIMATOLOGY
    COMFORT
    CORROSION
    DEHUMIDIFICATION
    DROP SIZE
    DRY HEAT
    ENVIRONMENTS
    HYGRAL PROPERTIES
    HYGROMETERS
    LAPSE RATE
    METEOROLOGICAL PARAMETERS
    METEOROLOGY
    MOISTURE
    MOISTURE CONTENT
    MOISTURE METERS
    PERSPIRATION
    PRECIPITATION (METEOROLOGY)
    PSYCHOLOGICAL EFFECTS
    PSYCHROMETERS
    REFRIGERATING
    TEMPERATURE
    VAPOR PRESSURE
    WATER
    WATER VAPOR
    WEATHER FORECASTING

HUMIDITY MEASUREMENT
RT   HYGROMETERS
   ∞MEASUREMENT
    METEOROLOGICAL INSTRUMENTS
    MOISTURE METERS
    PSYCHROMETERS

HUMMINGBIRD AIRCRAFT
USE   XV-4 AIRCRAFT

HUNGARY
GS   NATIONS
   . **HUNGARY**
RT   CENTRAL EUROPE
    EUROPE

HUNTER F-2 AIRCRAFT
USE   F-2 AIRCRAFT

HUNTING H-126 AIRCRAFT
USE   H-126 AIRCRAFT

HUNTING P-84 AIRCRAFT
USE   JET PROVOST AIRCRAFT

HURRICANES
GS   STORMS
   . CYCLONES
   . . **HURRICANES**
   . . . ANNA HURRICANE
   . STORMS (METEOROLOGY)
   . . TROPICAL STORMS
   . . . **HURRICANES**
   . . . . ANNA HURRICANE
RT   CLIMATOLOGY
    METEOROLOGY
    STORM DAMAGE
    STORM SURGES
    TORNADOES
    TYPHOONS

HUS-1 HELICOPTER
USE   UH-34 HELICOPTER

HUSKIE HELICOPTER
USE   HH-43 HELICOPTER

HUSTLER AIRCRAFT
USE   B-58 AIRCRAFT

HUYGENS PRINCIPLE
RT   DIFFRACTION
   ∞OPTICS
    POINT SOURCES
    REFRACTION
    SCATTERING
    SCHELKUNOFF PRINCIPLE
    SPHERICAL WAVES
    WAVE FRONTS
    WAVE PROPAGATION

HU2K-1 HELICOPTER
USE   UH-2 HELICOPTER

HVITTIS CHONDRITE
GS   CELESTIAL BODIES
   . METEORITES
   . . STONY METEORITES
   . . . CHONDRITES
   . . . . **HVITTIS CHONDRITE**

HYBRID CIRCUITS
GS   CIRCUITS
   . **HYBRID CIRCUITS**
RT   ELECTRONIC PACKAGING
    PRINTED CIRCUITS
    SEMICONDUCTOR DEVICES
    TRANSISTOR CIRCUITS

HYBRID COMBUSTION
USE   HYBRID PROPELLANT ROCKET ENGINES

HYBRID COMPUTERS
GS   DATA PROCESSING EQUIPMENT
   . COMPUTERS
   . . **HYBRID COMPUTERS**
RT   ANALOG COMPUTERS
    DIGITAL COMPUTERS

HYBRID NAVIGATION SYSTEMS
GS   NAVIGATION
   . **HYBRID NAVIGATION SYSTEMS**
RT   NAVIGATION AIDS
    NAVIGATION INSTRUMENTS
   ∞SYSTEMS

HYBRID PROPELLANT ROCKET ENGINES
UF   HYBRID COMBUSTION
GS   ENGINES
   . ROCKET ENGINES
   . . **HYBRID PROPELLANT ROCKET
       ENGINES**
   . . . LITHERGOL ROCKET ENGINES
RT   BOOSTER ROCKET ENGINES
    INTERNAL COMBUSTION ENGINES
    JET ENGINES
    LIQUID PROPELLANT ROCKET ENGINES
    RESTARTABLE ROCKET ENGINES
    SOLID PROPELLANT ROCKET ENGINES
    SUSTAINER ROCKET ENGINES
    VERNIER ENGINES

HYBRID PROPELLANTS
UF   LITHERGOLIC PROPELLANTS
GS   PROPELLANTS
   . **HYBRID PROPELLANTS**
RT   CASE BONDED PROPELLANTS
    CHEMICAL FUELS
    CRYOGENIC ROCKET PROPELLANTS
    GASEOUS ROCKET PROPELLANTS
    GELLED ROCKET PROPELLANTS
    HIGH ENERGY FUELS
    HIGH ENERGY PROPELLANTS
    HYPERGOLIC ROCKET PROPELLANTS
    LIQUID ROCKET PROPELLANTS
    METAL FUELS
    METAL PROPELLANTS
    SOLID PROPELLANT IGNITION
    SOLID PROPELLANTS
    SOLID ROCKET PROPELLANTS

HYBRID PROPULSION
UF   DUAL MODE PROPULSION
GS   PROPULSION
   . CHEMICAL PROPULSION
   . . **HYBRID PROPULSION**
RT   JET ENGINES
    LASER PROPULSION
    ROCKET ENGINES

∞ **HYBRID ROCKET ENGINES**
SN   *(USE OF A MORE SPECIFIC TERM IS
       RECOMMENDED--CONSULT THE TERMS
       LISTED BELOW)*

HYBRID ROCKET ENGINES-*(CONT.)*
RT   DUCTED ROCKET ENGINES

HYBRID STRUCTURES
RT   COMPOSITE STRUCTURES
    ELASTIC PROPERTIES
    FLEXIBLE BODIES
    RIGID STRUCTURES
    STRUCTURAL STABILITY
   ∞STRUCTURES

HYBRIDS (BIOLOGY)
USE   GENETIC ENGINEERING

HYDRATES
RT   AQUEOUS SOLUTIONS
    WATER

HYDRATION
RT   CHEMICAL REACTIONS
    DEHYDRATION
    HYDROLYSIS

HYDRAULIC ACTUATORS
USE   ACTUATORS
    HYDRAULIC EQUIPMENT

HYDRAULIC ANALOGIES
GS   ANALOGIES
   . **HYDRAULIC ANALOGIES**
RT   COMPUTERIZED SIMULATION
    FLOW VISUALIZATION
    FLUERICS
    FLUIDICS
    GAS FLOW
    NUMERICAL FLOW VISUALIZATION
    WAVE PROPAGATION

HYDRAULIC CONTROL
UF   ELECTROHYDRAULIC CONTROL
RT   AUTOMATIC CONTROL
   ∞CONTROL
    ELECTRONIC CONTROL
    ENGINE CONTROL
    FLUID POWER
    FLUIDICS
   ∞HYDRAULICS
    PNEUMATIC CONTROL
    REMOTE CONTROL
    SOLENOID VALVES

HYDRAULIC EQUIPMENT
UF   HYDRAULIC ACTUATORS
    HYDRAULIC HEATING SOURCES
    HYDRAULIC PUMPS
    HYDRAULIC SYSTEMS
    HYDRAULIC VALVES
GS   **HYDRAULIC EQUIPMENT**
   . AIRCRAFT HYDRAULIC SYSTEMS
RT   AIRBORNE EQUIPMENT
    AUTOMATIC CONTROL VALVES
    COCKS
    CUSHIONS
   ∞EQUIPMENT
    FLUID AMPLIFIERS
    FLUID POWER
    FLUID SWITCHING ELEMENTS
    FLY BY TUBE CONTROL
    GATES (OPENINGS)
   ∞HYDRAULICS
    MOTORS
    NETWORK ANALYSIS
    NETWORK SYNTHESIS
   ∞NETWORKS
    PUMPS
    RELIEF VALVES
    SERVOCONTROL
    SERVOMECHANISMS
    SHOCK ABSORBERS
   ∞SYSTEMS
    TURBINE WHEELS
    VALVES
    WATER HAMMER
    WHEEL BRAKES

HYDRAULIC FLUIDS
GS   LIQUIDS
   . **HYDRAULIC FLUIDS**
   . . SKYDROL (TRADEMARK)
RT   FLUID PRESSURE
    FLUID TRANSMISSION LINES
   ∞FLUIDS
    HIGH TEMPERATURE FLUIDS
   ∞HYDRAULICS

HYDRAULIC FLUIDS-(CONT.)
    OILS
    PATCH TESTS
    TRANSMISSION FLUIDS
    WORKING FLUIDS

HYDRAULIC HEATING SOURCES
    USE   HEAT SOURCES
           HYDRAULIC EQUIPMENT

HYDRAULIC JETS
    UF   WATER JETS
    GS   FLUID JETS
           . HYDRAULIC JETS
    RT   JET FLOW

HYDRAULIC PUMPS
    USE   HYDRAULIC EQUIPMENT
           PUMPS

HYDRAULIC SHOCK
    GS   MECHANICAL SHOCK
           . HYDRAULIC SHOCK

HYDRAULIC SYSTEMS
    USE   HYDRAULIC EQUIPMENT

HYDRAULIC TEST TUNNELS
    UF   WATER TUNNELS
    GS   TEST FACILITIES
           . HYDRAULIC TEST TUNNELS
    RT   ∞TUNNELS

HYDRAULIC VALVES
    USE   HYDRAULIC EQUIPMENT
           VALVES

∞ HYDRAULICS
    SN   (USE OF A MORE SPECIFIC TERM IS
           RECOMMENDED--CONSULT THE TERMS
           LISTED BELOW)
    RT   FLUID DYNAMICS
           FLUID FLOW
           FLUID MECHANICS
           FLUID POWER
           HYDRAULIC CONTROL
           HYDRAULIC EQUIPMENT
           HYDRAULIC FLUIDS
           HYDRODYNAMIC RAM EFFECT
           HYDRODYNAMICS
           HYDROLOGY
           HYDROMECHANICS
           HYDROSTATICS
           IMPEDANCE
           INFLUENCE COEFFICIENT
           LIMNOLOGY
           ∞MECHANICS (PHYSICS)
           PIPES (TUBES)
           PNEUMATICS
           PRESSURE HEADS
           THERMOHYDRAULICS
           WATER
           WATER FLOW
           WATER PRESSURE

HYDRAZIDES
    GS   ALIPHATIC COMPOUNDS
           . HYDRAZIDES
    RT   HYDRAZINES

HYDRAZINE BORANE
    GS   ALIPHATIC COMPOUNDS
           . HYDRAZINES
           . . HYDRAZINE BORANE
           BORON COMPOUNDS
           . BORON HYDRIDES
           . . BORANES
           . . . HYDRAZINE BORANE
           HYDROGEN COMPOUNDS
           . HYDRIDES
           . . BORON HYDRIDES
           . . . BORANES
           . . . . HYDRAZINE BORANE

HYDRAZINE ENGINES
    UF   NIMPHE (ENGINE)
    GS   ENGINES
           . ROCKET ENGINES
           . . LIQUID PROPELLANT ROCKET
               ENGINES
           . . . HYDRAZINE ENGINES
    RT   TURBOROCKET ENGINES

HYDRAZINE NITRATE
    GS   NITROGEN COMPOUNDS
           . NITRATES
           . . INORGANIC NITRATES
           . . . HYDRAZINE NITRATE

HYDRAZINE NITROFORM
    GS   ALIPHATIC COMPOUNDS
           . HYDRAZINE NITROFORM
           ESTERS
           . ORGANIC NITRATES
           . . NITROFORMS
           . . . HYDRAZINE NITROFORM
           EXPLOSIVES
           . HYDRAZINE NITROFORM
           NITROGEN COMPOUNDS
           . NITRATES
           . . ORGANIC NITRATES
           . . . NITROFORMS
           . . . . HYDRAZINE NITROFORM
           PROPELLANTS
           . HYDRAZINE NITROFORM

HYDRAZINE PERCHLORATES
    GS   ALIPHATIC COMPOUNDS
           . HYDRAZINES
           . . HYDRAZINE PERCHLORATES
           HALOGEN COMPOUNDS
           . CHLORINE COMPOUNDS
           . . PERCHLORATES
           . . . HYDRAZINE PERCHLORATES

HYDRAZINES
    GS   ALIPHATIC COMPOUNDS
           . HYDRAZINES
           . . CHLORPROMAZINE
           . . DIHYDRAZINE
           . . DIMETHYLHYDRAZINES
           . . ETHYLENE DIHYDRAZINE
           . . HYDRAZINE BORANE
           . . HYDRAZINE PERCHLORATES
           . . METHYLHYDRAZINE
           . . TETRAFLUOROHYDRAZINE
    RT   AEROZINE
           AMINES
           HYDRAZIDES
           HYDRAZONES
           LIQUID ROCKET PROPELLANTS
           ROCKET PROPELLANTS

HYDRAZINIUM COMPOUNDS
    GS   NITROGEN COMPOUNDS
           . HYDRAZINIUM COMPOUNDS
    RT   ∞CHEMICAL COMPOUNDS

HYDRAZOIC ACID
    GS   ACIDS
           . HYDRAZOIC ACID
           NITROGEN COMPOUNDS
           . HYDRAZOIC ACID
    RT   NITROGEN HYDRIDES

HYDRAZONES
    GS   NITROGEN COMPOUNDS
           . HYDRAZONES
    RT   HYDRAZINES

HYDRAZONIUM COMPOUNDS
    RT   ∞CHEMICAL COMPOUNDS

HYDRIDES
    GS   HYDROGEN COMPOUNDS
           . HYDRIDES
           . . BOROHYDRIDES
           . . . ALUMINUM BOROHYDRIDES
           . . . BERYLLIUM BOROHYDRIDES
           . . BORON HYDRIDES
           . . . ALUMINUM BOROHYDRIDES
           . . . BERYLLIUM BOROHYDRIDES
           . . . BORANES
           . . . . CARBORANE
           . . . . HYDRAZINE BORANE
           . . . . PENTABORANES
           . . DIBORANE
           . . DIHYDRIDES
           . . METAL HYDRIDES
           . . . ALUMINUM HYDRIDES
           . . . . ALUMINUM BOROHYDRIDES
           . . . BERYLLIUM HYDRIDES
           . . . CESIUM HYDRIDES
           . . . LITHIUM HYDRIDES
           . . . . LITHIUM ALUMINUM HYDRIDES
           . . . POTASSIUM HYDRIDES
           . . . SODIUM HYDRIDES
           . . NITROGEN HYDRIDES

HYDRIDES-(CONT.)
           . . PHOSPHINES
           . . SILANES
           . . . CHLOROSILANES
           . . . . METHYL CHLOROSILANES
           . . ZIRCONIUM HYDRIDES
    RT   DEUTERIDES
           HYDROGEN PRODUCTION

HYDROACOUSTICS
    USE   UNDERWATER ACOUSTICS

HYDROAEROMECHANICS
    USE   AERODYNAMICS

HYDROBALLISTICS
    GS   BALLISTICS
           . HYDROBALLISTICS
    RT   BALLISTIC RANGES
           HYDRODYNAMICS
           TORPEDOES
           UNDERWATER EXPLOSIONS
           UNDERWATER TRAJECTORIES

HYDROBAROPHONES
    USE   HYDROPHONES

HYDROBORATION
    GS   CHEMICAL REACTIONS
           . HYDROBORATION

HYDROBROMIC ACID
    GS   ACIDS
           . HYDROBROMIC ACID
           HALOGEN COMPOUNDS
           . BROMINE COMPOUNDS
           . . BROMIDES
           . . . HYDROBROMIC ACID
           . HALIDES
           . . BROMIDES
           . . . HYDROBROMIC ACID

HYDROBROMIDES
    GS   HALOGEN COMPOUNDS
           . BROMINE COMPOUNDS
           . . BROMIDES
           . . . HYDROBROMIDES
           . HALIDES
           . . BROMIDES
           . . . HYDROBROMIDES
           HYDROGEN COMPOUNDS
           . HYDROBROMIDES

HYDROCARBON COMBUSTION
    GS   COMBUSTION
           . HYDROCARBON COMBUSTION
    RT   EXPLOSIONS
           FUEL COMBUSTION
           OXIDATION
           PROPELLANT COMBUSTION
           SMOG

HYDROCARBON FUEL PRODUCTION
    GS   HYDROCARBON FUEL PRODUCTION
           . ATMOSPHERIC ENERGY SOURCES
    RT   AGRICULTURE
           BIOCONVERSION
           BIOMASS ENERGY PRODUCTION
           ENERGY TECHNOLOGY
           HYDROGEN FUELS
           LIGNITE
           SOLVENT REFINED COAL
           WASTE UTILIZATION

HYDROCARBON FUELS
    GS   FUELS
           . CHEMICAL FUELS
           . . HYDROCARBON FUELS
           . . . DIESEL FUELS
           . . . GASOLINE
           . . . JET ENGINE FUELS
           . . . . JP-4 JET FUEL
           . . . . JP-5 JET FUEL
           . . . . JP-6 JET FUEL
           . . . . JP-8 JET FUEL
           . . . RP-1 ROCKET PROPELLANTS
           . . . SYNTHANE
    RT   ACETYLENE
           AIRCRAFT FUELS
           ALKANES
           AMINES
           AUTOMOBILE FUELS
           BUTADIENE
           CLEAN FUELS

HYDROCARBON FUELS-*(CONT.)*
    COAL GASIFICATION
    COAL LIQUEFACTION
    COAL UTILIZATION
    CRUDE OIL
    ENDOTHERMIC FUELS
    ENERGY POLICY
    ENVIRONMENTAL CHEMISTRY
    ETHANE
    ETHYLENE
    FUEL PRODUCTION
    HEPTANES
    HEXENES
    HIGH ENERGY FUELS
    HYDROGEN FUELS
    HYDROGEN-BASED ENERGY
    HYPERGOLIC ROCKET PROPELLANTS
    KEROGEN
    KEROSENE
    LIGNITE
    LIQUEFIED NATURAL GAS
    METHANATION
    METHANE
    NATURAL GAS
    PARAFFINS
    PROPANE
    RETORT PROCESSING
    ROCKET PROPELLANTS
    SHALE OIL
    STORABLE PROPELLANTS
    SYNTHETIC FUELS

HYDROCARBON POISONING
RT   BENZENE POISONING
    INDUSTRIAL SAFETY
    ∞POISONING
    SMOG
    TOXICITY AND SAFETY HAZARD
    TOXICOLOGY

HYDROCARBONS
GS   **HYDROCARBONS**
    . ACETYLENE
    . ALKANES
    . . BUTANES
    . . CETANE
    . . ETHANE
    . . HEPTANES
    . . HEXENES
    . . METHANE
    . . NITROPROPANE
    . . NONANES
    . . OCTANES
    . . PARAFFINS
    . . . CERESIN
    . . PENTANES
    . . . NEOPENTANE
    . . PROPANE
    . ALKENES
    . . BUTADIENE
    . . BUTENES
    . . ETHYLENE
    . . . VINYLIDENE
    . . HEXENES
    . . PROPYLENE
    . . TRIENES
    . ALKYNES
    . . OXYACETYLENE
    . BENZENE
    . CAROTENE
    . CHLOROBENZENES
    . CUBANE
    . CYCLIC HYDROCARBONS
    . . ANTHRACENE
    . . COLCHICINE
    . . CYCLOBUTANE
    . . CYCLOPROPANE
    . . MENTHOL
    . . NAPHTHENES
    . CYCLOHEXANE
    . DIENES
    . . BUTADIENE
    . . HEPTADIENE
    . . HEXADIENE
    . DIPHENYL COMPOUNDS
    . . DIPHENYL HYDANTOIN
    . DURENE
    . INDENE
    . KEROGEN
    . MESITYLENE
    . METHYLENE
    . NAPHTHALENE
    . NATURAL GAS
    . . LIQUEFIED NATURAL GAS
    . PHENANTHRENE

HYDROCARBONS-*(CONT.)*
    . PYRENES
    . QUINOXALINES
    . STILBENE
    . TOLUENE
    . TRIPHENYLS
    . XYLENE
RT   ALIPHATIC COMPOUNDS
    ∞AROMATIC COMPOUNDS
    CARBON COMPOUNDS
    CRACKING (CHEMICAL ENGINEERING)
    ORGANIC PEROXIDES

HYDROCHLORIC ACID
GS   ACIDS
    . **HYDROCHLORIC ACID**
    HALOGEN COMPOUNDS
    . HALIDES
    . . CHLORIDES
    . . . HYDROGEN CHLORIDES
    . . . . **HYDROCHLORIC ACID**

HYDROCHLORIDES
GS   HALOGEN COMPOUNDS
    . CHLORINE COMPOUNDS
    . . CHLORIDES
    . . . **HYDROCHLORIDES**
    . HALIDES
    . . CHLORIDES
    . . . **HYDROCHLORIDES**
RT   HYDROGEN CHLORIDES

HYDROCLIMATOLOGY
RT   AGROCLIMATOLOGY
    CLIMATOLOGY
    HYDROGRAPHY
    HYDROLOGY
    METEOROLOGY
    OCEANOGRAPHY

HYDROCRACKING
GS   FRACTIONATION
    . **HYDROCRACKING**
RT   COAL GASIFICATION
    COAL LIQUEFACTION

HYDROCYANIC ACID
UF   HYDROGEN CYANIDES
    PRUSSIC ACID
GS   ACIDS
    . **HYDROCYANIC ACID**
    HYDROGEN COMPOUNDS
    . **HYDROCYANIC ACID**
    NITROGEN COMPOUNDS
    . **HYDROCYANIC ACID**
RT   CN EMISSION
    HCN LASERS

HYDRODYNAMIC COEFFICIENTS
RT   COMPUTATIONAL FLUID DYNAMICS
    ∞DRAG COEFFICIENTS
    FLOW DISTRIBUTION
    FLOW VELOCITY
    LIQUID FLOW
    SEA ROUGHNESS
    SHIP HULLS
    STEADY FLOW
    UNSTEADY FLOW
    WATER WAVES

HYDRODYNAMIC EQUATIONS
GS   EQUATIONS OF MOTION
    . KINETIC EQUATIONS
    . . **HYDRODYNAMIC EQUATIONS**
    . . . HELMHOLTZ VORTICITY EQUATION
RT   BOLTZMANN TRANSPORT EQUATION
    ∞EQUATIONS
    FLOW STABILITY
    FLOW THEORY
    FLUID MECHANICS
    GAS DYNAMICS
    HYDRODYNAMICS
    METEOROLOGY
    PLASMA DYNAMICS

HYDRODYNAMIC RAM EFFECT
RT   ∞EFFECTS
    FLUID FILLED SHELLS
    ∞HYDRAULICS
    HYPERVELOCITY IMPACT
    IMPACT
    KINETIC ENERGY
    LIQUID FILLED SHELLS
    MOMENTUM TRANSFER
    PENETRATION

HYDRODYNAMIC STABILITY
USE   FLOW STABILITY

HYDRODYNAMIC TUNNELS
USE   PLASMA JET WIND TUNNELS

HYDRODYNAMICS
GS   FLUID MECHANICS
    . FLUID DYNAMICS
    . . **HYDRODYNAMICS**
    . . . ELASTOHYDRODYNAMICS
    . . . ELECTROHYDRODYNAMICS
    . . . MAGNETOHYDRODYNAMICS
    . . . . CYLINDRICAL PLASMAS
    . HYDROMECHANICS
    . **HYDRODYNAMICS**
    . . . ELASTOHYDRODYNAMICS
    . . . ELECTROHYDRODYNAMICS
    . . . MAGNETOHYDRODYNAMICS
RT   BALLAST (MASS)
    ∞DYNAMICS
    EULER EQUATIONS OF MOTION
    FLOW THEORY
    FLUID FLOW
    FLUID POWER
    GAS DYNAMICS
    ∞HYDRAULICS
    HYDROBALLISTICS
    HYDRODYNAMIC EQUATIONS
    HYDROSTATICS
    KROOK EQUATION
    LAGRANGE COORDINATES
    ∞MECHANICS (PHYSICS)
    OCEAN DYNAMICS
    PRESSURE GRADIENTS
    PRESSURE HEADS
    SEEPAGE
    SHIP HULLS
    THERMOHYDRAULICS
    WATER
    WATER FLOW
    WATER HAMMER
    WATER PRESSURE

HYDROELASTICITY
GS   MECHANICAL PROPERTIES
    . ELASTIC PROPERTIES
    . . **HYDROELASTICITY**
RT   COMPRESSIBILITY
    COMPRESSIBLE FLUIDS
    MODULUS OF ELASTICITY
    THERMOELASTICITY
    VISCOELASTICITY

HYDROELECTRIC POWER STATIONS
UF   HYDROPOWER STATIONS
GS   STATIONS
    . **HYDROELECTRIC POWER STATIONS**
RT   ELECTRIC POWER PLANTS
    ELECTRIC POWER TRANSMISSION
    HYDROELECTRICITY
    ∞POWER PLANTS
    ∞POWER TRANSMISSION
    TURBOGENERATORS
    WATER WHEELS

HYDROELECTRICITY
GS   ELECTRICITY
    . **HYDROELECTRICITY**
RT   DAMS
    ELECTRIC CURRENT
    ∞ELECTRIC POWER
    HYDROELECTRIC POWER STATIONS
    ∞POWER PLANTS
    TURBOGENERATORS

HYDROFLUORIC ACID
UF   HYDROGEN FLUORIDES
GS   ACIDS
    . **HYDROFLUORIC ACID**
    HALOGEN COMPOUNDS
    . FLUORINE COMPOUNDS
    . . FLUORIDES
    . . . **HYDROFLUORIC ACID**
    . HALIDES
    . . FLUORIDES
    . . . **HYDROFLUORIC ACID**

HYDROFOIL BOATS
USE   HYDROFOIL CRAFT

HYDROFOIL CRAFT
UF   HYDROFOIL BOATS
RT   CAPTURED AIR BUBBLE VEHICLES
    HYDROFOILS

HYDROFOIL CRAFT-*(CONT.)*
    HYDROPLANES (VEHICLES)
    SHIPS

HYDROFOIL OSCILLATIONS
GS   OSCILLATIONS
    . **HYDROFOIL OSCILLATIONS**
RT   FLOW STABILITY
    FLUTTER
    HYDROFOILS
    SUPERCAVITATING FLOW

HYDROFOILS
GS   **HYDROFOILS**
    . KEELS
RT   AIRFOILS
    ∞BLADES
    ELEVATORS (CONTROL SURFACES)
    FINS
    ∞FOILS
    FOILS (MATERIALS)
    GUIDE VANES
    HULLS (STRUCTURES)
    HYDROFOIL CRAFT
    HYDROFOIL OSCILLATIONS
    HYDROPLANES (SURFACES)
    HYDROPLANING
    LANDING GEAR
    MARINE RUDDERS
    SHIPS
    SKIS
    STREAMLINING
    TAIL ASSEMBLIES

HYDROFORMING
GS   METAL WORKING
    . **HYDROFORMING**
RT   DEHYDROGENATION

HYDROGEN
GS   CHEMICAL ELEMENTS
    . **HYDROGEN**
    . . HYDROGEN ATOMS
    . . HYDROGEN IONS
    . . HYDROGEN ISOTOPES
    . . . DEUTERIUM
    . . . HYDROGEN 4
    . . . METALLIC HYDROGEN
    . . . TRITIUM
    . . HYDROGEN PLASMA
    . . . DEUTERIUM PLASMA
    . . LIQUID HYDROGEN
    . . ORTHO HYDROGEN
    . . PARA HYDROGEN
    GASES
    . **HYDROGEN**
    . . HYDROGEN ATOMS
    . . HYDROGEN IONS
    . . HYDROGEN ISOTOPES
    . . . DEUTERIUM
    . . . HYDROGEN 4
    . . . TRITIUM
    . . HYDROGEN PLASMA
    . . . DEUTERIUM PLASMA
    . . LIQUID HYDROGEN
    . . ORTHO HYDROGEN
    . . PARA HYDROGEN
RT   BALMER SERIES
    FUELS
    HYDROGENATION
    HYDROGENOLYSIS
    HYDRONIUM IONS
    METALLICITY
    NEPTUNE ATMOSPHERE
    PASCHEN SERIES
    RYDBERG SERIES
    SYNTHANE
    URANUS ATMOSPHERE

HYDROGEN AIR FUEL CELLS
USE   HYDROGEN OXYGEN FUEL CELLS

HYDROGEN ATOMS
GS   ATOMS
    . **HYDROGEN ATOMS**
    CHEMICAL ELEMENTS
    . HYDROGEN
    . . **HYDROGEN ATOMS**
    GASES
    . HYDROGEN
    . . **HYDROGEN ATOMS**

HYDROGEN AZIDES
GS   EXPLOSIVES
    . **HYDROGEN AZIDES**

HYDROGEN AZIDES-*(CONT.)*
    NITROGEN COMPOUNDS
    . AZIDES (INORGANIC)
    . . **HYDROGEN AZIDES**
    PROPELLANTS
    . **HYDROGEN AZIDES**

HYDROGEN BOMBS
USE   FUSION WEAPONS

HYDROGEN BONDS
GS   CHEMICAL BONDS
    . **HYDROGEN BONDS**

HYDROGEN CHLORIDES
GS   HALOGEN COMPOUNDS
    . HALIDES
    . . CHLORIDES
    . . . **HYDROGEN CHLORIDES**
    . . . . HYDROCHLORIC ACID
RT   HYDROCHLORIDES

HYDROGEN CLOUDS
GS   CLOUDS
    . **HYDROGEN CLOUDS**
RT   CLOUDS (METEOROLOGY)
    DROP SIZE
    GALACTIC ROTATION
    GASES
    MOLECULAR CLOUDS
    PLASMA CLOUDS
    SPIN TEMPERATURE
    VAPOR PHASES
    VAPORS

HYDROGEN COMPOUNDS
GS   **HYDROGEN COMPOUNDS**
    . DEUTERIUM COMPOUNDS
    . . DEUTERIDES
    . . DEUTERIUM FLUORIDES
    . HEAVY WATER
    . HYDRIDES
    . . BOROHYDRIDES
    . . . ALUMINUM BOROHYDRIDES
    . . . BERYLLIUM BOROHYDRIDES
    . . BORON HYDRIDES
    . . . ALUMINUM BOROHYDRIDES
    . . . BERYLLIUM BOROHYDRIDES
    . . . BORANES
    . . . . CARBORANE
    . . . . HYDRAZINE BORANE
    . . . . PENTABORANES
    . . . DIBORANE
    . . DIHYDRIDES
    . . METAL HYDRIDES
    . . . ALUMINUM HYDRIDES
    . . . . ALUMINUM BOROHYDRIDES
    . . . BERYLLIUM HYDRIDES
    . . . CESIUM HYDRIDES
    . . . LITHIUM HYDRIDES
    . . . . LITHIUM ALUMINUM HYDRIDES
    . . . POTASSIUM HYDRIDES
    . . . SODIUM HYDRIDES
    . . NITROGEN HYDRIDES
    . . PHOSPHINES
    . . SILANES
    . . . CHLOROSILANES
    . . . METHYL CHLOROSILANES
    . . ZIRCONIUM HYDRIDES
    . HYDROBROMIDES
    . HYDROCYANIC ACID
    . HYDROGEN PEROXIDE
    . HYDROGEN SULFIDE
    . HYDROSULFITES
    . LIGHT WATER
RT   ACIDS
    ∞CHEMICAL COMPOUNDS
    WATER

HYDROGEN CYANIDES
USE   HYDROCYANIC ACID

HYDROGEN DEUTERIUM OXIDE
USE   HEAVY WATER

HYDROGEN EMBRITTLEMENT
RT   CHEMISORPTION
    GAS-METAL INTERACTIONS
    IRON
    STEELS

HYDROGEN ENGINES
GS   ENGINES
    . INTERNAL COMBUSTION ENGINES

HYDROGEN ENGINES-*(CONT.)*
    . . GAS TURBINE ENGINES
    . . . **HYDROGEN ENGINES**
RT   AIRCRAFT ENGINES
    AUTOMOBILES

HYDROGEN FLUORIDES
USE   HYDROFLUORIC ACID

HYDROGEN FUELS
GS   FUELS
    . CHEMICAL FUELS
    . . LIQUID FUELS
    . . . **HYDROGEN FUELS**
RT   CRYOGENIC ROCKET PROPELLANTS
    DEUTERIUM
    FUEL CELLS
    FUEL PRODUCTION
    GASEOUS ROCKET PROPELLANTS
    GELLED PROPELLANTS
    HYDROCARBON FUEL PRODUCTION
    HYDROCARBON FUELS
    HYDROGEN-BASED ENERGY
    LIQUID HYDROGEN
    LIQUID ROCKET PROPELLANTS
    RAMJET ENGINES

HYDROGEN IONS
GS   CHEMICAL ELEMENTS
    . HYDROGEN
    . . **HYDROGEN IONS**
    GASES
    . HYDROGEN
    . . **HYDROGEN IONS**
    IONS
    . **HYDROGEN IONS**
RT   ACIDITY
    HYDRONIUM IONS
    PH
    PH FACTOR
    POSITIVE IONS
    PROTONS

HYDROGEN ISOTOPES
GS   CHEMICAL ELEMENTS
    . HYDROGEN
    . . **HYDROGEN ISOTOPES**
    . . . DEUTERIUM
    . . . HYDROGEN 4
    . . . METALLIC HYDROGEN
    . . . TRITIUM
    . NUCLIDES
    . ISOTOPES
    . . **HYDROGEN ISOTOPES**
    . . . DEUTERIUM
    . . . HYDROGEN 4
    . . . TRITIUM
    GASES
    . HYDROGEN
    . . **HYDROGEN ISOTOPES**
    . . . DEUTERIUM
    . . . HYDROGEN 4
    . . . TRITIUM

HYDROGEN MASERS
GS   STIMULATED EMISSION DEVICES
    . MASERS
    . . GAS MASERS
    . . . **HYDROGEN MASERS**

HYDROGEN METABOLISM
GS   METABOLISM
    . **HYDROGEN METABOLISM**
RT   CARBOHYDRATE METABOLISM
    NITROGEN METABOLISM
    OXYGEN METABOLISM
    RESPIRATION
    SECRETIONS

HYDROGEN OXYGEN ENGINES
UF   HYDROX ENGINES
    LOX-HYDROGEN ENGINES
GS   ENGINES
    . ROCKET ENGINES
    . . LIQUID PROPELLANT ROCKET
      ENGINES
    . . . **HYDROGEN OXYGEN ENGINES**
    . . . . J-2 ENGINE
    . . . . M-1 ENGINE
    . . . . RL-10-A-1 ENGINE
    . . . . RL-10-A-3 ENGINE
RT   AUXILIARY PROPULSION
    LIQUID AIR CYCLE ENGINES
    TURBOROCKET ENGINES

HYDROGEN OXYGEN FUEL CELLS
UF   HYDROGEN AIR FUEL CELLS
GS   ELECTRIC GENERATORS
. DIRECT POWER GENERATORS
. . FUEL CELLS
. . . **HYDROGEN OXYGEN FUEL CELLS**
ELECTROCHEMICAL CELLS
. FUEL CELLS
. . **HYDROGEN OXYGEN FUEL CELLS**
RT   PHOSPHORIC ACID FUEL CELLS

HYDROGEN PERCHLORATE
GS   HALOGEN COMPOUNDS
. CHLORINE COMPOUNDS
. . PERCHLORATES
. . . **HYDROGEN PERCHLORATE**

HYDROGEN PEROXIDE
GS   CHALCOGENIDES
. OXIDES
. . DIOXIDES
. . . **HYDROGEN PEROXIDE**
HYDROGEN COMPOUNDS
. **HYDROGEN PEROXIDE**
RT   ROCKET OXIDIZERS

HYDROGEN PLASMA
GS   CHEMICAL ELEMENTS
. HYDROGEN
. . **HYDROGEN PLASMA**
. . . DEUTERIUM PLASMA
GASES
. HYDROGEN
. . **HYDROGEN PLASMA**
. . . DEUTERIUM PLASMA
RT   ARGON PLASMA
DEUTERIUM
HELIUM PLASMA
OXYGEN PLASMA
SOLAR WIND
STARK EFFECT

HYDROGEN PRODUCTION
RT   ELECTROLYSIS
ENERGY CONVERSION
FUELS
HYDRIDES
HYDROGEN-BASED ENERGY
HYDROLYSIS
LIGNITE
SOLAR ENERGY CONVERSION
THERMAL DISSOCIATION

HYDROGEN RECOMBINATIONS
GS   RECOMBINATION REACTIONS
. **HYDROGEN RECOMBINATIONS**

HYDROGEN SULFIDE
GS   CHALCOGENIDES
. SULFIDES
. . INORGANIC SULFIDES
. . . **HYDROGEN SULFIDE**
HYDROGEN COMPOUNDS
. **HYDROGEN SULFIDE**
SULFUR COMPOUNDS
. SULFIDES
. . INORGANIC SULFIDES
. . . **HYDROGEN SULFIDE**

HYDROGEN 2
USE   DEUTERIUM

HYDROGEN 3
USE   TRITIUM

HYDROGEN 4
GS   CHEMICAL ELEMENTS
. HYDROGEN
. . HYDROGEN ISOTOPES
. . . **HYDROGEN 4**
. NUCLIDES
. . ISOTOPES
. . . HYDROGEN ISOTOPES
. . . . **HYDROGEN 4**
GASES
. HYDROGEN
. . HYDROGEN ISOTOPES
. . . **HYDROGEN 4**

HYDROGEN-BASED ENERGY
RT   ∞ENERGY
ENERGY TECHNOLOGY
FUEL CELLS
GAS MIXTURES

HYDROGEN-BASED ENERGY-(CONT.)
HYDROCARBON FUELS
HYDROGEN FUELS
HYDROGEN PRODUCTION
LIQUID HYDROGEN
NICKEL HYDROGEN BATTERIES

HYDROGENATION
GS   CHEMICAL REACTIONS
. REDUCTION (CHEMISTRY)
. . **HYDROGENATION**
RT   ASPHALTENES
CYCLOHEXANE
DEHYDROGENATION
HYDROGEN
REFINING

HYDROGENOLYSIS
GS   CHEMICAL REACTIONS
. **HYDROGENOLYSIS**
DECOMPOSITION
. **HYDROGENOLYSIS**
RT   CRACKING (CHEMICAL ENGINEERING)
DEHYDROGENATION
HYDROGEN
∞REDUCTION

HYDROGENOMONAS
GS   AUTOTROPHS
. **HYDROGENOMONAS**
MICROORGANISMS
. BACTERIA
. . **HYDROGENOMONAS**

HYDROGEOLOGY
UF   MARINE GEOLOGY
GS   GEOLOGY
. **HYDROGEOLOGY**
HYDROLOGY
. **HYDROGEOLOGY**
RT   AQUIFERS
CORE SAMPLING
EROSION
FLOOD PLAINS
FLOOD PREDICTIONS
GEYSERS
GLACIOLOGY
HYDROLOGY MODELS
HYDROSTATICS
∞SCIENCE
SOIL EROSION
STRATIGRAPHY
WATERSHEDS

HYDROGRAPHY
RT   GEOPHYSICS
HYDROCLIMATOLOGY
HYDROLOGY
HYDROMETEOROLOGY
ICE MAPPING
LIMNOLOGY
METEOROLOGY
OCEAN CURRENTS
OCEAN SURFACE
OCEANOGRAPHY

HYDROKINETICS
USE   HYDROMECHANICS

HYDROLOGY
GS   **HYDROLOGY**
. HYDROGEOLOGY
RT   ALLUVIUM
AQUIFERS
CLIMATOLOGY
DRAINAGE
DRAINAGE PATTERNS
DROUGHT
EARTH HYDROSPHERE
EARTH PLANETARY STRUCTURE
FLOOD CONTROL
FLOOD DAMAGE
FLOOD PLAINS
FLOOD PREDICTIONS
FLOODS
GEOCHEMISTRY
GEOPHYSICS
GREAT SALT LAKE (UT)
∞HYDRAULICS
HYDROCLIMATOLOGY
HYDROGRAPHY
HYDROMETEOROLOGY
ICE MAPPING
INTERNATIONAL HYDROLOGICAL
DECADE

HYDROLOGY-(CONT.)
LAKE ERIE
LAKE HURON
LAKE MICHIGAN
LAKE ONTARIO
LAKE SUPERIOR
LIMNOLOGY
MARINE CHEMISTRY
METEOROLOGICAL PARAMETERS
METEOROLOGY
OCEANOGRAPHY
POLAR METEOROLOGY
PRECIPITATION (METEOROLOGY)
RAIN
STREAMS
STRUCTURAL PROPERTIES (GEOLOGY)
WATER
WATER MANAGEMENT
WATER RESOURCES
WATERSHEDS

HYDROLOGY MODELS
GS   MODELS
. **HYDROLOGY MODELS**
RT   DRAINAGE
FLOODS
HYDROGEOLOGY
PRECIPITATION (METEOROLOGY)
RAIN
STREAMS
WATER FLOW

HYDROLYSIS
GS   CHEMICAL REACTIONS
. **HYDROLYSIS**
RT   AMMONOLYSIS
CRACKING (CHEMICAL ENGINEERING)
EXTRACTION
HYDRATION
HYDROGEN PRODUCTION

HYDROMAGNETIC FLOW
USE   MAGNETOHYDRODYNAMIC FLOW

HYDROMAGNETIC STABILITY
USE   MAGNETOHYDRODYNAMIC STABILITY

HYDROMAGNETIC WAVES
USE   MAGNETOHYDRODYNAMIC WAVES

HYDROMAGNETICS
USE   MAGNETOHYDRODYNAMICS

HYDROMAGNETISM
USE   MAGNETOHYDRODYNAMICS

HYDROMECHANICS
UF   HYDROKINETICS
GS   FLUID MECHANICS
. **HYDROMECHANICS**
. . HYDRODYNAMICS
. . . ELASTOHYDRODYNAMICS
. . . ELECTROHYDRODYNAMICS
. . . MAGNETOHYDRODYNAMICS
. . HYDROSTATICS
. . . MAGNETOHYDROSTATICS
RT   FLUID DYNAMICS
. FLUID FLOW
∞HYDRAULICS
KINETICS
∞SCIENCE
WATER

HYDROMETALLURGY
RT   CHLORINATION
ELECTRODIALYSIS
FILTRATION
ION EXCHANGING
LEACHING
∞METALLURGY
∞PRECIPITATION
PRECIPITATION (CHEMISTRY)
REFINING
SULFATION

HYDROMETEOROLOGY
GS   METEOROLOGY
. **HYDROMETEOROLOGY**
. . MARINE METEOROLOGY
RT   AGROMETEOROLOGY
HYDROGRAPHY
HYDROLOGY
PRECIPITATION (CHEMISTRY)
PRECIPITATION (METEOROLOGY)

HYDROMETEOROLOGY-(CONT.)
    WATER BALANCE

HYDROMETERS
  GS   MEASURING INSTRUMENTS
     . **HYDROMETERS**
  RT   CHEMICAL ANALYSIS
     DENSITY (MASS/VOLUME)
     DENSITY MEASUREMENT
     WEIGHT MEASUREMENT

HYDRONIUM IONS
  GS   IONS
     . **HYDRONIUM IONS**
  RT   HYDROGEN
     HYDROGEN IONS
     POSITIVE IONS

HYDROPHONES
  UF   HYDROBAROPHONES
  GS   TRANSDUCERS
     . SOUND TRANSDUCERS
     .. ELECTROACOUSTIC TRANSDUCERS
     ... **HYDROPHONES**
  RT   MICROPHONES
     SONAR
     SONOBUOYS

HYDROPLANES (SURFACES)
  UF   HYDROSKIS
  RT   HYDROFOILS
     HYDROPLANING
     SKIS
     ∞SYSTEMS

HYDROPLANES (VEHICLES)
  RT   HYDROFOIL CRAFT
     HYDROPLANING
     ∞VEHICLES

HYDROPLANING
  RT   HYDROFOILS
     HYDROPLANES (SURFACES)
     HYDROPLANES (VEHICLES)
     SKID LANDINGS
     SKIDDING
     WATER LANDING

HYDROPONICS
  RT   AGRICULTURE
     AQUATIC PLANTS
     AQUICULTURE
     PLANTS (BOTANY)
     VEGETATION GROWTH

HYDROPOWER STATIONS
  USE HYDROELECTRIC POWER STATIONS

HYDROPYROLYSIS
  GS   GASIFICATION
     . COAL GASIFICATION
     .. **HYDROPYROLYSIS**
  RT   COAL
     COAL LIQUEFACTION
     LIGNITE
     METHANATION
     METHANE

HYDROSKIS
  USE HYDROPLANES (SURFACES)

HYDROSPHERE (EARTH)
  USE EARTH HYDROSPHERE

HYDROSPINNING
  GS   FORMING TECHNIQUES
     . METAL SPINNING
     .. **HYDROSPINNING**
     METAL WORKING
     . METAL SPINNING
     .. **HYDROSPINNING**
     SPIN
     . METAL SPINNING
     .. **HYDROSPINNING**

HYDROSTATIC PRESSURE
  GS   PRESSURE
     . STATIC PRESSURE
     .. **HYDROSTATIC PRESSURE**
  RT   CENTER OF PRESSURE
     ELEVATION
     HEAD (FLUID MECHANICS)
     HYDROSTATICS
     ISOSTATIC PRESSURE

HYDROSTATIC PRESSURE-(CONT.)
     PRESSURE DEPENDENCE
     PRESSURE HEADS
     TRANSITION PRESSURE
     WATER PRESSURE

HYDROSTATICS
  GS   FLUID MECHANICS
     . HYDROMECHANICS
     .. **HYDROSTATICS**
     ... MAGNETOHYDROSTATICS
     STATICS
     . **HYDROSTATICS**
     .. MAGNETOHYDROSTATICS
  RT   AEROSTATICS
     ELEVATION
     HEAD (FLUID MECHANICS)
     ∞HYDRAULICS
     HYDRODYNAMICS
     HYDROGEOLOGY
     HYDROSTATIC PRESSURE
     ISOSTASY
     PRESSURE GRADIENTS
     PRESSURE HEADS
     WATER
     WATER PRESSURE

HYDROSULFITES
  GS   HYDROGEN COMPOUNDS
     . **HYDROSULFITES**
     SULFUR COMPOUNDS
     . SULFITES
     .. **HYDROSULFITES**

HYDROTHERMAL CRYSTAL GROWTH
  GS   GROWTH
     . CRYSTAL GROWTH
     .. **HYDROTHERMAL CRYSTAL GROWTH**
  RT   ELECTROEPITAXY

HYDROTHERMAL STRESS ANALYSIS
  RT   HYDROTHERMAL SYSTEMS
     HYGRAL PROPERTIES
     HYGROSCOPICITY
     MOISTURE CONTENT
     MOISTURE RESISTANCE

HYDROTHERMAL SYSTEMS
  RT   AQUIFERS
     ENERGY CONVERSION
     GEOTHERMAL RESOURCES
     GEYSERS
     HEATING
     HYDROTHERMAL STRESS ANALYSIS
     SOLAR HEATING
     ∞SYSTEMS

HYDROX ENGINES
  USE HYDROGEN OXYGEN ENGINES

HYDROXIDES
  GS   **HYDROXIDES**
     . LITHIUM HYDROXIDES
     . POTASSIUM HYDROXIDES
     . SODIUM HYDROXIDES
  RT   ALKALIES

HYDROXYCORTICOSTEROID
  GS   SECRETIONS
     . ENDOCRINE SECRETIONS
     .. HORMONES
     ... **HYDROXYCORTICOSTEROID**
     STEROIDS
     . CORTICOSTEROIDS
     .. **HYDROXYCORTICOSTEROID**
     ... CORTISONE
  RT   ADRENAL METABOLISM

HYDROXYL COMPOUNDS
  GS   **HYDROXYL COMPOUNDS**
     . ALCOHOLS
     .. ETHYL ALCOHOL
     .. GLYCOLS
     .. ISOPROPYL ALCOHOL
     .. METHYL ALCOHOLS
     .. PHENOLS
     ... BISPHENOLS
     ... CRESOLS
     ... PHLOROGLUCINOL
     ... THYMOL
     .. POLYVINYL ALCOHOL
     .. TRIOLS
     ... CYANURIC ACID
  RT   ALIPHATIC COMPOUNDS

HYDROXYL COMPOUNDS-(CONT.)
     ∞CHEMICAL COMPOUNDS

HYDROXYL EMISSION
  GS   DECAY
     . EMISSION
     .. RADIO EMISSION
     ... **HYDROXYL EMISSION**
     ELECTROMAGNETIC RADIATION
     . RADIO WAVES
     .. RADIO EMISSION
     ... **HYDROXYL EMISSION**
  RT   EMISSION SPECTRA
     RADIO SOURCES (ASTRONOMY)

HYDROXYL RADICALS
  GS   RADICALS
     . FREE RADICALS
     .. **HYDROXYL RADICALS**
  RT   ALCOHOLS
     ∞CHEMISTRY
     FORMYL IONS
     GLYCOLS
     IONS

HYDROXYLAMINE SULFATE
  GS   AMINES
     . **HYDROXYLAMINE SULFATE**
     SULFUR COMPOUNDS
     . SULFATES
     .. **HYDROXYLAMINE SULFATE**

HYDROXYLAMMONIUM PERCHLORATES
  GS   AMMONIUM COMPOUNDS
     . **HYDROXYLAMMONIUM**
       PERCHLORATES
     HALOGEN COMPOUNDS
     . CHLORINE COMPOUNDS
     .. PERCHLORATES
     ... **HYDROXYLAMMONIUM**
        PERCHLORATES

HYGIENE
  GS   **HYGIENE**
     . ORAL HYGIENE
  RT   BATHING
     CLEANLINESS
     CONSUMABLES (SPACECREW SUPPLIES)
     HEALTH
     HOUSEKEEPING (SPACECRAFT)
     PUBLIC HEALTH
     SANITATION

HYGRAL PROPERTIES
  RT   HUMIDITY
     HYDROTHERMAL STRESS ANALYSIS
     MOISTURE
     POROSITY
     ∞PROPERTIES

HYGROMETERS
  GS   MEASURING INSTRUMENTS
     . MOISTURE METERS
     .. **HYGROMETERS**
     ... PSYCHROMETERS
  RT   CHEMICAL ANALYSIS
     DEW POINT
     HUMIDITY
     HUMIDITY MEASUREMENT
     METEOROLOGICAL INSTRUMENTS

HYGROSCOPICITY
  RT   CHEMICAL PROPERTIES
     HYDROTHERMAL STRESS ANALYSIS
     MATERIAL ABSORPTION
     MOISTURE CONTENT
     MOISTURE RESISTANCE
     ∞PHYSICAL PROPERTIES
     SOLUBILITY
     WETTABILITY

HYLA-STAR ROCKET VEHICLE
  GS   LAUNCH VEHICLES
     . **HYLA-STAR ROCKET VEHICLE**
     ROCKET VEHICLES
     . SINGLE STAGE ROCKET VEHICLES
     .. **HYLA-STAR ROCKET VEHICLE**
  RT   LIQUID PROPELLANT ROCKET ENGINES
     TITAN 2 ICBM

HYLLERAAS COORDINATES
  GS   COORDINATES
     . **HYLLERAAS COORDINATES**
  RT   QUANTUM MECHANICS

HYLLERAAS COORDINATES-*(CONT.)*
   TWO BODY PROBLEM

HYOSCINE
UF   SCOPOLAMINE
GS   AMINES
   . **HYOSCINE**
   EPOXY COMPOUNDS
   . **HYOSCINE**
   HETEROCYCLIC COMPOUNDS
   . ALKALOIDS
   . . **HYOSCINE**
   NITROGEN COMPOUNDS
   . ALKALOIDS
   . . **HYOSCINE**

HYPERBARIC CHAMBERS
GS   COMPARTMENTS
   . TEST CHAMBERS
   . . PRESSURE CHAMBERS
   . . . **HYPERBARIC CHAMBERS**
RT   ∞CHAMBERS
   HIGH PRESSURE
   VACUUM CHAMBERS

HYPERBOLAS
GS   GEOMETRY
   . EUCLIDEAN GEOMETRY
   . . ANALYTIC GEOMETRY
   . . . CONICS
   . . . . **HYPERBOLAS**
RT   HYPERBOLIC TRAJECTORIES

HYPERBOLIC COORDINATES
UF   HYPERBOLIC SPACE
GS   COORDINATES
   . **HYPERBOLIC COORDINATES**

HYPERBOLIC DIFFERENTIAL EQUATIONS
GS   ANALYSIS (MATHEMATICS)
   . REAL VARIABLES
   . . DIFFERENTIAL EQUATIONS
   . . . **HYPERBOLIC DIFFERENTIAL
         EQUATIONS**
RT   DIRICHLET PROBLEM
   ∞EQUATIONS
   WAVE EQUATIONS

HYPERBOLIC FUNCTIONS
GS   ANALYSIS (MATHEMATICS)
   . COMPLEX VARIABLES
   . . **HYPERBOLIC FUNCTIONS**
   . REAL VARIABLES
   . . **HYPERBOLIC FUNCTIONS**
   FUNCTIONS (MATHEMATICS)
   . **HYPERBOLIC FUNCTIONS**
RT   EXPONENTIAL FUNCTIONS
   ∞HYPERBOLIC SYSTEMS
   METHOD OF CHARACTERISTICS
   ORTHOGONAL FUNCTIONS
   RIEMANN WAVES
   RIESZ THEOREM

HYPERBOLIC NAVIGATION
GS   NAVIGATION
   . RADIO NAVIGATION
   . . **HYPERBOLIC NAVIGATION**
   . . . DECCA NAVIGATION
   . . . LORAC NAVIGATION SYSTEM
   . . . LORAN
   . . . . LORAN C
   . . . . LORAN D
   . . . SHORAN
RT   AIR NAVIGATION
   ∞HYPERBOLIC SYSTEMS
   INERTIAL NAVIGATION
   SURFACE NAVIGATION

HYPERBOLIC REENTRY
GS   ATMOSPHERIC ENTRY
   . REENTRY
   . . **HYPERBOLIC REENTRY**
   SPACE FLIGHT
   . **HYPERBOLIC REENTRY**
RT   REENTRY TRAJECTORIES

HYPERBOLIC SPACE
USE   HYPERBOLIC COORDINATES

∞ **HYPERBOLIC SYSTEMS**
   SN   *(USE OF A MORE SPECIFIC TERM IS
         RECOMMENDED--CONSULT THE TERMS
         LISTED BELOW)*
   RT   HYPERBOLIC FUNCTIONS

HYPERBOLIC SYSTEMS-*(CONT.)*
   HYPERBOLIC NAVIGATION
   ∞SYSTEMS

HYPERBOLIC TRAJECTORIES
GS   TRAJECTORIES
   . **HYPERBOLIC TRAJECTORIES**
RT   CELESTIAL MECHANICS
   ESCAPE VELOCITY
   HYPERBOLAS
   SPACECRAFT TRAJECTORIES

HYPERCAPNIA
GS   CARBON DIOXIDE TENSION
   . **HYPERCAPNIA**
RT   BLOOD
   ∞BREATHING
   RESPIRATORY RATE
   RESPIRATORY SYSTEM

HYPERFINE STRUCTURE
RT   ATOMIC STRUCTURE
   FINE STRUCTURE
   LINE SPECTRA
   MUON SPIN ROTATION
   SPECTRUM ANALYSIS
   ∞STRUCTURES

HYPERGEOMETRIC FUNCTIONS
UF   JACOBI POLYNOMIALS
GS   ANALYSIS (MATHEMATICS)
   . COMPLEX VARIABLES
   . . **HYPERGEOMETRIC FUNCTIONS**
   FUNCTIONS (MATHEMATICS)
   . **HYPERGEOMETRIC FUNCTIONS**
RT   BESSEL FUNCTIONS
   GEOMETRY
   HYPERSPACES

HYPERGEOMETRY
USE   HYPERSPACES

HYPERGLYCEMIA
GS   METABOLISM
   . CARBOHYDRATE METABOLISM
   . . **HYPERGLYCEMIA**

HYPERGOLIC ROCKET PROPELLANTS
GS   LIQUIDS
   . LIQUID FUELS
   . . LIQUID ROCKET PROPELLANTS
   . . . **HYPERGOLIC ROCKET
         PROPELLANTS**
   PROPELLANTS
   . ROCKET PROPELLANTS
   . . LIQUID ROCKET PROPELLANTS
   . . . **HYPERGOLIC ROCKET
         PROPELLANTS**
RT   CRYOGENIC ROCKET PROPELLANTS
   GELLED ROCKET PROPELLANTS
   HYBRID PROPELLANTS
   HYDROCARBON FUELS
   PYROPHORIC MATERIALS
   SOLID PROPELLANT IGNITION
   SPONTANEOUS COMBUSTION
   STORABLE PROPELLANTS

HYPERION
GS   SATELLITES
   . NATURAL SATELLITES
   . . SATURN SATELLITES
   . . . **HYPERION**
RT   SATURN (PLANET)

HYPERKINESIA
GS   PHYSICAL EXERCISE
   . **HYPERKINESIA**
RT   EXHAUSTION
   FATIGUE (BIOLOGY)
   HYPOKINESIA
   STRESS (PHYSIOLOGY)
   WORK CAPACITY

HYPERNEA
RT   MENTAL PERFORMANCE

HYPERNUCLEI
GS   GASES
   . IONIZED GASES
   . . CHARGED PARTICLES
   . . . NUCLEI (NUCLEAR PHYSICS)
   . . . . **HYPERNUCLEI**
RT   ELEMENTARY PARTICLES
   RADIOACTIVE DECAY

HYPERONS
GS   PARTICLES
   . ELEMENTARY PARTICLES
   . . FERMIONS
   . . . BARYONS
   . . . . **HYPERONS**
   . . . . . XI HYPERONS
RT   ANTIPARTICLES
   BARYON RESONANCE
   CHARGED PARTICLES
   MESON RESONANCE
   NUCLEONS
   STRANGENESS

HYPEROPIA
GS   ACUITY
   . VISUAL ACUITY
   . . **HYPEROPIA**
RT   VISION

HYPEROXIA
UF   OXYGEN TOXICITY
RT   HYPERVENTILATION
   OXIMETRY
   OXYGEN CONSUMPTION
   TOXIC DISEASES
   TOXICITY

HYPERPLANES
GS   ANALYSIS (MATHEMATICS)
   . REAL VARIABLES
   . . **HYPERPLANES**
RT   HYPERSPACES
   POLYTOPES
   SET THEORY

HYPERPNEA
RT   ∞BREATHING
   RESPIRATORY RATE

HYPERSOMNIA
GS   SLEEP
   . **HYPERSOMNIA**

HYPERSONIC AIRCRAFT
SN   (AIRCRAFT DESIGNED TO FLY AT
      SPEEDS OF MACH 5 OR GREATER)
GS   HYPERSONIC VEHICLES
   . **HYPERSONIC AIRCRAFT**
   . . HYPERSONIC GLIDERS
   . . . X-20 AIRCRAFT
RT   AEROSPACEPLANES
   ∞AIRCRAFT
   ASTROPLANE
   BOOSTGLIDE VEHICLES
   HYPERSONICS
   JET AIRCRAFT
   ∞LOW WING AIRCRAFT
   RESEARCH AIRCRAFT
   SUPERSONIC AIRCRAFT
   SWEPTBACK TAIL SURFACES
   SWEPTBACK WINGS
   TRAPEZOIDAL TAIL SURFACES

HYPERSONIC BOUNDARY LAYER
GS   BOUNDARY LAYERS
   . **HYPERSONIC BOUNDARY LAYER**
RT   LAMINAR BOUNDARY LAYER
   THERMAL BOUNDARY LAYER
   TURBULENT BOUNDARY LAYER

HYPERSONIC COMBUSTION
GS   COMBUSTION
   . **HYPERSONIC COMBUSTION**
RT   EROSIVE BURNING
   FUEL COMBUSTION

HYPERSONIC FLIGHT
RT   AERODYNAMICS
   ∞FLIGHT
   HYPERSONICS
   MISSILES
   ROCKET FLIGHT
   SUPERSONIC FLIGHT

HYPERSONIC FLOW
GS   FLUID FLOW
   . **HYPERSONIC FLOW**
RT   AERODYNAMICS
   CASCADE WIND TUNNELS
   COMPRESSIBLE FLOW
   FLOW VELOCITY
   GAS FLOW
   HYPERSONICS

HYPERSONIC FLOW-*(CONT.)*
    HYPERVELOCITY WIND TUNNELS
    LIGHTHILL GAS MODEL
    SHOCK TUBES
    SHOCK TUNNELS
    SHOCK WAVES
    SUPERSONIC FLOW
    WIND TUNNELS

HYPERSONIC FORCES
  GS   AERODYNAMIC FORCES
     . **HYPERSONIC FORCES**
  RT   AERODYNAMIC DRAG
     HYPERSONICS
     LIFT

HYPERSONIC GLIDERS
  GS   GLIDERS
     . **HYPERSONIC GLIDERS**
     . . X-20 AIRCRAFT
     HYPERSONIC VEHICLES
     . HYPERSONIC AIRCRAFT
     . . **HYPERSONIC GLIDERS**
     . . . X-20 AIRCRAFT
  RT   AEROSPACEPLANES
     ∞AIRCRAFT
     ASSET GLIDERS
     ASTROPLANE
     BOOSTGLIDE VEHICLES
     HL-10 REENTRY VEHICLE
     HLD-35 REENTRY VEHICLE
     LIFTING REENTRY VEHICLES
     PARAGLIDERS

HYPERSONIC HEAT TRANSFER
  GS   TRANSMISSION
     . HEAT TRANSMISSION
     . . HEAT TRANSFER
     . . . AERODYNAMIC HEAT TRANSFER
     . . . . **HYPERSONIC HEAT TRANSFER**
  RT   AEROTHERMODYNAMICS
     HYPERSONICS
     SUPERSONIC HEAT TRANSFER

HYPERSONIC INLETS
  GS   INTAKE SYSTEMS
     . AIR INTAKES
     . . **HYPERSONIC INLETS**
  RT   BYPASS RATIO
     ∞DIFFUSERS
     ENGINE INLETS
     INLET AIRFRAME CONFIGURATIONS
     NOSE INLETS
     SIDE INLETS
     SUPERSONIC INLETS

HYPERSONIC NOZZLES
  RT   CONICAL NOZZLES
     ∞NOZZLES
     ROCKET NOZZLES
     SUPERSONIC NOZZLES
     TRANSONIC NOZZLES
     WIND TUNNEL NOZZLES

HYPERSONIC REENTRY
  GS   ATMOSPHERIC ENTRY
     . REENTRY
     . . **HYPERSONIC REENTRY**
     . . . UNCONTROLLED REENTRY
       (SPACECRAFT)
     SPACE FLIGHT
     . **HYPERSONIC REENTRY**
  RT   AERODYNAMIC HEATING
     AEROTHERMODYNAMICS
     BERENICE ROCKET VEHICLE
     BOUNDARY LAYER PLASMAS
     ENTRY GUIDANCE (STS)
     REENTRY EFFECTS
     REENTRY PHYSICS
     SPACECRAFT REENTRY

HYPERSONIC SHOCK
  RT   HYPERSONICS
     MACH CONES
     NOISE (SOUND)
     SHOCK WAVES

HYPERSONIC SPEED
  GS   RATES (PER TIME)
     . **HYPERSONIC SPEED**
     VELOCITY
     . **HYPERSONIC SPEED**
  RT   HIGH SPEED
     HYPERSONICS
     ∞HYPERVELOCITY

HYPERSONIC SPEED-*(CONT.)*
    SUPERSONIC SPEEDS

HYPERSONIC TEST APPARATUS
  RT   HYPERSONICS
     HYPERVELOCITY WIND TUNNELS
     MISSILE RANGES
     SUPERSONIC TEST APPARATUS
     ∞TEST EQUIPMENT

HYPERSONIC VEHICLES
  GS   **HYPERSONIC VEHICLES**
     . HYPERSONIC AIRCRAFT
     . . HYPERSONIC GLIDERS
     . . . X-20 AIRCRAFT
     . . LIFTING REENTRY VEHICLES
     . . HLD-35 REENTRY VEHICLE
     . . JANUS SPACECRAFT
     . . M-2F2 LIFTING BODY
     . . X-20 AIRCRAFT
  RT   ∞FLIGHT VEHICLES
     HYPERSONICS
     ∞INSULATED STRUCTURES
     RECOVERABLE SPACECRAFT
     REENTRY VEHICLES
     ∞SPACECRAFT
     TEST VEHICLES
     ∞VEHICLES
     ∞WINGED VEHICLES

HYPERSONIC WAKES
  GS   WAKES
     . **HYPERSONIC WAKES**
  RT   AIRCRAFT WAKES
     BOW WAVES
     HYPERSONICS
     SHOCK WAVES
     SUPERSONIC WAKES

HYPERSONIC WIND TUNNELS
  SN   (MACH 5 TO 10)
  GS   TEST FACILITIES
     . WIND TUNNELS
     . . **HYPERSONIC WIND TUNNELS**
     . . . CASCADE WIND TUNNELS
     . . . HOTSHOT WIND TUNNELS
     . . . PLASMA JET WIND TUNNELS
     . . . SHOCK TUNNELS
  RT   BLOWDOWN WIND TUNNELS
     COMBUSTION WIND TUNNELS
     HYPERVELOCITY WIND TUNNELS
     LOW DENSITY WIND TUNNELS
     MAGNETIC PISTONS
     SHOCK TUBES
     SUBSONIC WIND TUNNELS
     SUPERSONIC WIND TUNNELS
     TRANSONIC WIND TUNNELS

HYPERSONICS
  GS   FLUID MECHANICS
     . FLUID DYNAMICS
     . . GAS DYNAMICS
     . . . AERODYNAMICS
     . . . . **HYPERSONICS**
  RT   AEROTHERMODYNAMICS
     HYPERSONIC AIRCRAFT
     HYPERSONIC FLIGHT
     HYPERSONIC FLOW
     HYPERSONIC FORCES
     HYPERSONIC HEAT TRANSFER
     HYPERSONIC SHOCK
     HYPERSONIC SPEED
     HYPERSONIC TEST APPARATUS
     HYPERSONIC VEHICLES
     HYPERSONIC WAKES
     SUPERSONIC SPEEDS
     SUPERSONICS

HYPERSPACES
  UF   HYPERGEOMETRY
  RT   HYPERGEOMETRIC FUNCTIONS
     HYPERPLANES
     HYPERSPHERES
     PHASE-SPACE INTEGRAL
     ∞SPACE

HYPERSPHERES
  RT   GEOMETRY
     HYPERSPACES
     REAL VARIABLES

HYPERTENSIN
  GS   ACIDS
     . AMINO ACIDS
     . . PEPTIDES

HYPERTENSIN-*(CONT.)*
    . . . **HYPERTENSIN**
    DRUGS
    . VASOCONSTRICTOR DRUGS
    . . **HYPERTENSIN**
    ORGANIC COMPOUNDS
    . AMINO ACIDS
    . . PEPTIDES
    . . . **HYPERTENSIN**
    . . . . POLYPEPTIDES
    PROTEINS
    . PEPTIDES
    . . **HYPERTENSIN**
    . . . POLYPEPTIDES

HYPERTENSION
  GS   PRESSURE
     . BLOOD PRESSURE
     . . **HYPERTENSION**
  RT   MYOCARDIAL INFARCTION
     TRANQUILIZERS

HYPERTHERMIA
  RT   BODY TEMPERATURE
     FEVER
     HEAT STROKE
     SKIN TEMPERATURE (BIOLOGY)
     THERMOREGULATION

HYPERTONIA
  USE   OSMOSIS

HYPERTROPHY
  USE   GROWTH

∞ HYPERVELOCITY
  SN   *(USE OF A MORE SPECIFIC TERM IS*
     RECOMMENDED--CONSULT THE TERMS
     LISTED BELOW)
  RT   ESCAPE VELOCITY
     HYPERSONIC SPEED
     ORBITAL VELOCITY
     RELATIVISTIC VELOCITY

HYPERVELOCITY ACCELERATORS
  USE   HYPERVELOCITY GUNS

HYPERVELOCITY CRATERING
  USE   HYPERVELOCITY PROJECTILES
     PROJECTILE CRATERING

HYPERVELOCITY FLOW
  GS   FLUID FLOW
     . **HYPERVELOCITY FLOW**
  RT   FLOW VELOCITY
     SUPERSONIC FLOW

HYPERVELOCITY GUNS
  UF   HYPERVELOCITY ACCELERATORS
  RT   ∞ACCELERATORS
     BALLISTICS
     GAS GUNS
     ∞GUNS
     GUNS (ORDNANCE)
     RAILGUN ACCELERATORS

HYPERVELOCITY IMPACT
  GS   IMPACT
     . **HYPERVELOCITY IMPACT**
  RT   HYDRODYNAMIC RAM EFFECT
     IMPACT MELTS
     MECHANICAL SHOCK
     METEORITE COLLISIONS
     METEORITIC DAMAGE
     POINT IMPACT
     PROJECTILE CRATERING

HYPERVELOCITY LAUNCHERS
  GS   LAUNCHERS
     . **HYPERVELOCITY LAUNCHERS**
  RT   GUN LAUNCHERS
     RAILGUN ACCELERATORS

HYPERVELOCITY PROJECTILES
  UF   HYPERVELOCITY CRATERING
  GS   PROJECTILES
     . **HYPERVELOCITY PROJECTILES**
  RT   ∞BOMBARDMENT
     LIGHT GAS GUNS
     METEOROIDS
     MICROMETEORITES
     PROJECTILE CRATERING
     SIMULATION

HYPERVELOCITY WIND TUNNELS
SN   (ABOVE MACH 10)
GS   TEST FACILITIES
      . WIND TUNNELS
      . . **HYPERVELOCITY WIND TUNNELS**
      . . . CASCADE WIND TUNNELS
      . . . HOTSHOT WIND TUNNELS
      . . . PLASMA JET WIND TUNNELS
      . . . SHOCK TUNNELS
RT   BLOWDOWN WIND TUNNELS
      COMBUSTION WIND TUNNELS
      HYPERSONIC FLOW
      HYPERSONIC TEST APPARATUS
      HYPERSONIC WIND TUNNELS
      LOW DENSITY WIND TUNNELS
      MAGNETIC PISTONS
      SHOCK TUBES
      SUPERSONIC WIND TUNNELS

HYPERVENTILATION
RT   ACIDOSIS
      ALKALOSIS
      HYPEROXIA

HYPERVOLEMIA
RT   BLOOD CIRCULATION
      BLOOD VOLUME
      CIRCULATORY SYSTEM

HYPNOSIS
GS   SLEEP
      . **HYPNOSIS**
RT   ANESTHESIA
      SUGGESTION

HYPOBARIC ATMOSPHERES
RT   ALTITUDE SIMULATION
      ALTITUDE TOLERANCE
      ∞ATMOSPHERES
      HIGH ALTITUDE BREATHING
      HIGH ALTITUDE ENVIRONMENTS
      HIGH ALTITUDE PRESSURE
      LOW PRESSURE
      VACUUM TESTS

HYPOCAPNIA
GS   CARBON DIOXIDE TENSION
      . **HYPOCAPNIA**
RT   BLOOD

HYPODERMIS
GS   TISSUES (BIOLOGY)
      . **HYPODERMIS**

HYPODYNAMIA
RT   MUSCLES
      MUSCULAR FUNCTION

HYPOELASTICITY
GS   MECHANICAL PROPERTIES
      . ELASTIC PROPERTIES
      . . **HYPOELASTICITY**

HYPOGLYCEMIA
GS   METABOLISM
      . CARBOHYDRATE METABOLISM
      . . **HYPOGLYCEMIA**

HYPOKINESIA
RT   HYPERKINESIA
      MUSCULAR FUNCTION
      MUSCULOSKELETAL SYSTEM
      PHYSICAL EXERCISE

HYPOMETABOLISM
GS   METABOLISM
      . **HYPOMETABOLISM**
RT   THYROID GLAND

HYPOTENSION
GS   PRESSURE
      . BLOOD PRESSURE
      . . **HYPOTENSION**
RT   HEMORRHAGES

HYPOTHALAMUS
RT   BRAIN
      CEREBRAL CORTEX

HYPOTHERMIA
RT   BODY TEMPERATURE
      SKIN TEMPERATURE (BIOLOGY)
      THERMOREGULATION

HYPOTHESES
GS   **HYPOTHESES**
      . EXPECTANCY HYPOTHESIS
      . INTERMITTENCY HYPOTHESIS
      . LAGRANGE SIMILARITY HYPOTHESIS
      . NULL HYPOTHESIS
      . VORTICITY TRANSPORT HYPOTHESIS
RT   ASSUMPTIONS
      INFERENCE
      MATHEMATICAL LOGIC
      QUALITY CONTROL
      THEOREMS
      ∞THEORIES
      THESES

HYPOTONIA
GS   MUSCULAR TONUS
      . **HYPOTONIA**
RT   MUSCULAR FUNCTION

HYPOVENTILATION
GS   RATES (PER TIME)
      . RESPIRATORY RATE
      . . **HYPOVENTILATION**

HYPOVOLEMIA
RT   BLOOD CIRCULATION
      BLOOD VOLUME

HYPOXEMIA
GS   PRESSURE
      . PARTIAL PRESSURE
      . . OXYGEN TENSION
      . . . **HYPOXEMIA**
RT   HYPOXIA

HYPOXIA
UF   OXYGEN DEFICIENCY
RT   ANOXIA
      FASTING
      HYPOXEMIA
      OXIMETRY
      OXYGEN CONSUMPTION
      STRESS (PHYSIOLOGY)

HYPSOGRAPHY
GS   GEOGRAPHY
      . **HYPSOGRAPHY**
RT   CONTOURS
      DATUM (ELEVATION)
      ELEVATION
      MAPPING
      MAPS
      RELIEF MAPS
      TOPOGRAPHY

HYPSOMETERS
GS   MEASURING INSTRUMENTS
      . **HYPSOMETERS**
RT   ALTIMETERS
      BAROMETERS
      METEOROLOGICAL INSTRUMENTS
      PRESSURE GAGES

HYSTERESIS
RT   ACCURACY
      ANTIFERROELECTRICITY
      ANTIFERROMAGNETISM
      DAMPING
      DYNAMIC CHARACTERISTICS
      EDDY CURRENTS
      ELECTRICAL PROPERTIES
      ERRORS
      INTERNAL FRICTION
      MAGNETIC PERMEABILITY
      MAGNETIC PROPERTIES
      MECHANICAL PROPERTIES
      OPTICAL BISTABILITY
      ∞PHYSICAL PROPERTIES
      PRECISION
      RETARDING
      SHEAR PROPERTIES
      TENSILE STRENGTH
      TIME LAG
      TOLERANCES (MECHANICS)
      VISCOELASTICITY
      VISCOPLASTICITY

I

I BEAMS
GS   STRUCTURAL MEMBERS
      . BEAMS (SUPPORTS)
      . . **I BEAMS**
RT   CANTILEVER BEAMS
      CURVED BEAMS
      TRUSSES

IAPETUS
GS   CELESTIAL BODIES
      . NATURAL SATELLITES
      . . **IAPETUS**
      SATELLITES
      . NATURAL SATELLITES
      . . SATURN SATELLITES
      . . . **IAPETUS**
RT   CHARON
      SATURN (PLANET)

IBM COMPUTERS
GS   DATA PROCESSING EQUIPMENT
      . COMPUTERS
      . . **IBM COMPUTERS**
      . . . IBM 360 COMPUTER
      . . . IBM 370 COMPUTER
      . . . IBM 650 COMPUTER
      . . . IBM 704 COMPUTER
      . . . IBM 709 COMPUTER
      . . . IBM 1130 COMPUTER
      . . . IBM 1401 COMPUTER
      . . . IBM 1410 COMPUTER
      . . . IBM 1620 COMPUTER
      . . . IBM 2250 COMPUTER
      . . . IBM 7000 SERIES COMPUTERS
      . . . . IBM 7030 COMPUTER
      . . . . IBM 7040 COMPUTER
      . . . . IBM 7044 COMPUTER
      . . . . IBM 7070 COMPUTER
      . . . . IBM 7074 COMPUTER
      . . . . IBM 7090 COMPUTER
      . . . . IBM 7094 COMPUTER
RT   DIGITAL COMPUTERS

IBM 360 COMPUTER
GS   DATA PROCESSING EQUIPMENT
      . COMPUTERS
      . . DIGITAL COMPUTERS
      . . . **IBM 360 COMPUTER**
      . . IBM COMPUTERS
      . . . **IBM 360 COMPUTER**

IBM 370 COMPUTER
GS   DATA PROCESSING EQUIPMENT
      . COMPUTERS
      . . DIGITAL COMPUTERS
      . . . **IBM 370 COMPUTER**
      . . IBM COMPUTERS
      . . . **IBM 370 COMPUTER**

IBM 650 COMPUTER
GS   DATA PROCESSING EQUIPMENT
      . COMPUTERS
      . . DIGITAL COMPUTERS
      . . . **IBM 650 COMPUTER**
      . . IBM COMPUTERS
      . . . **IBM 650 COMPUTER**

IBM 704 COMPUTER
GS   DATA PROCESSING EQUIPMENT
      . COMPUTERS
      . . DIGITAL COMPUTERS
      . . . **IBM 704 COMPUTER**
      . . IBM COMPUTERS
      . . . **IBM 704 COMPUTER**

IBM 709 COMPUTER
GS   DATA PROCESSING EQUIPMENT
      . COMPUTERS
      . . DIGITAL COMPUTERS
      . . . **IBM 709 COMPUTER**
      . . IBM COMPUTERS
      . . . **IBM 709 COMPUTER**

IBM 1130 COMPUTER
GS   DATA PROCESSING EQUIPMENT
      . COMPUTERS
      . . DIGITAL COMPUTERS
      . . . **IBM 1130 COMPUTER**
      . . IBM COMPUTERS
      . . . **IBM 1130 COMPUTER**

IBM 1401 COMPUTER
GS   DATA PROCESSING EQUIPMENT
. COMPUTERS
. . DIGITAL COMPUTERS
. . . **IBM 1401 COMPUTER**
. . IBM COMPUTERS
. . . **IBM 1401 COMPUTER**

IBM 1410 COMPUTER
GS   DATA PROCESSING EQUIPMENT
. COMPUTERS
. . DIGITAL COMPUTERS
. . . **IBM 1410 COMPUTER**
. . IBM COMPUTERS
. . . **IBM 1410 COMPUTER**

IBM 1620 COMPUTER
GS   DATA PROCESSING EQUIPMENT
. COMPUTERS
. . DIGITAL COMPUTERS
. . . **IBM 1620 COMPUTER**
. . IBM COMPUTERS
. . . **IBM 1620 COMPUTER**

IBM 2250 COMPUTER
GS   DATA PROCESSING EQUIPMENT
. COMPUTERS
. . DIGITAL COMPUTERS
. . . **IBM 2250 COMPUTER**
. . IBM COMPUTERS
. . . **IBM 2250 COMPUTER**

IBM 7000 SERIES COMPUTERS
GS   DATA PROCESSING EQUIPMENT
. COMPUTERS
. . DIGITAL COMPUTERS
. . . **IBM 7000 SERIES COMPUTERS**
. . . . IBM 7030 COMPUTER
. . . . IBM 7040 COMPUTER
. . . . IBM 7044 COMPUTER
. . . . IBM 7070 COMPUTER
. . . . IBM 7074 COMPUTER
. . . . IBM 7090 COMPUTER
. . . . IBM 7094 COMPUTER
. . IBM COMPUTERS
. . . **IBM 7000 SERIES COMPUTERS**
. . . . IBM 7030 COMPUTER
. . . . IBM 7040 COMPUTER
. . . . IBM 7044 COMPUTER
. . . . IBM 7070 COMPUTER
. . . . IBM 7074 COMPUTER
. . . . IBM 7090 COMPUTER
. . . . IBM 7094 COMPUTER

IBM 7030 COMPUTER
GS   DATA PROCESSING EQUIPMENT
. COMPUTERS
. . DIGITAL COMPUTERS
. . . IBM 7000 SERIES COMPUTERS
. . . . **IBM 7030 COMPUTER**
. . IBM COMPUTERS
. . . IBM 7000 SERIES COMPUTERS
. . . . **IBM 7030 COMPUTER**

IBM 7040 COMPUTER
GS   DATA PROCESSING EQUIPMENT
. COMPUTERS
. . DIGITAL COMPUTERS
. . . IBM 7000 SERIES COMPUTERS
. . . . **IBM 7040 COMPUTER**
. . IBM COMPUTERS
. . . IBM 7000 SERIES COMPUTERS
. . . . **IBM 7040 COMPUTER**

IBM 7044 COMPUTER
GS   DATA PROCESSING EQUIPMENT
. COMPUTERS
. . DIGITAL COMPUTERS
. . . IBM 7000 SERIES COMPUTERS
. . . . **IBM 7044 COMPUTER**
. . IBM COMPUTERS
. . . IBM 7000 SERIES COMPUTERS
. . . . **IBM 7044 COMPUTER**

IBM 7070 COMPUTER
GS   DATA PROCESSING EQUIPMENT
. COMPUTERS
. . DIGITAL COMPUTERS
. . . IBM 7000 SERIES COMPUTERS
. . . . **IBM 7070 COMPUTER**
. . IBM COMPUTERS
. . . IBM 7000 SERIES COMPUTERS
. . . . **IBM 7070 COMPUTER**

IBM 7074 COMPUTER
GS   DATA PROCESSING EQUIPMENT
. COMPUTERS
. . DIGITAL COMPUTERS
. . . IBM 7000 SERIES COMPUTERS
. . . . **IBM 7074 COMPUTER**
. . IBM COMPUTERS
. . . IBM 7000 SERIES COMPUTERS
. . . . **IBM 7074 COMPUTER**

IBM 7090 COMPUTER
GS   DATA PROCESSING EQUIPMENT
. COMPUTERS
. . DIGITAL COMPUTERS
. . . IBM 7000 SERIES COMPUTERS
. . . . **IBM 7090 COMPUTER**
. . IBM COMPUTERS
. . . IBM 7000 SERIES COMPUTERS
. . . . **IBM 7090 COMPUTER**

IBM 7094 COMPUTER
GS   DATA PROCESSING EQUIPMENT
. COMPUTERS
. . DIGITAL COMPUTERS
. . . IBM 7000 SERIES COMPUTERS
. . . . **IBM 7094 COMPUTER**
. . IBM COMPUTERS
. . . IBM 7000 SERIES COMPUTERS
. . . . **IBM 7094 COMPUTER**

ICARUS ASTEROID
GS   CELESTIAL BODIES
. ASTEROID BELTS
. . ASTEROIDS
. . . **ICARUS ASTEROID**

ICBM (MISSILES)
USE   INTERCONTINENTAL BALLISTIC MISSILES

ICE
GS   **ICE**
. BAY ICE
. GLACIERS
. LAKE ICE
. . ICE FLOES
. LAND ICE
. SEA ICE
. . ICE FLOES
. . ICEBERGS
. . PRESSURE ICE
RT   AUFEIS (ICE)
CIRQUES (LANDFORMS)
FROST
POLAR CAPS
REFRIGERANTS
RUNWAY CONDITIONS
SLUSH
STORMS (METEOROLOGY)
WATER

ICE ENVIRONMENTS
UF   ANTARCTIC ENVIRONMENT
ARCTIC ENVIRONMENTS
GS   ENVIRONMENTS
. **ICE ENVIRONMENTS**
RT   ENVIRONMENT EFFECTS
MARINE ENVIRONMENTS
SEA ICE

ICE FLOES
GS   ICE
. LAKE ICE
. . **ICE FLOES**
. SEA ICE
. . **ICE FLOES**
RT   OCEANOGRAPHY

ICE FORMATION
UF   ICING
GS   **ICE FORMATION**
. CLOUD GLACIATION
RT   BAY ICE
FOULING
FREEZING
HAIL
LAKE ICE
LOW TEMPERATURE
PRESSURE ICE
SEA ICE
SNOW

ICE MAPPING
GS   MAPPING
. **ICE MAPPING**

ICE MAPPING-*(CONT.)*
RT   AERIAL PHOTOGRAPHY
BAY ICE
EARTH RESOURCES
HYDROGRAPHY
HYDROLOGY
INFRARED PHOTOGRAPHY
OCEANOGRAPHY
PHOTOGEOLOGY
PHOTOGRAPHY
PHOTOMAPPING
SEA ICE
SPACE SURVEILLANCE (SPACEBORNE)
SURVEILLANCE

ICE NUCLEI
RT   AITKEN NUCLEI
CLOUD GLACIATION
CONDENSATION NUCLEI
FREEZING
NUCLEATION
∞ NUCLEI

ICE OBSERVATION
USE   ICE REPORTING

ICE PACKS
USE   SEA ICE

ICE PREVENTION
GS   PREVENTION
. **ICE PREVENTION**
RT   ANTIICING ADDITIVES
DEFROSTING
DEICERS
DEICING
HEAT TAPES
HEATING
MELTING
STORM SUPPRESSION

ICE REPORTING
UF   ICE OBSERVATION
RT   BAY ICE
ICEBERGS
METEOROLOGICAL FLIGHT
POLAR METEOROLOGY
SPACE SURVEILLANCE (SPACEBORNE)
SURVEILLANCE

ICE SHELVES
USE   LAND ICE

ICEBERGS
GS   ICE
. SEA ICE
. . **ICEBERGS**
RESOURCES
. EARTH RESOURCES
. . **ICEBERGS**
RT   ICE REPORTING
LAND ICE

ICELAND
GS   LANDFORMS
. ISLANDS
. . **ICELAND**
NATIONS
. **ICELAND**

ICHTHYOLOGY
RT   FISHES
SCHOOLS (FISH)

ICING
USE   ICE FORMATION

ICL COMPUTERS
UF   INTERNATIONAL COMPUTERS LIMITED
GS   DATA PROCESSING EQUIPMENT
. COMPUTERS
. . DIGITAL COMPUTERS
. . . **ICL COMPUTERS**
RT   EUROPEAN SPACE AGENCY

ICOSAHEDRONS
GS   GEOMETRY
. EUCLIDEAN GEOMETRY
. . POLYHEDRONS
. . . **ICOSAHEDRONS**

IDAHO
GS   NATIONS
. UNITED STATES

IDAHO-(CONT.)
. . IDAHO
RT COLUMBIA RIVER BASIN (ID-OR-WA)
YELLOWSTONE NATIONAL PARK
(ID-MT-WY)

IDEAL FLUIDS
RT COMPRESSIBLE FLUIDS
EQUATIONS OF STATE
∞ FLUIDS
INCOMPRESSIBLE FLUIDS
MOLLIER DIAGRAM

IDEAL GAS
UF PERFECT GAS
GS GASES
. IDEAL GAS
RT DALTON LAW
EQUATIONS OF STATE
GAS DENSITY
KINETIC THEORY
KINETICS
REAL GASES

IDENTIFY FRIEND OR FOE
USE IFF SYSTEMS (IDENTIFICATION)

IDENTIFYING
GS IDENTIFYING
. CROP IDENTIFICATION
. IFF SYSTEMS (IDENTIFICATION)
. PARAMETER IDENTIFICATION
. RAPID BALLISTICS IDENTIFICATION
. SYSTEM IDENTIFICATION
. TIMBER IDENTIFICATION
RT CHEMICAL ANALYSIS
CODING
COGNITION
DETECTION
GAS DETECTORS
INSPECTION
MARKING
∞ MEASUREMENT
MISSILE DETECTION
PARTICULATE SAMPLING
PERCEPTION
RECOGNITION
SPECTRAL SIGNATURES
TRACKING (POSITION)
ULTRASONIC FLAW DETECTION
WISWESSER NOTATIONS

IDENTITIES
RT CONGRUENCES
∞ EQUATIONS

IDEP (DATA EXCHANGE)
USE INTERSERVICE DATA EXCHANGE
PROGRAM

IDLERS
RT BEARINGS
GEARS
PULLEYS
ROLLERS
VEHICULAR TRACKS

IFF SYSTEMS (IDENTIFICATION)
UF IDENTIFY FRIEND OR FOE
GS IDENTIFYING
. IFF SYSTEMS (IDENTIFICATION)
RT AIRCRAFT DETECTION
COGNITION
INTERROGATION
RECOGNITION
∞ SYSTEMS

IFR (RULES)
USE INSTRUMENT FLIGHT RULES

IGFET
USE FIELD EFFECT TRANSISTORS

IGNEOUS ROCKS
UF IGNIMBRITE
GS ROCKS
. IGNEOUS ROCKS
. . ANORTHOSITE
. . BASALT
. . DIORITE
. . DUNITE
. . ECLOGITE
. . FELSITE
. . GRANITE

IGNEOUS ROCKS-(CONT.)
. . OBSIDIAN
. . . MOLDAVITE
. . PERIDOTITE
. . PUMICE
. . SYENITE
. . TRACHYTE
RT ANDESITE
BATHOLITHS
BRECCIA
EFFUSIVES
ENSTATITE
FELDSPARS
ILMENITE
LAVA
MAGMA
MICA
MINERALS
OLIVINE
PYROXENES
QUARTZ
REGOLITH
ROCK INTRUSIONS
SEDIMENTARY ROCKS
SOILS
SPINEL
TOURMALINE

IGNIMBRITE
USE IGNEOUS ROCKS

IGNITERS
GS IGNITERS
. INITIATORS (EXPLOSIVES)
. . BOOSTERS (EXPLOSIVES)
. . CAPS (EXPLOSIVES)
. . DETONATORS
. . PRIMERS (EXPLOSIVES)
. SQUIBS
RT AMMUNITION
ELECTRIC IGNITION
EXPLOSIVE DEVICES
IGNITION
IGNITION SYSTEMS
INCENDIARY AMMUNITION
PYROPHORIC MATERIALS
SOLID PROPELLANT IGNITION
SPARK PLUGS

IGNITION
UF REIGNITION
GS IGNITION
. ELECTRIC IGNITION
. SOLID PROPELLANT IGNITION
. SPARK IGNITION
RT COMBUSTION
COMBUSTION PHYSICS
FIRING (IGNITING)
FLAME PROPAGATION
FLAMMABILITY
FLASH POINT
FUEL COMBUSTION
IGNITERS
PREMIXING
PROPELLANT COMBUSTION
ROASTING
SPARKS
SPONTANEOUS COMBUSTION
STARTING

IGNITION LIMITS
RT COMBUSTION
FLAME RETARDANTS
FLAMMABILITY
FUEL-AIR RATIO
GAS MIXTURES
∞ LIMITS

IGNITION SYSTEMS
RT AUTOMOBILES
DISTRIBUTORS
DWELL
ELECTRIC COILS
ELECTRIC IGNITION
ENGINES
IGNITERS
INTERNAL COMBUSTION ENGINES
ROCKET ENGINES
SPARK PLUGS
SQUIBS
STARTERS
∞ SYSTEMS

IGNITION TEMPERATURE
GS TEMPERATURE

IGNITION TEMPERATURE-(CONT.)
. IGNITION TEMPERATURE
. . FLASH POINT
RT COMBUSTION TEMPERATURE
FLAMMABILITY
PROPELLANT SENSITIVITY
PYROPHORIC MATERIALS
SOLID PROPELLANT IGNITION
SPONTANEOUS COMBUSTION
THERMITES

IGNITRONS
GS ELECTRON TUBES
. VACUUM TUBES
. . VACUUM TUBE OSCILLATORS
. . . GAS DISCHARGE TUBES
. . . . IGNITRONS
. . . MICROWAVE TUBES
. . . . PLANOTRONS
. . . . . IGNITRONS
MICROWAVE EQUIPMENT
. MICROWAVE TUBES
. . PLANOTRONS
. . . IGNITRONS
RECTIFIERS
. IGNITRONS

IGOSS
USE INTEGRATED GLOBAL OCEAN STATION
SYSTEMS

IGY (GEOPHYSICAL YEAR)
USE INTERNATIONAL GEOPHYSICAL YEAR

IL-14 AIRCRAFT
UF ILYUSHIN IL-14 AIRCRAFT
GS ILYUSHIN AIRCRAFT
. IL-14 AIRCRAFT
MONOPLANES
. IL-14 AIRCRAFT
TRANSPORT AIRCRAFT
. IL-14 AIRCRAFT
RT ∞ AIRCRAFT

IL-62 AIRCRAFT
UF CLASSIC AIRCRAFT
ILYUSHIN IL-62 AIRCRAFT
GS COMMERCIAL AIRCRAFT
. IL-62 AIRCRAFT
ILYUSHIN AIRCRAFT
. IL-62 AIRCRAFT
JET AIRCRAFT
. TURBOFAN AIRCRAFT
. . IL-62 AIRCRAFT
MONOPLANES
. IL-62 AIRCRAFT
PASSENGER AIRCRAFT
. IL-62 AIRCRAFT
RT ∞ AIRCRAFT

ILLIAC COMPUTERS
GS DATA PROCESSING EQUIPMENT
. COMPUTERS
. . DIGITAL COMPUTERS
. . . ILLIAC COMPUTERS
. . . . ILLIAC 3 COMPUTER
. . . . ILLIAC 4 COMPUTER

ILLIAC 3 COMPUTER
GS DATA PROCESSING EQUIPMENT
. COMPUTERS
. . DIGITAL COMPUTERS
. . . ILLIAC COMPUTERS
. . . . ILLIAC 3 COMPUTER
RT ANALOG TO DIGITAL CONVERTERS
PARALLEL PROCESSING (COMPUTERS)

ILLIAC 4 COMPUTER
GS DATA PROCESSING EQUIPMENT
. COMPUTERS
. . DIGITAL COMPUTERS
. . . ILLIAC COMPUTERS
. . . . ILLIAC 4 COMPUTER
RT ANALOG TO DIGITAL CONVERTERS
PARALLEL PROCESSING (COMPUTERS)

ILLINOIS
GS NATIONS
. UNITED STATES
. . ILLINOIS
RT OHIO RIVER (US)
WABASH RIVER BASIN (IL-IN-OH)

ILLITE
```
GS    CLAYS
      . ILLITE
      MINERALS
      . ILLITE
RT    SOILS
```

ILLUMINANCE
```
SN    (DETECTION RATE PER UNIT AREA OF
      VISIBLE RADIATION--EQUALS LIGHT
      PRESSURE TIMES SPEED OF LIGHT)
UF    LIGHT PRESSURE
GS    PRESSURE
      . RADIATION PRESSURE
      . . LUMINOUS INTENSITY
      . . . ILLUMINANCE
      RATES (PER TIME)
      . FLUX DENSITY
      . . RADIANT FLUX DENSITY
      . . . IRRADIANCE
      . . . . ILLUMINANCE
      . . . LUMINOUS INTENSITY
      . . . . ILLUMINANCE
RT    BRIGHTNESS
      ILLUMINATING
      ∞ILLUMINATION
      LIGHT (VISIBLE RADIATION)
      LUMINANCE
      LUMINOSITY
      RADIANCY
      SOLAR CONSTANT
      SOLAR FLUX DENSITY
      VISIBILITY
```

ILLUMINATING
```
UF    LIGHTING
RT    ARCHITECTURE
      BRIGHTNESS
      COMFORT
      DARKNESS
      ENVIRONMENTAL ENGINEERING
      ∞FLARES
      GLARE
      HUMAN FACTORS ENGINEERING
      ILLUMINANCE
      ∞ILLUMINATION
      LIGHT SOURCES
      LIGHT TRANSMISSION
      LIGHTING EQUIPMENT
      LUMINAIRES
      LUMINANCE
      PHOTOMETRY
      ∞PROJECTION
      PROJECTORS
      PYROTECHNICS
      SHADOWS
```

∞ ILLUMINATION
```
SN    (USE OF A MORE SPECIFIC TERM IS
      RECOMMENDED--CONSULT THE TERMS
      LISTED BELOW)
RT    BRIGHTNESS DISCRIMINATION
      DARKENING
      DARKNESS
      ∞DIFFUSERS
      ILLUMINANCE
      ILLUMINATING
      ILLUMINATORS
      ISOPHOTES
      LIGHT TRANSMISSION
      LUMINESCENCE
      PHOTOMETRY
```

ILLUMINATORS
```
GS    LIGHT SOURCES
      . ILLUMINATORS
      LIGHTING EQUIPMENT
      . ILLUMINATORS
RT    ∞ILLUMINATION
      INCANDESCENCE
      LUMINESCENCE
```

ILLUSIONS
```
GS    PSYCHOLOGICAL EFFECTS
      . ILLUSIONS
      . . HALLUCINATIONS
      . . MOON ILLUSION
      . . OCULOGRAVIC ILLUSIONS
      . . OPTICAL ILLUSION
      . . . ELEVATOR ILLUSION
RT    AFTERIMAGES
      IMAGES
      PERCEPTION
      VISION
```

ILMENITE
```
GS    CHALCOGENIDES
      . OXIDES
      . . METAL OXIDES
      . . . IRON OXIDES
      . . . . ILMENITE
      . . . TITANIUM OXIDES
      . . . . ILMENITE
      IRON COMPOUNDS
      . IRON OXIDES
      . . ILMENITE
      MINERALS
      . ILMENITE
      TITANIUM COMPOUNDS
      . TITANATES
      . . ILMENITE
      . TITANIUM OXIDES
      . . ILMENITE
RT    IGNEOUS ROCKS
      SANDS
```

ILS (LANDING SYSTEMS)
```
USE   INSTRUMENT LANDING SYSTEMS
```

ILYUSHIN AIRCRAFT
```
GS    ILYUSHIN AIRCRAFT
      . IL-14 AIRCRAFT
      . IL-62 AIRCRAFT
RT    ∞AIRCRAFT
```

ILYUSHIN IL-14 AIRCRAFT
```
USE   IL-14 AIRCRAFT
```

ILYUSHIN IL-62 AIRCRAFT
```
USE   IL-62 AIRCRAFT
```

IMAGE ANALYSIS
```
RT    CLUSTER ANALYSIS
      IMAGE ENHANCEMENT
      IMAGE PROCESSING
      IMAGE RESOLUTION
      PATTERN RECOGNITION
      RADAR IMAGERY
      REMOTE SENSING
      SATELLITE IMAGERY
      SCENE ANALYSIS
```

IMAGE CONTRAST
```
GS    CONTRAST
      . IMAGE CONTRAST
RT    FOCUSING
      GRAY SCALE
      PATTERN REGISTRATION
      RESOLUTION
      SELF FOCUSING
      SIGNAL TO NOISE RATIOS
      SMEAR
      VISIBILITY
```

IMAGE CONVERTERS
```
GS    OPTICAL EQUIPMENT
      . IMAGE CONVERTERS
      . . CELESCOPES
      . . IMAGE TUBES
      . . . THERMICONS
RT    CAMERA TUBES
      ∞CONVERTERS
      LALLEMAND CAMERAS
      LIGHT AMPLIFIERS
      MICROCHANNELS
      PHOTOCATHODES
```

IMAGE CORRELATORS
```
UF    SIMICOR (IMAGE CORRELATOR)
      SIMULTANEOUS IMAGE CORRELATOR
GS    CORRELATORS
      . IMAGE CORRELATORS
RT    HOLOGRAPHY
      IMAGING TECHNIQUES
      MAP MATCHING GUIDANCE
      PATTERN REGISTRATION
      VIDEO LANDMARK ACQUISITION AND
        TRACKING
```

IMAGE DISSECTOR TUBES
```
GS    ELECTRON TUBES
      . CAMERA TUBES
      . . IMAGE DISSECTOR TUBES
      TELEVISION EQUIPMENT
      . IMAGE DISSECTOR TUBES
RT    GUIDANCE SENSORS
      SATELLITE ORIENTATION
```

IMAGE ENHANCEMENT
```
RT    BAND RATIOING
      FOCUSING
      GEOMETRIC RECTIFICATION (IMAGERY)
      GRAY SCALE
      IMAGE ANALYSIS
      IMAGES
      IMAGING TECHNIQUES
      LIGHT AMPLIFIERS
      RADIOMETRIC CORRECTION
      RESOLUTION
      SIGNAL TO NOISE RATIOS
      TOMOGRAPHY
      VEGETATIVE INDEX
```

IMAGE FILTERS
```
RT    ∞FILTERS
      IMAGING TECHNIQUES
```

IMAGE FURNACES
```
GS    HEATING EQUIPMENT
      . FURNACES
      . . IMAGE FURNACES
      LABORATORY EQUIPMENT
      . IMAGE FURNACES
RT    ARC HEATING
      CARBON ARCS
```

IMAGE INTENSIFIERS
```
UF    INTENSIFIER TUBES
GS    INTENSIFIERS
      . IMAGE INTENSIFIERS
      . . IMAGE ORTHICONS
RT    AMPLIFIERS
      IMAGING TECHNIQUES
      LALLEMAND CAMERAS
      LIGHT AMPLIFIERS
      NIGHT VISION
      ORTHICONS
      PHOSPHORS
      PHOTOCATHODES
```

IMAGE MOTION COMPENSATION
```
RT    AERIAL PHOTOGRAPHY
      ∞COMPENSATION
      IMAGING TECHNIQUES
      PATTERN REGISTRATION
```

IMAGE ORTHICONS
```
GS    ELECTRON TUBES
      . CAMERA TUBES
      . . ORTHICONS
      . . . IMAGE ORTHICONS
      INTENSIFIERS
      . IMAGE INTENSIFIERS
      . . IMAGE ORTHICONS
      MICROWAVE EQUIPMENT
      . MICROWAVE TUBES
      . . ORTHICONS
      . . . IMAGE ORTHICONS
RT    PHOTOCATHODES
```

IMAGE PROCESSING
```
GS    IMAGE PROCESSING
      . BAND RATIOING
      . GEOMETRIC RECTIFICATION (IMAGERY)
RT    ATMOSPHERIC CORRECTION
      CHANGE DETECTION
      CLUSTER ANALYSIS
      COMPUTER AIDED TOMOGRAPHY
      DATA PROCESSING
      FEATURE IDENTIFICATION AND
        LOCATION EXPER
      FRAMES (DATA PROCESSING)
      GEOMETRIC ACCURACY
      GRAY SCALE
      IMAGE ANALYSIS
      IMAGERY
      IMAGING TECHNIQUES
      MULTISENSOR APPLICATIONS
      NAP-OF-THE-EARTH NAVIGATION
      ONBOARD DATA PROCESSING
      OPTICAL DATA PROCESSING
      POINT SPREAD FUNCTIONS
      PREPROCESSING
      PRINCIPAL COMPONENTS ANALYSIS
      ∞PROCESSING
      PUSHBROOM SENSOR MODES
      SPATIAL RESOLUTION
```

IMAGE RECONSTRUCTION
```
RT    DISPLAY DEVICES
      HOLOGRAPHY
      IMAGING TECHNIQUES
      PATTERN REGISTRATION
```

IMAGE RESOLUTION
GS   RESOLUTION
        . **IMAGE RESOLUTION**
RT   ELECTRO-OPTICAL PHOTOGRAPHY
        GEOMETRIC ACCURACY
        IMAGE ANALYSIS
        IMAGERY
        MATCHING
        MULTISPECTRAL PHOTOGRAPHY
        PATTERN REGISTRATION
        SPATIAL RESOLUTION

IMAGE ROTATION
RT   IMAGING TECHNIQUES
        ROTATION

IMAGE TRANSDUCERS
GS   TRANSDUCERS
        . **IMAGE TRANSDUCERS**
RT   CAMERA TUBES
        IMAGING TECHNIQUES
        LALLEMAND CAMERAS

IMAGE TUBES
GS   MICROWAVE EQUIPMENT
        . MICROWAVE TUBES
        . . **IMAGE TUBES**
        . . . THERMICONS
        OPTICAL EQUIPMENT
        . IMAGE CONVERTERS
        . . **IMAGE TUBES**
        . . . THERMICONS
RT   CATHODE RAY TUBES
        DISPLAY DEVICES
        FLYING SPOT SCANNERS
        HEAD-UP DISPLAYS
        MONOSCOPES

IMAGE VELOCITY SENSORS
RT   IMAGES
        IMAGING TECHNIQUES
        ∞SENSORS

IMAGERY
GS   **IMAGERY**
        . ACOUSTICAL HOLOGRAPHY
        . AERIAL PHOTOGRAPHY
        . ALL SKY PHOTOGRAPHY
        . ASTRONOMICAL PHOTOGRAPHY
        . BLACK AND WHITE PHOTOGRAPHY
        . CHRONOPHOTOGRAPHY
        . CINEMATOGRAPHY
        . CLOUD PHOTOGRAPHY
        . COLOR PHOTOGRAPHY
        . ELECTRO-OPTICAL PHOTOGRAPHY
        . ELECTRON PHOTOGRAPHY
        . HOLOGRAPHY
        . . MICROWAVE HOLOGRAPHY
        . . WHITE LIGHT HOLOGRAPHY
        . INFRARED IMAGERY
        . INFRARED PHOTOGRAPHY
        . . COLOR INFRARED PHOTOGRAPHY
        . KINOFORM
        . LUNAR PHOTOGRAPHY
        . MICROWAVE IMAGERY
        . MICROWAVE PHOTOGRAPHY
        . PHOTOMICROGRAPHY
        . PHOTORECONNAISSANCE
        . RADAR IMAGERY
        . RADAR PHOTOGRAPHY
        . RADIOGRAPHY
        . . ANGIOGRAPHY
        . . AUTORADIOGRAPHY
        . . NEUTRON RADIOGRAPHY
        . . TOMOGRAPHY
        . . . COMPUTER AIDED TOMOGRAPHY
        . . UROGRAPHY
        . REPRODUCTION (COPYING)
        . XEROGRAPHY
        . ROCKET-BORNE PHOTOGRAPHY
        . SATELLITE IMAGERY
        . SHADOWGRAPH PHOTOGRAPHY
        . . SCHLIEREN PHOTOGRAPHY
        . SPACEBORNE PHOTOGRAPHY
        . . SATELLITE-BORNE PHOTOGRAPHY
        . SPECTROHELIOGRAPHS
        . SPECTROPHOTOGRAPHY
        . STEREOSCOPY
        . . STEREOPHOTOGRAPHY
        . ULTRAVIOLET PHOTOMETRY
        . X RAY IMAGERY
RT   ACOUSTO-OPTICS
        APPEARANCE
        ATMOSPHERIC & OCEANOGRAPHIC
           INFORM SYS

IMAGERY-(CONT.)
        CHANGE DETECTION
        CONTOUR SENSORS
        DISPLAY DEVICES
        EARTH RESOURCES
        GEOGRAPHIC INFORMATION SYSTEMS
        GEOMETRIC RECTIFICATION (IMAGERY)
        GRAPHIC ARTS
        GROUND TRUTH
        IMAGE PROCESSING
        IMAGE RESOLUTION
        MICROWAVE SOUNDING
        MULTISPECTRAL PHOTOGRAPHY
        MULTISPECTRAL RADAR
        PHOTOGRAPHY
        RADAR SIGNATURES
        SCENE ANALYSIS
        SEA TRUTH
        SIGNATURE ANALYSIS

IMAGES
UF   OPTICAL IMAGES
GS   **IMAGES**
        . AFTERIMAGES
        . RETINAL IMAGES
RT   CONTOUR SENSORS
        DISPLAY DEVICES
        EVAPOROGRAPHY
        HALOS
        HELMET MOUNTED DISPLAYS
        ILLUSIONS
        IMAGE ENHANCEMENT
        IMAGE VELOCITY SENSORS
        ∞OPTICS
        PERCEPTION
        PHOTOGRAPHS
        REPRESENTATIONS
        SPATIAL FILTERING
        VISION

IMAGING RADAR
GS   RADAR
        IMAGING RADAR
RT   RADAR IMAGERY
        REMOTE SENSORS
        SIDE-LOOKING RADAR
        SYNTHETIC APERTURE RADAR

IMAGING TECHNIQUES
RT   ACOUSTIC MICROSCOPES
        ACOUSTICAL HOLOGRAPHY
        ADAPTIVE OPTICS
        CHARGE INJECTION DEVICES
        CROP IDENTIFICATION
        GRAY SCALE
        HOLOGRAMMETRY
        IMAGE CORRELATORS
        IMAGE ENHANCEMENT
        IMAGE FILTERS
        IMAGE INTENSIFIERS
        IMAGE MOTION COMPENSATION
        IMAGE PROCESSING
        IMAGE RECONSTRUCTION
        IMAGE ROTATION
        IMAGE TRANSDUCERS
        IMAGE VELOCITY SENSORS
        ∞METHODOLOGY
        MICROWAVE HOLOGRAPHY
        MODULATION TRANSFER FUNCTION
        MULTISENSOR APPLICATIONS
        MULTISPECTRAL BAND SCANNERS
        MULTISPECTRAL PHOTOGRAPHY
        MULTISPECTRAL RADAR
        OPTICAL RELAY SYSTEMS
        OPTICAL TRANSFER FUNCTION
        ∞OPTICS
        PATTERN REGISTRATION
        PHOTOGRAPHY
        PRINCIPAL COMPONENTS ANALYSIS
        RADAR IMAGERY
        RAPID BALLISTICS IDENTIFICATION
        RESOLUTION CELL
        SATELLITE IMAGERY
        SCENE ANALYSIS
        SPATIAL RESOLUTION
        STREAK PHOTOGRAPHY
        SYNTHETIC APERTURES
        TELEVISION SYSTEMS
        ULTRASONIC SCANNERS
        VEGETATIVE INDEX
        X RAY IMAGERY

IMBEDDINGS
∞
SN   *(USE OF A MORE SPECIFIC TERM IS*
        RECOMMENDED--CONSULT THE TERMS
        LISTED BELOW)
RT   ENCAPSULATING
        IMBEDDINGS (MATHEMATICS)
        INVARIANT IMBEDDINGS
        ∞MATRICES

IMBEDDINGS (MATHEMATICS)
GS   GEOMETRY
        . TOPOLOGY
        . . **IMBEDDINGS (MATHEMATICS)**
        . . . INVARIANT IMBEDDINGS
RT   ∞IMBEDDINGS
        STRANGE ATTRACTORS

IMBLMS
SN   (INTEGRATED MEDICAL AND
        BEHAVIORAL LABORATORY
        MEASUREMENT SYSTEM)
UF   INTEG MED AND BEHAVIORAL LAB
           MEASUR SYSTEM
RT   BIOINSTRUMENTATION
        BIOMEDICAL DATA
        MEASURING INSTRUMENTS
        MEDICAL EQUIPMENT

IMCC (CONTROL CENTER)
USE   INTEGRATED MISSION CONTROL
           CENTER

IME SATELLITE
USE   INTERNATIONAL MAGNETOSPHERIC
           EXPLORER

IMIDES
GS   NITROGEN COMPOUNDS
        . **IMIDES**
        . . SUCCINIMIDES
RT   AMIDES

IMINES
UF   SCHIFF BASES
GS   NITROGEN COMPOUNDS
        . **IMINES**
RT   AMINES

IMLSS
SN   (INTEGRATED MANEUVERING AND LIFE
        SUPPORT SYSTEM)
UF   INTEGRATED MANEUVERING LIFE
           SUPPORT SYS
GS   SELF MANEUVERING UNITS
        . **IMLSS**
RT   ASTRONAUT MANEUVERING EQUIPMENT
        EXTRAVEHICULAR ACTIVITY
        EXTRAVEHICULAR MOBILITY UNITS
        ∞SYSTEMS

IMMERSION
USE   SUBMERGING

IMMISCIBILITY
USE   SOLUBILITY

IMMITTANCE
USE   ELECTRICAL IMPEDANCE

IMMOBILIZATION
RT   DAMAGE
        IMPAIRMENT
        ∞MOTION

IMMUNITY
RT   INFECTIOUS DISEASES
        INOCULATION
        ∞RESISTANCE
        TOXINS AND ANTITOXINS

IMMUNOASSAY
UF   PLASMA RENIN ACTIVITY
GS   **IMMUNOASSAY**
        . RADIOIMMUNOASSAY
RT   ANTIGENS
        ASSAYING
        BIOCHEMISTRY
        IMMUNOLOGY
        RADIOBIOLOGY

IMMUNOLOGY
GS   MEDICAL SCIENCE
        . **IMMUNOLOGY**
RT   ALLERGIC DISEASES

IMMUNOLOGY-*(CONT.)*
    ANAPHYLAXIS
    ANTIBODIES
    ANTIGENS
    ANTISERUMS
    BIOCOMPATIBILITY
  ∞BIOLOGY
    IMMUNOASSAY
    PROPHYLAXIS
    RADIOIMMUNOASSAY
    VETERINARY MEDICINE

IMP
UF   INTERPLANETARY MONITORING
        PLATFORM
GS   SATELLITES
    . ARTIFICIAL SATELLITES
    . . EXPLORER SATELLITES
    . . . **IMP**
    . . LUNAR SATELLITES
    . . . **IMP**

IMP-A
USE   EXPLORER 18 SATELLITE

IMP-B
USE   EXPLORER 21 SATELLITE

IMP-C
USE   EXPLORER 28 SATELLITE

IMP-D
USE   EXPLORER 33 SATELLITE

IMP-E
USE   EXPLORER 35 SATELLITE

IMP-F
USE   EXPLORER 34 SATELLITE

IMP-G
USE   EXPLORER 41 SATELLITE

IMP-H
USE   EXPLORER 47 SATELLITE

IMP-I
USE   EXPLORER 43 SATELLITE

IMP-J
USE   EXPLORER 50 SATELLITE

IMP-1
USE   EXPLORER 18 SATELLITE

IMP-2
USE   EXPLORER 21 SATELLITE

IMP-3
USE   EXPLORER 28 SATELLITE

IMP-4
USE   EXPLORER 34 SATELLITE

IMP-5
USE   EXPLORER 41 SATELLITE

IMP-6
USE   EXPLORER 43 SATELLITE

IMP-7
USE   EXPLORER 47 SATELLITE

IMP-8
USE   EXPLORER 50 SATELLITE

IMPACT
GS   **IMPACT**
    . ECONOMIC IMPACT
    . ELECTRON IMPACT
    . HYPERVELOCITY IMPACT
    . ION IMPACT
    . POINT IMPACT
    . PROTON IMPACT
RT   DECELERATION
    HYDRODYNAMIC RAM EFFECT
    IMPINGEMENT
    MECHANICAL SHOCK
    PENETRATION
    PERCUSSION
    PRESSURE
    SHOCK ABSORBERS
    SHOCK RESISTANCE

IMPACT-*(CONT.)*
    SHOCK WAVES
    STRESSES

IMPACT ACCELERATION
UF   IMPACT DECELERATION
GS   RATES (PER TIME)
    . ACCELERATION (PHYSICS)
    . . **IMPACT ACCELERATION**
RT   ∞ACCELERATION
    DECELERATION
    MECHANICAL SHOCK
    PHYSIOLOGICAL ACCELERATION
    RAILROAD HUMPING TESTS
    SHOCK ABSORBERS

IMPACT DAMAGE
GS   DAMAGE
    . **IMPACT DAMAGE**
    . . METEORITIC DAMAGE
    . . RAIN IMPACT DAMAGE
RT   CRATERING
    CRATERS
    EJECTA
    IMPACT TOLERANCES
    MARS CRATERS
    METEOROID PROTECTION
    PLANETARY CRATERS

IMPACT DECELERATION
USE   DECELERATION
    IMPACT ACCELERATION

IMPACT FUSION
GS   INERTIAL CONFINEMENT FUSION
    . **IMPACT FUSION**
RT   FUSION REACTORS

IMPACT LOADS
UF   IMPACT PRESSURES
GS   LOADS (FORCES)
    . COMPRESSION LOADS
    . . **IMPACT LOADS**
    . DYNAMIC LOADS
    . . TRANSIENT LOADS
    . . . **IMPACT LOADS**
    PRESSURE
    . **IMPACT LOADS**
RT   BLAST LOADS
    DYNAMIC PRESSURE
    FIBER ORIENTATION
    LANDING LOADS
    LOADING RATE
    PRESSURE
    RANDOM LOADS
    SHOCK LOADS
    STRUCTURAL DESIGN CRITERIA

IMPACT MELTS
GS   MELTS (CRYSTAL GROWTH)
    . **IMPACT MELTS**
RT   CELESTIAL BODIES
    HYPERVELOCITY IMPACT
    LUNAR ROCKS
    MELTING
    METEORITES
    MINERALS
    PETROLOGY

IMPACT PREDICTION
UF   ARIP (IMPACT PREDICTION)
    AUTOMATIC ROCKET IMPACT
        PREDICTORS
GS   PREDICTIONS
    . **IMPACT PREDICTION**
RT   BALLISTIC TRAJECTORIES
    DOWNRANGE
    GUIDANCE (MOTION)
    LASER GUIDANCE
    MISSILE TRAJECTORIES
    RANGE SAFETY
    REENTRY
    TRAJECTORY ANALYSIS

IMPACT PRESSURES
USE   IMPACT LOADS

IMPACT RESISTANCE
UF   IMPACT SENSITIVITY
GS   SENSITIVITY
    . **IMPACT RESISTANCE**
    SHOCK RESISTANCE
    . **IMPACT RESISTANCE**
RT   CRASHWORTHINESS

IMPACT RESISTANCE-*(CONT.)*
    PROPELLANT SENSITIVITY
  ∞RESISTANCE
    TOLERANCES (PHYSIOLOGY)

IMPACT SENSITIVITY
USE   IMPACT RESISTANCE

IMPACT STRENGTH
GS   MECHANICAL PROPERTIES
    . **IMPACT STRENGTH**
RT   BRITTLE MATERIALS
    BRITTLENESS
    DUCTILITY
    EARTHQUAKE RESISTANCE
    HARDNESS
    NOTCH SENSITIVITY
  ∞RESISTANCE
    SHEAR PROPERTIES
  ∞STRENGTH
    STRESS CONCENTRATION
    WAVE RESISTANCE

IMPACT TESTING MACHINES
RT   DROP TESTS
    FATIGUE TESTS
  ∞MACHINERY
  ∞TEST EQUIPMENT

IMPACT TESTS
GS   **IMPACT TESTS**
    . CHARPY IMPACT TEST
RT   BRITTLENESS
    COMPRESSION TESTS
    DESTRUCTIVE TESTS
    DROP TESTS
    FATIGUE TESTS
    HARDNESS TESTS
    IMPACTORS
    LOAD TESTS
  ∞MATERIALS TESTS
    NOTCH SENSITIVITY
    NOTCH STRENGTH
    NOTCH TESTS
    SHOCK TESTS
    STRAIN RATE
    STRESS CONCENTRATION
  ∞TESTS
    TOUGHNESS

IMPACT TOLERANCES
GS   TOLERANCES (MECHANICS)
    . **IMPACT TOLERANCES**
RT   IMPACT DAMAGE

IMPACTORS
RT   CRUSHERS
    GRINDING MILLS
    HAMMERS
    IMPACT TESTS

IMPAIRMENT
RT   DAMAGE
    IMMOBILIZATION
    INJURIES
    LOSSES

IMPATT DIODES
USE   AVALANCHE DIODES

IMPEDANCE
UF   DUMMY LOADS
GS   **IMPEDANCE**
    . ACOUSTIC IMPEDANCE
    . ELECTRICAL IMPEDANCE
    . . ELECTRICAL RESISTANCE
    . . . CONTACT RESISTANCE
    . . . SKIN RESISTANCE
    . . REACTANCE
    . MECHANICAL IMPEDANCE
    . RESPIRATORY IMPEDANCE
RT   ATTENUATION COEFFICIENTS
    BANDWIDTH
    CHOKES (RESTRICTIONS)
  ∞CONDUCTIVITY
    CONSTRICTIONS
    DAMPING
    DIFFUSIVITY
    DYNAMIC CHARACTERISTICS
    DYNAMIC RESPONSE
    ELECTRIC COILS
    ELECTRICAL PROPERTIES
  ∞HYDRAULICS
    MECHANICAL PROPERTIES

IMPEDANCE-*(CONT.)*
 ∞PHYSICAL PROPERTIES
 ∞RESISTANCE
 RESONANT FREQUENCIES
 SMITH CHART
 TIME CONSTANT
 TRANSIENT RESPONSE

IMPEDANCE MATCHING
RT ANTENNA COUPLERS
 COUPLERS
 COUPLING CIRCUITS
 DIRECTIONAL COUPLERS
 ELECTRIC NETWORKS
 ELECTRICAL IMPEDANCE
 ITERATIVE NETWORKS
 MATCHING
 MODE TRANSFORMERS
 TRANSFER FUNCTIONS
 TRANSMISSION LINES
 WAVEGUIDE TUNERS
 WAVEGUIDE WINDOWS

IMPEDANCE MEASUREMENT
RT ELECTRICAL IMPEDANCE
 ELECTRICAL MEASUREMENT
 MECHANICAL IMPEDANCE
 MISMATCH (ELECTRICAL)
 RADIO FREQUENCY IMPEDANCE
  PROBES

IMPEDANCE PROBES
GS MEASURING INSTRUMENTS
 . **IMPEDANCE PROBES**
 . . RADIO FREQUENCY IMPEDANCE
  PROBES
RT RESONANCE PROBES

IMPELLER BLADES
USE ROTOR BLADES (TURBOMACHINERY)

IMPELLERS
GS ROTATING BODIES
 . ROTORS
 . . **IMPELLERS**
 . . . PUMP IMPELLERS
RT BLOWERS
 CENTRIFUGAL COMPRESSORS
 CENTRIFUGAL PUMPS
 COMPRESSOR ROTORS
 PUMPS
 ROTOR BLADES (TURBOMACHINERY)
 STATORS
 TURBINE WHEELS
 TURBINES
 TURBOMACHINE BLADES
 VANES

IMPERFECTIONS
USE DEFECTS

IMPERIAL VALLEY (CA)
GS VALLEYS
 . **IMPERIAL VALLEY (CA)**
RT CALIFORNIA
 DESERTS
 MEXICO

IMPINGEMENT
GS **IMPINGEMENT**
 . JET IMPINGEMENT
RT ABLATION
 ATTENUATION
 CAVITATION FLOW
 CORROSION
 EROSION
 GAS-SOLID INTERACTIONS
 IMPACT
 INCIDENCE
 REFLECTION
 SCATTERING

IMPLANTATION
GS **IMPLANTATION**
 . HEART IMPLANTATION
 . ION IMPLANTATION
RT GRAFTING
 INJECTION
 INSERTION

IMPLANTED ELECTRODES (BIOLOGY)
GS ELECTRODES
 . **IMPLANTED ELECTRODES (BIOLOGY)**
RT ∞BIOLOGY

IMPLICATION
RT INFERENCE

IMPLOSIONS
RT BURSTS
 EXPLOSIONS
 EXPLOSIVE DECOMPRESSION
 PROPELLANT EXPLOSIONS
 SHOCK WAVES

IMPREGNATING
RT CHEMICAL ATTACK
 COATINGS
 FINISHES
 INSERTION
 LUBRICATION
 PERMEATING
 POROSITY
 PRESERVING
 SELF LUBRICATING MATERIALS
 SELF LUBRICATION

IMPROVED TIROS OPERATIONAL SATELLITES
GS SATELLITES
 . ARTIFICIAL SATELLITES
 . . METEOROLOGICAL SATELLITES
 . . . TIROS SATELLITES
 . . . . **IMPROVED TIROS OPERATIONAL**
   SATELLITES
 . . . . . ITOS 1
 . . . . . ITOS 2
 . . . . . ITOS 3
 . . . . . ITOS 4
RT TIROS M
 TIROS N SERIES SATELLITES

IMPROVEMENT
RT CORRECTION
 PUBLIC RELATIONS
 UPGRADING

IMPULSE GENERATORS
RT ∞GENERATORS
 PULSE GENERATORS
 TURBINES

IMPULSES
RT HIGH IMPULSE
 SPECIFIC IMPULSE

IMPURITIES
GS **IMPURITIES**
 . PSEUDOPOTENTIALS
RT CONTAMINANTS
 CRYSTAL DEFECTS
 DIRT
 HETEROGENEITY
 INCLUSIONS
 POINT DEFECTS
 PSEUDOPOTENTIALS
 QUALITY
 TRACE CONTAMINANTS
 ULTRAPURE METALS
 WASTES

IMS
USE INTERNATIONAL MAGNETOSPHERIC
 STUDY

IN-FLIGHT MONITORING
RT CREW PROCEDURES (INFLIGHT)
 CREW PROCEDURES (PREFLIGHT)
 FLIGHT CONTROL
 FLIGHT TESTS
 MONITORS
 TELEMETRY

INACTIVATION
USE DEACTIVATION

INCANDESCENCE
GS DECAY
 . EMISSION
 . . LIGHT EMISSION
 . . . **INCANDESCENCE**
RT BRIGHTNESS
 COLOR
 EMISSIVITY
 ILLUMINATORS
 LIGHT (VISIBLE RADIATION)
 LUMINESCENCE
 LUMINOSITY
 LUMINOUS INTENSITY
 RADIANCE

INCANDESCENCE-*(CONT.)*
 SPECTRAL EMISSION
 THERMAL EMISSION

INCENDIARY AMMUNITION
GS AMMUNITION
 . **INCENDIARY AMMUNITION**
RT BOMBS (ORDNANCE)
 COMBUSTION
 EXOTHERMIC REACTIONS
 GRENADES
 GUNS (ORDNANCE)
 IGNITERS
 MISSILES
 PROJECTILES
 PROPELLANTS
 PYROTECHNICS
 ∞ROCKETS

INCENTIVE TECHNIQUES
RT COST INCENTIVES
 COST REDUCTION
 EFFICIENCY
 MANAGEMENT
 VALUE ENGINEERING

INCENTIVES
GS **INCENTIVES**
 . CONTRACT INCENTIVES
RT INCOME
 MANAGEMENT
 MANAGEMENT METHODS
 MOTIVATION
 PERSONNEL

INCIDENCE
GS **INCIDENCE**
 . GRAZING INCIDENCE
RT ANGLES (GEOMETRY)
 IMPINGEMENT

INCIDENT RADIATION
RT BISTATIC REFLECTIVITY
 CORPUSCULAR RADIATION
 ELECTROMAGNETIC RADIATION
 OBLIQUENESS
 OPTICAL REFLECTION
 PHOTON BEAMS
 ∞RADIATION
 REFLECTED WAVES
 REFRACTED WAVES
 RETROREFLECTION
 SCATTERING
 STOKES LAW OF RADIATION
 WAVE INCIDENCE CONTROL

INCINERATION
USE INCINERATORS

INCINERATORS
UF INCINERATION
RT BURNERS
 FURNACES
 WASTE DISPOSAL
 WASTE ENERGY UTILIZATION

∞ INCLINATION
SN *(USE OF A MORE SPECIFIC TERM IS*
 RECOMMENDED--CONSULT THE TERMS
 LISTED BELOW)
RT GEOMAGNETISM
 MAGNETIC EQUATOR
 ORBITS
 SLOPES
 TENDENCIES

INCLUSIONS
GS DEFECTS
 . **INCLUSIONS**
RT CASTING
 CASTINGS
 CLATHRATES
 HETEROGENEITY
 IMPURITIES
 METALLOGRAPHY
 VOIDS

INCOHERENCE
RT DISCONTINUITY
 NONSYNCHRONIZATION

INCOHERENT SCATTER RADAR
GS RADAR
 . **INCOHERENT SCATTER RADAR**

INCOHERENT SCATTER RADAR-*(CONT.)*
　　.. EISCAT RADAR SYSTEM (EUROPE)
RT　INCOHERENT SCATTERING
　　RADAR SCATTERING

INCOHERENT SCATTERING
GS　SCATTERING
　　. **INCOHERENT SCATTERING**
RT　COHERENT SCATTERING
　　EISCAT RADAR SYSTEM (EUROPE)
　　ELECTROMAGNETIC RADIATION
　　ELECTROMAGNETIC WAVE
　　　TRANSMISSION
　　INCOHERENT SCATTER RADAR
　　NUCLEAR SCATTERING
　　RADAR SCATTERING

INCOME
RT　BUDGETING
　　ECONOMICS
　　INCENTIVES

INCOMPATIBILITY
RT　ABILITIES
　　CORROSION
　　HAZARDS
　　∞ INTERFERENCE
　　SOLUBILITY

INCOMPRESSIBILITY
RT　COMPRESSIBILITY
　　FLUID MECHANICS

INCOMPRESSIBLE BOUNDARY LAYER
GS　BOUNDARY LAYERS
　　. **INCOMPRESSIBLE BOUNDARY LAYER**
RT　LAMINAR BOUNDARY LAYER
　　TURBULENT BOUNDARY LAYER

INCOMPRESSIBLE FLOW
GS　FLUID FLOW
　　. **INCOMPRESSIBLE FLOW**
　　.. STOKES FLOW
RT　AERODYNAMICS
　　BELTRAMI FLOW
　　COMPRESSIBLE FLOW
　　GAS FLOW
　　MILNE-THOMSON METHOD
　　NAVIER-STOKES EQUATION
　　REYNOLDS STRESS
　　STREAM FUNCTIONS (FLUIDS)
　　SUBSONIC FLOW

INCOMPRESSIBLE FLUIDS
GS　**INCOMPRESSIBLE FLUIDS**
　　. MICROPOLAR FLUIDS
RT　BOUSSINESQ APPROXIMATION
　　CHANNEL FLOW
　　COMPRESSIBLE FLUIDS
　　FLUID POWER
　　∞ FLUIDS
　　IDEAL FLUIDS
　　NAVIER-STOKES EQUATION
　　OSEEN APPROXIMATION
　　SUPERFLUIDITY

INCONEL (TRADEMARK)
GS　ALLOYS
　　. NICKEL ALLOYS
　　.. **INCONEL (TRADEMARK)**
RT　CHROMIUM ALLOYS
　　IRON ALLOYS

INCREASING
RT　ACCUMULATIONS
　　AUGMENTATION
　　GROWTH
　　MAGNIFICATION
　　PROMOTION
　　SWELLING

INDENE
GS　HYDROCARBONS
　　. **INDENE**

INDENTATION
RT　DEFORMATION
　　HARDNESS

INDEPENDENT VARIABLES
UF　ARGUMENTS (MATHEMATICS)
　　PARAMETERS
GS　**INDEPENDENT VARIABLES**
　　. LATTICE PARAMETERS

INDEPENDENT VARIABLES-*(CONT.)*
RT　DEPENDENT VARIABLES
　　DISTRIBUTED PARAMETER SYSTEMS
　　OBSERVABILITY (SYSTEMS)
　　PARAMETER IDENTIFICATION
　　∞ VARIABLE

∞ INDEXES
SN　*(USE OF A MORE SPECIFIC TERM IS*
　　RECOMMENDED--CONSULT THE TERMS
　　LISTED BELOW)
RT　INDEXES (DOCUMENTATION)
　　INDEXES (RATIOS)
　　KP INDEX
　　KWIC INDEXES

INDEXES (DOCUMENTATION)
GS　CLASSIFICATIONS
　　. **INDEXES (DOCUMENTATION)**
　　.. KWIC INDEXES
RT　ABSTRACTS
　　BIBLIOGRAPHIES
　　∞ CATALOGS
　　DOCUMENTATION
　　DOCUMENTS
　　HANDBOOKS
　　∞ INDEXES
　　INFORMATION DISSEMINATION
　　INFORMATION RETRIEVAL
　　LISTS
　　LITERATURE
　　∞ REFERENCE SYSTEMS
　　SELECTIVE DISSEMINATION OF
　　　INFORMATION
　　SPACE GLOSSARIES
　　SUMMARIES
　　SUPPLEMENTS
　　THESAURI

INDEXES (RATIOS)
GS　RATIOS
　　. **INDEXES (RATIOS)**
　　.. KP INDEX
　　.. MORPHOLOGICAL INDEXES
RT　EFFICIENCY
　　∞ INDEXES
　　MASS TO LIGHT RATIOS

INDIA
GS　NATIONS
　　. **INDIA**
RT　ASIA
　　BANGLADESH
　　BHUTAN
　　HIMALAYAS
　　INDIAN SPACECRAFT
　　ISRO
　　SIKKIM

INDIAN OCEAN
GS　OCEANS
　　. **INDIAN OCEAN**
RT　ARABIAN SEA
　　INDONESIA
　　MALAGASY REPUBLIC

INDIAN SPACE PROGRAM
GS　PROGRAMS
　　. SPACE PROGRAMS
　　.. **INDIAN SPACE PROGRAM**
RT　COMMUNICATION SATELLITES
　　MANNED SPACE FLIGHT
　　∞ RESEARCH PROJECTS
　　SATELLITE DESIGN
　　SPACE MISSIONS
　　∞ SPACECRAFT
　　SPACECRAFT DESIGN
　　TECHNOLOGY UTILIZATION

INDIAN SPACE RESEARCH ORGANIZATION
USE　ISRO

INDIAN SPACECRAFT
UF　ARYABHATA
　　INSAT SATELLITES
　　IRS (INDIAN SPACECRAFT)
　　SEO (INDIAN SPACECRAFT)
RT　INDIA
　　∞ SPACECRAFT

INDIANA
GS　NATIONS
　　. UNITED STATES
　　.. **INDIANA**

INDIANA-*(CONT.)*
RT　OHIO RIVER (US)
　　WABASH RIVER BASIN (IL-IN-OH)

INDICATING INSTRUMENTS
UF　TEMPERATURE INDICATORS
GS　MEASURING INSTRUMENTS
　　. **INDICATING INSTRUMENTS**
　　.. APPROACH INDICATORS
　　.. ATTITUDE INDICATORS
　　... GYRO HORIZONS
　　.. CLOUD HEIGHT INDICATORS
　　.. FLOW DIRECTION INDICATORS
　　... WIND VANES
　　.. GYROCOMPASSES
　　.. MICROWAVE SENSORS
　　.. POSITION INDICATORS
　　... ASTROLABES
　　... PLAN POSITION INDICATORS
　　... RADIO DIRECTION FINDERS
　　... SPACECRAFT POSITION
　　　　INDICATORS
　　.. SMOKE DETECTORS
　　.. SPEED INDICATORS
　　... ANEMOMETERS
　　.... DRAG FORCE ANEMOMETERS
　　.... HOT-FILM ANEMOMETERS
　　.... HOT-WIRE ANEMOMETERS
　　.... LASER ANEMOMETERS
　　.... SONIC ANEMOMETERS
　　... TACHOMETERS
　　.. WEIGHT INDICATORS
　　... MICROBALANCES
　　... STRAIN GAGE BALANCES
　　... THERMOBALANCES
RT　AIRCRAFT INSTRUMENTS
　　CONTROL MOMENT GYROSCOPES
　　∞ DETECTORS
　　DIALS
　　DISPLAY DEVICES
　　GAS DETECTORS
　　HEAD-UP DISPLAYS
　　HELMET MOUNTED DISPLAYS
　　∞ INDICATION
　　∞ INDICATORS
　　INSTRUMENT RECEIVERS
　　∞ INSTRUMENTS
　　RADARSCOPES
　　RECORDING INSTRUMENTS
　　THERMOCOUPLES
　　THERMOPILES

∞ INDICATION
SN　*(USE OF A MORE SPECIFIC TERM IS*
　　RECOMMENDED--CONSULT THE TERMS
　　LISTED BELOW)
RT　EVALUATION
　　INDICATING INSTRUMENTS
　　PROBABILITY THEORY
　　SIGNS AND SYMPTOMS

∞ INDICATORS
SN　*(USE OF A MORE SPECIFIC TERM IS*
　　RECOMMENDED--CONSULT THE TERMS
　　LISTED BELOW)
RT　AIRCRAFT GUIDANCE
　　CHEMICAL INDICATORS
　　INDICATING INSTRUMENTS
　　METHYLENE BLUE
　　RATE OF CLIMB INDICATORS

INDIUM
GS　CHEMICAL ELEMENTS
　　. **INDIUM**
　　METALS
　　. **INDIUM**

INDIUM ALLOYS
GS　ALLOYS
　　. **INDIUM ALLOYS**

INDIUM ANTIMONIDES
GS　ANTIMONY COMPOUNDS
　　. ANTIMONIDES
　　.. **INDIUM ANTIMONIDES**
　　INDIUM COMPOUNDS
　　. **INDIUM ANTIMONIDES**
RT　SEMICONDUCTORS (MATERIALS)

INDIUM ARSENIDES
GS　ARSENIC COMPOUNDS
　　. ARSENIDES
　　.. **INDIUM ARSENIDES**
　　INDIUM COMPOUNDS
　　. **INDIUM ARSENIDES**

INDIUM COMPOUNDS
GS   **INDIUM COMPOUNDS**
    . INDIUM ANTIMONIDES
    . INDIUM ARSENIDES
    . INDIUM PHOSPHATES
    . INDIUM PHOSPHIDES
    . INDIUM SULFIDES
    . INDIUM TELLURIDES
RT   ∞CHEMICAL COMPOUNDS
    ∞GROUP 3A COMPOUNDS
    ∞METAL COMPOUNDS

INDIUM ISOTOPES
GS   CHEMICAL ELEMENTS
    . NUCLIDES
    . . ISOTOPES
    . . . RADIOACTIVE ISOTOPES
    . . . . **INDIUM ISOTOPES**
    METALS
    . **INDIUM ISOTOPES**

INDIUM PHOSPHATES
GS   INDIUM COMPOUNDS
    . **INDIUM PHOSPHATES**
    PHOSPHORUS COMPOUNDS
    . PHOSPHATES
    . . **INDIUM PHOSPHATES**

INDIUM PHOSPHIDES
GS   INDIUM COMPOUNDS
    . **INDIUM PHOSPHIDES**
    PHOSPHORUS COMPOUNDS
    . PHOSPHIDES
    . . **INDIUM PHOSPHIDES**
RT   TRANSFERRED ELECTRON DEVICES

INDIUM SULFIDES
GS   CHALCOGENIDES
    . SULFIDES
    . . INORGANIC SULFIDES
    . . . **INDIUM SULFIDES**
    INDIUM COMPOUNDS
    . **INDIUM SULFIDES**
    SULFUR COMPOUNDS
    . SULFIDES
    . . INORGANIC SULFIDES
    . . . **INDIUM SULFIDES**

INDIUM TELLURIDES
GS   CHALCOGENIDES
    . TELLURIDES
    . . **INDIUM TELLURIDES**
    INDIUM COMPOUNDS
    . **INDIUM TELLURIDES**
    TELLURIUM COMPOUNDS
    . TELLURIDES
    . . **INDIUM TELLURIDES**
† RT   SEMICONDUCTORS (MATERIALS)

INDOLES
GS   HETEROCYCLIC COMPOUNDS
    . AZOLES
    . . PYRROLES
    . . . **INDOLES**
    . . . . TRYPTOPHAN
RT   METHOXY SYSTEMS

INDONESIA
GS   LANDFORMS
    . ISLANDS
    . . **INDONESIA**
    NATIONS
    . **INDONESIA**
RT   INDIAN OCEAN
    INDONESIAN SPACE PROGRAM
    PACIFIC OCEAN

INDONESIAN SPACE PROGRAM
GS   PROGRAMS
    . SPACE PROGRAMS
    . . **INDONESIAN SPACE PROGRAM**
RT   INDONESIA
    PALAPA SATELLITES
    PALAPA 2 SATELLITE

INDOOR AIR POLLUTION
GS   POLLUTION
    . ENVIRONMENT POLLUTION
    . . AIR POLLUTION
    . . . **INDOOR AIR POLLUTION**
RT   AIR QUALITY
    AIR SAMPLING
    ∞BUILDINGS

INDUCED FLUID FLOW
USE   FLUID FLOW

INDUCTANCE
GS   ELECTROMAGNETIC PROPERTIES
    . ELECTRICAL PROPERTIES
    . . **INDUCTANCE**
    . . . PROXIMITY EFFECT (ELECTRICITY)
RT   CAPACITANCE
    ELECTRICAL IMPEDANCE
    LC CIRCUITS
    MAGNETIC INDUCTION
    MAGNETIC PROPERTIES
    REACTANCE
    RL CIRCUITS
    TRANSFORMERS

∞ INDUCTION
SN   *(USE OF A MORE SPECIFIC TERM IS
    RECOMMENDED--CONSULT THE TERMS
    LISTED BELOW)*
RT   ARC GENERATORS
    DERIVATION
    INDUCTION (MATHEMATICS)
    INFERENCE
    INITIATION
    MAGNETIC INDUCTION
    NUMBER THEORY

INDUCTION (MATHEMATICS)
GS   NUMBER THEORY
    . **INDUCTION (MATHEMATICS)**
RT   ∞INDUCTION
    MATHEMATICAL LOGIC

INDUCTION HEATING
GS   HEATING
    . **INDUCTION HEATING**
RT   FURNACES
    MAGNETIC INDUCTION
    MAGNETIC PUMPING
    MELTING
    PLASMA HEATING
    RADIO FREQUENCY HEATING
    VACUUM MELTING

INDUCTION MOTORS
GS   MOTORS
    . ELECTRIC MOTORS
    . . **INDUCTION MOTORS**
RT   ALTERNATING CURRENT
    ARMATURES
    ASYNCHRONOUS MOTORS
    ∞ELECTRIC POWER
    ELECTRIC POWER SUPPLIES
    POWER FACTOR CONTROLLERS
    ∞ROTATING ELECTRICAL MACHINES
    SYNCHRONOUS MOTORS

INDUCTION SYSTEMS
USE   INTAKE SYSTEMS

INDUCTORS
RT   ARC GENERATORS
    BALLASTS (IMPEDANCES)
    CIRCUITS
    ∞COILS
    ELECTRIC ENERGY STORAGE
    ENERGY STORAGE
    MAGNET COILS
    TOROIDS

INDUSTRIAL AREAS
RT   CITIES
    COMMERCE
    CONSTRUCTION INDUSTRY
    ∞FACILITIES
    INDUSTRIES
    LAND USE
    MARKETING
    MEGALOPOLISES
    REGIONAL PLANNING
    SHIPYARDS
    SITE SELECTION
    URBAN DEVELOPMENT
    URBAN TRANSPORTATION

INDUSTRIAL ENERGY
RT   ALLOCATIONS
    COMMERCIAL ENERGY
    DISTRIBUTING
    DOMESTIC ENERGY
    ECONOMIC FACTORS
    ∞ENERGY

INDUSTRIAL ENERGY-*(CONT.)*
    ENERGY CONSUMPTION
    ENERGY CONVERSION
    TRANSPORTATION ENERGY

INDUSTRIAL MANAGEMENT
UF   BUSINESS MANAGEMENT
GS   MANAGEMENT
    . **INDUSTRIAL MANAGEMENT**
    . . ENGINEERING MANAGEMENT
    . . INVENTORY MANAGEMENT
    . . . INVENTORY CONTROLS
    . . PERSONNEL MANAGEMENT
RT   PRODUCTION MANAGEMENT
    RESEARCH MANAGEMENT
    SYSTEMS MANAGEMENT

INDUSTRIAL PLANTS
UF   FACTORIES
    PLANTS (INDUSTRIES)
GS   **INDUSTRIAL PLANTS**
    . FOUNDRIES
RT   CONSTRUCTION INDUSTRY
    ∞FACILITIES
    PILOT PLANTS

INDUSTRIAL SAFETY
GS   SAFETY
    . **INDUSTRIAL SAFETY**
RT   ACCIDENTS
    BENZENE POISONING
    BERYLLIUM POISONING
    CARBON TETRACHLORIDE POISONING
    HEALTH PHYSICS
    HYDROCARBON POISONING
    INDUSTRIES
    OCCUPATION
    REACTOR SAFETY

INDUSTRIAL WASTES
GS   WASTES
    . **INDUSTRIAL WASTES**
RT   INDUSTRIES
    LANDFILLS
    LIQUID WASTES
    SOLID WASTES
    WASTE DISPOSAL
    WASTE UTILIZATION
    WASTE WATER

INDUSTRIES
GS   **INDUSTRIES**
    . AEROSPACE INDUSTRY
    . . AIRCRAFT INDUSTRY
    . CONSTRUCTION INDUSTRY
    . DEFENSE INDUSTRY
    . . WEAPONS INDUSTRY
RT   AIRCRAFT PRODUCTION COSTS
    COMMERCE
    COMMERCIAL SPACECRAFT
    CONTRACT NEGOTIATION
    CONTRACTORS
    ECONOMIC DEVELOPMENT
    ECONOMIC IMPACT
    GROSS NATIONAL PRODUCT
    INDUSTRIAL AREAS
    INDUSTRIAL SAFETY
    INDUSTRIAL WASTES
    MANUFACTURING
    PERSONNEL SUBSYSTEMS
    RETIREMENT
    SHIPYARDS
    SPACE INDUSTRIALIZATION
    SPACE MANUFACTURING
    TECHNOLOGIES
    TECHNOLOGY ASSESSMENT
    TECHNOLOGY UTILIZATION
    UTILITIES

INELASTIC BODIES
USE   RIGID STRUCTURES

INELASTIC COLLISIONS
GS   COLLISIONS
    . **INELASTIC COLLISIONS**
RT   SCATTERING

INELASTIC SCATTERING
GS   SCATTERING
    . **INELASTIC SCATTERING**
RT   COHERENT SCATTERING
    COMPTON EFFECT
    ELASTIC SCATTERING
    ELECTRON SCATTERING
    MANDELSTAM REPRESENTATION

INELASTIC SCATTERING-*(CONT.)*
    NUCLEAR SCATTERING
    QUARK PARTON MODEL

INELASTIC STRESS
RT   CYCLIC LOADS
    MATHEMATICAL MODELS
    METAL FATIGUE
    STRESS ANALYSIS
    STRESS-STRAIN DIAGRAMS

INEQUALITIES
GS   **INEQUALITIES**
    . SCHWARTZ INEQUALITY
RT   ∞MATHEMATICS

INERT ATMOSPHERE
GS   CONTROLLED ATMOSPHERES
    . **INERT ATMOSPHERE**

INERT GASES
USE   RARE GASES

INERTIA
UF   INERTIAL FORCES
GS   **INERTIA**
    . INERTIA PRINCIPLE
    . . MACH INERTIA PRINCIPLE
RT   ∞FORCE
    FROUDE NUMBER
    MASS
    MOMENTS OF INERTIA
    ∞MOTION
    SIMILITUDE LAW

INERTIA BONDING
GS   BONDING
    . **INERTIA BONDING**
RT   ∞JOINING
    METAL-METAL BONDING

INERTIA MOMENTS
USE   MOMENTS OF INERTIA

INERTIA PRINCIPLE
GS   INERTIA
    . **INERTIA PRINCIPLE**
    . . MACH INERTIA PRINCIPLE
RT   EQUATIONS OF MOTION
    MOMENTS OF INERTIA

INERTIA WHEELS
USE   COUNTER-ROTATING WHEELS
    REACTION WHEELS

INERTIAL CONFINEMENT FUSION
GS   **INERTIAL CONFINEMENT FUSION**
    . IMPACT FUSION
RT   STRONGLY COUPLED PLASMAS

INERTIAL COORDINATES
GS   COORDINATES
    . **INERTIAL COORDINATES**
RT   ASTROGUIDE NAVIGATION SYSTEM
    GEOCENTRIC COORDINATES
    INERTIALESS STEERABLE ANTENNAS

INERTIAL FORCES
USE   INERTIA

INERTIAL FUSION (REACTOR)
RT   ∞FUSION
    FUSION REACTORS
    ION BEAMS
    LASER FUSION
    LASER PLASMAS
    NUCLEAR FUELS
    PLASMA COMPRESSION
    PULSED LASERS
    RELATIVISTIC ELECTRON BEAMS

INERTIAL GUIDANCE
GS   GUIDANCE (MOTION)
    . **INERTIAL GUIDANCE**
    . . STRAPDOWN INERTIAL GUIDANCE
RT   INJECTION GUIDANCE
    MIDCOURSE GUIDANCE
    REENTRY GUIDANCE
    SATELLITE GUIDANCE
    SPACECRAFT GUIDANCE
    STABILIZED PLATFORMS
    TERMINAL GUIDANCE

INERTIAL MEASURING UNITS
USE   INERTIAL PLATFORMS

INERTIAL NAVIGATION
GS   NAVIGATION
    . **INERTIAL NAVIGATION**
    . . ASTROGUIDE NAVIGATION SYSTEM
    . . GIMBALLESS INERTIAL NAVIGATION
RT   AIR NAVIGATION
    ALL-WEATHER AIR NAVIGATION
    CELESTIAL NAVIGATION
    DEAD RECKONING
    DIGITAL NAVIGATION
    HYPERBOLIC NAVIGATION
    NAVIGATION AIDS
    POLAR NAVIGATION
    RADAR NAVIGATION
    RADIO NAVIGATION
    SCHULER TUNING
    SPACE NAVIGATION
    STAR TRACKERS
    STRAPDOWN INERTIAL GUIDANCE
    SURFACE NAVIGATION

INERTIAL PLATFORMS
UF   INERTIAL MEASURING UNITS
RT   GIMBALLESS INERTIAL NAVIGATION
    GYROSCOPIC STABILITY
    KALMAN-SCHMIDT FILTERING
    NAVIGATION INSTRUMENTS
    ∞PLATFORMS
    THREE AXIS STABILIZATION

INERTIAL REFERENCE SYSTEMS
RT   CELESTIAL REFERENCE SYSTEMS
    ∞REFERENCE SYSTEMS
    RELATIVITY
    ∞SYSTEMS

INERTIAL UPPER STAGE
UF   INTERIM UPPER STAGE (STS)
    IUS
GS   INTERIM STAGES (SPACECRAFT)
    . **INERTIAL UPPER STAGE**
RT   ORBIT TRANSFER VEHICLES
    RECOVERABLE SPACECRAFT
    REUSABLE SPACECRAFT
    SPACE SHUTTLE ORBITERS
    SPACE SHUTTLES
    SPACE TRANSPORTATION
    SPACE TRANSPORTATION SYSTEM
    SPACE TUGS
    ULYSSES MISSION
    UPPER STAGE ROCKET ENGINES

INERTIALESS STEERABLE ANTENNAS
GS   ANTENNAS
    . DIRECTIONAL ANTENNAS
    . . STEERABLE ANTENNAS
    . . . **INERTIALESS STEERABLE**
        ANTENNAS
    ARRAYS
    . ANTENNA ARRAYS
    . . STEERABLE ANTENNAS
    . . . **INERTIALESS STEERABLE**
        ANTENNAS
RT   COMMUNICATION EQUIPMENT
    INERTIAL COORDINATES

INFARCTION
GS   DISEASES
    . **INFARCTION**
    . . MYOCARDIAL INFARCTION
RT   EMBOLISMS
    THROMBOSIS
    TISSUES (BIOLOGY)

INFECTIONS
USE   INFECTIOUS DISEASES

INFECTIOUS DISEASES
UF   INFECTIONS
GS   DISEASES
    . **INFECTIOUS DISEASES**
    . . AIRBORNE INFECTION
    . . CHOLERA
    . . CONJUNCTIVITIS
    . . DERMATITIS
    . . . CONTACT DERMATITIS
    . . DIPHTHERIA
    . . HEPATITIS
    . . INFLUENZA
    . . MENINGITIS
    . . POLIOMYELITIS
    . . SMALLPOX

INFECTIOUS DISEASES-*(CONT.)*
    . . SYPHILIS
    . . TUBERCULOSIS
    . . TYPHOID
    . . TYPHUS
RT   ANTISEPTICS
    ASPERGILLUS
    ∞BLISTERS
    EPIDEMIOLOGY
    IMMUNITY
    LEUKOPENIA
    PARASITIC DISEASES

INFERENCE
RT   ASSUMPTIONS
    DEDUCTION
    HYPOTHESES
    IMPLICATION
    ∞INDUCTION

INFESTATION
UF   INSECT DAMAGE
RT   BEETLES
    BOLL WEEVILS
    BOLLWORMS
    CHIRONOMUS FLIES
    FLATWORMS
    INSECTS
    LARVAE
    LOCUSTS
    MOTHS
    PARASITES
    PLANTS (BOTANY)
    SILKWORMS
    WORMS

INFILTRATION
RT   AIR CONDITIONING
    PERMEABILITY
    POROSITY
    VOIDS
    WARFARE

INFINITE SPAN WINGS
GS   AIRFOILS
    . THIN AIRFOILS
    . . THIN WINGS
    . . . **INFINITE SPAN WINGS**
    . WINGS
    . . SLENDER WINGS
    . . . **INFINITE SPAN WINGS**
    . . THIN WINGS
    . . . **INFINITE SPAN WINGS**
    . . UNSWEPT WINGS
    . . . **INFINITE SPAN WINGS**
    PLANFORMS
    . WING PLANFORMS
    . . **INFINITE SPAN WINGS**
RT   FLEXIBLE WINGS

INFINITY
RT   FUNCTIONS (MATHEMATICS)
    GEOMETRY
    NUMBER THEORY
    PROBABILITY THEORY
    REAL VARIABLES
    SERIES (MATHEMATICS)

INFLATABLE DEVICES
USE   INFLATABLE STRUCTURES

INFLATABLE GLIDERS
GS   EXPANDABLE STRUCTURES
    . INFLATABLE STRUCTURES
    . . **INFLATABLE GLIDERS**
    GLIDERS
    . PARAGLIDERS
    . . **INFLATABLE GLIDERS**
RT   ∞AIRCRAFT

INFLATABLE SPACECRAFT
GS   EXPANDABLE STRUCTURES
    . INFLATABLE STRUCTURES
    . . **INFLATABLE SPACECRAFT**
    . . . BEACON SATELLITES
    . . . . BEACON EXPLORER A
    . . . . EXPLORER 22 SATELLITE
    SPACE ERECTABLE STRUCTURES
    . **INFLATABLE SPACECRAFT**
    . . BEACON SATELLITES
    . . . EXPLORER 22 SATELLITE
RT   ARTIFICIAL SATELLITES
    ORBITAL ASSEMBLY
    SELF ERECTING DEVICES
    ∞SPACECRAFT

INFLATABLE SPACECRAFT-(CONT.)
    UNMANNED SPACECRAFT

INFLATABLE STRUCTURES
UF   DEFLATING
      INFLATABLE DEVICES
GS   EXPANDABLE STRUCTURES
      . **INFLATABLE STRUCTURES**
      . . BALLOONS
      . . . HIGH ALTITUDE BALLOONS
      . . . . JIMSPHERE BALLOONS
      . . . . SKYHOOK BALLOONS
      . . . . SUPERPRESSURE BALLOONS
      . . . METEOROLOGICAL BALLOONS
      . . . . JIMSPHERE BALLOONS
      . . . . ROBIN BALLOONS
      . . . MICROBALLOONS
      . . . TETHERED BALLOONS
      . . BALLUTES
      . . GAS BAGS
      . . INFLATABLE GLIDERS
      . . INFLATABLE SPACECRAFT
      . . . BEACON SATELLITES
      . . . . BEACON EXPLORER A
      . . . . EXPLORER 22 SATELLITE
      . . PARAVULCOONS
RT   AIRSHIPS
      BOATS
      ∞ BUILDINGS
      FLEXIBLE BODIES
      FLEXIBLE WINGS
      FLOATS
      FOLDING STRUCTURES
      INFLATING
      LIFE RAFTS
      LUNAR SHELTERS
      PNEUMATIC EQUIPMENT
      PRESSURE SUITS
      RADOMES
      SELF ERECTING DEVICES
      SPACE ERECTABLE STRUCTURES
      SPACE STATIONS
      ∞ STRUCTURES
      TIRES
      VARIABLE GEOMETRY STRUCTURES

INFLATING
SN   (EXCLUDES ECONOMIC INFLATION)
RT   EXPANSION
      GAS INJECTION
      GROWTH
      INFLATABLE STRUCTURES
      PRESSURE REDUCTION
      PRESSURIZING
      SWELLING

INFLECTION POINTS
GS   GEOMETRY
      . EUCLIDEAN GEOMETRY
      . . POINTS (MATHEMATICS)
      . . . **INFLECTION POINTS**
RT   CURVES (GEOMETRY)
      FUNCTIONS (MATHEMATICS)
      LINE SHAPE
      REAL VARIABLES

INFLUENCE COEFFICIENT
GS   COEFFICIENTS
      . **INFLUENCE COEFFICIENT**
      . . STRUCTURAL INFLUENCE
           COEFFICIENTS
RT   AEROELASTICITY
      DISCHARGE COEFFICIENT
      ELASTIC PROPERTIES
      FLUTTER
      FORCE DISTRIBUTION
      ∞ HYDRAULICS
      MOMENT DISTRIBUTION
      NOZZLE THRUST COEFFICIENTS
      PLASTIC PROPERTIES
      PRESSURE DISTRIBUTION
      STRESS ANALYSIS
      STRUCTURAL ANALYSIS

INFLUENZA
GS   DISEASES
      . INFECTIOUS DISEASES
      . . **INFLUENZA**
      . RESPIRATORY DISEASES
      . . **INFLUENZA**

INFORMATION
RT   ANNOTATIONS
      COMMUNICATING
      COMMUNICATION

INFORMATION-(CONT.)
      ∞ DATA
      DOCUMENTATION
      MATHEMATICAL TABLES
      NEWS MEDIA
      PRESENTATION
      PRIVACY
      REPORTS

INFORMATION ADAPTIVE SYSTEM
GS   INFORMATION SYSTEMS
      . **INFORMATION ADAPTIVE SYSTEM**
RT   COMMUNICATION EQUIPMENT
      ∞ SYSTEMS

INFORMATION DISSEMINATION
GS   COMMUNICATING
      . **INFORMATION DISSEMINATION**
      . . MESSAGES
      . . SELECTIVE DISSEMINATION OF
           INFORMATION
RT   BIBLIOGRAPHIES
      CATALOGS (PUBLICATIONS)
      DOCUMENTATION
      EXTRATERRESTRIAL COMMUNICATION
      INDEXES (DOCUMENTATION)
      INFORMATION TRANSFER
      INTEGRATED LIBRARY SYSTEMS
      LIBRARIES
      PRIVACY
      REPORTS
      SUMMARIES

INFORMATION FLOW
RT   AEROSPACE TECHNOLOGY TRANSFER
      COMMUNICATING
      COMMUNICATION
      ∞ FLOW
      INFORMATION TRANSFER
      MANAGEMENT
      MESSAGE PROCESSING
      SELECTIVE DISSEMINATION OF
           INFORMATION
      TECHNOLOGY TRANSFER

INFORMATION MANAGEMENT
GS   MANAGEMENT
      . **INFORMATION MANAGEMENT**
RT   COMMUNICATING
      COMMUNICATION
      DATA BASE MANAGEMENT SYSTEMS
      DATA RETRIEVAL
      DATA STORAGE
      INFORMATION TRANSFER
      INTEGRATED LIBRARY SYSTEMS
      TECHNOLOGY TRANSFER

INFORMATION RETRIEVAL
GS   RETRIEVAL
      . **INFORMATION RETRIEVAL**
RT   ABSTRACTS
      BIBLIOGRAPHIES
      COMMAND LANGUAGES
      COMPUTERS
      DATA PROCESSING
      DATA RETRIEVAL
      DOCUMENTATION
      DOCUMENTS
      INDEXES (DOCUMENTATION)
      INFORMATION TRANSFER
      INTEGRATED LIBRARY SYSTEMS
      INTERSERVICE DATA EXCHANGE
           PROGRAM
      LIBRARIES
      MANAGEMENT INFORMATION SYSTEMS
      NUMERICAL DATA BASES
      ON-LINE SYSTEMS
      QUERY LANGUAGES
      SEARCH PROFILES
      SELECTIVE DISSEMINATION OF
           INFORMATION
      SPACE GLOSSARIES
      STARSITE PROGRAM
      SUBJECTS
      THESAURI

INFORMATION SYSTEMS
GS   **INFORMATION SYSTEMS**
      . ATMOSPHERIC & OCEANOGRAPHIC
           INFORM SYS
      . EARTH RESOURCES INFORMATION
           SYSTEM
      . GEOGRAPHIC INFORMATION SYSTEMS
      . INFORMATION ADAPTIVE SYSTEM
      . INTEGRATED LIBRARY SYSTEMS

INFORMATION SYSTEMS-(CONT.)
      . MANAGEMENT INFORMATION
           SYSTEMS
      . NUMERICAL DATA BASES
RT   INFORMATION TRANSFER
      LIBRARIES
      MANAGEMENT SYSTEMS
      ON-LINE SYSTEMS
      SELECTIVE DISSEMINATION OF
           INFORMATION
      ∞ SYSTEMS
      SYSTEMS MANAGEMENT

INFORMATION THEORY
UF   SHANNON INFORMATION THEORY
RT   ∞ APPLICATIONS OF MATHEMATICS
      AUTOMATA THEORY
      ∞ AUTOMATION
      BCH CODES
      CODING
      COMBINATORIAL ANALYSIS
      COMMUNICATION THEORY
      COMPUTERS
      CORRECTION
      CORRELATION
      CRYPTOGRAPHY
      CYBERNETICS
      DATA PROCESSING
      DATA TRANSMISSION
      DECISION THEORY
      ERGODIC PROCESS
      ERROR DETECTION CODES
      FOURIER ANALYSIS
      GAME THEORY
      ∞ LOGIC
      MACHINE TRANSLATION
      MANAGEMENT INFORMATION SYSTEMS
      ∞ MATHEMATICS
      MAXIMUM ENTROPY METHOD
      MESSAGES
      ∞ NOISE
      OPERATIONS RESEARCH
      PARITY
      PETRI NETS
      PHASE SHIFT KEYING
      PROBABILITY THEORY
      QUANTUM AMPLIFIERS
      RANDOM PROCESSES
      REDUNDANCY
      SHANNON-WIENER MEASURE
      STATISTICAL ANALYSIS
      ∞ STATISTICS
      STOCHASTIC PROCESSES
      SYSTEMS ENGINEERING
      TELECOMMUNICATION
      TERMS
      ∞ THEORIES

INFORMATION TRANSFER
RT   INFORMATION DISSEMINATION
      INFORMATION FLOW
      INFORMATION MANAGEMENT
      INFORMATION RETRIEVAL
      INFORMATION SYSTEMS
      INTERNATIONAL COOPERATION
      TECHNOLOGY TRANSFER
      TECHNOLOGY UTILIZATION

INFORMATION TRANSMISSION
USE   DATA TRANSMISSION

INFRARED ABSORPTION
GS   ELECTROMAGNETIC PROPERTIES
      . ELECTROMAGNETIC ABSORPTION
      . . **INFRARED ABSORPTION**
      ENERGY ABSORPTION
      . ELECTROMAGNETIC ABSORPTION
      . . **INFRARED ABSORPTION**
      RADIATION ABSORPTION
      . ELECTROMAGNETIC ABSORPTION
      . . **INFRARED ABSORPTION**
RT   ∞ ABSORPTION
      ATMOSPHERIC ATTENUATION
      ATMOSPHERIC OPTICS
      INFRARED RADIATION
      LIGHT SCATTERING
      OPTICAL PROPERTIES
      THERMAL EMISSION
      TRANSMITTANCE
      WAVE ATTENUATION

INFRARED ASTRONOMY
GS   ASTRONOMY
      . **INFRARED ASTRONOMY**
RT   ASTRONOMICAL PHOTOGRAPHY

INFRARED ASTRONOMY-*(CONT.)*
    INFRARED ASTRONOMY SATELLITE
    INFRARED PHOTOMETRY
    SPACE INFRARED TELESCOPE FACILITY

INFRARED ASTRONOMY SATELLITE
UF   IRAS
GS   OBSERVATORIES
    . ASTRONOMICAL OBSERVATORIES
    . . ASTRONOMICAL SATELLITES
    . . . **INFRARED ASTRONOMY SATELLITE**
RT   GEOSYNCHRONOUS ORBITS
    INFRARED ASTRONOMY
    IRAS-ARAKI-ALCOCK COMET

INFRARED DETECTORS
GS   MEASURING INSTRUMENTS
    . RADIATION MEASURING INSTRUMENTS
    . . ACTINOMETERS
    . . . RADIOMETERS
    . . . . **INFRARED DETECTORS**
    . . INFRARED INSTRUMENTS
    . . . **INFRARED DETECTORS**
    . . . . FLIR DETECTORS
RT   BOLOMETERS
    ∞DETECTORS
    ELECTROMAGNETIC MEASUREMENT
    FOREST FIRE DETECTION
    INFRARED SIGNATURES
    MERCURY CADMIUM TELLURIDES

INFRARED FILTERS
GS   ELECTROMAGNETIC WAVE FILTERS
    . OPTICAL FILTERS
    . . **INFRARED FILTERS**
RT   ELECTRIC FILTERS
    ULTRAVIOLET FILTERS

INFRARED HORIZON SCANNERS
USE   HORIZON SCANNERS
    INFRARED SCANNERS

INFRARED IMAGERY
GS   IMAGERY
    . **INFRARED IMAGERY**
    PHOTOGRAPHY
    . **INFRARED IMAGERY**
RT   COLOR INFRARED PHOTOGRAPHY
    LUNAR EQUATOR
    THERMOGRAPHY
    X RAY IMAGERY

INFRARED INSPECTION
GS   INSPECTION
    . **INFRARED INSPECTION**
RT   QUALITY CONTROL

INFRARED INSTRUMENTS
GS   MEASURING INSTRUMENTS
    . RADIATION MEASURING INSTRUMENTS
    . . **INFRARED INSTRUMENTS**
    . . . INFRARED DETECTORS
    . . . . FLIR DETECTORS
    . . . INFRARED SCANNERS
    . . . INFRARED SPECTROMETERS
    . . . INFRARED SPECTROPHOTOMETERS
RT   FOREST FIRE DETECTION

INFRARED INTERFEROMETERS
GS   MEASURING INSTRUMENTS
    . INTERFEROMETERS
    . . **INFRARED INTERFEROMETERS**
RT   INTERFEROMETRY
    OPTICAL EQUIPMENT
    OPTICAL MEASUREMENT
    OPTICAL MEASURING INSTRUMENTS

INFRARED LASERS
UF   INFRARED MASERS
    IRASERS
GS   STIMULATED EMISSION DEVICES
    . LASERS
    . . **INFRARED LASERS**
RT   ARGON LASERS
    CARBON DIOXIDE LASERS
    CARBON LASERS
    CARBON MONOXIDE LASERS
    CHEMICAL LASERS
    DYE LASERS
    GAS LASERS
    HF LASERS
    LIQUID LASERS
    ORGANIC LASERS
    SOLID STATE LASERS

INFRARED LASERS-*(CONT.)*
    WAVEGUIDE LASERS

INFRARED MASERS
USE   INFRARED LASERS

INFRARED PHOTOGRAPHY
GS   IMAGERY
    . **INFRARED PHOTOGRAPHY**
    . . COLOR INFRARED PHOTOGRAPHY
    PHOTOGRAPHY
    . MULTISPECTRAL PHOTOGRAPHY
    . . **INFRARED PHOTOGRAPHY**
    . . . COLOR INFRARED PHOTOGRAPHY
RT   AERIAL PHOTOGRAPHY
    ASTRONOMICAL PHOTOGRAPHY
    BLACK AND WHITE PHOTOGRAPHY
    CINEMATOGRAPHY
    FAINT OBJECT CAMERA
    FOREST FIRE DETECTION
    GEOGRAPHIC INFORMATION SYSTEMS
    ICE MAPPING
    LUNAR PHOTOGRAPHY
    METEOROLOGICAL SATELLITES
    METEOSAT SATELLITE
    MULTISPECTRAL BAND CAMERAS
    NIMBUS SATELLITES
    RADIOMETERS
    SATELLITE-BORNE PHOTOGRAPHY
    TIMBER INVENTORY
    ULTRAVIOLET PHOTOGRAPHY

INFRARED PHOTOMETRY
GS   OPTICAL MEASUREMENT
    . PHOTOMETRY
    . . **INFRARED PHOTOMETRY**
RT   ASTRONOMICAL PHOTOMETRY
    INFRARED ASTRONOMY
    INFRARED SPECTRA
    NEAR INFRARED RADIATION
    STELLAR SPECTROPHOTOMETRY

INFRARED RADAR
GS   RADAR
    . **INFRARED RADAR**
RT   FLIR DETECTORS
    RADAR IMAGERY

INFRARED RADIATION
GS   ELECTROMAGNETIC RADIATION
    . **INFRARED RADIATION**
    . . FAR INFRARED RADIATION
    . . NEAR INFRARED RADIATION
RT   BEAMS (RADIATION)
    BLACK BODY RADIATION
    COHERENT ELECTROMAGNETIC
      RADIATION
    ENERGY ABSORPTION
    EVAPOROGRAPHY
    EXHAUST EMISSION
    HEAT
    INFRARED ABSORPTION
    INFRARED SIGNATURES
    LIGHT (VISIBLE RADIATION)
    MICROWAVES
    MONOCHROMATIC RADIATION
    PLANETARY RADIATION
    POLARIZED ELECTROMAGNETIC
      RADIATION
    ∞RADIATION
    SEYFERT GALAXIES
    SOLAR RADIATION
    SUNLIGHT
    TERRESTRIAL RADIATION
    THERMAL RADIATION
    WAVELENGTHS
    XENON LAMPS

INFRARED RADIOMETERS
GS   MONITORS
    . **INFRARED RADIOMETERS**
RT   AERIAL RECONNAISSANCE
    ATMOSPHERIC CORRECTION
    DATA ACQUISITION
    EARTH RESOURCES PROGRAM
    ENVIRONMENTAL MONITORING
    FOREST FIRE DETECTION
    PRESSURE MODULATOR RADIOMETERS
    RADIOMETRIC CORRECTION
    SATELLITE-BORNE INSTRUMENTS
    THERMAL MAPPING
    VISIBLE INFRARED SPIN SCAN
      RADIOMETER

INFRARED REFLECTION
GS   REFLECTION
    . **INFRARED REFLECTION**
RT   OPTICAL REFLECTION
    RADIATIVE HEAT TRANSFER
    RADIO ECHOES
    SPREAD REFLECTION
    ULTRAVIOLET REFLECTION

INFRARED SCANNERS
UF   INFRARED HORIZON SCANNERS
GS   MEASURING INSTRUMENTS
    . RADIATION MEASURING INSTRUMENTS
    . . ACTINOMETERS
    . . . RADIOMETERS
    . . . . **INFRARED SCANNERS**
    . . INFRARED INSTRUMENTS
    . . . **INFRARED SCANNERS**
    SCANNERS
    . **INFRARED SCANNERS**
RT   FOREST FIRE DETECTION
    HORIZON SCANNERS
    MULTISPECTRAL BAND SCANNERS
    OPTICAL EQUIPMENT
    SCANNER PROJECT
    THERMAL MAPPING

INFRARED SIGNATURES
GS   SIGNATURES
    . **INFRARED SIGNATURES**
RT   INFRARED DETECTORS
    INFRARED RADIATION
    INFRARED SPECTRA
    SIGNATURE ANALYSIS

INFRARED SPECTRA
GS   SPECTRA
    . RADIATION SPECTRA
    . . ELECTROMAGNETIC SPECTRA
    . . . **INFRARED SPECTRA**
RT   ∞ABSORPTION
    EMISSION SPECTRA
    INFRARED PHOTOMETRY
    INFRARED SIGNATURES
    LINE SPECTRA
    MICROWAVE SPECTRA
    MOLECULAR SPECTRA
    SOLAR SPECTRA
    STELLAR SPECTRA

INFRARED SPECTROMETERS
GS   MEASURING INSTRUMENTS
    . OPTICAL MEASURING INSTRUMENTS
    . . **INFRARED SPECTROMETERS**
    . . . FILTER WHEEL INFRARED
      SPECTROMETERS
    . RADIATION MEASURING INSTRUMENTS
    . . ACTINOMETERS
    . . . **INFRARED SPECTROMETERS**
    . . INFRARED INSTRUMENTS
    . . . **INFRARED SPECTROMETERS**
    . SPECTROMETERS
    . . **INFRARED SPECTROMETERS**
    . . . FILTER WHEEL INFRARED
      SPECTROMETERS
    OPTICAL EQUIPMENT
    . OPTICAL MEASURING INSTRUMENTS
    . . **INFRARED SPECTROMETERS**
    . . . FILTER WHEEL INFRARED
      SPECTROMETERS
RT   EBERT SPECTROMETERS
    SOLAR SPECTROMETERS

INFRARED SPECTROPHOTOMETERS
GS   MEASURING INSTRUMENTS
    . OPTICAL MEASURING INSTRUMENTS
    . . SPECTROPHOTOMETERS
    . . . **INFRARED SPECTROPHOTOMETERS**
    . RADIATION MEASURING INSTRUMENTS
    . . ACTINOMETERS
    . . . SPECTROPHOTOMETERS
    . . . . **INFRARED
      SPECTROPHOTOMETERS**
    . . INFRARED INSTRUMENTS
    . . . **INFRARED SPECTROPHOTOMETERS**
    OPTICAL EQUIPMENT
    . OPTICAL MEASURING INSTRUMENTS
    . . SPECTROPHOTOMETERS
    . . . **INFRARED SPECTROPHOTOMETERS**
RT   CHEMICAL ANALYSIS
    FILTER WHEEL INFRARED
      SPECTROMETERS
    PHOTOMETERS

INFRARED SPECTROSCOPY
- GS   SPECTROSCOPY
  - . **INFRARED SPECTROSCOPY**
- RT   ABSORPTION SPECTROSCOPY
  - ASTRONOMICAL SPECTROSCOPY
  - CHEMICAL ANALYSIS
  - ELECTRON SPECTROSCOPY
  - LASER SPECTROMETERS
  - MOLECULAR SPECTROSCOPY
  - MOLECULAR STRUCTURE
  - OPTOGALVANIC SPECTROSCOPY
  - RAMAN SPECTROSCOPY
  - SPECTROMETERS
  - SPECTROSCOPIC ANALYSIS
  - VACUUM SPECTROSCOPY

INFRARED STARS
- GS   CELESTIAL BODIES
  - . STARS
  - . . **INFRARED STARS**
- RT   HERBIG-HARO OBJECTS

INFRARED SUPPRESSION
- RT   AFTERBURNING
  - AIRCRAFT DETECTION
  - AIRCRAFT ENGINES
  - COOLING SYSTEMS
  - EXHAUST GASES
  - EXHAUST NOZZLES
  - HEAT SHIELDING
  - JET ENGINES
  - JET EXHAUST
  - REACTION PRODUCTS
  - SUPPRESSORS
  - TEMPERATURE CONTROL

INFRARED TELESCOPES
- GS   TELESCOPES
  - . ASTRONOMICAL TELESCOPES
  - . . **INFRARED TELESCOPES**
  - . . . SPACE INFRARED TELESCOPE FACILITY
- RT   ASTRONOMY
  - SPACE INFRARED TELESCOPE FACILITY

INFRARED TRACKING
- GS   TRACKING (POSITION)
  - . **INFRARED TRACKING**
- RT   ANTIMISSILE MISSILES
  - COMPENSATORY TRACKING
  - HOMING DEVICES
  - MISSILE TRACKING
  - OPTICAL TRACKING
  - PURSUIT TRACKING
  - RADIOMETERS

INFRARED WINDOWS
- RT   APERTURES
  - LASERS
  - ∞ WINDOWS

INFRASONIC FREQUENCIES
- GS   FREQUENCIES
  - . **INFRASONIC FREQUENCIES**
- RT   ACOUSTICS

∞ **INGESTION**
- SN   *(USE OF A MORE SPECIFIC TERM IS RECOMMENDED--CONSULT THE TERMS LISTED BELOW.*
- RT   INGESTION (BIOLOGY)
  - INGESTION (ENGINES)

INGESTION (BIOLOGY)
- GS   **INGESTION (BIOLOGY)**
  - . DRINKING
  - . EATING
  - . GRAZING
- RT   ∞ INGESTION
  - SWALLOWING

INGESTION (ENGINES)
- RT   ENGINE FAILURE
  - ∞ INGESTION

INGOTS
- GS   CASTINGS
  - . **INGOTS**
- RT   BILLETS
  - CASTING
  - MOLDS
  - SOLIDIFICATION

INGREDIENTS
- RT   ADMIXTURES
  - ∞ COMPONENTS
  - ∞ COMPOSITION
  - CONTENT
  - FORMULATIONS
  - MIXTURES

INGRESS (SPACECRAFT PASSAGEWAY)
- RT   AIR LOCKS
  - DOORS
  - EGRESS
  - HATCHES
  - OPENINGS

INHABITANTS
- GS   COMMUNITIES
  - . **INHABITANTS**
  - . . MOUNTAIN INHABITANTS
- RT   ABORIGINES
  - CITIES
  - DEMOGRAPHY
  - PERSONNEL
  - RESIDENTIAL AREAS

INHALATION
- USE   RESPIRATION

∞ **INHIBITION**
- SN   *(USE OF A MORE SPECIFIC TERM IS RECOMMENDED--CONSULT THE TERMS LISTED BELOW)*
- RT   ATTENUATION
  - CORROSION PREVENTION
  - DETACHMENT
  - DITHERS
  - FRUSTRATION
  - HANDICAPS
  - INHIBITION (PSYCHOLOGY)
  - INHIBITORS
  - INTROVERSION
  - LETHARGY
  - PASSIVITY
  - PREVENTION
  - ∞ REDUCTION
  - STOPPING

INHIBITION (PSYCHOLOGY)
- RT   CONDITIONING (LEARNING)
  - ∞ INHIBITION

INHIBITORS
- GS   **INHIBITORS**
  - . WEAR INHIBITORS
- RT   ADDITIVES
  - ANTIDOTES
  - ANTIFOULING
  - ANTIICING ADDITIVES
  - ANTIOXIDANTS
  - CASE BONDED PROPELLANTS
  - CATALYSTS
  - COATINGS
  - CORROSION PREVENTION
  - ∞ INHIBITION
  - NEUTRALIZERS
  - PACKAGING
  - PASSIVITY
  - PROPELLANT ADDITIVES
  - PROPELLANT DECOMPOSITION
  - PROPELLANT STORABILITY
  - RETARDANTS
  - SILENCERS
  - SOLID PROPELLANT IGNITION
  - SOLID PROPELLANTS
  - SUPPRESSORS

INHOMOGENEITY
- UF   NONHOMOGENEITY
- RT   DEFECTS
  - HETEROGENEITY
  - NONUNIFORMITY

INHOUR EQUATION
- RT   ∞ EQUATIONS
  - NUCLEAR REACTIONS
  - NUCLEAR REACTORS
  - REACTIVITY
  - REACTOR PHYSICS

INITIAL VALUE PROBLEMS
- USE   BOUNDARY VALUE PROBLEMS

INITIATION
- RT   ACTIVATION

INITIATION-*(CONT.)*
- ACTUATION
- DETONATION
- ∞ GENERATION
- ∞ INDUCTION
- INOCULATION
- NUCLEATION
- ∞ PRIMING
- REACTOR STARTUP TESTS
- STARTING
- STIMULATION

∞ **INITIATORS**
- SN   *(USE OF A MORE SPECIFIC TERM IS RECOMMENDED--CONSULT THE TERMS LISTED BELOW)*
- RT   CATALYSTS
  - INITIATORS (EXPLOSIVES)
  - STYPHNATES

INITIATORS (EXPLOSIVES)
- UF   ELECTROEXPLOSIVE DEVICES
- GS   EXPLOSIVE DEVICES
  - . **INITIATORS (EXPLOSIVES)**
  - . . BOOSTERS (EXPLOSIVES)
  - . . CAPS (EXPLOSIVES)
  - . . DETONATORS
  - . . EXPLODING WIRES
  - . . PRIMERS (EXPLOSIVES)
  - IGNITERS
  - . **INITIATORS (EXPLOSIVES)**
  - . . BOOSTERS (EXPLOSIVES)
  - . . CAPS (EXPLOSIVES)
  - . . DETONATORS
  - . . PRIMERS (EXPLOSIVES)
- RT   FUSES (ORDNANCE)
  - ∞ INITIATORS
  - PYROTECHNICS
  - STYPHNATES

INJECTION
- GS   **INJECTION**
  - . CARRIER INJECTION
  - . FLUID INJECTION
  - . . GAS INJECTION
  - . . LIQUID INJECTION
  - . . . DEEP WELL INJECTION (WASTES)
  - . . . WATER INJECTION
  - . FUEL INJECTION
  - . ION INJECTION
  - . SECONDARY INJECTION
  - . TRANSEARTH INJECTION
  - . TRANSLUNAR INJECTION
- RT   BARRITT DIODES
  - BLOWING
  - BOUNDARY LAYER SEPARATION
  - ∞ CHARGING
  - FEEDING (SUPPLYING)
  - FILLING
  - FLUID FLOW
  - IMPLANTATION
  - INJECTORS
  - INPUT
  - PERFORATING
  - SUPPLYING

INJECTION CARBURETORS
- USE   CARBURETORS
  - FUEL INJECTION

INJECTION GUIDANCE
- GS   GUIDANCE (MOTION)
  - . **INJECTION GUIDANCE**
- RT   ASCENT TRAJECTORIES
  - CELESTIAL NAVIGATION
  - COMMAND GUIDANCE
  - INERTIAL GUIDANCE
  - MIDCOURSE GUIDANCE
  - RENDEZVOUS GUIDANCE
  - SATELLITE GUIDANCE
  - SPACECRAFT GUIDANCE
  - TRANSEARTH INJECTION
  - TRANSLUNAR INJECTION

INJECTION LASERS
- GS   STIMULATED EMISSION DEVICES
  - . LASERS
  - . . **INJECTION LASERS**
- RT   GALLIUM ARSENIDE LASERS
  - GALLIUM ARSENIDES
  - INJECTION LOCKING
  - SEMICONDUCTOR LASERS

INJECTION LOCKING
- RT   CARRIER INJECTION

INJECTION LOCKING-*(CONT.)*
    INJECTION LASERS
    LASER MODE LOCKING

INJECTION MOLDING
GS   FORMING TECHNIQUES
    . **INJECTION MOLDING**
RT   CERAMICS
    DIES
    EXTRUDING
    MELTING
    MOLDING MATERIALS
    MOLDS
    PLASTICS

INJECTORS
GS   **INJECTORS**
    . VORTEX INJECTORS
RT   BLOWERS
    CARBURETORS
    EJECTORS
    FEEDERS
    FUEL INJECTION
    FUEL SYSTEMS
    INJECTION
    JET FLOW
    JET MIXING FLOW
    ∞JET NOZZLES
    ∞JETS
    NOZZLE FLOW
    ∞NOZZLES
    ORIFICES
    PUMPS
    SPRAY NOZZLES

INJUN EXPLORER
USE EXPLORER 25 SATELLITE

INJUN SATELLITES
GS   SATELLITES
    . ARTIFICIAL SATELLITES
    . . **INJUN SATELLITES**
    . . . EXPLORER 25 SATELLITE
    . . . INJUN 1 SATELLITE
    . . . INJUN 3 SATELLITE
    . . . INJUN 4 SATELLITE

INJUN 1 SATELLITE
GS   SATELLITES
    . ARTIFICIAL SATELLITES
    . . INJUN SATELLITES
    . . . **INJUN 1 SATELLITE**

INJUN 3 SATELLITE
GS   SATELLITES
    . ARTIFICIAL SATELLITES
    . . INJUN SATELLITES
    . . . **INJUN 3 SATELLITE**

INJUN 4 SATELLITE
GS   SATELLITES
    . ARTIFICIAL SATELLITES
    . . INJUN SATELLITES
    . . . **INJUN 4 SATELLITE**

INJUN 5 SATELLITE
USE EXPLORER 40 SATELLITE

INJURIES
GS   **INJURIES**
    . BACK INJURIES
    . BAROTRAUMA
    . BRAIN DAMAGE
    . BURNS (INJURIES)
    . CRASH INJURIES
    . EJECTION INJURIES
    . FROSTBITE
    . LESIONS
    . . PULMONARY LESIONS
    . NOISE INJURIES
    . PARACHUTING INJURY
    . PARALYSIS
    . RADIATION INJURIES
    . WHIPLASH INJURIES
RT   ACCIDENTS
    ∞BLISTERS
    CHEMICAL DEFENSE
    DAMAGE
    DEATH
    DIAGNOSIS
    FOREIGN BODIES
    HAZARDS
    HEMORRHAGES
    IMPAIRMENT

INJURIES-*(CONT.)*
    SABOTAGE
    VETERINARY MEDICINE
    WOUND HEALING

INKS
RT   DRAWINGS
    GRAPHIC ARTS
    PIGMENTS
    PRINTING

INLAND WATERS
GS   WATER
    . **INLAND WATERS**
    . . GROUND WATER
RT   GREAT LAKES (NORTH AMERICA)
    GREAT SALT LAKE (UT)
    LAKES
    RIVERS
    SPRINGS (WATER)
    WATER POLLUTION
    WATER RESOURCES
    WATER RUNOFF

INLET AIRFRAME CONFIGURATIONS
GS   INTAKE SYSTEMS
    . AIR INTAKES
    . . **INLET AIRFRAME CONFIGURATIONS**
RT   BYPASS RATIO
    ENGINE INLETS
    FLOW GEOMETRY
    HYPERSONIC INLETS
    INLET FLOW
    INLET NOZZLES
    NOSE INLETS
    SIDE INLETS
    SUPERSONIC INLETS

INLET FLOW
GS   FLUID FLOW
    . **INLET FLOW**
RT   BYPASS RATIO
    ∞DIFFUSERS
    FLOW GEOMETRY
    FLUID INJECTION
    HEAD FLOW
    INLET AIRFRAME CONFIGURATIONS
    INTAKE SYSTEMS
    ∞PRESSURE DROP
    SUPERSONIC INLETS
    VORTEX GENERATORS

INLET NOZZLES
RT   AIR INTAKES
    ANNULAR NOZZLES
    BYPASS RATIO
    CONICAL NOZZLES
    ∞DIFFUSERS
    ENGINE INLETS
    EXHAUST NOZZLES
    INLET AIRFRAME CONFIGURATIONS
    INTAKE SYSTEMS
    INTERNAL COMPRESSION INLETS
    ∞NOZZLES
    OPENINGS
    PIPE NOZZLES

INLET PRESSURE
GS   PRESSURE
    . **INLET PRESSURE**
RT   PRESSURE GRADIENTS
    PRESSURE RECOVERY
    STAGNATION PRESSURE
    WATER PRESSURE

INLET TEMPERATURE
GS   TEMPERATURE
    . **INLET TEMPERATURE**
RT   AIR INTAKES
    ENGINE INLETS
    FUEL SYSTEMS
    GAS TEMPERATURE
    INTAKE SYSTEMS

INLETS (DEVICES)
USE INTAKE SYSTEMS

INLETS (TOPOGRAPHY)
GS   LANDFORMS
    . **INLETS (TOPOGRAPHY)**
    . . BAYOUS
    . . COOK INLET (AK)
RT   BAYS (TOPOGRAPHIC FEATURES)
    DELAWARE BAY (US)

INLETS (TOPOGRAPHY)-*(CONT.)*
    FIORDS
    GULFS
    LAGOONS
    PERSIAN GULF
    SAGINAW BAY (MI)
    SOUNDS (TOPOGRAPHIC FEATURES)

INLIERS (LANDFORMS)
GS   LANDFORMS
    . **INLIERS (LANDFORMS)**
RT   EROSION
    GEOLOGICAL FAULTS
    GEOLOGY
    PETROGRAPHY
    PETROLOGY
    ROCK INTRUSIONS
    ROCKS
    STRUCTURAL PROPERTIES (GEOLOGY)

INNER RADIATION BELT
GS   EARTH ATMOSPHERE
    . RADIATION BELTS
    . . **INNER RADIATION BELT**
    ENVIRONMENTS
    . **INNER RADIATION BELT**
    PARTICLES
    . CHARGED PARTICLES
    . . MAGNETICALLY TRAPPED PARTICLES
    . . . RADIATION BELTS
    . . . . **INNER RADIATION BELT**
    . TRAPPED PARTICLES
    . . MAGNETICALLY TRAPPED PARTICLES
    . . . RADIATION BELTS
    . . . . **INNER RADIATION BELT**
RT   ARTIFICIAL RADIATION BELTS
    OUTER RADIATION BELT
    PROTON BELTS
    ∞RADIATION
    SINGLE EVENT UPSETS

INOCULATION
UF   SEEDING (INOCULATION)
RT   CRYSTAL GROWTH
    CRYSTALLIZATION
    IMMUNITY
    INITIATION
    NUCLEATION
    VACCINES

INOCULUM
GS   SERUMS
    . **INOCULUM**
    VACCINES
    . **INOCULUM**
RT   ANTIBODIES
    ANTIGENS
    PHYSIOLOGICAL DEFENSES

INORGANIC CHEMISTRY
RT   ANALYTICAL CHEMISTRY
    ∞CHEMISTRY

INORGANIC COATINGS
GS   COATINGS
    . **INORGANIC COATINGS**
    . . ANODIC COATINGS
    . . CERAMIC COATINGS
RT   ANTIRADAR COATINGS
    PROTECTIVE COATINGS

INORGANIC COMPOUNDS
GS   **INORGANIC COMPOUNDS**
    . AMMONIA
    . . LIQUID AMMONIA
RT   ACIDS
    ∞BASES
    ∞CHEMICAL COMPOUNDS
    FLAME RETARDANTS
    INTERMETALLICS
    MOLTEN SALTS
    ∞SALTS

∞ **INORGANIC MATERIALS**
SN   *(USE OF A MORE SPECIFIC TERM IS RECOMMENDED CONSULT THE TERMS LISTED BELOW)*
RT   ∞MATERIALS
    NONFLAMMABLE MATERIALS
    REFRACTORY MATERIALS
    THERMOCHROMATIC MATERIALS
    VITREOUS MATERIALS

INORGANIC NITRATES
GS   NITROGEN COMPOUNDS
. NITRATES
. . **INORGANIC NITRATES**
. . . AMMONIUM NITRATES
. . . HYDRAZINE NITRATE
. . . POTASSIUM NITRATES
. . . SILVER NITRATES
. . . SODIUM NITRATES

INORGANIC PEROXIDES
UF   SUPEROXIDES
GS   CHALCOGENIDES
. OXIDES
. . ANHYDRIDES
. . . PEROXIDES
. . . . **INORGANIC PEROXIDES**
RT   ORGANIC PEROXIDES

INORGANIC SULFIDES
GS   CHALCOGENIDES
. SULFIDES
. . **INORGANIC SULFIDES**
. . . BARIUM SULFIDES
. . . BISMUTH SULFIDES
. . . CADMIUM SULFIDES
. . . CALCIUM SULFIDES
. . . COPPER SULFIDES
. . . . ENARGITE
. . . HYDROGEN SULFIDE
. . . INDIUM SULFIDES
. . . LEAD SULFIDES
. . . MOLYBDENUM SULFIDES
. . . . MOLYBDENUM DISULFIDES
. . . POLYSULFIDES
. . . STRONTIUM SULFIDES
. . . ZINC SULFIDES
. . . . WURTZITE
. . . . ZINCBLENDE
SULFUR COMPOUNDS
. SULFIDES
. . **INORGANIC SULFIDES**
. . . BARIUM SULFIDES
. . . BISMUTH SULFIDES
. . . CADMIUM SULFIDES
. . . CALCIUM SULFIDES
. . . COPPER SULFIDES
. . . . ENARGITE
. . . HYDROGEN SULFIDE
. . . INDIUM SULFIDES
. . . LEAD SULFIDES
. . . MOLYBDENUM SULFIDES
. . . . MOLYBDENUM DISULFIDES
. . . POLYSULFIDES
. . . STRONTIUM SULFIDES
. . . ZINC SULFIDES
. . . . WURTZITE
. . . . ZINCBLENDE

INOSITOLS
GS   ALIPHATIC COMPOUNDS
. SUGARS
. . **INOSITOLS**
CARBOHYDRATES
. SUGARS
. . **INOSITOLS**

INPUT
RT   ACCUMULATIONS
COLLECTION
FEEDING (SUPPLYING)
FILLING
INJECTION
∞ LOADING
OUTPUT
READING
REPLENISHMENT
SUPPLYING

INPUT/OUTPUT ROUTINES
GS   COMPUTER PROGRAMS
. COMPUTER SYSTEMS PROGRAMS
. . **INPUT/OUTPUT ROUTINES**
RT   OPERATING SYSTEMS (COMPUTERS)
RANDOM ACCESS
∞ ROUTINES

INSAT SATELLITES
USE   INDIAN SPACECRAFT

INSECT DAMAGE
USE   INFESTATION

INSECTICIDES
GS   POISONS

INSECTICIDES-*(CONT.)*
. PESTICIDES
. . **INSECTICIDES**
. . . DIELDRIN
RT   ENDRIN
ENTOMOLOGY
TOXICOLOGY

INSECTS
GS   ANIMALS
. INVERTEBRATES
. . ARTHROPODS
. . . **INSECTS**
. . . . BEES
. . . . BOLL WEEVILS
. . . . CHIRONOMUS FLIES
. . . . COCKROACHES
. . . . COLEOPTERA
. . . . CRICKETS
. . . . . BEETLES
. . . . . . TRIBOLIA
. . . . DROSOPHILA
. . . . FIREFLIES
. . . . GRASSHOPPERS
. . . . LARVAE
. . . . . BOLLWORMS
. . . . . SILKWORMS
. . . . LOCUSTS
. . . . MOTHS
. . . . . SILKWORMS
RT   ENTOMOLOGY
INFESTATION
PUPA

INSENSITIVITY
USE   SENSITIVITY

INSERTION
RT   COLLATING
EMBEDDING
GRAFTING
IMPLANTATION
IMPREGNATING
INSERTS
NETWORK ANALYSIS
TRANSMISSION LOSS

INSERTION LOSS
RT   ENERGY DISSIPATION
LOSSES
TRANSMISSION LOSS

INSERTS
GS   **INSERTS**
. NOZZLE INSERTS
RT   ACCESSORIES
BUSHINGS
FASTENERS
FITTINGS
INSERTION
LININGS
SPACERS
SPOOLS
WASHERS (SPACERS)

INSHORE ZONES
USE   BEACHES

INSOLATION
RT   GREENHOUSE EFFECT
METEOROLOGY
SOLAR HEATING
SOLAR RADIATION
SUNLIGHT

INSOMNIA
GS   SLEEP
. **INSOMNIA**
RT   SLEEP DEPRIVATION

INSPECTION
GS   **INSPECTION**
. INFRARED INSPECTION
. X RAY INSPECTION
RT   ACCEPTABILITY
CHECKOUT
CHEMICAL TESTS
CONSTRUCTION
DETECTION
ENDOSCOPES
EVALUATION
EXAMINATION
IDENTIFYING
NONDESTRUCTIVE TESTS

INSPECTION-*(CONT.)*
PERFORMANCE TESTS
QUALITY CONTROL
SAMPLING
SPECIFICATIONS
STANDARDS
STATIC TESTS
STATISTICAL ANALYSIS
SURVEILLANCE
TOLERANCES (MECHANICS)
ULTRASONIC FLAW DETECTION

INSPECTOR SATELLITE
GS   MILITARY SPACECRAFT
. RECONNAISSANCE SPACECRAFT
. . **INSPECTOR SATELLITE**
SATELLITES
. ARTIFICIAL SATELLITES
. . **INSPECTOR SATELLITE**

INSPIRATION
RT   INTELLECT
MENTAL PERFORMANCE
PSYCHOLOGY

INSTABILITY
USE   STABILITY

INSTALLATION
USE   INSTALLING

INSTALLATION MANUALS
GS   DOCUMENTS
. MANUALS
. . **INSTALLATION MANUALS**

INSTALLING
UF   INSTALLATION
RT   ASSEMBLING
CONSTRUCTION
LOOK ANGLES (ELECTRONICS)
MAINTENANCE
RELOCATION
REPLACING
RETROFITTING

INSTANTONS
GS   FIELD THEORY (PHYSICS)
. QUANTUM CHROMODYNAMICS
. . **INSTANTONS**
RT   ELEMENTARY PARTICLES
PLASMA PHYSICS
QUARKS

INSTITUTIONS
GS   **INSTITUTIONS**
. BUREAUS (ORGANIZATIONS)
RT   FEDERATIONS
TEAMS

INSTRUCTION SETS (COMPUTERS)
RT   ALPHANUMERIC CHARACTERS
BOOLEAN ALGEBRA
COMPUTER PROGRAMS
MATHEMATICAL LOGIC

INSTRUCTIONS
USE   EDUCATION

INSTRUCTORS
GS   PERSONNEL
. **INSTRUCTORS**
RT   EDUCATION
LEARNING
SCHOOLS
STUDENTS
TRAINING EVALUATION
UNIVERSITIES

INSTRUMENT APPROACH
GS   APPROACH
. **INSTRUMENT APPROACH**
RT   AIRCRAFT APPROACH SPACING
AIRCRAFT INSTRUMENTS
APPROACH CONTROL
APPROACH INDICATORS
BLIND LANDING
FLIGHT CONTROL
FLIGHT INSTRUMENTS
GLIDE PATHS
LANDING AIDS
LANDING RADAR
NIGHT FLIGHTS (AIRCRAFT)

INSTRUMENT COMPENSATION
GS   **INSTRUMENT COMPENSATION**
.   TEMPERATURE COMPENSATION
RT   ADAPTIVE OPTICS
     CALIBRATING
     ∞COMPENSATION
     ERROR CORRECTING DEVICES

INSTRUMENT DRIFT
USE   DRIFT (INSTRUMENTATION)

INSTRUMENT ERRORS
GS   ERRORS
.   **INSTRUMENT ERRORS**
RT   BIAS
     BORESIGHT ERROR
     CALIBRATING
     DRIFT (INSTRUMENTATION)
     LINEARITY
     OPTICAL CORRECTION PROCEDURE
     SPECTRAL SENSITIVITY

INSTRUMENT FLIGHT RULES
UF   IFR (RULES)
GS   RULES
.   FLIGHT RULES
. . **INSTRUMENT FLIGHT RULES**
RT   AIR NAVIGATION
     AIR TRAFFIC CONTROL
     APPROACH CONTROL
     BEACONS
     BLIND LANDING
     FLIGHT CONDITIONS
     FLIGHT INSTRUMENTS
     FLIGHT PLANS
     LANDING
     LOW VISIBILITY

INSTRUMENT LANDING SYSTEMS
UF   ILS (LANDING SYSTEMS)
GS   LANDING AIDS
.   **INSTRUMENT LANDING SYSTEMS**
. . ALL-WEATHER LANDING SYSTEMS
. . AUTOMATIC LANDING CONTROL
RT   AIR TRAFFIC CONTROL
     AIRCRAFT GUIDANCE
     AIRCRAFT INSTRUMENTS
     AIRCRAFT LANDING
     AIRPORTS
     APPROACH CONTROL
     APPROACH INDICATORS
     BLIND LANDING
     DISPLAY DEVICES
     FLIGHT CONTROL
     FLIGHT INSTRUMENTS
     GLIDE PATHS
     GROUND BASED CONTROL
     ∞INSTRUMENTS
     LANDING
     LANDING INSTRUMENTS
     NIGHT FLIGHTS (AIRCRAFT)
     RADAR
     RADAR APPROACH CONTROL
     RADIO ALTIMETERS
     RADIO BEACONS
     ∞SYSTEMS
     TRACKING (POSITION)

INSTRUMENT ORIENTATION
RT   ALIGNMENT
     ATTITUDE (INCLINATION)
     BEARING (DIRECTION)
     DIRECTIVITY
     LOOK ANGLES (ELECTRONICS)
     ∞ORIENTATION
     POSITIONING

INSTRUMENT PACKAGES
GS   PACKAGES
.   **INSTRUMENT PACKAGES**
. . EASEP
. . EREP
RT   AMPS (SATELLITE PAYLOAD)
     AUTOMATIC WEATHER STATIONS
     DATA COLLECTION PLATFORMS
     LOCAL SCIENTIFIC SURVEY MODULE
     MODULES
     MOLECULAR SHIELDS
     OCEAN DATA ACQUISITIONS SYSTEMS
     ORBITING FROG OTOLITH
     PAYLOAD ASSIST MODULE
     PAYLOADS
     SATELLITE-BORNE INSTRUMENTS
     SIM
     SPACECRAFT INSTRUMENTS

INSTRUMENT PACKAGES-(CONT.)
     WEATHER STATIONS

INSTRUMENT RECEIVERS
RT   CONTROLLERS
     ∞DETECTORS
     DISPLAY DEVICES
     INDICATING INSTRUMENTS
     ∞INSTRUMENTS
     ISOTROPIC TURBULENCE
     MEASURING INSTRUMENTS
     RECEIVERS
     RECORDING INSTRUMENTS
     TRANSDUCERS

INSTRUMENT TRANSFORMERS
GS   TRANSFORMERS
.   **INSTRUMENT TRANSFORMERS**
RT   ∞CONVERTERS
     RESOLVERS

INSTRUMENT TRANSMITTERS
GS   TRANSMITTERS
.   **INSTRUMENT TRANSMITTERS**
RT   CONTROLLERS
     ∞INSTRUMENTS
     MEASURING INSTRUMENTS
     RECORDING INSTRUMENTS
     TRANSDUCERS

INSTRUMENTAL ANALYSIS
USE   ANALYZING
      AUTOMATION

INSTRUMENTATION
USE   INSTRUMENTS

∞ INSTRUMENTS
SN   *(USE OF A MORE SPECIFIC TERM IS
     RECOMMENDED--CONSULT THE TERMS
     LISTED BELOW)*
UF   INSTRUMENTATION
RT   ACTUATORS
     ADVANCED RANGE INSTRUMENTATION
       SHIP
     AIRCRAFT INSTRUMENTS
     APOLLO LUNAR SURFACE EXPERIMENTS
       PACKAGE
     AUTOMATIC CONTROL
     BIOINSTRUMENTATION
     BUBBLE TECHNIQUE
     CONTROLLERS
     DENSIMETERS
     DISPLAY DEVICES
     DRAG FORCE ANEMOMETERS
     EASEP
     EREP
     FLIGHT INSTRUMENTS
     FORCE VECTOR RECORDERS
     HAPLOSCOPES
     HELIOSTATS
     HELMET MOUNTED DISPLAYS
     INDICATING INSTRUMENTS
     INSTRUMENT LANDING SYSTEMS
     INSTRUMENT RECEIVERS
     INSTRUMENT TRANSMITTERS
     LASER ALTIMETERS
     MEASURING INSTRUMENTS
     METEOROLOGICAL INSTRUMENTS
     MICROWAVE SENSORS
     MONITORS
     NAVIGATION INSTRUMENTS
     OCULOMETERS
     PACKAGES
     PROPELLANT ACTUATED INSTRUMENTS
     RECORDING INSTRUMENTS
     REMOTE CONTROL
     ROCKET-BORNE INSTRUMENTS
     SATELLITE INSTRUMENTS
     SATELLITE-BORNE INSTRUMENTS
     SCATTEROMETERS
     SIM
     SODAR
     SOUND DETECTING AND RANGING
     SPACECRAFT INSTRUMENTS
     SURGICAL INSTRUMENTS
     TRANSDUCERS
     TRANSMITTERS
     TURBINE INSTRUMENTS
     ULTRASONIC DENSIMETERS

∞ INSULATED STRUCTURES
SN   *(USE OF A MORE SPECIFIC TERM IS
     RECOMMENDED--CONSULT THE TERMS
     LISTED BELOW)*

INSULATED STRUCTURES-(CONT.)
RT   DIELECTRICS
     ELECTRICAL INSULATION
     HEAT RADIATORS
     HEAT SHIELDING
     HYPERSONIC VEHICLES
     INSULATION
     RADIATION SHIELDING
     REENTRY SHIELDING
     REENTRY VEHICLES
     SPACECRAFT SHIELDING

INSULATING MATERIALS
USE   INSULATION

INSULATION
SN   (MATERIAL)
UF   INSULATING MATERIALS
GS   **INSULATION**
.   ELECTRICAL INSULATION
.   MULTILAYER INSULATION
.   THERMAL INSULATION
RT   ABSORBERS (MATERIALS)
     ASBESTOS
     CEILINGS (ARCHITECTURE)
     COMPOSITE MATERIALS
     CONCRETES
     ∞CONSTRUCTION MATERIALS
     DAMPING
     HONEYCOMB STRUCTURES
     ∞INSULATED STRUCTURES
     INSULATORS
     INTERLAYERS
     ISOLATION
     ISOLATORS
     JACKETS
     LINING PROCESSES
     LININGS
     ∞MATERIALS
     MICARTA
     OXIDES
     POTTING COMPOUNDS
     PROTECTION
     SUPPRESSORS
     VERMICULITE
     WATERPROOFING

INSULATORS
SN   (EXCLUDES THERMAL
     INSULATION--LIMITED TO DEVICES
     COMPOSED OF ELECTRICALLY
     INSULATIVE MATERIALS)
RT   ATTENUATORS
     DIELECTRICS
     ELECTRIC CONDUCTORS
     ELECTRICAL INSULATION
     INSULATION
     TRANSMISSION LINES

INSULIN
GS   DRUGS
.   **INSULIN**
     SECRETIONS
.   ENDOCRINE SECRETIONS
. . **INSULIN**
RT   DIABETES MELLITUS

INTAKE SYSTEMS
UF   INDUCTION SYSTEMS
     INLETS (DEVICES)
GS   **INTAKE SYSTEMS**
.   AIR INTAKES
. . ENGINE INLETS
. . HYPERSONIC INLETS
. . INLET AIRFRAME CONFIGURATIONS
. . SUPERSONIC INLETS
.   CONICAL INLETS
.   HELICAL INDUCERS
.   INTERNAL COMPRESSION INLETS
.   NOSE INLETS
.   SIDE INLETS
RT   AERODYNAMIC CONFIGURATIONS
     ANNULAR DUCTS
     BYPASS RATIO
     COOLING SYSTEMS
     DUCT GEOMETRY
     DUCTED BODIES
     DUCTS
     ENTRANCES
     EXHAUST SYSTEMS
     FEED SYSTEMS
     FEEDERS
     FUEL SYSTEMS
     INLET FLOW
     INLET NOZZLES

INTAKE SYSTEMS-*(CONT.)*
    INLET TEMPERATURE
    MANIFOLDS
    OPENINGS
    PIPE NOZZLES
    PLENUM CHAMBERS
    RAMPS (STRUCTURES)
    SCOOPS
    ∞SYSTEMS
    ∞WATER INTAKES

INTASAT SATELLITE
GS   SATELLITES
    . ARTIFICIAL SATELLITES
    .. SCIENTIFIC SATELLITES
    ... ENVIRONMENTAL RESEARCH
        SATELLITES
    .... **INTASAT SATELLITE**
RT   IONOSPHERE
    LOWER ATMOSPHERE
    MAGNETIC FIELDS
    TROPOSPHERE

**INTEG MED AND BEHAVIORAL LAB MEASUR
SYSTEM**
USE   IMBLMS

INTEG PROGRAM FOR AEROSPACE VEH DESIGN
USE   IPAD

INTEGERS
GS   NUMBER THEORY
    . **INTEGERS**
    REAL NUMBERS
    . **INTEGERS**
RT   ARITHMETIC
    COMPLEX NUMBERS
    CONGRUENCES
    DIGITS
    ∞NUMBERS

INTEGRAL CALCULUS
GS   ANALYSIS (MATHEMATICS)
    . CALCULUS
    .. **INTEGRAL CALCULUS**
    . REAL VARIABLES
    .. MEASURE AND INTEGRATION
    ... **INTEGRAL CALCULUS**
RT   AREA
    DIFFERENTIAL CALCULUS
    INTEGRALS
    J INTEGRAL
    NUMERICAL INTEGRATION
    OPERATIONAL CALCULUS

INTEGRAL EQUATIONS
UF   INTEGRODIFFERENTIAL EQUATIONS
GS   ANALYSIS (MATHEMATICS)
    . FUNCTIONAL ANALYSIS
    .. **INTEGRAL EQUATIONS**
    ... FREDHOLM EQUATIONS
    ... J INTEGRAL
    ... SINGULAR INTEGRAL EQUATIONS
    ... VOLTERRA EQUATIONS
    ... WIENER HOPF EQUATIONS
RT   ASYMPTOTIC PROPERTIES
    CALCULUS OF VARIATIONS
    DIFFERENTIAL EQUATIONS
    DISTRIBUTED PARAMETER SYSTEMS
    ∞EQUATIONS
    MELLIN TRANSFORMS
    METHOD OF MOMENTS
    NONLINEAR EQUATIONS
    PERCUS METHOD
    RANGE (EXTREMES)
    SCHMIDT METHOD
    TRANSPORT THEORY

INTEGRAL FUNCTIONS
USE   ENTIRE FUNCTIONS

INTEGRAL ROCKET RAMJETS
GS   ENGINES
    . AIR BREATHING ENGINES
    .. GAS TURBINE ENGINES
    ... JET ENGINES
    .... RAMJET ENGINES
    ..... **INTEGRAL ROCKET RAMJETS**
    . INTERNAL COMBUSTION ENGINES
    .. GAS TURBINE ENGINES
    ... JET ENGINES
    .... RAMJET ENGINES
    ..... **INTEGRAL ROCKET RAMJETS**
RT   SOLID PROPELLANT ROCKET ENGINES
    TURBINE ENGINES

INTEGRAL TRANSFORMATIONS
UF   TRANSFORM INTEGRALS
GS   ANALYSIS (MATHEMATICS)
    . FUNCTIONAL ANALYSIS
    .. **INTEGRAL TRANSFORMATIONS**
    ... CONVOLUTION INTEGRALS
    ... FOURIER TRANSFORMATION
    ... HILBERT TRANSFORMATION
    ... LAPLACE TRANSFORMATION
    TRANSFORMATIONS (MATHEMATICS)
    . **INTEGRAL TRANSFORMATIONS**
    .. FAST FOURIER TRANSFORMATIONS
    .. LAPLACE TRANSFORMATION
RT   LIGHTHILL METHOD
    OPERATORS (MATHEMATICS)

INTEGRALS
RT   DIFFERENTIAL EQUATIONS
    FUNCTIONALS
    INTEGRAL CALCULUS
    ∞MATHEMATICS

INTEGRATED CIRCUITS
UF   MONOLITHIC CIRCUITS
GS   CIRCUITS
    . **INTEGRATED CIRCUITS**
    .. DTL INTEGRATED CIRCUITS
    .. ENCAPSULATED MICROCIRCUITS
    .. LARGE SCALE INTEGRATION
    .. LINEAR INTEGRATED CIRCUITS
    .. TTL INTEGRATED CIRCUITS
    .. VERY LARGE SCALE INTEGRATION
    .. VHSIC (CIRCUITS)
RT   BURN-IN
    CHARGE FLOW DEVICES
    CHIPS (ELECTRONICS)
    CHIPS (MEMORY DEVICES)
    ELECTRONIC PACKAGING
    INTEGRATED OPTICS
    ION IMPLANTATION
    LATCH-UP
    MEDIUM SCALE INTEGRATION
    MICROCHANNEL PLATES
    MICROMINIATURIZATION
    MICROPROCESSORS
    MOLECULAR ELECTRONICS
    PHOTOMASKS
    PRINTED CIRCUITS
    THICK FILMS
    THIN FILMS
    TRANSISTOR CIRCUITS

INTEGRATED ENERGY SYSTEMS
RT   COMMUNITIES
    ELECTRIC POWER PLANTS
    ENERGY CONVERSION
    ENERGY DISTRIBUTION
    HEATING
    ∞SYSTEMS
    TOTAL ENERGY SYSTEMS
    UTILITIES

**INTEGRATED GLOBAL OCEAN STATION
SYSTEMS**
UF   IGOSS
RT   DATA COLLECTION PLATFORMS
    GLOBAL ATMOSPHERIC RESEARCH
        PROGRAM
    GROUND STATIONS
    INTERNATIONAL COOPERATION
    OCEANOGRAPHIC PARAMETERS
    ∞SYSTEMS
    WEATHER STATIONS

INTEGRATED LIBRARY SYSTEMS
GS   INFORMATION SYSTEMS
    . **INTEGRATED LIBRARY SYSTEMS**
RT   INFORMATION DISSEMINATION
    INFORMATION MANAGEMENT
    INFORMATION RETRIEVAL
    LIBRARIES
    ON-LINE SYSTEMS

INTEGRATED MANEUVERING LIFE SUPPORT SYS
USE   IMLSS

INTEGRATED MISSION CONTROL CENTER
UF   IMCC (CONTROL CENTER)
GS   STATIONS
    . GROUND STATIONS
    .. **INTEGRATED MISSION CONTROL
        CENTER**
RT   ∞CONTROL
    GEMINI PROJECT
    GROUND BASED CONTROL

INTEGRATED MISSION CONTROL-*(CONT.)*
    REAL TIME OPERATION

INTEGRATED OPTICS
RT   ELECTRO-OPTICS
    INTEGRATED CIRCUITS
    LENSES
    LIGHT TRANSMISSION
    MONOMOLECULAR FILMS
    OPTICAL BISTABILITY
    OPTICAL WAVEGUIDES
    ∞OPTICS
    THIN FILMS

INTEGRATION (REAL VARIABLES)
USE   MEASURE AND INTEGRATION

INTEGRATORS
GS   **INTEGRATORS**
    . DIGITAL INTEGRATORS
RT   CIRCUITS
    DIFFERENTIATORS
    ∞NETWORKS
    SOLIONS

INTEGRITY
GS   **INTEGRITY**
    . COMPUTER PROGRAM INTEGRITY
RT   COMPLETENESS
    PRIVACY
    SECURITY
    VULNERABILITY

INTEGRODIFFERENTIAL EQUATIONS
USE   DIFFERENTIAL EQUATIONS
    INTEGRAL EQUATIONS

INTEL 8080 MICROPROCESSOR
GS   DATA PROCESSING EQUIPMENT
    . MICROPROCESSORS
    .. **INTEL 8080 MICROPROCESSOR**
    PERIPHERAL EQUIPMENT (COMPUTERS)
    . **INTEL 8080 MICROPROCESSOR**
RT   COMPUTERS

INTELLECT
GS   INTELLIGENCE
    . **INTELLECT**
RT   ARTIFICIAL INTELLIGENCE
    INSPIRATION
    MENTAL PERFORMANCE
    PSYCHOLOGY

INTELLIGENCE
GS   **INTELLIGENCE**
    . ARTIFICIAL INTELLIGENCE
    .. EXPERT SYSTEMS
    . EXTRATERRESTRIAL INTELLIGENCE
    . INTELLECT
RT   DETECTION
    MENTAL HEALTH
    MENTAL PERFORMANCE
    MILITARY TECHNOLOGY
    RECONNAISSANCE
    SURVEILLANCE

INTELLIGIBILITY
GS   **INTELLIGIBILITY**
    . SPEECH RECOGNITION
RT   AMBIGUITY
    ∞COHERENCE
    COMMUNICATION THEORY
    ∞INTERPRETATION
    MESSAGES
    ORTHOGRAPHY
    PHONEMICS
    PHONETICS
    PSYCHOLINGUISTICS
    SCRAMBLING (COMMUNICATION)

INTELSAT SATELLITES
GS   SATELLITES
    . ARTIFICIAL SATELLITES
    .. COMMUNICATION SATELLITES
    ... **INTELSAT SATELLITES**

INTENSIFICATION
USE   AMPLIFICATION

INTENSIFIER TUBES
USE   IMAGE INTENSIFIERS

INTENSIFIERS
GS   **INTENSIFIERS**

INTENSIFIERS-*(CONT.)*
. IMAGE INTENSIFIERS
. . IMAGE ORTHICONS
RT AMPLIFIERS

∞ INTENSITY
 SN *(USE OF A MORE SPECIFIC TERM IS*
  RECOMMENDED--CONSULT THE TERMS
  LISTED BELOW)
 RT AMPLITUDES
  BRIGHTNESS
  FLUX (RATE)
  FLUX DENSITY
  LEVEL (QUANTITY)
  LOUDNESS
  LUMINANCE
  LUMINOUS INTENSITY
  MAGNITUDE
  NOISE INTENSITY
  RADIANCE
  STELLAR MAGNITUDE

INTERACTIONAL AERODYNAMICS
 GS FLUID MECHANICS
  . FLUID DYNAMICS
  . . GAS DYNAMICS
  . . . **INTERACTIONAL AERODYNAMICS**
 RT AIRFOILS
  COMPUTATIONAL FLUID DYNAMICS
  ∞FLOW
  LAMINAR BOUNDARY LAYER

∞ INTERACTIONS
 SN *(USE OF A MORE SPECIFIC TERM IS*
  RECOMMENDED--CONSULT THE TERMS
  LISTED BELOW)
 RT AIR LAND INTERACTIONS
  AIR SEA ICE INTERACTIONS
  AIR WATER INTERACTIONS
  ATOMIC COLLISIONS
  ATOMIC INTERACTIONS
  BEAM INTERACTIONS
  CONFIGURATION INTERACTION
  ELECTROMAGNETIC INTERACTIONS
  ELECTRON PHONON INTERACTIONS
  ELEMENTARY PARTICLE INTERACTIONS
  FLUID-SOLID INTERACTIONS
  GAS-GAS INTERACTIONS
  GAS-ION INTERACTIONS
  GAS-LIQUID INTERACTIONS
  GAS-METAL INTERACTIONS
  GAS-SOLID INTERACTIONS
  HIGH ENERGY INTERACTIONS
  ION ATOM INTERACTIONS
  LASER PLASMA INTERACTIONS
  LASER TARGET INTERACTIONS
  MAN ENVIRONMENT INTERACTIONS
  MESON-MESON INTERACTIONS
  MESON-NUCLEON INTERACTIONS
  MOLECULAR COLLISIONS
  MOLECULAR INTERACTIONS
  NUCLEAR CAPTURE
  NUCLEAR INTERACTIONS
  NUCLEAR REACTIONS
  NUCLEON-NUCLEON INTERACTIONS
  PARTICLE INTERACTIONS
  PARTICLE THEORY
  PHOTON-ELECTRON INTERACTION
  PLASMA INTERACTION EXPERIMENT
  PLASMA INTERACTIONS
  PLASMA-ELECTROMAGNETIC
   INTERACTION
  PLASMA-PARTICLE INTERACTIONS
  PROTON-PROTON REACTIONS
  SHOCK WAVE INTERACTION
  SOLAR TERRESTRIAL INTERACTIONS
  SOUND-SOUND INTERACTIONS
  SPIN-ORBIT INTERACTIONS
  STRONG INTERACTIONS (FIELD
   THEORY)
  WAVE INTERACTION
  WEAK ENERGY INTERACTIONS
  WEAK INTERACTIONS (FIELD THEORY)

INTERACTIVE CONTROL
 RT ACTIVE CONTROL
  ∞CONTROL
  CONTROL THEORY
  NUMERICAL CONTROL

INTERACTIVE GRAPHICS
 USE COMPUTER GRAPHICS

INTERATOMIC FORCES
 RT ATOMIC STRUCTURE

INTERATOMIC FORCES-*(CONT.)*
  VAN DER WAAL FORCES

INTERCALATION
 GS STRATIFICATION
  . **INTERCALATION**
 RT ∞CHEMICAL COMPOUNDS
  GRAPHITE
  INTERLAYERS
  ∞LAYERS

INTERCEPTION
 RT RENDEZVOUS
  SPACECRAFT DOCKING

INTERCEPTOR AIRCRAFT
 USE FIGHTER AIRCRAFT

∞ INTERCEPTORS
 SN *(USE OF A MORE SPECIFIC TERM IS*
  RECOMMENDED--CONSULT THE TERMS
  LISTED BELOW)
 RT FIGHTER AIRCRAFT
  SATELLITE INTERCEPTORS
  YF-12 AIRCRAFT

INTERCONNECTION
 USE JOINING

INTERCONTINENTAL BALLISTIC MISSILES
 UF ICBM (MISSILES)
 GS MISSILES
  . BALLISTIC MISSILES
  . . **INTERCONTINENTAL BALLISTIC**
   MISSILES
  . . . ATLAS ICBM
  . . . . ATLAS D ICBM
  . . . . ATLAS E ICBM
  . . . . ATLAS F ICBM
  . . . MINUTEMAN ICBM
  . . . TITAN ICBM
  . . . . TITAN 1 ICBM
  . . . . TITAN 2 ICBM
  . SURFACE TO SURFACE MISSILES
  . . **INTERCONTINENTAL BALLISTIC**
   MISSILES
  . . . ATLAS ICBM
  . . . . ATLAS D ICBM
  . . . . ATLAS E ICBM
  . . . . ATLAS F ICBM
  . . . MINUTEMAN ICBM
  . . . MX MISSILE
  . . . TITAN ICBM
  . . . . TITAN 1 ICBM
  . . . . TITAN 2 ICBM
 RT FLEET BALLISTIC MISSILES
  INTERMEDIATE RANGE BALLISTIC
   MISSILES
  MARK 1 REENTRY BODY
  MARK 2 REENTRY BODY
  MARK 3 REENTRY BODY
  MARK 4 REENTRY BODY
  MARK 5 REENTRY BODY
  MARK 6 REENTRY BODY
  MARK 11 REENTRY BODY
  MARK 12 REENTRY BODY
  MARK 17 REENTRY BODY
  TRANSOCEANIC SYSTEMS

INTERCOSMOS SATELLITES
 GS SATELLITES
  . ARTIFICIAL SATELLITES
  . . GEOPHYSICAL SATELLITES
  . . . COSMOS SATELLITES
  . . . . **INTERCOSMOS SATELLITES**
  . . SOVIET SATELLITES
  . . . COSMOS SATELLITES
  . . . . **INTERCOSMOS SATELLITES**

INTERCRANIAL CIRCULATION
 GS CIRCULATION
  . BLOOD CIRCULATION
  . . **INTERCRANIAL CIRCULATION**
 RT CRANIUM
  SKULL

INTERDIGITAL TRANSDUCERS
 GS TRANSDUCERS
  . **INTERDIGITAL TRANSDUCERS**
 RT DIGITAL TRANSDUCERS
  ELECTROACOUSTIC TRANSDUCERS
  PIEZOELECTRIC TRANSDUCERS
  SURFACE ACOUSTIC WAVE DEVICES

INTERFACE STABILITY
 RT FLUID BOUNDARIES
  GAS-SOLID INTERFACES
  INTERFACES
  LIQUID SLOSHING
  LIQUID-LIQUID INTERFACES
  LIQUID-SOLID INTERFACES
  LIQUID-VAPOR INTERFACES
  TAYLOR INSTABILITY
  ULLAGE

INTERFACES
 GS **INTERFACES**
  . FLUID BOUNDARIES
  . . GAS-SOLID INTERFACES
  . . JET BOUNDARIES
  . . LIQUID-LIQUID INTERFACES
  . . LIQUID-SOLID INTERFACES
  . . LIQUID-VAPOR INTERFACES
  . SOLID-SOLID INTERFACES
 RT BOUNDARIES
  COORDINATION
  DATA PROCESSING EQUIPMENT
  FREE BOUNDARIES
  INTERFACE STABILITY
  MANAGEMENT PLANNING
  PROJECT MANAGEMENT
  SURFACE PROPERTIES
  SURFACE REACTIONS
  ∞SURFACES
  TELECOMMUNICATION

INTERFACIAL ENERGY
 RT ADHESION
  ELECTRON ENERGY
  ∞ENERGY
  LIQUID-LIQUID INTERFACES
  SHEAR STRENGTH
  SURFACE ENERGY

INTERFACIAL STRAIN
 USE INTERFACIAL TENSION

INTERFACIAL TENSION
 UF INTERFACIAL STRAIN
  SURFACE TENSION
 GS SURFACE PROPERTIES
  . **INTERFACIAL TENSION**
 RT CAPILLARY WAVES
  GAS-LIQUID INTERACTIONS
  GIBBS ADSORPTION EQUATION
  GLOBULES
  LIQUID SURFACES
  LIQUID-LIQUID INTERFACES
  MARANGONI CONVECTION
  MECHANICAL PROPERTIES
  RIPPLES
  SLIDING
  SPREADING
  SURFACE ENERGY
  SURFACE STABILITY
  ∞SURFACES
  ∞TENSION
  TRIBOLOGY
  VAPOR PRESSURE
  WETTING

∞ INTERFERENCE
 SN *(USE OF A MORE SPECIFIC TERM IS*
  RECOMMENDED--CONSULT THE TERMS
  LISTED BELOW)
 RT AERODYNAMIC INTERFERENCE
  COHERENCE COEFFICIENT
  CROSSTALK
  DISRUPTING
  ELECTROMAGNETIC COMPATIBILITY
  ELECTROMAGNETIC INTERFERENCE
  HUM
  INCOMPATIBILITY
  INTERFERENCE FACTOR TABLE
  INTERFERENCE GRATING
  INTERSYMBOLIC INTERFERENCE
  JAMMING
  NONSYNCHRONIZATION
  RADIO FREQUENCY INTERFERENCE
  RAMSAUER EFFECT
  SUPPORT INTERFERENCE
  WAVE DIFFRACTION
  WAVE FRONT DEFORMATION

INTERFERENCE DRAG
 GS AERODYNAMIC CHARACTERISTICS
  . **INTERFERENCE DRAG**
  DYNAMIC CHARACTERISTICS
  . DRAG

INTERFERENCE DRAG-(CONT.)
```
      . . PRESSURE DRAG
      . . . WAVE DRAG
      . . . . INTERFERENCE DRAG
RT    PROPELLER SLIPSTREAMS
      SUPERSONIC DRAG
      UPWASH
```

INTERFERENCE FACTOR TABLE
```
GS    TABLES (DATA)
      . INTERFERENCE FACTOR TABLE
RT    ∞ INTERFERENCE
      MODULATION
      MULTICHANNEL COMMUNICATION
```
†

INTERFERENCE GRATING
```
RT    FRINGE MULTIPLICATION
      ∞ GRATINGS
      ∞ INTERFERENCE
      MOIRE EFFECTS
      MOIRE FRINGES
      RADIO FILTERS
      RADIO FREQUENCY INTERFERENCE
```

INTERFERENCE IMMUNITY
```
RT    ELECTROMAGNETIC INTERFERENCE
      NOISE REDUCTION
      RADIO FREQUENCY INTERFERENCE
      SIGNAL PROCESSING
      SIGNAL TO NOISE RATIOS
```

INTERFERENCE LIFT
```
GS    AERODYNAMIC CHARACTERISTICS
      . LIFT
      . . INTERFERENCE LIFT
      AERODYNAMIC FORCES
      . LIFT
      . . INTERFERENCE LIFT
      DISTRIBUTION (PROPERTY)
      . INTERFERENCE LIFT
      DYNAMIC CHARACTERISTICS
      . LIFT
      . . INTERFERENCE LIFT
RT    UPWASH
```

INTERFERENCE MONOCHROMATIZATION
```
USE   DIFFRACTION
      MONOCHROMATIZATION
```

INTERFEROGRAMS
```
USE   INTERFEROMETRY
```

INTERFEROMETERS
```
GS    MEASURING INSTRUMENTS
      . INTERFEROMETERS
      . . FABRY-PEROT INTERFEROMETERS
      . . INFRARED INTERFEROMETERS
      . . MACH-ZEHNDER INTERFEROMETERS
      . . MICHELSON INTERFEROMETERS
      . . MICROWAVE INTERFEROMETERS
      . . PHASE SWITCHING
              INTERFEROMETERS
      . . RADIO INTERFEROMETERS
RT    DIFFRACTOMETERS
      FLATNESS
      GONIOMETERS
      OPTICAL EQUIPMENT
      OPTICAL MEASUREMENT
      OPTICAL MEASURING INSTRUMENTS
      PHOTOGONIOMETERS
      RONCHI TEST
      SAGNAC EFFECT
      VERY LONG BASE INTERFEROMETRY
```

INTERFEROMETRY
```
UF    INTERFEROGRAMS
GS    INTERFEROMETRY
      . DIFFERENTIAL INTERFEROMETRY
      . HOLOGRAPHIC INTERFEROMETRY
      . LASER INTERFEROMETRY
      . MOIRE INTERFEROMETRY
      . RONCHI TEST
      . VERY LONG BASE INTERFEROMETRY
RT    DIFFRACTION PATTERNS
      FRESNEL DIFFRACTION
      FRESNEL REFLECTORS
      INFRARED INTERFEROMETERS
      ISOCHROMATICS
      NULL ZONES
      PLASMA FLUX MEASUREMENT
      SAGNAC EFFECT
      SCATTER PLATES (OPTICS)
```

INTERFERON
```
RT    BACTERIOPHAGES
      BIOCHEMISTRY
      ∞ BIOLOGY
      PHYSIOLOGICAL DEFENSES
      VIRUSES
```

INTERGALACTIC MEDIA
```
UF    EXTRAGALACTIC MEDIA
GS    MEDIA
      . INTERGALACTIC MEDIA
RT    COSMIC DUST
      COSMIC GASES
      COSMIC PLASMA
      MASS DISTRIBUTION
      STELLAR WINDS
```

INTERGRANULAR CORROSION
```
GS    CHEMICAL ATTACK
      . INTERGRANULAR CORROSION
      CORROSION
      . INTERGRANULAR CORROSION
RT    GRAIN BOUNDARIES
      STRESS CORROSION
      TRANSGRANULAR CORROSION
```

INTERIM STAGES (SPACECRAFT)
```
GS    INTERIM STAGES (SPACECRAFT)
      . INERTIAL UPPER STAGE
RT    MULTISTAGE ROCKET VEHICLES
      RECOVERABLE SPACECRAFT
      REUSABLE SPACECRAFT
      SPACE SHUTTLES
      STAGE SEPARATION
```

INTERIM UPPER STAGE (STS)
```
USE   INERTIAL UPPER STAGE
```

INTERIOR BALLISTICS
```
GS    BALLISTICS
      . INTERIOR BALLISTICS
RT    PROPELLANT TESTS
```

INTERLACING DRAINAGE
```
USE   DRAINAGE PATTERNS
```

INTERLAYERS
```
GS    INTERLAYERS
      . MULTILAYER INSULATION
RT    BARRIER LAYERS
      FABRICS
      INSULATION
      INTERCALATION
      LAMINATES
      ∞ LAYERS
      PLY ORIENTATION
      SANDWICH STRUCTURES
      ∞ TRANSITION LAYERS
```

INTERLOCKING
```
USE   LOCKING
```

INTERMEDIATE FREQUENCIES
```
GS    FREQUENCIES
      . INTERMEDIATE FREQUENCIES
RT    HIGH FREQUENCIES
      LOW FREQUENCIES
      RADIO FREQUENCIES
```

INTERMEDIATE FREQUENCY AMPLIFIERS
```
GS    AMPLIFIERS
      . INTERMEDIATE FREQUENCY
          AMPLIFIERS
RT    BEAT FREQUENCIES
      CRYSTAL FILTERS
      HETERODYNING
      LOGARITHMIC RECEIVERS
      PREAMPLIFIERS
      RADIO RECEIVERS
      TRANSISTOR AMPLIFIERS
```

INTERMEDIATE RANGE BALLISTIC MISSILES
```
UF    IRBM (MISSILES)
GS    MISSILES
      . BALLISTIC MISSILES
      . . INTERMEDIATE RANGE BALLISTIC
          MISSILES
      . . . BLUE STREAK MISSILE
      . . . JUPITER MISSILE
      . . . POLARIS MISSILES
      . . . . POLARIS A1 MISSILE
      . . . . POLARIS A2 MISSILE
      . . . . POLARIS A3 MISSILE
      . SURFACE TO SURFACE MISSILES
```

INTERMEDIATE RANGE BALLISTIC-(CONT.)
```
      . . INTERMEDIATE RANGE BALLISTIC
          MISSILES
      . . . BLUE STREAK MISSILE
      . . . JUPITER MISSILE
      . . . POLARIS MISSILES
      . . . . POLARIS A1 MISSILE
      . . . . POLARIS A2 MISSILE
      . . . . POLARIS A3 MISSILE
RT    FIELD ARMY BALLISTIC MISSILES
      FLEET BALLISTIC MISSILES
      INTERCONTINENTAL BALLISTIC MISSILES
      MARK 1 REENTRY BODY
      MARK 2 REENTRY BODY
      MARK 3 REENTRY BODY
      SHORT RANGE BALLISTIC MISSILES
```

INTERMETALLICS
```
SN    (COMPOUNDS CONSISTING OF ONLY
      METALLIC ELEMENTS)
UF    ELECTRON COMPOUNDS
RT    ALLOYS
      AMMINES
      ARSENIDES
      BORIDES
      INORGANIC COMPOUNDS
      METALLOIDS
      METALS
      PHASE DIAGRAMS
      PHOTOELECTROMAGNETIC EFFECTS
      SEMICONDUCTORS (MATERIALS)
      SILICIDES
      TELLURIDES
```

INTERMITTENCY
```
RT    CYCLES
      PULSES
      RANDOM PROCESSES
```

INTERMITTENCY HYPOTHESIS
```
GS    HYPOTHESES
      . INTERMITTENCY HYPOTHESIS
RT    PHOTOGRAPHIC RECORDING
```

INTERMODULATION
```
GS    MODULATION
      . INTERMODULATION
RT    DEMODULATION
      DISCRIMINATORS
      FREQUENCY ANALYZERS
      FREQUENCY MODULATION
      REMODULATION
      SOUND-SOUND INTERACTIONS
      WAVE INTERACTION
```

INTERMOLECULAR FORCES
```
GS    LEVEL (QUANTITY)
      . ENERGY LEVELS
      . . MOLECULAR ENERGY LEVELS
      . . . INTERMOLECULAR FORCES
RT    CONFIGURATION INTERACTION
      EXCIMERS
      LENNARD-JONES POTENTIAL
      MOLECULAR INTERACTIONS
      MOLECULAR STRUCTURE
      VAN DER WAAL FORCES
      VIRIAL COEFFICIENTS
```

INTERMONTANE FLOORS
```
USE   VALLEYS
```

INTERNAL COMBUSTION ENGINES
```
SN    (EXCLUDES ROCKET ENGINES)
GS    ENGINES
      . INTERNAL COMBUSTION ENGINES
      . . DIESEL ENGINES
      . . GAS TURBINE ENGINES
      . . . HYDROGEN ENGINES
      . . . JET ENGINES
      . . . . RAMJET ENGINES
      . . . . . INTEGRAL ROCKET RAMJETS
      . . . . . LOW VOLUME RAMJET ENGINES
      . . . . . PULSEJET ENGINES
      . . . . . SUPERSONIC COMBUSTION
              RAMJET ENGINES
      . . . . . TURBORAMJET ENGINES
      . . . . T-63 ENGINE
      . . . . T-76 ENGINE
      . . . . TURBOJET ENGINES
      . . . . . BRISTOL-SIDDELEY OLYMPUS
              593 ENGINE
      . . . . . BRISTOL-SIDDELEY VIPER
              ENGINE
      . . . . . DUCTED FAN ENGINES
      . . . . . J-33 ENGINE
```

INTERNAL COMBUSTION ENGINES-*(CONT.)*
- - - - - J-34 ENGINE
- - - - - J-47 ENGINE
- - - - - J-52 ENGINE
- - - - - J-57 ENGINE
- - - - - J-57-P-20 ENGINE
- - - - - J-65 ENGINE
- - - - - J-69-T-25 ENGINE
- - - - - J-71 ENGINE
- - - - - J-73 ENGINE
- - - - - J-75 ENGINE
- - - - - J-79 ENGINE
- - - - - J-85 ENGINE
- - - - - J-93 ENGINE
- - - - - RA-28 ENGINE
- - - - - TURBOFAN ENGINES
- - - - - - BRISTOL-SIDDELEY BS 53
     ENGINE
- - - - - - CF-700 ENGINE
- - - - - - J-97 ENGINE
- - - - - - TF-30 ENGINE
- - - - - - TF-41 ENGINE
- - - - - TURBOPROP ENGINES
- - - - - - T-34 ENGINE
- - - - - - T-38 ENGINE
- - - - - - T-53 ENGINE
- - - - - - T-56 ENGINE
- - - - - - T-64 ENGINE
- - - - - - T-74 ENGINE
- - - - - - T-78 ENGINE
- - - - - TURBORAMJET ENGINES
- - - T-58 ENGINE
- - - T-58-GE-8B ENGINE
- - HELICOPTER ENGINES
- - ROTARY ENGINES
- - - WANKEL ENGINES
- RT   AFTERBURNING
    AIRCRAFT ENGINES
    AUTOMOBILE ENGINES
    AUTOMOBILE FUELS
  ∞ BEARING
    BEARINGS
    BOOSTER ROCKET ENGINES
    CAMS
    CARBURETORS
    COMBUSTION
    COMBUSTION CHAMBERS
    DIESEL FUELS
    DISTRIBUTORS
    DUCTED ROCKET ENGINES
    ELECTRIC HYBRID VEHICLES
    ENGINE PARTS
    ENGINE PRIMERS
    ENGINE STARTERS
    EXHAUST SYSTEMS
    EXTERNAL COMBUSTION ENGINES
    FUEL CONSUMPTION
    FUEL INJECTION
    FUEL PUMPS
    FUEL SYSTEMS
    GAS TURBINES
    HYBRID PROPELLANT ROCKET ENGINES
    IGNITION SYSTEMS
    LIQUID PROPELLANT ROCKET ENGINES
    LUBRICATION SYSTEMS
    PISTON ENGINES
    PISTONS
    RETROROCKET ENGINES
    ROCKET ENGINES
    SOLID PROPELLANT ROCKET ENGINES
    SPARK PLUGS
    SUPERCHARGERS
    SUSTAINER ROCKET ENGINES
    THERMODYNAMIC CYCLES
    THERMODYNAMIC EFFICIENCY
    TORPEDO ENGINES
    VERNIER ENGINES

INTERNAL COMPRESSION INLETS
- GS   INTAKE SYSTEMS
    . **INTERNAL COMPRESSION INLETS**
- RT   AIR INTAKES
    COMPRESSING
    ENGINE INLETS
    INLET NOZZLES
    SUPERSONIC INLETS

INTERNAL CONVERSION
- RT   ∞ CONVERSION
    NUCLEAR REACTIONS

INTERNAL ENERGY
- RT   CHEMICAL ENERGY
    ∞ ENERGY
    FREE ENERGY

INTERNAL ENERGY-*(CONT.)*
    GIBBS-HELMHOLTZ EQUATIONS
    KINETIC ENERGY
  ∞ LEVEL
    MOLECULAR ENERGY LEVELS
    PARTICLE ENERGY
    POTENTIAL ENERGY
    THERMAL ENERGY
    THERMODYNAMICS

INTERNAL FRICTION
- GS   FRICTION
    . **INTERNAL FRICTION**
- RT   ANELASTICITY
    ATTENUATION
    COHESION
    DAMPING
    DENSITY (MASS/VOLUME)
    EDDY VISCOSITY
    HYSTERESIS
    MECHANICAL PROPERTIES
  ∞ PHYSICAL PROPERTIES
    PLASTIC FLOW
    VISCOSITY

INTERNAL PRESSURE
- SN   (PRESSURE INSIDE A PORTION OF
    MATTER) DUE TO ATTRACTION
    BETWEEN MOLECULES)
- GS   PRESSURE
    . **INTERNAL PRESSURE**
- RT   ADHESION
    COHESION
    GAS PRESSURE
    PARTIAL PRESSURE
    PRESSURE DISTRIBUTION
    SPREADING
    TEMPERATURE INVERSIONS

INTERNAL STRESS
- USE   RESIDUAL STRESS

INTERNAL WAVES
- GS   INTERNAL WAVES
    . PLANETARY WAVES
- RT   SURFACE WAVES
    ∞ WAVES

†

INTERNATIONAL COMPUTERS LIMITED
- USE   ICL COMPUTERS

INTERNATIONAL COOPERATION
- GS   FOREIGN POLICY
    . INTERNATIONAL RELATIONS
    . . **INTERNATIONAL COOPERATION**
    . . . OUTER SPACE TREATY
- RT   A-300 AIRCRAFT
    A-310 AIRCRAFT
    A-320 AIRCRAFT
    ANIK SATELLITES
    ANIK 1
    ANIK 2
    ANIK 3
    APOLLO SOYUZ TEST PROJECT
    ARABSAT
    ARCOMSAT
    AZUR SATELLITE
    COMMITTEE ON SPACE RESEARCH
    COMMUNICATIONS TECHNOLOGY
      SATELLITE
    CONVENTIONS
  ∞ COOPERATION
    COSMOS 782 SATELLITE
    COSMOS 936 SATELLITE
    COSMOS 1129 SATELLITE
    DISARMAMENT
    EISCAT RADAR SYSTEM (EUROPE)
    ESA SATELLITES
    EUROPEAN AIRBUS
    FEDERATIONS
    FRENCH SPACE PROGRAMS
    INFORMATION TRANSFER
    INTEGRATED GLOBAL OCEAN STATION
      SYSTEMS
    INTERNATIONAL HYDROLOGICAL
      DECADE
    INTERNATIONAL SATELLITE GEODESY
      EXPERIMENT
    NORTH ATLANTIC TREATY
      ORGANIZATION (NATO)
    ORBITING FROG OTOLITH
    PALAPA SATELLITES
    PALAPA 2 SATELLITE
    PEACETIME
    POLITICS

INTERNATIONAL COOPERATION-*(CONT.)*
    ROSAT MISSION
    SEA LAW
    SOVEREIGNTY
    SYMPHONIE SATELLITES
    U.S.S.R. SPACE PROGRAM
    UNITED NATIONS
    USER REQUIREMENTS
    VEGA PROJECT
    WORLD METEOROLOGICAL
      ORGANIZATION

INTERNATIONAL FIELD YEAR FOR GREAT LAKES
- RT   CANADA
    GREAT LAKES (NORTH AMERICA)
    UNITED STATES

INTERNATIONAL GEOPHYSICAL YEAR
- UF   IGY (GEOPHYSICAL YEAR)
- RT   GEOPHYSICS
    VANGUARD SATELLITES
    WORLD DATA CENTERS

**INTERNATIONAL GEOSPHERE-BIOSPHERE
PROGRAM**
- GS   PROGRAMS
    . **INTERNATIONAL
      GEOSPHERE-BIOSPHERE PROGRAM**
- RT   BIOGEOCHEMISTRY
    BIOSPHERE
    EARTH OBSERVATIONS (FROM SPACE)
    GEOPHYSICS
    MAN ENVIRONMENT INTERACTIONS
    SOLAR TERRESTRIAL INTERACTIONS

INTERNATIONAL HYDROLOGICAL DECADE
- RT   CANADA
    FOREIGN POLICY
    HYDROLOGY
    INTERNATIONAL COOPERATION
    INTERNATIONAL RELATIONS
    PRECIPITATION (METEOROLOGY)
    RIVER BASINS
    STREAMS
    UNITED STATES
    WATER RESOURCES
    WATERSHEDS

INTERNATIONAL LAW
- GS   LAW (JURISPRUDENCE)
    . **INTERNATIONAL LAW**
    . . AIR LAW
    . . SEA LAW
    . . SPACE LAW
- RT   CONVENTIONS
    LEGAL LIABILITY
    NATIONS
    OUTER SPACE TREATY
    PEACETIME
    POLITICS
    SOVEREIGNTY
    UNITED NATIONS
    WARFARE

INTERNATIONAL MAGNETOSPHERIC EXPLORER
- UF   IME SATELLITE
- GS   SATELLITES
    . ARTIFICIAL SATELLITES
    . . EXPLORER SATELLITES
    . . . **INTERNATIONAL MAGNETOSPHERIC
      EXPLORER**
- RT   DELTA LAUNCH VEHICLE
    MAGNETOSPHERE

INTERNATIONAL MAGNETOSPHERIC STUDY
- UF   IMS
- GS   INVESTIGATION
    . **INTERNATIONAL MAGNETOSPHERIC
      STUDY**
- RT   ATMOSPHERIC PHYSICS
    EUROPEAN SPACE PROGRAMS
    GEOMAGNETISM
    INTERPLANETARY MAGNETIC FIELDS
    MAGNETOSPHERE

INTERNATIONAL PRACTICAL TEMPERATURE
- USE   TEMPERATURE SCALES

INTERNATIONAL QUIET SUN YEAR
- UF   IQSY (INTERNATIONAL YEAR)
- RT   SOLAR ACTIVITY
    SOLAR CYCLES
    SOLAR PHYSICS

INTERNATIONAL RELATIONS
GS   FOREIGN POLICY
. **INTERNATIONAL RELATIONS**
. . INTERNATIONAL COOPERATION
. . . OUTER SPACE TREATY
RT   APOLLO SOYUZ TEST PROJECT
INTERNATIONAL HYDROLOGICAL
    DECADE
U.S.S.R. SPACE PROGRAM

**INTERNATIONAL SATELLITE GEODESY
EXPERIMENT**
UF   ISAGEX
RT   CELESTIAL GEODESY
EUROPEAN SPACE PROGRAMS
GEODETIC COORDINATES
INTERNATIONAL COOPERATION
SATELLITE TRACKING
U.S.S.R. SPACE PROGRAM

INTERNATIONAL SATS FOR IONOSPHERIC STUDY
USE   ISIS SATELLITES

INTERNATIONAL SOLAR POLAR MISSION
USE   ULYSSES MISSION

INTERNATIONAL SUN EARTH EXPLORER 1
GS   SATELLITES
. ARTIFICIAL SATELLITES
. . EXPLORER SATELLITES
. . . INTERNATIONAL SUN EARTH
    EXPLORERS
. . . . **INTERNATIONAL SUN EARTH
    EXPLORER 1**

INTERNATIONAL SUN EARTH EXPLORER 2
GS   SATELLITES
. ARTIFICIAL SATELLITES
. . EXPLORER SATELLITES
. . . INTERNATIONAL SUN EARTH
    EXPLORERS
. . . . **INTERNATIONAL SUN EARTH
    EXPLORER 2**

INTERNATIONAL SUN EARTH EXPLORER 3
GS   SATELLITES
. ARTIFICIAL SATELLITES
. . EXPLORER SATELLITES
. . . INTERNATIONAL SUN EARTH
    EXPLORERS
. . . . **INTERNATIONAL SUN EARTH
    EXPLORER 3**

INTERNATIONAL SUN EARTH EXPLORERS
UF   ISEE
GS   SATELLITES
. ARTIFICIAL SATELLITES
. . EXPLORER SATELLITES
. . . **INTERNATIONAL SUN EARTH
    EXPLORERS**
. . . . INTERNATIONAL SUN EARTH
    EXPLORER 1
. . . . INTERNATIONAL SUN EARTH
    EXPLORER 2
. . . . INTERNATIONAL SUN EARTH
    EXPLORER 3

INTERNATIONAL SYSTEM OF UNITS
UF   METRIC SYSTEM
SI
GS   UNITS OF MEASUREMENT
. **INTERNATIONAL SYSTEM OF UNITS**
RT   CONVERSION TABLES
∞ MEASUREMENT
MEASURING INSTRUMENTS
METRICATION
METROLOGY
∞ SYSTEMS

INTERNATIONAL TRADE
UF   EXPORTS
RT   ECONOMICS
REVENUE

INTERNATIONAL ULTRAVIOLET EXPLORER
USE   IUE

INTERNUCLEAR PROPERTIES
RT   MOLECULAR INTERACTIONS
∞ MOLECULAR PHYSICS

INTERORBITAL TRAJECTORIES
GS   TRAJECTORIES
. **INTERORBITAL TRAJECTORIES**

INTERORBITAL TRAJECTORIES-*(CONT.)*
RT   INTERPLANETARY TRAJECTORIES
ROUND TRIP TRAJECTORIES
SPACECRAFT TRAJECTORIES

INTERPERSONAL RELATIONS
USE   HUMAN RELATIONS

INTERPHONES
GS   COMMUNICATION EQUIPMENT
. **INTERPHONES**
RT   EARPHONES
MICROPHONES
TELECOMMUNICATION

INTERPLANETARY COMMUNICATION
GS   TELECOMMUNICATION
. SPACE COMMUNICATION
. . **INTERPLANETARY COMMUNICATION**
RT   CIRCUMLUNAR COMMUNICATION
EXTRATERRESTRIAL COMMUNICATION
FACSIMILE COMMUNICATION
LASERS
LUNAR COMMUNICATION
OPTICAL COMMUNICATION
RADIO COMMUNICATION
SPACECRAFT COMMUNICATION

INTERPLANETARY DUST
GS   DUST
. COSMIC DUST
. . **INTERPLANETARY DUST**
. . . METEOROID DUST CLOUDS
. . . . ZODIACAL DUST
MEDIA
. INTERPLANETARY MEDIUM
. . **INTERPLANETARY DUST**
. . . METEOROID DUST CLOUDS
. . . . ZODIACAL DUST
RT   METEOROIDS
MICROMETEOROIDS

INTERPLANETARY EXPLORER
USE   EXPLORER 18 SATELLITE

INTERPLANETARY FLIGHT
UF   PLANETARY SPACE FLIGHT
GS   SPACE FLIGHT
. **INTERPLANETARY FLIGHT**
. . LONG DURATION SPACE FLIGHT
RT   ASTRODYNAMICS
EARTH-VENUS TRAJECTORIES
FLYBY MISSIONS
INTERSTELLAR SPACECRAFT
MANNED SPACE FLIGHT
MARINER JUPITER-SATURN FLYBY
MARINER JUPITER-URANUS FLYBY
MARINER MARK 2 SPACECRAFT
ORBITS
OUTER PLANETS EXPLORERS
PLANETARY LANDING
RETURN TO EARTH SPACE FLIGHT
ROUND TRIP TRAJECTORIES
SPACE EXPLORATION
SPACE NAVIGATION
SPACECRAFT GUIDANCE
TOPS (SPACECRAFT)

INTERPLANETARY GAS
GS   EXTRATERRESTRIAL MATTER
. COSMIC GASES
. . **INTERPLANETARY GAS**
GASES
. RAREFIED GASES
. . COSMIC GASES
. . . **INTERPLANETARY GAS**
MEDIA
. INTERPLANETARY MEDIUM
. . **INTERPLANETARY GAS**
RT   COSMIC PLASMA
INTERSTELLAR GAS
NEUTRAL GASES
SOLAR WIND

INTERPLANETARY MAGNETIC FIELDS
GS   MAGNETIC FIELDS
. **INTERPLANETARY MAGNETIC FIELDS**
RT   CHAPMAN-FERRARO PROBLEM
INTERNATIONAL MAGNETOSPHERIC
    STUDY
SOLAR MAGNETIC FIELD

INTERPLANETARY MEDIUM
GS   MEDIA

INTERPLANETARY MEDIUM-*(CONT.)*
. **INTERPLANETARY MEDIUM**
. . INTERPLANETARY DUST
. . . METEOROID DUST CLOUDS
. . . . ZODIACAL DUST
. . INTERPLANETARY GAS
RT   MASS DISTRIBUTION
METEOROIDS
PLASMA CLOUDS
SOLAR WIND

INTERPLANETARY MONITORING PLATFORM
USE   IMP

INTERPLANETARY NAVIGATION
GS   NAVIGATION
. SPACE NAVIGATION
. . **INTERPLANETARY NAVIGATION**
RT   ASTRONAVIGATION
CELESTIAL NAVIGATION
CELESTIAL REFERENCE SYSTEMS
RADAR NAVIGATION
RADIO NAVIGATION

INTERPLANETARY PROPULSION
USE   INTERPLANETARY SPACECRAFT
ROCKET ENGINES

INTERPLANETARY SPACE
UF   TRANSLUNAR SPACE
GS   ENVIRONMENTS
. AEROSPACE ENVIRONMENTS
. DEEP SPACE
. . **INTERPLANETARY SPACE**
. EXTRATERRESTRIAL ENVIRONMENTS
. DEEP SPACE
. . . **INTERPLANETARY SPACE**
RT   CISLUNAR SPACE
HELIOSPHERE
INTERSTELLAR SPACE
POLAR CUSPS

INTERPLANETARY SPACECRAFT
UF   INTERPLANETARY PROPULSION
PLANETARY SPACECRAFT
GS   **INTERPLANETARY SPACECRAFT**
. EXPLORER 18 SATELLITE
. JUPITER PROBES
. . GALILEO PROBE
. GALILEO SPACECRAFT
. MARINER SPACE PROBES
. . MARINER R 2 SPACE PROBE
. MARINER VENUS-MERCURY 1973
. . MARINER 1 SPACE PROBE
. . MARINER 2 SPACE PROBE
. . MARINER 3 SPACE PROBE
. . MARINER 4 SPACE PROBE
. . MARINER 5 SPACE PROBE
. . MARINER 6 SPACE PROBE
. . MARINER 7 SPACE PROBE
. . MARINER 8 SPACE PROBE
. . MARINER 9 SPACE PROBE
. . MARINER 10 SPACE PROBE
. . MARINER 11 SPACE PROBE
. MARINER SPACECRAFT
. . MARINER C SPACECRAFT
. . MARINER VENUS 67 SPACECRAFT
. MARS PROBES
. . ADVANCED RECONN ELECTRIC
    SPACECRAFT
. . MARINER 3 SPACE PROBE
. . MARINER 4 SPACE PROBE
. . MARINER 6 SPACE PROBE
. . MARINER 7 SPACE PROBE
. . MARINER 8 SPACE PROBE
. . MARINER 10 SPACE PROBE
. . MARS 1 SPACECRAFT
. . MARS 2 SPACECRAFT
. . MARS 3 SPACECRAFT
. . MARS 4 SPACECRAFT
. . MARS 5 SPACECRAFT
. . MARS 6 SPACECRAFT
. . MARS 7 SPACECRAFT
. . VIKING SPACECRAFT
. . . VIKING 1 SPACECRAFT
. . . . VIKING LANDER SPACECRAFT
. . . . . VIKING LANDER 1
. . . . VIKING ORBITER SPACECRAFT
. . . . . VIKING ORBITER 1
. . . VIKING 2 SPACECRAFT
. . . . VIKING LANDER SPACECRAFT
. . . . . VIKING LANDER 2
. . . . VIKING ORBITER SPACECRAFT
. . . . . VIKING ORBITER 2
. . VIKING 75 ENTRY VEHICLE

INTERPLANETARY SPACECRAFT-(CONT.)
```
. . ZOND 2 SPACE PROBE
. PIONEER SPACE PROBES
. . PIONEER VENUS 2 ENTRY PROBES
. . . PIONEER VENUS 2 NIGHT PROBE
. . . PIONEER VENUS 2 SOUNDER
        PROBE
. . PIONEER 1 SPACE PROBE
. . PIONEER 2 SPACE PROBE
. . PIONEER 3 SPACE PROBE
. . PIONEER 4 SPACE PROBE
. . PIONEER 5 SPACE PROBE
. . PIONEER 6 SPACE PROBE
. . PIONEER 7 SPACE PROBE
. . PIONEER 8 SPACE PROBE
. . PIONEER 9 SPACE PROBE
. . PIONEER 10 SPACE PROBE
. PIONEER VENUS SPACECRAFT
. . PIONEER VENUS 1 SPACECRAFT
. . PIONEER VENUS 2 SPACECRAFT
. . . PIONEER VENUS 2 TRANSPORTER
        BUS
. TOPS (SPACECRAFT)
. VENUS PROBES
. . MARINER 1 SPACE PROBE
. . MARINER 2 SPACE PROBE
. . MARINER 5 SPACE PROBE
. PIONEER VENUS 2 SPACECRAFT
. . . PIONEER VENUS 2 TRANSPORTER
        BUS
. . VENERA SATELLITES
. . . VENERA 2 SATELLITE
. . . VENERA 3 SATELLITE
. . . VENERA 8 SATELLITE
. . . VENERA 9 SATELLITE
. . . VENERA 10 SATELLITE
. . . VENERA 11 SATELLITE
. . . VENERA 12 SATELLITE
. . ZOND 1 SPACE PROBE
. . ZOND 3 SPACE PROBE
. . ZOND 4 SPACE PROBE
. . ZOND 5 SPACE PROBE
. . ZOND 6 SPACE PROBE
. . ZOND 7 SPACE PROBE
. . ZOND 8 SPACE PROBE
. VOYAGER 1 SPACECRAFT
. VOYAGER 2 SPACECRAFT
. ZOND SPACE PROBES
. . ZOND 1 SPACE PROBE
. . ZOND 2 SPACE PROBE
. . ZOND 3 SPACE PROBE
. . ZOND 4 SPACE PROBE
. . ZOND 5 SPACE PROBE
. . ZOND 6 SPACE PROBE
. . ZOND 7 SPACE PROBE
. . ZOND 8 SPACE PROBE
```
```
RT    ARTIFICIAL SATELLITES
      INTERSTELLAR SPACECRAFT
      LANDING MODULES
      MANEUVERABLE SPACECRAFT
      MANNED SPACECRAFT
      RENDEZVOUS SPACECRAFT
      REUSABLE SPACECRAFT
      SATELLITES
      SPACE CAPSULES
      SPACE EXPLORATION
      SPACE PROBES
   ∞ SPACECRAFT
      UNMANNED SPACECRAFT
      VOYAGER 1977 MISSION
```

INTERPLANETARY TRAJECTORIES
```
GS    TRAJECTORIES
      . SPACECRAFT TRAJECTORIES
      . . INTERPLANETARY TRAJECTORIES
      . . . EARTH-MARS TRAJECTORIES
      . . . EARTH-MERCURY TRAJECTORIES
RT    EARTH-MOON TRAJECTORIES
      EARTH-VENUS TRAJECTORIES
      GODDARD TRAJECTORY
        DETERMINATION SYSTEM
      INTERORBITAL TRAJECTORIES
      ORBITAL LAUNCHING
      ORBITAL MECHANICS
      PARKING ORBITS
      PLANETARY ORBITS
      RENDEZVOUS TRAJECTORIES
      ROUND TRIP TRAJECTORIES
      SOLAR ORBITS
      SPACE NAVIGATION
      SPACECRAFT GUIDANCE
      TRANSFER ORBITS
      VIKING LANDER SPACECRAFT
      VIKING LANDER 1
      VIKING LANDER 2
      VIKING ORBITER SPACECRAFT
```

INTERPLANETARY TRAJECTORIES-(CONT.)
```
      VIKING ORBITER 1
      VIKING ORBITER 2
      VIKING 1 SPACECRAFT
      VIKING 2 SPACECRAFT
```

INTERPLANETARY TRANSFER ORBITS
```
GS    ORBITS
      . ELLIPTICAL ORBITS
      . . TRANSFER ORBITS
      . . . INTERPLANETARY TRANSFER
            ORBITS
      . SPACECRAFT ORBITS
      . . TRANSFER ORBITS
      . . . INTERPLANETARY TRANSFER
            ORBITS
RT    AEROASSIST
      AEROBRAKING
      AEROCAPTURE
      AEROMANEUVERING
      ORBITAL MECHANICS
      SWINGBY TECHNIQUE
```

INTERPOLATION
```
GS    ANALYSIS (MATHEMATICS)
      . NUMERICAL ANALYSIS
      . . INTERPOLATION
RT    COMMUTATION
      COMPUTATION
      EXTRAPOLATION
      FINITE DIFFERENCE THEORY
      STATISTICAL ANALYSIS
```

INTERPOLATORS
```
USE   REPEATERS
```

∞ INTERPRETATION
```
SN    (USE OF A MORE SPECIFIC TERM IS
      RECOMMENDED--CONSULT THE TERMS
      LISTED BELOW)
RT    DECODING
      INTELLIGIBILITY
      PERCEPTION
      PHOTOINTERPRETATION
      READING
      RECOGNITION
      SYNTAX
      TRANSLATING
```

INTERPROCESSOR COMMUNICATION
```
RT    COMPUTER NETWORKS
      COMPUTER SYSTEMS DESIGN
      DINING PHILOSOPHERS PROBLEM
      MULTIPROCESSING (COMPUTERS)
      PARALLEL PROCESSING (COMPUTERS)
```

INTERRELATIONSHIPS
```
USE   RELATIONSHIPS
```

INTERROGATION
```
RT    DATA PROCESSING
      IFF SYSTEMS (IDENTIFICATION)
      SECONDARY RADAR
      TRANSMITTER RECEIVERS
      TRANSPONDERS
```

INTERRUPTION
```
RT    ELECTRIC RELAYS
      PACKET SWITCHING
      SEQUENCING
      SWITCHES
      SWITCHING
```

INTERSECTIONS
```
SN    (EXCLUDES BOOLEAN LOGICAL
      PRODUCTS)
RT    CROSSINGS
      CROSSOVERS
      HIGHWAYS
   ∞ JUNCTIONS
      RAMPS (STRUCTURES)
      ROADS
      STREETS
      TRANSPORTATION NETWORKS
```

INTERSERVICE DATA EXCHANGE PROGRAM
```
UF    IDEP (DATA EXCHANGE)
RT    ∞ DATA
      DATA RETRIEVAL
      DATA STORAGE
      INFORMATION RETRIEVAL
      LIBRARIES
      MILITARY TECHNOLOGY
      RESEARCH
```

INTERSTELLAR CHEMISTRY
```
GS    PARTICLE INTERACTIONS
      . MOLECULAR INTERACTIONS
      . . INTERSTELLAR CHEMISTRY
RT    ASSOCIATION REACTIONS
      CHEMICAL REACTIONS
   ∞ CHEMISTRY
      FORMYL IONS
      MOLECULAR CLOUDS
      REACTION KINETICS
```

INTERSTELLAR COMMUNICATION
```
GS    COMMUNICATING
      . INTERSTELLAR COMMUNICATION
RT    EXTRATERRESTRIAL INTELLIGENCE
      RADIO COMMUNICATION
      SPACE COMMUNICATION
```

INTERSTELLAR EXTINCTION
```
UF    INTERSTELLAR REDDENING
GS    EXTINCTION
      . INTERSTELLAR EXTINCTION
RT    ASTROPHYSICS
      EVOLUTION (DEVELOPMENT)
      INTERSTELLAR GAS
      RADIATION ABSORPTION
      STELLAR EVOLUTION
      STELLAR RADIATION
```

INTERSTELLAR GAS
```
GS    EXTRATERRESTRIAL MATTER
      . COSMIC GASES
      . . INTERSTELLAR GAS
      . . . NEUTRAL GASES
      GASES
      . RAREFIED GASES
      . . COSMIC GASES
      . . . INTERSTELLAR GAS
      . . . . NEUTRAL GASES
RT    HELIOSPHERE
      INTERPLANETARY GAS
      INTERSTELLAR EXTINCTION
      MOLECULAR CLOUDS
      OPHIUCHI CLOUDS
      ORION NEBULA
      SPIN TEMPERATURE
      STELLAR MASS ACCRETION
      STELLAR WINDS
```

INTERSTELLAR MAGNETIC FIELDS
```
UF    GALACTIC MAGNETIC FIELDS
GS    MAGNETIC FIELDS
      . INTERSTELLAR MAGNETIC FIELDS
RT    STELLAR MAGNETIC FIELDS
```

INTERSTELLAR MASERS
```
GS    STIMULATED EMISSION DEVICES
      . MASERS
      . . INTERSTELLAR MASERS
RT    COHERENT ELECTROMAGNETIC
        RADIATION
      GAS MASERS
      LASERS
      MICROWAVE AMPLIFIERS
      MOLECULAR CLOUDS
      RADIATION SOURCES
      STIMULATED EMISSION
      WATER MASERS
```

INTERSTELLAR MATTER
```
RT    CELESTIAL BODIES
      COSMIC DUST
      FORMYL IONS
      MASS DISTRIBUTION
      METALLICITY
      MOLECULAR CLOUDS
      NEBULAE
      OPHIUCHI CLOUDS
      ORION NEBULA
      REFLECTION NEBULAE
      SPIN TEMPERATURE
      STELLAR MASS ACCRETION
```

INTERSTELLAR MICROWAVE SPECTRA
```
USE   INTERSTELLAR RADIATION
      MICROWAVE SPECTRA
```

INTERSTELLAR RADIATION
```
UF    INTERSTELLAR MICROWAVE SPECTRA
GS    EXTRATERRESTRIAL RADIATION
      . INTERSTELLAR RADIATION
RT    CORPUSCULAR RADIATION
      COSMIC NOISE
      COSMIC RAYS
      ELECTROMAGNETIC RADIATION
```

INTERSTELLAR RADIATION-(CONT.)
  GALACTIC RADIATION
  GAMMA RAY BURSTS
  ∞RADIATION
  RADIATIVE TRANSFER
  STELLAR RADIATION

INTERSTELLAR REDDENING
 USE INTERSTELLAR EXTINCTION

INTERSTELLAR SPACE
 GS ENVIRONMENTS
  . AEROSPACE ENVIRONMENTS
  . . DEEP SPACE
  . . . **INTERSTELLAR SPACE**
  . EXTRATERRESTRIAL ENVIRONMENTS
  . . DEEP SPACE
  . . . **INTERSTELLAR SPACE**
 RT INTERPLANETARY SPACE

INTERSTELLAR SPACECRAFT
 RT INTERPLANETARY FLIGHT
  INTERPLANETARY SPACECRAFT
  INTERSTELLAR TRAVEL
  SATELLITES
  SPACE EXPLORATION

INTERSTELLAR TRAVEL
 GS SPACE FLIGHT
  . **INTERSTELLAR TRAVEL**
 RT ASTRONAVIGATION
  CELESTIAL REFERENCE SYSTEMS
  EXTRATERRESTRIAL INTELLIGENCE
  INTERSTELLAR SPACECRAFT
  LONG DURATION SPACE FLIGHT
  MANNED SPACE FLIGHT

INTERSTICES
 RT CAVITIES
  CRACKS
  GRAIN BOUNDARIES
  PERCOLATION
  PERMEABILITY
  PINHOLES
  POROSITY
  POROUS MATERIALS
  VOIDS

INTERSTITIALS
 RT ADDITIVES
  CRYSTAL DEFECTS
  CRYSTAL STRUCTURE
  GRAIN BOUNDARIES

INTERSYMBOLIC INTERFERENCE
 RT DATA TRANSMISSION
  ∞INTERFERENCE
  SIGNAL DISTORTION
  TRANSMISSION EFFICIENCY

INTERTROPICAL CONVERGENT ZONES
 GS REGIONS
  . **INTERTROPICAL CONVERGENT ZONES**
 RT ATMOSPHERIC CIRCULATION
  FRONTS (METEOROLOGY)
  GARP ATLANTIC TROPICAL EXPERIMENT
  TROPICAL METEOROLOGY
  TROPICAL REGIONS

INTERVALS
 RT ALTERNATIONS
  CONSECUTIVE EVENTS
  SPACING
  STEP FUNCTIONS
  TIME
  TOPOLOGY

INTERVEHICLE SPACECREW TRANSFER
 USE SPACECREW TRANSFER

INTERVERTEBRAL DISKS
 GS DISKS (SHAPES)
  . **INTERVERTEBRAL DISKS**
 RT ∞DISKS
  MUSCULOSKELETAL SYSTEM
  VERTEBRAE

INTESTINES
 GS ANATOMY
  . DIGESTIVE SYSTEM
  . . GASTROINTESTINAL SYSTEM
  . . . **INTESTINES**
  . . . . RECTUM
  VISCERA

INTESTINES-(CONT.)
  . . **INTESTINES**
  . RECTUM
 RT ABDOMEN
  APPENDIX (ANATOMY)
  COLIC

INTOXICATION
 RT ∞POISONING
  TOXICITY AND SAFETY HAZARD
  TOXICOLOGY

INTRACRANIAL CAVITY
 GS ANATOMY
  . HEAD (ANATOMY)
  . . SKULL
  . . . CRANIUM
  . . . . **INTRACRANIAL CAVITY**
  . MUSCULOSKELETAL SYSTEM
  . . BONES
  . . . SKULL
  . . . . CRANIUM
  . . . . . **INTRACRANIAL CAVITY**

INTRACRANIAL PRESSURE
 GS PRESSURE
  . **INTRACRANIAL PRESSURE**
 RT BRAIN

INTRAMOLECULAR STRUCTURES
 RT MOLECULAR STRUCTURE
  ∞STRUCTURES

INTRAOCULAR PRESSURE
 UF TONOMETRY
 GS PRESSURE
  . **INTRAOCULAR PRESSURE**
 RT GLAUCOMA

INTRAORBIT TRANSFER VEHICLES
 RT LARGE SPACE STRUCTURES
  SPACE PLATFORMS
  SPACE SHUTTLES
  ∞VEHICLES

INTRAVASCULAR SYSTEM
 GS CIRCULATION
  . BLOOD CIRCULATION
  . . **INTRAVASCULAR SYSTEM**
 RT ∞SYSTEMS

INTRAVEHICULAR ACTIVITY
 RT ∞ACTIVITY
  ASTRONAUT LOCOMOTION
  ASTRONAUT MANEUVERING EQUIPMENT
  ASTRONAUT PERFORMANCE
  EXTRAVEHICULAR ACTIVITY
  HUMAN PERFORMANCE
  MANNED SPACE FLIGHT
  PILOT PERFORMANCE
  SPACECRAFT ENVIRONMENTS
  WEIGHTLESSNESS

INTRAVENOUS PROCEDURES
 RT CATHETERIZATION
  MEDICAL SERVICES

INTROVERSION
 RT ∞DEPRESSION
  DETACHMENT
  HUMAN BEHAVIOR
  ∞INHIBITION
  PSYCHOLOGY

INTRUDER AIRCRAFT
 USE A-6 AIRCRAFT

INTRUSION
 RT CONTAMINATION
  EXTRUDING
  LEAKAGE
  SEEPAGE

INVADER AIRCRAFT
 USE B-26 AIRCRAFT

INVALIDITY
 USE ERRORS

INVARIANCE
 GS **INVARIANCE**
  . GAUGE INVARIANCE
 RT ∞CONSTANT
  LORENTZ TRANSFORMATIONS

INVARIANT IMBEDDINGS
 GS GEOMETRY
  . TOPOLOGY
  . . IMBEDDINGS (MATHEMATICS)
  . . . **INVARIANT IMBEDDINGS**
 RT ANISOTROPIC FLUIDS
  CALCULUS OF VARIATIONS
  CONFORMAL MAPPING
  COORDINATE TRANSFORMATIONS
  DIFFERENTIAL GEOMETRY
  ∞IMBEDDINGS
  ISOTROPIC TURBULENCE

INVENTIONS
 RT PATENT APPLICATIONS
  PATENT POLICY
  PATENTS
  PRODUCT DEVELOPMENT

INVENTORIES
 GS **INVENTORIES**
  . CROP INVENTORIES
  . TIMBER INVENTORY
 RT LARGE AREA CROP INVENTORY
   EXPERIMENT
  RESERVES
  ∞STORAGE

INVENTORY CONTROLS
 GS MANAGEMENT
  . INDUSTRIAL MANAGEMENT
  . . INVENTORY MANAGEMENT
  . . . **INVENTORY CONTROLS**
  . LOGISTICS MANAGEMENT
  . . INVENTORY MANAGEMENT
  . . . **INVENTORY CONTROLS**
 RT ∞CONTROL
  DISTRIBUTING
  MATHEMATICAL MODELS
  OPTIMAL CONTROL
  RESERVES
  RISK
  ∞STORAGE
  TIME LAG

INVENTORY MANAGEMENT
 GS MANAGEMENT
  . INDUSTRIAL MANAGEMENT
  . . **INVENTORY MANAGEMENT**
  . . . INVENTORY CONTROLS
  . LOGISTICS MANAGEMENT
  . . **INVENTORY MANAGEMENT**
  . . . INVENTORY CONTROLS
 RT DOWNTIME
  LOGISTICS
  PROCUREMENT MANAGEMENT
  RESOURCES
  RETIREMENT FOR CAUSE
  SERVICES
  SPARE PARTS
  STOCKPILING
  ∞STORAGE

INVERSE SCATTERING
 GS SCATTERING
  . **INVERSE SCATTERING**
 RT FORWARD SCATTERING
  RESONANCE SCATTERING

INVERSIONS
 GS **INVERSIONS**
  . CENTRIFUGING STRESS
  . MAGNETIC FIELD INVERSIONS
  . POPULATION INVERSION
  . TEMPERATURE INVERSIONS

INVERTEBRATES
 GS ANIMALS
  . **INVERTEBRATES**
  . . ARTHROPODS
  . . . ARTEMIA
  . . . CEPHALOPODS
  . . . . MOLLUSKS
  . . . . OCTOPUSES
  . . . . SNAILS
  . . . CRABS
  . . . INSECTS
  . . . . BEES
  . . . . BOLL WEEVILS
  . . . . CHIRONOMUS FLIES
  . . . . COCKROACHES
  . . . . COLEOPTERA
  . . . . CRICKETS
  . . . . . BEETLES
  . . . . . . TRIBOLIA

INVERTEBRATES-(CONT.)
```
    . . . . DROSOPHILA
    . . . . FIREFLIES
    . . . . GRASSHOPPERS
    . . . . LARVAE
    . . . . . BOLLWORMS
    . . . . . SILKWORMS
    . . . . LOCUSTS
    . . . . MOTHS
    . . . . . SILKWORMS
    . . . PUPA
    . . SPIDERS
    . . PROTOZOA
    . . . FLAGELLATA
    . . . . TRYPANOSOME
    . . . PARAMECIA
    . . . ROTIFERA
    . . PELOMYXA
    . . SEA URCHINS
    . . SPORES
    . . . MICROSPORES
    . . WORMS
    . . . FLATWORMS
```
RT   BACTERIA
     HEMOCYTES
     MICROORGANISMS

INVERTED CONVERTERS (DC TO AC)
RT   ALTERNATING CURRENT
     ∞ CONVERTERS
     CURRENT CONVERTERS (AC TO DC)
     DIRECT CURRENT
     ELECTRIC CURRENT

INVERTERS
SN   (EXCLUDES AC TO DC INVERTERS)
GS   **INVERTERS**
     . STATIC INVERTERS
RT   ATTENUATORS
     OSCILLATORS

INVESTIGATION
UF   STUDIES
GS   **INVESTIGATION**
     . ACCIDENT INVESTIGATION
     . . AIRCRAFT ACCIDENT INVESTIGATION
     . INTERNATIONAL MAGNETOSPHERIC
       STUDY
RT   EXAMINATION
     EXPERIMENTATION
     EXPLORATION
     GEOPHYSICAL FLUID FLOW CELLS
     OSS-1 PAYLOAD
     PROGRAMS
     RESEARCH
     RESEARCH AND DEVELOPMENT
     SAMPLING
     UNIVERSITY PROGRAM

∞ INVESTMENT
SN   (USE OF A MORE SPECIFIC TERM IS
     RECOMMENDED--CONSULT THE TERMS
     LISTED BELOW)
RT   COMMERCE
     INVESTMENT CASTING
     INVESTMENTS

INVESTMENT CASTING
UF   LOST WAX PROCESS
GS   FORMING TECHNIQUES
     . CASTING
     . . **INVESTMENT CASTING**
RT   CENTRIFUGAL CASTING
     ∞ INVESTMENT

INVESTMENTS
RT   DEPRECIATION
     ECONOMIC IMPACT
     ECONOMICS
     FINANCE
     ∞ INVESTMENT

INVISCID FLOW
UF   NONVISCOUS FLOW
GS   FLUID FLOW
     . **INVISCID FLOW**
     . . STAGNATION FLOW
RT   AERODYNAMICS
     CROCCO METHOD
     CROCCO-LEE THEORY
     ∞ FLOW
     FLOW CHARACTERISTICS
     GAS FLOW
     LAMINAR FLOW
     POTENTIAL FLOW

INVISCID FLOW-(CONT.)
     PRANDTL NUMBER
     REYNOLDS NUMBER
     STAGNATION TEMPERATURE
     TURBULENT FLOW
     VISCOUS FLOW

INVISIBILITY
USE   VISIBILITY

INVOLUNTARINESS
USE   INVOLUNTARY ACTIONS

INVOLUNTARY ACTIONS
UF   INVOLUNTARINESS
RT   AUTONOMIC NERVOUS SYSTEM
     SNEEZING
     SPASMS
     TWITCHING

IO
GS   CELESTIAL BODIES
     . NATURAL SATELLITES
     . . JUPITER SATELLITES
     . . . GALILEAN SATELLITES
     . . . . **IO**
     SATELLITES
     . NATURAL SATELLITES
     . . JUPITER SATELLITES
     . . . GALILEAN SATELLITES
     . . . . **IO**
RT   CALLISTO
     CHARON
     GANYMEDE
     JUPITER (PLANET)

IODATES
GS   HALOGEN COMPOUNDS
     . IODINE COMPOUNDS
     . . **IODATES**
     . . . LITHIUM IODATES

IODIDES
GS   HALOGEN COMPOUNDS
     . IODINE COMPOUNDS
     . . **IODIDES**
     . . . CESIUM IODIDES
     . . . GALLAMINE TRIETHIODIDE
     . . . HAFNIUM IODIDES
     . . NIOBIUM IODIDES
     . . . POTASSIUM IODIDES
     . . . SILVER IODIDES
     . . . SODIUM IODIDES
     . . . ZIRCONIUM IODIDES

IODIMETRY
GS   CHEMICAL TESTS
     . CHEMICAL ANALYSIS
     . . **IODIMETRY**
RT   QUANTITATIVE ANALYSIS
     ∞ REDUCTION
     TITRATION

IODINE
GS   CHEMICAL ELEMENTS
     . HALOGENS
     . . **IODINE**
     . . . IODINE ISOTOPES
     . . . . IODINE 125
     . . . . IODINE 131
     . . . . IODINE 132

IODINE COMPOUNDS
GS   HALOGEN COMPOUNDS
     . **IODINE COMPOUNDS**
     . . IODATES
     . . . LITHIUM IODATES
     . . IODIDES
     . . . CESIUM IODIDES
     . . . GALLAMINE TRIETHIODIDE
     . . . HAFNIUM IODIDES
     . . . NIOBIUM IODIDES
     . . . POTASSIUM IODIDES
     . . . SILVER IODIDES
     . . . SODIUM IODIDES
     . . . ZIRCONIUM IODIDES
     . . IODOACETIC ACID
RT   ∞ CHEMICAL COMPOUNDS
     HALOCARBONS

IODINE ISOTOPES
GS   CHEMICAL ELEMENTS
     . HALOGENS
     . . IODINE

IODINE ISOTOPES-(CONT.)
     . . . **IODINE ISOTOPES**
     . . . . IODINE 125
     . . . . IODINE 131
     . . . . IODINE 132
     . NUCLIDES
     . . ISOTOPES
     . . . **IODINE ISOTOPES**
     . . . . IODINE 125
     . . . . IODINE 131
     . . . . IODINE 132

IODINE LASERS
GS   STIMULATED EMISSION DEVICES
     . LASERS
     . . **IODINE LASERS**

IODINE 125
GS   CHEMICAL ELEMENTS
     . HALOGENS
     . . IODINE
     . . . IODINE ISOTOPES
     . . . . **IODINE 125**
     . NUCLIDES
     . . ISOTOPES
     . . . IODINE ISOTOPES
     . . . . **IODINE 125**
     . . . RADIOACTIVE ISOTOPES
     . . . . **IODINE 125**

IODINE 131
GS   CHEMICAL ELEMENTS
     . HALOGENS
     . . IODINE
     . . . IODINE ISOTOPES
     . . . . **IODINE 131**
     . NUCLIDES
     . . ISOTOPES
     . . . IODINE ISOTOPES
     . . . . **IODINE 131**
     . . . RADIOACTIVE ISOTOPES
     . . . . **IODINE 131**

IODINE 132
GS   CHEMICAL ELEMENTS
     . HALOGENS
     . . IODINE
     . . . IODINE ISOTOPES
     . . . . **IODINE 132**
     . NUCLIDES
     . . ISOTOPES
     . . . IODINE ISOTOPES
     . . . . **IODINE 132**
     . . . RADIOACTIVE ISOTOPES
     . . . . **IODINE 132**

IODOACETIC ACID
GS   ACIDS
     . FATTY ACIDS
     . . CARBOXYLIC ACIDS
     . . . ACETIC ACID
     . . . . **IODOACETIC ACID**
     ALIPHATIC COMPOUNDS
     . FATTY ACIDS
     . . ACETIC ACID
     . . . **IODOACETIC ACID**
     HALOGEN COMPOUNDS
     . IODINE COMPOUNDS
     . . **IODOACETIC ACID**
     LIPIDS
     . **IODOACETIC ACID**
     ORGANIC COMPOUNDS
     . FATTY ACIDS
     . . ACETIC ACID
     . . . **IODOACETIC ACID**

ION ACCELERATORS
GS   PARTICLE ACCELERATORS
     . **ION ACCELERATORS**
RT   ∞ ACCELERATORS
     SYNCHROTRONS

ION ACOUSTIC WAVES
GS   ELASTIC WAVES
     . SOUND WAVES
     . . **ION ACOUSTIC WAVES**
RT   PLASMA OSCILLATIONS
     PLASMA WAVES
     WAVE PROPAGATION

ION ATOM INTERACTIONS
GS   PARTICLE INTERACTIONS
     . **ION ATOM INTERACTIONS**
RT   ATOMIC INTERACTIONS
     CHARGE EXCHANGE

ION ATOM INTERACTIONS-*(CONT.)*
    ELEMENTARY PARTICLE INTERACTIONS
    ∞ INTERACTIONS

ION BEAMS
  GS   BEAMS (RADIATION)
     . PARTICLE BEAMS
     .. **ION BEAMS**
     ION CURRENTS
     . **ION BEAMS**
  RT   ATOMIC BEAMS
     BEAM INJECTION
     BEAM NEUTRALIZATION
     INERTIAL FUSION (REACTOR)
     MOLECULAR BEAMS

ION CHAMBERS
  USE   IONIZATION CHAMBERS

ION CHARGE
  GS   ELECTRIC CHARGE
     . **ION CHARGE**
  RT   CHARGE EXCHANGE
     CHARGED PARTICLES
     IONIZATION
     VALENCE

ION CONCENTRATION
  RT   ACIDITY
     ∞ BASES
     IONOSPHERE
     PH FACTOR
     TITRATION

ION CURRENTS
  UF   IONIC CONDUCTIVITY
  GS   **ION CURRENTS**
     . ION BEAMS
  RT   SOLIONS

ION CYCLOTRON RADIATION
  GS   PARTICLES
     . CORPUSCULAR RADIATION
     .. CYCLOTRON RADIATION
     ... **ION CYCLOTRON RADIATION**
  RT   CYCLOTRON RESONANCE
     IONIC WAVES
     MAGNETIC PUMPING
     PLASMA RADIATION
     PLASMA WAVES
     ∞ RADIATION

ION DENSITY (CONCENTRATION)
  GS   DENSITY (NUMBER/VOLUME)
     . PARTICLE DENSITY (CONCENTRATION)
     .. **ION DENSITY (CONCENTRATION)**
     ... IONOSPHERIC ION DENSITY
     ... MAGNETOSPHERIC ION DENSITY
     .... MAGNETOSPHERIC PROTON
          DENSITY
     ... PROTON DENSITY
          (CONCENTRATION)
     .... MAGNETOSPHERIC PROTON
          DENSITY
  RT   ATMOSPHERIC DENSITY
     ATOM CONCENTRATION
     COSMIC RAYS
     ELECTRON DENSITY (CONCENTRATION)
     GERDIEN CONDENSERS
     IONIZATION
     IONOGRAMS
     IONOSPHERE
     PLASMA DENSITY
     POSITIVE IONS
     SAHA EQUATIONS
     SPACE DENSITY

ION DISTRIBUTION
  GS   DISTRIBUTION (PROPERTY)
     . **ION DISTRIBUTION**
  RT   CHARGE DISTRIBUTION
     CURRENT DISTRIBUTION
     ELECTROHYDRODYNAMICS
     IONIC MOBILITY
     SPATIAL DISTRIBUTION
     VERTICAL DISTRIBUTION

ION EMISSION
  GS   DECAY
     . EMISSION
     .. PARTICLE EMISSION
     ... **ION EMISSION**
  RT   IONIZATION
     THERMIONIC EMISSION

ION EMISSION-*(CONT.)*
    THERMIONICS

ION ENGINES
  UF   IONIC PROPELLANTS
     THERMIONIC REACTORS
  GS   ENGINES
     . ROCKET ENGINES
     .. ELECTRIC ROCKET ENGINES
     ... **ION ENGINES**
     .... CESIUM ENGINES
     .... MERCURY ION ENGINES
     .... RIT ENGINES
  RT   ARC JET ENGINES
     BEAM SWITCHING
     ELECTROSTATIC ENGINES
     ELECTROTHERMAL ENGINES
     IONIZERS
     NUCLEAR ROCKET ENGINES
     PLASMA ENGINES
     RESTARTABLE ROCKET ENGINES
     SUSTAINER ROCKET ENGINES
     ∞ THRUSTORS

ION EXCHANGE MEMBRANE ELECTROLYTES
  GS   CONDUCTORS
     . ELECTROLYTES
     .. **ION EXCHANGE MEMBRANE
          ELECTROLYTES**
     MEMBRANES
     . **ION EXCHANGE MEMBRANE
          ELECTROLYTES**
  RT   FUEL CELLS
     SEPARATORS

ION EXCHANGE RESINS
  GS   RESINS
     . **ION EXCHANGE RESINS**
  RT   PLASTICS
     ZEOLITES

ION EXCHANGING
  GS   EXCHANGING
     . **ION EXCHANGING**
  RT   BEDS (PROCESS ENGINEERING)
     CHARGE TRANSFER
     DEMINERALIZING
     GLASS ELECTRODES
     HYDROMETALLURGY
     ISOTOPE SEPARATION
     KAOLINITE
     METATHESIS
     ∞ SEPARATION
     SOFTENING
     WATER TREATMENT

ION EXTRACTION
  GS   EXTRACTION
     . **ION EXTRACTION**
  RT   ISOTOPE SEPARATION
     ∞ SEPARATION
     SOLVENT EXTRACTION

ION GAGES
  USE   IONIZATION GAGES

ION IMPACT
  GS   IMPACT
     . **ION IMPACT**
  RT   ELECTRON IMPACT
     POINT IMPACT
     RECOIL IONS
     TOROIDS
     TOWNSEND AVALANCHE
     TOWNSEND DISCHARGE

ION IMPLANTATION
  GS   IMPLANTATION
     . **ION IMPLANTATION**
  RT   AVALANCHE DIODES
     CARRIER MOBILITY
     DIODES
     FIELD EFFECT TRANSISTORS
     INTEGRATED CIRCUITS
     IONS
     JUNCTION TRANSISTORS
     METAL IONS
     METAL OXIDE SEMICONDUCTORS
     MICROELECTRONICS
     MOM (SEMICONDUCTORS)
     PHOTODIODES
     SEMICONDUCTOR DEVICES
     TRANSISTORS

ION INJECTION
  GS   INJECTION
     . **ION INJECTION**
  RT   CARRIER INJECTION
     IONIC MOBILITY
     PLASMA ACCELERATORS
     PLASMA GENERATORS
     PLASMA JETS

ION IRRADIATION
  GS   IRRADIATION
     . **ION IRRADIATION**
     .. DEUTERON IRRADIATION
     .. PROTON IRRADIATION
  RT   AURORAL IRRADIATION
     ELECTRON IRRADIATION
     IONIC COLLISIONS
     NEUTRON IRRADIATION

ION MICROSCOPES
  GS   MICROSCOPES
     . **ION MICROSCOPES**
  RT   ELECTRON MICROSCOPES
     ELECTRON MICROSCOPY

ION MOTION
  RT   IONIC WAVES
     ∞ MOTION
     PENNING DISCHARGE
     PLASMA COMPOSITION
     PLASMA DIFFUSION

ION OSCILLATION
  USE   PLASMA OSCILLATIONS

ION PLATING
  GS   PLATING
     . **ION PLATING**
  RT   IONS
     METAL COATINGS
     METAL IONS
     SPUTTERING
     THIN FILMS
     VACUUM DEPOSITION

ION PROBES
  GS   MEASURING INSTRUMENTS
     . **ION PROBES**
  RT   IONOSONDES
     RADIO FREQUENCY IMPEDANCE
         PROBES

ION PRODUCTION RATES
  GS   IONIZATION
     . **ION PRODUCTION RATES**
     RATES (PER TIME)
     . **ION PRODUCTION RATES**
  RT   AVALANCHES
     CHARGE EXCHANGE
     RECOIL IONS
     THERMIONIC CONVERTERS

ION PROPULSION
  GS   PROPULSION
     . ELECTRIC PROPULSION
     .. ELECTROSTATIC PROPULSION
     ... **ION PROPULSION**
     . LOW THRUST PROPULSION
     .. ELECTROSTATIC PROPULSION
     ... **ION PROPULSION**
     . SPACECRAFT PROPULSION
     .. ELECTROSTATIC PROPULSION
     ... **ION PROPULSION**
  RT   DUOPLASMATRONS
     ELECTROMAGNETIC PROPULSION
     HIGH TEMPERATURE PROPELLANTS
     NUCLEAR ELECTRIC PROPULSION
     PLASMA PROPULSION

ION PUMPS
  GS   PUMPS
     . VACUUM PUMPS
     .. **ION PUMPS**
     VACUUM APPARATUS
     . VACUUM PUMPS
     .. **ION PUMPS**
  RT   GETTERS

ION RECOMBINATION
  GS   CHEMICAL REACTIONS
     . **ION RECOMBINATION**
     RECOMBINATION REACTIONS
     . **ION RECOMBINATION**
  RT   ATOMIC RECOMBINATION

ION RECOMBINATION-*(CONT.)*
　　　DEIONIZATION
　　　ELECTRON RECOMBINATION
　　　ELECTRON-ION RECOMBINATION
　　　RECOMBINATION COEFFICIENT

ION SCATTERING
　GS　SCATTERING
　　　. **ION SCATTERING**
　RT　ELECTRON SCATTERING
　　　IONIC COLLISIONS
　　　IONIC DIFFUSION
　　　PROTON SCATTERING
　　　RECOIL IONS

ION SELECTIVE ELECTRODES
　GS　ELECTRODES
　　　. **ION SELECTIVE ELECTRODES**
　RT　CHEMICAL ANALYSIS

ION SHEATHS
　GS　SHEATHS
　　　. **ION SHEATHS**
　RT　PLASMA CLOUDS
　　　PLASMA PROBES
　　　PLASMA SHEATHS

ION SOURCES
　GS　**ION SOURCES**
　　　. PLASMATRONS
　　　. . DUOPLASMATRONS
　RT　ELECTRON SOURCES
　　　IONIZATION
　　　IONIZING RADIATION
　　　LINEAR ACCELERATORS
　　　PARTICLE ACCELERATORS
　　　PLASMA GENERATORS
　　　RADIATION SOURCES
　　　∞ SOURCES
　　　SPUTTERING

ION SPECTROMETERS
　USE　MASS SPECTROMETERS

ION STORAGE
　RT　FREQUENCY STANDARDS
　　　∞ STORAGE
　　　TRAPPING

ION STRIPPING
　RT　HEAVY IONS
　　　PARTICLE BEAMS
　　　PARTICLE DENSITY (CONCENTRATION)
　　　PARTICLES
　　　∞ SEPARATION
　　　∞ STRIPPING

ION TEMPERATURE
　GS　TEMPERATURE
　　　. **ION TEMPERATURE**
　RT　AURORAL TEMPERATURE
　　　IONOSPHERIC TEMPERATURE
　　　PLASMA TEMPERATURE
　　　SPACE TEMPERATURE
　　　SPECIFIC HEAT

ION TRAPS (INSTRUMENTATION)
　GS　MEASURING INSTRUMENTS
　　　. **ION TRAPS (INSTRUMENTATION)**
　　　TRAPS
　　　. **ION TRAPS (INSTRUMENTATION)**
　RT　RADIATION COUNTERS
　　　VAPOR TRAPS

ION-GAS INTERACTIONS
　USE　GAS-ION INTERACTIONS

IONIC COLLISIONS
　GS　COLLISIONS
　　　. **IONIC COLLISIONS**
　RT　ATOMIC COLLISIONS
　　　ION IRRADIATION
　　　ION SCATTERING
　　　RECOIL IONS

IONIC CONDUCTIVITY
　USE　ION CURRENTS

IONIC CRYSTALS
　GS　CRYSTALS
　　　. **IONIC CRYSTALS**
　RT　CHEMICAL BONDS
　　　CRYSTAL LATTICES
　　　EXCITONS

IONIC CRYSTALS-*(CONT.)*
　　　POLARONS

IONIC DIFFUSION
　GS　DIFFUSION
　　　. PARTICLE DIFFUSION
　　　. . **IONIC DIFFUSION**
　RT　AMBIPOLAR DIFFUSION
　　　DIFFUSION WAVES
　　　ELECTRON DIFFUSION
　　　ION SCATTERING
　　　PLASMA DIFFUSION
　　　SELF DIFFUSION (SOLID STATE)

IONIC MOBILITY
　GS　MOBILITY
　　　. **IONIC MOBILITY**
　　　TRANSPORT PROPERTIES
　　　. **IONIC MOBILITY**
　RT　AMBIPOLAR DIFFUSION
　　　ANIONS
　　　ATOMIC MOBILITIES
　　　CATIONS
　　　ELECTROHYDRODYNAMICS
　　　ELECTROLYSIS
　　　ELECTROLYTIC CELLS
　　　ELECTROMIGRATION
　　　ION DISTRIBUTION
　　　ION INJECTION
　　　IONS
　　　∞ MOTION
　　　NDM SEMICONDUCTOR DEVICES
　　　NEGATIVE IONS
　　　POSITIVE IONS

IONIC PROPELLANTS
　USE　ION ENGINES

IONIC REACTIONS
　RT　CHARGE TRANSFER
　　　MOLECULAR INTERACTIONS

IONIC WAVES
　GS　ELASTIC WAVES
　　　. **IONIC WAVES**
　RT　COLLISIONLESS PLASMAS
　　　ELECTROSTATIC WAVES
　　　ION CYCLOTRON RADIATION
　　　ION MOTION
　　　IONOSPHERIC CONDUCTIVITY
　　　IONOSPHERIC PROPAGATION
　　　PLASMA WAVES
　　　∞ WAVES

IONIZATION
　UF　ELECTRON IONIZATION
　GS　**IONIZATION**
　　　. AUTOIONIZATION
　　　. GAS IONIZATION
　　　. . ATMOSPHERIC IONIZATION
　　　. . . AURORAL IONIZATION
　　　. . FLAME IONIZATION
　　　. ION PRODUCTION RATES
　　　. NONEQUILIBRIUM IONIZATION
　　　. PHOTOIONIZATION
　　　. SURFACE IONIZATION
　RT　ATOMIC COLLISIONS
　　　ATOMIC EXCITATIONS
　　　CORONAS
　　　DISINTEGRATION
　　　DISSOCIATION
　　　ELECTRIC ARCS
　　　ELECTRIC CORONA
　　　ELECTRIC DISCHARGES
　　　ELECTRIC SPARKS
　　　ELECTRON ATTACHMENT
　　　EXCITATION
　　　ION CHARGE
　　　ION DENSITY (CONCENTRATION)
　　　ION EMISSION
　　　ION SOURCES
　　　IONOSPHERIC COMPOSITION
　　　MAGNETOHYDRODYNAMICS
　　　MOLECULAR EXCITATION
　　　OXYGEN RECOMBINATION
　　　SCHWARZSCHILD METRIC
　　　SINGLE EVENT UPSETS
　　　STELLAR CORONAS
　　　THERMAL DISSOCIATION

IONIZATION CHAMBERS
　UF　ION CHAMBERS
　　　IONIZATION COUNTERS
　GS　**IONIZATION CHAMBERS**
　　　. BUBBLE CHAMBERS

IONIZATION CHAMBERS-*(CONT.)*
　　　. CLOUD CHAMBERS
　　　. GEIGER COUNTERS
　　　. PROPORTIONAL COUNTERS
　　　. SPARK CHAMBERS
　RT　∞ CHAMBERS
　　　COUNTERS
　　　DOSIMETERS
　　　ELECTRON COUNTERS
　　　IONIZERS
　　　NEUTRON COUNTERS
　　　RADIATION COUNTERS
　　　RADIATION MEASURING INSTRUMENTS
　　　THRESHOLD DETECTORS (DOSIMETERS)

IONIZATION COEFFICIENTS
　GS　COEFFICIENTS
　　　. **IONIZATION COEFFICIENTS**

IONIZATION COUNTERS
　USE　IONIZATION CHAMBERS
　　　RADIATION COUNTERS

IONIZATION CROSS SECTIONS
　RT　ABSORPTION CROSS SECTIONS
　　　∞ CROSS SECTIONS
　　　NONADIABATIC THEORY
　　　SCATTERING CROSS SECTIONS

IONIZATION FREQUENCIES
　GS　FREQUENCIES
　　　. **IONIZATION FREQUENCIES**

IONIZATION GAGES
　UF　ION GAGES
　GS　MEASURING INSTRUMENTS
　　　. PRESSURE GAGES
　　　. . VACUUM GAGES
　　　. . . **IONIZATION GAGES**
　　　. . . . ALPHATRONS
　　　. . . . BAYARD-ALPERT IONIZATION
　　　　　　　GAGES
　　　. . . . PENNING GAGES
　　　. . . . PHILIPS IONIZATION GAGES
　　　VACUUM APPARATUS
　　　. VACUUM GAGES
　　　. . **IONIZATION GAGES**
　　　. . . ALPHATRONS
　　　. . . BAYARD-ALPERT IONIZATION
　　　　　　GAGES
　　　. . . PENNING GAGES
　　　. . . PHILIPS IONIZATION GAGES
　RT　HOT CATHODES
　　　KNUDSEN GAGES
　　　MCLEOD GAGES
　　　ORBITRONS
　　　PIRANI GAGES
　　　PRESSURE MEASUREMENT

IONIZATION POTENTIALS
　GS　POTENTIAL ENERGY
　　　. **IONIZATION POTENTIALS**
　RT　ACTIVATION
　　　ELECTRIC POTENTIAL
　　　NUCLEAR BINDING ENERGY
　　　∞ POTENTIAL
　　　SAHA EQUATIONS
　　　WORK FUNCTIONS

IONIZED GASES
　GS　GASES
　　　. **IONIZED GASES**
　　　. . CHARGED PARTICLES
　　　. . . ARGON PLASMA
　　　. . . BOUNDARY LAYER PLASMAS
　　　. . . COLD PLASMAS
　　　. . . COLLISIONAL PLASMAS
　　　. . . . STRONGLY COUPLED PLASMAS
　　　. . . COLLISIONLESS PLASMAS
　　　. . . CONDUCTION ELECTRONS
　　　. . . DENSE PLASMAS
　　　. . . . STRONGLY COUPLED PLASMAS
　　　. . . ELECTRON PLASMA
　　　. . . ELLIPTICAL PLASMAS
　　　. . . HELIUM PLASMA
　　　. . . HIGH TEMPERATURE PLASMAS
　　　. . . LASER PLASMAS
　　　. . . METALLIC PLASMAS
　　　. . . . CESIUM PLASMA
　　　. . . MICROPLASMAS
　　　. . . NITROGEN PLASMA
　　　. . . NUCLEI (NUCLEAR PHYSICS)
　　　. . . . EVEN-EVEN NUCLEI
　　　. . . . HEAVY NUCLEI
　　　. . . . HYPERNUCLEI

IONIZED GASES-(CONT.)
.... ODD-EVEN NUCLEI
.... ODD-ODD NUCLEI
... RAREFIED PLASMAS
... RELATIVISTIC PLASMAS
... ROTATING PLASMAS
... SOLAR WIND
... STELLAR WINDS
... THERMAL PLASMAS
RT   COSMIC GASES
ELECTRON GAS
FOKKER-PLANCK EQUATION
GAS IONIZATION
GAS TEMPERATURE
HIGH TEMPERATURE GASES
LORENTZ GAS
PLASMA COOLING
RECOMBINATION COEFFICIENT

IONIZED PLASMAS
USE   PLASMAS (PHYSICS)

IONIZERS
RT   ∞FILAMENTS
GAS IONIZATION
∞GRIDS
ION ENGINES
IONIZATION CHAMBERS
SURFACE IONIZATION
TUBE GRIDS

IONIZING RADIATION
GS   **IONIZING RADIATION**
. ALPHA PARTICLES
. BETA PARTICLES
. COSMIC RAYS
.. COSMIC RAY SHOWERS
.. GALACTIC COSMIC RAYS
.. GAMMA RAY BURSTS
.. PRIMARY COSMIC RAYS
... SOLAR COSMIC RAYS
.. SECONDARY COSMIC RAYS
. GAMMA RAY BEAMS
. GAMMA RAYS
.. GAMMA RAY BURSTS
. ULTRAVIOLET RADIATION
.. EXTREME ULTRAVIOLET RADIATION
.. FAR ULTRAVIOLET RADIATION
... LYMAN ALPHA RADIATION
... LYMAN BETA RADIATION
.. NEAR ULTRAVIOLET RADIATION
. X RAYS
.. COSMIC X RAYS
.. SOLAR X-RAYS
RT   ABSORPTION SPECTRA
AVALANCHES
BEAMS (RADIATION)
COHERENT ELECTROMAGNETIC
   RADIATION
CORPUSCULAR RADIATION
ELECTROMAGNETIC RADIATION
ELECTRON BEAMS
ELEMENTARY PARTICLES
EMISSION
FLUENCE
GAMMA RAY ABSORPTION
GAMMA RAY SPECTRA
ION SOURCES
IRRADIATION
LINEAR ENERGY TRANSFER (LET)
MONOCHROMATIC RADIATION
NUCLEAR RADIATION
PARTICLE TRAJECTORIES
∞RADIATION
RADIATION BELTS
RADIATION COUNTERS
RADIATION DAMAGE
RADIATION HAZARDS
RADIOACTIVE MATERIALS
RADIOACTIVITY
RADIOCHEMISTRY
RELATIVISTIC ELECTRON BEAMS
SOLAR RADIATION
STERILIZATION
SYSTEM GENERATED
   ELECTROMAGNETIC PULSES

IONOGRAMS
RT   ION DENSITY (CONCENTRATION)
IONOSONDES
RIOMETERS

IONOPAUSE
SN   (EXCLUDES PLASMAPAUSE)
RT   COMETARY ATMOSPHERES

IONOPAUSE-(CONT.)
PLANETARY ATMOSPHERES
PLASMAPAUSE
SPACE PLASMAS
VENUS ATMOSPHERE

IONOSONDES
GS   MEASURING INSTRUMENTS
. METEOROLOGICAL INSTRUMENTS
.. RADIOSONDES
... **IONOSONDES**
. SONDES
.. RADIOSONDES
... **IONOSONDES**
RADIO EQUIPMENT
. RADIO TRANSMITTERS
. RADIOSONDES
... **IONOSONDES**
TRANSMITTERS
. RADIO TRANSMITTERS
.. RADIOSONDES
... **IONOSONDES**
RT   ION PROBES
IONOGRAMS
IONOSPHERIC SOUNDING
RIOMETERS
SATELLITE SOUNDING
SOUNDING ROCKETS

IONOSPHERE
GS   EARTH ATMOSPHERE
. UPPER ATMOSPHERE
.. **IONOSPHERE**
... E REGION
.... E-1 LAYER
.... E-2 LAYER
.... SPORADIC E LAYER
... LOWER IONOSPHERE
... D REGION
... UPPER IONOSPHERE
... F REGION
..... F 1 REGION
..... F 2 REGION
ENVIRONMENTS
. **IONOSPHERE**
.. LOWER IONOSPHERE
RT   ATMOSPHERIC IONIZATION
CHEMOSPHERE
ELECTROJETS
EXOSPHERE
HETEROSPHERE
HOMOSPHERE
INTASAT SATELLITE
ION CONCENTRATION
ION DENSITY (CONCENTRATION)
IONOSPHERIC PROPAGATION
IONOSPHERIC STORMS
∞LAYERS
MAGNETOSPHERE
MESOSPHERE
MIDLATITUDE ATMOSPHERE
REGIONS
SATELLITE ATMOSPHERES
SHEAR LAYERS
THERMOSPHERE

IONOSPHERE EXPLORER A
USE   EXPLORER 20 SATELLITE

IONOSPHERIC ABSORPTION
USE   ELECTROMAGNETIC ABSORPTION
IONOSPHERIC PROPAGATION

IONOSPHERIC BLACKOUT
USE   BLACKOUT (PROPAGATION)

IONOSPHERIC COMPOSITION
GS   COMPOSITION (PROPERTY)
. ATMOSPHERIC COMPOSITION
.. **IONOSPHERIC COMPOSITION**
RT   ATOM CONCENTRATION
CHEMICAL COMPOSITION
GAS COMPOSITION
IONIZATION
PARTICLE DENSITY (CONCENTRATION)
PLASMA COMPOSITION
SATELLITE ATMOSPHERES

IONOSPHERIC CONDUCTIVITY
GS   TRANSPORT PROPERTIES
. ATMOSPHERIC CONDUCTIVITY
.. **IONOSPHERIC CONDUCTIVITY**
. ELECTRICAL RESISTIVITY
.. **IONOSPHERIC CONDUCTIVITY**
RT   ∞CONDUCTIVITY

IONOSPHERIC CONDUCTIVITY-(CONT.)
ELECTROJETS
IONIC WAVES
PLASMA CONDUCTIVITY

IONOSPHERIC CROSS MODULATION
GS   ELECTROMAGNETIC INTERFERENCE
. CROSSTALK
.. **IONOSPHERIC CROSS MODULATION**
. RADIO FREQUENCY INTERFERENCE
.. BLACKOUT (PROPAGATION)
... **IONOSPHERIC CROSS MODULATION**
MODULATION
. **IONOSPHERIC CROSS MODULATION**
RT   LUXEMBOURG EFFECT

IONOSPHERIC CURRENTS
GS   ELECTRIC CURRENT
. **IONOSPHERIC CURRENTS**
.. ELECTROJETS
... AURORAL ELECTROJETS
... EQUATORIAL ELECTROJET
ELECTRICITY
. ATMOSPHERIC ELECTRICITY
. **IONOSPHERIC CURRENTS**
.. ELECTROJETS
... AURORAL ELECTROJETS
.... EQUATORIAL ELECTROJET
RT   PLASMA CURRENTS
TRAVELING IONOSPHERIC
   DISTURBANCES

IONOSPHERIC DISTURBANCES
GS   **IONOSPHERIC DISTURBANCES**
. IONOSPHERIC STORMS
.. SUDDEN IONOSPHERIC
   DISTURBANCES
. TRAVELING IONOSPHERIC
   DISTURBANCES
RT   BLACKOUT (PROPAGATION)
∞DISTURBANCES
MAGNETIC VARIATIONS

IONOSPHERIC DRIFT
RT   ∞DRIFT
DRIFT RATE
ELECTROJETS
MAGNETIC RIGIDITY
POLARIZATION (CHARGE SEPARATION)
RADIATION BELTS

IONOSPHERIC ELECTRON DENSITY
GS   DENSITY (NUMBER/VOLUME)
. PARTICLE DENSITY (CONCENTRATION)
.. ELECTRON DENSITY
   (CONCENTRATION)
... **IONOSPHERIC ELECTRON DENSITY**
RT   ARIEL 4 SATELLITE
MAGNETOSPHERIC ELECTRON DENSITY

IONOSPHERIC F-SCATTER PROPAGATION
GS   SCATTERING
. WAVE SCATTERING
.. ELECTROMAGNETIC SCATTERING
... **IONOSPHERIC F-SCATTER
   PROPAGATION**
TRANSMISSION
. ELECTROMAGNETIC WAVE
   TRANSMISSION
.. RADIO TRANSMISSION
... IONOSPHERIC PROPAGATION
.... **IONOSPHERIC F-SCATTER
   PROPAGATION**
.. SCATTER PROPAGATION
... **IONOSPHERIC F-SCATTER
   PROPAGATION**
. SIGNAL TRANSMISSION
.. RADIO TRANSMISSION
... IONOSPHERIC PROPAGATION
.... **IONOSPHERIC F-SCATTER
   PROPAGATION**
. WAVE PROPAGATION
.. IONOSPHERIC PROPAGATION
... **IONOSPHERIC F-SCATTER
   PROPAGATION**
.. SCATTER PROPAGATION
... **IONOSPHERIC F-SCATTER
   PROPAGATION**

IONOSPHERIC HEATING
GS   HEATING
. **IONOSPHERIC HEATING**
RT   ATMOSPHERIC RADIATION
PLASMA HEATING

IONOSPHERIC ION DENSITY
GS   DENSITY (NUMBER/VOLUME)
    . PARTICLE DENSITY (CONCENTRATION)
    . . ION DENSITY (CONCENTRATION)
    . . . **IONOSPHERIC ION DENSITY**
RT   MAGNETOSPHERIC ION DENSITY
    POSITIVE IONS

IONOSPHERIC NOISE
GS   ATMOSPHERIC RADIATION
    . **IONOSPHERIC NOISE**
    . . WHISTLERS
    ELECTROMAGNETIC INTERFERENCE
    . RADIO FREQUENCY INTERFERENCE
    . . BLACKOUT (PROPAGATION)
    . . . ELECTROMAGNETIC NOISE
    . . . . **IONOSPHERIC NOISE**
    . . . . . WHISTLERS
RT   BACKGROUND NOISE
    BACKGROUND RADIATION
    RIOMETERS
    SKY WAVES

IONOSPHERIC PROPAGATION
UF   IONOSPHERIC ABSORPTION
    IONOSPHERIC REFLECTION
GS   TRANSMISSION
    . ELECTROMAGNETIC WAVE
      TRANSMISSION
    . . RADIO TRANSMISSION
    . . . **IONOSPHERIC PROPAGATION**
    . . . . IONOSPHERIC F-SCATTER
        PROPAGATION
    . SIGNAL TRANSMISSION
    . . RADIO TRANSMISSION
    . . . **IONOSPHERIC PROPAGATION**
    . . . . IONOSPHERIC F-SCATTER
        PROPAGATION
    . WAVE PROPAGATION
    . . **IONOSPHERIC PROPAGATION**
    . . . IONOSPHERIC F-SCATTER
        PROPAGATION
RT   ANTIPODES
    EISCAT RADAR SYSTEM (EUROPE)
    IONIC WAVES
    IONOSPHERE
    LOSSY MEDIA
    LUXEMBOURG EFFECT
    MAGNETOIONICS
    ORBIS
    ORBIS CAL SATELLITE
    POLAR RADIO BLACKOUT
    RIOMETERS
    SCATTER PROPAGATION
    TRAVELING IONOSPHERIC
      DISTURBANCES

IONOSPHERIC REFLECTION
USE   IONOSPHERIC PROPAGATION

IONOSPHERIC SOUNDING
GS   SOUNDING
    . **IONOSPHERIC SOUNDING**
RT   ALOUETTE PROJECT
    ALOUETTE 1 SATELLITE
    ALOUETTE 2 SATELLITE
    ARIEL 4 SATELLITE
    ATMOSPHERIC SOUNDING
    IONOSONDES
    ORBIS
    ORBIS CAL SATELLITE
    ROCKET SOUNDING
    SATELLITE SOUNDING

IONOSPHERIC STORMS
GS   IONOSPHERIC DISTURBANCES
    . **IONOSPHERIC STORMS**
    . . SUDDEN IONOSPHERIC
        DISTURBANCES
    STORMS
    . **IONOSPHERIC STORMS**
    . . SUDDEN IONOSPHERIC
        DISTURBANCES
RT   ∞DISTURBANCES
    IONOSPHERE
    IONOSPHERICS
    NOISE STORMS
    SOLAR STORMS
    SPREAD F
    TRAVELING IONOSPHERIC
      DISTURBANCES

IONOSPHERIC TEMPERATURE
GS   TEMPERATURE
    . ATMOSPHERIC TEMPERATURE

IONOSPHERIC TEMPERATURE-(CONT.)
    . . **IONOSPHERIC TEMPERATURE**
RT   AURORAL TEMPERATURE
    ELECTRON ENERGY
    ION TEMPERATURE

IONOSPHERIC TILTS
RT   TRAVELING IONOSPHERIC
      DISTURBANCES

IONOSPHERICS
GS   ELECTROMAGNETIC INTERFERENCE
    . RADIO FREQUENCY INTERFERENCE
    . . BLACKOUT (PROPAGATION)
    . . . ELECTROMAGNETIC NOISE
    . . . . ATMOSPHERICS
    . . . . . **IONOSPHERICS**
    . . . . . . DAWN CHORUS
    . . . . . . HISS
RT   IONOSPHERIC STORMS
    RADIO AURORAS

IONS
GS   **IONS**
    . ANIONS
    . ANTIPROTONS
    . CATIONS
    . . METAL IONS
    . . . FERRIC IONS
    . . . MANGANESE IONS
    . . . VANADYL RADICAL
    . CESIUM IONS
    . DEUTERONS
    . FORMYL IONS
    . HEAVY IONS
    . HELIUM IONS
    . HYDROGEN IONS
    . HYDRONIUM IONS
    . LIGHT IONS
    . MOLECULAR IONS
    . NEGATIVE IONS
    . NITROGEN IONS
    . OXYGEN IONS
    . POSITIVE IONS
    . PROTONS
    . . SOLAR PROTONS
    . RECOIL IONS
    . TRITONS
    . TRIVALENT IONS
RT   ALPHA PARTICLES
    ATOMS
    CHEMICAL ELEMENTS
    CORPUSCULAR RADIATION
    ELECTROLYTES
    FREE RADICALS
    HYDROXYL RADICALS
    ION IMPLANTATION
    ION PLATING
    IONIC MOBILITY
    MOLECULES
    MONATOMIC MOLECULES
    NUCLEI (NUCLEAR PHYSICS)
    PLASMAS (PHYSICS)
    POLYATOMIC MOLECULES
    VALENCE

IOWA
GS   NATIONS
    . UNITED STATES
    . . **IOWA**
RT   CEDAR RAPIDS (IA)
    MISSOURI RIVER (US)

IP (IMPACT PREDICTION)
USE   COMPUTERIZED SIMULATION

IPAD
SN   (INTEGRATED PROGRAM FOR
    AEROSPACE VEHICLE DESIGN)
UF   INTEG PROGRAM FOR AEROSPACE VEH
    DESIGN
GS   COMPUTER TECHNIQUES
    . COMPUTER AIDED DESIGN
    . . **IPAD**
    SPACECRAFT DESIGN
    . **IPAD**
RT   ∞DESIGN

IQSY (INTERNATIONAL YEAR)
USE   INTERNATIONAL QUIET SUN YEAR

IRAN
GS   NATIONS
    . **IRAN**
RT   ASIA

IRAQ
GS   NATIONS
    . **IRAQ**
RT   ASIA

IRAS
USE   INFRARED ASTRONOMY SATELLITE

IRAS-ARAKI-ALCOCK COMET
GS   CELESTIAL BODIES
    . COMETS
    . . **IRAS-ARAKI-ALCOCK COMET**
RT   INFRARED ASTRONOMY SATELLITE
    SOLAR SYSTEM

IRASERS
USE   INFRARED LASERS

IRBM (MISSILES)
USE   INTERMEDIATE RANGE BALLISTIC
      MISSILES

IRELAND
GS   LANDFORMS
    . ISLANDS
    . . **IRELAND**
    NATIONS
    . **IRELAND**

IRIDESCENCE
GS   ELECTROMAGNETIC PROPERTIES
    . OPTICAL PROPERTIES
    . . COLOR
    . . . **IRIDESCENCE**
RT   OPALESCENCE

IRIDIUM
GS   CHEMICAL ELEMENTS
    . REFRACTORY METALS
    . . **IRIDIUM**
    . . . IRIDIUM ISOTOPES
    METALS
    . TRANSITION METALS
    . . REFRACTORY METALS
    . . . **IRIDIUM**
    . . . . IRIDIUM ISOTOPES
    REFRACTORY MATERIALS
    . REFRACTORY METALS
    . . **IRIDIUM**
    . . . IRIDIUM ISOTOPES

IRIDIUM ISOTOPES
GS   CHEMICAL ELEMENTS
    . NUCLIDES
    . . ISOTOPES
    . . . **IRIDIUM ISOTOPES**
    . REFRACTORY METALS
    . . IRIDIUM
    . . . **IRIDIUM ISOTOPES**
    METALS
    . TRANSITION METALS
    . . REFRACTORY METALS
    . . . IRIDIUM
    . . . . **IRIDIUM ISOTOPES**
    REFRACTORY MATERIALS
    . REFRACTORY METALS
    . . IRIDIUM
    . . . **IRIDIUM ISOTOPES**

IRIS SATELLITES
GS   SATELLITES
    . ARTIFICIAL SATELLITES
    . . **IRIS SATELLITES**
RT   EUROPEAN SPACE PROGRAMS
    SATELLITE OBSERVATION
    SOLAR ACTIVITY
    SOLAR CYCLES
    SOLAR ENERGY
    SOLAR FLARES
    SOLAR RADIATION
    SOLAR SENSORS

IRISES (MECHANICAL APERTURES)
GS   OPENINGS
    . APERTURES
    . . **IRISES (MECHANICAL APERTURES)**
RT   CAMERA SHUTTERS
    WAVEGUIDE WINDOWS
    WAVEGUIDES

IRON
GS   CHEMICAL ELEMENTS
    . **IRON**
    . . IRON ISOTOPES

IRON--(CONT.)
```
        . . . IRON 57
        . . . IRON 58
        . . . IRON 59
      METALS
      . TRANSITION METALS
      . . IRON
        . . . IRON ISOTOPES
        . . . . IRON 57
        . . . . IRON 58
        . . . . IRON 59
RT    FERROUS METALS
      HYDROGEN EMBRITTLEMENT
      LOW CARBON STEELS
```

IRON ALLOYS
```
UF    FERROALLOYS
GS    ALLOYS
      . IRON ALLOYS
      . . STEELS
      . . . BAINITIC STEEL
      . . . CARBON STEELS
      . . . . LOW CARBON STEELS
      . . . CHROMIUM STEELS
      . . . CROLOY
      . . . HIGH STRENGTH STEELS
      . . . . MARAGING STEELS
      . . . NICKEL STEELS
      . . . STAINLESS STEELS
      . . . . AUSTENITIC STAINLESS STEELS
      . . . . FERRITIC STAINLESS STEELS
      . . . . MARTENSITIC STAINLESS STEELS
RT    AUSTENITE
      BAINITE
      BEARING ALLOYS
      CEMENTITE
      FERRITES
      HASTELLOY (TRADEMARK)
      INCONEL (TRADEMARK)
      KAMACITE
      MARTENSITE
      NIMONIC ALLOYS
      PEARLITE
      PERMALLOYS (TRADEMARK)
      ZIRCALOYS (TRADEMARK)
```

IRON CHLORIDES
```
GS    HALOGEN COMPOUNDS
      . CHLORINE COMPOUNDS
      . . CHLORIDES
      . . . IRON CHLORIDES
      . HALIDES
      . . CHLORIDES
      . . . IRON CHLORIDES
      . . METAL HALIDES
      . . . IRON CHLORIDES
      IRON COMPOUNDS
      . IRON CHLORIDES
```

IRON COMPOUNDS
```
GS    IRON COMPOUNDS
      . COHENITE
      . CORDIERITE
      . FAYALITE
      . FERRATES
      . . BARIUM FERRATES
      . FERRITES
      . FERROCENES
      . . ALKYLFERROCENE
      . IRON CHLORIDES
      . IRON CYANIDES
      . IRON OXIDES
      . . CHROMITES
      . . HEMATITE
      . . ILMENITE
      . MAGNETITE
      . LIMONITE
      . PYRITES
      . PYRRHOTITE
      . . TROILITE
      . SCHREIBERSITE
      . SIDERITES
RT    ∞CHEMICAL COMPOUNDS
      ∞GROUP 8 COMPOUNDS
      ∞METAL COMPOUNDS
```

IRON CYANIDES
```
GS    CYANIDES
      . IRON CYANIDES
      IRON COMPOUNDS
      . IRON CYANIDES
```

IRON ISOTOPES
```
GS    CHEMICAL ELEMENTS
      . IRON
```

IRON ISOTOPES--(CONT.)
```
      . . IRON ISOTOPES
      . . . IRON 57
      . . . IRON 58
      . . . IRON 59
      . NUCLIDES
      . . ISOTOPES
      . . . IRON ISOTOPES
      . . . . IRON 57
      . . . . IRON 58
      . . . . IRON 59
      METALS
      . TRANSITION METALS
      . . IRON
      . . . IRON ISOTOPES
      . . . . IRON 57
      . . . . IRON 58
      . . . . IRON 59
RT    FERROUS METALS
```

IRON METEORITES
```
UF    SIDERITE METEORITES
GS    CELESTIAL BODIES
      . METEORITES
      . . IRON METEORITES
      . . . AROOS METEORITE
      . . . ODESSA METEORITE
      . . . SIKHOTE-ALIN METEORITE
RT    ACHONDRITES
      HARLETON METEORITE
      KAMACITE
      LAZAREV METEORITE
      METEORITIC COMPOSITION
      METEORITIC MICROSTRUCTURES
      OKHANSK METEORITE
      SCHREIBERSITE
      STONY METEORITES
      TROILITE
      WIDMANSTATTEN STRUCTURE
```

IRON ORES
```
GS    MINERALS
      . IRON ORES
      . . HEMATITE
```

IRON OXIDES
```
GS    CHALCOGENIDES
      . OXIDES
      . . METAL OXIDES
      . . . IRON OXIDES
      . . . . CHROMITES
      . . . . HEMATITE
      . . . . ILMENITE
      . . . . MAGNETITE
      IRON COMPOUNDS
      . IRON OXIDES
      . . CHROMITES
      . . HEMATITE
      . . ILMENITE
      . . MAGNETITE
```

IRON 57
```
GS    CHEMICAL ELEMENTS
      . IRON
      . . IRON ISOTOPES
      . . . IRON 57
      . NUCLIDES
      . . ISOTOPES
      . . . IRON ISOTOPES
      . . . . IRON 57
      METALS
      . TRANSITION METALS
      . . IRON
      . . . IRON ISOTOPES
      . . . . IRON 57
```

IRON 58
```
GS    CHEMICAL ELEMENTS
      . IRON
      . . IRON ISOTOPES
      . . . IRON 58
      . NUCLIDES
      . . ISOTOPES
      . . . IRON ISOTOPES
      . . . . IRON 58
      METALS
      . TRANSITION METALS
      . . IRON
      . . . IRON ISOTOPES
      . . . . IRON 58
```

IRON 59
```
GS    CHEMICAL ELEMENTS
      . IRON
      . . IRON ISOTOPES
```

IRON 59--(CONT.)
```
      . . . IRON 59
      . NUCLIDES
      . . ISOTOPES
      . . . IRON ISOTOPES
      . . . . IRON 59
      . . . RADIOACTIVE ISOTOPES
      . . . . IRON 59
      METALS
      . TRANSITION METALS
      . . IRON
      . . . IRON ISOTOPES
      . . . . IRON 59
```

IROQUOIS HELICOPTER
```
USE   UH-1 HELICOPTER
```

IRRADIANCE
```
SN    (DETECTION RATE PER UNIT AREA OF
      RADIATION)
GS    RATES (PER TIME)
      . FLUX DENSITY
      . . RADIANT FLUX DENSITY
      . . . IRRADIANCE
      . . . . ILLUMINANCE
      . . . . SOLAR CONSTANT
RT    ELECTRON FLUX DENSITY
      LUMINANCE
      LUMINOUS INTENSITY
      NEUTRON FLUX DENSITY
      PROTON FLUX DENSITY
      RADIANCE
      SOLAR BACKSCATTER UV
        SPECTROMETER
      SOLAR FLUX DENSITY
```

IRRADIATION
```
GS    IRRADIATION
      . AURORAL IRRADIATION
      . ELECTRON IRRADIATION
      . ION IRRADIATION
      . . DEUTERON IRRADIATION
      . . PROTON IRRADIATION
      . NEUTRON IRRADIATION
      . X RAY IRRADIATION
RT    ACTIVATION
      BEAMS (RADIATION)
      ∞BOMBARDMENT
      DOSIMETERS
      ELECTROMAGNETIC ABSORPTION
      ELECTRON PROBES
      EMISSION
      EXCITATION
      EXPOSURE
      FLUX DENSITY
      IONIZING RADIATION
      NUCLEAR CAPTURE
      NUCLEAR FUSION
      NUCLEAR RADIATION
      PRESERVING
      ∞RADIATION
      RADIATION DOSAGE
      RADIATION EFFECTS
      RADIATION MEASUREMENT
      RADIATION TOLERANCE
      RADIOBIOLOGY
      RADIOGRAPHY
      TARGETS
```

IRRATIONALITY
```
RT    DISORIENTATION
      DITHERS
      MENTAL PERFORMANCE
      PREJUDICES
      PSYCHOSES
†     SCHIZOPHRENIA
```

IRREGULARITIES
```
RT    ABNORMALITIES
      DEFECTS
      DEVIATION
      NONUNIFORMITY
```

IRREVERSIBLE PROCESSES
```
RT    NONEQUILIBRIUM THERMODYNAMICS
      ONSAGER RELATIONSHIP
      REACTION KINETICS
      THERMODYNAMICS
      THERMOVISCOELASTICITY
      VARIATIONAL PRINCIPLES
```

IRRIGATION
```
RT    AGRICULTURE
      ALFALFA
      BARLEY
```

IRRIGATION-(CONT.)
- CANALS
- CITRUS TREES
- CORN
- CROP VIGOR
- DITCHES
- DRAINAGE
- DRAINAGE PATTERNS
- FARM CROPS
- FARMLANDS
- OATS
- ORCHARDS
- PONDS
- SEEPAGE
- SUGAR BEETS
- SUGAR CANE
- TROUGHS
- VEGETATION GROWTH
- VINEYARDS
- WATER CONSUMPTION

IRRITATION
- GS   **IRRITATION**
-    . TOXICITY AND SAFETY HAZARD
- RT   ∞REACTION

IRROTATIONAL FLOW
- USE   POTENTIAL FLOW

IRS (INDIAN SPACECRAFT)
- USE   INDIAN SPACECRAFT

ISAGEX
- USE   INTERNATIONAL SATELLITE GEODESY
  EXPERIMENT

ISCHEMIA
- GS   CIRCULATION
-    . BLOOD CIRCULATION
-    . . **ISCHEMIA**
- RT   CONGESTION
  VASOCONSTRICTION

ISEE
- USE   INTERNATIONAL SUN EARTH
  EXPLORERS

ISENTROPE
- RT   ADIABATIC CONDITIONS
  ISENTROPIC PROCESSES
  MOLLIER DIAGRAM
  POISSON EQUATION

ISENTROPIC PROCESSES
- GS   **ISENTROPIC PROCESSES**
-    . NONISENTROPICITY
- RT   BERNOULLI THEOREM
  ISENTROPE
  ISOENERGETIC PROCESSES
  ISOPYCNIC PROCESSES
  ∞PROCESSES
  THERMODYNAMIC EQUILIBRIUM

ISING MODEL
- USE   FERROMAGNETISM
  MATHEMATICAL MODELS

ISIS SATELLITES
- SN   (INTERNATIONAL SATELLITES FOR
  IONOSPHERIC STUDY)
- UF   INTERNATIONAL SATS FOR
  IONOSPHERIC STUDY
- GS   SATELLITES
-    . ARTIFICIAL SATELLITES
-    . . **ISIS SATELLITES**
-    . . . ALOUETTE 2 SATELLITE
-    . . . ISIS-A
-    . . . ISIS-B
-    . . . ISIS-X
- RT   ALOUETTE SATELLITES

ISIS-A
- GS   SATELLITES
-    . ARTIFICIAL SATELLITES
-    . . ISIS SATELLITES
-    . . . **ISIS-A**
- RT   ALOUETTE PROJECT

ISIS-B
- GS   SATELLITES
-    . ARTIFICIAL SATELLITES
-    . . ISIS SATELLITES
-    . . . **ISIS-B**

ISIS-X
- GS   SATELLITES
-    . ARTIFICIAL SATELLITES
-    . . ISIS SATELLITES
-    . . . **ISIS-X**
- RT   ALOUETTE B SATELLITE

ISKRA AIRCRAFT
- USE   TS-11 AIRCRAFT

ISLAND ARCS
- GS   LANDFORMS
-    . **ISLAND ARCS**
- RT   ALEUTIAN ISLANDS (US)
  ∞ARCS
  BARRIERS (LANDFORMS)
  ISLANDS
  KEYS (ISLANDS)
  LAGOONS
  REEFS

ISLANDS
- GS   LANDFORMS
-    . **ISLANDS**
-    . . ALEUTIAN ISLANDS (US)
-    . . ASSATEAGUE ISLAND (MD-VA)
-    . ATOLLS
-    . . BAHRAIN
-    . . BERMUDA
-    . CYPRUS
-    . . GREENLAND
-    . . HAWAII
-    . . ICELAND
-    . . INDONESIA
-    . IRELAND
-    . . KEYS (ISLANDS)
-    . . LONG ISLAND (NY)
-    . . MALAGASY REPUBLIC
-    . . MALDIVE ISLANDS
-    . . MALTA
-    . . MERRITT ISLAND (FL)
-    . NEWFOUNDLAND
-    . . NUNATAKS
-    . . PACIFIC ISLANDS
-    . . . GUAM
-    . . . JAPAN
-    . . . JOHNSTON ISLAND
-    . . . KURILE ISLANDS
-    . . . NEW GUINEA (ISLAND)
-    . . . NEW ZEALAND
-    . . . PHILIPPINES
-    . . SAMOA
-    . . PRINCE EDWARD ISLAND
-    . SICILY
-    . . TASMANIA
-    . . WALLOPS ISLAND
-    . . WEST INDIES
-    . . . BAHAMAS
-    . . . BARBADOS
-    . . . CUBA
-    . . . DOMINICA
-    . . . GUADELOUPE
-    . . . HAITI
-    . . . JAMAICA
-    . . . LESSER ANTILLES
-    . . . MARTINIQUE
-    . . . PUERTO RICO
-    . . . TRINIDAD AND TOBAGO
-    . . . VIRGIN ISLANDS
- RT   ARCHIPELAGOES
  CAPE VERDE
  CORAL REEFS
  ISLAND ARCS
  LAGOONS
  OUTER BANKS (NC)
  REEFS
  SEAMOUNTS

∞ ISOBARS
- SN   (USE OF A MORE SPECIFIC TERM IS
  RECOMMENDED--CONSULT THE TERMS
  LISTED BELOW)
- RT   BAROCLINITY
  BAROTROPISM
  ISOBARS (PRESSURE)
  NUCLEAR ISOBARS
  POLYTROPIC PROCESSES

ISOBARS (PRESSURE)
- RT   ATMOSPHERIC PRESSURE
  GEOSTROPHIC WIND
  GRADIENTS
  ∞ISOBARS
  ISOCHORIC PROCESSES
  ISOPYCNIC PROCESSES

ISOBARS (PRESSURE)-(CONT.)
- ISOTHERMAL PROCESSES
  METEOROLOGICAL CHARTS
  PRESSURE
  PRESSURE DISTRIBUTION
  PRESSURE GRADIENTS

ISOBUTANE
- USE   BUTANES

ISOBUTYLENE
- USE   BUTENES

ISOCHORIC PROCESSES
- RT   ISOBARS (PRESSURE)
  ISOPYCNIC PROCESSES
  THERMODYNAMIC EQUILIBRIUM
  VOLUME

ISOCHROMATICS
- RT   COLOR
  DICHROISM
  DIFFRACTION
  INTERFEROMETRY
  REFRACTION

ISOCYANATES
- GS   ESTERS
-    . **ISOCYANATES**
-    . . DIISOCYANATES
-    . . FULMINATES
-    NITROGEN COMPOUNDS
-    . CYANO COMPOUNDS
-    . . **ISOCYANATES**
-    . . . DIISOCYANATES
-    . . . FULMINATES

ISOELECTRONIC SEQUENCE
- RT   ATOMIC STRUCTURE
  SPECTRA
  SPECTROSCOPY

ISOENERGETIC PROCESSES
- RT   ADIABATIC CONDITIONS
  ISENTROPIC PROCESSES
  THERMODYNAMIC EQUILIBRIUM

ISOLATION
- GS   **ISOLATION**
-    . SOCIAL ISOLATION
- RT   CONFINEMENT
  CONFINING
  DEPRIVATION
  DISPOSAL
  EXCLUSION
  GNOTOBIOTICS
  INSULATION
  ISOLATORS
  ∞SEPARATION
  SPACING

ISOLATORS
- GS   **ISOLATORS**
-    . VIBRATION ISOLATORS
- RT   ATTENUATORS
  INSULATION
  ISOLATION
  NOISE REDUCTION
  SHOCK ABSORBERS
  SPACERS
  SUPPRESSORS
  VIBRATION

ISOMERIZATION
- GS   **ISOMERIZATION**
-    . ORTHO PARA CONVERSION
- RT   ∞CONVERSION
  REFINING

ISOMERS
- GS   **ISOMERS**
-    . ISOPROPYL ALCOHOL
-    . PHENANTHRENE
- RT   ATOMS
  NUCLEAR CHEMISTRY
  STEREOCHEMISTRY
  TAUTOMERS

ISOMORPHISM
- UF   MORPHOTROPISM
- GS   MORPHOLOGY
-    . **ISOMORPHISM**
- RT   CRYSTAL LATTICES
  CRYSTAL STRUCTURE

ISOMORPHISM-(CONT.)
DUALITY THEOREM
HOMOMORPHISMS

ISOPARAMETRIC FINITE ELEMENTS
RT   CONFORMAL MAPPING
COORDINATE TRANSFORMATIONS
∞ELEMENTS
FINITE ELEMENT METHOD
FRACTURE MECHANICS
NUMERICAL ANALYSIS
STRESS ANALYSIS

ISOPERIMETRIC PROBLEM
RT   ANALYTIC FUNCTIONS
CONTINUITY (MATHEMATICS)
LAGRANGE MULTIPLIERS
MATRICES (MATHEMATICS)
∞PROBLEMS
TOPOLOGY

ISOPHOTES
RT   ∞ILLUMINATION

ISOPLETHS
USE   NOMOGRAPHS

ISOPROPYL ALCOHOL
GS   ALIPHATIC COMPOUNDS
. ISOPROPYL COMPOUNDS
. . **ISOPROPYL ALCOHOL**
HYDROXYL COMPOUNDS
. ALCOHOLS
. . **ISOPROPYL ALCOHOL**
ISOMERS
. **ISOPROPYL ALCOHOL**

ISOPROPYL COMPOUNDS
GS   ALIPHATIC COMPOUNDS
. **ISOPROPYL COMPOUNDS**
. . ISOPROPYL ALCOHOL
RT   ∞CHEMICAL COMPOUNDS

ISOPROPYL NITRATE
GS   ALIPHATIC COMPOUNDS
. ALKYL COMPOUNDS
. . **ISOPROPYL NITRATE**
. NITRATE ESTERS
. . **ISOPROPYL NITRATE**
ESTERS
. NITRATE ESTERS
. . **ISOPROPYL NITRATE**
NITROGEN COMPOUNDS
. NITRATE ESTERS
. . **ISOPROPYL NITRATE**

ISOPYCNIC PROCESSES
UF   ISOSTERIC PROCESSES
RT   DENSITY (MASS/VOLUME)
ISENTROPIC PROCESSES
ISOBARS (PRESSURE)
ISOCHORIC PROCESSES

ISOSTASY
RT   ∞EQUILIBRIUM
GEOMORPHOLOGY
GEOPHYSICS
GLACIOLOGY
GRAVITATION
HYDROSTATICS
OROGRAPHY
SEISMOLOGY
SUBSIDENCE

ISOSTATIC PRESSURE
GS   PRESSURE
. **ISOSTATIC PRESSURE**
RT   ATMOSPHERIC PRESSURE
HYDROSTATIC PRESSURE
STATIC PRESSURE

ISOSTERIC PROCESSES
USE   ISOPYCNIC PROCESSES

ISOTENSOID STRUCTURES
RT   COMPOSITE WRAPPING
FILAMENT WINDING
PRESSURE VESSELS
PRESTRESSING
SHELLS (STRUCTURAL FORMS)
SPIRAL WRAPPING
∞STRUCTURES

ISOTHERMAL FLOW
GS   FLUID FLOW
. **ISOTHERMAL FLOW**
RT   FLOW DISTRIBUTION
ISOTHERMS
TEMPERATURE DISTRIBUTION

ISOTHERMAL LAYERS
RT   ISOTHERMS
LAMINAR BOUNDARY LAYER
STRATOSPHERE
TEMPERATURE DISTRIBUTION
TEMPERATURE GRADIENTS
THERMAL MAPPING
TROPOPAUSE

ISOTHERMAL PROCESSES
RT   ADIABATIC CONDITIONS
ISOBARS (PRESSURE)
ISOTHERMS
THERMODYNAMIC EQUILIBRIUM

ISOTHERMS
RT   ATMOSPHERIC & OCEANOGRAPHIC
     INFORM SYS
ATMOSPHERIC TEMPERATURE
     GRADIENTS
ISOTHERMAL FLOW
ISOTHERMAL LAYERS
ISOTHERMAL PROCESSES
METEOROLOGICAL PARAMETERS
METEOROLOGY
TEMPERATURE
TEMPERATURE DISTRIBUTION
TEMPERATURE GRADIENTS
THERMAL MAPPING
THERMODYNAMICS

ISOTONICITY
RT   BODY FLUIDS
OSMOSIS

ISOTOPE EFFECT
UF   ISOTOPE SHIFT
RT   ∞EFFECTS

ISOTOPE SEPARATION
RT   ATOMS
HEAVY IONS
ION EXCHANGING
ION EXTRACTION
ISOTOPES
JET MEMBRANE PROCESS

ISOTOPE SHIFT
USE   ISOTOPE EFFECT

ISOTOPES
GS   CHEMICAL ELEMENTS
. NUCLIDES
. . **ISOTOPES**
. . . ALUMINUM ISOTOPES
. . . . ALUMINUM 26
. . . . ALUMINUM 27
. . . ANTIMONY ISOTOPES
. . . ARGON ISOTOPES
. . . BARIUM ISOTOPES
. . . BERYLLIUM ISOTOPES
. . . . BERYLLIUM 7
. . . . BERYLLIUM 9
. . . . BERYLLIUM 10
. . . BISMUTH ISOTOPES
. . . BORON ISOTOPES
. . . . BORON 10
. . . BROMINE ISOTOPES
. . . CADMIUM ISOTOPES
. . . CALCIUM ISOTOPES
. . . CARBON ISOTOPES
. . . . CARBON 12
. . . . CARBON 13
. . . . CARBON 14
. . . CERIUM ISOTOPES
. . . . CERIUM 137
. . . . CERIUM 144
. . . CESIUM ISOTOPES
. . . . CESIUM 133
. . . . CESIUM 134
. . . . CESIUM 137
. . . . CESIUM 144
. . . CESIUM VAPOR
. . . CHROMIUM ISOTOPES
. . . COBALT ISOTOPES
. . . . COBALT 58
. . . . COBALT 60
. . . DYSPROSIUM ISOTOPES

ISOTOPES-(CONT.)
. . . ERBIUM ISOTOPES
. . . EUROPIUM ISOTOPES
. . . FLUORINE ISOTOPES
. . . GADOLINIUM ISOTOPES
. . . GALLIUM ISOTOPES
. . . GERMANIUM ISOTOPES
. . . HAFNIUM ISOTOPES
. . . HELIUM ISOTOPES
. . . HOLMIUM ISOTOPES
. . . HYDROGEN ISOTOPES
. . . . DEUTERIUM
. . . . HYDROGEN 4
. . . . TRITIUM
. . . IODINE ISOTOPES
. . . . IODINE 125
. . . . IODINE 131
. . . . IODINE 132
. . . IRIDIUM ISOTOPES
. . . IRON ISOTOPES
. . . . IRON 57
. . . . IRON 58
. . . . IRON 59
. . . KRYPTON ISOTOPES
. . . . KRYPTON 85
. . . LANTHANUM ISOTOPES
. . . LEAD ISOTOPES
. . . LITHIUM ISOTOPES
. . . LUTETIUM
. . . LUTETIUM ISOTOPES
. . . MANGANESE ISOTOPES
. . . MERCURY ISOTOPES
. . . NEODYMIUM ISOTOPES
. . . NEON ISOTOPES
. . . NICKEL ISOTOPES
. . . NIOBIUM ISOTOPES
. . . . NIOBIUM 95
. . . NITROGEN ISOTOPES
. . . . NITROGEN 15
. . . . NITROGEN 16
. . . OXYGEN ISOTOPES
. . . . OXYGEN 18
. . . PHOSPHORUS ISOTOPES
. . . . PHOSPHORUS 32
. . . PLATINUM ISOTOPES
. . . POLONIUM ISOTOPES
. . . . POLONIUM 208
. . . . POLONIUM 209
. . . . POLONIUM 210
. . . POTASSIUM ISOTOPES
. . . . POTASSIUM 38
. . . . POTASSIUM 39
. . . . POTASSIUM 40
. . . PRASEODYMIUM ISOTOPES
. . . PROMETHIUM ISOTOPES
. . . PROTACTINIUM ISOTOPES
. . . RADIOACTIVE ISOTOPES
. . . . ARSENIC ISOTOPES
. . . . ASTATINE ISOTOPES
. . . . BERYLLIUM 7
. . . . BERYLLIUM 9
. . . . BERYLLIUM 10
. . . . CARBON 14
. . . . CERIUM 137
. . . . CERIUM 144
. . . . CESIUM 134
. . . . CESIUM 137
. . . . CESIUM 144
. . . . COBALT 58
. . . . COBALT 60
. . . . GOLD ISOTOPES
. . . . . GOLD 198
. . . . INDIUM ISOTOPES
. . . . IODINE 125
. . . . IODINE 131
. . . . IODINE 132
. . . . IRON 59
. . . . KRYPTON 85
. . . . NIOBIUM 95
. . . . NITROGEN 16
. . . . PHOSPHORUS 32
. . . . POLONIUM 208
. . . . POLONIUM 209
. . . . POLONIUM 210
. . . . POTASSIUM 38
. . . . POTASSIUM 40
. . . . RUBIDIUM 86
. . . . SODIUM 22
. . . . SODIUM 24
. . . . STRONTIUM 85
. . . . STRONTIUM 88
. . . . STRONTIUM 89
. . . . STRONTIUM 90
. . . . TRANSURANIUM ELEMENTS
. . . . . AMERICIUM
. . . . . . AMERICIUM ISOTOPES

ISOTOPES-*(CONT.)*
```
. . . . . BERKELIUM
. . . . . CALIFORNIUM
. . . . . . CALIFORNIUM ISOTOPES
. . . . . CURIUM
. . . . . . CURIUM ISOTOPES
. . . . . EINSTEINIUM
. . . . . FERMIUM
. . . . . LAWRENCIUM
. . . . . MENDELEVIUM
. . . . . NEPTUNIUM
. . . . . . NEPTUNIUM ISOTOPES
. . . . . NOBELIUM
. . . . . PLUTONIUM
. . . . . . PLUTONIUM ISOTOPES
. . . . . SERGENIUM
. . . . TRITIUM
. . . . URANIUM 232
. . . . URANIUM 233
. . . . URANIUM 238
. . . . XENON 133
. . . . XENON 135
. . . . ZIRCONIUM 95
. . . RADIUM ISOTOPES
. . . . RADIUM 226
. . . RADON ISOTOPES
. . . RHODIUM ISOTOPES
. . . RUBIDIUM ISOTOPES
. . . . RUBIDIUM 86
. . . RUTHENIUM ISOTOPES
. . . SCANDIUM ISOTOPES
. . . SILVER ISOTOPES
. . . SODIUM ISOTOPES
. . . . SODIUM 22
. . . . SODIUM 24
. . . STRONTIUM ISOTOPES
. . . . STRONTIUM 85
. . . . STRONTIUM 87
. . . . STRONTIUM 89
. . . . STRONTIUM 90
. . . TANTALUM ISOTOPES
. . . TELLURIUM
. . . TELLURIUM ISOTOPES
. . . TERBIUM ISOTOPES
. . . THORIUM ISOTOPES
. . . THULIUM ISOTOPES
. . . TIN ISOTOPES
. . . TITANIUM ISOTOPES
. . . URANIUM ISOTOPES
. . . . URANIUM 232
. . . . URANIUM 233
. . . . URANIUM 234
. . . . URANIUM 235
. . . . URANIUM 238
. . . VANADIUM ISOTOPES
. . . XENON ISOTOPES
. . . . XENON 129
. . . . XENON 133
. . . . XENON 135
. . . YTTRIUM ISOTOPES
. . . ZINC ISOTOPES
. . . ZIRCONIUM ISOTOPES
. . . . ZIRCONIUM 95
```
RT   ATOMS
       HEAVY IONS
       ISOTOPE SEPARATION
       ISOTOPIC ENRICHMENT
       ISOTOPIC LABELING
       JET MEMBRANE PROCESS
       METALS
       NUCLEAR ISOBARS
       NUCLEI (NUCLEAR PHYSICS)
       RADIOACTIVE MATERIALS

ISOTOPIC ENRICHMENT
GS   ENRICHMENT
       . **ISOTOPIC ENRICHMENT**
       . . JET MEMBRANE PROCESS
RT   BENEFICIATION
       CHEMICAL ELEMENTS
      ∞CONCENTRATION
       ISOTOPES
       NUCLIDES

ISOTOPIC LABELING
GS   MARKING
       . **ISOTOPIC LABELING**
RT   CHEMICAL ANALYSIS
       ISOTOPES
       RADIOACTIVE ISOTOPES
       RADIOCHEMISTRY
       TRACE ELEMENTS
      ∞TRACERS

ISOTOPIC SPIN
GS   SPIN
       . PARTICLE SPIN
       . . **ISOTOPIC SPIN**

ISOTROPIC MEDIA
GS   ISOTROPY
       . **ISOTROPIC MEDIA**
RT   ANISOTROPIC MEDIA

ISOTROPIC TURBULENCE
GS   TURBULENCE
       . **ISOTROPIC TURBULENCE**
RT   ATMOSPHERIC TURBULENCE
       COORDINATE TRANSFORMATIONS
       HOMOGENEOUS TURBULENCE
       INSTRUMENT RECEIVERS
       INVARIANT IMBEDDINGS
       KOLMOGOROFF THEORY
       MAGNETOHYDRODYNAMIC TURBULENCE
       TURBULENT FLOW

ISOTROPISM
RT   REFRACTIVITY
       SYMMETRY

ISOTROPY
UF   SPATIAL ISOTROPY
GS   **ISOTROPY**
       . ISOTROPIC MEDIA
RT   ANISOTROPIC FLUIDS
       ANISOTROPY
       BRAGG ANGLE
       CRYSTAL STRUCTURE
       CRYSTALLOGRAPHY
       CRYSTALS
       DENDRITIC CRYSTALS
       DIRECTIVITY
       FIELD STRENGTH
       MECHANICAL PROPERTIES
       METALLOGRAPHY
       OPTICAL PROPERTIES
      ∞ORIENTATION
      ∞PHYSICAL PROPERTIES

ISRAEL
GS   NATIONS
       . **ISRAEL**
RT   ASIA

ISRO
UF   INDIAN SPACE RESEARCH
       ORGANIZATION
GS   ORGANIZATIONS
       . **ISRO**
RT   INDIA
       SPACE PROGRAMS

ISTHMUSES
GS   LAND
       . **ISTHMUSES**
       LANDFORMS
       . **ISTHMUSES**
RT   GEOLOGY
       OCEANOGRAPHY
       PENINSULAS
       TOPOGRAPHY
       WATER

ITALIAN SPACE PROGRAM
GS   PROGRAMS
       . SPACE PROGRAMS
       . . **ITALIAN SPACE PROGRAM**
RT   EUROPEAN SPACE PROGRAMS
       ITALY
       ORBITING FROG OTOLITH
       SIRIO SATELLITE

ITALY
GS   NATIONS
       ITALY
RT   ADRIATIC SEA
       ALPS MOUNTAINS (EUROPE)
       EUROPE
       ITALIAN SPACE PROGRAM
       SAN MARINO
       SICILY
       SIRIO SATELLITE
       VATICAN CITY

ITCHING
RT   CONTACT DERMATITIS
       DERMATITIS
       HISTAMINES

ITCHING-*(CONT.)*
       SENSITIVITY
       SENSORY PERCEPTION

ITERATION
GS   ANALYSIS (MATHEMATICS)
       . NUMERICAL ANALYSIS
       . . **ITERATION**
       . . . CONJUGATE GRADIENT METHOD
RT   PROBABILITY THEORY
       PROBLEM SOLVING

ITERATIVE NETWORKS
GS   CIRCUITS
       . **ITERATIVE NETWORKS**
RT   IMPEDANCE MATCHING

ITERATIVE SOLUTION
GS   PROBLEM SOLVING
       . **ITERATIVE SOLUTION**
RT   ASYMPTOTIC METHODS
       CHOLESKY FACTORIZATION
       CONJUGATE GRADIENT METHOD
       FINITE ELEMENT METHOD
       STRANGE ATTRACTORS

ITOS SATELLITES
GS   SATELLITES
       . ARTIFICIAL SATELLITES
       . . METEOROLOGICAL SATELLITES
       . . . TIROS SATELLITES
       . . . . **ITOS SATELLITES**
       . . . . . ITOS 1
       . . . . . ITOS 2
       . . . . . ITOS 3
       . . . . . ITOS 4
RT   TIROS M

ITOS 1
GS   SATELLITES
       . ARTIFICIAL SATELLITES
       . . METEOROLOGICAL SATELLITES
       . . . TIROS SATELLITES
       . . . . IMPROVED TIROS OPERATIONAL
       SATELLITES
       . . . . . **ITOS 1**
       . . . ITOS SATELLITES
       . . . . . **ITOS 1**
RT   TIROS M
       TIROS N SERIES SATELLITES
       TIROS OPERATIONAL SATELLITE
       SYSTEM

ITOS 2
GS   SATELLITES
       . ARTIFICIAL SATELLITES
       . . METEOROLOGICAL SATELLITES
       . . TIROS SATELLITES
       . . . IMPROVED TIROS OPERATIONAL
       SATELLITES
       . . . . . **ITOS 2**
       . . . . ITOS SATELLITES
       . . . . . **ITOS 2**
RT   TIROS M
       TIROS N SERIES SATELLITES
       TIROS OPERATIONAL SATELLITE
       SYSTEM

ITOS 3
GS   SATELLITES
       . ARTIFICIAL SATELLITES
       . . METEOROLOGICAL SATELLITES
       . . . TIROS SATELLITES
       . . . . IMPROVED TIROS OPERATIONAL
       SATELLITES
       . . . . . **ITOS 3**
       . . . . ITOS SATELLITES
       . . . . . **ITOS 3**
RT   TIROS M
       TIROS N SERIES SATELLITES
       TIROS OPERATIONAL SATELLITE
       SYSTEM

ITOS 4
GS   SATELLITES
       . ARTIFICIAL SATELLITES
       . . METEOROLOGICAL SATELLITES
       . . . TIROS SATELLITES
       . . . . IMPROVED TIROS OPERATIONAL
       SATELLITES
       . . . . . **ITOS 4**
       . . . . ITOS SATELLITES
       . . . . . **ITOS 4**
RT   TIROS M
       TIROS N SERIES SATELLITES

ITOS 4-(CONT.)
         TIROS OPERATIONAL SATELLITE
            SYSTEM

IUE
   UF   INTERNATIONAL ULTRAVIOLET
            EXPLORER
         SAS-D
   GS   OBSERVATORIES
         . ASTRONOMICAL OBSERVATORIES
         . . ASTRONOMICAL SATELLITES
         . . . **IUE**
   RT   EXPLORER SATELLITES
         EXTREME ULTRAVIOLET EXPLORER
            SATELLITE
         RADIO ASTRONOMY
         SPACEBORNE ASTRONOMY
         ULTRAVIOLET RADIATION

IUS
   USE INERTIAL UPPER STAGE

IVORY COAST
   GS   NATIONS
         . **IVORY COAST**
   RT   AFRICA

IVUNA METEORITE
   GS   CELESTIAL BODIES
         . METEORITES
         . . STONY METEORITES
         . . . CHONDRITES
         . . . . CARBONACEOUS METEORITES
         . . . . . **IVUNA METEORITE**

IZSAK ELLIPSOID
   USE ELLIPSOIDS
         GEODESY

I2S CAMERAS
   GS   OPTICAL EQUIPMENT
         . CAMERAS
         . . **I2S CAMERAS**
         PHOTOGRAPHIC EQUIPMENT
         . CAMERAS
         . . **I2S CAMERAS**
   RT   AIRCRAFT INSTRUMENTS
         MULTISPECTRAL PHOTOGRAPHY
         SPACECRAFT INSTRUMENTS

J

J INTEGRAL
   GS   ANALYSIS (MATHEMATICS)
         . FUNCTIONAL ANALYSIS
         . . INTEGRAL EQUATIONS
         . . . **J INTEGRAL**
         . REAL VARIABLES
         . . MEASURE AND INTEGRATION
         . . . **J INTEGRAL**
   RT   CRACK INITIATION
         CRACK PROPAGATION
         CRACKING (FRACTURING)
         CREEP RUPTURE STRENGTH
         ELASTOPLASTICITY
         FRACTURE MECHANICS
         FRACTURE STRENGTH
         INTEGRAL CALCULUS
         MECHANICAL PROPERTIES
         PLASTIC DEFORMATION
         STRUCTURAL ANALYSIS
         TOUGHNESS
         YIELD STRENGTH

J-2 ENGINE
   GS   ENGINES
         . ROCKET ENGINES
         . . LIQUID PROPELLANT ROCKET
               ENGINES
         . . . HYDROGEN OXYGEN ENGINES
         . . . . **J-2 ENGINE**
   RT   NOVA LAUNCH VEHICLES
         SATURN 1B LAUNCH VEHICLES
         SATURN 5 LAUNCH VEHICLES

J-33 ENGINE
   GS   ENGINES
         . AIR BREATHING ENGINES
         . . GAS TURBINE ENGINES
         . . . JET ENGINES
         . . . . TURBOJET ENGINES

J-33 ENGINE-(CONT.)
         . . . . . **J-33 ENGINE**
         . INTERNAL COMBUSTION ENGINES
         . . GAS TURBINE ENGINES
         . . . JET ENGINES
         . . . . TURBOJET ENGINES
         . . . . . **J-33 ENGINE**
         . TURBINE ENGINES
         . . GAS TURBINE ENGINES
         . . . JET ENGINES
         . . . . TURBOJET ENGINES
         . . . . . **J-33 ENGINE**
         TURBOMACHINERY
         . **J-33 ENGINE**
   RT   MACE MISSILES

J-34 ENGINE
   UF   XJ-34-WE-32 ENGINE
   GS   ENGINES
         . AIR BREATHING ENGINES
         . . GAS TURBINE ENGINES
         . . . JET ENGINES
         . . . . TURBOJET ENGINES
         . . . . . **J-34 ENGINE**
         . INTERNAL COMBUSTION ENGINES
         . . GAS TURBINE ENGINES
         . . . JET ENGINES
         . . . . TURBOJET ENGINES
         . . . . . **J-34 ENGINE**
         . TURBINE ENGINES
         . . GAS TURBINE ENGINES
         . . . JET ENGINES
         . . . . TURBOJET ENGINES
         . . . . . **J-34 ENGINE**

J-47 ENGINE
   GS   ENGINES
         . AIR BREATHING ENGINES
         . . GAS TURBINE ENGINES
         . . . JET ENGINES
         . . . . TURBOJET ENGINES
         . . . . . **J-47 ENGINE**
         . INTERNAL COMBUSTION ENGINES
         . . GAS TURBINE ENGINES
         . . . JET ENGINES
         . . . . TURBOJET ENGINES
         . . . . . **J-47 ENGINE**
         . TURBINE ENGINES
         . . GAS TURBINE ENGINES
         . . . JET ENGINES
         . . . . TURBOJET ENGINES
         . . . . . **J-47 ENGINE**

J-52 ENGINE
   GS   AIRCRAFT ENGINES
         . **J-52 ENGINE**
         ENGINES
         . INTERNAL COMBUSTION ENGINES
         . . GAS TURBINE ENGINES
         . . . JET ENGINES
         . . . . TURBOJET ENGINES
         . . . . . **J-52 ENGINE**
         . TURBINE ENGINES
         . . GAS TURBINE ENGINES
         . . . JET ENGINES
         . . . . TURBOJET ENGINES
         . . . . . **J-52 ENGINE**

J-57 ENGINE
   GS   ENGINES
         . AIR BREATHING ENGINES
         . . GAS TURBINE ENGINES
         . . . JET ENGINES
         . . . . TURBOJET ENGINES
         . . . . . **J-57 ENGINE**
         . INTERNAL COMBUSTION ENGINES
         . . GAS TURBINE ENGINES
         . . . JET ENGINES
         . . . . TURBOJET ENGINES
         . . . . . **J-57 ENGINE**
         . TURBINE ENGINES
         . . GAS TURBINE ENGINES
         . . . JET ENGINES
         . . . . TURBOJET ENGINES
         . . . . . **J-57 ENGINE**
   RT   AFTERBURNING

J-57-P-20 ENGINE
   GS   ENGINES
         . AIR BREATHING ENGINES
         . . GAS TURBINE ENGINES
         . . . JET ENGINES
         . . . . TURBOJET ENGINES
         . . . . . **J-57-P-20 ENGINE**
         . INTERNAL COMBUSTION ENGINES

J-57-P-20 ENGINE-(CONT.)
         . . GAS TURBINE ENGINES
         . . . JET ENGINES
         . . . . TURBOJET ENGINES
         . . . . . **J-57-P-20 ENGINE**
         . TURBINE ENGINES
         . . GAS TURBINE ENGINES
         . . . JET ENGINES
         . . . . TURBOJET ENGINES
         . . . . . **J-57-P-20 ENGINE**
   RT   AFTERBURNING

J-58 ENGINE
   GS   AIRCRAFT ENGINES
         . **J-58 ENGINE**

J-65 ENGINE
   GS   ENGINES
         . AIR BREATHING ENGINES
         . . GAS TURBINE ENGINES
         . . . JET ENGINES
         . . . . TURBOJET ENGINES
         . . . . . **J-65 ENGINE**
         . INTERNAL COMBUSTION ENGINES
         . . GAS TURBINE ENGINES
         . . . JET ENGINES
         . . . . TURBOJET ENGINES
         . . . . . **J-65 ENGINE**
         . TURBINE ENGINES
         . . GAS TURBINE ENGINES
         . . . JET ENGINES
         . . . . TURBOJET ENGINES
         . . . . . **J-65 ENGINE**
   RT   A-4 AIRCRAFT

J-69-T-25 ENGINE
   UF   MARBORE 2 ENGINE
   GS   ENGINES
         . AIR BREATHING ENGINES
         . . GAS TURBINE ENGINES
         . . . JET ENGINES
         . . . . TURBOJET ENGINES
         . . . . . **J-69-T-25 ENGINE**
         . INTERNAL COMBUSTION ENGINES
         . . GAS TURBINE ENGINES
         . . . JET ENGINES
         . . . . TURBOJET ENGINES
         . . . . . **J-69-T-25 ENGINE**
         . TURBINE ENGINES
         . . GAS TURBINE ENGINES
         . . . JET ENGINES
         . . . . TURBOJET ENGINES
         . . . . . **J-69-T-25 ENGINE**

J-71 ENGINE
   GS   ENGINES
         . AIR BREATHING ENGINES
         . . GAS TURBINE ENGINES
         . . . JET ENGINES
         . . . . TURBOJET ENGINES
         . . . . . **J-71 ENGINE**
         . INTERNAL COMBUSTION ENGINES
         . . GAS TURBINE ENGINES
         . . . JET ENGINES
         . . . . TURBOJET ENGINES
         . . . . . **J-71 ENGINE**
         . TURBINE ENGINES
         . . GAS TURBINE ENGINES
         . . . JET ENGINES
         . . . . TURBOJET ENGINES
         . . . . . **J-71 ENGINE**

J-73 ENGINE
   UF   YJ-73-GE-3 ENGINE
         YJ73 TURBOJET ENGINE
   GS   ENGINES
         . AIR BREATHING ENGINES
         . . GAS TURBINE ENGINES
         . . . JET ENGINES
         . . . . TURBOJET ENGINES
         . . . . . **J-73 ENGINE**
         . INTERNAL COMBUSTION ENGINES
         . . GAS TURBINE ENGINES
         . . . JET ENGINES
         . . . . TURBOJET ENGINES
         . . . . . **J-73 ENGINE**
         . TURBINE ENGINES
         . . GAS TURBINE ENGINES
         . . . JET ENGINES
         . . . . TURBOJET ENGINES
         . . . . . **J-73 ENGINE**

J-75 ENGINE
   GS   ENGINES
         . AIR BREATHING ENGINES

J-75 ENGINE-*(CONT.)*
    . . GAS TURBINE ENGINES
    . . . JET ENGINES
    . . . . TURBOJET ENGINES
    . . . . . **J-75 ENGINE**
    . INTERNAL COMBUSTION ENGINES
    . . GAS TURBINE ENGINES
    . . . JET ENGINES
    . . . . TURBOJET ENGINES
    . . . . . **J-75 ENGINE**
    . TURBINE ENGINES
    . . GAS TURBINE ENGINES
    . . . JET ENGINES
    . . . . TURBOJET ENGINES
    . . . . . **J-75 ENGINE**

J-79 ENGINE
UF   XJ-79-GE-1 ENGINE
       YJ-79 ENGINE
GS   ENGINES
    . AIR BREATHING ENGINES
    . . GAS TURBINE ENGINES
    . . . JET ENGINES
    . . . . TURBOJET ENGINES
    . . . . . **J-79 ENGINE**
    . INTERNAL COMBUSTION ENGINES
    . . GAS TURBINE ENGINES
    . . . JET ENGINES
    . . . . TURBOJET ENGINES
    . . . . . **J-79 ENGINE**
    . TURBINE ENGINES
    . . GAS TURBINE ENGINES
    . . . JET ENGINES
    . . . . TURBOJET ENGINES
    . . . . . **J-79 ENGINE**
RT   F-4 AIRCRAFT

J-85 ENGINE
UF   YJ-85 ENGINE
GS   ENGINES
    . AIR BREATHING ENGINES
    . . GAS TURBINE ENGINES
    . . . JET ENGINES
    . . . . TURBOJET ENGINES
    . . . . . **J-85 ENGINE**
    . INTERNAL COMBUSTION ENGINES
    . . GAS TURBINE ENGINES
    . . . JET ENGINES
    . . . . TURBOJET ENGINES
    . . . . . **J-85 ENGINE**
    . TURBINE ENGINES
    . . GAS TURBINE ENGINES
    . . . JET ENGINES
    . . . . TURBOJET ENGINES
    . . . . . **J-85 ENGINE**
RT   BLUE GOOSE MISSILE
       OSPREY MISSILE

J-93 ENGINE
UF   J93-MJ252H ENGINE
       J93-MJ280G ENGINE
       YJ-93 ENGINE
       YJ-93-GE-3 ENGINE
GS   ENGINES
    . AIR BREATHING ENGINES
    . . GAS TURBINE ENGINES
    . . . JET ENGINES
    . . . . TURBOJET ENGINES
    . . . . . **J-93 ENGINE**
    . INTERNAL COMBUSTION ENGINES
    . . GAS TURBINE ENGINES
    . . . JET ENGINES
    . . . . TURBOJET ENGINES
    . . . . . **J-93 ENGINE**
    . TURBINE ENGINES
    . . GAS TURBINE ENGINES
    . . . JET ENGINES
    . . . . TURBOJET ENGINES
    . . . . . **J-93 ENGINE**

J-97 ENGINE
GS   AIRCRAFT ENGINES
    . **J-97 ENGINE**
    ENGINES
    . AIR BREATHING ENGINES
    . . GAS TURBINE ENGINES
    . . . JET ENGINES
    . . . . TURBOJET ENGINES
    . . . . . TURBOFAN ENGINES
    . . . . . . **J-97 ENGINE**
    . INTERNAL COMBUSTION ENGINES
    . . GAS TURBINE ENGINES
    . . . JET ENGINES
    . . . . TURBOJET ENGINES
    . . . . . TURBOFAN ENGINES

J-97 ENGINE-*(CONT.)*
    . . . . . . **J-97 ENGINE**
    . TURBINE ENGINES
    . . GAS TURBINE ENGINES
    . . . JET ENGINES
    . . . . TURBOJET ENGINES
    . . . . . TURBOFAN ENGINES
    . . . . . . **J-97 ENGINE**

JABIRU ROCKET VEHICLE
USE   JAGUAR ROCKET VEHICLE

JACKETS
SN   (EXCLUDES CLOTHING)
RT   ABSORBERS (MATERIALS)
    ∞CASING
       COOLING
       COVERINGS
       HEATING
       INSULATION
       LININGS
       SHEATHS

JACKING EQUIPMENT
USE   JACKS (LIFTS)

∞ **JACKS**
SN   (USE OF A MORE SPECIFIC TERM IS
       RECOMMENDED--CONSULT THE TERMS
       LISTED BELOW)
RT   ELECTRIC CONNECTORS
       ELEVATORS (LIFTS)
       JACKS (LIFTS)

JACKS (ELECTRICAL)
USE   ELECTRIC CONNECTORS

JACKS (LIFTS)
UF   JACKING EQUIPMENT
RT   ∞JACKS
       ∞LIFTS
       POSITIONING DEVICES (MACHINERY)
       TUNNELING (EXCAVATION)

JACOBI INTEGRAL
GS   ANALYSIS (MATHEMATICS)
    . REAL VARIABLES
    . . **JACOBI INTEGRAL**
RT   CONFORMAL MAPPING
       DIFFUSION THEORY
       ELLIPTIC FUNCTIONS
       GREEN'S FUNCTIONS
       POTENTIAL THEORY
       WEIERSTRASS FUNCTIONS

JACOBI MATRIX METHOD
GS   ANALYSIS (MATHEMATICS)
    . REAL VARIABLES
    . . **JACOBI MATRIX METHOD**
RT   CALCULUS OF VARIATIONS
       EIGENVALUES
       EIGENVECTORS
       HERMITIAN POLYNOMIAL
       ∞METHODOLOGY

JACOBI POLYNOMIALS
USE   HYPERGEOMETRIC FUNCTIONS

JAGUAR AIRCRAFT
GS   ATTACK AIRCRAFT
    . FIGHTER AIRCRAFT
    . . **JAGUAR AIRCRAFT**
    SUPERSONIC AIRCRAFT
    . **JAGUAR AIRCRAFT**
    TRAINING AIRCRAFT
    . **JAGUAR AIRCRAFT**
RT   ∞AIRCRAFT
       BREGUET AIRCRAFT
       ∞MILITARY AIRCRAFT

JAGUAR ROCKET VEHICLE
UF   JABIRU ROCKET VEHICLE
GS   ROCKET VEHICLES
    . MULTISTAGE ROCKET VEHICLES
    . . **JAGUAR ROCKET VEHICLE**
    . SOUNDING ROCKETS
    . . **JAGUAR ROCKET VEHICLE**
RT   SOLID PROPELLANT ROCKET ENGINES

JAHN-TELLER EFFECT
RT   ∞EFFECTS
       ELECTRON PARAMAGNETIC RESONANCE
       ELECTRON TRANSITIONS
       ORBITALS

JAMAICA
GS   LANDFORMS
    . ISLANDS
    . . WEST INDIES
    . . . **JAMAICA**
    NATIONS
    . **JAMAICA**
RT   CARIBBEAN REGION

JAMMERS
RT   AIR DEFENSE
       JAMMING
       RADAR EQUIPMENT
       RADIO EQUIPMENT

JAMMING
GS   COUNTERMEASURES
    . **JAMMING**
    ELECTROMAGNETIC INTERFERENCE
    . **JAMMING**
RT   CLUTTER
       ELECTRONIC COUNTERMEASURES
       ELECTRONIC WARFARE
       FREQUENCY HOPPING
       ∞INTERFERENCE
       JAMMERS
       RADIO FREQUENCY INTERFERENCE
       WHITE NOISE

JANUS
GS   SATELLITES
    . NATURAL SATELLITES
    . . SATURN SATELLITES
    . . . **JANUS**
RT   SATURN (PLANET)

JANUS REACTOR
GS   NUCLEAR REACTORS
    . NUCLEAR RESEARCH AND TEST
        REACTORS
    . . **JANUS REACTOR**

JANUS SPACECRAFT
GS   GLIDERS
    . **JANUS SPACECRAFT**
    HYPERSONIC VEHICLES
    . LIFTING REENTRY VEHICLES
    . . **JANUS SPACECRAFT**
    LIFTING BODIES
    . LIFTING REENTRY VEHICLES
    . . **JANUS SPACECRAFT**
    MANEUVERABLE SPACECRAFT
    . LIFTING REENTRY VEHICLES
    . . **JANUS SPACECRAFT**
    MANNED SPACECRAFT
    . **JANUS SPACECRAFT**
    REENTRY VEHICLES
    . LIFTING REENTRY VEHICLES
    . . **JANUS SPACECRAFT**
    SOFT LANDING SPACECRAFT
    . **JANUS SPACECRAFT**

JAPAN
GS   LANDFORMS
    . ISLANDS
    . . PACIFIC ISLANDS
    . . . **JAPAN**
    NATIONS
    . **JAPAN**
RT   ASIA
       JAPANESE SPACECRAFT

JAPANESE SPACE PROGRAM
GS   PROGRAMS
    . SPACE PROGRAMS
    . . **JAPANESE SPACE PROGRAM**
RT   METEOROLOGICAL SATELLITES
       ∞RESEARCH PROJECTS
       SATELLITE DESIGN
       SPACE MISSIONS
       SPACE TRANSPORTATION
       ∞SPACECRAFT
       SPACECRAFT DESIGN

JAPANESE SPACECRAFT
UF   MOS (JAPANESE SPACECRAFT)
RT   JAPAN
       ∞SPACECRAFT

JARRING
USE   MECHANICAL SHOCK

JATO ENGINES
UF   JET ASSISTED TAKEOFF

JATO ENGINES-(CONT.)
GS   ENGINES
    . **JATO ENGINES**
    LAUNCHERS
    . AIRCRAFT LAUNCHING DEVICES
    . . **JATO ENGINES**
RT   SHORT TAKEOFF AIRCRAFT
    SOLID PROPELLANT ROCKET ENGINES
    TAKEOFF

JAVELIN AIRCRAFT
USE   GA-5 AIRCRAFT

JAVELIN ROCKET VEHICLE
GS   ROCKET VEHICLES
    . MULTISTAGE ROCKET VEHICLES
    . . **JAVELIN ROCKET VEHICLE**
RT   ARGO ROCKET VEHICLES
    ROCKET PROPELLED SLEDS
    SOLID PROPELLANT ROCKET ENGINES
    SOUNDING ROCKETS

JC-130 AIRCRAFT
USE   C-130 AIRCRAFT

JEANS THEORY
RT   ∞THEORIES

JEEPS
USE   AUTOMOBILES

JERBOAS
GS   ANIMALS
    . VERTEBRATES
    . . MAMMALS
    . . . RODENTS
    . . . . MICE
    . . . . . **JERBOAS**

JET AIRCRAFT
UF   JET FLIGHT
    TURBOJET AIRCRAFT
GS   **JET AIRCRAFT**
    . A-2 AIRCRAFT
    . A-3 AIRCRAFT
    . A-4 AIRCRAFT
    . A-5 AIRCRAFT
    . A-6 AIRCRAFT
    . ALPHA JET AIRCRAFT
    . AN-2 AIRCRAFT
    . AVRO 707 AIRCRAFT
    . B-47 AIRCRAFT
    . B-50 AIRCRAFT
    . B-52 AIRCRAFT
    . B-57 AIRCRAFT
    . B-58 AIRCRAFT
    . B-66 AIRCRAFT
    . B-70 AIRCRAFT
    . BOEING 747 AIRCRAFT
    . BOEING 2707 AIRCRAFT
    . BUCCANEER AIRCRAFT
    . C-5 AIRCRAFT
    . C-8A AUGMENTOR WING AIRCRAFT
    . C-9 AIRCRAFT
    . C-119 AIRCRAFT
    . C-135 AIRCRAFT
    . C-140 AIRCRAFT
    . CANBERRA AIRCRAFT
    . CL-41 AIRCRAFT
    . CL-823 AIRCRAFT
    . COMET 4 AIRCRAFT
    . CV-880 AIRCRAFT
    . D-558 AIRCRAFT
    . DC 9 AIRCRAFT
    . DC 10 AIRCRAFT
    . DH 112 AIRCRAFT
    . DH 115 AIRCRAFT
    . DH 125 AIRCRAFT
    . DHC 2 AIRCRAFT
    . EUROPEAN AIRBUS
    . . A-300 AIRCRAFT
    . . A-310 AIRCRAFT
    . . A-320 AIRCRAFT
    . F-2 AIRCRAFT
    . F-5 AIRCRAFT
    . F-8 AIRCRAFT
    . F-9 AIRCRAFT
    . F-14 AIRCRAFT
    . F-15 AIRCRAFT
    . F-16 AIRCRAFT
    . F-17 AIRCRAFT
    . F-18 AIRCRAFT
    . F-84 AIRCRAFT
    . F-86 AIRCRAFT
    . F-89 AIRCRAFT

JET AIRCRAFT-(CONT.)
    . F-94 AIRCRAFT
    . F-100 AIRCRAFT
    . F-101 AIRCRAFT
    . F-102 AIRCRAFT
    . F-104 AIRCRAFT
    . F-105 AIRCRAFT
    . F-106 AIRCRAFT
    . FD 2 AIRCRAFT
    . G-91 AIRCRAFT
    . G-95/4 AIRCRAFT
    . GA-5 AIRCRAFT
    . H-17 HELICOPTER
    . H-126 AIRCRAFT
    . HFB-320 AIRCRAFT
    . HP-115 AIRCRAFT
    . HS-801 AIRCRAFT
    . JET PROVOST AIRCRAFT
    . JETSTREAM AIRCRAFT
    . JINDIVIK TARGET AIRCRAFT
    . L-29 JET TRAINER
    . L-1011 AIRCRAFT
    . L-2000 AIRCRAFT
    . LEAR JET AIRCRAFT
    . MIRAGE AIRCRAFT
    . . MIRAGE 3 AIRCRAFT
    . NORD 1500 AIRCRAFT
    . P-3 AIRCRAFT
    . P-308 AIRCRAFT
    . P-1052 AIRCRAFT
    . PD-808 AIRCRAFT
    . PHANTOM AIRCRAFT
    . . F-4 AIRCRAFT
    . SC-1 AIRCRAFT
    . SCIMITAR AIRCRAFT
    . T-2 AIRCRAFT
    . T-33 AIRCRAFT
    . T-37 AIRCRAFT
    . T-38 AIRCRAFT
    . T-39 AIRCRAFT
    . TS-11 AIRCRAFT
    . TSR-2 AIRCRAFT
    . TU-104 AIRCRAFT
    . TU-124 AIRCRAFT
    . TURBOFAN AIRCRAFT
    . . A-7 AIRCRAFT
    . . BAC 111 AIRCRAFT
    . . BOEING 707 AIRCRAFT
    . . BOEING 720 AIRCRAFT
    . . BOEING 727 AIRCRAFT
    . . BOEING 733 AIRCRAFT
    . . BOEING 737 AIRCRAFT
    . . BOEING 757 AIRCRAFT
    . . BOEING 767 AIRCRAFT
    . . C-141 AIRCRAFT
    . . CONCORDE AIRCRAFT
    . . CV-990 AIRCRAFT
    . . DC 8 AIRCRAFT
    . . DH 121 AIRCRAFT
    . . DO-31 AIRCRAFT
    . . F-28 TRANSPORT AIRCRAFT
    . . F-111 AIRCRAFT
    . . IL-62 AIRCRAFT
    . . MYSTERE 20 AIRCRAFT
    . . P-1127 AIRCRAFT
    . . P-1154 AIRCRAFT
    . . SAAB 37 AIRCRAFT
    . . SAAB 105 AIRCRAFT
    . . SE-210 AIRCRAFT
    . . TU-134 AIRCRAFT
    . . TU-144 AIRCRAFT
    . TURBOPROP AIRCRAFT
    . . AN-22 AIRCRAFT
    . . AN-24 AIRCRAFT
    . . ARGOSY MK-1 AIRCRAFT
    . . BREGUET 941 AIRCRAFT
    . . BREGUET 1150 AIRCRAFT
    . . C-2 AIRCRAFT
    . . C-133 AIRCRAFT
    . . C-160 AIRCRAFT
    . . CL-44 AIRCRAFT
    . . CL-84 AIRCRAFT
    . . DHC 5 AIRCRAFT
    . . ELECTRA AIRCRAFT
    . . F-27 AIRCRAFT
    . . G-222 AIRCRAFT
    . . HS-748 AIRCRAFT
    . . MH-262 AIRCRAFT
    . . OV-1 AIRCRAFT
    . . OV-10 AIRCRAFT
    . . SC-5 AIRCRAFT
    . . VISCOUNT AIRCRAFT
    . . YS-11 AIRCRAFT
    . U-2 AIRCRAFT
    . VALIANT AIRCRAFT
    . VAMPIRE MK 35 AIRCRAFT

JET AIRCRAFT-(CONT.)
    . VC-10 AIRCRAFT
    . VICTOR MK-1 AIRCRAFT
    . VJ-101 AIRCRAFT
    . VULCAN AIRCRAFT
    . X-3 AIRCRAFT
    . X-5 AIRCRAFT
    . X-13 AIRCRAFT
    . X-14 AIRCRAFT
    . X-21 AIRCRAFT
    . X-21A AIRCRAFT
    . XC-142 AIRCRAFT
    . XV-4 AIRCRAFT
    . XV-5 AIRCRAFT
    . XV-9A AIRCRAFT
    . YAK 40 AIRCRAFT
RT   ∞AIRCRAFT
    AIRCRAFT NOISE
    ATTACK AIRCRAFT
    B-1 AIRCRAFT
    BOMBER AIRCRAFT
    CARGO AIRCRAFT
    COMMERCIAL AIRCRAFT
    FIGHTER AIRCRAFT
    FLYING PLATFORMS
    GENERAL AVIATION AIRCRAFT
    HYPERSONIC AIRCRAFT
    ∞JETS
    ∞LOW WING AIRCRAFT
    ∞MILITARY AIRCRAFT
    MYSTERE 50 AIRCRAFT
    PASSENGER AIRCRAFT
    RECONNAISSANCE AIRCRAFT
    RESEARCH AIRCRAFT
    SHORT TAKEOFF AIRCRAFT
    ∞SUBSONIC AIRCRAFT
    SUPERSONIC AIRCRAFT
    TAILLESS AIRCRAFT
    TANDEM WING AIRCRAFT
    TERRAIN FOLLOWING AIRCRAFT
    TRAINING AIRCRAFT
    TRANSPORT AIRCRAFT
    TU-154 AIRCRAFT
    TURBOJET ENGINES
    V/STOL AIRCRAFT
    ∞WINGED VEHICLES
    YF 12 AIRCRAFT

JET AIRCRAFT NOISE
UF   JET NOISE
GS   ELASTIC WAVES
    . SOUND WAVES
    . . NOISE (SOUND)
    . . . AIRCRAFT NOISE
    . . . . **JET AIRCRAFT NOISE**
RT   ACOUSTIC RETROFITTING
    AERODYNAMIC NOISE
    ∞AIRCRAFT
    AIRCRAFT RUNUP
    ENGINE NOISE
    MUFFLERS
    NOISE MEASUREMENT
    NOISE REDUCTION
    QUIET ENGINE PROGRAM
    SONIC BOOMS

JET AIRSTREAMS
USE   JET STREAMS (METEOROLOGY)

JET AMPLIFIERS
UF   FLUID JET AMPLIFIERS
GS   AMPLIFIERS
    . FLUID AMPLIFIERS
    . . **JET AMPLIFIERS**
RT   COANDA EFFECT
    FLUID JETS
    ∞JET NOZZLES
    NOZZLE WALLS

JET ASSISTED TAKEOFF
USE   JATO ENGINES

JET AUGMENTED WING FLAPS
USE   JET FLAPS
    WING FLAPS

JET BLAST EFFECTS
RT   ∞BLASTS
    ∞EFFECTS
    EXHAUST GASES
    GROUND EFFECT (AERODYNAMICS)
    NOISE (SOUND)
    PRESSURE EFFECTS
    TEMPERATURE EFFECTS

JET BOUNDARIES
GS   BOUNDARIES
  . FLUID BOUNDARIES
  . . **JET BOUNDARIES**
  INTERFACES
  . FLUID BOUNDARIES
  . . **JET BOUNDARIES**
RT   FREE JETS
  LIQUID SURFACES
  WALL JETS

JET CONDENSERS
GS   CONDENSERS (LIQUEFIERS)
  . **JET CONDENSERS**
RT   ∞CONDENSERS
  LIQUEFACTION
  NUCLEATION
  SPRAY CONDENSERS
  WORKING FLUIDS

JET CONTROL
RT   AUTOMATIC CONTROL
  BOUNDARY LAYER CONTROL
  ∞CONTROL
  DIRECTIONAL CONTROL
  SATELLITE ATTITUDE CONTROL
  SATELLITE CONTROL
  THRUST CONTROL
  VARIABLE THRUST

JET DAMPING
USE   DAMPING
  SPIN REDUCTION

JET DRAGON AIRCRAFT
USE   DH 125 AIRCRAFT

JET DRIVE
USE   JET PROPULSION

JET ENGINE FUELS
UF   JET FUELS
GS   FUELS
  . CHEMICAL FUELS
  . . HYDROCARBON FUELS
  . . . **JET ENGINE FUELS**
  . . . . JP-4 JET FUEL
  . . . . JP-5 JET FUEL
  . . . . JP-6 JET FUEL
  . . . . JP-8 JET FUEL
  . . LIQUID FUELS
  . . . **JET ENGINE FUELS**
  . . . . JP-4 JET FUEL
  . . . . JP-5 JET FUEL
  . . . . JP-6 JET FUEL
  . . . . JP-8 JET FUEL
RT   AIRCRAFT FUELS
  ANTIMISTING FUELS
  GASOLINE
  KEROSENE
  TURBINES

JET ENGINES
SN   (EXCLUDES HYDROJET ENGINES)
GS   ENGINES
  . AIR BREATHING ENGINES
  . . GAS TURBINE ENGINES
  . . . **JET ENGINES**
  . . . . RAMJET ENGINES
  . . . . . INTEGRAL ROCKET RAMJETS
  . . . . . LOW VOLUME RAMJET ENGINES
  . . . . PULSEJET ENGINES
  . . . . . SUPERSONIC COMBUSTION
     RAMJET ENGINES
  . . . . . TURBORAMJET ENGINES
  . . . . TURBOJET ENGINES
  . . . . . BRISTOL-SIDDELEY OLYMPUS
     593 ENGINE
  . . . . . BRISTOL-SIDDELEY VIPER
     ENGINE
  . . . . . DUCTED FAN ENGINES
  . . . . . J-33 ENGINE
  . . . . . J-34 ENGINE
  . . . . . J-47 ENGINE
  . . . . . J-57 ENGINE
  . . . . . J-57-P-20 ENGINE
  . . . . . J-65 ENGINE
  . . . . . J-69-T-25 ENGINE
  . . . . . J-71 ENGINE
  . . . . . J-73 ENGINE
  . . . . . J-75 ENGINE
  . . . . . J-79 ENGINE
  . . . . . J-85 ENGINE
  . . . . . J-93 ENGINE
  . . . . . RA-28 ENGINE

JET ENGINES-(CONT.)
  . . . . TURBOFAN ENGINES
  . . . . . BRISTOL-SIDDELEY BS 53
     ENGINE
  . . . . . CF-700 ENGINE
  . . . . . J-97 ENGINE
  . . . . . TF-41 ENGINE
  . . . . TURBOPROP ENGINES
  . . . . . T-53 ENGINE
  . . . . . T-56 ENGINE
  . . . . . T-64 ENGINE
  . . . . . T-74 ENGINE
  . . . . TURBORAMJET ENGINES
  . INTERNAL COMBUSTION ENGINES
  . . GAS TURBINE ENGINES
  . . . **JET ENGINES**
  . . . . RAMJET ENGINES
  . . . . . INTEGRAL ROCKET RAMJETS
  . . . . . LOW VOLUME RAMJET ENGINES
  . . . . . PULSEJET ENGINES
  . . . . . SUPERSONIC COMBUSTION
     RAMJET ENGINES
  . . . . . TURBORAMJET ENGINES
  . . . T-63 ENGINE
  . . . T-76 ENGINE
  . . . TURBOJET ENGINES
  . . . . BRISTOL-SIDDELEY OLYMPUS
     593 ENGINE
  . . . . BRISTOL-SIDDELEY VIPER
     ENGINE
  . . . . DUCTED FAN ENGINES
  . . . . J-33 ENGINE
  . . . . J-34 ENGINE
  . . . . J-47 ENGINE
  . . . . J-52 ENGINE
  . . . . J-57 ENGINE
  . . . . J-57-P-20 ENGINE
  . . . . J-65 ENGINE
  . . . . J-69-T-25 ENGINE
  . . . . J-71 ENGINE
  . . . . J-73 ENGINE
  . . . . J-75 ENGINE
  . . . . J-79 ENGINE
  . . . . J-85 ENGINE
  . . . . J-93 ENGINE
  . . . . RA-28 ENGINE
  . . . . TURBOFAN ENGINES
  . . . . . BRISTOL-SIDDELEY BS 53
     ENGINE
  . . . . . CF-700 ENGINE
  . . . . . J-97 ENGINE
  . . . . . TF-30 ENGINE
  . . . . . TF-41 ENGINE
  . . . . TURBOPROP ENGINES
  . . . . . T-34 ENGINE
  . . . . . T-38 ENGINE
  . . . . . T-53 ENGINE
  . . . . . T-56 ENGINE
  . . . . . T-64 ENGINE
  . . . . . T-74 ENGINE
  . . . . . T-78 ENGINE
  . . . . TURBORAMJET ENGINES
  . TURBINE ENGINES
  . . GAS TURBINE ENGINES
  . . **JET ENGINES**
  . . . RAMJET ENGINES
  . . . . LOW VOLUME RAMJET ENGINES
  . . . . PULSEJET ENGINES
  . . . . SUPERSONIC COMBUSTION
     RAMJET ENGINES
  . . . . TURBORAMJET ENGINES
  . . . T-63 ENGINE
  . . . T-76 ENGINE
  . . . TURBOJET ENGINES
  . . . . BRISTOL-SIDDELEY OLYMPUS
     593 ENGINE
  . . . . BRISTOL-SIDDELEY VIPER
     ENGINE
  . . . . DUCTED FAN ENGINES
  . . . . J-33 ENGINE
  . . . . J-34 ENGINE
  . . . . J-47 ENGINE
  . . . . J-52 ENGINE
  . . . . J-57 ENGINE
  . . . . J-57-P-20 ENGINE
  . . . . J-65 ENGINE
  . . . . J-69-T-25 ENGINE
  . . . . J-71 ENGINE
  . . . . J-73 ENGINE
  . . . . J-75 ENGINE
  . . . . J-79 ENGINE
  . . . . J-85 ENGINE
  . . . . J-93 ENGINE
  . . . . RA-28 ENGINE
  . . . . TURBOFAN ENGINES

JET ENGINES-(CONT.)
  . . . . . . BRISTOL-SIDDELEY BS 53
     ENGINE
  . . . . . . CF-700 ENGINE
  . . . . . . J-97 ENGINE
  . . . . . . TF-30 ENGINE
  . . . . . . TF-41 ENGINE
  . . . . . TURBOPROP ENGINES
  . . . . . . T-34 ENGINE
  . . . . . . T-38 ENGINE
  . . . . . . T-53 ENGINE
  . . . . . . T-56 ENGINE
  . . . . . . T-64 ENGINE
  . . . . . . T-74 ENGINE
  . . . . . . T-78 ENGINE
  . . . . . TURBORAMJET ENGINES
RT   AFTERBURNING
  AIRCRAFT ENGINES
  COMBUSTION CHAMBERS
  EJECTORS
  ENGINE STARTERS
  EXHAUST NOZZLES
  FLAMEOUT
  FLYING EJECTION SEATS
  FUEL INJECTION
  FUEL PUMPS
  HELICOPTER ENGINES
  HYBRID PROPELLANT ROCKET ENGINES
  HYBRID PROPULSION
  INFRARED SUPPRESSION
  ∞JET NOZZLES
  QUIET ENGINE PROGRAM
  REACTION PRODUCTS
  ROCKET ENGINES
  THRUST

JET EXHAUST
UF   HOT JET EXHAUST
RT   BASE HEATING
  EXHAUST EMISSION
  EXHAUST GASES
  INFRARED SUPPRESSION
  ROCKET EXHAUST

JET FLAMES
USE   FLAMES
  JET FLOW

JET FLAPS
UF   JET AUGMENTED WING FLAPS
GS   AIRFOILS
  . FLAPS (CONTROL SURFACES)
  . . **JET FLAPS**
  CONTROL SURFACES
  . FLAPS (CONTROL SURFACES)
  . . **JET FLAPS**
RT   EXTERNALLY BLOWN FLAPS
  H-126 AIRCRAFT
  SHORT TAKEOFF AIRCRAFT
  SPLIT FLAPS
  TRAILING EDGE FLAPS
  VORTEX FLAPS
  WING FLAPS

JET FLIGHT
USE   JET AIRCRAFT

JET FLOW
UF   HOT JETS
  JET FLAMES
  LAMINAR JETS
  REACTION JETS
GS   FLUID FLOW
  . **JET FLOW**
  . . AIR JETS
  . PERIPHERAL JET FLOW
  . . SUPERSONIC JET FLOW
RT   FLUID JETS
  FREE BOUNDARIES
  FREE JETS
  GAS STREAMS
  HYDRAULIC JETS
  INJECTORS
  JET MEMBRANE PROCESS
  ∞JETS
  NOZZLE FLOW
  PARTICLE LADEN JETS
  SPANWISE BLOWING
  TWO DIMENSIONAL JETS
  VAPOR JETS
  WALL JETS

JET FUELS
USE   JET ENGINE FUELS

JET IMPINGEMENT
GS　　IMPINGEMENT
　　　. **JET IMPINGEMENT**
RT　　ABLATION
　　　BASE HEATING

JET LAG
GS　　BIOLOGICAL EFFECTS
　　　. **JET LAG**
　　　DISORIENTATION
　　　. **JET LAG**
　　　PSYCHOLOGICAL EFFECTS
　　　. **JET LAG**
RT　　DESYNCHRONIZATION (BIOLOGY)
　　　DISORDERS
　　　FLIGHT STRESS (BIOLOGY)
　　　RHYTHM (BIOLOGY)
　　　SUPERSONIC FLIGHT

JET LIFT
GS　　AERODYNAMIC CHARACTERISTICS
　　　. LIFT
　　　. . **JET LIFT**
　　　AERODYNAMIC FORCES
　　　. LIFT
　　　. . **JET LIFT**
　　　DYNAMIC CHARACTERISTICS
　　　. LIFT
　　　. . **JET LIFT**
RT　　DISTRIBUTION (PROPERTY)

JET MEMBRANE PROCESS
GS　　ENRICHMENT
　　　. ISOTOPIC ENRICHMENT
　　　. . **JET MEMBRANE PROCESS**
RT　　GAS DYNAMICS
　　　ISOTOPE SEPARATION
　　　ISOTOPES
　　　JET FLOW
　　　MEMBRANES
　　　∞PROCESSES
　　　URANIUM

JET MIXING FLOW
GS　　FLUID FLOW
　　　. **JET MIXING FLOW**
RT　　FLUID JETS
　　　FREE BOUNDARIES
　　　FUEL INJECTION
　　　INJECTORS
　　　∞JETS
　　　MIXING
　　　PREMIXING
　　　TWO DIMENSIONAL JETS

JET NOISE
USE　　JET AIRCRAFT NOISE

∞ **JET NOZZLES**
SN　　(USE OF A MORE SPECIFIC TERM IS
　　　RECOMMENDED--CONSULT THE TERMS
　　　LISTED BELOW)
RT　　CARBURETORS
　　　CONICAL NOZZLES
　　　EXHAUST DIFFUSERS
　　　EXHAUST NOZZLES
　　　FUEL INJECTION
　　　INJECTORS
　　　JET AMPLIFIERS
　　　JET ENGINES
　　　SKIRTS

JET PILOTS
USE　　AIRCRAFT PILOTS

JET PROPULSION
UF　　JET DRIVE
GS　　PROPULSION
　　　. **JET PROPULSION**
RT　　AIRCRAFT ENGINES
　　　CHEMICAL PROPULSION
　　　HELICOPTER PROPELLER DRIVE
　　　MARINE PROPULSION
　　　ROCKET ENGINES
　　　SQUID PROJECT
　　　TURBINES

JET PROVOST AIRCRAFT
UF　　HUNTING P-84 AIRCRAFT
　　　P-84 AIRCRAFT
GS　　ATTACK AIRCRAFT
　　　. FIGHTER AIRCRAFT
　　　. . **JET PROVOST AIRCRAFT**
　　　BAC AIRCRAFT

JET PROVOST AIRCRAFT-*(CONT.)*
　　　. **JET PROVOST AIRCRAFT**
　　　JET AIRCRAFT
　　　. **JET PROVOST AIRCRAFT**
　　　MONOPLANES
　　　. **JET PROVOST AIRCRAFT**
　　　TRAINING AIRCRAFT
　　　. **JET PROVOST AIRCRAFT**
RT　　∞AIRCRAFT

JET PUMPS
SN　　(EXCLUDES DEVICES USING A LIQUID
　　　OR GAS TO INDUCE MOVEMENT OF A
　　　GAS SUCH AS AIR EJECTORS)
GS　　PUMPS
　　　. **JET PUMPS**
RT　　EJECTORS
　　　FUEL PUMPS
　　　∞JETS
　　　∞PUMPING
　　　TURBINE PUMPS
　　　VACUUM PUMPS

JET STAR AIRCRAFT
USE　　C-140 AIRCRAFT

JET STREAMS (METEOROLOGY)
UF　　JET AIRSTREAMS
GS　　FLUID FLOW
　　　. GAS FLOW
　　　. . AIR FLOW
　　　. . . AIR CURRENTS
　　　. . . . **JET STREAMS (METEOROLOGY)**
　　　WIND (METEOROLOGY)
　　　. WINDS ALOFT
　　　. . **JET STREAMS (METEOROLOGY)**
RT　　AIR JETS
　　　ATMOSPHERIC CIRCULATION
　　　CIRCUMPOLAR WESTERLIES
　　　CLEAR AIR TURBULENCE
　　　COANDA EFFECT
　　　FLUID JETS
　　　FREE BOUNDARIES
　　　TURBULENT JETS

JET THRUST
UF　　REACTION JETS
GS　　THRUST
　　　. **JET THRUST**
RT　　COLD GAS
　　　HIGH THRUST
　　　LOW THRUST
　　　MICROTHRUST
　　　ROCKET THRUST
　　　STATIC THRUST
　　　THRUST LOADS
　　　VARIABLE THRUST

JET VANES
GS　　CONTROL SURFACES
　　　. GUIDE VANES
　　　. . **JET VANES**
　　　VANES
　　　. GUIDE VANES
　　　. . **JET VANES**
RT　　AIRFOILS
　　　THRUST VECTOR CONTROL
　　　WALL JETS

JETAVATORS
USE　　GUIDE VANES

∞ **JETS**
SN　　(USE OF A MORE SPECIFIC TERM IS
　　　RECOMMENDED--CONSULT THE TERMS
　　　LISTED BELOW)
RT　　AIR JETS
　　　FLUID JETS
　　　FREE JETS
　　　GAS JETS
　　　INJECTORS
　　　JET AIRCRAFT
　　　JET FLOW
　　　JET MIXING FLOW
　　　JET PUMPS
　　　PLASMA JETS
　　　SPRAYERS
　　　TURBULENT JETS
　　　TWO DIMENSIONAL JETS
　　　WALL JETS

JETSTREAM AIRCRAFT
GS　　COMMERCIAL AIRCRAFT
　　　. **JETSTREAM AIRCRAFT**
　　　GRUMMAN AIRCRAFT

JETSTREAM AIRCRAFT-*(CONT.)*
　　　. **JETSTREAM AIRCRAFT**
　　　JET AIRCRAFT
　　　. **JETSTREAM AIRCRAFT**
　　　PASSENGER AIRCRAFT
　　　. **JETSTREAM AIRCRAFT**
RT　　∞AIRCRAFT

JETTIES
USE　　BREAKWATERS

JETTISON SYSTEMS
RT　　BAILOUT
　　　EJECTION
　　　EJECTION SEATS
　　　ESCAPE (ABANDONMENT)
　　　ESCAPE SYSTEMS
　　　JETTISONING
　　　∞SYSTEMS
　　　WING TANKS

JETTISONING
RT　　BAILOUT
　　　DISPOSAL
　　　DUMPING
　　　EJECTION
　　　EMPTYING
　　　ESCAPE (ABANDONMENT)
　　　EXPULSION
　　　JETTISON SYSTEMS
　　　SPILLING

JF 101 AIRCRAFT
USE　　F-101 AIRCRAFT

JFET
UF　　JUNCTION FIELD EFFECT TRANSISTORS
GS　　ELECTRONIC EQUIPMENT
　　　. SOLID STATE DEVICES
　　　. . SEMICONDUCTOR DEVICES
　　　. . . TRANSISTORS
　　　. . . . FIELD EFFECT TRANSISTORS
　　　. . . . . **JFET**
　　　. . . . JUNCTION TRANSISTORS
　　　. . . . . **JFET**
RT　　BARRIER LAYERS
　　　∞JUNCTIONS

JIGS
GS　　POSITIONING DEVICES (MACHINERY)
　　　. **JIGS**
RT　　CLAMPS
　　　FIXTURES
　　　HOLDERS
　　　MECHANICAL DEVICES
　　　TOOLS

JIMSPHERE BALLOONS
GS　　EXPANDABLE STRUCTURES
　　　. INFLATABLE STRUCTURES
　　　. . BALLOONS
　　　. . . HIGH ALTITUDE BALLOONS
　　　. . . . **JIMSPHERE BALLOONS**
　　　. . . METEOROLOGICAL BALLOONS
　　　. . . . **JIMSPHERE BALLOONS**
RT　　WIND (METEOROLOGY)

JINDIVIK TARGET AIRCRAFT
GS　　DRONE VEHICLES
　　　. DRONE AIRCRAFT
　　　. . TARGET DRONE AIRCRAFT
　　　. . . **JINDIVIK TARGET AIRCRAFT**
　　　JET AIRCRAFT
　　　. **JINDIVIK TARGET AIRCRAFT**
　　　MONOPLANES
　　　. **JINDIVIK TARGET AIRCRAFT**
　　　PILOTLESS AIRCRAFT
　　　. DRONE AIRCRAFT
　　　. . TARGET DRONE AIRCRAFT
　　　. . . **JINDIVIK TARGET AIRCRAFT**
　　　TARGETS
　　　. **JINDIVIK TARGET AIRCRAFT**
RT　　∞AIRCRAFT
　　　REMOTELY PILOTED VEHICLES

JITTER
USE　　VIBRATION

JOBS
USE　　TASKS

JODRELL BANK OBSERVATORY
GS　　OBSERVATORIES
　　　. **JODRELL BANK OBSERVATORY**

JODRELL BANK OBSERVATORY-*(CONT.)*
RT   ASTRONOMICAL OBSERVATORIES
GROUND STATIONS
RADIO TELESCOPES
TRACKING STATIONS

JOHNSTON ISLAND
GS   LANDFORMS
. ISLANDS
. . PACIFIC ISLANDS
. . . **JOHNSTON ISLAND**

∞ **JOINING**
SN   *(USE OF A MORE SPECIFIC TERM IS*
RECOMMENDED--CONSULT THE TERMS
LISTED BELOW)
UF   INTERCONNECTION
LINKING
RT   ADHESION
ADHESIVE BONDING
ASSEMBLING
BEAM LEADS
BINDING
BONDING
BRAZING
COLD WORKING
COUPLINGS
CROSSLINKING
FITTING
FUSION (MELTING)
INERTIA BONDING
JOINTS (JUNCTIONS)
LOCKING
MOORING
MOUNTING
POSITIONING
RETAINING
RIVETING
SEALING
SEWING
SOLDERING
SPLICING
ULTRASONIC SOLDERING
WELDING
YOKES

JOINT EUROPEAN TORUS
GS   NUCLEAR REACTORS
. TOKAMAK DEVICES
. . **JOINT EUROPEAN TORUS**
PLASMA GENERATORS
. TOKAMAK DEVICES
. . **JOINT EUROPEAN TORUS**
RT   CONTROLLED FUSION
REACTOR TECHNOLOGY

JOINTS (ANATOMY)
GS   ANATOMY
. MUSCULOSKELETAL SYSTEM
. . **JOINTS (ANATOMY)**
. . . ELBOW (ANATOMY)
. . . KNEE (ANATOMY)
. . . WRIST
RT   ARTHRITIS
BONES
CONNECTIVE TISSUE
FLEXORS
LIGAMENTS
SHOULDERS

JOINTS (JUNCTIONS)
UF   CONNECTIONS
SHANKS
GS   **JOINTS (JUNCTIONS)**
. BEAM LEADS
. BUTT JOINTS
. LAP JOINTS
. METAL JOINTS
. . SOLDERED JOINTS
. . WELDED JOINTS
. . . SPOT WELDS
. RIVETED JOINTS
. SEAMS (JOINTS)
RT   ADAPTERS
ADHESIVES
BALLS
BARRIER LAYERS
BELLOWS
BONDING
CLOSURES
CONNECTORS
CORNERS
COUPLINGS
FASTENERS
FILLETS

JOINTS (JUNCTIONS)-*(CONT.)*
FITTINGS
∞ JOINING
∞ JUNCTIONS
LINKAGES
METAL BONDING
SLEEVES
STRUCTURAL MEMBERS
SWIVELS
UNIONS (CONNECTORS)

JORDAN
GS   NATIONS
. **JORDAN**

JORDAN FORM
GS   ALGEBRA
. VECTOR SPACES
. . MATRICES (MATHEMATICS)
. . . **JORDAN FORM**
RT   EIGENVALUES
LINEAR TRANSFORMATIONS
TENSORS

JOSEPHSON JUNCTIONS
RT   SQUID (DETECTORS)
SUPERCONDUCTIVITY

JOUKOWSKI TRANSFORMATION
RT   AIRFOIL PROFILES
COMPLEX VARIABLES
COORDINATE TRANSFORMATIONS
KUTTA-JOUKOWSKI CONDITION
THEODORSEN TRANSFORMATION

JOULE HEATING
USE   OHMIC DISSIPATION
RESISTANCE HEATING

JOULE-THOMSON EFFECT
RT   CRYOGENICS
∞ EFFECTS
ENTHALPY
GAS EXPANSION
GAS FLOW
KINETIC THEORY
OHMIC DISSIPATION
THERMODYNAMIC PROPERTIES
THERMODYNAMICS
THROTTLING

JOURNAL BEARINGS
GS   BEARINGS
. **JOURNAL BEARINGS**
RT   ANTIFRICTION BEARINGS

∞ **JOURNALS**
SN   *(USE OF A MORE SPECIFIC TERM IS*
RECOMMENDED--CONSULT THE TERMS
LISTED BELOW)
RT   NEWS MEDIA
PERIODICALS
SHAFTS (MACHINE ELEMENTS)

JOURNALS (DOCUMENTS)
USE   PERIODICALS

JOURNALS (SHAFTS)
USE   SHAFTS (MACHINE ELEMENTS)

JP-4 JET FUEL
GS   FUELS
. CHEMICAL FUELS
. . HYDROCARBON FUELS
. . . JET ENGINE FUELS
. . . . **JP-4 JET FUEL**
. . LIQUID FUELS
. . . JET ENGINE FUELS
. . . . **JP-4 JET FUEL**
RT   JP-6 JET FUEL
JP-8 JET FUEL
RP-1 ROCKET PROPELLANTS

JP-5 JET FUEL
GS   FUELS
. CHEMICAL FUELS
. . HYDROCARBON FUELS
. . . JET ENGINE FUELS
. . . . **JP-5 JET FUEL**
. . LIQUID FUELS
. . . JET ENGINE FUELS
. . . . **JP-5 JET FUEL**

JP-6 JET FUEL
GS   FUELS
. CHEMICAL FUELS
. . HYDROCARBON FUELS
. . . JET ENGINE FUELS
. . . . **JP-6 JET FUEL**
. . LIQUID FUELS
. . . JET ENGINE FUELS
. . . . **JP-6 JET FUEL**
RT   JP-4 JET FUEL
JP-8 JET FUEL

JP-8 JET FUEL
GS   FUELS
. CHEMICAL FUELS
. . HYDROCARBON FUELS
. . . JET ENGINE FUELS
. . . . **JP-8 JET FUEL**
. . LIQUID FUELS
. . . JET ENGINE FUELS
. . . . **JP-8 JET FUEL**
RT   JP-4 JET FUEL
JP-6 JET FUEL
KEROSENE

JUDGMENTS
RT   DECISION MAKING
DECISIONS
LEGAL LIABILITY
PENALTIES

JUDI-DART ROCKET
GS   MEASURING INSTRUMENTS
. SONDES
. . **JUDI-DART ROCKET**
ROCKET VEHICLES
. SOUNDING ROCKETS
. . **JUDI-DART ROCKET**
RT   ROCKET SOUNDING

JUICES
GS   LIQUIDS
. **JUICES**
RT   CREATINE

JUMPERS
GS   CONDUCTORS
. ELECTROLYTES
. . **JUMPERS**
RT   CONNECTORS
SHORT CIRCUITS
∞ TERMINALS
WIRE

JUNCTION DIODES
GS   ELECTRONIC EQUIPMENT
. DIODES
. . SEMICONDUCTOR DIODES
. . . **JUNCTION DIODES**
. SOLID STATE DEVICES
. . SEMICONDUCTOR DEVICES
. . . **JUNCTION DIODES**
. . . . MIM DIODES
. . . . STEP RECOVERY DIODES
RT   BARRIER LAYERS
BARRITT DIODES
GERMANIUM DIODES
HETEROJUNCTION DEVICES
HETEROJUNCTIONS
TUNNEL DIODES
VARACTOR DIODES

JUNCTION FIELD EFFECT TRANSISTORS
USE   JFET

JUNCTION TRANSISTORS
GS   ELECTRONIC EQUIPMENT
. SOLID STATE DEVICES
. . SEMICONDUCTOR DEVICES
. . . TRANSISTORS
. . . . **JUNCTION TRANSISTORS**
. . . . . JFET
RT   BARRIER LAYERS
EPITAXY
ION IMPLANTATION
∞ JUNCTIONS
MBM JUNCTIONS
PHOTOTRANSISTORS
THYRISTORS

∞ **JUNCTIONS**
SN   *(USE OF A MORE SPECIFIC TERM IS*
RECOMMENDED--CONSULT THE TERMS
LISTED BELOW)

JUNCTIONS-*(CONT.)*
RT   CONNECTORS
      INTERSECTIONS
      JFET
      JOINTS (JUNCTIONS)
      JUNCTION TRANSISTORS
      P-N-P-N JUNCTIONS
      SEMICONDUCTOR DEVICES
      SEMICONDUCTOR JUNCTIONS

JUNGLES
USE   TROPICAL REGIONS

JUNO LAUNCH VEHICLES
GS   LAUNCH VEHICLES
     . **JUNO LAUNCH VEHICLES**
     . . JUNO 1 LAUNCH VEHICLE
     . . JUNO 2 LAUNCH VEHICLE
RT   LIQUID PROPELLANT ROCKET ENGINES
      SOLID PROPELLANT ROCKET ENGINES
      ∞VEHICLES

JUNO 1 LAUNCH VEHICLE
GS   LAUNCH VEHICLES
     . JUNO LAUNCH VEHICLES
     . . **JUNO 1 LAUNCH VEHICLE**
     ROCKET VEHICLES
     . MULTISTAGE ROCKET VEHICLES
     . . **JUNO 1 LAUNCH VEHICLE**
RT   EXPLORER SATELLITES
      JUPITER C ROCKET VEHICLE
      LIQUID PROPELLANT ROCKET ENGINES
      SERGEANT MISSILES
      SOLID PROPELLANT ROCKET ENGINES

JUNO 2 LAUNCH VEHICLE
GS   LAUNCH VEHICLES
     . JUNO LAUNCH VEHICLES
     . . **JUNO 2 LAUNCH VEHICLE**
     ROCKET VEHICLES
     . MULTISTAGE ROCKET VEHICLES
     . . **JUNO 2 LAUNCH VEHICLE**
RT   EXPLORER 11 SATELLITE
      JUPITER MISSILE
      LIQUID PROPELLANT ROCKET ENGINES
      PIONEER SPACE PROBES
      PIONEER 3 SPACE PROBE
      PIONEER 4 SPACE PROBE
      PIONEER 6 SPACE PROBE
      PIONEER 7 SPACE PROBE
      PIONEER 8 SPACE PROBE
      SERGEANT MISSILES
      SOLID PROPELLANT ROCKET ENGINES

JUPITER (PLANET)
GS   CELESTIAL BODIES
     . PLANETS
     . . GAS GIANT PLANETS
     . . . **JUPITER (PLANET)**
RT   AMALTHEA
      AMOR ASTEROID
      APOLLO ASTEROIDS
      CALLISTO
      EUROPA
      GALILEAN SATELLITES
      GALILEO PROBE
      GALILEO SPACECRAFT
      GANYMEDE
      IO
      JUPITER ATMOSPHERE
      JUPITER PROBES
      JUPITER RED SPOT
      JUPITER RINGS
      JUPITER SATELLITES
      VOYAGER 1 SPACECRAFT
      VOYAGER 2 SPACECRAFT
      VOYAGER 1977 MISSION

JUPITER ATMOSPHERE
GS   ENVIRONMENTS
     . EXTRATERRESTRIAL ENVIRONMENTS
     . . PLANETARY ENVIRONMENTS
     . . . PLANETARY ATMOSPHERES
     . . . . **JUPITER ATMOSPHERE**
RT   AEROSPACE ENVIRONMENTS
      GALILEO PROJECT
      GEOPHYSICAL FLUID FLOW CELLS
      JUPITER (PLANET)
      JUPITER RINGS

JUPITER C ROCKET VEHICLE
GS   ROCKET VEHICLES
     . MULTISTAGE ROCKET VEHICLES
     . . **JUPITER C ROCKET VEHICLE**
RT   EXPLORER SATELLITES

JUPITER C ROCKET VEHICLE-*(CONT.)*
      JUNO 1 LAUNCH VEHICLE
      JUPITER MISSILE
      LAUNCH VEHICLES
      LIQUID PROPELLANT ROCKET ENGINES
      SERGEANT MISSILES
      SOLID PROPELLANT ROCKET ENGINES

JUPITER MISSILE
GS   MISSILES
     . BALLISTIC MISSILES
     . . INTERMEDIATE RANGE BALLISTIC
        MISSILES
     . . . **JUPITER MISSILE**
     . SURFACE TO SURFACE MISSILES
     . . INTERMEDIATE RANGE BALLISTIC
        MISSILES
     . . . **JUPITER MISSILE**
RT   JUNO 2 LAUNCH VEHICLE
      JUPITER C ROCKET VEHICLE
      LIQUID PROPELLANT ROCKET ENGINES

JUPITER PROBES
GS   INTERPLANETARY SPACECRAFT
     . **JUPITER PROBES**
     . . GALILEO PROBE
     . . GALILEO SPACECRAFT
     UNMANNED SPACECRAFT
     . SPACE PROBES
     . . **JUPITER PROBES**
     . . . GALILEO PROBE
     . . . GALILEO SPACECRAFT
RT   GALILEO PROJECT
      JUPITER (PLANET)
      VOYAGER 1 SPACECRAFT
      VOYAGER 2 SPACECRAFT
      VOYAGER 1977 MISSION

JUPITER PROJECT
GS   PROGRAMS
     . NASA PROGRAMS
     . . **JUPITER PROJECT**
     . NASA SPACE PROGRAMS
     . . **JUPITER PROJECT**
     . PROJECTS
     . . **JUPITER PROJECT**
RT   LAUNCH VEHICLES

JUPITER RED SPOT
RT   GAS GIANT PLANETS
      JUPITER (PLANET)
      PLANETARY SURFACES
      PLANETS
      SURFACE PROPERTIES
      TOPOGRAPHY

JUPITER RINGS
GS   CELESTIAL BODIES
     . PLANETARY RINGS
     . . **JUPITER RINGS**
RT   JUPITER (PLANET)
      JUPITER ATMOSPHERE
      JUPITER SATELLITES
      PLANETARY COMPOSITION
      PLANETARY STRUCTURE
      PLANETOLOGY
      ∞RINGS
      SATURN RINGS
      SPACE EXPLORATION
      URANUS RINGS
      VOYAGER 1 SPACECRAFT

JUPITER SATELLITES
GS   CELESTIAL BODIES
     . NATURAL SATELLITES
     . . **JUPITER SATELLITES**
     . . . GALILEAN SATELLITES
     . . . . CALLISTO
     . . . . EUROPA
     . . . . GANYMEDE
     . . . . IO
     SATELLITES
     . NATURAL SATELLITES
     . . **JUPITER SATELLITES**
     . . . AMALTHEA
     . . . GALILEAN SATELLITES
     . . . . CALLISTO
     . . . . EUROPA
     . . . . GANYMEDE
     . . . . IO
RT   JUPITER (PLANET)
      JUPITER RINGS
      SOLAR SYSTEM

J93-MJ252H ENGINE
USE   J-93 ENGINE

J93-MJ280G ENGINE
USE   J-93 ENGINE

K

K BAND
USE   EXTREMELY HIGH FREQUENCIES

K LINES
GS   SPECTRA
     . RADIATION SPECTRA
     . . ELECTROMAGNETIC SPECTRA
     . . . LINE SPECTRA
     . . . . **K LINES**
RT   ABSORPTION SPECTRA
      EMISSION SPECTRA
      H LINES

K-MESONS
USE   KAONS

KA BAND
USE   EXTREMELY HIGH FREQUENCIES

KA-6 SAILPLANES
UF   SCHLEICHER KA-6 SAILPLANE
GS   GLIDERS
     . **KA-6 SAILPLANES**
RT   ∞AIRCRAFT
      GLIDING
      SAILWINGS
      ∞WINGED VEHICLES

KAKUTANI THEOREM
GS   THEOREMS
     . **KAKUTANI THEOREM**
RT   LATTICES (MATHEMATICS)
      STOCHASTIC PROCESSES
      VECTOR SPACES

KALIHARI BASIN (AFRICA)
GS   LANDFORMS
     . STRUCTURAL BASINS
     . . **KALIHARI BASIN (AFRICA)**
RT   AFRICA
      DESERTS
      REPUBLIC OF SOUTH AFRICA

KALMAN FILTERS
GS   LINEAR FILTERS
     . **KALMAN FILTERS**
RT   ELECTRIC FILTERS
      ∞FILTERS
      NAVIGATION AIDS
      OPTIMIZATION
      REDUCED ORDER FILTERS
      STATE ESTIMATION

KALMAN-SCHMIDT FILTERING
RT   ∞APPLICATIONS OF MATHEMATICS
      FEEDBACK CONTROL
      INERTIAL PLATFORMS
      NAVIGATION INSTRUMENTS
      OPTIMAL CONTROL
      OPTIMIZATION
      REMOTE CONTROL
      STOCHASTIC PROCESSES
      TIME SERIES ANALYSIS

KAMACITE
GS   ALLOYS
     . NICKEL ALLOYS
     . . **KAMACITE**
     MINERALS
     . **KAMACITE**
RT   IRON ALLOYS
      IRON METEORITES
      METEORITIC COMPOSITION

KAMAN AIRCRAFT
GS   **KAMAN AIRCRAFT**
     . H-43 HELICOPTER
     . HH-43 HELICOPTER
     . UH-2 HELICOPTER
RT   ∞AIRCRAFT

KAMAN UH-2A HELICOPTER
USE   UH-2 HELICOPTER

KAMPUCHEA
 USE   CAMBODIA

KANSAS
 GS   NATIONS
        . UNITED STATES
        . . **KANSAS**
 RT   MISSOURI RIVER (US)

KAOLINITE
 GS   ALUMINUM COMPOUNDS
        . ALUMINUM SILICATES
        . . **KAOLINITE**
       CLAYS
        . **KAOLINITE**
       MINERALS
        . **KAOLINITE**
       SILICON COMPOUNDS
        . ALUMINUM SILICATES
        . . **KAOLINITE**
        . SILICATES
        . . **KAOLINITE**
 RT   ALUMINUM OXIDES
       ION EXCHANGING
       SOILS

KAON PRODUCTION
 GS   PARTICLE PRODUCTION
        . **KAON PRODUCTION**
 RT   KAONS
       PARTICLE ACCELERATORS

KAONS
 UF   K-MESONS
 GS   PARTICLES
        . ELEMENTARY PARTICLES
        . . BOSONS
        . . . MESONS
        . . . . **KAONS**
        . . HADRONS
        . . . MESONS
        . . . . **KAONS**
        . NUCLEAR PARTICLES
        . . BOSONS
        . . . MESONS
        . . . . **KAONS**
 RT   BARYONS
       CHARGED PARTICLES
       KAON PRODUCTION
       KAONS
       KAONS
       PIONS
       POMERANCHUK THEOREM

KAPITZA RESISTANCE
 RT   ∞RESISTANCE

KAPOETA ACHONDRITE
 GS   CELESTIAL BODIES
        . METEORITES
        . . STONY METEORITES
        . . . ACHONDRITES
        . . . . **KAPOETA ACHONDRITE**

KAPPA ROCKET VEHICLES
 GS   ROCKET VEHICLES
        . MULTISTAGE ROCKET VEHICLES
        . . **KAPPA ROCKET VEHICLES**
        . . . KAPPA 8 ROCKET VEHICLE
        . . . KAPPA 9 ROCKET VEHICLE
        . SOUNDING ROCKETS
        . . **KAPPA ROCKET VEHICLES**
        . . . KAPPA 8 ROCKET VEHICLE
        . . . KAPPA 9 ROCKET VEHICLE
 RT   SOLID PROPELLANT ROCKET ENGINES
       ∞VEHICLES

KAPPA 8 ROCKET VEHICLE
 GS   ROCKET VEHICLES
        . MULTISTAGE ROCKET VEHICLES
        . . KAPPA ROCKET VEHICLES
        . . . **KAPPA 8 ROCKET VEHICLE**
        . SOUNDING ROCKETS
        . . KAPPA ROCKET VEHICLES
        . . . **KAPPA 8 ROCKET VEHICLE**
 RT   SOLID PROPELLANT ROCKET ENGINES

KAPPA 9 ROCKET VEHICLE
 GS   ROCKET VEHICLES
        . MULTISTAGE ROCKET VEHICLES
        . . KAPPA ROCKET VEHICLES
        . . . **KAPPA 9 ROCKET VEHICLE**
        . SOUNDING ROCKETS
        . . KAPPA ROCKET VEHICLES

KAPPA 9 ROCKET VEHICLE-*(CONT.)*
        . . . **KAPPA 9 ROCKET VEHICLE**
 RT   SOLID PROPELLANT ROCKET ENGINES

KAPTON (TRADEMARK)
 GS   POLYMERIC FILMS
        . . **KAPTON (TRADEMARK)**
 RT   ∞FILMS
       PLASTICS
       ∞POLYMERS

KARHUNEN-LOEVE EXPANSION
 GS   DATA PROCESSING
        . **KARHUNEN-LOEVE EXPANSION**
       EXPANSION
        . **KARHUNEN-LOEVE EXPANSION**
 RT   PRINCIPAL COMPONENTS ANALYSIS

KARL FISCHER REAGENT
 GS   CHEMICAL TESTS
        . CHEMICAL ANALYSIS
        . . **KARL FISCHER REAGENT**
 RT   DIOXIDES
       METHYL ALCOHOLS
       PYRIDINES
       QUANTITATIVE ANALYSIS

KARMAN VORTEX STREET
 GS   VORTEX STREETS
        . **KARMAN VORTEX STREET**
 RT   AEOLIAN TONES
       SUBSONIC FLOW
       VON KARMAN EQUATION
       VORTICITY EQUATIONS

KARMAN-BODEWADT FLOW
 GS   FLUID FLOW
        . AXISYMMETRIC FLOW
        . . **KARMAN-BODEWADT FLOW**
        . VISCOUS FLOW
        . . **KARMAN-BODEWADT FLOW**
       TRANSLATIONAL MOTION
        . **KARMAN-BODEWADT FLOW**
 RT   ROTATING DISKS
       ROTATING FLUIDS

KARST
 GS   LANDFORMS
        . STRUCTURAL BASINS
        . . **KARST**
        . . . SINKHOLES
 RT   CAVES
       CAVITIES
       KETTLES (GEOLOGY)
       ∞RIDGES
       ROCKS

KAWASAKI AIRCRAFT
 RT   ∞AIRCRAFT

KC-130 AIRCRAFT
 USE   C-130 AIRCRAFT

KC-135 AIRCRAFT
 USE   C-135 AIRCRAFT

KEELS
 GS   HYDROFOILS
        . **KEELS**
 RT   BOATS
       HULLS (STRUCTURES)
       LONGERONS
       SHIPS
       STABILIZERS (FLUID DYNAMICS)

KEL-F
 GS   HALOGEN COMPOUNDS
        . FLUORINE COMPOUNDS
        . . FLUORO COMPOUNDS
        . . . FLUORINE ORGANIC COMPOUNDS
        . . . . **KEL-F**
       ORGANIC COMPOUNDS
        . FLUORINE ORGANIC COMPOUNDS
        . . **KEL-F**
 RT   COPOLYMERS
       ∞POLYMERS

KELP
 USE   SEAWEEDS

KELVIN-HELMHOLTZ INSTABILITY
 RT   COLLISIONLESS PLASMAS
       FLOW STABILITY
       ∞HELMHOLTZ EQUATIONS

KELVIN-HELMHOLTZ INSTABILITY-*(CONT.)*
       MAGNETOHYDRODYNAMIC FLOW
       MAGNETOHYDRODYNAMIC STABILITY
       MASS FLOW
       NONUNIFORM PLASMAS
       PLASMAS (PHYSICS)
       SUPERFLUIDITY

KENTUCKY
 GS   NATIONS
        . UNITED STATES
        . . **KENTUCKY**
 RT   OHIO RIVER (US)
       TENNESSEE VALLEY (AL-KY-TN)

KENYA
 GS   NATIONS
        . **KENYA**
 RT   AFRICA

KEPLER LAWS
 GS   CLASSICAL MECHANICS
        . SPACE MECHANICS
        . . ORBITAL MECHANICS
        . . . **KEPLER LAWS**
       LAWS
        . **KEPLER LAWS**

KERATINS
 GS   PROTEINS
        . **KERATINS**
 RT   HAIR
       WOOL

KERATITIS
 GS   DISEASES
        . EYE DISEASES
        . . **KERATITIS**
 RT   CONJUNCTIVA
       CORNEA

KERNEL FUNCTIONS
 GS   ANALYSIS (MATHEMATICS)
        . REAL VARIABLES
        . . **KERNEL FUNCTIONS**
       FUNCTIONS (MATHEMATICS)
        . **KERNEL FUNCTIONS**
 RT   MELLIN TRANSFORMS

KEROGEN
 GS   HYDROCARBONS
        . **KEROGEN**
 RT   EARTH RESOURCES
       FUEL OILS
       FUELS
       GASOLINE
       GREASES
       HYDROCARBON FUELS
       KEROSENE
       LUBRICANTS
       OILS
       SHALE OIL

KEROSENE
 GS   FUELS
        . CHEMICAL FUELS
        . . LIQUID FUELS
        . . . **KEROSENE**
 RT   ANTIMISTING FUELS
       DIESEL FUELS
       FUEL OILS
       GASOLINE
       HYDROCARBON FUELS
       JET ENGINE FUELS
       JP-8 JET FUEL
       KEROGEN
       PARAFFINS
       RP-1 ROCKET PROPELLANTS
       SHALE OIL

KERR CELLS
 RT   CAMERA SHUTTERS
       ∞CELLS
       ∞ELECTRIC CELLS
       POLARIZED ELECTROMAGNETIC
         RADIATION
       POLARIZERS

∞ **KERR EFFECTS**
 SN   *(USE OF A MORE SPECIFIC TERM IS*
       RECOMMENDED--CONSULT THE TERMS
       LISTED BELOW)
 RT   ∞EFFECTS
       ELECTRO-OPTICAL EFFECT

KERR EFFECTS-*(CONT.)*
KERR ELECTROOPTICAL EFFECT
KERR MAGNETOOPTICAL EFFECT
MAGNETIC FIELDS

KERR ELECTROOPTICAL EFFECT
RT   ∞EFFECTS
ELECTRO-OPTICS
ELECTROMAGNETIC RADIATION
∞KERR EFFECTS
KERR MAGNETOOPTICAL EFFECT
LASERS
LIGHT MODULATION
OPTICAL PROPERTIES
POLARIZATION (WAVES)

KERR MAGNETOOPTICAL EFFECT
GS   ELECTROMAGNETIC PROPERTIES
. **KERR MAGNETOOPTICAL EFFECT**
RT   ∞EFFECTS
FARADAY EFFECT
∞KERR EFFECTS
KERR ELECTROOPTICAL EFFECT
MAGNETO-OPTICS
OPTICAL PROPERTIES
POLARIZATION (WAVES)
POLARIZED LIGHT

KESTREL AIRCRAFT
USE   P-1127 AIRCRAFT

KETENES
GS   ALIPHATIC COMPOUNDS
. **KETENES**
RT   KETONES

KETONES
GS   ALIPHATIC COMPOUNDS
. **KETONES**
. . ACETONE
. . ACETYLACETONE
. . ANTHRAQUINONES
. . CAMPHOR
. . NEMBUTAL (TRADEMARK)
. . PENTANONE
. . TRIMETHADIONE
RT   KETENES

KETTLES (GEOLOGY)
GS   GEOLOGY
. **KETTLES (GEOLOGY)**
LANDFORMS
. STRUCTURAL BASINS
. . **KETTLES (GEOLOGY)**
RT   CAVES
CAVITIES
EARTH RESOURCES
GLACIAL DRIFT
KARST
LAKES
SINKHOLES

KEVLAR (TRADEMARK)
GS   PLASTICS
. SYNTHETIC RESINS
. . THERMOSETTING RESINS
. . . FURAN RESINS
. . . . POLYAMIDE RESINS
. . . . . **KEVLAR (TRADEMARK)**
RESINS
. SYNTHETIC RESINS
. . THERMOSETTING RESINS
. . . FURAN RESINS
. . . . POLYAMIDE RESINS
. . . . . **KEVLAR (TRADEMARK)**
RT   NONFLAMMABLE MATERIALS
SYNTHETIC FIBERS

KEYING
GS   **KEYING**
. FREQUENCY SHIFT KEYING
. PHASE SHIFT KEYING
RT   MORSE CODE
RADIO TELEGRAPHY
TELEPRINTERS
TELETYPEWRITERS

KEYS (ISLANDS)
UF   CAYS
GS   LANDFORMS
. ISLANDS
. . **KEYS (ISLANDS)**
RT   CORAL REEFS
EARTH RESOURCES

KEYS (ISLANDS)-*(CONT.)*
ISLAND ARCS
OCEANS

KIDNEY DISEASES
GS   DISEASES
. **KIDNEY DISEASES**
. . NEPHRITIS
RT   CHOLERA

KIDNEYS
GS   ANATOMY
. ORGANS
. . **KIDNEYS**
VISCERA
. ORGANS
. . **KIDNEYS**
RT   RENAL FUNCTION
URINE
UROLITHIASIS
UROLOGY

KILOMETER WAVE ORBITING TELESCOPE
GS   TELESCOPES
. ASTRONOMICAL TELESCOPES
. . **KILOMETER WAVE ORBITING
TELESCOPE**
. RADIO TELESCOPES
. . **KILOMETER WAVE ORBITING
TELESCOPE**

KILOMETRIC WAVES
GS   ELECTROMAGNETIC RADIATION
· . **KILOMETRIC WAVES**
RT   ∞WAVES

KIMBERLITE
USE   BIOTITE
PERIDOTITE

KINEMATIC EQUATIONS
GS   EQUATIONS OF MOTION
. KINETIC EQUATIONS
. . **KINEMATIC EQUATIONS**
RT   ∞EQUATIONS

KINEMATICS
GS   **KINEMATICS**
. BODY KINEMATICS
RT   ACCELERATION (PHYSICS)
∞DYNAMICS
EQUATIONS OF MOTION
HODOGRAPHS
KINETICS
∞MECHANICS (PHYSICS)
MICROWAVE REFLECTOMETERS
∞MOTION
NUTATION
VELOCITY

KINESCOPES
USE   PICTURE TUBES

KINESTHESIA
GS   PERCEPTION
. SENSORY PERCEPTION
. . **KINESTHESIA**
RT   PROPRIOCEPTION

KINESTHESIS
USE   PROPRIOCEPTION

KINETIC ENERGY
UF   MOMENTUM ENERGY
GS   KINETICS
. **KINETIC ENERGY**
RT   CHEMICAL ENERGY
ELECTRON ENERGY
∞ENERGY
EQUIPARTITION THEOREM
FROUDE NUMBER
HYDRODYNAMIC RAM EFFECT
INTERNAL ENERGY
LAGRANGE SIMILARITY HYPOTHESIS
PARTICLE ENERGY
POTENTIAL ENERGY
PROTON ENERGY
THERMAL ENERGY
VIRIAL THEOREM
WORK
ZERO POINT ENERGY

KINETIC EQUATIONS
GS   EQUATIONS OF MOTION

KINETIC EQUATIONS-*(CONT.)*
. **KINETIC EQUATIONS**
. . HYDRODYNAMIC EQUATIONS
. . . HELMHOLTZ VORTICITY EQUATION
. . KINEMATIC EQUATIONS
RT   BBGKY HIERARCHY
BETHE-SALPETER EQUATION
EINSTEIN EQUATIONS
∞EQUATIONS
PARTIAL DIFFERENTIAL EQUATIONS
VIRIAL THEOREM

KINETIC FRICTION
GS   FRICTION
. **KINETIC FRICTION**
. . SLIDING FRICTION
RT   COEFFICIENT OF FRICTION
DRY FRICTION
FRICTION MEASUREMENT
STATIC FRICTION

KINETIC HEATING
GS   HEATING
. **KINETIC HEATING**
. . SHOCK HEATING
RT   GAS HEATING
MAGNETIC PUMPING
PLASMA HEATING

KINETIC THEORY
GS   **KINETIC THEORY**
. CHAPMAN-ENSKOG THEORY
. TRANSPORT THEORY
. . EYRING THEORY
. . MIXING LENGTH FLOW THEORY
RT   BINARY FLUIDS
BOLTZMANN DISTRIBUTION
BOLTZMANN TRANSPORT EQUATION
DIFFUSION
DIFFUSION THEORY
DIFFUSION WAVES
DYNAMIC PRESSURE
EQUATIONS OF STATE
FREE MOLECULAR FLOW
GAS TRANSPORT
GASEOUS SELF-DIFFUSION
IDEAL GAS
JOULE-THOMSON EFFECT
KNUDSEN FLOW
KROOK EQUATION
LORENTZ GAS
MASS FLOW
MAXWELL-BOLTZMANN DENSITY
FUNCTION
MOBILITY
MOMENTUM TRANSFER
MORSE POTENTIAL
REAL GASES
∞THEORIES
TRANSPORT PROPERTIES

KINETICS
GS   **KINETICS**
. ELECTROKINETICS
. KINETIC ENERGY
. NEWTON SECOND LAW
. NEWTON THEORY
. REACTION KINETICS
. VARIABLE MASS SYSTEMS
RT   ACCELERATION (PHYSICS)
ANGULAR MOMENTUM
BODY KINEMATICS
∞DYNAMICS
FLUID DYNAMICS
FLUID MECHANICS
∞FORCE
GAS DYNAMICS
HYDROMECHANICS
IDEAL GAS
KINEMATICS
∞MECHANICS (PHYSICS)
MOMENTUM TRANSFER
MOTION AFTEREFFECTS
NEWTON
PARTICLE COLLISIONS
∞PHYSICS
VELOCITY

KINOFORM
GS   DISPLAY DEVICES
. **KINOFORM**
IMAGERY
. **KINOFORM**
RT   COMPUTER PROGRAMMING
HOLOGRAPHY

∞ **KIRCHHOFF LAW**
SN   (USE OF A MORE SPECIFIC TERM IS
      RECOMMENDED--CONSULT THE TERMS
      LISTED BELOW)
RT   KIRCHHOFF LAW OF NETWORKS
      KIRCHHOFF LAW OF RADIATION

KIRCHHOFF LAW OF NETWORKS
RT   CIRCUITS
      ELECTRIC CURRENT
      ELECTRIC POTENTIAL
      ∞KIRCHHOFF LAW
      NETWORK ANALYSIS
      NETWORK SYNTHESIS

KIRCHHOFF LAW OF RADIATION
GS   LAWS
      . RADIATION LAWS
      . . **KIRCHHOFF LAW OF RADIATION**
RT   ABSORPTIVITY
      BLACK BODY RADIATION
      ∞KIRCHHOFF LAW
      ∞RADIATION
      STEFAN-BOLTZMANN LAW
      THERMODYNAMICS

KIRCHHOFF-HELMHOLTZ FLOW
USE   PIPE FLOW

KIRCHHOFF-HUYGENS PRINCIPLE
USE   DIFFRACTION
       WAVE PROPAGATION

KIRKENDALL EFFECT
RT   DIFFUSION THEORY
      DIFFUSION WELDING
      DIFFUSIVITY
      ∞EFFECTS
      THERMAL DIFFUSION

KITE BALLOONS
USE   TETHERED BALLOONS

KITS
RT   FIRST AID
      SURVIVAL
      TOOLS

KIWI B REACTORS
GS   NUCLEAR ELECTRIC POWER
        GENERATION
      . NUCLEAR POWER REACTORS
      . . KIWI REACTORS
      . . **KIWI B REACTORS**
      . . . KIWI B-1 REACTOR
      . . . KIWI B-4 REACTOR
      NUCLEAR REACTORS
      . GAS COOLED REACTORS
      . . KIWI REACTORS
      . . . **KIWI B REACTORS**
      . . . . KIWI B-1 REACTOR
      . . . . KIWI B-4 REACTOR
      . NUCLEAR POWER REACTORS
      . . KIWI REACTORS
      . . . **KIWI B REACTORS**
      . . . . KIWI B-1 REACTOR
      . . . . KIWI B-4 REACTOR
      . NUCLEAR RESEARCH AND TEST
        REACTORS
      . . KIWI REACTORS
      . . . **KIWI B REACTORS**
      . . . . KIWI B-1 REACTOR
      . . . . KIWI B-4 REACTOR

KIWI B-1 REACTOR
GS   NUCLEAR ELECTRIC POWER
        GENERATION
      . NUCLEAR POWER REACTORS
      . . KIWI REACTORS
      . . . KIWI B REACTORS
      . . . . **KIWI B-1 REACTOR**
      NUCLEAR REACTORS
      . GAS COOLED REACTORS
      . . KIWI REACTORS
      . . . KIWI B REACTORS
      . . . . **KIWI B-1 REACTOR**
      . NUCLEAR POWER REACTORS
      . . KIWI REACTORS
      . . . KIWI B REACTORS
      . . . . **KIWI B-1 REACTOR**
      . NUCLEAR RESEARCH AND TEST
        REACTORS
      . . KIWI REACTORS
      . . . KIWI B REACTORS

KIWI B-1 REACTOR-(CONT.)
      . . . . **KIWI B-1 REACTOR**

KIWI B-4 REACTOR
GS   NUCLEAR ELECTRIC POWER
        GENERATION
      . NUCLEAR POWER REACTORS
      . . KIWI REACTORS
      . . . KIWI B REACTORS
      . . . . **KIWI B-4 REACTOR**
      NUCLEAR REACTORS
      . GAS COOLED REACTORS
      . . KIWI REACTORS
      . . . KIWI B REACTORS
      . . . . **KIWI B-4 REACTOR**
      . NUCLEAR POWER REACTORS
      . . KIWI REACTORS
      . . . KIWI B REACTORS
      . . . . **KIWI B-4 REACTOR**
      . NUCLEAR RESEARCH AND TEST
        REACTORS
      . . KIWI REACTORS
      . . . KIWI B REACTORS
      . . . . **KIWI B-4 REACTOR**

KIWI REACTORS
UF   KIWI ROCKET REACTORS
GS   NUCLEAR ELECTRIC POWER
        GENERATION
      . NUCLEAR POWER REACTORS
      . . **KIWI REACTORS**
      . . . KIWI B REACTORS
      . . . . KIWI B-1 REACTOR
      . . . . KIWI B-4 REACTOR
      NUCLEAR REACTORS
      . GAS COOLED REACTORS
      . . **KIWI REACTORS**
      . . . KIWI B REACTORS
      . . . . KIWI B-1 REACTOR
      . . . . KIWI B-4 REACTOR
      . NUCLEAR POWER REACTORS
      . . **KIWI REACTORS**
      . . . KIWI B REACTORS
      . . . . KIWI B-1 REACTOR
      . . . . KIWI B-4 REACTOR
      . NUCLEAR RESEARCH AND TEST
        REACTORS
      . . **KIWI REACTORS**
      . . . KIWI B REACTORS
      . . . . KIWI B-1 REACTOR
      . . . . KIWI B-4 REACTOR
RT   NRX REACTORS
      NUCLEAR ENGINE FOR ROCKET
        VEHICLES
      PHOEBUS NUCLEAR REACTOR
      ROVER PROJECT

KIWI ROCKET REACTORS
USE   KIWI REACTORS

KJELDAHL METHOD
GS   CHEMICAL TESTS
      . CHEMICAL ANALYSIS
      . . QUANTITATIVE ANALYSIS
      . . . **KJELDAHL METHOD**
RT   AMMONIA
      ∞METHODOLOGY
      NITROGEN
      TITRATION

KLEBSIELLA
GS   MICROORGANISMS
      . BACTERIA
      . . **KLEBSIELLA**

KLEIN-DUNHAM POTENTIAL
RT   ∞POTENTIAL
      QUANTUM THEORY

KLEIN-GORDON EQUATION
GS   WAVE EQUATIONS
      . **KLEIN-GORDON EQUATION**
RT   DIRAC EQUATION
      ∞EQUATIONS

KLIPPEN
USE   OUTLIERS (LANDFORMS)

KLYSTRONS
GS   ELECTRON TUBES
      . VACUUM TUBES
      . . VACUUM TUBE OSCILLATORS
      . . . MICROWAVE TUBES
      . . . . **KLYSTRONS**

KLYSTRONS-(CONT.)
      MICROWAVE EQUIPMENT
      . MICROWAVE TUBES
      . . **KLYSTRONS**
RT   AMPLIFIERS
      CATCHERS
      CAVITY RESONATORS
      CYCLOTRON RESONANCE DEVICES
      ELECTRON BUNCHING
      ELECTRON CYCLOTRON HEATING
      ELECTROSTATIC GENERATORS
      MAGNETRONS

KNEE (ANATOMY)
GS   ANATOMY
      . LIMBS (ANATOMY)
      . . LEG (ANATOMY)
      . . . **KNEE (ANATOMY)**
      . MUSCULOSKELETAL SYSTEM
      . . JOINTS (ANATOMY)
      . . . **KNEE (ANATOMY)**
      APPENDAGES
      . LEG (ANATOMY)
      . . **KNEE (ANATOMY)**
RT   FEMUR

KNIGHT SHIFT
USE   NUCLEAR MAGNETIC RESONANCE

KNOBS
RT   HANDLES
      LEVERS
      MANUAL CONTROL

KNOOP HARDNESS
GS   MECHANICAL PROPERTIES
      . HARDNESS
      . . **KNOOP HARDNESS**
RT   HARDNESS TESTS
      MICROHARDNESS

KNOWLEDGE
GS   **KNOWLEDGE**
      . PHILOSOPHY
      . . PARADOXES
RT   AXIOMS
      DOCUMENTATION
      EDUCATION
      LEARNING
      LITERATURE
      PERCEPTION
      TEXTBOOKS
      TRAINING EVALUATION

KNOWLEDGE ENGINEERING
USE   EXPERT SYSTEMS

KNUDSEN CELLS
USE   KNUDSEN GAGES

KNUDSEN FLOW
UF   KNUDSEN NUMBER
GS   FLUID FLOW
      . DUCTED FLOW
      . . **KNUDSEN FLOW**
      . GAS FLOW
      . . **KNUDSEN FLOW**
RT   BOUNDARY LAYER TRANSITION
      FREE MOLECULAR FLOW
      KINETIC THEORY
      MEAN FREE PATH
      MOLECULAR FLOW
      PRESSURE GRADIENTS
      RAREFIED GAS DYNAMICS
      TRANSITION POINTS
      VACUUM
      VISCOUS FLOW

KNUDSEN GAGES
UF   KNUDSEN CELLS
GS   MEASURING INSTRUMENTS
      . PRESSURE GAGES
      . . VACUUM GAGES
      . . . **KNUDSEN GAGES**
      VACUUM APPARATUS
      . VACUUM GAGES
      . . **KNUDSEN GAGES**
RT   IONIZATION GAGES
      MCLEOD GAGES
      PIRANI GAGES
      PRESSURE MEASUREMENT
      RADIOMETERS

KNUDSEN NUMBER
USE   KNUDSEN FLOW

KNURLING
RT   GROOVING
      MACHINING
      METAL CUTTING

KOHOUTEK COMET
GS   CELESTIAL BODIES
      . COMETS
      . . KOHOUTEK COMET
RT   BESSEL-BREDICHIN THEORY
      ∞COMA
      RADIATION PRESSURE
      SOLAR SYSTEM

KOLMOGOROFF THEORY
RT   ISOTROPIC TURBULENCE
      LAGRANGE SIMILARITY HYPOTHESIS
      SHEAR FLOW
      ∞THEORIES
      TURBULENT FLOW
      VORTICES

KOLMOGOROFF-SMIRNOFF TEST
GS   STATISTICAL ANALYSIS
      . STATISTICAL TESTS
      . . KOLMOGOROFF-SMIRNOFF TEST

KONDO EFFECT
GS   TRANSPORT PROPERTIES
      . ELECTRICAL RESISTIVITY
      . . SUPERCONDUCTIVITY
      . . . KONDO EFFECT
RT   ALLOYS
      ∞EFFECTS
      LOW TEMPERATURE PHYSICS
      MAGNETIC MATERIALS
      NUCLEAR SPIN
      TRANSITION TEMPERATURE

∞ **KOREA**
SN   (USE OF A MORE SPECIFIC TERM IS
      RECOMMENDED--CONSULT THE TERMS
      LISTED BELOW)
RT   NORTH KOREA
      SOUTH KOREA

KORTEWEG-DEVRIES EQUATION
GS   WAVE EQUATIONS
      . KORTEWEG-DEVRIES EQUATION
RT   ∞EQUATIONS

KOSSEL PATTERN
GS   DISTRIBUTION (PROPERTY)
      . RADIATION DISTRIBUTION
      . . DIFFRACTION PATTERNS
      . . . KOSSEL PATTERN
RT   CRYSTAL LATTICES

KOVAR (TRADEMARK)
GS   ALLOYS
      . KOVAR (TRADEMARK)
RT   COBALT ALLOYS

KP INDEX
GS   RATIOS
      . INDEXES (RATIOS)
      . . KP INDEX
RT   GEOMAGNETIC PULSATIONS
      GEOMAGNETISM
      ∞INDEXES
      MAGNETIC DISTURBANCES
      MAGNETIC PROPERTIES
      MAGNETIC VARIATIONS
      MAGNETOSPHERE

KRAFT PROCESS (WOODPULP)
RT   MANUFACTURING
      PAPER (MATERIAL)
      ∞PROCESSES

KRAMERS-KRONIG FORMULA
RT   ∞DISPERSION
      ∞FORMULAS
      OPACITY
      SPECTRUM ANALYSIS

KREBS CYCLE
RT   CELLS (BIOLOGY)
      METABOLISM

KREEP
GS   MINERALS
      . KREEP
RT   GEOLOGY
      LUNAR SOIL
      PHOSPHATES
      POTASSIUM
      RARE EARTH ELEMENTS
      ROCKS

KRIGING
RT   VARIANCE (STATISTICS)

KRONECKER PRODUCT
USE   ORTHOGONALITY

KROOK EQUATION
RT   ∞EQUATIONS
      HYDRODYNAMICS
      KINETIC THEORY
      SHEAR FLOW
      SHOCK WAVE PROFILES

KRYPTON
GS   CHEMICAL ELEMENTS
      . RARE GASES
      . . KRYPTON
      GASES
      . RARE GASES
      . . KRYPTON

KRYPTON FLUORIDE LASERS
GS   STIMULATED EMISSION DEVICES
      . LASERS
      . . GAS LASERS
      . . . KRYPTON FLUORIDE LASERS
      . . RARE GAS-HALIDE LASERS
      . . . KRYPTON FLUORIDE LASERS
RT   COHERENT ELECTROMAGNETIC
      RADIATION
      LASING
      MASERS
      OPTICAL PUMPING

KRYPTON ISOTOPES
GS   CHEMICAL ELEMENTS
      . NUCLIDES
      . . ISOTOPES
      . . . KRYPTON ISOTOPES
      . . . . KRYPTON 85

KRYPTON 85
GS   CHEMICAL ELEMENTS
      . NUCLIDES
      . . ISOTOPES
      . . . KRYPTON ISOTOPES
      . . . . KRYPTON 85
      . . . RADIOACTIVE ISOTOPES
      . . . . KRYPTON 85

KU BAND
USE   SUPERHIGH FREQUENCIES

KUIPER AIRBORNE OBSERVATORY
USE   C-141 AIRCRAFT

KURILE ISLANDS
GS   LANDFORMS
      . ISLANDS
      . . PACIFIC ISLANDS
      . . . KURILE ISLANDS
RT   U.S.S.R.

KURTOSIS
GS   DISTRIBUTION (PROPERTY)
      . FREQUENCY DISTRIBUTION
      . . KURTOSIS
RT   ∞DISTRIBUTION
      FOURIER ANALYSIS
      ∞PATTERNS
      STATISTICAL DISTRIBUTIONS

KUTTA-JOUKOWSKI CONDITION
GS   CONDITIONS
      . KUTTA-JOUKOWSKI CONDITION
RT   AIRFOIL PROFILES
      BOUNDARY LAYER SEPARATION
      JOUKOWSKI TRANSFORMATION

KUWAIT
GS   NATIONS
      . KUWAIT
RT   ASIA

KWIC INDEXES
GS   CLASSIFICATIONS
      . INDEXES (DOCUMENTATION)
      . . KWIC INDEXES
RT   ∞INDEXES
      THESAURI

L

L BAND
USE   ULTRAHIGH FREQUENCIES

L-SAT
UF   EUROPEAN LARGE TELECOMM
      SATELLITE
GS   ESA SPACECRAFT
      . ESA SATELLITES
      . . L-SAT
      SATELLITES
      . ARTIFICIAL SATELLITES
      . . COMMUNICATION SATELLITES
      . . . L-SAT
      . . ESA SATELLITES
      . . . L-SAT
RT   SATELLITE NETWORKS

L-28 AIRCRAFT
USE   U-10 AIRCRAFT

L-29 AIRCRAFT
USE   L-29 JET TRAINER

L-29 JET TRAINER
UF   DELFIN AIRCRAFT
      L-29 AIRCRAFT
      OMNIPOL L-29 AIRCRAFT
GS   JET AIRCRAFT
      . L-29 JET TRAINER
      MONOPLANES
      . L-29 JET TRAINER
      TILT WING AIRCRAFT
      . L-29 JET TRAINER
      TRAINING AIRCRAFT
      . L-29 JET TRAINER
      V/STOL AIRCRAFT
      . L-29 JET TRAINER

L-1011 AIRCRAFT
GS   COMMERCIAL AIRCRAFT
      . L-1011 AIRCRAFT
      JET AIRCRAFT
      . L-1011 AIRCRAFT
      LOCKHEED AIRCRAFT
      . L-1011 AIRCRAFT
      PASSENGER AIRCRAFT
      . L-1011 AIRCRAFT
      TRANSPORT AIRCRAFT
      . L-1011 AIRCRAFT
RT   ∞AIRCRAFT
      TURBOFAN ENGINES

L-2000 AIRCRAFT
UF   LOCKHEED L-2000 AIRCRAFT
GS   JET AIRCRAFT
      . L-2000 AIRCRAFT
      LOCKHEED AIRCRAFT
      . L-2000 AIRCRAFT
      PASSENGER AIRCRAFT
      . L-2000 AIRCRAFT
      SUPERSONIC AIRCRAFT
      . SUPERSONIC TRANSPORTS
      . . L-2000 AIRCRAFT
      TRANSPORT AIRCRAFT
      . L-2000 AIRCRAFT
RT   ∞AIRCRAFT

LABELING (MARKING)
USE   MARKING

LABOR
RT   MANPOWER
      MEDIATION
      PERSONNEL SELECTION

LABORATORIES
GS   LABORATORIES
      . ADVANCED TECHNOLOGY
      LABORATORY
      . ENGINE TESTING LABORATORIES
      . ENVIRONMENTAL LABORATORIES
      . HUMAN FACTORS LABORATORIES

LABORATORIES-*(CONT.)*
    . LONG DURATION EXPOSURE FACILITY
    . LUNAR MOBILE LABORATORIES
    . LUNAR RECEIVING LABORATORY
    . SPACE LABORATORIES
    . . ATMOSPHERIC CLOUD PHYSICS LAB
       (SPACELAB)
    . . EARTH VIEWING APPLICATIONS
       LABORATORY
    . . MANNED ORBITAL LABORATORIES
    . . . MANNED ORBITAL RESEARCH
       LABORATORIES
    . UNDERWATER RESEARCH
       LABORATORIES
RT   EXPERIMENT DESIGN
      EXPERIMENTATION
      NUCLEAR RESEARCH
      RESEARCH FACILITIES
      SAIL PROJECT
      TEST FACILITIES
    ∞TESTS

LABORATORY EQUIPMENT
GS   **LABORATORY EQUIPMENT**
    . IMAGE FURNACES
    . SYRINGES
RT   AMPOULES
    ∞EQUIPMENT
      GLASSWARE
      MEASURING INSTRUMENTS
      PIPETTES

LABRADOR
RT   CANADA

LABYRINTH
GS   ANATOMY
    . SENSE ORGANS
    . . EAR
    . . . **LABYRINTH**
    . . . . COCHLEA
    . . . . VESTIBULES
RT   SEMICIRCULAR CANALS

LABYRINTH SEALS
GS   SEALS (STOPPERS)
    . **LABYRINTH SEALS**
RT   FLUID FLOW
      GASKETS
      GLANDS (SEALS)
      HERMETIC SEALS
      LEAKAGE
      O RING SEALS
      PACKINGS (SEALS)
      PLUGS
      PUMP SEALS
      ROTOR SPEED

LABYRINTHECTOMY
GS   SURGERY
    . **LABYRINTHECTOMY**
RT   EAR

LACATE (EXPERIMENT)
UF   LOWER ATMOSPHERIC COMPOSITION
      EXPERIMENT
RT   ATMOSPHERIC COMPOSITION
      ATMOSPHERIC TEMPERATURE
      LOWER ATMOSPHERE

LACE (ENGINE)
USE   LIQUID AIR CYCLE ENGINES

LACQUERS
GS   COATINGS
    . **LACQUERS**
      FINISHES
    . **LACQUERS**
RT   METAL COATINGS
      PRIMERS (COATINGS)
      PROTECTIVE COATINGS
      SPRAYED COATINGS

LACTATES
GS   ALIPHATIC COMPOUNDS
    . **LACTATES**
      ESTERS
    . **LACTATES**

LACTIC ACID
GS   ALIPHATIC COMPOUNDS
    . **LACTIC ACID**

LACTOSE
GS   ALIPHATIC COMPOUNDS
    . SUGARS
    . . **LACTOSE**
      CARBOHYDRATES
    . SUGARS
    . . **LACTOSE**

LACUNAS
RT   LICHENS
      PLANTS (BOTANY)

LADDERS
RT   ESCALATORS
      STAIRWAYS

LAG (DELAY)
USE   TIME LAG

LAGEOS (SATELLITE)
UF   LASER GEODYNAMIC SATELLITE
GS   SATELLITES
    . ARTIFICIAL SATELLITES
    . . PASSIVE SATELLITES
    . . . **LAGEOS (SATELLITE)**
      UNMANNED SPACECRAFT
    . PASSIVE SATELLITES
    . . **LAGEOS (SATELLITE)**
RT   LASER RANGE FINDERS
      RETROREFLECTION

LAGOONS
GS   LANDFORMS
    . **LAGOONS**
RT   ATOLLS
      BARS (LANDFORMS)
      BEACHES
      COASTS
      DUNES
      INLETS (TOPOGRAPHY)
      ISLAND ARCS
      ISLANDS
      LAKES
      PONDS
      RESERVOIRS
      TOPOGRAPHY

LAGRANGE COORDINATES
GS   COORDINATES
    . **LAGRANGE COORDINATES**
RT   CLASSICAL MECHANICS
      HYDRODYNAMICS
      LIBRATIONAL MOTION

LAGRANGE EQUATIONS OF MOTION
USE   EULER-LAGRANGE EQUATION

LAGRANGE MULTIPLIERS
RT   CHIRAL DYNAMICS
      DIFFERENTIAL EQUATIONS
      ISOPERIMETRIC PROBLEM
      MULTIPLIERS
      OPERATIONS RESEARCH
      OPTIMIZATION

LAGRANGE SIMILARITY HYPOTHESIS
GS   HYPOTHESES
    . **LAGRANGE SIMILARITY HYPOTHESIS**
      THEOREMS
    . SIMILARITY THEOREM
    . . **LAGRANGE SIMILARITY HYPOTHESIS**
RT   ENERGY DISSIPATION
      ENERGY TRANSFER
      KINETIC ENERGY
      KOLMOGOROFF THEORY
      TURBULENT FLOW

LAGRANGIAN EQUILIBRIUM POINTS
GS   GRAVITATIONAL EFFECTS
    . **LAGRANGIAN EQUILIBRIUM POINTS**
RT   CELESTIAL MECHANICS
      GRAVITATIONAL FIELDS
      ORBITAL MECHANICS

LAGUERRE FUNCTIONS
GS   ANALYSIS (MATHEMATICS)
    . COMPLEX VARIABLES
    . . **LAGUERRE FUNCTIONS**
      FUNCTIONS (MATHEMATICS)
    . **LAGUERRE FUNCTIONS**
RT   ORTHOGONAL FUNCTIONS

LAKE BEDS
USE   BEDS (GEOLOGY)

LAKE CHAMPLAIN BASIN (NY-VT)
GS   LANDFORMS
    . STRUCTURAL BASINS
    . . **LAKE CHAMPLAIN BASIN (NY-VT)**
RT   CANADA
      LAKES
      NEW YORK
      VERMONT

LAKE ERIE
GS   LAKES
    . GREAT LAKES (NORTH AMERICA)
    . . **LAKE ERIE**
RT   HYDROLOGY
      RIVERS
      STREAMS
      WATER
      WATER MANAGEMENT

LAKE HURON
GS   LAKES
    . GREAT LAKES (NORTH AMERICA)
    . . **LAKE HURON**
RT   HYDROLOGY
      RIVERS
      SAGINAW BAY (MI)
      STREAMS
      WATER
      WATER MANAGEMENT

LAKE ICE
GS   ICE
    . **LAKE ICE**
    . . ICE FLOES
RT   BAY ICE
      ICE FORMATION
      LAKES
      LAND ICE
      SEA ICE
      WATER

LAKE MICHIGAN
GS   LAKES
    . GREAT LAKES (NORTH AMERICA)
    . . **LAKE MICHIGAN**
RT   HYDROLOGY
      RIVERS
      STREAMS
      WATER
      WATER MANAGEMENT

LAKE ONTARIO
GS   LAKES
    . GREAT LAKES (NORTH AMERICA)
    . . **LAKE ONTARIO**
RT   HYDROLOGY
      RIVERS
      STREAMS
      WATER
      WATER MANAGEMENT

LAKE PONTCHARTRAIN (LA)
GS   LAKES
    . **LAKE PONTCHARTRAIN (LA)**
RT   LOUISIANA

LAKE SUPERIOR
GS   LAKES
    . GREAT LAKES (NORTH AMERICA)
    . . **LAKE SUPERIOR**
RT   HYDROLOGY
      RIVERS
      STREAMS
      WATER
      WATER MANAGEMENT

LAKE TAHOE (CA-NV)
GS   LAKES
    . **LAKE TAHOE (CA-NV)**
RT   CALIFORNIA
      NEVADA

LAKE TEXOMA (OK-TX)
GS   LAKES
    . **LAKE TEXOMA (OK-TX)**
RT   LIMNOLOGY
      OKLAHOMA
      RESERVOIRS
      TEXAS
      VADOSE WATER

LAKES
GS   LAKES
    . GREAT LAKES (NORTH AMERICA)

LAKES-(CONT.)
    . . LAKE ERIE
    . . LAKE HURON
    . . LAKE MICHIGAN
    . . LAKE ONTARIO
    . . LAKE SUPERIOR
    . GREAT SALT LAKE (UT)
    . LAKE PONTCHARTRAIN (LA)
    . LAKE TAHOE (CA-NV)
    . LAKE TEXOMA (OK-TX)
    . PYRAMID LAKE (NV)
RT   AQUIFERS
      BAYOUS
      BEACHES
      COASTS
      EARTH HYDROSPHERE
      EUTROPHICATION
      INLAND WATERS
      KETTLES (GEOLOGY)
      LAGOONS
      LAKE CHAMPLAIN BASIN (NY-VT)
      LAKE ICE
      LIMNOLOGY
      PLAYAS
      PONDS
      REGIONAL PLANNING
      RESERVOIRS
      RIVER BASINS
      SHOALS
      SHORELINES
      SPRINGS (WATER)
      STRAITS
      SURFACE WATER
      THERMAL POLLUTION
      WATER CIRCULATION
      WATER COLOR
      WATER DEPTH
      WATER RESOURCES
      WATERWAYS

LALLEMAND CAMERAS
GS   OPTICAL EQUIPMENT
    . CAMERAS
    . . **LALLEMAND CAMERAS**
    PHOTOGRAPHIC EQUIPMENT
    . CAMERAS
    . . **LALLEMAND CAMERAS**
RT   ASTRONOMICAL PHOTOGRAPHY
      ELECTRO-OPTICAL PHOTOGRAPHY
      IMAGE CONVERTERS
      IMAGE INTENSIFIERS
      IMAGE TRANSDUCERS
      LIGHT AMPLIFIERS
      SPECTROSCOPY
      TELEVISION CAMERAS

LAMB WAVES
GS   ELASTIC WAVES
    . SOUND WAVES
    . . **LAMB WAVES**
RT   ACOUSTIC PROPERTIES
      ACOUSTICS
      STURM-LIOUVILLE THEORY
      ULTRASONIC TESTS

LAMBDA ROCKET VEHICLES
GS   ROCKET VEHICLES
    . MULTISTAGE ROCKET VEHICLES
    . . **LAMBDA ROCKET VEHICLES**
    . SOUNDING ROCKETS
    . . **LAMBDA ROCKET VEHICLES**
RT   SOLID PROPELLANT ROCKET ENGINES
    ∞VEHICLES

LAMBDA TAURI STARS
GS   CELESTIAL BODIES
    . STARS
    . . **LAMBDA TAURI STARS**
RT   VARIABLE STARS

LAMBERT LAW
USE   BOUGUER LAW

LAMBERT SURFACE
RT   BOUGUER LAW
      CONFORMAL MAPPING
      COORDINATE TRANSFORMATIONS
      SURFACE DISTORTION
      ∞SURFACE GEOMETRY
      ∞SURFACES

LAME FUNCTIONS
GS   FUNCTIONS (MATHEMATICS)
    . **LAME FUNCTIONS**
RT   BOUNDARY VALUE PROBLEMS

LAME FUNCTIONS-(CONT.)
    DIFFERENTIAL EQUATIONS

LAME WAVE EQUATIONS
GS   ANALYSIS (MATHEMATICS)
    . REAL VARIABLES
    . . DIFFERENTIAL EQUATIONS
    . . . **LAME WAVE EQUATIONS**
    WAVE EQUATIONS
    . **LAME WAVE EQUATIONS**
RT   ACOUSTICS
      ELASTIC WAVES
      ∞EQUATIONS
      STURM-LIOUVILLE THEORY
      WAVE PROPAGATION

LAMELLA
GS   PHYSIOLOGY
    . **LAMELLA**
RT   BONES

LAMELLA (METALLURGY)
RT   ALUMINUM ALLOYS
      COPPER ALLOYS
      CRYSTALLOGRAPHY
      EUTECTIC ALLOYS
      MICROSTRUCTURE

LAMINA
USE   LAYERS

LAMINAR BOUNDARY LAYER
UF   LAMINAR BOUNDARY LAYER
      SEPARATION
      LAMINAR FLOW CONTROL
GS   BOUNDARY LAYERS
    . **LAMINAR BOUNDARY LAYER**
RT   BOUNDARY LAYER COMBUSTION
      BOUNDARY LAYER TRANSITION
      COMPRESSIBLE BOUNDARY LAYER
      GOERTLER INSTABILITY
      HYPERSONIC BOUNDARY LAYER
      INCOMPRESSIBLE BOUNDARY LAYER
      INTERACTIONAL AERODYNAMICS
      ISOTHERMAL LAYERS
      POHLHAUSEN METHOD
      SUPERSONIC BOUNDARY LAYERS
      THERMAL BOUNDARY LAYER
      THREE DIMENSIONAL BOUNDARY LAYER
      TURBULENT BOUNDARY LAYER
      TWO DIMENSIONAL BOUNDARY LAYER
      X-21 AIRCRAFT

LAMINAR BOUNDARY LAYER SEPARATION
USE   LAMINAR BOUNDARY LAYER

LAMINAR FLAMES
USE   FLAMES
      LAMINAR FLOW

LAMINAR FLOW
UF   LAMINAR FLAMES
      LAMINAR JETS
      POISEUILLE FLOW
      STREAMLINE FLOW
GS   FLUID FLOW
    . **LAMINAR FLOW**
    . . BLASIUS FLOW
    . . HARTMANN FLOW
    . . STRATIFIED FLOW
RT   AERODYNAMICS
      ATMOSPHERIC TURBULENCE
      BOUNDARY LAYER TRANSITION
      CAPILLARY FLOW
      CRITICAL FLOW
      FALKNER-SKAN EQUATION
      FLOW CHARACTERISTICS
      FLOW GEOMETRY
      FLOW STABILITY
      FORCED CONVECTION
      FREE CONVECTION
      GAS FLOW
      GAS STREAMS
      INVISCID FLOW
      LIQUID FLOW
      MASS FLOW
      MULTIPHASE FLOW
      NEWTON PRESSURE LAW
      OPEN CHANNEL FLOW
      ORIFICE FLOW
      PARALLEL FLOW
      PIPE FLOW
      PRANDTL-MEYER EXPANSION
      RAYLEIGH-BENARD CONVECTION
      REYNOLDS NUMBER

LAMINAR FLOW-(CONT.)
    ROSHKO PREDICTION
    SINGLE-PHASE FLOW
    STEADY FLOW
    STEAM FLOW
    TOLLMEIN-SCHLICHTING WAVES
    ∞TRANSITION LAYERS
    TURBULENT FLOW
    TWO PHASE FLOW
    UNIFORM FLOW
    UNSTEADY FLOW
    VISCOUS DRAG
    VISCOUS FLOW
    WEDGE FLOW
    X-21A AIRCRAFT

LAMINAR FLOW AIRFOILS
GS   AIRFOILS
    . **LAMINAR FLOW AIRFOILS**

LAMINAR FLOW CONTROL
USE   BOUNDARY LAYER CONTROL
      LAMINAR BOUNDARY LAYER

LAMINAR HEAT TRANSFER
GS   TRANSMISSION
    . HEAT TRANSMISSION
    . . HEAT TRANSFER
    . . . **LAMINAR HEAT TRANSFER**
RT   CONDUCTIVE HEAT TRANSFER
      CONVECTIVE HEAT TRANSFER
      THERMOHYDRAULICS
      TURBULENT HEAT TRANSFER

LAMINAR JETS
USE   JET FLOW
      LAMINAR FLOW

LAMINAR MIXING
GS   MIXING
    . **LAMINAR MIXING**
RT   FLUID INJECTION
      GAS MIXTURES
      TURBULENT MIXING

LAMINAR WAKES
GS   WAKES
    . **LAMINAR WAKES**
RT   AIRCRAFT WAKES
      TURBULENT WAKES

LAMINATED MATERIALS
USE   LAMINATES

LAMINATES
UF   LAMINATED MATERIALS
      LAMINATIONS
      MULTILAYER STRUCTURES
GS   COMPOSITE MATERIALS
    . **LAMINATES**
    . . BORAL
    . . PLYWOOD
    COMPOSITE STRUCTURES
    . **LAMINATES**
    . . BORAL
    . . PLYWOOD
RT   BONDING
      BORON-EPOXY COMPOUNDS
      CLADDING
      COATINGS
      FABRICS
      FIBER COMPOSITES
      FILAMENT WINDING
      ∞FILMS
      FORMICA
      GLASS FIBER REINFORCED PLASTICS
      HONEYCOMB STRUCTURES
      INTERLAYERS
      LAY-UP
      ∞LAYERS
      MAGNETIC CORES
      ∞MATERIALS
      MATRIX MATERIALS
      METAL BONDING
      METALLIZING
      MULTILAYER INSULATION
      PAPERS
      PLATING
      PLY ORIENTATION
      POLYMER MATRIX COMPOSITES
      PREPREGS
      REINFORCED PLASTICS
      REINFORCED PLATES
      SANDWICH STRUCTURES
      ∞SHEETS

LAMINATES-*(CONT.)*
    SUBSTRATES
    THERMOSETTING RESINS
    VENEERS

LAMINATIONS
    USE   LAMINATES

LAMPS
    USE   LUMINAIRES

LAMPS PROGRAM
    USE   LIGHT AIRBORNE MULTIPURPOSE
           SYSTEM

LANCE MISSILE
    GS   MISSILES
           . SURFACE TO SURFACE MISSILES
           . . **LANCE MISSILE**
    RT   LIQUID PROPELLANT ROCKET ENGINES
           TRAILBLAZER 1 REENTRY VEHICLE
           TX-77 ENGINE

LAND
    GS   **LAND**
           . ALLEGHENY PLATEAU (US)
           . ARID LANDS
           . BADLANDS
           . BARREN LAND
           . CASCADE RANGE (CA-OR-WA)
           . COLORADO PLATEAU (US)
           . DESERTS
           . . GOBI DESERT
           . . LIBYAN DESERT
           . . MOJAVE DESERT (CA)
           . . SAHARA DESERT (AFRICA)
           . FARMLANDS
           . GRASSLANDS
           . LLANOS ORIENTALES (COLOMBIA)
           . ISTHMUSES
           . MARSHLANDS
           . PARKS
           . . NATIONAL PARKS
           . . . YELLOWSTONE NATIONAL PARK
                 (ID-MT-WY)
           . PLAINS
           . . COASTAL PLAINS
           . . FLOOD PLAINS
           . . LLANOS ORIENTALES (COLOMBIA)
           . . PAMPAS
           . . PENEPLAINS
           . . PLAYAS
           . . TUNDRA
           . RANGELANDS
           . WETLANDS
    RT   CAPES (LANDFORMS)
           DESERTIFICATION
           DESERTLINE
           PENINSULAS
           RESIDENTIAL AREAS
           RURAL AREAS
           RURAL LAND USE
           SITES
           SOD
           SOILS
           TOPOGRAPHY

LAND ICE
    UF   ICE SHELVES
    GS   ICE
           . **LAND ICE**
           RESOURCES
           . EARTH RESOURCES
           . . **LAND ICE**
    RT   ANTARCTIC REGIONS
           GLACIAL DRIFT
           GLACIERS
           ICEBERGS
           LAKE ICE
           SEA ICE

LAND MANAGEMENT
    GS   MANAGEMENT
           . RESOURCES MANAGEMENT
           . . **LAND MANAGEMENT**
    RT   ENVIRONMENT MANAGEMENT
           REGIONAL PLANNING
           RURAL LAND USE
           URBAN PLANNING
           WILDERNESS

LAND MOBILE SATELLITE SERVICE
    GS   MOBILE COMMUNICATION SYSTEMS
           . **LAND MOBILE SATELLITE SERVICE**
    RT   COMMUNICATION SATELLITES

LAND MOBILE SATELLITE SERVICE-*(CONT.)*
    GROUND STATIONS
    MSAT
    RADIO COMMUNICATION

LAND USE
    GS   **LAND USE**
           . RURAL LAND USE
    RT   AGRISTARS PROJECT
           AIRPORT PLANNING
           BARREN LAND
           CHANGE DETECTION
           CONSERVATION
           DESERTIFICATION
           ∞DEVELOPMENT
           EARTH RESOURCES
           ECONOMIC DEVELOPMENT
           ENERGY POLICY
           ENVIRONMENT MANAGEMENT
           EXPLOITATION
           ∞FACILITIES
           FARMLANDS
           FOREST MANAGEMENT
           GRASSLANDS
           INDUSTRIAL AREAS
           LANDFILLS
           LEASING
           RESIDENTIAL AREAS
           SITE SELECTION
           SPOT (FRENCH SATELLITE)
           STARSITE PROGRAM
           SUBURBAN AREAS
           URBAN DEVELOPMENT
           URBAN PLANNING
           URBAN RESEARCH

LANDAU DAMPING
    GS   DAMPING
           . **LANDAU DAMPING**
    RT   ELECTRON PLASMA
           LANDAU FACTOR
           PHASE VELOCITY
           PLASMA WAVES
           SPACE CHARGE

LANDAU FACTOR
    RT   ATOMIC ENERGY LEVELS
           ATOMIC THEORY
           LANDAU DAMPING
           PLASMAS (PHYSICS)
           SUPERCONDUCTIVITY

LANDAU-GINZBURG EQUATIONS
    RT   ∞EQUATIONS
           QUANTUM ELECTRODYNAMICS
           SUPERCONDUCTIVITY

LANDFILLS
    RT   INDUSTRIAL WASTES
           LAND USE
           METHANE
           SOLID WASTES
           WASTE DISPOSAL
           WASTE UTILIZATION
           WATER POLLUTION

LANDFORMS
    GS   **LANDFORMS**
           . ARROYOS
           . BARRIERS (LANDFORMS)
           . . OUTER BANKS (NC)
           . . REEFS
           . BARS (LANDFORMS)
           . BEDS (GEOLOGY)
           . . SALT BEDS
           . BRIDGES (LANDFORMS)
           . CALDERAS
           . CANALS
           . CANYONS
           . . GRAND CANYON (AZ)
           . CAPES (LANDFORMS)
           . . CAPE HATTERAS (NC)
           . CONES (VOLCANOES)
           . CUSPS (LANDFORMS)
           . DELTAS
           . . MISSISSIPPI DELTA (LA)
           . . RHONE DELTA (FRANCE)
           . DIVIDES (LANDFORMS)
           . DUNES
           . ESCARPMENTS
           . FANS (LANDFORMS)
           . FIORDS
           . FLATS (LANDFORMS)
           . . TIDAL FLATS
           . GAPS (GEOLOGY)

LANDFORMS-*(CONT.)*
    . INLETS (TOPOGRAPHY)
    . . BAYOUS
    . . COOK INLET (AK)
    . INLIERS (LANDFORMS)
    . ISLAND ARCS
    . ISLANDS
    . . ALEUTIAN ISLANDS (US)
    . . ASSATEAGUE ISLAND (MD-VA)
    . . ATOLLS
    . . BAHRAIN
    . . BERMUDA
    . . CYPRUS
    . . GREENLAND
    . . HAWAII
    . . ICELAND
    . . INDONESIA
    . . IRELAND
    . . KEYS (ISLANDS)
    . . LONG ISLAND (NY)
    . . MALAGASY REPUBLIC
    . . MALDIVE ISLANDS
    . . MALTA
    . . MERRITT ISLAND (FL)
    . . NEWFOUNDLAND
    . . NUNATAKS
    . . PACIFIC ISLANDS
    . . . GUAM
    . . . JAPAN
    . . . JOHNSTON ISLAND
    . . . KURILE ISLANDS
    . . . NEW GUINEA (ISLAND)
    . . . NEW ZEALAND
    . . . PHILIPPINES
    . . . SAMOA
    . . PRINCE EDWARD ISLAND
    . . SICILY
    . . TASMANIA
    . . WALLOPS ISLAND
    . . WEST INDIES
    . . . BAHAMAS
    . . . BARBADOS
    . . . CUBA
    . . . DOMINICA
    . . . GUADELOUPE
    . . . HAITI
    . . . JAMAICA
    . . . LESSER ANTILLES
    . . . MARTINIQUE
    . . . PUERTO RICO
    . . . TRINIDAD AND TOBAGO
    . . . VIRGIN ISLANDS
    . ISTHMUSES
    . LAGOONS
    . MAGDALENA-CAUCA VALLEY
      (COLOMBIA)
    . MASSIFS
    . MOUNTAINS
    . . ADIRONDACK MOUNTAINS (NY)
    . . ALPS MOUNTAINS (EUROPE)
    . . ANDES MOUNTAINS (SOUTH
        AMERICA)
    . . APPALACHIAN MOUNTAINS (NORTH
        AMERICA)
    . . BIGHORN MOUNTAINS (MT-WY)
    . . BLACK HILLS (SD-WY)
    . . CARPATHIAN MOUNTAINS (EUROPE)
    . . CASCADE RANGE (CA-OR-WA)
    . . CAUCASUS MOUNTAINS (U.S.S.R.)
    . . COASTAL RANGES (CA)
    . . GREAT SMOKY MOUNTAINS (NC-TN)
    . . HIMALAYAS
    . . PENINSULAR RANGES (CA)
    . . PYRENEES MOUNTAINS (EUROPE)
    . . ROCKY MOUNTAINS (NORTH
        AMERICA)
    . . SAN JUAN MOUNTAINS (CO)
    . . SIERRA NEVADA MOUNTAINS (CA)
    . . WIND RIVER RANGE (WY)
    . . WRANGELL MOUNTAINS (AK)
    . MUSKEGS
    . OUTLIERS (LANDFORMS)
    . PANAMA CANAL ZONE
    . PEAKS (LANDFORMS)
    . PIKE'S PEAK (CO)
    . PENINSULAS
    . DELMARVA PENINSULA (DE-MD-VA)
    . PHOENIX QUADRANGLE (AZ)
    . PLAYAS
    . RAVINES
    . ST LAWRENCE VALLEY (NORTH
      AMERICA)
    . STEPPES
    . STRUCTURAL BASINS
    . . CIRQUES (LANDFORMS)
    . . GREAT BASIN (US)

LANDFORMS-*(CONT.)*
- . . KALIHARI BASIN (AFRICA)
- . . KARST
- . . . SINKHOLES
- . . KETTLES (GEOLOGY)
- . . LAKE CHAMPLAIN BASIN (NY-VT)
- . . RIVER BASINS
- . . . ATCHAFALAYA RIVER BASIN (LA)
- . . . CHENA RIVER BASIN (AK)
- . . . COLUMBIA RIVER BASIN (ID-OR-WA)
- . . . DELAWARE RIVER BASIN (US)
- . . . FEATHER RIVER BASIN (CA)
- . . . MISSOURI RIVER BASIN (US)
- . . . SUSQUEHANNA RIVER BASIN
       (MD-NY-PA)
- . . . WABASH RIVER BASIN (IL-IN-OH)
- . . WADIS
- . . WATERSHEDS
- . . WILLISTON BASIN (NORTH AMERICA)
- . TERRACES (LANDFORMS)
- . . PLATEAUS
- . . . ALLEGHENY PLATEAU (US)
- . . . COLORADO PLATEAU (US)
- . . . MESAS
- . . . . BUTTES
- . . . PIEDMONTS
- . . . . CENTRAL PIEDMONT (US)
- . TUNDRA
- . VOLCANOES
- . . MARS VOLCANOES

RT   ARCHIPELAGOES
    CROSSBEDDING (GEOLOGY)
    DEATH VALLEY (CA)
    DITCHES
    EARTHQUAKE RESISTANCE
   ∞FAULTS
    GEOLOGICAL FAULTS
    GLACIAL DRIFT
    LANDMARKS
    LANDSLIDES
    PLAINS
   ∞PLATFORMS
   ∞RIDGES
    SEAMOUNTS
    SINKHOLES
    SLOPES
    STRUCTURAL PROPERTIES (GEOLOGY)
    TERRAIN
    TOPOGRAPHY

LANDING
GS   **LANDING**
- . AIRCRAFT LANDING
- . . CRASH LANDING
- . . . DITCHING (LANDING)
- . . SKID LANDINGS
- . BLIND LANDING
- . GLIDE LANDINGS
- . . HORIZONTAL SPACECRAFT LANDING
- . HARD LANDING
- . SOFT LANDING
- . SPACECRAFT LANDING
- . . HORIZONTAL SPACECRAFT LANDING
- . . LUNAR LANDING
- . . MARS LANDING
- . PLANETARY LANDING
- . TOUCHDOWN
- . VERTICAL LANDING
- . WATER LANDING
- . . DITCHING (LANDING)

RT   AIR TRAFFIC CONTROL
    APPROACH
    APPROACH AND LANDING TESTS (STS)
    ARRIVALS
    GUIDANCE (MOTION)
    INSTRUMENT FLIGHT RULES
    INSTRUMENT LANDING SYSTEMS
    MANEUVERS
    RUNWAYS
    TAKEOFF
    VISUAL FLIGHT

LANDING AIDS
UF   LANDING SYSTEMS
GS   **LANDING AIDS**
- . AIRPORT BEACONS
- . . DISCRETE ADDRESS BEACON
     SYSTEM
- . AIRPORT LIGHTS
- . . RUNWAY LIGHTS
- . ARRESTING GEAR
- . INSTRUMENT LANDING SYSTEMS
- . . ALL-WEATHER LANDING SYSTEMS
- . . AUTOMATIC LANDING CONTROL
- . LANDING INSTRUMENTS

LANDING AIDS-*(CONT.)*
- . . APPROACH INDICATORS
- . LANDING RADAR
- . MICROVISION LANDING AID
- . MICROWAVE LANDING SYSTEMS
- . . MICROWAVE SCANNING BEAM
     LANDING SYSTEM

RT ∞AIDS
    AIR TRAFFIC CONTROL
    AIR TRAFFIC CONTROLLERS
    (PERSONNEL)
    AIRBORNE RADAR APPROACH
    AIRCRAFT EQUIPMENT
    AIRCRAFT INSTRUMENTS
    AIRCRAFT LANDING
    AIRCRAFT SAFETY
    AIRPORT TOWERS
    AIRPORTS
    ANTISKID DEVICES
    APPROACH
    APPROACH CONTROL
    AUTOMATIC PILOTS
    DELAYED FLAP APPROACH
    GROUND BASED CONTROL
    GROUND SUPPORT EQUIPMENT
    HEAD-UP DISPLAYS
    HELIPORTS
    INSTRUMENT APPROACH
    MILITARY AIR FACILITIES
    NATIONAL AVIATION SYSTEM
    NAVIGATION AIDS
    PLAT SYSTEM
    RADAR APPROACH CONTROL
    RADIO BEACONS
    RUNWAYS
    SAFETY DEVICES
    SOLAR COMPASSES

LANDING GEAR
UF   RETRACTABLE LANDING GEAR
RT   AIRCRAFT PARTS
    AIRCRAFT TIRES
    AIRFRAMES
   ∞BICYCLE
    BRAKES (FOR ARRESTING MOTION)
    CARRIAGES
    FAIRINGS
    FLOATS
   ∞GEAR
    HYDROFOILS
    NOSE WHEELS
    RETRACTABLE EQUIPMENT
    SELF ALIGNMENT
    SHOCK ABSORBERS
    SKIDDING
    SKIS
    SPRAY INGESTION
    TIRES
    UNDERCARRIAGES
    VEHICLE WHEELS
    WHEEL BRAKES
    WHEELS

LANDING INSTRUMENTS
GS   LANDING AIDS
- . **LANDING INSTRUMENTS**
- . . APPROACH INDICATORS

RT   AIR TRAFFIC CONTROL
    AIRCRAFT EQUIPMENT
    AIRCRAFT INSTRUMENTS
    ALTIMETERS
    AUTOMATIC CONTROL
    BLIND LANDING
    FLIGHT INSTRUMENTS
    INSTRUMENT LANDING SYSTEMS
    MANUAL CONTROL
    MEASURING INSTRUMENTS
    RADAR APPROACH CONTROL
    SPEED INDICATORS

LANDING LOADS
GS   LOADS (FORCES)
- . DYNAMIC LOADS
- . . TRANSIENT LOADS
- . . . **LANDING LOADS**

RT   DECELERATION
    IMPACT LOADS
    SHOCK LOADS
    STRUCTURAL DESIGN CRITERIA

LANDING MATS
RT   AIRCRAFT LANDING
    AIRPORTS
    MILITARY AIR FACILITIES
    RUNWAYS

LANDING MODULES
GS   MODULES
- . SPACECRAFT MODULES
- . . **LANDING MODULES**
- . . . LUNAR LANDING MODULES
- . . . . LUNAR MODULE
- . . . . . LSSM
- . . . MARS EXCURSION MODULE
   SOFT LANDING SPACECRAFT
- . **LANDING MODULES**
- . . LUNAR LANDING MODULES
- . . . LUNAR MODULE
- . . . . APOLLO LUNAR EXPERIMENT
       MODULE
- . . . . LSSM
- . . MARS EXCURSION MODULE

RT   APOLLO SPACECRAFT
    INTERPLANETARY SPACECRAFT
    LAUNCH VEHICLES
    MANEUVERABLE SPACECRAFT
    MANNED SPACECRAFT
    REENTRY VEHICLES
    REUSABLE SPACECRAFT
    SPACE CAPSULES
    SPACECRAFT DOCKING MODULES

LANDING RADAR
GS   LANDING AIDS
- . **LANDING RADAR**
   RADAR
- . **LANDING RADAR**

RT   AIR TRAFFIC CONTROL
    AIRCRAFT LANDING
    AIRCRAFT SAFETY
    APPROACH CONTROL
    INSTRUMENT APPROACH
    RADAR APPROACH CONTROL

LANDING SIMULATION
GS   SIMULATION
- . **LANDING SIMULATION**

RT   ALTITUDE SIMULATION
    ATMOSPHERIC ENTRY SIMULATION
    COMPUTERIZED SIMULATION
    FLIGHT SIMULATION
    SPACECRAFT LANDING
    TRAINING SIMULATORS

LANDING SITES
GS   SITES
- . **LANDING SITES**
- . . LUNAR LANDING SITES

RT   HELIPORTS
    RECOVERY ZONES
    RUNWAYS
    TRAJECTORY CONTROL

LANDING SPEED
GS   RATES (PER TIME)
- . **LANDING SPEED**
   VELOCITY
- . **LANDING SPEED**

RT   HIGH SPEED
    LOW SPEED

LANDING SYSTEMS
USE LANDING AIDS

LANDMARKS
RT   LANDFORMS
    TERRAIN
    TOPOGRAPHY

LANDSAT E
UF   EARTH RESOURCES TECHNOLOGY
    SATELLITE E
    EOS-A
    ERTS-E
GS   SATELLITES
- . ARTIFICIAL SATELLITES
- . . LANDSAT SATELLITES
- . . . **LANDSAT E**

LANDSAT F
UF   EARTH RESOURCES TECHNOLOGY
    SATELLITE F
    EOS-B
    ERTS-F
GS   SATELLITES
- . ARTIFICIAL SATELLITES
- . . LANDSAT SATELLITES
- . . . **LANDSAT F**

LANDSAT FOLLOW-ON MISSIONS
- UF   LFO
- RT   ∞ MISSIONS
         MULTIMISSION MODULAR SPACECRAFT

LANDSAT SATELLITES
- UF   EARTH RESOURCES TECHNOLOGY
             SATELLITES
         EOS
         ERTS
- GS   SATELLITES
         . ARTIFICIAL SATELLITES
         . . **LANDSAT SATELLITES**
         . . . LANDSAT E
         . . . LANDSAT F
         . . . LANDSAT 1
         . . . LANDSAT 2
         . . . LANDSAT 3
         . . . LANDSAT 4
         . . . LANDSAT 5
- RT   AGRISTARS PROJECT
         EARTH OBSERVATIONS (FROM SPACE)
         EARTHNET
         MAPSAT
         NASA PROGRAMS
         OCEANOGRAPHY
         SATELLITE OBSERVATION
         SEASAT PROGRAM
         SEASAT SATELLITES
         SEASAT 1
         SEASAT-B SATELLITE
         SYNCHRONOUS EARTH OBSERVATORY
             SATELLITE

LANDSAT 1
- UF   EARTH RESOURCES TECHNOLOGY
             SATELLITE 1
         ERTS-A
- GS   SATELLITES
         . ARTIFICIAL SATELLITES
         . . LANDSAT SATELLITES
         . . . **LANDSAT 1**

LANDSAT 2
- UF   EARTH RESOURCES TECHNOLOGY
             SATELLITE B
         ERTS-B
- GS   SATELLITES
         . ARTIFICIAL SATELLITES
         . . LANDSAT SATELLITES
         . . . **LANDSAT 2**

LANDSAT 3
- UF   EARTH RESOURCES TECHNOLOGY
             SATELLITE C
         ERTS-C
- GS   SATELLITES
         . ARTIFICIAL SATELLITES
         . . LANDSAT SATELLITES
         . . . **LANDSAT 3**
- RT   PLASMA INTERACTION EXPERIMENT

LANDSAT 4
- UF   EARTH RESOURCES TECHNOLOGY
             SATELLITE D
         ERTS-D
- GS   SATELLITES
         . ARTIFICIAL SATELLITES
         . . LANDSAT SATELLITES
         . . . **LANDSAT 4**

LANDSAT 5
- GS   SATELLITES
         . ARTIFICIAL SATELLITES
         . . LANDSAT SATELLITES
         . . . **LANDSAT 5**

LANDSCAPE
- USE   TERRAIN
         TOPOGRAPHY

LANDSLIDES
- GS   EARTH MOVEMENTS
         . **LANDSLIDES**
- RT   CLIFFS
         FLOOD DAMAGE
         LANDFORMS
         RAIN EROSION
         ROCKS
         SLOPES
         SOIL EROSION
         SOILS
         STORM DAMAGE

LANES
- USE   PATHS

LANGEVIN FORMULA
- RT   DISPERSING
         FERROMAGNETISM
         MAGNETIC MOMENTS

LANGLEY COMPLEX COORDINATOR
- GS   SIMULATORS
         . ENVIRONMENT SIMULATORS
         . . SPACE SIMULATORS
         . . . **LANGLEY COMPLEX COORDINATOR**
- RT   FLIGHT SIMULATORS
         GRAVITATIONAL EFFECTS
         ROTATING ENVIRONMENTS
         SPACE ENVIRONMENT SIMULATION
         SPACECRAFT ENVIRONMENTS
         WEIGHTLESSNESS SIMULATION

LANGMUIR PROBES
- USE   ELECTROSTATIC PROBES

LANGUAGE PROGRAMMING
- GS   SOFTWARE ENGINEERING
         . COMPUTER PROGRAMMING
         . . **LANGUAGE PROGRAMMING**
- RT   COMPUTER ASSISTED INSTRUCTION
         DATA PROCESSING
         HIGH LEVEL LANGUAGES
         LANGUAGES
         MACHINE ORIENTED LANGUAGES
         MACHINE TRANSLATION
         SYMBOLIC PROGRAMMING
         ∞ TRANSLATORS

LANGUAGES
- GS   **LANGUAGES**
         . ADA (PROGRAMMING LANGUAGE)
         . COMMAND LANGUAGES
         . QUERY LANGUAGES
         . ENGLISH LANGUAGE
         . ORTHOGRAPHY
         . PROGRAMMING LANGUAGES
         . . ALGOL
         . . APL (PROGRAMMING LANGUAGE)
         . . ASSEMBLY LANGUAGE
         . . . AUTOCODERS
         . . . COMPASS (PROGRAMMING
             LANGUAGE)
         . . . MAP (PROGRAMMING LANGUAGE)
         . . BASIC (PROGRAMMING LANGUAGE)
         . . COBOL
         . . COGO (PROGRAMMING LANGUAGE)
         . . CONTEXT FREE LANGUAGES
         . . FORTRAN
         . . HAL/S (LANGUAGE)
         . . HIGH LEVEL LANGUAGES
         . . LISP (PROGRAMMING LANGUAGE)
         . . MACHINE ORIENTED LANGUAGES
         . . . MARVS (PROGRAMMING
             LANGUAGE)
         . . . SLEUTH (PROGRAMMING
             LANGUAGE)
         . . NATURAL LANGUAGE (COMPUTERS)
         . . PASCAL (PROGRAMMING LANGUAGE)
         . . PL/1
         . SENTENCES
         . . WORDS (LANGUAGE)
         . . . SYLLABLES
- RT   ALPHABETS
         ARTICULATION
         CODING
         COMMUNICATION THEORY
         GRAMMARS
         LANGUAGE PROGRAMMING
         LINGUISTICS
         MACHINE TRANSLATION
         PHONEMES
         PHONEMICS
         PHONETICS
         SEMANTICS
         SPEECH
         SYMBOLS
         SYNTAX
         TRANSLATING
         VERBAL COMMUNICATION
         VOWELS

LANTHANIDE SERIES METALS
- USE   RARE EARTH ELEMENTS

LANTHANUM
- GS   CHEMICAL ELEMENTS
         . RARE EARTH ELEMENTS

LANTHANUM-*(CONT.)*
         . . **LANTHANUM**
         . . . LANTHANUM ISOTOPES
         METALS
         . RARE EARTH ELEMENTS
         . . **LANTHANUM**
         . . . LANTHANUM ISOTOPES
- RT   DIDYMIUM

LANTHANUM ALLOYS
- GS   ALLOYS
         . RARE EARTH ALLOYS
         . . **LANTHANUM ALLOYS**

LANTHANUM CHLORIDES
- GS   HALOGEN COMPOUNDS
         . CHLORINE COMPOUNDS
         . . CHLORIDES
         . . . **LANTHANUM CHLORIDES**
         . HALIDES
         . . CHLORIDES
         . . . **LANTHANUM CHLORIDES**
         . . METAL HALIDES
         . . . **LANTHANUM CHLORIDES**
         LANTHANUM COMPOUNDS
         . **LANTHANUM CHLORIDES**

LANTHANUM COMPOUNDS
- GS   **LANTHANUM COMPOUNDS**
         . LANTHANUM CHLORIDES
         . LANTHANUM FLUORIDES
         . LANTHANUM OXIDES
         . LANTHANUM TELLURIDES
- RT   ∞ CHEMICAL COMPOUNDS
         ∞ METAL COMPOUNDS

LANTHANUM FLUORIDES
- GS   HALOGEN COMPOUNDS
         . FLUORINE COMPOUNDS
         . . FLUORIDES
         . . . METAL FLUORIDES
         . . . . **LANTHANUM FLUORIDES**
         . HALIDES
         . . METAL HALIDES
         . . . **LANTHANUM FLUORIDES**
         LANTHANUM COMPOUNDS
         . **LANTHANUM FLUORIDES**

LANTHANUM ISOTOPES
- UF   LANTHANUM 140
- GS   CHEMICAL ELEMENTS
         . NUCLIDES
         . . ISOTOPES
         . . . **LANTHANUM ISOTOPES**
         . RARE EARTH ELEMENTS
         . . LANTHANUM
         . . . **LANTHANUM ISOTOPES**
         METALS
         . RARE EARTH ELEMENTS
         . . LANTHANUM
         . . . **LANTHANUM ISOTOPES**

LANTHANUM OXIDES
- GS   CHALCOGENIDES
         . OXIDES
         . . METAL OXIDES
         . . . **LANTHANUM OXIDES**
         LANTHANUM COMPOUNDS
         . **LANTHANUM OXIDES**

LANTHANUM TELLURIDES
- GS   CHALCOGENIDES
         . TELLURIDES
         . . **LANTHANUM TELLURIDES**
         LANTHANUM COMPOUNDS
         . **LANTHANUM TELLURIDES**
         RARE EARTH COMPOUNDS
         . **LANTHANUM TELLURIDES**
         TELLURIUM COMPOUNDS
         . TELLURIDES
         . . **LANTHANUM TELLURIDES**

LANTHANUM 140
- USE   LANTHANUM ISOTOPES

LAOS
- GS   NATIONS
         . **LAOS**
- RT   ASIA

LAP JOINTS
- GS   JOINTS (JUNCTIONS)
         . **LAP JOINTS**
- RT   BUTT JOINTS

LAP JOINTS-*(CONT.)*
METAL JOINTS
RIVETED JOINTS
SOLDERED JOINTS
WELDED JOINTS

LAPLACE EQUATION
RT   ∝ EQUATIONS
HARMONIC FUNCTIONS
PARTIAL DIFFERENTIAL EQUATIONS
POISSON EQUATION
STOKES-BELTRAMI EQUATION

LAPLACE OPERATORS
USE   LAPLACE TRANSFORMATION

LAPLACE TRANSFORMATION
UF   LAPLACE OPERATORS
GS   ANALYSIS (MATHEMATICS)
. FUNCTIONAL ANALYSIS
. . INTEGRAL TRANSFORMATIONS
. . . **LAPLACE TRANSFORMATION**
FUNCTIONS (MATHEMATICS)
. **LAPLACE TRANSFORMATION**
TRANSFORMATIONS (MATHEMATICS)
. INTEGRAL TRANSFORMATIONS
. . **LAPLACE TRANSFORMATION**
RT   DIFFERENTIAL EQUATIONS
OPERATORS (MATHEMATICS)

LAPSE RATE
RT   HUMIDITY
TEMPERATURE
TEMPERATURE INVERSIONS
TEPHIGRAMS

LARA AIRCRAFT
USE   COIN AIRCRAFT

LARGE APERTURE SEISMIC ARRAY
GS   ARRAYS
. **LARGE APERTURE SEISMIC ARRAY**
RT   EARTH MOVEMENTS
EARTHQUAKES
MEASURING INSTRUMENTS
SEISMIC WAVES
SEISMOLOGY

LARGE AREA CROP INVENTORY EXPERIMENT
RT   AGRICULTURE
AGROPHYSICAL UNITS
CROP GROWTH
CROP INVENTORIES
∝ CROPS
EARTH RESOURCES
EARTH RESOURCES PROGRAM
FARM CROPS
INVENTORIES

LARGE INFRARED TELESCOPE ON SPACELAB
USE   LIRTS (TELESCOPE)

LARGE SCALE INTEGRATION
UF   LSI
GS   CIRCUITS
. INTEGRATED CIRCUITS
. . **LARGE SCALE INTEGRATION**
. PRINTED CIRCUITS
. . **LARGE SCALE INTEGRATION**
MICROELECTRONICS
. **LARGE SCALE INTEGRATION**
RT   CHIPS (ELECTRONICS)
DTL INTEGRATED CIRCUITS
ELECTRONIC PACKAGING
LINEAR INTEGRATED CIRCUITS
MEDIUM SCALE INTEGRATION
MICROPROCESSORS
TTL INTEGRATED CIRCUITS
VERY LARGE SCALE INTEGRATION
VHSIC (CIRCUITS)

LARGE SPACE STRUCTURES
RT   CONTINUUM MODELING
EXPANDABLE STRUCTURES
FLEXIBLE SPACECRAFT
HOOP COLUMN ANTENNAS
INTRAORBIT TRANSFER VEHICLES
LASER GYROSCOPES
MAYPOLE ANTENNAS
MEGAMECHANICS
ORBITAL SERVICING
ORBITAL SPACE STATIONS
ORBITAL SPACE TESTS
SELF SHADOWING

LARGE SPACE STRUCTURES-*(CONT.)*
SHAPE CONTROL
SOLAR POWER SATELLITES
SPACE ERECTABLE STRUCTURES
SPACE OPERATIONS CENTER (NASA)
SPACE TECHNOLOGY EXPERIMENTS
∝ STRUCTURES

LARGE SPACE TELESCOPE
USE   HUBBLE SPACE TELESCOPE

LARGOS SATELLITE
GS   SATELLITES
. ARTIFICIAL SATELLITES
. . GEODETIC SATELLITES
. . . **LARGOS SATELLITE**
UNMANNED SPACECRAFT
. GEODETIC SATELLITES
. . **LARGOS SATELLITE**
RT   EXPLORER 29 SATELLITE
EXPLORER 36 SATELLITE
GEOS 1 SATELLITE
GEOS 2 SATELLITE
GEOS 3 SATELLITE

LARMOR PRECESSION
GS   GYRATION
. PRECESSION
. . **LARMOR PRECESSION**
RT   CYCLOTRON FREQUENCY
CYCLOTRON RADIATION

LARMOR RADIUS
GS   DIMENSIONS
. RADII
. . **LARMOR RADIUS**
RT   CYCLOTRON RADIATION
GYROMAGNETISM
PLASMA PHYSICS
PRECESSION

LARVAE
GS   ANIMALS
. INVERTEBRATES
. . ARTHROPODS
. . . INSECTS
. . . . **LARVAE**
. . . . . BOLLWORMS
. . . . . SILKWORMS
RT   INFESTATION
PUPA
WORMS

LARYNX
GS   **LARYNX**
. GLOTTIS
RT   CARTILAGE
VOCAL CORDS

LASER ALTIMETERS
GS   NAVIGATION AIDS
. **LASER ALTIMETERS**
SPACECRAFT INSTRUMENTS
. **LASER ALTIMETERS**
RT   ALTITUDE CONTROL
∝ INSTRUMENTS
LASERS
OPTICAL RADAR

LASER ANEMOMETERS
GS   MEASURING INSTRUMENTS
. INDICATING INSTRUMENTS
. . SPEED INDICATORS
. . . ANEMOMETERS
. . . . **LASER ANEMOMETERS**
RT   FLOW VELOCITY

LASER ANNEALING
GS   HEAT TREATMENT
. ANNEALING
. . **LASER ANNEALING**
RT   HEATING
LASER CUTTING
LASERS
NORMALIZING (HEAT TREATMENT)
RECRYSTALLIZATION
TEMPERING

LASER APPLICATIONS
GS   UTILIZATION
. **LASER APPLICATIONS**
. . LASER CUTTING
. . LASER FUSION
RT   AIRBORNE LASERS

LASER APPLICATIONS-*(CONT.)*
DATA TRANSMISSION
HOLE BURNING
LASER HEATING
LASERS
METAL WORKING
OPTICAL DATA STORAGE MATERIALS
OPTICAL DISKS
OPTICAL RELAY SYSTEMS
PLASMADYNAMIC LASERS
PRODUCTION ENGINEERING
RAPID BALLISTICS IDENTIFICATION
SPACEBORNE LASERS
TECHNOLOGY UTILIZATION
ULTRASHORT PULSED LASERS

LASER BEAM DEFOCUSING
USE   THERMAL BLOOMING

LASER CAVITIES
UF   OPTICAL GENERATORS
RT   AMPLIFIERS
LASER STABILITY
LASERS
LIGHT AMPLIFIERS
∝ OPTICS
PULSE GENERATORS
SEMICONDUCTOR LASERS
SOLID STATE DEVICES
SOLID STATE LASERS
STIMULATED EMISSION DEVICES

LASER COMMUNICATION
USE   OPTICAL COMMUNICATION

LASER CUTTING
GS   CUTTING
. **LASER CUTTING**
UTILIZATION
. LASER APPLICATIONS
. . **LASER CUTTING**
RT   BLANKING (CUTTING)
CUTTERS
FOCUSING
FORMING TECHNIQUES
LASER ANNEALING
LASER DRILLING
LASER HEATING
LASER OUTPUTS
LASER TARGET INTERACTIONS
MACHINING
∝ MATERIALS SCIENCE
METAL CUTTING
SPLITTING
THERMAL BLOOMING

LASER DAMAGE
GS   DAMAGE
. RADIATION DAMAGE
. . **LASER DAMAGE**
RADIATION EFFECTS
. **LASER DAMAGE**
RT   BURNS (INJURIES)
PULSED RADIATION
RADIATION HAZARDS

LASER DOPPLER VELOCIMETERS
GS   MEASURING INSTRUMENTS
. **LASER DOPPLER VELOCIMETERS**
OPTICAL EQUIPMENT
. **LASER DOPPLER VELOCIMETERS**
RT   FLOW MEASUREMENT
OPTICAL MEASURING INSTRUMENTS
VELOCITY MEASUREMENT

LASER DRILLING
GS   DRILLING
. **LASER DRILLING**
RT   FOCUSING
LASER CUTTING

LASER FUSION
GS   UTILIZATION
. LASER APPLICATIONS
. . **LASER FUSION**
RT   ∝ FUSION
GLASS LASERS
HIGH POWER LASERS
INERTIAL FUSION (REACTOR)
NOVA LASER SYSTEM
PLASMAS (PHYSICS)
SHIVA LASER SYSTEM

LASER GEODYNAMIC SATELLITE
USE   LAGEOS (SATELLITE)

LASER GUIDANCE
GS   GUIDANCE (MOTION)
    . TERMINAL GUIDANCE
    . . **LASER GUIDANCE**
RT   COMPUTER PROGRAMS
    HOMING DEVICES
    IMPACT PREDICTION
    MISSILE CONTROL

LASER GYROSCOPES
GS   GYROSCOPES
    . **LASER GYROSCOPES**
RT   LARGE SPACE STRUCTURES
    OPTICAL GYROSCOPES
    SAGNAC EFFECT
    ∞SENSORS
    SPACECRAFT GUIDANCE
    STABILIZATION

LASER HEATING
GS   HEATING
    . **LASER HEATING**
RT   HEAT SOURCES
    LASER APPLICATIONS
    LASER CUTTING
    PULSE HEATING
    PULSED LASERS
    THERMAL BLOOMING
†   YAG LASERS

LASER INTERFEROMETRY
GS   INTERFEROMETRY
    . **LASER INTERFEROMETRY**
RT   SAGNAC EFFECT

LASER MATERIALS
RT   ∞MATERIALS
    METAL VAPOR LASERS
    XENON CHLORIDE LASERS
    XENON FLUORIDE LASERS
    YAG LASERS

LASER MICROSCOPY
GS   MICROSCOPY
    . **LASER MICROSCOPY**
RT   ELECTRO-OPTICS
    LIGHT AMPLIFIERS
    METAL VAPOR LASERS
    MICROELECTRONICS

LASER MODE LOCKING
GS   LOCKING
    . **LASER MODE LOCKING**
RT   INJECTION LOCKING
    LASERS
    OPTICAL COUPLING

LASER MODES
GS   MODES
    . **LASER MODES**
RT   AXIAL MODES
    HELIUM-NEON LASERS
    LASER STABILITY
    OPTICAL RESONATORS
    TEA LASERS
    WAVEGUIDE LASERS
    WAVELENGTHS

LASER OUTPUTS
GS   OUTPUT
    . **LASER OUTPUTS**
RT   ATMOSPHERIC LASERS
    ∞COHERENCE
    COHERENT LIGHT
    DISTRIBUTED FEEDBACK LASERS
    DYE LASERS
    EXCIMER LASERS
    GASDYNAMIC LASERS
    GLASS LASERS
    HELIUM-NEON LASERS
    HIGH POWER LASERS
    HOLOGRAPHIC INTERFEROMETRY
    LASER CUTTING
    LASER STABILITY
    MASER OUTPUTS
    NOVA LASER SYSTEM
    OPTICAL RESONATORS
    OPTICAL WAVEGUIDES
    PHASE MATCHING
    PHOTON BEAMS
    PICOSECOND PULSES

LASER OUTPUTS-(CONT.)
    PULSE DURATION
    QUANTUM EFFICIENCY
    RADIANT FLUX DENSITY
    SHIVA LASER SYSTEM
    SPECKLE PATTERNS
    THERMAL BLOOMING
    TUBE LASERS
    TWO-WAVELENGTH LASERS
    ULTRAVIOLET LASERS
    VOLUMETRIC EFFICIENCY
    WAVEGUIDE LASERS
    WAVELENGTHS
    X RAY LASERS
    XENON CHLORIDE LASERS
    XENON FLUORIDE LASERS
    YAG LASERS

LASER PLASMA INTERACTIONS
GS   ELECTROMAGNETIC INTERACTIONS
    . PLASMA-ELECTROMAGNETIC
        INTERACTION
    . . **LASER PLASMA INTERACTIONS**
    PLASMA INTERACTIONS
    . PLASMA-ELECTROMAGNETIC
        INTERACTION
    . . **LASER PLASMA INTERACTIONS**
RT   BACKSCATTERING
    GLASS LASERS
    ∞INTERACTIONS
    PLASMAS (PHYSICS)
    THETA PINCH

LASER PLASMAS
SN   (EXCLUDES LASER OUTPUTS)
GS   GASES
    . IONIZED GASES
    . . CHARGED PARTICLES
    . . . **LASER PLASMAS**
    PARTICLES
    . CHARGED PARTICLES
    . . ENERGETIC PARTICLES
    . . . PLASMAS (PHYSICS)
    . . . . **LASER PLASMAS**
RT   INERTIAL FUSION (REACTOR)
    LASERS

LASER PROPULSION
GS   PROPULSION
    . ELECTRIC PROPULSION
    . . **LASER PROPULSION**
RT   AIRCRAFT ENGINES
    HYBRID PROPULSION
    OPTICAL PUMPING
    PROPULSION SYSTEM CONFIGURATIONS
    PROPULSIVE EFFICIENCY
    RANKINE CYCLE
    ROCKET ENGINES
    SPACECRAFT PROPULSION
    THERMODYNAMIC CYCLES

LASER PUMPING
GS   OPTICAL PUMPING
    . **LASER PUMPING**
RT   LASERS
    ∞PUMPING
    RARE GAS-HALIDE LASERS
    SOLAR-PUMPED LASERS
    STIMULATED EMISSION DEVICES
    WIGGLER MAGNETS

LASER RADAR
USE   OPTICAL RADAR

LASER RANGE FINDERS
GS   MEASURING INSTRUMENTS
    . DISTANCE MEASURING EQUIPMENT
    . . RANGE FINDERS
    . . . OPTICAL RANGE FINDERS
    . . . . **LASER RANGE FINDERS**
RT   LAGEOS (SATELLITE)
    LUNAR RANGEFINDING
    LUNAR RETROREFLECTORS
    NAVIGATION AIDS
    NAVIGATION INSTRUMENTS

LASER RANGER/TRACKER
RT   AIRBORNE LASERS
    RANGE FINDERS
    RANGEFINDING
    TRACKING (POSITION)

LASER SPECTROMETERS
GS   MEASURING INSTRUMENTS
    . SPECTROMETERS

LASER SPECTROMETERS-(CONT.)
    . . **LASER SPECTROMETERS**
RT   ABSORPTION SPECTRA
    INFRARED SPECTROSCOPY
    LASER SPECTROSCOPY

LASER SPECTROSCOPY
GS   SPECTROSCOPY
    . OPTICAL EMISSION SPECTROSCOPY
    . . **LASER SPECTROSCOPY**
RT   LASER SPECTROMETERS
    SPECTROSCOPIC ANALYSIS
    SPECTRUM ANALYSIS

LASER STABILITY
RT   CONTINUOUS WAVE LASERS
    FREQUENCY STABILITY
    LASER CAVITIES
    LASER MODES
    LASER OUTPUTS

LASER TARGET DESIGNATORS
RT   ∞DETECTORS
    MISSILE TRACKING
    SATELLITE TRACKING
    TARGET RECOGNITION
    TARGETS

LASER TARGET INTERACTIONS
RT   ∞INTERACTIONS
    LASER CUTTING
    LASERS
    PULSED LASERS
    TARGETS

LASER TARGETS
GS   TARGETS
    . **LASER TARGETS**
RT   GLASS LASERS
    LASERS

LASER WEAPONS
GS   WEAPON SYSTEMS
    . **LASER WEAPONS**
    WEAPONS
    . SPACE WEAPONS
    . . **LASER WEAPONS**
RT   FUSION WEAPONS
    LASERS
    MILITARY TECHNOLOGY
    STIMULATED EMISSION DEVICES

LASER WELDING
GS   WELDING
    . **LASER WELDING**
    . . FUSION WELDING
    . . . GAS WELDING
    . . . . BRAZING
    . . . . . LOW TEMPERATURE BRAZING
RT   BONDING
    HEATING
    PULSED LASERS
    SOLDERING

LASER WINDOWS
GS   WINDOWS (INTERVALS)
    . **LASER WINDOWS**
RT   BANDWIDTH
    ENERGY BANDS
    LASERS

LASERS
UF   FABRY-PEROT LASERS
    NATURAL LASERS
    OPTICAL MASERS
GS   STIMULATED EMISSION DEVICES
    . **LASERS**
    . . AIRBORNE LASERS
    . . ARGON LASERS
    . . ATMOSPHERIC LASERS
    . . CARBON LASERS
    . . CHEMICAL LASERS
    . . . HCL LASERS
    . . CONTINUOUS WAVE LASERS
    . . DISTRIBUTED FEEDBACK LASERS
    . . FREE ELECTRON LASERS
    . . GAMMA RAY LASERS
    . . GAS LASERS
    . . . CARBON DIOXIDE LASERS
    . . . CARBON MONOXIDE LASERS
    . . . DF LASERS
    . . . EXCIMER LASERS
    . . . HCL LASERS
    . . . . HCL ARGON LASERS

LASERS-(CONT.)
```
. . . HCN LASERS
. . . HELIUM-NEON LASERS
. . . HF LASERS
. . . KRYPTON FLUORIDE LASERS
. . . NITROGEN LASERS
. . . TEA LASERS
. . . ULTRAVIOLET LASERS
. . . XENON CHLORIDE LASERS
. . . XENON FLUORIDE LASERS
. . GASDYNAMIC LASERS
. . GLASS LASERS
. . HIGH POWER LASERS
. . . NOVA LASER SYSTEM
. . . SHIVA LASER SYSTEM
. . INFRARED LASERS
. . INJECTION LASERS
. . IODINE LASERS
. . LIQUID LASERS
. . METAL VAPOR LASERS
. . NEODYMIUM LASERS
. . NUCLEAR PUMPED LASERS
. . ORGANIC LASERS
. . . DYE LASERS
. . PLASMADYNAMIC LASERS
. . PULSED LASERS
. . . Q SWITCHED LASERS
. . . ULTRASHORT PULSED LASERS
. . . ULTRAVIOLET LASERS
. . RAMAN LASERS
. . RARE GAS-HALIDE LASERS
. . . KRYPTON FLUORIDE LASERS
. . . XENON CHLORIDE LASERS
. . . XENON FLUORIDE LASERS
. . RING LASERS
. . SEMICONDUCTOR LASERS
. . . GALLIUM ARSENIDE LASERS
. . SOLAR-PUMPED LASERS
. . SOLID STATE LASERS
. . . GALLIUM ARSENIDE LASERS
. . . RUBY LASERS
. . . YAG LASERS
. . SPACEBORNE LASERS
. . TUNABLE LASERS
. . TWO-WAVELENGTH LASERS
. . WAVEGUIDE LASERS
. . X RAY LASERS
```
```
RT    ALKALI VAPOR LAMPS
      AMPLIFIERS
      BEAM SWITCHING
   ∞COHERENCE
      COHERENT ELECTROMAGNETIC
        RADIATION
      COHERENT LIGHT
      ELECTRON PUMPING
      GARNETS
      HOLE BURNING
      HOLOGRAPHY
      INFRARED WINDOWS
      INTERPLANETARY COMMUNICATION
      INTERSTELLAR MASERS
      KERR ELECTROOPTICAL EFFECT
      LASER ALTIMETERS
      LASER ANNEALING
      LASER APPLICATIONS
      LASER CAVITIES
      LASER MODE LOCKING
      LASER PLASMAS
      LASER PUMPING
      LASER TARGET INTERACTIONS
      LASER TARGETS
      LASER WEAPONS
      LASER WINDOWS
      LASING
      LIGHT AMPLIFIERS
      LIGHT BEAMS
      LIGHT MODULATION
      LIGHT SOURCES
      LIGHT TRANSMISSION
      LUNAR COMMUNICATION
      MASERS
      MICROBALLOONS
      MOLECULAR OSCILLATORS
      NUCLEAR PUMPING
      OPTICAL COMMUNICATION
      OPTICAL DATA PROCESSING
      OPTICAL MEMORY (DATA STORAGE)
      OPTICAL PUMPING
   ∞OPTICS
      PHASE MATCHING
      PHOTODIODES
      PHOTONICS
      POWER TRANSMISSION (LASERS)
      PULSE GENERATORS
      PULSED RADIATION
      QUANTUM AMPLIFIERS
```

LASERS-(CONT.)
```
      QUANTUM ELECTRONICS
      RAPID BALLISTICS IDENTIFICATION
      SENARMONT POLARISCOPES
      SOLID STATE DEVICES
      SPACE COMMUNICATION
      STIMULATED EMISSION
      THERMAL BLOOMING
      THRESHOLD CURRENTS
      TRANSIENT OSCILLATIONS
      TRAVELING WAVE MODULATION
```

LASING
```
RT    DISTRIBUTED FEEDBACK LASERS
      ELECTRON TRANSITIONS
      EXCIMER LASERS
      HOLE BURNING
      KRYPTON FLUORIDE LASERS
      LASERS
      NITROGEN LASERS
      OPTICAL TRANSITION
      RARE GAS-HALIDE LASERS
      STIMULATED EMISSION DEVICES
```

LASV
```
USE   F-111 AIRCRAFT
```

LATCH-UP
```
RT    CMOS
      ELECTRICAL IMPEDANCE
      INTEGRATED CIRCUITS
      P-N-P-N JUNCTIONS
      SWITCHING CIRCUITS
```

LATCHES
```
RT    FASTENERS
      HOLDERS
      LINKAGES
      PINS
```

LATE STARS
```
GS    CELESTIAL BODIES
      . STARS
      . . LATE STARS
RT    COOL STARS
      DWARF STARS
      EARLY STARS
      GIANT STARS
      MAIN SEQUENCE STARS
      RED DWARF STARS
      RED GIANT STARS
      STELLAR EVOLUTION
      SUBGIANT STARS
```

LATENESS
```
RT    DELAY
      SCHEDULING
```

† LATENT HEAT OF FUSION
```
USE   HEAT OF FUSION
```

LATERAL CONTROL
```
UF    LATERALIZATION
      ROLL CONTROL
GS    ATTITUDE CONTROL
      . LATERAL CONTROL
RT    AILERONS
      AIRCRAFT CONTROL
      ALTITUDE CONTROL
      AUTOMATIC CONTROL
   ∞CONTROL
      DIRECTIONAL CONTROL
      ELEVONS
      HELICOPTER CONTROL
      LONGITUDINAL CONTROL
      MANUAL CONTROL
      MISSILE CONTROL
      ROLL
      SATELLITE ATTITUDE CONTROL
      SATELLITE CONTROL
```

LATERAL OSCILLATION
```
UF    SNAKING
RT    DIRECTIONAL STABILITY
      ROLL
      STABILITY AUGMENTATION
      TRANSVERSE OSCILLATION
      TURNING FLIGHT
      YAW
      YAWING MOMENTS
```

LATERAL STABILITY
```
UF    DIHEDRAL EFFECT
      LATERALITY
```

LATERAL STABILITY-(CONT.)
```
GS    DYNAMIC CHARACTERISTICS
      . DYNAMIC STABILITY
      . . MOTION STABILITY
      . . . ATTITUDE STABILITY
      . . . . LATERAL STABILITY
      STABILITY
      . DYNAMIC STABILITY
      . . MOTION STABILITY
      . . . ATTITUDE STABILITY
      . . . . LATERAL STABILITY
RT    AERODYNAMIC STABILITY
      AIRCRAFT STABILITY
      DIHEDRAL ANGLE
      DIRECTIONAL STABILITY
      FLOW STABILITY
      HANDEDNESS
      HOVERING STABILITY
      LONGITUDINAL STABILITY
      ROLL
      ROLLING MOMENTS
      ROTARY STABILITY
      SPACECRAFT STABILITY
      TURNING FLIGHT
      VERTICAL ORIENTATION
```

LATERALITY
```
USE   LATERAL STABILITY
```

LATERALIZATION
```
USE   LATERAL CONTROL
```

LATERITES
```
GS    SOILS
      . LATERITES
RT    DECOMPOSITION
      ROCKS
      TROPICAL REGIONS
      WATER
```

LATEX
```
GS    RUBBER
      . LATEX
RT    ACRYLIC RESINS
      ELASTOMERS
      SYNTHETIC RUBBERS
```

LATHES
```
GS    TOOLS
      . MACHINE TOOLS
      . . LATHES
      . . . TURRET LATHES
RT    ∞CONSTRUCTION MATERIALS
      GRINDING MACHINES
```

LATIN SQUARE METHOD
```
RT    ∞MATHEMATICS
      ∞METHODOLOGY
      ∞VARIABLE
```

LATITUDE
```
GS    LATITUDE
      . GEOMAGNETIC LATITUDE
RT    COORDINATES
      GEODETIC COORDINATES
      LONGITUDE
      POSITION (LOCATION)
```

LATITUDE MEASUREMENT
```
RT    LONGITUDE MEASUREMENT
      ∞MEASUREMENT
      NAVIGATION
      POSITIONING
```

LATTICE IMPERFECTIONS
```
USE   CRYSTAL DEFECTS
```

LATTICE PARAMETERS
```
GS    INDEPENDENT VARIABLES
      . LATTICE PARAMETERS
RT    CRYSTAL LATTICES
      CRYSTALLOGRAPHY
      PATTERSON MAP
      SUPERLATTICES
      X RAY ANALYSIS
```

LATTICE VIBRATIONS
```
GS    VIBRATION
      . LATTICE VIBRATIONS
RT    CRYSTAL DEFECTS
      CRYSTAL LATTICES
      FORBIDDEN BANDS
      PARTICLE MOTION
      PHONONS
```

LATTICE VIBRATIONS-(CONT.)
    RANDOM VIBRATION
    SPIN-LATTICE RELAXATION
    THERMAL ENERGY

∞ **LATTICES**
  SN *(USE OF A MORE SPECIFIC TERM IS*
     RECOMMENDED--CONSULT THE TERMS
     LISTED BELOW)
  RT CRYSTAL LATTICES
     LATTICES (MATHEMATICS)

LATTICES (MATHEMATICS)
  UF SUBLATTICES
  GS MATHEMATICAL LOGIC
    . **LATTICES (MATHEMATICS)**
    . . BOOLEAN ALGEBRA
    . . . BOOLEAN FUNCTIONS
  RT COMMUNICATION THEORY
     KAKUTANI THEOREM
     ∞LATTICES
     ∞MATHEMATICS
     ∞MATRICES
     SET THEORY

LATVIA
  RT BALTIC SEA
     EUROPE
     NATIONS

LAUE METHOD
  GS X RAY ANALYSIS
    . **LAUE METHOD**
  RT CRYSTAL LATTICES
     CRYSTALLOGRAPHY
     DIFFRACTION
     ∞METHODOLOGY
     X RAY DIFFRACTION

LAUGHING
  RT EMOTIONS
     HUMAN REACTIONS

LAUNCH COMPLEXES
  USE LAUNCHING BASES

LAUNCH DATES
  RT LAUNCHING
     SPACECRAFT LAUNCHING
     TIME
     TURNAROUND (STS)

LAUNCH ESCAPE SYSTEMS
  UF LES (ESCAPE SYSTEMS)
  GS ESCAPE SYSTEMS
    . **LAUNCH ESCAPE SYSTEMS**
  RT ESCAPE CAPSULES
     ESCAPE ROCKETS
     ∞SYSTEMS

LAUNCH TIME
  USE LAUNCH WINDOWS

LAUNCH VEHICLE CONFIGURATIONS
  RT AERODYNAMIC CONFIGURATIONS
     ∞CONFIGURATIONS
     MISSILE CONFIGURATIONS
     PROPULSION SYSTEM CONFIGURATIONS
     RECOVERABLE LAUNCH VEHICLES
     SPACECRAFT CONFIGURATIONS

LAUNCH VEHICLES
  UF CARRIER ROCKETS
  GS **LAUNCH VEHICLES**
    . ABLESTAR LAUNCH VEHICLE
    . ARIANE LAUNCH VEHICLE
    . ATLAS LAUNCH VEHICLES
    . . ATLAS ABLE 5 LAUNCH VEHICLE
    . . ATLAS AGENA B LAUNCH VEHICLE
    . . ATLAS AGENA LAUNCH VEHICLES
    . . ATLAS CENTAUR LAUNCH VEHICLE
    . . ATLAS SLV-3 LAUNCH VEHICLE
    . BLUE SCOUT ROCKET VEHICLE
    . BLUE STREAK LAUNCH VEHICLE
    . CENTAUR LAUNCH VEHICLE
    . . ATLAS CENTAUR LAUNCH VEHICLE
    . DELTA LAUNCH VEHICLE
    . DIAMANT LAUNCH VEHICLE
    . ELDO LAUNCH VEHICLE
    . EUROPA LAUNCH VEHICLES
    . . EUROPA 1 LAUNCH VEHICLE
    . . EUROPA 2 LAUNCH VEHICLE
    . . EUROPA 3 LAUNCH VEHICLE
    . . EUROPA 4 LAUNCH VEHICLE

LAUNCH VEHICLES-(CONT.)
    . HEAVY LIFT LAUNCH VEHICLES
    . HYLA-STAR ROCKET VEHICLE
    . JUNO LAUNCH VEHICLES
    . . JUNO 1 LAUNCH VEHICLE
    . . JUNO 2 LAUNCH VEHICLE
    . LITTLE JOE 2 LAUNCH VEHICLE
    . NOMAD LAUNCH VEHICLE
    . NOVA LAUNCH VEHICLES
    . RAM B LAUNCH VEHICLE
    . RECOVERABLE LAUNCH VEHICLES
    . REUSABLE LAUNCH VEHICLES
    . SINGLE STAGE TO ORBIT VEHICLES
    . SATURN LAUNCH VEHICLES
    . . SATURN D LAUNCH VEHICLE
    . . SATURN 1 LAUNCH VEHICLES
    . . . SATURN 1 SA-1 LAUNCH VEHICLE
    . . . SATURN 1 SA-2 LAUNCH VEHICLE
    . . . SATURN 1 SA-3 LAUNCH VEHICLE
    . . . SATURN 1 SA-4 LAUNCH VEHICLE
    . . . SATURN 1 SA-5 LAUNCH VEHICLE
    . . . SATURN 1 SA-6 LAUNCH VEHICLE
    . . . SATURN 1 SA-7 LAUNCH VEHICLE
    . . . SATURN 1 SA-8 LAUNCH VEHICLE
    . . . SATURN 1 SA-9 LAUNCH VEHICLE
    . . . SATURN 1 SA-10 LAUNCH VEHICLE
    . . SATURN 1B LAUNCH VEHICLES
    . . SATURN 2 LAUNCH VEHICLES
    . . SATURN 5 LAUNCH VEHICLES
    . STANDARD LAUNCH VEHICLES
    . . ATLAS SLV-3 LAUNCH VEHICLE
    . . STANDARD LAUNCH VEHICLE 5
    . THOR LAUNCH VEHICLES
    . . THOR ABLE ROCKET VEHICLE
    . . THOR AGENA LAUNCH VEHICLE
    . . THOR DELTA LAUNCH VEHICLE
    . THORAD LAUNCH VEHICLES
    . . THOR ABLE ROCKET VEHICLE
    . . THOR AGENA LAUNCH VEHICLE
    . . THOR DELTA LAUNCH VEHICLE
    . TITAN CENTAUR LAUNCH VEHICLE
    . TITAN LAUNCH VEHICLES
    . . TITAN 3 LAUNCH VEHICLE
    . VANGUARD 2 LAUNCH VEHICLE
    . VEGA LAUNCH VEHICLE
  RT AEROSPACEPLANES
     ASTROPLANE
     BOOSTER ROCKET ENGINES
     ∞BOOSTER ROCKETS
     ∞BOOSTERS
     CENTAUR PROJECT
     FLIGHT TEST VEHICLES
     JUPITER C ROCKET VEHICLE
     JUPITER PROJECT
     LANDING MODULES
     LAUNCHERS
     LAUNCHING
     MISSILE LAUNCHERS
     MISSILES
     MULTIENGINE VEHICLES
     MULTISTAGE ROCKET VEHICLES
     NATIONAL LAUNCH VEHICLE PROGRAM
     ROCKET CATAPULTS
     ROCKET ENGINES
     ROCKET LAUNCHERS
     ROCKET LAUNCHING
     ROCKET VEHICLES
     ∞ROCKETS
     SATURN PROJECT
     SCOUT PROJECT
     SPACE PROCESSING APPLICATIONS
      ROCKET
     ∞SPACECRAFT
     SPACECRAFT LAUNCHING
     SUSTAINER ROCKET ENGINES
     TEST VEHICLES
     TITAN PROJECT
     ∞VEHICLES
     VERNIER ENGINES
     ∞WINGED VEHICLES

LAUNCH WINDOWS
  UF LAUNCH TIME
  GS WINDOWS (INTERVALS)
    . **LAUNCH WINDOWS**
  RT LAUNCHING
     ROCKET LAUNCHING
     SPACECRAFT LAUNCHING

LAUNCHERS
  UF LAUNCHING DEVICES
  GS **LAUNCHERS**
    . AIRCRAFT LAUNCHING DEVICES
    . . JATO ENGINES

LAUNCHERS-(CONT.)
    . CATAPULTS
    . . ROCKET CATAPULTS
    . GUN LAUNCHERS
    . HYPERVELOCITY LAUNCHERS
    . MISSILE LAUNCHERS
    . . MOBILE MISSILE LAUNCHERS
    . ROCKET LAUNCHERS
    . . ROCKET CATAPULTS
  RT LAUNCH VEHICLES
     LAUNCHING
     LAUNCHING PADS
     LAUNCHING SITES
     NATIONAL LAUNCH VEHICLE PROGRAM
     ROCKET LAUNCHING
     TITAN PROJECT

LAUNCHING
  GS **LAUNCHING**
    . AIR LAUNCHING
    . ROCKET LAUNCHING
    . . LUNAR LAUNCH
    . . ORBITAL LAUNCHING
    . SEA LAUNCHING
    . SPACECRAFT LAUNCHING
  RT COUNTDOWN
     LAUNCH DATES
     LAUNCH VEHICLES
     LAUNCH WINDOWS
     LAUNCHERS
     MISSILE LAUNCHERS
     NATIONAL LAUNCH VEHICLE PROGRAM
     PRELAUNCH TESTS
     ROCKET LAUNCHERS
     ∞SHOT
     STARTING
     TITAN PROJECT

LAUNCHING BASES
  UF LAUNCH COMPLEXES
  GS **LAUNCHING BASES**
    . CAPE KENNEDY LAUNCH COMPLEX
  RT ∞FACILITIES
     GROUND SUPPORT EQUIPMENT

LAUNCHING DEVICES
  USE LAUNCHERS

LAUNCHING PADS
  GS SITES
    . LAUNCHING SITES
    . . **LAUNCHING PADS**
  RT FLAME DEFLECTORS
     GANTRY CRANES
     GROUND SUPPORT EQUIPMENT
     LAUNCHERS
     ∞PAD
     ∞PLATFORMS
     SPACECRAFT LAUNCHING
     UMBILICAL TOWERS

LAUNCHING SITES
  GS SITES
    . **LAUNCHING SITES**
    . . LAUNCHING PADS
  RT GANTRY CRANES
     GROUND SUPPORT EQUIPMENT
     LAUNCHERS
     MISSILE LAUNCHERS
     MISSILE SILOS
     MISSILES
     NATIONAL LAUNCH VEHICLE PROGRAM
     ROCKET CATAPULTS
     ROCKET LAUNCHERS

LAVA
  GS EFFUSIVES
    . **LAVA**
  RT AGGREGATES
     CALDERAS
     CONES (VOLCANOES)
     EARTH RESOURCES
     IGNEOUS ROCKS
     MAGMA
     MARIA
     MARS VOLCANOES
     MINERALS
     REGOLITH
     ROCKS
     SOILS
     VOLCANOES
     VOLCANOLOGY

LAVAL NUMBER
  GS RATIOS

LAVAL NUMBER-(CONT.)
. DIMENSIONLESS NUMBERS
. . **LAVAL NUMBER**

∞ **LAW**
SN (USE OF A MORE SPECIFIC TERM IS
RECOMMENDED--CONSULT THE TERMS
LISTED BELOW)
RT CLAIMING
LAW (JURISPRUDENCE)
LAWS

LAW (JURISPRUDENCE)
UF FORENSIC SCIENCES
GS **LAW (JURISPRUDENCE)**
. INTERNATIONAL LAW
. . AIR LAW
. . SEA LAW
. . SPACE LAW
. PUBLIC LAW
. LIABILITIES
. . LEGAL LIABILITY
. . PENALTIES
RT CONSTITUTION
CRIME
∞ LAW
POLITICS
REGULATIONS
VOTING

LAWRENCIUM
GS CHEMICAL ELEMENTS
. ACTINIDE SERIES
. . TRANSURANIUM ELEMENTS
. . . **LAWRENCIUM**
. NUCLIDES
. . ISOTOPES
. . . RADIOACTIVE ISOTOPES
. . . . TRANSURANIUM ELEMENTS
. . . . . **LAWRENCIUM**
HEAVY ELEMENTS
. TRANSURANIUM ELEMENTS
. . **LAWRENCIUM**
METALS
. ACTINIDE SERIES
. . TRANSURANIUM ELEMENTS
. . . **LAWRENCIUM**

LAWS
GS **LAWS**
. CHILD-LANGMUIR LAW
. CLOSURE LAW
. COFFIN-MANSON LAW
. CONSERVATION LAWS
. FOURIER LAW
. HOOKES LAW
. KEPLER LAWS
. NEWTON PRESSURE LAW
. NEWTON SECOND LAW
. NEWTON-BUSEMANN LAW
. OHMS LAW
. RADIATION LAWS
. . KIRCHHOFF LAW OF RADIATION
. . STEFAN-BOLTZMANN LAW
. . STOKES LAW OF RADIATION
. SCALING LAWS
. SIMILITUDE LAW
. SNELLS LAW
. TAFEL LAW
. WEBER-FECHNER LAW
RT ∞ LAW
RULES
∞ STOKES LAW

LAY-UP
RT CARBON FIBER REINFORCED PLASTICS
COMPOSITE MATERIALS
COMPOSITE STRUCTURES
EPOXY RESINS
FIBER ORIENTATION
LAMINATES
REINFORCING FIBERS

∞ **LAYERS**
SN (USE OF A MORE SPECIFIC TERM
RECOMMENDED--CONSULT THE TERMS
LISTED BELOW)
UF LAMINA
PLIES
RT ANTICLINES
ATMOSPHERIC BOUNDARY LAYER
BARRIER LAYERS
BOUNDARY LAYERS
COATINGS
DEEP SCATTERING LAYERS

LAYERS-(CONT.)
EKMAN LAYER
FLAT LAYERS
FOLDS (GEOLOGY)
FORMATIONS
GEOSYNCLINES
INTERCALATION
INTERLAYERS
IONOSPHERE
LAMINATES
MEMBRANES
MONOMOLECULAR FILMS
MULTILAYER INSULATION
PLASMA LAYERS
PLY ORIENTATION
REGIONS
SEDIMENTARY ROCKS
SHEAR LAYERS
SHOCK LAYERS
STRATA
STRATIFICATION
SUBSTRATES
SURFACE LAYERS
SYNCLINES
THREE DIMENSIONAL BOUNDARY LAYER
TURBULENT BOUNDARY LAYER

LAYOUTS
RT BLUEPRINTS
CIRCUIT DIAGRAMS
CONSTRUCTION
DESCRIPTIVE GEOMETRY
∞ DESIGN
∞ DRAWING
DRAWINGS
ENGINEERING DRAWINGS
MODELS
∞ PLANS
SURVEYS

LAZAREV METEORITE
GS CELESTIAL BODIES
. METEORITES
. . **LAZAREV METEORITE**
RT IRON METEORITES
STONY METEORITES

LC CIRCUITS
GS CIRCUITS
. **LC CIRCUITS**
RT INDUCTANCE
NETWORK ANALYSIS
NETWORK SYNTHESIS
PARAMETRIC AMPLIFIERS
RC CIRCUITS
RL CIRCUITS
RLC CIRCUITS
TIME CONSTANT

LCRE REACTOR
USE LITHIUM COOLED REACTOR
EXPERIMENT

LDEF
USE LONG DURATION EXPOSURE FACILITY

LEACHING
RT AUTOCLAVING
BENEFICIATION
DISSOLVING
ELUTION
EXTRACTION
FLUSHING
HYDROMETALLURGY
PERCOLATION
PERMEABILITY
∞ SEPARATION

LEAD (METAL)
GS CHEMICAL ELEMENTS
. **LEAD (METAL)**
. . LEAD ISOTOPES
METALS
. **LEAD (METAL)**
. . LEAD ISOTOPES

LEAD ACETATES
GS ACETATES
. **LEAD ACETATES**
ESTERS
. **LEAD ACETATES**
LEAD COMPOUNDS
. **LEAD ACETATES**
LEAD ORGANIC COMPOUNDS
. **LEAD ACETATES**

LEAD ACID BATTERIES
GS ELECTROCHEMICAL CELLS
. ELECTRIC BATTERIES
. . STORAGE BATTERIES
. . . **LEAD ACID BATTERIES**
RT CHEMICAL AUXILIARY POWER UNITS
∞ ELECTRIC CELLS
ELECTROLYTIC CELLS
ENERGY STORAGE
NICKEL IRON BATTERIES
∞ POWER SUPPLIES

LEAD ALLOYS
GS ALLOYS
. **LEAD ALLOYS**
RT BEARING ALLOYS
SOLDERS

LEAD CHLORIDES
GS HALOGEN COMPOUNDS
. CHLORINE COMPOUNDS
. . CHLORIDES
. . . **LEAD CHLORIDES**
. HALIDES
. . CHLORIDES
. . . **LEAD CHLORIDES**
. . METAL HALIDES
. . . **LEAD CHLORIDES**
LEAD COMPOUNDS
. **LEAD CHLORIDES**

LEAD COMPOUNDS
UF PLUMBANE
GS **LEAD COMPOUNDS**
. LEAD ACETATES
. LEAD CHLORIDES
. LEAD MOLYBDATES
. LEAD OXIDES
. LEAD SELENIDES
. LEAD SULFIDES
. LEAD TELLURIDES
. LEAD TITANATES
. LEAD TUNGSTATES
RT ∞ CHEMICAL COMPOUNDS
∞ GROUP 4A COMPOUNDS
∞ METAL COMPOUNDS

LEAD ISOTOPES
GS CHEMICAL ELEMENTS
. LEAD (METAL)
. . **LEAD ISOTOPES**
. NUCLIDES
. . ISOTOPES
. . . **LEAD ISOTOPES**
METALS
. LEAD (METAL)
. . **LEAD ISOTOPES**

LEAD MOLYBDATES
GS LEAD COMPOUNDS
. **LEAD MOLYBDATES**
MOLYBDENUM COMPOUNDS
. **LEAD MOLYBDATES**

LEAD ORGANIC COMPOUNDS
GS **LEAD ORGANIC COMPOUNDS**
. LEAD ACETATES
RT ∞ CHEMICAL COMPOUNDS
∞ METAL COMPOUNDS

LEAD OXIDES
GS CHALCOGENIDES
. OXIDES
. . METAL OXIDES
. . . **LEAD OXIDES**
LEAD COMPOUNDS
. **LEAD OXIDES**

LEAD POISONING
GS DISEASES
. TOXIC DISEASES
. . **LEAD POISONING**
TOXICITY
. **LEAD POISONING**
RT ∞ POISONING
SMOG

LEAD SELENIDES
GS CHALCOGENIDES
. SELENIDES
. . **LEAD SELENIDES**
LEAD COMPOUNDS
. **LEAD SELENIDES**
SELENIUM COMPOUNDS

LEAD SELENIDES-*(CONT.)*
. SELENIDES
. . **LEAD SELENIDES**

LEAD SULFIDES
GS   CHALCOGENIDES
. SULFIDES
. . INORGANIC SULFIDES
. . . **LEAD SULFIDES**
LEAD COMPOUNDS
. **LEAD SULFIDES**
SULFUR COMPOUNDS
. SULFIDES
. . INORGANIC SULFIDES
. . . **LEAD SULFIDES**

LEAD TELLURIDES
GS   CHALCOGENIDES
. TELLURIDES
. . **LEAD TELLURIDES**
LEAD COMPOUNDS
. **LEAD TELLURIDES**
TELLURIUM COMPOUNDS
. TELLURIDES
. . **LEAD TELLURIDES**

LEAD TITANATES
GS   LEAD COMPOUNDS
. **LEAD TITANATES**
TITANIUM COMPOUNDS
. TITANATES
. . **LEAD TITANATES**

LEAD TUNGSTATES
GS   LEAD COMPOUNDS
. **LEAD TUNGSTATES**
TUNGSTEN COMPOUNDS
. TUNGSTATES
. . **LEAD TUNGSTATES**

LEAD ZIRCONATE TITANATES
GS   TITANIUM COMPOUNDS
. TITANATES
. . PIEZOELECTRIC CERAMICS
. . . **LEAD ZIRCONATE TITANATES**

LEADERSHIP
RT   MORALE
PERSONNEL MANAGEMENT

LEADING EDGE FLAPS
GS   CONTROL SURFACES
. FLAPS (CONTROL SURFACES)
. . WING FLAPS
. . . **LEADING EDGE FLAPS**
RT   AIRCRAFT STRUCTURES
AIRFOILS
TRAILING-EDGE FLAPS
VORTEX FLAPS
∞WINGED VEHICLES
WINGS

LEADING EDGE SLATS
UF   WING SLATS
GS   AIRFOILS
. FLAPS (CONTROL SURFACES)
. . WING FLAPS
. . . **LEADING EDGE SLATS**
BRAKES (FOR ARRESTING MOTION)
. AERODYNAMIC BRAKES
. . WING FLAPS
. . . **LEADING EDGE SLATS**
. AIRCRAFT BRAKES
. . WING FLAPS
. . . **LEADING EDGE SLATS**
CONTROL SURFACES
. FLAPS (CONTROL SURFACES)
. . WING FLAPS
. . . **LEADING EDGE SLATS**
DRAG DEVICES
. AERODYNAMIC BRAKES
. . WING FLAPS
. . . **LEADING EDGE SLATS**
RT   BOUNDARY LAYER CONTROL
SPLIT FLAPS
SPOILERS
TRAILING EDGE FLAPS
WING SLOTS

LEADING EDGE SWEEP
GS   GEOMETRY
. EUCLIDEAN GEOMETRY
. . ANGLES (GEOMETRY)
. . . SWEEP ANGLE

LEADING EDGE SWEEP-*(CONT.)*
. . . . SWEEPBACK
. . . . . **LEADING EDGE SWEEP**

LEADING EDGE THRUST
GS   THRUST
. **LEADING EDGE THRUST**
RT   AERODYNAMIC FORCES
AIRFOILS
LEADING EDGES
LIFT
WING LOADING

LEADING EDGES
GS   EDGES
. **LEADING EDGES**
. . SHARP LEADING EDGES
RT   AIRFOILS
FOREBODIES
LEADING EDGE THRUST
THRUST DISTRIBUTION
TRAILING EDGES
VORTEX FLAPS

LEAKAGE
RT   CAVITIES
CRACKS
DEFECTS
∞ESCAPE
FLUID FLOW
INTRUSION
LABYRINTH SEALS
LOSS OF COOLANT
LOSSES
PERMEABILITY
PINHOLES
POROSITY
∞REDUCTION
SEEPAGE
WASTES

LEAR JET AIRCRAFT
GS   COMMERCIAL AIRCRAFT
. **LEAR JET AIRCRAFT**
JET AIRCRAFT
. **LEAR JET AIRCRAFT**
RT   ∞AIRCRAFT

LEARNING
GS   **LEARNING**
. ASTRONAUT TRAINING
. CONDITIONING (LEARNING)
. HABITUATION (LEARNING)
. MAZE LEARNING
. TRANSFER OF TRAINING
RT   ACHIEVEMENT
APTITUDE
BEHAVIOR
CHILD DEVICE
DECONDITIONING
EDUCATION
EDUCATIONAL TELEVISION
HABITS
INSTRUCTORS
KNOWLEDGE
MEMORY
MOTIVATION
REINFORCEMENT (PSYCHOLOGY)
RESPONSES
RETENTION (PSYCHOLOGY)
STUDENTS
TEACHING MACHINES
TEXTBOOKS
TRAINING ANALYSIS
TRAINING EVALUATION
UNIVERSITIES

LEARNING CURVES
RT   ASYMPTOTIC METHODS
∞CURVES

LEARNING MACHINES
UF   MACHINE LEARNING
GS   AUTOMATIC CONTROL
. ADAPTIVE CONTROL
. . **LEARNING MACHINES**
RT   ARTIFICIAL INTELLIGENCE
AUTOMATA THEORY
CYBERNETICS
FEEDBACK CONTROL
∞MACHINERY
SELF ORGANIZING SYSTEMS
TEACHING MACHINES

LEARNING THEORY
RT   CHILD DEVICE
EDUCATION
PROBLEM SOLVING
∞THEORIES

LEASING
GS   PROCUREMENT
. **LEASING**
RT   CONTRACTS
LAND USE
NASA PROGRAMS
RESOURCES MANAGEMENT
SITE SELECTION

LEAST SQUARES METHOD
GS   ANALYSIS (MATHEMATICS)
. NUMERICAL ANALYSIS
. . APPROXIMATION
. . . **LEAST SQUARES METHOD**
RT   CORRELATION
CURVE FITTING
GAUSS-MARKOV THEOREM
MEAN SQUARE VALUES
∞METHODOLOGY
OPTIMIZATION
PARAMETER IDENTIFICATION
QUALITY CONTROL
REGRESSION ANALYSIS
SIMULTANEOUS EQUATIONS

LEATHER
GS   SKIN (ANATOMY)
. **LEATHER**
RT   CLOTHING
COLLAGENS
SHOES

LEAVES
GS   PLANTS (BOTANY)
. **LEAVES**
RT   BROWN WAVE EFFECT
CANOPIES (VEGETATION)
DECIDUOUS TREES
DEFOLIANTS
DEFOLIATION
FOLIAGE
GREEN WAVE EFFECT
HERBICIDES

LEBANON
GS   NATIONS
. **LEBANON**
RT   ASIA

LEBESGUE THEOREM
GS   ANALYSIS (MATHEMATICS)
. REAL VARIABLES
. . MEASURE AND INTEGRATION
. . . **LEBESGUE THEOREM**
THEOREMS
. **LEBESGUE THEOREM**
RT   SET THEORY

LECTURES
UF   SPEECHES
RT   EDUCATION
PUBLIC SPEAKING
SPEECH
VERBAL COMMUNICATION

LED (DIODES)
USE   LIGHT EMITTING DIODES

LEDGES
RT   CLIFFS
ROCKS
TOPOGRAPHY

LEE WAVES
RT   AIR CURRENTS
BAROTROPIC FLOW
SURFACE WAVES
TROPOSPHERIC WAVES
VERTICAL AIR CURRENTS

LEG (ANATOMY)
GS   ANATOMY
. LIMBS (ANATOMY)
. . **LEG (ANATOMY)**
. . . FEET (ANATOMY)
. . . KNEE (ANATOMY)
APPENDAGES
. **LEG (ANATOMY)**

LEG (ANATOMY)-*(CONT.)*
```
          . . KNEE (ANATOMY)
RT      FEET (ANATOMY)
          FEMUR
          THIGH
          TIBIA
```

LEGAL LIABILITY
```
GS      LAW (JURISPRUDENCE)
          . PUBLIC LAW
          . . LIABILITIES
          . . . LEGAL LIABILITY
RT      AIR LAW
          CONTRACTS
          INTERNATIONAL LAW
          JUDGMENTS
          LOSSES
          PENALTIES
          PROHIBITION
```

LEGENDRE CODE
```
USE     COMPUTER PROGRAMMING
          NEUTRON SCATTERING
```

LEGENDRE FUNCTIONS
```
UF      LEGENDRE POLYNOMIALS
          LEGENDRE TRANSFORMATION
GS      ANALYSIS (MATHEMATICS)
          . COMPLEX VARIABLES
          . . LEGENDRE FUNCTIONS
          FUNCTIONS (MATHEMATICS)
          . LEGENDRE FUNCTIONS
RT      ORTHOGONAL FUNCTIONS
          SPHERICAL HARMONICS
```

LEGENDRE POLYNOMIALS
```
USE     LEGENDRE FUNCTIONS
```

LEGENDRE TRANSFORMATION
```
USE     LEGENDRE FUNCTIONS
```

LEGIBILITY
```
RT      CHARACTER RECOGNITION
          CONTRAST
          PERCEPTION
          PRINTING
          READING
          RESOLUTION
          SYMBOLS
          VISIBILITY
          VISION
```

LEGUMINOUS PLANTS
```
GS      FARM CROPS
          . LEGUMINOUS PLANTS
RT      AGRICULTURE
          BOTANY
          EARTH RESOURCES
          ∞ FOOD
          HAY
          NITROGENATION
          NODULES
          VEGETABLES
```

LEIDENFROST PHENOMENON
```
GS      PHASE TRANSFORMATIONS
          . VAPORIZING
          . . BOILING
          . . . NUCLEATE BOILING
          . . . . LEIDENFROST PHENOMENON
RT      FILM BOILING
          HEAT TRANSFER
```

LEM (LUNAR MODULE)
```
USE     LUNAR MODULE
```

LEMMAS
```
USE     THEOREMS
```

LENGTH
```
GS      DIMENSIONS
          . LENGTH
RT      DISTANCE
          THICKNESS
```

LENNARD-JONES GAS
```
RT      BINARY FLUIDS
          GAS VISCOSITY
          LENNARD-JONES POTENTIAL
```

LENNARD-JONES POTENTIAL
```
RT      COMPUTERIZED SIMULATION
          INTERMOLECULAR FORCES
          LENNARD-JONES GAS
```

LENNARD-JONES POTENTIAL-*(CONT.)*
```
          MOLECULAR INTERACTIONS
          POTENTIAL THEORY
```

LENS ANTENNAS
```
GS      ANTENNAS
          . DIRECTIONAL ANTENNAS
          . . LENS ANTENNAS
          MICROWAVE EQUIPMENT
          . MICROWAVE ANTENNAS
          . . LENS ANTENNAS
RT      ANTENNA DESIGN
          DIPOLE ANTENNAS
          HORN ANTENNAS
          LENSES
          RADAR ANTENNAS
          WAVEGUIDE ANTENNAS
          WIRE GRID LENSES
```

LENS DESIGN
```
RT      ANTIREFLECTION COATINGS
          COMPUTER AIDED DESIGN
          ∞ DESIGN
          GRADIENT INDEX OPTICS
          LENSES
          OPTICAL CORRECTION PROCEDURE
          ∞ OPTICS
          PRODUCT DEVELOPMENT
          STIGMATISM
          ZOOM LENSES
```

LENSES
```
GS      LENSES
          . CONTACT LENSES
          . FRESNEL LENSES
          . GRAVITATIONAL LENSES
          . MAGNETIC LENSES
          . WIDE ANGLE LENSES
          . WIRE GRID LENSES
          . ZOOM LENSES
RT      ASTIGMATISM
          CAMERAS
          CATARACTS
          CIRCUMSOLAR TELESCOPES
          EYE (ANATOMY)
          EYEPIECES
          FOCUSING
          GRADIENT INDEX OPTICS
          INTEGRATED OPTICS
          LENS ANTENNAS
          LENS DESIGN
          MAGNIFICATION
          OPTICAL EQUIPMENT
          OPTICAL FILTERS
          ∞ OPTICS
          PANORAMIC CAMERAS
          PHOTOGRAPHIC EQUIPMENT
          REFRACTING TELESCOPES
          REFRACTION
          RETICLES
          STIGMATISM
          STREAK CAMERAS
          TELESCOPES
          VIGNETTING
```

LENTICULAR BODIES
```
GS      SYMMETRICAL BODIES
          . LENTICULAR BODIES
RT      AXISYMMETRIC BODIES
          ∞ BODIES
          CONVEXITY
```

† **LEON-QUERETARO AREA (MEXICO)**
```
RT      MEXICO
```

LEONID METEOROIDS
```
GS      CELESTIAL BODIES
          . METEOROID SHOWERS
          . . LEONID METEOROIDS
          . METEOROIDS
          . . LEONID METEOROIDS
```

LEPTONS
```
GS      PARTICLES
          . ELEMENTARY PARTICLES
          . . FERMIONS
          . . . LEPTONS
          . . . . ANTINEUTRINOS
          . . . . MUONS
          . . . . NEUTRINOS
          . . . . . SOLAR NEUTRINOS
RT      CHARGED PARTICLES
          CHARM (PARTICLE PHYSICS)
          GLUONS
          MESONS
```

LEPTONS-*(CONT.)*
```
          PARTONS
          QUANTUM CHROMODYNAMICS
          QUARK PARTON MODEL
```

LES (ESCAPE SYSTEMS)
```
USE     LAUNCH ESCAPE SYSTEMS
```

LES (SATELLITES)
```
USE     LINCOLN EXPERIMENTAL SATELLITES
```

LESA (LUNAR EXPLORATION SYSTEM)
```
USE     LUNAR EXPLORATION SYSTEM FOR
            APOLLO
```

LESIONS
```
GS      INJURIES
          . LESIONS
          . . PULMONARY LESIONS
RT      ABRASION
          BURNS (INJURIES)
          MILIARIA
```

LESOTHO
```
GS      NATIONS
          . LESOTHO
RT      AFRICA
          REPUBLIC OF SOUTH AFRICA
```

LESSER ANTILLES
```
GS      LANDFORMS
          . ISLANDS
          . . WEST INDIES
          . . . LESSER ANTILLES
RT      ATLANTIC OCEAN
```

LETHALITY
```
RT      CARBON MONOXIDE POISONING
          DAMAGE
          DESTRUCTION
```

LETHARGY
```
RT      BOREDOM
          ∞ DEPRESSION
          DETACHMENT
          FRUSTRATION
          HUMAN BEHAVIOR
          ∞ INHIBITION
          MONOTONY
```

LETTERS (SYMBOLS)
```
USE     SYMBOLS
```

LEUCINE
```
GS      ACIDS
          . AMINO ACIDS
          . . LEUCINE
          . . . NORLEUCINE
          ORGANIC COMPOUNDS
          . AMINO ACIDS
          . . LEUCINE
          . . . NORLEUCINE
```

LEUKEMIAS
```
GS      DISEASES
          . TUMORS
          . . NEOPLASMS
          . . . CANCER
          . . . . LEUKEMIAS
RT      BONE MARROW
```

LEUKOCYTES
```
GS      ANATOMY
          . CARDIOVASCULAR SYSTEM
          . . LEUKOCYTES
          . . . EOSINOPHILS
          . . . LYMPHOCYTES
          BODY FLUIDS
          . BLOOD
          . . LEUKOCYTES
          . . . EOSINOPHILS
          . . . LYMPHOCYTES
          CELLS (BIOLOGY)
          . LEUKOCYTES
          . . EOSINOPHILS
          . . LYMPHOCYTES
RT      BIOCOMPATIBILITY
          ERYTHROCYTES
          WHITE BLOOD CELLS
```

LEUKOPENIA
```
GS      SIGNS AND SYMPTOMS
          . LEUKOPENIA
RT      INFECTIOUS DISEASES
```

∞ LEVEL
SN   (USE OF A MORE SPECIFIC TERM IS
      RECOMMENDED--CONSULT THE TERMS
      LISTED BELOW)
RT   CHEMICAL ENERGY
      FREE ENERGY
      HEIGHT
      INTERNAL ENERGY
      LEVEL (HORIZONTAL)
      LEVEL (QUANTITY)

LEVEL (HORIZONTAL)
GS   LEVEL (HORIZONTAL)
      . LIQUID LEVELS
RT   ∞ GRADE
      GRADIENTS
      ∞ LEVEL
      SLOPES

LEVEL (QUANTITY)
GS   LEVEL (QUANTITY)
      . EFFECTIVE PERCEIVED NOISE LEVELS
      . ENERGY LEVELS
      . . ATOMIC ENERGY LEVELS
      . . ELECTRON STATES
      . . GROUND STATE
      . . MOLECULAR ENERGY LEVELS
      . . . INTERMOLECULAR FORCES
RT   AMPLITUDES
      DISPLACEMENT
      ∞ FLUX
      FLUX (RATE)
      FLUX DENSITY
      ∞ INTENSITY
      ∞ LEVEL
      LOUDNESS
      MAGNITUDE
      VALUE

LEVELING
SN   (EXCLUDES METAL WORKING)
RT   ADJUSTING
      CONSISTENCY
      DATUM (ELEVATION)
      FLATTENING
      METAL WORKING
      ∞ ROLLING
      SMOOTHING
      WINDING

LEVERS
RT   CANTILEVER MEMBERS
      HANDLES
      KNOBS
      ∞ MACHINERY
      MANUAL CONTROL
      MECHANICAL DEVICES
      PEDALS

LEVITATION
GS   LEVITATION
      . ACOUSTIC LEVITATION
RT   BUOYANCY
      ELECTROSTATIC GYROSCOPES
      FLOTATION
      FRICTIONLESS ENVIRONMENTS
      LEVITATION MELTING
      MAGNETIC BEARINGS
      MAGNETIC LEVITATION VEHICLES
      SUSPENSION SYSTEMS (VEHICLES)
      VACUUM MELTING

LEVITATION MELTING
GS   PHASE TRANSFORMATIONS
      . MELTING
      . . LEVITATION MELTING
RT   ELECTRIC CURRENT
      EXTERNAL SURFACE CURRENTS
      LEVITATION
      LIQUID METALS
      LOW GRAVITY MANUFACTURING
      MAGNETIC SUSPENSION
      ∞ METALLURGY
      OHMIC DISSIPATION
      RESISTANCE HEATING
      SPACE MANUFACTURING
      SPACE PROCESSING

LEWIS BASE
RT   AMINES
      ELECTRONS

LEWIS NUMBERS
GS   RATIOS
      . DIMENSIONLESS NUMBERS

LEWIS NUMBERS-(CONT.)
      . . LEWIS NUMBERS
RT   DENSITY (MASS/VOLUME)
      DIFFUSION COEFFICIENT
      FLUID FLOW
      HEAT TRANSFER
      MASS FLOW
      MASS TRANSFER
      SPECIFIC HEAT
      THERMAL CONDUCTIVITY

LEXAN (TRADEMARK)
GS   CARBON COMPOUNDS
      . CARBONATES
      . . POLYCARBONATES
      . . . LEXAN (TRADEMARK)
      ESTERS
      . POLYCARBONATES
      . . LEXAN (TRADEMARK)
RT   ∞ POLYMERS
      RESINS

LFO
USE   LANDSAT FOLLOW-ON MISSIONS

LIABILITIES
GS   LAW (JURISPRUDENCE)
      . PUBLIC LAW
      . . LIABILITIES
      . . . LEGAL LIABILITY
RT   AIR LAW
      COMMERCE
      DISCIPLINING
      LOSSES
      PENALTIES
      REGULATIONS

LIAPUNOV FUNCTIONS
UF   LYAPUNOV FUNCTIONS
GS   ANALYSIS (MATHEMATICS)
      . REAL VARIABLES
      . . LIAPUNOV FUNCTIONS
      FUNCTIONS (MATHEMATICS)
      . LIAPUNOV FUNCTIONS
RT   DIFFERENTIAL EQUATIONS

LIBERIA
GS   NATIONS
      . LIBERIA
RT   AFRICA

LIBRARIES
RT   BIBLIOGRAPHIES
      CATALOGS (PUBLICATIONS)
      DATA RETRIEVAL
      DOCUMENTATION
      DOCUMENTS
      INFORMATION DISSEMINATION
      INFORMATION RETRIEVAL
      INFORMATION SYSTEMS
      INTEGRATED LIBRARY SYSTEMS
      INTERSERVICE DATA EXCHANGE
        PROGRAM
      LITERATURE
      MUSEUMS
      ∞ REFERENCE SYSTEMS
      SELECTIVE DISSEMINATION OF
        INFORMATION
      TEXTBOOKS
      WORLD DATA CENTERS

LIBRATION
RT   EARTH LIMB
      LISSAJOUS FIGURES
      LUNAR FAR SIDE
      LUNAR LIMB
      ∞ MOTION
      NUTATION
      PRECESSION
      ROTATION

LIBRATIONAL MOTION
RT   LAGRANGE COORDINATES
      ∞ MOTION
      NUTATION

LIBYA
GS   NATIONS
      . LIBYA
RT   AFRICA

LIBYAN DESERT
GS   LAND
      . DESERTS

LIBYAN DESERT-(CONT.)
      . . LIBYAN DESERT
RT   AFRICA

LICENSING
RT   COPYRIGHTS
      PATENT APPLICATIONS
      POLICIES
      REGULATIONS

LICHENS
GS   PLANTS (BOTANY)
      . LICHENS
RT   ALGAE
      FUNGI
      LACUNAS
      SYMBIOSIS

LIDAR
USE   OPTICAL RADAR

LIE GROUPS
GS   ALGEBRA
      . LIE GROUPS
      . . SPINOR GROUPS
      GEOMETRY
      . DIFFERENTIAL GEOMETRY
      . . LIE GROUPS
      . . . SPINOR GROUPS
RT   GROUP THEORY

LIECHTENSTEIN
GS   NATIONS
      . LIECHTENSTEIN
RT   EUROPE

LIENARD POTENTIAL
GS   POTENTIAL ENERGY
      . ELECTRIC POTENTIAL
      . . LIENARD POTENTIAL
RT   ELECTRIC CURRENT
      ELECTRIC FIELDS

LIES
RT   GALVANIC SKIN RESPONSE
†   MENTAL PERFORMANCE

LIFE (BIOLOGY)
USE   LIFE SCIENCES

LIFE (DURABILITY)
UF   LIFETIME (DURABILITY)
GS   LIFE (DURABILITY)
      . CARRIER LIFETIME
      . FATIGUE LIFE
      . HALF LIFE
      . PLASMA LIFETIME
      . SATELLITE LIFETIME
      . SERVICE LIFE
      . STORAGE STABILITY
RT   ACCELERATED LIFE TESTS
      AIRCRAFT SURVIVABILITY
      DEPLETION
      DEPRECIATION
      DURABILITY
      FAILURE ANALYSIS
      LONG TERM EFFECTS
      ∞ LONGEVITY
      MILLS RATIO
      MTBF
      ∞ RESISTANCE
      RETIREMENT FOR CAUSE
      VULNERABILITY

LIFE CYCLE COSTS
GS   COSTS
      . LIFE CYCLE COSTS
RT   COST ANALYSIS
      COST EFFECTIVENESS
      DESIGN TO COST
      FINANCIAL MANAGEMENT
      MANAGEMENT PLANNING
      PRODUCTION COSTS
      SYSTEMS ENGINEERING
      VALUE ENGINEERING

LIFE DETECTORS
RT   BIOSATELLITES
      ∞ DETECTORS
      EXTRATERRESTRIAL LIFE

LIFE RAFTS
GS   RAFTS
      . LIFE RAFTS

LIFE RAFTS-(CONT.)
RT   FLOATS
        INFLATABLE STRUCTURES
        LIFEBOATS

LIFE SCIENCES
UF   LIFE (BIOLOGY)
GS   **LIFE SCIENCES**
      . EXTRATERRESTRIAL LIFE
      . MOLECULAR BIOLOGY
RT   ABIOGENESIS
        AGING (BIOLOGY)
        BIOLOGICAL EVOLUTION
      ∞BIOLOGY
        CHEMICAL EVOLUTION
        ENVIRONMENTAL ENGINEERING
        NEUROLOGY
      ∞PHYSICAL SCIENCES
        PSYCHOPHARMACOLOGY
      ∞SCIENCE

LIFE SPAN
SN   (LIMITED TO THE LIFE SCIENCES)
RT   AGE FACTOR
        AGING (BIOLOGY)
        DEATH
        EXISTENCE
        GERONTOLOGY
      ∞LONGEVITY
        MORTALITY
      ∞SPAN

LIFE SUPPORT SYSTEMS
GS   SUPPORT SYSTEMS
      . **LIFE SUPPORT SYSTEMS**
      . . BIOPAKS
      . . CLOSED ECOLOGICAL SYSTEMS
      . . EMERGENCY LIFE SUSTAINING
          SYSTEMS
      . . . AEPS
      . . PORTABLE LIFE SUPPORT SYSTEMS
      . . . AEPS
RT   AEROSPACE ENVIRONMENTS
        AIR CONDITIONING
        ARTIFICIAL GRAVITY
        ASTRONAUT LOCOMOTION
      ∞ATMOSPHERES
        BIOSATELLITES
        BREATHING APPARATUS
        ENVIRONMENTAL ENGINEERING
        ENVIRONMENTS
        EXOBIOLOGY
        HUMAN FACTORS ENGINEERING
        LONG TERM EFFECTS
        LUNAR ENVIRONMENT
        LUNAR LOGISTICS
        LUNAR SHELTERS
        MANNED MANEUVERING UNITS
      ∞NUTRIENTS
        ONBOARD EQUIPMENT
        OXYGEN MASKS
        OXYGEN SUPPLY EQUIPMENT
        PLANETARY ENVIRONMENTS
        PRESSURE SUITS
        PRESSURIZED CABINS
        PROVISIONING
        REBREATHING
        SPACE FLIGHT FEEDING
        SPACE HABITATS
        SPACECRAFT ENVIRONMENTS
        SURVIVAL
      ∞SUSTAINING
      ∞SYSTEMS
        THERMAL ENVIRONMENTS
        UNDERWATER BREATHING APPARATUS
        VAPOR BARRIER CLOTHING
        VENTILATION
        WATER
        WEIGHTLESSNESS

LIFEBOATS
GS   SURFACE VEHICLES
      . BOATS
      . . **LIFEBOATS**
      WATER VEHICLES
      . BOATS
      . . **LIFEBOATS**
RT   LIFE RAFTS
        RAFTS
        SURVIVAL EQUIPMENT

LIFETIME (DURABILITY)
USE   LIFE (DURABILITY)

LIFT
UF   AERODYNAMIC LIFT
        LIFT COEFFICIENTS
        LIFT DISTRIBUTION
        LIFT FORCES
        VARIABLE LIFT
GS   AERODYNAMIC CHARACTERISTICS
      . **LIFT**
      . . INTERFERENCE LIFT
      . . JET LIFT
      . . ROTOR LIFT
      . . ZERO LIFT
      AERODYNAMIC FORCES
      . **LIFT**
      . . INTERFERENCE LIFT
      . . JET LIFT
      . . ROTOR LIFT
      . . ZERO LIFT
      DYNAMIC CHARACTERISTICS
      . **LIFT**
      . . INTERFERENCE LIFT
      . . JET LIFT
      . . ROTOR LIFT
      . . ZERO LIFT
RT   AERODYNAMIC COEFFICIENTS
        AERODYNAMIC CONFIGURATIONS
        AERODYNAMIC DRAG
        AERODYNAMICS
        AIRFOILS
        ANGLE OF ATTACK
        ASPECT RATIO
        CAMBER
        DISTRIBUTION (PROPERTY)
        DRAG
        EXTERNALLY BLOWN FLAPS
        GLIDING
        GROUND EFFECT (AERODYNAMICS)
        HYPERSONIC FORCES
        LEADING EDGE THRUST
        PRESSURE DISTRIBUTION
        SWEEP EFFECT
        UNDER SURFACE BLOWING
        UPPER SURFACE BLOWING

LIFT AUGMENTATION
RT   BOUNDARY LAYER CONTROL
        CIRCULATION CONTROL AIRFOILS
        DOWNWASH
        PERIPHERAL JET FLOW
        SPANWISE BLOWING
        UPPER SURFACE BLOWN FLAPS
        VORTEX FLAPS

LIFT COEFFICIENTS
USE   AERODYNAMIC COEFFICIENTS
        LIFT

LIFT DEVICES
UF   LIFTING SURFACES
RT   BOUNDARY LAYER CONTROL
      ∞DEVICES
        DIRECT LIFT CONTROLS
        DRAG DEVICES
        EXTERNALLY BLOWN FLAPS
        FLAPS (CONTROL SURFACES)
        MAGNETIC LEVITATION VEHICLES
        SLOTS
        UPPER SURFACE BLOWN FLAPS

LIFT DISTRIBUTION
USE   FORCE DISTRIBUTION
        LIFT

LIFT DRAG RATIO
UF   DRAG BALANCE
GS   RATIOS
      . **LIFT DRAG RATIO**
RT   AERODYNAMIC BALANCE
        AERODYNAMIC COEFFICIENTS
        AERODYNAMIC DRAG
        AERODYNAMIC STALLING
        BOUNDARY LAYER SEPARATION
        DRAG REDUCTION
        FORCE DISTRIBUTION
        PRESSURE RATIO

LIFT FANS
UF   FANLIFT DEVICES
RT   DUCTED FANS
        FAN IN WING AIRCRAFT
        LIFTING ROTORS
        PROPELLER FANS
        ROTARY WINGS
        SHORT TAKEOFF AIRCRAFT
        TURBOFANS

LIFT FANS-(CONT.)
        VERTICAL TAKEOFF AIRCRAFT
        XV-11A AIRCRAFT

LIFT FORCES
USE   LIFT

LIFTING BODIES
UF   LIFTING SURFACES
GS   **LIFTING BODIES**
      . LIFTING REENTRY VEHICLES
      . . HL-10 REENTRY VEHICLE
      . . HLD-35 REENTRY VEHICLE
      . . JANUS SPACECRAFT
      . . M-2 LIFTING BODY
      . . . M-2F2 LIFTING BODY
      . . X-20 AIRCRAFT
      . . X-24 AIRCRAFT
      . M-2F3 LIFTING BODY
RT   AERODYNAMIC CONFIGURATIONS
        AIRFOILS
        BLUFF BODIES
      ∞BODIES
      ∞DEVICES
        LUNAR FLYING VEHICLES
        REENTRY VEHICLES
        TOWED BODIES

LIFTING REENTRY VEHICLES
UF   REENTRY GLIDERS
        SPACE GLIDERS
GS   HYPERSONIC VEHICLES
      . **LIFTING REENTRY VEHICLES**
      . . HLD-35 REENTRY VEHICLE
      . . JANUS SPACECRAFT
      . . M-2F2 LIFTING BODY
      . . X-20 AIRCRAFT
      LIFTING BODIES
      . **LIFTING REENTRY VEHICLES**
      . . HL-10 REENTRY VEHICLE
      . . HLD-35 REENTRY VEHICLE
      . . JANUS SPACECRAFT
      . . M-2 LIFTING BODY
      . . . M-2F2 LIFTING BODY
      . . X-20 AIRCRAFT
      . . X-24 AIRCRAFT
      MANEUVERABLE SPACECRAFT
      . **LIFTING REENTRY VEHICLES**
      . . HL-10 REENTRY VEHICLE
      . . HLD-35 REENTRY VEHICLE
      . . JANUS SPACECRAFT
      . . M-2 LIFTING BODY
      . . M-2F2 LIFTING BODY
      . . X-20 AIRCRAFT
      REENTRY VEHICLES
      . **LIFTING REENTRY VEHICLES**
      . . FDL-5 REENTRY VEHICLE
      . . HL-10 REENTRY VEHICLE
      . . HLD-35 REENTRY VEHICLE
      . . JANUS SPACECRAFT
      . . M-2 LIFTING BODY
      . . . M-2F2 LIFTING BODY
      . . X-20 AIRCRAFT
RT   AEROSPACEPLANES
      ∞AIRCRAFT
        ASSET GLIDERS
        ASTRO VEHICLE
        ASTROPLANE
        BOOSTGLIDE VEHICLES
        GLIDERS
        HYPERSONIC GLIDERS
        MANNED REENTRY
        MANNED SPACECRAFT
        RECOVERABLE SPACECRAFT
        REENTRY
        SPACECRAFT REENTRY

LIFTING ROTORS
GS   AIRFOILS
      . WINGS
      . . ROTARY WINGS
      . . . **LIFTING ROTORS**
      . . . . BEARINGLESS ROTORS
      ROTATING BODIES
      . ROTORS
      . . ROTARY WINGS
      . . . **LIFTING ROTORS**
      . . . . BEARINGLESS ROTORS
RT   GROUND EFFECT MACHINES
        LIFT FANS
        ROTARY WING AIRCRAFT
        SHORT TAKEOFF AIRCRAFT
        VERTICAL TAKEOFF AIRCRAFT

LIFTING SURFACES
USE　LIFT DEVICES
　　　LIFTING BODIES
　　　SURFACES

∞ **LIFTS**
　SN　(USE OF A MORE SPECIFIC TERM IS
　　　　RECOMMENDED--CONSULT THE TERMS
　　　　LISTED BELOW)
　RT　CONVEYORS
　　　CRANES
　　　ELEVATORS (LIFTS)
　　　ESCALATORS
　　　JACKS (LIFTS)
　　　WINCHES

LIGAMENTS
　RT　CONNECTIVE TISSUE
　　　JOINTS (ANATOMY)

LIGANDS
　RT　CHEMICAL BONDS
　　　CHEMICAL COMPOSITION

LIGHT (VISIBLE RADIATION)
　UF　EXTRAGALACTIC LIGHT
　　　OPTICAL SPECTRUM
　　　VISIBLE RADIATION
　GS　ELECTROMAGNETIC RADIATION
　　　. **LIGHT (VISIBLE RADIATION)**
　　　. . COHERENT LIGHT
　　　. . GEGENSCHEIN
　　　. . LIGHT BEAMS
　　　. . POLARIZED LIGHT
　　　. . SKY RADIATION
　　　. . . AIRGLOW
　　　. . . . GEOCORONAL EMISSIONS
　　　. . . . NIGHTGLOW
　　　. . . . TWILIGHT GLOW
　　　. . . DAYGLOW
　　　. . SUNLIGHT
　　　. . ZODIACAL LIGHT
　RT　ATMOSPHERIC RADIATION
　　　ATTENUATION
　　　BEAMS (RADIATION)
　　　BLACK BODY RADIATION
　　　BRIGHTNESS
　　　CERENKOV RADIATION
　　　COHERENT ELECTROMAGNETIC
　　　　RADIATION
　　　COHERENT RADIATION
　　　COLOR
　　　CRITICAL FREQUENCIES
　　　DARKNESS
　　　DICHROISM
　　　ELECTROMAGNETIC SPECTRA
　　　ENERGY ABSORPTION
　　　EXCITONS
　　　EXTRATERRESTRIAL RADIATION
　∞FLASH
　　　GEOMETRICAL OPTICS
　　　GLARE
　　　ILLUMINANCE
　　　INCANDESCENCE
　　　INFRARED RADIATION
　　　LIGHT CURVE
　　　LIGHTING EQUIPMENT
　　　LINE SPECTRA
　　　LUMENS
　　　LUMINAIRES
　　　LUMINANCE
　　　LUMINESCENCE
　　　LUMINOSITY
　　　LUMINOUS INTENSITY
　　　MONOCHROMATIC RADIATION
　　　NEAR INFRARED RADIATION
　　　NEAR ULTRAVIOLET RADIATION
　　　OPACITY
　　　OPTICAL DEPOLARIZATION
　　　OPTICAL EMISSION SPECTROSCOPY
　　　OPTICAL MEASUREMENT
　　　OPTICAL PROPERTIES
　∞OPTICS
　　　PHOTICS
　　　PHOTOMETRY
　　　PHOTONS
　　　PHOTONUCLEAR REACTIONS
　　　PHOTOPHILIC PLANTS
　　　PHOTOPHORESIS
　　　PHOTOSENSITIVITY
　　　PLANETARY RADIATION
　　　POLARIZED ELECTROMAGNETIC
　　　　RADIATION
　　　POLARIZERS
　∞RADIATION

LIGHT (VISIBLE RADIATION)-(CONT.)
　　　RAMAN SPECTRA
　　　REFLECTION
　　　REFRACTION
　　　REFRACTIVITY
　　　SHADOWS
　　　SKY BRIGHTNESS
　　　SOLAR RADIATION
　　　THERMAL RADIATION
　　　TRANSMITTANCE
　　　ULTRAVIOLET SPECTRA
　　　VISIBILITY
　　　VISIBLE SPECTRUM

LIGHT ABSORPTION
USE　ELECTROMAGNETIC ABSORPTION

LIGHT ADAPTATION
　GS　ADAPTATION
　　　. RETINAL ADAPTATION
　　　. . **LIGHT ADAPTATION**
　　　SENSITIVITY
　　　. PHOTOSENSITIVITY
　　　. . **LIGHT ADAPTATION**
　RT　FLASH BLINDNESS
　　　NIGHT VISION
　　　PUPILLOMETRY
　　　THRESHOLDS (PERCEPTION)
　　　VISION

LIGHT AIRBORNE MULTIPURPOSE SYSTEM
　UF　LAMPS PROGRAM
　GS　NAVIGATION AIDS
　　　. **LIGHT AIRBORNE MULTIPURPOSE**
　　　　SYSTEM
　　　ONBOARD EQUIPMENT
　　　. AIRBORNE EQUIPMENT
　　　. . **LIGHT AIRBORNE MULTIPURPOSE**
　　　　　SYSTEM
　RT　AIRCRAFT EQUIPMENT
　　　FLIGHT INSTRUMENTS
　　　NAVIGATION INSTRUMENTS
　∞SYSTEMS

LIGHT AIRCRAFT
　GS　**LIGHT AIRCRAFT**
　　　. BEECH 99 AIRCRAFT
　　　. . BEECHCRAFT 18 AIRCRAFT
　　　. . C-33 AIRCRAFT
　　　. . C-35 AIRCRAFT
　　　. CESSNA L-19 AIRCRAFT
　　　. CESSNA 172 AIRCRAFT
　　　. CESSNA 205 AIRCRAFT
　　　. CESSNA 210 AIRCRAFT
　　　. CESSNA 402B AIRCRAFT
　　　. DH 125 AIRCRAFT
　　　. DO-27 AIRCRAFT
　　　. DO-28 AIRCRAFT
　　　. F-28 HELICOPTER
　　　. FIREBEE 2 TARGET DRONE AIRCRAFT
　　　. G-1 AIRCRAFT
　　　. G-91 AIRCRAFT
　　　. LIGHT INTRATHEATER TRANSPORT
　　　. MH-262 AIRCRAFT
　　　. MYSTERE 20 AIRCRAFT
　　　. OH-4 HELICOPTER
　　　. OH-5 HELICOPTER
　　　. OH-6 HELICOPTER
　　　. OH-13 HELICOPTER
　　　. OH-23 HELICOPTER
　　　. OH-58 HELICOPTER
　　　. P-166 AIRCRAFT
　　　. PD-808 AIRCRAFT
　　　. PIPER AIRCRAFT
　　　. . PA-34 SENECA AIRCRAFT
　　　. SAAB 105 AIRCRAFT
　　　. SC-7 AIRCRAFT
　　　. U-10 AIRCRAFT
　　　. VZ-8 AIRCRAFT
　　　. YAK 40 AIRCRAFT
　RT　AGRICULTURAL AIRCRAFT
　∞AIRCRAFT
　　　BIPLANES
　　　DRONE AIRCRAFT
　　　GENERAL AVIATION AIRCRAFT
　∞LOW WING AIRCRAFT
　∞MILITARY AIRCRAFT
　　　MYSTERE 50 AIRCRAFT
　　　OBSERVATION AIRCRAFT
　　　PASSENGER AIRCRAFT
　　　PILOTLESS AIRCRAFT
　　　RECONNAISSANCE AIRCRAFT
　　　SUBMERSIBLE AIRCRAFT
　∞SUBSONIC AIRCRAFT
　　　TERRAIN FOLLOWING AIRCRAFT

LIGHT AIRCRAFT-(CONT.)
　　　TRAINING AIRCRAFT
　　　TRANSPORT AIRCRAFT
　　　ULTRALIGHT AIRCRAFT
　　　UTILITY AIRCRAFT
　　　WATER TAKEOFF AND LANDING
　　　　AIRCRAFT

LIGHT ALLOYS
　GS　ALLOYS
　　　. **LIGHT ALLOYS**
　　　. . ALUMINUM ALLOYS
　　　. . BERYLLIUM ALLOYS
　　　. . MAGNESIUM ALLOYS
　RT　∞METALLURGY
　　　METALS

LIGHT AMPLIFIERS
　UF　OPTICAL AMPLIFIERS
　GS　AMPLIFIERS
　　　. **LIGHT AMPLIFIERS**
　RT　HCN LASERS
　　　IMAGE CONVERTERS
　　　IMAGE ENHANCEMENT
　　　IMAGE INTENSIFIERS
　　　LALLEMAND CAMERAS
　　　LASER CAVITIES
　　　LASER MICROSCOPY
　　　LASERS
　　　MICROCHANNELS
　∞OPTICS
　　　PHOTOCATHODES
　　　ULTRASHORT PULSED LASERS
　　　ULTRAVIOLET LASERS

LIGHT ARMED RECONNAISSANCE AIRCRAFT
USE　COIN AIRCRAFT

LIGHT BEAMS
　UF　LIGHT PROBES
　GS　BEAMS (RADIATION)
　　　. PHOTON BEAMS
　　　. . **LIGHT BEAMS**
　　　ELECTROMAGNETIC RADIATION
　　　. LIGHT (VISIBLE RADIATION)
　　　. . **LIGHT BEAMS**
　　　. PHOTON BEAMS
　　　. . **LIGHT BEAMS**
　　　PARTICLES
　　　. ELEMENTARY PARTICLES
　　　. . BOSONS
　　　. . . PHOTONS
　　　. . . . **LIGHT BEAMS**
　RT　LASERS
　　　OPTICAL WAVEGUIDES
　　　RARE GAS-HALIDE LASERS

LIGHT BULBS
USE　LUMINAIRES

LIGHT COMMUNICATION
USE　OPTICAL COMMUNICATION

LIGHT CURVE
　RT　∞CURVES
　　　LIGHT (VISIBLE RADIATION)
　　　STELLAR RADIATION

LIGHT DURATION
USE　FLASH
　　　PULSE DURATION

LIGHT ELEMENTS
　GS　CHEMICAL ELEMENTS
　　　. **LIGHT ELEMENTS**
　RT　∞ELEMENTS
　　　LOW DENSITY MATERIALS

LIGHT EMISSION
　UF　OPTICAL EMISSION
　GS　DECAY
　　　. EMISSION
　　　. . **LIGHT EMISSION**
　　　. . . INCANDESCENCE
　　　. . . LUMINESCENCE
　　　. . . . BIOLUMINESCENCE
　　　. . . . CATHODE GLOW
　　　. . . . CATHODOLUMINESCENCE
　　　. . . . CHEMILUMINESCENCE
　　　. . . . ELECTROLUMINESCENCE
　　　. . . . FLUORESCENCE
　　　. . . . . PHOSPHORESCENCE
　　　. . . . . RESONANCE FLUORESCENCE
　　　. . . . . X RAY FLUORESCENCE

LIGHT EMISSION-*(CONT.)*
```
        . . . . LUNAR LUMINESCENCE
        . . . . OPTICAL RESONANCE
        . . . . PHOTOLUMINESCENCE
        . . . . . TRIBOLUMINESCENCE
        . . . . . X RAY FLUORESCENCE
        . . . . SHOCK WAVE LUMINESCENCE
        . . . . SONOLUMINESCENCE
        . . . . THERMOLUMINESCENCE
RT      AIRGLOW
        AURORAL ABSORPTION
        AURORAL IONIZATION
        AURORAL SPECTROSCOPY
        AURORAS
        DIMMING
        ELECTROMAGNETIC RADIATION
        LINEAR POLARIZATION
        ∞OPTICS
        SELF SUSTAINED EMISSION
        SKY BRIGHTNESS
        SPECTRAL EMISSION
        STIMULATED EMISSION
        WHITE HOLES (ASTRONOMY)
```

LIGHT EMITTING DIODES
```
UF      LED (DIODES)
GS      ELECTRONIC EQUIPMENT
        . DIODES
        . . SEMICONDUCTOR DIODES
        . . . LIGHT EMITTING DIODES
        . SOLID STATE DEVICES
        . . SEMICONDUCTOR DEVICES
        . . . LIGHT EMITTING DIODES
RT      AIRCRAFT INSTRUMENTS
        ALPHANUMERIC CHARACTERS
        DISPLAY DEVICES
        ELECTROLUMINESCENCE
        LUMINESCENCE
        PHOTONICS
```

LIGHT GAS GUNS
```
GS      GAS GUNS
        . LIGHT GAS GUNS
RT      HYPERVELOCITY PROJECTILES
```

LIGHT INTENSITY
```
USE     LUMINOUS INTENSITY
```

LIGHT INTRATHEATER TRANSPORT
```
GS      LIGHT AIRCRAFT
        . LIGHT INTRATHEATER TRANSPORT
        TRANSPORT AIRCRAFT
        . LIGHT INTRATHEATER TRANSPORT
RT      ∞AIRCRAFT
        COIN AIRCRAFT
```

LIGHT IONS
```
GS      IONS
        . LIGHT IONS
RT      CHEMICAL ELEMENTS
        HEAVY IONS
        PLASMAS (PHYSICS)
```

LIGHT MODULATION
```
UF      OPTICAL MASER MODULATION
        OPTICAL MODULATION
GS      MODULATION
        . LIGHT MODULATION
        . . MIROS SYSTEM
        . . ULTRASONIC LIGHT MODULATION
RT      AMPLITUDE MODULATION
        ELECTRO-OPTICAL EFFECT
        FREQUENCY MODULATION
        KERR ELECTROOPTICAL EFFECT
        LASERS
        LIGHT VALVES
        MODULATORS
        OPTICAL HETERODYNING
        ∞OPTICS
        PULSE MODULATION
        TRAVELING WAVE MODULATION
        TUNABLE LASERS
```

LIGHT PRESSURE
```
USE     ILLUMINANCE
```

LIGHT PROBES
```
USE     LIGHT BEAMS
```

LIGHT SCATTERING
```
GS      SCATTERING
        . WAVE SCATTERING
        . . ELECTROMAGNETIC SCATTERING
        . . . LIGHT SCATTERING
```

LIGHT SCATTERING-*(CONT.)*
```
        . . . . HALOS
        TRANSMISSION
        . ELECTROMAGNETIC WAVE
            TRANSMISSION
        . . LIGHT TRANSMISSION
        . . . LIGHT SCATTERING
        . . . . HALOS
        . WAVE PROPAGATION
        . . LIGHT SCATTERING
        . . . HALOS
RT      AFTERGLOWS
        ATMOSPHERIC SCATTERING
        BRILLOUIN EFFECT
        CIRCUMSOLAR RADIATION
        DIFFUSE RADIATION
        ELECTROMAGNETIC ABSORPTION
        FORWARD SCATTERING
        GRAVITATIONAL LENSES
        INFRARED ABSORPTION
        RAYLEIGH SCATTERING
        REFLECTION NEBULAE
        SCATTER PLATES (OPTICS)
        SPECKLE PATTERNS
        TRANSMISSIVITY
        TROPOSPHERIC SCATTERING
        UMKEHR EFFECT
```

LIGHT SCATTERING METERS
```
GS      MEASURING INSTRUMENTS
        . OPTICAL MEASURING INSTRUMENTS
        . . LIGHT SCATTERING METERS
        OPTICAL EQUIPMENT
        . OPTICAL MEASURING INSTRUMENTS
        . . LIGHT SCATTERING METERS
RT      METEOROLOGICAL INSTRUMENTS
```

LIGHT SOURCES
```
GS      LIGHT SOURCES
        . ILLUMINATORS
RT      ARC LAMPS
        CATHODOLUMINESCENCE
        DAYGLOW
        DUOCHROMATORS
        ELECTRIC ARCS
        ELECTROLUMINESCENCE
        FLASH LAMPS
        GLOW DISCHARGES
        HCN LASERS
        HEAT SOURCES
        ILLUMINATING
        LASERS
        LIGHTING EQUIPMENT
        LUMINAIRES
        MERCURY LAMPS
        MONOCHROMATORS
        MOON
        PLASMA DISPLAY DEVICES
        POINT SOURCES
        RADIATION SOURCES
        SUN
```

LIGHT SPEED
```
GS      RATES (PER TIME)
        . LIGHT SPEED
        VELOCITY
        . LIGHT SPEED
RT      HIGH SPEED
        RELATIVISTIC VELOCITY
        SCHWARZSCHILD METRIC
```

LIGHT TRANSMISSION
```
UF      OPTICAL ABSORPTION
GS      TRANSMISSION
        . ELECTROMAGNETIC WAVE
            TRANSMISSION
        . . LIGHT TRANSMISSION
        . . . LIGHT SCATTERING
        . . . . HALOS
RT      ABSORPTANCE
        ATMOSPHERIC OPTICS
        ATMOSPHERIC REFRACTION
        FERMAT PRINCIPLE
        FIBER OPTICS
        FLICKER
        GAMMA RAY LASERS
        GEOMETRICAL OPTICS
        HAZE
        ILLUMINATING
        ∞ILLUMINATION
        INTEGRATED OPTICS
        LASERS
        LOW VISIBILITY
        MOLECULAR ABSORPTION
        OPACITY
```

LIGHT TRANSMISSION-*(CONT.)*
```
        OPTICAL BISTABILITY
        OPTICAL COUPLING
        OPTICAL PROPERTIES
        OPTICAL REFLECTION
        OPTICAL WAVEGUIDES
        RAINBOWS
        SAGNAC EFFECT
        STIMULATED EMISSION DEVICES
        TRANSLUCENCE
        TRANSPARENCE
        TURBIDITY
        ULTRAVIOLET LASERS
        VISIBILITY
        WAVE DISPERSION
        WAVE PROPAGATION
```

LIGHT TRANSPORT AIRCRAFT
```
GS      COMMERCIAL AIRCRAFT
        . LIGHT TRANSPORT AIRCRAFT
        TRANSPORT AIRCRAFT
        . LIGHT TRANSPORT AIRCRAFT
RT      ∞AIRCRAFT
        MULTIENGINE VEHICLES
        PASSENGER AIRCRAFT
```

LIGHT VALVES
```
RT      ELECTRO-OPTICS
        LIGHT MODULATION
        LIQUID CRYSTALS
        OPTICAL DATA PROCESSING
```

LIGHT WATER
```
UF      PROTIUM
GS      HYDROGEN COMPOUNDS
        . LIGHT WATER
        WATER
        . LIGHT WATER
```

LIGHT WATER BREEDER REACTORS
```
GS      NUCLEAR REACTORS
        . BREEDER REACTORS
        . . LIGHT WATER BREEDER REACTORS
RT      HEAVY WATER REACTORS
```

LIGHT WATER REACTORS
```
GS      NUCLEAR REACTORS
        . LIQUID COOLED REACTORS
        . . WATER COOLED REACTORS
        . . . LIGHT WATER REACTORS
RT      WATER MODERATED REACTORS
```

LIGHT-CONE EXPANSION
```
GS      ELECTRODYNAMICS
        . QUANTUM ELECTRODYNAMICS
        . . LIGHT-CONE EXPANSION
        NUCLEAR REACTIONS
        . HIGH ENERGY INTERACTIONS
        . . LIGHT-CONE EXPANSION
        QUANTUM MECHANICS
        . QUANTUM ELECTRODYNAMICS
        . . LIGHT-CONE EXPANSION
```

LIGHTHILL GAS MODEL
```
GS      MODELS
        . LIGHTHILL GAS MODEL
RT      BOUNDARY LAYER FLOW
        GAS MIXTURES
        GAS TRANSPORT
        HYPERSONIC FLOW
        MOLECULAR THEORY
        TRANSPORT PROPERTIES
```

LIGHTHILL METHOD
```
RT      AIRFOIL PROFILES
        AIRFOILS
        CONFORMAL MAPPING
        FLOW THEORY
        INTEGRAL TRANSFORMATIONS
        ∞METHODOLOGY
```

LIGHTING
```
USE     ILLUMINATING
```

LIGHTING EQUIPMENT
```
GS      LIGHTING EQUIPMENT
        . AIRCRAFT LIGHTS
        . ILLUMINATORS
        . LUMINAIRES
        . . AIRPORT LIGHTS
        . . . RUNWAY LIGHTS
        . . ARC LAMPS
        . . . FLASH LAMPS
        . . . ALKALI VAPOR LAMPS
```

LIGHTING EQUIPMENT-(CONT.)
- .. MERCURY LAMPS
- .. QUARTZ LAMPS
- .. SEARCHLIGHTS
- .. XENON LAMPS
- RT   ∞ELECTRIC EQUIPMENT
- ∞EQUIPMENT
- ∞FLARES
- ILLUMINATING
- LIGHT (VISIBLE RADIATION)
- LIGHT SOURCES
- ONBOARD EQUIPMENT
- WASTE ENERGY UTILIZATION

LIGHTNING
- GS   ELECTRIC CURRENT
- . ELECTRIC DISCHARGES
- .. **LIGHTNING**
- ... BALL LIGHTNING
- RT   ∞ARRESTERS
- ATMOSPHERIC ELECTRICITY
- ELECTRIC ARCS
- ELECTRIC SPARKS
- ELECTRICITY
- ELECTRODELESS DISCHARGES
- GAS DISCHARGES
- LIGHTNING SUPPRESSION
- NITROGENATION
- RADIATIVE RECOMBINATION
- STATIC ELECTRICITY
- THUNDERSTORMS
- WHISTLERS

LIGHTNING SUPPRESSION
- GS   WEATHER MODIFICATION
- . **LIGHTNING SUPPRESSION**
- RT   ATMOSPHERIC ELECTRICITY
- CLIMATOLOGY
- ELECTRIC DISCHARGES
- LIGHTNING
- THUNDERSTORMS

LIGHTS
- USE   LUMINAIRES

LIGNIN
- RT   CELLULOSE
- ∞POLYMERS

LIGNITE
- GS   RESOURCES
- . EARTH RESOURCES
- .. FOSSIL FUELS
- ... COAL
- .... **LIGNITE**
- ROCKS
- . COAL
- .. **LIGNITE**
- SEDIMENTARY ROCKS
- . CARBONACEOUS ROCKS
- .. COAL
- ... **LIGNITE**
- RT   ASHES
- BITUMENS
- CARBONACEOUS MATERIALS
- COAL GASIFICATION
- COAL LIQUEFACTION
- COAL UTILIZATION
- COKE
- ENERGY CONVERSION
- ENERGY POLICY
- ENERGY TECHNOLOGY
- GASEOUS FUELS
- HYDROCARBON FUEL PRODUCTION
- HYDROCARBON FUELS
- HYDROGEN PRODUCTION
- HYDROPYROLYSIS
- SYNTHANE

LIKELIHOOD RATIO
- GS   STATISTICAL ANALYSIS
- . **LIKELIHOOD RATIO**
- RT   ESTIMATES
- MATHEMATICAL MODELS
- MAXIMUM LIKELIHOOD ESTIMATES
- PROBABILITY THEORY
- STATISTICAL TESTS

LIMB BRIGHTENING
- RT   B STARS
- BRIGHTNESS
- BRIGHTNESS TEMPERATURE
- ∞LIMBS
- SOLAR FLUX
- SOLAR FLUX DENSITY

LIMB BRIGHTENING-(CONT.)
- SOLAR GRANULATION
- SOLAR LIMB
- STELLAR ATMOSPHERES
- STELLAR LUMINOSITY

LIMB DARKENING
- GS   DARKENING
- . **LIMB DARKENING**
- RT   B STARS
- BINARY STARS
- ∞LIMBS
- SOLAR LIMB
- STELLAR ATMOSPHERES
- STELLAR LUMINOSITY

∞ LIMBS
- SN   *(USE OF A MORE SPECIFIC TERM IS RECOMMENDED--CONSULT THE TERMS LISTED BELOW)*
- RT   EARTH LIMB
- LIMB BRIGHTENING
- LIMB DARKENING
- LIMBS (ANATOMY)
- LUNAR LIMB
- PLANETARY LIMB
- SOLAR LIMB

LIMBS (ANATOMY)
- GS   ANATOMY
- . **LIMBS (ANATOMY)**
- .. ARM (ANATOMY)
- ... ELBOW (ANATOMY)
- ... FOREARM
- .. HAND (ANATOMY)
- ... FINGERS
- .. LEG (ANATOMY)
- ... FEET (ANATOMY)
- ... KNEE (ANATOMY)
- RT   APPENDAGES
- FEET (ANATOMY)
- HUMAN BODY
- ∞LIMBS

LIME
- USE   CALCIUM OXIDES

LIMEN
- RT   PSYCHOLOGICAL TESTS
- THRESHOLDS (PERCEPTION)

LIMESTONE
- GS   ROCKS
- . **LIMESTONE**
- SEDIMENTARY ROCKS
- . **LIMESTONE**
- RT   AGGREGATES
- CALCIUM CARBONATES
- DOLOMITE (MINERAL)
- EARTH RESOURCES
- FLUXES
- MINERALS
- SCHIST
- SOILS

LIMITATIONS
- USE   CONSTRAINTS

LIMITER AMPLIFIERS
- GS   AMPLIFIERS
- . **LIMITER AMPLIFIERS**

LIMITER CIRCUITS
- GS   CIRCUITS
- . **LIMITER CIRCUITS**
- .. CLIPPER CIRCUITS
- RT   CIRCULATORS (PHASE SHIFT CIRCUITS)
- CLAMPING CIRCUITS
- CURRENT REGULATORS
- POWER LIMITERS

LIMITERS (FUSION REACTORS)
- RT   BLANKETS (FUSION REACTORS)
- CONTROLLED FUSION
- FUSION REACTORS
- MODERATION (ENERGY ABSORPTION)
- MODERATORS
- PLASMA CONTROL
- PLASMA LOSS
- REACTOR DESIGN
- REACTOR MATERIALS
- TOKAMAK DEVICES
- TOROIDAL PLASMAS
- WALLS

∞ LIMITS
- SN   *(USE OF A MORE SPECIFIC TERM IS RECOMMENDED--CONSULT THE TERMS LISTED BELOW)*
- RT   IGNITION LIMITS
- LIMITS (MATHEMATICS)
- RANGE (EXTREMES)

LIMITS (MATHEMATICS)
- GS   ANALYSIS (MATHEMATICS)
- . CALCULUS
- .. **LIMITS (MATHEMATICS)**
- . REAL VARIABLES
- .. EXTREMUM VALUES
- ... **LIMITS (MATHEMATICS)**
- RT   DIFFERENTIAL CALCULUS
- ∞ENVELOPES
- ∞LIMITS

LIMNOLOGY
- RT   AQUIFERS
- ARROYOS
- EARTH HYDROSPHERE
- FRESH WATER
- GEOCHEMISTRY
- GEOPHYSICS
- GROUND WATER
- ∞HYDRAULICS
- HYDROGRAPHY
- HYDROLOGY
- LAKE TEXOMA (OK-TX)
- LAKES
- MARINE BIOLOGY
- MARINE CHEMISTRY
- PONDS
- POTABLE WATER
- RAIN
- STREAMS
- WATER
- WATER MANAGEMENT
- WATER POLLUTION
- WATER RESOURCES
- WELLS

LIMONITE
- GS   IRON COMPOUNDS
- . **LIMONITE**
- MINERALS
- . **LIMONITE**

LINCOLN EXPERIMENTAL SATELLITES
- UF   LES (SATELLITES)
- GS   SATELLITES
- . ARTIFICIAL SATELLITES
- .. **LINCOLN EXPERIMENTAL SATELLITES**

LINE CURRENT
- GS   ELECTRIC CURRENT
- . **LINE CURRENT**
- RT   ELECTRIC POWER SUPPLIES
- ELECTRODYNAMICS
- MAGNETOHYDRODYNAMIC FLOW
- PLASMA CURRENTS
- ∞POWER SUPPLIES

LINE OF SIGHT
- RT   AREA
- COORDINATES
- ∞DIRECTION
- LOCI
- TARGETS

LINE OF SIGHT COMMUNICATION
- GS   TELECOMMUNICATION
- . COMMUNICATION
- .. **LINE OF SIGHT COMMUNICATION**
- RT   BORESIGHT ERROR
- FREQUENCY MODULATION
- SPACE COMMUNICATION
- TELEVISION TRANSMISSION

LINE SHAPE
- GS   SHAPES
- . **LINE SHAPE**
- RT   CURVES (GEOMETRY)
- INFLECTION POINTS
- ∞PROFILES

LINE SPECTRA
- UF   SPECTRAL LINES
- GS   SPECTRA
- . RADIATION SPECTRA
- .. ELECTROMAGNETIC SPECTRA

LINE SPECTRA-(CONT.)
. . . **LINE SPECTRA**
. . . . BALMER SERIES
. . . . D LINES
. . . . ELECTRONIC SPECTRA
. . . . FRAUNHOFER LINES
. . . . H LINES
. . . . . H ALPHA LINE
. . . . . H BETA LINE
. . . . . H GAMMA LINE
. . . . K LINES
. . . . LYMAN SPECTRA
. . . . PASCHEN SERIES
. . . . RYDBERG SERIES
. . . . TELLURIC LINES
RT ABSORPTION SPECTRA
ATOMIC ENERGY LEVELS
BOHR THEORY
EMISSION SPECTRA
FINE STRUCTURE
FLAME SPECTROSCOPY
FREQUENCIES
HYPERFINE STRUCTURE
INFRARED SPECTRA
LIGHT (VISIBLE RADIATION)
∞LINES
MOLECULAR SPECTROSCOPY
OSCILLATOR STRENGTHS
PRESSURE BROADENING
RAMAN SPECTRA
RAMAN SPECTROSCOPY
RESONANCE LINES
SEYFERT GALAXIES
SOLAR SPECTRA
SPECTRAL BANDS
SPECTRAL EMISSION
SPECTRAL ENERGY DISTRIBUTION
SPECTRAL LINE WIDTH
SPECTRAL RESOLUTION
SPECTROGRAMS
SPECTRUM ANALYSIS
STARK EFFECT
STELLAR SPECTRA
ULTRAVIOLET SPECTRA
VISIBLE SPECTRUM

LINEAMENT
USE STRUCTURAL PROPERTIES (GEOLOGY)

LINEAR ACCELERATORS
GS PARTICLE ACCELERATORS
. **LINEAR ACCELERATORS**
RT ∞ACCELERATORS
ELECTRON ACCELERATORS
ION SOURCES
MULTIPACTOR DISCHARGES
NEUTRON SOURCES

LINEAR AMPLIFIERS
GS AMPLIFIERS
. **LINEAR AMPLIFIERS**

LINEAR ARRAYS
GS ARRAYS
. ANTENNA ARRAYS
. **LINEAR ARRAYS**
. . . ENDFIRE ARRAYS
. . . . YAGI ANTENNAS
RT DIPOLE ANTENNAS
MULTIPLE BEAM INTERVAL SCANNERS
PHASED ARRAYS
PUSHBROOM SENSOR MODES

LINEAR CIRCUITS
GS CIRCUITS
. **LINEAR CIRCUITS**
RT AMPLIFIERS
DISTRIBUTED PARAMETER SYSTEMS
ELECTRICAL RESISTANCE
SUPERPOSITION (MATHEMATICS)
VOLT-AMPERE CHARACTERISTICS

LINEAR ENERGY TRANSFER (LET)
GS ENERGY TRANSFER
. **LINEAR ENERGY TRANSFER (LET)**
RT IONIZING RADIATION

LINEAR EQUATIONS
GS ALGEBRA
. **LINEAR EQUATIONS**
. . LINEAR EVOLUTION EQUATIONS
ANALYSIS (MATHEMATICS)
. REAL VARIABLES
. . **LINEAR EQUATIONS**
. . . LINEAR EVOLUTION EQUATIONS

LINEAR EQUATIONS-(CONT.)
RT DETERMINANTS
∞EQUATIONS
FLOQUET THEOREM
MATRICES (MATHEMATICS)
OPERATIONAL CALCULUS
POLYNOMIALS

LINEAR EVOLUTION EQUATIONS
GS ALGEBRA
. LINEAR EQUATIONS
. . **LINEAR EVOLUTION EQUATIONS**
ANALYSIS (MATHEMATICS)
. REAL VARIABLES
. LINEAR EQUATIONS
. . **LINEAR EVOLUTION EQUATIONS**
RT DIFFERENCE EQUATIONS
∞EQUATIONS

LINEAR FILTERS
GS **LINEAR FILTERS**
. KALMAN FILTERS
. REDUCED ORDER FILTERS
RT ADAPTIVE FILTERS
ELECTRIC FILTERS
ELECTROMAGNETIC WAVE FILTERS
∞FILTERS
∞FREQUENCY RESPONSE
NONLINEAR FILTERS

LINEAR INTEGRATED CIRCUITS
GS CIRCUITS
. INTEGRATED CIRCUITS
. . **LINEAR INTEGRATED CIRCUITS**
RT ELECTRONIC PACKAGING
LARGE SCALE INTEGRATION
MICROMINIATURIZATION
MOLECULAR ELECTRONICS
TRANSISTOR CIRCUITS

LINEAR POLARIZATION
GS POLARIZATION (WAVES)
. **LINEAR POLARIZATION**
RT LIGHT EMISSION
MICROWAVE EMISSION
OPTICAL POLARIZATION
∞POLARIZATION
POLARIZED ELECTROMAGNETIC
RADIATION
POLARIZED RADIATION
RADIO ASTRONOMY

LINEAR PREDICTION
GS PREDICTIONS
. **LINEAR PREDICTION**
RT COMPUTATION
DIFFERENTIAL PULSE CODE
MODULATION
MATHEMATICAL MODELS
OPERATIONS RESEARCH
QUALITY CONTROL
STATISTICAL ANALYSIS

LINEAR PROGRAMMING
GS OPERATIONS RESEARCH
. MATHEMATICAL PROGRAMMING
. . **LINEAR PROGRAMMING**
RESEARCH
. MATHEMATICAL PROGRAMMING
. . **LINEAR PROGRAMMING**
RT ∞APPLICATIONS OF MATHEMATICS
COMPUTER PROGRAMMING
CONSTRAINTS
DYNAMIC PROGRAMMING
FORMALISM
GAME THEORY
MATRICES (MATHEMATICS)
NONLINEAR PROGRAMMING
NUMERICAL ANALYSIS
OPERATIONS RESEARCH
OPTIMIZATION
∞PROGRAMMING
SIMPLEX METHOD

LINEAR RECEIVERS
GS RECEIVERS
. **LINEAR RECEIVERS**
RT ∞FREQUENCY RESPONSE
NYQUIST FREQUENCIES

LINEAR SYSTEMS
RT DISTRIBUTED PARAMETER SYSTEMS
NONLINEAR SYSTEMS
ROBUSTNESS (MATHEMATICS)
STATE ESTIMATION

LINEAR SYSTEMS-(CONT.)
∞SYSTEMS
TRACKING PROBLEM

LINEAR TRANSFORMATIONS
GS ALGEBRA
. **LINEAR TRANSFORMATIONS**
FUNCTIONS (MATHEMATICS)
. **LINEAR TRANSFORMATIONS**
RT FOURIER ANALYSIS
JORDAN FORM
MATRICES (MATHEMATICS)
ORTHOGONAL FUNCTIONS
SCHWARTZ INEQUALITY
VECTOR SPACES

LINEAR VIBRATION
GS VIBRATION
. STRUCTURAL VIBRATION
. . **LINEAR VIBRATION**
RT FREE VIBRATION
MISSILE VIBRATION
RANDOM VIBRATION
VIBRATION MODE

LINEARITY
GS **LINEARITY**
. COLLINEARITY
RT ACCURACY
CONSISTENCY
DIFFERENTIAL EQUATIONS
DYNAMIC CHARACTERISTICS
ERRORS
FUNCTIONS (MATHEMATICS)
INSTRUMENT ERRORS
LINEARIZATION
NONLINEARITY
TOLERANCES (MECHANICS)
VARIABILITY

LINEARIZATION
RT BERNOULLI THEOREM
∞EQUATIONS
GALERKIN METHOD
LINEARITY
SIMPLIFICATION

LINEN
GS FABRICS
. **LINEN**
FIBERS
. **LINEN**
TEXTILES
. **LINEN**
RT ORGANIC MATERIALS

LINERS
USE LININGS

∞ **LINES**
SN *(USE OF A MORE SPECIFIC TERM IS RECOMMENDED--CONSULT THE TERMS LISTED BELOW)*
RT DELAY LINES
LINE SPECTRA
LINES OF FORCE
PIPELINES
TERMINATOR LINES
TETHERLINES
TRANSMISSION LINES
UNDERGROUND TRANSMISSION LINES

LINES (GEOMETRY)
GS GEOMETRY
. EUCLIDEAN GEOMETRY
. . **LINES (GEOMETRY)**
. . . CHORDS (GEOMETRY)
. . . GEODESIC LINES
RT RADII
RECIPROCAL THEOREMS
SEGMENTS

LINES OF FORCE
RT BARIUM ION CLOUDS
CONJUGATE POINTS
FLUX PINNING
∞FORCE
∞LINES
MAGNETIC CIRCUITS
MAGNETIC DOMAINS
MAGNETIC FIELDS
MAGNETIC FLUX
MAGNETIC MIRRORS
MAGNETIC PROPERTIES

LINES OF FORCE-*(CONT.)*
MAGNETOSTATIC FIELDS
NONUNIFORM MAGNETIC FIELDS
POLAR CUSPS

LING-TEMCO-VOUGHT AIRCRAFT
UF   LTV AIRCRAFT
GS   **LING-TEMCO-VOUGHT AIRCRAFT**
. A-7 AIRCRAFT
. F-8 AIRCRAFT
. XC-142 AIRCRAFT
RT   ∞AIRCRAFT

LINGUISTICS
GS   **LINGUISTICS**
. MACHINE TRANSLATION
. PHONEMES
. PHONEMICS
. PSYCHOLINGUISTICS
. SEMANTICS
. SYNTAX
.. SENTENCES
... WORDS (LANGUAGE)
.... SYLLABLES
RT   LANGUAGES
ORTHOGRAPHY
SPEECH

LINING PROCESSES
RT   COATING
COATINGS
INSULATION
LININGS
SEALING
TUNNELING (EXCAVATION)

LININGS
UF   LINERS
GS   **LININGS**
. ROCKET LININGS
RT   BUSHINGS
∞CASING
COATINGS
INSERTS
INSULATION
JACKETS
LINING PROCESSES
SHEATHS
SHIELDING
∞TUBES

LINKAGES
RT   CAMS
CONNECTORS
COUPLING
COUPLINGS
ECCENTRICS
FASTENERS
FITTINGS
JOINTS (JUNCTIONS)
LATCHES
∞LINKS
MECHANICAL DEVICES
UNIONS (CONNECTORS)
YOKES

LINKING
USE   JOINING

∞ LINKS
SN   *(USE OF A MORE SPECIFIC TERM IS RECOMMENDED--CONSULT THE TERMS LISTED BELOW)*
RT   CHAINS
LINKAGES
LINKS (MATHEMATICS)

LINKS (MATHEMATICS)
GS   GEOMETRY
. TOPOLOGY
.. **LINKS (MATHEMATICS)**
RT   ∞LINKS

LIOUVILLE EQUATIONS
GS   ANALYSIS (MATHEMATICS)
. REAL VARIABLES
.. DIFFERENTIAL EQUATIONS
... PARTIAL DIFFERENTIAL EQUATIONS
.... **LIOUVILLE EQUATIONS**
RT   ∞EQUATIONS
PLASMA PHYSICS
PLASMAS (PHYSICS)
STATISTICAL MECHANICS

LIOUVILLE THEOREM
GS   ANALYSIS (MATHEMATICS)
. COMPLEX VARIABLES
.. **LIOUVILLE THEOREM**
THEOREMS
. **LIOUVILLE THEOREM**

LIP READING
GS   COMMUNICATING
. **LIP READING**
READING
. **LIP READING**

LIPID METABOLISM
GS   METABOLISM
. PROTEIN METABOLISM
.. **LIPID METABOLISM**
RT   ∞NUTRIENTS
OILS

LIPIDS
GS   **LIPIDS**
. CASTOR OIL
. IODOACETIC ACID
. LIPOIC ACID
. LIPOPROTEINS
. OLEIC ACID
. PROPIONIC ACID
. SEBACIC ACID
. VALERIC ACID
RT   ESTERS
GLYCEROLS
MYELIN
∞NUTRIENTS

LIPOIC ACID
GS   ACIDS
. FATTY ACIDS
.. **LIPOIC ACID**
ALIPHATIC COMPOUNDS
. FATTY ACIDS
.. **LIPOIC ACID**
LIPIDS
. **LIPOIC ACID**
ORGANIC COMPOUNDS
. FATTY ACIDS
.. **LIPOIC ACID**

LIPOPROTEINS
GS   LIPIDS
. **LIPOPROTEINS**
PROTEINS
. **LIPOPROTEINS**

LIPS (ANATOMY)
RT   FACE (ANATOMY)
HEAD (ANATOMY)

LIPSCHITZ CONDITION
GS   ANALYSIS (MATHEMATICS)
. REAL VARIABLES
.. **LIPSCHITZ CONDITION**
CONDITIONS
. **LIPSCHITZ CONDITION**
RT   DIFFERENTIAL EQUATIONS

LIQUEFACTION
GS   PHASE TRANSFORMATIONS
. **LIQUEFACTION**
.. COAL LIQUEFACTION
RT   ∞CONDENSATION
∞CONVERSION
JET CONDENSERS
MELTING
NONCONDENSABLE GASES
THIXOTROPY

LIQUEFIED GASES
GS   GASES
. **LIQUEFIED GASES**
.. LIQUEFIED NATURAL GAS
.. LIQUID AIR
.. LIQUID AMMONIA
.. LIQUID FLUORINE
.. LIQUID HELIUM
... LIQUID HELIUM 2
.. LIQUID HYDROGEN
.. LIQUID NEON
.. LIQUID NITROGEN
.. LIQUID OXYGEN
LIQUIDS
. **LIQUEFIED GASES**
.. LIQUID AMMONIA
.. LIQUID HELIUM

LIQUEFIED GASES-*(CONT.)*
.. LIQUID HYDROGEN
.. LIQUID NEON
.. LIQUID NITROGEN
.. LIQUID OXYGEN
RT   CONDENSATES
CONDENSERS (LIQUEFIERS)
CRYOGENIC ROCKET PROPELLANTS
CRYOGENICS
GAS MIXTURES
LIQUID ROCKET PROPELLANTS
SOLID CRYOGEN COOLING

LIQUEFIED NATURAL GAS
UF   LNG
GS   FUELS
. **LIQUEFIED NATURAL GAS**
GASES
. FLAMMABLE GASES
.. **LIQUEFIED NATURAL GAS**
. LIQUEFIED GASES
.. **LIQUEFIED NATURAL GAS**
. NATURAL GAS
.. **LIQUEFIED NATURAL GAS**
HYDROCARBONS
. NATURAL GAS
.. **LIQUEFIED NATURAL GAS**
RT   HYDROCARBON FUELS
METHANE

LIQUID AIR
GS   GASES
. AIR
.. **LIQUID AIR**
. GAS MIXTURES
.. **LIQUID AIR**
. LIQUEFIED GASES
.. **LIQUID AIR**

LIQUID AIR CYCLE ENGINES
UF   LACE (ENGINE)
GS   ENGINES
. ROCKET ENGINES
.. LIQUID PROPELLANT ROCKET ENGINES
... **LIQUID AIR CYCLE ENGINES**
RT   AEROSPACEPLANES
ASTROPLANE
HYDROGEN OXYGEN ENGINES
SUSTAINER ROCKET ENGINES
TURBOROCKET ENGINES

LIQUID ALLOYS
GS   ALLOYS
. **LIQUID ALLOYS**
RT   METALS

LIQUID AMMONIA
GS   GASES
. LIQUEFIED GASES
.. **LIQUID AMMONIA**
INORGANIC COMPOUNDS
. AMMONIA
.. **LIQUID AMMONIA**
LIQUIDS
. LIQUEFIED GASES
.. **LIQUID AMMONIA**
NITROGEN COMPOUNDS
. AMMONIA
.. **LIQUID AMMONIA**
RT   FUELS
LIQUID FUELS

LIQUID ATOMIZATION
GS   ATOMIZING
. **LIQUID ATOMIZATION**
RT   GAS ATOMIZATION
SPRAYING

LIQUID BEARINGS
GS   BEARINGS
. **LIQUID BEARINGS**
RT   LUBRICATION

LIQUID BREATHING
GS   RESPIRATION
. **LIQUID BREATHING**
RT   ACCLIMATIZATION
PRESSURE BREATHING
RESUSCITATION

LIQUID CHROMATOGRAPHY
UF   GEL PERMEATION CHROMATOGRAPHY
GS   CHROMATOGRAPHY

LIQUID CHROMATOGRAPHY-(CONT.)
. **LIQUID CHROMATOGRAPHY**
RT   CHEMICAL ANALYSIS
     CHEMICAL TESTS
     COLORIMETRY
     PAPER CHROMATOGRAPHY
     SORPTION

LIQUID COOLED REACTORS
GS   NUCLEAR REACTORS
    . **LIQUID COOLED REACTORS**
    . . LIQUID METAL COOLED REACTORS
    . . . ADVANCED SODIUM COOLED
        REACTOR
    . . . EXPERIMENTAL BREEDER REACTOR
        1
    . . . EXPERIMENTAL BREEDER REACTOR
        2
    . . . LITHIUM COOLED REACTOR
        EXPERIMENT
    . . . LOS ALAMOS MOLTEN PLUTONIUM
        REACTOR
    . . . MILITARY COMPACT REACTORS
    . . . SODIUM GRAPHITE REACTORS
    . . . SODIUM REACTOR EXPERIMENT
    . . ORGANIC COOLED REACTORS
    . . . EXPERIMENTAL ORGANIC COOLED
        REACTORS
    . . WATER COOLED REACTORS
    . . . BOILING WATER REACTORS
    . . . . EXPERIMENTAL BOILING WATER
        REACTORS
    . . . . HALDEN BOILING WATER
        REACTOR
    . . . . LOS ALAMOS WATER BOILER
        REACTOR
    . . . . PATHFINDER NUCLEAR REACTOR
    . . . . SPERT REACTORS
    . . . HEAVY WATER REACTORS
    . . . . HEAVY WATER COMPONENTS
        TEST REACTORS
    . . . . PLUTONIUM RECYCLE TEST
        REACTOR
    . . . . ZERO POWER REACTOR 2
    . . . LIGHT WATER REACTORS
    . . . NRX REACTORS
    . . . PLUM BROOK REACTOR
    . . . PRESSURIZED WATER REACTORS
    . . . SPECTRAL SHIFT CONTROL
        REACTOR
    . . . SWIMMING POOL REACTORS
    . . . ZERO POWER REACTORS
    . . . . ZERO POWER REACTOR 2
    . . . . ZERO POWER REACTOR 3
    . . . . ZERO POWER REACTOR 6
    . . . . ZERO POWER REACTOR 9
RT   SODIUM COOLING

LIQUID COOLING
SN   (COOLING WITH LIQUIDS)
UF   WATER COOLING
GS   COOLING
    . **LIQUID COOLING**
    . . FILM COOLING
RT   AIR COOLING
     COOLANTS
     COOLING SYSTEMS
     SODIUM COOLING
     SPACE COOLING (BUILDINGS)
     SWEAT COOLING
     THERMAL POLLUTION
     WATER IMMERSION

LIQUID CRYSTALS
GS   CRYSTALS
    . **LIQUID CRYSTALS**
RT   ANISOTROPIC FLUIDS
     CHOLESTEROL
     LIGHT VALVES

LIQUID DROPS
USE   DROPS (LIQUIDS)

LIQUID FILLED SHELLS
GS   SHELLS (STRUCTURAL FORMS)
    . FLUID FILLED SHELLS
    . . **LIQUID FILLED SHELLS**
RT   HYDRODYNAMIC RAM EFFECT
     PROPELLANT TANKS
     REINFORCED SHELLS
     SHELL STABILITY
    ∝ STORAGE
     TANKS (CONTAINERS)
    ∝ VESSELS

LIQUID FLOW
GS   FLUID FLOW
    . **LIQUID FLOW**
    . . OPEN CHANNEL FLOW
    . . WATER FLOW
RT   CRITICAL FLOW
     GAS FLOW
     HEAD (FLUID MECHANICS)
     HEAD FLOW
     HYDRODYNAMIC COEFFICIENTS
     LAMINAR FLOW
     MASS FLOW
     MULTIPHASE FLOW
     NONNEWTONIAN FLOW
     ORIFICE FLOW
     PIPE FLOW
     PRESSURE GRADIENTS
     PRESSURE HEADS
     RHEOLOGY
     SINGLE-PHASE FLOW
     SORET COEFFICIENT
     STEADY FLOW
     SUBCRITICAL FLOW
     SUPERCRITICAL FLOW
     TURBULENT FLOW
     TWO PHASE FLOW
     UNIFORM FLOW
     UNSTEADY FLOW

LIQUID FLUORINE
GS   GASES
    . LIQUEFIED GASES
    . . **LIQUID FLUORINE**

LIQUID FUELS
GS   FUELS
    . CHEMICAL FUELS
    . . **LIQUID FUELS**
    . . . AIRCRAFT FUELS
    . . . AUTOMOBILE FUELS
    . . . DIESEL FUELS
    . . . GASOLINE
    . . . HYDROGEN FUELS
    . . . JET ENGINE FUELS
    . . . . JP-4 JET FUEL
    . . . . JP-5 JET FUEL
    . . . . JP-6 JET FUEL
    . . . . JP-8 JET FUEL
    . . . KEROSENE
    . . LIQUIDS
    . **LIQUID FUELS**
    . . LIQUID ROCKET PROPELLANTS
    . . . CRYOGENIC ROCKET PROPELLANTS
    . . . HYPERGOLIC ROCKET
        PROPELLANTS
    . . . MONOPROPELLANTS
    . . . . AEROZINE
    . . . RP-1 ROCKET PROPELLANTS
    . . . SLURRY PROPELLANTS
RT   FUEL PRODUCTION
     GASEOUS FUELS
     GELLED ROCKET PROPELLANTS
     LIQUID AMMONIA
     LIQUID HYDROGEN
     ROCKET PROPELLANTS
     SYNTHETIC FUELS

LIQUID HELIUM
UF   HELIUM 2
GS   CHEMICAL ELEMENTS
    . RARE GASES
    . . HELIUM
    . . . **LIQUID HELIUM**
    . . . . LIQUID HELIUM 2
    . GASES
    . LIQUEFIED GASES
    . . **LIQUID HELIUM**
    . . . LIQUID HELIUM 2
    . RARE GASES
    . . HELIUM
    . . . **LIQUID HELIUM**
    . . . . LIQUID HELIUM 2
    . LIQUIDS
    . CRYOGENIC FLUIDS
    . . **LIQUID HELIUM**
    . LIQUEFIED GASES
    . . **LIQUID HELIUM**
RT   CRYOSTATS
     SUPERFLUIDITY
     TWO FLUID MODELS

LIQUID HELIUM 2
GS   CHEMICAL ELEMENTS
    . RARE GASES
    . . HELIUM

LIQUID HELIUM 2-(CONT.)
    . . . LIQUID HELIUM
    . . . . **LIQUID HELIUM 2**
    . GASES
    . LIQUEFIED GASES
    . . LIQUID HELIUM
    . . . **LIQUID HELIUM 2**
    . RARE GASES
    . . HELIUM
    . . . LIQUID HELIUM
    . . . . **LIQUID HELIUM 2**
RT   CRYOSTATS
     SUPERFLUIDITY

LIQUID HYDROGEN
GS   CHEMICAL ELEMENTS
    . HYDROGEN
    . . **LIQUID HYDROGEN**
    . GASES
    . HYDROGEN
    . . **LIQUID HYDROGEN**
    . LIQUEFIED GASES
    . . **LIQUID HYDROGEN**
    . LIQUIDS
    . CRYOGENIC FLUIDS
    . . **LIQUID HYDROGEN**
    . LIQUEFIED GASES
    . . **LIQUID HYDROGEN**
RT   CRYOGENIC ROCKET PROPELLANTS
     FUELS
     HYDROGEN FUELS
     HYDROGEN-BASED ENERGY
     LIQUID FUELS
     TOPPING CYCLE ENGINES

LIQUID INJECTION
GS   INJECTION
    . FLUID INJECTION
    . . **LIQUID INJECTION**
    . . . DEEP WELL INJECTION (WASTES)
    . . . WATER INJECTION
RT   FILM COOLING
     FUEL INJECTION
     FUEL SPRAYS
     MIXING
     PROPELLANT SPRAYS
     THRUST VECTOR CONTROL

LIQUID LASERS
GS   STIMULATED EMISSION DEVICES
    . LASERS
    . . **LIQUID LASERS**
RT   CARBON LASERS
     CHEMICAL LASERS
     DYE LASERS
     INFRARED LASERS
     ORGANIC LASERS

LIQUID LEVELS
GS   LEVEL (HORIZONTAL)
    . **LIQUID LEVELS**
RT   FLUID BOUNDARIES

LIQUID LITHIUM
GS   CHEMICAL ELEMENTS
    . ALKALI METALS
    . . LITHIUM
    . . . **LIQUID LITHIUM**
    . LIQUIDS
    . LIQUID METALS
    . . **LIQUID LITHIUM**
    . METALS
    . ALKALI METALS
    . . LITHIUM
    . . . **LIQUID LITHIUM**
    . LIQUID METALS
    . . **LIQUID LITHIUM**

LIQUID MERCURY
USE   MERCURY (METAL)

LIQUID METAL COOLED REACTORS
UF   LMCR (REACTORS)
GS   NUCLEAR REACTORS
    . LIQUID COOLED REACTORS
    . . **LIQUID METAL COOLED REACTORS**
    . . . ADVANCED SODIUM COOLED
        REACTOR
    . . . EXPERIMENTAL BREEDER REACTOR
        1
    . . . EXPERIMENTAL BREEDER REACTOR
        2
    . . . LITHIUM COOLED REACTOR
        EXPERIMENT

LIQUID METAL COOLED REACTORS-(CONT.)
- . . . LOS ALAMOS MOLTEN PLUTONIUM
      REACTOR
- . . . MILITARY COMPACT REACTORS
- . . . SODIUM GRAPHITE REACTORS
- . . . SODIUM REACTOR EXPERIMENT
- RT   ENRICO FERMI ATOMIC POWER PLANT
       SODIUM
       SODIUM REACTOR EXPERIMENT

LIQUID METAL FAST BREEDER REACTORS
- UF   LMFBR
- GS   NUCLEAR REACTORS
- . BREEDER REACTORS
- . . **LIQUID METAL FAST BREEDER
      REACTORS**
- . FAST NUCLEAR REACTORS
- . . **LIQUID METAL FAST BREEDER
      REACTORS**
- RT   NUCLEAR POWER REACTORS

LIQUID METALS
- GS   LIQUIDS
- . **LIQUID METALS**
- . . LIQUID LITHIUM
- . . LIQUID POTASSIUM
- . . LIQUID SODIUM
- . . MERCURY (METAL)
- . . . MERCURY VAPOR
- METALS
- . **LIQUID METALS**
- . . LIQUID LITHIUM
- . . LIQUID POTASSIUM
- . . LIQUID SODIUM
- . . MERCURY (METAL)
- . . . MERCURY VAPOR
- RT   CASTING
       LEVITATION MELTING
       LUBRICANTS
       MELTING
       METAL VAPORS

LIQUID NEON
- GS   CHEMICAL ELEMENTS
- . RARE GASES
- . . NEON
- . . . **LIQUID NEON**
- GASES
- . LIQUEFIED GASES
- . . **LIQUID NEON**
- . RARE GASES
- . . NEON
- . . . **LIQUID NEON**
- LIQUIDS
- . LIQUEFIED GASES
- . . **LIQUID NEON**

LIQUID NITROGEN
- GS   CHEMICAL ELEMENTS
- . NITROGEN
- . . **LIQUID NITROGEN**
- GASES
- . LIQUEFIED GASES
- . . **LIQUID NITROGEN**
- . NITROGEN
- . . **LIQUID NITROGEN**
- LIQUIDS
- . CRYOGENIC FLUIDS
- . . **LIQUID NITROGEN**
- . LIQUEFIED GASES
- . . **LIQUID NITROGEN**
- RT   SOLID CRYOGENS

LIQUID OXIDIZERS
- GS   LIQUIDS
- . **LIQUID OXIDIZERS**
- OXIDIZERS
- . **LIQUID OXIDIZERS**
- RT   ROCKET OXIDIZERS

LIQUID OXYGEN
- UF   LOX (OXYGEN)
- GS   GASES
- . LIQUEFIED GASES
- . . **LIQUID OXYGEN**
- . OXYGEN
- . . **LIQUID OXYGEN**
- LIQUIDS
- . CRYOGENIC FLUIDS
- . . **LIQUID OXYGEN**
- . LIQUEFIED GASES
- . . **LIQUID OXYGEN**
- OXIDIZERS
- . **LIQUID OXYGEN**
- RT   CRYOGENIC ROCKET PROPELLANTS

LIQUID OXYGEN-(CONT.)
- FLOX
- ROCKET OXIDIZERS

LIQUID PHASE EPITAXY
- GS   GROWTH
- . CRYSTAL GROWTH
- . . EPITAXY
- . . . **LIQUID PHASE EPITAXY**
- RT   CRYSTAL STRUCTURE
       LIQUID PHASES
       VAPOR PHASE EPITAXY

LIQUID PHASES
- RT   ALLOYS
       CRITICAL PRESSURE
       ELECTROEPITAXY
       EUTECTICS
       LIQUID PHASE EPITAXY
       LIQUIDS
       LIQUIDUS
       MELTING POINTS
       PHASE DIAGRAMS
- ∞ PHASES
       SOLID PHASES
       SOLID SOLUTIONS
       SOLIDUS
       SOLUBILITY
       SUPERCRITICAL PRESSURES
       SYNTECTIC ALLOYS
       TRANSITION TEMPERATURE
       VAPOR PHASE EPITAXY
       VAPOR PHASES

LIQUID PLUS SOLID ZONES
- USE   MUSHY ZONES

LIQUID POTASSIUM
- GS   CHEMICAL ELEMENTS
- . ALKALI METALS
- . . POTASSIUM
- . . . **LIQUID POTASSIUM**
- LIQUIDS
- . LIQUID METALS
- . . **LIQUID POTASSIUM**
- METALS
- . ALKALI METALS
- . . POTASSIUM
- . . . **LIQUID POTASSIUM**
- . LIQUID METALS
- . . **LIQUID POTASSIUM**

LIQUID PROPELLANT ROCKET ENGINES
- GS   ENGINES
- . ROCKET ENGINES
- . . **LIQUID PROPELLANT ROCKET
      ENGINES**
- . . . AJ-10 ENGINE
- . . . F-1 ROCKET ENGINE
- . . . H-1 ENGINE
- . . . HYDRAZINE ENGINES
- . . . HYDROGEN OXYGEN ENGINES
- . . . . J-2 ENGINE
- . . . . M-1 ENGINE
- . . . . RL-10-A-1 ENGINE
- . . . . RL-10-A-3 ENGINE
- . . . LIQUID AIR CYCLE ENGINES
- . . . LR-62-RM-2 ENGINE
- . . . LR-87-AJ-5 ENGINE
- . . . LR-91-AJ-5 ENGINE
- . . . LR-99 ENGINE
- . . . MA-2 ENGINE
- . . . MA-3 ENGINE
- . . . MA-5 ENGINE
- . . . RL-10 ENGINES
- . . . . RL-10-A-1 ENGINE
- . . . . RL-10-A-3 ENGINE
- . . . SPACE SHUTTLE MAIN ENGINE
- . . . X-405 ENGINE
- . . . XLR-99 ENGINE
- . . . YLR-91-AJ-1 ENGINE
- RT   ABLESTAR LAUNCH VEHICLE
       ATLAS SLV-3 LAUNCH VEHICLE
       BLACK KNIGHT ROCKET VEHICLE
       BLUE STEEL MISSILE
       BLUE STREAK LAUNCH VEHICLE
       BLUE STREAK MISSILE
       BOMARC A MISSILE
       BOMARC B MISSILE
       BOOSTER ROCKET ENGINES
       CENTAUR LAUNCH VEHICLE
       CORPORAL MISSILE
       CORVUS MISSILE
       DIAMANT LAUNCH VEHICLE

LIQUID PROPELLANT ROCKET-(CONT.)
- DORNIER PARAGLIDER ROCKET
  VEHICLE
- DUCTED ROCKET ENGINES
- HYBRID PROPELLANT ROCKET ENGINES
- HYLA-STAR ROCKET VEHICLE
- INTERNAL COMBUSTION ENGINES
- JUNO LAUNCH VEHICLES
- JUNO 1 LAUNCH VEHICLE
- JUNO 2 LAUNCH VEHICLE
- JUPITER C ROCKET VEHICLE
- JUPITER MISSILE
- LANCE MISSILE
- METEOR 1 ROCKET VEHICLE
- NAVAHO MISSILE
- NIKE-AJAX MISSILE
- NOMAD LAUNCH VEHICLE
- NOVA LAUNCH VEHICLES
- PROPELLANT TANKS
- RESTARTABLE ROCKET ENGINES
- RETROROCKET ENGINES
- SATURN S-1 STAGE
- SATURN S-1B STAGE
- SATURN S-1C STAGE
- SATURN S-2 STAGE
- SATURN S-4 STAGE
- SATURN S-4B STAGE
- SATURN STAGES
- SOLID PROPELLANT ROCKET ENGINES
- SPARROW 3 MISSILE
- SUSTAINER ROCKET ENGINES
- TALOS MISSILE
- THOR ABLE ROCKET VEHICLE
- THOR AGENA LAUNCH VEHICLE
- THOR DELTA LAUNCH VEHICLE
- THOR LAUNCH VEHICLES
- THORAD LAUNCH VEHICLES
- TITAN ICBM
- TITAN LAUNCH VEHICLES
- V-1 MISSILE
- V-2 MISSILE
- VANGUARD 2 LAUNCH VEHICLE
- VEGA LAUNCH VEHICLE
- VERNIER ENGINES
- VERONIQUE ROCKET VEHICLES
- VIKING ROCKET VEHICLE

LIQUID ROCKET PROPELLANTS
- UF   BIPROPELLANTS
       TRIPROPELLANTS
- GS   LIQUIDS
- . LIQUID FUELS
- . . **LIQUID ROCKET PROPELLANTS**
- . . . CRYOGENIC ROCKET PROPELLANTS
- . . . HYPERGOLIC ROCKET
          PROPELLANTS
- . . . MONOPROPELLANTS
- . . . . AEROZINE
- . . . RP-1 ROCKET PROPELLANTS
- . . . SLURRY PROPELLANTS
- PROPELLANTS
- . ROCKET PROPELLANTS
- . . **LIQUID ROCKET PROPELLANTS**
- . . . CRYOGENIC ROCKET PROPELLANTS
- . . . GELLED ROCKET PROPELLANTS
- . . . HYPERGOLIC ROCKET
          PROPELLANTS
- . . . MONOPROPELLANTS
- . . . . AEROZINE
- . . . RP-1 ROCKET PROPELLANTS
- . . . SLURRY PROPELLANTS
- RT   AIRCRAFT FUELS
       CHLORINE FLUORIDES
       FUEL TANK PRESSURIZATION
       GASEOUS ROCKET PROPELLANTS
       HIGH ENERGY PROPELLANTS
       HYBRID PROPELLANTS
       HYDRAZINES
       HYDROGEN FUELS
       LIQUEFIED GASES
       NITROGEN TETROXIDE
       PROPELLANT SPRAYS
       SOLID ROCKET PROPELLANTS
       STORABLE PROPELLANTS

LIQUID ROTATION
- USE   ROTATING LIQUIDS

LIQUID SLOSHING
- UF   SLOSHING
- RT   AERODYNAMIC STABILITY
       AIRCRAFT STABILITY
       BAFFLES
       CONTROLLABILITY
       FUEL CONTROL

LIQUID SLOSHING-*(CONT.)*
    FUEL TANKS
    INTERFACE STABILITY
    PROPELLANT TANKS
    PROPELLANT TRANSFER
    ROTATING FLUIDS
    SPACECRAFT STABILITY
    STORAGE STABILITY
    TANK GEOMETRY
    ULLAGE

LIQUID SODIUM
GS   CHEMICAL ELEMENTS
    . ALKALI METALS
    . . SODIUM
    . . . **LIQUID SODIUM**
    LIQUIDS
    . LIQUID METALS
    . . **LIQUID SODIUM**
    METALS
    . ALKALI METALS
    . . SODIUM
    . . . **LIQUID SODIUM**
    . LIQUID METALS
    . . **LIQUID SODIUM**

LIQUID SURFACES
GS   **LIQUID SURFACES**
    . MENISCI
RT   FLUID BOUNDARIES
    FREE BOUNDARIES
    INTERFACIAL TENSION
    JET BOUNDARIES
    SOLID SURFACES
    SURFACE WAVES
    ∞SURFACES

LIQUID WASTES
GS   WASTES
    . **LIQUID WASTES**
    . . URINE
    . . WASTE WATER
RT   DRAINAGE
    EFFLUENTS
    HUMAN WASTES
    INDUSTRIAL WASTES
    METABOLIC WASTES
    PONDS
    SEWAGE
    SLUDGE
    SOLID WASTES

LIQUID-GAS MIXTURES
GS   MIXTURES
    . DISPERSIONS
    . . **LIQUID-GAS MIXTURES**
    . . . AEROSOLS
    . . . . FOG
RT   AIR WATER INTERACTIONS
    BINARY MIXTURES
    GAS MIXTURES
    MENISCI
    SOLUBILITY
    VAPOR PHASES
    VAPOR PRESSURE

LIQUID-LIQUID INTERFACES
GS   BOUNDARIES
    . FLUID BOUNDARIES
    . . **LIQUID-LIQUID INTERFACES**
    INTERFACES
    . FLUID BOUNDARIES
    . . **LIQUID-LIQUID INTERFACES**
RT   BOUNDARY LAYERS
    FREE BOUNDARIES
    HEAT TRANSFER
    INTERFACE STABILITY
    INTERFACIAL ENERGY
    INTERFACIAL TENSION
    PRESSURE GRADIENTS
    SOLUBILITY

LIQUID-SOLID INTERFACES
GS   BOUNDARIES
    . FLUID BOUNDARIES
    . . **LIQUID-SOLID INTERFACES**
    INTERFACES
    . FLUID BOUNDARIES
    . . **LIQUID-SOLID INTERFACES**
RT   BOUNDARY LAYERS
    FLUID FILMS
    FLUID-SOLID INTERACTIONS
    ∞FUSION
    HEAT TRANSFER
    INTERFACE STABILITY

LIQUID-SOLID INTERFACES-*(CONT.)*
    MELTING
    MENISCI
    METAL SURFACES
    PHASE CHANGE MATERIALS
    SOLID PHASES
    SOLID-SOLID INTERFACES
    SQUEEZE FILMS

LIQUID-VAPOR EQUILIBRIUM
UF   VAPOR LIQUID EQUILIBRIUM
RT   ∞EQUILIBRIUM
    THERMODYNAMIC EQUILIBRIUM
    VAPORS

LIQUID-VAPOR INTERFACES
GS   BOUNDARIES
    . FLUID BOUNDARIES
    . . **LIQUID-VAPOR INTERFACES**
    INTERFACES
    . FLUID BOUNDARIES
    . . **LIQUID-VAPOR INTERFACES**
RT   AIR WATER INTERACTIONS
    EVAPORATION
    FREE BOUNDARIES
    HEAT TRANSFER
    INTERFACE STABILITY
    MENISCI
    PRESSURE GRADIENTS
    SOLUBILITY
    VAPOR PHASES
    VAPOR PRESSURE

LIQUIDS
GS   **LIQUIDS**
    . CRYOGENIC FLUIDS
    . . FERMI LIQUIDS
    . . FLOX
    . . LIQUID HELIUM
    . . LIQUID HYDROGEN
    . . LIQUID NITROGEN
    . . LIQUID OXYGEN
    . FERROFLUIDS
    . HYDRAULIC FLUIDS
    . . SKYDROL (TRADEMARK)
    . JUICES
    . LIQUEFIED GASES
    . . LIQUID AMMONIA
    . . LIQUID HELIUM
    . . LIQUID HYDROGEN
    . . LIQUID NEON
    . . LIQUID NITROGEN
    . . LIQUID OXYGEN
    . LIQUID FUELS
    . LIQUID ROCKET PROPELLANTS
    . . CRYOGENIC ROCKET PROPELLANTS
    . . HYPERGOLIC ROCKET
        PROPELLANTS
    . . MONOPROPELLANTS
    . . . AEROZINE
    . . RP-1 ROCKET PROPELLANTS
    . . SLURRY PROPELLANTS
    . LIQUID METALS
    . . LIQUID LITHIUM
    . . LIQUID POTASSIUM
    . . LIQUID SODIUM
    . . MELTING
    . . MERCURY (METAL)
    . . . MERCURY VAPOR
    . LIQUID OXIDIZERS
    . ORGANIC LIQUIDS
    . POTABLE LIQUIDS
    . BEVERAGES
    . . . WINES
    . ROTATING LIQUIDS
RT   ∞FLUIDS
    GLOBULES
    GLYCEROLS
    LIQUID PHASES
    NONPOINT SOURCES
    PHASE DIAGRAMS
    VAPOR PHASES
    WATER

LIQUIDUS
RT   CRYSTALLIZATION
    LIQUID PHASES
    MELTING POINTS
    PHASE DIAGRAMS
    SOLID PHASES
    SOLID SOLUTIONS
    SOLIDUS

LIRTS (TELESCOPE)
UF   LARGE INFRARED TELESCOPE ON
    SPACELAB

LIRTS (TELESCOPE)-*(CONT.)*
GS   TELESCOPES
    . SPACEBORNE TELESCOPES
    . . **LIRTS (TELESCOPE)**
RT   EUROPEAN SPACE AGENCY
    PAYLOADS
    SPACE SHUTTLES
    SPACELAB

LISP (PROGRAMMING LANGUAGE)
GS   LANGUAGES
    . PROGRAMMING LANGUAGES
    . . **LISP (PROGRAMMING LANGUAGE)**
RT   COMPUTER PROGRAMMING
    RECURSIVE FUNCTIONS

LISSAJOUS FIGURES
RT   ECCENTRIC ORBITS
    EQUATIONS OF MOTION
    LIBRATION
    LUNAR ORBITS
    SATELLITE ORBITS

LISTS
GS   **LISTS**
    . HARDWARE UTILIZATION LISTS
RT   ∞CATALOGS
    DISPLAY DEVICES
    ENUMERATION
    INDEXES (DOCUMENTATION)
    PRINTOUTS

LITERATURE
GS   **LITERATURE**
    . BIOGRAPHY
    . DOCUMENTATION
RT   BIBLIOGRAPHIES
    DOCUMENTS
    INDEXES (DOCUMENTATION)
    KNOWLEDGE
    LIBRARIES
    PAPERS
    PHILOSOPHY

LITHERGOL ROCKET ENGINES
GS   ENGINES
    . ROCKET ENGINES
    . . HYBRID PROPELLANT ROCKET
        ENGINES
    . . . **LITHERGOL ROCKET ENGINES**

LITHERGOLIC PROPELLANTS
USE   HYBRID PROPELLANTS

LITHIASIS
GS   DISEASES
    . **LITHIASIS**
RT   CALCULI
    DENTAL CALCULI

LITHIUM
GS   CHEMICAL ELEMENTS
    . ALKALI METALS
    . . **LITHIUM**
    . . . LIQUID LITHIUM
    . . . LITHIUM ISOTOPES
    METALS
    . ALKALI METALS
    . . **LITHIUM**
    . . . LIQUID LITHIUM
    . . . LITHIUM ISOTOPES

LITHIUM ALLOYS
GS   ALLOYS
    . **LITHIUM ALLOYS**

LITHIUM ALUMINUM HYDRIDES
GS   ALUMINUM COMPOUNDS
    . **LITHIUM ALUMINUM HYDRIDES**
    HYDROGEN COMPOUNDS
    . HYDRIDES
    . . METAL HYDRIDES
    . . . LITHIUM HYDRIDES
    . . . . **LITHIUM ALUMINUM HYDRIDES**
    LITHIUM COMPOUNDS
    . **LITHIUM ALUMINUM HYDRIDES**
RT   POWDERED ALUMINUM

LITHIUM BORATES
GS   BORON COMPOUNDS
    . BORATES
    . . **LITHIUM BORATES**
    LITHIUM COMPOUNDS
    . **LITHIUM BORATES**

LITHIUM CHLORIDES
GS   HALOGEN COMPOUNDS
    . CHLORINE COMPOUNDS
    . . CHLORIDES
    . . . **LITHIUM CHLORIDES**
    . HALIDES
    . . CHLORIDES
    . . . **LITHIUM CHLORIDES**
    . . METAL HALIDES
    . . . **LITHIUM CHLORIDES**
    LITHIUM COMPOUNDS
    . **LITHIUM CHLORIDES**

LITHIUM COMPOUNDS
GS   **LITHIUM COMPOUNDS**
    . LITHIUM ALUMINUM HYDRIDES
    . LITHIUM BORATES
    . LITHIUM CHLORIDES
    . LITHIUM FLUORIDES
    . LITHIUM HYDRIDES
    . LITHIUM HYDROXIDES
    . LITHIUM IODATES
    . LITHIUM NIOBATES
    . LITHIUM OXIDES
    . LITHIUM PERCHLORATES
    . LITHIUM SULFATES
    . ORGANIC LITHIUM COMPOUNDS
    . SPODUMENE
RT   ∞ALKALI METAL COMPOUNDS
    ∞CHEMICAL COMPOUNDS
    ∞METAL COMPOUNDS
    METAL FUELS

LITHIUM COOLED REACTOR EXPERIMENT
UF   LCRE REACTOR
GS   NUCLEAR REACTORS
    . LIQUID COOLED REACTORS
    . . LIQUID METAL COOLED REACTORS
    . . . **LITHIUM COOLED REACTOR**
           EXPERIMENT

LITHIUM FLUORIDES
GS   HALOGEN COMPOUNDS
    . FLUORINE COMPOUNDS
    . . FLUORIDES
    . . . METAL FLUORIDES
    . . . . **LITHIUM FLUORIDES**
    . HALIDES
    . . METAL HALIDES
    . . . **LITHIUM FLUORIDES**
    LITHIUM COMPOUNDS
    . **LITHIUM FLUORIDES**

LITHIUM HYDRIDES
GS   HYDROGEN COMPOUNDS
    . HYDRIDES
    . . METAL HYDRIDES
    . . . **LITHIUM HYDRIDES**
    . . . . LITHIUM ALUMINUM HYDRIDES
    LITHIUM COMPOUNDS
    . **LITHIUM HYDRIDES**

LITHIUM HYDROXIDES
GS   ALKALIES
    . **LITHIUM HYDROXIDES**
    HYDROXIDES
    . **LITHIUM HYDROXIDES**
    LITHIUM COMPOUNDS
    . **LITHIUM HYDROXIDES**

LITHIUM IODATES
GS   HALOGEN COMPOUNDS
    . IODINE COMPOUNDS
    . . IODATES
    . . . **LITHIUM IODATES**
    LITHIUM COMPOUNDS
    . **LITHIUM IODATES**
RT   ∞METAL COMPOUNDS

LITHIUM ISOTOPES
UF   LITHIUM 4
    LITHIUM 6
GS   CHEMICAL ELEMENTS
    . ALKALI METALS
    . . LITHIUM
    . . . **LITHIUM ISOTOPES**
    . NUCLIDES
    . . ISOTOPES
    . . . **LITHIUM ISOTOPES**
    METALS
    . ALKALI METALS
    . . LITHIUM
    . . . **LITHIUM ISOTOPES**

LITHIUM NIOBATES
GS   LITHIUM COMPOUNDS
    . **LITHIUM NIOBATES**

LITHIUM OXIDES
GS   CHALCOGENIDES
    . OXIDES
    . . METAL OXIDES
    . . . **LITHIUM OXIDES**
    LITHIUM COMPOUNDS
    . **LITHIUM OXIDES**

LITHIUM PERCHLORATES
GS   HALOGEN COMPOUNDS
    . CHLORINE COMPOUNDS
    . . PERCHLORATES
    . . . **LITHIUM PERCHLORATES**
    LITHIUM COMPOUNDS
    . **LITHIUM PERCHLORATES**

LITHIUM SULFATES
GS   LITHIUM COMPOUNDS
    . **LITHIUM SULFATES**
    SULFUR COMPOUNDS
    . SULFATES
    . . **LITHIUM SULFATES**

LITHIUM SULFUR BATTERIES
GS   ELECTROCHEMICAL CELLS
    . **LITHIUM SULFUR BATTERIES**
RT   ∞CELLS
    ∞ELECTRIC CELLS
    ∞ENERGY SOURCES
    ∞POWER SUPPLIES

LITHIUM 4
USE   LITHIUM ISOTOPES

LITHIUM 6
USE   LITHIUM ISOTOPES

LITHOGRAPHY
GS   PRINTING
    . **LITHOGRAPHY**
    . . PHOTOLITHOGRAPHY
RT   PHOTOMECHANICAL EFFECT
    REPRODUCTION (COPYING)

LITHOLOGY
GS   GEOLOGY
    . **LITHOLOGY**
RT   REGOLITH
    ROCKS

LITHOSPHERE
GS   **LITHOSPHERE**
    . EARTH CORE
    . EARTH CRUST
    . EARTH MANTLE
    . EARTH SURFACE
RT   EARTH PLANETARY STRUCTURE
    PLANETARY MANTLES
    PLATES (TECTONICS)
    SUBDUCTION (GEOLOGY)

LITHUANIA
RT   EUROPE
    NATIONS

LITTLE JOE 2 LAUNCH VEHICLE
GS   LAUNCH VEHICLES
    . **LITTLE JOE 2 LAUNCH VEHICLE**
    ROCKET VEHICLES
    . MULTISTAGE ROCKET VEHICLES
    . . **LITTLE JOE 2 LAUNCH VEHICLE**
RT   ALGOL ENGINE
    MERCURY PROJECT
    SERGEANT MISSILES
    SOLID PROPELLANT ROCKET ENGINES
    TX-354 ENGINE
    XM-33 ENGINE

LITTLE JOHN ROCKET VEHICLE
GS   ROCKET VEHICLES
    . SINGLE STAGE ROCKET VEHICLES
    . . **LITTLE JOHN ROCKET VEHICLE**
    . SURFACE TO SURFACE ROCKETS
    . . **LITTLE JOHN ROCKET VEHICLE**
RT   HERCULES ENGINE
    SOLID PROPELLANT ROCKET ENGINES

LITTORAL CURRENTS
USE   COASTAL CURRENTS

LITTORAL DRIFT
RT   BARS (LANDFORMS)
    BEACHES
    BREAKWATERS
    COASTS
    OCEAN CURRENTS
    SANDS
    SEDIMENTS

LITTORAL TRANSPORT
RT   BREAKWATERS
    OCEAN CURRENTS
    SANDS
    WATER WAVES
    ∞WAVES

LIVER
GS   ANATOMY
    . ORGANS
    . . **LIVER**
    VISCERA
    . ORGANS
    . . **LIVER**
RT   GASTROINTESTINAL SYSTEM
    GLANDS (ANATOMY)
    HEPATITIS
    TYROSINE

LIVERMORE POOL TYPE REACTOR
UF   LPTR REACTOR
GS   NUCLEAR REACTORS
    . NUCLEAR RESEARCH AND TEST
        REACTORS
    . . **LIVERMORE POOL TYPE REACTOR**

LIVESTOCK
GS   ANIMALS
    . **LIVESTOCK**
RT   CALVES
    CATTLE
    DEER
    GOATS
    HORSES
    RANGELANDS
    SHEEP
    SWINE
    TURKEYS

LIXISCOPES
UF   LOW INTENSITY X RAY IMAGING SCOPE
GS   MEDICAL EQUIPMENT
    . X RAY APPARATUS
    . . **LIXISCOPES**
RT   PORTABLE EQUIPMENT
    RADIOGRAPHY
    X RAY ASTRONOMY
    X RAY IMAGERY

LIZARDS
GS   ANIMALS
    . POIKILOTHERMIA
    . . REPTILES
    . . . **LIZARDS**
    . VERTEBRATES
    . REPTILES
    . . . **LIZARDS**

LLANOS ORIENTALES (COLOMBIA)
GS   LAND
    . GRASSLANDS
    . . **LLANOS ORIENTALES (COLOMBIA)**
    . PLAINS
    . . **LLANOS ORIENTALES (COLOMBIA)**
RT   COLOMBIA

LMCR (REACTORS)
USE   LIQUID METAL COOLED REACTORS

LMFBR
USE   LIQUID METAL FAST BREEDER
        REACTORS

LNG
USE   LIQUEFIED NATURAL GAS

LOAD DISTRIBUTION (FORCES)
GS   DISTRIBUTION (PROPERTY)
    . **LOAD DISTRIBUTION (FORCES)**
RT   ∞DISTRIBUTION

LOAD FACTORS
USE   LOADS (FORCES)

LOAD TESTING MACHINES
RT   ∞ MACHINERY
      ∞ TEST EQUIPMENT

LOAD TESTS
RT   COMPRESSION TESTS
      CREEP TESTS
      DESTRUCTIVE TESTS
      FATIGUE TESTS
      IMPACT TESTS
      LOADING RATE
      NONDESTRUCTIVE TESTS
      SHOCK TESTS
      SPECIMEN GEOMETRY
      SPIN TESTS
      STATIC TESTS
      TENSILE TESTS
      ∞ TESTS

∞ **LOADING**
SN   (USE OF A MORE SPECIFIC TERM IS
      RECOMMENDED--CONSULT THE TERMS
      LISTED BELOW)
UF   DUMMY LOADS
RT   FEEDING (SUPPLYING)
      FILLING
      INPUT
      LOADING OPERATIONS
      LOADS (FORCES)
      PAYLOADS
      REFILLING
      REPLENISHMENT
      SHAFTS (MACHINE ELEMENTS)
      SWEEP EFFECT

LOADING FORCES
USE   LOADS (FORCES)

LOADING MOMENTS
GS   MOMENTS
      . LOADING MOMENTS
RT   AERODYNAMIC LOADS
      BENDING MOMENTS
      FLEXING
      LOADS (FORCES)
      MASS DISTRIBUTION
      MOMENT DISTRIBUTION
      PRESSURE DISTRIBUTION
      STATIC LOADS
      STRUCTURAL ANALYSIS
      TORQUE

LOADING OPERATIONS
RT   FEEDERS
      ∞ LOADING
      MATERIALS HANDLING
      ∞ OPERATIONS
      UNLOADING

LOADING RATE
GS   RATES (PER TIME)
      . LOADING RATE
RT   IMPACT LOADS
      LOAD TESTS
      LOADS (FORCES)
      STRAIN RATE
      VELOCITY

LOADING WAVES
USE   ELASTIC WAVES
       LOADS (FORCES)

LOADS (FORCES)
UF   LOAD FACTORS
      LOADING FORCES
      LOADING WAVES
GS   **LOADS (FORCES)**
      . AXIAL LOADS
      . . AXIAL COMPRESSION LOADS
      . COMPRESSION LOADS
      . . AXIAL COMPRESSION LOADS
      . . IMPACT LOADS
      . CRITICAL LOADING
      . DYNAMIC LOADS
      . . AERODYNAMIC LOADS
      . . . BLAST LOADS
      . . . GUST LOADS
      . . CYCLIC LOADS
      . . ROLLING CONTACT LOADS
      . . THRUST LOADS
      . . TRANSIENT LOADS
      . . . GUST LOADS
      . . . IMPACT LOADS
      . . LANDING LOADS
      . . SHOCK LOADS

LOADS (FORCES)-(CONT.)
      . . . . BLAST LOADS
      . . VIBRATORY LOADS
      . . WING LOADING
      . EDGE LOADING
      . RANDOM LOADS
      . . GUST LOADS
      . . STATIC LOADS
RT   BALLAST (MASS)
      ∞ EQUILIBRIUM
      ∞ FORCE
      FORCE DISTRIBUTION
      HUGONIOT EQUATION OF STATE
      ∞ LOADING
      LOADING MOMENTS
      LOADING RATE
      MASS DISTRIBUTION
      ∞ MECHANICS (PHYSICS)
      MOMENT DISTRIBUTION
      PAYLOADS
      PRESSURE
      PRESSURE DISTRIBUTION
      PRESSURE EFFECTS
      SHAFTS (MACHINE ELEMENTS)
      SHEARING
      STRESS CONCENTRATION
      STRESS INTENSITY FACTORS
      STRESSES
      STRUCTURAL DESIGN CRITERIA
      WEIGHT (MASS)
      WIND PRESSURE

∞ **LOBES**
SN   (USE OF A MORE SPECIFIC TERM IS
      RECOMMENDED--CONSULT THE TERMS
      LISTED BELOW)
RT   ANTENNA DESIGN
      ANTENNA RADIATION PATTERNS
      BACKLOBES
      SIDELOBES

LOCAL GROUP (ASTRONOMY)
GS   CELESTIAL BODIES
      . GALAXIES
      . . GALACTIC CLUSTERS
      . . . LOCAL GROUP (ASTRONOMY)
      . . . . ANDROMEDA GALAXIES
RT   BARRED GALAXIES
      COSMOLOGY
      DISK GALAXIES
      DWARF GALAXIES
      ELLIPTICAL GALAXIES
      SPIRAL GALAXIES
      VIRGO GALACTIC CLUSTER

LOCAL SCIENTIFIC SURVEY MODULE
GS   MODULES
      . LOCAL SCIENTIFIC SURVEY MODULE
RT   INSTRUMENT PACKAGES
      LUNAR EXPLORATION
      MEASURING INSTRUMENTS

LOCALIZATION
USE   POSITION (LOCATION)

LOCATES SYSTEM
UF   LOCATION OF AIR TRAFFIC SATELLITES
RT   AIR TRAFFIC CONTROL
      BEACON SATELLITES
      NAVIGATION SATELLITES
      SATELLITE GUIDANCE
      ∞ SYSTEMS

LOCATION
USE   POSITION (LOCATION)

LOCATION OF AIR TRAFFIC SATELLITES
USE   LOCATES SYSTEM

LOCI
GS   GEOMETRY
      . EUCLIDEAN GEOMETRY
      . . ANALYTIC GEOMETRY
      . . . LOCI
RT   ∞ CENTERS
      CONICS
      FOCI
      LINE OF SIGHT
      POINTS (MATHEMATICS)
      RESOLUTION

LOCKHEED AIRCRAFT
GS   **LOCKHEED AIRCRAFT**
      . C-5 AIRCRAFT

LOCKHEED AIRCRAFT-(CONT.)
      . C-121 AIRCRAFT
      . C-130 AIRCRAFT
      . C-140 AIRCRAFT
      . C-141 AIRCRAFT
      . CL-823 AIRCRAFT
      . EC-121 AIRCRAFT
      . ELECTRA AIRCRAFT
      . F-94 AIRCRAFT
      . F-104 AIRCRAFT
      . L-1011 AIRCRAFT
      . L-2000 AIRCRAFT
      . LOCKHEED MODEL 18 AIRCRAFT
      . P-3 AIRCRAFT
      . T-33 AIRCRAFT
      . U-2 AIRCRAFT
      . XH-51 HELICOPTER
      . XV-4 AIRCRAFT
RT   ∞ AIRCRAFT

LOCKHEED C-5 AIRCRAFT
USE   C-5 AIRCRAFT

LOCKHEED CL-595 HELICOPTER
USE   XH-51 HELICOPTER

LOCKHEED CL-823 AIRCRAFT
USE   CL-823 AIRCRAFT

LOCKHEED CONSTELLATION AIRCRAFT
USE   C-121 AIRCRAFT

LOCKHEED L-2000 AIRCRAFT
USE   L-2000 AIRCRAFT

LOCKHEED MODEL 18 AIRCRAFT
GS   LOCKHEED AIRCRAFT
      . LOCKHEED MODEL 18 AIRCRAFT
      MONOPLANES
      . LOCKHEED MODEL 18 AIRCRAFT
      TRANSPORT AIRCRAFT
      . LOCKHEED MODEL 18 AIRCRAFT
RT   ∞ AIRCRAFT

LOCKHEED U-2 AIRCRAFT
USE   U-2 AIRCRAFT

LOCKHEED XV-4A AIRCRAFT
USE   XV-4 AIRCRAFT

LOCKHEED 186 HELICOPTER
USE   XH-51 HELICOPTER

LOCKING
UF   INTERLOCKING
GS   **LOCKING**
      . LASER MODE LOCKING
RT   FASTENERS
      ∞ JOINING
      LOCKS (FASTENERS)
      RETAINING

∞ **LOCKS**
SN   (USE OF A MORE SPECIFIC TERM IS
      RECOMMENDED--CONSULT THE TERMS
      LISTED BELOW)
RT   AIR LOCKS
      LOCKS (FASTENERS)

LOCKS (FASTENERS)
GS   FASTENERS
      . LOCKS (FASTENERS)
RT   LOCKING
      ∞ LOCKS

LOCOMOTION
UF   MOTILITY
GS   **LOCOMOTION**
      . ASTRONAUT LOCOMOTION
      . WALKING
RT   EXERCISE PHYSIOLOGY
      NAVIGATION
      PHYSIOLOGY
      PROPULSION
      WHEELCHAIRS

LOCOMOTIVES
RT   DIESEL ENGINES
      HANDLING EQUIPMENT
      RAIL TRANSPORTATION
      WINDSHIELDS

LOCUSTS
GS   ANIMALS

LOCUSTS-*(CONT.)*
    . INVERTEBRATES
    . . ARTHROPODS
    . . . INSECTS
    . . . . **LOCUSTS**
RT   FARM CROPS
      FOLIAGE
      INFESTATION
      VEGETATION

LOFAR
RT   SONAR
      UNDERWATER ACOUSTICS

LOFTI SATELLITES
USE LOW FREQUENCY TRANSIONOSPHERIC
      SATELLITES

LOFTING
RT   AIRCRAFT DESIGN
      ASCENT TRAJECTORIES
      COMPUTER AIDED DESIGN
      DIFFERENTIAL GEOMETRY
      ENGINEERING DRAWINGS
      MATHEMATICAL MODELS
      SPACECRAFT DESIGN
      STRUCTURAL DESIGN
      ∞ SURFACE GEOMETRY
      TEMPLATES

LOG PERIODIC ANTENNAS
GS   ANTENNAS
      . DIRECTIONAL ANTENNAS
      . . **LOG PERIODIC ANTENNAS**
RT   ANTENNA ARRAYS
      ANTENNA DESIGN
      BROADBAND
      DIPOLE ANTENNAS
      ∞ FREQUENCY RESPONSE

LOG SPIRAL ANTENNAS
GS   ANTENNAS
      . SPIRAL ANTENNAS
      . . **LOG SPIRAL ANTENNAS**
RT   DIPOLE ANTENNAS

LOGARITHMIC RECEIVERS
GS   RECEIVERS
      . **LOGARITHMIC RECEIVERS**
RT   COMMUNICATION EQUIPMENT
      ∞ FREQUENCY RESPONSE
      INTERMEDIATE FREQUENCY AMPLIFIERS
      TRANSFER FUNCTIONS

LOGARITHMS
GS   ANALYSIS (MATHEMATICS)
      . COMPLEX VARIABLES
      . . EXPONENTIAL FUNCTIONS
      . . . **LOGARITHMS**
      FUNCTIONS (MATHEMATICS)
      . TRANSCENDENTAL FUNCTIONS
      . EXPONENTIAL FUNCTIONS
      . . . **LOGARITHMS**
RT   EXPONENTS

LOGGING (INDUSTRY)
RT   FORESTS
      TREES (PLANTS)

∞ **LOGIC**
SN   *(USE OF A MORE SPECIFIC TERM IS*
      RECOMMENDED--CONSULT THE TERMS
      LISTED BELOW)
RT   ARTIFICIAL INTELLIGENCE
      AXIOMS
      BOOLEAN ALGEBRA
      BRANCHING (MATHEMATICS)
      COMPLEMENTS (MATHEMATICS)
      EXPERT SYSTEMS
      FLUID LOGIC
      FLUIDICS
      FORMALISM
      INFORMATION THEORY
      LOGIC CIRCUITS
      LOGIC DESIGN
      MATHEMATICAL LOGIC
      PARADOXES
      PHILOSOPHY
      ∞ PRINCIPLES
      THRESHOLD LOGIC
      TRANSISTOR LOGIC

LOGIC CIRCUITS
UF   LOGIC NETWORKS

LOGIC CIRCUITS-*(CONT.)*
GS   CIRCUITS
      . **LOGIC CIRCUITS**
      . . THRESHOLD GATES
RT   ADDING CIRCUITS
      ARCHITECTURE (COMPUTERS)
      CENTRAL PROCESSING UNITS
      COMPUTERS
      COUNTING CIRCUITS
      DECISIONS
      DIGITAL COMPUTERS
      FLUID LOGIC
      GATES (CIRCUITS)
      ∞ LOGIC
      LOGICAL ELEMENTS
      MATRICES (CIRCUITS)
      MULTIPLIERS
      MULTIVIBRATORS
      NEURAL NETS
      ∞ RELAY
      SWITCHING CIRCUITS
      THRESHOLD LOGIC
      TRANSISTOR CIRCUITS
      TRANSISTOR LOGIC

LOGIC DESIGN
RT   AMPLIFIER DESIGN
      ARCHITECTURE (COMPUTERS)
      COMPUTER AIDED DESIGN
      COMPUTER DESIGN
      COMPUTER PROGRAMMING
      ∞ DESIGN
      DESIGN ANALYSIS
      ∞ LOGIC
      LOGIC PROGRAMMING
      LOGICAL ELEMENTS
      SWITCHING THEORY
      TRANSISTOR LOGIC

LOGIC NETWORKS
USE LOGIC CIRCUITS

LOGIC PROGRAMMING
GS   SOFTWARE ENGINEERING
      . COMPUTER PROGRAMMING
      . . **LOGIC PROGRAMMING**
RT   ARTIFICIAL INTELLIGENCE
      EXPERT SYSTEMS
      LOGIC DESIGN

LOGICAL ELEMENTS
UF   DECISION ELEMENTS
RT   COMPUTER COMPONENTS
      ∞ ELEMENTS
      GATES (CIRCUITS)
      LOGIC CIRCUITS
      LOGIC DESIGN

LOGISTICS
GS   **LOGISTICS**
      . LUNAR LOGISTICS
      . SPACE LOGISTICS
RT   AIRCRAFT MAINTENANCE
      ARMY-NAVY INSTRUMENTATION
         PROGRAM
      COMMAND AND CONTROL
      CONSUMABLES (SPACECRAFT)
      CRANES
      DEPLOYMENT
      DOWNTIME
      ∞ ELECTRIC EQUIPMENT
      ENERGY POLICY
      ∞ FACILITIES
      INVENTORY MANAGEMENT
      MAINTENANCE
      MATRIX MANAGEMENT
      PORTABLE EQUIPMENT
      RAPID TRANSIT SYSTEMS
      RESOURCE ALLOCATION
      RESOURCES
      SERVICES
      SHIPYARDS
      SITE SELECTION
      STOCKPILING
      ∞ STORAGE
      STOWAGE (ONBOARD EQUIPMENT)
      TRANSPORTATION
      ∞ TRAVEL
      UTILITIES

LOGISTICS MANAGEMENT
GS   MANAGEMENT
      . **LOGISTICS MANAGEMENT**
      . . INVENTORY MANAGEMENT
      . . . INVENTORY CONTROLS

LOGISTICS MANAGEMENT-*(CONT.)*
RT   ∞ FACILITIES
      MAINTENANCE
      RESOURCES
      SERVICES
      SPARE PARTS
      ∞ STORAGE

LOGISTICS OVER THE SHORE (LOTS) CARRIER
RT   MILITARY TECHNOLOGY

LOH HELICOPTER
USE OH-6 HELICOPTER

LOKI ROCKET VEHICLE
GS   ROCKET VEHICLES
      . SINGLE STAGE ROCKET VEHICLES
      . . **LOKI ROCKET VEHICLE**
      . SOUNDING ROCKETS
      . . **LOKI ROCKET VEHICLE**
RT   SOLID PROPELLANT ROCKET ENGINES
      WASP SOUNDING ROCKET

LOLA (SIMULATOR)
USE LUNAR ORBIT AND LANDING
      SIMULATORS

LOMONOSOV CURRENT
GS   CIRCULATION
      . WATER CIRCULATION
      . . WATER CURRENTS
      . . . OCEAN CURRENTS
      . . . . **LOMONOSOV CURRENT**
RT   ATLANTIC OCEAN
      GULF STREAM
      TROPICAL REGIONS

LONG DURATION EXPOSURE FACILITY
UF   LDEF
GS   LABORATORIES
      . **LONG DURATION EXPOSURE FACILITY**
      MANNED SPACECRAFT
      . SPACE STATIONS
      . . ORBITAL SPACE STATIONS
      . . . **LONG DURATION EXPOSURE**
          FACILITY
      SATELLITES
      . ARTIFICIAL SATELLITES
      . . ORBITAL SPACE STATIONS
      . . . **LONG DURATION EXPOSURE**
          FACILITY
      STATIONS
      . SPACE STATIONS
      . . ORBITAL SPACE STATIONS
      . . . **LONG DURATION EXPOSURE**
          FACILITY
RT   SPACE LABORATORIES

LONG DURATION SPACE FLIGHT
UF   EXTENDED DURATION SPACE FLIGHT
GS   SPACE FLIGHT
      . INTERPLANETARY FLIGHT
      . . **LONG DURATION SPACE FLIGHT**
RT   DEEP SPACE
      EXTRATERRESTRIAL ENVIRONMENTS
      ∞ FLIGHT
      FLYBY MISSIONS
      INTERSTELLAR TRAVEL
      MANNED SPACE FLIGHT
      ∞ MISSIONS
      PLANETARY ENVIRONMENTS
      SPACE ADAPTATION SYNDROME

LONG ISLAND (NY)
GS   LANDFORMS
      . ISLANDS
      . . **LONG ISLAND (NY)**
RT   ATLANTIC OCEAN
      NEW YORK

LONG RANGE NAVIGATION
USE LORAN
      LORAN D

LONG RANGE WEATHER FORECASTING
GS   FORECASTING
      . WEATHER FORECASTING
      . . **LONG RANGE WEATHER**
          FORECASTING
      METEOROLOGY
      . WEATHER FORECASTING
      . . **LONG RANGE WEATHER**
          FORECASTING
RT   NUMERICAL WEATHER FORECASTING

LONG RANGE WEATHER-*(CONT.)*
    STATISTICAL WEATHER FORECASTING

LONG TERM EFFECTS
UF	SECULAR PERTURBATION
RT	CELESTIAL MECHANICS
	CLIMATE
	CLOSED ECOLOGICAL SYSTEMS
	CYCLES
	DURABILITY
	∞EFFECTS
	LIFE (DURABILITY)
	LIFE SUPPORT SYSTEMS
	ORBIT PERTURBATION
	∞PERFORMANCE
	PERIODIC VARIATIONS
	PERTURBATION
	STORAGE STABILITY
	TIME TEMPERATURE PARAMETER
	WEATHER

LONG TERM ZONAL EARTH ENERGY EXPERIMENT
USE	LZEEBE SATELLITE

LONG WAVE RADIATION
GS	ELECTROMAGNETIC RADIATION
	. RADIO WAVES
	. . **LONG WAVE RADIATION**
RT	FAR INFRARED RADIATION
	MONOCHROMATIC RADIATION
	∞RADIATION
	SHORT WAVE RADIATION
	SOLAR RADIATION

LONG WAVES (METEOROLOGY)
USE	PLANETARY WAVES

LONGERONS
UF	ASTROMASTS
GS	STRUCTURAL MEMBERS
	. **LONGERONS**
RT	KEELS
	REINFORCEMENT (STRUCTURES)
	RIBS (SUPPORTS)
	STRAKES
	STRINGERS
	STRUCTURAL STABILITY

∞ **LONGEVITY**
SN	*(USE OF A MORE SPECIFIC TERM IS*
	RECOMMENDED--CONSULT THE TERMS
	LISTED BELOW)
RT	LIFE (DURABILITY)
	LIFE SPAN

LONGITUDE
GS	**LONGITUDE**
	. SOLAR LONGITUDE
RT	COORDINATES
	GEODETIC COORDINATES
	LATITUDE
	POSITION (LOCATION)

LONGITUDE MEASUREMENT
RT	LATITUDE MEASUREMENT
	∞MEASUREMENT
	NAVIGATION
	POSITIONING

LONGITUDINAL CONTROL
UF	PITCH ATTITUDE CONTROL
GS	ATTITUDE CONTROL
	. **LONGITUDINAL CONTROL**
RT	AIRCRAFT CONTROL
	ALTITUDE CONTROL
	AUTOMATIC CONTROL
	∞CONTROL
	DIRECTIONAL CONTROL
	HELICOPTER CONTROL
	LATERAL CONTROL
	MANUAL CONTROL
	MISSILE CONTROL
	PILOT INDUCED OSCILLATION
	PITCH (INCLINATION)
	SATELLITE ATTITUDE CONTROL
	SATELLITE CONTROL

LONGITUDINAL STABILITY
GS	DYNAMIC CHARACTERISTICS
	. DYNAMIC STABILITY
	. . MOTION STABILITY
	. . . ATTITUDE STABILITY
	. . . . **LONGITUDINAL STABILITY**

LONGITUDINAL STABILITY-*(CONT.)*
	STABILITY
	. DYNAMIC STABILITY
	. . MOTION STABILITY
	. . . ATTITUDE STABILITY
	. . . . **LONGITUDINAL STABILITY**
RT	AERODYNAMIC STABILITY
	AIRCRAFT STABILITY
	DIRECTIONAL STABILITY
	FLOW STABILITY
	HOVERING STABILITY
	LATERAL STABILITY
	PITCH (INCLINATION)
	PITCHING MOMENTS
	POGO EFFECTS
	ROTARY STABILITY
	SPACECRAFT STABILITY

LONGITUDINAL WAVES
GS	**LONGITUDINAL WAVES**
	. PLANE WAVES
RT	BEAMS (RADIATION)
	DILATATIONAL WAVES
	ELASTIC WAVES
	ELECTROSTATIC WAVES
	FREQUENCIES
	NORMAL SHOCK WAVES
	∞RADIATION
	SEISMIC WAVES
	SHOCK WAVES
	SOLAR RADIATION
	SOUND WAVES
	TRANSVERSE WAVES
	WAVE PACKETS
	WAVELENGTHS
	∞WAVES

LONGSHORE CURRENTS
USE	COASTAL CURRENTS

LOOK ANGLES (ELECTRONICS)
GS	GEOMETRY
	. EUCLIDEAN GEOMETRY
	. . ANGLES (GEOMETRY)
	. . . **LOOK ANGLES (ELECTRONICS)**
RT	ALIGNMENT
	DIRECTIVITY
	INSTALLING
	INSTRUMENT ORIENTATION
	OPTICAL EQUIPMENT
	POSITIONING
	RADAR EQUIPMENT

LOOK ANGLES (TRACKING)
GS	GEOMETRY
	. EUCLIDEAN GEOMETRY
	. . ANGLES (GEOMETRY)
	. . . **LOOK ANGLES (TRACKING)**
RT	AZIMUTH
	ELEVATION ANGLE
	FIELD OF VIEW

LOOP ANTENNAS
GS	ANTENNAS
	. DIRECTIONAL ANTENNAS
	. . **LOOP ANTENNAS**
RT	AIRCRAFT ANTENNAS
	LOOPS
	MONOPOLE ANTENNAS

LOOPS
GS	**LOOPS**
	. CORROSION TEST LOOPS
RT	CIRCUITS
	CLOSED CYCLES
	LOOP ANTENNAS
	TORUSES
	TRUSSES

LOR (RENDEZVOUS)
USE	LUNAR ORBITAL RENDEZVOUS

LORAC NAVIGATION SYSTEM
GS	NAVIGATION
	. RADIO NAVIGATION
	. . HYPERBOLIC NAVIGATION
	. . . **LORAC NAVIGATION SYSTEM**
RT	DISTANCE MEASURING EQUIPMENT
	NAVIGATION AIDS
	NAVIGATION INSTRUMENTS
	SURFACE NAVIGATION
	∞SYSTEMS

LORAN
UF	LONG RANGE NAVIGATION
GS	NAVIGATION
	. RADIO NAVIGATION
	. . HYPERBOLIC NAVIGATION
	. . . **LORAN**
	. . . . LORAN C
	. . . . LORAN D
RT	AIR NAVIGATION
	DECCA NAVIGATION
	DISTANCE MEASURING EQUIPMENT
	NAVIGATION AIDS
	NAVIGATION INSTRUMENTS
	POLAR NAVIGATION
	SOLAR COMPASSES
	SURFACE NAVIGATION
	SURVEYS

LORAN C
GS	NAVIGATION
	. RADIO NAVIGATION
	. . HYPERBOLIC NAVIGATION
	. . . LORAN
	. . . . **LORAN C**
RT	AIR NAVIGATION
	DECCA NAVIGATION
	NAVIGATION AIDS

LORAN D
UF	LONG RANGE NAVIGATION
GS	NAVIGATION
	. RADIO NAVIGATION
	. . HYPERBOLIC NAVIGATION
	. . . LORAN
	. . . . **LORAN D**
RT	AIR NAVIGATION
	DECCA NAVIGATION
	NAVIGATION AIDS

LORENTZ CONTRACTION
UF	FITZGERALD-LORENTZ CONTRACTION
RT	RELATIVITY

LORENTZ FORCE
RT	CHARGED PARTICLES
	∞FORCE
	MAGNETIC FIELDS
	PONDEROMOTIVE FORCES

LORENTZ GAS
GS	GASES
	. **LORENTZ GAS**
RT	GAS DYNAMICS
	IONIZED GASES
	KINETIC THEORY

LORENTZ TRANSFORMATIONS
GS	FUNCTIONS (MATHEMATICS)
	. **LORENTZ TRANSFORMATIONS**
RT	DIRAC EQUATION
	INVARIANCE
	MANDELSTAM REPRESENTATION

LORV
USE	LOW OBSERVABLE REENTRY VEHICLES

LOS ALAMOS MOLTEN PLUTONIUM REACTOR
GS	NUCLEAR REACTORS
	. LIQUID COOLED REACTORS
	. . LIQUID METAL COOLED REACTORS
	. . . **LOS ALAMOS MOLTEN PLUTONIUM REACTOR**
	. NUCLEAR RESEARCH AND TEST REACTORS
	. . **LOS ALAMOS MOLTEN PLUTONIUM REACTOR**

LOS ALAMOS TURRET REACTOR
USE	HIGH TEMPERATURE NUCLEAR REACTORS

LOS ALAMOS WATER BOILER REACTOR
GS	NUCLEAR REACTORS
	. LIQUID COOLED REACTORS
	. . WATER COOLED REACTORS
	. . . BOILING WATER REACTORS
	. . . . **LOS ALAMOS WATER BOILER REACTOR**

LOSS OF COOLANT
UF	COOLANT LOSS
GS	ACCIDENTS
	. **LOSS OF COOLANT**
RT	COOLANTS

LOSS OF COOLANT-(CONT.)
    LEAKAGE
    LOSSES
    NUCLEAR REACTORS
    REACTOR MATERIALS

LOSSES
  RT   AUDITORY DEFECTS
        COMMERCE
        DAMAGE
        DEPLETION
        EDDY CURRENTS
        ENERGY DISSIPATION
        IMPAIRMENT
        INSERTION LOSS
        LEAKAGE
        LEGAL LIABILITY
        LIABILITIES
        LOSS OF COOLANT
        OHMIC DISSIPATION
        PLASMA LOSS
        SEEPAGE
        TRANSMISSION LOSS
        WASTES
        WATER LOSS
        YIELD

LOSSLESS EQUIPMENT
  RT   LOSSLESS MATERIALS

LOSSLESS MATERIALS
  SN   (DIELECTRIC MATERIALS THAT DO NOT
        DISSIPATE ENERGY OR THAT DO NOT
        DAMPEN OSCILLATIONS)
  GS   DIELECTRICS
        . **LOSSLESS MATERIALS**
  RT   LOSSLESS EQUIPMENT
        ∞MATERIALS

LOSSY MEDIA
  RT   ELECTROMAGNETIC WAVE
          TRANSMISSION
        IONOSPHERIC PROPAGATION
        TRANSMISSION LOSS
        WAVE PROPAGATION

LOST WAX PROCESS
  USE   INVESTMENT CASTING

LOTS CARGO SHIPS
  USE   CARGO SHIPS

LOUDNESS
  RT   ACOUSTICS
        EFFECTIVE PERCEIVED NOISE LEVELS
        FLUX DENSITY
        HEARING
        ∞INTENSITY
        LEVEL (QUANTITY)
        NOISE (SOUND)
        NOISE MEASUREMENT
        NOISE REDUCTION
        POWER SPECTRA
        SOUND INTENSITY
        SOUND PRESSURE
        SOUND WAVES

LOUDSPEAKERS
  GS   AUDIO EQUIPMENT
        . **LOUDSPEAKERS**
        TRANSDUCERS
        . SOUND TRANSDUCERS
        . . ELECTROACOUSTIC TRANSDUCERS
        . . . **LOUDSPEAKERS**
  RT   MONAURAL SIGNALS
        RADIO RECEIVERS
        SOUND GENERATORS

LOUISIANA
  GS   NATIONS
        . UNITED STATES
        . . **LOUISIANA**
  RT   ATCHAFALAYA RIVER BASIN (LA)
        GULF OF MEXICO
        LAKE PONTCHARTRAIN (LA)
        MISSISSIPPI DELTA (LA)

∞ **LOUNGES**
  SN   *(USE OF A MORE SPECIFIC TERM IS*
        RECOMMENDED--CONSULT THE TERMS
        LISTED BELOW)
  RT   MOBILE LOUNGES
        ROOMS
        SEATS

LOUVERS
  RT   APERTURES
        BAFFLES
        ∞DIFFUSERS
        ∞SCREENING
        SHADES
        SHIELDING
        ∞SHUTTERS
        SLOTS
        VENTS

LOVE WAVES
  GS   ELASTIC WAVES
        . SEISMIC WAVES
        . . **LOVE WAVES**
  RT   SURFACE WAVES

LOW ALLOY STEELS
  USE   HIGH STRENGTH STEELS

LOW ALTITUDE
  GS   ALTITUDE
        . **LOW ALTITUDE**
  RT   ELEVATION
        LOWER ATMOSPHERE
        MIDALTITUDE
        NAP-OF-THE-EARTH NAVIGATION

LOW ASPECT RATIO
  GS   RATIOS
        . ASPECT RATIO
        . . **LOW ASPECT RATIO**

LOW ASPECT RATIO WINGS
  UF   DIAMOND WINGS
  GS   AIRFOILS
        . WINGS
        . . **LOW ASPECT RATIO WINGS**
        . . . DELTA WINGS
        . . . TRAPEZOIDAL WINGS
  RT   CRUCIFORM WINGS
        FIXED WINGS
        RIGID WINGS
        WING PLANFORMS

LOW CARBON STEELS
  GS   ALLOYS
        . IRON ALLOYS
        . . STEELS
        . . . CARBON STEELS
        . . . . **LOW CARBON STEELS**
  RT   IRON

LOW CONCENTRATIONS
  GS   COMPOSITION (PROPERTY)
        . CONCENTRATION (COMPOSITION)
        . . **LOW CONCENTRATIONS**
  RT   DILUTION

LOW CONDUCTIVITY
  RT   ELECTRIC CURRENT
        ELECTRICAL RESISTIVITY

LOW COST
  GS   COSTS
        . **LOW COST**
  RT   ECONOMY

LOW CURRENTS
  GS   ELECTRIC CURRENT
        . **LOW CURRENTS**
  RT   LOW VOLTAGE
        PLASMA CURRENTS

LOW DENSITY FLOW
  RT   ∞FLOW
        FLUID DYNAMICS
        MOLECULAR FLOW
        RAREFIED GAS DYNAMICS
        RAREFIED GASES

LOW DENSITY GASES
  USE   RAREFIED GASES

LOW DENSITY MATERIALS
  RT   ABSORBENTS
        ABSORBERS (MATERIALS)
        FOAMS
        GRANULAR MATERIALS
        HONEYCOMB CORES
        HONEYCOMB STRUCTURES
        LIGHT ELEMENTS
        ∞MATERIALS
        POLYURETHANE FOAM

LOW DENSITY MATERIALS-(CONT.)
        POROUS MATERIALS
        POROUS PLATES
        POWDER METALLURGY

LOW DENSITY RESEARCH
  GS   RESEARCH
        . **LOW DENSITY RESEARCH**
  RT   BLOWDOWN WIND TUNNELS
        COLLISIONLESS PLASMAS
        COMPOSITE MATERIALS
        EPOXY MATRIX COMPOSITES
        NONUNIFORM PLASMAS
        PLASMAS (PHYSICS)
        RAREFIED GASES
        SHOCK TUBES
        SHOCK TUNNELS
        SHOCK WAVE LUMINESCENCE
        ULTRAHIGH VACUUM
        VACUUM APPARATUS

LOW DENSITY WIND TUNNELS
  GS   TEST FACILITIES
        . WIND TUNNELS
        . . **LOW DENSITY WIND TUNNELS**
  RT   HYPERSONIC WIND TUNNELS
        HYPERVELOCITY WIND TUNNELS
        PLASMA JETS
        RAREFIED GAS DYNAMICS
        SHOCK TUBES
        SHOCK TUNNELS
        SLIP FLOW
        SUPERSONIC WIND TUNNELS

        †

LOW FREQUENCIES
  GS   FREQUENCIES
        . RADIO FREQUENCIES
        . . **LOW FREQUENCIES**
        . . . SUBAUDIBLE FREQUENCIES
        . . . . VERY LOW FREQUENCIES
  RT   EXTREMELY LOW FREQUENCIES
        INTERMEDIATE FREQUENCIES

LOW FREQUENCY BANDS
  GS   FREQUENCIES
        . RADIO FREQUENCIES
        . . **LOW FREQUENCY BANDS**
        . . . VERY LOW FREQUENCIES
  RT   ∞BANDS
        HIGH FREQUENCIES
        ULTRAHIGH FREQUENCIES
        VERY HIGH FREQUENCIES

**LOW FREQUENCY TRANSIONOSPHERIC
SATELLITES**
  UF   LOFTI SATELLITES
  GS   SATELLITES
        . ARTIFICIAL SATELLITES
        . . COMMUNICATION SATELLITES
        . . . **LOW FREQUENCY
             TRANSIONOSPHERIC
             SATELLITES**

LOW GRAVITY
  USE   REDUCED GRAVITY

LOW GRAVITY MANUFACTURING
  GS   MANUFACTURING
        . **LOW GRAVITY MANUFACTURING**
  RT   CONTAINERLESS MELTS
        DROP TOWERS
        FABRICATION
        LEVITATION MELTING
        MARANGONI CONVECTION
        METAL FOAMS
        ∞MICROGRAVITY APPLICATIONS
        REDUCED GRAVITY
        SPACE MANUFACTURING
        SPACE PROCESSING
        SPACE TOOLS
        TECHNOLOGIES

LOW INTENSITY X RAY IMAGING SCOPE
  USE   LIXISCOPES

LOW LATITUDES
  USE   TROPICAL REGIONS

LOW LEVEL TURBULENCE
  GS   TURBULENCE
        . ATMOSPHERIC TURBULENCE
        . . **LOW LEVEL TURBULENCE**
  RT   HOMOGENEOUS TURBULENCE

LOW MASS
USE   MASS

LOW MOLECULAR WEIGHTS
GS   MOLECULAR WEIGHT
    . **LOW MOLECULAR WEIGHTS**
RT   DIATOMIC MOLECULES
    MOLECULES
    MONATOMIC MOLECULES
    WEIGHT (MASS)

LOW NOISE
RT   PREAMPLIFIERS
    SIGNAL TO NOISE RATIOS

LOW OBSERVABLE REENTRY VEHICLES
UF   LORV
GS   REENTRY VEHICLES
    . **LOW OBSERVABLE REENTRY**
      VEHICLES
RT   RADAR CROSS SECTIONS
    REENTRY
    REENTRY PHYSICS
   ∞VEHICLES

LOW PASS FILTERS
RT   BANDSTOP FILTERS
    ELECTRIC FILTERS
    ELECTROMAGNETIC WAVE FILTERS
   ∞FILTERS
    MICROWAVE FILTERS
    OPTICAL FILTERS

LOW PRESSURE
GS   PRESSURE
    . **LOW PRESSURE**
    . . HIGH ALTITUDE PRESSURE
RT   ALTITUDE TOLERANCE
    CYCLONES
   ∞DEPRESSION
    HIGH ALTITUDE ENVIRONMENTS
    HIGH PRESSURE
    HYPOBARIC ATMOSPHERES
    TROUGHS
    VACUUM

LOW PRESSURE CHAMBERS
USE   VACUUM CHAMBERS

∞ **LOW RESISTANCE**
SN   *(USE OF A MORE SPECIFIC TERM IS*
    RECOMMENDED--CONSULT THE TERMS
    LISTED BELOW)
RT   CHEMICAL PROPERTIES
    ELECTRICAL RESISTANCE
    FLOW RESISTANCE
    MECHANICAL PROPERTIES
   ∞RESISTANCE
    THERMAL RESISTANCE

LOW REYNOLDS NUMBER
SN   (RN BELOW 2,000)
GS   RATIOS
    . DIMENSIONLESS NUMBERS
    . . REYNOLDS NUMBER
    . . . LOW REYNOLDS NUMBER
RT   HIGH REYNOLDS NUMBER

LOW SPEED
UF   LOW VELOCITY
GS   RATES (PER TIME)
    . **LOW SPEED**
    VELOCITY
    . **LOW SPEED**
RT   AIRSPEED
    FLOW VELOCITY
    GROUND SPEED
    LANDING SPEED
    SUBSONIC SPEED

LOW SPEED STABILITY
GS   DYNAMIC CHARACTERISTICS
    . DYNAMIC STABILITY
    . . MOTION STABILITY
    . . . **LOW SPEED STABILITY**
    STABILITY
    . DYNAMIC STABILITY
    . . MOTION STABILITY
    . . . **LOW SPEED STABILITY**
RT   AERODYNAMIC STABILITY
    AERODYNAMIC STALLING
    AIRCRAFT STABILITY

LOW SPEED STABILITY *-(CONT.)*
    ATTITUDE STABILITY
    CONTROLLABILITY
    DYNAMIC TESTS
    FLIGHT CHARACTERISTICS
    FLOW STABILITY
    HOVERING STABILITY
    SPACECRAFT STABILITY

LOW SPEED WIND TUNNELS
GS   TEST FACILITIES
    . WIND TUNNELS
    . . **LOW SPEED WIND TUNNELS**
    . . . SUBSONIC WIND TUNNELS
RT   BLOWDOWN WIND TUNNELS

LOW TEMPERATURE
GS   TEMPERATURE
    . **LOW TEMPERATURE**
    . . ULTRALOW TEMPERATURES
RT   BAY ICE
    COOLING
    CRYOGENICS
    FREEZING
    FROST
    FROST DAMAGE
    ICE FORMATION
    MAGNETIC COOLING
    PRESSURE ICE
    REFRIGERATING

LOW TEMPERATURE BRAZING
GS   WELDING
    . LASER WELDING
    . . FUSION WELDING
    . . . GAS WELDING
    . . . . BRAZING
    . . . . . **LOW TEMPERATURE BRAZING**
RT   SOLDERING

LOW TEMPERATURE ENVIRONMENTS
GS   ENVIRONMENTS
    . **LOW TEMPERATURE ENVIRONMENTS**
RT   COLD STRENGTH
    COLD WEATHER
    HIGH ALTITUDE ENVIRONMENTS
    LUNAR TEMPERATURE
    MAGNETIC COOLING
    MOUNTAIN INHABITANTS
    THERMAL ENVIRONMENTS

LOW TEMPERATURE PHYSICS
RT   CRYOCHEMISTRY
    CRYOGENICS
    KONDO EFFECT
   ∞PHYSICS
   ∞SCIENCE
    SOLIDIFIED GASES
    SUPERCONDUCTING POWER
      TRANSMISSION
    SUPERCONDUCTIVITY

LOW TEMPERATURE PLASMAS
USE   COLD PLASMAS

LOW TEMPERATURE TESTS
GS   ENVIRONMENTAL TESTS
    . **LOW TEMPERATURE TESTS**
RT   CHEMICAL TESTS
    COLD STRENGTH
    COLD WEATHER TESTS
    CRYOSTATS
    HARDNESS TESTS
    LUBRICANT TESTS
    MELTING POINTS
    NONDESTRUCTIVE TESTS
    QUALITY CONTROL
    TEMPERATURE CONTROL
   ∞TESTS
    THERMAL EXPANSION
    THERMAL STABILITY

LOW THRUST
GS   THRUST
    . **LOW THRUST**
    . . MICROTHRUST
RT   HIGH THRUST
    JET THRUST
    ROCKET THRUST
    VARIABLE THRUST

LOW THRUST PROPULSION
GS   PROPULSION
    . **LOW THRUST PROPULSION**
    . . ELECTROMAGNETIC PROPULSION
    . . ELECTROSTATIC PROPULSION
    . . . ION PROPULSION
    . . MAN OPERATED PROPULSION
      SYSTEMS
    . . PHOTONIC PROPULSION

LOW THRUST PROPULSION-*(CONT.)*
    . . PLASMA PROPULSION
    . . SOLAR PROPULSION
    . . . SOLAR ELECTRIC PROPULSION
    . . . SOLAR THERMAL PROPULSION
RT   ELECTRIC PROPULSION
    MICROTHRUST
    ROCKET THRUST
    SPACECRAFT PROPULSION
    VARIABLE THRUST

LOW TURBULENCE
GS   TURBULENCE
    . **LOW TURBULENCE**
RT   STEADY FLOW

LOW VACUUM
SN   (PRESSURES BETWEEN 3.001 AND 1.0
      TORR)
GS   PRESSURE
    . VACUUM
    . . **LOW VACUUM**
RT   HIGH VACUUM

LOW VELOCITY
USE   LOW SPEED

LOW VISIBILITY
GS   VISIBILITY
    . **LOW VISIBILITY**
RT   AIRCRAFT LANDING
    ALL-WEATHER LANDING SYSTEMS
    HAZARDS
    HAZE
    INSTRUMENT FLIGHT RULES
    LIGHT TRANSMISSION

LOW VOLTAGE
GS   POTENTIAL ENERGY
    . ELECTRIC POTENTIAL
    . . **LOW VOLTAGE**
RT   LOW CURRENTS

LOW VOLUME RAMJET ENGINES
GS   ENGINES
    . AIR BREATHING ENGINES
    . . GAS TURBINE ENGINES
    . . . JET ENGINES
    . . . . RAMJET ENGINES
    . . . . . **LOW VOLUME RAMJET ENGINES**
    . INTERNAL COMBUSTION ENGINES
    . . GAS TURBINE ENGINES
    . . . JET ENGINES
    . . . . RAMJET ENGINES
    . . . . . **LOW VOLUME RAMJET ENGINES**
    . TURBINE ENGINES
    . . GAS TURBINE ENGINES
    . . . JET ENGINES
    . . . . RAMJET ENGINES
    . . . . . **LOW VOLUME RAMJET ENGINES**

LOW WEIGHT
RT   GRAVITATION
    REDUCED GRAVITY
    WEIGHTLESSNESS

∞ **LOW WING AIRCRAFT**
SN   *(USE OF A MORE SPECIFIC TERM IS*
    RECOMMENDED--CONSULT THE TERMS
    LISTED BELOW)
RT   ∞AIRCRAFT
    AIRCRAFT CONFIGURATIONS
    BEECH 99 AIRCRAFT
    GENERAL AVIATION AIRCRAFT
    HYPERSONIC AIRCRAFT
    JET AIRCRAFT
    LIGHT AIRCRAFT
    MONOPLANES
    PASSENGER AIRCRAFT
    TAILLESS AIRCRAFT
    TRANSPORT AIRCRAFT
    TURBOFAN AIRCRAFT
    TURBOPROP AIRCRAFT

LOWER ATMOSPHERE
SN   (ALTITUDE BELOW ABOUT 50 KM)
GS   EARTH ATMOSPHERE
    . **LOWER ATMOSPHERE**
    . . BIOSPHERE
    . . OZONOSPHERE
    . . TROPOSPHERE
RT   CHEMOSPHERE
    HETEROSPHERE
    HOMOSPHERE
    INTASAT SATELLITE
    LACATE (EXPERIMENT)
    LOW ALTITUDE
    MESOMETEOROLOGY
    MIDDLE ATMOSPHERE

LOWER ATMOSPHERE-*(CONT.)*
    TROPOPAUSE

**LOWER ATMOSPHERIC COMPOSITION
EXPERIMENT**
USE   LACATE (EXPERIMENT)

LOWER BODY NEGATIVE PRESSURE
GS   HEMODYNAMICS
    . **LOWER BODY NEGATIVE PRESSURE**
RT   ACCELERATION STRESSES
        (PHYSIOLOGY)
    ARTIFICIAL GRAVITY
    CARDIOVASCULAR SYSTEM
    GRAVITATIONAL EFFECTS
    ORTHOSTATIC TOLERANCE
    SPACE FLIGHT STRESS
    STRESS (PHYSIOLOGY)
    WEIGHTLESSNESS

LOWER CALIFORNIA (MEXICO)
UF   BAJA CALIFORNIA
RT   MEXICO
    NORTH AMERICA

LOWER IONOSPHERE
GS   EARTH ATMOSPHERE
    . UPPER ATMOSPHERE
    . . IONOSPHERE
    . . . **LOWER IONOSPHERE**
    . . . . D REGION
    ENVIRONMENTS
    . IONOSPHERE
    . . **LOWER IONOSPHERE**
RT   E REGION

LOX (OXYGEN)
USE   LIQUID OXYGEN

LOX-HYDROGEN ENGINES
USE   HYDROGEN OXYGEN ENGINES

LPTR REACTOR
USE   LIVERMORE POOL TYPE REACTOR

LR CIRCUITS
USE   RL CIRCUITS

LR-62-RM-2 ENGINE
GS   ENGINES
    . ROCKET ENGINES
    . . LIQUID PROPELLANT ROCKET
        ENGINES
    . . . **LR-62-RM-2 ENGINE**
RT   BULLPUP B MISSILE
    BULLPUP MISSILES

LR-87-AJ-5 ENGINE
GS   ENGINES
    . ROCKET ENGINES
    . . BOOSTER ROCKET ENGINES
    . . . **LR-87-AJ-5 ENGINE**
    . . LIQUID PROPELLANT ROCKET
        ENGINES
    . . . **LR-87-AJ-5 ENGINE**
RT   TITAN 1 ICBM

LR-91-AJ-5 ENGINE
UF   XLR-91-AJ-5 ENGINE
GS   ENGINES
    . ROCKET ENGINES
    . . LIQUID PROPELLANT ROCKET
        ENGINES
    . . . **LR-91-AJ-5 ENGINE**
RT   TITAN ICBM

LR-99 ENGINE
UF   YLR-99-RM-1 ENGINE
GS   ENGINES
    . ROCKET ENGINES
    . . LIQUID PROPELLANT ROCKET
        ENGINES
    . . . **LR-99 ENGINE**
RT   X-15 AIRCRAFT

LRC CIRCUITS
USE   RLC CIRCUITS

LRV (VEHICLE)
USE   LUNAR ROVING VEHICLES

LSI
USE   LARGE SCALE INTEGRATION

LSSM
UF   LUNAR SURFACE SCIENTIFIC MODULES
GS   LUNAR SPACECRAFT
    . LUNAR LANDING MODULES
    . . LUNAR MODULE
    . . . **LSSM**
    MANNED SPACECRAFT
    . LUNAR MODULE
    . . **LSSM**
    MODULES
    . SPACECRAFT MODULES
    . . LANDING MODULES
    . . . LUNAR LANDING MODULES
    . . . . LUNAR MODULE
    . . . . . **LSSM**
    SOFT LANDING SPACECRAFT
    . LANDING MODULES
    . . LUNAR LANDING MODULES
    . . . LUNAR MODULE
    . . . . **LSSM**
RT   APOLLO PROJECT
    ∞SURFACES

LST
USE   HUBBLE SPACE TELESCOPE

LTV AIRCRAFT
USE   LING-TEMCO-VOUGHT AIRCRAFT

LUBRICANT TESTS
RT   ENGINE TESTS
    HIGH TEMPERATURE TESTS
    LOW TEMPERATURE TESTS
    ∞MATERIALS TESTS
    ∞TESTS

LUBRICANTS
GS   **LUBRICANTS**
    . GAS LUBRICANTS
    . HIGH TEMPERATURE LUBRICANTS
    . LUBRICATING OILS
    . SOLID LUBRICANTS
RT   ADDITIVES
    BOUNDARY LUBRICATION
    GRAPHITE
    GREASES
    KEROGEN
    LIQUID METALS
    LUBRICATION
    LUBRICATION SYSTEMS
    MAINTENANCE
    OILS
    PETROLEUM PRODUCTS
    SQUEEZE FILMS

LUBRICATING OILS
GS   LUBRICANTS
    . **LUBRICATING OILS**
    OILS
    . **LUBRICATING OILS**
RT   DETERGENTS
    LUBRICATION
    MINERAL OILS
    SHALE OIL

LUBRICATION
GS   **LUBRICATION**
    . BOUNDARY LUBRICATION
    . SELF LUBRICATION
    . SPACECRAFT LUBRICATION
RT   BEARINGS
    ELASTOHYDRODYNAMICS
    ENGINES
    FRICTION REDUCTION
    GEARS
    IMPREGNATING
    LIQUID BEARINGS
    LUBRICANTS
    LUBRICATING OILS
    LUBRICATION SYSTEMS
    MAINTENANCE
    SELF LUBRICATING MATERIALS
    SLIDING
    TRIBOLOGY

LUBRICATION SYSTEMS
RT   AUTOMOBILES
    COOLING SYSTEMS
    INTERNAL COMBUSTION ENGINES
    LUBRICANTS
    LUBRICATION
    PUMPS
    ∞SYSTEMS

LUCITE (TRADEMARK)
USE   POLYMETHYL METHACRYLATE

LUDER BANDS
USE   PLASTIC DEFORMATION
    YIELD POINT

LUDOX (TRADEMARK)
GS   REFRACTORY MATERIALS
    . **LUDOX (TRADEMARK)**
RT   DENSIFICATION
    HEAT SHIELDING
    REENTRY SHIELDING
    SPACECRAFT CONSTRUCTION
        MATERIALS
    THERMAL PROTECTION
    TILES

LUGS
RT   FASTENERS
    HOLDERS
    STUDS (STRUCTURAL MEMBERS)
    SUPPORTS

LUMBAR REGION
GS   REGIONS
    . **LUMBAR REGION**
RT   ANATOMY
    HUMAN BODY

LUMBERING AREAS
USE   FORESTS

LUMENS
GS   PRESSURE
    . RADIATION PRESSURE
    . . **LUMENS**
    RATES (PER TIME)
    . FLUX DENSITY
    . . RADIANT FLUX DENSITY
    . . . **LUMENS**
RT   LIGHT (VISIBLE RADIATION)
    LUMINANCE
    LUMINESCENCE
    LUMINOSITY
    OPTICAL PROPERTIES
    RADIANCE

LUMINAIRES
UF   ELECTROLUMINESCENT LAMPS
    LAMPS
    LIGHT BULBS
    LIGHTS
GS   LIGHTING EQUIPMENT
    . **LUMINAIRES**
    . . AIRPORT LIGHTS
    . . . RUNWAY LIGHTS
    . . ARC LAMPS
    . . FLASH LAMPS
    . . . ALKALI VAPOR LAMPS
    . . MERCURY LAMPS
    . . QUARTZ LAMPS
    . . SEARCHLIGHTS
    . . XENON LAMPS
RT   BALLASTS (IMPEDANCES)
    BULBS
    FIXTURES
    ∞FLARES
    ∞GLOBES
    ILLUMINATING
    LIGHT (VISIBLE RADIATION)
    LIGHT SOURCES
    PROJECTORS
    VISUAL SIGNALS

LUMINANCE
SN   (EMISSION RATE PER UNIT AREA OF OF
        VISIBLE RADIATION)
GS   PRESSURE
    . RADIATION PRESSURE
    . . LUMINOUS INTENSITY
    . . . **LUMINANCE**
    RATES (PER TIME)
    . FLUX DENSITY
    . . RADIANT FLUX DENSITY
    . . . LUMINOUS INTENSITY
    . . . . **LUMINANCE**
RT   BRIGHTNESS
    GLARE
    ILLUMINANCE
    ILLUMINATING
    ∞INTENSITY
    IRRADIANCE
    LIGHT (VISIBLE RADIATION)
    LUMENS

LUMINANCE-*(CONT.)*
```
        OPTICAL PROPERTIES
        PHOTOMETRY
        SKY BRIGHTNESS
        SOLAR FLUX DENSITY
        STELLAR MAGNITUDE
```

LUMINESCENCE
```
UF      GLOW
        NOCTILUCENCE
GS      DECAY
        . EMISSION
        . . LIGHT EMISSION
        . . . LUMINESCENCE
        . . . . BIOLUMINESCENCE
        . . . . CATHODE GLOW
        . . . . CATHODOLUMINESCENCE
        . . . . CHEMILUMINESCENCE
        . . . . ELECTROLUMINESCENCE
        . . . . FLUORESCENCE
        . . . . . PHOSPHORESCENCE
        . . . . . RESONANCE FLUORESCENCE
        . . . . . X RAY FLUORESCENCE
        . . . . LUNAR LUMINESCENCE
        . . . . OPTICAL RESONANCE
        . . . . PHOTOLUMINESCENCE
        . . . . . TRIBOLUMINESCENCE
        . . . . . X RAY FLUORESCENCE
        . . . . SHOCK WAVE LUMINESCENCE
        . . . . SONOLUMINESCENCE
        . . . . THERMOLUMINESCENCE
RT      AFTERGLOWS
        ALKALI VAPOR LAMPS
        BRIGHTNESS
        ELECTRON-HOLE DROPS
        FRAUNHOFER LINE DISCRIMINATORS
        ∞ILLUMINATION
        ILLUMINATORS
        INCANDESCENCE
        LIGHT (VISIBLE RADIATION)
        LIGHT EMITTING DIODES
        LUMENS
        LUMINOSITY
        LUMINOUS INTENSITY
        NOCTILUCENT CLOUDS
        OPTICAL TRANSITION
        PLASMA RADIATION
        STELLAR LUMINOSITY
        STOKES LAW OF RADIATION
        VISIBILITY
```

LUMINESCENT INTENSITY
```
USE     LUMINOUS INTENSITY
```

LUMINOSITY
```
GS      ELECTROMAGNETIC PROPERTIES
        . OPTICAL PROPERTIES
-       . . LUMINOSITY
        . . . STELLAR LUMINOSITY
RT      BRIGHTNESS
        EMISSIVITY
        EMITTANCE
        ILLUMINATION
        INCANDESCENCE
        LIGHT (VISIBLE RADIATION)
        LUMENS
        LUMINESCENCE
        MASS TO LIGHT RATIOS
        PHOSPHENE
        RADIANCE
        RADIANT FLUX DENSITY
        VISIBILITY
```

LUMINOUS FLUX DENSITY
```
USE     LUMINOUS INTENSITY
```

LUMINOUS INTENSITY
```
SN      (EMISSION OR DETECTION RATE PER
        UNIT AREA OF VISIBLE RADIATION)
UF      LIGHT INTENSITY
        LUMINESCENT INTENSITY
        LUMINOUS FLUX DENSITY
GS      PRESSURE
        . RADIATION PRESSURE
        . . LUMINOUS INTENSITY
        . . . ILLUMINANCE
        . . . LUMINANCE
        RATES (PER TIME)
        . FLUX DENSITY
        . . RADIANT FLUX DENSITY
        . . . LUMINOUS INTENSITY
        . . . . ILLUMINANCE
        . . . . LUMINANCE
RT      BL LACERTAE OBJECTS
        BRIGHTNESS
```

LUMINOUS INTENSITY-*(CONT.)*
```
        EMITTANCE
        FLUX (RATE)
        INCANDESCENCE
∞       INTENSITY
        IRRADIANCE
        LIGHT (VISIBLE RADIATION)
        LUMINESCENCE
        MASS TO LIGHT RATIOS
        RADIANCY
        SEYFERT GALAXIES
        SOLAR FLUX DENSITY
        STELLAR MAGNITUDE
```

LUMPED PARAMETER SYSTEMS
```
RT      LUMPING
        MATHEMATICAL MODELS
        MATRICES (MATHEMATICS)
        ∞SYSTEMS
```

LUMPING
```
RT      AGGLOMERATION
        COAGULATION
        COLLECTION
        COMPOSITION (PROPERTY)
        LUMPED PARAMETER SYSTEMS
```

LUNA LUNAR PROBES
```
USE     LUNIK LUNAR PROBES
```

LUNAR ALBEDO
```
GS      ALBEDO
        . LUNAR ALBEDO
RT      ABSORPTANCE
        COSMIC RAY ALBEDO
        EARTH ALBEDO
        OPTICAL PROPERTIES
        SURFACE PROPERTIES
```

LUNAR ATMOSPHERE
```
UF      LUNAR IONOSPHERE
GS      ENVIRONMENTS
        . EXTRATERRESTRIAL ENVIRONMENTS
        . . LUNAR ENVIRONMENT
        . . . LUNAR ATMOSPHERE
        . , SATELLITE ATMOSPHERES
        . . . LUNAR ATMOSPHERE
RT      MOON
        PLANETARY ATMOSPHERES
```

LUNAR BASES
```
RT      AEPS
        ∞ASTRONAUTICS
        ∞BASES
        MOON
        SPACE COLONIES
        STATIONS
```

LUNAR CINEMATOGRAPHY
```
USE     LUNAR PHOTOGRAPHY
```

LUNAR COMMUNICATION
```
GS      TELECOMMUNICATION
        . SPACE COMMUNICATION
        . . LUNAR COMMUNICATION
        . . . CIRCUMLUNAR COMMUNICATION
RT      FACSIMILE COMMUNICATION
        INTERPLANETARY COMMUNICATION
        LASERS
        MOON
        OPTICAL COMMUNICATION
        RADAR
        RADIO COMMUNICATION
        SPACECRAFT COMMUNICATION
```

LUNAR COMPOSITION
```
GS      COMPOSITION (PROPERTY)
        . LUNAR COMPOSITION
RT      MOON
        PRE-IMBRIAN PERIOD
        SELENOLOGY
```

LUNAR CORE
```
GS      CORES
        . LUNAR CORE
        SELENOLOGY
        . LUNAR CORE
RT      STELLAR CORES
```

LUNAR CRATERS
```
GS      CRATERS
        . LUNAR CRATERS
        . . PTOLEMAEUS CRATER
        . . TYCHO CRATER
```

LUNAR CRATERS-*(CONT.)*
```
RT      METEORITE CRATERS
        MOON
        PRE-IMBRIAN PERIOD
        SELENOGRAPHY
        SELENOLOGY
```

LUNAR CRUST
```
GS      CRUSTS
        . LUNAR CRUST
RT      EARTH CRUST
        MOON
        SELENOGRAPHY
        SELENOLOGY
```

LUNAR DUST
```
GS      DUST
        . LUNAR DUST
        SOILS
        . LUNAR SOIL
        . . LUNAR DUST
RT      MOON
        SELENOLOGY
```

LUNAR ECHOES
```
GS      ECHOES
        . LUNAR ECHOES
        . . LUNAR RADAR ECHOES
RT      RADIO ECHOES
```

LUNAR ECLIPSES
```
GS      ECLIPSES
        . LUNAR ECLIPSES
RT      MOON
```

LUNAR EFFECTS
```
UF      LUNAR PERTURBATION
GS      LUNAR EFFECTS
        . LUNAR GRAVITATIONAL EFFECTS
        . LUNAR TIDES
RT      ∞EFFECTS
        ORBIT PERTURBATION
```

LUNAR ENVIRONMENT
```
GS      ENVIRONMENTS
        . EXTRATERRESTRIAL ENVIRONMENTS
        . . LUNAR ENVIRONMENT
        . . . LUNAR ATMOSPHERE
RT      AEROSPACE ENVIRONMENTS
        BIOASTRONAUTICS
        EXOBIOLOGY
        LIFE SUPPORT SYSTEMS
        MOON
        PLANETARY ENVIRONMENTS
        THERMAL ENVIRONMENTS
```

LUNAR EQUATOR
```
GS      EQUATORS
        . LUNAR EQUATOR
RT      INFRARED IMAGERY
        RADAR IMAGERY
```

LUNAR ESCAPE DEVICES
```
RT      ESCAPE CAPSULES
        ESCAPE ROCKETS
```

LUNAR EVOLUTION
```
GS      EVOLUTION (DEVELOPMENT)
        . LUNAR EVOLUTION
RT      MOON
        PRE-IMBRIAN PERIOD
        SELENOLOGY
```

LUNAR EXPLORATION
```
GS      EXPLORATION
        . LUNAR EXPLORATION
RT      APOLLO LUNAR EXPERIMENT MODULE
        APOLLO LUNAR SURFACE EXPERIMENTS
            PACKAGE
        APOLLO PROJECT
        APOLLO 5 FLIGHT
        APOLLO 6 FLIGHT
        APOLLO 7 FLIGHT
        APOLLO 8 FLIGHT
        APOLLO 9 FLIGHT
        APOLLO 10 FLIGHT
        APOLLO 11 FLIGHT
        APOLLO 12 FLIGHT
        APOLLO 13 FLIGHT
        APOLLO 14 FLIGHT
        APOLLO 15 FLIGHT
        APOLLO 16 FLIGHT
        APOLLO 17 FLIGHT
        EASEP
```

LUNAR EXPLORATION-*(CONT.)*
  EXTRATERRESTRIAL RESOURCES
  LOCAL SCIENTIFIC SURVEY MODULE
  MOON
  SPACE EXPLORATION

LUNAR EXPLORATION SYSTEM FOR APOLLO
UF LESA (LUNAR EXPLORATION SYSTEM)
RT APOLLO PROJECT
  APOLLO 5 FLIGHT
  APOLLO 6 FLIGHT
  APOLLO 7 FLIGHT
  APOLLO 8 FLIGHT
  APOLLO 9 FLIGHT
  APOLLO 10 FLIGHT
  APOLLO 11 FLIGHT
  APOLLO 12 FLIGHT
  APOLLO 13 FLIGHT
  APOLLO 14 FLIGHT
  APOLLO 15 FLIGHT
  APOLLO 16 FLIGHT
  APOLLO 17 FLIGHT
  ∞ SYSTEMS

LUNAR FAR SIDE
RT LIBRATION
  MOON

LUNAR FIGURE
RT SELENOLOGY

LUNAR FLIGHT
GS SPACE FLIGHT
  . LUNAR FLIGHT
RT APOLLO 5 FLIGHT
  APOLLO 6 FLIGHT
  APOLLO 7 FLIGHT
  APOLLO 8 FLIGHT
  APOLLO 9 FLIGHT
  APOLLO 10 FLIGHT
  APOLLO 11 FLIGHT
  APOLLO 12 FLIGHT
  APOLLO 13 FLIGHT
  APOLLO 14 FLIGHT
  APOLLO 15 FLIGHT
  APOLLO 16 FLIGHT
  APOLLO 17 FLIGHT
  CIRCUMLUNAR TRAJECTORIES
  CISLUNAR SPACE
  EARTH-MOON TRAJECTORIES
  ∞ FLIGHT
  FLYBY MISSIONS
  MOON-EARTH TRAJECTORIES
  ORBITS

LUNAR FLYING VEHICLES
RT ∞ FLIGHT VEHICLES
  LIFTING BODIES
  ∞ VEHICLES

LUNAR GEOLOGY
GS GEOLOGY
  . LUNAR GEOLOGY
RT GEOMORPHOLOGY
  MOON
  MOONQUAKES
  PRE-IMBRIAN PERIOD
  REGOLITH
  SEISMOLOGY
  SELENOLOGY

LUNAR GRAVITATION
GS GRAVITATION
  . LUNAR GRAVITATION
RT MOON
  PLANETARY GRAVITATION

LUNAR GRAVITATIONAL EFFECTS
GS GRAVITATION
  . LUNAR GRAVITATIONAL EFFECTS
  GRAVITATIONAL EFFECTS
  . LUNAR GRAVITATIONAL EFFECTS
  LUNAR EFFECTS
  . LUNAR GRAVITATIONAL EFFECTS
RT ∞ EFFECTS

LUNAR GRAVITY SIMULATOR
GS SIMULATORS
  . ENVIRONMENT SIMULATORS
  .. LUNAR GRAVITY SIMULATOR
RT GRAVITATION

LUNAR IONOSPHERE
USE LUNAR ATMOSPHERE

LUNAR LANDING
GS LANDING
  . SPACECRAFT LANDING
  .. LUNAR LANDING
RT APOLLO LUNAR EXPERIMENT MODULE
  APOLLO 5 FLIGHT
  APOLLO 6 FLIGHT
  APOLLO 7 FLIGHT
  APOLLO 8 FLIGHT
  APOLLO 9 FLIGHT
  APOLLO 10 FLIGHT
  APOLLO 11 FLIGHT
  APOLLO 12 FLIGHT
  APOLLO 13 FLIGHT
  APOLLO 14 FLIGHT
  APOLLO 15 FLIGHT
  APOLLO 16 FLIGHT
  APOLLO 17 FLIGHT
  CRASH LANDING
  HARD LANDING
  PLANETARY LANDING
  SOFT LANDING
  SURVEYOR PROJECT

LUNAR LANDING MODULES
GS LUNAR SPACECRAFT
  . LUNAR LANDING MODULES
  .. LUNAR MODULE
  ... LSSM
  MODULES
  . SPACECRAFT MODULES
  .. LANDING MODULES
  ... LUNAR LANDING MODULES
  .... LUNAR MODULE
  ..... LSSM
  SOFT LANDING SPACECRAFT
  . LANDING MODULES
  .. LUNAR LANDING MODULES
  ... LUNAR MODULE
  .... APOLLO LUNAR EXPERIMENT
   MODULE
  .... LSSM
RT APOLLO EXTENSION SYSTEM
  MANEUVERABLE SPACECRAFT
  MANNED SPACECRAFT
  REUSABLE SPACECRAFT
  UNMANNED SPACECRAFT

LUNAR LANDING SITES
GS SITES
  . LANDING SITES
  .. LUNAR LANDING SITES
RT MOON
  SELENOGRAPHY

LUNAR LAUNCH
GS LAUNCHING
  . ROCKET LAUNCHING
  .. LUNAR LAUNCH
RT APOLLO 5 FLIGHT
  APOLLO 6 FLIGHT
  APOLLO 7 FLIGHT
  APOLLO 8 FLIGHT
  APOLLO 9 FLIGHT
  APOLLO 10 FLIGHT
  APOLLO 11 FLIGHT
  APOLLO 12 FLIGHT
  APOLLO 13 FLIGHT
  APOLLO 14 FLIGHT
  APOLLO 15 FLIGHT
  APOLLO 16 FLIGHT
  APOLLO 17 FLIGHT
  ORBITAL LAUNCHING
  SATURN PROJECT

LUNAR LIMB
RT LIBRATION
  ∞ LIMBS
  MOON
  PLANETARY LIMB

LUNAR LOGISTICS
GS LOGISTICS
  . LUNAR LOGISTICS
RT LIFE SUPPORT SYSTEMS
  MANNED LUNAR SURFACE VEHICLES
  MATERIALS HANDLING

LUNAR LUMINESCENCE
GS DECAY
  . EMISSION
  .. LIGHT EMISSION
  ... LUMINESCENCE
  .... LUNAR LUMINESCENCE
RT MOON

LUNAR MAGNETIC FIELDS
GS MAGNETIC FIELDS
  . LUNAR MAGNETIC FIELDS
RT MOON

LUNAR MANTLE
RT CRUSTS
  EARTH MANTLE
  PLANETARY MANTLES
  PLANETARY STRUCTURE
  REGOLITH
  SELENOLOGY

LUNAR MAPS
GS MAPS
  . LUNAR MAPS
RT ASTRONOMICAL MAPS
  MOON
  SELENOGRAPHY

LUNAR MARIA
GS MARIA
  . LUNAR MARIA

LUNAR MOBILE LABORATORIES
UF MOLABS
GS LABORATORIES
  . LUNAR MOBILE LABORATORIES
  SURFACE VEHICLES
  . LUNAR SURFACE VEHICLES
  .. LUNAR MOBILE LABORATORIES
RT APOLLO PROJECT
  MANNED LUNAR SURFACE VEHICLES
  SELENOGRAPHY

LUNAR MODULE
UF LEM (LUNAR MODULE)
GS LUNAR SPACECRAFT
  . LUNAR LANDING MODULES
  .. LUNAR MODULE
  ... LSSM
  MANNED SPACECRAFT
  . LUNAR MODULE
  .. APOLLO LUNAR EXPERIMENT
   MODULE
  .. LSSM
  .. LUNAR MODULE 5
  .. LUNAR MODULE 7
  MODULES
  . SPACECRAFT MODULES
  .. LANDING MODULES
  ... LUNAR LANDING MODULES
  .... LUNAR MODULE
  ..... LSSM
  SOFT LANDING SPACECRAFT
  . LANDING MODULES
  .. LUNAR LANDING MODULES
  ... LUNAR MODULE
  .... APOLLO LUNAR EXPERIMENT
   MODULE
  .... LSSM
RT APOLLO SPACECRAFT
  APOLLO 5 FLIGHT
  APOLLO 6 FLIGHT
  APOLLO 7 FLIGHT
  APOLLO 8 FLIGHT
  APOLLO 9 FLIGHT
  APOLLO 10 FLIGHT
  APOLLO 11 FLIGHT
  APOLLO 12 FLIGHT
  APOLLO 13 FLIGHT
  APOLLO 14 FLIGHT
  APOLLO 15 FLIGHT
  APOLLO 16 FLIGHT
  APOLLO 17 FLIGHT
  ASCENT PROPULSION SYSTEMS

LUNAR MODULE ASCENT STAGE
RT ASCENT
  ASCENT TRAJECTORIES
  ROCKET ENGINES
  STAGE SEPARATION

LUNAR MODULE 5
GS MANNED SPACECRAFT
  . LUNAR MODULE
  .. LUNAR MODULE 5
RT APOLLO SPACECRAFT

LUNAR MODULE 7
GS MANNED SPACECRAFT
  . LUNAR MODULE
  .. LUNAR MODULE 7
RT APOLLO SPACECRAFT

LUNAR OBSERVATORIES
GS   OBSERVATORIES
   . **LUNAR OBSERVATORIES**
RT   ASTRONOMICAL OBSERVATORIES

LUNAR OCCULTATION
GS   OCCULTATION
   . **LUNAR OCCULTATION**
   .. SOLAR ECLIPSES
RT   EXOSAT SATELLITE
   MOON
   STELLAR OCCULTATION

LUNAR ORBIT AND LANDING SIMULATORS
UF   LOLA (SIMULATOR)
GS   SIMULATORS
   . **LUNAR ORBIT AND LANDING**
     SIMULATORS
RT   FLIGHT SIMULATORS
   TRAINING SIMULATORS

LUNAR ORBITAL RENDEZVOUS
UF   LOR (RENDEZVOUS)
GS   MANEUVERS
   . ORBITAL RENDEZVOUS
   .. **LUNAR ORBITAL RENDEZVOUS**
   RENDEZVOUS
   . SPACE RENDEZVOUS
   .. ORBITAL RENDEZVOUS
   ... **LUNAR ORBITAL RENDEZVOUS**
RT   EARTH ORBITAL RENDEZVOUS
   ORBITAL MECHANICS
   SPACECRAFT TRAJECTORIES

LUNAR ORBITER
GS   LUNAR SPACECRAFT
   . LUNAR SATELLITES
   .. **LUNAR ORBITER**
   ... LUNAR ORBITER 1
   ... LUNAR ORBITER 2
   ... LUNAR ORBITER 3
   ... LUNAR ORBITER 4
   ... LUNAR ORBITER 5
   SATELLITES
   . ARTIFICIAL SATELLITES
   .. LUNAR SATELLITES
   ... **LUNAR ORBITER**
   .... LUNAR ORBITER 1
   .... LUNAR ORBITER 2
   .... LUNAR ORBITER 3
   .... LUNAR ORBITER 4
   .... LUNAR ORBITER 5
RT   ORBITER PROJECT

LUNAR ORBITER A
USE   LUNAR ORBITER 1

LUNAR ORBITER B
USE   LUNAR ORBITER 2

LUNAR ORBITER C
USE   LUNAR ORBITER 3

LUNAR ORBITER D
USE   LUNAR ORBITER 4

LUNAR ORBITER E
USE   LUNAR ORBITER 5

LUNAR ORBITER 1
UF   LUNAR ORBITER A
GS   LUNAR SPACECRAFT
   . LUNAR SATELLITES
   .. LUNAR ORBITER
   ... **LUNAR ORBITER 1**
   SATELLITES
   . ARTIFICIAL SATELLITES
   .. LUNAR SATELLITES
   ... LUNAR ORBITER
   .... **LUNAR ORBITER 1**
RT   ORBITER PROJECT

LUNAR ORBITER 2
UF   LUNAR ORBITER B
GS   LUNAR SPACECRAFT
   . LUNAR SATELLITES
   .. LUNAR ORBITER
   ... **LUNAR ORBITER 2**
   SATELLITES
   . ARTIFICIAL SATELLITES
   .. LUNAR SATELLITES
   ... LUNAR ORBITER
   .... **LUNAR ORBITER 2**
RT   ORBITER PROJECT

LUNAR ORBITER 3
UF   LUNAR ORBITER C
GS   LUNAR SPACECRAFT
   . LUNAR SATELLITES
   .. LUNAR ORBITER
   ... **LUNAR ORBITER 3**
   SATELLITES
   . ARTIFICIAL SATELLITES
   .. LUNAR SATELLITES
   ... LUNAR ORBITER
   .... **LUNAR ORBITER 3**
RT   ORBITER PROJECT

LUNAR ORBITER 4
UF   LUNAR ORBITER D
GS   LUNAR SPACECRAFT
   . LUNAR SATELLITES
   .. LUNAR ORBITER
   ... **LUNAR ORBITER 4**
   SATELLITES
   . ARTIFICIAL SATELLITES
   .. LUNAR SATELLITES
   ... LUNAR ORBITER
   .... **LUNAR ORBITER 4**
RT   ORBITER PROJECT

LUNAR ORBITER 5
UF   LUNAR ORBITER E
GS   LUNAR SPACECRAFT
   . LUNAR SATELLITES
   .. LUNAR ORBITER
   ... **LUNAR ORBITER 5**
   SATELLITES
   . ARTIFICIAL SATELLITES
   .. LUNAR SATELLITES
   ... LUNAR ORBITER
   .... **LUNAR ORBITER 5**
RT   ORBITER PROJECT

LUNAR ORBITS
UF   EVECTION
GS   ORBITS
   . **LUNAR ORBITS**
RT   ARTIFICIAL SATELLITES
   CIRCULAR ORBITS
   CIRCUMLUNAR TRAJECTORIES
   CISLUNAR SPACE
   COMMAND SERVICE MODULES
   EARTH ORBITS
   EARTH-MOON TRAJECTORIES
   ELLIPTICAL ORBITS
   EQUATORIAL ORBITS
   LISSAJOUS FIGURES
   MOON
   ORBITAL MECHANICS
   PARKING ORBITS
   PERILUNES
   POLAR ORBITS
   SATELLITE ORBITS
   SPACECRAFT ORBITS
   TRANSFER ORBITS

LUNAR PERTURBATION
USE   LUNAR EFFECTS

LUNAR PHASES
RT   MOON
   ∞ PHASES
   TERMINATOR LINES

LUNAR PHOTOGRAPHS
GS   PHOTOGRAPHS
   . **LUNAR PHOTOGRAPHS**
RT   ASTRONOMICAL PHOTOGRAPHY
   PHOTOGRAPHY
   RANGER PROJECT
   SPACEBORNE PHOTOGRAPHY

LUNAR PHOTOGRAPHY
UF   LUNAR CINEMATOGRAPHY
GS   IMAGERY
   . **LUNAR PHOTOGRAPHY**
   PHOTOGRAPHY
   . **LUNAR PHOTOGRAPHY**
RT   ASTRONOMICAL PHOTOGRAPHY
   BLACK AND WHITE PHOTOGRAPHY
   INFRARED PHOTOGRAPHY
   MOON
   RANGER PROJECT
   SPACEBORNE PHOTOGRAPHY

LUNAR PROBES
GS   LUNAR SPACECRAFT
   . **LUNAR PROBES**
   .. LUNIK LUNAR PROBES

LUNAR PROBES-(CONT.)
   ... LUNIK 2 LUNAR PROBE
   ... LUNIK 3 LUNAR PROBE
   ... LUNIK 9 LUNAR PROBE
   ... LUNIK 10 LUNAR PROBE
   ... LUNIK 11 LUNAR PROBE
   ... LUNIK 12 LUNAR PROBE
   ... LUNIK 13 LUNAR PROBE
   ... LUNIK 14 LUNAR PROBE
   ... LUNIK 16 LUNAR PROBE
   ... LUNIK 17 LUNAR PROBE
   ... LUNIK 19 LUNAR PROBE
   ... LUNIK 20 LUNAR PROBE
   ... LUNIK 22 LUNAR PROBE
   .. RANGER LUNAR PROBES
   ... RANGER LUNAR LANDING
     VEHICLES
   ... RANGER 1 LUNAR PROBE
   ... RANGER 2 LUNAR PROBE
   ... RANGER 3 LUNAR PROBE
   ... RANGER 4 LUNAR PROBE
   ... RANGER 5 LUNAR PROBE
   ... RANGER 6 LUNAR PROBE
   ... RANGER 7 LUNAR PROBE
   ... RANGER 8 LUNAR PROBE
   ... RANGER 9 LUNAR PROBE
   .. SURVEYOR LUNAR PROBES
   ... SURVEYOR 1 LUNAR PROBE
   ... SURVEYOR 2 LUNAR PROBE
   ... SURVEYOR 3 LUNAR PROBE
   ... SURVEYOR 4 LUNAR PROBE
   ... SURVEYOR 5 LUNAR PROBE
   ... SURVEYOR 6 LUNAR PROBE
   ... SURVEYOR 7 LUNAR PROBE
   UNMANNED SPACECRAFT
   . SPACE PROBES
   .. **LUNAR PROBES**
   ... LUNIK LUNAR PROBES
   .... LUNIK 2 LUNAR PROBE
   .... LUNIK 3 LUNAR PROBE
   .... LUNIK 9 LUNAR PROBE
   .... LUNIK 10 LUNAR PROBE
   .... LUNIK 11 LUNAR PROBE
   .... LUNIK 12 LUNAR PROBE
   .... LUNIK 13 LUNAR PROBE
   .... LUNIK 14 LUNAR PROBE
   .... LUNIK 16 LUNAR PROBE
   .... LUNIK 17 LUNAR PROBE
   .... LUNIK 19 LUNAR PROBE
   .... LUNIK 20 LUNAR PROBE
   .... LUNIK 22 LUNAR PROBE
   ... RANGER LUNAR PROBES
   .... RANGER LUNAR LANDING
     VEHICLES
   .... RANGER 1 LUNAR PROBE
   .... RANGER 2 LUNAR PROBE
   .... RANGER 3 LUNAR PROBE
   .... RANGER 4 LUNAR PROBE
   .... RANGER 5 LUNAR PROBE
   .... RANGER 6 LUNAR PROBE
   .... RANGER 7 LUNAR PROBE
   .... RANGER 8 LUNAR PROBE
   .... RANGER 9 LUNAR PROBE
   ... SURVEYOR LUNAR PROBES
   .... SURVEYOR 1 LUNAR PROBE
   .... SURVEYOR 2 LUNAR PROBE
   .... SURVEYOR 3 LUNAR PROBE
   .... SURVEYOR 4 LUNAR PROBE
   .... SURVEYOR 5 LUNAR PROBE
   .... SURVEYOR 6 LUNAR PROBE
   .... SURVEYOR 7 LUNAR PROBE
RT   APOLLO PROJECT
   ATLAS ABLE 5 LAUNCH VEHICLE
   MANEUVERABLE SPACECRAFT
   PIONEER PROJECT
   RANGER PROJECT
   SOFT LANDING SPACECRAFT
   SURVEYOR PROJECT

LUNAR PROGRAMS
GS   PROGRAMS
   . **LUNAR PROGRAMS**
   .. APOLLO PROJECT
   .. SURVEYOR PROJECT

LUNAR RADAR ECHOES
UF   LUNAR SCATTERING
GS   ECHOES
   . LUNAR ECHOES
   .. **LUNAR RADAR ECHOES**
   . RADAR ECHOES
   .. **LUNAR RADAR ECHOES**

LUNAR RADIATION
GS   EXTRATERRESTRIAL RADIATION

LUNAR RADIATION-(CONT.)
. **LUNAR RADIATION**
RT ∞ RADIATION

LUNAR RANGEFINDING
GS RANGEFINDING
. **LUNAR RANGEFINDING**
RT DISTANCE MEASURING EQUIPMENT
LASER RANGE FINDERS
MEASURING INSTRUMENTS
OPTICAL RANGE FINDERS
RANGE FINDERS

LUNAR RAYS
SN (EXCLUDES RADIATION)
RT METEORITE CRATERS
MOON
∞ RAYS
SELENOGRAPHY

LUNAR RECEIVING LABORATORY
GS LABORATORIES
. **LUNAR RECEIVING LABORATORY**

LUNAR RETROREFLECTORS
RT APOLLO LUNAR SURFACE EXPERIMENTS
PACKAGE
EARTH-MOON SYSTEM
GEODESY
LASER RANGE FINDERS
RETROREFLECTION
U.S.S.R. SPACE PROGRAM

LUNAR ROCKS
GS ROCKS
. **LUNAR ROCKS**
RT IMPACT MELTS
PARTICLE TRACKS
PRE-IMBRIAN PERIOD
REGOLITH
SELENOGRAPHY
SELENOLOGY

LUNAR ROTATION
GS ROTATING BODIES
. **LUNAR ROTATION**
RT CENTER OF GRAVITY
SPIN DYNAMICS

LUNAR ROVING VEHICLES
UF LRV (VEHICLE)
GS SURFACE VEHICLES
. LUNAR SURFACE VEHICLES
. . **LUNAR ROVING VEHICLES**
. . . LUNOKHOD LUNAR ROVING
VEHICLES
. . . MANNED LUNAR SURFACE
VEHICLES
. ROVING VEHICLES
. . **LUNAR ROVING VEHICLES**
. . . LUNOKHOD LUNAR ROVING
VEHICLES
RT PROVING
RESEARCH VEHICLES
∞ VEHICLES

LUNAR SATELLITES
GS LUNAR SPACECRAFT
. **LUNAR SATELLITES**
. . LUNAR ORBITER
. . . LUNAR ORBITER 1
. . . LUNAR ORBITER 2
. . . LUNAR ORBITER 3
. . . LUNAR ORBITER 4
. . . LUNAR ORBITER 5
SATELLITES
. ARTIFICIAL SATELLITES
. . **LUNAR SATELLITES**
. . . EXPLORER 18 SATELLITE
. . . EXPLORER 28 SATELLITE
. . . IMP
. . . LUNAR ORBITER
. . . . LUNAR ORBITER 1
. . . . LUNAR ORBITER 2
. . . . LUNAR ORBITER 3
. . . . LUNAR ORBITER 4
. . . . LUNAR ORBITER 5
RT MANEUVERABLE SPACECRAFT
MANNED SPACECRAFT
PERILUNES
POLAR ORBITS
UNMANNED SPACECRAFT

LUNAR SCATTERING
USE DIFFUSE RADIATION
LUNAR RADAR ECHOES

LUNAR SEISMOGRAPHS
GS MEASURING INSTRUMENTS
. VIBRATION METERS
. . SEISMOGRAPHS
. . . **LUNAR SEISMOGRAPHS**
RECORDING INSTRUMENTS
. SEISMOGRAPHS
. . **LUNAR SEISMOGRAPHS**

LUNAR SHADOW
GS SHADOWS
. **LUNAR SHADOW**
RT ECLIPSES
MOON
SOLAR ECLIPSES

LUNAR SHELTERS
GS SHELTERS
. **LUNAR SHELTERS**
RT INFLATABLE STRUCTURES
LIFE SUPPORT SYSTEMS
SPACE COLONIES
SURVIVAL
∞ TUNNELS

LUNAR SOIL
GS SOILS
. **LUNAR SOIL**
. . LUNAR DUST
RT KREEP
MINERALS
MOON
PENETROMETERS

LUNAR SPACECRAFT
GS **LUNAR SPACECRAFT**
. APOLLO LUNAR EXPERIMENT MODULE
. APOLLO SPACECRAFT
. LUNAR LANDING MODULES
. . LUNAR MODULE
. . . LSSM
. LUNAR PROBES
. . LUNIK LUNAR PROBES
. . . LUNIK 2 LUNAR PROBE
. . . LUNIK 3 LUNAR PROBE
. . . LUNIK 9 LUNAR PROBE
. . . LUNIK 10 LUNAR PROBE
. . . LUNIK 11 LUNAR PROBE
. . . LUNIK 12 LUNAR PROBE
. . . LUNIK 13 LUNAR PROBE
. . . LUNIK 14 LUNAR PROBE
. . . LUNIK 16 LUNAR PROBE
. . . LUNIK 17 LUNAR PROBE
. . . LUNIK 19 LUNAR PROBE
. . . LUNIK 20 LUNAR PROBE
. . . LUNIK 22 LUNAR PROBE
. . RANGER LUNAR PROBES
. . RANGER LUNAR LANDING
VEHICLES
. . . RANGER 1 LUNAR PROBE
. . . RANGER 2 LUNAR PROBE
. . . RANGER 3 LUNAR PROBE
. . . RANGER 4 LUNAR PROBE
. . . RANGER 5 LUNAR PROBE
. . . RANGER 6 LUNAR PROBE
. . . RANGER 7 LUNAR PROBE
. . . RANGER 8 LUNAR PROBE
. . . RANGER 9 LUNAR PROBE
. . SURVEYOR LUNAR PROBES
. . . SURVEYOR 1 LUNAR PROBE
. . . SURVEYOR 2 LUNAR PROBE
. . . SURVEYOR 3 LUNAR PROBE
. . . SURVEYOR 4 LUNAR PROBE
. . . SURVEYOR 5 LUNAR PROBE
. . . SURVEYOR 6 LUNAR PROBE
. . . SURVEYOR 7 LUNAR PROBE
. LUNAR SATELLITES
. . LUNAR ORBITER
. . . LUNAR ORBITER 1
. . . LUNAR ORBITER 2
. . . LUNAR ORBITER 3
. . . LUNAR ORBITER 4
. . . LUNAR ORBITER 5
RT APOLLO 5 FLIGHT
APOLLO 6 FLIGHT
ARTIFICIAL SATELLITES
HALO ORBIT SPACE STATION
MANNED SPACECRAFT
ORBITING LUNAR STATIONS
RENDEZVOUS SPACECRAFT
SATELLITES

LUNAR SPACECRAFT-(CONT.)
SPACE CAPSULES
∞ SPACECRAFT
SURVEYOR PROJECT
UNMANNED SPACECRAFT

LUNAR SURFACE
RT SELENOLOGY
SURFACE LAYERS
SURFACE PROPERTIES
∞ SURFACES

LUNAR SURFACE SCIENTIFIC MODULES
USE LSSM

LUNAR SURFACE VEHICLES
GS SURFACE VEHICLES
. **LUNAR SURFACE VEHICLES**
. . LUNAR MOBILE LABORATORIES
. . LUNAR ROVING VEHICLES
. . . LUNOKHOD LUNAR ROVING
VEHICLES
. . . MANNED LUNAR SURFACE
VEHICLES
RT CRAWLER TRACTORS
∞ SURFACES
∞ VEHICLES
WALKING MACHINES

LUNAR TEMPERATURE
GS TEMPERATURE
. **LUNAR TEMPERATURE**
RT HIGH TEMPERATURE ENVIRONMENTS
LOW TEMPERATURE ENVIRONMENTS
MOON

LUNAR TIDES
GS LUNAR EFFECTS
. **LUNAR TIDES**
TIDES
. **LUNAR TIDES**
RT ATMOSPHERIC TIDES
EARTH TIDES
MOONQUAKES

LUNAR TOPOGRAPHY
GS TOPOGRAPHY
. **LUNAR TOPOGRAPHY**
RT MOON
SELENOGRAPHY
SELENOLOGY
SURFACE PROPERTIES
SURFACE ROUGHNESS

LUNAR TRAJECTORIES
GS TRAJECTORIES
. SPACECRAFT TRAJECTORIES
. . **LUNAR TRAJECTORIES**
. . . CIRCUMLUNAR TRAJECTORIES
. . . EARTH-MOON TRAJECTORIES
. . . MOON-EARTH TRAJECTORIES
RT PARKING ORBITS
TRANSFER ORBITS

LUNATION
USE MONTH

LUNEBERG LENSES
USE RADAR CORNER REFLECTORS

LUNG MORPHOLOGY
GS MORPHOLOGY
. **LUNG MORPHOLOGY**
RT ALVEOLI
PULMONARY LESIONS
RESPIRATORY DISEASES

LUNGS
GS ANATOMY
. ORGANS
. . **LUNGS**
. RESPIRATORY SYSTEM
. . **LUNGS**
VISCERA
. ORGANS
. . **LUNGS**
RT ALVEOLAR AIR
ALVEOLI
ATELECTASIS
BRONCHI
PLEURAE
PNEUMOGRAPHY
PNEUMOTHORAX
PULMONARY CIRCULATION

LUNGS-(CONT.)
    PULMONARY FUNCTIONS
    PULMONARY LESIONS
    SPIROMETERS

LUNIK LUNAR PROBES
UF   LUNA LUNAR PROBES
GS   LUNAR SPACECRAFT
    . LUNAR PROBES
    . . **LUNIK LUNAR PROBES**
    . . . LUNIK 2 LUNAR PROBE
    . . . LUNIK 3 LUNAR PROBE
    . . . LUNIK 9 LUNAR PROBE
    . . . LUNIK 10 LUNAR PROBE
    . . . LUNIK 11 LUNAR PROBE
    . . . LUNIK 12 LUNAR PROBE
    . . . LUNIK 13 LUNAR PROBE
    . . . LUNIK 14 LUNAR PROBE
    . . . LUNIK 16 LUNAR PROBE
    . . . LUNIK 17 LUNAR PROBE
    . . . LUNIK 19 LUNAR PROBE
    . . . LUNIK 20 LUNAR PROBE
    . . . LUNIK 22 LUNAR PROBE
    SOVIET SPACECRAFT
    . **LUNIK LUNAR PROBES**
    . . LUNIK 2 LUNAR PROBE
    . . LUNIK 3 LUNAR PROBE
    . . LUNIK 9 LUNAR PROBE
    . . LUNIK 10 LUNAR PROBE
    . . LUNIK 11 LUNAR PROBE
    . . LUNIK 12 LUNAR PROBE
    . . LUNIK 13 LUNAR PROBE
    . . LUNIK 14 LUNAR PROBE
    . . LUNIK 16 LUNAR PROBE
    . . LUNIK 17 LUNAR PROBE
    . . LUNIK 19 LUNAR PROBE
    . . LUNIK 20 LUNAR PROBE
    . . LUNIK 22 LUNAR PROBE
    UNMANNED SPACECRAFT
    . SPACE PROBES
    . . LUNAR PROBES
    . . . **LUNIK LUNAR PROBES**
    . . . . LUNIK 2 LUNAR PROBE
    . . . . LUNIK 3 LUNAR PROBE
    . . . . LUNIK 9 LUNAR PROBE
    . . . . LUNIK 10 LUNAR PROBE
    . . . . LUNIK 11 LUNAR PROBE
    . . . . LUNIK 12 LUNAR PROBE
    . . . . LUNIK 13 LUNAR PROBE
    . . . . LUNIK 14 LUNAR PROBE
    . . . . LUNIK 16 LUNAR PROBE
    . . . . LUNIK 17 LUNAR PROBE
    . . . . LUNIK 19 LUNAR PROBE
    . . . . LUNIK 20 LUNAR PROBE
    . . . . LUNIK 22 LUNAR PROBE
RT   LUNOKHOD LUNAR ROVING VEHICLES
    U.S.S.R. SPACE PROGRAM

LUNIK 2 LUNAR PROBE
GS   LUNAR SPACECRAFT
    . LUNAR PROBES
    . . LUNIK LUNAR PROBES
    . . . **LUNIK 2 LUNAR PROBE**
    SOVIET SPACECRAFT
    . LUNIK LUNAR PROBES
    . . **LUNIK 2 LUNAR PROBE**
    UNMANNED SPACECRAFT
    . SPACE PROBES
    . . LUNAR PROBES
    . . . LUNIK LUNAR PROBES
    . . . . **LUNIK 2 LUNAR PROBE**

LUNIK 3 LUNAR PROBE
GS   LUNAR SPACECRAFT
    . LUNAR PROBES
    . . LUNIK LUNAR PROBES
    . . . **LUNIK 3 LUNAR PROBE**
    SOVIET SPACECRAFT
    . LUNIK LUNAR PROBES
    . . **LUNIK 3 LUNAR PROBE**
    UNMANNED SPACECRAFT
    . SPACE PROBES
    . . LUNAR PROBES
    . . . LUNIK LUNAR PROBES
    . . . . **LUNIK 3 LUNAR PROBE**

LUNIK 9 LUNAR PROBE
GS   LUNAR SPACECRAFT
    . LUNAR PROBES
    . . LUNIK LUNAR PROBES
    . . . **LUNIK 9 LUNAR PROBE**
    SOVIET SPACECRAFT
    . LUNIK LUNAR PROBES
    . . **LUNIK 9 LUNAR PROBE**
    UNMANNED SPACECRAFT

LUNIK 9 LUNAR PROBE-(CONT.)
    . SPACE PROBES
    . . LUNAR PROBES
    . . . LUNIK LUNAR PROBES
    . . . . **LUNIK 9 LUNAR PROBE**

LUNIK 10 LUNAR PROBE
GS   LUNAR SPACECRAFT
    . LUNAR PROBES
    . . LUNIK LUNAR PROBES
    . . . **LUNIK 10 LUNAR PROBE**
    SOVIET SPACECRAFT
    . LUNIK LUNAR PROBES
    . . **LUNIK 10 LUNAR PROBE**
    UNMANNED SPACECRAFT
    . SPACE PROBES
    . . LUNAR PROBES
    . . . LUNIK LUNAR PROBES
    . . . . **LUNIK 10 LUNAR PROBE**

LUNIK 11 LUNAR PROBE
GS   LUNAR SPACECRAFT
    . LUNAR PROBES
    . . LUNIK LUNAR PROBES
    . . . **LUNIK 11 LUNAR PROBE**
    SOVIET SPACECRAFT
    . LUNIK LUNAR PROBES
    . . **LUNIK 11 LUNAR PROBE**
    UNMANNED SPACECRAFT
    . SPACE PROBES
    . . LUNAR PROBES
    . . . LUNIK LUNAR PROBES
    . . . . **LUNIK 11 LUNAR PROBE**

LUNIK 12 LUNAR PROBE
GS   LUNAR SPACECRAFT
    . LUNAR PROBES
    . . LUNIK LUNAR PROBES
    . . . **LUNIK 12 LUNAR PROBE**
    SOVIET SPACECRAFT
    . LUNIK LUNAR PROBES
    . . **LUNIK 12 LUNAR PROBE**
    UNMANNED SPACECRAFT
    . SPACE PROBES
    . . LUNAR PROBES
    . . . LUNIK LUNAR PROBES
    . . . . **LUNIK 12 LUNAR PROBE**

LUNIK 13 LUNAR PROBE
GS   LUNAR SPACECRAFT
    . LUNAR PROBES
    . . LUNIK LUNAR PROBES
    . . . **LUNIK 13 LUNAR PROBE**
    SOVIET SPACECRAFT
    . LUNIK LUNAR PROBES
    . . **LUNIK 13 LUNAR PROBE**
    UNMANNED SPACECRAFT
    . SPACE PROBES
    . . LUNAR PROBES
    . . . LUNIK LUNAR PROBES
    . . . . **LUNIK 13 LUNAR PROBE**

LUNIK 14 LUNAR PROBE
GS   LUNAR SPACECRAFT
    . LUNAR PROBES
    . . LUNIK LUNAR PROBES
    . . . **LUNIK 14 LUNAR PROBE**
    SOVIET SPACECRAFT
    . LUNIK LUNAR PROBES
    . . **LUNIK 14 LUNAR PROBE**
    UNMANNED SPACECRAFT
    . SPACE PROBES
    . . LUNAR PROBES
    . . . LUNIK LUNAR PROBES
    . . . . **LUNIK 14 LUNAR PROBE**

LUNIK 16 LUNAR PROBE
GS   LUNAR SPACECRAFT
    . LUNAR PROBES
    . . LUNIK LUNAR PROBES
    . . . **LUNIK 16 LUNAR PROBE**
    SOVIET SPACECRAFT
    . LUNIK LUNAR PROBES
    . . **LUNIK 16 LUNAR PROBE**
    UNMANNED SPACECRAFT
    . SPACE PROBES
    . . LUNAR PROBES
    . . . LUNIK LUNAR PROBES
    . . . . **LUNIK 16 LUNAR PROBE**

LUNIK 17 LUNAR PROBE
GS   LUNAR SPACECRAFT
    . LUNAR PROBES
    . . LUNIK LUNAR PROBES
    . . . **LUNIK 17 LUNAR PROBE**

LUNIK 17 LUNAR PROBE-(CONT.)
    SOVIET SPACECRAFT
    . LUNIK LUNAR PROBES
    . . **LUNIK 17 LUNAR PROBE**
    UNMANNED SPACECRAFT
    . SPACE PROBES
    . . LUNAR PROBES
    . . . LUNIK LUNAR PROBES
    . . . . **LUNIK 17 LUNAR PROBE**

LUNIK 19 LUNAR PROBE
GS   LUNAR SPACECRAFT
    . LUNAR PROBES
    . . LUNIK LUNAR PROBES
    . . . **LUNIK 19 LUNAR PROBE**
    SOVIET SPACECRAFT
    . LUNIK LUNAR PROBES
    . . **LUNIK 19 LUNAR PROBE**
    UNMANNED SPACECRAFT
    . SPACE PROBES
    . . LUNAR PROBES
    . . . LUNIK LUNAR PROBES
    . . . . **LUNIK 19 LUNAR PROBE**
RT   U.S.S.R. SPACE PROGRAM

LUNIK 20 LUNAR PROBE
GS   LUNAR SPACECRAFT
    . LUNAR PROBES
    . . LUNIK LUNAR PROBES
    . . . **LUNIK 20 LUNAR PROBE**
    SOVIET SPACECRAFT
    . LUNIK LUNAR PROBES
    . . **LUNIK 20 LUNAR PROBE**
    UNMANNED SPACECRAFT
    . SPACE PROBES
    . . LUNAR PROBES
    . . . LUNIK LUNAR PROBES
    . . . . **LUNIK 20 LUNAR PROBE**

LUNIK 22 LUNAR PROBE
GS   LUNAR SPACECRAFT
    . LUNAR PROBES
    . . LUNIK LUNAR PROBES
    . . . **LUNIK 22 LUNAR PROBE**
    SOVIET SPACECRAFT
    . LUNIK LUNAR PROBES
    . . **LUNIK 22 LUNAR PROBE**
    UNMANNED SPACECRAFT
    . SPACE PROBES
    . . LUNAR PROBES
    . . . LUNIK LUNAR PROBES
    . . . . **LUNIK 22 LUNAR PROBE**
RT   U.S.S.R. SPACE PROGRAM

LUNOKHOD LUNAR ROVING VEHICLES
GS   SURFACE VEHICLES
    . LUNAR SURFACE VEHICLES
    . . LUNAR ROVING VEHICLES
    . . . **LUNOKHOD LUNAR ROVING VEHICLES**
    . ROVING VEHICLES
    . . LUNAR ROVING VEHICLES
    . . . **LUNOKHOD LUNAR ROVING VEHICLES**
RT   LUNIK LUNAR PROBES
    U.S.S.R. SPACE PROGRAM
    ∞ VEHICLES

LUSTER
UF   DULLNESS
RT   BRIGHTNESS
    FINISHES
    GLARE
    REFLECTANCE

LUTETIUM
GS   CHEMICAL ELEMENTS
    . NUCLIDES
    . . ISOTOPES
    . . . **LUTETIUM**
    . . . . LUTETIUM ISOTOPES
    . RARE EARTH ELEMENTS
    . . **LUTETIUM**
    . . . LUTETIUM ISOTOPES
    METALS
    . RARE EARTH ELEMENTS
    . . **LUTETIUM**
    . . . LUTETIUM ISOTOPES

LUTETIUM COMPOUNDS
GS   RARE EARTH COMPOUNDS
    . **LUTETIUM COMPOUNDS**
RT   ∞ CHEMICAL COMPOUNDS
    ∞ METAL COMPOUNDS

LUTETIUM ISOTOPES
UF   LUTETIUM 176
GS   CHEMICAL ELEMENTS
   . NUCLIDES
   . . ISOTOPES
   . . . LUTETIUM
   . . . . **LUTETIUM ISOTOPES**
   . RARE EARTH ELEMENTS
   . . LUTETIUM
   . . . **LUTETIUM ISOTOPES**
   METALS
   . RARE EARTH ELEMENTS
   . . LUTETIUM
   . . . **LUTETIUM ISOTOPES**

LUTETIUM 176
USE   LUTETIUM ISOTOPES

LUXEMBOURG
GS   NATIONS
   . **LUXEMBOURG**
RT   EUROPE

LUXEMBOURG EFFECT
RT   ∞EFFECTS
       IONOSPHERIC CROSS MODULATION
       IONOSPHERIC PROPAGATION

LYAPUNOV FUNCTIONS
USE   LIAPUNOV FUNCTIONS

LYMAN ALPHA RADIATION
GS   ELECTROMAGNETIC RADIATION
   . ULTRAVIOLET RADIATION
   . . FAR ULTRAVIOLET RADIATION
   . . . **LYMAN ALPHA RADIATION**
   IONIZING RADIATION
   . ULTRAVIOLET RADIATION
   . . FAR ULTRAVIOLET RADIATION
   . . . **LYMAN ALPHA RADIATION**
RT   ATOMIC SPECTRA
       EXTRATERRESTRIAL RADIATION
       POLARIZED ELECTROMAGNETIC
         RADIATION
       ∞RADIATION
       ULTRAVIOLET ASTRONOMY

LYMAN BETA RADIATION
GS   ELECTROMAGNETIC RADIATION
   . ULTRAVIOLET RADIATION
   . . FAR ULTRAVIOLET RADIATION
   . . . **LYMAN BETA RADIATION**
   IONIZING RADIATION
   . ULTRAVIOLET RADIATION
   . . FAR ULTRAVIOLET RADIATION
   . . . **LYMAN BETA RADIATION**
RT   ATOMIC SPECTRA
       EXTRATERRESTRIAL RADIATION
       POLARIZED ELECTROMAGNETIC
         RADIATION
       ∞RADIATION
       ULTRAVIOLET ASTRONOMY

LYMAN SPECTRA
GS   SPECTRA
   . RADIATION SPECTRA
   . . ELECTROMAGNETIC SPECTRA
   . . . LINE SPECTRA
   . . . . **LYMAN SPECTRA**
RT   ATOMIC SPECTRA
       ELECTRONIC SPECTRA
       EMISSION SPECTRA
       H LINES
       SOLAR SPECTRA
       SPECTRAL THEORY
       ULTRAVIOLET SPECTRA

LYMPH
GS   BODY FLUIDS
   . **LYMPH**
   . . LYMPHOCYTES
RT   CORPUSCLES

LYMPHOCYTES
GS   ANATOMY
   . CARDIOVASCULAR SYSTEM
   . . LEUKOCYTES
   . . . **LYMPHOCYTES**
   BODY FLUIDS
   . BLOOD
   . . LEUKOCYTES
   . . . **LYMPHOCYTES**
   . LYMPH
   . . **LYMPHOCYTES**

LYMPHOCYTES-*(CONT.)*
   CELLS (BIOLOGY)
   . LEUKOCYTES
   . . **LYMPHOCYTES**
RT   CORPUSCLES

LYOPHILIZATION
USE   COLLOIDING

LYOPHILS
USE   COLLOIDS

LYRA CONSTELLATION
GS   CONSTELLATIONS
   . **LYRA CONSTELLATION**
RT   CELESTIAL BODIES
       CELESTIAL SPHERE
       STARS

LYSERGAMIDE
GS   NITROGEN COMPOUNDS
   . AMIDES
   . . **LYSERGAMIDE**
RT   DRUGS

LYSERGINE
GS   HETEROCYCLIC COMPOUNDS
   . ALKALOIDS
   . . **LYSERGINE**
   NITROGEN COMPOUNDS
   . ALKALOIDS
   . . **LYSERGINE**

LYSIMETERS
RT   GROUND WATER
       MOISTURE CONTENT
       PERCOLATION
       SOIL MOISTURE
       SOILS
       WATER BALANCE
       WATER POLLUTION

LYSINE
GS   ACIDS
   . AMINO ACIDS
   . . **LYSINE**
   ORGANIC COMPOUNDS
   . AMINO ACIDS
   . . **LYSINE**
RT   DIGESTING
       LYSOGENESIS

LYSOGENESIS
RT   DISINTEGRATION
       LYSINE

LYSOZYME
GS   ENZYMES
   . **LYSOZYME**
RT   BODY FLUIDS

LZEEBE SATELLITE
UF   EARTH ENERGY BUDGET EXPERIMENT
       LONG TERM ZONAL EARTH ENERGY
         EXPERIMENT
       ZONAL EARTH ENERGY BUDGET
         EXPERIMENT
GS   SATELLITES
   . ARTIFICIAL SATELLITES
   . . SCIENTIFIC SATELLITES
   . . . **LZEEBE SATELLITE**

M

M REGION
GS   REGIONS
   . **M REGION**
RT   GEOMAGNETISM
       SOLAR ATMOSPHERE
       SOLAR CORPUSCULAR RADIATION
       SOLAR WIND

M STARS
GS   CELESTIAL BODIES
   . STARS
   . . **M STARS**
   . . . FLARE STARS
RT   GIANT STARS
       MAIN SEQUENCE STARS
       RED GIANT STARS
       S STARS

M STARS-*(CONT.)*
       SUBGIANT STARS
       SUPERGIANT STARS
       SYMBIOTIC STARS

M WINGS
USE   VARIABLE SWEEP WINGS

M-1 ENGINE
UF   AJ-1000 ENGINE
GS   ENGINES
   . ROCKET ENGINES
   . . BOOSTER ROCKET ENGINES
   . . . **M-1 ENGINE**
   . . LIQUID PROPELLANT ROCKET
       ENGINES
   . . . HYDROGEN OXYGEN ENGINES
   . . . . **M-1 ENGINE**
RT   NOVA LAUNCH VEHICLES
       SATURN 1 LAUNCH VEHICLES
       SATURN 1B LAUNCH VEHICLES

M-2 LIFTING BODY
GS   LIFTING BODIES
   . LIFTING REENTRY VEHICLES
   . . **M-2 LIFTING BODY**
   . . . M-2F2 LIFTING BODY
   MANEUVERABLE SPACECRAFT
   . LIFTING REENTRY VEHICLES
   . . **M-2 LIFTING BODY**
   . . . M-2F2 LIFTING BODY
   REENTRY VEHICLES
   . LIFTING REENTRY VEHICLES
   . . **M-2 LIFTING BODY**
   . . . M-2F2 LIFTING BODY

M-2F2 LIFTING BODY
GS   HYPERSONIC VEHICLES
   . LIFTING REENTRY VEHICLES
   . . **M-2F2 LIFTING BODY**
   LIFTING BODIES
   . LIFTING REENTRY VEHICLES
   . . M-2 LIFTING BODY
   . . . **M-2F2 LIFTING BODY**
   MANEUVERABLE SPACECRAFT
   . LIFTING REENTRY VEHICLES
   . . M-2 LIFTING BODY
   . . . **M-2F2 LIFTING BODY**
   REENTRY VEHICLES
   . LIFTING REENTRY VEHICLES
   . . M-2 LIFTING BODY
   . . . **M-2F2 LIFTING BODY**

M-2F3 LIFTING BODY
GS   LIFTING BODIES
   . **M-2F3 LIFTING BODY**

M-46 ENGINE
GS   ENGINES
   . ROCKET ENGINES
   . . SOLID PROPELLANT ROCKET
       ENGINES
   . . . **M-46 ENGINE**
RT   FALCON MISSILE

M-55 ENGINE
GS   ENGINES
   . ROCKET ENGINES
   . . BOOSTER ROCKET ENGINES
   . . . **M-55 ENGINE**
   . . SOLID PROPELLANT ROCKET
       ENGINES
   . . . **M-55 ENGINE**
RT   MINUTEMAN ICBM

M-56 ENGINE
GS   ENGINES
   . ROCKET ENGINES
   . . SOLID PROPELLANT ROCKET
       ENGINES
   . . . **M-56 ENGINE**
RT   MINUTEMAN ICBM

M-57 ENGINE
GS   ENGINES
   . ROCKET ENGINES
   . . SOLID PROPELLANT ROCKET
       ENGINES
   . . . **M-57 ENGINE**
RT   MINUTEMAN ICBM

M-100 ENGINE
GS   ENGINES
   . ROCKET ENGINES

M-100 ENGINE-*(CONT.)*
   . . **M-100 ENGINE**

MA-2 ENGINE
GS   ENGINES
   . ROCKET ENGINES
   . . BOOSTER ROCKET ENGINES
   . . . **MA-2 ENGINE**
   . . LIQUID PROPELLANT ROCKET
        ENGINES
   . . . **MA-2 ENGINE**
RT   ATLAS ICBM
   VERNIER ENGINES

MA-2 MISSION
USE   MERCURY MA-2 FLIGHT

MA-3 ENGINE
GS   ENGINES
   . ROCKET ENGINES
   . . BOOSTER ROCKET ENGINES
   . . . **MA-3 ENGINE**
   . . LIQUID PROPELLANT ROCKET
        ENGINES
   . . . **MA-3 ENGINE**
RT   ATLAS ICBM
   VERNIER ENGINES

MA-3 FLIGHT
USE   MERCURY MA-3 FLIGHT

MA-4 FLIGHT
USE   MERCURY MA-4 FLIGHT

MA-5 ENGINE
GS   ENGINES
   . ROCKET ENGINES
   . . BOOSTER ROCKET ENGINES
   . . . **MA-5 ENGINE**
   . . LIQUID PROPELLANT ROCKET
        ENGINES
   . . . **MA-5 ENGINE**
RT   ATLAS LAUNCH VEHICLES
   ATLAS SLV-3 LAUNCH VEHICLE
   VERNIER ENGINES

MA-5 FLIGHT
USE   MERCURY MA-5 FLIGHT

MA-8 FLIGHT
USE   MERCURY MA-8 FLIGHT

MA-9 FLIGHT
USE   MERCURY MA-9 FLIGHT

MAARS
USE   CRATERS

MACE MISSILES
GS   MISSILES
   . SURFACE TO SURFACE MISSILES
   . . **MACE MISSILES**
RT   BOOSTER ROCKET ENGINES
   J-33 ENGINE
   SOLID PROPELLANT ROCKET ENGINES
   TURBOJET ENGINES

MACH CONES
GS   ELASTIC WAVES
   . SHOCK WAVES
   . . **MACH CONES**
RT   ACOUSTIC VELOCITY
   BOW WAVES
   CONES
   HYPERSONIC SHOCK
   SOUND WAVES
   SUPERSONIC FLIGHT
   SUPERSONIC FLOW
   SUPERSONICS

MACH INERTIA PRINCIPLE
GS   INERTIA
   . INERTIA PRINCIPLE
   . . **MACH INERTIA PRINCIPLE**
RT   EQUATIONS OF MOTION
   MOMENTS OF INERTIA

MACH NUMBER
UF   CRITICAL MACH NUMBER
   GLAUERT COEFFICIENT
GS   RATIOS
   . DIMENSIONLESS NUMBERS
   . . **MACH NUMBER**
RT   ACOUSTIC VELOCITY

MACH NUMBER-*(CONT.)*
   AERODYNAMICS
   AIRSPEED
   SHOCK WAVES
   SUPERHARMONICS
   SWEEP ANGLE

MACH REFLECTION
GS   REFLECTION
   . WAVE REFLECTION
   . . **MACH REFLECTION**
RT   SHOCK WAVES

MACH-ZEHNDER INTERFEROMETERS
GS   MEASURING INSTRUMENTS
   . INTERFEROMETERS
   . . **MACH-ZEHNDER INTERFEROMETERS**
RT   AERODYNAMICS
   ARGON LASERS
   CARBON DIOXIDE LASERS
   DIFFRACTOMETERS
   GAS LASERS
   GONIOMETERS
   OPTICAL EQUIPMENT
   OPTICAL MEASURING INSTRUMENTS
   SCHLIEREN PHOTOGRAPHY

MACHINE LEARNING
USE   LEARNING MACHINES

MACHINE LIFE
USE   SERVICE LIFE

MACHINE ORIENTED LANGUAGES
GS   LANGUAGES
   . PROGRAMMING LANGUAGES
   . . **MACHINE ORIENTED LANGUAGES**
   . . . MARVS (PROGRAMMING
        LANGUAGE)
   . . . SLEUTH (PROGRAMMING
        LANGUAGE)
RT   ALGOL
   ASSEMBLY LANGUAGE
   AUTOCODERS
   LANGUAGE PROGRAMMING
   PL/1

MACHINE RECOGNITION
USE   ARTIFICIAL INTELLIGENCE

MACHINE STORAGE
USE   COMPUTER STORAGE DEVICES
   CORE STORAGE

MACHINE TOOLS
GS   TOOLS
   . . **MACHINE TOOLS**
   . . GRINDING MACHINES
   . LATHES
   . . . TURRET LATHES
   . . MILLING MACHINES
   . . SHAPERS
RT   CUTTERS
   DIES
   DRILLS
   ∞MACHINERY
   MACHINING
   MANDRELS
   MECHANICAL DEVICES
   MECHANICAL ENGINEERING
   METAL CUTTING
   NUMERICAL CONTROL
   PRESSES
   PUNCHES
   SAWS
   SHEARS
   TAPS
   ULTRASONIC CLEANING

MACHINE TRANSLATION
GS   LINGUISTICS
   . **MACHINE TRANSLATION**
   TRANSLATING
   . **MACHINE TRANSLATION**
RT   COMPUTER PROGRAMS
   INFORMATION THEORY
   LANGUAGE PROGRAMMING
   LANGUAGES

MACHINE-INDEPENDENT PROGRAMS
GS   COMPUTER PROGRAMS
   . **MACHINE-INDEPENDENT PROGRAMS**
   PROGRAMS
   . **MACHINE-INDEPENDENT PROGRAMS**

MACHINE-INDEPENDENT PROGRAMS-*(CONT.)*
RT   COMPUTER PROGRAMMING
   COMPUTERS
   MULTIPROGRAMMING

∞ **MACHINERY**
SN   *(USE OF A MORE SPECIFIC TERM IS*
   RECOMMENDED--CONSULT THE TERMS
   LISTED BELOW)
RT   BORING MACHINES
   COMPUTERS
   DRAFTING MACHINES
   ENGINES
   FATIGUE TESTING MACHINES
   GRINDING MACHINES
   GROUND EFFECT MACHINES
   IMPACT TESTING MACHINES
   LEARNING MACHINES
   LEVERS
   LOAD TESTING MACHINES
   MACHINE TOOLS
   MECHANICAL ENGINEERING
   ∞MECHANISM
   MECHANIZATION
   MILLING MACHINES
   POSITIONING DEVICES (MACHINERY)
   REFRIGERATING MACHINERY
   ∞ROTATING ELECTRICAL MACHINES
   SELF FOCUSING
   TEACHING MACHINES
   TIDE POWERED MACHINES
   TOOLS
   TURBOMACHINERY
   TURING MACHINES
   VIBRATION SIMULATORS
   WALKING MACHINES
   WATERWAVE POWERED MACHINES
   WELDING MACHINES
   WINDMILLS (WINDPOWERED MACHINES)

MACHINING
UF   MATERIAL REMOVAL (MACHINING)
GS   **MACHINING**
   . CHEMICAL MACHINING
   . . ELECTROCHEMICAL MACHINING
   . HOT MACHINING
   . MILLING (MACHINING)
   . SPARK MACHINING
   . ULTRASONIC MACHINING
RT   ∞CUT-OFF
   CUTTING
   DRILLING
   FINISHES
   FORMING TECHNIQUES
   GRINDING (MATERIAL REMOVAL)
   GROOVING
   KNURLING
   LASER CUTTING
   MACHINE TOOLS
   METAL CUTTING
   METAL WORKING
   PLANING
   RESIDUAL STRESS
   SETUPS
   SURFACE FINISHING
   SURFACE ROUGHNESS
   TOOLING
   V GROOVES

MACLAURIN SERIES
UF   MCLAURIN SERIES
GS   ANALYSIS (MATHEMATICS)
   . CALCULUS
   . . SERIES (MATHEMATICS)
   . . . POWER SERIES
   . . . . TAYLOR SERIES
   . . . . . **MACLAURIN SERIES**
   . REAL VARIABLES
   . . SERIES (MATHEMATICS)
   . . . POWER SERIES
   . . . . TAYLOR SERIES
   . . . . . **MACLAURIN SERIES**

MACROCLIMATE
USE   CLIMATE

MACROMOLECULES
USE   MOLECULES

MACROPHAGES
GS   CELLS (BIOLOGY)
   . **MACROPHAGES**
RT   TISSUES (BIOLOGY)

MACROSCOPIC EQUATIONS
RT   ∞EQUATIONS
      ∞MEASUREMENT
      ∞PROPERTIES
      STATISTICAL MECHANICS

MACULAR VISION
USE   VISION

MADAGASCAR
USE   MALAGASY REPUBLIC

MAFFEI GALAXIES
GS   CELESTIAL BODIES
      . GALAXIES
      . . **MAFFEI GALAXIES**
RT   NEBULAE
      RADIO ASTRONOMY
      RADIO GALAXIES
      RADIO SOURCES (ASTRONOMY)
      SPIRAL GALAXIES

MAGAZINES (SUPPLY CHAMBERS)
RT   AMMUNITION
      PHOTOGRAPHIC FILM
      SPOOLS

MAGDALENA-CAUCA VALLEY (COLOMBIA)
GS   LANDFORMS
      . **MAGDALENA-CAUCA VALLEY
        (COLOMBIA)**
      VALLEYS
      . **MAGDALENA-CAUCA VALLEY
        (COLOMBIA)**
RT   COLOMBIA
      SOUTH AMERICA

MAGELLAN MISSION
GS   OBSERVATORIES
      . ASTRONOMICAL OBSERVATORIES
      . . ASTRONOMICAL SATELLITES
      . . . **MAGELLAN MISSION**
      SATELLITES
      . ARTIFICIAL SATELLITES
      . . ESA SATELLITES
      . . . **MAGELLAN MISSION**
RT   EXTREME ULTRAVIOLET RADIATION
      FAR ULTRAVIOLET RADIATION
      SPACEBORNE ASTRONOMY

MAGELLANIC CLOUDS
GS   CLOUDS
      . **MAGELLANIC CLOUDS**
RT   NEBULAE
      ORION NEBULA
      STAR CLUSTERS
      STARS

MAGIC TEES
RT   DUPLEXERS

MAGMA
RT   IGNEOUS ROCKS
      LAVA
      REGOLITH
      ROCKS
      SOILS

MAGNESIUM
GS   CHEMICAL ELEMENTS
      . **MAGNESIUM**
      . . MAGNESIUM ISOTOPES
      METALS
      . **MAGNESIUM**
      . . MAGNESIUM ISOTOPES

MAGNESIUM ALLOYS
GS   ALLOYS
      . LIGHT ALLOYS
      . . **MAGNESIUM ALLOYS**

MAGNESIUM BROMIDES
GS   HALOGEN COMPOUNDS
      . BROMINE COMPOUNDS
      . . BROMIDES
      . . . **MAGNESIUM BROMIDES**
      . HALIDES
      . . BROMIDES
      . . . **MAGNESIUM BROMIDES**
      . . METAL HALIDES
      . . . **MAGNESIUM BROMIDES**
      MAGNESIUM COMPOUNDS
      . **MAGNESIUM BROMIDES**

MAGNESIUM CELLS
GS   ELECTRIC GENERATORS
      . DIRECT POWER GENERATORS
      . . PRIMARY BATTERIES
      . . . DRY CELLS
      . . . . **MAGNESIUM CELLS**
      ELECTROCHEMICAL CELLS
      . ELECTRIC BATTERIES
      . . PRIMARY BATTERIES
      . . . DRY CELLS
      . . . . **MAGNESIUM CELLS**
RT   CHEMICAL AUXILIARY POWER UNITS
      ELECTROLYTIC POLARIZATION

MAGNESIUM CHLORIDES
GS   HALOGEN COMPOUNDS
      . CHLORINE COMPOUNDS
      . . CHLORIDES
      . . . **MAGNESIUM CHLORIDES**
      . HALIDES
      . . CHLORIDES
      . . . **MAGNESIUM CHLORIDES**
      MAGNESIUM COMPOUNDS
      . **MAGNESIUM CHLORIDES**

MAGNESIUM COMPOUNDS
GS   **MAGNESIUM COMPOUNDS**
      . AKERMANITE
      . BRUCITE
      . CHLOROPHYLLS
      . CORDIERITE
      . DOLOMITE (MINERAL)
      . ENSTATITE
      . FORSTERITE
      . GEHLENITE
      . MAGNESIUM BROMIDES
      . MAGNESIUM CHLORIDES
      . MAGNESIUM FLUORIDES
      . MAGNESIUM GERMANATES
      . MAGNESIUM GERMANIDES
      . MAGNESIUM OXIDES
      . . PERICLASE
      . MAGNESIUM PERCHLORATES
      . MAGNESIUM SULFATES
      . . HEXAHEDRITE
      . MAGNESIUM TITANATES
      . MERWINITE
      . MONTICELLITE
      . TALC
RT   ∞ALKALINE EARTH COMPOUNDS
      ∞CHEMICAL COMPOUNDS
      ∞METAL COMPOUNDS

MAGNESIUM FLUORIDES
GS   HALOGEN COMPOUNDS
      . FLUORINE COMPOUNDS
      . . FLUORIDES
      . . . METAL FLUORIDES
      . . . . **MAGNESIUM FLUORIDES**
      . HALIDES
      . . METAL HALIDES
      . . . **MAGNESIUM FLUORIDES**
      MAGNESIUM COMPOUNDS
      . **MAGNESIUM FLUORIDES**

MAGNESIUM GERMANATES
GS   GERMANIUM COMPOUNDS
      . GERMANATES
      . . **MAGNESIUM GERMANATES**
      MAGNESIUM COMPOUNDS
      . **MAGNESIUM GERMANATES**

MAGNESIUM GERMANIDES
GS   GERMANIUM COMPOUNDS
      . GERMANIDES
      . . **MAGNESIUM GERMANIDES**
      MAGNESIUM COMPOUNDS
      . **MAGNESIUM GERMANIDES**

MAGNESIUM ISOTOPES
GS   CHEMICAL ELEMENTS
      . MAGNESIUM
      . . **MAGNESIUM ISOTOPES**
      METALS
      . MAGNESIUM
      . . **MAGNESIUM ISOTOPES**

MAGNESIUM OXIDES
GS   CHALCOGENIDES
      . OXIDES
      . . METAL OXIDES
      . . . ALKALINE EARTH OXIDES
      . . . . **MAGNESIUM OXIDES**
      . . . . . PERICLASE
      MAGNESIUM COMPOUNDS

MAGNESIUM OXIDES-(CONT.)
      . **MAGNESIUM OXIDES**
      . . PERICLASE

MAGNESIUM PERCHLORATES
GS   HALOGEN COMPOUNDS
      . CHLORINE COMPOUNDS
      . . PERCHLORATES
      . . . **MAGNESIUM PERCHLORATES**
      MAGNESIUM COMPOUNDS
      . **MAGNESIUM PERCHLORATES**

MAGNESIUM SULFATES
GS   MAGNESIUM COMPOUNDS
      . **MAGNESIUM SULFATES**
      . . HEXAHEDRITE
      SULFUR COMPOUNDS
      . SULFATES
      . . **MAGNESIUM SULFATES**
      . . . HEXAHEDRITE
RT   BLOEDITE

MAGNESIUM TITANATES
GS   MAGNESIUM COMPOUNDS
      . **MAGNESIUM TITANATES**
      TITANIUM COMPOUNDS
      . TITANATES
      . . **MAGNESIUM TITANATES**

MAGNESYN (TRADEMARK)
USE   SERVOMOTORS

MAGNET COILS
GS   **MAGNET COILS**
      . FIELD COILS
RT   ∞COILS
      ELECTROMAGNETISM
      ELECTROMAGNETS
      INDUCTORS
      MAGNETIC CIRCUITS
      MAGNETIC CORES
      MAGNETIC ENERGY STORAGE
      MAGNETS
      SATURABLE REACTORS
      SOLENOIDS
      SUPERCONDUCTING MAGNETS
      TOROIDS
      TRANSFORMERS
      WIRE WINDING
      YOKES

MAGNETIC ABSORPTION
USE   ELECTROMAGNETIC ABSORPTION

MAGNETIC AMPLIFIERS
GS   AMPLIFIERS
      . **MAGNETIC AMPLIFIERS**
RT   MAGNETOSTATIC AMPLIFIERS
      NONLINEARITY
      POWER AMPLIFIERS
      SATURABLE REACTORS
      VOLTAGE AMPLIFIERS

MAGNETIC ANNULAR ARC
RT   ∞ARCS
      CURRENT DISTRIBUTION
      PLASMA ACCELERATORS
      PLASMA CONTROL
      PLASMA PROPULSION

MAGNETIC ANNULAR SHOCK TUBES
UF   MAST SHOCK TUBES
GS   SHOCK WAVE GENERATORS
      . SHOCK TUBES
      . . **MAGNETIC ANNULAR SHOCK TUBES**

MAGNETIC ANOMALIES
UF   GEOMAGNETIC ANOMALIES
GS   ANOMALIES
      . **MAGNETIC ANOMALIES**
      . . GEOMAGNETIC HOLLOW
RT   AEROMAGNETISM
      GEOMAGNETISM
      NONUNIFORM MAGNETIC FIELDS

MAGNETIC BEARINGS
GS   BEARINGS
      . **MAGNETIC BEARINGS**
RT   LEVITATION
      MAGNETIC SUSPENSION

MAGNETIC CHARGE DENSITY
UF   SCALAR MAGNETIC CHARGE
GS   DIVERGENCE

MAGNETIC CHARGE DENSITY-*(CONT.)*
. **MAGNETIC CHARGE DENSITY**
RT ∞CHARGING
CONSTITUTIVE EQUATIONS
MAXWELL EQUATION

MAGNETIC CIRCUITS
GS CIRCUITS
. **MAGNETIC CIRCUITS**
RT FLUX (RATE)
LINES OF FORCE
MAGNET COILS
∞NETWORKS
SATURABLE REACTORS
TRANSFORMERS

MAGNETIC COILS
GS ELECTRIC COILS
. **MAGNETIC COILS**
RT ∞COILS
ELECTROMAGNETIC HAMMERS
ELECTROMAGNETISM
FLUX PUMPS

MAGNETIC COMPASSES
GS AIRCRAFT INSTRUMENTS
. COMPASSES
. . **MAGNETIC COMPASSES**
NAVIGATION AIDS
. NAVIGATION INSTRUMENTS
. COMPASSES
. . . **MAGNETIC COMPASSES**
RT GYROCOMPASSES
SOLAR COMPASSES

MAGNETIC COMPRESSION
RT COMPRESSING
CONFINEMENT
PLASMA CONTROL
PLASMAS (PHYSICS)

MAGNETIC CONTROL
RT ATTITUDE CONTROL
∞CONTROL

MAGNETIC COOLING
GS COOLING
. **MAGNETIC COOLING**
RT ABSORPTION COOLING
ADIABATIC DEMAGNETIZATION COOLING
LOW TEMPERATURE
LOW TEMPERATURE ENVIRONMENTS
REFRIGERATING

MAGNETIC CORES
GS CORES
. **MAGNETIC CORES**
RT BUBBLE MEMORY DEVICES
ELECTRIC COILS
FERRITES
FERROMAGNETISM
LAMINATES
MAGNET COILS
MAGNETS
PARAMETRONS
SATURABLE REACTORS
TOROIDS
TRANSFORMERS

MAGNETIC DIFFUSION
SN (DIFFUSION VIA A MAGNETIC FIELD)
GS DIFFUSION
. **MAGNETIC DIFFUSION**
RT FIELD STRENGTH

MAGNETIC DIPOLES
RT ∞DIPOLES
ELECTRIC DIPOLES
∞PHYSICAL PROPERTIES
∞POLES

MAGNETIC DISKS
GS MAGNETIC STORAGE
. **MAGNETIC DISKS**
RT CORE STORAGE
∞DISKS
MEMORY (COMPUTERS)
VIDEO DISKS

MAGNETIC DISPERSION
RT ∞DISPERSION
ELECTROMAGNETIC SCATTERING
FERROMAGNETISM
MAGNETIZATION

MAGNETIC DISPERSION-*(CONT.)*
WAVE SCATTERING

MAGNETIC DISTURBANCES
GS **MAGNETIC DISTURBANCES**
. MAGNETIC STORMS
RT AURORAS
∞DISTURBANCES
GEOMAGNETISM
KP INDEX
NONADIABATIC THEORY
SOLAR ACTIVITY
SOLAR ACTIVITY EFFECTS
SOLAR FLARES
SOLAR PLANETARY INTERACTIONS
SOLAR TERRESTRIAL INTERACTIONS
SOLAR WIND VELOCITY
STARSPOTS
STELLAR ACTIVITY
SUDDEN IONOSPHERIC DISTURBANCES
SUDDEN STORM COMMENCEMENTS
SUNSPOTS

MAGNETIC DOMAINS
GS DOMAINS
. **MAGNETIC DOMAINS**
RT BUBBLE MEMORY DEVICES
BUBBLE TECHNIQUE
DIPOLE MOMENTS
DOMAIN WALL
LINES OF FORCE

MAGNETIC DRUMS
GS MAGNETIC STORAGE
. **MAGNETIC DRUMS**
RT CORE STORAGE
∞DRUMS

MAGNETIC EFFECTS
UF GEOMAGNETIC EFFECTS
GS MAGNETIC PROPERTIES
. **MAGNETIC EFFECTS**
. . MAGNETIC RIGIDITY
RT ∞EFFECTS
GEOMAGNETISM
MAGNETOACTIVITY
PLASMA COMPRESSION
TEMPERATURE EFFECTS

MAGNETIC ENERGY STORAGE
GS ENERGY STORAGE
. **MAGNETIC ENERGY STORAGE**
RT ENERGY TECHNOLOGY
MAGNET COILS
MAGNETIC FIELDS
SUPERCONDUCTING MAGNETS

MAGNETIC EQUATOR
UF GEOMAGNETIC EQUATOR
GS EQUATORS
. **MAGNETIC EQUATOR**
RT GEOMAGNETISM
∞INCLINATION

MAGNETIC FIELD CONFIGURATIONS
RT ASTROPHYSICS
ELECTROMAGNETIC FIELDS
PLASMA COMPRESSION
PLASMA CONTROL
PLASMA PHYSICS
POLAR CUSPS
POLOIDAL FLUX
SPHEROMAKS
STELLAR MAGNETIC FIELDS

MAGNETIC FIELD INTENSITY
USE MAGNETIC FLUX

MAGNETIC FIELD INVERSIONS
GS INVERSIONS
. **MAGNETIC FIELD INVERSIONS**
RT ELECTROMAGNETIC FIELDS
ELECTROMAGNETISM
ELECTROMECHANICS
FIELD THEORY (PHYSICS)

† **MAGNETIC FIELDS**
GS **MAGNETIC FIELDS**
. BIOMAGNETISM
. FORCE-FREE MAGNETIC FIELDS
. GEOMAGNETISM
. INTERPLANETARY MAGNETIC FIELDS
. INTERSTELLAR MAGNETIC FIELDS
. LUNAR MAGNETIC FIELDS

MAGNETIC FIELDS-*(CONT.)*
. MAGNETOSTATIC FIELDS
. NONUNIFORM MAGNETIC FIELDS
. PALEOMAGNETISM
. PLANETARY MAGNETIC FIELDS
. STELLAR MAGNETIC FIELDS
. . SOLAR MAGNETIC FIELD
. TRAPPED MAGNETIC FIELDS
RT BERNSTEIN ENERGY PRINCIPLE
BETA FACTOR
CONJUGATE POINTS
CONSTITUTIVE EQUATIONS
CROSSED FIELDS
DEMAGNETIZATION
ELECTRIC FIELDS
ELECTROMAGNETIC ACCELERATION
ELECTROMAGNETIC FIELDS
ELECTROMAGNETISM
ELECTROMECHANICS
ELECTRON-HOLE DROPS
FERROMAGNETIC RESONANCE
FIELD EMISSION
FIELD STRENGTH
FIELD THEORY (PHYSICS)
∞FIELDS
FLUX PUMPS
GEOMAGNETIC TAIL
HELIOS SATELLITES
INTASAT SATELLITE
∞KERR EFFECTS
LINES OF FORCE
LORENTZ FORCE
MAGNETIC ENERGY STORAGE
MAGNETIZATION
MAGNETO-OPTICS
MAGNETOACTIVITY
MAGNETOPLASMADYNAMICS
MAGNETORESISTIVITY
MAGNETOSPHERE
MAGNETOSTATICS
MAGNETS
MULTIPOLAR FIELDS
PARTICLE ACCELERATION
PINCH EFFECT
POLAR CUSPS
POLARITY
RACETRACKS (PARTICLE
ACCELERATORS)
RADIATION BELTS
SCREW PINCH
SCYLLA
SELF CONSISTENT FIELDS
SQUARE WELLS
SUHL EFFECT
ZEEMAN EFFECT

MAGNETIC FILMS
GS COATINGS
. **MAGNETIC FILMS**
RT ∞FILMS

MAGNETIC FLUX
UF MAGNETIC FIELD INTENSITY
GS FIELD STRENGTH
. **MAGNETIC FLUX**
RATES (PER TIME)
. FLUX (RATE)
. . **MAGNETIC FLUX**
RT BETA FACTOR
CONSTITUTIVE EQUATIONS
CURRENT SHEETS
FLUX PINNING
LINES OF FORCE
PINNING

MAGNETIC FORMING
GS FORMING TECHNIQUES
. **MAGNETIC FORMING**
METAL WORKING
. **MAGNETIC FORMING**
RT BULGING
COLD WORKING
DEEP DRAWING
ELECTROMAGNETIC HAMMERS
METAL DRAWING

MAGNETIC INDUCTION
UF ELECTROMAGNETIC DEDUCTION
GS MAGNETIC PROPERTIES
. **MAGNETIC INDUCTION**
RT COUPLING COEFFICIENTS
FLUX (RATE)
INDUCTANCE
∞INDUCTION
INDUCTION HEATING

MAGNETIC INDUCTION PROBES
USE   MAGNETIC PROBES

MAGNETIC LENSES
UF   QUADRUPOLE LENSES
GS   LENSES
       . **MAGNETIC LENSES**
RT   CATHODE RAY TUBES
       ELECTRON BEAMS
       ELECTRON GUNS
       ELECTRON MICROSCOPES
       ELECTRON MICROSCOPY
       PLASMA GUNS
       PLASMA JETS
       WIRE GRID LENSES

MAGNETIC LEVITATION VEHICLES
GS   SURFACE VEHICLES
       . **MAGNETIC LEVITATION VEHICLES**
RT   LEVITATION
       LIFT DEVICES
       MASS DRIVERS (PAYLOAD DELIVERY)
       RAIL TRANSPORTATION
       SUSPENSION SYSTEMS (VEHICLES)
       ∞ VEHICLES

MAGNETIC MATERIALS
UF   MAGNETIC METALS
GS   **MAGNETIC MATERIALS**
       . FERRIMAGNETIC MATERIALS
       . FERROMAGNETIC MATERIALS
       .. FERROFLUIDS
       .. FERROMAGNETIC FILMS
       .. MAGNETITE
       .. PERMALLOYS (TRADEMARK)
RT   KONDO EFFECT
       MAGNETS
       ∞ MATERIALS

MAGNETIC MEASUREMENT
SN   (MEASUREMENT OF MAGNETIC
       PROPERTIES, QUANTITIES OR
       CONDITIONS)
UF   FLUXMETERS
       MAGNETOMETRY
RT   ELECTROMAGNETIC MEASUREMENT
       MAGNETOMETERS
       ∞ MATERIALS TESTS
       ∞ MEASUREMENT
       SQUID (DETECTORS)

MAGNETIC MEMORIES
USE   MAGNETIC STORAGE

MAGNETIC METALS
USE   MAGNETIC MATERIALS
       METALS

MAGNETIC MIRRORS
GS   MIRRORS
       . **MAGNETIC MIRRORS**
RT   LINES OF FORCE
       MIRROR FUSION
       MIRROR POINT
       NONUNIFORM MAGNETIC FIELDS
       NUCLEAR FUSION
       PLASMA CONTROL
       PLASMA EQUILIBRIUM
       Q DEVICES
       SCYLLA
       SPHEROMAKS

MAGNETIC MOMENTS
GS   MAGNETIC PROPERTIES
       . **MAGNETIC MOMENTS**
       MOMENTS
       . DIPOLE MOMENTS
       .. **MAGNETIC MOMENTS**
RT   BOHR MAGNETON
       ELECTRIC MOMENTS
       LANGEVIN FORMULA
       QUENCHING (ATOMIC PHYSICS)

MAGNETIC MONOPOLES
GS   MONOPOLES
       . **MAGNETIC MONOPOLES**
       PARTICLES
       . ELEMENTARY PARTICLES
       .. **MAGNETIC MONOPOLES**
RT   QUANTUM THEORY

MAGNETIC PERMEABILITY
UF   MAGNETIC SUSCEPTIBILITY
       SUSCEPTIBILITY (MAGNETISM)

MAGNETIC PERMEABILITY-(CONT.)
GS   MAGNETIC PROPERTIES
       . **MAGNETIC PERMEABILITY**
RT   CURIE-WEISS LAW
       DIELECTRIC PERMEABILITY
       HYSTERESIS
       NEEL TEMPERATURE
       RELUCTANCE

MAGNETIC PISTONS
GS   PISTONS
       . **MAGNETIC PISTONS**
RT   HYPERSONIC WIND TUNNELS
       HYPERVELOCITY WIND TUNNELS
       SHOCK TUBES
       SHOCK WAVE GENERATORS

MAGNETIC POLES
RT   AURORAL ZONES
       ∞ DIPOLES
       GEOMAGNETISM
       POLARITY
       ∞ POLES

MAGNETIC PROBES
UF   MAGNETIC INDUCTION PROBES
GS   MEASURING INSTRUMENTS
       . **MAGNETIC PROBES**
RT   MAGNETOMETERS
       RESONANCE PROBES
       SPACE PROBES

MAGNETIC PROPERTIES
GS   **MAGNETIC PROPERTIES**
       . ANTIFERROMAGNETISM
       . BIOMAGNETISM
       . CURIE TEMPERATURE
       . DIAMAGNETISM
       . FERRIMAGNETISM
       . FERROMAGNETISM
       . GEOMAGNETISM
       . GYROMAGNETISM
       .. GYROFREQUENCY
       . MAGNETIC EFFECTS
       .. MAGNETIC RIGIDITY
       . MAGNETIC INDUCTION
       . MAGNETIC MOMENTS
       . MAGNETIC PERMEABILITY
       . MAGNETIC RELAXATION
       .. SPIN-LATTICE RELAXATION
       . MAGNETIC SUSPENSION
       . MAGNETOACOUSTICS
       . MAGNETOACTIVITY
       .. MAGNETORESISTIVITY
       . MAGNETOSTRICTION
       . PALEOMAGNETISM
       . PARAMAGNETISM
       . POLARIZATION CHARACTERISTICS
       . RELUCTANCE
       . REMANENCE
       . THERMOMAGNETIC EFFECTS
RT   COERCIVITY
       CURIE-WEISS LAW
       DIPOLE MOMENTS
       EDDY CURRENTS
       ELECTRICAL PROPERTIES
       ELECTROMAGNETIC PROPERTIES
       ELECTROMAGNETISM
       FERRITIC STAINLESS STEELS
       FIELD STRENGTH
       HYSTERESIS
       INDUCTANCE
       KP INDEX
       LINES OF FORCE
       MAGNETIZATION
       MAGNETOMECHANICS (PHYSICS)
       MAGNETS
       MAXWELL EQUATION
       ∞ PHYSICAL PROPERTIES
       POLARIZATION (SPIN ALIGNMENT)
       ∞ PROPERTIES
       ∞ SOLID STATE PHYSICS
       SPIN GLASS

MAGNETIC PUMPING
RT   ELECTRON CYCLOTRON HEATING
       INDUCTION HEATING
       ION CYCLOTRON RADIATION
       KINETIC HEATING
       PLASMA HEATING
       ∞ PUMPING

MAGNETIC RECORDING
UF   MAGNETIC TAPE RECORDERS
GS   RECORDING

MAGNETIC RECORDING-(CONT.)
       . **MAGNETIC RECORDING**
RT   BUBBLE MEMORY DEVICES
       DATA RECORDING
       RECORDING HEADS

MAGNETIC RELAXATION
GS   MAGNETIC PROPERTIES
       . **MAGNETIC RELAXATION**
       .. SPIN-LATTICE RELAXATION
RT   RELAXATION (MECHANICS)

MAGNETIC RESONANCE
GS   RESONANCE
       . **MAGNETIC RESONANCE**
       .. FERROMAGNETIC RESONANCE
       .. NUCLEAR MAGNETIC RESONANCE
       ... PROTON MAGNETIC RESONANCE
       ... PROTON RESONANCE
       .. PARAMAGNETIC RESONANCE
       ... ELECTRON PARAMAGNETIC
           RESONANCE
RT   NUCLEAR SPIN
       OVERHAUSER EFFECT
       SPECTRUM ANALYSIS

MAGNETIC RIGIDITY
UF   GYROINTERACTION
GS   MAGNETIC PROPERTIES
       . MAGNETIC EFFECTS
       .. **MAGNETIC RIGIDITY**
RT   ELECTRON TRAJECTORIES
       IONOSPHERIC DRIFT
       PARTICLE MASS
       PARTICLE MOTION
       ∞ RIGIDITY

MAGNETIC SHIELDING
GS   SHIELDING
       . **MAGNETIC SHIELDING**
RT   ELECTROMAGNETIC SHIELDING
       MAGNETOMETERS
       RADIATION SHIELDING

MAGNETIC SIGNALS
RT   NUCLEAR MAGNETIC RESONANCE
       SIGNAL MIXING
       ∞ SIGNALS

MAGNETIC SIGNATURES
UF   MAGNETOGRAMS
GS   SIGNATURES
       . **MAGNETIC SIGNATURES**
RT   PATTERN REGISTRATION

MAGNETIC SPECTROSCOPY
GS   SPECTROSCOPY
       . **MAGNETIC SPECTROSCOPY**
RT   GAS SPECTROSCOPY
       MASS SPECTROSCOPY
       SPECTROSCOPIC ANALYSIS
       VACUUM SPECTROSCOPY

MAGNETIC STARS
GS   CELESTIAL BODIES
       . STARS
       .. **MAGNETIC STARS**
RT   PECULIAR STARS

MAGNETIC STORAGE
UF   MAGNETIC MEMORIES
GS   **MAGNETIC STORAGE**
       . BUBBLE MEMORY DEVICES
       . CORE STORAGE
       . MAGNETIC DISKS
       . MAGNETIC DRUMS
RT   COMPUTER STORAGE DEVICES
       DATA RECORDING
       DATA STORAGE
       ∞ DRUMS
       PARAMETRONS
       ∞ STORAGE
       VIRTUAL MEMORY SYSTEMS

MAGNETIC STORMS
UF   GEOMAGNETIC STORMS
       MAGNETIC SUBSTORMS
GS   MAGNETIC DISTURBANCES
       . **MAGNETIC STORMS**
       STORMS
       . **MAGNETIC STORMS**
RT   DAWN CHORUS
       FORBUSH DECREASES
       NOISE STORMS

MAGNETIC STORMS-*(CONT.)*
    SOLAR STORMS
    SOLAR TERRESTRIAL INTERACTIONS
    SPREAD F
    SUDDEN IONOSPHERIC DISTURBANCES
    SUDDEN STORM COMMENCEMENTS

MAGNETIC SUBSTORMS
  USE   MAGNETIC STORMS

MAGNETIC SURVEYS
  UF   MAGNETOTELLURIC PROFILING
  RT   AEROMAGNETISM
       GEOMAGNETISM

MAGNETIC SUSCEPTIBILITY
  USE   MAGNETIC PERMEABILITY

MAGNETIC SUSPENSION
  GS   MAGNETIC PROPERTIES
      . **MAGNETIC SUSPENSION**
      SUSPENDING (HANGING)
      . **MAGNETIC SUSPENSION**
  RT   ANNULAR SUSPENSION AND POINTING
       SYSTEM
      LEVITATION MELTING
      MAGNETIC BEARINGS

MAGNETIC SWITCHING
  GS   SWITCHING
      . **MAGNETIC SWITCHING**
  RT   ANTIFERROMAGNETISM
      BEAM SWITCHING
      BUBBLE MEMORY DEVICES
      SATURABLE REACTORS

MAGNETIC TAPE RECORDERS
  USE   MAGNETIC RECORDING
       TAPE RECORDERS

MAGNETIC TAPE TRANSPORTS
  GS   MECHANICAL DRIVES
      . **MAGNETIC TAPE TRANSPORTS**
  RT   TAPE RECORDERS

MAGNETIC TAPES
  GS   PERIPHERAL EQUIPMENT (COMPUTERS)
      . COMPUTER STORAGE DEVICES
      . . **MAGNETIC TAPES**
      . . . COMPUTER COMPATIBLE TAPES
  RT   PLASTIC TAPES
      PLAYBACKS
      PUNCHED TAPES
      READERS
      RECORDING HEADS
      REELS
      TAPE RECORDERS
      ∞TAPES

MAGNETIC TRANSDUCERS
  GS   TRANSDUCERS
      . **MAGNETIC TRANSDUCERS**
  RT   ELECTROMAGNETIC MEASUREMENT
      ELECTRONIC TRANSDUCERS
      MICROPHONES

MAGNETIC VARIATIONS
  GS   VARIATIONS
      . **MAGNETIC VARIATIONS**
      . . GEOMAGNETIC PULSATIONS
      . . . GEOMAGNETIC MICROPULSATIONS
      . . NOCTURNAL VARIATIONS
  RT   AEROMAGNETISM
      ANNUAL VARIATIONS
      DIURNAL VARIATIONS
      IONOSPHERIC DISTURBANCES
      KP INDEX
      SCYLLA
      TRAVELING IONOSPHERIC
       DISTURBANCES

MAGNETICALLY TRAPPED PARTICLES
  GS   PARTICLES
      . CHARGED PARTICLES
      . . **MAGNETICALLY TRAPPED**
         PARTICLES
      . . . RADIATION BELTS
      . . . . ARTIFICIAL RADIATION BELTS
      . . . . INNER RADIATION BELT
      . . . . OUTER RADIATION BELT
      . . . . PROTON BELTS
      . TRAPPED PARTICLES
      . . **MAGNETICALLY TRAPPED**
         PARTICLES

MAGNETICALLY TRAPPED PARTICLES-*(CONT.)*
      . . . RADIATION BELTS
      . . . . ARTIFICIAL RADIATION BELTS
      . . . . INNER RADIATION BELT
      . . . . OUTER RADIATION BELT
      . . . . PROTON BELTS
  RT   PLASMA CONTROL
      TRAPPED MAGNETIC FIELDS

MAGNETITE
  GS   CHALCOGENIDES
      . OXIDES
      . . METAL OXIDES
      . . . IRON OXIDES
      . . . . **MAGNETITE**
      IRON COMPOUNDS
      . IRON OXIDES
      . . **MAGNETITE**
      MAGNETIC MATERIALS
      . FERROMAGNETIC MATERIALS
      . . **MAGNETITE**
      MINERALS
      . **MAGNETITE**

MAGNETIZATION
  UF   REMAGNETIZATION
  RT   COERCIVITY
      MAGNETIC DISPERSION
      MAGNETIC FIELDS
      MAGNETIC PROPERTIES
      MAGNETOMECHANICS (PHYSICS)
      MAGNETS
      MAGNONS
      ∞POLARIZATION
      POLARIZATION (CHARGE SEPARATION)
      POLARIZATION (SPIN ALIGNMENT)

MAGNETO-OPTICS
  RT   ACOUSTO-OPTICS
      ELECTRO-OPTICS
      ELECTROMAGNETIC RADIATION
      FARADAY EFFECT
      KERR MAGNETOOPTICAL EFFECT
      MAGNETIC FIELDS
      ∞OPTICS
      POLARIZATION (WAVES)
      POLARIZED ELECTROMAGNETIC
       RADIATION

MAGNETOACOUSTIC WAVES
  GS   ELASTIC WAVES
      . MAGNETOELASTIC WAVES
      . . **MAGNETOACOUSTIC WAVES**
  RT   MAGNETOHYDRODYNAMIC WAVES
      PLASMA WAVES

MAGNETOACOUSTICS
  GS   ACOUSTICS
      . **MAGNETOACOUSTICS**
      MAGNETIC PROPERTIES
      . **MAGNETOACOUSTICS**

MAGNETOACTIVITY
  GS   MAGNETIC PROPERTIES
      . **MAGNETOACTIVITY**
      . . MAGNETORESISTIVITY
  RT   MAGNETIC EFFECTS
      MAGNETIC FIELDS

MAGNETOCARDIOGRAPHY
  GS   BIOENGINEERING
      . BIOMETRICS
      . . CARDIOGRAPHY
      . . . **MAGNETOCARDIOGRAPHY**
  RT   BIOINSTRUMENTATION

MAGNETOELASTIC VIBRATIONS
  USE   MAGNETOELASTIC WAVES

MAGNETOELASTIC WAVES
  UF   MAGNETOELASTIC VIBRATIONS
  GS   ELASTIC WAVES
      . **MAGNETOELASTIC WAVES**
      . . MAGNETOACOUSTIC WAVES
  RT   ELECTROSTATIC WAVES
      MAGNETOSONIC RESONANCE
      MAGNETOSPHERIC INSTABILITY
      MAGNETOSTRICTION
      PLASMA WAVES
      SOUND WAVES
      ULTRASONIC RADIATION

MAGNETOELASTICITY
  USE   MAGNETOSTRICTION

MAGNETOELECTRIC MEDIA
  RT   DIELECTRICS
      MAGNETOIONICS
      MAXWELL EQUATION
      MECHANICAL DRIVES

MAGNETOGASDYNAMICS
  USE   MAGNETOHYDRODYNAMICS

MAGNETOGRAMS
  USE   MAGNETIC SIGNATURES

MAGNETOHYDRODYNAMIC ACCELERATION
  USE   PLASMA ACCELERATION

MAGNETOHYDRODYNAMIC FLOW
  UF   HYDROMAGNETIC FLOW
      PLASMA FLOW
  GS   FLUID FLOW
      . **MAGNETOHYDRODYNAMIC FLOW**
  RT   COMPRESSIBLE FLOW
      CORE FLOW
      GAS FLOW
      GEOMAGNETIC HOLLOW
      HARTMANN FLOW
      KELVIN-HELMHOLTZ INSTABILITY
      LINE CURRENT
      MAGNETOHYDRODYNAMICS
      PLASMA FLUX MEASUREMENT
      PLASMA TURBULENCE
      PLASMAS (PHYSICS)
      REVERSE FIELD PINCH
      SCREW PINCH
      SOLAR WIND VELOCITY
      TRANSVERSE WAVES
      TWO FLUID MODELS

MAGNETOHYDRODYNAMIC GENERATORS
  GS   ELECTRIC GENERATORS
      . DIRECT POWER GENERATORS
      . . **MAGNETOHYDRODYNAMIC**
         GENERATORS
  RT   FUEL CELLS
      ∞GENERATORS
      MAGNETOHYDRODYNAMICS
      PLASMA ACCELERATORS
      PLASMA GENERATORS
      THERMIONIC CONVERTERS
      THERMOELECTRIC GENERATORS

MAGNETOHYDRODYNAMIC SHEAR HEATING
  GS   HEATING
      . **MAGNETOHYDRODYNAMIC SHEAR**
        HEATING
  RT   PLASMA HEATING
      PLASMA SHEATHS
      SHOCK HEATING
      VISCOUS FLOW

MAGNETOHYDRODYNAMIC STABILITY
  UF   HYDROMAGNETIC STABILITY
      PLASMA INSTABILITY
      PLASMA STABILITY
  GS   DYNAMIC CHARACTERISTICS
      . DYNAMIC STABILITY
      . . MOTION STABILITY
      . . . FLOW STABILITY
      . . . . **MAGNETOHYDRODYNAMIC**
         STABILITY
      . . . . . WEIBEL INSTABILITY
      . FLOW CHARACTERISTICS
      . FLOW STABILITY
      . . . **MAGNETOHYDRODYNAMIC**
         STABILITY
      . . . . WEIBEL INSTABILITY
      STABILITY
      . DYNAMIC STABILITY
      . . MOTION STABILITY
      . . . FLOW STABILITY
      . . . . **MAGNETOHYDRODYNAMIC**
         STABILITY
  RT   BALLOONING MODES
      BETA FACTOR
      ELASTIC WAVES
      ELLIPTICAL PLASMAS
      HELICAL FLOW
      KELVIN-HELMHOLTZ INSTABILITY
      MAGNETOHYDRODYNAMICS
      MAGNETOHYDROSTATICS
      NONEQUILIBRIUM PLASMAS
      NONUNIFORM PLASMAS
      PLASMA CONDUCTIVITY
      PLASMA COOLING
      PLASMA DECAY
      PLASMA DRIFT

MAGNETOHYDRODYNAMIC STABILITY-*(CONT.)*
    PLASMA EQUILIBRIUM
    PLASMA LIFETIME
    PLASMA LOSS
    PLASMA PINCH
    PLASMA POTENTIALS
    PLASMA SLABS
    PLASMA TEMPERATURE
    PLASMA TURBULENCE
    PLASMAS (PHYSICS)
    PLASMONS
    SPACE PLASMAS
    STRONGLY COUPLED PLASMAS
    THERMAL INSTABILITY
    ZETA PINCH

MAGNETOHYDRODYNAMIC TURBULENCE
GS   TURBULENCE
    . **MAGNETOHYDRODYNAMIC TURBULENCE**
    . . PLASMA TURBULENCE
RT   HOMOGENEOUS TURBULENCE
    ISOTROPIC TURBULENCE

MAGNETOHYDRODYNAMIC WAVES
UF   ALFVEN WAVES
    HYDROMAGNETIC WAVES
    PLASMA SOUND WAVES
GS   ELASTIC WAVES
    . **MAGNETOHYDRODYNAMIC WAVES**
    . . PLASMA WAVES
    . . . ELECTROSTATIC WAVES
RT   MAGNETOACOUSTIC WAVES
    MAGNETOHYDRODYNAMICS
    NORMAL SHOCK WAVES
    OBLIQUE SHOCK WAVES
    SHOCK WAVES

MAGNETOHYDRODYNAMICS
UF   GEOMETRICAL HYDROMAGNETICS
    HYDROMAGNETICS
    HYDROMAGNETISM
    MAGNETOGASDYNAMICS
GS   FLUID MECHANICS
    . FLUID DYNAMICS
    . . HYDRODYNAMICS
    . . . **MAGNETOHYDRODYNAMICS**
    . . . . CYLINDRICAL PLASMAS
    . HYDROMECHANICS
    . . HYDRODYNAMICS
    . . . **MAGNETOHYDRODYNAMICS**
RT   ALPHA PLASMA DEVICES
    CONDUCTING FLUIDS
    ∞DYNAMICS
    ELECTRIC ARCS
    ELECTROHYDRODYNAMICS
    GAS DYNAMICS
    GAS TRANSPORT
    HALL ACCELERATORS
    HALL EFFECT
    HARTMANN FLOW
    HARTMANN NUMBER
    IONIZATION
    MAGNETOHYDRODYNAMIC FLOW
    MAGNETOHYDRODYNAMIC GENERATORS
    MAGNETOHYDRODYNAMIC STABILITY
    MAGNETOHYDRODYNAMIC WAVES
    MAGNETOHYDROSTATICS
    MAGNETOIONICS
    MAGNETOSONIC RESONANCE
    PINCH EFFECT
    PLASMA CURRENTS
    PLASMA DYNAMICS
    PLASMA PHYSICS
    PLASMA PROPULSION
    PLASMAS (PHYSICS)
    SPACE CHARGE
    SPACE MECHANICS
    SPACE PLASMAS
    STELLAR ACTIVITY
    STELLARATORS
    THERMONUCLEAR REACTIONS
    URANIUM PLASMAS

MAGNETOHYDROSTATICS
GS   FLUID MECHANICS
    . HYDROMECHANICS
    . . HYDROSTATICS
    . . . **MAGNETOHYDROSTATICS**
    STATICS
    . HYDROSTATICS
    . . **MAGNETOHYDROSTATICS**
RT   MAGNETOHYDRODYNAMIC STABILITY
    MAGNETOHYDRODYNAMICS
    MAGNETOIONICS

MAGNETOHYDROSTATICS-*(CONT.)*
    PLASMA PHYSICS
    STATIC STABILITY

MAGNETOIONIC PLASMA
USE   PLASMAS (PHYSICS)

MAGNETOIONICS
RT   ELECTROMAGNETIC WAVE TRANSMISSION
    ELLIPTICAL POLARIZATION
    GEOMAGNETISM
    GYROFREQUENCY
    IONOSPHERIC PROPAGATION
    MAGNETOELECTRIC MEDIA
    MAGNETOHYDRODYNAMICS
    MAGNETOHYDROSTATICS
    PLASMAS (PHYSICS)
    RADIO TRANSMISSION

MAGNETOMECHANICS (PHYSICS)
RT   MAGNETIC PROPERTIES
    MAGNETIZATION
    ∞PHYSICS

MAGNETOMETERS
UF   GAUSSMETERS
    GRADIOMETERS
GS   MEASURING INSTRUMENTS
    . **MAGNETOMETERS**
    . . VARIOMETERS
RT   ELECTRICAL MEASUREMENT
    FIELD INTENSITY METERS
    GEOMAGNETISM
    MAGNETIC MEASUREMENT
    MAGNETIC PROBES
    MAGNETIC SHIELDING
    MAGSAT A SATELLITE
    MAGSAT B SATELLITE
    MAGSAT SATELLITES
    MAGSAT 1 SATELLITE
    NUCLEAR MAGNETIC RESONANCE
    PROTON MASERS

MAGNETOMETRY
USE   MAGNETIC MEASUREMENT

MAGNETOPAUSE
GS   EARTH ATMOSPHERE
    . UPPER ATMOSPHERE
    . . MAGNETOSPHERE
    . . . **MAGNETOPAUSE**
    ENVIRONMENTS
    . MAGNETOSPHERE
    . . **MAGNETOPAUSE**
RT   CHAPMAN-FERRARO PROBLEM
    MAGNETOSPHERIC INSTABILITY
    POLAR CUSPS
    SATELLITE ATMOSPHERES
    SOLAR WIND

MAGNETOPLASMADYNAMICS
RT   MAGNETIC FIELDS
    PLASMA DENSITY
    PLASMA PROPULSION
    ROCKET ENGINES
    SPACECRAFT PROPULSION

MAGNETOPLASMAS
USE   PLASMAS (PHYSICS)

MAGNETORESISTIVITY
GS   MAGNETIC PROPERTIES
    . MAGNETOACTIVITY
    . . **MAGNETORESISTIVITY**
    TRANSPORT PROPERTIES
    . ELECTRICAL RESISTIVITY
    . . **MAGNETORESISTIVITY**
RT   ∞CONDUCTIVITY
    ELECTROMAGNETISM
    FERMI SURFACES
    MAGNETIC FIELDS
    RELUCTANCE
    ∞RESISTANCE

MAGNETOSONIC RESONANCE
GS   RESONANCE
    . **MAGNETOSONIC RESONANCE**
RT   MAGNETOELASTIC WAVES
    MAGNETOHYDRODYNAMICS

MAGNETOSPHERE
GS   EARTH ATMOSPHERE
    . UPPER ATMOSPHERE

MAGNETOSPHERE-*(CONT.)*
    . . **MAGNETOSPHERE**
    . . . GEOMAGNETIC TAIL
    . . . MAGNETOPAUSE
    ENVIRONMENTS
    . **MAGNETOSPHERE**
    . . GEOMAGNETIC TAIL
    . . MAGNETOPAUSE
RT   AMPTE (SATELLITES)
    BARIUM ION CLOUDS
    CHAPMAN-FERRARO PROBLEM
    COROTATION
    EXOSPHERE
    GEOMAGNETIC HOLLOW
    GEOMAGNETISM
    GEOS SATELLITES (ESA)
    HETEROSPHERE
    INTERNATIONAL MAGNETOSPHERIC EXPLORER
    INTERNATIONAL MAGNETOSPHERIC STUDY
    IONOSPHERE
    KP INDEX
    MAGNETIC FIELDS
    NEUTRAL SHEETS
    OPEN PROJECT
    PLASMA CLOUDS
    PLASMAPAUSE
    PLASMASPHERE
    POLAR CUSPS
    RADIATION BELTS
    RADIATION TRAPPING
    SATELLITE ATMOSPHERES
    SCREEN EFFECT
    SOLAR PLANETARY INTERACTIONS
    SOLAR TERRESTRIAL INTERACTIONS
    SOLAR WIND VELOCITY
    SPACE PLASMAS
    THERMOSPHERE

MAGNETOSPHERIC ELECTRON DENSITY
GS   DENSITY (NUMBER/VOLUME)
    . PARTICLE DENSITY (CONCENTRATION)
    . . ELECTRON DENSITY (CONCENTRATION)
    . . . **MAGNETOSPHERIC ELECTRON DENSITY**
RT   ATMOSPHERIC DENSITY
    IONOSPHERIC ELECTRON DENSITY
    PLASMA DENSITY

MAGNETOSPHERIC INSTABILITY
GS   STABILITY
    . **MAGNETOSPHERIC INSTABILITY**
RT   GEOMAGNETIC PULSATIONS
    MAGNETOELASTIC WAVES
    MAGNETOPAUSE

MAGNETOSPHERIC ION DENSITY
GS   DENSITY (NUMBER/VOLUME)
    . PARTICLE DENSITY (CONCENTRATION)
    . . ION DENSITY (CONCENTRATION)
    . . . **MAGNETOSPHERIC ION DENSITY**
    . . . . MAGNETOSPHERIC PROTON DENSITY
RT   ATMOSPHERIC DENSITY
    IONOSPHERIC ION DENSITY
    PLASMA DENSITY
    POSITIVE IONS

MAGNETOSPHERIC PROTON DENSITY
GS   DENSITY (NUMBER/VOLUME)
    . PARTICLE DENSITY (CONCENTRATION)
    . . ION DENSITY (CONCENTRATION)
    . . . MAGNETOSPHERIC ION DENSITY
    . . . . **MAGNETOSPHERIC PROTON DENSITY**
    . . . PROTON DENSITY (CONCENTRATION)
    . . . . **MAGNETOSPHERIC PROTON DENSITY**
RT   ATMOSPHERIC DENSITY
    PLASMA DENSITY

MAGNETOSTATIC AMPLIFIERS
GS   AMPLIFIERS
    . **MAGNETOSTATIC AMPLIFIERS**
RT   MAGNETIC AMPLIFIERS
    PARAMETRIC AMPLIFIERS
    TRAVELING WAVE TUBES
    YTTRIUM-ALUMINUM GARNET
    YTTRIUM-IRON GARNET

MAGNETOSTATIC FIELDS
GS   MAGNETIC FIELDS

MAGNETOSTATIC FIELDS-*(CONT.)*
. **MAGNETOSTATIC FIELDS**
RT   FIELD THEORY (PHYSICS)
     LINES OF FORCE

MAGNETOSTATICS
GS   ELECTROMAGNETISM
     . **MAGNETOSTATICS**
RT   ELECTROSTATICS
     FLUX (RATE)
     MAGNETIC FIELDS

MAGNETOSTRICTION
UF   MAGNETOELASTICITY
GS   MAGNETIC PROPERTIES
     . **MAGNETOSTRICTION**
     MECHANICAL PROPERTIES
     . ELASTIC PROPERTIES
     . . **MAGNETOSTRICTION**
RT   ELECTROSTRICTION
     MAGNETOELASTIC WAVES

MAGNETOTELLURIC PROFILING
USE   MAGNETIC SURVEYS

MAGNETOVARIOGRAPHS
USE   VARIOMETERS

MAGNETRON SPUTTERING
GS   SPUTTERING
     . **MAGNETRON SPUTTERING**
RT   DEPOSITION
     METAL COATINGS

MAGNETRONS
GS   ELECTRON TUBES
     . VACUUM TUBES
     . . VACUUM TUBE OSCILLATORS
     . . . MICROWAVE TUBES
     . . . . **MAGNETRONS**
     . . . . . NIGOTRONS
     MICROWAVE EQUIPMENT
     . MICROWAVE TUBES
     . . **MAGNETRONS**
     . . . NIGOTRONS
     OSCILLATORS
     . VACUUM TUBE OSCILLATORS
     . . MICROWAVE TUBES
     . . . **MAGNETRONS**
     . . . . NIGOTRONS
RT   CAVITY RESONATORS
     CROSSED FIELD AMPLIFIERS
     CROSSED FIELDS
     ELECTROSTATIC GENERATORS
     KLYSTRONS
     MULTIMODE RESONATORS
     PLANOTRONS
     RESONATORS
     TRAVELING WAVE TUBES

MAGNETS
GS   **MAGNETS**
     . CRYOGENIC MAGNETS
     . ELECTROMAGNETS
     . . HIGH FIELD MAGNETS
     . . . SUPERCONDUCTING MAGNETS
     . FERRIMAGNETS
     . WIGGLER MAGNETS
RT   ELECTRETS
     FERROMAGNETIC MATERIALS
     FERROMAGNETISM
     MAGNET COILS
     MAGNETIC CORES
     MAGNETIC FIELDS
     MAGNETIC MATERIALS
     MAGNETIC PROPERTIES
     MAGNETIZATION
     PERMALLOYS (TRADEMARK)

MAGNIFICATION
UF   MAGNIFIERS
RT   AMPLIFICATION
     INCREASING
     LENSES
     ∞ PROJECTION

MAGNIFIERS
USE   MAGNIFICATION

MAGNITUDE
GS   **MAGNITUDE**
     . STELLAR MAGNITUDE
RT   AMPLITUDES
     DIMENSIONS

MAGNITUDE-*(CONT.)*
     DISPLACEMENT
     ∞ INTENSITY
     LEVEL (QUANTITY)

MAGNONS
UF   SPIN WAVES
GS   ELEMENTARY EXCITATIONS
     . **MAGNONS**
RT   ANTIFERROMAGNETISM
     FERRIMAGNETISM
     FERROMAGNETISM
     MAGNETIZATION
     PLASMONS

MAGNUS EFFECT
RT   BERNOULLI THEOREM
     BOUNDARY LAYER FLOW
     ∞ EFFECTS
     FLUID DYNAMICS
     FLUID FLOW
     MISSILE DESIGN
     ROTATING CYLINDERS

MAGSAT A SATELLITE
GS   SATELLITES
     . ARTIFICIAL SATELLITES
     . . SCIENTIFIC SATELLITES
     . . . **MAGSAT A SATELLITE**
RT   GEOMAGNETISM
     MAGNETOMETERS

MAGSAT B SATELLITE
GS   SATELLITES
     . ARTIFICIAL SATELLITES
     . . SCIENTIFIC SATELLITES
     . . . **MAGSAT B SATELLITE**
RT   GEOMAGNETISM
     MAGNETOMETERS

MAGSAT SATELLITES
GS   SATELLITES
     . ARTIFICIAL SATELLITES
     . . SCIENTIFIC SATELLITES
     . . . **MAGSAT SATELLITES**
RT   GEOMAGNETISM
     MAGNETOMETERS

MAGSAT 1 SATELLITE
GS   SATELLITES
     . ARTIFICIAL SATELLITES
     . . SCIENTIFIC SATELLITES
     . . . **MAGSAT 1 SATELLITE**
RT   GEOMAGNETISM
     MAGNETOMETERS

MAIN SEQUENCE STARS
GS   CELESTIAL BODIES
     . STARS
     . . **MAIN SEQUENCE STARS**
     . . . PRE-MAIN SEQUENCE STARS
RT   COLOR-MAGNITUDE DIAGRAM
     DWARF STARS
     EARLY STARS
     GIANT STARS
     LATE STARS
     M STARS
     RED DWARF STARS
     STELLAR EVOLUTION
     STELLAR MASS
     SUBDWARF STARS
     SUBGIANT STARS

MAINE
GS   NATIONS
     . UNITED STATES
     . . **MAINE**
RT   ST LAWRENCE VALLEY (NORTH
       AMERICA)

MAINTAINABILITY
RT   DESIGN ANALYSIS
     MAINTENANCE
     RELIABILITY

MAINTENANCE
UF   REPAIRING
     TROUBLESHOOTING
GS   **MAINTENANCE**
     . AIRCRAFT MAINTENANCE
     . FILE MAINTENANCE (COMPUTERS)
     . SPACE MAINTENANCE
     . SPACECRAFT MAINTENANCE
RT   CHECKOUT

MAINTENANCE-*(CONT.)*
     CONSTRUCTION
     DAMAGE ASSESSMENT
     DOWNTIME
     EQUIPMENT SPECIFICATIONS
     ∞ FIXING
     GROUND CREWS
     GROUND SUPPORT EQUIPMENT
     INSTALLING
     LOGISTICS
     LOGISTICS MANAGEMENT
     LUBRICANTS
     LUBRICATION
     MAINTAINABILITY
     MANUALS
     MECHANICAL ENGINEERING
     OPERATING COSTS
     RELIABILITY
     REPLACING
     SELF REPAIRING DEVICES
     SERVICE LIFE
     SHIPYARDS
     SHOPS
     SPARE PARTS
     SPECIFICATIONS

MAINTENANCE TRAINING
GS   EDUCATION
     . **MAINTENANCE TRAINING**

MAJORITY CARRIERS
GS   CHARGE CARRIERS
     . **MAJORITY CARRIERS**
RT   ADDITIVES
     BIPOLAR TRANSISTORS
     CARRIER INJECTION
     ELECTRON MOBILITY
     ELECTRONS
     HOLES (ELECTRON DEFICIENCIES)
     SEMICONDUCTORS (MATERIALS)

MALAGASY REPUBLIC
UF   MADAGASCAR
GS   LANDFORMS
     . ISLANDS
     . **MALAGASY REPUBLIC**
     NATIONS
     . **MALAGASY REPUBLIC**
RT   AFRICA
     INDIAN OCEAN

MALAWI
GS   NATIONS
     . **MALAWI**
RT   AFRICA

MALAYA
USE   MALAYSIA

MALAYSIA
UF   MALAYA
GS   NATIONS
     . MALAYSIA

MALDIVE ISLANDS
GS   LANDFORMS
     . ISLANDS
     NATIONS
     . MALDIVE ISLANDS

MALEATES
GS   ALIPHATIC COMPOUNDS
     . **MALEATES**
     ESTERS
     . **MALEATES**

MALES
RT   CHILDREN
     FEMALES
     SEX
     SEX FACTOR

MALFUNCTIONS
RT   ABORTED MISSIONS
     AIRCRAFT ACCIDENTS
     AIRCRAFT HAZARDS
     DOWNTIME
     ERRORS
     FAILURE
     SYSTEM FAILURES

MALI
GS   NATIONS
     . **MALI**

MALI-*(CONT.)*
RT   AFRICA

MALKUS THEORY
RT   STATISTICAL MECHANICS
      ∞ THEORIES

MALLEABILITY
GS   MECHANICAL PROPERTIES
      . **MALLEABILITY**
RT   DUCTILITY
      METAL WORKING

MALONONITRILE
GS   CYANIDES
      . **MALONONITRILE**
      NITRILES
      . **MALONONITRILE**

MALTA
GS   LANDFORMS
      . ISLANDS
      . . **MALTA**
RT   MEDITERRANEAN SEA

MAMMALS
GS   ANIMALS
      . VERTEBRATES
      . . **MAMMALS**
      . . . BATS
      . . . BEARS
      . . . CATS
      . . . CATTLE
      . . . . CALVES
      . . . DEER
      . . . . CARIBOUS
      . . . DOGS
      . . . DOLPHINS
      . . . GOATS
      . . . HORSES
      . . . MANATEES
      . . . MARINE MAMMALS
      . . . PORPOISES
      . . . PRIMATES
      . . . . APES
      . . . . . CHIMPANZEES
      . . . . BABOONS
      . . . . HUMAN BEINGS
      . . . MONKEYS
      . . . RODENTS
      . . . . GUINEA PIGS
      . . . . HAMSTERS
      . . . . MICE
      . . . . . JERBOAS
      . . . . POCKET MICE
      . . . . RABBITS
      . . . . RATS
      . . . . SQUIRRELS
      . . . . . GROUND SQUIRRELS
      . . . SEALS (ANIMALS)
      . . . SHEEP
      . . . SHREWS
      . . . SWINE
      . . . WHALES
      . . . WOLVES
RT   EARTH RESOURCES
      MAMMARY GLANDS

MAMMARY GLANDS
GS   ANATOMY
      . GLANDS (ANATOMY)
      . . **MAMMARY GLANDS**
RT   MAMMALS

MAN
USE   HUMAN BEINGS

MAN ENVIRONMENT INTERACTIONS
RT   DESERTIFICATION
      ENVIRONMENT EFFECTS
      ENVIRONMENT MANAGEMENT
      HUMAN BEINGS
      ∞ INTERACTIONS
      INTERNATIONAL
         GEOSPHERE-BIOSPHERE PROGRAM
      RESOURCES

MAN MACHINE SYSTEMS
RT   ASTRONAUT PERFORMANCE
      ∞ AUTOMATION
      BALANCING
      BIONICS
      BIOTECHNOLOGY
      COMPUTER SYSTEMS DESIGN

MAN MACHINE SYSTEMS-*(CONT.)*
      CONSOLES
      CYBERNETICS
      DATA PROCESSING TERMINALS
      DEPERSONALIZATION
      DISPLAY DEVICES
      ∞ ENGINEERING
      HUMAN FACTORS ENGINEERING
      MANAGEMENT
      MECHANIZATION
      PILOT INDUCED OSCILLATION
      ROBOTICS
      ∞ SYSTEMS
      SYSTEMS ANALYSIS
      SYSTEMS ENGINEERING
      SYSTEMS MANAGEMENT
      TELEOPERATORS
      WORKSTATIONS

MAN OPERATED PROPULSION SYSTEMS
UF   MOPS (PROPULSION SYSTEMS)
GS   PROPULSION
      . LOW THRUST PROPULSION
      . . **MAN OPERATED PROPULSION
          SYSTEMS**
RT   ASTRONAUT LOCOMOTION
      COMPRESSED AIR
      EXTRAVEHICULAR ACTIVITY
      GASEOUS ROCKET PROPELLANTS
      MANNED SPACE FLIGHT
      PILOT PERFORMANCE
      RETROROCKET ENGINES
      ∞ SYSTEMS

MAN POWERED AIRCRAFT
RT   ∞ AIRCRAFT
      HANG GLIDERS
      SOARING
      ULTRALIGHT AIRCRAFT
      ∞ WINGED VEHICLES

MANAGEMENT
UF   ADMINISTRATION
GS   **MANAGEMENT**
      . CONFIGURATION MANAGEMENT
      . CONTRACT MANAGEMENT
      . DATA MANAGEMENT
      . ENVIRONMENT MANAGEMENT
      . FINANCIAL MANAGEMENT
      . INDUSTRIAL MANAGEMENT
      . . ENGINEERING MANAGEMENT
      . . INVENTORY MANAGEMENT
      . . . INVENTORY CONTROLS
      . . PERSONNEL MANAGEMENT
      . INFORMATION MANAGEMENT
      . LOGISTICS MANAGEMENT
      . . INVENTORY MANAGEMENT
      . . . INVENTORY CONTROLS
      . MATRIX MANAGEMENT
      . PROCUREMENT MANAGEMENT
      . PRODUCTION MANAGEMENT
      . PROJECT MANAGEMENT
      . RESEARCH MANAGEMENT
      . RESOURCES MANAGEMENT
      . . FOREST MANAGEMENT
      . . . REFORESTATION
      . . LAND MANAGEMENT
      . SAFETY MANAGEMENT
      . SYSTEMS MANAGEMENT
      . TERMINAL AREA ENERGY
          MANAGEMENT
      . WATER MANAGEMENT
      . WEAPON SYSTEM MANAGEMENT
RT   AUTONOMY
      CENTRAL ELECTRONIC MANAGEMENT
         SYSTEM
      COMMAND AND CONTROL
      CONTRACT NEGOTIATION
      COST ANALYSIS
      COST ESTIMATES
      COST INCENTIVES
      COST REDUCTION
      CYBERNETICS
      DECISION MAKING
      DECISIONS
      ∞ DIRECTION
      ECONOMIC ANALYSIS
      ECONOMIC FACTORS
      EVALUATION
      FIDUCIARIES
      FORECASTING
      GERT
      INCENTIVE TECHNIQUES
      INCENTIVES
      INFORMATION FLOW

MANAGEMENT-*(CONT.)*
      MAN MACHINE SYSTEMS
      MARKETING
      MISSION PLANNING
      OPERATIONS RESEARCH
      PERFORMANCE PREDICTION
      PERSONNEL DEVELOPMENT
      PREJUDICES
      PROBLEM SOLVING
      PROCUREMENT POLICY
      PRODUCT DEVELOPMENT
      PRODUCTION ENGINEERING
      PROGRESS
      PROJECT PLANNING
      ∞ RESEARCH PROJECTS
      STATISTICAL ANALYSIS
      SYSTEMS ENGINEERING

MANAGEMENT ANALYSIS
RT   ∞ ANALYZING
      COST ANALYSIS
      GERT
      PERT
      TRADEOFFS

MANAGEMENT INFORMATION SYSTEMS
GS   INFORMATION SYSTEMS
      . **MANAGEMENT INFORMATION
          SYSTEMS**
      MANAGEMENT SYSTEMS
      . **MANAGEMENT INFORMATION
          SYSTEMS**
RT   COMPUTER TECHNIQUES
      DATA BASE MANAGEMENT SYSTEMS
      DATA RETRIEVAL
      DATA STORAGE
      DATA SYSTEMS
      INFORMATION RETRIEVAL
      INFORMATION THEORY
      ∞ SYSTEMS

MANAGEMENT METHODS
GS   **MANAGEMENT METHODS**
      . DELPHI METHOD (FORECASTING)
      . PATTERN METHOD (FORECASTING)
      . PROBE METHOD (FORECASTING)
      . PROFILE METHOD (FORECASTING)
RT   COMPUTER TECHNIQUES
      COST REDUCTION
      CRITICAL PATH METHOD
      DECISION MAKING
      ESTIMATES
      FORECASTING
      GERT
      INCENTIVES
      MATRIX MANAGEMENT
      ∞ METHODOLOGY
      NASA INTERACTIVE PLANNING SYSTEM
      OPERATIONS RESEARCH
      PERT
      RETRAINING
      STARSITE PROGRAM
      SYSTEMS MANAGEMENT

MANAGEMENT PLANNING
GS   PLANNING
      . **MANAGEMENT PLANNING**
      . . PRODUCTION PLANNING
      . . PROJECT PLANNING
RT   CONSULTING
      COST ANALYSIS
      COST REDUCTION
      DECISION MAKING
      ∞ DEVELOPMENT
      ECONOMY
      ESTIMATES
      FEASIBILITY ANALYSIS
      FINANCE
      FORECASTING
      GERT
      INTERFACES
      LIFE CYCLE COSTS
      MEDIATION
      MISSION PLANNING
      OPERATIONS RESEARCH
      PERSONNEL MANAGEMENT
      PERT
      PROGRAM TREND LINE ANALYSIS
      PROJECT MANAGEMENT
      RESEARCH AND DEVELOPMENT
      SELECTIVE DISSEMINATION OF
          INFORMATION
      SYSTEMS ENGINEERING
      TRADEOFFS
      VALUE ENGINEERING

MANAGEMENT SYSTEMS
GS   **MANAGEMENT SYSTEMS**
. FLIGHT MANAGEMENT SYSTEMS
. MANAGEMENT INFORMATION
   SYSTEMS
RT   COMPUTER TECHNIQUES
INFORMATION SYSTEMS
PROJECT MANAGEMENT
∞ SYSTEMS

MANATEES
GS   ANIMALS
. VERTEBRATES
. . MAMMALS
. . . **MANATEES**

MANDELSTAM REPRESENTATION
GS   ATTENUATION
. WAVE ATTENUATION
. . RADIO ATTENUATION
. . . **MANDELSTAM REPRESENTATION**
RELATIVISTIC THEORY
. **MANDELSTAM REPRESENTATION**
RT   ELECTROMAGNETIC ABSORPTION
INELASTIC SCATTERING
LORENTZ TRANSFORMATIONS
MICROWAVE SCATTERING
NUCLEAR SCATTERING
QUANTUM THEORY
RADAR ATTENUATION

MANDRELS
RT   CORES
MACHINE TOOLS
MOLDS
SHAFTS (MACHINE ELEMENTS)

MANEUVERABILITY
RT   AIR SLEW MISSILES
AIRCRAFT CONTROL
AIRCRAFT MANEUVERS
AIRCRAFT PERFORMANCE
CONTROLLABILITY
FLIGHT CHARACTERISTICS
FLIGHT CONTROL
HELICOPTER PERFORMANCE
MANEUVERS
SPACECRAFT MANEUVERS

MANEUVERABLE REENTRY BODIES
GS   REENTRY VEHICLES
. **MANEUVERABLE REENTRY BODIES**
RT   ∞ BODIES

MANEUVERABLE SPACECRAFT
GS   **MANEUVERABLE SPACECRAFT**
. AEROSPACEPLANES
. . ASTROPLANE
. APOLLO SPACECRAFT
. . APOLLO LUNAR EXPERIMENT
   MODULE
. ASTRO VEHICLE
. EVASIVE SATELLITES
. FERRY SPACECRAFT
. LIFTING REENTRY VEHICLES
. . HL-10 REENTRY VEHICLE
. . HLD-35 REENTRY VEHICLE
. . JANUS SPACECRAFT
. . M-2 LIFTING BODY
. . . M-2F2 LIFTING BODY
. . X-20 AIRCRAFT
. RENDEZVOUS SPACECRAFT
RT   ARTIFICIAL SATELLITES
INTERPLANETARY SPACECRAFT
LANDING MODULES
LUNAR LANDING MODULES
LUNAR PROBES
LUNAR SATELLITES
MANNED SPACECRAFT
MARS (MANNED REUSABLE
   SPACECRAFT)
RECOVERABLE SPACECRAFT
REENTRY VEHICLES
SPACE PROBES
∞ SPACECRAFT
SPACECRAFT MANEUVERS
THRUST VECTOR CONTROL

MANEUVERS
GS   **MANEUVERS**
. AIRCRAFT MANEUVERS
. EVASIVE ACTIONS
. HOVERING
. ORBITAL RENDEZVOUS
. . EARTH ORBITAL RENDEZVOUS

MANEUVERS-_(CONT.)_
. . LUNAR ORBITAL RENDEZVOUS
. SIDESLIP
. SPACECRAFT DOCKING
. SPACECRAFT MANEUVERS
. . ORBITAL MANEUVERS
RT   ACROBATICS
AIRCRAFT SPIN
FLIGHT CONTROL
LANDING
MANEUVERABILITY
MINOR CIRCLE TURNING FLIGHT
SELF MANEUVERING UNITS
TAKEOFF
TURNING FLIGHT

MANGANESE
GS   CHEMICAL ELEMENTS
. **MANGANESE**
. . MANGANESE ISOTOPES
METALS
. TRANSITION METALS
. . **MANGANESE**
. . . MANGANESE ISOTOPES
RT   STRATEGIC MATERIALS

MANGANESE ALLOYS
GS   ALLOYS
. **MANGANESE ALLOYS**
. . MANGANIN (TRADEMARK)

MANGANESE COMPOUNDS
GS   **MANGANESE COMPOUNDS**
. HOPCALITE (TRADEMARK)
. MANGANESE OXIDES
. MANGANESE PHOSPHIDES
. PERMANGANATES
RT   ∞ CHEMICAL COMPOUNDS
∞ GROUP 7B COMPOUNDS
∞ METAL COMPOUNDS

MANGANESE IONS
GS   IONS
. CATIONS
. . METAL IONS
. . . **MANGANESE IONS**
PARTICLES
. CHARGED PARTICLES
. . CATIONS
. . . METAL IONS
. . . . **MANGANESE IONS**
RT   PERMANGANATES

MANGANESE ISOTOPES
UF   MANGANESE 53
MANGANESE 54
MANGANESE 56
GS   CHEMICAL ELEMENTS
. MANGANESE
. . **MANGANESE ISOTOPES**
. NUCLIDES
. . ISOTOPES
. . . **MANGANESE ISOTOPES**
METALS
. TRANSITION METALS
. . MANGANESE
. . . **MANGANESE ISOTOPES**

MANGANESE OXIDES
GS   CHALCOGENIDES
. OXIDES
. . METAL OXIDES
. . . **MANGANESE OXIDES**
. . . . HOPCALITE (TRADEMARK)
MANGANESE COMPOUNDS
. **MANGANESE OXIDES**

MANGANESE PHOSPHIDES
GS   MANGANESE COMPOUNDS
. **MANGANESE PHOSPHIDES**
PHOSPHORUS COMPOUNDS
. PHOSPHIDES
. . **MANGANESE PHOSPHIDES**

MANGANESE 53
USE   MANGANESE ISOTOPES

MANGANESE 54
USE   MANGANESE ISOTOPES

MANGANESE 56
USE   MANGANESE ISOTOPES

MANGANIN (TRADEMARK)
GS   ALLOYS
. COPPER ALLOYS
. . **MANGANIN (TRADEMARK)**
. MANGANESE ALLOYS
. . **MANGANIN (TRADEMARK)**
RT   ELECTRICAL RESISTANCE
THERMOCOUPLES

MANIFOLDS
RT   AIR INTAKES
EXHAUST SYSTEMS
FUEL SYSTEMS
INTAKE SYSTEMS
PIPES (TUBES)
PLENUM CHAMBERS
∞ TUBES
∞ WATER INTAKES

MANIFOLDS (MATHEMATICS)
GS   **MANIFOLDS (MATHEMATICS)**
. RIEMANN MANIFOLD
RT   COORDINATES
CURVES (GEOMETRY)
FIBERS (MATHEMATICS)
FIXED POINTS (MATHEMATICS)
TOPOLOGY

MANIPULATION
USE   MANIPULATORS

MANIPULATORS
SN   (MECHANICAL DEVICES FOR REMOTE
   HANDLING)
UF   MANIPULATION
GS   **MANIPULATORS**
. REMOTE MANIPULATOR SYSTEM
RT   CONTROL EQUIPMENT
PAYLOAD DEPLOYMENT & RETRIEVAL
   SYSTEM
REMOTE CONTROL
REMOTE HANDLING
ROBOTICS
SERVOCONTROL
SHIELDING
TELEOPERATORS

MANITOBA
GS   NATIONS
. CANADA
. . **MANITOBA**

MANITOU (CO)
GS   CITIES
. **MANITOU (CO)**
RT   COLORADO

MANN-WHITNEY-WILCOXON U TEST
GS   STATISTICAL ANALYSIS
. STATISTICAL TESTS
. . **MANN-WHITNEY-WILCOXON U TEST**
RT   QUALITY CONTROL

MANNED AERODYNAMIC REUSABLE SPACESHIP
USE   MARS (MANNED REUSABLE
   SPACECRAFT)

MANNED LUNAR SURFACE VEHICLES
GS   SURFACE VEHICLES
. LUNAR SURFACE VEHICLES
. . LUNAR ROVING VEHICLES
. . . **MANNED LUNAR SURFACE
     VEHICLES**
RT   CRAWLER TRACTORS
LUNAR LOGISTICS
LUNAR MOBILE LABORATORIES
∞ SURFACES
∞ VEHICLES
WALKING MACHINES

MANNED MANEUVERING UNITS
GS   EXTRAVEHICULAR MOBILITY UNITS
. ASTRONAUT MANEUVERING
   EQUIPMENT
. . **MANNED MANEUVERING UNITS**
RT   ASTRONAUT LOCOMOTION
EXTRAVEHICULAR ACTIVITY
LIFE SUPPORT SYSTEMS
ORBITAL SERVICING
SELF MANEUVERING UNITS
SPACE SUITS

MANNED ORBITAL LABORATORIES
UF   MOL (ORBITAL LABORATORIES)

MANNED ORBITAL LABORATORIES-*(CONT.)*
GS   LABORATORIES
  . SPACE LABORATORIES
  . . **MANNED ORBITAL LABORATORIES**
  . . . MANNED ORBITAL RESEARCH
         LABORATORIES
    MANNED SPACECRAFT
  . **MANNED ORBITAL LABORATORIES**
  . . MANNED ORBITAL RESEARCH
       LABORATORIES
RT   APOLLO SPACECRAFT
    ORBITAL SPACE STATIONS
    ORBITAL WORKSHOPS
    RECONNAISSANCE SPACECRAFT
    TITAN 3 LAUNCH VEHICLE

MANNED ORBITAL RESEARCH LABORATORIES
UF   MORL
GS   LABORATORIES
  . SPACE LABORATORIES
  . . MANNED ORBITAL LABORATORIES
  . . . **MANNED ORBITAL RESEARCH
         LABORATORIES**
    MANNED SPACECRAFT
  . MANNED ORBITAL LABORATORIES
  . . **MANNED ORBITAL RESEARCH
       LABORATORIES**
RT   APOLLO SPACECRAFT
    ORBITAL SPACE STATIONS
  ∞SPACECRAFT

MANNED ORBITAL SPACE STATIONS
USE   ORBITAL SPACE STATIONS

MANNED ORBITAL TELESCOPES
UF   MOT (ORBITAL TELESCOPES)
GS   TELESCOPES
  . **MANNED ORBITAL TELESCOPES**
RT   ASTRONOMICAL TELESCOPES
    OAO

MANNED REENTRY
GS   ATMOSPHERIC ENTRY
  . REENTRY
  . . **MANNED REENTRY**
    SPACE FLIGHT
  . MANNED SPACE FLIGHT
  . . **MANNED REENTRY**
RT   DESCENT TRAJECTORIES
    ENVIRONMENTAL CONTROL
    LIFTING REENTRY VEHICLES
    REENTRY COMMUNICATION
    SPACECRAFT REENTRY

MANNED SPACE FLIGHT
GS   SPACE FLIGHT
  . **MANNED SPACE FLIGHT**
  . . APOLLO FLIGHTS
  . . . APOLLO 5 FLIGHT
  . . . APOLLO 6 FLIGHT
  . . . APOLLO 7 FLIGHT
  . . . APOLLO 8 FLIGHT
  . . . APOLLO 9 FLIGHT
  . . . APOLLO 10 FLIGHT
  . . . APOLLO 11 FLIGHT
  . . . APOLLO 12 FLIGHT
  . . . APOLLO 13 FLIGHT
  . . . APOLLO 14 FLIGHT
  . . . APOLLO 15 FLIGHT
  . . . APOLLO 16 FLIGHT
  . . . APOLLO 17 FLIGHT
  . . GEMINI FLIGHTS
  . . . GEMINI 3 FLIGHT
  . . . GEMINI 4 FLIGHT
  . . . GEMINI 5 FLIGHT
  . . . GEMINI 6 FLIGHT
  . . . GEMINI 7 FLIGHT
  . . . GEMINI 8 FLIGHT
  . . . GEMINI 9 FLIGHT
  . . . GEMINI 10 FLIGHT
  . . . GEMINI 11 FLIGHT
  . . . GEMINI 12 FLIGHT
  . . MANNED REENTRY
  . . MERCURY FLIGHTS
  . . . MERCURY MA-1 FLIGHT
  . . . MERCURY MA-2 FLIGHT
  . . . MERCURY MA-3 FLIGHT
  . . . MERCURY MA-4 FLIGHT
  . . . MERCURY MA-5 FLIGHT
  . . . MERCURY MA-6 FLIGHT
  . . . MERCURY MA-7 FLIGHT
  . . . MERCURY MA-8 FLIGHT
  . . . MERCURY MA-9 FLIGHT
  . . . MERCURY MR-1 FLIGHT
  . . . MERCURY MR-2 FLIGHT

MANNED SPACE FLIGHT-*(CONT.)*
  . . . MERCURY MR-3 FLIGHT
  . . . MERCURY MR-4 FLIGHT
  . . SPACE SHUTTLE MISSION 31-A
  . . SPACE SHUTTLE MISSION 31-B
  . . SPACE SHUTTLE MISSION 31-C
  . . SPACE SHUTTLE MISSION 31-D
  . . SPACE SHUTTLE MISSION 41-A
  . . SPACE SHUTTLE MISSION 41-B
  . . SPACE SHUTTLE MISSION 41-C
  . . SPACE SHUTTLE MISSION 41-D
  . . SPACE SHUTTLE MISSION 41-G
  . . SPACE SHUTTLE MISSION 51-A
  . . SPACE SHUTTLE MISSION 51-B
  . . SPACE SHUTTLE MISSION 51-C
  . . SPACE SHUTTLE MISSION 51-D
  . . SPACE SHUTTLE MISSION 51-E
  . . SPACE SHUTTLE MISSION 51-F
  . . SPACE SHUTTLE MISSION 51-G
  . . SPACE SHUTTLE MISSION 51-H
  . . SPACE SHUTTLE MISSION 51-I
  . . SPACE SHUTTLE MISSION 51-J
  . . SPACE SHUTTLE MISSION 51-L
  . . SPACE SHUTTLE MISSION 61-A
  . . SPACE SHUTTLE MISSION 61-B
  . . SPACE SHUTTLE MISSION 61-C
  . . SPACE SHUTTLE MISSION 61-E
RT   AEROSPACE ENVIRONMENTS
    APOLLO EXTENSION SYSTEM
    ATLANTIS (ORBITER)
    COLUMBIA (ORBITER)
    DISCOVERY (ORBITER)
    ENTERPRISE (ORBITER)
    EXTRAVEHICULAR ACTIVITY
    GEMINI (GT-1) SPACECRAFT
    GEMINI B SPACECRAFT
    GEMINI SPACECRAFT
    GEMINI 2 SPACECRAFT
    HUMAN FACTORS ENGINEERING
    INDIAN SPACE PROGRAM
    INTERPLANETARY FLIGHT
    INTERSTELLAR TRAVEL
    INTRAVEHICULAR ACTIVITY
    LONG DURATION SPACE FLIGHT
    MAN OPERATED PROPULSION SYSTEMS
    MERCURY PROJECT
    SPACE ADAPTATION SYNDROME
    SPACE COMMUNICATION
    SPACE EXPLORATION
    SPACE FLIGHT STRESS
    SPACE LOGISTICS
    SPACE PROGRAMS
    SPACE PSYCHOLOGY
    SPACE SHUTTLE ORBITERS
    SPACE SHUTTLES
    SPACECREW TRANSFER
    SUBORBITAL FLIGHT

MANNED SPACE FLIGHT NETWORK
GS   TRACKING NETWORKS
  . **MANNED SPACE FLIGHT NETWORK**
RT   ADVANCED RANGE INSTRUMENTATION
      SHIP
    UNIFIED S BAND

MANNED SPACECRAFT
GS   **MANNED SPACECRAFT**
  . AEROSPACEPLANES
  . . ASTROPLANE
  . APOLLO SPACECRAFT
  . . APOLLO LUNAR EXPERIMENT
      MODULE
  . ASTRO VEHICLE
  . FERRY SPACECRAFT
  . GEMINI B SPACECRAFT
  . GEMINI SPACECRAFT
  . GEMINI (GT-1) SPACECRAFT
  . . GEMINI 2 SPACECRAFT
  . JANUS SPACECRAFT
  . LUNAR MODULE
  . . APOLLO LUNAR EXPERIMENT
      MODULE
  . . LSSM
  . . LUNAR MODULE 5
  . . LUNAR MODULE 7
  . MANNED ORBITAL LABORATORIES
  . . MANNED ORBITAL RESEARCH
      LABORATORIES
  . MARS (MANNED REUSABLE
     SPACECRAFT)
  . MERCURY SPACECRAFT
  . . AURORA 7
  . . FAITH 7
  . . FRIENDSHIP 7
  . . SIGMA 7

MANNED SPACECRAFT-*(CONT.)*
  . SHUTTLE DERIVED VEHICLES
  . SOYUZ SPACECRAFT
  . SPACE SHUTTLES
  . . HERMES MANNED SPACEPLANE
  . SPACE STATIONS
  . . ORBITAL SPACE STATIONS
  . . . LONG DURATION EXPOSURE
        FACILITY
  . . ORBITAL WORKSHOPS
  . . . SALYUT SPACE STATION
  . . . SKYLAB 1
  . . . SKYLAB 2
  . . . SKYLAB 3
  . . . SKYLAB 4
  . . . SPACE OPERATIONS CENTER
        (NASA)
  . SPACE BASE COMMAND CENTER
  . VOSKHOD MANNED SPACECRAFT
  . . VOSKHOD 1 SPACECRAFT
  . . VOSKHOD 2 SPACECRAFT
  . VOSTOK SPACECRAFT
  . . VOSTOK 1 SPACECRAFT
  . . VOSTOK 2 SPACECRAFT
  . . VOSTOK 3 SPACECRAFT
  . . VOSTOK 4 SPACECRAFT
  . . VOSTOK 5 SPACECRAFT
  . . VOSTOK 6 SPACECRAFT
RT   APOLLO PROJECT
    APOLLO SOYUZ TEST PROJECT
    APOLLO 7 FLIGHT
    APOLLO 8 FLIGHT
    APOLLO 10 FLIGHT
    APOLLO 11 FLIGHT
    APOLLO 12 FLIGHT
    APOLLO 13 FLIGHT
    APOLLO 14 FLIGHT
    APOLLO 15 FLIGHT
    APOLLO 16 FLIGHT
    APOLLO 17 FLIGHT
    APPROACH AND LANDING TESTS (STS)
    ARTIFICIAL SATELLITES
    BIOSATELLITES
    BOOSTGLIDE VEHICLES
    CHALLENGER (ORBITER)
    COMMAND SERVICE MODULES
    ENVIRONMENTAL CONTROL
    GRAVITY GRADIENT SATELLITES
    INTERPLANETARY SPACECRAFT
    LANDING MODULES
    LIFTING REENTRY VEHICLES
    LUNAR LANDING MODULES
    LUNAR SATELLITES
    LUNAR SPACECRAFT
    MANEUVERABLE SPACECRAFT
    MERCURY FLIGHTS
    MERCURY PROJECT
    MILITARY SPACECRAFT
    RECONNAISSANCE SPACECRAFT
    RECOVERABLE SPACECRAFT
    RENDEZVOUS SPACECRAFT
    REUSABLE SPACECRAFT
    SATELLITES
    SPACE CAPSULES
    SPACE NAVIGATION
    SPACE SHUTTLE BOOSTERS
  ∞SPACECRAFT
    SPACECRAFT CABIN SIMULATORS
    UNMANNED SPACECRAFT
    X-20 AIRCRAFT

MANNING THEORY
RT   FLUID FLOW
  ∞THEORIES
    WALL FLOW

MANNITOL
GS   ALIPHATIC COMPOUNDS
  . SUGARS
  . . **MANNITOL**
    CARBOHYDRATES
  . SUGARS
  . . **MANNITOL**

MANOMETERS
UF   MICROMANOMETERS
    U TUBES
GS   MEASURING INSTRUMENTS
  . PRESSURE GAGES
  . . **MANOMETERS**
RT   BAROMETERS
    BLOOD PRESSURE
    FLAME PROBES
    PRESSURE DISTRIBUTION
    PRESSURE MEASUREMENT

MANOMETERS-*(CONT.)*
    VACUUM GAGES

MANPOWER
GS   **MANPOWER**
      . SCIENTISTS
RT   ENGINEERING MANAGEMENT
      HUMAN RESOURCES
      LABOR
      PERSONNEL
      RESEARCH MANAGEMENT
      RESOURCES
      RETRAINING

MANTLE (EARTH STRUCTURE)
USE   EARTH MANTLE

∞ **MANUAL**
SN   *(USE OF A MORE SPECIFIC TERM IS*
      RECOMMENDED--CONSULT THE TERMS
      LISTED BELOW)
RT   MANUAL CONTROL
      MANUALS

MANUAL CONTROL
GS   **MANUAL CONTROL**
      . VISUAL CONTROL
RT   AIRCRAFT CONTROL
      ATTITUDE CONTROL
      AUTOMATIC CONTROL
      ∞ BUTTONS
      CONSOLES
      ∞ CONTROL
      CONTROL BOARDS
      CONTROL EQUIPMENT
      DIRECTIONAL CONTROL
      ENGINE CONTROL
      GUIDANCE (MOTION)
      HANDLES
      HELICOPTER CONTROL
      HUMAN FACTORS ENGINEERING
      KNOBS
      LANDING INSTRUMENTS
      LATERAL CONTROL
      LEVERS
      LONGITUDINAL CONTROL
      ∞ MANUAL
      PEDALS
      REENTRY GUIDANCE
      REMOTE CONTROL
      SATELLITE CONTROL
      SATELLITE GUIDANCE
      SERVOCONTROL
      SPACECRAFT CONTROL
      SPACECRAFT GUIDANCE
      SPEED CONTROL
      TEMPERATURE CONTROL

MANUALS
GS   DOCUMENTS
      . **MANUALS**
      . . INSTALLATION MANUALS
      . . USER MANUALS (COMPUTER
          PROGRAMS)
RT   DIRECTORIES
      HANDBOOKS
      MAINTENANCE
      ∞ MANUAL
      TEXTBOOKS

MANUFACTURING
GS   **MANUFACTURING**
      . COMPUTER AIDED MANUFACTURING
      . LOW GRAVITY MANUFACTURING
      . SPACE MANUFACTURING
RT   AIRCRAFT PRODUCTION COSTS
      COMMERCE
      COMMODITIES
      CONTAINERLESS MELTS
      CONTRACT NEGOTIATION
      ECONOMIC DEVELOPMENT
      FABRICATION
      INDUSTRIES
      KRAFT PROCESS (WOODPULP)
      ∞ PROCESSING
      PRODUCTION MANAGEMENT
      PRODUCTS
      SPACE INDUSTRIALIZATION
      TECHNOLOGIES
      TECHNOLOGY ASSESSMENT
      TECHNOLOGY UTILIZATION

MANURES
GS   WASTES
      . **MANURES**

MANURES-*(CONT.)*
RT   BIOMASS ENERGY PRODUCTION
      METABOLIC WASTES
      WASTE DISPOSAL
      WASTE UTILIZATION

MANY BODY PROBLEM
UF   MANY PARTICLE THEORY
RT   BCS THEORY
      CELESTIAL MECHANICS
      ELEMENTARY EXCITATIONS
      FIELD THEORY (PHYSICS)
      FOUR BODY PROBLEM
      GREEN'S FUNCTIONS
      HARTREE APPROXIMATION
      ORBITAL MECHANICS
      ORBITS
      PARTICLE THEORY
      PERTURBATION
      PERTURBATION THEORY
      ∞ PROBLEMS
      QUANTUM STATISTICS
      STATISTICAL MECHANICS
      SUPERFLUIDITY
      THREE BODY PROBLEM
      TROJAN ORBITS
      TWO BODY PROBLEM

MANY ELECTRON EFFECTS
RT   AUTOIONIZATION
      ELECTRON CAPTURE
      ELECTRON SCATTERING
      ELECTRON STATES
      ELECTRON TRANSITIONS

MANY PARTICLE THEORY
USE   MANY BODY PROBLEM

MAP (PROGRAMMING LANGUAGE)
GS   LANGUAGES
      . PROGRAMMING LANGUAGES
      . . ASSEMBLY LANGUAGE
      . . . **MAP (PROGRAMMING LANGUAGE)**
RT   COMPUTER PROGRAMMING

MAP MATCHING GUIDANCE
GS   GUIDANCE (MOTION)
      . **MAP MATCHING GUIDANCE**
RT   AIRBORNE EQUIPMENT
      DISPLAY DEVICES
      IMAGE CORRELATORS
      RADAR MAPS
      RADAR NAVIGATION
      TERCOM
      VIDEO LANDMARK ACQUISITION AND
          TRACKING

MAPPING
SN   (EXCLUDES CONFORMAL MAPPING)
UF   CARTOGRAPHY
      FLUX MAPPING
GS   **MAPPING**
      . CADASTRAL MAPPING
      . COMPUTER AIDED MAPPING
      . ICE MAPPING
      . PHOTOMAPPING
      . PLANETARY MAPPING
      . SOIL MAPPING
      . THEMATIC MAPPING
      . THERMAL MAPPING
RT   ASTROGRAPHY
      BONNE PROJECTION
      CONTOURS
      DECLINATION
      FIXED POINTS (MATHEMATICS)
      FUNCTIONS (MATHEMATICS)
      GEOGRAPHIC APPLICATIONS PROGRAM
      GEOGRAPHY
      HEAT CAPACITY MAPPING MISSION
      HYPSOGRAPHY
      MAPS
      MAPSAT
      ORTHOPHOTOGRAPHY
      PHOENIX QUADRANGLE (AZ)
      PHOTOGRAMMETRY
      PHOTOGRAPHY
      SCALE (RATIO)
      SPOT (FRENCH SATELLITE)
      SURVEYS
      TERRAIN ANALYSIS
      TOPOGRAPHY
      TOPOLOGY
      TRIANGULATION

MAPS
GS   **MAPS**
      . ASTRONOMICAL MAPS
      . . PLANISPHERES
      . LUNAR MAPS
      . METEOROLOGICAL CHARTS
      . PHOTOMAPS
      . RADAR CLUTTER MAPS
      . RADAR MAPS
      . RELIEF MAPS
RT   BONNE PROJECTION
      CADASTRAL MAPPING
      CHARTS
      COMPUTER AIDED MAPPING
      COORDINATES
      DATUM (ELEVATION)
      GEOGRAPHY
      ∞ GLOBES
      HYPSOGRAPHY
      MAPPING
      MERCATOR PROJECTION
      NAVIGATION AIDS
      PHOTOMAPPING
      SOIL MAPPING
      SURVEYS
      THEMATIC MAPPING

MAPSAT
GS   SATELLITES
      ARTIFICIAL SATELLITES
      . . **MAPSAT**
RT   LANDSAT SATELLITES
      MAPPING
      REMOTE SENSING
      STEREOPHOTOGRAPHY

MARAGING
GS   HARDENING (MATERIALS)
      . PRECIPITATION HARDENING
      . . **MARAGING**
      HEAT TREATMENT
      . **MARAGING**

MARAGING STEELS
GS   ALLOYS
      . HIGH STRENGTH ALLOYS
      . . HIGH STRENGTH STEELS
      . . . **MARAGING STEELS**
      . IRON ALLOYS
      . . STEELS
      . . . HIGH STRENGTH STEELS
      . . . . **MARAGING STEELS**
RT   MARTENSITIC STAINLESS STEELS
      STAINLESS STEELS

MARANGONI CONVECTION
GS   CONVECTION
      . **MARANGONI CONVECTION**
RT   CONVECTIVE FLOW
      FREE CONVECTION
      INTERFACIAL TENSION
      LOW GRAVITY MANUFACTURING
      MELTS (CRYSTAL GROWTH)
      REDUCED GRAVITY
      SPACE PROCESSING

MARBORE 2 ENGINE
USE   J-69-T-25 ENGINE

MARECS MARITIME SATELLITES
GS   ESA SPACECRAFT
      . ESA SATELLITES
      . . **MARECS MARITIME SATELLITES**
      SATELLITES
      . ARTIFICIAL SATELLITES
      . . COMMUNICATION SATELLITES
      . . . **MARECS MARITIME SATELLITES**
      . . ESA SATELLITES
      . . . **MARECS MARITIME SATELLITES**
      . . MARITIME SATELLITES
      . . . **MARECS MARITIME SATELLITES**
RT   EUROPEAN SPACE PROGRAMS
      SATELLITE NETWORKS

MARGINS
RT   BORDERS
      EDGES
      RIMS

MARIA
GS   **MARIA**
      . LUNAR MARIA
RT   LAVA
      METEORITE CRATERS
      TOPOGRAPHY

MARIJUANA
GS   DRUGS
. CENTRAL NERVOUS SYSTEM
    DEPRESSANTS
. . **MARIJUANA**
RT   ALKALOIDS

MARINE BIOLOGY
RT   ALGAE
    AQUATIC PLANTS
    AQUICULTURE
    ∞BIOLOGY
    ENVIRONMENT EFFECTS
    ENVIRONMENTAL QUALITY
    FISHERIES
    LIMNOLOGY
    OCEANOGRAPHY
    ∞SCIENCE
    SEA GRASSES
    SEALS (ANIMALS)
    SEAWEEDS
    SHELLFISH
    THERMAL POLLUTION
    WATERFOWL
    WETLANDS

MARINE CHEMISTRY
RT   BIOCHEMISTRY
    ∞CHEMISTRY
    GEOCHEMISTRY
    HYDROLOGY
    LIMNOLOGY
    OCEAN BOTTOM
    ∞SCIENCE
    SEDIMENTS

MARINE ENVIRONMENTS
GS   ENVIRONMENTS
. **MARINE ENVIRONMENTS**
RT   AQUICULTURE
    BEACHES
    COASTAL ECOLOGY
    COASTS
    ENVIRONMENT EFFECTS
    ICE ENVIRONMENTS
    NEARSHORE WATER
    OCEAN MODELS
    OCEANOGRAPHY
    RED TIDE
    SEA BREEZE
    SHELLFISH
    WATERFOWL
    WETLANDS

MARINE GEOLOGY
USE HYDROGEOLOGY

MARINE MAMMALS
GS   ANIMALS
. VERTEBRATES
. . MAMMALS
. . . **MARINE MAMMALS**
RT   DOLPHINS
    PORPOISES
    SEALS (ANIMALS)
    WHALES

MARINE METEOROLOGY
GS   METEOROLOGY
. HYDROMETEOROLOGY
. . **MARINE METEOROLOGY**
RT   FRONTS (METEOROLOGY)
    OCEANOGRAPHY
    ∞SCIENCE
    TYPHOONS
    WIND (METEOROLOGY)

MARINE NAVIGATION
USE SURFACE NAVIGATION

MARINE PROPULSION
GS   PROPULSION
. **MARINE PROPULSION**
. . UNDERWATER PROPULSION
. . . SUBMARINE PROPULSION
RT   CHEMICAL PROPULSION
    ELECTRIC PROPULSION
    JET PROPULSION
    NUCLEAR ELECTRIC PROPULSION
    NUCLEAR PROPULSION
    PROPELLER DRIVE
    SAVANNAH NUCLEAR SHIP

MARINE RESOURCES
GS   RESOURCES
. EARTH RESOURCES
. . **MARINE RESOURCES**
RT   AQUICULTURE
    COASTAL ECOLOGY
    FISHERIES
    OCEANOGRAPHY
    OCEANS
    SEA WATER
    SHELLFISH
    UNDERWATER RESOURCES
    WATER POLLUTION
    WETLANDS

MARINE RUDDERS
GS   CONTROL SURFACES
. RUDDERS
. . **MARINE RUDDERS**
RT   AERIAL RUDDERS
    HYDROFOILS
    TAIL ASSEMBLIES

MARINE TECHNOLOGY
GS   TECHNOLOGIES
. **MARINE TECHNOLOGY**
RT   AQUICULTURE
    ARTIFICIAL HARBORS
    DEEPWATER TERMINALS
    OCEANOGRAPHY
    OFFSHORE DOCKING
    OFFSHORE ENERGY SOURCES
    OFFSHORE PLATFORMS
    TANKER TERMINALS
    WHARVES

MARINE TRANSPORTATION
GS   TRANSPORTATION
. **MARINE TRANSPORTATION**
RT   AIR TRANSPORTATION
    DEEPWATER TERMINALS
    HARBORS
    OFFSHORE DOCKING
    RAIL TRANSPORTATION
    SHIPS
    TANKER SHIPS
    WATER VEHICLES

MARINER C SPACECRAFT
GS   INTERPLANETARY SPACECRAFT
. MARINER SPACECRAFT
. . **MARINER C SPACECRAFT**
    UNMANNED SPACECRAFT
. SPACE PROBES
. . MARINER SPACECRAFT
. . . **MARINER C SPACECRAFT**

MARINER JUPITER-SATURN FLYBY
GS   FLYBY MISSIONS
. GRAND TOURS
. . **MARINER JUPITER-SATURN FLYBY**
RT   INTERPLANETARY FLIGHT
    ∞MISSIONS
    SPACE FLIGHT
    SPACE MISSIONS

MARINER JUPITER-URANUS FLYBY
GS   FLYBY MISSIONS
. GRAND TOURS
. . **MARINER JUPITER-URANUS FLYBY**
RT   INTERPLANETARY FLIGHT
    ∞MISSIONS
    SPACE FLIGHT
    SPACE MISSIONS

MARINER MARK 2 SPACECRAFT
RT   FLYBY MISSIONS
    INTERPLANETARY FLIGHT
    ∞SPACECRAFT

MARINER PROGRAM
GS   PROGRAMS
. NASA PROGRAMS
. . **MARINER PROGRAM**
. NASA SPACE PROGRAMS
. . **MARINER PROGRAM**
RT   AGENA B ROCKET VEHICLE
    AGENA ROCKET VEHICLES
    ATLAS AGENA LAUNCH VEHICLES
    ATLAS LAUNCH VEHICLES
    CENTAUR PROJECT
    FLYBY MISSIONS
    MARS PROBES
    SPACE PROBES
    UNMANNED SPACECRAFT

MARINER PROGRAM-*(CONT.)*
    VENUS PROBES

MARINER R 2 SPACE PROBE
GS   INTERPLANETARY SPACECRAFT
. MARINER SPACE PROBES
. . **MARINER R 2 SPACE PROBE**
    UNMANNED SPACECRAFT
. SPACE PROBES
. . MARINER SPACE PROBES
. . . **MARINER R 2 SPACE PROBE**

MARINER SPACE PROBES
GS   INTERPLANETARY SPACECRAFT
. **MARINER SPACE PROBES**
. . MARINER R 2 SPACE PROBE
. . MARINER VENUS-MERCURY 1973
. . MARINER 1 SPACE PROBE
. . MARINER 2 SPACE PROBE
. . MARINER 3 SPACE PROBE
. . MARINER 4 SPACE PROBE
. . MARINER 5 SPACE PROBE
. . MARINER 6 SPACE PROBE
. . MARINER 7 SPACE PROBE
. . MARINER 8 SPACE PROBE
. . MARINER 9 SPACE PROBE
. . MARINER 10 SPACE PROBE
. . MARINER 11 SPACE PROBE
    UNMANNED SPACECRAFT
. SPACE PROBES
. . **MARINER SPACE PROBES**
. . . MARINER R 2 SPACE PROBE
. . . MARINER VENUS-MERCURY 1973
. . . MARINER 1 SPACE PROBE
. . . MARINER 2 SPACE PROBE
. . . MARINER 3 SPACE PROBE
. . . MARINER 4 SPACE PROBE
. . . MARINER 5 SPACE PROBE
. . . MARINER 6 SPACE PROBE
. . . MARINER 7 SPACE PROBE
. . . MARINER 8 SPACE PROBE
. . . MARINER 9 SPACE PROBE
. . . MARINER 10 SPACE PROBE
. . . MARINER 11 SPACE PROBE
. . . MARINER-MERCURY 1973

MARINER SPACECRAFT
GS   INTERPLANETARY SPACECRAFT
. **MARINER SPACECRAFT**
. . MARINER C SPACECRAFT
. . MARINER VENUS 67 SPACECRAFT
    UNMANNED SPACECRAFT
. SPACE PROBES
. . **MARINER SPACECRAFT**
. . . MARINER C SPACECRAFT
. . . MARINER VENUS 67 SPACECRAFT

MARINER VENUS 67 SPACECRAFT
GS   INTERPLANETARY SPACECRAFT
. MARINER SPACECRAFT
. . **MARINER VENUS 67 SPACECRAFT**
    UNMANNED SPACECRAFT
. SPACE PROBES
. . MARINER SPACECRAFT
. . . **MARINER VENUS 67 SPACECRAFT**
RT   VENUS PROBES

MARINER VENUS-MERCURY 1973
GS   FLYBY MISSIONS
. **MARINER VENUS-MERCURY 1973**
    INTERPLANETARY SPACECRAFT
. MARINER SPACE PROBES
. . **MARINER VENUS-MERCURY 1973**
    UNMANNED SPACECRAFT
. SPACE PROBES
. . MARINER SPACE PROBES
. . . **MARINER VENUS-MERCURY 1973**
RT   MARINER-MERCURY 1973

MARINER 1 SPACE PROBE
GS   INTERPLANETARY SPACECRAFT
. MARINER SPACE PROBES
. . **MARINER 1 SPACE PROBE**
. VENUS PROBES
. . **MARINER 1 SPACE PROBE**
    UNMANNED SPACECRAFT
. SPACE PROBES
. . MARINER SPACE PROBES
. . . **MARINER 1 SPACE PROBE**
. VENUS PROBES
. . . **MARINER 1 SPACE PROBE**

MARINER 2 SPACE PROBE
GS   INTERPLANETARY SPACECRAFT
. MARINER SPACE PROBES

MARINER 2 SPACE PROBE-*(CONT.)*
  . . **MARINER 2 SPACE PROBE**
  . VENUS PROBES
  . . **MARINER 2 SPACE PROBE**
  UNMANNED SPACECRAFT
  . SPACE PROBES
  . . MARINER SPACE PROBES
  . . . **MARINER 2 SPACE PROBE**
  . . VENUS PROBES
  . . . **MARINER 2 SPACE PROBE**
RT ATLAS AGENA B LAUNCH VEHICLE

MARINER 3 SPACE PROBE
GS INTERPLANETARY SPACECRAFT
  . MARINER SPACE PROBES
  . . **MARINER 3 SPACE PROBE**
  . MARS PROBES
  . . **MARINER 3 SPACE PROBE**
  UNMANNED SPACECRAFT
  . SPACE PROBES
  . . MARINER SPACE PROBES
  . . . **MARINER 3 SPACE PROBE**
  . . MARS PROBES
  . . . **MARINER 3 SPACE PROBE**

MARINER 4 SPACE PROBE
GS INTERPLANETARY SPACECRAFT
  . MARINER SPACE PROBES
  . . **MARINER 4 SPACE PROBE**
  . MARS PROBES
  . . **MARINER 4 SPACE PROBE**
  UNMANNED SPACECRAFT
  . SPACE PROBES
  . . MARINER SPACE PROBES
  . . . **MARINER 4 SPACE PROBE**
  . . MARS PROBES
  . . . **MARINER 4 SPACE PROBE**

MARINER 5 SPACE PROBE
GS INTERPLANETARY SPACECRAFT
  . MARINER SPACE PROBES
  . . **MARINER 5 SPACE PROBE**
  . VENUS PROBES
  . . **MARINER 5 SPACE PROBE**
  UNMANNED SPACECRAFT
  . SPACE PROBES
  . . MARINER SPACE PROBES
  . . . **MARINER 5 SPACE PROBE**
  . . VENUS PROBES
  . . . **MARINER 5 SPACE PROBE**
RT ATLAS AGENA LAUNCH VEHICLES

MARINER 6 SPACE PROBE
GS INTERPLANETARY SPACECRAFT
  . MARINER SPACE PROBES
  . . **MARINER 6 SPACE PROBE**
  . MARS PROBES
  . . **MARINER 6 SPACE PROBE**
  UNMANNED SPACECRAFT
  . SPACE PROBES
  . . MARINER SPACE PROBES
  . . . **MARINER 6 SPACE PROBE**
  . . MARS PROBES
  . . . **MARINER 6 SPACE PROBE**
RT ATLAS AGENA LAUNCH VEHICLES
     MARS 69 PROJECT

MARINER 7 SPACE PROBE
GS INTERPLANETARY SPACECRAFT
  . MARINER SPACE PROBES
  . . **MARINER 7 SPACE PROBE**
  . MARS PROBES
  . . **MARINER 7 SPACE PROBE**
  UNMANNED SPACECRAFT
  . SPACE PROBES
  . . MARINER SPACE PROBES
  . . . **MARINER 7 SPACE PROBE**
  . . MARS PROBES
  . . . **MARINER 7 SPACE PROBE**
RT MARS 69 PROJECT

MARINER 8 SPACE PROBE
GS INTERPLANETARY SPACECRAFT
  . MARINER SPACE PROBES
  . . **MARINER 8 SPACE PROBE**
  . MARS PROBES
  . . **MARINER 8 SPACE PROBE**
  UNMANNED SPACECRAFT
  . SPACE PROBES
  . . MARINER SPACE PROBES
  . . . **MARINER 8 SPACE PROBE**
  . . MARS PROBES
  . . . **MARINER 8 SPACE PROBE**
RT MARS 71 PROJECT

MARINER 9 SPACE PROBE
GS INTERPLANETARY SPACECRAFT
  . MARINER SPACE PROBES
  . . **MARINER 9 SPACE PROBE**
  UNMANNED SPACECRAFT
  . SPACE PROBES
  . . MARINER SPACE PROBES
  . . . **MARINER 9 SPACE PROBE**

MARINER 10 SPACE PROBE
GS INTERPLANETARY SPACECRAFT
  . MARINER SPACE PROBES
  . . **MARINER 10 SPACE PROBE**
  . MARS PROBES
  . . **MARINER 10 SPACE PROBE**
  UNMANNED SPACECRAFT
  . SPACE PROBES
  . . MARINER SPACE PROBES
  . . . **MARINER 10 SPACE PROBE**
  . . MARS PROBES
  . . . **MARINER 10 SPACE PROBE**

MARINER 11 SPACE PROBE
GS INTERPLANETARY SPACECRAFT
  . MARINER SPACE PROBES
  . . **MARINER 11 SPACE PROBE**
  UNMANNED SPACECRAFT
  . SPACE PROBES
  . . MARINER SPACE PROBES
  . . . **MARINER 11 SPACE PROBE**

MARINER-MERCURY 1973
GS FLYBY MISSIONS
  . **MARINER-MERCURY 1973**
  UNMANNED SPACECRAFT
  . SPACE PROBES
  . . MARINER SPACE PROBES
  . . . **MARINER-MERCURY 1973**
RT MARINER VENUS-MERCURY 1973

MARISAT SATELLITES
GS SATELLITES
  . ARTIFICIAL SATELLITES
  . . **MARISAT SATELLITES**
  . . . MARISAT 1 SATELLITE
HI COMMUNICATION
     FLEET SATELLITE COMMUNICATION
      SYSTEM
     MAROTS (ESA)
     RADIO COMMUNICATION

MARISAT 1 SATELLITE
GS SATELLITES
  . ARTIFICIAL SATELLITES
  . . MARISAT SATELLITES
  . . . **MARISAT 1 SATELLITE**
RT RADIO COMMUNICATION

MARITIME COMMUNICATION SATELLITE (ESA)
USE MAROTS (ESA)

MARITIME ORBITAL TEST SATELLITE
USE MAROTS (ESA)

MARITIME SATELLITES
GS SATELLITES
  . ARTIFICIAL SATELLITES
  . . **MARITIME SATELLITES**
  . . . ERS-1 (ESA SATELLITE)
  . . . MARECS MARITIME SATELLITES
  . . . MAROTS (ESA)
RT MSAT
     NATIONAL OCEANIC SATELLITE SYSTEM
     TOPEX

MARK 1 REENTRY BODY
GS REENTRY VEHICLES
  . **MARK 1 REENTRY BODY**
RT INTERCONTINENTAL BALLISTIC MISSILES
     INTERMEDIATE RANGE BALLISTIC
      MISSILES

MARK 1 SPACECRAFT
RT SATELLITES
  ∞SPACECRAFT

MARK 2 REENTRY BODY
GS REENTRY VEHICLES
  . **MARK 2 REENTRY BODY**
RT INTERCONTINENTAL BALLISTIC MISSILES
     INTERMEDIATE RANGE BALLISTIC
      MISSILES

MARK 3 REENTRY BODY
GS REENTRY VEHICLES
  . **MARK 3 REENTRY BODY**
RT INTERCONTINENTAL BALLISTIC MISSILES
     INTERMEDIATE RANGE BALLISTIC
      MISSILES

MARK 4 REENTRY BODY
GS REENTRY VEHICLES
  . **MARK 4 REENTRY BODY**
RT INTERCONTINENTAL BALLISTIC MISSILES

MARK 5 REENTRY BODY
GS REENTRY VEHICLES
  . **MARK 5 REENTRY BODY**
RT INTERCONTINENTAL BALLISTIC MISSILES

MARK 6 REENTRY BODY
GS REENTRY VEHICLES
  . **MARK 6 REENTRY BODY**
RT INTERCONTINENTAL BALLISTIC MISSILES

MARK 11 REENTRY BODY
GS REENTRY VEHICLES
  . **MARK 11 REENTRY BODY**
RT INTERCONTINENTAL BALLISTIC MISSILES

MARK 12 REENTRY BODY
GS REENTRY VEHICLES
  . **MARK 12 REENTRY BODY**
RT INTERCONTINENTAL BALLISTIC MISSILES

MARK 17 REENTRY BODY
GS REENTRY VEHICLES
  . **MARK 17 REENTRY BODY**
RT INTERCONTINENTAL BALLISTIC MISSILES

∞ **MARKERS**
SN *(USE OF A MORE SPECIFIC TERM IS*
    RECOMMENDED--CONSULT THE TERM
    LISTED BELOW)
RT BEACONS
     BUOYS
     CRAYONS
     DYES
     RADIO BEACONS
     RUNWAY LIGHTS
     SMOKE

MARKET RESEARCH
GS RESEARCH
  . **MARKET RESEARCH**
RT COMMERCE
     COMMODITIES
     CONSUMERS
     MARKETING
     PRODUCT DEVELOPMENT

MARKETING
RT COMMERCE
     CONSUMERS
     FINANCE
     INDUSTRIAL AREAS
     MANAGEMENT
     MARKET RESEARCH
     PRODUCT DEVELOPMENT
     SUPPLYING

MARKING
UF LABELING (MARKING)
     TAGGING
GS **MARKING**
  . ISOTOPIC LABELING
RT DETECTION
     IDENTIFYING
     MATERIALS HANDLING
     PACKAGING
     STAINING
  ∞TRACERS

MARKOV CHAINS
GS STOCHASTIC PROCESSES
  . **MARKOV CHAINS**
RT MONTE CARLO METHOD
     RANDOM WALK

MARKOV PROCESSES
GS STOCHASTIC PROCESSES
  . **MARKOV PROCESSES**
RT RANDOM PROCESSES

MAROTS (ESA)
UF   MARITIME COMMUNICATION SATELLITE
          (ESA)
        MARITIME ORBITAL TEST SATELLITE
GS   ESA SPACECRAFT
        . ESA SATELLITES
        . . **MAROTS (ESA)**
        SATELLITES
        . ARTIFICIAL SATELLITES
        . . COMMUNICATION SATELLITES
        . . . **MAROTS (ESA)**
        . ESA SATELLITES
        . . **MAROTS (ESA)**
        . . MARITIME SATELLITES
        . . . **MAROTS (ESA)**
RT   EUROPEAN SPACE AGENCY
        MARISAT SATELLITES
        RANGEFINDING
        RESCUE OPERATIONS
        SHIP TERMINALS

MARQUARDT R4D ENGINE
GS   ENGINES
        . **MARQUARDT R4D ENGINE**
RT   APOLLO PROJECT
        AUXILIARY PROPULSION
        COMMAND MODULES
        ∞REACTION CONTROL
        SATELLITE ATTITUDE CONTROL
        SPACECRAFT CONTROL

MARROW
GS   ANATOMY
        . MUSCULOSKELETAL SYSTEM
        . . BONES
        . . . **MARROW**
        . . CONNECTIVE TISSUE
        . . . **MARROW**

∞ **MARS**
SN   *(USE OF A MORE SPECIFIC TERM IS
         RECOMMENDED--CONSULT THE TERMS
         LISTED BELOW)*
RT   MARS (MANNED REUSABLE
          SPACECRAFT)
        MARS (PLANET)
        NAVIGATION AIDS
        TRACKING STATIONS

MARS (MANNED REUSABLE SPACECRAFT)
SN   (NOT RESTRICTED ONLY TO
         SPACECRAFT FOR FLIGHT TO PLANET
         MARS)
UF   MANNED AERODYNAMIC REUSABLE
          SPACESHIP
GS   MANNED SPACECRAFT
        . **MARS (MANNED REUSABLE
          SPACECRAFT)**
        REENTRY VEHICLES
        . RECOVERABLE SPACECRAFT
        . . REUSABLE SPACECRAFT
        . . . **MARS (MANNED REUSABLE
          SPACECRAFT)**
RT   FERRY SPACECRAFT
        MANEUVERABLE SPACECRAFT
        ∞MARS

MARS (PLANET)
GS   CELESTIAL BODIES
        . PLANETS
        . . TERRESTRIAL PLANETS
        . . . **MARS (PLANET)**
RT   AMOR ASTEROID
        APOLLO ASTEROIDS
        DEIMOS
        DUST STORMS
        ∞MARS
        MARS ATMOSPHERE
        MARS ENVIRONMENT
        MARS SURFACE
        MARS VOLCANOES
        PHOBOS
        PLANETARY CRATERS
        POLAR CAPS

MARS ATMOSPHERE
GS   ENVIRONMENTS
        . EXTRATERRESTRIAL ENVIRONMENTS
        . . PLANETARY ENVIRONMENTS
        . . . MARS ENVIRONMENT
        . . . . **MARS ATMOSPHERE**
        . . . PLANETARY ATMOSPHERES
        . . . . **MARS ATMOSPHERE**
RT   AEROSPACE ENVIRONMENTS
        MARS (PLANET)

MARS ATMOSPHERE-*(CONT.)*
        MARS VOLCANOES

MARS CRATERS
GS   CRATERS
        . PLANETARY CRATERS
        . . **MARS CRATERS**
RT   CRATERING
        EJECTA
        IMPACT DAMAGE
        METEORITE CRATERS
        METEORITIC DAMAGE

MARS ENVIRONMENT
GS   ENVIRONMENTS
        . EXTRATERRESTRIAL ENVIRONMENTS
        . . PLANETARY ENVIRONMENTS
        . . . **MARS ENVIRONMENT**
        . . . . MARS ATMOSPHERE
RT   DUST STORMS
        MARS (PLANET)
        MARS VOLCANOES

MARS EXCURSION MODULE
UF   MEM (EXCURSION MODULE)
GS   MODULES
        . SPACECRAFT MODULES
        . . LANDING MODULES
        . . . **MARS EXCURSION MODULE**
        SOFT LANDING SPACECRAFT
        . LANDING MODULES
        . . **MARS EXCURSION MODULE**

MARS LANDING
GS   LANDING
        . SPACECRAFT LANDING
        . . **MARS LANDING**
RT   AEPS
        PLANETARY LANDING
        SOFT LANDING
        VIKING 75 ENTRY VEHICLE

MARS PHOTOGRAPHS
GS   PHOTOGRAPHS
        . **MARS PHOTOGRAPHS**
RT   PHOTOGRAPHY
        SATELLITE-BORNE PHOTOGRAPHY
        SPACEBORNE PHOTOGRAPHY

MARS PROBES
GS   INTERPLANETARY SPACECRAFT
        . **MARS PROBES**
        . . ADVANCED RECONN ELECTRIC
          SPACECRAFT
        . . MARINER 3 SPACE PROBE
        . . MARINER 4 SPACE PROBE
        . . MARINER 6 SPACE PROBE
        . . MARINER 7 SPACE PROBE
        . . MARINER 8 SPACE PROBE
        . . MARINER 10 SPACE PROBE
        . . MARS 1 SPACECRAFT
        . . MARS 2 SPACECRAFT
        . . MARS 3 SPACECRAFT
        . . MARS 4 SPACECRAFT
        . . MARS 5 SPACECRAFT
        . . MARS 6 SPACECRAFT
        . . MARS 7 SPACECRAFT
        . . VIKING SPACECRAFT
        . . . VIKING 1 SPACECRAFT
        . . . . VIKING LANDER SPACECRAFT
        . . . . . VIKING LANDER 1
        . . . . VIKING ORBITER SPACECRAFT
        . . . . . VIKING ORBITER 1
        . . . VIKING 2 SPACECRAFT
        . . . . VIKING LANDER SPACECRAFT
        . . . . . VIKING LANDER 2
        . . . . VIKING ORBITER SPACECRAFT
        . . . . . VIKING ORBITER 2
        . . VIKING 75 ENTRY VEHICLE
        . . ZOND 2 SPACE PROBE
        UNMANNED SPACECRAFT
        . SPACE PROBES
        . . **MARS PROBES**
        . . . ADVANCED RECONN ELECTRIC
          SPACECRAFT
        . . . MARINER 3 SPACE PROBE
        . . . MARINER 4 SPACE PROBE
        . . . MARINER 6 SPACE PROBE
        . . . MARINER 7 SPACE PROBE
        . . . MARINER 8 SPACE PROBE
        . . . MARINER 10 SPACE PROBE
        . . . MARS 1 SPACECRAFT
        . . . MARS 2 SPACECRAFT
        . . . MARS 3 SPACECRAFT
        . . . MARS 4 SPACECRAFT

MARS PROBES-*(CONT.)*
        . . . MARS 5 SPACECRAFT
        . . . MARS 6 SPACECRAFT
        . . . MARS 7 SPACECRAFT
        . . . VIKING ORBITER 1975
        . . . ZOND 2 SPACE PROBE
RT   MARINER PROGRAM
        OUTER PLANETS EXPLORERS
        VENUS PROBES
        VOYAGER PROJECT
        ZOND SPACE PROBES

MARS SURFACE
GS   PLANETARY SURFACES
        . **MARS SURFACE**
RT   CANALS
        DUST STORMS
        MARS (PLANET)
        MARS VOLCANOES
        METEORITE CRATERS
        PLANETARY CRATERS
        ∞SURFACES
        TOPOGRAPHY

MARS SURFACE SAMPLES
GS   SAMPLES
        . **MARS SURFACE SAMPLES**
RT   ASSAYING
        CHEMICAL ANALYSIS
        SPECIMENS
        ∞SURFACES
        VIKING LANDER 1
        VIKING LANDER 2

MARS VOLCANOES
GS   GEOLOGY
        . VOLCANOES
        . . **MARS VOLCANOES**
        LANDFORMS
        . VOLCANOES
        . . **MARS VOLCANOES**
        PLANETARY GEOLOGY
        . **MARS VOLCANOES**
RT   BASALT
        CALDERAS
        CONES (VOLCANOES)
        EFFUSIVES
        LAVA
        MARS (PLANET)
        MARS ATMOSPHERE
        MARS ENVIRONMENT
        MARS SURFACE
        MOUNTAINS
        OROGRAPHY
        PALEOMAGNETISM
        PETROLOGY
        ROUSE BELTS
        VOLCANOLOGY

MARS 1 SPACECRAFT
GS   INTERPLANETARY SPACECRAFT
        . MARS PROBES
        . . **MARS 1 SPACECRAFT**
        SOVIET SPACECRAFT
        . **MARS 1 SPACECRAFT**
        UNMANNED SPACECRAFT
        . SPACE PROBES
        . . MARS PROBES
        . . . **MARS 1 SPACECRAFT**

MARS 2 SPACECRAFT
GS   INTERPLANETARY SPACECRAFT
        . MARS PROBES
        . . **MARS 2 SPACECRAFT**
        SOVIET SPACECRAFT
        . **MARS 2 SPACECRAFT**
        UNMANNED SPACECRAFT
        . SPACE PROBES
        . . MARS PROBES
        . . . **MARS 2 SPACECRAFT**
RT   U.S.S.R. SPACE PROGRAM

MARS 3 SPACECRAFT
GS   INTERPLANETARY SPACECRAFT
        . MARS PROBES
        . . **MARS 3 SPACECRAFT**
        SOVIET SPACECRAFT
        . **MARS 3 SPACECRAFT**
        SPACECRAFT DESIGN
        . **MARS 3 SPACECRAFT**
        SPACECRAFT STRUCTURES
        . **MARS 3 SPACECRAFT**
        UNMANNED SPACECRAFT
        . SPACE PROBES
        . . MARS PROBES

MARS 3 SPACECRAFT-*(CONT.)*
      . . . **MARS 3 SPACECRAFT**
RT   SPACECRAFT DESIGN
       SPACECRAFT STRUCTURES
       U.S.S.R. SPACE PROGRAM

MARS 4 SPACECRAFT
GS   INTERPLANETARY SPACECRAFT
      . MARS PROBES
      . . **MARS 4 SPACECRAFT**
      SOVIET SPACECRAFT
      . **MARS 4 SPACECRAFT**
      UNMANNED SPACECRAFT
      . SPACE PROBES
      . . MARS PROBES
      . . . **MARS 4 SPACECRAFT**
RT   U.S.S.R. SPACE PROGRAM

MARS 5 SPACECRAFT
GS   INTERPLANETARY SPACECRAFT
      . MARS PROBES
      . . **MARS 5 SPACECRAFT**
      SOVIET SPACECRAFT
      . **MARS 5 SPACECRAFT**
      UNMANNED SPACECRAFT
      . SPACE PROBES
      . . MARS PROBES
      . . . **MARS 5 SPACECRAFT**
RT   U.S.S.R. SPACE PROGRAM

MARS 6 SPACECRAFT
GS   INTERPLANETARY SPACECRAFT
      . MARS PROBES
      . . **MARS 6 SPACECRAFT**
      SOVIET SPACECRAFT
      . **MARS 6 SPACECRAFT**
      UNMANNED SPACECRAFT
      . SPACE PROBES
      . . MARS PROBES
      . . . **MARS 6 SPACECRAFT**
RT   U.S.S.R. SPACE PROGRAM

MARS 7 SPACECRAFT
GS   INTERPLANETARY SPACECRAFT
      . MARS PROBES
      MARS 7 SPACECRAFT
      SOVIET SPACECRAFT
      . **MARS 7 SPACECRAFT**
      UNMANNED SPACECRAFT
      . SPACE PROBES
      . . MARS PROBES
      . . . **MARS 7 SPACECRAFT**
RT   U.S.S.R. SPACE PROGRAM

MARS 69 PROJECT
GS   PROGRAMS
      . NASA PROGRAMS
      . . **MARS 69 PROJECT**
      . NASA SPACE PROGRAMS
      . . **MARS 69 PROJECT**
      . PROJECTS
      . . **MARS 69 PROJECT**
RT   MARINER 6 SPACE PROBE
      MARINER 7 SPACE PROBE
      SPACE EXPLORATION

MARS 71 PROJECT
GS   PROGRAMS
      . NASA PROGRAMS
      . . **MARS 71 PROJECT**
      . NASA SPACE PROGRAMS
      . . **MARS 71 PROJECT**
      . PROJECTS
      . . **MARS 71 PROJECT**
RT   MARINER 8 SPACE PROBE
      SPACE EXPLORATION

MARSHES
USE   MARSHLANDS

MARSHLANDS
UF   BOGS
      COASTAL MARSHLANDS
      MARSHES
      SWAMPS
GS   LAND
      . **MARSHLANDS**
RT   BAYOUS
      EARTH RESOURCES
      EARTH SURFACE
      FLATS (LANDFORMS)
      MUSKEGS
      OCEANOGRAPHY
      TIDAL FLATS
      WATERFOWL

MARSHLANDS-*(CONT.)*
      WETLANDS

MARTENSITE
RT   AUSTENITE
      HARDENING (MATERIALS)
      HEAT TREATMENT
      IRON ALLOYS
      MARTENSITIC STAINLESS STEELS
      MICROSTRUCTURE
      PHASE TRANSFORMATIONS
      STEELS

MARTENSITIC STAINLESS STEELS
GS   ALLOYS
      . IRON ALLOYS
      . . STEELS
      . . . STAINLESS STEELS
      . . . . **MARTENSITIC STAINLESS STEELS**
RT   AUSTENITIC STAINLESS STEELS
      MARAGING STEELS
      MARTENSITE

MARTENSITIC TRANSFORMATION
GS   PHASE TRANSFORMATIONS
      . **MARTENSITIC TRANSFORMATION**
RT   AUSTENITE

MARTIN AIRCRAFT
GS   **MARTIN AIRCRAFT**
      . B-26 AIRCRAFT
      . B-57 AIRCRAFT
RT   ∞ AIRCRAFT

MARTINGALES
RT   DECISION THEORY
      GAME THEORY
      ∞ MATHEMATICS
      PROBABILITY THEORY
      STOCHASTIC PROCESSES

MARTINIQUE
GS   LANDFORMS
      . ISLANDS
      . . WEST INDIES
      . . . **MARTINIQUE**
      NATIONS
      . FRANCE
      . . **MARTINIQUE**
RT   CARIBBEAN REGION

MARVS (PROGRAMMING LANGUAGE)
GS   LANGUAGES
      . PROGRAMMING LANGUAGES
      . . MACHINE ORIENTED LANGUAGES
      . . . **MARVS (PROGRAMMING
         LANGUAGE)**

MARYLAND
GS   NATIONS
      . UNITED STATES
      . . **MARYLAND**
RT   ALLEGHENY PLATEAU (US)
      ASSATEAGUE ISLAND (MD-VA)
      CHESAPEAKE BAY (US)
      DELMARVA PENINSULA (DE-MD-VA)
      POTOMAC RIVER VALLEY (MD-VA-WV)
      SUSQUEHANNA RIVER BASIN
        (MD-NY-PA)

MASCONS
GS   COMPOSITION (PROPERTY)
      . CONCENTRATION (COMPOSITION)
      . . **MASCONS**
RT   CENTER OF MASS
      GRAVITY ANOMALIES
      MASS
      WEIGHT (MASS)

MASER OUTPUTS
GS   OUTPUT
      . **MASER OUTPUTS**
RT   ∞ COHERENCE
      LASER OUTPUTS
      PULSE DURATION
      RADIANT FLUX DENSITY
      WATER MASERS
      WAVELENGTHS

MASER RESONATORS
USE   MASERS

MASERS
UF   MASER RESONATORS
      PARAMAGNETIC AMPLIFIERS
      RASERS
GS   STIMULATED EMISSION DEVICES
      . **MASERS**
      . . GAS MASERS
      . . . HYDROGEN MASERS
      . . INTERSTELLAR MASERS
      . . PROTON MASERS
      . . TRAVELING WAVE MASERS
      . . WATER MASERS
RT   AMPLIFIERS
      ATOMIC CLOCKS
      COHERENT ELECTROMAGNETIC
        RADIATION
      CROSS RELAXATION
      FREQUENCY STANDARDS
      KRYPTON FLUORIDE LASERS
      LASERS
      MICROWAVE AMPLIFIERS
      MOLECULAR OSCILLATORS
      RESONATORS
      STIMULATED EMISSION
      TRANSIENT OSCILLATIONS
      TWO-WAVELENGTH LASERS
      ULTRAVIOLET LASERS

MASKING
GS   **MASKING**
      . TARGET MASKING
RT   AUDIOMETRY
      CHEMISORPTION
      COVERINGS
      PHOTOMASKS

MASKS
GS   **MASKS**
      . OXYGEN MASKS
RT   CHEMICAL DEFENSE
      PROTECTIVE CLOTHING

MASONITE (TRADEMARK)
RT   CELLULOSE
      ∞ CONSTRUCTION MATERIALS
      TREES (PLANTS)
      WOOD

MASONRY
GS   **MASONRY**
      . BRICKS
RT   CEMENTS
      CERAMICS
      CLAYS
      CONCRETES
      CONSTRUCTION
      ∞ CONSTRUCTION MATERIALS
      MORTARS (MATERIAL)
      STRUCTURAL MEMBERS
      TILES
      VENEERS

MASS
UF   LOW MASS
GS   **MASS**
      . CENTER OF MASS
      . CRITICAL MASS
      . MISSING MASS (ASTROPHYSICS)
      . PARTICLE MASS
      . . ELECTRON MASS
      . PLANETARY MASS
      . STELLAR MASS
      . SUBCRITICAL MASS
RT   CENTER OF GRAVITY
      DE BROGLIE WAVELENGTHS
      INERTIA
      MASCONS
      MASS TO LIGHT RATIOS
      MOMENTS OF INERTIA
      RELATIVISTIC EFFECTS
      WEIGHT (MASS)

∞ MASS BALANCE
SN   *(USE OF A MORE SPECIFIC TERM IS
      RECOMMENDED--CONSULT THE TERMS
      LISTED BELOW)*
RT   BALANCE
      MASS DISTRIBUTION
      MATERIAL BALANCE
      VARIABLE MASS SYSTEMS

MASS DISTRIBUTION
GS   DISTRIBUTION (PROPERTY)
      . **MASS DISTRIBUTION**
RT   AERODYNAMIC BALANCE

MASS DISTRIBUTION-*(CONT.)*
    AERODYNAMIC STABILITY
    ANGULAR DISTRIBUTION
    BALANCE
    BALLAST (MASS)
    CHARGE DISTRIBUTION
    COSMOLOGY
    COUNTERBALANCES
    DENSITY WAVE MODEL
    ∞ DISTRIBUTION
    FLUX DENSITY
    FORCE DISTRIBUTION
    INTERGALACTIC MEDIA
    INTERPLANETARY MEDIUM
    INTERSTELLAR MATTER
    LOADING MOMENTS
    LOADS (FORCES)
    ∞ MASS BALANCE
    MASS TO LIGHT RATIOS
    METEOROID CONCENTRATION
    MISSING MASS (ASTROPHYSICS)
    MOMENT DISTRIBUTION
    MOMENTS OF INERTIA
    PRESSURE DISTRIBUTION
    SIZE DISTRIBUTION
    STAR DISTRIBUTION
    STATIC LOADS
    STRUCTURAL DESIGN CRITERIA
    VARIABLE MASS SYSTEMS

MASS DRIVERS (PAYLOAD DELIVERY)
  GS PROPULSION
    . SPACECRAFT PROPULSION
    . . ELECTROMAGNETIC PROPULSION
    . . . **MASS DRIVERS (PAYLOAD
        DELIVERY)**
  RT MAGNETIC LEVITATION VEHICLES
    MOON-EARTH TRAJECTORIES
    REMOTE MANIPULATOR SYSTEM

MASS FILTERS
  USE FLUID FILTERS

MASS FLOW
  GS FLUID FLOW
    . **MASS FLOW**
  RT CROCCO-LEE THEORY
    ∞ FLOW
    FLOW THEORY
    GAS FLOW
    KELVIN-HELMHOLTZ INSTABILITY
    KINETIC THEORY
    LAMINAR FLOW
    LEWIS NUMBERS
    LIQUID FLOW
    MOLECULAR INTERACTIONS
    MULTIPHASE FLOW
    PIPE FLOW
    SEDIMENT TRANSPORT
    SINGLE-PHASE FLOW
    SLIDING
    SLUMPING
    SOLIDS FLOW
    STEADY FLOW
    STEAM FLOW
    TURBULENT FLOW
    UNIFORM FLOW
    UNSTEADY FLOW

MASS FLOW FACTORS
  RT DISCHARGE COEFFICIENT
    FLOW COEFFICIENTS
    HEAT TRANSFER COEFFICIENTS
    HEAT TRANSMISSION
    NOZZLE GEOMETRY

MASS FLOW RATE
  GS RATES (PER TIME)
    . **MASS FLOW RATE**
  RT CONVECTIVE FLOW
    DIFFUSION COEFFICIENT
    FLOW VELOCITY
    PNEUMATIC PROBES
    SPECIFIC IMPULSE
    TRANSIENT PRESSURES

MASS RATIOS
  GS RATIOS
    . **MASS RATIOS**
    . . MASS TO LIGHT RATIOS
    . . PAYLOAD MASS RATIO
    . . PROPELLANT MASS RATIO
  RT METALLICITY
    PRESSURE RATIO
    STRUCTURAL WEIGHT

MASS RATIOS-*(CONT.)*
    THRUST-WEIGHT RATIO

MASS SPECTRA
  GS SPECTRA
    . **MASS SPECTRA**
  RT ENERGY SPECTRA
    MOLECULAR SPECTRA
    RADIATION SPECTRA

MASS SPECTROMETERS
  UF ION SPECTROMETERS
    RETARDING ION MASS
        SPECTROMETERS
  GS MEASURING INSTRUMENTS
    . SPECTROMETERS
    . . **MASS SPECTROMETERS**
  RT CHEMICAL ANALYSIS
    GAS ANALYSIS
    MICROANALYSIS
    NEUTRON ACTIVATION ANALYSIS
    QUALITATIVE ANALYSIS

MASS SPECTROMETRY
  USE MASS SPECTROSCOPY

MASS SPECTROSCOPY
  UF MASS SPECTROMETRY
  GS SPECTROSCOPY
    . **MASS SPECTROSCOPY**
  RT CHEMICAL ANALYSIS
    GAS SPECTROSCOPY
    MAGNETIC SPECTROSCOPY
    NUCLEAR RADIATION SPECTROSCOPY
    SPECTROSCOPIC ANALYSIS
    VACUUM SPECTROSCOPY

MASS TO LIGHT RATIOS
  GS RATIOS
    . MASS RATIOS
    . . **MASS TO LIGHT RATIOS**
  RT ASTRONOMY
    ASTROPHYSICS
    GALACTIC RADIATION
    INDEXES (RATIOS)
    LUMINOSITY
    LUMINOUS INTENSITY
    MASS
    MASS DISTRIBUTION
    MISSING MASS (ASTROPHYSICS)
    RADIANT FLUX DENSITY
    STELLAR LUMINOSITY
    STELLAR MASS

MASS TRANSFER
  RT ABLATION
    CHARGE TRANSFER
    CONVECTIVE FLOW
    CONVECTIVE HEAT TRANSFER
    ENERGY TRANSFER
    GAS TRANSPORT
    GAS-LIQUID INTERACTIONS
    HEAT TRANSFER
    LEWIS NUMBERS
    POROUS BOUNDARY LAYER CONTROL
    SEDIMENT TRANSPORT
    TRANSFERRING
    TRANSPIRATION

MASSACHUSETTS
  GS NATIONS
    . UNITED STATES
    . . **MASSACHUSETTS**

MASSAGING
  GS THERAPY
    . **MASSAGING**
  RT FATIGUE (BIOLOGY)
    RELAXATION (PHYSIOLOGY)

MASSIFS
  GS LANDFORMS
    . **MASSIFS**
  RT EARTH CRUST
    ∞ FAULTS
    GEOLOGICAL FAULTS
    GEOLOGY
    MOUNTAINS

MAST SHOCK TUBES
  USE MAGNETIC ANNULAR SHOCK TUBES

MASTICATION
  UF CHEWING

MASTICATION-*(CONT.)*
  RT DIGESTING
    EATING
    TEETH

MASTOIDS
  GS ANATOMY
    . MUSCULOSKELETAL SYSTEM
    . . BONES
    . . . **MASTOIDS**
    . SENSE ORGANS
    . . EAR
    . . . **MASTOIDS**

MATCHED FILTERS
  GS ELECTROMAGNETIC WAVE FILTERS
    . **MATCHED FILTERS**
  RT COMMUNICATION EQUIPMENT
    DEMODULATORS
    ∞ FILTERS
    MODULATORS
    SIGNAL TO NOISE RATIOS

MATCHING
  RT ADJUSTING
    COMPARISON
    FITTING
    HOMOLOGY
    IMAGE RESOLUTION
    IMPEDANCE MATCHING
    MISMATCH (ELECTRICAL)
    PATTERN REGISTRATION

MATERIAL ABSORPTION
  RT ABSORBENTS
    ABSORBERS (EQUIPMENT)
    ∞ ABSORPTION
    ASSIMILATION
    EXTRACTION
    HYGROSCOPICITY
    RADIATION ABSORPTION
    SORPTION
    WATER TREATMENT

MATERIAL BALANCE
  GS **MATERIAL BALANCE**
    . WATER BALANCE
  RT BALANCE
    HEAT BALANCE
    ∞ MASS BALANCE
    STOICHIOMETRY

MATERIAL REMOVAL (MACHINING)
  USE MACHINING

∞ **MATERIALS**
  SN *(USE OF A MORE SPECIFIC TERM IS
    RECOMMENDED--CONSULT THE TERMS
    LISTED BELOW)*
  UF SUBSTANCES
  RT ABLATIVE MATERIALS
    ABSORBENTS
    ABSORBERS (MATERIALS)
    ACCEPTOR MATERIALS
    AGING (MATERIALS)
    AIRCRAFT CONSTRUCTION MATERIALS
    AIRFRAME MATERIALS
    AMORPHOUS MATERIALS
    ANISOTROPIC MEDIA
    BINARY SYSTEMS (MATERIALS)
    BINDERS (MATERIALS)
    BITUMENS
    BORON REINFORCED MATERIALS
    BORSIC (TRADENAME)
    BRITTLE MATERIALS
    CARBONACEOUS MATERIALS
    COMPOSITE MATERIALS
    CONCRETE STRUCTURES
    ∞ CONSTRUCTION MATERIALS
    CONTAMINANTS
    CORK (MATERIALS)
    CURL (MATERIALS)
    DISLOCATIONS (MATERIALS)
    DONOR MATERIALS
    DREDGED MATERIALS
    ELECTRIC FURNACES
    ELECTRONS
    EPOXY MATRIX COMPOSITES
    FATIGUE (MATERIALS)
    FERRIMAGNETIC MATERIALS
    FERROMAGNETIC MATERIALS
    FISSIONABLE MATERIALS
    FOAMS
    FOILS (MATERIALS)
    FRACTURES (MATERIALS)

MATERIALS-(CONT.)
 GLASS
 GLASSY CARBON
 GRANULAR MATERIALS
 GRAPHITE-EPOXY COMPOSITES
 HOLES (ELECTRON DEFICIENCIES)
 ∞ INORGANIC MATERIALS
 INSULATION
 LAMINATES
 LASER MATERIALS
 LOSSLESS MATERIALS
 LOW DENSITY MATERIALS
 MAGNETIC MATERIALS
 MATERIALS HANDLING
 MATERIALS RECOVERY
 ∞ MATERIALS SCIENCE
 ∞ MATERIALS TESTS
 MATRIX MATERIALS
 MECHANICAL PROPERTIES
 METAL MATRIX COMPOSITES
 MOLDING MATERIALS
 NONFLAMMABLE MATERIALS
 ORGANIC MATERIALS
 PAPER (MATERIAL)
 PHASE CHANGE MATERIALS
 PHOTOELASTIC MATERIALS
 PHOTOELECTRIC MATERIALS
 POLYMER MATRIX COMPOSITES
 POROUS MATERIALS
 PYROLYTIC MATERIALS
 PYROPHORIC MATERIALS
 RADIOACTIVE MATERIALS
 RADOME MATERIALS
 REACTOR MATERIALS
 REFRACTORY MATERIALS
 REINFORCING MATERIALS
 RESERVES
 RESOURCES
 SELF LUBRICATING MATERIALS
 SEMICONDUCTORS (MATERIALS)
 SIZING MATERIALS
 SOLIDS
 SPACECRAFT CONSTRUCTION
  MATERIALS
 SPONGES (MATERIALS)
 STRATEGIC MATERIALS
 SUPERHYBRID MATERIALS
 THERMOCHROMATIC MATERIALS
 THERMOELECTRIC MATERIALS
 THICKENERS (MATERIALS)
 THREE DIMENSIONAL COMPOSITES
 VITREOUS MATERIALS
 VYCOR

MATERIALS HANDLING
 GS   **MATERIALS HANDLING**
    . GROUND HANDLING
    . PROPELLANT TRANSFER
    . REMOTE HANDLING
 RT   AIRFIELD SURFACE MOVEMENTS
    ∞ AUTOMATION
    BLOWERS
    CANALS
    CARGO
    CARGO AIRCRAFT
    CARTS
    CHEMICAL ENGINEERING
    CHUTES
    ∞ CONTAINERS
    CONTINGENCY
    CONVEYORS
    CRANES
    DELIVERY
    DISPENSERS
    DISPOSAL
    DISTRIBUTING
    ∞ DISTRIBUTION
    DISTRIBUTORS
    DOLLIES
    DUMPING
    EJECTION
    EJECTORS
    EMPTYING
    ENCAPSULATING
    EXCAVATION
    FEEDERS
    FEEDING (SUPPLYING)
    FLUID FLOW
    FUEL PUMPS
    HAULING
    HEAVY LIFT AIRSHIPS
    HOPPERS
    LOADING OPERATIONS
    LUNAR LOGISTICS
    MARKING
    ∞ MATERIALS

MATERIALS HANDLING-(CONT.)
    MECHANICAL ENGINEERING
    MINES (EXCAVATIONS)
    MOORING
    PACKAGING
    PIPELINES
    ∞ PUMPING
    PUMPS
    RAILROAD HUMPING TESTS
    RELEASING
    RIGGING
    SERVICES
    SIPHONS
    SPRAYERS
    SPREADING
    STACKS
    ∞ STORAGE
    TANKS (CONTAINERS)
    TRACTORS
    TRANSFERRING
    TRANSPORTATION
    TRUCKS
    UNLOADING
    VACUUM PUMPS
    WASTE DISPOSAL
    WHARVES

MATERIALS RECOVERY
 SN   (TREATMENT OF A MATERIAL TO
     RECLAIM ONE OR MORE OF ITS
     COMPONENTS)
 GS   RECLAMATION
    . **MATERIALS RECOVERY**
    . . NUCLEAR FUEL REPROCESSING
    . . SOLVOLYSIS
    . . WATER RECLAMATION
 RT   ∞ ABSORPTION
    BY-PRODUCTS
    CENTRIFUGING
    CRYSTALLIZATION
    DISPOSAL
    DISTILLATION
    EXTRACTION
    FILTRATION
    ∞ MATERIALS
    ∞ PRECIPITATION
    PRECIPITATION (CHEMISTRY)
    ∞ PROCESSING
    ∞ RECOVERY
    RECYCLING
    REFINING
    REMOVAL
    ∞ SEPARATION

∞ **MATERIALS SCIENCE**
 SN   *(USE OF A MORE SPECIFIC TERM IS
     RECOMMENDED--CONSULT THE TERMS
     LISTED BELOW)*
 RT   CERAMICS
    LASER CUTTING
    ∞ MATERIALS
    METAL FOAMS
    PLASTICS
    ∞ PROPERTIES
    ∞ SCIENCE

MATERIALS TESTING REACTORS
 USE   NUCLEAR RESEARCH AND TEST
     REACTORS

∞ **MATERIALS TESTS**
 SN   *(USE OF A MORE SPECIFIC TERM IS
     RECOMMENDED--CONSULT THE TERM
     LISTED BELOW)*
 RT   BEND TESTS
    BURST TESTS
    CHARPY IMPACT TEST
    CHEMICAL ANALYSIS
    COMPRESSION TESTS
    CORROSION TESTS
    DESTRUCTIVE TESTS
    ELECTROPHOTOMETRY
    ENVIRONMENTAL TESTS
    FATIGUE TESTS
    FUEL TESTS
    GAS ANALYSIS
    HARDNESS TESTS
    HIGH TEMPERATURE TESTS
    IMPACT TESTS
    LUBRICANT TESTS
    MAGNETIC MEASUREMENT
    ∞ MATERIALS
    MECHANICAL PROPERTIES
    METALLOGRAPHY
    MICROANALYSIS

MATERIALS TESTS-(CONT.)
    NEUTRON RADIOGRAPHY
    NONDESTRUCTIVE TESTS
    PROPELLANT TESTS
    QUALITY
    QUALITY CONTROL
    RADIOGRAPHY
    SPECIFICATIONS
    STATIC TESTS
    ∞ TESTS
    ULTRASONIC TESTS
    WEAR TESTS
    X RAY ANALYSIS
    X RAY SPECTROSCOPY

MATHEMATICAL ANALYSIS
 USE   APPLICATIONS OF MATHEMATICS

MATHEMATICAL LOGIC
 GS   **MATHEMATICAL LOGIC**
    . ALGORITHMS
    . . PARSING ALGORITHMS
    . . SIMPLEX METHOD
    . AXIOMS
    . FORMULAS (MATHEMATICS)
    . . BETHE-HEITLER FORMULA
    . LATTICES (MATHEMATICS)
    . . BOOLEAN ALGEBRA
    . . . BOOLEAN FUNCTIONS
    . SET THEORY
    . . BOREL SETS
    . . EQUIVALENCE
    . . THRESHOLD LOGIC
 RT   BRANCHING (MATHEMATICS)
    FUNCTIONS (MATHEMATICS)
    HYPOTHESES
    INDUCTION (MATHEMATICS)
    INSTRUCTION SETS (COMPUTERS)
    ∞ LOGIC
    PHILOSOPHY
    PROVING
    THEOREMS
    TURING MACHINES
    VENN DIAGRAMS

MATHEMATICAL MODELS
 UF   ISING MODEL
 GS   MODELS
    . **MATHEMATICAL MODELS**
    . . ANALOG SIMULATION
    . . BIOLOGICAL MODELS (MATHEMATICS)
    . . DIGITAL SIMULATION
    . . THOMAS-FERMI MODEL
    . . VENEZIANO MODEL
 RT   AIRCRAFT MODELS
    ∞ APPLICATIONS OF MATHEMATICS
    ASTRONOMICAL MODELS
    ASYMPTOTIC PROPERTIES
    ATMOSPHERIC MODELS
    BOND GRAPHS
    BROKEN SYMMETRY
    CHAOS
    COMPUTATIONAL ASTROPHYSICS
    COMPUTATIONAL GRIDS
    COMPUTER SYSTEMS SIMULATION
    COMPUTERIZED SIMULATION
    CONTINUUM MODELING
    CONTROL SYSTEMS DESIGN
    DECISION THEORY
    DYNAMIC MODELS
    DYNAMIC PROGRAMMING
    DYNAMICAL SYSTEMS
    EXHAUST FLOW SIMULATION
    EXPERIMENT DESIGN
    FACTORIAL DESIGN
    FLOW CHARTS
    FOOTPRINTS
    FORECASTING
    FUNCTIONS (MATHEMATICS)
    GAME THEORY
    GOODNESS OF FIT
    GRAPH THEORY
    INELASTIC STRESS
    INVENTORY CONTROLS
    LIKELIHOOD RATIO
    LINEAR PREDICTION
    LOFTING
    LUMPED PARAMETER SYSTEMS
    METHOD OF MOMENTS
    ∞ MISSILE SIMULATORS
    MONTE CARLO METHOD
    NUMERICAL WEATHER FORECASTING
    OCEAN MODELS
    OPERATIONS RESEARCH
    OUTLIERS (STATISTICS)

MATHEMATICAL MODELS-(CONT.)

PARAMETER IDENTIFICATION
PARAMETERIZATION
QUANTILES
QUEUEING THEORY
REGRESSION COEFFICIENTS
RISK
ROBUSTNESS (MATHEMATICS)
SCHEDULING
SIMILARITY THEOREM
SIMULATION
SPACECRAFT MODELS
STATISTICAL DISTRIBUTIONS
STOCHASTIC PROCESSES
SYSTEM IDENTIFICATION
SYSTEMS ANALYSIS
SYSTEMS ENGINEERING
SYSTEMS SIMULATION
TRAJECTORY ANALYSIS
TWO DIMENSIONAL BODIES
VALIDITY
WAR GAMES

MATHEMATICAL PROGRAMMING

GS   OPERATIONS RESEARCH
    . **MATHEMATICAL PROGRAMMING**
    . . LINEAR PROGRAMMING
    . . NONLINEAR PROGRAMMING
    . . QUADRATIC PROGRAMMING
    RESEARCH
    . **MATHEMATICAL PROGRAMMING**
    . . DYNAMIC PROGRAMMING
    . . LINEAR PROGRAMMING
    . . NONLINEAR PROGRAMMING
    . . QUADRATIC PROGRAMMING
RT   COMPUTER PROGRAMMING
    GAME THEORY
    OPTIMIZATION
    ∞ PROGRAMMING
    SIMPLEX METHOD

MATHEMATICAL TABLES

GS   TABLES (DATA)
    . **MATHEMATICAL TABLES**
RT   INFORMATION
    NUMERICAL ANALYSIS
    RANDOM NUMBERS

∞ MATHEMATICS

SN   *(USE OF A MORE SPECIFIC TERM IS*
    RECOMMENDED--CONSULT THE TERMS
    LISTED BELOW)
RT   ALGEBRA
    ANALYSIS (MATHEMATICS)
    AXIOMS
    BOND GRAPHS
    CALCULUS
    CURRENT ALGEBRA
    DUALITY THEOREM
    FORMULAS (MATHEMATICS)
    FRACTALS
    FUNCTIONS (MATHEMATICS)
    GEOMETRY
    INEQUALITIES
    INFORMATION THEORY
    INTEGRALS
    LATIN SQUARE METHOD
    LATTICES (MATHEMATICS)
    MARTINGALES
    MORPHOLOGY
    NUMBER THEORY
    NUMERICAL ANALYSIS
    PRIMITIVE EQUATIONS
    ∞ PRINCIPLES
    PROBABILITY THEORY
    RINGS (MATHEMATICS)
    ∞ SCIENCE
    SERIES EXPANSION
    STARS (MATHEMATICS)
    STATISTICAL ANALYSIS
    SUPERPOSITION (MATHEMATICS)
    SYMBOLS
    THEOREMS

MATHIEU EQUATION

USE   MATHIEU FUNCTION

MATHIEU FUNCTION

UF   MATHIEU EQUATION
GS   ANALYSIS (MATHEMATICS)
    . COMPLEX VARIABLES
    . . **MATHIEU FUNCTION**
    FUNCTIONS (MATHEMATICS)
    . **MATHIEU FUNCTION**
RT   BOUNDARY VALUE PROBLEMS

MATHIEU FUNCTION-(CONT.)

DIFFERENTIAL EQUATIONS
EIGENVECTORS
∞ EQUATIONS
HILL DETERMINANT
ORTHOGONAL FUNCTIONS

MATRA MISSILE

GS   MISSILES
    . AIR TO AIR MISSILES
    . . **MATRA MISSILE**
RT   SOLID PROPELLANT ROCKET ENGINES

∞ MATRICES

SN   *(USE OF A MORE SPECIFIC TERM IS*
    RECOMMENDED--CONSULT THE TERMS
    LISTED BELOW)
RT   COMPOSITE MATERIALS
    EPOXY MATRIX COMPOSITES
    EUTECTIC COMPOSITES
    ∞ GRIDS
    ∞ IMBEDDINGS
    LATTICES (MATHEMATICS)
    MATRICES (CIRCUITS)
    MATRICES (MATHEMATICS)
    METAL MATRIX COMPOSITES
    MONOTECTIC ALLOYS
    POLYMER MATRIX COMPOSITES

MATRICES (CIRCUITS)

GS   CIRCUITS
    . **MATRICES (CIRCUITS)**
RT   LOGIC CIRCUITS
    ∞ MATRICES
    SWITCHING CIRCUITS

MATRICES (MATHEMATICS)

UF   DIFFERENTIAL ALGEBRA
    MATRIX ANALYSIS
GS   ALGEBRA
    . VECTOR SPACES
    . . **MATRICES (MATHEMATICS)**
    . . . ADJOINTS
    . . . CANONICAL FORMS
    . . . EIGENVALUES
    . . . EIGENVECTORS
    . . . JORDAN FORM
    . . . STIFFNESS MATRIX
RT   ARRAYS
    CHIRAL DYNAMICS
    DETERMINANTS
    ENERGY METHODS
    FACTOR ANALYSIS
    FINITE ELEMENT METHOD
    HERMITIAN POLYNOMIAL
    HESSIAN MATRICES
    ISOPERIMETRIC PROBLEM
    LINEAR EQUATIONS
    LINEAR PROGRAMMING
    LINEAR TRANSFORMATIONS
    LUMPED PARAMETER SYSTEMS
    ∞ MATRICES
    METHOD OF MOMENTS
    ROOTS OF EQUATIONS
    SIMPLEX METHOD
    SIMULTANEOUS EQUATIONS
    SUBGROUPS
    U SPIN SPACE
    WALSH FUNCTION

MATRIX ANALYSIS

USE   MATRICES (MATHEMATICS)

MATRIX MANAGEMENT

GS   MANAGEMENT
    . **MATRIX MANAGEMENT**
RT   ALLOCATIONS
    LOGISTICS
    MANAGEMENT METHODS
    ∞ METHODOLOGY
    OPERATIONS RESEARCH
    PRODUCTIVITY
    PROJECT PLANNING
    SCHEDULING
    TASKS

MATRIX MATERIALS

RT   CERAMIC MATRIX COMPOSITES
    COMPOSITE MATERIALS
    EPOXY MATRIX COMPOSITES
    FIBER REINFORCED COMPOSITES
    LAMINATES
    ∞ MATERIALS
    METAL MATRIX COMPOSITES
    POLYMER MATRIX COMPOSITES

MATRIX MATERIALS-(CONT.)

REINFORCING MATERIALS
RESIN MATRIX COMPOSITES

MATRIX METHODS

SN   (LIMITED TO METHODS FOR
    STRUCTURAL ANALYSIS)
UF   MATRIX STRESS CALCULATION
GS   STRUCTURAL ANALYSIS
    . **MATRIX METHODS**
RT   EQUILIBRIUM METHODS
    ∞ METHODOLOGY
    NASTRAN
    SPLINE FUNCTIONS

MATRIX STRESS CALCULATION

USE   MATRIX METHODS

MATRIX THEORY

RT   OPERATORS (MATHEMATICS)
    ∞ THEORIES

MATTER (PHYSICS)

RT   ANTIMATTER
    EXTRATERRESTRIAL MATTER
    ∞ PHYSICS
    ROTATING MATTER

MATTS (SYSTEMS)

UF   MULTIPLE TARGET TRAJECTORY
    SYSTEMS
GS   TRACKING NETWORKS
    . **MATTS (SYSTEMS)**
RT   ABORT TRAJECTORIES
    AIRBORNE EQUIPMENT
    ANGULAR CORRELATION
    TARGET ACQUISITION

MATURING

USE   GROWTH

MAULER MISSILE

GS   MISSILES
    . ANTIAIRCRAFT MISSILES
    . . **MAULER MISSILE**
    . ANTIMISSILE MISSILES
    . . **MAULER MISSILE**
    . SURFACE TO AIR MISSILES
    . . **MAULER MISSILE**
RT   SINGLE STAGE ROCKET VEHICLES
    SOLID PROPELLANT ROCKET ENGINES

MAURITANIA

GS   NATIONS
    . **MAURITANIA**
RT   AFRICA

MAVERICK MISSILES

GS   MISSILES
    . AIR TO SURFACE MISSILES
    . . **MAVERICK MISSILES**

MAX HOLSTE MH-262 AIRCRAFT

USE   MH-262 AIRCRAFT

MAXIMA

GS   ANALYSIS (MATHEMATICS)
    . REAL VARIABLES
    . . EXTREMUM VALUES
    . . . **MAXIMA**
RT   APEXES
    CALCULUS OF VARIATIONS
    CUSPS (MATHEMATICS)
    MINIMA
    OPTIMIZATION
    ∞ PEAKS
    PENALTY FUNCTION
    RANGE (EXTREMES)
    ZENITH

MAXIMUM ENTROPY METHOD

GS   ENTROPY (STATISTICS)
    . **MAXIMUM ENTROPY METHOD**
    SPECTRUM ANALYSIS
    . **MAXIMUM ENTROPY METHOD**
RT   DISTRIBUTION FUNCTIONS
    ENTROPY
    FOURIER TRANSFORMATION
    INFORMATION THEORY
    ∞ METHODOLOGY
    POWER SPECTRA
    SIGNAL PROCESSING
    SIGNAL TO NOISE RATIOS
    STATISTICAL ANALYSIS

MAXIMUM ENTROPY METHOD-*(CONT.)*
TIME SERIES ANALYSIS

MAXIMUM LIKELIHOOD ESTIMATES
RT   CONFIDENCE LIMITS
       FORECASTING
       GOODNESS OF FIT
       LIKELIHOOD RATIO
       PARAMETER IDENTIFICATION
       PREDICTIONS
       RELIABILITY
       RISK
       SYSTEM IDENTIFICATION

MAXIMUM PRINCIPLE
RT   COMPLEX VARIABLES
       DIFFERENTIAL EQUATIONS
       ELLIPTIC DIFFERENTIAL EQUATIONS
       HARMONIC FUNCTIONS
       PONTRYAGIN PRINCIPLE
       REAL VARIABLES

MAXIMUM USABLE FREQUENCY
GS   FREQUENCIES
       . **MAXIMUM USABLE FREQUENCY**
RT   FREQUENCY ASSIGNMENT
       FREQUENCY REUSE
       HIGH FREQUENCIES
       VERY HIGH FREQUENCIES

MAXWELL BODIES
RT   CLASSICAL MECHANICS
       CONTINUUM MECHANICS
       HOOKES LAW
       OSCILLATION DAMPERS
       RELAXATION TIME

MAXWELL EQUATION
RT   BOLTZMANN-VLASOV EQUATION
       BORN-INFELD THEORY
       ELECTRICITY
       ELECTRODYNAMICS
       ELECTROMECHANICS
       ∞EQUATIONS
       GAUSS EQUATION
       MAGNETIC CHARGE DENSITY
       MAGNETIC PROPERTIES
       MAGNETOELECTRIC MEDIA
       POYNTING THEOREM
       ∞STOKES LAW

MAXWELL FLUIDS
RT   COMPRESSIBLE FLUIDS
       FLUID MECHANICS
       ∞FLUIDS
       RHEOLOGY
       VISCOELASTICITY
       VISCOUS FLOW
       VISCOUS FLUIDS

MAXWELL-BOLTZMANN DENSITY FUNCTION
UF   MAXWELLIAN DISTRIBUTION (DENSITY)
GS   FUNCTIONS (MATHEMATICS)
       . **MAXWELL-BOLTZMANN DENSITY**
         FUNCTION
       STATISTICAL ANALYSIS
       . **MAXWELL-BOLTZMANN DENSITY**
         FUNCTION
RT   DENSITY DISTRIBUTION
       KINETIC THEORY
       PROBABILITY THEORY
       STATISTICAL MECHANICS

MAXWELL-MOHR METHOD
RT   DEFLECTION
       ∞EQUILIBRIUM
       ∞METHODOLOGY
       STATIC DEFORMATION
       TRUSSES

MAXWELLIAN DISTRIBUTION (DENSITY)
USE   MAXWELL-BOLTZMANN DENSITY
         FUNCTION

MAYER PROBLEM
RT   ∞CONDENSATION
       CRITICAL POINT
       GIBBS FREE ENERGY
       ∞MOLECULAR PHYSICS
       ∞PROBLEMS
       SUPERSATURATION

MAYPOLE ANTENNAS
RT   ANTENNA DESIGN

MAYPOLE ANTENNAS-*(CONT.)*
       LARGE SPACE STRUCTURES
       SPACE ERECTABLE STRUCTURES

MAZE LEARNING
GS   LEARNING
       . **MAZE LEARNING**
RT   PROBLEM SOLVING

MB-1 ROCKET VEHICLE
USE   GENIE ROCKET VEHICLE

MBM JUNCTIONS
UF   METAL-BARRIER-METAL JUNCTIONS
GS   SEMICONDUCTOR JUNCTIONS
       . **MBM JUNCTIONS**
RT   BARRIER LAYERS
       ∞BARRIERS
       JUNCTION TRANSISTORS

MCDONNELL AIRCRAFT
GS   MCDONNELL DOUGLAS AIRCRAFT
       . **MCDONNELL AIRCRAFT**
       . . C-9 AIRCRAFT
       . . DC 10 AIRCRAFT
       . . F-101 AIRCRAFT
       . . PHANTOM AIRCRAFT
       . . . F-4 AIRCRAFT
       . . . RF-4 AIRCRAFT
RT   ∞AIRCRAFT

MCDONNELL DOUGLAS AIRCRAFT
GS   **MCDONNELL DOUGLAS AIRCRAFT**
       . A-1 AIRCRAFT
       . DOUGLAS AIRCRAFT
       . . A-3 AIRCRAFT
       . . A-4 AIRCRAFT
       . B-66 AIRCRAFT
       . . C-9 AIRCRAFT
       . . C-47 AIRCRAFT
       . . C-54 AIRCRAFT
       . . C-118 AIRCRAFT
       . . C-124 AIRCRAFT
       . . C-133 AIRCRAFT
       . D-558 AIRCRAFT
       . . DC 3 AIRCRAFT
       . . DC 7 AIRCRAFT
       . . DC 8 AIRCRAFT
       . . DC 9 AIRCRAFT
       . . DC 10 AIRCRAFT
       . PD-808 AIRCRAFT
       . . X-3 AIRCRAFT
       . F-18 AIRCRAFT
       . MCDONNELL AIRCRAFT
       . . C-9 AIRCRAFT
       . . DC 10 AIRCRAFT
       . . F-101 AIRCRAFT
       . . PHANTOM AIRCRAFT
       . . . F-4 AIRCRAFT
       . . . RF-4 AIRCRAFT
RT   ∞AIRCRAFT

MCLAURIN SERIES
USE   MACLAURIN SERIES

MCLEOD GAGES
GS   MEASURING INSTRUMENTS
       . PRESSURE GAGES
       . . VACUUM GAGES
       . . . **MCLEOD GAGES**
       VACUUM APPARATUS
       . VACUUM GAGES
       . . **MCLEOD GAGES**
RT   IONIZATION GAGES
       KNUDSEN GAGES
       PIRANI GAGES
       PRESSURE MEASUREMENT

MCMURDO SOUND
GS   REGIONS
       . POLAR REGIONS
       . . ANTARCTIC REGIONS
       . . . **MCMURDO SOUND**
       SOUTHERN HEMISPHERE
       . ANTARCTIC REGIONS
       . . **MCMURDO SOUND**
RT   ROSS ICE SHELF

MCR REACTORS
USE   MILITARY COMPACT REACTORS

MDA
USE   MULTIPLE DOCKING ADAPTERS

ME P-160 AIRCRAFT
USE   P-160 AIRCRAFT

ME P-308 AIRCRAFT
USE   P-308 AIRCRAFT

MEADOWLANDS
USE   GRASSLANDS

MEAN
GS   AVERAGE
       . **MEAN**
       MOMENTS
       . DISTRIBUTION MOMENTS
       . . **MEAN**
RT   MEDIAN (STATISTICS)
       MODE (STATISTICS)
       NORMALITY
       QUALITY CONTROL
       RANGE (EXTREMES)
       STATISTICAL ANALYSIS
       VARIANCE (STATISTICS)

MEAN FREE PATH
RT   COLLISION PARAMETERS
       ∞CROSS SECTIONS
       KNUDSEN FLOW
       PARTICLE COLLISIONS
       PARTICLE MOTION
       ∞PATHS
       SCATTERING
       VACUUM

MEAN SQUARE VALUES
GS   ANALYSIS (MATHEMATICS)
       . NUMERICAL ANALYSIS
       . . APPROXIMATION
       . . . **MEAN SQUARE VALUES**
RT   ALGORITHMS
       ERROR ANALYSIS
       LEAST SQUARES METHOD

MEAN TIME BETWEEN FAILURES
USE   MTBF

MEANDERS
RT   OPEN CHANNEL FLOW
       RAPIDS
       RIVER BASINS
       RIVERS
       STREAMS
       TOPOGRAPHY
       VALLEYS

MEASURE AND INTEGRATION
UF   INTEGRATION (REAL VARIABLES)
       MEASURE THEORY
       RIEMANN INTEGRAL
GS   ANALYSIS (MATHEMATICS)
       . REAL VARIABLES
       . . **MEASURE AND INTEGRATION**
       . . . BINARY INTEGRATION
       . . . BOREL SETS
       . . . FUNCTIONAL INTEGRATION
       . . . INTEGRAL CALCULUS
       . . . J INTEGRAL
       . . . LEBESGUE THEOREM
       . . . NUMERICAL INTEGRATION
       . . . . RUNGE-KUTTA METHOD
       . . . STIELTJES INTEGRAL
       . . . WEIGHTING FUNCTIONS
RT   FOURIER ANALYSIS

MEASURE THEORY
USE   MEASURE AND INTEGRATION

∞ MEASUREMENT
SN   *(USE OF A MORE SPECIFIC TERM IS*
       RECOMMENDED--CONSULT THE TERMS
       LISTED BELOW)
UF   DETERMINATION
       MEASURING
       QUANTIZATION
RT   ACCURACY
       ACOUSTIC MEASUREMENT
       AIRBORNE RANGE AND ORBIT
         DETERMINATION
       AIRCRAFT INSTRUMENTS
       ANALOG DATA
       ASTROMETRY
       AUDIOMETRY
       CHEMICAL ANALYSIS
       CONFIDENCE LIMITS
       CONSISTENCY

MEASUREMENT-(CONT.)

COUNTING
∞ DATA
∞ DEFINITION
DENSIMETERS
DENSITY MEASUREMENT
DEPTH MEASUREMENT
DETECTION
DILATOMETRY
DIMENSIONAL MEASUREMENT
DOWNRANGE ANTIMISSILE
   MEASUREMENT PROGRAM
DOWNRANGE MEASUREMENT
DRAG MEASUREMENT
EARTH TERMINAL MEASUREMENT
   SYSTEM
ELECTRICAL MEASUREMENT
ELECTROMAGNETIC MEASUREMENT
ELECTROMAGNETIC NOISE
   MEASUREMENT
ESTIMATING
EVALUATION
EXAMINATION
FLOW MEASUREMENT
FREQUENCY MEASUREMENT
FRICTION MEASUREMENT
GAMMA RAY ABSORPTIOMETRY
GEOMETRY
GRAVIMETRY
HEAT MEASUREMENT
HIGH ALT TARGET AND BACKGROUND
   MEASUREMENT
HUMIDITY MEASUREMENT
IDENTIFYING
INTERNATIONAL SYSTEM OF UNITS
LATITUDE MEASUREMENT
LONGITUDE MEASUREMENT
MACROSCOPIC EQUATIONS
MAGNETIC MEASUREMENT
∞ MEASURES
MEASURING INSTRUMENTS
MECHANICAL MEASUREMENT
METROLOGY
MONITORS
NOISE MEASUREMENT
OPTICAL MEASUREMENT
OPTOMETRY
PHOTOGRAPHIC MEASUREMENT
PNEUMOGRAPHY
PRESSURE MEASUREMENT
PROVING
PUPILLOMETRY
RADAR MEASUREMENT
RADIATION MEASUREMENT
RADIOACTIVE AGE DETERMINATION
RANGEFINDING
SIGNAL MEASUREMENT
SIZE DETERMINATION
SOUNDING
SPHYGMOGRAPHY
STANDARDS
STRAIN MEASUREMENT
SYNOPTIC MEASUREMENT
TEMPERATURE MEASUREMENT
THRUST MEASUREMENT
TIME MEASUREMENT
TRAJECTORY MEASUREMENT
ULTRASONIC DENSIMETERS
UNITS OF MEASUREMENT
VELOCITY MEASUREMENT
VIBRATION MEASUREMENT
WEIGHT MEASUREMENT
WIND MEASUREMENT

∞ MEASURES

SN   (USE OF A MORE SPECIFIC TERM IS
    RECOMMENDED--CONSULT THE TERMS
    LISTED BELOW)
RT   CRITERIA
    ∞ MEASUREMENT
    STANDARDS

MEASURING

USE   MEASUREMENT

MEASURING INSTRUMENTS

UF   FLUXMETERS
    GAGES
    METERS
    RATE METERS
GS   MEASURING INSTRUMENTS
   . ACCELEROMETERS
   . . STRAIN GAGE ACCELEROMETERS
   . AMMETERS
   . . MICROMILLIAMMETERS

MEASURING INSTRUMENTS-(CONT.)

. . THERMOELEMENT AMMETERS
. ANALYZERS
. . ENGINE ANALYZERS
. . SIGNAL ANALYZERS
. BALLOON-BORNE INSTRUMENTS
. BATHYMETERS
. BURETTES
. CALORIMETERS
. . BOMB CALORIMETERS
. . DROP CALORIMETERS
. . FLAME CALORIMETERS
. COMPARATORS
. COMPASSES
. . GYROCOMPASSES
. . ELECTRICAL CONDUCTIVITY METERS
. COULOMETERS
. COUNTERS
. . RADIATION COUNTERS
. . . CERENKOV COUNTERS
. . . ELECTRON COUNTERS
. . . GEIGER COUNTERS
. . . NEUTRON COUNTERS
. . . . NEUTRON SPECTROMETERS
. . PARTICLE TELESCOPES
. . PROPORTIONAL COUNTERS
. . . QUANTUM COUNTERS
. . SCINTILLATION COUNTERS
. . . SPARK CHAMBERS
. DEFORMETERS
. DENSIMETERS
. . ULTRASONIC DENSIMETERS
. DENSITOMETERS
. . MICRODENSITOMETERS
. DISTANCE MEASURING EQUIPMENT
. ALTIMETERS
. . RADIO ALTIMETERS
. . GEODIMETERS
. RANGE FINDERS
. . OPTICAL RANGE FINDERS
. . . LASER RANGE FINDERS
. . STADIMETERS
. . TELLUROMETERS
. DYNAMOMETERS
. ELASTOMETERS
. ELECTROMETERS
. ELECTRON PROBES
. ENGINE MONITORING INSTRUMENTS
. ERGOMETERS
. EUDIOMETERS
. EXTENSOMETERS
. FIELD INTENSITY METERS
. FLAME PROBES
. FLIGHT LOAD RECORDERS
. FLIGHT RECORDERS
. FLOWMETERS
. . GAS METERS
. . HOT-WIRE FLOWMETERS
. . RHEOMETERS
. FORCE VECTOR RECORDERS
. FUEL GAGES
. . CAPACITIVE FUEL GAGES
. GALVANOMETERS
. GERDIEN CONDENSERS
. GONIOMETERS
. . PHOTOGONIOMETERS
. . RADIOGONIOMETERS
. GRAVIMETERS
. GRAVITY GRADIOMETERS
. HELIOMETERS
. . PYROHELIOMETERS
. HYDROMETERS
. HYPSOMETERS
. IMPEDANCE PROBES
. . RADIO FREQUENCY IMPEDANCE
   PROBES
. INDICATING INSTRUMENTS
. . APPROACH INDICATORS
. . ATTITUDE INDICATORS
. . . GYRO HORIZONS
. . CLOUD HEIGHT INDICATORS
. . FLOW DIRECTION INDICATORS
. . . WIND VANES
. . GYROCOMPASSES
. . MICROWAVE SENSORS
. . POSITION INDICATORS
. . . ASTROLABES
. . . PLAN POSITION INDICATORS
. . . RADIO DIRECTION FINDERS
. . . SPACECRAFT POSITION
   INDICATORS
. . SMOKE DETECTORS
. . SPEED INDICATORS
. ANEMOMETERS
. . . DRAG FORCE ANEMOMETERS

MEASURING INSTRUMENTS-(CONT.)

. . . . HOT-FILM ANEMOMETERS
. . . . HOT-WIRE ANEMOMETERS
. . . . LASER ANEMOMETERS
. . . . SONIC ANEMOMETERS
. . . TACHOMETERS
. . WEIGHT INDICATORS
. . . MICROBALANCES
. . . STRAIN GAGE BALANCES
. . . THERMOBALANCES
. INTERFEROMETERS
. . FABRY-PEROT INTERFEROMETERS
. . INFRARED INTERFEROMETERS
. . MACH-ZEHNDER INTERFEROMETERS
. . MICHELSON INTERFEROMETERS
. . MICROWAVE INTERFEROMETERS
. . PHASE SWITCHING
   INTERFEROMETERS
. . RADIO INTERFEROMETERS
. ION PROBES
. ION TRAPS (INSTRUMENTATION)
. LASER DOPPLER VELOCIMETERS
. MAGNETIC PROBES
. MAGNETOMETERS
. . VARIOMETERS
. MECHANOGRAMS
. METEOROLOGICAL INSTRUMENTS
. . BAROMETERS
. . CLOUD HEIGHT INDICATORS
. . DROPSONDES
. . RADIOMETEOROGRAPHS
. . RADIOSONDES
. . . ENDORADIOSONDES
. . . IONOSONDES
. . . RAWINSONDES
. . RAIN GAGES
. . WEATHER DATA RECORDERS
. . WIND VANES
. MICROMETERS
. MICROWAVE PROBES
. . MICROWAVE PLASMA PROBES
. MOISTURE METERS
. . HYGROMETERS
. . . PSYCHROMETERS
. MONOCHROMATORS
. NOISE METERS
. OHMMETERS
. OMEGA NAVIGATION SYSTEM
. OPTICAL MEASURING INSTRUMENTS
. . CATHETOMETERS
. . DIFFRACTOMETERS
. . EBERT SPECTROMETERS
. . ELLIPSOMETERS
. . GEODIMETERS
. . HAPLOSCOPES
. . INFRARED SPECTROMETERS
. . . FILTER WHEEL INFRARED
   SPECTROMETERS
. . LIGHT SCATTERING METERS
. . MICRODENSITOMETERS
. . NEPHELOMETERS
. . OCULOMETERS
. . OPTICAL PYROMETERS
. . OPTICAL RANGE FINDERS
. . OPTICAL SCANNERS
. . . MULTISPECTRAL BAND SCANNERS
. . PHOTOGONIOMETERS
. . PHOTOMETERS
. . . ELECTROPHOTOMETERS
. . . ULTRAVIOLET SPECTROMETERS
. . . ULTRAVIOLET
   SPECTROPHOTOMETERS
. . POLARIMETERS
. . REFLECTOMETERS
. . MICROWAVE REFLECTOMETERS
. . REFRACTOMETERS
. . SEXTANTS
. . SPECTROPHOTOMETERS
. . . INFRARED SPECTROPHOTOMETERS
. . . ULTRAVIOLET
   SPECTROPHOTOMETERS
. . TRANSITS
. . . THEODOLITES
. . . . CINETHEODOLITES
. . TRANSMISSOMETERS
. OSCILLOGRAPHS
. OXYGEN ANALYZERS
. PENETROMETERS
. PLASMA PROBES
. . ELECTROSTATIC PROBES
. POLARISCOPES
. . SENARMONT POLARISCOPES
. POTENTIOMETERS (INSTRUMENTS)
. PRESSURE GAGES
. . BAROMETERS
. . MANOMETERS

MEASURING INSTRUMENTS-*(CONT.)*
- . . OSMOMETERS
- . . PIEZOELECTRIC GAGES
- . . PIEZOMETERS
- . . VACUUM GAGES
- . . . IONIZATION GAGES
- . . . . ALPHATRONS
- . . . . BAYARD-ALPERT IONIZATION
          GAGES
- . . . . PENNING GAGES
- . . . . PHILIPS IONIZATION GAGES
- . . . KNUDSEN GAGES
- . . . MCLEOD GAGES
- . . . PIRANI GAGES
- . PROFILOMETERS
- . PROTRACTORS
- . RADIATION MEASURING INSTRUMENTS
- . . ACTINOMETERS
- . . . INFRARED SPECTROMETERS
- . . . PYRANOMETERS
- . . . RADIOMETERS
- . . . . DICKE RADIOMETERS
- . . . . INFRARED DETECTORS
- . . . . INFRARED SCANNERS
- . . . . MICROWAVE RADIOMETERS
- . . . . PASSIVE L-BAND RADIOMETERS
- . . . . PRESSURE MODULATOR
            RADIOMETERS
            SPECTRORADIOMETERS
- . . . SOLAR SPECTROMETERS
- . . . SPECTROHELIOGRAPHS
- . . . SPECTROPHOTOMETERS
- . . . . INFRARED
            SPECTROPHOTOMETERS
- . . . . ULTRAVIOLET
            SPECTROPHOTOMETERS
- . . . ULTRAVIOLET SPECTROMETERS
- . . BOLOMETERS
- . . EBERT SPECTROMETERS
- . . ELECTROSTATIC PROBES
- . . FABRY-PEROT SPECTROMETERS
- . . HODOSCOPES
- . . INFRARED INSTRUMENTS
- . . . INFRARED DETECTORS
- . . . . FLIR DETECTORS
- . . . INFRARED SCANNERS
- . . . INFRARED SPECTROMETERS
- . . . INFRARED SPECTROPHOTOMETERS
- . . PHOTOMETERS
- . . . ELECTROPHOTOMETERS
- . . . ULTRAVIOLET SPECTROMETERS
- . . . ULTRAVIOLET
          SPECTROPHOTOMETERS
- . . RADIATION COUNTERS
- . . . CERENKOV COUNTERS
- . . . ELECTRON COUNTERS
- . . . GEIGER COUNTERS
- . . . NEUTRON COUNTERS
- . . . . NEUTRON SPECTROMETERS
- . . . PARTICLE TELESCOPES
- . . . PROPORTIONAL COUNTERS
- . . . QUANTUM COUNTERS
- . . . SCINTILLATION COUNTERS
- . . . SPARK CHAMBERS
- . . RADIATION DETECTORS
- . . . DOSIMETERS
- . . . . THRESHOLD DETECTORS
            (DOSIMETERS)
- . . . GOLAY DETECTOR CELLS
- . . . SILICON RADIATION DETECTORS
- . RIOMETERS
- . RATIOMETERS
- . RESONANCE PROBES
- . . CYCLOTRON RESONANCE DEVICES
- . RESPIROMETERS
- . SATELLITE-BORNE INSTRUMENTS
- . . AMPS (SATELLITE PAYLOAD)
- . SCATTEROMETERS
- . SHOCK MEASURING INSTRUMENTS
- . SONDES
- . . DROPSONDES
- . . JUDI-DART ROCKET
- . . RADIOSONDES
- . . . ENDORADIOSONDES
- . . . IONOSONDES
- . . . RAWINSONDES
- . SPECTROMETERS
- . . EBERT SPECTROMETERS
- . . FABRY-PEROT SPECTROMETERS
- . . GAMMA RAY SPECTROMETERS
- . . INFRARED SPECTROMETERS
- . . . FILTER WHEEL INFRARED
          SPECTROMETERS
- . . LASER SPECTROMETERS
- . . MASS SPECTROMETERS
- . . MICROWAVE SPECTROMETERS

MEASURING INSTRUMENTS-*(CONT.)*
- . . NEUTRON SPECTROMETERS
- . . SOLAR BACKSCATTER UV
        SPECTROMETER
- . . SOLAR SPECTROMETERS
- . . SPECTROHELIOGRAPHS
- . . TIME OF FLIGHT SPECTROMETERS
- . . ULTRAVIOLET SPECTROMETERS
- . . . HIGH DISPERSION
          SPECTROGRAPHS
- . SPUTTERING GAGES
- . STRAIN GAGES
- . TEMPERATURE MEASURING
      INSTRUMENTS
- . . BATHYTHERMOGRAPHS
- . . OPTICAL PYROMETERS
- . . PYROMETERS
- . . . RADIATION PYROMETERS
- . . . THERMOCOUPLE PYROMETERS
- . . TEMPERATURE PROBES
- . . . PNEUMATIC PROBES
- . . THERMOMETERS
- . . . RESISTANCE THERMOMETERS
- . TENSIOMETERS
- . TENSOMETERS
- . THERMAL CONDUCTIVITY GAGES
- . TILTMETERS
- . TIME MEASURING INSTRUMENTS
- . . CLOCKS
- . . . ATOMIC CLOCKS
- . . . AUTONOMOUS SPACECRAFT
          CLOCKS
- . . . CHRONOMETERS
- . . TIMING DEVICES
- . . TITRIMETERS
- . TORQUEMETERS
- . TURBULENCE METERS
- . VIBRATION METERS
- . . SEISMOGRAPHS
- . . . LUNAR SEISMOGRAPHS
- . VISCOMETERS
- . VOLTMETERS
- . . MILLIVOLTMETERS
- . WATTMETERS

RT   AIRCRAFT INSTRUMENTS
       AUTOMATIC CONTROL
       AUTOMATIC TEST EQUIPMENT
       BIOINSTRUMENTATION
       CALIBRATING
       CIRCUMSOLAR TELESCOPES
       CONTROL MOMENT GYROSCOPES
       CONTROLLERS
   ∞ DETECTORS
       DRAG MEASUREMENT
       DUOCHROMATORS
       ELECTRIC BRIDGES
       ELECTRICAL MEASUREMENT
       FLIGHT INSTRUMENTS
       FOREST FIRE DETECTION
       FRAUNHOFER LINE DISCRIMINATORS
       IMBLMS
       INSTRUMENT RECEIVERS
       INSTRUMENT TRANSMITTERS
   ∞ INSTRUMENTS
       INTERNATIONAL SYSTEM OF UNITS
       LABORATORY EQUIPMENT
       LANDING INSTRUMENTS
       LARGE APERTURE SEISMIC ARRAY
       LOCAL SCIENTIFIC SURVEY MODULE
       LUNAR RANGEFINDING
   ∞ MEASUREMENT
       METROLOGY
       MICROINSTRUMENTATION
       MONITORS
       NAVIGATION INSTRUMENTS
   ∞ PROBES
       PROPELLANT ACTUATED INSTRUMENTS
       RADIO PROBING
       RADIO TELEMETRY
       RAPID BALLISTICS IDENTIFICATION
       RECORDING INSTRUMENTS
       REMOTE SENSORS
       ROCKET-BORNE INSTRUMENTS
       RONCHI TEST
       SATELLITE INSTRUMENTS
   ∞ SENSORS
       SODAR
       SOUND DETECTING AND RANGING
       SPACECRAFT INSTRUMENTS
       SYNCHROSCOPES
       TELEMETRY
   ∞ TEST EQUIPMENT
       TRANSDUCERS
       ULTRASONIC SCANNERS
       VENTURI TUBES
       WHEATSTONE BRIDGES

MEASURING INSTRUMENTS-*(CONT.)*
       WIND TUNNEL CALIBRATION

MECAMYLAMINE
GS   AMINES
       . **MECAMYLAMINE**
       TERPENES
       . **MECAMYLAMINE**

MECHANICAL DEVICES
RT   CAMS
       CLAMPS
       CLIPS
       CLUTCHES
   ∞ DEVICES
   ∞ EQUIPMENT
       HOLDERS
       JIGS
       LEVERS
       LINKAGES
       MACHINE TOOLS
   ∞ MECHANISM
       MECHANIZATION
       TOOLS

MECHANICAL DRAWINGS
USE   ENGINEERING DRAWINGS

MECHANICAL DRIVES
UF   ROTARY DRIVES
GS   **MECHANICAL DRIVES**
       . HELICOPTER PROPELLER DRIVE
       . MAGNETIC TAPE TRANSPORTS
       . PROPELLER DRIVE
       . TRANSMISSIONS (MACHINE ELEMENTS)
RT   CLUTCHES
       COUNTER-ROTATING WHEELS
       COUPLING
       COUPLINGS
   ∞ DRIVES
   ∞ GEAR
       GEAR TEETH
       GEARS
       MAGNETOELECTRIC MEDIA
   ∞ POWER TRANSMISSION
       SHAFTS (MACHINE ELEMENTS)
       VEHICLE WHEELS
       WIND TUNNEL DRIVES
       WINDMILLS (WINDPOWERED MACHINES)

MECHANICAL ENGINEERING
RT   AERONAUTICAL ENGINEERING
       AEROSPACE ENGINEERING
   ∞ ENGINEERING
       FLUID FLOW
       FLYWHEELS
       FURNACES
       HEAT TRANSFER
       MACHINE TOOLS
   ∞ MACHINERY
       MAINTENANCE
       MATERIALS HANDLING
       STRESS ANALYSIS
       THERMODYNAMICS
       VIBRATION TESTS
††

MECHANICAL IMPEDANCE
GS   IMPEDANCE
       . **MECHANICAL IMPEDANCE**
RT   ATTENUATION
       DAMPING
       FRICTION
       IMPEDANCE MEASUREMENT

MECHANICAL MEASUREMENT
SN   (MEASUREMENT OF MECHANICAL
       PROPERTIES, QUANTITIES OR
       CONDITIONS)
GS   **MECHANICAL MEASUREMENT**
       . DISPLACEMENT MEASUREMENT
       . DRAG MEASUREMENT
       . FLOW MEASUREMENT
       . FRICTION MEASUREMENT
       . PRESSURE MEASUREMENT
       . STRESS MEASUREMENT
       . . X RAY STRESS MEASUREMENT
       . THRUST MEASUREMENT
       . VELOCITY MEASUREMENT
       . . WIND VELOCITY MEASUREMENT
       . VIBRATION MEASUREMENT
       . WIND MEASUREMENT
       . . WIND VELOCITY MEASUREMENT
RT   ACCELEROMETERS
       ACOUSTIC MEASUREMENT
       DEFORMETERS

MECHANICAL MEASUREMENT-*(CONT.)*
    DENSITY MEASUREMENT
    DEPTH MEASUREMENT
    DYNAMOMETERS
    EXTENSOMETERS
    FLOWMETERS
    ∞MEASUREMENT
    STRAIN GAGES
    TENSIOMETERS
    TORQUEMETERS
    WEIGHT INDICATORS

MECHANICAL OSCILLATORS
GS   OSCILLATORS
    . **MECHANICAL OSCILLATORS**
    . . PENDULUMS
    . . . GYROSCOPIC PENDULUMS
RT   ELECTRIC CHOPPERS
    HARMONIC OSCILLATORS
    RECIPROCATION
    RESONANT VIBRATION
    VIBRATION

MECHANICAL PROPERTIES
UF   METEORITE COMPRESSION TESTS
    STRENGTH OF MATERIALS
GS   **MECHANICAL PROPERTIES**
    . ABRASION RESISTANCE
    . BRITTLENESS
    . BULK MODULUS
    . COLD STRENGTH
    . COMPRESSIBILITY
    . COMPRESSIVE STRENGTH
    . CREEP PROPERTIES
    . . SHEAR CREEP
    . . STEADY STATE CREEP
    . . TENSILE CREEP
    . CREEP RUPTURE STRENGTH
    . CREEP STRENGTH
    . DIMENSIONAL STABILITY
    . . STRUCTURAL STABILITY
    . . . SHELL STABILITY
    . DUCTILITY
    . EARTHQUAKE RESISTANCE
    . ELASTIC PROPERTIES
    . . AEROELASTICITY
    . . . AEROTHERMOELASTICITY
    . . ANELASTICITY
    . . ELASTOPLASTICITY
    . . HYDROELASTICITY
    . . HYPOELASTICITY
    . . MAGNETOSTRICTION
    . . MODULUS OF ELASTICITY
    . . . DYNAMIC MODULUS OF ELASTICITY
    . . PHOTOELASTICITY
    . . PHOTOVISCOELASTICITY
    . . PROPORTIONAL LIMIT
    . . THERMOELASTICITY
    . . AEROTHERMOELASTICITY
    . . VISCOELASTICITY
    . . . PHOTOVISCOELASTICITY
    . . . THERMOVISCOELASTICITY
    . ELECTROSTRICTION
    . FATIGUE LIFE
    . FIBER STRENGTH
    . FLEXIBILITY
    . FRACTURE STRENGTH
    . HARDNESS
    . . KNOOP HARDNESS
    . . MICROHARDNESS
    . . ROCKWELL HARDNESS
    . HIGH STRENGTH
    . IMPACT STRENGTH
    . MALLEABILITY
    . MODULAR RATIOS
    . NOTCH STRENGTH
    . PIEZOELECTRICITY
    . PLASTIC PROPERTIES
    . . ELASTOPLASTICITY
    . . PHOTOPLASTICITY
    . . SUPERPLASTICITY
    . . THERMOPLASTICITY
    . . VISCOPLASTICITY
    . . YIELD POINT
    . POISSON RATIO
    . RESILIENCE
    . SET
    . SHEAR PROPERTIES
    . . SHEAR STRENGTH
    . STIFFNESS
    . STRESS CYCLES
    . STRESS RATIO
    . STRESS RELAXATION
    . TENSILE PROPERTIES
    . TENSILE STRENGTH

MECHANICAL PROPERTIES-*(CONT.)*
    . THERMAL RESISTANCE
    . TOUGHNESS
    . . NOTCH SENSITIVITY
    . WELD STRENGTH
    . YIELD STRENGTH
    . . MICROYIELD STRENGTH
RT   ACOUSTIC PROPERTIES
    AGING (MATERIALS)
    ANISOTROPY
    BUOYANCY
    CAST ALLOYS
    COEFFICIENTS
    COMPRESSING
    COMPRESSION LOADS
    DEFORMATION
    DURABILITY
    ELASTIC PROPERTIES
    ELONGATION
    FATIGUE (MATERIALS)
    FERRITIC STAINLESS STEELS
    FIBER ORIENTATION
    FLATNESS
    ∞HIGH RESISTANCE
    HYSTERESIS
    IMPEDANCE
    INTERFACIAL TENSION
    INTERNAL FRICTION
    ISOTROPY
    J INTEGRAL
    ∞LOW RESISTANCE
    ∞MATERIALS
    ∞MATERIALS TESTS
    ∞METALLURGY
    MICROMECHANICS
    MICROPOROSITY
    PEELING
    PERMEABILITY
    ∞PHYSICAL PROPERTIES
    PROPELLANT PROPERTIES
    ∞PROPERTIES
    RADIATION EFFECTS
    RELIABILITY
    ∞RIGIDITY
    ROUGHNESS
    RUGGEDNESS
    SHEAR STRAIN
    SHEAR STRESS
    SHOCK RESISTANCE
    SOLID MECHANICS
    SPECIFICATIONS
    SPECIMEN GEOMETRY
    STRAIN RATE
    ∞STRENGTH
    STRESS CONCENTRATION
    STRESSES
    STRUCTURAL FAILURE
    SUPERCOOLING
    SURFACE DEFECTS
    SURFACE PROPERTIES
    SURFACE ROUGHNESS
    TEARING
    TEMPERATURE INVERSIONS
    TEXTURES
    TOLERANCES (MECHANICS)
    TRIAXIAL STRESSES
    TRIBOLUMINESCENCE
    WEATHERING

MECHANICAL RESONANCE
USE   RESONANT VIBRATION

MECHANICAL SHOCK
UF   JARRING
GS   **MECHANICAL SHOCK**
    . HYDRAULIC SHOCK
RT   ACCELERATION (PHYSICS)
    HIGH ACCELERATION
    HYPERVELOCITY IMPACT
    IMPACT
    IMPACT ACCELERATION
    ∞SHOCK
    SHOCK ABSORBERS
    SHOCK RESISTANCE
    SHOCK SPECTRA
    SHOCK WAVES
    VIBRATION

MECHANICAL TWINNING
GS   TWINNING
    . **MECHANICAL TWINNING**
RT   CRYSTAL DEFECTS
    CRYSTAL GROWTH
    CRYSTAL STRUCTURE
    WORK HARDENING

∞ MECHANICS (PHYSICS)
SN   *(USE OF A MORE SPECIFIC TERM IS
    RECOMMENDED--CONSULT THE TERMS
    LISTED BELOW)*
RT   CELESTIAL MECHANICS
    CLASSICAL MECHANICS
    CONTINUUM MECHANICS
    ∞DYNAMICS
    ELECTROMECHANICS
    FLIGHT MECHANICS
    FLUID DYNAMICS
    FLUID MECHANICS
    FRACTURE MECHANICS
    GAS DYNAMICS
    ∞HYDRAULICS
    HYDRODYNAMICS
    KINEMATICS
    KINETICS
    LOADS (FORCES)
    ∞MECHANISM
    ORBITAL MECHANICS
    ∞PHYSICS
    QUANTUM MECHANICS
    ∞SCIENCE
    SOLID MECHANICS
    STATICS
    STATISTICAL MECHANICS
    VIRIAL THEOREM
    WIGNER COEFFICIENT

∞ MECHANISM
SN   *(USE OF A MORE SPECIFIC TERM IS
    RECOMMENDED--CONSULT THE TERMS
    LISTED BELOW)*
RT   ∞MACHINERY
    MECHANICAL DEVICES
    ∞MECHANICS (PHYSICS)
    ∞METHODOLOGY

MECHANIZATION
RT   ∞AUTOMATION
    DATA PROCESSING
    DEPERSONALIZATION
    ∞MACHINERY
    MAN MACHINE SYSTEMS
    MECHANICAL DEVICES
    ∞OPERATIONS
    SYSTEMS ENGINEERING
    TOOLING
    TOOLS

MECHANOGRAMS
GS   MEASURING INSTRUMENTS
    . **MECHANOGRAMS**
    MEDICAL EQUIPMENT
    . **MECHANOGRAMS**
    RECORDING INSTRUMENTS
    . **MECHANOGRAMS**
RT   MUSCULAR FUNCTION

MECHANORECEPTORS
GS   ANATOMY
    . SENSE ORGANS
    . . **MECHANORECEPTORS**
    RECEPTORS (PHYSIOLOGY)
    . **MECHANORECEPTORS**
RT   SENSITOMETRY

MECLIZINE
GS   HALOGEN COMPOUNDS
    . CHLORINE COMPOUNDS
    . . **MECLIZINE**
    HETEROCYCLIC COMPOUNDS
    . AZINES
    . **MECLIZINE**
    PYRAZINES
    . AZINES
    . . **MECLIZINE**

MEDIA
SN   (EXCLUDES COMMUNICATION
    TECHNIQUES)
GS   **MEDIA**
    . ANISOTROPIC MEDIA
    . . ANISOTROPIC FLUIDS
    . ELASTIC MEDIA
    . INTERGALACTIC MEDIA
    . INTERPLANETARY MEDIUM
    . . INTERPLANETARY DUST
    . . . METEOROID DUST CLOUDS
    . . . . ZODIACAL DUST
    . . INTERPLANETARY GAS
RT   ∞CHANNELS

MEDIAN (STATISTICS)
RT　AVERAGE
　　DISTRIBUTION MOMENTS
　　ERRORS
　　MEAN
　　MODE (STATISTICS)
　　NORMALITY
　　NORMS
　　QUALITY CONTROL
　　STATISTICAL ANALYSIS
∞ TESTS

MEDIASTINUM
RT　SEPTUM
　　TISSUES (BIOLOGY)

MEDIATION
RT　LABOR
　　MANAGEMENT PLANNING

MEDICAL ELECTRONICS
RT　ECHOENCEPHALOGRAPHY
　　ELECTROCARDIOGRAPHY
　　ELECTROENCEPHALOGRAPHY
　　ELECTROMYOGRAPHY
∞ ELECTRONICS
　　ELECTROPLETHYSMOGRAPHY
　　ELECTRORETINOGRAPHY

MEDICAL EQUIPMENT
GS　MEDICAL EQUIPMENT
　. ARTIFICIAL CARDIAC PACEMAKER
　. ARTIFICIAL HEART VALVES
　. BLOOD PUMPS
　. CARDIOTACHOMETERS
　. ENDOSCOPES
　. MECHANOGRAMS
　. PROSTHETIC DEVICES
　.. ARTIFICIAL EARS
　. RESPIRATORS
　. STETHOSCOPES
　. STRETCHERS
　. SURGICAL INSTRUMENTS
　. SYRINGES
　. TOURNIQUETS
　. X RAY APPARATUS
　.. LIXISCOPES
　.. X RAY TUBES
RT　CARDIOGRAPHY
　　DENTISTRY
　　DIAGNOSIS
　　ECHOENCEPHALOGRAPHY
　　ELECTROENCEPHALOGRAPHY
　　EMERGENCY LIFE SUSTAINING
　　　SYSTEMS
∞ EQUIPMENT
　　FIRST AID
　　FLUOROSCOPY
　　HOSPITALS
　　IMBLMS
∞ MEDICINE
　　MICROTOMY
　　MOBILE QUARANTINE FACILITY
　　THERAPY

MEDICAL PERSONNEL
GS　PERSONNEL
　. MEDICAL PERSONNEL
　.. FLIGHT NURSES
　.. PHYSICIANS
　.. SURGEONS
　... FLIGHT SURGEONS
RT　∞ MEDICINE

MEDICAL PHENOMENA
GS　PHENOMENOLOGY
　. MEDICAL PHENOMENA
RT　DIVING (UNDERWATER)

MEDICAL SCIENCE
GS　MEDICAL SCIENCE
　. ANESTHESIOLOGY
　. DENTISTRY
　. DERMATOLOGY
　. ENDOCRINOLOGY
　. EPIDEMIOLOGY
　. GERIATRICS
　. GYNECOLOGY
　. HISTOLOGY
　. IMMUNOLOGY
　. NEUROLOGY
　. NEUROPSYCHIATRY
　. NUCLEAR MEDICINE
　.. RADIOBIOLOGY
　. OPHTHALMOLOGY

MEDICAL SCIENCE-(CONT.)
　. ORTHOPEDICS
　. OTOLARYNGOLOGY
　. OTOLOGY
　. PATHOLOGY
　.. HUMAN PATHOLOGY
　. PSYCHIATRY
　.. SOCIAL PSYCHIATRY
　. RADIOLOGY
　. RADIOPATHOLOGY
　. SYMPTOMOLOGY
　. TOOTH DISEASES
　. UROLOGY
RT　AEROSPACE MEDICINE
∞ BIOLOGY
　　CLINICAL MEDICINE
　　DIAGNOSIS
　　DISEASES
　　FIRST AID
∞ MEDICINE
　　OPTOMETRY
　　PHARMACOLOGY
　　PHENOMENOLOGY
　　PNEUMOTHORAX
　　PSYCHOPHARMACOLOGY
　　RADIATION THERAPY
∞ SCIENCE
　　SPORTS MEDICINE
　　TRANSFUSION

MEDICAL SERVICES
GS　SERVICES
　. MEDICAL SERVICES
RT　AMBULANCES
　　FIRST AID
　　INTRAVENOUS PROCEDURES
　　MOBILE QUARANTINE FACILITY
　　PUBLIC HEALTH

∞ MEDICINE
SN　(USE OF A MORE SPECIFIC TERM IS
　　RECOMMENDED--CONSULT THE TERMS
　　LISTED BELOW)
RT　AEROSPACE MEDICINE
　　CLINICAL MEDICINE
　　MEDICAL EQUIPMENT
　　MEDICAL PERSONNEL
　　MEDICAL SCIENCE
　　NEUROPSYCHIATRY
　　PHARMACOLOGY
　　PSYCHOPHARMACOLOGY
　　RADIOBIOLOGY
　　RADIOLOGY
　　VETERINARY MEDICINE

MEDITERRANEAN SEA
GS　SEAS
　. MEDITERRANEAN SEA
　.. ADRIATIC SEA
RT　MALTA
　　RHONE DELTA (FRANCE)
　　SICILY

MEDIUM SCALE INTEGRATION
GS　CIRCUITS
　. PRINTED CIRCUITS
　.. MEDIUM SCALE INTEGRATION
　　MICROELECTRONICS
　. MEDIUM SCALE INTEGRATION
RT　ELECTRONIC PACKAGING
　　INTEGRATED CIRCUITS
　　LARGE SCALE INTEGRATION

MEETINGS
USE　CONFERENCES

MEGALOPOLISES
RT　CITIES
　　COMMUNITIES
　　DEMOGRAPHY
　　INDUSTRIAL AREAS
　　REGIONAL PLANNING
　　RESIDENTIAL AREAS
　　RURAL AREAS
　　SUBURBAN AREAS
　　URBAN DEVELOPMENT
　　URBAN TRANSPORTATION

MEGAMECHANICS
RT　LARGE SPACE STRUCTURES
　　STRUCTURAL ANALYSIS
　　STRUCTURAL ENGINEERING
　　TRUSSES

MEISSNER EFFECT
USE　DIAMAGNETISM
　　SUPERCONDUCTIVITY

MELAMINE
GS　AMINES
　. MELAMINE
RT　RESINS

MELANIN
GS　PIGMENTS
　. MELANIN
　　PROTEINS
　. MELANIN
RT　DOPA
　　SKIN (ANATOMY)

MELANOIDIN
GS　ACIDS
　. AMINO ACIDS
　.. MELANOIDIN
　　ORGANIC COMPOUNDS
　. AMINO ACIDS
　.. MELANOIDIN

MELLIN TRANSFORMS
GS　FUNCTIONS (MATHEMATICS)
　. MELLIN TRANSFORMS
RT　INTEGRAL EQUATIONS
　　KERNEL FUNCTIONS

MELT SPINNING
GS　CRYSTALLIZATION
　. MELT SPINNING
　　MELTS (CRYSTAL GROWTH)
　. MELT SPINNING
　　SOLIDIFICATION
　. MELT SPINNING
RT　∞ METALLURGY
　　PHASE TRANSFORMATIONS

MELTING
UF　REMELTING
　　THAWING
GS　PHASE TRANSFORMATIONS
　. MELTING
　.. FUSION (MELTING)
　.. LEVITATION MELTING
　.. VACUUM MELTING
RT　ABLATION
　　AUFEIS (ICE)
　　BURNTHROUGH (FAILURE)
　　CASTING
　　COAL LIQUEFACTION
　　CONTAINERLESS MELTS
　　COOLING
　　DEFROSTING
　　DEICING
　　DROP TRANSFER
　　EXTRACTION
　　FREEZING
　　FURNACES
　　HEAT OF FUSION
　　HEATING
　　ICE PREVENTION
　　IMPACT MELTS
　　INDUCTION HEATING
　　INJECTION MOLDING
　　LIQUEFACTION
　　LIQUID METALS
　　LIQUID-SOLID INTERFACES
　　MELTS (CRYSTAL GROWTH)
　　METAL CUTTING
　　METAL FOAMS
∞ METALLURGY
　　MOLDS
　　PHASE CHANGE MATERIALS
　　PSEUDOPOTENTIALS
∞ SEPARATION
　　SMELTING
　　SOLAR FURNACES
　　SPIKING
　　ZONE MELTING

MELTING POINTS
UF　FREEZING POINTS
GS　THERMODYNAMIC PROPERTIES
　. THERMOPHYSICAL PROPERTIES
　.. MELTING POINTS
RT　HIGH TEMPERATURE TESTS
　　LIQUID PHASES
　　LIQUIDUS
　　LOW TEMPERATURE TESTS
　　PHASE DIAGRAMS
　　REACTION BONDING

MELTING POINTS-*(CONT.)*
    SOLID SOLUTIONS
    SOLID STATE
    SOLIDIFICATION
    SOLIDIFIED GASES
    SPECIFIC HEAT
    TEMPERATURE
    TRANSITION TEMPERATURE

MELTS (CRYSTAL GROWTH)
GS   **MELTS (CRYSTAL GROWTH)**
   . CONTAINERLESS MELTS
   . IMPACT MELTS
   . MELT SPINNING
RT   ATOMIC STRUCTURE
   CRYSTAL GROWTH
   CRYSTALLIZATION
   FLOAT ZONES
   MARANGONI CONVECTION
   MELTING
   SEMICONDUCTORS (MATERIALS)

MEM (EXCURSION MODULE)
USE MARS EXCURSION MODULE

MEMBRANE ANALOGY
USE MEMBRANE STRUCTURES
     STRUCTURAL ANALYSIS

MEMBRANE STRUCTURES
UF   MEMBRANE ANALOGY
GS   MEMBRANES
   . **MEMBRANE STRUCTURES**
   . . SKIN (STRUCTURAL MEMBER)
   STRUCTURAL MEMBERS
   . **MEMBRANE STRUCTURES**
   . . SKIN (STRUCTURAL MEMBER)
RT   DIAPHRAGMS (MECHANICS)
   METAL SHELLS
   PERFORATED SHELLS
   SCOTCHLITE (TRADEMARK)
   ∞SHEETS
   SHELLS (STRUCTURAL FORMS)
   ∞STRUCTURES
   THIN WALLED SHELLS
   WEBS (SUPPORTS)

MEMBRANE THEORY
USE STRUCTURAL ANALYSIS

MEMBRANES
UF   WEBS (MEMBRANES)
GS   **MEMBRANES**
   . CHOROID MEMBRANES
   . CONJUNCTIVA
   . ION EXCHANGE MEMBRANE
     ELECTROLYTES
   . MEMBRANE STRUCTURES
   . . SKIN (STRUCTURAL MEMBER)
   . PLEURAE
RT   ∞DIAPHRAGMS
   DIAPHRAGMS (MECHANICS)
   ∞FILMS
   JET MEMBRANE PROCESS
   ∞LAYERS
   OSMOSIS
   PERITONEUM
   REVERSE OSMOSIS
   SEPTUM
   ∞SHEETS
   SHELLS (STRUCTURAL FORMS)
   SKIN (ANATOMY)
   ∞WEBS
   WEBS (SHEETS)
   WEBS (SUPPORTS)

MEMORY
SN   (LIMITED TO SENTIENT ORGANISMS--
     EXCLUDES COMPUTER STORAGE
     DEVICES AND PLASTIC MEMORY)
RT   EDUCATION
   LEARNING
   MNEMONICS
   RECOGNITION
   RETENTION (PSYCHOLOGY)

MEMORY (COMPUTERS)
RT   ARCHITECTURE (COMPUTERS)
   COMPUTER DESIGN
   COMPUTER STORAGE DEVICES
   COMPUTERS
   HOLE BURNING
   MAGNETIC DISKS
   VIDEO DISKS

MENDELEVIUM
GS   CHEMICAL ELEMENTS
   . ACTINIDE SERIES
   . . TRANSURANIUM ELEMENTS
   . . . **MENDELEVIUM**
   . NUCLIDES
   . . ISOTOPES
   . . . RADIOACTIVE ISOTOPES
   . . . . TRANSURANIUM ELEMENTS
   . . . . . **MENDELEVIUM**
   HEAVY ELEMENTS
   . TRANSURANIUM ELEMENTS
   . . **MENDELEVIUM**
   METALS
   . ACTINIDE SERIES
   . . TRANSURANIUM ELEMENTS
   . . . **MENDELEVIUM**

MENINGITIS
GS   DISEASES
   . INFECTIOUS DISEASES
   . . **MENINGITIS**

MENISCI
GS   LIQUID SURFACES
   . **MENISCI**
RT   CURVES (GEOMETRY)
   LIQUID-GAS MIXTURES
   LIQUID-SOLID INTERFACES
   LIQUID-VAPOR INTERFACES
   ∞SURFACES

MENSTRUATION
GS   PHYSIOLOGY
   . **MENSTRUATION**
RT   FEMALES
   OVARIES

MENTAL HEALTH
GS   HEALTH
   . **MENTAL HEALTH**
RT   HUMAN PERFORMANCE
   INTELLIGENCE
   NEUROPSYCHIATRY
   PSYCHOTHERAPY
   RORSCHACH TESTS
   SCHIZOPHRENIA

MENTAL PERFORMANCE
RT   CONSCIOUSNESS
   HUMAN PERFORMANCE
   HYPERNEA
   INSPIRATION
   INTELLECT
   INTELLIGENCE
   IRRATIONALITY
   LIES
   OPERATOR PERFORMANCE
   ∞PERFORMANCE
   PSYCHOMOTOR PERFORMANCE
   STRESS (PSYCHOLOGY)
   WORKLOADS (PSYCHOPHYSIOLOGY)

MENTAL STRESS
USE STRESS (PSYCHOLOGY)

MENTHOL
GS   ALIPHATIC COMPOUNDS
   . CYCLIC HYDROCARBONS
   . . **MENTHOL**
   HYDROCARBONS
   . CYCLIC HYDROCARBONS
   . . **MENTHOL**
   TERPENES
   . **MENTHOL**

MEPROBAMATE
GS   ALIPHATIC COMPOUNDS
   . **MEPROBAMATE**
   ESTERS
   . **MEPROBAMATE**

MERCAPTAN
USE THIOLS

MERCAPTO COMPOUNDS
USE THIOLS

MERCATOR PROJECTION
GS   GEOMETRY
   . EUCLIDEAN GEOMETRY
   . . ANALYTIC GEOMETRY
   . . . **MERCATOR PROJECTION**
   . . PROJECTIVE GEOMETRY

MERCATOR PROJECTION-*(CONT.)*
   . . . **MERCATOR PROJECTION**
RT   MAPS

MERCURE AIRCRAFT
GS   TRANSPORT AIRCRAFT
   . SHORT HAUL AIRCRAFT
   . . **MERCURE AIRCRAFT**
RT   ∞AIRCRAFT
   CARGO AIRCRAFT
   PASSENGER AIRCRAFT

MERCURY (METAL)
UF   LIQUID MERCURY
GS   CHEMICAL ELEMENTS
   . **MERCURY (METAL)**
   . . MERCURY ISOTOPES
   . . MERCURY VAPOR
   LIQUIDS
   . LIQUID METALS
   . . **MERCURY (METAL)**
   . . . MERCURY VAPOR
   METALS
   . LIQUID METALS
   . . **MERCURY (METAL)**
   . . . MERCURY VAPOR

MERCURY (PLANET)
GS   CELESTIAL BODIES
   . PLANETS
   . . TERRESTRIAL PLANETS
   . . . **MERCURY (PLANET)**
RT   PLANETARY CRATERS

MERCURY ALLOYS
GS   ALLOYS
   . **MERCURY ALLOYS**
   . . MERCURY AMALGAMS

MERCURY AMALGAMS
UF   AMALGAMS
GS   ALLOYS
   . MERCURY ALLOYS
   . . **MERCURY AMALGAMS**

MERCURY ARCS
GS   ELECTRIC CURRENT
   . ELECTRIC DISCHARGES
   . . ELECTRIC ARCS
   . . . **MERCURY ARCS**
RT   ARC LAMPS
   METALLIC PLASMAS
   RECTIFIERS

MERCURY CADMIUM TELLURIDES
UF   CADMIUM MERCURY TELLURIDES
RT   INFRARED DETECTORS
   PHOTOCONDUCTIVITY
   PHOTOCONDUCTORS
   PHOTODIODES

MERCURY COMPOUNDS
GS   **MERCURY COMPOUNDS**
   . MERCURY OXIDES
   . MERCURY TELLURIDES
RT   ∞CHEMICAL COMPOUNDS
   ∞GROUP 2B COMPOUNDS
   ∞METAL COMPOUNDS

MERCURY FLIGHTS
GS   SPACE FLIGHT
   . MANNED SPACE FLIGHT
   . . **MERCURY FLIGHTS**
   . . . MERCURY MA-1 FLIGHT
   . . . MERCURY MA-2 FLIGHT
   . . . MERCURY MA-3 FLIGHT
   . . . MERCURY MA-4 FLIGHT
   . . . MERCURY MA-5 FLIGHT
   . . . MERCURY MA-6 FLIGHT
   . . . MERCURY MA-7 FLIGHT
   . . . MERCURY MA-8 FLIGHT
   . . . MERCURY MA-9 FLIGHT
   . . . MERCURY MR-1 FLIGHT
   . . . MERCURY MR-2 FLIGHT
   . . . MERCURY MR-3 FLIGHT
   . . . MERCURY MR-4 FLIGHT
RT   ATLAS LAUNCH VEHICLES
   MANNED SPACECRAFT
   SPACE CAPSULES

MERCURY ION ENGINES
GS   ENGINES
   . ROCKET ENGINES
   . . ELECTRIC ROCKET ENGINES

MERCURY ION ENGINES-*(CONT.)*
```
    . . . ION ENGINES
    . . . . MERCURY ION ENGINES
RT    ELECTROSTATIC ENGINES
      NUCLEAR PROPULSION
      NUCLEAR ROCKET ENGINES
      PLASMA ENGINES
```

MERCURY ISOTOPES
```
GS    CHEMICAL ELEMENTS
      . MERCURY (METAL)
      . . MERCURY ISOTOPES
      . NUCLIDES
      . . ISOTOPES
      . . . MERCURY ISOTOPES
```

MERCURY LAMPS
```
GS    LIGHTING EQUIPMENT
      . LUMINAIRES
      . . MERCURY LAMPS
RT    LIGHT SOURCES
      PHOSPHORS
      STERILIZATION
      XENON LAMPS
```

MERCURY MA-1 FLIGHT
```
GS    SPACE FLIGHT
      . MANNED SPACE FLIGHT
      . . MERCURY FLIGHTS
      . . . MERCURY MA-1 FLIGHT
RT    ATLAS LAUNCH VEHICLES
```

MERCURY MA-2 FLIGHT
```
UF    MA-2 MISSION
GS    SPACE FLIGHT
      . MANNED SPACE FLIGHT
      . . MERCURY FLIGHTS
      . . . MERCURY MA-2 FLIGHT
RT    ATLAS LAUNCH VEHICLES
      FAITH 7
```

MERCURY MA-3 FLIGHT
```
UF    MA-3 FLIGHT
GS    SPACE FLIGHT
      MANNED SPACE FLIGHT
      . . MERCURY FLIGHTS
      . . . MERCURY MA-3 FLIGHT
RT    ATLAS LAUNCH VEHICLES
```

MERCURY MA-4 FLIGHT
```
UF    MA-4 FLIGHT
GS    SPACE FLIGHT
      . MANNED SPACE FLIGHT
      . . MERCURY FLIGHTS
      . . . MERCURY MA-4 FLIGHT
RT    ATLAS LAUNCH VEHICLES
```

MERCURY MA-5 FLIGHT
```
UF    MA-5 FLIGHT
GS    SPACE FLIGHT
      . MANNED SPACE FLIGHT
      . . MERCURY FLIGHTS
      . . . MERCURY MA-5 FLIGHT
RT    ATLAS LAUNCH VEHICLES
```

MERCURY MA-6 FLIGHT
```
GS    SPACE FLIGHT
      . MANNED SPACE FLIGHT
      . . MERCURY FLIGHTS
      . . . MERCURY MA-6 FLIGHT
RT    ATLAS LAUNCH VEHICLES
      FRIENDSHIP 7
```

MERCURY MA-7 FLIGHT
```
GS    SPACE FLIGHT
      . MANNED SPACE FLIGHT
      . . MERCURY FLIGHTS
      . . . MERCURY MA-7 FLIGHT
RT    ATLAS LAUNCH VEHICLES
      AURORA 7
```

MERCURY MA-8 FLIGHT
```
UF    MA-8 FLIGHT
GS    SPACE FLIGHT
      . MANNED SPACE FLIGHT
      . . MERCURY FLIGHTS
      . . . MERCURY MA-8 FLIGHT
RT    ATLAS LAUNCH VEHICLES
      SIGMA 7
```

MERCURY MA-9 FLIGHT
```
UF    MA-9 FLIGHT
GS    SPACE FLIGHT
      . MANNED SPACE FLIGHT
```

MERCURY MA-9 FLIGHT-*(CONT.)*
```
      . . MERCURY FLIGHTS
      . . . MERCURY MA-9 FLIGHT
RT    ATLAS LAUNCH VEHICLES
      FAITH 7
```

MERCURY MR-1 FLIGHT
```
GS    SPACE FLIGHT
      . MANNED SPACE FLIGHT
      . . MERCURY FLIGHTS
      . . . MERCURY MR-1 FLIGHT
```

MERCURY MR-2 FLIGHT
```
GS    SPACE FLIGHT
      . MANNED SPACE FLIGHT
      . . MERCURY FLIGHTS
      . . . MERCURY MR-2 FLIGHT
```

MERCURY MR-3 FLIGHT
```
UF    MR-3 FLIGHT
GS    SPACE FLIGHT
      . MANNED SPACE FLIGHT
      . . MERCURY FLIGHTS
      . . . MERCURY MR-3 FLIGHT
```

MERCURY MR-4 FLIGHT
```
GS    SPACE FLIGHT
      . MANNED SPACE FLIGHT
      . . MERCURY FLIGHTS
      . . . MERCURY MR-4 FLIGHT
```

MERCURY OXIDES
```
GS    CHALCOGENIDES
      . OXIDES
      . . METAL OXIDES
      . . . MERCURY OXIDES
      MERCURY COMPOUNDS
      . MERCURY OXIDES
```

MERCURY PROJECT
```
GS    PROGRAMS
      . NASA PROGRAMS
      . . MERCURY PROJECT
      . NASA SPACE PROGRAMS
      . . MERCURY PROJECT
      . PROJECTS
      . . MERCURY PROJECT
RT    APOLLO PROJECT
      ATLAS LAUNCH VEHICLES
      GEMINI PROJECT
      LITTLE JOE 2 LAUNCH VEHICLE
      MANNED SPACE FLIGHT
      MANNED SPACECRAFT
```

MERCURY SPACECRAFT
```
GS    MANNED SPACECRAFT
      . MERCURY SPACECRAFT
      . . AURORA 7
      . . FAITH 7
      . . FRIENDSHIP 7
      . . SIGMA 7
      REENTRY VEHICLES
      . RECOVERABLE SPACECRAFT
      . . MERCURY SPACECRAFT
      . . . AURORA 7
      . . . FAITH 7
      . . . FRIENDSHIP 7
      . . . SIGMA 7
      SOFT LANDING SPACECRAFT
      . MERCURY SPACECRAFT
      . . AURORA 7
      . . FAITH 7
      . . FRIENDSHIP 7
      . . SIGMA 7
      SPACE CAPSULES
      . MERCURY SPACECRAFT
      . . AURORA 7
      . . FAITH 7
      . . FRIENDSHIP 7
      . . SIGMA 7
```

MERCURY TELLURIDES
```
GS    CHALCOGENIDES
      . TELLURIDES
      . . MERCURY TELLURIDES
      MERCURY COMPOUNDS
      . MERCURY TELLURIDES
      TELLURIUM COMPOUNDS
      . TELLURIDES
      . . MERCURY TELLURIDES
```

MERCURY VAPOR
```
GS    CHEMICAL ELEMENTS
      . MERCURY (METAL)
```

MERCURY VAPOR-*(CONT.)*
```
      . MERCURY VAPOR
      LIQUIDS
      . LIQUID METALS
      . . MERCURY (METAL)
      . . . MERCURY VAPOR
      METALS
      . LIQUID METALS
      . . MERCURY (METAL)
      . . . MERCURY VAPOR
      . METAL VAPORS
      . . MERCURY VAPOR
      VAPORS
      . METAL VAPORS
      . . MERCURY VAPOR
RT    CESIUM VAPOR
      SODIUM VAPOR
```

MERGING ROUTINES
```
GS    COMPUTER PROGRAMS
      . MERGING ROUTINES
RT    ∞ ROUTINES
```

MERIDIONAL FLOW
```
GS    FLUID FLOW
      . GAS FLOW
      . . AIR FLOW
      . . . AIR CURRENTS
      . . . . MERIDIONAL FLOW
RT    ATMOSPHERIC CIRCULATION
      FLOW GEOMETRY
      WIND (METEOROLOGY)
      WIND DIRECTION
```

MEROMORPHIC FUNCTIONS
```
GS    ANALYSIS (MATHEMATICS)
      . COMPLEX VARIABLES
      . . MEROMORPHIC FUNCTIONS
      . . . ELLIPTIC FUNCTIONS
      . . . RATIONAL FUNCTIONS
      FUNCTIONS (MATHEMATICS)
      . MEROMORPHIC FUNCTIONS
      . . ELLIPTIC FUNCTIONS
      . . RATIONAL FUNCTIONS
```

MERRITT ISLAND (FL)
```
GS    LANDFORMS
      . ISLANDS
      . . MERRITT ISLAND (FL)
RT    FLORIDA
```

MERWINITE
```
GS    CALCIUM COMPOUNDS
      . MERWINITE
      MAGNESIUM COMPOUNDS
      . MERWINITE
      MINERALS
      . MERWINITE
      SILICON COMPOUNDS
      . SILICATES
      . . MERWINITE
```

MESAS
```
GS    LANDFORMS
      . TERRACES (LANDFORMS)
      . . PLATEAUS
      . . . MESAS
      . . . . BUTTES
RT    FLATS (LANDFORMS)
      HIGHLANDS
      MOUNTAINS
```

MESFETS
```
USE   FIELD EFFECT TRANSISTORS
```

MESH
```
RT    FABRICS
      ∞ GRIDS
      STRANDS
      WEBBING
      ∞ WEBS
```

MESH (MATHEMATICS)
```
USE   COMPUTATIONAL GRIDS
```

MESITYLENE
```
GS    HYDROCARBONS
      . MESITYLENE
```

MESOMETEOROLOGY
```
GS    METEOROLOGY
      . MESOMETEOROLOGY
RT    AERONOMY
      LOWER ATMOSPHERE
```

MESOMETEOROLOGY-*(CONT.)*
  MICROMETEOROLOGY

MESON RESONANCE
  GS   PARTICLES
        . ELEMENTARY PARTICLES
        . . BOSONS
        . . . MESONS
        . . . . **MESON RESONANCE**
        . . . . . X MESONS
        . . FERMIONS
        . . . **MESON RESONANCE**
        . NUCLEAR PARTICLES
        . . BOSONS
        . . . MESONS
        . . . . **MESON RESONANCE**
        . . . . . X MESONS
        RESONANCE
        . **MESON RESONANCE**
        . . X MESONS
  RT   BARYONS
        HYPERONS

MESON-MESON INTERACTIONS
  GS   PARTICLE INTERACTIONS
        . ELEMENTARY PARTICLE
          INTERACTIONS
        . . **MESON-MESON INTERACTIONS**
  RT   ELECTROMAGNETIC INTERACTIONS
        ∞ INTERACTIONS

MESON-NUCLEON INTERACTIONS
  GS   PARTICLE INTERACTIONS
        . ELEMENTARY PARTICLE
          INTERACTIONS
        . . **MESON-NUCLEON INTERACTIONS**
  RT   CHARGED PARTICLES
        ∞ INTERACTIONS
        MESONS
        YUKAWA POTENTIAL

MESONS
  GS   PARTICLES
        . ELEMENTARY PARTICLES
        . . BOSONS
        . . . **MESONS**
        . . . . ETA-MESONS
        . . . . KAONS
        . . . . MESON RESONANCE
        . . . . . X MESONS
        . . . . MUONS
        . . . . PIONS
        . . . . VECTOR MESONS
        . . . . . RHO-MESONS
        . . . . . SIGMA-MESONS
        . . HADRONS
        . . . **MESONS**
        . . . . KAONS
        . . . . MUONS
        . . . . OMEGA-MESONS
        . . . . VECTOR MESONS
        . . . . . RHO-MESONS
        . . . . . SIGMA-MESONS
        . NUCLEAR PARTICLES
        . . BOSONS
        . . . **MESONS**
        . . . . ETA-MESONS
        . . . . KAONS
        . . . . MESON RESONANCE
        . . . . . X MESONS
        . . . . MUONS
        . . . . PIONS
        . . . . VECTOR MESONS
        . . . . . RHO-MESONS
        . . . . . SIGMA-MESONS
  RT   BARYONS
        BOSON FIELDS
        CHARGED PARTICLES
        CORPUSCULAR RADIATION
        COSMIC RAYS
        GLUONS
        LEPTONS
        MESON-NUCLEON INTERACTIONS
        MUONIUM
        POMERANCHUK THEOREM
        STRANGENESS

MESOPAUSE
  SN   (ALTITUDE APPROXIMATELY 90 KM)
  GS   EARTH ATMOSPHERE
        . UPPER ATMOSPHERE
        . . **MESOPAUSE**
        ENVIRONMENTS
        . **MESOPAUSE**
  RT   MESOSPHERE

MESOPAUSE-*(CONT.)*
        STRATOPAUSE

MESOPHILES
  GS   MICROORGANISMS
        . **MESOPHILES**
  RT   PSYCHROPHILES
        THERMOPHILES

MESOSCALE PHENOMENA
  GS   PHENOMENOLOGY
        . **MESOSCALE PHENOMENA**
  RT   METEOROLOGY
        WIND (METEOROLOGY)

MESOSPHERE
  GS   EARTH ATMOSPHERE
        . UPPER ATMOSPHERE
        . . **MESOSPHERE**
        ENVIRONMENTS
        . **MESOSPHERE**
  RT   CHEMOSPHERE
        HOMOSPHERE
        IONOSPHERE
        MESOPAUSE
        SOLAR MESOSPHERE EXPLORER
        STRATOPAUSE

MESSAGE PROCESSING
  RT   COMMUNICATING
        COMMUNICATION
        CRYPTOGRAPHY
        INFORMATION FLOW
        MESSAGES
        PACKET TRANSMISSION
        ∞ PROCESSING
        SEMANTICS
        SIGNAL PROCESSING
        SIGNAL TRANSMISSION
        SYMBOLS

MESSAGES
  GS   COMMUNICATING
        . INFORMATION DISSEMINATION
        . . **MESSAGES**
  RT   COMMUNICATION THEORY
        INFORMATION THEORY
        INTELLIGIBILITY
        MESSAGE PROCESSING
        SEMANTICS
        SENTENCES
        SIGNAL TRANSMISSION
        ∞ SIGNALS
        SYLLABLES
        SYMBOLS
        VOCODERS
        WORDS (LANGUAGE)

MESSERSCHMITT ME P-160 AIRCRAFT
  USE   P-160 AIRCRAFT

MESSERSCHMITT ME P-308 AIRCRAFT
  USE   P-308 AIRCRAFT

METABOLIC DISEASES
  GS   DISEASES
        . **METABOLIC DISEASES**
  RT   METABOLISM

METABOLIC WASTES
  GS   WASTES
        . **METABOLIC WASTES**
        . . HUMAN WASTES
        . . . URINE
  RT   ACTIVATED SLUDGE
        AIR POLLUTION
        CARBON DIOXIDE
        COMPOSTING
        ENVIRONMENT EFFECTS
        ENVIRONMENT POLLUTION
        ENVIRONMENTAL SURVEYS
        EXPIRED AIR
        LIQUID WASTES
        MANURES
        METABOLITES
        ORGANIC WASTES (FUEL CONVERSION)
        POLLUTION
        SEWAGE
        SEWERS
        SOLID WASTES
        WASTE DISPOSAL

METABOLISM
  GS   **METABOLISM**

METABOLISM-*(CONT.)*
        . ADRENAL METABOLISM
        . ASCORBIC ACID METABOLISM
        . CALCIUM METABOLISM
        . CARBOHYDRATE METABOLISM
        . . HYPERGLYCEMIA
        . . HYPOGLYCEMIA
        . CATABOLISM
        . ELECTROLYTE METABOLISM
        . ENZYME ACTIVITY
        . . FERMENTATION
        . HORMONE METABOLISMS
        . HYDROGEN METABOLISM
        . HYPOMETABOLISM
        . MINERAL METABOLISM
        . NITROGEN METABOLISM
        . OXYGEN METABOLISM
        . PHOSPHORUS METABOLISM
        . PROTEIN METABOLISM
        . . LIPID METABOLISM
  RT   CALORIC REQUIREMENTS
        ENZYMOLOGY
        HETEROTROPHS
        HOMEOSTASIS
        KREBS CYCLE
        METABOLIC DISEASES
        METABOLITES
        NUTRITION
        OBESITY
        OSTEOPOROSIS
        OXYGEN CONSUMPTION
        RESPIRATION
        SECRETIONS
        THERMOREGULATION

METABOLITES
  RT   BIOCHEMISTRY
        BIOSYNTHESIS
        METABOLIC WASTES
        METABOLISM
        ORGANIC COMPOUNDS

METAGALAXY
  USE   UNIVERSE

METAL AIR BATTERIES
  GS   ELECTRIC GENERATORS
        . DIRECT POWER GENERATORS
        . . PRIMARY BATTERIES
        . . . **METAL AIR BATTERIES**
        . . . . ZINC-OXYGEN BATTERIES
        ELECTROCHEMICAL CELLS
        . ELECTRIC BATTERIES
        . . PRIMARY BATTERIES
        . . . **METAL AIR BATTERIES**
        . . . . ZINC-OXYGEN BATTERIES
  RT   DRY CELLS
        STORAGE BATTERIES

METAL BONDING
  GS   BONDING
        . **METAL BONDING**
        . . METAL-METAL BONDING
  RT   ADHESION
        ADHESIVE BONDING
        BIMETALS
        BRAZING
        DIFFUSION WELDING
        EXPLOSIVE WELDING
        JOINTS (JUNCTIONS)
        LAMINATES
        RESIN BONDING
        SOLDERING
        WELDING

METAL COATINGS
  SN   (COATINGS CONSISTING OF METAL)
  GS   COATINGS
        . **METAL COATINGS**
        . . GOLD COATINGS
        . . NICKEL COATINGS
        . . ZINC COATINGS
        METALS
        . **METAL COATINGS**
        . . ALUMINUM COATINGS
        . . GOLD COATINGS
        . . NICKEL COATINGS
        . . ZINC COATINGS
  RT   ANODIC STRIPPING
        ANTIRADAR COATINGS
        CERAMIC COATINGS
        CLADDING
        CORROSION
        CORROSION PREVENTION
        DEPOSITION

METAL COATINGS-(CONT.)
ELECTROLESS DEPOSITION
HOT CORROSION
ION PLATING
LACQUERS
MAGNETRON SPUTTERING
METALLIZING
∞METALLURGY
OXIDES
PAINTS
∞PLATES
PLATING
PRIMERS (COATINGS)
PROTECTIVE COATINGS
SPRAYED COATINGS

METAL COMBUSTION
GS   COMBUSTION
. **METAL COMBUSTION**
RT   FUEL COMBUSTION
GAS-METAL INTERACTIONS
METALS
OXIDATION
PROPELLANT COMBUSTION
PYROPHORIC MATERIALS
SOLID PROPELLANT COMBUSTION
SOLID PROPELLANT IGNITION

∞ **METAL COMPOUNDS**
SN   *(USE OF A MORE SPECIFIC TERM IS*
RECOMMENDED--CONSULT THE TERMS
LISTED BELOW)
RT   ∞ALKALI METAL COMPOUNDS
ALUMINUM COMPOUNDS
AMMINES
ANTIMONY COMPOUNDS
BARIUM COMPOUNDS
BERYLLIUM COMPOUNDS
BISMUTH COMPOUNDS
CADMIUM COMPOUNDS
CALCIUM COMPOUNDS
CERIUM COMPOUNDS
CESIUM COMPOUNDS
∞CHEMICAL COMPOUNDS
CHROMIUM COMPOUNDS
COBALT COMPOUNDS
COPPER COMPOUNDS
DYSPROSIUM COMPOUNDS
ERBIUM COMPOUNDS
EUROPIUM COMPOUNDS
GALLIUM COMPOUNDS
GERMANIUM COMPOUNDS
HAFNIUM COMPOUNDS
INDIUM COMPOUNDS
IRON COMPOUNDS
LANTHANUM COMPOUNDS
LEAD COMPOUNDS
LEAD ORGANIC COMPOUNDS
LITHIUM COMPOUNDS
LITHIUM IODATES
LUTETIUM COMPOUNDS
MAGNESIUM COMPOUNDS
MANGANESE COMPOUNDS
MERCURY COMPOUNDS
METAL FLUORIDES
METAL HALIDES
METAL HYDRIDES
METAL OXIDES
METALS
MOLYBDENUM COMPOUNDS
NEODYMIUM COMPOUNDS
NEPTUNIUM COMPOUNDS
NICKEL COMPOUNDS
NIOBIUM COMPOUNDS
ORGANIC ALUMINUM COMPOUNDS
ORGANIC GERMANIUM COMPOUNDS
ORGANIC LITHIUM COMPOUNDS
ORGANIC TIN COMPOUNDS
ORGANOMETALLIC COMPOUNDS
OSMIUM COMPOUNDS
PLATINUM COMPOUNDS
PLUTONIUM COMPOUNDS
POTASSIUM COMPOUNDS
PROTACTINIUM COMPOUNDS
RARE EARTH COMPOUNDS
REFRACTORY MATERIALS
RHENIUM COMPOUNDS
RUBIDIUM COMPOUNDS
RUTHENIUM COMPOUNDS
SAMARIUM COMPOUNDS
SCANDIUM COMPOUNDS
SILVER COMPOUNDS
SODIUM COMPOUNDS
STRONTIUM COMPOUNDS
TANTALUM COMPOUNDS

METAL COMPOUNDS-(CONT.)
TECHNETIUM COMPOUNDS
THALLIUM COMPOUNDS
THORIUM COMPOUNDS
THULIUM COMPOUNDS
TIN COMPOUNDS
TITANIUM COMPOUNDS
TUNGSTEN COMPOUNDS
URANIUM COMPOUNDS
VANADIUM COMPOUNDS
VANADYL COMPOUNDS
YTTERBIUM COMPOUNDS
YTTRIUM COMPOUNDS
ZINC COMPOUNDS
ZIRCONIUM COMPOUNDS

METAL CORROSION
USE   CORROSION

METAL CRYSTALS
GS   CRYSTALS
. **METAL CRYSTALS**
METALS
. **METAL CRYSTALS**
RT   CRYSTAL LATTICES
CRYSTAL STRUCTURE
METALLOGRAPHY

METAL CUTTING
GS   CUTTING
. **METAL CUTTING**
RT   COUNTERSINKING
GRINDING (MATERIAL REMOVAL)
KNURLING
LASER CUTTING
MACHINE TOOLS
MACHINING
MELTING
MILLING (MACHINING)
PERFORATING
PLANING
PLASMA ARC CUTTING
SCARFING
SHEARING
SLICING
SPARK MACHINING
SPIKING

METAL DRAWING
GS   FORMING TECHNIQUES
. **METAL DRAWING**
METAL WORKING
. **METAL DRAWING**
RT   BULGING
BUNDLE DRAWING
COLD DRAWING
COLD WORKING
∞DRAWING
DUCTILITY
HOT WORKING
MAGNETIC FORMING
STRETCH FORMING

METAL FATIGUE
GS   FATIGUE (MATERIALS)
. **METAL FATIGUE**
RT   BENDING FATIGUE
COFFIN-MANSON LAW
CRACK CLOSURE
CRACK INITIATION
CRACK PROPAGATION
FRACTURING
GRAIN SIZE
INELASTIC STRESS
RETIREMENT FOR CAUSE
RUPTURING
S-N DIAGRAMS
SEGRE CHARACTERISTIC
STRESS CORROSION
STRESS CORROSION CRACKING
THERMAL FATIGUE
TRANSGRANULAR CORROSION

METAL FIBERS
GS   FIBERS
. **METAL FIBERS**
RT   BORSIC (TRADENAME)
FIBER COMPOSITES
FILAMENT WINDING
REINFORCING FIBERS

METAL FILMS
GS   METALS
. **METAL FILMS**
RT   COATINGS

METAL FILMS-(CONT.)
∞ FILMS
METALLIZING
NICKEL COATINGS
PHOTOTHERMAL CONVERSION
PICKLING (METALLURGY)
SPUTTERING GAGES
THIN FILMS

METAL FINISHING
GS   **METAL FINISHING**
. ELECTROPOLISHING
. PEENING
. . SHOT PEENING
RT   CLEANING
COATING
COATINGS
DESCALING
PICKLING (METALLURGY)
PLATING
SURFACE FINISHING

METAL FLUORIDES
GS   HALOGEN COMPOUNDS
. FLUORINE COMPOUNDS
. . FLUORIDES
. . . **METAL FLUORIDES**
. . . . ALUMINUM FLUORIDES
. . . . BERYLLIUM FLUORIDES
. . . . CADMIUM FLUORIDES
. . . . CALCIUM FLUORIDES
. . . . CESIUM FLUORIDES
. . . . CHROMIUM FLUORIDES
. . . . COBALT FLUORIDES
. . . . COPPER FLUORIDES
. . . . LANTHANUM FLUORIDES
. . . . LITHIUM FLUORIDES
. . . . MAGNESIUM FLUORIDES
. . . . NICKEL FLUORIDES
. . . . PLUTONIUM FLUORIDES
. . . . PROTACTINIUM FLUORIDES
. . . . SODIUM FLUORIDES
. . . . STRONTIUM FLUORIDES
. . . . THORIUM FLUORIDES
. . . . TUNGSTEN FLUORIDES
. . . . URANIUM FLUORIDES
. . . . ZINC FLUORIDES
RT   ∞METAL COMPOUNDS

METAL FOAMS
GS   FOAMS
. **METAL FOAMS**
RT   BUBBLES
FOAMING
LOW GRAVITY MANUFACTURING
∞MATERIALS SCIENCE
MELTING
∞METALLURGY
SPACE PROCESSING APPLICATIONS
ROCKET

METAL FOILS
GS   FOILS (MATERIALS)
. **METAL FOILS**
METALS
. **METAL FOILS**
RT   HONEYCOMB STRUCTURES
MULTILAYER INSULATION
∞SHEETS

METAL FORGING
USE   FORGING

METAL FORMING
USE   FORMING TECHNIQUES
METAL WORKING

METAL FUELS
GS   FUELS
. CHEMICAL FUELS
. . **METAL FUELS**
RT   ALUMINUM COMPOUNDS
BERYLLIUM COMPOUNDS
BORON COMPOUNDS
CESIUM COMPOUNDS
GELLED PROPELLANTS
HYBRID PROPELLANTS
LITHIUM COMPOUNDS
METALS
SLURRY PROPELLANTS
SOLID PROPELLANTS

METAL GRINDING
GS   GRINDING (MATERIAL REMOVAL)
. **METAL GRINDING**

METAL GRINDING-*(CONT.)*
RT   FORMING TECHNIQUES
      GRINDING MACHINES
      SURFACE FINISHING

METAL HALIDES
GS   HALOGEN COMPOUNDS
      . HALIDES
      . . **METAL HALIDES**
      . . . ALKALI HALIDES
      . . . . CESIUM HALIDES
      . . . . . CESIUM BROMIDES
      . . . . . CESIUM FLUORIDES
      . . . . . CESIUM IODIDES
      . . . POTASSIUM IODIDES
      . . . SODIUM BROMIDES
      . . . SODIUM CHLORIDES
      . . . SODIUM FLUORIDES
      . . . SODIUM IODIDES
      . . . ALUMINUM CHLORIDES
      . . . ALUMINUM FLUORIDES
      . . BARIUM FLUORIDES
      . . BERYLLIUM CHLORIDES
      . . BERYLLIUM FLUORIDES
      . . CADMIUM CHLORIDES
      . . CADMIUM FLUORIDES
      . . CALCIUM CHLORIDES
      . . CALCIUM FLUORIDES
      . . . FLUORSPAR
      . . CHROMIUM BROMIDES
      . . CHROMIUM FLUORIDES
      . . COBALT FLUORIDES
      . . COPPER CHLORIDES
      . . COPPER FLUORIDES
      . . HAFNIUM IODIDES
      . . IRON CHLORIDES
      . . LANTHANUM CHLORIDES
      . . LANTHANUM FLUORIDES
      . . LEAD CHLORIDES
      . . LITHIUM CHLORIDES
      . . LITHIUM FLUORIDES
      . . MAGNESIUM BROMIDES
      . . MAGNESIUM FLUORIDES
      . . NICKEL FLUORIDES
      . . NIOBIUM IODIDES
      . . PLUTONIUM FLUORIDES
      . . POTASSIUM BROMIDES
      . . POTASSIUM CHLORIDES
      . . PROTACTINIUM FLUORIDES
      . . SILVER HALIDES
      . . . SILVER BROMIDES
      . . . SILVER CHLORIDES
      . . . SILVER IODIDES
      . . STRONTIUM BROMIDES
      . . STRONTIUM FLUORIDES
      . . TECHNETIUM FLUORIDES
      . . THORIUM FLUORIDES
      . . TITANIUM CHLORIDES
      . . TUNGSTEN HALIDES
      . . . TUNGSTEN CHLORIDES
      . . . TUNGSTEN FLUORIDES
      . . URANIUM FLUORIDES
      . . ZINC CHLORIDES
      . . ZINC FLUORIDES
      . . ZIRCONIUM IODIDES
RT   ∞METAL COMPOUNDS

METAL HARDENING
USE   HARDENING (MATERIALS)

METAL HYDRIDES
UF   PLUMBANE
GS   HYDROGEN COMPOUNDS
      . HYDRIDES
      . . **METAL HYDRIDES**
      . . . ALUMINUM HYDRIDES
      . . . . ALUMINUM BOROHYDRIDES
      . . . BERYLLIUM HYDRIDES
      . . . CESIUM HYDRIDES
      . . . LITHIUM HYDRIDES
      . . . . LITHIUM ALUMINUM HYDRIDES
      . . . POTASSIUM HYDRIDES
      . . . SODIUM HYDRIDES
RT   ∞METAL COMPOUNDS

METAL INSULATOR SEMICONDUCTORS
USE   MIS (SEMICONDUCTORS)

METAL IONS
GS   IONS
      . CATIONS
      . . **METAL IONS**
      . . . FERRIC IONS
      . . . MANGANESE IONS
      . . . VANADYL RADICAL

METAL IONS-*(CONT.)*
      PARTICLES
      . CHARGED PARTICLES
      . . CATIONS
      . . . **METAL IONS**
      . . . . FERRIC IONS
      . . . . MANGANESE IONS
RT   BARIUM ION CLOUDS
      ION IMPLANTATION
      ION PLATING
      POSITIVE IONS

METAL JOINTS
GS   JOINTS (JUNCTIONS)
      . **METAL JOINTS**
      . . SOLDERED JOINTS
      . . WELDED JOINTS
      . . . SPOT WELDS
RT   BUTT JOINTS
      EXPLOSIVE WELDING
      LAP JOINTS
      RIVETED JOINTS
      SEAMS (JOINTS)

METAL MATRIX COMPOSITES
GS   COMPOSITE MATERIALS
      . **METAL MATRIX COMPOSITES**
      . . BORSIC (TRADENAME)
      . . EUTECTIC COMPOSITES
      METALS
      . **METAL MATRIX COMPOSITES**
      MIXTURES
      . **METAL MATRIX COMPOSITES**
      . . BORSIC (TRADENAME)
RT   BORON FIBERS
      ELECTRODEPOSITION
      FIBER COMPOSITES
      FIBERS
      ∞MATERIALS
      ∞MATRICES
      MATRIX MATERIALS
      MONOTECTIC ALLOYS
      PLASMA SPRAYING
      POWDER METALLURGY
      REINFORCING FIBERS
      RESIN MATRIX COMPOSITES
      WHISKER COMPOSITES

METAL NITRIDES
GS   NITROGEN COMPOUNDS
      . NITRIDES
      . . **METAL NITRIDES**
RT   ALUMINUM NITRIDES
      BERYLLIUM NITRIDES
      TANTALUM NITRIDES
      TITANIUM NITRIDES
      TRANSITION METALS
      ZIRCONIUM NITRIDES

METAL OXIDE SEMICONDUCTORS
UF   MOS (SEMICONDUCTORS)
GS   ELECTRONIC EQUIPMENT
      . SOLID STATE DEVICES
      . . SEMICONDUCTOR DEVICES
      . . . **METAL OXIDE SEMICONDUCTORS**
      . . . . CHARGE TRANSFER DEVICES
      . . . . . BUCKET BRIGADE DEVICES
      . . . . . CHARGE COUPLED DEVICES
      . . . . . CHARGE INJECTION DEVICES
      . . . . CMOS
      . . . . SOS (SEMICONDUCTORS)
      SEMICONDUCTORS (MATERIALS)
      . **METAL OXIDE SEMICONDUCTORS**
      . . CHARGE TRANSFER DEVICES
      . . CMOS
      . . SOS (SEMICONDUCTORS)
RT   CAPACITANCE-VOLTAGE
          CHARACTERISTICS
      ION IMPLANTATION
      RECTIFIERS

METAL OXIDES
GS   CHALCOGENIDES
      . OXIDES
      . . **METAL OXIDES**
      . . . ALKALINE EARTH OXIDES
      . . . . BARIUM OXIDES
      . . . . BERYLLIUM OXIDES
      . . . . CALCIUM OXIDES
      . . . . MAGNESIUM OXIDES
      . . . . . PERICLASE
      . . . ALUMINUM OXIDES
      . . . . SAPPHIRE
      . . . BISMUTH OXIDES
      . . . CESIUM OXIDES

METAL OXIDES-*(CONT.)*
      . . . CHROMIUM OXIDES
      . . . . CHROMITES
      . . . COBALT OXIDES
      . . . COPPER OXIDES
      . . . HAFNIUM OXIDES
      . . . IRON OXIDES
      . . . . CHROMITES
      . . . . HEMATITE
      . . . . ILMENITE
      . . . . MAGNETITE
      . . . LANTHANUM OXIDES
      . . . LEAD OXIDES
      . . . LITHIUM OXIDES
      . . . MANGANESE OXIDES
      . . . . HOPCALITE (TRADEMARK)
      . . . MERCURY OXIDES
      . . . MIXED OXIDES
      . . . MOLYBDENUM OXIDES
      . . . NICKEL OXIDES
      . . . NIOBIUM OXIDES
      . . . PLATINUM OXIDES
      . . . PLUTONIUM OXIDES
      . . . POTASSIUM OXIDES
      . . . SCANDIUM OXIDES
      . . . SILVER OXIDES
      . . . SODIUM PEROXIDES
      . . . TANTALUM OXIDES
      . . . THORIUM OXIDES
      . . . TIN OXIDES
      . . . TITANIUM OXIDES
      . . . . ANATASE
      . . . . ILMENITE
      . . . . RUTILE
      . . . TUNGSTEN OXIDES
      . . . . SCHEELITE
      . . . URANIUM OXIDES
      . . . VANADIUM OXIDES
      . . . YTTRIUM OXIDES
      . . . ZINC OXIDES
      . . . ZIRCONIUM OXIDES
RT   CATHODIC COATINGS
      ∞METAL COMPOUNDS
      OXIDE FILMS
      VANADATES

METAL PARTICLES
GS   PARTICLES
      . **METAL PARTICLES**
      . . PLATINUM BLACK
      . . POWDERED ALUMINUM
RT   POWDER METALLURGY
      SCRAP
      SPUTTERING

METAL PLATES
UF   PLATE (METAL)
GS   STRUCTURAL MEMBERS
      . PLATES (STRUCTURAL MEMBERS)
      . . **METAL PLATES**
      . . . BOILER PLATE
RT   ARMOR
      BARS
      BILLETS
      FLANGES
      FLAT PLATES
      GIRDER WEBS
      PARALLEL PLATES
      ∞PLATES
      RECTANGULAR PLATES
      SLABS
      THICK PLATES
      THIN PLATES

METAL POLISHING
UF   POLISHED METALS
GS   POLISHING
      . **METAL POLISHING**
      . . ELECTROPOLISHING
RT   CLEANING
      SURFACE FINISHING

METAL POWDER
UF   POWDERED METALS
GS   METALS
      . **METAL POWDER**
      . . PLATINUM BLACK
      . . POWDERED ALUMINUM
RT   ATOMIZING
      BEARING ALLOYS
      COMMINUTION
      COMPRESSIBILITY
      COMPRESSING
      ELECTRODEPOSITION
      FLAKES

METAL POWDER-(CONT.)
MIXING
PARTICLES
POROUS MATERIALS
POWDER METALLURGY
REDUCTION (CHEMISTRY)
SINTERING
SIZE SEPARATION

METAL PROPELLANTS
GS   FUELS
. SOLID ROCKET PROPELLANTS
. . **METAL PROPELLANTS**
PROPELLANTS
. ROCKET PROPELLANTS
. . SOLID ROCKET PROPELLANTS
. . . **METAL PROPELLANTS**
. SOLID PROPELLANTS
. . SOLID ROCKET PROPELLANTS
. . . **METAL PROPELLANTS**
RT   ALUMINUM COMPOUNDS
BERYLLIUM COMPOUNDS
BORON COMPOUNDS
GELLED PROPELLANTS
GELLED ROCKET PROPELLANTS
HYBRID PROPELLANTS
MONOPROPELLANTS
SLURRY PROPELLANTS

METAL SHEETS
UF   SHEET METAL
RT   PLATES (STRUCTURAL MEMBERS)
∞ SHEETS

METAL SHELLS
GS   SHELLS (STRUCTURAL FORMS)
. **METAL SHELLS**
RT   CIRCULAR SHELLS
CYLINDRICAL SHELLS
HEMISPHERICAL SHELLS
HULLS (STRUCTURES)
MEMBRANE STRUCTURES
ORTHOTROPIC SHELLS
REINFORCED SHELLS
SKIN (STRUCTURAL MEMBER)
SPHERICAL SHELLS
THIN WALLED SHELLS
TOROIDAL SHELLS

METAL SPINNING
UF   SPIN FORGING
SPINNING (METALLURGY)
GS   FORMING TECHNIQUES
. **METAL SPINNING**
. . HYDROSPINNING
METAL WORKING
. **METAL SPINNING**
. . HYDROSPINNING
SPIN
. **METAL SPINNING**
. . HYDROSPINNING
RT   COLD WORKING
EXTRUDING
HOT WORKING

METAL SPRAYING
GS   SPRAYING
. **METAL SPRAYING**
RT   ARC SPRAYING
COATING
COATINGS
FLAME SPRAYING
METALLIZING
SURFACE FINISHING

METAL STRIPS
RT   BILLETS
RIBBONS
STRAKES
∞ STRIP

METAL SURFACES
RT   CRACK INITIATION
CRYSTAL SURFACES
EROSION
GAS-SOLID INTERFACES
LIQUID-SOLID INTERFACES
OXIDE FILMS
SOLID SURFACES
SURFACE FINISHING
SURFACE PROPERTIES
SURFACE REACTIONS
∞ SURFACES

METAL VAPOR LASERS
GS   STIMULATED EMISSION DEVICES
. LASERS
. . **METAL VAPOR LASERS**
RT   LASER MATERIALS
LASER MICROSCOPY
OPTICAL PUMPING

METAL VAPORS
GS   METALS
. **METAL VAPORS**
. . MERCURY VAPOR
. . SODIUM VAPOR
VAPORS
. **METAL VAPORS**
. . MERCURY VAPOR
. . SODIUM VAPOR
RT   ALKALI METALS
ALKALI VAPOR LAMPS
GAS-METAL INTERACTIONS
HEAT TRANSFER
LIQUID METALS
VAPOR DEPOSITION

METAL WHISKER REINFORCEMENT
USE   WHISKER COMPOSITES

METAL WORKING
SN   (METAL DEFORMATION FOR CHANGING
SHAPE AND FOR
PROPERTIES--EXCLUDES CASTING,
CUTTING, DEPOSITION PROCESS AND
MACHINING)
UF   METAL FORMING
GS   **METAL WORKING**
. AUSFORMING
. BULGING
. CLADDING
. COINING
. EXPLOSIVE FORMING
. FORGING
. HYDROFORMING
. MAGNETIC FORMING
. METAL DRAWING
. METAL SPINNING
. . HYDROSPINNING
. SIZING (SHAPING)
RT   BRAKES (FORMING OR BENDING)
∞ BREAKDOWN
CASTING
COLD PRESSING
COLD ROLLING
COLD WORKING
DECARBURIZATION
DEEP DRAWING
DIMPLING
ELECTROHYDRAULIC FORMING
ELECTROMAGNETIC HAMMERS
EXPLOSIVE WELDING
EXTRUDING
FLATTENING
FORMING TECHNIQUES
HARDENING (MATERIALS)
HOT PRESSING
HOT WORKING
LASER APPLICATIONS
LEVELING
MACHINING
MALLEABILITY
∞ METALLURGY
PEENING
PERFORATING
PIERCING
PLASMA ARC CUTTING
PRESSING (FORMING)
PYROMETALLURGY
∞ REDUCTION
ROLL FORMING
∞ ROLLING
SHEARING
SHOT PEENING
STAMPING
STRETCH FORMING
STRETCHING
SWAGING
TEMPERING
WINDING
WORK HARDENING

METAL-BARRIER-METAL JUNCTIONS
USE   MBM JUNCTIONS

METAL-GAS SYSTEMS
RT   GAS LUBRICANTS
GAS-METAL INTERACTIONS

METAL-GAS SYSTEMS-(CONT.)
GASES
METALS
∞ SYSTEMS
VAPOR PHASES

METAL-INSULATOR-METAL DIODES
USE   MIM DIODES

METAL-INSULATOR-METAL SEMICONDUCTORS
USE   MIM (SEMICONDUCTORS)

METAL-METAL BONDING
GS   BONDING
. METAL BONDING
. . **METAL-METAL BONDING**
RT   ADHESIVE BONDING
ADHESIVES
DIFFUSION WELDING
EXPLOSIVE WELDING
INERTIA BONDING
RESIN BONDING
SOLDERING
WELDING

METAL-NITRIDE-OXIDE-SEMICONDUCTORS
GS   ELECTRONIC EQUIPMENT
. SOLID STATE DEVICES
. . **METAL-NITRIDE-OXIDE-SEMICONDUCT
ORS**
SEMICONDUCTORS (MATERIALS)
. **METAL-NITRIDE-OXIDE-SEMICONDUCT
ORS**
RT   CHIPS (MEMORY DEVICES)

METAL-NITRIDE-OXIDE-SILICON
UF   MNOS
GS   SEMICONDUCTORS (MATERIALS)
. **METAL-NITRIDE-OXIDE-SILICON**

METAL-OXIDE-METAL SEMICONDUCTORS
USE   MOM (SEMICONDUCTORS)

† ## METAL-WATER REACTIONS
GS   CHEMICAL REACTIONS
. **METAL-WATER REACTIONS**
RT   CORROSION
ELECTROCHEMICAL CORROSION
EROSION
PITTING
RUSTING
SURFACE REACTIONS

METALLIC GLASSES
GS   GLASS
. **METALLIC GLASSES**
RT   GLASS COATINGS
GLASS FIBERS
OPTICAL PROPERTIES
SILICON DIOXIDE
SPIN GLASS
VITREOUS MATERIALS

METALLIC HYDROGEN
GS   CHEMICAL ELEMENTS
. HYDROGEN
. . HYDROGEN ISOTOPES
. . . **METALLIC HYDROGEN**
RT   CRITICAL TEMPERATURE
SOLID PHASES
SOLID STATE
SOLIDIFIED GASES
SOLIDS

METALLIC PLASMAS
GS   GASES
. IONIZED GASES
. . CHARGED PARTICLES
. . . **METALLIC PLASMAS**
. . . . CESIUM PLASMA
PARTICLES
. CHARGED PARTICLES
. . ENERGETIC PARTICLES
. . . PLASMAS (PHYSICS)
. . . . **METALLIC PLASMAS**
. . . . . CESIUM PLASMA
RT   ELECTRON PLASMA
MERCURY ARCS
PLASMA SHEATHS

METALLIC STARS
GS   CELESTIAL BODIES
. STARS
. . **METALLIC STARS**

METALLIC STARS-(CONT.)
RT   ABUNDANCE
      CHEMICAL COMPOSITION
      METALLICITY
      STELLAR ATMOSPHERES
      STELLAR STRUCTURE

METALLICITY
RT   ABUNDANCE
      CHEMICAL ANALYSIS
      CHEMICAL COMPOSITION
      GALACTIC CLUSTERS
      GALAXIES
      GLOBULAR CLUSTERS
      HYDROGEN
      INTERSTELLAR MATTER
      MASS RATIOS
      METALLIC STARS
      METALS
      SPECTROSCOPIC ANALYSIS
      STAR CLUSTERS
      STARS

METALLIZING
GS   COATING
      . **METALLIZING**
      COATINGS
      . **METALLIZING**
RT   CLADDING
      ELECTROPLATING
      FINISHES
      FLAME SPRAYING
      LAMINATES
      METAL COATINGS
      METAL FILMS
      METAL SPRAYING
      PLATING
      SPRAYING
      SUBSTRATES
      VAPOR DEPOSITION

METALLOGRAPHY
RT   ABRASION
      ALLOYS
      ANISOTROPY
      CRYSTAL LATTICES
      CRYSTALLOGRAPHY
      ELECTROPOLISHING
      ETCHING
      FERROGRAPHY
      INCLUSIONS
      ISOTROPY
      ∞MATERIALS TESTS
      METAL CRYSTALS
      ∞METALLURGY
      METALS
      MICROPOROSITY
      MICROSCOPES
      MICROSTRUCTURE
      MUSHY ZONES
      ORDER-DISORDER TRANSFORMATIONS
      PHOTOMICROGRAPHY
      POLISHING
      RADIOGRAPHY
      REPLICAS
      SOLID SUSPENSIONS
      TIME TEMPERATURE PARAMETER
      VIBRATORY POLISHING
      WIDMANSTATTEN STRUCTURE
      X RAY DIFFRACTION

METALLOIDS
UF   SEMIMETALS
GS   CHEMICAL ELEMENTS
      . **METALLOIDS**
      . . ANTIMONY
      . . ANTIMONY ISOTOPES
      . . ARSENIC
      . . . ARSENIC ISOTOPES
      . . BORON
      . . . BORON ISOTOPES
      . . . . BORON 10
      . . GERMANIUM
      . . . GERMANIUM ISOTOPES
      . . POLONIUM
      . . . POLONIUM ISOTOPES
      . . . . POLONIUM 208
      . . . . POLONIUM 209
      . . . . POLONIUM 210
      . . SILICON
      . . . SILICON ISOTOPES
      . . TELLURIUM
      . . . TELLURIUM ISOTOPES
RT   ALLOYS
      ARSENIC ALLOYS

METALLOIDS-(CONT.)
      BORON ALLOYS
      INTERMETALLICS
      METALS
      ORGANOMETALLIC COMPOUNDS
      SEMICONDUCTORS (MATERIALS)

METALLORGANIC COMPOUNDS
USE   ORGANOMETALLIC COMPOUNDS

METALLOSILOXANE POLYMER
RT   ORGANOMETALLIC COMPOUNDS
      ∞POLYMERS

METALLOXANE POLYMER
GS   ORGANOMETALLIC COMPOUNDS
      . **METALLOXANE POLYMER**
RT   ∞POLYMERS

∞ **METALLURGY**
SN   *(USE OF A MORE SPECIFIC TERM IS*
      RECOMMENDED--CONSULT THE TERMS
      LISTED BELOW)
RT   ALLOYS
      BENEFICIATION
      CASTING
      COATING
      COATINGS
      CORROSION
      CRYSTALLOGRAPHY
      DUPLEX OPERATION
      FERROUS METALS
      FOUNDRIES
      FRACTOGRAPHY
      FURNACES
      HARDENING (MATERIALS)
      HEAT TREATMENT
      HYDROMETALLURGY
      LEVITATION MELTING
      LIGHT ALLOYS
      MECHANICAL PROPERTIES
      MELT SPINNING
      MELTING
      METAL COATINGS
      METAL FOAMS
      METAL WORKING
      METALLOGRAPHY
      METALS
      NONFERROUS METALS
      ∞PHYSICAL SCIENCES
      POWDER METALLURGY
      PYROMETALLURGY
      RAPID QUENCHING (METALLURGY)
      RECRYSTALLIZATION
      ∞SCIENCE
      SMELTING
      THERMOMECHANICAL TREATMENT

METALS
UF   MAGNETIC METALS
GS   **METALS**
      . ACTINIDE SERIES
      . . ACTINIUM
      . . RADIUM
      . . . RADIUM ISOTOPES
      . . . . RADIUM 226
      . . THORIUM
      . . . THORIUM ISOTOPES
      . . TRANSURANIUM ELEMENTS
      . . . AMERICIUM
      . . . . AMERICIUM ISOTOPES
      . . . . . AMERICIUM 241
      . . . BERKELIUM
      . . . CALIFORNIUM
      . . . . CALIFORNIUM ISOTOPES
      . . . CURIUM
      . . . . CURIUM ISOTOPES
      . . . . . CURIUM 242
      . . . . . CURIUM 244
      . . . EINSTEINIUM
      . . . FERMIUM
      . . . LAWRENCIUM
      . . . MENDELEVIUM
      . . . NEPTUNIUM
      . . . . NEPTUNIUM ISOTOPES
      . . . NOBELIUM
      . . . PLUTONIUM
      . . . . PLUTONIUM ISOTOPES
      . . . . . PLUTONIUM 238
      . . . . . PLUTONIUM 239
      . . . . . PLUTONIUM 240
      . . . . . PLUTONIUM 241
      . . . . . PLUTONIUM 244
      . . . SERGENIUM
      . . URANIUM

METALS-(CONT.)
      . . . URANIUM ISOTOPES
      . . . . URANIUM 232
      . . . . URANIUM 233
      . . . . URANIUM 234
      . . . . URANIUM 235
      . . . . URANIUM 238
      . . . URANIUM PLASMAS
      . ALKALI METALS
      . . CESIUM
      . . . CESIUM ISOTOPES
      . . . . CESIUM 133
      . . . . CESIUM 134
      . . . . CESIUM 137
      . . . . CESIUM 144
      . . . CESIUM VAPOR
      . . FRANCIUM
      . . LITHIUM
      . . . LIQUID LITHIUM
      . . . LITHIUM ISOTOPES
      . . POTASSIUM
      . . . LIQUID POTASSIUM
      . . . POTASSIUM ISOTOPES
      . . . . POTASSIUM 38
      . . . . POTASSIUM 39
      . . . . POTASSIUM 40
      . . RUBIDIUM
      . . . RUBIDIUM ISOTOPES
      . . . . RUBIDIUM 86
      . . SODIUM
      . . . LIQUID SODIUM
      . . . SODIUM ISOTOPES
      . . . . SODIUM 22
      . . . . SODIUM 24
      . . . SODIUM VAPOR
      . ALKALINE EARTH METALS
      . . BARIUM ISOTOPES
      . ALUMINUM
      . . ALUMINUM ISOTOPES
      . . . ALUMINUM 26
      . . . ALUMINUM 27
      . . POWDERED ALUMINUM
      . . SINTERED ALUMINUM POWDER
      . ANTIMONY ISOTOPES
      . ASTATINE
      . ASTATINE ISOTOPES
      . BARIUM
      . . BARIUM ISOTOPES
      . BERYLLIUM
      . . BERYLLIUM ISOTOPES
      . . . BERYLLIUM 7
      . . . BERYLLIUM 9
      . . . BERYLLIUM 10
      . BISMUTH
      . . BISMUTH ISOTOPES
      . CALCIUM
      . . CALCIUM ISOTOPES
      . COPPER ISOTOPES
      . FERROUS METALS
      . GALLIUM
      . . GALLIUM ISOTOPES
      . INDIUM
      . INDIUM ISOTOPES
      . LEAD (METAL)
      . . LEAD ISOTOPES
      . LIQUID METALS
      . . LIQUID LITHIUM
      . . LIQUID POTASSIUM
      . . LIQUID SODIUM
      . . MERCURY (METAL)
      . . . MERCURY VAPOR
      . MAGNESIUM
      . . MAGNESIUM ISOTOPES
      . METAL COATINGS
      . . ALUMINUM COATINGS
      . . GOLD COATINGS
      . . NICKEL COATINGS
      . . ZINC COATINGS
      . METAL CRYSTALS
      . METAL FILMS
      . METAL FOILS
      . METAL MATRIX COMPOSITES
      . METAL POWDER
      . . PLATINUM BLACK
      . . POWDERED ALUMINUM
      . METAL VAPORS
      . . MERCURY VAPOR
      . . SODIUM VAPOR
      . NOBLE METALS
      . . GOLD
      . . . GOLD ISOTOPES
      . . . . GOLD 198
      . . RUTHENIUM
      . . . RUTHENIUM ISOTOPES
      . . SILVER
      . . . SILVER ISOTOPES

METEORITES-(CONT.)
      FOREIGN BODIES
      IMPACT MELTS
      METEORITE CRATERS
      METEORITIC COMPOSITION
      METEORITIC MICROSTRUCTURES
      METEOROID SHOWERS
      METEOROIDS
      MICROMETEOROIDS
      MOLDAVITE

METEORITIC COMPOSITION
GS   COMPOSITION (PROPERTY)
      . **METEORITIC COMPOSITION**
RT   CARBONACEOUS METEORITES
      COSMOCHEMISTRY
      IRON METEORITES
      KAMACITE
      METEORITES
      SCHREIBERSITE
      STONY METEORITES
      TEKTITES
      TROILITE

METEORITIC DAMAGE
GS   DAMAGE
      . IMPACT DAMAGE
      . . **METEORITIC DAMAGE**
RT   ∞BOMBARDMENT
      CRATERING
      EJECTA
      HYPERVELOCITY IMPACT
      MARS CRATERS
      METEORITE COLLISIONS
      METEORITE CRATERS
      METEOROID HAZARDS
      METEOROID PROTECTION
      PROJECTILE CRATERING

METEORITIC DIAMONDS
GS   DIAMONDS
      . **METEORITIC DIAMONDS**

METEORITIC DUST
USE   MICROMETEOROIDS

METEORITIC IONIZATION
USE   ATMOSPHERIC IONIZATION
       METEOR TRAILS

METEORITIC MICROSTRUCTURES
GS   MICROSTRUCTURE
      . **METEORITIC MICROSTRUCTURES**
RT   IRON METEORITES
      METEORITES
      STONY METEORITES
      TEKTITES
      WIDMANSTATTEN STRUCTURE

METEOROID CONCENTRATION
GS   COMPOSITION (PROPERTY)
      . CONCENTRATION (COMPOSITION)
      . . **METEOROID CONCENTRATION**
      DENSITY (NUMBER/VOLUME)
      . **METEOROID CONCENTRATION**
RT   FLUX DENSITY
      MASS DISTRIBUTION
      SPATIAL DISTRIBUTION
      SPORADIC METEOROIDS

METEOROID CRATERS
USE   METEORITE CRATERS

METEOROID DUST CLOUDS
GS   CELESTIAL BODIES
      . METEOROIDS
      . . MICROMETEOROIDS
      . . . **METEOROID DUST CLOUDS**
      . . . . ZODIACAL DUST
      DUST
      . COSMIC DUST
      . . INTERPLANETARY DUST
      . . . **METEOROID DUST CLOUDS**
      . . . . ZODIACAL DUST
      MEDIA
      . INTERPLANETARY MEDIUM
      . . INTERPLANETARY DUST
      . . . **METEOROID DUST CLOUDS**
      . . . . ZODIACAL DUST
RT   CLOUDS
      EXPLORER SATELLITES
      TERRESTRIAL DUST BELT

METEOROID HAZARDS
UF   METEOR HAZARDS
GS   HAZARDS
      . FLIGHT HAZARDS
      . . **METEOROID HAZARDS**
RT   METEORITE COLLISIONS
      METEORITIC DAMAGE
      METEOROIDS
      OPERATIONAL HAZARDS
      PROJECTILE CRATERING

METEOROID PROTECTION
GS   PROTECTION
      . **METEOROID PROTECTION**
RT   BUMPERS
      IMPACT DAMAGE
      METEORITIC DAMAGE
      SPACECRAFT SHIELDING
      SPACECRAFT STRUCTURES

METEOROID SHOWERS
UF   METEOR BURSTS
GS   CELESTIAL BODIES
      . **METEOROID SHOWERS**
      . . AQUARID METEOROIDS
      . . ARIETID METEOROIDS
      . . CYRILLID METEOROIDS
      . . DRACONID METEOROIDS
      . . GEMINID METEOROIDS
      . . LEONID METEOROIDS
      . . ORIONID METEOROIDS
      . . PERSEID METEOROIDS
      . . QUADRANTID METEOROIDS
      . . TAURID METEOROIDS
RT   ASTRONOMY
      BOLIDES
      COMETS
      METEOR TRAILS
      METEORITES
      METEOROIDS
      ∞SHOWERS

METEOROID TECHNOLOGY SATELLITE
USE   EXPLORER 46 SATELLITE

METEOROIDS
SN   (SOLID OBJECTS IN SPACE, MUCH
       SMALLER THAN AN ASTEROID AND
       MUCH LARGER THAN A MOLECULE)
UF   METEORS
GS   CELESTIAL BODIES
      . **METEOROIDS**
      . . AQUARID METEOROIDS
      . . ARIETID METEOROIDS
      . . BOLIDES
      . . . CYRILLID METEOROIDS
      . . DRACONID METEOROIDS
      . . GEMINID METEOROIDS
      . . LEONID METEOROIDS
      . . MICROMETEOROIDS
      . . . METEOROID DUST CLOUDS
      . . . . ZODIACAL DUST
      . . ORIONID METEOROIDS
      . . PERSEID METEOROIDS
      . . QUADRANTID METEOROIDS
      . . RADIO METEORS
      . . SPORADIC METEOROIDS
      . . TAURID METEOROIDS
RT   ASTEROID BELTS
      ASTEROIDS
      BUMPERS
      CHIRON
      COMETS
      COSMIC DUST
      HYPERVELOCITY PROJECTILES
      INTERPLANETARY DUST
      INTERPLANETARY MEDIUM
      METEOR TRAILS
      METEORITES
      METEOROID HAZARDS
      METEOROID SHOWERS
      MICROMETEORITES
      NATURAL SATELLITES
      PARTICLE TRACKS
      RADIATION METEOROID SPACECRAFT
      SOLAR SYSTEM
      SPACE DEBRIS
      TEMPEL 2 COMET
      TORO ASTEROID
      VESTA ASTEROID

METEOROLOGICAL BALLOONS
GS   EXPANDABLE STRUCTURES
      . INFLATABLE STRUCTURES
      . . BALLOONS

METEOROLOGICAL BALLOONS-(CONT.)
      . . . **METEOROLOGICAL BALLOONS**
      . . . . JIMSPHERE BALLOONS
      . . . . ROBIN BALLOONS
RT   DROPSONDES
      HIGH ALTITUDE BALLOONS
      RADIOSONDES
      RAWINSONDES
      ROCKOONS
      SKYHOOK BALLOONS
      SOUNDING
      SUPERPRESSURE BALLOONS
      TETHERED BALLOONS
      UPPER ATMOSPHERE
      WEATHER FORECASTING

METEOROLOGICAL CHARTS
UF   WEATHER CHARTS
      WEATHER MAPS
GS   CHARTS
      . **METEOROLOGICAL CHARTS**
      MAPS
      . **METEOROLOGICAL CHARTS**
RT   ISOBARS (PRESSURE)
      RADAR MAPS
      SYNOPTIC METEOROLOGY

METEOROLOGICAL FLIGHT
RT   AERIAL RECONNAISSANCE
      BALLOON FLIGHT
      ∞FLIGHT
      ICE REPORTING
      ROCKET FLIGHT
      SIRS B SATELLITE
      SOUNDING
      SPACE FLIGHT
      WEATHER FORECASTING

METEOROLOGICAL INSTRUMENTS
GS   MEASURING INSTRUMENTS
      . **METEOROLOGICAL INSTRUMENTS**
      . . BAROMETERS
      . . CLOUD HEIGHT INDICATORS
      . . DROPSONDES
      . . RADIOMETEOROGRAPHS
      . . RADIOSONDES
      . . . ENDORADIOSONDES
      . . . IONOSONDES
      . . . RAWINSONDES
      . . RAIN GAGES
      . . WEATHER DATA RECORDERS
      . . WIND VANES
RT   ANEMOMETERS
      BALLOON-BORNE INSTRUMENTS
      HOT-FILM ANEMOMETERS
      HOT-WIRE ANEMOMETERS
      HUMIDITY MEASUREMENT
      HYGROMETERS
      HYPSOMETERS
      ∞INSTRUMENTS
      LIGHT SCATTERING METERS
      METEOROLOGY
      NEPHANALYSIS
      PSYCHROMETERS
      RECORDING INSTRUMENTS
      ROCKET-BORNE INSTRUMENTS
      SIRS B SATELLITE
      SODAR
      SOUND DETECTING AND RANGING
      SOUNDING ROCKETS
      TRANSDUCERS
      WEATHER RECONNAISSANCE AIRCRAFT
      WEATHER STATIONS

METEOROLOGICAL PARAMETERS
GS   CONSTRAINTS
      . **METEOROLOGICAL PARAMETERS**
      . . BRUNT-VAISALA FREQUENCY
RT   AEROLOGY
      AGROCLIMATOLOGY
      ANNUAL VARIATIONS
      ATMOSPHERIC & OCEANOGRAPHIC
       INFORM SYS
      ATMOSPHERIC CLOUD PHYSICS LAB
       (SPACELAB)
      ATMOSPHERIC TURBULENCE
      CEILINGS (METEOROLOGY)
      CLOUD COVER
      COLD FRONTS
      EQUATORIAL ATMOSPHERE
      FRONTS (METEOROLOGY)
      HUMIDITY
      HYDROLOGY
      ISOTHERMS
      MOISTURE

METEOROLOGICAL PARAMETERS-*(CONT.)*
OCEAN DATA ACQUISITIONS SYSTEMS
OCEANOGRAPHIC PARAMETERS
PRECIPITATION (METEOROLOGY)
STORMS (METEOROLOGY)
TEMPERATURE INVERSIONS
TROPICAL METEOROLOGY
WARM FRONTS
WEATHER
WIND MEASUREMENT

METEOROLOGICAL PROBES
USE   SONDES

METEOROLOGICAL RADAR
UF   WEATHER RADAR
GS   RADAR
      . **METEOROLOGICAL RADAR**
RT   PRECIPITATION PARTICLE
          MEASUREMENT
      PULSE RADAR
      RADAR SCANNING
      RADAR TRACKING
      RADIO METEOROLOGY
      SURVEILLANCE RADAR
      WEATHER FORECASTING

METEOROLOGICAL RESEARCH AIRCRAFT
RT   ∞AIRCRAFT
      DATA ACQUISITION
      RESEARCH AIRCRAFT

METEOROLOGICAL ROCKETS
USE   SOUNDING ROCKETS

METEOROLOGICAL SATELLITES
GS   SATELLITES
      . ARTIFICIAL SATELLITES
      . . **METEOROLOGICAL SATELLITES**
      . . . AEROS SATELLITE
      . . . COSMOS 144 SATELLITE
      . . . D-2 SATELLITES
      . . . ELEKTRON SATELLITES
      . . . . ELEKTRON 1 SATELLITE
      . . . . ELEKTRON 2 SATELLITE
      . . . . ELEKTRON 4 SATELLITE
      . . . EOLE SATELLITES
      . . . ESSA SATELLITES
      . . . . ESSA 1 SATELLITE
      . . . . ESSA 2 SATELLITE
      . . . . ESSA 3 SATELLITE
      . . . . ESSA 4 SATELLITE
      . . . . ESSA 5 SATELLITE
      . . . . ESSA 6 SATELLITE
      . . . . ESSA 7 SATELLITE
      . . . . ESSA 8 SATELLITE
      . . . . ESSA 9 SATELLITE
      . . . EXPLORER 9 SATELLITE
      . . . EXPLORER 17 SATELLITE
      . . . EXPLORER 19 SATELLITE
      . . . GEOLE SATELLITES
      . . . METEOSAT SATELLITE
      . . . NIMBUS SATELLITES
      . . . . NIMBUS 1 SATELLITE
      . . . . NIMBUS 2 SATELLITE
      . . . . NIMBUS 3 SATELLITE
      . . . . NIMBUS 4 SATELLITE
      . . . . NIMBUS 5 SATELLITE
      . . . . NIMBUS 6 SATELLITE
      . . . . NIMBUS 7 SATELLITE
      . . . NOAA SATELLITES
      . . . . NOAA 2 SATELLITE
      . . . . NOAA 3 SATELLITE
      . . . . NOAA 4 SATELLITE
      . . . . NOAA 5 SATELLITE
      . . . . NOAA 6 SATELLITE
      . . . . NOAA 7 SATELLITE
      . . . . NOAA 8 SATELLITE
      . . . SAN MARCO SATELLITES
      . . . . SAN MARCO 1 SATELLITE
      . . . . SAN MARCO 2 SATELLITE
      . . . . SAN MARCO 3 SATELLITE
      . . . SEOCS (SATELLITE)
      . . . SIRS B SATELLITE
      . . . SPUTNIK 1 SATELLITE
      . . . SPUTNIK 2 SATELLITE
      . . . SPUTNIK 3 SATELLITE
      . . . SRET SATELLITES
      . . . . SRET 1 SATELLITE
      . . . . SRET 2 SATELLITE
      . . . SYNCHRONOUS EARTH
              OBSERVATORY SATELLITE
      . . . . SMS 1
      . . . . SMS 2

METEOROLOGICAL SATELLITES-*(CONT.)*
. . . SYNCHRONOUS METEOROLOGICAL
        SATELLITE
. . . . SMS 1
. . . . SMS 2
. . . TIROS SATELLITES
. . . IMPROVED TIROS OPERATIONAL
        SATELLITES
. . . . . ITOS 1
. . . . . ITOS 2
. . . . . ITOS 3
. . . . . ITOS 4
. . . ITOS SATELLITES
. . . . . ITOS 1
. . . . . ITOS 2
. . . . . ITOS 3
. . . . . ITOS 4
. . . . TIROS M
. . . . TIROS N SERIES SATELLITES
. . . . TIROS 1 SATELLITE
. . . . TIROS 2 SATELLITE
. . . . TIROS 3 SATELLITE
. . . . TIROS 4 SATELLITE
. . . . TIROS 5 SATELLITE
. . . . TIROS 6 SATELLITE
. . . . TIROS 7 SATELLITE
. . . . TIROS 8 SATELLITE
. . . . TIROS 9 SATELLITE
. . . . TIROS 10 SATELLITE
. . . VANGUARD 2 SATELLITE
RT   AGRISTARS PROJECT
      ATS
      CLOUD PHOTOGRAPHY
      GEOPHYSICAL SATELLITES
      GOES 1
      GOES 2
      GOES 3
      GOES 4
      GOES 5
      INFRARED PHOTOGRAPHY
      JAPANESE SPACE PROGRAM
      METEOROLOGY
      MILITARY SPACECRAFT
      NAVIGATION SATELLITES
      NIMBUS PROJECT
      NOESS
      SATELLITE OBSERVATION
      SATELLITE SOUNDING
      SATELLITE TELEVISION
      SOUNDING ROCKETS
      SPACE PROBES
      TIROS PROJECT
      UNMANNED SPACECRAFT
      VANGUARD SATELLITES
      WEATHER FORECASTING
      WEATHER STATIONS

METEOROLOGICAL SERVICES
GS   SERVICES
      . **METEOROLOGICAL SERVICES**
RT   AUTOMATIC WEATHER STATIONS
      FLIGHT CONDITIONS
      WEATHER FORECASTING
      WEATHER STATIONS

METEOROLOGICAL SOLENOIDS
RT   BAROCLINITY
      VORTICES

METEOROLOGICAL STATIONS
USE   WEATHER STATIONS

METEOROLOGY
UF   ATMOSPHERIC CONDITIONS
GS   **METEOROLOGY**
      . AEROLOGY
      . AGROMETEOROLOGY
      . ALPINE METEOROLOGY
      . BIOMETEOROLOGY
      . HYDROMETEOROLOGY
      . . MARINE METEOROLOGY
      . MESOMETEOROLOGY
      . MICROMETEOROLOGY
      . NUCLEAR METEOROLOGY
      . POLAR METEOROLOGY
      . RADIO METEOROLOGY
      . SYNOPTIC METEOROLOGY
      . TROPICAL METEOROLOGY
      . WEATHER FORECASTING
      . . LONG RANGE WEATHER
            FORECASTING
      . . NOWCASTING
      . . NUMERICAL WEATHER FORECASTING
      . . STATISTICAL WEATHER
            FORECASTING

METEOROLOGY-*(CONT.)*
RT   ACID RAIN
      ACOUSTIC SOUNDING
      AERONOMY
      AGROCLIMATOLOGY
      AIR LAND INTERACTIONS
      AIR MASSES
      ANNUAL VARIATIONS
      ANTICYCLONES
      ANVIL CLOUDS
      ARC CLOUDS
      ∞ATMOSPHERES
      ATMOSPHERIC & OCEANOGRAPHIC
          INFORM SYS
      ATMOSPHERIC DENSITY
      ATMOSPHERIC PHYSICS
      ATMOSPHERIC TURBULENCE
      BAROCLINIC INSTABILITY
      BRIGHTNESS TEMPERATURE
      CAP CLOUDS
      CEILINGS (METEOROLOGY)
      CIRROCUMULUS CLOUDS
      CIRROSTRATUS CLOUDS
      CIRRUS SHIELDS
      CLIMATE
      CLIMATOLOGY
      CLOUD COVER
      CLOUDS (METEOROLOGY)
      COLD FRONTS
      CONDENSATION NUCLEI
      CONVECTION
      CONVECTION CLOUDS
      CORIOLIS EFFECT
      CYCLONES
      DMSP SATELLITES
      ENVIRONMENTAL ENGINEERING
      ENVIRONMENTAL MONITORING
      FRONTS (METEOROLOGY)
      GARP ATLANTIC TROPICAL EXPERIMENT
      GEOLOGY
      GEOPHYSICS
      GLOBAL ATMOSPHERIC RESEARCH
          PROGRAM
      HAILSTORMS
      HUMIDITY
      HURRICANES
      HYDROCLIMATOLOGY
      HYDRODYNAMIC EQUATIONS
      HYDROGRAPHY
      HYDROLOGY
      INSOLATION
      ISOTHERMS
      MESOSCALE PHENOMENA
      METEOROLOGICAL INSTRUMENTS
      METEOROLOGICAL SATELLITES
      METEOSAT SATELLITE
      METHOD OF CHARACTERISTICS
      MOISTURE
      NATIONAL SEVERE STORMS PROJECT
      NEPHANALYSIS
      NOESS
      OCEANOGRAPHY
      ∞PHYSICAL SCIENCES
      PRECIPITATION (METEOROLOGY)
      ∞SCIENCE
      SEA BREEZE
      SEASONS
      SODAR
      SOUND DETECTING AND RANGING
      STORMS (METEOROLOGY)
      TEMPERATURE
      TEMPERATURE INVERSIONS
      TROPICAL REGIONS
      TROPICAL STORMS
      TYPHOONS
      WARM FRONTS
      WEATHER
      WIND (METEOROLOGY)
      WIND MEASUREMENT
      WORLD METEOROLOGICAL
          ORGANIZATION

METEORS
USE   METEOROIDS

METEOSAT SATELLITE
GS   ESA SPACECRAFT
      . ESA SATELLITES
      . . **METEOSAT SATELLITE**
      SATELLITES
      . ARTIFICIAL SATELLITES
      . . ESA SATELLITES
      . . . **METEOSAT SATELLITE**
      . . METEOROLOGICAL SATELLITES
      . . . **METEOSAT SATELLITE**

METEOSAT SATELLITE-*(CONT.)*
RT CLOUD COVER
    CLOUD PHOTOGRAPHY
    EUROPEAN SPACE AGENCY
    EUROPEAN SPACE PROGRAMS
    FRENCH SATELLITES
    FRENCH SPACE PROGRAMS
    INFRARED PHOTOGRAPHY
    METEOROLOGY
    SATELLITE OBSERVATION
    WEATHER

METERS
USE MEASURING INSTRUMENTS

METHACRYLATE RESINS
USE ACRYLIC RESINS

METHAMPHETAMINE
GS AMINES
    . AMPHETAMINES
    . . **METHAMPHETAMINE**
    DRUGS
    . **METHAMPHETAMINE**

METHANATION
GS CHEMICAL REACTIONS
    . **METHANATION**
RT BIOMASS ENERGY PRODUCTION
    COAL GASIFICATION
    HYDROCARBON FUELS
    HYDROPYROLYSIS

METHANE
GS ALIPHATIC COMPOUNDS
    . ALKANES
    . . **METHANE**
    HYDROCARBONS
    . ALKANES
    . . **METHANE**
RT BIOCONVERSION
    COAL DERIVED GASES
    HYDROCARBON FUELS
    HYDROPYROLYSIS
    LANDFILLS
    LIQUEFIED NATURAL GAS
    NATURAL GAS
    NATURAL GAS EXPLORATION
    NEPTUNE ATMOSPHERE
    OIL FIELDS
    PETROLEUM PRODUCTS
    SYNTHANE
    URANUS ATMOSPHERE

METHIONINE
GS ACIDS
    . AMINO ACIDS
    . . **METHIONINE**
    ORGANIC COMPOUNDS
    . AMINO ACIDS
    . . **METHIONINE**

METHOD OF CHARACTERISTICS
UF CHARACTERISTIC METHOD
RT ∞ CHARACTERISTICS
    COMPRESSIBLE FLUIDS
    FLOW DISTRIBUTION
    HYPERBOLIC FUNCTIONS
    METEOROLOGY
    ∞ METHODOLOGY
    PARTIAL DIFFERENTIAL EQUATIONS
    PLASTIC PROPERTIES
    PRANDTL-MEYER EXPANSION
    STEADY FLOW
    UNSTEADY FLOW

METHOD OF MOMENTS
RT DISTRIBUTION MOMENTS
    INTEGRAL EQUATIONS
    MATHEMATICAL MODELS
    MATRICES (MATHEMATICS)
    ∞ METHODOLOGY
    MOMENT DISTRIBUTION
    MOMENTS
    NUMERICAL ANALYSIS

∞ METHODOLOGY
SN *(USE OF A MORE SPECIFIC TERM IS*
    RECOMMENDED--CONSULT THE TERMS
    LISTED BELOW)
UF METHODS
    TECHNIQUES
RT APPROXIMATION
    ASYMPTOTIC METHODS

METHODOLOGY-*(CONT.)*
    BIOT METHOD
    BOUNDARY INTEGRAL METHOD
    BRIDGMAN METHOD
    CRITICAL PATH METHOD
    CROCCO METHOD
    CZOCHRALSKI METHOD
    DEBYE-SCHERRER METHOD
    DELPHI METHOD (FORECASTING)
    DIGITAL TECHNIQUES
    EMERGENCY BREATHING TECHNIQUES
    ENCKE METHOD
    ENERGY METHODS
    EQUILIBRIUM METHODS
    ETHICS
    FINITE ELEMENT METHOD
    FINITE VOLUME METHOD
    FUJITA METHOD
    GALERKIN METHOD
    GERT
    GLIMM METHOD
    HALPHEN METHOD
    HARTREE-FOCK-SLATER METHOD
    HEURISTIC METHODS
    HILL METHOD
    IMAGING TECHNIQUES
    JACOBI MATRIX METHOD
    KJELDAHL METHOD
    LATIN SQUARE METHOD
    LAUE METHOD
    LEAST SQUARES METHOD
    LIGHTHILL METHOD
    MANAGEMENT METHODS
    MATRIX MANAGEMENT
    MATRIX METHODS
    MAXIMUM ENTROPY METHOD
    MAXWELL-MOHR METHOD
∞ MECHANISM
    METHOD OF CHARACTERISTICS
    METHOD OF MOMENTS
    MILNE METHOD
    MILNE-THOMSON METHOD
    MINIMUM ENTROPY METHOD
    MOIRE EFFECTS
    MONTE CARLO METHOD
    NEWTON-RAPHSON METHOD
    PANEL METHOD (FLUID DYNAMICS)
    PARTICLE IN CELL TECHNIQUE
    PATTERN METHOD (FORECASTING)
    PERCUS METHOD
    POHLHAUSEN METHOD
    PROBE METHOD (FORECASTING)
    PROBLEM SOLVING
    PROFILE METHOD (FORECASTING)
    RAYLEIGH-RITZ METHOD
    RELAXATION METHOD (MATHEMATICS)
    RITZ AVERAGING METHOD
    RULER METHOD
    RUNGE-KUTTA METHOD
    SCHMIDT METHOD
    SCHWARTZ METHOD
    SIMPLEX METHOD
    STEEPEST DESCENT METHOD
    STRAIN ENERGY METHODS
    TRAVELING SOLVENT METHOD
    VAN SLYKE METHOD
    VON ZEIPEL METHOD
    WENTZEL-KRAMER-BRILLOUIN METHOD
    WING FLOW METHOD TESTS

METHODS
USE METHODOLOGY
    PROCEDURES

METHOXY SYSTEMS
RT ALCOHOLS
    ∞ CHEMICAL COMPOUNDS
    INDOLES
    ORGANIC CHEMISTRY
    ORGANIC COMPOUNDS
    PYRROLES
    ∞ SYSTEMS

METHYL ALCOHOLS
GS ALIPHATIC COMPOUNDS
    . **METHYL ALCOHOLS**
    HYDROXYL COMPOUNDS
    . ALCOHOLS
    . . **METHYL ALCOHOLS**
RT KARL FISCHER REAGENT

METHYL CHLORIDE
GS DRUGS
    . ANESTHETICS
    . . **METHYL CHLORIDE**

METHYL CHLORIDE-*(CONT.)*
RT CHLORIDES

METHYL CHLOROSILANES
GS ALIPHATIC COMPOUNDS
    . METHYL COMPOUNDS
    . . **METHYL CHLOROSILANES**
    HYDROGEN COMPOUNDS
    . HYDRIDES
    . . SILANES
    . . . **METHYL CHLOROSILANES**
    SILICON COMPOUNDS
    . SILANES
    . . **METHYL CHLOROSILANES**

METHYL COMPOUNDS
GS ALIPHATIC COMPOUNDS
    . **METHYL COMPOUNDS**
    . . METHYL CHLOROSILANES
    . . METHYL NITRATE
RT ∞ CHEMICAL COMPOUNDS
    TRIMETHYL COMPOUNDS

METHYL NITRATE
GS ALIPHATIC COMPOUNDS
    . ALKYL COMPOUNDS
    . . **METHYL NITRATE**
    . METHYL COMPOUNDS
    . . **METHYL NITRATE**
    NITROGEN COMPOUNDS
    . NITRATES
    . . **METHYL NITRATE**

METHYL POLYSILOXANE
GS SILICON POLYMERS
    . **METHYL POLYSILOXANE**
RT ∞ POLYMERS
    SILICON COMPOUNDS

METHYLATION
GS CHEMICAL REACTIONS
    . **METHYLATION**
RT ALKYLATION

METHYLENE
GS HYDROCARBONS
    . **METHYLENE**
RT DYES
    STAINING

METHYLENE BLUE
GS DYES
    . **METHYLENE BLUE**
    HETEROCYCLIC COMPOUNDS
    . AZINES
    . . **METHYLENE BLUE**
    PYRAZINES
    . AZINES
    . . **METHYLENE BLUE**
RT CHEMICAL ANALYSIS
    CHEMICAL INDICATORS
    ∞ INDICATORS
    STAINING

METHYLENE DIAMINE
GS ALIPHATIC COMPOUNDS
    . **METHYLENE DIAMINE**
    AMINES
    . **METHYLENE DIAMINE**

METHYLHYDRAZINE
GS ALIPHATIC COMPOUNDS
    . HYDRAZINES
    . . **METHYLHYDRAZINE**
RT DIMETHYLHYDRAZINES

METRAZOL
GS DRUGS
    . **METRAZOL**

METRIC CONVERSION
USE METRICATION

METRIC PHOTOGRAPHY
GS PHOTOGRAPHY
    . **METRIC PHOTOGRAPHY**

METRIC SPACE
GS GEOMETRY
    . TOPOLOGY
    . . **METRIC SPACE**
RT BANACH SPACE
    BIMETRIC THEORIES

METRIC SYSTEM
USE   INTERNATIONAL SYSTEM OF UNITS

METRICATION
UF   METRIC CONVERSION
RT   ∞CONVERSION
        INTERNATIONAL SYSTEM OF UNITS
        METROLOGY
        STANDARDIZATION
        UNITS OF MEASUREMENT

METROLOGY
RT   INTERNATIONAL SYSTEM OF UNITS
        ∞MEASUREMENT
        MEASURING INSTRUMENTS
        METRICATION
        STANDARDS
        UNITS OF MEASUREMENT

METROPOLITAN AIRCRAFT
USE   CV-440 AIRCRAFT

METROPOLITAN AREAS
USE   CITIES

MEXICO
GS   NATIONS
        . MEXICO
RT   CHIAPAS (MEXICO)
        COLORADO RIVER (NORTH AMERICA)
        GULF OF CALIFORNIA (MEXICO)
        GULF OF MEXICO
        IMPERIAL VALLEY (CA)
        LEON-QUERETARO AREA (MEXICO)
        LOWER CALIFORNIA (MEXICO)
        NORTH AMERICA
        RIO GRANDE (NORTH AMERICA)
        SAN ANDREAS FAULT
        SOUTHERN CALIFORNIA

MH-262 AIRCRAFT
UF   MAX HOLSTE MH-262 AIRCRAFT
        NORD 262 AIRCRAFT
GS   JET AIRCRAFT
        . TURBOPROP AIRCRAFT
        . . MH-262 AIRCRAFT
        LIGHT AIRCRAFT
        . MH-262 AIRCRAFT
        MONOPLANES
        . MH-262 AIRCRAFT
        NORD AIRCRAFT
        . MH-262 AIRCRAFT
        TRANSPORT AIRCRAFT
        . MH-262 AIRCRAFT
RT   ∞AIRCRAFT
        CARGO AIRCRAFT
        PASSENGER AIRCRAFT

MICA
UF   FLUOROMICA
GS   MINERALS
        . MICA
        . . BIOTITE
        . . FLUOROPHLOGOPITE
        . . MUSCOVITE
RT   IGNEOUS ROCKS
        VERMICULITE

MICARTA
GS   COMPOSITE MATERIALS
        . REINFORCED PLASTICS
        . . MICARTA
        PLASTICS
        . REINFORCED PLASTICS
        . . MICARTA
        SYNTHETIC RESINS
        . . THERMOSETTING RESINS
        . . . PHENOLIC RESINS
        . . . . MICARTA
        RESINS
        . SYNTHETIC RESINS
        . . THERMOSETTING RESINS
        . . . PHENOLIC RESINS
        . . . . MICARTA
RT   FABRICS
        FIBER COMPOSITES
        INSULATION
        ∞POLYMERS

MICE
GS   ANIMALS
        . VERTEBRATES
        . . MAMMALS
        . . . RODENTS

MICE-(CONT.)
        . . . . MICE
        . . . . . JERBOAS
        . . . . . POCKET MICE
RT   RATS

MICHAEL REACTION
GS   CHEMICAL REACTIONS
        . MICHAEL REACTION

MICHAELIS THEORY
RT   ∞THEORIES

MICHELL THEOREM
GS   THEOREMS
        . MICHELL THEOREM
RT   STRESS ANALYSIS
        STRUCTURAL ANALYSIS

MICHELSON INTERFEROMETERS
GS   MEASURING INSTRUMENTS
        . INTERFEROMETERS
        . . MICHELSON INTERFEROMETERS
RT   ASTROPHYSICS
        RADIO ASTRONOMY
        SPECTROMETERS

MICHIGAN
GS   NATIONS
        . UNITED STATES
        . . MICHIGAN
RT   PONTIAC (MI)
        SAGINAW BAY (MI)

MICROANALYSIS
GS   CHEMICAL TESTS
        . CHEMICAL ANALYSIS
        . . MICROANALYSIS
RT   ELECTRON MICROSCOPES
        ELECTRON MICROSCOPY
        ELECTROPHOTOMETRY
        MASS SPECTROMETERS
        ∞MATERIALS TESTS
        NEUTRON ACTIVATION ANALYSIS
        QUALITATIVE ANALYSIS
        QUANTITATIVE ANALYSIS
        SPECTROSCOPIC ANALYSIS
        X RAY ANALYSIS

MICROBALANCES
UF   MICROSCALES
GS   MEASURING INSTRUMENTS
        . INDICATING INSTRUMENTS
        . . WEIGHT INDICATORS
        . . . MICROBALANCES

MICROBALLOONS
GS   EXPANDABLE STRUCTURES
        . INFLATABLE STRUCTURES
        . . BALLOONS
        . . . MICROBALLOONS
RT   GLOBULES
        LASERS
        SPHERES
        TARGETS

MICROBE
USE   MICROORGANISMS

MICROBEAMS
GS   BEAMS (RADIATION)
        . MICROBEAMS
RT   COLLIMATION
        CRYSTALLOGRAPHY
        X RAY ANALYSIS

MICROBIOLOGY
GS   MICROBIOLOGY
        . BACTERIOLOGY
RT   ∞BIOLOGY
        CULTURE TECHNIQUES
        GNOTOBIOTICS

MICROCALORIMETERS
USE   CALORIMETERS

MICROCHANNEL PLATES
UF   MULTICHANNEL PLATES
RT   CHANNEL MULTIPLIERS
        INTEGRATED CIRCUITS
        MICROCHANNELS
        MICROWAVE EQUIPMENT
        PHOTOMULTIPLIER TUBES
        ∞PLATES

MICROCHANNEL PLATES-(CONT.)
        THIN FILMS

MICROCHANNELS
RT   FREQUENCIES
        IMAGE CONVERTERS
        LIGHT AMPLIFIERS
        MICROCHANNEL PLATES
        MULTI-ANODE MICROCHANNEL ARRAYS
        NIGHT VISION
        OPTICAL EQUIPMENT
        PHOTOCATHODES
        ULTRAVIOLET RADIATION

MICROCIRCUITS
USE   MICROELECTRONICS

MICROCLIMATOLOGY
GS   CLIMATOLOGY
        . MICROCLIMATOLOGY
RT   AGROCLIMATOLOGY
        BIOMETEOROLOGY
        MICROMETEOROLOGY

MICROCOMPUTERS
GS   DATA PROCESSING EQUIPMENT
        . COMPUTERS
        . . DIGITAL COMPUTERS
        . . . MICROCOMPUTERS
        . . . . PERSONAL COMPUTERS
RT   MICROPROCESSORS
        MINICOMPUTERS

MICROCRACKS
GS   CRACKS
        . MICROCRACKS
RT   CRACK CLOSURE
        CRACK GEOMETRY
        CRACK INITIATION
        ELBER EQUATION
        SURFACE CRACKS

MICROCRYSTALS
GS   CRYSTALS
        . MICROCRYSTALS
RT   CRYSTALLITES
        SPHERULITES

MICROCYSTIS
GS   ALGAE
        . MICROCYSTIS
        MICROORGANISMS
        . MICROCYSTIS
RT   POLLUTION

MICRODENSITOMETERS
GS   MEASURING INSTRUMENTS
        . DENSITOMETERS
        . . MICRODENSITOMETERS
        . OPTICAL MEASURING INSTRUMENTS
        . . MICRODENSITOMETERS
        OPTICAL EQUIPMENT
        . OPTICAL MEASURING INSTRUMENTS
        . . MICRODENSITOMETERS
RT   GRAVIMETERS
        OPTICAL DENSITY
        OPTICAL MEASUREMENT
        PHOTOMETERS

MICROELECTRONICS
UF   MICROCIRCUITS
GS   MICROELECTRONICS
        . LARGE SCALE INTEGRATION
        . MEDIUM SCALE INTEGRATION
RT   BEAM LEADS
        CIRCUITS
        ∞ELECTRONICS
        ENCAPSULATED MICROCIRCUITS
        ION IMPLANTATION
        LASER MICROSCOPY
        MICROINSTRUMENTATION
        MICROMINIATURIZATION
        MICROMINIATURIZED ELECTRONIC
          DEVICES
        MICROMODULES
        MOLECULAR ELECTRONICS
        PHOTOLITHOGRAPHY
        PHOTOMASKS
        SINGLE EVENT UPSETS
        TRANSISTOR CIRCUITS
        WAFERS

MICROFIBERS
GS   FIBERS

MICROFIBERS-*(CONT.)*
. **MICROFIBERS**

MICROFILMS
GS   PHOTOGRAPHIC FILM
. **MICROFILMS**
RT   DATA RETRIEVAL
DATA STORAGE
MICROPHOTOGRAPHS
READERS
REPRODUCTION (COPYING)

MICROGRAPHY
USE   PHOTOMICROGRAPHY

MICROGRAVITY
USE   REDUCED GRAVITY

∞ **MICROGRAVITY APPLICATIONS**
SN   *(USE OF A MORE SPECIFIC TERM IS
RECOMMENDED--CONSULT THE TERMS
LISTED BELOW)*
RT   BIOPROCESSING
COMMERCE LAB
ELECTROPHORESIS
LOW GRAVITY MANUFACTURING
SPACE COMMERCIALIZATION
SPACE MANUFACTURING
SPACE PROCESSING

MICROHARDNESS
UF   MICROINDENTATION
GS   MECHANICAL PROPERTIES
. HARDNESS
. . **MICROHARDNESS**
RT   KNOOP HARDNESS
ROCKWELL HARDNESS

MICROINDENTATION
USE   MICROHARDNESS

MICROINSTRUMENTATION
RT   MEASURING INSTRUMENTS
MICROELECTRONICS
MICROMINIATURIZATION

MICROMANOMETERS
USE   MANOMETERS

MICROMECHANICS
RT   COMPOSITE MATERIALS
CRACK PROPAGATION
FRACTURE MECHANICS
MECHANICAL PROPERTIES
MICROSTRUCTURE
REINFORCING FIBERS
STRESS CONCENTRATION

MICROMETEORITES
GS   CELESTIAL BODIES
. METEORITES
. . **MICROMETEORITES**
RT   COSMIC DUST
HYPERVELOCITY PROJECTILES
METEOROIDS
MICROMETEOROIDS
TEKTITES
ZODIACAL DUST

MICROMETEOROID EXPLORER SATELLITES
GS   SATELLITES
. ARTIFICIAL SATELLITES
. . EXPLORER SATELLITES
. . . **MICROMETEOROID EXPLORER
SATELLITES**

MICROMETEOROIDS
UF   METEORITIC DUST
MICROMETEORS
GS   CELESTIAL BODIES
. METEOROIDS
. . **MICROMETEOROIDS**
. . . METEOROID DUST CLOUDS
. . . . ZODIACAL DUST
RT   COSMIC DUST
EXPLORER SATELLITES
INTERPLANETARY DUST
METEOR TRAILS
METEORITES
MICROMETEORITES
POYNTING-ROBERTSON EFFECT
SPACE DEBRIS
TERRESTRIAL DUST BELT
ZODIACAL LIGHT

MICROMETEOROLOGY
GS   METEOROLOGY
. **MICROMETEOROLOGY**
RT   AGROMETEOROLOGY
MESOMETEOROLOGY
MICROCLIMATOLOGY
TURBULENCE

MICROMETEORS
USE   MICROMETEOROIDS

MICROMETERS
GS   MEASURING INSTRUMENTS
. **MICROMETERS**
RT   DIMENSIONAL MEASUREMENT
DISTANCE MEASURING EQUIPMENT

MICROMILLIAMMETERS
GS   MEASURING INSTRUMENTS
. AMMETERS
. . **MICROMILLIAMMETERS**
RT   ELECTRIC CURRENT
ELECTRICAL MEASUREMENT
GALVANOMETERS

MICROMINIATURIZATION
GS   MINIATURIZATION
. **MICROMINIATURIZATION**
RT   CIRCUITS
DTL INTEGRATED CIRCUITS
INTEGRATED CIRCUITS
LINEAR INTEGRATED CIRCUITS
MICROELECTRONICS
MICROINSTRUMENTATION
MICROMINIATURIZED ELECTRONIC
DEVICES
MINIATURE ELECTRONIC EQUIPMENT
MOLECULAR ELECTRONICS
SEMICONDUCTOR DEVICES
SUBMINIATURIZATION
THICK FILMS
THIN FILMS
TTL INTEGRATED CIRCUITS
WAFERS

MICROMINIATURIZED ELECTRONIC DEVICES
GS   **MICROMINIATURIZED ELECTRONIC
DEVICES**
. MICROMODULES
RT   MICROELECTRONICS
MICROMINIATURIZATION
MINIATURE ELECTRONIC EQUIPMENT

MICROMODULES
GS   ELECTRONIC EQUIPMENT
. ELECTRONIC MODULES
. . **MICROMODULES**
MICROMINIATURIZED ELECTRONIC
DEVICES
. **MICROMODULES**
MODULES
. ELECTRONIC MODULES
. . **MICROMODULES**
RT   BEAM LEADS
∞CONTAINERS
ELECTRONIC PACKAGING
MICROELECTRONICS
MICROPROCESSORS
MINIATURE ELECTRONIC EQUIPMENT
PHOTOLITHOGRAPHY

MICROMOTORS
SN   (EXCLUDES ROCKET ENGINES)
GS   MOTORS
. ELECTRIC MOTORS
. . **MICROMOTORS**

MICROORGANISMS
UF   MICROBE
GS   **MICROORGANISMS**
. AEROBES
. ANAEROBES
. BACTERIA
. . ACTINOMYCETES
. . AZOTOBACTER
. . BACILLUS
. . CLOSTRIDIUM BOTULINUM
. . ESCHERICHIA
. . HYDROGENOMONAS
. . KLEBSIELLA
. . NITROBACTER
. . PSEUDOMONAS
. . SALMONELLA
. . SARCINA
. . SERRATIA

MICROORGANISMS-*(CONT.)*
. . STAPHYLOCOCCUS
. . STEAROTHERMOPHILUS
. . STREPTOCOCCUS
. . STREPTOMYCETES
. MESOPHILES
. MICROCYSTIS
. PROTOZOA
. . AMOEBA
. . . PELOMYXA
. . FLAGELLATA
. . . TRYPANOSOME
. . PARAMECIA
. . . ROTIFERA
. PSYCHROPHILES
. SAPROPHYTES
. SPORES
. . MICROSPORES
. VIRUSES
. . ADENOVIRUSES
. . BACTERIOPHAGES
RT   ANIMALS
ANTIBIOTICS
GNOTOBIOTICS
INVERTEBRATES
MICROPARTICLES
PLANTS (BOTANY)
POLLUTION
RED TIDE
VIRULENCE

MICROPARTICLES
GS   PARTICLES
. **MICROPARTICLES**
RT   CONDENSATION NUCLEI
FERROFLUIDS
MICROORGANISMS

MICROPHONES
GS   AUDIO EQUIPMENT
. **MICROPHONES**
TRANSDUCERS
. SOUND TRANSDUCERS
. . ELECTROACOUSTIC TRANSDUCERS
. . . **MICROPHONES**
RT   HYDROPHONES
INTERPHONES
MAGNETIC TRANSDUCERS
MONAURAL SIGNALS
TRANSMITTERS
ULTRASONIC WAVE TRANSDUCERS

MICROPHOTOGRAPHS
GS   PHOTOGRAPHS
. **MICROPHOTOGRAPHS**
RT   DATA STORAGE
MICROFILMS
PHOTOGRAPHY
PHOTOMASKS

MICROPHOTOMETERS
USE   PHOTOMETERS

MICROPLASMAS
GS   GASES
. IONIZED GASES
. . CHARGED PARTICLES
. . . **MICROPLASMAS**
PARTICLES
. CHARGED PARTICLES
. . ENERGETIC PARTICLES
. . . PLASMAS (PHYSICS)
. . . . **MICROPLASMAS**

MICROPOLAR FLUIDS
GS   INCOMPRESSIBLE FLUIDS
. **MICROPOLAR FLUIDS**
RT   FLUID MECHANICS
∞FLUIDS
MICROSTRUCTURE

MICROPOROSITY
GS   POROSITY
. **MICROPOROSITY**
RT   MECHANICAL PROPERTIES
METALLOGRAPHY
MICROSTRUCTURE

MICROPROCESSORS
GS   DATA PROCESSING EQUIPMENT
. **MICROPROCESSORS**
. . INTEL 8080 MICROPROCESSOR
RT   COMPUTER DESIGN
COMPUTER STORAGE DEVICES
COMPUTER TECHNIQUES

MICROPROCESSORS-*(CONT.)*
                DATA PROCESSING
                DISTRIBUTED PROCESSING
                FIRMWARE
                INTEGRATED CIRCUITS
                LARGE SCALE INTEGRATION
                MICROCOMPUTERS
                MICROMODULES
                ONBOARD DATA PROCESSING

MICROPROGRAMMING
GS   SOFTWARE ENGINEERING
          . COMPUTER PROGRAMMING
          . . **MICROPROGRAMMING**
RT   FIRMWARE
          ∞ PROGRAMMING

MICROPULSATIONS
GS   PULSES
          . **MICROPULSATIONS**
          . . GEOMAGNETIC MICROPULSATIONS
RT   VARIATIONS

MICROROCKET ENGINES
GS   ENGINES
          . ROCKET ENGINES
          . . **MICROROCKET ENGINES**
          . . . ORBIT MANEUVERING ENGINE
                (SPACE SHUTTLE)
RT   ELECTRIC ROCKET ENGINES
          ELECTROSTATIC ENGINES
          MICROTHRUST
          VERNIER ENGINES

MICROSCALES
USE   MICROBALANCES

MICROSCOPES
GS   **MICROSCOPES**
          . ACOUSTIC MICROSCOPES
          . ELECTRON MICROSCOPES
          . ION MICROSCOPES
          . OPTICAL MICROSCOPES
RT   BINOCULARS
          EYEPIECES
          METALLOGRAPHY
          MICROSCOPY
          OPTICAL EQUIPMENT
          OPTICAL MEASURING INSTRUMENTS
          PHOTOMICROGRAPHY
          ULTRAVIOLET MICROSCOPY

MICROSCOPY
GS   **MICROSCOPY**
          . ELECTRON MICROSCOPY
          . LASER MICROSCOPY
          . ULTRAVIOLET MICROSCOPY
RT   MICROSCOPES
          MICROTOMY
          PHASE CONTRAST
          PHOTOMICROGRAPHY
          SLIDES (MICROSCOPY)

MICROSEISMS
GS   ELASTIC WAVES
          . SEISMIC WAVES
          . . **MICROSEISMS**
RT   CRUSTAL FRACTURES
          EARTHQUAKE DAMAGE
          EARTHQUAKES

MICROSONICS
GS   ACOUSTICS
          . **MICROSONICS**
RT   ELASTIC PROPERTIES
          PIEZOELECTRIC CRYSTALS
          SOUND FIELDS
          SOUND WAVES
          SURFACE WAVES

MICROSPORES
GS   ANIMALS
          . INVERTEBRATES
          . . SPORES
          . . . **MICROSPORES**
          FUNGI
          . SPORES
          . . **MICROSPORES**
          MICROORGANISMS
          . SPORES
          . . **MICROSPORES**
RT   PLANTS (BOTANY)

MICROSTRIP TRANSMISSION LINES
UF   FLAT COAXIAL TRANSMISSION LINES
          PARALLEL STRIP LINES
GS   TRANSMISSION LINES
          . STRIP TRANSMISSION LINES
          . . **MICROSTRIP TRANSMISSION LINES**
RT   DIRECTIONAL COUPLERS
          MICROWAVE TRANSMISSION

MICROSTRUCTURE
GS   **MICROSTRUCTURE**
          . METEORITIC MICROSTRUCTURES
          . WIDMANSTATTEN STRUCTURE
RT   AGING (MATERIALS)
          AGING (METALLURGY)
          AUSTENITE
          BAINITE
          BAUSCHINGER EFFECT
          CAST ALLOYS
          CASTING
          CASTINGS
          CEMENTITE
          CRYSTAL STRUCTURE
          CRYSTALLOGRAPHY
          FERRITES
          GRAIN SIZE
          HARDENING (MATERIALS)
          HEAT TREATMENT
          LAMELLA (METALLURGY)
          MARTENSITE
          METALLOGRAPHY
          MICROMECHANICS
          MICROPOLAR FLUIDS
          MICROPOROSITY
          ORDER-DISORDER TRANSFORMATIONS
          PEARLITE
          PHOTOMICROGRAPHY
          QUENCHING (COOLING)
          SHAPE MEMORY ALLOYS
          SPHERULITES
          ∞ STRUCTURES
          THERMOMECHANICAL TREATMENT
          WORK SOFTENING

MICROTHRUST
GS   THRUST
          . LOW THRUST
          . . **MICROTHRUST**
RT   JET THRUST
          LOW THRUST PROPULSION
          MICROROCKET ENGINES
          ROCKET THRUST
          VARIABLE THRUST

MICROTOMY
RT   MEDICAL EQUIPMENT
          MICROSCOPY

MICROTRONS
GS   PARTICLE ACCELERATORS
          . CYCLOTRONS
          . . **MICROTRONS**
RT   BETATRONS
          SYNCHROTRONS

MICROVISION LANDING AID
GS   DISPLAY DEVICES
          . **MICROVISION LANDING AID**
          LANDING AIDS
          . **MICROVISION LANDING AID**

MICROWAVE AMPLIFIERS
GS   AMPLIFIERS
          . **MICROWAVE AMPLIFIERS**
          MICROWAVE EQUIPMENT
          . **MICROWAVE AMPLIFIERS**
RT   CROSSED FIELD AMPLIFIERS
          INTERSTELLAR MASERS
          MASERS
          PARAMETRIC AMPLIFIERS
          TRANSFERRED ELECTRON DEVICES

MICROWAVE ANTENNAS
GS   ANTENNAS
          . RADIO ANTENNAS
          . . **MICROWAVE ANTENNAS**
          . . . SPACETENNAS
          MICROWAVE EQUIPMENT
          . **MICROWAVE ANTENNAS**
          . . HORN ANTENNAS
          . . LENS ANTENNAS
          . . RECTENNAS
          . . SLOT ANTENNAS
          . . SPACETENNAS
RT   AIRCRAFT ANTENNAS

MICROWAVE ANTENNAS-*(CONT.)*
                ANTENNA ARRAYS
                DIRECTIONAL ANTENNAS
                GREGORIAN ANTENNAS
                HELICAL ANTENNAS
                MISSILE ANTENNAS
                OMNIDIRECTIONAL ANTENNAS
                PARABOLIC ANTENNAS
                PARABOLIC REFLECTORS
                RADAR ANTENNAS
                WAVEGUIDE ANTENNAS

MICROWAVE ATTENUATION
GS   ATTENUATION
          . **MICROWAVE ATTENUATION**
          TRANSMISSION
          . ELECTROMAGNETIC WAVE
                TRANSMISSION
          . . RADIO TRANSMISSION
          . . . **MICROWAVE ATTENUATION**
          . SIGNAL TRANSMISSION
          . . RADIO TRANSMISSION
          . . . **MICROWAVE ATTENUATION**
RT   WAVE PROPAGATION

MICROWAVE CIRCUITS
GS   CIRCUITS
          . **MICROWAVE CIRCUITS**

MICROWAVE COUPLING
GS   COUPLING
          . **MICROWAVE COUPLING**
RT   ANTENNA COUPLERS
          CROSS COUPLING
          DIRECTIONAL ANTENNAS
          DIRECTIONAL COUPLERS
          OPTICAL COUPLING

MICROWAVE EMISSION
GS   DECAY
          . EMISSION
          . . **MICROWAVE EMISSION**
          ELECTROMAGNETIC RADIATION
          . RADIO WAVES
          . . SHORT WAVE RADIATION
          . . . MICROWAVES
          . . . . **MICROWAVE EMISSION**
RT   COSMIC NOISE
          EXTRATERRESTRIAL RADIATION
          EXTRATERRESTRIAL RADIO WAVES
          LINEAR POLARIZATION
          STELLAR RADIATION

MICROWAVE EQUIPMENT
GS   **MICROWAVE EQUIPMENT**
          . GYRATORS
          . . MICROWAVE FILTERS
          . MICROWAVE AMPLIFIERS
          . MICROWAVE ANTENNAS
          . . HORN ANTENNAS
          . . LENS ANTENNAS
          . . RECTENNAS
          . . SLOT ANTENNAS
          . . SPACETENNAS
          . MICROWAVE INTERFEROMETERS
          . MICROWAVE PROBES
          . . MICROWAVE PLASMA PROBES
          . MICROWAVE RADIOMETERS
          . MICROWAVE SCANNING BEAM
                LANDING SYSTEM
          . MICROWAVE TUBES
          . . COLD CATHODE TUBES
          . . . PHOTOTUBES
          . . . . PHOTOMULTIPLIER TUBES
          . . IMAGE TUBES
          . . . THERMICONS
          . . KLYSTRONS
          . . MAGNETRONS
          . . . NIGOTRONS
          . . MICROWAVE OSCILLATORS
          . . . BACKWARD WAVE TUBES
          . . MONOSCOPES
          . . ORTHICONS
          . . . IMAGE ORTHICONS
          . . PLANOTRONS
          . . . CARCINOTRONS
          . . . CELESCOPES
          . . . HELITRONS
          . . . IGNITRONS
          . . . THERMICONS
          . . . THERMIONIC DIODES
          . . . THYRATRONS
          . . TRAVELING WAVE TUBES
          . . . CARCINOTRONS
          . . . HELITRONS

MICROWAVE EQUIPMENT-*(CONT.)*
. VIDICONS
. . THERMICONS
RT   GAS DISCHARGE TUBES
     MICROCHANNEL PLATES

MICROWAVE FILTERS
GS   ELECTROMAGNETIC WAVE FILTERS
. ELECTRIC FILTERS
. . **MICROWAVE FILTERS**
    MICROWAVE EQUIPMENT
. GYRATORS
. . **MICROWAVE FILTERS**
RT   BANDPASS FILTERS
    BANDSTOP FILTERS
    DIGITAL FILTERS
    FIR FILTERS
    HIGH PASS FILTERS
    LOW PASS FILTERS
    RADAR FILTERS
    RADIO FILTERS
    RECTANGULAR WAVEGUIDES
    WAVEGUIDE FILTERS

MICROWAVE FREQUENCIES
GS   FREQUENCIES
. RADIO FREQUENCIES
. . **MICROWAVE FREQUENCIES**
. . . C BAND
. . . EXTREMELY HIGH FREQUENCIES
. . . P BAND
. . . SUPERHIGH FREQUENCIES
RT   ACOUSTIC MICROSCOPES
    CENTIMETER WAVES
    MICROWAVES
    PASSIVE L-BAND RADIOMETERS
    PRAETERSONIC DEVICES

MICROWAVE HOLOGRAPHY
GS   IMAGERY
. HOLOGRAPHY
. . **MICROWAVE HOLOGRAPHY**
    PHOTOGRAPHY
. HOLOGRAPHY
. . **MICROWAVE HOLOGRAPHY**
RT   IMAGING TECHNIQUES
    MICROWAVES
    WAVE FRONT RECONSTRUCTION

MICROWAVE IMAGERY
GS   IMAGERY
. **MICROWAVE IMAGERY**
RT   RADARSCOPES
    SYNTHETIC APERTURE RADAR
    X RAY IMAGERY

MICROWAVE INTERFEROMETERS
GS   MEASURING INSTRUMENTS
. INTERFEROMETERS
. . **MICROWAVE INTERFEROMETERS**
    MICROWAVE EQUIPMENT
. **MICROWAVE INTERFEROMETERS**
RT   FABRY-PEROT INTERFEROMETERS
    PLASMA DIAGNOSTICS

MICROWAVE LANDING SYSTEMS
GS   LANDING AIDS
. **MICROWAVE LANDING SYSTEMS**
. . MICROWAVE SCANNING BEAM
      LANDING SYSTEM
RT   AIR TRAFFIC CONTROL
    AIRCRAFT LANDING
    AIRCRAFT SAFETY
    APPROACH CONTROL
    AUTOMATED EN ROUTE ATC
    AUTOMATIC LANDING CONTROL
  ∞SYSTEMS

MICROWAVE OSCILLATORS
GS   ELECTRON TUBES
. VACUUM TUBES
. . VACUUM TUBE OSCILLATORS
. . . MICROWAVE TUBES
. . . . **MICROWAVE OSCILLATORS**
    MICROWAVE EQUIPMENT
. MICROWAVE TUBES
. . **MICROWAVE OSCILLATORS**
. . . BACKWARD WAVE TUBES
    OSCILLATORS
. VACUUM TUBE OSCILLATORS
. MICROWAVE TUBES
. . . **MICROWAVE OSCILLATORS**
RT   BARRITT DIODES
    GAS DISCHARGE TUBES
    TRANSFERRED ELECTRON DEVICES

MICROWAVE OSCILLATORS-*(CONT.)*
    VOLTAGE CONTROLLED OSCILLATORS

MICROWAVE PHOTOGRAPHY
GS   IMAGERY
. **MICROWAVE PHOTOGRAPHY**
    PHOTOGRAPHY
. **MICROWAVE PHOTOGRAPHY**
RT   RADAR DATA
    RADAR PHOTOGRAPHY
    RADARSCOPES

MICROWAVE PLASMA PROBES
GS   MEASURING INSTRUMENTS
. MICROWAVE PROBES
. . **MICROWAVE PLASMA PROBES**
    MICROWAVE EQUIPMENT
. MICROWAVE PROBES
. . **MICROWAVE PLASMA PROBES**
RT   ELECTRON PROBES
    PLASMA FLUX MEASUREMENT
    PLASMAGUIDES
    PLASMAS (PHYSICS)
    RESONANCE PROBES

MICROWAVE PROBES
GS   MEASURING INSTRUMENTS
. **MICROWAVE PROBES**
. . MICROWAVE PLASMA PROBES
    MICROWAVE EQUIPMENT
. **MICROWAVE PROBES**
. . MICROWAVE PLASMA PROBES
RT   RADIO FREQUENCY IMPEDANCE
     PROBES

MICROWAVE RADIATION
USE MICROWAVES

MICROWAVE RADIOMETERS
GS   MEASURING INSTRUMENTS
. RADIATION MEASURING INSTRUMENTS
. . ACTINOMETERS
. . . RADIOMETERS
. . . . **MICROWAVE RADIOMETERS**
    MICROWAVE EQUIPMENT
. **MICROWAVE RADIOMETERS**

MICROWAVE REFLECTOMETERS
GS   MEASURING INSTRUMENTS
. OPTICAL MEASURING INSTRUMENTS
. . REFLECTOMETERS
. . . **MICROWAVE REFLECTOMETERS**
    OPTICAL EQUIPMENT
. OPTICAL MEASURING INSTRUMENTS
. . REFLECTOMETERS
. . . **MICROWAVE REFLECTOMETERS**
RT   KINEMATICS

MICROWAVE RESONANCE
GS   RESONANCE
. **MICROWAVE RESONANCE**
RT   CAVITY RESONATORS
    HARMONIC ANALYSIS
    NONRESONANCE

MICROWAVE SCANNING BEAM LANDING SYSTEM
UF   MSBLS
GS   LANDING AIDS
. MICROWAVE LANDING SYSTEMS
. . **MICROWAVE SCANNING BEAM
      LANDING SYSTEM**
    MICROWAVE EQUIPMENT
. **MICROWAVE SCANNING BEAM
     LANDING SYSTEM**
    NAVIGATION AIDS
. **MICROWAVE SCANNING BEAM
     LANDING SYSTEM**
RT   APPROACH INDICATORS
    SPACE SHUTTLE ORBITERS
  ∞SYSTEMS

MICROWAVE SCATTERING
GS   SCATTERING
. WAVE SCATTERING
. . ELECTROMAGNETIC SCATTERING
. . . **MICROWAVE SCATTERING**
RT   ATMOSPHERIC SCATTERING
    MANDELSTAM REPRESENTATION
    SCATTEROMETERS

MICROWAVE SENSORS
GS   MEASURING INSTRUMENTS
. INDICATING INSTRUMENTS
. . **MICROWAVE SENSORS**

MICROWAVE SENSORS-*(CONT.)*
RT ∞INSTRUMENTS
    RADAR RECEIVERS
  ∞SENSORS
    SIGNAL DETECTORS
    SYNTHETIC APERTURE RADAR

MICROWAVE SOUNDING
GS   SOUNDING
. **MICROWAVE SOUNDING**
RT   IMAGERY
    MICROWAVES
    ROCKET SOUNDING

MICROWAVE SPECTRA
UF   INTERSTELLAR MICROWAVE SPECTRA
GS   SPECTRA
. RADIATION SPECTRA
. . ELECTROMAGNETIC SPECTRA
. . . RADIO SPECTRA
. . . . **MICROWAVE SPECTRA**
RT   ABSORPTION SPECTRA
    INFRARED SPECTRA
    MOLECULAR ROTATION
    MOLECULAR SPECTRA
    MOLECULAR SPECTROSCOPY

MICROWAVE SPECTROMETERS
GS   MEASURING INSTRUMENTS
. SPECTROMETERS
. . **MICROWAVE SPECTROMETERS**

MICROWAVE SWITCHING
GS   SWITCHING
. **MICROWAVE SWITCHING**
RT   FERROELECTRICITY
    GYRATORS
    PACKET SWITCHING
    PHASE SHIFT
    SWITCHING CIRCUITS
    WAVEGUIDES

MICROWAVE TRANSMISSION
GS   TRANSMISSION
. ELECTROMAGNETIC WAVE
     TRANSMISSION
. . RADIO TRANSMISSION
. . . **MICROWAVE TRANSMISSION**
. SIGNAL TRANSMISSION
. RADIO TRANSMISSION
. . **MICROWAVE TRANSMISSION**
RT   CIRCULAR WAVEGUIDES
    DOMESTIC SATELLITE COMMUNICATIONS
     SYSTEMS
    DOWNLINKING
    FLEET SATELLITE COMMUNICATION
     SYSTEM
    FREQUENCY REUSE
    MICROSTRIP TRANSMISSION LINES
    SATELLITE SOLAR ENERGY
     CONVERSION
    SATELLITE SOLAR POWER STATIONS
    SPACETENNAS
    TELETYPEWRITER SYSTEMS
    UPLINKING
    WAVE PROPAGATION

MICROWAVE TUBES
GS   ELECTRON TUBES
. VACUUM TUBES
. . VACUUM TUBE OSCILLATORS
. . . **MICROWAVE TUBES**
. . . . KLYSTRONS
. . . . MAGNETRONS
. . . . . NIGOTRONS
. . . . MICROWAVE OSCILLATORS
. . . . PLANOTRONS
. . . . . CARCINOTRONS
. . . . . CELESCOPES
. . . . . HELITRONS
. . . . . IGNITRONS
. . . . . THERMICONS
. . . . . THERMIONIC DIODES
. . . . . THYRATRONS
. . . . TRAVELING WAVE TUBES
. . . . . CARCINOTRONS
. . . . . HELITRONS
    MICROWAVE EQUIPMENT
. **MICROWAVE TUBES**
. . COLD CATHODE TUBES
. . PHOTOTUBES
. . . PHOTOMULTIPLIER TUBES
. . IMAGE TUBES
. . . THERMICONS
. . KLYSTRONS

MICROWAVE TUBES-*(CONT.)*
```
       . . MAGNETRONS
       . . . NIGOTRONS
       . . MICROWAVE OSCILLATORS
       . . . BACKWARD WAVE TUBES
       . . MONOSCOPES
       . . ORTHICONS
       . . . IMAGE ORTHICONS
       . . PLANOTRONS
       . . . CARCINOTRONS
       . . . CELESCOPES
       . . . HELITRONS
       . . IGNITRONS
       . . . THERMICONS
       . . . THERMIONIC DIODES
       . . THYRATRONS
       . . TRAVELING WAVE TUBES
       . . . CARCINOTRONS
       . . . HELITRONS
       OSCILLATORS
       . VACUUM TUBE OSCILLATORS
       . . MICROWAVE TUBES
       . . . MAGNETRONS
       . . . . NIGOTRONS
       . . . MICROWAVE OSCILLATORS
       . . . PLANOTRONS
       . . . . CARCINOTRONS
       . . . . HELITRONS
RT     GAS DISCHARGE TUBES
       TRIODES
       ∞TUBES
```

MICROWAVES
```
UF     MICROWAVE RADIATION
GS     ELECTROMAGNETIC RADIATION
       . RADIO WAVES
       . . SHORT WAVE RADIATION
       . . . MICROWAVES
       . . . . CENTIMETER WAVES
       . . . . DECIMETER WAVES
       . . . . MICROWAVE EMISSION
       . . . . MILLIMETER WAVES
RT     COSMIC NOISE
       ELECTROMAGNETIC NOISE
       EXTRATERRESTRIAL RADIO WAVES
       INFRARED RADIATION
       MICROWAVE FREQUENCIES
       MICROWAVE HOLOGRAPHY
       MICROWAVE SOUNDING
       ∞RADIATION
       SATELLITE SOLAR ENERGY
          CONVERSION
       SATELLITE SOLAR POWER STATIONS
       SCATTEROMETERS
       SUBMILLIMETER WAVES
       WHISTLERS
```

MICROWEIGHING
```
USE    WEIGHT MEASUREMENT
```

MICROYIELD STRENGTH
```
GS     MECHANICAL PROPERTIES
       . YIELD STRENGTH
       . . MICROYIELD STRENGTH
RT     ELASTIC PROPERTIES
       ∞STRENGTH
       STRESSES
       YIELD POINT
```

MICTURITION
```
USE    URINATION
```

MIDAIR COLLISIONS
```
GS     COLLISIONS
       . MIDAIR COLLISIONS
       . . BIRD-AIRCRAFT COLLISIONS
RT     AIR TRAFFIC CONTROL
       AIRCRAFT ACCIDENTS
       AIRCRAFT HAZARDS
       AIRCRAFT SAFETY
       BEACON COLLISION AVOIDANCE
          SYSTEM
       COLLISION AVOIDANCE
       CRASHES
       FLIGHT HAZARDS
       FLIGHT SAFETY
       PILOT ERROR
       THREAT EVALUATION
```

MIDALTITUDE
```
GS     ALTITUDE
       . MIDALTITUDE
RT     FLIGHT ALTITUDE
       HIGH ALTITUDE
       LOW ALTITUDE
```

MIDAS SATELLITES
```
GS     MILITARY SPACECRAFT
       . RECONNAISSANCE SPACECRAFT
       . . MIDAS SATELLITES
       . . . MIDAS 2 SATELLITE
       . . . MIDAS 3 SATELLITE
       . . . MIDAS 4 SATELLITE
       . . . MIDAS 5 SATELLITE
       . . . MIDAS 6 SATELLITE
       . . . MIDAS 7 SATELLITE
       SATELLITES
       . ARTIFICIAL SATELLITES
       . . MIDAS SATELLITES
       . . . MIDAS 2 SATELLITE
       . . . MIDAS 3 SATELLITE
       . . . MIDAS 4 SATELLITE
       . . . MIDAS 5 SATELLITE
       . . . MIDAS 6 SATELLITE
       . . . MIDAS 7 SATELLITE
RT     ATLAS AGENA B LAUNCH VEHICLE
```

MIDAS 2 SATELLITE
```
GS     MILITARY SPACECRAFT
       . RECONNAISSANCE SPACECRAFT
       . . MIDAS SATELLITES
       . . . MIDAS 2 SATELLITE
       SATELLITES
       . ARTIFICIAL SATELLITES
       . . MIDAS SATELLITES
       . . . MIDAS 2 SATELLITE
```

MIDAS 3 SATELLITE
```
GS     MILITARY SPACECRAFT
       . RECONNAISSANCE SPACECRAFT
       . . MIDAS SATELLITES
       . . . MIDAS 3 SATELLITE
       SATELLITES
       . ARTIFICIAL SATELLITES
       . . MIDAS SATELLITES
       . . . MIDAS 3 SATELLITE
```

MIDAS 4 SATELLITE
```
GS     MILITARY SPACECRAFT
       . RECONNAISSANCE SPACECRAFT
       . . MIDAS SATELLITES
       . . . MIDAS 4 SATELLITE
       SATELLITES
       . ARTIFICIAL SATELLITES
       . . MIDAS SATELLITES
       . . . MIDAS 4 SATELLITE
```

MIDAS 5 SATELLITE
```
GS     MILITARY SPACECRAFT
       . RECONNAISSANCE SPACECRAFT
       . . MIDAS SATELLITES
       . . . MIDAS 5 SATELLITE
       SATELLITES
       . ARTIFICIAL SATELLITES
       . . MIDAS SATELLITES
       . . . MIDAS 5 SATELLITE
```

MIDAS 6 SATELLITE
```
GS     MILITARY SPACECRAFT
       . RECONNAISSANCE SPACECRAFT
       . . MIDAS SATELLITES
       . . . MIDAS 6 SATELLITE
       SATELLITES
       . ARTIFICIAL SATELLITES
       . . MIDAS SATELLITES
       . . . MIDAS 6 SATELLITE
```

MIDAS 7 SATELLITE
```
GS     MILITARY SPACECRAFT
       . RECONNAISSANCE SPACECRAFT
       . . MIDAS SATELLITES
       . . . MIDAS 7 SATELLITE
       SATELLITES
       . ARTIFICIAL SATELLITES
       . . MIDAS SATELLITES
       . . . MIDAS 7 SATELLITE
```

MIDCOURSE GUIDANCE
```
GS     GUIDANCE (MOTION)
       . MIDCOURSE GUIDANCE
RT     COMMAND GUIDANCE
       INERTIAL GUIDANCE
       INJECTION GUIDANCE
       RENDEZVOUS GUIDANCE
       SPACECRAFT GUIDANCE
       TERMINAL GUIDANCE
       TRANSEARTH INJECTION
       TRANSLUNAR INJECTION
```

MIDCOURSE TRAJECTORIES
```
GS     TRAJECTORIES
```

MIDCOURSE TRAJECTORIES-*(CONT.)*
```
       . MIDCOURSE TRAJECTORIES
RT     ASCENT TRAJECTORIES
       BALLISTIC TRAJECTORIES
       COASTING FLIGHT
       DESCENT TRAJECTORIES
       PARABOLIC FLIGHT
```

MIDDLE ATMOSPHERE
```
GS     EARTH ATMOSPHERE
       . HOMOSPHERE
       . . MIDDLE ATMOSPHERE
       . . . CHEMOSPHERE
       . . . STRATOSPHERE
RT     AIR
       AIR POLLUTION
       ∞ATMOSPHERES
       ATMOSPHERIC CHEMISTRY
       ATMOSPHERIC CIRCULATION
       ATMOSPHERIC COMPOSITION
       CLIMATOLOGY
       EQUATORIAL ATMOSPHERE
       FREE ATMOSPHERE
       HETEROSPHERE
       LOWER ATMOSPHERE
       MIDLATITUDE ATMOSPHERE
       STRATOPAUSE
       TROPOPAUSE
       UPPER ATMOSPHERE
```

MIDDLE EAR
```
GS     ANATOMY
       . SENSE ORGANS
       . . EAR
       . . . MIDDLE EAR
RT     SEMICIRCULAR CANALS
```

MIDDLE EAR PRESSURE
```
GS     PRESSURE
       . MIDDLE EAR PRESSURE
RT     EAR PRESSURE TEST
       EARDRUMS
```

MIDLATITUDE ATMOSPHERE
```
GS     EARTH ATMOSPHERE
       . MIDLATITUDE ATMOSPHERE
       ENVIRONMENTS
       . MIDLATITUDE ATMOSPHERE
RT     IONOSPHERE
       MIDDLE ATMOSPHERE
       SPORADIC E LAYER
```

MIDLATITUDES
```
USE    TEMPERATE REGIONS
```

MIE SCATTERING
```
UF     MIE THEORY
GS     SCATTERING
       . WAVE SCATTERING
       . . ELECTROMAGNETIC SCATTERING
       . . . MIE SCATTERING
       . . . . RAYLEIGH SCATTERING
```

MIE THEORY
```
USE    MIE SCATTERING
```

MIG AIRCRAFT
```
GS     ATTACK AIRCRAFT
       . FIGHTER AIRCRAFT
       . . MIG AIRCRAFT
       SUPERSONIC AIRCRAFT
       . MIG AIRCRAFT
RT     ∞AIRCRAFT
```

MIGRATION
```
RT     BEHAVIOR
       PHENOLOGY
       WATERFOWL
```

MIL AIRCRAFT
```
RT     ∞AIRCRAFT
```

MILANKOVITCH THEORY
```
USE    CLIMATOLOGY
```

MILIARIA
```
GS     DISEASES
       . MILIARIA
RT     LESIONS
       SWEAT
```

MILITARY AIR FACILITIES
```
UF     AIRCRAFT BASES
RT     AIR TRAFFIC CONTROL
```

MILITARY AIR FACILITIES-(CONT.)
∞ AIRCRAFT
   AIRCRAFT CARRIERS
   AIRPORTS
∞ FACILITIES
∞ FIELDS
   HANGARS
   HELIPORTS
   LANDING AIDS
   LANDING MATS
   NAVIGATION AIDS
   STATIONS

∞ MILITARY AIRCRAFT
SN *(USE OF A MORE SPECIFIC TERM IS RECOMMENDED--CONSULT THE TERMS LISTED BELOW)*
UF BOEING MILITARY AIRCRAFT
   CESSNA MILITARY AIRCRAFT
   CHANCE-VOUGHT MILITARY AIRCRAFT
   CONVAIR MILITARY AIRCRAFT
   CURTISS-WRIGHT MILITARY AIRCRAFT
   DOUGLAS MILITARY AIRCRAFT
   FAIRCHILD MILITARY AIRCRAFT
   GENERAL DYNAMICS MILITARY AIRCRAFT
   HILLER MILITARY AIRCRAFT
   HUGHES MILITARY AIRCRAFT
   REPUBLIC MILITARY AIRCRAFT
RT A-37 AIRCRAFT
   AH-1G HELICOPTER
   AH-63 HELICOPTER
   AH-64 HELICOPTER
∞ AIRCRAFT
   AIRCRAFT CARRIERS
   AIRCRAFT SURVIVABILITY
   AIRSHIPS
   ALPHA JET AIRCRAFT
   ANTISUBMARINE WARFARE AIRCRAFT
   ARMED FORCES
   ARMED FORCES (FOREIGN)
   ARMED FORCES (UNITED STATES)
   ATTACK AIRCRAFT
   ATTACKING (ASSAULTING)
   AWACS AIRCRAFT
   B-1 AIRCRAFT
   BOMBER AIRCRAFT
   C-1A AIRCRAFT
   CARGO AIRCRAFT
   CH-62 HELICOPTER
   CL-600 CHALLENGER AIRCRAFT
   DRONE AIRCRAFT
   DRONE VEHICLES
   E-2 AIRCRAFT
   E-3A AIRCRAFT
   E-4A AIRCRAFT
   FIGHTER AIRCRAFT
   FIREBEE 2 TARGET DRONE AIRCRAFT
   FV-12A AIRCRAFT
   GLIDERS
   GROUND EFFECT MACHINES
   H-60 HELICOPTER
   HARRIER AIRCRAFT
   HELICOPTERS
   JAGUAR AIRCRAFT
   JET AIRCRAFT
   LIGHT AIRCRAFT
∞ MILITARY AVIATION
   MILITARY HELICOPTERS
   MRCA AIRCRAFT
   NUCLEAR PROPELLED AIRCRAFT
   OBSERVATION AIRCRAFT
   PANAVIA MILITARY AIRCRAFT
   PASSENGER AIRCRAFT
   PILOTLESS AIRCRAFT
   RECONNAISSANCE AIRCRAFT
   RESEARCH AIRCRAFT
   ROTARY WING AIRCRAFT
   S-3 AIRCRAFT
   SHORT TAKEOFF AIRCRAFT
   SUBMERSIBLE AIRCRAFT
   TAILLESS AIRCRAFT
   TANKER AIRCRAFT
   TARGET DRONE AIRCRAFT
   TERRAIN FOLLOWING AIRCRAFT
   TRAINING AIRCRAFT
   TRANSPORT AIRCRAFT
   UTILITY AIRCRAFT
   V/STOL AIRCRAFT
   VERTICAL TAKEOFF AIRCRAFT
   WEAPON SYSTEMS
   YC-14 AIRCRAFT
   YF-12 AIRCRAFT
   YF-16 AIRCRAFT

∞ MILITARY AVIATION
SN *(USE OF A MORE SPECIFIC TERM IS RECOMMENDED--CONSULT THE TERMS LISTED BELOW)*
RT ∞ AERONAUTICS
   AIR LAW
   ARMED FORCES
   BOMBER AIRCRAFT
   FIGHTER AIRCRAFT
∞ MILITARY AIRCRAFT
   RECONNAISSANCE AIRCRAFT

MILITARY COMPACT REACTORS
UF MCR REACTORS
GS NUCLEAR REACTORS
  . LIQUID COOLED REACTORS
  .. LIQUID METAL COOLED REACTORS
  ... **MILITARY COMPACT REACTORS**
  . NUCLEAR RESEARCH AND TEST REACTORS
  .. **MILITARY COMPACT REACTORS**

MILITARY HELICOPTERS
GS V/STOL AIRCRAFT
  . ROTARY WING AIRCRAFT
  .. HELICOPTERS
  ... **MILITARY HELICOPTERS**
  .... AH-1G HELICOPTER
  .... AH-64 HELICOPTER
  .... CH-3 HELICOPTER
  .... CH-34 HELICOPTER
  .... CH-46 HELICOPTER
  .... CH-47 HELICOPTER
  .... CH-54 HELICOPTER
  .... H-43 HELICOPTER
  .... H-53 HELICOPTER
  .... H-56 HELICOPTER
  .... H-60 HELICOPTER
  .... HEAVY LIFT HELICOPTERS
  ..... CH-62 HELICOPTER
  .... HH-43 HELICOPTER
  .... OH-4 HELICOPTER
  .... OH-5 HELICOPTER
  .... OH-6 HELICOPTER
  .... OH-13 HELICOPTER
  .... OH-23 HELICOPTER
  .... OH-58 HELICOPTER
  .... QH-50 HELICOPTER
  .... S-58 HELICOPTER
  .... S-61 HELICOPTER
  .... SA-330 HELICOPTER
  .... UH-1 HELICOPTER
  .... UH-2 HELICOPTER
  .... UH-34 HELICOPTER
  .... UH-60A HELICOPTER
  .... UH-61A HELICOPTER
RT ∞ AIRCRAFT
   ATTACK AIRCRAFT
∞ MILITARY AIRCRAFT
   RECONNAISSANCE AIRCRAFT

MILITARY OPERATIONS
GS **MILITARY OPERATIONS**
  . COMBAT
  . ELECTRONIC WARFARE
RT DEPLOYMENT
   TACTICS
   TANKS (COMBAT VEHICLES)

MILITARY PSYCHIATRY
USE MILITARY PSYCHOLOGY

MILITARY PSYCHOLOGY
UF MILITARY PSYCHIATRY
GS PSYCHOLOGY
  . **MILITARY PSYCHOLOGY**
RT AVIATION PSYCHOLOGY
   PSYCHIATRY
   PSYCHOLOGICAL EFFECTS
   PSYCHOLOGICAL TESTS
   PSYCHOMETRICS
   SPACE PSYCHOLOGY

MILITARY SPACECRAFT
GS **MILITARY SPACECRAFT**
  . DMSP SATELLITES
  . RECONNAISSANCE SPACECRAFT
  .. INSPECTOR SATELLITE
  .. MIDAS SATELLITES
  ... MIDAS 2 SATELLITE
  ... MIDAS 3 SATELLITE
  ... MIDAS 4 SATELLITE
  ... MIDAS 5 SATELLITE
  ... MIDAS 6 SATELLITE
  ... MIDAS 7 SATELLITE

MILITARY SPACECRAFT-(CONT.)
  .. PHOTO RECONNAISSANCE SPACECRAFT
  . VELA SATELLITES
RT AEROSPACEPLANES
   ARMED FORCES
   ARTIFICIAL SATELLITES
   ASTROPLANE
   EVASIVE SATELLITES
   MANNED SPACECRAFT
   METEOROLOGICAL SATELLITES
∞ MILITARY VEHICLES
   NAVIGATION SATELLITES
   RECOVERABLE SPACECRAFT
   RENDEZVOUS SPACECRAFT
   SATELLITE NETWORKS
   SATELLITES
   SPACE STATIONS
   SPACE SURVEILLANCE (SPACEBORNE)
∞ SPACECRAFT
   SYNCHRONOUS SATELLITES
   UNMANNED SPACECRAFT
   WEAPON SYSTEMS

MILITARY TECHNOLOGY
GS TECHNOLOGIES
  . **MILITARY TECHNOLOGY**
RT ANTIMISSILE DEFENSE
   ANTIRADIATION MISSILES
   ANTISUBMARINE WARFARE
   ARMED FORCES (FOREIGN)
   ARMED FORCES (UNITED STATES)
   ARMY-NAVY INSTRUMENTATION PROGRAM
   AWACS AIRCRAFT
   BALLISTIC MISSILE EARLY WARNING SYSTEM
   DEFENSE COMMUNICATIONS SYSTEM (DCS)
   DEFENSE INDUSTRY
   DEFENSE PROGRAM
   DEPLOYMENT
   E-2 AIRCRAFT
   E-3A AIRCRAFT
   E-4A AIRCRAFT
   FLEET SATELLITE COMMUNICATION SYSTEM
   INTELLIGENCE
   INTERSERVICE DATA EXCHANGE PROGRAM
   LASER WEAPONS
   LOGISTICS OVER THE SHORE (LOTS) CARRIER
   MISSILE DEFENSE
   OPTICAL COUNTERMEASURES
   RADAR HOMING MISSILES
   SAFEGUARD SYSTEM
   TACTICS
   WEAPONS
   WEAPONS DELIVERY
   WEAPONS INDUSTRY

∞ MILITARY VEHICLES
SN *(USE OF A MORE SPECIFIC TERM IS RECOMMENDED--CONSULT THE TERMS LISTED BELOW)*
RT AEROQUATIC VEHICLES
   AIRCRAFT CARRIERS
   AMBULANCES
   AMPHIBIOUS VEHICLES
   ARMED FORCES
   ARMED FORCES (FOREIGN)
   ARMED FORCES (UNITED STATES)
   AUTOMOBILES
   BOATS
   MILITARY SPACECRAFT
   RECOVERY VEHICLES
   RESEARCH VEHICLES
   SHIPS
   SUBMARINES
   TANKS (COMBAT VEHICLES)
   TRUCKS
   UNDERWATER VEHICLES
∞ VEHICLES
   WATER VEHICLES

MILK
RT BEVERAGES
∞ FOOD

MILKY WAY GALAXY
GS CELESTIAL BODIES
  . GALAXIES
  .. SPIRAL GALAXIES
  ... **MILKY WAY GALAXY**

MILKY WAY GALAXY-_(CONT.)_
  RT   ORION NEBULA
        RADIO SOURCES (ASTRONOMY)
        STARS

MILLET
  GS   FARM CROPS
        . MILLET
  RT   EARTH RESOURCES
        FLOUR (FOOD)
        ∞ FOOD
        GRAINS (FOOD)
        GRASSES

MILLIMETER WAVES
  GS   ELECTROMAGNETIC RADIATION
        . RADIO WAVES
        . . SHORT WAVE RADIATION
        . . . MICROWAVES
        . . . . **MILLIMETER WAVES**
  RT   BEAM PLASMA AMPLIFIERS
        C BAND
        CN EMISSION
        CYCLOTRON RESONANCE DEVICES
        DECIMETER WAVES
        ELECTROMAGNETIC NOISE
        EXTRATERRESTRIAL RADIO WAVES
        EXTREMELY HIGH FREQUENCIES
        FREQUENCIES
        SOLAR RADIO EMISSION
        SUBMILLIMETER WAVES
        WAVELENGTHS

∞ **MILLING**
  SN   _(USE OF A MORE SPECIFIC TERM IS_
        RECOMMENDED--CONSULT THE TERMS
        LISTED BELOW)
  RT   COMMINUTION
        COMPOUNDING
        MILLING (MACHINING)

MILLING (MACHINING)
  GS   CUTTING
        . **MILLING (MACHINING)**
        MACHINING
        . **MILLING (MACHINING)**
  RT   CHEMICAL MACHINING
        GROOVING
        METAL CUTTING
        MILLING MACHINES
        PLANING

MILLING (MIXING)
  USE   COMPOUNDING

MILLING MACHINES
  GS   TOOLS
        . MACHINE TOOLS
        . . **MILLING MACHINES**
  RT   GRINDING MACHINES
        ∞ MACHINERY
        SHAPERS

MILLIVOLTMETERS
  GS   MEASURING INSTRUMENTS
        . VOLTMETERS
        . . **MILLIVOLTMETERS**
  RT   GALVANOMETERS

MILLS RATIO
  GS   RATIOS
        . **MILLS RATIO**
  RT   FAILURE
        FAILURE ANALYSIS
        LIFE (DURABILITY)
        MORTALITY
        PROBABILITY DENSITY FUNCTIONS
        STATISTICAL ANALYSIS

MILNE METHOD
  GS   ANALYSIS (MATHEMATICS)
        . NUMERICAL ANALYSIS
        . . APPROXIMATION
        . . . **MILNE METHOD**
  RT   DIFFERENTIAL EQUATIONS
        ∞ METHODOLOGY

MILNE-THOMSON METHOD
  RT   INCOMPRESSIBLE FLOW
        ∞ METHODOLOGY
        NAVIER-STOKES EQUATION
        VISCOUS FLOW

MIM (SEMICONDUCTORS)
  UF   METAL-INSULATOR-METAL
        SEMICONDUCTORS
  GS   ELECTRONIC EQUIPMENT
        . SOLID STATE DEVICES
        . . SEMICONDUCTOR DEVICES
        . . . **MIM (SEMICONDUCTORS)**
        SEMICONDUCTORS (MATERIALS)
        . **MIM (SEMICONDUCTORS)**
  RT   SIS (SEMICONDUCTORS)

MIM DIODES
  UF   METAL-INSULATOR-METAL DIODES
  GS   ELECTRONIC EQUIPMENT
        . SOLID STATE DEVICES
        . . SEMICONDUCTOR DEVICES
        . . . JUNCTION DIODES
        . . . . **MIM DIODES**
  RT   ELECTRON TUNNELING
        NEGATIVE RESISTANCE DEVICES
        SEMICONDUCTOR DIODES
        TUNNEL DIODES

MIMAS
  GS   SATELLITES
        . NATURAL SATELLITES
        . . SATURN SATELLITES
        . . . **MIMAS**
  RT   SATURN (PLANET)

MINE DETECTORS
  GS   WARNING SYSTEMS
        . **MINE DETECTORS**
  RT   ∞ DETECTORS
        WARNING

MINER RULE
  USE   PALMGREN-MINER RULE

MINERAL DEPOSITS
  RT   CONTACTS (GEOLOGY)
        DREDGING
        EARTH RESOURCES
        EXCAVATION
        GEOLOGY
        MINERALOGY
        MINERALS
        MINES (EXCAVATIONS)
        MINING
        RESERVES
        STRIP MINING
        UNDERWATER RESOURCES

MINERAL EXPLORATION
  GS   EXPLORATION
        . **MINERAL EXPLORATION**
  RT   EXCAVATION
        MINERALS
        MINES (EXCAVATIONS)
        MINING

MINERAL METABOLISM
  GS   METABOLISM
        . **MINERAL METABOLISM**
  RT   BODY FLUIDS
        CALORIC REQUIREMENTS
        ENDOCRINE SYSTEMS
        SECRETIONS

MINERAL OILS
  GS   OILS
        . **MINERAL OILS**
  RT   LUBRICATING OILS

MINERALOGY
  RT   CRYSTALLOGRAPHY
        GEOCHEMISTRY
        GEOLOGY
        MINERAL DEPOSITS
        MINERALS
        PETROLOGY
        ∞ PHYSICAL SCIENCES

MINERALS
  UF   APATITES
        ORES
  GS   **MINERALS**
        . AMPHIBOLES
        . ANATASE
        . ARAGONITE
        . ASBESTOS
        . BARITE
        . BASTNASITE
        . BERYL

MINERALS-_(CONT.)_
        . BLOEDITE
        . BRUCITE
        . CALCITE
        . CHROMITES
        . COHENITE
        . CORDIERITE
        . CRYOLITE
        . DAWSONITE
        . DOLOMITE (MINERAL)
        . EUXENITE
        . FAYALITE
        . FELDSPARS
        . FLUORITE
        . FLUORSPAR
        . GARNETS
        . GEHLENITE
        . AKERMANITE
        . GRAPHITE
        . GYPSUM
        . HEXAHEDRITE
        . ILLITE
        . ILMENITE
        . IRON ORES
        . . HEMATITE
        . KAMACITE
        . KAOLINITE
        . KREEP
        . LIMONITE
        . MAGNETITE
        . MERWINITE
        . MICA
        . . BIOTITE
        . . FLUOROPHLOGOPITE
        . . MUSCOVITE
        . MONTICELLITE
        . MONTMORILLONITE
        . NEPHELINE
        . NEPHELITE
        . OLIVINE
        . . FORSTERITE
        . PEROVSKITES
        . PROUSTITE
        . PYRITES
        . PYROPHYLLITE
        . PYROXENES
        . . ENSTATITE
        . PYRRHOTITE
        . . TROILITE
        . QUARTZ
        . . COESITE
        . . STISHOVITE
        . SCHEELITE
        . SCHREIBERSITE
        . SERPENTINE
        . SIDERITES
        . SPINEL
        . SPODUMENE
        . TALC
        . TOURMALINE
        . VERMICULITE
        . WURTZITE
        . ZINCBLENDE
  RT   ALUMINUM SILICATES
        ANDESITE
        BAUXITE
        BENEFICIATION
        BIOGEOCHEMISTRY
        BONE MINERAL CONTENT
        BOREHOLES
        CALCIUM SILICATES
        CRYSTALLITES
        DIORITE
        DUNITE
        EARTH RESOURCES
        FELSITE
        FLUOROSILICATES
        GEOLOGY
        IGNEOUS ROCKS
        IMPACT MELTS
        LAVA
        LIMESTONE
        LUNAR SOIL
        MINERAL DEPOSITS
        MINERAL EXPLORATION
        MINERALOGY
        MONAZITE SANDS
        MULLITES
        ∞ NUTRIENTS
        OBSIDIAN
        POTASSIUM SILICATES
        ROCKS
        RUTILE
        SHALES
        SILICATES
        SODIUM SILICATES

MINERALS-*(CONT.)*
　　　SOILS
　　　UNDERGROUND ACOUSTICS
　　　YTTRIUM-ALUMINUM GARNET
　　　YTTRIUM-IRON GARNET
　　　ZEOLITES

∞ MINES
　SN　*(USE OF A MORE SPECIFIC TERM IS*
　　　RECOMMENDED--CONSULT THE TERMS
　　　LISTED BELOW)
　RT　MINES (EXCAVATIONS)
　　　MINES (ORDNANCE)

MINES (EXCAVATIONS)
　UF　QUARRIES
　RT　CORE SAMPLING
　　　DRAINAGE
　　　EXPLOITATION
　　　EXPLORATION
　　　MATERIALS HANDLING
　　　MINERAL DEPOSITS
　　　MINERAL EXPLORATION
　　　∞MINES
　　　MINING
　　　PITS (EXCAVATIONS)
　　　RESERVES
　　　STRATIGRAPHY
　　　STRIP MINING
　　　SUBSIDENCE
　　　UNDERGROUND EXPLOSIONS
　　　UNDERGROUND STORAGE
　　　UNDERGROUND STRUCTURES
　　　WASTE DISPOSAL

MINES (ORDNANCE)
　GS　WEAPONS
　　　. **MINES (ORDNANCE)**
　RT　AMMUNITION
　　　∞MINES

MINIATURE ELECTRONIC EQUIPMENT
　GS　ELECTRONIC EQUIPMENT
　　　. **MINIATURE ELECTRONIC EQUIPMENT**
　RT　CIRCUITS
　　　∞ELECTRIC EQUIPMENT
　　　ELECTRONIC MODULES
　　　∞EQUIPMENT
　　　MICROMINIATURIZATION
　　　MICROMINIATURIZED ELECTRONIC
　　　　DEVICES
　　　MICROMODULES
　　　MINIATURIZATION
　　　MOLECULAR ELECTRONICS
　　　PRINTED CIRCUITS
　　　SOLID STATE DEVICES
　　　SUBMINIATURIZATION
　　　THIN FILMS

MINIATURIZATION
　GS　**MINIATURIZATION**
　　　. MICROMINIATURIZATION
　　　. SUBMINIATURIZATION
　RT　CIRCUITS
　　　MINIATURE ELECTRONIC EQUIPMENT
　　　PRINTED CIRCUITS
　　　PRINTED RESISTORS
　　　TRANSISTORS
　　　WAFERS

MINICOMPUTERS
　GS　DATA PROCESSING EQUIPMENT
　　　. COMPUTERS
　　　. . DIGITAL COMPUTERS
　　　. . . **MINICOMPUTERS**
　　　. . . . NOVA COMPUTERS
　RT　AIRBORNE/SPACEBORNE COMPUTERS
　　　ATMOSPHERIC & OCEANOGRAPHIC
　　　　INFORM SYS
　　　MICROCOMPUTERS

MINIMA
　GS　ANALYSIS (MATHEMATICS)
　　　. REAL VARIABLES
　　　. . EXTREMUM VALUES
　　　. . . **MINIMA**
　RT　CUSPS (MATHEMATICS)
　　　DIFFERENTIAL CALCULUS
　　　MAXIMA
　　　OPERATIONS RESEARCH
　　　OPTIMIZATION
　　　PENALTY FUNCTION
　　　RANGE (EXTREMES)
　　　STEEPEST DESCENT METHOD

MINIMAL SURFACES
　RT　BOUNDARY VALUE PROBLEMS
　　　CONFORMAL MAPPING
　　　FINITE ELEMENT METHOD
　　　∞SURFACES

MINIMAX TECHNIQUE
　RT　APPROXIMATION
　　　CURVE FITTING
　　　GAME THEORY
　　　OPERATIONS RESEARCH
　　　RESEARCH
　　　SADDLE POINTS

MINIMIZATION
　USE　OPTIMIZATION

MINIMUM DRAG
　GS　DYNAMIC CHARACTERISTICS
　　　. DRAG
　　　. . **MINIMUM DRAG**
　RT　AIRCRAFT PERFORMANCE
　　　FRICTION DRAG

MINIMUM ENTROPY METHOD
　GS　ENTROPY (STATISTICS)
　　　. **MINIMUM ENTROPY METHOD**
　RT　∞METHODOLOGY

MINIMUM VARIANCE ORBIT DETERMINATION
　UF　MINIVAR ORBIT DETERMINATION
　GS　CLASSICAL MECHANICS
　　　. SPACE MECHANICS
　　　. . ORBITAL MECHANICS
　　　. . . **MINIMUM VARIANCE ORBIT**
　　　　　　DETERMINATION
　　　COMPUTATION
　　　. ORBIT CALCULATION
　　　. . **MINIMUM VARIANCE ORBIT**
　　　　　DETERMINATION
　RT　STATISTICAL ANALYSIS

MINING
　GS　**MINING**
　　　. STRIP MINING
　RT　CLAYS
　　　DREDGING
　　　ENERGY POLICY
　　　EXCAVATION
　　　EXPLOITATION
　　　MINERAL DEPOSITS
　　　MINERAL EXPLORATION
　　　MINES (EXCAVATIONS)
　　　UNDERGROUND STRUCTURES

MINITRACK OPTICAL TRACKING SYSTEM
　USE　MINITRACK SYSTEM

MINITRACK SYSTEM
　UF　MINITRACK OPTICAL TRACKING SYSTEM
　　　MOTS (TRACKING SYSTEM)
　RT　GLOBAL TRACKING NETWORK
　　　OPTICAL TRACKING
　　　SATELLITE TRACKING
　　　SPACE DETECTION AND TRACKING
　　　　SYSTEM
　　　SPACE SURVEILLANCE (GROUND
　　　　BASED)
　　　SPACECRAFT TRACKING
　　　STDN (NETWORK)
　　　∞SYSTEMS
　　　TRACKING NETWORKS
　　　TRACKING STATIONS
　　　∞TRACKS

MINIVAR ORBIT DETERMINATION
　USE　MINIMUM VARIANCE ORBIT
　　　　DETERMINATION

MINKOWSKI SPACE
　RT　FEYNMAN DIAGRAMS
　　　PROBABILITY THEORY
　　　SPACE-TIME FUNCTIONS

MINNESOTA
　GS　NATIONS
　　　. UNITED STATES
　　　. . **MINNESOTA**

MINOR CIRCLE TURNING FLIGHT
　GS　TURNING FLIGHT
　　　. **MINOR CIRCLE TURNING FLIGHT**
　RT　AIRCRAFT CONTROL
　　　MANEUVERS

MINOR PLANET 1221
　USE　AMOR ASTEROID

MINOR PLANET 2060
　USE　CHIRON

MINORITIES
　RT　AMERICAN INDIANS
　　　ANTHROPOLOGY
　　　COMMUNITIES
　　　CULTURE (SOCIAL SCIENCES)
　　　NATIONS
　　　RACES
　　　SOCIOLOGY
　　　VOTING

MINORITY CARRIERS
　GS　CHARGE CARRIERS
　　　. **MINORITY CARRIERS**
　RT　ADDITIVES
　　　BIPOLAR TRANSISTORS
　　　CARRIER INJECTION
　　　CARRIER LIFETIME
　　　ELECTRON MOBILITY
　　　ELECTRONS
　　　HOLES (ELECTRON DEFICIENCIES)
　　　SEMICONDUCTORS (MATERIALS)

MINOS COMPUTER
　GS　DATA PROCESSING EQUIPMENT
　　　. COMPUTERS
　　　. . **MINOS COMPUTER**

MINUTEMAN ICBM
　UF　MINUTEMAN MISSILES
　GS　MISSILES
　　　. BALLISTIC MISSILES
　　　. . INTERCONTINENTAL BALLISTIC
　　　　　MISSILES
　　　. . . **MINUTEMAN ICBM**
　　　. SURFACE TO SURFACE MISSILES
　　　. . INTERCONTINENTAL BALLISTIC
　　　　　MISSILES
　　　. . . **MINUTEMAN ICBM**
　RT　M-55 ENGINE
　　　M-56 ENGINE
　　　M-57 ENGINE
　　　MULTISTAGE ROCKET VEHICLES
　　　SOLID PROPELLANT ROCKET ENGINES
　　　SPACE WEAPONS

MINUTEMAN MISSILES
　USE　MINUTEMAN ICBM

MIOSIS
　RT　EYE (ANATOMY)
　　　OPHTHALMOLOGY
　　　TETRAD THEORY
　　　VISION

MIRAGE AIRCRAFT
　GS　ATTACK AIRCRAFT
　　　. FIGHTER AIRCRAFT
　　　. . **MIRAGE AIRCRAFT**
　　　. . . MIRAGE 3 AIRCRAFT
　　　DASSAULT AIRCRAFT
　　　. **MIRAGE AIRCRAFT**
　　　. . MIRAGE 3 AIRCRAFT
　　　JET AIRCRAFT
　　　. **MIRAGE AIRCRAFT**
　　　. . MIRAGE 3 AIRCRAFT
　　　MONOPLANES
　　　. **MIRAGE AIRCRAFT**
　　　. . MIRAGE 3 AIRCRAFT
　　　OBSERVATION AIRCRAFT
　　　. **MIRAGE AIRCRAFT**
　　　. . MIRAGE 3 AIRCRAFT
　　　RECONNAISSANCE AIRCRAFT
　　　. **MIRAGE AIRCRAFT**
　　　. . MIRAGE 3 AIRCRAFT
　　　SUPERSONIC AIRCRAFT
　　　. **MIRAGE AIRCRAFT**
　　　. . MIRAGE 3 AIRCRAFT

MIRAGE 3 AIRCRAFT
　UF　DASSAULT MIRAGE 3 AIRCRAFT
　GS　ATTACK AIRCRAFT
　　　. FIGHTER AIRCRAFT
　　　. . MIRAGE AIRCRAFT
　　　. . . **MIRAGE 3 AIRCRAFT**
　　　DASSAULT AIRCRAFT
　　　. MIRAGE AIRCRAFT
　　　. . **MIRAGE 3 AIRCRAFT**
　　　JET AIRCRAFT

MIRAGE 3 AIRCRAFT-*(CONT.)*
　. MIRAGE AIRCRAFT
　. . **MIRAGE 3 AIRCRAFT**
　MONOPLANES
　. MIRAGE AIRCRAFT
　. . **MIRAGE 3 AIRCRAFT**
　OBSERVATION AIRCRAFT
　. MIRAGE AIRCRAFT
　. . **MIRAGE 3 AIRCRAFT**
　RECONNAISSANCE AIRCRAFT
　. MIRAGE AIRCRAFT
　. . **MIRAGE 3 AIRCRAFT**
　SUPERSONIC AIRCRAFT
　. MIRAGE AIRCRAFT
　. . **MIRAGE 3 AIRCRAFT**
　TAILLESS AIRCRAFT
　. **MIRAGE 3 AIRCRAFT**

†

MIRANDA SATELLITE
GS　SATELLITES
　. ARTIFICIAL SATELLITES
　. . SYNCHRONOUS SATELLITES
　. . . **MIRANDA SATELLITE**
RT　ATTITUDE CONTROL

MIROS SYSTEM
UF　MODULATING RETRODIRECTIVE OPTICS
GS　MODULATION
　. LIGHT MODULATION
　. . **MIROS SYSTEM**
RT　OPTICAL MEASURING INSTRUMENTS
　∞SYSTEMS

MIRROR FUSION
RT　FUSION REACTORS
　MAGNETIC MIRRORS
　NUCLEAR FUSION
　PLASMA CONTROL

MIRROR POINT
RT　MAGNETIC MIRRORS
　RADIATION BELTS

MIRRORS
GS　**MIRRORS**
　. CELESCOPES
　. FRESNEL REFLECTORS
　. MAGNETIC MIRRORS
　. PARABOLOID MIRRORS
　. ROTATING MIRRORS
　. SOLETTAS
RT　CASSEGRAIN OPTICS
　CIRCUMSOLAR TELESCOPES
　COLLIMATORS
　HELIOSTATS
　OPTICAL EQUIPMENT
　OPTICAL RESONATORS
　∞OPTICS
　REFLECTING TELESCOPES
　REFLECTORS
　SOLAR COLLECTORS
　SOLAR REFLECTORS
　SPECULAR REFLECTION
　TELESCOPES

MIS (SEMICONDUCTORS)
UF　METAL INSULATOR SEMICONDUCTORS
GS　ELECTRONIC EQUIPMENT
　. SOLID STATE DEVICES
　. . SEMICONDUCTOR DEVICES
　. . . **MIS (SEMICONDUCTORS)**
　SEMICONDUCTORS (MATERIALS)
　. **MIS (SEMICONDUCTORS)**
RT　SIS (SEMICONDUCTORS)

MISALIGNMENT
SN　(EXCLUDES PSYCHOLOGICAL
　DISORIENTATION)
UF　MISORIENTATION
RT　ATTITUDE (INCLINATION)
　DISORIENTATION
　POSITION (LOCATION)

MISCIBILITY
USE　SOLUBILITY

MISFETS
USE　FIELD EFFECT TRANSISTORS

MISMATCH (ELECTRICAL)
RT　ELECTRICAL MEASUREMENT
　IMPEDANCE MEASUREMENT
　MATCHING

MISORIENTATION
USE　MISALIGNMENT

MISS DISTANCE
GS　DISTANCE
　. **MISS DISTANCE**
RT　ACCURACY
　AIR TO SURFACE MISSILES

MISSILE ANTENNAS
GS　ANTENNAS
　. **MISSILE ANTENNAS**
　MISSILE COMPONENTS
　. **MISSILE ANTENNAS**
RT　AIRCRAFT ANTENNAS
　DIRECTIONAL ANTENNAS
　MICROWAVE ANTENNAS

MISSILE BODIES
UF　MISSILE CASES
GS　MISSILE COMPONENTS
　. **MISSILE BODIES**
RT　AIRFRAMES
　AXISYMMETRIC BODIES
　BLUNT BODIES
　∞BODIES
　CASES (CONTAINERS)
　FINNED BODIES
　ROCKET ENGINE CASES
　SLENDER BODIES
　STREAMLINED BODIES

MISSILE CASES
USE　MISSILE BODIES

MISSILE COMPONENTS
GS　**MISSILE COMPONENTS**
　. MISSILE ANTENNAS
　. MISSILE BODIES
RT　∞COMPONENTS
　ENGINES
　FINS
　NOSE CONES
　WARHEADS
　WINGS

MISSILE CONFIGURATIONS
GS　**MISSILE CONFIGURATIONS**
　. SANDPIPER TARGET MISSILE
RT　AERODYNAMIC CONFIGURATIONS
　AIRCRAFT CONFIGURATIONS
　∞CONFIGURATIONS
　HAMMERHEAD CONFIGURATION
　LAUNCH VEHICLE CONFIGURATIONS
　MISSILES
　MULTIENGINE VEHICLES
　PATRIOT MISSILE
　PROPULSION SYSTEM CONFIGURATIONS
　ROCKET ENGINES
　ROCKET VEHICLES

MISSILE CONSTRUCTION
USE　MISSILE STRUCTURES

MISSILE CONTROL
UF　MISSILE GUIDANCE
　MISSILE STABILIZATION
RT　ACTUATORS
　ANALOG COMPUTERS
　ATTITUDE CONTROL
　AUTOMATIC CONTROL
　AUTOMATIC FLIGHT CONTROL
　BEAM RIDER GUIDANCE
　∞CONTROL
　DIRECTIONAL CONTROL
　FLIGHT CONTROL
　GROUND BASED CONTROL
　HOMING
　LASER GUIDANCE
　LATERAL CONTROL
　LONGITUDINAL CONTROL
　MISSILES
　RADAR HOMING MISSILES
　RADIO CONTROL
　REMOTE CONTROL
　ROCKET ENGINE CONTROL
　SPACECRAFT CONTROL
　STAR TRACKERS
　THRUST VECTOR CONTROL
　VISUAL CONTROL

MISSILE DEFENSE
SN　(SYSTEMS DESIGNED TO PROTECT
　MISSILES AGAINST ATTACK)

MISSILE DEFENSE-*(CONT.)*
RT　ANTIMISSILE DEFENSE
　ANTIMISSILE MISSILES
　ANTIRADIATION MISSILES
　BALLISTIC MISSILE DECOYS
　∞DEFENSE
　DEFENSE INDUSTRY
　DEFENSE PROGRAM
　HARDENING (SYSTEMS)
　MILITARY TECHNOLOGY
　MISSILES
　OPTICAL COUNTERMEASURES
　REENTRY DECOYS
　SAFEGUARD SYSTEM
　WEAPONS DELIVERY

MISSILE DESIGN
RT　AEROSPACE ENGINEERING
　AIRCRAFT DESIGN
　COMPUTER AIDED DESIGN
　∞DESIGN
　∞DEVELOPMENT
　ENGINE DESIGN
　FLIGHT TESTS
　FUNCTIONAL DESIGN SPECIFICATIONS
　MAGNUS EFFECT
　RELIABILITY
　STRUCTURAL DESIGN
　SYSTEMS ENGINEERING

MISSILE DETECTION
GS　DETECTION
　. **MISSILE DETECTION**
　. . RADAR DETECTION
RT　EARLY WARNING SYSTEMS
　ELECTRONIC WARFARE
　IDENTIFYING
　TARGET ACQUISITION
　TARGET RECOGNITION

MISSILE ENGINE CASES
USE　ROCKET ENGINE CASES

MISSILE GUIDANCE
USE　MISSILE CONTROL

MISSILE LAUNCHERS
GS　LAUNCHERS
　. **MISSILE LAUNCHERS**
　. . MOBILE MISSILE LAUNCHERS
RT　BALLISTIC MISSILE SUBMARINES
　CATAPULTS
　GROUND SUPPORT EQUIPMENT
　GUN LAUNCHERS
　LAUNCH VEHICLES
　LAUNCHING
　LAUNCHING SITES
　MISSILES
　ROCKET LAUNCHERS
　SEA LAUNCHING
　WEAPON SYSTEMS

MISSILE RANGES
SN　(EXCLUDES DISTANCE OF MISSILE
　TRAVEL)
GS　RANGES (FACILITIES)
　. TEST RANGES
　. . **MISSILE RANGES**
　TEST FACILITIES
　. TEST RANGES
　. . **MISSILE RANGES**
RT　BALLISTIC RANGES
　DOWNRANGE
　HYPERSONIC TEST APPARATUS
　MISSILES
　RANGE SAFETY
　REENTRY RANGE

MISSILE SIGNATURES
GS　SIGNATURES
　. **MISSILE SIGNATURES**
RT　DETECTION
　SIGNATURE ANALYSIS
　TARGET RECOGNITION

MISSILE SILOS
UF　SILOS (MISSILE STORAGE)
RT　∞BUILDINGS
　LAUNCHING SITES
　MX MISSILE
　∞STORAGE

∞ **MISSILE SIMULATORS**
    SN   *(USE OF A MORE SPECIFIC TERM IS*
         RECOMMENDED--CONSULT THE TERMS
         LISTED BELOW)
    RT   COMPUTERIZED SIMULATION
         FLIGHT SIMULATORS
         MATHEMATICAL MODELS
         MISSILES
         SIMULATORS
         TRAINING SIMULATORS
         WIND TUNNEL MODELS

MISSILE STABILIZATION
    USE MISSILE CONTROL
         STABILIZATION

MISSILE STORAGE
    RT   GROUND SUPPORT EQUIPMENT
         MOBILE MISSILE LAUNCHERS
         PROPELLANT STORAGE
       ∞STORAGE
         UNDERGROUND STORAGE

MISSILE STRUCTURES
    UF   MISSILE CONSTRUCTION
    RT   AIRFRAMES
       ∞STRUCTURES
         TAIL ASSEMBLIES

MISSILE SYSTEMS
    GS   WEAPON SYSTEMS
         . **MISSILE SYSTEMS**
         . . NIKE X SYSTEMS
         . . SAFEGUARD SYSTEM
    RT   AEROSPACE SYSTEMS
         BEAM RIDER GUIDANCE
         MOBILE MISSILE LAUNCHERS
         RADAR HOMING MISSILES
       ∞SYSTEMS

MISSILE TESTS
    RT   CAPTIVE TESTS
         ENGINE TESTS
         FLIGHT TESTS
         FUEL TESTS
         GROUND TESTS
         MISSILES
         PRELAUNCH TESTS
         PROPELLANT TESTS
         STABILITY TESTS
         STATIC TESTS
         TEST FIRING
         TEST VEHICLES
       ∞TESTS
         WIND TUNNEL STABILITY TESTS

MISSILE TRACKING
    GS   TRACKING (POSITION)
         . **MISSILE TRACKING**
    RT   INFRARED TRACKING
         LASER TARGET DESIGNATORS
         POLYSTATION DOPPLER TRACKING
           SYSTEM
         RANGE AND RANGE RATE TRACKING
         SPACE DETECTION AND TRACKING
           SYSTEM
         SPACECRAFT TRACKING
         TRACKING NETWORKS
         TRACKING STATIONS

MISSILE TRAJECTORIES
    GS   TRAJECTORIES
         . **MISSILE TRAJECTORIES**
    RT   ASCENT TRAJECTORIES
         BALLISTIC TRAJECTORIES
         COBRA DANE (RADAR)
         DESCENT TRAJECTORIES
         FLIGHT MECHANICS
         FLIGHT PATHS
         IMPACT PREDICTION
         PARABOLIC FLIGHT
         REENTRY TRAJECTORIES
         SPINNING UNGUIDED ROCKET
           TRAJECTORY
         UNDERWATER TRAJECTORIES

MISSILE VIBRATION
    GS   VIBRATION
         . STRUCTURAL VIBRATION
         . . **MISSILE VIBRATION**
    RT   BENDING VIBRATION
         BREATHING VIBRATION
         FLUTTER
         LINEAR VIBRATION
         RANDOM VIBRATION

MISSILE VIBRATION-*(CONT.)*
         SELF INDUCED VIBRATION
         SUPERSONIC FLUTTER
         TORSIONAL VIBRATION
         TRANSONIC FLUTTER

MISSILES
    GS   **MISSILES**
         . AIR SLEW MISSILES
         . AIR TO AIR MISSILES
         . . FALCON MISSILE
         . . MATRA MISSILE
         . . SIDEWINDER MISSILES
         . . SPARROW MISSILES
         . . . SPARROW 2 MISSILE
         . . . SPARROW 3 MISSILE
         . AIR TO SURFACE MISSILES
         . . BULLPUP MISSILES
         . . CONDOR MISSILE
         . . HARPOON MISSILE
         . . HOUND DOG MISSILE
         . . MAVERICK MISSILES
         . . QUAIL MISSILE
         . . SHRIKE MISSILE
         . ANTELOPE MISSILE
         . ANTIAIRCRAFT MISSILES
         . . BOMARC MISSILES
         . . FALCON MISSILE
         . . MAULER MISSILE
         . . NIKE-AJAX MISSILE
         . . NIKE-HERCULES MISSILE
         . . REDEYE MISSILE
         . . SIAM MISSILES
         . . SIDEWINDER MISSILES
         . . TARTAR MISSILE
         . . TERRIER MISSILE
         . ANTIMISSILE MISSILES
         . . MAULER MISSILE
         . . NIKE-ZEUS MISSILE
         . . SPARTAN MISSILE
         . . SPRINT MISSILE
         . ANTIRADIATION MISSILES
         . ANTISHIP MISSILES
         . BALLISTIC MISSILES
         . FIELD ARMY BALLISTIC MISSILES
         . . SUBROC MISSILE
         . . INTERCONTINENTAL BALLISTIC
           MISSILES
         . . . ATLAS ICBM
         . . . . ATLAS D ICBM
         . . . . ATLAS E ICBM
         . . . . ATLAS F ICBM
         . . . MINUTEMAN ICBM
         . . . TITAN ICBM
         . . . . TITAN 1 ICBM
         . . . . TITAN 2 ICBM
         . . INTERMEDIATE RANGE BALLISTIC
           MISSILES
         . . . BLUE STREAK MISSILE
         . . . JUPITER MISSILE
         . . . POLARIS MISSILES
         . . . . POLARIS A1 MISSILE
         . . . . POLARIS A2 MISSILE
         . . . . POLARIS A3 MISSILE
         . . PERSHING MISSILE
         . . POSEIDON MISSILES
         . SHORT RANGE BALLISTIC MISSILES
         . SKYBOLT MISSILE
         . . V-2 MISSILE
         . BLUE STEEL MISSILE
         . BUMBLEBEE PROJECT
         . CORVUS MISSILE
         . OSPREY MISSILE
         . PRECISION GUIDED PROJECTILES
         . RADAR HOMING MISSILES
         . RAMJET MISSILES
         . . NAVAHO MISSILE
         . . SUPERSONIC LOW ALTITUDE MISSILE
         . SANDPIPER TARGET MISSILE
         . SS-11 MISSILE
         . SURFACE TO AIR MISSILES
         . . BLUE GOOSE MISSILE
         . . BOMARC MISSILES
         . . . BOMARC A MISSILE
         . . . BOMARC B MISSILE
         . . CHAPARRAL MISSILE
         . . HAWK MISSILE
         . . MAULER MISSILE
         . . NIKE MISSILES
         . . . NIKE-AJAX MISSILE
         . . . NIKE-HERCULES MISSILE
         . . . NIKE-ZEUS MISSILE
         . . PATRIOT MISSILE
         . . REDEYE MISSILE
         . . SPRINT MISSILE
         . . TALOS MISSILE

MISSILES-*(CONT.)*
         . . TARTAR MISSILE
         . . TERRIER MISSILE
         . SURFACE TO SURFACE MISSILES
         . . ANTITANK MISSILES
         . . . SHILLELAGH MISSILES
         . . . TOW MISSILES
         . . CORPORAL MISSILE
         . . CRUISE MISSILES
         . . . NAVAHO MISSILE
         . . . TOMAHAWK MISSILES
         . . FLEET BALLISTIC MISSILES
         . . . POLARIS A1 MISSILE
         . . . POLARIS A2 MISSILE
         . . . POLARIS A3 MISSILE
         . . . POSEIDON MISSILES
         . . . SUBROC MISSILE
         . . INTERCONTINENTAL BALLISTIC
           MISSILES
         . . . ATLAS ICBM
         . . . . ATLAS D ICBM
         . . . . ATLAS E ICBM
         . . . . ATLAS F ICBM
         . . . MINUTEMAN ICBM
         . . . MX MISSILE
         . . . TITAN ICBM
         . . . . TITAN 1 ICBM
         . . . . TITAN 2 ICBM
         . . INTERMEDIATE RANGE BALLISTIC
           MISSILES
         . . . BLUE STREAK MISSILE
         . . . JUPITER MISSILE
         . . . POLARIS MISSILES
         . . . . POLARIS A1 MISSILE
         . . . . POLARIS A2 MISSILE
         . . . . POLARIS A3 MISSILE
         . . LANCE MISSILE
         . . MACE MISSILES
         . . PERSHING MISSILE
         . . REGULUS MISSILE
         . . SERGEANT MISSILES
         . . SHORT RANGE BALLISTIC MISSILES
         . . SUPERSONIC LOW ALTITUDE MISSILE
         . . V-1 MISSILE
         . . UNDERWATER TO SURFACE MISSILES
         . . SUBROC MISSILE
    RT   AMMUNITION
         ANTIMISSILE DEFENSE
         ANTISHIP WARFARE
         ARTILLERY
         ASCENT PROPULSION SYSTEMS
         AUXILIARY PROPULSION
         BOMBS (ORDNANCE)
         FLIGHT TEST VEHICLES
       ∞FLIGHT VEHICLES
         GROUND SUPPORT EQUIPMENT
         GUIDANCE (MOTION)
         HOMING DEVICES
         HYPERSONIC FLIGHT
         INCENDIARY AMMUNITION
         LAUNCH VEHICLES
         LAUNCHING SITES
         MISSILE CONFIGURATIONS
         MISSILE CONTROL
         MISSILE DEFENSE
         MISSILE LAUNCHERS
         MISSILE RANGES
       ∞MISSILE SIMULATORS
         MISSILE TESTS
         MULTIENGINE VEHICLES
         NIKE X SYSTEMS
         NUCLEAR WEAPONS
         PLASMA SHEATHS
         PROPULSION
         REENTRY
         REENTRY VEHICLES
         ROCKET CATAPULTS
         ROCKET ENGINES
         ROCKET PROPELLANTS
       ∞ROCKETS
       ∞SCRAM
         SPACECRAFT LAUNCHING
         SPIN STABILIZATION
         STAGE SEPARATION
         SUPERSONIC COMBUSTION RAMJET
           ENGINES
         SUPERSONIC FLIGHT
         TERMINAL BALLISTICS
         TEST VEHICLES
         TORPEDOES
         TRAJECTORIES
         TRANSPORTATION
       ∞VEHICLES
         WARHEADS
         WEAPON SYSTEMS
         WEAPONS

MISSILES-(CONT.)
  ∞WINGED VEHICLES

MISSING MASS (ASTROPHYSICS)
  GS   COSMOLOGY
        . **MISSING MASS (ASTROPHYSICS)**
        MASS
        . **MISSING MASS (ASTROPHYSICS)**
  RT   ASTRONOMY
        ASTROPHYSICS
        DYNAMIC STABILITY
        GALACTIC CLUSTERS
        GALACTIC STRUCTURE
        MASS DISTRIBUTION
        MASS TO LIGHT RATIOS
        VIRIAL THEOREM

MISSION PLANNING
  GS   PLANNING
        . **MISSION PLANNING**
  RT   BUDGETING
        COMMERCE LAB
        CRITICAL PATH METHOD
        ESTIMATING
        FORECASTING
        MANAGEMENT
        MANAGEMENT PLANNING
        ∞MISSIONS
        ∞OPERATIONS
        OPERATIONS RESEARCH
        PAYLOAD INTEGRATION
        ∞PLANS
        PREDICTIONS
        PRELAUNCH SUMMARIES
        PROGRAMS
        PROJECT MANAGEMENT
        SCHEDULING
        ULYSSES MISSION

∞ **MISSIONS**
  SN   *(USE OF A MORE SPECIFIC TERM IS*
        RECOMMENDED--CONSULT THE TERMS
        LISTED BELOW)
  RT   ABORTED MISSIONS
        ASTEROID MISSIONS
        EARTH-VENUS TRAJECTORIES
        EXPEDITIONS
        FLYBY MISSIONS
        GALILEO SPACECRAFT
        GRAND TOURS
        HEAT CAPACITY MAPPING MISSION
        LANDSAT FOLLOW-ON MISSIONS
        LONG DURATION SPACE FLIGHT
        MARINER JUPITER-SATURN FLYBY
        MARINER JUPITER-URANUS FLYBY
        MISSION PLANNING
        PLANNING
        PROGRAMS
        PROJECT PLANNING
        PROJECTS
        SOLAR MAXIMUM MISSION
        SOLAR MAXIMUM MISSION-A
        SPACE FLIGHT
        SPACE MISSIONS
        TARGETS
        ULYSSES MISSION
        VOYAGER 1977 MISSION

MISSISSIPPI
  GS   NATIONS
        . UNITED STATES
        . . **MISSISSIPPI**
  RT   GULF OF MEXICO

MISSISSIPPI DELTA (LA)
  GS   LANDFORMS
        . DELTAS
        . . **MISSISSIPPI DELTA (LA)**
  RT   LOUISIANA
        RIVERS

MISSISSIPPI RIVER (US)
  GS   RIVERS
        . **MISSISSIPPI RIVER (US)**
  RT   DRAINAGE PATTERNS
        EARTH RESOURCES
        FLOODS
        RESOURCES
        RIVER BASINS

MISSOURI
  GS   NATIONS
        . UNITED STATES
        . . **MISSOURI**
  RT   MISSOURI RIVER (US)

MISSOURI-(CONT.)
        ST LOUIS-KANSAS CITY CORRIDOR (MO)

MISSOURI RIVER (US)
  GS   RIVERS
        . **MISSOURI RIVER (US)**
  RT   IOWA
        KANSAS
        MISSOURI
        MONTANA
        NEBRASKA
        NORTH DAKOTA
        RIVER BASINS
        SOUTH DAKOTA
        UNITED STATES
        VALLEYS

MISSOURI RIVER BASIN (US)
  GS   LANDFORMS
        . STRUCTURAL BASINS
        . . RIVER BASINS
        . . . **MISSOURI RIVER BASIN (US)**
  RT   RIVERS
        WATERSHEDS

MIST
  SN   (ATMOSPHERIC WATER)
  GS   PARTICLES
        . **MIST**
  RT   AEROSOLS
        DISPERSIONS
        FOG
        FOG DISPERSAL
        HAZE
        HAZE DETECTION
        PRECIPITATION (METEOROLOGY)

MITOCHONDRIA
  GS   CELLS (BIOLOGY)
        . **MITOCHONDRIA**
        PYRIMIDINES
        . **MITOCHONDRIA**
  RT   CYTOLOGY

MITOSIS
  RT   CELLS (BIOLOGY)
        CHROMOSOMES
        CYTOLOGY
        CYTOPLASM
        MUTATIONS
        PHYSIOLOGY
        ∞REPRODUCTION

MITRA
  RT   FUNGI
        PLANTS (BOTANY)

MIUS
  USE   MODULAR INTEGRATED UTILITY SYSTEM

MIXED CRYSTALS
  GS   CRYSTALS
        . **MIXED CRYSTALS**
  RT   POWDER METALLURGY
        SINTERING

MIXED FLOW
  USE   MULTIPHASE FLOW

MIXED OXIDES
  GS   CHALCOGENIDES
        . OXIDES
        . . METAL OXIDES
        . . . **MIXED OXIDES**
  RT   NUCLEAR FUELS
        PLUTONIUM OXIDES
        URANIUM OXIDES

MIXERS
  SN   (EXCLUDES MIXING CIRCUITS)
  RT   ADMIXTURES
        AERATION
        AEROSOLS
        AGITATION
        BAFFLES
        BLOWERS
        CARBURETORS
        COALESCING
        CONTACTORS
        ∞DIFFUSERS
        ∞DISPERSION
        FEEDERS
        GRINDING MILLS
        MIXING

MIXERS-(CONT.)
        MIXTURES
        PADDLES
        PLOWS
        PLUNGERS
        SEPARATORS
        SHAKERS
        SPRAYERS
        STIRRING
        TUMBLING MOTION

MIXING
  GS   **MIXING**
        . COLLOIDING
        . COMPOUNDING
        . DISSOLVING
        . HOMOGENIZING
        . LAMINAR MIXING
        . PREMIXING
        . SIGNAL MIXING
        . SUSPENDING (MIXING)
        . TURBULENT MIXING
  RT   AERATION
        AGITATION
        BLOWING
        CHOKES
        DIFFUSION
        DILUTION
        GRINDING (COMMINUTION)
        JET MIXING FLOW
        LIQUID INJECTION
        METAL POWDER
        MIXERS
        MIXTURES
        PREMIXED FLAMES
        ∞SEPARATION
        SHAKING
        SPRAYING
        SWIRLING
        TANGLING
        TRAPPED VORTEXES
        TURBULENCE
        VORTICES

MIXING CIRCUITS
  GS   CIRCUITS
        . **MIXING CIRCUITS**
  RT   FREQUENCY CONVERTERS
        FREQUENCY SYNTHESIZERS
        HETERODYNING
        PREAMPLIFIERS

MIXING DEPTH
  USE   MIXING HEIGHT

MIXING HEIGHT
  UF   MIXING DEPTH
  RT   AIR POLLUTION
        ATMOSPHERIC CIRCULATION
        CONVECTION
        CONVECTION CURRENTS
        VERTICAL AIR CURRENTS
        WIND (METEOROLOGY)

MIXING LENGTH FLOW THEORY
  GS   FLOW THEORY
        . **MIXING LENGTH FLOW THEORY**
        KINETIC THEORY
        . TRANSPORT THEORY
        . . **MIXING LENGTH FLOW THEORY**
  RT   SHEAR FLOW
        ∞THEORIES
        TURBULENT FLOW
        TURBULENT MIXING
        VORTICITY TRANSPORT HYPOTHESIS

MIXTURES
  UF   BLENDS
  GS   **MIXTURES**
        . ADMIXTURES
        . BINARY MIXTURES
        . . BINARY FLUIDS
        . . EUTECTICS
        . . . EUTECTIC ALLOYS
        . DISPERSIONS
        . . COLLOIDS
        . . . AEROSOLS
        . . . . FOG
        . . . COLLOIDAL PROPELLANTS
        . . EMULSIONS
        . . . PHOTOGRAPHIC EMULSIONS
        . . . . NUCLEAR EMULSIONS
        . . LIQUID-GAS MIXTURES
        . . . AEROSOLS
        . . . . FOG

MIXTURES-(CONT.)
.. PLASTISOLS
... SMOKE
. GADOLINIUM ALLOYS
. METAL MATRIX COMPOSITES
.. BORSIC (TRADENAME)
. SIALON
. SLURRIES
. SOLID SUSPENSIONS
. SOLUTIONS
.. AQUEOUS SOLUTIONS
. GAS MIXTURES
... DETONABLE GAS MIXTURES
.. PHOTOGRAPHIC EMULSIONS
... NUCLEAR EMULSIONS
.. SOLID SOLUTIONS
RT   ALLOYS
      AZEOTROPES
      ∞COMBINATION
      COMPOSITE MATERIALS
      COMPOSITION (PROPERTY)
      DISSOLVED GASES
      EUTECTIC COMPOSITES
      FORMULATIONS
      GADOLINIUM ALLOYS
      INGREDIENTS
      MIXERS
      MIXING
      MONOTECTIC ALLOYS
      PASTE (CONSISTENCY)
      SOLUBILITY
      SYNTECTIC ALLOYS

ML-1 NUCLEAR POWER PLANT
GS   ELECTRIC POWER PLANTS
      . NUCLEAR POWER PLANTS
      .. **ML-1 NUCLEAR POWER PLANT**
      NUCLEAR ELECTRIC POWER
        GENERATION
      . NUCLEAR POWER PLANTS
      .. **ML-1 NUCLEAR POWER PLANT**
RT   ∞POWER PLANTS

MLA
USE   MULTISPECTRAL LINEAR ARRAYS

MMS
USE   MULTIMISSION MODULAR SPACECRAFT

MNEMONICS
RT   MEMORY
      NOMENCLATURES
      SYMBOLIC PROGRAMMING
      SYMBOLS

MNOS
USE   METAL-NITRIDE-OXIDE-SILICON

MOBILE COMMUNICATION SYSTEMS
GS   **MOBILE COMMUNICATION SYSTEMS**
      . LAND MOBILE SATELLITE SERVICE
RT   COMMUNICATION SATELLITES
      MSAT
      RADIO COMMUNICATION

MOBILE LOUNGES
RT   AIRFIELD SURFACE MOVEMENTS
      AIRPORTS
      GROUND HANDLING
      ∞LOUNGES

MOBILE MISSILE LAUNCHERS
GS   LAUNCHERS
      . MISSILE LAUNCHERS
      .. **MOBILE MISSILE LAUNCHERS**
RT   BALLISTIC MISSILE SUBMARINES
      MISSILE STORAGE
      MISSILE SYSTEMS
      WEAPON SYSTEMS

MOBILE QUARANTINE FACILITY
RT   AEROSPACE MEDICINE
      EVACUATING (TRANSPORTATION)
      ∞FACILITIES
      MEDICAL EQUIPMENT
      MEDICAL SERVICES
      PHYSICAL EXAMINATIONS
      PHYSIOLOGICAL TESTS

MOBILITY
SN   (EXCLUDES CONSIDERATIONS OF
      MANNED AND UNMANNED CRAFT)
GS   **MOBILITY**
      . ATOMIC MOBILITIES

MOBILITY-(CONT.)
. CARRIER MOBILITY
.. ELECTRON MOBILITY
.. HOLE MOBILITY
.. IONIC MOBILITY
RT   ∞CONDUCTIVITY
      DIFFUSIVITY
      DRIFT RATE
      HALL EFFECT
      KINETIC THEORY
      PORTABLE EQUIPMENT
      TRANSPORT PROPERTIES

MODAL RESPONSE
UF   MODE SHAPES
GS   RESPONSES
      . **MODAL RESPONSE**
RT   DYNAMIC RESPONSE
      STROKING TESTS

MODCOMP II COMPUTER
GS   DATA PROCESSING EQUIPMENT
      . COMPUTERS
      .. DIGITAL COMPUTERS
      ... **MODCOMP II COMPUTER**

MODCOMP IV COMPUTER
GS   DATA PROCESSING EQUIPMENT
      . COMPUTERS
      .. DIGITAL COMPUTERS
      ... **MODCOMP IV COMPUTER**

∞ **MODE**
SN   (USE OF A MORE SPECIFIC TERM IS
      RECOMMENDED--CONSULT THE TERMS
      LISTED BELOW)
RT   FAILURE MODES
      MODE (STATISTICS)
      MODES

MODE (STATISTICS)
RT   AVERAGE
      DISTRIBUTION MOMENTS
      FAILURE MODES
      MEAN
      MEDIAN (STATISTICS)
      ∞MODE
      MODES
      MOMENTS
      QUALITY CONTROL

MODE COUPLING
USE   COUPLED MODES

MODE OF VIBRATION
USE   VIBRATION MODE

MODE SHAPES
USE   MODAL RESPONSE

MODE TRANSFORMERS
GS   TRANSDUCERS
      . **MODE TRANSFORMERS**
      TRANSFORMERS
      . **MODE TRANSFORMERS**
RT   IMPEDANCE MATCHING
      PROPAGATION MODES
      TRANSMISSION LINES
      VIBRATION MODE
      WAVEGUIDE TUNERS

MODELS
GS   **MODELS**
      . AIRCRAFT MODELS
      . ASTRONOMICAL MODELS
      . DENSITY WAVE MODEL
      .. STELLAR MODELS
      . ATMOSPHERIC MODELS
      . REFERENCE ATMOSPHERES
      . BREADBOARD MODELS
      . DYNAMIC MODELS
      . ENVIRONMENT MODELS
      . GUTENBERG ZONE
      . HYDROLOGY MODELS
      . LIGHTHILL GAS MODEL
      . MATHEMATICAL MODELS
      .. ANALOG SIMULATION
      .. BIOLOGICAL MODELS (MATHEMATICS)
      .. DIGITAL SIMULATION
      .. THOMAS-FERMI MODEL
      .. VENEZIANO MODEL
      . NUCLEAR MODELS
      . OCEAN MODELS
      . QUARK PARTON MODEL

MODELS-(CONT.)
. SCALE MODELS
. SEMISPAN MODELS
. SPACECRAFT MODELS
. STATIC MODELS
. VECTOR DOMINANCE MODEL
. WIND TUNNEL MODELS
.. POWERED MODELS
RT   ANALOGS
      DUMMIES
      LAYOUTS
      PILOT PLANTS
      REPLICAS
      SIMULATORS
      TEST FACILITIES

MODEMS
UF   MODULATORS-DEMODULATORS
GS   DEMODULATORS
      . **MODEMS**
      MODULATORS
      . **MODEMS**
RT   DATA TRANSMISSION
      PHASE DEMODULATORS
      PHASE MODULATION
      PULSE AMPLITUDE MODULATION
      PULSE COMMUNICATION
      PULSE DURATION MODULATION
      PULSE FREQUENCY MODULATION
      PULSE MODULATION
      PULSE POSITION MODULATION

MODERATION (ENERGY ABSORPTION)
GS   ENERGY ABSORPTION
      . **MODERATION (ENERGY ABSORPTION)**
      .. THERMALIZATION (ENERGY
         ABSORPTION)
      ... NEUTRON THERMALIZATION
RT   ∞ABSORPTION
      LIMITERS (FUSION REACTORS)
      MODERATORS

MODERATORS
RT   BERYLLIUM
      BLANKETS (FUSION REACTORS)
      GRAPHITE
      HEAVY WATER
      LIMITERS (FUSION REACTORS)
      MODERATION (ENERGY ABSORPTION)
      NEUTRON ABSORBERS
      NUCLEAR REACTORS
      REACTOR MATERIALS
      WATER

MODES
GS   **MODES**
      . AXIAL MODES
      . BALLOONING MODES
      . COUPLED MODES
      . FAILURE MODES
      . LASER MODES
      . MODES (STANDING WAVES)
      . PROPAGATION MODES
      . PUSHBROOM SENSOR MODES
      . VIBRATION MODE
      .. UNCOUPLED MODES
RT   ∞MODE
      MODE (STATISTICS)
      TEARING MODES (PLASMAS)

MODES (STANDING WAVES)
GS   MODES
      . **MODES (STANDING WAVES)**
RT   UNCOUPLED MODES
      VIBRATION

MODIFICATION
USE   REVISIONS

MODULAR INTEGRATED UTILITY SYSTEM
UF   MIUS
RT   AIR CONDITIONING
      COMMUNITIES
      ELECTRIC POWER PLANTS
      HEATING
      POTABLE WATER
      SEWAGE TREATMENT
      ∞SYSTEMS
      UTILITIES
      WASTE DISPOSAL

MODULAR RATIOS
GS   MECHANICAL PROPERTIES
      . **MODULAR RATIOS**
      RATIOS

MODULAR RATIOS-*(CONT.)*
. **MODULAR RATIOS**
RT   COMPOSITE MATERIALS
      STRESS RATIO
      STRUCTURAL ANALYSIS
      STRUCTURAL ENGINEERING

MODULARITY
RT   ARCHITECTURE (COMPUTERS)
      AVIONICS
      COMPUTER PROGRAMS
      ELECTRONIC MODULES
      SYSTEMS ENGINEERING

MODULATED CONTINUOUS RADIATION
GS   CONTINUOUS RADIATION
      . **MODULATED CONTINUOUS RADIATION**
      ELECTROMAGNETIC RADIATION
      . **MODULATED CONTINUOUS RADIATION**
RT   COHERENT ELECTROMAGNETIC
        RADIATION
      PHASE DEVIATION
      ∞ RADIATION

MODULATING RETRODIRECTIVE OPTICS
USE   MIROS SYSTEM

MODULATION
UF   CARRIER MODULATION
GS   **MODULATION**
      . AMPLITUDE MODULATION
      . FREQUENCY MODULATION
      . . FEEDBACK FREQUENCY MODULATION
      . . FM/PM (MODULATION)
      . . FREQUENCY SHIFT KEYING
      . . PULSE FREQUENCY MODULATION
      . . PULSE FREQUENCY MODULATION
          TELEMETRY
      . INTERMODULATION
      . IONOSPHERIC CROSS MODULATION
      . LIGHT MODULATION
      . . MIROS SYSTEM
      . . ULTRASONIC LIGHT MODULATION
      . PHASE MODULATION
      . . FM/PM (MODULATION)
      . . PHASE SHIFT KEYING
      . PULSE MODULATION
      . . PULSE AMPLITUDE MODULATION
      . . PULSE CODE MODULATION
      . . . DELTA MODULATION
      . . . DIFFERENTIAL PULSE CODE
            MODULATION
      . . PULSE FREQUENCY MODULATION
      . . PULSE FREQUENCY MODULATION
          TELEMETRY
      . . PULSE TIME MODULATION
      . . . PULSE DURATION MODULATION
      . . . PULSE POSITION MODULATION
      . TRAVELING WAVE MODULATION
      . VELOCITY MODULATION
RT   CARRIER FREQUENCIES
      CARRIER WAVES
      COMPANDING
      CRYSTALLIZATION
      DEMODULATION
      DEMODULATORS
      DOUBLE SIDEBAND TRANSMISSION
      INTERFERENCE FACTOR TABLE
      MODULATORS
      P.A.C.M. TELEMETRY
      RADIO TRANSMISSION
      REMODULATION
      SELECTIVE FADING
      TELECOMMUNICATION
      WAVE INTERACTION

MODULATION TRANSFER FUNCTION
UF   MTF
GS   FUNCTIONS (MATHEMATICS)
      . TRANSFER FUNCTIONS
      . . **MODULATION TRANSFER FUNCTION**
RT   FIGURE OF MERIT
      IMAGING TECHNIQUES
      OPTICAL MEASUREMENT
      OPTICAL TRANSFER FUNCTION
      ∞ PERFORMANCE
      SYSTEM EFFECTIVENESS
      SYSTEMS ANALYSIS

MODULATORS
GS   **MODULATORS**
      . MODEMS
RT   AMPLIFIERS
      AMPLITUDE MODULATION
      DEMODULATORS

MODULATORS-*(CONT.)*
      ELECTRON TUBES
      FREQUENCY MODULATION
      LIGHT MODULATION
      MATCHED FILTERS
      MODULATION
      PHASE MODULATION
      PULSE MODULATION

MODULATORS-DEMODULATORS
USE   MODEMS

MODULES
GS   **MODULES**
      . AIRLOCK MODULES
      . CHEMICAL RELEASE MODULES
      . ELECTRONIC MODULES
      . . MICROMODULES
      . LOCAL SCIENTIFIC SURVEY MODULE
      . PAYLOAD ASSIST MODULE
      . POWER MODULES (STS)
      . SERVICE MODULES
      . SPACECRAFT DOCKING MODULES
      . SPACECRAFT MODULES
      . . COMMAND MODULES
      . . COMMAND SERVICE MODULES
      . . LANDING MODULES
      . . . LUNAR LANDING MODULES
      . . . . LUNAR MODULE
      . . . . . LSSM
      . . . MARS EXCURSION MODULE
      . . . SIM
RT   CIRCUITS
      COMPARTMENTS
      ∞ COMPONENTS
      INSTRUMENT PACKAGES
      SPACE TUGS
      SPARE PARTS

MODULUS OF ELASTICITY
UF   COMPLIANCE (ELASTICITY)
      ELASTIC MODULUS
      YOUNG MODULUS
GS   MECHANICAL PROPERTIES
      . ELASTIC PROPERTIES
      . . **MODULUS OF ELASTICITY**
      . . . DYNAMIC MODULUS OF ELASTICITY
RT   ANELASTICITY
      BENDING
      HOOKES LAW
      HYDROELASTICITY
      POISSON RATIO
      PROPORTIONAL LIMIT
      ∞ RIGIDITY
      SHEAR PROPERTIES
      STIFFNESS
      STRESS-STRAIN DIAGRAMS

MOHAWK AIRCRAFT
USE   OV-1 AIRCRAFT

MOHR CIRCLES
USE   FRACTURE MECHANICS

MOIRE EFFECTS
RT   BEAT FREQUENCIES
      BIREFRINGENCE
      DIFFRACTION
      ∞ EFFECTS
      FRINGE MULTIPLICATION
      INTERFERENCE GRATING
      ∞ METHODOLOGY
      MOIRE FRINGES
      MOIRE INTERFEROMETRY
      PHOTOELASTIC ANALYSIS
      SCHLIEREN PHOTOGRAPHY

MOIRE FRINGES
RT   DIFFRACTION PATTERNS
      FRINGE MULTIPLICATION
      INTERFERENCE GRATING
      MOIRE EFFECTS
      STRESS ANALYSIS
      STRESS CONCENTRATION

MOIRE INTERFEROMETRY
GS   INTERFEROMETRY
      . **MOIRE INTERFEROMETRY**
RT   DIFFRACTION PATTERNS
      HOLOGRAPHIC INTERFEROMETRY
      MOIRE EFFECTS

MOISTURE
GS   **MOISTURE**

MOISTURE-*(CONT.)*
      . SOIL MOISTURE
RT   HUMIDITY
      HYGRAL PROPERTIES
      METEOROLOGICAL PARAMETERS
      METEOROLOGY
      WATER
      WATER VAPOR

MOISTURE CONTENT
UF   DAMPNESS
      WATER CONTENT
      WETNESS
GS   COMPOSITION (PROPERTY)
      . CONCENTRATION (COMPOSITION)
      . . **MOISTURE CONTENT**
      . . . ATMOSPHERIC MOISTURE
RT   ATMOSPHERIC COMPOSITION
      CHEMICAL PROPERTIES
      HUMIDITY
      HYDROTHERMAL STRESS ANALYSIS
      HYGROSCOPICITY
      LYSIMETERS
      SOIL MOISTURE
      WATER
      WATER VAPOR

MOISTURE DETECTORS
USE   MOISTURE METERS

MOISTURE METERS
UF   MOISTURE DETECTORS
GS   MEASURING INSTRUMENTS
      . **MOISTURE METERS**
      . . HYGROMETERS
      . . . PSYCHROMETERS
RT   CHEMICAL ANALYSIS
      HUMIDITY
      HUMIDITY MEASUREMENT

MOISTURE RESISTANCE
RT   CAULKING
      COATINGS
      HYDROTHERMAL STRESS ANALYSIS
      HYGROSCOPICITY
      POROSITY
      ∞ RESISTANCE
      SEALING
      WATERPROOFING
      WEATHERPROOFING

MOJAVE DESERT (CA)
GS   LAND
      . DESERTS
      . . **MOJAVE DESERT (CA)**
RT   ARID LANDS
      CALIFORNIA
      DESERTIFICATION
      REMOTE REGIONS

MOL (ORBITAL LABORATORIES)
USE   MANNED ORBITAL LABORATORIES

MOLABS
USE   LUNAR MOBILE LABORATORIES

∞ **MOLD**
SN   *(USE OF A MORE SPECIFIC TERM IS*
      RECOMMENDED--CONSULT THE TERMS
      LISTED BELOW)
RT   ASPERGILLUS
      FUNGI
      MOLDS
      RHIZOPUS
      RUST FUNGI

MOLDAVITE
GS   ROCKS
      . IGNEOUS ROCKS
      . . OBSIDIAN
      . . . **MOLDAVITE**
RT   GLASS
      METEORITES
      SOILS

MOLDING MATERIALS
RT   BINDERS (MATERIALS)
      CASTING
      CLAYS
      CORES
      INJECTION MOLDING
      ∞ MATERIALS
      MOLDS
      PLASTERS

MOLDING MATERIALS-(CONT.)
  PLASTICS
  SAND CASTING
  SANDS
  TENITE

MOLDS
  SN   (EXCLUDES ORGANISMS)
  RT   CASTING
        CASTINGS
        DIES
        FLAT PATTERNS
        FOUNDRIES
        INGOTS
        INJECTION MOLDING
        MANDRELS
        MELTING
      ∞ MOLD
        MOLDING MATERIALS
        ORGANIC MATERIALS
      ∞ PATTERNS
        PREFORMS
        PRESSING (FORMING)
        PUNCHES
        TABLETS
        TEMPLATES

MOLECULAR ABSORPTION
  GS   ENERGY ABSORPTION
        . **MOLECULAR ABSORPTION**
        RADIATION ABSORPTION
        . **MOLECULAR ABSORPTION**
  RT ∞ ABSORPTION
        ATMOSPHERIC ATTENUATION
        BEER LAW
        ELECTROMAGNETIC ABSORPTION
        LIGHT TRANSMISSION

MOLECULAR BEAM EPITAXY
  GS   GROWTH
        . CRYSTAL GROWTH
        . . EPITAXY
        . . . **MOLECULAR BEAM EPITAXY**

MOLECULAR BEAMS
  GS   BEAMS (RADIATION)
        . PARTICLE BEAMS
        . . NEUTRAL BEAMS
        . . . **MOLECULAR BEAMS**
  RT   ATOMIC BEAMS
        ATOMIC CLOCKS
        FREE MOLECULAR FLOW
        ION BEAMS
        MOLECULES
        RAREFIED GAS DYNAMICS

MOLECULAR BIOLOGY
  GS   LIFE SCIENCES
        . **MOLECULAR BIOLOGY**
  RT   BIOCHEMISTRY
      ∞ BIOLOGY
        PHYSIOCHEMISTRY

MOLECULAR BONDS
  USE   CHEMICAL BONDS

MOLECULAR CHAINS
  RT   CHAINS
        CRYSTAL LATTICES
        MONOMERS

MOLECULAR CLOUDS
  GS   CLOUDS
        . **MOLECULAR CLOUDS**
  RT   ASTRONOMICAL MODELS
        COSMIC DUST
        HYDROGEN CLOUDS
        INTERSTELLAR CHEMISTRY
        INTERSTELLAR GAS
        INTERSTELLAR MASERS
        INTERSTELLAR MATTER

MOLECULAR COLLISIONS
  GS   COLLISIONS
        . **MOLECULAR COLLISIONS**
        PARTICLE INTERACTIONS
        . MOLECULAR INTERACTIONS
        . . **MOLECULAR COLLISIONS**
  RT   ATOMIC COLLISIONS
      ∞ INTERACTIONS
        PARTICLE COLLISIONS
        RIGID ROTORS (PLASMA PHYSICS)

MOLECULAR DIFFUSION
  GS   DIFFUSION
        . **MOLECULAR DIFFUSION**
  RT   ATMOSPHERIC DIFFUSION
        DIFFUSION COEFFICIENT
        DIFFUSION WAVES
        DISSOCIATION
        GASEOUS DIFFUSION
        GASEOUS SELF-DIFFUSION
        PARTICLE DIFFUSION
        SELF DIFFUSION (SOLID STATE)
        SURFACE DIFFUSION

MOLECULAR DISSOCIATION
  USE   DISSOCIATION

MOLECULAR ELECTRONICS
  GS   ELECTROPHYSICS
        . **MOLECULAR ELECTRONICS**
  RT   DTL INTEGRATED CIRCUITS
      ∞ ELECTRONICS
        INTEGRATED CIRCUITS
        LINEAR INTEGRATED CIRCUITS
        MICROELECTRONICS
        MICROMINIATURIZATION
        MINIATURE ELECTRONIC EQUIPMENT
        MONOMOLECULAR FILMS
        PI-ELECTRONS
        SEMICONDUCTOR DEVICES
        THIN FILMS
        TTL INTEGRATED CIRCUITS

MOLECULAR ENERGY LEVELS
  GS   LEVEL (QUANTITY)
        . ENERGY LEVELS
        . . **MOLECULAR ENERGY LEVELS**
        . . . INTERMOLECULAR FORCES
  RT   CHEMICAL ENERGY
      ∞ ENERGY
        ENERGY OF FORMATION
        EXCIMERS
        FREE ENERGY
        HEAT OF SOLUTION
        INTERNAL ENERGY
      ∞ NUCLEAR ENERGY

MOLECULAR EXCITATION
  GS   EXCITATION
        . **MOLECULAR EXCITATION**
  RT   ATOMIC EXCITATIONS
        ENERGY LEVELS
        IONIZATION
        PARTICLE COLLISIONS

MOLECULAR FLOW
  SN   (FLOW WITH KNUDSEN NUMBERS
        GREATER THAN 0.01--FOR SPECIFIC
        FLOWS IN THIS RANGE USE
        NARROWER TERMS--FOR DUCTED
        MOLECULAR FLOW USE KNUDSEN
        FLOW)
  GS   FLUID FLOW
        . GAS FLOW
        . . **MOLECULAR FLOW**
        . . . SLIP FLOW
        . . . TRANSITION FLOW
  RT   BOUNDARY LAYER TRANSITION
        CONTINUUM FLOW
        KNUDSEN FLOW
        LOW DENSITY FLOW
        RAREFIED GAS DYNAMICS
        TRANSPIRATION

MOLECULAR GASES
  GS   GASES
        . **MOLECULAR GASES**
        . . POLAR GASES
        . . POLYATOMIC GASES
        . . . DIATOMIC GASES
  RT   ASSOCIATION REACTIONS
        GAS DYNAMICS
        MONATOMIC GASES
        NONPOLAR GASES
        RAREFIED GASES
        REAL GASES

MOLECULAR INTERACTIONS
  GS   PARTICLE INTERACTIONS
        . **MOLECULAR INTERACTIONS**
        . . INTERSTELLAR CHEMISTRY
        . . MOLECULAR COLLISIONS
  RT   CONFIGURATION INTERACTION
        DISSOCIATION
      ∞ INTERACTIONS
        INTERMOLECULAR FORCES

MOLECULAR INTERACTIONS-(CONT.)
        INTERNUCLEAR PROPERTIES
        IONIC REACTIONS
        LENNARD-JONES POTENTIAL
        MASS FLOW
      ∞ MOLECULAR PHYSICS
        TRANSPORT THEORY

MOLECULAR IONS
  GS   IONS
        . **MOLECULAR IONS**
  RT   FORMYL IONS
      ∞ MOLECULAR PHYSICS
        POSITIVE IONS

MOLECULAR ORBITALS
  GS   ORBITALS
        . **MOLECULAR ORBITALS**
        WAVE FUNCTIONS
        . **MOLECULAR ORBITALS**
  RT   QUANTUM THEORY
        SELF CONSISTENT FIELDS

MOLECULAR OSCILLATIONS
  GS   OSCILLATIONS
        . **MOLECULAR OSCILLATIONS**
  RT   ARGON LASERS
        CARBON DIOXIDE LASERS
        CARBON MONOXIDE LASERS
        GAS LASERS
        OSCILLATOR STRENGTHS

MOLECULAR OSCILLATORS
  GS   OSCILLATORS
        . **MOLECULAR OSCILLATORS**
  RT   LASERS
        MASERS
        OSCILLATOR STRENGTHS
        TWO-WAVELENGTH LASERS
        ULTRAVIOLET LASERS

∞ **MOLECULAR PHYSICS**
  SN   (USE OF A MORE SPECIFIC TERM IS
        RECOMMENDED--CONSULT THE TERMS
        LISTED BELOW)
  RT   INTERNUCLEAR PROPERTIES
        MAYER PROBLEM
        MOLECULAR INTERACTIONS
        MOLECULAR IONS
        MONATOMIC MOLECULES
      ∞ PHYSICS
      ∞ SCIENCE

MOLECULAR PUMPS
  GS   PUMPS
        . VACUUM PUMPS
        . . **MOLECULAR PUMPS**
        VACUUM APPARATUS
        . VACUUM PUMPS
        . . **MOLECULAR PUMPS**
  RT   PLASMA PUMPING
        PUMP SEALS

MOLECULAR RELAXATION
  UF   CHEMICAL RELAXATION
        VIBRATIONAL RELAXATION
  RT   GAS FLOW
        POPULATION INVERSION
      ∞ RELAXATION
        RELAXATION (MECHANICS)
        RELAXATION TIME
        SHOCK WAVES
        THERMODYNAMICS
        VIBRATION DAMPING
        VIBRATIONAL SPECTRA

MOLECULAR ROTATION
  GS   GYRATION
        . ROTATION
        . . **MOLECULAR ROTATION**
  RT   MICROWAVE SPECTRA
        RAMAN SPECTRA
        RIGID ROTORS (PLASMA PHYSICS)

MOLECULAR SHIELDS
  RT   CONTAMINATION
        HIGH VACUUM
        INSTRUMENT PACKAGES
        SPACEBORNE EXPERIMENTS

MOLECULAR SIEVES
  USE   ABSORBENTS

MOLECULAR SPECTRA
GS   SPECTRA
     . **MOLECULAR SPECTRA**
     . . ELECTRONIC SPECTRA
     . . RAMAN SPECTRA
     . . VIBRATIONAL SPECTRA
RT   ABSORPTION SPECTRA
     ELECTROMAGNETIC SPECTRA
     EMISSION SPECTRA
     ENERGY SPECTRA
     INFRARED SPECTRA
     MASS SPECTRA
     MICROWAVE SPECTRA
     OXYGEN SPECTRA
     SOLAR SPECTRA
     STELLAR SPECTRA
     SWAN BANDS
     ULTRAVIOLET SPECTRA
     VEGARD-KAPLAN BANDS
     VISIBLE SPECTRUM

MOLECULAR SPECTROSCOPY
GS   SPECTROSCOPY
     . **MOLECULAR SPECTROSCOPY**
     . . RAMAN SPECTROSCOPY
RT   ABSORPTION SPECTRA
     ELECTRON SPECTROSCOPY
     EMISSION SPECTROSCOPY
     INFRARED SPECTROSCOPY
     LINE SPECTRA
     MICROWAVE SPECTRA
     OPTOGALVANIC SPECTROSCOPY
     SPECTROSCOPIC ANALYSIS
     ULTRAVIOLET SPECTROSCOPY
     VACUUM SPECTROSCOPY
     X RAY SPECTROSCOPY

MOLECULAR STRUCTURE
RT   ATOMIC INTERACTIONS
     ATOMIC STRUCTURE
     COMPLEX COMPOUNDS
     CONFIGURATION INTERACTION
     CRYSTAL LATTICES
     INFRARED SPECTROSCOPY
     INTERMOLECULAR FORCES
     INTRAMOLECULAR STRUCTURES
     MOLECULES
     MONATOMIC MOLECULES
     NUCLEAR MAGNETIC RESONANCE
     NUCLEAR MODELS
     ORDER-DISORDER TRANSFORMATIONS
     POLYATOMIC MOLECULES
     POLYWATER
     ∞STRUCTURES
     UNIMOLECULAR STRUCTURES
     WISWESSER NOTATIONS

MOLECULAR THEORY
RT   LIGHTHILL GAS MODEL
     ∞THEORIES

MOLECULAR TRAJECTORIES
GS   TRAJECTORIES
     . **MOLECULAR TRAJECTORIES**
RT   GAS FLOW

MOLECULAR WEIGHT
GS   **MOLECULAR WEIGHT**
     . LOW MOLECULAR WEIGHTS
RT   MOLECULES
     MONATOMIC MOLECULES
     POLYATOMIC MOLECULES
     WEIGHT (MASS)

MOLECULES
UF   MACROMOLECULES
GS   **MOLECULES**
     . DIATOMIC MOLECULES
     . MONATOMIC MOLECULES
     . POLYATOMIC MOLECULES
     . . TRIATOMIC MOLECULES
RT   ATOMS
     CHEMICAL BONDS
     ∞CHEMICAL COMPOUNDS
     IONS
     LOW MOLECULAR WEIGHTS
     MOLECULAR BEAMS
     MOLECULAR STRUCTURE
     MOLECULAR WEIGHT

MOLES
GS   ANIMALS
     . **MOLES**

MOLIERE FORMULA
USE   COSMIC RAY SHOWERS
     SECONDARY COSMIC RAYS
     SPATIAL DISTRIBUTION

MOLLIER DIAGRAM
UF   ENTHALPY-ENTROPY DIAGRAMS
GS   CHARTS
     . GRAPHS (CHARTS)
     . . **MOLLIER DIAGRAM**
     DIAGRAMS
     . **MOLLIER DIAGRAM**
RT   ENTHALPY
     ENTROPY
     EQUATIONS OF STATE
     IDEAL FLUIDS
     ISENTROPE
     THERMODYNAMICS

MOLLUSKS
GS   ANIMALS
     . INVERTEBRATES
     . . ARTHROPODS
     . . . CEPHALOPODS
     . . . . **MOLLUSKS**
RT   SHELLFISH

MOLNIYA SATELLITES
GS   SATELLITES
     . ARTIFICIAL SATELLITES
     . . COMMUNICATION SATELLITES
     . . . **MOLNIYA SATELLITES**
     . SOVIET SATELLITES
     . . . **MOLNIYA SATELLITES**
RT   ELECTRIC DISCHARGES
     RADIO RELAY SYSTEMS
     SATELLITE NETWORKS
     TELECOMMUNICATION
     TELEVISION TRANSMISSION
     U.S.S.R. SPACE PROGRAM

MOLTEN SALT ELECTROLYTES
GS   CONDUCTORS
     . ELECTROLYTES
     . . **MOLTEN SALT ELECTROLYTES**

MOLTEN SALT NUCLEAR REACTORS
UF   MSRE REACTORS
GS   NUCLEAR REACTORS
     . **MOLTEN SALT NUCLEAR REACTORS**
RT   ∞REACTORS

MOLTEN SALTS
RT   HALIDES
     HALITES
     INORGANIC COMPOUNDS
     NITRIDES
     SALT BATHS
     ∞SALTS
     SODIUM CHLORIDES

MOLTING
RT   PHENOLOGY
     SHEDDING

MOLYBDATES
GS   MOLYBDENUM COMPOUNDS
     . **MOLYBDATES**

MOLYBDENUM
GS   CHEMICAL ELEMENTS
     . REFRACTORY METALS
     . . **MOLYBDENUM**
     METALS
     . TRANSITION METALS
     . . REFRACTORY METALS
     . . . **MOLYBDENUM**
     REFRACTORY MATERIALS
     . REFRACTORY METALS
     . . **MOLYBDENUM**

MOLYBDENUM ALLOYS
GS   ALLOYS
     . HEAT RESISTANT ALLOYS
     . . REFRACTORY METAL ALLOYS
     . . . **MOLYBDENUM ALLOYS**
     . . . . RENE 41
     . . . . RENE 63
     . . . . RENE 77
     REFRACTORY MATERIALS
     . REFRACTORY METAL ALLOYS
     . . **MOLYBDENUM ALLOYS**
     . . . RENE 41
     . . . RENE 63

MOLYBDENUM ALLOYS-(CONT.)
     . . . RENE 77
RT   HASTELLOY (TRADEMARK)
     MULBERRY (ALLOY)
     PERMALLOYS (TRADEMARK)
     STAINLESS STEELS

MOLYBDENUM CARBIDES
GS   CARBON COMPOUNDS
     . CARBIDES
     . . **MOLYBDENUM CARBIDES**

MOLYBDENUM COMPOUNDS
GS   **MOLYBDENUM COMPOUNDS**
     . LEAD MOLYBDATES
     . MOLYBDATES
     . MOLYBDENUM DISULFIDES
     . MOLYBDENUM OXIDES
RT   ∞CHEMICAL COMPOUNDS
     ∞GROUP 6B COMPOUNDS
     ∞METAL COMPOUNDS

MOLYBDENUM DISULFIDES
GS   CHALCOGENIDES
     . SULFIDES
     . . INORGANIC SULFIDES
     . . . MOLYBDENUM SULFIDES
     . . . . **MOLYBDENUM DISULFIDES**
     MOLYBDENUM COMPOUNDS
     . **MOLYBDENUM DISULFIDES**
     SULFUR COMPOUNDS
     . SULFIDES
     . . INORGANIC SULFIDES
     . . . MOLYBDENUM SULFIDES
     . . . . **MOLYBDENUM DISULFIDES**

MOLYBDENUM OXIDES
GS   CHALCOGENIDES
     . OXIDES
     . . METAL OXIDES
     . . . **MOLYBDENUM OXIDES**
     MOLYBDENUM COMPOUNDS
     . **MOLYBDENUM OXIDES**

MOLYBDENUM SULFIDES
GS   CHALCOGENIDES
     . SULFIDES
     . . INORGANIC SULFIDES
     . . . **MOLYBDENUM SULFIDES**
     . . . . MOLYBDENUM DISULFIDES
     SULFUR COMPOUNDS
     . SULFIDES
     . . INORGANIC SULFIDES
     . . . **MOLYBDENUM SULFIDES**
     . . . . MOLYBDENUM DISULFIDES

MOM (SEMICONDUCTORS)
UF   METAL-OXIDE-METAL SEMICONDUCTORS
GS   SEMICONDUCTORS (MATERIALS)
     . **MOM (SEMICONDUCTORS)**
RT   ION IMPLANTATION

MOMENT DISTRIBUTION
GS   DISTRIBUTION (PROPERTY)
     . **MOMENT DISTRIBUTION**
RT   ANGULAR DISTRIBUTION
     FORCE DISTRIBUTION
     INFLUENCE COEFFICIENT
     LOADING MOMENTS
     LOADS (FORCES)
     MASS DISTRIBUTION
     METHOD OF MOMENTS
     MOMENTS
     MOMENTS OF INERTIA
     PRESSURE DISTRIBUTION
     STATIC LOADS
     STRESS CONCENTRATION
     STRUCTURAL ANALYSIS
     STRUCTURAL DESIGN CRITERIA

MOMENTS
GS   **MOMENTS**
     . BENDING MOMENTS
     . DIPOLE MOMENTS
     . . ELECTRIC MOMENTS
     . . MAGNETIC MOMENTS
     . DISTRIBUTION MOMENTS
     . . MEAN
     . . ORTHOGONALITY
     . . STANDARD DEVIATION
     . LOADING MOMENTS
     . MOMENTS OF INERTIA
     . STABILITY DERIVATIVES
     . . PITCHING MOMENTS
     . . ROLLING MOMENTS

MOMENTS-*(CONT.)*
```
       . . YAWING MOMENTS
       . TORQUE
RT     METHOD OF MOMENTS
       MODE (STATISTICS)
       MOMENT DISTRIBUTION
       MOMENTUM
       SKEWNESS
       TORSION
       VARIANCE (STATISTICS)
```

MOMENTS OF INERTIA
```
UF     INERTIA MOMENTS
GS     MOMENTS
       . MOMENTS OF INERTIA
RT     ANGULAR MOMENTUM
       CENTER OF GRAVITY
       CENTER OF PRESSURE
       CENTROIDS
       EQUATIONS OF MOTION
       EULER EQUATIONS OF MOTION
       INERTIA
       INERTIA PRINCIPLE
       MACH INERTIA PRINCIPLE
       MASS
       MASS DISTRIBUTION
       MOMENT DISTRIBUTION
       PITCHING MOMENTS
       ROLLING MOMENTS
       STABILITY DERIVATIVES
       STRESS ANALYSIS
       STRUCTURAL STRAIN
       TORQUE
       YAWING MOMENTS
```

MOMENTUM
```
GS     MOMENTUM
       . ANGULAR MOMENTUM
RT     CLASSICAL MECHANICS
       DE BROGLIE WAVELENGTHS
       ∞ DYNAMICS
       MOMENTS
       ∞ MOTION
       MOTION AFTEREFFECTS
       PENDULUMS
       TURNING FLIGHT
```

MOMENTUM ENERGY
```
USE    KINETIC ENERGY
```

MOMENTUM THEORY
```
RT     CONSERVATION LAWS
       NEWTON SECOND LAW
       ∞ THEORIES
```

MOMENTUM TRANSFER
```
RT     ∞ DYNAMICS
       ENERGY TRANSFER
       GAS-LIQUID INTERACTIONS
       HYDRODYNAMIC RAM EFFECT
       KINETIC THEORY
       KINETICS
       PRANDTL NUMBER
       TRANSFERRING
```

MONACO
```
GS     NATIONS
       . MONACO
RT     EUROPE
```

MONATOMIC GASES
```
UF     ATOMIC GASES
GS     GASES
       . MONATOMIC GASES
RT     CHAPMAN-ENSKOG THEORY
       MOLECULAR GASES
       RARE GASES
       REAL GASES
```

MONATOMIC MOLECULES
```
GS     MOLECULES
       . MONATOMIC MOLECULES
RT     ATOMS
       CHEMICAL BONDS
       ∞ CHEMICAL COMPOUNDS
       IONS
       LOW MOLECULAR WEIGHTS
       ∞ MOLECULAR PHYSICS
       MOLECULAR STRUCTURE
       MOLECULAR WEIGHT
       POSITIVE IONS
```

MONAURAL SIGNALS
```
RT     AUDIO EQUIPMENT
```

MONAURAL SIGNALS-*(CONT.)*
```
       AUDIO FREQUENCIES
       AUDITORY PERCEPTION
       AUDITORY SIGNALS
       LOUDSPEAKERS
       MICROPHONES
       SOUND TRANSMISSION
```

MONAZITE SANDS
```
GS     PHOSPHORUS COMPOUNDS
       . PHOSPHATES
       . . MONAZITE SANDS
       SEDIMENTS
       . SANDS
       . . MONAZITE SANDS
       SOILS
       . SANDS
       . . MONAZITE SANDS
RT     MINERALS
       SEDIMENTARY ROCKS
```

MONEL (TRADEMARK)
```
GS     ALLOYS
       . NICKEL ALLOYS
       . . MONEL (TRADEMARK)
```

MONGE-AMPERE EQUATION
```
GS     ALGEBRA
       . NONLINEAR EQUATIONS
       . . MONGE-AMPERE EQUATION
       ANALYSIS (MATHEMATICS)
       . REAL VARIABLES
       . . DIFFERENTIAL EQUATIONS
       . . . PARTIAL DIFFERENTIAL EQUATIONS
       . . . . ELLIPTIC DIFFERENTIAL
                EQUATIONS
       . . . . . MONGE-AMPERE EQUATION
       . NONLINEAR EQUATIONS
       . . MONGE-AMPERE EQUATION
RT     BOUNDARY VALUE PROBLEMS
       ∞ EQUATIONS
```

MONGOLIA
```
GS     NATIONS
       . MONGOLIA
RT     ASIA
```

MONITORS
```
GS     MONITORS
       . INFRARED RADIOMETERS
RT     AIRCRAFT INSTRUMENTS
       ALARM PROJECT
       ANALYZERS
       CONICAL SCANNING
       COUNTERS
       DATA RECORDERS
       ∞ DETECTORS
       DISPLAY DEVICES
       ENVIRONMENTAL MONITORING
       GAS DETECTORS
       HELMET MOUNTED DISPLAYS
       IN-FLIGHT MONITORING
       ∞ INSTRUMENTS
       ∞ MEASUREMENT
       MEASURING INSTRUMENTS
       OPTICAL SCANNERS
       POLLUTION MONITORING
       RADIATION MEASURING INSTRUMENTS
       SCANNING
       WARNING
       WARNING SYSTEMS
```

MONKEYS
```
GS     ANIMALS
       . VERTEBRATES
       . . MAMMALS
       . . . PRIMATES
       . . . . MONKEYS
       . WILDLIFE
       . . MONKEYS
```

MONOCHROMATIC RADIATION
```
SN     (LIMITED TO ELECTROMAGNETIC
        RADIATION)
GS     ELECTROMAGNETIC RADIATION
       . MONOCHROMATIC RADIATION
RT     BEAMS (RADIATION)
       BRILLOUIN EFFECT
       COHERENT ELECTROMAGNETIC
        RADIATION
       COHERENT LIGHT
       ∞ FILTERS
       GAMMA RAYS
       INFRARED RADIATION
       IONIZING RADIATION
```

MONOCHROMATIC RADIATION-*(CONT.)*
```
       LIGHT (VISIBLE RADIATION)
       LONG WAVE RADIATION
       MONOCHROMATIZATION
       MONOCHROMATORS
       POLARIZED ELECTROMAGNETIC
        RADIATION
       POLARIZED LIGHT
       ∞ RADIATION
       RADIO WAVES
       SHORT WAVE RADIATION
       ULTRAVIOLET RADIATION
       X RAYS
```

MONOCHROMATIZATION
```
UF     INTERFERENCE MONOCHROMATIZATION
RT     MONOCHROMATIC RADIATION
       PARTICLE ENERGY
       POLARIZATION (WAVES)
```

MONOCHROMATORS
```
GS     MEASURING INSTRUMENTS
       . MONOCHROMATORS
       RADIATION SOURCES
       . MONOCHROMATORS
RT     COMPARATORS
       DUOCHROMATORS
       GONIOMETERS
       LIGHT SOURCES
       MONOCHROMATIC RADIATION
       OPTICAL EQUIPMENT
       OPTICAL MEASURING INSTRUMENTS
       PHOTOGONIOMETERS
       SPECTROPHOTOMETERS
```

MONOCOQUE STRUCTURES
```
RT     ∞ CYLINDERS
       SHELLS (STRUCTURAL FORMS)
       STRESSED-SKIN STRUCTURES
       ∞ STRUCTURES
```

MONOCRYSTALS
```
USE    SINGLE CRYSTALS
```

MONOCULAR VISION
```
GS     VISION
       . MONOCULAR VISION
RT     HUMAN FACTORS ENGINEERING
       MOTION PERCEPTION
       PERCEPTION
       SPACE PERCEPTION
```

MONOETHANOLAMINE (MEA)
```
GS     ALIPHATIC COMPOUNDS
       . MONOETHANOLAMINE (MEA)
       AMINES
       . MONOETHANOLAMINE (MEA)
```

MONOIDS
```
GS     ALGEBRA
       . GROUP THEORY
       . . HOMOMORPHISMS
       . . . MONOIDS
```

MONOLITHIC CIRCUITS
```
USE    INTEGRATED CIRCUITS
```

MONOMERS
```
RT     DIBASIC COMPOUNDS
       DIMERS
       MOLECULAR CHAINS
       ∞ POLYMERS
       PREPOLYMERS
       TRIMERS
```

MONOMOLECULAR FILMS
```
GS     SURFACE LAYERS
       . MONOMOLECULAR FILMS
       THIN FILMS
       . MONOMOLECULAR FILMS
RT     ENERGY ABSORPTION FILMS
       ∞ FILMS
       INTEGRATED OPTICS
       ∞ LAYERS
       MOLECULAR ELECTRONICS
       SURFACTANTS
       THIN LAYER CHROMATOGRAPHY
```

MONOPLANES
```
GS     MONOPLANES
       . A-1 AIRCRAFT
       . A-2 AIRCRAFT
       . A-3 AIRCRAFT
       . A-4 AIRCRAFT
```

MONOPLANES—*(CONT.)*
- . A-5 AIRCRAFT
- . A-6 AIRCRAFT
- . A-7 AIRCRAFT
- . A-37 AIRCRAFT
- . AN-2 AIRCRAFT
- . AN-22 AIRCRAFT
- . AN-24 AIRCRAFT
- . ARGOSY MK 1 AIRCRAFT
- . AVRO 707 AIRCRAFT
- . B-26 AIRCRAFT
- . B-47 AIRCRAFT
- . B-50 AIRCRAFT
- . B-52 AIRCRAFT
- . B-57 AIRCRAFT
- . B-58 AIRCRAFT
- . B-66 AIRCRAFT
- . B-70 AIRCRAFT
- . BAC 111 AIRCRAFT
- . BEECHCRAFT 18 AIRCRAFT
- . BOEING 707 AIRCRAFT
- . BOEING 720 AIRCRAFT
- . BOEING 733 AIRCRAFT
- . BOEING 737 AIRCRAFT
- . BOEING 757 AIRCRAFT
- . BOEING 767 AIRCRAFT
- . BREGUET 940 AIRCRAFT
- . BREGUET 941 AIRCRAFT
- . BREGUET 1150 AIRCRAFT
- . BUCCANEER AIRCRAFT
- . C-2 AIRCRAFT
- . C-33 AIRCRAFT
- . C-35 AIRCRAFT
- . C-46 AIRCRAFT
- . C-47 AIRCRAFT
- . C-54 AIRCRAFT
- . C-118 AIRCRAFT
- . C-121 AIRCRAFT
- . C-123 AIRCRAFT
- . C-124 AIRCRAFT
- . C-130 AIRCRAFT
- . C-131 AIRCRAFT
- . C-133 AIRCRAFT
- . C-135 AIRCRAFT
- . C-140 AIRCRAFT
- . C-141 AIRCRAFT
- . C-160 AIRCRAFT
- . CANBERRA AIRCRAFT
- . CESSNA L-19 AIRCRAFT
- . CESSNA 172 AIRCRAFT
- . CESSNA 205 AIRCRAFT
- . CESSNA 210 AIRCRAFT
- . CESSNA 402B AIRCRAFT
- . CL-41 AIRCRAFT
- . CL-44 AIRCRAFT
- . COMET 4 AIRCRAFT
- . CV-340 AIRCRAFT
- . CV-440 AIRCRAFT
- . CV-880 AIRCRAFT
- . CV-990 AIRCRAFT
- . D-558 AIRCRAFT
- . DC 3 AIRCRAFT
- . DC 7 AIRCRAFT
- . DC 8 AIRCRAFT
- . DH 112 AIRCRAFT
- . DH 115 AIRCRAFT
- . DH 121 AIRCRAFT
- . DH 125 AIRCRAFT
- . DHC 2 AIRCRAFT
- . DHC 4 AIRCRAFT
- . DHC 5 AIRCRAFT
- . DO-27 AIRCRAFT
- . DO-28 AIRCRAFT
- . DO-31 AIRCRAFT
- . EC-121 AIRCRAFT
- . ELECTRA AIRCRAFT
- . F-2 AIRCRAFT
- . F-5 AIRCRAFT
- . F-8 AIRCRAFT
- . F-9 AIRCRAFT
- . F-17 AIRCRAFT
- . F-27 AIRCRAFT
- . F-28 TRANSPORT AIRCRAFT
- . F-84 AIRCRAFT
- . F-86 AIRCRAFT
- . F-89 AIRCRAFT
- . F-94 AIRCRAFT
- . F-100 AIRCRAFT
- . F-102 AIRCRAFT
- . F-104 AIRCRAFT
- . F-105 AIRCRAFT
- . F-106 AIRCRAFT
- . FD 2 AIRCRAFT
- . G-1 AIRCRAFT
- . G-91 AIRCRAFT
- . G-95/4 AIRCRAFT

MONOPLANES—*(CONT.)*
- . G-222 AIRCRAFT
- . GA-5 AIRCRAFT
- . H-126 AIRCRAFT
- . HFB-320 AIRCRAFT
- . HP-115 AIRCRAFT
- . HS-748 AIRCRAFT
- . IL-14 AIRCRAFT
- . IL-62 AIRCRAFT
- . JET PROVOST AIRCRAFT
- . JINDIVIK TARGET AIRCRAFT
- . L-29 JET TRAINER
- . LOCKHEED MODEL 18 AIRCRAFT
- . MH-262 AIRCRAFT
- . MIRAGE AIRCRAFT
- . MIRAGE 3 AIRCRAFT
- . MYSTERE 20 AIRCRAFT
- . NORD 1500 AIRCRAFT
- . OV-1 AIRCRAFT
- . OV-10 AIRCRAFT
- . P-3 AIRCRAFT
- . P-51 AIRCRAFT
- . P-166 AIRCRAFT
- . P-308 AIRCRAFT
- . P-1052 AIRCRAFT
- . P-1127 AIRCRAFT
- . P-1154 AIRCRAFT
- . PD-808 AIRCRAFT
- . PHANTOM AIRCRAFT
- . RB-50 AIRCRAFT
- . RF-4 AIRCRAFT
- . S-2 AIRCRAFT
- . SAAB 105 AIRCRAFT
- . SC-1 AIRCRAFT
- . SC-5 AIRCRAFT
- . SC-7 AIRCRAFT
- . SCIMITAR AIRCRAFT
- . SE-210 AIRCRAFT
- . SHACKLETON BOMBER
- . T-2 AIRCRAFT
- . T-28 AIRCRAFT
- . T-33 AIRCRAFT
- . T-37 AIRCRAFT
- . T-38 AIRCRAFT
- . T-39 AIRCRAFT
- . TS-11 AIRCRAFT
- . TSR-2 AIRCRAFT
- . TU-104 AIRCRAFT
- . TU-124 AIRCRAFT
- . TU-134 AIRCRAFT
- . U-2 AIRCRAFT
- . U-10 AIRCRAFT
- . VALIANT AIRCRAFT
- . VC-10 AIRCRAFT
- . VICTOR MK-1 AIRCRAFT
- . VISCOUNT AIRCRAFT
- . VJ-101 AIRCRAFT
- . X-1 AIRCRAFT
- . X-2 AIRCRAFT
- . X-3 AIRCRAFT
- . X-5 AIRCRAFT
- . X-13 AIRCRAFT
- . X-14 AIRCRAFT
- . X-21 AIRCRAFT
- . X-21A AIRCRAFT
- . XC-142 AIRCRAFT
- . XV-4 AIRCRAFT
- . XV-5 AIRCRAFT
- . YS-11 AIRCRAFT
- . Z-37 AIRCRAFT
- RT   AERODYNAMIC CONFIGURATIONS
- ∞AIRCRAFT
-   AIRFOILS
-   BIPLANES
-   CARGO AIRCRAFT
-   GLIDERS
- ∞LOW WING AIRCRAFT
-   SEAPLANES
-   TAILLESS AIRCRAFT
-   WATER TAKEOFF AND LANDING AIRCRAFT
-   WING PLANFORMS
-   WING PROFILES
- ∞WINGED VEHICLES

MONOPOLE ANTENNAS
- UF   SPIKE ANTENNAS
- GS   ANTENNAS
- . OMNIDIRECTIONAL ANTENNAS
- . . **MONOPOLE ANTENNAS**
- . . . WHIP ANTENNAS
- RT   ANTENNA DESIGN
-   DIPOLE ANTENNAS
-   LOOP ANTENNAS
-   MONOPOLES
- ∞SPIKES

MONOPOLES
- GS   **MONOPOLES**
- . MAGNETIC MONOPOLES
- RT   ∞DIPOLES
-   MONOPOLE ANTENNAS
-   MULTIPOLES
- ∞POLES

MONOPROPELLANTS
- GS   LIQUIDS
- . LIQUID FUELS
- . . LIQUID ROCKET PROPELLANTS
- . . . **MONOPROPELLANTS**
- . . . . AEROZINE
-   PROPELLANTS
- . ROCKET PROPELLANTS
- . . LIQUID ROCKET PROPELLANTS
- . . . **MONOPROPELLANTS**
- . . . . AEROZINE
- RT   AIRCRAFT FUELS
-   CHEMICAL FUELS
-   GASEOUS ROCKET PROPELLANTS
-   GELLED ROCKET PROPELLANTS
-   METAL PROPELLANTS
-   PLASTIC PROPELLANTS
-   PROPELLANT DECOMPOSITION
-   SLURRY PROPELLANTS
-   SOLID ROCKET PROPELLANTS

MONOPULSE ANTENNAS
- GS   ANTENNAS
- . **MONOPULSE ANTENNAS**
- RT   DIRECTIONAL ANTENNAS
-   PHASED ARRAYS
-   WAVEGUIDE ANTENNAS

MONOPULSE RADAR
- GS   RADAR
- . PULSE RADAR
- . . PULSE DOPPLER RADAR
- . . . **MONOPULSE RADAR**
- RT   DOPPLER RADAR
-   DUPLEXERS
-   RADAR TRACKING
-   TRACKING RADAR

MONOSACCHARIDES
- GS   ALIPHATIC COMPOUNDS
- . SUGARS
- . . **MONOSACCHARIDES**
- . . . RIBOSE
- . . . XYLOSE
-   CARBOHYDRATES
- . SUGARS
- . . **MONOSACCHARIDES**
- . . . RIBOSE
- . . . XYLOSE

MONOSCOPES
- GS   MICROWAVE EQUIPMENT
- . MICROWAVE TUBES
- . . **MONOSCOPES**
-   TELEVISION EQUIPMENT
- . **MONOSCOPES**
- RT   CAMERA TUBES
-   ELECTRON BEAMS
-   IMAGE TUBES
-   SECONDARY EMISSION
- ∞TEST EQUIPMENT

MONOSTABLE MULTIVIBRATORS
- GS   CIRCUITS
- . MULTIVIBRATORS
- . . **MONOSTABLE MULTIVIBRATORS**

MONOTECTIC ALLOYS
- GS   ALLOYS
- . **MONOTECTIC ALLOYS**
- RT   COMPOSITE MATERIALS
- ∞MATRICES
-   METAL MATRIX COMPOSITES
-   METALS
-   MIXTURES

MONOTONE FUNCTIONS
- GS   FUNCTIONS (MATHEMATICS)
- . **MONOTONE FUNCTIONS**
- RT   ANALYSIS (MATHEMATICS)
-   CALCULUS
-   REAL VARIABLES

MONOTONY
- RT   BOREDOM
-   LETHARGY

MONOTONY-*(CONT.)*
SENSORY DEPRIVATION

MONSOONS
- GS   WIND (METEOROLOGY)
- . **MONSOONS**
- RT   ANNUAL VARIATIONS
   ATMOSPHERIC CIRCULATION
   GROUND WIND
   PRECIPITATION (METEOROLOGY)
   SEA BREEZE

MONTANA
- GS   NATIONS
- . UNITED STATES
- . . **MONTANA**
- RT   BIGHORN MOUNTAINS (MT-WY)
   MISSOURI RIVER (US)
   WILLISTON BASIN (NORTH AMERICA)
   YELLOWSTONE NATIONAL PARK
      (ID-MT-WY)

MONTE CARLO METHOD
- GS   ANALYSIS (MATHEMATICS)
- . NUMERICAL ANALYSIS
- . . **MONTE CARLO METHOD**
- RT   DIFFUSION THEORY
   EXPECTANCY HYPOTHESIS
   GAME THEORY
   MARKOV CHAINS
   MATHEMATICAL MODELS
   ∞METHODOLOGY
   PROBABILITY THEORY
   RANDOM PROCESSES
   RANDOM WALK
   SIMULATION
   STATISTICAL ANALYSIS
   STOCHASTIC PROCESSES
   TRANSPORT THEORY

MONTEREY BAY (CA)
- GS   BAYS (TOPOGRAPHIC FEATURES)
- . **MONTEREY BAY (CA)**
- RT   CALIFORNIA
   PACIFIC OCEAN

MONTH
- UF   LUNATION
- RT   CALENDARS
   TIME
   UNITS OF MEASUREMENT

MONTICELLITE
- GS   CALCIUM COMPOUNDS
- . **MONTICELLITE**
   MAGNESIUM COMPOUNDS
- . **MONTICELLITE**
   MINERALS
- . **MONTICELLITE**
   SILICON COMPOUNDS
- . SILICATES
- . . **MONTICELLITE**
- RT   OLIVINE

MONTMORILLONITE
- GS   ALUMINUM COMPOUNDS
- . ALUMINUM SILICATES
- . . **MONTMORILLONITE**
   CLAYS
- . **MONTMORILLONITE**
   MINERALS
- . **MONTMORILLONITE**
   SILICON COMPOUNDS
- . ALUMINUM SILICATES
- . . **MONTMORILLONITE**
- . SILICATES
- . . **MONTMORILLONITE**
- RT   BENTONITE

MOODS
- RT   EMOTIONAL FACTORS
   EMOTIONS
   PSYCHOLOGICAL EFFECTS
   PSYCHOLOGICAL FACTORS
   SENSORY FEEDBACK

MOON
- GS   CELESTIAL BODIES
- . NATURAL SATELLITES
- . . **MOON**
   SATELLITES
- . NATURAL SATELLITES
- . . **MOON**
- RT   EARTH-MOON SYSTEM

MOON-*(CONT.)*
LIGHT SOURCES
LUNAR ATMOSPHERE
LUNAR BASES
LUNAR COMMUNICATION
LUNAR COMPOSITION
LUNAR CRATERS
LUNAR CRUST
LUNAR DUST
LUNAR ECLIPSES
LUNAR ENVIRONMENT
LUNAR EVOLUTION
LUNAR EXPLORATION
LUNAR FAR SIDE
LUNAR GEOLOGY
LUNAR GRAVITATION
LUNAR LANDING SITES
LUNAR LIMB
LUNAR LUMINESCENCE
LUNAR MAGNETIC FIELDS
LUNAR MAPS
LUNAR OCCULTATION
LUNAR ORBITS
LUNAR PHASES
LUNAR PHOTOGRAPHY
LUNAR RAYS
LUNAR SHADOW
LUNAR SOIL
LUNAR TEMPERATURE
LUNAR TOPOGRAPHY
SELENOGRAPHY
SELENOLOGY

MOON ILLUSION
- GS   PSYCHOLOGICAL EFFECTS
- . ILLUSIONS
- . . **MOON ILLUSION**
- RT   OPTICAL ILLUSION
   SENSORY FEEDBACK

MOON-EARTH TRAJECTORIES
- GS   TRAJECTORIES
- . SPACECRAFT TRAJECTORIES
- . . LUNAR TRAJECTORIES
- . . . **MOON-EARTH TRAJECTORIES**
- RT   APOLLO 5 FLIGHT
   APOLLO 6 FLIGHT
   APOLLO 7 FLIGHT
   APOLLO 8 FLIGHT
   APOLLO 10 FLIGHT
   APOLLO 11 FLIGHT
   APOLLO 12 FLIGHT
   APOLLO 13 FLIGHT
   APOLLO 14 FLIGHT
   APOLLO 15 FLIGHT
   APOLLO 16 FLIGHT
   APOLLO 17 FLIGHT
   CIRCUMLUNAR TRAJECTORIES
   EARTH-MOON TRAJECTORIES
   GODDARD TRAJECTORY
      DETERMINATION SYSTEM
   LUNAR FLIGHT
   MASS DRIVERS (PAYLOAD DELIVERY)
   ORBITAL MECHANICS
   REENTRY TRAJECTORIES
   ROUND TRIP TRAJECTORIES
   TRANSFER ORBITS

MOONQUAKES
- GS   SEISMOLOGY
- . **MOONQUAKES**
- RT   LUNAR GEOLOGY
   LUNAR TIDES
   PLANETARY QUAKES
   SELENOLOGY

MOORING
- UF   MOORINGS
- RT   AIRPORTS
   ANCHORS (FASTENERS)
   FASTENERS
   ∞JOINING
   MATERIALS HANDLING
   MULTIPLE DOCKING ADAPTERS
   SPACECRAFT DOCKING

MOORINGS
- USE   MOORING

MOPS (PROPULSION SYSTEMS)
- USE   MAN OPERATED PROPULSION SYSTEMS

MORAINES
- USE   GLACIAL DRIFT

MORALE
- RT   CREATIVITY
   DISCIPLINING
   LEADERSHIP
   MOTIVATION
   PRODUCTIVITY
   PSYCHOLOGY
   RECREATION

MOREHOUSE COMET
- GS   CELESTIAL BODIES
- . COMETS
- . . **MOREHOUSE COMET**

MORL
- USE   MANNED ORBITAL RESEARCH
      LABORATORIES

MORNING
- RT   DAYTIME
   SUNRISE

MOROCCO
- GS   NATIONS
- . **MOROCCO**
- RT   AFRICA

MORPHINE
- GS   DRUGS
- . NARCOTICS
- . . **MORPHINE**
   HETEROCYCLIC COMPOUNDS
- . ALKALOIDS
- . . **MORPHINE**
   NITROGEN COMPOUNDS
- . ALKALOIDS
- . . **MORPHINE**

MORPHOLOGICAL INDEXES
- GS   RATIOS
- . INDEXES (RATIOS)
- . . **MORPHOLOGICAL INDEXES**
- RT   MORPHOLOGY

MORPHOLOGY
- GS   **MORPHOLOGY**
- . GEOMORPHOLOGY
- . ISOMORPHISM
- . LUNG MORPHOLOGY
- . POLYMORPHISM
- RT   ANATOMY
   ∞BIOLOGY
   DIFFERENTIATION (BIOLOGY)
   GEOLOGY
   HISTOLOGY
   ∞MATHEMATICS
   MORPHOLOGICAL INDEXES
   SHAPES
   VESTIBULES

MORPHOTROPISM
- USE   ISOMORPHISM

MORSE CODE
- RT   ∞CODES
   COMMUNICATING
   KEYING
   RADIO TELEGRAPHY
   TELECOMMUNICATION

MORSE POTENTIAL
- RT   DIATOMIC MOLECULES
   KINETIC THEORY
   POTENTIAL ENERGY

MORTALITY
- RT   AGING (BIOLOGY)
   DEATH
   EXPIRATION
   LIFE SPAN
   MILLS RATIO

MORTARS (MATERIAL)
- RT   ADMIXTURES
   BRICKS
   CEMENTS
   CERAMICS
   CONCRETES
   GROUT
   MASONRY
   PLASTERS
   REFRACTORIES

MOS (JAPANESE SPACECRAFT)
USE   JAPANESE SPACECRAFT

MOS (SEMICONDUCTORS)
USE   METAL OXIDE SEMICONDUCTORS

MOSAICS
RT   ASSEMBLIES
        DIFFRACTION
        ∞NETWORKS
        PHOTOGRAPHS

MOSCOW
GS   CITIES
        . MOSCOW
RT   U.S.S.R.

MOSFET
USE   FIELD EFFECT TRANSISTORS

MOSS (SPACE STATIONS)
USE   ORBITAL SPACE STATIONS

MOSSBAUER EFFECT
RT   CRYSTAL LATTICES
        ∞EFFECTS
        ELECTROMAGNETIC ABSORPTION
        FLUORESCENCE
        GAMMA RAYS
        RESONANCE SCATTERING
        RESONANT FREQUENCIES

MOT (ORBITAL TELESCOPES)
USE   MANNED ORBITAL TELESCOPES

MOTHS
GS   ANIMALS
        . INVERTEBRATES
        . . ARTHROPODS
        . . . INSECTS
        . . . . MOTHS
        . . . . . SILKWORMS
RT   BOLLWORMS
        INFESTATION

MOTILITY
USE   LOCOMOTION

∞ **MOTION**
SN   (USE OF A MORE SPECIFIC TERM IS
        RECOMMENDED--CONSULT THE TERMS
        LISTED BELOW)
UF   ANIMATION
        MOVEMENT
RT   ACCELERATION (PHYSICS)
        ATTITUDE (INCLINATION)
        BROWNIAN MOVEMENTS
        DISPLACEMENT
        DOMAIN WALL
        GLIDING
        GYRATION
        HARMONIC MOTION
        HEAD MOVEMENT
        HEAVING
        HIGH ACCELERATION
        IMMOBILIZATION
        INERTIA
        ION MOTION
        IONIC MOBILITY
        KINEMATICS
        LIBRATION
        MOMENTUM
        NUTATION
        ORBITS
        OSCILLATIONS
        OSCILLATORS
        PARTICLE MOTION
        PARTICLE TRAJECTORIES
        PITCH (INCLINATION)
        ROTATION
        SACCADIC EYE MOVEMENTS
        SOLAR ORBITS
        SPACECRAFT MOTION
        SPACECRAFT TRAJECTORIES
        STELLAR MOTIONS
        SWARMING
        TEETERING
        TRANSIT TIME
        TRANSLATIONAL MOTION
        TUMBLING MOTION
        TURBULENCE
        VELOCITY
        VERTICAL MOTION
        VERTICAL MOTION SIMULATORS

MOTION-*(CONT.)*
        VIBRATION
        VISCOSITY
        YAW

MOTION AFTEREFFECTS
RT   EQUATIONS OF MOTION
        KINETICS
        MOMENTUM

MOTION EQUATIONS
USE   EQUATIONS OF MOTION

MOTION PERCEPTION
GS   PERCEPTION
        . MOTION PERCEPTION
RT   BINOCULAR VISION
        MONOCULAR VISION
        VISUAL PERCEPTION

MOTION PICTURES
UF   CINEFLUOROGRAPHY
        CINERADIOGRAPHY
GS   PHOTOGRAPHS
        . MOTION PICTURES
RT   CHRONOPHOTOGRAPHY
        CINEMATOGRAPHY
        GRAPHIC ARTS
        PROJECTORS
        SUPPLEMENTS
        VIDEO EQUIPMENT

MOTION SICKNESS
UF   AIR SICKNESS
RT   ACCELERATION STRESSES
            (PHYSIOLOGY)
        AEROSPACE MEDICINE
        NAUSEA
        SPACE ADAPTATION SYNDROME
        VOMITING

MOTION SICKNESS DRUGS
GS   DRUGS
        . MOTION SICKNESS DRUGS
RT   PHARMACOLOGY

MOTION SIMULATION
GS   HUMAN FACTORS ENGINEERING
        . MOTION SIMULATION
        SIMULATION
        . MOTION SIMULATION
RT   FLIGHT SIMULATION
        FLIGHT SIMULATORS
        MOTION SIMULATORS

MOTION SIMULATORS
GS   SIMULATORS
        . MOTION SIMULATORS
RT   COMPUTERIZED SIMULATION
        CONTROL SIMULATION
        FLIGHT SIMULATORS
        MOTION SIMULATION
        SPACE ENVIRONMENT SIMULATION
        TEST FACILITIES

MOTION STABILITY
GS   DYNAMIC CHARACTERISTICS
        . DYNAMIC STABILITY
        . . MOTION STABILITY
        . . . AERODYNAMIC STABILITY
        . . . AIRCRAFT STABILITY
        . . . . HOVERING STABILITY
        . . . ATTITUDE STABILITY
        . . . . DIRECTIONAL STABILITY
        . . . . . GYROSCOPIC STABILITY
        . . . . LATERAL STABILITY
        . . . . LONGITUDINAL STABILITY
        . . . FLOW STABILITY
        . . . . BOUNDARY LAYER STABILITY
        . . . . FLAME STABILITY
        . . . . MAGNETOHYDRODYNAMIC
            STABILITY
        . . . . . WEIBEL INSTABILITY
        . . . LOW SPEED STABILITY
        . . . ROTARY STABILITY
        . . . . GYROSCOPIC STABILITY
        . . . SPACECRAFT STABILITY
        STABILITY
        . DYNAMIC STABILITY
        . . MOTION STABILITY
        . . . AERODYNAMIC STABILITY
        . . . AIRCRAFT STABILITY
        . . . . HOVERING STABILITY
        . . . ATTITUDE STABILITY

MOTION STABILITY-*(CONT.)*
        . . . . DIRECTIONAL STABILITY
        . . . . . GYROSCOPIC STABILITY
        . . . . LATERAL STABILITY
        . . . . LONGITUDINAL STABILITY
        . . . FLOW STABILITY
        . . . . BOUNDARY LAYER STABILITY
        . . . . FLAME STABILITY
        . . . . MAGNETOHYDRODYNAMIC
            STABILITY
        . . . LOW SPEED STABILITY
        . . . ROTARY STABILITY
        . . . . GYROSCOPIC STABILITY
        . . . SPACECRAFT STABILITY
RT   COMBUSTION STABILITY
        CONTROL STABILITY
        DYNAMIC TESTS
        ROUGHNESS
        SEA KEEPING
        SPACECRAFT MOTION
        STABLE OSCILLATIONS
        SURFACE STABILITY

MOTIVATION
GS   MOTIVATION
        . CONTRACT INCENTIVES
RT   ∞DRIVES
        INCENTIVES
        LEARNING
        MORALE
        REINFORCEMENT (PSYCHOLOGY)
        SELF STIMULATION
        ∞STIMULI

MOTOR SYSTEMS (BIOLOGY)
USE   EFFERENT NERVOUS SYSTEMS

MOTOR VEHICLES
GS   SURFACE VEHICLES
        . MOTOR VEHICLES
        . . AUTOMATED MIXED TRAFFIC
            VEHICLES
        . . AUTOMOBILES
        . . . ELECTRIC AUTOMOBILES
        . . ELECTRIC MOTOR VEHICLES
        . . TRACTORS
        . . . CRAWLER TRACTORS
        . . . TRACKED VEHICLES
        . . TRUCKS
        . . . TANK TRUCKS
RT   TRANSPORTATION
        ∞VEHICLES

MOTORS
SN   (MACHINES SUPPLIED WITH EXTERNAL
        ENERGY WHICH IS CONVERTED INTO
        FORCE AND/OR MOTION)
GS   MOTORS
        . ELECTRIC MOTORS
        . . ASYNCHRONOUS MOTORS
        . . INDUCTION MOTORS
        . . MICROMOTORS
        . STEPPING MOTORS
        . . SYNCHRONOUS MOTORS
        . TORQUE MOTORS
        . SERVOMOTORS
RT   APOGEE BOOST MOTORS
        ∞ELECTRIC EQUIPMENT
        ENERGY CONVERSION EFFICIENCY
        ENGINES
        ∞GENERATORS
        HYDRAULIC EQUIPMENT
        STATORS

MOTS (TRACKING SYSTEM)
USE   MINITRACK SYSTEM

MOUNTAIN INHABITANTS
GS   COMMUNITIES
        . INHABITANTS
        . . MOUNTAIN INHABITANTS
RT   ALTITUDE ACCLIMATIZATION
        HIGH ALTITUDE ENVIRONMENTS
        LOW TEMPERATURE ENVIRONMENTS

MOUNTAINS
GS   LANDFORMS
        . MOUNTAINS
        . . ADIRONDACK MOUNTAINS (NY)
        . . ALPS MOUNTAINS (EUROPE)
        . . ANDES MOUNTAINS (SOUTH
            AMERICA)
        . . APPALACHIAN MOUNTAINS (NORTH
            AMERICA)
        . . BIGHORN MOUNTAINS (MT-WY)

MOUNTAINS-(CONT.)
- . . BLACK HILLS (SD-WY)
- . . CARPATHIAN MOUNTAINS (EUROPE)
- . . CASCADE RANGE (CA-OR-WA)
- . . CAUCASUS MOUNTAINS (U.S.S.R.)
- . . COASTAL RANGES (CA)
- . . GREAT SMOKY MOUNTAINS (NC-TN)
- . . HIMALAYAS
- . . PENINSULAR RANGES (CA)
- . . PYRENEES MOUNTAINS (EUROPE)
- . . ROCKY MOUNTAINS (NORTH AMERICA)
- . . SAN JUAN MOUNTAINS (CO)
- . . SIERRA NEVADA MOUNTAINS (CA)
- . . WIND RIVER RANGE (WY)
- . . WRANGELL MOUNTAINS (AK)

RT CENTRAL PIEDMONT (US)
- CIRQUES (LANDFORMS)
- CONES (VOLCANOES)
- CONTINENTS
- DIVIDES (LANDFORMS)
- FORMATIONS
- GAPS (GEOLOGY)
- GEOMORPHOLOGY
- HIGHLANDS
- MARS VOLCANOES
- MASSIFS
- MESAS
- OROGRAPHY
- ∞PEAKS
- PEAKS (LANDFORMS)
- PIEDMONTS
- PIKE'S PEAK (CO)
- ∞RIDGES
- TERRACES (LANDFORMS)
- VOLCANOES
- VOLCANOLOGY
- WATERSHEDS

MOUNTING
GS **MOUNTING**
- . RIGID MOUNTING
RT ASSEMBLING
- ∞ATTACHMENT
- BRACKETS
- ∞JOINING
- SUSPENDING (HANGING)

MOUNTS
USE SUPPORTS

MOUTH
GS ANATOMY
- . DIGESTIVE SYSTEM
- . . **MOUTH**
RT SALIVARY GLANDS
- TEETH
- TONGUE

MOVEMENT
USE MOTION

MOVING TARGET INDICATORS
UF MTI RADAR
GS RADAR
- . **MOVING TARGET INDICATORS**
RT CANCELLATION CIRCUITS
- COHERENT RADAR
- DOPPLER RADAR
- OVER-THE-HORIZON RADAR
- RADAR CROSS SECTIONS
- RADAR TRACKING
- TARGET ACQUISITION

MOZAMBIQUE
GS NATIONS
- . **MOZAMBIQUE**
RT AFRICA

MR-3 FLIGHT
USE MERCURY MR-3 FLIGHT

MRCA AIRCRAFT
UF MULTI-ROLE COMBAT AIRCRAFT
- TORNADO AIRCRAFT
RT ∞AIRCRAFT
- ATTACK AIRCRAFT
- FIGHTER AIRCRAFT
- ∞MILITARY AIRCRAFT

MRKOS COMET
GS CELESTIAL BODIES
- . COMETS
- . . **MRKOS COMET**

MSAT
RT COMMUNICATION SATELLITES
- GROUND STATIONS
- LAND MOBILE SATELLITE SERVICE
- MARITIME SATELLITES
- MOBILE COMMUNICATION SYSTEMS
- RADIO COMMUNICATION
- RADIO RELAY SYSTEMS
- SATELLITE TRANSMISSION

MSBLS
USE MICROWAVE SCANNING BEAM LANDING
     SYSTEM

† **MSRE REACTORS**
USE MOLTEN SALT NUCLEAR REACTORS

MTBF
UF MEAN TIME BETWEEN FAILURES
GS TIME
- . **MTBF**
RT DOWNTIME
- FAILURE ANALYSIS
- FAILURE MODES
- LIFE (DURABILITY)
- RATES (PER TIME)
- RELIABILITY
- STATISTICAL ANALYSIS

MTF
USE MODULATION TRANSFER FUNCTION

MTI RADAR
USE MOVING TARGET INDICATORS

MUBIS (SCANNERS)
USE MULTIPLE BEAM INTERVAL SCANNERS

MUCOCELES
GS CYSTS
- . **MUCOCELES**
RT ∞BLISTERS

MUCUS
GS BODY FLUIDS
- . **MUCUS**
RT SALIVA

MUD
GS SEDIMENTS
- . **MUD**
- SOILS
- . **MUD**
RT ALLUVIUM
- CLAYS
- FANS (LANDFORMS)
- GROUT
- OCEAN BOTTOM
- RAIN EROSION
- SLUDGE
- TIDAL FLATS

MUFFLERS
RT ACOUSTIC RETROFITTING
- AIRCRAFT NOISE
- ATTENUATORS
- BAFFLES
- DAMPING
- ∞DIFFUSERS
- EXHAUST SYSTEMS
- FURNACES
- JET AIRCRAFT NOISE
- NOISE (SOUND)
- NOISE REDUCTION
- ROCKET ENGINE NOISE
- SILENCERS
- SUPPRESSORS

MULBERRY (ALLOY)
GS ALLOYS
- . **MULBERRY (ALLOY)**
RT ANTIMONY ALLOYS
- MOLYBDENUM ALLOYS

MULLITES
RT ALUMINUM SILICATES
- MINERALS

MULTI-ANODE MICROCHANNEL ARRAYS
RT ANODES
- ARRAYS
- ASTRONOMICAL TELESCOPES
- MICROCHANNELS
- RADIATION DETECTORS

MULTI-ANODE MICROCHANNEL-(CONT.)
- SPACEBORNE TELESCOPES

MULTI-ROLE COMBAT AIRCRAFT
USE MRCA AIRCRAFT

MULTIBEAM ANTENNAS
RT BEAMS (RADIATION)
- SATELLITE ANTENNAS

MULTICHANNEL COMMUNICATION
GS TELECOMMUNICATION
- . **MULTICHANNEL COMMUNICATION**
RT CODE DIVISION MULTIPLE ACCESS
- INTERFERENCE FACTOR TABLE
- MULTIPLE ACCESS
- RADIO TRANSMITTERS
- TELECONFERENCING
- TIME DIVISION MULTIPLE ACCESS

MULTICHANNEL PLATES
USE MICROCHANNEL PLATES

MULTIENGINE VEHICLES
RT ∞AIRCRAFT
- B-1 AIRCRAFT
- LAUNCH VEHICLES
- LIGHT TRANSPORT AIRCRAFT
- MISSILE CONFIGURATIONS
- MISSILES
- MULTISTAGE ROCKET VEHICLES
- RECOVERABLE LAUNCH VEHICLES
- ROCKET VEHICLES
- ∞VEHICLES

MULTILAYER INSULATION
GS INSULATION
- . **MULTILAYER INSULATION**
- INTERLAYERS
- . **MULTILAYER INSULATION**
RT COMPOSITE MATERIALS
- CRYOGENIC FLUID STORAGE
- FABRICS
- ∞FOILS
- FOILS (MATERIALS)
- LAMINATES
- ∞LAYERS
- METAL FOILS
- PLY ORIENTATION
- SANDWICH STRUCTURES
- ∞SHEETS

MULTILAYER STRUCTURES
USE LAMINATES

MULTILOOP SYSTEMS
USE CASCADE CONTROL

MULTIMISSION MODULAR SPACECRAFT
UF MMS
RT BESS (SATELLITE)
- LANDSAT FOLLOW-ON MISSIONS
- SATELLITE NETWORKS
- SOLAR MAXIMUM MISSION

MULTIMODE RESONATORS
GS RESONATORS
- . **MULTIMODE RESONATORS**
RT CAVITY RESONATORS
- MAGNETRONS
- PROPAGATION MODES

MULTIPACTOR DISCHARGES
GS ELECTRIC CURRENT
- . ELECTRIC DISCHARGES
- . . **MULTIPACTOR DISCHARGES**
RT LINEAR ACCELERATORS
- PHOTOMULTIPLIER TUBES
- SECONDARY EMISSION
- SPARK GAPS

MULTIPATH TRANSMISSION
GS TRANSMISSION
- . ELECTROMAGNETIC WAVE
     TRANSMISSION
- . . RADIO TRANSMISSION
- . . . **MULTIPATH TRANSMISSION**
- . SIGNAL TRANSMISSION
- . . RADIO TRANSMISSION
- . . . **MULTIPATH TRANSMISSION**
RT CEPSTRAL ANALYSIS
- DIFFRACTION PATHS
- FERMAT PRINCIPLE
- MULTISTATIC RADAR

MULTIPATH TRANSMISSION-*(CONT.)*
```
        OPTICAL PATHS
     ∞PATHS
        RADIO WAVES
        SOUND TRANSMISSION
        WAVE PROPAGATION
```

MULTIPHASE FLOW
```
UF      MIXED FLOW
GS      FLUID FLOW
        . MULTIPHASE FLOW
        . . TWO PHASE FLOW
RT      CONICAL FLOW
        CRITICAL FLOW
        CROCCO-LEE THEORY
        FLOW DISTORTION
        FLOW MEASUREMENT
        GAS FLOW
        LAMINAR FLOW
        LIQUID FLOW
        MASS FLOW
        ORIFICE FLOW
        PIPE FLOW
        PRESSURE GRADIENTS
        SINGLE-PHASE FLOW
        SOLIDS FLOW
        STEADY FLOW
        STEAM FLOW
        SUBCRITICAL FLOW
        SUPERCRITICAL FLOW
        TURBULENT FLOW
        UNIFORM FLOW
        UNSTEADY FLOW
```

MULTIPHOTON ABSORPTION
```
GS      ENERGY ABSORPTION
        . ELECTROMAGNETIC ABSORPTION
        . . MULTIPHOTON ABSORPTION
        RADIATION ABSORPTION
        . ELECTROMAGNETIC ABSORPTION
        . . MULTIPHOTON ABSORPTION
RT      ∞ABSORPTION
        PHOTON ABSORPTIOMETRY
```

MULTIPLE ACCESS
```
GS      TELECOMMUNICATION
        . MULTIPLE ACCESS
        . . ALOHA SYSTEM
        . . CODE DIVISION MULTIPLE ACCESS
        . . DEMAND ASSIGNMENT MULTIPLE
              ACCESS
        . . FREQUENCY DIVISION MULTIPLE
              ACCESS
        . . TIME DIVISION MULTIPLE ACCESS
        TRANSMISSION
        . SIGNAL TRANSMISSION
        . . DATA TRANSMISSION
        . . . MULTIPLE ACCESS
        . . . . CODE DIVISION MULTIPLE ACCESS
        . . . . FREQUENCY DIVISION MULTIPLE
              ACCESS
RT      ACCESS CONTROL
        FREQUENCY DIVISION MULTIPLEXING
        MULTICHANNEL COMMUNICATION
        PACKET SWITCHING
        PULSE COMMUNICATION
        WIDEBAND COMMUNICATION
```

MULTIPLE BEAM INTERVAL SCANNERS
```
UF      MUBIS (SCANNERS)
GS      ANTENNAS
        . MULTIPLE BEAM INTERVAL
              SCANNERS
RT      LINEAR ARRAYS
        RADAR SCANNING
```

MULTIPLE DOCKING ADAPTERS
```
UF      MDA
GS      ADAPTERS
        . MULTIPLE DOCKING ADAPTERS
RT      AIRLOCK MODULES
        MOORING
        ORBITAL RENDEZVOUS
        SATURN WORKSHOPS
        SATURN 1 WORKSHOP
        SATURN 5 WORKSHOP
        SKYLAB 1
        SKYLAB 2
        SKYLAB 3
        SKYLAB 4
        SPACECRAFT DOCKING
```

MULTIPLE OUTPUT PROGRAMS
```
GS      COMPUTER PROGRAMS
        . MULTIPLE OUTPUT PROGRAMS
```

MULTIPLE OUTPUT PROGRAMS-*(CONT.)*
```
RT      MULTIPROGRAMMING
        READOUT
        TIME SHARING
```

MULTIPLE TARGET TRAJECTORY SYSTEMS
```
USE     MATTS (SYSTEMS)
```

MULTIPLETS
```
USE     FINE STRUCTURE
```

MULTIPLEX TRANSMISSION
```
USE     MULTIPLEXING
```

MULTIPLEXERS
```
USE     MULTIPLEXING
```

MULTIPLEXING
```
UF      MULTIPLEX TRANSMISSION
        MULTIPLEXERS
GS      TRANSMISSION
        . MULTIPLEXING
        . . CODE DIVISION MULTIPLEXING
        . . FREQUENCY DIVISION MULTIPLEXING
        . . TIME DIVISION MULTIPLEXING
        . . WAVELENGTH DIVISION
              MULTIPLEXING
ПT      ACCESS CONTROL
        CARRIER FREQUENCIES
        CODE DIVISION MULTIPLE ACCESS
        DATA TRANSMISSION
        DEMULTIPLEXING
        FREQUENCY DIVISION MULTIPLE
              ACCESS
        PULSE COMMUNICATION
        RADIO TRANSMISSION
        SATELLITE TRANSMISSION
        SIGNAL TRANSMISSION
```

MULTIPLICATION
```
GS      NUMBER THEORY
        . MULTIPLICATION
RT      ARITHMETIC
        COMPUTATION
        FRINGE MULTIPLICATION
```

MULTIPLIER PHOTOTUBES
```
USE     PHOTOMULTIPLIER TUBES
```

MULTIPLIERS
```
GS      MULTIPLIERS
        . CHANNEL MULTIPLIERS
RT      LAGRANGE MULTIPLIERS
        LOGIC CIRCUITS
        PHOTOMULTIPLIER TUBES
```

MULTIPOLAR FIELDS
```
RT      CONTINUUM MECHANICS
        FIELD THEORY (PHYSICS)
        GRAVITATIONAL FIELDS
        MAGNETIC FIELDS
        MULTIPOLES
```

MULTIPOLES
```
RT      MONOPOLES
        MULTIPOLAR FIELDS
```

MULTIPROCESSING (COMPUTERS)
```
GS      DATA PROCESSING
        . MULTIPROCESSING (COMPUTERS)
        . . PIPELINING (COMPUTERS)
RT      ASSOCIATIVE PROCESSING
              (COMPUTERS)
        COMPUTERS
        CONCURRENT PROCESSING
        DATA PROCESSING EQUIPMENT
        INTERPROCESSOR COMMUNICATION
        MULTIPROGRAMMING
        PARALLEL PROGRAMMING
        REAL TIME OPERATION
        TIME SHARING
```

MULTIPROGRAMMING
```
GS      SOFTWARE ENGINEERING
        . COMPUTER PROGRAMMING
        . . MULTIPROGRAMMING
RT      MACHINE-INDEPENDENT PROGRAMS
        MULTIPLE OUTPUT PROGRAMS
        MULTIPROCESSING (COMPUTERS)
        PIPELINING (COMPUTERS)
        ∞PROGRAMMING
        TIME SHARING
```

MULTIPROPELLANTS
```
USE     ROCKET PROPELLANTS
```

MULTIRADAR TRACKING
```
USE     RADAR NETWORKS
```

MULTISENSOR APPLICATIONS
```
RT      IMAGE PROCESSING
        IMAGING TECHNIQUES
        PATTERN RECOGNITION
        REMOTE SENSING
        REMOTE SENSORS
```

MULTISPECTRAL BAND CAMERAS
```
GS      OPTICAL EQUIPMENT
        . CAMERAS
        . . MULTISPECTRAL BAND CAMERAS
        PHOTOGRAPHIC EQUIPMENT
        . CAMERAS
        . . MULTISPECTRAL BAND CAMERAS
RT      INFRARED PHOTOGRAPHY
        PHOTOGRAPHY
```

MULTISPECTRAL BAND SCANNERS
```
GS      MEASURING INSTRUMENTS
        . OPTICAL MEASURING INSTRUMENTS
        . . OPTICAL SCANNERS
        . . . MULTISPECTRAL BAND SCANNERS
        OPTICAL EQUIPMENT
        . OPTICAL MEASURING INSTRUMENTS
        . . OPTICAL SCANNERS
        . . . MULTISPECTRAL BAND SCANNERS
        SCANNERS
        . OPTICAL SCANNERS
        . . MULTISPECTRAL BAND SCANNERS
RT      BAND RATIOING
        CHANGE DETECTION
        EARTH OBSERVATIONS (FROM SPACE)
        IMAGING TECHNIQUES
        INFRARED SCANNERS
        OCEAN COLOR SCANNER
        PANORAMIC SCANNING
        PHOTOGRAPHY
        RADIOMETRIC CORRECTION
        RADIOMETRIC RESOLUTION
        SCANNING
        SPACEBORNE PHOTOGRAPHY
        SPECTRAL RECONNAISSANCE
        VEGETATIVE INDEX
```

MULTISPECTRAL LINEAR ARRAYS
```
UF      MLA
GS      ELECTRONIC EQUIPMENT
        . SOLID STATE DEVICES
        . . MULTISPECTRAL LINEAR ARRAYS
        SPACECRAFT INSTRUMENTS
        . SATELLITE INSTRUMENTS
        . . MULTISPECTRAL LINEAR ARRAYS
RT      ∞DETECTORS
        ∞SENSORS
```

MULTISPECTRAL PHOTOGRAPHY
```
GS      PHOTOGRAPHY
        . MULTISPECTRAL PHOTOGRAPHY
        . . INFRARED PHOTOGRAPHY
        . . . COLOR INFRARED PHOTOGRAPHY
        . . RADAR PHOTOGRAPHY
RT      CHANGE DETECTION
        CROP IDENTIFICATION
        EARTH OBSERVATIONS (FROM SPACE)
        IMAGE RESOLUTION
        IMAGERY
        IMAGING TECHNIQUES
        I2S CAMERAS
        SPECTRAL RECONNAISSANCE
```

MULTISPECTRAL RADAR
```
GS      RADAR
        . MULTISPECTRAL RADAR
RT      IMAGERY
        IMAGING TECHNIQUES
        SPECTRAL RECONNAISSANCE
```

MULTISPECTRAL RESOURCE SAMPLER
```
GS      SATELLITES
        . ARTIFICIAL SATELLITES
        . . MULTISPECTRAL RESOURCE
              SAMPLER
RT      REMOTE SENSING
```

MULTISPECTRAL TRACKING TELESCOPES
```
GS      OPTICAL EQUIPMENT
        . ASTRONOMICAL TELESCOPES
        . . SPECTROSCOPIC TELESCOPES
```

MULTISPECTRAL TRACKING-(CONT.)
```
   . . . MULTISPECTRAL TRACKING
             TELESCOPES
   TELESCOPES
   . ASTRONOMICAL TELESCOPES
   . . SPECTROSCOPIC TELESCOPES
   . . . MULTISPECTRAL TRACKING
             TELESCOPES
RT OPTICAL MEASURING INSTRUMENTS
   OPTICAL TRACKING
   TRACKING (POSITION)
```

MULTISTAGE COMPRESSORS
```
USE   TURBOCOMPRESSORS
```

MULTISTAGE ROCKET VEHICLES
```
GS  ROCKET VEHICLES
    . MULTISTAGE ROCKET VEHICLES
    . . ANTARES ROCKET VEHICLE
    . . ARGO ROCKET VEHICLES
    . . ARIANE LAUNCH VEHICLE
    . . ASTROBEE ROCKET VEHICLES
    . . ASTROBEE 1500 ROCKET VEHICLE
    . . ATHENA ROCKET VEHICLE
    . . ATLAS LAUNCH VEHICLES
    . . . ATLAS ABLE 5 LAUNCH VEHICLE
    . . . ATLAS AGENA B LAUNCH VEHICLE
    . . . ATLAS AGENA LAUNCH VEHICLES
    . . . ATLAS CENTAUR LAUNCH VEHICLE
    . . . ATLAS SLV-3 LAUNCH VEHICLE
    . . BERENICE ROCKET VEHICLE
    . . BLACK KNIGHT ROCKET VEHICLE
    . . BLUE SCOUT ROCKET VEHICLE
    . . DIAMANT LAUNCH VEHICLE
    . . ELDO LAUNCH VEHICLE
    . . EXOS SOUNDING ROCKET
    . . JAGUAR ROCKET VEHICLE
    . . JAVELIN ROCKET VEHICLE
    . . JUNO 1 LAUNCH VEHICLE
    . . JUNO 2 LAUNCH VEHICLE
    . . JUPITER C ROCKET VEHICLE
    . . KAPPA ROCKET VEHICLES
    . . . KAPPA 8 ROCKET VEHICLE
    . . . KAPPA 9 ROCKET VEHICLE
    . . LAMBDA ROCKET VEHICLES
    . . LITTLE JOE 2 LAUNCH VEHICLE
    . . NIKE ROCKET VEHICLES
    . . . NIKE-APACHE ROCKET VEHICLE
    . . . NIKE-CAJUN ROCKET VEHICLE
    . . . NIKE-HYDAC ROCKET VEHICLE
    . . . NIKE-IROQUOIS ROCKET VEHICLE
    . . . NIKE-JAVELIN ROCKET VEHICLE
    . . . NIKE-TOMAHAWK ROCKET VEHICLE
    . . NOVA LAUNCH VEHICLES
    . . PHOENIX SOUNDING ROCKET
    . . RAM B LAUNCH VEHICLE
    . . RUBIS ROCKET VEHICLE
    . . SATURN LAUNCH VEHICLES
    . . . SATURN D LAUNCH VEHICLE
    . . . SATURN 1 LAUNCH VEHICLES
    . . . . SATURN 1 SA-1 LAUNCH VEHICLE
    . . . . SATURN 1 SA-2 LAUNCH VEHICLE
    . . . . SATURN 1 SA-3 LAUNCH VEHICLE
    . . . . SATURN 1 SA-4 LAUNCH VEHICLE
    . . . . SATURN 1 SA-5 LAUNCH VEHICLE
    . . . . SATURN 1 SA-6 LAUNCH VEHICLE
    . . . . SATURN 1 SA-7 LAUNCH VEHICLE
    . . . . SATURN 1 SA-8 LAUNCH VEHICLE
    . . . . SATURN 1 SA-9 LAUNCH VEHICLE
    . . . . SATURN 1 SA-10 LAUNCH
              VEHICLE
    . . . SATURN 1B LAUNCH VEHICLES
    . . . SATURN 2 LAUNCH VEHICLES
    . . . SATURN 5 LAUNCH VEHICLES
    . . SCOUT LAUNCH VEHICLE
    . . SKYLARK ROCKET VEHICLE
    . . THOR LAUNCH VEHICLES
    . . . THOR ABLE LAUNCH VEHICLE
    . . . THOR AGENA LAUNCH VEHICLE
    . . . THOR DELTA LAUNCH VEHICLE
    . . TITAN LAUNCH VEHICLES
    . . . TITAN 3 LAUNCH VEHICLE
    . . VANGUARD 2 LAUNCH VEHICLE
    . . VEGA LAUNCH VEHICLE
    . . WASP SOUNDING ROCKET
RT  AIR LAUNCHING
    EXPENDABLE STAGES (SPACECRAFT)
    INTERIM STAGES (SPACECRAFT)
    LAUNCH VEHICLES
    MINUTEMAN ICBM
    MULTIENGINE VEHICLES
    NAVAHO MISSILE
    PAYLOAD MASS RATIO
    PERSHING MISSILE
    PIGGYBACK SYSTEMS
```

MULTISTAGE ROCKET VEHICLES-(CONT.)
```
    POLARIS MISSILES
    PROPULSIVE EFFICIENCY
    ROCKET ENGINES
    SS-11 MISSILE
    STAGE SEPARATION
    SUNBLAZER SPACE PROBE
    TALOS MISSILE
    TERRIER MISSILE
    TITAN ICBM
    TRAILBLAZER 1 REENTRY VEHICLE
    TRAILBLAZER 2 REENTRY VEHICLE
    UPPER STAGE ROCKET ENGINES
∞ VEHICLES
```

MULTISTATIC RADAR
```
UF   BISTATIC RADAR
GS   RADAR
     . DOPPLER RADAR
     . . MULTISTATIC RADAR
     . SURVEILLANCE RADAR
     . . MULTISTATIC RADAR
RT   MULTIPATH TRANSMISSION
     PULSE RADAR
     RADAR DETECTION
     TARGET RECOGNITION
```

MULTITEMPORAL ANALYSIS
```
USE   TEMPORAL RESOLUTION
```

MULTIVARIATE STATISTICAL ANALYSIS
```
GS   STATISTICAL ANALYSIS
     . VARIANCE (STATISTICS)
     . . MULTIVARIATE STATISTICAL
           ANALYSIS
     . . . BIVARIATE ANALYSIS
     . . . COVARIANCE
     . . . REGRESSION ANALYSIS
RT   ∞ ANALYZING
     CORRELATION
     DISCRIMINANT ANALYSIS (STATISTICS)
     ∞ VARIANCE
```

MULTIVIBRATORS
```
GS   CIRCUITS
     . MULTIVIBRATORS
     . . FLIP-FLOPS
     . . MONOSTABLE MULTIVIBRATORS
RT   AMPLIFIERS
     BISTABLE CIRCUITS
     LOGIC CIRCUITS
     OSCILLATORS
     POSITIVE FEEDBACK
     SWITCHING CIRCUITS
     TRIGGER CIRCUITS
```

MUON SPIN ROTATION
```
GS   GYRATION
     . ROTATION
     . . MUON SPIN ROTATION
RT   CHARGED PARTICLES
     HYPERFINE STRUCTURE
     MUONS
     PARTICLE DIFFUSION
     PARTICLE SPIN
     PRECESSION
```

MUONIUM
```
RT   ELECTRONS
     MESONS
```

MUONS
```
GS   PARTICLES
     . ELEMENTARY PARTICLES
     . . BOSONS
     . . . MESONS
     . . . . MUONS
     . . FERMIONS
     . . . LEPTONS
     . . . . MUONS
     . . HADRONS
     . . . MESONS
     . . . . MUONS
     . NUCLEAR PARTICLES
     . . BOSONS
     . . . MESONS
     . . . . MUONS
RT   BARYONS
     CHARGED PARTICLES
     ELECTRON DECAY RATE
     MUON SPIN ROTATION
```

MURCHISON METEORITE
```
GS   CELESTIAL BODIES
     . METEORITES
```

MURCHISON METEORITE-(CONT.)
```
     . . STONY METEORITES
     . . . CHONDRITES
     . . . . CARBONACEOUS CHONDRITES
     . . . . . MURCHISON METEORITE
```

MURRAY METEORITE
```
GS   CELESTIAL BODIES
     . METEORITES
     . . STONY METEORITES
     . . . CHONDRITES
     . . . . CARBONACEOUS METEORITES
     . . . . . MURRAY METEORITE
```

MUSCLE RELAXANTS
```
GS   DRUGS
     . MUSCLE RELAXANTS
```

MUSCLES
```
GS   MUSCLES
     . MYOCARDIUM
RT   ALDOLASE
     ATAXIA
     CONGENERS
     CONVULSIONS
     DIAPHRAGM (ANATOMY)
     ELECTROCARDIOGRAPHY
     FIBRILLATION
     HYPODYNAMIA
     MUSCULAR FATIGUE
     MUSCULAR STRENGTH
     MUSCULAR TONUS
     MYOELECTRIC POTENTIALS
     MYOELECTRICITY
     SPASMS
     TWITCHING
```

MUSCOVITE
```
GS   ALUMINUM COMPOUNDS
     . MUSCOVITE
     CHALCOGENIDES
     . OXIDES
     . . SILICON OXIDES
     . . . MUSCOVITE
     MINERALS
     . MICA
     . . MUSCOVITE
     SILICON COMPOUNDS
     . SILICON OXIDES
     . . MUSCOVITE
```

MUSCULAR FATIGUE
```
GS   FATIGUE (BIOLOGY)
     . MUSCULAR FATIGUE
RT   MUSCLES
     MUSCULOSKELETAL SYSTEM
     STRESS (PHYSIOLOGY)
```

MUSCULAR FUNCTION
```
GS   MUSCULAR FUNCTION
     . SPASMS
RT   CRAMPS
     ∞ FUNCTIONS
     HYPODYNAMIA
     HYPOKINESIA
     HYPOTONIA
     MECHANOGRAMS
     TWITCHING
```

MUSCULAR STRENGTH
```
RT   MUSCLES
     MUSCULOSKELETAL SYSTEM
     ∞ STRENGTH
```

MUSCULAR TONUS
```
UF   TONUS
GS   MUSCULAR TONUS
     . HYPOTONIA
RT   EXERCISE PHYSIOLOGY
     MUSCLES
```

MUSCULOSKELETAL SYSTEM
```
UF   SKELETON
GS   ANATOMY
     . MUSCULOSKELETAL SYSTEM
     . . BONES
     . . . CARTILAGE
     . . . CEREBRUM
     . . . FEMUR
     . . . MARROW
     . . . MASTOIDS
     . . . PELVIS
     . . . SCAPULA
     . . . SCIATIC REGION
```

MUSCULOSKELETAL SYSTEM-*(CONT.)*
. . . SKULL
. . . . CRANIUM
. . . . . INTRACRANIAL CAVITY
. . . STERNUM
. . . TIBIA
. . . ULNA
. . . VERTEBRAE
. . CHIN
. . CONNECTIVE TISSUE
. . . CARTILAGE
. . . COLLAGENS
. . . CONGENERS
. . . MARROW
. . CONSTRICTORS
. . FLEXORS
. . JOINTS (ANATOMY)
. . . ELBOW (ANATOMY)
. . . KNEE (ANATOMY)
. . . WRIST
. . VERTEBRAL COLUMN
RT   EXOSKELETONS
HYPOKINESIA
INTERVERTEBRAL DISKS
MUSCULAR FATIGUE
MUSCULAR STRENGTH
SPINE
STRIATION
∞ SYSTEMS

MUSEUMS
RT   ANTHROPOLOGY
ARTIFACTS
∞ BUILDINGS
COLLECTION
HISTORIES
LIBRARIES

MUSHY ZONES
UF   LIQUID PLUS SOLID ZONES
RT   CASTING
COOLING
METALLOGRAPHY
PHASE TRANSFORMATIONS
SOLIDIFICATION

MUSIC
RT   ARTS
OCTAVES

MUSKEGS
GS   LANDFORMS
. **MUSKEGS**
RT   ARCTIC REGIONS
MARSHLANDS
SOILS
TOPOGRAPHY
WATER

MUSTANG AIRCRAFT
USE   P-51 AIRCRAFT

MUTAGENS
RT   AIR POLLUTION
BIOCHEMISTRY
BIOLOGICAL EVOLUTION
CELLS (BIOLOGY)
CHEMICAL ANALYSIS
GENETICS
MUTATIONS

MUTATIONS
GS   GENETICS
. **MUTATIONS**
RT   BIOLOGICAL EVOLUTION
CELLS (BIOLOGY)
CHROMOSOMES
MITOSIS
MUTAGENS
NUCLEOGENESIS
RADIATION HAZARDS

MX MISSILE
GS   MISSILES
. SURFACE TO SURFACE MISSILES
. . INTERCONTINENTAL BALLISTIC
MISSILES
. . . **MX MISSILE**
RT   MISSILE SILOS

MYELIN
GS   NERVOUS SYSTEM
. **MYELIN**
RT   FATS

MYELIN-*(CONT.)*
LIPIDS

MYLAR (TRADEMARK)
GS   POLYMERIC FILMS
. **MYLAR (TRADEMARK)**
RT   POLYETHYLENE TEREPHTHALATE
∞ POLYMERS

MYOCARDIAL INFARCTION
GS   DISEASES
. INFARCTION
. . **MYOCARDIAL INFARCTION**
RT   ARTERIOSCLEROSIS
BLOOD COAGULATION
CORONARY ARTERY DISEASE
HEART DISEASES
HYPERTENSION
THROMBOSIS

MYOCARDIUM
GS   ANATOMY
. CARDIOVASCULAR SYSTEM
. . HEART
. . . **MYOCARDIUM**
MUSCLES
. **MYOCARDIUM**
RT   ANGINA PECTORIS

MYOELECTRIC POTENTIALS
GS   MYOELECTRICITY
. **MYOELECTRIC POTENTIALS**
RT   MUSCLES
∞ POTENTIAL

MYOELECTRICITY
GS   **MYOELECTRICITY**
. MYOELECTRIC POTENTIALS
RT   ELECTROMYOGRAPHY
MUSCLES

MYOGLOBIN
RT   PIGMENTS
PROTEINS

MYOPIA
RT   VISION

MYSTERE 20 AIRCRAFT
UF   DASSAULT MYSTERE 20 AIRCRAFT
GS   DASSAULT AIRCRAFT
. **MYSTERE 20 AIRCRAFT**
JET AIRCRAFT
. TURBOFAN AIRCRAFT
. . **MYSTERE 20 AIRCRAFT**
LIGHT AIRCRAFT
. **MYSTERE 20 AIRCRAFT**
MONOPLANES
. **MYSTERE 20 AIRCRAFT**
PASSENGER AIRCRAFT
. **MYSTERE 20 AIRCRAFT**
TRANSPORT AIRCRAFT
. **MYSTERE 20 AIRCRAFT**

MYSTERE 50 AIRCRAFT
UF   DASSAULT MYSTERE 50 AIRCRAFT
RT   DASSAULT AIRCRAFT
JET AIRCRAFT
LIGHT AIRCRAFT
PASSENGER AIRCRAFT
TRANSPORT AIRCRAFT
TURBOFAN AIRCRAFT

N

N ELECTRONS
GS   PARTICLES
. CHARGED PARTICLES
. . ENERGETIC PARTICLES
. . . ELECTRONS
. . . . **N ELECTRONS**
RT   BETA PARTICLES

N-N JUNCTIONS
GS   SEMICONDUCTOR JUNCTIONS
. **N-N JUNCTIONS**

N-P JUNCTIONS
USE   P-N JUNCTIONS

N-P-N JUNCTIONS
GS   SEMICONDUCTOR JUNCTIONS
. **N-P-N JUNCTIONS**
RT   BIPOLAR TRANSISTORS

N-TYPE SEMICONDUCTORS
GS   SEMICONDUCTORS (MATERIALS)
. **N-TYPE SEMICONDUCTORS**
RT   ELECTRONS
SCHOTTKY DIODES
SEMICONDUCTOR JUNCTIONS
SUHL EFFECT

N-156 AIRCRAFT
USE   F-5 AIRCRAFT

NA-300 AIRCRAFT
USE   OV-10 AIRCRAFT

NACELLES
RT   AERODYNAMIC CONFIGURATIONS
AIR INTAKES
AIRFRAMES
COWLINGS
DUCTED BODIES
ENGINE INLETS
EXTERNAL STORE SEPARATION
EXTERNAL STORES
EXTERNAL TANKS
FAIRINGS
HOUSINGS
NOSE INLETS
PERFORATED SHELLS
PODS (EXTERNAL STORES)
PROTUBERANCES
SHELLS (STRUCTURAL FORMS)
WING-FUSELAGE STORES

NAKED SINGULARITIES
GS   ANALYSIS (MATHEMATICS)
. COMPLEX VARIABLES
. . SINGULARITY (MATHEMATICS)
. . . **NAKED SINGULARITIES**
RT   ASTROPHYSICS
BLACK HOLES (ASTRONOMY)
COSMOLOGY
GRAVITATIONAL COLLAPSE
POINTS (MATHEMATICS)
RELATIVITY
SPACE-TIME FUNCTIONS
THEORETICAL PHYSICS
WHITE HOLES (ASTRONOMY)

NAMC AIRCRAFT
USE   NIHON AIRCRAFT

NAMIBIA
UF   SOUTH WEST AFRICA
RT   AFRICA
NATIONS
REPUBLIC OF SOUTH AFRICA

NAMING
GS   **NAMING**
. NORMS
RT   SPECIFICATIONS
STANDARDIZATION

NAP-OF-THE-EARTH NAVIGATION
UF   NOE NAVIGATION
GS   NAVIGATION
. AIR NAVIGATION
. . **NAP-OF-THE-EARTH NAVIGATION**
RT   HELICOPTERS
IMAGE PROCESSING
LOW ALTITUDE
NIGHT FLIGHTS (AIRCRAFT)
NIGHT VISION
TARGET RECOGNITION
TERRAIN ANALYSIS
TERRAIN FOLLOWING AIRCRAFT

NAPHTHALENE
GS   HYDROCARBONS
. **NAPHTHALENE**

NAPHTHENES
GS   ALIPHATIC COMPOUNDS
. CYCLIC HYDROCARBONS
. . **NAPHTHENES**
HYDROCARBONS
. CYCLIC HYDROCARBONS
. . **NAPHTHENES**

NAPPES
USE   FOLDS (GEOLOGY)

NARCOLEPSY
GS   DISEASES
        . **NARCOLEPSY**

NARCOSIS
GS   UNCONSCIOUSNESS
        . **NARCOSIS**
RT   ∞POISONING

NARCOTICS
GS   DRUGS
        . **NARCOTICS**
        . . MORPHINE
RT   PENTOBARBITAL
        PHENOBARBITAL
        PSYCHOTROPIC DRUGS

NARROWBAND
GS   BANDWIDTH
        . **NARROWBAND**
RT   ∞BANDS
        BROADBAND
        FREQUENCIES

NASA COMMUNICATION NETWORK
USE   NASCOM NETWORK

NASA END-TO-END DATA SYSTEM
USE   NEEDS (DATA SYSTEM)

NASA INTERACTIVE PLANNING SYSTEM
UF   NIPS (SYSTEM)
RT   COMPUTER PROGRAMS
        EARTH RESOURCES
        MANAGEMENT METHODS
        NASA PROGRAMS
        PROJECT PLANNING
        RESOURCE ALLOCATION
        RESOURCES MANAGEMENT
        ∞SYSTEMS

NASA PROGRAMS
GS   PROGRAMS
        . **NASA PROGRAMS**
        . . APOLLO APPLICATIONS PROGRAM
        . . APOLLO PROJECT
        . . ASSESS PROGRAM
        . . ATLIT PROJECT
        . . BIOASTRONAUTICAL ORBITAL SPACE
              SYSTEM
        . . CENTAUR PROJECT
        . . DAST PROGRAM
        . . EARTH & OCEAN PHYSICS
              APPLICATIONS PROGRAM
        . . EARTH RESOURCES PROGRAM
        . . . EARTH RESOURCES SURVEY
              PROGRAM
        . . . . SEASAT PROGRAM
        . . ECHO PROJECT
        . . GALILEO PROJECT
        . . GEMINI PROJECT
        . . HELIOS PROJECT
        . . JUPITER PROJECT
        . . MARINER PROGRAM
        . . MARS 69 PROJECT
        . . MARS 71 PROJECT
        . . MERCURY PROJECT
        . . NATIONAL LAUNCH VEHICLE
              PROGRAM
        . . NEW MOONS PROJECT
        . . NIMBUS PROJECT
        . . PIONEER PROJECT
        . . PROJECT SETI
        . . QUIET ENGINE PROGRAM
        . . RANGER PROJECT
        . . . AGENA B RANGER PROGRAM
        . . ROVER PROJECT
        . . SAIL PROJECT
        . . SATURN PROJECT
        . . SCOUT PROJECT
        . . SKYLAB PROGRAM
        . . SUPERSONIC CRUISE AIRCRAFT
              RESEARCH
        . . SURVEYOR PROJECT
        . . SYNCHRONOUS COMMUNICATIONS
              SATELLITE PROJ
        . . TACT PROGRAM
        . . TEKTITE PROJECT
        . . TERMINAL CONFIGURED VEHICLE
              PROGRAM
        . . TILT ROTOR RESEARCH AIRCRAFT
              PROGRAM

NASA PROGRAMS-(CONT.)
        . . TIROS PROJECT
        . . TITAN PROJECT
        . . VANGUARD PROJECT
        . . VOYAGER PROJECT
RT   AGRISTARS PROJECT
        APOLLO EXTENSION SYSTEM
        CANADIAN SPACE PROGRAM
        COMMITTEE ON SPACE RESEARCH
        COMMUNICATIONS TECHNOLOGY
          SATELLITE
        DELAYED FLAP APPROACH
        EARTH RESOURCES INFORMATION
          SYSTEM
        GARP ATLANTIC TROPICAL EXPERIMENT
        GEOGRAPHIC APPLICATIONS PROGRAM
        GLOBAL ATMOSPHERIC RESEARCH
          PROGRAM
        GRANTS
        GRAVITY PROBE B
        LANDSAT SATELLITES
        LEASING
        NASA INTERACTIVE PLANNING SYSTEM
        NASA SPACE PROGRAMS
        NOESS
        OSS-1 PAYLOAD
        PAYLOAD DEPLOYMENT & RETRIEVAL
          SYSTEM
        QUASAT
        QUESTOL
        ∞RESEARCH PROJECTS
        ROTOR SYSTEMS RESEARCH AIRCRAFT
        SEASAT SATELLITES
        SINGLE STAGE TO ORBIT VEHICLES
        SPACE PROGRAMS
        SPACE TRANSPORTATION SYSTEM
        SPACELAB
        STARSITE PROGRAM
        STORMSAT SATELLITE
        SYNCHRONOUS EARTH OBSERVATORY
          SATELLITE
        TECHNOLOGY UTILIZATION
        TRANSIT NAVIGATION SYSTEM
        UNIVERSITY PROGRAM
        VIKING MARS PROGRAM

NASA SPACE PROGRAMS
GS   PROGRAMS
        . **NASA SPACE PROGRAMS**
        . . APOLLO APPLICATIONS PROGRAM
        . . APOLLO PROJECT
        . . CENTAUR PROJECT
        . . EARTH & OCEAN PHYSICS
              APPLICATIONS PROGRAM
        . . EARTH RESOURCES PROGRAM
        . . . EARTH RESOURCES SURVEY
              PROGRAM
        . . . . SEASAT PROGRAM
        . . ECHO PROJECT
        . . GALILEO PROJECT
        . . GEMINI PROJECT
        . . HELIOS PROJECT
        . . JUPITER PROJECT
        . . MARINER PROGRAM
        . . MARS 69 PROJECT
        . . MARS 71 PROJECT
        . . MERCURY PROJECT
        . . NATIONAL LAUNCH VEHICLE
              PROGRAM
        . . NEW MOONS PROJECT
        . . NIMBUS PROJECT
        . . PIONEER PROJECT
        . . PROJECT SETI
        . . QUIET ENGINE PROGRAM
        . . RANGER PROJECT
        . . . AGENA B RANGER PROGRAM
        . . ROVER PROJECT
        . . SAIL PROJECT
        . . SATURN PROJECT
        . . SCOUT PROJECT
        . . SKYLAB PROGRAM
        . . SURVEYOR PROJECT
        . . SYNCHRONOUS COMMUNICATIONS
              SATELLITE PROJ
        . . TEKTITE PROJECT
        . . TIROS PROJECT
        . . TITAN PROJECT
        . . VANGUARD PROJECT
        . . VOYAGER PROJECT
RT   NASA PROGRAMS

NASA STRUCTURAL ANALYSIS PROGRAM
USE   NASTRAN

NASARR
USE   NORTH AMERICAN SEARCH AND
        RANGING RADAR

NASCOM NETWORK
UF   NASA COMMUNICATION NETWORK
GS   COMMUNICATING
        . POINT TO POINT COMMUNICATION
        . . **NASCOM NETWORK**
        COMMUNICATION NETWORKS
        . **NASCOM NETWORK**
RT   FLEET SATELLITE COMMUNICATION
          SYSTEM
        GLOBAL TRACKING NETWORK
        ∞NETWORKS
        RADIO COMMUNICATION
        TELECOMMUNICATION

NASTRAN
UF   NASA STRUCTURAL ANALYSIS
          PROGRAM
GS   COMPUTER PROGRAMS
        . **NASTRAN**
RT   BENDING MOMENTS
        COMPUTER TECHNIQUES
        DYNAMIC LOADS
        FINITE ELEMENT METHOD
        MATRIX METHODS
        STRESS ANALYSIS
        STRUCTURAL ANALYSIS

NATIONAL AIRSPACE SYSTEM
RT   AIR TRAFFIC CONTROL
        AIRCRAFT SAFETY
        AIRPORTS
        AIRSPACE
        NATIONAL AIRSPACE UTILIZATION
          SYSTEM
        NATIONAL AVIATION SYSTEM

NATIONAL AIRSPACE UTILIZATION SYSTEM
RT   AIR LAW
        AIR NAVIGATION
        AIR TRAFFIC
        AIR TRAFFIC CONTROL
        AIRCRAFT APPROACH SPACING
        AIRSPACE
        COLLISION AVOIDANCE
        FLIGHT PATHS
        FLIGHT PLANS
        FLIGHT RULES
        NATIONAL AIRSPACE SYSTEM
        ∞SYSTEMS

NATIONAL AVIATION SYSTEM
RT   AIR TRAFFIC
        AIR TRAFFIC CONTROL
        AIR TRANSPORTATION
        AIRCRAFT APPROACH SPACING
        FLIGHT RULES
        LANDING AIDS
        NATIONAL AIRSPACE SYSTEM
        ∞SYSTEMS
        TRAFFIC CONTROL

NATIONAL LAUNCH VEHICLE PROGRAM
GS   PROGRAMS
        . NASA PROGRAMS
        . . **NATIONAL LAUNCH VEHICLE
              PROGRAM**
        . NASA SPACE PROGRAMS
        . . **NATIONAL LAUNCH VEHICLE
              PROGRAM**
RT   LAUNCH VEHICLES
        LAUNCHERS
        LAUNCHING
        LAUNCHING SITES
        SPACE PROGRAMS

NATIONAL OCEANIC SATELLITE SYSTEM
RT   ARTIFICIAL SATELLITES
        MARITIME SATELLITES
        SATELLITES
        ∞SYSTEMS

**NATIONAL OPERATIONAL ENVIRONMENTAL SAT
SYS**
USE   NOESS

NATIONAL PARKS
GS   LAND
        . PARKS
        . . **NATIONAL PARKS**

NATIONAL PARKS-*(CONT.)*
. . . YELLOWSTONE NATIONAL PARK
    (ID-MT-WY)

NATIONAL SEVERE STORMS PROJECT
RT   METEOROLOGY
        TORNADOES
        WARNING SYSTEMS

NATIONS
GS   **NATIONS**
. AFGHANISTAN
. ALBANIA
. ALGERIA
. ANDORRA
. ANGOLA
. ARGENTINA
. AUSTRALIA
. AUSTRIA
. BAHAMAS
. BAHRAIN
. BANGLADESH
. BARBADOS
. BELGIUM
. BELIZE
. BENIN
. BOLIVIA
. BOTSWANA
. BRAZIL
. BRUNEI
. BULGARIA
. BURKINA
. BURMA
. BURUNDI
. CAMBODIA
. CAMEROON
. CANADA
. . ALBERTA
. . BRITISH COLUMBIA
. . MANITOBA
. . NEW BRUNSWICK
. . NEWFOUNDLAND
. . NORTHWEST TERRITORIES
. . NOVA SCOTIA
. . ONTARIO
. . PRINCE EDWARD ISLAND
. . QUEBEC
. . SASKATCHEWAN
. . YUKON TERRITORY
. CAPE VERDE
. CENTRAL AFRICAN REPUBLIC
. CHAD
. CHILE
. CHINA
. COLOMBIA
. CONGO (BRAZZAVILLE)
. COSTA RICA
. CUBA
. CYPRUS
. CZECHOSLOVAKIA
. DENMARK
. DOMINICA
. DOMINICAN REPUBLIC
. EAST GERMANY
. ECUADOR
. EGYPT
. EL SALVADOR
. ETHIOPIA
. FINLAND
. FRANCE
. . FRENCH GUIANA
. . GUADELOUPE
. . MARTINIQUE
. GABON
. GAMBIA
. GHANA
. GREECE
. GUATEMALA
. GUINEA
. GUYANA
. HAITI
. HONDURAS
. HUNGARY
. ICELAND
. INDIA
. INDONESIA
. IRAN
. IRAQ
. IRELAND
. ISRAEL
. ITALY
. IVORY COAST
. JAMAICA
. JAPAN
. JORDAN

NATIONS-*(CONT.)*
. KENYA
. KUWAIT
. LAOS
. LEBANON
. LESOTHO
. LIBERIA
. LIBYA
. LIECHTENSTEIN
. LUXEMBOURG
. MALAGASY REPUBLIC
. MALAWI
. MALAYSIA
. MALDIVE ISLANDS
. MALI
. MAURITANIA
. MEXICO
. MONACO
. MONGOLIA
. MOROCCO
. MOZAMBIQUE
. NEPAL
. NETHERLANDS
. NEW ZEALAND
. NICARAGUA
. NIGER
. NIGERIA
. NORTH KOREA
. NORWAY
. OMAN
. PAKISTAN
. PANAMA
. PARAGUAY
. PERU
. PHILIPPINES
. POLAND
. PORTUGAL
. REPUBLIC OF SOUTH AFRICA
. ROMANIA
. RWANDA
. SAN MARINO
. SAUDI ARABIA
. SENEGAL
. SIERRA LEONE
. SIKKIM
. SINGAPORE
. SOMALIA
. SOUTH KOREA
. SOUTHERN YEMEN
. SPAIN
. SRI LANKA
. SUDAN
. SURINAM
. SWAZILAND
. SWEDEN
. SWITZERLAND
. SYRIA
. TAIWAN
. TANZANIA
. THAILAND
. TIBET
. TOGO
. TRINIDAD AND TOBAGO
. TUNISIA
. TURKEY
. UGANDA
. UNITED ARAB EMIRATES
. UNITED KINGDOM
. . ENGLAND
. . SCOTLAND
. UNITED STATES
. . ALABAMA
. . ALASKA
. . ARIZONA
. . ARKANSAS
. . CALIFORNIA
. . COLORADO
. . CONNECTICUT
. . DELAWARE
. . FLORIDA
. . GEORGIA
. . HAWAII
. . IDAHO
. . ILLINOIS
. . INDIANA
. . IOWA
. . KANSAS
. . KENTUCKY
. . LOUISIANA
. . MAINE
. . MARYLAND
. . MASSACHUSETTS
. . MICHIGAN
. . MINNESOTA
. . MISSISSIPPI
. . MISSOURI

NATIONS-*(CONT.)*
. . MONTANA
. . NEBRASKA
. . NEVADA
. . NEW HAMPSHIRE
. . NEW JERSEY
. . NEW MEXICO
. . NEW YORK
. . NORTH CAROLINA
. . NORTH DAKOTA
. . OHIO
. . OKLAHOMA
. . OREGON
. . PENNSYLVANIA
. . RHODE ISLAND
. . SOUTH CAROLINA
. . SOUTH DAKOTA
. . TENNESSEE
. . TEXAS
. . UTAH
. . VERMONT
. . VIRGINIA
. . WASHINGTON
. . WEST VIRGINIA
. . WISCONSIN
. . WYOMING
. URUGUAY
. U.S.S.R.
. VATICAN CITY
. VENEZUELA
. VIETNAM
. WEST GERMANY
. YEMEN
. YUGOSLAVIA
. ZAIRE
. ZAMBIA
. ZIMBABWE
RT   AFRICA
        ASIA
        CITIES
        COMMUNITIES
        DEMOGRAPHY
        DEVELOPING NATIONS
        ESTONIA
        EUROPE
        FEDERATIONS
        HONG KONG
        INTERNATIONAL LAW
        LATVIA
        LITHUANIA
        MINORITIES
        NAMIBIA
        POLITICS
        REGIMES
        SPANISH SAHARA
        UNITED NATIONS

NATO 3B SATELLITE
GS   SATELLITES
. ARTIFICIAL SATELLITES
. . COMMUNICATION SATELLITES
. . . COMMUNICATIONS TECHNOLOGY
        SATELLITE
. . . . **NATO 3B SATELLITE**

NATURAL FREQUENCIES
USE   RESONANT FREQUENCIES

NATURAL GAS
GS   GASES
        NATURAL GAS
. . LIQUEFIED NATURAL GAS
        HYDROCARBONS
        NATURAL GAS
. . LIQUEFIED NATURAL GAS
RT   HYDROCARBON FUELS
        METHANE
        NATURAL GAS EXPLORATION
        OIL FIELDS

NATURAL GAS EXPLORATION
RT   DRILLING
        METHANE
        NATURAL GAS
        OIL EXPLORATION
        PHOTOGEOLOGY

NATURAL LANGUAGE (COMPUTERS)
GS   LANGUAGES
. PROGRAMMING LANGUAGES
. . **NATURAL LANGUAGE (COMPUTERS)**
RT   COMPUTER PROGRAMMING
        CONTEXT
        DATA PROCESSING

NATURAL LASERS
USE   LASERS

NATURAL SATELLITES
SN   (EXCLUDES PLANETS)
UF   PLANETARY SATELLITES
GS   CELESTIAL BODIES
      . **NATURAL SATELLITES**
      . . CHARON
      . . DEIMOS
      . . IAPETUS
      . . JUPITER SATELLITES
      . . . GALILEAN SATELLITES
      . . . . CALLISTO
      . . . . EUROPA
      . . . . GANYMEDE
      . . . . IO
      . . MOON
      . . RHEA (ASTRONOMY)
      . . TITAN
      . . TRITON
      . . URANUS SATELLITES
      SATELLITES
      . **NATURAL SATELLITES**
      . . CHARON
      . . DEIMOS
      . . JUPITER SATELLITES
      . . . AMALTHEA
      . . . GALILEAN SATELLITES
      . . . . CALLISTO
      . . . . EUROPA
      . . . . GANYMEDE
      . . . . IO
      . . MOON
      . . SATURN SATELLITES
      . . . DIONE
      . . . ENCELADUS
      . . . HYPERION
      . . . IAPETUS
      . . . JANUS
      . . . MIMAS
      . . . RHEA (ASTRONOMY)
      . . . TETHYS
      . . . TITAN
      . . TRITON
RT   ARTIFICIAL SATELLITES
      CYRILLID METEOROIDS
      EARTH-MOON SYSTEM
      METEOROIDS
      PLANETS
      ROCHE LIMIT
      SATELLITE ATMOSPHERES
      SATELLITE SURFACES
      SATURN RINGS
      SOLAR SYSTEM
      TEKTITES
      URANUS RINGS

NAUSEA
RT   ANTIEMETICS AND ANTINAUSEANTS
      MOTION SICKNESS
      VOMITING

NAUTICAL CHARTS
GS   CHARTS
      . **NAUTICAL CHARTS**
RT   NAVIGATION AIDS
      NAVIGATION SATELLITES
      SURFACE NAVIGATION

NAVAHO MISSILE
GS   MISSILES
      . RAMJET MISSILES
      . . **NAVAHO MISSILE**
      . SURFACE TO SURFACE MISSILES
      . . CRUISE MISSILES
      . . . **NAVAHO MISSILE**
RT   LIQUID PROPELLANT ROCKET ENGINES
      MULTISTAGE ROCKET VEHICLES
      RAMJET ENGINES

NAVIER-STOKES EQUATION
GS   EQUATIONS OF MOTION
      . **NAVIER-STOKES EQUATION**
RT   BURGER EQUATION
      COMPUTATIONAL FLUID DYNAMICS
      ∞ EQUATIONS
      FLOW THEORY
      INCOMPRESSIBLE FLOW
      INCOMPRESSIBLE FLUIDS
      MILNE-THOMSON METHOD
      NEWTONIAN FLUIDS
      OSEEN APPROXIMATION
      REYNOLDS EQUATION
      REYNOLDS STRESS

NAVIER-STOKES EQUATION-(CONT.)
      VISCOUS FLOW
      VISCOUS FLUIDS

NAVIGATION
GS   **NAVIGATION**
      . AIR NAVIGATION
      . . ALL-WEATHER AIR NAVIGATION
      . . AREA NAVIGATION
      . . NAP-OF-THE-EARTH NAVIGATION
      . CELESTIAL NAVIGATION
      . . ASTROGUIDE NAVIGATION SYSTEM
      . . ASTRONAVIGATION
      . DEAD RECKONING
      . DIGITAL NAVIGATION
      . DOPPLER NAVIGATION
      . HYBRID NAVIGATION SYSTEMS
      . INERTIAL NAVIGATION
      . . ASTROGUIDE NAVIGATION SYSTEM
      . . GIMBALLESS INERTIAL NAVIGATION
      . OMEGA NAVIGATION SYSTEM
      . POLAR NAVIGATION
      . RADAR NAVIGATION
      . RADIO NAVIGATION
      . . HYPERBOLIC NAVIGATION
      . . . DECCA NAVIGATION
      . . . LORAC NAVIGATION SYSTEM
      . . . LORAN
      . . . . LORAN C
      . . . . LORAN D
      . . . SHORAN
      . . TACAN
      . . VHF OMNIRANGE NAVIGATION
      . SPACE NAVIGATION
      . . INTERPLANETARY NAVIGATION
      . SURFACE NAVIGATION
RT   AUTOMATIC FLIGHT CONTROL
      AZIMUTH
      BAY ICE
      DECLINATION
      DISTANCE MEASURING EQUIPMENT
      ∞ FIXING
      FLIGHT CONTROL
      FLIGHT PATHS
      GLOBAL POSITIONING SYSTEM
      GUIDANCE (MOTION)
      GYROSCOPIC COUPLING
      HOMING DEVICES
      LATITUDE MEASUREMENT
      LOCOMOTION
      LONGITUDE MEASUREMENT
      ORBITAL POSITION ESTIMATION
      PLOTTING
      POSITION (LOCATION)
      POSITION ERRORS
      POSITIONING
      STAR TRACKERS
      STATIONKEEPING
      ∞ SYSTEMS
      TRIANGULATION

NAVIGATION AIDS
GS   **NAVIGATION AIDS**
      . BEACONS
      . . AIRPORT BEACONS
      . . . DISCRETE ADDRESS BEACON
            SYSTEM
      . . RADAR BEACONS
      . . . DISCRETE ADDRESS BEACON
            SYSTEM
      . . RADIO BEACONS
      . . . OMNIDIRECTIONAL RADIO RANGES
      . . . . SELF CALIBRATING OMNIRANGE
      . . RADIO DIRECTION FINDERS
      . LASER ALTIMETERS
      . LIGHT AIRBORNE MULTIPURPOSE
        SYSTEM
      . MICROWAVE SCANNING BEAM
        LANDING SYSTEM
      . NAVIGATION INSTRUMENTS
      . . COMPASSES
      . . . GYROCOMPASSES
      . . . MAGNETIC COMPASSES
      . . . SOLAR COMPASSES
      . . RADIO DIRECTION FINDERS
      . TERCOM
RT   ∞ AIDS
      AIR NAVIGATION
      AIR TRAFFIC CONTROL
      AIRCRAFT EQUIPMENT
      AIRCRAFT INSTRUMENTS
      AIRCRAFT SAFETY
      AIRPORTS
      ALL-WEATHER AIR NAVIGATION
      ALTIMETERS

NAVIGATION AIDS-(CONT.)
      APPROACH INDICATORS
      ATTITUDE INDICATORS
      AUTOMATIC FLIGHT CONTROL
      AUTOMATIC PILOTS
      AUTOMATIC TRAFFIC ADVISORY AND
        RESOLUTION
      AUTONOMOUS NAVIGATION
      BUOYS
      CHARTS
      DECCA NAVIGATION
      DISPLAY DEVICES
      DISTANCE MEASURING EQUIPMENT
      ECHO SOUNDING
      FLIGHT CONTROL
      FLIGHT MANAGEMENT SYSTEMS
      FLIGHT PATHS
      GYROSTABILIZERS
      HEAD-UP DISPLAYS
      HELIPORTS
      HOMING DEVICES
      HYBRID NAVIGATION SYSTEMS
      INERTIAL NAVIGATION
      KALMAN FILTERS
      LANDING AIDS
      LASER RANGE FINDERS
      LORAC NAVIGATION SYSTEM
      LORAN
      LORAN C
      LORAN D
      MAPS
      ∞ MARS
      MILITARY AIR FACILITIES
      NAUTICAL CHARTS
      PLOTTERS
      POSITION INDICATORS
      RADIO NAVIGATION
      RANGE FINDERS
      REDUCED ORDER FILTERS
      REFERENCE STARS
      SEXTANTS
      SHORAN
      SOLAR SENSORS
      SONAR
      STAR TRACKERS
      SURFACE NAVIGATION
      TACAN
      VHF OMNIRANGE NAVIGATION
      WEATHER

NAVIGATION INSTRUMENTS
GS   NAVIGATION AIDS
      . **NAVIGATION INSTRUMENTS**
      . . COMPASSES
      . . . GYROCOMPASSES
      . . . MAGNETIC COMPASSES
      . . . SOLAR COMPASSES
      . . RADIO DIRECTION FINDERS
RT   AIRCRAFT EQUIPMENT
      AIRCRAFT INSTRUMENTS
      ALTIMETERS
      AUTONOMOUS NAVIGATION
      BORESIGHT ERROR
      FLIGHT CONTROL
      FLIGHT INSTRUMENTS
      GIMBALLESS INERTIAL NAVIGATION
      HORIZON SCANNERS
      HYBRID NAVIGATION SYSTEMS
      INERTIAL PLATFORMS
      ∞ INSTRUMENTS
      KALMAN-SCHMIDT FILTERING
      LASER RANGE FINDERS
      LIGHT AIRBORNE MULTIPURPOSE
        SYSTEM
      LORAC NAVIGATION SYSTEM
      LORAN
      MEASURING INSTRUMENTS
      POSITION INDICATORS
      RADAR
      SOLAR SENSORS
      STAR TRACKERS
      TERCOM

NAVIGATION SATELLITES
GS   SATELLITES
      . ARTIFICIAL SATELLITES
      . . **NAVIGATION SATELLITES**
      . . . AEROSAT SATELLITES
      . . . EXPLORER 22 SATELLITE
      . . . NAVIGATION TECHNOLOGY
            SATELLITES
      . . . NAVSTAR SATELLITES
      . . . NOVA SATELLITES
      . . . REFSAT

NAVIGATION SATELLITES-*(CONT.)*
```
        . . . TRANSIT ATTITUDE CONTROL
                     SATELLITE
        . . . TRANSIT SATELLITES
        UNMANNED SPACECRAFT
        . NAVIGATION SATELLITES
        . EXPLORER 22 SATELLITE
        . . NAVSTAR SATELLITES
        . . NOVA SATELLITES
        . . REFSAT
        . . TRANSIT ATTITUDE CONTROL
                     SATELLITE
        . . TRANSIT SATELLITES
RT      ACTIVE SATELLITES
        ATS
        GEODETIC SATELLITES
        LOCATES SYSTEM
        METEOROLOGICAL SATELLITES
        MILITARY SPACECRAFT
        NAUTICAL CHARTS
        PASSIVE SATELLITES
        SATELLITE NAVIGATION SYSTEMS
        SATELLITE NETWORKS
        SYNCHRONOUS SATELLITES
        TRANSIT NAVIGATION SYSTEM
```

NAVIGATION TECHNOLOGY SATELLITES
```
UF      NTS
GS      SATELLITES
        . ARTIFICIAL SATELLITES
        . . NAVIGATION SATELLITES
        . . . NAVIGATION TECHNOLOGY
                     SATELLITES
RT      NAVSTAR SATELLITES
```

NAVIGATORS
```
GS      PERSONNEL
        . NAVIGATORS
RT      FLIGHT CREWS
        FLYING PERSONNEL
```

NAVION AIRCRAFT
```
GS      NAVION AIRCRAFT
        . G-1 AIRCRAFT
RT      ∞AIRCRAFT
```

NAVION G-1 AIRCRAFT
```
USE     G-1 AIRCRAFT
```

NAVION RANGEMASTER AIRCRAFT
```
USE     G-1 AIRCRAFT
```

NAVSTAR SATELLITES
```
GS      SATELLITES
        . ARTIFICIAL SATELLITES
        . . NAVIGATION SATELLITES
        . . . NAVSTAR SATELLITES
        UNMANNED SPACECRAFT
        . NAVIGATION SATELLITES
        . . NAVSTAR SATELLITES
RT      ACTIVE SATELLITES
        ATS
        GEODETIC SATELLITES
        NAVIGATION TECHNOLOGY SATELLITES
        REFSAT
        SATELLITE NETWORKS
```

NAVY
```
GS      ARMED FORCES
        . NAVY
RT      AIRCRAFT CARRIERS
        BALLISTIC MISSILE SUBMARINES
        FLEET SATELLITE COMMUNICATION
                     SYSTEM
        NUCLEAR POWERED SHIPS
        SHIPS
        SUBMARINES
        TRIDENT SUBMARINE
        ∞VESSELS
```

NC-130 AIRCRAFT
```
USE     C-130 AIRCRAFT
```

NDM SEMICONDUCTOR DEVICES
```
SN      (NEGATIVE DIFFERENTIAL MOBILITY
                     SEMICONDUCTOR DEVICES)
UF      NEGATIVE DIFF MOBILITY
                     SEMICONDUCTORS
GS      ELECTRONIC EQUIPMENT
        . SOLID STATE DEVICES
        . . SEMICONDUCTOR DEVICES
        . . . NDM SEMICONDUCTOR DEVICES
RT      CONDUCTION BANDS
        ∞DEVICES
```

NDM SEMICONDUCTOR DEVICES-*(CONT.)*
```
        DIFFUSIVITY
        ELECTRON MOBILITY
        IONIC MOBILITY
```

NEAR FIELDS
```
GS      ELECTROMAGNETIC FIELDS
        . NEAR FIELDS
RT      ANTENNA RADIATION PATTERNS
        ANTENNAS
        ELECTROMAGNETIC RADIATION
        FAR FIELDS
        RADIO EQUIPMENT
        SIDELOBES
```

NEAR INFRARED RADIATION
```
SN      (0.75 TO 3 MICRONS)
GS      ELECTROMAGNETIC RADIATION
        . INFRARED RADIATION
        . . NEAR INFRARED RADIATION
RT      FAR INFRARED RADIATION
        INFRARED PHOTOMETRY
        LIGHT (VISIBLE RADIATION)
        ∞RADIATION
        RADIATIVE HEAT TRANSFER
        RADIATIVE TRANSFER
        TERRESTRIAL RADIATION
        THERMAL RADIATION
```

NEAR ULTRAVIOLET RADIATION
```
SN      (2000 TO 4000 ANGSTROMS)
GS      ELECTROMAGNETIC RADIATION
        . ULTRAVIOLET RADIATION
        . . NEAR ULTRAVIOLET RADIATION
        IONIZING RADIATION
        . ULTRAVIOLET RADIATION
        . . NEAR ULTRAVIOLET RADIATION
RT      FAR ULTRAVIOLET RADIATION
        LIGHT (VISIBLE RADIATION)
        ∞RADIATION
```

NEAR WAKES
```
GS      WAKES
        . NEAR WAKES
```

NEARSHORE WATER
```
GS      WATER
        . NEARSHORE WATER
        . . COASTAL WATER
RT      MARINE ENVIRONMENTS
        OCEANS
        SEA WATER
        VADOSE WATER
        WATER DEPTH
        WETLANDS
```

NEBRASKA
```
GS      NATIONS
        . UNITED STATES
        . . NEBRASKA
RT      MISSOURI RIVER (US)
        SAND HILLS REGION (NE)
```

NEBULAE
```
GS      CELESTIAL BODIES
        . NEBULAE
        . . CASSIOPEIA A
        . . CRAB NEBULA
        . . GUM NEBULA
        . . ORION NEBULA
        . . PLANETARY NEBULAE
        . . REFLECTION NEBULAE
RT      GALAXIES
        INTERSTELLAR MATTER
        MAFFEI GALAXIES
        MAGELLANIC CLOUDS
        NORTH POLAR SPUR (ASTRONOMY)
        OPHIUCHI CLOUDS
        OPIK THEORY
        SOLAR CORONA
        SUPERNOVAE
```

NECK (ANATOMY)
```
GS      ANATOMY
        . NECK (ANATOMY)
RT      VERTEBRAE
```

NEEDLE BEARINGS
```
GS      BEARINGS
        . NEEDLE BEARINGS
RT      ANTIFRICTION BEARINGS
        BALL BEARINGS
        ROLLER BEARINGS
```

NEEDLES
```
RT      DENDRITIC CRYSTALS
        SEWING
        SINGLE CRYSTALS
        SURGICAL INSTRUMENTS
```

NEEDS (DATA SYSTEM)
```
UF      NASA END-TO-END DATA SYSTEM
GS      DATA SYSTEMS
        . NEEDS (DATA SYSTEM)
        END-TO-END DATA SYSTEMS
        . NEEDS (DATA SYSTEM)
RT      DATA ACQUISITION
        DATA PROCESSING
        SATELLITE INSTRUMENTS
        ∞SYSTEMS
```

NEEL TEMPERATURE
```
GS      TEMPERATURE
        . NEEL TEMPERATURE
RT      ANTIFERROMAGNETISM
        MAGNETIC PERMEABILITY
        PHASE TRANSFORMATIONS
        SPECIFIC HEAT
        THERMAL EXPANSION
```

NEGATIVE CONDUCTANCE
```
RT      AVALANCHE DIODES
        GALLIUM ARSENIDES
        GUNN EFFECT
        TUNNEL DIODES
```

NEGATIVE DIFF MOBILITY SEMICONDUCTORS
```
USE     NDM SEMICONDUCTOR DEVICES
```

NEGATIVE FEEDBACK
```
UF      DEGENERATIVE FEEDBACK
GS      FEEDBACK
        . NEGATIVE FEEDBACK
RT      AUTOMATIC CONTROL
        DAMPING
        DEGENERATION
        FEEDBACK CONTROL
        NONLINEAR FEEDBACK
        OSCILLATORS
        TRANSFER FUNCTIONS
```

NEGATIVE IONS
```
GS      IONS
        . NEGATIVE IONS
        PARTICLES
        . CHARGED PARTICLES
        . . NEGATIVE IONS
RT      ANIONS
        FORMYL IONS
        FREE RADICALS
        IONIC MOBILITY
        NITROGEN IONS
        OXYGEN IONS
        PLASMA PHYSICS
```

NEGATIVE RESISTANCE CIRCUITS
```
GS      CIRCUITS
        . NEGATIVE RESISTANCE CIRCUITS
RT      ∞RESISTANCE
        TUNNEL DIODES
```

NEGATIVE RESISTANCE DEVICES
```
RT      ALUMINUM GALLIUM ARSENIDES
        GALLIUM ARSENIDES
        GUNN DIODES
        GUNN EFFECT
        MIM DIODES
        NEGATRONS
        PARAMETRIC AMPLIFIERS
        RAMSAUER EFFECT
        ∞RESISTANCE
```

NEGATRONS
```
GS      PARTICLES
        . CHARGED PARTICLES
        . . ENERGETIC PARTICLES
        . . . ELECTRONS
        . . . . NEGATRONS
RT      NEGATIVE RESISTANCE DEVICES
```

NEMBUTAL (TRADEMARK)
```
GS      ALIPHATIC COMPOUNDS
        . KETONES
        . . NEMBUTAL (TRADEMARK)
        DRUGS
        . NEMBUTAL (TRADEMARK)
        SODIUM COMPOUNDS
        . NEMBUTAL (TRADEMARK)
```

NEMBUTAL (TRADEMARK)-*(CONT.)*
RT   PENTOBARBITAL SODIUM

NEODYMIUM
GS   CHEMICAL ELEMENTS
    . RARE EARTH ELEMENTS
    . **NEODYMIUM**
    METALS
    . RARE EARTH ELEMENTS
    . . **NEODYMIUM**
RT   DIDYMIUM

NEODYMIUM ALLOYS
GS   ALLOYS
    . RARE EARTH ALLOYS
    . . **NEODYMIUM ALLOYS**

NEODYMIUM COMPOUNDS
GS   RARE EARTH COMPOUNDS
    . **NEODYMIUM COMPOUNDS**
RT   ∞CHEMICAL COMPOUNDS
    ∞METAL COMPOUNDS

NEODYMIUM ISOTOPES
GS   CHEMICAL ELEMENTS
    . NUCLIDES
    . . ISOTOPES
    . . . **NEODYMIUM ISOTOPES**
    . RARE EARTH ELEMENTS
    . **NEODYMIUM ISOTOPES**
    METALS
    . RARE EARTH ELEMENTS
    . . **NEODYMIUM ISOTOPES**

NEODYMIUM LASERS
GS   STIMULATED EMISSION DEVICES
    . LASERS
    . . **NEODYMIUM LASERS**
RT   COHERENT LIGHT
    GLASS LASERS
    OPTICAL PUMPING
    RARE EARTH ELEMENTS

NEON
GS   CHEMICAL ELEMENTS
    . RARE GASES
    . . **NEON**
    . . . LIQUID NEON
    . . . NEON ISOTOPES
    GASES
    . RARE GASES
    . . **NEON**
    . . . LIQUID NEON
    . . . NEON ISOTOPES

NEON ISOTOPES
UF   NEON 19
GS   CHEMICAL ELEMENTS
    . NUCLIDES
    . . ISOTOPES
    . . . **NEON ISOTOPES**
    . RARE GASES
    . . **NEON**
    . . . **NEON ISOTOPES**
    GASES
    . RARE GASES
    . . **NEON**
    . . . **NEON ISOTOPES**

NEON 19
USE   NEON ISOTOPES

NEOPENTANE
GS   ALIPHATIC COMPOUNDS
    . ALKANES
    . . PENTANES
    . . . **NEOPENTANE**
    HYDROCARBONS
    . ALKANES
    . . PENTANES
    . . . **NEOPENTANE**

NEOPLASMS
GS   DISEASES
    . TUMORS
    . . **NEOPLASMS**
    . . . CANCER
    . . . . LEUKEMIAS
RT   CARCINOGENS
    CYSTS

NEOPRENES
USE   CHLOROPRENE RESINS

NEPAL
GS   NATIONS
    . **NEPAL**
RT   ASIA

NEPHANALYSIS
RT   ALPINE METEOROLOGY
    ANVIL CLOUDS
    ATMOSPHERIC CLOUD PHYSICS LAB
      (SPACELAB)
    CAP CLOUDS
    CHEMICAL ANALYSIS
    CIRROCUMULUS CLOUDS
    CIRROSTRATUS CLOUDS
    CLOUD COVER
    CLOUD PHYSICS
    CLOUDS (METEOROLOGY)
    CONVECTION CLOUDS
    METEOROLOGICAL INSTRUMENTS
    METEOROLOGY
    NEPHELOMETERS
    PRECIPITATION (METEOROLOGY)
    SYNOPTIC MEASUREMENT
    SYNOPTIC METEOROLOGY
    WEATHER FORECASTING

NEPHELINE
GS   ALUMINUM COMPOUNDS
    . **NEPHELINE**
    MINERALS
    . **NEPHELINE**
    POTASSIUM COMPOUNDS
    . **NEPHELINE**
    SILICON COMPOUNDS
    . SILICATES
    . . **NEPHELINE**
    SODIUM COMPOUNDS
    . **NEPHELINE**
RT   NEPHELITE

NEPHELITE
GS   ALUMINUM COMPOUNDS
    . **NEPHELITE**
    CHALCOGENIDES
    . OXIDES
    . . SILICON OXIDES
    . . . **NEPHELITE**
    MINERALS
    . **NEPHELITE**
    SILICON COMPOUNDS
    . SILICON OXIDES
    . . **NEPHELITE**
RT   NEPHELINE

NEPHELOMETERS
GS   MEASURING INSTRUMENTS
    . OPTICAL MEASURING INSTRUMENTS
    . . **NEPHELOMETERS**
    OPTICAL EQUIPMENT
    . OPTICAL MEASURING INSTRUMENTS
    . . **NEPHELOMETERS**
RT   NEPHANALYSIS
    OPTICAL MEASUREMENT
    PHOTOMETERS

NEPHRITIS
GS   DISEASES
    . KIDNEY DISEASES
    . . **NEPHRITIS**

NEPTUNE (PLANET)
GS   CELESTIAL BODIES
    . PLANETS
    . . GAS GIANT PLANETS
    . . . **NEPTUNE (PLANET)**
RT   NEPTUNE ATMOSPHERE
    TRITON

NEPTUNE ATMOSPHERE
GS   ENVIRONMENTS
    . EXTRATERRESTRIAL ENVIRONMENTS
    . . PLANETARY ENVIRONMENTS
    . . . PLANETARY ATMOSPHERES
    . . . . **NEPTUNE ATMOSPHERE**
RT   AEROSPACE ENVIRONMENTS
    ∞ATMOSPHERES
    GAS GIANT PLANETS
    HYDROGEN
    METHANE
    NEPTUNE (PLANET)
    TRITON

NEPTUNIUM
GS   CHEMICAL ELEMENTS
    . ACTINIDE SERIES

NEPTUNIUM-*(CONT.)*
    . . TRANSURANIUM ELEMENTS
    . . . **NEPTUNIUM**
    . . . . NEPTUNIUM ISOTOPES
    . . NUCLIDES
    . . ISOTOPES
    . . RADIOACTIVE ISOTOPES
    . . . . TRANSURANIUM ELEMENTS
    . . . . . **NEPTUNIUM**
    . . . . . . NEPTUNIUM ISOTOPES
    HEAVY ELEMENTS
    . TRANSURANIUM ELEMENTS
    . . **NEPTUNIUM**
    . . . NEPTUNIUM ISOTOPES
    METALS
    . ACTINIDE SERIES
    . . TRANSURANIUM ELEMENTS
    . . . **NEPTUNIUM**
    . . . . NEPTUNIUM ISOTOPES

NEPTUNIUM COMPOUNDS
GS   ACTINIDE SERIES COMPOUNDS
    . **NEPTUNIUM COMPOUNDS**
RT   ∞CHEMICAL COMPOUNDS
    ∞METAL COMPOUNDS

NEPTUNIUM ISOTOPES
GS   CHEMICAL ELEMENTS
    . ACTINIDE SERIES
    . . TRANSURANIUM ELEMENTS
    . . . NEPTUNIUM
    . . . . **NEPTUNIUM ISOTOPES**
    . NUCLIDES
    . . ISOTOPES
    . . . RADIOACTIVE ISOTOPES
    . . . . TRANSURANIUM ELEMENTS
    . . . . . NEPTUNIUM
    . . . . . . **NEPTUNIUM ISOTOPES**
    HEAVY ELEMENTS
    . TRANSURANIUM ELEMENTS
    . . NEPTUNIUM
    . . . **NEPTUNIUM ISOTOPES**
    METALS
    . ACTINIDE SERIES
    . . TRANSURANIUM ELEMENTS
    . . . NEPTUNIUM
    . . . . **NEPTUNIUM ISOTOPES**

NERNST GENERATORS
USE   THERMOMAGNETIC COOLING

NERNST HEAT THEOREM
USE   NERNST-ETTINGSHAUSEN EFFECT

NERNST-ETTINGSHAUSEN EFFECT
UF   NERNST HEAT THEOREM
GS   GALVANOMAGNETIC EFFECTS
    . **NERNST-ETTINGSHAUSEN EFFECT**
RT   ∞EFFECTS
    TEMPERATURE EFFECTS
    THERMOMAGNETIC EFFECTS

NERVA (ENGINE)
USE   NUCLEAR ENGINE FOR ROCKET
      VEHICLES

NERVES
GS   NERVOUS SYSTEM
    . GANGLIA
    . . **NERVES**
RT   CAROTID SINUS BODY
    CAROTID SINUS REFLEX
    HIS BUNDLE
    NEURITIS

NERVOUS SYSTEM
UF   VASOMOTOR NERVOUS SYSTEM
GS   **NERVOUS SYSTEM**
    . AFFERENT NERVOUS SYSTEMS
    . AUTONOMIC NERVOUS SYSTEM
    . . SYMPATHETIC NERVOUS SYSTEM
    . AXONS
    . CENTRAL NERVOUS SYSTEM
    . . BRAIN
    . . . BRAIN STEM
    . . . CEREBELLUM
    . . . CEREBRAL CORTEX
    . . . CEREBRAL VENTRICLES
    . . . CEREBRUM
    . . . HIPPOCAMPUS
    . . . SPINAL CORD
    . . . SPINE
    . . THALAMUS
    . DIENCEPHALON
    . EFFERENT NERVOUS SYSTEMS

NERVOUS SYSTEM-*(CONT.)*
　. GANGLIA
　. . NERVES
　. . . NEURONS
　. MYELIN
　. NEUROBLASTS
　. NEUROGLIA
　. PERIPHERAL NERVOUS SYSTEM
　. SYNAPSES
RT　ELECTROPHYSIOLOGY
　　HOMEOSTASIS
　　NEURASTHENIA
　　NEURITIS
　　NEUROPSYCHIATRY
　　NEUROTRANSMITTERS
　　PROPRIOCEPTORS
　　PSYCHOPHARMACOLOGY
　　SENSE ORGANS
　　∞SYSTEMS

NETHERLANDS
UF　HOLLAND
GS　NATIONS
　. **NETHERLANDS**
RT　ASTRONOMICAL NETHERLANDS
　　　SATELLITE
　　EUROPE
　　SURINAM

NETS
GS　**NETS**
　. NEURAL NETS
　. PETRI NETS
RT　∞NETWORKS

NETWORK ANALYSIS
UF　TELLEGEN THEORY
GS　**NETWORK ANALYSIS**
　. CRITICAL PATH METHOD
　. SNEAK CIRCUIT ANALYSIS
RT　∞ANALYZING
　　CIRCUITS
　　DATA FLOW ANALYSIS
　　DISTRIBUTED PARAMETER SYSTEMS
　　DUALITY PRINCIPLE
　　EQUIVALENT CIRCUITS
　　FOSTER THEORY
　　GYRATORS
　　HYDRAULIC EQUIPMENT
　　INSERTION
　　KIRCHHOFF LAW OF NETWORKS
　　LC CIRCUITS
　　∞NETWORKS
　　∞PATHS
　　RC CIRCUITS
　　RL CIRCUITS
　　RLC CIRCUITS
　　SIGNAL FLOW GRAPHS
　　SUPERPOSITION (MATHEMATICS)

NETWORK CONTROL
RT　COMMUNICATION NETWORKS
　　COMMUNICATION SATELLITES
　　COMPUTER NETWORKS
　　∞CONTROL
　　PACKET SWITCHING
　　SATELLITE NETWORKS
　　TRANSMISSION EFFICIENCY

NETWORK SYNTHESIS
UF　TELLEGEN THEORY
RT　COMMUNICATION THEORY
　　EQUIVALENT CIRCUITS
　　HYDRAULIC EQUIPMENT
　　KIRCHHOFF LAW OF NETWORKS
　　LC CIRCUITS
　　∞NETWORKS
　　RC CIRCUITS
　　RICHARDS THEOREM
　　RL CIRCUITS
　　RLC CIRCUITS
　　SUPERPOSITION (MATHEMATICS)
　　SWITCHING THEORY
　　∞SYNTHESIS
　　TOPOLOGY

∞ **NETWORKS**
SN　*(USE OF A MORE SPECIFIC TERM IS*
　　RECOMMENDED--CONSULT THE TERMS
　　LISTED BELOW)
RT　ARPA COMPUTER NETWORK
　　BOND GRAPHS
　　CIRCUITS
　　COMPUTER NETWORKS
　　COUPLING CIRCUITS

NETWORKS-*(CONT.)*
　　DEEP SPACE NETWORK
　　DIFFERENTIATORS
　　ELECTRIC NETWORKS
　　ELECTRIC POWER TRANSMISSION
　　GRAPHS (CHARTS)
　　GYRATORS
　　HYDRAULIC EQUIPMENT
　　INTEGRATORS
　　MAGNETIC CIRCUITS
　　MOSAICS
　　NASCOM NETWORK
　　NETS
　　NETWORK ANALYSIS
　　NETWORK SYNTHESIS
　　NEURAL NETS
　　PNEUMATIC EQUIPMENT
　　QUADRUPOLE NETWORKS
　　RADAR NETWORKS
　　SATELLITE NETWORKS
　　SIGNAL FLOW GRAPHS
　　TOPOLOGY
　　TRACKING NETWORKS
　　TRANSMISSION LINES
　　TRANSPORTATION NETWORKS
　　UNDERGROUND TRANSMISSION LINES

NEUMANN PROBLEM
GS　ANALYSIS (MATHEMATICS)
　. REAL VARIABLES
　. . **NEUMANN PROBLEM**
　　BOUNDARY VALUE PROBLEMS
　. **NEUMANN PROBLEM**
RT　DIFFERENTIAL EQUATIONS
　　PARTIAL DIFFERENTIAL EQUATIONS
　　∞PROBLEMS

NEURAL NETS
GS　NETS
　. **NEURAL NETS**
RT　CYBERNETICS
　　LOGIC CIRCUITS
　　∞NETWORKS

NEURASTHENIA
GS　DISEASES
　. **NEURASTHENIA**
RT　NERVOUS SYSTEM

NEURISTORS
GS　ELECTRONIC EQUIPMENT
　. SOLID STATE DEVICES
　. . SEMICONDUCTOR DEVICES
　. . . **NEURISTORS**
RT　BIONICS

NEURITIS
GS　DISEASES
　. **NEURITIS**
RT　NERVES
　　NERVOUS SYSTEM

NEUROBLASTS
GS　CELLS (BIOLOGY)
　. **NEUROBLASTS**
　　NERVOUS SYSTEM
　. **NEUROBLASTS**

NEUROGLIA
GS　NERVOUS SYSTEM
　. **NEUROGLIA**
　　TISSUES (BIOLOGY)
　. **NEUROGLIA**
RT　CELLS (BIOLOGY)

NEUROLOGY
UF　NEUROSCIENCE
GS　MEDICAL SCIENCE
　. **NEUROLOGY**
RT　BRAIN
　　CHEMICAL DEFENSE
　　ELECTROPHYSIOLOGY
　　LIFE SCIENCES
　　NEUROPSYCHIATRY
　　THRESHOLDS (PERCEPTION)

NEUROMUSCULAR TRANSMISSION
RT　BIOELECTRICITY
　　CHOLINESTERASE
　　NEUROTRANSMITTERS
　　PERIPHERAL NERVOUS SYSTEM
　　SYNAPSES

NEURON TRANSMISSION
USE　BIOELECTRICITY

NEURONS
GS　CELLS (BIOLOGY)
　. **NEURONS**
　　NERVOUS SYSTEM
　. GANGLIA
　. . **NEURONS**
RT　BLOOD-BRAIN BARRIER
　　SYNCODERS

NEUROPHYSIOLOGY
GS　PHYSIOLOGY
　. **NEUROPHYSIOLOGY**
RT　GANGLIA
　　ONTOGENY
　　PSYCHOTROPIC DRUGS
　　∞SCIENCE

NEUROPSYCHIATRY
GS　MEDICAL SCIENCE
　. **NEUROPSYCHIATRY**
RT　HUMAN BEHAVIOR
　　∞MEDICINE
　　MENTAL HEALTH
　　NERVOUS SYSTEM
　　NEUROLOGY
　　PSYCHOTHERAPY

NEUROSCIENCE
USE　NEUROLOGY

NEUROSES
GS　**NEUROSES**
　. NEUROTIC DEPRESSION
RT　FEAR
　　FEAR OF FLYING
　　PSYCHOSES

NEUROSPORA
GS　FUNGI
　. **NEUROSPORA**
RT　GENETICS

NEUROTIC DEPRESSION
GS　NEUROSES
　. **NEUROTIC DEPRESSION**
RT　∞DEPRESSION
　　PSYCHOTIC DEPRESSION

NEUROTRANSMITTERS
RT　AXONS
　　CELLS (BIOLOGY)
　　NERVOUS SYSTEM
　　NEUROMUSCULAR TRANSMISSION

NEUROTROPISM
GS　TROPISM
　. **NEUROTROPISM**

NEUTRAL ATMOSPHERES
RT　∞ATMOSPHERES

NEUTRAL ATOMS
GS　ATOMS
　. **NEUTRAL ATOMS**
RT　ATOMIC BEAMS
　　CHARGE DISTRIBUTION
　　∞ELEMENTS
　　NEUTRAL BEAMS

NEUTRAL BEAMS
GS　BEAMS (RADIATION)
　. PARTICLE BEAMS
　. . **NEUTRAL BEAMS**
　. . . MOLECULAR BEAMS
　. . . NEUTRON BEAMS
RT　ATOMIC BEAMS
　　BEAM INJECTION
　　BEAM NEUTRALIZATION
　　NEUTRAL ATOMS
　　PARTICLES
　　PION BEAMS

NEUTRAL CURRENTS
RT　CURRENT DISTRIBUTION
　　GRAVITATIONAL COLLAPSE
　　NEUTRAL PARTICLES
　　NEUTRINOS
　　NEUTRON STARS
　　PARTICLE INTERACTIONS
　　STELLAR EVOLUTION

NEUTRAL GASES
GS   EXTRATERRESTRIAL MATTER
   . COSMIC GASES
   . . INTERSTELLAR GAS
   . . . **NEUTRAL GASES**
   GASES
   . RAREFIED GASES
   . . COSMIC GASES
   . . . INTERSTELLAR GAS
   . . . . **NEUTRAL GASES**
RT   INTERPLANETARY GAS

NEUTRAL PARTICLES
GS   PARTICLES
   . **NEUTRAL PARTICLES**
   . . GRAVITINOS
   . . NEUTRONS
   . . . COLD NEUTRONS
   . . . FAST NEUTRONS
   . . . PHOTONEUTRONS
   . . . THERMAL NEUTRONS
RT   ELECTRON RECOMBINATION
   NEUTRAL CURRENTS

NEUTRAL SHEETS
RT   ATMOSPHERIC PHYSICS
   CHARGED PARTICLES
   MAGNETOSPHERE
   PARTICLE MOTION
   PLASMA PHYSICS
   ∞SHEETS

NEUTRALIZERS
RT   ADDITIVES
   ∞AGENTS
   BUFFERS (CHEMISTRY)
   DISCHARGERS
   INHIBITORS
   PRESERVATIVES
   RETARDANTS
   STABILIZERS (AGENTS)
   SUPPRESSORS

NEUTRINO BEAMS
GS   BEAMS (RADIATION)
   . PARTICLE BEAMS
   . . **NEUTRINO BEAMS**

NEUTRINOS
GS   PARTICLES
   . ELEMENTARY PARTICLES
   . . FERMIONS
   . . . LEPTONS
   . . . . **NEUTRINOS**
   . . . . . SOLAR NEUTRINOS
RT   ANTINEUTRINOS
   GRAVITINOS
   NEUTRAL CURRENTS

NEUTRON ABSORBERS
RT   ABSORBERS (MATERIALS)
   CONTROL RODS
   MODERATORS
   POISONING (REACTION INHIBITION)
   RADIATION ABSORPTION
   RADIATION SHIELDING

NEUTRON ACTIVATION ANALYSIS
GS   ACTIVATION ANALYSIS
   . **NEUTRON ACTIVATION ANALYSIS**
   CHEMICAL TESTS
   . CHEMICAL ANALYSIS
   . . **NEUTRON ACTIVATION ANALYSIS**
RT   MASS SPECTROMETERS
   MICROANALYSIS
   QUALITATIVE ANALYSIS
   QUANTITATIVE ANALYSIS
   SPECTROSCOPIC ANALYSIS

NEUTRON BEAMS
GS   BEAMS (RADIATION)
   . PARTICLE BEAMS
   . . NEUTRAL BEAMS
   . . . **NEUTRON BEAMS**
   NUCLEAR RADIATION
   . **NEUTRON BEAMS**
RT   ATOMIC BEAMS
   PARTICLES
   PION BEAMS
   PROTON BEAMS

NEUTRON COUNTERS
UF   NEUTRON DETECTORS
GS   MEASURING INSTRUMENTS

NEUTRON COUNTERS-*(CONT.)*
   . COUNTERS
   . . RADIATION COUNTERS
   . . . **NEUTRON COUNTERS**
   . . . . NEUTRON SPECTROMETERS
   . RADIATION MEASURING INSTRUMENTS
   . . RADIATION COUNTERS
   . . . **NEUTRON COUNTERS**
   . . . . NEUTRON SPECTROMETERS
RT   DOSIMETERS
   GEIGER COUNTERS
   IONIZATION CHAMBERS
   PROPORTIONAL COUNTERS
   SCINTILLATION COUNTERS
   SPARK CHAMBERS

NEUTRON CROSS SECTIONS
RT   ABSORPTION CROSS SECTIONS
   ∞CROSS SECTIONS
   NUCLEAR PARTICLES
   SCATTERING CROSS SECTIONS
   STOPPING POWER

NEUTRON DECAY
GS   DECAY
   . **NEUTRON DECAY**
RT   HOT ATOMS

NEUTRON DETECTORS
USE   NEUTRON COUNTERS

NEUTRON DIFFRACTION
GS   DIFFRACTION
   . **NEUTRON DIFFRACTION**
RT   CRYSTALLOGRAPHY

NEUTRON DISTRIBUTION
GS   DISTRIBUTION (PROPERTY)
   . **NEUTRON DISTRIBUTION**
RT   NUCLEAR PARTICLES

NEUTRON EMISSION
GS   DECAY
   . EMISSION
   . . PARTICLE EMISSION
   . . . **NEUTRON EMISSION**
   . RADIOACTIVE DECAY
   . . **NEUTRON EMISSION**
   NUCLEAR REACTIONS
   . RADIOACTIVE DECAY
   . . **NEUTRON EMISSION**
RT   NEUTRONS
   SELECTION RULES (NUCLEAR PHYSICS)

NEUTRON FLUX
USE   FLUX (RATE)

NEUTRON FLUX DENSITY
SN   (NEUTRON EMISSION OR DETECTION
   RATE PER UNIT AREA)
GS   RATES (PER TIME)
   . FLUX DENSITY
   . . RADIANT FLUX DENSITY
   . . . PARTICLE FLUX DENSITY
   . . . . **NEUTRON FLUX DENSITY**
RT   HIGH FLUX ISOTOPE REACTORS
   IRRADIANCE
   NUCLEAR FISSION
   RADIANCE
   RADIANCY
   RADIATION SHIELDING

NEUTRON IRRADIATION
GS   IRRADIATION
   . **NEUTRON IRRADIATION**
RT   ION IRRADIATION
   TRANSMUTATION

NEUTRON PHYSICS
RT   ∞PHYSICS
   ∞SCIENCE

NEUTRON RADIOGRAPHY
GS   IMAGERY
   . RADIOGRAPHY
   . . **NEUTRON RADIOGRAPHY**
   NONDESTRUCTIVE TESTS
   . **NEUTRON RADIOGRAPHY**
RT   ∞MATERIALS TESTS

NEUTRON SCATTERING
UF   LEGENDRE CODE
GS   NUCLEAR REACTIONS
   . NUCLEAR SCATTERING

NEUTRON SCATTERING-*(CONT.)*
   . . **NEUTRON SCATTERING**
   SCATTERING
   . NUCLEAR SCATTERING
   . . **NEUTRON SCATTERING**
RT   ELEMENTARY PARTICLES
   NUCLEAR PARTICLES
   RESONANCE SCATTERING

NEUTRON SOURCES
GS   RADIATION SOURCES
   . **NEUTRON SOURCES**
RT   LINEAR ACCELERATORS
   NUCLEAR FUELS
   NUCLEAR RESEARCH AND TEST
   REACTORS
   PARTICLE ACCELERATORS
   SPENT FUELS

NEUTRON SPECTRA
GS   SPECTRA
   . ENERGY SPECTRA
   . . **NEUTRON SPECTRA**

NEUTRON SPECTROMETERS
UF   TRIPLE AXIS SPECTROMETERS
GS   MEASURING INSTRUMENTS
   . COUNTERS
   . . RADIATION COUNTERS
   . . . NEUTRON COUNTERS
   . . . . **NEUTRON SPECTROMETERS**
   . RADIATION MEASURING INSTRUMENTS
   . . RADIATION COUNTERS
   . . . NEUTRON COUNTERS
   . . . . **NEUTRON SPECTROMETERS**
   . SPECTROMETERS
   . . **NEUTRON SPECTROMETERS**

NEUTRON STARS
SN   (EXCLUDES TRACKS OF PARTICLES
   EMANATING FROM A NUCLEAR
   COLLISION)
GS   CELESTIAL BODIES
   . STARS
   . . **NEUTRON STARS**
   . . . PULSARS
RT   EXTARS
   GRAVITATIONAL LENSES
   NEUTRAL CURRENTS
   SUPERNOVA REMNANTS
   X RAY BINARIES

NEUTRON THERMALIZATION
GS   ENERGY ABSORPTION
   . MODERATION (ENERGY ABSORPTION)
   . . THERMALIZATION (ENERGY
   ABSORPTION)
   . . . **NEUTRON THERMALIZATION**

NEUTRON TRANSMUTATION
USE   NUCLEAR REACTIONS

NEUTRONS
GS   PARTICLES
   . ELEMENTARY PARTICLES
   . . FERMIONS
   . . . **NEUTRONS**
   . . . . COLD NEUTRONS
   . . . . FAST NEUTRONS
   . . . . PHOTONEUTRONS
   . . . . THERMAL NEUTRONS
   . NEUTRAL PARTICLES
   . . **NEUTRONS**
   . . . COLD NEUTRONS
   . . . FAST NEUTRONS
   . . . PHOTONEUTRONS
   . . . THERMAL NEUTRONS
RT   BARYONS
   CHARGED PARTICLES
   CORPUSCULAR RADIATION
   COSMIC RAYS
   NEUTRON EMISSION
   NUCLEAR RADIATION
   NUCLEI (NUCLEAR PHYSICS)
   NUCLEON POTENTIAL
   NUCLEONS
   RADIATION EFFECTS
   RADIATION SHIELDING

NEVADA
GS   NATIONS
   . UNITED STATES
   . . **NEVADA**
RT   GREAT BASIN (US)
   LAKE TAHOE (CA-NV)

NEVADA-(CONT.)
    PYRAMID LAKE (NV)
    SOUTHERN CALIFORNIA

NEW BRUNSWICK
  GS   NATIONS
       . CANADA
       . . **NEW BRUNSWICK**

NEW ENGLAND (US)
  GS   REGIONS
       . **NEW ENGLAND (US)**
  RT   UNITED STATES

NEW GUINEA (ISLAND)
  GS   LANDFORMS
       . ISLANDS
       . . PACIFIC ISLANDS
       . . . **NEW GUINEA (ISLAND)**
  RT   TORRES STRAIT

NEW HAMPSHIRE
  GS   NATIONS
       . UNITED STATES
       . . **NEW HAMPSHIRE**
  RT   ST LAWRENCE VALLEY (NORTH
         AMERICA)

NEW HAVEN (CT)
  GS   CITIES
       . **NEW HAVEN (CT)**
  RT   CONNECTICUT

NEW JERSEY
  GS   NATIONS
       . UNITED STATES
       . . **NEW JERSEY**
  RT   DELAWARE BAY (US)
       DELAWARE RIVER BASIN (US)
       HUDSON RIVER (NY-NJ)

NEW MEXICO
  GS   NATIONS
       . UNITED STATES
       . . **NEW MEXICO**
  RT   COLORADO PLATEAU (US)
       RIO GRANDE (NORTH AMERICA)

NEW MOONS PROJECT
  GS   PROGRAMS
       . NASA PROGRAMS
       . . **NEW MOONS PROJECT**
       . NASA SPACE PROGRAMS
       . . **NEW MOONS PROJECT**
       . PROJECTS
       . . **NEW MOONS PROJECT**
  RT   NUCLEAR PROPULSION
       STRUCTURAL WEIGHT
       WEIGHT ANALYSIS

NEW YORK
  GS   NATIONS
       . UNITED STATES
       . . **NEW YORK**
  RT   ADIRONDACK MOUNTAINS (NY)
       DELAWARE RIVER BASIN (US)
       HUDSON RIVER (NY-NJ)
       LAKE CHAMPLAIN BASIN (NY-VT)
       LONG ISLAND (NY)
       NEW YORK CITY (NY)
       ST LAWRENCE VALLEY (NORTH
         AMERICA)
       SUSQUEHANNA RIVER BASIN
         (MD-NY-PA)

NEW YORK CITY (NY)
  GS   CITIES
       . **NEW YORK CITY (NY)**
  RT   NEW YORK

NEW ZEALAND
  GS   LANDFORMS
       . ISLANDS
       . . PACIFIC ISLANDS
       . . . **NEW ZEALAND**
       NATIONS
       . **NEW ZEALAND**

NEWFOUNDLAND
  GS   LANDFORMS
       . ISLANDS
       . . **NEWFOUNDLAND**
       NATIONS
       . CANADA

NEWFOUNDLAND-(CONT.)
    . . **NEWFOUNDLAND**

NEWS
  RT   DOCUMENTATION

NEWS MEDIA
  RT   DATA ACQUISITION
       INFORMATION
       ∞JOURNALS

NEWTON
  RT   ∞FORCE
       KINETICS
       NEWTONIAN FLUIDS
       NONNEWTONIAN FLUIDS

NEWTON PRESSURE LAW
  GS   LAWS
       . **NEWTON PRESSURE LAW**
  RT   COMPRESSIBLE FLOW
       LAMINAR FLOW
       NEWTONIAN FLUIDS
       PRANDTL-MEYER EXPANSION
       PRESSURE
       PRESSURE DISTRIBUTION

NEWTON SECOND LAW
  GS   KINETICS
       . **NEWTON SECOND LAW**
       LAWS
       . **NEWTON SECOND LAW**
  RT   CONSERVATION
       MOMENTUM THEORY

NEWTON THEORY
  GS   KINETICS
       . **NEWTON THEORY**
       THEORETICAL PHYSICS
       . **NEWTON THEORY**
  RT   CONSERVATION LAWS
       NEWTONIAN FLUIDS
       NONNEWTONIAN FLUIDS
       NONRELATIVISTIC MECHANICS
       ∞THEORIES

NEWTON-BUSEMANN LAW
  GS   LAWS
       . **NEWTON-BUSEMANN LAW**

NEWTON-RAPHSON METHOD
  GS   ANALYSIS (MATHEMATICS)
       . NUMERICAL ANALYSIS
       . . APPROXIMATION
       . . . **NEWTON-RAPHSON METHOD**
  RT   ∞METHODOLOGY

NEWTONIAN FLUIDS
  RT   ANISOTROPIC FLUIDS
       ∞FLUIDS
       NAVIER-STOKES EQUATION
       NEWTON
       NEWTON PRESSURE LAW
       NEWTON THEORY
       NONNEWTONIAN FLUIDS
       STRESS-STRAIN-TIME RELATIONS
       VISCOUS FLUIDS

NICARAGUA
  GS   NATIONS
       . **NICARAGUA**
  RT   CENTRAL AMERICA

NICHROME (TRADEMARK)
  GS   ALLOYS
       . NICKEL ALLOYS
       . . **NICHROME (TRADEMARK)**

NICKEL
  GS   CHEMICAL ELEMENTS
       . **NICKEL**
       . . NICKEL ISOTOPES
       METALS
       . TRANSITION METALS
       . . **NICKEL**
       . . . NICKEL ISOTOPES
  RT   CONSTANTAN

NICKEL ALLOYS
  GS   ALLOYS
       . **NICKEL ALLOYS**
       . . ASTROLOY (TRADEMARK)
       . . HASTELLOY (TRADEMARK)
       . . INCONEL (TRADEMARK)

NICKEL ALLOYS-(CONT.)
    . . KAMACITE
    . . MONEL (TRADEMARK)
    . . NICHROME (TRADEMARK)
    . . NITINOL ALLOYS
    . . RENE 41
    . . RENE 63
    . . RENE 77
    . . RENE 95
    . . UDIMET ALLOYS
    . . WASPALOY
  RT   HEAT RESISTANT ALLOYS
       NIMONIC ALLOYS
       PERMALLOYS (TRADEMARK)
       SHAPE MEMORY ALLOYS
       STAINLESS STEELS
       SULFIDATION

NICKEL CADMIUM BATTERIES
  UF   CADMIUM NICKEL BATTERIES
  GS   ELECTROCHEMICAL CELLS
       . ELECTRIC BATTERIES
       . . STORAGE BATTERIES
       . . . **NICKEL CADMIUM BATTERIES**
  RT   DRY CELLS
       SILVER CADMIUM BATTERIES

NICKEL COATINGS
  GS   COATINGS
       METAL COATINGS
       . . **NICKEL COATINGS**
       METALS
       . METAL COATINGS
       . . **NICKEL COATINGS**
  RT   CORROSION PREVENTION
       METAL FILMS
       PROTECTIVE COATINGS

NICKEL COMPOUNDS
  GS   **NICKEL COMPOUNDS**
       . COHENITE
       . NICKEL FLUORIDES
       . NICKEL OXIDES
       . SCHREIBERSITE
  RT   ∞CHEMICAL COMPOUNDS
       ∞GROUP 8 COMPOUNDS
       ∞METAL COMPOUNDS

NICKEL FLUORIDES
  GS   HALOGEN COMPOUNDS
       . FLUORINE COMPOUNDS
       . . FLUORIDES
       . . . METAL FLUORIDES
       . . . . **NICKEL FLUORIDES**
       . HALIDES
       . . METAL HALIDES
       . . . **NICKEL FLUORIDES**
       NICKEL COMPOUNDS
       . **NICKEL FLUORIDES**

NICKEL HYDROGEN BATTERIES
  GS   ELECTROCHEMICAL CELLS
       . ELECTRIC BATTERIES
       . . STORAGE BATTERIES
       . . . **NICKEL HYDROGEN BATTERIES**
  RT   ENERGY STORAGE
       HYDROGEN-BASED ENERGY
       SPACECRAFT POWER SUPPLIES

NICKEL IRON BATTERIES
  GS   ELECTROCHEMICAL CELLS
       . ELECTRIC BATTERIES
       . . **NICKEL IRON BATTERIES**
  RT   LEAD ACID BATTERIES
       NICKEL ZINC BATTERIES
       STORAGE BATTERIES

NICKEL ISOTOPES
  GS   CHEMICAL ELEMENTS
       . NICKEL
       . . **NICKEL ISOTOPES**
       . NUCLIDES
       . . ISOTOPES
       . . . **NICKEL ISOTOPES**
       METALS
       . TRANSITION METALS
       . . NICKEL
       . . . **NICKEL ISOTOPES**

NICKEL OXIDES
  GS   CHALCOGENIDES
       . OXIDES
       . . METAL OXIDES
       . . . **NICKEL OXIDES**
       NICKEL COMPOUNDS

NICKEL OXIDES-*(CONT.)*
. NICKEL OXIDES

NICKEL PLATE
GS   PLATING
. **NICKEL PLATE**
RT   ELECTROPLATING
GOLD COATINGS

NICKEL STEELS
GS   ALLOYS
. IRON ALLOYS
. . STEELS
. . . **NICKEL STEELS**
RT   STAINLESS STEELS

NICKEL ZINC BATTERIES
UF   ZINC NICKEL BATTERIES
GS   ELECTRIC GENERATORS
. DIRECT POWER GENERATORS
. . PRIMARY BATTERIES
. . . DRY CELLS
. . . . **NICKEL ZINC BATTERIES**
ELECTROCHEMICAL CELLS
. ELECTRIC BATTERIES
. . PRIMARY BATTERIES
. . . DRY CELLS
. . . . **NICKEL ZINC BATTERIES**
. . STORAGE BATTERIES
. . . **NICKEL ZINC BATTERIES**
RT   NICKEL IRON BATTERIES

NICOTINAMIDE
GS   HETEROCYCLIC COMPOUNDS
. ALKALOIDS
. . **NICOTINAMIDE**
NITROGEN COMPOUNDS
. ALKALOIDS
. . **NICOTINAMIDE**
. AMIDES
. . **NICOTINAMIDE**
VITAMINS
. **NICOTINAMIDE**

NICOTINE
GS   HETEROCYCLIC COMPOUNDS
. ALKALOIDS
. . **NICOTINE**
NITROGEN COMPOUNDS
. ALKALOIDS
. . **NICOTINE**
RT   TOBACCO

NICOTINIC ACID
GS   HETEROCYCLIC COMPOUNDS
. **NICOTINIC ACID**
PYRIDINES
. **NICOTINIC ACID**
VITAMINS
. **NICOTINIC ACID**

NIGELLA
GS   PLANTS (BOTANY)
. **NIGELLA**
RT   BOTANY

NIGER
GS   NATIONS
. **NIGER**
RT   AFRICA

NIGERIA
GS   NATIONS
. **NIGERIA**
RT   AFRICA

NIGHT
RT   DARKENING
DARKNESS
DAYTIME
DIURNAL VARIATIONS
EVENING
SHADOWS
SKY BRIGHTNESS
TWILIGHT GLOW

NIGHT AIRGLOW
USE   NIGHTGLOW

NIGHT E LAYER
USE   E REGION
NIGHT SKY

NIGHT F LAYER
USE   F REGION
NIGHT SKY

NIGHT FLIGHTS (AIRCRAFT)
RT   ∞AIRCRAFT
APPROACH CONTROL
BLIND LANDING
DARKNESS
FLIGHT INSTRUMENTS
INSTRUMENT APPROACH
INSTRUMENT LANDING SYSTEMS
NAP-OF-THE-EARTH NAVIGATION
RADAR
RADIO BEACONS
VISIBILITY

NIGHT SKY
UF   NIGHT E LAYER
NIGHT F LAYER
GS   SKY
. **NIGHT SKY**
RT   AIRGLOW
AURORAS
GEGENSCHEIN
NIGHTGLOW
SKY BRIGHTNESS
TWILIGHT GLOW
ZODIACAL LIGHT

NIGHT VISION
GS   VISION
. **NIGHT VISION**
RT   DARK ADAPTATION
IMAGE INTENSIFIERS
LIGHT ADAPTATION
MICROCHANNELS
NAP-OF-THE-EARTH NAVIGATION

NIGHTGLOW
UF   NIGHT AIRGLOW
GS   ATMOSPHERIC RADIATION
. SKY RADIATION
. . AIRGLOW
. . . **NIGHTGLOW**
ELECTROMAGNETIC RADIATION
. LIGHT (VISIBLE RADIATION)
. . SKY RADIATION
. . . AIRGLOW
. . . . **NIGHTGLOW**
RT   BIOMETEOROLOGY
NIGHT SKY
RADIO AURORAS
SKY BRIGHTNESS

NIGOTRONS
GS   ELECTRON TUBES
. VACUUM TUBES
. . VACUUM TUBE OSCILLATORS
. . . MICROWAVE TUBES
. . . . MAGNETRONS
. . . . . **NIGOTRONS**
MICROWAVE EQUIPMENT
. MICROWAVE TUBES
. . MAGNETRONS
. . . **NIGOTRONS**
OSCILLATORS
. VACUUM TUBE OSCILLATORS
. . MICROWAVE TUBES
. . . MAGNETRONS
. . . . **NIGOTRONS**

NIHON AIRCRAFT
UF   NAMC AIRCRAFT
GS   **NIHON AIRCRAFT**
. YS-11 AIRCRAFT
RT   ∞AIRCRAFT

NIHON YS-11 AIRCRAFT
USE   YS-11 AIRCRAFT

NIKE BOOSTER ROCKET ENGINES
GS   ENGINES
. ROCKET ENGINES
. . BOOSTER ROCKET ENGINES
. . . **NIKE BOOSTER ROCKET ENGINES**
. . SOLID PROPELLANT ROCKET
ENGINES
. . . **NIKE BOOSTER ROCKET ENGINES**
RT   ∞NIKE ROCKETS

NIKE MISSILES
GS   MISSILES
. SURFACE TO AIR MISSILES

NIKE MISSILES-*(CONT.)*
. . NIKE MISSILES
. . . NIKE-AJAX MISSILE
. . . NIKE-HERCULES MISSILE
. . . NIKE-ZEUS MISSILE
RT   ANTIAIRCRAFT MISSILES
ANTIMISSILE MISSILES
∞NIKE ROCKETS
SENTINEL SYSTEM

NIKE PROJECT
GS   PROGRAMS
. PROJECTS
. . **NIKE PROJECT**
RT   ∞NIKE ROCKETS

NIKE ROCKET VEHICLES
GS   ROCKET VEHICLES
. MULTISTAGE ROCKET VEHICLES
. . **NIKE ROCKET VEHICLES**
. . . NIKE-APACHE ROCKET VEHICLE
. . . NIKE-CAJUN ROCKET VEHICLE
. . . NIKE-HYDAC ROCKET VEHICLE
. . . NIKE-IROQUOIS ROCKET VEHICLE
. . . NIKE-JAVELIN ROCKET VEHICLE
. . . NIKE-TOMAHAWK ROCKET VEHICLE
RT   ∞NIKE ROCKETS
∞VEHICLES

∞ NIKE ROCKETS
SN   *(USE OF A MORE SPECIFIC TERM IS
RECOMMENDED--CONSULT THE TERMS
LISTED BELOW)*
RT   NIKE BOOSTER ROCKET ENGINES
NIKE MISSILES
NIKE PROJECT
NIKE ROCKET VEHICLES

NIKE X SYSTEMS
GS   WEAPON SYSTEMS
. MISSILE SYSTEMS
. . **NIKE X SYSTEMS**
RT   ANTIMISSILE MISSILES
MISSILES
SURFACE TO AIR MISSILES
∞SYSTEMS

NIKE-AJAX MISSILE
GS   MISSILES
. ANTIAIRCRAFT MISSILES
. . **NIKE-AJAX MISSILE**
. SURFACE TO AIR MISSILES
. . NIKE MISSILES
. . . **NIKE-AJAX MISSILE**
RT   ARGO ROCKET VEHICLES
EXOS SOUNDING ROCKET
LIQUID PROPELLANT ROCKET ENGINES
SOLID PROPELLANT ROCKET ENGINES
TRAILBLAZER 1 REENTRY VEHICLE

NIKE-APACHE ROCKET VEHICLE
GS   ROCKET VEHICLES
. MULTISTAGE ROCKET VEHICLES
. . NIKE ROCKET VEHICLES
. . . **NIKE-APACHE ROCKET VEHICLE**
RT   SOLID PROPELLANT ROCKET ENGINES

NIKE-ASP ROCKET
USE   ASP ROCKET VEHICLE

NIKE-CAJUN ROCKET VEHICLE
GS   ROCKET VEHICLES
. MULTISTAGE ROCKET VEHICLES
. . NIKE ROCKET VEHICLES
. . . **NIKE-CAJUN ROCKET VEHICLE**
RT   CAJUN ROCKET VEHICLE
SOLID PROPELLANT ROCKET ENGINES

NIKE-HERCULES MISSILE
GS   MISSILES
. ANTIAIRCRAFT MISSILES
. . **NIKE-HERCULES MISSILE**
. SURFACE TO AIR MISSILES
. . NIKE MISSILES
. . . **NIKE-HERCULES MISSILE**
RT   SOLID PROPELLANT ROCKET ENGINES

NIKE-HYDAC ROCKET VEHICLE
GS   ROCKET VEHICLES
. MULTISTAGE ROCKET VEHICLES
. . NIKE ROCKET VEHICLES
. . . **NIKE-HYDAC ROCKET VEHICLE**
RT   ∞VEHICLES

NIKE-IROQUOIS ROCKET VEHICLE
GS ROCKET VEHICLES
  . MULTISTAGE ROCKET VEHICLES
  . . NIKE ROCKET VEHICLES
  . . . **NIKE-IROQUOIS ROCKET VEHICLE**
RT ∞VEHICLES

NIKE-JAVELIN ROCKET VEHICLE
GS ROCKET VEHICLES
  . MULTISTAGE ROCKET VEHICLES
  . . NIKE ROCKET VEHICLES
  . . . **NIKE-JAVELIN ROCKET VEHICLE**
RT SOLID PROPELLANT ROCKET ENGINES
  SOUNDING ROCKETS

NIKE-TOMAHAWK ROCKET VEHICLE
GS ROCKET VEHICLES
  . MULTISTAGE ROCKET VEHICLES
  . . NIKE ROCKET VEHICLES
  . . . **NIKE-TOMAHAWK ROCKET
    VEHICLE**
RT SOLID PROPELLANT ROCKET ENGINES

NIKE-ZEUS MISSILE
UF ZEUS MISSILE
GS MISSILES
  . ANTIMISSILE MISSILES
  . . **NIKE-ZEUS MISSILE**
  . SURFACE TO AIR MISSILES
  . . NIKE MISSILES
  . . . **NIKE-ZEUS MISSILE**
RT SOLID PROPELLANT ROCKET ENGINES
  SPARTAN MISSILE
  SPRINT MISSILE

NIMBOSTRATUS CLOUDS
UF NIMBUS CLOUDS
GS CLOUDS
  . CLOUDS (METEOROLOGY)
  . . CONVECTION CLOUDS
  . . . **NIMBOSTRATUS CLOUDS**
RT CUMULONIMBUS CLOUDS
  PRECIPITATION (METEOROLOGY)
  STRATUS CLOUDS

NIMBUS CLOUDS
USE NIMBOSTRATUS CLOUDS

NIMBUS PROJECT
GS PROGRAMS
  . NASA PROGRAMS
  . . **NIMBUS PROJECT**
  . NASA SPACE PROGRAMS
  . . **NIMBUS PROJECT**
  . PROJECTS
  . . **NIMBUS PROJECT**
RT CLOUD PHOTOGRAPHY
  METEOROLOGICAL SATELLITES
  SATELLITE OBSERVATION

NIMBUS SATELLITES
GS SATELLITES
  . ARTIFICIAL SATELLITES
  . . METEOROLOGICAL SATELLITES
  . . . **NIMBUS SATELLITES**
  . . . . NIMBUS 1 SATELLITE
  . . . . NIMBUS 2 SATELLITE
  . . . . NIMBUS 3 SATELLITE
  . . . . NIMBUS 4 SATELLITE
  . . . . NIMBUS 5 SATELLITE
  . . . . NIMBUS 6 SATELLITE
  . . . . NIMBUS 7 SATELLITE
RT CLOUD PHOTOGRAPHY
  ESSA SATELLITES
  INFRARED PHOTOGRAPHY
  SATELLITE OBSERVATION
  THOR AGENA LAUNCH VEHICLE

NIMBUS 1 SATELLITE
GS SATELLITES
  . ARTIFICIAL SATELLITES
  . . METEOROLOGICAL SATELLITES
  . . . NIMBUS SATELLITES
  . . . . **NIMBUS 1 SATELLITE**
RT CLOUD PHOTOGRAPHY
  THOR AGENA LAUNCH VEHICLE

NIMBUS 2 SATELLITE
GS SATELLITES
  . ARTIFICIAL SATELLITES
  . . METEOROLOGICAL SATELLITES
  . . . NIMBUS SATELLITES
  . . . . **NIMBUS 2 SATELLITE**
RT CLOUD PHOTOGRAPHY

NIMBUS 2 SATELLITE-*(CONT.)*
  THOR AGENA LAUNCH VEHICLE

NIMBUS 3 SATELLITE
GS SATELLITES
  . ARTIFICIAL SATELLITES
  . . METEOROLOGICAL SATELLITES
  . . . NIMBUS SATELLITES
  . . . . **NIMBUS 3 SATELLITE**

NIMBUS 4 SATELLITE
GS SATELLITES
  . ARTIFICIAL SATELLITES
  . . METEOROLOGICAL SATELLITES
  . . . NIMBUS SATELLITES
  . . . . **NIMBUS 4 SATELLITE**

NIMBUS 5 SATELLITE
GS SATELLITES
  . ARTIFICIAL SATELLITES
  . . METEOROLOGICAL SATELLITES
  . . . NIMBUS SATELLITES
  . . . . **NIMBUS 5 SATELLITE**

NIMBUS 6 SATELLITE
GS SATELLITES
  . ARTIFICIAL SATELLITES
  . . METEOROLOGICAL SATELLITES
  . . . NIMBUS SATELLITES
  . . . . **NIMBUS 6 SATELLITE**

NIMBUS 7 SATELLITE
GS SATELLITES
  . ARTIFICIAL SATELLITES
  . . METEOROLOGICAL SATELLITES
  . . . NIMBUS SATELLITES
  . . . . **NIMBUS 7 SATELLITE**

NIMONIC ALLOYS
GS ALLOYS
  . HEAT RESISTANT ALLOYS
  . . **NIMONIC ALLOYS**
RT IRON ALLOYS
  NICKEL ALLOYS

NIMPHE (ENGINE)
USE HYDRAZINE ENGINES

NIMROD ACCELERATOR
GS PARTICLE ACCELERATORS
  . **NIMROD ACCELERATOR**
RT ∞ACCELERATORS

NIOBATES
GS NIOBIUM COMPOUNDS
  . **NIOBATES**
RT EUXENITE
  OXIDES
  ∞OXYGEN COMPOUNDS

NIOBIUM
UF COLUMBIUM
GS CHEMICAL ELEMENTS
  . REFRACTORY METALS
  . . **NIOBIUM**
  . . . NIOBIUM ISOTOPES
  . . . . NIOBIUM 95
  METALS
  . TRANSITION METALS
  . . REFRACTORY METALS
  . . . **NIOBIUM**
  . . . . NIOBIUM ISOTOPES
  . . . . . NIOBIUM 95
  REFRACTORY MATERIALS
  . REFRACTORY METALS
  . . **NIOBIUM**
  . . . NIOBIUM ISOTOPES
  . . . . NIOBIUM 95

NIOBIUM ALLOYS
GS ALLOYS
  . HEAT RESISTANT ALLOYS
  . . REFRACTORY METAL ALLOYS
  . . . **NIOBIUM ALLOYS**
  REFRACTORY MATERIALS
  . REFRACTORY METAL ALLOYS
  . . **NIOBIUM ALLOYS**

NIOBIUM CARBIDES
GS CARBON COMPOUNDS
  . CARBIDES
  . . **NIOBIUM CARBIDES**
  NIOBIUM COMPOUNDS
  . **NIOBIUM CARBIDES**

NIOBIUM COMPOUNDS
GS **NIOBIUM COMPOUNDS**
  . NIOBATES
  . NIOBIUM CARBIDES
  . NIOBIUM IODIDES
  . NIOBIUM OXIDES
  . NIOBIUM STANNIDES
RT ∞CHEMICAL COMPOUNDS
  ∞GROUP 5B COMPOUNDS
  ∞METAL COMPOUNDS

NIOBIUM IODIDES
GS HALOGEN COMPOUNDS
  . HALIDES
  . . METAL HALIDES
  . . . **NIOBIUM IODIDES**
  . IODINE COMPOUNDS
  . . IODIDES
  . . . **NIOBIUM IODIDES**
  NIOBIUM COMPOUNDS
  . **NIOBIUM IODIDES**

NIOBIUM ISOTOPES
GS CHEMICAL ELEMENTS
  . NUCLIDES
  . . ISOTOPES
  . . . **NIOBIUM ISOTOPES**
  . . . . NIOBIUM 95
  . REFRACTORY METALS
  . . NIOBIUM
  . . . **NIOBIUM ISOTOPES**
  . . . . NIOBIUM 95
  METALS
  . TRANSITION METALS
  . . REFRACTORY METALS
  . . . NIOBIUM
  . . . . **NIOBIUM ISOTOPES**
  . . . . . NIOBIUM 95
  REFRACTORY MATERIALS
  . REFRACTORY METALS
  . . NIOBIUM
  . . . **NIOBIUM ISOTOPES**
  . . . . NIOBIUM 95

NIOBIUM OXIDES
GS CHALCOGENIDES
  . OXIDES
  . . METAL OXIDES
  . . . **NIOBIUM OXIDES**
  NIOBIUM COMPOUNDS
  . **NIOBIUM OXIDES**

NIOBIUM STANNIDES
GS NIOBIUM COMPOUNDS
  . **NIOBIUM STANNIDES**
  TIN COMPOUNDS
  . STANNIDES
  . . **NIOBIUM STANNIDES**

NIOBIUM 95
GS CHEMICAL ELEMENTS
  . NUCLIDES
  . . ISOTOPES
  . . . NIOBIUM ISOTOPES
  . . . . **NIOBIUM 95**
  . . . RADIOACTIVE ISOTOPES
  . . . . **NIOBIUM 95**
  . REFRACTORY METALS
  . . NIOBIUM
  . . . NIOBIUM ISOTOPES
  . . . . **NIOBIUM 95**
  METALS
  . TRANSITION METALS
  . . REFRACTORY METALS
  . . . NIOBIUM
  . . . . NIOBIUM ISOTOPES
  . . . . . **NIOBIUM 95**
  REFRACTORY MATERIALS
  . REFRACTORY METALS
  . . NIOBIUM
  . . . NIOBIUM ISOTOPES
  . . . . **NIOBIUM 95**

NIPS (SYSTEM)
USE NASA INTERACTIVE PLANNING SYSTEM

NITINOL ALLOYS
GS ALLOYS
  . NICKEL ALLOYS
  . . **NITINOL ALLOYS**
  . SHAPE MEMORY ALLOYS
  . . **NITINOL ALLOYS**
  . TITANIUM ALLOYS
  . . **NITINOL ALLOYS**

NITRAMINE PROPELLANTS
GS   PROPELLANTS
    . ROCKET PROPELLANTS
    . . **NITRAMINE PROPELLANTS**
    . SOLID PROPELLANTS
    . . **NITRAMINE PROPELLANTS**
RT   OXIDIZERS

NITRASOL EXPLOSIVES
GS   EXPLOSIVES
    . **NITRASOL EXPLOSIVES**
    PROPELLANTS
    . **NITRASOL EXPLOSIVES**

NITRATE ESTERS
GS   ALIPHATIC COMPOUNDS
    . **NITRATE ESTERS**
    . . ISOPROPYL NITRATE
    . . PROPYL NITRATE
    ESTERS
    . **NITRATE ESTERS**
    . . ISOPROPYL NITRATE
    . . PROPYL NITRATE
    NITROGEN COMPOUNDS
    . **NITRATE ESTERS**
    . . ISOPROPYL NITRATE
    . . PROPYL NITRATE

NITRATES
GS   NITROGEN COMPOUNDS
    . **NITRATES**
    . . DINITRATES
    . . INORGANIC NITRATES
    . . . AMMONIUM NITRATES
    . . . HYDRAZINE NITRATE
    . . . POTASSIUM NITRATES
    . . . SILVER NITRATES
    . . . SODIUM NITRATES
    . . METHYL NITRATE
    . . ORGANIC NITRATES
    . . . CELLULOSE NITRATE
    . . . NITROFORMS
    . . . . HYDRAZINE NITROFORM
    . . . NITROGLYCERIN
    . . . PETN

NITRATION
GS   CHEMICAL REACTIONS
    . **NITRATION**
RT   DENITROGENATION

NITRIC ACID
GS   ACIDS
    . **NITRIC ACID**
    NITROGEN COMPOUNDS
    . **NITRIC ACID**
RT   NITROUS ACID

NITRIC OXIDE
GS   CHALCOGENIDES
    . OXIDES
    . . NITROGEN OXIDES
    . . . **NITRIC OXIDE**
    NITROGEN COMPOUNDS
    . NITROGEN OXIDES
    . . **NITRIC OXIDE**
RT   NITROSYLS

NITRIDES
GS   NITROGEN COMPOUNDS
    . **NITRIDES**
    . . ALUMINUM NITRIDES
    . . BERYLLIUM NITRIDES
    . . BORON NITRIDES
    . . METAL NITRIDES
    . . OXYNITRIDES
    . . SILICON NITRIDES
    . . TANTALUM NITRIDES
    . . TITANIUM NITRIDES
    . . ZIRCONIUM NITRIDES
RT   CERAMIC NUCLEAR FUELS
    MOLTEN SALTS

NITRIDING
GS   CHEMICAL REACTIONS
    . **NITRIDING**
    HARDENING (MATERIALS)
    . **NITRIDING**
    HEAT TREATMENT
    . **NITRIDING**

NITRILES
GS   **NITRILES**
    . ACRYLONITRILES

NITRILES-(CONT.)
    . MALONONITRILE
    . PHOSPHONITRILES
RT   CYANO COMPOUNDS

NITRITES
GS   NITROGEN COMPOUNDS
    . **NITRITES**
RT   ALIPHATIC COMPOUNDS

NITRO COMPOUNDS
GS   NITROGEN COMPOUNDS
    . **NITRO COMPOUNDS**
    . . NITROBENZENES
    . . . TRINITROTOLUENE
    . . NITROGLYCERIN
    . . NITROGUANIDINE
    . . NITROMETHANE
    . . NITROPROPANE
    . . PICRATES
    . . . AMMONIUM PICRATES
    . . POLYBUTADIENE TETRANITRAMINE
    . . TETRYL
    . . TRINITRO COMPOUNDS
RT   ∞CHEMICAL COMPOUNDS

NITROAMINES
GS   ALIPHATIC COMPOUNDS
    . **NITROAMINES**
    AMINES
    . **NITROAMINES**
    NITROGEN COMPOUNDS
    . **NITROAMINES**

NITROBACTER
GS   MICROORGANISMS
    . BACTERIA
    . . **NITROBACTER**

NITROBENZENES
GS   NITROGEN COMPOUNDS
    . NITRO COMPOUNDS
    . . **NITROBENZENES**
    . . . TRINITROTOLUENE

NITROCELLULOSE
USE   CELLULOSE NITRATE

NITROFLUORAMINES
GS   ALIPHATIC COMPOUNDS
    . FLUOROAMINES
    . . **NITROFLUORAMINES**
    AMINES
    . FLUOROAMINES
    . . **NITROFLUORAMINES**
    HALOGEN COMPOUNDS
    . FLUORINE COMPOUNDS
    . . FLUORO COMPOUNDS
    . . . FLUORINE ORGANIC COMPOUNDS
    . . . . FLUOROAMINES
    . . . . . **NITROFLUORAMINES**
    NITROGEN COMPOUNDS
    . **NITROFLUORAMINES**
    ORGANIC COMPOUNDS
    . FLUORINE ORGANIC COMPOUNDS
    . . FLUOROAMINES
    . . . **NITROFLUORAMINES**

NITROFORMATES
GS   FORMATES
    . **NITROFORMATES**
    NITROGEN COMPOUNDS
    . **NITROFORMATES**

NITROFORMS
GS   ESTERS
    . ORGANIC NITRATES
    . . **NITROFORMS**
    . . . HYDRAZINE NITROFORM
    NITROGEN COMPOUNDS
    . NITRATES
    . . ORGANIC NITRATES
    . . . **NITROFORMS**
    . . . . HYDRAZINE NITROFORM

NITROGEN
GS   CHEMICAL ELEMENTS
    . **NITROGEN**
    . . LIQUID NITROGEN
    . . NITROGEN ATOMS
    . . NITROGEN IONS
    . . NITROGEN ISOTOPES
    . . . NITROGEN 15
    . . . NITROGEN 16

NITROGEN-(CONT.)
    . . SOLID NITROGEN
    GASES
    . **NITROGEN**
    . . LIQUID NITROGEN
    . . NITROGEN IONS
RT   KJELDAHL METHOD
    NITROGEN LASERS
    NITROGENATION
    NITROLYSIS
    REACTION BONDING
    SIALON
    VEGARD-KAPLAN BANDS
    WOLF-RAYET STARS

NITROGEN ATOMS
GS   ATOMS
    . **NITROGEN ATOMS**
    CHEMICAL ELEMENTS
    . NITROGEN
    . . **NITROGEN ATOMS**

NITROGEN COMPOUNDS
GS   **NITROGEN COMPOUNDS**
    . ACETANILIDE
    . ALKALOIDS
    . . ATROPINE
    . . BETAINES
    . . COLCHICINE
    . . ERGOTAMINE
    . . HYOSCINE
    . . LYSERGINE
    . . MORPHINE
    . . NICOTINAMIDE
    . . NICOTINE
    . . PILOCARPINE
    . . QUINOLINE
    . . RESERPINE
    . . TROPYL COMPOUNDS
    . AMIDES
    . . CARBAMIDES
    . . CYANAMIDES
    . . FORMHYDROXAMIC ACID
    . . LYSERGAMIDE
    . . NICOTINAMIDE
    . . OXAMIC ACIDS
    . . POLYIMIDES
    . . SUCCINIMIDES
    . . UREAS
    . . . DIFLUOROUREA
    . . . THIOUREAS
    . . . THIURONIUM
    . AMMONIA
    . . LIQUID AMMONIA
    . AZIDES (INORGANIC)
    . . HYDROGEN AZIDES
    . . SODIUM AZIDES
    . AZIDES (ORGANIC)
    . . SODIUM AZIDES
    . . TRIAMINOGUANIDINIUM AZIDE
    . AZO COMPOUNDS
    . RDX
    . CYANO COMPOUNDS
    . . CYANAMIDES
    . . ISOCYANATES
    . . . DIISOCYANATES
    . . FULMINATES
    . FOLIC ACID
    . HYDRAZINIUM COMPOUNDS
    . HYDRAZOIC ACID
    . HYDRAZONES
    . HYDROCYANIC ACID
    . IMIDES
    . . SUCCINIMIDES
    . IMINES
    . NITRATE ESTERS
    . . ISOPROPYL NITRATE
    . . PROPYL NITRATE
    . NITRATES
    . . DINITRATES
    . . INORGANIC NITRATES
    . . . AMMONIUM NITRATES
    . . . HYDRAZINE NITRATE
    . . . POTASSIUM NITRATES
    . . . SILVER NITRATES
    . . . SODIUM NITRATES
    . . METHYL NITRATE
    . . ORGANIC NITRATES
    . . . CELLULOSE NITRATE
    . . . NITROFORMS
    . . . . HYDRAZINE NITROFORM
    . . . NITROGLYCERIN
    . . . PETN
    . NITRIC ACID
    . NITRIDES

NITROGEN COMPOUNDS-(CONT.)
```
. . ALUMINUM NITRIDES
. . BERYLLIUM NITRIDES
. . BORON NITRIDES
. . METAL NITRIDES
. . OXYNITRIDES
. . SILICON NITRIDES
. . TANTALUM NITRIDES
. . TITANIUM NITRIDES
. . ZIRCONIUM NITRIDES
. NITRITES
. NITRO COMPOUNDS
. . NITROBENZENES
. . . TRINITROTOLUENE
. . NITROGLYCERIN
. . NITROGUANIDINE
. . NITROMETHANE
. . NITROPROPANE
. . PICRATES
. . . AMMONIUM PICRATES
. . POLYBUTADIENE TETRANITRAMINE
. . TETRYL
. . TRINITRO COMPOUNDS
. NITROAMINES
. NITROFLUORAMINES
. NITROFORMATES
. NITROGEN FLUORIDES
. NITROGEN HYDRIDES
. NITROGEN OXIDES
. . NITRIC OXIDE
. . NITROGEN DIOXIDE
. . NITROGEN TETROXIDE
. . NITROUS OXIDES
. NITROGEN POLYMERS
. NITROSAMINE
. NITROSO COMPOUNDS
. NITROSYLS
. NITROXYCHLORIDES
. NITRYL CHLORIDES
. NITRYL FLUORIDES
. PHOSPHONITRILES
. THIAZINE (TRADEMARK)
. THYMINE
. TRINITRAMINE
. TRYPTOPHAN
. URACIL
. XANTHINES
. . CAFFEINE
. . GUANINES
. . URIC ACID
```
RT   ∞CHEMICAL COMPOUNDS
      CYANIDES
      ∞GROUP 5A COMPOUNDS
      PHOSPHAZENE

NITROGEN DIOXIDE
GS   CHALCOGENIDES
```
. OXIDES
. . NITROGEN OXIDES
. . . NITROGEN DIOXIDE
NITROGEN COMPOUNDS
. NITROGEN OXIDES
. . NITROGEN DIOXIDE
```

NITROGEN FIXATION
USE   NITROGENATION

NITROGEN FLUORIDES
GS   HALOGEN COMPOUNDS
```
. FLUORINE COMPOUNDS
. . FLUORIDES
. . . NITROGEN FLUORIDES
. HALIDES
. . FLUORIDES
. . . NITROGEN FLUORIDES
NITROGEN COMPOUNDS
. NITROGEN FLUORIDES
```

NITROGEN HYDRIDES
GS   HYDROGEN COMPOUNDS
```
. HYDRIDES
. . NITROGEN HYDRIDES
NITROGEN COMPOUNDS
. NITROGEN HYDRIDES
```
RT   AMMONIA
      HYDRAZOIC ACID

NITROGEN IONS
GS   CHEMICAL ELEMENTS
```
. NITROGEN
. . NITROGEN IONS
GASES
. NITROGEN
. . NITROGEN IONS
IONS
```

NITROGEN IONS-(CONT.)
```
. . NITROGEN IONS
```
RT   NEGATIVE IONS

NITROGEN ISOTOPES
GS   CHEMICAL ELEMENTS
```
. NITROGEN
. . NITROGEN ISOTOPES
. . . NITROGEN 15
. . . NITROGEN 16
. NUCLIDES
. . ISOTOPES
. . . NITROGEN ISOTOPES
. . . . NITROGEN 15
. . . . NITROGEN 16
```

NITROGEN LASERS
GS   STIMULATED EMISSION DEVICES
```
. LASERS
. . GAS LASERS
. . . NITROGEN LASERS
```
RT   LASING
      NITROGEN
      POPULATION INVERSION
      PULSED LASERS
      ULTRAVIOLET LASERS

NITROGEN METABOLISM
GS   METABOLISM
```
. NITROGEN METABOLISM
```
RT   BIOCHEMISTRY
    ∞BIOLOGY
      ENZYMOLOGY
      HYDROGEN METABOLISM
      NUTRITION

NITROGEN OXIDES
GS   CHALCOGENIDES
```
. OXIDES
. . NITROGEN OXIDES
. . . NITRIC OXIDE
. . . NITROGEN DIOXIDE
. . . NITROGEN TETROXIDE
. . . NITROUS OXIDES
NITROGEN COMPOUNDS
. NITROGEN OXIDES
. . NITRIC OXIDE
. . NITROGEN DIOXIDE
. . NITROGEN TETROXIDE
. . NITROUS OXIDES
```
RT   NITROSYLS
      NITROUS ACID
      PHOTOCHEMICAL OXIDANTS

NITROGEN PLASMA
GS   GASES
```
. IONIZED GASES
. . CHARGED PARTICLES
. . . NITROGEN PLASMA
PARTICLES
. CHARGED PARTICLES
. . ENERGETIC PARTICLES
. . . PLASMAS (PHYSICS)
. . . . NITROGEN PLASMA
```

NITROGEN POLYMERS
GS   NITROGEN COMPOUNDS
```
. NITROGEN POLYMERS
```
RT   ∞POLYMERS

NITROGEN TETROXIDE
GS   CHALCOGENIDES
```
. OXIDES
. . NITROGEN OXIDES
. . . NITROGEN TETROXIDE
NITROGEN COMPOUNDS
. NITROGEN OXIDES
. . NITROGEN TETROXIDE
```
RT   LIQUID ROCKET PROPELLANTS
      ROCKET OXIDIZERS

NITROGEN 15
GS   CHEMICAL ELEMENTS
```
. NITROGEN
. . NITROGEN ISOTOPES
. . . NITROGEN 15
. NUCLIDES
. . ISOTOPES
. . . NITROGEN ISOTOPES
. . . . NITROGEN 15
```

NITROGEN 16
GS   CHEMICAL ELEMENTS
```
. NITROGEN
```

NITROGEN 16-(CONT.)
```
. . NITROGEN ISOTOPES
. . . NITROGEN 16
. NUCLIDES
. . ISOTOPES
. . . NITROGEN ISOTOPES
. . . . NITROGEN 16
. . . . RADIOACTIVE ISOTOPES
. . . . . NITROGEN 16
```

NITROGENATION
UF   NITROGEN FIXATION
GS   CHEMICAL REACTIONS
```
. NITROGENATION
```
RT   LEGUMINOUS PLANTS
      LIGHTNING
      NITROGEN

NITROGLYCERIN
GS   ESTERS
```
. ORGANIC NITRATES
. . NITROGLYCERIN
NITROGEN COMPOUNDS
. NITRATES
. . ORGANIC NITRATES
. . . NITROGLYCERIN
. NITRO COMPOUNDS
. . NITROGLYCERIN
```
RT   DOUBLE BASE PROPELLANTS
      DOUBLE BASE ROCKET PROPELLANTS
      DYNAMITE
      EXPLOSIVES
      GLYCERIDES
      GLYCEROLS

NITROGUANIDINE
UF   HBNQ
GS   NITROGEN COMPOUNDS
```
. NITRO COMPOUNDS
. . NITROGUANIDINE
```
RT   EXPLOSIVES
      SOLID PROPELLANTS

NITROLYSIS
GS   CHEMICAL REACTIONS
```
. NITROLYSIS
DECOMPOSITION
. NITROLYSIS
```
RT   CRACKING (CHEMICAL ENGINEERING)
      NITROGEN

NITROMETHANE
GS   NITROGEN COMPOUNDS
```
. NITRO COMPOUNDS
. . NITROMETHANE
```
RT   BSX
      EXPLOSIVES

NITRONIUM COMPOUNDS
GS   NITRONIUM COMPOUNDS
```
. NITRONIUM PERCHLORATE
```
RT   ∞CHEMICAL COMPOUNDS

NITRONIUM PERCHLORATE
GS   HALOGEN COMPOUNDS
```
. CHLORINE COMPOUNDS
. PERCHLORATES
. . . NITRONIUM PERCHLORATE
NITRONIUM COMPOUNDS
. NITRONIUM PERCHLORATE
```
RT   ROCKET OXIDIZERS

NITROPROPANE
GS   ALIPHATIC COMPOUNDS
```
. ALKANES
. . NITROPROPANE
HYDROCARBONS
. ALKANES
. . NITROPROPANE
NITROGEN COMPOUNDS
. NITRO COMPOUNDS
. . NITROPROPANE
```
RT   PROPANE

NITROSAMINE
GS   ALIPHATIC COMPOUNDS
```
. NITROSAMINE
AMINES
. NITROSAMINE
NITROGEN COMPOUNDS
. NITROSAMINE
```

NITROSO COMPOUNDS
GS   NITROGEN COMPOUNDS

NITROSO COMPOUNDS-*(CONT.)*
. **NITROSO COMPOUNDS**
. . NITROSYLS
RT   ∞CHEMICAL COMPOUNDS
ORGANIC COMPOUNDS

NITROSYL CHLORIDES
GS   HALOGEN COMPOUNDS
. CHLORINE COMPOUNDS
. . CHLORIDES
. . . **NITROSYL CHLORIDES**
. HALIDES
. . CHLORIDES
. . . **NITROSYL CHLORIDES**
. NITROSYLS
. . **NITROSYL CHLORIDES**

NITROSYLS
GS   HALOGEN COMPOUNDS
. **NITROSYLS**
. . NITROSYL CHLORIDES
NITROGEN COMPOUNDS
. NITROSO COMPOUNDS
. . **NITROSYLS**
RT   ALIPHATIC COMPOUNDS
AMINES
ESTERS
HALIDES
NITRIC OXIDE
NITROGEN OXIDES

NITROUS ACID
RT   AIR POLLUTION
ATMOSPHERIC CHEMISTRY
NITRIC ACID
NITROGEN OXIDES
REACTION KINETICS

NITROUS OXIDES
GS   CHALCOGENIDES
. OXIDES
. . NITROGEN OXIDES
. . . **NITROUS OXIDES**
NITROGEN COMPOUNDS
. NITROGEN OXIDES
. . **NITROUS OXIDES**

NITROXYCHLORIDES
GS   HALOGEN COMPOUNDS
. CHLORINE COMPOUNDS
. . CHLORIDES
. . . **NITROXYCHLORIDES**
. HALIDES
. . CHLORIDES
. . . **NITROXYCHLORIDES**
NITROGEN COMPOUNDS
. **NITROXYCHLORIDES**

NITRYL CHLORIDES
GS   HALOGEN COMPOUNDS
. CHLORINE COMPOUNDS
. . CHLORIDES
. . . **NITRYL CHLORIDES**
. HALIDES
. . CHLORIDES
. . . **NITRYL CHLORIDES**
NITROGEN COMPOUNDS
. **NITRYL CHLORIDES**

NITRYL FLUORIDES
GS   NITROGEN COMPOUNDS
. **NITRYL FLUORIDES**

NMR
USE   NUCLEAR MAGNETIC RESONANCE

NOAA E
USE   NOAA 8 SATELLITE

NOAA SATELLITES
GS   SATELLITES
. ARTIFICIAL SATELLITES
. . METEOROLOGICAL SATELLITES
. . . **NOAA SATELLITES**
. . . . NOAA 2 SATELLITE
. . . . NOAA 3 SATELLITE
. . . . NOAA 4 SATELLITE
. . . . NOAA 5 SATELLITE
. . . . NOAA 6 SATELLITE
. . . . NOAA 7 SATELLITE
. . . . NOAA 8 SATELLITE
RT   SMS 1
SMS 2

NOAA 2 SATELLITE
GS   SATELLITES
. ARTIFICIAL SATELLITES
. . METEOROLOGICAL SATELLITES
. . . NOAA SATELLITES
. . . . **NOAA 2 SATELLITE**

NOAA 3 SATELLITE
GS   SATELLITES
. ARTIFICIAL SATELLITES
. . METEOROLOGICAL SATELLITES
. . . NOAA SATELLITES
. . . . **NOAA 3 SATELLITE**

NOAA 4 SATELLITE
GS   SATELLITES
. ARTIFICIAL SATELLITES
. . METEOROLOGICAL SATELLITES
. . . NOAA SATELLITES
. . . . **NOAA 4 SATELLITE**

NOAA 5 SATELLITE
GS   SATELLITES
. ARTIFICIAL SATELLITES
. . METEOROLOGICAL SATELLITES
. . . NOAA SATELLITES
. . . . **NOAA 5 SATELLITE**

NOAA 6 SATELLITE
GS   SATELLITES
. ARTIFICIAL SATELLITES
. . METEOROLOGICAL SATELLITES
. . . NOAA SATELLITES
. . . . **NOAA 6 SATELLITE**
RT   TIROS N SERIES SATELLITES

NOAA 7 SATELLITE
GS   SATELLITES
. ARTIFICIAL SATELLITES
. . METEOROLOGICAL SATELLITES
. . . NOAA SATELLITES
. . . . **NOAA 7 SATELLITE**
RT   TIROS N SERIES SATELLITES

NOAA 8 SATELLITE
UF   NOAA E
GS   SATELLITES
. ARTIFICIAL SATELLITES
. . METEOROLOGICAL SATELLITES
. . . NOAA SATELLITES
. . . . **NOAA 8 SATELLITE**
RT   SARSAT

NOBELIUM
GS   CHEMICAL ELEMENTS
. ACTINIDE SERIES
. . TRANSURANIUM ELEMENTS
. . . **NOBELIUM**
. NUCLIDES
. . ISOTOPES
. . . RADIOACTIVE ISOTOPES
. . . . TRANSURANIUM ELEMENTS
. . . . . **NOBELIUM**
HEAVY ELEMENTS
. TRANSURANIUM ELEMENTS
. . **NOBELIUM**
METALS
. ACTINIDE SERIES
. . TRANSURANIUM ELEMENTS
. . . **NOBELIUM**

NOBLE GASES
USE   RARE GASES

NOBLE METALS
UF   PRECIOUS METALS
GS   METALS
. **NOBLE METALS**
. . GOLD
. . . GOLD ISOTOPES
. . . . GOLD 198
. . RUTHENIUM
. . . RUTHENIUM ISOTOPES
. . SILVER
. . . SILVER ISOTOPES
RT   ∞GROUP 1B COMPOUNDS

NOCTILUCENCE
USE   LUMINESCENCE

NOCTILUCENT CLOUDS
GS   CLOUDS
. CLOUDS (METEOROLOGY)
. . **NOCTILUCENT CLOUDS**

NOCTILUCENT CLOUDS-*(CONT.)*
RT   LUMINESCENCE

NOCTURNAL VARIATIONS
GS   VARIATIONS
. MAGNETIC VARIATIONS
. . **NOCTURNAL VARIATIONS**
. PERIODIC VARIATIONS
. . **NOCTURNAL VARIATIONS**
RT   DIURNAL VARIATIONS
GEOMAGNETIC MICROPULSATIONS
GEOMAGNETIC PULSATIONS

NODES (STANDING WAVES)
RT   ANTINODES
HARMONICS
RESONANT FREQUENCIES
STANDING WAVES
VIBRATION
WAVELENGTHS
∞WAVES

NODULES
RT   LEGUMINOUS PLANTS
PARTICLES
SPHERES
SPHERULITES

NOE NAVIGATION
USE   NAP-OF-THE-EARTH NAVIGATION

NOESS
SN   (NATIONAL OPERATIONAL
ENVIRONMENTAL SATELLITE SYSTEM)
UF   NATIONAL OPERATIONAL
ENVIRONMENTAL SAT SYS
RT   METEOROLOGICAL SATELLITES
METEOROLOGY
NASA PROGRAMS
OBSERVATION
SATELLITES
∞SYSTEMS

∞ **NOISE**
SN   *(USE OF A MORE SPECIFIC TERM IS
RECOMMENDED--CONSULT THE TERMS
LISTED BELOW)*
RT   BACKGROUND NOISE
CONTINUOUS NOISE
EFFECTIVE PERCEIVED NOISE LEVELS
ELECTROMAGNETIC NOISE
HUM
INFORMATION THEORY
NOISE (SOUND)
NOISE PROPAGATION
NOISE SPECTRA
RANDOM NOISE
SIGNAL TO NOISE RATIOS
SPATIAL FILTERING
WHITE NOISE

NOISE (SOUND)
UF   NOISE HAZARDS
GS   ELASTIC WAVES
. SOUND WAVES
. . **NOISE (SOUND)**
. . . AERODYNAMIC NOISE
. . . AIRCRAFT NOISE
. . . . JET AIRCRAFT NOISE
. . . . SONIC BOOMS
. . . ENGINE NOISE
. . . . ROCKET ENGINE NOISE
. . . THERMAL NOISE
RT   ACOUSTICS
AEOLIAN TONES
AIRCRAFT HAZARDS
AUDITORY STIMULI
AUDITORY TASKS
BACKGROUND NOISE
ECHOES
EFFECTIVE PERCEIVED NOISE LEVELS
FLIGHT HAZARDS
HUMAN FACTORS ENGINEERING
HYPERSONIC SHOCK
JET BLAST EFFECTS
LOUDNESS
MUFFLERS
NOISE INJURIES
OPERATIONAL HAZARDS
RANDOM NOISE
RANDOM VIBRATION
REVERBERATION
SHOCK WAVES
SOUND PRESSURE
UNDERWATER ACOUSTICS

NOISE (SOUND)-*(CONT.)*
    WHITE NOISE

NOISE ATTENUATION
  USE   NOISE REDUCTION

NOISE ELIMINATION
  USE   NOISE REDUCTION

NOISE GENERATORS
  RT   ELECTROMAGNETIC NOISE
       ∞GENERATORS
       RADIO FREQUENCY INTERFERENCE
       RANDOM NOISE
       SOUND GENERATORS
       SOUND PROPAGATION

NOISE HAZARDS
  USE   HAZARDS
       NOISE (SOUND)

NOISE INJURIES
  GS   INJURIES
      . **NOISE INJURIES**
  RT   EAR PROTECTORS

NOISE INTENSITY
  RT   AIRCRAFT NOISE
       AUDITORY STIMULI
       EFFECTIVE PERCEIVED NOISE LEVELS
       ELECTROMAGNETIC NOISE
       ∞INTENSITY
       PSYCHOACOUSTICS
       SIRENS
       SOUND INTENSITY

NOISE MEASUREMENT
  GS   ACOUSTIC MEASUREMENT
      . **NOISE MEASUREMENT**
  RT   AERODYNAMIC NOISE
       AIRCRAFT NOISE
       BACKGROUND NOISE
       JET AIRCRAFT NOISE
       LOUDNESS
       ∞MEASUREMENT
       NOISE (SOUND)
       SOUND INTENSITY

NOISE METERS
  SN   (LIMITED TO ACOUSTIC NOISE)
  GS   MEASURING INSTRUMENTS
      . **NOISE METERS**
  RT   ACOUSTIC MEASUREMENT
       FIELD INTENSITY METERS
       PRESSURE MEASUREMENT

NOISE POLLUTION
  GS   POLLUTION
      . **NOISE POLLUTION**
  RT   ACOUSTICS
       AUDIO FREQUENCIES
       ENVIRONMENT EFFECTS
       ENVIRONMENT POLLUTION
       ENVIRONMENTAL QUALITY
       HUMAN REACTIONS
       HUMAN TOLERANCES
       PHYSIOLOGICAL EFFECTS
       PHYSIOLOGICAL FACTORS
       SOUND WAVES

NOISE PREDICTION
  GS   PREDICTIONS
      . **NOISE PREDICTION**

NOISE PREDICTION (AIRCRAFT)
  UF   AIRCRAFT NOISE PREDICTION
  GS   PREDICTIONS
      . **NOISE PREDICTION (AIRCRAFT)**
  RT   AEROACOUSTICS
       ∞AIRCRAFT
       AIRCRAFT NOISE
       ESTIMATES
       FORECASTING
       SOUND WAVES

NOISE PROPAGATION
  RT   ACOUSTICS
       COHERENCE COEFFICIENT
       CONTINUOUS NOISE
       FAR FIELDS
       NOISE SPECTRA
       SIGNAL TO NOISE RATIOS
       SOUND PROPAGATION

NOISE REDUCTION
  UF   NOISE ATTENUATION
       NOISE ELIMINATION
       NOISE SUPPRESSORS
  RT   ACOUSTIC ATTENUATION
       ACOUSTIC DUCTS
       ACOUSTIC RETROFITTING
       ACOUSTICS
       AERODYNAMIC NOISE
       AIRCRAFT NOISE
       COAXIAL NOZZLES
       DELAYED FLAP APPROACH
       EAR PROTECTORS
       ECHO SUPPRESSORS
       EFFECTIVE PERCEIVED NOISE LEVELS
       ELECTRICAL GROUNDING
       ELECTROMAGNETIC INTERFERENCE
       ELECTROMAGNETIC NOISE
       FLIGHT RULES
       GRAZING FLOW
       HELMHOLTZ RESONATORS
       INTERFERENCE IMMUNITY
       ISOLATORS
       JET AIRCRAFT NOISE
       LOUDNESS
       MUFFLERS
       QUIET ENGINE PROGRAM
      ∞REDUCTION
       SHOCK WAVE ATTENUATION
       SILENCE
       SQUELCH CIRCUITS
       SUPPRESSORS
       SYNCHROPHASING
       VIBRATION ISOLATORS

NOISE SPECTRA
  GS   SPECTRA
      . **NOISE SPECTRA**
  RT   BACKGROUND NOISE
       CHANNEL NOISE
       ELECTROMAGNETIC COMPATIBILITY
       ELECTROMAGNETIC NOISE
       ELECTROMAGNETIC SPECTRA
       NOISE (SOUND)
       RADIATION SPECTRA
       RANDOM NOISE
       RANDOM SIGNALS
       SHOCK SPECTRA
       SIGNAL TO NOISE RATIOS
       WHITE NOISE

NOISE STORMS
  GS   STORMS
      . **NOISE STORMS**
  RT   COSMIC NOISE
       ELECTROMAGNETIC NOISE
       IONOSPHERIC STORMS
       MAGNETIC STORMS
       RADIO FREQUENCY INTERFERENCE
       SOLAR STORMS

NOISE SUPPRESSORS
  USE   NOISE REDUCTION

NOISE TEMPERATURE
  GS   TEMPERATURE
      . **NOISE TEMPERATURE**
  RT   ELECTROMAGNETIC NOISE
       ELECTRON ENERGY
       ELECTRON STATES
       TEMPERATURE MEASUREMENT
       THERMAL NOISE

NOISE THRESHOLD
  RT   AUDITORY FATIGUE
       AUDITORY PERCEPTION
       BACKGROUND NOISE
       SIGNAL TO NOISE RATIOS
      ∞THRESHOLDS

NOISE TOLERANCE
  RT   HAZARDS
       HUMAN TOLERANCES
       TOLERANCES (PHYSIOLOGY)

NOMAD LAUNCH VEHICLE
  GS   LAUNCH VEHICLES
      . **NOMAD LAUNCH VEHICLE**
      ROCKET VEHICLES
      . SINGLE STAGE ROCKET VEHICLES
      . . **NOMAD LAUNCH VEHICLE**
  RT   ATLAS LAUNCH VEHICLES
       LIQUID PROPELLANT ROCKET ENGINES

NOMENCLATURES
  RT   ∞DEFINITION
       DESCRIPTIONS
       DICTIONARIES
       MNEMONICS
       SEMANTICS
       SYMBOLS
       TERMINOLOGY
       THESAURI

NOMINAL VALUES
  USE   APPROXIMATION

NOMOGRAMS
  USE   NOMOGRAPHS

NOMOGRAPHS
  UF   ISOPLETHS
       NOMOGRAMS
  GS   ANALYSIS (MATHEMATICS)
      . NUMERICAL ANALYSIS
      . . **NOMOGRAPHS**
  RT   CHARTS
       GRAPHS (CHARTS)

NONADIABATIC CONDITIONS
  GS   CONDITIONS
      . **NONADIABATIC CONDITIONS**
  RT   ENERGY TRANSFER
       HEAT TRANSFER
       NONISOTHERMAL PROCESSES
       THERMODYNAMICS

NONADIABATIC PROCESSES
  USE   HEAT TRANSFER

NONADIABATIC THEORY
  RT   ADIABATIC EQUATIONS
       CHARGED PARTICLES
       ENERGY DISSIPATION
       IONIZATION CROSS SECTIONS
       MAGNETIC DISTURBANCES
      ∞THEORIES
       WAVE PROPAGATION

NONANES
  GS   ALIPHATIC COMPOUNDS
      . ALKANES
      . . **NONANES**
      HYDROCARBONS
      . ALKANES
      . . **NONANES**

NONAQUEOUS ELECTROLYTES
  GS   CONDUCTORS
      . ELECTROLYTES
      . . **NONAQUEOUS ELECTROLYTES**
  RT   ELECTRIC BATTERIES
      ∞ELECTRIC CELLS
       ELECTROCHEMISTRY
       ELECTROLYTIC CELLS
       PRIMARY BATTERIES
       STORAGE BATTERIES
       WET CELLS

NONCONDENSABLE GASES
  GS   GASES
      . **NONCONDENSABLE GASES**
  RT   CRITICAL TEMPERATURE
       GAS-LIQUID INTERACTIONS
       LIQUEFACTION

NONCONDUCTORS
  USE   ELECTRICAL INSULATION

NONCONSERVATIVE FORCES
  RT   CONSERVATION
       CONSERVATION EQUATIONS
       CONSERVATION LAWS
       CONTINUITY EQUATION
      ∞FORCE

NONDESTRUCTIVE TESTS
  UF   FLAW DETECTION
  GS   **NONDESTRUCTIVE TESTS**
      . NEUTRON RADIOGRAPHY
      . PRELAUNCH TESTS
      . . STATIC FIRING
  RT   ADHESION TESTS
       CHEMICAL TESTS
       DESTRUCTIVE TESTS
       ELECTRONIC EQUIPMENT TESTS
       ENGINE TESTS
       HARDNESS TESTS

NONDESTRUCTIVE TESTS-*(CONT.)*
    HIGH TEMPERATURE TESTS
    INSPECTION
    LOAD TESTS
    LOW TEMPERATURE TESTS
    ∞ MATERIALS TESTS
    QUALITY CONTROL
    RADIOGRAPHY
    RELIABILITY
    STATIC TESTS
    ∞ TESTS
    THERMOGRAPHY
    TOLERANCES (MECHANICS)
    ULTRASONIC FLAW DETECTION
    ULTRASONIC SPECTROSCOPY
    ULTRASONIC TESTS
    X RAY INSPECTION

NONELECTROLYTES
RT   ELECTROLYTES

NONEQUILIBRIUM CONDITIONS
GS   CONDITIONS
    . **NONEQUILIBRIUM CONDITIONS**
RT   ∞ EQUILIBRIUM
    UNSTEADY STATE

NONEQUILIBRIUM DRAG
USE   FRICTION DRAG

NONEQUILIBRIUM FLOW
GS   FLUID FLOW
    . GAS FLOW
    . . **NONEQUILIBRIUM FLOW**
RT   EQUILIBRIUM FLOW
    ∞ FLUIDS
    HEAT TRANSMISSION
    OSCILLATING FLOW
    QUASI-STEADY STATES
    UNSTEADY FLOW

NONEQUILIBRIUM IONIZATION
GS   IONIZATION
    . **NONEQUILIBRIUM IONIZATION**

NONEQUILIBRIUM PLASMAS
GS   PARTICLES
    . CHARGED PARTICLES
    . . ENERGETIC PARTICLES
    . . . PLASMAS (PHYSICS)
    . . . . **NONEQUILIBRIUM PLASMAS**
RT   MAGNETOHYDRODYNAMIC STABILITY
    NONUNIFORM PLASMAS
    PLASMA COMPOSITION
    PLASMA POTENTIALS
    PLASMA RADIATION
    PLASMA SHEATHS
    ROTATING PLASMAS

NONEQUILIBRIUM RADIATION
GS   ELECTROMAGNETIC RADIATION
    . **NONEQUILIBRIUM RADIATION**
RT   SHOCK WAVE PROPAGATION

NONEQUILIBRIUM THERMODYNAMICS
GS   THERMODYNAMICS
    . **NONEQUILIBRIUM THERMODYNAMICS**
RT   IRREVERSIBLE PROCESSES

NONEUCLIDIAN GEOMETRY
USE   DIFFERENTIAL GEOMETRY

NONFERROUS METALS
GS   METALS
    . **NONFERROUS METALS**
RT   CHEMICAL ELEMENTS
    CONDUCTORS
    ∞ METALLURGY

NONFLAMMABLE MATERIALS
RT   ASBESTOS
    FIREPROOFING
    ∞ INORGANIC MATERIALS
    KEVLAR (TRADEMARK)
    ∞ MATERIALS
    OXIDES
    REFRACTORY MATERIALS

NONGRAY ATMOSPHERES
RT   ∞ ATMOSPHERES
    BLACK BODY RADIATION
    EMISSIVITY
    GRAY GAS
    PLANETARY ATMOSPHERES

NONGRAY GAS
GS   GASES
    . **NONGRAY GAS**
RT   ∞ ATMOSPHERES
    BLACK BODY RADIATION
    EMISSIVITY
    HEAT TRANSFER
    SPECTRAL EMISSION
    THERMAL RADIATION
    THERMODYNAMICS

NONHOLONOMIC EQUATIONS
GS   ANALYSIS (MATHEMATICS)
    . COMPLEX VARIABLES
    . . **NONHOLONOMIC EQUATIONS**
RT   ANALYTIC FUNCTIONS
    ∞ EQUATIONS

NONHOMOGENEITY
USE   INHOMOGENEITY

NONISENTROPICITY
GS   ISENTROPIC PROCESSES
    . **NONISENTROPICITY**
RT   ENTROPY
    ∞ PROCESSES

NONISOTHERMAL PROCESSES
RT   ENERGY TRANSFER
    HEAT TRANSFER
    NONADIABATIC CONDITIONS
    PRESSURE EFFECTS
    ∞ PROCESSES
    TEMPERATURE GRADIENTS
    THERMODYNAMICS

NONISOTROPIC PLATES
USE   ANISOTROPIC PLATES

NONISOTROPY
USE   ANISOTROPY

NONLIFTING VEHICLES
USE   BALLISTIC VEHICLES

NONLINEAR EQUATIONS
GS   ALGEBRA
    . **NONLINEAR EQUATIONS**
    . . CUBIC EQUATIONS
    . . DUFFING DIFFERENTIAL EQUATION
    . . MONGE-AMPERE EQUATION
    . . NONLINEAR EVOLUTION EQUATIONS
    . . QUADRATIC EQUATIONS
    . . QUARTIC EQUATIONS
    ANALYSIS (MATHEMATICS)
    . REAL VARIABLES
    . . **NONLINEAR EQUATIONS**
    . . . CUBIC EQUATIONS
    . . . DUFFING DIFFERENTIAL EQUATION
    . . . MONGE-AMPERE EQUATION
    . . . NONLINEAR EVOLUTION
        EQUATIONS
    . . . QUADRATIC EQUATIONS
    . . . QUARTIC EQUATIONS
RT   BORN-INFELD THEORY
    DIFFERENTIAL EQUATIONS
    ∞ EQUATIONS
    FIELD THEORY (ALGEBRA)
    INTEGRAL EQUATIONS
    POLYNOMIALS
    ROOTS OF EQUATIONS

NONLINEAR EVOLUTION EQUATIONS
GS   ALGEBRA
    . NONLINEAR EQUATIONS
    . . **NONLINEAR EVOLUTION EQUATIONS**
    ANALYSIS (MATHEMATICS)
    . REAL VARIABLES
    . . NONLINEAR EQUATIONS
    . . . **NONLINEAR EVOLUTION
        EQUATIONS**
RT   DIFFERENCE EQUATIONS
    ∞ EQUATIONS

NONLINEAR FEEDBACK
GS   FEEDBACK
    . **NONLINEAR FEEDBACK**
RT   FEEDBACK AMPLIFIERS
    FEEDBACK CONTROL
    NEGATIVE FEEDBACK
    POSITIVE FEEDBACK
    SENSORY FEEDBACK
    TRANSFER FUNCTIONS

NONLINEAR FILTERS
RT   ELECTRIC FILTERS
    ELECTROMAGNETIC WAVE FILTERS
    ∞ FILTERS
    LINEAR FILTERS

NONLINEAR OPTICS
RT   BIREFRINGENCE
    ELECTRO-OPTICAL EFFECT
    ELECTROMAGNETIC RADIATION
    GEOMETRICAL OPTICS
    GRADIENT INDEX OPTICS
    OPTICAL BISTABILITY
    ∞ OPTICS
    RAMAN SPECTRA
    SAGNAC EFFECT

NONLINEAR PROGRAMMING
GS   OPERATIONS RESEARCH
    . MATHEMATICAL PROGRAMMING
    . . **NONLINEAR PROGRAMMING**
    RESEARCH
    . MATHEMATICAL PROGRAMMING
    . . **NONLINEAR PROGRAMMING**
RT   ∞ APPLICATIONS OF MATHEMATICS
    CONSTRAINTS
    FORMALISM
    LINEAR PROGRAMMING
    OPERATIONS RESEARCH
    ∞ PROGRAMMING

NONLINEAR SYSTEMS
SN   (DYNAMIC SYSTEMS HAVING
    NONLINEAR RESPONSES)
RT   CHAOS
    CONTROL EQUIPMENT
    DISTRIBUTED PARAMETER SYSTEMS
    DYNAMIC PROGRAMMING
    DYNAMICAL SYSTEMS
    LINEAR SYSTEMS
    STRANGE ATTRACTORS
    ∞ SYSTEMS
    TRACKING PROBLEM

NONLINEARITY
UF   QUASILINEARITY
RT   DIFFERENTIAL EQUATIONS
    FUNCTIONS (MATHEMATICS)
    LINEARITY
    MAGNETIC AMPLIFIERS
    VARIABILITY
    VOLTERRA EQUATIONS

NONNEWTONIAN FLOW
GS   FLUID FLOW
    . **NONNEWTONIAN FLOW**
RT   LIQUID FLOW
    STEADY FLOW
    THIXOTROPY
    UNSTEADY FLOW
    VISCOELASTICITY
    VISCOPLASTICITY

NONNEWTONIAN FLUIDS
RT   COLLOIDS
    ∞ FLUIDS
    GELATINS
    GELS
    NEWTON
    NEWTON THEORY
    NEWTONIAN FLUIDS
    RHEOLOGY
    VISCOELASTICITY
    VISCOPLASTICITY
    VISCOUS FLUIDS

NONOHMIC EFFECT
RT   BARRIER LAYERS
    CONTACT RESISTANCE
    ∞ EFFECTS
    SPACE CHARGE

NONOSCILLATORY ACTION
GS   OSCILLATIONS
    . **NONOSCILLATORY ACTION**
RT   OSCILLATION DAMPERS
    OSCILLATORS
    VIBRATION DAMPING

NONPARAMETRIC STATISTICS
GS   STATISTICAL ANALYSIS
    . **NONPARAMETRIC STATISTICS**
RT   ∞ STATISTICS

NONPOINT SOURCES
RT CONTAMINANTS
  CONTAMINATION
  DIFFUSION
  ENVIRONMENTS
  GASES
  LIQUIDS
  PARTICLES
 ∞POINTS
  POISONS
  POLLUTION
  PUBLIC HEALTH
  RADIOACTIVE WASTES
 ∞SOURCES
  TOXICOLOGY
  WASTES

NONPOLAR GASES
GS GASES
 . **NONPOLAR GASES**
RT MOLECULAR GASES
  RARE GASES

NONREFLECTION
USE ENERGY ABSORPTION

NONRELATIVISTIC ELECTRONS
USE ELECTRONS

NONRELATIVISTIC MECHANICS
RT NEWTON THEORY
  RELATIVITY

NONRESONANCE
RT MICROWAVE RESONANCE
  TRANSMISSION LINES
  TRAVELING WAVES

NONRIGIDITY
USE FLEXIBILITY

NONSTABILIZED OSCILLATION
GS OSCILLATIONS
 . **NONSTABILIZED OSCILLATION**
RT OSCILLATION DAMPERS
  OSCILLATORS
  PILOT INDUCED OSCILLATION
  STABLE OSCILLATIONS
  VIBRATION DAMPING

NONSYNCHRONIZATION
RT DEVIATION
  INCOHERENCE
 ∞INTERFERENCE
  NONUNIFORMITY

NONUNIFORM FLOW
GS FLUID FLOW
 . **NONUNIFORM FLOW**
RT FLOW CHARACTERISTICS
  GAS FLOW
  TURBULENT FLOW
  UNIFORM FLOW
  UNSTEADY FLOW

NONUNIFORM MAGNETIC FIELDS
GS MAGNETIC FIELDS
 . **NONUNIFORM MAGNETIC FIELDS**
RT LINES OF FORCE
  MAGNETIC ANOMALIES
  MAGNETIC MIRRORS

NONUNIFORM PLASMAS
GS PARTICLES
 . CHARGED PARTICLES
 . . ENERGETIC PARTICLES
 . . . PLASMAS (PHYSICS)
 . . . . **NONUNIFORM PLASMAS**
RT KELVIN-HELMHOLTZ INSTABILITY
  LOW DENSITY RESEARCH
  MAGNETOHYDRODYNAMIC STABILITY
  NONEQUILIBRIUM PLASMAS
  PLASMA COMPOSITION
  PLASMA OSCILLATIONS
  PLASMA WAVES
  RAREFIED PLASMAS

NONUNIFORMITY
RT FLEXIBILITY
  INHOMOGENEITY
  IRREGULARITIES
  NONSYNCHRONIZATION
  OSCILLATIONS
  TURBULENCE

NONVISCOUS FLOW
USE INVISCID FLOW

NOON
RT DAYTIME
  ZENITH

NORADRENALINE
GS DRUGS
 . STIMULANTS
 . . **NORADRENALINE**
RT NOREPINEPHRINE

NORD AIRCRAFT
GS **NORD AIRCRAFT**
 . C-160 AIRCRAFT
 . MH-262 AIRCRAFT
 . NORD 1500 AIRCRAFT
RT ∞AIRCRAFT

NORD 262 AIRCRAFT
USE MH-262 AIRCRAFT

NORD 1500 AIRCRAFT
UF GRIFFON AIRCRAFT
GS JET AIRCRAFT
 . **NORD 1500 AIRCRAFT**
 MONOPLANES
 . **NORD 1500 AIRCRAFT**
 NORD AIRCRAFT
 . **NORD 1500 AIRCRAFT**
 RESEARCH AIRCRAFT
 . **NORD 1500 AIRCRAFT**
 SUPERSONIC AIRCRAFT
 . **NORD 1500 AIRCRAFT**
RT ∞AIRCRAFT

NOREPINEPHRINE
GS DRUGS
 . EPINEPHRINE
 . . **NOREPINEPHRINE**
 . STIMULANTS
 . . **NOREPINEPHRINE**
RT NORADRENALINE

NORLEUCINE
GS ACIDS
 . AMINO ACIDS
 . . LEUCINE
 . . . **NORLEUCINE**
 ORGANIC COMPOUNDS
 . AMINO ACIDS
 . . LEUCINE
 . . . **NORLEUCINE**

NORMAL DENSITY FUNCTIONS
UF GAUSSIAN DISTRIBUTIONS
 NORMAL DISTRIBUTIONS
GS FUNCTIONS (MATHEMATICS)
 . PROBABILITY DENSITY FUNCTIONS
 . . **NORMAL DENSITY FUNCTIONS**
 STATISTICAL ANALYSIS
 . PROBABILITY DENSITY FUNCTIONS
 . . **NORMAL DENSITY FUNCTIONS**
RT CONTINUITY (MATHEMATICS)
  DISCRETE FUNCTIONS
  HISTOGRAMS

NORMAL DISTRIBUTIONS
USE NORMAL DENSITY FUNCTIONS

NORMAL FORCE DISTRIBUTION
USE FORCE DISTRIBUTION

NORMAL SHOCK WAVES
GS ELASTIC WAVES
 . SHOCK WAVES
 . . **NORMAL SHOCK WAVES**
RT LONGITUDINAL WAVES
  MAGNETOHYDRODYNAMIC WAVES
  OBLIQUE SHOCK WAVES
  PLANE WAVES
  SHOCK LAYERS

NORMALITY
RT ASYMPTOTIC PROPERTIES
  AVERAGE
  MEAN
  MEDIAN (STATISTICS)
  NORMS
  STATISTICAL TESTS

∞ **NORMALIZING**
SN *(USE OF A MORE SPECIFIC TERM IS RECOMMENDED--CONSULT THE TERMS LISTED BELOW)*
RT NORMALIZING (HEAT TREATMENT)
  NORMALIZING (STATISTICS)

NORMALIZING (HEAT TREATMENT)
GS HEAT TREATMENT
 . **NORMALIZING (HEAT TREATMENT)**
RT ANNEALING
  HARDENING (MATERIALS)
  LASER ANNEALING
 ∞NORMALIZING
  TEMPERING

NORMALIZING (STATISTICS)
RT EVALUATION
 ∞NORMALIZING
  QUALITY CONTROL
  RATINGS

NORMS
GS NAMING
 . **NORMS**
RT AVERAGE
  ETHICS
  MEDIAN (STATISTICS)
  NORMALITY
  PSYCHOMETRICS
  VALUE

NORTH AMERICA
GS CONTINENTS
 . **NORTH AMERICA**
RT APPALACHIAN MOUNTAINS (NORTH AMERICA)
  CANADA
  CENTRAL AMERICA
  LOWER CALIFORNIA (MEXICO)
  MEXICO
  TUNDRA
  UNITED STATES
  WILLISTON BASIN (NORTH AMERICA)

NORTH AMERICAN AIRCRAFT
GS **NORTH AMERICAN AIRCRAFT**
 . A-2 AIRCRAFT
 . A-5 AIRCRAFT
 . B-1 AIRCRAFT
 . B-70 AIRCRAFT
 . F-86 AIRCRAFT
 . F-100 AIRCRAFT
 . OV-10 AIRCRAFT
 . P-51 AIRCRAFT
 . T-2 AIRCRAFT
 . T-28 AIRCRAFT
 . T-39 AIRCRAFT
 . X-15 AIRCRAFT
RT ∞AIRCRAFT

NORTH AMERICAN SEARCH AND RANGING RADAR
UF NASARR
GS RADAR
 . SEARCH RADAR
 . . **NORTH AMERICAN SEARCH AND RANGING RADAR**
RT RANGE AND RANGE RATE TRACKING

NORTH ATLANTIC TREATY ORGANIZATION (NATO)
GS ORGANIZATIONS
 . **NORTH ATLANTIC TREATY ORGANIZATION (NATO)**
RT INTERNATIONAL COOPERATION

NORTH CAROLINA
GS NATIONS
 . UNITED STATES
 . . **NORTH CAROLINA**
RT CAPE HATTERAS (NC)
  GREAT SMOKY MOUNTAINS (NC-TN)
  OUTER BANKS (NC)
  SAND HILLS REGION (GA-NC-SC)

NORTH DAKOTA
GS NATIONS
 . UNITED STATES
 . . **NORTH DAKOTA**
RT MISSOURI RIVER (US)
  WILLISTON BASIN (NORTH AMERICA)

NORTH KOREA
- UF   DEMOCRATIC PEOPLES REPUBLIC OF KOREA
- GS   NATIONS
  - . **NORTH KOREA**
- RT   ASIA
  - ∞KOREA
  - SOUTH KOREA

NORTH POLAR SPUR (ASTRONOMY)
- GS   EXTRATERRESTRIAL RADIATION
  - . EXTRATERRESTRIAL RADIO WAVES
  - . . GALACTIC RADIO WAVES
  - . . . **NORTH POLAR SPUR (ASTRONOMY)**
  - . GALACTIC RADIATION
  - . . GALACTIC RADIO WAVES
  - . . . **NORTH POLAR SPUR (ASTRONOMY)**
- RT   NEBULAE
  - SUPERNOVA REMNANTS
  - X RAY SPECTRA

NORTH SEA
- GS   SEAS
  - . **NORTH SEA**
- RT   ENGLISH CHANNEL

NORTH VIETNAM
- USE   VIETNAM

NORTHERN HEMISPHERE
- GS   **NORTHERN HEMISPHERE**
  - . ARCTIC REGIONS
- RT   ∞HEMISPHERES
  - NORTHERN SKY
  - SOUTHERN HEMISPHERE

NORTHERN SKY
- RT   ASTRONOMICAL CATALOGS
  - ASTRONOMICAL COORDINATES
  - ASTRONOMICAL OBSERVATORIES
  - NORTHERN HEMISPHERE
  - SOUTHERN SKY

NORTHROP AIRCRAFT
- GS   **NORTHROP AIRCRAFT**
  - . A-9 AIRCRAFT
  - . F-5 AIRCRAFT
  - . F-18 AIRCRAFT
  - . F-89 AIRCRAFT
  - . T-38 AIRCRAFT
  - . X-21 AIRCRAFT
  - . X-21A AIRCRAFT
- RT   ∞AIRCRAFT

NORTHWEST TERRITORIES
- GS   NATIONS
  - . CANADA
  - . . **NORTHWEST TERRITORIES**

NORTON COUNTY ACHONDRITE
- GS   CELESTIAL BODIES
  - . METEORITES
  - . . STONY METEORITES
  - . . . ACHONDRITES
  - . . . . **NORTON COUNTY ACHONDRITE**

NORWAY
- GS   NATIONS
  - . **NORWAY**
- RT   EUROPE
  - FIORDS
  - SCANDINAVIA

∞ **NOSE**
- SN   *(USE OF A MORE SPECIFIC TERM IS RECOMMENDED--CONSULT THE TERMS LISTED BELOW)*
- RT   NOSE (ANATOMY)
  - NOSES (FOREBODIES)

NOSE (ANATOMY)
- GS   ANATOMY
  - . RESPIRATORY SYSTEM
  - . . **NOSE (ANATOMY)**
  - FACE (ANATOMY)
  - . **NOSE (ANATOMY)**
- RT   HEAD (ANATOMY)
  - ∞NOSE
  - PARANASAL SINUSES
  - SINUSES

NOSE CAPS
- USE   NOSE CONES

NOSE CONES
- UF   NOSE CAPS
- GS   CONES
  - . **NOSE CONES**
  - . . ABLATIVE NOSE CONES
  - . . ROCKET NOSE CONES
- RT   ABLATIVE MATERIALS
  - BLUNT BODIES
  - ∞CAPS
  - CIRCULAR CONES
  - HALF CONES
  - MISSILE COMPONENTS
  - OGIVES
  - REENTRY SHIELDING
  - REENTRY VEHICLES
  - SPACECRAFT COMPONENTS
  - SPACECRAFT SHIELDING
  - SPHERICAL CAPS
  - WARHEADS

NOSE FINS
- GS   FINS
  - . **NOSE FINS**
- RT   CONTROL SURFACES
  - FINNED BODIES
  - NOSES (FOREBODIES)
  - VANES

NOSE INLETS
- GS   INTAKE SYSTEMS
  - . **NOSE INLETS**
- RT   AIR INTAKES
  - ANNULAR DUCTS
  - BYPASS RATIO
  - DUCTED BODIES
  - DUCTS
  - HYPERSONIC INLETS
  - INLET AIRFRAME CONFIGURATIONS
  - NACELLES
  - NOSES (FOREBODIES)
  - SCOOPS
  - SIDE INLETS
  - SUPERSONIC INLETS
  - ∞WATER INTAKES

NOSE TIPS
- GS   TIPS
  - . **NOSE TIPS**
- RT   AERODYNAMIC CONFIGURATIONS
  - AIRFOIL PROFILES
  - NOSES (FOREBODIES)

NOSE WHEELS
- GS   WHEELS
  - . VEHICLE WHEELS
  - . . **NOSE WHEELS**
- RT   BRAKES (FOR ARRESTING MOTION)
  - LANDING GEAR

NOSES (FOREBODIES)
- GS   AIRCRAFT STRUCTURES
  - . FOREBODIES
  - . . **NOSES (FOREBODIES)**
- RT   ∞NOSE
  - NOSE FINS
  - NOSE INLETS
  - NOSE TIPS

NOSTOC
- GS   ALGAE
  - . BLUE GREEN ALGAE
  - . . **NOSTOC**

NOTATION
- USE   CODING

NOTCH SENSITIVITY
- GS   MECHANICAL PROPERTIES
  - . TOUGHNESS
  - . . **NOTCH SENSITIVITY**
  - SENSITIVITY
  - . **NOTCH SENSITIVITY**
- RT   CHARPY IMPACT TEST
  - FATIGUE (MATERIALS)
  - IMPACT STRENGTH
  - IMPACT TESTS

NOTCH STRENGTH
- GS   MECHANICAL PROPERTIES
  - . **NOTCH STRENGTH**
- RT   BRITTLENESS

NOTCH STRENGTH-*(CONT.)*
- DUCTILITY
- FATIGUE TESTS
- IMPACT TESTS
- ∞STRENGTH
- STRESS CONCENTRATION
- STRESS INTENSITY FACTORS

NOTCH TESTS
- UF   NOTCHED METALS
- GS   **NOTCH TESTS**
  - . CHARPY IMPACT TEST
- RT   BRITTLENESS
  - DROP TESTS
  - FATIGUE TESTS
  - HARDNESS
  - IMPACT TESTS
  - STRESS CONCENTRATION
  - ∞TESTS

NOTCHED METALS
- USE   NOTCH TESTS

NOTCHES
- RT   PASSAGEWAYS
  - V GROOVES

∞ **NOVA**
- SN   *(USE OF A MORE SPECIFIC TERM IS RECOMMENDED--CONSULT THE TERMS LISTED BELOW)*
- RT   NOVA LAUNCH VEHICLES
  - NOVAE

NOVA COMPUTERS
- GS   DATA PROCESSING EQUIPMENT
  - . COMPUTERS
  - . . DIGITAL COMPUTERS
  - . . . MINICOMPUTERS
  - . . . . **NOVA COMPUTERS**

NOVA LASER SYSTEM
- GS   STIMULATED EMISSION DEVICES
  - . LASERS
  - . . HIGH POWER LASERS
  - . . . **NOVA LASER SYSTEM**
- RT   LASER FUSION
  - LASER OUTPUTS
  - SHIVA LASER SYSTEM
  - ∞SYSTEMS

NOVA LAUNCH VEHICLES
- GS   LAUNCH VEHICLES
  - . **NOVA LAUNCH VEHICLES**
  - ROCKET VEHICLES
  - . MULTISTAGE ROCKET VEHICLES
  - . . **NOVA LAUNCH VEHICLES**
- RT   F-1 ROCKET ENGINE
  - J-2 ENGINE
  - LIQUID PROPELLANT ROCKET ENGINES
  - M-1 ENGINE
  - ∞NOVA
  - ∞VEHICLES

NOVA SATELLITES
- GS   SATELLITES
  - . ARTIFICIAL SATELLITES
  - . . NAVIGATION SATELLITES
  - . . . **NOVA SATELLITES**
  - UNMANNED SPACECRAFT
  - . NAVIGATION SATELLITES
  - . . **NOVA SATELLITES**
- RT   DISCOS (SATELLITE ATTITUDE CONTROL)
  - TRANSIT NAVIGATION SYSTEM

NOVA SCOTIA
- GS   NATIONS
  - . CANADA
  - . . **NOVA SCOTIA**

NOVAE
- GS   CELESTIAL BODIES
  - . STARS
  - . . VARIABLE STARS
  - . . . **NOVAE**
  - . . . . DWARF NOVAE
  - . . . . HERCULES NOVA
- RT   CATACLYSMIC VARIABLES
  - ∞NOVA
  - SHOCK WAVES
  - STELLAR MASS
  - STELLAR MASS EJECTION
  - SUPERNOVAE

NOVAE-*(CONT.)*
   SYMBIOTIC STARS

NOVOCAIN
   GS   DRUGS
         . ANESTHETICS
         . . **NOVOCAIN**

NOWCASTING
   GS   FORECASTING
         . WEATHER FORECASTING
         . . **NOWCASTING**
         METEOROLOGY
         . WEATHER FORECASTING
         . . **NOWCASTING**

NOXIOUS MATERIALS
   USE   CONTAMINANTS

NOZZLE COEFFICIENT
   USE   NOZZLE FLOW

NOZZLE DESIGN
   RT   ∞DESIGN
         ENGINE DESIGN
         ∞NOZZLES

NOZZLE EFFICIENCY
   GS   EFFICIENCY
         . **NOZZLE EFFICIENCY**
   RT   ∞NOZZLES
         POWER EFFICIENCY
         PROPULSIVE EFFICIENCY
         THERMODYNAMIC EFFICIENCY

NOZZLE FLOW
   UF   NOZZLE COEFFICIENT
   GS   FLUID FLOW
         . **NOZZLE FLOW**
   RT   AEROTHERMOCHEMISTRY
         ANNULAR FLOW
         CORNER FLOW
         DISCHARGE COEFFICIENT
         EXHAUST GASES
         EXHAUST NOZZLES
         FISSIONABLE MATERIALS
         FLOW GEOMETRY
         FLUID INJECTION
         INJECTORS
         JET FLOW
         OUTLET FLOW
         PNEUMATIC PROBES
         SUPERSONIC JET FLOW
         TRANSONIC FLOW

NOZZLE GEOMETRY
   GS   GEOMETRY
         . **NOZZLE GEOMETRY**
   RT   COAXIAL NOZZLES
         CONICAL NOZZLES
         CONVERGENT NOZZLES
         CONVERGENT-DIVERGENT NOZZLES
         DISCHARGE COEFFICIENT
         DIVERGENT NOZZLES
         MASS FLOW FACTORS
         ∞NOZZLES
         PIPE NOZZLES
         PLUG NOZZLES
         SHROUDED NOZZLES
         SPIKE NOZZLES
         THROATS

NOZZLE INSERTS
   GS   INSERTS
         . **NOZZLE INSERTS**
   RT   ABLATIVE MATERIALS
         CHOKES
         CONICAL NOZZLES
         CONVERGENT-DIVERGENT NOZZLES
         EXHAUST NOZZLES
         ∞NOZZLES
         ROCKET NOZZLES
         THROATS

NOZZLE THRUST COEFFICIENTS
   GS   COEFFICIENTS
         . **NOZZLE THRUST COEFFICIENTS**
   RT   DISCHARGE COEFFICIENT
         FLOW COEFFICIENTS
         INFLUENCE COEFFICIENT
         THRUST
         THRUST VECTOR CONTROL

NOZZLE WALLS
   GS   WALLS
         . **NOZZLE WALLS**
   RT   CONICAL NOZZLES
         CONVERGENT NOZZLES
         DIVERGENT NOZZLES
         JET AMPLIFIERS
         ∞NOZZLES
         REFRACTORY MATERIALS
         SHROUDED NOZZLES
         THROATS

NOZZLELESS ROCKET ENGINES
   GS   ENGINES
         . ROCKET ENGINES
         . . **NOZZLELESS ROCKET ENGINES**
   RT   ROCKET NOZZLES

∞ **NOZZLES**
   SN   *(USE OF A MORE SPECIFIC TERM IS*
         RECOMMENDED--CONSULT THE TERMS
         LISTED BELOW)
   RT   ACOUSTIC NOZZLES
         ANNULAR NOZZLES
         ATOMIZERS
         BLOWERS
         CHOKES
         CHOKES (RESTRICTIONS)
         COAXIAL NOZZLES
         CONICAL NOZZLES
         CONVERGENT NOZZLES
         CONVERGENT-DIVERGENT NOZZLES
         ∞DIFFUSERS
         DIVERGENT NOZZLES
         DUAL THRUST NOZZLES
         EXHAUST NOZZLES
         FLOW MEASUREMENT
         FUNNELS
         HYPERSONIC NOZZLES
         INJECTORS
         INLET NOZZLES
         NOZZLE DESIGN
         NOZZLE EFFICIENCY
         NOZZLE GEOMETRY
         NOZZLE INSERTS
         NOZZLE WALLS
         ORIFICES
         OUTLETS
         PIPE NOZZLES
         PLUG NOZZLES
         ROCKET NOZZLES
         SHROUDED NOZZLES
         SONIC NOZZLES
         SPIKE NOZZLES
         SPRAY NOZZLES
         SPRAYERS
         SUPERSONIC NOZZLES
         TRANSONIC NOZZLES
         TURBINES
         VENTS
         WIND TUNNEL NOZZLES

NRX REACTORS
   GS   NUCLEAR REACTORS
         . LIQUID COOLED REACTORS
         . . WATER COOLED REACTORS
         . . . **NRX REACTORS**
         . NUCLEAR RESEARCH AND TEST
           REACTORS
         . . **NRX REACTORS**
   RT   KIWI REACTORS
         NUCLEAR ENGINE FOR ROCKET
           VEHICLES

NTS
   USE   NAVIGATION TECHNOLOGY SATELLITES

NU FACTOR
   RT   POISSON RATIO

NUCLEAR AUXILIARY POWER UNITS
   GS   AUXILIARY POWER SOURCES
         . **NUCLEAR AUXILIARY POWER UNITS**
         . . SNAP
         . . . FISSION ELECTRIC CELLS
         . . . . SNAP 2
         . . . . SNAP 4
         . . . . SNAP 8
         . . . . SNAP 10A
         . . . SNAP 1
         . . . SNAP 3
         . . . SNAP 7
         . . . SNAP 9A
         . . . SNAP 11
         . . . SNAP 13

NUCLEAR AUXILIARY POWER UNITS-*(CONT.)*
         . . . SNAP 15
         . . . SNAP 17
         . . . SNAP 19
         . . . SNAP 21
         . . . SNAP 23
         . . . SNAP 27
         . . . SNAP 29
         . . . SNAP 50
         . . SPACE POWER REACTORS
         . . . FISSION ELECTRIC CELLS
         . . . . SNAP 2
         . . . . SNAP 4
         . . . . SNAP 8
         . . . . SNAP 10A
         . . . . SNAP 50
         . . . SPACE POWER UNIT REACTORS
         NUCLEAR ELECTRIC POWER
           GENERATION
         . **NUCLEAR AUXILIARY POWER UNITS**
         . . SNAP
         . . . FISSION ELECTRIC CELLS
         . . . . SNAP 2
         . . . . SNAP 4
         . . . . SNAP 8
         . . . . SNAP 10A
         . . . SNAP 1
         . . . SNAP 3
         . . . SNAP 7
         . . . SNAP 9A
         . . . SNAP 11
         . . . SNAP 13
         . . . SNAP 15
         . . . SNAP 17
         . . . SNAP 19
         . . . SNAP 21
         . . . SNAP 23
         . . . SNAP 27
         . . . SNAP 29
         . . . SNAP 50
         . . SPACE POWER REACTORS
         . . . FISSION ELECTRIC CELLS
         . . . . SNAP 2
         . . . . SNAP 4
         . . . . SNAP 8
         . . . . SNAP 10A
         . . . . SNAP 50
         . . . SPACE POWER UNIT REACTORS
   RT   ∞POWER SUPPLIES
         RADIOISOTOPE BATTERIES
         SPACECRAFT POWER SUPPLIES
         THERMOELECTRIC GENERATORS
         THERMOELECTRIC POWER GENERATION

NUCLEAR BINDING ENERGY
   RT   ACTIVATION ENERGY
         ∞ENERGY
         IONIZATION POTENTIALS

NUCLEAR CAPTURE
   GS   NUCLEAR REACTIONS
         . NUCLEAR INTERACTIONS
         . . **NUCLEAR CAPTURE**
         . . . ELECTRON CAPTURE
         PARTICLE INTERACTIONS
         . ELEMENTARY PARTICLE
           INTERACTIONS
         . . **NUCLEAR CAPTURE**
         . . . ELECTRON CAPTURE
         . NUCLEAR INTERACTIONS
         . . **NUCLEAR CAPTURE**
         . . . ELECTRON CAPTURE
   RT   ACTIVATION ENERGY
         CAPTURE EFFECT
         ELECTRON TRANSITIONS
         ENERGY LEVELS
         ∞INTERACTIONS
         IRRADIATION
         SPIN
         TRANSITION PROBABILITIES

NUCLEAR CHEMISTRY
   RT   ATOMIC STRUCTURE
         ∞CHEMISTRY
         ISOMERS
         ∞NUCLEAR ENERGY
         PHYSICAL CHEMISTRY
         PLASMA CHEMISTRY
         QUANTUM CHEMISTRY
         RADIOCHEMISTRY

NUCLEAR DEFORMATION
   GS   DEFORMATION
         . **NUCLEAR DEFORMATION**

NUCLEAR DEVICES
GS   EXPLOSIVE DEVICES
    . **NUCLEAR DEVICES**
RT   ∞DEVICES
    FISSION WEAPONS
    THERMONUCLEAR EXPLOSIONS
    WARHEADS

NUCLEAR ELECTRIC POWER GENERATION
UF   NUCLEAR POWER GENERATION
GS   **NUCLEAR ELECTRIC POWER
       GENERATION**
    . NUCLEAR AUXILIARY POWER UNITS
    . . SNAP
    . . . FISSION ELECTRIC CELLS
    . . . . SNAP 2
    . . . . SNAP 4
    . . . . SNAP 8
    . . . . SNAP 10A
    . . . SNAP 1
    . . . SNAP 3
    . . . SNAP 7
    . . . SNAP 9A
    . . . SNAP 11
    . . . SNAP 13
    . . . SNAP 15
    . . . SNAP 17
    . . . SNAP 19
    . . . SNAP 21
    . . . SNAP 23
    . . . SNAP 27
    . . . SNAP 29
    . . . SNAP 50
    . . SPACE POWER REACTORS
    . . . FISSION ELECTRIC CELLS
    . . . . SNAP 2
    . . . . SNAP 4
    . . . . SNAP 8
    . . . . SNAP 10A
    . . . SNAP 50
    . . SPACE POWER UNIT REACTORS
    . NUCLEAR POWER PLANTS
    . . ENRICO FERMI ATOMIC POWER
       PLANT
    . . HALLAM NUCLEAR POWER FACILITY
    . . ML-1 NUCLEAR POWER PLANT
    . NUCLEAR POWER REACTORS
    . . KIWI REACTORS
    . . . KIWI B REACTORS
    . . . . KIWI B-1 REACTOR
    . . . . KIWI B-4 REACTOR
    . PATHFINDER NUCLEAR REACTOR
    . PLUTONIUM RECYCLE TEST
       REACTOR
    . . SPACE POWER REACTORS
    . . . FISSION ELECTRIC CELLS
    . . . . SNAP 2
    . . . . SNAP 4
    . . . . SNAP 8
    . . . . SNAP 10A
    . . . SPACE POWER UNIT REACTORS
    . . TORY 2 REACTOR
    . . TORY 2-A REACTOR
    . . TORY 2-C REACTOR
    . THERMONUCLEAR POWER
       GENERATION
RT   ELECTRIC GENERATORS
    ∞NUCLEAR ENERGY

NUCLEAR ELECTRIC PROPULSION
GS   PROPULSION
    . NUCLEAR PROPULSION
    . . **NUCLEAR ELECTRIC PROPULSION**
RT   ELECTRIC PROPULSION
    ELECTROTHERMAL ENGINES
    ION PROPULSION
    MARINE PROPULSION
    PLASMA PROPULSION
    SPACECRAFT PROPULSION

NUCLEAR EMULSIONS
GS   MIXTURES
    . DISPERSIONS
    . . EMULSIONS
    . . . PHOTOGRAPHIC EMULSIONS
    . . . . **NUCLEAR EMULSIONS**
    . SOLUTIONS
    . . PHOTOGRAPHIC EMULSIONS
    . . . **NUCLEAR EMULSIONS**
RT   DOSIMETERS
    RADIATION COUNTERS
    RADIATION MEASURING INSTRUMENTS

∞ **NUCLEAR ENERGY**
SN   *(USE OF A MORE SPECIFIC TERM IS
    RECOMMENDED--CONSULT THE TERMS
    LISTED BELOW)*
UF   ATOMIC ENERGY
RT   ANNULAR CORE PULSE REACTORS
    ATOMIC THEORY
    CHEMICAL ENERGY
    ENERGY POLICY
    ENERGY STORAGE
    MOLECULAR ENERGY LEVELS
    NUCLEAR CHEMISTRY
    NUCLEAR ELECTRIC POWER
       GENERATION
    NUCLEAR FISSION
    NUCLEAR FUELS
    NUCLEAR FUSION
    NUCLEAR HEAT
    NUCLEAR PHYSICS
    NUCLEAR PROPULSION
    NUCLEAR REACTORS
    NUCLEAR RESEARCH
    NUCLEAR WARFARE
    NUCLEAR WEAPONS
    NUCLEONICS
    SPENT FUELS

NUCLEAR ENGINE FOR ROCKET VEHICLES
UF   NERVA (ENGINE)
GS   ENGINES
    . ROCKET ENGINES
    . . **NUCLEAR ENGINE FOR ROCKET
       VEHICLES**
RT   BOOSTER ROCKET ENGINES
    KIWI REACTORS
    NRX REACTORS
    PHOEBUS NUCLEAR REACTOR
    ROVER PROJECT
    SUSTAINER ROCKET ENGINES
    ∞VEHICLES
    WATER COOLED REACTORS

NUCLEAR EXPLOSION EFFECT
RT   ∞EFFECTS
    FALLOUT
    RADIATION EFFECTS
    RADIATION HAZARDS

NUCLEAR EXPLOSIONS
UF   ATOMIC EXPLOSIONS
GS   EXPLOSIONS
    . **NUCLEAR EXPLOSIONS**
    . . THERMONUCLEAR EXPLOSIONS
RT   AERIAL EXPLOSIONS
    ARTIFICIAL RADIATION BELTS
    CIVIL DEFENSE
    CRATERING
    FALLOUT
    ∞FIREBALLS
    FISHBOWL OPERATION
    HIGH ENERGY INTERACTIONS
    RADIATION HAZARDS
    UNDERGROUND EXPLOSIONS
    UNDERWATER EXPLOSIONS
    VELA SATELLITES

NUCLEAR FISSION
GS   DECAY
    . **NUCLEAR FISSION**
    NUCLEAR REACTIONS
    . **NUCLEAR FISSION**
RT   CRITICAL EXPERIMENTS
    CRITICAL MASS
    FISSION PRODUCTS
    FUSION REACTORS
    FUSION-FISSION HYBRID REACTORS
    HIGH ENERGY INTERACTIONS
    NEUTRON FLUX DENSITY
    ∞NUCLEAR ENERGY
    RADIOACTIVE DECAY
    RADIOACTIVE MATERIALS
    SUBCRITICAL MASS

NUCLEAR FUEL BURNUP
GS   COMBUSTION
    . FUEL COMBUSTION
    . . **NUCLEAR FUEL BURNUP**
RT   CRITICAL MASS
    REACTOR PHYSICS
    REACTOR TECHNOLOGY

NUCLEAR FUEL ELEMENTS
UF   FUEL ELEMENTS (NUCLEAR REACTORS)
RT   ANNULAR CORE PULSE REACTORS
    ∞ELEMENTS

NUCLEAR FUEL ELEMENTS-*(CONT.)*
    FUELS
    PLUTONIUM ALLOYS
    REACTOR CORES
    REACTOR MATERIALS
    URANIUM ALLOYS
    URANIUM CARBIDES

NUCLEAR FUEL REPROCESSING
GS   RECLAMATION
    . MATERIALS RECOVERY
    . . **NUCLEAR FUEL REPROCESSING**
RT   NUCLEAR FUELS
    ∞PROCESSING
    ∞RECOVERY
    RECYCLING
    SPENT FUELS

NUCLEAR FUELS
UF   REACTOR FUELS
GS   FUELS
    . **NUCLEAR FUELS**
    . . CERAMIC NUCLEAR FUELS
    . . FISSIUM
    . . SPENT FUELS
RT   ANNULAR CORE PULSE REACTORS
    DEUTERIUM
    ENERGY POLICY
    FISSILE FUELS
    FISSION
    FISSIONABLE MATERIALS
    FUEL CAPSULES
    FUEL PRODUCTION
    INERTIAL FUSION (REACTOR)
    MIXED OXIDES
    NEUTRON SOURCES
    ∞NUCLEAR ENERGY
    NUCLEAR FUEL REPROCESSING
    PELLETS
    PLUTONIUM
    PLUTONIUM ALLOYS
    PLUTONIUM COMPOUNDS
    REACTOR CORES
    REACTOR MATERIALS
    REACTOR STARTUP TESTS
    SOL-GEL PROCESSES
    THORIUM
    THORIUM ALLOYS
    THORIUM COMPOUNDS
    TRITIUM
    URANIUM
    URANIUM ALLOYS
    URANIUM CARBIDES
    URANIUM COMPOUNDS
    URANIUM OXIDES
    URANIUM 233
    URANIUM 235
    URANIUM 238

NUCLEAR FUSION
UF   NUCLEOSYNTHESIS
GS   NUCLEAR REACTIONS
    . THERMONUCLEAR REACTIONS
    . . **NUCLEAR FUSION**
    . . . CONTROLLED FUSION
RT   COLLISIONAL PLASMAS
    DENSE PLASMAS
    DEUTERON IRRADIATION
    ∞FUSION
    FUSION REACTORS
    FUSION WEAPONS
    FUSION-FISSION HYBRID REACTORS
    HIGH ENERGY INTERACTIONS
    IRRADIATION
    MAGNETIC MIRRORS
    MIRROR FUSION
    ∞NUCLEAR ENERGY
    PLASMA FOCUS
    RAILGUN ACCELERATORS
    STELLAR PHYSICS
    ∞SYNTHESIS
    TOKAMAK DEVICES

NUCLEAR GYROSCOPES
GS   GYROSCOPES
    . **NUCLEAR GYROSCOPES**

NUCLEAR HEAT
GS   HEAT
    . **NUCLEAR HEAT**
RT   ∞NUCLEAR ENERGY

NUCLEAR INTERACTIONS
GS   NUCLEAR REACTIONS
    . **NUCLEAR INTERACTIONS**

NUCLEAR INTERACTIONS-*(CONT.)*
```
      . . NUCLEAR CAPTURE
      . . . ELECTRON CAPTURE
      . . SPIN-ORBIT INTERACTIONS
      . . . ELECTRON CAPTURE
      . PARTICLE INTERACTIONS
      . NUCLEAR INTERACTIONS
      . . NUCLEAR CAPTURE
      . . . ELECTRON CAPTURE
      . . SPIN-ORBIT INTERACTIONS
      . . . ELECTRON CAPTURE
      . . STRONG INTERACTIONS (FIELD
            THEORY)
      . . WEAK INTERACTIONS (FIELD
            THEORY)
RT    COLLISION PARAMETERS
      ELEMENTARY PARTICLES
      ∞ INTERACTIONS
```

NUCLEAR ISOBARS
```
RT    CHEMICAL ELEMENTS
      ∞ ISOBARS
      ISOTOPES
      NUCLEI (NUCLEAR PHYSICS)
      NUCLIDES
```

NUCLEAR LIGHTBULB ENGINES
```
GS    ENGINES
      . ROCKET ENGINES
      . . NUCLEAR ROCKET ENGINES
      . . . NUCLEAR LIGHTBULB ENGINES
RT    GASEOUS FISSION REACTORS
```

NUCLEAR MAGNETIC RESONANCE
```
UF    KNIGHT SHIFT
      NMR
GS    RESONANCE
      . MAGNETIC RESONANCE
      . . NUCLEAR MAGNETIC RESONANCE
      . . . PROTON MAGNETIC RESONANCE
      . . . PROTON RESONANCE
RT    MAGNETIC SIGNALS
      MAGNETOMETERS
      MOLECULAR STRUCTURE
      PARAMAGNETIC RESONANCE
      PARTICLE SPIN
      PLANCKS CONSTANT
      SPIN RESONANCE
      SPIN-LATTICE RELAXATION
```

NUCLEAR MEDICINE
```
UF    RADIATION MEDICINE
GS    MEDICAL SCIENCE
      . NUCLEAR MEDICINE
      . . RADIOBIOLOGY
RT    ANTIRADIATION DRUGS
      HEALTH PHYSICS
      ∞ RADIATION
      RADIOPATHOLOGY
```

NUCLEAR METEOROLOGY
```
GS    METEOROLOGY
      . NUCLEAR METEOROLOGY
RT    FALLOUT
```

NUCLEAR MODELS
```
GS    MODELS
      . NUCLEAR MODELS
RT    ATOMIC STRUCTURE
      ENERGY LEVELS
      MOLECULAR STRUCTURE
      QUARK PARTON MODEL
```

NUCLEAR PARTICLES
```
GS    PARTICLES
      . NUCLEAR PARTICLES
      . . ANTIPARTICLES
      . . . ANTINEUTRINOS
      . . . ANTINUCLEONS
      . . . ANTIPROTONS
      . . . POSITRONS
      . . BETA PARTICLES
      . . BOSONS
      . . . ALPHA PARTICLES
      . . . MESONS
      . . . . ETA-MESONS
      . . . . KAONS
      . . . . MESON RESONANCE
      . . . . . X MESONS
      . . . . MUONS
      . . . . PIONS
      . . . . VECTOR MESONS
      . . . . . RHO-MESONS
      . . . . . SIGMA-MESONS
      . . . PHOTONS
```

NUCLEAR PARTICLES-*(CONT.)*
```
      . . . XI HYPERONS
      . . NUCLEONS
      . . PHOTOELECTRONS
RT    CORPUSCULAR RADIATION
      COSMIC RAYS
      ELEMENTARY PARTICLES
      FISSION PRODUCTS
      GAMMA RAY BURSTS
      NEUTRON CROSS SECTIONS
      NEUTRON DISTRIBUTION
      NEUTRON SCATTERING
      NUCLEON POTENTIAL
      NUCLEON-NUCLEON SCATTERING
      PARTICLE ACCELERATORS
      PARTICLE TRACKS
      PHOTONEUTRONS
      PI-ELECTRONS
      POSITRON ANNIHILATION
      PROTON RESONANCE
      PROTONS
```

NUCLEAR PHYSICS
```
GS    NUCLEAR PHYSICS
      . REACTOR PHYSICS
RT    ATOMIC STRUCTURE
      CURRENT ALGEBRA
      ELECTROMAGNETIC ABSORPTION
      FIELD THEORY (PHYSICS)
      HEALTH PHYSICS
      ∞ NUCLEAR ENERGY
      NUCLEONICS
      PARITY
      PARTICLE SPIN
      ∞ PHYSICS
      QUANTUM THEORY
      ∞ SCIENCE
      THEORETICAL PHYSICS
```

NUCLEAR POTENTIAL
```
GS    POTENTIAL ENERGY
      . NUCLEAR POTENTIAL
RT    NUCLEON POTENTIAL
```

NUCLEAR POWER GENERATION
```
USE   NUCLEAR ELECTRIC POWER
            GENERATION
```

NUCLEAR POWER PLANTS
```
GS    ELECTRIC POWER PLANTS
      . NUCLEAR POWER PLANTS
      . . ENRICO FERMI ATOMIC POWER
            PLANT
      . . HALLAM NUCLEAR POWER FACILITY
      . . ML-1 NUCLEAR POWER PLANT
      NUCLEAR ELECTRIC POWER
            GENERATION
      . NUCLEAR POWER PLANTS
      . . ENRICO FERMI ATOMIC POWER
            PLANT
      . . HALLAM NUCLEAR POWER FACILITY
      . . ML-1 NUCLEAR POWER PLANT
RT    PLASMA CORE REACTORS
```

NUCLEAR POWER REACTORS
```
GS    NUCLEAR ELECTRIC POWER
            GENERATION
      . NUCLEAR POWER REACTORS
      . . KIWI REACTORS
      . . . KIWI B REACTORS
      . . . . KIWI B-1 REACTOR
      . . . . KIWI B-4 REACTOR
      . . PATHFINDER NUCLEAR REACTOR
      . . PLUTONIUM RECYCLE TEST
            REACTOR
      . . SPACE POWER REACTORS
      . . . FISSION ELECTRIC CELLS
      . . . . SNAP 2
      . . . . SNAP 4
      . . . . SNAP 8
      . . . . SNAP 10A
      . . . SPACE POWER UNIT REACTORS
      . . TORY 2 REACTOR
      . . TORY 2-A REACTOR
      . . TORY 2-C REACTOR
      NUCLEAR REACTORS
      . NUCLEAR POWER REACTORS
      . . KIWI REACTORS
      . . . KIWI B REACTORS
      . . . . KIWI B-1 REACTOR
      . . . . KIWI B-4 REACTOR
      . . PATHFINDER NUCLEAR REACTOR
      . . PLUTONIUM RECYCLE TEST
            REACTOR
      . . SPACE POWER REACTORS
```

NUCLEAR POWER REACTORS-*(CONT.)*
```
      . . . FISSION ELECTRIC CELLS
      . . . . SNAP 2
      . . . . SNAP 4
      . . . . SNAP 8
      . . . . SNAP 10A
      . . SPACE POWER UNIT REACTORS
      . . TORY 2 REACTOR
      . . TORY 2-A REACTOR
      . . TORY 2-C REACTOR
RT    BOILING WATER REACTORS
      BREEDER REACTORS
      FAST NUCLEAR REACTORS
      FAST OXIDE REACTORS
      HIGH TEMPERATURE GAS COOLED
            REACTORS
      LIQUID METAL FAST BREEDER
            REACTORS
      POWER REACTORS
      PRESSURIZED WATER REACTORS
      SNAP
      SODIUM GRAPHITE REACTORS
```

NUCLEAR POWERED SHIPS
```
GS    SURFACE VEHICLES
      . NUCLEAR POWERED SHIPS
      . . SAVANNAH NUCLEAR SHIP
      WATER VEHICLES
      . SHIPS
      . NUCLEAR POWERED SHIPS
      . . . SAVANNAH NUCLEAR SHIP
RT    AIRCRAFT CARRIERS
      CARGO SHIPS
      NAVY
      SUBMARINES
```

NUCLEAR PROPELLED AIRCRAFT
```
RT    ∞ AIRCRAFT
      ∞ MILITARY AIRCRAFT
      RESEARCH AIRCRAFT
```

NUCLEAR PROPULSION
```
UF    CHEMONUCLEAR PROPULSION
      THERMONUCLEAR PROPULSION
GS    PROPULSION
      . NUCLEAR PROPULSION
      . . NUCLEAR ELECTRIC PROPULSION
RT    AIRCRAFT ENGINES
      GASEOUS FISSION REACTORS
      HIGH TEMPERATURE NUCLEAR
            REACTORS
      HIGH TEMPERATURE PROPELLANTS
      MARINE PROPULSION
      MERCURY ION ENGINES
      NEW MOONS PROJECT
      ∞ NUCLEAR ENERGY
      ROVER PROJECT
      SAVANNAH NUCLEAR SHIP
      SPACECRAFT PROPULSION
      TRIDENT SUBMARINE
      UNDERWATER PROPULSION
```

NUCLEAR PUMPED LASERS
```
GS    STIMULATED EMISSION DEVICES
      . LASERS
      . . NUCLEAR PUMPED LASERS
RT    OPTICAL PUMPING
      OPTICAL RESONANCE
```

NUCLEAR PUMPING
```
RT    ELECTRON PUMPING
      ENERGY TRANSFER
      FISSION PRODUCTS
      GAS LASERS
      LASERS
      OPTICAL PUMPING
      POPULATION INVERSION
      ∞ PUMPING
      STIMULATED EMISSION
      STIMULATED EMISSION DEVICES
```

NUCLEAR QUADRUPOLE RESONANCE
```
GS    RESONANCE
      . NUCLEAR QUADRUPOLE RESONANCE
RT    ENERGY LEVELS
      QUADRUPOLES
```

NUCLEAR RADIATION
```
GS    NUCLEAR RADIATION
      . BETA PARTICLES
      . FAST NEUTRONS
      . GAMMA RAY BEAMS
      . GAMMA RAYS
      . . GAMMA RAY BURSTS
      . NEUTRON BEAMS
```

NUCLEAR RADIATION-*(CONT.)*
. PHOTONEUTRONS
. POST-BLAST NUCLEAR RADIATION
. SPALLATION
. THERMAL NEUTRONS
RT   ALPHA PARTICLES
     BREMSSTRAHLUNG
     CERENKOV RADIATION
     CORPUSCULAR RADIATION
     ELECTROMAGNETIC RADIATION
     ELECTRON PUMPING
     ELECTRON RADIATION
     ELECTRONS
     ELEMENTARY PARTICLES
     EMISSION SPECTRA
     FISSION PRODUCTS
     GAMMA RAY ABSORPTION
     HEALTH PHYSICS
     HIGH ENERGY INTERACTIONS
     IONIZING RADIATION
     IRRADIATION
     NEUTRONS
     PARTICLE PRODUCTION
     PHOTONS
∞RADIATION
     RADIATION EFFECTS
     RADIATION HAZARDS
     RADIOACTIVE CONTAMINANTS
     RADIOACTIVE DECAY
     RADIOACTIVE MATERIALS
     RADIOACTIVITY
     RADIOBIOLOGY
     RADIOCHEMISTRY
     VELA SATELLITES

NUCLEAR RADIATION SPECTROSCOPY
GS   SPECTROSCOPY
. **NUCLEAR RADIATION SPECTROSCOPY**
RT   MASS SPECTROSCOPY
     SPECTROSCOPIC ANALYSIS
     VACUUM SPECTROSCOPY

NUCLEAR RAMJET ENGINES
GS   ENGINES
. ROCKET ENGINES
.. **NUCLEAR RAMJET ENGINES**
RT   PLUTO REACTORS
∞ROCKETS
     SUPERSONIC LOW ALTITUDE MISSILE

NUCLEAR REACTIONS
UF   NEUTRON TRANSMUTATION
GS   **NUCLEAR REACTIONS**
. ALPHA DECAY
. ANNIHILATION REACTIONS
.. POSITRON ANNIHILATION
. ELECTRON SCATTERING
. HIGH ENERGY INTERACTIONS
.. LIGHT-CONE EXPANSION
. NUCLEAR FISSION
. NUCLEAR INTERACTIONS
.. NUCLEAR CAPTURE
... ELECTRON CAPTURE
.. SPIN-ORBIT INTERACTIONS
... ELECTRON CAPTURE
.. NUCLEAR SCATTERING
.. NEUTRON SCATTERING
.. RESONANCE SCATTERING
. NUCLEAR TRANSFORMATIONS
. PHOTONUCLEAR REACTIONS
. PHOTOPRODUCTION
. PROTON SCATTERING
. PROTON-PROTON REACTIONS
. RADIOACTIVE DECAY
.. NEUTRON EMISSION
. SPALLATION
. THERMONUCLEAR REACTIONS
.. NUCLEAR FUSION
... CONTROLLED FUSION
RT   BRAGG CURVE
     COMPTON EFFECT
     CRITICAL EXPERIMENTS
     CRITICAL MASS
     EMISSION
     HALF LIFE
     INHOUR EQUATION
∞INTERACTIONS
     INTERNAL CONVERSION
     PAIR PRODUCTION
     PARTICLE INTERACTIONS
     PARTICLE PRODUCTION
     PHOTONEUTRONS
     POISONING (REACTION INHIBITION)
     POMERONS
     RADIATION ABSORPTION

NUCLEAR REACTIONS-*(CONT.)*
     RADIOGENIC MATERIALS
∞REACTION
     REACTION KINETICS
     REACTIVITY
     SOLAR NEUTRINOS
     SPALLING
     SUBCRITICAL MASS
     TRANSMUTATION

NUCLEAR REACTOR CONTROL
RT   CONFINEMENT
∞CONTROL
     CONTROL RODS
∞REACTION CONTROL
     REACTOR SAFETY

NUCLEAR REACTORS
GS   **NUCLEAR REACTORS**
. ANNULAR CORE PULSE REACTORS
. ASTRON THERMONUCLEAR REACTOR
. BREEDER REACTORS
.. EXPERIMENTAL BREEDER REACTOR
    1
.. EXPERIMENTAL BREEDER REACTOR
    2
.. LIGHT WATER BREEDER REACTORS
.. LIQUID METAL FAST BREEDER
    REACTORS
. ENGINEERING TEST REACTORS
. FAST NUCLEAR REACTORS
. EXPERIMENTAL BREEDER REACTOR
    1
. EXPERIMENTAL BREEDER REACTOR
    2
.. FAST OXIDE REACTORS
.. FAST TEST REACTORS
.. GAS COOLED FAST REACTORS
.. LIQUID METAL FAST BREEDER
    REACTORS
. FUSION REACTORS
. SPHEROMAKS
. FUSION-FISSION HYBRID REACTORS
. GAS COOLED REACTORS
.. EXPERIMENTAL GAS COOLED
    REACTORS
.. GAS COOLED FAST REACTORS
.. HIGH TEMPERATURE NUCLEAR
    REACTORS
... HIGH TEMPERATURE GAS COOLED
    REACTORS
.. KIWI REACTORS
... KIWI B REACTORS
.... KIWI B-1 REACTOR
.... KIWI B-4 REACTOR
.. TORY 2 REACTOR
.. TORY 2-A REACTOR
.. TORY 2-C REACTOR
. GASEOUS FISSION REACTORS
. HANFORD REACTORS
. HIGH FLUX ISOTOPE REACTORS
. LIQUID COOLED REACTORS
.. LIQUID METAL COOLED REACTORS
... ADVANCED SODIUM COOLED
    REACTOR
... EXPERIMENTAL BREEDER REACTOR
    1
... EXPERIMENTAL BREEDER REACTOR
    2
... LITHIUM COOLED REACTOR
    EXPERIMENT
... LOS ALAMOS MOLTEN PLUTONIUM
    REACTOR
... MILITARY COMPACT REACTORS
... SODIUM GRAPHITE REACTORS
... SODIUM REACTOR EXPERIMENT
.. ORGANIC COOLED REACTORS
... EXPERIMENTAL ORGANIC COOLED
    REACTORS
.. WATER COOLED REACTORS
... BOILING WATER REACTORS
.... EXPERIMENTAL BOILING WATER
    REACTORS
.... HALDEN BOILING WATER
    REACTOR
.... LOS ALAMOS WATER BOILER
    REACTOR
.... PATHFINDER NUCLEAR REACTOR
.... SPERT REACTORS
... HEAVY WATER REACTORS
.... HEAVY WATER COMPONENTS
    TEST REACTORS
.... PLUTONIUM RECYCLE TEST
    REACTOR
.... ZERO POWER REACTOR 2

NUCLEAR REACTORS-*(CONT.)*
... LIGHT WATER REACTORS
... NRX REACTORS
... PLUM BROOK REACTOR
... PRESSURIZED WATER REACTORS
.... SPECTRAL SHIFT CONTROL
    REACTOR
... SWIMMING POOL REACTORS
... ZERO POWER REACTORS
.... ZERO POWER REACTOR 2
.... ZERO POWER REACTOR 3
.... ZERO POWER REACTOR 6
.... ZERO POWER REACTOR 9
. MOLTEN SALT NUCLEAR REACTORS
. NUCLEAR POWER REACTORS
.. KIWI REACTORS
... KIWI B REACTORS
.... KIWI B-1 REACTOR
.... KIWI B-4 REACTOR
. PATHFINDER NUCLEAR REACTOR
. PLUTONIUM RECYCLE TEST
    REACTOR
. SPACE POWER REACTORS
.. FISSION ELECTRIC CELLS
.... SNAP 2
.... SNAP 4
.... SNAP 8
.... SNAP 10A
.. SPACE POWER UNIT REACTORS
. TORY 2 REACTOR
. TORY 2-A REACTOR
. TORY 2-C REACTOR
. NUCLEAR RESEARCH AND TEST
    REACTORS
.. ADVANCED TEST REACTORS
.. EXPERIMENTAL BOILING WATER
    REACTORS
.. EXPERIMENTAL BREEDER REACTOR
    1
.. EXPERIMENTAL BREEDER REACTOR
    2
.. EXPERIMENTAL GAS COOLED
    REACTORS
.. EXPERIMENTAL ORGANIC COOLED
    REACTORS
.. HEALTH PHYSICS RESEARCH
    REACTOR
.. HEAVY WATER COMPONENTS TEST
    REACTORS
.. HERO REACTOR
.. HIGH TEMPERATURE NUCLEAR
    REACTORS
.. JANUS REACTOR
.. KIWI REACTORS
... KIWI B REACTORS
.... KIWI B-1 REACTOR
.... KIWI B-4 REACTOR
.. LIVERMORE POOL TYPE REACTOR
.. LOS ALAMOS MOLTEN PLUTONIUM
    REACTOR
.. MILITARY COMPACT REACTORS
.. NRX REACTORS
.. PLUM BROOK REACTOR
.. PLUTONIUM RECYCLE TEST
    REACTOR
.. SODIUM REACTOR EXPERIMENT
.. SPERT REACTORS
.. TORY 2 REACTOR
.. TORY 2-A REACTOR
.. TORY 2-C REACTOR
.. TOWER SHIELDING REACTOR 2
.. ZERO POWER REACTOR 2
.. ZERO POWER REACTOR 3
.. ZERO POWER REACTOR 6
.. ZERO POWER REACTOR 9
. ORGANIC MODERATED REACTORS
.. EXPERIMENTAL ORGANIC COOLED
    REACTORS
. PEBBLE BED REACTORS
. PHOEBUS NUCLEAR REACTOR
. PLASMA CORE REACTORS
. PLUTO REACTORS
. THERMAL REACTORS
. TOKAMAK DEVICES
.. JOINT EUROPEAN TORUS
. WATER MODERATED REACTORS
.. EXPERIMENTAL BOILING WATER
    REACTORS
.. HEAVY WATER COMPONENTS TEST
    REACTORS
.. PLUTONIUM RECYCLE TEST
    REACTOR
RT   CLOSED CYCLES
     CONTROL RODS
     COOLANTS
     FISSILE FUELS

NUCLEAR REACTORS-*(CONT.)*
    HIGH FLUX BEAM REACTORS
    INHOUR EQUATION
    LOSS OF COOLANT
    MODERATORS
   ∞NUCLEAR ENERGY
   ∞PILES
    RADIATION SHIELDING
    REACTOR CORES
    REACTOR DESIGN
    REACTOR MATERIALS
    REACTOR PHYSICS
    REACTOR SAFETY
    REACTOR STARTUP TESTS
    REACTOR TECHNOLOGY
   ∞REACTORS
    ROVER PROJECT
    STELLARATORS
    THERMAL NEUTRONS
    THERMAL POLLUTION

NUCLEAR RELAXATION
  RT   RELAXATION (MECHANICS)

NUCLEAR RESEARCH
  GS   RESEARCH
    . **NUCLEAR RESEARCH**
  RT   HIGH ENERGY INTERACTIONS
    LABORATORIES
   ∞NUCLEAR ENERGY
    PLASMA CORE REACTORS
    RADIOCHEMISTRY

NUCLEAR RESEARCH AND TEST REACTORS
  UF   MATERIALS TESTING REACTORS
    NUCLEAR TEST REACTORS
    PHYSICAL CONSTANTS TESTING
      REACTOR
  GS   NUCLEAR REACTORS
    . **NUCLEAR RESEARCH AND TEST**
      REACTORS
    . . ADVANCED TEST REACTORS
    . . EXPERIMENTAL BOILING WATER
      REACTORS
    . . EXPERIMENTAL BREEDER REACTOR
      1
    . . EXPERIMENTAL BREEDER REACTOR
      2
    . . EXPERIMENTAL GAS COOLED
      REACTORS
    . . EXPERIMENTAL ORGANIC COOLED
      REACTORS
    . . HEALTH PHYSICS RESEARCH
      REACTOR
    . . HEAVY WATER COMPONENTS TEST
      REACTORS
    . . HERO REACTOR
    . . HIGH TEMPERATURE NUCLEAR
      REACTORS
    . . JANUS REACTOR
    . . KIWI REACTORS
    . . . KIWI B REACTORS
    . . . . KIWI B-1 REACTOR
    . . . . KIWI B-4 REACTOR
    . . LIVERMORE POOL TYPE REACTOR
    . . LOS ALAMOS MOLTEN PLUTONIUM
      REACTOR
    . . MILITARY COMPACT REACTORS
    . . NRX REACTORS
    . . PLUM BROOK REACTOR
    . . PLUTONIUM RECYCLE TEST
      REACTOR
    . . SODIUM REACTOR EXPERIMENT
    . . SPERT REACTORS
    . . TORY 2 REACTOR
    . . TORY 2-A REACTOR
    . . TORY 2-C REACTOR
    . . TOWER SHIELDING REACTOR 2
    . . ZERO POWER REACTOR 2
    . . ZERO POWER REACTOR 3
    . . ZERO POWER REACTOR 6
    . . ZERO POWER REACTOR 9
  RT   BOILING WATER REACTORS
    NEUTRON SOURCES
    REACTOR DESIGN
    REACTOR TECHNOLOGY
   ∞REACTORS
    TRANSIENT REACTOR TEST FACILITY

NUCLEAR ROCKET ENGINES
  UF   THERMIONIC REACTORS
  GS   ENGINES
    . ROCKET ENGINES
    . . **NUCLEAR ROCKET ENGINES**
    . . . NUCLEAR LIGHTBULB ENGINES

NUCLEAR ROCKET ENGINES-*(CONT.)*
  RT   BOOSTER ROCKET ENGINES
    ION ENGINES
    MERCURY ION ENGINES
    PHOEBUS NUCLEAR REACTOR
    PLUTO REACTORS
    RESTARTABLE ROCKET ENGINES
   ∞ROCKETS
    SUSTAINER ROCKET ENGINES

NUCLEAR SCATTERING
  SN   (SCATTERING CAUSED BY NUCLEUS
      AND NOT BY ORBITAL ELECTRONS)
  GS   NUCLEAR REACTIONS
    . **NUCLEAR SCATTERING**
    . . NEUTRON SCATTERING
    . . RESONANCE SCATTERING
    SCATTERING
    . **NUCLEAR SCATTERING**
    . . NEUTRON SCATTERING
    . . RESONANCE SCATTERING
  RT   ANGULAR DISTRIBUTION
    BACKSCATTERING
    COHERENT SCATTERING
    ELASTIC SCATTERING
    ELECTRON SCATTERING
    FORWARD SCATTERING
    INCOHERENT SCATTERING
    INELASTIC SCATTERING
    MANDELSTAM REPRESENTATION

NUCLEAR SHIELDING
  USE   RADIATION SHIELDING

NUCLEAR SPIN
  GS   SPIN
    . PARTICLE SPIN
    . . **NUCLEAR SPIN**
  RT   ELECTRON SPIN
    ENERGY LEVELS
    KONDO EFFECT
    MAGNETIC RESONANCE
    OVERHAUSER EFFECT
    QUANTUM NUMBERS
    QUANTUM THEORY

NUCLEAR STRUCTURE
  RT   ENERGY LEVELS
    EVEN-EVEN NUCLEI
    ODD-ODD NUCLEI

NUCLEAR TEST REACTORS
  USE   NUCLEAR RESEARCH AND TEST
      REACTORS

NUCLEAR TRANSFORMATIONS
  GS   NUCLEAR REACTIONS
    . **NUCLEAR TRANSFORMATIONS**

NUCLEAR VULNERABILITY
  GS   VULNERABILITY
    . **NUCLEAR VULNERABILITY**
  RT   PENETRATION
    RADIATION EFFECTS
    THERMONUCLEAR EXPLOSIONS

NUCLEAR WARFARE
  GS   WARFARE
    . **NUCLEAR WARFARE**
  RT   CIVIL DEFENSE
    HARDENING (SYSTEMS)
   ∞NUCLEAR ENERGY

NUCLEAR WARHEADS
  GS   WEAPONS
    . WARHEADS
    . . **NUCLEAR WARHEADS**

NUCLEAR WASTES
  USE   RADIOACTIVE WASTES

NUCLEAR WEAPONS
  GS   WEAPONS
    . **NUCLEAR WEAPONS**
    . . FISSION WEAPONS
  RT   BOMBS (ORDNANCE)
    EXPLOSIVES
    MISSILES
   ∞NUCLEAR ENERGY
    PROJECTILES
   ∞ROCKETS
    SPACE WEAPONS
    TORPEDOES
    WARHEADS

NUCLEAR WEAPONS-*(CONT.)*
    WEAPON SYSTEMS
    WEAPONS DELIVERY

NUCLEASE
  GS   ALIPHATIC COMPOUNDS
    . **NUCLEASE**
    ENZYMES
    . **NUCLEASE**
    ORGANIC COMPOUNDS
    . **NUCLEASE**
    PROTEINS
    . **NUCLEASE**

NUCLEATE BOILING
  GS   PHASE TRANSFORMATIONS
    . VAPORIZING
    . . BOILING
    . . . **NUCLEATE BOILING**
    . . . . LEIDENFROST PHENOMENON
  RT   FILM BOILING
    HEAT TRANSFER
    HEAT TRANSFER COEFFICIENTS
    NUCLEATION

NUCLEATION
  GS   **NUCLEATION**
    . CLOUD SEEDING
  RT   ACCUMULATIONS
    AITKEN NUCLEI
    CONDENSATION NUCLEI
    CONDENSING
    CRYSTAL GROWTH
    CRYSTALLIZATION
    DROP SIZE
   ∞FORMATION
    HEAT TREATMENT
    ICE NUCLEI
    INITIATION
    INOCULATION
    JET CONDENSERS
    NUCLEATE BOILING
   ∞NUCLEI
    RECRYSTALLIZATION
    SUPERCOOLING

∞ **NUCLEI**
  SN   *(USE OF A MORE SPECIFIC TERM IS*
      RECOMMENDED--CONSULT THE TERMS
      LISTED BELOW)
  RT   AITKEN NUCLEI
    CELLS (BIOLOGY)
    CHARGED PARTICLES
    CHROMOSOMES
    CONDENSATION NUCLEI
    ICE NUCLEI
    NUCLEATION
    NUCLEI (NUCLEAR PHYSICS)
    NUCLEOGENESIS
    ODD-ODD NUCLEI

NUCLEI (NUCLEAR PHYSICS)
  GS   GASES
    . IONIZED GASES
    . . CHARGED PARTICLES
    . . . **NUCLEI (NUCLEAR PHYSICS)**
    . . . . EVEN-EVEN NUCLEI
    . . . . HEAVY NUCLEI
    . . . . HYPERNUCLEI
    . . . . ODD-EVEN NUCLEI
    . . . . ODD-ODD NUCLEI
    PARTICLES
    . CHARGED PARTICLES
    . . ENERGETIC PARTICLES
    . . . **NUCLEI (NUCLEAR PHYSICS)**
  RT   ATOMS
    CORPUSCULAR RADIATION
    COSMIC RAYS
    ELEMENTARY PARTICLES
    IONS
    ISOTOPES
    NEUTRONS
    NUCLEAR ISOBARS
   ∞NUCLEI
    NUCLEONS
   ∞PHYSICS
    PROTONS

NUCLEIC ACIDS
  GS   ACIDS
    . **NUCLEIC ACIDS**
    . . DEOXYRIBONUCLEIC ACID
    . . GUANOSINES
    . . RIBONUCLEIC ACIDS
    . . URIDYLIC ACID

NUCLEOGENESIS
RT   GENETICS
      MUTATIONS
      ∞ NUCLEI

NUCLEON POTENTIAL
RT   NEUTRONS
      NUCLEAR PARTICLES
      NUCLEAR POTENTIAL
      ∞ POTENTIAL
      PROTONS
      ∞ RADIATION

NUCLEON-NUCLEON INTERACTIONS
GS   PARTICLE INTERACTIONS
     . ELEMENTARY PARTICLE
        INTERACTIONS
     . . **NUCLEON-NUCLEON INTERACTIONS**
RT   CHARGED PARTICLES
      ∞ INTERACTIONS

NUCLEON-NUCLEON SCATTERING
GS   SCATTERING
     . **NUCLEON-NUCLEON SCATTERING**
RT   NUCLEAR PARTICLES
      PARTICLE COLLISIONS
      POMERANCHUK THEOREM

NUCLEONICS
RT   ∞ ELECTRONICS
      ∞ NUCLEAR ENERGY
      NUCLEAR PHYSICS
      TECHNOLOGIES

NUCLEONS
GS   PARTICLES
     . ELEMENTARY PARTICLES
     . . **NUCLEONS**
     . NUCLEAR PARTICLES
     . . **NUCLEONS**
RT   ALPHA PARTICLES
      ANTINUCLEONS
      BARYONS
      CHARGED PARTICLES
      FAST NEUTRONS
      HYPERONS
      NEUTRONS
      NUCLEI (NUCLEAR PHYSICS)'
      PROTONS
      VECTOR DOMINANCE MODEL

NUCLEOPHILES
GS   ELECTRON ATTACHMENT
     . **NUCLEOPHILES**

NUCLEOSIDES
GS   ALIPHATIC COMPOUNDS
     . **NUCLEOSIDES**
     . . ADENINES
     . . ADENOSINES
     . . . ADENOSINE DIPHOSPHATE
     . . . ADENOSINE TRIPHOSPHATE
     CARBOHYDRATES
     . **NUCLEOSIDES**
     . . ADENOSINES
     . . . ADENOSINE DIPHOSPHATE
     . . . ADENOSINE TRIPHOSPHATE
RT   GLUCOSIDES
      RIBOSE
      THYMIDINE

NUCLEOSYNTHESIS
USE   NUCLEAR FUSION

NUCLEOTIDES
GS   ORGANIC COMPOUNDS
     . **NUCLEOTIDES**
     . . ADENINES
     . . ADENOSINES
     . . . ADENOSINE DIPHOSPHATE
     . . . ADENOSINE TRIPHOSPHATE
     . . CYCLIC AMP
     . . POLYNUCLEOTIDES
     . . PYRIDINE NUCLEOTIDES
     . . URIDYLIC ACID
     PROTEINS
     . **NUCLEOTIDES**
     . . ADENINES
     . . ADENOSINES
     . . . ADENOSINE DIPHOSPHATE
     . . . ADENOSINE TRIPHOSPHATE
     . . . CYCLIC AMP
     . . POLYNUCLEOTIDES
     . . PYRIDINE NUCLEOTIDES

NUCLEOTIDES-*(CONT.)*
     . . URIDYLIC ACID

NUCLIDES
GS   CHEMICAL ELEMENTS
     . **NUCLIDES**
     . . ISOTOPES
     . . . ALUMINUM ISOTOPES
     . . . . ALUMINUM 26
     . . . . ALUMINUM 27
     . . . ANTIMONY ISOTOPES
     . . . ARGON ISOTOPES
     . . . BARIUM ISOTOPES
     . . . BERYLLIUM ISOTOPES
     . . . . BERYLLIUM 7
     . . . . BERYLLIUM 9
     . . . . BERYLLIUM 10
     . . . BISMUTH ISOTOPES
     . . . BORON ISOTOPES
     . . . . BORON 10
     . . . BROMINE ISOTOPES
     . . . CADMIUM ISOTOPES
     . . . CALCIUM ISOTOPES
     . . . CARBON ISOTOPES
     . . . . CARBON 12
     . . . . CARBON 13
     . . . . CARBON 14
     . . . CERIUM ISOTOPES
     . . . . CERIUM 137
     . . . . CERIUM 144
     . . . CESIUM ISOTOPES
     . . . . CESIUM 133
     . . . . CESIUM 134
     . . . . CESIUM 137
     . . . . CESIUM 144
     . . . CESIUM VAPOR
     . . . CHROMIUM ISOTOPES
     . . . COBALT ISOTOPES
     . . . . COBALT 58
     . . . . COBALT 60
     . . . DYSPROSIUM ISOTOPES
     . . . ERBIUM ISOTOPES
     . . . EUROPIUM ISOTOPES
     . . . FLUORINE ISOTOPES
     . . . GADOLINIUM ISOTOPES
     . . . GALLIUM ISOTOPES
     . . . GERMANIUM ISOTOPES
     . . . HAFNIUM ISOTOPES
     . . . HELIUM ISOTOPES
     . . . HOLMIUM ISOTOPES
     . . . HYDROGEN ISOTOPES
     . . . . DEUTERIUM
     . . . . HYDROGEN 4
     . . . . TRITIUM
     . . . IODINE ISOTOPES
     . . . . IODINE 125
     . . . . IODINE 131
     . . . . IODINE 132
     . . . IRIDIUM ISOTOPES
     . . . IRON ISOTOPES
     . . . . IRON 57
     . . . . IRON 58
     . . . . IRON 59
     . . . KRYPTON ISOTOPES
     . . . . KRYPTON 85
     . . . LANTHANUM ISOTOPES
     . . . LEAD ISOTOPES
     . . . LITHIUM ISOTOPES
     . . . LUTETIUM
     . . . . LUTETIUM ISOTOPES
     . . . MANGANESE ISOTOPES
     . . . MERCURY ISOTOPES
     . . . NEODYMIUM ISOTOPES
     . . . NEON ISOTOPES
     . . . NICKEL ISOTOPES
     . . . NIOBIUM ISOTOPES
     . . . . NIOBIUM 95
     . . . NITROGEN ISOTOPES
     . . . . NITROGEN 15
     . . . . NITROGEN 16
     . . . OXYGEN ISOTOPES
     . . . . OXYGEN 18
     . . . PHOSPHORUS ISOTOPES
     . . . . PHOSPHORUS 32
     . . . PLATINUM ISOTOPES
     . . . POLONIUM ISOTOPES
     . . . . POLONIUM 208
     . . . . POLONIUM 209
     . . . . POLONIUM 210
     . . . POTASSIUM ISOTOPES
     . . . . POTASSIUM 38
     . . . . POTASSIUM 39
     . . . . POTASSIUM 40
     . . . PRASEODYMIUM ISOTOPES
     . . . PROMETHIUM ISOTOPES
     . . . PROTACTINIUM ISOTOPES

NUCLIDES-*(CONT.)*
     . . . RADIOACTIVE ISOTOPES
     . . . . ARSENIC ISOTOPES
     . . . . . ASTATINE ISOTOPES
     . . . . BERYLLIUM 7
     . . . . BERYLLIUM 9
     . . . . BERYLLIUM 10
     . . . . CARBON 14
     . . . . CERIUM 137
     . . . . CERIUM 144
     . . . . CESIUM 134
     . . . . CESIUM 137
     . . . . CESIUM 144
     . . . . COBALT 58
     . . . . COBALT 60
     . . . . GOLD ISOTOPES
     . . . . . GOLD 198
     . . . . INDIUM ISOTOPES
     . . . . IODINE 125
     . . . . IODINE 131
     . . . . IODINE 132
     . . . . IRON 59
     . . . . KRYPTON 85
     . . . . NIOBIUM 95
     . . . . NITROGEN 16
     . . . . PHOSPHORUS 32
     . . . . POLONIUM 208
     . . . . POLONIUM 209
     . . . . POLONIUM 210
     . . . . POTASSIUM 38
     . . . . POTASSIUM 40
     . . . . RUBIDIUM 86
     . . . . SODIUM 22
     . . . . SODIUM 24
     . . . . STRONTIUM 85
     . . . . STRONTIUM 88
     . . . . STRONTIUM 89
     . . . . STRONTIUM 90
     . . . . TRANSURANIUM ELEMENTS
     . . . . . AMERICIUM
     . . . . . . AMERICIUM ISOTOPES
     . . . . . BERKELIUM
     . . . . . CALIFORNIUM
     . . . . . . CALIFORNIUM ISOTOPES
     . . . . . CURIUM
     . . . . . . CURIUM ISOTOPES
     . . . . . EINSTEINIUM
     . . . . . FERMIUM
     . . . . . LAWRENCIUM
     . . . . . MENDELEVIUM
     . . . . . NEPTUNIUM
     . . . . . . NEPTUNIUM ISOTOPES
     . . . . . NOBELIUM
     . . . . . PLUTONIUM
     . . . . . . PLUTONIUM ISOTOPES
     . . . . . SERGENIUM
     . . . . TRITIUM
     . . . . URANIUM 232
     . . . . URANIUM 233
     . . . . URANIUM 238
     . . . . XENON 133
     . . . . XENON 135
     . . . . ZIRCONIUM 95
     . . . RADIUM ISOTOPES
     . . . . RADIUM 226
     . . . RADON ISOTOPES
     . . . RHODIUM ISOTOPES
     . . . RUBIDIUM ISOTOPES
     . . . . RUBIDIUM 86
     . . . RUTHENIUM ISOTOPES
     . . . SCANDIUM ISOTOPES
     . . SILVER ISOTOPES
     . . SODIUM ISOTOPES
     . . . SODIUM 22
     . . . SODIUM 24
     . . STRONTIUM ISOTOPES
     . . . STRONTIUM 85
     . . . STRONTIUM 87
     . . . STRONTIUM 89
     . . . STRONTIUM 90
     . . TANTALUM ISOTOPES
     . . TELLURIUM
     . . . TELLURIUM ISOTOPES
     . . TERBIUM ISOTOPES
     . . THORIUM ISOTOPES
     . . THULIUM ISOTOPES
     . . TIN ISOTOPES
     . . TITANIUM ISOTOPES
     . . URANIUM ISOTOPES
     . . . URANIUM 232
     . . . URANIUM 233
     . . . URANIUM 234
     . . . URANIUM 235
     . . . URANIUM 238
     . . VANADIUM ISOTOPES
     . . XENON ISOTOPES

NUCLIDES-*(CONT.)*
. . . . XENON 129
. . . . XENON 133
. . . . XENON 135
. . . YTTRIUM ISOTOPES
. . . ZINC ISOTOPES
. . . ZIRCONIUM ISOTOPES
. . . . ZIRCONIUM 95
RT   ISOTOPIC ENRICHMENT
        NUCLEAR ISOBARS
        PARTICLE MASS

NULL HYPOTHESIS
GS   HYPOTHESES
        . **NULL HYPOTHESIS**
RT   CONFIDENCE LIMITS
        DEGREES OF FREEDOM
        SIGNIFICANCE
        STATISTICAL TESTS

NULL ZONES
GS   REGIONS
        . **NULL ZONES**
RT   DIFFRACTION PATTERNS
        FIELD THEORY (PHYSICS)
        ∞ FORCE
        INTERFEROMETRY
        RADIATION DISTRIBUTION
        VERY LONG BASE INTERFEROMETRY

NUMBER THEORY
GS   **NUMBER THEORY**
        . ADDITION THEOREM
        . ARITHMETIC
        . . DOUBLE PRECISION ARITHMETIC
        . . FIXED POINT ARITHMETIC
        . . FLOATING POINT ARITHMETIC
        . CONGRUENCES
        . DIOPHANTINE EQUATION
        . DIVIDING (MATHEMATICS)
        . EXPONENTS
        . INDUCTION (MATHEMATICS)
        . INTEGERS
        . MULTIPLICATION
        . SUBTRACTION
RT   ADDITION
        COMBINATORIAL ANALYSIS
        DECIMALS
        DIGITS
        ∞ DIVISION
        ENUMERATION
        FIBONACCI NUMBERS
        FUNCTIONS (MATHEMATICS)
        ∞ INDUCTION
        INFINITY
        ∞ MATHEMATICS
        ∞ NUMBERS
        QUATERNIONS
        SUBGROUPS
        ∞ THEORIES
        UNIQUENESS THEOREM

∞ **NUMBERS**
SN   *(USE OF A MORE SPECIFIC TERM IS
        RECOMMENDED--CONSULT THE TERMS
        LISTED BELOW)*
RT   ALPHANUMERIC CHARACTERS
        BIOT NUMBER
        COMPLEX NUMBERS
        COUNTING
        DAMKOHLER NUMBER
        DECIMALS
        DIGITS
        DIMENSIONLESS NUMBERS
        DOUBLE PRECISION ARITHMETIC
        FIBONACCI NUMBERS
        INTEGERS
        NUMBER THEORY
        QUANTUM NUMBERS
        RANDOM NUMBERS
        REAL NUMBERS

NUMERICAL ANALYSIS
GS   ANALYSIS (MATHEMATICS)
        . **NUMERICAL ANALYSIS**
        . . APPROXIMATION
        . . . BORN APPROXIMATION
        . . . BORN-OPPENHEIMER
                APPROXIMATION
        . . . CHEBYSHEV APPROXIMATION
        . . . EDDINGTON APPROXIMATION
        . . . FINITE DIFFERENCE THEORY
        . . . FINITE ELEMENT METHOD
        . . . HARTREE APPROXIMATION
        . . . LEAST SQUARES METHOD

NUMERICAL ANALYSIS-*(CONT.)*
. . . MEAN SQUARE VALUES
. . . MILNE METHOD
. . . NEWTON-RAPHSON METHOD
. . . NUMERICAL DIFFERENTIATION
. . . OSEEN APPROXIMATION
. . . PADE APPROXIMATION
. . . PARTICLE IN CELL TECHNIQUE
. . . POHLHAUSEN METHOD
. . . RAYLEIGH-RITZ METHOD
. . . RELAXATION METHOD
            (MATHEMATICS)
. . . RITZ AVERAGING METHOD
. . . SOMMERFELD APPROXIMATION
. . COMPUTATIONAL CHEMISTRY
. . COMPUTATIONAL FLUID DYNAMICS
. . DIFFERENCE EQUATIONS
. . ERROR ANALYSIS
. . GRAEFF CALCULUS
. . INTERPOLATION
. . ITERATION
. . . CONJUGATE GRADIENT METHOD
. . MONTE CARLO METHOD
. . NOMOGRAPHS
. . NUMERICAL INTEGRATION
. . . RUNGE-KUTTA METHOD
RT   ADJOINTS
        ALGORITHMS
        ∞ ANALYZING
        ∞ APPLICATIONS OF MATHEMATICS
        ASYMPTOTES
        COMPUTATIONAL GRIDS
        COMPUTER PROGRAMMING
        CRANK-NICHOLSON METHOD
        DIFFERENTIAL EQUATIONS
        ISOPARAMETRIC FINITE ELEMENTS
        LINEAR PROGRAMMING
        MATHEMATICAL TABLES
        ∞ MATHEMATICS
        METHOD OF MOMENTS
        SIGNIFICANCE
        SPATIAL MARCHING
        TIME MARCHING
        TRAJECTORY ANALYSIS

NUMERICAL CONTROL
UF   COMPUTERIZED CONTROL
GS   AUTOMATIC CONTROL
        . **NUMERICAL CONTROL**
RT   ∞ AUTOMATION
        COMPUTER PROGRAMS
        ∞ CONTROL
        CONTROL SYSTEMS DESIGN
        DIGITAL COMMAND SYSTEMS
        DIGITAL TECHNIQUES
        ELECTRIC CONTROL
        INTERACTIVE CONTROL
        MACHINE TOOLS
        PRODUCTION ENGINEERING
        SEQUENTIAL CONTROL
        STANDARDIZATION

NUMERICAL DATA BASES
GS   INFORMATION SYSTEMS
        . **NUMERICAL DATA BASES**
RT   INFORMATION RETRIEVAL
        ON-LINE SYSTEMS

NUMERICAL DIFFERENTIATION
GS   ANALYSIS (MATHEMATICS)
        . NUMERICAL ANALYSIS
        . . APPROXIMATION
        . . . **NUMERICAL DIFFERENTIATION**
        . REAL VARIABLES
        . . **NUMERICAL DIFFERENTIATION**
RT   ALGORITHMS
        COMPUTER TECHNIQUES
        DIFFERENTIAL CALCULUS
        DIFFERENTIAL EQUATIONS
        ESTIMATING
        FUNCTIONS (MATHEMATICS)
        ∞ THEORIES

NUMERICAL FLOW VISUALIZATION
GS   FLOW VISUALIZATION
        . **NUMERICAL FLOW VISUALIZATION**
RT   FLOW DISTRIBUTION
        HYDRAULIC ANALOGIES

NUMERICAL INTEGRATION
UF   COWELL METHOD
GS   ANALYSIS (MATHEMATICS)
        . NUMERICAL ANALYSIS
        . . **NUMERICAL INTEGRATION**
        . . . RUNGE-KUTTA METHOD

NUMERICAL INTEGRATION-*(CONT.)*
. REAL VARIABLES
. . MEASURE AND INTEGRATION
. . . **NUMERICAL INTEGRATION**
. . . . RUNGE-KUTTA METHOD
RT   DIFFERENTIAL EQUATIONS
        DIGITAL INTEGRATORS
        INTEGRAL CALCULUS

NUMERICAL STABILITY
RT   APPROXIMATION
        BACKWARD DIFFERENCING
        DIFFERENCE EQUATIONS
        DIFFERENTIAL EQUATIONS
        STRANGE ATTRACTORS

NUMERICAL WEATHER FORECASTING
GS   FORECASTING
        . WEATHER FORECASTING
        . . **NUMERICAL WEATHER
                FORECASTING**
        METEOROLOGY
        . WEATHER FORECASTING
        . . **NUMERICAL WEATHER
                FORECASTING**
RT   ATMOSPHERIC MODELS
        COMPUTERIZED SIMULATION
        LONG RANGE WEATHER FORECASTING
        MATHEMATICAL MODELS
        STATISTICAL WEATHER FORECASTING

NUNATAKS
GS   LANDFORMS
        . ISLANDS
        . . **NUNATAKS**
RT   ARCTIC REGIONS
        ROCKS
        SEA ICE

NUSSELT NUMBER
GS   RATIOS
        . DIMENSIONLESS NUMBERS
        . . **NUSSELT NUMBER**
RT   CONVECTIVE HEAT TRANSFER
        HEAT TRANSFER
        PRANDTL NUMBER
        SCHMIDT NUMBER

NUTATION
UF   NUTATIONAL OSCILLATION
RT   ACTUATION
        DISPLACEMENT
        ∞ DYNAMICS
        KINEMATICS
        LIBRATION
        ∞ MOTION
        PERTURBATION
        POLAR WANDERING (GEOLOGY)
        PRECESSION
        ROTATION
        VIBRATION

NUTATION DAMPERS
RT   CONTROL MOMENT GYROSCOPES
        ∞ DAMPERS
        OSCILLATION DAMPERS
        SPACECRAFT STABILITY

NUTATIONAL OSCILLATION
USE   NUTATION

∞ **NUTRIENTS**
SN   *(USE OF A MORE SPECIFIC TERM IS
        RECOMMENDED--CONSULT THE TERMS
        LISTED BELOW)*
RT   AQUICULTURE
        CALORIC REQUIREMENTS
        CARBOHYDRATES
        EUTROPHICATION
        FATS
        FATTY ACIDS
        FISHES
        LIFE SUPPORT SYSTEMS
        LIPID METABOLISM
        LIPIDS
        MINERALS
        NUTRITION
        NUTRITIONAL REQUIREMENTS
        PROTEINS
        TRACE ELEMENTS
        VITAMINS

NUTRITION
RT   BIOCHEMISTRY

NUTRITION-(CONT.)
  BROTHS
  CALORIC REQUIREMENTS
  DIETS
  ∞FOOD
  METABOLISM
  NITROGEN METABOLISM
  ∞NUTRIENTS
  NUTRITIONAL REQUIREMENTS
  PHYSIOLOGY
  SPACE FLIGHT FEEDING

NUTRITIONAL REQUIREMENTS
  GS **NUTRITIONAL REQUIREMENTS**
    . CALORIC REQUIREMENTS
  RT ATROPHY
    ∞NUTRIENTS
    NUTRITION
    SPACE FLIGHT FEEDING
    SYNTHETIC FOOD

NUTS (FASTENERS)
  GS FASTENERS
    . **NUTS (FASTENERS)**
  RT ANCHORS (FASTENERS)
    BOLTS
    HOLDERS
    SCREWS
    THREADS

NUTS (FRUITS)
  RT ANGIOSPERMS
    ORCHARDS
    SEEDS

NYLON (TRADEMARK)
  GS FIBERS
    . SYNTHETIC FIBERS
    . . **NYLON (TRADEMARK)**
  RT POLYMERIC FILMS
    ∞POLYMERS

NYLON RESINS
  USE POLYAMIDE RESINS

NYQUIST DIAGRAM
  GS DIAGRAMS
    . **NYQUIST DIAGRAM**
  RT CONTROL STABILITY
    TRANSFER FUNCTIONS

NYQUIST FREQUENCIES
  GS FREQUENCIES
    . **NYQUIST FREQUENCIES**
  RT LINEAR RECEIVERS

NYSTAGMUS
  GS EYE MOVEMENTS
    . **NYSTAGMUS**
  RT ELECTRONYSTAGMOGRAPHY

O

O RING SEALS
  GS SEALS (STOPPERS)
    . **O RING SEALS**
  RT GASKETS
    GLANDS (SEALS)
    HERMETIC SEALS
    LABYRINTH SEALS
    ∞RINGS

O STARS
  GS CELESTIAL BODIES
    . STARS
    . . HOT STARS
    . . . **O STARS**
  RT BLUE STARS
    WOLF-RAYET STARS

OAK RIDGE ISOCHRONOUS CYCLOTRON
  UF ORIC CYCLOTRON
  GS PARTICLE ACCELERATORS
    . CYCLOTRONS
    . . **OAK RIDGE ISOCHRONOUS
        CYCLOTRON**

OAO
  UF ORBITING ASTRONOMICAL
      OBSERVATORY
    S-18 SATELLITE

OAO-(CONT.)
  GS OBSERVATORIES
    . ASTRONOMICAL OBSERVATORIES
    . . ASTRONOMICAL SATELLITES
    . . . **OAO**
    . . . . OAO 1
    . . . . OAO 2
    . . . . OAO 3
  RT AGENA B ROCKET VEHICLE
    ATLAS LAUNCH VEHICLES
    HEAO
    HEAO 1
    HEAO 2
    HEAO 3
    MANNED ORBITAL TELESCOPES

OAO 1
  UF OAO-A
  GS OBSERVATORIES
    . ASTRONOMICAL OBSERVATORIES
    . . ASTRONOMICAL SATELLITES
    . . . OAO
    . . . . **OAO 1**
  RT ATLAS CENTAUR LAUNCH VEHICLE

OAO 2
  UF OAO-A2
  GS OBSERVATORIES
    . ASTRONOMICAL OBSERVATORIES
    . . ASTRONOMICAL SATELLITES
    . . . OAO
    . . . . **OAO 2**
  RT ATLAS CENTAUR LAUNCH VEHICLE

OAO 3
  UF COPERNICUS SPACECRAFT
    OAO-C
  GS OBSERVATORIES
    . ASTRONOMICAL OBSERVATORIES
    . . ASTRONOMICAL SATELLITES
    . . . OAO
    . . . . **OAO 3**
  RT ATLAS CENTAUR LAUNCH VEHICLE

OAO-A
  USE OAO 1

OAO-A2
  USE OAO 2

OAO-C
  USE OAO 3

OASES
  RT AQUIFERS
    ARID LANDS
    DESERTIFICATION
    DESERTS
    POTABLE WATER
    SPRINGS (WATER)
    VEGETATION
    WELLS

OATS
  GS FARM CROPS
    . GRAINS (FOOD)
    . . **OATS**
    PLANTS (BOTANY)
    . **OATS**
  RT AGRICULTURE
    BOTANY
    CROP GROWTH
    CROP VIGOR
    CURING
    EARTH RESOURCES
    ∞FOOD
    GRASSES
    IRRIGATION
    SEEDS

†

OBESITY
  RT BODY FLUIDS
    BODY MEASUREMENT (BIOLOGY)
    BODY SIZE (BIOLOGY)
    BODY VOLUME (BIOLOGY)
    BODY WEIGHT
    METABOLISM

OBJECT PROGRAMS
  GS COMPUTER PROGRAMS
    . **OBJECT PROGRAMS**

OBLATE SPHEROIDS
  GS GEOMETRY

OBLATE SPHEROIDS-(CONT.)
    . EUCLIDEAN GEOMETRY
    . . ANALYTIC GEOMETRY
    . . . SPHEROIDS
    . . . . **OBLATE SPHEROIDS**
  RT ELLIPTICITY
    FINENESS RATIO
    FLATTENING
    GEODESY
    GEOIDS
    PROLATE SPHEROIDS
    SHAPES
    SOLAR OBLATENESS

OBLIQUE COORDINATES
  GS COORDINATES
    . **OBLIQUE COORDINATES**
  RT CARTESIAN COORDINATES

OBLIQUE SHOCK WAVES
  GS ELASTIC WAVES
    . SHOCK WAVES
    . . **OBLIQUE SHOCK WAVES**
  RT MAGNETOHYDRODYNAMIC WAVES
    NORMAL SHOCK WAVES
    SHOCK LAYERS
    SUPERSONIC COMPRESSORS

OBLIQUE WINGS
  GS AIRFOILS
    . WINGS
    . . **OBLIQUE WINGS**
  RT AERODYNAMIC CONFIGURATIONS
    AIRCRAFT PARTS
    AIRCRAFT STRUCTURES
    DRONE AIRCRAFT
    PILOTLESS AIRCRAFT
    REMOTELY PILOTED VEHICLES
    WING PLANFORMS

OBLIQUENESS
  RT ANGLES (GEOMETRY)
    INCIDENT RADIATION

OBSCURATION
  USE OCCULTATION

OBSERVABILITY (SYSTEMS)
  RT BOUNDARY VALUE PROBLEMS
    CONTROL THEORY
    DEPENDENT VARIABLES
    FEEDBACK CONTROL
    INDEPENDENT VARIABLES
    OBSERVATION
    PARAMETER IDENTIFICATION
    STATE VECTORS
    SYSTEM IDENTIFICATION
    ∞SYSTEMS
    SYSTEMS ANALYSIS
    SYSTEMS ENGINEERING

OBSERVATION
  GS **OBSERVATION**
    . EARTH OBSERVATIONS (FROM SPACE)
    . . SATELLITE OBSERVATION
    . . SPOT (FRENCH SATELLITE)
    . SPACE OBSERVATIONS (FROM EARTH)
    . VISUAL OBSERVATION
  RT COUNTING
    DATA ACQUISITION
    DETECTION
    EVALUATION
    EXAMINATION
    FOREST FIRE DETECTION
    OBSERVABILITY (SYSTEMS)
    ∞PERFORMANCE
    RADIO OBSERVATION
    RECONNAISSANCE

OBSERVATION AIRCRAFT
  GS **OBSERVATION AIRCRAFT**
    . A-2 AIRCRAFT
    . BREGUET 1150 AIRCRAFT
    . CESSNA L-19 AIRCRAFT
    . CL-84 AIRCRAFT
    . E-2 AIRCRAFT
    . F-5 AIRCRAFT
    . G-91 AIRCRAFT
    . G-95/4 AIRCRAFT
    . MIRAGE AIRCRAFT
    . . MIRAGE 3 AIRCRAFT
    . OH-4 HELICOPTER
    . OH-5 HELICOPTER
    . OH-6 HELICOPTER
    . OV-1 AIRCRAFT

OBSERVATION AIRCRAFT-*(CONT.)*
```
      . OV-10 AIRCRAFT
      . P-1127 AIRCRAFT
      . P-1154 AIRCRAFT
      . RB-50 AIRCRAFT
      . RF-4 AIRCRAFT
      . TSR-2 AIRCRAFT
      . U-2 AIRCRAFT
   RT  ∞AIRCRAFT
      ANTISUBMARINE WARFARE AIRCRAFT
      ARC CLOUDS
      BALLOONS
      FLYING PLATFORMS
      GLIDERS
      HS-801 AIRCRAFT
      LIGHT AIRCRAFT
      ∞MILITARY AIRCRAFT
      RECONNAISSANCE AIRCRAFT
      TERRAIN FOLLOWING AIRCRAFT
      UTILITY AIRCRAFT
      WEATHER RECONNAISSANCE AIRCRAFT
```

OBSERVATORIES
```
   GS  OBSERVATORIES
      . AOSO
      . ASTRONOMICAL OBSERVATORIES
      .. ASTRONOMICAL SATELLITES
      ... ASTRONOMICAL NETHERLANDS
               SATELLITE
      ... HEAO
      .... HEAO 1
      .... HEAO 2
      .... HEAO 3
      ... HUBBLE SPACE TELESCOPE
      ... INFRARED ASTRONOMY SATELLITE
      ... IUE
      ... MAGELLAN MISSION
      ... OAO
      .... OAO 1
      .... OAO 2
      .... OAO 3
      ... OSO
      .... OSO-1
      .... OSO-2
           OSO-3
      .... OSO-4
      .... OSO-5
      .... OSO-6
      .... OSO-7
      .... OSO-8
      ... QUASAT
      ... SAS
      .... SAS-1
      .... SAS-2
      .... SAS-3
      ... SPACE INFRARED TELESCOPE
              FACILITY
      ... SPARTAN SATELLITES
      .. GAMMA RAY OBSERVATORY
      . ROSAT MISSION
      . GEOPHYSICAL OBSERVATORIES
      .. OGO
      ... EGO
      ... OGO-A
      ... OGO-3
      ... OGO-5
      ... POGO
      .... OGO-C
      .... OGO-4
      .... OGO-6
      .. OSO
      ... OSO-C
      ... OSO-1
      ... OSO-2
      ... OSO-3
      ... OSO-4
      ... OSO-5
      ... OSO-6
      ... OSO-7
      ... OSO-8
      . JODRELL BANK OBSERVATORY
      . LUNAR OBSERVATORIES
      . SOLAR OBSERVATORIES
      .. OSO
      ... OSO-C
      ... OSO-1
      ... OSO-2
      ... OSO-3
      ... OSO-4
      ... OSO-5
      ... OSO-6
      ... OSO-7
      ... OSO-8
   RT  ARTIFICIAL SATELLITES
```

OBSIDIAN
```
   GS  ROCKS
      . IGNEOUS ROCKS
      .. OBSIDIAN
      ... MOLDAVITE
   RT  GLASS
      MINERALS
      POWDER (PARTICLES)
      PUMICE
      SILICON DIOXIDE
      SOILS
```

OBSIDIAN GLASS
```
   GS  GLASS
      . OBSIDIAN GLASS
```

OBSTACLE AVOIDANCE
```
   RT  AIRCRAFT MANEUVERS
      EVASIVE ACTIONS
      TACTICS
      VULNERABILITY
```

OBSTACLES
```
   USE  BARRIERS
```

OBSTRUCTING
```
   USE  BLOCKING
```

OCCIPITAL LOBES
```
   GS  ANATOMY
      . HEAD (ANATOMY)
      .. OCCIPITAL LOBES
```

OCCLUSION
```
   RT  DEGASSING
      GAS-METAL INTERACTIONS
      GAS-SOLID INTERFACES
      SOLIDIFICATION
```

OCCULTATION
```
   UF  OBSCURATION
   GS  OCCULTATION
      . LUNAR OCCULTATION
      .. SOLAR ECLIPSES
      . RADIO OCCULTATION
      . STELLAR OCCULTATION
   RT  ∞CONJUNCTION
      ECLIPSES
      EXTINGUISHING
      ∞TRANSIT
```

OCCUPATION
```
   RT  INDUSTRIAL SAFETY
      PERSONNEL
      WORK
```

OCCURRENCES
```
   RT  EVENTS
```

OCEAN BOTTOM
```
   RT  BEDS (GEOLOGY)
      CONTINENTAL SHELVES
      CORE SAMPLING
      CRATONS
      GEOLOGY
      MARINE CHEMISTRY
      MUD
      OCEANOGRAPHY
      SEAMOUNTS
      SEDIMENTS
      SLUDGE
      UNDERWATER RESOURCES
```

OCEAN COLOR SCANNER
```
   RT  CHLOROPHYLLS
      COASTAL WATER
      COLORIMETRY
      MULTISPECTRAL BAND SCANNERS
      OCEAN DATA ACQUISITIONS SYSTEMS
      OCEANOGRAPHIC PARAMETERS
      OCEANOGRAPHY
      PHOTOMAPPING
      REMOTE SENSORS
      WATER COLOR
```

OCEAN CURRENTS
```
   GS  CIRCULATION
      . WATER CIRCULATION
      .. WATER CURRENTS
      ... OCEAN CURRENTS
      .... COASTAL CURRENTS
      .... EL NINO
      .... GULF STREAM
      .... LOMONOSOV CURRENT
```

OCEAN CURRENTS-*(CONT.)*
```
   RT  CORE SAMPLING
      ∞CURRENTS
      FLUID FLOW
      FRONTAL WAVES
      GYRES
      HYDROGRAPHY
      LITTORAL DRIFT
      LITTORAL TRANSPORT
      OCEAN DYNAMICS
      OCEANOGRAPHY
      OCEANS
      PRESSURE ICE
      SALINITY
      SPITSBERGEN (NORWAY)
      TIDAL WAVES
      TIDE POWERED GENERATORS
      TIDE POWERED MACHINES
      TIDEPOWER
      TIDES
      TOPEX
      UPWELLING WATER
      WATERWAVE ENERGY CONVERSION
      WATERWAVE POWERED MACHINES
```

OCEAN DATA ACQUISITIONS SYSTEMS
```
   UF  OCEAN DATA PLATFORMS
      OCEAN DATA STATIONS
      ODAS
   RT  AUTOMATIC WEATHER STATIONS
      BUOYS
      ∞DATA
      DATA ACQUISITION
      GROUND STATIONS
      INSTRUMENT PACKAGES
      METEOROLOGICAL PARAMETERS
      OCEAN COLOR SCANNER
      OCEANOGRAPHIC PARAMETERS
      SHIPS
      TRANSOCEANIC SYSTEMS
      UNDERWATER RESEARCH
         LABORATORIES
      WEATHER STATIONS
```

OCEAN DATA PLATFORMS
```
   USE  OCEAN DATA ACQUISITIONS SYSTEMS
```

OCEAN DATA STATIONS
```
   USE  OCEAN DATA ACQUISITIONS SYSTEMS
```

OCEAN DYNAMICS
```
   RT  AIR WATER INTERACTIONS
      DYNAMIC CHARACTERISTICS
      ∞DYNAMICS
      FLUID DYNAMICS
      HYDRODYNAMICS
      OCEAN CURRENTS
      OCEAN MODELS
      OCEAN SURFACE
      OCEANOGRAPHY
      WATER WAVES
```

OCEAN MODELS
```
   GS  MODELS
      . OCEAN MODELS
   RT  AIR WATER INTERACTIONS
      ATMOSPHERIC MODELS
      DYNAMIC MODELS
      MARINE ENVIRONMENTS
      MATHEMATICAL MODELS
      OCEAN DYNAMICS
      OCEANOGRAPHY
      SARGASSO SEA
      SEA ROUGHNESS
      SEA STATES
```

OCEAN SURFACE
```
   RT  EARTH SURFACE
      FLUID FLOW
      HYDROGRAPHY
      OCEAN DYNAMICS
      OCEANOGRAPHIC PARAMETERS
      OCEANOGRAPHY
      SARGASSO SEA
      SEA LEVEL
      SEA ROUGHNESS
      SEA STATES
      SEA SURFACE TEMPERATURE
      SEA TRUTH
      SEA WATER
      STORM SURGES
      ∞SURFACES
      TIDAL WAVES
      TIDE POWERED GENERATORS
      TIDE POWERED MACHINES
```

OCEAN SURFACE-(CONT.)
    TIDEPOWER
    TIDES
    TOPEX
    WATERWAVE ENERGY CONVERSION
    WATERWAVE POWERED MACHINES

OCEAN TEMPERATURE
GS   OCEANOGRAPHIC PARAMETERS
    . **OCEAN TEMPERATURE**
    TEMPERATURE
    TEMPERATURE
    . WATER TEMPERATURE
    . . **OCEAN TEMPERATURE**
    . . . SEA SURFACE TEMPERATURE
RT   EL NINO
    OCEANOGRAPHY
    OCEANS
    OFFSHORE ENERGY SOURCES
    SEA STATES
    SEA TRUTH
    SEA WATER
    SEAS
    SOLAR SEA POWER PLANTS
    SURFACE TEMPERATURE
    TEMPERATURE DISTRIBUTION
    TEMPERATURE GRADIENTS
    THERMAL POLLUTION

OCEAN THERMAL ENERGY CONVERSION
GS   ENERGY CONVERSION
    . **OCEAN THERMAL ENERGY
      CONVERSION**
RT   ∞CONVERSION
    ∞ENERGY SOURCES
    GEOTHERMAL ENERGY CONVERSION
    GEOTHERMAL TECHNOLOGY
    SOLAR SEA POWER PLANTS
    TEMPERATURE

OCEANOGRAPHIC PARAMETERS
GS   **OCEANOGRAPHIC PARAMETERS**
    . OCEAN TEMPERATURE
RT   ATMOSPHERIC & OCEANOGRAPHIC
      INFORM SYS
    INTEGRATED GLOBAL OCEAN STATION
      SYSTEMS
    METEOROLOGICAL PARAMETERS
    OCEAN COLOR SCANNER
    OCEAN DATA ACQUISITIONS SYSTEMS
    OCEAN SURFACE
    SALINITY
    SEA STATES

OCEANOGRAPHY
RT   ARTIFICIAL HARBORS
    ATMOSPHERIC & OCEANOGRAPHIC
      INFORM SYS
    BATHYMETERS
    BAY ICE
    BREAKWATERS
    COASTAL CURRENTS
    CORE SAMPLING
    DEEP SCATTERING LAYERS
    DEEPWATER TERMINALS
    EARTH & OCEAN PHYSICS
      APPLICATIONS PROGRAM
    EARTH PLANETARY STRUCTURE
    EARTH RESOURCES
    ENVIRONMENTAL MONITORING
    EROS (SATELLITES)
    ESTUARIES
    FIORDS
    FRONTAL WAVES
    GARP ATLANTIC TROPICAL EXPERIMENT
    GEOGRAPHY
    GEOLOGY
    GEOPHYSICS
    GYRES
    HARBORS
    HYDROCLIMATOLOGY
    HYDROGRAPHY
    HYDROLOGY
    ICE FLOES
    ICE MAPPING
    ISTHMUSES
    LANDSAT SATELLITES
    MARINE BIOLOGY
    MARINE ENVIRONMENTS
    MARINE METEOROLOGY
    MARINE RESOURCES
    MARINE TECHNOLOGY
    MARSHLANDS
    METEOROLOGY
    OCEAN BOTTOM
    OCEAN COLOR SCANNER

OCEANOGRAPHY-(CONT.)
    OCEAN CURRENTS
    OCEAN DYNAMICS
    OCEAN MODELS
    OCEAN SURFACE
    OCEAN TEMPERATURE
    OCEANS
    OFFSHORE DOCKING
    OFFSHORE PLATFORMS
    OIL SLICKS
    PELAGIC ZONE
∞ PHYSICAL SCIENCES
    RED TIDE
    REEFS
    SARGASSO SEA
∞ SCIENCE
    SEA GRASSES
    SEA ICE
    SEA LEVEL
    SEA ROUGHNESS
    SEA STATES
    SEA SURFACE TEMPERATURE
    SEA WATER
    SEAS
    SEASAT PROGRAM
    SEASAT SATELLITES
    SEASAT 1
    SEASAT-B SATELLITE
    SEAWEEDS
    SHALLOW WATER
    SHIPYARDS
    SHOALS
    SHORELINES
    STORM SURGES
    TANKER TERMINALS
    THERMOCLINES
    TIDAL WAVES
    TIDE POWERED GENERATORS
    TIDEPOWER
    TIDES
    TOPEX
    TOPOGRAPHY
    UNDERWATER RESEARCH
      LABORATORIES
    UNDERWATER RESOURCES
    WATER CIRCULATION
    WATER CURRENTS
    WATERFOWL
    WATERWAVE ENERGY
    WATERWAVE ENERGY CONVERSION
    WETLANDS

OCEANS
GS   **OCEANS**
    . ARCTIC OCEAN
    . ATLANTIC OCEAN
    . INDIAN OCEAN
    . PACIFIC OCEAN
RT   COASTAL CURRENTS
    COASTAL WATER
    COASTS
    EARTH HYDROSPHERE
    GEOGRAPHY
    KEYS (ISLANDS)
    MARINE RESOURCES
    NEARSHORE WATER
    OCEAN CURRENTS
    OCEAN TEMPERATURE
    OCEANOGRAPHY
    SEAS
    SEAWEEDS
    SHALLOW WATER
    SHOALS
    SHORELINES
    SOUNDS (TOPOGRAPHIC FEATURES)
    THERMAL POLLUTION
    TIDAL FLATS
    TIDE POWERED GENERATORS
    WATER COLOR
    WATER DEPTH
    WATER RESOURCES
    WATERWAVE ENERGY CONVERSION

OCTAHEDRAL RESEARCH SATELLITES
USE   ENVIRONMENTAL RESEARCH
      SATELLITES

OCTAHEDRITE
USE   ANATASE

OCTAHEDRONS
GS   GEOMETRY
    . EUCLIDEAN GEOMETRY
    . . POLYHEDRONS
    . . . **OCTAHEDRONS**

∞ OCTANE
SN   *(USE OF A MORE SPECIFIC TERM IS
      RECOMMENDED--CONSULT THE TERMS
      LISTED BELOW)*
RT   ANTIKNOCK ADDITIVES
    OCTANES

OCTANE NUMBER
RT   GASOLINE

OCTANES
SN   (ACYCLIC HYDROCARBONS)
GS   ALIPHATIC COMPOUNDS
    . ALKANES
    . . **OCTANES**
    HYDROCARBONS
    . ALKANES
    . . **OCTANES**
RT   ANTIKNOCK ADDITIVES
∞ OCTANE

OCTAVES
GS   RANGE (EXTREMES)
    . FREQUENCY RANGES
    . . **OCTAVES**
RT   ACOUSTICS
    MUSIC

OCTETS
GS   VALENCE
    . **OCTETS**
RT   ATOMIC STRUCTURE
    CHEMICAL BONDS

OCTOATES
GS   ALIPHATIC COMPOUNDS
    . **OCTOATES**
    ESTERS
    . **OCTOATES**

OCTOL (EXPLOSIVE)
GS   EXPLOSIVES
    . **OCTOL (EXPLOSIVE)**

OCTOPUSES
GS   ANIMALS
    . INVERTEBRATES
    . . ARTHROPODS
    . . . CEPHALOPODS
    . . . . **OCTOPUSES**

OCULAR CIRCULATION
GS   CIRCULATION
    . BLOOD CIRCULATION
    . . **OCULAR CIRCULATION**

OCULOGRAVIC ILLUSIONS
GS   PSYCHOLOGICAL EFFECTS
    . ILLUSIONS
    . . **OCULOGRAVIC ILLUSIONS**
RT   GRAVIRECEPTORS
    OTOLITH ORGANS
    VERTICAL PERCEPTION

OCULOMETERS
GS   MEASURING INSTRUMENTS
    . OPTICAL MEASURING INSTRUMENTS
    . . **OCULOMETERS**
RT   EYE MOVEMENTS
    ∞INSTRUMENTS
    OPTICAL TRACKING

OCULOMOTOR NERVES
GS   ANATOMY
    . SENSE ORGANS
    . . EYE (ANATOMY)
    . . . **OCULOMOTOR NERVES**
RT   VISION

ODAS
USE   OCEAN DATA ACQUISITIONS SYSTEMS

ODD-EVEN NUCLEI
GS   GASES
    . IONIZED GASES
    . . CHARGED PARTICLES
    . . . NUCLEI (NUCLEAR PHYSICS)
    . . . . **ODD-EVEN NUCLEI**
RT   EVEN-EVEN NUCLEI
    ODD-ODD NUCLEI

ODD-ODD NUCLEI
GS   GASES
    . IONIZED GASES

ODD-ODD NUCLEI-*(CONT.)*
. . CHARGED PARTICLES
. . . NUCLEI (NUCLEAR PHYSICS)
. . . . **ODD-ODD NUCLEI**
RT   EVEN-EVEN NUCLEI
NUCLEAR STRUCTURE
∞NUCLEI
ODD-EVEN NUCLEI

ODESSA METEORITE
GS   CELESTIAL BODIES
. METEORITES
. . IRON METEORITES
. . . **ODESSA METEORITE**

ODORS
RT   AIR POLLUTION
COMBUSTION PRODUCTS
EXHAUST GASES
GASES

OFF-ON CONTROL
UF   BANG-BANG CONTROL
GS   AUTOMATIC CONTROL
. **OFF-ON CONTROL**
RT   ∞CONTROL
CONTROL EQUIPMENT
CONTROL THEORY
PROPORTIONAL CONTROL
SERVOCONTROL
SOLENOID VALVES

OFFGASSING
RT   DEGASSING
VACUUM
VACUUM EFFECTS

OFFICE OF SPACE & TERRESTR APPLIC PAYLOADS
USE OSTA-1 PAYLOAD
OSTA-2 PAYLOAD

OFFSHORE DOCKING
RT   ARTIFICIAL HARBORS
CARGO SHIPS
DEEPWATER TERMINALS
MARINE TECHNOLOGY
MARINE TRANSPORTATION
OCEANOGRAPHY
SHIP TERMINALS
TANKER SHIPS
TANKER TERMINALS
∞TANKERS
TERMINAL FACILITIES
TRANSPORTATION

OFFSHORE ENERGY SOURCES
RT   CRUDE OIL
DEEPWATER TERMINALS
DRILLING
ENERGY TECHNOLOGY
MARINE TECHNOLOGY
OCEAN TEMPERATURE
OIL EXPLORATION
OIL FIELDS
SEA BREEZE
SEEPAGE

OFFSHORE PLATFORMS
RT   ARTIFICIAL HARBORS
CARGO SHIPS
DEEPWATER TERMINALS
MARINE TECHNOLOGY
OCEANOGRAPHY
∞PLATFORMS
TANKER SHIPS
TANKER TERMINALS
∞TANKERS
TERMINAL FACILITIES
TRANSPORTATION

OFFSHORE REACTOR SITES
GS   SITES
. **OFFSHORE REACTOR SITES**
RT   REACTOR DESIGN
REACTOR SAFETY
REACTOR TECHNOLOGY
REMOTE REGIONS

OFT
USE SPACE TRANSPORTATION SYSTEM FLIGHTS

OFT 1
USE SPACE TRANSPORTATION SYSTEM 1 FLIGHT

OFT 2
USE SPACE TRANSPORTATION SYSTEM 2 FLIGHT

OFT 3
USE SPACE TRANSPORTATION SYSTEM 3 FLIGHT

OFT 4
USE SPACE TRANSPORTATION SYSTEM 4 FLIGHT

OGEE SHAPE
GS   SHAPES
. **OGEE SHAPE**
RT   VARIABLE SWEEP WINGS

OGEE WINGS
USE VARIABLE SWEEP WINGS

OGIVES
RT   BODIES OF REVOLUTION
ELLIPSOIDS
FAIRINGS
NOSE CONES
SPHERES
STREAMLINED BODIES
SYMMETRICAL BODIES

OGO
UF   ORBITING GEOPHYSICAL OBSERVATORY
GS   OBSERVATORIES
. GEOPHYSICAL OBSERVATORIES
. . **OGO**
. . . EGO
. . . OGO-A
. . . OGO-3
. . . OGO-5
. . . POGO
. . . . OGO-C
. . . . OGO-4
. . . . OGO-6
SATELLITES
. ARTIFICIAL SATELLITES
. . GEOPHYSICAL SATELLITES
. . . **OGO**
. . . . EGO
. . . . OGO-A
. . . . OGO-3
. . . . OGO-5
. . . . POGO
. . . . . OGO-C
. . . . . OGO-4
. . . . . OGO-6
RT   GAMMA RAY OBSERVATORY

OGO-A
UF   S-49 SATELLITE
GS   OBSERVATORIES
. GEOPHYSICAL OBSERVATORIES
. . OGO
. . . **OGO-A**
SATELLITES
. ARTIFICIAL SATELLITES
. . GEOPHYSICAL SATELLITES
. . . OGO
. . . . **OGO-A**
RT   ATLAS AGENA LAUNCH VEHICLES

OGO-B
USE OGO-3

OGO-C
UF   S-50 SATELLITE
GS   OBSERVATORIES
. GEOPHYSICAL OBSERVATORIES
. . OGO
. . . POGO
. . . . **OGO-C**
SATELLITES
. ARTIFICIAL SATELLITES
. . GEOPHYSICAL SATELLITES
. . . OGO
. . . . POGO
. . . . . **OGO-C**

OGO-D
USE OGO-4

OGO-E
USE OGO-5

OGO-F
USE OGO-6

OGO-3
UF   OGO-B
GS   OBSERVATORIES
. GEOPHYSICAL OBSERVATORIES
. . OGO
. . . **OGO-3**
SATELLITES
. ARTIFICIAL SATELLITES
. . GEOPHYSICAL SATELLITES
. . . OGO
. . . . **OGO-3**
RT   THOR AGENA LAUNCH VEHICLE

OGO-4
UF   OGO-D
GS   OBSERVATORIES
. GEOPHYSICAL OBSERVATORIES
. . OGO
. . . POGO
. . . . **OGO-4**
SATELLITES
. ARTIFICIAL SATELLITES
. . GEOPHYSICAL SATELLITES
. . . OGO
. . . . POGO
. . . . . **OGO-4**
RT   GEODESY

OGO-5
UF   OGO-E
GS   OBSERVATORIES
. GEOPHYSICAL OBSERVATORIES
. . OGO
. . . **OGO-5**
SATELLITES
. ARTIFICIAL SATELLITES
. . GEOPHYSICAL SATELLITES
. . . OGO
. . . . **OGO-5**
RT   GEODESY

OGO-6
UF   OGO-F
GS   OBSERVATORIES
. GEOPHYSICAL OBSERVATORIES
. . OGO
. . . POGO
. . . . **OGO-6**
SATELLITES
. ARTIFICIAL SATELLITES
. . GEOPHYSICAL SATELLITES
. . . OGO
. . . . POGO
. . . . . **OGO-6**

OH-4 HELICOPTER
UF   HO-4 HELICOPTER
GS   BELL AIRCRAFT
. **OH-4 HELICOPTER**
LIGHT AIRCRAFT
. **OH-4 HELICOPTER**
OBSERVATION AIRCRAFT
. **OH-4 HELICOPTER**
V/STOL AIRCRAFT
. ROTARY WING AIRCRAFT
. . HELICOPTERS
. . . MILITARY HELICOPTERS
. . . . **OH-4 HELICOPTER**

OH-5 HELICOPTER
UF   FH-1100 HELICOPTER
HO-5 HELICOPTER
GS   FAIRCHILD-HILLER AIRCRAFT
. **OH-5 HELICOPTER**
HILLER AIRCRAFT
. **OH-5 HELICOPTER**
LIGHT AIRCRAFT
. **OH-5 HELICOPTER**
OBSERVATION AIRCRAFT
. **OH-5 HELICOPTER**
PASSENGER AIRCRAFT
. **OH-5 HELICOPTER**
V/STOL AIRCRAFT
. ROTARY WING AIRCRAFT
. . HELICOPTERS
. . . MILITARY HELICOPTERS
. . . . **OH-5 HELICOPTER**
RT   RIGID ROTOR HELICOPTERS

OH-6 HELICOPTER
- UF   HO-6 HELICOPTER
-       LOH HELICOPTER
- GS   HUGHES AIRCRAFT
-       . **OH-6 HELICOPTER**
-       LIGHT AIRCRAFT
-       . **OH-6 HELICOPTER**
-       OBSERVATION AIRCRAFT
-       . **OH-6 HELICOPTER**
-       V/STOL AIRCRAFT
-       . ROTARY WING AIRCRAFT
-       . . HELICOPTERS
-       . . . MILITARY HELICOPTERS
-       . . . . **OH-6 HELICOPTER**

OH-13 HELICOPTER
- UF   H-13 HELICOPTER
-       SIOUX HELICOPTER
-       UH-13 HELICOPTER
- GS   BELL AIRCRAFT
-       . **OH-13 HELICOPTER**
-       LIGHT AIRCRAFT
-       . **OH-13 HELICOPTER**
-       UTILITY AIRCRAFT
-       . **OH-13 HELICOPTER**
-       V/STOL AIRCRAFT
-       . ROTARY WING AIRCRAFT
-       . . HELICOPTERS
-       . . . MILITARY HELICOPTERS
-       . . . . **OH-13 HELICOPTER**

OH-23 HELICOPTER
- UF   H-23 HELICOPTER
-       RAVEN HELICOPTER
-       UH-12 HELICOPTER
- GS   FAIRCHILD-HILLER AIRCRAFT
-       . **OH-23 HELICOPTER**
-       LIGHT AIRCRAFT
-       . **OH-23 HELICOPTER**
-       UTILITY AIRCRAFT
-       . **OH-23 HELICOPTER**
-       V/STOL AIRCRAFT
-       . ROTARY WING AIRCRAFT
-       . . HELICOPTERS
-       . . . MILITARY HELICOPTERS
-       . . . . **OH-23 HELICOPTER**

OH-58 HELICOPTER
- GS   LIGHT AIRCRAFT
-       . **OH-58 HELICOPTER**
-       V/STOL AIRCRAFT
-       . ROTARY WING AIRCRAFT
-       . . HELICOPTERS
-       . . . MILITARY HELICOPTERS
-       . . . . **OH-58 HELICOPTER**

OHIO
- GS   NATIONS
-       . UNITED STATES
-       . . **OHIO**
- RT   OHIO RIVER (US)
-       WABASH RIVER BASIN (IL-IN-OH)

OHIO RIVER (US)
- GS   RIVERS
-       . **OHIO RIVER (US)**
- RT   ILLINOIS
-       INDIANA
-       KENTUCKY
-       OHIO
-       PENNSYLVANIA
-       WEST VIRGINIA

OHMIC DISSIPATION
- UF   JOULE HEATING
- GS   DISSIPATION
-       . **OHMIC DISSIPATION**
- RT   JOULE-THOMSON EFFECT
-       LEVITATION MELTING
-       LOSSES

OHMMETERS
- GS   MEASURING INSTRUMENTS
-       . **OHMMETERS**
- RT   ELECTRICAL CONDUCTIVITY METERS
-       ELECTRICAL IMPEDANCE
-       ELECTRICAL MEASUREMENT
-       ELECTRICAL RESISTANCE
-       RESISTANCE THERMOMETERS
-       WHEATSTONE BRIDGES

OHMS LAW
- GS   CIRCUITS
-       . **OHMS LAW**
-       LAWS

OHMS LAW -*(CONT.)*
-       . **OHMS LAW**
- RT   ∞ CONDUCTIVITY
-       ELECTRIC CURRENT
-       ELECTRICAL RESISTANCE
-       ELECTRICITY
-       ELECTROMOTIVE FORCES
-       VOLT-AMPERE CHARACTERISTICS

OIL ADDITIVES
- GS   ADDITIVES
-       . **OIL ADDITIVES**

OIL EXPLORATION
- GS   EXPLORATION
-       . **OIL EXPLORATION**
- RT   CRUDE OIL
-       DRILLING
-       ENERGY POLICY
-       GEOLOGY
-       NATURAL GAS EXPLORATION
-       OFFSHORE ENERGY SOURCES
-       TAR SANDS
-       UNDERWATER RESOURCES

OIL FIELDS
- GS   RESOURCES
-       . EARTH RESOURCES
-       . . **OIL FIELDS**
- RT   CRUDE OIL
-       DRILLING
-       METHANE
-       NATURAL GAS
-       OFFSHORE ENERGY SOURCES
-       OILS
-       TAR SANDS

OIL POLLUTION
- GS   POLLUTION
-       . ENVIRONMENT POLLUTION
-       . . WATER POLLUTION
-       . . . **OIL POLLUTION**
- RT   COASTAL ECOLOGY
-       WETLANDS

OIL RECOVERY
- RT   ENERGY TECHNOLOGY
-       RECLAMATION
-       ∞ RECOVERY
-       REUSE

OIL SLICKS
- UF   SLICKS
- RT   DUMPING
-       ENVIRONMENT POLLUTION
-       OCEANOGRAPHY
-       POLLUTION
-       SPILLING
-       WATER POLLUTION

OILS
- GS   **OILS**
-       . CASTOR OIL
-       . CRUDE OIL
-       . FUEL OILS
-       . LUBRICATING OILS
-       . MINERAL OILS
-       . SHALE OIL
- RT   ENERGY POLICY
-       FATS
-       FUELS
-       GREASES
-       HYDRAULIC FLUIDS
-       KEROGEN
-       LIPID METABOLISM
-       LUBRICANTS
-       OIL FIELDS
-       PETROLEUM PRODUCTS
-       PITCH (MATERIAL)
-       RETORT PROCESSING
-       TAR SANDS

OKHANSK METEORITE
- GS   CELESTIAL BODIES
-       . METEORITES
-       . . **OKHANSK METEORITE**
- RT   IRON METEORITES
-       STONY METEORITES

OKLAHOMA
- GS   NATIONS
-       . UNITED STATES
-       . . **OKLAHOMA**
- RT   LAKE TEXOMA (OK-TX)

OLEFINS
- USE   ALKENES

OLEIC ACID
- GS   ACIDS
-       . FATTY ACIDS
-       . . CARBOXYLIC ACIDS
-       . . . **OLEIC ACID**
-       ALIPHATIC COMPOUNDS
-       . FATTY ACIDS
-       . . **OLEIC ACID**
-       LIPIDS
-       . **OLEIC ACID**
-       ORGANIC COMPOUNDS
-       . FATTY ACIDS
-       . . **OLEIC ACID**

OLFACTORY PERCEPTION
- UF   SMELL
- GS   PERCEPTION
-       . SENSORY PERCEPTION
-       . . **OLFACTORY PERCEPTION**
- RT   CHEMORECEPTORS
-       SENSE ORGANS

OLIVINE
- GS   MINERALS
-       . **OLIVINE**
-       . . FORSTERITE
- RT   DUNITE
-       IGNEOUS ROCKS
-       MONTICELLITE
-       PERIDOTITE
-       REGOLITH
-       ROCKS
-       SOILS

OMAN
- GS   NATIONS
-       . **OMAN**

OME
- USE   ORBIT MANEUVERING ENGINE (SPACE SHUTTLE)

OMEGA NAVIGATION SYSTEM
- GS   MEASURING INSTRUMENTS
-       . **OMEGA NAVIGATION SYSTEM**
-       NAVIGATION
-       . **OMEGA NAVIGATION SYSTEM**
- RT   AIR NAVIGATION
-       ∞ SYSTEMS

OMEGA-MESONS
- GS   PARTICLES
-       . ELEMENTARY PARTICLES
-       . . FERMIONS
-       . . . BARYONS
-       . . . . **OMEGA-MESONS**
-       . . HADRONS
-       . . . BARYONS
-       . . . . **OMEGA-MESONS**
-       . . . MESONS
-       . . . . **OMEGA-MESONS**
- RT   CHARGED PARTICLES
-       ETA-MESONS

OMEGATRONS
- GS   PARTICLE ACCELERATORS
-       . CYCLOTRONS
-       . . **OMEGATRONS**

OMICRON CETI STAR
- GS   CELESTIAL BODIES
-       . STARS
-       . . **OMICRON CETI STAR**
- RT   VARIABLE STARS

OMNIDIRECTIONAL ANTENNAS
- GS   ANTENNAS
-       . **OMNIDIRECTIONAL ANTENNAS**
-       . . MONOPOLE ANTENNAS
-       . . . WHIP ANTENNAS
-       . . TURNSTILE ANTENNAS
- RT   DIPOLE ANTENNAS
-       DIRECTIONAL ANTENNAS
-       MICROWAVE ANTENNAS
-       RADIO ANTENNAS

OMNIDIRECTIONAL RADIO RANGES
- GS   NAVIGATION AIDS
-       . BEACONS
-       . . RADIO BEACONS
-       . . . **OMNIDIRECTIONAL RADIO RANGES**

OMNIDIRECTIONAL RADIO RANGES-_(CONT.)_
```
      . . . . SELF CALIBRATING OMNIRANGE
        RADIO EQUIPMENT
      . RADIO TRANSMITTERS
      . . RADIO BEACONS
      . . . OMNIDIRECTIONAL RADIO RANGES
      . . . . SELF CALIBRATING OMNIRANGE
        TRANSMITTERS
      . RADIO TRANSMITTERS
      . . RADIO BEACONS
      . . . OMNIDIRECTIONAL RADIO RANGES
      . . . . SELF CALIBRATING OMNIRANGE
RT    DISTANCE MEASURING EQUIPMENT
      RADIO NAVIGATION
      SOLAR COMPASSES
```

OMNIPOL HC-3 HELICOPTER
```
USE   HC-3 HELICOPTER
```

OMNIPOL L-29 AIRCRAFT
```
USE   L-29 JET TRAINER
```

OMNIPOL Z-37 AIRCRAFT
```
USE   Z-37 AIRCRAFT
```

OMNIRANGE NAVIGATION
```
USE   VHF OMNIRANGE NAVIGATION
```

ON-LINE PROGRAMMING
```
GS    SOFTWARE ENGINEERING
      . COMPUTER PROGRAMMING
      . . ON-LINE PROGRAMMING
```

ON-LINE SYSTEMS
```
RT    COMPUTER PROGRAMS
      COMPUTER TECHNIQUES
      DATA MANAGEMENT
      DATA PROCESSING
      INFORMATION RETRIEVAL
      INFORMATION SYSTEMS
      INTEGRATED LIBRARY SYSTEMS
      NUMERICAL DATA BASES
      ∞ SYSTEMS
```

ONBOARD COMPUTERS
```
USE   AIRBORNE/SPACEBORNE COMPUTERS
```

ONBOARD DATA PROCESSING
```
RT    AIRBORNE/SPACEBORNE COMPUTERS
      ∞ DATA
      FLIGHT MANAGEMENT SYSTEMS
      IMAGE PROCESSING
      MICROPROCESSORS
      REAL TIME OPERATION
      SIGNAL PROCESSING
```

ONBOARD EQUIPMENT
```
GS    ONBOARD EQUIPMENT
      . AIRBORNE EQUIPMENT
      . . AIRBORNE/SPACEBORNE
        COMPUTERS
      . . LIGHT AIRBORNE MULTIPURPOSE
        SYSTEM
      . . TERCOM
      . AIRBORNE LASERS
      . AIRCRAFT EQUIPMENT
      . . BOMBING EQUIPMENT
      . . FLYING EJECTION SEATS
      . . TERCOM
      . SPACECRAFT EQUIPMENT
      . . SPACECRAFT ELECTRONIC
        EQUIPMENT
RT    AIRBORNE SURVEILLANCE RADAR
      ∞ AIRCRAFT
      BUBBLE TECHNIQUE
      CREW PROCEDURES (PREFLIGHT)
      ∞ EQUIPMENT
      FLIGHT INSTRUMENTS
      FLIGHT OPERATIONS
      FLIGHT SAFETY
      HEATING EQUIPMENT
      LIFE SUPPORT SYSTEMS
      LIGHTING EQUIPMENT
      RADAR EQUIPMENT
      RADIO EQUIPMENT
      SPACECRAFT INSTRUMENTS
      STOWAGE (ONBOARD EQUIPMENT)
      SURVIVAL EQUIPMENT
      TELECOMMUNICATION
      ∞ TEST EQUIPMENT
      TRAINING DEVICES
```

ONE DIMENSIONAL FLOW
```
GS    FLUID FLOW
```

ONE DIMENSIONAL FLOW-_(CONT.)_
```
      . ONE DIMENSIONAL FLOW
RT    ANNULAR FLOW
      AXIAL FLOW
      CORE FLOW
      FLOW GEOMETRY
      HUGONIOT EQUATION OF STATE
      THREE DIMENSIONAL FLOW
      TWO DIMENSIONAL FLOW
```

ONE-PHASE FLOW
```
USE   SINGLE-PHASE FLOW
```

ONISOTROPY
```
USE   ANISOTROPY
```

ONSAGER PHENOMENOLOGICAL COEFFICIENT
```
GS    COEFFICIENTS
      . ONSAGER PHENOMENOLOGICAL
        COEFFICIENT
RT    FLUX DENSITY
      PLASMAS (PHYSICS)
      STATISTICAL MECHANICS
      VARIATIONAL PRINCIPLES
```

ONSAGER RELATIONSHIP
```
RT    ∞ EQUILIBRIUM
      IRREVERSIBLE PROCESSES
      THERMODYNAMICS
```

ONTARIO
```
GS    NATIONS
      . CANADA
      . . ONTARIO
```

ONTOGENESIS
```
USE   ONTOGENY
```

ONTOGENY
```
UF    ONTOGENESIS
RT    BIOGENY
      EVOLUTION (DEVELOPMENT)
      GROWTH
      NEUROPHYSIOLOGY
```

OOCYTES
```
USE   GAMETOCYTES
```

OPACIFIERS
```
GS    ADDITIVES
      . OPACIFIERS
RT    ∞ AGENTS
      FILLERS
```

OPACITY
```
GS    ELECTROMAGNETIC PROPERTIES
      . OPTICAL PROPERTIES
      . . OPACITY
RT    ABSORPTANCE
      ABSORPTIVITY
      ACOUSTICS
      ATMOSPHERIC OPTICS
      ATTENUATION COEFFICIENTS
      CLARITY
      DENSITY (MASS/VOLUME)
      ELECTROMAGNETIC ABSORPTION
      HAZE
      KRAMERS-KRONIG FORMULA
      LIGHT (VISIBLE RADIATION)
      LIGHT TRANSMISSION
      REFRACTIVITY
      TRANSLUCENCE
      TRANSMISSION EFFICIENCY
      TRANSMISSIVITY
      TRANSPARENCE
      TURBIDITY
      UNDERWATER OPTICS
      VISIBILITY
```

OPALESCENCE
```
RT    IRIDESCENCE
      OPTICAL PROPERTIES
```

OPEN CHANNEL FLOW
```
GS    FLUID FLOW
      . CHANNEL FLOW
      . . OPEN CHANNEL FLOW
      . LIQUID FLOW
      . . OPEN CHANNEL FLOW
RT    LAMINAR FLOW
      MEANDERS
      PIPE FLOW
      TURBULENT FLOW
      WATER FLOW
```

OPEN CIRCUIT VOLTAGE
```
GS    POTENTIAL ENERGY
      . ELECTRIC POTENTIAL
      . . OPEN CIRCUIT VOLTAGE
RT    BIAS
      CAPACITANCE
      ELECTRICAL PROPERTIES
      ELECTRICAL RESISTIVITY
      ELECTROMOTIVE FORCES
      ENERGY CONVERSION EFFICIENCY
      OVERVOLTAGE
      ∞ POTENTIAL
      POWER GAIN
      SHORT CIRCUIT CURRENTS
      SOLAR CELLS
      STATIC ELECTRICITY
      VOLT-AMPERE CHARACTERISTICS
```

OPEN PROJECT
```
UF    ORIGIN OF PLASMAS IN EARTH
        NEIGHBORHOOD
RT    EARTH ATMOSPHERE
      MAGNETOSPHERE
      PLASMA DIAGNOSTICS
      PLASMA PHYSICS
      PLASMASPHERE
      SATELLITE-BORNE INSTRUMENTS
      SPACE PLASMAS
```

OPENINGS
```
UF    CUT-OUTS
GS    OPENINGS
      . APERTURES
      . . IRISES (MECHANICAL APERTURES)
      . . SYNTHETIC APERTURES
      . PORTS (OPENINGS)
      . SLITS
RT    ANNULAR DUCTS
      CAVITIES
      CRACKS
      CURTAINS
      DOORS
      DUCT GEOMETRY
      DUCTS
      EGRESS
      EXHAUST NOZZLES
      EXHAUST SYSTEMS
      GAPS
      GATES (OPENINGS)
      INGRESS (SPACECRAFT PASSAGEWAY)
      INLET NOZZLES
      INTAKE SYSTEMS
      ORIFICES
      OUTLETS
      PASSAGEWAYS
      PERFORATED PLATES
      PIPE NOZZLES
      SLOTS
      VENTS
      WINDOWS (APERTURES)
```

OPERATING COSTS
```
GS    COSTS
      . OPERATING COSTS
RT    AIRLINE OPERATIONS
      ECONOMIC ANALYSIS
      ENERGY POLICY
      FUELS
      MAINTENANCE
      PRODUCTION COSTS
      SYSTEMS ANALYSIS
```

OPERATING SYSTEMS (COMPUTERS)
```
SN    (COMPUTER PROGRAMS FOR
        EXPEDITING,CONTROLLING AND/OR
        RECORDING COMPUTER USE BY OTHER
        PROGRAMS)
GS    COMPUTER PROGRAMS
      . COMPUTER SYSTEMS PROGRAMS
      . . OPERATING SYSTEMS (COMPUTERS)
RT    ASSEMBLER ROUTINES
      COMPILERS
      COMPUTER INFORMATION SECURITY
      COMPUTER SYSTEMS DESIGN
      INPUT/OUTPUT ROUTINES
      ∞ ROUTINES
      ∞ SYSTEMS
```

OPERATING TEMPERATURE
```
GS    TEMPERATURE
      . OPERATING TEMPERATURE
RT    AMBIENT TEMPERATURE
      COMBUSTION TEMPERATURE
      ROOM TEMPERATURE
      WALL TEMPERATURE
```

OPERATIONAL AMPLIFIERS
GS   AMPLIFIERS
     . **OPERATIONAL AMPLIFIERS**
RT   AMPLIFIER DESIGN
     ANALOG CIRCUITS
     ANALOG COMPUTERS
     DIFFERENTIAL AMPLIFIERS
     TRANSISTOR AMPLIFIERS

OPERATIONAL CALCULUS
RT   ∞APPLICATIONS OF MATHEMATICS
     CALCULUS
     CALCULUS OF VARIATIONS
     DIFFERENTIAL EQUATIONS
     FOURIER ANALYSIS
     INTEGRAL CALCULUS
     LINEAR EQUATIONS

OPERATIONAL HAZARDS
GS   HAZARDS
     . **OPERATIONAL HAZARDS**
RT   AIR PIRACY
     AIRCRAFT HAZARDS
     CUMULATIVE DAMAGE
     FLIGHT HAZARDS
     METEOROID HAZARDS
     NOISE (SOUND)
     RADIATION HAZARDS

OPERATIONAL PROBLEMS
RT   ∞PROBLEMS

∞ **OPERATIONS**
SN   *(USE OF A MORE SPECIFIC TERM IS*
     RECOMMENDED--CONSULT THE TERMS
     LISTED BELOW)
RT   AIR DROP OPERATIONS
     AIR TRAFFIC CONTROL
     AIRLINE OPERATIONS
     CHEMICAL ENGINEERING
     CHEMICAL REACTIONS
     CLINICAL MEDICINE
     DEPLOYMENT
     FISHBOWL OPERATION
     LOADING OPERATIONS
     MECHANIZATION
     MISSION PLANNING
     OPERATIONS RESEARCH
     ORIFICES
     PREFLIGHT OPERATIONS
     PREMATURE OPERATION
     PRODUCTION ENGINEERING
     PROGRAMS
     PROJECTS
     RESCUE OPERATIONS
     RUNNING
     SEQUENCING
     STRATEGY
     SURGERY
     SYSTEMS ENGINEERING

OPERATIONS RESEARCH
GS   **OPERATIONS RESEARCH**
     . DYNAMIC PROGRAMMING
     . MATHEMATICAL PROGRAMMING
     .. LINEAR PROGRAMMING
     .. NONLINEAR PROGRAMMING
     .. QUADRATIC PROGRAMMING
RT   ∞APPLICATIONS OF MATHEMATICS
     COMPUTER SYSTEMS SIMULATION
     COMPUTERIZED SIMULATION
     CONSTRAINTS
     CONTROL SYSTEMS DESIGN
     CRITICAL PATH METHOD
     DECISION THEORY
     DELPHI METHOD (FORECASTING)
     DYNAMIC PROGRAMMING
     EXPERIMENT DESIGN
     FORECASTING
     FUNCTIONS (MATHEMATICS)
     GAME THEORY
     INFORMATION THEORY
     LAGRANGE MULTIPLIERS
     LINEAR PREDICTION
     LINEAR PROGRAMMING
     MANAGEMENT
     MANAGEMENT METHODS
     MANAGEMENT PLANNING
     MATHEMATICAL MODELS
     MATRIX MANAGEMENT
     MINIMA
     MINIMAX TECHNIQUE
     MISSION PLANNING
     NONLINEAR PROGRAMMING
     ∞OPERATIONS

OPERATIONS RESEARCH-*(CONT.)*
     OPTIMIZATION
     ∞PATHS
     PATTERN METHOD (FORECASTING)
     PROBABILITY THEORY
     PROBE METHOD (FORECASTING)
     PROFILE METHOD (FORECASTING)
     PROJECT PLANNING
     QUALITY CONTROL
     QUEUEING THEORY
     RAND PROJECT
     RAYLEIGH DISTRIBUTION
     RESEARCH AND DEVELOPMENT
     RESEARCH MANAGEMENT
     RISK
     SADDLE POINTS (GAME THEORY)
     SEQUENCING
     SIMULATION
     STATISTICAL ANALYSIS
     STOCHASTIC PROCESSES
     STRATEGY
     ∞SYNTHESIS
     SYNTHESIS (CHEMISTRY)
     SYSTEMS ANALYSIS
     SYSTEMS ENGINEERING
     SYSTEMS MANAGEMENT
     SYSTEMS SIMULATION
     TRAVELING SALESMAN PROBLEM
     URBAN DEVELOPMENT
     WAR GAMES

OPERATOR PERFORMANCE
GS   HUMAN PERFORMANCE
     . **OPERATOR PERFORMANCE**
RT   ASTRONAUT PERFORMANCE
     COMPUTER SYSTEMS PERFORMANCE
     MENTAL PERFORMANCE
     ∞PERFORMANCE
     PILOT PERFORMANCE
     PSYCHOMOTOR PERFORMANCE

∞ **OPERATORS**
SN   *(USE OF A MORE SPECIFIC TERM IS*
     RECOMMENDED--CONSULT THE TERMS
     LISTED BELOW)
RT   OPERATORS (MATHEMATICS)
     OPERATORS (PERSONNEL)
     REACTOR CORES

OPERATORS (MATHEMATICS)
UF   DIFFERENTIAL OPERATORS
     FREDHOLM OPERATORS
GS   **OPERATORS (MATHEMATICS)**
     . BERGMAN OPERATOR
RT   FUNCTIONS (MATHEMATICS)
     INTEGRAL TRANSFORMATIONS
     LAPLACE TRANSFORMATION
     MATRIX THEORY
     ∞OPERATORS
     PERTURBATION THEORY
     S MATRIX THEORY

OPERATORS (PERSONNEL)
GS   PERSONNEL
     . **OPERATORS (PERSONNEL)**
     .. PILOTS (PERSONNEL)
     ... AIRCRAFT PILOTS
     .... TEST PILOTS
RT   ∞OPERATORS

OPHIUCHI CLOUDS
RT   CLOUD PHYSICS
     INTERSTELLAR GAS
     INTERSTELLAR MATTER
     NEBULAE

OPHTHALMODYNAMOMETRY
RT   BLOOD PRESSURE
     EYE (ANATOMY)
     VISION

OPHTHALMOLOGY
GS   MEDICAL SCIENCE
     . **OPHTHALMOLOGY**
RT   ELECTRONYSTAGMOGRAPHY
     EYE (ANATOMY)
     EYE DISEASES
     MIOSIS
     VESTIBULAR NYSTAGMUS

OPIK THEORY
RT   NEBULAE
     ORION CONSTELLATION
     ORION NEBULA
     SUPERNOVAE

OPIK THEORY-*(CONT.)*
     ∞THEORIES

OPTICAL ABSORPTION
USE ELECTROMAGNETIC ABSORPTION
     LIGHT TRANSMISSION

OPTICAL ACTIVITY
RT   BIOCHEMISTRY
     CARBOHYDRATES
     ∞OPTICS
     ORGANIC CHEMISTRY
     POLARIMETRY
     POLARIZED LIGHT
     STEREOCHEMISTRY

OPTICAL AMPLIFIERS
USE LIGHT AMPLIFIERS

OPTICAL BISTABILITY
GS   ELECTROMAGNETIC PROPERTIES
     . OPTICAL PROPERTIES
     .. **OPTICAL BISTABILITY**
RT   HYSTERESIS
     INTEGRATED OPTICS
     LIGHT TRANSMISSION
     NONLINEAR OPTICS
     OPTICAL DATA STORAGE MATERIALS
     OPTICAL EQUIPMENT
     OPTICAL MEASURING INSTRUMENTS
     OPTICAL MEMORY (DATA STORAGE)
     OPTICAL WAVEGUIDES
     SWITCHING CIRCUITS

OPTICAL COMMUNICATION
UF   LASER COMMUNICATION
     LIGHT COMMUNICATION
     OPTICAL SIGNALS
GS   TELECOMMUNICATION
     . COMMUNICATION
     .. **OPTICAL COMMUNICATION**
RT   DYE LASERS
     GROUND-AIR-GROUND COMMUNICATION
     HIGH POWER LASERS
     INTERPLANETARY COMMUNICATION
     LASERS
     LUNAR COMMUNICATION
     ∞OPTICS
     SPACE COMMUNICATION
     SPACECRAFT COMMUNICATION
     TUNABLE LASERS
     VISUAL SIGNALS
     WIRELESS COMMUNICATION

OPTICAL COMPUTERS
GS   DATA PROCESSING EQUIPMENT
     . COMPUTERS
     .. **OPTICAL COMPUTERS**
RT   COHERENT LIGHT
     COMPUTER DESIGN
     ELECTRO-OPTICS
     OPTICAL EQUIPMENT
     OPTICAL MEMORY (DATA STORAGE)

OPTICAL CORRECTION PROCEDURE
GS   CORRECTION
     . **OPTICAL CORRECTION PROCEDURE**
     PROCEDURES
     . **OPTICAL CORRECTION PROCEDURE**
RT   ADAPTIVE OPTICS
     ADJUSTING
     ERRORS
     INSTRUMENT ERRORS
     LENS DESIGN
     ∞OPTICS
     PHOTOGRAPHIC MEASUREMENT
     PHOTOGRAPHS
     POSITION ERRORS
     SELF FOCUSING

OPTICAL COUNTERMEASURES
GS   COUNTERMEASURES
     . **OPTICAL COUNTERMEASURES**
RT   AIR DEFENSE
     ANTIMISSILE DEFENSE
     DECEPTION
     ELECTRONIC COUNTERMEASURES
     MILITARY TECHNOLOGY
     MISSILE DEFENSE
     OPTICAL RADAR
     ∞OPTICS
     SPACE SURVEILLANCE (SPACEBORNE)

OPTICAL COUPLING
GS   COUPLING
      . **OPTICAL COUPLING**
RT   COUPLES
      CROSS COUPLING
      CROSS POLARIZATION
      LASER MODE LOCKING
      LIGHT TRANSMISSION
      MICROWAVE COUPLING
      ∞OPTICS
      PHASE LOCKED SYSTEMS
      POLARIZATION (WAVES)

OPTICAL DATA PROCESSING
GS   DATA PROCESSING
      . **OPTICAL DATA PROCESSING**
      . . SCENE ANALYSIS
RT   CHARACTER RECOGNITION
      ∞DATA
      DATA ACQUISITION
      DATA PROCESSING EQUIPMENT
      GRAY SCALE
      IMAGE PROCESSING
      LASERS
      LIGHT VALVES
      OPTICAL DISKS
      OPTICAL RELAY SYSTEMS
      ∞OPTICS
      PHOTONICS
      ∞PROCESSING
      READERS
      SCANNERS
      TOMOGRAPHY

OPTICAL DATA STORAGE MATERIALS
RT   ∞DATA
      DATA RECORDING
      DATA STORAGE
      LASER APPLICATIONS
      OPTICAL BISTABILITY
      OPTICAL MEMORY (DATA STORAGE)
      ∞OPTICS
      PHOTOGRAPHIC FILM
      VIDEO DISKS

OPTICAL DENSITY
RT   ∞DENSITY
      MICRODENSITOMETERS
      ∞OPTICS
      TRANSLUCENCE
      TRANSMITTANCE
      TRANSPARENCE
      TURBIDITY
      UNDERWATER OPTICS

OPTICAL DEPOLARIZATION
RT   LIGHT (VISIBLE RADIATION)
      ∞OPTICS
      POLARIZED LIGHT

OPTICAL DISKS
GS   PERIPHERAL EQUIPMENT (COMPUTERS)
      . COMPUTER STORAGE DEVICES
      . . **OPTICAL DISKS**
RT   DATA STORAGE
      LASER APPLICATIONS
      OPTICAL DATA PROCESSING
      OPTICAL EQUIPMENT
      OPTICAL MEMORY (DATA STORAGE)
      VIDEO DISKS

OPTICAL EMISSION
USE   LIGHT EMISSION

OPTICAL EMISSION SPECTROSCOPY
GS   SPECTROSCOPY
      . **OPTICAL EMISSION SPECTROSCOPY**
      . . LASER SPECTROSCOPY
RT   AURORAL SPECTROSCOPY
      ELECTRON SPECTROSCOPY
      EMISSION SPECTRA
      LIGHT (VISIBLE RADIATION)
      ∞OPTICS

OPTICAL EQUIPMENT
GS   **OPTICAL EQUIPMENT**
      . ASTRONOMICAL TELESCOPES
      . . HELIOMETERS
      . . . PYROHELIOMETERS
      . . SPECTROSCOPIC TELESCOPES
      . . . MULTISPECTRAL TRACKING
            TELESCOPES
      . . . STRATOSCOPE TELESCOPES
      . . X RAY TELESCOPES
      . BINOCULARS

OPTICAL EQUIPMENT-*(CONT.)*
      . CAMERAS
      . . BAKER-NUNN CAMERA
      . . BALLISTIC CAMERAS
      . . DELFT CAMERA
      . . DIFFRACTION LIMITED CAMERAS
      . . FAINT OBJECT CAMERA
      . . HIGH SPEED CAMERAS
      . . . FRAMING CAMERAS
      . . I2S CAMERAS
      . . LALLEMAND CAMERAS
      . . MULTISPECTRAL BAND CAMERAS
      . . PANORAMIC CAMERAS
      . . PINHOLE CAMERAS
      . . SCHMIDT CAMERAS
      . . STREAK CAMERAS
      . . TELEVISION CAMERAS
      . COLLIMATORS
      . ENDOSCOPES
      . EYEPIECES
      . IMAGE CONVERTERS
      . . CELESCOPES
      . . IMAGE TUBES
      . . . THERMICONS
      . LASER DOPPLER VELOCIMETERS
      . OPTICAL GYROSCOPES
      . OPTICAL MEASURING INSTRUMENTS
      . . CATHETOMETERS
      . . DIFFRACTOMETERS
      . . EBERT SPECTROMETERS
      . . ELLIPSOMETERS
      . . GEODIMETERS
      . . INFRARED SPECTROMETERS
      . . . FILTER WHEEL INFRARED
            SPECTROMETERS
      . . LIGHT SCATTERING METERS
      . . MICRODENSITOMETERS
      . . NEPHELOMETERS
      . . OPTICAL PYROMETERS
      . . OPTICAL RANGE FINDERS
      . . OPTICAL SCANNERS
      . . . MULTISPECTRAL BAND SCANNERS
      . . PHOTOGONIOMETERS
      . . PHOTOMETERS
      . . . ELECTROPHOTOMETERS
      . . . ULTRAVIOLET SPECTROMETERS
      . . . ULTRAVIOLET
            SPECTROPHOTOMETERS
      . . POLARIMETERS
      . . REFLECTOMETERS
      . . MICROWAVE REFLECTOMETERS
      . . REFRACTOMETERS
      . . SEXTANTS
      . . SPECTROPHOTOMETERS
      . . . INFRARED SPECTROPHOTOMETERS
      . . . ULTRAVIOLET
            SPECTROPHOTOMETERS
      . . TRANSITS
      . . . THEODOLITES
      . . . . CINETHEODOLITES
      . . . TRANSMISSOMETERS
      . OPTICAL MICROSCOPES
      . OPTICAL RADAR
      . PERISCOPES
      . PHOTOGRAPHIC RECTIFIERS
      . POLARISCOPES
      . . SENARMONT POLARISCOPES
      . PRISMS
      . . PRISMATIC BARS
      . SCATTER PLATES (OPTICS)
      . SPECTROHELIOGRAPHS
      . STROBOSCOPES
      . WIDE ANGLE LENSES
RT   ABSORPTION SPECTROSCOPY
      ACOUSTIC MICROSCOPES
      CIRCUMSOLAR TELESCOPES
      DENSITOMETERS
      FABRY-PEROT SPECTROMETERS
      GEOMETRICAL OPTICS
      HORIZON SCANNERS
      INFRARED INTERFEROMETERS
      INFRARED SCANNERS
      INTERFEROMETERS
      LENSES
      LOOK ANGLES (ELECTRONICS)
      MACH-ZEHNDER INTERFEROMETERS
      MICROCHANNELS
      MICROSCOPES
      MIRRORS
      MONOCHROMATORS
      OPTICAL BISTABILITY
      OPTICAL COMPUTERS
      OPTICAL DISKS
      ∞OPTICS
      OPTOGALVANIC SPECTROSCOPY
      PHOTOGRAPHIC EQUIPMENT

OPTICAL EQUIPMENT-*(CONT.)*
      RADIO TELESCOPES
      REFLECTING TELESCOPES
      REFRACTING TELESCOPES
      RETICLES
      SCANNER PROJECT
      SCANNERS
      SPECTROMETERS
      TELESCOPES
      TRIPODS
      VIDEO EQUIPMENT

OPTICAL FILTERS
GS   ELECTROMAGNETIC WAVE FILTERS
      . **OPTICAL FILTERS**
      . . BIREFRINGENT FILTERS
      . . INFRARED FILTERS
      . . ULTRAVIOLET FILTERS
RT   ADAPTIVE FILTERS
      BANDPASS FILTERS
      BANDSTOP FILTERS
      DIAPHRAGMS (MECHANICS)
      DIDYMIUM
      ELECTRIC FILTERS
      ∞FILTERS
      ∞GRATINGS
      GRATINGS (SPECTRA)
      HIGH PASS FILTERS
      LENSES
      LOW PASS FILTERS
      OPTICAL RELAY SYSTEMS
      ∞OPTICS
      PHOTOGRAPHIC EQUIPMENT
      PHOTOGRAPHIC FILM
      ROWLAND CIRCLES
      SUNGLASSES
      TRANSMISSION

OPTICAL GENERATORS
USE   LASER CAVITIES

OPTICAL GYROSCOPES
GS   GYROSCOPES
      . **OPTICAL GYROSCOPES**
      OPTICAL EQUIPMENT
      . **OPTICAL GYROSCOPES**
RT   LASER GYROSCOPES
      ∞OPTICS
      SAGNAC EFFECT

OPTICAL HETERODYNING
GS   HETERODYNING
      . **OPTICAL HETERODYNING**
RT   DOPPLER EFFECT
      LIGHT MODULATION
      ∞OPTICS

OPTICAL ILLUSION
GS   PSYCHOLOGICAL EFFECTS
      . ILLUSIONS
      . . **OPTICAL ILLUSION** .
      . . . ELEVATOR ILLUSION
RT   MOON ILLUSION
      ∞OPTICS

OPTICAL IMAGES
USE   IMAGES

OPTICAL MASER MODULATION
USE   LIGHT MODULATION

OPTICAL MASERS
USE   LASERS

OPTICAL MEASUREMENT
SN   (MEASUREMENTS OF OPTICAL
      PROPERTIES, QUANTITIES OR
      CONDITIONS)
GS   **OPTICAL MEASUREMENT**
      . ASTRONOMICAL PHOTOMETRY
      . . STELLAR SPECTROPHOTOMETRY
      . COLORIMETRY
      . OPTOMETRY
      . PHOTOMETRY
      . . ELECTROPHOTOMETRY
      . . INFRARED PHOTOMETRY
      . . SPECTROPHOTOMETRY
      . . . STELLAR SPECTROPHOTOMETRY
      . . TELEPHOTOMETRY
      . . ULTRAVIOLET PHOTOMETRY
      . . VISUAL PHOTOMETRY
      . POLARIMETRY
RT   CHEMICAL ANALYSIS
      COLLIMATORS

OPTICAL MEASUREMENT-*(CONT.)*
DENSITOMETERS
DIFFRACTOMETERS
ELECTRO-OPTICAL PHOTOGRAPHY
EMISSIVITY
FARADAY EFFECT
GAMMA RAY SPECTROMETERS
GEODIMETERS
GEOMETRICAL OPTICS
GRAZING INCIDENCE
INFRARED INTERFEROMETERS
INTERFEROMETERS
LIGHT (VISIBLE RADIATION)
∞MEASUREMENT
MICRODENSITOMETERS
MODULATION TRANSFER FUNCTION
NEPHELOMETERS
∞OPTICS
PHASE CONTRAST
PHOTOGRAPHIC MEASUREMENT
PHOTOMETERS
POLARIMETERS
RAY TRACING
REFLECTANCE
REFLECTOMETERS
REFRACTOMETERS
RONCHI TEST
SPECTRAL SIGNATURES
SPECTROMETERS
SPECTROPHOTOMETERS
STROBOSCOPES

OPTICAL MEASURING INSTRUMENTS
SN   (INSTRUMENTS UTILIZING OPTICAL
         PRINCIPLES FOR MEASUREMENT)
UF   OPTICAL SENSORS
GS   MEASURING INSTRUMENTS
      . **OPTICAL MEASURING INSTRUMENTS**
      . . CATHETOMETERS
      . . DIFFRACTOMETERS
      . . EBERT SPECTROMETERS
      . . ELLIPSOMETERS
      . . GEODIMETERS
      . . HAPLOSCOPES
      . . INFRARED SPECTROMETERS
      . . . FILTER WHEEL INFRARED
            SPECTROMETERS
      . . LIGHT SCATTERING METERS
      . . MICRODENSITOMETERS
      . . NEPHELOMETERS
      . . OCULOMETERS
      . . OPTICAL PYROMETERS
      . . OPTICAL RANGE FINDERS
      . . OPTICAL SCANNERS
      . . . MULTISPECTRAL BAND SCANNERS
      . . PHOTOGONIOMETERS
      . . PHOTOMETERS
      . . . ELECTROPHOTOMETERS
      . . . ULTRAVIOLET SPECTROMETERS
      . . . ULTRAVIOLET
            SPECTROPHOTOMETERS
      . . POLARIMETERS
      . . REFLECTOMETERS
      . . . MICROWAVE REFLECTOMETERS
      . . REFRACTOMETERS
      . . SEXTANTS
      . . SPECTROPHOTOMETERS
      . . . INFRARED SPECTROPHOTOMETERS
      . . . ULTRAVIOLET
            SPECTROPHOTOMETERS
      . . TRANSITS
      . . . THEODOLITES
      . . . . CINETHEODOLITES
      . . TRANSMISSOMETERS
OPTICAL EQUIPMENT
. **OPTICAL MEASURING INSTRUMENTS**
. . CATHETOMETERS
. . DIFFRACTOMETERS
. . EBERT SPECTROMETERS
. . ELLIPSOMETERS
. . GEODIMETERS
. . INFRARED SPECTROMETERS
. . . FILTER WHEEL INFRARED
      SPECTROMETERS
. . LIGHT SCATTERING METERS
. . MICRODENSITOMETERS
. . NEPHELOMETERS
. . OPTICAL PYROMETERS
. . OPTICAL RANGE FINDERS
. . OPTICAL SCANNERS
. . . MULTISPECTRAL BAND SCANNERS
. . PHOTOGONIOMETERS
. . PHOTOMETERS
. . . ELECTROPHOTOMETERS
. . . ULTRAVIOLET SPECTROMETERS

OPTICAL MEASURING INSTRUMENTS-*(CONT.)*
. . . ULTRAVIOLET
      SPECTROPHOTOMETERS
. . POLARIMETERS
. . REFLECTOMETERS
. . . MICROWAVE REFLECTOMETERS
. . REFRACTOMETERS
. . SEXTANTS
. . SPECTROPHOTOMETERS
. . . INFRARED SPECTROPHOTOMETERS
. . . ULTRAVIOLET
      SPECTROPHOTOMETERS
. . TRANSITS
. . . THEODOLITES
. . . . CINETHEODOLITES
. . TRANSMISSOMETERS
RT   ABSORPTION SPECTROSCOPY
      CINESPECTROGRAPHS
      COLORIMETRY
      DENSITOMETERS
      FABRY-PEROT SPECTROMETERS
      FAINT OBJECT CAMERA
      GONIOMETERS
      GUIDANCE SENSORS
      INFRARED INTERFEROMETERS
      INTERFEROMETERS
      LASER DOPPLER VELOCIMETERS
      MACH-ZEHNDER INTERFEROMETERS
      MICROSCOPES
      MIROS SYSTEM
      MONOCHROMATORS
      MULTISPECTRAL TRACKING
         TELESCOPES
      OPTICAL BISTABILITY
∞OPTICS
      OPTOGALVANIC SPECTROSCOPY
      PERISCOPES
      POLARIMETRY
      POLARISCOPES
      RADIATION MEASURING INSTRUMENTS
      REFLECTING TELESCOPES
      REFRACTING TELESCOPES
      SELF FOCUSING
      SENARMONT POLARISCOPES
      SOLAR INSTRUMENTS
      TELEPHOTOMETRY
      TELESCOPES

OPTICAL MEMORY (DATA STORAGE)
RT   COHERENT LIGHT
      COMPUTER STORAGE DEVICES
∞DATA
      HOLOGRAPHY
      LASERS
      OPTICAL BISTABILITY
      OPTICAL COMPUTERS
      OPTICAL DATA STORAGE MATERIALS
      OPTICAL DISKS
∞OPTICS
      VIDEO DISKS

OPTICAL METHODS
USE   OPTICS

OPTICAL MICROSCOPES
GS   MICROSCOPES
      . **OPTICAL MICROSCOPES**
      OPTICAL EQUIPMENT
      . **OPTICAL MICROSCOPES**
RT   ELECTRON MICROSCOPES
∞OPTICS

OPTICAL MODULATION
USE   LIGHT MODULATION

OPTICAL PATHS
RT   DIFFRACTION PATHS
      GEOMETRICAL OPTICS
      MULTIPATH TRANSMISSION
∞OPTICS
∞PATHS
      PHASE CONTRAST
      PHOTON BEAMS
      SAGNAC EFFECT
      UNDERWATER OPTICS
      VOIGT EFFECT
      WAVE DISPERSION

OPTICAL POLARIZATION
RT   CIRCULAR POLARIZATION
      LINEAR POLARIZATION
∞OPTICS
      POLARIZED LIGHT
      POLARIZERS
      POLAROGRAPHY

OPTICAL PROPERTIES
SN   (INCLUDES PROPERTIES AND EFFECTS
         OF VISIBLE, INFRARED AND
         ULTRAVIOLET ELECTROMAGNETIC
         WAVES)
GS   ELECTROMAGNETIC PROPERTIES
      . **OPTICAL PROPERTIES**
      . . ABSORPTIVITY
      . . BIREFRINGENCE
      . . BRIGHTNESS
      . . . SOLAR GRANULATION
      . . BRIGHTNESS DISTRIBUTION
      . . COLOR
      . . . IRIDESCENCE
      . . . WATER COLOR
      . . DICHROISM
      . . LUMINOSITY
      . . . STELLAR LUMINOSITY
      . . OPACITY
      . . OPTICAL BISTABILITY
      . . OPTICAL REFLECTION
      . . PHOSPHORESCENCE
      . . PHOTOCONDUCTIVITY
      . . PHOTOELECTRIC EFFECT
      . . . PHOTOIONIZATION
      . . PHOTOELECTRIC EMISSION
      . . PHOTOVISCOELASTICITY
      . . PHOTOVOLTAIC EFFECT
      . . RADIANCE
      . . REFLECTANCE
      . . REFRACTIVITY
      . . SKY BRIGHTNESS
      . . STIGMATISM
      . . TRANSLUCENCE
      . . TRANSMISSIVITY
      . . TRANSMITTANCE
      . . TRANSPARENCE
      . . TURBIDITY
RT   ACOUSTO-OPTICS
      ALBEDO
      BIREFRINGENT FILTERS
      CLARITY
      COEFFICIENTS
      COHERENT RADIATION
      CROSS POLARIZATION
      DARKNESS
      DIFFRACTION
      ELECTRICAL PROPERTIES
      ELECTROMAGNETIC ABSORPTION
      EMITTANCE
      EXCITONS
      FARADAY EFFECT
      GEOMETRICAL OPTICS
      GLARE
      GLASS
      GRADIENT INDEX OPTICS
      HAZE
      INFRARED ABSORPTION
      ISOTROPY
      KERR ELECTROOPTICAL EFFECT
      KERR MAGNETOOPTICAL EFFECT
      LIGHT (VISIBLE RADIATION)
      LIGHT TRANSMISSION
      LUMENS
      LUMINANCE
      LUNAR ALBEDO
      METALLIC GLASSES
      OPALESCENCE
∞OPTICS
∞ORIENTATION
      PHOTOELECTRICITY
      PHOTONS
      PHOTOTROPISM
      PHYSICAL OPTICS
∞PHYSICAL PROPERTIES
      POLARIZATION (WAVES)
∞PROPERTIES
∞SOLID STATE PHYSICS
      SURFACE PROPERTIES
      THERMOCHROMATIC MATERIALS
      THERMODYNAMIC PROPERTIES
      VISIBILITY
      WAVE DISPERSION

OPTICAL PUMPING
GS   **OPTICAL PUMPING**
      . LASER PUMPING
RT   ELECTRON PUMPING
      ELECTRON-HOLE DROPS
      EXCIMER LASERS
      GAMMA RAY LASERS
      GLASS LASERS
      HCN LASERS
      KRYPTON FLUORIDE LASERS
      LASER PROPULSION
      LASERS

OPTICAL PUMPING-*(CONT.)*
　　　　METAL VAPOR LASERS
　　　　NEODYMIUM LASERS
　　　　NUCLEAR PUMPED LASERS
　　　　NUCLEAR PUMPING
　　∞OPTICS
　　　　PULSE REPETITION RATE
　　∞PUMPING
　　　　RARE GAS-HALIDE LASERS
　　　　SOLAR-PUMPED LASERS
　　　　STIMULATED EMISSION
　　　　STIMULATED EMISSION DEVICES

OPTICAL PYROMETERS
　GS　MEASURING INSTRUMENTS
　　　. OPTICAL MEASURING INSTRUMENTS
　　　.. **OPTICAL PYROMETERS**
　　　. TEMPERATURE MEASURING
　　　　　INSTRUMENTS
　　　.. **OPTICAL PYROMETERS**
　　　OPTICAL EQUIPMENT
　　　. OPTICAL MEASURING INSTRUMENTS
　　　.. **OPTICAL PYROMETERS**
　RT　∞OPTICS
　　　RADIATION PYROMETERS

OPTICAL RADAR
　UF　LASER RADAR
　　　LIDAR
　GS　OPTICAL EQUIPMENT
　　　. **OPTICAL RADAR**
　　　RADAR
　　　. **OPTICAL RADAR**
　RT　LASER ALTIMETERS
　　　OPTICAL COUNTERMEASURES
　　∞OPTICS
　　　OVER-THE-HORIZON RADAR
　　　RADAR DETECTION

OPTICAL RANGE FINDERS
　GS　MEASURING INSTRUMENTS
　　　. DISTANCE MEASURING EQUIPMENT
　　　.. RANGE FINDERS
　　　... **OPTICAL RANGE FINDERS**
　　　.... LASER RANGE FINDERS
　　　. OPTICAL MEASURING INSTRUMENTS
　　　.. **OPTICAL RANGE FINDERS**
　　　OPTICAL EQUIPMENT
　　　. OPTICAL MEASURING INSTRUMENTS
　　　.. **OPTICAL RANGE FINDERS**
　RT　LUNAR RANGEFINDING
　　∞OPTICS

OPTICAL REFLECTION
　GS　ELECTROMAGNETIC PROPERTIES
　　　. OPTICAL PROPERTIES
　　　.. **OPTICAL REFLECTION**
　　　REFLECTION
　　　. **OPTICAL REFLECTION**
　RT　ANTIREFLECTION COATINGS
　　　ELECTROMAGNETIC ABSORPTION
　　　GEOMETRICAL OPTICS
　　　INCIDENT RADIATION
　　　INFRARED REFLECTION
　　　LIGHT TRANSMISSION
　　∞OPTICS
　　　REFLECTANCE
　　　REFLECTED WAVES
　　　SPREAD REFLECTION

OPTICAL RELAY SYSTEMS
　RT　ELECTRO-OPTICS
　　　IMAGING TECHNIQUES
　　　LASER APPLICATIONS
　　　OPTICAL DATA PROCESSING
　　　OPTICAL FILTERS
　　　OPTICAL RESONATORS
　　∞OPTICS
　　　PATTERN RECOGNITION
　　∞SYSTEMS

OPTICAL RESONANCE
　GS　DECAY
　　　. EMISSION
　　　.. LIGHT EMISSION
　　　... LUMINESCENCE
　　　.... **OPTICAL RESONANCE**
　　　RESONANCE
　　　. **OPTICAL RESONANCE**
　RT　NUCLEAR PUMPED LASERS
　　∞OPTICS
　　　PLASMA RADIATION
　　　PLASMA SPECTRA
　　　RESONANCE LINES
　　　SPECTRUM ANALYSIS

OPTICAL RESONATORS
　GS　RESONATORS
　　　. **OPTICAL RESONATORS**
　RT　LASER MODES
　　　LASER OUTPUTS
　　　MIRRORS
　　　OPTICAL RELAY SYSTEMS
　　∞OPTICS

OPTICAL SATELLITE TRACKING PROGRAM
　RT　∞OPTICS
　　　SATELLITE TRACKING

OPTICAL SCANNERS
　GS　MEASURING INSTRUMENTS
　　　. OPTICAL MEASURING INSTRUMENTS
　　　.. **OPTICAL SCANNERS**
　　　... MULTISPECTRAL BAND SCANNERS
　　　OPTICAL EQUIPMENT
　　　. OPTICAL MEASURING INSTRUMENTS
　　　.. **OPTICAL SCANNERS**
　　　... MULTISPECTRAL BAND SCANNERS
　　　SCANNERS
　　　. **OPTICAL SCANNERS**
　　　.. FLYING SPOT SCANNERS
　　　.. MULTISPECTRAL BAND SCANNERS
　RT　CHARACTER RECOGNITION
　　　DATA ACQUISITION
　　　MONITORS
　　∞OPTICS
　　　PHOTON BEAMS
　　　READERS
　　　TELEVISION CAMERAS

OPTICAL SENSORS
　USE　OPTICAL MEASURING INSTRUMENTS

OPTICAL SIGNALS
　USE　OPTICAL COMMUNICATION

OPTICAL SLANT RANGE
　GS　DISTANCE
　　　. **OPTICAL SLANT RANGE**
　RT　∞OPTICS
　　　RADAR RANGE

OPTICAL SPECTRUM
　USE　LIGHT (VISIBLE RADIATION)
　　　SPECTRA

OPTICAL THICKNESS
　RT　ANTIREFLECTION COATINGS
　　　FERMAT PRINCIPLE
　　∞OPTICS
　　　REFRACTIVITY
　　　THICKNESS

OPTICAL TRACKING
　UF　VISUAL TRACKING
　GS　TRACKING (POSITION)
　　　. **OPTICAL TRACKING**
　RT　BALLISTIC CAMERAS
　　　BORESIGHT ERROR
　　　BORESIGHTS
　　　COMPENSATORY TRACKING
　　　GLOBAL TRACKING NETWORK
　　　INFRARED TRACKING
　　　MINITRACK SYSTEM
　　　MULTISPECTRAL TRACKING
　　　TELESCOPES
　　　OCULOMETERS
　　∞OPTICS
　　　PHOTOGRAPHIC TRACKING
　　　RANGE AND RANGE RATE TRACKING
　　　SPACE DETECTION AND TRACKING
　　　　SYSTEM
　　　SPACECRAFT TRACKING
　　　STDN (NETWORK)

OPTICAL TRANSFER FUNCTION
　UF　OTF
　GS　FUNCTIONS (MATHEMATICS)
　　　. TRANSFER FUNCTIONS
　　　.. **OPTICAL TRANSFER FUNCTION**
　RT　ADAPTIVE OPTICS
　　　ASTRONOMICAL TELESCOPES
　　　COST ANALYSIS
　　　FIGURE OF MERIT
　　　IMAGING TECHNIQUES
　　　MODULATION TRANSFER FUNCTION
　　∞OPTICS
　　∞PERFORMANCE
　　　SPACEBORNE TELESCOPES
　　　SYSTEM EFFECTIVENESS

OPTICAL TRANSFER FUNCTION-*(CONT.)*
　　　SYSTEMS ANALYSIS
　　　SYSTEMS ENGINEERING
　　　TELESCOPES

OPTICAL TRANSITION
　RT　ELECTRON TRANSITIONS
　　　EMISSION SPECTRA
　　　FRANCK-CONDON PRINCIPLE
　　　LASING
　　　LUMINESCENCE
　　∞OPTICS

OPTICAL WAVEGUIDES
　GS　TRANSMISSION LINES
　　　. COMMUNICATION CABLES
　　　.. WAVEGUIDES
　　　　OPTICAL WAVEGUIDES
　RT　INTEGRATED OPTICS
　　　LASER OUTPUTS
　　　LIGHT BEAMS
　　　LIGHT TRANSMISSION
　　　OPTICAL BISTABILITY
　　∞OPTICS
　　　PHOTONICS
　　　WAVEGUIDE LASERS

∞ **OPTICS**
　SN　*(USE OF A MORE SPECIFIC TERM IS*
　　　RECOMMENDED--CONSULT THE TERMS
　　　LISTED BELOW)
　UF　OPTICAL METHODS
　RT　ACOUSTO-OPTICS
　　　ADAPTIVE OPTICS
　　　ANGULAR RESOLUTION
　　　ASPHERICITY
　　　ASTIGMATISM
　　　ATMOSPHERIC OPTICS
　　　CASSEGRAIN OPTICS
　　　CAUSTICS (OPTICS)
　　　CRYSTAL OPTICS
　　　DEFOCUSING
　　　DIFFRACTION PATTERNS
　　　DIFFRACTION PROPAGATION
　　　ELECTRO-OPTICAL EFFECT
　　　ELECTRO-OPTICAL PHOTOGRAPHY
　　　ELECTRO-OPTICS
　　　ELECTRON OPTICS
　　　FIBER OPTICS
　　　FOCI
　　　FRESNEL LENSES
　　　GEOMETRICAL OPTICS
　　　GRADIENT INDEX OPTICS
　　　HUYGENS PRINCIPLE
　　　IMAGES
　　　IMAGING TECHNIQUES
　　　INTEGRATED OPTICS
　　　LASER CAVITIES
　　　LASERS
　　　LENS DESIGN
　　　LENSES
　　　LIGHT (VISIBLE RADIATION)
　　　LIGHT AMPLIFIERS
　　　LIGHT EMISSION
　　　LIGHT MODULATION
　　　MAGNETO-OPTICS
　　　MIRRORS
　　　NONLINEAR OPTICS
　　　OPTICAL ACTIVITY
　　　OPTICAL COMMUNICATION
　　　OPTICAL CORRECTION PROCEDURE
　　　OPTICAL COUNTERMEASURES
　　　OPTICAL COUPLING
　　　OPTICAL DATA PROCESSING
　　　OPTICAL DATA STORAGE MATERIALS
　　　OPTICAL DENSITY
　　　OPTICAL DEPOLARIZATION
　　　OPTICAL EMISSION SPECTROSCOPY
　　　OPTICAL EQUIPMENT
　　　OPTICAL FILTERS
　　　OPTICAL GYROSCOPES
　　　OPTICAL HETERODYNING
　　　OPTICAL ILLUSION
　　　OPTICAL MEASUREMENT
　　　OPTICAL MEASURING INSTRUMENTS
　　　OPTICAL MEMORY (DATA STORAGE)
　　　OPTICAL MICROSCOPES
　　　OPTICAL PATHS
　　　OPTICAL POLARIZATION
　　　OPTICAL PROPERTIES
　　　OPTICAL PUMPING
　　　OPTICAL PYROMETERS
　　　OPTICAL RADAR
　　　OPTICAL RANGE FINDERS
　　　OPTICAL REFLECTION

OPTICS-*(CONT.)*
      OPTICAL RELAY SYSTEMS
      OPTICAL RESONANCE
      OPTICAL RESONATORS
      OPTICAL SATELLITE TRACKING
        PROGRAM
      OPTICAL SCANNERS
      OPTICAL SLANT RANGE
      OPTICAL THICKNESS
      OPTICAL TRACKING
      OPTICAL TRANSFER FUNCTION
      OPTICAL TRANSITION
      OPTICAL WAVEGUIDES
      PARALLAX
      PHOTICS
      PHOTOELASTIC ANALYSIS
      PHYSICAL OPTICS
      PREFOCUSING
      REFLECTION
      RESOLUTION
      SCATTER PLATES (OPTICS)
      ∞SCIENCE
      SNELLS LAW
      SPECTROSCOPY
      STARLAB
      UNDERWATER OPTICS

OPTIMAL CONTROL
UF   OPTIMUM CONTROL
GS   AUTOMATIC CONTROL
     . **OPTIMAL CONTROL**
     . . TIME OPTIMAL CONTROL
     OPTIMIZATION
     . **OPTIMAL CONTROL**
     . . TIME OPTIMAL CONTROL
RT   ADAPTIVE CONTROL
     ∞CONTROL
     CONTROL SYSTEMS DESIGN
     CONTROL THEORY
     FEEDBACK CONTROL
     FEEDFORWARD CONTROL
     INVENTORY CONTROLS
     KALMAN-SCHMIDT FILTERING
     PARAMETER IDENTIFICATION
     TRACKING PROBLEM
     TRAJECTORY CONTROL

OPTIMIZATION
UF   MINIMIZATION
     REDUCTION (MATHEMATICS)
GS   **OPTIMIZATION**
     . FLIGHT OPTIMIZATION
     . OPTIMAL CONTROL
     . . TIME OPTIMAL CONTROL
     . TRAJECTORY OPTIMIZATION
RT   ∞APPLICATIONS OF MATHEMATICS
     BELLMAN THEORY
     BOLZA PROBLEMS
     CONSTRAINTS
     CORRELATION
     ∞DESIGN
     DESIGN ANALYSIS
     DIFFERENTIAL CALCULUS
     EFFICIENCY
     EXTREMUM VALUES
     GRADIENTS
     HESSIAN MATRICES
     KALMAN FILTERS
     KALMAN-SCHMIDT FILTERING
     LAGRANGE MULTIPLIERS
     LEAST SQUARES METHOD
     LINEAR PROGRAMMING
     MATHEMATICAL PROGRAMMING
     MAXIMA
     MINIMA
     OPERATIONS RESEARCH
     PARAMETER IDENTIFICATION
     PENALTY FUNCTION
     PLANNING
     PONTRYAGIN PRINCIPLE
     QUALITY CONTROL
     RANGE (EXTREMES)
     ∞REDUCTION
     SCHEDULING
     SIMPLEX METHOD
     STATIC MODELS
     STEEPEST DESCENT METHOD
     STOPPING
     SYSTEM IDENTIFICATION
     TRAJECTORY CONTROL
     WIENER FILTERING

OPTIMUM CONTROL
USE   OPTIMAL CONTROL

OPTIMUM THRUST PROGRAMMING
USE   THRUST PROGRAMMING

OPTIONS
RT   ALTERNATIVES
     CONTRACTS
     SELECTION
     SITE SELECTION
     SUBCONTRACTS

OPTOGALVANIC SPECTROSCOPY
GS   SPECTROSCOPY
     . ABSORPTION SPECTROSCOPY
     . . **OPTOGALVANIC SPECTROSCOPY**
RT   FLAME SPECTROSCOPY
     FRAUNHOFER LINES
     GAS SPECTROSCOPY
     INFRARED SPECTROSCOPY
     MOLECULAR SPECTROSCOPY
     OPTICAL EQUIPMENT
     OPTICAL MEASURING INSTRUMENTS
     RAMAN SPECTROSCOPY
     ULTRAVIOLET SPECTROSCOPY
     VOLT-AMPERE CHARACTERISTICS

OPTOMETRY
GS   OPTICAL MEASUREMENT
     . **OPTOMETRY**
RT   ANASTIGMATISM
     BLINDNESS
     EYE (ANATOMY)
     HAPLOSCOPES
     ∞MEASUREMENT
     MEDICAL SCIENCE
     VISION

OR-GATES
USE   GATES (CIRCUITS)

ORAL HYGIENE
GS   HYGIENE
     . **ORAL HYGIENE**
RT   CLEANLINESS
     DENTISTRY
     HEALTH
     PUBLIC HEALTH
     TEETH
     TOOTH DISEASES

ORATORY
USE   PUBLIC SPEAKING

ORBIS
UF   ORBITING RADIO BEACON IONOSPHERIC
     SOUNDER
GS   SATELLITES
     . ARTIFICIAL SATELLITES
     . . SCIENTIFIC SATELLITES
     . . . **ORBIS**
     . . . . ORBIS CAL SATELLITE
RT   IONOSPHERIC PROPAGATION
     IONOSPHERIC SOUNDING
     RADIO BEACONS

ORBIS CAL SATELLITE
GS   SATELLITES
     . ARTIFICIAL SATELLITES
     . . GRAVITY GRADIENT SATELLITES
     . . . **ORBIS CAL SATELLITE**
     . . SCIENTIFIC SATELLITES
     . . . ORBIS
     . . . . **ORBIS CAL SATELLITE**
RT   IONOSPHERIC PROPAGATION
     IONOSPHERIC SOUNDING
     RADIO BEACONS

ORBIT CALCULATION
UF   SATELLITE ORBIT CALCULATION
GS   COMPUTATION
     . **ORBIT CALCULATION**
     . . MINIMUM VARIANCE ORBIT
       DETERMINATION
RT   FLIGHT MECHANICS
     GODDARD TRAJECTORY
       DETERMINATION SYSTEM
     ORBITAL ELEMENTS
     ORBITAL MECHANICS
     ORBITAL POSITION ESTIMATION
     QUADRATURES

ORBIT DECAY
RT   AERODYNAMIC DRAG
     ATMOSPHERIC ENTRY
     FLIGHT MECHANICS

ORBIT DECAY-*(CONT.)*
     ORBITAL MECHANICS
     SATELLITE LIFETIME

ORBIT EQUATIONS
USE   ORBITAL MECHANICS

ORBIT MANEUVERING ENGINE (SPACE SHUTTLE)
UF   OME
GS   ENGINES
     . ROCKET ENGINES
     . . MICROROCKET ENGINES
     . . . **ORBIT MANEUVERING ENGINE**
       (SPACE SHUTTLE)
RT   AEROMANEUVERING ORBIT TO ORBIT
     SHUTTLE
     ORBITAL MANEUVERS
     SPACE SHUTTLES

ORBIT PERTURBATION
UF   EVECTION
GS   PERTURBATION
     . **ORBIT PERTURBATION**
     . . SATELLITE PERTURBATION
RT   DRIFT RATE
     LONG TERM EFFECTS
     LUNAR EFFECTS
     ORBITAL ELEMENTS
     ORBITAL MECHANICS
     PERTURBATION THEORY
     SCHACH EFFECT
     VINTI THEORY

ORBIT SPECTRUM UTILIZATION
RT   COMMUNICATION SATELLITES
     FREQUENCY ASSIGNMENT
     RADIO RELAY SYSTEMS
     SATELLITE ORBITS
     SYSTEMS ENGINEERING
     TELEVISION SYSTEMS

ORBIT TRANSFER VEHICLES
UF   OTV
RT   INERTIAL UPPER STAGE
     ORBITAL MANEUVERING VEHICLES
     ORBITAL SERVICING
     PAYLOAD DELIVERY (STS)
     PAYLOAD DEPLOYMENT & RETRIEVAL
       SYSTEM
     PAYLOAD RETRIEVAL (STS)
     SPACE SHUTTLES
     SPACE TRANSPORTATION
     SPACE TUGS
     ∞SPACECRAFT
     ∞VEHICLES

ORBITAL ASSEMBLY
UF   CONSTRUCTION IN SPACE
     SPACECRAFT ORBITAL ASSEMBLY
GS   ASSEMBLING
     . **ORBITAL ASSEMBLY**
RT   EXPANDABLE STRUCTURES
     INFLATABLE SPACECRAFT
     SELF ERECTING DEVICES
     SPACE ERECTABLE STRUCTURES
     SPACE OPERATIONS CENTER (NASA)
     SPACECRAFT MODULES
     SPACECRAFT STRUCTURES

ORBITAL ELEMENTS
RT   ∞ELEMENTS
     ORBIT CALCULATION
     ORBIT PERTURBATION
     PERTURBATION THEORY
     SLATER ORBITALS

ORBITAL FLIGHT TEST 1 (SHUTTLE)
USE   SPACE TRANSPORTATION SYSTEM 1
     FLIGHT

ORBITAL FLIGHT TEST 2 (SHUTTLE)
USE   SPACE TRANSPORTATION SYSTEM 2
     FLIGHT

ORBITAL FLIGHT TEST 3 (SHUTTLE)
USE   SPACE TRANSPORTATION SYSTEM 3
     FLIGHT

ORBITAL FLIGHT TEST 4 (SHUTTLE)
USE   SPACE TRANSPORTATION SYSTEM 4
     FLIGHT

ORBITAL FLIGHT TESTS (SHUTTLE)
USE   SPACE TRANSPORTATION SYSTEM
        FLIGHTS

ORBITAL LAUNCHING
SN   (LAUNCHING FROM AN
        ORBIT--EXCLUDES LAUNCHING INTO
        ORBIT FROM GROUND)
GS   LAUNCHING
        . ROCKET LAUNCHING
        . . **ORBITAL LAUNCHING**
RT   INTERPLANETARY TRAJECTORIES
        LUNAR LAUNCH
        PAYLOAD DELIVERY (STS)
        SPACECRAFT LAUNCHING
        TRANSFER ORBITS

ORBITAL LIFETIME
RT   ATTITUDE CONTROL
        EARTH ORBITS

ORBITAL MANEUVERING VEHICLES
RT   ORBIT TRANSFER VEHICLES
        ORBITAL SERVICING
        POWER MODULES (STS)
        REMOTELY PILOTED VEHICLES
        SATELLITES
        ∞ SPACECRAFT

ORBITAL MANEUVERS
GS   MANEUVERS
        . SPACECRAFT MANEUVERS
        . . **ORBITAL MANEUVERS**
RT   ORBIT MANEUVERING ENGINE (SPACE
        SHUTTLE)
        SPACE NAVIGATION
        SPACE SHUTTLES

ORBITAL MECHANICS
UF   ORBIT EQUATIONS
GS   CLASSICAL MECHANICS
        . SPACE MECHANICS
        . . **ORBITAL MECHANICS**
        . . . KEPLER LAWS
        . . . MINIMUM VARIANCE ORBIT
            DETERMINATION
RT   AEROMANEUVERING ORBIT TO ORBIT
        SHUTTLE
        APSIDES
        ASTRODYNAMICS
        CELESTIAL MECHANICS
        CIRCULAR ORBITS
        DRIFT RATE
        EARTH ORBITAL RENDEZVOUS
        EARTH ORBITS
        EARTH-MARS TRAJECTORIES
        EARTH-MERCURY TRAJECTORIES
        EARTH-MOON SYSTEM
        ELLIPTICAL ORBITS
        EQUATORIAL ORBITS
        FLIGHT MECHANICS
        FLIGHT OPTIMIZATION
        GODDARD TRAJECTORY
            DETERMINATION SYSTEM
        HANSEN LUNAR THEORY
        HILL LUNAR THEORY
        HILL METHOD
        INTERPLANETARY TRAJECTORIES
        INTERPLANETARY TRANSFER ORBITS
        LAGRANGIAN EQUILIBRIUM POINTS
        LUNAR ORBITAL RENDEZVOUS
        LUNAR ORBITS
        MANY BODY PROBLEM
    ∞ MECHANICS (PHYSICS)
        MOON-EARTH TRAJECTORIES
        ORBIT CALCULATION
        ORBIT DECAY
        ORBIT PERTURBATION
        ORBITS
        PARKING ORBITS
        PERTURBATION
        PLANETARY LANDING
        POYNTING-ROBERTSON EFFECT
        QUADRATURES
        RENDEZVOUS
        RENDEZVOUS TRAJECTORIES
        ROUND TRIP TRAJECTORIES
        SATELLITE ORBITS
        SATELLITE PERTURBATION
        SATELLITES
        SPACE NAVIGATION
        SPACECRAFT ORBITS
        STATIONKEEPING
        SWINGBY TECHNIQUE
        THRUST PROGRAMMING

ORBITAL MECHANICS-(CONT.)
        TRAJECTORY ANALYSIS
        TRANSEARTH INJECTION
        TRANSFER ORBITS
        TRANSLUNAR INJECTION
        TWENTY-FOUR HOUR ORBITS
        TWO BODY PROBLEM

ORBITAL MOTION
USE   ORBITS

ORBITAL POSITION ESTIMATION
GS   ESTIMATING
        . **ORBITAL POSITION ESTIMATION**
RT   CELESTIAL SPHERE
        GODDARD TRAJECTORY
            DETERMINATION SYSTEM
        NAVIGATION
        ORBIT CALCULATION
    ∞ ORIENTATION
        POSITION (LOCATION)
        POSITION ERRORS
    ∞ RANGE
        SATELLITE ORBITS
        SPACECRAFT ORBITS
        SPACECRAFT POSITION INDICATORS
        STATE ESTIMATION

ORBITAL RENDEZVOUS
UF   SATELLITE RENDEZVOUS
GS   MANEUVERS
        . **ORBITAL RENDEZVOUS**
        . . EARTH ORBITAL RENDEZVOUS
        . . LUNAR ORBITAL RENDEZVOUS
        RENDEZVOUS
        . SPACE RENDEZVOUS
        . . **ORBITAL RENDEZVOUS**
        . . . LUNAR ORBITAL RENDEZVOUS
RT   ATLAS LAUNCH VEHICLES
        MULTIPLE DOCKING ADAPTERS
        PAYLOAD RETRIEVAL (STS)
        RENDEZVOUS GUIDANCE
        RENDEZVOUS SPACECRAFT
        RENDEZVOUS TRAJECTORIES
        SPACECRAFT DOCKING
        SPACECRAFT TRAJECTORIES
        TETHERING

ORBITAL SERVICING
RT   LARGE SPACE STRUCTURES
        MANNED MANEUVERING UNITS
        ORBIT TRANSFER VEHICLES
        ORBITAL MANEUVERING VEHICLES
        ORBITAL SPACE STATIONS
        PAYLOAD TRANSFER
        SPACE OPERATIONS CENTER (NASA)
        SPACE PLATFORMS
        SPACE SHUTTLE PAYLOADS
        SPACE TRANSPORTATION SYSTEM
        SPACE TUGS

ORBITAL SHOTS
RT   ∞ SHOT
        SPACECRAFT LAUNCHING

ORBITAL SIMULATORS
USE   SPACE SIMULATORS

ORBITAL SPACE STATIONS
UF   MANNED ORBITAL SPACE STATIONS
        MOSS (SPACE STATIONS)
GS   MANNED SPACECRAFT
        . SPACE STATIONS
        . . **ORBITAL SPACE STATIONS**
        . . . LONG DURATION EXPOSURE
            FACILITY
        . . . ORBITAL WORKSHOPS
        . . . SALYUT SPACE STATION
        . . . SKYLAB 1
        . . . SKYLAB 2
        . . . SKYLAB 3
        . . . SKYLAB 4
        . . . SPACE OPERATIONS CENTER
            (NASA)
        SATELLITES
        . ARTIFICIAL SATELLITES
        . . **ORBITAL SPACE STATIONS**
        . . . EOSS
        . . . LONG DURATION EXPOSURE
            FACILITY
        . . . ORBITAL WORKSHOPS
        STATIONS
        . SPACE STATIONS
        . . **ORBITAL SPACE STATIONS**
        . . . HALO ORBIT SPACE STATION

ORBITAL SPACE STATIONS-(CONT.)
        . . . LONG DURATION EXPOSURE
            FACILITY
        . . . ORBITAL WORKSHOPS
        . . . ORBITING LUNAR STATIONS
        . . . SALYUT SPACE STATION
        . . . SPACE OPERATIONS CENTER
            (NASA)
RT   LARGE SPACE STRUCTURES
        MANNED ORBITAL LABORATORIES
        MANNED ORBITAL RESEARCH
            LABORATORIES
        ORBITAL SERVICING
        SPACE BASES
        SPACE COLONIES
        SPACE MECHANICS
        SPACE STATIONS
        SPACECRAFT DOCKING
        STRUCTURAL ANALYSIS
    ∞ TESTS

ORBITAL SPACE TESTS
RT   LARGE SPACE STRUCTURES
        SPACE MECHANICS
        SPACE STATIONS
        STRUCTURAL ANALYSIS
    ∞ TESTS

ORBITAL TEST SATELLITE (ESA)
USE   OTS (ESA)

ORBITAL TRANSFER
USE   TRANSFER ORBITS

ORBITAL VELOCITY
GS   RATES (PER TIME)
        . **ORBITAL VELOCITY**
        VELOCITY
        . **ORBITAL VELOCITY**
RT   ANGULAR VELOCITY
        ESCAPE VELOCITY
    ∞ HYPERVELOCITY
        VELOCITY ERRORS

ORBITAL WORKERS
GS   PERSONNEL
        . FLYING PERSONNEL
        . . ASTRONAUTS
        . . . **ORBITAL WORKERS**
RT   ASTRONAUT LOCOMOTION
        EXTRAVEHICULAR ACTIVITY
        SPACE MAINTENANCE
        SPACE TOOLS
        WORK CAPACITY

ORBITAL WORKSHOPS
GS   MANNED SPACECRAFT
        . SPACE STATIONS
        . . ORBITAL SPACE STATIONS
        . . . **ORBITAL WORKSHOPS**
        SATELLITES
        . ARTIFICIAL SATELLITES
        . . ORBITAL SPACE STATIONS
        . . . **ORBITAL WORKSHOPS**
        STATIONS
        . SPACE STATIONS
        . . ORBITAL SPACE STATIONS
        . . . **ORBITAL WORKSHOPS**
RT   APOLLO EXTENSION SYSTEM
        CONTAINERLESS MELTS
        MANNED ORBITAL LABORATORIES
        SKYLAB PROGRAM
        SKYLAB 1
        SKYLAB 2
        SKYLAB 3
        SKYLAB 4
        SPACE PROCESSING

ORBITALS
GS   **ORBITALS**
        . ELECTRON ORBITALS
        . MOLECULAR ORBITALS
        . SLATER ORBITALS
RT   JAHN-TELLER EFFECT
        ORBITS
        SCHWARZSCHILD METRIC

ORBITER PROJECT
GS   PROGRAMS
        . PROJECTS
        . . **ORBITER PROJECT**
RT   LUNAR ORBITER
        LUNAR ORBITER 1
        LUNAR ORBITER 2
        LUNAR ORBITER 3

ORBITER PROJECT-(CONT.)
    LUNAR ORBITER 4
    LUNAR ORBITER 5

ORBITING ASTRONOMICAL OBSERVATORY
  USE   OAO

ORBITING DIPOLES
  GS   ELECTRIC CHARGE
    . ELECTRIC DIPOLES
    . . **ORBITING DIPOLES**
  RT   COMMUNICATION EQUIPMENT
    ∞ DIPOLES

ORBITING FROG OTOLITH
  GS   SATELLITES
    . ARTIFICIAL SATELLITES
    . . BIOSATELLITES
    . . . **ORBITING FROG OTOLITH**
  RT   BIOLOGICAL EFFECTS
    BIOMETRICS
    BIOTELEMETRY
    INSTRUMENT PACKAGES
    INTERNATIONAL COOPERATION
    ITALIAN SPACE PROGRAM

ORBITING GEOPHYSICAL OBSERVATORY
  USE   OGO

ORBITING LUNAR STATIONS
  GS   STATIONS
    . SPACE STATIONS
    . . ORBITAL SPACE STATIONS
    . . . **ORBITING LUNAR STATIONS**
  RT   EOSS
    LUNAR SPACECRAFT
    ∞ SPACECRAFT

ORBITING RADIO BEACON IONOSPHERIC SOUNDER
  USE   ORBIS

ORBITING SATELLITES
  USE   ARTIFICIAL SATELLITES

ORBITING SOLAR OBSERVATORY
  USE   OSO

ORBITRONS
  RT   ELECTRON CLOUDS
    ELECTRON TUBES
    IONIZATION GAGES
    SPACE CHARGE
    VACUUM GAGES

ORBITS
  UF   ORBITAL MOTION
    PERIODIC ORBITS
  GS   **ORBITS**
    . APHELIONS
    . CIRCULAR ORBITS
    . . STATIONARY ORBITS
    . EARTH ORBITS
    . . APOGEES
    . . PERIGEES
    . ECCENTRIC ORBITS
    . ELLIPTICAL ORBITS
    . . APOGEES
    . . PERIGEES
    . . PERIHELIONS
    . . TRANSFER ORBITS
    . . . INTERPLANETARY TRANSFER ORBITS
    . EQUATORIAL ORBITS
    . . STATIONARY ORBITS
    . LUNAR ORBITS
    . PLANETARY ORBITS
    . SOLAR ORBITS
    . . PERIHELIONS
    . SPACECRAFT ORBITS
    . . SATELLITE ORBITS
    . . . GEOSYNCHRONOUS ORBITS
    . . . PARKING ORBITS
    . . . POLAR ORBITS
    . . . STATIONARY ORBITS
    . . . TWENTY-FOUR HOUR ORBITS
    . . . TRANSFER ORBITS
    . . . INTERPLANETARY TRANSFER ORBITS
    . . TROJAN ORBITS
    . STELLAR ORBITS
  RT   AIRBORNE RANGE AND ORBIT DETERMINATION
    APEXES

ORBITS-(CONT.)
    ARTIFICIAL SATELLITES
    ASTRODYNAMICS
    CELESTIAL BODIES
    CELESTIAL MECHANICS
    ∞ CONJUNCTION
    EARTH-VENUS TRAJECTORIES
    EPHEMERIDES
    FLIGHT OPTIMIZATION
    FLIGHT PATHS
    FOUR BODY PROBLEM
    GROUND TRACKS
    ∞ INCLINATION
    INTERPLANETARY FLIGHT
    LUNAR FLIGHT
    MANY BODY PROBLEM
    ∞ MOTION
    ORBITAL MECHANICS
    ORBITALS
    ∞ PATHS
    QUADRATURES
    ROCHE LIMIT
    SATELLITE GROUND TRACKS
    SATELLITES
    SCHWARZSCHILD METRIC
    SPACE FLIGHT
    SPACE NAVIGATION
    SPACECRAFT GUIDANCE
    STATIONKEEPING
    SUBORBITAL FLIGHT
    THREE BODY PROBLEM
    TRAJECTORIES
    TWO BODY PROBLEM

ORCHARDS
  GS   PLANTS (BOTANY)
    . **ORCHARDS**
  RT   AGRICULTURE
    BLIGHT
    CITRUS TREES
    CROP GROWTH
    CROP VIGOR
    ∞ CROPS
    CURING
    FARM CROPS
    ∞ FOOD
    FROST DAMAGE
    FRUITS
    IRRIGATION
    NUTS (FRUITS)
    RURAL LAND USE
    SILVICULTURE
    TREES (PLANTS)

ORDER-DISORDER TRANSFORMATIONS
  RT   ATOMIC STRUCTURE
    CRYSTAL DEFECTS
    CRYSTAL LATTICES
    CRYSTAL STRUCTURE
    CRYSTALLOGRAPHY
    HOLES (ELECTRON DEFICIENCIES)
    METALLOGRAPHY
    MICROSTRUCTURE
    MOLECULAR STRUCTURE
    PHASE TRANSFORMATIONS
    SOLID SOLUTIONS
    ∞ TRANSFORMATIONS

ORDNANCE
  RT   AIR TO SURFACE MISSILES
    AMMUNITION
    ARMOR
    BALLISTICS
    EXPLOSIVES
    GROUND SUPPORT EQUIPMENT
    PYROTECHNICS
    TANKS (COMBAT VEHICLES)
    TRAJECTORIES
    WARFARE
    WEAPON SYSTEMS
    WEAPONS

OREGON
  GS   NATIONS
    . UNITED STATES
    . . **OREGON**
  RT   CASCADE RANGE (CA-OR-WA)
    COLUMBIA RIVER BASIN (ID-OR-WA)

ORES
  USE   MINERALS

ORGAN WEIGHT
  GS   WEIGHT (MASS)
    . **ORGAN WEIGHT**

ORGANIC ALUMINUM COMPOUNDS
  GS   ALUMINUM COMPOUNDS
    . **ORGANIC ALUMINUM COMPOUNDS**
    ORGANOMETALLIC COMPOUNDS
    . **ORGANIC ALUMINUM COMPOUNDS**
  RT   ∞ CHEMICAL COMPOUNDS
    ∞ METAL COMPOUNDS

ORGANIC BORON COMPOUNDS
  GS   BORON COMPOUNDS
    . **ORGANIC BORON COMPOUNDS**
  RT   ∞ CHEMICAL COMPOUNDS

ORGANIC CHARGE TRANSFER SALTS
  GS   ORGANIC COMPOUNDS
    . **ORGANIC CHARGE TRANSFER SALTS**
  RT   CHARGE TRANSFER DEVICES
    ∞ SALTS
    SEMICONDUCTORS (MATERIALS)

ORGANIC CHEMISTRY
  RT   BIOCHEMISTRY
    ∞ CHEMISTRY
    CRACKING (CHEMICAL ENGINEERING)
    CYCLIC COMPOUNDS
    DIELS-ALDER REACTIONS
    HISTOCHEMICAL ANALYSIS
    METHOXY SYSTEMS
    OPTICAL ACTIVITY
    PHYSIOCHEMISTRY

ORGANIC COMPOUNDS
  GS   **ORGANIC COMPOUNDS**
    . AMIDASE
    . AMINO ACIDS
    . . ALANINE
    . . . PHENYLALANINE
    . . ASPARTIC ACID
    . . CYSTEINE
    . FOLIC ACID
    . . GLUTAMIC ACID
    . . GLUTAMINE
    . . GLUTATHIONE
    . . GLYCINE
    . . HIPPURIC ACID
    . . HISTIDINE
    . . LEUCINE
    . . . NORLEUCINE
    . . LYSINE
    . . MELANOIDIN
    . . METHIONINE
    . . PAPAIN
    . . PEPTIDES
    . . . HYPERTENSIN
    . . . . POLYPEPTIDES
    . . PROTOPROTEINS
    . . PYRIDINE NUCLEOTIDES
    . . THYROXINE
    . . TRYPTOPHAN
    . . TYROSINE
    . URIDYLIC ACID
    . ASPARTATES
    . CASTOR OIL
    . CYCLIC COMPOUNDS
    . . CYCLIC AMP
    . DOPA
    . FATS
    . . CHOLINE
    . FATTY ACIDS
    . . ACETIC ACID
    . . . ETHYLENEDIAMINETETRAACETIC ACIDS
    . . . IODOACETIC ACID
    . . LIPOIC ACID
    . . OLEIC ACID
    . . PROPIONIC ACID
    . . SEBACIC ACID
    . . VALERIC ACID
    . FLUORINE ORGANIC COMPOUNDS
    . . FLUOROAMINES
    . . . NITROFLUORAMINES
    . . . TRIFLUOROAMINE OXIDE
    . . FLUOROCARBONS
    . . FLUOROHYDROCARBONS
    . . CARBON TETRAFLUORIDE
    . KEL-F
    . PERFLUOROALKANE
    . PERFLUOROGUANIDINE
    . FURANS
    . TETRAHYDROFURAN
    . NUCLEASE
    . NUCLEOTIDES
    . . ADENINES
    . . ADENOSINES
    . . . ADENOSINE DIPHOSPHATE

ORGANIC COMPOUNDS-*(CONT.)*
- . . . ADENOSINE TRIPHOSPHATE
- . . . CYCLIC AMP
- . . POLYNUCLEOTIDES
- . . PYRIDINE NUCLEOTIDES
- . . URIDYLIC ACID
- . ORGANIC CHARGE TRANSFER SALTS
- . ORGANIC LIQUIDS
- . ORGANIC SULFUR COMPOUNDS
- . OXIDASE
- . PENTANONE
- . PHOSPHAZENE
- . POLYNUCLEAR ORGANIC COMPOUNDS
- . PROPARGYL GROUPS
- . QUINOLINE
- . QUINOXALINES
- . SEROTONIN
- RT   ACETYL COMPOUNDS
-      ACIDS
-      ∞AROMATIC COMPOUNDS
-      ∞CHEMICAL COMPOUNDS
-      CHEMICAL EVOLUTION
-      METABOLITES
-      METHOXY SYSTEMS
-      NITROSO COMPOUNDS
-      ORGANIC WASTES (FUEL CONVERSION)
-      ORGANOMETALLIC COMPOUNDS
-      PROSTAGLANDINS
-      ∞SALTS

ORGANIC COOLANTS
- GS   COOLANTS
-      . **ORGANIC COOLANTS**

ORGANIC COOLED REACTORS
- UF   ORGEL REACTOR
- GS   NUCLEAR REACTORS
-      . LIQUID COOLED REACTORS
-      . . **ORGANIC COOLED REACTORS**
-      . . . EXPERIMENTAL ORGANIC COOLED
-           REACTORS
- RT   REACTOR DESIGN
-      REACTOR TECHNOLOGY

ORGANIC FLUORINE COMPOUNDS
- USE   FLUORINE ORGANIC COMPOUNDS

ORGANIC GERMANIUM COMPOUNDS
- GS   GERMANIUM COMPOUNDS
-      . **ORGANIC GERMANIUM COMPOUNDS**
-      ORGANOMETALLIC COMPOUNDS
-      . **ORGANIC GERMANIUM COMPOUNDS**
- RT   ∞CHEMICAL COMPOUNDS
-      ∞METAL COMPOUNDS

ORGANIC LASERS
- GS   STIMULATED EMISSION DEVICES
-      . LASERS
-      . . **ORGANIC LASERS**
-      . . . DYE LASERS
- RT   CARBON DIOXIDE LASERS
-      CARBON LASERS
-      CHEMICAL LASERS
-      GAS LASERS
-      INFRARED LASERS
-      LIQUID LASERS

ORGANIC LIQUIDS
- GS   LIQUIDS
-      . **ORGANIC LIQUIDS**
-      ORGANIC COMPOUNDS
-      . **ORGANIC LIQUIDS**
- RT   PYRUVATES
-      XANTHIC ACIDS

ORGANIC LITHIUM COMPOUNDS
- GS   LITHIUM COMPOUNDS
-      . **ORGANIC LITHIUM COMPOUNDS**
-      ORGANOMETALLIC COMPOUNDS
-      . **ORGANIC LITHIUM COMPOUNDS**
- RT   ∞CHEMICAL COMPOUNDS
-      ∞METAL COMPOUNDS

ORGANIC MATERIALS
- GS   **ORGANIC MATERIALS**
-      . PEAT
- RT   BIODEGRADABILITY
-      CARBONACEOUS MATERIALS
-      CORK (MATERIALS)
-      COTTON FIBERS
-      ELASTOMERS
-      LINEN
-      ∞MATERIALS
-      MOLDS
-      PAPER (MATERIAL)

ORGANIC MATERIALS-*(CONT.)*
- PHASE CHANGE MATERIALS
- PLASTICS
- ∞POLYMERS
- ROSIN
- RUBBER
- SILK
- THERMOCHROMATIC MATERIALS
- WOOD
- WOOL

ORGANIC MODERATED REACTORS
- GS   NUCLEAR REACTORS
-      . **ORGANIC MODERATED REACTORS**
-      . . EXPERIMENTAL ORGANIC COOLED
-          REACTORS

ORGANIC NITRATES
- GS   ESTERS
-      . **ORGANIC NITRATES**
-      . . CELLULOSE NITRATE
-      . . NITROFORMS
-      . . . HYDRAZINE NITROFORM
-      . . NITROGLYCERIN
-      . . PETN
-      NITROGEN COMPOUNDS
-      . NITRATES
-      . **ORGANIC NITRATES**
-      . . CELLULOSE NITRATE
-      . . . NITROFORMS
-      . . . . HYDRAZINE NITROFORM
-      . . . NITROGLYCERIN
-      . . . PETN

ORGANIC PEROXIDES
- GS   CHALCOGENIDES
-      . OXIDES
-      . . ANHYDRIDES
-      . . . PEROXIDES
-      . . . . **ORGANIC PEROXIDES**
- RT   AIR POLLUTION
-      HYDROCARBONS
-      INORGANIC PEROXIDES

ORGANIC PHOSPHORUS COMPOUNDS
- GS   PHOSPHORUS COMPOUNDS
-      . **ORGANIC PHOSPHORUS COMPOUNDS**
-      . . PHOSPHONITRILES
-      . . URIDYLIC ACID
- RT   ∞CHEMICAL COMPOUNDS

ORGANIC SEMICONDUCTORS
- GS   SEMICONDUCTORS (MATERIALS)
-      . **ORGANIC SEMICONDUCTORS**
- RT   ∞CHEMICAL COMPOUNDS
-      CONDUCTORS
-      SEMICONDUCTOR DEVICES

ORGANIC SILICON COMPOUNDS
- GS   SILICON COMPOUNDS
-      . **ORGANIC SILICON COMPOUNDS**
-      . . TRIPHENYL SILICON
- RT   ∞CHEMICAL COMPOUNDS

ORGANIC SOLIDS
- RT   ASTRONOMICAL SPECTROSCOPY
-      COSMIC DUST
-      PLANETARY ATMOSPHERES
-      SOLIDS

ORGANIC SULFUR COMPOUNDS
- GS   ORGANIC COMPOUNDS
-      . **ORGANIC SULFUR COMPOUNDS**
-      SULFUR COMPOUNDS
-      . **ORGANIC SULFUR COMPOUNDS**
- RT   ∞CHEMICAL COMPOUNDS

ORGANIC TIN COMPOUNDS
- GS   ORGANOMETALLIC COMPOUNDS
-      . **ORGANIC TIN COMPOUNDS**
-      TIN COMPOUNDS
-      . **ORGANIC TIN COMPOUNDS**
- RT   ∞CHEMICAL COMPOUNDS
-      ∞METAL COMPOUNDS

ORGANIC WASTES (FUEL CONVERSION)
- RT   ∞CONVERSION
-      ENERGY CONVERSION
-      GARBAGE
-      HUMAN WASTES
-      METABOLIC WASTES
-      ORGANIC COMPOUNDS
-      RESIDUES
-      SEWAGE

ORGANIC WASTES (FUEL-*(CONT.)*
- SLUDGE
- WASTES

ORGANISMS
- RT   ANIMALS
-      BIOMASS
-      CARBON CYCLE
-      DEEP SCATTERING LAYERS
-      PHYSIOLOGY
-      PLANTS (BOTANY)

ORGANIZATIONS
- UF   ASSOCIATIONS
- GS   **ORGANIZATIONS**
-      . EUROPEAN SPACE AGENCY
-      . FEDERATIONS
-      . . BUREAUS (ORGANIZATIONS)
-      . ISRO
-      . NORTH ATLANTIC TREATY
-          ORGANIZATION (NATO)
-      . WORLD METEOROLOGICAL
-          ORGANIZATION
- RT   TEAMS
-      UNITED NATIONS

ORGANIZING
- RT   PERSONNEL
-      UNIONIZATION

ORGANOMETALLIC COMPOUNDS
- UF   METALLORGANIC COMPOUNDS
- GS   **ORGANOMETALLIC COMPOUNDS**
-      . ALKYLFERROCENE
-      . CHLOROPHYLLS
-      . FERROCENES
-      . HEMOGLOBIN
-      . . CARBOXYHEMOGLOBIN
-      . . OXYHEMOGLOBIN
-      . METALLOXANE POLYMER
-      . ORGANIC ALUMINUM COMPOUNDS
-      . ORGANIC GERMANIUM COMPOUNDS
-      . ORGANIC LITHIUM COMPOUNDS
-      . ORGANIC TIN COMPOUNDS
-      . PORPHINES
- RT   CHELATES
-      ∞CHEMICAL COMPOUNDS
-      ∞METAL COMPOUNDS
-      METALLOIDS
-      METALLOSILOXANE POLYMER
-      ORGANIC COMPOUNDS
-      SYNTHETIC METALS

ORGANOMETALLIC POLYMERS
- RT   ∞POLYMERS

ORGANS
- GS   ANATOMY
-      . **ORGANS**
-      . . BLADDER
-      . . ESOPHAGUS
-      . . KIDNEYS
-      . . LIVER
-      . . LUNGS
-      . . OVARIES
-      . . PITUITARY GLAND
-      . . SPLEEN
-      . . STOMACH
-      . . TESTES
-      VISCERA
-      . **ORGANS**
-      . . BLADDER
-      . . ESOPHAGUS
-      . . KIDNEYS
-      . . LIVER
-      . . LUNGS
-      . . OVARIES
-      . . PITUITARY GLAND
-      . . SPLEEN
-      . . STOMACH
-      . . TESTES
- RT   DIGESTIVE SYSTEM
-      GASTROINTESTINAL SYSTEM
-      GLANDS (ANATOMY)
-      PNEUMOTHORAX
-      PULMONARY LESIONS

ORGEL REACTOR
- USE   ORGANIC COOLED REACTORS

ORGUEIL METEORITE
- GS   CELESTIAL BODIES
-      . METEORITES
-      . . STONY METEORITES
-      . . . CHONDRITES

ORGUEIL METEORITE-_(CONT.)_
..... CARBONACEOUS METEORITES
..... **ORGUEIL METEORITE**

ORIC CYCLOTRON
USE OAK RIDGE ISOCHRONOUS CYCLOTRON

∞ **ORIENTATION**
SN _(USE OF A MORE SPECIFIC TERM IS_
RECOMMENDED--CONSULT THE TERMS
LISTED BELOW)
RT ALIGNMENT
ATTITUDE (INCLINATION)
AZIMUTH
BEARING (DIRECTION)
BRAGG ANGLE
COLLIMATION
CRYSTALLOGRAPHY
DIRECTIVITY
EDUCATION
FIBER ORIENTATION
FIELD STRENGTH
HORIZONTAL ORIENTATION
INSTRUMENT ORIENTATION
ISOTROPY
OPTICAL PROPERTIES
ORBITAL POSITION ESTIMATION
PLY ORIENTATION
POLARIZATION (SPIN ALIGNMENT)
POLARIZATION (WAVES)
POSITION (LOCATION)
POSITIONING
SOUND LOCALIZATION
VERTICAL ORIENTATION
VERTICAL PERCEPTION
VISUAL PERCEPTION

ORIFICE FLOW
GS FLUID FLOW
. **ORIFICE FLOW**
RT CRITICAL FLOW
∞FLOW
GAS FLOW
GRAZING FLOW
LAMINAR FLOW
LIQUID FLOW
MULTIPHASE FLOW
ORIFICES
PIPE FLOW
PRESSURE GRADIENTS
SINGLE-PHASE FLOW
STEADY FLOW
STEAM FLOW
SUBCRITICAL FLOW
SUPERCRITICAL FLOW
TURBULENT FLOW
UNSTEADY FLOW

ORIFICES
RT ANNULAR DUCTS
APERTURES
CAVITIES
CHOKES (FUEL SYSTEMS)
CHOKES (RESTRICTIONS)
DUCTS
FLOW MEASUREMENT
FLOWMETERS
GAPS
INJECTORS
∞NOZZLES
OPENINGS
∞OPERATIONS
ORIFICE FLOW
PORTS (OPENINGS)
SPRAY NOZZLES
THROATS
VENTURI TUBES

ORIGIN OF PLASMAS IN EARTH NEIGHBORHOOD
USE OPEN PROJECT

∞ **ORIGINS**
SN _(USE OF A MORE SPECIFIC TERM IS_
RECOMMENDED-CONSULT TERMS
LISTED BELOW)
RT CAUSES
COORDINATES
DERIVATION
GRAPHS (CHARTS)

ORION (RADIO INTERFEROMETRY NETWORK)
RT RADIO INTERFEROMETERS
RADIO RECEIVERS
TRACKING NETWORKS

ORION AIRCRAFT
USE P-3 AIRCRAFT

ORION CONSTELLATION
GS CONSTELLATIONS
. **ORION CONSTELLATION**
RT OPIK THEORY
SIGMA ORIONIS

ORION NEBULA
GS CELESTIAL BODIES
. NEBULAE
.. **ORION NEBULA**
RT ASTROPHYSICS
CASSIOPEIA A
CRAB NEBULA
GALAXIES
GUM NEBULA
INTERSTELLAR GAS
INTERSTELLAR MATTER
MAGELLANIC CLOUDS
MILKY WAY GALAXY
OPIK THEORY
PLANETARY NEBULAE
STELLAR CORONAS
SUPERNOVAE

ORIONID METEOROIDS
GS CELESTIAL BODIES
. METEOROID SHOWERS
.. **ORIONID METEOROIDS**
. METEOROIDS
.. **ORIONID METEOROIDS**
RT AQUARID METEOROIDS

ORLICZ SPACE
RT SET THEORY

ORNITHOPTER AIRCRAFT
USE RESEARCH AIRCRAFT

ORNSTEIN-UHLENBECK PROCESS
RT ∞PROCESSES

OROGRAPHIC CLOUDS
USE CAP CLOUDS

OROGRAPHY
GS GEOGRAPHY
. **OROGRAPHY**
GEOLOGY
. **OROGRAPHY**
RT CONES (VOLCANOES)
GEOMORPHOLOGY
ISOSTASY
MARS VOLCANOES
MOUNTAINS
PEAKS (LANDFORMS)
VOLCANOES
VOLCANOLOGY

ORR-SOMMERFELD EQUATIONS
RT ∞EQUATIONS
FLOW DISTORTION
FLOW STABILITY
FLOW THEORY
QUANTUM MECHANICS
VELOCITY DISTRIBUTION

ORRERIES
USE ASTRONOMICAL MODELS

ORTHICONS
GS ELECTRON TUBES
. CAMERA TUBES
.. **ORTHICONS**
... IMAGE ORTHICONS
MICROWAVE EQUIPMENT
. MICROWAVE TUBES
.. **ORTHICONS**
... IMAGE ORTHICONS
RT IMAGE INTENSIFIERS
PHOTOCATHODES
TELEVISION CAMERAS
TELEVISION EQUIPMENT

ORTHO HYDROGEN
GS CHEMICAL ELEMENTS
. HYDROGEN
.. **ORTHO HYDROGEN**
GASES
. HYDROGEN
.. **ORTHO HYDROGEN**

ORTHO PARA CONVERSION
GS ISOMERIZATION
. **ORTHO PARA CONVERSION**
RT ∞CONVERSION
PARA HYDROGEN

ORTHOGONAL FUNCTIONS
GS ANALYSIS (MATHEMATICS)
. COMPLEX VARIABLES
.. **ORTHOGONAL FUNCTIONS**
FUNCTIONS (MATHEMATICS)
. **ORTHOGONAL FUNCTIONS**
.. WALSH FUNCTION
RT BESSEL FUNCTIONS
EXPONENTIAL FUNCTIONS
FUNCTION SPACE
HANKEL FUNCTIONS
HYPERBOLIC FUNCTIONS
LAGUERRE FUNCTIONS
LEGENDRE FUNCTIONS
LINEAR TRANSFORMATIONS
MATHIEU FUNCTION
ORTHOGONALITY
ORTHONORMAL FUNCTIONS
QUALITY CONTROL

ORTHOGONAL MULTIPLEXING THEORY
RT PULSE COMMUNICATION
SIGNAL TRANSMISSION
∞THEORIES
WAVE INTERACTION
WAVELENGTH DIVISION MULTIPLEXING

ORTHOGONALITY
UF KRONECKER PRODUCT
GS MOMENTS
. DISTRIBUTION MOMENTS
.. **ORTHOGONALITY**
RT COVARIANCE
EXPERIMENT DESIGN
FACTOR ANALYSIS
ORTHOGONAL FUNCTIONS
QUALITY CONTROL

ORTHOGRAPHY
GS LANGUAGES
. **ORTHOGRAPHY**
RT HANDWRITING
INTELLIGIBILITY
LINGUISTICS
SEMANTICS
SYNTAX
WORDS (LANGUAGE)

ORTHONORMAL FUNCTIONS
GS FUNCTIONS (MATHEMATICS)
. **ORTHONORMAL FUNCTIONS**
RT ORTHOGONAL FUNCTIONS

ORTHOPEDICS
GS MEDICAL SCIENCE
. **ORTHOPEDICS**

ORTHOPHOTOGRAPHY
GS PHOTOGRAPHY
. **ORTHOPHOTOGRAPHY**
RT AERIAL PHOTOGRAPHY
COLOR PHOTOGRAPHY
MAPPING

ORTHOSTATIC TOLERANCE
RT BED REST
BLOOD PRESSURE
HUMAN TOLERANCES
LOWER BODY NEGATIVE PRESSURE
POSTURE
TOLERANCES (PHYSIOLOGY)

ORTHOTROPIC CYLINDERS
RT ∞CYLINDERS
CYLINDRICAL BODIES
CYLINDRICAL SHELLS
ROCKET ENGINE CASES

ORTHOTROPIC PLATES
GS STRUCTURAL MEMBERS
. PLATES (STRUCTURAL MEMBERS)
.. **ORTHOTROPIC PLATES**

ORTHOTROPIC SHELLS
GS SHELLS (STRUCTURAL FORMS)
. **ORTHOTROPIC SHELLS**
RT CYLINDRICAL SHELLS
METAL SHELLS

ORTHOTROPIC SHELLS-*(CONT.)*
    REINFORCED SHELLS
    SHELL STABILITY
    THIN WALLED SHELLS

ORTHOTROPISM
RT   PLATES (STRUCTURAL MEMBERS)

OSCILLATING CYLINDERS
RT   ∞CYLINDERS
    CYLINDRICAL BODIES
    CYLINDRICAL SHELLS
    OSCILLATIONS
    VIBRATION

OSCILLATING FLOW
GS   FLUID FLOW
    . UNSTEADY FLOW
    . . **OSCILLATING FLOW**
RT   BUFFETING
    COMPRESSIBILITY EFFECTS
    FLOW DISTORTION
    NONEQUILIBRIUM FLOW
    SMALL PERTURBATION FLOW
    STROUHAL NUMBER

OSCILLATION DAMPERS
RT   ∞ABSORBERS
    ∞DAMPERS
    MAXWELL BODIES
    NONOSCILLATORY ACTION
    NONSTABILIZED OSCILLATION
    NUTATION DAMPERS
    SPRINGS (ELASTIC)
    VIBRATION ISOLATORS

OSCILLATIONS
UF   PHUGOID OSCILLATIONS
GS   **OSCILLATIONS**
    . ELECTRON OSCILLATIONS
    . HARMONIC OSCILLATION
    . HYDROFOIL OSCILLATIONS
    . MOLECULAR OSCILLATIONS
    . NONOSCILLATORY ACTION
    . NONSTABILIZED OSCILLATION
    . PLASMA OSCILLATIONS
    . PRESSURE OSCILLATIONS
    . SELF OSCILLATION
    . STABLE OSCILLATIONS
    . STELLAR OSCILLATIONS
    . . SOLAR OSCILLATIONS
    . TRANSIENT OSCILLATIONS
    . TRANSVERSE OSCILLATION
    . . H WAVES
    . UNDAMPED OSCILLATIONS
    . WING OSCILLATIONS
RT   AMPLITUDES
    BRUNT-VAISALA FREQUENCY
    CRYSTAL OSCILLATORS
    DAMPING
    FEEDBACK
    ∞MOTION
    NONUNIFORMITY
    OSCILLATING CYLINDERS
    OSCILLATORS
    PENDULUMS
    PERIODIC VARIATIONS
    PERTURBATION
    RESONANCE
    RESONANT VIBRATION
    ∞RHYTHM
    SPACECRAFT MOTION
    SPRINGS (ELASTIC)
    SYNTONY
    TRAVELING WAVE TUBES
    VIBRATION
    VIBRATION TESTS

OSCILLATOR STRENGTHS
RT   ABSORPTION SPECTRA
    ABSORPTIVITY
    ELECTRON OSCILLATIONS
    ELECTRON TRANSITIONS
    LINE SPECTRA
    MOLECULAR OSCILLATIONS
    MOLECULAR OSCILLATORS
    OSCILLATORS
    SPECTRAL LINE WIDTH

OSCILLATORS
UF   PHUGOID OSCILLATIONS
    WAVE OSCILLATORS
GS   **OSCILLATORS**
    . AUTODYNES
    . CRYSTAL OSCILLATORS

OSCILLATORS-*(CONT.)*
    . HARMONIC OSCILLATORS
    . MECHANICAL OSCILLATORS
    . . PENDULUMS
    . . . GYROSCOPIC PENDULUMS
    . MOLECULAR OSCILLATORS
    . RELAXATION OSCILLATORS
    . . PHANTASTRONS
    . SYNCHRONIZED OSCILLATORS
    . VACUUM TUBE OSCILLATORS
    . . CATHODE RAY TUBES
    . . . PICTURE TUBES
    . . GAS DISCHARGE TUBES
    . . MICROWAVE TUBES
    . . . MAGNETRONS
    . . . . NIGOTRONS
    . . . MICROWAVE OSCILLATORS
    . . . PLANOTRONS
    . . . . CARCINOTRONS
    . . . . HELITRONS
    . VOLTAGE CONTROLLED OSCILLATORS
RT   AMPLIFIERS
    AUTOMATIC FREQUENCY CONTROL
    CAVITY RESONATORS
    CIRCUITS
    ELECTRON TUBES
    FEEDBACK
    FEEDBACK AMPLIFIERS
    FLIP-FLOPS
    FREQUENCY STABILITY
    FREQUENCY SYNTHESIZERS
    HARMONIC GENERATORS
    INVERTERS
    ∞MOTION
    MULTIVIBRATORS
    NEGATIVE FEEDBACK
    NONOSCILLATORY ACTION
    NONSTABILIZED OSCILLATION
    OSCILLATIONS
    OSCILLATOR STRENGTHS
    PARAMETRONS
    PERIODIC VARIATIONS
    PERTURBATION
    POSITIVE FEEDBACK
    RESONANT FREQUENCIES
    RESONATORS
    SELF EXCITATION
    SEMICONDUCTOR DEVICES
    SIGNAL GENERATORS
    SOLID STATE DEVICES
    SUBHARMONIC GENERATORS
    TRANSFORMERS
    VIBRATION

OSCILLOGRAMS
USE   OSCILLOGRAPHS

OSCILLOGRAPHS
UF   OSCILLOGRAMS
GS   MEASURING INSTRUMENTS
    . **OSCILLOGRAPHS**
    RECORDING INSTRUMENTS
    . **OSCILLOGRAPHS**
RT   BARKHAUSEN EFFECT
    ELECTRICAL MEASUREMENT
    OSCILLOSCOPES
    TIME MEASUREMENT

OSCILLOSCOPES
RT   CATHODE RAY TUBES
    ELECTRONIC EQUIPMENT TESTS
    FLYING SPOT SCANNERS
    FREQUENCY ANALYZERS
    OSCILLOGRAPHS
    SWEEP CIRCUITS
    SWEEP FREQUENCY
    SYNCHROSCOPES
    VIDEO EQUIPMENT

OSCULATIONS
USE   DOUBLE CUSPS

OSEEN APPROXIMATION
GS   ANALYSIS (MATHEMATICS)
    . NUMERICAL ANALYSIS
    . . APPROXIMATION
    . . . **OSEEN APPROXIMATION**
RT   INCOMPRESSIBLE FLUIDS
    NAVIER-STOKES EQUATION
    ROSHKO PREDICTION
    STOKES FLOW
    VISCOUS FLUIDS

OSMIUM
GS   CHEMICAL ELEMENTS

OSMIUM-*(CONT.)*
    . REFRACTORY METALS
    . . **OSMIUM**
    . . . OSMIUM ISOTOPES
    METALS
    . TRANSITION METALS
    . . REFRACTORY METALS
    . . . **OSMIUM**
    . . . . OSMIUM ISOTOPES
    REFRACTORY MATERIALS
    . REFRACTORY METALS
    . . **OSMIUM**
    . . . OSMIUM ISOTOPES

OSMIUM ALLOYS
GS   ALLOYS
    . HEAT RESISTANT ALLOYS
    . . REFRACTORY METAL ALLOYS
    . . . **OSMIUM ALLOYS**
    REFRACTORY MATERIALS
    . REFRACTORY METAL ALLOYS
    . . **OSMIUM ALLOYS**

OSMIUM COMPOUNDS
RT   ∞CHEMICAL COMPOUNDS
    ∞GROUP 8 COMPOUNDS
    ∞METAL COMPOUNDS

OSMIUM ISOTOPES
GS   CHEMICAL ELEMENTS
    . REFRACTORY METALS
    . . OSMIUM
    . . . **OSMIUM ISOTOPES**
    METALS
    . TRANSITION METALS
    . . REFRACTORY METALS
    . . . OSMIUM
    . . . . **OSMIUM ISOTOPES**
    REFRACTORY MATERIALS
    . REFRACTORY METALS
    . . OSMIUM
    . . . **OSMIUM ISOTOPES**

OSMOMETERS
GS   MEASURING INSTRUMENTS
    . PRESSURE GAGES
    . . **OSMOMETERS**

OSMOSIS
UF   HYPERTONIA
    OSMOTIC PRESSURE
GS   **OSMOSIS**
    . REVERSE OSMOSIS
RT   DEMINERALIZING
    DESALINIZATION
    DIAPHRAGMS (MECHANICS)
    DIFFUSION
    EXTRACTION
    HOMEOSTASIS
    ISOTONICITY
    MEMBRANES
    PERMEATING
    PRESSURE
    ∞SEPARATION
    WATER BALANCE

OSMOTIC PRESSURE
USE   OSMOSIS

OSO
UF   ORBITING SOLAR OBSERVATORY
GS   OBSERVATORIES
    . ASTRONOMICAL OBSERVATORIES
    . . ASTRONOMICAL SATELLITES
    . . . **OSO**
    . . . . OSO-1
    . . . . OSO-2
    . . . . OSO-3
    . . . . OSO-4
    . . . . OSO-5
    . . . . OSO-6
    . . . . OSO-7
    . . . . OSO-8
    . GEOPHYSICAL OBSERVATORIES
    . . **OSO**
    . . . OSO-C
    . . . OSO-1
    . . . OSO-2
    . . . OSO-3
    . . . OSO-4
    . . . OSO-5
    . . . OSO-6
    . . . OSO-7
    . . . OSO-8
    . SOLAR OBSERVATORIES

oso-(CONT.)

```
. . OSO
. . . OSO-C
. . . OSO-1
. . . OSO-2
. . . OSO-3
. . . OSO-4
. . . OSO-5
. . . OSO-6
. . . OSO-7
. . . OSO-8
SATELLITES
. ARTIFICIAL SATELLITES
. . GEOPHYSICAL SATELLITES
. . . OSO
. . . . OSO-C
. . . . OSO-1
. . . . OSO-2
. . . . OSO-3
. . . . OSO-4
. . . . OSO-5
. . . . OSO-6
. . . . OSO-7
. . . . OSO-8
UNMANNED SPACECRAFT
. SOLAR OBSERVATORIES
. . OSO
. . . AOSO
. . . OSO-C
. . . OSO-1
. . . OSO-2
. . . OSO-3
. . . OSO-4
. . . OSO-5
. . . OSO-6
. . . OSO-7
. . . OSO-8
```
RT   SUN
      THOR DELTA LAUNCH VEHICLE

OSO-A
USE   OSO-1

OSO-B
USE   OSO-2

OSO-C
UF   S-57 SATELLITE
GS   OBSERVATORIES
```
. GEOPHYSICAL OBSERVATORIES
. . OSO
. . . OSO-C
. SOLAR OBSERVATORIES
. . OSO
. . . OSO-C
SATELLITES
. ARTIFICIAL SATELLITES
. . GEOPHYSICAL SATELLITES
. . . OSO
. . . . OSO-C
UNMANNED SPACECRAFT
. SOLAR OBSERVATORIES
. . OSO
. . . OSO-C
```
RT   DELTA LAUNCH VEHICLE

OSO-D
USE   OSO-4

OSO-E
USE   OSO-3

OSO-F
USE   OSO-5

OSO-G
USE   OSO-6

OSO-H
USE   OSO-7

OSO-J
USE   OSO-8

OSO-1
UF   OSO-A
      S-16 SATELLITE
GS   OBSERVATORIES
```
. ASTRONOMICAL OBSERVATORIES
. . ASTRONOMICAL SATELLITES
. . . OSO
. . . . OSO-1
. GEOPHYSICAL OBSERVATORIES
. . OSO
```

oso-1-(CONT.)
```
. . . OSO-1
. SOLAR OBSERVATORIES
. . OSO
. . . OSO-1
SATELLITES
. ARTIFICIAL SATELLITES
. . GEOPHYSICAL SATELLITES
. . . OSO
. . . . OSO-1
UNMANNED SPACECRAFT
. SOLAR OBSERVATORIES
. . OSO
. . . OSO-1
```
RT   DELTA LAUNCH VEHICLE

OSO-2
UF   OSO-B
      S-17 SATELLITE
GS   OBSERVATORIES
```
. ASTRONOMICAL OBSERVATORIES
. . ASTRONOMICAL SATELLITES
. . . OSO
. . . . OSO-2
. GEOPHYSICAL OBSERVATORIES
. . OSO
. . . OSO-2
. SOLAR OBSERVATORIES
. . OSO
. . . OSO-2
SATELLITES
. ARTIFICIAL SATELLITES
. . GEOPHYSICAL SATELLITES
. . . OSO
. . . . OSO-2
UNMANNED SPACECRAFT
. SOLAR OBSERVATORIES
. . OSO
. . . OSO-2
```
RT   DELTA LAUNCH VEHICLE

OSO-3
UF   OSO-E
GS   OBSERVATORIES
```
. ASTRONOMICAL OBSERVATORIES
. . ASTRONOMICAL SATELLITES
. . . OSO
. . . . OSO-3
. GEOPHYSICAL OBSERVATORIES
. . OSO
. . . OSO-3
. SOLAR OBSERVATORIES
. . OSO
. . . OSO-3
SATELLITES
. ARTIFICIAL SATELLITES
. . GEOPHYSICAL SATELLITES
. . . OSO
. . . . OSO-3
UNMANNED SPACECRAFT
. SOLAR OBSERVATORIES
. . OSO
. . . OSO-3
```

OSO-4
UF   OSO-D
GS   OBSERVATORIES
```
. ASTRONOMICAL OBSERVATORIES
. . ASTRONOMICAL SATELLITES
. . . OSO
. . . . OSO-4
. GEOPHYSICAL OBSERVATORIES
. . OSO
. . . OSO-4
. SOLAR OBSERVATORIES
. . OSO
. . . OSO-4
SATELLITES
. ARTIFICIAL SATELLITES
. . GEOPHYSICAL SATELLITES
. . . OSO
. . . . OSO-4
UNMANNED SPACECRAFT
. SOLAR OBSERVATORIES
. . OSO
. . . OSO-4
```
RT   DELTA LAUNCH VEHICLE

OSO-5
UF   OSO-F
GS   OBSERVATORIES
```
. ASTRONOMICAL OBSERVATORIES
. . ASTRONOMICAL SATELLITES
. . . OSO
. . . . OSO-5
```

oso-5-(CONT.)
```
. GEOPHYSICAL OBSERVATORIES
. . OSO
. . . OSO-5
. SOLAR OBSERVATORIES
. . OSO
. . . OSO-5
SATELLITES
. ARTIFICIAL SATELLITES
. . GEOPHYSICAL SATELLITES
. . . OSO
. . . . OSO-5
UNMANNED SPACECRAFT
. SOLAR OBSERVATORIES
. . OSO
. . . OSO-5
```

OSO-6
UF   OSO-G
GS   OBSERVATORIES
```
. ASTRONOMICAL OBSERVATORIES
. . ASTRONOMICAL SATELLITES
. . . OSO
. . . . OSO-6
. GEOPHYSICAL OBSERVATORIES
. . OSO
. . . OSO-6
. SOLAR OBSERVATORIES
. . OSO
. . . OSO-6
SATELLITES
. ARTIFICIAL SATELLITES
. . GEOPHYSICAL SATELLITES
. . . OSO
. . . . OSO-6
UNMANNED SPACECRAFT
. SOLAR OBSERVATORIES
. . OSO
. . . OSO-6
```

OSO-7
UF   OSO-H
GS   OBSERVATORIES
```
. ASTRONOMICAL OBSERVATORIES
. . ASTRONOMICAL SATELLITES
. . . OSO
. . . . OSO-7
. GEOPHYSICAL OBSERVATORIES
. . OSO
. . . OSO-7
. SOLAR OBSERVATORIES
. . OSO
. . . OSO-7
SATELLITES
. ARTIFICIAL SATELLITES
. . GEOPHYSICAL SATELLITES
. . . OSO
. . . . OSO-7
UNMANNED SPACECRAFT
. SOLAR OBSERVATORIES
. . OSO
. . . OSO-7
```
RT   DUAL SPIN SPACECRAFT

OSO-8
UF   OSO-J
GS   OBSERVATORIES
```
. ASTRONOMICAL OBSERVATORIES
. . ASTRONOMICAL SATELLITES
. . . OSO
. . . . OSO-8
. GEOPHYSICAL OBSERVATORIES
. . OSO
. . . OSO-8
. SOLAR OBSERVATORIES
. . OSO
. . . OSO-8
SATELLITES
. ARTIFICIAL SATELLITES
. . GEOPHYSICAL SATELLITES
. . . OSO
. . . . OSO-8
UNMANNED SPACECRAFT
. SOLAR OBSERVATORIES
. . OSO
. . . OSO-8
```

OSPREY MISSILE
GS   MISSILES
      . OSPREY MISSILE
RT   J-85 ENGINE

OSS-1 PAYLOAD
GS   PAYLOADS
      . SPACE SHUTTLE PAYLOADS

OSS-1 PAYLOAD-(CONT.)
. . **OSS-1 PAYLOAD**
RT   EXPLORATION
      INVESTIGATION
      NASA PROGRAMS
      SPACE TRANSPORTATION SYSTEM
      SPACEBORNE EXPERIMENTS

OSTA-1 PAYLOAD
SN   (OFFICE OF SPACE & TERRESTRIAL
      APPLICATIONS PAYLOADS)
UF   OFFICE OF SPACE & TERRESTR APPLIC
      PAYLOADS
GS   PAYLOADS
      . SPACE SHUTTLE PAYLOADS
      . . **OSTA-1 PAYLOAD**
RT   SPACE TRANSPORTATION SYSTEM
      SPACEBORNE EXPERIMENTS

OSTA-2 PAYLOAD
SN   (OFFICE OF SPACE & TERRESTRIAL
      APPLICATIONS PAYLOADS)
UF   OFFICE OF SPACE & TERRESTR APPLIC
      PAYLOADS
GS   PAYLOADS
      . **OSTA-2 PAYLOAD**
RT   SPACE TRANSPORTATION SYSTEM
      SPACELAB

OSTEOPOROSIS
GS   DISEASES
      . **OSTEOPOROSIS**
RT   BONE DEMINERALIZATION
      BONE MINERAL CONTENT
      BONES
      CALCIUM METABOLISM
      METABOLISM

OT-2
USE   ESSA 2 SATELLITE

OT-3
USE   ESSA 1 SATELLITE

OTF
USE   OPTICAL TRANSFER FUNCTION

OTOLARYNGOLOGY
GS   MEDICAL SCIENCE
      . **OTOLARYNGOLOGY**
RT   EAR

OTOLITH ORGANS
GS   ANATOMY
      . SENSE ORGANS
      . . GRAVIRECEPTORS
      . . . **OTOLITH ORGANS**
RT   OCULOGRAVIC ILLUSIONS
      VERTICAL PERCEPTION

OTOLOGY
GS   MEDICAL SCIENCE
      . **OTOLOGY**
RT   EAR

OTS (ESA)
UF   ORBITAL TEST SATELLITE (ESA)
GS   ESA SPACECRAFT
      . ESA SATELLITES
      . . **OTS (ESA)**
      SATELLITES
      . ARTIFICIAL SATELLITES
      . . ESA SATELLITES
      . . . **OTS (ESA)**
RT   EUROPEAN COMMUNICATIONS
      SATELLITE
      EUROPEAN SPACE PROGRAMS

OTTO CYCLE
GS   CYCLES
      . THERMODYNAMIC CYCLES
      . . **OTTO CYCLE**
RT   RANKINE CYCLE

OTV
USE   ORBIT TRANSFER VEHICLES

OUTCROPS
RT   FOLDS (GEOLOGY)
      FORMATIONS
      GEOLOGY

OUTER BANKS (NC)
GS   LANDFORMS

OUTER BANKS (NC)-(CONT.)
      . BARRIERS (LANDFORMS)
      . . **OUTER BANKS (NC)**
RT   ATLANTIC OCEAN
      ISLANDS
      NORTH CAROLINA

OUTER PLANET MISSIONS
USE   GRAND TOURS

OUTER PLANET SPACECRAFT
USE   OUTER PLANETS EXPLORERS

OUTER PLANETS EXPLORERS
UF   OUTER PLANET SPACECRAFT
      PLANETARY EXPLORER
RT   DELTA LAUNCH VEHICLE
      EXPLORER SATELLITES
      FLYBY MISSIONS
      GRAND TOURS
      INTERPLANETARY FLIGHT
      MARS PROBES
      ∞ SPACECRAFT
      TOPS (SPACECRAFT)
      VENUS PROBES

OUTER RADIATION BELT
GS   EARTH ATMOSPHERE
      . RADIATION BELTS
      . . **OUTER RADIATION BELT**
      PARTICLES
      . CHARGED PARTICLES
      . . MAGNETICALLY TRAPPED PARTICLES
      . . . RADIATION BELTS
      . . . . **OUTER RADIATION BELT**
      . TRAPPED PARTICLES
      . . MAGNETICALLY TRAPPED PARTICLES
      . . . RADIATION BELTS
      . . . . **OUTER RADIATION BELT**
RT   ARTIFICIAL RADIATION BELTS
      INNER RADIATION BELT
      PROTON BELTS
      ∞ RADIATION

OUTER SPACE TREATY
GS   FOREIGN POLICY
      . INTERNATIONAL RELATIONS
      . . INTERNATIONAL COOPERATION
      . . . **OUTER SPACE TREATY**
RT   CONVENTIONS
      INTERNATIONAL LAW
      RESEARCH AND DEVELOPMENT
      RESOURCE ALLOCATION
      SPACE LAW

OUTGASSING
RT   DEGASSING
      DESORPTION
      EVOLUTION (LIBERATION)
      GAS EVOLUTION
      PURGING
      RESIDUAL GAS
      TRANSPIRATION
      VACUUM
      VACUUM PUMPS

OUTLET FLOW
GS   FLUID FLOW
      . **OUTLET FLOW**
RT   CASCADE FLOW
      CHANNEL FLOW
      ∞ FLOW
      FLOW CHARACTERISTICS
      NOZZLE FLOW

OUTLETS
GS   **OUTLETS**
      . VENTS
RT   APERTURES
      CAVITIES
      ∞ DISCHARGE
      DOORS
      DUCTS
      EGRESS
      EXHAUST NOZZLES
      EXHAUST SYSTEMS
      GATES (OPENINGS)
      ∞ NOZZLES
      OPENINGS
      OUTPUT
      PIPE NOZZLES
      PLUGS
      PORTS (OPENINGS)
      ∞ TERMINALS

OUTLETS (GEOLOGY)
USE   ESTUARIES

OUTLIERS (LANDFORMS)
UF   KLIPPEN
GS   LANDFORMS
      . **OUTLIERS (LANDFORMS)**
RT   FORMATIONS
      ROCKS
      SOIL EROSION

OUTLIERS (STATISTICS)
RT   MATHEMATICAL MODELS
      PROBABILITY THEORY
      STATISTICAL ANALYSIS
      STATISTICAL DISTRIBUTIONS
      STATISTICAL TESTS

OUTPUT
UF   DUMMY LOADS
GS   **OUTPUT**
      . LASER OUTPUTS
      . MASER OUTPUTS
RT   ∞ CAPACITY
      CATCHERS
      COMPUTER SYSTEMS PERFORMANCE
      DELIVERY
      EFFLUX
      INPUT
      OUTLETS
      ∞ PERFORMANCE
      POWER CONDITIONING
      PRINTOUTS
      ∞ PRODUCTION
      PRODUCTS
      READOUT
      SUPPLYING
      TRACKING PROBLEM
      TRANSFER FUNCTIONS
      TRANSMISSION
      YIELD

OV-1 AIRCRAFT
UF   AO-1 AIRCRAFT
      GRUMMAN OV-1C AIRCRAFT
      MOHAWK AIRCRAFT
GS   GRUMMAN AIRCRAFT
      . **OV-1 AIRCRAFT**
      JET AIRCRAFT
      . TURBOPROP AIRCRAFT
      . . **OV-1 AIRCRAFT**
      MONOPLANES
      . **OV-1 AIRCRAFT**
      OBSERVATION AIRCRAFT
      . **OV-1 AIRCRAFT**
RT   ∞ AIRCRAFT

OV-1 SATELLITES
GS   SATELLITES
      . ARTIFICIAL SATELLITES
      . . SCIENTIFIC SATELLITES
      . . . **OV-1 SATELLITES**
RT   GRAVITY GRADIENT SATELLITES
      SPIN STABILIZATION

OV-2 SATELLITES
GS   SATELLITES
      . ARTIFICIAL SATELLITES
      . . SCIENTIFIC SATELLITES
      . . . **OV-2 SATELLITES**
RT   GRAVITY GRADIENT SATELLITES
      SPIN STABILIZATION

OV-3 SATELLITES
GS   SATELLITES
      . ARTIFICIAL SATELLITES
      . . SCIENTIFIC SATELLITES
      . . . **OV-3 SATELLITES**
RT   GRAVITY GRADIENT SATELLITES
      SPIN STABILIZATION

OV-4 SATELLITES
GS   SATELLITES
      . ARTIFICIAL SATELLITES
      . . SCIENTIFIC SATELLITES
      . . . **OV-4 SATELLITES**
RT   GRAVITY GRADIENT SATELLITES
      SPIN STABILIZATION

OV-5 SATELLITES
GS   SATELLITES
      . ARTIFICIAL SATELLITES
      . . SCIENTIFIC SATELLITES
      . . . **OV-5 SATELLITES**

OV-5 SATELLITES-*(CONT.)*
RT   GRAVITY GRADIENT SATELLITES
        SPIN STABILIZATION

OV-10 AIRCRAFT
UF   NA-300 AIRCRAFT
GS   ATTACK AIRCRAFT
      . **OV-10 AIRCRAFT**
      COIN AIRCRAFT
      . **OV-10 AIRCRAFT**
      JET AIRCRAFT
      . TURBOPROP AIRCRAFT
      . . **OV-10 AIRCRAFT**
      MONOPLANES
      . **OV-10 AIRCRAFT**
      NORTH AMERICAN AIRCRAFT
      . **OV-10 AIRCRAFT**
      OBSERVATION AIRCRAFT
      . **OV-10 AIRCRAFT**
      RECONNAISSANCE AIRCRAFT
      . **OV-10 AIRCRAFT**
RT   ∞AIRCRAFT

OVARIES
GS   ANATOMY
      . GENITOURINARY SYSTEM
      . **OVARIES**
      . GLANDS (ANATOMY)
      . . ENDOCRINE GLANDS
      . . . **OVARIES**
      . . SEX GLANDS
      . . . **OVARIES**
      . ORGANS
      . . **OVARIES**
      REPRODUCTIVE SYSTEMS
      . **OVARIES**
      VISCERA
      . ENDOCRINE GLANDS
      . . **OVARIES**
      . ORGANS
      . . **OVARIES**
      . SEX GLANDS
      . . **OVARIES**
RT   EGGS
        MENSTRUATION
        ∞REPRODUCTION

OVENS
GS   HEATING EQUIPMENT
      . **OVENS**
RT   BAKING
        DRY HEAT
        FURNACES
        WASTE ENERGY UTILIZATION

OVER-THE-HORIZON RADAR
GS   RADAR
      . SEARCH RADAR
      . . **OVER-THE-HORIZON RADAR**
RT   EARLY WARNING SYSTEMS
        MOVING TARGET INDICATORS
        OPTICAL RADAR
        RADAR DETECTION
        RADAR RANGE

OVERCAST
USE   CLOUD COVER

OVERCOMPRESSION
USE   OVERCONSOLIDATION

OVERCONSOLIDATION
UF   OVERCOMPRESSION
RT   CONSOLIDATION
        FOUNDATIONS

OVERHAUSER EFFECT
RT   ∞EFFECTS
        MAGNETIC RESONANCE
        NUCLEAR SPIN
        ∞POLARIZATION
        RESONANCE

OVERPRESSURE
GS   PRESSURE
      . **OVERPRESSURE**
RT   BLAST LOADS
        DYNAMIC PRESSURE

OVERTONES
USE   HARMONICS

OVERVOLTAGE
RT   CIRCUIT PROTECTION

OVERVOLTAGE-*(CONT.)*
        DECOMPOSITION
        ELECTRIC POTENTIAL
        GEIGER COUNTERS
        OPEN CIRCUIT VOLTAGE
        POLARIZATION (CHARGE SEPARATION)
        SURGES

OXALATES
GS   **OXALATES**
      . COBALT OXALATES
RT   OXALIC ACID
        ∞OXYGEN COMPOUNDS

OXALIC ACID
GS   ALIPHATIC COMPOUNDS
      . **OXALIC ACID**
RT   OXALATES

OXAMIC ACIDS
GS   ALIPHATIC COMPOUNDS
      . **OXAMIC ACIDS**
      NITROGEN COMPOUNDS
      . AMIDES
      . . **OXAMIC ACIDS**

OXAZOLE
GS   HETEROCYCLIC COMPOUNDS
      . AZOLES
      . . **OXAZOLE**

OXIDASE
GS   ACIDS
      . **OXIDASE**
      ENZYMES
      . **OXIDASE**
      ORGANIC COMPOUNDS
      . **OXIDASE**
      PROTEINS
      . **OXIDASE**

OXIDATION
GS   CHEMICAL REACTIONS
      . **OXIDATION**
      . . ELECTROCHEMICAL OXIDATION
      . . PHOTOOXIDATION
      . . RUSTING
RT   ASSOCIATION REACTIONS
        CHARRING
        CHEMICAL ATTACK
        COMBUSTION
        CORROSION
        DEGRADATION
        DEHYDROGENATION
        DOPA
        ELECTRON TRANSFER
        EPOXIDATION
        EROSIVE BURNING
        FUEL COMBUSTION
        HOT CORROSION
        HYDROCARBON COMBUSTION
        METAL COMBUSTION
        OXIDIZERS
        OXYGENATION
        PASSIVITY
        REDUCTION (CHEMISTRY)
        ROASTING
        THERMAL RESISTANCE

OXIDATION RESISTANCE
GS   CORROSION RESISTANCE
      . **OXIDATION RESISTANCE**
RT   PASSIVITY
        ∞RESISTANCE
        RUSTING
        SILICONIZING
        THERMAL RESISTANCE

OXIDATION-REDUCTION REACTIONS
GS   CHEMICAL REACTIONS
      . **OXIDATION-REDUCTION REACTIONS**
RT   ELECTROCHEMISTRY
        REDUCTION (CHEMISTRY)

OXIDE FILMS
RT   CATHODIC COATINGS
        ∞FILMS
        METAL OXIDES
        METAL SURFACES
        SURFACE LAYERS
        THIN FILMS

OXIDES
GS   CHALCOGENIDES

OXIDES-*(CONT.)*
    . **OXIDES**
    . . ANHYDRIDES
    . . . PEROXIDES
    . . . . INORGANIC PEROXIDES
    . . . . ORGANIC PEROXIDES
    . . . . SODIUM PEROXIDES
    . . BORON OXIDES
    . . BRUCITE
    . . CARBON MONOXIDE
    . . CARBON SUBOXIDES
    . . CHLORINE OXIDES
    . . DIOXIDES
    . . . CARBON DIOXIDE
    . . . FLINT
    . . . HYDROGEN PEROXIDE
    . . . SILICON DIOXIDE
    . . . . QUARTZ
    . . . . . COESITE
    . . . SULFUR DIOXIDES
    . . GERMANIUM OXIDES
    . . HEAVY WATER
    . . METAL OXIDES
    . . . ALKALINE EARTH OXIDES
    . . . . BARIUM OXIDES
    . . . . BERYLLIUM OXIDES
    . . . . CALCIUM OXIDES
    . . . . MAGNESIUM OXIDES
    . . . . . PERICLASE
    . . . ALUMINUM OXIDES
    . . . . SAPPHIRE
    . . . BISMUTH OXIDES
    . . . CESIUM OXIDES
    . . CHROMIUM OXIDES
    . . . CHROMITES
    . . COBALT OXIDES
    . . COPPER OXIDES
    . . HAFNIUM OXIDES
    . . IRON OXIDES
    . . . CHROMITES
    . . . HEMATITE
    . . . . ILMENITE
    . . . MAGNETITE
    . . . LANTHANUM OXIDES
    . . LEAD OXIDES
    . . LITHIUM OXIDES
    . . MANGANESE OXIDES
    . . . HOPCALITE (TRADEMARK)
    . . MERCURY OXIDES
    . . MIXED OXIDES
    . . MOLYBDENUM OXIDES
    . . NICKEL OXIDES
    . . NIOBIUM OXIDES
    . . PLATINUM OXIDES
    . . PLUTONIUM OXIDES
    . . POTASSIUM OXIDES
    . . SCANDIUM OXIDES
    . . SILVER OXIDES
    . . SODIUM PEROXIDES
    . . TANTALUM OXIDES
    . . THORIUM OXIDES
    . . TIN OXIDES
    . . TITANIUM OXIDES
    . . . ANATASE
    . . . . ILMENITE
    . . . RUTILE
    . . TUNGSTEN OXIDES
    . . . SCHEELITE
    . . URANIUM OXIDES
    . . VANADIUM OXIDES
    . . YTTRIUM OXIDES
    . . ZINC OXIDES
    . . ZIRCONIUM OXIDES
    . NITROGEN OXIDES
    . . NITRIC OXIDE
    . . NITROGEN DIOXIDE
    . . NITROGEN TETROXIDE
    . . NITROUS OXIDES
    . PHOSPHORUS OXIDES
    . PYROXENES
    . . ENSTATITE
    . SELENIUM OXIDES
    . SILICON OXIDES
    . . MUSCOVITE
    . . NEPHELITE
    . . SILICON DIOXIDE
    . . . QUARTZ
    . . . . COESITE
    . . . SPODUMENE
    . . SULFUR OXIDES
    . . . SULFUR DIOXIDES
RT   ANODIC COATINGS
        CATHODIC COATINGS
        EUXENITE
        INSULATION
        METAL COATINGS

OXIDES-(CONT.)
    NIOBATES
    NONFLAMMABLE MATERIALS
    ∞OXYGEN COMPOUNDS
    WATER

OXIDIZERS
GS   **OXIDIZERS**
    . HIGH ENERGY OXIDIZERS
    . LIQUID OXIDIZERS
    . LIQUID OXYGEN
    . PHOTOCHEMICAL OXIDANTS
    . ROCKET OXIDIZERS
    . . FLOX
    . . TAGN
RT   ∞AGENTS
    AIR POLLUTION
    FLUORINE
    FUELS
    NITRAMINE PROPELLANTS
    OXIDATION

OXIMETRY
RT   BIOCHEMICAL OXYGEN DEMAND
    BLOOD
    HYPEROXIA
    HYPOXIA
    OXYGEN CONSUMPTION

OXYACETYLENE
GS   ALIPHATIC COMPOUNDS
    . ALKYNES
    . . **OXYACETYLENE**
    HYDROCARBONS
    . ALKYNES
    . . **OXYACETYLENE**
RT   ACETYLENE
    DETONABLE GAS MIXTURES
    ∞OXYGEN COMPOUNDS

OXYALKYLATION
USE   ALKYLATION

OXYFLUORIDES
GS   HALOGEN COMPOUNDS
    . FLUORINE COMPOUNDS
    . . FLUORIDES
    . . . **OXYFLUORIDES**
    . HALIDES
    . . FLUORIDES
    . . . **OXYFLUORIDES**
RT   ∞OXYGEN COMPOUNDS

OXYGEN
GS   CHEMICAL ELEMENTS
    . **OXYGEN**
    . . OXYGEN ATOMS
    . . OXYGEN IONS
    . . OXYGEN ISOTOPES
    . . . OXYGEN 17
    . . . OXYGEN 18
    . . OXYGEN PLASMA
    GASES
    . **OXYGEN**
    . . HIGH PRESSURE OXYGEN
    . . LIQUID OXYGEN
    . . OXYGEN ATOMS
    . . OXYGEN IONS
    . . OXYGEN ISOTOPES
    . . . OXYGEN 18
    . . OXYGEN PLASMA
    . . OZONE
    . . . OZONATES
    . . . OZONIDES
RT   REACTION BONDING
    SCHUMANN-RUNGE BANDS
    SIALON

OXYGEN AFTERGLOW
GS   AFTERGLOWS
    . **OXYGEN AFTERGLOW**

OXYGEN ANALYZERS
UF   OXYGEN DETECTORS
GS   MEASURING INSTRUMENTS
    . **OXYGEN ANALYZERS**
RT   GAS ANALYSIS

OXYGEN ATOMS
GS   ATOMS
    . **OXYGEN ATOMS**
    CHEMICAL ELEMENTS
    . OXYGEN
    . . **OXYGEN ATOMS**

OXYGEN ATOMS-(CONT.)
    GASES
    . OXYGEN
    . . **OXYGEN ATOMS**

OXYGEN BREATHING
RT   ∞BREATHING

∞ **OXYGEN COMPOUNDS**
SN   (USE OF A MORE SPECIFIC TERM IS
    RECOMMENDED--CONSULT THE TERMS
    LISTED BELOW)
RT   ACIDS
    ALUMINATES
    ARSENATES
    BORATES
    BROMATES
    CARBOHYDRATES
    CARBON SUBOXIDES
    CARBONATES
    ∞CHEMICAL COMPOUNDS
    CHLORATES
    CHROMATES
    GERMANIUM OXIDES
    NIOBATES
    OXALATES
    OXIDES
    OXYACETYLENE
    OXYFLUORIDES
    OXYGEN FLUORIDES
    OZONATES
    OZONE FLUORIDE
    OZONIDES
    STANNATES

OXYGEN CONSUMPTION
GS   CONSUMPTION
    . **OXYGEN CONSUMPTION**
RT   BIOCHEMICAL OXYGEN DEMAND
    HYPEROXIA
    HYPOXIA
    METABOLISM
    OXIMETRY

OXYGEN DEFICIENCY
USE   HYPOXIA

OXYGEN DETECTORS
USE   OXYGEN ANALYZERS

OXYGEN FLUORIDES
GS   HALOGEN COMPOUNDS
    . FLUORINE COMPOUNDS
    . . FLUORIDES
    . . . **OXYGEN FLUORIDES**
    . HALIDES
    . . FLUORIDES
    . . . **OXYGEN FLUORIDES**
RT   ∞OXYGEN COMPOUNDS

OXYGEN IONS
GS   CHEMICAL ELEMENTS
    . OXYGEN
    . . **OXYGEN IONS**
    GASES
    . OXYGEN
    . . **OXYGEN IONS**
    IONS
    . **OXYGEN IONS**
RT   FREE RADICALS
    NEGATIVE IONS

OXYGEN ISOTOPES
GS   CHEMICAL ELEMENTS
    . NUCLIDES
    . . ISOTOPES
    . . . **OXYGEN ISOTOPES**
    . . . . OXYGEN 18
    . OXYGEN
    . . **OXYGEN ISOTOPES**
    . . . OXYGEN 17
    . . . OXYGEN 18
    GASES
    . OXYGEN
    . . **OXYGEN ISOTOPES**
    . . . OXYGEN 18

OXYGEN MASKS
GS   BREATHING APPARATUS
    . **OXYGEN MASKS**
    MASKS
    . **OXYGEN MASKS**
    OXYGEN SUPPLY EQUIPMENT
    . **OXYGEN MASKS**

OXYGEN MASKS-(CONT.)
RT   HIGH ALTITUDE BREATHING
    LIFE SUPPORT SYSTEMS
    PORTABLE LIFE SUPPORT SYSTEMS

OXYGEN METABOLISM
GS   METABOLISM
    . **OXYGEN METABOLISM**
RT   HYDROGEN METABOLISM
    RESPIRATION

OXYGEN PLASMA
GS   CHEMICAL ELEMENTS
    . OXYGEN
    . . **OXYGEN PLASMA**
    GASES
    . OXYGEN
    . . **OXYGEN PLASMA**
RT   ARGON PLASMA
    HELIUM PLASMA
    HYDROGEN PLASMA

OXYGEN PRODUCTION
RT   CLOSED ECOLOGICAL SYSTEMS
    GAS EXCHANGE

OXYGEN RECOMBINATION
GS   CHEMICAL REACTIONS
    . ATOMIC RECOMBINATION
    . . **OXYGEN RECOMBINATION**
    RECOMBINATION REACTIONS
    . ATOMIC RECOMBINATION
    . . **OXYGEN RECOMBINATION**
RT   IONIZATION

OXYGEN REGULATORS
GS   REGULATORS
    . **OXYGEN REGULATORS**
RT   FLOW REGULATORS
    PRESSURE REGULATORS

OXYGEN SPECTRA
GS   SPECTRA
    . **OXYGEN SPECTRA**
RT   AIRGLOW
    HERZBERG BANDS
    MOLECULAR SPECTRA
    SOLAR SPECTRA

OXYGEN SUPPLY EQUIPMENT
UF   OXYGEN SYSTEMS
GS   **OXYGEN SUPPLY EQUIPMENT**
    . OXYGEN MASKS
RT   AEPS
    AIR CONDITIONING EQUIPMENT
    BREATHING APPARATUS
    CABIN ATMOSPHERES
    COMPRESSED AIR
    CONTROLLED ATMOSPHERES
    EMERGENCY LIFE SUSTAINING
      SYSTEMS
    LIFE SUPPORT SYSTEMS
    PRESSURIZED CABINS
    SURVIVAL EQUIPMENT

OXYGEN SYSTEMS
USE   OXYGEN SUPPLY EQUIPMENT

OXYGEN TENSION
GS   PRESSURE
    . PARTIAL PRESSURE
    . . **OXYGEN TENSION**
    . . . HYPOXEMIA

OXYGEN TOXICITY
USE   HYPEROXIA

OXYGEN 17
GS   CHEMICAL ELEMENTS
    . OXYGEN
    . . OXYGEN ISOTOPES
    . . . **OXYGEN 17**

OXYGEN 18
GS   CHEMICAL ELEMENTS
    . NUCLIDES
    . . ISOTOPES
    . . . OXYGEN ISOTOPES
    . . . . **OXYGEN 18**
    . OXYGEN
    . . OXYGEN ISOTOPES
    . . . **OXYGEN 18**
    GASES
    . OXYGEN

P

OXYGEN 18-*(CONT.)*
```
      . . OXYGEN ISOTOPES
      . . . OXYGEN 18
```

OXYGENATION
```
GS    CHEMICAL REACTIONS
      . OXYGENATION
RT    AERATION
      DISSOLVED GASES
      OXIDATION
```

OXYHALIDES
```
GS    HALOGEN COMPOUNDS
      . HALIDES
      . . OXYHALIDES
```

OXYHEMOGLOBIN
```
GS    CELLS (BIOLOGY)
      . HEMOGLOBIN
      . . OXYHEMOGLOBIN
      ORGANOMETALLIC COMPOUNDS
      . HEMOGLOBIN
      . . OXYHEMOGLOBIN
      PROTEINS
      . OXYHEMOGLOBIN
RT    ERYTHROCYTES
```

OXYNITRIDES
```
GS    NITROGEN COMPOUNDS
      . NITRIDES
      . . OXYNITRIDES
RT    ∞CHEMICAL COMPOUNDS
      ∞GROUP 5A COMPOUNDS
```

OZONATES
```
GS    GASES
      . OXYGEN
      . . OZONE
      . . . OZONATES
RT    ∞OXYGEN COMPOUNDS
```

OZONE
```
GS    GASES
      . OXYGEN
      . . OZONE
      . . . OZONATES
      . . . OZONIDES
RT    HALOGEN OCCULTATION EXPERIMENT
      OZONE FLUORIDE
      OZONOMETRY
      PHOTOCHEMICAL OXIDANTS
      SAGE SATELLITE
      SOLAR MESOSPHERE EXPLORER
```

OZONE FLUORIDE
```
GS    HALOGEN COMPOUNDS
      . FLUORINE COMPOUNDS
      . . FLUORIDES
      . . . OZONE FLUORIDE
      . HALIDES
      . . FLUORIDES
      . . . OZONE FLUORIDE
RT    ∞OXYGEN COMPOUNDS
      OZONE
```

OZONIDES
```
GS    GASES
      . OXYGEN
      . . OZONE
      . . . OZONIDES
RT    ∞OXYGEN COMPOUNDS
```

OZONOMETRY
```
GS    CHEMICAL TESTS
      . CHEMICAL ANALYSIS
      . . GAS ANALYSIS
      . . . OZONOMETRY
RT    OZONE
```

OZONOSPHERE
```
GS    EARTH ATMOSPHERE
      . LOWER ATMOSPHERE
      . . OZONOSPHERE
RT    CHEMOSPHERE
      HOMOSPHERE
      UMKEHR EFFECT
      UPPER ATMOSPHERE
```

P BAND
```
GS    FREQUENCIES
      . RADIO FREQUENCIES
      . . MICROWAVE FREQUENCIES
      . . . P BAND
      . . ULTRAHIGH FREQUENCIES
      . . . P BAND
      . . VERY HIGH FREQUENCIES
      . . . P BAND
```

P WAVES
```
GS    ELASTIC WAVES
      . P WAVES
RT    COMPRESSIBLE FLUIDS
      COMPRESSION WAVES
      CRUSTAL FRACTURES
      DILATATIONAL WAVES
      S WAVES
      SEISMIC WAVES
      SURFACE WAVES
```

P.A.C.M. TELEMETRY
```
SN    (PULSE AMPLITUDE CODE MODULATION)
GS    TELECOMMUNICATION
      . TELEMETRY
      . . P.A.C.M. TELEMETRY
RT    AMPLITUDE MODULATION
      COMMUNICATION EQUIPMENT
      DIFFERENTIAL PULSE CODE
        MODULATION
      MODULATION
      PULSE AMPLITUDE MODULATION
      PULSE CODE MODULATION
```

P-I-N DIODES
```
USE   DIODES
      P-I-N JUNCTIONS
```

P-I-N JUNCTIONS
```
UF    P-I-N DIODES
GS    SEMICONDUCTOR JUNCTIONS
      . P-I-N JUNCTIONS
```

P-N JUNCTIONS
```
UF    N-P JUNCTIONS
GS    SEMICONDUCTOR JUNCTIONS
      . P-N JUNCTIONS
RT    SIS (SEMICONDUCTORS)
```

P-N-P JUNCTIONS
```
GS    SEMICONDUCTOR JUNCTIONS
      . P-N-P JUNCTIONS
```

P-N-P-N JUNCTIONS
```
GS    SEMICONDUCTOR JUNCTIONS
      . P-N-P-N JUNCTIONS
RT    ∞JUNCTIONS
      LATCH-UP
      THYRISTORS
```

P-TYPE SEMICONDUCTORS
```
GS    SEMICONDUCTORS (MATERIALS)
      . P-TYPE SEMICONDUCTORS
RT    HOLES (ELECTRON DEFICIENCIES)
      SEMICONDUCTOR JUNCTIONS
```

P-1 ENGINE
```
GS    ENGINES
      . ROCKET ENGINES
      . . BOOSTER ROCKET ENGINES
      . . . P-1 ENGINE
      . . SOLID PROPELLANT ROCKET
        ENGINES
      . . . P-1 ENGINE
```

P-3 AIRCRAFT
```
UF    ORION AIRCRAFT
      P3V AIRCRAFT
GS    ANTISUBMARINE WARFARE AIRCRAFT
      . P-3 AIRCRAFT
      JET AIRCRAFT
      . P-3 AIRCRAFT
      LOCKHEED AIRCRAFT
      . P-3 AIRCRAFT
      MONOPLANES
      . P-3 AIRCRAFT
RT    ∞AIRCRAFT
      TURBOPROP ENGINES
```

P-51 AIRCRAFT
```
UF    MUSTANG AIRCRAFT
GS    ATTACK AIRCRAFT
```

P-51 AIRCRAFT-*(CONT.)*
```
      . FIGHTER AIRCRAFT
      . . P-51 AIRCRAFT
      MONOPLANES
      . P-51 AIRCRAFT
      NORTH AMERICAN AIRCRAFT
      . P-51 AIRCRAFT
RT    ∞AIRCRAFT
```

P-84 AIRCRAFT
```
USE   JET PROVOST AIRCRAFT
```

P-160 AIRCRAFT
```
UF    ME P-160 AIRCRAFT
      MESSERSCHMITT ME P-160 AIRCRAFT
GS    COMMERCIAL AIRCRAFT
      . P-160 AIRCRAFT
      PASSENGER AIRCRAFT
      . P-160 AIRCRAFT
RT    ∞AIRCRAFT
```

P-166 AIRCRAFT
```
UF    PIAGGIO P-166 AIRCRAFT
GS    LIGHT AIRCRAFT
      . P-166 AIRCRAFT
      MONOPLANES
      . P-166 AIRCRAFT
      PASSENGER AIRCRAFT
      . P-166 AIRCRAFT
      PIAGGIO AIRCRAFT
      . P-166 AIRCRAFT
      TRANSPORT AIRCRAFT
      . CARGO AIRCRAFT
      . . P-166 AIRCRAFT
RT    ∞AIRCRAFT
```

P-308 AIRCRAFT
```
UF    ME P-308 AIRCRAFT
      MESSERSCHMITT ME P-308 AIRCRAFT
GS    ATTACK AIRCRAFT
      . P-308 AIRCRAFT
      JET AIRCRAFT
      . P-308 AIRCRAFT
      MONOPLANES
      . P-308 AIRCRAFT
RT    ∞AIRCRAFT
```

P-531 HELICOPTER
```
UF    SCOUT HELICOPTER
      WESTLAND P-531 HELICOPTER
GS    UTILITY AIRCRAFT
      . P-531 HELICOPTER
      V/STOL AIRCRAFT
      . ROTARY WING AIRCRAFT
      . . HELICOPTERS
      . . . P-531 HELICOPTER
      WESTLAND AIRCRAFT
      . P-531 HELICOPTER
RT    ANTISUBMARINE WARFARE AIRCRAFT
      PASSENGER AIRCRAFT
```

P-1052 AIRCRAFT
```
UF    HAWKER P-1052 AIRCRAFT
GS    HAWKER SIDDELEY AIRCRAFT
      . P-1052 AIRCRAFT
      JET AIRCRAFT
      . P-1052 AIRCRAFT
      MONOPLANES
      . P-1052 AIRCRAFT
      RESEARCH AIRCRAFT
      . P-1052 AIRCRAFT
RT    ∞AIRCRAFT
```

P-1127 AIRCRAFT
```
UF    HAWKER P-1127 AIRCRAFT
      KESTREL AIRCRAFT
      VZ-12 AIRCRAFT
      XV-6A AIRCRAFT
GS    ATTACK AIRCRAFT
      . FIGHTER AIRCRAFT
      . . P-1127 AIRCRAFT
      HAWKER SIDDELEY AIRCRAFT
      . P-1127 AIRCRAFT
      JET AIRCRAFT
      . TURBOFAN AIRCRAFT
      . . P-1127 AIRCRAFT
      MONOPLANES
      . P-1127 AIRCRAFT
      OBSERVATION AIRCRAFT
      . P-1127 AIRCRAFT
      RECONNAISSANCE AIRCRAFT
      . P-1127 AIRCRAFT
      V/STOL AIRCRAFT
      . P-1127 AIRCRAFT
RT    ∞AIRCRAFT
```

P-1127 AIRCRAFT-*(CONT.)*
    BRISTOL-SIDDELEY BS 53 ENGINE
    HARRIER AIRCRAFT
    TURBOFAN ENGINES

P-1154 AIRCRAFT
UF   HAWKER P-1154 AIRCRAFT
GS   ATTACK AIRCRAFT
    . FIGHTER AIRCRAFT
    . . **P-1154 AIRCRAFT**
    HAWKER SIDDELEY AIRCRAFT
    . **P-1154 AIRCRAFT**
    JET AIRCRAFT
    . TURBOFAN AIRCRAFT
    . . **P-1154 AIRCRAFT**
    MONOPLANES
    . **P-1154 AIRCRAFT**
    OBSERVATION AIRCRAFT
    . **P-1154 AIRCRAFT**
    RECONNAISSANCE AIRCRAFT
    . **P-1154 AIRCRAFT**
    SUPERSONIC AIRCRAFT
    . **P-1154 AIRCRAFT**
    V/STOL AIRCRAFT
    . **P-1154 AIRCRAFT**
RT   ∞AIRCRAFT
    TURBOFAN ENGINES

PA-34 SENECA AIRCRAFT
UF   SENECA AIRCRAFT
GS   LIGHT AIRCRAFT
    . PIPER AIRCRAFT
    . . **PA-34 SENECA AIRCRAFT**
RT   ∞AIRCRAFT
    ATLIT PROJECT
    GAW-1 AIRFOIL
    GENERAL DYNAMICS AIRCRAFT

PACE
USE   PHYSICS AND CHEMISTRY EXPERIMENT
       IN SPACE

PACIFIC ISLANDS
GS   LANDFORMS
    . ISLANDS
    . . **PACIFIC ISLANDS**
    . . . GUAM
    . . . JAPAN
    . . . JOHNSTON ISLAND
    . . . KURILE ISLANDS
    . . . NEW GUINEA (ISLAND)
    . . . NEW ZEALAND
    . . . PHILIPPINES
    . . . SAMOA

PACIFIC NORTHWEST (US)
GS   REGIONS
    . **PACIFIC NORTHWEST (US)**
RT   CANADA
    UNITED STATES

PACIFIC OCEAN
GS   OCEANS
    . **PACIFIC OCEAN**
RT   BERING SEA
    COASTAL RANGES (CA)
    EL NINO
    GULF OF ALASKA
    GULF OF CALIFORNIA (MEXICO)
    INDONESIA
    MONTEREY BAY (CA)
    SALTON SEA (CA)
    SAN FRANCISCO BAY (CA)
    SEA OF OKHOTSK
    SOUTHERN CALIFORNIA

PACKAGES
GS   **PACKAGES**
    . APOLLO LUNAR SURFACE
       EXPERIMENTS PACKAGE
    . INSTRUMENT PACKAGES
    . . EASEP
    . . EREP
RT   BAGS
    BOXES (CONTAINERS)
    BUNDLES
    CARTRIDGES
    CASES (CONTAINERS)
    ∞CONTAINERS
    ∞INSTRUMENTS
    PACKAGING

PACKAGING
GS   **PACKAGING**
    . ELECTRONIC PACKAGING

PACKAGING-*(CONT.)*
RT   ∞CONTAINERS
    CORROSION PREVENTION
    ENCAPSULATING
    ENCLOSURE
    HAULING
    HOPPERS
    INHIBITORS
    MARKING
    MATERIALS HANDLING
    PACKAGES
    ∞PACKING
    PRESERVING
    SEALERS
    SPIRAL WRAPPING
    ∞STORAGE
    TRANSPORTATION
    VERMICULITE
    WEATHERPROOFING
    ∞WRAP

PACKET SWITCHING
GS   SWITCHING
    . PACKET TRANSMISSION
    . . **PACKET SWITCHING**
RT   BEAM SWITCHING
    COMMUNICATION NETWORKS
    DATA TRANSMISSION
    INTERRUPTION
    MICROWAVE SWITCHING
    MULTIPLE ACCESS
    NETWORK CONTROL
    PACKETS (COMMUNICATION)
    PROTOCOL (COMPUTERS)
    RADIO TRANSMISSION
    SEQUENCING
    SIGNAL TRANSMISSION
    SWITCHING CIRCUITS
    SWITCHING THEORY
    TELECOMMUNICATION
    TIME DIVISION MULTIPLE ACCESS

PACKET TRANSMISSION
GS   SWITCHING
    . **PACKET TRANSMISSION**
    . . PACKET SWITCHING
RT   ALOHA SYSTEM
    CHANNEL CAPACITY
    DATA TRANSMISSION
    MESSAGE PROCESSING
    PACKETS (COMMUNICATION)
    SPACECRAFT COMMUNICATION
    TRANSMISSION EFFICIENCY

PACKETS (COMMUNICATION)
RT   ALOHA SYSTEM
    COMMUNICATION NETWORKS
    DATA TRANSMISSION
    PACKET SWITCHING
    PACKET TRANSMISSION
    TRANSMISSION EFFICIENCY
    WAVE PACKETS

∞ **PACKING**
SN   *(USE OF A MORE SPECIFIC TERM IS*
    RECOMMENDED--CONSULT THE TERMS
    LISTED BELOW)
RT   PACKAGING
    PACKING DENSITY
    PACKINGS (SEALS)
    SEALING

PACKING DENSITY
GS   DENSITY (NUMBER/VOLUME)
    . **PACKING DENSITY**
RT   BRAVAIS CRYSTALS
    CRYSTAL STRUCTURE
    CRYSTALS
    ∞PACKING
    VOID RATIO

PACKINGS (SEALS)
GS   SEALS (STOPPERS)
    . **PACKINGS (SEALS)**
RT   BEARINGS
    GLANDS (SEALS)
    LABYRINTH SEALS
    ∞PACKING
    PUMPS
    SEALERS
    SEALING
    SHAFTS (MACHINE ELEMENTS)
    VALVES

∞ **PAD**
SN   *(USE OF A MORE SPECIFIC TERM IS*
    RECOMMENDED--CONSULT THE TERMS
    LISTED BELOW)
RT   CUSHIONS
    FOUNDATIONS
    LAUNCHING PADS

PADDLES
RT   FOLDING STRUCTURES
    MIXERS
    SOLAR GENERATORS
    TURBOMACHINE BLADES

PADE APPROXIMATION
GS   ANALYSIS (MATHEMATICS)
    . CALCULUS
    . . SERIES (MATHEMATICS)
    . . . **PADE APPROXIMATION**
    . NUMERICAL ANALYSIS
    . . APPROXIMATION
    . . . **PADE APPROXIMATION**
    . REAL VARIABLES
    . . SERIES (MATHEMATICS)
    . . . **PADE APPROXIMATION**

PAGEOS SATELLITE
GS   SATELLITES
    ARTIFICIAL SATELLITES
    . . GEODETIC SATELLITES
    . . . **PAGEOS SATELLITE**
    . . PASSIVE SATELLITES
    . . . **PAGEOS SATELLITE**
    UNMANNED SPACECRAFT
    . GEODETIC SATELLITES
    . . **PAGEOS SATELLITE**
    . PASSIVE SATELLITES
    . . **PAGEOS SATELLITE**
RT   EXPLORER 29 SATELLITE
    EXPLORER 36 SATELLITE
    GEOS 1 SATELLITE
    GEOS 2 SATELLITE
    GEOS 3 SATELLITE

PAIN
GS   PERCEPTION
    . SENSORY PERCEPTION
    . . **PAIN**
RT   ANALGESIA

PAIN SENSITIVITY
GS   PERCEPTION
    . SENSORY PERCEPTION
    . . **PAIN SENSITIVITY**
    SENSITIVITY
    . **PAIN SENSITIVITY**

PAINTS
GS   COATINGS
    . **PAINTS**
RT   FILLERS
    FINISHES
    METAL COATINGS
    PIGMENTS
    PRIMERS (COATINGS)
    PROTECTIVE COATINGS
    RUBBER COATINGS
    SEALERS
    SPRAYED COATINGS
    TURPENTINE
    VARNISHES

PAIR PRODUCTION
GS   PARTICLE PRODUCTION
    . **PAIR PRODUCTION**
RT   ELECTROMAGNETIC ABSORPTION
    ELECTRON EMISSION
    ELECTRON PHOTON CASCADES
    EMISSION
    HIGH ENERGY INTERACTIONS
    NUCLEAR REACTIONS
    PHOTOPRODUCTION
    POSITRON ANNIHILATION
    POSITRONS

PAKISTAN
GS   NATIONS
    . **PAKISTAN**
RT   ASIA
    BANGLADESH
    HIMALAYAS

PALAPA B SATELLITE
USE   PALAPA 2 SATELLITE

PALAPA SATELLITES
```
GS    SATELLITES
      . ARTIFICIAL SATELLITES
      . . COMMUNICATION SATELLITES
      . . . PALAPA SATELLITES
      . . . . PALAPA 2 SATELLITE
RT    INDONESIAN SPACE PROGRAM
      INTERNATIONAL COOPERATION
```

PALAPA 2 SATELLITE
```
UF    PALAPA B SATELLITE
GS    SATELLITES
      . ARTIFICIAL SATELLITES
      . . COMMUNICATION SATELLITES
      . . . PALAPA SATELLITES
      . . . . PALAPA 2 SATELLITE
RT    INDONESIAN SPACE PROGRAM
      INTERNATIONAL COOPERATION
```

PALEOBIOLOGY
```
RT    ∞ BIOLOGY
      FOSSILS
      GEOCHEMISTRY
      GEOCHRONOLOGY
      PALEONTOLOGY
```

PALEOMAGNETISM
```
GS    MAGNETIC FIELDS
      . PALEOMAGNETISM
      MAGNETIC PROPERTIES
      . PALEOMAGNETISM
RT    ARCHAEOLOGY
      CONES (VOLCANOES)
      CONTINENTAL DRIFT
      GEOLOGY
      GEOMAGNETISM
      GEOPHYSICS
      MARS VOLCANOES
      REMANENCE
      ROCKS
      VOLCANOES
      VOLCANOLOGY
```

PALEONTOLOGY
```
RT    FORMATIONS
      FOSSILS
      GEOCHEMISTRY
      GEOCHRONOLOGY
      GEOLOGICAL SURVEYS
      GEOLOGY
      HISTORIES
      PALEOBIOLOGY
      PRECAMBRIAN PERIOD
      STRATIGRAPHY
```

PALLADIUM
```
GS    CHEMICAL ELEMENTS
      . PALLADIUM
      METALS
      . TRANSITION METALS
      . . PALLADIUM
```

PALLADIUM ALLOYS
```
GS    ALLOYS
      . PALLADIUM ALLOYS
```

PALLADIUM COMPOUNDS
```
RT    ∞ CHEMICAL COMPOUNDS
      METALS
      TRANSITION METALS
```

PALMAR SWEAT INDEX
```
RT    PERSPIRATION
      STRESS (PHYSIOLOGY)
      STRESS (PSYCHOLOGY)
```

PALMGREN-MINER RULE
```
UF    MINER RULE
GS    RULES
      . PALMGREN-MINER RULE
RT    FATIGUE LIFE
```

PALMITIC ACID
```
GS    ACIDS
      . FATTY ACIDS
      . . PALMITIC ACID
      ALIPHATIC COMPOUNDS
      . FATTY ACIDS
      . . PALMITIC ACID
RT    FATS
```

PALO VERDE VALLEY (CA)
```
GS    VALLEYS
```

PALO VERDE VALLEY (CA)-(CONT.)
```
      . PALO VERDE VALLEY (CA)
RT    CALIFORNIA
      DESERTS
```

PAM (MODULATION)
```
USE   PULSE AMPLITUDE MODULATION
```

PAMPAS
```
GS    LAND
      . PLAINS
      . . PAMPAS
```

PANAMA
```
GS    NATIONS
      . PANAMA
RT    CANALS
      CENTRAL AMERICA
```

PANAMA CANAL ZONE
```
GS    LANDFORMS
      . PANAMA CANAL ZONE
      REGIONS
      . PANAMA CANAL ZONE
RT    CARIBBEAN SEA
      CENTRAL AMERICA
      UNITED STATES
```

PANAVIA MILITARY AIRCRAFT
```
RT    ∞ AIRCRAFT
      AIRCRAFT DESIGN
      FIGHTER AIRCRAFT
      ∞ MILITARY AIRCRAFT
      VARIABLE SWEEP WINGS
      WEAPON SYSTEMS
```

PANCREAS
```
GS    ANATOMY
      . DIGESTIVE SYSTEM
      . . PANCREAS
      . GLANDS (ANATOMY)
      . . ENDOCRINE GLANDS
      . . . PANCREAS
      VISCERA
      . ENDOCRINE GLANDS
      . . PANCREAS
RT    DIABETES MELLITUS
      GASTROINTESTINAL SYSTEM
      TRYPSIN
```

PANEL FLUTTER
```
GS    VIBRATION
      . STRUCTURAL VIBRATION
      . . FLUTTER
      . . . PANEL FLUTTER
      . . SELF INDUCED VIBRATION
      . . . PANEL FLUTTER
RT    AERODYNAMIC NOISE
      AEROELASTICITY
      BENDING VIBRATION
```

PANEL METHOD (FLUID DYNAMICS)
```
GS    PROCEDURES
      . PANEL METHOD (FLUID DYNAMICS)
RT    BERNOULLI THEOREM
      BOUNDARY LAYERS
      COMPUTATIONAL FLUID DYNAMICS
      FINITE ELEMENT METHOD
      ∞ FLOW
      FLOW THEORY
      FLUID DYNAMICS
      GAS-SOLID INTERACTIONS
      ∞ METHODOLOGY
      TURBULENCE
```

PANELS
```
SN    (EXCLUDES GROUPS OF PEOPLE)
GS    PANELS
      . CURVED PANELS
      . RECTANGULAR PANELS
      . WING PANELS
RT    BAFFLES
      CEILINGS (ARCHITECTURE)
      ∞ CONSTRUCTION MATERIALS
      DIVIDERS
      FLAT PLATES
      ∞ PLATES
      ∞ SHEETS
      SHIELDING
      THIN PLATES
      WALLS
```

PANIC
```
RT    EMOTIONAL FACTORS
```

PANIC-(CONT.)
```
      EMOTIONS
      FEAR
      HUMAN BEHAVIOR
```

PANORAMIC CAMERAS
```
GS    OPTICAL EQUIPMENT
      . CAMERAS
      . . PANORAMIC CAMERAS
      PHOTOGRAPHIC EQUIPMENT
      . CAMERAS
      . . PANORAMIC CAMERAS
RT    CAMERA SHUTTERS
      FOCUSING
      LENSES
      PHOTOGRAPHY
      WIDE ANGLE LENSES
```

PANORAMIC SCANNING
```
GS    SCANNING
      . PANORAMIC SCANNING
RT    CONICAL SCANNING
      FREQUENCY SCANNING
      MULTISPECTRAL BAND SCANNERS
      RADAR SCANNING
      SCANNERS
      SEARCHING
      SURVEILLANCE
```

PANSPERMIA
```
RT    ABIOGENESIS
      AEROSPACE ENVIRONMENTS
      BACTERIA
      BIOLOGICAL EVOLUTION
      EXOBIOLOGY
      EXTRATERRESTRIAL LIFE
      FUNGI
```

PANT PROGRAM
```
UF    ABLATED NOSETIPS
      PASSIVE NOSETIP TECHNOLOGY
GS    PROGRAMS
      . PANT PROGRAM
```

PANTAR CHONDRITES
```
GS    CELESTIAL BODIES
      . METEORITES
      . . STONY METEORITES
      . . . CHONDRITES
      . . . . PANTAR CHONDRITES
```

PANTHER AIRCRAFT
```
USE   F-9 AIRCRAFT
```

PAPAIN
```
GS    ACIDS
      . AMINO ACIDS
      . . PAPAIN
      ENZYMES
      . PAPAIN
      ORGANIC COMPOUNDS
      . AMINO ACIDS
      . . PAPAIN
RT    PEPSIN
```

PAPER (MATERIAL)
```
RT    BOARDS (PAPER)
      KRAFT PROCESS (WOODPULP)
      ∞ MATERIALS
      ORGANIC MATERIALS
      WEBS (SHEETS)
      WOOD
```

PAPER CHROMATOGRAPHY
```
GS    CHEMICAL TESTS
      . CHEMICAL ANALYSIS
      . . PAPER CHROMATOGRAPHY
      CHROMATOGRAPHY
      . PAPER CHROMATOGRAPHY
RT    GAS CHROMATOGRAPHY
      LIQUID CHROMATOGRAPHY
```

PAPERS
```
GS    DOCUMENTS
      . PAPERS
RT    BOARDS (PAPER)
      CONFERENCES
      FIBERS
      ∞ FILMS
      LAMINATES
      LITERATURE
      PRESIDENTIAL REPORTS
      PRIVACY
      REPORTS
```

PAPERS-*(CONT.)*
 ∞SHEETS
 WEBS (SHEETS)

PAPILLAE
RT PROTUBERANCES

† PARA HYDROGEN
GS CHEMICAL ELEMENTS
 . HYDROGEN
 . . **PARA HYDROGEN**
 GASES
 . HYDROGEN
 . . **PARA HYDROGEN**
RT ORTHO PARA CONVERSION

PARABOLAS
GS GEOMETRY
 . EUCLIDEAN GEOMETRY
 . . ANALYTIC GEOMETRY
 . . . CONICS
 . . . . **PARABOLAS**

PARABOLIC ANTENNAS
GS ANTENNAS
 . DIRECTIONAL ANTENNAS
 . . **PARABOLIC ANTENNAS**
RT ANTENNA DESIGN
 CASSEGRAIN ANTENNAS
 HORN ANTENNAS
 MICROWAVE ANTENNAS
 RADAR ANTENNAS
 RADAR EQUIPMENT
 RADAR REFLECTORS

PARABOLIC BODIES
UF PARABOLOIDS
GS SYMMETRICAL BODIES
 . BODIES OF REVOLUTION
 . . **PARABOLIC BODIES**

PARABOLIC DIFFERENTIAL EQUATIONS
GS ANALYSIS (MATHEMATICS)
 . REAL VARIABLES
 . . DIFFERENTIAL EQUATIONS
 . . . PARTIAL DIFFERENTIAL EQUATIONS
 . . . . **PARABOLIC DIFFERENTIAL
     EQUATIONS**
RT ∞EQUATIONS

PARABOLIC FLIGHT
RT ASCENT TRAJECTORIES
 BALLISTIC TRAJECTORIES
 CLIMBING FLIGHT
 COASTING FLIGHT
 DESCENT TRAJECTORIES
 ∞FLIGHT
 MIDCOURSE TRAJECTORIES
 MISSILE TRAJECTORIES
 SUBORBITAL FLIGHT
 TRAJECTORIES
 WEIGHTLESSNESS
 WEIGHTLESSNESS SIMULATION

PARABOLIC REFLECTORS
UF DISHES
GS REFLECTORS
 . **PARABOLIC REFLECTORS**
 . . PARABOLOID MIRRORS
RT MICROWAVE ANTENNAS
 RADAR REFLECTORS
 SCHWARZSCHILD ANTENNAS
 SOLAR REFLECTORS

PARABOLIC VELOCITY
USE ESCAPE VELOCITY

PARABOLOID MIRRORS
GS MIRRORS
 . **PARABOLOID MIRRORS**
 REFLECTORS
 . PARABOLIC REFLECTORS
 . . **PARABOLOID MIRRORS**
RT REFLECTING TELESCOPES
 SOLAR REFLECTORS

PARABOLOIDS
USE PARABOLIC BODIES

PARACHUTE DESCENT
UF PARACHUTING
GS DESCENT
 . **PARACHUTE DESCENT**
RT BAILOUT

PARACHUTE DESCENT-*(CONT.)*
 EJECTION
 EJECTION TRAINING
 ESCAPE (ABANDONMENT)
 FREE FALL
 PARACHUTES

PARACHUTE FABRICS
GS FABRICS
 . **PARACHUTE FABRICS**
RT FORTISAN (TRADEMARK)
 GORES
 PARACHUTES

PARACHUTES
GS **PARACHUTES**
 . DRAG CHUTES
 . RECOVERY PARACHUTES
 . RIBBON PARACHUTES
 . ROTOCHUTES
RT AIR DROP OPERATIONS
 AIRDROPS
 BALLUTES
 BRAKES (FOR ARRESTING MOTION)
 FOLDING STRUCTURES
 PARACHUTE DESCENT
 PARACHUTE FABRICS
 PARACONE
 PARAVULCOONS
 PARAWINGS
 TOWED BODIES
 WHIRL TOWERS

PARACHUTING
USE PARACHUTE DESCENT

PARACHUTING INJURY
GS INJURIES
 . **PARACHUTING INJURY**

PARACONE
RT ESCAPE CAPSULES
 ESCAPE SYSTEMS
 PARACHUTES

PARADOXES
GS KNOWLEDGE
 . PHILOSOPHY
 . . **PARADOXES**
RT ∞LOGIC
 RELATIVITY

PARAFFINS
GS ALIPHATIC COMPOUNDS
 . ALKANES
 . . **PARAFFINS**
 . . . CERESIN
 HYDROCARBONS
 . ALKANES
 . . **PARAFFINS**
 . . . CERESIN
RT HYDROCARBON FUELS
 KEROSENE
 SHALE OIL

PARAGLIDERS
GS GLIDERS
 . **PARAGLIDERS**
 . . INFLATABLE GLIDERS
 . . PARAWINGS
RT FOLDING STRUCTURES
 HYPERSONIC GLIDERS
 ∞SUBSONIC AIRCRAFT

PARAGUAY
GS NATIONS
 . **PARAGUAY**
RT SOUTH AMERICA

PARALLAX
GS **PARALLAX**
 . SOLAR PARALLAX
 . STELLAR PARALLAX
RT ASTROMETRY
 COMPANION STARS
 ∞OPTICS

PARALLEL COMPUTERS
GS DATA PROCESSING EQUIPMENT
 . COMPUTERS
 . . DIGITAL COMPUTERS
 . . . **PARALLEL COMPUTERS**

PARALLEL FLOW
GS FLUID FLOW
 . **PARALLEL FLOW**
 . . PIPE FLOW
 . . THREE DIMENSIONAL FLOW
RT FLOW GEOMETRY
 FLOW VELOCITY
 FLUID DYNAMICS
 LAMINAR FLOW
 STEADY FLOW

PARALLEL PLATES
RT CAPACITORS
 ∞CHANNELS
 FLAT PLATES
 METAL PLATES
 ∞PLATES
 THIN PLATES
 WAVEGUIDES

PARALLEL PROCESSING (COMPUTERS)
GS DATA PROCESSING
 . **PARALLEL PROCESSING (COMPUTERS)**
RT ASSOCIATIVE PROCESSING
  (COMPUTERS)
 CONCURRENT PROCESSING
 ILLIAC 3 COMPUTER
 ILLIAC 4 COMPUTER
 INTERPROCESSOR COMMUNICATION

PARALLEL PROGRAMMING
GS SOFTWARE ENGINEERING
 . COMPUTER PROGRAMMING
 . . **PARALLEL PROGRAMMING**
RT MULTIPROCESSING (COMPUTERS)
 PIPELINING (COMPUTERS)

PARALLEL STRIP LINES
USE MICROSTRIP TRANSMISSION LINES

PARALLELEPIPEDS
GS GEOMETRY
 . EUCLIDEAN GEOMETRY
 . . POLYHEDRONS
 . . . **PARALLELEPIPEDS**

PARALLELOGRAMS
GS GEOMETRY
 . EUCLIDEAN GEOMETRY
 . . POLYGONS
 . . . TETRAGONS
 . . . . **PARALLELOGRAMS**
 . . . . . RHOMBOIDS

PARALYSIS
GS DISEASES
 . **PARALYSIS**
 INJURIES
 . **PARALYSIS**
RT TREMORS

PARAMAGNETIC AMPLIFIERS
USE MASERS

PARAMAGNETIC RESONANCE
GS RESONANCE
 . MAGNETIC RESONANCE
 . . **PARAMAGNETIC RESONANCE**
 . . . ELECTRON PARAMAGNETIC
    RESONANCE
RT ABSORPTION SPECTRA
 FERROMAGNETIC RESONANCE
 NUCLEAR MAGNETIC RESONANCE
 PARAMAGNETISM

PARAMAGNETISM
GS MAGNETIC PROPERTIES
 . **PARAMAGNETISM**
RT ANTIFERROMAGNETISM
 CURIE-WEISS LAW
 DIAMAGNETISM
 PARAMAGNETIC RESONANCE

PARAMECIA
GS ANIMALS
 . INVERTEBRATES
 . . PROTOZOA
 . . . **PARAMECIA**
 . . . . ROTIFERA
 MICROORGANISMS
 . PROTOZOA
 . . **PARAMECIA**
 . . . ROTIFERA

PARAMETER IDENTIFICATION
GS   ESTIMATING
. **PARAMETER IDENTIFICATION**
IDENTIFYING
. **PARAMETER IDENTIFICATION**
PARAMETERIZATION
. **PARAMETER IDENTIFICATION**
RT   COMPLEX SYSTEMS
CONTROL SYSTEMS DESIGN
DYNAMIC RESPONSE
ESTIMATES
INDEPENDENT VARIABLES
LEAST SQUARES METHOD
MATHEMATICAL MODELS
MAXIMUM LIKELIHOOD ESTIMATES
OBSERVABILITY (SYSTEMS)
OPTIMAL CONTROL
OPTIMIZATION
PREDICTION ANALYSIS TECHNIQUES
PROBABILITY THEORY
STATISTICAL ANALYSIS
STEEPEST DESCENT METHOD
SYSTEM IDENTIFICATION
SYSTEMS ANALYSIS
SYSTEMS ENGINEERING

PARAMETERIZATION
GS   **PARAMETERIZATION**
. PARAMETER IDENTIFICATION
RT   ALGORITHMS
∞APPLICATIONS OF MATHEMATICS
DEPENDENT VARIABLES
DERIVATION
DIMENSIONAL ANALYSIS
∞ESTIMATORS
FORMALISM
FORMULATIONS
MATHEMATICAL MODELS
SCALE EFFECT
SEMIEMPIRICAL EQUATIONS
SYSTEM IDENTIFICATION
UNITS OF MEASUREMENT

PARAMETERS
USE   INDEPENDENT VARIABLES

PARAMETRIC AMPLIFIERS
UF   PARAMETRIC OSCILLATORS
GS   AMPLIFIERS
. **PARAMETRIC AMPLIFIERS**
RT   FREQUENCY CONVERTERS
LC CIRCUITS
MAGNETOSTATIC AMPLIFIERS
MICROWAVE AMPLIFIERS
NEGATIVE RESISTANCE DEVICES
POWER AMPLIFIERS
SEMICONDUCTOR DEVICES

PARAMETRIC DIODES
GS   ELECTRONIC EQUIPMENT
. DIODES
. . SEMICONDUCTOR DIODES
. . . **PARAMETRIC DIODES**
. SOLID STATE DEVICES
. . SEMICONDUCTOR DEVICES
. . . **PARAMETRIC DIODES**
RT   VARACTOR DIODES

PARAMETRIC FREQUENCY CONVERTERS
GS   FREQUENCY CONVERTERS
. **PARAMETRIC FREQUENCY
   CONVERTERS**
RT   ∞CONVERTERS
PHASE MODULATION
UP-CONVERTERS

PARAMETRIC OSCILLATORS
USE   PARAMETRIC AMPLIFIERS

PARAMETRONS
RT   COMPUTER STORAGE DEVICES
MAGNETIC CORES
MAGNETIC STORAGE
OSCILLATORS
PHASE LOCK DEMODULATORS
THIN FILMS

PARANASAL SINUSES
GS   SINUSES
. **PARANASAL SINUSES**
RT   NOSE (ANATOMY)

PARAPLASTS
GS   PLASTERS

PARAPLASTS-*(CONT.)*
. **PARAPLASTS**
RT   RESINS

PARAPSYCHOLOGY
USE   EXTRASENSORY PERCEPTION

PARASITES
RT   ANIMALS
BLIGHT
INFESTATION

PARASITIC DISEASES
GS   DISEASES
. **PARASITIC DISEASES**
. . BLIGHT
RT   AIRBORNE INFECTION
AMOEBA
CHOLERA
FUNGI
INFECTIOUS DISEASES
RUST FUNGI
TRYPANOSOME

PARATHYROID GLAND
GS   ANATOMY
. GLANDS (ANATOMY)
. . ENDOCRINE GLANDS
. . . **PARATHYROID GLAND**
VISCERA
. ENDOCRINE GLANDS
. . **PARATHYROID GLAND**
RT   CALCIUM METABOLISM

PARAVULCOONS
GS   BRAKES (FOR ARRESTING MOTION)
. AERODYNAMIC BRAKES
. . **PARAVULCOONS**
DRAG DEVICES
. AERODYNAMIC BRAKES
. . **PARAVULCOONS**
EXPANDABLE STRUCTURES
. INFLATABLE STRUCTURES
. . **PARAVULCOONS**
RT   AIR DROP OPERATIONS
BALLOONS
FOLDING STRUCTURES
PARACHUTES

PARAWINGS
GS   AIRFOILS
. WINGS
. . FLEXIBLE WINGS
. . . **PARAWINGS**
GLIDERS
. PARAGLIDERS
. . **PARAWINGS**
RT   AIR DROP OPERATIONS
FOLDING STRUCTURES
HANG GLIDERS
PARACHUTES

PARENTERAL FUNCTIONS
RT   ∞FUNCTIONS

PARENTS
RT   CHILDREN
HUMAN BEINGS

PARITY
RT   BCH CODES
CODING
CONSERVATION
CORRECTION
EQUIVALENCE
ERROR DETECTION CODES
INFORMATION THEORY
NUCLEAR PHYSICS
PARTICLE SPIN
QUANTUM NUMBERS
QUANTUM THEORY
STRANGENESS
VECTOR CURRENTS

∞ **PARKING**
SN   *(USE OF A MORE SPECIFIC TERM IS
RECOMMENDED--CONSULT THE TERMS
LISTED BELOW)*
RT   PARKING ORBITS
RAMPS (STRUCTURES)

PARKING ORBITS
GS   ORBITS
. SPACECRAFT ORBITS

PARKING ORBITS-*(CONT.)*
. . SATELLITE ORBITS
. . . **PARKING ORBITS**
RT   EARTH ORBITS
EARTH-MOON TRAJECTORIES
FLIGHT OPTIMIZATION
INTERPLANETARY TRAJECTORIES
LUNAR ORBITS
LUNAR TRAJECTORIES
ORBITAL MECHANICS
∞PARKING
PLANETARY ORBITS
THRUST PROGRAMMING
TRANSFER ORBITS

PARKINSON DISEASE
GS   DISEASES
. **PARKINSON DISEASE**
RT   TREMORS

PARKS
GS   LAND
. **PARKS**
. . NATIONAL PARKS
. . . YELLOWSTONE NATIONAL PARK
    (ID-MT-WY)
RT   RECREATION
REGIONAL PLANNING
URBAN DEVELOPMENT
URBAN PLANNING

PAROTID GLAND
USE   SALIVARY GLANDS

PARSING ALGORITHMS
GS   MATHEMATICAL LOGIC
. ALGORITHMS
. . **PARSING ALGORITHMS**
RT   COMPILERS
COMPUTER TECHNIQUES
GRAMMARS
SEMANTICS
SUBROUTINES
SYNTAX

PARTIAL DIFFERENTIAL EQUATIONS
GS   ANALYSIS (MATHEMATICS)
. REAL VARIABLES
. . DIFFERENTIAL EQUATIONS
. . . **PARTIAL DIFFERENTIAL
      EQUATIONS**
. . . . BIHARMONIC EQUATIONS
. . . . BURGER EQUATION
. . . . ELLIPTIC DIFFERENTIAL
       EQUATIONS
. . . . . MONGE-AMPERE EQUATION
. . . . EULER-CAUCHY EQUATIONS
. . . . FOKKER-PLANCK EQUATION
. . . . GAUSS EQUATION
. . . . HELMHOLTZ VORTICITY EQUATION
. . . . LIOUVILLE EQUATIONS
. . . . PARABOLIC DIFFERENTIAL
       EQUATIONS
. . . . . VLASOV EQUATIONS
RT   BOLTZMANN-VLASOV EQUATION
CAUCHY-RIEMANN EQUATIONS
∞EQUATIONS
FUNCTIONAL INTEGRATION
KINETIC EQUATIONS
LAPLACE EQUATION
METHOD OF CHARACTERISTICS
NEUMANN PROBLEM
POISSON EQUATION
WAVE EQUATIONS

PARTIAL PRESSURE
GS   PRESSURE
. **PARTIAL PRESSURE**
. . OXYGEN TENSION
. . . HYPOXEMIA
RT   DALTON LAW
GAS PRESSURE
HENRY LAW
INTERNAL PRESSURE
RAOULT LAW
RESIDUAL GAS
∞TENSION
VAPOR PRESSURE

PARTICLE ACCELERATION
GS   RATES (PER TIME)
. ACCELERATION (PHYSICS)
. . **PARTICLE ACCELERATION**
RT   ∞ACCELERATION
ELECTROMAGNETIC ACCELERATION

PARTICLE ACCELERATION-*(CONT.)*
    MAGNETIC FIELDS
    PLASMA ACCELERATION
    RACETRACKS (PARTICLE
        ACCELERATORS)

PARTICLE ACCELERATOR TARGETS
GS   TARGETS
    . **PARTICLE ACCELERATOR TARGETS**
RT   ∞ACCELERATORS
      TARGET THICKNESS

PARTICLE ACCELERATORS
GS   **PARTICLE ACCELERATORS**
    . CYCLIC ACCELERATORS
    . . BETATRONS
    . . SYNCHROCYCLOTRONS
    . . SYNCHROTRONS
    . . . BEVATRON
    . . . STORAGE RINGS (PARTICLE
        ACCELERATORS)
    . CYCLOTRONS
    . . GEOCYCLOTRONS
    . . MICROTRONS
    . . OAK RIDGE ISOCHRONOUS
        CYCLOTRON
    . . OMEGATRONS
    . . SYNCHROCYCLOTRONS
    . ELECTRON ACCELERATORS
    . . BETATRONS
    . ION ACCELERATORS
    . LINEAR ACCELERATORS
    . NIMROD ACCELERATOR
    . SYNCHROPHASOTRONS
    . VAN DE GRAAFF ACCELERATORS
RT   ∞ACCELERATORS
      BEAM SPLITTERS
      ELECTRON GUNS
      ELEMENTARY PARTICLES
      ION SOURCES
      KAON PRODUCTION
      NEUTRON SOURCES
      NUCLEAR PARTICLES
      RACETRACKS (PARTICLE
        ACCELERATORS)
      RAILGUN ACCELERATORS
      SEPAC (PAYLOAD)

PARTICLE BEAMS
GS   BEAMS (RADIATION)
    . **PARTICLE BEAMS**
    . . ATOMIC BEAMS
    . . ELECTRON BEAMS
    . . . RELATIVISTIC ELECTRON BEAMS
    . . ION BEAMS
    . . NEUTRAL BEAMS
    . . . MOLECULAR BEAMS
    . . . NEUTRON BEAMS
    . . . NEUTRINO BEAMS
    . . PION BEAMS
    . . PROTON BEAMS
RT   BEAM SPLITTERS
      ELECTRON ACCELERATION
      ELECTRON BOMBARDMENT
      FLUX (RATE)
      ION STRIPPING
      PHONON BEAMS

PARTICLE CHARGING
RT   CHARGED PARTICLES

PARTICLE COLLISIONS
GS   COLLISIONS
    . **PARTICLE COLLISIONS**
RT   ATOMIC COLLISIONS
      ATOMIC EXCITATIONS
      DENSE PLASMAS
      FADDEEV EQUATIONS
      KINETICS
      MEAN FREE PATH
      MOLECULAR COLLISIONS
      MOLECULAR EXCITATION
      NUCLEON-NUCLEON SCATTERING
      SCATTERING

PARTICLE COUNTERS
USE   RADIATION COUNTERS

PARTICLE DECAY
USE   RADIOACTIVE DECAY

PARTICLE DENSITY (CONCENTRATION)
GS   DENSITY (NUMBER/VOLUME)
    . **PARTICLE DENSITY (CONCENTRATION)**

PARTICLE DENSITY (CONCENTRATION)-*(CONT.)*
    . . ELECTRON DENSITY
        (CONCENTRATION)
    . . . CARRIER DENSITY (SOLID STATE)
    . . . ELECTRON DENSITY PROFILES
    . . . IONOSPHERIC ELECTRON DENSITY
    . . . MAGNETOSPHERIC ELECTRON
        DENSITY
    . . ELECTRON DISTRIBUTION
    . . . ELECTRON DENSITY PROFILES
    . . ION DENSITY (CONCENTRATION)
    . . . IONOSPHERIC ION DENSITY
    . . . MAGNETOSPHERIC ION DENSITY
    . . . . MAGNETOSPHERIC PROTON
        DENSITY
    . . . PROTON DENSITY
        (CONCENTRATION)
    . . . . MAGNETOSPHERIC PROTON
        DENSITY
    . . PLASMA DENSITY
RT   ATMOSPHERIC DENSITY
      ESRO 4 SATELLITE
      ION STRIPPING
      IONOSPHERIC COMPOSITION
      SPACE DENSITY
      SPATIAL DISTRIBUTION

PARTICLE DETECTORS
USE   RADIATION COUNTERS

PARTICLE DIFFUSION
GS   DIFFUSION
    . **PARTICLE DIFFUSION**
    . . ELECTRON DIFFUSION
    . . IONIC DIFFUSION
RT   ATOMIC BEAMS
      BOLTZMANN TRANSPORT EQUATION
      DIFFUSION COEFFICIENT
      DROP SIZE
      FLUX (RATE)
      GASEOUS SELF-DIFFUSION
      MOLECULAR DIFFUSION
      MUON SPIN ROTATION
      SELF DIFFUSION (SOLID STATE)

PARTICLE EMISSION
GS   DECAY
    . EMISSION
    . . **PARTICLE EMISSION**
    . . . ELECTRON EMISSION
    . . . . FIELD EMISSION
    . . . . PHOTOELECTRIC EMISSION
    . . . . SECONDARY EMISSION
    . . . ION EMISSION
    . . . NEUTRON EMISSION
    . . . THERMIONIC EMISSION
RT   EXHAUST EMISSION
      EXPULSION
      SELF SUSTAINED EMISSION
      STIMULATED EMISSION

PARTICLE ENERGY
GS   **PARTICLE ENERGY**
    . ELECTRON ENERGY
    . . ELECTRON STATES
    . PROTON ENERGY
RT   ∞ENERGY
      INTERNAL ENERGY
      KINETIC ENERGY
      MONOCHROMATIZATION
     ∞PARTICLE INTENSITY

PARTICLE FLUX
USE   FLUX (RATE)

PARTICLE FLUX DENSITY
SN   (PARTICLE EMISSION OR DETECTION
      RATE PER UNIT AREA)
GS   RATES (PER TIME)
    . FLUX DENSITY
    . . RADIANT FLUX DENSITY
    . . . **PARTICLE FLUX DENSITY**
    . . . . ELECTRON FLUX DENSITY
    . . . . NEUTRON FLUX DENSITY
    . . . . PROTON FLUX DENSITY
RT   HELIOS SATELLITES
     ∞PARTICLE INTENSITY
      RADIANCY
      RADIATION COUNTERS
      RADIATION PRESSURE
      SOLAR CONSTANT
      SOLAR FLUX DENSITY

PARTICLE IN CELL TECHNIQUE
GS   ANALYSIS (MATHEMATICS)

PARTICLE IN CELL TECHNIQUE-*(CONT.)*
    . NUMERICAL ANALYSIS
    . . APPROXIMATION
    . . . **PARTICLE IN CELL TECHNIQUE**
RT   ∞CELLS
      CRYSTAL LATTICES
      FLOW EQUATIONS
     ∞METHODOLOGY

∞ **PARTICLE INTENSITY**
SN   *(USE OF A MORE SPECIFIC TERM IS*
      RECOMMENDED--CONSULT THE TERMS
      LISTED BELOW)
RT   PARTICLE ENERGY
      PARTICLE FLUX DENSITY

PARTICLE INTERACTIONS
GS   **PARTICLE INTERACTIONS**
    . CONFIGURATION INTERACTION
    . ELEMENTARY PARTICLE
        INTERACTIONS
    . . MESON-MESON INTERACTIONS
    . . MESON-NUCLEON INTERACTIONS
    . . NUCLEAR CAPTURE
    . . . ELECTRON CAPTURE
    . . NUCLEON-NUCLEON INTERACTIONS
    . ION ATOM INTERACTIONS
    . MOLECULAR INTERACTIONS
        INTERSTELLAR CHEMISTRY
    . . MOLECULAR COLLISIONS
    . NUCLEAR INTERACTIONS
    . . NUCLEAR CAPTURE
    . . . ELECTRON CAPTURE
    . . SPIN-ORBIT INTERACTIONS
    . . . ELECTRON CAPTURE
    . . STRONG INTERACTIONS (FIELD
        THEORY)
    . . WEAK INTERACTIONS (FIELD
        THEORY)
    . PLASMA-PARTICLE INTERACTIONS
    . WEAK ENERGY INTERACTIONS
    . . WEAK INTERACTIONS (FIELD
        THEORY)
RT   BRAGG CURVE
      CHARM (PARTICLE PHYSICS)
      CHEMICAL REACTIONS
      COLLISION PARAMETERS
      ELECTRON PHONON INTERACTIONS
      ELECTRON SCATTERING
      FEYNMAN DIAGRAMS
      FLAVOR (PARTICLE PHYSICS)
      HIGH ENERGY INTERACTIONS
     ∞INTERACTIONS
      NEUTRAL CURRENTS
      NUCLEAR REACTIONS
      PHOTONUCLEAR REACTIONS
      PHOTOPHORESIS
      QUANTUM CHROMODYNAMICS
      SPHINX

PARTICLE LADEN JETS
RT   FUEL FLOW
      JET FLOW
      PARTICLES
      TURBULENT FLOW

PARTICLE MASS
GS   MASS
    . **PARTICLE MASS**
    . . ELECTRON MASS
RT   GRAVITINOS
      MAGNETIC RIGIDITY
      NUCLIDES

PARTICLE MOTION
RT   ELECTROPHORESIS
      FALLING
      LATTICE VIBRATIONS
      MAGNETIC RIGIDITY
      MEAN FREE PATH
     ∞MOTION
      NEUTRAL SHEETS
      PHOTOPHORESIS
      RECOILINGS
      RELATIVISTIC VELOCITY
      SETTLING

PARTICLE PRECIPITATION
GS   **PARTICLE PRECIPITATION**
    . ELECTRON PRECIPITATION
    . PROTON PRECIPITATION
RT   ATOMIC STRUCTURE
      CHARGED PARTICLES
     ∞PRECIPITATION

PARTICLE PRODUCTION
GS   **PARTICLE PRODUCTION**
. KAON PRODUCTION
. PAIR PRODUCTION
RT   COMMINUTION
CORPUSCULAR RADIATION
HIGH ENERGY INTERACTIONS
NUCLEAR RADIATION
NUCLEAR REACTIONS
PARTICLES
RADIOACTIVITY
SPALLATION

PARTICLE SIZE DISTRIBUTION
GS   SIZE DISTRIBUTION
. **PARTICLE SIZE DISTRIBUTION**
RT   DIMENSIONS
DROP SIZE
FINENESS
FINES
FRACTIONS
GRAIN SIZE
PARTICLES
PRECIPITATION PARTICLE
MEASUREMENT
SIZE DETERMINATION
SIZE SEPARATION
SOLIDS FLOW

PARTICLE SPIN
GS   SPIN
. **PARTICLE SPIN**
. . ELECTRON SPIN
. . ISOTOPIC SPIN
. . NUCLEAR SPIN
RT   ANGULAR MOMENTUM
MUON SPIN ROTATION
NUCLEAR MAGNETIC RESONANCE
NUCLEAR PHYSICS
PARITY
QUENCHING (ATOMIC PHYSICS)
SPIN RESONANCE

PARTICLE TELESCOPES
UF   ELECTRON TELESCOPES
GEP TELESCOPES
GODDARD EXPERIMENT PACKAGE
TELESCOPE
PROTON TELESCOPES
GS   MEASURING INSTRUMENTS
. COUNTERS
. . RADIATION COUNTERS
. . . **PARTICLE TELESCOPES**
. RADIATION MEASURING INSTRUMENTS
. . RADIATION COUNTERS
. . . **PARTICLE TELESCOPES**
TELESCOPES
. **PARTICLE TELESCOPES**
RT   ASTRONOMICAL TELESCOPES
GEIGER COUNTERS
SATELLITE-BORNE INSTRUMENTS
SCINTILLATION COUNTERS

PARTICLE THEORY
RT   BODY KINEMATICS
CHARM (PARTICLE PHYSICS)
COLLISION PARAMETERS
FLAVOR (PARTICLE PHYSICS)
∞ INTERACTIONS
MANY BODY PROBLEM
PLASMA-PARTICLE INTERACTIONS
∞ THEORIES
UNIFIED FIELD THEORY
WEAK ENERGY INTERACTIONS

PARTICLE TRACKS
RT   CHEMICAL ANALYSIS
CORE SAMPLING
COSMIC RAYS
FOSSILS
GEOCHRONOLOGY
LUNAR ROCKS
METEOROIDS
NUCLEAR PARTICLES
RADIATION EFFECTS
STRATIGRAPHY
TRACE ELEMENTS
∞ TRACKS

PARTICLE TRAJECTORIES
GS   TRAJECTORIES
. **PARTICLE TRAJECTORIES**
. . ELECTRON TRAJECTORIES
RT   BUBBLE CHAMBERS
CHARGED PARTICLES

PARTICLE TRAJECTORIES-*(CONT.)*
ELECTRON OPTICS
IONIZING RADIATION
∞ MOTION
∞ PATHS
RACETRACKS (PARTICLE
ACCELERATORS)
∞ TRACKS

PARTICLES
GS   **PARTICLES**
. AEROSOLS
. . FOG
. CHARGED PARTICLES
. . ANIONS
. . ANTIPROTONS
. . CATIONS
. . . METAL IONS
. . . . FERRIC IONS
. . . . MANGANESE IONS
. . ENERGETIC PARTICLES
. . ELECTRONS
. . . CONDUCTION ELECTRONS
. . . HIGH ENERGY ELECTRONS
. . . HOT ELECTRONS
. . . N ELECTRONS
. . . NEGATRONS
. . . PI-ELECTRONS
. . NUCLEI (NUCLEAR PHYSICS)
. . PLASMAS (PHYSICS)
. . . ARGON PLASMA
. . . BETA PARTICLES
. . . BOUNDARY LAYER PLASMAS
. . . COLD PLASMAS
. . . COLLISIONAL PLASMAS
. . . . STRONGLY COUPLED PLASMAS
. . . COLLISIONLESS PLASMAS
. . . COSMIC PLASMA
. . . CYLINDRICAL PLASMAS
. . . DENSE PLASMAS
. . . . PLASMA FOCUS
. . . . STRONGLY COUPLED PLASMAS
. . . ELECTRON PLASMA
. . . ELLIPTICAL PLASMAS
. . . HELIUM PLASMA
. . . HIGH TEMPERATURE PLASMAS
. . . LASER PLASMAS
. . . METALLIC PLASMAS
. . . . CESIUM PLASMA
. . . MICROPLASMAS
. . . NITROGEN PLASMA
. . . NONEQUILIBRIUM PLASMAS
. . . NONUNIFORM PLASMAS
. . . RAREFIED PLASMAS
. . . RELATIVISTIC PLASMAS
. . . ROTATING PLASMAS
. . . SEMICONDUCTOR PLASMAS
. . . SOLAR WIND
. . . SPACE PLASMAS
. . . SPHERICAL PLASMAS
. . . STELLAR WINDS
. . . THERMAL PLASMAS
. . . TOROIDAL PLASMAS
. . MAGNETICALLY TRAPPED PARTICLES
. . RADIATION BELTS
. . . ARTIFICIAL RADIATION BELTS
. . . INNER RADIATION BELT
. . . OUTER RADIATION BELT
. . . PROTON BELTS
. . NEGATIVE IONS
. . PARTONS
. . PLASMA CLOUDS
. . PLASMA LAYERS
. . PLASMA SHEATHS
. . PLASMA SLABS
. . POSITRONS
. . PROTONS
. . . RECOIL PROTONS
. . . SOLAR PROTONS
. CORPUSCULAR RADIATION
. . CYCLOTRON RADIATION
. . . ION CYCLOTRON RADIATION
. . ELECTRON PRECIPITATION
. . ELECTRON RADIATION
. . . BETA PARTICLES
. . . ELECTRON BEAMS
. . . . RELATIVISTIC ELECTRON BEAMS
. . PRIMARY COSMIC RAYS
. . SOLAR COSMIC RAYS
. . RADIATION BELTS
. . SOLAR CORPUSCULAR RADIATION
. . SOLAR ELECTRONS
. . SOLAR PROTONS
. DROPS (LIQUIDS)
. . RAINDROPS
. ELEMENTARY PARTICLES

PARTICLES-*(CONT.)*
. . ANTIPARTICLES
. . . ANTINEUTRINOS
. . . ANTINUCLEONS
. . . ANTIPROTONS
. . . POSITRONS
. . BETA PARTICLES
. . BOSONS
. . . ALPHA PARTICLES
. . MESONS
. . . . ETA-MESONS
. . . . KAONS
. . . MESON RESONANCE
. . . . . X MESONS
. . . MUONS
. . . PIONS
. . . VECTOR MESONS
. . . . RHO-MESONS
. . . . SIGMA-MESONS
. . PHOTONS
. . . LIGHT BEAMS
. . XI HYPERONS
. DEUTERONS
. FERMIONS
. . BARYONS
. . . HYPERONS
. . . . XI HYPERONS
. . . OMEGA-MESONS
. . . RHO-MESONS
. . . SIGMA-MESONS
. . ETA-MESONS
. . LEPTONS
. . . ANTINEUTRINOS
. . . MUONS
. . . NEUTRINOS
. . . . SOLAR NEUTRINOS
. . MESON RESONANCE
. . NEUTRONS
. . . COLD NEUTRONS
. . . FAST NEUTRONS
. . . PHOTONEUTRONS
. . . THERMAL NEUTRONS
. . PROTONS
. . . RECOIL PROTONS
. . . SOLAR PROTONS
. GLUONS
. GRAVITINOS
. GRAVITONS
. HADRONS
. . BARYONS
. . . OMEGA-MESONS
. . . RHO-MESONS
. . . SIGMA-MESONS
. . MESONS
. . . KAONS
. . . MUONS
. . . OMEGA-MESONS
. . . VECTOR MESONS
. . . . RHO-MESONS
. . . . SIGMA-MESONS
. . MAGNETIC MONOPOLES
. . NUCLEONS
. . PARTONS
. QUARKS
. TACHYONS
. FLAKES
. METAL PARTICLES
. . PLATINUM BLACK
. . POWDERED ALUMINUM
. MICROPARTICLES
. MIST
. NEUTRAL PARTICLES
. . GRAVITINOS
. . NEUTRONS
. . . COLD NEUTRONS
. . . FAST NEUTRONS
. . . PHOTONEUTRONS
. . . THERMAL NEUTRONS
. NUCLEAR PARTICLES
. . ANTIPARTICLES
. . . ANTINEUTRINOS
. . . ANTINUCLEONS
. . . ANTIPROTONS
. . . POSITRONS
. . BETA PARTICLES
. . BOSONS
. . . ALPHA PARTICLES
. . . MESONS
. . . . ETA-MESONS
. . . . KAONS
. . . . MESON RESONANCE
. . . . . X MESONS
. . . . MUONS
. . . . PIONS
. . . . VECTOR MESONS
. . . . . RHO-MESONS

PARTICLES-(CONT.)

```
. . . . . SIGMA-MESONS
. . . . PHOTONS
. . . XI HYPERONS
. . NUCLEONS
. . PHOTOELECTRONS
. PLASMA JETS
. POLLEN
. POWDER (PARTICLES)
. . FINES
. . PLATINUM BLACK
. . POWDERED ALUMINUM
. RELATIVISTIC PARTICLES
. . RELATIVISTIC ELECTRON BEAMS
. SOOT
. TRAPPED PARTICLES
. . MAGNETICALLY TRAPPED PARTICLES
. . . RADIATION BELTS
. . . . ARTIFICIAL RADIATION BELTS
. . . . INNER RADIATION BELT
. . . . OUTER RADIATION BELT
. . . . PROTON BELTS
```
```
RT    AIR POLLUTION
      CHEMICAL CLOUDS
      CLOUDS
      COLLOIDS
      DEUTERON IRRADIATION
      DIRT
      DISPERSIONS
      DUST
      GAS ATOMIZATION
      GRANULAR MATERIALS
      GRIT
      ION STRIPPING
      METAL POWDER
      NEUTRAL BEAMS
      NEUTRON BEAMS
      NODULES
      NONPOINT SOURCES
      PARTICLE LADEN JETS
      PARTICLE PRODUCTION
      PARTICLE SIZE DISTRIBUTION
      POSITRON ANNIHILATION
      PRECIPITATION PARTICLE
        MEASUREMENT
      PROTON PRECIPITATION
      SMOKE
```

PARTICULATE FILTERS
```
USE   FLUID FILTERS
```

PARTICULATE SAMPLING
```
GS    SAMPLING
      . PARTICULATE SAMPLING
RT    ASSAYING
      CHEMICAL ANALYSIS
      CONCENTRATION (COMPOSITION)
      IDENTIFYING
```

∞ PARTITIONS
```
SN    (USE OF A MORE SPECIFIC TERM IS
      RECOMMENDED--CONSULT THE TERMS
      LISTED BELOW)
RT    CURTAINS
      PARTITIONS (MATHEMATICS)
      PARTITIONS (STRUCTURES)
      SEPTUM
```

PARTITIONS (MATHEMATICS)
```
GS    ANALYSIS (MATHEMATICS)
      . COMBINATORIAL ANALYSIS
      . . PARTITIONS (MATHEMATICS)
RT    COMBINATIONS (MATHEMATICS)
      EQUIVALENCE
      ∞ PARTITIONS
      PERMUTATIONS
```

PARTITIONS (STRUCTURES)
```
RT    BULKHEADS
      CURTAINS
      ∞ PARTITIONS
      THIN WALLS
      WALLS
```

PARTONS
```
GS    PARTICLES
      . CHARGED PARTICLES
      . . PARTONS
      . ELEMENTARY PARTICLES
      . . PARTONS
RT    HADRONS
      LEPTONS
      QUARK PARTON MODEL
      QUARKS
```

PARTS
```
USE   COMPONENTS
```

PAS
```
UF    PERIGEE-APOGEE SATELLITES
GS    SATELLITES
      . ARTIFICIAL SATELLITES
      . . PAS
RT    ELLIPTICAL ORBITS
      TWENTY-FOUR HOUR ORBITS
```

PASCAL (PROGRAMMING LANGUAGE)
```
GS    LANGUAGES
      . PROGRAMMING LANGUAGES
      . . PASCAL (PROGRAMMING LANGUAGE)
RT    COMPILERS
      COMPUTER PROGRAMMING
```

PASCHEN SERIES
```
GS    SPECTRA
      . RADIATION SPECTRA
      . . ELECTROMAGNETIC SPECTRA
      . . . LINE SPECTRA
      . . . . PASCHEN SERIES
RT    ABSORPTION SPECTRA
      ATOMIC SPECTRA
      ELECTRON TRANSITIONS
      EMISSION SPECTRA
      H LINES
      HYDROGEN
```

PASSAGEWAYS
```
GS    PASSAGEWAYS
      . STRAITS
      . . TORRES STRAIT
      . TRANSFER TUNNELS
RT    APPROACH
      CAVITIES
      CORRIDORS
      GAPS
      NOTCHES
      OPENINGS
      ∞ PATHS
      ROADS
      ∞ TUNNELS
      UNDERGROUND STRUCTURES
      VESTIBULES
```

PASSENGER AIRCRAFT
```
UF    EXECUTIVE AIRCRAFT
GS    PASSENGER AIRCRAFT
      . BAC 111 AIRCRAFT
      . BO-105 HELICOPTER
      . BOEING 707 AIRCRAFT
      . BOEING 720 AIRCRAFT
      . BOEING 727 AIRCRAFT
      . BOEING 737 AIRCRAFT
      . BOEING 747 AIRCRAFT
      . BOEING 757 AIRCRAFT
      . BOEING 767 AIRCRAFT
      . BOEING 2707 AIRCRAFT
      . BREGUET 941 AIRCRAFT
      . C-33 AIRCRAFT
      . C-35 AIRCRAFT
      . C-46 AIRCRAFT
      . CESSNA 172 AIRCRAFT
      . CESSNA 205 AIRCRAFT
      . CESSNA 210 AIRCRAFT
      . CESSNA 402B AIRCRAFT
      . CH-3 HELICOPTER
      . CH-46 HELICOPTER
      . CH-47 HELICOPTER
      . CH-54 HELICOPTER
      . COMET 4 AIRCRAFT
      . CV-340 AIRCRAFT
      . CV-440 AIRCRAFT
      . CV-880 AIRCRAFT
      . CV-990 AIRCRAFT
      . DC 8 AIRCRAFT
      . DC 10 AIRCRAFT
      . DH 121 AIRCRAFT
      . DH 125 AIRCRAFT
      . DO-27 AIRCRAFT
      . DO-28 AIRCRAFT
      . ELECTRA AIRCRAFT
      . EUROPEAN AIRBUS
      . . A-300 AIRCRAFT
      . . A-310 AIRCRAFT
      . . A-320 AIRCRAFT
      . F-27 AIRCRAFT
      . F-28 HELICOPTER
      . F-28 TRANSPORT AIRCRAFT
      . G-1 AIRCRAFT
      . G-222 AIRCRAFT
      . H-19 HELICOPTER
```

PASSENGER AIRCRAFT-(CONT.)
```
      . H-53 HELICOPTER
      . H-56 HELICOPTER
      . HFB-320 AIRCRAFT
      . HS-748 AIRCRAFT
      . IL-62 AIRCRAFT
      . JETSTREAM AIRCRAFT
      . L-1011 AIRCRAFT
      . L-2000 AIRCRAFT
      . MYSTERE 20 AIRCRAFT
      . OH-5 HELICOPTER
      . P-160 AIRCRAFT
      . P-166 AIRCRAFT
      . SE-210 AIRCRAFT
      . T-39 AIRCRAFT
      . TU-104 AIRCRAFT
      . TU-124 AIRCRAFT
      . TU-134 AIRCRAFT
      . TU-144 AIRCRAFT
      . U-10 AIRCRAFT
      . VC-10 AIRCRAFT
      . VISCOUNT AIRCRAFT
      . YAK 40 AIRCRAFT
      . YS-11 AIRCRAFT
RT    ∞ AIRCRAFT
      AN-22 AIRCRAFT
      AN-24 AIRCRAFT
      CARGO AIRCRAFT
      CIVIL AVIATION
      COMMERCIAL AIRCRAFT
      DC 7 AIRCRAFT
      E-2 AIRCRAFT
      GENERAL AVIATION AIRCRAFT
      GROUND EFFECT MACHINES
      HC-3 HELICOPTER
      JET AIRCRAFT
      LIGHT AIRCRAFT
      LIGHT TRANSPORT AIRCRAFT
      ∞ LOW WING AIRCRAFT
      MERCURE AIRCRAFT
      MH-262 AIRCRAFT
      ∞ MILITARY AIRCRAFT
      MYSTERE 50 AIRCRAFT
      P-531 HELICOPTER
      PD-808 AIRCRAFT
      ROTARY WING AIRCRAFT
      SAAB 105 AIRCRAFT
      SC-7 AIRCRAFT
      SHORT HAUL AIRCRAFT
      ∞ SUBSONIC AIRCRAFT
      SUPERSONIC AIRCRAFT
      SUPERSONIC TRANSPORTS
      TRANSPORT AIRCRAFT
      TU-154 AIRCRAFT
      TURBOFAN AIRCRAFT
      TURBOPROP AIRCRAFT
      V/STOL AIRCRAFT
      WATER TAKEOFF AND LANDING
        AIRCRAFT
```

PASSENGERS
```
RT    AIRLINE OPERATIONS
      AUTOMATED GUIDEWAY TRANSIT
        VEHICLES
      AUTOMATED MIXED TRAFFIC VEHICLES
      AUTOMATED TRANSIT VEHICLES
      PAYLOADS
      RAPID TRANSIT SYSTEMS
      RIDING QUALITY
      TRANSPORTATION
```

PASSES
```
USE   GAPS (GEOLOGY)
```

PASSIVATION
```
USE   PASSIVITY
```

PASSIVE L-BAND RADIOMETERS
```
GS    MEASURING INSTRUMENTS
      . RADIATION MEASURING INSTRUMENTS
      . . ACTINOMETERS
      . . . RADIOMETERS
      . . . . PASSIVE L-BAND RADIOMETERS
RT    MICROWAVE FREQUENCIES
      ULTRAHIGH FREQUENCIES
```

PASSIVE NOSETIP TECHNOLOGY
```
USE   PANT PROGRAM
```

PASSIVE SATELLITES
```
UF    REFLECTOR SATELLITES
GS    SATELLITES
      . ARTIFICIAL SATELLITES
      . . PASSIVE SATELLITES
      . . . BEACON SATELLITES
```

PASSIVE SATELLITES-*(CONT.)*
```
        . . . . BEACON EXPLORER A
        . . . . EXPLORER 22 SATELLITE
        . . . ECHO SATELLITES
        . . . . ECHO 1 SATELLITE
        . . . . ECHO 2 SATELLITE
        . . LAGEOS (SATELLITE)
        . . . PAGEOS SATELLITE
        UNMANNED SPACECRAFT
        . PASSIVE SATELLITES
        . . BEACON SATELLITES
        . . . BEACON EXPLORER A
        . . EXPLORER 22 SATELLITE
        . . ECHO SATELLITES
        . . . ECHO 1 SATELLITE
        . . . ECHO 2 SATELLITE
        . . LAGEOS (SATELLITE)
        . . PAGEOS SATELLITE
RT      ACTIVE SATELLITES
        COMMUNICATION SATELLITES
        ECHO PROJECT
        GEODETIC SATELLITES
        NAVIGATION SATELLITES
        SYNCHRONOUS SATELLITES
```

PASSIVITY
```
UF      PASSIVATION
RT      ANODIZING
        CHEMICAL ATTACK
        CHEMICAL PROPERTIES
        COATINGS
        CORROSION
        CORROSION PREVENTION
        CORROSION RESISTANCE
        DEACTIVATION
        ELECTROLYSIS
     ∞ INHIBITION
        INHIBITORS
        OXIDATION
        OXIDATION RESISTANCE
        RUSTING
        SILICONIZING
```

PASTE (CONSISTENCY)
```
RT      MIXTURES
```

PASTES
```
GS      _ADHESIVES
        . PASTES
RT      GLUES
        PLASTERS
```

PASTEURIZING
```
GS      HEATING
        . PASTEURIZING
RT      PURIFICATION
        STERILIZATION
```

PATCH TESTS
```
RT      CONTACT DERMATITIS
        HYDRAULIC FLUIDS
     ∞ TESTS
        TUBERCULOSIS
        WEAR TESTS
        WELD STRENGTH
```

PATENT APPLICATIONS
```
RT      COPYRIGHTS
        INVENTIONS
        LICENSING
        PATENTS
        PRODUCT DEVELOPMENT
        TECHNOLOGY UTILIZATION
```

PATENT POLICY
```
GS      POLICIES
        . PATENT POLICY
RT      INVENTIONS
        PATENTS
        PRODUCT DEVELOPMENT
        REGULATIONS
        RULES
```

PATENTS
```
RT      CLAIMING
        GRANTS
        INVENTIONS
        PATENT APPLICATIONS
        PATENT POLICY
```

PATHFINDER NUCLEAR REACTOR
```
GS      NUCLEAR ELECTRIC POWER
          GENERATION
        . NUCLEAR POWER REACTORS
```

PATHFINDER NUCLEAR REACTOR-*(CONT.)*
```
        . . PATHFINDER NUCLEAR REACTOR
        NUCLEAR REACTORS
        . LIQUID COOLED REACTORS
        . WATER COOLED REACTORS
        . . BOILING WATER REACTORS
        . . . PATHFINDER NUCLEAR REACTOR
        . NUCLEAR POWER REACTORS
        . . PATHFINDER NUCLEAR REACTOR
```

PATHOGENESIS
```
RT      CHOLERA
        DISEASES
        PATHOGENS
```

PATHOGENS
```
RT      BACTERIA
        CLOSTRIDIUM BOTULINUM
        PATHOGENESIS
```

PATHOLOGICAL EFFECTS
```
RT      BIOLOGICAL EFFECTS
        CARBON MONOXIDE POISONING
        CHOLERA
        DISEASES
     ∞ EFFECTS
        PHYSIOLOGICAL RESPONSES
     ∞ STRESS (BIOLOGY)
```

PATHOLOGY
```
GS      MEDICAL SCIENCE
        . PATHOLOGY
        . . HUMAN PATHOLOGY
RT      AUTOPSIES
        DIAGNOSIS
        DISSECTION
        HEMORRHAGES
        RADIATION THERAPY
        VETERINARY MEDICINE
```

∞ PATHS
```
SN      (USE OF A MORE SPECIFIC TERM IS
        RECOMMENDED--CONSULT THE TERMS
        LISTED BELOW)
UF      COURSES
        LANES
RT      CRITICAL PATH METHOD
        DIFFRACTION PATHS
        DUALITY PRINCIPLE
        FLIGHT PATHS
        GROUND TRACKS
        MEAN FREE PATH
        METEOR TRAILS
        MULTIPATH TRANSMISSION
        NETWORK ANALYSIS
        OPERATIONS RESEARCH
        OPTICAL PATHS
        ORBITS
        PARTICLE TRAJECTORIES
        PASSAGEWAYS
        PERT
        ROUTES
        SOUND TRANSMISSION
        THERMODYNAMICS
        TRAJECTORIES
```

PATIENTS
```
RT      HUMAN BEINGS
        HUMAN PATHOLOGY
        THERAPY
```

PATRIOT MISSILE
```
GS      MISSILES
        . SURFACE TO AIR MISSILES
        . . PATRIOT MISSILE
RT      MISSILE CONFIGURATIONS
     ∞ ROCKETS
        WEAPONS
```

PATROLS
```
RT      RECONNAISSANCE
```

PATTERN DISTRIBUTION
```
USE     DISTRIBUTION (PROPERTY)
```

PATTERN METHOD (FORECASTING)
```
GS      FORECASTING
        . TECHNOLOGICAL FORECASTING
        . . PATTERN METHOD (FORECASTING)
        MANAGEMENT METHODS
        . PATTERN METHOD (FORECASTING)
RT      DELPHI METHOD (FORECASTING)
        ESTIMATING
     ∞ METHODOLOGY
```

PATTERN METHOD (FORECASTING)-*(CONT.)*
```
        OPERATIONS RESEARCH
        PLANNING
        PREDICTIONS
        PROBE METHOD (FORECASTING)
        TECHNOLOGY ASSESSMENT
```

PATTERN RECOGNITION
```
UF      AUTOMATIC PATTERN RECOGNITION
        FEATURE EXTRACTION
GS      RECOGNITION
        . PATTERN RECOGNITION
        . . CHARACTER RECOGNITION
        . . GRAPHOLOGY
RT      CHANGE DETECTION
        CLUMPS
        CLUSTER ANALYSIS
        COMPUTER VISION
        CONTEXT
        FEATURE IDENTIFICATION AND
          LOCATION EXPER
        GRAY SCALE
        IMAGE ANALYSIS
        MULTISENSOR APPLICATIONS
        OPTICAL RELAY SYSTEMS
        PRINCIPAL COMPONENTS ANALYSIS
        READERS
        REPETITION
```

PATTERN REGISTRATION
```
RT      COMPARISON
        IMAGE CONTRAST
        IMAGE CORRELATORS
        IMAGE MOTION COMPENSATION
        IMAGE RECONSTRUCTION
        IMAGE RESOLUTION
        IMAGING TECHNIQUES
        MAGNETIC SIGNATURES
        MATCHING
```

∞ PATTERNS
```
SN      (USE OF A MORE SPECIFIC TERM IS
        RECOMMENDED--CONSULT THE TERMS
        LISTED BELOW)
RT      DIFFRACTION PATTERNS
        DISTRIBUTION (PROPERTY)
        DRAINAGE PATTERNS
        KURTOSIS
        MOLDS
        PHOTOMASKS
        PROTOTYPES
        RADIATION DISTRIBUTION
        RESINS
        SPECKLE PATTERNS
        SYNTHETIC ARRAYS
        TEMPLATES
        TEST PATTERN GENERATORS
        WIDMANSTATTEN STRUCTURE
```

PATTERSON MAP
```
GS      CHARTS
        . GRAPHS (CHARTS)
        . . PATTERSON MAP
RT      CRYSTAL LATTICES
        CRYSTAL STRUCTURE
        LATTICE PARAMETERS
```

PAULI EXCLUSION PRINCIPLE
```
GS      QUANTUM MECHANICS
        . PAULI EXCLUSION PRINCIPLE
        WAVE FUNCTIONS
        . PAULI EXCLUSION PRINCIPLE
RT      ATOMIC STRUCTURE
        EXCLUSION
        FERMIONS
```

PAVEMENTS
```
RT      ASPHALT
        COATINGS
        CONCRETES
        FOUNDATIONS
        HIGHWAYS
        ROADS
        RUNWAYS
        STREETS
```

PAYLOAD ASSIST MODULE
```
GS      MODULES
        . PAYLOAD ASSIST MODULE
        ROCKET VEHICLES
        . PAYLOAD ASSIST MODULE
RT      INSTRUMENT PACKAGES
        PAYLOADS
        SPACE SHUTTLE PAYLOADS
        SPACE TRANSPORTATION SYSTEM
```

PAYLOAD ASSIST MODULE-*(CONT.)*
  SPACEBORNE EXPERIMENTS

PAYLOAD CONTROL
  RT   ∞CONTROL
        PAYLOADS
        SOUNDING ROCKETS
        SPACE SHUTTLES

PAYLOAD DELIVERY (STS)
  GS   DELIVERY
        . **PAYLOAD DELIVERY (STS)**
  RT   ELECTRIC POWER SUPPLIES
        ORBIT TRANSFER VEHICLES
        ORBITAL LAUNCHING
        PAYLOADS
        POWER MODULES (STS)
        SOLAR ARRAYS
        SPACE TRANSPORTATION SYSTEM

PAYLOAD DEPLOYMENT & RETRIEVAL SYSTEM
  GS   **PAYLOAD DEPLOYMENT & RETRIEVAL
        SYSTEM**
        . REMOTE MANIPULATOR SYSTEM
  RT   MANIPULATORS
        NASA PROGRAMS
        ORBIT TRANSFER VEHICLES
        PAYLOADS
        REMOTE HANDLING
        SPACE SHUTTLES
        ∞SYSTEMS

PAYLOAD INTEGRATION
  RT   MISSION PLANNING
        PAYLOAD INTEGRATION PLAN
        PAYLOADS
        SPACE SHUTTLE PAYLOADS

PAYLOAD INTEGRATION PLAN
  RT   PAYLOAD INTEGRATION
        PAYLOADS
        ∞PLANS
        SPACE SHUTTLE ORBITERS
        SPACE SHUTTLE PAYLOADS
        SPACE TRANSPORTATION SYSTEM
        SPACEBORNE EXPERIMENTS

PAYLOAD MASS RATIO
  GS   RATIOS
        . MASS RATIOS
        . . **PAYLOAD MASS RATIO**
  RT   MULTISTAGE ROCKET VEHICLES
        PIGGYBACK SYSTEMS
        PRESSURE RATIO
        PROPELLANT MASS RATIO

PAYLOAD RETRIEVAL (STS)
  GS   RETRIEVAL
        . **PAYLOAD RETRIEVAL (STS)**
  RT   ORBIT TRANSFER VEHICLES
        ORBITAL RENDEZVOUS
        PAYLOAD TRANSFER
        REMOTE MANIPULATOR SYSTEM
        SPACE SHUTTLES
        SPACE TRANSPORTATION SYSTEM
        STATIONKEEPING

PAYLOAD STATIONS
  GS   STATIONS
        . **PAYLOAD STATIONS**
  RT   PAYLOADS
        SPACE TRANSPORTATION

PAYLOAD TRANSFER
  RT   ORBITAL SERVICING
        PAYLOAD RETRIEVAL (STS)
        SPACE MAINTENANCE

PAYLOADS
  GS   **PAYLOADS**
        . AMPS (SATELLITE PAYLOAD)
        . EARTH VIEWING APPLICATIONS
          LABORATORY
        . EXPOS (SPACELAB PAYLOAD)
        . OSTA-2 PAYLOAD
        . SEPAC (PAYLOAD)
        . SHUTTLE IMAGING RADAR
        . SORTIE SYSTEMS
        . SPACE SHUTTLE PAYLOADS
        . . ELECTROMAGNETIC ENVIRONMENT
            EXPERIMENT
        . . HALOGEN OCCULTATION
            EXPERIMENT
        . . OSS-1 PAYLOAD

PAYLOADS-*(CONT.)*
        . . OSTA-1 PAYLOAD
        . . SPACEBORNE EXPERIMENTS
        . . . ATMOSPHERIC GENERAL
              CIRCULATION EXPERIMENT
        . . . EARTH RADIATION BUDGET
              EXPERIMENT
        . . . PHYSICS AND CHEMISTRY
              EXPERIMENT IN SPACE
        . . . PLASMA INTERACTION EXPERIMENT
        . . SPACELAB
        . . X RAY ASTROPHYSICS FACILITY
        . SPACELAB PAYLOADS
        . . ATMOSPHERIC GENERAL
            CIRCULATION EXPERIMENT
        . . GEOPHYSICAL FLUID FLOW CELLS
        . . SOLAR CELL CALIBRATION FACILITY
        . STARLAB
  RT   AIRCRAFT PERFORMANCE
        AIRCRAFT SPECIFICATIONS
        ANNULAR SUSPENSION AND POINTING
          SYSTEM
        APOLLO LUNAR SURFACE EXPERIMENTS
          PACKAGE
        ASTEROID CAPTURE
        EASEP
        INSTRUMENT PACKAGES
        LIRTS (TELESCOPE)
        ∞LOADING
        LOADS (FORCES)
        PASSENGERS
        PAYLOAD ASSIST MODULE
        PAYLOAD CONTROL
        PAYLOAD DELIVERY (STS)
        PAYLOAD DEPLOYMENT & RETRIEVAL
          SYSTEM
        PAYLOAD INTEGRATION
        PAYLOAD INTEGRATION PLAN
        PAYLOAD STATIONS
        PIGGYBACK SYSTEMS
        SPACE PROCESSING APPLICATIONS
          ROCKET
        SPACE TRANSPORTATION
        SPACE TUGS
        SPACEBORNE EXPERIMENTS
        VERTICAL 8 ROCKET
        WARHEADS
        ∞WEIGHT
        WEIGHT (MASS)

PBB
  USE   POLYBROMINATED BIPHENYLS

PBRE (REACTORS)
  USE   PEBBLE BED REACTORS

PCB
  USE   POLYCHLORINATED BIPHENYLS

PCM (MATERIALS)
  USE   PHASE CHANGE MATERIALS

PCM (MODULATION)
  USE   PULSE CODE MODULATION

PCM TELEMETRY
  GS   TELECOMMUNICATION
        . TELEMETRY
        . . **PCM TELEMETRY**
        TRANSMISSION
        . SIGNAL TRANSMISSION
        . . **PCM TELEMETRY**
  RT   DIFFERENTIAL PULSE CODE
          MODULATION
        PULSE CODE MODULATION

PD-808 AIRCRAFT
  UF   DOUGLAS PD-808 AIRCRAFT
        PIAGGIO-DOUGLAS PD-808 AIRCRAFT
  GS   JET AIRCRAFT
        . **PD-808 AIRCRAFT**
        LIGHT AIRCRAFT
        . **PD-808 AIRCRAFT**
        MCDONNELL DOUGLAS AIRCRAFT
        . DOUGLAS AIRCRAFT
        . . **PD-808 AIRCRAFT**
        MONOPLANES
        . **PD-808 AIRCRAFT**
        PIAGGIO AIRCRAFT
        . **PD-808 AIRCRAFT**
        UTILITY AIRCRAFT
        . **PD-808 AIRCRAFT**
  RT   ∞AIRCRAFT
        PASSENGER AIRCRAFT

PDM (MODULATION)
  USE   PULSE DURATION MODULATION

PDP COMPUTERS
  GS   DATA PROCESSING EQUIPMENT
        . COMPUTERS
        . . DIGITAL COMPUTERS
        . . . **PDP COMPUTERS**
        . . . . PDP 7 COMPUTER
        . . . . PDP 8 COMPUTER
        . . . . PDP 9 COMPUTER
        . . . . PDP 10 COMPUTER
        . . . . PDP 11 COMPUTER
        . . . . PDP 11/20 COMPUTER
        . . . . PDP 11/40 COMPUTER
        . . . . PDP 11/45 COMPUTER
        . . . . PDP 11/50 COMPUTER
        . . . . PDP 11/70 COMPUTER
        . . . . PDP 12 COMPUTER

PDP 7 COMPUTER
  GS   DATA PROCESSING EQUIPMENT
        . COMPUTERS
        . . DIGITAL COMPUTERS
        . . . PDP COMPUTERS
        . . . . **PDP 7 COMPUTER**
  RT   PDP 9 COMPUTER

PDP 8 COMPUTER
  GS   DATA PROCESSING EQUIPMENT
        . COMPUTERS
        . . DIGITAL COMPUTERS
        . . . PDP COMPUTERS
        . . . . **PDP 8 COMPUTER**

PDP 9 COMPUTER
  GS   DATA PROCESSING EQUIPMENT
        . COMPUTERS
        . . DIGITAL COMPUTERS
        . . . PDP COMPUTERS
        . . . . **PDP 9 COMPUTER**
  RT   PDP 7 COMPUTER

PDP 10 COMPUTER
  UF   SYSTEM 10 COMPUTER
  GS   DATA PROCESSING EQUIPMENT
        . COMPUTERS
        . . DIGITAL COMPUTERS
        . . . PDP COMPUTERS
        . . . . **PDP 10 COMPUTER**

PDP 11 COMPUTER
  GS   DATA PROCESSING EQUIPMENT
        . COMPUTERS
        . . DIGITAL COMPUTERS
        . . . PDP COMPUTERS
        . . . . **PDP 11 COMPUTER**

PDP 11/20 COMPUTER
  GS   DATA PROCESSING EQUIPMENT
        . COMPUTERS
        . . DIGITAL COMPUTERS
        . . . PDP COMPUTERS
        . . . . **PDP 11/20 COMPUTER**

PDP 11/40 COMPUTER
  GS   DATA PROCESSING EQUIPMENT
        . COMPUTERS
        . . DIGITAL COMPUTERS
        . . . PDP COMPUTERS
        . . . . **PDP 11/40 COMPUTER**

PDP 11/45 COMPUTER
  GS   DATA PROCESSING EQUIPMENT
        . COMPUTERS
        . . DIGITAL COMPUTERS
        . . . PDP COMPUTERS
        . . . . **PDP 11/45 COMPUTER**

PDP 11/50 COMPUTER
  GS   DATA PROCESSING EQUIPMENT
        . COMPUTERS
        . . DIGITAL COMPUTERS
        . . . PDP COMPUTERS
        . . . . **PDP 11/50 COMPUTER**

PDP 11/70 COMPUTER
  GS   DATA PROCESSING EQUIPMENT
        . COMPUTERS
        . . DIGITAL COMPUTERS
        . . . PDP COMPUTERS
        . . . . **PDP 11/70 COMPUTER**

PDP 12 COMPUTER
GS   DATA PROCESSING EQUIPMENT
   . COMPUTERS
   . . DIGITAL COMPUTERS
   . . . PDP COMPUTERS
   . . . . **PDP 12 COMPUTER**

PDP 15 COMPUTER
GS   DATA PROCESSING EQUIPMENT
   . COMPUTERS
   . . DIGITAL COMPUTERS
   . . . **PDP 15 COMPUTER**

PEACETIME
RT   ELECTRONIC WARFARE
   HISTORIES
   INTERNATIONAL COOPERATION
   INTERNATIONAL LAW
   WARFARE

∞ **PEAKS**
SN   *(USE OF A MORE SPECIFIC TERM IS RECOMMENDED--CONSULT THE TERMS LISTED BELOW)*
RT   APEXES
   EXTREMUM VALUES
   MAXIMA
   MOUNTAINS
   PLATEAUS

PEAKS (LANDFORMS)
UF   PINNACLES
GS   LANDFORMS
   . **PEAKS (LANDFORMS)**
   . . PIKE'S PEAK (CO)
RT   MOUNTAINS
   OROGRAPHY
   TOPOGRAPHY

PEARLITE
RT   CEMENTITE
   FERRITES
   IRON ALLOYS
   MICROSTRUCTURE
   STEELS

PEARSON DISTRIBUTIONS
GS   FUNCTIONS (MATHEMATICS)
   . PROBABILITY DENSITY FUNCTIONS
   . . **PEARSON DISTRIBUTIONS**
   STATISTICAL ANALYSIS
   . PROBABILITY DENSITY FUNCTIONS
   . . **PEARSON DISTRIBUTIONS**
   STATISTICAL DISTRIBUTIONS
   . **PEARSON DISTRIBUTIONS**

PEAT
GS   CARBONACEOUS MATERIALS
   . **PEAT**
   ORGANIC MATERIALS
   . **PEAT**
   SEDIMENTARY ROCKS
   . **PEAT**
RT   COAL

PEBBLE BED REACTORS
UF   PBRE (REACTORS)
GS   NUCLEAR REACTORS
   . **PEBBLE BED REACTORS**
RT   REACTOR DESIGN
   REACTOR TECHNOLOGY

PECLET NUMBER
GS   RATIOS
   . DIMENSIONLESS NUMBERS
   . . **PECLET NUMBER**
RT   ADVECTION
   HEAT TRANSFER
   PRANDTL NUMBER
   REYNOLDS NUMBER
   THERMAL DIFFUSION

PECULIAR STARS
GS   CELESTIAL BODIES
   . STARS
   . . **PECULIAR STARS**
   . . . SYMBIOTIC STARS
RT   A STARS
   B STARS
   HOT STARS
   MAGNETIC STARS
   STELLAR SPECTRA
   STELLAR SPECTROPHOTOMETRY
   STELLAR STRUCTURE

PEDALS
RT   LEVERS
   MANUAL CONTROL

PEDIMENTS
USE   PIEDMONTS

PEDIPLAINS
USE   PIEDMONTS

PEDOLOGY
USE   SOIL SCIENCE

PEELING
RT   ADHESION
   CUTTING
   DELAMINATING
   FLAKING
   MECHANICAL PROPERTIES
   SHEDDING
   ∞ STRIPPING

PEENING
GS   METAL FINISHING
   . **PEENING**
   . . SHOT PEENING
RT   COLD WORKING
   HARDENING (MATERIALS)
   METAL WORKING
   WORK HARDENING

PEGASUS COMPUTER
GS   DATA PROCESSING EQUIPMENT
   . COMPUTERS
   . . **PEGASUS COMPUTER**

PEGASUS ENGINE
USE   BRISTOL-SIDDELEY BS 53 ENGINE

PEGASUS SATELLITES
GS   SATELLITES
   . ARTIFICIAL SATELLITES
   . . **PEGASUS SATELLITES**
RT   SATURN PROJECT

PELAGIC ZONE
GS   REGIONS
   . **PELAGIC ZONE**
RT   OCEANOGRAPHY

PELLETS
RT   BRIQUETS
   FUEL CAPSULES
   GRANULAR MATERIALS
   NUCLEAR FUELS
   ∞ SHOT

PELLICLE
RT   THIN FILMS

PELOMYXA
GS   ANIMALS
   . AMOEBA
   . . **PELOMYXA**
   . INVERTEBRATES
   . . PROTOZOA
   . . . **PELOMYXA**
   MICROORGANISMS
   . PROTOZOA
   . . AMOEBA
   . . . **PELOMYXA**

PELTIER EFFECTS
RT   ∞ EFFECTS
   SEEBECK EFFECT
   TEMPERATURE EFFECTS
   THERMOCOUPLES
   THERMOELECTRIC COOLING
   THERMOELECTRICITY
   THERMOPHYSICAL PROPERTIES

PELVIS
GS   ANATOMY
   . MUSCULOSKELETAL SYSTEM
   . . BONES
   . . . **PELVIS**
RT   GIRDLES

PENALTIES
GS   LAW (JURISPRUDENCE)
   . PUBLIC LAW
   . . **PENALTIES**
RT   AIR LAW
   DISCIPLINING

PENALTIES-*(CONT.)*
   JUDGMENTS
   LEGAL LIABILITY
   LIABILITIES
   PROHIBITION
   REGULATIONS

PENALTY FUNCTION
GS   FUNCTIONS (MATHEMATICS)
   . **PENALTY FUNCTION**
RT   CONSTRAINTS
   ∞ FUNCTIONS
   MAXIMA
   MINIMA
   OPTIMIZATION

PENDULOUS GYROSCOPES
USE   GYROSCOPIC PENDULUMS

PENDULUMS
GS   OSCILLATORS
   . MECHANICAL OSCILLATORS
   . . **PENDULUMS**
   . . . GYROSCOPIC PENDULUMS
RT   ACCELEROMETERS
   GRAVITATION
   MOMENTUM
   OSCILLATIONS
   TIMING DEVICES

PENEPLAINS
GS   LAND
   . PLAINS
   . . **PENEPLAINS**
RT   EROSION
   GEOLOGY
   SOIL EROSION
   TOPOGRAPHY
   WIND EROSION

PENETRANTS
RT   ∞ AGENTS
   PRESERVATIVES
   RETARDANTS

PENETRATING PARTICLES
USE   CORPUSCULAR RADIATION

PENETRATION
RT   DIFFUSION
   DRILLING
   FRAGMENTATION
   HYDRODYNAMIC RAM EFFECT
   IMPACT
   NUCLEAR VULNERABILITY
   PERCOLATION
   PERFORATING
   PERMEABILITY
   PERMEATING
   PIERCING
   ∞ SATURATION
   SEEPAGE
   TERMINAL BALLISTICS
   VULNERABILITY

PENETRATION BALLISTICS
USE   TERMINAL BALLISTICS

PENETROMETERS
GS   MEASURING INSTRUMENTS
   . **PENETROMETERS**
RT   LUNAR SOIL

PENICILLIN
GS   DRUGS
   . ANTIBIOTICS
   . . **PENICILLIN**
   STEROIDS
   . **PENICILLIN**

PENINSULAR RANGES (CA)
GS   LANDFORMS
   . MOUNTAINS
   . . **PENINSULAR RANGES (CA)**
RT   CALIFORNIA

PENINSULAS
GS   LANDFORMS
   . **PENINSULAS**
   . . DELMARVA PENINSULA (DE-MD-VA)
RT   ISTHMUSES
   LAND
   WATER

PENNING DISCHARGE
GS   ELECTRIC CURRENT
    . ELECTRIC DISCHARGES
    . . **PENNING DISCHARGE**
RT   ELECTRODELESS DISCHARGES
    GAS IONIZATION
    ION MOTION
    PLASMA GENERATORS

PENNING EFFECT
RT   ∞EFFECTS
    GAS IONIZATION
    METASTABLE ATOMS

PENNING GAGES
GS   MEASURING INSTRUMENTS
    . PRESSURE GAGES
    . . VACUUM GAGES
    . . . IONIZATION GAGES
    . . . . **PENNING GAGES**
    VACUUM APPARATUS
    . VACUUM GAGES
    . . IONIZATION GAGES
    . . . **PENNING GAGES**

PENNSYLVANIA
GS   NATIONS
    . UNITED STATES
    . . **PENNSYLVANIA**
RT   ALLEGHENY PLATEAU (US)
    DELAWARE BAY (US)
    DELAWARE RIVER BASIN (US)
    OHIO RIVER (US)
    SUSQUEHANNA RIVER BASIN
    (MD-NY-PA)

∞ **PENS**
SN   *(USE OF A MORE SPECIFIC TERM IS
    RECOMMENDED--CONSULT THE TERMS
    LISTED BELOW)*
UF   STYLUSES
RT   ENCLOSURES
    RECORDING INSTRUMENTS

PENTABORANES
GS   BORON COMPOUNDS
    . BORON HYDRIDES
    . . BORANES
    . . . **PENTABORANES**
    HYDROGEN COMPOUNDS
    . HYDRIDES
    . . BORON HYDRIDES
    . . . BORANES
    . . . . **PENTABORANES**

PENTACHLORIDES
USE   CHLORIDES

PENTAERYTHRITOL TETRANITRATE
USE   PETN

PENTANES
GS   ALIPHATIC COMPOUNDS
    . ALKANES
    . . **PENTANES**
    . . . NEOPENTANE
    HYDROCARBONS
    . ALKANES
    . . **PENTANES**
    . . . NEOPENTANE

PENTANONE
GS   ALIPHATIC COMPOUNDS
    . KETONES
    . . **PENTANONE**
    ORGANIC COMPOUNDS
    . **PENTANONE**
RT   ACETONE
    ACETYLACETONE

PENTOBARBITAL
RT   DRUGS
    NARCOTICS
    SEDATIVES

PENTOBARBITAL SODIUM
GS   DRUGS
    . **PENTOBARBITAL SODIUM**
    . . RESERPINE
RT   NEMBUTAL (TRADEMARK)

PENTODES
RT   ELECTRON TUBES
    SEMICONDUCTOR DEVICES

PENTODES-*(CONT.)*
    TETRODES
    TRANSISTORS
    VACUUM TUBES

PENTOLITE
GS   EXPLOSIVES
    . **PENTOLITE**
    PROPELLANTS
    . **PENTOLITE**

PENTOSE
GS   ALIPHATIC COMPOUNDS
    . SUGARS
    . . **PENTOSE**
    . . . RIBOSE
    . . . XYLOSE
    CARBOHYDRATES
    . SUGARS
    . . **PENTOSE**
    . . . RIBOSE
    . . . XYLOSE

PENUMBRAS
GS   SHADOWS
    . **PENUMBRAS**
RT   ECLIPSES
    UMBRAS

PEOLE SATELLITES
GS   SATELLITES
    . ARTIFICIAL SATELLITES
    . . FRENCH SATELLITES
    . . . **PEOLE SATELLITES**
RT   GEOPHYSICAL SATELLITES

PEOPLES DEMOCRATIC REPUBLIC OF GERMANY
USE   EAST GERMANY

PEPPERS
RT   ∞FOOD

PEPSIN
GS   ENZYMES
    . **PEPSIN**
RT   PAPAIN

PEPTIDES
GS   ACIDS
    . AMINO ACIDS
    . . **PEPTIDES**
    . . . HYPERTENSIN
    ORGANIC COMPOUNDS
    . AMINO ACIDS
    . . **PEPTIDES**
    . . . HYPERTENSIN
    . . . . POLYPEPTIDES
    PROTEINS
    . **PEPTIDES**
    . . HYPERTENSIN
    . . . POLYPEPTIDES
RT   ASPARTIC ACID

PERCENTAGE
USE   RATIOS

PERCEPTION
GS   **PERCEPTION**
    . AUDITORY PERCEPTION
    . BINAURAL HEARING
    . MOTION PERCEPTION
    . SENSORY PERCEPTION
    . . CONSCIOUSNESS
    . . EXTRASENSORY PERCEPTION
    . . KINESTHESIA
    . . OLFACTORY PERCEPTION
    . . PAIN
    . . PAIN SENSITIVITY
    . . PROPRIOCEPTION
    . . AUTOKINESIS
    . . TASTE
    . . TOUCH
    . . . TACTILE DISCRIMINATION
    . . VERTICAL PERCEPTION
    . . VIBRATION PERCEPTION
    . . VISUAL PERCEPTION
    . . . CRITICAL FLICKER FUSION
    . . . SPACE PERCEPTION
    . . . . AUTOKINESIS
    . . . VISUAL DISCRIMINATION
    . SOUND LOCALIZATION
RT   ACUITY
    ADAPTATION
    ARTIFICIAL INTELLIGENCE

PERCEPTION-*(CONT.)*
    CHARACTER RECOGNITION
    COGNITION
    COLOR
    CONTRAST
    ELECTROCUTANEOUS COMMUNICATION
    ∞FREQUENCY RESPONSE
    IDENTIFYING
    ILLUSIONS
    IMAGES
    ∞INTERPRETATION
    KNOWLEDGE
    LEGIBILITY
    MONOCULAR VISION
    PERCEPTUAL TIME CONSTANT
    READING
    RESOLUTION
    RETINAL ADAPTATION
    SENSITIVITY
    SENSORY DEPRIVATION
    SENSORY FEEDBACK
    SYMBOLS
    THRESHOLDS (PERCEPTION)
    VISIBILITY
    VISION

PERCEPTRONS
USE   SELF ORGANIZING SYSTEMS

PERCEPTUAL ERRORS
RT   DISPLAY DEVICES
    VISUAL PERCEPTION
    VISUAL STIMULI

PERCEPTUAL TIME CONSTANT
GS   CONSTANTS
    . TIME CONSTANT
    . . **PERCEPTUAL TIME CONSTANT**
RT   PERCEPTION
    REACTION TIME
    SENSE ORGANS
    SENSORIMOTOR PERFORMANCE
    VELOCITY

PERCHLORATES
GS   HALOGEN COMPOUNDS
    . CHLORINE COMPOUNDS
    . . **PERCHLORATES**
    . . . ALUMINUM PERCHLORATES
    . . . AMMONIUM PERCHLORATES
    . . . HYDRAZINE PERCHLORATES
    . . . HYDROGEN PERCHLORATE
    . . . HYDROXYLAMMONIUM
        PERCHLORATES
    . . . LITHIUM PERCHLORATES
    . . . MAGNESIUM PERCHLORATES
    . . . NITRONIUM PERCHLORATE
    . . . POTASSIUM PERCHLORATES
RT   CHLORATES
    PERCHLORIC ACID

PERCHLORIC ACID
GS   ACIDS
    . **PERCHLORIC ACID**
RT   PERCHLORATES

PERCHLORYL FLUORIDES
GS   HALOGEN COMPOUNDS
    . FLUORINE COMPOUNDS
    . . FLUORIDES
    . . . **PERCHLORYL FLUORIDES**
    . HALIDES
    . . FLUORIDES
    . . . **PERCHLORYL FLUORIDES**

PERCOLATION
RT   BEDS (PROCESS ENGINEERING)
    CONCENTRATING
    DIFFUSION
    EXTRACTION
    FILTRATION
    INTERSTICES
    LEACHING
    LYSIMETERS
    PENETRATION
    PERMEABILITY
    PERMEATING
    SEEPAGE
    ∞SEPARATION
    VOIDS

PERCUS METHOD
RT   FLOW EQUATIONS
    INTEGRAL EQUATIONS
    ∞METHODOLOGY

PERCUSSION
RT　DETONATION
　　IMPACT
　　PHYSICAL EXAMINATIONS
　　PRIMERS (EXPLOSIVES)

PERFECT GAS
USE　IDEAL GAS

PERFLUORO COMPOUNDS
GS　**PERFLUORO COMPOUNDS**
　．PERFLUOROALKANE
　．PERFLUOROGUANIDINE

PERFLUOROALKANE
GS　HALOGEN COMPOUNDS
　．FLUORINE COMPOUNDS
　．．FLUORO COMPOUNDS
　．．．DIFLUORO COMPOUNDS
　．．．．**PERFLUOROALKANE**
　．．．FLUORINE ORGANIC COMPOUNDS
　．．．．**PERFLUOROALKANE**
　ORGANIC COMPOUNDS
　．FLUORINE ORGANIC COMPOUNDS
　．．**PERFLUOROALKANE**
　PERFLUORO COMPOUNDS
　．**PERFLUOROALKANE**

PERFLUOROGUANIDINE
GS　HALOGEN COMPOUNDS
　．FLUORINE COMPOUNDS
　．．FLUORO COMPOUNDS
　．．．FLUORINE ORGANIC COMPOUNDS
　．．．．**PERFLUOROGUANIDINE**
　ORGANIC COMPOUNDS
　．FLUORINE ORGANIC COMPOUNDS
　．．**PERFLUOROGUANIDINE**
　PERFLUORO COMPOUNDS
　．**PERFLUOROGUANIDINE**
RT　GUANIDINES

PERFORATED PLATES
GS　STRUCTURAL MEMBERS
　．PLATES (STRUCTURAL MEMBERS)
　．．**PERFORATED PLATES**
RT　ANISOTROPIC PLATES
　　CAVITIES
　　HOLE GEOMETRY (MECHANICS)
　　OPENINGS
　∞PERFORATION
　　STRESS CONCENTRATION

PERFORATED SHELLS
GS　SHELLS (STRUCTURAL FORMS)
　．**PERFORATED SHELLS**
RT　ARCHES
　　CAVITIES
　　ENCLOSURES
　　FAIRINGS
　　HOLE DISTRIBUTION (MECHANICS)
　　HOLE GEOMETRY (MECHANICS)
　　HOUSINGS
　　HULLS (STRUCTURES)
　　MEMBRANE STRUCTURES
　　NACELLES
　∞PERFORATION
　　PRESSURE VESSEL DESIGN
　　ROCKET ENGINE CASES
　　SHELL THEORY
　　STRESS CONCENTRATION

PERFORATING
RT　BURNTHROUGH (FAILURE)
　　CUTTING
　　DRILLING
　　FORMATIONS
　　FRACTURING
　　GAS INJECTION
　　INJECTION
　　METAL CUTTING
　　METAL WORKING
　　PENETRATION
　∞PERFORATION
　　PIERCING
　　WATER INJECTION

∞ **PERFORATION**
SN　(USE OF A MORE SPECIFIC TERM IS
　　RECOMMENDED--CONSULT THE TERMS
　　LISTED BELOW)
RT　CAVITIES
　　PERFORATED PLATES
　　PERFORATED SHELLS
　　PERFORATING
　　PIERCING

∞ **PERFORMANCE**
SN　(USE OF A MORE SPECIFIC TERM IS
　　RECOMMENDED--CONSULT THE TERMS
　　LISTED BELOW)
RT　AIRCRAFT PERFORMANCE
　　ASTRONAUT PERFORMANCE
　　COMFORT
　　COMPLEXITY
　　COMPUTER SYSTEMS PERFORMANCE
　　CONSISTENCY
　　EFFICIENCY
　　EFFORT
　　ENVIRONMENTS
　　EVALUATION
　　EXAMINATION
　　FATIGUE (BIOLOGY)
　　FIGURE OF MERIT
　　FLIGHT CHARACTERISTICS
　　HUMAN FACTORS ENGINEERING
　　HUMAN PERFORMANCE
　　LONG TERM EFFECTS
　　MENTAL PERFORMANCE
　　MODULATION TRANSFER FUNCTION
　　OBSERVATION
　　OPERATOR PERFORMANCE
　　OPTICAL TRANSFER FUNCTION
　　OUTPUT
　　PERFORMANCE TESTS
　　PILOT PERFORMANCE
　　POSTFLIGHT ANALYSIS
　　PROPULSION SYSTEM PERFORMANCE
　　QUALITY
　　RATINGS
　　RELIABILITY
　　SPACECRAFT PERFORMANCE
　　STANDARDS
　　TASK COMPLEXITY
　　TRAINING EVALUATION

PERFORMANCE PREDICTION
GS　FORECASTING
　．**PERFORMANCE PREDICTION**
　．．PREDICTION ANALYSIS TECHNIQUES
　PREDICTIONS
　．**PERFORMANCE PREDICTION**
RT　EVALUATION
　　MANAGEMENT
　　RELIABILITY
　　RELIABILITY ANALYSIS
　　RELIABILITY ENGINEERING

PERFORMANCE TESTS
SN　(APPLY ONLY TO OPERATING
　　EQUIPMENT)
RT　ACCELERATED LIFE TESTS
　　ACCEPTABILITY
　　CERTIFICATION
　　CHECKOUT
　　COMPUTER SYSTEMS PERFORMANCE
　　INSPECTION
　∞PERFORMANCE
　　SPACE VEHICLE CHECKOUT PROGRAM
　　SPECIFICATIONS
　　STANDARDS
　∞TESTS

PERFUSION
USE　DIFFUSION

PERICLASE
GS　CHALCOGENIDES
　．OXIDES
　．．METAL OXIDES
　．．．ALKALINE EARTH OXIDES
　．．．．MAGNESIUM OXIDES
　．．．．．**PERICLASE**
　MAGNESIUM COMPOUNDS
　．MAGNESIUM OXIDES
　．．**PERICLASE**

PERIDOTITE
UF　KIMBERLITE
GS　ROCKS
　．IGNEOUS ROCKS
　．．**PERIDOTITE**
RT　CHROMITES
　　DUNITE
　　OLIVINE
　　REGOLITH
　　SOILS

PERIGEE-APOGEE SATELLITES
USE　PAS

PERIGEES
GS　APSIDES
　．**PERIGEES**
　ORBITS
　．EARTH ORBITS
　．．**PERIGEES**
　．ELLIPTICAL ORBITS
　．．**PERIGEES**
RT　APOGEES
　　PERILUNES

PERIHELIONS
SN　(PERIASTRONS IN THE SOLAR SYSTEM)
GS　APSIDES
　．**PERIHELIONS**
　ORBITS
　．ELLIPTICAL ORBITS
　．．**PERIHELIONS**
　．SOLAR ORBITS
　．．**PERIHELIONS**
RT　APHELIONS

PERILUNES
GS　APSIDES
　．**PERILUNES**
RT　LUNAR ORBITS
　　LUNAR SATELLITES
　　PERIGEES

PERIOD EQUATIONS
USE　PERIODIC FUNCTIONS

PERIODIC FUNCTIONS
UF　PERIOD EQUATIONS
GS　ANALYSIS (MATHEMATICS)
　．REAL VARIABLES
　．．**PERIODIC FUNCTIONS**
　．．．TRIGONOMETRIC FUNCTIONS
　．．．．COSINE SERIES
　．．．．SINE SERIES
　．．．．TANGENTS
　FUNCTIONS (MATHEMATICS)
　．TRANSCENDENTAL FUNCTIONS
　．．**PERIODIC FUNCTIONS**
　．．．TRIGONOMETRIC FUNCTIONS
　．．．．COSINE SERIES
　．．．．SINE SERIES
　．．．．TANGENTS
RT　FLOQUET THEOREM
　　FOURIER ANALYSIS

PERIODIC ORBITS
USE　ORBITS

PERIODIC PROCESSES
USE　CYCLES

PERIODIC VARIATIONS
UF　PERIODICITY
GS　VARIATIONS
　．**PERIODIC VARIATIONS**
　．．ANNUAL VARIATIONS
　．．DIURNAL VARIATIONS
　．．NOCTURNAL VARIATIONS
　．．SECULAR VARIATIONS
RT　AUTOCORRELATION
　　CATACLYSMIC VARIABLES
　　CLIMATOLOGY
　　CYCLES
　　DENDROCHRONOLOGY
　　EL NINO
　　EXTRAPOLATION
　　FOURIER ANALYSIS
　　LONG TERM EFFECTS
　　OSCILLATIONS
　　OSCILLATORS
　　POLAR WANDERING (GEOLOGY)
　∞RHYTHM
　　TRENDS
　　VARIABILITY
　　VARIABLE STARS

PERIODICALS
UF　JOURNALS (DOCUMENTS)
GS　DOCUMENTS
　．**PERIODICALS**
RT　∞JOURNALS
　　RECORDS

PERIODICITY
USE　PERIODIC VARIATIONS

PERIODICITY (BIOLOGY)
USE　RHYTHM (BIOLOGY)

PERIPHERAL CIRCULATION
GS   CIRCULATION
. BLOOD CIRCULATION
. . **PERIPHERAL CIRCULATION**

PERIPHERAL EQUIPMENT (COMPUTERS)
GS   **PERIPHERAL EQUIPMENT (COMPUTERS)**
. ACCUMULATORS (COMPUTERS)
. COMPUTER STORAGE DEVICES
. . BUFFER STORAGE
. . CRYOGENIC COMPUTER STORAGE
. . MAGNETIC TAPES
. . . COMPUTER COMPATIBLE TAPES
. . OPTICAL DISKS
. . RANDOM ACCESS MEMORY
. . READ-ONLY MEMORY DEVICES
. . REGISTERS (COMPUTERS)
. CONSOLES
. . REMOTE CONSOLES
. INTEL 8080 MICROPROCESSOR
RT   ∞EQUIPMENT

PERIPHERAL JET FLOW
GS   FLUID FLOW
. JET FLOW
. . **PERIPHERAL JET FLOW**
RT   DOWNWASH
GROUND EFFECT (AERODYNAMICS)
GROUND EFFECT MACHINES
LIFT AUGMENTATION

PERIPHERAL NERVOUS SYSTEM
GS   NERVOUS SYSTEM
. **PERIPHERAL NERVOUS SYSTEM**
RT   NEUROMUSCULAR TRANSMISSION
∞SYSTEMS

PERIPHERAL VISION
GS   VISION
. **PERIPHERAL VISION**
RT   SPACE PERCEPTION
VISUAL ACUITY
VISUAL FIELDS

PERIPHERIES
USE   BOUNDARIES

PERISCOPES
GS   OPTICAL EQUIPMENT
. **PERISCOPES**
RT   BINOCULARS
EYEPIECES
OPTICAL MEASURING INSTRUMENTS
TELESCOPES
VIEWING

PERITONEUM
GS   TISSUES (BIOLOGY)
. **PERITONEUM**
RT   ABDOMEN
MEMBRANES

PERMAFROST
UF   FROZEN SOILS
GS   SOILS
. **PERMAFROST**
RT   AUFEIS (ICE)
POLAR REGIONS

PERMALLOYS (TRADEMARK)
GS   ALLOYS
. **PERMALLOYS (TRADEMARK)**
MAGNETIC MATERIALS
. FERROMAGNETIC MATERIALS
. . **PERMALLOYS (TRADEMARK)**
RT   IRON ALLOYS
MAGNETS
MOLYBDENUM ALLOYS
NICKEL ALLOYS

PERMANGANATES
GS   MANGANESE COMPOUNDS
. **PERMANGANATES**
RT   MANGANESE IONS

PERMEABILITY
SN   (EXCLUDES MAGNETIC PERMEABILITY)
GS   **PERMEABILITY**
. DIELECTRIC PERMEABILITY
RT   AQUIFERS
DENSITY (MASS/VOLUME)
DIFFUSION
DIFFUSIVITY
DRAINAGE

PERMEABILITY-(CONT.)
FORMATIONS
INFILTRATION
INTERSTICES
LEACHING
LEAKAGE
MECHANICAL PROPERTIES
PENETRATION
PERCOLATION
PERMEATING
∞PHYSICAL PROPERTIES
POROSITY
∞RESISTANCE
SEEPAGE
SURFACE PROPERTIES
VOID RATIO
VOIDS
WETTABILITY

PERMEATING
RT   ∞ABSORPTION
DESORPTION
DIALYSIS
DIFFUSION
DISPERSING
IMPREGNATING
OSMOSIS
PENETRATION
PERCOLATION
PERMEABILITY
POROSITY
REVERSE OSMOSIS
∞SATURATION
SORPTION
TRANSPIRATION

PERMISSIVITY
RT   COMPATIBILITY
PSYCHOLOGICAL FACTORS

PERMITTIVITY
UF   DIELECTRIC CONSTANT
GS   ELECTROMAGNETIC PROPERTIES
. ELECTRICAL PROPERTIES
. . DIELECTRIC PROPERTIES
. . . **PERMITTIVITY**
RT   ELECTRIC FIELDS
FIELD STRENGTH

PERMUTATIONS
GS   ANALYSIS (MATHEMATICS)
. COMBINATORIAL ANALYSIS
. . **PERMUTATIONS**
RT   ∞COMBINATION
COMBINATIONS (MATHEMATICS)
PARTITIONS (MATHEMATICS)
SET THEORY

PEROVSKITES
GS   CALCIUM COMPOUNDS
. **PEROVSKITES**
MINERALS
. **PEROVSKITES**
TITANIUM COMPOUNDS
. TITANATES
. . **PEROVSKITES**

PEROXIDES
GS   CHALCOGENIDES
. OXIDES
. . ANHYDRIDES
. . . **PEROXIDES**
. . . . INORGANIC PEROXIDES
. . . . ORGANIC PEROXIDES
. . . . SODIUM PEROXIDES
RT   DIOXIDES

PERSEID METEOROIDS
GS   CELESTIAL BODIES
. METEOROID SHOWERS
. . **PERSEID METEOROIDS**
. METEOROIDS
. . **PERSEID METEOROIDS**

PERSHING MISSILE
GS   MISSILES
. BALLISTIC MISSILES
. . **PERSHING MISSILE**
. SURFACE TO SURFACE MISSILES
. . **PERSHING MISSILE**
RT   MULTISTAGE ROCKET VEHICLES
SOLID PROPELLANT ROCKET ENGINES

PERSIAN GULF
GS   GULFS
. **PERSIAN GULF**
RT   INLETS (TOPOGRAPHY)

PERSONAL COMPUTERS
GS   DATA PROCESSING EQUIPMENT
. COMPUTERS
. . DIGITAL COMPUTERS
. . . MICROCOMPUTERS
. . . . **PERSONAL COMPUTERS**
RT   COMPUTER TECHNIQUES

PERSONALITY
RT   DEPERSONALIZATION
PERSONNEL SELECTION

PERSONALITY TESTS
RT   PSYCHOLOGICAL TESTS
PSYCHOMETRICS
QUALIFICATIONS
∞TESTS

PERSONNEL
GS   **PERSONNEL**
. AIR TRAFFIC CONTROLLERS
(PERSONNEL)
. CREWS
. . FLIGHT CREWS
. . . SPACECREWS
. ENEMY PERSONNEL
. FLYING PERSONNEL
. ASTRONAUTS
. . . ORBITAL WORKERS
. . COSMONAUTS
. . FLIGHT CREWS
. . . SPACECREWS
. . PILOTS (PERSONNEL)
. . . AIRCRAFT PILOTS
. . . . TEST PILOTS
. GROUND CREWS
. INSTRUCTORS
. MEDICAL PERSONNEL
. . FLIGHT NURSES
. . PHYSICIANS
. . SURGEONS
. . . FLIGHT SURGEONS
. NAVIGATORS
. OPERATORS (PERSONNEL)
. PILOTS (PERSONNEL)
. . AIRCRAFT PILOTS
. . . TEST PILOTS
. POLICE
. PROGRAMMERS
. SCIENTISTS
RT   ∞COMPLEMENT
CONSULTING
CREW EXPERIMENT STATIONS
CREW OBSERVATION STATIONS
CREW STATIONS
CREW WORKSTATIONS
DEPERSONALIZATION
DEPLOYMENT
EMPLOYEE RELATIONS
∞ESTIMATORS
HELMET MOUNTED DISPLAYS
INCENTIVES
INHABITANTS
MANPOWER
OCCUPATION
ORGANIZING
POSITION (TITLE)
QUALIFICATIONS
RESEARCH MANAGEMENT
RESOURCES
RETIREMENT
RETRAINING
SERVICES
UNIONIZATION
WAGE SURVEYS

PERSONNEL DEVELOPMENT
RT   ∞DEVELOPMENT
EMPLOYEE RELATIONS
MANAGEMENT
RESOURCES
TRAINING ANALYSIS

PERSONNEL MANAGEMENT
GS   MANAGEMENT
. INDUSTRIAL MANAGEMENT
. . **PERSONNEL MANAGEMENT**
RT   EMPLOYEE RELATIONS
HUMAN RELATIONS
LEADERSHIP

PERSONNEL MANAGEMENT-*(CONT.)*
    MANAGEMENT PLANNING

PERSONNEL PROPULSION SYSTEMS
  USE   SELF MANEUVERING UNITS

PERSONNEL SELECTION
  GS   SELECTION
      . **PERSONNEL SELECTION**
      . . PILOT SELECTION
  RT   APTITUDE
      EMPLOYMENT
      LABOR
      PERSONALITY
      PHYSICAL EXAMINATIONS
      PHYSIOLOGICAL TESTS

PERSONNEL SUBSYSTEMS
  RT   INDUSTRIES

PERSPEX (TRADEMARK)
  GS   PLASTICS
      . **PERSPEX (TRADEMARK)**

PERSPIRATION
  UF   SWEATING
  RT   BODY FLUIDS
      BODY TEMPERATURE
      EVAPORATION
      EXCRETION
      FECES
      HEAT ACCLIMATIZATION
      HUMIDITY
      PALMAR SWEAT INDEX
      SKIN (ANATOMY)
      SWEAT
      TRANSPIRATION

PERT
  RT   COMMERCE
      CONTRACT MANAGEMENT
      CRITICAL PATH METHOD
      GERT
      MANAGEMENT ANALYSIS
      MANAGEMENT METHODS
      MANAGEMENT PLANNING
      ∞PATHS
      PROGRAM TREND LINE ANALYSIS
      PROJECT MANAGEMENT

PERTURBATION
  GS   **PERTURBATION**
      . ORBIT PERTURBATION
      . . SATELLITE PERTURBATION
  RT   ∞DISTURBANCES
      FOUR BODY PROBLEM
      GEODESY
      LONG TERM EFFECTS
      MANY BODY PROBLEM
      NUTATION
      ORBITAL MECHANICS
      OSCILLATIONS
      OSCILLATORS
      RADIATION PRESSURE
      SCHACH EFFECT
      THREE BODY PROBLEM
      TWO BODY PROBLEM
      VARIATIONS

PERTURBATION THEORY
  UF   DISTURBANCE THEORY
  GS   **PERTURBATION THEORY**
      . VINTI THEORY
  RT   BOUSSINESQ APPROXIMATION
      CELESTIAL MECHANICS
      DISTURBING FUNCTIONS
      HANSEN LUNAR THEORY
      HARTREE APPROXIMATION
      HILL LUNAR THEORY
      HILL METHOD
      MANY BODY PROBLEM
      OPERATORS (MATHEMATICS)
      ORBIT PERTURBATION
      ORBITAL ELEMENTS
      QUANTUM THEORY
      STRANGE ATTRACTORS
      TAYLOR INSTABILITY
      ∞THEORIES
      VON ZEIPEL METHOD
      WAVE FUNCTIONS
      WENTZEL-KRAMER-BRILLOUIN METHOD
      YANG-MILLS FIELDS
      YANG-MILLS THEORY

PERU
  GS   NATIONS
      . **PERU**
  RT   SOUTH AMERICA

PERVEANCE
  GS   RATIOS
      . **PERVEANCE**
  RT   CHILD-LANGMUIR LAW
      SPACE CHARGE
      THERMIONIC DIODES
      VACUUM TUBES
      WORK FUNCTIONS

PESTICIDES
  GS   POISONS
      . **PESTICIDES**
      . . INSECTICIDES
      . . . DIELDRIN
  RT   CROP DUSTING
      ENVIRONMENTAL CHEMISTRY
      TOXICOLOGY

PETALS
  RT   PLANTS (BOTANY)

PETECHIA
  GS   HEMORRHAGES
      . **PETECHIA**
  RT   SKIN (ANATOMY)

PETN
  UF   PENTAERYTHRITOL TETRANITRATE
  GS   ESTERS
      . ORGANIC NITRATES
      . . **PETN**
      NITROGEN COMPOUNDS
      . NITRATES
      . . ORGANIC NITRATES
      . . . **PETN**
  RT   EXPLOSIVES

PETREL SOUNDING ROCKET
  GS   ROCKET VEHICLES
      . SOUNDING ROCKETS
      . . **PETREL SOUNDING ROCKET**
  RT   ∞ROCKETS

PETRI NETS
  GS   NETS
      . **PETRI NETS**
  RT   CONSECUTIVE EVENTS
      DYNAMIC MODELS
      GRAPHS (CHARTS)
      INFORMATION THEORY
      SEQUENCING
      TREES (MATHEMATICS)

PETROGRAPHY
  GS   GEOLOGY
      . PETROLOGY
      . . **PETROGRAPHY**
  RT   INLIERS (LANDFORMS)
      ROCKS
      SEDIMENTARY ROCKS

PETROLEUM
  USE   CRUDE OIL

PETROLEUM PRODUCTS
  GS   PRODUCTS
      . **PETROLEUM PRODUCTS**
      . . ASPHALT
      . . GREASES
      . . TARS
  RT   CRUDE OIL
      LUBRICANTS
      METHANE
      OILS
      POLYNUCLEAR ORGANIC COMPOUNDS

PETROLOGY
  GS   GEOLOGY
      . **PETROLOGY**
      . . PETROGRAPHY
  RT   CONES (VOLCANOES)
      FORMATIONS
      GEOCHEMISTRY
      GEOLOGICAL SURVEYS
      GEOPHYSICS
      IMPACT MELTS
      INLIERS (LANDFORMS)
      MARS VOLCANOES
      MINERALOGY

PETROLOGY-*(CONT.)*
      ROCKS
      STRATIGRAPHY
      VOLCANOES
      VOLCANOLOGY

PFAFF EQUATION
  GS   ANALYSIS (MATHEMATICS)
      . **PFAFF EQUATION**
  RT   DIFFERENTIAL EQUATIONS
      ∞EQUATIONS
      THERMODYNAMICS

PFM (MODULATION)
  USE   PULSE FREQUENCY MODULATION

PH
  RT   ACID BASE EQUILIBRIUM
      ACID RAIN
      ACIDITY
      ACIDOSIS
      ALKALINITY
      ALKALOSIS
      BUFFERS (CHEMISTRY)
      HYDROGEN IONS

PH FACTOR
  RT   ACID BASE EQUILIBRIUM
      ACIDOSIS
      ALKALOSIS
      HYDROGEN IONS
      ION CONCENTRATION

PHANTASTRONS
  GS   CIRCUITS
      . DELAY CIRCUITS
      . . **PHANTASTRONS**
      OSCILLATORS
      . RELAXATION OSCILLATORS
      . . **PHANTASTRONS**
  RT   FEEDBACK AMPLIFIERS

PHANTOM AIRCRAFT
  GS   JET AIRCRAFT
      . **PHANTOM AIRCRAFT**
      . . F-4 AIRCRAFT
      MCDONNELL DOUGLAS AIRCRAFT
      . MCDONNELL AIRCRAFT
      . . **PHANTOM AIRCRAFT**
      . . . F-4 AIRCRAFT
      MONOPLANES
      . **PHANTOM AIRCRAFT**
      SUPERSONIC AIRCRAFT
      . **PHANTOM AIRCRAFT**
      . . F-4 AIRCRAFT
  RT   ∞AIRCRAFT

PHARMACOLOGY
  GS   **PHARMACOLOGY**
      . PSYCHOPHARMACOLOGY
  RT   ANESTHESIOLOGY
      ANTIRADIATION DRUGS
      BIOPROCESSING
      CYCLIC AMP
      DRUGS
      MEDICAL SCIENCE
      ∞MEDICINE
      MOTION SICKNESS DRUGS
      VASOCONSTRICTOR DRUGS
      VETERINARY MEDICINE

PHARYNX
  GS   ANATOMY
      . RESPIRATORY SYSTEM
      . . BRONCHIAL TUBE
      . . . **PHARYNX**

PHASE ANGLE
  USE   PHASE SHIFT

PHASE CHANGE MATERIALS
  UF   PCM (MATERIALS)
  RT   CERESIN
      CONDENSING
      FUSION (MELTING)
      HEAT OF FUSION
      HEAT STORAGE
      HEAT TRANSFER
      LIQUID-SOLID INTERFACES
      ∞MATERIALS
      MELTING
      ORGANIC MATERIALS
      PHASE TRANSFORMATIONS
      SOLAR ENERGY

PHASE CHANGE MATERIALS-*(CONT.)*
   SOLAR ENERGY CONVERSION
   SOLAR HEATING
   SUBLIMATION
   TROMBE WALLS
   WAXES
   WORKING FLUIDS

PHASE COHERENCE
 RT ∞COHERENCE
   COHERENCE COEFFICIENT
   COHERENT LIGHT
   WAVE FRONTS

PHASE CONJUGATION
 GS CONJUGATION
   . **PHASE CONJUGATION**

PHASE CONTRAST
 GS CONTRAST
   . **PHASE CONTRAST**
 RT DIFFRACTION PATTERNS
   ELECTRON MICROSCOPY
   MICROSCOPY
   OPTICAL MEASUREMENT
   OPTICAL PATHS

PHASE CONTROL
 RT CIRCUIT PROTECTION
   ∞CONTROL
   PHASE LOCKED SYSTEMS
   TRANSFORMERS

PHASE DEMODULATORS
 GS DEMODULATORS
   . **PHASE DEMODULATORS**
 RT MODEMS

PHASE DETECTORS
 GS CIRCUITS
   . **PHASE DETECTORS**
   . . SYNCHROSCOPES
 RT ∞DETECTORS
   PHASE LOCKED SYSTEMS
   SIGNAL DETECTION
   SYNCHRONISM

PHASE DEVIATION
 RT MODULATED CONTINUOUS RADIATION

PHASE DIAGRAMS
 UF CONSTITUTIONAL DIAGRAMS
   EQUILIBRIUM DIAGRAMS
   EUTECTIC DIAGRAMS
 GS DIAGRAMS
   . **PHASE DIAGRAMS**
 RT ALLOYS
   BINARY SYSTEMS (MATERIALS)
   CRITICAL TEMPERATURE
   EUTECTICS
   HEAT OF FUSION
   HEAT TREATMENT
   INTERMETALLICS
   LIQUID PHASES
   LIQUIDS
   LIQUIDUS
   MELTING POINTS
   SOLID PHASES
   SOLID SOLUTIONS
   SOLID SUSPENSIONS
   SOLUBILITY
   STOICHIOMETRY
   TRANSITION POINTS
   TRANSITION TEMPERATURE
   VAPOR PHASES

PHASE ERROR
 GS ERRORS
   . **PHASE ERROR**
 RT CIRCUIT PROTECTION
   ERROR SIGNALS

PHASE LOCK DEMODULATORS
 GS DEMODULATORS
   . **PHASE LOCK DEMODULATORS**
 RT CORRELATION DETECTION
   PARAMETRONS

PHASE LOCKED SYSTEMS
 RT FEEDBACK FREQUENCY MODULATION
   OPTICAL COUPLING
   PHASE CONTROL
   PHASE DETECTORS
   SYNCHRONIZED OSCILLATORS

PHASE LOCKED SYSTEMS-*(CONT.)*
   ∞SYSTEMS
   TRACKING FILTERS

PHASE MATCHING
 RT CRYSTAL OPTICS
   CRYSTALS
   FREQUENCY MULTIPLIERS
   HARMONIC GENERATIONS
   LASER OUTPUTS
   LASERS

PHASE MODULATION
 GS MODULATION
   . **PHASE MODULATION**
   . . FM/PM (MODULATION)
   . . PHASE SHIFT KEYING
 RT AMPLITUDE MODULATION
   DEMODULATION
   DEMODULATORS
   FREQUENCY MODULATION
   MODEMS
   MODULATORS
   PARAMETRIC FREQUENCY CONVERTERS
   PULSE MODULATION
   PUSH-PULL AMPLIFIERS

PHASE RULE
 GS RULES
   . **PHASE RULE**
 RT CHEMICAL EQUILIBRIUM
   DEGREES OF FREEDOM
   ∞GIBBS EQUATIONS

PHASE SHIFT
 UF PHASE ANGLE
 GS **PHASE SHIFT**
   . SAGNAC EFFECT
 RT ANGLES (GEOMETRY)
   EQUALIZERS (CIRCUITS)
   MICROWAVE SWITCHING
   ∞PHASES
   ∞SHIFT

PHASE SHIFT CIRCUITS
 GS CIRCUITS
   . **PHASE SHIFT CIRCUITS**
   . . CIRCULATORS (PHASE SHIFT
    CIRCUITS)
 RT DELAY CIRCUITS
   DUPLEX OPERATION
   GYRATORS

PHASE SHIFT KEYING
 GS KEYING
   . **PHASE SHIFT KEYING**
   MODULATION
   . PHASE MODULATION
   . . **PHASE SHIFT KEYING**
 RT INFORMATION THEORY

PHASE SWITCHING INTERFEROMETERS
 GS MEASURING INSTRUMENTS
   . INTERFEROMETERS
   . . **PHASE SWITCHING**
     INTERFEROMETERS
 RT RADIO ASTRONOMY
   RADIO TELESCOPES

PHASE TRANSFORMATIONS
 GS **PHASE TRANSFORMATIONS**
   . ARC MELTING
   . FREEZING
   . . VIBRATIONAL FREEZING
   . . ZONE MELTING
   . LIQUEFACTION
   . . COAL LIQUEFACTION
   . MARTENSITIC TRANSFORMATION
   . MELTING
   . . FUSION (MELTING)
   . . LEVITATION MELTING
   . . VACUUM MELTING
   . VAPORIZING
   . . BOILING
   . . . FILM BOILING
   . . . NUCLEATE BOILING
   . . . . LEIDENFROST PHENOMENON
   . . EVAPORATION
   . . . EVAPOTRANSPIRATION
   . . . PROPELLANT EVAPORATION
   . . . TRANSPIRATION
   . . FLASHING (VAPORIZING)
   . . PREVAPORIZATION
   . . SUBLIMATION
 RT COLD HARDENING

PHASE TRANSFORMATIONS-*(CONT.)*
   CONDENSING
   CRITICAL TEMPERATURE
   CRYSTALLIZATION
   DIRECTIONAL SOLIDIFICATION
    (CRYSTALS)
   ELECTRON-HOLE DROPS
   HEAT OF FUSION
   MARTENSITE
   MELT SPINNING
   METAMORPHISM (GEOLOGY)
   MUSHY ZONES
   NEEL TEMPERATURE
   ORDER-DISORDER TRANSFORMATIONS
   PHASE CHANGE MATERIALS
   ∞PHASES
   SHAPE MEMORY ALLOYS
   SOLIDIFICATION
   SOLIDS
   SYNTECTIC ALLOYS
   ∞TRANSFORMATIONS
   ∞TRANSITION
   TRANSITION PRESSURE
   TRANSITION TEMPERATURE

PHASE VELOCITY
 GS RATES (PER TIME)
   . **PHASE VELOCITY**
   VELOCITY
   . **PHASE VELOCITY**
 RT ELECTROMAGNETIC RADIATION
   GROUP VELOCITY
   LANDAU DAMPING
   PROPAGATION VELOCITY
   QUANTUM MECHANICS
   TRAVELING WAVES
   WAVE FRONTS
   WAVE PROPAGATION

PHASE-SPACE INTEGRAL
 GS ANALYSIS (MATHEMATICS)
   . **PHASE-SPACE INTEGRAL**
 RT CLASSICAL MECHANICS
   EUCLIDEAN GEOMETRY
   HYPERSPACES
   STATE VECTORS

PHASED ARRAYS
 GS ARRAYS
   . **PHASED ARRAYS**
 RT ANTENNA ARRAYS
   LINEAR ARRAYS
   MONOPULSE ANTENNAS
   SEISMOGRAPHS
   STEERABLE ANTENNAS

∞ PHASES
 SN *(USE OF A MORE SPECIFIC TERM IS*
   RECOMMENDED--CONSULT THE TERMS
   LISTED BELOW)
 RT CYCLES
   LIQUID PHASES
   LUNAR PHASES
   PHASE SHIFT
   PHASE TRANSFORMATIONS
   SOLID PHASES
   TERMINATOR LINES
   VAPOR PHASES

PHENACETIN
 USE ACETANILIDE

PHENANTHRENE
 GS HYDROCARBONS
   . **PHENANTHRENE**
   ISOMERS
   . **PHENANTHRENE**
 RT ANTHRACENE
   DYES

PHENOBARBITAL
 RT DRUGS
   NARCOTICS
   SEDATIVES

PHENOL FORMALDEHYDE
 RT FORMALDEHYDE
   PHENOLIC RESINS
   RESINS

PHENOLIC EPOXY RESINS
 GS RESINS
   . SYNTHETIC RESINS
   . . THERMOSETTING RESINS

PHENOLIC EPOXY RESINS-*(CONT.)*
```
        . . . PHENOLIC RESINS
        . . . . PHENOLIC EPOXY RESINS
RT      ADHESIVES
        AMINES
        CROSSLINKING
```

PHENOLIC RESINS
```
GS      PLASTICS
        . SYNTHETIC RESINS
        . . THERMOSETTING RESINS
        . . . PHENOLIC RESINS
        . . . . MICARTA
        RESINS
        . SYNTHETIC RESINS
        . . THERMOSETTING RESINS
        . . . PHENOLIC RESINS
        . . . . MICARTA
        . . . . PHENOLIC EPOXY RESINS
RT      PHENOL FORMALDEHYDE
```

PHENOLOGY
```
GS      PHENOMENOLOGY
        . PHENOLOGY
RT      ACTIVITY CYCLES (BIOLOGY)
        BIOMETEOROLOGY
        CLIMATOLOGY
        COASTAL ECOLOGY
        ECOLOGY
        MIGRATION
        MOLTING
        RHYTHM (BIOLOGY)
```

PHENOLS
```
GS      HYDROXYL COMPOUNDS
        . ALCOHOLS
        . . PHENOLS
        . . . BISPHENOLS
        . . . CRESOLS
        . . . PHLOROGLUCINOL
        . . . THYMOL
RT      THIOLS
```

PHENOMENOLOGY
```
GS      PHENOMENOLOGY
        . MEDICAL PHENOMENA
        . MESOSCALE PHENOMENA
        . PHENOLOGY
RT      CASE HISTORIES
        MEDICAL SCIENCE
```

PHENOTHIAZINES
```
GS      HETEROCYCLIC COMPOUNDS
        . AZINES
        . . PHENOTHIAZINES
        PYRAZINES
        . AZINES
        . . PHENOTHIAZINES
```

PHENYLALANINE
```
GS      ACIDS
        . AMINO ACIDS
        . . ALANINE
        . . . PHENYLALANINE
        ORGANIC COMPOUNDS
        . AMINO ACIDS
        . . ALANINE
        . . . PHENYLALANINE
```

PHENYLS
```
GS      PHENYLS
        . POLYCHLORINATED BIPHENYLS
        . POLYPHENYLS
        . . TETRAPHENYLS
        . TERPHENYLS
RT      PROPARGYL GROUPS
```

PHILCO 2000 COMPUTER
```
GS      DATA PROCESSING EQUIPMENT
        . COMPUTERS
        . . DIGITAL COMPUTERS
        . . . PHILCO 2000 COMPUTER
```

PHILIPPINES
```
GS      LANDFORMS
        . ISLANDS
        . . PACIFIC ISLANDS
        . . . PHILIPPINES
        NATIONS
        . PHILIPPINES
```

PHILIPS IONIZATION GAGES
```
GS      MEASURING INSTRUMENTS
        . PRESSURE GAGES
```

PHILIPS IONIZATION GAGES-*(CONT.)*
```
        . . VACUUM GAGES
        . . . IONIZATION GAGES
        . . . . PHILIPS IONIZATION GAGES
        VACUUM APPARATUS
        . VACUUM GAGES
        . . IONIZATION GAGES
        . . . PHILIPS IONIZATION GAGES
RT      PRESSURE MEASUREMENT
```

PHILOSOPHY
```
GS      KNOWLEDGE
        . PHILOSOPHY
        . . PARADOXES
RT      LITERATURE
        ∞ LOGIC
        MATHEMATICAL LOGIC
```

PHLOROGLUCINOL
```
GS      HYDROXYL COMPOUNDS
        . ALCOHOLS
        . . PHENOLS
        . . . PHLOROGLUCINOL
RT      CHEMICAL INDICATORS
        RESINS
```

PHOBIAS
```
GS      PHOBIAS
        . FEAR
        . . FEAR OF FLYING
RT      ANXIETY
        EMOTIONAL FACTORS
```

PHOBOS
```
GS      CELESTIAL BODIES
        . PHOBOS
RT      DEIMOS
        MARS (PLANET)
        SATURN (PLANET)
```

PHOEBUS NUCLEAR REACTOR
```
GS      NUCLEAR REACTORS
        . PHOEBUS NUCLEAR REACTOR
RT      KIWI REACTORS
        NUCLEAR ENGINE FOR ROCKET
          VEHICLES
        NUCLEAR ROCKET ENGINES
```

PHOENIX (AZ)
```
GS      CITIES
        . PHOENIX (AZ)
RT      ARIZONA
```

PHOENIX QUADRANGLE (AZ)
```
GS      LANDFORMS
        . PHOENIX QUADRANGLE (AZ)
RT      ARIZONA
        GEODETIC SURVEYS
        MAPPING
```

PHOENIX SOUNDING ROCKET
```
GS      ROCKET VEHICLES
        . MULTISTAGE ROCKET VEHICLES
        . . PHOENIX SOUNDING ROCKET
        . SOUNDING ROCKETS
        . . PHOENIX SOUNDING ROCKET
RT      SOLID PROPELLANT ROCKET ENGINES
```

PHONEMES
```
GS      LINGUISTICS
        . PHONEMES
        SPEECH
        . PHONEMES
RT      LANGUAGES
        PHONEMICS
        PHONETICS
        PSYCHOLINGUISTICS
        SPEECH RECOGNITION
        WORDS (LANGUAGE)
```

PHONEMICS
```
GS      LINGUISTICS
        . PHONEMICS
RT      INTELLIGIBILITY
        LANGUAGES
        PHONEMES
        PHONETICS
        PSYCHOLINGUISTICS
        SPEECH
        SPEECH DEFECTS
        SPEECH RECOGNITION
        WORDS (LANGUAGE)
```

PHONETICS
```
GS      SPEECH
        . PHONETICS
RT      ACOUSTICS
        INTELLIGIBILITY
        LANGUAGES
        PHONEMES
        PHONEMICS
        SPEECH DEFECTS
        SPEECH RECOGNITION
        VERBAL COMMUNICATION
        WORDS (LANGUAGE)
```

PHONOARTERIOGRAPHY
```
RT      ARTERIES
        BLOOD CIRCULATION
        PHONOCARDIOGRAPHY
```

PHONOCARDIOGRAMS
```
USE     PHONOCARDIOGRAPHY
```

PHONOCARDIOGRAPHY
```
UF      PHONOCARDIOGRAMS
        VIBROCARDIOGRAPHY
GS      BIOENGINEERING
        . BIOMETRICS
        . . CARDIOGRAPHY
        . . . PHONOCARDIOGRAPHY
        . . . . ECHOCARDIOGRAPHY
RT      BALLISTOCARDIOGRAPHY
        ELECTROCARDIOGRAPHY
        HEART
        HEART DISEASES
        PHONOARTERIOGRAPHY
        VECTORCARDIOGRAPHY
```

PHONON BEAMS
```
GS      BEAMS (RADIATION)
        . PHONON BEAMS
        ELASTIC WAVES
        . PHONONS
        . . PHONON BEAMS
        ELECTROMAGNETIC RADIATION
        . THERMAL RADIATION
        . . PHONON BEAMS
        ELEMENTARY EXCITATIONS
        . PHONONS
        . . PHONON BEAMS
RT      CORPUSCULAR RADIATION
        PARTICLE BEAMS
        PHOTON BEAMS
        UMKLAPP PROCESS
```

PHONONS
```
GS      ELASTIC WAVES
        . PHONONS
        . . PHONON BEAMS
        ELEMENTARY EXCITATIONS
        . PHONONS
        . . PHONON BEAMS
RT      CRYSTAL STRUCTURE
        LATTICE VIBRATIONS
        PLASMONS
        POLARONS
        SOUND WAVES
        UMKLAPP PROCESS
```

PHORIA
```
GS      DISEASES
        . EYE DISEASES
        . . PHORIA
```

PHOSGENE
```
GS      ALIPHATIC COMPOUNDS
        . PHOSGENE
        GASES
        . PHOSGENE
        HALOGEN COMPOUNDS
        . CHLORINE COMPOUNDS
        . . CHLORIDES
        . . . PHOSGENE
        . HALIDES
        . . CHLORIDES
        . . . PHOSGENE
        POISONS
        . PHOSGENE
RT      ∞ CHEMICAL COMPOUNDS
```

PHOSPHATES
```
GS      PHOSPHORUS COMPOUNDS
        . PHOSPHATES
        . . ADENOSINE TRIPHOSPHATE
        . . AMMONIUM PHOSPHATES
        . . CALCIUM PHOSPHATES
        . . CYCLIC AMP
```

PHOSPHATES-(CONT.)
```
        . . DIPHOSPHATES
        . . . ADENOSINE DIPHOSPHATE
        . . INDIUM PHOSPHATES
        . . MONAZITE SANDS
        . . POLYNUCLEOTIDES
        . . POTASSIUM PHOSPHATES
        . . PYRIDINE NUCLEOTIDES
        . . URIDYLIC ACID
   RT   KREEP
        PHOSPHORIC ACID
        SKYDROL (TRADEMARK)
```

PHOSPHAZENE
```
   GS   ORGANIC COMPOUNDS
        . PHOSPHAZENE
   RT   NITROGEN COMPOUNDS
        PHOSPHINES
        PHOSPHONITRILES
        POLYMER CHEMISTRY
```

PHOSPHENE
```
   RT   LUMINOSITY
        RETINA
        VISION
```

PHOSPHIDES
```
   GS   PHOSPHORUS COMPOUNDS
        . PHOSPHIDES
        . . BORON PHOSPHIDES
        . . GALLIUM PHOSPHIDES
        . . INDIUM PHOSPHIDES
        . . MANGANESE PHOSPHIDES
        . . SCHREIBERSITE
```

PHOSPHINES
```
   GS   HYDROGEN COMPOUNDS
        . HYDRIDES
        . . PHOSPHINES
        PHOSPHORUS COMPOUNDS
        . PHOSPHINES
   RT   PHOSPHAZENE
```

PHOSPHONITRILES
```
   GS   NITRILES
        . PHOSPHONITRILES
        NITROGEN COMPOUNDS
        . PHOSPHONITRILES
        PHOSPHORUS COMPOUNDS
        . ORGANIC PHOSPHORUS COMPOUNDS
        . . PHOSPHONITRILES
   RT   PHOSPHAZENE
```

PHOSPHONIUM COMPOUNDS
```
   GS   PHOSPHORUS COMPOUNDS
        . PHOSPHONIUM COMPOUNDS
   RT   ∞CHEMICAL COMPOUNDS
```

PHOSPHORESCENCE
```
   GS   DECAY
        . EMISSION
        . . LIGHT EMISSION
        . . . LUMINESCENCE
        . . . . FLUORESCENCE
        . . . . . PHOSPHORESCENCE
        ELECTROMAGNETIC PROPERTIES
        . OPTICAL PROPERTIES
        . . PHOSPHORESCENCE
   RT   AFTERGLOWS
        BIOLUMINESCENCE
        CHEMILUMINESCENCE
        PHOSPHORS
        PLASMA RADIATION
        SCINTILLATION
        TRAPPING
```

PHOSPHORIC ACID
```
   GS   ACIDS
        . PHOSPHORIC ACID
        PHOSPHORUS COMPOUNDS
        . PHOSPHORIC ACID
   RT   PHOSPHATES
```

PHOSPHORIC ACID FUEL CELLS
```
   GS   ELECTRIC GENERATORS
        . DIRECT POWER GENERATORS
        . . FUEL CELLS
        . . . PHOSPHORIC ACID FUEL CELLS
        ELECTROCHEMICAL CELLS
        . FUEL CELLS
        . . PHOSPHORIC ACID FUEL CELLS
   RT   BIOCHEMICAL FUEL CELLS
        ELECTROLYTIC CELLS
        ENERGY TECHNOLOGY
```

PHOSPHORIC ACID FUEL CELLS-(CONT.)
```
        HYDROGEN OXYGEN FUEL CELLS
        REGENERATIVE FUEL CELLS
        TOTAL ENERGY SYSTEMS
```

PHOSPHORS
```
   GS   PHOSPHORS
        . RADIOPHOSPHORS
   RT   FLUORESCENCE
        IMAGE INTENSIFIERS
        MERCURY LAMPS
        PHOSPHORESCENCE
        PHOTOGRAPHIC FILM
```

PHOSPHORUS
```
   GS   CHEMICAL ELEMENTS
        . PHOSPHORUS
        . . PHOSPHORUS ISOTOPES
        . . . PHOSPHORUS 32
```

PHOSPHORUS COMPOUNDS
```
   GS   PHOSPHORUS COMPOUNDS
        . ADENINES
        . DIETHYL HYDROGEN PHOSPHITE
          (DEHP)
        . ORGANIC PHOSPHORUS COMPOUNDS
        . . PHOSPHONITRILES
        . . URIDYLIC ACID
        . PHOSPHATES
        . . ADENOSINE TRIPHOSPHATE
        . . AMMONIUM PHOSPHATES
        . . CALCIUM PHOSPHATES
        . . CYCLIC AMP
        . . DIPHOSPHATES
        . . . ADENOSINE DIPHOSPHATE
        . . INDIUM PHOSPHATES
        . . MONAZITE SANDS
        . . POLYNUCLEOTIDES
        . . POTASSIUM PHOSPHATES
        . . PYRIDINE NUCLEOTIDES
        . . URIDYLIC ACID
        . PHOSPHIDES
        . . BORON PHOSPHIDES
        . . GALLIUM PHOSPHIDES
        . . INDIUM PHOSPHIDES
        . . MANGANESE PHOSPHIDES
        . . SCHREIBERSITE
        . PHOSPHINES
        . PHOSPHONIUM COMPOUNDS
        . PHOSPHORIC ACID
        . PHOSPHORUS OXIDES
        . PHOSPHORUS POLYMERS
   RT   ALIPHATIC COMPOUNDS
        ∞CHEMICAL COMPOUNDS
        ∞GROUP 5A COMPOUNDS
```

PHOSPHORUS ISOTOPES
```
   GS   CHEMICAL ELEMENTS
        . NUCLIDES
        . ISOTOPES
        . . . PHOSPHORUS ISOTOPES
        . . . . PHOSPHORUS 32
        . PHOSPHORUS
        . . PHOSPHORUS ISOTOPES
        . . . PHOSPHORUS 32
```

PHOSPHORUS METABOLISM
```
   GS   METABOLISM
        . PHOSPHORUS METABOLISM
```

PHOSPHORUS OXIDES
```
   GS   CHALCOGENIDES
        . OXIDES
        . . PHOSPHORUS OXIDES
        PHOSPHORUS COMPOUNDS
        . PHOSPHORUS OXIDES
```

PHOSPHORUS POLYMERS
```
   GS   PHOSPHORUS COMPOUNDS
        . PHOSPHORUS POLYMERS
   RT   ∞POLYMERS
```

PHOSPHORUS 32
```
   GS   CHEMICAL ELEMENTS
        . NUCLIDES
        . . ISOTOPES
        . . . PHOSPHORUS ISOTOPES
        . . . . PHOSPHORUS 32
        . . . RADIOACTIVE ISOTOPES
        . . . . PHOSPHORUS 32
        . PHOSPHORUS
        . . PHOSPHORUS ISOTOPES
        . . . PHOSPHORUS 32
```

PHOSPHORYLATION
```
   GS   CHEMICAL REACTIONS
        . PHOSPHORYLATION
```

PHOTICS
```
   RT   LIGHT (VISIBLE RADIATION)
        ∞OPTICS
```

PHOTO RECONNAISSANCE SPACECRAFT
```
   GS   MILITARY SPACECRAFT
        . RECONNAISSANCE SPACECRAFT
        . . PHOTO RECONNAISSANCE
            SPACECRAFT
   RT   ∞SPACECRAFT
```

PHOTOABSORPTION
```
   GS   ENERGY ABSORPTION
        . ELECTROMAGNETIC ABSORPTION
        . . PHOTOABSORPTION
        RADIATION ABSORPTION
        . ELECTROMAGNETIC ABSORPTION
        . . PHOTOABSORPTION
   RT   ∞ABSORPTION
```

PHOTOACOUSTIC MICROSCOPY

PHOTOACOUSTIC SPECTROSCOPY
```
   GS   SPECTROSCOPY
        . PHOTOACOUSTIC SPECTROSCOPY
```

PHOTOCATHODES
```
   GS   ELECTRODES
        . CATHODES
        . . TUBE CATHODES
        . . . PHOTOCATHODES
   RT   ELECTRODE MATERIALS
        IMAGE CONVERTERS
        IMAGE INTENSIFIERS
        IMAGE ORTHICONS
        LIGHT AMPLIFIERS
        MICROCHANNELS
        ORTHICONS
        PHOTOELECTRIC CELLS
        PHOTOELECTRIC EMISSION
        PHOTOELECTRIC MATERIALS
        PHOTOMULTIPLIER TUBES
```

PHOTOCELLS
```
   USE   PHOTOELECTRIC CELLS
```

PHOTOCHEMICAL OXIDANTS
```
   GS   OXIDIZERS
        . PHOTOCHEMICAL OXIDANTS
   RT   AIR POLLUTION
        ATMOSPHERIC CHEMISTRY
        NITROGEN OXIDES
        OZONE
        PHOTOOXIDATION
```

PHOTOCHEMICAL REACTIONS
```
   UF   PHOTOCHEMISTRY
        PHOTOREDUCTION
   GS   CHEMICAL REACTIONS
        . PHOTOCHEMICAL REACTIONS
        . . PHOTOCHROMISM
        . . PHOTODECOMPOSITION
        . . PHOTOLYSIS
        . . . RADIOLYSIS
        . . PHOTOSYNTHESIS
   RT   ASSOCIATION REACTIONS
        ATMOSPHERIC CHEMISTRY
        CHARGE TRANSFER
        SODALITE
```

PHOTOCHEMISTRY
```
   USE   PHOTOCHEMICAL REACTIONS
```

PHOTOCHROMISM
```
   GS   CHEMICAL REACTIONS
        . PHOTOCHEMICAL REACTIONS
        . . PHOTOCHROMISM
   RT   COLOR PHOTOGRAPHY
        SODALITE
```

PHOTOCLINOMETRY
```
   USE   PHOTOGRAMMETRY
```

PHOTOCONDUCTIVE CELLS
```
   GS   PHOTOELECTRIC CELLS
        . PHOTOCONDUCTIVE CELLS
   RT   ∞CELLS
        PHOTOVOLTAIC CELLS
```

PHOTOCONDUCTIVITY
UF   PHOTORESISTIVITY
GS   ELECTROMAGNETIC PROPERTIES
. ELECTRICAL PROPERTIES
. . **PHOTOCONDUCTIVITY**
. OPTICAL PROPERTIES
. . **PHOTOCONDUCTIVITY**
TRANSPORT PROPERTIES
. ELECTRICAL RESISTIVITY
. . **PHOTOCONDUCTIVITY**
RT   ∞CONDUCTIVITY
MERCURY CADMIUM TELLURIDES
PHOTOCONDUCTORS
PHOTOELECTRICITY
SQUARE WELLS

PHOTOCONDUCTORS
UF   PHOTORESISTORS
GS   CONDUCTORS
. **PHOTOCONDUCTORS**
SEMICONDUCTORS (MATERIALS)
. **PHOTOCONDUCTORS**
RT   MERCURY CADMIUM TELLURIDES
PHOTOCONDUCTIVITY
PHOTODIODES
PHOTOELECTRIC CELLS
PHOTOELECTRIC MATERIALS
PHOTOMETERS
PHOTOTRANSISTORS
RESISTORS

PHOTOCURRENTS
USE   ELECTRIC CURRENT
PHOTOELECTRIC EMISSION

PHOTODECOMPOSITION
GS   CHEMICAL REACTIONS
. PHOTOCHEMICAL REACTIONS
. . **PHOTODECOMPOSITION**
DECOMPOSITION
. **PHOTODECOMPOSITION**
RADIATION CHEMISTRY
. **PHOTODECOMPOSITION**
RT   ELECTROMAGNETIC ABSORPTION
PHOTODETACHMENT
PHOTOLYSIS

PHOTODETACHMENT
RT   PHOTODECOMPOSITION
PHOTOIONIZATION

PHOTODETECTORS
USE   PHOTOMETERS

PHOTODIODES
GS   ELECTRONIC EQUIPMENT
. DIODES
. . SEMICONDUCTOR DIODES
. . . **PHOTODIODES**
. SOLID STATE DEVICES
. . SEMICONDUCTOR DEVICES
. . . **PHOTODIODES**
RT   ION IMPLANTATION
LASERS
MERCURY CADMIUM TELLURIDES
PHOTOCONDUCTORS
PHOTOELECTRIC CELLS
PHOTOELECTRIC MATERIALS
PHOTOTRANSISTORS
PHOTOTUBES
PUSHBROOM SENSOR MODES
SIS (SEMICONDUCTORS)
SOLAR CELLS

PHOTODISSOCIATION
GS   DECOMPOSITION
. **PHOTODISSOCIATION**
DISSOCIATION
. **PHOTODISSOCIATION**
RADIATION CHEMISTRY
. **PHOTODISSOCIATION**
RT   ELECTROMAGNETIC ABSORPTION
PHOTOLYSIS

PHOTOELASTIC ANALYSIS
UF   PHOTOELASTIC STRESS MEASUREMENT
RT   FRINGE MULTIPLICATION
MOIRE EFFECTS
∞OPTICS
PHOTOGRAPHIC MEASUREMENT
∞POLARIZATION
POLARIZATION (WAVES)
STRESS ANALYSIS
STRESS MEASUREMENT
TEMPERATURE INVERSIONS

PHOTOELASTIC MATERIALS
RT   ∞MATERIALS
PHOTOELASTICITY

PHOTOELASTIC STRESS MEASUREMENT
USE   PHOTOELASTIC ANALYSIS

PHOTOELASTICITY
GS   ELECTROMAGNETIC PROPERTIES
. **PHOTOELASTICITY**
. . PHOTOVISCOELASTICITY
MECHANICAL PROPERTIES
. ELASTIC PROPERTIES
. . **PHOTOELASTICITY**
. . . PHOTOVISCOELASTICITY
RT   BIREFRINGENCE
DICHROISM
PHOTOELASTIC MATERIALS
POLARIZED LIGHT
PRISMS
REFRACTION
STRESS ANALYSIS

PHOTOELECTRIC CELLS
UF   PHOTOCELLS
GS   **PHOTOELECTRIC CELLS**
. PHOTOCONDUCTIVE CELLS
. PHOTOVOLTAIC CELLS
RT   ∞CELLS
DIRECT POWER GENERATORS
∞ELECTRIC CELLS
ELECTROCHEMICAL CELLS
ELECTRODE MATERIALS
ENERGY ABSORPTION FILMS
PHOTOCATHODES
PHOTOCONDUCTORS
PHOTODIODES
PHOTOMETERS
PHOTOMULTIPLIER TUBES
PHOTOTRANSISTORS
PHOTOTUBES
SOLAR CELLS
SOLAR GENERATORS
TRANSDUCERS

PHOTOELECTRIC EFFECT
GS   DECAY
. EMISSION
. . **PHOTOELECTRIC EFFECT**
. . . PHOTOIONIZATION
ELECTROMAGNETIC PROPERTIES
. OPTICAL PROPERTIES
. . **PHOTOELECTRIC EFFECT**
. . . PHOTOIONIZATION
RT   ∞EFFECTS
PHOTOELECTRICITY
PHOTOELECTRONS
PHOTOTHERMAL CONVERSION

PHOTOELECTRIC EMISSION
UF   PHOTOCURRENTS
PHOTOEMISSION
PHOTOEMISSIVITY
GS   DECAY
. EMISSION
. . PARTICLE EMISSION
. . . ELECTRON EMISSION
. . . . **PHOTOELECTRIC EMISSION**
ELECTROMAGNETIC PROPERTIES
. OPTICAL PROPERTIES
. . **PHOTOELECTRIC EMISSION**
RT   ELECTRICAL PROPERTIES
EXTERNAL SURFACE CURRENTS
PHOTOCATHODES
PHOTOELECTRICITY
PHOTOIONIZATION
PHOTOPEAK
STIMULATED EMISSION
WORK FUNCTIONS

PHOTOELECTRIC GENERATORS
GS   ELECTRIC GENERATORS
. DIRECT POWER GENERATORS
. . **PHOTOELECTRIC GENERATORS**
RT   ∞GENERATORS
PHOTOELECTROCHEMICAL DEVICES
SOLAR ENERGY CONVERSION
SOLAR GENERATORS
THERMOELECTRIC GENERATORS

PHOTOELECTRIC MATERIALS
UF   PHOTOEMITTERS
RT   ELECTRODE MATERIALS
ELECTRON EMISSION
∞MATERIALS

PHOTOELECTRIC MATERIALS-*(CONT.)*
PHOTOCATHODES
PHOTOCONDUCTORS
PHOTODIODES
PHOTOELECTRICITY
PHOTOELECTRONS
PHOTOTRANSISTORS
PHOTOTUBES
PHOTOVOLTAIC CELLS

PHOTOELECTRIC PHOTOMETERS
USE   ELECTROPHOTOMETERS

PHOTOELECTRICITY
UF   PHOTOELECTRONICS
PHOTOSENSORS
RT   COMPTON EFFECT
ELECTRICITY
OPTICAL PROPERTIES
PHOTOCONDUCTIVITY
PHOTOELECTRIC EFFECT
PHOTOELECTRIC EMISSION
PHOTOELECTRIC MATERIALS
PHOTOELECTRONS
PHOTOVOLTAGES
PHOTOVOLTAIC EFFECT

PHOTOELECTROCHEMICAL DEVICES
RT   ∞DEVICES
ELECTROCHEMICAL CELLS
ELECTRODE MATERIALS
ENERGY TECHNOLOGY
PHOTOELECTRIC GENERATORS
PHOTOELECTROCHEMISTRY
PHOTON BEAMS
SOLAR ENERGY CONVERSION

PHOTOELECTROCHEMISTRY
GS   ELECTROCHEMISTRY
. **PHOTOELECTROCHEMISTRY**
RT   ∞CHEMISTRY
PHOTOELECTROCHEMICAL DEVICES

PHOTOELECTROMAGNETIC DETECTORS
USE   PHOTOELECTROMAGNETIC EFFECTS
RADIATION MEASURING INSTRUMENTS

PHOTOELECTROMAGNETIC EFFECTS
UF   PHOTOELECTROMAGNETIC DETECTORS
RT   ∞EFFECTS
EXCITONS
INTERMETALLICS

PHOTOELECTRON SPECTROSCOPY
GS   SPECTROSCOPY
. **PHOTOELECTRON SPECTROSCOPY**
RT   ELECTRON EMISSION
SPECTROSCOPIC ANALYSIS

PHOTOELECTRONICS
USE   ELECTRONICS
PHOTOELECTRICITY

PHOTOELECTRONS
GS   PARTICLES
. NUCLEAR PARTICLES
. . **PHOTOELECTRONS**
RT   ELECTRON EMISSION
PHOTOELECTRIC EFFECT
PHOTOELECTRIC MATERIALS
PHOTOELECTRICITY
PHOTOIONIZATION
PHOTOMAGNETIC EFFECTS
PHOTONEUTRONS
PHOTONUCLEAR REACTIONS
PHOTOVOLTAIC EFFECT

PHOTOEMISSION
USE   PHOTOELECTRIC EMISSION

PHOTOEMISSIVITY
USE   EMISSIVITY
PHOTOELECTRIC EMISSION

PHOTOEMITTERS
USE   PHOTOELECTRIC MATERIALS

PHOTOENGRAVING
RT   PHOTOMECHANICAL EFFECT
PRINTING

PHOTOGEOLOGY
GS   GEOLOGY
. **PHOTOGEOLOGY**

PHOTOGEOLOGY-(CONT.)
RT   AERIAL PHOTOGRAPHY
       EARTH RESOURCES SURVEY AIRCRAFT
       GEOLOGICAL SURVEYS
       GEOMORPHOLOGY
       ICE MAPPING
       NATURAL GAS EXPLORATION
       PHOTOGRAMMETRY
       PHOTOMAPPING
       RECONNAISSANCE
       THEMATIC MAPPING

PHOTOGONIOMETERS
GS   MEASURING INSTRUMENTS
     . GONIOMETERS
     . . **PHOTOGONIOMETERS**
     . OPTICAL MEASURING INSTRUMENTS
     . . **PHOTOGONIOMETERS**
     OPTICAL EQUIPMENT
     . OPTICAL MEASURING INSTRUMENTS
     . . **PHOTOGONIOMETERS**
RT   ANGLES (GEOMETRY)
       DIFFRACTOMETERS
       INTERFEROMETERS
       MONOCHROMATORS
       SPECTROMETERS

PHOTOGRAMMETRY
UF   PHOTOCLINOMETRY
GS   PHOTOGRAPHIC MEASUREMENT
     . **PHOTOGRAMMETRY**
RT   AERIAL PHOTOGRAPHY
       MAPPING
       PHOTOGEOLOGY
       PHOTORECONNAISSANCE
       PROJECTORS
       RELIEF MAPS
       STEREOPHOTOGRAPHY
       SURVEYS
       TERRAIN ANALYSIS

PHOTOGRAPH INTERPRETATION
USE PHOTOINTERPRETATION

PHOTOGRAPHIC DEVELOPERS
UF   DEVELOPERS (PHOTOGRAPHY)
RT   ∞DEVELOPMENT
       PHOTOGRAPHS
       PHOTOGRAPHY

PHOTOGRAPHIC EMULSIONS
GS   MIXTURES
     . DISPERSIONS
     . . EMULSIONS
     . . . **PHOTOGRAPHIC EMULSIONS**
     . . . . NUCLEAR EMULSIONS
     . SOLUTIONS
     . . **PHOTOGRAPHIC EMULSIONS**
     . . . NUCLEAR EMULSIONS
RT   PHOTOGRAPHY
       PHOTOSENSITIVITY

PHOTOGRAPHIC EQUIPMENT
GS   **PHOTOGRAPHIC EQUIPMENT**
     . CAMERAS
     . . BAKER-NUNN CAMERA
     . . BALLISTIC CAMERAS
     . . DELFT CAMERA
     . . DIFFRACTION LIMITED CAMERAS
     . . FAINT OBJECT CAMERA
     . . HIGH SPEED CAMERAS
     . . . FRAMING CAMERAS
     . . I2S CAMERAS
     . . LALLEMAND CAMERAS
     . . MULTISPECTRAL BAND CAMERAS
     . . PANORAMIC CAMERAS
     . . PINHOLE CAMERAS
     . . SCHMIDT CAMERAS
     . . STREAK CAMERAS
     . . TELEVISION CAMERAS
     . PHOTOGRAPHIC PROCESSING
       EQUIPMENT
     . PHOTOGRAPHIC RECTIFIERS
RT   ∞EQUIPMENT
       LENSES
       OPTICAL EQUIPMENT
       OPTICAL FILTERS
       PHOTOGRAPHY
       PHOTOMETERS
       PROJECTORS

PHOTOGRAPHIC FILM
GS   **PHOTOGRAPHIC FILM**
     . MICROFILMS
RT   ∞FILMS

PHOTOGRAPHIC FILM-(CONT.)
       MAGAZINES (SUPPLY CHAMBERS)
       OPTICAL DATA STORAGE MATERIALS
       OPTICAL FILTERS
       PHOSPHORS
       PHOTOGRAPHS
       PHOTOGRAPHY
       POLYMERIC FILMS
       SABATIER REACTION

PHOTOGRAPHIC MEASUREMENT
GS   **PHOTOGRAPHIC MEASUREMENT**
     . PHOTOGRAMMETRY
RT   DOSIMETERS
     ∞MEASUREMENT
       OPTICAL CORRECTION PROCEDURE
       OPTICAL MEASUREMENT
       PHOTOELASTIC ANALYSIS
       PHOTOGRAPHY
       PHOTOINTERPRETATION
       PHOTOMETRY
       PHOTORECONNAISSANCE
       SPECTROMETERS

PHOTOGRAPHIC PLATES
RT   GLASS
       PHOTOGRAPHIC PROCESSING
       PHOTOGRAPHS
       PHOTOGRAPHY
     ∞PLATES

PHOTOGRAPHIC PROCESSING
RT   DARKROOMS
       PHOTOGRAPHIC PLATES
       PHOTOGRAPHS
       PHOTOGRAPHY
       PRINTING
     ∞PROCESSING

PHOTOGRAPHIC PROCESSING EQUIPMENT
GS   PHOTOGRAPHIC EQUIPMENT
     . **PHOTOGRAPHIC PROCESSING**
       EQUIPMENT
RT   DARKROOMS
       PHOTOGRAPHY
       REPRODUCTION (COPYING)

PHOTOGRAPHIC RECORDING
GS   RECORDING
     . **PHOTOGRAPHIC RECORDING**
RT   DATA RECORDING
       HIGH SPEED PHOTOGRAPHY
       HOLOGRAMMETRY
       INTERMITTENCY HYPOTHESIS
       PHOTOGRAPHS
       PHOTOGRAPHY
       RECORDING INSTRUMENTS

PHOTOGRAPHIC RECTIFIERS
GS   OPTICAL EQUIPMENT
     . **PHOTOGRAPHIC RECTIFIERS**
     PHOTOGRAPHIC EQUIPMENT
     . **PHOTOGRAPHIC RECTIFIERS**
RT   ∞CONDENSERS
       PHOTOGRAPHY

PHOTOGRAPHIC TRACKING
GS   TRACKING (POSITION)
     . **PHOTOGRAPHIC TRACKING**
RT   CINETHEODOLITES
       OPTICAL TRACKING
       PHOTOGRAPHY
       SATELLITE TRACKING
       SPACE DETECTION AND TRACKING
        SYSTEM

PHOTOGRAPHS
GS   **PHOTOGRAPHS**
     . CLOUD PHOTOGRAPHS
     . LUNAR PHOTOGRAPHS
     . MARS PHOTOGRAPHS
     . MICROPHOTOGRAPHS
     . MOTION PICTURES
     . PHOTOMICROGRAPHS
RT   DISPLAY DEVICES
       IMAGES
       MOSAICS
       OPTICAL CORRECTION PROCEDURE
       PHOTOGRAPHIC DEVELOPERS
       PHOTOGRAPHIC FILM
       PHOTOGRAPHIC PLATES
       PHOTOGRAPHIC PROCESSING
       PHOTOGRAPHIC RECORDING
       PHOTOGRAPHY
       REPRESENTATIONS

PHOTOGRAPHS-(CONT.)
       SPATIAL FILTERING
       VISUAL AIDS
       XEROGRAPHY

PHOTOGRAPHY
GS   **PHOTOGRAPHY**
     . AERIAL PHOTOGRAPHY
     . ALL SKY PHOTOGRAPHY
     . ASTRONOMICAL PHOTOGRAPHY
     . AUTORADIOGRAPHY
     . BLACK AND WHITE PHOTOGRAPHY
     . CHRONOPHOTOGRAPHY
     . CINEMATOGRAPHY
     . CLOUD PHOTOGRAPHY
     . COLOR PHOTOGRAPHY
     . ELECTRO-OPTICAL PHOTOGRAPHY
     . ELECTRON PHOTOGRAPHY
     . FRACTOGRAPHY
     . FRAME PHOTOGRAPHY
     . HIGH SPEED PHOTOGRAPHY
     . HOLOGRAPHY
     . . ACOUSTICAL HOLOGRAPHY
     . . MICROWAVE HOLOGRAPHY
     . . WHITE LIGHT HOLOGRAPHY
     . INFRARED IMAGERY
     . LUNAR PHOTOGRAPHY
     . METRIC PHOTOGRAPHY
     . MICROWAVE PHOTOGRAPHY
     . MULTISPECTRAL PHOTOGRAPHY
     . . INFRARED PHOTOGRAPHY
     . . . COLOR INFRARED PHOTOGRAPHY
     . . RADAR PHOTOGRAPHY
     . ORTHOPHOTOGRAPHY
     . PHOTOMICROGRAPHY
     . ROCKET-BORNE PHOTOGRAPHY
     . SHADOWGRAPH PHOTOGRAPHY
     . SCHLIEREN PHOTOGRAPHY
     . SPACEBORNE PHOTOGRAPHY
     . . SATELLITE-BORNE PHOTOGRAPHY
     . SPECTROPHOTOGRAPHY
     . STEREOSCOPY
     . . STEREOPHOTOGRAPHY
     . STREAK PHOTOGRAPHY
     . ULTRAVIOLET PHOTOGRAPHY
     . ULTRAVIOLET PHOTOMETRY
     . UNDERWATER PHOTOGRAPHY
     . UROGRAPHY
RT   BRIGHTNESS DISTRIBUTION
       BRIGHTNESS TEMPERATURE
       CAMERAS
       CLOUD PHOTOGRAPHS
       DARKROOMS
       EARTH OBSERVATIONS (FROM SPACE)
       EARTH RESOURCES
       EVAPOROGRAPHY
       EXPOSURE
       GRAPHIC ARTS
       HS-801 AIRCRAFT
       ICE MAPPING
       IMAGERY
       IMAGING TECHNIQUES
       LUNAR PHOTOGRAPHS
       MAPPING
       MARS PHOTOGRAPHS
       MICROPHOTOGRAPHS
       MULTISPECTRAL BAND CAMERAS
       MULTISPECTRAL BAND SCANNERS
       PANORAMIC CAMERAS
       PHOTOGRAPHIC DEVELOPERS
       PHOTOGRAPHIC EMULSIONS
       PHOTOGRAPHIC EQUIPMENT
       PHOTOGRAPHIC FILM
       PHOTOGRAPHIC MEASUREMENT
       PHOTOGRAPHIC PLATES
       PHOTOGRAPHIC PROCESSING
       PHOTOGRAPHIC PROCESSING
        EQUIPMENT
       PHOTOGRAPHIC RECORDING
       PHOTOGRAPHIC RECTIFIERS
       PHOTOGRAPHIC TRACKING
       PHOTOGRAPHS
       PHOTOINTERPRETATION
       PHOTOLITHOGRAPHY
       PHOTOMAPPING
       PHOTOMAPS
       PHOTOMASKS
       PHOTOMECHANICAL EFFECT
       PHOTOMICROGRAPHY
       PHOTORECONNAISSANCE
       PINHOLE CAMERAS
       PROJECTORS
       RADIOGRAPHY
       RAPID BALLISTICS IDENTIFICATION
       REPRODUCTION (COPYING)
       TIMBER INVENTORY

PHOTOGRAPHY-*(CONT.)*
    WAVE FRONT RECONSTRUCTION
    XEROGRAPHY

PHOTOINTERPRETATION
UF   PHOTOGRAPH INTERPRETATION
RT   AERIAL PHOTOGRAPHY
    ∞ANALYZING
    CHANGE DETECTION
    GROUND TRUTH
    ∞INTERPRETATION
    PHOTOGRAPHIC MEASUREMENT
    PHOTOGRAPHY
    PHOTOMAPPING
    PHOTORECONNAISSANCE
    SEA TRUTH
    SPATIAL FILTERING

PHOTOIONIZATION
GS   DECAY
    . EMISSION
    . . PHOTOELECTRIC EFFECT
    . . . **PHOTOIONIZATION**
    ELECTROMAGNETIC PROPERTIES
    . OPTICAL PROPERTIES
    . . PHOTOELECTRIC EFFECT
    . . . **PHOTOIONIZATION**
    IONIZATION
    . **PHOTOIONIZATION**
RT   ATMOSPHERIC IONIZATION
    AURORAL IONIZATION
    AURORAL IRRADIATION
    ELECTRON EMISSION
    GAS IONIZATION
    PHOTODETACHMENT
    PHOTOELECTRIC EMISSION
    PHOTOELECTRONS

PHOTOLITHOGRAPHY
GS   PRINTING
    . LITHOGRAPHY
    . . **PHOTOLITHOGRAPHY**
RT   MICROELECTRONICS
    MICROMODULES
    PHOTOGRAPHY

PHOTOLUMINESCENCE
GS   DECAY
    . EMISSION
    . . LIGHT EMISSION
    . . . LUMINESCENCE
    . . . . **PHOTOLUMINESCENCE**
    . . . . . TRIBOLUMINESCENCE
    . . . . . X RAY FLUORESCENCE
RT   FLUORESCENCE
    PHOTOLUMINESCENT BANDS

PHOTOLUMINESCENT BANDS
GS   SPECTRA
    . SPECTRAL BANDS
    . . **PHOTOLUMINESCENT BANDS**
RT   ABSORPTION SPECTRA
    ∞BANDS
    EMISSION SPECTRA
    PHOTOLUMINESCENCE
    TRIBOLUMINESCENCE

PHOTOLYSIS
GS   CHEMICAL REACTIONS
    . PHOTOCHEMICAL REACTIONS
    . . **PHOTOLYSIS**
    . . . RADIOLYSIS
    DECOMPOSITION
    . **PHOTOLYSIS**
    . . RADIOLYSIS
    RADIATION CHEMISTRY
    . **PHOTOLYSIS**
RT   CRACKING (CHEMICAL ENGINEERING)
    ELECTROLYSIS
    PHOTODECOMPOSITION
    PHOTODISSOCIATION

PHOTOMAGNETIC EFFECTS
RT   DEUTERONS
    ∞EFFECTS
    GAMMA RAYS
    PHOTOELECTRONS
    SPIN DECOUPLING

PHOTOMAPPING
GS   MAPPING
    . **PHOTOMAPPING**
RT   AERIAL PHOTOGRAPHY
    COLOR PHOTOGRAPHY
    DMSP SATELLITES

PHOTOMAPPING-*(CONT.)*
    EARTH RESOURCES
    GEODESY
    GEOLOGY
    GNOMONIC PROJECTION
    HOLOGRAMMETRY
    ICE MAPPING
    MAPS
    OCEAN COLOR SCANNER
    PHOTOGEOLOGY
    PHOTOGRAPHY
    PHOTOINTERPRETATION
    ROCKET-BORNE PHOTOGRAPHY
    SATELLITE-BORNE PHOTOGRAPHY
    SOIL MAPPING
    SPACEBORNE PHOTOGRAPHY
    THEMATIC MAPPING
    THERMAL MAPPING
    TOPOGRAPHY

PHOTOMAPS
GS   MAPS
    . **PHOTOMAPS**
RT   AERIAL PHOTOGRAPHY
    PHOTOGRAPHY
    RELIEF MAPS
    SATELLITE-BORNE PHOTOGRAPHY
    SPACEBORNE PHOTOGRAPHY
    THEMATIC MAPPING

PHOTOMASKS
RT   ARRAYS
    CIRCUIT DIAGRAMS
    INTEGRATED CIRCUITS
    MASKING
    MICROELECTRONICS
    MICROPHOTOGRAPHS
    ∞PATTERNS
    PHOTOGRAPHY
    PRINTED CIRCUITS
    SOLID STATE DEVICES
    SUBSTRATES
    WAFERS

PHOTOMECHANICAL EFFECT
RT   ∞EFFECTS
    LITHOGRAPHY
    PHOTOENGRAVING
    PHOTOGRAPHY
    PRINTING

PHOTOMETERS
UF   MICROPHOTOMETERS
    PHOTODETECTORS
GS   MEASURING INSTRUMENTS
    . OPTICAL MEASURING INSTRUMENTS
    . . **PHOTOMETERS**
    . . . ELECTROPHOTOMETERS
    . . . ULTRAVIOLET SPECTROMETERS
    . . . ULTRAVIOLET
          SPECTROPHOTOMETERS
    . RADIATION MEASURING INSTRUMENTS
    . . **PHOTOMETERS**
    . . . ELECTROPHOTOMETERS
    . . . ULTRAVIOLET SPECTROMETERS
    . . . ULTRAVIOLET
          SPECTROPHOTOMETERS
    OPTICAL EQUIPMENT
    . OPTICAL MEASURING INSTRUMENTS
    . . **PHOTOMETERS**
    . . . ELECTROPHOTOMETERS
    . . . ULTRAVIOLET SPECTROMETERS
    . . . ULTRAVIOLET
          SPECTROPHOTOMETERS
RT   BOLOMETERS
    DENSITOMETERS
    ELECTROPHOTOMETRY
    ELLIPSOMETERS
    HORIZON SCANNERS
    INFRARED SPECTROPHOTOMETERS
    MICRODENSITOMETERS
    NEPHELOMETERS
    OPTICAL MEASUREMENT
    PHOTOCONDUCTORS
    PHOTOELECTRIC CELLS
    PHOTOGRAPHIC EQUIPMENT
    PHOTOMETRY
    PHOTOTRANSISTORS
    POLARIMETERS
    PYRANOMETERS
    RADIOMETERS
    REFLECTOMETERS
    SPECTROMETERS
    SPECTROPHOTOMETERS
    TELEPHOTOMETRY

PHOTOMETERS-*(CONT.)*
    TRANSMISSOMETERS

PHOTOMETRY
GS   OPTICAL MEASUREMENT
    . **PHOTOMETRY**
    . . ELECTROPHOTOMETRY
    . . INFRARED PHOTOMETRY
    . . SPECTROPHOTOMETRY
    . . . STELLAR SPECTROPHOTOMETRY
    . . TELEPHOTOMETRY
    . . ULTRAVIOLET PHOTOMETRY
    . . VISUAL PHOTOMETRY
RT   ASTRONOMICAL PHOTOMETRY
    CHEMICAL ANALYSIS
    COLORIMETRY
    ILLUMINATING
    ∞ILLUMINATION
    LIGHT (VISIBLE RADIATION)
    LUMINANCE
    PHOTOGRAPHIC MEASUREMENT
    PHOTOMETERS
    POLARIMETRY
    REFLECTANCE
    SPECTROSCOPY
    TRANSMITTANCE

PHOTOMICROGRAPHS
GS   PHOTOGRAPHS
    . **PHOTOMICROGRAPHS**
RT   PHOTOMICROGRAPHY

PHOTOMICROGRAPHY
UF   MICROGRAPHY
GS   IMAGERY
    . **PHOTOMICROGRAPHY**
    PHOTOGRAPHY
    . **PHOTOMICROGRAPHY**
RT   BLACK AND WHITE PHOTOGRAPHY
    ELECTRON MICROSCOPES
    METALLOGRAPHY
    MICROSCOPES
    MICROSCOPY
    MICROSTRUCTURE
    PHOTOGRAPHY
    PHOTOMICROGRAPHS

PHOTOMULTIPLIER TUBES
UF   ELECTRON MULTIPLIERS
    MULTIPLIER PHOTOTUBES
GS   AMPLIFIERS
    . CURRENT AMPLIFIERS
    . . **PHOTOMULTIPLIER TUBES**
    . . . FREQUENCY MODULATION
          PHOTOMULTIPLIERS
    MICROWAVE EQUIPMENT
    . MICROWAVE TUBES
    . . COLD CATHODE TUBES
    . . . PHOTOTUBES
    . . . . **PHOTOMULTIPLIER TUBES**
RT   CATHODES
    CHANNEL MULTIPLIERS
    DYNODES
    ELECTRODES
    MICROCHANNEL PLATES
    MULTIPACTOR DISCHARGES
    MULTIPLIERS
    PHOTOCATHODES
    PHOTOELECTRIC CELLS
    SCINTILLATION COUNTERS
    SECONDARY EMISSION

PHOTON ABSORPTIOMETRY
GS   DENSITY MEASUREMENT
    . **PHOTON ABSORPTIOMETRY**
RT   ABSORPTION SPECTRA
    DENSITOMETERS
    ELECTROMAGNETIC ABSORPTION
    ENERGY ABSORPTION
    GAMMA RAY ABSORPTIOMETRY
    GAMMA RAY ABSORPTION
    MULTIPHOTON ABSORPTION
    RADIATION ABSORPTION

PHOTON BEAMS
GS   BEAMS (RADIATION)
    . **PHOTON BEAMS**
    . . LIGHT BEAMS
    ELECTROMAGNETIC RADIATION
    . **PHOTON BEAMS**
    . . LIGHT BEAMS
RT   BEAM WAVEGUIDES
    GAMMA RAY BEAMS
    INCIDENT RADIATION
    LASER OUTPUTS

PHOTON BEAMS-*(CONT.)*
    OPTICAL PATHS
    OPTICAL SCANNERS
    PHONON BEAMS
    PHOTOELECTROCHEMICAL DEVICES
    PHOTONIC PROPULSION
    PHOTONS
    REFLECTED WAVES
    REFRACTED WAVES
    THERMAL BLOOMING

PHOTON DENSITY
GS   RATES (PER TIME)
    . FLUX DENSITY
    . . **PHOTON DENSITY**

PHOTON-ELECTRON INTERACTION
RT   ELASTIC SCATTERING
    ELECTRON SCATTERING
    ELEMENTARY PARTICLE INTERACTIONS
    ∞INTERACTIONS
    UMKLAPP PROCESS

PHOTONEUTRONS
GS   NUCLEAR RADIATION
    . **PHOTONEUTRONS**
    PARTICLES
    . ELEMENTARY PARTICLES
    . . FERMIONS
    . . . NEUTRONS
    . . . . **PHOTONEUTRONS**
    . NEUTRAL PARTICLES
    . . NEUTRONS
    . . . **PHOTONEUTRONS**
RT   BARYONS
    NUCLEAR PARTICLES
    NUCLEAR REACTIONS
    PHOTOELECTRONS
    PHOTONUCLEAR REACTIONS
    VECTOR DOMINANCE MODEL

PHOTONIC PROPULSION
GS   PROPULSION
    . LOW THRUST PROPULSION
    . . **PHOTONIC PROPULSION**
    . SPACECRAFT PROPULSION
    . . **PHOTONIC PROPULSION**
RT   ELECTROMAGNETIC PROPULSION
    PHOTON BEAMS

PHOTONICS
RT   ELECTRO-OPTICS
    FIBER OPTICS
    LASERS
    LIGHT EMITTING DIODES
    OPTICAL DATA PROCESSING
    OPTICAL WAVEGUIDES
    PHOTONS

PHOTONS
GS   PARTICLES
    . ELEMENTARY PARTICLES
    . . BOSONS
    . . . **PHOTONS**
    . . . . LIGHT BEAMS
    . NUCLEAR PARTICLES
    . . BOSONS
    . . . **PHOTONS**
RT   ANNIHILATION REACTIONS
    COSMIC RAYS
    ELECTROMAGNETIC RADIATION
    GAMMA RAYS
    LIGHT (VISIBLE RADIATION)
    NUCLEAR RADIATION
    OPTICAL PROPERTIES
    PHOTON BEAMS
    PHOTONICS
    PHOTONUCLEAR REACTIONS
    PLANCKS CONSTANT
    QUANTUM THEORY
    ∞RADIATION
    ROTONS

PHOTONUCLEAR REACTIONS
GS   NUCLEAR REACTIONS
    . **PHOTONUCLEAR REACTIONS**
RT   LIGHT (VISIBLE RADIATION)
    PARTICLE INTERACTIONS
    PHOTOELECTRONS
    PHOTONEUTRONS
    PHOTONS

PHOTOOXIDATION
GS   CHEMICAL REACTIONS
    . OXIDATION

PHOTOOXIDATION-*(CONT.)*
    . . PHOTOOXIDATION
RT   ASSOCIATION REACTIONS
    PHOTOCHEMICAL OXIDANTS

PHOTOPEAK
RT   AMPLITUDE DISTRIBUTION ANALYSIS
    PHOTOELECTRIC EMISSION
    PULSE AMPLITUDE
    SCINTILLATION COUNTERS

PHOTOPHILIC PLANTS
GS   PLANTS (BOTANY)
    . **PHOTOPHILIC PLANTS**
RT   LIGHT (VISIBLE RADIATION)
    PHOTOSENSITIVITY

PHOTOPHORESIS
RT   AEROSOLS
    LIGHT (VISIBLE RADIATION)
    PARTICLE INTERACTIONS
    PARTICLE MOTION
    RADIATION PRESSURE

PHOTOPLASTICITY
GS   MECHANICAL PROPERTIES
    . PLASTIC PROPERTIES
    . . **PHOTOPLASTICITY**

PHOTOPRODUCTION
GS   DECAY
    . **PHOTOPRODUCTION**
    NUCLEAR REACTIONS
    . **PHOTOPRODUCTION**
RT   ELECTROMAGNETIC ABSORPTION
    PAIR PRODUCTION
    RADIOACTIVE DECAY
    VECTOR DOMINANCE MODEL

PHOTORECEPTORS
GS   ANATOMY
    . SENSE ORGANS
    . . **PHOTORECEPTORS**
    RECEPTORS (PHYSIOLOGY)
    . **PHOTORECEPTORS**
RT   PHOTOSENSITIVITY
    RETINA
    SENSITOMETRY
    VISUAL PIGMENTS
    YOUNG-HELMHOLTZ THEORY

PHOTORECONNAISSANCE
GS   IMAGERY
    . **PHOTORECONNAISSANCE**
    RECONNAISSANCE
    . **PHOTORECONNAISSANCE**
RT   AERIAL RECONNAISSANCE
    AIRBORNE INTEGRATED
      RECONNAISSANCE SYSTEM
    BLACK AND WHITE PHOTOGRAPHY
    DMSP SATELLITES
    EARTH RESOURCES SURVEY AIRCRAFT
    GROUND TRUTH
    HS-801 AIRCRAFT
    PHOTOGRAMMETRY
    PHOTOGRAPHIC MEASUREMENT
    PHOTOGRAPHY
    PHOTOINTERPRETATION
    SPECTRAL RECONNAISSANCE

PHOTOREDUCTION
USE PHOTOCHEMICAL REACTIONS

PHOTORESISTIVITY
USE PHOTOCONDUCTIVITY

PHOTORESISTORS
USE PHOTOCONDUCTORS

PHOTOSENSITIVITY
GS   SENSITIVITY
    . **PHOTOSENSITIVITY**
    . . LIGHT ADAPTATION
    . . PHOTOTROPISM
RT   LIGHT (VISIBLE RADIATION)
    PHOTOGRAPHIC EMULSIONS
    PHOTOPHILIC PLANTS
    PHOTORECEPTORS
    SENSITOMETRY
    THRESHOLDS (PERCEPTION)
    VISUAL PIGMENTS

PHOTOSENSORS
USE PHOTOELECTRICITY
    RADIATION MEASURING INSTRUMENTS

PHOTOSPHERE
GS   **PHOTOSPHERE**
    . SOLAR GRANULATION
RT   CHROMOSPHERE
    FACULAE
    SOLAR ATMOSPHERE
    SOLAR PHYSICS
    SPICULES
    STARSPOTS
    STELLAR ACTIVITY
    SUN
    SUNSPOTS

PHOTOSTRESSES
GS   STRESSES
    . **PHOTOSTRESSES**

PHOTOSYNTHESIS
GS   CHEMICAL REACTIONS
    . PHOTOCHEMICAL REACTIONS
    . . **PHOTOSYNTHESIS**
RT   ALGAE
    CARBOHYDRATES
    CHLOROPHYLLS
    CHLOROPLASTS
    CROP GROWTH
    RESPIRATION

PHOTOTHERMAL CONVERSION
GS   ENERGY CONVERSION
    . **PHOTOTHERMAL CONVERSION**
RT   ∞CONVERSION
    ENERGY ABSORPTION FILMS
    ENERGY CONVERSION EFFICIENCY
    ENERGY TECHNOLOGY
    METAL FILMS
    PHOTOELECTRIC EFFECT
    SELECTIVITY
    SEMICONDUCTING FILMS
    SOLAR ENERGY ABSORBERS
    SOLAR ENERGY CONVERSION
    SPECTRAL SENSITIVITY
    THERMAL ENERGY
    THERMODYNAMICS
    THIN FILMS

PHOTOTHERMOTROPISM
USE ANISOTROPY
    PHOTOTROPISM
    TEMPERATURE EFFECTS

PHOTOTRANSISTORS
GS   ELECTRONIC EQUIPMENT
    . SOLID STATE DEVICES
    . . SEMICONDUCTOR DEVICES
    . . . TRANSISTORS
    . . . . **PHOTOTRANSISTORS**
RT   JUNCTION TRANSISTORS
    PHOTOCONDUCTORS
    PHOTODIODES
    PHOTOELECTRIC CELLS
    PHOTOELECTRIC MATERIALS
    PHOTOMETERS
    PHOTOTUBES

PHOTOTROPISM
UF   PHOTOTHERMOTROPISM
GS   SENSITIVITY
    . PHOTOSENSITIVITY
    . . **PHOTOTROPISM**
RT   COLOR
    CROP VIGOR
    OPTICAL PROPERTIES
    PLANTS (BOTANY)

PHOTOTUBES
GS   MICROWAVE EQUIPMENT
    . MICROWAVE TUBES
    . . COLD CATHODE TUBES
    . . . **PHOTOTUBES**
    . . . . PHOTOMULTIPLIER TUBES
RT   CATHODES
    ELECTRODES
    FLYING SPOT SCANNERS
    GAS DISCHARGE TUBES
    PHOTODIODES
    PHOTOELECTRIC CELLS
    PHOTOELECTRIC MATERIALS
    PHOTOTRANSISTORS

PHOTOVISCOELASTICITY
- GS   ELECTROMAGNETIC PROPERTIES
- . OPTICAL PROPERTIES
- . . **PHOTOVISCOELASTICITY**
- . PHOTOELASTICITY
- . . **PHOTOVISCOELASTICITY**
- MECHANICAL PROPERTIES
- . ELASTIC PROPERTIES
- . . PHOTOELASTICITY
- . . . **PHOTOVISCOELASTICITY**
- . . VISCOELASTICITY
- . . . **PHOTOVISCOELASTICITY**

PHOTOVOLTAGES
- GS   POTENTIAL ENERGY
- . ELECTRIC POTENTIAL
- . . **PHOTOVOLTAGES**
- RT   PHOTOELECTRICITY
- PHOTOVOLTAIC EFFECT

PHOTOVOLTAIC CELLS
- GS   ELECTRONIC EQUIPMENT
- . SOLID STATE DEVICES
- . . SEMICONDUCTOR DEVICES
- . . . **PHOTOVOLTAIC CELLS**
- PHOTOELECTRIC CELLS
- . **PHOTOVOLTAIC CELLS**
- VOLTAGE GENERATORS
- . **PHOTOVOLTAIC CELLS**
- RT   ∞CELLS
- ELECTROCHEMICAL CELLS
- PHOTOCONDUCTIVE CELLS
- PHOTOELECTRIC MATERIALS
- SHORT CIRCUIT CURRENTS
- SIS (SEMICONDUCTORS)
- SOLAR CELLS
- SOLAR GENERATORS

PHOTOVOLTAIC CONVERSION
- RT   ∞CONVERSION
- ENERGY ABSORPTION FILMS

PHOTOVOLTAIC EFFECT
- GS   ELECTROMAGNETIC PROPERTIES
- . ELECTRICAL PROPERTIES
- . . **PHOTOVOLTAIC EFFECT**
- . OPTICAL PROPERTIES
- . . **PHOTOVOLTAIC EFFECT**
- RT   ∞EFFECTS
- ELECTRON EMISSION
- PHOTOELECTRICITY
- PHOTOELECTRONS
- PHOTOVOLTAGES

PHREATOPHYTES
- GS   PLANTS (BOTANY)
- . **PHREATOPHYTES**
- RT   TREES (PLANTS)

PHTHALATES
- GS   ESTERS
- . **PHTHALATES**

PHTHALOCYANIN
- GS   HETEROCYCLIC COMPOUNDS
- . **PHTHALOCYANIN**
- RT   PIGMENTS

PHUGOID OSCILLATIONS
- USE   OSCILLATIONS
- OSCILLATORS
- PITCH (INCLINATION)

PHYLLOQUINONE
- UF   VITAMIN K
- GS   HETEROCYCLIC COMPOUNDS
- . **PHYLLOQUINONE**
- VITAMINS
- . **PHYLLOQUINONE**

PHYSICAL CHEMISTRY
- GS   **PHYSICAL CHEMISTRY**
- . CRYOCHEMISTRY
- . QUANTUM CHEMISTRY
- RT   AEROTHERMOCHEMISTRY
- ATMOSPHERIC CHEMISTRY
- CHEMICAL ANALYSIS
- ∞CHEMISTRY
- COMPUTATIONAL CHEMISTRY
- NUCLEAR CHEMISTRY
- ∞PHYSICS
- THERMOCHEMISTRY
- THERMODYNAMICS

PHYSICAL CONSTANTS TESTING REACTOR
- USE   NUCLEAR RESEARCH AND TEST
- REACTORS
- WATER COOLED REACTORS

PHYSICAL ENDURANCE
- USE   PHYSICAL FITNESS

PHYSICAL EXAMINATIONS
- RT   FLIGHT FITNESS
- MOBILE QUARANTINE FACILITY
- PERCUSSION
- PERSONNEL SELECTION

PHYSICAL EXERCISE
- UF   EXERCISE
- GYMNASTICS
- GS   **PHYSICAL EXERCISE**
- . HYPERKINESIA
- RT   ANGINA PECTORIS
- ATHLETES
- ATROPHY
- FATIGUE (BIOLOGY)
- HYPOKINESIA
- PHYSICAL FITNESS
- RUNNING
- SWIMMING
- TREADMILLS
- WALKING

PHYSICAL FACTORS
- GS   PHYSIOLOGICAL FACTORS
- . **PHYSICAL FACTORS**
- RT   ∞PHYSICS
- WORK

PHYSICAL FITNESS
- UF   PHYSICAL ENDURANCE
- GS   FITNESS
- . **PHYSICAL FITNESS**
- RT   ATHLETES
- COMPETITION
- EXERCISE PHYSIOLOGY
- FLIGHT FITNESS
- PHYSICAL EXERCISE
- PHYSIOLOGICAL TESTS
- POSTURE
- SPORTS MEDICINE
- SWIMMING
- TREADMILLS
- WORK CAPACITY

PHYSICAL OPTICS
- RT   CRYSTAL OPTICS
- FIBER OPTICS
- GEOMETRICAL OPTICS
- GRADIENT INDEX OPTICS
- OPTICAL PROPERTIES
- ∞OPTICS
- QUANTUM THEORY
- ∞THEORIES

∞ **PHYSICAL PROPERTIES**
- SN   *(USE OF A MORE SPECIFIC TERM IS*
- *RECOMMENDED-CONSULT THE TERMS*
- *LISTED BELOW)*
- RT   ACOUSTIC PROPERTIES
- ADSORPTIVITY
- BRAGG ANGLE
- BUOYANCY
- CHEMICAL PROPERTIES
- COLOR
- DENSITY (MASS/VOLUME)
- DIFFUSIVITY
- DURABILITY
- EDDY CURRENTS
- ELASTIC PROPERTIES
- ELECTRICAL PROPERTIES
- ELECTROMAGNETIC PROPERTIES
- FUSIBILITY
- ∞HIGH RESISTANCE
- HYGROSCOPICITY
- HYSTERESIS
- IMPEDANCE
- INTERNAL FRICTION
- ISOTROPY
- MAGNETIC DIPOLES
- MAGNETIC PROPERTIES
- MECHANICAL PROPERTIES
- OPTICAL PROPERTIES
- PERMEABILITY
- POLYMORPHISM
- PROPELLANT PROPERTIES
- ∞PROPERTIES
- SURFACE PROPERTIES

PHYSICAL PROPERTIES-(*CONT.*)
- THERMAL EXPANSION
- THERMODYNAMIC PROPERTIES
- THIXOTROPY
- TRANSMISSIVITY
- TRANSPORT PROPERTIES
- VIRTUAL PROPERTIES
- VISCOSITY

∞ **PHYSICAL SCIENCES**
- SN   *(USE OF A MORE SPECIFIC TERM IS*
- *RECOMMENDED--CONSULT THE TERMS*
- *LISTED BELOW)*
- RT   ASTRONOMY
- ∞CHEMISTRY
- GEOLOGY
- LIFE SCIENCES
- ∞METALLURGY
- METEOROLOGY
- MINERALOGY
- OCEANOGRAPHY
- ∞PHYSICS
- ∞SCIENCE

PHYSICAL WORK
- UF   EXERTION
- GS   WORK
- . **PHYSICAL WORK**
- RT   EFFORT
- HORSEPOWER
- TASKS
- TREADMILLS
- WORK CAPACITY
- WORKLOADS (PSYCHOPHYSIOLOGY)

PHYSICIANS
- GS   PERSONNEL
- . MEDICAL PERSONNEL
- . . **PHYSICIANS**
- RT   STETHOSCOPES

∞ **PHYSICS**
- SN   *(USE OF A MORE SPECIFIC TERM IS*
- *RECOMMENDED--CONSULT THE TERMS*
- *LISTED BELOW)*
- RT   ASTROPHYSICS
- ATMOSPHERIC PHYSICS
- ATOMIC PHYSICS
- BIOPHYSICS
- BRANCHING (PHYSICS)
- CHARM (PARTICLE PHYSICS)
- CLOUD PHYSICS
- COMBUSTION
- COMBUSTION PHYSICS
- ELECTROPHYSICS
- FIELD THEORY (PHYSICS)
- GEOPHYSICS
- HEALTH PHYSICS
- HEALTH PHYSICS RESEARCH REACTOR
- KINETICS
- LOW TEMPERATURE PHYSICS
- MAGNETOMECHANICS (PHYSICS)
- MATTER (PHYSICS)
- ∞MECHANICS (PHYSICS)
- ∞MOLECULAR PHYSICS
- NEUTRON PHYSICS
- NUCLEAR PHYSICS
- NUCLEI (NUCLEAR PHYSICS)
- PHYSICAL CHEMISTRY
- PHYSICAL FACTORS
- ∞PHYSICAL SCIENCES
- PLASMA PHYSICS
- PLASMAS (PHYSICS)
- POLYMER PHYSICS
- PSYCHOPHYSICS
- QUENCHING (ATOMIC PHYSICS)
- RADIO PHYSICS
- REACTOR PHYSICS
- REENTRY PHYSICS
- RIGID ROTORS (PLASMA PHYSICS)
- ∞SCIENCE
- SELECTION RULES (NUCLEAR PHYSICS)
- SOLAR PHYSICS
- ∞SOLID STATE PHYSICS
- STRANGE ATTRACTORS
- THEORETICAL PHYSICS

PHYSICS AND CHEMISTRY EXPERIMENT IN SPACE
- UF   PACE
- GS   EXPERIMENTATION
- . **PHYSICS AND CHEMISTRY EXPERIMENT IN SPACE**
- PAYLOADS
- . SPACE SHUTTLE PAYLOADS

PHYSICS AND CHEMISTRY-*(CONT.)*
　. . SPACEBORNE EXPERIMENTS
　. . . **PHYSICS AND CHEMISTRY**
　　　　EXPERIMENT IN SPACE
RT　∞CHEMISTRY
　　SPACE FLIGHT

PHYSIOCHEMISTRY
GS　ENVIRONMENTAL CHEMISTRY
　. BIOCHEMISTRY
　. . **PHYSIOCHEMISTRY**
RT　∞CHEMISTRY
　　EXERCISE PHYSIOLOGY
　　MOLECULAR BIOLOGY
　　ORGANIC CHEMISTRY
　　PHYSIOLOGY
　　PSYCHOTROPIC DRUGS
　　∞SCIENCE

PHYSIOGRAPHY
USE　GEOMORPHOLOGY

PHYSIOLOGICAL ACCELERATION
GS　RATES (PER TIME)
　. **PHYSIOLOGICAL ACCELERATION**
RT　∞ACCELERATION
　　ACCELERATION (PHYSICS)
　　ACCELERATION STRESSES
　　　(PHYSIOLOGY)
　　DECELERATION
　　GRAVITATIONAL PHYSIOLOGY
　　IMPACT ACCELERATION

PHYSIOLOGICAL DEFENSES
RT　ANTIBODIES
　　ANTIGENS
　　BIOCOMPATIBILITY
　　∞DEFENSE
　　INOCULUM
　　INTERFERON

PHYSIOLOGICAL EFFECTS
GS　**PHYSIOLOGICAL EFFECTS**
　. PHYSIOLOGICAL RESPONSES
　. . HEMODYNAMIC RESPONSES
RT　BIOLOGICAL EFFECTS
　　BONE DEMINERALIZATION
　　CHOLERA
　　COMFORT
　　DIVING (UNDERWATER)
　　∞EFFECTS
　　ENVIRONMENTAL ENGINEERING
　　ENVIRONMENTS
　　EXERCISE PHYSIOLOGY
　　GEOTROPISM
　　GONADS
　　GRAVITATIONAL PHYSIOLOGY
　　HEAT ACCLIMATIZATION
　　HEAT STROKE
　　HEMATOPOIESIS
　　HEMATOPOIETIC SYSTEM
　　HUMAN REACTIONS
　　NOISE POLLUTION
　　RADIATION EFFECTS
　　REACTION TIME
　　RELATIVE BIOLOGICAL EFFECTIVENESS
　　　(RBE)
　　SHOCK (PHYSIOLOGY)
　　SPACE ADAPTATION SYNDROME
　　SPORTS MEDICINE

PHYSIOLOGICAL FACTORS
GS　**PHYSIOLOGICAL FACTORS**
　. PHYSICAL FACTORS
RT　ASTRONAUT PERFORMANCE
　　CHEMICAL DEFENSE
　　CHEMICAL WARFARE
　　FLIGHT STRESS (BIOLOGY)
　　NOISE POLLUTION
　　SEX FACTOR

PHYSIOLOGICAL RESPONSES
GS　PHYSIOLOGICAL EFFECTS
　. **PHYSIOLOGICAL RESPONSES**
　. . HEMODYNAMIC RESPONSES
　RESPONSES
　. **PHYSIOLOGICAL RESPONSES**
　. . HEMODYNAMIC RESPONSES
RT　DESYNCHRONIZATION (BIOLOGY)
　　EVOKED RESPONSE
　　　(PSYCHOPHYSIOLOGY)
　　GRAVITATIONAL PHYSIOLOGY
　　PATHOLOGICAL EFFECTS

PHYSIOLOGICAL TELEMETRY
USE　BIOTELEMETRY

PHYSIOLOGICAL TESTS
GS　**PHYSIOLOGICAL TESTS**
　. BODY SWAY TEST
　. CARBOXYHEMOGLOBIN TEST
　. EAR PRESSURE TEST
　. ELECTRONYSTAGMOGRAPHY
　. VESTIBULAR TESTS
　. WEBER TEST
RT　CARDIOGRAPHY
　　CERTIFICATION
　　ENVIRONMENTAL INDEX
　　ENVIRONMENTAL TESTS
　　MOBILE QUARANTINE FACILITY
　　PERSONNEL SELECTION
　　PHYSICAL FITNESS
　　PILOT SELECTION
　　PSYCHOMOTOR PERFORMANCE
　　SENSORIMOTOR PERFORMANCE
　　TAYLOR MANIFEST ANXIETY SCALE
　　∞TESTS
　　TREADMILLS
　　URINALYSIS

PHYSIOLOGY
GS　**PHYSIOLOGY**
　. AUDIOLOGY
　. BODY COMPOSITION (BIOLOGY)
　. ELECTROPHYSIOLOGY
　. EXERCISE PHYSIOLOGY
　. GRAVITATIONAL PHYSIOLOGY
　. HEMATOPOIETIC SYSTEM
　. LAMELLA
　. MENSTRUATION
　. NEUROPHYSIOLOGY
　. PSYCHOPHYSIOLOGY
　. . WORKLOADS (PSYCHOPHYSIOLOGY)
　. RESPIRATORY PHYSIOLOGY
　. UNDERWATER PHYSIOLOGY
RT　CATABOLISM
　　CHRONIC CONDITIONS
　　DIFFERENTIATION (BIOLOGY)
　　ELECTRIC STIMULI
　　EXOSKELETONS
　　FEET (ANATOMY)
　　HOMEOSTASIS
　　HUMAN BODY
　　LOCOMOTION
　　MITOSIS
　　NUTRITION
　　ORGANISMS
　　PHYSIOCHEMISTRY
　　REGENERATION (PHYSIOLOGY)
　　REGULARITY
　　RUGGEDNESS
　　∞SCIENCE
　　SEX FACTOR
　　STRESS (PHYSIOLOGY)

PHYTOTRONS
UF　GERMINATORS
　　GROWTH CHAMBERS
RT　GERMINATION
　　GREENHOUSES
　　GROWTH
　　PLANTS (BOTANY)

PI-ELECTRONS
GS　PARTICLES
　. CHARGED PARTICLES
　. . ENERGETIC PARTICLES
　. . . ELECTRONS
　. . . . **PI-ELECTRONS**
RT　MOLECULAR ELECTRONICS
　　NUCLEAR PARTICLES

PIAGGIO AIRCRAFT
GS　**PIAGGIO AIRCRAFT**
　. P-166 AIRCRAFT
　. PD-808 AIRCRAFT
RT　∞AIRCRAFT

PIAGGIO P-166 AIRCRAFT
USE　P-166 AIRCRAFT

PIAGGIO-DOUGLAS PD-808 AIRCRAFT
USE　PD-808 AIRCRAFT

PIASECKI AIRCRAFT
GS　**PIASECKI AIRCRAFT**
　. VZ-8 AIRCRAFT
RT　∞AIRCRAFT

PICKLING (METALLURGY)
GS　CHEMICAL CLEANING
　. **PICKLING (METALLURGY)**
RT　DESCALING
　　METAL FILMS
　　METAL FINISHING
　　SCALE (CORROSION)

PICKOFFS
USE　SENSORS

PICKUPS
USE　SENSORS

PICOSECOND PULSES
GS　PULSES
　. **PICOSECOND PULSES**
RT　AMPLITUDES
　　ELECTROMAGNETIC PULSES
　　LASER OUTPUTS
　　PULSE RATE
　　PULSED RADIATION
　　TIME SIGNALS

PICRATES
GS　NITROGEN COMPOUNDS
　. NITRO COMPOUNDS
　. . **PICRATES**
　. . . AMMONIUM PICRATES

PICTURE TUBES
UF　KINESCOPES
GS　ELECTRON TUBES
　. VACUUM TUBES
　. . VACUUM TUBE OSCILLATORS
　. . . CATHODE RAY TUBES
　. . . . **PICTURE TUBES**
　　OSCILLATORS
　. VACUUM TUBE OSCILLATORS
　. . CATHODE RAY TUBES
　. . . **PICTURE TUBES**
　　VIDEO EQUIPMENT
　. **PICTURE TUBES**
RT　DISPLAY DEVICES
　　FLYING SPOT SCANNERS
　　TELEVISION EQUIPMENT

PIEDMONTS
UF　PEDIMENTS
　　PEDIPLAINS
GS　LANDFORMS
　. TERRACES (LANDFORMS)
　. . PLATEAUS
　. . . **PIEDMONTS**
　. . . . CENTRAL PIEDMONT (US)
RT　COASTAL PLAINS
　　MOUNTAINS

PIERCING
UF　PUNCTURING
RT　CUTTING
　　DRILLING
　　EXTRUDING
　　FORGING
　　METAL WORKING
　　PENETRATION
　　PERFORATING
　　∞PERFORATION
　　SPARK MACHINING
　　VULNERABILITY

PIERS
USE　WHARVES

PIEZOELECTRIC CERAMICS
GS　TITANIUM COMPOUNDS
　. TITANATES
　. . **PIEZOELECTRIC CERAMICS**
　. . . LEAD ZIRCONATE TITANATES

PIEZOELECTRIC CRYSTALS
GS　CRYSTALS
　. CRYSTAL OSCILLATORS
　. . **PIEZOELECTRIC CRYSTALS**
RT　MICROSONICS
　　PIEZORESISTIVE TRANSDUCERS
　　QUARTZ TRANSDUCERS
　　SINGLE CRYSTALS

PIEZOELECTRIC GAGES
GS　MEASURING INSTRUMENTS
　. PRESSURE GAGES
　. . **PIEZOELECTRIC GAGES**
　　TRANSDUCERS

PIEZOELECTRIC GAGES-*(CONT.)*
. PIEZOELECTRIC TRANSDUCERS
. . **PIEZOELECTRIC GAGES**
. PIEZORESISTIVE TRANSDUCERS
. . **PIEZOELECTRIC GAGES**
RT PRESSURE SENSORS
STRAIN GAGES

PIEZOELECTRIC TRANSDUCERS
GS TRANSDUCERS
. **PIEZOELECTRIC TRANSDUCERS**
. . PIEZOELECTRIC GAGES
RT INTERDIGITAL TRANSDUCERS
PRAETERSONIC DEVICES
ULTRASONIC CLEANING

PIEZOELECTRICITY
GS MECHANICAL PROPERTIES
. **PIEZOELECTRICITY**
RT CRYSTAL OSCILLATORS
ELASTIC PROPERTIES
ELECTRICITY
ELECTROSTRICTION
PIEZORESISTIVE TRANSDUCERS
PYROELECTRICITY

PIEZOMETERS
GS MEASURING INSTRUMENTS
. PRESSURE GAGES
. . **PIEZOMETERS**
RT SEEPAGE

PIEZORESISTIVE TRANSDUCERS
GS TRANSDUCERS
. **PIEZORESISTIVE TRANSDUCERS**
. . PIEZOELECTRIC GAGES
RT PIEZOELECTRIC CRYSTALS
PIEZOELECTRICITY

PIGEONS
GS ANIMALS
. VERTEBRATES
. . BIRDS
. . . **PIGEONS**

PIGGYBACK SYSTEMS
RT AIR LAUNCHING
MULTISTAGE ROCKET VEHICLES
PAYLOAD MASS RATIO
PAYLOADS
∞SYSTEMS

PIGMENTS
GS **PIGMENTS**
. CAROTENE
. CHLOROPHYLLS
. CYTOCHROMES
. MELANIN
. VISUAL PIGMENTS
RT ADDITIVES
ALBINISM
ANATASE
DOPA
FILLERS
INKS
MYOGLOBIN
PAINTS
PHTHALOCYANIN
RUTILE
SKIN (ANATOMY)

PIGS (SWINE)
USE SWINE

PIKE'S PEAK (CO)
GS LANDFORMS
. PEAKS (LANDFORMS)
. . **PIKE'S PEAK (CO)**
RT COLORADO
MOUNTAINS

PILE FOUNDATIONS
GS FOUNDATIONS
. **PILE FOUNDATIONS**
RT ∞PILES

∞ PILES
SN *(USE OF A MORE SPECIFIC TERM IS
RECOMMENDED--CONSULT THE TERMS
LISTED BELOW)*
RT NUCLEAR REACTORS
PILE FOUNDATIONS

PILLOWS
RT COUCHES
CUSHIONS
SEATS

PILOCARPINE
GS HETEROCYCLIC COMPOUNDS
. ALKALOIDS
. . **PILOCARPINE**
NITROGEN COMPOUNDS
. ALKALOIDS
. . **PILOCARPINE**

PILOT ERROR
UF FLIGHT TECHNICAL ERROR
GS ERRORS
. **PILOT ERROR**
RT AIRCRAFT ACCIDENTS
COLLISIONS
CRASH LANDING
CRASHES
HUMAN FACTORS ENGINEERING
HUMAN PERFORMANCE
MIDAIR COLLISIONS
PILOT INDUCED OSCILLATION

PILOT INDUCED OSCILLATION
RT AERODYNAMIC STABILITY
AIRCRAFT CONTROL
AIRCRAFT STABILITY
CONTROL STABILITY
HIGH GAIN
LONGITUDINAL CONTROL
MAN MACHINE SYSTEMS
NONSTABILIZED OSCILLATION
PILOT ERROR
PILOT PERFORMANCE
SELF INDUCED VIBRATION
STABLE OSCILLATIONS
TRANSIENT OSCILLATIONS

PILOT LANDING AID TELEVISION SYSTEM
USE PLAT SYSTEM

PILOT PERFORMANCE
GS HUMAN PERFORMANCE
. **PILOT PERFORMANCE**
RT AIRCRAFT PERFORMANCE
ASTRONAUT PERFORMANCE
INTRAVEHICULAR ACTIVITY
MAN OPERATED PROPULSION SYSTEMS
OPERATOR PERFORMANCE
∞PERFORMANCE
PILOT INDUCED OSCILLATION
PSYCHOMOTOR PERFORMANCE
SENSORIMOTOR PERFORMANCE

PILOT PLANTS
RT INDUSTRIAL PLANTS
MODELS
PRODUCT DEVELOPMENT
PROTOTYPES

PILOT SELECTION
GS SELECTION
. PERSONNEL SELECTION
. . **PILOT SELECTION**
RT PHYSIOLOGICAL TESTS
PSYCHOLOGICAL TESTS

PILOT TRAINING
GS EDUCATION
. **PILOT TRAINING**
RT ASTRONAUT TRAINING
AVIATION PSYCHOLOGY
EJECTION INJURIES
EJECTION TRAINING
FLIGHT SIMULATORS
FLIGHT TRAINING
SPACE FLIGHT TRAINING
TRAINING SIMULATORS

PILOTED CENTRIFUGES
USE HUMAN CENTRIFUGES

PILOTLESS AIRCRAFT
SN (CONVENTIONAL AIRCRAFT CONVERTED
FOR REMOTE CONTROL)
GS **PILOTLESS AIRCRAFT**
. DRONE AIRCRAFT
. . TARGET DRONE AIRCRAFT
. . . FIREBEE 2 TARGET DRONE
AIRCRAFT
. . . JINDIVIK TARGET AIRCRAFT

PILOTLESS AIRCRAFT-*(CONT.)*
RT ∞AIRCRAFT
BALLOONS
DRONE VEHICLES
LIGHT AIRCRAFT
∞MILITARY AIRCRAFT
OBLIQUE WINGS
RECONNAISSANCE AIRCRAFT
REMOTELY PILOTED VEHICLES

∞ PILOTS
SN *(USE OF A MORE SPECIFIC TERM IS
RECOMMENDED--CONSULT THE TERMS
LISTED BELOW)*
RT AIRCRAFT PILOTS
AUTOMATIC PILOTS
PILOTS (PERSONNEL)
TEST PILOTS

PILOTS (PERSONNEL)
GS PERSONNEL
. FLYING PERSONNEL
. . **PILOTS (PERSONNEL)**
. . . AIRCRAFT PILOTS
. . . . TEST PILOTS
. OPERATORS (PERSONNEL)
. . **PILOTS (PERSONNEL)**
. . . AIRCRAFT PILOTS
. . . . TEST PILOTS
RT ASTRONAUTS
COSMONAUTS
CREWS
FLIGHT CREWS
∞PILOTS

PINCH EFFECT
GS **PINCH EFFECT**
. PLASMA PINCH
. . SCREW PINCH
. . THETA PINCH
. . ZETA PINCH
. REVERSE FIELD PINCH
RT CYLINDRICAL PLASMAS
∞EFFECTS
MAGNETIC FIELDS
MAGNETOHYDRODYNAMICS
PLASMA COMPRESSION
PLASMA CONTROL
RELATIVISTIC PLASMAS
STELLARATORS
THERMONUCLEAR POWER GENERATION
THERMONUCLEAR REACTIONS
ZETA THERMONUCLEAR REACTOR

PINEAL GLAND
GS ANATOMY
. GLANDS (ANATOMY)
. . ENDOCRINE GLANDS
. . . **PINEAL GLAND**
VISCERA
. ENDOCRINE GLANDS
. . **PINEAL GLAND**
RT BRAIN

PINHOLE CAMERAS
GS OPTICAL EQUIPMENT
. CAMERAS
. . **PINHOLE CAMERAS**
PHOTOGRAPHIC EQUIPMENT
. CAMERAS
. . **PINHOLE CAMERAS**
RT APERTURES
PHOTOGRAPHY
PINHOLES

PINHOLES
RT CASTING
CASTINGS
DEFECTS
INTERSTICES
LEAKAGE
PINHOLE CAMERAS
POROSITY

PINNACLES
USE PEAKS (LANDFORMS)

PINNING
SN (LIMITED TO ELECTRONICS)
GS **PINNING**
. FLUX PINNING
RT CRYSTAL DEFECTS
CRYSTAL DISLOCATIONS
CURRENT DENSITY
MAGNETIC FLUX

PINNING-*(CONT.)*
    SUPERCONDUCTORS

PINS
  GS   FASTENERS
     . **PINS**
  RT   COUPLINGS
      HOLDERS
      LATCHES
      RIVETS
    ∞SPIKES
      STUDS (STRUCTURAL MEMBERS)

PINTLES
  RT   PIVOTS
      RUDDERS
      SHAFTS (MACHINE ELEMENTS)

PION BEAMS
  GS   BEAMS (RADIATION)
     . PARTICLE BEAMS
     . . **PION BEAMS**
  RT   NEUTRAL BEAMS
      NEUTRON BEAMS

PIONEER F SPACE PROBE
  USE   PIONEER 10 SPACE PROBE

PIONEER G SPACE PROBE
  USE   PIONEER 11 SPACE PROBE

PIONEER PROJECT
  GS   PROGRAMS
     . NASA PROGRAMS
     . . **PIONEER PROJECT**
     . NASA SPACE PROGRAMS
     . . **PIONEER PROJECT**
     . PROJECTS
     . . **PIONEER PROJECT**
  RT   LUNAR PROBES
      PIONEER SPACE PROBES
      SPACE PROBES

PIONEER SATURN SPACECRAFT
  USE   PIONEER 11 SPACE PROBE

PIONEER SPACE PROBES
  GS   INTERPLANETARY SPACECRAFT
     . **PIONEER SPACE PROBES**
     . . PIONEER VENUS 2 ENTRY PROBES
     . . . PIONEER VENUS 2 NIGHT PROBE
     . . . PIONEER VENUS 2 SOUNDER
          PROBE
     . . PIONEER 1 SPACE PROBE
     . . PIONEER 2 SPACE PROBE
     . . PIONEER 3 SPACE PROBE
     . . PIONEER 4 SPACE PROBE
     . . PIONEER 5 SPACE PROBE
     . . PIONEER 6 SPACE PROBE
     . . PIONEER 7 SPACE PROBE
     . . PIONEER 8 SPACE PROBE
     . . PIONEER 9 SPACE PROBE
     . . PIONEER 10 SPACE PROBE
     UNMANNED SPACECRAFT
     . SPACE PROBES
     . . **PIONEER SPACE PROBES**
     . . . PIONEER VENUS 2 ENTRY PROBES
     . . . . PIONEER VENUS 2 NIGHT PROBE
     . . . PIONEER 1 SPACE PROBE
     . . . PIONEER 2 SPACE PROBE
     . . . PIONEER 3 SPACE PROBE
     . . . PIONEER 4 SPACE PROBE
     . . . PIONEER 5 SPACE PROBE
     . . . PIONEER 6 SPACE PROBE
     . . . PIONEER 7 SPACE PROBE
     . . . PIONEER 8 SPACE PROBE
     . . . PIONEER 9 SPACE PROBE
     . . . PIONEER 10 SPACE PROBE
     . . . PIONEER 11 SPACE PROBE
  RT   JUNO 2 LAUNCH VEHICLE
      PIONEER PROJECT
      PIONEER VENUS SPACECRAFT
      PIONEER VENUS 1 SPACECRAFT
      PIONEER VENUS 2 SPACECRAFT
      SOLAR PROBES

PIONEER VENUS SPACECRAFT
  UF   PIONEER 12 SPACE PROBE
  GS   INTERPLANETARY SPACECRAFT
     . **PIONEER VENUS SPACECRAFT**
     . . PIONEER VENUS 1 SPACECRAFT
     . . PIONEER VENUS 2 SPACECRAFT
     . . . PIONEER VENUS 2 TRANSPORTER
          BUS

PIONEER VENUS SPACECRAFT-*(CONT.)*
    UNMANNED SPACECRAFT
    . **PIONEER VENUS SPACECRAFT**
    . . PIONEER VENUS 1 SPACECRAFT
    . . PIONEER VENUS 2 SPACECRAFT
    . . . PIONEER VENUS 2 TRANSPORTER
        BUS
  RT   PIONEER SPACE PROBES
    ∞PROBES
      SPACE PROBES

PIONEER VENUS 1 SPACECRAFT
  GS   INTERPLANETARY SPACECRAFT
     . PIONEER VENUS SPACECRAFT
     . . **PIONEER VENUS 1 SPACECRAFT**
     UNMANNED SPACECRAFT
     . PIONEER VENUS SPACECRAFT
     . . **PIONEER VENUS 1 SPACECRAFT**
  RT   PIONEER SPACE PROBES
    ∞PROBES
      SPACE PROBES

PIONEER VENUS 2 ENTRY PROBES
  GS   INTERPLANETARY SPACECRAFT
     . PIONEER SPACE PROBES
     . . **PIONEER VENUS 2 ENTRY PROBES**
     . . . PIONEER VENUS 2 NIGHT PROBE
     . . . PIONEER VENUS 2 SOUNDER
         PROBE
     UNMANNED SPACECRAFT
     . SPACE PROBES
     . . PIONEER SPACE PROBES
     . . . **PIONEER VENUS 2 ENTRY PROBES**
     . . . . PIONEER VENUS 2 NIGHT PROBE
  RT   ∞PROBES

PIONEER VENUS 2 MULTIPROBE SPACECRAFT
  USE   PIONEER VENUS 2 SPACECRAFT

PIONEER VENUS 2 NIGHT PROBE
  GS   INTERPLANETARY SPACECRAFT
     . PIONEER SPACE PROBES
     . . PIONEER VENUS 2 ENTRY PROBES
     . . . **PIONEER VENUS 2 NIGHT PROBE**
     UNMANNED SPACECRAFT
     . SPACE PROBES
     . . PIONEER SPACE PROBES
     . . . PIONEER VENUS 2 ENTRY PROBES
     . . . . **PIONEER VENUS 2 NIGHT PROBE**
  RT   ∞PROBES

PIONEER VENUS 2 SOUNDER PROBE
  GS   INTERPLANETARY SPACECRAFT
     . PIONEER SPACE PROBES
     . . PIONEER VENUS 2 ENTRY PROBES
     . . . **PIONEER VENUS 2 SOUNDER**
         PROBE

PIONEER VENUS 2 SPACECRAFT
  UF   PIONEER VENUS 2 MULTIPROBE
      SPACECRAFT
  GS   INTERPLANETARY SPACECRAFT
     . PIONEER VENUS SPACECRAFT
     . . **PIONEER VENUS 2 SPACECRAFT**
     . . . PIONEER VENUS 2 TRANSPORTER
         BUS
     . VENUS PROBES
     . . **PIONEER VENUS 2 SPACECRAFT**
     . . . PIONEER VENUS 2 TRANSPORTER
         BUS
     UNMANNED SPACECRAFT
     . PIONEER VENUS SPACECRAFT
     . . **PIONEER VENUS 2 SPACECRAFT**
     . . . PIONEER VENUS 2 TRANSPORTER
         BUS
  RT   PIONEER SPACE PROBES
    ∞PROBES
    ∞SPACECRAFT

PIONEER VENUS 2 TRANSPORTER BUS
  GS   INTERPLANETARY SPACECRAFT
     . PIONEER VENUS SPACECRAFT
     . . PIONEER VENUS 2 SPACECRAFT
     . . . **PIONEER VENUS 2 TRANSPORTER**
         BUS
     . VENUS PROBES
     . . PIONEER VENUS 2 SPACECRAFT
     . . . **PIONEER VENUS 2 TRANSPORTER**
         BUS
     UNMANNED SPACECRAFT
     . PIONEER VENUS SPACECRAFT
     . . PIONEER VENUS 2 SPACECRAFT
     . . . **PIONEER VENUS 2 TRANSPORTER**
         BUS
  RT   ∞PROBES

PIONEER 1 SPACE PROBE
  GS   INTERPLANETARY SPACECRAFT
     . PIONEER SPACE PROBES
     . . **PIONEER 1 SPACE PROBE**
     UNMANNED SPACECRAFT
     . SPACE PROBES
     . . PIONEER SPACE PROBES
     . . . **PIONEER 1 SPACE PROBE**
  RT   THOR ABLE ROCKET VEHICLE

PIONEER 2 SPACE PROBE
  GS   INTERPLANETARY SPACECRAFT
     . PIONEER SPACE PROBES
     . . **PIONEER 2 SPACE PROBE**
     UNMANNED SPACECRAFT
     . SPACE PROBES
     . . PIONEER SPACE PROBES
     . . . **PIONEER 2 SPACE PROBE**

PIONEER 3 SPACE PROBE
  GS   INTERPLANETARY SPACECRAFT
     . PIONEER SPACE PROBES
     . . **PIONEER 3 SPACE PROBE**
     UNMANNED SPACECRAFT
     . SPACE PROBES
     . . PIONEER SPACE PROBES
     . . . **PIONEER 3 SPACE PROBE**
  RT   JUNO 2 LAUNCH VEHICLE

PIONEER 4 LUNAR PROBE
  USE   PIONEER 4 SPACE PROBE

PIONEER 4 SPACE PROBE
  UF   PIONEER 4 LUNAR PROBE
  GS   INTERPLANETARY SPACECRAFT
     . PIONEER SPACE PROBES
     . . **PIONEER 4 SPACE PROBE**
     UNMANNED SPACECRAFT
     . SPACE PROBES
     . . PIONEER SPACE PROBES
     . . . **PIONEER 4 SPACE PROBE**
  RT   JUNO 2 LAUNCH VEHICLE

PIONEER 5 SPACE PROBE
  GS   INTERPLANETARY SPACECRAFT
     . PIONEER SPACE PROBES
     . . **PIONEER 5 SPACE PROBE**
     UNMANNED SPACECRAFT
     . SPACE PROBES
     . . PIONEER SPACE PROBES
     . . . **PIONEER 5 SPACE PROBE**
  RT   THOR ABLE ROCKET VEHICLE

PIONEER 6 SPACE PROBE
  GS   INTERPLANETARY SPACECRAFT
     . PIONEER SPACE PROBES
     . . **PIONEER 6 SPACE PROBE**
     UNMANNED SPACECRAFT
     . SPACE PROBES
     . . PIONEER SPACE PROBES
     . . . **PIONEER 6 SPACE PROBE**
  RT   DELTA LAUNCH VEHICLE
      JUNO 2 LAUNCH VEHICLE

PIONEER 7 SPACE PROBE
  GS   INTERPLANETARY SPACECRAFT
     . PIONEER SPACE PROBES
     . . **PIONEER 7 SPACE PROBE**
     UNMANNED SPACECRAFT
     . SPACE PROBES
     . . PIONEER SPACE PROBES
     . . . **PIONEER 7 SPACE PROBE**
  RT   DELTA LAUNCH VEHICLE
      JUNO 2 LAUNCH VEHICLE

PIONEER 8 SPACE PROBE
  GS   INTERPLANETARY SPACECRAFT
     . PIONEER SPACE PROBES
     . . **PIONEER 8 SPACE PROBE**
     UNMANNED SPACECRAFT
     . SPACE PROBES
     . . PIONEER SPACE PROBES
     . . . **PIONEER 8 SPACE PROBE**
  RT   JUNO 2 LAUNCH VEHICLE
    ∞PROBES

PIONEER 9 SPACE PROBE
  GS   INTERPLANETARY SPACECRAFT
     . PIONEER SPACE PROBES
     . . **PIONEER 9 SPACE PROBE**
     UNMANNED SPACECRAFT
     . SPACE PROBES
     . . PIONEER SPACE PROBES
     . . . **PIONEER 9 SPACE PROBE**

PIONEER 9 SPACE PROBE-(CONT.)
RT ∞PROBES

PIONEER 10 SPACE PROBE
UF   PIONEER F SPACE PROBE
GS   INTERPLANETARY SPACECRAFT
. PIONEER SPACE PROBES
. . **PIONEER 10 SPACE PROBE**
UNMANNED SPACECRAFT
. SPACE PROBES
. . PIONEER SPACE PROBES
. . . **PIONEER 10 SPACE PROBE**
RT   ∞PROBES

PIONEER 11 SPACE PROBE
UF   PIONEER G SPACE PROBE
PIONEER SATURN SPACECRAFT
GS   UNMANNED SPACECRAFT
. SPACE PROBES
. . PIONEER SPACE PROBES
. . . **PIONEER 11 SPACE PROBE**
RT   ∞PROBES

PIONEER 12 SPACE PROBE
USE   PIONEER VENUS SPACECRAFT

PIONS
GS   PARTICLES
. ELEMENTARY PARTICLES
. . BOSONS
. . . MESONS
. . . . **PIONS**
. NUCLEAR PARTICLES
. . BOSONS
. . . MESONS
. . . . **PIONS**
RT   BARYONS
CHARGED PARTICLES
KAONS

PIPE FLOW
UF   KIRCHHOFF-HELMHOLTZ FLOW
GS   FLUID FLOW
. GAS FLOW
. . **PIPE FLOW**
. PARALLEL FLOW
. . **PIPE FLOW**
RT   CHANNEL FLOW
CRITICAL FLOW
LAMINAR FLOW
LIQUID FLOW
MASS FLOW
MULTIPHASE FLOW
OPEN CHANNEL FLOW
ORIFICE FLOW
PIPES (TUBES)
PRESSURE GRADIENTS
SINGLE-PHASE FLOW
STEADY FLOW
STEAM FLOW
SUBCRITICAL FLOW
SUPERCRITICAL FLOW
TURBULENT FLOW
UNIFORM FLOW
UNSTEADY FLOW
WATER FLOW
WATER HAMMER
WATER PRESSURE

PIPE NOZZLES
RT   EXHAUST SYSTEMS
INLET NOZZLES
INTAKE SYSTEMS
NOZZLE GEOMETRY
∞NOZZLES
OPENINGS
OUTLETS
TANKS (CONTAINERS)

PIPELINES
GS   **PIPELINES**
. SEWERS
RT   CROSSINGS
∞LINES
MATERIALS HANDLING
PIPES (TUBES)
PUMPS
SIPHONS
STEAM FLOW
∞STORAGE
STORAGE TANKS
TRANSPORTATION
WASTE DISPOSAL
WATER HAMMER

PIPELINING (COMPUTERS)
GS   DATA PROCESSING
. MULTIPROCESSING (COMPUTERS)
. . **PIPELINING (COMPUTERS)**
RT   ASSOCIATIVE PROCESSING
(COMPUTERS)
DATA PROCESSING EQUIPMENT
MULTIPROGRAMMING
PARALLEL PROGRAMMING
TIME SHARING

PIPER AIRCRAFT
GS   LIGHT AIRCRAFT
. **PIPER AIRCRAFT**
. . PA-34 SENECA AIRCRAFT
RT   ∞AIRCRAFT
GENERAL AVIATION AIRCRAFT

PIPERIDINE
GS   HETEROCYCLIC COMPOUNDS
. **PIPERIDINE**
PYRIDINES
. **PIPERIDINE**

PIPES (TUBES)
UF   TUBING
GS   **PIPES (TUBES)**
. GAS PIPES
. U BENDS
RT   ∞CASING
CIRCULAR TUBES
DUCTS
FLUID FLOW
∞HEADERS
HOSES
∞HYDRAULICS
MANIFOLDS
PIPE FLOW
PIPELINES
RISERS
SIPHONS
SYRINGES
∞TUBES

PIPETTES
RT   BURETTES
GLASSWARE
LABORATORY EQUIPMENT

PIRANI GAGES
GS   MEASURING INSTRUMENTS
. PRESSURE GAGES
. . VACUUM GAGES
. . . **PIRANI GAGES**
VACUUM APPARATUS
. VACUUM GAGES
. . **PIRANI GAGES**
RT   HOT-WIRE FLOWMETERS
IONIZATION GAGES
KNUDSEN GAGES
MCLEOD GAGES
PRESSURE MEASUREMENT

PISTON ENGINES
UF   RECIPROCATING ENGINES
GS   ENGINES
. **PISTON ENGINES**
. . DIESEL ENGINES
RT   AIRCRAFT ENGINES
AUTOMOBILE ENGINES
EXTERNAL COMBUSTION ENGINES
FUEL INJECTION
INTERNAL COMBUSTION ENGINES
PISTONS
RECIPROCATION
ROTARY ENGINES
WANKEL ENGINES

PISTON THEORY
RT   COMPRESSING
FLUID DYNAMICS
PISTONS
∞THEORIES

PISTONS
GS   **PISTONS**
. MAGNETIC PISTONS
RT   COMBUSTION CHAMBERS
ENGINE PARTS
INTERNAL COMBUSTION ENGINES
PISTON ENGINES
PISTON THEORY
PLUNGERS
RECIPROCATION

∞ **PITCH**
SN   *(USE OF A MORE SPECIFIC TERM IS
RECOMMENDED--CONSULT THE TERMS
LISTED BELOW)*
UF   TONE
RT   FREQUENCIES
PITCH (INCLINATION)
PITCH (MATERIAL)

PITCH (INCLINATION)
UF   DAMPING IN PITCH
PHUGOID OSCILLATIONS
PITCH ANGLES
GS   ATTITUDE (INCLINATION)
. **PITCH (INCLINATION)**
RT   ANGLES (GEOMETRY)
HEAVING
LONGITUDINAL CONTROL
LONGITUDINAL STABILITY
∞MOTION
∞PITCH
ROLL
ROTATION
SLOPES
STABILITY AUGMENTATION
VARIABLE PITCH PROPELLERS
YAW

PITCH (MATERIAL)
RT   ASPHALT
OILS
∞PITCH
TARS

PITCH ANGLES
USE   PITCH (INCLINATION)

PITCH ATTITUDE CONTROL
USE   LONGITUDINAL CONTROL

PITCHING MOMENTS
GS   MOMENTS
. STABILITY DERIVATIVES
. . **PITCHING MOMENTS**
RT   AERODYNAMIC COEFFICIENTS
LONGITUDINAL STABILITY
MOMENTS OF INERTIA
ROLLING MOMENTS
TORQUE
YAWING MOMENTS

PITOT TUBES
UF   PRESTON TUBES
RT   FLOW MEASUREMENT
FLOWMETERS
PRESSURE MEASUREMENT
PROTUBERANCES
SPEED INDICATORS
STATIC PRESSURE
∞TUBES
VELOCITY MEASUREMENT
VENTURI TUBES

∞ **PITS**
SN   *(USE OF A MORE SPECIFIC TERM IS
RECOMMENDED--CONSULT THE TERMS
LISTED BELOW)*
RT   PITS (EXCAVATIONS)
PITTING

PITS (EXCAVATIONS)
RT   BOREHOLES
EXCAVATION
MINES (EXCAVATIONS)
∞PITS
SUMPS

PITTING
RT   CHEMICAL ATTACK
CHIPPING
CORROSION
CORROSION RESISTANCE
CORROSION TESTS
DEGRADATION
EROSION
EROSIVE BURNING
ETCHING
HOT CORROSION
METAL-WATER REACTIONS
∞PITS
SCORING

PITUITARY GLAND
GS   ANATOMY

PITUITARY GLAND-(CONT.)
. GLANDS (ANATOMY)
. . ENDOCRINE GLANDS
. . . **PITUITARY GLAND**
. ORGANS
. . **PITUITARY GLAND**
VISCERA
. ENDOCRINE GLANDS
. . **PITUITARY GLAND**
. ORGANS
. . **PITUITARY GLAND**

PITUITARY HORMONES
GS   SECRETIONS
. ENDOCRINE SECRETIONS
. . HORMONES
. . . **PITUITARY HORMONES**
. . . . ADRENOCORTICOTROPIN (ACTH)

PIVOTED WING AIRCRAFT
USE   TILT WING AIRCRAFT

PIVOTS
UF   TROCHOIDS
RT   BEARINGS
GIMBALS
HINGES
PINTLES
SHAFTS (MACHINE ELEMENTS)
SUPPORTS
SWIVELS

PIX
USE   PLASMA INTERACTION EXPERIMENT

PL/1
GS   LANGUAGES
. PROGRAMMING LANGUAGES
. . **PL/1**
RT   COBOL
COMPILERS
COMPUTER PROGRAMMING
FORTRAN
MACHINE ORIENTED LANGUAGES

PLAGES (FACULAE)
USE   FACULAE

PLAINS
GS   LAND
. **PLAINS**
. . COASTAL PLAINS
. . FLOOD PLAINS
. . LLANOS ORIENTALES (COLOMBIA)
. . PAMPAS
. . PENEPLAINS
. . PLAYAS
. . TUNDRA
RT   FARMLANDS
FLATS (LANDFORMS)
GEOGRAPHY
GRASSLANDS
GREAT PLAINS CORRIDOR (NORTH
AMERICA)
LANDFORMS
PLATEAUS
STEPPES
TOPOGRAPHY
WILDERNESS

PLAN POSITION INDICATORS
UF   PPI (POSITION INDICATORS)
GS   AIRCRAFT INSTRUMENTS
. POSITION INDICATORS
. . **PLAN POSITION INDICATORS**
DISPLAY DEVICES
. POSITION INDICATORS
. . **PLAN POSITION INDICATORS**
. RADARSCOPES
. . **PLAN POSITION INDICATORS**
MEASURING INSTRUMENTS
. INDICATING INSTRUMENTS
. . POSITION INDICATORS
. . . **PLAN POSITION INDICATORS**
RADAR EQUIPMENT
. RADARSCOPES
. . **PLAN POSITION INDICATORS**

PLANAR STRUCTURES
RT   FLAT LAYERS
FLAT PLATES
FLAT SURFACES
FLATNESS
∞ STRUCTURES

PLANAR STRUCTURES-(CONT.)
SURFACE PROPERTIES

PLANCKS CONSTANT
GS   CONSTANTS
. **PLANCKS CONSTANT**
RT   BLACK BODY RADIATION
DE BROGLIE WAVELENGTHS
ELECTROMAGNETIC RADIATION
NUCLEAR MAGNETIC RESONANCE
PHOTONS
QUANTUM THEORY
THERMAL RADIATION
WENTZEL-KRAMER-BRILLOUIN METHOD

PLANE STRAIN
RT   CRACK PROPAGATION
ELASTIC DEFORMATION
FRACTURE MECHANICS
PLASTIC DEFORMATION
STRESS INTENSITY FACTORS
STRESS-STRAIN RELATIONSHIPS

PLANE WAVES
GS   LONGITUDINAL WAVES
. **PLANE WAVES**
RT   BEAMS (RADIATION)
CYLINDRICAL WAVES
ELASTIC WAVES
NORMAL SHOCK WAVES
∞ RADIATION
SHOCK WAVES
SOLITARY WAVES
SOUND WAVES
SPATIAL FILTERING
SPHERICAL WAVES
TRANSVERSE WAVES
TRAVELING WAVES
∞ WAVES

PLANET EPHEMERIDES
GS   EPHEMERIDES
. **PLANET EPHEMERIDES**
RT   GEOCENTRIC COORDINATES
PLANETS

PLANET ORIGINS
USE   PLANETARY EVOLUTION

PLANETARIUMS
RT   ASTRONOMICAL MODELS
DISPLAY DEVICES

PLANETARY ATMOSPHERES
SN   (EXCLUDES EARTH ATMOSPHERE)
GS   ENVIRONMENTS
. EXTRATERRESTRIAL ENVIRONMENTS
. . PLANETARY ENVIRONMENTS
. . . **PLANETARY ATMOSPHERES**
. . . . HELIUM HYDROGEN
ATMOSPHERES
. . . . JUPITER ATMOSPHERE
. . . . MARS ATMOSPHERE
. . . . NEPTUNE ATMOSPHERE
. . . . SATURN ATMOSPHERE
. . . . URANUS ATMOSPHERE
. . . . VENUS ATMOSPHERE
RT   ∞ ABSORPTION
∞ ATMOSPHERES
ATMOSPHERIC ATTENUATION
ATMOSPHERIC COMPOSITION
ATMOSPHERIC DENSITY
ATMOSPHERIC TEMPERATURE
EARTH ATMOSPHERE
IONOPAUSE
LUNAR ATMOSPHERE
NONGRAY ATMOSPHERES
ORGANIC SOLIDS
PLANETARY RINGS
PRIMITIVE EARTH ATMOSPHERE
RADIATIVE TRANSFER
RADIO OCCULTATION
SATELLITE ATMOSPHERES
SATURN RINGS
SOLAR PLANETARY INTERACTIONS

PLANETARY BASES
RT   EXTRATERRESTRIAL RESOURCES
SPACE EXPLORATION
STATIONS

PLANETARY BOUNDARY LAYER
GS   BOUNDARY LAYERS
. **PLANETARY BOUNDARY LAYER**

PLANETARY BOUNDARY LAYER-(CONT.)
RT   ATMOSPHERIC BOUNDARY LAYER

PLANETARY COMPOSITION
GS   COMPOSITION (PROPERTY)
. **PLANETARY COMPOSITION**
RT   EARTH PLANETARY STRUCTURE
GAS GIANT PLANETS
JUPITER RINGS
SATURN RINGS
SPACE EXPLORATION
STRUCTURAL PROPERTIES (GEOLOGY)

PLANETARY CORES
GS   CORES
. **PLANETARY CORES**
. . EARTH CORE
RT   PLANETS
STELLAR CORES

PLANETARY CRATERS
GS   CRATERS
. **PLANETARY CRATERS**
. . MARS CRATERS
RT   EARTH (PLANET)
IMPACT DAMAGE
MARS (PLANET)
MARS SURFACE
MERCURY (PLANET)
METEORITE CRATERS
PLANETS
VENUS (PLANET)
VENUS SURFACE

PLANETARY ENTRY
USE   ATMOSPHERIC ENTRY

PLANETARY ENVIRONMENTS
SN   (EXCLUDES EARTH)
GS   ENVIRONMENTS
. EXTRATERRESTRIAL ENVIRONMENTS
. . **PLANETARY ENVIRONMENTS**
. . . MARS ENVIRONMENT
. . . . MARS ATMOSPHERE
. . . PLANETARY ATMOSPHERES
. . . . HELIUM HYDROGEN
ATMOSPHERES
. . . . JUPITER ATMOSPHERE
. . . . MARS ATMOSPHERE
. . . . NEPTUNE ATMOSPHERE
. . . . SATURN ATMOSPHERE
. . . . URANUS ATMOSPHERE
. . . . VENUS ATMOSPHERE
RT   AEROSPACE ENVIRONMENTS
BIOASTRONAUTICS
EXOBIOLOGY
LIFE SUPPORT SYSTEMS
LONG DURATION SPACE FLIGHT
LUNAR ENVIRONMENT
PLANETS
PROTOPLANETS
TERRESTRIAL PLANETS
THERMAL ENVIRONMENTS

PLANETARY EVOLUTION
UF   PLANET ORIGINS
GS   EVOLUTION (DEVELOPMENT)
. **PLANETARY EVOLUTION**
RT   COSMOLOGY
PROTOPLANETS
STELLAR EVOLUTION

PLANETARY EXPLORATION
USE   SPACE EXPLORATION

PLANETARY EXPLORER
USE   OUTER PLANETS EXPLORERS

PLANETARY GEOLOGY
GS   **PLANETARY GEOLOGY**
. MARS VOLCANOES

PLANETARY GRAVITATION
GS   GRAVITATION
. **PLANETARY GRAVITATION**
RT   ESCAPE VELOCITY
LUNAR GRAVITATION

PLANETARY LANDING
SN   (EXCLUDES LANDING ON THE PLANET
EARTH)
GS   LANDING
. SPACECRAFT LANDING
. . **PLANETARY LANDING**

PLANETARY LANDING-(CONT.)
RT   CRASH LANDING
      GLIDE LANDINGS
      HARD LANDING
      HORIZONTAL SPACECRAFT LANDING
      INTERPLANETARY FLIGHT
      LUNAR LANDING
      MARS LANDING
      ORBITAL MECHANICS
      ROVING VEHICLES
      SOFT LANDING
      WATER LANDING

PLANETARY LIMB
RT   EARTH LIMB
   ∞ LIMBS
      LUNAR LIMB
      SOLAR LIMB

PLANETARY MAGNETIC FIELDS
GS   MAGNETIC FIELDS
    . PLANETARY MAGNETIC FIELDS
RT   GEOMAGNETIC TAIL
      GEOMAGNETISM
      POLAR CUSPS
      SOLAR PLANETARY INTERACTIONS

PLANETARY MANTLES
GS   PLANETARY MANTLES
    . EARTH MANTLE
RT   CRUSTS
      LITHOSPHERE
      LUNAR MANTLE

PLANETARY MAPPING
GS   MAPPING
    . PLANETARY MAPPING
RT   ASTROGRAPHY
      HEAT CAPACITY MAPPING MISSION
      THERMAL MAPPING

PLANETARY MASS
GS   MASS
    . PLANETARY MASS
RT   PROTOPLANETS

PLANETARY MOTION
USE   SOLAR ORBITS

PLANETARY NEBULAE
GS   CELESTIAL BODIES
    . NEBULAE
    . . PLANETARY NEBULAE
RT   ORION NEBULA

PLANETARY ORBITS
GS   ORBITS
    . PLANETARY ORBITS
RT   AMOR ASTEROID
      APOLLO ASTEROIDS
      CHARON
      CIRCULAR ORBITS
      EARTH ORBITS
      ELLIPTICAL ORBITS
      EQUATORIAL ORBITS
      INTERPLANETARY TRAJECTORIES
      PARKING ORBITS
      POLAR ORBITS
      SATELLITE ORBITS
      SPACECRAFT ORBITS
      SWINGBY TECHNIQUE
      TRANSFER ORBITS
      TWENTY-FOUR HOUR ORBITS
      VIKING ORBITER SPACECRAFT

PLANETARY QUAKES
RT   EARTHQUAKES
      GEODYNAMICS
      MOONQUAKES
      SEISMIC WAVES
      SHOCK WAVES

PLANETARY QUARANTINE
RT   SPACECRAFT STERILIZATION

PLANETARY RADIATION
SN   (EXCLUDES TERRESTRIAL RADIATION)
GS   ELECTROMAGNETIC RADIATION
    . PLANETARY RADIATION
      EXTRATERRESTRIAL RADIATION
    . PLANETARY RADIATION
RT   ALBEDO
      DECIMETER WAVES
      INFRARED RADIATION

PLANETARY RADIATION-(CONT.)
      LIGHT (VISIBLE RADIATION)
   ∞ RADIATION
      RADIO WAVES
      SATURN ATMOSPHERE
      TERRESTRIAL RADIATION
      THERMAL RADIATION
      VLF EMISSION RECORDERS

PLANETARY RINGS
GS   CELESTIAL BODIES
    . PLANETARY RINGS
    . . JUPITER RINGS
    . . SATURN RINGS
    . . URANUS RINGS
RT   PLANETARY ATMOSPHERES
      PLANETS
   ∞ RINGS

PLANETARY ROTATION
GS   GYRATION
    . ROTATION
    . . PLANETARY ROTATION
RT   ASTROPHYSICS
      PLANETOLOGY
      ROTATING BODIES
      STELLAR ROTATION

PLANETARY SATELLITES
USE   NATURAL SATELLITES

PLANETARY SPACE FLIGHT
USE   INTERPLANETARY FLIGHT

PLANETARY SPACECRAFT
USE   INTERPLANETARY SPACECRAFT

PLANETARY STRUCTURE
RT   CHEMICAL COMPOSITION
      EARTH PLANETARY STRUCTURE
      JUPITER RINGS
      LUNAR MANTLE
      PLANETOLOGY
      URANUS RINGS

PLANETARY SURFACES
GS   PLANETARY SURFACES
    . MARS SURFACE
    . VENUS SURFACE
RT   EARTH SURFACE
      JUPITER RED SPOT
      ROVING VEHICLES
      SATURN RINGS
      SURFACE PROPERTIES
   ∞ SURFACES
      TOPOGRAPHY

PLANETARY TEMPERATURE
GS   TEMPERATURE
    . PLANETARY TEMPERATURE
RT   ATMOSPHERIC TEMPERATURE
      SATURN RINGS

PLANETARY WAVES
UF   LONG WAVES (METEOROLOGY)
      ROSSBY WAVES
GS   BAROTROPISM
    . PLANETARY WAVES
      INTERNAL WAVES
    . PLANETARY WAVES
      TROPOSPHERIC WAVES
    . PLANETARY WAVES
RT   ATMOSPHERIC CIRCULATION
      BAROTROPIC FLOW
      CORIOLIS EFFECT
      FLUID FLOW
      GRAVITY WAVES
      ROSSBY REGIMES
      ROTATING FLUIDS
      ROTATING LIQUIDS
      VORTICES
   ∞ WAVES

PLANETISMALS
USE   PROTOPLANETS

PLANETOCENTRIC COORDINATES
GS   COORDINATES
    . PLANETOCENTRIC COORDINATES
    . . GEOCENTRIC COORDINATES
RT   ASTRONOMICAL COORDINATES
      CELESTIAL REFERENCE SYSTEMS
      SPHERICAL COORDINATES

PLANETOLOGY
RT   JUPITER RINGS
      PLANETARY ROTATION
      PLANETARY STRUCTURE
      SATURN RINGS
      TERRESTRIAL PLANETS

PLANETS
GS   CELESTIAL BODIES
    . PLANETS
    . . EXTRASOLAR PLANETS
    . . GAS GIANT PLANETS
    . . . JUPITER (PLANET)
    . . . NEPTUNE (PLANET)
    . . . SATURN (PLANET)
    . . . URANUS (PLANET)
    . . . . URANUS RINGS
    . . . . URANUS SATELLITES
    . . PLUTO (PLANET)
    . . TERRESTRIAL PLANETS
    . . . EARTH (PLANET)
    . . . MARS (PLANET)
    . . . MERCURY (PLANET)
    . . . VENUS (PLANET)
RT   CELESTIAL MECHANICS
      CHIRON
      ECLIPTIC
      JUPITER RED SPOT
      NATURAL SATELLITES
      PLANET EPHEMERIDES
      PLANETARY CORES
      PLANETARY CRATERS
      PLANETARY ENVIRONMENTS
      PLANETARY RINGS
      PROTOPLANETS
      SATURN RINGS
      SOLAR SYSTEM
      SUN

PLANFORMS
GS   PLANFORMS
    . CARET WINGS
    . RECTANGULAR PLANFORMS
    . . RECTANGULAR PANELS
    . . RECTANGULAR PLATES
    . . RECTANGULAR WINGS
    . SWEPTBACK TAIL SURFACES
    . TRAPEZOIDAL TAIL SURFACES
    . WING PLANFORMS
    . . CHANNEL WINGS
    . . INFINITE SPAN WINGS
    . . SWEPT FORWARD WINGS
    . . . TRAPEZOIDAL WINGS
    . . SWEPTBACK WINGS
    . . . ARROW WINGS
    . . . DELTA WINGS
    . . . TRAPEZOIDAL WINGS
    . . VARIABLE SWEEP WINGS
RT ∞ BODIES
    ∞ CROSS SECTIONS
      GEOMETRY
    ∞ PROFILES
      SHAPES
    ∞ SURFACE GEOMETRY

PLANIGRAPHY
USE   TOMOGRAPHY

PLANING
SN   (EXCLUDES MOTION INVOLVING
      DYNAMIC SUPPORTING FORCES)
GS   CUTTING
    . PLANING
RT   GRINDING (MATERIAL REMOVAL)
      MACHINING
      METAL CUTTING
      MILLING (MACHINING)
      SLICING
      SMOOTHING

PLANISPHERES
GS   MAPS
    . ASTRONOMICAL MAPS
    . . PLANISPHERES
RT   ASTRONOMICAL COORDINATES
      CELESTIAL SPHERE
      CONSTELLATIONS
      POLAR COORDINATES

PLANKTON
UF   PLANKTON BLOOM
RT   ALGAE
      ANIMALS
      PLANTS (BOTANY)
      RED TIDE

PLANKTON-*(CONT.)*
    THERMAL POLLUTION

PLANKTON BLOOM
  USE   PLANKTON

PLANNING
  GS   **PLANNING**
     . AIRPORT PLANNING
     . MANAGEMENT PLANNING
     . . PRODUCTION PLANNING
     . . PROJECT PLANNING
     . MISSION PLANNING
     . REGIONAL PLANNING
     . . URBAN PLANNING
  RT   BUDGETING
     CRITICAL PATH METHOD
     DELPHI METHOD (FORECASTING)
     ∞DESIGN
     FORECASTING
     ∞MISSIONS
     OPTIMIZATION
     PATTERN METHOD (FORECASTING)
     PROBE METHOD (FORECASTING)
     PRODUCTION ENGINEERING
     PROFILE METHOD (FORECASTING)
     PROGRESS
     SEQUENCING
     SLICING
     STARSITE PROGRAM
     TRAINING ANALYSIS
     URBAN DEVELOPMENT

PLANOTRONS
  UF   AMPLITRONS (TRADEMARK)
  GS   ELECTRON TUBES
     . VACUUM TUBES
     . . VACUUM TUBE OSCILLATORS
     . . . MICROWAVE TUBES
     . . . . **PLANOTRONS**
     . . . . . CARCINOTRONS
     . . . . . CELESCOPES
     . . . . . HELITRONS
     . . . . . IGNITRONS
     . . . . . THERMICONS
     . . . . . THERMIONIC DIODES
     . . . . . THYRATRONS
     MICROWAVE EQUIPMENT
     . MICROWAVE TUBES
     . . **PLANOTRONS**
     . . . CARCINOTRONS
     . . . CELESCOPES
     . . . HELITRONS
     . . . IGNITRONS
     . . . THERMICONS
     . . . THERMIONIC DIODES
     . . . THYRATRONS
     OSCILLATORS
     . VACUUM TUBE OSCILLATORS
     . . MICROWAVE TUBES
     . . . **PLANOTRONS**
     . . . . CARCINOTRONS
     . . . . HELITRONS
  RT   CAMERA TUBES
     ELECTRIC ARCS
     MAGNETRONS

∞ PLANS
  SN   *(USE OF A MORE SPECIFIC TERM IS*
     RECOMMENDED--CONSULT THE TERMS
     LISTED BELOW)
  RT   DRAWINGS
     FLIGHT PLANS
     LAYOUTS
     MISSION PLANNING
     PAYLOAD INTEGRATION PLAN
     URBAN DEVELOPMENT

PLANT DESIGN
  SN   (EXCLUDES BIOLOGICAL PLANTS)
  RT   ARCHITECTURE
     ∞DESIGN
     STRUCTURAL DESIGN

PLANT ROOTS
  RT   BULBS
     PLANTS (BOTANY)
     ∞ROOTS
     VEGETATION GROWTH

PLANT STRESS
  RT   AERIAL PHOTOGRAPHY
     AGRICULTURE
     CROP GROWTH
     CROP VIGOR

PLANT STRESS-*(CONT.)*
     EARTH RESOURCES PROGRAM
     REMOTE SENSING
     SOIL MOISTURE
     SPECTRAL REFLECTANCE
     VEGETATION GROWTH

PLANTAR TISSUES
  GS   TISSUES (BIOLOGY)
     . **PLANTAR TISSUES**

PLANTING
  RT   AGRICULTURE
     ∞CROPS
     CULTIVATION
     FARM CROPS
     FARMLANDS
     FERTILIZERS
     PLANTS (BOTANY)
     PLOWING
     PLOWS
     SEEDS
     SILVICULTURE
     SOILS
     TRACTORS
     VEGETABLES

PLANTS (BOTANY)
  UF   FLORA
  GS   **PLANTS (BOTANY)**
     . ALFALFA
     . AQUATIC PLANTS
     . BARLEY
     . BRUSH (BOTANY)
     . . CHAPARRAL
     . BRYOPHYTES
     . CORN
     . FOLIAGE
     . GRASSES
     . . HAY
     . . REEDS (PLANTS)
     . . SEA GRASSES
     . . SORGHUM
     . GUAYULE
     . LEAVES
     . LICHENS
     . NIGELLA
     . OATS
     . ORCHARDS
     . PHOTOPHILIC PLANTS
     . PHREATOPHYTES
     . SAPROPHYTES
     . SUGAR CANE
     . SUNFLOWERS
     . THERMOPHILIC PLANTS
     . TOBACCO
     . TRAGACANTH
     . TREES (PLANTS)
     . . CITRUS TREES
     . . CONIFERS
     . . DECIDUOUS TREES
  RT   AGRICULTURE
     ANGIOSPERMS
     ANIMALS
     BIOCHEMICAL OXYGEN DEMAND
     BIOGEOCHEMISTRY
     BIOMASS
     BLIGHT
     BOTANY
     CANOPIES (VEGETATION)
     CARBON CYCLE
     CHLOROPHYLLS
     CORTEXES (BOTANY)
     COTTON
     CROP GROWTH
     CROP VIGOR
     DEFOLIANTS
     DEFOLIATION
     EARTH RESOURCES
     ENVIRONMENTS
     FOOD CHAIN
     FORESTS
     FROST DAMAGE
     GEOBOTANY
     GEOTROPISM
     GRAVITROPISM
     GREENHOUSES
     HALOPHILES
     HERBICIDES
     HETEROTROPHS
     HYDROPONICS
     INFESTATION
     LACUNAS
     MICROORGANISMS
     MICROSPORES

PLANTS (BOTANY)-*(CONT.)*
     MITRA
     ORGANISMS
     PETALS
     PHOTOTROPISM
     PHYTOTRONS
     PLANKTON
     PLANT ROOTS
     PLANTING
     PLOWING
     POLLEN
     POTATOES
     RAIN FORESTS
     SEEDS
     SPINACH
     SPORES
     STEMS
     TRADESCANTIA
     UTRICLE
     VEGETATION
     VIABILITY
     VINEYARDS
     WOOD

PLANTS (INDUSTRIES)
  USE   INDUSTRIAL PLANTS

PLASMA ACCELERATION
  UF   MAGNETOHYDRODYNAMIC
     ACCELERATION
  GS   RATES (PER TIME)
     . ACCELERATION (PHYSICS)
     . . **PLASMA ACCELERATION**
  RT   ∞ACCELERATION
     PARTICLE ACCELERATION
     PLASMAS (PHYSICS)
     WAVE PROPAGATION

PLASMA ACCELERATORS
  GS   **PLASMA ACCELERATORS**
     . ALPHA PLASMA DEVICES
     . COAXIAL PLASMA ACCELERATORS
     . CYCLOPS PLASMA ACCELERATOR
  RT   ∞ACCELERATORS
     ELECTROMAGNETIC ACCELERATION
     ION INJECTION
     MAGNETIC ANNULAR ARC
     MAGNETOHYDRODYNAMIC GENERATORS
     PLASMAS (PHYSICS)

PLASMA ANTENNAS
  GS   ANTENNAS
     . **PLASMA ANTENNAS**
  RT   ANTENNA DESIGN
     ANTENNA RADIATION PATTERNS
     PLASMA CYLINDERS
     SPACECRAFT COMMUNICATION

PLASMA ARC CUTTING
  RT   METAL CUTTING
     METAL WORKING
     PLASMA ARC WELDING
     PLASMA TORCHES
     PLASMAS (PHYSICS)

PLASMA ARC SPRAYING
  USE   ARC SPRAYING

PLASMA ARC WELDING
  GS   WELDING
     . ELECTRIC WELDING
     . . ARC WELDING
     . . . **PLASMA ARC WELDING**
  RT   PLASMA ARC CUTTING
     PLASMA TORCHES
     PLASMAS (PHYSICS)

PLASMA ARCS
  USE   PLASMA JETS

PLASMA BUBBLES
  RT   F REGION
     PLASMA DENSITY

PLASMA CHEMISTRY
  RT   ∞CHEMISTRY
     NUCLEAR CHEMISTRY
     PLASMAS (PHYSICS)

PLASMA CLOUDS
  GS   CLOUDS
     . **PLASMA CLOUDS**
     PARTICLES
     . CHARGED PARTICLES

PLASMA CLOUDS-*(CONT.)*
 . . **PLASMA CLOUDS**
RT CHEMICAL CLOUDS
  CLOUDS (METEOROLOGY)
  COSMIC PLASMA
  GEOMAGNETIC HOLLOW
  HYDROGEN CLOUDS
  INTERPLANETARY MEDIUM
  ION SHEATHS
  MAGNETOSPHERE
  PLASMAPAUSE
  PLASMAS (PHYSICS)

PLASMA COMPOSITION
GS COMPOSITION (PROPERTY)
 . **PLASMA COMPOSITION**
RT ATOM CONCENTRATION
  GAS COMPOSITION
  ION MOTION
  IONOSPHERIC COMPOSITION
  NONEQUILIBRIUM PLASMAS
  NONUNIFORM PLASMAS
  PLASMAS (PHYSICS)
  THOMAS-FERMI MODEL
  URANIUM PLASMAS

PLASMA COMPRESSION
GS COMPRESSING
 . **PLASMA COMPRESSION**
RT CONTROLLED FUSION
  DENSE PLASMAS
  INERTIAL FUSION (REACTOR)
  MAGNETIC EFFECTS
  MAGNETIC FIELD CONFIGURATIONS
  PINCH EFFECT
  PLASMA FOCUS
  PLASMAS (PHYSICS)
  STRONGLY COUPLED PLASMAS
  THETA PINCH
  TOKAMAK DEVICES
  ZETA PINCH

PLASMA CONDUCTIVITY
GS TRANSPORT PROPERTIES
 . ELECTRICAL RESISTIVITY
 . . **PLASMA CONDUCTIVITY**
RT COLLISIONAL PLASMAS
  ∞CONDUCTIVITY
  IONOSPHERIC CONDUCTIVITY
  MAGNETOHYDRODYNAMIC STABILITY
  PLASMAS (PHYSICS)
  STRONGLY COUPLED PLASMAS

PLASMA CONFINEMENT
USE PLASMA CONTROL

PLASMA CONTROL
UF PLASMA CONFINEMENT
RT BALLOONING MODES
  BETA FACTOR
  BUMPY TORUSES
  CONFINEMENT
  ∞CONTROL
  CROSSED FIELD GUNS
  CROSSED FIELDS
  ELECTRON-ION RECOMBINATION
  ELLIPTICAL PLASMAS
  HELICAL INDUCERS
  LIMITERS (FUSION REACTORS)
  MAGNETIC ANNULAR ARC
  MAGNETIC COMPRESSION
  MAGNETIC FIELD CONFIGURATIONS
  MAGNETIC MIRRORS
  MAGNETICALLY TRAPPED PARTICLES
  MIRROR FUSION
  PINCH EFFECT
  PLASMAS (PHYSICS)
  REVERSE FIELD PINCH
  RIGID ROTORS (PLASMA PHYSICS)
  SCREW PINCH
  TOKAMAK DEVICES
  TOROIDAL PLASMAS
  TRANSFORMERS
  TRAP PROGRAM
  TRAPPED MAGNETIC FIELDS
  ZETA PINCH

PLASMA COOLING
GS COOLING
 . **PLASMA COOLING**
RT CONTROLLED FUSION
  IONIZED GASES
  MAGNETOHYDRODYNAMIC STABILITY
  PLASMAS (PHYSICS)
  TEMPERATURE CONTROL

PLASMA CORE REACTORS
GS NUCLEAR REACTORS
 . **PLASMA CORE REACTORS**
RT CRITICAL MASS
  NUCLEAR POWER PLANTS
  NUCLEAR RESEARCH
  PLASMAS (PHYSICS)
  RADIOACTIVE WASTES
  REACTOR CORES
  ∞REACTORS
  REFLECTORS
  WASTE DISPOSAL

PLASMA CURRENTS
GS ELECTRIC CURRENT
 . **PLASMA CURRENTS**
RT BEAM CURRENTS
  CONTROLLED FUSION
  EDDY CURRENTS
  ELECTRIC DISCHARGES
  ELECTRICAL RESISTIVITY
  HIGH CURRENT
  IONOSPHERIC CURRENTS
  LINE CURRENT
  LOW CURRENTS
  MAGNETOHYDRODYNAMICS
  PLASMA-PARTICLE INTERACTIONS
  PLASMAS (PHYSICS)
  RING CURRENTS
  SPHEROMAKS
  TOROIDAL PLASMAS

PLASMA CYLINDERS
GS **PLASMA CYLINDERS**
 . CYLINDRICAL PLASMAS
RT ∞CYLINDERS
  CYLINDRICAL BODIES
  CYLINDRICAL SHELLS
  PLASMA ANTENNAS
  PLASMAGUIDES
  PLASMAS (PHYSICS)

PLASMA DECAY
GS DECAY
 . **PLASMA DECAY**
RT AFTERGLOWS
  ELECTROMAGNETIC WAVE
   TRANSMISSION
  HELIUM AFTERGLOW
  MAGNETOHYDRODYNAMIC STABILITY
  PLASMAS (PHYSICS)

PLASMA DENSITY
GS DENSITY (NUMBER/VOLUME)
 . PARTICLE DENSITY (CONCENTRATION)
 . . **PLASMA DENSITY**
RT ATMOSPHERIC DENSITY
  ATOM CONCENTRATION
  CAVITONS
  COLLISIONAL PLASMAS
  ELECTRON DENSITY (CONCENTRATION)
  ELECTRON-HOLE DROPS
  ION DENSITY (CONCENTRATION)
  MAGNETOPLASMADYNAMICS
  MAGNETOSPHERIC ELECTRON DENSITY
  MAGNETOSPHERIC ION DENSITY
  MAGNETOSPHERIC PROTON DENSITY
  PLASMA BUBBLES
  PLASMA DRIFT
  PLASMAS (PHYSICS)
  PROTON DENSITY (CONCENTRATION)
  SPACE DENSITY
  SPACE PLASMAS
  STRONGLY COUPLED PLASMAS

PLASMA DIAGNOSTICS
RT FABRY-PEROT INTERFEROMETERS
  MICROWAVE INTERFEROMETERS
  OPEN PROJECT
  PLASMAS (PHYSICS)
  RESONANCE PROBES
  SPACE PLASMAS

PLASMA DIFFUSION
UF PLASMA DISPERSION
GS DIFFUSION
 . **PLASMA DIFFUSION**
RT AMBIPOLAR DIFFUSION
  COLLOIDAL GENERATORS
  DIFFUSION WAVES
  ELECTRON DIFFUSION
  GASEOUS SELF-DIFFUSION
  ION MOTION
  IONIC DIFFUSION
  PLASMAS (PHYSICS)

PLASMA DIODES
GS ELECTRONIC EQUIPMENT
 . DIODES
 . . **PLASMA DIODES**
RT CESIUM DIODES
  PLASMAS (PHYSICS)

PLASMA DISCHARGES
USE PLASMA JETS

PLASMA DISPERSION
USE PLASMA DIFFUSION

PLASMA DISPLAY DEVICES
GS DISPLAY DEVICES
 . **PLASMA DISPLAY DEVICES**
RT ∞DEVICES
  GAS IONIZATION
  GLOW DISCHARGES
  LIGHT SOURCES
  PLASMAS (PHYSICS)

PLASMA DRIFT
RT MAGNETOHYDRODYNAMIC STABILITY
  PLASMA DENSITY
  PLASMA WAVES
  PLASMAS (PHYSICS)

PLASMA DYNAMICS
UF SNOWPLOW EFFECT
RT ∞DYNAMICS
  HYDRODYNAMIC EQUATIONS
  MAGNETOHYDRODYNAMICS
  PLASMAS (PHYSICS)

PLASMA ELECTRODES
GS ELECTRODES
 . **PLASMA ELECTRODES**
RT HOT-WIRE FLOWMETERS
  PLASMAGUIDES
  PLASMATRONS
  ZETA PINCH

PLASMA ENGINES
GS PLASMA POWER SOURCES
 . **PLASMA ENGINES**
 . . TWO STAGE PLASMA ENGINES
RT ARC JET ENGINES
  COAXIAL PLASMA ACCELERATORS
  HIGH TEMPERATURE PROPELLANTS
  ION ENGINES
  MERCURY ION ENGINES
  PLASMAS (PHYSICS)
  PULSED JET ENGINES
  RESISTOJET ENGINES
  RIT ENGINES

PLASMA EQUILIBRIUM
RT BALLOONING MODES
  BETA FACTOR
  CONFINEMENT
  ELECTRON-HOLE DROPS
  ∞EQUILIBRIUM
  EQUILIBRIUM FLOW
  MAGNETIC MIRRORS
  MAGNETOHYDRODYNAMIC STABILITY
  PLASMAS (PHYSICS)
  STRONGLY COUPLED PLASMAS

PLASMA ETCHING
RT PLASMAS (PHYSICS)
  SPUTTERING

PLASMA FLOW
USE MAGNETOHYDRODYNAMIC FLOW

PLASMA FLUX MEASUREMENT
RT INTERFEROMETRY
  MAGNETOHYDRODYNAMIC FLOW
  MICROWAVE PLASMA PROBES
  PLASMAS (PHYSICS)
  ROTATING PLASMAS

PLASMA FOCUS
GS FOCI
 . **PLASMA FOCUS**
  PARTICLES
 . CHARGED PARTICLES
 . . ENERGETIC PARTICLES
 . . . PLASMAS (PHYSICS)
 . . . . DENSE PLASMAS
 . . . . . **PLASMA FOCUS**
RT NUCLEAR FUSION
  PLASMA COMPRESSION

PLASMA PUMPING-*(CONT.)*
    PLASMAS (PHYSICS)
    ∞PUMPING

PLASMA RADIATION
RT   ELECTRON RADIATION
    FLUORESCENCE
    GLOW DISCHARGES
    ION CYCLOTRON RADIATION
    LUMINESCENCE
    NONEQUILIBRIUM PLASMAS
    OPTICAL RESONANCE
    PHOSPHORESCENCE
    PLASMAS (PHYSICS)
    POLARIZED RADIATION
    ∞RADIATION
    RELATIVISTIC PLASMAS

PLASMA RENIN ACTIVITY
USE   IMMUNOASSAY

PLASMA RESONANCE
GS   RESONANCE
    . **PLASMA RESONANCE**
RT   CAVITONS
    CYCLOTRON RESONANCE
    ELECTROMAGNETIC INTERACTIONS
    PLASMAS (PHYSICS)
    RESONANCE LINES
    RESONANCE PROBES

PLASMA RINGS
USE   TOROIDAL PLASMAS

PLASMA SHEATHS
GS   PARTICLES
    . CHARGED PARTICLES
    . . PLASMA LAYERS
    . . . **PLASMA SHEATHS**
    SHEATHS
    . **PLASMA SHEATHS**
RT   BLACKOUT (PROPAGATION)
    BOUNDARY LAYER PLASMAS
    ION SHEATHS
    MAGNETOHYDRODYNAMIC SHEAR
       HEATING
    METALLIC PLASMAS
    MISSILES
    NONEQUILIBRIUM PLASMAS
    PLASMAS (PHYSICS)
    REENTRY COMMUNICATION
    REENTRY EFFECTS
    REENTRY PHYSICS
    SYSTEM GENERATED
       ELECTROMAGNETIC PULSES
    UNCONTROLLED REENTRY
       (SPACECRAFT)

PLASMA SLABS
GS   PARTICLES
    . CHARGED PARTICLES
    . . **PLASMA SLABS**
RT   MAGNETOHYDRODYNAMIC STABILITY
    PLASMAS (PHYSICS)

PLASMA SOUND WAVES
USE   MAGNETOHYDRODYNAMIC WAVES
    PLASMA WAVES

PLASMA SPECTRA
GS   SPECTRA
    . **PLASMA SPECTRA**
RT   EMISSION SPECTRA
    ENERGY SPECTRA
    OPTICAL RESONANCE
    PLASMAS (PHYSICS)
    RADIATION SPECTRA

PLASMA SPRAYING
GS   SPRAYING
    . **PLASMA SPRAYING**
RT   COATING
    COATINGS
    FLAME SPRAYING
    METAL MATRIX COMPOSITES
    PLASMAS (PHYSICS)
    SPRAYED COATINGS

PLASMA STABILITY
USE   MAGNETOHYDRODYNAMIC STABILITY

PLASMA TEMPERATURE
GS   TEMPERATURE
    . **PLASMA TEMPERATURE**

PLASMA TEMPERATURE-*(CONT.)*
RT   ION TEMPERATURE
    MAGNETOHYDRODYNAMIC STABILITY
    PLASMA HEATING
    PLASMAS (PHYSICS)
    THERMAL PLASMAS

PLASMA THEORY
USE   PLASMA PHYSICS

PLASMA TORCHES
RT   PLASMA ARC CUTTING
    PLASMA ARC WELDING
    PLASMA JETS
    PLASMAS (PHYSICS)

PLASMA TURBULENCE
GS   TURBULENCE
    . MAGNETOHYDRODYNAMIC
       TURBULENCE
    . . **PLASMA TURBULENCE**
RT   MAGNETOHYDRODYNAMIC FLOW
    MAGNETOHYDRODYNAMIC STABILITY
    PLASMAS (PHYSICS)

PLASMA WAVES
UF   PLASMA SOUND WAVES
GS   ELASTIC WAVES
    . MAGNETOHYDRODYNAMIC WAVES
    . . **PLASMA WAVES**
    . . . ELECTROSTATIC WAVES
RT   COLLISIONAL PLASMAS
    DIFFUSION WAVES
    ELECTROACOUSTIC WAVES
    ELECTRON PLASMA
    ION ACOUSTIC WAVES
    ION CYCLOTRON RADIATION
    IONIC WAVES
    LANDAU DAMPING
    MAGNETOACOUSTIC WAVES
    MAGNETOELASTIC WAVES
    NONUNIFORM PLASMAS
    PLASMA DRIFT
    PLASMAS (PHYSICS)
    SHOCK WAVES
    SPACE PLASMAS
    WAVE PACKETS

PLASMA-ELECTROMAGNETIC INTERACTION
GS   ELECTROMAGNETIC INTERACTIONS
    . **PLASMA-ELECTROMAGNETIC
       INTERACTION**
    . . LASER PLASMA INTERACTIONS
    PLASMA INTERACTIONS
    . **PLASMA-ELECTROMAGNETIC
       INTERACTION**
    . . LASER PLASMA INTERACTIONS
RT   ∞INTERACTIONS
    PLASMAS (PHYSICS)
    SPACE PLASMAS

PLASMA-PARTICLE INTERACTIONS
GS   PARTICLE INTERACTIONS
    . **PLASMA-PARTICLE INTERACTIONS**
    PLASMA INTERACTIONS
    . **PLASMA-PARTICLE INTERACTIONS**
RT   BEAM INJECTION
    BEAM PLASMA AMPLIFIERS
    CHARGE EXCHANGE
    ELECTROMAGNETIC INTERACTIONS
    ELECTRON PHONON INTERACTIONS
    ELECTRON PLASMA
    ∞INTERACTIONS
    PARTICLE THEORY
    PLASMA CURRENTS
    PLASMAS (PHYSICS)
    RELATIVISTIC ELECTRON BEAMS
    RELATIVISTIC PLASMAS
    SPHINX

PLASMADYNAMIC LASERS
GS   STIMULATED EMISSION DEVICES
    . LASERS
    . . **PLASMADYNAMIC LASERS**
RT   COHERENT LIGHT
    GASDYNAMIC LASERS
    LASER APPLICATIONS
    PLASMAS (PHYSICS)

PLASMAGUIDES
GS   TRANSMISSION LINES
    . COMMUNICATION CABLES
    . . WAVEGUIDES
    . . . **PLASMAGUIDES**
RT   BEAM WAVEGUIDES

PLASMAGUIDES-*(CONT.)*
    ELECTROMAGNETIC WAVE
       TRANSMISSION
    MICROWAVE PLASMA PROBES
    PLASMA CYLINDERS
    PLASMA ELECTRODES
    PLASMA PROBES
    PLASMAS (PHYSICS)
    WAVE PROPAGATION

PLASMAPAUSE
SN   (LIMITED TO EARTH'S ATMOSPHERE)
RT   COSMIC PLASMA
    IONOPAUSE
    MAGNETOSPHERE
    PLASMA CLOUDS
    PLASMA OSCILLATIONS
    PLASMA PHYSICS
    PLASMAS (PHYSICS)
    SOLAR WIND

PLASMAS (PHYSICS)
UF   ELECTROSTATIC PLASMA
    IONIZED PLASMAS
    MAGNETOIONIC PLASMA
    MAGNETOPLASMAS
    PLASMOIDS
GS   PARTICLES
    . CHARGED PARTICLES
    . . ENERGETIC PARTICLES
    . . . **PLASMAS (PHYSICS)**
    . . . . ARGON PLASMA
    . . . . BETA PARTICLES
    . . . . BOUNDARY LAYER PLASMAS
    . . . . COLD PLASMAS
    . . . . COLLISIONAL PLASMAS
    . . . . . STRONGLY COUPLED PLASMAS
    . . . . COLLISIONLESS PLASMAS
    . . . . COSMIC PLASMA
    . . . . CYLINDRICAL PLASMAS
    . . . . DENSE PLASMAS
    . . . . . PLASMA FOCUS
    . . . . . STRONGLY COUPLED PLASMAS
    . . . . ELECTRON PLASMA
    . . . . ELLIPTICAL PLASMAS
    . . . . HELIUM PLASMA
    . . . . HIGH TEMPERATURE PLASMAS
    . . . . LASER PLASMAS
    . . . . METALLIC PLASMAS
    . . . . . CESIUM PLASMA
    . . . . MICROPLASMAS
    . . . . NITROGEN PLASMA
    . . . . NONEQUILIBRIUM PLASMAS
    . . . . NONUNIFORM PLASMAS
    . . . . RAREFIED PLASMAS
    . . . . RELATIVISTIC PLASMAS
    . . . . ROTATING PLASMAS
    . . . . SEMICONDUCTOR PLASMAS
    . . . . SOLAR WIND
    . . . . SPACE PLASMAS
    . . . . SPHERICAL PLASMAS
    . . . . STELLAR WINDS
    . . . . THERMAL PLASMAS
    . . . . TOROIDAL PLASMAS
RT   ALPHA PLASMA DEVICES
    BEAM PLASMA AMPLIFIERS
    BLACKOUT (PROPAGATION)
    COMBUSTION PHYSICS
    CORE FLOW
    CYCLOPS PLASMA ACCELERATOR
    DEBYE LENGTH
    DEUTERON IRRADIATION
    DEUTERONS
    DUOPLASMATRONS
    ELECTRIC ARCS
    ELECTRON ENERGY
    GASES
    HIGH TEMPERATURE FLUIDS
    IONS
    KELVIN-HELMHOLTZ INSTABILITY
    LANDAU FACTOR
    LASER FUSION
    LASER PLASMA INTERACTIONS
    LIGHT IONS
    LIOUVILLE EQUATIONS
    LOW DENSITY RESEARCH
    MAGNETIC COMPRESSION
    MAGNETOHYDRODYNAMIC FLOW
    MAGNETOHYDRODYNAMIC STABILITY
    MAGNETOHYDRODYNAMICS
    MAGNETOIONICS
    MICROWAVE PLASMA PROBES
    ONSAGER PHENOMENOLOGICAL
       COEFFICIENT
    ∞PHYSICS

PLASMAS (PHYSICS)-*(CONT.)*
PLASMA ACCELERATION
PLASMA ACCELERATORS
PLASMA ARC CUTTING
PLASMA ARC WELDING
PLASMA CHEMISTRY
PLASMA CLOUDS
PLASMA COMPOSITION
PLASMA COMPRESSION
PLASMA CONDUCTIVITY
PLASMA CONTROL
PLASMA COOLING
PLASMA CORE REACTORS
PLASMA CURRENTS
PLASMA CYLINDERS
PLASMA DECAY
PLASMA DENSITY
PLASMA DIAGNOSTICS
PLASMA DIFFUSION
PLASMA DIODES
PLASMA DISPLAY DEVICES
PLASMA DRIFT
PLASMA DYNAMICS
PLASMA ENGINES
PLASMA EQUILIBRIUM
PLASMA ETCHING
PLASMA FLUX MEASUREMENT
PLASMA FREQUENCIES
PLASMA GENERATORS
PLASMA GUNS
PLASMA HEATING
PLASMA INTERACTION EXPERIMENT
PLASMA INTERACTIONS
PLASMA JET SYNTHESIS
PLASMA JET WIND TUNNELS
PLASMA JETS
PLASMA LAYERS
PLASMA LIFETIME
PLASMA LOSS
PLASMA OSCILLATIONS
PLASMA PHYSICS
PLASMA PINCH
PLASMA POTENTIALS
PLASMA POWER SOURCES
PLASMA PROBES
PLASMA PROPULSION
PLASMA PUMPING
PLASMA RADIATION
PLASMA RESONANCE
PLASMA SHEATHS
PLASMA SLABS
PLASMA SPECTRA
PLASMA SPRAYING
PLASMA TEMPERATURE
PLASMA TORCHES
PLASMA TURBULENCE
PLASMA WAVES
PLASMA-ELECTROMAGNETIC
   INTERACTION
PLASMA-PARTICLE INTERACTIONS
PLASMADYNAMIC LASERS
PLASMAGUIDES
PLASMAPAUSE
PLASMASPHERE
PLASMATRONS
RADIATION BELTS
RAREFIED GAS DYNAMICS
SCYLLA
SOLAR PHYSICS
SOLAR WIND
SPACE CHARGE
SPHINX
STELLAR MAGNETIC FIELDS
TEARING MODES (PLASMAS)
THERMAL DISSOCIATION
THERMODYNAMICS
THERMONUCLEAR REACTIONS
TWO STAGE PLASMA ENGINES

PLASMAS-IN-SPACE PAYLOAD
USE   AMPS (SATELLITE PAYLOAD)

PLASMASPHERE
RT   ATMOSPHERIC IONIZATION
   CHEMOSPHERE
   EARTH ATMOSPHERE
   MAGNETOSPHERE
   OPEN PROJECT
   PLASMAS (PHYSICS)
   UPPER ATMOSPHERE

PLASMATRONS
GS   ION SOURCES
   . **PLASMATRONS**
   . . DUOPLASMATRONS

PLASMATRONS-*(CONT.)*
   PLASMA GENERATORS
   . **PLASMATRONS**
   . . DUOPLASMATRONS
RT   PLASMA ELECTRODES
   PLASMA GUNS
   PLASMA JETS
   PLASMA PROPULSION
   PLASMAS (PHYSICS)

PLASMOIDS
USE   PLASMAS (PHYSICS)

PLASMOLYSIS
RT   CELLS (BIOLOGY)
   CYTOLOGY
   DEHYDRATION

PLASMONS
SN   (EXCLUDES ORGANIC CYTOPLASMIC
   CONDITIONS)
GS   ELECTROMAGNETIC RADIATION
   . **PLASMONS**
   ELEMENTARY EXCITATIONS
   . **PLASMONS**
   POLARITONS
   . **PLASMONS**
RT   ELECTRON GAS
   EXCITONS
   MAGNETOHYDRODYNAMIC STABILITY
   MAGNONS
   PHONONS
   PLASMA FREQUENCIES
   PLASMA OSCILLATIONS
   POLARONS

PLASTERS
GS   **PLASTERS**
   . GYPSUM
   . PARAPLASTS
RT   CASTS
   GROUT
   MOLDING MATERIALS
   MORTARS (MATERIAL)
   PASTES

PLASTIC AIRCRAFT STRUCTURES
GS   AIRCRAFT STRUCTURES
   . **PLASTIC AIRCRAFT STRUCTURES**
RT   AIRCRAFT CONSTRUCTION MATERIALS
   AIRCRAFT SURVIVABILITY
   BORON-EPOXY COMPOUNDS
   GLASS FIBER REINFORCED PLASTICS
   PLASTICS

PLASTIC ANISOTROPY
GS   ANISOTROPY
   . **PLASTIC ANISOTROPY**
   . . ELASTIC ANISOTROPY
RT   VISCOPLASTICITY

PLASTIC COATINGS
GS   COATINGS
   . **PLASTIC COATINGS**
RT   ANTIRADAR COATINGS
   ENCAPSULATING
   POLYMERIC FILMS
   PROTECTIVE COATINGS
   SPRAYED COATINGS

PLASTIC DEFORMATION
UF   LUDER BANDS
   PLASTIC YIELDING
   STRAIN SOFTENING
GS   DEFORMATION
   . **PLASTIC DEFORMATION**
RT   ∞BANDS
   BENDING
   BORDONI PEAKS
   CREEP PROPERTIES
   CREEP TESTS
   ELASTIC DEFORMATION
   ELONGATION
   J INTEGRAL
   PLANE STRAIN
   SAINT VENANT PRINCIPLE
   SHEAR CREEP
   ∞SLIP
   STRESS PROPAGATION
   STRESS RELAXATION
   STRESS-STRAIN RELATIONSHIPS
   STRETCHING
   STRUCTURAL STRAIN
   SUPERPLASTICITY
   TEMPERATURE INVERSIONS

PLASTIC DEFORMATION-*(CONT.)*
TENSILE CREEP
TENSILE DEFORMATION
THERMOMECHANICAL TREATMENT
WARPAGE
WORK SOFTENING
YIELD STRENGTH

PLASTIC FILMS
USE   POLYMERIC FILMS

PLASTIC FLOW
GS   FLUID FLOW
   . **PLASTIC FLOW**
   . . TRESCA FLOW
RT   CREEP PROPERTIES
   ∞FLOW
   INTERNAL FRICTION
   RHEOLOGY
   SHEAR FLOW
   STEADY STATE CREEP
   STRESS RELAXATION
   SUPERPLASTICITY
   VISCOELASTICITY
   VISCOPLASTICITY

PLASTIC MATERIALS
USE   PLASTICS

PLASTIC MEMORY
RT   SHAPE MEMORY ALLOYS
   STRESS RELAXATION

PLASTIC PROPELLANTS
GS   PROPELLANTS
   . SOLID PROPELLANTS
   . . **PLASTIC PROPELLANTS**
RT   CHEMICAL FUELS
   COMPOSITE PROPELLANTS
   EXPLOSIVES
   GELLED PROPELLANTS
   HTPB PROPELLANTS
   MONOPROPELLANTS
   PLASTICIZERS
   POLYBUTADIENE TETRANITRAMINE
   PYROTECHNICS

PLASTIC PROPERTIES
UF   PLASTICITY
GS   MECHANICAL PROPERTIES
   . **PLASTIC PROPERTIES**
   . . ELASTOPLASTICITY
   . . PHOTOPLASTICITY
   . . SUPERPLASTICITY
   . . THERMOPLASTICITY
   . . VISCOPLASTICITY
   . . YIELD POINT
RT   COHESION
   COLD FLOW TESTS
   DUCTILITY
   ELASTIC PROPERTIES
   FATIGUE (MATERIALS)
   FLEXIBILITY
   HARDNESS
   INFLUENCE COEFFICIENT
   METHOD OF CHARACTERISTICS
   ∞PROPERTIES
   RHEOLOGY
   SEMISOLIDS
   STRESS RELAXATION
   STRESS TENSORS
   STRUCTURAL STABILITY

PLASTIC TAPES
RT   ADHESIVES
   MAGNETIC TAPES
   ∞TAPES

PLASTIC YIELDING
USE   PLASTIC DEFORMATION

PLASTICITY
USE   PLASTIC PROPERTIES

PLASTICIZERS
UF   CASTING SOLVENTS
   ELASTICIZERS
GS   ADDITIVES
   . **PLASTICIZERS**
RT   CASE BONDED PROPELLANTS
   COATINGS
   DOMINO PROPELLANTS
   ESTERS
   PLASTIC PROPELLANTS

PLASTICIZERS-*(CONT.)*
　　　PROPELLANT ADDITIVES
　　　SKYDROL (TRADEMARK)
　　　SOLID PROPELLANTS
　　　SOLID ROCKET BINDERS
　　　SURFACTANTS
　　　TRIACETIN

PLASTICS
　UF　　PLASTIC MATERIALS
　GS　　**PLASTICS**
　　　. CARBON FIBER REINFORCED
　　　　　PLASTICS
　　　. DELRIN (TRADEMARK)
　　　. PERSPEX (TRADEMARK)
　　　. POLYBUTADIENE
　　　. POLYETHYLENES
　　　. . POLYETHYLENE TEREPHTHALATE
　　　. POLYISOBUTYLENE
　　　. POLYPROPYLENE
　　　. POLYSTYRENE
　　　. POLYTETRAFLUOROETHYLENE
　　　. POLYVINYL ALCOHOL
　　　. POLYVINYL CHLORIDE
　　　. POLYVINYL FLUORIDE
　　　. REINFORCED PLASTICS
　　　. . MICARTA
　　　. SYNTHETIC RESINS
　　　. . ADDITION RESINS
　　　. . . ACRYLIC RESINS
　　　. . . VINYL COPOLYMERS
　　　. . POLYESTER RESINS
　　　. . POLYETHER RESINS
　　　. . . POLYMETHYL METHACRYLATE
　　　. . THERMOPLASTIC RESINS
　　　. . . QUINOXALINES
　　　. . . THERMOPLASTIC FILMS
　　　. . THERMOSETTING RESINS
　　　. . . EPOXY RESINS
　　　. . . FURAN RESINS
　　　. . . POLYAMIDE RESINS
　　　. . . . KEVLAR (TRADEMARK)
　　　. . . PHENOLIC RESINS
　　　. . . . MICARTA
　　　. TEFLON (TRADEMARK)
　　　. VITON
　RT　　ACRYLONITRILES
　　　BORON REINFORCED MATERIALS
　　　∞CONSTRUCTION MATERIALS
　　　ELASTOMERS
　　　FLUOROPOLYMERS
　　　FURANS
　　　INJECTION MOLDING
　　　ION EXCHANGE RESINS
　　　KAPTON (TRADEMARK)
　　　∞MATERIALS SCIENCE
　　　MOLDING MATERIALS
　　　ORGANIC MATERIALS
　　　PLASTIC AIRCRAFT STRUCTURES
　　　∞POLYMERS
　　　STYROFOAM (TRADEMARK)
　　　TETRAHYDROFURAN
　　　THIOPLASTICS

PLASTISOLS
　GS　　MIXTURES
　　　. DISPERSIONS
　　　. . **PLASTISOLS**
　　　. . . SMOKE
　RT　　COLLOIDS
　　　COMPOSITE PROPELLANTS
　　　DOUBLE BASE PROPELLANTS
　　　RESINS

PLAT SYSTEM
　UF　　PILOT LANDING AID TELEVISION
　　　　　SYSTEM
　GS　　COMMUNICATION EQUIPMENT
　　　. **PLAT SYSTEM**
　　　TELECOMMUNICATION
　　　. **PLAT SYSTEM**
　　　TELEVISION SYSTEMS
　　　. **PLAT SYSTEM**
　RT　　LANDING AIDS
　　　∞SYSTEMS

PLATE (METAL)
　USE　　METAL PLATES

PLATE THEORY
　RT　　FLAT PLATES
　　　STRUCTURAL ANALYSIS
　　　∞THEORIES

PLATEAUS
　GS　　LANDFORMS
　　　. TERRACES (LANDFORMS)
　　　. . **PLATEAUS**
　　　. . . ALLEGHENY PLATEAU (US)
　　　. . . COLORADO PLATEAU (US)
　　　. . . MESAS
　　　. . . . BUTTES
　　　. . . PIEDMONTS
　　　. . . . CENTRAL PIEDMONT (US)
　RT　　APEXES
　　　EROSION
　　　HIGHLANDS
　　　∞PEAKS
　　　PLAINS
　　　STRATIGRAPHY

PLATELETS
　RT　　BLOOD COAGULATION
　　　BLOOD GROUPS
　　　HISTOLOGY
　　　THROMBOPLASTIN

PLATENS
　RT　　∞PLATES
　　　PRESSES
　　　∞PRESSING
　　　PUNCHES
　　　RAMS (PRESSES)
　　　ROLLERS
　　　TOOLS

∞ **PLATES**
　SN　　*(USE OF A MORE SPECIFIC TERM IS*
　　　　RECOMMENDED--CONSULT THE TERMS
　　　　LISTED BELOW)
　RT　　CORRUGATING
　　　DISKS (SHAPES)
　　　FLAT PLATES
　　　METAL COATINGS
　　　METAL PLATES
　　　MICROCHANNEL PLATES
　　　PANELS
　　　PARALLEL PLATES
　　　PHOTOGRAPHIC PLATES
　　　PLATENS
　　　PLATES (STRUCTURAL MEMBERS)
　　　PLATING
　　　RECTANGULAR PLATES
　　　SCATTER PLATES (OPTICS)
　　　THICK PLATES
　　　THIN PLATES
　　　TRAYS

PLATES (STRUCTURAL MEMBERS)
　GS　　STRUCTURAL MEMBERS
　　　. **PLATES (STRUCTURAL MEMBERS)**
　　　. . ANISOTROPIC PLATES
　　　. . ANNULAR PLATES
　　　. . CANTILEVER PLATES
　　　. . CIRCULAR PLATES
　　　. . CORRUGATED PLATES
　　　. . ELASTIC PLATES
　　　. . END PLATES
　　　. . GIRDER WEBS
　　　. . METAL PLATES
　　　. . . BOILER PLATE
　　　. . ORTHOTROPIC PLATES
　　　. . PERFORATED PLATES
　　　. . POROUS PLATES
　　　. . REINFORCED PLATES
　RT　　FLAT PLATES
　　　GIRDERS
　　　METAL SHEETS
　　　ORTHOTROPISM
　　　∞PLATES
　　　REISSNER THEORY
　　　SLABS
　　　THICK PLATES
　　　THIN PLATES

PLATES (TECTONICS)
　RT　　EARTH CRUST
　　　EARTH MANTLE
　　　EARTH PLANETARY STRUCTURE
　　　EARTHQUAKES
　　　GEOLOGICAL FAULTS
　　　GEOPHYSICS
　　　LITHOSPHERE
　　　STRUCTURAL PROPERTIES (GEOLOGY)
　　　SUBDUCTION (GEOLOGY)
　　　TECTONICS

∞ **PLATFORMS**
　SN　　*(USE OF A MORE SPECIFIC TERM IS*
　　　　RECOMMENDED--CONSULT THE TERMS
　　　　LISTED BELOW)
　RT　　FLIGHT MECHANICS
　　　FLOORS
　　　FLYING PLATFORMS
　　　GUIDANCE (MOTION)
　　　INERTIAL PLATFORMS
　　　LANDFORMS
　　　LAUNCHING PADS
　　　OFFSHORE PLATFORMS
　　　SLABS
　　　SOLETTAS
　　　SPACE STATIONS
　　　STABILIZED PLATFORMS
　　　SUPPORTS
　　　SYNCHRONOUS PLATFORMS

PLATING
　GS　　**PLATING**
　　　. ELECTROPLATING
　　　. FLAME PLATING
　　　. ION PLATING
　　　. NICKEL PLATE
　RT　　ANODIC STRIPPING
　　　CATHODIC COATINGS
　　　CLADDING
　　　DEPOSITION
　　　DEPOSITS
　　　ELECTRODEPOSITION
　　　ELECTROLESS DEPOSITION
　　　FINISHES
　　　LAMINATES
　　　METAL COATINGS
　　　METAL FINISHING
　　　METALLIZING
　　　∞PLATES
　　　PROTECTIVE COATINGS
　　　SUBSTRATES
　　　THIN FILMS

PLATINUM
　GS　　CHEMICAL ELEMENTS
　　　. **PLATINUM**
　　　. . PLATINUM BLACK
　　　. . PLATINUM ISOTOPES
　　　METALS
　　　. TRANSITION METALS
　　　. . **PLATINUM**
　　　. . . PLATINUM BLACK
　　　. . . PLATINUM ISOTOPES

PLATINUM ALLOYS
　GS　　ALLOYS
　　　. **PLATINUM ALLOYS**
　RT　　RHODIUM ALLOYS

PLATINUM BLACK
　GS　　CHEMICAL ELEMENTS
　　　. PLATINUM
　　　. . **PLATINUM BLACK**
　　　METALS
　　　. METAL POWDER
　　　. . **PLATINUM BLACK**
　　　. TRANSITION METALS
　　　. . PLATINUM
　　　. . . **PLATINUM BLACK**
　　　PARTICLES
　　　. METAL PARTICLES
　　　. . **PLATINUM BLACK**
　　　. POWDER (PARTICLES)
　　　. . **PLATINUM BLACK**
　RT　　CATALYSTS

PLATINUM COMPOUNDS
　GS　　**PLATINUM COMPOUNDS**
　　　. PLATINUM OXIDES
　RT　　∞CHEMICAL COMPOUNDS
　　　∞GROUP 8 COMPOUNDS
　　　∞METAL COMPOUNDS

PLATINUM ISOTOPES
　GS　　CHEMICAL ELEMENTS
　　　. NUCLIDES
　　　. . ISOTOPES
　　　. . . **PLATINUM ISOTOPES**
　　　. PLATINUM
　　　. . **PLATINUM ISOTOPES**
　　　METALS
　　　. TRANSITION METALS
　　　. . PLATINUM
　　　. . . **PLATINUM ISOTOPES**

PLATINUM OXIDES
GS   CHALCOGENIDES
. OXIDES
. . METAL OXIDES
. . . **PLATINUM OXIDES**
PLATINUM COMPOUNDS
. **PLATINUM OXIDES**

PLAYAS
GS   LAND
. PLAINS
. . **PLAYAS**
LANDFORMS
. **PLAYAS**
RT   DESERTS
LAKES

PLAYBACKS
RT   MAGNETIC TAPES
∞RECORDERS
RECORDING
RECORDS
∞TAPES
VIDEO DISKS

PLENUM CHAMBERS
RT   AIR INTAKES
∞CHAMBERS
DUCTS
EXHAUST SYSTEMS
FUEL SYSTEMS
INTAKE SYSTEMS
MANIFOLDS
∞WATER INTAKES

PLETHYSMOGRAPHY
GS   BIOENGINEERING
. BIOMETRICS
. . **PLETHYSMOGRAPHY**
. . . ELECTROPLETHYSMOGRAPHY

PLEURAE
GS   MEMBRANES
. **PLEURAE**
RT   LUNGS

PLEUROTIN
GS   DRUGS
. ANTIBIOTICS
. . **PLEUROTIN**
RT   STAPHYLOCOCCUS

PLEXIGLASS (TRADEMARK)
USE   POLYMETHYL METHACRYLATE

PLIES
USE   LAYERS

∞ **PLOTS**
SN   *(USE OF A MORE SPECIFIC TERM IS
RECOMMENDED--CONSULT THE TERMS
LISTED BELOW)*
RT   CHARTS
DISPLAY DEVICES
PLOTTERS
PLOTTING
SITES

PLOTTERS
UF   PLOTTING INSTRUMENTS
GS   RECORDING INSTRUMENTS
. **PLOTTERS**
. . X-Y PLOTTERS
RT   COMPUTER GRAPHICS
DIGITAL TO ANALOG CONVERTERS
DISPLAY DEVICES
NAVIGATION AIDS
∞PLOTS
PLOTTING
POSITION INDICATORS
PRINTERS
REMOTE CONSOLES

PLOTTING
RT   ANALOG TO DIGITAL CONVERTERS
DISPLAY DEVICES
NAVIGATION
∞PLOTS
PLOTTERS
PRINTING
RECORDING

PLOTTING INSTRUMENTS
USE   PLOTTERS

PLOWED FIELDS
USE   FARMLANDS

PLOWING
GS   CULTIVATION
. **PLOWING**
RT   AGRICULTURE
FARM CROPS
FARMLANDS
GRASSLANDS
PLANTING
PLANTS (BOTANY)
PLOWS
SOD
TRACTORS

PLOWS
RT   AGRICULTURE
MIXERS
PLANTING
PLOWING

PLSS
USE   PORTABLE LIFE SUPPORT SYSTEMS

PLUG NOZZLES
GS   EXHAUST NOZZLES
. **PLUG NOZZLES**
RT   ANNULAR NOZZLES
CONICAL NOZZLES
NOZZLE GEOMETRY
∞NOZZLES
ROCKET NOZZLES
SPIKE NOZZLES

PLUGGING
UF   CLOGGING
RT   AGGLOMERATION
BLOCKING
CAULKING
CLOSING
CLOSURES
CONSTRICTIONS
FOULING
PLUGS
SEALING
SEALS (STOPPERS)

PLUGS
SN   (EXCLUDES SPARKPLUGS OR
ELECTRICAL CONNECTORS)
GS   SEALS (STOPPERS)
. **PLUGS**
RT   BLOCKING
CLOSURES
LABYRINTH SEALS
OUTLETS
PLUGGING
STOPPING

PLUM BROOK REACTOR
GS   NUCLEAR REACTORS
. LIQUID COOLED REACTORS
. . WATER COOLED REACTORS
. . . **PLUM BROOK REACTOR**
. NUCLEAR RESEARCH AND TEST
REACTORS
. . **PLUM BROOK REACTOR**

PLUMAGE
RT   BIRDS

PLUMBANE
USE   LEAD COMPOUNDS
METAL HYDRIDES

PLUMES
GS   **PLUMES**
. ROCKET EXHAUST
RT   CHIMNEYS
CONDENSATES
POLLUTION TRANSPORT
SHOCK WAVES

PLUNGERS
RT   MIXERS
PISTONS
RAMS (PRESSES)
RAMS (PUMPS)

PLUTO (PLANET)
GS   CELESTIAL BODIES
. PLANETS
. . **PLUTO (PLANET)**

PLUTO (PLANET)-*(CONT.)*
RT   CHARON

PLUTO REACTORS
GS   NUCLEAR REACTORS
. **PLUTO REACTORS**
RT   NUCLEAR RAMJET ENGINES
NUCLEAR ROCKET ENGINES
SUPERSONIC LOW ALTITUDE MISSILE

PLUTONIUM
GS   CHEMICAL ELEMENTS
. ACTINIDE SERIES
. . TRANSURANIUM ELEMENTS
. . . **PLUTONIUM**
. . . . PLUTONIUM ISOTOPES
. . . . . PLUTONIUM 238
. . . . . PLUTONIUM 239
. . . . . PLUTONIUM 240
. . . . . PLUTONIUM 241
. . . . . PLUTONIUM 244
. NUCLIDES
. . ISOTOPES
. . . RADIOACTIVE ISOTOPES
. . . . TRANSURANIUM ELEMENTS
. . . . . **PLUTONIUM**
. . . . . . PLUTONIUM ISOTOPES
HEAVY ELEMENTS
. TRANSURANIUM ELEMENTS
. . **PLUTONIUM**
. . . PLUTONIUM ISOTOPES
. . . . PLUTONIUM 238
. . . . PLUTONIUM 239
. . . . PLUTONIUM 240
. . . . PLUTONIUM 241
. . . . PLUTONIUM 244
METALS
. ACTINIDE SERIES
. . TRANSURANIUM ELEMENTS
. . . **PLUTONIUM**
. . . . PLUTONIUM ISOTOPES
. . . . . PLUTONIUM 238
. . . . . PLUTONIUM 239
. . . . . PLUTONIUM 240
. . . . . PLUTONIUM 241
. . . . . PLUTONIUM 244
RT   FISSIONABLE MATERIALS
NUCLEAR FUELS

PLUTONIUM ALLOYS
GS   ALLOYS
. **PLUTONIUM ALLOYS**
RT   NUCLEAR FUEL ELEMENTS
NUCLEAR FUELS

PLUTONIUM CARBIDES
USE   PLUTONIUM COMPOUNDS

PLUTONIUM COMPOUNDS
UF   PLUTONIUM CARBIDES
GS   ACTINIDE SERIES COMPOUNDS
. **PLUTONIUM COMPOUNDS**
. . PLUTONIUM FLUORIDES
. . PLUTONIUM OXIDES
RT   CERAMIC NUCLEAR FUELS
∞CHEMICAL COMPOUNDS
∞METAL COMPOUNDS
NUCLEAR FUELS

PLUTONIUM FLUORIDES
GS   ACTINIDE SERIES COMPOUNDS
. PLUTONIUM COMPOUNDS
. . **PLUTONIUM FLUORIDES**
HALOGEN COMPOUNDS
. FLUORINE COMPOUNDS
. . FLUORIDES
. . . METAL FLUORIDES
. . . . **PLUTONIUM FLUORIDES**
. HALIDES
. . METAL HALIDES
. . . **PLUTONIUM FLUORIDES**

PLUTONIUM ISOTOPES
GS   CHEMICAL ELEMENTS
. ACTINIDE SERIES
. . TRANSURANIUM ELEMENTS
. . . PLUTONIUM
. . . . **PLUTONIUM ISOTOPES**
. . . . . PLUTONIUM 238
. . . . . PLUTONIUM 239
. . . . . PLUTONIUM 240
. . . . . PLUTONIUM 241
. . . . . PLUTONIUM 244
. NUCLIDES
. . ISOTOPES

PLUTONIUM ISOTOPES-(CONT.)
```
. . . RADIOACTIVE ISOTOPES
. . . . TRANSURANIUM ELEMENTS
. . . . . PLUTONIUM
. . . . . . PLUTONIUM ISOTOPES
HEAVY ELEMENTS
. TRANSURANIUM ELEMENTS
. . PLUTONIUM
. . . PLUTONIUM ISOTOPES
. . . . PLUTONIUM 238
. . . . PLUTONIUM 239
. . . . PLUTONIUM 240
. . . . PLUTONIUM 241
. . . . PLUTONIUM 244
METALS
. ACTINIDE SERIES
. . TRANSURANIUM ELEMENTS
. . PLUTONIUM
. . . PLUTONIUM ISOTOPES
. . . . PLUTONIUM 238
. . . . PLUTONIUM 239
. . . . PLUTONIUM 240
. . . . PLUTONIUM 241
. . . . PLUTONIUM 244
```

PLUTONIUM OXIDES
```
GS    ACTINIDE SERIES COMPOUNDS
      . PLUTONIUM COMPOUNDS
      . . PLUTONIUM OXIDES
      CHALCOGENIDES
      . OXIDES
      . . METAL OXIDES
      . . . PLUTONIUM OXIDES
RT    CERAMIC NUCLEAR FUELS
      MIXED OXIDES
```

PLUTONIUM RECYCLE TEST REACTOR
```
UF    PRTR (REACTOR)
GS    NUCLEAR ELECTRIC POWER
         GENERATION
      . NUCLEAR POWER REACTORS
      . . PLUTONIUM RECYCLE TEST
            REACTOR
      NUCLEAR REACTORS
      . LIQUID COOLED REACTORS
      . . WATER COOLED REACTORS
      . . . HEAVY WATER REACTORS
      . . . . PLUTONIUM RECYCLE TEST
                REACTOR
      . NUCLEAR POWER-REACTORS
      . . PLUTONIUM RECYCLE TEST
            REACTOR
      . NUCLEAR RESEARCH AND TEST
            REACTORS
      . PLUTONIUM RECYCLE TEST
          REACTOR
      . WATER MODERATED REACTORS
      . . PLUTONIUM RECYCLE TEST
            REACTOR
```

PLUTONIUM 238
```
GS    CHEMICAL ELEMENTS
      . ACTINIDE SERIES
      . . TRANSURANIUM ELEMENTS
      . . . PLUTONIUM
      . . . . PLUTONIUM ISOTOPES
      . . . . . PLUTONIUM 238
      . NUCLIDES
      . . ISOTOPES
      . . . RADIOACTIVE ISOTOPES
      . . . . TRANSURANIUM ELEMENTS
      . . . . . PLUTONIUM
      . . . . . . PLUTONIUM ISOTOPES
      HEAVY ELEMENTS
      . TRANSURANIUM ELEMENTS
      . . PLUTONIUM
      . . . PLUTONIUM ISOTOPES
      . . . . PLUTONIUM 238
      METALS
      . ACTINIDE SERIES
      . . TRANSURANIUM ELEMENTS
      . . . PLUTONIUM
      . . . . PLUTONIUM ISOTOPES
      . . . . . PLUTONIUM 238
```

PLUTONIUM 239
```
GS    CHEMICAL ELEMENTS
      . ACTINIDE SERIES
      . . TRANSURANIUM ELEMENTS
      . . . PLUTONIUM
      . . . . PLUTONIUM ISOTOPES
      . . . . . PLUTONIUM 239
      . NUCLIDES
      . . ISOTOPES
      . . . RADIOACTIVE ISOTOPES
```

PLUTONIUM 239-(CONT.)
```
. . . . TRANSURANIUM ELEMENTS
. . . . . PLUTONIUM
. . . . . . PLUTONIUM ISOTOPES
HEAVY ELEMENTS
. TRANSURANIUM ELEMENTS
. . PLUTONIUM
. . . PLUTONIUM ISOTOPES
. . . . PLUTONIUM 239
METALS
. ACTINIDE SERIES
. . TRANSURANIUM ELEMENTS
. . . PLUTONIUM
. . . . PLUTONIUM ISOTOPES
. . . . . PLUTONIUM 239
```

PLUTONIUM 240
```
GS    CHEMICAL ELEMENTS
      . ACTINIDE SERIES
      . . TRANSURANIUM ELEMENTS
      . . . PLUTONIUM
      . . . . PLUTONIUM ISOTOPES
      . . . . . PLUTONIUM 240
      . NUCLIDES
      . . ISOTOPES
      . . . RADIOACTIVE ISOTOPES
      . . . . TRANSURANIUM ELEMENTS
      . . . . . PLUTONIUM
      . . . . . . PLUTONIUM ISOTOPES
      HEAVY ELEMENTS
      . TRANSURANIUM ELEMENTS
      . . PLUTONIUM
      . . . PLUTONIUM ISOTOPES
      . . . . PLUTONIUM 240
      METALS
      . ACTINIDE SERIES
      . . TRANSURANIUM ELEMENTS
      . . . PLUTONIUM
      . . . . PLUTONIUM ISOTOPES
      . . . . . PLUTONIUM 240
```

PLUTONIUM 241
```
GS    CHEMICAL ELEMENTS
      . ACTINIDE SERIES
      . . TRANSURANIUM ELEMENTS
      . . . PLUTONIUM
      . . . . PLUTONIUM ISOTOPES
      . . . . . PLUTONIUM 241
      . NUCLIDES
      . . ISOTOPES
      . . . RADIOACTIVE ISOTOPES
      . . . . TRANSURANIUM ELEMENTS
      . . . . . PLUTONIUM
      . . . . . . PLUTONIUM ISOTOPES
      HEAVY ELEMENTS
      . TRANSURANIUM ELEMENTS
      . . PLUTONIUM
      . . . PLUTONIUM ISOTOPES
      . . . . PLUTONIUM 241
      METALS
      . ACTINIDE SERIES
      . . TRANSURANIUM ELEMENTS
      . . . PLUTONIUM
      . . . . PLUTONIUM ISOTOPES
      . . . . . PLUTONIUM 241
```

PLUTONIUM 244
```
GS    CHEMICAL ELEMENTS
      . ACTINIDE SERIES
      . . TRANSURANIUM ELEMENTS
      . . . PLUTONIUM
      . . . . PLUTONIUM ISOTOPES
      . . . . . PLUTONIUM 244
      . NUCLIDES
      . . ISOTOPES
      . . . RADIOACTIVE ISOTOPES
      . . . . TRANSURANIUM ELEMENTS
      . . . . . PLUTONIUM
      . . . . . . PLUTONIUM ISOTOPES
      HEAVY ELEMENTS
      . TRANSURANIUM ELEMENTS
      . . PLUTONIUM
      . . . PLUTONIUM ISOTOPES
      . . . . PLUTONIUM 244
      METALS
      . ACTINIDE SERIES
      . . TRANSURANIUM ELEMENTS
      . . . PLUTONIUM
      . . . . PLUTONIUM ISOTOPES
      . . . . . PLUTONIUM 244
```

PLUVIOGRAPHS
```
USE   RAIN GAGES
      RECORDING INSTRUMENTS
```

PLY ORIENTATION
```
RT    ALIGNMENT
      COMPOSITE MATERIALS
      INTERLAYERS
      LAMINATES
      ∞LAYERS
      MULTILAYER INSULATION
      ∞ORIENTATION
      PLYWOOD
      POSITIONING
      SANDWICH STRUCTURES
      SUBSTRATES
```

PLYWOOD
```
GS    COMPOSITE MATERIALS
      . LAMINATES
      . . PLYWOOD
      COMPOSITE STRUCTURES
      . LAMINATES
      . . PLYWOOD
      WOOD
      . PLYWOOD
RT    PLY ORIENTATION
      TREES (PLANTS)
      WOODEN STRUCTURES
```

PNEUMATIC CIRCUITS
```
GS    CIRCUITS
      . PNEUMATIC CIRCUITS
      PNEUMATIC EQUIPMENT
      . PNEUMATIC CIRCUITS
RT    FLUIDICS
      VALVES
```

PNEUMATIC CONTROL
```
UF    PNEUMATIC RESET
RT    AUTOMATIC CONTROL
      AUTOMATIC CONTROL VALVES
      COMPRESSED GAS
      ∞CONTROL
      CONTROL EQUIPMENT
      CONTROL VALVES
      CONTROLLERS
      ELECTRONIC CONTROL
      ENGINE CONTROL
      FLUID POWER
      FLUIDICS
      HYDRAULIC CONTROL
      REMOTE CONTROL
```

PNEUMATIC EQUIPMENT
```
GS    PNEUMATIC EQUIPMENT
      . GAS VALVES
      . PNEUMATIC CIRCUITS
RT    AIR BAG RESTRAINT DEVICES
      COMPRESSED AIR
      CUSHIONS
      ∞EQUIPMENT
      FLUID AMPLIFIERS
      FLUID POWER
      FLUID SWITCHING ELEMENTS
      FLUIDICS
      GAS GENERATORS
      GOLAY DETECTOR CELLS
      INFLATABLE STRUCTURES
      ∞NETWORKS
      SERVOCONTROL
      SERVOMECHANISMS
      SHOCK ABSORBERS
      ∞SYSTEMS
      VALVES
```

PNEUMATIC PROBES
```
GS    MEASURING INSTRUMENTS
      . TEMPERATURE MEASURING
          INSTRUMENTS
      . . TEMPERATURE PROBES
      . . . PNEUMATIC PROBES
RT    FLOW MEASUREMENT
      HIGH TEMPERATURE GASES
      MASS FLOW RATE
      NOZZLE FLOW
      PRESSURE MEASUREMENT
```

PNEUMATIC RESET
```
USE   PNEUMATIC CONTROL
```

PNEUMATICS
```
GS    FLUID MECHANICS
      . PNEUMATICS
RT    FLOW THEORY
      FLUID POWER
      FLUIDICS
      GASES
      ∞HYDRAULICS
```

PNEUMOGRAPHS
USE   PNEUMOGRAPHY

PNEUMOGRAPHY
UF   PNEUMOGRAPHS
RT   BIOTELEMETRY
      LUNGS
      ∞MEASUREMENT
      RADIOGRAPHY

PNEUMONIA
GS   DISEASES
      . RESPIRATORY DISEASES
      .. **PNEUMONIA**
RT   CONGESTION

PNEUMOTHORAX
RT   DISEASES
      LUNGS
      MEDICAL SCIENCE
      ORGANS

PNICTIDES
USE   GROUP 5A COMPOUNDS

POCKELS EFFECT
USE   BIREFRINGENCE

POCKET MICE
GS   ANIMALS
      . VERTEBRATES
      .. MAMMALS
      ... RODENTS
      .... MICE
      ..... **POCKET MICE**
RT   RATS

PODS (EXTERNAL STORES)
GS   EXTERNAL STORES
      . **PODS (EXTERNAL STORES)**
RT   COWLINGS
      EXTERNAL STORE SEPARATION
      FUEL TANKS
      NACELLES
      WING-FUSELAGE STORES

POGO
UF   POLAR ORBIT GEOPHYSICAL
          OBSERVATORY
GS   OBSERVATORIES
      . GEOPHYSICAL OBSERVATORIES
      .. OGO
      ... **POGO**
      .... OGO-C
      .... OGO-4
      .... OGO-6
      SATELLITES
      . ARTIFICIAL SATELLITES
      .. GEOPHYSICAL SATELLITES
      ... OGO
      .... **POGO**
      ..... OGO-C
      ..... OGO-4
      ..... OGO-6
RT   AGENA B ROCKET VEHICLE
      EGO

POGO EFFECTS
GS   VIBRATION
      . **POGO EFFECTS**
      VIBRATION EFFECTS
      . **POGO EFFECTS**
RT   ∞EFFECTS
      LONGITUDINAL STABILITY

POHLHAUSEN METHOD
UF   POHLHAUSEN SOLUTION
GS   ANALYSIS (MATHEMATICS)
      . NUMERICAL ANALYSIS
      .. APPROXIMATION
      ... **POHLHAUSEN METHOD**
RT   LAMINAR BOUNDARY LAYER
      ∞METHODOLOGY
      VELOCITY DISTRIBUTION
      VISCOUS FLOW

POHLHAUSEN SOLUTION
USE   POHLHAUSEN METHOD

POIKILOTHERMIA
GS   ANIMALS
      . **POIKILOTHERMIA**
      .. AMPHIBIA
      ... FROGS

POIKILOTHERMIA-(CONT.)
      .. REPTILES
      ... LIZARDS
      ... SNAKES
      ... TURTLES
RT   BODY TEMPERATURE

POINCARE PROBLEM
RT   ∞PROBLEMS

POINCARE SPHERES
GS   SYMMETRICAL BODIES
      . BODIES OF REVOLUTION
      .. SPHERES
      ... **POINCARE SPHERES**
RT   GEOMETRY

POINT DEFECTS
GS   DEFECTS
      . CRYSTAL DEFECTS
      .. **POINT DEFECTS**
      ... VACANCIES (CRYSTAL DEFECTS)
      .... FRENKEL DEFECTS
RT   CRYSTAL DISLOCATIONS
      IMPURITIES
      SURFACE DEFECTS

POINT IMPACT
GS   IMPACT
      . **POINT IMPACT**
RT   ELECTRON IMPACT
      HYPERVELOCITY IMPACT
      ION IMPACT
      PROTON IMPACT

POINT MATCHING METHOD (MATHEMATICS)
USE   BOUNDARY VALUE PROBLEMS

POINT SOURCES
GS   RADIATION SOURCES
      . **POINT SOURCES**
RT   DIFFUSE RADIATION
      ∞ENERGY SOURCES
      HUYGENS PRINCIPLE
      LIGHT SOURCES
      SPHERICAL WAVES

POINT SPREAD FUNCTIONS
GS   FUNCTIONS (MATHEMATICS)
      . **POINT SPREAD FUNCTIONS**
RT   IMAGE PROCESSING

POINT TO POINT COMMUNICATION
GS   COMMUNICATING
      . **POINT TO POINT COMMUNICATION**
      .. NASCOM NETWORK
RT   RADIO COMMUNICATION
      TELECOMMUNICATION
      WESTAR SATELLITES
      WIDEBAND COMMUNICATION

POINTERS
USE   DIALS

POINTING CONTROL SYSTEMS
GS   FLIGHT CONTROL
      . **POINTING CONTROL SYSTEMS**
      .. ANNULAR SUSPENSION AND
          POINTING SYSTEM
RT   ∞CONTROL
      ENTRY GUIDANCE (STS)
      GUIDANCE (MOTION)
      SPACE FLIGHT
      SPACECRAFT CONTROL
      ∞SYSTEMS

∞ POINTS
SN   (USE OF A MORE SPECIFIC TERM IS
      RECOMMENDED--CONSULT THE TERMS
      LISTED BELOW)
RT   NONPOINT SOURCES
      POINTS (MATHEMATICS)
      POSITION (LOCATION)

POINTS (MATHEMATICS)
GS   GEOMETRY
      . EUCLIDEAN GEOMETRY
      .. **POINTS (MATHEMATICS)**
      ... FIXED POINTS (MATHEMATICS)
      ... INFLECTION POINTS
RT   FOCI
      LOCI
      NAKED SINGULARITIES
      ∞POINTS

POINTS (MATHEMATICS)-(CONT.)
      RECIPROCAL THEOREMS
      SINGULARITY (MATHEMATICS)

POISEUILLE FLOW
USE   LAMINAR FLOW

∞ POISONING
SN   (USE OF A MORE SPECIFIC TERM IS
      RECOMMENDED--CONSULT THE TERMS
      LISTED BELOW)
RT   BENZENE POISONING
      BERYLLIUM POISONING
      CARBON MONOXIDE POISONING
      CARBON TETRACHLORIDE POISONING
      CURARE
      HYDROCARBON POISONING
      INTOXICATION
      LEAD POISONING
      NARCOSIS
      POISONING (REACTION INHIBITION)
      TOXIC DISEASES
      TOXIC HAZARDS

POISONING (REACTION INHIBITION)
RT   CONTROL RODS
      NEUTRON ABSORBERS
      NUCLEAR REACTIONS
      ∞POISONING
      RADIOACTIVE WASTES

POISONING (TOXICOLOGY)
USE   TOXIC DISEASES

POISONS
GS   POISONS
      . CARBAMATES (TRADENAME)
      .. URETHANES
      . CURARE
      . ENDOTOXINS
      . PESTICIDES
      .. INSECTICIDES
      ... DIELDRIN
      . PHOSGENE
      . STRYCHNINE
RT   ENVIRONMENT EFFECTS
      ENVIRONMENT POLLUTION
      ENVIRONMENTAL SURVEYS
      NONPOINT SOURCES
      POLLUTION
      TOXICITY

POISSON DENSITY FUNCTIONS
UF   POISSON PROCESS
GS   FUNCTIONS (MATHEMATICS)
      . **POISSON DENSITY FUNCTIONS**
      STATISTICAL ANALYSIS
      . **POISSON DENSITY FUNCTIONS**
RT   CONTINUITY (MATHEMATICS)
      DISCRETE FUNCTIONS
      EXPONENTIAL FUNCTIONS

POISSON EQUATION
GS   ANALYSIS (MATHEMATICS)
      . REAL VARIABLES
      .. DIFFERENTIAL EQUATIONS
      ... **POISSON EQUATION**
RT   CLASSICAL MECHANICS
      ELECTROSTATICS
      ∞EQUATIONS
      ISENTROPE
      LAPLACE EQUATION
      PARTIAL DIFFERENTIAL EQUATIONS

POISSON PROCESS
USE   POISSON DENSITY FUNCTIONS
      STOCHASTIC PROCESSES

POISSON RATIO
GS   MECHANICAL PROPERTIES
      . **POISSON RATIO**
      RATIOS
      . **POISSON RATIO**
RT   AIRY FUNCTION
      COMPRESSIVE STRENGTH
      ELASTIC PROPERTIES
      FIBER STRENGTH
      MODULUS OF ELASTICITY
      NU FACTOR
      STRESS-STRAIN DIAGRAMS
      TENSILE STRENGTH

POLAIRE SATELLITE
USE   D-2 SATELLITES

POLAND
GS   NATIONS
    . **POLAND**
RT   CENTRAL EUROPE
    EUROPE

POLAR AURORAS
USE   AURORAS

POLAR CAP ABSORPTION
GS   ENERGY ABSORPTION
    . ELECTROMAGNETIC ABSORPTION
    . . **POLAR CAP ABSORPTION**
    . THERMAL ABSORPTION
    . . **POLAR CAP ABSORPTION**
    RADIATION ABSORPTION
    . ELECTROMAGNETIC ABSORPTION
    . . **POLAR CAP ABSORPTION**
RT   ∞ABSORPTION

POLAR CAPS
RT   ANTARCTIC REGIONS
    ARCTIC REGIONS
    ∞CAPS
    EARTH (PLANET)
    ICE
    MARS (PLANET)

POLAR COORDINATES
GS   COORDINATES
    . **POLAR COORDINATES**
RT   ASTRONOMICAL COORDINATES
    PLANISPHERES
    SMITH CHART
    SPHERICAL COORDINATES

POLAR CUSPS
RT   AERONOMY
    ∞CUSPS
    GEOMAGNETIC LATITUDE
    GEOMAGNETIC TAIL
    GEOMAGNETISM
    GEOPHYSICS
    INTERPLANETARY SPACE
    LINES OF FORCE
    MAGNETIC FIELD CONFIGURATIONS
    MAGNETIC FIELDS
    MAGNETOPAUSE
    MAGNETOSPHERE
    PLANETARY MAGNETIC FIELDS
    POLAR REGIONS
    SPACE PLASMAS

POLAR GASES
GS   GASES
    . MOLECULAR GASES
    . . **POLAR GASES**
RT   CARBON DIOXIDE LASERS
    GAS COMPOSITION
    GAS DISCHARGES
    GAS DYNAMICS
    GAS LASERS
    GAS MASERS
    POLARIZATION (CHARGE SEPARATION)

POLAR IONOSPHERE BEACON
USE   BEACON SATELLITES

POLAR METEOROLOGY
GS   METEOROLOGY
    . **POLAR METEOROLOGY**
RT   AEROLOGY
    CLIMATOLOGY
    HYDROLOGY
    ICE REPORTING

POLAR NAVIGATION
GS   NAVIGATION
    . **POLAR NAVIGATION**
RT   AIR NAVIGATION
    CELESTIAL NAVIGATION
    DEAD RECKONING
    DIGITAL NAVIGATION
    INERTIAL NAVIGATION
    LORAN

POLAR ORBIT GEOPHYSICAL OBSERVATORY
USE   POGO

POLAR ORBITS
GS   ORBITS
    . SPACECRAFT ORBITS
    . . SATELLITE ORBITS
    . . . **POLAR ORBITS**

POLAR ORBITS-*(CONT.)*
RT   CIRCULAR ORBITS
    EARTH ORBITS
    ELLIPTICAL ORBITS
    EQUATORIAL ORBITS
    LUNAR ORBITS
    LUNAR SATELLITES
    PLANETARY ORBITS
    TIROS SATELLITES
    TWENTY-FOUR HOUR ORBITS

POLAR RADIO BLACKOUT
GS   ELECTROMAGNETIC INTERFERENCE
    . RADIO FREQUENCY INTERFERENCE
    . . BLACKOUT (PROPAGATION)
    . . . **POLAR RADIO BLACKOUT**
RT   AURORAL ZONES
    IONOSPHERIC PROPAGATION

POLAR REGIONS
UF   HIGH LATITUDES
GS   REGIONS
    . **POLAR REGIONS**
    . . ANTARCTIC REGIONS
    . . . MCMURDO SOUND
    . . . ROSS ICE SHELF
    . . ARCTIC REGIONS
RT   AURORAL ZONES
    CLIMATOLOGY
    GEOGRAPHY
    PERMAFROST
    POLAR CUSPS
    TEMPERATE REGIONS
    TIMBERLINE

POLAR SUBSTORMS
GS   STORMS
    . STORMS (METEOROLOGY)
    . . **POLAR SUBSTORMS**

POLAR WANDERING (GEOLOGY)
UF   CHANDLER MOTION
RT   EARTH AXIS
    ∞EARTH MOTION
    GEODESY
    NUTATION
    PERIODIC VARIATIONS
    PRECESSION

POLARIMETERS
UF   SPECTROPOLARIMETERS
GS   MEASURING INSTRUMENTS
    . OPTICAL MEASURING INSTRUMENTS
    . . **POLARIMETERS**
    OPTICAL EQUIPMENT
    . OPTICAL MEASURING INSTRUMENTS
    . . **POLARIMETERS**
RT   CHEMICAL ANALYSIS
    ELLIPSOMETERS
    OPTICAL MEASUREMENT
    PHOTOMETERS
    POLARIMETRY
    POLARISCOPES
    POLARIZERS
    POLAROGRAPHY
    SOLAR MAXIMUM MISSION

POLARIMETRY
GS   OPTICAL MEASUREMENT
    . **POLARIMETRY**
RT   OPTICAL ACTIVITY
    OPTICAL MEASURING INSTRUMENTS
    PHOTOMETRY
    POLARIMETERS
    POLARIZATION (WAVES)

POLARIS A1 MISSILE
GS   MISSILES
    . BALLISTIC MISSILES
    . . INTERMEDIATE RANGE BALLISTIC
       MISSILES
    . . . POLARIS MISSILES
    . . . . **POLARIS A1 MISSILE**
    . SURFACE TO SURFACE MISSILES
    . . FLEET BALLISTIC MISSILES
    . . . **POLARIS A1 MISSILE**
    . . INTERMEDIATE RANGE BALLISTIC
       MISSILES
    . . . POLARIS MISSILES
    . . . . **POLARIS A1 MISSILE**

POLARIS A2 MISSILE
GS   MISSILES
    . BALLISTIC MISSILES

POLARIS A2 MISSILE-*(CONT.)*
    . . INTERMEDIATE RANGE BALLISTIC
       MISSILES
    . . . POLARIS MISSILES
    . . . . **POLARIS A2 MISSILE**
    . SURFACE TO SURFACE MISSILES
    . . FLEET BALLISTIC MISSILES
    . . . **POLARIS A2 MISSILE**
    . . INTERMEDIATE RANGE BALLISTIC
       MISSILES
    . . . POLARIS MISSILES
    . . . . **POLARIS A2 MISSILE**

POLARIS A3 MISSILE
GS   MISSILES
    . BALLISTIC MISSILES
    . . INTERMEDIATE RANGE BALLISTIC
       MISSILES
    . . . POLARIS MISSILES
    . . . . **POLARIS A3 MISSILE**
    . SURFACE TO SURFACE MISSILES
    . . FLEET BALLISTIC MISSILES
    . . . **POLARIS A3 MISSILE**
    . . INTERMEDIATE RANGE BALLISTIC
       MISSILES
    . . . POLARIS MISSILES
    . . . . **POLARIS A3 MISSILE**

POLARIS MISSILES
GS   MISSILES
    . BALLISTIC MISSILES
    . . INTERMEDIATE RANGE BALLISTIC
       MISSILES
    . . . **POLARIS MISSILES**
    . . . . POLARIS A1 MISSILE
    . . . . POLARIS A2 MISSILE
    . . . . POLARIS A3 MISSILE
    . SURFACE TO SURFACE MISSILES
    . . INTERMEDIATE RANGE BALLISTIC
       MISSILES
    . . . **POLARIS MISSILES**
    . . . . POLARIS A1 MISSILE
    . . . . POLARIS A2 MISSILE
    . . . . POLARIS A3 MISSILE
RT   MULTISTAGE ROCKET VEHICLES
    SOLID PROPELLANT ROCKET ENGINES
    XM-33 ENGINE

POLARIS SUBMARINES
USE   GUIDED MISSILE SUBMARINES

POLARISCOPES
GS   MEASURING INSTRUMENTS
    . **POLARISCOPES**
    . . SENARMONT POLARISCOPES
    OPTICAL EQUIPMENT
    . **POLARISCOPES**
    . . SENARMONT POLARISCOPES
RT   OPTICAL MEASURING INSTRUMENTS
    POLARIMETERS
    POLARIZATION (WAVES)
    POLARIZERS

POLARITONS
GS   **POLARITONS**
    . PLASMONS

POLARITY
RT   ∞DIPOLES
    ELECTRIC CHARGE
    ELECTRIC FIELDS
    MAGNETIC FIELDS
    MAGNETIC POLES
    POLARIZATION (CHARGE SEPARATION)
    POLARIZATION (SPIN ALIGNMENT)
    QUADRUPOLES

∞ POLARIZATION
SN   *(USE OF A MORE SPECIFIC TERM IS*
    RECOMMENDED--CONSULT THE TERMS
    LISTED BELOW)
RT   ANTIFERROELECTRICITY
    BIPOLARITY
    LINEAR POLARIZATION
    MAGNETIZATION
    OVERHAUSER EFFECT
    PHOTOELASTIC ANALYSIS
    POLARIZATION (CHARGE SEPARATION)
    POLARIZATION (SPIN ALIGNMENT)
    POLARIZATION (WAVES)
    POLARIZED RADIATION

POLARIZATION (CHARGE SEPARATION)
UF   CHARGE SEPARATION
GS   **POLARIZATION (CHARGE SEPARATION)**

POLARIZATION (CHARGE SEPARATION) -*(CONT.)*
. DIELECTRIC POLARIZATION
. ELECTROLYTIC POLARIZATION
RT   CHARGE DISTRIBUTION
CHARGE TRANSFER
DEACTIVATION
DEPOLARIZATION
ELECTRETS
ELECTRIC CHARGE
ELECTRIC MOMENTS
ELECTRODE FILM BARRIERS
ELECTROMIGRATION
HALL EFFECT
IONOSPHERIC DRIFT
MAGNETIZATION
OVERVOLTAGE
POLAR GASES
POLARITY
∞POLARIZATION
PYROELECTRICITY
∞SEPARATION
TAFEL LAW

POLARIZATION (SPIN ALIGNMENT)
RT   ALIGNMENT
ANISOTROPY
DEACTIVATION
MAGNETIC PROPERTIES
MAGNETIZATION
∞ORIENTATION
POLARITY
∞POLARIZATION
ROTATION
SPIN TESTS

POLARIZATION (WAVES)
UF   POLARIZATION CHARTS
GS   **POLARIZATION (WAVES)**
. CIRCULAR POLARIZATION
. CROSS POLARIZATION
. ELLIPTICAL POLARIZATION
. LINEAR POLARIZATION
RT   ANISOTROPIC MEDIA
ANISOTROPY
BIREFRINGENCE
BI LACERTAE OBJECTS
COLLIMATION
FARADAY EFFECT
KERR ELECTROOPTICAL EFFECT
KERR MAGNETOOPTICAL EFFECT
MAGNETO-OPTICS
MONOCHROMATIZATION
OPTICAL COUPLING
OPTICAL PROPERTIES
∞ORIENTATION
PHOTOELASTIC ANALYSIS
POLARIMETRY
POLARISCOPES
∞POLARIZATION
POLARIZED ELECTROMAGNETIC
RADIATION
POLARIZERS
POLARONS
REFRACTIVITY
ROTATION

POLARIZATION CHARACTERISTICS
GS   MAGNETIC PROPERTIES
. **POLARIZATION CHARACTERISTICS**
RT   BREWSTER ANGLE
∞CHARACTERISTICS
POLARIZED RADIATION

POLARIZATION CHARTS
USE   GRAPHS (CHARTS)
POLARIZATION (WAVES)

POLARIZED ELASTIC WAVES
GS   ELASTIC WAVES
. **POLARIZED ELASTIC WAVES**
POLARIZED RADIATION
. **POLARIZED ELASTIC WAVES**
RT   S WAVES
SEISMIC WAVES
SOUND WAVES

POLARIZED ELECTROMAGNETIC RADIATION
GS   ELECTROMAGNETIC RADIATION
. **POLARIZED ELECTROMAGNETIC
RADIATION**
. . POLARIZED LIGHT
. . SYNCHROTRON RADIATION
POLARIZED RADIATION
. **POLARIZED ELECTROMAGNETIC
RADIATION**

POLARIZED ELECTROMAGNETIC-*(CONT.)*
. . POLARIZED LIGHT
. . SYNCHROTRON RADIATION
RT   CROSS POLARIZATION
EXTRATERRESTRIAL RADIATION
FARADAY EFFECT
INFRARED RADIATION
KERR CELLS
LIGHT (VISIBLE RADIATION)
LINEAR POLARIZATION
LYMAN ALPHA RADIATION
LYMAN BETA RADIATION
MAGNETO-OPTICS
MONOCHROMATIC RADIATION
POLARIZATION (WAVES)
POLARIZERS
∞RADIATION
RADIATIVE TRANSFER
RADIO WAVES
STELLAR RADIATION
ULTRAVIOLET RADIATION

POLARIZED LIGHT
GS   ELECTROMAGNETIC RADIATION
. LIGHT (VISIBLE RADIATION)
. . **POLARIZED LIGHT**
. POLARIZED ELECTROMAGNETIC
RADIATION
. . **POLARIZED LIGHT**
POLARIZED RADIATION
. POLARIZED ELECTROMAGNETIC
RADIATION
. . **POLARIZED LIGHT**
RT   GEGENSCHEIN
KERR MAGNETOOPTICAL EFFECT
MONOCHROMATIC RADIATION
OPTICAL ACTIVITY
OPTICAL DEPOLARIZATION
OPTICAL POLARIZATION
PHOTOELASTICITY
ZODIACAL LIGHT

POLARIZED RADIATION
GS   **POLARIZED RADIATION**
. POLARIZED ELASTIC WAVES
. POLARIZED ELECTROMAGNETIC
RADIATION
. . POLARIZED LIGHT
. . SYNCHROTRON RADIATION
RT   CAUSTICS (OPTICS)
ELASTIC WAVES
ELECTROMAGNETIC RADIATION
EXTRATERRESTRIAL RADIATION
LINEAR POLARIZATION
PLASMA RADIATION
∞POLARIZATION
POLARIZATION CHARACTERISTICS
∞RADIATION
∞RAYS

POLARIZERS
RT   KERR CELLS
LIGHT (VISIBLE RADIATION)
OPTICAL POLARIZATION
POLARIMETERS
POLARISCOPES
POLARIZATION (WAVES)
POLARIZED ELECTROMAGNETIC
RADIATION

POLAROGRAPHS
USE   POLAROGRAPHY

POLAROGRAPHY
UF   POLAROGRAPHS
GS   ELECTRICAL MEASUREMENT
. **POLAROGRAPHY**
RT   CHEMICAL ANALYSIS
OPTICAL POLARIZATION
POLARIMETERS
QUANTITATIVE ANALYSIS

POLARONS
GS   ELEMENTARY EXCITATIONS
. **POLARONS**
RT   CONDUCTION BANDS
CROSS POLARIZATION
ELECTRON PHONON INTERACTIONS
IONIC CRYSTALS
PHONONS
PLASMONS
POLARIZATION (WAVES)

∞ **POLES**
SN   *(USE OF A MORE SPECIFIC TERM IS
RECOMMENDED--CONSULT THE TERMS
LISTED BELOW)*
RT   ∞DIPOLES
MAGNETIC DIPOLES
MAGNETIC POLES
MONOPOLES
POLES (SUPPORTS)
REGGE POLES

POLES (SUPPORTS)
RT   ELECTRIC POWER TRANSMISSION
∞POLES

POLICE
GS   PERSONNEL
. **POLICE**
RT   COMMUNITIES
CRIME
REGULATIONS
SECURITY
SOCIAL FACTORS
VIOLENCE

POLICIES
GS   **POLICIES**
. ENERGY POLICY
. PATENT POLICY
. PROCUREMENT POLICY
RT   COPYRIGHTS
GOVERNMENTS
LICENSING
PROHIBITION
REGULATIONS
RULES

POLIOMYELITIS
GS   DISEASES
. INFECTIOUS DISEASES
. . **POLIOMYELITIS**

POLISH TS-11 AIRCRAFT
USE   TS-11 AIRCRAFT

POLISHED METALS
USE   METAL POLISHING

POLISHING
GS   **POLISHING**
. METAL POLISHING
. . ELECTROPOLISHING
. VIBRATORY POLISHING
RT   ABRASION
CLEANING
FINISHES
GRINDING (MATERIAL REMOVAL)
METALLOGRAPHY
SMOOTHING
SURFACE FINISHING
ULTRASONIC CLEANING

POLITICS
RT   AIR LAW
COMMUNITIES
CULTURE (SOCIAL SCIENCES)
GOVERNMENTS
INTERNATIONAL COOPERATION
INTERNATIONAL LAW
LAW (JURISPRUDENCE)
NATIONS
REGIMES
SOCIOLOGY
SOVEREIGNTY
UNITED NATIONS
VOTING
WARFARE

POLLEN
GS   PARTICLES
. . **POLLEN**
RT   AEROBIOLOGY
AIR POLLUTION
DUST
PLANTS (BOTANY)
∞REPRODUCTION

POLLUTANTS
USE   CONTAMINANTS

POLLUTION
GS   **POLLUTION**
. ENVIRONMENT POLLUTION
. . AIR POLLUTION

POLLUTION-(CONT.)
```
        . . . GLOBAL AIR POLLUTION
        . . . INDOOR AIR POLLUTION
        . . WATER POLLUTION
        . . . OIL POLLUTION
        . NOISE POLLUTION
        . THERMAL POLLUTION
RT      CLEAN FUELS
        CONTAMINANTS
        CONTAMINATION
        DEBRIS
        DECONTAMINATION
        DISSIPATION
        ELIMINATION
        ENDANGERED SPECIES
        ENERGY POLICY
        ENVIRONMENT EFFECTS
        ENVIRONMENT PROTECTION
        ENVIRONMENTAL QUALITY
        ENVIRONMENTAL SURVEYS
        HUMAN WASTES
        METABOLIC WASTES
        MICROCYSTIS
        MICROORGANISMS
        NONPOINT SOURCES
        OIL SLICKS
        POISONS
        PREVENTION
        PUBLIC HEALTH
        PURITY
        QUALITY
        RADIOACTIVE WASTES
        SMOKE ABATEMENT
        SOLID WASTES
        TOXICOLOGY
        WASTE DISPOSAL
        WASTES
        WATER
        WATER RECLAMATION
        WATER TREATMENT
```

POLLUTION CONTROL
```
RT      AIR QUALITY
        BIOCHEMICAL OXYGEN DEMAND
     ∞ CONTROL
        DEWATERING
        ENVIRONMENTAL SURVEYS
        FLUE GASES
        FLY ASH
```

POLLUTION MONITORING
```
RT      AIR POLLUTION
        AIR QUALITY
        ENVIRONMENT POLLUTION
        GLOBAL AIR POLLUTION
        GROUND STATIONS
        MONITORS
        WARNING SYSTEMS
        WATER POLLUTION
```

POLLUTION TRANSPORT
```
UF      ATMOSPHERIC LOADING
RT      AEROSOLS
        AIR POLLUTION
        ATMOSPHERIC CIRCULATION
        ATMOSPHERIC DIFFUSION
        COMBUSTION PRODUCTS
        DISPERSING
        ENVIRONMENT POLLUTION
        EXHAUST EMISSION
        EXHAUST GASES
        GAS TRANSPORT
        GASEOUS DIFFUSION
        GLOBAL AIR POLLUTION
        PLUMES
        THERMAL POLLUTION
        TRACE CONTAMINANTS
        TRANSPORT PROPERTIES
        TRANSPORT THEORY
        WATER CIRCULATION
        WATER POLLUTION
```

POLOIDAL FLUX
```
RT      MAGNETIC FIELD CONFIGURATIONS
        TOKAMAK DEVICES
        TOROIDAL PLASMAS
```

POLONIUM
```
GS      CHEMICAL ELEMENTS
        . METALLOIDS
        . . POLONIUM
        . . . POLONIUM ISOTOPES
        . . . . POLONIUM 208
        . . . . POLONIUM 209
        . . . . POLONIUM 210
```

POLONIUM-(CONT.)
```
RT      METALS
```

POLONIUM COMPOUNDS
```
RT     ∞ CHEMICAL COMPOUNDS
       ∞ GROUP 6A COMPOUNDS
```

POLONIUM ISOTOPES
```
GS      CHEMICAL ELEMENTS
        . METALLOIDS
        . . POLONIUM
        . . . POLONIUM ISOTOPES
        . . . . POLONIUM 208
        . . . . POLONIUM 209
        . . . . POLONIUM 210
        . NUCLIDES
        . . ISOTOPES
        . . . POLONIUM ISOTOPES
        . . . . POLONIUM 208
        . . . . POLONIUM 209
        . . . . POLONIUM 210
RT      METALS
```

POLONIUM 208
```
GS      CHEMICAL ELEMENTS
        . METALLOIDS
        . . POLONIUM
        . . . POLONIUM ISOTOPES
        . . . . POLONIUM 208
        . NUCLIDES
        . . ISOTOPES
        . . . POLONIUM ISOTOPES
        . . . . POLONIUM 208
        . . . RADIOACTIVE ISOTOPES
        . . . . POLONIUM 208
RT      METALS
```

POLONIUM 209
```
GS      CHEMICAL ELEMENTS
        . METALLOIDS
        . . POLONIUM
        . . . POLONIUM ISOTOPES
        . . . . POLONIUM 209
        . NUCLIDES
        . . ISOTOPES
        . . . POLONIUM ISOTOPES
        . . . . POLONIUM 209
        . . . RADIOACTIVE ISOTOPES
        . . . . POLONIUM 209
RT      METALS
```

POLONIUM 210
```
GS      CHEMICAL ELEMENTS
        . METALLOIDS
        . . POLONIUM
        . . . POLONIUM ISOTOPES
        . . . . POLONIUM 210
        . NUCLIDES
        . . ISOTOPES
        . . . POLONIUM ISOTOPES
        . . . . POLONIUM 210
        . . . RADIOACTIVE ISOTOPES
        . . . . POLONIUM 210
RT      METALS
```

POLYACETYLENE
```
RT      SEMICONDUCTORS (MATERIALS)
```

POLYACRYLATES
```
USE     ACRYLIC RESINS
```

POLYAMIDE RESINS
```
UF      NYLON RESINS
GS      PLASTICS
        . SYNTHETIC RESINS
        . . THERMOSETTING RESINS
        . . . FURAN RESINS
        . . . . POLYAMIDE RESINS
        . . . . . KEVLAR (TRADEMARK)
        RESINS
        . SYNTHETIC RESINS
        . . THERMOSETTING RESINS
        . . . FURAN RESINS
        . . . . POLYAMIDE RESINS
        . . . . . KEVLAR (TRADEMARK)
```

POLYATOMIC GASES
```
GS      GASES
        . MOLECULAR GASES
        . . POLYATOMIC GASES
        . . . DIATOMIC GASES
```

POLYATOMIC MOLECULES
```
GS      MOLECULES
```

POLYATOMIC MOLECULES-(CONT.)
```
        . POLYATOMIC MOLECULES
        . . TRIATOMIC MOLECULES
RT      ATOMS
        CHEMICAL BONDS
       ∞ CHEMICAL COMPOUNDS
        IONS
        MOLECULAR STRUCTURE
        MOLECULAR WEIGHT
        POSITIVE IONS
```

POLYBENZIMIDAZOLE
```
RT      SYNTHETIC FIBERS
```

POLYBROMINATED BIPHENYLS
```
UF      PBB
GS      TOXINS AND ANTITOXINS
        . POLYBROMINATED BIPHENYLS
RT      FLAME RETARDANTS
        POLYCHLORINATED BIPHENYLS
```

POLYBUTADIENE
```
GS      PLASTICS
        . POLYBUTADIENE
RT      ADDITION RESINS
        BUTADIENE
        HTPB PROPELLANTS
        SYNTHETIC RUBBERS
```

POLYBUTADIENE TETRANITRAMINE
```
GS      NITROGEN COMPOUNDS
        . NITRO COMPOUNDS
        . . POLYBUTADIENE TETRANITRAMINE
RT      PLASTIC PROPELLANTS
```

POLYCARBONATES
```
GS      CARBON COMPOUNDS
        . CARBONATES
        . . POLYCARBONATES
        . . . LEXAN (TRADEMARK)
        ESTERS
        . POLYCARBONATES
        . . LEXAN (TRADEMARK)
RT     ∞ POLYMERS
```

POLYCHLORINATED BIPHENYLS
```
UF      PCB
GS      PHENYLS
        . POLYCHLORINATED BIPHENYLS
RT      POLYBROMINATED BIPHENYLS
```

POLYCRYSTALS
```
GS      CRYSTALS
        . POLYCRYSTALS
RT      BICRYSTALS
        CRYSTAL STRUCTURE
        SINGLE CRYSTALS
```

POLYCYTHEMIA
```
RT      HEMOGLOBIN
        HEMOLYSIS
        HEMORRHAGES
        SPLEEN
```

POLYESTER RESINS
```
GS      PLASTICS
        . SYNTHETIC RESINS
        . . POLYESTER RESINS
        RESINS
        . SYNTHETIC RESINS
        . . POLYESTER RESINS
RT      DACRON (TRADEMARK)
        THERMOSETTING RESINS
```

POLYESTERS
```
GS      ESTERS
        . POLYESTERS
RT     ∞ POLYMERS
        SYNTHETIC FIBERS
```

POLYETHER RESINS
```
GS      PLASTICS
        . SYNTHETIC RESINS
        . . POLYETHER RESINS
        . . . POLYMETHYL METHACRYLATE
        RESINS
        . SYNTHETIC RESINS
        . . POLYETHER RESINS
        . . . POLYMETHYL METHACRYLATE
RT      VULCANIZED ELASTOMERS
```

POLYETHYLENE TEREPHTHALATE
```
GS      ESTERS
        . POLYETHYLENE TEREPHTHALATE
```

POLYETHYLENE TEREPHTHALATE-*(CONT.)*
    PLASTICS
    . POLYETHYLENES
    . . **POLYETHYLENE TEREPHTHALATE**
RT   ADDITION RESINS
    MYLAR (TRADEMARK)
    ∞POLYMERS

POLYETHYLENES
GS   PLASTICS
    . **POLYETHYLENES**
    . . POLYETHYLENE TEREPHTHALATE
RT   ADDITION RESINS
    ETHYLENE
    SYNTHETIC RESINS
    THERMOPLASTIC RESINS

POLYGONIZATION
RT   CRYSTAL DEFECTS
    CRYSTAL GROWTH
    RECRYSTALLIZATION

POLYGONS
GS   GEOMETRY
    . EUCLIDEAN GEOMETRY
    . . **POLYGONS**
    . . . HEXAGONS
    . . . TETRAGONS
    . . . . PARALLELOGRAMS
    . . . . . RHOMBOIDS
    . . . . RECTANGLES
    . . . . . SQUARES (MATHEMATICS)
    . . . . TRAPEZOIDS
    . . . TRIANGLES
RT   POLYTOPES

POLYHEDRONS
GS   GEOMETRY
    . EUCLIDEAN GEOMETRY
    . . **POLYHEDRONS**
    . . . CUBES (MATHEMATICS)
    . . . ICOSAHEDRONS
    . . . OCTAHEDRONS
    . . . PARALLELEPIPEDS
    . . . PYRAMIDS
    . . . RHOMBOHEDRONS
    . . . TETRAHEDRONS
RT   POLYTOPES

POLYIMIDE RESINS
GS   RESINS
    . **POLYIMIDE RESINS**
RT   POLYIMIDES
    RESIN MATRIX COMPOSITES

POLYIMIDES
GS   NITROGEN COMPOUNDS
    . AMIDES
    . . **POLYIMIDES**
RT   POLYIMIDE RESINS

POLYISOBUTYLENE
GS   PLASTICS
    . **POLYISOBUTYLENE**
RT   ADDITION RESINS
    SYNTHETIC RUBBERS

POLYISOPRENES
RT   ∞POLYMERS
    RUBBER
    SYNTHETIC RUBBERS

POLYMER CHEMISTRY
RT   ∞CHEMISTRY
    PHOSPHAZENE
    POLYMER PHYSICS
    POLYWATER

POLYMER MATRIX COMPOSITES
GS   COMPOSITE MATERIALS
    . **POLYMER MATRIX COMPOSITES**
RT   BORON FIBERS
    ∞CONSTRUCTION MATERIALS
    FIBER COMPOSITES
    LAMINATES
    ∞MATERIALS
    ∞MATRICES
    MATRIX MATERIALS
    ∞POLYMERS
    PULTRUSION
    REINFORCED PLASTICS
    RESIN MATRIX COMPOSITES

POLYMER PHYSICS
RT   ∞PHYSICS
    POLYMER CHEMISTRY
    POLYWATER
    ∞SCIENCE

POLYMERIC FILMS
UF   PLASTIC FILMS
GS   **POLYMERIC FILMS**
    . KAPTON (TRADEMARK)
    . MYLAR (TRADEMARK)
RT   CASTING
    FIBERS
    ∞FILMS
    NYLON (TRADEMARK)
    PHOTOGRAPHIC FILM
    PLASTIC COATINGS
    ∞SHEETS

POLYMERIZATION
GS   **POLYMERIZATION**
    . COPOLYMERIZATION
    . DIMERIZATION
    . VINYL COPOLYMERS
RT   CHEMICAL REACTIONS
    COUPLED MODES
    DEPOLYMERIZATION
    QUINOXALINES
    REFINING
    ∞SETTING
    ZIEGLER CATALYST

∞ **POLYMERS**
SN   *(USE OF A MORE SPECIFIC TERM IS*
    RECOMMENDED--CONSULT THE TERMS
    LISTED BELOW)
RT   CELLOPHANE
    COORDINATION POLYMERS
    COPOLYMERIZATION
    COPOLYMERS
    ELASTOMERS
    FLUOROPOLYMERS
    FORMICA
    HIGH POLYMERS
    KAPTON (TRADEMARK)
    KEL-F
    LEXAN (TRADEMARK)
    LIGNIN
    METALLOSILOXANE POLYMER
    METALLOXANE POLYMER
    METHYL POLYSILOXANE
    MICARTA
    MONOMERS
    MYLAR (TRADEMARK)
    NITROGEN POLYMERS
    NYLON (TRADEMARK)
    ORGANIC MATERIALS
    ORGANOMETALLIC POLYMERS
    PHOSPHORUS POLYMERS
    PLASTICS
    POLYCARBONATES
    POLYESTERS
    POLYETHYLENE TEREPHTHALATE
    POLYISOPRENES
    POLYMER MATRIX COMPOSITES
    POLYQUINOXALINES
    POLYTETRAFLUOROETHYLENE
    POLYURETHANE FOAM
    POLYVINYL FLUORIDE
    PREPOLYMERS
    PYRRONES (TRADEMARK)
    SILICON POLYMERS
    SILICONES
    SILOXANES
    SOLITHANES
    STYROFOAM (TRADEMARK)
    SYNTHETIC RESINS
    TEFLON (TRADEMARK)
    VINYL COPOLYMERS
    VINYL POLYMERS

POLYMETHYL METHACRYLATE
UF   LUCITE (TRADEMARK)
    PLEXIGLASS (TRADEMARK)
GS   PLASTICS
    . SYNTHETIC RESINS
    . . POLYETHER RESINS
    . . . **POLYMETHYL METHACRYLATE**
    RESINS
    . SYNTHETIC RESINS
    . . POLYETHER RESINS
    . . . **POLYMETHYL METHACRYLATE**

POLYMORPHISM
GS   MORPHOLOGY

POLYMORPHISM-*(CONT.)*
    . **POLYMORPHISM**
RT   ALLOTROPY
    CRYSTAL LATTICES
    CRYSTAL STRUCTURE
    ∞PHYSICAL PROPERTIES

POLYNOMIALS
GS   ALGEBRA
    . **POLYNOMIALS**
    . . BINOMIALS
    . . DYADICS
    . . . HERMITIAN POLYNOMIAL
RT   COEFFICIENTS
    CUBIC EQUATIONS
    EIGENVALUES
    EIGENVECTORS
    ∞EQUATIONS
    LINEAR EQUATIONS
    NONLINEAR EQUATIONS
    QUADRATIC EQUATIONS
    ROOTS OF EQUATIONS

POLYNUCLEAR ORGANIC COMPOUNDS
GS   ORGANIC COMPOUNDS
    . **POLYNUCLEAR ORGANIC COMPOUNDS**
RT   AIR POLLUTION
    ∞CHEMICAL COMPOUNDS
    PETROLEUM PRODUCTS
    PURIFICATION

POLYNUCLEOTIDES
GS   ORGANIC COMPOUNDS
    . NUCLEOTIDES
    . . **POLYNUCLEOTIDES**
    PHOSPHORUS COMPOUNDS
    . PHOSPHATES
    . **POLYNUCLEOTIDES**
    PROTEINS
    . NUCLEOTIDES
    . . **POLYNUCLEOTIDES**
RT   RIBONUCLEIC ACIDS

POLYOT SATELLITES
GS   SATELLITES
    . ARTIFICIAL SATELLITES
    . **POLYOT SATELLITES**
    UNMANNED SPACECRAFT
    . **POLYOT SATELLITES**

POLYPEPTIDES
GS   ORGANIC COMPOUNDS
    . AMINO ACIDS
    . . PEPTIDES
    . . . HYPERTENSIN
    . . . . **POLYPEPTIDES**
    PROTEINS
    . PEPTIDES
    . . HYPERTENSIN
    . . . **POLYPEPTIDES**

POLYPHENYL ETHER
GS   ETHERS
    . **POLYPHENYL ETHER**

POLYPHENYLS
GS   PHENYLS
    . **POLYPHENYLS**
    . . TETRAPHENYLS

POLYPROPYLENE
GS   PLASTICS
    . **POLYPROPYLENE**
RT   ADDITION RESINS
    SYNTHETIC RESINS

POLYQUINOXALINES
RT   ∞CHEMICAL COMPOUNDS
    ∞POLYMERS

POLYSACCHARIDES
GS   ALIPHATIC COMPOUNDS
    . **POLYSACCHARIDES**
    . . CELLULOSE
    . . . FORTISAN (TRADEMARK)
    . . CHITIN
    . . DEXTRANS
    . . GLYCOGENS
    . . STARCHES
    CARBOHYDRATES
    . **POLYSACCHARIDES**
    . . CELLULOSE
    . . . FORTISAN (TRADEMARK)
    . . CHITIN

POLYSACCHARIDES-*(CONT.)*
. . DEXTRANS
. . GLYCOGENS
. . STARCHES
RT   GUMS (SUBSTANCES)

POLYSLIPS
RT   ∞SLIP

POLYSTATION DOPPLER TRACKING SYSTEM
GS   STATIONS
. GROUND STATIONS
. . **POLYSTATION DOPPLER TRACKING
    SYSTEM**
. TRACKING STATIONS
. . **POLYSTATION DOPPLER TRACKING
    SYSTEM**
TRACKING (POSITION)
. **POLYSTATION DOPPLER TRACKING
   SYSTEM**
TRACKING NETWORKS
. **POLYSTATION DOPPLER TRACKING
   SYSTEM**
RT   DOPPLER RADAR
MISSILE TRACKING
PULSE RADAR
RADAR NETWORKS
SATELLITE DOPPLER POSITIONING
SPACE DETECTION AND TRACKING
   SYSTEM
SPACECRAFT TRACKING
∞SYSTEMS

POLYSTYRENE
GS   PLASTICS
. **POLYSTYRENE**
STYRENES
. **POLYSTYRENE**
. . STYROFOAM (TRADEMARK)
RT   ADDITION RESINS
SANTOWAX (TRADEMARK)
SYNTHETIC RESINS
THERMOPLASTIC RESINS

POLYSULFIDES
GS   CHALCOGENIDES
. SULFIDES
. . INORGANIC SULFIDES
. . . **POLYSULFIDES**
SULFUR COMPOUNDS
. SULFIDES
. . INORGANIC SULFIDES
. . . **POLYSULFIDES**
RT   COMPOSITE PROPELLANTS

POLYTETRAFLUOROETHYLENE
GS   HALOGEN COMPOUNDS
. FLUORINE COMPOUNDS
. . FLUORO COMPOUNDS
. . . DIFLUORO COMPOUNDS
. . . . **POLYTETRAFLUOROETHYLENE**
PLASTICS
. **POLYTETRAFLUOROETHYLENE**
RT   ∞POLYMERS
SYNTHETIC RESINS
TEFLON (TRADEMARK)

POLYTOPES
RT   ANALYTIC GEOMETRY
EUCLIDEAN GEOMETRY
HYPERPLANES
POLYGONS
POLYHEDRONS

POLYTROPIC PROCESSES
RT   ADIABATIC CONDITIONS
∞ISOBARS
THERMODYNAMICS

POLYURETHANE FOAM
RT   FOAMS
LOW DENSITY MATERIALS
∞POLYMERS
SOILS
SPONGES (MATERIALS)

POLYURETHANE RESINS
GS   RESINS
. **POLYURETHANE RESINS**
RT   COMPOSITE PROPELLANTS

POLYVINYL ALCOHOL
GS   HYDROXYL COMPOUNDS
. ALCOHOLS

POLYVINYL ALCOHOL-*(CONT.)*
. . POLYVINYL ALCOHOL
PLASTICS
. **POLYVINYL ALCOHOL**
RT   ADDITION RESINS
SYNTHETIC RESINS
VINYL POLYMERS

POLYVINYL CHLORIDE
UF   GEON (TRADEMARK)
GS   PLASTICS
. **POLYVINYL CHLORIDE**
RT   ADDITION RESINS
CHLORIDES
SYNTHETIC RESINS
TETRAHYDROFURAN
VINYL POLYMERS

POLYVINYL FLUORIDE
GS   PLASTICS
. **POLYVINYL FLUORIDE**
VINYL POLYMERS
. **POLYVINYL FLUORIDE**
RT   ∞POLYMERS

POLYWATER
GS   WATER
. **POLYWATER**
RT   ATOMIC STRUCTURE
CHEMICAL BONDS
MOLECULAR STRUCTURE
POLYMER CHEMISTRY
POLYMER PHYSICS

POMERANCHUK THEOREM
GS   THEOREMS
. **POMERANCHUK THEOREM**
RT   ANTIPARTICLES
DEUTERONS
DIFFRACTION PATTERNS
EIKONAL EQUATION
ELASTIC SCATTERING
ELECTRONS
ELEMENTARY PARTICLES
FIELD THEORY (PHYSICS)
FREDHOLM EQUATIONS
GLAUBER THEORY
HIGH ENERGY INTERACTIONS
KAONS
MESONS
NUCLEON-NUCLEON SCATTERING
POMERONS
REGGE POLES
SCATTERING CROSS SECTIONS

POMERONS
RT   NUCLEAR REACTIONS
POMERANCHUK THEOREM
PROTON-PROTON REACTIONS
REGGE POLES
SCATTERING CROSS SECTIONS

PONDEROMOTIVE FORCES
GS   ELECTROMOTIVE FORCES
. **PONDEROMOTIVE FORCES**
RT   ELECTRODYNAMICS
∞FORCE
LORENTZ FORCE
RELATIVISTIC PLASMAS
RELATIVITY

PONDS
RT   AQUIFERS
GREAT SALT LAKE (UT)
IRRIGATION
LAGOONS
LAKES
LIMNOLOGY
LIQUID WASTES
RESERVOIRS
SOLAR PONDS (HEAT STORAGE)
SURFACE WATER
WASTE DISPOSAL
WATER RESOURCES
WATERSHEDS
WINDPOWERED PUMPS

PONTIAC (MI)
GS   CITIES
. **PONTIAC (MI)**
RT   MICHIGAN

PONTRYAGIN PRINCIPLE
RT   CALCULUS OF VARIATIONS

PONTRYAGIN PRINCIPLE-*(CONT.)*
MAXIMUM PRINCIPLE
OPTIMIZATION
REACTION TIME

POPULATION INVERSION
GS   INVERSIONS
. **POPULATION INVERSION**
RT   ELECTRON PUMPING
ENERGY LEVELS
MOLECULAR RELAXATION
NITROGEN LASERS
NUCLEAR PUMPING
POPULATIONS
STIMULATED EMISSION

POPULATION THEORY
RT   POPULATIONS
PROBABILITY THEORY
∞THEORIES

POPULATIONS
RT   BIOMASS
DISCRIMINANT ANALYSIS (STATISTICS)
POPULATION INVERSION
POPULATION THEORY
PREDATORS
∞STATISTICS

PORCELAIN
GS   CERAMICS
. **PORCELAIN**
REFRACTORY MATERIALS
. **PORCELAIN**
RT   CERAMIC COATINGS
ENAMELS
GLASS
GLAZES
SILICON DIOXIDE
VITREOUS MATERIALS
VITRIFICATION

PORES
USE   POROSITY

POROSITY
UF   PORES
GS   **POROSITY**
. MICROPOROSITY
RT   AQUIFERS
BUOYANCY
COMPRESSIBILITY
DEFECTS
∞DENSITY
DENSITY (MASS/VOLUME)
FORMATIONS
GAS INJECTION
HOLE DISTRIBUTION (MECHANICS)
HYGRAL PROPERTIES
IMPREGNATING
INFILTRATION
INTERSTICES
LEAKAGE
MOISTURE RESISTANCE
PERMEABILITY
PERMEATING
PINHOLES
POROUS BOUNDARY LAYER CONTROL
POROUS MATERIALS
POROUS PLATES
∞PROPERTIES
SINTERING
TEXTURES
VOID RATIO
VOIDS
WETTABILITY

POROUS BOUNDARY LAYER CONTROL
GS   BOUNDARY LAYER CONTROL
. **POROUS BOUNDARY LAYER CONTROL**
RT   ∞CONTROL
CONVECTIVE FLOW
EKMAN LAYER
FREE CONVECTION
MASS TRANSFER
POROSITY

POROUS MATERIALS
RT   BRITTLE MATERIALS
∞CELLS
HONEYCOMB STRUCTURES
INTERSTICES
LOW DENSITY MATERIALS
∞MATERIALS
METAL POWDER

POROUS MATERIALS-*(CONT.)*
    POROSITY
    POWDER METALLURGY
    SANDS
    SOILS
    SPONGES (MATERIALS)

POROUS PLATES
GS   STRUCTURAL MEMBERS
    . PLATES (STRUCTURAL MEMBERS)
    . . **POROUS PLATES**
RT   LOW DENSITY MATERIALS
    POROSITY

POROUS WALLS
GS   WALLS
    . **POROUS WALLS**
RT   ∞DIFFUSERS

PORPHINES
GS   ORGANOMETALLIC COMPOUNDS
    . **PORPHINES**
RT   CHLOROPHYLLS
    HEMOGLOBIN

PORPHYRA
GS   ALGAE
    . **PORPHYRA**

PORPHYRINS
GS   **PORPHYRINS**
    . CHLOROPHYLLS
RT   HEMOGLOBIN

PORPOISES
GS   ANIMALS
    . VERTEBRATES
    . . MAMMALS
    . . . **PORPOISES**
RT   MARINE MAMMALS

PORTABLE EQUIPMENT
RT   ∞EQUIPMENT
    LIXISCOPES
    LOGISTICS
    MOBILITY
    STOWAGE (ONBOARD EQUIPMENT)

PORTABLE LIFE SUPPORT SYSTEMS
UF   PLSS
GS   SUPPORT SYSTEMS
    . LIFE SUPPORT SYSTEMS
    . . **PORTABLE LIFE SUPPORT SYSTEMS**
    . . . AEPS
RT   ARGON-OXYGEN ATMOSPHERES
    BIOPAKS
    BREATHING APPARATUS
    EMERGENCY LIFE SUSTAINING
      SYSTEMS
    HELIUM-OXYGEN ATMOSPHERES
    OXYGEN MASKS
    PRESSURE SUITS
    ∞SYSTEMS

∞ PORTS
SN   *(USE OF A MORE SPECIFIC TERM IS*
    RECOMMENDED--CONSULT THE TERMS
    LISTED BELOW)
RT   AIRPORTS
    DRYDOCKS
    HARBORS
    PORTS (OPENINGS)
    SHIPYARDS
    WHARVES

PORTS (OPENINGS)
GS   OPENINGS
    . **PORTS (OPENINGS)**
RT   APERTURES
    CAVITIES
    DUCTS
    EXHAUST SYSTEMS
    ORIFICES
    OUTLETS
    ∞PORTS
    VENTS
    ∞WINDOWS
    WINDOWS (APERTURES)

PORTUGAL
GS   NATIONS
    . **PORTUGAL**
RT   EUROPE

POSEIDON MISSILES
GS   MISSILES
    . BALLISTIC MISSILES
    . . **POSEIDON MISSILES**
    . SURFACE TO SURFACE MISSILES
    . . FLEET BALLISTIC MISSILES
    . . . **POSEIDON MISSILES**
RT   BALLISTIC MISSILE SUBMARINES
    GUIDED MISSILE SUBMARINES
    SEA LAUNCHING

POSEIDON SATELLITE
GS   SATELLITES
    . ARTIFICIAL SATELLITES
    . . FRENCH SATELLITES
    . . . **POSEIDON SATELLITE**

∞ POSITION
SN   *(USE OF A MORE SPECIFIC TERM IS*
    RECOMMENDED--CONSULT THE TERMS
    LISTED BELOW)
RT   ATTITUDE (INCLINATION)
    POSITION (LOCATION)
    POSITION (TITLE)

POSITION (LOCATION)
UF   LOCALIZATION
    LOCATION
GS   **POSITION (LOCATION)**
    . SOLAR POSITION
RT   ALTITUDE
    ASTROLABES
    AZIMUTH
    BEARING (DIRECTION)
    COLLATING
    COLLOCATION
    COORDINATES
    DETECTION
    DISTANCE
    EPHEMERIDES
    EXPOSURE
    ∞FIXING
    GEOMETRY
    LATITUDE
    LONGITUDE
    MISALIGNMENT
    NAVIGATION
    ORBITAL POSITION ESTIMATION
    ∞ORIENTATION
    ∞POINTS
    POSITION ERRORS
    POSITION SENSING
    POSITIONING
    RADAR BEACONS
    SITES
    SOUND RANGING
    SPATIAL DISTRIBUTION
    SPHERICAL COORDINATES
    STATIONS
    SURVEYS
    TRACKING (POSITION)

POSITION (TITLE)
RT   EMPLOYEE RELATIONS
    EVALUATION
    ∞GRADE
    PERSONNEL
    ∞POSITION
    RATINGS

POSITION ERRORS
GS   ERRORS
    . **POSITION ERRORS**
RT   ASTROLABES
    ERROR SIGNALS
    NAVIGATION
    OPTICAL CORRECTION PROCEDURE
    ORBITAL POSITION ESTIMATION
    POSITIONING
    VELOCITY ERRORS

POSITION INDICATORS
GS   AIRCRAFT INSTRUMENTS
    . **POSITION INDICATORS**
    . . PLAN POSITION INDICATORS
    . . RADIO DIRECTION FINDERS
    . . SPACECRAFT POSITION INDICATORS
    DISPLAY DEVICES
    . **POSITION INDICATORS**
    . . PLAN POSITION INDICATORS
    . . RADIO DIRECTION FINDERS
    . . SPACECRAFT POSITION INDICATORS
    MEASURING INSTRUMENTS
    . INDICATING INSTRUMENTS
    . . **POSITION INDICATORS**

POSITION INDICATORS-*(CONT.)*
    . . . ASTROLABES
    . . . PLAN POSITION INDICATORS
    . . . RADIO DIRECTION FINDERS
    . . . SPACECRAFT POSITION
        INDICATORS
RT   ALTIMETERS
    BEACONS
    DISTANCE MEASURING EQUIPMENT
    FLIGHT INSTRUMENTS
    GLOBAL POSITIONING SYSTEM
    HEAD-UP DISPLAYS
    NAVIGATION AIDS
    NAVIGATION INSTRUMENTS
    PLOTTERS
    POSITION SENSING
    RANGE FINDERS
    ROCKET-BORNE INSTRUMENTS
    SEXTANTS
    SOLAR COMPASSES
    SOUND LOCALIZATION

POSITION SENSING
RT   COMPUTER VISION
    ELECTRO-OPTICS
    POSITION (LOCATION)
    POSITION INDICATORS
    ROBOTICS

POSITIONING
RT   ADJUSTING
    ALIGNMENT
    AMBIGUITY
    COLLATING
    COLLOCATION
    DISPLACEMENT
    DISTRIBUTING
    ∞DISTRIBUTION
    EXPOSURE
    FITTING
    ∞FIXING
    INSTRUMENT ORIENTATION
    ∞JOINING
    LATITUDE MEASUREMENT
    LONGITUDE MEASUREMENT
    LOOK ANGLES (ELECTRONICS)
    NAVIGATION
    ∞ORIENTATION
    PLY ORIENTATION
    POSITION (LOCATION)
    POSITION ERRORS
    RADIO NAVIGATION
    RELOCATION
    SATELLITE DOPPLER POSITIONING
    ∞SETTING
    SPACING
    STATIONKEEPING

POSITIONING DEVICES (MACHINERY)
GS   **POSITIONING DEVICES (MACHINERY)**
    . BOOMS (EQUIPMENT)
    . CAMS
    . JIGS
RT   ∞DEVICES
    HOLDERS
    JACKS (LIFTS)
    ∞MACHINERY
    SLEWING

POSITIVE FEEDBACK
UF   REGENERATIVE FEEDBACK
GS   FEEDBACK
    . **POSITIVE FEEDBACK**
RT   AMPLIFICATION
    FEEDBACK AMPLIFIERS
    MULTIVIBRATORS
    NONLINEAR FEEDBACK
    OSCILLATORS
    REGENERATION (ENGINEERING)
    SELF OSCILLATION
    TRANSFER FUNCTIONS

POSITIVE IONS
GS   IONS
    . **POSITIVE IONS**
RT   ATOMS
    CATIONS
    FORMYL IONS
    HYDROGEN IONS
    HYDRONIUM IONS
    ION DENSITY (CONCENTRATION)
    IONIC MOBILITY
    IONOSPHERIC ION DENSITY
    MAGNETOSPHERIC ION DENSITY
    METAL IONS

POSITIVE IONS-*(CONT.)*
    MOLECULAR IONS
    MONATOMIC MOLECULES
    POLYATOMIC MOLECULES
    PROTONS
    TRIVALENT IONS
    VALENCE

POSITRON ANNIHILATION
GS   NUCLEAR REACTIONS
    . ANNIHILATION REACTIONS
    . . **POSITRON ANNIHILATION**
RT   ANTIPARTICLES
    ELEMENTARY PARTICLES
    NUCLEAR PARTICLES
    PAIR PRODUCTION
    PARTICLES

POSITRONIUM
RT   ATOMS
    EXCITONS

POSITRONS
GS   ANTIMATTER
    . ANTIPARTICLES
    . . **POSITRONS**
    PARTICLES
    . CHARGED PARTICLES
    . . **POSITRONS**
    . ELEMENTARY PARTICLES
    . . ANTIPARTICLES
    . . . **POSITRONS**
    . NUCLEAR PARTICLES
    . . ANTIPARTICLES
    . . . **POSITRONS**
RT   PAIR PRODUCTION

POST BOOST PROPULSION SYSTEM
RT   ASCENT TRAJECTORIES
    PROPULSION
    PROPULSION SYSTEM CONFIGURATIONS
    ROCKET ENGINES
    SPACECRAFT PROPULSION
    ∞SYSTEMS
    TRAJECTORY CONTROL

POST-BLAST NUCLEAR RADIATION
GS   NUCLEAR RADIATION
    . **POST-BLAST NUCLEAR RADIATION**
RT   FALLOUT
    HALF LIFE
    RADIANT FLUX DENSITY
    ∞RADIATION
    RADIATION EFFECTS
    RADIOACTIVE DECAY
    RADIOACTIVITY
    VELA SATELLITES

POSTAMPLIFIERS
GS   AMPLIFIERS
    . **POSTAMPLIFIERS**
RT   PREAMPLIFIERS

POSTERIOR SECTIONS
RT   ANATOMY
    DORSAL SECTIONS

POSTFLIGHT ANALYSIS
RT   ∞ANALYZING
    ∞PERFORMANCE
    POSTMISSION ANALYSIS (SPACECRAFT)

POSTLAUNCH REPORTS
GS   DOCUMENTS
    . **POSTLAUNCH REPORTS**
    REPORTS
    . **POSTLAUNCH REPORTS**
RT   PRELAUNCH SUMMARIES
    SPACECRAFT LAUNCHING
    SPACECRAFT PERFORMANCE
    SUMMARIES

POSTMISSION ANALYSIS (SPACECRAFT)
RT   FLIGHT TESTS
    POSTFLIGHT ANALYSIS

POSTULATES
USE   AXIOMS

POSTURE
RT   HUMAN BODY
    ORTHOSTATIC TOLERANCE
    PHYSICAL FITNESS

POTABLE LIQUIDS
GS   LIQUIDS
    . **POTABLE LIQUIDS**
    . . BEVERAGES
    . . . WINES
RT   PURITY

POTABLE WATER
GS   WATER
    . **POTABLE WATER**
RT   COLD WATER
    CONSERVATION
    DROUGHT
    FRESH WATER
    GROUND WATER
    LIMNOLOGY
    MODULAR INTEGRATED UTILITY SYSTEM
    OASES
    PURIFICATION
    SANITATION
    SPRINGS (WATER)
    WATER MANAGEMENT
    WATER RESOURCES
    WATER TABLES
    WATER TREATMENT

POTASSIUM
GS   CHEMICAL ELEMENTS
    . ALKALI METALS
    . . **POTASSIUM**
    . . . LIQUID POTASSIUM
    . . . POTASSIUM ISOTOPES
    . . . . POTASSIUM 38
    . . . . POTASSIUM 39
    . . . . POTASSIUM 40
    METALS
    . ALKALI METALS
    . . **POTASSIUM**
    . . . LIQUID POTASSIUM
    . . . POTASSIUM ISOTOPES
    . . . . POTASSIUM 38
    . . . . POTASSIUM 39
    . . . . POTASSIUM 40
RT   KREEP

POTASSIUM ALLOYS
GS   ALLOYS
    . **POTASSIUM ALLOYS**

POTASSIUM BROMIDES
GS   HALOGEN COMPOUNDS
    . BROMINE COMPOUNDS
    . . BROMIDES
    . . . **POTASSIUM BROMIDES**
    . HALIDES
    . . BROMIDES
    . . . **POTASSIUM BROMIDES**
    . METAL HALIDES
    . . . **POTASSIUM BROMIDES**
    POTASSIUM COMPOUNDS
    . **POTASSIUM BROMIDES**

POTASSIUM CHLORIDES
GS   HALOGEN COMPOUNDS
    . CHLORINE COMPOUNDS
    . . CHLORIDES
    . . . **POTASSIUM CHLORIDES**
    . HALIDES
    . . CHLORIDES
    . . . **POTASSIUM CHLORIDES**
    . . METAL HALIDES
    . . . **POTASSIUM CHLORIDES**
    POTASSIUM COMPOUNDS
    . **POTASSIUM CHLORIDES**

POTASSIUM CHROMATES
GS   CHROMIUM COMPOUNDS
    . CHROMATES
    . . **POTASSIUM CHROMATES**
    POTASSIUM COMPOUNDS
    . **POTASSIUM CHROMATES**

POTASSIUM COMPOUNDS
GS   **POTASSIUM COMPOUNDS**
    . ALUM
    . NEPHELINE
    . POTASSIUM BROMIDES
    . POTASSIUM CHLORIDES
    . POTASSIUM CHROMATES
    . POTASSIUM HYDRIDES
    . POTASSIUM HYDROXIDES
    . POTASSIUM IODIDES
    . POTASSIUM NITRATES
    . POTASSIUM OXIDES
    . POTASSIUM PERCHLORATES

POTASSIUM COMPOUNDS-*(CONT.)*
    . POTASSIUM PEROXIDES
    . POTASSIUM PHOSPHATES
    . POTASSIUM SILICATES
RT   ∞ALKALI METAL COMPOUNDS
    ∞CHEMICAL COMPOUNDS
    ∞METAL COMPOUNDS

POTASSIUM HYDRIDES
GS   HYDROGEN COMPOUNDS
    . HYDRIDES
    . . METAL HYDRIDES
    . . . **POTASSIUM HYDRIDES**
    POTASSIUM COMPOUNDS
    . **POTASSIUM HYDRIDES**

POTASSIUM HYDROXIDES
GS   ALKALIES
    . **POTASSIUM HYDROXIDES**
    HYDROXIDES
    . **POTASSIUM HYDROXIDES**
    POTASSIUM COMPOUNDS
    . **POTASSIUM HYDROXIDES**

POTASSIUM IODIDES
GS   HALOGEN COMPOUNDS
    . HALIDES
    . . METAL HALIDES
    . . . ALKALI HALIDES
    . . . . **POTASSIUM IODIDES**
    . IODINE COMPOUNDS
    . . IODIDES
    . . . **POTASSIUM IODIDES**
    POTASSIUM COMPOUNDS
    . **POTASSIUM IODIDES**

POTASSIUM ISOTOPES
GS   CHEMICAL ELEMENTS
    . ALKALI METALS
    . . POTASSIUM
    . . . **POTASSIUM ISOTOPES**
    . . . . POTASSIUM 38
    . . . . POTASSIUM 39
    . . . . POTASSIUM 40
    . NUCLIDES
    . . ISOTOPES
    . . . **POTASSIUM ISOTOPES**
    . . . . POTASSIUM 38
    . . . . POTASSIUM 39
    . . . . POTASSIUM 40
    METALS
    . ALKALI METALS
    . . POTASSIUM
    . . . **POTASSIUM ISOTOPES**
    . . . . POTASSIUM 38
    . . . . POTASSIUM 39
    . . . . POTASSIUM 40

POTASSIUM NITRATES
GS   NITROGEN COMPOUNDS
    . NITRATES
    . . INORGANIC NITRATES
    . . . **POTASSIUM NITRATES**
    POTASSIUM COMPOUNDS
    . **POTASSIUM NITRATES**

POTASSIUM OXIDES
GS   CHALCOGENIDES
    . OXIDES
    . . METAL OXIDES
    . . . **POTASSIUM OXIDES**
    POTASSIUM COMPOUNDS
    . **POTASSIUM OXIDES**

POTASSIUM PERCHLORATES
GS   HALOGEN COMPOUNDS
    . CHLORINE COMPOUNDS
    . . PERCHLORATES
    . . . **POTASSIUM PERCHLORATES**
    POTASSIUM COMPOUNDS
    . **POTASSIUM PERCHLORATES**
RT   EXPLOSIVES
    SOLID ROCKET PROPELLANTS

POTASSIUM PEROXIDES
GS   POTASSIUM COMPOUNDS
    . **POTASSIUM PEROXIDES**

POTASSIUM PHOSPHATES
GS   PHOSPHORUS COMPOUNDS
    . PHOSPHATES
    . . **POTASSIUM PHOSPHATES**
    POTASSIUM COMPOUNDS
    . **POTASSIUM PHOSPHATES**

POTASSIUM SILICATES
GS   POTASSIUM COMPOUNDS
.   **POTASSIUM SILICATES**
         SILICON COMPOUNDS
.   SILICATES
. .   **POTASSIUM SILICATES**
RT   MINERALS

POTASSIUM 38
GS   CHEMICAL ELEMENTS
.   ALKALI METALS
. .   POTASSIUM
. . .   POTASSIUM ISOTOPES
. . . . **POTASSIUM 38**
.   NUCLIDES
.   ISOTOPES
. . .   POTASSIUM ISOTOPES
. . . . **POTASSIUM 38**
. . .   RADIOACTIVE ISOTOPES
. . . . **POTASSIUM 38**
         METALS
.   ALKALI METALS
. .   POTASSIUM
. . .   POTASSIUM ISOTOPES
. . . . **POTASSIUM 38**

POTASSIUM 39
GS   CHEMICAL ELEMENTS
.   ALKALI METALS
. .   POTASSIUM
. . .   POTASSIUM ISOTOPES
. . . . **POTASSIUM 39**
.   NUCLIDES
. .   ISOTOPES
. . .   POTASSIUM ISOTOPES
. . . . **POTASSIUM 39**
         METALS
.   ALKALI METALS
. .   POTASSIUM
. . .   POTASSIUM ISOTOPES
. . . . **POTASSIUM 39**

POTASSIUM 40
GS   CHEMICAL ELEMENTS
.   ALKALI METALS
. .   POTASSIUM
. . .   POTASSIUM ISOTOPES
. . . . **POTASSIUM 40**
.   NUCLIDES
. .   ISOTOPES
. . .   POTASSIUM ISOTOPES
. . . . **POTASSIUM 40**
. . .   RADIOACTIVE ISOTOPES
. . . . **POTASSIUM 40**
         METALS
.   ALKALI METALS
. .   POTASSIUM
. . .   POTASSIUM ISOTOPES
. . . . **POTASSIUM 40**

POTATOES
GS   FARM CROPS
.   **POTATOES**
         VEGETABLES
.   **POTATOES**
RT   ∞FOOD
         PLANTS (BOTANY)

∞ **POTENTIAL**
SN   (USE OF A MORE SPECIFIC TERM IS
         RECOMMENDED--CONSULT THE TERMS
         LISTED BELOW)
RT   COULOMB POTENTIAL
         ELECTRIC POTENTIAL
         GEOPOTENTIAL
         IONIZATION POTENTIALS
         KLEIN-DUNHAM POTENTIAL
         MYOELECTRIC POTENTIALS
         NUCLEON POTENTIAL
         OPEN CIRCUIT VOLTAGE
         PLASMA POTENTIALS
         POTENTIAL ENERGY
         POTENTIAL FIELDS
         POTENTIAL THEORY
         YUKAWA POTENTIAL

POTENTIAL ENERGY
GS   **POTENTIAL ENERGY**
.   ELECTRIC POTENTIAL
. .   BIOELECTRIC POTENTIAL
. .   CONTACT POTENTIALS
. .   COULOMB POTENTIAL
. .   LIENARD POTENTIAL
. .   LOW VOLTAGE
. .   OPEN CIRCUIT VOLTAGE

POTENTIAL ENERGY-(CONT.)
. .   PHOTOVOLTAGES
. .   QUANTUM WELLS
. .   SPIKE POTENTIALS
.   GEOPOTENTIAL HEIGHT
.   IONIZATION POTENTIALS
.   NUCLEAR POTENTIAL
.   PLASMA POTENTIALS
RT   CHEMICAL ENERGY
         ELECTRIC ENERGY STORAGE
         ∞ENERGY
         ENERGY STORAGE
         FROUDE NUMBER
         GEOPOTENTIAL
         INTERNAL ENERGY
         KINETIC ENERGY
         MORSE POTENTIAL
         ∞POTENTIAL

POTENTIAL FIELDS
RT   FIELD THEORY (PHYSICS)
         ∞POTENTIAL

POTENTIAL FLOW
UF   IRROTATIONAL FLOW
GS   FLUID FLOW
.   **POTENTIAL FLOW**
. .   EQUIPOTENTIALS
RT   CARTAN SPACE
         HEAT TRANSMISSION
         INVISCID FLOW
         VORTICITY

POTENTIAL GRADIENTS
GS   GRADIENTS
.   **POTENTIAL GRADIENTS**
RT   PRESSURE GRADIENTS
         SPARK GAPS
         TEMPERATURE GRADIENTS

POTENTIAL THEORY
RT   DIFFERENTIAL EQUATIONS
         JACOBI INTEGRAL
         LENNARD-JONES POTENTIAL
         ∞POTENTIAL
         STREAM FUNCTIONS (FLUIDS)
         ∞THEORIES

∞ **POTENTIOMETERS**
SN   (USE OF A MORE SPECIFIC TERM IS
         RECOMMENDED--CONSULT THE TERMS
         LISTED BELOW)
RT   POTENTIOMETERS (INSTRUMENTS)
         POTENTIOMETERS (RESISTORS)

POTENTIOMETERS (INSTRUMENTS)
GS   MEASURING INSTRUMENTS
.   **POTENTIOMETERS (INSTRUMENTS)**
RT   BOLOMETERS
         ELECTRIC POTENTIAL
         ELECTRICAL MEASUREMENT
         ELECTROMETERS
         ∞POTENTIOMETERS
         THERMOCOUPLE PYROMETERS
         THERMOCOUPLES
         VOLTMETERS

POTENTIOMETERS (RESISTORS)
GS   ATTENUATORS
.   RESISTORS
. .   **POTENTIOMETERS (RESISTORS)**
RT   ∞POTENTIOMETERS

POTENTIOMETRIC ANALYSIS
UF   POTENTIOMETRY
GS   CHEMICAL TESTS
.   CHEMICAL ANALYSIS
. .   **POTENTIOMETRIC ANALYSIS**

POTENTIOMETRY
USE   POTENTIOMETRIC ANALYSIS

POTEZ AIRCRAFT
RT   ∞AIRCRAFT

POTOMAC RIVER VALLEY (MD-VA-WV)
GS   VALLEYS
.   **POTOMAC RIVER VALLEY (MD-VA-WV)**
RT   DISTRICT OF COLUMBIA
         MARYLAND
         VIRGINIA
         WEST VIRGINIA

POTTING COMPOUNDS
RT   ∞COMPOUNDS
         ENCAPSULATING
         INSULATION

POURING
RT   CASTING
         CASTINGS

POWDER (PARTICLES)
GS   PARTICLES
.   **POWDER (PARTICLES)**
. .   FINES
. .   PLATINUM BLACK
. .   POWDERED ALUMINUM
RT   COMPRESSIBILITY
         CROP DUSTING
         DUST
         EXPLOSIVES
         FLAKES
         FLOUR (FOOD)
         GRANULAR MATERIALS
         OBSIDIAN
         PUMICE
         SIZE SEPARATION

POWDER METALLURGY
RT   ALLOYS
         AUTOCLAVING
         CERMETS
         COMMINUTION
         COMPACTING
         COMPOSITE MATERIALS
         ELECTRODEPOSITION
         LOW DENSITY MATERIALS
         METAL MATRIX COMPOSITES
         METAL PARTICLES
         METAL POWDER
         ∞METALLURGY
         MIXED CRYSTALS
         POROUS MATERIALS
         PREFORMS
         REACTION BONDING
         SINTERED ALUMINUM POWDER
         SINTERING
         VACUUM MELTING

POWDERED ALUMINUM
GS   CHEMICAL ELEMENTS
.   ALUMINUM
. .   **POWDERED ALUMINUM**
         METALS
.   ALUMINUM
. .   **POWDERED ALUMINUM**
.   METAL POWDER
. .   **POWDERED ALUMINUM**
         PARTICLES
.   METAL PARTICLES
. .   **POWDERED ALUMINUM**
.   POWDER (PARTICLES)
. .   **POWDERED ALUMINUM**
RT   LITHIUM ALUMINUM HYDRIDES
         SINTERED ALUMINUM POWDER

POWDERED METALS
USE   METAL POWDER

∞ **POWER**
SN   (USE OF A MORE SPECIFIC TERM IS
         RECOMMENDED--CONSULT THE TERMS
         LISTED BELOW)
RT   ELECTRIC GENERATORS
         FLUID POWER
         FLUX (RATE)
         FLUX DENSITY
         HORSEPOWER
         RESOLUTION
         THRUST

POWER AMPLIFIERS
GS   AMPLIFIERS
.   **POWER AMPLIFIERS**
RT   AMPLIDYNES
         CYCLOTRON RESONANCE DEVICES
         FEEDBACK AMPLIFIERS
         MAGNETIC AMPLIFIERS
         PARAMETRIC AMPLIFIERS
         PUSH-PULL AMPLIFIERS
         TRANSISTOR AMPLIFIERS

POWER CONDITIONING
UF   POWER PROCESSING SYSTEMS
RT   ∞CONDITIONING
         ELECTRIC CURRENT
         ELECTRIC GENERATORS

POWER CONDITIONING-*(CONT.)*
    ELECTRIC POTENTIAL
    ENERGY CONVERSION
    ENERGY CONVERSION EFFICIENCY
    OUTPUT
    SATELLITE SOLAR ENERGY
        CONVERSION
    SATELLITE SOLAR POWER STATIONS

POWER CONVERTERS
  RT   ∞CONVERTERS
        TORQUE CONVERTERS

POWER DENSITY (ELECTROMAGNETIC)
  USE   RADIANT FLUX DENSITY

POWER EFFICIENCY
  GS   EFFICIENCY
        . **POWER EFFICIENCY**
  RT   COMBUSTION EFFICIENCY
        COMPRESSOR EFFICIENCY
        HORSEPOWER
        NOZZLE EFFICIENCY
        POWER FACTOR CONTROLLERS
        ∞POWER LOSS
        PROPELLER EFFICIENCY
        PROPULSION SYSTEM PERFORMANCE
        PROPULSIVE EFFICIENCY
        THERMODYNAMIC EFFICIENCY
        TRANSMISSION EFFICIENCY

POWER FACTOR CONTROLLERS
  GS   CONTROLLERS
        . **POWER FACTOR CONTROLLERS**
  RT   CURRENT REGULATORS
        ELECTRIC MOTORS
        ENERGY CONSERVATION
        ENERGY CONVERSION EFFICIENCY
        INDUCTION MOTORS
        POWER EFFICIENCY
        VOLTAGE REGULATORS

POWER GAIN
  GS   AMPLIFICATION
        . **POWER GAIN**
  RT   CATT DEVICES
        HIGH GAIN
        OPEN CIRCUIT VOLTAGE

POWER GENERATORS
  USE   ELECTRIC GENERATORS

POWER LIMITED SPACECRAFT
  RT   ∞SPACECRAFT

POWER LIMITERS
  RT   ATTENUATORS
        CLAMPING CIRCUITS
        CLIPPER CIRCUITS
        LIMITER CIRCUITS

POWER LINES
  GS   TRANSMISSION LINES
        . **POWER LINES**
  RT   BUS CONDUCTORS
        ∞CABLES
        COAXIAL CABLES
        ELECTRIC POWER TRANSMISSION
        ELECTRIC WIRE
        SUBMARINE CABLES
        SUPERCONDUCTING POWER
            TRANSMISSION
        UNDERGROUND TRANSMISSION LINES

∞ **POWER LOSS**
  SN   *(USE OF A MORE SPECIFIC TERM IS*
        RECOMMENDED--CONSULT THE TERMS
        LISTED BELOW)
  RT   ENERGY DISSIPATION
        POWER EFFICIENCY

POWER MODULES (STS)
  GS   MODULES
        . **POWER MODULES (STS)**
  RT   ORBITAL MANEUVERING VEHICLES
        PAYLOAD DELIVERY (STS)
        SOLAR ARRAYS
        SPACE TRANSPORTATION SYSTEM
        SPACECRAFT POWER SUPPLIES

∞ **POWER PLANTS**
  SN   *(USE OF A MORE SPECIFIC TERM IS*
        RECOMMENDED--CONSULT THE TERMS
        LISTED BELOW)

POWER PLANTS-*(CONT.)*
  RT   COGENERATION
        ELECTRIC POWER PLANTS
        ENGINES
        ENRICO FERMI ATOMIC POWER PLANT
        GEOTHERMAL ENERGY UTILIZATION
        HALLAM NUCLEAR POWER FACILITY
        HYDROELECTRIC POWER STATIONS
        HYDROELECTRICITY
        ML-1 NUCLEAR POWER PLANT
        SOLAR SEA POWER PLANTS
        SOLAR THERMAL ELECTRIC POWER
            PLANTS

POWER PROCESSING SYSTEMS
  USE   POWER CONDITIONING

POWER REACTORS
  RT   NUCLEAR POWER REACTORS
        ∞REACTORS
        SATURABLE REACTORS

POWER SERIES
  GS   ANALYSIS (MATHEMATICS)
        . CALCULUS
        . . SERIES (MATHEMATICS)
        . . . **POWER SERIES**
        . . . . TAYLOR SERIES
        . . . . . MACLAURIN SERIES
        . REAL VARIABLES
        . SERIES (MATHEMATICS)
        . . . **POWER SERIES**
        . . . . TAYLOR SERIES
        . . . . . MACLAURIN SERIES
  RT   ANALYTIC FUNCTIONS
        BESSEL FUNCTIONS

POWER SPECTRA
  GS   SPECTRA
        . **POWER SPECTRA**
        . . CEPSTRA
  RT   ACOUSTICS
        CEPSTRAL ANALYSIS
        ENERGY SPECTRA
        FLUX DENSITY
        LOUDNESS
        MAXIMUM ENTROPY METHOD

∞ **POWER SUPPLIES**
  SN   *(USE OF A MORE SPECIFIC TERM IS*
        RECOMMENDED--CONSULT THE TERMS
        LISTED BELOW)
  RT   AIRCRAFT ENGINES
        AIRCRAFT POWER SUPPLIES
        AUXILIARY POWER SOURCES
        ELECTRIC BATTERIES
        ELECTRIC GENERATORS
        ELECTRIC POWER SUPPLIES
        ELECTRON SOURCES
        ENERGY REQUIREMENTS
        HEAT SOURCES
        LEAD ACID BATTERIES
        LINE CURRENT
        LITHIUM SULFUR BATTERIES
        NUCLEAR AUXILIARY POWER UNITS
        PLASMA POWER SOURCES
        PROPELLANTS
        RECTIFIERS
        SOLAR GENERATORS
        SPACECRAFT POWER SUPPLIES
        VOLTAGE CONVERTERS (AC TO AC)
        VOLTAGE CONVERTERS (DC TO DC)

POWER SUPPLY CIRCUITS
  GS   CIRCUITS
        . **POWER SUPPLY CIRCUITS**
  RT   CURRENT REGULATORS
        RECTIFIERS
        TRANSFORMERS
        VOLTAGE CONVERTERS (DC TO DC)
        VOLTAGE REGULATORS

∞ **POWER TRANSMISSION**
  SN   *(USE OF A MORE SPECIFIC TERM IS*
        RECOMMENDED--CONSULT THE TERMS
        LISTED BELOW)
  RT   BUS CONDUCTORS
        ELECTRIC OUTLETS
        ELECTRIC POWER TRANSMISSION
        ELECTRICAL ENGINEERING
        ELECTRIFICATION
        HYDROELECTRIC POWER STATIONS
        MECHANICAL DRIVES
        WINDMILLS (WINDPOWERED MACHINES)

POWER TRANSMISSION (LASERS)
  RT   CARBON DIOXIDE LASERS
        CARBON MONOXIDE LASERS
        ELECTRIC POWER TRANSMISSION
        GAS LASERS
        LASERS
        SOLAR POWER SATELLITES

POWERED LIFT AIRCRAFT
  RT   ∞AIRCRAFT
        EXTERNALLY BLOWN FLAPS
        SHORT TAKEOFF AIRCRAFT
        VERTICAL TAKEOFF AIRCRAFT

POWERED MODELS
  SN   (LIMITED TO TEST FACILITIES)
  GS   MODELS
        . WIND TUNNEL MODELS
        . . **POWERED MODELS**
  RT   AIRCRAFT MODELS
        DYNAMIC MODELS

POYNTING THEOREM
  GS   THEOREMS
        . **POYNTING THEOREM**
  RT   ∞ELECTRIC POWER
        ELECTROMAGNETIC RADIATION
        ENERGY TRANSFER
        MAXWELL EQUATION
        VECTOR ANALYSIS

POYNTING-ROBERTSON EFFECT
  RT   ∞EFFECTS
        MICROMETEOROIDS
        ORBITAL MECHANICS
        RADIATION EFFECTS
        ZODIACAL DUST
        ZODIACAL LIGHT

PPI (POSITION INDICATORS)
  USE   PLAN POSITION INDICATORS

PPM (MODULATION)
  USE   PULSE POSITION MODULATION

PRACTICES
  USE   PROCEDURES

PRAESEPE STAR CLUSTERS
  GS   CELESTIAL BODIES
        . STAR CLUSTERS
        . . **PRAESEPE STAR CLUSTERS**
        . STARS
        . . **PRAESEPE STAR CLUSTERS**

PRAETERSONIC DEVICES
  RT   MICROWAVE FREQUENCIES
        PIEZOELECTRIC TRANSDUCERS
        THIN FILMS
        ULTRAHIGH FREQUENCIES

PRAIRIES
  USE   GRASSLANDS

PRANDTL NUMBER
  GS   RATIOS
        . DIMENSIONLESS NUMBERS
        . . **PRANDTL NUMBER**
  RT   FORCED CONVECTION
        HEAT TRANSFER
        INVISCID FLOW
        MOMENTUM TRANSFER
        NUSSELT NUMBER
        PECLET NUMBER
        REYNOLDS NUMBER
        SCHMIDT NUMBER
        THERMODYNAMIC PROPERTIES
        VISCOUS FLOW

PRANDTL-MEYER EXPANSION
  GS   EXPANSION
        . **PRANDTL-MEYER EXPANSION**
  RT   BLASIUS EQUATION
        FALKNER-SKAN EQUATION
        LAMINAR FLOW
        METHOD OF CHARACTERISTICS
        NEWTON PRESSURE LAW
        SUPERSONIC FLOW
        TWO DIMENSIONAL FLOW

PRASEODYMIUM
  GS   CHEMICAL ELEMENTS
        . RARE EARTH ELEMENTS
        . . **PRASEODYMIUM**

PRASEODYMIUM-(CONT.)
```
        . . . PRASEODYMIUM ISOTOPES
        METALS
        . RARE EARTH ELEMENTS
        . PRASEODYMIUM
        . . . PRASEODYMIUM ISOTOPES
RT      DIDYMIUM
```

PRASEODYMIUM ISOTOPES
```
UF      PRASEODYMIUM 144
GS      CHEMICAL ELEMENTS
        . NUCLIDES
        . . ISOTOPES
        . . . PRASEODYMIUM ISOTOPES
        . RARE EARTH ELEMENTS
        . . PRASEODYMIUM
        . . . PRASEODYMIUM ISOTOPES
        METALS
        . RARE EARTH ELEMENTS
        . . PRASEODYMIUM
        . . . PRASEODYMIUM ISOTOPES
```

PRASEODYMIUM 144
```
USE     PRASEODYMIUM ISOTOPES
```

PRE-IMBRIAN PERIOD
```
RT      LUNAR COMPOSITION
        LUNAR CRATERS
        LUNAR EVOLUTION
        LUNAR GEOLOGY
        LUNAR ROCKS
```

PRE-MAIN SEQUENCE STARS
```
GS      CELESTIAL BODIES
        . STARS
        . . MAIN SEQUENCE STARS
        . . . PRE-MAIN SEQUENCE STARS
RT      STELLAR EVOLUTION
```

PREAMPLIFIERS
```
UF      PRESELECTORS
GS      AMPLIFIERS
        . PREAMPLIFIERS
RT      INTERMEDIATE FREQUENCY AMPLIFIERS
        LOW NOISE
        MIXING CIRCUITS
        POSTAMPLIFIERS
        SIGNAL DETECTION
        SIGNAL DETECTORS
        SIGNAL RECEPTION
        TRANSISTOR AMPLIFIERS
        VOLTAGE AMPLIFIERS
```

PREBURNERS
```
GS      PRESSURE VESSELS
        . PREBURNERS
RT      PUMPS
        TURBINE PUMPS
```

PRECAMBRIAN PERIOD
```
RT      BALTIC SHIELD (EUROPE)
        CANADIAN SHIELD
        GEOLOGY
        PALEONTOLOGY
```

PRECAUTIONS
```
USE     ACCIDENT PREVENTION
```

PRECESSION
```
GS      GYRATION
        . PRECESSION
        . . LARMOR PRECESSION
        . . PROTON PRECESSION
        . . QUENCHING (ATOMIC PHYSICS)
RT      GYROSCOPES
        GYROSCOPIC STABILITY
        LARMOR RADIUS
        LIBRATION
        MUON SPIN ROTATION
        NUTATION
        POLAR WANDERING (GEOLOGY)
        ROTATION
        VORTEX PRECESSION
```

PRECIOUS METALS
```
USE     NOBLE METALS
```

∞ PRECIPITATION
```
SN      (USE OF A MORE SPECIFIC TERM IS
        RECOMMENDED-CONSULT THE TERM
        LISTED BELOW)
RT      DROP SIZE
        ELECTRON PRECIPITATION
        FALLING
```

PRECIPITATION-(CONT.)
```
        HYDROMETALLURGY
        MATERIALS RECOVERY
        PARTICLE PRECIPITATION
        PRECIPITATION (CHEMISTRY)
        PRECIPITATION (METEOROLOGY)
        PROTON PRECIPITATION
```

PRECIPITATION (CHEMISTRY)
```
RT      AGGLOMERATION
        CEMENTATION
      ∞ CHEMISTRY
        COAGULATION
        COLLOIDING
        CONCENTRATING
        CRYSTALLIZATION
        DEPOSITION
        DISSOLVING
        FILTRATION
        FLOCCULATING
        HYDROMETALLURGY
        HYDROMETEOROLOGY
        MATERIALS RECOVERY
      ∞ PRECIPITATION
        PRECIPITATORS
      ∞ SATURATION
        SATURATION (CHEMISTRY)
      ∞ SEPARATION
        SETTLING
        SOLUBILITY
        SUPERSATURATION
        UNSATURATION (CHEMISTRY)
```

PRECIPITATION (METEOROLOGY)
```
GS      PRECIPITATION (METEOROLOGY)
        . DEW
        . HAIL
        . RAIN
        . . ACID RAIN
        . SNOW
        . SNOW COVER
RT      ALPINE METEOROLOGY
        ANVIL CLOUDS
        ATMOSPHERIC MOISTURE
        CAP CLOUDS
        CIRROCUMULUS CLOUDS
        CIRROSTRATUS CLOUDS
        CLIMATOLOGY
        CLOUD PHYSICS
        CLOUD SEEDING
        CLOUDS (METEOROLOGY)
        CUMULONIMBUS CLOUDS
        CYCLONES
        DRAINAGE PATTERNS
        DROUGHT
        FLOOD DAMAGE
        FLOOD PREDICTIONS
        FLOODS
        FOG
        FOG DISPERSAL
        HAILSTORMS
        HUMIDITY
        HYDROLOGY
        HYDROLOGY MODELS
        HYDROMETEOROLOGY
        INTERNATIONAL HYDROLOGICAL
            DECADE
        METEOROLOGICAL PARAMETERS
        METEOROLOGY
        MIST
        MONSOONS
        NEPHANALYSIS
        NIMBOSTRATUS CLOUDS
      ∞ PRECIPITATION
        RAINMAKING
        RAINSTORMS
      ∞ SATURATION
        SNOWSTORMS
        STORM DAMAGE
        STORM ENHANCEMENT
        STORM SUPPRESSION
        STORMS
        STORMS (METEOROLOGY)
        WATER
        WATER RESOURCES
        WATERSHEDS
        WEATHER
        WEATHER FORECASTING
```

PRECIPITATION HARDENING
```
UF      AGE HARDENING
        DISPERSION PRECIPITATION HARDENING
        STRAIN AGING
GS      HARDENING (MATERIALS)
        . PRECIPITATION HARDENING
```

PRECIPITATION HARDENING-(CONT.)
```
        . . MARAGING
RT      COLD HARDENING
        EUTECTIC COMPOSITES
        HEAT TREATMENT
        SOLID SOLUTIONS
        STRAIN HARDENING
        SUPERSATURATION
        TIME TEMPERATURE PARAMETER
```

PRECIPITATION PARTICLE MEASUREMENT
```
GS      SIZE DETERMINATION
        . PRECIPITATION PARTICLE
            MEASUREMENT
RT      DROP SIZE
        METEOROLOGICAL RADAR
        PARTICLE SIZE DISTRIBUTION
        PARTICLES
```

PRECIPITATORS
```
GS      SEPARATORS
        . PRECIPITATORS
        . . ELECTROSTATIC PRECIPITATORS
RT      AIR FILTERS
        CONCENTRATORS
        DUST COLLECTORS
        PRECIPITATION (CHEMISTRY)
        THICKENERS (EQUIPMENT)
```

PRECISION
```
UF      EXACTNESS
RT      ACCURACY
        ALLOWANCES
        CONFIDENCE LIMITS
        CONSISTENCY
      ∞ DEFINITION
        DYNAMIC CHARACTERISTICS
        ERRORS
        GEOMETRIC DILUTION OF PRECISION
        HIGH RESOLUTION
        HYSTERESIS
        QUALITY
        QUALITY CONTROL
        RELIABILITY
        RESOLUTION
        SCHEDULES
        SENSITIVITY
      ∞ SHARPNESS
        TOLERANCES (MECHANICS)
        TRUNCATION ERRORS
        VALIDITY
```

PRECISION GUIDED PROJECTILES
```
GS      MISSILES
        . PRECISION GUIDED PROJECTILES
        PROJECTILES
        . PRECISION GUIDED PROJECTILES
        WEAPONS
        . ARTILLERY
        . . PRECISION GUIDED PROJECTILES
        . GUNS (ORDNANCE)
        . . PRECISION GUIDED PROJECTILES
        . WARHEADS
        . . PRECISION GUIDED PROJECTILES
RT      ∞ BOMBS
```

PRECONDITIONING
```
GS      PREPARATION
        . PRECONDITIONING
RT      ∞ CONDITIONING
```

PRECOOLING
```
GS      COOLING
        . PRECOOLING
RT      REGENERATIVE COOLING
```

PREDATORS
```
RT      ANIMALS
        ECOLOGY
        ECOSYSTEMS
        POPULATIONS
```

PREDICTION ANALYSIS TECHNIQUES
```
GS      FORECASTING
        . PERFORMANCE PREDICTION
        . . PREDICTION ANALYSIS TECHNIQUES
        SCHEDULING
        . PREDICTION ANALYSIS TECHNIQUES
RT      ∞ ANALYZING
        PARAMETER IDENTIFICATION
        SYSTEM IDENTIFICATION
```

PREDICTION RECORDING
```
GS      RECORDING
```

PREDICTION RECORDING-*(CONT.)*
. **PREDICTION RECORDING**
RT   PREDICTIONS

PREDICTIONS
UF   PREDICTORS
GS   **PREDICTIONS**
. FLOOD PREDICTIONS
. IMPACT PREDICTION
. LINEAR PREDICTION
. NOISE PREDICTION
. NOISE PREDICTION (AIRCRAFT)
. PERFORMANCE PREDICTION
. ROSHKO PREDICTION
RT   CATASTROPHE THEORY
CONFIDENCE LIMITS
CONTINGENCY
DELPHI METHOD (FORECASTING)
ESTIMATES
FORECASTING
MAXIMUM LIKELIHOOD ESTIMATES
MISSION PLANNING
PATTERN METHOD (FORECASTING)
PREDICTION RECORDING
PROBE METHOD (FORECASTING)
PROFILE METHOD (FORECASTING)
∞PROJECTION
RISK
SCHEDULES
TECHNOLOGICAL FORECASTING

PREDICTORS
USE   PREDICTIONS

PREEMPTING
RT   CLAIMING
PREVENTION

PREFIRING TESTS
GS   ENGINE TESTS
. **PREFIRING TESTS**
RT   CAPTIVE TESTS
CHECKOUT
GROUND TESTS
PREFLIGHT ANALYSIS
PRELAUNCH TESTS
ROCKET ENGINE DESIGN
SPACE VEHICLE CHECKOUT PROGRAM
STATIC TESTS
TEST FIRING
TEST STANDS
∞TESTS

PREFLIGHT ANALYSIS
RT   ∞ANALYZING
PREFIRING TESTS
SYSTEMS ANALYSIS
∞TESTS
TRAJECTORY ANALYSIS
WEIGHT ANALYSIS

PREFLIGHT OPERATIONS
GS   **PREFLIGHT OPERATIONS**
. AIRCRAFT RUNUP
. COUNTDOWN
RT   CREW PROCEDURES (PREFLIGHT)
GROUND TESTS
∞OPERATIONS
PRELAUNCH TESTS
REFUELING

PREFOCUSING
GS   FOCUSING
. **PREFOCUSING**
RT   ∞OPTICS

PREFORMS
RT   BLANKS
COMPOSITE MATERIALS
MOLDS
POWDER METALLURGY

PREGNANCY
RT   BIRTH

PREHEATERS
USE   HEATING EQUIPMENT

PREHEATING
USE   HEATING

PREIMPREGNATION
RT   FILAMENT WINDING
PULTRUSION

PREJUDICES
RT   ECONOMICS
IRRATIONALITY
MANAGEMENT
∞PROPERTIES
PSYCHOLOGY

PRELAUNCH PROBLEMS
RT   COUNTDOWN
∞PROBLEMS
RELIABILITY
SPACECRAFT RELIABILITY

PRELAUNCH SUMMARIES
GS   SUMMARIES
. **PRELAUNCH SUMMARIES**
RT   MISSION PLANNING
POSTLAUNCH REPORTS
SPACECRAFT LAUNCHING

PRELAUNCH TESTS
GS   GROUND TESTS
. **PRELAUNCH TESTS**
. . STATIC FIRING
NONDESTRUCTIVE TESTS
. **PRELAUNCH TESTS**
. . STATIC FIRING
RT   CAPTIVE TESTS
COLD FLOW TESTS
COUNTDOWN
CREW PROCEDURES (PREFLIGHT)
ENGINE TESTS
LAUNCHING
MISSILE TESTS
PREFIRING TESTS
PREFLIGHT OPERATIONS
SPACECRAFT MAINTENANCE
STATIC TESTS
TEST FIRING
TEST STANDS
∞TESTS

PRELOADING
USE   PRESTRESSING

PREMATURE OPERATION
RT   ∞OPERATIONS
PREPARATION

PREMIXED FLAMES
GS   FLAMES
. **PREMIXED FLAMES**
RT   CARBURETORS
FLAME PROPAGATION
GAS MIXTURES
MIXING
PREMIXING

PREMIXING
GS   MIXING
. **PREMIXING**
RT   FUEL-AIR RATIO
FUELS
GAS MIXTURES
HOMOGENIZING
IGNITION
JET MIXING FLOW
PREMIXED FLAMES
SPRAYING

PREPARATION
GS   **PREPARATION**
. PRECONDITIONING
. PRETREATMENT
. . PRESTRESSING
. PREWHIRLING
. PREWHITENING
RT   ASSEMBLING
PREMATURE OPERATION
∞PRIMING

PREPOLYMERS
GS   **PREPOLYMERS**
. DIMERS
. TRIMERS
RT   MONOMERS
∞POLYMERS

PREPREGS
RT   COMPOSITE MATERIALS
EPOXY RESINS
LAMINATES
RESIN MATRIX COMPOSITES

PREPROCESSING
RT   DATA PROCESSING
DATA REDUCTION
IMAGE PROCESSING

PRESBYOPIA
RT   VISION

PRESELECTORS
USE   PREAMPLIFIERS

PRESENTATION
RT   INFORMATION

PRESERVATIVES
RT   ADDITIVES
∞AGENTS
ANTICOAGULANTS
ANTIOXIDANTS
NEUTRALIZERS
PENETRANTS
PRESERVING
RETARDANTS
STABILIZERS (AGENTS)

PRESERVING
GS   FOOD PROCESSING
. **PRESERVING**
RT   BIOPAKS
∞CONTAINERS
CORROSION PREVENTION
COVERINGS
CURING
DEGRADATION
DEHYDRATED FOOD
∞FOOD
FREEZE DRYING
FREEZING
FROZEN FOODS
IMPREGNATING
IRRADIATION
PACKAGING
PRESERVATIVES
RADIATION EFFECTS
REFRIGERATING
∞STORAGE
WEATHERPROOFING

PRESIDENTIAL REPORTS
GS   DOCUMENTS
. **PRESIDENTIAL REPORTS**
REPORTS
. **PRESIDENTIAL REPORTS**
RT   CONGRESSIONAL REPORTS
PAPERS
RECORDS

PRESINTERING
USE   SINTERING

PRESSES
GS   **PRESSES**
. RAMS (PRESSES)
RT   COMPACTING
HAMMERS
MACHINE TOOLS
PLATENS
∞PRESSING
PRESSING (FORMING)
PUNCHES
TOOLS

∞ **PRESSING**
SN   *(USE OF A MORE SPECIFIC TERM IS*
RECOMMENDED--CONSULT THE TERMS
LISTED BELOW)
RT   COLD PRESSING
COMPACTING
COMPRESSING
HOT PRESSING
PLATENS
PRESSES
PRESSING (FORMING)

PRESSING (FORMING)
GS   FORMING TECHNIQUES
. **PRESSING (FORMING)**
. . BLANKING (CUTTING)
. . COINING
. . STAMPING
RT   COLD PRESSING
COMPACTING
EXTRUDING
FORGING

PRESSING (FORMING)-*(CONT.)*
. HOT PRESSING
. METAL WORKING
. MOLDS
. PRESSES
∞ PRESSING
. PULTRUSION
. SIZING (SHAPING)
. UPSETTING

PRESSORS
USE   VASOCONSTRICTOR DRUGS

PRESSURE
UF   SURFACE PRESSURE
GS   **PRESSURE**
. ATMOSPHERIC PRESSURE
. BASE PRESSURE
. BLOOD PRESSURE
. . DIASTOLIC PRESSURE
. . HYPERTENSION
. . HYPOTENSION
. . SYSTOLIC PRESSURE
. CRITICAL PRESSURE
. DENSIFICATION
. DIFFERENTIAL PRESSURE
. DYNAMIC PRESSURE
. FLUID PRESSURE
. . WATER PRESSURE
. GAS PRESSURE
. GEOPRESSURE
. HIGH PRESSURE
. IMPACT LOADS
. INLET PRESSURE
. INTERNAL PRESSURE
. INTRACRANIAL PRESSURE
. INTRAOCULAR PRESSURE
. ISOSTATIC PRESSURE
. LOW PRESSURE
. . HIGH ALTITUDE PRESSURE
. MIDDLE EAR PRESSURE
. OVERPRESSURE
. PARTIAL PRESSURE
. . OXYGEN TENSION
. . . HYPOXEMIA
. RADIATION PRESSURE
. . ELECTRON PRESSURE
. . LUMENS
. . LUMINOUS INTENSITY
. . . ILLUMINANCE
. . . LUMINANCE
. . SOUND PRESSURE
. STAGNATION PRESSURE
. STATIC PRESSURE
. . HYDROSTATIC PRESSURE
. SUPERCRITICAL PRESSURES
. THRUST CHAMBER PRESSURE
. TRANSIENT PRESSURES
. TRANSITION PRESSURE
. VACUUM
. . HIGH VACUUM
. . LOW VACUUM
. . ULTRAHIGH VACUUM
. VAPOR PRESSURE
. WALL PRESSURE
. WIND PRESSURE
RT   BARORECEPTORS
BLAST LOADS
CENTER OF PRESSURE
COMPRESSING
EAR PRESSURE TEST
ELASTIC WAVES
ENVIRONMENTS
∞ FORCE
FUEL TANK PRESSURIZATION
GIBBS-HELMHOLTZ EQUATIONS
HEAD (FLUID MECHANICS)
HIGH PRESSURE OXYGEN
IMPACT
IMPACT LOADS
ISOBARS (PRESSURE)
LOADS (FORCES)
NEWTON PRESSURE LAW
OSMOSIS
PRESSURE BREATHING
PRESSURE BROADENING
PRESSURE CHAMBERS
PRESSURE DISTRIBUTION
PRESSURE DRAG
∞ PRESSURE DROP
PRESSURE EFFECTS
PRESSURE GAGES
PRESSURE GRADIENTS
PRESSURE HEADS
PRESSURE ICE

PRESSURE-*(CONT.)*
PRESSURE MEASUREMENT
PRESSURE MODULATOR RADIOMETERS
PRESSURE OSCILLATIONS
PRESSURE PULSES
PRESSURE RATIO
PRESSURE RECORDERS
PRESSURE RECOVERY
PRESSURE REDUCTION
PRESSURE SENSORS
PRESSURE SUITS
PRESSURE VESSEL DESIGN
PRESSURE VESSELS
PRESSURE WELDING
PRESSURIZED CABINS
PRESSURIZED WATER REACTORS
PRESSURIZING
TEMPERATURE INVERSIONS
TRANSIENT PRESSURES
VACUUM CHAMBERS
WEIGHT (MASS)

PRESSURE BREATHING
GS   RESPIRATION
. **PRESSURE BREATHING**
RT   EMERGENCY BREATHING TECHNIQUES
LIQUID BREATHING
PRESSURE
STRESS (PHYSIOLOGY)

PRESSURE BROADENING
RT   LINE SPECTRA
PRESSURE
SPECTROSCOPY

PRESSURE CABINS
USE   PRESSURIZED CABINS

PRESSURE CHAMBERS
GS   COMPARTMENTS
. TEST CHAMBERS
. . **PRESSURE CHAMBERS**
. . . HYPERBARIC CHAMBERS
. . . VACUUM CHAMBERS
RT   AIR LOCKS
∞ CHAMBERS
ENCLOSURES
PRESSURE
PRESSURIZED CABINS
WIND TUNNEL DRIVES

PRESSURE DEPENDENCE
RT   BURNING RATE
HYDROSTATIC PRESSURE
REACTION KINETICS

PRESSURE DISTRIBUTION
UF   PRESSURE FIELDS
GS   DISTRIBUTION (PROPERTY)
. **PRESSURE DISTRIBUTION**
RT   AERODYNAMIC COEFFICIENTS
AERODYNAMIC LOADS
AERODYNAMIC STABILITY
CENTER OF PRESSURE
DIFFERENTIAL PRESSURE
∞ DISTRIBUTION
FIELD THEORY (PHYSICS)
INFLUENCE COEFFICIENT
INTERNAL PRESSURE
ISOBARS (PRESSURE)
LIFT
LOADING MOMENTS
LOADS (FORCES)
MANOMETERS
MASS DISTRIBUTION
MOMENT DISTRIBUTION
NEWTON PRESSURE LAW
PRESSURE
∞ PRESSURE DROP
SHOCK WAVE PROFILES
SPANWISE BLOWING
STATIC LOADS
STRUCTURAL DESIGN CRITERIA
THEODORSEN TRANSFORMATION
THRUST DISTRIBUTION
VELOCITY DISTRIBUTION
VERTICAL DISTRIBUTION
WALL PRESSURE

PRESSURE DRAG
GS   DYNAMIC CHARACTERISTICS
. DRAG
. . **PRESSURE DRAG**
. . . SUPERSONIC DRAG
. . . WAVE DRAG

PRESSURE DRAG-*(CONT.)*
. . . . INTERFERENCE DRAG
RT   AERODYNAMIC DRAG
FRICTION DRAG
PRESSURE

∞ **PRESSURE DROP**
SN   *(USE OF A MORE SPECIFIC TERM IS*
RECOMMENDED--CONSULT THE TERMS
LISTED BELOW)
RT   FLUID POWER
FRICTION
GAS FLOW
HEAD FLOW
INLET FLOW
PRESSURE
PRESSURE DISTRIBUTION
PRESSURE GRADIENTS
PRESSURE OSCILLATIONS
PRESSURE REDUCTION
TWO PHASE FLOW

PRESSURE EFFECTS
RT   BETA FACTOR
COMPRESSIBILITY EFFECTS
∞ EFFECTS
JET BLAST EFFECTS
LOADS (FORCES)
NONISOTHERMAL PROCESSES
PRESSURE
SUCTION
TEMPERATURE EFFECTS
TEMPERATURE INVERSIONS
TRANSITION PRESSURE
VACUUM EFFECTS
WIND EFFECTS

PRESSURE FIELDS
USE   PRESSURE DISTRIBUTION

PRESSURE GAGES
UF   BOMBS (PRESSURE GAGES)
GS   MEASURING INSTRUMENTS
. **PRESSURE GAGES**
. . BAROMETERS
. . MANOMETERS
. . OSMOMETERS
. . PIEZOELECTRIC GAGES
. . PIEZOMETERS
. . VACUUM GAGES
. . . IONIZATION GAGES
. . . . ALPHATRONS
. . . . BAYARD-ALPERT IONIZATION
GAGES
. . . . PENNING GAGES
. . . . PHILIPS IONIZATION GAGES
. . . KNUDSEN GAGES
. . . MCLEOD GAGES
. . . PIRANI GAGES
RT   ∞ BOMBS
BOURDON TUBES
FLOWMETERS
HYPSOMETERS
PRESSURE
SHOCK MEASURING INSTRUMENTS
STRAIN GAGE ACCELEROMETERS
STRAIN GAGE BALANCES
STRAIN GAGES
WEIGHT INDICATORS

PRESSURE GRADIENTS
GS   GRADIENTS
. **PRESSURE GRADIENTS**
RT   ATMOSPHERIC PRESSURE
BATHYTHERMOGRAPHS
CRITICAL FLOW
DIFFERENTIAL PRESSURE
FLUID BOUNDARIES
FLUID FLOW
FRICTION FACTOR
GEOPRESSURE
HYDRODYNAMICS
HYDROSTATICS
INLET PRESSURE
ISOBARS (PRESSURE)
KNUDSEN FLOW
LIQUID FLOW
LIQUID-LIQUID INTERFACES
LIQUID-VAPOR INTERFACES
MULTIPHASE FLOW
ORIFICE FLOW
PIPE FLOW
POTENTIAL GRADIENTS
PRESSURE
∞ PRESSURE DROP

PRESSURE GRADIENTS-*(CONT.)*
      RANKINE-HUGONIOT RELATION
      STEADY FLOW
      STEAM FLOW
      SUBCRITICAL FLOW
      SUCTION
      SUPERCRITICAL FLOW
      UNIFORM FLOW
      UNSTEADY FLOW
      VENTURI TUBES

PRESSURE HEADS
UF   HEAD (PRESSURE)
GS   FLUID FLOW
      . HEAD (FLUID MECHANICS)
      . . **PRESSURE HEADS**
RT   CENTER OF PRESSURE
      ELEVATION
      FLUID FLOW
      ∞HYDRAULICS
      HYDRODYNAMICS
      HYDROSTATIC PRESSURE
      HYDROSTATICS
      LIQUID FLOW
      PRESSURE

PRESSURE ICE
UF   PRESSURE RIDGES
GS   ICE
      . SEA ICE
      . . **PRESSURE ICE**
RT   COLD WEATHER
      FREEZING
      ICE FORMATION
      LOW TEMPERATURE
      OCEAN CURRENTS
      PRESSURE
      TIDES
      WIND (METEOROLOGY)
      WINTER

PRESSURE MEASUREMENT
UF   TONOMETRY
GS   MECHANICAL MEASUREMENT
      . **PRESSURE MEASUREMENT**
RT   BAROMETERS
      BOURDON TUBES
      DIFFERENTIAL PRESSURE
      FLOW MEASUREMENT
      FLOWMETERS
      IONIZATION GAGES
      KNUDSEN GAGES
      MANOMETERS
      MCLEOD GAGES
      ∞MEASUREMENT
      NOISE METERS
      PHILIPS IONIZATION GAGES
      PIRANI GAGES
      PITOT TUBES
      PNEUMATIC PROBES
      PRESSURE
      VACUUM
      VACUUM GAGES
      VELOCITY
      VELOCITY MEASUREMENT
      VENTURI TUBES
      WEIGHT INDICATORS
      WIND TUNNEL CALIBRATION
      WIND TUNNEL TESTS

PRESSURE MODULATOR RADIOMETERS
GS   MEASURING INSTRUMENTS
      . RADIATION MEASURING INSTRUMENTS
      . . ACTINOMETERS
      . . . RADIOMETERS
      . . . . **PRESSURE MODULATOR
            RADIOMETERS**
RT   INFRARED RADIOMETERS
      PRESSURE

PRESSURE OSCILLATIONS
GS   OSCILLATIONS
      . **PRESSURE OSCILLATIONS**
RT   COMBUSTION STABILITY
      FLAME PROPAGATION
      PRESSURE
      ∞PRESSURE DROP
      TURBULENT FLOW

PRESSURE PROBES
USE   PRESSURE SENSORS

PRESSURE PULSES
GS   PULSES
      . **PRESSURE PULSES**

PRESSURE PULSES-*(CONT.)*
RT   BLAST LOADS
      FLAME PROPAGATION
      PRESSURE
      SHOCK WAVES

PRESSURE RATIO
GS   RATIOS
      . **PRESSURE RATIO**
RT   FUEL-AIR RATIO
      LIFT DRAG RATIO
      MASS RATIOS
      PAYLOAD MASS RATIO
      PRESSURE
      PROPELLANT MASS RATIO
      STRESS RATIO
      THRUST-WEIGHT RATIO

PRESSURE RECORDERS
GS   RECORDING INSTRUMENTS
      . **PRESSURE RECORDERS**
RT   PRESSURE

PRESSURE RECOVERY
RT   ∞DIFFUSERS
      EXPLOSIVE DECOMPRESSION
      FLUID AMPLIFIERS
      INLET PRESSURE
      PRESSURE
      ∞RECOVERY

PRESSURE REDUCTION
UF   BLEED-OFF
      DECOMPRESSION
      DEFLATING
      DEPRESSURIZATION
GS   **PRESSURE REDUCTION**
      . EXPLOSIVE DECOMPRESSION
RT   ∞BLEEDING
      COMPRESSING
      GAS EXPANSION
      INFLATING
      PRESSURE
      ∞PRESSURE DROP
      ∞REDUCTION

PRESSURE REGULATORS
GS   CONTROL EQUIPMENT
      . **PRESSURE REGULATORS**
      REGULATORS
      . **PRESSURE REGULATORS**
      VALVES
      . AUTOMATIC CONTROL VALVES
      . . **PRESSURE REGULATORS**
RT   CONTROLLERS
      FLOW REGULATORS
      FUEL TANK PRESSURIZATION
      OXYGEN REGULATORS
      PRESSURIZING
      RELIEF VALVES

PRESSURE RIDGES
USE   PRESSURE ICE

PRESSURE SENSORS
UF   PRESSURE PROBES
      PRESSURE TRANSDUCERS
GS   TRANSDUCERS
      . **PRESSURE SENSORS**
      . . BOURDON TUBES
RT   ELECTROACOUSTIC WAVES
      PIEZOELECTRIC GAGES
      PRESSURE
      QUARTZ TRANSDUCERS
      ∞RAKES
      SHOCK WAVE GENERATORS
      TRANSIENT PRESSURES
      TRANSIENT RESPONSE
      ULTRASONIC WAVE TRANSDUCERS

PRESSURE SUITS
GS   CLOTHING
      . PROTECTIVE CLOTHING
      . . **PRESSURE SUITS**
      . . . SPACE SUITS
      . SUITS
      . . **PRESSURE SUITS**
      . . . SPACE SUITS
RT   FLIGHT CLOTHING
      HELMETS
      INFLATABLE STRUCTURES
      LIFE SUPPORT SYSTEMS
      PORTABLE LIFE SUPPORT SYSTEMS
      PRESSURE
      SAFETY DEVICES

PRESSURE SWITCHES
GS   CONTROL EQUIPMENT
      . **PRESSURE SWITCHES**
      SWITCHES
      . **PRESSURE SWITCHES**
RT   ELECTRIC SWITCHES

PRESSURE TRANSDUCERS
USE   PRESSURE SENSORS

PRESSURE VESSEL DESIGN
GS   STRUCTURAL DESIGN
      . **PRESSURE VESSEL DESIGN**
RT   ∞DESIGN
      PERFORATED SHELLS
      PRESSURE
      SHELLS (STRUCTURAL FORMS)

PRESSURE VESSELS
GS   **PRESSURE VESSELS**
      . PREBURNERS
RT   ACCUMULATORS
      AUTOCLAVES
      BELLS
      BOILERS
      BULBS
      BURST TESTS
      ∞CONTAINERS
      DOMES (STRUCTURAL FORMS)
      FUEL TANK PRESSURIZATION
      FUEL TANKS
      HEMISPHERE CYLINDER BODIES
      ISOTENSOID STRUCTURES
      PRESSURE
      PROPELLANT TANKS
      REACTOR MATERIALS
      SHALLOW SHELL EQUATIONS
      SPHERICAL TANKS
      STORAGE TANKS
      TANKS (CONTAINERS)
      ∞VESSELS
      WALL PRESSURE
      WIND TUNNEL WALLS

PRESSURE WAVES
USE   ELASTIC WAVES

PRESSURE WELDING
GS   WELDING
      . **PRESSURE WELDING**
      . . DIFFUSION WELDING
      . . EXPLOSIVE WELDING
      . . ULTRASONIC WELDING
RT   ARC WELDING
      ELECTRIC WELDING
      FLASH WELDING
      FRICTION WELDING
      FUSION WELDING
      GAS WELDING
      PRESSURE
      SPOT WELDS

PRESSURIZED CABINS
UF   PRESSURE CABINS
GS   COMPARTMENTS
      . **PRESSURIZED CABINS**
RT   AIRCRAFT COMPARTMENTS
      CABIN ATMOSPHERES
      ∞CABINS
      COCKPITS
      EMERGENCY LIFE SUSTAINING
        SYSTEMS
      ENVIRONMENTAL CONTROL
      ESCAPE CAPSULES
      EXPLOSIVE DECOMPRESSION
      LIFE SUPPORT SYSTEMS
      OXYGEN SUPPLY EQUIPMENT
      PRESSURE
      PRESSURE CHAMBERS
      SPACECRAFT CABIN ATMOSPHERES
      SPACECRAFT CABINS

PRESSURIZED WATER REACTORS
GS   NUCLEAR REACTORS
      . LIQUID COOLED REACTORS
      . . WATER COOLED REACTORS
      . . . **PRESSURIZED WATER REACTORS**
      . . . . SPECTRAL SHIFT CONTROL
        REACTOR
RT   NUCLEAR POWER REACTORS
      PRESSURE

PRESSURIZING
GS   **PRESSURIZING**
      . FUEL TANK PRESSURIZATION

PRESSURIZING-(CONT.)
RT   ACCUMULATORS
       DENSIFICATION
       EXPULSION
       EXPULSION BLADDERS
       GAS GENERATORS
       GAS INJECTION
       INFLATING
       PRESSURE
       PRESSURE REGULATORS
       STIMULATION

PRESTON TUBES
USE   PITOT TUBES
       SPEED INDICATORS

PRESTRAINING
USE   PRESTRESSING

PRESTRESSING
UF   PRELOADING
       PRESTRAINING
       PRETWISTING
GS   PREPARATION
       . PRETREATMENT
       . . PRESTRESSING
RT   ELASTIC DEFORMATION
       ISOTENSOID STRUCTURES
       STRESSES
       STRUCTURAL STRAIN

PRETESTS
USE   TESTS

PRETREATMENT
GS   PREPARATION
       . PRETREATMENT
       . . PRESTRESSING
RT   PREVENTION
       ∞PRIMING

PRETWISTING
USE   PRESTRESSING
       TWISTING

PREVAPORIZATION
GS   PHASE TRANSFORMATIONS
       . VAPORIZING
       . . PREVAPORIZATION
RT   FLASHING (VAPORIZING)
       GASES
       VAPOR PHASES
       VAPORS
       VOLATILITY

PREVENTION
GS   PREVENTION
       . ACCIDENT PREVENTION
       . CORROSION PREVENTION
       . FIRE PREVENTION
       . ICE PREVENTION
RT   BLOCKING
       ETIOLOGY
       ∞INHIBITION
       POLLUTION
       PREEMPTING
       PRETREATMENT
       PROTECTION
       ∞REDUCTION
       ∞RESISTANCE
       RETARDING
       SABOTAGE
       SAFETY
       STOPPING

PREWHIRLING
GS   PREPARATION
       . PREWHIRLING

PREWHITENING
GS   PREPARATION
       . PREWHITENING
RT   COLOR
       ∞TREATMENT

PRIBRAM METEORITE
GS   CELESTIAL BODIES
       . METEORITES
       . . STONY METEORITES
       . . . CHONDRITES
       . . . . PRIBRAM METEORITE
RT   BOLIDES
       METEOR TRAILS

PRIMARY BATTERIES
SN   (NON-RECHARGEABLE BATTERIES)
GS   ELECTRIC GENERATORS
       . DIRECT POWER GENERATORS
       . . PRIMARY BATTERIES
       . . . ALKALINE BATTERIES
       . . . DRY CELLS
       . . . . MAGNESIUM CELLS
       . . . . NICKEL ZINC BATTERIES
       . . . METAL AIR BATTERIES
       . . . . ZINC-OXYGEN BATTERIES
       . . . THERMAL BATTERIES
       ELECTROCHEMICAL CELLS
       . ELECTRIC BATTERIES
       . . PRIMARY BATTERIES
       . . . DRY CELLS
       . . . . MAGNESIUM CELLS
       . . . . NICKEL ZINC BATTERIES
       . . . METAL AIR BATTERIES
       . . . . ZINC-OXYGEN BATTERIES
       . . . SODIUM SULFUR BATTERIES
       . . . THERMAL BATTERIES
RT   CHARGE EFFICIENCY
       ELECTROLYTES
       NONAQUEOUS ELECTROLYTES
       STORAGE BATTERIES
       WET CELLS

PRIMARY COSMIC RAYS
UF   HEAVY COSMIC RAY PRIMARIES
GS   EXTRATERRESTRIAL RADIATION
       . PRIMARY COSMIC RAYS
       . . SOLAR COSMIC RAYS
       IONIZING RADIATION
       . COSMIC RAYS
       . . PRIMARY COSMIC RAYS
       . . . SOLAR COSMIC RAYS
       PARTICLES
       . CORPUSCULAR RADIATION
       . . PRIMARY COSMIC RAYS
       . . . SOLAR COSMIC RAYS
RT   COSMIC RAY ALBEDO
       HEAVY NUCLEI
       SECONDARY COSMIC RAYS

PRIMATES
GS   ANIMALS
       . VERTEBRATES
       . . MAMMALS
       . . . PRIMATES
       . . . . APES
       . . . . . CHIMPANZEES
       . . . . BABOONS
       . . . . HUMAN BEINGS
       . . . . MONKEYS

∞ PRIMERS
SN   (USE OF A MORE SPECIFIC TERM IS
       RECOMMENDED--CONSULT THE TERMS
       LISTED BELOW)
RT   ENGINE PRIMERS
       PRIMERS (COATINGS)
       PRIMERS (EXPLOSIVES)

PRIMERS (COATINGS)
GS   COATINGS
       . PROTECTIVE COATINGS
       . . PRIMERS (COATINGS)
RT   DOPES
       FILLERS
       FINISHES
       LACQUERS
       METAL COATINGS
       PAINTS
       ∞PRIMERS
       SPRAYED COATINGS
       SUBSTRATES
       VARNISHES

PRIMERS (EXPLOSIVES)
GS   EXPLOSIVE DEVICES
       . INITIATORS (EXPLOSIVES)
       . . PRIMERS (EXPLOSIVES)
       IGNITERS
       . INITIATORS (EXPLOSIVES)
       . . PRIMERS (EXPLOSIVES)
RT   CAPS (EXPLOSIVES)
       DETONATION
       DETONATORS
       EXPLODING WIRES
       PERCUSSION
       ∞PRIMERS
       SQUIBS

∞ PRIMING
SN   (USE OF A MORE SPECIFIC TERM IS
       RECOMMENDED--CONSULT THE TERMS
       LISTED BELOW)
RT   COATING
       COATINGS
       INITIATION
       PREPARATION
       PRETREATMENT
       STARTING

PRIMITIVE EARTH ATMOSPHERE
GS   EARTH ATMOSPHERE
       . PRIMITIVE EARTH ATMOSPHERE
RT   ∞ATMOSPHERES
       ATMOSPHERIC COMPOSITION
       ATMOSPHERIC ELECTRICITY
       ATMOSPHERIC MODELS
       EARTH PLANETARY STRUCTURE
       FREE ATMOSPHERE
       PLANETARY ATMOSPHERES

PRIMITIVE EQUATIONS
RT   ATMOSPHERIC BOUNDARY LAYER
       CLIMATOLOGY
       ∞EQUATIONS
       EULER EQUATIONS OF MOTION
       FLUID DYNAMICS
       ∞MATHEMATICS

PRINCE EDWARD ISLAND
GS   LANDFORMS
       . ISLANDS
       . . PRINCE EDWARD ISLAND
       NATIONS
       . CANADA
       . . PRINCE EDWARD ISLAND

PRINCE WILLIAM SOUND (AK)
GS   SOUNDS (TOPOGRAPHIC FEATURES)
       . PRINCE WILLIAM SOUND (AK)
RT   ALASKA

PRINCETON SAILWINGS
USE   SAILWINGS

PRINCIPAL COMPONENTS ANALYSIS
RT   IMAGE PROCESSING
       IMAGING TECHNIQUES
       KARHUNEN-LOEVE EXPANSION
       PATTERN RECOGNITION

∞ PRINCIPLES
SN   (USE OF A MORE SPECIFIC TERM IS
       RECOMMENDED--CONSULT THE TERMS
       LISTER BELOW)
RT   DUALITY PRINCIPLE
       ∞LOGIC
       ∞MATHEMATICS

PRINTED CIRCUITS
GS   CIRCUITS
       . PRINTED CIRCUITS
       . . LARGE SCALE INTEGRATION
       . . MEDIUM SCALE INTEGRATION
RT   BREADBOARD MODELS
       CIRCUIT BOARDS
       ELECTRONIC PACKAGING
       HYBRID CIRCUITS
       INTEGRATED CIRCUITS
       MINIATURE ELECTRONIC EQUIPMENT
       MINIATURIZATION
       PHOTOMASKS
       SUBMINIATURIZATION
       THICK FILMS
       TRANSISTOR CIRCUITS

PRINTED RESISTORS
GS   ATTENUATORS
       . RESISTORS
       . . PRINTED RESISTORS
RT   MINIATURIZATION

PRINTERS
GS   PRINTERS
       . PRINTERS (DATA PROCESSING)
       . TELEPRINTERS
RT   CATHODE RAY TUBES
       DATA PROCESSING EQUIPMENT
       PLOTTERS
       PRINTING
       PROJECTORS
       TYPEWRITERS

PRINTERS (DATA PROCESSING)
- GS   DATA PROCESSING EQUIPMENT
- . AUXILIARY EQUIPMENT (COMPUTERS)
- . . **PRINTERS (DATA PROCESSING)**
-       PRINTERS
- . **PRINTERS (DATA PROCESSING)**
- RT   AUTOMATIC TYPEWRITERS
-       COMPUTERS
-       ∞ DATA
-       DISPLAY DEVICES
-       PRINTOUTS
-       READOUT
-       TELEPRINTERS

PRINTING
- GS   **PRINTING**
- . LITHOGRAPHY
- . . PHOTOLITHOGRAPHY
- RT   BINDING
-       CONTRAST
-       ELECTRONOGRAPHY
-       ENGRAVING
-       INKS
-       LEGIBILITY
-       PHOTOENGRAVING
-       PHOTOGRAPHIC PROCESSING
-       PHOTOMECHANICAL EFFECT
-       PLOTTING
-       PRINTERS
-       READING
-       REPRODUCTION (COPYING)
-       STENCIL PROCESSES

PRINTOUTS
- RT   FORMAT
-       LISTS
-       OUTPUT
-       PRINTERS (DATA PROCESSING)
-       READOUT
-       TABLES (DATA)

PRIORITIES
- RT   ENGINEERING MANAGEMENT
-       PROJECT PLANNING
-       RESEARCH MANAGEMENT
-       RESOURCE ALLOCATION
-       SEQUENCING

PRISMATIC BARS
- GS   BARS
- . **PRISMATIC BARS**
-       OPTICAL EQUIPMENT
- . PRISMS
- . . **PRISMATIC BARS**

PRISMS
- GS   OPTICAL EQUIPMENT
- . **PRISMS**
- . . PRISMATIC BARS
- RT   PHOTOELASTICITY
-       REFRACTION

PRIVACY
- RT   COMPUTER INFORMATION SECURITY
-       INFORMATION
-       INFORMATION DISSEMINATION
-       INTEGRITY
-       PAPERS
-       RECORDING
-       RECORDS
-       SECURITY

PRIVATE AIRCRAFT
- USE   GENERAL AVIATION AIRCRAFT

PROBABILITY
- USE   PROBABILITY THEORY

PROBABILITY DENSITY FUNCTIONS
- GS   FUNCTIONS (MATHEMATICS)
- . **PROBABILITY DENSITY FUNCTIONS**
- . . NORMAL DENSITY FUNCTIONS
- . . PEARSON DISTRIBUTIONS
- . . RAYLEIGH DISTRIBUTION
- . . WEIBULL DENSITY FUNCTIONS
-       STATISTICAL ANALYSIS
- . **PROBABILITY DENSITY FUNCTIONS**
- . . NORMAL DENSITY FUNCTIONS
- . . PEARSON DISTRIBUTIONS
- . . RAYLEIGH DISTRIBUTION
- . . WEIBULL DENSITY FUNCTIONS
- RT   BINOMIAL THEOREM
-       CENSORED DATA (MATHEMATICS)
-       CONTINUITY (MATHEMATICS)

PROBABILITY DENSITY FUNCTIONS-(CONT.)
-       DISCRETE FUNCTIONS
-       EVENTS
-       EXPECTANCY HYPOTHESIS
-       EXPONENTIAL FUNCTIONS
-       FAILURE ANALYSIS
-       GAS DENSITY
-       MILLS RATIO
-       QUARTILES

PROBABILITY DISTRIBUTION FUNCTIONS
- GS   FUNCTIONS (MATHEMATICS)
- . **PROBABILITY DISTRIBUTION**
    FUNCTIONS
-       STATISTICAL ANALYSIS
- . **PROBABILITY DISTRIBUTION**
    FUNCTIONS
-       STATISTICAL DISTRIBUTIONS
- . **PROBABILITY DISTRIBUTION**
    FUNCTIONS
- RT   DISCRETE FUNCTIONS
-       GOODNESS OF FIT

PROBABILITY THEORY
- UF   PROBABILITY
-       STATISTICAL PROBABILITY
- RT   ∞ APPLICATIONS OF MATHEMATICS
-       BINOMIAL THEOREM
-       BOREL SETS
-       COMBINATORIAL ANALYSIS
-       CONFIDENCE
-       ∞ CONJUNCTION
-       CONSECUTIVE EVENTS
-       CONTINUUMS
-       CORRELATION
-       DECISION THEORY
-       DISTRIBUTION FUNCTIONS
-       DUFFING DIFFERENTIAL EQUATION
-       EINSTEIN EQUATIONS
-       ERGODIC PROCESS
-       ERROR ANALYSIS
-       EVENTS
-       EXTREMUM VALUES
-       FORECASTING
-       FUZZY SYSTEMS
-       GAME THEORY
-       GOODNESS OF FIT
-       ∞ INDICATION
-       INFINITY
-       INFORMATION THEORY
-       ITERATION
-       LIKELIHOOD RATIO
-       MARTINGALES
-       ∞ MATHEMATICS
-       MAXWELL-BOLTZMANN DENSITY
           FUNCTION
-       MINKOWSKI SPACE
-       MONTE CARLO METHOD
-       OPERATIONS RESEARCH
-       OUTLIERS (STATISTICS)
-       PARAMETER IDENTIFICATION
-       POPULATION THEORY
-       QUALITY CONTROL
-       RANDOM ERRORS
-       RANDOM NOISE
-       RELIABILITY
-       SAMPLING
-       STATISTICAL ANALYSIS
-       STATISTICAL DISTRIBUTIONS
-       ∞ STATISTICS
-       STIELTJES INTEGRAL
-       STOCHASTIC PROCESSES
-       SUBGROUPS
-       SYSTEM IDENTIFICATION
-       ∞ THEORIES
-       TRANSITION PROBABILITIES
-       TRAVELING SALESMAN PROBLEM
-       UNIQUENESS THEOREM

PROBE METHOD (FORECASTING)
- GS   FORECASTING
- . TECHNOLOGICAL FORECASTING
- . . **PROBE METHOD (FORECASTING)**
-       MANAGEMENT METHODS
- . **PROBE METHOD (FORECASTING)**
- RT   DELPHI METHOD (FORECASTING)
-       ESTIMATING
-       ∞ METHODOLOGY
-       OPERATIONS RESEARCH
-       PATTERN METHOD (FORECASTING)
-       PLANNING
-       PREDICTIONS
-       PROFILE METHOD (FORECASTING)
-       TECHNOLOGY ASSESSMENT

∞ PROBES
- SN   *(USE OF A MORE SPECIFIC TERM IS*
    RECOMMENDED--CONSULT THE TERMS
    LISTED BELOW)
- RT   GALILEO PROBE
-       GAS DETECTORS
-       MEASURING INSTRUMENTS
-       PIONEER VENUS SPACECRAFT
-       PIONEER VENUS 1 SPACECRAFT
-       PIONEER VENUS 2 ENTRY PROBES
-       PIONEER VENUS 2 NIGHT PROBE
-       PIONEER VENUS 2 SPACECRAFT
-       PIONEER VENUS 2 TRANSPORTER BUS
-       PIONEER 8 SPACE PROBE
-       PIONEER 9 SPACE PROBE
-       PIONEER 10 SPACE PROBE
-       PIONEER 11 SPACE PROBE
-       RADIO PROBING
-       REMOTE SENSORS
-       SONDES
-       SPACE PROBES
-       TRANSDUCERS

PROBLEM SOLVING
- GS   **PROBLEM SOLVING**
- . ASYMPTOTIC METHODS
- . ITERATIVE SOLUTION
- . THEOREM PROVING
- RT   APPROXIMATION
-       BACKWARD DIFFERENCING
-       COMPUTATIONAL GRIDS
-       CRANK-NICHOLSON METHOD
-       DECISION MAKING
-       DINING PHILOSOPHERS PROBLEM
-       EXISTENCE THEOREMS
-       GROUP DYNAMICS
-       HOMOTROPY
-       HOUSEHOLDER TRANSFORMATIONS
-       ITERATION
-       LEARNING THEORY
-       MANAGEMENT
-       MAZE LEARNING
-       ∞ METHODOLOGY
-       SIMPLEX METHOD
-       ∞ SOLUTION

∞ PROBLEMS
- SN   *(USE OF A MORE SPECIFIC TERM IS*
    RECOMMENDED--CONSULT THE TERMS
    LISTED BELOW)
- RT   BOLZA PROBLEMS
-       BOUNDARY VALUE PROBLEMS
-       CAUCHY PROBLEM
-       CHAPMAN-FERRARO PROBLEM
-       DIRICHLET PROBLEM
-       FOUR BODY PROBLEM
-       ISOPERIMETRIC PROBLEM
-       MANY BODY PROBLEM
-       MAYER PROBLEM
-       NEUMANN PROBLEM
-       OPERATIONAL PROBLEMS
-       POINCARE PROBLEM
-       PRELAUNCH PROBLEMS
-       THREE BODY PROBLEM
-       TRACKING PROBLEM
-       TRAVELING SALESMAN PROBLEM
-       TWO BODY PROBLEM

PROCEDURES
- UF   METHODS
-       PRACTICES
- GS   **PROCEDURES**
- . BOUNDARY INTEGRAL METHOD
- . CREW PROCEDURES (INFLIGHT)
- . CREW PROCEDURES (PREFLIGHT)
- . FINITE ELEMENT METHOD
- . FINITE VOLUME METHOD
- . GLIMM METHOD
- . OPTICAL CORRECTION PROCEDURE
- . PANEL METHOD (FLUID DYNAMICS)
- RT   PROCUREMENT POLICY
-       SYSTEMS ANALYSIS

PROCEEDINGS
- USE   CONFERENCES

PROCESS CONTROL (INDUSTRY)
- RT   COMPONENT RELIABILITY
-       ∞ CONTROL
-       PRODUCT DEVELOPMENT
-       QUALITY CONTROL
-       SAMPLING
-       SPECIFICATIONS

PROCESS HEAT
GS   HEAT
      . **PROCESS HEAT**
RT   HEAT GENERATION

∞ **PROCESSES**
SN   *(USE OF A MORE SPECIFIC TERM IS*
      RECOMMENDED--CONSULT THE TERMS
      LISTED BELOW)
RT   AUTOREGRESSIVE PROCESSES
      ISENTROPIC PROCESSES
      JET MEMBRANE PROCESS
      KRAFT PROCESS (WOODPULP)
      NONISENTROPICITY
      NONISOTHERMAL PROCESSES
      ORNSTEIN-UHLENBECK PROCESS
      PRODUCT DEVELOPMENT
      QUALITY CONTROL
      SOL-GEL PROCESSES
      SPACE INDUSTRIALIZATION
      UMKLAPP PROCESS

∞ **PROCESSING**
SN   *(USE OF A MORE SPECIFIC TERM IS*
      RECOMMENDED--CONSULT THE TERMS
      LISTED BELOW)
RT   ASSOCIATIVE PROCESSING
      (COMPUTERS)
      BATCH PROCESSING
      DATA PROCESSING
      FOOD PROCESSING
      IMAGE PROCESSING
      MANUFACTURING
      MATERIALS RECOVERY
      MESSAGE PROCESSING
      NUCLEAR FUEL REPROCESSING
      OPTICAL DATA PROCESSING
      PHOTOGRAPHIC PROCESSING
      PRODUCTION ENGINEERING
      RECYCLING
      REFINING
      RETORT PROCESSING
      SETTLING
      SIGNAL PROCESSING
      WET SPINNING

PROCESSORS (COMPUTERS)
USE   CENTRAL PROCESSING UNITS

PROCUREMENT
GS   **PROCUREMENT**
      . GOVERNMENT PROCUREMENT
      . LEASING
RT   CONTRACTS
      EQUIPMENT SPECIFICATIONS
      GOVERNMENT/INDUSTRY RELATIONS
      ∞ RECEIVING
      SERVICES
      SPECIFICATIONS
      SUBCONTRACTS

PROCUREMENT MANAGEMENT
GS   MANAGEMENT
      . **PROCUREMENT MANAGEMENT**
RT   ALLOCATIONS
      BUDGETING
      ∞ BUDGETS
      COMMODITIES
      FEDERAL BUDGETS
      FINANCIAL MANAGEMENT
      INVENTORY MANAGEMENT
      PRODUCTS
      SERVICES

PROCUREMENT POLICY
GS   POLICIES
      . **PROCUREMENT POLICY**
RT   DECISIONS
      MANAGEMENT
      PROCEDURES
      REGULATIONS
      RULES

PRODUCT DEVELOPMENT
UF   ENGINEERING DEVELOPMENT
GS   **PRODUCT DEVELOPMENT**
      . WEAPONS DEVELOPMENT
RT   AIRCRAFT DESIGN
      AIRCRAFT PRODUCTION
      AMPLIFIER DESIGN
      ANTENNA DESIGN
      BREADBOARD MODELS
      COMMERCE
      COMPUTER DESIGN
      CONSUMERS

PRODUCT DEVELOPMENT-*(CONT.)*
      ∞ DESIGN
      ∞ DEVELOPMENT
      ENGINE DESIGN
      FUNCTIONAL DESIGN SPECIFICATIONS
      HELICOPTER DESIGN
      INVENTIONS
      LENS DESIGN
      MANAGEMENT
      MARKET RESEARCH
      MARKETING
      PATENT APPLICATIONS
      PATENT POLICY
      PILOT PLANTS
      PROCESS CONTROL (INDUSTRY)
      ∞ PROCESSES
      ∞ PRODUCTION
      PRODUCTION ENGINEERING
      QUALITY
      QUALITY CONTROL
      REACTOR DESIGN
      RELIABILITY
      SATELLITE DESIGN
      SOLVENT REFINED COAL
      SPACE INDUSTRIALIZATION
      SPACECRAFT DESIGN
      STANDARDIZATION
      STRUCTURAL DESIGN

∞ **PRODUCTION**
SN   *(USE OF A MORE SPECIFIC TERM IS*
      RECOMMENDED--CONSULT THE TERMS
      LISTED BELOW)
RT   AIRCRAFT PRODUCTION
      FUEL PRODUCTION
      OUTPUT
      PRODUCT DEVELOPMENT
      PRODUCTION ENGINEERING
      PRODUCTION PLANNING
      RESERVES
      TOOLS

PRODUCTION COSTS
GS   **PRODUCTION COSTS**
      . AIRCRAFT PRODUCTION COSTS
RT   COST ANALYSIS
      COST ESTIMATES
      DESIGN TO COST
      LIFE CYCLE COSTS
      OPERATING COSTS

PRODUCTION ENGINEERING
UF   PRODUCTION METHODS
GS   **PRODUCTION ENGINEERING**
      . PRODUCTION PLANNING
RT   AIRCRAFT PRODUCTION
      AIRCRAFT PRODUCTION COSTS
      ∞ CAPACITY
      ∞ ENGINEERING
      HUMAN FACTORS ENGINEERING
      LASER APPLICATIONS
      MANAGEMENT
      NUMERICAL CONTROL
      ∞ OPERATIONS
      PLANNING
      ∞ PROCESSING
      PRODUCT DEVELOPMENT
      ∞ PRODUCTION
      PRODUCTIVITY
      PRODUCTS
      SCHEDULING
      STANDARDIZATION

PRODUCTION MANAGEMENT
GS   MANAGEMENT
      . **PRODUCTION MANAGEMENT**
RT   AIRCRAFT PRODUCTION COSTS
      EMPLOYEE RELATIONS
      ESTIMATES
      FABRICATION
      INDUSTRIAL MANAGEMENT
      MANUFACTURING
      QUALITY CONTROL
      RELIABILITY
      RESOURCES
      SAFETY MANAGEMENT

PRODUCTION METHODS
USE   PRODUCTION ENGINEERING

PRODUCTION PLANNING
GS   PLANNING
      . MANAGEMENT PLANNING
      . . **PRODUCTION PLANNING**
      PRODUCTION ENGINEERING

PRODUCTION PLANNING-*(CONT.)*
      . **PRODUCTION PLANNING**
RT   SCHEDULES

PRODUCTIVITY
RT   AIRCRAFT PRODUCTION COSTS
      ALLOWANCES
      EFFICIENCY
      MATRIX MANAGEMENT
      MORALE
      PRODUCTION ENGINEERING
      RELIABILITY
      WORKSTATIONS

PRODUCTS
GS   **PRODUCTS**
      . GROSS NATIONAL PRODUCT
      . PETROLEUM PRODUCTS
      . . ASPHALT
      . . GREASES
      . . TARS
      . REACTION PRODUCTS
RT   BY-PRODUCTS
      COMBUSTION PRODUCTS
      COMMODITIES
      FISSION PRODUCTS
      MANUFACTURING
      OUTPUT
      PROCUREMENT MANAGEMENT
      PRODUCTION ENGINEERING
      QUALITY CONTROL
      RESOURCE ALLOCATION
      SERVICES
      SPACE INDUSTRIALIZATION

PROFICIENCY
USE   ABILITIES

PROFILE METHOD (FORECASTING)
GS   FORECASTING
      . TECHNOLOGICAL FORECASTING
      . . **PROFILE METHOD (FORECASTING)**
      MANAGEMENT METHODS
      . **PROFILE METHOD (FORECASTING)**
RT   DELPHI METHOD (FORECASTING)
      ESTIMATING
      ∞ METHODOLOGY
      OPERATIONS RESEARCH
      PLANNING
      PREDICTIONS
      PROBE METHOD (FORECASTING)
      TECHNOLOGY ASSESSMENT

∞ **PROFILES**
SN   *(USE OF A MORE SPECIFIC TERM IS*
      RECOMMENDED--CONSULT THE TERMS
      LISTED BELOW)
RT   AIRFOIL PROFILES
      ANGLES (GEOMETRY)
      CURVATURE
      DELINEATION
      DISTRIBUTION (PROPERTY)
      GEOMETRY
      GRADIENTS
      LINE SHAPE
      PLANFORMS
      PROFILOMETERS
      SEARCH PROFILES
      SHAPES
      SHOCK WAVE PROFILES
      SLOPES
      STREAMLINING
      TOPOGRAPHY
      WIND PROFILES

PROFILOMETERS
GS   MEASURING INSTRUMENTS
      . **PROFILOMETERS**
RT   ∞ PROFILES
      ROUGHNESS
      SHAPES
      SURFACE PROPERTIES
      SURFACE ROUGHNESS

PROGENY
RT   CHILDREN
      REPRODUCTION (BIOLOGY)

PROGNOSIS
RT   DIAGNOSIS

PROGNOZ SATELLITES
GS   SATELLITES
      . ARTIFICIAL SATELLITES

PROGNOZ SATELLITES-*(CONT.)*
. . SOVIET SATELLITES
. . . **PROGNOZ SATELLITES**

PROGRAM MANAGEMENT
USE   PROJECT MANAGEMENT

PROGRAM TREND LINE ANALYSIS
RT   CRITICAL PATH METHOD
         MANAGEMENT PLANNING
         PERT
         PROGRAMS
         PROJECT MANAGEMENT

PROGRAM VERIFICATION (COMPUTERS)
RT   CHECKOUT
         COMPUTER PROGRAMMING
         FILE MAINTENANCE (COMPUTERS)
         PROVING
         SOFTWARE TOOLS
         SYSTEMS ANALYSIS
         ∞ TESTS

PROGRAMMED INSTRUCTION
GS   **PROGRAMMED INSTRUCTION**
         . COMPUTER ASSISTED INSTRUCTION
RT   COMPILERS
         COMPUTER PROGRAMMING
         COMPUTER PROGRAMS

PROGRAMMERS
GS   PERSONNEL
         . **PROGRAMMERS**
RT   CODERS
         COMPUTER PROGRAMMING
         FILE MAINTENANCE (COMPUTERS)

∞ **PROGRAMMING**
SN   *(USE OF A MORE SPECIFIC TERMS IS*
         RECOMMENDED--CONSULT THE TERMS
         LISTED BELOW)
RT   COMPUTER PROGRAMMING
         DYNAMIC PROGRAMMING
         FILE MAINTENANCE (COMPUTERS)
         LINEAR PROGRAMMING
         MATHEMATICAL PROGRAMMING
         MICROPROGRAMMING
         MULTIPROGRAMMING
         NONLINEAR PROGRAMMING
         QUADRATIC PROGRAMMING

PROGRAMMING (SCHEDULING)
GS   **PROGRAMMING (SCHEDULING)**
         . THRUST PROGRAMMING
RT   CRITICAL PATH METHOD
         ∞ STEPS

PROGRAMMING LANGUAGES
GS   LANGUAGES
         . **PROGRAMMING LANGUAGES**
         . . ALGOL
         . . APL (PROGRAMMING LANGUAGE)
         . . ASSEMBLY LANGUAGE
         . . . AUTOCODERS
         . . . COMPASS (PROGRAMMING
                  LANGUAGE)
         . . . MAP (PROGRAMMING LANGUAGE)
         . . BASIC (PROGRAMMING LANGUAGE)
         . . COBOL
         . . COGO (PROGRAMMING LANGUAGE)
         . . CONTEXT FREE LANGUAGES
         . . FORTRAN
         . . HAL/S (LANGUAGE)
         . . HIGH LEVEL LANGUAGES
         . . LISP (PROGRAMMING LANGUAGE)
         . . MACHINE ORIENTED LANGUAGES
         . . . MARVS (PROGRAMMING
                  LANGUAGE)
         . . . SLEUTH (PROGRAMMING
                  LANGUAGE)
         . . NATURAL LANGUAGE (COMPUTERS)
         . . PASCAL (PROGRAMMING LANGUAGE)
         . . PL/1
RT   COMPUTER PROGRAMMING

PROGRAMS
GS   **PROGRAMS**
         . ACEE PROGRAM
         . ARMY-NAVY INSTRUMENTATION
            PROGRAM
         . COMSAT PROGRAM
         . DEFENSE PROGRAM
         . DOWNRANGE ANTIMISSILE
            MEASUREMENT PROGRAM

PROGRAMS-*(CONT.)*
. . GLOBAL ATMOSPHERIC RESEARCH
      PROGRAM
. . GARP ATLANTIC TROPICAL
      EXPERIMENT
. GULLIVER PROGRAM
. INTERNATIONAL
      GEOSPHERE-BIOSPHERE PROGRAM
. LUNAR PROGRAMS
. . APOLLO PROJECT
. . SURVEYOR PROJECT
. MACHINE-INDEPENDENT PROGRAMS
. NASA PROGRAMS
. . APOLLO APPLICATIONS PROGRAM
. . APOLLO PROJECT
. . ASSESS PROGRAM
. . ATLIT PROJECT
. . BIOASTRONAUTICAL ORBITAL SPACE
      SYSTEM
. . CENTAUR PROJECT
. . DAST PROGRAM
. . EARTH & OCEAN PHYSICS
      APPLICATIONS PROGRAM
. . EARTH RESOURCES PROGRAM
. . . EARTH RESOURCES SURVEY
         PROGRAM
. . . . SEASAT PROGRAM
. . ECHO PROJECT
. . GALILEO PROJECT
. . GEMINI PROJECT
. . HELIOS PROJECT
. . JUPITER PROJECT
. . MARINER PROGRAM
. . MARS 69 PROJECT
. . MARS 71 PROJECT
. . MERCURY PROJECT
. . NATIONAL LAUNCH VEHICLE
      PROGRAM
. . NEW MOONS PROJECT
. . NIMBUS PROJECT
. . PIONEER PROJECT
. . PROJECT SETI
. . QUIET ENGINE PROGRAM
. . RANGER PROJECT
. . . AGENA B RANGER PROGRAM
. . ROVER PROJECT
. . SAIL PROJECT
. . SATURN PROJECT
. . SCOUT PROJECT
. . SKYLAB PROGRAM
. . SUPERSONIC CRUISE AIRCRAFT
      RESEARCH
. . SURVEYOR PROJECT
. . SYNCHRONOUS COMMUNICATIONS
      SATELLITE PROJ
. . TACT PROGRAM
. . TEKTITE PROJECT
. . TERMINAL CONFIGURED VEHICLE
      PROGRAM
. . TILT ROTOR RESEARCH AIRCRAFT
      PROGRAM
. . TIROS PROJECT
. . TITAN PROJECT
. . VANGUARD PROJECT
. . VOYAGER PROJECT
. NASA SPACE PROGRAMS
. . APOLLO APPLICATIONS PROGRAM
. . APOLLO PROJECT
. . CENTAUR PROJECT
. . EARTH & OCEAN PHYSICS
      APPLICATIONS PROGRAM
. . EARTH RESOURCES PROGRAM
. . . EARTH RESOURCES SURVEY
         PROGRAM
. . . . SEASAT PROGRAM
. . ECHO PROJECT
. . GALILEO PROJECT
. . GEMINI PROJECT
. . HELIOS PROJECT
. . JUPITER PROJECT
. . MARINER PROGRAM
. . MARS 69 PROJECT
. . MARS 71 PROJECT
. . MERCURY PROJECT
. . NATIONAL LAUNCH VEHICLE
      PROGRAM
. . NEW MOONS PROJECT
. . NIMBUS PROJECT
. . PIONEER PROJECT
. . PROJECT SETI
. . QUIET ENGINE PROGRAM
. . RANGER PROJECT
. . . AGENA B RANGER PROGRAM
. . ROVER PROJECT
. . SAIL PROJECT
. . SATURN PROJECT

PROGRAMS-*(CONT.)*
. . SCOUT PROJECT
. . SKYLAB PROGRAM
. . SURVEYOR PROJECT
. . SYNCHRONOUS COMMUNICATIONS
      SATELLITE PROJ
. . TEKTITE PROJECT
. . TIROS PROJECT
. . TITAN PROJECT
. . VANGUARD PROJECT
. . VOYAGER PROJECT
. PANT PROGRAM
. PROJECTS
. . ADVENT PROJECT
. . AGRISTARS PROJECT
. . ALARM PROJECT
. . APOLLO PROJECT
. . APOLLO SOYUZ TEST PROJECT
. . ARGUS PROJECT
. . ASSET PROJECT
. . ATLIT PROJECT
. . BIG SHOT PROJECT
. . BIOS PROJECT
. . BUMBLEBEE PROJECT
. . CENTAUR PROJECT
. . DEFENDER PROJECT
. . EARTH & OCEAN PHYSICS
      APPLICATIONS PROGRAM
. . ECHO PROJECT
. . ECLIPSE PROJECT
. . EXPERIMENTAL REFLECTOR ORBITAL
      SHOT PROJ
. . GALILEO PROJECT
. . GEMINI PROJECT
. . GEOSARI PROJECT
. . HARVARD RADIO METEOR PROJECT
. . HELIOS PROJECT
. . JUPITER PROJECT
. . MARS 69 PROJECT
. . MARS 71 PROJECT
. . MERCURY PROJECT
. . NEW MOONS PROJECT
. . NIKE PROJECT
. . NIMBUS PROJECT
. . ORBITER PROJECT
. . PIONEER PROJECT
. . PROJECT SETI
. . RADIO ATTENUATION MEASUREMENT
      PROJECT
. . RAND PROJECT
. . RANGER PROJECT
. . . AGENA B RANGER PROGRAM
. . ROVER PROJECT
. . SAIL PROJECT
. . SATURN PROJECT
. . SCANNER PROJECT
. . SCOUT PROJECT
. . SEAFARER PROJECT
. . SQUID PROJECT
. . SUBMARINE INTEGRATED CONTROL
      PROJECT
. . SUCCESS PROJECT
. . SURVEYOR PROJECT
. . SYNCHRONOUS COMMUNICATIONS
      SATELLITE PROJ
. . TELSTAR PROJECT
. . THEMIS PROJECT
. . TIROS PROJECT
. . TITAN PROJECT
. . VANGUARD PROJECT
. . VOYAGER PROJECT
. . WEST FORD PROJECT
. RADAR TARGET SCATTER SITE
      PROGRAM
. SOURCE PROGRAMS
. SPACE PROGRAMS
. . BRAZILIAN SPACE PROGRAM
. . CANADIAN SPACE PROGRAM
. . CHINESE SPACE PROGRAM
. . EUROPEAN SPACE PROGRAMS
. . FRENCH SPACE PROGRAMS
. . GEOGRAPHIC APPLICATIONS
      PROGRAM
. . INDIAN SPACE PROGRAM
. . INDONESIAN SPACE PROGRAM
. . ITALIAN SPACE PROGRAM
. . JAPANESE SPACE PROGRAM
. . SAUDI ARABIAN SPACE PROGRAM
. . SWEDISH SPACE PROGRAM
. . SWISS SPACE PROGRAM
. . U.S.S.R. SPACE PROGRAM
. . UK SPACE PROGRAM
. STARSITE PROGRAM
. TRAP PROGRAM
. UNIVERSITY PROGRAM
RT   BUREAUS (ORGANIZATIONS)

PROGRAMS-(CONT.)
                COMMITTEE ON SPACE RESEARCH
                COMPUTER PROGRAM INTEGRITY
                COMPUTER PROGRAMS
                DAST PROGRAM
                EARTH RESOURCES INFORMATION
                    SYSTEM
                INVESTIGATION
                MISSION PLANNING
        ∞MISSIONS
        ∞OPERATIONS
                PROGRAM TREND LINE ANALYSIS
                RESEARCH AND DEVELOPMENT
        ∞RESEARCH PROJECTS
                SEASAT SATELLITES
                SEASAT 1
                SEASAT-B SATELLITE
                SOLAR MAXIMUM MISSION
                SYNCHRONOUS EARTH OBSERVATORY
                    SATELLITE
                USER MANUALS (COMPUTER
                    PROGRAMS)

PROGRESS
RT   ECONOMICS
        MANAGEMENT
        PLANNING
        ∞PROPERTIES

PROGRESSIONS
GS   ANALYSIS (MATHEMATICS)
        . CALCULUS
        . . SERIES (MATHEMATICS)
        . . . **PROGRESSIONS**
        . REAL VARIABLES
        . . SERIES (MATHEMATICS)
        . . . **PROGRESSIONS**

PROHIBITION
RT   LEGAL LIABILITY
        PENALTIES
        POLICIES
        REGULATIONS

PROJECT MANAGEMENT
UF   PROGRAM MANAGEMENT
GS   MANAGEMENT
        . **PROJECT MANAGEMENT**
RT   COMMERCE
        CONTRACT MANAGEMENT
        CRITICAL PATH METHOD
        GERT
        INTERFACES
        MANAGEMENT PLANNING
        MANAGEMENT SYSTEMS
        MISSION PLANNING
        PERT
        PROGRAM TREND LINE ANALYSIS
        PROJECTS
        RESEARCH AND DEVELOPMENT
        ∞RESEARCH PROJECTS
        WEAPON SYSTEM MANAGEMENT

PROJECT PLANNING
GS   PLANNING
        . MANAGEMENT PLANNING
        . . **PROJECT PLANNING**
RT   ALLOCATIONS
        BUDGETING
        DECISIONS
        ESTIMATES
        FORECASTING
        GOALS
        MANAGEMENT
        MATRIX MANAGEMENT
        ∞MISSIONS
        NASA INTERACTIVE PLANNING SYSTEM
        OPERATIONS RESEARCH
        PRIORITIES
        PROJECTS

PROJECT SETI
UF   SEARCH FOR EXTRATERRESTRIAL
            INTELLIGENCE
        SETI
GS   PROGRAMS
        . NASA PROGRAMS
        . . **PROJECT SETI**
        . NASA SPACE PROGRAMS
        . . **PROJECT SETI**
        . PROJECTS
        . . **PROJECT SETI**
RT   EXTRATERRESTRIAL INTELLIGENCE
        RADIO COMMUNICATION
        RADIO SIGNALS

PROJECTILE CRATERING
UF   HYPERVELOCITY CRATERING
GS   CRATERING
        . **PROJECTILE CRATERING**
RT   EJECTA
        HYPERVELOCITY IMPACT
        HYPERVELOCITY PROJECTILES
        METEORITE CRATERS
        METEORITIC DAMAGE
        METEOROID HAZARDS

PROJECTILE PENETRATION
USE   TERMINAL BALLISTICS

PROJECTILES
GS   **PROJECTILES**
        . HYPERVELOCITY PROJECTILES
        . PRECISION GUIDED PROJECTILES
        . SABOT PROJECTILES
RT   AMMUNITION
        BALLISTICS
        BOMBS (ORDNANCE)
        CARTRIDGES
        FINNED BODIES
        GUNFIRE
        GUNS (ORDNANCE)
        INCENDIARY AMMUNITION
        NUCLEAR WEAPONS
        PYROTECHNICS
        SHAPED CHARGES
        SHRAPNEL
        TERMINAL BALLISTICS
        TERRADYNAMICS
        WARHEADS
        WEAPONS

∞ **PROJECTION**
SN   (USE OF A MORE SPECIFIC TERM IS
        RECOMMENDED--CONSULT THE TERMS
        LISTED BELOW)
RT   BONNE PROJECTION
        DESCRIPTIVE GEOMETRY
        DRAWINGS
        FORECASTING
        GNOMONIC PROJECTION
        GRAPHIC ARTS
        ILLUMINATING
        MAGNIFICATION
        PREDICTIONS
        PROJECTORS
        TRENDS

PROJECTIVE GEOMETRY
GS   GEOMETRY
        . EUCLIDEAN GEOMETRY
        . . **PROJECTIVE GEOMETRY**
        . . . MERCATOR PROJECTION
RT   ANALYTIC GEOMETRY
        DESCRIPTIVE GEOMETRY
        GNOMONIC PROJECTION
        RECIPROCAL THEOREMS

PROJECTORS
SN   (LIGHT AND IMAGE)
RT   BEACONS
        ILLUMINATING
        LUMINAIRES
        MOTION PICTURES
        PHOTOGRAMMETRY
        PHOTOGRAPHIC EQUIPMENT
        PHOTOGRAPHY
        PRINTERS
        ∞PROJECTION
        SEARCHLIGHTS

PROJECTS
GS   PROGRAMS
        . **PROJECTS**
        . . ADVENT PROJECT
        . . AGRISTARS PROJECT
        . . ALARM PROJECT
        . . APOLLO PROJECT
        . . APOLLO SOYUZ TEST PROJECT
        . . ARGUS PROJECT
        . . ASSET PROJECT
        . . ATLIT PROJECT
        . . BIG SHOT PROJECT
        . . BIOS PROJECT
        . . BUMBLEBEE PROJECT
        . . CENTAUR PROJECT
        . . DEFENDER PROJECT
        . . EARTH & OCEAN PHYSICS
            APPLICATIONS PROGRAM
        . . ECHO PROJECT
        . . ECLIPSE PROJECT

PROJECTS-(CONT.)
        . . EXPERIMENTAL REFLECTOR ORBITAL
            SHOT PROJ
        . . GALILEO PROJECT
        . . GEMINI PROJECT
        . . GEOSARI PROJECT
        . . HARVARD RADIO METEOR PROJECT
        . . HELIOS PROJECT
        . . JUPITER PROJECT
        . . MARS 69 PROJECT
        . . MARS 71 PROJECT
        . . MERCURY PROJECT
        . . NEW MOONS PROJECT
        . . NIKE PROJECT
        . . NIMBUS PROJECT
        . . ORBITER PROJECT
        . . PIONEER PROJECT
        . . PROJECT SETI
        . . RADIO ATTENUATION MEASUREMENT
            PROJECT
        . . RAND PROJECT
        . . RANGER PROJECT
        . . . AGENA B RANGER PROGRAM
        . . ROVER PROJECT
        . . SAIL PROJECT
        . . SATURN PROJECT
        . . SCANNER PROJECT
        . . SCOUT PROJECT
        . . SEAFARER PROJECT
        . . SQUID PROJECT
        . . SUBMARINE INTEGRATED CONTROL
            PROJECT
        . . SUCCESS PROJECT
        . . SURVEYOR PROJECT
        . . SYNCHRONOUS COMMUNICATIONS
            SATELLITE PROJ
        . . TELSTAR PROJECT
        . . THEMIS PROJECT
        . . TIROS PROJECT
        . . TITAN PROJECT
        . . VANGUARD PROJECT
        . . VOYAGER PROJECT
        . . WEST FORD PROJECT
RT   BUREAUS (ORGANIZATIONS)
        CONTRACTS
        ESTIMATING
        ∞MISSIONS
        ∞OPERATIONS
        PROJECT MANAGEMENT
        PROJECT PLANNING
        ∞RESEARCH PROJECTS
        TASKS
        TEAMS

PROLATE SPHEROIDS
GS   GEOMETRY
        . EUCLIDEAN GEOMETRY
        . . ANALYTIC GEOMETRY
        . . . SPHEROIDS
        . . . . **PROLATE SPHEROIDS**
RT   OBLATE SPHEROIDS

PROLATENESS
RT   SHAPES

PROLONGATION
GS   EXTENSIONS
        . **PROLONGATION**
RT   TIME

PROMETHAZINE
GS   AMINES
        . **PROMETHAZINE**
        DRUGS
        . ANTIHISTAMINICS
        . . **PROMETHAZINE**
        HETEROCYCLIC COMPOUNDS
        . **PROMETHAZINE**

PROMETHIUM
GS   CHEMICAL ELEMENTS
        . RARE EARTH ELEMENTS
        . . **PROMETHIUM**
        . . . PROMETHIUM ISOTOPES
        METALS
        . RARE EARTH ELEMENTS
        . . **PROMETHIUM**
        . . . PROMETHIUM ISOTOPES

PROMETHIUM ISOTOPES
UF   PROMETHIUM 146
GS   CHEMICAL ELEMENTS
        . NUCLIDES
        . . ISOTOPES
        . . . **PROMETHIUM ISOTOPES**

PROMETHIUM ISOTOPES-*(CONT.)*
. RARE EARTH ELEMENTS
. . PROMETHIUM
. . . **PROMETHIUM ISOTOPES**
METALS
. RARE EARTH ELEMENTS
. . PROMETHIUM
. . . **PROMETHIUM ISOTOPES**

PROMETHIUM 146
USE   PROMETHIUM ISOTOPES

PROMINENCES
GS   **PROMINENCES**
. SOLAR PROMINENCES
RT   SOLAR ACTIVITY

PROMOTION
RT   DISPLAY DEVICES
INCREASING
PUBLIC RELATIONS
UPGRADING

PRONE POSITION
RT   REST
SITTING POSITION
SUPINE POSITION

PRONY SERIES
GS   ANALYSIS (MATHEMATICS)
. CALCULUS
. . SERIES (MATHEMATICS)
. . . **PRONY SERIES**
. REAL VARIABLES
. . SERIES (MATHEMATICS)
. . . **PRONY SERIES**

PROOFS
USE   PROVING

PROP-FAN TECHNOLOGY
RT   PROPELLER BLADES
PROPELLER EFFICIENCY
PROPELLER FANS
TURBOPROP ENGINES

∞ PROPAGATION
SN   *(USE OF A MORE SPECIFIC TERM IS
RECOMMENDED--CONSULT THE TERMS
LISTED BELOW)*
UF   PROPAGATORS
RT   ACOUSTIC PROPAGATION
ATTENUATION -
CRACK PROPAGATION
DIFFRACTION PROPAGATION
DIFFUSION
ELECTROMAGNETIC RADIATION
FLAME PROPAGATION
PROPAGATION (EXTENSION)
SELF PROPAGATION
STRESS PROPAGATION
TRANSEQUATORIAL PROPAGATION
TRANSMISSION
WAVE PROPAGATION

PROPAGATION (EXTENSION)
GS   **PROPAGATION (EXTENSION)**
. CRACK PROPAGATION
. FLAME PROPAGATION
RT   ∞ PROPAGATION

PROPAGATION MODES
GS   MODES
. **PROPAGATION MODES**
RT   ANTIPODES
CIRCULAR WAVEGUIDES
ELECTROMAGNETIC SURFACE WAVES
MODE TRANSFORMERS
MULTIMODE RESONATORS
PROPAGATION VELOCITY
SHOCK WAVE INTERACTION
WAVE INTERACTION
WAVE PROPAGATION
WAVEGUIDES

PROPAGATION VELOCITY
GS   RATES (PER TIME)
. **PROPAGATION VELOCITY**
VELOCITY
. **PROPAGATION VELOCITY**
RT   ELECTROMAGNETIC RADIATION
GROUP VELOCITY
PHASE VELOCITY
PROPAGATION MODES

PROPAGATION VELOCITY-*(CONT.)*
WAVE PROPAGATION

PROPAGATORS
USE   PROPAGATION

PROPANE
GS   ALIPHATIC COMPOUNDS
. ALKANES
. . **PROPANE**
HYDROCARBONS
. ALKANES
. . **PROPANE**
RT   CYCLOPROPANE
HYDROCARBON FUELS
NITROPROPANE

PROPARGYL GROUPS
GS   ORGANIC COMPOUNDS
. **PROPARGYL GROUPS**
RT   ETHERS
GETTERS
PHENYLS

∞ PROPELLANT ACTUATED DEVICES
SN   *(USE OF A MORE SPECIFIC TERM IS
RECOMMENDED--CONSULT THE TERMS
LISTED BELOW)*
RT   EJECTION SEATS
EXPLOSIVE DEVICES
PROPELLANT ACTUATED INSTRUMENTS
ROCKET ENGINES

PROPELLANT ACTUATED INSTRUMENTS
RT   ACTUATORS
CONTROLLERS
∞ INSTRUMENTS
MEASURING INSTRUMENTS
∞ PROPELLANT ACTUATED DEVICES

PROPELLANT ADDITIVES
GS   ADDITIVES
. **PROPELLANT ADDITIVES**
. . PROPELLANT BINDERS
. . . SOLID ROCKET BINDERS
RT   ANTIICING ADDITIVES
ANTIOXIDANTS
CATALYSTS
COMPOSITE PROPELLANTS
CORROSION PREVENTION
GELLED PROPELLANTS
INHIBITORS
PLASTICIZERS
STORABLE PROPELLANTS

PROPELLANT BINDERS
GS   ADDITIVES
. PROPELLANT ADDITIVES
. . **PROPELLANT BINDERS**
. . . SOLID ROCKET BINDERS
BINDERS (MATERIALS)
. **PROPELLANT BINDERS**
. . SOLID ROCKET BINDERS
RT   COMPOSITE PROPELLANTS
ROCKET PROPELLANTS
SOLID PROPELLANTS

PROPELLANT CASTING
GS   CASTINGS
. **PROPELLANT CASTING**
FORMING TECHNIQUES
. CASTING
. . **PROPELLANT CASTING**

PROPELLANT CHEMISTRY
RT   ∞ CHEMISTRY
SOLID PROPELLANT COMBUSTION
THERMOCHEMISTRY

PROPELLANT COMBUSTION
GS   COMBUSTION
. **PROPELLANT COMBUSTION**
. . SOLID PROPELLANT COMBUSTION
. . . SOLID PROPELLANT IGNITION
RT   AXIAL MODES
COMBUSTION EFFICIENCY
COMBUSTION STABILITY
EROSIVE BURNING
FUEL COMBUSTION
HYDROCARBON COMBUSTION
IGNITION
METAL COMBUSTION
VELOCITY COUPLING

PROPELLANT DECOMPOSITION
GS   DECOMPOSITION
. **PROPELLANT DECOMPOSITION**
RT   ENDOTHERMIC FUELS
FUEL CORROSION
INHIBITORS
MONOPROPELLANTS
STORABLE PROPELLANTS

PROPELLANT EVAPORATION
GS   PHASE TRANSFORMATIONS
. VAPORIZING
. . EVAPORATION
. . . **PROPELLANT EVAPORATION**
RT   EVAPORATIVE COOLING
STORABLE PROPELLANTS

PROPELLANT EXPLOSIONS
RT   DETONATION
IMPLOSIONS
ROCKET ENGINES

PROPELLANT GRAINS
RT   BURNING RATE
∞ GRAINS
SOLID PROPELLANTS
SOLID ROCKET PROPELLANTS

PROPELLANT MASS RATIO
GS   RATIOS
. MASS RATIOS
. . **PROPELLANT MASS RATIO**
RT   PAYLOAD MASS RATIO
PRESSURE RATIO
PROPULSION SYSTEM PERFORMANCE
PROPULSIVE EFFICIENCY
SPECIFIC IMPULSE
STAGE SEPARATION

PROPELLANT OXIDIZERS
USE   ROCKET OXIDIZERS

PROPELLANT PROPERTIES
GS   **PROPELLANT PROPERTIES**
. PROPELLANT SENSITIVITY
. PROPELLANT STORABILITY
RT   CHEMICAL PROPERTIES
ELASTIC PROPERTIES
MECHANICAL PROPERTIES
∞ PHYSICAL PROPERTIES
∞ PROPERTIES
THERMODYNAMIC PROPERTIES

PROPELLANT SENSITIVITY
GS   PROPELLANT PROPERTIES
. **PROPELLANT SENSITIVITY**
SENSITIVITY
. **PROPELLANT SENSITIVITY**
RT   IGNITION TEMPERATURE
IMPACT RESISTANCE
SHOCK RESISTANCE
SPONTANEOUS COMBUSTION
STORABLE PROPELLANTS

PROPELLANT SPRAYS
RT   FUEL INJECTION
FUEL SPRAYS
LIQUID INJECTION
LIQUID ROCKET PROPELLANTS
SPRAYERS

PROPELLANT STORABILITY
GS   PROPELLANT PROPERTIES
. **PROPELLANT STORABILITY**
RT   FUEL CORROSION
INHIBITORS
STORABLE PROPELLANTS

PROPELLANT STORAGE
GS   CONSUMABLES (SPACECRAFT)
. **PROPELLANT STORAGE**
RT   EXPULSION BLADDERS
FUEL TANK PRESSURIZATION
FUEL TANKS
GROUND SUPPORT EQUIPMENT
HANDLING EQUIPMENT
MISSILE STORAGE
ROCKET PROPELLANTS
SPACE STORAGE
STORABLE PROPELLANTS
∞ STORAGE
UNDERGROUND STORAGE

PROPELLANT TANKS
UF   ROCKET PROPELLANT TANKS
GS   TANKS (CONTAINERS)
. **PROPELLANT TANKS**
RT   CYLINDRICAL TANKS
      EXPULSION BLADDERS
      EXTERNAL TANKS
      FLUID FILLED SHELLS
      FUEL TANK PRESSURIZATION
      FUEL TANKS
      LIQUID FILLED SHELLS
      LIQUID PROPELLANT ROCKET ENGINES
      LIQUID SLOSHING
      PRESSURE VESSELS
      SPHERICAL TANKS
      STORAGE TANKS
      TANK GEOMETRY
      ULLAGE

PROPELLANT TESTS
RT   COLD FLOW TESTS
      CORROSION TESTS
      ENGINE TESTS
      FUEL TESTS
      INTERIOR BALLISTICS
    ∞MATERIALS TESTS
      MISSILE TESTS
      PROPULSIVE EFFICIENCY
      STABILITY TESTS
    ∞TESTS

PROPELLANT TRANSFER
GS   FLUID FLOW
. FUEL FLOW
. . **PROPELLANT TRANSFER**
      MATERIALS HANDLING
. **PROPELLANT TRANSFER**
RT   FUEL CONTROL
      FUEL SYSTEMS
      LIQUID SLOSHING
      REFUELING

PROPELLANTS
GS   **PROPELLANTS**
. COLLOIDAL PROPELLANTS
. DOUBLE BASE PROPELLANTS
. . DOUBLE BASE ROCKET
        PROPELLANTS
. GELLED PROPELLANTS
. . GELLED ROCKET PROPELLANTS
. GUN PROPELLANTS
. HIGH ENERGY PROPELLANTS
. DOMINO PROPELLANTS
. HIGH TEMPERATURE PROPELLANTS
. HYBRID PROPELLANTS
. HYDRAZINE NITROFORM
. HYDROGEN AZIDES
. NITRASOL EXPLOSIVES
. PENTOLITE
. RDX
. ROCKET PROPELLANTS
. . GASEOUS ROCKET PROPELLANTS
. . LIQUID ROCKET PROPELLANTS
. . . CRYOGENIC ROCKET PROPELLANTS
. . . GELLED ROCKET PROPELLANTS
. . . HYPERGOLIC ROCKET
          PROPELLANTS
. . . MONOPROPELLANTS
. . . . AEROZINE
. . . RP-1 ROCKET PROPELLANTS
. . . SLURRY PROPELLANTS
. . NITRAMINE PROPELLANTS
. . SOLID ROCKET PROPELLANTS
. . . DOUBLE BASE ROCKET
          PROPELLANTS
. . . METAL PROPELLANTS
. . TAGN
. . TATB
. SOLID PROPELLANTS
. . CASE BONDED PROPELLANTS
. . COMPOSITE PROPELLANTS
. . NITRAMINE PROPELLANTS
. . PLASTIC PROPELLANTS
. . SOLID ROCKET PROPELLANTS
. . . DOUBLE BASE ROCKET
          PROPELLANTS
. . . HTPB PROPELLANTS
. . . METAL PROPELLANTS
. STORABLE PROPELLANTS
. . RP-1 ROCKET PROPELLANTS
. TETRYL
RT   AMMUNITION
      ASCENT PROPULSION SYSTEMS
      AUXILIARY PROPULSION
      BALLISTICS

PROPELLANTS-*(CONT.)*
      BURNING RATE
      CARTRIDGES
    ∞ENERGY SOURCES
      EXPLOSIVES
      FUEL TANKS
      FUELS
      FULMINATES
      GUNS (ORDNANCE)
      INCENDIARY AMMUNITION
    ∞POWER SUPPLIES
      PROPULSION
      SPACECRAFT POWER SUPPLIES
      SPACECRAFT PROPULSION
      SPECIFIC IMPULSE
      TORPEDOES

PROPELLER BLADES
GS   AIRFOILS
. **PROPELLER BLADES**
RT   BLADE TIPS
    ∞BLADES
      FAN BLADES
      FEATHERING
      PROP-FAN TECHNOLOGY
      PROPELLERS
      ROTARY WINGS
      SYNCHROPHASING

PROPELLER DRIVE
GS   MECHANICAL DRIVES
. **PROPELLER DRIVE**
RT   CONTRAROTATING PROPELLERS
      MARINE PROPULSION
      PROPELLERS
      UNDERWATER PROPULSION

PROPELLER EFFICIENCY
GS   EFFICIENCY
. PROPULSIVE EFFICIENCY
. . **PROPELLER EFFICIENCY**
RT   CONTRAROTATING PROPELLERS
      POWER EFFICIENCY
      PROP-FAN TECHNOLOGY
      PROPELLERS

PROPELLER FANS
GS   PROPELLERS
. **PROPELLER FANS**
RT   DUCTED FANS
    ∞FANS
      LIFT FANS
      PROP-FAN TECHNOLOGY

PROPELLER SLIPSTREAMS
GS   WAKES
. AIRCRAFT WAKES
. . SLIPSTREAMS
. . . **PROPELLER SLIPSTREAMS**
. TURBULENT WAKES
. SLIPSTREAMS
. . . **PROPELLER SLIPSTREAMS**
RT   INTERFERENCE DRAG

PROPELLERS
GS   **PROPELLERS**
. CONTRAROTATING PROPELLERS
. PROPELLER FANS
. SHROUDED PROPELLERS
. TILTED PROPELLERS
. VARIABLE PITCH PROPELLERS
RT   ACTUATOR DISKS
      FEATHERING
      PROPELLER BLADES
      PROPELLER DRIVE
      PROPELLER EFFICIENCY
      SHIPS

∞ **PROPERTIES**
SN   *(USE OF A MORE SPECIFIC TERM IS*
      RECOMMENDED--CONSULT THE TERMS
      LISTED BELOW)
UF   ATTRIBUTES
RT   ACOUSTIC PROPERTIES
      BIODEGRADABILITY
      CHEMICAL PROPERTIES
      CREEP PROPERTIES
      DIELECTRIC PROPERTIES
      DYNAMIC CHARACTERISTICS
      ELECTRICAL PROPERTIES
      ELECTROMAGNETIC PROPERTIES
      HYGRAL PROPERTIES
      MACROSCOPIC EQUATIONS
      MAGNETIC PROPERTIES
    ∞MATERIALS SCIENCE

PROPERTIES-*(CONT.)*
      MECHANICAL PROPERTIES
      OPTICAL PROPERTIES
    ∞PHYSICAL PROPERTIES
      PLASTIC PROPERTIES
      POROSITY
      PREJUDICES
      PROGRESS
      PROPELLANT PROPERTIES
      PROXIMITY
      RECOVERABILITY
      REGULARITY
      SHEAR PROPERTIES
      STRUCTURAL PROPERTIES (GEOLOGY)
      SURFACE PROPERTIES
      TENSILE PROPERTIES
      THERMOCHEMICAL PROPERTIES
      THERMODYNAMIC PROPERTIES
      THERMOPHYSICAL PROPERTIES
      TRANSPORT PROPERTIES
      TURBIDITY
      VIRTUAL PROPERTIES

PROPHYLAXIS
RT   DISEASES
      IMMUNOLOGY

PROPIONIC ACID
GS   ACIDS
. FATTY ACIDS
. . CARBOXYLIC ACIDS
. . . **PROPIONIC ACID**
      ALIPHATIC COMPOUNDS
. FATTY ACIDS
. **PROPIONIC ACID**
      LIPIDS
. **PROPIONIC ACID**
      ORGANIC COMPOUNDS
. FATTY ACIDS
. . **PROPIONIC ACID**

PROPORTION
RT   DISTRIBUTING
      RATIOS

PROPORTIONAL CONTROL
GS   AUTOMATIC CONTROL
. **PROPORTIONAL CONTROL**
RT   ∞CONTROL
      CONTROL EQUIPMENT
      FEEDBACK CONTROL
      OFF-ON CONTROL
      SERVOCONTROL

PROPORTIONAL COUNTERS
GS   IONIZATION CHAMBERS
. **PROPORTIONAL COUNTERS**
      MEASURING INSTRUMENTS
. COUNTERS
. . RADIATION COUNTERS
. . . **PROPORTIONAL COUNTERS**
. RADIATION MEASURING INSTRUMENTS
. . RADIATION COUNTERS
. . . **PROPORTIONAL COUNTERS**
RT   DOSIMETERS
      GEIGER COUNTERS
      NEUTRON COUNTERS

PROPORTIONAL LIMIT
UF   ELASTIC STRENGTH
GS   MECHANICAL PROPERTIES
. ELASTIC PROPERTIES
. . **PROPORTIONAL LIMIT**
      RANGE (EXTREMES)
. **PROPORTIONAL LIMIT**
RT   CRITICAL LOADING
      MODULUS OF ELASTICITY
      STRESS-STRAIN DIAGRAMS

PROPRIOCEPTION
UF   KINESTHESIS
GS   PERCEPTION
. SENSORY PERCEPTION
. . **PROPRIOCEPTION**
. . . AUTOKINESIS
RT   KINESTHESIA

PROPRIOCEPTORS
GS   ANATOMY
. SENSE ORGANS
. . **PROPRIOCEPTORS**
      RECEPTORS (PHYSIOLOGY)
. **PROPRIOCEPTORS**
RT   BARORECEPTORS
      NERVOUS SYSTEM

PROPRIOCEPTORS-*(CONT.)*
   SENSITOMETRY

PROPULSION
  GS **PROPULSION**
   . ASCENT PROPULSION SYSTEMS
   . AUXILIARY PROPULSION
   . CHEMICAL PROPULSION
   . . HYBRID PROPULSION
   . DESCENT PROPULSION SYSTEMS
   . ELECTRIC PROPULSION
   . . ELECTROMAGNETIC PROPULSION
   . . ELECTROSTATIC PROPULSION
   . . . ION PROPULSION
   . . LASER PROPULSION
   . . PLASMA PROPULSION
   . . SOLAR ELECTRIC PROPULSION
   . JET PROPULSION
   . LOW THRUST PROPULSION
   . . ELECTROMAGNETIC PROPULSION
   . . ELECTROSTATIC PROPULSION
   . . . ION PROPULSION
   . . MAN OPERATED PROPULSION
      SYSTEMS
   . . PHOTONIC PROPULSION
   . . PLASMA PROPULSION
   . . SOLAR PROPULSION
   . . . SOLAR ELECTRIC PROPULSION
   . . . SOLAR THERMAL PROPULSION
   . MARINE PROPULSION
   . . UNDERWATER PROPULSION
   . . SUBMARINE PROPULSION
   . NUCLEAR PROPULSION
   . . NUCLEAR ELECTRIC PROPULSION
   . SPACECRAFT PROPULSION
   . . ELECTROMAGNETIC PROPULSION
   . . . MASS DRIVERS (PAYLOAD
      DELIVERY)
   . . ELECTROSTATIC PROPULSION
   . . . ION PROPULSION
   . . PHOTONIC PROPULSION
   . . PLASMA PROPULSION
   . . SOLAR PROPULSION
   . . . SOLAR ELECTRIC PROPULSION
   . . . SOLAR THERMAL PROPULSION
  RT AERONAUTICAL ENGINEERING
     ∞AIRCRAFT
     ∞ASTRONAUTICS
     ∞DRIVES
     ENGINES
     EXHAUST GASES
     FUEL TANK PRESSURIZATION
     FUEL TANKS
     HIGH IMPULSE
     LOCOMOTION
     MISSILES
     POST BOOST PROPULSION SYSTEM
     PROPELLANTS
     PROPULSIVE EFFICIENCY
     PULLING
     PUSHING
     ROCKET PROPELLANTS
     SOLAR SAILS
     SPACE FLIGHT
     SPACE SHUTTLE MAIN ENGINE
     SPACE TUGS
     THRUST

PROPULSION SYSTEM CONFIGURATIONS
  GS **PROPULSION SYSTEM
      CONFIGURATIONS**
   . ASCENT PROPULSION SYSTEMS
   . DESCENT PROPULSION SYSTEMS
  RT AERODYNAMIC CONFIGURATIONS
     AIRCRAFT CONFIGURATIONS
     AUXILIARY PROPULSION
     ∞CONFIGURATIONS
     LASER PROPULSION
     LAUNCH VEHICLE CONFIGURATIONS
     MISSILE CONFIGURATIONS
     POST BOOST PROPULSION SYSTEM
     SPACECRAFT CONFIGURATIONS
     ∞SYSTEMS
     TOPPING CYCLE ENGINES

PROPULSION SYSTEM PERFORMANCE
  RT COLD FLOW TESTS
     COMBUSTION EFFICIENCY
     ∞PERFORMANCE
     POWER EFFICIENCY
     PROPELLANT MASS RATIO
     PROPULSIVE EFFICIENCY
     ROCKET THRUST
     SOLAR THERMAL PROPULSION
     SPECIFIC IMPULSE

PROPULSION SYSTEM PERFORMANCE-*(CONT.)*
     ∞SYSTEMS
     THERMODYNAMIC EFFICIENCY

PROPULSIVE EFFICIENCY
  GS EFFICIENCY
   . **PROPULSIVE EFFICIENCY**
   . . PROPELLER EFFICIENCY
  RT COMBUSTION EFFICIENCY
     ENGINE TESTS
     LASER PROPULSION
     MULTISTAGE ROCKET VEHICLES
     NOZZLE EFFICIENCY
     POWER EFFICIENCY
     PROPELLANT MASS RATIO
     PROPELLANT TESTS
     PROPULSION
     PROPULSION SYSTEM PERFORMANCE
     SPECIFIC IMPULSE
     THERMODYNAMIC EFFICIENCY
     THRUST PROGRAMMING

PROPYL COMPOUNDS
  GS ALIPHATIC COMPOUNDS
   . **PROPYL COMPOUNDS**
  RT ∞CHEMICAL COMPOUNDS

PROPYL NITRATE
  GS ALIPHATIC COMPOUNDS
   . ALKYL COMPOUNDS
   . . **PROPYL NITRATE**
   . NITRATE ESTERS
   . . **PROPYL NITRATE**
   ESTERS
   . NITRATE ESTERS
   . . **PROPYL NITRATE**
   NITROGEN COMPOUNDS
   . NITRATE ESTERS
   . . **PROPYL NITRATE**

PROPYLENE
  GS ALIPHATIC COMPOUNDS
   . ALKENES
   . . **PROPYLENE**
   HYDROCARBONS
   . ALKENES
   . . **PROPYLENE**

PROPYLENE OXIDE
  GS EPOXY COMPOUNDS
   . **PROPYLENE OXIDE**

PROSPECTING
  USE EXPLORATION

PROSTAGLANDINS
  GS SECRETIONS
   . ENDOCRINE SECRETIONS
   . . HORMONES
   . . . **PROSTAGLANDINS**
  RT BIOSYNTHESIS
     ORGANIC COMPOUNDS
     PROSTATE GLAND

PROSTATE GLAND
  GS ANATOMY
   . GENITOURINARY SYSTEM
   . . **PROSTATE GLAND**
   . GLANDS (ANATOMY)
   . . ENDOCRINE GLANDS
   . . . **PROSTATE GLAND**
   . . SEX GLANDS
   . . . **PROSTATE GLAND**
   VISCERA
   . ENDOCRINE GLANDS
   . . **PROSTATE GLAND**
   . SEX GLANDS
   . . **PROSTATE GLAND**
  RT BLADDER
     PROSTAGLANDINS

PROSTHETIC DEVICES
  GS MEDICAL EQUIPMENT
   . **PROSTHETIC DEVICES**
   . . ARTIFICIAL EARS
  RT WALKING MACHINES

PROTACTINIUM
  GS CHEMICAL ELEMENTS
   . **PROTACTINIUM**
   . . PROTACTINIUM ISOTOPES
   METALS
   . **PROTACTINIUM**
   . . PROTACTINIUM ISOTOPES

PROTACTINIUM COMPOUNDS
  GS **PROTACTINIUM COMPOUNDS**
   . PROTACTINIUM FLUORIDES
  RT ∞CHEMICAL COMPOUNDS
     ∞METAL COMPOUNDS

PROTACTINIUM FLUORIDES
  GS HALOGEN COMPOUNDS
   . FLUORINE COMPOUNDS
   . . FLUORIDES
   . . . METAL FLUORIDES
   . . . . **PROTACTINIUM FLUORIDES**
   . HALIDES
   . . METAL HALIDES
   . . . **PROTACTINIUM FLUORIDES**
   PROTACTINIUM COMPOUNDS
   . **PROTACTINIUM FLUORIDES**

PROTACTINIUM ISOTOPES
  UF PROTACTINIUM 234
  GS CHEMICAL ELEMENTS
   . NUCLIDES
   . . ISOTOPES
   . . . **PROTACTINIUM ISOTOPES**
   . PROTACTINIUM
   . . **PROTACTINIUM ISOTOPES**
   METALS
   . PROTACTINIUM
   . . **PROTACTINIUM ISOTOPES**

PROTACTINIUM 234
  USE PROTACTINIUM ISOTOPES

PROTEASE
  GS ENZYMES
   . **PROTEASE**

PROTECTION
  GS **PROTECTION**
   . ACCELERATION PROTECTION
   . CIRCUIT PROTECTION
   . CORROSION PREVENTION
   . ENVIRONMENT PROTECTION
   . EYE PROTECTION
   . METEOROID PROTECTION
   . RADIATION PROTECTION
   . . RADIATION SHIELDING
   . . . SOLAR RADIATION SHIELDING
   . THERMAL PROTECTION
  RT ACCIDENT PREVENTION
     AIRPORT SECURITY
     CIVIL DEFENSE
     COATINGS
     COUNTERMEASURES
     FLYING EJECTION SEATS
     HAZARDS
     HOUSINGS
     INSULATION
     PREVENTION
     PROTECTORS
     ∞RESISTANCE
     SAFETY
     SAFETY DEVICES
     SHIELDING
     WARNING
     WARNING SYSTEMS

PROTECTIVE CLOTHING
  GS CLOTHING
   . **PROTECTIVE CLOTHING**
   . . HELMETS
   . . PRESSURE SUITS
   . . . SPACE SUITS
   . . VAPOR BARRIER CLOTHING
  RT ARMOR
     CHEMICAL DEFENSE
     COVERALLS
     EMERGENCY LIFE SUSTAINING
      SYSTEMS
     FLIGHT CLOTHING
     GLOVES
     GOGGLES
     MASKS
     SAFETY DEVICES
     SHOES

PROTECTIVE COATINGS
  UF CERAMAL PROTECTIVE COATINGS
     SPRAYED PROTECTIVE COATINGS
  GS COATINGS
   . **PROTECTIVE COATINGS**
   . . ANODIC COATINGS
   . . CERAMIC COATINGS
   . . PRIMERS (COATINGS)
   . . REFRACTORY COATINGS

PROTECTIVE COATINGS-(CONT.)
RT   ALKYD RESINS
      ANODIZING
      CLADDING
    ∞CONSTRUCTION MATERIALS
      CORROSION
      DESENSITIZING
      ELECTROPLATING
      ENCAPSULATING
      FINISHES
      GLASS COATINGS
      GLAZES
      GOLD COATINGS
      INORGANIC COATINGS
      LACQUERS
      METAL COATINGS
      NICKEL COATINGS
      PAINTS
      PLASTIC COATINGS
      PLATING
      RUBBER COATINGS
      SPRAYED COATINGS
      SURFACE FINISHING
      TRANSGRANULAR CORROSION
      VARNISHES
      WATERPROOFING
      ZINC COATINGS

PROTECTORS
GS   **PROTECTORS**
    . EAR PROTECTORS
RT   BUMPERS
    ∞CONTAINERS
      ENCLOSURES
      FAIRINGS
      HOUSINGS
      PROTECTION
      SAFETY DEVICES
    ∞SCREENS
      SHEATHS
      SHIELDING

PROTEIN METABOLISM
GS   METABOLISM
    . **PROTEIN METABOLISM**
    . . LIPID METABOLISM
RT   PROTEIN SYNTHESIS
      SYNTHETIC FOOD

PROTEIN SYNTHESIS
RT   BIOLOGICAL EVOLUTION
      CHEMICAL EVOLUTION
      PROTEIN METABOLISM

PROTEINOIDS
GS   PROTEINS
    . **PROTEINOIDS**

PROTEINS
GS   **PROTEINS**
    . ALBUMINS
    . AMIDASE
    . ASPARTATES
    . CARBONIC ANHYDRASE
    . CARBOXYHEMOGLOBIN
    . COENZYMES
    . . GLUTATHIONE
    . ELASTIN
    . FIBRIN
    . FIBRINOGEN
    . GLOBULINS
    . . GAMMA GLOBULIN
    . GUANINES
    . KERATINS
    . LIPOPROTEINS
    . MELANIN
    . NUCLEASE
    . NUCLEOTIDES
    . . ADENINES
    . . ADENOSINES
    . . . ADENOSINE DIPHOSPHATE
    . . . ADENOSINE TRIPHOSPHATE
    . . CYCLIC AMP
    . . POLYNUCLEOTIDES
    . . PYRIDINE NUCLEOTIDES
    . . URIDYLIC ACID
    . OXIDASE
    . OXYHEMOGLOBIN
    . PEPTIDES
    . . HYPERTENSIN
    . . . POLYPEPTIDES
    . PROTEINOIDS
    . PROTOPROTEINS
    . THROMBIN
RT   ADRENOCORTICOTROPIN (ACTH)

PROTEINS-(CONT.)
    ALANINE
    CATALASE
    COLLAGENS
    CYSTEAMINE
    CYSTEINE
    ∞FOOD
    MYOGLOBIN
    ∞NUTRIENTS
    PROTOPLASM
    SERUMS
    SYNTHETIC FOOD

PROTHROMBIN
GS   ENZYMES
    . **PROTHROMBIN**
RT   THROMBIN

PROTIUM
USE   LIGHT WATER

PROTOBIOLOGY
RT   ∞BIOLOGY
      VIRUSES

PROTOCOL (COMPUTERS)
RT   CHANNELS (DATA TRANSMISSION)
      COMMUNICATION NETWORKS
      COMPUTER NETWORKS
      DATA LINKS
      DATA PROCESSING
      DATA TRANSMISSION
      PACKET SWITCHING

PROTON BEAMS
GS   BEAMS (RADIATION)
    . PARTICLE BEAMS
    . . **PROTON BEAMS**
RT   NEUTRON BEAMS

PROTON BELTS
GS   EARTH ATMOSPHERE
    . RADIATION BELTS
    . . **PROTON BELTS**
    PARTICLES
    . CHARGED PARTICLES
    . . MAGNETICALLY TRAPPED PARTICLES
    . . . RADIATION BELTS
    . . . . **PROTON BELTS**
    . TRAPPED PARTICLES
    . . MAGNETICALLY TRAPPED PARTICLES
    . . . RADIATION BELTS
    . . . . **PROTON BELTS**
RT   ∞BELTS
      INNER RADIATION BELT
      OUTER RADIATION BELT

PROTON DAMAGE
GS   DAMAGE
    . **PROTON DAMAGE**

PROTON DENSITY (CONCENTRATION)
GS   DENSITY (NUMBER/VOLUME)
    . PARTICLE DENSITY (CONCENTRATION)
    . . ION DENSITY (CONCENTRATION)
    . . . **PROTON DENSITY**
          (CONCENTRATION)
    . . . . MAGNETOSPHERIC PROTON
          DENSITY
RT   ATMOSPHERIC DENSITY
      ATOM CONCENTRATION
      PLASMA DENSITY
      SPACE DENSITY

PROTON ENERGY
GS   PARTICLE ENERGY
    . **PROTON ENERGY**
RT   ACTIVATION ENERGY
      ELECTRON ENERGY
    ∞ENERGY
      KINETIC ENERGY
      SURFACE ENERGY

PROTON FLUX DENSITY
SN   (PROTON EMISSION OR DETECTION
      RATE PER UNIT AREA)
GS   RATES (PER TIME)
    . FLUX DENSITY
    . . RADIANT FLUX DENSITY
    . . . PARTICLE FLUX DENSITY
    . . . . **PROTON FLUX DENSITY**
RT   IRRADIANCE
      RADIANCY
      RADIATION COUNTERS

PROTON FLUX DENSITY-(CONT.)
    SOLAR FLUX DENSITY

PROTON IMPACT
GS   IMPACT
    . **PROTON IMPACT**
RT   ELECTRON IMPACT
      POINT IMPACT

PROTON IRRADIATION
GS   IRRADIATION
    . ION IRRADIATION
    . . **PROTON IRRADIATION**
RT   DEUTERON IRRADIATION
      ELECTRON RADIATION

PROTON MAGNETIC RESONANCE
GS   RESONANCE
    . MAGNETIC RESONANCE
    . . NUCLEAR MAGNETIC RESONANCE
    . . . **PROTON MAGNETIC RESONANCE**

PROTON MASERS
GS   STIMULATED EMISSION DEVICES
    . MASERS
    . . **PROTON MASERS**
RT   MAGNETOMETERS

PROTON PRECESSION
GS   GYRATION
    . PRECESSION
    . . **PROTON PRECESSION**
RT   FREE VIBRATION

PROTON PRECIPITATION
GS   PARTICLE PRECIPITATION
    . **PROTON PRECIPITATION**
RT   AURORAS
      ELECTRON PRECIPITATION
      PARTICLES
    ∞PRECIPITATION
      RADIATION BELTS
      TRAPPED PARTICLES
      UPPER ATMOSPHERE

PROTON PROTUBERANCES
GS   PROTUBERANCES
    . **PROTON PROTUBERANCES**

PROTON RESONANCE
GS   RESONANCE
    . MAGNETIC RESONANCE
    . . NUCLEAR MAGNETIC RESONANCE
    . . . **PROTON RESONANCE**
RT   NUCLEAR PARTICLES

PROTON SATELLITES
GS   SATELLITES
    . ARTIFICIAL SATELLITES
    . . SOVIET SATELLITES
    . . . **PROTON SATELLITES**
    . . . . PROTON 1 SATELLITE
    . . . . PROTON 2 SATELLITE
    . . . . PROTON 3 SATELLITE
    . . . . PROTON 4 SATELLITE
RT   U.S.S.R. SPACE PROGRAM

PROTON SCATTERING
GS   NUCLEAR REACTIONS
    . **PROTON SCATTERING**
    SCATTERING
    . **PROTON SCATTERING**
RT   ION SCATTERING

PROTON TELESCOPES
USE   PARTICLE TELESCOPES

PROTON 1 SATELLITE
GS   SATELLITES
    . ARTIFICIAL SATELLITES
    . . SOVIET SATELLITES
    . . . PROTON SATELLITES
    . . . . **PROTON 1 SATELLITE**

PROTON 2 SATELLITE
GS   SATELLITES
    . ARTIFICIAL SATELLITES
    . . SOVIET SATELLITES
    . . . PROTON SATELLITES
    . . . . **PROTON 2 SATELLITE**

PROTON 3 SATELLITE
GS   SATELLITES
    . ARTIFICIAL SATELLITES

PROTON 3 SATELLITE-*(CONT.)*
. . SOVIET SATELLITES
. . . PROTON SATELLITES
. . . . **PROTON 3 SATELLITE**

PROTON 4 SATELLITE
GS   SATELLITES
. ARTIFICIAL SATELLITES
. . SOVIET SATELLITES
. . . PROTON SATELLITES
. . . . **PROTON 4 SATELLITE**

PROTON-PROTON REACTIONS
GS   NUCLEAR REACTIONS
. **PROTON-PROTON REACTIONS**
RT   ANNIHILATION REACTIONS
∞ INTERACTIONS
POMERONS
THERMONUCLEAR REACTIONS

PROTONS
GS   IONS
. **PROTONS**
. . SOLAR PROTONS
PARTICLES
. CHARGED PARTICLES
. . **PROTONS**
. . . RECOIL PROTONS
. . . SOLAR PROTONS
. ELEMENTARY PARTICLES
. . FERMIONS
. . . **PROTONS**
. . . . RECOIL PROTONS
. . . . SOLAR PROTONS
RT   ALPHA PARTICLES
ANTIPROTONS
BARYONS
COSMIC RAYS
DEUTERONS
FLUX DENSITY
HYDROGEN IONS
NUCLEAR PARTICLES
NUCLEI (NUCLEAR PHYSICS)
NUCLEON POTENTIAL
NUCLEONS
POSITIVE IONS
RADIATION BELTS
RADIATION SHIELDING
TRITONS

PROTOPLANETS
UF   PLANETISMALS
GS   CELESTIAL BODIES
. **PROTOPLANETS**
RT   COSMOLOGY
PLANETARY ENVIRONMENTS
PLANETARY EVOLUTION
PLANETARY MASS
PLANETS
SOLAR ORBITS
SOLAR SYSTEM
STELLAR EVOLUTION

PROTOPLASM
RT   PROTEINS

PROTOPLASTS
GS   CELLS (BIOLOGY)
. **PROTOPLASTS**

PROTOPROTEINS
GS   ACIDS
. AMINO ACIDS
. . **PROTOPROTEINS**
ORGANIC COMPOUNDS
. AMINO ACIDS
. . **PROTOPROTEINS**
PROTEINS
. **PROTOPROTEINS**

PROTOSTARS
GS   CELESTIAL BODIES
. STARS
. . EARLY STARS
. . . **PROTOSTARS**
. . . . T TAURI STARS
RT   STELLAR EVOLUTION
STELLAR MASS ACCRETION

PROTOTYPES
RT   BREADBOARD MODELS
∞ PATTERNS
PILOT PLANTS

PROTOZOA
GS   ANIMALS
. INVERTEBRATES
. . **PROTOZOA**
. . . FLAGELLATA
. . . . TRYPANOSOME
. . . PARAMECIA
. . . . ROTIFERA
. . . PELOMYXA
MICROORGANISMS
. **PROTOZOA**
. . AMOEBA
. . PELOMYXA
. . FLAGELLATA
. . . TRYPANOSOME
. . PARAMECIA
. . . ROTIFERA

PROTRACTORS
GS   MEASURING INSTRUMENTS
. **PROTRACTORS**
RT   ANGLES (GEOMETRY)

PROTUBERANCES
SN   (COMPONENTS MOUNTED EXTERNAL TO
THE STRUCTURE)
GS   **PROTUBERANCES**
. PROTON PROTUBERANCES
RT   AERODYNAMIC CONFIGURATIONS
AERODYNAMIC INTERFERENCE
AIRCRAFT ANTENNAS
AIRCRAFT PARTS
AIRFRAMES
∞ BLISTERS
COWLINGS
DOMES (STRUCTURAL FORMS)
EXTERNAL STORE SEPARATION
EXTERNAL STORES
FAIRINGS
FUEL TANKS
HOUSINGS
NACELLES
PAPILLAE
PITOT TUBES
RADOMES
∞ RIDGES
SHELLS (STRUCTURAL FORMS)
VORTEX ALLEVIATION
WING-FUSELAGE STORES
WINGLETS

PROUSTITE
GS   ARSENIC COMPOUNDS
. ARSENIDES
. . **PROUSTITE**
MINERALS
. **PROUSTITE**

PROVIDER AIRCRAFT
USE   C-123 AIRCRAFT

PROVING
UF   CONFIRMATION
DEMONSTRATION
PROOFS
VALIDATION
VERIFICATION (PROVING)
GS   **PROVING**
. THEOREM PROVING
RT   ACCEPTABILITY
APPROACH AND LANDING TESTS (STS)
ERROR DETECTION CODES
EVALUATION
EXAMINATION
LUNAR ROVING VEHICLES
MATHEMATICAL LOGIC
∞ MEASUREMENT
PROGRAM VERIFICATION (COMPUTERS)
∞ ROVINGS
∞ TESTS

PROVISIONING
RT   CONSUMABLES (SPACECREW SUPPLIES)
∞ FOOD
LIFE SUPPORT SYSTEMS
SPACE RATIONS
STOWAGE (ONBOARD EQUIPMENT)

PROXIMITY
RT   DISTANCE
∞ PROPERTIES
TIGHTNESS

PROXIMITY EFFECT (ELECTRICITY)
GS   ELECTROMAGNETIC PROPERTIES

PROXIMITY EFFECT (ELECTRICITY)-*(CONT.)*
. ELECTRICAL PROPERTIES
. . INDUCTANCE
. . . **PROXIMITY EFFECT (ELECTRICITY)**
RT   ∞ EFFECTS
ELECTRICITY
SUPERCONDUCTORS
VOLT-AMPERE CHARACTERISTICS

PRTR (REACTOR)
USE   PLUTONIUM RECYCLE TEST REACTOR

PRUSSIC ACID
USE   HYDROCYANIC ACID

PSEUDOMONAS
GS   MICROORGANISMS
. BACTERIA
. . **PSEUDOMONAS**

PSEUDONOISE
RT   RANDOM NOISE

PSEUDOPOTENTIALS
GS   IMPURITIES
. **PSEUDOPOTENTIALS**
RT   IMPURITIES
MELTING
SEMICONDUCTORS (MATERIALS)

PSEUDORANDOM SEQUENCES
RT   RANDOM NUMBERS
∞ SIGNALS

PSYCHIATRY
GS   MEDICAL SCIENCE
. **PSYCHIATRY**
. . SOCIAL PSYCHIATRY
RT   BRAIN
MILITARY PSYCHOLOGY
PSYCHOLOGY
PSYCHOTHERAPY

PSYCHOACOUSTICS
GS   ACOUSTICS
. **PSYCHOACOUSTICS**
PSYCHOLOGY
. PSYCHOPHYSICS
. . **PSYCHOACOUSTICS**
RT   AUDITORY PERCEPTION
AUDITORY SIGNALS
BELLS
BIOACOUSTICS
NOISE INTENSITY
PSYCHOLOGICAL EFFECTS
PSYCHOLOGICAL FACTORS

PSYCHOLINGUISTICS
GS   LINGUISTICS
. **PSYCHOLINGUISTICS**
RT   INTELLIGIBILITY
PHONEMES
PHONEMICS
ROBOTS
SEMANTICS
SYLLABLES
SYNTAX

PSYCHOLOGICAL EFFECTS
GS   **PSYCHOLOGICAL EFFECTS**
. DESYNCHRONIZATION (BIOLOGY)
. ILLUSIONS
. . HALLUCINATIONS
. . MOON ILLUSION
. . OCULOGRAVIC ILLUSIONS
. . OPTICAL ILLUSION
. . . ELEVATOR ILLUSION
. JET LAG
RT   AFTERIMAGES
AVIATION PSYCHOLOGY
BIOLOGICAL EFFECTS
BOREDOM
COMFORT
CONFIDENCE
DISORIENTATION
∞ EFFECTS
EMOTIONS
ENVIRONMENTAL ENGINEERING
ENVIRONMENTS
FRUSTRATION
HUMAN FACTORS ENGINEERING
HUMAN REACTIONS
HUMIDITY
MILITARY PSYCHOLOGY

PSYCHOLOGICAL EFFECTS-*(CONT.)*
- MOODS
- PSYCHOACOUSTICS
- REACTION TIME
- SPACE ADAPTATION SYNDROME
- SPACE PSYCHOLOGY
- STRESS (PSYCHOLOGY)
- TAYLOR MANIFEST ANXIETY SCALE

PSYCHOLOGICAL FACTORS
- RT ASTRONAUT PERFORMANCE
- AVIATION PSYCHOLOGY
- EMOTIONAL FACTORS
- FLIGHT STRESS (BIOLOGY)
- HABITS
- MOODS
- PERMISSIVITY
- PSYCHOACOUSTICS
- PSYCHOSOMATICS
- REWARD (PSYCHOLOGY)
- SEX FACTOR
- SPACE PSYCHOLOGY
- ∞STIMULI
- STRESS (PSYCHOLOGY)

PSYCHOLOGICAL INDEXES
- USE PSYCHOLOGICAL TESTS

PSYCHOLOGICAL SETS
- GS PSYCHOLOGY
- . **PSYCHOLOGICAL SETS**

PSYCHOLOGICAL TESTS
- UF PSYCHOLOGICAL INDEXES
- GS **PSYCHOLOGICAL TESTS**
- . RORSCHACH TESTS
- RT CERTIFICATION
- ENVIRONMENTAL TESTS
- LIMEN
- MILITARY PSYCHOLOGY
- PERSONALITY TESTS
- PILOT SELECTION
- PSYCHOMETRICS
- RATIOS
- SKINNER BOXES
- TAYLOR MANIFEST ANXIETY SCALE
- ∞TESTS

PSYCHOLOGY
- GS **PSYCHOLOGY**
- . AVIATION PSYCHOLOGY
- . COGNITIVE PSYCHOLOGY
- . MILITARY PSYCHOLOGY
- . PSYCHOLOGICAL SETS
- . PSYCHOPHYSICS
- . . PSYCHOACOUSTICS
- . SPACE PSYCHOLOGY
- RT BIOFEEDBACK
- BOREDOM
- BRAIN
- CYBERNETICS
- DETACHMENT
- DIAGNOSIS
- DISORDERS
- DISORIENTATION
- EMOTIONAL FACTORS
- EMOTIONS
- EXTROVERSION
- FRUSTRATION
- INSPIRATION
- INTELLECT
- INTROVERSION
- MORALE
- PREJUDICES
- PSYCHIATRY
- PSYCHOMETRICS
- PSYCHOTHERAPY
- RORSCHACH TESTS
- STRESS (PSYCHOLOGY)
- SUBLIMINAL STIMULI

PSYCHOMETRICS
- RT DIAGNOSIS
- EDUCATION
- MILITARY PSYCHOLOGY
- NORMS
- PERSONALITY TESTS
- PSYCHOLOGICAL TESTS
- PSYCHOLOGY
- PSYCHOMOTOR PERFORMANCE
- PSYCHOPHYSICS
- PSYCHOSOMATICS
- SKINNER BOXES

PSYCHOMOTOR PERFORMANCE
- GS SENSORIMOTOR PERFORMANCE
- . **PSYCHOMOTOR PERFORMANCE**
- . . PSYCHOSOMATICS
- RT BIOCONTROL SYSTEMS
- HUMAN PERFORMANCE
- HUMAN REACTIONS
- MENTAL PERFORMANCE
- OPERATOR PERFORMANCE
- PHYSIOLOGICAL TESTS
- PILOT PERFORMANCE
- PSYCHOMETRICS
- REACTION TIME
- WORKLOADS (PSYCHOPHYSIOLOGY)

PSYCHOPHARMACOLOGY
- GS PHARMACOLOGY
- . **PSYCHOPHARMACOLOGY**
- RT CENTRAL NERVOUS SYSTEM
- CENTRAL NERVOUS SYSTEM
  DEPRESSANTS
- CENTRAL NERVOUS SYSTEM
  STIMULANTS
- LIFE SCIENCES
- MEDICAL SCIENCE
- ∞MEDICINE
- NERVOUS SYSTEM
- PSYCHOTROPIC DRUGS

PSYCHOPHYSICS
- GS PSYCHOLOGY
- . **PSYCHOPHYSICS**
- . . PSYCHOACOUSTICS
- RT ∞PHYSICS
- PSYCHOMETRICS
- ∞SCIENCE

PSYCHOPHYSIOLOGY
- GS PHYSIOLOGY
- . **PSYCHOPHYSIOLOGY**
- . . WORKLOADS (PSYCHOPHYSIOLOGY)
- RT EVOKED RESPONSE
  (PSYCHOPHYSIOLOGY)
- ∞SCIENCE

PSYCHOSES
- GS **PSYCHOSES**
- . PSYCHOTIC DEPRESSION
- . SCHIZOPHRENIA
- RT DETACHMENT
- DISORDERS
- FEAR
- IRRATIONALITY
- NEUROSES

PSYCHOSOMATICS
- GS SENSORIMOTOR PERFORMANCE
- . PSYCHOMOTOR PERFORMANCE
- . . **PSYCHOSOMATICS**
- RT PSYCHOLOGICAL FACTORS
- PSYCHOMETRICS

PSYCHOTHERAPY
- GS THERAPY
- . **PSYCHOTHERAPY**
- RT CONVULSIONS
- GESTALT THEORY
- HEALTH
- MENTAL HEALTH
- NEUROPSYCHIATRY
- PSYCHIATRY
- PSYCHOLOGY
- PSYCHOTROPIC DRUGS

PSYCHOTIC DEPRESSION
- GS PSYCHOSES
- . **PSYCHOTIC DEPRESSION**
- RT ∞DEPRESSION
- NEUROTIC DEPRESSION

PSYCHOTROPIC DRUGS
- GS DRUGS
- . **PSYCHOTROPIC DRUGS**
- RT CENTRAL NERVOUS SYSTEM
- NARCOTICS
- NEUROPHYSIOLOGY
- PHYSIOCHEMISTRY
- PSYCHOPHARMACOLOGY
- PSYCHOTHERAPY
- SEDATIVES

PSYCHROMETERS
- GS MEASURING INSTRUMENTS
- . MOISTURE METERS

PSYCHROMETERS-*(CONT.)*
- . . HYGROMETERS
- . . **PSYCHROMETERS**
- RT ATMOSPHERIC MOISTURE
- CHEMICAL ANALYSIS
- HUMIDITY
- HUMIDITY MEASUREMENT
- METEOROLOGICAL INSTRUMENTS

PSYCHROPHILES
- GS MICROORGANISMS
- . **PSYCHROPHILES**
- RT MESOPHILES
- THERMOPHILES

PTM (MODULATION)
- USE PULSE TIME MODULATION

PTOLEMAEUS CRATER
- GS CRATERS
- . LUNAR CRATERS
- . . **PTOLEMAEUS CRATER**
- RT METEORITE CRATERS

PUBLIC ADDRESS SYSTEMS
- RT ∞SYSTEMS
- WARNING SYSTEMS

PUBLIC HEALTH
- GS BIOPHYSICS
- . HEALTH PHYSICS
- . . **PUBLIC HEALTH**
- HEALTH
- . HEALTH PHYSICS
- . . **PUBLIC HEALTH**
- RT HAZARDOUS MATERIAL DISPOSAL (IN
  SPACE)
- HYGIENE
- MEDICAL SERVICES
- NONPOINT SOURCES
- ORAL HYGIENE
- POLLUTION
- SANITATION
- URBAN PLANNING

PUBLIC LAW
- GS LAW (JURISPRUDENCE)
- . **PUBLIC LAW**
- . . LIABILITIES
- . . . LEGAL LIABILITY
- . . PENALTIES
- RT AIR LAW

PUBLIC RELATIONS
- RT ∞COOPERATION
- IMPROVEMENT
- PROMOTION
- UPGRADING

PUBLIC SPEAKING
- UF ORATORY
- RT LECTURES
- SPEECH

PUBLICATIONS
- USE DOCUMENTS

PUERTO RICO
- GS LANDFORMS
- . ISLANDS
- . . WEST INDIES
- . . . **PUERTO RICO**
- RT UNITED STATES

PULLEYS
- RT ∞BELTS
- BLOCKS
- IDLERS
- ROLLERS
- WHEELS
- WINCHES

PULLING
- RT ∞DRAWING
- ∞FORCE
- PROPULSION
- TRACTION

PULMONARY CIRCULATION
- GS CIRCULATION
- . BLOOD CIRCULATION
- . . **PULMONARY CIRCULATION**
- RT ALVEOLI
- ARTIFICIAL CARDIAC PACEMAKER

PULMONARY CIRCULATION-(CONT.)
    BLOOD PUMPS
    HEART IMPLANTATION
    LUNGS
    RESPIRATORY SYSTEM

PULMONARY FUNCTIONS
  RT   ALVEOLI
      ∞FUNCTIONS
      LUNGS

PULMONARY LESIONS
  GS   DISEASES
     . **PULMONARY LESIONS**
     INJURIES
     . LESIONS
     . . **PULMONARY LESIONS**
  RT   LUNG MORPHOLOGY
     LUNGS
     ORGANS
     RESPIRATORY DISEASES

PULSARS
  GS   CELESTIAL BODIES
     . RADIO SOURCES (ASTRONOMY)
     . . RADIO STARS
     . . . **PULSARS**
     . STARS
     . . NEUTRON STARS
     . . . **PULSARS**
     . . RADIO STARS
     . . . **PULSARS**
  RT   QUASARS
     RADIATION SOURCES
     RADIO ASTRONOMY
     RADIO BURSTS
     SUPERNOVA REMNANTS

PULSATING FLOW
  USE UNSTEADY FLOW

PULSE AMPLITUDE
  UF   PULSE HEIGHT
  GS   AMPLITUDES
     . **PULSE AMPLITUDE**
     WAVEFORMS
     . **PULSE AMPLITUDE**
  RT   AMPLITUDE DISTRIBUTION ANALYSIS
     ELECTRIC PULSES
     PHOTOPEAK
     PULSED RADIATION
     SAWTOOTH WAVEFORMS
     SQUARE WAVES

PULSE AMPLITUDE MODULATION
  UF   PAM (MODULATION)
  GS   MODULATION
     . PULSE MODULATION
     . . **PULSE AMPLITUDE MODULATION**
  RT   MODEMS
     P.A.C.M. TELEMETRY

PULSE CHARGING
  RT   BATTERY CHARGERS
     ELECTRIC BATTERIES
     ELECTRIC CHARGE
     STORAGE BATTERIES

PULSE CODE MODULATION
  UF   PCM (MODULATION)
  GS   MODULATION
     . PULSE MODULATION
     . . **PULSE CODE MODULATION**
     . . . DELTA MODULATION
     . . . DIFFERENTIAL PULSE CODE
        MODULATION
  RT   BITERNARY CODE
     DECOMMUTATORS
     P.A.C.M. TELEMETRY
     PCM TELEMETRY
     UNIFIED S BAND

PULSE COMMUNICATION
  UF   DIGITAL COMMUNICATION
  GS   TELECOMMUNICATION
     . **PULSE COMMUNICATION**
     . . DIGITAL SPACECRAFT TELEVISION
  RT   BIT ERROR RATE
     COMMUNICATION NETWORKS
     DATA TRANSMISSION
     DELTA MODULATION
     DIFFERENTIAL PULSE CODE
        MODULATION
     DIGITAL TELEVISION

PULSE COMMUNICATION-(CONT.)
    ELECTROMAGNETIC PULSES
    FREQUENCY DIVISION MULTIPLEXING
    MODEMS
    MULTIPLE ACCESS
    MULTIPLEXING
    ORTHOGONAL MULTIPLEXING THEORY
    RADIO COMMUNICATION
    RADIO TRANSMISSION
    SATELLITE TRANSMISSION
    SIGNAL TRANSMISSION
    SPACE COMMUNICATION
    TELEGRAPH SYSTEMS
    TELEMETRY
    TIME DIVISION MULTIPLE ACCESS

PULSE COMPRESSION
  RT   CODING
     COMPRESSING
     RADAR

PULSE DIFFRACTION
  GS   DIFFRACTION
     . **PULSE DIFFRACTION**
  RT   PLASMA JETS
     PULSED RADIATION
     WAVE PROPAGATION

PULSE DOPPLER RADAR
  GS   RADAR
     . PULSE RADAR
     . . **PULSE DOPPLER RADAR**
     . . . EARTH RESOURCES SHUTTLE
        IMAGING RADAR
     . . . MONOPULSE RADAR
  RT   CANCELLATION CIRCUITS
     COHERENT RADAR

PULSE DURATION
  UF   LIGHT DURATION
     PULSE WIDTH
  GS   WAVEFORMS
     . **PULSE DURATION**
  RT   ELECTRIC PULSES
     LASER OUTPUTS
     MASER OUTPUTS
     PULSE REPETITION RATE
     PULSED RADIATION
     SAWTOOTH WAVEFORMS
     SQUARE WAVES
     TIME SIGNALS
     ULTRASHORT PULSED LASERS

PULSE DURATION MODULATION
  UF   PDM (MODULATION)
     PULSE WIDTH MODULATION
     PWM (MODULATION)
  GS   MODULATION
     . PULSE MODULATION
     . . PULSE TIME MODULATION
     . . . **PULSE DURATION MODULATION**
  RT   MODEMS

PULSE FREQUENCY MODULATION
  UF   PFM (MODULATION)
  GS   MODULATION
     . FREQUENCY MODULATION
     . . **PULSE FREQUENCY MODULATION**
     . PULSE MODULATION
     . . **PULSE FREQUENCY MODULATION**
  RT   COMMUNICATION EQUIPMENT
     DIFFERENTIAL PULSE CODE
        MODULATION
     MODEMS

PULSE FREQUENCY MODULATION TELEMETRY
  GS   MODULATION
     . FREQUENCY MODULATION
     . . **PULSE FREQUENCY MODULATION**
        TELEMETRY
     . PULSE MODULATION
     . . **PULSE FREQUENCY MODULATION**
        TELEMETRY
     TELECOMMUNICATION
     . RADIO COMMUNICATION
     . . RADIO TELEMETRY
     . . . **PULSE FREQUENCY MODULATION**
        TELEMETRY
     . TELEMETRY
     . . RADIO TELEMETRY
     . . . **PULSE FREQUENCY MODULATION**
        TELEMETRY
  RT   COMMUNICATION EQUIPMENT
     RADIO TRANSMISSION

PULSE GENERATORS
  RT   COMPULSATORS
     ELECTRIC PULSES
     FUNCTION GENERATORS
     ∞GENERATORS
     IMPULSE GENERATORS
     LASER CAVITIES
     LASERS
     PLASMA GENERATORS
     PULSE REPETITION RATE
     PULSED RADIATION
     SHOCK WAVE GENERATORS

PULSE HEATING
  GS   HARDENING (MATERIALS)
     . **PULSE HEATING**
     HEAT TREATMENT
     . ANNEALING
     . . **PULSE HEATING**
     HEATING
     . TRANSIENT HEATING
     . . **PULSE HEATING**
  RT   LASER HEATING

PULSE HEIGHT
  USE PULSE AMPLITUDE

PULSE MODULATION
  GS   MODULATION
     . **PULSE MODULATION**
     . . PULSE AMPLITUDE MODULATION
     . . PULSE CODE MODULATION
     . . . DELTA MODULATION
     . . . DIFFERENTIAL PULSE CODE
        MODULATION
     . . PULSE FREQUENCY MODULATION
     . . PULSE FREQUENCY MODULATION
        TELEMETRY
     . . PULSE TIME MODULATION
     . . . PULSE DURATION MODULATION
     . . . PULSE POSITION MODULATION
  RT   AMPLITUDE MODULATION
     DEMODULATION
     DEMODULATORS
     ELECTRIC PULSES
     ELECTROMAGNETIC PULSES
     FREQUENCY MODULATION
     LIGHT MODULATION
     MODEMS
     MODULATORS
     PHASE MODULATION
     PULSED RADIATION
     RADIO TELEMETRY
     TIME DIVISION MULTIPLEXING
     TRIGATRONS

PULSE POSITION MODULATION
  UF   PPM (MODULATION)
  GS   MODULATION
     . PULSE MODULATION
     . . PULSE TIME MODULATION
     . . . **PULSE POSITION MODULATION**
  RT   MODEMS

PULSE RADAR
  GS   RADAR
     . **PULSE RADAR**
     . . PULSE DOPPLER RADAR
     . . . EARTH RESOURCES SHUTTLE
        IMAGING RADAR
     . . . MONOPULSE RADAR
  RT   COHERENT RADAR
     CONTINUOUS WAVE RADAR
     DOPPLER RADAR
     ECHO SUPPRESSORS
     ELECTROMAGNETIC PULSES
     METEOROLOGICAL RADAR
     MULTISTATIC RADAR
     POLYSTATION DOPPLER TRACKING
        SYSTEM
     SEARCH RADAR
     SURVEILLANCE RADAR
     SYNCHRONIZERS
     TRACKING RADAR

PULSE RATE
  UF   CHRONOTRONS
  GS   RATES (PER TIME)
     . **PULSE RATE**
     . . PULSE REPETITION RATE
  RT   ELECTRIC PULSES
     PICOSECOND PULSES
     PULSED RADIATION

PULSE RECORDERS
USE   COUNTERS

PULSE REPETITION RATE
GS   RATES (PER TIME)
      . PULSE RATE
      . . **PULSE REPETITION RATE**
RT   ∞FREQUENCY RESPONSE
      OPTICAL PUMPING
      PULSE DURATION
      PULSE GENERATORS
      PULSED LASERS

PULSE TIME MODULATION
UF   PTM (MODULATION)
GS   MODULATION
      . PULSE MODULATION
      . . **PULSE TIME MODULATION**
      . . . PULSE DURATION MODULATION
      . . . PULSE POSITION MODULATION

PULSE WIDTH
USE   PULSE DURATION

PULSE WIDTH AMPLITUDE CONVERTERS
RT   ∞CONVERTERS
      FREQUENCY CONVERTERS

PULSE WIDTH MODULATION
USE   PULSE DURATION MODULATION

PULSED JET ENGINES
GS   ENGINES
      . **PULSED JET ENGINES**
RT   ELECTRIC ROCKET ENGINES
      ELECTROTHERMAL ENGINES
      PLASMA ENGINES
      RESISTOJET ENGINES

PULSED LASERS
GS   STIMULATED EMISSION DEVICES
      . LASERS
      . . **PULSED LASERS**
      . . . Q SWITCHED LASERS
      . . . ULTRASHORT PULSED LASERS
      . . . ULTRAVIOLET LASERS
RT   ARGON LASERS
      CARBON DIOXIDE LASERS
      GAS LASERS
      GLASS LASERS
      INERTIAL FUSION (REACTOR)
      LASER HEATING
      LASER TARGET INTERACTIONS
      LASER WELDING
      NITROGEN LASERS
      PULSE REPETITION RATE
      RUBY LASERS
      SEMICONDUCTOR LASERS
      TEA LASERS
      TUBE LASERS
      WAVEGUIDE LASERS

PULSED RADIATION
GS   **PULSED RADIATION**
      . ELECTROMAGNETIC PULSES
      . . SYSTEM GENERATED
              ELECTROMAGNETIC PULSES
RT   CONTINUOUS RADIATION
      CORPUSCULAR RADIATION
      ELASTIC WAVES
      ELECTROMAGNETIC RADIATION
      GAMMA RAY LASERS
      LASER DAMAGE
      LASERS
      PICOSECOND PULSES
      PULSE AMPLITUDE
      PULSE DIFFRACTION
      PULSE DURATION
      PULSE GENERATORS
      PULSE MODULATION
      PULSE RATE
      ∞RADIATION
      ∞RAYS

PULSEJET ENGINES
GS   ENGINES
      . AIR BREATHING ENGINES
      . . GAS TURBINE ENGINES
      . . . JET ENGINES
      . . . . RAMJET ENGINES
      . . . . . **PULSEJET ENGINES**
      . INTERNAL COMBUSTION ENGINES
      . . GAS TURBINE ENGINES
      . . . JET ENGINES

PULSEJET ENGINES-*(CONT.)*
      . . . . RAMJET ENGINES
      . . . . . **PULSEJET ENGINES**
      . TURBINE ENGINES
      . . GAS TURBINE ENGINES
      . . . JET ENGINES
      . . . . RAMJET ENGINES
      . . . . . **PULSEJET ENGINES**
RT   V-1 MISSILE

PULSES
GS   **PULSES**
      . ELECTRIC PULSES
      . ELECTROMAGNETIC PULSES
      . . SYSTEM GENERATED
              ELECTROMAGNETIC PULSES
      . GEOMAGNETIC PULSATIONS
      . . GEOMAGNETIC MICROPULSATIONS
      . MICROPULSATIONS
      . . GEOMAGNETIC MICROPULSATIONS
      . PICOSECOND PULSES
      . PRESSURE PULSES
RT   AMPLITUDES
      INTERMITTENCY
      SOLITARY WAVES

PULTRUSION
GS   FORMING TECHNIQUES
      . EXTRUDING
      . . **PULTRUSION**
RT   CASTING
      COMPOSITE STRUCTURES
      DIES
      EPOXY MATRIX COMPOSITES
      FIBER COMPOSITES
      GLASS FIBER REINFORCED PLASTICS
      HOT WORKING
      POLYMER MATRIX COMPOSITES
      PREIMPREGNATION
      PRESSING (FORMING)
      RESIN MATRIX COMPOSITES

PULVERIZING
USE   GRINDING (COMMINUTION)

PUMICE
GS   ROCKS
      . IGNEOUS ROCKS
      . . **PUMICE**
RT   ABRASIVES
      OBSIDIAN
      POWDER (PARTICLES)
      SOILS

PUMP IMPELLERS
GS   ROTATING BODIES
      . ROTORS
      . . IMPELLERS
      . . . **PUMP IMPELLERS**
RT   CENTRIFUGAL COMPRESSORS
      CENTRIFUGAL PUMPS

PUMP SEALS
GS   SEALS (STOPPERS)
      . **PUMP SEALS**
RT   GASKETS
      ∞GLANDS
      GLANDS (SEALS)
      HERMETIC SEALS
      LABYRINTH SEALS
      MOLECULAR PUMPS

∞ **PUMPING**
SN   *(USE OF A MORE SPECIFIC TERM IS
      RECOMMENDED--CONSULT THE TERMS
      LISTED BELOW)*
RT   BLOWING
      COMPRESSED AIR
      COMPRESSING
      CRYOPUMPING
      DRAINAGE
      ELECTRON PUMPING
      JET PUMPS
      LASER PUMPING
      MAGNETIC PUMPING
      MATERIALS HANDLING
      NUCLEAR PUMPING
      OPTICAL PUMPING
      PLASMA PUMPING
      PUMPS
      PURGING
      WINDMILLS (WINDPOWERED MACHINES)
      WINDPOWERED PUMPS

PUMPS
SN   (LIMITED TO PUMPS FOR MATERIALS--
      EXCLUDES HEAT PUMPS)
UF   HYDRAULIC PUMPS
GS   **PUMPS**
      . AXIAL FLOW PUMPS
      . . TURBINE PUMPS
      . BLOOD PUMPS
      . CENTRIFUGAL PUMPS
      . DIFFUSION PUMPS
      . ELECTROMAGNETIC PUMPS
      . FUEL PUMPS
      . JET PUMPS
      . RAMS (PUMPS)
      . VACUUM PUMPS
      . . CONDENSATION PUMPS
      . . ION PUMPS
      . . MOLECULAR PUMPS
      . VISCOPUMPS
      . WINDPOWERED PUMPS
RT   BELLOWS
      CENTRIFUGAL COMPRESSORS
      EJECTORS
      FEED SYSTEMS
      HEAT PUMPS
      HYDRAULIC EQUIPMENT
      IMPELLERS
      INJECTORS
      LUBRICATION SYSTEMS
      MATERIALS HANDLING
      PACKINGS (SEALS)
      PIPELINES
      PREBURNERS
      ∞PUMPING
      SIPHONS
      STATORS
      TURBOMACHINERY
      VANELESS DIFFUSERS

PUNCHED CARDS
GS   CARDS
      . **PUNCHED CARDS**
RT   COMPUTER STORAGE DEVICES
      DATA RECORDERS
      DATA RECORDING
      DATA STORAGE
      READERS

PUNCHED TAPES
RT   AUTOMATIC TYPEWRITERS
      COMPUTER STORAGE DEVICES
      DATA RECORDING
      MAGNETIC TAPES
      READERS
      ∞TAPES

PUNCHES
RT   DIES
      MACHINE TOOLS
      MOLDS
      PLATENS
      PRESSES
      STAMPING

PUNCTURING
USE   PIERCING

PUPA
GS   ANIMALS
      . INVERTEBRATES
      . . ARTHROPODS
      . . . **PUPA**
RT   INSECTS
      LARVAE

PUPIL SIZE
RT   PUPILLOMETRY

PUPILLOMETRY
RT   BIOMETRICS
      DARK ADAPTATION
      LIGHT ADAPTATION
      ∞MEASUREMENT
      PUPIL SIZE

PUPILS
GS   ANATOMY
      . SENSE ORGANS
      . . EYE (ANATOMY)
      . . . **PUPILS**
RT   VISION

PURGING
RT   CIRCULATION

PURGING-*(CONT.)*
    CLEARING
    DECONTAMINATION
    DEGASSING
    DISTILLATION
    EVACUATING (VACUUM)
    FLUSHING
    OUTGASSING
    ∞PUMPING
    PURIFICATION
    RELIEVING
    ∞SEPARATION
    VENTING

PURIFICATION
UF   PURIFIERS
GS   **PURIFICATION**
    . AIR PURIFICATION
RT   AERATION
    ANTISEPTICS
    BENEFICIATION
    CHEMICAL STERILIZATION
    CLEANING
    CRYSTALLIZATION
    DECONTAMINATION
    DEMINERALIZING
    DESALINIZATION
    DISSIPATION
    DISTILLATION
    ELIMINATION
    ELUTION
    ENRICHMENT
    FLUSHING
    GETTERS
    PASTEURIZING
    POLYNUCLEAR ORGANIC COMPOUNDS
    POTABLE WATER
    PURGING
    PURITY
    RECTIFICATION
    ∞REDUCTION
    REDUCTION (CHEMISTRY)
    REFINING
    SCAVENGING
    ∞SEPARATION
    SEWAGE TREATMENT
    SOLVENT EXTRACTION
    SPACECRAFT STERILIZATION
    STERILIZATION
    SUBLIMATION
    ULTRAPURE METALS
    UPGRADING
    WASHING
    WATER TREATMENT
    ZONE MELTING

PURIFIERS
USE PURIFICATION

PURINES
GS   **PURINES**
    . ADENINES
    . XANTHINES
    . . CAFFEINE
    . GUANINES
    . . URIC ACID

PURITY
RT   CLARITY
    CONCENTRATION (COMPOSITION)
    CONTAMINANTS
    CONTAMINATION
    DECONTAMINATION
    DILUTION
    FINENESS
    POLLUTION
    POTABLE LIQUIDS
    PURIFICATION
    QUALITY
    TRACE CONTAMINANTS
    ULTRAPURE METALS
    WATER POLLUTION

PURPOSES
RT   GOALS

PURSUIT TRACKING
GS   TRACKING (POSITION)
    . **PURSUIT TRACKING**
RT   INFRARED TRACKING
    RADAR TRACKING
    SATELLITE INTERCEPTORS

PUSH-PULL AMPLIFIERS
UF   BALANCED AMPLIFIERS

PUSH-PULL AMPLIFIERS-*(CONT.)*
GS   AMPLIFIERS
    . **PUSH-PULL AMPLIFIERS**
RT   PHASE MODULATION
    POWER AMPLIFIERS

PUSHBROOM SENSOR MODES
GS   MODES
    . **PUSHBROOM SENSOR MODES**
RT   ARRAYS
    ELECTRO-OPTICS
    IMAGE PROCESSING
    LINEAR ARRAYS
    PHOTODIODES

PUSHING
RT   ∞FORCE
    PROPULSION

PWM (MODULATION)
USE PULSE DURATION MODULATION

PYCNOMETERS
RT   DENSITY (MASS/VOLUME)

PYLON MOUNTING
RT   AERODYNAMIC CONFIGURATIONS
    AIRCRAFT STRUCTURES
    COLUMNS (SUPPORTS)
    RIGID MOUNTING
    STRUCTURAL MEMBERS
    SUPPORTS
    WIND TUNNEL MODELS

PYLONS
GS   SUPPORTS
    . **PYLONS**
RT   COLUMNS (SUPPORTS)
    STRUCTURAL MEMBERS
    STRUTS
    TOWERS

PYRAMID LAKE (NV)
GS   LAKES
    . **PYRAMID LAKE (NV)**
RT   NEVADA
    WATER MANAGEMENT
    WATER RESOURCES

PYRAMIDAL BODIES
RT   ∞BODIES
    PYRAMIDS
    REENTRY VEHICLES

PYRAMIDS
GS   GEOMETRY
    . EUCLIDEAN GEOMETRY
    . . POLYHEDRONS
    . . . **PYRAMIDS**
RT   FRUSTUMS
    PYRAMIDAL BODIES

PYRANOMETERS
GS   MEASURING INSTRUMENTS
    . RADIATION MEASURING INSTRUMENTS
    . . ACTINOMETERS
    . . . **PYRANOMETERS**
RT   PHOTOMETERS
    RADIOMETERS
    SKY RADIATION

PYRAZINES
GS   **PYRAZINES**
    . AZINES
    . . CYANURATES
    . . CYANURIC ACID
    . . MECLIZINE
    . . METHYLENE BLUE
    . . PHENOTHIAZINES

PYRENEES MOUNTAINS (EUROPE)
GS   LANDFORMS
    . MOUNTAINS
    . . **PYRENEES MOUNTAINS (EUROPE)**
RT   ANDORRA
    FRANCE
    SPAIN

PYRENES
GS   HYDROCARBONS
    . **PYRENES**

PYREX (TRADEMARK)
USE BOROSILICATE GLASS

PYRIDINE NUCLEOTIDES
GS   ACIDS
    . AMINO ACIDS
    . . **PYRIDINE NUCLEOTIDES**
    ORGANIC COMPOUNDS
    . AMINO ACIDS
    . . **PYRIDINE NUCLEOTIDES**
    . NUCLEOTIDES
    . . **PYRIDINE NUCLEOTIDES**
    PHOSPHORUS COMPOUNDS
    . PHOSPHATES
    . **PYRIDINE NUCLEOTIDES**
    PROTEINS
    . NUCLEOTIDES
    . . **PYRIDINE NUCLEOTIDES**

PYRIDINES
GS   **PYRIDINES**
    . NICOTINIC ACID
    . PIPERIDINE
RT   KARL FISCHER REAGENT

PYRIDOXINE
UF   VITAMIN B 6
GS   HETEROCYCLIC COMPOUNDS
    . **PYRIDOXINE**
    VITAMINS
    . **PYRIDOXINE**

PYRIMIDINES
GS   **PYRIMIDINES**
    . ALLOXAN
    . MITOCHONDRIA
    . THYMIDINE
    . THYMINE
    . URACIL

PYRITES
GS   CHALCOGENIDES
    . SULFIDES
    . . **PYRITES**
    IRON COMPOUNDS
    . **PYRITES**
    MINERALS
    . **PYRITES**
    SULFUR COMPOUNDS
    . SULFIDES
    . . **PYRITES**

PYROCERAM (TRADEMARK)
GS   CERAMICS
    . **PYROCERAM (TRADEMARK)**
    GLASS
    . **PYROCERAM (TRADEMARK)**

PYROELECTRICITY
GS   THERMODYNAMIC PROPERTIES
    . THERMOPHYSICAL PROPERTIES
    . . **PYROELECTRICITY**
RT   PIEZOELECTRICITY
    POLARIZATION (CHARGE SEPARATION)

PYROGEN
GS   GASES
    . FLAMMABLE GASES
    . . **PYROGEN**
RT   TORCHES

PYROGRAPHALLOY
USE COMPOSITE MATERIALS
    PYROLYTIC GRAPHITE
    REFRACTORY MATERIALS

PYROHELIOMETERS
UF   HELIOMETRY
GS   MEASURING INSTRUMENTS
    . HELIOMETERS
    . . **PYROHELIOMETERS**
    OPTICAL EQUIPMENT
    . ASTRONOMICAL TELESCOPES
    . . HELIOMETERS
    . . . **PYROHELIOMETERS**
    TELESCOPES
    . ASTRONOMICAL TELESCOPES
    . . HELIOMETERS
    . . . **PYROHELIOMETERS**

PYROHYDROLYSIS
GS   CHEMICAL REACTIONS
    . **PYROHYDROLYSIS**
RT   PYROLYSIS

PYROLYSIS
- SN (TRANSFORMATION OF A CHEMICAL COMPOUND INTO ONE OR MORE NEW COMPOUNDS SOLELY THROUGH THE APPLICATION OF HEAT)
- GS CHEMICAL REACTIONS
  - . CRACKING (CHEMICAL ENGINEERING)
  - . . **PYROLYSIS**
  - . THERMAL DECOMPOSITION
  - . . **PYROLYSIS**
- RT ABLATION
  - ENDOTHERMIC REACTIONS
  - EXOTHERMIC REACTIONS
  - PYROHYDROLYSIS
  - THERMAL ABSORPTION
  - THERMAL DEGRADATION
  - THERMAL INSTABILITY
  - THERMOGRAVIMETRY

PYROLYTIC GRAPHITE
- UF PYROGRAPHALLOY
- GS PYROLYTIC MATERIALS
  - . **PYROLYTIC GRAPHITE**
- RT HEAT SHIELDING

PYROLYTIC MATERIALS
- GS **PYROLYTIC MATERIALS**
  - . PYROLYTIC GRAPHITE
- RT ABLATIVE MATERIALS
  - CERAMICS
  - ∞MATERIALS
  - REFRACTORY COATINGS
  - REFRACTORY MATERIALS

PYROMETALLURGY
- RT AEROTHERMOCHEMISTRY
  - CHLORINATION
  - ∞CONVERTERS
  - HEAT BALANCE
  - METAL WORKING
  - ∞METALLURGY
  - REFINING
  - SINTERING
  - SUBLIMATION
  - THERMOCHEMISTRY

PYROMETERS
- GS MEASURING INSTRUMENTS
  - . TEMPERATURE MEASURING INSTRUMENTS
  - . . **PYROMETERS**
  - . . . RADIATION PYROMETERS
  - . . . THERMOCOUPLE PYROMETERS
- RT TEMPERATURE MEASUREMENT

PYROMETRY
- USE TEMPERATURE MEASUREMENT

PYROPHORIC MATERIALS
- RT EXPLOSIVES
  - FLAMMABILITY
  - HYPERGOLIC ROCKET PROPELLANTS
  - IGNITERS
  - IGNITION TEMPERATURE
  - ∞MATERIALS
  - METAL COMBUSTION
  - SOLID PROPELLANT IGNITION
  - SPONTANEOUS COMBUSTION

PYROPHYLLITE
- GS ALUMINUM COMPOUNDS
  - . ALUMINUM SILICATES
  - . . **PYROPHYLLITE**
  - MINERALS
  - . **PYROPHYLLITE**
  - SILICON COMPOUNDS
  - . ALUMINUM SILICATES
  - . . **PYROPHYLLITE**
  - . SILICATES
  - . . **PYROPHYLLITE**
- RT ALUMINUM OXIDES

PYROTECHNICS
- UF FIREWORKS
- RT AMMUNITION
  - BOMBS (ORDNANCE)
  - CHEMICAL FUELS
  - DOUBLE BASE PROPELLANTS
  - EXPLOSIVES
  - ∞FLARES
  - GRENADES
  - ILLUMINATING
  - INCENDIARY AMMUNITION
  - INITIATORS (EXPLOSIVES)
  - ORDNANCE

PYROTECHNICS-*(CONT.)*
- PLASTIC PROPELLANTS
- PROJECTILES
- RDX
- ∞ROCKETS
- ∞SIGNALS
- THERMITES

PYROXENES
- GS CHALCOGENIDES
  - . OXIDES
  - . . **PYROXENES**
  - . . . ENSTATITE
  - MINERALS
  - . **PYROXENES**
  - . . ENSTATITE
  - SILICON COMPOUNDS
  - . SILICATES
  - . . **PYROXENES**
  - . . . ENSTATITE
- RT ECLOGITE
  - IGNEOUS ROCKS
  - REGOLITH
  - ROCKS
  - SOILS

PYROXYLIN
- USE CELLULOSE NITRATE

PYRRHOTITE
- GS CHALCOGENIDES
  - . SULFIDES
  - . . **PYRRHOTITE**
  - . . . TROILITE
  - IRON COMPOUNDS
  - . **PYRRHOTITE**
  - . . TROILITE
  - MINERALS
  - . **PYRRHOTITE**
  - . . TROILITE
  - SULFUR COMPOUNDS
  - . SULFIDES
  - . . **PYRRHOTITE**
  - . . . TROILITE

PYRROLES
- GS HETEROCYCLIC COMPOUNDS
  - . AZOLES
  - . . **PYRROLES**
  - . . . CARBAZOLES
  - . . . INDOLES
  - . . . . TRYPTOPHAN
- RT METHOXY SYSTEMS

PYRRONES (TRADEMARK)
- RT ∞POLYMERS

PYRUVATES
- GS ACIDS
  - . AMINO ACIDS
  - . . **PYRUVATES**
- RT ORGANIC LIQUIDS

P3V AIRCRAFT
- USE P-3 AIRCRAFT

P78-2 SATELLITE
- USE SCATHA SATELLITE

Q

Q DEVICES
- RT MAGNETIC MIRRORS
  - PLASMA PINCH
  - ZETA PINCH

Q FACTORS
- UF HIGH Q
  - QUALITY FACTORS
- RT FIGURE OF MERIT
  - RESONANT VIBRATION
  - SPECTRAL RESOLUTION
  - TUNING

Q SWITCHED LASERS
- GS STIMULATED EMISSION DEVICES
  - . LASERS
  - . . PULSED LASERS
  - . . . **Q SWITCHED LASERS**
- RT ARGON LASERS
  - CARBON DIOXIDE LASERS

Q SWITCHED LASERS-*(CONT.)*
- CHEMICAL LASERS
- GAS LASERS
- RUBY LASERS
- SEMICONDUCTOR LASERS
- SOLID STATE LASERS

Q VALUES
- GS VALUE
  - . **Q VALUES**

QCD
- USE QUANTUM CHROMODYNAMICS

QH-50 HELICOPTER
- UF DASH HELICOPTER
  - DSN HELICOPTER
  - GYRODYNE DSN-3 HELICOPTER
  - GYRODYNE MILITARY AIRCRAFT
- GS GYRODYNE AIRCRAFT
  - . **QH-50 HELICOPTER**
  - V/STOL AIRCRAFT
  - . ROTARY WING AIRCRAFT
  - . . HELICOPTERS
  - . . . MILITARY HELICOPTERS
  - . . . . **QH-50 HELICOPTER**

QSO (RADIO SOURCES)
- USE QUASARS

QUADRANTID METEOROIDS
- GS CELESTIAL BODIES
  - . METEOROID SHOWERS
  - . . **QUADRANTID METEOROIDS**
  - . METEOROIDS
  - . . **QUADRANTID METEOROIDS**

QUADRANTS
- GS GEOMETRY
  - . EUCLIDEAN GEOMETRY
  - . . ANALYTIC GEOMETRY
  - . . . **QUADRANTS**

QUADRATIC EQUATIONS
- GS ALGEBRA
  - . NONLINEAR EQUATIONS
  - . . **QUADRATIC EQUATIONS**
  - ANALYSIS (MATHEMATICS)
  - . REAL VARIABLES
  - . . NONLINEAR EQUATIONS
  - . . . **QUADRATIC EQUATIONS**
  - FIELD THEORY (ALGEBRA)
  - . **QUADRATIC EQUATIONS**
- RT ∞EQUATIONS
  - POLYNOMIALS

QUADRATIC PROGRAMMING
- GS OPERATIONS RESEARCH
  - . MATHEMATICAL PROGRAMMING
  - . . **QUADRATIC PROGRAMMING**
  - RESEARCH
  - . MATHEMATICAL PROGRAMMING
  - . . **QUADRATIC PROGRAMMING**
- RT ∞PROGRAMMING

QUADRATURE APPROXIMATION
- USE QUADRATURES

QUADRATURES
- UF QUADRATURE APPROXIMATION
- RT CIRCULAR ORBITS
  - ORBIT CALCULATION
  - ORBITAL MECHANICS
  - ORBITS
  - SPACE MECHANICS

QUADRUPOLE LENSES
- USE MAGNETIC LENSES

QUADRUPOLE NETWORKS
- RT ∞NETWORKS

QUADRUPOLES
- RT ∞DIPOLES
  - NUCLEAR QUADRUPOLE RESONANCE
  - POLARITY

QUAIL MISSILE
- GS DECOYS
  - . **QUAIL MISSILE**
  - MISSILES
  - . AIR TO SURFACE MISSILES
  - . . **QUAIL MISSILE**
- RT COUNTERMEASURES

QUAIL MISSILE-*(CONT.)*
    TURBOJET ENGINES

QUALIFICATIONS
    RT   CERTIFICATION
          CONTRACTORS
          EDUCATION
          EXPERIENCE
          FITNESS
          PERSONALITY TESTS
          PERSONNEL
        ∞TESTS

QUALITATIVE ANALYSIS
    GS   CHEMICAL TESTS
          . CHEMICAL ANALYSIS
          . . **QUALITATIVE ANALYSIS**
    RT   ANALYTICAL CHEMISTRY
          ELECTROPHOTOMETRY
          FLAME SPECTROSCOPY
          GAS ANALYSIS
          MASS SPECTROMETERS
          MICROANALYSIS
          NEUTRON ACTIVATION ANALYSIS
          QUANTITATIVE ANALYSIS
          SPECTROSCOPIC ANALYSIS

QUALITY
    GS   **QUALITY**
          . AIR QUALITY
          . ENVIRONMENTAL QUALITY
          . . WATER QUALITY
          . RIDING QUALITY
    RT   ACCURACY
          ADEQUACY
          APPEARANCE
          COMPUTER SYSTEMS PERFORMANCE
          CONCENTRATION (COMPOSITION)
          CONSISTENCY
          CONTAMINANTS
          CONTROLLABILITY
          DURABILITY
          EVALUATION
          FIGURE OF MERIT
          FINENESS
          FLIGHT CHARACTERISTICS
        ∞GRADE
          IMPURITIES
        ∞MATERIALS TESTS
        ∞PERFORMANCE
          POLLUTION
          PRECISION
          PRODUCT DEVELOPMENT
          PURITY
          RELIABILITY
        ∞RESISTANCE
          SPECIFICATIONS
          STABILITY
        ∞TESTS
          UPGRADING
          VALIDITY
          VARIABILITY

QUALITY CONTROL
    UF   RELIABILITY CONTROL
    RT   ACCELERATED LIFE TESTS
          ACCEPTABILITY
          AIRCRAFT RELIABILITY
          ASSURANCE
          AVERAGE
          BAYES THEOREM
          BURN-IN
          CERTIFICATION
          CHEMICAL TESTS
          CIRCUIT RELIABILITY
          COMPONENT RELIABILITY
          CONFIDENCE
          CONFIDENCE LIMITS
          CONSTRUCTION
        ∞CONTROL
          CORRELATION
          CORRELATION COEFFICIENTS
          COVARIANCE
          DATA SAMPLING
          DEGREES OF FREEDOM
          ELECTRICAL PROPERTIES
          ELECTRONIC EQUIPMENT TESTS
          ERROR DETECTION CODES
          ERRORS
          ESTIMATES
          ESTIMATING
          EXPERIMENT DESIGN
          EXTRAPOLATION
          HYPOTHESES
          INFRARED INSPECTION

QUALITY CONTROL-*(CONT.)*
          INSPECTION
          LEAST SQUARES METHOD
          LINEAR PREDICTION
          LOW TEMPERATURE TESTS
          MANN-WHITNEY-WILCOXON U TEST
        ∞MATERIALS TESTS
          MEAN
          MEDIAN (STATISTICS)
          MODE (STATISTICS)
          NONDESTRUCTIVE TESTS
          NORMALIZING (STATISTICS)
          OPERATIONS RESEARCH
          OPTIMIZATION
          ORTHOGONAL FUNCTIONS
          ORTHOGONALITY
          PRECISION
          PROBABILITY THEORY
          PROCESS CONTROL (INDUSTRY)
        ∞PROCESSES
          PRODUCT DEVELOPMENT
          PRODUCTION MANAGEMENT
          PRODUCTS
          RANDOM ERRORS
          RANDOM SAMPLING
          RANGE (EXTREMES)
          REGRESSION ANALYSIS
          REGRESSION COEFFICIENTS
          RELIABILITY
          RELIABILITY ENGINEERING
          SAMPLING
          SCHEDULING
          SEQUENTIAL ANALYSIS
          SPACECRAFT RELIABILITY
          SPECIFICATIONS
          STANDARD DEVIATION
          STANDARDIZATION
          STANDARDS
          STATIC TESTS
          STATISTICAL ANALYSIS
          STATISTICAL CORRELATION
          STATISTICAL DISTRIBUTIONS
          STATISTICAL TESTS
          STRUCTURAL RELIABILITY
        ∞SYSTEMS
          TASK COMPLEXITY
          TASKS
        ∞TESTS
          TOLERANCES (MECHANICS)
          ULTRASONIC FLAW DETECTION
          VALUE ENGINEERING
          VARIABILITY
          VARIANCE (STATISTICS)
          WEAR TESTS

QUALITY FACTORS
    USE   Q FACTORS

QUANTILES
    GS   STATISTICAL ANALYSIS
          . **QUANTILES**
    RT   MATHEMATICAL MODELS
          STATISTICAL DISTRIBUTIONS
        ∞STATISTICS
          SYMMETRY

QUANTITATIVE ANALYSIS
    GS   CHEMICAL TESTS
          . CHEMICAL ANALYSIS
          . . **QUANTITATIVE ANALYSIS**
          . . . KJELDAHL METHOD
          . . . VAN SLYKE METHOD
    RT   ANALYTICAL CHEMISTRY
          CHROMATOGRAPHY
          ELECTROPHOTOMETRY
          GAS ANALYSIS
          GRAVIMETRY
          IODIMETRY
          KARL FISCHER REAGENT
          MICROANALYSIS
          NEUTRON ACTIVATION ANALYSIS
          POLAROGRAPHY
          QUALITATIVE ANALYSIS
          RADIOCHEMICAL SEPARATION
          SPECTROSCOPIC ANALYSIS
          VOLUMETRIC ANALYSIS

QUANTITY
    USE   AMOUNT

QUANTIZATION
    USE   MEASUREMENT

QUANTIZER
    USE   COUNTERS

QUANTUM AMPLIFIERS
    RT   INFORMATION THEORY
          LASERS
          TWO-WAVELENGTH LASERS
          ULTRASHORT PULSED LASERS
          ULTRAVIOLET LASERS

QUANTUM CHEMISTRY
    GS   PHYSICAL CHEMISTRY
          . **QUANTUM CHEMISTRY**
    RT   ∞CHEMISTRY
          NUCLEAR CHEMISTRY
          QUANTUM MECHANICS

QUANTUM CHROMODYNAMICS
    UF   COLOR (PARTICLE PHYSICS)
          QCD
    GS   FIELD THEORY (PHYSICS)
          . **QUANTUM CHROMODYNAMICS**
          . . INSTANTONS
    RT   ∞DYNAMICS
          GLUONS
          LEPTONS
          PARTICLE INTERACTIONS
          QUANTUM MECHANICS
          QUARKS
        ∞THEORIES

QUANTUM COUNTERS
    GS   MEASURING INSTRUMENTS
          . COUNTERS
          . . RADIATION COUNTERS
          . . . **QUANTUM COUNTERS**
          . RADIATION MEASURING INSTRUMENTS
          . . RADIATION COUNTERS
          . . . **QUANTUM COUNTERS**
    RT   SQUID (DETECTORS)

QUANTUM EFFICIENCY
    RT   ENERGY CONVERSION EFFICIENCY
          ENERGY TECHNOLOGY
          HETEROJUNCTION DEVICES
          LASER OUTPUTS
          SOLAR CELLS
          VOLT-AMPERE CHARACTERISTICS

QUANTUM ELECTRODYNAMICS
    GS   ELECTRODYNAMICS
          . **QUANTUM ELECTRODYNAMICS**
          . . LIGHT-CONE EXPANSION
          QUANTUM MECHANICS
          . **QUANTUM ELECTRODYNAMICS**
          . . LIGHT-CONE EXPANSION
    RT   ELECTROMAGNETIC FIELDS
          FEYNMAN DIAGRAMS
          FIELD THEORY (PHYSICS)
          LANDAU-GINZBURG EQUATIONS
          RESONANCE FLUORESCENCE
          SELF CONSISTENT FIELDS

QUANTUM ELECTRONICS
    RT   ∞ELECTRONICS
          LASERS
          QUANTUM MECHANICS
          QUANTUM THEORY

QUANTUM GENERATORS
    USE   STIMULATED EMISSION DEVICES

QUANTUM MECHANICS
    GS   **QUANTUM MECHANICS**
          . PAULI EXCLUSION PRINCIPLE
          . QUANTUM ELECTRODYNAMICS
          . . LIGHT-CONE EXPANSION
    RT   ATOMIC INTERACTIONS
          BORN APPROXIMATION
          DYSON THEORY
          ELECTROMAGNETIC INTERACTIONS
          ENERGY DISTRIBUTION
          FERMI-DIRAC STATISTICS
          FUNCTION SPACE
          GROUP VELOCITY
          HYLLERAAS COORDINATES
        ∞MECHANICS (PHYSICS)
          ORR-SOMMERFELD EQUATIONS
          PHASE VELOCITY
          QUANTUM CHEMISTRY
          QUANTUM CHROMODYNAMICS
          QUANTUM ELECTRONICS
          RACAH COEFFICIENT
          RELATIVITY
          STATISTICAL MECHANICS
          STRANGENESS
          U SPIN SPACE
          WAVE PACKETS

QUANTUM NUMBERS
RT   ANGULAR MOMENTUM
      ELECTRONS
      ENERGY LEVELS
      NUCLEAR SPIN
   ∞ NUMBERS
      PARITY
      SELECTION RULES (NUCLEAR PHYSICS)

QUANTUM STATISTICS
UF   BOSE-EINSTEIN STATISTICS
RT   BOSONS
      FERMI-DIRAC STATISTICS
      FERMIONS
      MANY BODY PROBLEM
   ∞ STATISTICS
      SUPERFLUIDITY
      THOMAS-FERMI MODEL

QUANTUM THEORY
UF   · WIGHTMAN THEORY
GS   THEORETICAL PHYSICS
      . **QUANTUM THEORY**
      . . BOHR THEORY
RT   ANGULAR MOMENTUM
      ATOMIC THEORY
      CHARM (PARTICLE PHYSICS)
      DE BROGLIE WAVELENGTHS
      DIRAC EQUATION
      ELEMENTARY PARTICLES
      EMISSION
      ENERGY LEVELS
      FIELD THEORY (PHYSICS)
      FLAVOR (PARTICLE PHYSICS)
      FORBIDDEN TRANSITIONS
      GROUND STATE
      HAMILTONIAN FUNCTIONS
      KLEIN-DUNHAM POTENTIAL
      MAGNETIC MONOPOLES
      MANDELSTAM REPRESENTATION
      MOLECULAR ORBITALS
      NUCLEAR PHYSICS
      NUCLEAR SPIN
      PARITY
      PERTURBATION THEORY
      PHOTONS
      PHYSICAL OPTICS
      PLANCKS CONSTANT
      QUANTUM ELECTRONICS
   ∞ RADIATION
      RADIATION LAWS
      SCHUMANN-RUNGE BANDS
      STATISTICAL DISTRIBUTIONS
      STATISTICAL MECHANICS
   ∞ THEORIES
      WAVE EQUATIONS

QUANTUM WELLS
GS   POTENTIAL ENERGY
      . ELECTRIC POTENTIAL
      . . **QUANTUM WELLS**
RT   BAND STRUCTURE OF SOLIDS
      CONDUCTION BANDS
      CONDUCTION ELECTRONS
      ENERGY BANDS
      ENERGY GAPS (SOLID STATE)
      GAPS
      HETEROJUNCTION DEVICES
      HETEROJUNCTIONS
      VALENCE

QUARK PARTON MODEL
GS   MODELS
      . **QUARK PARTON MODEL**
RT   HADRONS
      INELASTIC SCATTERING
      LEPTONS
      NUCLEAR MODELS
      PARTONS
      QUARKS

QUARKS
GS   PARTICLES
      . ELEMENTARY PARTICLES
      . . **QUARKS**
RT   FLAVOR (PARTICLE PHYSICS)
      GLUONS
      INSTANTONS
      PARTONS
      QUANTUM CHROMODYNAMICS
      QUARK PARTON MODEL

QUARRIES
USE   MINES (EXCAVATIONS)

QUARTIC EQUATIONS
GS   ALGEBRA
      . NONLINEAR EQUATIONS
      . . **QUARTIC EQUATIONS**
      ANALYSIS (MATHEMATICS)
      . REAL VARIABLES
      . . NONLINEAR EQUATIONS
      . . . **QUARTIC EQUATIONS**
RT   ∞ EQUATIONS

QUARTILES
RT   PROBABILITY DENSITY FUNCTIONS
      STATISTICAL ANALYSIS
      STATISTICAL DISTRIBUTIONS

QUARTZ
GS   CHALCOGENIDES
      . OXIDES
      . . DIOXIDES
      . . . SILICON DIOXIDE
      . . . . **QUARTZ**
      . . . . . COESITE
      . . SILICON OXIDES
      . . . SILICON DIOXIDE
      . . . . **QUARTZ**
      . . . . . COESITE
      MINERALS
      . **QUARTZ**
      . . COESITE
      . . STISHOVITE
      SILICON COMPOUNDS
      . SILICON OXIDES
      . . SILICON DIOXIDE
      . . . **QUARTZ**
      . . . . COESITE
RT   ABRASIVES
      FELSITE
      FLINT
      IGNEOUS ROCKS
      ROCKS
      SANDS
      SOILS

QUARTZ CRYSTALS
GS   CRYSTALS
      . **QUARTZ CRYSTALS**
RT   FREQUENCY CONTROL
      FREQUENCY STABILITY
      RADIO TRANSMITTERS
      SILICON DIOXIDE

QUARTZ LAMPS
GS   LIGHTING EQUIPMENT
      . LUMINAIRES
      . . **QUARTZ LAMPS**

QUARTZ TRANSDUCERS
GS   TRANSDUCERS
      . **QUARTZ TRANSDUCERS**
RT   PIEZOELECTRIC CRYSTALS
      PRESSURE SENSORS

QUASARS
UF   QSO (RADIO SOURCES)
      QUASI-STELLAR RADIO SOURCES
GS   CELESTIAL BODIES
      . RADIO SOURCES (ASTRONOMY)
      . . **QUASARS**
RT   EXTRAGALACTIC RADIO SOURCES
      GALAXIES
      GRAVITATIONAL COLLAPSE
      PULSARS
      RADIO ASTRONOMY
      RADIO BURSTS
      RADIO EMISSION
      RADIO GALAXIES
      RADIO STARS
      STARS
      X RAY SPECTRA

QUASAT
SN   (QUASAR SATELLITE)
GS   OBSERVATORIES
      . ASTRONOMICAL OBSERVATORIES
      . . ASTRONOMICAL SATELLITES
      . . . **QUASAT**
RT   EUROPEAN SPACE PROGRAMS
      NASA PROGRAMS
      RADIO ASTRONOMY
      RADIO TELESCOPES
      SPACEBORNE ASTRONOMY
      VERY LONG BASE INTERFEROMETRY

QUASI-PARTICLES
USE   ELEMENTARY EXCITATIONS

QUASI-STEADY STATES
RT   EQUILIBRIUM FLOW
      FLUID DYNAMICS
      NONEQUILIBRIUM FLOW
      STEADY FLOW
      STEADY STATE CREEP
      UNIFORM FLOW

QUASI-STELLAR RADIO SOURCES
USE   QUASARS

QUASILINEARITY
USE   NONLINEARITY

QUATERNARY ALLOYS
GS   ALLOYS
      . **QUATERNARY ALLOYS**

QUATERNIONS
RT   CLASSICAL MECHANICS
      NUMBER THEORY

QUEBEC
GS   NATIONS
      . CANADA
      . . **QUEBEC**

QUEFRENCIES
GS   FREQUENCIES
      . AUDIO FREQUENCIES
      . . **QUEFRENCIES**
RT   CEPSTRA

∞ **QUENCHING**
SN   (USE OF A MORE SPECIFIC TERM IS
      RECOMMENDED--CONSULT THE TERMS
      LISTED BELOW)
RT   EXTINGUISHING
      QUENCHING (COOLING)
      RAPID QUENCHING (METALLURGY)

QUENCHING (ATOMIC PHYSICS)
GS   GYRATION
      . PRECESSION
      . . **QUENCHING (ATOMIC PHYSICS)**
RT   ANGULAR MOMENTUM
      MAGNETIC MOMENTS
      PARTICLE SPIN
   ∞ PHYSICS

QUENCHING (COOLING)
UF   FLAME QUENCHING
GS   COOLING
      . **QUENCHING (COOLING)**
RT   BATHS
      COMBUSTION
      DIPPING
      EXTINGUISHING
      HARDENING (MATERIALS)
      HEAT TREATMENT
      MICROSTRUCTURE
   ∞ QUENCHING
      SUBMERGING
      SUPERCOOLING
      SUPERSATURATION
      THERMOMECHANICAL TREATMENT
      WATER IMMERSION

QUERY LANGUAGES
GS   LANGUAGES
      . COMMAND LANGUAGES
      . . **QUERY LANGUAGES**
RT   INFORMATION RETRIEVAL

QUESTOL
SN   (EXPERIMENTAL STOL TRANSPORT
      RESEARCH AIRPLANE)
UF   EXPERIMENTAL STOL TRANSPORT RSCH
      AIRPLANE
GS   RESEARCH AIRCRAFT
      . **QUESTOL**
      V/STOL AIRCRAFT
      . SHORT TAKEOFF AIRCRAFT
      . . **QUESTOL**
RT   ∞ AIRCRAFT
      NASA PROGRAMS

QUEUEING THEORY
RT   ARPA COMPUTER NETWORK
      BUNCHING
      MATHEMATICAL MODELS
      OPERATIONS RESEARCH
      STATISTICAL ANALYSIS
      STOCHASTIC PROCESSES

QUEUEING THEORY-(CONT.)
∞ THEORIES

QUIET ENGINE PROGRAM
GS   PROGRAMS
. NASA PROGRAMS
. . **QUIET ENGINE PROGRAM**
. NASA SPACE PROGRAMS
. . **QUIET ENGINE PROGRAM**
RT   AIRCRAFT ENGINES
ENGINE NOISE
JET AIRCRAFT NOISE
JET ENGINES
NOISE REDUCTION

QUINOLINE
GS   NITROGEN COMPOUNDS
. ALKALOIDS
. . **QUINOLINE**
ORGANIC COMPOUNDS
. **QUINOLINE**
RT   DRUGS

QUINOXALINES
GS   HYDROCARBONS
. **QUINOXALINES**
ORGANIC COMPOUNDS
. **QUINOXALINES**
PLASTICS
. SYNTHETIC RESINS
. . THERMOPLASTIC RESINS
. . . **QUINOXALINES**
RESINS
. SYNTHETIC RESINS
. . THERMOPLASTIC RESINS
. . . **QUINOXALINES**
RT   POLYMERIZATION

QUOTIENTS
RT   DIVIDING (MATHEMATICS)

R

RA-28 ENGINE
GS   ENGINES
. AIR BREATHING ENGINES
. . GAS TURBINE ENGINES
. . . JET ENGINES
. . . . TURBOJET ENGINES
. . . . . **RA-28 ENGINE**
. INTERNAL COMBUSTION ENGINES
. . GAS TURBINE ENGINES
. . . JET ENGINES
. . . . TURBOJET ENGINES
. . . . . **RA-28 ENGINE**
. TURBINE ENGINES
. . GAS TURBINE ENGINES
. . . JET ENGINES
. . . . TURBOJET ENGINES
. . . . . **RA-28 ENGINE**

RABBITS
GS   ANIMALS
. VERTEBRATES
. . MAMMALS
. . . RODENTS
. . . . **RABBITS**

RACAH COEFFICIENT
RT   ANGULAR MOMENTUM
COEFFICIENTS
COUPLING
QUANTUM MECHANICS
TRANSFORMATIONS (MATHEMATICS)

RACE FACTORS
RT   ANTHROPOLOGY
CULTURE (SOCIAL SCIENCES)
HUMAN BEINGS
HUMAN PERFORMANCE
RACES
SOCIAL FACTORS
SOCIOLOGY
∞ VARIABLE

RACES
RT   AMERICAN INDIANS
ANTHROPOLOGY
CULTURE (SOCIAL SCIENCES)
HUMAN BEINGS
MINORITIES

RACES-(CONT.)
RACE FACTORS

RACETRACKS (PARTICLE ACCELERATORS)
RT   ∞ ACCELERATORS
ELECTROMAGNETS
MAGNETIC FIELDS
PARTICLE ACCELERATION
PARTICLE ACCELERATORS
PARTICLE TRAJECTORIES

∞ RACKS
SN   *(USE OF A MORE SPECIFIC TERM IS RECOMMENDED CONSULT THE TERMS LISTED BLOW)*
RT   RACKS (FRAMES)
RACKS (GEARS)

RACKS (FRAMES)
RT   ∞ RACKS
∞ SHELVES
∞ STORAGE
SUPPORTS

RACKS (GEARS)
GS   GEARS
. **RACKS (GEARS)**
RT   ∞ RACKS
TRANSLATIONAL MOTION

RACON BEACONS
USE   RADAR BEACONS

RADANT
GS   ANTENNAS
. DIRECTIONAL ANTENNAS
. . RADAR ANTENNAS
. . . **RADANT**
RADAR EQUIPMENT
. RADAR ANTENNAS
. . **RADANT**
RT   RADAR FILTERS
RADOMES
SLOT ANTENNAS
WAVEGUIDE ANTENNAS

RADAR
GS   **RADAR**
. COHERENT RADAR
. CONTINUOUS WAVE RADAR
. DOPPLER RADAR
. . MULTISTATIC RADAR
. IMAGING RADAR
. INCOHERENT SCATTER RADAR
. . EISCAT RADAR SYSTEM (EUROPE)
. INFRARED RADAR
. LANDING RADAR
. METEOROLOGICAL RADAR
. MOVING TARGET INDICATORS
. MULTISPECTRAL RADAR
. OPTICAL RADAR
. PULSE RADAR
. . PULSE DOPPLER RADAR
. . . EARTH RESOURCES SHUTTLE IMAGING RADAR
. . . MONOPULSE RADAR
. RADAR MEASUREMENT
. RANGE AND RANGE RATE TRACKING
. SATELLITE-BORNE RADAR
. SEARCH RADAR
. . NORTH AMERICAN SEARCH AND RANGING RADAR
. . OVER-THE-HORIZON RADAR
. SECONDARY RADAR
. SPACE BASED RADAR
. SURVEILLANCE RADAR
. . AIRBORNE SURVEILLANCE RADAR
. . COBRA DANE (RADAR)
. . MULTISTATIC RADAR
. SYNTHETIC APERTURE RADAR
. . SIDE-LOOKING RADAR
. TRACKING RADAR
. . COBRA DANE (RADAR)
. TRADEX RADAR SYSTEM
. VENUS ORBITING IMAGING RADAR (SPACECRAFT)
RT   AIRCRAFT INSTRUMENTS
CANCELLATION CIRCUITS
CIRCUMLUNAR COMMUNICATION
COLLISION AVOIDANCE
DISPLAY DEVICES
DISTANCE MEASURING EQUIPMENT
ELECTROMAGNETIC RADIATION
FLIGHT INSTRUMENTS
INSTRUMENT LANDING SYSTEMS

RADAR-(CONT.)
LUNAR COMMUNICATION
NAVIGATION INSTRUMENTS
NIGHT FLIGHTS (AIRCRAFT)
PULSE COMPRESSION
RADAR DETECTION
SCATTEROMETERS

RADAR ABSORBERS
RT   ABSORBERS (MATERIALS)
ANTIRADAR COATINGS
COUNTERMEASURES
ELECTROMAGNETIC ABSORPTION

RADAR ABSORBING MATERIALS
USE   ANTIRADAR COATINGS

RADAR ALTIMETERS
USE   RADIO ALTIMETERS

RADAR ANTENNAS
GS   ANTENNAS
. DIRECTIONAL ANTENNAS
. . **RADAR ANTENNAS**
. . . RADANT
RADAR EQUIPMENT
. **RADAR ANTENNAS**
. . RADANT
RT   AIRCRAFT ANTENNAS
DIPLEXERS
DIPOLE ANTENNAS
DOGHOUSES (ELECTRONICS)
HIGH RESOLUTION COVERAGE ANTENNAS
HORN ANTENNAS
LENS ANTENNAS
MICROWAVE ANTENNAS
PARABOLIC ANTENNAS
RADOMES
SCHWARZSCHILD ANTENNAS
SIDELOBE REDUCTION
SLOT ANTENNAS
SPINNERS
STEERABLE ANTENNAS

RADAR APPROACH CONTROL
UF   RAPCON (CONTROL)
GS   APPROACH CONTROL
. **RADAR APPROACH CONTROL**
GROUND BASED CONTROL
. AIR TRAFFIC CONTROL
. . **RADAR APPROACH CONTROL**
TRAFFIC CONTROL
. AIR TRAFFIC CONTROL
. . **RADAR APPROACH CONTROL**
RT   AIRBORNE RADAR APPROACH
AIRCRAFT GUIDANCE
APPROACH INDICATORS
∞ CONTROL
INSTRUMENT LANDING SYSTEMS
LANDING AIDS
LANDING INSTRUMENTS
LANDING RADAR
RADARSCOPES
SURVEILLANCE RADAR

RADAR ASTRONOMY
GS   ASTRONOMY
. **RADAR ASTRONOMY**
RT   RADIO ASTRONOMY

RADAR ATTENUATION
GS   ATTENUATION
. WAVE ATTENUATION
. . **RADAR ATTENUATION**
RT   ATMOSPHERIC ATTENUATION
ELECTROMAGNETIC ABSORPTION
ELECTROMAGNETIC WAVE TRANSMISSION
MANDELSTAM REPRESENTATION
RADIO ATTENUATION
SIDELOBE REDUCTION
SIGNAL TRANSMISSION
TRANSMISSION
WAVE PROPAGATION

RADAR BEACONS
UF   RACON BEACONS
GS   NAVIGATION AIDS
. BEACONS
. . **RADAR BEACONS**
. . . DISCRETE ADDRESS BEACON SYSTEM
RADAR EQUIPMENT
. **RADAR BEACONS**

RADAR BEACONS-(CONT.)
RT   AIRCRAFT COMMUNICATION
       COMPASSES
       POSITION (LOCATION)
       SOLAR COMPASSES
       TRANSPONDERS

RADAR BEAMS
GS   BEAMS (RADIATION)
      . **RADAR BEAMS**

RADAR CLUTTER MAPS
GS   MAPS
      . **RADAR CLUTTER MAPS**

RADAR CORNER REFLECTORS
UF   LUNEBERG LENSES
GS   RADAR EQUIPMENT
      . RADAR REFLECTORS
      . . **RADAR CORNER REFLECTORS**
      REFLECTORS
      . RADAR REFLECTORS
      . . **RADAR CORNER REFLECTORS**
RT   RETROREFLECTORS

RADAR CROSS SECTIONS
RT   ANGELS
      ∞CROSS SECTIONS
      LOW OBSERVABLE REENTRY VEHICLES
      MOVING TARGET INDICATORS

RADAR DATA
RT   ∞DATA
      MICROWAVE PHOTOGRAPHY
      VIDEO DATA

RADAR DETECTION
GS   DETECTION
      . MISSILE DETECTION
      . . **RADAR DETECTION**
RT   COHERENT RADAR
      CONTINUOUS WAVE RADAR
      DIGITAL RADAR SYSTEMS
      DOPPLER RADAR
      ELECTRONIC COUNTERMEASURES
      ELECTRONIC WARFARE
      MULTISTATIC RADAR
      OPTICAL RADAR
      OVER-THE-HORIZON RADAR
      RADAR
      RADARSCOPES
      RESOLUTION CELL
      SATELLITE-BORNE RADAR
      SEARCH RADAR
      SIGNAL DETECTION

RADAR DIRECTION FINDERS
USE   RADIO DIRECTION FINDERS

RADAR DISPLAYS
USE   RADARSCOPES

RADAR ECHOES
UF   RADAR REFLECTIONS
GS   ECHOES
      . **RADAR ECHOES**
      . . CLUTTER
      . . LUNAR RADAR ECHOES
      . . SOLAR RADAR ECHOES
      . . VENUS RADAR ECHOES
RT   AURORAL ECHOES
      CHAFF
      GHOSTS
      GLINT
      TARGETS

RADAR EQUIPMENT
GS   **RADAR EQUIPMENT**
      . DISCRETE ADDRESS BEACON SYSTEM
      . DOGHOUSES (ELECTRONICS)
      . RADAR ANTENNAS
      . . RADANT
      . RADAR BEACONS
      . RADAR FILTERS
      . RADAR RECEIVERS
      . RADAR REFLECTORS
      . . RADAR CORNER REFLECTORS
      . RADAR TRANSMITTERS
      . RADARSCOPES
      . . PLAN POSITION INDICATORS
      . RETROREFLECTORS
RT   AIRBORNE EQUIPMENT
      AIRPORT SURFACE DETECTION
        EQUIPMENT

RADAR EQUIPMENT-(CONT.)
      AUTOMATED RADAR TERMINAL SYSTEM
      DIGITAL RADAR SYSTEMS
      DIPLEXERS
      DISTANCE MEASURING EQUIPMENT
      DOPPLER RADAR
    ∞EQUIPMENT
      JAMMERS
      LOOK ANGLES (ELECTRONICS)
      ONBOARD EQUIPMENT
      PARABOLIC ANTENNAS
      RADIO EQUIPMENT
      RADOMES
      RANGE FINDERS
      SERVOMECHANISMS
      SYNTHETIC APERTURE RADAR
      TRANSPONDERS

RADAR FILTERS
GS   ELECTROMAGNETIC WAVE FILTERS
      . ELECTRIC FILTERS
      . . **RADAR FILTERS**
      RADAR EQUIPMENT
      . **RADAR FILTERS**
RT   FIR FILTERS
      MICROWAVE FILTERS
      RADANT
      RADIO FILTERS
      WAVEGUIDE FILTERS

RADAR GEOLOGY
GS   GEOLOGY
      . **RADAR GEOLOGY**
RT   GEOLOGICAL SURVEYS
      SHUTTLE IMAGING RADAR

RADAR HOMING MISSILES
GS   MISSILES
      . **RADAR HOMING MISSILES**
RT   MILITARY TECHNOLOGY
      MISSILE CONTROL
      MISSILE SYSTEMS
      TARGET RECOGNITION

RADAR IMAGERY
GS   IMAGERY
      . **RADAR IMAGERY**
RT   CHANGE DETECTION
      EARTH RESOURCES SHUTTLE IMAGING
        RADAR
      IMAGE ANALYSIS
      IMAGING RADAR
      IMAGING TECHNIQUES
      INFRARED RADAR
      LUNAR EQUATOR
      RESOLUTION CELL
      SHUTTLE IMAGING RADAR
      X RAY IMAGERY

RADAR MAPS
GS   MAPS
      . **RADAR MAPS**
RT   MAP MATCHING GUIDANCE
      METEOROLOGICAL CHARTS

RADAR MEASUREMENT
GS   RADAR
      . **RADAR MEASUREMENT**
RT   DISTANCE MEASURING EQUIPMENT
      ∞MEASUREMENT
      RADIO ALTIMETERS
      RANGEFINDING

RADAR NAVIGATION
GS   NAVIGATION
      . **RADAR NAVIGATION**
RT   AIR NAVIGATION
      AIR TRAFFIC CONTROL
      AIRCRAFT GUIDANCE
      ALL-WEATHER AIR NAVIGATION
      AUTOMATIC FLIGHT CONTROL
      CELESTIAL NAVIGATION
      COLLISION AVOIDANCE
      DEAD RECKONING
      DISTANCE
      DISTANCE MEASURING EQUIPMENT
      DOPPLER NAVIGATION
      DOPPLER RADAR
      DOPPLER-FIZEAU EFFECT
      GROUND BASED CONTROL
      INERTIAL NAVIGATION
      INTERPLANETARY NAVIGATION
      MAP MATCHING GUIDANCE
      RADARSCOPES
      RADIO NAVIGATION

RADAR NAVIGATION-(CONT.)
      SATELLITE NAVIGATION SYSTEMS
      SPACE NAVIGATION
      SURFACE NAVIGATION
      TACAN

RADAR NETWORKS
UF   MULTIRADAR TRACKING
GS   TRACKING NETWORKS
      . **RADAR NETWORKS**
RT   DOPPLER RADAR
      ∞NETWORKS
      POLYSTATION DOPPLER TRACKING
        SYSTEM
      TRACKING STATIONS

RADAR OBSERVATION
USE   RADAR TRACKING

RADAR PHOTOGRAPHY
GS   IMAGERY
      . **RADAR PHOTOGRAPHY**
      PHOTOGRAPHY
      . MULTISPECTRAL PHOTOGRAPHY
      . . **RADAR PHOTOGRAPHY**
RT   BLACK AND WHITE PHOTOGRAPHY
      MICROWAVE PHOTOGRAPHY
      RADARSCOPES
      SPECTRAL RECONNAISSANCE
      ULTRAVIOLET PHOTOGRAPHY

RADAR RANGE
GS   DISTANCE
      . **RADAR RANGE**
RT   CONTINUOUS WAVE RADAR
      OPTICAL SLANT RANGE
      OVER-THE-HORIZON RADAR
      RADIO RANGE

RADAR RECEIVERS
GS   RADAR EQUIPMENT
      . **RADAR RECEIVERS**
      RECEIVERS
      . **RADAR RECEIVERS**
RT   DIGITAL RADAR SYSTEMS
      ELECTROMAGNETIC NOISE
      MICROWAVE SENSORS
      RADIO RECEIVERS
      REPEATERS

RADAR RECEPTION
RT   RADIO RECEPTION
      ∞RECEIVING
      SIDELOBE REDUCTION

RADAR REFLECTIONS
USE   RADAR ECHOES

RADAR REFLECTORS
GS   RADAR EQUIPMENT
      . **RADAR REFLECTORS**
      . . RADAR CORNER REFLECTORS
      REFLECTORS
      . **RADAR REFLECTORS**
      . . RADAR CORNER REFLECTORS
RT   PARABOLIC ANTENNAS
      PARABOLIC REFLECTORS
      RADIO ECHOES
      SIDELOBE REDUCTION

RADAR RESOLUTION
GS   RESOLUTION
      . **RADAR RESOLUTION**
RT   ANGULAR RESOLUTION
      DISPLAY DEVICES
      HIGH RESOLUTION COVERAGE
        ANTENNAS
      SIDELOBE REDUCTION
      SPECTRAL RESOLUTION

RADAR SCANNING
GS   SCANNING
      . **RADAR SCANNING**
      . . RADAR TARGET SCATTER SITE
        PROGRAM
RT   CONICAL SCANNING
      DIGITAL RADAR SYSTEMS
      FREQUENCY SCANNING
      METEOROLOGICAL RADAR
      MULTIPLE BEAM INTERVAL SCANNERS
      PANORAMIC SCANNING
      RADIO TRACKING
      SIDE-LOOKING RADAR
      SURVEILLANCE

RADAR SCATTERING
GS   SCATTERING
. **RADAR SCATTERING**
RT   EISCAT RADAR SYSTEM (EUROPE)
INCOHERENT SCATTER RADAR
INCOHERENT SCATTERING
SCATTEROMETERS

RADAR SIGNATURES
GS   SIGNATURES
. **RADAR SIGNATURES**
RT   COBRA DANE (RADAR)
DETECTION
IMAGERY
SIGNATURE ANALYSIS
TARGET RECOGNITION

RADAR TARGET SCATTER SITE PROGRAM
UF   RATSCAT PROGRAM
GS   PROGRAMS
. **RADAR TARGET SCATTER SITE PROGRAM**
SCANNING
. **RADAR SCANNING**
.. **RADAR TARGET SCATTER SITE PROGRAM**
RT   TARGETS

RADAR TARGETS
GS   TARGETS
. **RADAR TARGETS**
RT   DIGITAL RADAR SYSTEMS
EARLY WARNING SYSTEMS
RADIAL VELOCITY

RADAR TRACKING
UF   RADAR OBSERVATION
GS   TRACKING (POSITION)
. **RADAR TRACKING**
RT   AUTOMATED RADAR TERMINAL SYSTEM
BALLISTIC MISSILE EARLY WARNING SYSTEM
COMPENSATORY TRACKING
DIGITAL RADAR SYSTEMS
DOPPLER RADAR
EARLY WARNING SYSTEMS
METEOROLOGICAL RADAR
MONOPULSE RADAR
MOVING TARGET INDICATORS
PURSUIT TRACKING
RADIO TRACKING
RANGE AND RANGE RATE TRACKING
RANGEFINDING
RAWINSONDES
SEARCH RADAR
SLEWING
SPACECRAFT TRACKING
SURVEILLANCE RADAR
THREAT EVALUATION
TRACKING RADAR
TRANSPONDER CONTROL GROUP

RADAR TRANSMISSION
GS   TRANSMISSION
. ELECTROMAGNETIC WAVE TRANSMISSION
.. **RADAR TRANSMISSION**
. SIGNAL TRANSMISSION
.. **RADAR TRANSMISSION**
RT   ATMOSPHERIC ATTENUATION
DIGITAL RADAR SYSTEMS
EISCAT RADAR SYSTEM (EUROPE)
ELECTROMAGNETIC PULSES
RADIO TRANSMISSION
RADOME MATERIALS
WAVE ATTENUATION
WAVE PROPAGATION

RADAR TRANSMITTERS
GS   RADAR EQUIPMENT
. **RADAR TRANSMITTERS**
TRANSMITTERS
. **RADAR TRANSMITTERS**

RADARSAT
GS   CANADIAN SPACECRAFT
. **RADARSAT**
RT   CANADIAN SPACE PROGRAM
SYNTHETIC APERTURE RADAR

RADARSCOPES
UF   RADAR DISPLAYS
GS   DISPLAY DEVICES
. **RADARSCOPES**
.. PLAN POSITION INDICATORS

RADARSCOPES-*(CONT.)*
RADAR EQUIPMENT
. **RADARSCOPES**
.. PLAN POSITION INDICATORS
RT   AIRCRAFT GUIDANCE
INDICATING INSTRUMENTS
MICROWAVE IMAGERY
MICROWAVE PHOTOGRAPHY
RADAR APPROACH CONTROL
RADAR DETECTION
RADAR NAVIGATION
RADAR PHOTOGRAPHY
SURVEILLANCE RADAR

RADIAL DISTRIBUTION
GS   DISTRIBUTION (PROPERTY)
. **RADIAL DISTRIBUTION**
RT   RAYLEIGH DISTRIBUTION
STAR DISTRIBUTION
WIND PROFILES

RADIAL DRAINAGE PATTERNS
USE   DRAINAGE PATTERNS

RADIAL FLOW
GS   FLUID FLOW
. **RADIAL FLOW**
RT   AXIAL FLOW
CENTRIFUGAL COMPRESSORS
DIFFUSION
FLOW GEOMETRY
GAS FLOW
HEAT TRANSMISSION
TWO DIMENSIONAL FLOW

RADIAL VELOCITY
GS   RATES (PER TIME)
. **RADIAL VELOCITY**
VELOCITY
. **RADIAL VELOCITY**
RT   ASTRONOMICAL SPECTROSCOPY
DOPPLER EFFECT
RADAR TARGETS
RED SHIFT
VELOCITY MEASUREMENT

RADIANCE
SN   (DIRECTIONAL EMISSION RATE PER UNIT AREA OF RADIATION)
GS   ELECTROMAGNETIC PROPERTIES
. OPTICAL PROPERTIES
.. **RADIANCE**
RATES (PER TIME)
. FLUX DENSITY
.. RADIANT FLUX DENSITY
... **RADIANCE**
RT   BLACK BODY RADIATION
BRIGHTNESS
EMISSIVITY
EMITTANCE
GLARE
INCANDESCENCE
∞INTENSITY
IRRADIANCE
LUMENS
LUMINOSITY
NEUTRON FLUX DENSITY
SOLAR FLUX DENSITY
TRANSMISSOMETERS
VISIBILITY

RADIANCY
SN   (EMISSION RATE PER UNIT AREA OF RADIATION)
GS   RATES (PER TIME)
. FLUX DENSITY
.. RADIANT FLUX DENSITY
... **RADIANCY**
RT   ELECTRON FLUX DENSITY
ILLUMINANCE
LUMINOUS INTENSITY
NEUTRON FLUX DENSITY
PARTICLE FLUX DENSITY
PROTON FLUX DENSITY
SOLAR FLUX DENSITY

RADIANT COOLING
GS   COOLING
. **RADIANT COOLING**
RT   RADIATIVE HEAT TRANSFER
SURFACE COOLING

RADIANT ENERGY
USE   RADIATION

RADIANT FLUX DENSITY
SN   (DYNES/CM-SEC AS DISTINGUISHED FROM FROM RADIATION PRESSURE--DYNES/CM2)
UF   POWER DENSITY (ELECTROMAGNETIC)
RADIANT INTENSITY
RADIATION INTENSITY
GS   RATES (PER TIME)
. FLUX DENSITY
.. **RADIANT FLUX DENSITY**
... IRRADIANCE
.... ILLUMINANCE
.... SOLAR CONSTANT
... LUMENS
... LUMINOUS INTENSITY
.... ILLUMINANCE
.... LUMINANCE
.. PARTICLE FLUX DENSITY
.... ELECTRON FLUX DENSITY
.... NEUTRON FLUX DENSITY
.... PROTON FLUX DENSITY
... RADIANCE
... RADIANCY
... SOLAR FLUX DENSITY
.... SOLAR CONSTANT
RT   BL LACERTAE OBJECTS
BRIGHTNESS
BRIGHTNESS DISTRIBUTION
DOSIMETERS
EMISSIVITY
EMITTANCE
FAR FIELDS
FLUX (RATE)
GAMMA RAY BURSTS
LASER OUTPUTS
LUMINOSITY
MASER OUTPUTS
MASS TO LIGHT RATIOS
POST-BLAST NUCLEAR RADIATION
∞RADIATION
RADIATION COUNTERS
RADIATION PRESSURE
RADIO SPECTRA
SCATTERING FUNCTIONS
SOLAR REFLECTORS
SOUND INTENSITY
VIEW EFFECTS

RADIANT HEATING
UF   RADIATION HEATING
GS   HEATING
. **RADIANT HEATING**
RT   ∞ENERGY
GAS HEATING
∞RADIATION
RADIATIVE HEAT TRANSFER
RADIATIVE TRANSFER
SOLAR HEATING

RADIANT INTENSITY
USE   RADIANT FLUX DENSITY

∞ RADIATION
SN   *(USE OF A MORE SPECIFIC TERM IS RECOMMENDED--CONSULT THE TERMS LISTED BELOW)*
UF   RADIANT ENERGY
RADIATION EMISSION
RT   ALPHA PARTICLES
ANTENNA RADIATION PATTERNS
ARTIFICIAL RADIATION BELTS
ATMOSPHERIC RADIATION
BACKGROUND NOISE
BACKGROUND RADIATION
BASE HEATING
BEAMS (RADIATION)
BLACK BODY RADIATION
CERENKOV RADIATION
CIRCUMSOLAR RADIATION
COHERENT ACOUSTIC RADIATION
COHERENT ELECTROMAGNETIC RADIATION
COHERENT RADIATION
CONTINUOUS RADIATION
CORPUSCULAR RADIATION
COSMIC RAYS
CYCLOTRON RADIATION
DIFFUSE RADIATION
EARTH RADIATION BUDGET EXPERIMENT
ELASTIC WAVES
ELECTROMAGNETIC NOISE
ELECTROMAGNETIC RADIATION
ELECTRON RADIATION
EMISSION

RADIATION-*(CONT.)*
 EXTRATERRESTRIAL RADIATION
 EXTREME ULTRAVIOLET RADIATION
 FALLOUT
 FAR INFRARED RADIATION
 FAR ULTRAVIOLET RADIATION
 FLUX (RATE)
 FLUX DENSITY
 GALACTIC RADIATION
 GAMMA RAYS
 GEOPHYSICS
 GRAVITATIONAL WAVES
 HARMONIC RADIATION
 HEATING
 INCIDENT RADIATION
 INFRARED RADIATION
 INNER RADIATION BELT
 INTERSTELLAR RADIATION
 ION CYCLOTRON RADIATION
 IONIZING RADIATION
 IRRADIATION
 KIRCHHOFF LAW OF RADIATION
 LIGHT (VISIBLE RADIATION)
 LONG WAVE RADIATION
 LONGITUDINAL WAVES
 LUNAR RADIATION
 LYMAN ALPHA RADIATION
 LYMAN BETA RADIATION
 MICROWAVES
 MODULATED CONTINUOUS RADIATION
 MONOCHROMATIC RADIATION
 NEAR INFRARED RADIATION
 NEAR ULTRAVIOLET RADIATION
 NUCLEAR MEDICINE
 NUCLEAR RADIATION
 NUCLEON POTENTIAL
 OUTER RADIATION BELT
 PHOTONS
 PLANE WAVES
 PLANETARY RADIATION
 PLASMA RADIATION
 POLARIZED ELECTROMAGNETIC
  RADIATION
 POLARIZED RADIATION
 POST-BLAST NUCLEAR RADIATION
 PULSED RADIATION
 QUANTUM THEORY
 RADIANT FLUX DENSITY
 RADIANT HEATING
 RADIATION ABSORPTION
 RADIATION BELTS
 RADIATION CHEMISTRY
 RADIATION COUNTERS
 RADIATION DAMAGE
 RADIATION DETECTORS
 RADIATION DISTRIBUTION
 RADIATION DOSAGE
 RADIATION EFFECTS
 RADIATION HARDENING
 RADIATION HAZARDS
 RADIATION INJURIES
 RADIATION LAWS
 RADIATION MEASUREMENT
 RADIATION MEASURING INSTRUMENTS
 RADIATION METEOROID SPACECRAFT
 RADIATION PRESSURE
 RADIATION PROTECTION
 RADIATION PYROMETERS
 RADIATION SHIELDING
 RADIATION SICKNESS
 RADIATION SOURCES
 RADIATION SPECTRA
 RADIATION THERAPY
 RADIATION TOLERANCE
 RADIATION TRANSPORT
 RADIATION TRAPPING
 RADIATIVE TRANSFER
 RADIOACTIVITY
 RADIOLOGY
 REFLECTED WAVES
 RELIC RADIATION
 RESONANCE FLUORESCENCE
 SELF ABSORPTION
 SHORT WAVE RADIATION
 SILICON RADIATION DETECTORS
 SKY RADIATION
 SOLAR CORPUSCULAR RADIATION
 SOLAR RADIATION
 SOLAR RADIATION SHIELDING
 SOLAR RADIATION 1 SATELLITE
 SOLAR RADIATION 3 SATELLITE
 SOLAR WIND
 SOUND WAVES
 SPECTRAL EMISSION
 STANDING WAVES
 STELLAR RADIATION

RADIATION-*(CONT.)*
 STOKES LAW OF RADIATION
 STRATOSPHERE RADIATION
 SYNCHROTRON RADIATION
 TEMPERATURE EFFECTS
 TERRESTRIAL RADIATION
 THERMAL RADIATION
 TRAP PROGRAM
 TROPOSPHERIC RADIATION
 ULTRASONIC RADIATION
 ULTRAVIOLET RADIATION
 VOLTERRA EQUATIONS
 X RAY SOURCES

RADIATION ABSORPTION
 GS **RADIATION ABSORPTION**
  . ATMOSPHERIC ATTENUATION
  . . AURORAL ABSORPTION
  . . ELECTROMAGNETIC ABSORPTION
  . . AURORAL ABSORPTION
  . . GAMMA RAY ABSORPTION
  . . INFRARED ABSORPTION
  . . MULTIPHOTON ABSORPTION
  . . PHOTOABSORPTION
  . . POLAR CAP ABSORPTION
  . . ULTRAVIOLET ABSORPTION
  . . X RAY ABSORPTION
  . MOLECULAR ABSORPTION
  . SELF ABSORPTION
 RT ∞ABSORPTION
  ABSORPTION CROSS SECTIONS
  GAMMA RAY ABSORPTIOMETRY
  GOLAY DETECTOR CELLS
  GRAY GAS
  INTERSTELLAR EXTINCTION
  MATERIAL ABSORPTION
  NEUTRON ABSORBERS
  NUCLEAR REACTIONS
  PHOTON ABSORPTIOMETRY
  ∞RADIATION
  RADIATION CHEMISTRY
  STOPPING POWER

RADIATION AND METEOROID SATELLITE
 GS SATELLITES
  . ARTIFICIAL SATELLITES
  . . GEOPHYSICAL SATELLITES
  . . . **RADIATION AND METEOROID
    SATELLITE**

RADIATION BELTS
 UF GEOMAGNETICALLY TRAPPED
   PARTICLES
  VAN ALLEN RADIATION BELTS
 GS EARTH ATMOSPHERE
  . **RADIATION BELTS**
  . . ARTIFICIAL RADIATION BELTS
  . . INNER RADIATION BELT
  . . OUTER RADIATION BELT
  . . PROTON BELTS
  PARTICLES
  . CHARGED PARTICLES
  . . MAGNETICALLY TRAPPED PARTICLES
  . . . **RADIATION BELTS**
  . . . . ARTIFICIAL RADIATION BELTS
  . . . . INNER RADIATION BELT
  . . . . OUTER RADIATION BELT
  . . . . PROTON BELTS
  . CORPUSCULAR RADIATION
  . **RADIATION BELTS**
  . TRAPPED PARTICLES
  . . MAGNETICALLY TRAPPED PARTICLES
  . . . **RADIATION BELTS**
  . . . . ARTIFICIAL RADIATION BELTS
  . . . . INNER RADIATION BELT
  . . . . OUTER RADIATION BELT
  . . . . PROTON BELTS
 RT AEROSPACE ENVIRONMENTS
  ∞BELTS
  COSMIC RAYS
  ELECTRON DENSITY (CONCENTRATION)
  ELECTRON PRECIPITATION
  ELECTRON TRAJECTORIES
  ELECTRONS
  ELEMENTARY PARTICLES
  ENTRAPMENT
  EXOSPHERE
  EXTRATERRESTRIAL RADIATION
  IONIZING RADIATION
  IONOSPHERIC DRIFT
  MAGNETIC FIELDS
  MAGNETOSPHERE
  MIRROR POINT
  PLASMAS (PHYSICS)
  PROTON PRECIPITATION

RADIATION BELTS-*(CONT.)*
  PROTONS
  ∞RADIATION
  SOLAR RADIATION
  TRAPPING
  UPPER ATMOSPHERE

RADIATION CHEMISTRY
 GS **RADIATION CHEMISTRY**
  . PHOTODECOMPOSITION
  . PHOTODISSOCIATION
  . PHOTOLYSIS
 RT ∞CHEMISTRY
  ELECTROMAGNETIC RADIATION
  ∞RADIATION
  RADIATION ABSORPTION

RADIATION COUNTERS
 UF IONIZATION COUNTERS
  PARTICLE COUNTERS
  PARTICLE DETECTORS
 GS MEASURING INSTRUMENTS
  . COUNTERS
  . . **RADIATION COUNTERS**
  . . . CERENKOV COUNTERS
  . . . ELECTRON COUNTERS
  . . . GEIGER COUNTERS
  . . . NEUTRON COUNTERS
  . . . . NEUTRON SPECTROMETERS
  . . . PARTICLE TELESCOPES
  . . . PROPORTIONAL COUNTERS
  . . . QUANTUM COUNTERS
  . . . SCINTILLATION COUNTERS
  . . . SPARK CHAMBERS
  . RADIATION MEASURING INSTRUMENTS
  . . **RADIATION COUNTERS**
  . . . CERENKOV COUNTERS
  . . . ELECTRON COUNTERS
  . . . GEIGER COUNTERS
  . . . NEUTRON COUNTERS
  . . . . NEUTRON SPECTROMETERS
  . . . PARTICLE TELESCOPES
  . . . PROPORTIONAL COUNTERS
  . . . QUANTUM COUNTERS
  . . . SCINTILLATION COUNTERS
  . . . SPARK CHAMBERS
 RT BUBBLE CHAMBERS
  CHANNEL MULTIPLIERS
  CLOUD CHAMBERS
  COINCIDENCE CIRCUITS
  DOSIMETERS
  ELECTROSTATIC PROBES
  FLUENCE
  GAS DISCHARGE TUBES
  HODOSCOPES
  ION TRAPS (INSTRUMENTATION)
  IONIZATION CHAMBERS
  IONIZING RADIATION
  NUCLEAR EMULSIONS
  PARTICLE FLUX DENSITY
  PROTON FLUX DENSITY
  RADIANT FLUX DENSITY
  ∞RADIATION
  SPECTROMETERS

RADIATION DAMAGE
 GS DAMAGE
  . **RADIATION DAMAGE**
  . . LASER DAMAGE
  RADIATION EFFECTS
  . **RADIATION DAMAGE**
 RT IONIZING RADIATION
  ∞RADIATION
  SINGLE EVENT UPSETS

RADIATION DETECTORS
 GS MEASURING INSTRUMENTS
  . RADIATION MEASURING INSTRUMENTS
  . . **RADIATION DETECTORS**
  . . . DOSIMETERS
  . . . . THRESHOLD DETECTORS
    (DOSIMETERS)
  . . . GOLAY DETECTOR CELLS
  . . . SILICON RADIATION DETECTORS
 RT ∞DETECTORS
  GEIGER COUNTERS
  HEALTH PHYSICS
  MULTI-ANODE MICROCHANNEL ARRAYS
  ∞RADIATION
  SATELLITE-BORNE INSTRUMENTS
  VELA SATELLITES

RADIATION DISTRIBUTION
 UF RADIATION FIELDS
 GS DISTRIBUTION (PROPERTY)

RADIATION DISTRIBUTION-(CONT.)
- . RADIATION DISTRIBUTION
- . . ANTENNA RADIATION PATTERNS
- . . . SIDELOBES
- . . DIFFRACTION PATTERNS
- . . . KOSSEL PATTERN
- . . . RAINBOWS
- RT   CORPUSCULAR RADIATION
-     ELASTIC WAVES
-     ELECTROMAGNETIC RADIATION
-     FIELD THEORY (PHYSICS)
-     FLUX DENSITY
-     NULL ZONES
-     ∞PATTERNS
-     ∞RADIATION
-     VERTICAL DISTRIBUTION
-     WAVE DISPERSION

RADIATION DOSAGE
- UF   RADIATION EXPOSURE
- GS   DOSAGE
- . RADIATION DOSAGE
- RT   BIOLOGICAL EFFECTS
-     DOSIMETERS
-     EXPOSURE
-     HEALTH PHYSICS
-     IRRADIATION
-     ∞RADIATION
-     SINGLE EVENT UPSETS

RADIATION EFFECTS
- GS   RADIATION EFFECTS
- . LASER DAMAGE
- . RADIATION DAMAGE
- . RADIATION INJURIES
- . RADIOLYSIS
- . SINGLE EVENT UPSETS
- RT   BIOLOGICAL EFFECTS
-     BLACKOUT (PROPAGATION)
-     BRAGG CURVE
-     DAMAGE
-     DOSIMETERS
-     ∞EFFECTS
-     ELECTRON RADIATION
-     FALLOUT
-     GAMMA RAYS
-     HEALTH PHYSICS
-     HEMATOPOIESIS
-     IRRADIATION
-     MECHANICAL PROPERTIES
-     NEUTRONS
-     NUCLEAR EXPLOSION EFFECT
-     NUCLEAR RADIATION
-     NUCLEAR VULNERABILITY
-     PARTICLE TRACKS
-     PHYSIOLOGICAL EFFECTS
-     POST-BLAST NUCLEAR RADIATION
-     POYNTING-ROBERTSON EFFECT
-     PRESERVING
-     ∞RADIATION
-     RADIOACTIVE CONTAMINANTS
-     SPACE BASED RADAR

RADIATION EMISSION
- USE   RADIATION

RADIATION EXPOSURE
- USE   RADIATION DOSAGE

RADIATION FIELDS
- USE   RADIATION DISTRIBUTION

RADIATION HARDENING
- RT   ANTENNAS
-     ELECTRONIC EQUIPMENT
-     ∞RADIATION

RADIATION HAZARDS
- GS   HAZARDS
- . RADIATION HAZARDS
- RT   DERMATITIS
-     DOSIMETERS
-     ELECTROMAGNETIC RADIATION
-     FALLOUT
-     FLUX DENSITY
-     HEALTH PHYSICS
-     IONIZING RADIATION
-     LASER DAMAGE
-     MUTATIONS
-     NUCLEAR EXPLOSION EFFECT
-     NUCLEAR EXPLOSIONS
-     NUCLEAR RADIATION
-     OPERATIONAL HAZARDS
-     ∞RADIATION
-     RADIOACTIVE CONTAMINANTS

RADIATION HAZARDS-(CONT.)
-     RADIOACTIVE MATERIALS
-     RADIOACTIVE WASTES
-     RADIOACTIVITY
-     REACTOR SAFETY

RADIATION HEATING
- USE   RADIANT HEATING

RADIATION INJURIES
- GS   INJURIES
- . RADIATION INJURIES
-     RADIATION EFFECTS
- . RADIATION INJURIES
- RT   BURNS (INJURIES)
-     HEALTH PHYSICS
-     ∞RADIATION

RADIATION INTENSITY
- USE   RADIANT FLUX DENSITY

RADIATION LAWS
- GS   LAWS
- . RADIATION LAWS
- . . KIRCHHOFF LAW OF RADIATION
- . . STEFAN-BOLTZMANN LAW
- . . STOKES LAW OF RADIATION
- RT   ELECTROMAGNETIC RADIATION
-     QUANTUM THEORY
-     ∞RADIATION

RADIATION MEASUREMENT
- RT   DOSAGE
-     DOSIMETERS
-     IRRADIATION
-     ∞MEASUREMENT
-     ∞RADIATION
-     SENSITOMETRY

RADIATION MEASURING INSTRUMENTS
- UF   PHOTOELECTROMAGNETIC DETECTORS
-     PHOTOSENSORS
-     RADIATION METERS
- GS   MEASURING INSTRUMENTS
- . RADIATION MEASURING
-     INSTRUMENTS
- . . ACTINOMETERS
- . . . INFRARED SPECTROMETERS
- . . . PYRANOMETERS
- . . RADIOMETERS
- . . . . DICKE RADIOMETERS
- . . . . INFRARED DETECTORS
- . . . . INFRARED SCANNERS
- . . . . MICROWAVE RADIOMETERS
- . . . . PASSIVE L-BAND RADIOMETERS
- . . . PRESSURE MODULATOR
-     RADIOMETERS
- . . . SPECTRORADIOMETERS
- . . SOLAR SPECTROMETERS
- . . SPECTROHELIOGRAPHS
- . . SPECTROPHOTOMETERS
- . . . INFRARED
-     SPECTROPHOTOMETERS
- . . . ULTRAVIOLET
-     SPECTROPHOTOMETERS
- . . . ULTRAVIOLET SPECTROMETERS
- . . BOLOMETERS
- . . EBERT SPECTROMETERS
- . . ELECTROSTATIC PROBES
- . FABRY-PEROT SPECTROMETERS
- . . HODOSCOPES
- . . INFRARED INSTRUMENTS
- . . . INFRARED DETECTORS
- . . . . FLIR DETECTORS
- . . . INFRARED SCANNERS
- . . . INFRARED SPECTROMETERS
- . . . INFRARED SPECTROPHOTOMETERS
- . . PHOTOMETERS
- . . . ELECTROPHOTOMETERS
- . . . ULTRAVIOLET SPECTROMETERS
- . . . ULTRAVIOLET
-     SPECTROPHOTOMETERS
- . . RADIATION COUNTERS
- . . . CERENKOV COUNTERS
- . . . ELECTRON COUNTERS
- . . . GEIGER COUNTERS
- . . . NEUTRON COUNTERS
- . . . NEUTRON SPECTROMETERS
- . . . PARTICLE TELESCOPES
- . . . PROPORTIONAL COUNTERS
- . . . QUANTUM COUNTERS
- . . . SCINTILLATION COUNTERS
- . . SPARK CHAMBERS
- . . RADIATION DETECTORS
- . . . DOSIMETERS

RADIATION MEASURING INSTRUMENTS-(CONT.)
- . . . . THRESHOLD DETECTORS
-     (DOSIMETERS)
- . . . GOLAY DETECTOR CELLS
- . . . SILICON RADIATION DETECTORS
- . . RIOMETERS
- RT   ∞DETECTORS
-     EARTH RADIATION BUDGET
-     EXPERIMENT
-     HEALTH PHYSICS
-     IONIZATION CHAMBERS
-     MONITORS
-     NUCLEAR EMULSIONS
-     OPTICAL MEASURING INSTRUMENTS
-     ∞RADIATION
-     SAFETY DEVICES
-     SOLAR INSTRUMENTS
-     VELA SATELLITES
-     VIEW EFFECTS

RADIATION MEDICINE
- USE   NUCLEAR MEDICINE

RADIATION METEOROID SPACECRAFT
- RT   METEOROIDS
-     ∞RADIATION
-     ∞SPACECRAFT
-     SPACECRAFT CONFIGURATIONS

RADIATION METERS
- USE   RADIATION MEASURING INSTRUMENTS

RADIATION NOISE
- USE   ELECTROMAGNETIC NOISE

RADIATION PRESSURE
- SN   (DYNES PER CM SQUARED, AS
-     DISTINGUISHED FROM RADIANT FLUX
-     DENSITY--DYNES PER CM-SEC)
- GS   PRESSURE
- . RADIATION PRESSURE
- . . ELECTRON PRESSURE
- . . LUMENS
- . . LUMINOUS INTENSITY
- . . . ILLUMINANCE
- . . . LUMINANCE
- . SOUND PRESSURE
- RT   BAROCLINIC WAVES
-     BESSEL-BREDICHIN THEORY
-     COMET TAILS
-     CORPUSCULAR RADIATION
-     ELASTIC WAVES
-     ELECTROMAGNETIC RADIATION
-     KOHOUTEK COMET
-     PARTICLE FLUX DENSITY
-     PERTURBATION
-     PHOTOPHORESIS
-     RADIANT FLUX DENSITY
-     ∞RADIATION
-     SOLAR FLUX DENSITY
-     SOLAR RADIATION
-     SOLAR WIND
-     STELLAR WINDS

RADIATION PROTECTION
- GS   PROTECTION
- . RADIATION PROTECTION
- . . RADIATION SHIELDING
- . . . SOLAR RADIATION SHIELDING
- RT   ANTIRADIATION DRUGS
-     HEALTH PHYSICS
-     ∞RADIATION
-     SYNCHROTRON RADIATION
-     THERMAL PROTECTION
-     VISORS

RADIATION PYROMETERS
- GS   MEASURING INSTRUMENTS
- . TEMPERATURE MEASURING
-     INSTRUMENTS
- . . PYROMETERS
- . . . RADIATION PYROMETERS
- RT   BOLOMETERS
-     CIRCUMSOLAR TELESCOPES
-     OPTICAL PYROMETERS
-     ∞RADIATION
-     TEMPERATURE MEASUREMENT
-     THERMOCOUPLE PYROMETERS

RADIATION RESISTANCE
- USE   RADIATION TOLERANCE

RADIATION SHIELDING
- UF   NUCLEAR SHIELDING

RADIATION SHIELDING-(CONT.)
GS   PROTECTION
      . RADIATION PROTECTION
      . . RADIATION SHIELDING
      . . . SOLAR RADIATION SHIELDING
      SHIELDING
      . RADIATION SHIELDING
      . . SOLAR RADIATION SHIELDING
RT   ABSORBERS (MATERIALS)
      ATTENUATORS
      BORAL
      ELECTROMAGNETIC ABSORPTION
      ELECTROMAGNETIC SHIELDING
      GAMMA RAYS
      ∞INSULATED STRUCTURES
      MAGNETIC SHIELDING
      NEUTRON ABSORBERS
      NEUTRON FLUX DENSITY
      NEUTRONS
      NUCLEAR REACTORS
      PROTONS
      ∞RADIATION
      RADIO FREQUENCY SHIELDING
      REACTOR MATERIALS
      REFLECTORS
      SAFETY DEVICES
      SPACE BASED RADAR
      SPACECRAFT SHIELDING
      STOPPING POWER
      TOWER SHIELDING REACTOR 2

RADIATION SICKNESS
GS   DISEASES
      . RADIATION SICKNESS
RT   ANTIRADIATION DRUGS
      DERMATITIS
      HEALTH PHYSICS
      ∞RADIATION
      RADIOPATHOLOGY

RADIATION SOURCES
UF   COHERENT SOURCES
GS   RADIATION SOURCES
      . MONOCHROMATORS
      . NEUTRON SOURCES
      . POINT SOURCES
RT   CORPUSCULAR RADIATION
      DUOCHROMATORS
      ELECTROMAGNETIC RADIATION
      ELECTRON SOURCES
      EXTARS
      EXTRAGALACTIC RADIO SOURCES
      ∞GENERATORS
      HEAT SOURCES
      INTERSTELLAR MASERS
      ION SOURCES
      LIGHT SOURCES
      PULSARS
      ∞RADIATION
      RADIO SOURCES (ASTRONOMY)
      RADIOACTIVE MATERIALS
      SOUND GENERATORS
      ∞SOURCES

RADIATION SPECTRA
GS   SPECTRA
      . RADIATION SPECTRA
      . . ELECTROMAGNETIC SPECTRA
      . . . GAMMA RAY SPECTRA
      . . . INFRARED SPECTRA
      . . . LINE SPECTRA
      . . . . BALMER SERIES
      . . . . D LINES
      . . . . ELECTRONIC SPECTRA
      . . . . FRAUNHOFER LINES
      . . . . H LINES
      . . . . . H ALPHA LINE
      . . . . . H BETA LINE
      . . . . . H GAMMA LINE
      . . . . K LINES
      . . . . LYMAN SPECTRA
      . . . . PASCHEN SERIES
      . . . . RYDBERG SERIES
      . . . . TELLURIC LINES
      . . . RADIO SPECTRA
      . . . . MICROWAVE SPECTRA
      . . . RAMAN SPECTRA
      . . . STELLAR SPECTRA
      . . . . SOLAR SPECTRA
      . . . UBV SPECTRA
      . . . ULTRAVIOLET SPECTRA
      . . . VIBRATIONAL SPECTRA
      . . . VISIBLE SPECTRUM
      . . . X RAY SPECTRA
      . . EMISSION SPECTRA

RADIATION SPECTRA-(CONT.)
      . . HERZBERG BANDS
RT   ASTRONOMICAL SPECTROSCOPY
      COSMIC BACKGROUND EXPLORER
        SATELLITE
      ENERGY SPECTRA
      GAMMA RAY SPECTRA
      MASS SPECTRA
      NOISE SPECTRA
      PLASMA SPECTRA
      ∞RADIATION

RADIATION THERAPY
UF   RADIOTHERAPY
GS   THERAPY
      . RADIATION THERAPY
RT   CANCER
      MEDICAL SCIENCE
      PATHOLOGY
      ∞RADIATION

RADIATION TOLERANCE
UF   RADIATION RESISTANCE
      RADIOSENSITIVITY
GS   SENSITIVITY
      . RADIATION TOLERANCE
      TOLERANCES (PHYSIOLOGY)
      . RADIATION TOLERANCE
RT   HUMAN TOLERANCES
      IRRADIATION
      ∞RADIATION
      ∞RESISTANCE
      TOLERANCES (MECHANICS)

RADIATION TRANSPORT
RT   EXPLODING WIRES
      ∞RADIATION
      RADIATIVE TRANSFER
      TRANSPORT PROPERTIES

RADIATION TRAPPING
RT   ARGON
      EXCITATION
      MAGNETOSPHERE
      METASTABLE STATE
      PLASMA PHYSICS
      ∞RADIATION

RADIATIVE HEAT TRANSFER
GS   RADIATIVE TRANSFER
      . RADIATIVE HEAT TRANSFER
      TRANSMISSION
      . HEAT TRANSMISSION
      . . HEAT TRANSFER
      . . . RADIATIVE HEAT TRANSFER
RT   CONCENTRATORS
      CONVECTIVE HEAT TRANSFER
      COOLING FINS
      HEAT RADIATORS
      INFRARED REFLECTION
      NEAR INFRARED RADIATION
      RADIANT COOLING
      RADIANT HEATING
      SATELLITE TEMPERATURE
      SPACECRAFT RADIATORS
      STEFAN-BOLTZMANN LAW
      SURFACE COOLING
      THERMOHYDRAULICS
      TROMBE WALLS
      VIEW EFFECTS

RADIATIVE LIFETIME
RT   DECAY
      HALF LIFE

RADIATIVE RECOMBINATION
GS   RECOMBINATION REACTIONS
      . ELECTRON RECOMBINATION
      . . RADIATIVE RECOMBINATION
      . ELECTRON-ION RECOMBINATION
      . . RADIATIVE RECOMBINATION
RT   AIRGLOW
      ATOMIC RECOMBINATION
      CARRIER INJECTION
      DEIONIZATION
      LIGHTNING

RADIATIVE TRANSFER
GS   RADIATIVE TRANSFER
      . RADIATIVE HEAT TRANSFER
RT   ATMOSPHERIC CORRECTION
      COSMIC RAYS
      ELECTROMAGNETIC RADIATION
      ENERGY TRANSFER
      EXTRATERRESTRIAL RADIATION

RADIATIVE TRANSFER-(CONT.)
      GALACTIC RADIATION
      HEAT TRANSFER
      HEAT TRANSMISSION
      INTERSTELLAR RADIATION
      NEAR INFRARED RADIATION
      PLANETARY ATMOSPHERES
      POLARIZED ELECTROMAGNETIC
        RADIATION
      RADIANT HEATING
      ∞RADIATION
      RADIATION TRANSPORT
      RADIO BURSTS
      RADIO STARS
      SOLAR RADIATION
      STELLAR ATMOSPHERES
      STELLAR RADIATION

∞ **RADIATORS**
SN   (USE OF A MORE SPECIFIC TERM IS
      RECOMMENDED--CONSULT THE TERMS
      LISTED BELOW)
RT   ANTENNAS
      HEAT RADIATORS
      SOUND TRANSDUCERS
      THERMOSIPHONS

RADICALS
GS   RADICALS
      . FORMYL IONS
      . FREE RADICALS
      . . HYDROXYL RADICALS
      . VANADYL RADICAL
      . VINYL RADICAL
RT   ∞ROOTS

RADII
UF   RADIUS
GS   DIMENSIONS
      . RADII
      . . LARMOR RADIUS
      GEOMETRY
      . EUCLIDEAN GEOMETRY
      . . RADII
RT   CIRCLES (GEOMETRY)
      CIRCUMFERENCES
      DIAMETERS
      LINES (GEOMETRY)
      RADIO FREQUENCIES
      SEGMENTS

RADIO ALTIMETERS
UF   RADAR ALTIMETERS
GS   AIRCRAFT INSTRUMENTS
      . ALTIMETERS
      . . RADIO ALTIMETERS
      FLIGHT INSTRUMENTS
      . RADIO ALTIMETERS
      MEASURING INSTRUMENTS
      . DISTANCE MEASURING EQUIPMENT
      . . ALTIMETERS
      . . . RADIO ALTIMETERS
RT   AUTOMATIC PILOTS
      INSTRUMENT LANDING SYSTEMS
      RADAR MEASUREMENT

RADIO ANTENNAS
GS   ANTENNAS
      . RADIO ANTENNAS
      . . MICROWAVE ANTENNAS
      . . . SPACETENNAS
      RADIO EQUIPMENT
      . RADIO ANTENNAS
RT   AIRCRAFT ANTENNAS
      DIRECTIONAL ANTENNAS
      OMNIDIRECTIONAL ANTENNAS
      RECEPTION DIVERSITY
      RHOMBIC ANTENNAS
      SATELLITE ANTENNAS
      SCHWARZSCHILD ANTENNAS
      TWO REFLECTOR ANTENNAS
      WHIP ANTENNAS

RADIO ASTRONOMY
GS   ASTRONOMY
      . RADIO ASTRONOMY
RT   ASTRONOMICAL OBSERVATORIES
      ASTRONOMICAL SPECTROSCOPY
      BRIGHTNESS DISTRIBUTION
      BRIGHTNESS TEMPERATURE
      CORONAL HOLES
      EXTRAGALACTIC RADIO SOURCES
      EXTRATERRESTRIAL RADIO WAVES
      GAMMA RAY ASTRONOMY
      IUE

RADIO ASTRONOMY-*(CONT.)*
- LINEAR POLARIZATION
- MAFFEI GALAXIES
- MICHELSON INTERFEROMETERS
- PHASE SWITCHING INTERFEROMETERS
- PULSARS
- QUASARS
- QUASAT
- RADAR ASTRONOMY
- SAS-2
- SAS-3
- ∞ SCIENCE
- VERY HIGH FREQUENCY RADIO
  - EQUIPMENT
- VERY LONG BASE INTERFEROMETRY

RADIO ASTRONOMY EXPLORER B
- USE   EXPLORER 49 SATELLITE

RADIO ASTRONOMY EXPLORER SATELLITE
- GS   SATELLITES
- . ARTIFICIAL SATELLITES
- . . EXPLORER SATELLITES
- . . . **RADIO ASTRONOMY EXPLORER
  SATELLITE**

RADIO ASTRONOMY EXPLORER 2
- USE   EXPLORER 49 SATELLITE

RADIO ATTENUATION
- UF   RADIO SIGNAL ATTENUATION
- GS   ATTENUATION
- . WAVE ATTENUATION
- . . **RADIO ATTENUATION**
- . . . MANDELSTAM REPRESENTATION
- RT   ATMOSPHERIC ATTENUATION
- ELECTROMAGNETIC ABSORPTION
- ELECTROMAGNETIC WAVE
  - TRANSMISSION
- GROUND EFFECT (COMMUNICATIONS)
- RADAR ATTENUATION
- SIGNAL TRANSMISSION
- TRANSHORIZON RADIO PROPAGATION
- TRANSMISSION
- WAVE PROPAGATION

RADIO ATTENUATION MEASUREMENT PROJECT
- UF   RAM PROJECT
- GS   PROGRAMS
- . PROJECTS
- . . **RADIO ATTENUATION
  MEASUREMENT PROJECT**

RADIO AURORAS
- GS   ATMOSPHERIC RADIATION
- . AURORAS
- . . **RADIO AURORAS**
- RT   ∞ DISTURBANCES
- IONOSPHERICS
- NIGHTGLOW
- SOLAR ACTIVITY

RADIO BEACONS
- UF   RADIO RANGES
- GS   NAVIGATION AIDS
- . BEACONS
- . . **RADIO BEACONS**
- . . . OMNIDIRECTIONAL RADIO RANGES
- . . . . SELF CALIBRATING OMNIRANGE
- RADIO EQUIPMENT
- . RADIO TRANSMITTERS
- . . **RADIO BEACONS**
- . . . OMNIDIRECTIONAL RADIO RANGES
- . . . . SELF CALIBRATING OMNIRANGE
- TRANSMITTERS
- . RADIO TRANSMITTERS
- . . **RADIO BEACONS**
- . . . OMNIDIRECTIONAL RADIO RANGES
- . . . . SELF CALIBRATING OMNIRANGE
- RT   AIRPORT BEACONS
- BEACON COLLISION AVOIDANCE
  - SYSTEM
- HOMING DEVICES
- INSTRUMENT LANDING SYSTEMS
- LANDING AIDS
- ∞ MARKERS
- NIGHT FLIGHTS (AIRCRAFT)
- ORBIS
- ORBIS CAL SATELLITE
- RANGES (FACILITIES)
- SOLAR COMPASSES

RADIO BROADCASTING
- USE   BROADCASTING

RADIO BURSTS
- GS   BURSTS
- . **RADIO BURSTS**
- . . SOLAR RADIO BURSTS
- . . . TYPE 2 BURSTS
- . . . TYPE 3 BURSTS
- . . . TYPE 4 BURSTS
- . . . TYPE 5 BURSTS
- ELECTROMAGNETIC RADIATION
- . RADIO WAVES
- . . EXTRATERRESTRIAL RADIO WAVES
- . . . **RADIO BURSTS**
- . . . . SOLAR RADIO BURSTS
- . . . . . TYPE 2 BURSTS
- . . . . . TYPE 3 BURSTS
- . . . . . TYPE 4 BURSTS
- . . . . . TYPE 5 BURSTS
- . RADIO EMISSION
- . . . **RADIO BURSTS**
- . . . . SOLAR RADIO BURSTS
- . . . . . TYPE 2 BURSTS
- . . . . . TYPE 3 BURSTS
- . . . . . TYPE 4 BURSTS
- . . . . . TYPE 5 BURSTS
- EXTRATERRESTRIAL RADIATION
- . EXTRATERRESTRIAL RADIO WAVES
- . . **RADIO BURSTS**
- . . . SOLAR RADIO BURSTS
- . . . . TYPE 2 BURSTS
- . . . . TYPE 3 BURSTS
- . . . . TYPE 4 BURSTS
- . . . . TYPE 5 BURSTS
- RT   ∞ DISTURBANCES
- PULSARS
- QUASARS
- RADIATIVE TRANSFER
- SOLAR RADIO EMISSION
- STELLAR RADIATION

RADIO COMMUNICATION
- GS   TELECOMMUNICATION
- . **RADIO COMMUNICATION**
- . . RADIO RELAY SYSTEMS
- . . . CODE DIVISION MULTIPLE ACCESS
- . . . TIME DIVISION MULTIPLE ACCESS
- . . RADIO TELEGRAPHY
- . . RADIO TELEMETRY
- . . . PULSE FREQUENCY MODULATION
  - TELEMETRY
- . . TELEPHONY
- RT   ACCESS CONTROL
- AIRCRAFT COMMUNICATION
- BLACKOUT (PROPAGATION)
- BROADCASTING
- CIRCUMLUNAR COMMUNICATION
- CODE DIVISION MULTIPLEXING
- COMMUNICATION EQUIPMENT
- COMMUNICATION NETWORKS
- FLEET SATELLITE COMMUNICATION
  - SYSTEM
- FREQUENCY DIVISION MULTIPLE
  - ACCESS
- FREQUENCY DIVISION MULTIPLEXING
- GROUND-AIR-GROUND COMMUNICATION
- INTERPLANETARY COMMUNICATION
- INTERSTELLAR COMMUNICATION
- LAND MOBILE SATELLITE SERVICE
- LUNAR COMMUNICATION
- MARISAT SATELLITES
- MARISAT 1 SATELLITE
- MOBILE COMMUNICATION SYSTEMS
- MSAT
- NASCOM NETWORK
- POINT TO POINT COMMUNICATION
- PROJECT SETI
- PULSE COMMUNICATION
- RADIOTELEPHONES
- REENTRY COMMUNICATION
- SHIP TO SHORE COMMUNICATION
- SPACE COMMUNICATION
- SPACECRAFT COMMUNICATION
- TELEGRAPH SYSTEMS
- TELEMETRY
- TELEVISION SYSTEMS
- TRANSOCEANIC COMMUNICATION
- UNDERGROUND COMMUNICATION
- VOCODERS
- VOICE COMMUNICATION

RADIO CONTROL
- GS   REMOTE CONTROL
- . **RADIO CONTROL**
- RT   AIRCRAFT CONTROL
- AUTOMATIC CONTROL
- ∞ CONTROL

RADIO CONTROL-*(CONT.)*
- DEEP SPACE INSTRUMENTATION
  - FACILITY
- GROUND BASED CONTROL
- MISSILE CONTROL
- SPACECRAFT CONTROL

RADIO DIRECTION FINDERS
- UF   DIRECTION FINDERS (RADIO)
- RADAR DIRECTION FINDERS
- GS   AIRCRAFT INSTRUMENTS
- . POSITION INDICATORS
- . . **RADIO DIRECTION FINDERS**
- DISPLAY DEVICES
- . POSITION INDICATORS
- . . **RADIO DIRECTION FINDERS**
- MEASURING INSTRUMENTS
- . INDICATING INSTRUMENTS
- . . POSITION INDICATORS
- . . . **RADIO DIRECTION FINDERS**
- NAVIGATION AIDS
- . BEACONS
- . . **RADIO DIRECTION FINDERS**
- . NAVIGATION INSTRUMENTS
- . . **RADIO DIRECTION FINDERS**
- RT   AIRCRAFT EQUIPMENT
- COMPASSES
- DIRECTION FINDING
- FLIGHT INSTRUMENTS
- GYROCOMPASSES
- HOMING
- HOMING DEVICES
- RADIOGONIOMETERS
- VHF OMNIRANGE NAVIGATION

RADIO ECHOES
- UF   RADIO REFLECTION
- GS   ECHOES
- . **RADIO ECHOES**
- RT   ANGELS
- AURORAL ECHOES
- GHOSTS
- HARVARD RADIO METEOR PROJECT
- INFRARED REFLECTION
- LUNAR ECHOES
- RADAR REFLECTORS
- ULTRAVIOLET REFLECTION

RADIO ELECTRONICS
- RT   ∞ ELECTRONICS

RADIO EMISSION
- GS   DECAY
- . EMISSION
- . . **RADIO EMISSION**
- . . . CN EMISSION
- . . . HYDROXYL EMISSION
- . . . SOLAR RADIO EMISSION
- . . . . SOLAR RADIO BURSTS
- . . . . . TYPE 2 BURSTS
- . . . . . TYPE 3 BURSTS
- . . . . . TYPE 4 BURSTS
- . . . . . TYPE 5 BURSTS
- ELECTROMAGNETIC RADIATION
- . RADIO WAVES
- . . **RADIO EMISSION**
- . . . CN EMISSION
- . . . HYDROXYL EMISSION
- . . . RADIO BURSTS
- . . . . SOLAR RADIO BURSTS
- . . . . . TYPE 2 BURSTS
- . . . . . TYPE 3 BURSTS
- . . . . . TYPE 4 BURSTS
- . . . . . TYPE 5 BURSTS
- . . . SOLAR RADIO EMISSION
- . . . . SOLAR RADIO BURSTS
- . . . . . TYPE 2 BURSTS
- . . . . . TYPE 3 BURSTS
- . . . . . TYPE 4 BURSTS
- . . . . . TYPE 5 BURSTS
- RT   EXTRAGALACTIC RADIO SOURCES
- EXTRATERRESTRIAL RADIO WAVES
- QUASARS

RADIO EQUIPMENT
- GS   **RADIO EQUIPMENT**
- . RADIO ANTENNAS
- . RADIO FILTERS
- . RADIO RECEIVERS
- . . SUPERHETERODYNE RECEIVERS
- . . TRANSMITTER RECEIVERS
- . . WHISTLER RECORDERS
- . RADIO TELESCOPES
- . RADIO TRANSMITTERS
- . . RADIO BEACONS

RADIO EQUIPMENT-*(CONT.)*
  . . . OMNIDIRECTIONAL RADIO RANGES
  . . . . SELF CALIBRATING OMNIRANGE
  . . RADIOSONDES
  . . . ENDORADIOSONDES
  . . . IONOSONDES
  . . . RAWINSONDES
  . . RADIOTELEPHONES
  . . SONOBUOYS
  . . TRANSMITTER RECEIVERS
  . RECEPTION DIVERSITY
  . SPACECRAFT ANTENNAS
  . TRANSPONDERS
  . VERY HIGH FREQUENCY RADIO
     EQUIPMENT
RT   AIRBORNE EQUIPMENT
    ANTENNAS
    BROADCASTING
    COMMUNICATION EQUIPMENT
    CRYSTAL FILTERS
    CYLINDRICAL ANTENNAS
    JAMMERS
    NEAR FIELDS
    ONBOARD EQUIPMENT
    RADAR EQUIPMENT

RADIO FILTERS
GS   ELECTROMAGNETIC WAVE FILTERS
    . ELECTRIC FILTERS
    . . **RADIO FILTERS**
    RADIO EQUIPMENT
    . **RADIO FILTERS**
RT   CRYSTAL FILTERS
    ∞FILTERS
    INTERFERENCE GRATING
    MICROWAVE FILTERS
    RADAR FILTERS

RADIO FREQUENCIES
GS   FREQUENCIES
    . **RADIO FREQUENCIES**
    . . EXTREMELY LOW RADIO
      FREQUENCIES
    . . HIGH FREQUENCIES
    . . LOW FREQUENCIES
    . . . SUBAUDIBLE FREQUENCIES
    . . . VERY LOW FREQUENCIES
    . . LOW FREQUENCY BANDS
    . . . VERY LOW FREQUENCIES
    . . MICROWAVE FREQUENCIES
    . . . C BAND
    . . . EXTREMELY HIGH FREQUENCIES
    . . . P BAND
    . . . SUPERHIGH FREQUENCIES
    . . ULTRAHIGH FREQUENCIES
    . . . P BAND
    . . VERY HIGH FREQUENCIES
    . . . P BAND
RT   AUDIO FREQUENCIES
    CARRIER FREQUENCIES
    EXTREMELY LOW FREQUENCIES
    INTERMEDIATE FREQUENCIES
    RADII

RADIO FREQUENCY DISCHARGE
GS   ELECTRIC CURRENT
    . ELECTRIC DISCHARGES
    . . **RADIO FREQUENCY DISCHARGE**
RT   ELECTRODELESS DISCHARGES
    ELECTRON EMISSION
    RING DISCHARGE

RADIO FREQUENCY HEATING
GS   HEATING
    . **RADIO FREQUENCY HEATING**
RT   INDUCTION HEATING
    PLASMA HEATING

RADIO FREQUENCY IMPEDANCE PROBES
GS   MEASURING INSTRUMENTS
    . IMPEDANCE PROBES
    . . **RADIO FREQUENCY IMPEDANCE
      PROBES**
RT   IMPEDANCE MEASUREMENT
    ION PROBES
    MICROWAVE PROBES
    PLASMA PROBES

RADIO FREQUENCY INTERFERENCE
UF   RADIO INTERFERENCE
GS   ELECTROMAGNETIC INTERFERENCE
    . **RADIO FREQUENCY INTERFERENCE**
    . . BLACKOUT (PROPAGATION)
    . . . ELECTROMAGNETIC NOISE
    . . . . ATMOSPHERICS

RADIO FREQUENCY INTERFERENCE-*(CONT.)*
    . . . . . IONOSPHERICS
    . . . . . . DAWN CHORUS
    . . . . . . HISS
    . . . . . SUDDEN ENHANCEMENT OF
       ATMOSPHERICS
    . . . . . WHISTLERS
    . . . . COSMIC NOISE
    . . . IONOSPHERIC NOISE
    . . . . WHISTLERS
    . . . SHOT NOISE
    . . . . WHITE NOISE
    . . . . . THERMAL NOISE
    . . IONOSPHERIC CROSS MODULATION
    . . . POLAR RADIO BLACKOUT
    . . CHIRP
    . . . CHIRP SIGNALS
RT   CLUTTER
    CROSS COUPLING
    ELECTROMAGNETIC COMPATIBILITY
    ELECTRONIC COUNTERMEASURES
    ELECTRONIC WARFARE
    EXTRATERRESTRIAL RADIO WAVES
    ∞INTERFERENCE
    INTERFERENCE GRATING
    INTERFERENCE IMMUNITY
    JAMMING
    NOISE GENERATORS
    NOISE STORMS
    SIGNAL FADING

RADIO FREQUENCY ION THRUSTOR ENGINES
USE   RIT ENGINES

RADIO FREQUENCY NOISE
USE   ELECTROMAGNETIC NOISE

RADIO FREQUENCY RADIATION
USE   RADIO WAVES

RADIO FREQUENCY SHIELDING
GS   SHIELDING
    . ELECTROMAGNETIC SHIELDING
    . . **RADIO FREQUENCY SHIELDING**
RT   RADIATION SHIELDING
    SPACECRAFT SHIELDING

RADIO GALAXIES
GS   CELESTIAL BODIES
    . GALAXIES
    . . **RADIO GALAXIES**
    . RADIO SOURCES (ASTRONOMY)
    . . EXTRAGALACTIC RADIO SOURCES
    . . . **RADIO GALAXIES**
RT   DISK GALAXIES
    MAFFEI GALAXIES
    QUASARS

RADIO HORIZONS
GS   HORIZON
    . **RADIO HORIZONS**
RT   HORIZON SCANNERS

RADIO INTERFERENCE
USE   RADIO FREQUENCY INTERFERENCE

RADIO INTERFEROMETERS
GS   MEASURING INSTRUMENTS
    . INTERFEROMETERS
    . . **RADIO INTERFEROMETERS**
RT   ASTROPHYSICS
    ORION (RADIO INTERFEROMETRY
      NETWORK)
    VERY LONG BASE INTERFEROMETRY

†

RADIO METEOROLOGY
GS   METEOROLOGY
    . **RADIO METEOROLOGY**
RT   ATMOSPHERICS
    METEOROLOGICAL RADAR
    RADIOSONDES

RADIO METEORS
GS   CELESTIAL BODIES
    . METEOROIDS
    . . **RADIO METEORS**
RT   ATMOSPHERIC IONIZATION
    METEOR TRAILS

RADIO NAVIGATION
GS   NAVIGATION
    . **RADIO NAVIGATION**
    . . HYPERBOLIC NAVIGATION
    . . . DECCA NAVIGATION

RADIO NAVIGATION-*(CONT.)*
    . . . LORAC NAVIGATION SYSTEM
    . . . LORAN
    . . . . LORAN C
    . . . . LORAN D
    . . . SHORAN
    . . TACAN
    . . VHF OMNIRANGE NAVIGATION
RT   AIR NAVIGATION
    AIR TRAFFIC CONTROL
    AIRCRAFT GUIDANCE
    ALL-WEATHER AIR NAVIGATION
    ASTRONAVIGATION
    AUTOMATIC FLIGHT CONTROL
    CELESTIAL NAVIGATION
    COLLISION AVOIDANCE
    DEAD RECKONING
    DISTANCE MEASURING EQUIPMENT
    DOPPLER NAVIGATION
    FLIGHT CONTROL
    GUIDANCE (MOTION)
    HOMING DEVICES
    INERTIAL NAVIGATION
    INTERPLANETARY NAVIGATION
    NAVIGATION AIDS
    OMNIDIRECTIONAL RADIO RANGES
    POSITIONING
    RADAR NAVIGATION
    SATELLITE NAVIGATION SYSTEMS
    SOLAR COMPASSES
    SPACE NAVIGATION
    SURFACE NAVIGATION

RADIO OBSERVATION
RT   OBSERVATION
    SPACE OBSERVATIONS (FROM EARTH)

RADIO OCCULTATION
GS   OCCULTATION
    . **RADIO OCCULTATION**
RT   ATMOSPHERIC COMPOSITION
    ATMOSPHERIC PRESSURE
    ATMOSPHERIC TEMPERATURE
    PLANETARY ATMOSPHERES
    SPACE PROBES
    SPACECRAFT TRAJECTORIES

RADIO PHYSICS
RT   ∞PHYSICS
    ∞SCIENCE
    THEORETICAL PHYSICS

RADIO PROBING
RT   MEASURING INSTRUMENTS
    ∞PROBES

RADIO PROPAGATION
USE   RADIO TRANSMISSION

RADIO RANGE
SN   (EXCLUDES RADIO BEACONS)
GS   DISTANCE
    . **RADIO RANGE**
    RANGE (EXTREMES)
    . FREQUENCY RANGES
    . . **RADIO RANGE**
RT   RADAR RANGE

RADIO RANGES
USE   RADIO BEACONS

RADIO RECEIVERS
GS   COMMUNICATION EQUIPMENT
    . **RADIO RECEIVERS**
    . . SUPERHETERODYNE RECEIVERS
    . . TRANSMITTER RECEIVERS
    . . WHISTLER RECORDERS
    RADIO EQUIPMENT
    . **RADIO RECEIVERS**
    . . SUPERHETERODYNE RECEIVERS
    . . TRANSMITTER RECEIVERS
    . . WHISTLER RECORDERS
    RECEIVERS
    . **RADIO RECEIVERS**
    . . SUPERHETERODYNE RECEIVERS
RT   DATA LINKS
    DIRECTORS (ANTENNA ELEMENTS)
    ELECTROMAGNETIC NOISE
    INTERMEDIATE FREQUENCY AMPLIFIERS
    LOUDSPEAKERS
    ORION (RADIO INTERFEROMETRY
      NETWORK)
    RADAR RECEIVERS
    RECEPTION DIVERSITY
    TELEVISION RECEPTION

RADIO RECEIVERS-*(CONT.)*
    TRANSPONDERS
    TUNERS

RADIO RECEPTION
  RT   HOMODYNE RECEPTION
      RADAR RECEPTION
     ∞ RECEIVING
      RECEPTION DIVERSITY
      SCATTER PROPAGATION
      TELEVISION RECEPTION

RADIO REFLECTION
  USE RADIO ECHOES

RADIO RELAY SYSTEMS
  GS   TELECOMMUNICATION
     . RADIO COMMUNICATION
     . . **RADIO RELAY SYSTEMS**
     . . . CODE DIVISION MULTIPLE ACCESS
     . . . TIME DIVISION MULTIPLE ACCESS
  RT   COMMUNICATION EQUIPMENT
      COMMUNICATION SATELLITES
      DATA LINKS
      DEFENSE COMMUNICATIONS SATELLITE
        SYSTEM
      EARTH TERMINAL MEASUREMENT
        SYSTEM
      EARTH TERMINALS
      GLOBAL TRACKING NETWORK
      MOLNIYA SATELLITES
      MSAT
      ORBIT SPECTRUM UTILIZATION
     ∞ RELAY
     ∞ SYSTEMS
      TDR SATELLITES

RADIO SCATTERING
  RT   ATMOSPHERIC DIFFUSION
      ATMOSPHERIC SCATTERING
      SCATTER PROPAGATION
      SIGNAL FADING
      SIGNAL TRANSMISSION

RADIO SIGNAL ATTENUATION
  USE RADIO ATTENUATION

RADIO SIGNAL PROPAGATION
  USE RADIO TRANSMISSION

RADIO SIGNALS
  RT   BROADCASTING
      PROJECT SETI
      SIGNAL DISTORTION
      SIGNAL MIXING
     ∞ SIGNALS
      TRANSHORIZON RADIO PROPAGATION
      WHISTLERS

RADIO SOURCES (ASTRONOMY)
  SN   (LIMITED TO EXTRATERRESTRIAL RADIO
      SOURCES)
  GS   CELESTIAL BODIES
     . **RADIO SOURCES (ASTRONOMY)**
     . . CASSIOPEIA A
     . . EXTRAGALACTIC RADIO SOURCES
     . . . RADIO GALAXIES
     . . QUASARS
     . . RADIO STARS
     . . . PULSARS
  RT   BL LACERTAE OBJECTS
      CN EMISSION
      EXTRATERRESTRIAL RADIO WAVES
      GALACTIC NUCLEI
      GALAXIES
      HYDROXYL EMISSION
      MAFFEI GALAXIES
      MILKY WAY GALAXY
      RADIATION SOURCES
     ∞ SOURCES

RADIO SPECTRA
  GS   SPECTRA
     . RADIATION SPECTRA
     . . ELECTROMAGNETIC SPECTRA
     . . . **RADIO SPECTRA**
     . . . . MICROWAVE SPECTRA
  RT   CARRIER WAVES
      ELECTROMAGNETIC NOISE
      RADIANT FLUX DENSITY

RADIO SPECTROSCOPY
  GS   SPECTROSCOPY
     . **RADIO SPECTROSCOPY**

RADIO SPECTROSCOPY-*(CONT.)*
  RT   ASTRONOMICAL SPECTROSCOPY
      ULTRAVIOLET SPECTRA
      ULTRAVIOLET SPECTROSCOPY
      X RAY SPECTROSCOPY

RADIO STARS
  GS   CELESTIAL BODIES
     . RADIO SOURCES (ASTRONOMY)
     . . **RADIO STARS**
     . . . PULSARS
     . STARS
     . . **RADIO STARS**
     . . . PULSARS
  RT   QUASARS
      RADIATIVE TRANSFER
      STELLAR RADIATION

RADIO TELEGRAPHY
  GS   TELECOMMUNICATION
     . RADIO COMMUNICATION
     . . **RADIO TELEGRAPHY**
  RT   COMMUNICATION EQUIPMENT
      KEYING
      MORSE CODE

RADIO TELEMETRY
  GS   TELECOMMUNICATION
     . RADIO COMMUNICATION
     . . **RADIO TELEMETRY**
     . . . PULSE FREQUENCY MODULATION
        TELEMETRY
     . . TELEMETRY
     . . **RADIO TELEMETRY**
     . . . PULSE FREQUENCY MODULATION
        TELEMETRY
  RT   COMMUNICATION EQUIPMENT
      DATA TRANSMISSION
      EXTRATERRESTRIAL COMMUNICATION
      GROUND SUPPORT EQUIPMENT
      MEASURING INSTRUMENTS
      PULSE MODULATION
      RADIOMETEOROGRAPHS
      RADIOSONDES
      SPACE COMMUNICATION
      WIRELESS COMMUNICATION

RADIO TELESCOPES
  GS   RADIO EQUIPMENT
     . **RADIO TELESCOPES**
     TELESCOPES
     . **RADIO TELESCOPES**
     . . KILOMETER WAVE ORBITING
        TELESCOPE
  RT   ANTENNAS
      ASTRONOMICAL TELESCOPES
      JODRELL BANK OBSERVATORY
      OPTICAL EQUIPMENT
      PHASE SWITCHING INTERFEROMETERS
      QUASAT

RADIO TRACKING
  GS   TRACKING (POSITION)
     . **RADIO TRACKING**
     . . WILDLIFE RADIOLOCATION
  RT   RADAR SCANNING
      RADAR TRACKING
      RANGE AND RANGE RATE TRACKING
      RANGEFINDING
      RAWINSONDES
      SPACECRAFT TRACKING

RADIO TRANSMISSION
  UF   RADIO PROPAGATION
      RADIO SIGNAL PROPAGATION
  GS   TRANSMISSION
     . ELECTROMAGNETIC WAVE
        TRANSMISSION
     . . **RADIO TRANSMISSION**
     . . . DOUBLE SIDEBAND TRANSMISSION
     . . . IONOSPHERIC PROPAGATION
     . . . . IONOSPHERIC F-SCATTER
        PROPAGATION
     . . . MICROWAVE ATTENUATION
     . . . MICROWAVE TRANSMISSION
     . . . MULTIPATH TRANSMISSION
     . . . SHORT WAVE RADIO
        TRANSMISSION
     . . . SINGLE SIDEBAND TRANSMISSION
     . . . SPREAD SPECTRUM TRANSMISSION
     . . . TRANSEQUATORIAL PROPAGATION
     . . . TRANSHORIZON RADIO
        PROPAGATION
     . SIGNAL TRANSMISSION
     . . **RADIO TRANSMISSION**

RADIO TRANSMISSION-*(CONT.)*
     . . . DOUBLE SIDEBAND TRANSMISSION
     . . . IONOSPHERIC PROPAGATION
     . . . . IONOSPHERIC F-SCATTER
        PROPAGATION
     . . . MICROWAVE ATTENUATION
     . . . MICROWAVE TRANSMISSION
     . . . MULTIPATH TRANSMISSION
     . . . SHORT WAVE RADIO
        TRANSMISSION
     . . . SINGLE SIDEBAND TRANSMISSION
     . . . TRANSEQUATORIAL PROPAGATION
     . . . TRANSHORIZON RADIO
        PROPAGATION
  RT   ANTIPODES
      ATMOSPHERIC ATTENUATION
      BROADCASTING
      CODE DIVISION MULTIPLEXING
      COMPANDING
      DATA TRANSMISSION
      FREQUENCY REUSE
      FREQUENCY SHIFT KEYING
      MAGNETOIONICS
      MODULATION
      MULTIPLEXING
      PACKET SWITCHING
      PULSE COMMUNICATION
      PULSE FREQUENCY MODULATION
        TELEMETRY
      RADAR TRANSMISSION
      RADOME MATERIALS
      SATELLITE TRANSMISSION
      SCATTER PROPAGATION
      SEAFARER PROJECT
      SOMMERFELD APPROXIMATION
      SYMPHONIE SATELLITES
      VOICE OF AMERICA
      WAVE ATTENUATION
      WAVE PROPAGATION

RADIO TRANSMITTERS
  GS   RADIO EQUIPMENT
     . **RADIO TRANSMITTERS**
     . . RADIO BEACONS
     . . . OMNIDIRECTIONAL RADIO RANGES
     . . . . SELF CALIBRATING OMNIRANGE
     . . RADIOSONDES
     . . . ENDORADIOSONDES
     . . . IONOSONDES
     . . . RAWINSONDES
     . . RADIOTELEPHONES
     . . SONOBUOYS
     . . TRANSMITTER RECEIVERS
     TRANSMITTERS
     . **RADIO TRANSMITTERS**
     . . RADIO BEACONS
     . . . OMNIDIRECTIONAL RADIO RANGES
     . . . . SELF CALIBRATING OMNIRANGE
     . . RADIOMETEOROGRAPHS
     . . RADIOSONDES
     . . . ENDORADIOSONDES
     . . . IONOSONDES
     . . . RAWINSONDES
     . . RADIOTELEPHONES
     . . SONOBUOYS
     . . TRANSMITTER RECEIVERS
  RT   MULTICHANNEL COMMUNICATION
      QUARTZ CRYSTALS
      TELEVISION TRANSMISSION
      TRANSPONDERS
      WILDLIFE RADIOLOCATION

RADIO WAVE REFRACTION
  GS   REFRACTION
     . ATMOSPHERIC REFRACTION
     . . **RADIO WAVE REFRACTION**
  RT   WAVE DISPERSION

RADIO WAVES
  UF   RADIO FREQUENCY RADIATION
  GS   ELECTROMAGNETIC RADIATION
     . **RADIO WAVES**
     . . DECAMETRIC WAVES
     . . EXTRATERRESTRIAL RADIO WAVES
     . . . GALACTIC RADIO WAVES
     . . RADIO BURSTS
     . . . SOLAR RADIO BURSTS
     . . . . TYPE 2 BURSTS
     . . . . TYPE 3 BURSTS
     . . . . TYPE 4 BURSTS
     . . . . TYPE 5 BURSTS
     . . SOLAR RADIO EMISSION
     . . . SOLAR RADIO BURSTS
     . . . . TYPE 2 BURSTS
     . . . . TYPE 3 BURSTS

RADIO WAVES-(CONT.)
```
. . . . .   TYPE 4 BURSTS
. . . . .   TYPE 5 BURSTS
. .   LONG WAVE RADIATION
. .   RADIO EMISSION
. . .   CN EMISSION
. . .   HYDROXYL EMISSION
. . .   RADIO BURSTS
. . . .   SOLAR RADIO BURSTS
. . . . .   TYPE 2 BURSTS
. . . . .   TYPE 3 BURSTS
. . . . .   TYPE 4 BURSTS
. . . . .   TYPE 5 BURSTS
. . .   SOLAR RADIO EMISSION
. . . .   SOLAR RADIO BURSTS
. . . . .   TYPE 2 BURSTS
. . . . .   TYPE 3 BURSTS
. . . . .   TYPE 4 BURSTS
. . . . .   TYPE 5 BURSTS
. .   SHORT WAVE RADIATION
. .   MICROWAVES
. . . .   CENTIMETER WAVES
. . . .   DECIMETER WAVES
. . . .   MICROWAVE EMISSION
. . . .   MILLIMETER WAVES
. . .   SUBMILLIMETER WAVES
. .   SKY WAVES
```
RT   ATMOSPHERICS
     COHERENT ELECTROMAGNETIC
      RADIATION
     ELECTROMAGNETIC NOISE
     ELECTROMAGNETIC SURFACE WAVES
     EXTRATERRESTRIAL RADIATION
     FAR INFRARED RADIATION
     FREQUENCIES
     GROUND WAVE PROPAGATION
     MONOCHROMATIC RADIATION
     MULTIPATH TRANSMISSION
     PLANETARY RADIATION
     POLARIZED ELECTROMAGNETIC
      RADIATION
     SCATTER PROPAGATION
     SOLAR RADIATION
     SOLITARY WAVES
     THERMAL RADIATION
     TRANSVERSE WAVES
     TRAVELING WAVES
     TROPOSPHERIC WAVES

RADIOACTIVE AGE DETERMINATION
UF   RADIOACTIVE DATING
RT   ∞AGING
     FOSSILS
     GEOCHRONOLOGY
     HALF LIFE
     ∞MEASUREMENT
     RADIOCHEMISTRY
     RADIOGENIC MATERIALS
     TIME MEASUREMENT

RADIOACTIVE CONTAMINANTS
GS   CONTAMINANTS
    . **RADIOACTIVE CONTAMINANTS**
RT   ATMOSPHERIC COMPOSITION
     FALLOUT
     NUCLEAR RADIATION
     RADIATION EFFECTS
     RADIATION HAZARDS

RADIOACTIVE DATING
USE   RADIOACTIVE AGE DETERMINATION

∞ RADIOACTIVE DEBRIS
SN   *(USE OF A MORE SPECIFIC TERM IS*
     RECOMMENDED--CONSULT THE TERMS
     LISTED BELOW)
RT   DEBRIS
     FALLOUT
     RADIOACTIVE MATERIALS
     RADIOACTIVE WASTES
     RADIOGENIC MATERIALS

RADIOACTIVE DECAY
UF   PARTICLE DECAY
GS   DECAY
    . **RADIOACTIVE DECAY**
    . . ALPHA DECAY
    . . NEUTRON EMISSION
    NUCLEAR REACTIONS
    . **RADIOACTIVE DECAY**
    . . NEUTRON EMISSION
RT   EMISSION
     GAMMA RAY BEAMS
     GAMMA RAYS
     HALF LIFE

RADIOACTIVE DECAY-(CONT.)
    HYPERNUCLEI
    NUCLEAR FISSION
    NUCLEAR RADIATION
    PHOTOPRODUCTION
    POST-BLAST NUCLEAR RADIATION
    RADIOACTIVITY
    RADIOGENIC MATERIALS
    THERMONUCLEAR REACTIONS
    VECTOR CURRENTS

RADIOACTIVE ELEMENTS
USE   RADIOACTIVE ISOTOPES

RADIOACTIVE ISOTOPES
UF   RADIOACTIVE ELEMENTS
     RADIOACTIVE NUCLIDES
     RADIONUCLIDES
GS   CHEMICAL ELEMENTS
    . NUCLIDES
    . . ISOTOPES
    . . . **RADIOACTIVE ISOTOPES**
```
. . . .   ARSENIC ISOTOPES
. . . .   ASTATINE ISOTOPES
. . .   BERYLLIUM 7
. . .   BERYLLIUM 9
. . .   BERYLLIUM 10
. . .   CARBON 14
. . .   CERIUM 137
. . .   CERIUM 144
. . .   CESIUM 134
. . .   CESIUM 137
. . .   CESIUM 144
. . .   COBALT 58
. . .   COBALT 60
. . .   GOLD ISOTOPES
. . . .   GOLD 198
. . .   INDIUM ISOTOPES
. . .   IODINE 125
. . .   IODINE 131
. . .   IODINE 132
. . .   IRON 59
. . .   KRYPTON 85
. . .   NIOBIUM 95
. . .   NITROGEN 16
. . .   PHOSPHORUS 32
. . .   POLONIUM 208
. . .   POLONIUM 209
. . .   POLONIUM 210
. . .   POTASSIUM 38
. . .   POTASSIUM 40
. . .   RUBIDIUM 86
. . .   SODIUM 22
. . .   SODIUM 24
. . .   STRONTIUM 85
. . .   STRONTIUM 88
. . .   STRONTIUM 89
. . .   STRONTIUM 90
. . .   TRANSURANIUM ELEMENTS
. . . .   AMERICIUM
. . . . .   AMERICIUM ISOTOPES
. . . .   BERKELIUM
. . . .   CALIFORNIUM
. . . . .   CALIFORNIUM ISOTOPES
. . . .   CURIUM
. . . . .   CURIUM ISOTOPES
. . . .   EINSTEINIUM
. . . .   FERMIUM
. . . .   LAWRENCIUM
. . . .   MENDELEVIUM
. . . .   NEPTUNIUM
. . . . .   NEPTUNIUM ISOTOPES
. . . .   NOBELIUM
. . . .   PLUTONIUM
. . . . .   PLUTONIUM ISOTOPES
. . . .   SERGENIUM
. . . .   TRITIUM
. . .   URANIUM 232
. . .   URANIUM 233
. . .   URANIUM 238
. . .   XENON 133
. . .   XENON 135
. . .   ZIRCONIUM 95
```
RT   ACTINIDE SERIES
     ISOTOPIC LABELING
     RADIOISOTOPE BATTERIES
     RADIOPHOSPHORS

RADIOACTIVE MATERIALS
RT   ACTINIDE SERIES
     FISSILE FUELS
     FISSION PRODUCTS
     FISSIONABLE MATERIALS
     IONIZING RADIATION
     ISOTOPES

RADIOACTIVE MATERIALS-(CONT.)
    ∞MATERIALS
    NUCLEAR FISSION
    NUCLEAR RADIATION
    RADIATION HAZARDS
    RADIATION SOURCES
    ∞RADIOACTIVE DEBRIS
    RADIOACTIVITY
    RADIOBIOLOGY
    RADIOCHEMISTRY
    URANIUM PLASMAS

RADIOACTIVE NUCLIDES
USE   RADIOACTIVE ISOTOPES

RADIOACTIVE WASTES
UF   NUCLEAR WASTES
GS   WASTES
    . **RADIOACTIVE WASTES**
RT   CONTAMINATION
     DECOMMISSIONING
     ENVIRONMENT POLLUTION
     ENVIRONMENT PROTECTION
     ENVIRONMENTAL SURVEYS
     FISSION PRODUCTS
     HAZARDOUS MATERIAL DISPOSAL (IN
      SPACE)
     NONPOINT SOURCES
     PLASMA CORE REACTORS
     POISONING (REACTION INHIBITION)
     POLLUTION
     RADIATION HAZARDS
     ∞RADIOACTIVE DEBRIS
     RADIOCHEMISTRY
     RADIOGENIC MATERIALS
     SOLID WASTES
     WASTE DISPOSAL

RADIOACTIVITY
RT   ∞ACTIVITY
     ALPHA PARTICLES
     EMISSION
     FALLOUT
     FISSION PRODUCTS
     GAMMA RAYS
     GEOCHEMISTRY
     GEOPHYSICS
     HALF LIFE
     IONIZING RADIATION
     NUCLEAR RADIATION
     PARTICLE PRODUCTION
     POST-BLAST NUCLEAR RADIATION
     ∞RADIATION
     RADIATION HAZARDS
     RADIOACTIVE DECAY
     RADIOACTIVE MATERIALS
     RADIOCHEMISTRY

RADIOBIOLOGY
GS   MEDICAL SCIENCE
    . NUCLEAR MEDICINE
    . . **RADIOBIOLOGY**
RT   ANTIRADIATION DRUGS
     ∞BIOLOGY
     BIOMAGNETISM
     DOSIMETERS
     HEALTH PHYSICS
     IMMUNOASSAY
     IRRADIATION
     ∞MEDICINE
     NUCLEAR RADIATION
     RADIOACTIVE MATERIALS
     RADIOCHEMISTRY
     RADIOIMMUNOASSAY

RADIOCARDIOGRAPHY
GS   BIOENGINEERING
    . BIOMETRICS
    . . **RADIOCARDIOGRAPHY**
    CARDIOLOGY
    . **RADIOCARDIOGRAPHY**
RT   CARDIOLOGY

RADIOCHEMICAL SEPARATION
GS   RADIOCHEMISTRY
    . **RADIOCHEMICAL SEPARATION**
RT   CHEMICAL REACTIONS
     QUANTITATIVE ANALYSIS
     ∞SEPARATION

RADIOCHEMISTRY
UF   REACTOR CHEMISTRY
GS   **RADIOCHEMISTRY**
    . RADIOCHEMICAL SEPARATION
RT   CHEMICAL ANALYSIS

RADIOCHEMISTRY-(CONT.)
∞ CHEMISTRY
IONIZING RADIATION
ISOTOPIC LABELING
NUCLEAR CHEMISTRY
NUCLEAR RADIATION
NUCLEAR RESEARCH
RADIOACTIVE AGE DETERMINATION
RADIOACTIVE MATERIALS
RADIOACTIVE WASTES
RADIOACTIVITY
RADIOBIOLOGY

RADIOGENIC MATERIALS
RT   NUCLEAR REACTIONS
RADIOACTIVE AGE DETERMINATION
∞ RADIOACTIVE DEBRIS
RADIOACTIVE DECAY
RADIOACTIVE WASTES
TRANSMUTATION

RADIOGONIOMETERS
GS   MEASURING INSTRUMENTS
. GONIOMETERS
. . RADIOGONIOMETERS
RT   RADIO DIRECTION FINDERS

RADIOGRAPHY
UF   CINEFLUOROGRAPHY
CINERADIOGRAPHY
GS   IMAGERY
. RADIOGRAPHY
. . ANGIOGRAPHY
. . AUTORADIOGRAPHY
. . NEUTRON RADIOGRAPHY
. . TOMOGRAPHY
. . . COMPUTER AIDED TOMOGRAPHY
. . UROGRAPHY
RT   BRAGG ANGLE
CRYSTALLOGRAPHY
∞ FLASH
IRRADIATION
LIXISCOPES
∞ MATERIALS TESTS
METALLOGRAPHY
NONDESTRUCTIVE TESTS
PHOTOGRAPHY
PNEUMOGRAPHY
RADIOLOGY
X RAY ANALYSIS
X RAY APPARATUS
X RAY ASTRONOMY
X RAY DIFFRACTION
X RAY FLUORESCENCE
X RAY IMAGERY
X RAY INSPECTION
X RAY SPECTROSCOPY
X RAY TELESCOPES
X RAY TUBES
X RAYS

RADIOIMMUNOASSAY
GS   IMMUNOASSAY
. RADIOIMMUNOASSAY
RT   ANTIGENS
ASSAYING
BIOCHEMISTRY
IMMUNOLOGY
RADIOBIOLOGY

RADIOISOTOPE BATTERIES
UF   ATOMIC BATTERIES
GS   ELECTRIC GENERATORS
. DIRECT POWER GENERATORS
. . RADIOISOTOPE BATTERIES
. . . SNAP 7
. . . SNAP 9A
. . . SNAP 11
. . . SNAP 13
. . . SNAP 15
. . . SNAP 17
. . . SNAP 19
. . . SNAP 21
. . . SNAP 23
. . . SNAP 27
. . . SNAP 29
RT   ELECTRIC BATTERIES
FISSION ELECTRIC CELLS
NUCLEAR AUXILIARY POWER UNITS
RADIOACTIVE ISOTOPES
THERMIONIC CONVERTERS
THERMOELECTRIC GENERATORS

RADIOLOGY
GS   MEDICAL SCIENCE

RADIOLOGY-(CONT.)
. RADIOLOGY
RT   AEROSPACE MEDICINE
∞ MEDICINE
∞ RADIATION
RADIOGRAPHY
X RAY ANALYSIS
X RAYS

RADIOLYSIS
GS   CHEMICAL REACTIONS
. PHOTOCHEMICAL REACTIONS
. . PHOTOLYSIS
. . . RADIOLYSIS
DECOMPOSITION
. PHOTOLYSIS
. . RADIOLYSIS
. RADIOLYSIS
RADIATION EFFECTS
. RADIOLYSIS

RADIOMETEOROGRAPHS
GS   MEASURING INSTRUMENTS
. METEOROLOGICAL INSTRUMENTS
. . RADIOMETEOROGRAPHS
RECORDING INSTRUMENTS
. RADIOMETEOROGRAPHS
TRANSMITTERS
. RADIO TRANSMITTERS
. . RADIOMETEOROGRAPHS
RT   RADIO TELEMETRY
RADIOSONDES

RADIOMETERS
GS   MEASURING INSTRUMENTS
. RADIATION MEASURING INSTRUMENTS
. . ACTINOMETERS
. . RADIOMETERS
. . . . DICKE RADIOMETERS
. . . . INFRARED DETECTORS
. . . . INFRARED SCANNERS
. . . . MICROWAVE RADIOMETERS
. . . . PASSIVE L-BAND RADIOMETERS
. . . . PRESSURE MODULATOR
RADIOMETERS
. . . . SPECTRORADIOMETERS
RT   BOLOMETERS
FOREST FIRE DETECTION
HORIZON SCANNERS
INFRARED PHOTOGRAPHY
INFRARED TRACKING
KNUDSEN GAGES
PHOTOMETERS
PYRANOMETERS
RADIOMETRIC RESOLUTION
SPECTROPHOTOMETERS
THERMISTORS

RADIOMETRIC CORRECTION
UF   RADIOMETRIC RECTIFICATION
RT   IMAGE ENHANCEMENT
INFRARED RADIOMETERS
MULTISPECTRAL BAND SCANNERS
VEGETATIVE INDEX

RADIOMETRIC RECTIFICATION
USE   RADIOMETRIC CORRECTION

RADIOMETRIC RESOLUTION
GS   RESOLUTION
. RADIOMETRIC RESOLUTION
RT   MULTISPECTRAL BAND SCANNERS
RADIOMETERS
REMOTE SENSORS
SPECTRAL RESOLUTION

RADIONUCLIDES
USE   RADIOACTIVE ISOTOPES

RADIOPATHOLOGY
GS   MEDICAL SCIENCE
. RADIOPATHOLOGY
RT   ANTIRADIATION DRUGS
NUCLEAR MEDICINE
RADIATION SICKNESS

RADIOPHOSPHORS
GS   PHOSPHORS
. RADIOPHOSPHORS
RT   RADIOACTIVE ISOTOPES

RADIOPROTECTIVE AGENTS
USE   ANTIRADIATION DRUGS

RADIOSENSITIVITY
USE   RADIATION TOLERANCE

RADIOSONDES
GS   MEASURING INSTRUMENTS
. METEOROLOGICAL INSTRUMENTS
. . RADIOSONDES
. . . ENDORADIOSONDES
. . . IONOSONDES
. . . RAWINSONDES
. SONDES
. . RADIOSONDES
. . . ENDORADIOSONDES
. . . IONOSONDES
. . . RAWINSONDES
RADIO EQUIPMENT
. RADIO TRANSMITTERS
. . RADIOSONDES
. . . ENDORADIOSONDES
. . . IONOSONDES
. . . RAWINSONDES
TRANSMITTERS
. RADIO TRANSMITTERS
. . RADIOSONDES
. . . ENDORADIOSONDES
. . . IONOSONDES
. . . RAWINSONDES
RT   ARCAS ROCKET VEHICLES
BALLOON SOUNDING
BALLOON-BORNE INSTRUMENTS
DROPSONDES
METEOROLOGICAL BALLOONS
RADIO METEOROLOGY
RADIO TELEMETRY
RADIOMETEOROGRAPHS
ROBIN BALLOONS
SATELLITE SOUNDING
SOUNDING ROCKETS

RADIOTELEPHONES
GS   RADIO EQUIPMENT
. RADIO TRANSMITTERS
. . RADIOTELEPHONES
RECEIVERS
. RADIOTELEPHONES
TELECOMMUNICATION
. RADIOTELEPHONES
TELEPHONES
. RADIOTELEPHONES
TRANSMITTERS
. RADIO TRANSMITTERS
. . RADIOTELEPHONES
RT   ECHO SUPPRESSORS
RADIO COMMUNICATION
TELEPHONY
VOICE COMMUNICATION

RADIOTHERAPY
USE   RADIATION THERAPY

RADIUM
GS   CHEMICAL ELEMENTS
. ACTINIDE SERIES
. . RADIUM
. . . RADIUM ISOTOPES
. . . . RADIUM 226
METALS
. ACTINIDE SERIES
. . RADIUM
. . . RADIUM ISOTOPES
. . . . RADIUM 226

RADIUM ISOTOPES
GS   CHEMICAL ELEMENTS
. ACTINIDE SERIES
. . RADIUM
. . . RADIUM ISOTOPES
. . . . RADIUM 226
. NUCLIDES
. ISOTOPES
. . . RADIUM ISOTOPES
. . . . RADIUM 226
METALS
. ACTINIDE SERIES
. . RADIUM
. . . RADIUM ISOTOPES
. . . . RADIUM 226

RADIUM 226
GS   CHEMICAL ELEMENTS
. ACTINIDE SERIES
. . RADIUM
. . . RADIUM ISOTOPES
. . . . RADIUM 226
. NUCLIDES

RADIUM 226-*(CONT.)*
```
. . ISOTOPES
. . . RADIUM ISOTOPES
. . . . RADIUM 226
  METALS
. ACTINIDE SERIES
. . RADIUM
. . . RADIUM ISOTOPES
. . . . RADIUM 226
```

RADIUS
```
USE    RADII
```

RADOME MATERIALS
```
GS     DIELECTRICS
. RADOME MATERIALS
RT     ELECTROMAGNETIC WAVE
         TRANSMISSION
       ∞MATERIALS
       RADAR TRANSMISSION
       RADIO TRANSMISSION
       RADOMES
       TRANSPARENCE
```

RADOMES
```
GS     HOUSINGS
. RADOMES
  SHELLS (STRUCTURAL FORMS)
. DOMES (STRUCTURAL FORMS)
. . RADOMES
RT     INFLATABLE STRUCTURES
       PROTUBERANCES
       RADANT
       RADAR ANTENNAS
       RADAR EQUIPMENT
       RADOME MATERIALS
```

RADON
```
GS     CHEMICAL ELEMENTS
. RARE GASES
. . RADON
. . . RADON ISOTOPES
  GASES
. RARE GASES
. . RADON
. . . RADON ISOTOPES
```

RADON ISOTOPES
```
UF     THORON
GS     CHEMICAL ELEMENTS
. NUCLIDES
. . ISOTOPES
. . . RADON ISOTOPES
. RARE GASES
. . RADON
. . . RADON ISOTOPES
  GASES
. RARE GASES
. . RADON
. . . RADON ISOTOPES
```

RADUGA SATELLITE
```
GS     SATELLITES
. ARTIFICIAL SATELLITES
. . COMMUNICATION SATELLITES
. . . RADUGA SATELLITE
. . SOVIET SATELLITES
. . . RADUGA SATELLITE
```

RAE B
```
USE    EXPLORER 49 SATELLITE
```

RAE 1
```
USE    EXPLORER 49 SATELLITE
```

RAE 2
```
USE    EXPLORER 49 SATELLITE
```

RAE-1
```
USE    EXPLORER 38 SATELLITE
```

RAFTS
```
GS     RAFTS
. LIFE RAFTS
RT     FLOATS
       LIFEBOATS
       SURVIVAL EQUIPMENT
```

RAIL TRANSPORTATION
```
UF     RAILROADS
GS     TRANSPORTATION
. RAIL TRANSPORTATION
RT     AUTOMATED GUIDEWAY TRANSIT
         VEHICLES
```

RAIL TRANSPORTATION-*(CONT.)*
```
       AUTOMATED TRANSIT VEHICLES
       LOCOMOTIVES
       MAGNETIC LEVITATION VEHICLES
       MARINE TRANSPORTATION
       RAILS
       RAPID TRANSIT SYSTEMS
       SURFACE VEHICLES
       URBAN TRANSPORTATION
```

RAILGUN ACCELERATORS
```
RT     ∞ACCELERATORS
       HYPERVELOCITY GUNS
       HYPERVELOCITY LAUNCHERS
       NUCLEAR FUSION
       PARTICLE ACCELERATORS
       SPACECRAFT LAUNCHING
```

RAILROAD HUMPING TESTS
```
RT     CARGO
       IMPACT ACCELERATION
       MATERIALS HANDLING
       SHOCK TESTS
       ∞TESTS
```

RAILROADS
```
USE    RAIL TRANSPORTATION
```

RAILS
```
RT     RAIL TRANSPORTATION
       RAPID TRANSIT SYSTEMS
       SURFACE VEHICLES
```

RAIN
```
GS     PRECIPITATION (METEOROLOGY)
. RAIN
. . ACID RAIN
RT     AQUIFERS
       CLOUD SEEDING
       CONDENSATION NUCLEI
       FLOOD PREDICTIONS
       HYDROLOGY
       HYDROLOGY MODELS
       LIMNOLOGY
       RAINBOWS
       RAINDROPS
       RAINSTORMS
       ∞SHOWERS
       THUNDERSTORMS
       WATERSHEDS
```

RAIN EROSION
```
GS     EROSION
. RAIN EROSION
RT     LANDSLIDES
       MUD
       SANDS
       SOIL EROSION
```

RAIN FORESTS
```
GS     RESOURCES
. EARTH RESOURCES
. . FORESTS
. . . RAIN FORESTS
RT     CANOPIES (VEGETATION)
       GEOBOTANY
       PLANTS (BOTANY)
       ∞SHOWERS
       TROPICAL REGIONS
       VEGETATION
```

RAIN GAGES
```
UF     PLUVIOGRAPHS
GS     MEASURING INSTRUMENTS
. METEOROLOGICAL INSTRUMENTS
. . RAIN GAGES
```

RAIN IMPACT DAMAGE
```
GS     DAMAGE
. IMPACT DAMAGE
. . RAIN IMPACT DAMAGE
RT     ARROYOS
       EROSION
       SOIL EROSION
       WATER EROSION
```

RAINBOWS
```
GS     DISTRIBUTION (PROPERTY)
. RADIATION DISTRIBUTION
. . DIFFRACTION PATTERNS
. . . RAINBOWS
RT     HALOS
       LIGHT TRANSMISSION
       RAIN
```

RAINDROPS
```
GS     PARTICLES
. DROPS (LIQUIDS)
. . RAINDROPS
RT     DROP SIZE
       FALLING SPHERES
       RAIN
       RAINMAKING
```

RAINMAKING
```
GS     WEATHER MODIFICATION
. RAINMAKING
RT     CLIMATOLOGY
       CLOUD SEEDING
       PRECIPITATION (METEOROLOGY)
       RAINDROPS
       WATER RESOURCES
```

RAINSTORMS
```
GS     STORMS
. STORMS (METEOROLOGY)
. . RAINSTORMS
. . . THUNDERSTORMS
RT     ACID RAIN
       FLOOD CONTROL
       FLOOD PREDICTIONS
       HAILSTORMS
       PRECIPITATION (METEOROLOGY)
       RAIN
       ∞SHOWERS
       STORM DAMAGE
       STORM ENHANCEMENT
       STORM SUPPRESSION
       TORNADOES
```

∞ RAKES
```
SN     (USE OF A MORE SPECIFIC TERM IS
         RECOMMENDED--CONSULT THE TERMS
         LISTED BELOW)
RT     PRESSURE SENSORS
       SLOPES
```

∞ RAM
```
SN     (USE OF A MORE SPECIFIC TERM IS
         RECOMMENDED--CONSULT THE TERMS
         LISTED BELOW)
RT     ANTIRADAR COATINGS
       RAMS (PRESSES)
       RAMS (PUMPS)
```

RAM B LAUNCH VEHICLE
```
GS     LAUNCH VEHICLES
. RAM B LAUNCH VEHICLE
  ROCKET VEHICLES
. MULTISTAGE ROCKET VEHICLES
. . RAM B LAUNCH VEHICLE
RT     SOLID PROPELLANT ROCKET ENGINES
       TX-354 ENGINE
```

RAM PROJECT
```
USE    RADIO ATTENUATION MEASUREMENT
         PROJECT
```

RAMAN EFFECT
```
USE    RAMAN SPECTRA
```

RAMAN LASERS
```
GS     STIMULATED EMISSION DEVICES
. LASERS
. . RAMAN LASERS
```

RAMAN SCATTERING
```
USE    RAMAN SPECTRA
```

RAMAN SPECTRA
```
UF     RAMAN EFFECT
       RAMAN SCATTERING
GS     SCATTERING
. WAVE SCATTERING
. . ELECTROMAGNETIC SCATTERING
. . . RAMAN SPECTRA
  SPECTRA
. MOLECULAR SPECTRA
. . RAMAN SPECTRA
. RADIATION SPECTRA
. . ELECTROMAGNETIC SPECTRA
. . . RAMAN SPECTRA
RT     ABSORPTION SPECTRA
       EMISSION SPECTRA
       LIGHT (VISIBLE RADIATION)
       LINE SPECTRA
       MOLECULAR ROTATION
       NONLINEAR OPTICS
       VIBRATIONAL SPECTRA
```

RAMAN SPECTROSCOPY
UF   COHERENT ANTI-STOKES RAMAN
        SPECTROSCOPY
GS   SPECTROSCOPY
    . MOLECULAR SPECTROSCOPY
    .. **RAMAN SPECTROSCOPY**
RT   ASTRONOMICAL SPECTROSCOPY
    INFRARED SPECTROSCOPY
    LINE SPECTRA
    OPTOGALVANIC SPECTROSCOPY
    RAYLEIGH SCATTERING
    SPECTROSCOPIC ANALYSIS

RAMJET ENGINES
UF   ATHODYDS
GS   ENGINES
    . AIR BREATHING ENGINES
    .. GAS TURBINE ENGINES
    ... JET ENGINES
    .... **RAMJET ENGINES**
    ..... INTEGRAL ROCKET RAMJETS
    ..... LOW VOLUME RAMJET ENGINES
    ..... PULSEJET ENGINES
    ..... SUPERSONIC COMBUSTION
         RAMJET ENGINES
    ..... TURBORAMJET ENGINES
    . INTERNAL COMBUSTION ENGINES
    .. GAS TURBINE ENGINES
    ... JET ENGINES
    .... **RAMJET ENGINES**
    ..... INTEGRAL ROCKET RAMJETS
    ..... LOW VOLUME RAMJET ENGINES
    ..... PULSEJET ENGINES
    ..... SUPERSONIC COMBUSTION
         RAMJET ENGINES
    ..... TURBORAMJET ENGINES
    . TURBINE ENGINES
    .. GAS TURBINE ENGINES
    ... JET ENGINES
    .... **RAMJET ENGINES**
    ..... LOW VOLUME RAMJET ENGINES
    ..... PULSEJET ENGINES
    ..... SUPERSONIC COMBUSTION
         RAMJET ENGINES
    ..... TURBORAMJET ENGINES
RT   HYDROGEN FUELS
    METEOR 1 ROCKET VEHICLE
    NAVAHO MISSILE
    SUPERSONIC LOW ALTITUDE MISSILE
    TURBOJET ENGINES

RAMJET MISSILES
GS   MISSILES
    . **RAMJET MISSILES**
    .. NAVAHO MISSILE
    .. SUPERSONIC LOW ALTITUDE MISSILE
RT   AIR TO AIR MISSILES
    ANTIAIRCRAFT MISSILES
    SUPERSONIC COMBUSTION RAMJET
        ENGINES
    SURFACE TO AIR MISSILES
    SURFACE TO SURFACE MISSILES

RAMP FUNCTIONS
GS   FUNCTIONS (MATHEMATICS)
    . **RAMP FUNCTIONS**
RT   DYNAMIC RESPONSE
    ∞FREQUENCY RESPONSE
    ∞RAMPS
    REACTION TIME
    SLOPES
    STEP FUNCTIONS

∞ **RAMPS**
SN   *(USE OF A MORE SPECIFIC TERM IS*
    RECOMMENDED--CONSULT THE TERMS
    LISTED BELOW)
RT   RAMP FUNCTIONS
    RAMPS (STRUCTURES)

RAMPS (STRUCTURES)
RT   BRIDGES (STRUCTURES)
    CROSSINGS
    HIGHWAYS
    INTAKE SYSTEMS
    INTERSECTIONS
    ∞PARKING
    ∞RAMPS
    SLOPES
    WHEELCHAIRS

RAMS (PRESSES)
GS   PRESSES
    . **RAMS (PRESSES)**
RT   HAMMERS

RAMS (PRESSES)-*(CONT.)*
    PLATENS
    PLUNGERS
    ∞RAM

RAMS (PUMPS)
GS   PUMPS
    . **RAMS (PUMPS)**
RT   PLUNGERS
    ∞RAM
    WATER HAMMER

RAMSAUER EFFECT
RT   ∞EFFECTS
    ELECTRON SCATTERING
    ∞INTERFERENCE
    NEGATIVE RESISTANCE DEVICES
    RARE GASES
    SCATTERING CROSS SECTIONS

RAND PROJECT
GS   PROGRAMS
    . PROJECTS
    .. **RAND PROJECT**
RT   OPERATIONS RESEARCH

RANDOM ACCESS
GS   **RANDOM ACCESS**
    . ALOHA SYSTEM
RT   COMPUTER STORAGE DEVICES
    INPUT/OUTPUT ROUTINES
    RANDOM ACCESS MEMORY
    RANDOM PROCESSES

RANDOM ACCESS MEMORY
GS   PERIPHERAL EQUIPMENT (COMPUTERS)
    . COMPUTER STORAGE DEVICES
    .. **RANDOM ACCESS MEMORY**
RT   RANDOM ACCESS

RANDOM DISTRIBUTIONS
USE   STATISTICAL DISTRIBUTIONS

RANDOM ERRORS
GS   ERRORS
    . **RANDOM ERRORS**
RT   BCH CODES
    ∞DISPERSION
    PROBABILITY THEORY
    QUALITY CONTROL
    SAMPLING
    STOCHASTIC PROCESSES

RANDOM LOADS
GS   LOADS (FORCES)
    . **RANDOM LOADS**
    .. GUST LOADS
RT   DYNAMIC LOADS
    IMPACT LOADS
    STATIC LOADS
    STRUCTURAL DESIGN CRITERIA
    TRANSIENT LOADS

RANDOM NOISE
UF   GAUSSIAN NOISE
GS   **RANDOM NOISE**
    . RANDOM SIGNALS
RT   BACKGROUND NOISE
    CHANNEL NOISE
    COMMUNICATION THEORY
    ELECTROMAGNETIC NOISE
    ∞NOISE
    NOISE (SOUND)
    NOISE GENERATORS
    NOISE SPECTRA
    PROBABILITY THEORY
    PSEUDONOISE
    SIGNAL TO NOISE RATIOS
    STOCHASTIC PROCESSES
    WHITE NOISE

RANDOM NUMBERS
RT   MATHEMATICAL TABLES
    ∞NUMBERS
    PSEUDORANDOM SEQUENCES

RANDOM PROCESSES
GS   STOCHASTIC PROCESSES
    . **RANDOM PROCESSES**
    .. RANDOM WALK
RT   COMMUNICATION THEORY
    INFORMATION THEORY
    INTERMITTENCY
    MARKOV PROCESSES

RANDOM PROCESSES-*(CONT.)*
    MONTE CARLO METHOD
    RANDOM ACCESS
    STATISTICAL ANALYSIS

RANDOM SAMPLING
GS   SAMPLING
    . **RANDOM SAMPLING**
RT   QUALITY CONTROL

RANDOM SIGNALS
GS   RANDOM NOISE
    . **RANDOM SIGNALS**
RT   NOISE SPECTRA
    SIGNAL TO NOISE RATIOS
    ∞SIGNALS
    STOCHASTIC PROCESSES

RANDOM VARIABLES
RT   FUNCTIONS (MATHEMATICS)
    SHANNON-WIENER MEASURE
    ∞STATISTICS
    ∞VARIABLE

RANDOM VIBRATION
GS   VIBRATION
    . **RANDOM VIBRATION**
RT   BENDING VIBRATION
    FLUTTER
    FORCED VIBRATION
    LATTICE VIBRATIONS
    LINEAR VIBRATION
    MISSILE VIBRATION
    NOISE (SOUND)
    SELF INDUCED VIBRATION
    STRUCTURAL VIBRATION
    TORSIONAL VIBRATION

RANDOM WALK
GS   STOCHASTIC PROCESSES
    . RANDOM PROCESSES
    .. **RANDOM WALK**
RT   MARKOV CHAINS
    MONTE CARLO METHOD

∞ **RANGE**
SN   *(USE OF A MORE SPECIFIC TERM IS*
    RECOMMENDED--CONSULT THE TERMS
    LISTED BELOW)
RT   DISTANCE
    ORBITAL POSITION ESTIMATION
    RANGE (EXTREMES)
    RANGES (FACILITIES)

RANGE (EXTREMES)
UF   EXTREMA
    GUMBEL THEORY
GS   **RANGE (EXTREMES)**
    . FREQUENCY RANGES
    .. OCTAVES
    .. RADIO RANGE
    .. SUBAUDIBLE FREQUENCIES
    . PROPORTIONAL LIMIT
    . ROCHE LIMIT
RT   CONFIDENCE LIMITS
    CONSTRAINTS
    DISTANCE
    DOMAINS
    DYNAMIC CHARACTERISTICS
    ERRORS
    FUNCTIONS (MATHEMATICS)
    HETEROGENEITY
    HORIZON
    INTEGRAL EQUATIONS
    ∞LIMITS
    MAXIMA
    MEAN
    MINIMA
    OPTIMIZATION
    QUALITY CONTROL
    ∞RANGE
    SENSITIVITY
    STANDARD DEVIATION
    STATISTICAL TESTS
    TOLERANCES (MECHANICS)
    TOLERANCES (PHYSIOLOGY)
    ∞TRAVEL
    VARIABILITY
    VARIANCE (STATISTICS)

RANGE AND RANGE RATE TRACKING
GS   DISTANCE
    . **RANGE AND RANGE RATE TRACKING**
    RADAR
    . **RANGE AND RANGE RATE TRACKING**

RANGE AND RANGE RATE TRACKING-(CONT.)
TRACKING (POSITION)
. **RANGE AND RANGE RATE TRACKING**
RT   GLOBAL TRACKING NETWORK
     MISSILE TRACKING
     NORTH AMERICAN SEARCH AND
       RANGING RADAR
     OPTICAL TRACKING
     RADAR TRACKING
     RADIO TRACKING
     SATELLITE TRACKING
     STDN (NETWORK)

RANGE CONTROL
USE   TRAJECTORY CONTROL

RANGE ERRORS
SN   (EXCLUDES ERRORS IN DISTANCE
     TRAVELED--LIMITED TO ERRORS IN
     DISTANCE MEASUREMENT)
GS   ERRORS
    . **RANGE ERRORS**
RT   ACCURACY
     BORESIGHT ERROR
     DISTANCE MEASURING EQUIPMENT
     ERROR ANALYSIS
     ERROR SIGNALS

RANGE FINDERS
UF   RANGE INDICATORS
GS   MEASURING INSTRUMENTS
    . DISTANCE MEASURING EQUIPMENT
     . . **RANGE FINDERS**
     . . . OPTICAL RANGE FINDERS
     . . . . LASER RANGE FINDERS
RT   ALTIMETERS
     FIRE CONTROL
     GEODIMETERS
     LASER RANGER/TRACKER
     LUNAR RANGEFINDING
     NAVIGATION AIDS
     POSITION INDICATORS
     RADAR EQUIPMENT
     RANGEFINDING
     SOUND LOCALIZATION
     SPACE PERCEPTION
     STADIMETERS
     TELLUROMETERS

RANGE INDICATORS
USE   RANGE FINDERS

RANGE MEASUREMENT
USE   RANGEFINDING

RANGE RESOURCES
GS   RESOURCES
    . EARTH RESOURCES
    . . **RANGE RESOURCES**

RANGE SAFETY
GS   SAFETY
    . **RANGE SAFETY**
RT   AEROSPACE SAFETY
     IMPACT PREDICTION
     MISSILE RANGES
     TEST RANGES
     TRAJECTORY CONTROL

RANGEFINDING
UF   RANGE MEASUREMENT
     RANGING
GS   **RANGEFINDING**
    . AIRBORNE RANGE AND ORBIT
      DETERMINATION
    . LUNAR RANGEFINDING
    . SOUND RANGING
RT   BALLISTIC CAMERAS
     LASER RANGER/TRACKER
     MAROTS (ESA)
     ∞MEASUREMENT
     RADAR MEASUREMENT
     RADAR TRACKING
     RADIO TRACKING
     RANGE FINDERS
     TRACKING (POSITION)

RANGELANDS
GS   LAND
    . **RANGELANDS**
RT   CATTLE
     GRASSLANDS
     GRAZING
     LIVESTOCK

RANGELANDS-(CONT.)
RURAL AREAS
RURAL LAND USE

RANGEMASTER AIRCRAFT
USE   G-1 AIRCRAFT

RANGER BLOCK 3 TELEVISION SYSTEM
GS   COMMUNICATION EQUIPMENT
    . SPACECRAFT TELEVISION
    . . **RANGER BLOCK 3 TELEVISION
      SYSTEM**
     TELEVISION SYSTEMS
    . SPACECRAFT TELEVISION
    . . **RANGER BLOCK 3 TELEVISION
      SYSTEM**
RT   ∞SYSTEMS

RANGER LUNAR LANDING VEHICLES
GS   LUNAR SPACECRAFT
    . LUNAR PROBES
    . . RANGER LUNAR PROBES
    . . . **RANGER LUNAR LANDING
      VEHICLES**
     UNMANNED SPACECRAFT
    . SPACE PROBES
    . . LUNAR PROBES
    . . . RANGER LUNAR PROBES
    . . . . **RANGER LUNAR LANDING
      VEHICLES**
RT   BE-3 ENGINE
     ∞VEHICLES

RANGER LUNAR PROBES
UF   RANGER SATELLITES
GS   LUNAR SPACECRAFT
    . LUNAR PROBES
    . . **RANGER LUNAR PROBES**
    . . . RANGER LUNAR LANDING
      VEHICLES
    . . . RANGER 1 LUNAR PROBE
    . . . RANGER 2 LUNAR PROBE
    . . . RANGER 3 LUNAR PROBE
    . . . RANGER 4 LUNAR PROBE
    . . . RANGER 5 LUNAR PROBE
    . . . RANGER 6 LUNAR PROBE
    . . . RANGER 7 LUNAR PROBE
    . . . RANGER 8 LUNAR PROBE
    . . . RANGER 9 LUNAR PROBE
     UNMANNED SPACECRAFT
    . SPACE PROBES
    . . LUNAR PROBES
    . . **RANGER LUNAR PROBES**
    . . . . RANGER LUNAR LANDING
      VEHICLES
    . . . . RANGER 1 LUNAR PROBE
    . . . . RANGER 2 LUNAR PROBE
    . . . . RANGER 3 LUNAR PROBE
    . . . . RANGER 4 LUNAR PROBE
    . . . . RANGER 5 LUNAR PROBE
    . . . . RANGER 6 LUNAR PROBE
    . . . . RANGER 7 LUNAR PROBE
    . . . . RANGER 8 LUNAR PROBE
    . . . . RANGER 9 LUNAR PROBE
RT   ATLAS AGENA B LAUNCH VEHICLE

RANGER PROJECT
GS   PROGRAMS
    . NASA PROGRAMS
    . . **RANGER PROJECT**
    . . . AGENA B RANGER PROGRAM
    . NASA SPACE PROGRAMS
    . . **RANGER PROJECT**
    . . . AGENA B RANGER PROGRAM
    . PROJECTS
    . . **RANGER PROJECT**
    . . . AGENA B RANGER PROGRAM
RT   AGENA B ROCKET VEHICLE
     AGENA ROCKET VEHICLES
     ATLAS LAUNCH VEHICLES
     LUNAR PHOTOGRAPHS
     LUNAR PHOTOGRAPHY
     LUNAR PROBES

RANGER SATELLITES
USE   RANGER LUNAR PROBES

RANGER 1 LUNAR PROBE
GS   LUNAR SPACECRAFT
    . LUNAR PROBES
    . . RANGER LUNAR PROBES
    . . . **RANGER 1 LUNAR PROBE**
     UNMANNED SPACECRAFT
    . SPACE PROBES
    . . LUNAR PROBES

RANGER 1 LUNAR PROBE-(CONT.)
. . . RANGER LUNAR PROBES
. . . . **RANGER 1 LUNAR PROBE**

RANGER 2 LUNAR PROBE
GS   LUNAR SPACECRAFT
    . LUNAR PROBES
    . . RANGER LUNAR PROBES
    . . . **RANGER 2 LUNAR PROBE**
     UNMANNED SPACECRAFT
    . SPACE PROBES
    . . LUNAR PROBES
    . . . RANGER LUNAR PROBES
    . . . . **RANGER 2 LUNAR PROBE**

RANGER 3 LUNAR PROBE
GS   LUNAR SPACECRAFT
    . LUNAR PROBES
    . . RANGER LUNAR PROBES
    . . . **RANGER 3 LUNAR PROBE**
     UNMANNED SPACECRAFT
    . SPACE PROBES
    . . LUNAR PROBES
    . . . RANGER LUNAR PROBES
    . . . . **RANGER 3 LUNAR PROBE**

RANGER 4 LUNAR PROBE
GS   LUNAR SPACECRAFT
    . LUNAR PROBES
    . . RANGER LUNAR PROBES
    . . . **RANGER 4 LUNAR PROBE**
     UNMANNED SPACECRAFT
    . SPACE PROBES
    . . LUNAR PROBES
    . . . RANGER LUNAR PROBES
    . . . . **RANGER 4 LUNAR PROBE**
RT   ATLAS AGENA B LAUNCH VEHICLE

RANGER 5 LUNAR PROBE
GS   LUNAR SPACECRAFT
    . LUNAR PROBES
    . . RANGER LUNAR PROBES
    . . . **RANGER 5 LUNAR PROBE**
     UNMANNED SPACECRAFT
    . SPACE PROBES
    . . LUNAR PROBES
    . . . RANGER LUNAR PROBES
    . . . . **RANGER 5 LUNAR PROBE**

RANGER 6 LUNAR PROBE
GS   LUNAR SPACECRAFT
    . LUNAR PROBES
    . . RANGER LUNAR PROBES
    . . . **RANGER 6 LUNAR PROBE**
     UNMANNED SPACECRAFT
    . SPACE PROBES
    . . LUNAR PROBES
    . . . RANGER LUNAR PROBES
    . . . . **RANGER 6 LUNAR PROBE**

RANGER 7 LUNAR PROBE
GS   LUNAR SPACECRAFT
    . LUNAR PROBES
    . . RANGER LUNAR PROBES
    . . . **RANGER 7 LUNAR PROBE**
     UNMANNED SPACECRAFT
    . SPACE PROBES
    . . LUNAR PROBES
    . . . RANGER LUNAR PROBES
    . . . . **RANGER 7 LUNAR PROBE**

RANGER 8 LUNAR PROBE
GS   LUNAR SPACECRAFT
    . LUNAR PROBES
    . . RANGER LUNAR PROBES
    . . . **RANGER 8 LUNAR PROBE**
     UNMANNED SPACECRAFT
    . SPACE PROBES
    . . LUNAR PROBES
    . . . RANGER LUNAR PROBES
    . . . . **RANGER 8 LUNAR PROBE**

RANGER 9 LUNAR PROBE
GS   LUNAR SPACECRAFT
    . LUNAR PROBES
    . . RANGER LUNAR PROBES
    . . . **RANGER 9 LUNAR PROBE**
     UNMANNED SPACECRAFT
    . SPACE PROBES
    . . LUNAR PROBES
    . . . RANGER LUNAR PROBES
    . . . . **RANGER 9 LUNAR PROBE**

RANGES (FACILITIES)
GS   **RANGES (FACILITIES)**
. TEST RANGES
. . BALLISTIC RANGES
. . MISSILE RANGES
RT   RADIO BEACONS
∞RANGE

RANGING
USE   RANGEFINDING

RANK TESTS
GS   STATISTICAL ANALYSIS
. STATISTICAL TESTS
. . **RANK TESTS**
RT   ∞TESTS

RANKINE CYCLE
GS   CYCLES
. THERMODYNAMIC CYCLES
. . **RANKINE CYCLE**
RT   ASTEC SOLAR TURBOELECTRIC
GENERATOR
BRAYTON CYCLE
CARNOT CYCLE
LASER PROPULSION
OTTO CYCLE
SOLAR GENERATORS
THERMODYNAMICS

RANKINE-HUGONIOT RELATION
RT   AEROTHERMODYNAMICS
∞DENSITY
PRESSURE GRADIENTS
SHOCK WAVE PROPAGATION

RANKING
RT   ARRAYS
COMPARISON
EVALUATION
RATINGS
SELECTION
SEQUENCING
VALUE

RAOULT LAW
RT   COMPOSITION (PROPERTY)
HENRY LAW
PARTIAL PRESSURE
SOLUTIONS
VAPOR PRESSURE

RAPCON (CONTROL)
USE   RADAR APPROACH CONTROL

RAPID BALLISTICS IDENTIFICATION
GS   IDENTIFYING
. **RAPID BALLISTICS IDENTIFICATION**
RT   DISPLAY DEVICES
IMAGING TECHNIQUES
LASER APPLICATIONS
LASERS
MEASURING INSTRUMENTS
PHOTOGRAPHY
SCANNING
STIMULATED EMISSION
STIMULATED EMISSION DEVICES

RAPID EYE MOVEMENT STATE
UF   DESYNCHRONIZED SLEEP
REMS
RT   DREAMS
EYE MOVEMENTS
SLEEP

RAPID QUENCHING (METALLURGY)
RT   CRYSTAL GROWTH
CRYSTAL LATTICES
CRYSTAL STRUCTURE
∞METALLURGY
∞QUENCHING

RAPID TRANSIT SYSTEMS
UF   HIGH SPEED TRANSPORTATION
GS   TRANSPORTATION
. **RAPID TRANSIT SYSTEMS**
RT   AIR TRANSPORTATION
AUTOMATED GUIDEWAY TRANSIT
VEHICLES
AUTOMATED TRANSIT VEHICLES
CARGO
GROUND EFFECT MACHINES
HIGHWAYS
LOGISTICS

RAPID TRANSIT SYSTEMS-*(CONT.)*
PASSENGERS
RAIL TRANSPORTATION
RAILS
ROADS
∞SYSTEMS
∞TRANSPORT VEHICLES
TRANSPORTATION NETWORKS
URBAN TRANSPORTATION

RAPIDS
RT   MEANDERS
RIVER BASINS
RIVERS
STREAMS
WATER CURRENTS
WATER FLOW

RARE EARTH ALLOYS
GS   ALLOYS
. **RARE EARTH ALLOYS**
. . ERBIUM ALLOYS
. . GADOLINIUM ALLOYS
. . LANTHANUM ALLOYS
. . NEODYMIUM ALLOYS
RT   YTTRIUM ALLOYS

RARE EARTH COMPOUNDS
GS   **RARE EARTH COMPOUNDS**
. CERIUM COMPOUNDS
. . BASTNASITE
. . CERIUM OXIDES
. DYSPROSIUM COMPOUNDS
. ERBIUM COMPOUNDS
. EUROPIUM COMPOUNDS
. LANTHANUM TELLURIDES
. LUTETIUM COMPOUNDS
. NEODYMIUM COMPOUNDS
. SAMARIUM COMPOUNDS
. SCANDIUM COMPOUNDS
. . SCANDIUM OXIDES
. THULIUM COMPOUNDS
. YTTERBIUM COMPOUNDS
RT   ∞CHEMICAL COMPOUNDS
∞GROUP 3B COMPOUNDS
∞METAL COMPOUNDS

RARE EARTH ELEMENTS
UF   LANTHANIDE SERIES METALS
GS   CHEMICAL ELEMENTS
. **RARE EARTH ELEMENTS**
. . CERIUM
. . . CERIUM ISOTOPES
. . . . CERIUM 137
. . . . CERIUM 144
. . DYSPROSIUM
. . . DYSPROSIUM ISOTOPES
. . ERBIUM
. . . ERBIUM ISOTOPES
. . EUROPIUM
. . . EUROPIUM ISOTOPES
. . GADOLINIUM
. . . GADOLINIUM ISOTOPES
. . HOLMIUM
. . . HOLMIUM ISOTOPES
. . LANTHANUM
. . . LANTHANUM ISOTOPES
. . LUTETIUM
. . . LUTETIUM ISOTOPES
. . NEODYMIUM
. . . NEODYMIUM ISOTOPES
. . PRASEODYMIUM
. . . PRASEODYMIUM ISOTOPES
. . PROMETHIUM
. . . PROMETHIUM ISOTOPES
. . SAMARIUM
. . . SAMARIUM ISOTOPES
. . SCANDIUM
. . . SCANDIUM ISOTOPES
. . TERBIUM
. . . TERBIUM ISOTOPES
. . THULIUM
. . . THULIUM ISOTOPES
. . YTTERBIUM
. . . YTTERBIUM ISOTOPES
. . YTTRIUM
. . . YTTRIUM ISOTOPES
. METALS
. **RARE EARTH ELEMENTS**
. . CERIUM
. . . CERIUM ISOTOPES
. . . . CERIUM 137
. . . . CERIUM 144
. . DYSPROSIUM
. . . DYSPROSIUM ISOTOPES

RARE EARTH ELEMENTS-*(CONT.)*
. . ERBIUM
. . . ERBIUM ISOTOPES
. . EUROPIUM
. . . EUROPIUM ISOTOPES
. . GADOLINIUM
. . . GADOLINIUM ISOTOPES
. . HOLMIUM
. . . HOLMIUM ISOTOPES
. . LANTHANUM
. . . LANTHANUM ISOTOPES
. . LUTETIUM
. . . LUTETIUM ISOTOPES
. . NEODYMIUM
. . . NEODYMIUM ISOTOPES
. . PRASEODYMIUM
. . . PRASEODYMIUM ISOTOPES
. . PROMETHIUM
. . . PROMETHIUM ISOTOPES
. . SAMARIUM
. . . SAMARIUM ISOTOPES
. . SCANDIUM
. . . SCANDIUM ISOTOPES
. . TERBIUM
. . . TERBIUM ISOTOPES
. . THULIUM
. . . THULIUM ISOTOPES
. . YTTERBIUM
. . . YTTERBIUM ISOTOPES
. . YTTRIUM
. . . YTTRIUM ISOTOPES
RT   ALKALI VAPOR LAMPS
KREEP
NEODYMIUM LASERS
TRANSITION METALS

∞ **RARE GAS COMPOUNDS**
SN   *(USE OF A MORE SPECIFIC TERM IS
RECOMMENDED--CONSULT THE TERMS
LISTED BELOW)*
RT   ∞CHEMICAL COMPOUNDS
EXCIMER LASERS
EXCIMERS
HELIUM COMPOUNDS
XENON COMPOUNDS

RARE GAS-HALIDE LASERS
GS   STIMULATED EMISSION DEVICES
. LASERS
. . **RARE GAS-HALIDE LASERS**
. . . KRYPTON FLUORIDE LASERS
. . . XENON CHLORIDE LASERS
. . . XENON FLUORIDE LASERS
RT   COHERENT LIGHT
GAS LASERS
LASER PUMPING
LASING
LIGHT BEAMS
OPTICAL PUMPING
STIMULATED EMISSION

RARE GASES
UF   INERT GASES
NOBLE GASES
GS   CHEMICAL ELEMENTS
. **RARE GASES**
. . ARGON
. . . ARGON ISOTOPES
. . HELIUM
. . . HELIUM ATOMS
. . . HELIUM FILM
. . . HELIUM ISOTOPES
. . . LIQUID HELIUM
. . . . LIQUID HELIUM 2
. . KRYPTON
. . NEON
. . . LIQUID NEON
. . . NEON ISOTOPES
. . RADON
. . . RADON ISOTOPES
. . XENON
. . . XENON ISOTOPES
. . . . XENON 129
. . . . XENON 133
. . . . XENON 135
. GASES
. **RARE GASES**
. . ARGON
. . . ARGON ISOTOPES
. . HELIUM
. . . HELIUM ATOMS
. . . HELIUM FILM
. . . HELIUM ISOTOPES
. . . LIQUID HELIUM
. . . . LIQUID HELIUM 2

RARE GASES-*(CONT.)*
```
       . . KRYPTON
       . . NEON
       . . . LIQUID NEON
       . . . NEON ISOTOPES
       . . RADON
       . . . RADON ISOTOPES
       . . XENON
       . . . XENON ISOTOPES
       . . . . XENON 129
       . . . . XENON 133
       . . . . XENON 135
    RT   MONATOMIC GASES
         NONPOLAR GASES
         RAMSAUER EFFECT
```

RAREFACTION
```
    RT   ANTINODES
         COMPRESSING
         ELASTIC WAVES
         EXPANSION
         VACUUM
```

RAREFACTION WAVES
```
    USE  ELASTIC WAVES
```

RAREFIED GAS DYNAMICS
```
    GS   FLUID MECHANICS
         . FLUID DYNAMICS
         . . GAS DYNAMICS
         . . . RAREFIED GAS DYNAMICS
    RT   ATOMIC BEAMS
         CHAPMAN-ENSKOG THEORY
         CONTINUUM FLOW
       ∞DYNAMICS
         FREE MOLECULAR FLOW
         KNUDSEN FLOW
         LOW DENSITY FLOW
         LOW DENSITY WIND TUNNELS
         MOLECULAR BEAMS
         MOLECULAR FLOW
         PLASMAS (PHYSICS)
         SLIP FLOW
         TRANSITION FLOW
```

RAREFIED GASES
```
    UF   LOW DENSITY GASES
    GS   GASES
         . RAREFIED GASES
         . . COSMIC GASES
         . . . INTERPLANETARY GAS
         . . . INTERSTELLAR GAS
         . . . . NEUTRAL GASES
    RT   ELECTRON GAS
         FREE MOLECULAR FLOW
         GAS DENSITY
         GAS TEMPERATURE
         HIGH TEMPERATURE GASES
         LOW DENSITY FLOW
         LOW DENSITY RESEARCH
         MOLECULAR GASES
```

RAREFIED PLASMAS
```
    GS   GASES
         . IONIZED GASES
         . . CHARGED PARTICLES
         . . . RAREFIED PLASMAS
         PARTICLES
         . CHARGED PARTICLES
         . . ENERGETIC PARTICLES
         . . . PLASMAS (PHYSICS)
         . . . . RAREFIED PLASMAS
    RT   CATHODE GLOW
         COLD PLASMAS
         COLLISIONLESS PLASMAS
         ELECTRON PLASMA
         NONUNIFORM PLASMAS
```

RASERS
```
    USE  MASERS
```

RATE METERS
```
    USE  MEASURING INSTRUMENTS
```

RATE OF CLIMB INDICATORS
```
    GS   AIRCRAFT INSTRUMENTS
         . RATE OF CLIMB INDICATORS
    RT   ALTIMETERS
         FLIGHT INSTRUMENTS
       ∞INDICATORS
```

RATES (PER TIME)
```
    GS   RATES (PER TIME)
         . ACCELERATION (PHYSICS)
```

RATES (PER TIME)-*(CONT.)*
```
         . . ANGULAR ACCELERATION
         . . DECELERATION
         . . . SPIN REDUCTION
         . . ELECTRON ACCELERATION
         . . HIGH ACCELERATION
         . . HIGH GRAVITY ENVIRONMENTS
         . . IMPACT ACCELERATION
         . . PARTICLE ACCELERATION
         . . PLASMA ACCELERATION
         . . TRANSVERSE ACCELERATION
         . ACOUSTIC VELOCITY
         . AIRSPEED
         . ANGULAR VELOCITY
         . BIT ERROR RATE
         . BURNING RATE
         . COLLISION PARAMETERS
         . COLLISION RATES
         . CRITICAL VELOCITY
         . DECAY RATES
         . . ELECTRON DECAY RATE
         . DRIFT RATE
         . ESCAPE VELOCITY
         . EVAPORATION RATE
         . EXHAUST VELOCITY
         . FLOW VELOCITY
         . . SOLAR WIND VELOCITY
         . FLUX (RATE)
         . . HEAT FLUX
         . . MAGNETIC FLUX
         . . SOLAR FLUX
         . . FLUX DENSITY
         . . . CURRENT DENSITY
         . . . PHOTON DENSITY
         . . . RADIANT FLUX DENSITY
         . . . . IRRADIANCE
         . . . . . ILLUMINANCE
         . . . . . . SOLAR CONSTANT
         . . . . . LUMENS
         . . . . LUMINOUS INTENSITY
         . . . . . ILLUMINANCE
         . . . . . LUMINANCE
         . . . PARTICLE FLUX DENSITY
         . . . . ELECTRON FLUX DENSITY
         . . . . NEUTRON FLUX DENSITY
         . . . . PROTON FLUX DENSITY
         . . . RADIANCE
         . . . RADIANCY
         . . . SOLAR FLUX DENSITY
         . . . . SOLAR CONSTANT
         . . SOUND INTENSITY
         . . . ZERO SOUND
         . GROUND SPEED
         . GROUP VELOCITY
         . HEART RATE
         . . ARRHYTHMIA
         . . BRADYCARDIA
         . . SYSTOLE
         . . TACHYCARDIA
         . HIGH SPEED
         . HYPERSONIC SPEED
         . ION PRODUCTION RATES
         . LANDING SPEED
         . LIGHT SPEED
         . LOADING RATE
         . LOW SPEED
         . MASS FLOW RATE
         . ORBITAL VELOCITY
         . PHASE VELOCITY
         . PHYSIOLOGICAL ACCELERATION
         . PROPAGATION VELOCITY
         . PULSE RATE
         . . PULSE REPETITION RATE
         . RADIAL VELOCITY
         . RECOMBINATION COEFFICIENT
         . RELATIVISTIC VELOCITY
         . RESPIRATORY RATE
         . . DYSPNEA
         . . HYPOVENTILATION
         . . TACHYPNEA
         . ROTOR SPEED
         . SIGNAL FADING RATE
         . SOLAR VELOCITY
         . STRAIN RATE
         . SUBSONIC SPEED
         . SUPERSONIC SPEEDS
         . TERMINAL VELOCITY
         . TIP SPEED
         . TRANSONIC SPEED
         . WIND VELOCITY
         . . SOLAR WIND VELOCITY
    RT   ACCESS TIME
         MTBF
         SOLITARY WAVES
         TIME FUNCTIONS
         TIME MEASUREMENT
```

RATES (PER TIME)-*(CONT.)*
```
         VOLUME
```

RATINGS
```
    RT   ASSESSMENTS
         CONSISTENCY
         EVALUATION
         NORMALIZING (STATISTICS)
       ∞PERFORMANCE
         POSITION (TITLE)
         RANKING
```

RATIOMETERS
```
    GS   MEASURING INSTRUMENTS
         . RATIOMETERS
```

RATIONAL FUNCTIONS
```
    GS   ANALYSIS (MATHEMATICS)
         . COMPLEX VARIABLES
         . . MEROMORPHIC FUNCTIONS
         . . . RATIONAL FUNCTIONS
         FUNCTIONS (MATHEMATICS)
         . MEROMORPHIC FUNCTIONS
         . . RATIONAL FUNCTIONS
```

RATIONS
```
    GS   RATIONS
         . SPACE RATIONS
    RT   ∞FOOD
```

RATIOS
```
    UF   PERCENTAGE
    GS   RATIOS
         . ASPECT RATIO
         . . HIGH ASPECT RATIO
         . . LOW ASPECT RATIO
         . COMPRESSION RATIO
         . DIMENSIONLESS NUMBERS
         . . BIOT NUMBER
         . . FROUDE NUMBER
         . . GRASHOF NUMBER
         . . HARTMANN NUMBER
         . . LAVAL NUMBER
         . . LEWIS NUMBERS
         . . MACH NUMBER
         . . NUSSELT NUMBER
         . . PECLET NUMBER
         . . PRANDTL NUMBER
         . . RAYLEIGH NUMBER
         . . REYNOLDS NUMBER
         . . . HIGH REYNOLDS NUMBER
         . . . LOW REYNOLDS NUMBER
         . . RICHARDSON NUMBER
         . . SCHMIDT NUMBER
         . . SIMILARITY NUMBERS
         . . STANTON NUMBER
         . . STROUHAL NUMBER
         . FINENESS RATIO
         . FUEL-AIR RATIO
         . INDEXES (RATIOS)
         . . KP INDEX
         . . MORPHOLOGICAL INDEXES
         . LIFT DRAG RATIO
         . MASS RATIOS
         . . MASS TO LIGHT RATIOS
         . . PAYLOAD MASS RATIO
         . . PROPELLANT MASS RATIO
         . MILLS RATIO
         . MODULAR RATIOS
         . PERVEANCE
         . POISSON RATIO
         . PRESSURE RATIO
         . SCALE (RATIO)
         . SIGNAL TO NOISE RATIOS
         . STANDING WAVE RATIOS
         . STRESS RATIO
         . THICKNESS RATIO
         . THRUST-WEIGHT RATIO
         . VOID RATIO
    RT   EFFICIENCY
         FRACTALS
         FRACTIONS
         PROPORTION
         PSYCHOLOGICAL TESTS
         REFRACTIVITY
         TEMPERATURE RATIO
```

RATS
```
    GS   ANIMALS
         . VERTEBRATES
         . . MAMMALS
         . . . RODENTS
         . . . . RATS
    RT   MICE
         POCKET MICE
```

RATSCAT PROGRAM
USE   RADAR TARGET SCATTER SITE
        PROGRAM

RAVEN HELICOPTER
USE   OH-23 HELICOPTER

RAVINES
GS   LANDFORMS
     . **RAVINES**
RT   CANYONS
     EROSION
     RIVER BASINS
     TOPOGRAPHY
     VALLEYS
     WATER EROSION

RAWINSONDES
GS   MEASURING INSTRUMENTS
     . METEOROLOGICAL INSTRUMENTS
     . . RADIOSONDES
     . . . **RAWINSONDES**
     . SONDES
     . . RADIOSONDES
     . . . **RAWINSONDES**
     RADIO EQUIPMENT
     . RADIO TRANSMITTERS
     . . RADIOSONDES
     . . . **RAWINSONDES**
     TRANSMITTERS
     . RADIO TRANSMITTERS
     . . RADIOSONDES
     . . . **RAWINSONDES**
RT   DROPSONDES
     METEOROLOGICAL BALLOONS
     RADAR TRACKING
     RADIO TRACKING
     WIND MEASUREMENT

RAY ACOUSTICS
USE   GEOMETRICAL ACOUSTICS

RAY OPTICS
USE   GEOMETRICAL OPTICS

RAY TRACING
RT   DIFFRACTION
     GEOMETRICAL OPTICS
     GEOMETRICAL THEORY OF
       DIFFRACTION
     GRADIENT INDEX OPTICS
     GRAZING INCIDENCE
     OPTICAL MEASUREMENT
     REFLECTANCE
     TRACKING (POSITION)
     TRANSMITTANCE

RAYLEIGH DISTRIBUTION
GS   FUNCTIONS (MATHEMATICS)
     . PROBABILITY DENSITY FUNCTIONS
     . . **RAYLEIGH DISTRIBUTION**
     STATISTICAL ANALYSIS
     . PROBABILITY DENSITY FUNCTIONS
     . . **RAYLEIGH DISTRIBUTION**
     STATISTICAL DISTRIBUTIONS
     . **RAYLEIGH DISTRIBUTION**
RT   ERROR ANALYSIS
     OPERATIONS RESEARCH
     RADIAL DISTRIBUTION

RAYLEIGH EQUATIONS
RT   ∞ EQUATIONS
     HEAT TRANSFER
     THERMODYNAMICS

RAYLEIGH NUMBER
GS   RATIOS
     . DIMENSIONLESS NUMBERS
     . . **RAYLEIGH NUMBER**
RT   BENARD CELLS
     BUOYANCY
     RAYLEIGH-BENARD CONVECTION

RAYLEIGH SCATTERING
GS   SCATTERING
     . WAVE SCATTERING
     . . ELECTROMAGNETIC SCATTERING
     . . . MIE SCATTERING
     . . . . **RAYLEIGH SCATTERING**
RT   AIRGLOW
     GRAY GAS
     LIGHT SCATTERING
     RAMAN SPECTROSCOPY
     SKY

RAYLEIGH WAVES
GS   ELASTIC WAVES
     . SEISMIC WAVES
     . . **RAYLEIGH WAVES**
RT   BAROTROPIC FLOW
     FLUID FLOW
     S WAVES
     TWO DIMENSIONAL FLOW

RAYLEIGH-BENARD CONVECTION
GS   CONVECTION
     . FREE CONVECTION
     . . **RAYLEIGH-BENARD CONVECTION**
     FLUID FLOW
     . CONVECTIVE FLOW
     . . **RAYLEIGH-BENARD CONVECTION**
     . . . BENARD CELLS
RT   CONVECTION CURRENTS
     CONVECTIVE HEAT TRANSFER
     FORCED CONVECTION
     HOT SURFACES
     LAMINAR FLOW
     RAYLEIGH NUMBER
     THERMAL BOUNDARY LAYER

RAYLEIGH-RITZ METHOD
GS   ANALYSIS (MATHEMATICS)
     . NUMERICAL ANALYSIS
     . . APPROXIMATION
     . . . **RAYLEIGH-RITZ METHOD**
RT   ∞ METHODOLOGY
     VARIATIONAL PRINCIPLES

RAYON
GS   FIBERS
     . SYNTHETIC FIBERS
     . . **RAYON**
     TEXTILES
     . **RAYON**

∞ RAYS
SN   *(USE OF A MORE SPECIFIC TERM IS*
     RECOMMENDED--CONSULT THE TERMS
     LISTED BELOW)
RT   ATMOSPHERIC RADIATION
     BACKGROUND NOISE
     BEAMS (RADIATION)
     CAUSTICS (OPTICS)
     COHERENT RADIATION
     CONTINUOUS RADIATION
     CORPUSCULAR RADIATION
     ELECTROMAGNETIC RADIATION
     EXTRATERRESTRIAL RADIATION
     GAMMA RAYS
     LUNAR RAYS
     POLARIZED RADIATION
     PULSED RADIATION

RAYTHEON COMPUTERS
GS   DATA PROCESSING EQUIPMENT
     . COMPUTERS
     . . DIGITAL COMPUTERS
     . . . **RAYTHEON COMPUTERS**

RAZOR BLADES
GS   CUTTERS
     . BLADES (CUTTERS)
     . . **RAZOR BLADES**

RB-47 AIRCRAFT
USE   B-47 AIRCRAFT

RB-50 AIRCRAFT
UF   SUPER FORTRESS AIRCRAFT
GS   BOEING AIRCRAFT
     . **RB-50 AIRCRAFT**
     MONOPLANES
     . **RB-50 AIRCRAFT**
     OBSERVATION AIRCRAFT
     . **RB-50 AIRCRAFT**
     RECONNAISSANCE AIRCRAFT
     . **RB-50 AIRCRAFT**
RT   ∞ AIRCRAFT
     BOMBER AIRCRAFT
     WEATHER RECONNAISSANCE AIRCRAFT

RB-57 AIRCRAFT
USE   B-57 AIRCRAFT

RB-66 AIRCRAFT
USE   B-66 AIRCRAFT

RBE
USE   RELATIVE BIOLOGICAL EFFECTIVENESS
     (RBE)

RC CIRCUITS
UF   RC NETWORKS
GS   CIRCUITS
     . **RC CIRCUITS**
RT   CAPACITANCE
     COUPLING CIRCUITS
     DISCRIMINATORS
     ELECTRIC FILTERS
     ELECTRICAL RESISTANCE
     LC CIRCUITS
     NETWORK ANALYSIS
     NETWORK SYNTHESIS
     RLC CIRCUITS
     TIME CONSTANT

RC NETWORKS
USE   RC CIRCUITS

RCA COMPUTERS
GS   DATA PROCESSING EQUIPMENT
     . COMPUTERS
     . . **RCA COMPUTERS**
     . . . RCA SPECTRA 70 COMPUTER
     . . . RCA-110 COMPUTERS
RT   DATA PROCESSING

RCA SATCOM SATELLITES
GS   COMMERCIAL SPACECRAFT
     . **RCA SATCOM SATELLITES**
     SATELLITES
     . ARTIFICIAL SATELLITES
     . . COMMUNICATION SATELLITES
     . . . **RCA SATCOM SATELLITES**
RT   DELTA LAUNCH VEHICLE
     DOMESTIC SATELLITE COMMUNICATIONS
     SYSTEMS

RCA SPECTRA 70 COMPUTER
GS   DATA PROCESSING EQUIPMENT
     . COMPUTERS
     . . DIGITAL COMPUTERS
     . . . **RCA SPECTRA 70 COMPUTER**
     . . RCA COMPUTERS
     . . . **RCA SPECTRA 70 COMPUTER**

RCA-110 COMPUTERS
GS   DATA PROCESSING EQUIPMENT
     . COMPUTERS
     . . RCA COMPUTERS
     . . . **RCA-110 COMPUTERS**

RDX
UF   CYCLOTRIMETHYLENE TRINITRAMINE
     TRINITROTRIAZOCYCLOHEXANE
GS   EXPLOSIVES
     . **RDX**
     HETEROCYCLIC COMPOUNDS
     . **RDX**
     NITROGEN COMPOUNDS
     . AZO COMPOUNDS
     . . **RDX**
     PROPELLANTS
     . **RDX**
RT   PYROTECHNICS
     SOLID PROPELLANTS
     SOLID ROCKET PROPELLANTS

REACTANCE
GS   IMPEDANCE
     . ELECTRICAL IMPEDANCE
     . . **REACTANCE**
RT   CAPACITANCE
     ELECTRICAL RESISTANCE
     FOSTER THEORY
     INDUCTANCE
     SMITH CHART

∞ REACTION
SN   *(USE OF A MORE SPECIFIC TERM IS*
     RECOMMENDED--CONSULT THE TERMS
     LISTED BELOW)
RT   CHEMICAL REACTIONS
     HUMAN REACTIONS
     IRRITATION
     NUCLEAR REACTIONS
     THRUST

REACTION BONDING
GS   BONDING
     . **REACTION BONDING**

REACTION BONDING-(CONT.)
RT   ALUMINUM
       CERAMICS
       CHEMICAL REACTIONS
       MELTING POINTS
       NITROGEN
       OXYGEN
       POWDER METALLURGY
       SIALON
       SILICON
       SILICON NITRIDES
       SINTERING

∞ **REACTION CONTROL**
SN   *(USE OF A MORE SPECIFIC TERM IS RECOMMENDED--CONSULT THE TERMS LISTED BELOW)*
RT   CHEMICAL REACTION CONTROL
       COMBUSTION CONTROL
      ∞CONTROL
       DIRECTIONAL CONTROL
       MARQUARDT R4D ENGINE
       NUCLEAR REACTOR CONTROL
       REACTOR SAFETY
       THRUST CONTROL

REACTION JET BACKPACKS
USE   SELF MANEUVERING UNITS

REACTION JETS
USE   JET FLOW
       JET THRUST

REACTION KINETICS
UF   CHEMICAL KINETICS
       REACTION RATE
GS   KINETICS
      . **REACTION KINETICS**
RT   ASSOCIATION REACTIONS
       AUTOCATALYSIS
       CATALYSIS
       CHEMICAL EQUILIBRIUM
       CHEMICAL REACTIONS
       DAMKOHLER NUMBER
       FISCHER-TROPSCH PROCESS
       HALF LIFE
       HEAT OF DISSOCIATION
       INTERSTELLAR CHEMISTRY
       IRREVERSIBLE PROCESSES
       NITROUS ACID
       NUCLEAR REACTIONS
       PRESSURE DEPENDENCE
       REAGENTS
       SOLVATION

REACTION PRODUCTS
GS   PRODUCTS
      . **REACTION PRODUCTS**
RT   ASHES
       BY-PRODUCTS
       COMBUSTION PRODUCTS
       EFFLUENTS
       ENGINES
       EXHAUST GASES
       FUMES
       GASES
       INFRARED SUPPRESSION
       JET ENGINES
       RESIDUES
       SLAGS
       SLUDGE

REACTION RATE
USE   REACTION KINETICS

REACTION TIME
UF   REVERSE TIME
GS   TIME
      . **REACTION TIME**
      . . CHRONAXY
RT   ADAPTATION
       CONDITIONED REFLEXES
       DYNAMIC CHARACTERISTICS
       DYNAMIC RESPONSE
       HUMAN REACTIONS
       PERCEPTUAL TIME CONSTANT
       PHYSIOLOGICAL EFFECTS
       PONTRYAGIN PRINCIPLE
       PSYCHOLOGICAL EFFECTS
       PSYCHOMOTOR PERFORMANCE
       RAMP FUNCTIONS
       REFLEXES
       REFRACTORY PERIOD
       SENSITIVITY
       SENSORIMOTOR PERFORMANCE

REACTION TIME-(CONT.)
       STEP FUNCTIONS
       TIME CONSTANT
       TIME LAG

REACTION WHEELS
UF   INERTIA WHEELS
GS   WHEELS
      . **REACTION WHEELS**
RT   ATTITUDE CONTROL
       COUNTER-ROTATING WHEELS
       FLYWHEELS

REACTIVITY
RT   CHEMICAL REACTIONS
       INHOUR EQUATION
       NUCLEAR REACTIONS

REACTOR CHEMISTRY
USE   RADIOCHEMISTRY

REACTOR CORES
GS   CORES
      . **REACTOR CORES**
RT   ANNULAR CORE PULSE REACTORS
       CONTROL RODS
       NUCLEAR FUEL ELEMENTS
       NUCLEAR FUELS
       NUCLEAR REACTORS
      ∞OPERATORS
       PLASMA CORE REACTORS
       REFLECTOMETERS
       REFLECTORS
       VOID RATIO

REACTOR DESIGN
RT   ANNULAR CORE PULSE REACTORS
       BLANKETS (FISSION REACTORS)
       BLANKETS (FUSION REACTORS)
       CHEMICAL REACTORS
       COMPUTER AIDED DESIGN
      ∞DESIGN
       ENGINE DESIGN
       ENGINEERING TEST REACTORS
       HANFORD REACTORS
       HIGH TEMPERATURE NUCLEAR REACTORS
       LIMITERS (FUSION REACTORS)
       NUCLEAR REACTORS
       NUCLEAR RESEARCH AND TEST REACTORS
       OFFSHORE REACTOR SITES
       ORGANIC COOLED REACTORS
       PEBBLE BED REACTORS
       PRODUCT DEVELOPMENT

REACTOR FUELS
USE   NUCLEAR FUELS

REACTOR IN FLIGHT TEST PROGRAM
USE   RIFT (REACTOR IN FLIGHT TEST)

REACTOR MATERIALS
RT   ANNULAR CORE PULSE REACTORS
       BLANKETS (FISSION REACTORS)
       BLANKETS (FUSION REACTORS)
       CHEMICAL REACTORS
      ∞CONSTRUCTION MATERIALS
       COOLANTS
       LIMITERS (FUSION REACTORS)
       LOSS OF COOLANT
      ∞MATERIALS
       MODERATORS
       NUCLEAR FUEL ELEMENTS
       NUCLEAR FUELS
       NUCLEAR REACTORS
       PRESSURE VESSELS
       RADIATION SHIELDING
       SPENT FUELS

REACTOR PHYSICS
GS   NUCLEAR PHYSICS
      . **REACTOR PHYSICS**
RT   ANNULAR CORE PULSE REACTORS
       BETA FACTOR
       HANFORD REACTORS
       INHOUR EQUATION
       NUCLEAR FUEL BURNUP
       NUCLEAR REACTORS
      ∞PHYSICS
      ∞SCIENCE

REACTOR SAFETY
GS   SAFETY

REACTOR SAFETY-(CONT.)
      . **REACTOR SAFETY**
RT   ANNULAR CORE PULSE REACTORS
       CHEMICAL REACTORS
       CONTROL RODS
       EXPLOSIONS
       INDUSTRIAL SAFETY
       NUCLEAR REACTOR CONTROL
       NUCLEAR REACTORS
       OFFSHORE REACTOR SITES
       RADIATION HAZARDS
      ∞REACTION CONTROL
       RELIEF VALVES
       TRANSIENT REACTOR TEST FACILITY

REACTOR STARTUP TESTS
GS   FUEL TESTS
      . **REACTOR STARTUP TESTS**
RT   INITIATION
       NUCLEAR FUELS
       NUCLEAR REACTORS
       STARTING
      ∞TESTS

REACTOR TECHNOLOGY
GS   TECHNOLOGIES
      . **REACTOR TECHNOLOGY**
RT   ANNULAR CORE PULSE REACTORS
      ∞ENGINEERING
       ENGINEERING TEST REACTORS
       HANFORD REACTORS
       HIGH TEMPERATURE NUCLEAR REACTORS
       JOINT EUROPEAN TORUS
       NUCLEAR FUEL BURNUP
       NUCLEAR REACTORS
       NUCLEAR RESEARCH AND TEST REACTORS
       OFFSHORE REACTOR SITES
       ORGANIC COOLED REACTORS
       PEBBLE BED REACTORS
       REVERSE FIELD PINCH

∞ **REACTORS**
SN   *(USE OF A MORE SPECIFIC TERM IS RECOMMENDED--CONSULT THE TERMS LISTED BELOW)*
RT   ANNULAR CORE PULSE REACTORS
       CHEMICAL REACTORS
       ELECTRIC REACTORS
       FAST TEST REACTORS
       FUSION REACTORS
       FUSION-FISSION HYBRID REACTORS
       HIGH TEMPERATURE NUCLEAR REACTORS
       MOLTEN SALT NUCLEAR REACTORS
       NUCLEAR REACTORS
       NUCLEAR RESEARCH AND TEST REACTORS
       PLASMA CORE REACTORS
       POWER REACTORS
       RIFT (REACTOR IN FLIGHT TEST)
       SNAPTRAN REACTOR
       SPHEROMAKS
       SWIMMING POOL REACTORS
       THERMAL REACTORS
       TOKAMAK DEVICES
       WATER COOLED REACTORS

READ-ONLY MEMORY DEVICES
GS   PERIPHERAL EQUIPMENT (COMPUTERS)
      . COMPUTER STORAGE DEVICES
      . . **READ-ONLY MEMORY DEVICES**
RT   COMPUTER COMPONENTS
       COMPUTER DESIGN
       COMPUTER SYSTEMS DESIGN
       COMPUTERS

READERS
UF   READING MACHINES
RT   CHARACTER RECOGNITION
       CONICAL SCANNING
      ∞DETECTORS
       MAGNETIC TAPES
       MICROFILMS
       OPTICAL DATA PROCESSING
       OPTICAL SCANNERS
       PATTERN RECOGNITION
       PUNCHED CARDS
       PUNCHED TAPES
       READING

READING
GS   **READING**
      . LIP READING

READING-*(CONT.)*
RT   CHARACTER RECOGNITION
      CONICAL SCANNING
      DATA TRANSMISSION
      DISPLAY DEVICES
      INPUT
      ∞INTERPRETATION
      LEGIBILITY
      PERCEPTION
      PRINTING
      READERS
      SCANNERS
      SCANNING
      SYMBOLS
      VISIBILITY

READING MACHINES
USE   READERS

READJUSTMENT
USE   ADJUSTING

READOUT
RT   DISPLAY DEVICES
      MULTIPLE OUTPUT PROGRAMS
      OUTPUT
      PRINTERS (DATA PROCESSING)
      PRINTOUTS
      REMOTE CONSOLES

REAGENTS
RT   CATALYSTS
      CHEMICAL ANALYSIS
      CHEMICAL REACTIONS
      REACTION KINETICS

REAL GASES
GS   GASES
      . **REAL GASES**
RT   EQUATIONS OF STATE
      GAS DENSITY
      IDEAL GAS
      KINETIC THEORY
      MOLECULAR GASES
      MONATOMIC GASES

REAL NUMBERS
GS   **REAL NUMBERS**
      . INTEGERS
RT   COMPLEX NUMBERS
      ∞NUMBERS

REAL TIME OPERATION
RT   AUTOMATIC CONTROL
      COMPUTER PROGRAMMING
      COMPUTERS
      DISPLAY DEVICES
      INTEGRATED MISSION CONTROL
         CENTER
      MULTIPROCESSING (COMPUTERS)
      ONBOARD DATA PROCESSING

REAL VARIABLES
GS   ANALYSIS (MATHEMATICS)
      . **REAL VARIABLES**
      . . ABEL FUNCTION
      . . ASYMPTOTES
      . . BESSEL FUNCTIONS
      . . . HANKEL FUNCTIONS
      . . BETHE-SALPETER EQUATION
      . . CALCULUS OF VARIATIONS
      . . COMPOSITE FUNCTIONS
      . . DELTA FUNCTION
      . . DIFFERENTIAL EQUATIONS
      . . . BLASIUS EQUATION
      . . . CAUCHY-RIEMANN EQUATIONS
      . . . CHANDRASEKHAR EQUATION
      . . . COSINE SERIES
      . . . DUFFING DIFFERENTIAL EQUATION
      . . . FALKNER-SKAN EQUATION
      . . . HYPERBOLIC DIFFERENTIAL
            EQUATIONS
      . . . LAME WAVE EQUATIONS
      . . . PARTIAL DIFFERENTIAL EQUATIONS
      . . . . BIHARMONIC EQUATIONS
      . . . . BURGER EQUATION
      . . . . ELLIPTIC DIFFERENTIAL
               EQUATIONS
      . . . . . MONGE-AMPERE EQUATION
      . . . . EULER-CAUCHY EQUATIONS
      . . . . FOKKER-PLANCK EQUATION
      . . . . GAUSS EQUATION
      . . . . HELMHOLTZ VORTICITY EQUATION
      . . . . LIOUVILLE EQUATIONS

REAL VARIABLES-*(CONT.)*
      . . . . PARABOLIC DIFFERENTIAL
               EQUATIONS
      . . . . VLASOV EQUATIONS
      . . . POISSON EQUATION
      . . EINSTEIN EQUATIONS
      . . EXISTENCE THEOREMS
      . . EXTREMUM VALUES
      . . . LIMITS (MATHEMATICS)
      . . . MAXIMA
      . . . MINIMA
      . . FOURIER-BESSEL TRANSFORMATIONS
      . . GREEN'S FUNCTIONS
      . . HYPERBOLIC FUNCTIONS
      . . HYPERPLANES
      . . JACOBI INTEGRAL
      . . JACOBI MATRIX METHOD
      . . KERNEL FUNCTIONS
      . . LIAPUNOV FUNCTIONS
      . . LINEAR EQUATIONS
      . . . LINEAR EVOLUTION EQUATIONS
      . . LIPSCHITZ CONDITION
      . . MEASURE AND INTEGRATION
      . . . BINARY INTEGRATION
      . . . BOREL SETS
      . . . FUNCTIONAL INTEGRATION
      . . . INTEGRAL CALCULUS
      . . . J INTEGRAL
      . . . LEBESGUE THEOREM
      . . . NUMERICAL INTEGRATION
      . . . . RUNGE-KUTTA METHOD
      . . . STIELTJES INTEGRAL
      . . . WEIGHTING FUNCTIONS
      . . NEUMANN PROBLEM
      . . NONLINEAR EQUATIONS
      . . . CUBIC EQUATIONS
      . . . DUFFING DIFFERENTIAL EQUATION
      . . . MONGE-AMPERE EQUATION
      . . . NONLINEAR EVOLUTION
            EQUATIONS
      . . . QUADRATIC EQUATIONS
      . . . QUARTIC EQUATIONS
      . . NUMERICAL DIFFERENTIATION
      . . PERIODIC FUNCTIONS
      . . TRIGONOMETRIC FUNCTIONS
      . . . . COSINE SERIES
      . . . SINE SERIES
      . . . . TANGENTS
      . . SERIES (MATHEMATICS)
      . . . ASYMPTOTIC SERIES
      . . . CAMPBELL-HAUSDORFF SERIES
      . . . COSINE SERIES
      . . . FOURIER SERIES
      . . . PADE APPROXIMATION
      . . . POWER SERIES
      . . . . TAYLOR SERIES
      . . . . . MACLAURIN SERIES
      . . . PROGRESSIONS
      . . . PRONY SERIES
      . . . SINE SERIES
      . . STURM-LIOUVILLE THEORY
      . . VECTOR ANALYSIS
      . . . COLLINEARITY
      . . . COPLANARITY
      . . . CURL (VECTORS)
      . . . . VORTICITY
      . . WEIERSTRASS FUNCTIONS
      . . WHITTAKER FUNCTIONS
RT   APERIODIC FUNCTIONS
      CALCULUS
      CHOLESKY FACTORIZATION
      COMPLEX VARIABLES
      CONTINUUMS
      DEPENDENT VARIABLES
      DIFFERENTIAL CALCULUS
      FACTORIZATION
      FOURIER ANALYSIS
      HERMITIAN POLYNOMIAL
      HYPERSPHERES
      INFINITY
      INFLECTION POINTS
      MAXIMUM PRINCIPLE
      MONOTONE FUNCTIONS
      SCHMIDT METHOD
      STABILITY DERIVATIVES
      UNIQUENESS THEOREM
      ∞VARIABLE

REARWARD FACING STEPS
USE   BACKWARD FACING STEPS

REATTACHED FLOW
GS   FLUID FLOW
      . VISCOUS FLOW
      . . BOUNDARY LAYER FLOW

REATTACHED FLOW-*(CONT.)*
      . . . **REATTACHED FLOW**
RT   ∞ATTACHMENT
      BACKWARD FACING STEPS
      BOUNDARY LAYER SEPARATION
      COANDA EFFECT
      CROCCO-LEE THEORY
      FLOW CHARACTERISTICS
      FLOW DISTRIBUTION
      SEPARATED FLOW

REATTACHMENT
USE   ATTACHMENT

REB
USE   RELATIVISTIC ELECTRON BEAMS

REBREATHING
RT   AIR PURIFICATION
      CARBON DIOXIDE CONCENTRATION
      CARBON DIOXIDE REMOVAL
      EXPIRED AIR
      LIFE SUPPORT SYSTEMS
      SPACECRAFT CABIN ATMOSPHERES

RECEIVERS
UF   RECEIVING SYSTEMS
GS   **RECEIVERS**
      . LINEAR RECEIVERS
      . LOGARITHMIC RECEIVERS
      . RADAR RECEIVERS
      . RADIO RECEIVERS
      . . SUPERHETERODYNE RECEIVERS
      . RADIOTELEPHONES
      . TELEVISION RECEIVERS
RT   AMPLIFIERS
      ∞DETECTORS
      DISPLAY DEVICES
      DUPLEXERS
      ELECTRIC FILTERS
      INSTRUMENT RECEIVERS
      ∞RECEIVING
      REPEATERS
      TANKS (CONTAINERS)
      TELEPRINTERS
      TELETYPEWRITERS
      TRANSMITTERS

∞ **RECEIVING**
SN   (USE OF A MORE SPECIFIC TERM IS
      RECOMMENDED--CONSULT THE TERMS
      LISTED BELOW)
UF   RECEPTION
RT   ACQUISITION
      COLLECTION
      DELIVERY
      PROCUREMENT
      RADAR RECEPTION
      RADIO RECEPTION
      RECEIVERS
      RECOGNITION
      SIGNAL RECEPTION
      TELEVISION RECEPTION

RECEIVING SYSTEMS
USE   RECEIVERS

RECEPTACLES (CONTAINERS)
USE   CONTAINERS

RECEPTION
USE   RECEIVING

RECEPTION DIVERSITY
UF   SPACE DIVERSITY
GS   RADIO EQUIPMENT
      . **RECEPTION DIVERSITY**
RT   FADING
      RADIO ANTENNAS
      RADIO RECEIVERS
      RADIO RECEPTION
      SIGNAL FADING

RECEPTORS (PHYSIOLOGY)
GS   **RECEPTORS (PHYSIOLOGY)**
      . BARORECEPTORS
      . CHEMORECEPTORS
      . GRAVIRECEPTORS
      . MECHANORECEPTORS
      . PHOTORECEPTORS
      . PROPRIOCEPTORS
      . THERMORECEPTORS
RT   SENSE ORGANS
      SENSITOMETRY

RECESSES
RT   CAVITIES
       CREVASSES
       ∞ HOLLOW

RECESSION
RT   ∞ DEPRESSION
       ECONOMICS

RECHARGING
RT   CHARGE EFFICIENCY

RECIPROCAL THEOREMS
GS   THEOREMS
       . **RECIPROCAL THEOREMS**
RT   ANGLES (GEOMETRY)
       GEOMETRY
       LINES (GEOMETRY)
       POINTS (MATHEMATICS)
       PROJECTIVE GEOMETRY

RECIPROCATING ENGINES
USE   PISTON ENGINES

RECIPROCATION
RT   CYCLES
       MECHANICAL OSCILLATORS
       PISTON ENGINES
       PISTONS

RECIPROCITY THEOREM
GS   THEOREMS
       . **RECIPROCITY THEOREM**
RT   ACOUSTIC SCATTERING
       ELECTROMAGNETIC FIELDS
       ELECTROMAGNETIC SCATTERING
       WAVE SCATTERING

RECIRCULATION
USE   CIRCULATION

RECIRCULATIVE FLUID FLOW
GS   FLUID FLOW
       . **RECIRCULATIVE FLUID FLOW**
RT   BACKWARD FACING STEPS
       BOUNDARY LAYER FLOW
       BOUNDARY LAYER SEPARATION
       REVERSED FLOW
       TURBULENT FLOW
       TURBULENT MIXING
       VORTICES

RECLAMATION
GS   **RECLAMATION**
       . GAS RECOVERY
       . MATERIALS RECOVERY
       . . NUCLEAR FUEL REPROCESSING
       . . SOLVOLYSIS
       . . WATER RECLAMATION
RT   OIL RECOVERY
       ∞ RECOVERY
       RECYCLING
       REGENERATION (ENGINEERING)

RECOGNITION
GS   **RECOGNITION**
       . PATTERN RECOGNITION
       . . CHARACTER RECOGNITION
       . . GRAPHOLOGY
       . SPEECH RECOGNITION
       . TARGET RECOGNITION
       . TIMBER IDENTIFICATION
RT   ACQUISITION
       CONSCIOUSNESS
       CROP IDENTIFICATION
       IDENTIFYING
       IFF SYSTEMS (IDENTIFICATION)
       ∞ INTERPRETATION
       MEMORY
       ∞ RECEIVING
       REMOTE SENSING

RECOIL ATOMS
GS   ATOMS
       . **RECOIL ATOMS**
RT   RECOILINGS

RECOIL IONS
GS   IONS
       . **RECOIL IONS**
RT   ATOMIC COLLISIONS
       CHARGE EXCHANGE
       ELECTRON SCATTERING
       ION IMPACT

RECOIL IONS-*(CONT.)*
       ION PRODUCTION RATES
       ION SCATTERING
       IONIC COLLISIONS
       RECOILINGS

RECOIL PROTONS
GS   PARTICLES
       . CHARGED PARTICLES
       . . PROTONS
       . . . **RECOIL PROTONS**
       . ELEMENTARY PARTICLES
       . . FERMIONS
       . . . PROTONS
       . . . . **RECOIL PROTONS**
RT   BARYONS
       RECOILINGS

RECOILINGS
RT   COLLISIONS
       PARTICLE MOTION
       RECOIL ATOMS
       RECOIL IONS
       RECOIL PROTONS

RECOMBINATION COEFFICIENT
GS   COEFFICIENTS
       . **RECOMBINATION COEFFICIENT**
       RATES (PER TIME)
       . **RECOMBINATION COEFFICIENT**
RT   FREE ELECTRONS
       ION RECOMBINATION
       IONIZED GASES

RECOMBINATION REACTIONS
GS   **RECOMBINATION REACTIONS**
       . ATOMIC RECOMBINATION
       . . OXYGEN RECOMBINATION
       . ELECTRON RECOMBINATION
       . . RADIATIVE RECOMBINATION
       . ELECTRON-ION RECOMBINATION
       . . RADIATIVE RECOMBINATION
       . HYDROGEN RECOMBINATIONS
       . ION RECOMBINATION
RT   ATOMIC COLLISIONS
       CAPTURE EFFECT
       FERTILIZATION
       SUHL EFFECT

RECOMMENDATIONS
GS   **RECOMMENDATIONS**
       . SUGGESTION
RT   DECISION THEORY

RECOMPRESSION
USE   COMPRESSING

RECONNAISSANCE
GS   **RECONNAISSANCE**
       . AERIAL RECONNAISSANCE
       . . AIRBORNE INTEGRATED
               RECONNAISSANCE SYSTEM
       . PHOTORECONNAISSANCE
       . SPECTRAL RECONNAISSANCE
RT   COSPAS
       EARTH RESOURCES
       INTELLIGENCE
       OBSERVATION
       PATROLS
       PHOTOGEOLOGY
       SARSAT
       SEARCHING
       SPACE OBSERVATIONS (FROM EARTH)
       SPACE SURVEILLANCE (SPACEBORNE)
       SURVEILLANCE
       SURVEYS
       TERRAIN ANALYSIS

RECONNAISSANCE AIRCRAFT
GS   **RECONNAISSANCE AIRCRAFT**
       . A-9 AIRCRAFT
       . BREGUET 1150 AIRCRAFT
       . CESSNA L-19 AIRCRAFT
       . CL-84 AIRCRAFT
       . EARTH RESOURCES SURVEY
          AIRCRAFT
       . F-5 AIRCRAFT
       . G-91 AIRCRAFT
       . G-95/4 AIRCRAFT
       . HS-801 AIRCRAFT
       . MIRAGE AIRCRAFT
       . . MIRAGE 3 AIRCRAFT
       . OV-10 AIRCRAFT
       . P-1127 AIRCRAFT
       . P-1154 AIRCRAFT

RECONNAISSANCE AIRCRAFT-*(CONT.)*
       . RB-50 AIRCRAFT
       . TSR-2 AIRCRAFT
       . U-2 AIRCRAFT
       . VICTOR MK-1 AIRCRAFT
       . WEATHER RECONNAISSANCE
          AIRCRAFT
RT   AERIAL RECONNAISSANCE
       ∞ AIRCRAFT
       ANTISUBMARINE WARFARE AIRCRAFT
       FLYING PLATFORMS
       JET AIRCRAFT
       LIGHT AIRCRAFT
       ∞ MILITARY AIRCRAFT
       ∞ MILITARY AVIATION
       MILITARY HELICOPTERS
       OBSERVATION AIRCRAFT
       PILOTLESS AIRCRAFT
       SUBMERSIBLE AIRCRAFT
       SUPERSONIC AIRCRAFT
       UTILITY AIRCRAFT
       V/STOL AIRCRAFT
       VALIANT AIRCRAFT
       WATER TAKEOFF AND LANDING
          AIRCRAFT
       YF-12 AIRCRAFT

RECONNAISSANCE SPACECRAFT
GS   MILITARY SPACECRAFT
       . **RECONNAISSANCE SPACECRAFT**
       . . INSPECTOR SATELLITE
       . . MIDAS SATELLITES
       . . . MIDAS 2 SATELLITE
       . . . MIDAS 3 SATELLITE
       . . . MIDAS 4 SATELLITE
       . . . MIDAS 5 SATELLITE
       . . . MIDAS 6 SATELLITE
       . . . MIDAS 7 SATELLITE
       . . PHOTO RECONNAISSANCE
          SPACECRAFT
RT   AERIAL RECONNAISSANCE
       ARTIFICIAL SATELLITES
       MANNED ORBITAL LABORATORIES
       MANNED SPACECRAFT
       UNMANNED SPACECRAFT

RECONSTRUCTION
GS   **RECONSTRUCTION**
       . WAVE FRONT RECONSTRUCTION
RT   CONSTRUCTION
       RESTORATION

∞ **RECORDERS**
SN   *(USE OF A MORE SPECIFIC TERM IS*
       RECOMMENDED--CONSULT THE TERMS
       LISTED BELOW)
RT   CABLE FORCE RECORDERS
       DATA RECORDERS
       PLAYBACKS
       RECORDING INSTRUMENTS
       REGISTERS (COMPUTERS)
       TAPE RECORDERS
       VLF EMISSION RECORDERS

RECORDING
GS   **RECORDING**
       . DATA RECORDING
       . DATA SMOOTHING
       . MAGNETIC RECORDING
       . PHOTOGRAPHIC RECORDING
       . PREDICTION RECORDING
RT   PLAYBACKS
       PLOTTING
       PRIVACY
       ∞ STORAGE

RECORDING HEADS
RT   DATA RECORDING
       MAGNETIC RECORDING
       MAGNETIC TAPES
       RECORDING INSTRUMENTS
       TAPE RECORDERS
       VIDEO EQUIPMENT

RECORDING INSTRUMENTS
UF   EMISSOGRAPHS
       PLUVIOGRAPHS
       THERMOGRAMS
GS   **RECORDING INSTRUMENTS**
       . BATHYTHERMOGRAPHS
       . CABLE FORCE RECORDERS
       . FLIGHT LOAD RECORDERS
       . FLIGHT RECORDERS
       . FORCE VECTOR RECORDERS
       . MECHANOGRAMS

RECORDING INSTRUMENTS-(CONT.)
. OSCILLOGRAPHS
. PLOTTERS
. X-Y PLOTTERS
. PRESSURE RECORDERS
. RADIOMETEOROGRAPHS
. SEISMOGRAPHS
. . LUNAR SEISMOGRAPHS
. WEATHER DATA RECORDERS
. WHISTLER RECORDERS
RT AIRCRAFT INSTRUMENTS
AUTOMATIC CONTROL
BUBBLE TECHNIQUE
CONTROL EQUIPMENT
COUNTERS
DATA RECORDERS
ELECTRONIC RECORDING SYSTEMS
FLIGHT INSTRUMENTS
GRAPHS (CHARTS)
INDICATING INSTRUMENTS
INSTRUMENT RECEIVERS
INSTRUMENT TRANSMITTERS
∞INSTRUMENTS
MEASURING INSTRUMENTS
METEOROLOGICAL INSTRUMENTS
∞PENS
PHOTOGRAPHIC RECORDING
∞RECORDERS
RECORDING HEADS
SONOGRAMS
SPHYGMOGRAPHY
TAPE RECORDERS
TRANSDUCERS
VLF EMISSION RECORDERS

RECORDS
GS DOCUMENTS
. RECORDS
. . VIDEO DISKS
RT CASE HISTORIES
∞DATA
DATA PROCESSING
DATA RECORDING
DOCUMENTATION
∞DRAWING
FORMAT
HISTORIES
PERIODICALS
PLAYBACKS
PRESIDENTIAL REPORTS
PRIVACY
REPORTS
∞REPRODUCTION
SUPPLEMENTS
TECHNICAL WRITING
∞TESTS
TEXTS

RECOVERABILITY
RT DAMAGE ASSESSMENT
∞PROPERTIES
∞RECOVERY

RECOVERABLE LAUNCH VEHICLES
GS LAUNCH VEHICLES
. RECOVERABLE LAUNCH VEHICLES
RT BOOSTER RECOVERY
LAUNCH VEHICLE CONFIGURATIONS
MULTIENGINE VEHICLES
∞RECOVERY
RECOVERY PARACHUTES
REUSABLE LAUNCH VEHICLES
∞VEHICLES
∞WINGED VEHICLES

RECOVERABLE SATELLITES
USE RECOVERABLE SPACECRAFT

RECOVERABLE SPACECRAFT
UF RECOVERABLE SATELLITES
GS REENTRY VEHICLES
. RECOVERABLE SPACECRAFT
. . APOLLO SPACECRAFT
. . . APOLLO LUNAR EXPERIMENT
MODULE
. . ASTRO VEHICLE
. . GEMINI B SPACECRAFT
. . GEMINI SPACECRAFT
. . . GEMINI (GT-1) SPACECRAFT
. . . GEMINI 2 SPACECRAFT
. . MERCURY SPACECRAFT
. . . AURORA 7
. . . FAITH 7
. . . FRIENDSHIP 7
. . . SIGMA 7

RECOVERABLE SPACECRAFT-(CONT.)
. . REUSABLE SPACECRAFT
. . . ASTROPLANE
. . . ATLANTIS (ORBITER)
. . . CHALLENGER (ORBITER)
. . . COLUMBIA (ORBITER)
. . . DISCOVERY (ORBITER)
. . . ENTERPRISE (ORBITER)
. . . HERMES MANNED SPACEPLANE
. . . MARS (MANNED REUSABLE
SPACECRAFT)
. . VOSKHOD MANNED SPACECRAFT
. . . VOSKHOD 1 SPACECRAFT
. . . VOSKHOD 2 SPACECRAFT
. . VOSTOK SPACECRAFT
. . . VOSTOK 1 SPACECRAFT
. . . VOSTOK 2 SPACECRAFT
. . . VOSTOK 3 SPACECRAFT
. . . VOSTOK 4 SPACECRAFT
. . . VOSTOK 5 SPACECRAFT
. . . VOSTOK 6 SPACECRAFT
RT BOOSTER ROCKET ENGINES
BOOSTGLIDE VEHICLES
EXPENDABLE STAGES (SPACECRAFT)
HYPERSONIC VEHICLES
INERTIAL UPPER STAGE
INTERIM STAGES (SPACECRAFT)
LIFTING REENTRY VEHICLES
MANEUVERABLE SPACECRAFT
MANNED SPACECRAFT
MILITARY SPACECRAFT
RENDEZVOUS SPACECRAFT
SATELLITES
SPACE CAPSULES
SPACE SHUTTLE ORBITERS
∞SPACECRAFT
SPACECRAFT RECOVERY
UNMANNED SPACECRAFT
∞WINGED VEHICLES

∞ RECOVERY
SN (USE OF A MORE SPECIFIC TERM IS
RECOMMENDED--CONSULT THE TERMS
LISTED BELOW)
RT BOOSTER RECOVERY
GAS RECOVERY
MATERIALS RECOVERY
NUCLEAR FUEL REPROCESSING
OIL RECOVERY
PRESSURE RECOVERY
RECLAMATION
RECOVERABILITY
RECOVERABLE LAUNCH VEHICLES
RECOVERY PARACHUTES
RETRIEVAL
REUSE
SPACECRAFT RECOVERY
STRESS RELAXATION
STRESS RELIEVING
VISUAL DISCRIMINATION

RECOVERY PARACHUTES
GS PARACHUTES
. RECOVERY PARACHUTES
RT BOOSTER RECOVERY
DISCOVERER RECOVERY CAPSULES
RECOVERABLE LAUNCH VEHICLES
∞RECOVERY
RIBBON PARACHUTES
SPACECRAFT RECOVERY

RECOVERY VEHICLES
SN (EXCLUDES RECOVERABLE VEHICLES)
RT HELICOPTERS
∞MILITARY VEHICLES
TRUCKS
∞VEHICLES

RECOVERY ZONES
RT DOWNRANGE
LANDING SITES
REENTRY RANGE
REGIONS
SPACECRAFT RECOVERY

RECREATION
RT MORALE
PARKS
RELAXATION (PHYSIOLOGY)
REST
STARSITE PROGRAM
URBAN PLANNING
URBAN RESEARCH

RECRYSTALLIZATION
GS CRYSTALLIZATION
. RECRYSTALLIZATION
RT ANNEALING
HEAT TREATMENT
LASER ANNEALING
∞METALLURGY
NUCLEATION
POLYGONIZATION
∞SEPARATION

RECTANGLES
GS GEOMETRY
. EUCLIDEAN GEOMETRY
. . POLYGONS
. . . TETRAGONS
. . . . RECTANGLES
RT RECTANGULAR PLANFORMS

RECTANGULAR BEAMS
GS STRUCTURAL MEMBERS
. BEAMS (SUPPORTS)
. . RECTANGULAR BEAMS
RT BOX BEAMS

RECTANGULAR COORDINATES
USE CARTESIAN COORDINATES

RECTANGULAR DRAINAGE
USE DRAINAGE PATTERNS

RECTANGULAR PANELS
GS PANELS
. RECTANGULAR PANELS
PLANFORMS
. RECTANGULAR PLANFORMS
. . RECTANGULAR PANELS
RT STRAKES
STRUCTURAL MEMBERS
WING PANELS

RECTANGULAR PLANFORMS
GS PLANFORMS
. RECTANGULAR PLANFORMS
. . RECTANGULAR PANELS
. . RECTANGULAR PLATES
. . RECTANGULAR WINGS
RT RECTANGLES
WING PLANFORMS

RECTANGULAR PLATES
GS PLANFORMS
. RECTANGULAR PLANFORMS
. . RECTANGULAR PLATES
RT FLAT PLATES
METAL PLATES
∞PLATES

RECTANGULAR WAVEGUIDES
GS TRANSMISSION LINES
. COMMUNICATION CABLES
. . WAVEGUIDES
. . . RECTANGULAR WAVEGUIDES
RT BEAM WAVEGUIDES
MICROWAVE FILTERS

RECTANGULAR WIND TUNNELS
GS TEST FACILITIES
. WIND TUNNELS
. . RECTANGULAR WIND TUNNELS
RT SUBSONIC WIND TUNNELS

RECTANGULAR WINGS
UF STRAIGHT WINGS
GS AIRFOILS
. WINGS
. . UNSWEPT WINGS
. . . RECTANGULAR WINGS
PLANFORMS
. RECTANGULAR PLANFORMS
. . RECTANGULAR WINGS

RECTENNAS
UF RECTIFIER ANTENNAS
GS MICROWAVE EQUIPMENT
. MICROWAVE ANTENNAS
. . RECTENNAS
RT SATELLITE POWER TRANSMISSION (TO
EARTH)
SOLAR RADIATION
SPACETENNAS

RECTIFICATION
GS RECTIFICATION

RECTIFICATION-*(CONT.)*
. GEOMETRIC RECTIFICATION (IMAGERY)
RT   ∞CONDENSATION
       DISTILLATION
       PURIFICATION
       REFINING

RECTIFIER ANTENNAS
USE   RECTENNAS

RECTIFIERS
SN   (EXCLUDES PHOTOGRAPHIC RECTIFIER)
GS   **RECTIFIERS**
. AVALANCHE DIODES
. . CRYOSAR
. CRYSTAL RECTIFIERS
. GERMANIUM DIODES
. IGNITRONS
. THYRATRONS
. THYRISTORS
. . SILICON CONTROLLED RECTIFIERS
RT   BARRITT DIODES
       CURRENT CONVERTERS (AC TO DC)
       DIODES
       ELECTRON TUBES
       ∞ENERGY SOURCES
       FORM FACTORS
       MERCURY ARCS
       METAL OXIDE SEMICONDUCTORS
       ∞POWER SUPPLIES
       POWER SUPPLY CIRCUITS
       SEMICONDUCTOR DEVICES
       SOLID STATE DEVICES
       THIN FILMS

RECTUM
GS   ANATOMY
. DIGESTIVE SYSTEM
. . GASTROINTESTINAL SYSTEM
. . . INTESTINES
. . . . **RECTUM**
       VISCERA
. INTESTINES
. . **RECTUM**

RECUPERATORS
USE   REGENERATORS

RECURSION FORMULAS
USE   RECURSIVE FUNCTIONS

RECURSIVE FUNCTIONS
UF   RECURSION FORMULAS
GS   FUNCTIONS (MATHEMATICS)
. **RECURSIVE FUNCTIONS**
RT   FIR FILTERS
       LISP (PROGRAMMING LANGUAGE)
       STRANGE ATTRACTORS

RECYCLING
RT   ECONOMY
       EXTRACTION
       MATERIALS RECOVERY
       NUCLEAR FUEL REPROCESSING
       ∞PROCESSING
       RECLAMATION
       REFINING
       RESOURCES
       SOLVOLYSIS
       SPENT FUELS

RED ARCS
GS   ATMOSPHERIC RADIATION
. AURORAS
. . AURORAL ARCS
. . . **RED ARCS**
RT   ∞ARCS
       AURORAL IONIZATION

RED BLOOD CELLS
USE   ERYTHROCYTES

RED DWARF STARS
GS   CELESTIAL BODIES
. STARS
. . DWARF STARS
. . . **RED DWARF STARS**
RT   HOT STARS
       LATE STARS
       MAIN SEQUENCE STARS
       STELLAR LUMINOSITY
       STELLAR MAGNITUDE
       SUBDWARF STARS
       SUPERNOVA REMNANTS

RED DWARF STARS-*(CONT.)*
WHITE DWARF STARS

RED GIANT STARS
GS   CELESTIAL BODIES
. STARS
. . S STARS
. . . GIANT STARS
. . . . **RED GIANT STARS**
. . . . . CARBON STARS
RT   LATE STARS
       M STARS
       STELLAR EVOLUTION
       STELLAR LUMINOSITY

RED SEA
GS   SEAS
. **RED SEA**
RT   AFRICA
       ASIA

RED SHIFT
RT   COSMOLOGY
       DOPPLER EFFECT
       DOPPLER-FIZEAU EFFECT
       GALAXIES
       HUBBLE CONSTANT
       HUBBLE DIAGRAM
       RADIAL VELOCITY

RED TIDE
RT   FISHES
       MARINE ENVIRONMENTS
       MICROORGANISMS
       OCEANOGRAPHY
       PLANKTON
       SEA WATER
       TOXICOLOGY

REDEYE MISSILE
GS   MISSILES
. ANTIAIRCRAFT MISSILES
. . **REDEYE MISSILE**
. SURFACE TO AIR MISSILES
. . **REDEYE MISSILE**
RT   SOLID PROPELLANT ROCKET ENGINES

REDOX CELLS
GS   ELECTROCHEMICAL CELLS
. ELECTRIC BATTERIES
. . **REDOX CELLS**
RT   ELECTROCHEMISTRY
       ELECTROLYTES
       ENERGY CONVERSION EFFICIENCY
       ENERGY STORAGE

REDUCED GRAVITY
UF   LOW GRAVITY
       MICROGRAVITY
       SUBGRAVITY
GS   GRAVITATION
. **REDUCED GRAVITY**
RT   ANTIGRAVITY
       BIOPROCESSING
       DROP TOWERS
       FLUID MANAGEMENT
       HIGH GRAVITY ENVIRONMENTS
       LOW GRAVITY MANUFACTURING
       LOW WEIGHT
       MARANGONI CONVECTION

REDUCED ORDER FILTERS
GS   LINEAR FILTERS
. **REDUCED ORDER FILTERS**
RT   ELECTRIC FILTERS
       ∞FILTERS
       KALMAN FILTERS
       NAVIGATION AIDS

∞ REDUCTION
SN   *(USE OF A MORE SPECIFIC TERM IS
       RECOMMENDED--CONSULT THE TERMS
       LISTED BELOW)*
UF   DECREMENTING
       DIMINUTION
       SHORTENING
RT   ATTENUATION
       CLEANING
       COMMINUTION
       CONTRACTION
       DAMPING
       DATA REDUCTION
       DECELERATION
       DECONTAMINATION

REDUCTION-*(CONT.)*
DEMAGNETIZATION
DEOXYGENATION
DEPLETION
DEPOLARIZATION
DILUTION
DIMMING
DISPERSING
DISSIPATION
DRAG REDUCTION
ELIMINATION
FRICTION REDUCTION
HYDROGENOLYSIS
∞INHIBITION
IODIMETRY
LEAKAGE
METAL WORKING
NOISE REDUCTION
OPTIMIZATION
PRESSURE REDUCTION
PREVENTION
PURIFICATION
REDUCTION (CHEMISTRY)
REFINING
RELAXATION (MECHANICS)
REMOVAL
RETARDING
SHRINKAGE
SIDELOBE REDUCTION
SPIN REDUCTION
STOPPING
TAPERING

REDUCTION (CHEMISTRY)
GS   CHEMICAL REACTIONS
. **REDUCTION (CHEMISTRY)**
. . DEOXIDIZING
. . HYDROGENATION
RT   ∞CHEMISTRY
       DEHYDROGENATION
       ELECTRODEPOSITION
       ELECTROLYSIS
       METAL POWDER
       OXIDATION
       OXIDATION-REDUCTION REACTIONS
       PURIFICATION
       ∞REDUCTION
       ROASTING
       SMELTING

REDUCTION (MATHEMATICS)
USE   OPTIMIZATION

REDUNDANCY
RT   ASSURANCE
       COMMUNICATION THEORY
       COMPUTER PROGRAM INTEGRITY
       CORRECTION
       ERROR DETECTION CODES
       INFORMATION THEORY
       RELIABILITY

REDUNDANCY ENCODING
GS   CODING
. **REDUNDANCY ENCODING**
RT   CONCATENATED CODES
       DATA TRANSMISSION
       ERROR CORRECTING CODES
       ERROR CORRECTING DEVICES
       ERROR DETECTION CODES
       REPETITION
       SIGNAL ENCODING

REDUNDANT COMPONENTS
UF   REDUNDANT STRUCTURES
RT   BACKUPS
       ∞COMPONENTS
       RELIABILITY
       SPARE PARTS
       ∞STRUCTURES

REDUNDANT STRUCTURES
USE   REDUNDANT COMPONENTS

REEDS (PLANTS)
GS   PLANTS (BOTANY)
. GRASSES
. . **REEDS (PLANTS)**

REEFS
GS   LANDFORMS
. BARRIERS (LANDFORMS)
. . **REEFS**
RT   ATOLLS
       BARS (LANDFORMS)

REEFS-*(CONT.)*
    CORAL REEFS
    ISLAND ARCS
    ISLANDS
    OCEANOGRAPHY
    ROCKS
    SANDS
    SHALLOW WATER
   ∞SHELVES
    SHOALS

REELS
RT   CABLES (ROPES)
   ∞CONTAINERS
    MAGNETIC TAPES
    SPOOLS
    TETHERED BALLOONS
    TETHERED SATELLITES
    TETHERING

REENTRY
GS   ATMOSPHERIC ENTRY
   . **REENTRY**
   . . HYPERBOLIC REENTRY
   . . HYPERSONIC REENTRY
   . . . UNCONTROLLED REENTRY
     (SPACECRAFT)
   . . MANNED REENTRY
   . . SPACECRAFT REENTRY
   . . . UNCONTROLLED REENTRY
     (SPACECRAFT)
RT   ABLATION
    AERODYNAMIC HEATING
    AERODYNAMIC STABILITY
    AERODYNAMICS
    AEROTHERMODYNAMICS
    DESCENT
    DESCENT TRAJECTORIES
   ∞ENTRY
    FLIGHT PATHS
    IMPACT PREDICTION
    LIFTING REENTRY VEHICLES
    LOW OBSERVABLE REENTRY VEHICLES
    MISSILES
   ∞ROCKETS
    SPACE FLIGHT
    TERMINAL GUIDANCE

REENTRY BODIES
USE   REENTRY VEHICLES

REENTRY COMMUNICATION
GS   TELECOMMUNICATION
   . SPACE COMMUNICATION
   . . SPACECRAFT COMMUNICATION
   . . . **REENTRY COMMUNICATION**
RT   BLACKOUT (PROPAGATION)
    MANNED REENTRY
    PLASMA SHEATHS
    RADIO COMMUNICATION
    VOICE COMMUNICATION

REENTRY DECOYS
GS   COUNTERMEASURES
   . **REENTRY DECOYS**
    DECOYS
   . **REENTRY DECOYS**
    REENTRY VEHICLES
   . **REENTRY DECOYS**
RT   BALLISTIC MISSILE DECOYS
    MISSILE DEFENSE

REENTRY EFFECTS
RT   ABLATION
    AERODYNAMIC HEATING
    BLACKOUT (PROPAGATION)
   ∞EFFECTS
    HYPERSONIC REENTRY
    PLASMA SHEATHS
    TEMPERATURE EFFECTS

REENTRY GLIDERS
USE   LIFTING REENTRY VEHICLES

REENTRY GUIDANCE
GS   GUIDANCE (MOTION)
   . **REENTRY GUIDANCE**
RT   AUTOMATIC CONTROL
    DESCENT TRAJECTORIES
    INERTIAL GUIDANCE
    MANUAL CONTROL
    SATELLITE GUIDANCE
    SPACECRAFT GUIDANCE
    TERMINAL GUIDANCE

REENTRY PHYSICS
RT   ABLATION
    AEROTHERMOCHEMISTRY
    AEROTHERMODYNAMICS
    HYPERSONIC REENTRY
    LOW OBSERVABLE REENTRY VEHICLES
   ∞PHYSICS
    PLASMA SHEATHS
   ∞SCIENCE

REENTRY RANGE
GS   DISTANCE
   . **REENTRY RANGE**
RT   MISSILE RANGES
    RECOVERY ZONES

REENTRY SHIELDING
GS   SHIELDING
   . HEAT SHIELDING
   . . **REENTRY SHIELDING**
RT   ABLATION
    ABLATIVE NOSE CONES
    AERODYNAMIC HEATING
    AEROTHERMOCHEMISTRY
    HEAT SINKS
   ∞INSULATED STRUCTURES
    LUDOX (TRADEMARK)
    NOSE CONES
    REUSABLE HEAT SHIELDING
    SPACECRAFT SHIELDING
    THERMAL CONTROL COATINGS
    THERMAL INSULATION
    THERMAL PROTECTION

REENTRY TRAJECTORIES
GS   TRAJECTORIES
   . DESCENT TRAJECTORIES
   . . **REENTRY TRAJECTORIES**
RT   CIRCUMLUNAR TRAJECTORIES
    FLIGHT MECHANICS
    HYPERBOLIC REENTRY
    MISSILE TRAJECTORIES
    MOON-EARTH TRAJECTORIES
    SPACECRAFT TRAJECTORIES
    TERMINAL GUIDANCE

REENTRY VEHICLES
UF   REENTRY BODIES
GS   **REENTRY VEHICLES**
   . AEROSPACEPLANES
   . . ASTROPLANE
   . BOOSTGLIDE VEHICLES
   . . X-20 AIRCRAFT
   . LIFTING REENTRY VEHICLES
   . . FDL-5 REENTRY VEHICLE
   . . HL-10 REENTRY VEHICLE
   . . HLD-35 REENTRY VEHICLE
   . . JANUS SPACECRAFT
   . . M-2 LIFTING BODY
   . . . M-2F2 LIFTING BODY
   . . X-20 AIRCRAFT
   . LOW OBSERVABLE REENTRY
    VEHICLES
   . MANEUVERABLE REENTRY BODIES
   . MARK 1 REENTRY BODY
   . MARK 2 REENTRY BODY
   . MARK 3 REENTRY BODY
   . MARK 4 REENTRY BODY
   . MARK 5 REENTRY BODY
   . MARK 6 REENTRY BODY
   . MARK 11 REENTRY BODY
   . MARK 12 REENTRY BODY
   . MARK 17 REENTRY BODY
   . RECOVERABLE SPACECRAFT
   . . APOLLO SPACECRAFT
   . . . APOLLO LUNAR EXPERIMENT
     MODULE
   . . ASTRO VEHICLE
   . . GEMINI B SPACECRAFT
   . . GEMINI SPACECRAFT
   . . . GEMINI (GT-1) SPACECRAFT
   . . . GEMINI 2 SPACECRAFT
   . . MERCURY SPACECRAFT
   . . . AURORA 7
   . . . FAITH 7
   . . . FRIENDSHIP 7
   . . . SIGMA 7
   . . REUSABLE SPACECRAFT
   . . . ASTROPLANE
   . . . ATLANTIS (ORBITER)
   . . . CHALLENGER (ORBITER)
   . . . COLUMBIA (ORBITER)
   . . . DISCOVERY (ORBITER)
   . . . ENTERPRISE (ORBITER)
   . . . HERMES MANNED SPACEPLANE

REENTRY VEHICLES-*(CONT.)*
   . . . MARS (MANNED REUSABLE
     SPACECRAFT)
   . . VOSKHOD MANNED SPACECRAFT
   . . . VOSKHOD 1 SPACECRAFT
   . . . VOSKHOD 2 SPACECRAFT
   . . VOSTOK SPACECRAFT
   . . . VOSTOK 1 SPACECRAFT
   . . . VOSTOK 2 SPACECRAFT
   . . . VOSTOK 3 SPACECRAFT
   . . . VOSTOK 4 SPACECRAFT
   . . . VOSTOK 5 SPACECRAFT
   . . . VOSTOK 6 SPACECRAFT
   . REENTRY DECOYS
   . TRAILBLAZER 1 REENTRY VEHICLE
   . TRAILBLAZER 2 REENTRY VEHICLE
   . X-17 REENTRY VEHICLE
RT   ABLATIVE NOSE CONES
    AEROTHERMOCHEMISTRY
    ATHENA ROCKET VEHICLE
   ∞BALLISTIC VEHICLES
    BLUFF BODIES
   ∞BODIES
    FERRY SPACECRAFT
   ∞FLIGHT VEHICLES
    HYPERSONIC VEHICLES
   ∞INSULATED STRUCTURES
    LANDING MODULES
    LIFTING BODIES
    MANEUVERABLE SPACECRAFT
    MISSILES
    NOSE CONES
    PYRAMIDAL BODIES
   ∞ROCKETS
    SPACE CAPSULES
   ∞SPACECRAFT
    SPACECRAFT CONFIGURATIONS
    TERMINAL GUIDANCE
    TEST VEHICLES
   ∞VEHICLES
   ∞WINGED VEHICLES

REFERENCE ATMOSPHERES
UF   STANDARD ATMOSPHERES
GS   MODELS
   . ATMOSPHERIC MODELS
   . . **REFERENCE ATMOSPHERES**
    STANDARDS
   . **REFERENCE ATMOSPHERES**

REFERENCE STARS
GS   CELESTIAL BODIES
   . STARS
   . . **REFERENCE STARS**
RT   ASTRONOMICAL COORDINATES
    ASTRONOMICAL PHOTOGRAPHY
    CELESTIAL NAVIGATION
    NAVIGATION AIDS
    SPACE NAVIGATION

∞ **REFERENCE SYSTEMS**
SN   *(USE OF A MORE SPECIFIC TERM IS*
    RECOMMENDED--CONSULT THE TERMS
    LISTED BELOW)
RT   BIBLIOGRAPHIES
    CELESTIAL REFERENCE SYSTEMS
    COORDINATES
    DOCUMENTATION
    INDEXES (DOCUMENTATION)
    INERTIAL REFERENCE SYSTEMS
    LIBRARIES
    SPHERICAL COORDINATES
   ∞SYSTEMS
    WISWESSER NOTATIONS

REFERENCES (STANDARDS)
USE   STANDARDS

REFILLING
GS   FILLING
   . **REFILLING**
RT   ∞LOADING
    REPLENISHMENT

REFINING
GS   **REFINING**
   . ELECTROREFINING
   . ELECTROSLAG REFINING
RT   ALKYLATION
    BENEFICIATION
    CHEMICAL FRACTIONATION
    CLEAN FUELS
    CLEANING
   ∞CONVERSION
    CRYSTALLIZATION

REFINING-*(CONT.)*
    DESULFURIZING
    DEWAXING
    DISTILLATION
    DROP TRANSFER
    ENERGY POLICY
    ENRICHMENT
    EXTRACTION
    FRACTIONATION
    HYDROGENATION
    HYDROMETALLURGY
    ISOMERIZATION
    MATERIALS RECOVERY
    POLYMERIZATION
   ∞PROCESSING
    PURIFICATION
    PYROMETALLURGY
    RECTIFICATION
    RECYCLING
   ∞REDUCTION
   ∞SEPARATION
    SMELTING
    SUBLIMATION
    UPGRADING
    ZONE MELTING

REFLECTANCE
UF   REFLECTION COEFFICIENT
    REFLECTIVITY
GS   ELECTROMAGNETIC PROPERTIES
   . OPTICAL PROPERTIES
   . . **REFLECTANCE**
RT   ABSORPTANCE
    ALBEDO
    ATTENUATION COEFFICIENTS
    BIREFRINGENCE
    BISTATIC REFLECTIVITY
    BRIGHTNESS
    COARSENESS
    COSMIC RAY ALBEDO
    EARTH ALBEDO
    FLOW COEFFICIENTS
    GEOMETRICAL THEORY OF
      DIFFRACTION
    LUSTER
    OPTICAL MEASUREMENT
    OPTICAL REFLECTION
    PHOTOMETRY
    RAY TRACING
    REFLECTION
    REFLECTOMETERS
    SURFACE PROPERTIES
    SURFACE ROUGHNESS EFFECTS
    TRANSMITTANCE
    VEGETATIVE INDEX

REFLECTED RADIATION
USE   REFLECTED WAVES

REFLECTED RAYS
USE   REFLECTED WAVES

REFLECTED WAVES
UF   REFLECTED RADIATION
    REFLECTED RAYS
RT   CORPUSCULAR RADIATION
    ELASTIC WAVES
    ELECTROMAGNETIC RADIATION
    INCIDENT RADIATION
    OPTICAL REFLECTION
    PHOTON BEAMS
   ∞RADIATION
    REFRACTED WAVES
    RETROREFLECTION
    WAVE REFLECTION

REFLECTING TELESCOPES
GS   TELESCOPES
   . **REFLECTING TELESCOPES**
   . . STARSAT TELESCOPE
RT   ASTRONOMICAL TELESCOPES
    CASSEGRAIN OPTICS
    MIRRORS
    OPTICAL EQUIPMENT
    OPTICAL MEASURING INSTRUMENTS
    PARABOLOID MIRRORS
    REFLECTORS
    SCHMIDT TELESCOPES
    SPECTROSCOPIC TELESCOPES
    STRATOSCOPE TELESCOPES

REFLECTION
GS   **REFLECTION**
   . INFRARED REFLECTION
   . OPTICAL REFLECTION

REFLECTION-*(CONT.)*
   . RETROREFLECTION
   . SIGNAL REFLECTION
   . SPECULAR REFLECTION
   . SPREAD REFLECTION
   . ULTRAVIOLET REFLECTION
   . WAVE REFLECTION
   . . MACH REFLECTION
RT   BREWSTER ANGLE
    DEFLECTION
    DIFFUSION
    ECHELETTE GRATINGS
    ELECTROMAGNETIC ABSORPTION
    ELECTROMAGNETIC RADIATION
    IMPINGEMENT
    LIGHT (VISIBLE RADIATION)
   ∞OPTICS
    REFLECTANCE
    REFRACTION
    SCATTERING
    TRANSMISSION
    ZERO SOUND

REFLECTION COEFFICIENT
USE   REFLECTANCE

REFLECTION NEBULAE
GS   CELESTIAL BODIES
   . NEBULAE
   . . **REFLECTION NEBULAE**
RT   COSMIC DUST
    INTERSTELLAR MATTER
    LIGHT SCATTERING

REFLECTIVITY
USE   REFLECTANCE

REFLECTOMETERS
GS   MEASURING INSTRUMENTS
   . OPTICAL MEASURING INSTRUMENTS
   . . **REFLECTOMETERS**
   . . . MICROWAVE REFLECTOMETERS
    OPTICAL EQUIPMENT
   . OPTICAL MEASURING INSTRUMENTS
   . . **REFLECTOMETERS**
   . . . MICROWAVE REFLECTOMETERS
RT   COMPARATORS
    DIRECTORS (ANTENNA ELEMENTS)
    OPTICAL MEASUREMENT
    PHOTOMETERS
    REACTOR CORES
    REFLECTANCE
    SCHELKUNOFF PRINCIPLE
    TWO REFLECTOR ANTENNAS
    ULTRAVIOLET REFLECTION

† REFLECTOR SATELLITES
USE   PASSIVE SATELLITES

REFLECTORS
GS   **REFLECTORS**
   . FRESNEL REFLECTORS
   . PARABOLIC REFLECTORS
   . . PARABOLOID MIRRORS
   . RADAR REFLECTORS
   . . RADAR CORNER REFLECTORS
   . RETROREFLECTORS
   . SOLAR COLLECTORS
   . . SOLAR REFLECTORS
   . WIGGLER MAGNETS
RT   ANTENNAS
    ATTENUATORS
    BAFFLES
    CEILINGS (ARCHITECTURE)
    DEFLECTORS
    DIRECTORS (ANTENNA ELEMENTS)
    HELIOSTATS
    MIRRORS
    PLASMA CORE REACTORS
    RADIATION SHIELDING
    REACTOR CORES
    REFLECTING TELESCOPES
    SCHELKUNOFF PRINCIPLE
    SUBREFLECTORS
    TELESCOPES
    TWO REFLECTOR ANTENNAS

REFLEXES
GS   **REFLEXES**
   . CAROTID SINUS REFLEX
   . CONDITIONED REFLEXES
   . RESPIRATORY REFLEXES
   . . COUGH
   . . HERING-BREVER REFLEX
   . . SNEEZING

REFLEXES-*(CONT.)*
   . VESTIBULAR NYSTAGMUS
RT   DECONDITIONING
    REACTION TIME
    VASOCONSTRICTION
    VASODILATION

REFORESTATION
GS   MANAGEMENT
   . RESOURCES MANAGEMENT
   . . FOREST MANAGEMENT
   . . . **REFORESTATION**
RT   FORESTS
    TIMBER INVENTORY

REFRACTED RADIATION
USE   REFRACTED WAVES

REFRACTED RAYS
USE   REFRACTED WAVES

REFRACTED WAVES
UF   REFRACTED RADIATION
    REFRACTED RAYS
RT   CORPUSCULAR RADIATION
    EIKONAL EQUATION
    ELASTIC WAVES
    ELECTROMAGNETIC RADIATION
    INCIDENT RADIATION
    PHOTON BEAMS
    REFLECTED WAVES
    REFRACTION
   ∞WAVES

REFRACTING TELESCOPES
GS   TELESCOPES
   . **REFRACTING TELESCOPES**
RT   ASTRONOMICAL TELESCOPES
    LENSES
    OPTICAL EQUIPMENT
    OPTICAL MEASURING INSTRUMENTS
    SPECTROSCOPIC TELESCOPES
    STRATOSCOPE TELESCOPES

REFRACTION
GS   **REFRACTION**
   . ATMOSPHERIC REFRACTION
   . . RADIO WAVE REFRACTION
   . BIREFRINGENCE
RT   ASPHERICITY
    ASTIGMATISM
   ∞CONDUCTION
    DEFLECTION
    DIFFRACTION
    DISTORTION
    DIVERGENCE
    HUYGENS PRINCIPLE
    ISOCHROMATICS
    LENSES
    LIGHT (VISIBLE RADIATION)
    PHOTOELASTICITY
    PRISMS
    REFLECTION
    REFRACTED WAVES
    REFRACTIVITY
    SINKING
    SNELLS LAW
    TRANSMISSION
    VOIGT EFFECT
    WAVE DISPERSION
    WAVE PROPAGATION

REFRACTIVE INDEX
USE   REFRACTIVITY

REFRACTIVITY
UF   REFRACTIVE INDEX
GS   ELECTROMAGNETIC PROPERTIES
   . OPTICAL PROPERTIES
   . . **REFRACTIVITY**
RT   ATMOSPHERIC REFRACTION
    BIREFRINGENCE
    BIREFRINGENT COATINGS
    BIREFRINGENT FILTERS
    BREWSTER ANGLE
    GRADIENT INDEX OPTICS
    ISOTROPISM
    LIGHT (VISIBLE RADIATION)
    OPACITY
    OPTICAL THICKNESS
    POLARIZATION (WAVES)
    RATIOS
    REFRACTION
    REFRACTOMETERS
    SNELLS LAW

REFRACTIVITY-*(CONT.)*
UNDERWATER OPTICS

REFRACTOMETERS
GS   MEASURING INSTRUMENTS
. OPTICAL MEASURING INSTRUMENTS
. . **REFRACTOMETERS**
OPTICAL EQUIPMENT
. OPTICAL MEASURING INSTRUMENTS
. . **REFRACTOMETERS**
RT   GONIOMETERS
OPTICAL MEASUREMENT
REFRACTIVITY

REFRACTORIES
GS   REFRACTORY MATERIALS
. **REFRACTORIES**
RT   CERAMICS
CERMETS
COMBUSTION CHAMBERS
FORSTERITE
FURNACES
HEARTHS
MORTARS (MATERIAL)
REFRACTORY COATINGS
ROCKET ENGINES
ROCKET LININGS
THERMAL INSULATION
TURBINES

REFRACTORY COATINGS
GS   COATINGS
. PROTECTIVE COATINGS
. . **REFRACTORY COATINGS**
RT   CERAMICS
PYROLYTIC MATERIALS
REFRACTORIES
THERMAL INSULATION

REFRACTORY MATERIALS
UF   HIGH MELTING COMPOUNDS
HIGH TEMPERATURE MATERIALS
PYROGRAPHALLOY
GS   **REFRACTORY MATERIALS**
. LUDOX (TRADEMARK)
. PORCELAIN
. REFRACTORIES
. REFRACTORY METAL ALLOYS
. . MOLYBDENUM ALLOYS
. . . RENE 41
. . . RENE 63
. . . RENE 77
. . NIOBIUM ALLOYS
. . OSMIUM ALLOYS
. . RHENIUM ALLOYS
. . TANTALUM ALLOYS
. . TUNGSTEN ALLOYS
. REFRACTORY METALS
. . CHROMIUM
. . . CHROMIUM ISOTOPES
. . IRIDIUM
. . . IRIDIUM ISOTOPES
. . MOLYBDENUM
. . NIOBIUM
. . . NIOBIUM ISOTOPES
. . . . NIOBIUM 95
. . OSMIUM
. . . OSMIUM ISOTOPES
. . RHENIUM
. . . RHENIUM ISOTOPES
. . TANTALUM
. . . TANTALUM ISOTOPES
. . TUNGSTEN
. . . TUNGSTEN ISOTOPES
RT   ABLATIVE MATERIALS
CARBIDES
CARBORUNDUM (TRADEMARK)
CERAMICS
CERMETS
∞ CHEMICAL COMPOUNDS
CLAYS
HIGH TEMPERATURE RESEARCH
∞ INORGANIC MATERIALS
∞ MATERIALS
∞ METAL COMPOUNDS
NONFLAMMABLE MATERIALS
NOZZLE WALLS
PYROLYTIC MATERIALS
SCOTCHLITE (TRADEMARK)
SIALON

REFRACTORY METAL ALLOYS
GS   ALLOYS
. HEAT RESISTANT ALLOYS
. . **REFRACTORY METAL ALLOYS**

REFRACTORY METAL ALLOYS-*(CONT.)*
. . . MOLYBDENUM ALLOYS
. . . . RENE 41
. . . . RENE 63
. . . . RENE 77
. . . NIOBIUM ALLOYS
. . . OSMIUM ALLOYS
. . . RHENIUM ALLOYS
. . . TANTALUM ALLOYS
. . . TUNGSTEN ALLOYS
REFRACTORY MATERIALS
. **REFRACTORY METAL ALLOYS**
. . MOLYBDENUM ALLOYS
. . . RENE 41
. . . RENE 63
. . . RENE 77
. . NIOBIUM ALLOYS
. . OSMIUM ALLOYS
. . RHENIUM ALLOYS
. . TANTALUM ALLOYS
. . TUNGSTEN ALLOYS

REFRACTORY METALS
SN   (METALS MELTING ABOVE
APPROXIMATELY 2400 C)
GS   CHEMICAL ELEMENTS
. **REFRACTORY METALS**
. . CHROMIUM
. . . CHROMIUM ISOTOPES
. . IRIDIUM
. . . IRIDIUM ISOTOPES
. . MOLYBDENUM
. . NIOBIUM
. . . NIOBIUM ISOTOPES
. . . . NIOBIUM 95
. . OSMIUM
. . . OSMIUM ISOTOPES
. . TANTALUM
. . . TANTALUM ISOTOPES
. . TUNGSTEN
. . TUNGSTEN ISOTOPES
METALS
. TRANSITION METALS
. . **REFRACTORY METALS**
. . . CHROMIUM
. . . . CHROMIUM ISOTOPES
. . . IRIDIUM
. . . . IRIDIUM ISOTOPES
. . . MOLYBDENUM
. . . NIOBIUM
. . . . NIOBIUM ISOTOPES
. . . . . NIOBIUM 95
. . . OSMIUM
. . . . OSMIUM ISOTOPES
. . . RHENIUM
. . . . RHENIUM ISOTOPES
. . . TANTALUM
. . . . TANTALUM ISOTOPES
. . . TUNGSTEN
. . . TUNGSTEN ISOTOPES
REFRACTORY MATERIALS
. **REFRACTORY METALS**
. . CHROMIUM
. . . CHROMIUM ISOTOPES
. . IRIDIUM
. . . IRIDIUM ISOTOPES
. . MOLYBDENUM
. . NIOBIUM
. . . NIOBIUM ISOTOPES
. . . . NIOBIUM 95
. . OSMIUM
. . . OSMIUM ISOTOPES
. . RHENIUM
. . . RHENIUM ISOTOPES
. . TANTALUM
. . . TANTALUM ISOTOPES
. . TUNGSTEN
. . TUNGSTEN ISOTOPES
RT   HEAT RESISTANT ALLOYS

REFRACTORY PERIOD
RT   REACTION TIME
∞ RELAXATION
RESPONSES
TIME LAG

REFRASIL (TRADEMARK)
USE   FIBERS
SILICON DIOXIDE

REFRIGERANTS
RT   ABSORBERS (MATERIALS)
ABSORPTION COOLING
AIR CONDITIONING
AMMONIA

REFRIGERANTS-*(CONT.)*
BRINES
COOLANTS
COOLING SYSTEMS
FLUOROHYDROCARBONS
FREON
ICE
REFRIGERATING
REFRIGERATING MACHINERY
REFRIGERATORS
SOLID NITROGEN

REFRIGERATING
RT   AIR CONDITIONING
AIR COOLING
COLD TRAPS
CONDENSING
COOLERS
COOLING
COOLING SYSTEMS
CRYOGENIC COOLING
CRYOGENIC EQUIPMENT
CRYOGENICS
DEFROSTING
DEHUMIDIFICATION
FREEZING
FREON
FROZEN FOODS
HUMIDITY
LOW TEMPERATURE
MAGNETIC COOLING
PRESERVING
REFRIGERANTS
REFRIGERATORS
TEMPERATURE
TEMPERATURE CONTROL
TEMPERATURE DISTRIBUTION
THERMOELECTRIC COOLING
VENTILATION

REFRIGERATING MACHINERY
GS   **REFRIGERATING MACHINERY**
. REFRIGERATORS
RT   ABSORBERS (EQUIPMENT)
AIR CONDITIONING
AIR CONDITIONING EQUIPMENT
BLOWERS
COMPRESSORS
CONDENSERS (LIQUEFIERS)
COOLERS
COOLING SYSTEMS
CRYOGENIC EQUIPMENT
EVAPORATORS
HEAT PUMPS
∞ MACHINERY
REFRIGERANTS
TEMPERATURE CONTROL
THERMOELECTRIC COOLING

REFRIGERATORS
GS   REFRIGERATING MACHINERY
. **REFRIGERATORS**
RT   COOLERS
DEFROSTING
REFRIGERANTS
REFRIGERATING

REFSAT
GS   SATELLITES
. ARTIFICIAL SATELLITES
. . NAVIGATION SATELLITES
. . . **REFSAT**
UNMANNED SPACECRAFT
. NAVIGATION SATELLITES
. . **REFSAT**
RT   NAVSTAR SATELLITES
SYNCHRONOUS SATELLITES

REFUELING
UF   FUELING
GS   **REFUELING**
. AIR TO AIR REFUELING
RT   AIRCRAFT HAZARDS
FLIGHT OPERATIONS
FUEL CONSUMPTION
FUEL CONTAMINATION
FUEL CONTROL
FUEL SYSTEMS
GROUND SUPPORT EQUIPMENT
PREFLIGHT OPERATIONS
PROPELLANT TRANSFER
REPLENISHMENT
RETRACTABLE EQUIPMENT

∞ **REGENERATION**
SN   *(USE OF A MORE SPECIFIC TERM IS*
        RECOMMENDED--CONSULT THE TERMS
        LISTED BELOW)
RT   REGENERATION (ENGINEERING)
        REGENERATION (PHYSIOLOGY)

REGENERATION (ENGINEERING)
UF   REGENERATIVE CYCLES
RT   ∞GENERATION
        POSITIVE FEEDBACK
        RECLAMATION
        ∞REGENERATION

REGENERATION (PHYSIOLOGY)
UF   BIOREGENERATION
RT   PHYSIOLOGY
        ∞REGENERATION

REGENERATIVE COOLING
GS   COOLING
        . **REGENERATIVE COOLING**
RT   HEAT EXCHANGERS
        PRECOOLING
        REGENERATORS

REGENERATIVE CYCLES
USE   REGENERATION (ENGINEERING)

REGENERATIVE FEEDBACK
USE   POSITIVE FEEDBACK

REGENERATIVE FUEL CELLS
GS   ELECTRIC GENERATORS
        . DIRECT POWER GENERATORS
        . . FUEL CELLS
        . . . **REGENERATIVE FUEL CELLS**
        ELECTROCHEMICAL CELLS
        . FUEL CELLS
        . . **REGENERATIVE FUEL CELLS**
RT   BIOCHEMICAL FUEL CELLS
        PHOSPHORIC ACID FUEL CELLS
        STORAGE BATTERIES

REGENERATORS
UF   RECUPERATORS
GS   **REGENERATORS**
        . THERMOSIPHONS
RT   ENERGY STORAGE
        HEAT EXCHANGERS
        HEAT SINKS
        REGENERATIVE COOLING
        TUBE HEAT EXCHANGERS

REGGE POLES
RT   ANGULAR MOMENTUM
        ∞POLES
        POMERANCHUK THEOREM
        POMERONS
        SCATTERING CROSS SECTIONS

REGIMES
RT   COMMUNITIES
        CULTURE (SOCIAL SCIENCES)
        ENVIRONMENTS
        GOVERNMENTS
        NATIONS
        POLITICS

REGIONAL PLANNING
GS   PLANNING
        . **REGIONAL PLANNING**
        . . URBAN PLANNING
RT   CONSERVATION
        FARMLANDS
        FOREST MANAGEMENT
        FORESTS
        HARBORS
        HIGHWAYS
        INDUSTRIAL AREAS
        LAKES
        LAND MANAGEMENT
        MEGALOPOLISES
        PARKS
        RESIDENTIAL AREAS
        RURAL AREAS
        RURAL LAND USE
        ST LOUIS-KANSAS CITY CORRIDOR (MO)
        SUBURBAN AREAS
        URBAN DEVELOPMENT
        URBAN TRANSPORTATION

REGIONS
UF   ZONES

REGIONS-*(CONT.)*
GS   **REGIONS**
        . AURORAL ZONES
        . BRILLOUIN ZONES
        . CENTRAL ATLANTIC REGION (US)
        . D REGION
        . E REGION
        . EQUATORIAL REGIONS
        . F REGION
        . FRESNEL REGION
        . GUTENBERG ZONE
        . HABITATS
        . INTERTROPICAL CONVERGENT ZONES
        . LUMBAR REGION
        . M REGION
        . NEW ENGLAND (US)
        . NULL ZONES
        . PACIFIC NORTHWEST (US)
        . PANAMA CANAL ZONE
        . PELAGIC ZONE
        . POLAR REGIONS
        . . ANTARCTIC REGIONS
        . . . MCMURDO SOUND
        . . . ROSS ICE SHELF
        . . ARCTIC REGIONS
        . REMOTE REGIONS
        . . ANTARCTIC REGIONS
        . . ARCTIC REGIONS
        . SAND HILLS REGION (GA-NC-SC)
        . SAND HILLS REGION (NE)
        . SOUTHEAST ASIA
        . SOUTHERN CALIFORNIA
        . SUBARCTIC REGIONS
        . TEMPERATE REGIONS
        . TROPICAL REGIONS
        . . AMAZON REGION (SOUTH AMERICA)
RT   ASTEROID BELTS
        ∞BELTS
        BOUNDARIES
        CENTRAL AMERICA
        IONOSPHERE
        ∞LAYERS
        RECOVERY ZONES
        ∞SECTORS
        SITES

∞ **REGISTERS**
SN   *(USE OF A MORE SPECIFIC TERM IS*
        RECOMMENDED--CONSULT THE TERMS
        LISTED BELOW)
RT   REGISTERS (AIR CIRCULATION)
        REGISTERS (COMPUTERS)

REGISTERS (AIR CIRCULATION)
RT   COOLING SYSTEMS
        ∞REGISTERS

REGISTERS (COMPUTERS)
GS   PERIPHERAL EQUIPMENT (COMPUTERS)
        . COMPUTER STORAGE DEVICES
        . . **REGISTERS (COMPUTERS)**
RT   CENTRAL PROCESSING UNITS
        ∞RECORDERS
        ∞REGISTERS
        SHIFT REGISTERS

REGOLITH
GS   ROCKS
        . **REGOLITH**
RT   BASALT
        BEDROCK
        BRECCIA
        CARBONACEOUS ROCKS
        COAL
        EARTH MANTLE
        EARTH RESOURCES
        ENSTATITE
        GEOLOGY
        IGNEOUS ROCKS
        LAVA
        LITHOLOGY
        LUNAR GEOLOGY
        LUNAR MANTLE
        LUNAR ROCKS
        MAGMA
        OLIVINE
        PERIDOTITE
        PYROXENES
        ROCK INTRUSIONS
        SELENOLOGY
        STRATIGRAPHY

REGRESSION (STATISTICS)
USE   REGRESSION ANALYSIS

REGRESSION ANALYSIS
UF   REGRESSION (STATISTICS)
GS   STATISTICAL ANALYSIS
        . VARIANCE (STATISTICS)
        . . MULTIVARIATE STATISTICAL
            ANALYSIS
        . . . **REGRESSION ANALYSIS**
RT   AUTOREGRESSIVE PROCESSES
        CLUMPS
        CORRELATION
        COVARIANCE
        EXPERIMENT DESIGN
        FACTOR ANALYSIS
        FORECASTING
        LEAST SQUARES METHOD
        QUALITY CONTROL
        SIGNIFICANCE
        STATISTICAL TESTS
        VARIABILITY

REGRESSION COEFFICIENTS
GS   COEFFICIENTS
        . **REGRESSION COEFFICIENTS**
RT   CORRELATION
        FORECASTING
        MATHEMATICAL MODELS
        QUALITY CONTROL

REGULARITY
RT   PHYSIOLOGY
        ∞PROPERTIES

REGULATION
USE   CONTROL

REGULATIONS
RT   AIR LAW
        ALLOWANCES
        ∞CONTROL
        COPYRIGHTS
        CRIME
        LAW (JURISPRUDENCE)
        LIABILITIES
        LICENSING
        PATENT POLICY
        PENALTIES
        POLICE
        POLICIES
        PROCUREMENT POLICY
        PROHIBITION
        RULES

REGULATORS
SN   (EXTRACT ENERGY NEEDED FOR
        OPERATION FROM THE VARIABLE,
        MATERIAL, OR CONDITION BEING
        REGULATED)
GS   **REGULATORS**
        . AUTOMATIC FREQUENCY CONTROL
        . CURRENT REGULATORS
        . FLOW REGULATORS
        . . FUEL FLOW REGULATORS
        . FREQUENCY CONTROL
        . GIBBERELLINS
        . OXYGEN REGULATORS
        . PRESSURE REGULATORS
        . RELIEF VALVES
        . SPEED REGULATORS
        . THERMOSTATS
        . VOLTAGE REGULATORS
RT   ACTUATORS
        AUTOMATIC CONTROL
        AUTOMATIC CONTROL VALVES
        ∞CONTROL
        CONTROLLERS
        CRYOSTATS
        SPEED CONTROL

REGULUS MISSILE
GS   MISSILES
        . SURFACE TO SURFACE MISSILES
        . . **REGULUS MISSILE**
RT   SOLID PROPELLANT ROCKET ENGINES
        TURBOJET ENGINES

REHEATING
USE   HEATING

REIGNITION
USE   IGNITION

REINFORCED MATERIALS
USE   COMPOSITE MATERIALS

REINFORCED PLASTICS
GS   COMPOSITE MATERIALS
    . **REINFORCED PLASTICS**
    . . GLASS FIBER REINFORCED PLASTICS
    . . MICARTA
    PLASTICS
    . **REINFORCED PLASTICS**
    . . MICARTA
RT   AIRCRAFT SURVIVABILITY
    BORON FIBERS
    BORON REINFORCED MATERIALS
    FIBER COMPOSITES
    GRAPHITE-EPOXY COMPOSITES
    LAMINATES
    POLYMER MATRIX COMPOSITES
    REINFORCEMENT (STRUCTURES)
    THERMOSETTING RESINS

REINFORCED PLATES
GS   STRUCTURAL MEMBERS
    . PLATES (STRUCTURAL MEMBERS)
    . . **REINFORCED PLATES**
RT   ANISOTROPIC PLATES
    CORRUGATED PLATES
    LAMINATES
    REINFORCEMENT (STRUCTURES)

REINFORCED SHELLS
GS   SHELLS (STRUCTURAL FORMS)
    . **REINFORCED SHELLS**
RT   ANISOTROPIC SHELLS
    CORRUGATED SHELLS
    CYLINDRICAL SHELLS
    FLUID FILLED SHELLS
    HEMISPHERICAL SHELLS
    LIQUID FILLED SHELLS
    METAL SHELLS
    ORTHOTROPIC SHELLS
    REINFORCEMENT (STRUCTURES)
    REINFORCEMENT RINGS
    SHELL STABILITY
    SPHERICAL SHELLS
    THIN WALLED SHELLS
    TOROIDAL SHELLS
    WIND TUNNEL WALLS

∞ **REINFORCEMENT**
SN   *(USE OF A MORE SPECIFIC TERM IS*
    RECOMMENDED--CONSULT THE TERMS
    LISTED BELOW)
RT   REINFORCEMENT (PSYCHOLOGY)
    REINFORCEMENT (STRUCTURES)

REINFORCEMENT (PSYCHOLOGY)
GS   **REINFORCEMENT (PSYCHOLOGY)**
    . REWARD (PSYCHOLOGY)
RT   LEARNING
    MOTIVATION
    ∞ REINFORCEMENT
    SELF STIMULATION

REINFORCEMENT (STRUCTURES)
RT   BULKHEADS
    COMPOSITE MATERIALS
    FILLERS
    LONGERONS
    REINFORCED PLASTICS
    REINFORCED PLATES
    REINFORCED SHELLS
    ∞ REINFORCEMENT
    REINFORCEMENT RINGS
    RIBS (SUPPORTS)
    RIGID STRUCTURES
    RING STRUCTURES
    STIFFENING
    STRAKES
    STRINGERS
    STRUCTURAL MEMBERS
    STRUCTURAL STABILITY
    STRUCTURAL STRAIN
    SUPPORTS
    THICK WALLS
    WIRE
    WIRE CLOTH

REINFORCEMENT RINGS
GS   RING STRUCTURES
    . **REINFORCEMENT RINGS**
RT   REINFORCED SHELLS
    REINFORCEMENT (STRUCTURES)
    RIBS (SUPPORTS)
    ∞ RINGS

REINFORCING FIBERS
GS   FIBERS

REINFORCING FIBERS-*(CONT.)*
    . **REINFORCING FIBERS**
    . . BORON FIBERS
    . . CARBON FIBERS
RT   BORON REINFORCED MATERIALS
    CARBON FIBER REINFORCED PLASTICS
    CARBON-CARBON COMPOSITES
    CERAMIC MATRIX COMPOSITES
    COMPOSITE MATERIALS
    DACRON (TRADEMARK)
    FIBER COMPOSITES
    FIBER ORIENTATION
    GLASS FIBER REINFORCED PLASTICS
    GLASS FIBERS
    GRAPHITE-EPOXY COMPOSITES
    LAY-UP
    METAL FIBERS
    METAL MATRIX COMPOSITES
    MICROMECHANICS
    REINFORCING MATERIALS
    SUPERHYBRID MATERIALS
    SYNTHETIC FIBERS
    WHISKER COMPOSITES

REINFORCING MATERIALS
RT   COMPOSITE MATERIALS
    FABRICS
    FIBER REINFORCED COMPOSITES
    FIBERS
    ∞ FILAMENTS
    ∞ MATERIALS
    MATRIX MATERIALS
    REINFORCING FIBERS

REISSNER THEORY
RT   PLATES (STRUCTURAL MEMBERS)
    STRESS ANALYSIS
    ∞ THEORIES

REISSNER-NORDSTROM SOLUTION
RT   ASTRONOMICAL MODELS
    BLACK HOLES (ASTRONOMY)
    CHARGED PARTICLES
    GRAVITATIONAL EFFECTS
    RELATIVITY

REJECTION
RT   ACCEPTABILITY
    ELIMINATION
    EVALUATION
    EXCLUSION
    REMOVAL
    SELECTION

∞ **RELATIONSHIPS**
SN   *(USE OF A MORE SPECIFIC TERM IS*
    RECOMMENDED--CONSULT THE TERMS
    LISTED BELOW)
UF   INTERRELATIONSHIPS
RT   APPROXIMATION
    DUALITY THEOREM
    HOMOLOGY
    STRESS-STRAIN RELATIONSHIPS

RELATIVE BIOLOGICAL EFFECTIVENESS (RBE)
UF   RBE
GS   BIOLOGICAL EFFECTS
    . **RELATIVE BIOLOGICAL**
      EFFECTIVENESS (RBE)
RT   ∞ BIOLOGY
    PHYSIOLOGICAL EFFECTS

RELATIVISTIC EFFECTS
RT   DIMENSIONS
    ∞ EFFECTS
    GRAVITATIONAL LENSES
    MASS
    RELATIVITY
    TIME
    VELOCITY

RELATIVISTIC ELECTRON BEAMS
UF   REB
GS   BEAMS (RADIATION)
    . PARTICLE BEAMS
    . . ELECTRON BEAMS
    . . . **RELATIVISTIC ELECTRON BEAMS**
    PARTICLES
    . CORPUSCULAR RADIATION
    . . ELECTRON RADIATION
    . . . ELECTRON BEAMS
    . . . . **RELATIVISTIC ELECTRON BEAMS**
    . RELATIVISTIC PARTICLES
    . . **RELATIVISTIC ELECTRON BEAMS**
RT   BEAM PLASMA AMPLIFIERS

RELATIVISTIC ELECTRON BEAMS-*(CONT.)*
    BETA PARTICLES
    CONTROLLED FUSION
    ELECTRON BOMBARDMENT
    ELECTRON SCATTERING
    INERTIAL FUSION (REACTOR)
    IONIZING RADIATION
    PLASMA HEATING
    PLASMA JETS
    PLASMA-PARTICLE INTERACTIONS

RELATIVISTIC PARTICLES
GS   PARTICLES
    . **RELATIVISTIC PARTICLES**
    . . RELATIVISTIC ELECTRON BEAMS
RT   HAMILTON-JACOBI EQUATION

RELATIVISTIC PLASMAS
GS   GASES
    . IONIZED GASES
    . . CHARGED PARTICLES
    . . . **RELATIVISTIC PLASMAS**
    PARTICLES
    . CHARGED PARTICLES
    . . ENERGETIC PARTICLES
    . . . PLASMAS (PHYSICS)
    . . . . **RELATIVISTIC PLASMAS**
RT   ASTRON THERMONUCLEAR REACTOR
    BREMSSTRAHLUNG
    COSMIC PLASMA
    ELECTRON PLASMA
    GRAVITATIONAL COLLAPSE
    HIGH TEMPERATURE PLASMAS
    PINCH EFFECT
    PLASMA JETS
    PLASMA RADIATION
    PLASMA-PARTICLE INTERACTIONS
    PONDEROMOTIVE FORCES

RELATIVISTIC THEORY
UF   WIGHTMAN THEORY
GS   **RELATIVISTIC THEORY**
    . MANDELSTAM REPRESENTATION
RT   ∞ THEORIES

RELATIVISTIC VELOCITY
GS   RATES (PER TIME)
    . **RELATIVISTIC VELOCITY**
    VELOCITY
    . **RELATIVISTIC VELOCITY**
RT   HIGH SPEED
    ∞ HYPERVELOCITY
    LIGHT SPEED
    PARTICLE MOTION

RELATIVITY
UF   GEOMETRODYNAMICS
    SPACE-TIME CONTINUUM
RT   BIG BANG COSMOLOGY
    CONTINUUMS
    DIFFERENTIAL GEOMETRY
    FIELD THEORY (PHYSICS)
    GRAVITATIONAL LENSES
    GRAVITY PROBE B
    INERTIAL REFERENCE SYSTEMS
    LORENTZ CONTRACTION
    NAKED SINGULARITIES
    NONRELATIVISTIC MECHANICS
    PARADOXES
    PONDEROMOTIVE FORCES
    QUANTUM MECHANICS
    REISSNER-NORDSTROM SOLUTION
    RELATIVISTIC EFFECTS
    SCHWARZSCHILD METRIC
    SPACE-TIME FUNCTIONS
    TENSOR ANALYSIS
    UNIFIED FIELD THEORY

∞ **RELAXATION**
SN   *(USE OF A MORE SPECIFIC TERM IS*
    RECOMMENDED--CONSULT THE TERMS
    LISTED BELOW)
RT   CROSS RELAXATION
    MOLECULAR RELAXATION
    REFRACTORY PERIOD
    RELAXATION (MECHANICS)
    RELAXATION (PHYSIOLOGY)
    RELAXATION METHOD (MATHEMATICS)

RELAXATION (MECHANICS)
GS   **RELAXATION (MECHANICS)**
    . SPIN-LATTICE RELAXATION
    . STRESS RELAXATION
RT   ∞ EQUILIBRIUM
    EXPANSION

RELAXATION (MECHANICS)-*(CONT.)*
    MAGNETIC RELAXATION
    MOLECULAR RELAXATION
    NUCLEAR RELAXATION
    ∞REDUCTION
    RELAXATION METHOD (MATHEMATICS)
    RELAXATION TIME
    RESIDUAL STRESS
    VISCOELASTICITY
    VISCOPLASTICITY

RELAXATION (PHYSIOLOGY)
RT   COMPRESSIBILITY EFFECTS
    MASSAGING
    RECREATION
    ∞RELAXATION
    WORK-REST CYCLE

RELAXATION METHOD (MATHEMATICS)
GS   ANALYSIS (MATHEMATICS)
    . NUMERICAL ANALYSIS
    . . APPROXIMATION
    . . . **RELAXATION METHOD (MATHEMATICS)**
RT   COMPUTATIONAL FLUID DYNAMICS
    ∞METHODOLOGY
    ∞RELAXATION
    RELAXATION (MECHANICS)

RELAXATION OSCILLATORS
GS   OSCILLATORS
    . **RELAXATION OSCILLATORS**
    . . PHANTASTRONS

RELAXATION TIME
GS   TIME
    . **RELAXATION TIME**
RT   ∞EQUILIBRIUM
    EXCITATION
    MAXWELL BODIES
    MOLECULAR RELAXATION
    SPIN-LATTICE RELAXATION
    TIME CONSTANT

∞ **RELAY**
SN   *(USE OF A MORE SPECIFIC TERM IS RECOMMENDED--CONSULT THE TERMS LISTED BELOW)*
RT   DISCONNECT DEVICES
    ELECTRIC CONTACTS
    ELECTRIC RELAYS
    LOGIC CIRCUITS
    RADIO RELAY SYSTEMS
    REPEATERS

RELAY SATELLITES
GS   SATELLITES
    . ARTIFICIAL SATELLITES
    . . COMMUNICATION SATELLITES
    . . . **RELAY SATELLITES**
    . . . . RELAY 1 SATELLITE
    . . . . RELAY 2 SATELLITE
RT   ADVENT PROJECT
    THOR DELTA LAUNCH VEHICLE
    TRANSOCEANIC COMMUNICATION

RELAY 1 SATELLITE
GS   SATELLITES
    . ARTIFICIAL SATELLITES
    . . COMMUNICATION SATELLITES
    . . . RELAY SATELLITES
    . . . . **RELAY 1 SATELLITE**

RELAY 2 SATELLITE
GS   SATELLITES
    . ARTIFICIAL SATELLITES
    . . COMMUNICATION SATELLITES
    . . . RELAY SATELLITES
    . . . . **RELAY 2 SATELLITE**

RELEASING
GS   **RELEASING**
    . FIBER RELEASE
RT   DECOUPLING
    ∞DISCHARGE
    DISCONNECT DEVICES
    DISPERSING
    DUMPING
    EJECTION
    EMISSION
    EMPTYING
    EXHAUST EMISSION
    MATERIALS HANDLING
    RELIEVING

RELEASING-*(CONT.)*
    SCATTERING
    SPILLING
    UNLOADING
    VENTING

RELIABILITY
GS   **RELIABILITY**
    . AIRCRAFT RELIABILITY
    . CIRCUIT RELIABILITY
    . COMPONENT RELIABILITY
    . SPACECRAFT RELIABILITY
    . STRUCTURAL RELIABILITY
RT   ACCEPTABILITY
    ACCURACY
    AIRCRAFT SURVIVABILITY
    ALLOWANCES
    ASSURANCE
    CENSORED DATA (MATHEMATICS)
    COMPUTER SYSTEMS PERFORMANCE
    CONFIDENCE
    CONFIDENCE LIMITS
    CONSISTENCY
    CUMULATIVE DAMAGE
    ∞DESIGN
    DESIGN ANALYSIS
    DOWNTIME
    DURABILITY
    DYNAMIC CHARACTERISTICS
    ERRORS
    ESTIMATES
    EXPECTATION
    FAILURE ANALYSIS
    FORECASTING
    MAINTAINABILITY
    MAINTENANCE
    MAXIMUM LIKELIHOOD ESTIMATES
    MECHANICAL PROPERTIES
    MISSILE DESIGN
    MTBF
    NONDESTRUCTIVE TESTS
    ∞PERFORMANCE
    PERFORMANCE PREDICTION
    PRECISION
    PRELAUNCH PROBLEMS
    PROBABILITY THEORY
    PRODUCT DEVELOPMENT
    PRODUCTION MANAGEMENT
    PRODUCTIVITY
    QUALITY
    QUALITY CONTROL
    REDUNDANCY
    REDUNDANT COMPONENTS
    RISK
    SAFETY FACTORS
    SAMPLING
    SPECIFICATIONS
    STABILITY
    STANDARDS
    STATISTICAL ANALYSIS
    STATISTICAL DISTRIBUTIONS
    STATISTICAL TESTS
    ∞STATISTICS
    SYSTEM EFFECTIVENESS
    SYSTEMS COMPATIBILITY
    SYSTEMS ENGINEERING
    ∞TESTS
    TOLERANCES (MECHANICS)
    VALIDITY
    VARIABILITY
    VULNERABILITY

RELIABILITY ANALYSIS
RT   ∞ANALYZING
    DESIGN ANALYSIS
    PERFORMANCE PREDICTION

RELIABILITY CONTROL
USE   QUALITY CONTROL
    RELIABILITY ENGINEERING

RELIABILITY ENGINEERING
UF   RELIABILITY CONTROL
RT   COMPLEX SYSTEMS
    ∞ENGINEERING
    FAULT TOLERANCE
    PERFORMANCE PREDICTION
    QUALITY CONTROL
    SNEAK CIRCUIT ANALYSIS
    SYSTEM EFFECTIVENESS
    SYSTEM IDENTIFICATION
    SYSTEMS COMPATIBILITY
    SYSTEMS ENGINEERING
    VALUE ENGINEERING

RELIC RADIATION
RT   ASTRONOMY
    ASTROPHYSICS
    BACKGROUND RADIATION
    BIG BANG COSMOLOGY
    EXTRATERRESTRIAL RADIATION
    ∞RADIATION
    UNIVERSE

RELIEF MAPS
GS   MAPS
    . **RELIEF MAPS**
RT   HYPSOGRAPHY
    PHOTOGRAMMETRY
    PHOTOMAPS
    TOPOGRAPHY

RELIEF VALVES
GS   REGULATORS
    . **RELIEF VALVES**
    VALVES
    . AUTOMATIC CONTROL VALVES
    . . **RELIEF VALVES**
RT   AUTOMATIC CONTROL
    BYPASSES
    FUEL TANK PRESSURIZATION
    FUEL VALVES
    GAS VALVES
    HYDRAULIC EQUIPMENT
    PRESSURE REGULATORS
    REACTOR SAFETY
    VENTING
    VENTS

RELIEVING
GS   **RELIEVING**
    . STRESS RELIEVING
RT   ∞DISCHARGE
    EXHAUSTING
    PURGING
    RELEASING

RELOCATION
RT   INSTALLING
    POSITIONING
    REPLACING

RELUCTANCE
UF   RELUCTIVITY
GS   MAGNETIC PROPERTIES
    . **RELUCTANCE**
RT   MAGNETIC PERMEABILITY
    MAGNETORESISTIVITY

RELUCTIVITY
USE   RELUCTANCE

REMAGNETIZATION
USE   MAGNETIZATION

REMANENCE
GS   MAGNETIC PROPERTIES
    . **REMANENCE**
RT   FLUX DENSITY
    PALEOMAGNETISM

REMELTING
USE   MELTING

REMODULATION
RT   DEMODULATION
    INTERMODULATION
    MODULATION

REMOTE CONSOLES
GS   PERIPHERAL EQUIPMENT (COMPUTERS)
    . CONSOLES
    . . **REMOTE CONSOLES**
RT   COMPUTER COMPONENTS
    COMPUTER GRAPHICS
    DATA LINKS
    DATA PROCESSING EQUIPMENT
    DATA PROCESSING TERMINALS
    DISPLAY DEVICES
    PLOTTERS
    READOUT

REMOTE CONTROL
UF   ELECTROMAGNETIC CONTROL
GS   **REMOTE CONTROL**
    . RADIO CONTROL
RT   AIRCRAFT CONTROL
    ANTIRADIATION MISSILES
    ATTITUDE CONTROL

REMOTE CONTROL-*(CONT.)*
    AUTOMATIC CONTROL
∞ AUTOMATION
    CASCADE CONTROL
∞ CONTROL
    CONTROL BOARDS
    CONTROLLERS
    DIGITAL COMMAND SYSTEMS
    DYNAMIC CHARACTERISTICS
    ELECTRIC CONTROL
    ELECTRONIC CONTROL
    ENGINE CONTROL
    FLIGHT CONTROL
    GROUND BASED CONTROL
    GUIDANCE (MOTION)
    HYDRAULIC CONTROL
∞ INSTRUMENTS
    KALMAN-SCHMIDT FILTERING
    MANIPULATORS
    MANUAL CONTROL
    MISSILE CONTROL
    PNEUMATIC CONTROL
    REMOTE MANIPULATOR SYSTEM
    ROCKET ENGINE CONTROL
    SATELLITE CONTROL
    SERVOCONTROL
    SERVOMECHANISMS
    SPACECRAFT CONTROL
    TELEOPERATORS
    TEMPERATURE CONTROL
    TURBOJET ENGINE CONTROL
    VISUAL CONTROL

REMOTE HANDLING
UF TELECHIRICS
GS MATERIALS HANDLING
    . REMOTE HANDLING
RT MANIPULATORS
    PAYLOAD DEPLOYMENT & RETRIEVAL
        SYSTEM
    TELEOPERATORS

REMOTE MANIPULATOR SYSTEM
GS MANIPULATORS
    . REMOTE MANIPULATOR SYSTEM
    PAYLOAD DEPLOYMENT & RETRIEVAL
        SYSTEM
    . REMOTE MANIPULATOR SYSTEM
RT MASS DRIVERS (PAYLOAD DELIVERY)
    PAYLOAD RETRIEVAL (STS)
    REMOTE CONTROL
    SPACE MAINTENANCE
    SPACE TRANSPORTATION SYSTEM
∞ SYSTEMS

REMOTE REGIONS
GS REGIONS
    . REMOTE REGIONS
    . . ANTARCTIC REGIONS
    . . ARCTIC REGIONS
RT DESERTS
    MOJAVE DESERT (CA)
    OFFSHORE REACTOR SITES
    SAHARA DESERT (AFRICA)
    WILDERNESS

REMOTE SENSING
GS DETECTION
    . REMOTE SENSING
RT AEROMAGNETISM
    BAND RATIOING
    CHANGE DETECTION
    CLUSTER ANALYSIS
    DESERTIFICATION
    DMSP SATELLITES
    EARTH RESOURCES
    FEATURE IDENTIFICATION AND
        LOCATION EXPER
    GEOGRAPHIC INFORMATION SYSTEMS
    IMAGE ANALYSIS
    MAPSAT
    MULTISENSOR APPLICATIONS
    MULTISPECTRAL RESOURCE SAMPLER
    PLANT STRESS
    RECOGNITION
    SHUTTLE IMAGING RADAR
    SWATH WIDTH
    VEGETATIVE INDEX

REMOTE SENSORS
RT AGRISTARS PROJECT
    AIRBORNE LASERS
    AUTOMATIC WEATHER STATIONS
    CROP IDENTIFICATION
    CROP INVENTORIES

REMOTE SENSORS-*(CONT.)*
    DATA ACQUISITION
    DATA COLLECTION PLATFORMS
∞ DETECTORS
    EARTH RESOURCES
    EARTHNET
    EROS (SATELLITES)
    FEATURE IDENTIFICATION AND
        LOCATION EXPER
    GEOGRAPHIC APPLICATIONS PROGRAM
    HAZE DETECTION
    IMAGING RADAR
    MEASURING INSTRUMENTS
    MULTISENSOR APPLICATIONS
    OCEAN COLOR SCANNER
∞ PROBES
    RADIOMETRIC RESOLUTION
    SATELLITE-BORNE INSTRUMENTS
∞ SENSORS
    SPACEBORNE LASERS
    TRANSDUCERS
    WILDLIFE RADIOLOCATION

REMOTELY PILOTED VEHICLES
UF RPV
RT ∞ AIRCRAFT
    DAST PROGRAM
    DRONE AIRCRAFT
    HIGHLY MANEUVERABLE AIRCRAFT
    JINDIVIK TARGET AIRCRAFT
    OBLIQUE WINGS
    ORBITAL MANEUVERING VEHICLES
    PILOTLESS AIRCRAFT
    TARGET DRONE AIRCRAFT
    VATOL AIRCRAFT
∞ VEHICLES

REMOVAL
RT ANODIC STRIPPING
    CANCELLATION
    CARBON DIOXIDE REMOVAL
    CLEARING
    DELETION
    DEPLETION
    DISPOSAL
    DISSIPATION
    EJECTION
    EMPTYING
    EVACUATING (TRANSPORTATION)
    EVACUATING (VACUUM)
    EXHAUSTING
    EXPULSION
    EXTRACTION
    MATERIALS RECOVERY
∞ REDUCTION
    REJECTION
∞ SEPARATION
    UNLOADING
    WEAR

REMS
USE RAPID EYE MOVEMENT STATE

RENAL CALCULI
USE CALCULI

RENAL FUNCTION
RT ∞ FUNCTIONS
    KIDNEYS

RENDEZVOUS
GS **RENDEZVOUS**
    . EARTH ORBITAL RENDEZVOUS
    . SPACE RENDEZVOUS
    . . ORBITAL RENDEZVOUS
    . . . LUNAR ORBITAL RENDEZVOUS
RT APOLLO SOYUZ TEST PROJECT
    FLIGHT MECHANICS
    INTERCEPTION
    ORBITAL MECHANICS

RENDEZVOUS GUIDANCE
GS GUIDANCE (MOTION)
    . RENDEZVOUS GUIDANCE
RT COMMAND GUIDANCE
    HOMING DEVICES
    INJECTION GUIDANCE
    MIDCOURSE GUIDANCE
    ORBITAL RENDEZVOUS
    SATELLITE GUIDANCE
    SPACECRAFT GUIDANCE
    TERMINAL GUIDANCE

RENDEZVOUS SPACECRAFT
GS MANEUVERABLE SPACECRAFT

RENDEZVOUS SPACECRAFT-*(CONT.)*
    . RENDEZVOUS SPACECRAFT
RT COMMAND GUIDANCE
    FERRY SPACECRAFT
    INTERPLANETARY SPACECRAFT
    LUNAR SPACECRAFT
    MANNED SPACECRAFT
    MILITARY SPACECRAFT
    ORBITAL RENDEZVOUS
    RECOVERABLE SPACECRAFT
    SATELLITES
    SPACE CAPSULES
    SPACE STATIONS
    SPACECREW TRANSFER
    UNMANNED SPACECRAFT

RENDEZVOUS TRAJECTORIES
GS TRAJECTORIES
    . RENDEZVOUS TRAJECTORIES
RT ASCENT TRAJECTORIES
    CIRCUMLUNAR TRAJECTORIES
    EARTH ORBITAL RENDEZVOUS
    EARTH-MOON TRAJECTORIES
    FLIGHT MECHANICS
    INTERPLANETARY TRAJECTORIES
    ORBITAL MECHANICS
    ORBITAL RENDEZVOUS
    SPACE RENDEZVOUS
    SPACECRAFT DOCKING
    SPACECRAFT TRAJECTORIES

RENE 41
GS ALLOYS
    . CHROMIUM ALLOYS
    . . RENE 41
    . COBALT ALLOYS
    . . RENE 41
    . HEAT RESISTANT ALLOYS
    . . REFRACTORY METAL ALLOYS
    . . . MOLYBDENUM ALLOYS
    . . . . RENE 41
    . NICKEL ALLOYS
    . . RENE 41
    REFRACTORY MATERIALS
    . REFRACTORY METAL ALLOYS
    . . MOLYBDENUM ALLOYS
    . . . RENE 41
RT WROUGHT ALLOYS

RENE 63
GS ALLOYS
    . CHROMIUM ALLOYS
    . . RENE 63
    . COBALT ALLOYS
    . . RENE 63
    . HEAT RESISTANT ALLOYS
    . . REFRACTORY METAL ALLOYS
    . . . MOLYBDENUM ALLOYS
    . . . . RENE 63
    . NICKEL ALLOYS
    . . RENE 63
    REFRACTORY MATERIALS
    . REFRACTORY METAL ALLOYS
    . . MOLYBDENUM ALLOYS
    . . . RENE 63
RT WROUGHT ALLOYS

RENE 77
GS ALLOYS
    . CHROMIUM ALLOYS
    . . RENE 77
    . COBALT ALLOYS
    . . RENE 77
    . HEAT RESISTANT ALLOYS
    . . REFRACTORY METAL ALLOYS
    . . . MOLYBDENUM ALLOYS
    . . . . RENE 77
    . NICKEL ALLOYS
    . . RENE 77
    REFRACTORY MATERIALS
    . REFRACTORY METAL ALLOYS
    . . MOLYBDENUM ALLOYS
    . . . RENE 77
RT WROUGHT ALLOYS

RENE 95
GS ALLOYS
    . CHROMIUM ALLOYS
    . . RENE 95
    . COBALT ALLOYS
    . . RENE 95
    . NICKEL ALLOYS
    . . RENE 95

REORIENTATION
USE   RETRAINING

REPAIRING
USE   MAINTENANCE

REPEATERS
UF   INTERPOLATORS
GS   TRANSMITTERS
     . **REPEATERS**
RT   AMPLIFIERS
     RADAR RECEIVERS
     RECEIVERS
     ∞RELAY
     ∞TRANSLATORS

REPETITION
RT   COUNTING
     PATTERN RECOGNITION
     REDUNDANCY ENCODING

REPLACING
RT   DAMAGE ASSESSMENT
     INSTALLING
     MAINTENANCE
     RELOCATION
     REPLENISHMENT
     SUBSTITUTES

REPLENISHMENT
RT   FILLING
     INPUT
     ∞LOADING
     REFILLING
     REFUELING
     REPLACING

REPLICAS
RT   ELECTRON MICROSCOPES
     METALLOGRAPHY
     MODELS
     REPRODUCTION (COPYING)

REPORT GENERATORS
RT   COMPUTER PROGRAMS
     COMPUTER SYSTEMS PROGRAMS
     ∞GENERATORS
     USER MANUALS (COMPUTER
        PROGRAMS)

REPORTS
GS   **REPORTS**
     . CONGRESSIONAL REPORTS
     . POSTLAUNCH REPORTS
     . PRESIDENTIAL REPORTS
     . WAGE SURVEYS
RT   AEROSPACE TECHNOLOGY TRANSFER
     CONFERENCES
     ∞DISCUSSION
     DOCUMENTATION
     DOCUMENTS
     INFORMATION
     INFORMATION DISSEMINATION
     PAPERS
     RECORDS
     SUMMARIES
     SUPPLEMENTS
     TECHNOLOGY TRANSFER

REPRESENTATIONS
RT   CHARACTERIZATION
     DESCRIPTIONS
     DRAWINGS
     GRAPHS (CHARTS)
     IMAGES
     PHOTOGRAPHS
     SIGNATURES

∞ **REPRODUCTION**
SN   *(USE OF A MORE SPECIFIC TERM IS*
     RECOMMENDED--CONSULT THE TERMS
     LISTED BELOW)
RT   BIRTH
     BREEDING (REPRODUCTION)
     CELL DIVISION
     FERTILITY
     FERTILIZATION
     GONADS
     MITOSIS
     OVARIES
     POLLEN
     RECORDS
     REPRODUCTION (BIOLOGY)
     REPRODUCTION (COPYING)

REPRODUCTION-(CONT.)
     REPRODUCTIVE SYSTEMS
     SEX GLANDS
     SPERMATOZOA
     SPORES
     TESTES
     UTERUS

REPRODUCTION (BIOLOGY)
RT   ∞BIOLOGY
     BREEDING (REPRODUCTION)
     EMBRYOLOGY
     FERTILITY
     FETUSES
     PROGENY
     ∞REPRODUCTION

REPRODUCTION (COPYING)
UF   DUPLICATING
GS   IMAGERY
     . **REPRODUCTION (COPYING)**
     . . XEROGRAPHY
RT   BLUEPRINTS
     DOCUMENT STORAGE
     DRAWINGS
     ENGINEERING DRAWINGS
     LITHOGRAPHY
     MICROFILMS
     PHOTOGRAPHIC PROCESSING
        EQUIPMENT
     PHOTOGRAPHY
     PRINTING
     REPLICAS
     ∞REPRODUCTION
     STENCIL PROCESSES

REPRODUCTIVE SYSTEMS
GS   **REPRODUCTIVE SYSTEMS**
     . OVARIES
     . TESTES
     . UTERUS
RT   BIRTH
     CHROMOSOMES
     FERTILITY
     FETUSES
     ∞REPRODUCTION
     ∞SYSTEMS

REPTILES
GS   ANIMALS
     . POIKILOTHERMIA
     . . **REPTILES**
     . . . LIZARDS
     . . . SNAKES
     . . . TURTLES
     . VERTEBRATES
     . . **REPTILES**
     . . . LIZARDS
     . . . SNAKES
     . . . TURTLES

REPUBLIC AIRCRAFT
GS   **REPUBLIC AIRCRAFT**
     . A-10 AIRCRAFT
     . F-84 AIRCRAFT
     . F-105 AIRCRAFT
RT   ∞AIRCRAFT

REPUBLIC MILITARY AIRCRAFT
USE   MILITARY AIRCRAFT

REPUBLIC OF CHINA
USE   TAIWAN

REPUBLIC OF KOREA
USE   SOUTH KOREA

REPUBLIC OF SOUTH AFRICA
UF   SOUTH AFRICA
GS   NATIONS
     . **REPUBLIC OF SOUTH AFRICA**
RT   AFRICA
     BOTSWANA
     KALIHARI BASIN (AFRICA)
     LESOTHO
     NAMIBIA
     SWAZILAND

REPUBLIC OF VIETNAM
USE   VIETNAM

REPULSION
USE   FORCE

REQUIREMENTS
RT   SPECIFICATIONS
     USER REQUIREMENTS

RESCUE OPERATIONS
RT   AERONAUTICAL SATELLITES
     COSPAS
     MAROTS (ESA)
     ∞OPERATIONS
     SARSAT
     SPACECRAFT RECOVERY

RESEARCH
GS   **RESEARCH**
     . HIGH TEMPERATURE RESEARCH
     . LOW DENSITY RESEARCH
     . MARKET RESEARCH
     . MATHEMATICAL PROGRAMMING
     . . DYNAMIC PROGRAMMING
     . . LINEAR PROGRAMMING
     . . NONLINEAR PROGRAMMING
     . . QUADRATIC PROGRAMMING
     . NUCLEAR RESEARCH
RT   CRITICAL PATH METHOD
     ∞DESIGN
     ETHICS
     EXPLORATION
     INTERSERVICE DATA EXCHANGE
        PROGRAM
     INVESTIGATION
     MINIMAX TECHNIQUE
     ∞RESEARCH PROJECTS

RESEARCH AIRCRAFT
UF   ORNITHOPTER AIRCRAFT
GS   **RESEARCH AIRCRAFT**
     . AVIAN 2/180 AUTOGIRO
     . AVRO 707 AIRCRAFT
     . B-70 AIRCRAFT
     . BREGUET 940 AIRCRAFT
     . C-8A AUGMENTOR WING AIRCRAFT
     . D-558 AIRCRAFT
     . FD 2 AIRCRAFT
     . FIREBEE 2 TARGET DRONE AIRCRAFT
     . H-17 HELICOPTER
     . H-126 AIRCRAFT
     . HP-115 AIRCRAFT
     . NORD 1500 AIRCRAFT
     . P-1052 AIRCRAFT
     . QUESTOL
     . ROTOR SYSTEMS RESEARCH
        AIRCRAFT
     . SC-1 AIRCRAFT
     . U-2 AIRCRAFT
     . VZ-2 AIRCRAFT
     . VZ-8 AIRCRAFT
     . X-1 AIRCRAFT
     . X-2 AIRCRAFT
     . X-3 AIRCRAFT
     . X-5 AIRCRAFT
     . X-13 AIRCRAFT
     . X-14 AIRCRAFT
     . X-15 AIRCRAFT
     . X-19 AIRCRAFT
     . X-20 AIRCRAFT
     . X-21 AIRCRAFT
     . X-21A AIRCRAFT
     . X-22 AIRCRAFT
     . X-22A AIRCRAFT
     . X-24 AIRCRAFT
     . XH-51 HELICOPTER
     . XV-4 AIRCRAFT
     . XV-5 AIRCRAFT
     . XV-8A AIRCRAFT
     . XV-9A AIRCRAFT
     . XV-11A AIRCRAFT
RT   AEROSPACEPLANES
     ∞AIRCRAFT
     ASTROPLANE
     DRONE AIRCRAFT
     FAN IN WING AIRCRAFT
     FLIGHT TEST VEHICLES
     FLYING PLATFORMS
     GROUND EFFECT MACHINES
     HOVERCRAFT GROUND EFFECT
        MACHINES
     HYPERSONIC AIRCRAFT
     JET AIRCRAFT
     METEOROLOGICAL RESEARCH
        AIRCRAFT
     ∞MILITARY AIRCRAFT
     NUCLEAR PROPELLED AIRCRAFT
     ROCKET PLANES
     SUBMERSIBLE AIRCRAFT
     SUPERSONIC AIRCRAFT

RESEARCH AIRCRAFT-(CONT.)
  TAILLESS AIRCRAFT
  TANDEM WING AIRCRAFT
  TEST VEHICLES
  TILT WING AIRCRAFT
  V/STOL AIRCRAFT
  VERTICAL TAKEOFF AIRCRAFT
  ∞ WINGED VEHICLES
  YF-12 AIRCRAFT

RESEARCH AND DEVELOPMENT
  RT   ∞ DESIGN
       INVESTIGATION
       MANAGEMENT PLANNING
       OPERATIONS RESEARCH
       OUTER SPACE TREATY
       PROGRAMS
       PROJECT MANAGEMENT
       ∞ RESEARCH PROJECTS
       SYSTEMS ENGINEERING
       TECHNOLOGY UTILIZATION
       WEAPONS DEVELOPMENT

RESEARCH FACILITIES
  RT   ∞ FACILITIES
       LABORATORIES
       ∞ RESEARCH PROJECTS
       SPACE INDUSTRIALIZATION
       SPACE LABORATORIES
       TEST FACILITIES

RESEARCH MANAGEMENT
  GS   MANAGEMENT
       . RESEARCH MANAGEMENT
  RT   ALLOCATIONS
       BLOCK DIAGRAMS
       ∞ BUDGETS
       ENGINEERING MANAGEMENT
       FEASIBILITY ANALYSIS
       GOALS
       INDUSTRIAL MANAGEMENT
       MANPOWER
       OPERATIONS RESEARCH
       PERSONNEL
       PRIORITIES

∞ RESEARCH PROJECTS
  SN   (USE OF A MORE SPECIFIC TERM IS
       RECOMMENDED--CONSULT THE TERMS
       LISTED BELOW)
  RT   CHINESE SPACE PROGRAM
       EARTH & OCEAN PHYSICS
         APPLICATIONS PROGRAM
       FRENCH SPACE PROGRAMS
       INDIAN SPACE PROGRAM
       JAPANESE SPACE PROGRAM
       MANAGEMENT
       NASA PROGRAMS
       PROGRAMS
       PROJECT MANAGEMENT
       PROJECTS
       RESEARCH
       RESEARCH AND DEVELOPMENT
       RESEARCH FACILITIES
       SPACE PROGRAMS
       SPHINX

RESEARCH VEHICLES
  SN   (VEHICLES DESIGNED TO BE SUBJECTS
       OF RESEARCH - - NOT RESEARCH
       EQUIPMENT CONTAINERS)
  GS   RESEARCH VEHICLES
       . AUTOMATED MIXED TRAFFIC
           VEHICLES
       . UNDERWATER RESEARCH
           LABORATORIES
  RT   BOATS
       ELECTRIC MOTOR VEHICLES
       ∞ FLIGHT VEHICLES
       LUNAR ROVING VEHICLES
       ∞ MILITARY VEHICLES
       ROVING VEHICLES
       SHIPS
       SPACE LABORATORIES
       ∞ SPACECRAFT
       SURFACE EFFECT SHIPS
       UNDERWATER VEHICLES
       ∞ VEHICLES
       WATER VEHICLES

RESERPINE
  GS   DRUGS
       . PENTOBARBITAL SODIUM
       . . RESERPINE
       HETEROCYCLIC COMPOUNDS

RESERPINE-(CONT.)
       . ALKALOIDS
       . . RESERPINE
       NITROGEN COMPOUNDS
       . ALKALOIDS
       . . RESERPINE
  RT   ANTIHYPERTENSIVE AGENTS

RESERVES
  RT   ABUNDANCE
       AVAILABILITY
       BACKUPS
       CONTINGENCY
       CRUDE OIL
       ECONOMIC FACTORS
       ENERGY POLICY
       ESTIMATES
       ESTIMATING
       EVALUATION
       EXPLOITATION
       EXPLORATION
       FORECASTING
       INVENTORIES
       INVENTORY CONTROLS
       ∞ MATERIALS
       MINERAL DEPOSITS
       MINES (EXCAVATIONS)
       ∞ PRODUCTION
       RESOURCES
       STOCKPILING
       ∞ STORAGE

RESERVOIRS
  SN   (FOR SURFACE WATER STORAGE--NOT
       OIL OR GAS POOLS)
  RT   DAMS
       EVAPORATION
       FRESH WATER
       LAGOONS
       LAKE TEXOMA (OK-TX)
       LAKES
       PONDS
       RIVERS
       SOLAR PONDS (HEAT STORAGE)
       STREAMS
       WATER RESOURCES
       WINDPOWERED PUMPS

RESIDENTIAL AREAS
  RT   CITIES
       INHABITANTS
       LAND
       LAND USE
       MEGALOPOLISES
       REGIONAL PLANNING
       RURAL AREAS
       SUBURBAN AREAS
       URBAN DEVELOPMENT

RESIDENTIAL ENERGY
  RT   ENERGY CONSERVATION
       ENERGY TECHNOLOGY
       HEAT PUMPS
       SOLAR COOLING
       SOLAR HEATING
       SOLAR HOUSES
       SPACE COOLING (BUILDINGS)
       SPACE HEATING (BUILDINGS)
       WATER HEATING

RESIDUAL GAS
  GS   GASES
       . RESIDUAL GAS
  RT   GETTERS
       HIGH VACUUM
       OUTGASSING
       PARTIAL PRESSURE
       ULTRAHIGH VACUUM
       VACUUM APPARATUS
       VACUUM TUBES

RESIDUAL STRENGTH
  RT   CRACK PROPAGATION
       FATIGUE (MATERIALS)
       FRACTURE MECHANICS
       FRACTURE STRENGTH
       RESIDUAL STRESS
       ∞ STRENGTH
       TENSILE STRENGTH

RESIDUAL STRESS
  UF   INTERNAL STRESS
  GS   STRESSES
       . RESIDUAL STRESS
  RT   CREEP PROPERTIES

RESIDUAL STRESS-(CONT.)
       MACHINING
       RELAXATION (MECHANICS)
       RESIDUAL STRENGTH
       STRAIN HARDENING
       STRESS RELAXATION
       STRESS RELIEVING
       STRESS-STRAIN RELATIONSHIPS
       TEMPERATURE INVERSIONS

RESIDUES
  RT   ASHES
       ORGANIC WASTES (FUEL CONVERSION)
       REACTION PRODUCTS
       SOLID WASTES
       WASTE TREATMENT
       WASTE WATER
       WASTES

RESILIENCE
  GS   MECHANICAL PROPERTIES
       . RESILIENCE
  RT   COMPRESSIVE STRENGTH
       ELASTIC PROPERTIES
       SHEAR PROPERTIES
       SPRINGS (ELASTIC)
       TENSILE STRENGTH

RESIN BONDING
  GS   BONDING
       . RESIN BONDING
  RT   ADHESIVE BONDING
       METAL BONDING
       METAL-METAL BONDING

RESIN MATRIX COMPOSITES
  GS   COMPOSITE MATERIALS
       . RESIN MATRIX COMPOSITES
       . . GRAPHITE-EPOXY COMPOSITES
  RT   EPOXY RESINS
       MATRIX MATERIALS
       METAL MATRIX COMPOSITES
       POLYIMIDE RESINS
       POLYMER MATRIX COMPOSITES
       PREPREGS
       PULTRUSION
       RESINS

RESINS
  GS   RESINS
       . ACRYLIC RESINS
       . ALKYD RESINS
       . ION EXCHANGE RESINS
       . POLYIMIDE RESINS
       . POLYURETHANE RESINS
       . SILICONE RESINS
       . SYNTHETIC RESINS
       . . ADDITION RESINS
       . . . VINYL COPOLYMERS
       . . POLYESTER RESINS
       . . POLYETHER RESINS
       . . . POLYMETHYL METHACRYLATE
       . . THERMOPLASTIC RESINS
       . . . QUINOXALINES
       . . . THERMOPLASTIC FILMS
       . . THERMOSETTING RESINS
       . . . EPOXY RESINS
       . . . FURAN RESINS
       . . . . POLYAMIDE RESINS
       . . . . . KEVLAR (TRADEMARK)
       . . . PHENOLIC RESINS
       . . . . MICARTA
       . . . . PHENOLIC EPOXY RESINS
  RT   ACRYLATES
       BAKELITE (TRADEMARK)
       DELRIN (TRADEMARK)
       FILLERS
       LEXAN (TRADEMARK)
       MELAMINE
       PARAPLASTS
       ∞ PATTERNS
       PHENOL FORMALDEHYDE
       PHLOROGLUCINOL
       PLASTISOLS
       RESIN MATRIX COMPOSITES
       TEFLON (TRADEMARK)

∞ RESISTANCE
  SN   (USE OF A MORE SPECIFIC TERM IS
       RECOMMENDED--CONSULT THE TERMS
       LISTED BELOW)
  UF   CONDUCTANCE
       RESISTANCE COEFFICIENTS
  RT   ABRASION RESISTANCE
       ACCELERATION TOLERANCE

RESISTANCE-*(CONT.)*
- ACOUSTIC PROPERTIES
- AERODYNAMIC DRAG
- CHEMICAL PROPERTIES
- CONSTRICTIONS
- CONTACT RESISTANCE
- CORROSION RESISTANCE
- CRACK PROPAGATION
- CREEP STRENGTH
- DAMPING
- DIFFUSIVITY
- DURABILITY
- EARTHQUAKE RESISTANCE
- ELECTRICAL PROPERTIES
- ELECTRICAL RESISTANCE
- ELECTRICAL RESISTIVITY
- FLAMMABILITY
- FLOW RESISTANCE
- FRACTURE STRENGTH
- FUSIBILITY
- ∞HIGH RESISTANCE
- IMMUNITY
- IMPACT RESISTANCE
- IMPACT STRENGTH
- IMPEDANCE
- KAPITZA RESISTANCE
- LIFE (DURABILITY)
- ∞LOW RESISTANCE
- MAGNETORESISTIVITY
- MOISTURE RESISTANCE
- NEGATIVE RESISTANCE CIRCUITS
- NEGATIVE RESISTANCE DEVICES
- OXIDATION RESISTANCE
- PERMEABILITY
- PREVENTION
- PROTECTION
- QUALITY
- RADIATION TOLERANCE
- RESISTANCE THERMOMETERS
- RETARDING
- SENSITIVITY
- SHOCK RESISTANCE
- SKIN RESISTANCE
- STABILITY
- THERMAL RESISTANCE
- TOLERANCES (PHYSIOLOGY)
- VULNERABILITY
- WAVE RESISTANCE

RESISTANCE COEFFICIENTS
- USE   RESISTANCE

RESISTANCE HEATING
- UF   JOULE HEATING
- GS   HEATING
-     . **RESISTANCE HEATING**
- RT   ARC HEATING
-     ELECTROSLAG REFINING
-     GAS HEATING
-     LEVITATION MELTING

RESISTANCE THERMOMETERS
- GS   MEASURING INSTRUMENTS
-     . TEMPERATURE MEASURING INSTRUMENTS
-     . . THERMOMETERS
-     . . . **RESISTANCE THERMOMETERS**
- RT   BOLOMETERS
-     OHMMETERS
-     ∞RESISTANCE
-     TEMPERATURE MEASUREMENT
-     THERMOCOUPLE PYROMETERS

RESISTIVITY
- USE   ELECTRICAL RESISTIVITY

RESISTOJET ENGINES
- UF   RESISTOJETS
- GS   ENGINES
-     . ROCKET ENGINES
-     . . ELECTRIC ROCKET ENGINES
-     . . . ELECTROTHERMAL ENGINES
-     . . . . **RESISTOJET ENGINES**
- RT   ARC JET ENGINES
-     PLASMA ENGINES
-     PULSED JET ENGINES

RESISTOJETS
- USE   RESISTOJET ENGINES

RESISTORS
- UF   TUNNEL RESISTORS
- GS   ATTENUATORS
-     . **RESISTORS**
-     . . POTENTIOMETERS (RESISTORS)

RESISTORS-*(CONT.)*
-     . . PRINTED RESISTORS
-     . . THERMISTORS
- RT   BALLASTS (IMPEDANCES)
-     ELECTRIC CONDUCTORS
-     ELECTRIC FILTERS
-     ELECTRIC REACTORS
-     ∞FILAMENTS
-     PHOTOCONDUCTORS
-     SEMICONDUCTORS (MATERIALS)
-     SOLID STATE DEVICES
-     VARISTORS

RESOLUTION
- UF   RESOLVING POWER
- GS   **RESOLUTION**
-     . ANGULAR RESOLUTION
-     . HIGH RESOLUTION
-     . IMAGE RESOLUTION
-     . RADAR RESOLUTION
-     . RADIOMETRIC RESOLUTION
-     . SPATIAL RESOLUTION
-     . SPECTRAL RESOLUTION
-     . TEMPORAL RESOLUTION
- RT   ACCURACY
-     AUTOMATIC TRAFFIC ADVISORY AND RESOLUTION
-     BLURRING
-     CHARACTER RECOGNITION
-     CONTRAST
-     ∞DEFINITION
-     DYNAMIC CHARACTERISTICS
-     ERRORS
-     FOCI
-     HIGH RESOLUTION COVERAGE ANTENNAS
-     IMAGE CONTRAST
-     IMAGE ENHANCEMENT
-     LEGIBILITY
-     LOCI
-     ∞OPTICS
-     PERCEPTION
-     ∞POWER
-     PRECISION
-     RESOLUTION CELL
-     SENSITIVITY
-     SPATIAL FILTERING
-     STARK EFFECT
-     ∞THRESHOLDS
-     TOLERANCES (MECHANICS)
-     VISIBILITY
-     VISION

RESOLUTION CELL
- RT   ∞CELLS
-     IMAGING TECHNIQUES
-     RADAR DETECTION
-     RADAR IMAGERY
-     RESOLUTION

RESOLVERS
- RT   ANALOG COMPUTERS
-     INSTRUMENT TRANSFORMERS
-     TRANSFORMERS

RESOLVING POWER
- USE   RESOLUTION

RESONANCE
- GS   **RESONANCE**
-     . BARYON RESONANCE
-     . CYCLOTRON RESONANCE
-     . MAGNETIC RESONANCE
-     . . FERROMAGNETIC RESONANCE
-     . . NUCLEAR MAGNETIC RESONANCE
-     . . . PROTON MAGNETIC RESONANCE
-     . . . PROTON RESONANCE
-     . . PARAMAGNETIC RESONANCE
-     . . . ELECTRON PARAMAGNETIC RESONANCE
-     . MAGNETOSONIC RESONANCE
-     . MESON RESONANCE
-     . . X MESONS
-     . MICROWAVE RESONANCE
-     . NUCLEAR QUADRUPOLE RESONANCE
-     . OPTICAL RESONANCE
-     . PLASMA RESONANCE
-     . RESONANT VIBRATION
-     . SPIN RESONANCE
- RT   FOSTER THEORY
-     OSCILLATIONS
-     OVERHAUSER EFFECT
-     RESONANT FREQUENCIES
-     SYNTONY
-     TUNING

RESONANCE-*(CONT.)*
-     VIBRATION

RESONANCE CHARGE EXCHANGE
- GS   EXCHANGING
-     . CHARGE EXCHANGE
-     . . **RESONANCE CHARGE EXCHANGE**
- RT   SPIN EXCHANGE

RESONANCE FLUORESCENCE
- UF   RESONANCE RADIATION
- GS   DECAY
-     . EMISSION
-     . . LIGHT EMISSION
-     . . . LUMINESCENCE
-     . . . . FLUORESCENCE
-     . . . . . **RESONANCE FLUORESCENCE**
- RT   ATOMIC EXCITATIONS
-     ATOMIC PHYSICS
-     QUANTUM ELECTRODYNAMICS
-     ∞RADIATION

RESONANCE LINES
- UF   DIELECTRONIC SATELLITE LINES
- RT   LINE SPECTRA
-     OPTICAL RESONANCE
-     PLASMA RESONANCE

RESONANCE PROBES
- GS   MEASURING INSTRUMENTS
-     . **RESONANCE PROBES**
-     . . CYCLOTRON RESONANCE DEVICES
- RT   IMPEDANCE PROBES
-     MAGNETIC PROBES
-     MICROWAVE PLASMA PROBES
-     PLASMA DIAGNOSTICS
-     PLASMA RESONANCE
-     TUNERS

RESONANCE RADIATION
- USE   RESONANCE FLUORESCENCE

RESONANCE SCATTERING
- SN   (INTERACTION WITH THE INTERIOR OF THE NUCLEUS--EXCLUDES POTENTIAL SCATTERING)
- GS   NUCLEAR REACTIONS
-     . NUCLEAR SCATTERING
-     . . **RESONANCE SCATTERING**
-     SCATTERING
-     . NUCLEAR SCATTERING
-     . . **RESONANCE SCATTERING**
- RT   INVERSE SCATTERING
-     MOSSBAUER EFFECT
-     NEUTRON SCATTERING

RESONANCE TESTING
- RT   DAMPING TESTS
-     ELASTIC DAMPING
-     ELECTRONIC EQUIPMENT TESTS
-     FATIGUE TESTS
-     RESONANT FREQUENCIES
-     STABILITY TESTS
-     STATIC TESTS
-     STRUCTURAL STABILITY
-     ∞TESTS
-     VIBRATION TESTS
-     VISCOUS DAMPING

RESONANT CAVITIES
- USE   CAVITY RESONATORS

RESONANT FREQUENCIES
- UF   NATURAL FREQUENCIES
- GS   FREQUENCIES
-     . **RESONANT FREQUENCIES**
- RT   ANTINODES
-     BANDWIDTH
-     BEAT FREQUENCIES
-     BORDONI PEAKS
-     CAVITY RESONATORS
-     CRITICAL FREQUENCIES
-     CRITICAL VELOCITY
-     DAMPING
-     DYNAMIC CHARACTERISTICS
-     ∞DYNAMICS
-     ELECTROMAGNETIC ABSORPTION
-     HARMONICS
-     IMPEDANCE
-     MOSSBAUER EFFECT
-     NODES (STANDING WAVES)
-     OSCILLATORS
-     RESONANCE
-     RESONANCE TESTING

RESONANT FREQUENCIES-*(CONT.)*
　　RESONATORS
　　STANDING WAVES
　　TRANSIENT RESPONSE
　　TUNERS
　　TUNING

RESONANT VIBRATION
　UF　MECHANICAL RESONANCE
　GS　RESONANCE
　　　. **RESONANT VIBRATION**
　　　VIBRATION
　　　. **RESONANT VIBRATION**
　RT　DAMPING
　　　DYNAMIC STABILITY
　　∞DYNAMICS
　　　FLAPPING
　　　FLUTTER
　　　MECHANICAL OSCILLATORS
　　　OSCILLATIONS
　　　Q FACTORS
　　　STABLE OSCILLATIONS
　　　STRUCTURAL VIBRATION
　　　UNDAMPED OSCILLATIONS

RESONATORS
　GS　**RESONATORS**
　　　. CAVITY RESONATORS
　　　. HELMHOLTZ RESONATORS
　　　. MULTIMODE RESONATORS
　　　. OPTICAL RESONATORS
　RT　DELTA ANTENNAS
　　　ELECTRON TUBES
　　　FREQUENCY STANDARDS
　　　GRAZING FLOW
　　　MAGNETRONS
　　　MASERS
　　　OSCILLATORS
　　　RESONANT FREQUENCIES
　　　SELF EXCITATION
　　　TUNING
　　　TUNING FORK GYROSCOPES

RESOURCE ALLOCATION
　GS　ALLOCATIONS
　　　. **RESOURCE ALLOCATION**
　RT　DISTRIBUTING
　　　ENERGY CONSERVATION
　　　ENERGY POLICY
　　　ENGINEERING MANAGEMENT
　　　LOGISTICS
　　　NASA INTERACTIVE PLANNING SYSTEM
　　　OUTER SPACE TREATY
　　　PRIORITIES
　　　PRODUCTS
　　　RESOURCES

RESOURCES
　GS　**RESOURCES**
　　　. AQUIFERS
　　　. EARTH RESOURCES
　　　. . FORESTS
　　　. . . RAIN FORESTS
　　　. . FOSSIL FUELS
　　　. . . COAL
　　　. . . . LIGNITE
　　　. . . . SOLVENT REFINED COAL
　　　. . . CRUDE OIL
　　　. . GLACIERS
　　　. . ICEBERGS
　　　. . LAND ICE
　　　. . MARINE RESOURCES
　　　. . OIL FIELDS
　　　. . RANGE RESOURCES
　　　. . SPRINGS (WATER)
　　　. . TAR SANDS
　　　. . THERMAL RESOURCES
　　　. . . GEOTHERMAL RESOURCES
　　　. . . . GEYSERS
　　　. . UNDERWATER RESOURCES
　　　. . WATER RESOURCES
　　　. EXTRATERRESTRIAL RESOURCES
　RT　ABUNDANCE
　　　AVAILABILITY
　　　CONSULTING
　　　DEPLETION
　　　ECONOMIC DEVELOPMENT
　　　ECONOMIC FACTORS
　　　ECONOMIC IMPACT
　　　ECONOMICS
　　　ENERGY CONSERVATION
　　　ENERGY POLICY
　　　ENGINEERING MANAGEMENT
　　　GEOTHERMAL TECHNOLOGY
　　　GREAT LAKES (NORTH AMERICA)

RESOURCES-*(CONT.)*
　　INVENTORY MANAGEMENT
　　LOGISTICS
　　LOGISTICS MANAGEMENT
　　MAN ENVIRONMENT INTERACTIONS
　　MANPOWER
　∞MATERIALS
　　MISSISSIPPI RIVER (US)
　　PERSONNEL
　　PERSONNEL DEVELOPMENT
　　PRODUCTION MANAGEMENT
　　RECYCLING
　　RESERVES
　　RESOURCE ALLOCATION
　　SITE SELECTION
　　URBAN DEVELOPMENT
　　VEGETATION

RESOURCES MANAGEMENT
　GS　MANAGEMENT
　　　. **RESOURCES MANAGEMENT**
　　　. . FOREST MANAGEMENT
　　　. . . REFORESTATION
　　　. . LAND MANAGEMENT
　RT　EARTH RESOURCES
　　　ENVIRONMENT MANAGEMENT
　　　ENVIRONMENTAL CONTROL
　　　LEASING
　　　NASA INTERACTIVE PLANNING SYSTEM
　　　THERMAL RESOURCES
　　　WATER RUNOFF

RESPIRATION
　UF　APNEA
　　　INHALATION
　GS　**RESPIRATION**
　　　. HIGH ALTITUDE BREATHING
　　　. LIQUID BREATHING
　　　. PRESSURE BREATHING
　RT　ALVEOLI
　　　ASPHYXIA
　∞BREATHING
　　　EXPIRATION
　　　EXPIRED AIR
　　　HYDROGEN METABOLISM
　　　METABOLISM
　　　OXYGEN METABOLISM
　　　PHOTOSYNTHESIS
　　　RESPIRATORS
　　　RESPIRATORY SYSTEM
　　　RESUSCITATION
　　　SINUSES
　　　VALSALVA EXERCISE

RESPIRATORS
　GS　MEDICAL EQUIPMENT
　　　. **RESPIRATORS**
　RT　BREATHING APPARATUS
　　　EMERGENCY BREATHING TECHNIQUES
　　　RESPIRATION
　　　RESUSCITATION
　　　THERAPY

RESPIRATORY DISEASES
　GS　DISEASES
　　　. **RESPIRATORY DISEASES**
　　　. . AEROSINUSITIS
　　　. . ASTHMA
　　　. . EMPHYSEMA
　　　. . INFLUENZA
　　　. . PNEUMONIA
　　　. . TUBERCULOSIS
　RT　BERYLLIUM POISONING
　　　CONGESTION
　　　LUNG MORPHOLOGY
　　　PULMONARY LESIONS

RESPIRATORY IMPEDANCE
　GS　IMPEDANCE
　　　. **RESPIRATORY IMPEDANCE**

RESPIRATORY PHYSIOLOGY
　GS　PHYSIOLOGY
　　　. **RESPIRATORY PHYSIOLOGY**
　RT　EXERCISE PHYSIOLOGY
　∞SCIENCE

RESPIRATORY RATE
　GS　RATES (PER TIME)
　　　. **RESPIRATORY RATE**
　　　. . DYSPNEA
　　　. . HYPOVENTILATION
　　　. . TACHYPNEA
　RT　HYPERCAPNIA
　　　HYPERPNEA

RESPIRATORY RATE-*(CONT.)*
　　SPIROMETERS

RESPIRATORY REFLEXES
　GS　REFLEXES
　　　. **RESPIRATORY REFLEXES**
　　　. . COUGH
　　　. . HERING-BREVER REFLEX
　　　. . SNEEZING
　RT　∞BREATHING

RESPIRATORY SYSTEM
　GS　ANATOMY
　　　. **RESPIRATORY SYSTEM**
　　　. . BRONCHI
　　　. . BRONCHIAL TUBE
　　　. . . PHARYNX
　　　. . . TRACHEA
　　　. . DIAPHRAGM (ANATOMY)
　　　. . LUNGS
　　　. . NOSE (ANATOMY)
　RT　ALVEOLI
　　　EVAPORATION
　　　HOMEOSTASIS
　　　HYPERCAPNIA
　　　PULMONARY CIRCULATION
　　　RESPIRATION
　∞SYSTEMS

RESPIROMETERS
　GS　MEASURING INSTRUMENTS
　　　. **RESPIROMETERS**
　RT　BIOINSTRUMENTATION
　　　EXHALATION

RESPONDERS
　USE　TRANSPONDERS

RESPONSE BIAS
　GS　BIAS
　　　. **RESPONSE BIAS**
　RT　DYNAMIC RESPONSE
　　　ERRORS
　∞TIME RESPONSE
　　　TRANSIENT RESPONSE

RESPONSE TIME (COMPUTERS)
　GS　TIME
　　　. **RESPONSE TIME (COMPUTERS)**
　RT　COMPUTER PROGRAMMING
　　　COMPUTER SYSTEMS PERFORMANCE
　　　DATA PROCESSING

RESPONSES
　GS　**RESPONSES**
　　　. DYNAMIC RESPONSE
　　　. . TRANSIENT RESPONSE
　　　. GALVANIC SKIN RESPONSE
　　　. MODAL RESPONSE
　　　. PHYSIOLOGICAL RESPONSES
　　　. . HEMODYNAMIC RESPONSES
　RT　CHRONAXY
　∞FREQUENCY RESPONSE
　　　LEARNING
　　　REFRACTORY PERIOD
　∞THRESHOLDS
　　　TIME LAG
　∞TIME RESPONSE

REST
　GS　**REST**
　　　. BED REST
　RT　PRONE POSITION
　　　RECREATION
　　　SITTING POSITION
　　　SLEEP
　　　SUPINE POSITION

RESTARTABLE ROCKET ENGINES
　GS　ENGINES
　　　. ROCKET ENGINES
　　　. . **RESTARTABLE ROCKET ENGINES**
　RT　DUCTED ROCKET ENGINES
　　　ELECTRIC ROCKET ENGINES
　　　ELECTROSTATIC ENGINES
　　　ELECTROTHERMAL ENGINES
　　　HYBRID PROPELLANT ROCKET ENGINES
　　　ION ENGINES
　　　LIQUID PROPELLANT ROCKET ENGINES
　　　NUCLEAR ROCKET ENGINES
　　　RETROROCKET ENGINES
　　　SOLID PROPELLANT ROCKET ENGINES
　　　SUSTAINER ROCKET ENGINES
　　　TURBOROCKET ENGINES

RESTARTABLE ROCKET ENGINES-*(CONT.)*
VERNIER ENGINES

RESTORATION
RT   ADDITION
RECONSTRUCTION

RESTRAINTS
USE   CONSTRAINTS

RESTRICTIONS
USE   CONSTRICTIONS

RESULTANTS
RT   VECTOR ANALYSIS

RESUSCITATION
UF   ARTIFICIAL RESPIRATION
RT   EMERGENCY BREATHING TECHNIQUES
FIRST AID
LIQUID BREATHING
RESPIRATION
RESPIRATORS

RETAINING
RT   ASTEROID CAPTURE
CONSTRAINTS
CONTAINMENT
∞HOLDING
∞JOINING
LOCKING
∞RETENTION
SEALING
∞STORAGE

RETARDANTS
GS   **RETARDANTS**
. FLAME RETARDANTS
RT   ACCELERATING AGENTS
ADDITIVES
ANTIICING ADDITIVES
ANTIKNOCK ADDITIVES
ANTIOXIDANTS
CATALYSTS
EXPLOSION SUPPRESSION
INHIBITORS
NEUTRALIZERS
PENETRANTS
PRESERVATIVES
∞RETARDERS
RETARDING
STABILIZERS (AGENTS)
SUPPRESSORS
SURFACTANTS
WEAR INHIBITORS

∞ **RETARDERS**
SN   *(USE OF A MORE SPECIFIC TERM IS
RECOMMENDED--CONSULT THE TERMS
LISTED BELOW)*
RT   RETARDANTS
RETARDERS (DEVICES)

RETARDERS (DEVICES)
RT   BLOCKING
BRAKES (FOR ARRESTING MOTION)
BRAKING
CONSTRICTIONS
∞RETARDERS
RETARDING

RETARDING
UF   SUPPRESSION
RT   ATTENUATION
BLOCKING
BRAKING
DAMPING
DECELERATION
FOULING
HYSTERESIS
PREVENTION
∞REDUCTION
∞RESISTANCE
RETARDANTS
RETARDERS (DEVICES)
STOPPING

RETARDING ION MASS SPECTROMETERS
USE   MASS SPECTROMETERS

∞ **RETENTION**
SN   *(USE OF A MORE SPECIFIC TERM IS
RECOMMENDED--CONSULT THE TERMS
LISTED BELOW)*

RETENTION-*(CONT.)*
RT   RETAINING
RETENTION (PSYCHOLOGY)

RETENTION (PSYCHOLOGY)
RT   LEARNING
MEMORY
∞RETENTION

RETICLES
GS   **RETICLES**
. WIRE GRID LENSES
RT   CONTACT LENSES
EYEPIECES
∞GRIDS
LENSES
OPTICAL EQUIPMENT
SCALE (RATIO)

RETICULOCYTES
GS   ANATOMY
. CARDIOVASCULAR SYSTEM
. . ERYTHROCYTES
. . . **RETICULOCYTES**
BODY FLUIDS
. BLOOD
. . ERYTHROCYTES
. . . **RETICULOCYTES**
CELLS (BIOLOGY)
. ERYTHROCYTES
. . **RETICULOCYTES**
RT   HEMATOLOGY
HEMOGLOBIN
HEMOLYSIS
HEMORRHAGES

RETINA
GS   ANATOMY
. SENSE ORGANS
. . EYE (ANATOMY)
. . . **RETINA**
. . . . FOVEA
RT   ELECTRORETINOGRAPHY
PHOSPHENE
PHOTORECEPTORS
VISION
VISUAL FIELDS

RETINAL ADAPTATION
GS   ADAPTATION
. **RETINAL ADAPTATION**
. . DARK ADAPTATION
. . LIGHT ADAPTATION
RT   PERCEPTION
VISIBILITY
VISION

RETINAL IMAGES
GS   IMAGES
. **RETINAL IMAGES**
RT   VISION
VISUAL FIELDS

RETINENE
UF   VITAMIN A
GS   HETEROCYCLIC COMPOUNDS
. **RETINENE**
VITAMINS
. **RETINENE**
RT   ALDEHYDES
CAROTENE

RETIREMENT
RT   EMPLOYEE RELATIONS
INDUSTRIES
PERSONNEL
SOCIOLOGY

RETIREMENT FOR CAUSE
RT   COMPONENT RELIABILITY
ENGINE PARTS
FATIGUE LIFE
FRACTURE STRENGTH
INVENTORY MANAGEMENT
LIFE (DURABILITY)
METAL FATIGUE
SERVICE LIFE
SPARE PARTS

RETORC (TORPEDOES)
USE   TORPEDOES

RETORT PROCESSING
RT   FRACTIONATION

RETORT PROCESSING-*(CONT.)*
HYDROCARBON FUELS
OILS
∞PROCESSING
SHALE OIL

RETRACTABLE EQUIPMENT
UF   RETRACTABLE LANDING GEAR
RT   AERODYNAMIC BRAKES
LANDING GEAR
REFUELING

RETRACTABLE LANDING GEAR
USE   LANDING GEAR
RETRACTABLE EQUIPMENT

RETRAINING
UF   REORIENTATION
RT   ADAPTATION
EDUCATION
MANAGEMENT METHODS
MANPOWER
PERSONNEL
TASKS
TRAINING ANALYSIS

RETRIEVAL
GS   **RETRIEVAL**
. DATA RETRIEVAL
. INFORMATION RETRIEVAL
. PAYLOAD RETRIEVAL (STS)
RT   ∞RECOVERY
SEARCHING

RETROACTION
USE   RETROTHRUST

RETROFIRING
GS   FIRING (IGNITING)
. ROCKET FIRING
. . **RETROFIRING**
RT   DECELERATION
RETROROCKET ENGINES
RETROTHRUST

RETROFITTING
GS   **RETROFITTING**
. ACOUSTIC RETROFITTING
RT   INSTALLING

RETROREFLECTION
GS   REFLECTION
. **RETROREFLECTION**
RT   ANTENNA ARRAYS
INCIDENT RADIATION
LAGEOS (SATELLITE)
LUNAR RETROREFLECTORS
REFLECTED WAVES
RETROREFLECTORS

RETROREFLECTORS
GS   RADAR EQUIPMENT
. **RETROREFLECTORS**
REFLECTORS
. **RETROREFLECTORS**
RT   RADAR CORNER REFLECTORS
RETROREFLECTION

RETROROCKET ENGINES
GS   ENGINES
. ROCKET ENGINES
. . **RETROROCKET ENGINES**
. . . BE-3 ENGINE
RT   CONTROL ROCKETS
INTERNAL COMBUSTION ENGINES
LIQUID PROPELLANT ROCKET ENGINES
MAN OPERATED PROPULSION SYSTEMS
RESTARTABLE ROCKET ENGINES
RETROFIRING
RETROTHRUST
SOLID PROPELLANT ROCKET ENGINES

RETROTHRUST
UF   RETROACTION
GS   THRUST
. ROCKET THRUST
. . **RETROTHRUST**
RT   DECELERATION
RETROFIRING
RETROROCKET ENGINES

RETURN BEAM VIDICONS
GS   ELECTRON TUBES
. CAMERA TUBES

RETURN BEAM VIDICONS-(CONT.)
　　. . VIDICONS
　　. . . **RETURN BEAM VIDICONS**
　　. . . . THERMICONS
　RT　TELEVISION CAMERAS

RETURN TO EARTH SPACE FLIGHT
　GS　SPACE FLIGHT
　　. **RETURN TO EARTH SPACE FLIGHT**
　RT　INTERPLANETARY FLIGHT
　　　SPACECRAFT REENTRY

REUSABLE HEAT SHIELDING
　GS　SHIELDING
　　. HEAT SHIELDING
　　. . **REUSABLE HEAT SHIELDING**
　RT　COOLING
　　　HEAT TRANSFER
　　　REENTRY SHIELDING
　　　SPACECRAFT SHIELDING
　　　TEMPERATURE CONTROL
　　　THERMAL CONTROL COATINGS
　　　THERMAL PROTECTION

REUSABLE LAUNCH VEHICLES
　GS　LAUNCH VEHICLES
　　. **REUSABLE LAUNCH VEHICLES**
　　. . SINGLE STAGE TO ORBIT VEHICLES
　RT　AEROMANEUVERING ORBIT TO ORBIT
　　　　SHUTTLE
　　　RECOVERABLE LAUNCH VEHICLES
　　　ROCKET ENGINES
　　　SPACECRAFT LAUNCHING
　　　SPACECRAFT RECOVERY
　　　∞VEHICLES

REUSABLE ROCKET ENGINES
　GS　ENGINES
　　. ROCKET ENGINES
　　. . **REUSABLE ROCKET ENGINES**

REUSABLE SPACECRAFT
　GS　REENTRY VEHICLES
　　. RECOVERABLE SPACECRAFT
　　. . **REUSABLE SPACECRAFT**
　　. . . ASTROPLANE
　　. . . ATLANTIS (ORBITER)
　　. . . CHALLENGER (ORBITER)
　　. . . COLUMBIA (ORBITER)
　　. . . DISCOVERY (ORBITER)
　　. . . ENTERPRISE (ORBITER)
　　. . . HERMES MANNED SPACEPLANE
　　. . . MARS (MANNED REUSABLE
　　　　　　SPACECRAFT)
　RT　EXPENDABLE STAGES (SPACECRAFT)
　　　FERRY SPACECRAFT
　　　INERTIAL UPPER STAGE
　　　INTERIM STAGES (SPACECRAFT)
　　　INTERPLANETARY SPACECRAFT
　　　LANDING MODULES
　　　LUNAR LANDING MODULES
　　　MANNED SPACECRAFT
　　　SOFT LANDING SPACECRAFT
　　　SPACE SHUTTLE BOOSTERS
　　　SPACE SHUTTLE ORBITERS
　　　UNMANNED SPACECRAFT

REUSE
　GS　UTILIZATION
　　. **REUSE**
　RT　OIL RECOVERY
　　　∞RECOVERY

REVENUE
　RT　ALLOCATIONS
　　　ASSESSMENTS
　　　BUDGETING
　　　COSTS
　　　INTERNATIONAL TRADE

REVERBERATION
　GS　ACOUSTIC PROPERTIES
　　. ACOUSTIC SCATTERING
　　. . **REVERBERATION**
　　　SCATTERING
　　. WAVE SCATTERING
　　. . ACOUSTIC SCATTERING
　　. . . **REVERBERATION**
　RT　ECHOES
　　　NOISE (SOUND)
　　　SOUND WAVES

REVERSE FIELD PINCH
　GS　PINCH EFFECT

REVERSE FIELD PINCH-(CONT.)
　　. **REVERSE FIELD PINCH**
　RT　MAGNETOHYDRODYNAMIC FLOW
　　　PLASMA CONTROL
　　　REACTOR TECHNOLOGY
　　　SCREW PINCH
　　　TOROIDAL PLASMAS

REVERSE OSMOSIS
　GS　OSMOSIS
　　. **REVERSE OSMOSIS**
　RT　DEMINERALIZING
　　　DESALINIZATION
　　　MEMBRANES
　　　PERMEATING

REVERSE TIME
　USE　REACTION TIME

REVERSED FLOW
　GS　FLUID FLOW
　　. **REVERSED FLOW**
　RT　BOUNDARY LAYER SEPARATION
　　　RECIRCULATIVE FLUID FLOW
　　　REVERSING
　　　SEPARATED FLOW

REVERSING
　RT　∞DIRECTION
　　　REVERSED FLOW

REVIEWING
　RT　∞DISCUSSION
　　　EVALUATION
　　　EXAMINATION
　　　TRAINING EVALUATION

REVISIONS
　UF　ALTERATION
　　　MODIFICATION
　RT　ADJUSTING
　　　CONTRACTS
　　　CORRECTION
　　　EXTENSIONS
　　　VARIATIONS

REVOLUTION (MOTION)
　USE　REVOLVING

REVOLVING
　UF　REVOLUTION (MOTION)
　GS　GYRATION
　　. **REVOLVING**
　RT　ANGULAR VELOCITY
　　　CENTRIPETAL FORCE
　　　ROTATION

REWARD (PSYCHOLOGY)
　GS　REINFORCEMENT (PSYCHOLOGY)
　　. **REWARD (PSYCHOLOGY)**
　RT　COMFORT
　　　HUMAN REACTIONS
　　　PSYCHOLOGICAL FACTORS

REYNOLDS EQUATION
　UF　REYNOLDS LAW
　GS　EQUATIONS OF MOTION
　　. **REYNOLDS EQUATION**
　RT　AERODYNAMIC CONFIGURATIONS
　　　∞EQUATIONS
　　　NAVIER-STOKES EQUATION
　　　SCALE MODELS

REYNOLDS LAW
　USE　REYNOLDS EQUATION

REYNOLDS NUMBER
　UF　CRITICAL REYNOLDS NUMBER
　GS　RATIOS
　　. DIMENSIONLESS NUMBERS
　　. . **REYNOLDS NUMBER**
　　. . . HIGH REYNOLDS NUMBER
　　. . . LOW REYNOLDS NUMBER
　RT　BOUNDARY LAYER FLOW
　　　BOUNDARY LAYER STABILITY
　　　BOUNDARY LAYER TRANSITION
　　　FLUID FLOW
　　　FROUDE NUMBER
　　　GRASHOF NUMBER
　　　INVISCID FLOW
　　　LAMINAR FLOW
　　　PECLET NUMBER
　　　PRANDTL NUMBER
　　　RICHARDSON NUMBER
　　　SCALE EFFECT

REYNOLDS NUMBER-(CONT.)
　　　TRANSITION POINTS
　　　TURBULENT FLOW
　　　VISCOUS FLOW

REYNOLDS STRESS
　GS　STRESSES
　　. **REYNOLDS STRESS**
　RT　INCOMPRESSIBLE FLOW
　　　NAVIER-STOKES EQUATION
　　　TURBULENT BOUNDARY LAYER
　　　TURBULENT FLOW

RF-4 AIRCRAFT
　GS　MCDONNELL DOUGLAS AIRCRAFT
　　. MCDONNELL AIRCRAFT
　　. . **RF-4 AIRCRAFT**
　　　MONOPLANES
　　. **RF-4 AIRCRAFT**
　　　OBSERVATION AIRCRAFT
　　. **RF-4 AIRCRAFT**
　　　SUPERSONIC AIRCRAFT
　　. **RF-4 AIRCRAFT**
　RT　∞AIRCRAFT
　　　F-4 AIRCRAFT

RF-8 AIRCRAFT
　USE　F-8 AIRCRAFT

RH-2 HELICOPTER
　USE　UH-1 HELICOPTER

RHEA (ASTRONOMY)
　GS　CELESTIAL BODIES
　　. NATURAL SATELLITES
　　. . **RHEA (ASTRONOMY)**
　　　SATELLITES
　　. NATURAL SATELLITES
　　. . SATURN SATELLITES
　　. . . **RHEA (ASTRONOMY)**
　RT　SATURN (PLANET)
　　　SOLAR SYSTEM

RHENIUM
　GS　METALS
　　. TRANSITION METALS
　　. . REFRACTORY METALS
　　. . . **RHENIUM**
　　. . . . RHENIUM ISOTOPES
　　　REFRACTORY MATERIALS
　　. REFRACTORY METALS
　　. . **RHENIUM**
　　. . . RHENIUM ISOTOPES

RHENIUM ALLOYS
　GS　ALLOYS
　　. HEAT RESISTANT ALLOYS
　　. . REFRACTORY METAL ALLOYS
　　. . . **RHENIUM ALLOYS**
　　　REFRACTORY MATERIALS
　　. REFRACTORY METAL ALLOYS
　　. . **RHENIUM ALLOYS**

RHENIUM COMPOUNDS
　RT　∞CHEMICAL COMPOUNDS
　　　∞GROUP 7B COMPOUNDS
　　　∞METAL COMPOUNDS

RHENIUM ISOTOPES
　GS　METALS
　　. TRANSITION METALS
　　. . REFRACTORY METALS
　　. . . RHENIUM
　　. . . . **RHENIUM ISOTOPES**
　　　REFRACTORY MATERIALS
　　. REFRACTORY METALS
　　. . RHENIUM
　　. . . **RHENIUM ISOTOPES**

RHEOCASTING
　GS　FORMING TECHNIQUES
　　. CASTING
　　. . **RHEOCASTING**
　RT　ALLOYS
　　　CAST ALLOYS
　　　DIES
　　　FORGING
　　　SLURRIES
　　　SOLIDIFICATION

RHEOELECTRICAL SIMULATION
　GS　SIMULATION
　　. **RHEOELECTRICAL SIMULATION**
　RT　ANALOG CIRCUITS

RHEOELECTRICAL SIMULATION-*(CONT.)*
    ANALOG SIMULATION
    BIONICS
    FLOW DISTRIBUTION

RHEOENCEPHALOGRAPHY
  RT   BLOOD CIRCULATION
      BRAIN
      BRAIN CIRCULATION

RHEOLOGY
  RT   FLOW MEASUREMENT
      FLOW THEORY
      ∞FLUIDS
      LIQUID FLOW
      MAXWELL FLUIDS
      NONNEWTONIAN FLUIDS
      PLASTIC FLOW
      PLASTIC PROPERTIES
      VISCOSITY

RHEOMETERS
  GS   MEASURING INSTRUMENTS
      . FLOWMETERS
      . . **RHEOMETERS**
  RT   BLOOD CIRCULATION

RHESUS FACTOR
  RT   ANTIGENS
      BLOOD
      CONGENITAL ANOMALIES

RHEUMATIC DISEASES
  GS   DISEASES
      . **RHEUMATIC DISEASES**

RHIZOPUS
  GS   FUNGI
      . **RHIZOPUS**
  RT   BLIGHT
      ∞MOLD

RHO-MESONS
  GS   PARTICLES
      . ELEMENTARY PARTICLES
      . . BOSONS
      . . . MESONS
      . . . . VECTOR MESONS
      . . . . . **RHO-MESONS**
      . FERMIONS
      . . . BARYONS
      . . . . **RHO-MESONS**
      . . HADRONS
      . . . BARYONS
      . . . . **RHO-MESONS**
      . . . MESONS
      . . . . VECTOR MESONS
      . . . . . **RHO-MESONS**
      . NUCLEAR PARTICLES
      . . BOSONS
      . . . MESONS
      . . . . VECTOR MESONS
      . . . . . **RHO-MESONS**
  RT   CHARGED PARTICLES
      ETA-MESONS

RHODE ISLAND
  GS   NATIONS
      . UNITED STATES
      . . **RHODE ISLAND**
  RT   BLOCK ISLAND SOUND (RI)

RHODESIA
  USE   ZIMBABWE

RHODIUM
  GS   CHEMICAL ELEMENTS
      . **RHODIUM**
      . . RHODIUM ISOTOPES
      METALS
      . TRANSITION METALS
      . . **RHODIUM**
      . . . RHODIUM ISOTOPES

RHODIUM ALLOYS
  GS   ALLOYS
      . **RHODIUM ALLOYS**
  RT   PLATINUM ALLOYS

RHODIUM COMPOUNDS
  RT   ∞CHEMICAL COMPOUNDS
      ∞GROUP 8 COMPOUNDS

RHODIUM ISOTOPES
  UF   RHODIUM 102
      RHODIUM 106
  GS   CHEMICAL ELEMENTS
      . NUCLIDES
      . . ISOTOPES
      . . . **RHODIUM ISOTOPES**
      . RHODIUM
      . . **RHODIUM ISOTOPES**
      METALS
      . TRANSITION METALS
      . . RHODIUM
      . . . **RHODIUM ISOTOPES**

RHODIUM 102
  USE   RHODIUM ISOTOPES

RHODIUM 106
  USE   RHODIUM ISOTOPES

RHOMBIC ANTENNAS
  GS   ANTENNAS
      . DIRECTIONAL ANTENNAS
      . . **RHOMBIC ANTENNAS**
  RT   ANTENNA DESIGN
      RADIO ANTENNAS

RHOMBOHEDRONS
  GS   GEOMETRY
      . EUCLIDEAN GEOMETRY
      . . POLYHEDRONS
      . . . **RHOMBOHEDRONS**

RHOMBOIDS
  GS   GEOMETRY
      . EUCLIDEAN GEOMETRY
      . . POLYGONS
      . . . TETRAGONS
      . . . . PARALLELOGRAMS
      . . . . . **RHOMBOIDS**
  RT   GEOMETRY

RHONE DELTA (FRANCE)
  GS   LANDFORMS
      . DELTAS
      . . **RHONE DELTA (FRANCE)**
  RT   FRANCE
      MEDITERRANEAN SEA
      RIVERS

∞ RHYTHM
  SN   *(USE OF A MORE SPECIFIC TERM IS*
      RECOMMENDED--CONSULT THE TERMS
      LISTED BELOW)
  RT   OSCILLATIONS
      PERIODIC VARIATIONS
      RHYTHM (BIOLOGY)

RHYTHM (BIOLOGY)
  UF   BIOLOGICAL CLOCKS
      BIOLOGICAL RHYTHM
      PERIODICITY (BIOLOGY)
  GS   **RHYTHM (BIOLOGY)**
      . CIRCADIAN RHYTHMS
  RT   ACTIVITY CYCLES (BIOLOGY)
      ALTERNATIONS
      ∞BIOLOGY
      CYCLES
      DESYNCHRONIZATION (BIOLOGY)
      JET LAG
      PHENOLOGY
      ∞RHYTHM

RIBBON PARACHUTES
  GS   PARACHUTES
      . **RIBBON PARACHUTES**
  RT   DRAG CHUTES
      RECOVERY PARACHUTES

RIBBONS
  RT   CONVEYORS
      FABRICS
      FASTENERS
      METAL STRIPS
      ∞STRIP
      ∞TAPES

RIBOFLAVIN
  UF   VITAMIN B 2
      VITAMIN G
  GS   HETEROCYCLIC COMPOUNDS
      . **RIBOFLAVIN**
      VITAMINS
      . **RIBOFLAVIN**

RIBONUCLEIC ACIDS
  UF   RNA
  GS   ACIDS
      . NUCLEIC ACIDS
      . . **RIBONUCLEIC ACIDS**
  RT   GUANOSINES
      POLYNUCLEOTIDES

RIBOSE
  GS   ALIPHATIC COMPOUNDS
      . SUGARS
      . . MONOSACCHARIDES
      . . . **RIBOSE**
      . PENTOSE
      . . . **RIBOSE**
      CARBOHYDRATES
      . SUGARS
      . . MONOSACCHARIDES
      . . . **RIBOSE**
      . PENTOSE
      . . . **RIBOSE**
  RT   NUCLEOSIDES

RIBS (SUPPORTS)
  RT   LONGERONS
      REINFORCEMENT (STRUCTURES)
      REINFORCEMENT RINGS
      STIFFENING
      WEBS (SUPPORTS)

RICCATI EQUATION
  RT   DIFFERENTIAL EQUATIONS

RICE
  GS   FARM CROPS
      . GRAINS (FOOD)
      . . **RICE**
  RT   WHEAT

RICHARDS THEOREM
  GS   THEOREMS
      . **RICHARDS THEOREM**
  RT   NETWORK SYNTHESIS
      SIGNAL FLOW GRAPHS

RICHARDSON NUMBER
  GS   RATIOS
      . DIMENSIONLESS NUMBERS
      . . **RICHARDSON NUMBER**
  RT   AERODYNAMIC STABILITY
      REYNOLDS NUMBER
      SHEAR FLOW

RICHARDSON-DUSHMAN EQUATION
  USE   TEMPERATURE EFFECTS
      THERMIONIC EMISSION

∞ RIDGES
  SN   *(USE OF A MORE SPECIFIC TERM IS*
      RECOMMENDED--CONSULT THE TERMS
      LISTED BELOW
  UF   CUESTAS
      HOGBACKS
  RT   BRIDGES (LANDFORMS)
      BUCKLING
      CORRUGATING
      GAPS (GEOLOGY)
      KARST
      LANDFORMS
      MOUNTAINS
      PROTUBERANCES
      WRINKLING

RIDING QUALITY
  GS   QUALITY
      . **RIDING QUALITY**
  RT   COMFORT
      PASSENGERS
      SEATS
      SUSPENSION SYSTEMS (VEHICLES)
      TRANSPORTATION

RIEMANN INTEGRAL
  USE   MEASURE AND INTEGRATION

RIEMANN MANIFOLD
  UF   RIEMANN SPACE
      RIEMANN SPHERE
  GS   GEOMETRY
      . DIFFERENTIAL GEOMETRY
      . . **RIEMANN MANIFOLD**
      MANIFOLDS (MATHEMATICS)
      . **RIEMANN MANIFOLD**
  RT   EUCLIDEAN GEOMETRY

RIEMANN PROBLEM
USE   CAUCHY PROBLEM

RIEMANN SPACE
USE   RIEMANN MANIFOLD

RIEMANN SPHERE
USE   RIEMANN MANIFOLD

RIEMANN WAVES
GS   ELASTIC WAVES
   . SHOCK WAVES
   . . **RIEMANN WAVES**
RT   BLAST LOADS
   DIFFERENTIAL EQUATIONS
   DYNAMIC PRESSURE
   EXPLOSIONS
   HYPERBOLIC FUNCTIONS

RIESZ THEOREM
GS   THEOREMS
   . **RIESZ THEOREM**
RT   DIFFERENTIAL EQUATIONS
   HYPERBOLIC FUNCTIONS

RIFLES
GS   WEAPONS
   . GUNS (ORDNANCE)
   . . **RIFLES**
RT   ARTILLERY

RIFT (REACTOR IN FLIGHT TEST)
UF   REACTOR IN FLIGHT TEST PROGRAM
RT   ELECTRIC PROPULSION
   ∞REACTORS
   ROCKET ENGINES
   ROVER PROJECT
   SATURN PROJECT

RIFT VALLEYS
USE   VALLEYS

RIFTS
USE   GEOLOGICAL FAULTS

RIGGING
RT   ASSEMBLING
   CONSTRUCTION
   ∞EQUIPMENT
   MATERIALS HANDLING
   SHROUDS

RIGID BODIES
USE   RIGID STRUCTURES

RIGID MOUNTING
GS   MOUNTING
   . **RIGID MOUNTING**
RT   PYLON MOUNTING

RIGID ROTOR HELICOPTERS
GS   V/STOL AIRCRAFT
   . ROTARY WING AIRCRAFT
   . . HELICOPTERS
   . . . **RIGID ROTOR HELICOPTERS**
   . . . . CH-3 HELICOPTER
   . . . . F-28 HELICOPTER
   . . . . XH-51 HELICOPTER
RT   ∞AIRCRAFT
   OH-5 HELICOPTER

RIGID ROTORS
UF   HINGELESS ROTORS
GS   AIRFOILS
   . WINGS
   . . ROTARY WINGS
   . . . **RIGID ROTORS**
   RIGID STRUCTURES
   . **RIGID ROTORS**
   ROTATING BODIES
   . ROTORS
   . . ROTARY WINGS
   . . . **RIGID ROTORS**
RT   BEARINGLESS ROTORS

RIGID ROTORS (PLASMA PHYSICS)
GS   **RIGID ROTORS (PLASMA PHYSICS)**
   . PLASMA PHYSICS
RT   MOLECULAR COLLISIONS
   MOLECULAR ROTATION
   ∞PHYSICS
   PLASMA CONTROL

RIGID STRUCTURES
UF   INELASTIC BODIES
   RIGID BODIES
   STIFF STRUCTURES
GS   **RIGID STRUCTURES**
   . RIGID ROTORS
   . RIGID WINGS
RT   ARCHES
   COMPOSITE MATERIALS
   CONCRETE STRUCTURES
   EULER EQUATIONS OF MOTION
   HYBRID STRUCTURES
   REINFORCEMENT (STRUCTURES)
   SANDWICH STRUCTURES
   SPACE ERECTABLE STRUCTURES
   STEEL STRUCTURES
   ∞STRUCTURES
   TRANSLATIONAL MOTION
   WELDED STRUCTURES

RIGID WINGS
SN   (EXCLUDES RIGID ROTORS)
GS   AIRFOILS
   . WINGS
   . . **RIGID WINGS**
   RIGID STRUCTURES
   . **RIGID WINGS**
RT   AEROELASTICITY
   FIXED WINGS
   FLEXIBLE WINGS
   LOW ASPECT RATIO WINGS

∞ **RIGIDITY**
SN   *(USE OF A MORE SPECIFIC TERM IS*
   RECOMMENDED--CONSULT THE TERMS
   LISTED BELOW)
RT   FLEXIBILITY
   MAGNETIC RIGIDITY
   MECHANICAL PROPERTIES
   MODULUS OF ELASTICITY
   RUGGEDNESS
   STIFFNESS
   STRUCTURAL STABILITY

RILLS
USE   VALLEYS

RIMS
RT   ∞BLADES
   BORDERS
   EDGES
   MARGINS
   SIDES

RING CURRENTS
GS   ELECTRIC CURRENT
   . **RING CURRENTS**
RT   ATMOSPHERIC ELECTRICITY
   ELECTROJETS
   PLASMA CURRENTS

RING DISCHARGE
GS   ELECTRIC CURRENT
   . ELECTRIC DISCHARGES
   . . TOWNSEND DISCHARGE
   . . . GAS DISCHARGES
   . . . . TOROIDAL DISCHARGE
   . . . . . **RING DISCHARGE**
RT   ∞DISCHARGE
   ELECTRODELESS DISCHARGES
   GAS IONIZATION
   HIGH FREQUENCIES
   RADIO FREQUENCY DISCHARGE

RING LASERS
GS   STIMULATED EMISSION DEVICES
   . LASERS
   . . **RING LASERS**

RING STRUCTURES
GS   **RING STRUCTURES**
   . REINFORCEMENT RINGS
   . RING WINGS
RT   AERODYNAMIC CONFIGURATIONS
   ∞BANDS
   HOOPS
   REINFORCEMENT (STRUCTURES)
   ∞RINGS
   ∞STRUCTURES

RING WINGS
GS   AIRFOILS
   . WINGS
   . . UNCAMBERED WINGS

RING WINGS-*(CONT.)*
   . . . **RING WINGS**
   . . UNSWEPT WINGS
   . . . **RING WINGS**
   RING STRUCTURES
   . **RING WINGS**
RT   DUCTED FANS
   SHROUDED PROPELLERS
   TWISTED WINGS

∞ **RINGS**
SN   *(USE OF A MORE SPECIFIC TERM IS*
   RECOMMENDED--CONSULT THE TERMS
   LISTED BELOW)
RT   ANNULI
   BODIES OF REVOLUTION
   CIRCLES (GEOMETRY)
   JUPITER RINGS
   O RING SEALS
   PLANETARY RINGS
   REINFORCEMENT RINGS
   RING STRUCTURES
   RINGS (MATHEMATICS)
   SATURN RINGS
   STORAGE RINGS (PARTICLE
     ACCELERATORS)
   TOROIDAL PLASMAS
   TORUSES
   URANUS RINGS
   VORTEX RINGS

RINGS (MATHEMATICS)
RT   ∞MATHEMATICS
   ∞RINGS

RIO GRANDE (NORTH AMERICA)
GS   RIVERS
   . **RIO GRANDE (NORTH AMERICA)**
RT   GULF OF MEXICO
   MEXICO
   NEW MEXICO
   TEXAS

RIOMETERS
GS   MEASURING INSTRUMENTS
   . RADIATION MEASURING INSTRUMENTS
   . . **RIOMETERS**
RT   ATMOSPHERIC IONIZATION
   AURORAL ABSORPTION
   IONOGRAMS
   IONOSONDES
   IONOSPHERIC NOISE
   IONOSPHERIC PROPAGATION

RIPPLES
GS   ELASTIC WAVES
   . CAPILLARY WAVES
   . . **RIPPLES**
   SURFACE WAVES
   . CAPILLARY WAVES
   . . **RIPPLES**
RT   GRAVITY WAVES
   INTERFACIAL TENSION
   WATER WAVES
   WIND (METEOROLOGY)

RISERS
RT   CASTINGS
   PIPES (TUBES)

RISK
RT   ACCEPTABILITY
   ASSUMPTIONS
   ∞CAPACITY
   COMMERCE
   CONFIDENCE
   CONFIDENCE LIMITS
   CONTINGENCY
   DECISION THEORY
   ESTIMATES
   ESTIMATING
   FINANCE
   FORECASTING
   GAME THEORY
   HAZARDS
   INVENTORY CONTROLS
   MATHEMATICAL MODELS
   MAXIMUM LIKELIHOOD ESTIMATES
   OPERATIONS RESEARCH
   PREDICTIONS
   RELIABILITY
   STRATEGY

RIT ENGINES
UF   RADIO FREQUENCY ION THRUSTOR
       ENGINES
GS   ENGINES
    . ROCKET ENGINES
    . . ELECTRIC ROCKET ENGINES
    . . . ION ENGINES
    . . . . **RIT ENGINES**
RT   ELECTROSTATIC ENGINES
    PLASMA ENGINES

RITZ AVERAGING METHOD
GS   ANALYSIS (MATHEMATICS)
    . NUMERICAL ANALYSIS
    . . APPROXIMATION
    . . . **RITZ AVERAGING METHOD**
RT   ∞METHODOLOGY

RIVER BASINS
GS   LANDFORMS
    . STRUCTURAL BASINS
    . . **RIVER BASINS**
    . . . ATCHAFALAYA RIVER BASIN (LA)
    . . . CHENA RIVER BASIN (AK)
    . . . COLUMBIA RIVER BASIN (ID-OR-WA)
    . . . DELAWARE RIVER BASIN (US)
    . . . FEATHER RIVER BASIN (CA)
    . . . MISSOURI RIVER BASIN (US)
    . . . SUSQUEHANNA RIVER BASIN
       (MD-NY-PA)
    . . . WABASH RIVER BASIN (IL-IN-OH)
    . . . WADIS
RT   AMAZON REGION (SOUTH AMERICA)
    CHESAPEAKE BAY (US)
    DEATH VALLEY (CA)
    INTERNATIONAL HYDROLOGICAL
      DECADE
    LAKES
    MEANDERS
    MISSISSIPPI RIVER (US)
    MISSOURI RIVER (US)
    RAPIDS
    RAVINES
    RIVERS
    SACRAMENTO VALLEY (CA)
    SAGINAW BAY (MI)
    SAN JOAQUIN VALLEY (CA)
    SHENANDOAH VALLEY (VA)
    VADOSE WATER
    WATERSHEDS

RIVERS
GS   **RIVERS**
    . COLORADO RIVER (NORTH AMERICA)
    . HUDSON RIVER (NY-NJ)
    . MISSISSIPPI RIVER (US)
    . MISSOURI RIVER (US)
    . OHIO RIVER (US)
    . RIO GRANDE (NORTH AMERICA)
RT   ALLUVIUM
    AMAZON REGION (SOUTH AMERICA)
    ATCHAFALAYA RIVER BASIN (LA)
    AUFEIS (ICE)
    BAYOUS
    CANYONS
    COLUMBIA RIVER BASIN (ID-OR-WA)
    DELAWARE RIVER BASIN (US)
    DELTAS
    EARTH RESOURCES
    EROSION
    ESTUARIES
    FEATHER RIVER BASIN (CA)
    INLAND WATERS
    LAKE ERIE
    LAKE HURON
    LAKE MICHIGAN
    LAKE ONTARIO
    LAKE SUPERIOR
    MEANDERS
    MISSISSIPPI DELTA (LA)
    MISSOURI RIVER BASIN (US)
    RAPIDS
    RESERVOIRS
    RHONE DELTA (FRANCE)
    RIVER BASINS
    SHOALS
    SHORELINES
    SOUNDS (TOPOGRAPHIC FEATURES)
    STREAMS
    SURFACE WATER
    SUSQUEHANNA RIVER BASIN
      (MD-NY-PA)
    TRIBUTARIES
    VALLEYS
    WABASH RIVER BASIN (IL-IN-OH)

RIVERS-(CONT.)
    WADIS
    WATER COLOR
    WATER RUNOFF
    WATERSHEDS
    WATERWAYS
    WHARVES

RIVETED JOINTS
GS   JOINTS (JUNCTIONS)
    . **RIVETED JOINTS**
RT   BUTT JOINTS
    LAP JOINTS
    METAL JOINTS
    WELDED JOINTS

RIVETING
RT   ∞JOINING
    RIVETS
    SEALING

RIVETS
GS   FASTENERS
    . **RIVETS**
RT   COUPLINGS
    HOLDERS
    PINS
    RIVETING

RL CIRCUITS
UF   LR CIRCUITS
GS   CIRCUITS
    . **RL CIRCUITS**
    . . RLC CIRCUITS
RT   COUPLING CIRCUITS
    ELECTRICAL RESISTANCE
    INDUCTANCE
    LC CIRCUITS
    NETWORK ANALYSIS
    NETWORK SYNTHESIS
    TIME CONSTANT

RL-10 ENGINES
GS   ENGINES
    . ROCKET ENGINES
    . . LIQUID PROPELLANT ROCKET
       ENGINES
    . . . **RL-10 ENGINES**
    . . . . RL-10-A-1 ENGINE
    . . . . RL-10-A-3 ENGINE
RT   ATLAS CENTAUR LAUNCH VEHICLE
    CENTAUR PROJECT
    CRYOGENIC ROCKET PROPELLANTS
    SATURN LAUNCH VEHICLES

RL-10-A-1 ENGINE
GS   ENGINES
    . ROCKET ENGINES
    . . LIQUID PROPELLANT ROCKET
       ENGINES
    . . . HYDROGEN OXYGEN ENGINES
    . . . . **RL-10-A-1 ENGINE**
    . . . RL-10 ENGINES
    . . . . **RL-10-A-1 ENGINE**

RL-10-A-3 ENGINE
GS   ENGINES
    . ROCKET ENGINES
    . . LIQUID PROPELLANT ROCKET
       ENGINES
    . . . HYDROGEN OXYGEN ENGINES
    . . . . **RL-10-A-3 ENGINE**
    . . . RL-10 ENGINES
    . . . . **RL-10-A-3 ENGINE**
RT   SATURN D LAUNCH VEHICLE

RLC CIRCUITS
UF   LRC CIRCUITS
    RLC NETWORKS
GS   CIRCUITS
    . RL CIRCUITS
    . . **RLC CIRCUITS**
RT   CAPACITANCE
    CAPACITANCE SWITCHES
    ELECTRICAL RESISTANCE
    LC CIRCUITS
    NETWORK ANALYSIS
    NETWORK SYNTHESIS
    RC CIRCUITS
    TIME CONSTANT

RLC NETWORKS
USE   RLC CIRCUITS

RNA
USE   RIBONUCLEIC ACIDS

ROADS
GS   **ROADS**
    . HIGHWAYS
RT   ∞FACILITIES
    INTERSECTIONS
    PASSAGEWAYS
    PAVEMENTS
    RAPID TRANSIT SYSTEMS
    SITE SELECTION
    STREETS
    TRANSPORTATION
    TRANSPORTATION NETWORKS

ROADWAY POWERED VEHICLES
GS   SURFACE VEHICLES
    . **ROADWAY POWERED VEHICLES**
RT   ELECTRIC BATTERIES
    ELECTRIC MOTOR VEHICLES
    ENERGY STORAGE

ROASTING
UF   CALCINATION
RT   BAKING
    DESULFURIZING
    DRYING
    HEATING
    IGNITION
    OXIDATION
    REDUCTION (CHEMISTRY)
    SINTERING

ROBIN BALLOONS
GS   EXPANDABLE STRUCTURES
    . INFLATABLE STRUCTURES
    . . BALLOONS
    . . . METEOROLOGICAL BALLOONS
    . . . . **ROBIN BALLOONS**
RT   HIGH ALTITUDE BALLOONS
    RADIOSONDES
    ROCKOONS
    SKYHOOK BALLOONS
    SOUNDING

ROBOTICS
RT   ARTIFICIAL INTELLIGENCE
    AUTOMATA THEORY
    AUTOMATIC CONTROL
    ∞AUTOMATION
    COMPUTER AIDED DESIGN
    COMPUTER AIDED MANUFACTURING
    COMPUTER AIDED MAPPING
    COMPUTER VISION
    MAN MACHINE SYSTEMS
    MANIPULATORS
    POSITION SENSING
    ROBOTS
    TELEOPERATORS
    VOICE CONTROL

ROBOTS
RT   ARTIFICIAL INTELLIGENCE
    AUTOMATA THEORY
    BIONICS
    COMPUTER VISION
    PSYCHOLINGUISTICS
    ROBOTICS
    SERVOMECHANISMS
    VOICE CONTROL

ROBUSTNESS (MATHEMATICS)
RT   ALGORITHMS
    CONTROL STABILITY
    CONTROL THEORY
    FEEDBACK CONTROL
    LINEAR SYSTEMS
    MATHEMATICAL MODELS

ROCHE LIMIT
GS   RANGE (EXTREMES)
    . **ROCHE LIMIT**
RT   CELESTIAL MECHANICS
    DIMENSIONAL STABILITY
    GRAVITATION
    NATURAL SATELLITES
    ORBITS
    ROTATING BODIES
    TWO BODY PROBLEM

ROCK BOLTS
GS   FASTENERS
    . BOLTS

ROCK BOLTS-*(CONT.)*	**ROCKET ENGINES**-*(CONT.)*	**ROCKET ENGINES**-*(CONT.)*
. . **ROCK BOLTS**	. **ROCKET ENGINES**	. . TURBOROCKET ENGINES
	. . BOOSTER ROCKET ENGINES	. . ULLAGE ROCKET ENGINES
ROCK INTRUSIONS	. . . AJ-10 ENGINE	. . UPPER STAGE ROCKET ENGINES
UF DIKES (GEOLOGY)	. . . ALGOL ENGINE	. . VERNIER ENGINES
GS **ROCK INTRUSIONS**	. . . APOGEE BOOST MOTORS	. . . SYNCOM APOGEE ENGINES
. BATHOLITHS	. . . H-1 ENGINE	RT AIRCRAFT ENGINES
RT CONTACTS (GEOLOGY)	. . . LR-87-AJ-5 ENGINE	AXIAL MODES
IGNEOUS ROCKS	. . . M-1 ENGINE	BURNING TIME
INLIERS (LANDFORMS)	. . . M-55 ENGINE	EJECTORS
REGOLITH	. . . MA-2 ENGINE	EXHAUST NOZZLES
ROCKS	. . . MA-3 ENGINE	EXPENDABLE STAGES (SPACECRAFT)
	. . . MA-5 ENGINE	HEAVY LIFT LAUNCH VEHICLES
ROCK MECHANICS	. . . NIKE BOOSTER ROCKET ENGINES	HYBRID PROPULSION
RT FRACTURE MECHANICS	. . . P-1 ENGINE	IGNITION SYSTEMS
GEOLOGY	. . . ROCKET ENGINE 9KS-11000	INTERNAL COMBUSTION ENGINES
ROCKS	. . . X-405 ENGINE	JET ENGINES
SOIL MECHANICS	. . DUCTED ROCKET ENGINES	JET PROPULSION
STRUCTURAL PROPERTIES (GEOLOGY)	. . ELECTRIC ROCKET ENGINES	LASER PROPULSION
	. . . ELECTROSTATIC ENGINES	LAUNCH VEHICLES
ROCK SALT	. . . ELECTROTHERMAL ENGINES	LUNAR MODULE ASCENT STAGE
USE HALITES	. . . . ARC JET ENGINES	MAGNETOPLASMADYNAMICS
	. . . . RESISTOJET ENGINES	MISSILE CONFIGURATIONS
ROCKET BOOSTERS	. . . ION ENGINES	MISSILES
USE BOOSTER ROCKET ENGINES	. . . . CESIUM ENGINES	MULTISTAGE ROCKET VEHICLES
	. . . . MERCURY ION ENGINES	POST BOOST PROPULSION SYSTEM
ROCKET CATAPULTS	. . . . RIT ENGINES	∞PROPELLANT ACTUATED DEVICES
GS LAUNCHERS	. . HEUS ROCKET ENGINES	PROPELLANT EXPLOSIONS
. CATAPULTS	. . HOT WATER ROCKET ENGINES	REFRACTORIES
. . **ROCKET CATAPULTS**	. . HYBRID PROPELLANT ROCKET	REUSABLE LAUNCH VEHICLES
. ROCKET LAUNCHERS	ENGINES	RIFT (REACTOR IN FLIGHT TEST)
. . **ROCKET CATAPULTS**	. . LITHERGOL ROCKET ENGINES	∞ROCKETS
RT GUN LAUNCHERS	. . LIQUID PROPELLANT ROCKET	SINGLE STAGE ROCKET VEHICLES
LAUNCH VEHICLES	ENGINES	SOLID ROCKET PROPELLANTS
LAUNCHING SITES	. . . AJ-10 ENGINE	SPACECRAFT COMPONENTS
MISSILES	. . . F-1 ROCKET ENGINE	SPACECRAFT PROPULSION
∞ROCKETS	. . . H-1 ENGINE	SPACECRAFT STRUCTURES
	. . . HYDRAZINE ENGINES	THRUST
ROCKET CHAMBERS	. . . HYDROGEN OXYGEN ENGINES	THRUST VECTOR CONTROL
USE THRUST CHAMBERS	. . . . J-2 ENGINE	THRUST-WEIGHT RATIO
	. . . . M-1 ENGINE	∞THRUSTORS
ROCKET ENGINE CASES	. . . . RL-10-A-1 ENGINE	
UF MISSILE ENGINE CASES	. . . . RL-10-A-3 ENGINE	**ROCKET EXHAUST**
ROCKET MOTOR CASES	. . . LIQUID AIR CYCLE ENGINES	GS PLUMES
GS CASES (CONTAINERS)	. . . LR-62-RM-2 ENGINE	. **ROCKET EXHAUST**
. **ROCKET ENGINE CASES**	. . . LR-87-AJ-5 ENGINE	RT BASE HEATING
RT MISSILE BODIES	. . . LR-91-AJ-5 ENGINE	COMBUSTION PRODUCTS
ORTHOTROPIC CYLINDERS	. . . LR-99 ENGINE	EXHAUST GASES
PERFORATED SHELLS	. . . MA-2 ENGINE	EXHAUST SYSTEMS
SHELLS (STRUCTURAL FORMS)	. . . MA-3 ENGINE	JET EXHAUST
THRUST CHAMBERS	. . . MA-5 ENGINE	
	. . . RL-10 ENGINES	**ROCKET FIRING**
ROCKET ENGINE CONTROL	. . . . RL-10-A-1 ENGINE	GS FIRING (IGNITING)
GS ENGINE CONTROL	. . . . RL-10-A-3 ENGINE	. **ROCKET FIRING**
. **ROCKET ENGINE CONTROL**	. . . SPACE SHUTTLE MAIN ENGINE	. . RETROFIRING
RT ATTITUDE CONTROL	. . . X-405 ENGINE	RT BURNING TIME
AUTOMATIC CONTROL	. . . XLR-99 ENGINE	DETONATION
∞CONTROL	. . . YLR-91-AJ-1 ENGINE	STATIC FIRING
DIRECTIONAL CONTROL	. . M-100 ENGINE	TEST FIRING
FLIGHT CONTROL	. . MICROROCKET ENGINES	
FUEL CONTROL	. . . ORBIT MANEUVERING ENGINE	**ROCKET FLIGHT**
HEUS ROCKET ENGINES	(SPACE SHUTTLE)	RT CLIMBING FLIGHT
MISSILE CONTROL	. . NOZZLELESS ROCKET ENGINES	COASTING FLIGHT
REMOTE CONTROL	. . NUCLEAR ENGINE FOR ROCKET	∞FLIGHT
SERVOCONTROL	VEHICLES	FLIGHT PATHS
SPACECRAFT CONTROL	. . NUCLEAR RAMJET ENGINES	HORIZONTAL FLIGHT
THRUST CONTROL	. . NUCLEAR ROCKET ENGINES	HYPERSONIC FLIGHT
	. . . NUCLEAR LIGHTBULB ENGINES	METEOROLOGICAL FLIGHT
ROCKET ENGINE DESIGN	. . RESTARTABLE ROCKET ENGINES	SPACE FLIGHT
GS ENGINE DESIGN	. . RETROROCKET ENGINES	SUBORBITAL FLIGHT
. **ROCKET ENGINE DESIGN**	. . . BE-3 ENGINE	SUPERSONIC FLIGHT
RT COLD FLOW TESTS	. . REUSABLE ROCKET ENGINES	TRAJECTORIES
∞DESIGN	. . SOLID PROPELLANT ROCKET	TRANSONIC FLIGHT
ENGINE TESTS	ENGINES	VERTICAL FLIGHT
PREFIRING TESTS	. . . ALGOL ENGINE	
ROVER PROJECT	. . . APOGEE BOOST MOTORS	**ROCKET LAUNCHERS**
	. . . ASROC ENGINE	GS LAUNCHERS
ROCKET ENGINE NOISE	. . . HERCULES ENGINE	. **ROCKET LAUNCHERS**
GS ELASTIC WAVES	. . . M-46 ENGINE	. . ROCKET CATAPULTS
. SOUND WAVES	. . . M-55 ENGINE	RT GROUND SUPPORT EQUIPMENT
. . NOISE (SOUND)	. . . M-56 ENGINE	GUN LAUNCHERS
. . . ENGINE NOISE	. . . M-57 ENGINE	LAUNCH VEHICLES
. . . . **ROCKET ENGINE NOISE**	. . . NIKE BOOSTER ROCKET ENGINES	LAUNCHING
RT MUFFLERS	. . . P-1 ENGINE	LAUNCHING SITES
	. . . SL-3 ROCKET ENGINE	MISSILE LAUNCHERS
ROCKET ENGINE 9KS-11000	. . . SYNCOM APOGEE ENGINES	∞ROCKETS
GS ENGINES	. . . TU-121 ENGINE	ROCKOONS
. ROCKET ENGINES	. . . TX-77 ENGINE	SEA LAUNCHING
. . BOOSTER ROCKET ENGINES	. . . TX-354 ENGINE	
. . . **ROCKET ENGINE 9KS-11000**	. . . X-248 ENGINE	**ROCKET LAUNCHING**
	. . . X-254 ENGINE	UF BLASTOFF
ROCKET ENGINES	. . . X-258 ENGINES	GS LAUNCHING
UF INTERPLANETARY PROPULSION	. . . . X-258-B1 ENGINE	. **ROCKET LAUNCHING**
GS ENGINES	. . . X-259 ENGINE	. . LUNAR LAUNCH
	. . . XM-33 ENGINE	. . ORBITAL LAUNCHING
	. . SUSTAINER ROCKET ENGINES	

ROCKET LAUNCHING-*(CONT.)*
RT LAUNCH VEHICLES
    LAUNCH WINDOWS
    LAUNCHERS
    SPACECRAFT LAUNCHING

ROCKET LININGS
GS LININGS
    . **ROCKET LININGS**
RT ENGINE PARTS
    REFRACTORIES

ROCKET MOTOR CASES
USE ROCKET ENGINE CASES

ROCKET NOSE CONES
GS CONES
    . NOSE CONES
    . . **ROCKET NOSE CONES**
RT ABLATIVE NOSE CONES

ROCKET NOZZLES
GS **ROCKET NOZZLES**
    . DUAL THRUST NOZZLES
RT ANNULAR NOZZLES
    CONICAL NOZZLES
    CONVERGENT-DIVERGENT NOZZLES
    DIVERGENT NOZZLES
    HYPERSONIC NOZZLES
    NOZZLE INSERTS
    NOZZLELESS ROCKET ENGINES
    ∞NOZZLES
    PLUG NOZZLES
    SKIRTS
    SPIKE NOZZLES
    SUPERSONIC NOZZLES

ROCKET OXIDIZERS
UF PROPELLANT OXIDIZERS
GS OXIDIZERS
    . **ROCKET OXIDIZERS**
    . . FLOX
    . . TAGN
RT CRYOGENIC FLUIDS
    DOMINO PROPELLANTS
    HIGH ENERGY OXIDIZERS
    HYDROGEN PEROXIDE
    LIQUID OXIDIZERS
    LIQUID OXYGEN
    NITROGEN TETROXIDE
    NITRONIUM PERCHLORATE
    TATB

ROCKET PLANES
GS ROCKET VEHICLES
    . **ROCKET PLANES**
RT AEROSPACEPLANES
    ∞AIRCRAFT
    ASTROPLANE
    BOOSTGLIDE VEHICLES
    RESEARCH AIRCRAFT

ROCKET PROPELLANT TANKS
USE PROPELLANT TANKS

ROCKET PROPELLANTS
UF MULTIPROPELLANTS
GS PROPELLANTS
    . **ROCKET PROPELLANTS**
    . . GASEOUS ROCKET PROPELLANTS
    . . LIQUID ROCKET PROPELLANTS
    . . . CRYOGENIC ROCKET PROPELLANTS
    . . . GELLED ROCKET PROPELLANTS
    . . . HYPERGOLIC ROCKET
          PROPELLANTS
    . . MONOPROPELLANTS
    . . . . AEROZINE
    . . . RP-1 ROCKET PROPELLANTS
    . . SLURRY PROPELLANTS
    . . NITRAMINE PROPELLANTS
    . . SOLID ROCKET PROPELLANTS
    . . . DOUBLE BASE ROCKET
          PROPELLANTS
    . . . METAL PROPELLANTS
    . . TAGN
    . . TATB
RT ASCENT PROPULSION SYSTEMS
    AUXILIARY PROPULSION
    FUELS
    HYDRAZINES
    HYDROCARBON FUELS
    LIQUID FUELS
    MISSILES
    PROPELLANT BINDERS
    PROPELLANT STORAGE

ROCKET PROPELLANTS-*(CONT.)*
    PROPULSION
    SOLID PROPELLANTS
    SPECIFIC IMPULSE
    STORABLE PROPELLANTS
    THRUST
    TORPEDO ENGINES

ROCKET PROPELLED SLEDS
GS SURFACE VEHICLES
    . SLEDS
    . . **ROCKET PROPELLED SLEDS**
RT JAVELIN ROCKET VEHICLE
    ∞TEST EQUIPMENT

ROCKET SONDES
USE SOUNDING ROCKETS

ROCKET SOUNDING
GS SOUNDING
    . **ROCKET SOUNDING**
RT ACOUSTIC SOUNDING
    ATMOSPHERIC SOUNDING
    BARIUM ION CLOUDS
    IONOSPHERIC SOUNDING
    JUDI-DART ROCKET
    MICROWAVE SOUNDING
    SATELLITE SOUNDING
    SOUNDING ROCKETS

ROCKET TEST FACILITIES
GS TEST FACILITIES
    . **ROCKET TEST FACILITIES**
RT ENGINE TESTS
    TEST FIRING
    TEST RANGES
    TEST STANDS

ROCKET THRUST
GS THRUST
    . **ROCKET THRUST**
    . . RETROTHRUST
RT HIGH THRUST
    JET THRUST
    LOW THRUST
    LOW THRUST PROPULSION
    MICROTHRUST
    PROPULSION SYSTEM PERFORMANCE
    SPECIFIC IMPULSE
    STATIC THRUST
    THRUST LOADS
    THRUST TERMINATION
    VARIABLE THRUST

ROCKET VEHICLES
GS **ROCKET VEHICLES**
    . ABLESTAR LAUNCH VEHICLE
    . ARCON ROCKET VEHICLE
    . BLUE STREAK LAUNCH VEHICLE
    . BLUE STREAK MISSILE
    . CENTAUR LAUNCH VEHICLE
    . . ATLAS CENTAUR LAUNCH VEHICLE
    . FOLDING FIN AIRCRAFT ROCKET
      VEHICLE
    . HOVERING ROCKET VEHICLES
    . METEOR 1 ROCKET VEHICLE
    . MULTISTAGE ROCKET VEHICLES
    . . ANTARES ROCKET VEHICLE
    . . ARGO ROCKET VEHICLES
    . . ARIANE LAUNCH VEHICLE
    . . ASTROBEE ROCKET VEHICLES
    . . . ASTROBEE 1500 ROCKET VEHICLE
    . . ATHENA ROCKET VEHICLE
    . . ATLAS LAUNCH VEHICLES
    . . . ATLAS ABLE 5 LAUNCH VEHICLE
    . . . ATLAS AGENA B LAUNCH VEHICLE
    . . . ATLAS AGENA LAUNCH VEHICLES
    . . . ATLAS CENTAUR LAUNCH VEHICLE
    . . . ATLAS SLV-3 LAUNCH VEHICLE
    . . BERENICE ROCKET VEHICLE
    . . BLACK KNIGHT ROCKET VEHICLE
    . . BLUE SCOUT ROCKET VEHICLE
    . . DIAMANT LAUNCH VEHICLE
    . . ELDO LAUNCH VEHICLE
    . . EXOS SOUNDING ROCKET
    . . JAGUAR ROCKET VEHICLE
    . . JAVELIN ROCKET VEHICLE
    . . JUNO 1 LAUNCH VEHICLE
    . . JUNO 2 LAUNCH VEHICLE
    . . JUPITER C ROCKET VEHICLE
    . . KAPPA ROCKET VEHICLES
    . . . KAPPA 8 ROCKET VEHICLE
    . . . KAPPA 9 ROCKET VEHICLE
    . . LAMBDA ROCKET VEHICLES
    . . LITTLE JOE 2 LAUNCH VEHICLE

ROCKET VEHICLES-*(CONT.)*
    . . NIKE ROCKET VEHICLES
    . . . NIKE-APACHE ROCKET VEHICLE
    . . . NIKE-CAJUN ROCKET VEHICLE
    . . . NIKE-HYDAC ROCKET VEHICLE
    . . . NIKE-IROQUOIS ROCKET VEHICLE
    . . . NIKE-JAVELIN ROCKET VEHICLE
    . . . NIKE-TOMAHAWK ROCKET VEHICLE
    . . NOVA LAUNCH VEHICLES
    . PHOENIX SOUNDING ROCKET
    . RAM B LAUNCH VEHICLE
    . RUBIS ROCKET VEHICLE
    . SATURN LAUNCH VEHICLES
    . . SATURN D LAUNCH VEHICLE
    . . SATURN 1 LAUNCH VEHICLES
    . . . SATURN 1 SA-1 LAUNCH VEHICLE
    . . . SATURN 1 SA-2 LAUNCH VEHICLE
    . . . SATURN 1 SA-3 LAUNCH VEHICLE
    . . . SATURN 1 SA-4 LAUNCH VEHICLE
    . . . SATURN 1 SA-5 LAUNCH VEHICLE
    . . . SATURN 1 SA-6 LAUNCH VEHICLE
    . . . SATURN 1 SA-7 LAUNCH VEHICLE
    . . . SATURN 1 SA-8 LAUNCH VEHICLE
    . . . SATURN 1 SA-9 LAUNCH VEHICLE
    . . . SATURN 1 SA-10 LAUNCH
          VEHICLE
    . . SATURN 1B LAUNCH VEHICLES
    . . SATURN 2 LAUNCH VEHICLES
    . . SATURN 5 LAUNCH VEHICLES
    . SCOUT LAUNCH VEHICLE
    . SKYLARK ROCKET VEHICLE
    . THOR LAUNCH VEHICLES
    . . THOR ABLE ROCKET VEHICLE
    . . THOR AGENA LAUNCH VEHICLE
    . . THOR DELTA LAUNCH VEHICLE
    . . TITAN LAUNCH VEHICLES
    . . TITAN 3 LAUNCH VEHICLE
    . VANGUARD 2 LAUNCH VEHICLE
    . VEGA LAUNCH VEHICLE
    . WASP SOUNDING ROCKET
    . PAYLOAD ASSIST MODULE
    . ROCKET PLANES
    . SATURN STAGES
    . . SATURN S-1 STAGE
    . . SATURN S-1B STAGE
    . . SATURN S-1C STAGE
    . . SATURN S-2 STAGE
    . . SATURN S-4 STAGE
    . . SATURN S-4B STAGE
    . SINGLE STAGE ROCKET VEHICLES
    . AGENA ROCKET VEHICLES
    . . AGENA A ROCKET VEHICLE
    . . AGENA B ROCKET VEHICLE
    . . AGENA C ROCKET VEHICLE
    . . AGENA D ROCKET VEHICLE
    . . ARCAS ROCKET VEHICLES
    . . BLACK BRANT SOUNDING ROCKETS
    . . . BLACK BRANT 1 SOUNDING
          ROCKET
    . . . BLACK BRANT 2 SOUNDING
          ROCKET
    . . . BLACK BRANT 3 SOUNDING
          ROCKET
    . . . BLACK BRANT 4 SOUNDING
          ROCKET
    . . . BLACK BRANT 5 SOUNDING
          ROCKET
    . . BLACK KNIGHT ROCKET VEHICLE
    . . DORNIER PARAGLIDER ROCKET
        VEHICLE
    . . GENIE ROCKET VEHICLE
    . . HONEST JOHN ROCKET VEHICLE
    . . HYLA-STAR ROCKET VEHICLE
    . . LITTLE JOHN ROCKET VEHICLE
    . . LOKI ROCKET VEHICLE
    . . NOMAD LAUNCH VEHICLE
    . . VERONIQUE ROCKET VEHICLES
    . . VIKING ROCKET VEHICLE
    . . ZUNI ROCKET VEHICLE
    . SOUNDING ROCKETS
    . . AEROBEE ROCKET VEHICLE
    . . ANTARES ROCKET VEHICLE
    . . APACHE ROCKET VEHICLE
    . . ARCAS ROCKET VEHICLES
    . . ARIES SOUNDING ROCKET
    . . ASP ROCKET VEHICLE
    . . ASTROBEE ROCKET VEHICLES
    . . . ASTROBEE 1500 ROCKET VEHICLE
    . . BLACK BRANT SOUNDING ROCKETS
    . . . BLACK BRANT 1 SOUNDING
          ROCKET
    . . . BLACK BRANT 2 SOUNDING
          ROCKET
    . . . BLACK BRANT 3 SOUNDING
          ROCKET

ROCKET VEHICLES-(CONT.)
... BLACK BRANT 4 SOUNDING
    ROCKET
... BLACK BRANT 5 SOUNDING
    ROCKET
.. CAJUN ROCKET VEHICLE
.. DORNIER PARAGLIDER ROCKET
    VEHICLE
.. EXOS SOUNDING ROCKET
.. JAGUAR ROCKET VEHICLE
.. JUDI-DART ROCKET
.. KAPPA ROCKET VEHICLES
... KAPPA 8 ROCKET VEHICLE
... KAPPA 9 ROCKET VEHICLE
.. LAMBDA ROCKET VEHICLES
.. LOKI ROCKET VEHICLE
.. PETREL SOUNDING ROCKET
.. PHOENIX SOUNDING ROCKET
.. SKUA ROCKET VEHICLES
.. SKYLARK ROCKET VEHICLES
.. VENUS FLY TRAP ROCKET VEHICLE
.. VERONIQUE ROCKET VEHICLES
.. VERTICAL 8 ROCKET
.. WASP SOUNDING ROCKET
. STANDARD LAUNCH VEHICLES
. ATLAS SLV-3 LAUNCH VEHICLE
.. STANDARD LAUNCH VEHICLE 5
. SURFACE TO SURFACE ROCKETS
. HONEST JOHN ROCKET VEHICLE
.. LITTLE JOHN ROCKET VEHICLE
. THORAD LAUNCH VEHICLES
. THOR ABLE ROCKET VEHICLE
.. THOR AGENA LAUNCH VEHICLE
. THOR DELTA LAUNCH VEHICLE
. TITAN CENTAUR LAUNCH VEHICLE
RT   ACOUSTIC SOUNDING
    ∞BALLISTIC VEHICLES
    ∞FLIGHT VEHICLES
    HEUS ROCKET ENGINES
    LAUNCH VEHICLES
    MISSILE CONFIGURATIONS
    MULTIENGINE VEHICLES
    ∞ROCKETS
    SPACE PROCESSING APPLICATIONS
        ROCKET
    STAGE SEPARATION
    TEST VEHICLES
    TRAILBLAZER 1 REENTRY VEHICLE
    TRAILBLAZER 2 REENTRY VEHICLE
    ∞VEHICLES
    ∞WINGED VEHICLES
    X-1 AIRCRAFT
    X-2 AIRCRAFT
    X-15 AIRCRAFT
    X-17 REENTRY VEHICLE

ROCKET-BORNE INSTRUMENTS
RT   CONTROLLERS
    FLIGHT TEST INSTRUMENTS
    ∞INSTRUMENTS
    MEASURING INSTRUMENTS
    METEOROLOGICAL INSTRUMENTS
    POSITION INDICATORS

ROCKET-BORNE PHOTOGRAPHY
GS   IMAGERY
    . **ROCKET-BORNE PHOTOGRAPHY**
    PHOTOGRAPHY
    . **ROCKET-BORNE PHOTOGRAPHY**
RT   AERIAL PHOTOGRAPHY
    ASTRONOMICAL PHOTOGRAPHY
    BLACK AND WHITE PHOTOGRAPHY
    PHOTOMAPPING
    SATELLITE-BORNE PHOTOGRAPHY
    SPACEBORNE PHOTOGRAPHY

∞ ROCKETS
SN   *(USE OF A MORE SPECIFIC TERM IS*
    RECOMMENDED--CONSULT THE TERMS
    LISTED BELOW)
RT   AIR SLEW MISSILES
    AMMUNITION
    ESCAPE ROCKETS
    HEAVY LIFT LAUNCH VEHICLES
    INCENDIARY AMMUNITION
    LAUNCH VEHICLES
    MISSILES
    NUCLEAR RAMJET ENGINES
    NUCLEAR ROCKET ENGINES
    NUCLEAR WEAPONS
    PATRIOT MISSILE
    PETREL SOUNDING ROCKET
    PYROTECHNICS
    REENTRY
    REENTRY VEHICLES

ROCKETS-(CONT.)
    ROCKET CATAPULTS
    ROCKET ENGINES
    ROCKET LAUNCHERS
    ROCKET VEHICLES
    ROCKOONS
    SPACE FLIGHT
    SURFACE TO AIR MISSILES
    SURFACE TO SURFACE MISSILES
    SURFACE TO SURFACE ROCKETS
    TORPEDOES
    VERTICAL 8 ROCKET
    WARHEADS
    WEAPON SYSTEMS
    WEAPONS DELIVERY

ROCKOONS
RT   HIGH ALTITUDE BALLOONS
    METEOROLOGICAL BALLOONS
    ROBIN BALLOONS
    ROCKET LAUNCHERS
    ∞ROCKETS
    SKYHOOK BALLOONS

ROCKS
UF   STONES (ROCKS)
GS   **ROCKS**
    . ANDESITE
    . ATAXITE
    . BEDROCK
    . BALTIC SHIELD (EUROPE)
    . BRECCIA
    . COAL
    .. LIGNITE
    .. SOLVENT REFINED COAL
    . GNEISS
    . IGNEOUS ROCKS
    .. ANORTHOSITE
    .. BASALT
    .. DIORITE
    .. DUNITE
    .. ECLOGITE
    .. FELSITE
    .. GRANITE
    .. OBSIDIAN
    ... MOLDAVITE
    .. PERIDOTITE
    .. PUMICE
    .. SYENITE
    . TRACHYTE
    . LIMESTONE
    . LUNAR ROCKS
    . REGOLITH
    . SANDSTONES
    . SCHIST
    . SHALES
    . SHATTER CONES
RT   AGGREGATES
    BATHOLITHS
    BAUXITE
    BOREHOLES
    CARBONACEOUS ROCKS
    CLAYS
    CONTACTS (GEOLOGY)
    CROSSBEDDING (GEOLOGY)
    DIRT
    DOLOMITE (MINERAL)
    EARTH RESOURCES
    EFFUSIVES
    ENSTATITE
    FOLDS (GEOLOGY)
    FORMATIONS
    GEOLOGY
    GYPSUM
    INLIERS (LANDFORMS)
    KARST
    KREEP
    LANDSLIDES
    LATERITES
    LAVA
    LEDGES
    LITHOLOGY
    MAGMA
    METAMORPHISM (GEOLOGY)
    MINERALS
    NUNATAKS
    OLIVINE
    OUTLIERS (LANDFORMS)
    PALEOMAGNETISM
    PETROGRAPHY
    PETROLOGY
    PYROXENES
    QUARTZ
    REEFS
    ROCK INTRUSIONS

ROCKS-(CONT.)
    ROCK MECHANICS
    SEDIMENTARY ROCKS
    SERPENTINE
    SOILS
    STRATIGRAPHY
    TUNNELING (EXCAVATION)

ROCKWELL HARDNESS
GS   MECHANICAL PROPERTIES
    . HARDNESS
    .. **ROCKWELL HARDNESS**
RT   MICROHARDNESS

ROCKY MOUNTAINS (NORTH AMERICA)
GS   LANDFORMS
    . MOUNTAINS
    .. **ROCKY MOUNTAINS (NORTH**
        AMERICA)
RT   CANADA
    UNITED STATES

RODENTS
GS   ANIMALS
    . VERTEBRATES
    .. MAMMALS
    ... **RODENTS**
    .... GUINEA PIGS
    .... HAMSTERS
    .... MICE
    ..... JERBOAS
    ..... POCKET MICE
    .... RABBITS
    .... RATS
    .... SQUIRRELS
    ..... GROUND SQUIRRELS

RODS
GS   **RODS**
    . CONTROL RODS
RT   BARS
    BILLETS
    DIRECTORS (ANTENNA ELEMENTS)
    STRUCTURAL MEMBERS
    WIRE

ROENTGEN SATELLITE
USE   ROSAT MISSION

ROGALLO WINGS
USE   FLEXIBLE WINGS
    FOLDING STRUCTURES

ROLL
UF   DAMPING IN ROLL
GS   ATTITUDE (INCLINATION)
    . **ROLL**
RT   LATERAL CONTROL
    LATERAL OSCILLATION
    LATERAL STABILITY
    PITCH (INCLINATION)
    ROLLERS
    ∞ROLLING
    ROLLING MOMENTS
    ROTATION
    SIDESLIP
    TURNING FLIGHT
    YAW

ROLL CONTROL
USE   LATERAL CONTROL

ROLL FORMING
GS   FORMING TECHNIQUES
    . **ROLL FORMING**
RT   COLD WORKING
    METAL WORKING

ROLLER BEARINGS
GS   BEARINGS
    . ANTIFRICTION BEARINGS
    .. **ROLLER BEARINGS**
RT   BALL BEARINGS
    NEEDLE BEARINGS
    THRUST BEARINGS

ROLLERS
RT   CONVEYORS
    CYLINDRICAL BODIES
    DISPENSERS
    DISTRIBUTORS
    IDLERS
    PLATENS
    PULLEYS

ROLLERS-*(CONT.)*
- ROLL
- TIRES
- VEHICLE WHEELS
- WHEELS

∞ **ROLLING**
- SN   *(USE OF A MORE SPECIFIC TERM IS RECOMMENDED--CONSULT THE TERMS LISTED BELOW)*
- RT   AUSFORMING
- FLATTENING
- FORGING
- LEVELING
- METAL WORKING
- ROLL

ROLLING CONTACT LOADS
- GS   LOADS (FORCES)
- . DYNAMIC LOADS
- . . **ROLLING CONTACT LOADS**
- RT   ANTIFRICTION BEARINGS
- STRESSES
- STRUCTURAL DESIGN CRITERIA

ROLLING MOMENTS
- GS   MOMENTS
- . STABILITY DERIVATIVES
- . . **ROLLING MOMENTS**
- RT   AERODYNAMIC COEFFICIENTS
- LATERAL STABILITY
- MOMENTS OF INERTIA
- PITCHING MOMENTS
- ROLL
- TORQUE
- YAWING MOMENTS

ROLLUP SOLAR ARRAYS
- USE   SOLAR ARRAYS

ROMANIA
- UF   RUMANIA
- GS   NATIONS
- . **ROMANIA**
- RT   BLACK SEA
- CENTRAL EUROPE
- EUROPE

RONCHI TEST
- GS   INTERFEROMETRY
- . **RONCHI TEST**
- RT   ELECTROMAGNETIC RADIATION
- GRATINGS (SPECTRA)
- INTERFEROMETERS
- MEASURING INSTRUMENTS
- OPTICAL MEASUREMENT

ROOFS
- RT   ∞ BUILDINGS
- SHEATHS

ROOM TEMPERATURE
- GS   TEMPERATURE
- . **ROOM TEMPERATURE**
- RT   AMBIENT TEMPERATURE
- OPERATING TEMPERATURE

ROOMS
- GS   **ROOMS**
- . CLEAN ROOMS
- . DARKROOMS
- RT   COMPARTMENTS
- ENCLOSURES
- ∞ LOUNGES

ROOT-MEAN-SQUARE ERRORS
- GS   ERRORS
- . **ROOT-MEAN-SQUARE ERRORS**
- RT   ERROR ANALYSIS
- STATISTICAL ANALYSIS

∞ **ROOTS**
- SN   *(USE OF A MORE SPECIFIC TERM IS RECOMMENDED--CONSULT THE TERMS LISTED BELOW)*
- RT   PLANT ROOTS
- RADICALS
- ROOTS OF EQUATIONS
- WING ROOTS

ROOTS OF EQUATIONS
- UF   ZERO CROSSINGS
- RT   EIGENVALUES
- ∞ EQUATIONS

ROOTS OF EQUATIONS-*(CONT.)*
- EXISTENCE THEOREMS
- MATRICES (MATHEMATICS)
- NONLINEAR EQUATIONS
- POLYNOMIALS
- ∞ ROOTS

RORSCHACH TESTS
- GS   PSYCHOLOGICAL TESTS
- . **RORSCHACH TESTS**
- RT   MENTAL HEALTH
- PSYCHOLOGY
- ∞ TESTS

ROSAT MISSION
- UF   ROENTGEN SATELLITE
- GS   OBSERVATORIES
- . ASTRONOMICAL OBSERVATORIES
- . . **ROSAT MISSION**
- SATELLITES
- . ARTIFICIAL SATELLITES
- . . **ROSAT MISSION**
- RT   ASTRONOMICAL SATELLITES
- INTERNATIONAL COOPERATION
- SPACEBORNE ASTRONOMY
- SPACEBORNE TELESCOPES
- X RAY ASTRONOMY
- X RAY SOURCES
- X RAY TELESCOPES

ROSETTE SHAPES
- GS   SHAPES
- . **ROSETTE SHAPES**
- RT   ANTENNA RADIATION PATTERNS
- CRYSTALLITES
- SPHERULITES
- STRAIN GAGES

ROSHKO PREDICTION
- GS   PREDICTIONS
- . **ROSHKO PREDICTION**
- RT   BLUFF BODIES
- LAMINAR FLOW
- OSEEN APPROXIMATION
- THREE DIMENSIONAL FLOW

ROSIN
- GS   GUMS (SUBSTANCES)
- . **ROSIN**
- RT   ORGANIC MATERIALS

ROSS ICE SHELF
- GS   REGIONS
- . POLAR REGIONS
- . . ANTARCTIC REGIONS
- . . . **ROSS ICE SHELF**
- SOUTHERN HEMISPHERE
- . ANTARCTIC REGIONS
- . . **ROSS ICE SHELF**
- RT   MCMURDO SOUND

ROSSBY REGIMES
- RT   BAROTROPIC FLOW
- PLANETARY WAVES

ROSSBY WAVES
- USE   PLANETARY WAVES

ROTARY DRIVES
- USE   MECHANICAL DRIVES

ROTARY ENGINES
- GS   ENGINES
- . INTERNAL COMBUSTION ENGINES
- . . **ROTARY ENGINES**
- . . . WANKEL ENGINES
- RT   AIRCRAFT ENGINES
- AUTOMOBILE ENGINES
- PISTON ENGINES

ROTARY GYROSCOPES
- GS   GYROSCOPES
- . **ROTARY GYROSCOPES**
- . . FLUID ROTOR GYROSCOPES
- RT   GYROSCOPE FLUIDS
- GYROSCOPIC STABILITY
- ROTATING BODIES

ROTARY STABILITY
- UF   WHIRL INSTABILITY
- GS   DYNAMIC CHARACTERISTICS
- . DYNAMIC STABILITY
- . . MOTION STABILITY
- . . . **ROTARY STABILITY**

ROTARY STABILITY-*(CONT.)*
- . . . . GYROSCOPIC STABILITY
- STABILITY
- . DYNAMIC STABILITY
- . . MOTION STABILITY
- . . . **ROTARY STABILITY**
- . . . . GYROSCOPIC STABILITY
- RT   DIRECTIONAL STABILITY
- FLOW STABILITY
- LATERAL STABILITY
- LONGITUDINAL STABILITY
- ROTATING BODIES
- ROTATION

ROTARY WING AIRCRAFT
- UF   ROTORCRAFT
- GS   V/STOL AIRCRAFT
- . **ROTARY WING AIRCRAFT**
- . . AUTOGYROS
- . . . AVIAN 2/180 AUTOGIRO
- . . HELICOPTERS
- . . . ALOUETTE HELICOPTERS
- . . . . SA-330 HELICOPTER
- . . . . SE-3160 HELICOPTER
- . . . BELL 214A HELICOPTER
- . . . BO-105 HELICOPTER
- . . . CH-21 HELICOPTER
- . . . COMPOUND HELICOPTERS
- . . . H-17 HELICOPTER
- . . . H-54 HELICOPTER
- . . . HC-3 HELICOPTER
- . . . MILITARY HELICOPTERS
- . . . . AH-1G HELICOPTER
- . . . . AH-64 HELICOPTER
- . . . . CH-3 HELICOPTER
- . . . . CH-34 HELICOPTER
- . . . . CH-46 HELICOPTER
- . . . . CH-47 HELICOPTER
- . . . . CH-54 HELICOPTER
- . . . . H-43 HELICOPTER
- . . . . H-53 HELICOPTER
- . . . . H-56 HELICOPTER
- . . . . H-60 HELICOPTER
- . . . . HEAVY LIFT HELICOPTERS
- . . . . . CH-62 HELICOPTER
- . . . . HH-43 HELICOPTER
- . . . . OH-4 HELICOPTER
- . . . . OH-5 HELICOPTER
- . . . . OH-6 HELICOPTER
- . . . . OH-13 HELICOPTER
- . . . . OH-23 HELICOPTER
- . . . . OH-58 HELICOPTER
- . . . . QH-50 HELICOPTER
- . . . . S-58 HELICOPTER
- . . . . S-61 HELICOPTER
- . . . . SA-330 HELICOPTER
- . . . . UH-1 HELICOPTER
- . . . . UH-2 HELICOPTER
- . . . . UH-34 HELICOPTER
- . . . . UH-60A HELICOPTER
- . . . . UH-61A HELICOPTER
- . . . P-531 HELICOPTER
- . . RIGID ROTOR HELICOPTERS
- . . . . CH-3 HELICOPTER
- . . . . F-28 HELICOPTER
- . . . . XH-51 HELICOPTER
- . . SH-3 HELICOPTER
- . . SH-4 HELICOPTER
- . . . SIKORSKY WHIRLWIND HELICOPTER
- . . . TANDEM ROTOR HELICOPTERS
- . . . . CH-46 HELICOPTER
- . . . . CH-47 HELICOPTER
- . . . . H-25 HELICOPTER
- . . . . TH-55 HELICOPTER
- . . WESTLAND WHIRLWIND HELICOPTER
- . . XV-9A AIRCRAFT
- . . TILT ROTOR AIRCRAFT
- . . . XV-15 AIRCRAFT
- RT   ∞ AIRCRAFT
- AUTOROTATION
- COMMERCIAL AIRCRAFT
- LIFTING ROTORS
- ∞ MILITARY AIRCRAFT
- PASSENGER AIRCRAFT
- SHORT TAKEOFF AIRCRAFT
- ∞ SUBSONIC AIRCRAFT
- TRANSPORT AIRCRAFT
- VERTICAL TAKEOFF AIRCRAFT
- WESER AIRCRAFT
- WESTLAND AIRCRAFT

ROTARY WINGS
- UF   HELICOPTER ROTORS
- HINGED ROTOR BLADES

ROTARY WINGS-*(CONT.)*
GS   AIRFOILS
        . WINGS
        . . **ROTARY WINGS**
        . . . LIFTING ROTORS
        . . . . BEARINGLESS ROTORS
        . . . RIGID ROTORS
        . . . TILTING ROTORS
        . . . TIP DRIVEN ROTORS
        ROTATING BODIES
        . ROTORS
        . **ROTARY WINGS**
        . . . CIRCULATION CONTROL ROTORS
        . . . LIFTING ROTORS
        . . . . BEARINGLESS ROTORS
        . . . RIGID ROTORS
        . . . TILTING ROTORS
        . . . TIP DRIVEN ROTORS
        . . . X WING ROTORS
RT   BLADE TIPS
        ∞BLADES
        FAN BLADES
        FLAPPING HINGES
        FOLDING STRUCTURES
        GROUND RESONANCE
        HARMONIC CONTROL
        HELICOPTER PROPELLER DRIVE
        HELICOPTER TAIL ROTORS
        LIFT FANS
        PROPELLER BLADES
        ∞ROTOR BLADES
        TAIL ROTORS
        TILT ROTOR RESEARCH AIRCRAFT
            PROGRAM
        WHIRL TOWERS

ROTATING
USE   ROTATION

ROTATING BODIES
UF   ROTATING VEHICLES
        SOLID ROTATION
GS   **ROTATING BODIES**
        . LUNAR ROTATION
        . ROTATING CYLINDERS
        . ROTATING DISKS
        . ROTATING SPHERES
        . ROTORS
        . . COMPRESSOR ROTORS
        . . FLYWHEELS
        . . IMPELLERS
        . . PUMP IMPELLERS
        . . ROTARY WINGS
        . . . CIRCULATION CONTROL ROTORS
        . . . LIFTING ROTORS
        . . . . BEARINGLESS ROTORS
        . . . RIGID ROTORS
        . . . TILTING ROTORS
        . . . TIP DRIVEN ROTORS
        . . . X WING ROTORS
        . . TAIL ROTORS
        . . . HELICOPTER TAIL ROTORS
        . . TIP VANES
        . . TURBINE WHEELS
RT   AXES OF ROTATION
        ∞BODIES
        PLANETARY ROTATION
        ROCHE LIMIT
        ROTARY GYROSCOPES
        ROTARY STABILITY
        ROTATION
        SPINNING UNGUIDED ROCKET
            TRAJECTORY

ROTATING CYLINDERS
GS   ROTATING BODIES
        . **ROTATING CYLINDERS**
        SYMMETRICAL BODIES
        . BODIES OF REVOLUTION
        . . CYLINDRICAL BODIES
        . . . **ROTATING CYLINDERS**
RT   COUETTE FLOW
        ∞CYLINDERS
        CYLINDRICAL SHELLS
        ELASTOHYDRODYNAMICS
        MAGNUS EFFECT
        SHAFTS (MACHINE ELEMENTS)
        VISCOMETERS
        VISCOMETRY

ROTATING DISKS
GS   DISKS (SHAPES)
        . **ROTATING DISKS**
        ROTATING BODIES
        . **ROTATING DISKS**

ROTATING DISKS-*(CONT.)*
RT   ACCRETION DISKS
        COUNTER ROTATION
        KARMAN-BODEWADT FLOW

∞ **ROTATING ELECTRICAL MACHINES**
SN   *(USE OF A MORE SPECIFIC TERM IS
        RECOMMENDED--CONSULT THE TERMS
        LISTED BELOW)*
RT   ARMATURES
        COMMUTATORS
        ELECTRIC HYBRID VEHICLES
        ELECTRIC MOTORS
        INDUCTION MOTORS
        ∞MACHINERY
        ROTATING GENERATORS
        ROTORS
        SERVOMOTORS
        STATORS

ROTATING ENVIRONMENTS
GS   ENVIRONMENTS
        . **ROTATING ENVIRONMENTS**
RT   ARTIFICIAL GRAVITY
        BARANY CHAIR
        CORIOLIS EFFECT
        HIGH GRAVITY ENVIRONMENTS
        LANGLEY COMPLEX COORDINATOR
        SPACECRAFT ENVIRONMENTS
        TUMBLING MOTION

ROTATING FLUIDS
GS   **ROTATING FLUIDS**
        . ROTATING LIQUIDS
RT   COUNTER ROTATION
        ∞FLUIDS
        GOERTLER INSTABILITY
        KARMAN-BODEWADT FLOW
        LIQUID SLOSHING
        PLANETARY WAVES
        SUPERROTATION
        TAYLOR INSTABILITY
        TRAPPED VORTEXES
        TURBULENT FLOW
        VORTEX SHEETS
        VORTICES
        WING TIP VORTICES

ROTATING GENERATORS
UF   DYNAMOS
GS   ELECTRIC GENERATORS
        . **ROTATING GENERATORS**
        . . AMPLIDYNES
        . . DYNAMOMETERS
        . . HOMOPOLAR GENERATORS
        . . STATIC ALTERNATORS
        . . TURBOGENERATORS
        . . . ASTEC SOLAR TURBOELECTRIC
            GENERATOR
RT   COMMUTATORS
        ELECTROSTATIC GENERATORS
        ∞GENERATORS
        ∞ROTATING ELECTRICAL MACHINES
        TURBINES
        TURBOMACHINERY

ROTATING LIQUIDS
UF   LIQUID ROTATION
GS   LIQUIDS
        . **ROTATING LIQUIDS**
        ROTATING FLUIDS
        . **ROTATING LIQUIDS**
RT   GOERTLER INSTABILITY
        PLANETARY WAVES
        ROTATION
        TRAPPED VORTEXES
        VORTICES

ROTATING MATTER
RT   MATTER (PHYSICS)
        ROTATION
        SPIN DYNAMICS

ROTATING MIRRORS
GS   MIRRORS
        . **ROTATING MIRRORS**
RT   FRAMING CAMERAS
        HIGH SPEED CAMERAS

ROTATING PLASMAS
GS   GASES
        . IONIZED GASES
        . . CHARGED PARTICLES
        . . . **ROTATING PLASMAS**
        PARTICLES

ROTATING PLASMAS-*(CONT.)*
        . CHARGED PARTICLES
        . . ENERGETIC PARTICLES
        . . . PLASMAS (PHYSICS)
        . . . . **ROTATING PLASMAS**
RT   DRIFT RATE
        NONEQUILIBRIUM PLASMAS
        PLASMA FLUX MEASUREMENT
        THETA PINCH
        TOROIDAL PLASMAS
        TWO FLUID MODELS
        ZETA PINCH

ROTATING SHAFTS
GS   **ROTATING SHAFTS**
        . SHAFTS (MACHINE ELEMENTS)
        . . TURBOSHAFTS

ROTATING SPHERES
GS   ROTATING BODIES
        . **ROTATING SPHERES**
        SYMMETRICAL BODIES
        . BODIES OF REVOLUTION
        . . SPHERES
        . . . **ROTATING SPHERES**
RT   EQUATORS
        SPHERICAL SHELLS

ROTATING STALLS
RT   BOUNDARY LAYER SEPARATION
        COMPRESSOR BLADES
        ∞ROTOR BLADES
        TURBOCOMPRESSORS

ROTATING VEHICLES
USE   ROTATING BODIES
        VEHICLES

ROTATION
UF   ROTATING
        WHIRL
        WHIRLING
GS   GYRATION
        . **ROTATION**
        . . AUTOROTATION
        . . COROTATION
        . . COUNTER ROTATION
        . . EARTH ROTATION
        . . MOLECULAR ROTATION
        . . MUON SPIN ROTATION
        . . PLANETARY ROTATION
        . . SATELLITE ROTATION
        . . STELLAR ROTATION
        . . . SOLAR ROTATION
RT   ANGULAR ACCELERATION
        ANGULAR VELOCITY
        AXES OF ROTATION
        CIRCULATION
        CORIOLIS EFFECT
        CROSS POLARIZATION
        FARADAY EFFECT
        IMAGE ROTATION
        LIBRATION
        ∞MOTION
        NUTATION
        PITCH (INCLINATION)
        POLARIZATION (SPIN ALIGNMENT)
        POLARIZATION (WAVES)
        PRECESSION
        REVOLVING
        ROLL
        ROTARY STABILITY
        ROTATING BODIES
        ROTATING LIQUIDS
        ROTATING MATTER
        ROTONS
        TORQUE
        VORTEX AVOIDANCE
        VORTICES
        YAW

ROTATIONAL FLOW
USE   FLUID FLOW
        VORTICES

ROTIFERA
GS   ANIMALS
        . INVERTEBRATES
        . . PROTOZOA
        . . . PARAMECIA
        . . . . **ROTIFERA**
        MICROORGANISMS
        . PROTOZOA
        . . PARAMECIA
        . . . **ROTIFERA**

ROTOCHUTES
GS   PARACHUTES
      . **ROTOCHUTES**
RT   AUTOROTATION

ROTONS
GS   FLUID MECHANICS
      . FLUID DYNAMICS
      . . **ROTONS**
RT   ACTIVATION ENERGY
      EXCITATION
      PHOTONS
      ROTATION

ROTOR AERODYNAMICS
GS   FLUID MECHANICS
      . FLUID DYNAMICS
      . . GAS DYNAMICS
      . . . AERODYNAMICS
      . . . . **ROTOR AERODYNAMICS**
RT   FLAPPING
      FLAPPING HINGES
      GROUND RESONANCE
      ROTOR BODY INTERACTIONS
      ROTORS
      WHIRL TOWERS

∞ **ROTOR BLADES**
SN   (USE OF A MORE SPECIFIC TERM IS
      RECOMMENDED--CONSULT THE TERMS
      LISTED BELOW)
RT   HELICOPTER TAIL ROTORS
      ROTARY WINGS
      ROTATING STALLS
      ROTOR BLADES (TURBOMACHINERY)
      TAIL ROTORS
      X WING ROTORS

ROTOR BLADES (TURBOMACHINERY)
UF   IMPELLER BLADES
GS   TURBOMACHINE BLADES
      . **ROTOR BLADES (TURBOMACHINERY)**
RT   AIRFOILS
      BLADE TIPS
      ∞ BLADES
      COMPRESSOR BLADES
      COMPRESSOR ROTORS
      IMPELLERS
      ∞ ROTOR BLADES
      ROTORS
      STATOR BLADES
      TURBINE BLADES

ROTOR BODY INTERACTIONS
RT   AERODYNAMIC CHARACTERISTICS
      AERODYNAMIC CONFIGURATIONS
      HELICOPTER DESIGN
      ROTOR AERODYNAMICS

ROTOR DISKS
USE   TURBINE WHEELS

ROTOR HUBS
USE   HUBS
      ROTORS

ROTOR LIFT
GS   AERODYNAMIC CHARACTERISTICS
      . LIFT
      . . **ROTOR LIFT**
      AERODYNAMIC FORCES
      . LIFT
      . . **ROTOR LIFT**
      DYNAMIC CHARACTERISTICS
      . LIFT
      . . **ROTOR LIFT**
RT   DISTRIBUTION (PROPERTY)

ROTOR SPEED
GS   RATES (PER TIME)
      . **ROTOR SPEED**
      VELOCITY
      . **ROTOR SPEED**
RT   ANGULAR VELOCITY
      HIGH SPEED
      LABYRINTH SEALS
      TIP SPEED

ROTOR SYSTEMS RESEARCH AIRCRAFT
GS   RESEARCH AIRCRAFT
      . **ROTOR SYSTEMS RESEARCH
        AIRCRAFT**
RT   ∞ AIRCRAFT
      AIRCRAFT DESIGN

ROTOR SYSTEMS RESEARCH-(CONT.)
      HELICOPTERS
      NASA PROGRAMS
      ∞ SYSTEMS

ROTORCRAFT
USE   ROTARY WING AIRCRAFT

ROTORCRAFT AIRCRAFT
RT   ∞ AIRCRAFT

ROTORS
UF   ROTOR HUBS
GS   ROTATING BODIES
      . **ROTORS**
      . . COMPRESSOR ROTORS
      . . FLYWHEELS
      . . IMPELLERS
      . . PUMP IMPELLERS
      . . ROTARY WINGS
      . . . CIRCULATION CONTROL ROTORS
      . . . LIFTING ROTORS
      . . . . BEARINGLESS ROTORS
      . . . RIGID ROTORS
      . . . TILTING ROTORS
      . . . TIP DRIVEN ROTORS
      . . . X WING ROTORS
      . . TAIL ROTORS
      . . . HELICOPTER TAIL ROTORS
      . . TIP VANES
      . . TURBINE WHEELS
RT   AIRFOILS
      ARMATURES
      CENTRIFUGAL COMPRESSORS
      HEAVY LIFT AIRSHIPS
      ∞ ROTATING ELECTRICAL MACHINES
      ROTOR AERODYNAMICS
      ROTOR BLADES (TURBOMACHINERY)
      STATORS
      TURBINES
      TURBOCOMPRESSORS
      TURBOMACHINE BLADES
      TURBOSHAFTS
      WHEELS
      WINGS

ROUGHNESS
GS   **ROUGHNESS**
      . SEA ROUGHNESS
      . SURFACE ROUGHNESS
RT   COARSENESS
      CONTOURS
      FLATNESS
      MECHANICAL PROPERTIES
      MOTION STABILITY
      PROFILOMETERS
      SMOOTHING
      SURFACE PROPERTIES

ROUND TRIP TRAJECTORIES
GS   TRAJECTORIES
      . **ROUND TRIP TRAJECTORIES**
      . . CIRCUMLUNAR TRAJECTORIES
RT   EARTH-MOON TRAJECTORIES
      INTERORBITAL TRAJECTORIES
      INTERPLANETARY FLIGHT
      INTERPLANETARY TRAJECTORIES
      MOON-EARTH TRAJECTORIES
      ORBITAL MECHANICS
      SPACECRAFT TRAJECTORIES
      SWINGBY TECHNIQUE

ROUSE BELTS
RT   ∞ BELTS
      CONES (VOLCANOES)
      EARTHQUAKES
      GEOLOGICAL FAULTS
      MARS VOLCANOES
      SEISMOLOGY
      TREMORS
      VOLCANOES
      VOLCANOLOGY

ROUTES
RT   AIR TRAFFIC CONTROL
      FLIGHT PLANS
      ∞ PATHS
      SITE SELECTION
      TRANSPORTATION

∞ **ROUTINES**
SN   (USE OF A MORE SPECIFIC TERM IS
      RECOMMENDED--CONSULT THE TERMS
      LISTED BELOW)
RT   COMPUTER PROGRAMS

ROUTINES-(CONT.)
      COMPUTER SYSTEMS PROGRAMS
      DATA CONVERSION ROUTINES
      INPUT/OUTPUT ROUTINES
      MERGING ROUTINES
      OPERATING SYSTEMS (COMPUTERS)
      USER MANUALS (COMPUTER
        PROGRAMS)

ROVER PROJECT
GS   PROGRAMS
      . NASA PROGRAMS
      . . **ROVER PROJECT**
      . NASA SPACE PROGRAMS
      . . **ROVER PROJECT**
      . PROJECTS
      . . **ROVER PROJECT**
RT   KIWI REACTORS
      NUCLEAR ENGINE FOR ROCKET
        VEHICLES
      NUCLEAR PROPULSION
      NUCLEAR REACTORS
      RIFT (REACTOR IN FLIGHT TEST)
      ROCKET ENGINE DESIGN
      SPACECRAFT PROPULSION

ROVING VEHICLES
UF   EXTRATERRESTRIAL ROVING VEHICLES
GS   SURFACE VEHICLES
      . **ROVING VEHICLES**
      . . LUNAR ROVING VEHICLES
      . . . LUNOKHOD LUNAR ROVING
            VEHICLES
RT   PLANETARY LANDING
      PLANETARY SURFACES
      RESEARCH VEHICLES
      TOROIDAL WHEELS
      ∞ VEHICLES

∞ **ROVINGS**
SN   (USE OF A MORE SPECIFIC TERM IS
      RECOMMENDED--CONSULT THE TERMS
      LISTED BELOW)
RT   COMPOSITE MATERIALS
      PROVING
      WEBS (SHEETS)
      YARNS

ROWLAND CIRCLES
RT   GRATINGS (SPECTRA)
      OPTICAL FILTERS

RP-1 ROCKET PROPELLANTS
GS   FUELS
      . CHEMICAL FUELS
      . . HYDROCARBON FUELS
      . . . **RP-1 ROCKET PROPELLANTS**
      LIQUIDS
      . LIQUID FUELS
      . . LIQUID ROCKET PROPELLANTS
      . . . **RP-1 ROCKET PROPELLANTS**
      PROPELLANTS
      . ROCKET PROPELLANTS
      . . LIQUID ROCKET PROPELLANTS
      . . . **RP-1 ROCKET PROPELLANTS**
      . STORABLE PROPELLANTS
      . . **RP-1 ROCKET PROPELLANTS**
RT   JP-4 JET FUEL
      KEROSENE

RPV
USE   REMOTELY PILOTED VEHICLES

RTV-40 RUBBER (TRADEMARK)
GS   RUBBER
      . SILICONE RUBBER
      . . **RTV-40 RUBBER (TRADEMARK)**
      . SYNTHETIC RUBBERS
      . . **RTV-40 RUBBER (TRADEMARK)**

RTV-60 RUBBER (TRADEMARK)
GS   RUBBER
      . SILICONE RUBBER
      . . **RTV-60 RUBBER (TRADEMARK)**
      . SYNTHETIC RUBBERS
      . . **RTV-60 RUBBER (TRADEMARK)**

RUANDA-URUNDI
USE   BURUNDI
      RWANDA

RUBBER
GS   **RUBBER**
      . ADIPRENE (TRADEMARK)

RUBBER-*(CONT.)*
- . LATEX
- . SILICONE RUBBER
- . . RTV-40 RUBBER (TRADEMARK)
- . . RTV-60 RUBBER (TRADEMARK)
- . SYNTHETIC RUBBERS
- . . BUNA (TRADEMARK)
- . ELASTOMERS
- . . . CHLOROPRENE RESINS
- . . . THIOPLASTICS
- . . . VITON RUBBER (TRADEMARK)
- . . . VULCANIZED ELASTOMERS
- . . RTV-40 RUBBER (TRADEMARK)
- . . RTV-60 RUBBER (TRADEMARK)
- RT GUAYULE
- GUMS (SUBSTANCES)
- ORGANIC MATERIALS
- POLYISOPRENES

RUBBER COATINGS
- GS COATINGS
- . **RUBBER COATINGS**
- RT PAINTS
- PROTECTIVE COATINGS

RUBIDIUM
- GS CHEMICAL ELEMENTS
- . ALKALI METALS
- . . **RUBIDIUM**
- . . . RUBIDIUM ISOTOPES
- . . . . RUBIDIUM 86
- METALS
- . ALKALI METALS
- . . **RUBIDIUM**
- . . . RUBIDIUM ISOTOPES
- . . . . RUBIDIUM 86

RUBIDIUM COMPOUNDS
- RT ∞ALKALI METAL COMPOUNDS
- ∞CHEMICAL COMPOUNDS
- ∞METAL COMPOUNDS

RUBIDIUM ISOTOPES
- GS CHEMICAL ELEMENTS
- . ALKALI METALS
- . . RUBIDIUM
- . . . **RUBIDIUM ISOTOPES**
- . . . . RUBIDIUM 86
- . NUCLIDES
- . . ISOTOPES
- . . . **RUBIDIUM ISOTOPES**
- . . . . RUBIDIUM 86
- METALS
- . ALKALI METALS
- . . RUBIDIUM
- . . . **RUBIDIUM ISOTOPES**
- . . . . RUBIDIUM 86

RUBIDIUM 86
- GS CHEMICAL ELEMENTS
- . ALKALI METALS
- . . RUBIDIUM
- . . . RUBIDIUM ISOTOPES
- . . . . **RUBIDIUM 86**
- . NUCLIDES
- . . ISOTOPES
- . . . RADIOACTIVE ISOTOPES
- . . . . **RUBIDIUM 86**
- . . . RUBIDIUM ISOTOPES
- . . . . **RUBIDIUM 86**
- METALS
- . ALKALI METALS
- . . RUBIDIUM
- . . . RUBIDIUM ISOTOPES
- . . . . **RUBIDIUM 86**

RUBIS ROCKET VEHICLE
- GS ROCKET VEHICLES
- . MULTISTAGE ROCKET VEHICLES
- . . **RUBIS ROCKET VEHICLE**
- RT SOLID PROPELLANT ROCKET ENGINES

RUBY
- RT ALUMINUM OXIDES
- CRYSTALS

RUBY LASERS
- GS ELECTRONIC EQUIPMENT
- . SOLID STATE DEVICES
- . . SOLID STATE LASERS
- . . . **RUBY LASERS**
- STIMULATED EMISSION DEVICES
- . LASERS
- . . SOLID STATE LASERS
- . . . **RUBY LASERS**

RUBY LASERS-*(CONT.)*
- RT PULSED LASERS
- Q SWITCHED LASERS
- VERNEUIL PROCESS

RUDDERS
- GS CONTROL SURFACES
- . **RUDDERS**
- . . MARINE RUDDERS
- RT AIRFOILS
- FINS
- PINTLES
- STABILIZERS (FLUID DYNAMICS)
- SWEPTBACK TAIL SURFACES
- TABS (CONTROL SURFACES)
- TAIL ASSEMBLIES
- TAIL SURFACES
- TRAPEZOIDAL TAIL SURFACES

RUGGEDNESS
- RT DURABILITY
- ∞HIGH RESISTANCE
- MECHANICAL PROPERTIES
- PHYSIOLOGY
- ∞RIGIDITY

RULER METHOD
- RT ∞METHODOLOGY

RULES
- GS **RULES**
- . FLIGHT RULES
- . . INSTRUMENT FLIGHT RULES
- . . VISUAL FLIGHT RULES
- . PALMGREN-MINER RULE
- . PHASE RULE
- . SELECTION RULES (NUCLEAR PHYSICS)
- . SUM RULES
- . WHITHAM RULE
- RT LAWS
- PATENT POLICY
- POLICIES
- PROCUREMENT POLICY
- REGULATIONS
- SEA LAW

RUMANIA
- USE ROMANIA

RUN TIME (COMPUTERS)
- RT COMPUTER PROGRAMMING
- COMPUTERS
- TIME SHARING

RUNGE-KUTTA METHOD
- GS ANALYSIS (MATHEMATICS)
- . NUMERICAL ANALYSIS
- . . NUMERICAL INTEGRATION
- . . . **RUNGE-KUTTA METHOD**
- . REAL VARIABLES
- . . MEASURE AND INTEGRATION
- . . . NUMERICAL INTEGRATION
- . . . . **RUNGE-KUTTA METHOD**
- RT ∞METHODOLOGY

RUNNING
- RT ∞OPERATIONS
- PHYSICAL EXERCISE

RUNOFFS
- USE DRAINAGE

RUNWAY ALIGNMENT
- SN (ALIGNMENT WITH RUNWAYS--NOT ALIGNMENT OF RUNWAYS)
- RT AIRCRAFT LANDING
- TAKEOFF RUNS

RUNWAY CONDITIONS
- GS CONDITIONS
- . **RUNWAY CONDITIONS**
- RT ICE
- RUNWAYS
- SLUSH
- SURFACE ROUGHNESS
- WATER
- WEATHER

RUNWAY LIGHTS
- GS LANDING AIDS
- . AIRPORT LIGHTS
- . . **RUNWAY LIGHTS**
- LIGHTING EQUIPMENT

RUNWAY LIGHTS-*(CONT.)*
- . LUMINAIRES
- . . AIRPORT LIGHTS
- . . . **RUNWAY LIGHTS**
- RT APPROACH CONTROL
- ∞FLARES
- ∞MARKERS
- RUNWAYS
- SEARCHLIGHTS
- VISUAL CONTROL

RUNWAYS
- RT AIRFIELD SURFACE MOVEMENTS
- AIRPORTS
- LANDING
- LANDING AIDS
- LANDING MATS
- LANDING SITES
- PAVEMENTS
- RUNWAY CONDITIONS
- RUNWAY LIGHTS
- ∞STRIP
- TAKEOFF
- TAXIING

RUPTURING
- RT ∞BLISTERS
- BURSTS
- CRACKING (FRACTURING)
- DISRUPTING
- FAILURE
- FRACTURE MECHANICS
- METAL FATIGUE
- SELF SEALING
- STRUCTURAL STRAIN
- TEARING

RURAL AREAS
- RT AGRICULTURE
- FARMLANDS
- GRASSLANDS
- LAND
- MEGALOPOLISES
- RANGELANDS
- REGIONAL PLANNING
- RESIDENTIAL AREAS
- SUBURBAN AREAS
- WILDERNESS

RURAL LAND USE
- GS LAND USE
- . **RURAL LAND USE**
- RT AGRICULTURE
- CONSERVATION
- ∞DEVELOPMENT
- EARTH RESOURCES
- FARMLANDS
- GRASSLANDS
- GRAZING
- GREAT PLAINS CORRIDOR (NORTH AMERICA)
- LAND
- LAND MANAGEMENT
- ORCHARDS
- RANGELANDS
- REGIONAL PLANNING
- SITES

RUST FUNGI
- UF RUSTS (BOTANY)
- GS FUNGI
- . **RUST FUNGI**
- RT BLIGHT
- ∞MOLD
- PARASITIC DISEASES

RUSTING
- GS CHEMICAL REACTIONS
- . OXIDATION
- . . **RUSTING**
- CORROSION
- . **RUSTING**
- RT ATMOSPHERIC EFFECTS
- CHEMICAL ATTACK
- COATINGS
- CORROSION RESISTANCE
- DEGRADATION
- DESENSITIZING
- DETERIORATION
- HOT CORROSION
- METAL-WATER REACTIONS
- OXIDATION RESISTANCE
- PASSIVITY
- SCALE (CORROSION)
- WEATHERING

RUSTS (BOTANY)
USE   RUST FUNGI

RUTHENIUM
GS   CHEMICAL ELEMENTS
.   **RUTHENIUM**
. .   RUTHENIUM ISOTOPES
METALS
.   NOBLE METALS
. .   **RUTHENIUM**
. . .   RUTHENIUM ISOTOPES
.   TRANSITION METALS
. .   **RUTHENIUM**
. . .   RUTHENIUM ISOTOPES

RUTHENIUM ALLOYS
GS   ALLOYS
.   **RUTHENIUM ALLOYS**

RUTHENIUM COMPOUNDS
RT   ∞CHEMICAL COMPOUNDS
∞METAL COMPOUNDS
TRANSITION METALS

RUTHENIUM ISOTOPES
UF   RUTHENIUM 106
GS   CHEMICAL ELEMENTS
.   NUCLIDES
ISOTOPES
. . .   **RUTHENIUM ISOTOPES**
.   RUTHENIUM
. .   **RUTHENIUM ISOTOPES**
METALS
.   NOBLE METALS
. .   RUTHENIUM
. . .   **RUTHENIUM ISOTOPES**
.   TRANSITION METALS
. .   RUTHENIUM
. . .   **RUTHENIUM ISOTOPES**

RUTHENIUM 106
USE   RUTHENIUM ISOTOPES

RUTILE
GS   CHALCOGENIDES
.   OXIDES
. .   METAL OXIDES
. . .   TITANIUM OXIDES
. . . .   **RUTILE**
TITANIUM COMPOUNDS
.   TITANIUM OXIDES
. .   **RUTILE**
RT   ANATASE
COESITE
CROSS RELAXATION
MINERALS
PIGMENTS
STISHOVITE

RWANDA
UF   RUANDA-URUNDI
GS   NATIONS
.   **RWANDA**
RT   AFRICA
BURUNDI

RYAN AIRCRAFT
UF   RYAN MILITARY AIRCRAFT
GS   **RYAN AIRCRAFT**
.   FIREBEE 2 TARGET DRONE AIRCRAFT
.   X-13 AIRCRAFT
.   XC-142 AIRCRAFT
.   XV-5 AIRCRAFT
.   XV-8A AIRCRAFT
RT   ∞AIRCRAFT

RYAN MILITARY AIRCRAFT
USE   RYAN AIRCRAFT

RYDBERG SERIES
GS   SPECTRA
.   RADIATION SPECTRA
. .   ELECTROMAGNETIC SPECTRA
. . .   LINE SPECTRA
. . . .   **RYDBERG SERIES**
RT   ABSORPTION SPECTRA
ATOMIC SPECTRA
ELECTRON TRANSITIONS
EMISSION SPECTRA
H LINES
HYDROGEN

R5D AIRCRAFT
USE   C-54 AIRCRAFT

R7V AIRCRAFT
USE   C-121 AIRCRAFT
EC-121 AIRCRAFT

S

S BAND
USE   SUPERHIGH FREQUENCIES
ULTRAHIGH FREQUENCIES

S CURVES
GS   GEOMETRY
.   CURVES (GEOMETRY)
. .   **S CURVES**
. . .   GOMPERTZ CURVES
.   EUCLIDEAN GEOMETRY
. .   ANALYTIC GEOMETRY
. . .   **S CURVES**
. . . .   GOMPERTZ CURVES

S GLASS
GS   GLASS
.   E GLASS
. .   **S GLASS**
RT   COMPOSITE MATERIALS
GLASS FIBER REINFORCED PLASTICS
GLASS FIBERS
SILICON DIOXIDE

S MATRIX THEORY
UF   SCATTERING MATRIX
RT   OPERATORS (MATHEMATICS)
SCATTERING CROSS SECTIONS
∞THEORIES

S STARS
GS   CELESTIAL BODIES
.   STARS
. .   **S STARS**
. . .   GIANT STARS
. . . .   RED GIANT STARS
. . . . .   CARBON STARS
RT   M STARS

S WAVES
UF   SECONDARY WAVES
SHEAR DISTURBANCES
SHEAR WAVES
GS   ELASTIC WAVES
.   **S WAVES**
RT   CRUSTAL FRACTURES
DILATATIONAL WAVES
P WAVES
POLARIZED ELASTIC WAVES
RAYLEIGH WAVES
SEISMIC WAVES
SURFACE WAVES
TRANSVERSE WAVES

S-A-W DEVICES
USE   SURFACE ACOUSTIC WAVE DEVICES

S-N DIAGRAMS
UF   FATIGUE DIAGRAMS
GS   DIAGRAMS
.   **S-N DIAGRAMS**
RT   BENDING FATIGUE
CYCLIC LOADS
FATIGUE (MATERIALS)
FATIGUE LIFE
FATIGUE TESTS
METAL FATIGUE
STRESS ANALYSIS
STRESS CYCLES
STRESS MEASUREMENT
STRESS RATIO

S-2 AIRCRAFT
UF   SNOW AERIAL APPLICATOR AIRCRAFT
S-2B
SNOW S-2 AIRCRAFT
US-2A AIRCRAFT
GS   MONOPLANES
.   **S-2 AIRCRAFT**
SNOW AIRCRAFT
.   **S-2 AIRCRAFT**
UTILITY AIRCRAFT
.   **S-2 AIRCRAFT**
RT   ∞AIRCRAFT

S-3 AIRCRAFT
GS   ANTISUBMARINE WARFARE AIRCRAFT
.   **S-3 AIRCRAFT**
RT   ∞AIRCRAFT
∞MILITARY AIRCRAFT

S-3 SATELLITE
USE   EXPLORER 12 SATELLITE

S-6 SATELLITE
USE   EXPLORER 17 SATELLITE

S-16 SATELLITE
USE   OSO-1

S-17 SATELLITE
USE   OSO-2

S-18 SATELLITE
USE   OAO

S-27 SATELLITE
USE   ALOUETTE 1 SATELLITE

S-49 SATELLITE
USE   OGO-A

S-50 SATELLITE
USE   OGO-C

S-51 SATELLITE
USE   ARIEL 1 SATELLITE

S-52 SATELLITE
USE   ARIEL 2 SATELLITE

S-57 SATELLITE
USE   OSO-C

S-58 HELICOPTER
UF   SIKORSKY S-58 HELICOPTER
GS   SIKORSKY AIRCRAFT
.   **S-58 HELICOPTER**
TRANSPORT AIRCRAFT
.   **S-58 HELICOPTER**
V/STOL AIRCRAFT
.   ROTARY WING AIRCRAFT
. .   HELICOPTERS
. . .   MILITARY HELICOPTERS
. . . .   **S-58 HELICOPTER**
RT   CH-34 HELICOPTER
UH-34 HELICOPTER

S-61 HELICOPTER
UF   SIKORSKY S-61 HELICOPTER
GS   SIKORSKY AIRCRAFT
.   **S-61 HELICOPTER**
TRANSPORT AIRCRAFT
.   **S-61 HELICOPTER**
V/STOL AIRCRAFT
.   ROTARY WING AIRCRAFT
. .   HELICOPTERS
. . .   MILITARY HELICOPTERS
. . . .   **S-61 HELICOPTER**
RT   ANTISUBMARINE WARFARE AIRCRAFT
CH-3 HELICOPTER
SH-3 HELICOPTER
SH-4 HELICOPTER
WATER TAKEOFF AND LANDING
AIRCRAFT

S-64 HELICOPTER
USE   CH-54 HELICOPTER

S-66 SATELLITE
USE   BEACON EXPLORER A

S-67 HELICOPTER
UF   SIKORSKY S-67 HELICOPTER
GS   SIKORSKY AIRCRAFT
.   **S-67 HELICOPTER**

S-74 SATELLITE
USE   EXPLORER 18 SATELLITE

SA-321 HELICOPTER
UF   SUD AVIATION SA-321 HELICOPTER
GS   SUD AVIATION AIRCRAFT
.   **SA-321 HELICOPTER**

SA-330 HELICOPTER
UF   SUD AVIATION SA-330 HELICOPTER
GS   SUD AVIATION AIRCRAFT

SA-330 HELICOPTER-(CONT.)
. ALOUETTE HELICOPTERS
. . **SA-330 HELICOPTER**
TRANSPORT AIRCRAFT
. **SA-330 HELICOPTER**
V/STOL AIRCRAFT
. ROTARY WING AIRCRAFT
. . HELICOPTERS
. . . ALOUETTE HELICOPTERS
. . . . **SA-330 HELICOPTER**
. . . MILITARY HELICOPTERS
. . . . **SA-330 HELICOPTER**

SAAB AIRCRAFT
GS **SAAB AIRCRAFT**
. SAAB 37 AIRCRAFT
. SAAB 105 AIRCRAFT
RT ∞AIRCRAFT

SAAB 37 AIRCRAFT
GS ATTACK AIRCRAFT
. FIGHTER AIRCRAFT
. . **SAAB 37 AIRCRAFT**
JET AIRCRAFT
. TURBOFAN AIRCRAFT
. . **SAAB 37 AIRCRAFT**
SAAB AIRCRAFT
. **SAAB 37 AIRCRAFT**
SUPERSONIC AIRCRAFT
. **SAAB 37 AIRCRAFT**
RT ∞AIRCRAFT
CANARD CONFIGURATIONS
HARRIER AIRCRAFT

SAAB 105 AIRCRAFT
GS JET AIRCRAFT
. TURBOFAN AIRCRAFT
. . **SAAB 105 AIRCRAFT**
LIGHT AIRCRAFT
. **SAAB 105 AIRCRAFT**
MONOPLANES
. **SAAB 105 AIRCRAFT**
SAAB AIRCRAFT
. **SAAB 105 AIRCRAFT**
UTILITY AIRCRAFT
. **SAAB 105 AIRCRAFT**
RT ∞AIRCRAFT
PASSENGER AIRCRAFT

SABATIER REACTION
GS CHEMICAL REACTIONS
. **SABATIER REACTION**
RT PHOTOGRAPHIC FILM

SABOT PROJECTILES
GS PROJECTILES
. **SABOT PROJECTILES**
RT ARTILLERY
FRAGMENTATION
GUN LAUNCHERS
GUNS (ORDNANCE)

SABOTAGE
RT ACCIDENTS
AIR DEFENSE
DAMAGE
DEACTIVATION
DISASTERS
HAZARDS
INJURIES
PREVENTION
SAFETY
SPACE LAW
WRECKAGE

SABRE AIRCRAFT
USE F-86 AIRCRAFT

SABRELINER AIRCRAFT
USE T-39 AIRCRAFT

SACCADIC EYE MOVEMENTS
GS EYE MOVEMENTS
. **SACCADIC EYE MOVEMENTS**
RT FOVEA
∞MOTION
VISUAL FIELDS

SACCHARIDES
USE CARBOHYDRATES

SACCHAROMYCES
GS FUNGI
. **SACCHAROMYCES**

SACRAMENTO VALLEY (CA)
GS VALLEYS
. **SACRAMENTO VALLEY (CA)**
RT CALIFORNIA
RIVER BASINS

SADDLE POINTS
GS **SADDLE POINTS**
. SADDLE POINTS (GAME THEORY)
RT CURVE FITTING
GAME THEORY
MINIMAX TECHNIQUE

SADDLE POINTS (GAME THEORY)
GS GAME THEORY
. **SADDLE POINTS (GAME THEORY)**
SADDLE POINTS
. **SADDLE POINTS (GAME THEORY)**
RT OPERATIONS RESEARCH
SADDLES
∞THEORIES

SADDLES
RT SADDLE POINTS (GAME THEORY)

SADDLES (SUPPORTS)
GS STRUCTURAL MEMBERS
. **SADDLES (SUPPORTS)**
SUPPORTS
. **SADDLES (SUPPORTS)**

SAFEGUARD SYSTEM
GS WEAPON SYSTEMS
. MISSILE SYSTEMS
. . **SAFEGUARD SYSTEM**
RT ANTIMISSILE DEFENSE
BALLISTIC MISSILES
MILITARY TECHNOLOGY
MISSILE DEFENSE
SENTINEL SYSTEM
∞SYSTEMS

SAFETY
GS **SAFETY**
. AEROSPACE SAFETY
. AIRCRAFT SAFETY
. FLIGHT SAFETY
. INDUSTRIAL SAFETY
. RANGE SAFETY
. REACTOR SAFETY
RT ACCIDENT PREVENTION
ACCIDENTS
AIR BAG RESTRAINT DEVICES
CRASHES
∞DETECTORS
EMERGENCY LIFE SUSTAINING
SYSTEMS
ENERGY POLICY
EXPLOSIONS
FIRE PREVENTION
FIREPROOFING
FIRES
HAZARDS
PREVENTION
PROTECTION
SABOTAGE
∞STORAGE
VORTEX AVOIDANCE
WARNING
WARNING SYSTEMS

SAFETY DEVICES
GS **SAFETY DEVICES**
. ABORT APPARATUS
. AIR BAG RESTRAINT DEVICES
. ARRESTING GEAR
. EJECTION SEATS
. . FLYING EJECTION SEATS
. ESCAPE CAPSULES
. ESCAPE ROCKETS
. HELMETS
. SEAT BELTS
. SPACE SUITS
RT ACCIDENT PREVENTION
ACCIDENT PRONENESS
ACCIDENTS
AIRCRAFT SAFETY
AMBULANCES
ANTISKID DEVICES
AUTOMOBILE ACCIDENTS
∞BARRIERS
CHEMICAL DEFENSE
DEFLECTORS
∞DEVICES

SAFETY DEVICES-(CONT.)
EMERGENCY LIFE SUSTAINING
SYSTEMS
ENCLOSURES
∞EQUIPMENT
FAIL-SAFE SYSTEMS
FIRE PREVENTION
FLAME DEFLECTORS
FLIGHT SAFETY
GATES (OPENINGS)
GUARDS (SHIELDS)
HARNESSES
HAZARDS
HUMAN FACTORS ENGINEERING
LANDING AIDS
PRESSURE SUITS
PROTECTION
PROTECTIVE CLOTHING
PROTECTORS
RADIATION MEASURING INSTRUMENTS
RADIATION SHIELDING
SHIELDING
SMOKE DETECTORS
SPACECRAFT SHIELDING
WARNING
WARNING SYSTEMS

SAFETY FACTORS
RT ACCIDENT PRONENESS
AEROSPACE SAFETY
DESIGN ANALYSIS
ESCAPE SYSTEMS
HAZARDS
HEALTH PHYSICS
RELIABILITY
STABILITY

SAFETY MANAGEMENT
GS MANAGEMENT
. **SAFETY MANAGEMENT**
RT ACCIDENT PREVENTION
AEROSPACE SAFETY
EDUCATION
FAIL-SAFE SYSTEMS
FIRE PREVENTION
GUARDS (SHIELDS)
HAZARDS
HUMAN FACTORS ENGINEERING
PRODUCTION MANAGEMENT
WARNING SYSTEMS

SAGE AIR DEFENSE SYSTEM
GS AIR DEFENSE
. **SAGE AIR DEFENSE SYSTEM**
RT ∞SYSTEMS

SAGE SATELLITE
UF STRATOSPHERIC AEROSOL & GAS
EXPERIMENT
GS SATELLITES
. ARTIFICIAL SATELLITES
. . **SAGE SATELLITE**
RT AEROSOLS
OZONE

SAGINAW BAY (MI)
GS BAYS (TOPOGRAPHIC FEATURES)
. **SAGINAW BAY (MI)**
RT INLETS (TOPOGRAPHY)
LAKE HURON
MICHIGAN
RIVER BASINS

SAGITTARIUS CONSTELLATION
GS CONSTELLATIONS
. **SAGITTARIUS CONSTELLATION**

SAGNAC EFFECT
GS PHASE SHIFT
. **SAGNAC EFFECT**
RT ANGULAR VELOCITY
FIBER OPTICS
INTERFEROMETERS
INTERFEROMETRY
LASER GYROSCOPES
LASER INTERFEROMETRY
LIGHT TRANSMISSION
NONLINEAR OPTICS
OPTICAL GYROSCOPES
OPTICAL PATHS
WAVE PROPAGATION

SAHA EQUATIONS
RT ARC HEATING
ELECTRIC ARCS

SAHA EQUATIONS-*(CONT.)*
∞ EQUATIONS
   ION DENSITY (CONCENTRATION)
   IONIZATION POTENTIALS
   TEMPERATURE

SAHARA DESERT (AFRICA)
GS   LAND
     . DESERTS
     . . **SAHARA DESERT (AFRICA)**
RT   AFRICA
     ARID LANDS
     BARREN LAND
     DESERTIFICATION
     DUNES
     REMOTE REGIONS

SAIL PROJECT
UF   SHUTTLE AVIONICS INTEGRATION
        LABORATORY
GS   PROGRAMS
     . NASA PROGRAMS
     . . **SAIL PROJECT**
     . NASA SPACE PROGRAMS
     . . **SAIL PROJECT**
     . PROJECTS
     . . **SAIL PROJECT**
RT   EARTH VIEWING APPLICATIONS
        LABORATORY
     LABORATORIES
     SPACE LABORATORIES
     SPACE SHUTTLES

SAILPLANES
USE   GLIDERS

SAILS
GS   **SAILS**
     . SAILWINGS
     . SOLAR SAILS
RT   FINS
     GLIDERS
     TAIL ASSEMBLIES

SAILWINGS
UF   PRINCETON SAILWINGS
GS   FOLDING STRUCTURES
     . **SAILWINGS**
     SAILS
     . **SAILWINGS**
RT   GLIDERS
     HANG GLIDERS
     KA-6 SAILPLANES

SAINT ELMO FIRE
GS   ELECTRIC CURRENT
     . ELECTRIC DISCHARGES
     . . **SAINT ELMO FIRE**
RT   FIRES

SAINT VENANT FLEXURE PROBLEM
USE   SAINT VENANT PRINCIPLE

SAINT VENANT PRINCIPLE
UF   SAINT VENANT FLEXURE PROBLEM
     ST VENANT FLEXURE PROBLEM
RT   PLASTIC DEFORMATION
     STATIC DEFORMATION
     STATIC LOADS
     STRESS ANALYSIS
     STRESS CONCENTRATION
     TEMPERATURE INVERSIONS

SALICYLATES
GS   **SALICYLATES**
     . SODIUM SALICYLATES
RT   ACETYLSALICYLIC ACID
     DRUGS
     ESTERS

SALINITY
GS   CHEMICAL PROPERTIES
     . **SALINITY**
RT   ALKALINITY
     BRINES
     CORE SAMPLING
     DESALINIZATION
     OCEAN CURRENTS
     OCEANOGRAPHIC PARAMETERS
     SEA WATER

SALIVA
GS   BODY FLUIDS
     . **SALIVA**

SALIVA-*(CONT.)*
RT   DIGESTIVE SYSTEM
     MUCUS
     SALIVARY GLANDS

SALIVARY GLANDS
UF   PAROTID GLAND
GS   ANATOMY
     . GLANDS (ANATOMY)
     . . **SALIVARY GLANDS**
RT   MOUTH
     SALIVA

SALMONELLA
GS   MICROORGANISMS
     . BACTERIA
     . . **SALMONELLA**

SALT BATHS
GS   BATHS
     . **SALT BATHS**
RT   BRINES
     HEAT TREATMENT
     MOLTEN SALTS

SALT BEDS
GS   LANDFORMS
     . BEDS (GEOLOGY)
     . . **SALT BEDS**
RT   BRINES
     BROMIDES
     CHLORIDES
     FLATS (LANDFORMS)
     SODIUM CHLORIDES

SALT FLATS
USE   FLATS (LANDFORMS)

SALT SPRAY TESTS
GS   CHEMICAL TESTS
     . **SALT SPRAY TESTS**
     ENVIRONMENTAL TESTS
     . CORROSION TESTS
     . . **SALT SPRAY TESTS**
RT   CORROSION
     CORROSION RESISTANCE
     SPRAY INGESTION
     STRESS CORROSION
∞ TESTS

SALTON SEA (CA)
GS   SEAS
     . **SALTON SEA (CA)**
RT   CALIFORNIA
     PACIFIC OCEAN

∞ SALTS
SN   *(USE OF A MORE SPECIFIC TERM IS
      RECOMMENDED--CONSULT THE TERMS
      LISTED BELOW)*
RT   HALITES
     INORGANIC COMPOUNDS
     MOLTEN SALTS
     ORGANIC CHARGE TRANSFER SALTS
     ORGANIC COMPOUNDS
     SODIUM CHLORIDES
     SULFONATES

SALYUT SPACE STATION
GS   MANNED SPACECRAFT
     . SPACE STATIONS
     . . ORBITAL SPACE STATIONS
     . . . **SALYUT SPACE STATION**
     SOVIET SPACECRAFT
     . **SALYUT SPACE STATION**
     STATIONS
     . SPACE STATIONS
     . . ORBITAL SPACE STATIONS
     . . . **SALYUT SPACE STATION**
RT   SOYUZ SPACECRAFT
     SPACE BASES
     SPACE LABORATORIES
     SPACECRAFT DOCKING
     U.S.S.R. SPACE PROGRAM

SAMARITAN AIRCRAFT
USE   C-131 AIRCRAFT

SAMARIUM
GS   CHEMICAL ELEMENTS
     . RARE EARTH ELEMENTS
     . . **SAMARIUM**
     . . . SAMARIUM ISOTOPES
     METALS

SAMARIUM-*(CONT.)*
     . RARE EARTH ELEMENTS
     . . **SAMARIUM**
     . . . SAMARIUM ISOTOPES

SAMARIUM COMPOUNDS
GS   RARE EARTH COMPOUNDS
     . **SAMARIUM COMPOUNDS**
RT   ∞ CHEMICAL COMPOUNDS
     ∞ METAL COMPOUNDS

SAMARIUM ISOTOPES
GS   CHEMICAL ELEMENTS
     . RARE EARTH ELEMENTS
     . . SAMARIUM
     . . . **SAMARIUM ISOTOPES**
     METALS
     . RARE EARTH ELEMENTS
     . . SAMARIUM
     . . . **SAMARIUM ISOTOPES**

SAMOA
GS   LANDFORMS
     . ISLANDS
     . . PACIFIC ISLANDS
     . . . **SAMOA**

SAMOS
GS   SATELLITES
     . **SAMOS**

SAMPLED DATA
USE   DATA SAMPLING

SAMPLED DATA SYSTEMS
USE   DATA SAMPLING

SAMPLERS
UF   BOMBS (SAMPLERS)
     SAMPLING DEVICES
RT   ∞ BOMBS
     CORE SAMPLING
     SAMPLES
     SAMPLING
     SELECTORS
     ∞ TEST EQUIPMENT

SAMPLES
GS   **SAMPLES**
     . MARS SURFACE SAMPLES
RT   ACCEPTABILITY
     SAMPLERS
     SAMPLING
     SPECIMENS

SAMPLING
GS   **SAMPLING**
     . AIR SAMPLING
     . CORE SAMPLING
     . DATA SAMPLING
     . PARTICULATE SAMPLING
     . RANDOM SAMPLING
RT   ALLOWANCES
     ASSAYING
     BAYES THEOREM
     CENSORED DATA (MATHEMATICS)
     CHEMICAL ANALYSIS
     CHEMICAL TESTS
     COLLECTION
     CONCENTRATION (COMPOSITION)
     CONFIDENCE LIMITS
     COUNTING
     ESTIMATING
     EXPLORATION
     GLOBAL AIR SAMPLING PROGRAM
     HETEROGENEITY
     HOMOGENEITY
     INSPECTION
     INVESTIGATION
     PROBABILITY THEORY
     PROCESS CONTROL (INDUSTRY)
     QUALITY CONTROL
     RANDOM ERRORS
     RELIABILITY
     SAMPLERS
     SAMPLES
     SELECTION
     SEQUENTIAL ANALYSIS
     SPECIMENS
     STANDARDS
     STATISTICAL ANALYSIS
     ∞ STATISTICS
     SWEEP CIRCUITS
     ∞ TESTS

SAMPLING-(CONT.)
    VARIABILITY
    WEIBULL DENSITY FUNCTIONS

SAMPLING DEVICES
USE   SAMPLERS

SAN ANDREAS FAULT
GS   GEOLOGICAL FAULTS
    . SAN ANDREAS FAULT
RT   CALIFORNIA
    CRUSTAL FRACTURES
    EARTH CRUST
    EARTHQUAKES
    MEXICO

SAN ANDREAS FAULT EXPERIMENT
RT   EARTHQUAKES
    GEOLOGICAL FAULTS

SAN FRANCISCO (CA)
GS   CITIES
    . SAN FRANCISCO (CA)
RT   CALIFORNIA

SAN FRANCISCO BAY (CA)
GS   BAYS (TOPOGRAPHIC FEATURES)
    . SAN FRANCISCO BAY (CA)
RT   CALIFORNIA
    PACIFIC OCEAN
    SAN PABLO BAY (CA)

SAN JOAQUIN VALLEY (CA)
GS   VALLEYS
    . SAN JOAQUIN VALLEY (CA)
RT   CALIFORNIA
    RIVER BASINS

SAN JUAN MOUNTAINS (CO)
GS   LANDFORMS
    . MOUNTAINS
    .. SAN JUAN MOUNTAINS (CO)
RT   COLORADO

SAN MARCO SATELLITES
GS   SATELLITES
    . ARTIFICIAL SATELLITES
    .. METEOROLOGICAL SATELLITES
    ... SAN MARCO SATELLITES
    .... SAN MARCO 1 SATELLITE
    .... SAN MARCO 2 SATELLITE
    .... SAN MARCO 3 SATELLITE
RT   SCOUT LAUNCH VEHICLE

SAN MARCO 1 SATELLITE
GS   SATELLITES
    . ARTIFICIAL SATELLITES
    .. METEOROLOGICAL SATELLITES
    ... SAN MARCO SATELLITES
    .... SAN MARCO 1 SATELLITE

SAN MARCO 2 SATELLITE
GS   SATELLITES
    . ARTIFICIAL SATELLITES
    .. METEOROLOGICAL SATELLITES
    ... SAN MARCO SATELLITES
    .... SAN MARCO 2 SATELLITE

SAN MARCO 3 SATELLITE
GS   SATELLITES
    . ARTIFICIAL SATELLITES
    .. METEOROLOGICAL SATELLITES
    ... SAN MARCO SATELLITES
    .... SAN MARCO 3 SATELLITE

SAN MARINO
GS   NATIONS
    . SAN MARINO
RT   EUROPE
    ITALY

SAN PABLO BAY (CA)
GS   BAYS (TOPOGRAPHIC FEATURES)
    . SAN PABLO BAY (CA)
RT   CALIFORNIA
    SAN FRANCISCO BAY (CA)

SAND CASTING
GS   FORMING TECHNIQUES
    . CASTING
    .. SAND CASTING
RT   MOLDING MATERIALS
    SANDS

SAND DUNES
USE   DUNES

SAND HILLS REGION (GA-NC-SC)
GS   REGIONS
    . SAND HILLS REGION (GA-NC-SC)
RT   GEORGIA
    NORTH CAROLINA
    SOUTH CAROLINA

SAND HILLS REGION (NE)
GS   REGIONS
    . SAND HILLS REGION (NE)
RT   NEBRASKA

SANDPIPER TARGET MISSILE
GS   MISSILE CONFIGURATIONS
    . SANDPIPER TARGET MISSILE
    MISSILES
    . SANDPIPER TARGET MISSILE
RT   DRONE VEHICLES
    TARGETS

SANDS
GS   SEDIMENTS
    . SANDS
    .. MONAZITE SANDS
    .. TAR SANDS
    SOILS
    . SANDS
    .. MONAZITE SANDS
    .. TAR SANDS
RT   AGGREGATES
    ALLUVIUM
    AQUIFERS
    DELTAS
    DUNES
    DUST
    EARTH RESOURCES
    FANS (LANDFORMS)
    GRAVELS
    GRIT
    ILMENITE
    LITTORAL DRIFT
    LITTORAL TRANSPORT
    MOLDING MATERIALS
    POROUS MATERIALS
    QUARTZ
    RAIN EROSION
    REEFS
    SAND CASTING
    SANDSTONES
    SEDIMENTARY ROCKS
    SILICA GLASS
    SILICON DIOXIDE

SANDSTONES
GS   ROCKS
    . SANDSTONES
    SEDIMENTARY ROCKS
    . SANDSTONES
RT   EARTH RESOURCES
    SANDS
    SCHIST
    SOILS

SANDWICH CONSTRUCTION
USE   SANDWICH STRUCTURES

SANDWICH STRUCTURES
UF   SANDWICH CONSTRUCTION
RT   COMPOSITE MATERIALS
    EPOXY MATRIX COMPOSITES
    HONEYCOMB CORES
    HONEYCOMB STRUCTURES
    INTERLAYERS
    LAMINATES
    MULTILAYER INSULATION
    PLY ORIENTATION
    RIGID STRUCTURES
∞   STRUCTURES
    WALLS

SANITATION
RT   CONSUMABLES (SPACECREW SUPPLIES)
    HEALTH
    HOUSEKEEPING (SPACECRAFT)
    HYGIENE
    POTABLE WATER
    PUBLIC HEALTH
    SEWERS
    WARNING SYSTEMS
    WASTE DISPOSAL

SANTOWAX (TRADEMARK)
RT   POLYSTYRENE

SAPPHIRE
GS   ALUMINUM COMPOUNDS
    . ALUMINUM OXIDES
    .. SAPPHIRE
    CHALCOGENIDES
    . OXIDES
    .. METAL OXIDES
    ... ALUMINUM OXIDES
    .... SAPPHIRE

SAPROPHYTES
GS   MICROORGANISMS
    . SAPROPHYTES
    PLANTS (BOTANY)
    . SAPROPHYTES
RT   BACTERIA

SARCINA
GS   MICROORGANISMS
    . BACTERIA
    .. SARCINA

SARCOMA
USE   CANCER

SARGASSO SEA
RT   ATLANTIC OCEAN
    GULF STREAM
    OCEAN MODELS
    OCEAN SURFACE
    OCEANOGRAPHY
    SEAS

SARSAT
UF   SEARCH AND RESCUE SATELLITE
GS   SATELLITES
    . ARTIFICIAL SATELLITES
    .. SARSAT
RT   COSPAS
    NOAA 8 SATELLITE
    RECONNAISSANCE
    RESCUE OPERATIONS
    SEARCHING

SAS
UF   SMALL ASTRONOMY SATELLITES
GS   OBSERVATORIES
    . ASTRONOMICAL OBSERVATORIES
    .. ASTRONOMICAL SATELLITES
    ... SAS
    .... SAS-1
    .... SAS-2
    .... SAS-3
RT   EXPLORER 48 SATELLITE
    UHURU SATELLITE

SAS-D
USE   IUE

SAS-1
UF   SMALL ASTRONOMY SATELLITE 1
GS   OBSERVATORIES
    . ASTRONOMICAL OBSERVATORIES
    .. ASTRONOMICAL SATELLITES
    ... SAS
    .... SAS-1
RT   RADIO ASTRONOMY
    SPACEBORNE ASTRONOMY

SAS-2
UF   SMALL ASTRONOMY SATELLITE 2
GS   OBSERVATORIES
    . ASTRONOMICAL OBSERVATORIES
    .. ASTRONOMICAL SATELLITES
    ... SAS
    .... SAS-2
RT   EXPLORER 48 SATELLITE
    RADIO ASTRONOMY
    SPACEBORNE ASTRONOMY

SAS-3
UF   SMALL ASTRONOMY SATELLITE 3
GS   OBSERVATORIES
    . ASTRONOMICAL OBSERVATORIES
    .. ASTRONOMICAL SATELLITES
    ... SAS
    .... SAS-3
RT   EXPLORER 53 SATELLITE
    RADIO ASTRONOMY
    SPACEBORNE ASTRONOMY
    X RAY ASTRONOMY

SASKATCHEWAN
GS   NATIONS
. CANADA
. . **SASKATCHEWAN**

SATAN (SENSOR)
USE   TERRAIN ANALYSIS

SATELLITE ANTENNAS
GS   ANTENNAS
. **SATELLITE ANTENNAS**
RT   FURLABLE ANTENNAS
MULTIBEAM ANTENNAS
RADIO ANTENNAS
TELECOMMUNICATION

SATELLITE ATMOSPHERES
GS   ENVIRONMENTS
. EXTRATERRESTRIAL ENVIRONMENTS
. . **SATELLITE ATMOSPHERES**
. . . LUNAR ATMOSPHERE
RT   ∞ATMOSPHERES
ATMOSPHERIC CHEMISTRY
ATMOSPHERIC COMPOSITION
ATMOSPHERIC PHYSICS
EARTH ATMOSPHERE
IONOSPHERE
IONOSPHERIC COMPOSITION
MAGNETOPAUSE
MAGNETOSPHERE
NATURAL SATELLITES
PLANETARY ATMOSPHERES
STELLAR ATMOSPHERES
TITAN
TRITON
UPPER ATMOSPHERE

SATELLITE ATTITUDE CONTROL
GS   ATTITUDE CONTROL
. **SATELLITE ATTITUDE CONTROL**
SPACECRAFT CONTROL
. SATELLITE CONTROL
. . **SATELLITE ATTITUDE CONTROL**
RT   ATTITUDE STABILITY
AUTOMATIC CONTROL
∞CONTROL
DIRECTIONAL CONTROL
GRAVITY GRADIENT SATELLITES
JET CONTROL
LATERAL CONTROL
LONGITUDINAL CONTROL
MARQUARDT R4D ENGINE
THREE AXIS STABILIZATION
TRANSIT ATTITUDE CONTROL
SATELLITE

SATELLITE ATTITUDE DISTURBANCE
USE   ATTITUDE STABILITY
SPACECRAFT STABILITY

SATELLITE CAPTURE
USE   SPACECRAFT RECOVERY

SATELLITE COMMUNICATION
USE   SPACECRAFT COMMUNICATION

SATELLITE COMMUNICATIONS SHIPS
UF   USNS KINGSPORT
GS   SURFACE VEHICLES
. **SATELLITE COMMUNICATIONS SHIPS**
WATER VEHICLES
. SHIPS
. . **SATELLITE COMMUNICATIONS SHIPS**
RT   SPACECRAFT COMMUNICATION

SATELLITE CONFIGURATIONS
GS   SPACECRAFT CONFIGURATIONS
. **SATELLITE CONFIGURATIONS**
RT   AERODYNAMIC CONFIGURATIONS

SATELLITE CONTROL
GS   SPACECRAFT CONTROL
. **SATELLITE CONTROL**
. . SATELLITE ATTITUDE CONTROL
RT   ATTITUDE CONTROL
AUTOMATIC CONTROL
∞CONTROL
DIRECTIONAL CONTROL
FLEXIBLE SPACECRAFT
GRAVITY GRADIENT SATELLITES
JET CONTROL
LATERAL CONTROL
LONGITUDINAL CONTROL
MANUAL CONTROL

SATELLITE CONTROL-*(CONT.)*
REMOTE CONTROL
THRUST CONTROL

SATELLITE DEFENSE
USE   SPACECRAFT DEFENSE

SATELLITE DESIGN
GS   SPACECRAFT DESIGN
. **SATELLITE DESIGN**
RT   COMPUTER AIDED DESIGN
∞DESIGN
INDIAN SPACE PROGRAM
JAPANESE SPACE PROGRAM
PRODUCT DEVELOPMENT
SPACECRAFT STRUCTURES
STRUCTURAL DESIGN
SYSTEMS ENGINEERING

SATELLITE DOPPLER POSITIONING
RT   DOPPLER EFFECT
DOPPLER NAVIGATION
DOPPLER RADAR
GEODESY
GEODETIC ACCURACY
GEODETIC COORDINATES
GEODETIC SATELLITES
GEODETIC SURVEYS
POLYSTATION DOPPLER TRACKING
SYSTEM
POSITIONING
SATELLITE TRACKING
TRACKING (POSITION)

SATELLITE DRAG
GS   DYNAMIC CHARACTERISTICS
. DRAG
. . **SATELLITE DRAG**
RT   AERODYNAMIC DRAG
ELECTROSTATIC DRAG
FRICTION DRAG

SATELLITE GROUND SUPPORT
RT   GROUND SUPPORT EQUIPMENT
SPACECRAFT COMMUNICATION

SATELLITE GROUND TRACKS
GS   GROUND TRACKS
. **SATELLITE GROUND TRACKS**
RT   FLIGHT PATHS
ORBITS

SATELLITE GUIDANCE
GS   GUIDANCE (MOTION)
. SPACECRAFT GUIDANCE
. . **SATELLITE GUIDANCE**
RT   AUTOMATIC CONTROL
INERTIAL GUIDANCE
INJECTION GUIDANCE
LOCATES SYSTEM
MANUAL CONTROL
REENTRY GUIDANCE
RENDEZVOUS GUIDANCE
SPACE NAVIGATION

SATELLITE IMAGERY
GS   IMAGERY
. **SATELLITE IMAGERY**
RT   ATMOSPHERIC CORRECTION
IMAGE ANALYSIS
IMAGING TECHNIQUES
SATELLITE OBSERVATION
SATELLITE-BORNE PHOTOGRAPHY
VEGETATIVE INDEX

SATELLITE INSTRUMENTS
GS   SPACECRAFT INSTRUMENTS
. **SATELLITE INSTRUMENTS**
. . MULTISPECTRAL LINEAR ARRAYS
RT   FLIGHT INSTRUMENTS
∞INSTRUMENTS
MEASURING INSTRUMENTS
NEEDS (DATA SYSTEM)
WILDLIFE RADIOLOCATION

SATELLITE INTERCEPTORS
RT   ∞INTERCEPTORS
PURSUIT TRACKING

SATELLITE LAUNCHING
USE   SPACECRAFT LAUNCHING

SATELLITE LIFETIME
GS   LIFE (DURABILITY)

SATELLITE LIFETIME-*(CONT.)*
. **SATELLITE LIFETIME**
RT   ORBIT DECAY
SPACECRAFT REENTRY

SATELLITE MANEUVERS
USE   SPACECRAFT MANEUVERS

SATELLITE NAVIGATION SYSTEMS
GS   **SATELLITE NAVIGATION SYSTEMS**
. TRANSIT NAVIGATION SYSTEM
RT   AUTONOMOUS NAVIGATION
GLOBAL POSITIONING SYSTEM
NAVIGATION SATELLITES
RADAR NAVIGATION
RADIO NAVIGATION
SPACE NAVIGATION
∞SYSTEMS

SATELLITE NETWORKS
SN   (NETWORKS INCORPORATING
SATELLITES)
RT   AERONAUTICAL SATELLITES
AEROSAT SATELLITES
CODE DIVISION MULTIPLE ACCESS
COMMUNICATION NETWORKS
COMMUNICATION SATELLITES
COMSTAR SATELLITES
DEMAND ASSIGNMENT MULTIPLE
ACCESS
DOMESTIC SATELLITE COMMUNICATIONS
SYSTEMS
HET EXPERIMENT
L-SAT
MARECS MARITIME SATELLITES
MILITARY SPACECRAFT
MOLNIYA SATELLITES
MULTIMISSION MODULAR SPACECRAFT
NAVIGATION SATELLITES
NAVSTAR SATELLITES
NETWORK CONTROL
∞NETWORKS
SATELLITES
SKYNET SATELLITES
TDR SATELLITES
TELECONFERENCING
TIME DIVISION MULTIPLE ACCESS

SATELLITE OBSERVATION
GS   OBSERVATION
. EARTH OBSERVATIONS (FROM SPACE)
. . **SATELLITE OBSERVATION**
RT   ARC CLOUDS
EARTH RESOURCES PROGRAM
EROS (SATELLITES)
ESSA SATELLITES
IRIS SATELLITES
LANDSAT SATELLITES
METEOROLOGICAL SATELLITES
METEOSAT SATELLITE
NIMBUS PROJECT
NIMBUS SATELLITES
SATELLITE IMAGERY
SATELLITES
SIRS B SATELLITE
SPACEBORNE PHOTOGRAPHY
SWATH WIDTH
SYNCHRONOUS EARTH OBSERVATORY
SATELLITE
TIROS OPERATIONAL SATELLITE
SYSTEM
TIROS SATELLITES
TOPEX
UHURU SATELLITE
VEGETATIVE INDEX
VELA SATELLITES
WILDLIFE RADIOLOCATION

SATELLITE ORBIT CALCULATION
USE   ORBIT CALCULATION

SATELLITE ORBITS
SN   (LIMITED TO ORBITS OF ARTIFICIAL
SATELLITES)
GS   ORBITS
. SPACECRAFT ORBITS
. . **SATELLITE ORBITS**
. . . GEOSYNCHRONOUS ORBITS
. . . PARKING ORBITS
. . . POLAR ORBITS
. . . STATIONARY ORBITS
. . . . TWENTY-FOUR HOUR ORBITS
RT   CIRCULAR ORBITS
EARTH ORBITS
ELLIPTICAL ORBITS

SATELLITE ORBITS-(CONT.)
    EQUATORIAL ORBITS
    LISSAJOUS FIGURES
    LUNAR ORBITS
    ORBIT SPECTRUM UTILIZATION
    ORBITAL MECHANICS
    ORBITAL POSITION ESTIMATION
    PLANETARY ORBITS
    SATELLITES
    TRANSFER ORBITS

SATELLITE ORIENTATION
GS   ATTITUDE (INCLINATION)
    . **SATELLITE ORIENTATION**
RT   FLEXIBLE SPACECRAFT
    IMAGE DISSECTOR TUBES
    SPIN STABILIZATION
    THREE AXIS STABILIZATION

SATELLITE PERTURBATION
GS   PERTURBATION
    . ORBIT PERTURBATION
    . . **SATELLITE PERTURBATION**
RT   DISCOS (SATELLITE ATTITUDE
        CONTROL)
    GRAVITATIONAL FIELDS
    ORBITAL MECHANICS
    SCHACH EFFECT
    SPACECRAFT STABILITY
    TESSERAL HARMONICS

SATELLITE POWER TRANSMISSION (TO EARTH)
RT   RECTENNAS
    SOLAR ARRAYS
    SOLAR CELLS
    SOLAR POWER SATELLITES

SATELLITE RENDEZVOUS
USE   ORBITAL RENDEZVOUS

SATELLITE ROTATION
GS   GYRATION
    . ROTATION
    . . **SATELLITE ROTATION**
RT   FLEXIBLE SPACECRAFT
    SPIN REDUCTION
    SPIN STABILIZATION
    TUMBLING MOTION
    YO-YO DEVICES

SATELLITE SOLAR ENERGY CONVERSION
GS   ENERGY CONVERSION
    . **SATELLITE SOLAR ENERGY
        CONVERSION**
RT   ∞CONVERSION
    MICROWAVE TRANSMISSION
    MICROWAVES
    POWER CONDITIONING
    SOLAR CELLS
    SUN

SATELLITE SOLAR POWER STATIONS
RT   ENERGY CONVERSION
    MICROWAVE TRANSMISSION
    MICROWAVES
    POWER CONDITIONING
    SOLAR CELLS
    SUN

SATELLITE SOUNDING
GS   SOUNDING
    . **SATELLITE SOUNDING**
RT   ATMOSPHERIC SOUNDING
    IONOSONDES
    IONOSPHERIC SOUNDING
    METEOROLOGICAL SATELLITES
    RADIOSONDES
    ROCKET SOUNDING
    SATELLITES
    VISIBLE INFRARED SPIN SCAN
        RADIOMETER

SATELLITE SURFACES
SN   (RESTRICTED TO NATURAL SATELLITES)
RT   CRATERS
    NATURAL SATELLITES
    ∞SURFACES
    TERRAIN ANALYSIS

SATELLITE TELEVISION
GS   COMMUNICATION EQUIPMENT
    . SPACECRAFT TELEVISION
    . . **SATELLITE TELEVISION**
    TELECOMMUNICATION

SATELLITE TELEVISION-(CONT.)
    . SPACECRAFT TELEVISION
    . . **SATELLITE TELEVISION**
    TELEVISION SYSTEMS
    . SPACECRAFT TELEVISION
    . . **SATELLITE TELEVISION**
RT   COLOR TELEVISION
    METEOROLOGICAL SATELLITES
    SPACE PROBES
    STEREOTELEVISION
    SYMPHONIE SATELLITES
    TELEVISION CAMERAS
    TELEVISION TRANSMISSION

SATELLITE TEMPERATURE
GS   TEMPERATURE
    . **SATELLITE TEMPERATURE**
RT   AMBIENT TEMPERATURE
    RADIATIVE HEAT TRANSFER
    SOLAR RADIATION SHIELDING
    SPACECRAFT DESIGN
    SPACECRAFT ENVIRONMENTS
    SPACECRAFT TEMPERATURE
    TEMPERATURE DISTRIBUTION
    TEMPERATURE MEASUREMENT
    THERMAL ENVIRONMENTS

SATELLITE TRACKING
GS   TRACKING (POSITION)
    . SPACECRAFT TRACKING
    . . **SATELLITE TRACKING**
    . . . SATELLITE-TO-SATELLITE TRACKING
RT   CINETHEODOLITES
    GLOBAL TRACKING NETWORK
    INTERNATIONAL SATELLITE GEODESY
        EXPERIMENT
    LASER TARGET DESIGNATORS
    MINITRACK SYSTEM
    OPTICAL SATELLITE TRACKING
        PROGRAM
    PHOTOGRAPHIC TRACKING
    RANGE AND RANGE RATE TRACKING
    SATELLITE DOPPLER POSITIONING
    SATELLITES
    SPACE FLIGHT TRACKING AND DATA
        NETWORK
    STDN (NETWORK)
    TRACKING NETWORKS
    TRACKING STATIONS
    TRANSPONDER CONTROL GROUP

SATELLITE TRACKING AND DATA ACQ NETWORK
USE   STDN (NETWORK)

SATELLITE TRANSMISSION
GS   TRANSMISSION
    . SIGNAL TRANSMISSION
    . . **SATELLITE TRANSMISSION**
RT   ALOHA SYSTEM
    CODE DIVISION MULTIPLEXING
    DATA TRANSMISSION
    DOMESTIC SATELLITE COMMUNICATIONS
        SYSTEMS
    DOWNLINKING
    EARTH TERMINALS
    FREQUENCY DIVISION MULTIPLEXING
    FREQUENCY REUSE
    MSAT
    MULTIPLEXING
    PULSE COMMUNICATION
    RADIO TRANSMISSION
    SINGLE CHANNEL PER CARRIER
        TRANSMISSION
    SPACECRAFT TELEVISION
    TDR SATELLITES
    TELEVISION TRANSMISSION
    UPLINKING

SATELLITE-BORNE INSTRUMENTS
GS   MEASURING INSTRUMENTS
    . **SATELLITE-BORNE INSTRUMENTS**
    . . AMPS (SATELLITE PAYLOAD)
RT   AMPTE (SATELLITES)
    DIAL SATELLITE
    INFRARED RADIOMETERS
    INSTRUMENT PACKAGES
    ∞INSTRUMENTS
    OPEN PROJECT
    PARTICLE TELESCOPES
    RADIATION DETECTORS
    REMOTE SENSORS
    SINGLE EVENT UPSETS
    SOLAR BACKSCATTER UV
        SPECTROMETER

SATELLITE-BORNE INSTRUMENTS-(CONT.)
    VISIBLE INFRARED SPIN SCAN
    RADIOMETER

SATELLITE-BORNE PHOTOGRAPHY
GS   IMAGERY
    . SPACEBORNE PHOTOGRAPHY
    . . **SATELLITE-BORNE PHOTOGRAPHY**
    PHOTOGRAPHY
    . SPACEBORNE PHOTOGRAPHY
    . . **SATELLITE-BORNE PHOTOGRAPHY**
RT   AERIAL PHOTOGRAPHY
    ASTRONOMICAL PHOTOGRAPHY
    BLACK AND WHITE PHOTOGRAPHY
    DMSP SATELLITES
    FOREST FIRE DETECTION
    GEOGRAPHIC APPLICATIONS PROGRAM
    INFRARED PHOTOGRAPHY
    MARS PHOTOGRAPHS
    PHOTOMAPPING
    PHOTOMAPS
    ROCKET-BORNE PHOTOGRAPHY
    SATELLITE IMAGERY
    SPACE SURVEILLANCE (SPACEBORNE)
    SPECTRAL RECONNAISSANCE
    TIMBER INVENTORY

SATELLITE-BORNE RADAR
GS   RADAR
    . **SATELLITE-BORNE RADAR**
RT   RADAR DETECTION
    SEARCH RADAR
    SURVEILLANCE RADAR
    SYNTHETIC APERTURE RADAR
    TRACKING RADAR

SATELLITE-TO-SATELLITE TRACKING
GS   TRACKING (POSITION)
    . SPACECRAFT TRACKING
    . . SATELLITE TRACKING
    . . . **SATELLITE-TO-SATELLITE
        TRACKING**
RT   SPACE SURVEILLANCE (SPACEBORNE)
    TRACKING NETWORKS

SATELLITES
SN   (EXCLUDES PLANETS)
GS   **SATELLITES**
    . ACTIVE SATELLITES
    . . SYNCOM SATELLITES
    . . . EARLY BIRD SATELLITES
    . . . SYNCOM 1 SATELLITE
    . . . SYNCOM 2 SATELLITE
    . . . SYNCOM 3 SATELLITE
    . ARTIFICIAL SATELLITES
    . . ALOUETTE SATELLITES
    . . . ALOUETTE B SATELLITE
    . . . ALOUETTE 1 SATELLITE
    . . . ALOUETTE 2 SATELLITE
    . . ARABSAT
    . . ARIEL SATELLITES
    . . . ARIEL 1 SATELLITE
    . . . ARIEL 2 SATELLITE
    . . . ARIEL 3 SATELLITE
    . . . ARIEL 4 SATELLITE
    . . . ARIEL 5 SATELLITE
    . . BESS (SATELLITE)
    . . BIOSATELLITES
    . . . BIOSATELLITE 1
    . . . BIOSATELLITE 2
    . . . BIOSATELLITE 3
    . . . ORBITING FROG OTOLITH
    . . . SPUTNIK 2 SATELLITE
    . . COMMUNICATION SATELLITES
    . . . AERONAUTICAL SATELLITES
    . . . . AEROSAT SATELLITES
    . . . ARCOMSAT
    . . . COMMUNICATIONS TECHNOLOGY
        SATELLITE
    . . . . COMSTAR C
    . . . . NATO 3B SATELLITE
    . . . COMSTAR SATELLITES
    . . . EUROPEAN COMMUNICATIONS
        SATELLITE
    . . . INTELSAT SATELLITES
    . . . L-SAT
    . . . LOW FREQUENCY
        TRANSIONOSPHERIC SATELLITES
    . . . MARECS MARITIME SATELLITES
    . . . MAROTS (ESA)
    . . . MOLNIYA SATELLITES
    . . . PALAPA SATELLITES
    . . . PALAPA 2 SATELLITE
    . . . RADUGA SATELLITE
    . . . RCA SATCOM SATELLITES

SATELLITES-*(CONT.)*

. . . RELAY SATELLITES
. . . . RELAY 1 SATELLITE
. . . . RELAY 2 SATELLITE
. . . SYMPHONIE SATELLITES
. . . SYNCOM SATELLITES
. . . . EARLY BIRD SATELLITES
. . . . SYNCOM 1 SATELLITE
. . . . SYNCOM 2 SATELLITE
. . . . SYNCOM 3 SATELLITE
. . . . SYNCOM 4 SATELLITE
. . . WESTAR SATELLITES
. . COSPAS
. . COURIER SATELLITE
. . DIADEME SATELLITES
. . DISCOVERER SATELLITES
. . DODGE SATELLITE
. . EROS (SATELLITES)
. . ESA SATELLITES
. . . AEROSAT SATELLITES
. . . COS-B SATELLITE
. . . ERS-1 (ESA SATELLITE)
. . . ESRO 1 SATELLITE
. . . ESRO 2 SATELLITE
. . . ESRO 4 SATELLITE
. . . EUROPEAN COMMUNICATIONS
          SATELLITE
. . . EXOSAT SATELLITE
. . . GEOS SATELLITES (ESA)
. . . HEOS SATELLITES
. . . HEOS A SATELLITE
. . . HEOS B SATELLITE
. . . HIPPARCOS SATELLITE
. . . L-SAT
. . MAGELLAN MISSION
. . . MARECS MARITIME SATELLITES
. . . MAROTS (ESA)
. . . METEOSAT SATELLITE
. . . OTS (ESA)
. . TD SATELLITES
. . . . TD-1 SATELLITE
. EUROPEAN 1 SPACECRAFT
. EVASIVE SATELLITES
. EXPLORER SATELLITES
. . APPLICATIONS EXPLORER
          SATELLITES
. . . COSMIC BACKGROUND EXPLORER
          SATELLITE
. . . DUAL AIR DENSITY EXPLORER
. . . DYNAMICS EXPLORER SATELLITES
. . . . DYNAMICS EXPLORER 1
          SATELLITE
. . . . DYNAMICS EXPLORER 2
          SATELLITE
. . . EXPLORER 1 SATELLITE
. . . EXPLORER 2 SATELLITE
. . . EXPLORER 3 SATELLITE
. . . EXPLORER 4 SATELLITE
. . . EXPLORER 5 SATELLITE
. . . EXPLORER 6 SATELLITE
. . . EXPLORER 7 SATELLITE
. . . EXPLORER 8 SATELLITE
. . . EXPLORER 9 SATELLITE
. . . EXPLORER 10 SATELLITE
. . . EXPLORER 11 SATELLITE
. . . EXPLORER 12 SATELLITE
. . . EXPLORER 14 SATELLITE
. . . EXPLORER 15 SATELLITE
. . . EXPLORER 16 SATELLITE
. . . EXPLORER 17 SATELLITE
. . . EXPLORER 18 SATELLITE
. . . EXPLORER 19 SATELLITE
. . . EXPLORER 20 SATELLITE
. . . EXPLORER 21 SATELLITE
. . . EXPLORER 22 SATELLITE
. . . EXPLORER 23 SATELLITE
. . . EXPLORER 24 SATELLITE
. . . EXPLORER 25 SATELLITE
. . . EXPLORER 26 SATELLITE
. . . EXPLORER 27 SATELLITE
. . . EXPLORER 28 SATELLITE
. . . EXPLORER 29 SATELLITE
. . . EXPLORER 30 SATELLITE
. . . EXPLORER 31 SATELLITE
. . . EXPLORER 32 SATELLITE
. . . EXPLORER 33 SATELLITE
. . . EXPLORER 34 SATELLITE
. . . EXPLORER 35 SATELLITE
. . . EXPLORER 36 SATELLITE
. . . EXPLORER 37 SATELLITE
. . . EXPLORER 38 SATELLITE
. . . EXPLORER 39 SATELLITE
. . . EXPLORER 40 SATELLITE
. . . EXPLORER 41 SATELLITE
. . . EXPLORER 43 SATELLITE
. . . EXPLORER 44 SATELLITE

SATELLITES-*(CONT.)*

. . . EXPLORER 46 SATELLITE
. . . EXPLORER 47 SATELLITE
. . . EXPLORER 48 SATELLITE
. . . EXPLORER 49 SATELLITE
. . . EXPLORER 50 SATELLITE
. . . EXPLORER 51 SATELLITE
. . . EXPLORER 52 SATELLITE
. . . EXPLORER 53 SATELLITE
. . . EXPLORER 54 SATELLITE
. . . EXPLORER 55 SATELLITE
. . . EXTREME ULTRAVIOLET EXPLORER
          SATELLITE
. . . FAR UV SPECTROSCOPIC
          EXPLORER
. . . IMP
. . . INTERNATIONAL MAGNETOSPHERIC
          EXPLORER
. . . INTERNATIONAL SUN EARTH
          EXPLORERS
. . . . INTERNATIONAL SUN EARTH
          EXPLORER 1
. . . . INTERNATIONAL SUN EARTH
          EXPLORER 2
. . . . INTERNATIONAL SUN EARTH
          EXPLORER 3
. . . MICROMETEOROID EXPLORER
          SATELLITES
. . . RADIO ASTRONOMY EXPLORER
          SATELLITE
. . . SOLAR MESOSPHERE EXPLORER
. . . X RAY TIMING EXPLORER
. . FRENCH SATELLITES
. . . D-1 SATELLITE
. . . D-2 SATELLITES
. . . EOLE SATELLITES
. . . FR-1 SATELLITE
. . . GEOLE SATELLITES
. . . PEOLE SATELLITES
. . . POSEIDON SATELLITE
. . . SPOT (FRENCH SATELLITE)
. . . SRET SATELLITES
. . . . SRET 1 SATELLITE
. . . . SRET 2 SATELLITE
. . GEODETIC SATELLITES
. . . ANNA SATELLITES
. . . EXPLORER 29 SATELLITE
. . . EXPLORER 36 SATELLITE
. . . GEOLE SATELLITES
. . . GEOS 1 SATELLITE
. . . GEOS 2 SATELLITE
. . . GEOS 3 SATELLITE
. . . LARGOS SATELLITE
. . . PAGEOS SATELLITE
. . . VANGUARD 1 SATELLITE
. GEOPHYSICAL SATELLITES
. . COSMOS SATELLITES
. . . INTERCOSMOS SATELLITES
. . . EXPLORER 6 SATELLITE
. . . EXPLORER 10 SATELLITE
. . . EXPLORER 12 SATELLITE
. . . EXPLORER 45 SATELLITE
. . . OGO
. . . . EGO
. . . . OGO-A
. . . . OGO-3
. . . . OGO-5
. . . . POGO
. . . . . OGO-C
. . . . . OGO-4
. . . . . OGO-6
. . . OSO
. . . . OSO-C
. . . . OSO-1
. . . . OSO-2
. . . . OSO-3
. . . . OSO-4
. . . . OSO-5
. . . . OSO-6
. . . . OSO-7
. . . . OSO-8
. . . RADIATION AND METEOROID
          SATELLITE
. . SPUTNIK 3 SATELLITE
. . . VANGUARD 3 SATELLITE
. GEOS-D SATELLITE
. . GRAVITY GRADIENT SATELLITES
. . . ATS
. . . . ATS 1
. . . . ATS 2
. . . . ATS 3
. . . . ATS 4
. . . . ATS 5
. . . . ATS 6
. . . . ATS 7
. . . . ATS 8

SATELLITES-*(CONT.)*

. . . ORBIS CAL SATELLITE
. . GREB SATELLITES
. . HELIOS SATELLITES
. . . HELIOS A
. . . HELIOS B
. . . HELIOS 1
. . . HELIOS 2
. . INJUN SATELLITES
. . . EXPLORER 25 SATELLITE
. . . INJUN 1 SATELLITE
. . . INJUN 3 SATELLITE
. . . INJUN 4 SATELLITE
. . INSPECTOR SATELLITE
. . IRIS SATELLITES
. . ISIS SATELLITES
. . . ALOUETTE 2 SATELLITE
. . . ISIS-A
. . . ISIS-B
. . . ISIS-X
. . LANDSAT SATELLITES
. . . LANDSAT E
. . . LANDSAT F
. . . LANDSAT 1
. . . LANDSAT 2
. . . LANDSAT 3
. . . LANDSAT 4
. . . LANDSAT 5
. . LINCOLN EXPERIMENTAL SATELLITES
. . LUNAR SATELLITES
. . . EXPLORER 18 SATELLITE
. . . EXPLORER 28 SATELLITE
. . . IMP
. . . LUNAR ORBITER
. . . . LUNAR ORBITER 1
. . . . LUNAR ORBITER 2
. . . . LUNAR ORBITER 3
. . . . LUNAR ORBITER 4
. . . . LUNAR ORBITER 5
. . . MAPSAT
. . MARISAT SATELLITES
. . . MARISAT 1 SATELLITE
. . MARITIME SATELLITES
. . . ERS-1 (ESA SATELLITE)
. . . MARECS MARITIME SATELLITES
. . . MAROTS (ESA)
. . METEOROLOGICAL SATELLITES
. . . AEROS SATELLITE
. . . COSMOS 144 SATELLITE
. . . D-2 SATELLITES
. . . ELEKTRON SATELLITES
. . . . ELEKTRON 1 SATELLITE
. . . . ELEKTRON 2 SATELLITE
. . . . ELEKTRON 4 SATELLITE
. . . EOLE SATELLITES
. . . ESSA SATELLITES
. . . . ESSA 1 SATELLITE
. . . . ESSA 2 SATELLITE
. . . . ESSA 3 SATELLITE
. . . . ESSA 4 SATELLITE
. . . . ESSA 5 SATELLITE
. . . . ESSA 6 SATELLITE
. . . . ESSA 7 SATELLITE
. . . . ESSA 8 SATELLITE
. . . . ESSA 9 SATELLITE
. . . EXPLORER 9 SATELLITE
. . . EXPLORER 17 SATELLITE
. . . EXPLORER 19 SATELLITE
. . . GEOLE SATELLITES
. . . METEOSAT SATELLITE
. . . NIMBUS SATELLITES
. . . . NIMBUS 1 SATELLITE
. . . . NIMBUS 2 SATELLITE
. . . . NIMBUS 3 SATELLITE
. . . . NIMBUS 4 SATELLITE
. . . . NIMBUS 5 SATELLITE
. . . . NIMBUS 6 SATELLITE
. . . . NIMBUS 7 SATELLITE
. . . NOAA SATELLITES
. . . . NOAA 2 SATELLITE
. . . . NOAA 3 SATELLITE
. . . . NOAA 4 SATELLITE
. . . . NOAA 5 SATELLITE
. . . . NOAA 6 SATELLITE
. . . . NOAA 7 SATELLITE
. . . . NOAA 8 SATELLITE
. . . SAN MARCO SATELLITES
. . . . SAN MARCO 1 SATELLITE
. . . . SAN MARCO 2 SATELLITE
. . . . SAN MARCO 3 SATELLITE
. . . SEOCS (SATELLITE)
. . . SIRS B SATELLITE
. . . SPUTNIK 1 SATELLITE
. . . SPUTNIK 2 SATELLITE
. . . SPUTNIK 3 SATELLITE
. . . SRET SATELLITES

SATURABLE REACTORS-*(CONT.)*
    POWER REACTORS
    TRANSFORMERS

SATURATED HYDROCARBONS
  USE   ALKANES

∞ **SATURATION**
  SN   *(USE OF A MORE SPECIFIC TERM IS*
      RECOMMENDED--CONSULT THE TERMS
      LISTED BELOW)
  RT   CONCENTRATION (COMPOSITION)
      CONDENSING
      CROWDING
      DESATURATION
      PENETRATION
      PERMEATING
      PRECIPITATION (CHEMISTRY)
      PRECIPITATION (METEOROLOGY)
      UNSATURATION (CHEMISTRY)
      WETTING

SATURATION (CHEMISTRY)
  RT   CHEMICAL BONDS
     ∞ CHEMISTRY
      PRECIPITATION (CHEMISTRY)
      UNSATURATION (CHEMISTRY)

∞ **SATURN**
  SN   *(USE OF A MORE SPECIFIC TERM IS*
      RECOMMENDED--CONSULT THE TERMS
      LISTED BELOW)
  RT   SATURN (PLANET)
      SATURN PROJECT

SATURN (PLANET)
  GS   CELESTIAL BODIES
     . PLANETS
     . . GAS GIANT PLANETS
     . . . **SATURN (PLANET)**
  RT   DIONE
      ENCELADUS
      HYPERION
      IAPETUS
      JANUS
      MIMAS
      PHOBOS
      RHEA (ASTRONOMY)
     ∞ SATURN
      TETHYS
      TITAN

SATURN ATMOSPHERE
  GS   ENVIRONMENTS
     . EXTRATERRESTRIAL ENVIRONMENTS
     . . PLANETARY ENVIRONMENTS
     . . . PLANETARY ATMOSPHERES
     . . . . **SATURN ATMOSPHERE**
  RT   ATMOSPHERIC COMPOSITION
      PLANETARY RADIATION

SATURN D LAUNCH VEHICLE
  GS   LAUNCH VEHICLES
     . SATURN LAUNCH VEHICLES
     . . **SATURN D LAUNCH VEHICLE**
      ROCKET VEHICLES
     . MULTISTAGE ROCKET VEHICLES
     . . SATURN LAUNCH VEHICLES
     . . . **SATURN D LAUNCH VEHICLE**
  RT   RL-10-A-3 ENGINE

SATURN LAUNCH VEHICLES
  GS   LAUNCH VEHICLES
     . **SATURN LAUNCH VEHICLES**
     . . SATURN D LAUNCH VEHICLE
     . . SATURN 1 LAUNCH VEHICLES
     . . . SATURN 1 SA-1 LAUNCH VEHICLE
     . . . SATURN 1 SA-2 LAUNCH VEHICLE
     . . . SATURN 1 SA-3 LAUNCH VEHICLE
     . . . SATURN 1 SA-4 LAUNCH VEHICLE
     . . . SATURN 1 SA-5 LAUNCH VEHICLE
     . . . SATURN 1 SA-6 LAUNCH VEHICLE
     . . . SATURN 1 SA-7 LAUNCH VEHICLE
     . . . SATURN 1 SA-8 LAUNCH VEHICLE
     . . . SATURN 1 SA-9 LAUNCH VEHICLE
     . . . SATURN 1 SA-10 LAUNCH VEHICLE
     . . SATURN 1B LAUNCH VEHICLES
     . . SATURN 2 LAUNCH VEHICLES
     . . SATURN 5 LAUNCH VEHICLES
      ROCKET VEHICLES
     . MULTISTAGE ROCKET VEHICLES
     . . **SATURN LAUNCH VEHICLES**
     . . . SATURN D LAUNCH VEHICLE
     . . . SATURN 1 LAUNCH VEHICLES
     . . . . SATURN 1 SA-1 LAUNCH VEHICLE

SATURN LAUNCH VEHICLES-*(CONT.)*
     . . . . SATURN 1 SA-2 LAUNCH VEHICLE
     . . . . SATURN 1 SA-3 LAUNCH VEHICLE
     . . . . SATURN 1 SA-4 LAUNCH VEHICLE
     . . . . SATURN 1 SA-5 LAUNCH VEHICLE
     . . . . SATURN 1 SA-6 LAUNCH VEHICLE
     . . . . SATURN 1 SA-7 LAUNCH VEHICLE
     . . . . SATURN 1 SA-8 LAUNCH VEHICLE
     . . . . SATURN 1 SA-9 LAUNCH VEHICLE
     . . . . SATURN 1 SA-10 LAUNCH
         VEHICLE
     . . . SATURN 1B LAUNCH VEHICLES
     . . . SATURN 2 LAUNCH VEHICLES
     . . . SATURN 5 LAUNCH VEHICLES
  RT   APOLLO PROJECT
      F-1 ROCKET ENGINE
      RL-10 ENGINES
     ∞ VEHICLES

SATURN PROJECT
  GS   PROGRAMS
     . NASA PROGRAMS
     . . **SATURN PROJECT**
     . NASA SPACE PROGRAMS
     . . **SATURN PROJECT**
     . PROJECTS
     . . **SATURN PROJECT**
  RT   APOLLO APPLICATIONS PROGRAM
      APOLLO SPACECRAFT
      CENTAUR LAUNCH VEHICLE
      LAUNCH VEHICLES
      LUNAR LAUNCH
      PEGASUS SATELLITES
      RIFT (REACTOR IN FLIGHT TEST)
      VOYAGER PROJECT

SATURN RINGS
  GS   CELESTIAL BODIES
     . PLANETARY RINGS
     . . **SATURN RINGS**
  RT   GAS GIANT PLANETS
      JUPITER RINGS
      NATURAL SATELLITES
      PLANETARY ATMOSPHERES
      PLANETARY COMPOSITION
      PLANETARY SURFACES
      PLANETARY TEMPERATURE
      PLANETOLOGY
      PLANETS
     ∞ RINGS
      SOLAR SYSTEM
      URANUS RINGS

SATURN S-1 STAGE
  GS   ROCKET VEHICLES
     . SATURN STAGES
     . . **SATURN S-1 STAGE**
  RT   LIQUID PROPELLANT ROCKET ENGINES

SATURN S-1B STAGE
  GS   ROCKET VEHICLES
     . SATURN STAGES
     . . **SATURN S-1B STAGE**
  RT   LIQUID PROPELLANT ROCKET ENGINES

SATURN S-1C STAGE
  GS   ROCKET VEHICLES
     . SATURN STAGES
     . . **SATURN S-1C STAGE**
  RT   LIQUID PROPELLANT ROCKET ENGINES

SATURN S-2 STAGE
  GS   ROCKET VEHICLES
     . SATURN STAGES
     . . **SATURN S-2 STAGE**
  RT   LIQUID PROPELLANT ROCKET ENGINES

SATURN S-4 STAGE
  GS   ROCKET VEHICLES
     . SATURN STAGES
     . . **SATURN S-4 STAGE**
  RT   LIQUID PROPELLANT ROCKET ENGINES

SATURN S-4B STAGE
  GS   ROCKET VEHICLES
     . SATURN STAGES
     . . **SATURN S-4B STAGE**
  RT   LIQUID PROPELLANT ROCKET ENGINES

SATURN SATELLITES
  GS   SATELLITES
     . NATURAL SATELLITES
     . . **SATURN SATELLITES**
     . . . DIONE

SATURN SATELLITES-*(CONT.)*
     . . . ENCELADUS
     . . . HYPERION
     . . . IAPETUS
     . . . JANUS
     . . . MIMAS
     . . . RHEA (ASTRONOMY)
     . . . TETHYS
     . . . TITAN

SATURN STAGES
  GS   ROCKET VEHICLES
     . **SATURN STAGES**
     . . SATURN S-1 STAGE
     . . SATURN S-1B STAGE
     . . SATURN S-1C STAGE
     . . SATURN S-2 STAGE
     . . SATURN S-4 STAGE
     . . SATURN S-4B STAGE
  RT   LIQUID PROPELLANT ROCKET ENGINES

SATURN WORKSHOPS
  GS   **SATURN WORKSHOPS**
     . SATURN 1 WORKSHOP
     . SATURN 5 WORKSHOP
  RT   AIRLOCK MODULES
      APOLLO APPLICATIONS PROGRAM
      APOLLO PROJECT
      MULTIPLE DOCKING ADAPTERS
      SKYLAB PROGRAM
      SPACE STATIONS

SATURN 1 LAUNCH VEHICLES
  GS   LAUNCH VEHICLES
     . SATURN LAUNCH VEHICLES
     . . **SATURN 1 LAUNCH VEHICLES**
     . . . SATURN 1 SA-1 LAUNCH VEHICLE
     . . . SATURN 1 SA-2 LAUNCH VEHICLE
     . . . SATURN 1 SA-3 LAUNCH VEHICLE
     . . . SATURN 1 SA-4 LAUNCH VEHICLE
     . . . SATURN 1 SA-5 LAUNCH VEHICLE
     . . . SATURN 1 SA-6 LAUNCH VEHICLE
     . . . SATURN 1 SA-7 LAUNCH VEHICLE
     . . . SATURN 1 SA-8 LAUNCH VEHICLE
     . . . SATURN 1 SA-9 LAUNCH VEHICLE
     . . . SATURN 1 SA-10 LAUNCH VEHICLE
      ROCKET VEHICLES
     . MULTISTAGE ROCKET VEHICLES
     . . SATURN LAUNCH VEHICLES
     . . . **SATURN 1 LAUNCH VEHICLES**
     . . . . SATURN 1 SA-1 LAUNCH VEHICLE
     . . . . SATURN 1 SA-2 LAUNCH VEHICLE
     . . . . SATURN 1 SA-3 LAUNCH VEHICLE
     . . . . SATURN 1 SA-4 LAUNCH VEHICLE
     . . . . SATURN 1 SA-5 LAUNCH VEHICLE
     . . . . SATURN 1 SA-6 LAUNCH VEHICLE
     . . . . SATURN 1 SA-7 LAUNCH VEHICLE
     . . . . SATURN 1 SA-8 LAUNCH VEHICLE
     . . . . SATURN 1 SA-9 LAUNCH VEHICLE
     . . . . SATURN 1 SA-10 LAUNCH
         VEHICLE
  RT   H-1 ENGINE
      M-1 ENGINE

SATURN 1 SA-1 LAUNCH VEHICLE
  GS   LAUNCH VEHICLES
     . SATURN LAUNCH VEHICLES
     . . SATURN 1 LAUNCH VEHICLES
     . . . **SATURN 1 SA-1 LAUNCH VEHICLE**
      ROCKET VEHICLES
     . MULTISTAGE ROCKET VEHICLES
     . . SATURN LAUNCH VEHICLES
     . . . SATURN 1 LAUNCH VEHICLES
     . . . . **SATURN 1 SA-1 LAUNCH VEHICLE**

SATURN 1 SA-2 LAUNCH VEHICLE
  GS   LAUNCH VEHICLES
     . SATURN LAUNCH VEHICLES
     . . SATURN 1 LAUNCH VEHICLES
     . . . **SATURN 1 SA-2 LAUNCH VEHICLE**
      ROCKET VEHICLES
     . MULTISTAGE ROCKET VEHICLES
     . . SATURN LAUNCH VEHICLES
     . . . SATURN 1 LAUNCH VEHICLES
     . . . . **SATURN 1 SA-2 LAUNCH VEHICLE**

SATURN 1 SA-3 LAUNCH VEHICLE
  GS   LAUNCH VEHICLES
     . SATURN LAUNCH VEHICLES
     . . SATURN 1 LAUNCH VEHICLES
     . . . **SATURN 1 SA-3 LAUNCH VEHICLE**
      ROCKET VEHICLES
     . MULTISTAGE ROCKET VEHICLES
     . . SATURN LAUNCH VEHICLES
     . . . SATURN 1 LAUNCH VEHICLES

SATURN 1 SA-3 LAUNCH VEHICLE-(CONT.)
. . . . SATURN 1 SA-3 LAUNCH VEHICLE

SATURN 1 SA-4 LAUNCH VEHICLE
GS   LAUNCH VEHICLES
. SATURN LAUNCH VEHICLES
. . SATURN 1 LAUNCH VEHICLES
. . . SATURN 1 SA-4 LAUNCH VEHICLE
ROCKET VEHICLES
. MULTISTAGE ROCKET VEHICLES
. . SATURN LAUNCH VEHICLES
. . . SATURN 1 LAUNCH VEHICLES
. . . . SATURN 1 SA-4 LAUNCH VEHICLE

SATURN 1 SA-5 LAUNCH VEHICLE
GS   LAUNCH VEHICLES
. SATURN LAUNCH VEHICLES
. . SATURN 1 LAUNCH VEHICLES
. . . SATURN 1 SA-5 LAUNCH VEHICLE
ROCKET VEHICLES
. MULTISTAGE ROCKET VEHICLES
. . SATURN LAUNCH VEHICLES
. . . SATURN 1 LAUNCH VEHICLES
. . . . SATURN 1 SA-5 LAUNCH VEHICLE

SATURN 1 SA-6 LAUNCH VEHICLE
GS   LAUNCH VEHICLES
. SATURN LAUNCH VEHICLES
. . SATURN 1 LAUNCH VEHICLES
. . . SATURN 1 SA-6 LAUNCH VEHICLE
ROCKET VEHICLES
. MULTISTAGE ROCKET VEHICLES
. . SATURN LAUNCH VEHICLES
. . . SATURN 1 LAUNCH VEHICLES
. . . . SATURN 1 SA-6 LAUNCH VEHICLE

SATURN 1 SA-7 LAUNCH VEHICLE
GS   LAUNCH VEHICLES
. . SATURN LAUNCH VEHICLES
. . SATURN 1 LAUNCH VEHICLES
. . . SATURN 1 SA-7 LAUNCH VEHICLE
ROCKET VEHICLES
. MULTISTAGE ROCKET VEHICLES
. . SATURN LAUNCH VEHICLES
. . . SATURN 1 LAUNCH VEHICLES
. . . . SATURN 1 SA-7 LAUNCH VEHICLE

SATURN 1 SA-8 LAUNCH VEHICLE
GS   LAUNCH VEHICLES
. . SATURN LAUNCH VEHICLES
. . SATURN 1 LAUNCH VEHICLES
. . . SATURN 1 SA-8 LAUNCH VEHICLE
ROCKET VEHICLES
. MULTISTAGE ROCKET VEHICLES
. . SATURN LAUNCH VEHICLES
. . SATURN 1 LAUNCH VEHICLES
. . . . SATURN 1 SA-8 LAUNCH VEHICLE

SATURN 1 SA-9 LAUNCH VEHICLE
GS   LAUNCH VEHICLES
. . SATURN LAUNCH VEHICLES
. . SATURN 1 LAUNCH VEHICLES
. . . SATURN 1 SA-9 LAUNCH VEHICLE
ROCKET VEHICLES
. MULTISTAGE ROCKET VEHICLES
. . SATURN LAUNCH VEHICLES
. . SATURN 1 LAUNCH VEHICLES
. . . . SATURN 1 SA-9 LAUNCH VEHICLE

SATURN 1 SA-10 LAUNCH VEHICLE
GS   LAUNCH VEHICLES
. SATURN LAUNCH VEHICLES
. . SATURN 1 LAUNCH VEHICLES
. . . SATURN 1 SA-10 LAUNCH VEHICLE
ROCKET VEHICLES
. MULTISTAGE ROCKET VEHICLES
. . SATURN LAUNCH VEHICLES
. . . SATURN 1 LAUNCH VEHICLES
. . . . SATURN 1 SA-10 LAUNCH
VEHICLE

SATURN 1 WORKSHOP
GS   SATURN WORKSHOPS
. SATURN 1 WORKSHOP
RT   AIRLOCK MODULES
APOLLO APPLICATIONS PROGRAM
APOLLO PROJECT
MULTIPLE DOCKING ADAPTERS
SKYLAB PROGRAM
SPACE STATIONS

SATURN 1B LAUNCH VEHICLES
GS   LAUNCH VEHICLES
. SATURN LAUNCH VEHICLES

SATURN 1B LAUNCH VEHICLES-(CONT.)
. . SATURN 1B LAUNCH VEHICLES
ROCKET VEHICLES
. MULTISTAGE ROCKET VEHICLES
. . SATURN LAUNCH VEHICLES
. . . SATURN 1B LAUNCH VEHICLES
RT   H-1 ENGINE
J-2 ENGINE
M-1 ENGINE
SKYLAB 2
SKYLAB 3
SKYLAB 4

SATURN 2 LAUNCH VEHICLES
GS   LAUNCH VEHICLES
. SATURN LAUNCH VEHICLES
. . SATURN 2 LAUNCH VEHICLES
ROCKET VEHICLES
. MULTISTAGE ROCKET VEHICLES
. . SATURN LAUNCH VEHICLES
. . . SATURN 2 LAUNCH VEHICLES

SATURN 5 LAUNCH VEHICLES
GS   LAUNCH VEHICLES
. SATURN LAUNCH VEHICLES
. . SATURN 5 LAUNCH VEHICLES
ROCKET VEHICLES
. MULTISTAGE ROCKET VEHICLES
. . SATURN LAUNCH VEHICLES
. . . SATURN 5 LAUNCH VEHICLES
RT   J-2 ENGINE
SKYLAB 2
SKYLAB 3
SKYLAB 4

SATURN 5 WORKSHOP
GS   SATURN WORKSHOPS
. SATURN 5 WORKSHOP
RT   AIRLOCK MODULES
APOLLO APPLICATIONS PROGRAM
APOLLO PROJECT
MULTIPLE DOCKING ADAPTERS
SKYLAB PROGRAM
SPACE STATIONS

SAUDI ARABIA
GS   NATIONS
. SAUDI ARABIA
RT   ASIA
SAUDI ARABIAN SPACE PROGRAM

SAUDI ARABIAN SPACE PROGRAM
GS   PROGRAMS
. SPACE PROGRAMS
. . SAUDI ARABIAN SPACE PROGRAM
RT   ARABSAT
ARCOMSAT
SAUDI ARABIA
SPACE SHUTTLE MISSION 51-G

SAVAGE AIRCRAFT
USE   A-2 AIRCRAFT

SAVANNAH NUCLEAR SHIP
GS   SURFACE VEHICLES
. CARGO SHIPS
. . SAVANNAH NUCLEAR SHIP
. NUCLEAR POWERED SHIPS
. . SAVANNAH NUCLEAR SHIP
WATER VEHICLES
. SHIPS
. . CARGO SHIPS
. . . SAVANNAH NUCLEAR SHIP
. . NUCLEAR POWERED SHIPS
. . . SAVANNAH NUCLEAR SHIP
RT   MARINE PROPULSION
NUCLEAR PROPULSION

SAVANNAHS
USE   GRASSLANDS

SAWS
GS   CUTTERS
. SAWS
TOOLS
. SAWS
RT   MACHINE TOOLS
SHEARS

SAWTOOTH WAVEFORMS
GS   WAVEFORMS
. SAWTOOTH WAVEFORMS
RT   PULSE AMPLITUDE
PULSE DURATION

SAWTOOTH WAVEFORMS-(CONT.)
SQUARE WAVES

SC-1 AIRCRAFT
UF   SHORT SC-1 AIRCRAFT
GS   JET AIRCRAFT
. SC-1 AIRCRAFT
MONOPLANES
. SC-1 AIRCRAFT
RESEARCH AIRCRAFT
. SC-1 AIRCRAFT
TAILLESS AIRCRAFT
. SC-1 AIRCRAFT
V/STOL AIRCRAFT
. VERTICAL TAKEOFF AIRCRAFT
. . SC-1 AIRCRAFT
RT   ∞ AIRCRAFT

SC-5 AIRCRAFT
UF   BELFAST AIRCRAFT
SHORT BELFAST C MK-1 AIRCRAFT
SHORT SC-5 AIRCRAFT
GS   JET AIRCRAFT
. TURBOPROP AIRCRAFT
. . SC-5 AIRCRAFT
MONOPLANES
. SC-5 AIRCRAFT
TRANSPORT AIRCRAFT
. SC-5 AIRCRAFT
RT   ∞ AIRCRAFT

SC-7 AIRCRAFT
UF   SHORT SC-7 AIRCRAFT
SKYVAN AIRCRAFT
TURBO-SKYVAN AIRCRAFT
GS   LIGHT AIRCRAFT
. SC-7 AIRCRAFT
MONOPLANES
. SC-7 AIRCRAFT
TRANSPORT AIRCRAFT
. SC-7 AIRCRAFT
RT   ∞ AIRCRAFT
CARGO AIRCRAFT
PASSENGER AIRCRAFT

SCALAR MAGNETIC CHARGE
USE   MAGNETIC CHARGE DENSITY

SCALARS
RT   TENSOR ANALYSIS
TENSORS

∞ SCALE
SN   (USE OF A MORE SPECIFIC TERM IS
RECOMMENDED--CONSULT THE TERMS
LISTED BELOW)
RT   SCALE (CORROSION)
SCALE (RATIO)
TEMPERATURE SCALES
WEIGHT INDICATORS

SCALE (CORROSION)
GS   CORROSION
. SCALE (CORROSION)
RT   CHEMICAL ATTACK
DEGRADATION
DESCALING
HOT CORROSION
PICKLING (METALLURGY)
RUSTING
∞ SCALE
∞ SCALING

SCALE (RATIO)
GS   RATIOS
. SCALE (RATIO)
RT   MAPPING
RETICLES
∞ SCALE

SCALE EFFECT
RT   ∞ EFFECTS
FORCE DISTRIBUTION
PARAMETERIZATION
REYNOLDS NUMBER
∞ SCALING

SCALE HEIGHT *
GS   DIMENSIONS
. HEIGHT
. . SCALE HEIGHT
RT   EARTH ATMOSPHERE
GEOPOTENTIAL HEIGHT
HEAD (FLUID MECHANICS)

SCALE MODELS
- GS   MODELS
  - . SCALE MODELS
- RT   AERODYNAMIC CONFIGURATIONS
  - AIRCRAFT MODELS
  - REYNOLDS EQUATION
  - SCALING LAWS
  - SEMISPAN MODELS
  - SIMILARITY THEOREM
  - SIMILITUDE LAW
  - SPACECRAFT MODELS
  - WIND TUNNEL MODELS

SCALERS
- GS   CIRCUITS
  - . COUNTING CIRCUITS
  - . . SCALERS
- RT   TENSOR ANALYSIS

∞ SCALING
- SN   (USE OF A MORE SPECIFIC TERM IS RECOMMENDED--CONSULT THE TERMS LISTED BELOW)
- RT   CALIBRATING
  - ERRORS
  - SCALE (CORROSION)
  - SCALE EFFECT
  - SCALING LAWS

SCALING LAWS
- GS   LAWS
  - . SCALING LAWS
- RT   DIMENSIONAL ANALYSIS
  - DIMENSIONLESS NUMBERS
  - SCALE MODELS
  - ∞SCALING
  - SIMILARITY NUMBERS
  - WIND TUNNEL CALIBRATION

SCALLOPING
- RT   EDGES
  - ELECTRON BEAMS
  - TRAVELING WAVE TUBES

SCANDINAVIA
- RT   DENMARK
  - FINLAND
  - NORWAY
  - SWEDEN

SCANDIUM
- GS   CHEMICAL ELEMENTS
  - . RARE EARTH ELEMENTS
  - . . SCANDIUM
  - . . . SCANDIUM ISOTOPES
  - METALS
  - . RARE EARTH ELEMENTS
  - . . SCANDIUM
  - . . . SCANDIUM ISOTOPES
  - . TRANSITION METALS
  - . . SCANDIUM
  - . . . SCANDIUM ISOTOPES

SCANDIUM COMPOUNDS
- GS   RARE EARTH COMPOUNDS
  - . SCANDIUM COMPOUNDS
  - . . SCANDIUM OXIDES
- RT   ∞CHEMICAL COMPOUNDS
  - ∞GROUP 3B COMPOUNDS
  - ∞METAL COMPOUNDS

SCANDIUM ISOTOPES
- UF   SCANDIUM 46
- GS   CHEMICAL ELEMENTS
  - . NUCLIDES
  - . . ISOTOPES
  - . . . SCANDIUM ISOTOPES
  - . RARE EARTH ELEMENTS
  - . . SCANDIUM
  - . . . SCANDIUM ISOTOPES
  - METALS
  - . RARE EARTH ELEMENTS
  - . . SCANDIUM
  - . . . SCANDIUM ISOTOPES
  - . TRANSITION METALS
  - . . SCANDIUM
  - . . . SCANDIUM ISOTOPES

SCANDIUM OXIDES
- GS   CHALCOGENIDES
  - . OXIDES
  - . . METAL OXIDES
  - . . . SCANDIUM OXIDES
  - RARE EARTH COMPOUNDS

SCANDIUM OXIDES-(CONT.)
  - . SCANDIUM COMPOUNDS
  - . . SCANDIUM OXIDES

SCANDIUM 46
- USE   SCANDIUM ISOTOPES

SCANNER PROJECT
- GS   PROGRAMS
  - . PROJECTS
  - . . SCANNER PROJECT
- RT   HORIZON SCANNERS
  - INFRARED SCANNERS
  - OPTICAL EQUIPMENT

SCANNERS
- UF   SCANNING DEVICES
- GS   SCANNERS
  - . COASTAL ZONE COLOR SCANNER
  - . HORIZON SCANNERS
  - . INFRARED SCANNERS
  - . OPTICAL SCANNERS
  - . . FLYING SPOT SCANNERS
  - . . MULTISPECTRAL BAND SCANNERS
  - . ULTRASONIC SCANNERS
- RT   CONICAL SCANNING
  - OPTICAL DATA PROCESSING
  - OPTICAL EQUIPMENT
  - PANORAMIC SCANNING
  - READING
  - SCANNING
  - SUBREFLECTORS

SCANNING
- GS   SCANNING
  - . CONICAL SCANNING
  - . FREQUENCY SCANNING
  - . PANORAMIC SCANNING
  - . RADAR SCANNING
  - . . RADAR TARGET SCATTER SITE PROGRAM
- RT   EARTH RESOURCES
  - EROS (SATELLITES)
  - EXAMINATION
  - MONITORS
  - MULTISPECTRAL BAND SCANNERS
  - RAPID BALLISTICS IDENTIFICATION
  - READING
  - SCANNERS
  - SEARCHING
  - SURVEILLANCE
  - ULTRASONIC SCANNERS

SCANNING DEVICES
- USE   SCANNERS

SCANNING LASER ACOUSTIC MICROSCOPE (SLAM)
- USE   ACOUSTIC MICROSCOPES

SCAPULA
- GS   ANATOMY
  - . MUSCULOSKELETAL SYSTEM
  - . . BONES
  - . . . SCAPULA
- RT   ARM (ANATOMY)
  - SHOULDERS

SCAR PROGRAM
- USE   SUPERSONIC CRUISE AIRCRAFT RESEARCH

SCARFING
- GS   CUTTING
  - . SCARFING
- RT   CLEANING
  - GRINDING (MATERIAL REMOVAL)
  - METAL CUTTING
  - SLICING

SCARPS
- USE   ESCARPMENTS

SCARS
- GS   TISSUES (BIOLOGY)
  - . SCARS

SCARS (GEOLOGY)
- USE   EROSION

SCAT
- USE   SUPERSONIC COMMERCIAL AIR TRANSPORT

SCATHA SATELLITE
- UF   P78-2 SATELLITE
- GS   SATELLITES
  - . ARTIFICIAL SATELLITES
  - . . SCATHA SATELLITE
- RT   ∞CHARGING
  - ELECTRIC CHARGE
  - ELECTROMAGNETIC INTERFERENCE
  - ELECTROSTATIC CHARGE
  - ELECTROSTATIC PROBES
  - ELECTROSTATIC SHIELDING
  - HIGH ENERGY ELECTRONS

SCATTER PLATES (OPTICS)
- GS   OPTICAL EQUIPMENT
  - . SCATTER PLATES (OPTICS)
- RT   BEAM SPLITTERS
  - COHERENT LIGHT
  - HOLOGRAPHIC INTERFEROMETRY
  - HOLOGRAPHY
  - INTERFEROMETRY
  - LIGHT SCATTERING
  - ∞OPTICS
  - ∞PLATES

SCATTER PROPAGATION
- GS   TRANSMISSION
  - . ELECTROMAGNETIC WAVE TRANSMISSION
  - . . SCATTER PROPAGATION
  - . . . IONOSPHERIC F-SCATTER PROPAGATION
  - . WAVE PROPAGATION
  - . . SCATTER PROPAGATION
  - . . . IONOSPHERIC F-SCATTER PROPAGATION
- RT   BACKSCATTERING
  - FORWARD SCATTERING
  - IONOSPHERIC PROPAGATION
  - METEOR TRAILS
  - RADIO RECEPTION
  - RADIO SCATTERING
  - RADIO TRANSMISSION
  - RADIO WAVES

SCATTERERS
- USE   SCATTERING

SCATTERING
- UF   SCATTERERS
- GS   SCATTERING
  - . BACKSCATTERING
  - . COHERENT SCATTERING
  - . COMPTON EFFECT
  - . ELASTIC SCATTERING
  - . ELECTRON SCATTERING
  - . . CONFIGURATION INTERACTION
  - . . ELECTRON RUNAWAY (PLASMA PHYSICS)
  - . FORWARD SCATTERING
  - . INCOHERENT SCATTERING
  - . INELASTIC SCATTERING
  - . INVERSE SCATTERING
  - . ION SCATTERING
  - . NUCLEAR SCATTERING
  - . . NEUTRON SCATTERING
  - . . RESONANCE SCATTERING
  - . NUCLEON-NUCLEON SCATTERING
  - . PROTON SCATTERING
  - . RADAR SCATTERING
  - . WAVE SCATTERING
  - . . ACOUSTIC SCATTERING
  - . . . REVERBERATION
  - . . ATMOSPHERIC SCATTERING
  - . . . TROPOSPHERIC SCATTERING
  - . . ELECTROMAGNETIC SCATTERING
  - . . . IONOSPHERIC F-SCATTER PROPAGATION
  - . . . LIGHT SCATTERING
  - . . . . HALOS
  - . . . MICROWAVE SCATTERING
  - . . . MIE SCATTERING
  - . . . . RAYLEIGH SCATTERING
  - . . . . RAMAN SPECTRA
  - . . . THOMSON SCATTERING
  - . . . X RAY SCATTERING
- RT   ATOMIC COLLISIONS
  - BISTATIC REFLECTIVITY
  - CIRCUMSOLAR RADIATION
  - COLLISION PARAMETERS
  - COLLISIONS
  - DEEP SCATTERING LAYERS
  - DEFLECTION
  - DIFFUSION
  - DISPERSING

SCATTERING-*(CONT.)*
    ELECTROMAGNETIC RADIATION
    ENCOUNTERS
    HUYGENS PRINCIPLE
    IMPINGEMENT
    INCIDENT RADIATION
    INELASTIC COLLISIONS
    MEAN FREE PATH
    PARTICLE COLLISIONS
    POMERONS
    REFLECTION
    RELEASING
    SCATTEROMETERS
    SHOCK WAVE INTERACTION
    SPREAD REFLECTION
    SPREADING
    SPRINKLING
    STATISTICAL DISTRIBUTIONS
    TRANSMITTANCE
    WAVE DEGRADATION
    WAVE DISPERSION
    WAVE INTERACTION

SCATTERING AMPLITUDE
  GS   AMPLITUDES
    . SCATTERING AMPLITUDE
  RT   FADDEEV EQUATIONS
    WAVE SCATTERING

SCATTERING COEFFICIENTS
  GS   COEFFICIENTS
    . SCATTERING COEFFICIENTS
  RT   ABSORPTIVITY
    ATTENUATION COEFFICIENTS
    FORM FACTORS

SCATTERING CROSS SECTIONS
  RT   ABSORPTION CROSS SECTIONS
    BARYON RESONANCE
    BORN APPROXIMATION
    ∞CROSS SECTIONS
    ELECTRON RUNAWAY (PLASMA
      PHYSICS)
    IONIZATION CROSS SECTIONS
    NEUTRON CROSS SECTIONS
    POMERANCHUK THEOREM
    POMERONS
    RAMSAUER EFFECT
    REGGE POLES
    S MATRIX THEORY
    STOPPING POWER

SCATTERING FUNCTIONS
  RT   FLUX DENSITY
    ∞FUNCTIONS
    RADIANT FLUX DENSITY

SCATTERING MATRIX
  USE S MATRIX THEORY

SCATTEROMETERS
  GS   MEASURING INSTRUMENTS
    . SCATTEROMETERS
  RT   ∞INSTRUMENTS
    MICROWAVE SCATTERING
    MICROWAVES
    RADAR
    RADAR SCATTERING
    SCATTERING
    WAVE SCATTERING

SCAVENGING
  RT   CLEANING
    DEGASSING
    DEOXIDIZING
    PURIFICATION

SCCF
  USE SOLAR CELL CALIBRATION FACILITY

SCENE ANALYSIS
  GS   DATA PROCESSING
    . OPTICAL DATA PROCESSING
    .. SCENE ANALYSIS
  RT   CHANGE DETECTION
    CHARACTER RECOGNITION
    FEATURE IDENTIFICATION AND
      LOCATION EXPER
    IMAGE ANALYSIS
    IMAGERY
    IMAGING TECHNIQUES
    VIDEO LANDMARK ACQUISITION AND
      TRACKING

SCENEDESMUS
  GS   ALGAE
    . SCENEDESMUS

SCF
  USE SELF CONSISTENT FIELDS

SCHACH EFFECT
  RT   CELESTIAL MECHANICS
    ∞EFFECTS
    ORBIT PERTURBATION
    PERTURBATION
    SATELLITE PERTURBATION

SCHAUDER FIXPOINT THEOREM
  GS   THEOREMS
    . SCHAUDER FIXPOINT THEOREM
  RT   COMPLEX VARIABLES
    DIFFERENTIAL EQUATIONS

SCHEDULES
  GS   SCHEDULES
    . COUNTDOWN
  RT   CONTRACT MANAGEMENT
    PRECISION
    PREDICTIONS
    PRODUCTION PLANNING
    TIME
    TIME LAG
    TURNAROUND (STS)

SCHEDULING
  GS   SCHEDULING
    . PREDICTION ANALYSIS TECHNIQUES
  RT   CALENDARS
    CONSECUTIVE EVENTS
    CONTINUITY
    ∞CONTROL
    CROP CALENDARS
    DECISION THEORY
    FORECASTING
    FORMALISM
    LATENESS
    MATHEMATICAL MODELS
    MATRIX MANAGEMENT
    MISSION PLANNING
    OPTIMIZATION
    PRODUCTION ENGINEERING
    QUALITY CONTROL
    SEQUENCING
    TASK COMPLEXITY
    TASKS
    TIME SERIES ANALYSIS

SCHEELITE
  GS   CALCIUM COMPOUNDS
    . SCHEELITE
    CHALCOGENIDES
    . OXIDES
    .. METAL OXIDES
    ... TUNGSTEN OXIDES
    .... SCHEELITE
    MINERALS
    . SCHEELITE
    TUNGSTEN COMPOUNDS
    . TUNGSTEN OXIDES
    .. SCHEELITE

SCHELKUNOFF PRINCIPLE
  RT   ANTENNA RADIATION PATTERNS
    HORN ANTENNAS
    HUYGENS PRINCIPLE
    REFLECTOMETERS
    REFLECTORS

SCHEMATICS
  USE CIRCUIT DIAGRAMS

SCHIFF BASES
  USE IMINES

SCHIST
  GS   ROCKS
    . SCHIST
  RT   LIMESTONE
    SANDSTONES

SCHIZOPHRENIA
  GS   PSYCHOSES
    . SCHIZOPHRENIA
  RT   ∞DEPRESSION
    IRRATIONALITY
    MENTAL HEALTH

SCHLEICHER AIRCRAFT
  RT   ∞AIRCRAFT

SCHLEICHER KA-6 SAILPLANE
  USE KA-6 SAILPLANES

SCHLIEREN PHOTOGRAPHY
  GS   IMAGERY
    . SHADOWGRAPH PHOTOGRAPHY
    .. SCHLIEREN PHOTOGRAPHY
    PHOTOGRAPHY
    . SHADOWGRAPH PHOTOGRAPHY
    .. SCHLIEREN PHOTOGRAPHY
  RT   BLACK AND WHITE PHOTOGRAPHY
    FLOW VISUALIZATION
    MACH-ZEHNDER INTERFEROMETERS
    MOIRE EFFECTS

SCHMIDT CAMERAS
  GS   OPTICAL EQUIPMENT
    . CAMERAS
    .. SCHMIDT CAMERAS
    PHOTOGRAPHIC EQUIPMENT
    . CAMERAS
    .. SCHMIDT CAMERAS
    TELESCOPES
    . SCHMIDT CAMERAS
  RT   ASTRONOMICAL PHOTOGRAPHY
    ASTRONOMICAL TELESCOPES
    BAKER-NUNN CAMERA

SCHMIDT METHOD
  RT   DIFFERENTIAL EQUATIONS
    INTEGRAL EQUATIONS
    ∞METHODOLOGY
    REAL VARIABLES

SCHMIDT NUMBER
  GS   RATIOS
    . DIMENSIONLESS NUMBERS
    .. SCHMIDT NUMBER
  RT   NUSSELT NUMBER
    PRANDTL NUMBER

SCHMIDT TELESCOPES
  GS   TELESCOPES
    . SCHMIDT TELESCOPES
  RT   ASTRONOMICAL TELESCOPES
    REFLECTING TELESCOPES

SCHOOLS
  RT   EDUCATION
    INSTRUCTORS
    TRAINING EVALUATION
    UNIVERSITIES

SCHOOLS (FISH)
  GS   ANIMALS
    . VERTEBRATES
    .. FISHES
    ... SCHOOLS (FISH)
  RT   ICHTHYOLOGY

SCHOTTKY BARRIER DIODES
  USE SCHOTTKY DIODES

SCHOTTKY DIODES
  UF   SCHOTTKY BARRIER DIODES
  GS   ELECTRONIC EQUIPMENT
    . DIODES
    .. SEMICONDUCTOR DIODES
    ... SCHOTTKY DIODES
    . SOLID STATE DEVICES
    .. SEMICONDUCTOR DEVICES
    ... SCHOTTKY DIODES
  RT   ∞BARRIERS
    BARRITT DIODES
    GALLIUM ARSENIDES
    N-TYPE SEMICONDUCTORS
    SEMICONDUCTOR JUNCTIONS
    SILICON
    SIS (SEMICONDUCTORS)
    WORK FUNCTIONS
    ZINC SELENIDES

SCHOTTKY EFFECT
  USE WORK FUNCTIONS

SCHREIBERSITE
  GS   IRON COMPOUNDS
    . SCHREIBERSITE
    MINERALS
    . SCHREIBERSITE
    NICKEL COMPOUNDS

SCHREIBERSITE-*(CONT.)*
. **SCHREIBERSITE**
PHOSPHORUS COMPOUNDS
. PHOSPHIDES
. . **SCHREIBERSITE**
RT   IRON METEORITES
METEORITIC COMPOSITION
STONY METEORITES

SCHROEDINGER EQUATION
GS   WAVE EQUATIONS
. **SCHROEDINGER EQUATION**
RT   ∞EQUATIONS
WENTZEL-KRAMER-BRILLOUIN METHOD

SCHULER TUNING
GS   TUNING
. **SCHULER TUNING**
RT   GYROSCOPIC PENDULUMS
GYROSCOPIC STABILITY
INERTIAL NAVIGATION

SCHUMANN-RUNGE BANDS
GS   SPECTRA
. SPECTRAL BANDS
. . **SCHUMANN-RUNGE BANDS**
RT   ABSORPTION SPECTRA
∞BANDS
EMISSION SPECTRA
HERZBERG BANDS
OXYGEN
QUANTUM THEORY

SCHWARTZ INEQUALITY
GS   INEQUALITIES
. **SCHWARTZ INEQUALITY**
RT   ALGEBRA
LINEAR TRANSFORMATIONS
VECTORS (MATHEMATICS)

SCHWARTZ METHOD
GS   STRESS ANALYSIS
. **SCHWARTZ METHOD**
RT   ∞METHODOLOGY

SCHWARZ-CHRISTOFFEL TRANSFORMATION
GS   ANALYSIS (MATHEMATICS)
. COMPLEX VARIABLES
. . **SCHWARZ-CHRISTOFFEL**
TRANSFORMATION
FUNCTIONS (MATHEMATICS)
. **SCHWARZ-CHRISTOFFEL**
TRANSFORMATION
RT   CONFORMAL MAPPING

SCHWARZSCHILD ANTENNAS
GS   ANTENNAS
. **SCHWARZSCHILD ANTENNAS**
RT   HORNS
PARABOLIC REFLECTORS
RADAR ANTENNAS
RADIO ANTENNAS

SCHWARZSCHILD METRIC
RT   BIMETRIC THEORIES
COORDINATE TRANSFORMATIONS
ESCAPE VELOCITY
GRAVITATIONAL FIELDS
IONIZATION
LIGHT SPEED
ORBITALS
ORBITS
RELATIVITY

SCHWASSMANN-WACHMANN COMET
GS   CELESTIAL BODIES
. COMETS
. . **SCHWASSMANN-WACHMANN COMET**

SCIATIC REGION
GS   ANATOMY
. MUSCULOSKELETAL SYSTEM
. . BONES
. . . **SCIATIC REGION**

∞ **SCIENCE**
SN   *(USE OF A MORE SPECIFIC TERM IS*
RECOMMENDED--CONSULT THE TERMS
LISTED BELOW)
RT   ACOUSTICS
AEROACOUSTICS
AERODYNAMICS
∞AERONAUTICS
AEROSPACE MEDICINE

SCIENCE-*(CONT.)*
AEROTHERMODYNAMICS
ALGEBRA
ANTHROPOLOGY
ASTRODYNAMICS
ASTRONOMY
ASTROPHYSICS
ATMOSPHERIC PHYSICS
ATOMIC PHYSICS
BIOACOUSTICS
BIOASTRONAUTICS
BIODYNAMICS
∞BIOLOGY
BIOPHYSICS
BOTANY
CLOUD PHYSICS
COMBUSTION PHYSICS
COMPUTATIONAL ASTROPHYSICS
ELECTROPHYSICS
ELECTROPHYSIOLOGY
ENTOMOLOGY
FLUID DYNAMICS
FLUID MECHANICS
GEOLOGY
GEOMETRY
GEOPHYSICS
HEALTH PHYSICS
HELIOSEISMOLOGY
HYDROGEOLOGY
HYDROMECHANICS
LIFE SCIENCES
LOW TEMPERATURE PHYSICS
MARINE BIOLOGY
MARINE CHEMISTRY
MARINE METEOROLOGY
∞MATERIALS SCIENCE
∞MATHEMATICS
∞MECHANICS (PHYSICS)
MEDICAL SCIENCE
∞METALLURGY
METEOROLOGY
∞MOLECULAR PHYSICS
NEUROPHYSIOLOGY
NEUTRON PHYSICS
NUCLEAR PHYSICS
OCEANOGRAPHY
∞OPTICS
∞PHYSICAL SCIENCES
∞PHYSICS
PHYSIOCHEMISTRY
PHYSIOLOGY
PLASMA PHYSICS
POLYMER PHYSICS
PSYCHOPHYSICS
PSYCHOPHYSIOLOGY
RADIO ASTRONOMY
RADIO PHYSICS
REACTOR PHYSICS
REENTRY PHYSICS
RESPIRATORY PHYSIOLOGY
SEISMOLOGY
SOLAR DIAMETER
SOLAR PHYSICS
SOLID MECHANICS
∞SOLID STATE PHYSICS
STELLAR PHYSICS
SUNRISE
SUNSET
TAXONOMY
THEORETICAL PHYSICS
TRIGONOMETRY
UNDERWATER PHYSIOLOGY
∞ZOOLOGY

SCIENTIFIC INSTRUMENT MODULES
USE   SIM

SCIENTIFIC SATELLITES
GS   SATELLITES
. ARTIFICIAL SATELLITES
. . **SCIENTIFIC SATELLITES**
. . . AMPTE (SATELLITES)
. . . ASTRONOMICAL SATELLITES
. . . ATS
. . . . ATS 1
. . . . ATS 2
. . . . ATS 3
. . . . ATS 4
. . . . ATS 5
. . . . ATS 6
. . . . ATS 7
. . . . ATS 8
. . . AZUR SATELLITE
. . . CANNONBALL 2 SATELLITE
. . . DIAL SATELLITE

SCIENTIFIC SATELLITES-*(CONT.)*
. . . ENVIRONMENTAL RESEARCH
SATELLITES
. . . . ERS 17
. . . . ERS 18
. . . . INTASAT SATELLITE
. . . EXOSAT SATELLITE
. . . EXPLORER 45 SATELLITE
. . . GRAVSAT SATELLITE
. . . HAWKEYE SATELLITES
. . . LZEEBE SATELLITE
. . . MAGSAT A SATELLITE
. . . MAGSAT B SATELLITE
. . . MAGSAT SATELLITES
. . . MAGSAT 1 SATELLITE
. . . ORBIS
. . . . ORBIS CAL SATELLITE
. . . OV-1 SATELLITES
. . . OV-2 SATELLITES
. . . OV-3 SATELLITES
. . . OV-4 SATELLITES
. . . OV-5 SATELLITES
. . . SMALL SCIENTIFIC SATELLITES
. . . UK SATELLITES
. . . . UK 4 SATELLITE
RT   CANADIAN SPACE PROGRAM
ESRO 4 SATELLITE
EXPLORER 55 SATELLITE
TECHNOLOGY FEASIBILITY SPACECRAFT

SCIENTISTS
GS   MANPOWER
. **SCIENTISTS**
PERSONNEL
. **SCIENTISTS**
RT   AWARDS

SCIMITAR AIRCRAFT
UF   VICKERS SCIMITAR AIRCRAFT
GS   ATTACK AIRCRAFT
. FIGHTER AIRCRAFT
. . **SCIMITAR AIRCRAFT**
BAC AIRCRAFT
. **SCIMITAR AIRCRAFT**
JET AIRCRAFT
. **SCIMITAR AIRCRAFT**
MONOPLANES
. **SCIMITAR AIRCRAFT**
RT   ∞AIRCRAFT

SCINTILLATION
RT   GLINT
PHOSPHORESCENCE

SCINTILLATION COUNTERS
UF   SCINTILLATORS
SCINTILLOMETERS
GS   MEASURING INSTRUMENTS
. COUNTERS
. . RADIATION COUNTERS
. . . **SCINTILLATION COUNTERS**
. RADIATION MEASURING INSTRUMENTS
. . RADIATION COUNTERS
. . . **SCINTILLATION COUNTERS**
RT   CERENKOV COUNTERS
NEUTRON COUNTERS
PARTICLE TELESCOPES
PHOTOMULTIPLIER TUBES
PHOTOPEAK

SCINTILLATORS
USE   SCINTILLATION COUNTERS

SCINTILLOMETERS
USE   SCINTILLATION COUNTERS

SCISSION
USE   CLEAVAGE

SCOOPS
RT   AIR INTAKES
CONVEYORS
DUCTS
INTAKE SYSTEMS
NOSE INLETS
SIDE INLETS
∞WATER INTAKES

SCOPOLAMINE
USE   HYOSCINE

SCORE OMNIRANGE
USE   SELF CALIBRATING OMNIRANGE

SCORE SATELLITE
GS   SATELLITES
      . ARTIFICIAL SATELLITES
      . . **SCORE SATELLITE**

SCORING
UF   SCRIBING
RT   ABRASION
      DEFECTS
      FRICTION
      PITTING
      WEAR

SCORPIO CONSTELLATION
USE   SCORPIUS CONSTELLATION

SCORPIUS CONSTELLATION
UF   SCORPIO CONSTELLATION
GS   CONSTELLATIONS
      . **SCORPIUS CONSTELLATION**
RT   ZODIAC

SCOTCHLITE (TRADEMARK)
RT   MEMBRANE STRUCTURES
      REFRACTORY MATERIALS

SCOTLAND
GS   NATIONS
      . UNITED KINGDOM
      . . **SCOTLAND**

SCOUT HELICOPTER
USE   P-531 HELICOPTER

SCOUT LAUNCH VEHICLE
GS   LAUNCH VEHICLES
      . **SCOUT LAUNCH VEHICLE**
      ROCKET VEHICLES
      . MULTISTAGE ROCKET VEHICLES
      . . **SCOUT LAUNCH VEHICLE**
RT   ALGOL ENGINE
      EXPLORER 9 SATELLITE
      EXPLORER 16 SATELLITE
      EXPLORER 19 SATELLITE
      EXPLORER 20 SATELLITE
      EXPLORER 22 SATELLITE
      EXPLORER 23 SATELLITE
      EXPLORER 24 SATELLITE
      EXPLORER 25 SATELLITE
      EXPLORER 27 SATELLITE
      EXPLORER 30 SATELLITE
      EXPLORER 37 SATELLITE
      EXPLORER 39 SATELLITE
      EXPLORER 40 SATELLITE
      SAN MARCO SATELLITES
      SOLID PROPELLANT ROCKET ENGINES
      TX-354 ENGINE
      X-248 ENGINE
      X-254 ENGINE
      X-258 ENGINES
      X-259 ENGINE
      XM-33 ENGINE

SCOUT PROJECT
GS   PROGRAMS
      . NASA PROGRAMS
      . . **SCOUT PROJECT**
      . NASA SPACE PROGRAMS
      . . **SCOUT PROJECT**
      . PROJECTS
      . . **SCOUT PROJECT**
RT   ∞ BOOSTERS
      EXPLORER SATELLITES
      LAUNCH VEHICLES

SCPC TRANSMISSION
USE   SINGLE CHANNEL PER CARRIER
      TRANSMISSION

SCR (RECTIFIERS)
USE   SILICON CONTROLLED RECTIFIERS

∞ **SCRAM**
SN   (USE OF A MORE SPECIFIC TERM IS
      RECOMMENDED--CONSULT THE TERMS
      LISTED BELOW)
RT   MISSILES
      SHUTDOWNS
      SUPERSONIC COMBUSTION RAMJET
      ENGINES

SCRAMBLING (COMMUNICATION)
RT   INTELLIGIBILITY
      SECURITY

SCRAMBLING (COMMUNICATION)-(CONT.)
      SIGNAL DISTORTION
      SIGNAL ENCODING
      VOCODERS
      VOICE COMMUNICATION

SCRAMJET ENGINES
USE   SUPERSONIC COMBUSTION RAMJET
      ENGINES

SCRAMJETS
USE   SUPERSONIC COMBUSTION RAMJET
      ENGINES

SCRAP
RT   CHIPS
      DEBRIS
      METAL PARTICLES
      WASTES

SCRAPERS
RT   CUTTERS
      FILES (TOOLS)
      HONING
      ∞ SEPARATION

SCREEN EFFECT
RT   ∞ COMA
      DIELECTRICS
      ∞ EFFECTS
      ELECTROMAGNETIC WAVE FILTERS
      ELECTROMAGNETIC WAVE
      TRANSMISSION
      ELECTRON GAS
      MAGNETOSPHERE
      SEMICONDUCTORS (MATERIALS)
      WAVE PROPAGATION

∞ **SCREENING**
SN   (USE OF A MORE SPECIFIC TERM IS
      RECOMMENDED--CONSULT THE TERMS
      LISTED BELOW)
RT   FILTRATION
      FINES
      LOUVERS
      SELECTION
      WATER TREATMENT

∞ **SCREENS**
SN   (USE OF A MORE SPECIFIC TERM IS
      RECOMMENDED--CONSULT THE TERMS
      LISTED BELOW)
RT   CURTAINS
      DISPLAY DEVICES
      PROTECTORS
      SHIELDING
      SIZING SCREENS
      WIRE CLOTH

SCREW DISLOCATIONS
GS   DEFECTS
      . CRYSTAL DEFECTS
      . . CRYSTAL DISLOCATIONS
      . . . **SCREW DISLOCATIONS**
      DISLOCATIONS (MATERIALS)
      . CRYSTAL DISLOCATIONS
      . . **SCREW DISLOCATIONS**
RT   EDGE DISLOCATIONS

SCREW PINCH
GS   PINCH EFFECT
      . PLASMA PINCH
      . . **SCREW PINCH**
RT   MAGNETIC FIELDS
      MAGNETOHYDRODYNAMIC FLOW
      PLASMA CONTROL
      REVERSE FIELD PINCH
      THETA PINCH
      ZETA PINCH

SCREWS
SN   (EXCLUDES PROPELLERS AND CRYSTAL
      DEFECTS)
GS   FASTENERS
      . **SCREWS**
RT   ANCHORS (FASTENERS)
      BOLTS
      COUPLINGS
      HOLDERS
      NUTS (FASTENERS)
      STUDS (STRUCTURAL MEMBERS)
      THREADS

SCRIBING
USE   SCORING

SCRUBBERS
RT   CLEANING
      COLUMNS (PROCESS ENGINEERING)
      FLUE GASES
      WASHING

SCRUBBING
USE   WASHING

SCRUBS (BOTANY)
USE   BRUSH (BOTANY)

SCUTUM CONSTELLATION
GS   CONSTELLATIONS
      . **SCUTUM CONSTELLATION**
RT   ZODIAC

SCYLLA
GS   PLASMA GENERATORS
      . **SCYLLA**
RT   MAGNETIC FIELDS
      MAGNETIC MIRRORS
      MAGNETIC VARIATIONS
      PLASMAS (PHYSICS)
      THERMONUCLEAR REACTIONS

SDP (COMPUTERS)
USE   SITE DATA PROCESSORS

SDS 900 SERIES COMPUTERS
GS   DATA PROCESSING EQUIPMENT
      . COMPUTERS
      . . DIGITAL COMPUTERS
      . . . **SDS 900 SERIES COMPUTERS**
      . . . . SDS 930 COMPUTER

SDS 930 COMPUTER
GS   DATA PROCESSING EQUIPMENT
      . COMPUTERS
      . . DIGITAL COMPUTERS
      . . . SDS 900 SERIES COMPUTERS
      . . . . **SDS 930 COMPUTER**

SDS 9300 COMPUTER
GS   DATA PROCESSING EQUIPMENT
      . COMPUTERS
      . . DIGITAL COMPUTERS
      . . . **SDS 9300 COMPUTER**

SDV
USE   SHUTTLE DERIVED VEHICLES

SE-A
USE   EXPLORER 30 SATELLITE

SE-210 AIRCRAFT
UF   CARAVELLE AIRCRAFT
      SUD AVIATION SE-210 AIRCRAFT
GS   COMMERCIAL AIRCRAFT
      . **SE-210 AIRCRAFT**
      JET AIRCRAFT
      . TURBOFAN AIRCRAFT
      . . **SE-210 AIRCRAFT**
      MONOPLANES
      . **SE-210 AIRCRAFT**
      PASSENGER AIRCRAFT
      . **SE-210 AIRCRAFT**
      SUD AVIATION AIRCRAFT
      . **SE-210 AIRCRAFT**
RT   ∞ AIRCRAFT

SE-3160 HELICOPTER
UF   ALOUETTE 3 HELICOPTER
      SUD AVIATION SE-3160 HELICOPTER
GS   SUD AVIATION AIRCRAFT
      . ALOUETTE HELICOPTERS
      . . **SE-3160 HELICOPTER**
      V/STOL AIRCRAFT
      . ROTARY WING AIRCRAFT
      . . HELICOPTERS
      . . . ALOUETTE HELICOPTERS
      . . . . **SE-3160 HELICOPTER**

SEA BREEZE
GS   WIND (METEOROLOGY)
      . **SEA BREEZE**
RT   AEROLOGY
      AIR CURRENTS
      ATMOSPHERIC CIRCULATION
      BAROTROPIC FLOW
      CLIMATOLOGY

SEA BREEZE-*(CONT.)*
  GEOSTROPHIC WIND
  GUSTS
  MARINE ENVIRONMENTS
  METEOROLOGY
  MONSOONS
  OFFSHORE ENERGY SOURCES
  TIDAL WAVES
  WIND DIRECTION
  WIND EFFECTS
  WIND EROSION
  WIND MEASUREMENT
  WINDPOWER UTILIZATION
  WINDS ALOFT

SEA GRASSES
  GS   PLANTS (BOTANY)
       . GRASSES
       . . **SEA GRASSES**
  RT   MARINE BIOLOGY
       OCEANOGRAPHY
       SEAWEEDS
       VEGETATION
       WETLANDS

SEA ICE
  UF   ICE PACKS
  GS   ICE
       . **SEA ICE**
       . . ICE FLOES
       . . ICEBERGS
       . . PRESSURE ICE
  RT   AIR SEA ICE INTERACTIONS
       BAY ICE
       FREEZING
       GLACIAL DRIFT
       GLACIERS
       ICE ENVIRONMENTS
       ICE FORMATION
       ICE MAPPING
       LAKE ICE
       LAND ICE
       NUNATAKS
       OCEANOGRAPHY

SEA KEEPING
  RT   ATTITUDE GYROS
       DAMPING
       GYROSCOPIC STABILITY
       GYROSTABILIZERS
       MOTION STABILITY
       ∞STABILIZERS
       TORQUERS

SEA KING HELICOPTER
  USE   SH-3 HELICOPTER

SEA KNIGHT HELICOPTER
  USE   CH-46 HELICOPTER

SEA LAUNCHING
  GS   LAUNCHING
       . **SEA LAUNCHING**
  RT   ANTISHIP MISSILES
       ANTISHIP WARFARE
       BALLISTIC MISSILE SUBMARINES
       CATAPULTS
       DRYDOCKS
       FLEET BALLISTIC MISSILES
       MISSILE LAUNCHERS
       POSEIDON MISSILES
       ROCKET LAUNCHERS
       TORPEDOES
       WATER TAKEOFF AND LANDING
         AIRCRAFT

SEA LAW
  GS   LAW (JURISPRUDENCE)
       . INTERNATIONAL LAW
       . . **SEA LAW**
  RT   ∞COOPERATION
       INTERNATIONAL COOPERATION
       RULES
       UNITED NATIONS

SEA LEVEL
  GS   ALTITUDE
       . **SEA LEVEL**
  RT   OCEAN SURFACE
       OCEANOGRAPHY
       SEA STATES

SEA OF JAPAN
  GS   SEAS

SEA OF JAPAN-*(CONT.)*
       . **SEA OF JAPAN**
  RT   ASIA

SEA OF OKHOTSK
  GS   SEAS
       . **SEA OF OKHOTSK**
  RT   PACIFIC OCEAN
       U.S.S.R.

SEA ROUGHNESS
  GS   ROUGHNESS
       . **SEA ROUGHNESS**
  RT   HYDRODYNAMIC COEFFICIENTS
       OCEAN MODELS
       OCEAN SURFACE
       OCEANOGRAPHY
       SURFACE WAVES
       TIDAL WAVES
       TIDE POWERED GENERATORS
       TIDE POWERED MACHINES
       TIDEPOWER
       TIDES
       TURBULENCE
       WATER CURRENTS
       WATER WAVES
       WATERWAVE ENERGY CONVERSION
       WATERWAVE POWERED MACHINES
       WIND EFFECTS
       WIND VELOCITY

SEA STATES
  RT   OCEAN MODELS
       OCEAN SURFACE
       OCEAN TEMPERATURE
       OCEANOGRAPHIC PARAMETERS
       OCEANOGRAPHY
       SEA LEVEL
       TOPEX
       WATER CURRENTS
       WATER WAVES
       WIND EFFECTS

SEA SURFACE TEMPERATURE
  GS   TEMPERATURE
       . WATER TEMPERATURE
       . . OCEAN TEMPERATURE
       . . . **SEA SURFACE TEMPERATURE**
  RT   AIR WATER INTERACTIONS
       OCEAN SURFACE
       OCEANOGRAPHY
       SURFACE TEMPERATURE

SEA TRUTH
  RT   AERIAL PHOTOGRAPHY
       COASTAL CURRENTS
       IMAGERY
       OCEAN SURFACE
       OCEAN TEMPERATURE
       PHOTOINTERPRETATION

SEA URCHINS
  GS   ANIMALS
       . INVERTEBRATES
       . . **SEA URCHINS**

SEA WALLS
  USE   BREAKWATERS

SEA WATER
  GS   WATER
       . **SEA WATER**
  RT   BRINES
       COASTAL WATER
       FISHERIES
       MARINE RESOURCES
       NEARSHORE WATER
       OCEAN SURFACE
       OCEAN TEMPERATURE
       OCEANOGRAPHY
       RED TIDE
       SALINITY
       SEAWEEDS
       THERMOCLINES
       UNDERWATER PHOTOGRAPHY
       UNDERWATER RESOURCES
       WATER RESOURCES

SEAFARER PROJECT
  UF   GLOBAL COMMUNICATIONS ANTENNA
         GRID (NAVY)
       UNDERGROUND RADIO ANTENNA GRID
         (NAVY)
  GS   PROGRAMS

SEAFARER PROJECT-*(CONT.)*
       . PROJECTS
       . . **SEAFARER PROJECT**
  RT   EXTREMELY LOW FREQUENCIES
       RADIO TRANSMISSION
       SUBMARINES
       TELECOMMUNICATION
       UNDERWATER COMMUNICATION

SEAHORSE HELICOPTER
  USE   UH-34 HELICOPTER

SEALANTS
  USE   SEALERS

SEALERS
  UF   SEALANTS
  RT   ADHESIVES
       COATINGS
       DOPES
       FILLERS
       PACKAGING
       PACKINGS (SEALS)
       PAINTS
       SEALING
       SEALS (STOPPERS)
       SEAMS (JOINTS)
       SOLDERS
       VARNISHES

SEALING
  GS   **SEALING**
       . SELF SEALING
  RT   ADHESION
       ADHESIVE BONDING
       BINDING
       BLOCKING
       BLOWERS
       BONDING
       BRAZING
       CAULKING
       CEMENTS
       CLAMPS
       CLOSING
       COATING
       COATINGS
       CONTAINMENT
       COVERINGS
       ENCAPSULATING
       GLANDS (SEALS)
       ∞JOINING
       LINING PROCESSES
       MOISTURE RESISTANCE
       ∞PACKING
       PACKINGS (SEALS)
       PLUGGING
       RETAINING
       RIVETING
       SEALERS
       SOLDERING
       SPRAYING
       STOPPING
       WATERPROOFING
       WELDING

SEALS (ANIMALS)
  GS   ANIMALS
       . VERTEBRATES
       . . MAMMALS
       . . . **SEALS (ANIMALS)**
  RT   MARINE BIOLOGY
       MARINE MAMMALS

SEALS (STOPPERS)
  GS   **SEALS (STOPPERS)**
       . GASKETS
       . GLANDS (SEALS)
       . HERMETIC SEALS
       . LABYRINTH SEALS
       . O RING SEALS
       . PACKINGS (SEALS)
       . PLUGS
       . PUMP SEALS
  RT   AIR LOCKS
       BARRIER LAYERS
       ∞BARRIERS
       BLOCKING
       ∞CAPS
       CLOSURES
       CONSTRICTIONS
       CUFFS
       PLUGGING
       SEALERS
       SPHERICAL CAPS
       ∞TAPES

SEALS (STOPPERS)-_(CONT.)_
    VALVES

SEAMOUNTS
  RT   CONTINENTAL SHELVES
        CREVASSES
        ∞FAULTS
        FOLDS (GEOLOGY)
        ISLANDS
        LANDFORMS
        OCEAN BOTTOM
        STRUCTURAL BASINS

SEAMS (JOINTS)
  GS   JOINTS (JUNCTIONS)
        . **SEAMS (JOINTS)**
  RT   ADHESIVES
        FILLETS
        METAL JOINTS
        SEALERS

SEAPLANES
  GS   WATER TAKEOFF AND LANDING
          AIRCRAFT
        . **SEAPLANES**
  RT   AMPHIBIOUS AIRCRAFT
        AMPHIBIOUS VEHICLES
        HULLS (STRUCTURES)
        MONOPLANES

SEARCH AND RESCUE SATELLITE
  USE   SARSAT

**SEARCH FOR EXTRATERRESTRIAL
INTELLIGENCE**
  USE   PROJECT SETI

SEARCH PROFILES
  GS   SEARCHING
        . **SEARCH PROFILES**
  RT   DATA RETRIEVAL
        INFORMATION RETRIEVAL
        ∞PROFILES

SEARCH RADAR
  GS   RADAR
        . **SEARCH RADAR**
        . . NORTH AMERICAN SEARCH AND
            RANGING RADAR
        . . OVER-THE-HORIZON RADAR
  RT   AIRPORT SURFACE DETECTION
          EQUIPMENT
        COHERENT RADAR
        CONTINUOUS WAVE RADAR
        PULSE RADAR
        RADAR DETECTION
        RADAR TRACKING
        SATELLITE-BORNE RADAR
        SIDE-LOOKING RADAR
        SURVEILLANCE RADAR
        TRACKING RADAR
        TRADEX RADAR SYSTEM

SEARCHING
  GS   **SEARCHING**
        . SEARCH PROFILES
  RT   CONICAL SCANNING
        COSPAS
        PANORAMIC SCANNING
        RECONNAISSANCE
        RETRIEVAL
        SARSAT
        SCANNING
        SELECTION

SEARCHLIGHTS
  GS   LIGHTING EQUIPMENT
        . LUMINAIRES
        . . **SEARCHLIGHTS**
  RT   AIRPORT LIGHTS
        ARC LAMPS
        BEACONS
        PROJECTORS
        RUNWAY LIGHTS

SEAS
  GS   **SEAS**
        . ARABIAN SEA
        . BALTIC SEA
        . BARENTS SEA
        . BEAUFORT SEA (NORTH AMERICA)
        . BERING SEA
        . BLACK SEA
        . CARIBBEAN SEA

SEAS-_(CONT.)_
        . CASPIAN SEA
        . CHUCKCHI SEA
        . MEDITERRANEAN SEA
        . . ADRIATIC SEA
        . NORTH SEA
        . RED SEA
        . SALTON SEA (CA)
        . SEA OF JAPAN
        . SEA OF OKHOTSK
  RT   ADRIATIC SEA
        ARCHIPELAGOES
        COASTAL CURRENTS
        COASTS
        EARTH HYDROSPHERE
        OCEAN TEMPERATURE
        OCEANOGRAPHY
        OCEANS
        SARGASSO SEA
        SEAWEEDS
        SHALLOW WATER
        SHOALS
        STRAITS
        THERMAL POLLUTION
        UNDERWATER PHOTOGRAPHY

SEASAT PROGRAM
  GS   PROGRAMS
        . NASA PROGRAMS
        . . EARTH RESOURCES PROGRAM
        . . . EARTH RESOURCES SURVEY
            PROGRAM
        . . . . **SEASAT PROGRAM**
        . NASA SPACE PROGRAMS
        . . EARTH RESOURCES PROGRAM
        . . . EARTH RESOURCES SURVEY
            PROGRAM
        . . . . **SEASAT PROGRAM**
  RT   LANDSAT SATELLITES
        OCEANOGRAPHY

SEASAT SATELLITES
  GS   SATELLITES
        . ARTIFICIAL SATELLITES
        . . **SEASAT SATELLITES**
        . . . SEASAT 1
        . . . SEASAT-B SATELLITE
  RT   LANDSAT SATELLITES
        NASA PROGRAMS
        OCEANOGRAPHY
        PROGRAMS
        SYNCHRONOUS EARTH OBSERVATORY
          SATELLITE

SEASAT 1
  GS   SATELLITES
        . ARTIFICIAL SATELLITES
        . . SEASAT SATELLITES
        . . . **SEASAT 1**
  RT   LANDSAT SATELLITES
        OCEANOGRAPHY
        PROGRAMS
        SEASAT-B SATELLITE

SEASAT-B SATELLITE
  GS   SATELLITES
        . ARTIFICIAL SATELLITES
        . . SEASAT SATELLITES
        . . . **SEASAT-B SATELLITE**
  RT   LANDSAT SATELLITES
        OCEANOGRAPHY
        PROGRAMS
        SEASAT 1

SEASONAL VARIATIONS
  USE   ANNUAL VARIATIONS

SEASONS
  GS   **SEASONS**
        . AUTUMN
        . SPRING (SEASON)
        . SUMMER
        . WINTER
  RT   ANNUAL VARIATIONS
        CLIMATOLOGY
        CROP CALENDARS
        EQUINOXES
        METEOROLOGY
        SOLAR POSITION
        SOLSTICES
        WEATHER
        WIND VARIATIONS

SEASPRITE HELICOPTER
  USE   UH-2 HELICOPTER

SEAT BELTS
  GS   SAFETY DEVICES
        . **SEAT BELTS**
  RT   ∞BELTS
        HARNESSES
        SEATS

SEATS
  UF   BENCHES
        CHAIRS
  GS   **SEATS**
        . BARANY CHAIR
        . EJECTION SEATS
        . . FLYING EJECTION SEATS
  RT   COMFORT
        COUCHES
        CUSHIONS
        HARNESSES
        ∞LOUNGES
        PILLOWS
        RIDING QUALITY
        SEAT BELTS
        SITTING POSITION

SEAWEEDS
  UF   KELP
  RT   MARINE BIOLOGY
        OCEANOGRAPHY
        OCEANS
        SEA GRASSES
        SEA WATER
        SEAS

SEBACEOUS GLANDS
  GS   ANATOMY
        . GLANDS (ANATOMY)
        . . **SEBACEOUS GLANDS**

SEBACIC ACID
  GS   ACIDS
        . FATTY ACIDS
        . . CARBOXYLIC ACIDS
        . . . **SEBACIC ACID**
        ALIPHATIC COMPOUNDS
        . FATTY ACIDS
        . . **SEBACIC ACID**
        LIPIDS
        . **SEBACIC ACID**
        ORGANIC COMPOUNDS
        . FATTY ACIDS
        . . **SEBACIC ACID**

SECONDARY BATTERIES
  USE   STORAGE BATTERIES

SECONDARY COSMIC RAYS
  UF   MOLIERE FORMULA
  GS   IONIZING RADIATION
        . COSMIC RAYS
        . . **SECONDARY COSMIC RAYS**
  RT   ATMOSPHERIC RADIATION
        COSMIC RAY ALBEDO
        COSMIC RAY SHOWERS
        ELECTRON DECAY RATE
        ELECTRON PHOTON CASCADES
        ELECTRON PRECIPITATION
        PRIMARY COSMIC RAYS
        SINGLE EVENT UPSETS

SECONDARY EMISSION
  GS   DECAY
        . EMISSION
        . . PARTICLE EMISSION
        . . . ELECTRON EMISSION
        . . . . **SECONDARY EMISSION**
  RT   DYNODES
        ELECTRON IRRADIATION
        FIELD EMISSION
        MONOSCOPES
        MULTIPACTOR DISCHARGES
        PHOTOMULTIPLIER TUBES
        TOWNSEND AVALANCHE

SECONDARY FLOW
  GS   FLUID FLOW
        . VISCOUS FLOW
        . . BOUNDARY LAYER FLOW
        . . . **SECONDARY FLOW**
        TRANSLATIONAL MOTION
        . THREE DIMENSIONAL MOTION
        . . THREE DIMENSIONAL FLOW
        . . . **SECONDARY FLOW**
  RT   COMPRESSIBILITY EFFECTS
        CORNER FLOW
        THREE DIMENSIONAL BOUNDARY LAYER

SECONDARY FLOW-*(CONT.)*
    VORTICES
    VORTICITY

SECONDARY INJECTION
GS   INJECTION
    . **SECONDARY INJECTION**
RT   FLUID INJECTION
    SHOCK WAVE CONTROL
    SHOCK WAVE PROPAGATION
    SUPERSONIC FLOW
    THRUST AUGMENTATION
    THRUST VECTOR CONTROL

SECONDARY RADAR
GS   RADAR
    . **SECONDARY RADAR**
RT   DISCRETE ADDRESS BEACON SYSTEM
    INTERROGATION

SECONDARY WAVES
USE   S WAVES

SECRETIONS
GS   **SECRETIONS**
    . ENDOCRINE SECRETIONS
    . . HORMONES
    . . . ESTROGENS
    . . . HYDROXYCORTICOSTEROID
    . . . PITUITARY HORMONES
    . . . . ADRENOCORTICOTROPIN (ACTH)
    . . . PROSTAGLANDINS
    . . INSULIN
    . SWEAT
RT   BODY FLUIDS
    GALL
    GLANDS (ANATOMY)
    HYDROGEN METABOLISM
    METABOLISM
    MINERAL METABOLISM
    SKIN (ANATOMY)

∞ **SECTIONS**
SN   *(USE OF A MORE SPECIFIC TERM IS*
    RECOMMENDED--CONSULT THE TERMS
    LISTED BELOW)
RT   CATEGORIES
    CLASSES
    SUBDIVISIONS
    SUBSIDIARIES

∞ **SECTORS**
SN   *(USE OF A MORE SPECIFIC TERM IS*
    RECOMMENDED--CONSULT THE TERMS
    LISTED BELOW)
RT   AREA
    CIRCLES (GEOMETRY)
    REGIONS

SECULAR PERTURBATION
USE   LONG TERM EFFECTS

SECULAR VARIATIONS
GS   VARIATIONS
    . PERIODIC VARIATIONS
    . . **SECULAR VARIATIONS**
RT   ATMOSPHERIC PHYSICS
    SOLAR ACTIVITY EFFECTS
    SOLAR CYCLES

SECURITY
GS   **SECURITY**
    . AIRPORT SECURITY
    . COMPUTER INFORMATION SECURITY
RT   ∞ CLASSIFYING
    COMPUTER PROGRAM INTEGRITY
    CRIME
    INTEGRITY
    POLICE
    PRIVACY
    SCRAMBLING (COMMUNICATION)
    VULNERABILITY

SEDATIVES
GS   DRUGS
    . **SEDATIVES**
RT   PENTOBARBITAL
    PHENOBARBITAL
    PSYCHOTROPIC DRUGS
    TRANQUILIZERS

SEDIMENT TRANSPORT
RT   MASS FLOW
    MASS TRANSFER

SEDIMENT TRANSPORT-*(CONT.)*
    SEDIMENTS

SEDIMENTARY ROCKS
GS   **SEDIMENTARY ROCKS**
    . CARBONACEOUS ROCKS
    . . COAL
    . . . LIGNITE
    . LIMESTONE
    . PEAT
    . SANDSTONES
    . SHALES
RT   ALLUVIUM
    BRECCIA
    CLAYS
    DOLOMITE (MINERAL)
    GYPSUM
    IGNEOUS ROCKS
    ∞ LAYERS
    MONAZITE SANDS
    PETROGRAPHY
    ROCKS
    SANDS
    SEDIMENTS
    SHATTER CONES
    SOILS
    STRATIGRAPHY

SEDIMENTS
UF   SILTS
GS   **SEDIMENTS**
    . GRAVELS
    . MUD
    . SANDS
    . . MONAZITE SANDS
    . . TAR SANDS
RT   ALLUVIUM
    CLAYS
    DEPOSITION
    DEPOSITS
    DREDGED MATERIALS
    FANS (LANDFORMS)
    GLACIAL DRIFT
    GRIT
    LITTORAL DRIFT
    MARINE CHEMISTRY
    OCEAN BOTTOM
    SEDIMENT TRANSPORT
    SEDIMENTARY ROCKS
    SEWAGE TREATMENT
    SLUDGE

SEEBECK COEFFICIENT
USE   SEEBECK EFFECT

SEEBECK EFFECT
UF   SEEBECK COEFFICIENT
RT   ∞ EFFECTS
    PELTIER EFFECTS
    TEMPERATURE EFFECTS
    THERMOCOUPLES
    THERMODYNAMIC PROPERTIES
    THERMOELECTRICITY
    THERMOPHYSICAL PROPERTIES
    TRANSPORT PROPERTIES

SEEDING (INOCULATION)
USE   INOCULATION

SEEDS
RT   ALFALFA
    BARLEY
    CITRUS TREES
    CORN
    EMBRYOS
    FARM CROPS
    GRAINS (FOOD)
    NUTS (FRUITS)
    OATS
    PLANTING
    PLANTS (BOTANY)
    SUGAR BEETS
    SUGAR CANE
    UTRICLE
    VEGETABLES
    VIABILITY

SEEKERS
USE   HOMING DEVICES

SEEPAGE
RT   CANALS
    DRAINAGE
    FLOOD DAMAGE
    FLOW NETS

SEEPAGE-*(CONT.)*
    HYDRODYNAMICS
    INTRUSION
    IRRIGATION
    LEAKAGE
    LOSSES
    OFFSHORE ENERGY SOURCES
    PENETRATION
    PERCOLATION
    PERMEABILITY
    PIEZOMETERS
    WATER CONSUMPTION

SEGMENTS
RT   CIRCLES (GEOMETRY)
    ∞ COMPONENTS
    CURVES (GEOMETRY)
    LINES (GEOMETRY)
    RADII

SEGRE CHARACTERISTIC
RT   ∞ CHARACTERISTICS
    CRACK PROPAGATION
    METAL FATIGUE

SEGREGATION
USE   SEPARATION

SEISMIC ENERGY
RT   EARTHQUAKE DAMAGE
    ∞ ENERGY
    STRAIN ENERGY METHODS

SEISMIC WAVES
UF   ELECTROSEISMIC EFFECT
GS   ELASTIC WAVES
    . **SEISMIC WAVES**
    . . LOVE WAVES
    . . MICROSEISMS
    . . RAYLEIGH WAVES
RT   CRUSTAL FRACTURES
    DETONATION WAVES
    DILATATIONAL WAVES
    EARTH MOVEMENTS
    EARTHQUAKE DAMAGE
    EARTHQUAKE RESISTANCE
    EARTHQUAKE RESISTANT STRUCTURES
    EARTHQUAKES
    GUTENBERG ZONE
    LARGE APERTURE SEISMIC ARRAY
    LONGITUDINAL WAVES
    P WAVES
    PLANETARY QUAKES
    POLARIZED ELASTIC WAVES
    S WAVES
    SEISMOLOGY
    SHOCK WAVES
    SURFACE WAVES
    TSUNAMI WAVES
    UNDERGROUND EXPLOSIONS
    ∞ WAVES

SEISMOCARDIOGRAPHY
SN   (MEASUREMENT OF THE HIGH
    FREQUENCY VIBRATIONS OF THE
    HEART)
GS   BIOENGINEERING
    . BIOMETRICS
    . . CARDIOGRAPHY
    . . . **SEISMOCARDIOGRAPHY**
RT   BALLISTOCARDIOGRAPHY
    FIBRILLATION

SEISMOGRAMS
RT   SEISMOGRAPHS

SEISMOGRAPHS
UF   SEISMOMETERS
GS   MEASURING INSTRUMENTS
    . VIBRATION METERS
    . . **SEISMOGRAPHS**
    . . . LUNAR SEISMOGRAPHS
    RECORDING INSTRUMENTS
    . **SEISMOGRAPHS**
    . . LUNAR SEISMOGRAPHS
RT   ACCELEROMETERS
    ACOUSTIC MEASUREMENT
    PHASED ARRAYS
    SEISMOGRAMS
    SHOCK MEASURING INSTRUMENTS
    TILTMETERS

SEISMOLOGY
GS   **SEISMOLOGY**

SEISMOLOGY-*(CONT.)*
. HELIOSEISMOLOGY
. MOONQUAKES
RT   CRUSTAL FRACTURES
EARTH MOVEMENTS
EARTHQUAKE DAMAGE
EARTHQUAKES
GEOLOGY
GEOPHYSICS
ISOSTASY
LARGE APERTURE SEISMIC ARRAY
LUNAR GEOLOGY
ROUSE BELTS
∞ SCIENCE
SEISMIC WAVES
SUBDUCTION (GEOLOGY)
TIDAL WAVES

SEISMOMETERS
USE   SEISMOGRAPHS

SEIZURES
RT   CONVULSIONS
CRAMPS

SEL COMPUTERS
GS   DATA PROCESSING EQUIPMENT
. COMPUTERS
. . DIGITAL COMPUTERS
. . . **SEL COMPUTERS**

SELECTION
UF   CHOICE
GS   **SELECTION**
. PERSONNEL SELECTION
. . PILOT SELECTION
. SITE SELECTION
RT   CERTIFICATION
∞ CLASSIFYING
COLLECTION
DECISIONS
EVALUATION
FIGURE OF MERIT
OPTIONS
RANKING
REJECTION
SAMPLING
∞ SCREENING
SEARCHING
∞ TESTS

SELECTION RULES (NUCLEAR PHYSICS)
GS   RULES
. **SELECTION RULES (NUCLEAR PHYSICS)**
RT   ALPHA DECAY
EMISSION
FORBIDDEN TRANSITIONS
NEUTRON EMISSION
∞ PHYSICS
QUANTUM NUMBERS

SELECTIVE DISSEMINATION OF INFORMATION
GS   COMMUNICATING
. INFORMATION DISSEMINATION
. . **SELECTIVE DISSEMINATION OF INFORMATION**
TELECOMMUNICATION
. **SELECTIVE DISSEMINATION OF INFORMATION**
RT   DATA STORAGE
DOCUMENTATION
INDEXES (DOCUMENTATION)
INFORMATION FLOW
INFORMATION RETRIEVAL
INFORMATION SYSTEMS
LIBRARIES
MANAGEMENT PLANNING
TECHNOLOGY TRANSFER

SELECTIVE FADING
GS   FADING
. SIGNAL FADING
. . **SELECTIVE FADING**
RT   FREQUENCY ANALYZERS
GROUND WAVE PROPAGATION
MODULATION
SIDEBANDS
SIGNAL FADING RATE

SELECTIVE SURFACES
UF   SOLAR SELECTIVE COATINGS
RT   ENERGY ABSORPTION FILMS
SELECTIVITY
SOLAR COLLECTORS

SELECTIVE SURFACES-*(CONT.)*
SOLAR ENERGY ABSORBERS

SELECTIVITY
RT   DISCRIMINATION
PHOTOTHERMAL CONVERSION
SELECTIVE SURFACES

SELECTORS
RT   ANALYZERS
CIRCUITS
ELECTRIC RELAYS
SAMPLERS
SWITCHES
SWITCHING CIRCUITS

SELENIDES
GS   CHALCOGENIDES
. **SELENIDES**
. . CADMIUM SELENIDES
. . COPPER SELENIDES
. . GALLIUM SELENIDES
. . LEAD SELENIDES
. . ZINC SELENIDES
SELENIUM COMPOUNDS
. **SELENIDES**
. . CADMIUM SELENIDES
. . COPPER SELENIDES
. . GALLIUM SELENIDES
. . LEAD SELENIDES
. . ZINC SELENIDES

SELENIUM
GS   CHEMICAL ELEMENTS
. **SELENIUM**

SELENIUM ALLOYS
GS   ALLOYS
. **SELENIUM ALLOYS**
RT   COPPER

SELENIUM COMPOUNDS
GS   **SELENIUM COMPOUNDS**
. SELENIDES
. . CADMIUM SELENIDES
. . COPPER SELENIDES
. . GALLIUM SELENIDES
. . LEAD SELENIDES
. . ZINC SELENIDES
. SELENIUM OXIDES
RT   ∞ CHEMICAL COMPOUNDS
∞ GROUP 6A COMPOUNDS

SELENIUM OXIDES
GS   CHALCOGENIDES
. OXIDES
. . **SELENIUM OXIDES**
SELENIUM COMPOUNDS
. **SELENIUM OXIDES**

SELENOGRAPHY
RT   GEOGRAPHY
LUNAR CRATERS
LUNAR CRUST
LUNAR LANDING SITES
LUNAR MAPS
LUNAR MOBILE LABORATORIES
LUNAR RAYS
LUNAR ROCKS
LUNAR TOPOGRAPHY
MOON
SELENOLOGY
SURFACE PROPERTIES

SELENOLOGY
GS   **SELENOLOGY**
. LUNAR CORE
RT   ASTRONOMY
LUNAR COMPOSITION
LUNAR CRATERS
LUNAR CRUST
LUNAR DUST
LUNAR EVOLUTION
LUNAR FIGURE
LUNAR GEOLOGY
LUNAR MANTLE
LUNAR ROCKS
LUNAR SURFACE
LUNAR TOPOGRAPHY
MOON
MOONQUAKES
REGOLITH
SELENOGRAPHY

SELF ABSORPTION
GS   ENERGY ABSORPTION
. **SELF ABSORPTION**
RADIATION ABSORPTION
. **SELF ABSORPTION**
RT   ∞ ABSORPTION
ABSORPTION SPECTRA
ABSORPTIVITY
AUTOMATIC CONTROL
DIFFUSION
∞ RADIATION

SELF ADAPTIVE CONTROL SYSTEMS
GS   AUTOMATIC CONTROL
. ADAPTIVE CONTROL
. . **SELF ADAPTIVE CONTROL SYSTEMS**
RT   ACTIVE CONTROL
ADAPTIVE OPTICS
AUTOMATA THEORY
AUTONOMY
∞ CONTROL
∞ SYSTEMS

SELF ALIGNMENT
GS   ALIGNMENT
. **SELF ALIGNMENT**
AUTOMATIC CONTROL
. **SELF ALIGNMENT**
RT   ACTIVE CONTROL
ADAPTIVE CONTROL
LANDING GEAR
SERVOMECHANISMS

SELF CALIBRATING OMNIRANGE
UF   SCORE OMNIRANGE
GS   NAVIGATION AIDS
. BEACONS
. . RADIO BEACONS
. . . OMNIDIRECTIONAL RADIO RANGES
. . . . **SELF CALIBRATING OMNIRANGE**
RADIO EQUIPMENT
. RADIO TRANSMITTERS
. . RADIO BEACONS
. . . OMNIDIRECTIONAL RADIO RANGES
. . . . **SELF CALIBRATING OMNIRANGE**
TRANSMITTERS
. RADIO TRANSMITTERS
. . RADIO BEACONS
. . . OMNIDIRECTIONAL RADIO RANGES
. . . . **SELF CALIBRATING OMNIRANGE**
RT   SOLAR COMPASSES

SELF CONSISTENT FIELDS
UF   SCF
RT   FIELD THEORY (PHYSICS)
∞ FIELDS
HARTREE APPROXIMATION
MAGNETIC FIELDS
MOLECULAR ORBITALS
QUANTUM ELECTRODYNAMICS
SHELL THEORY

SELF DEPLOYING SPACE STATIONS
USE   SELF ERECTING DEVICES
SPACE STATIONS

SELF DIFFUSION (SOLID STATE)
GS   DIFFUSION
. **SELF DIFFUSION (SOLID STATE)**
RT   ATOMIC MOBILITIES
IONIC DIFFUSION
MOLECULAR DIFFUSION
PARTICLE DIFFUSION

SELF ERECTING DEVICES
UF   SELF DEPLOYING SPACE STATIONS
RT   ∞ AUTOMATION
∞ DEVICES
∞ EQUIPMENT
INFLATABLE SPACECRAFT
INFLATABLE STRUCTURES
ORBITAL ASSEMBLY
SPACE ERECTABLE STRUCTURES

SELF EXCITATION
GS   EXCITATION
. **SELF EXCITATION**
RT   FORCED VIBRATION
FREE VIBRATION
OSCILLATORS
RESONATORS

SELF FOCUSING
GS   FOCUSING

SELF FOCUSING-*(CONT.)*
- . **SELF FOCUSING**
- RT IMAGE CONTRAST
- ∞MACHINERY
- OPTICAL CORRECTION PROCEDURE
- OPTICAL MEASURING INSTRUMENTS

SELF INDUCED VIBRATION
- GS VIBRATION
- . STRUCTURAL VIBRATION
- . . **SELF INDUCED VIBRATION**
- . . . PANEL FLUTTER
- . . . SUBSONIC FLUTTER
- . . . SUPERSONIC FLUTTER
- . . . TRANSONIC FLUTTER
- RT BENDING VIBRATION
- FLUTTER
- FORCED VIBRATION
- FREE VIBRATION
- MISSILE VIBRATION
- PILOT INDUCED OSCILLATION
- RANDOM VIBRATION
- TORSIONAL VIBRATION

SELF INITIATED ANTIAIRCRAFT MISSILES
- USE SIAM MISSILES

SELF LUBRICATING MATERIALS
- RT IMPREGNATING
- LUBRICATION
- ∞MATERIALS
- SOLID LUBRICANTS
- SPACECRAFT LUBRICATION

SELF LUBRICATION
- GS LUBRICATION
- . **SELF LUBRICATION**
- RT IMPREGNATING

SELF MANEUVERING UNITS
- UF PERSONNEL PROPULSION SYSTEMS
- REACTION JET BACKPACKS
- SMU (MANEUVERING UNITS)
- SPACE SELF MANEUVERING UNITS
- GS **SELF MANEUVERING UNITS**
- . IMLSS
- RT ASTRONAUT MANEUVERING EQUIPMENT
- EXTRAVEHICULAR ACTIVITY
- EXTRAVEHICULAR MOBILITY UNITS
- MANEUVERS
- MANNED MANEUVERING UNITS

SELF ORGANIZING SYSTEMS
- UF PERCEPTRONS
- RT ARTIFICIAL INTELLIGENCE
- LEARNING MACHINES
- ∞SYSTEMS
- TURING MACHINES

SELF OSCILLATION
- GS OSCILLATIONS
- . **SELF OSCILLATION**
- RT FEEDBACK AMPLIFIERS
- POSITIVE FEEDBACK
- TRANSFER FUNCTIONS

SELF PROPAGATION
- GS DIFFUSION
- . **SELF PROPAGATION**
- TRANSMISSION
- . **SELF PROPAGATION**
- RT ∞PROPAGATION

SELF REGULATING
- USE AUTOMATIC CONTROL

SELF REPAIRING DEVICES
- RT ∞AUTOMATION
- ∞DEVICES
- MAINTENANCE

SELF SEALING
- GS SEALING
- . **SELF SEALING**
- RT FLIGHT SAFETY
- FUEL SYSTEMS
- RUPTURING
- SUPPORT SYSTEMS

SELF SHADOWING
- RT LARGE SPACE STRUCTURES
- SHADOWS
- SOLAR ARRAYS

SELF STIMULATION
- RT MOTIVATION
- REINFORCEMENT (PSYCHOLOGY)

SELF SUBTRACTION HOLOGRAPHY
- USE HOLOGRAPHIC SUBTRACTION

SELF SUSTAINED EMISSION
- GS DECAY
- . EMISSION
- . . **SELF SUSTAINED EMISSION**
- RT ELECTRON EMISSION
- LIGHT EMISSION
- PARTICLE EMISSION
- STIMULATED EMISSION

SELSYNS (TRADEMARK)
- USE SERVOMOTORS

SEMANTICS
- GS LINGUISTICS
- . **SEMANTICS**
- RT COMMUNICATION THEORY
- GRAMMARS
- LANGUAGES
- MESSAGE PROCESSING
- MESSAGES
- NOMENCLATURES
- ORTHOGRAPHY
- PARSING ALGORITHMS
- PSYCHOLINGUISTICS
- SENTENCES
- SPEECH
- SYLLABLES
- SYMBOLS
- SYNTAX
- WORDS (LANGUAGE)

SEMICIRCULAR CANALS
- GS ANATOMY
- . SENSE ORGANS
- . . EAR
- . . . **SEMICIRCULAR CANALS**
- RT EARDRUMS
- LABYRINTH
- MIDDLE EAR
- VESTIBULES

SEMICONDUCTING FILMS
- RT AMORPHOUS SEMICONDUCTORS
- ENERGY ABSORPTION FILMS
- ∞FILMS
- PHOTOTHERMAL CONVERSION
- THICK FILMS
- THIN FILMS

SEMICONDUCTOR DEVICES
- GS ELECTRONIC EQUIPMENT
- . SOLID STATE DEVICES
- . . **SEMICONDUCTOR DEVICES**
- . . . AVALANCHE DIODES
- . . . . CRYOSAR
- . . . BARRITT DIODES
- . . . GERMANIUM DIODES
- . . . HETEROJUNCTION DEVICES
- . . . . HIGH ELECTRON MOBILITY TRANSISTORS
- . . . JUNCTION DIODES
- . . . . MIM DIODES
- . . . . STEP RECOVERY DIODES
- . . . LIGHT EMITTING DIODES
- . . . METAL OXIDE SEMICONDUCTORS
- . . . . CHARGE TRANSFER DEVICES
- . . . . . BUCKET BRIGADE DEVICES
- . . . . . CHARGE COUPLED DEVICES
- . . . . . CHARGE INJECTION DEVICES
- . . . . CMOS
- . . . . SOS (SEMICONDUCTORS)
- . . . MIM (SEMICONDUCTORS)
- . . . MIS (SEMICONDUCTORS)
- . . . NDM SEMICONDUCTOR DEVICES
- . . . NEURISTORS
- . . . PARAMETRIC DIODES
- . . . PHOTODIODES
- . . . PHOTOVOLTAIC CELLS
- . . . SCHOTTKY DIODES
- . . . SEMICONDUCTOR LASERS
- . . . . GALLIUM ARSENIDE LASERS
- . . . THERMISTORS
- . . . THYRISTORS
- . . . . SILICON CONTROLLED RECTIFIERS
- . . . TRANSFERRED ELECTRON DEVICES
- . . . TRANSISTOR AMPLIFIERS
- . . . TRANSISTORS

SEMICONDUCTOR DEVICES-*(CONT.)*
- . . . . BIPOLAR TRANSISTORS
- . . . . FIELD EFFECT TRANSISTORS
- . . . . . CHARGE FLOW DEVICES
- . . . . . JFET
- . . . . HIGH ELECTRON MOBILITY TRANSISTORS
- . . . . JUNCTION TRANSISTORS
- . . . . . JFET
- . . . . PHOTOTRANSISTORS
- . . . SILICON TRANSISTORS
- . . . . SOS (SEMICONDUCTORS)
- . . TRAPATT DEVICES
- . . VARACTOR DIODES
- . . VARISTORS
- RT BARRIER LAYERS
- BUBBLE TECHNIQUE
- CHIPS (MEMORY DEVICES)
- CRYSTAL RECTIFIERS
- DIFFUSION ELECTRODES
- DIODES
- GUNN DIODES
- GUNN EFFECT
- HALL EFFECT
- HYBRID CIRCUITS
- ION IMPLANTATION
- ∞JUNCTIONS
- MICROMINIATURIZATION
- MOLECULAR ELECTRONICS
- ORGANIC SEMICONDUCTORS
- OSCILLATORS
- PARAMETRIC AMPLIFIERS
- PENTODES
- RECTIFIERS
- SEMICONDUCTORS (MATERIALS)
- SILICON FILMS
- SOLAR CELLS
- TETRODES
- TRIODES
- WAFERS

SEMICONDUCTOR DIODES
- GS ELECTRONIC EQUIPMENT
- . DIODES
- . . **SEMICONDUCTOR DIODES**
- . . . AVALANCHE DIODES
- . . . BARRITT DIODES
- . . . GERMANIUM DIODES
- . . . GUNN DIODES
- . . . JUNCTION DIODES
- . . . LIGHT EMITTING DIODES
- . . . PARAMETRIC DIODES
- . . . PHOTODIODES
- . . . SCHOTTKY DIODES
- . . . TUNNEL DIODES
- . . . VARACTOR DIODES
- RT MIM DIODES
- SIS (SEMICONDUCTORS)

SEMICONDUCTOR INSULATOR SEMICONDUCTORS
- USE SIS (SEMICONDUCTORS)

SEMICONDUCTOR JUNCTIONS
- GS **SEMICONDUCTOR JUNCTIONS**
- . MBM JUNCTIONS
- . N-N JUNCTIONS
- . N-P-N JUNCTIONS
- . P-I-N JUNCTIONS
- . P-N JUNCTIONS
- . P-N-P JUNCTIONS
- . P-N-P-N JUNCTIONS
- . SILICON JUNCTIONS
- RT BARRITT DIODES
- HETEROJUNCTION DEVICES
- HETEROJUNCTIONS
- HOMOJUNCTIONS
- ∞JUNCTIONS
- N-TYPE SEMICONDUCTORS
- P-TYPE SEMICONDUCTORS
- SCHOTTKY DIODES
- SIS (SEMICONDUCTORS)

SEMICONDUCTOR LASERS
- GS ELECTRONIC EQUIPMENT
- . SOLID STATE DEVICES
- . . SEMICONDUCTOR DEVICES
- . . . **SEMICONDUCTOR LASERS**
- . . . . GALLIUM ARSENIDE LASERS
- STIMULATED EMISSION DEVICES
- . LASERS
- . . **SEMICONDUCTOR LASERS**
- . . . GALLIUM ARSENIDE LASERS
- RT DISTRIBUTED FEEDBACK LASERS
- GALLIUM ARSENIDES
- GUNN EFFECT

SEMICONDUCTOR LASERS-*(CONT.)*
    INJECTION LASERS
    LASER CAVITIES
    PULSED LASERS
    Q SWITCHED LASERS
    SOLID STATE LASERS
    WAVEGUIDE LASERS

SEMICONDUCTOR PLASMAS
GS   PARTICLES
    . CHARGED PARTICLES
    . . ENERGETIC PARTICLES
    . . . PLASMAS (PHYSICS)
    . . . . **SEMICONDUCTOR PLASMAS**
RT   ELECTRON MOBILITY
    ELECTRON-HOLE DROPS
    HOLES (ELECTRON DEFICIENCIES)
    PLASMA PHYSICS
    SEMICONDUCTORS (MATERIALS)

SEMICONDUCTORS (MATERIALS)
GS   **SEMICONDUCTORS (MATERIALS)**
    . ACCEPTOR MATERIALS
    . AMORPHOUS SEMICONDUCTORS
    . DONOR MATERIALS
    . METAL OXIDE SEMICONDUCTORS
    . . CHARGE TRANSFER DEVICES
    . . CMOS
    . . SOS (SEMICONDUCTORS)
    . METAL-NITRIDE-OXIDE-SEMICONDUCTO
      RS
    . METAL-NITRIDE-OXIDE-SILICON
    . MIM (SEMICONDUCTORS)
    . MIS (SEMICONDUCTORS)
    . MOM (SEMICONDUCTORS)
    . N-TYPE SEMICONDUCTORS
    . ORGANIC SEMICONDUCTORS
    . P-TYPE SEMICONDUCTORS
    . PHOTOCONDUCTORS
    . SUPERLATTICES
    . VYCOR
RT   ALUMINUM ARSENIDES
    BIPOLAR TRANSISTORS
    BUCKET BRIGADE DEVICES
    CARRIER DENSITY (SOLID STATE)
    CARRIER INJECTION
    CHARGE COUPLED DEVICES
    CONDUCTION BANDS
    CONDUCTORS
    ELECTRIC CONDUCTORS
    ELECTRON DENSITY (CONCENTRATION)
    ELECTRON TUNNELING
    ELECTRONS
    EMITTERS
    EXCITONS
    GALLIUM NITRIDES
    HOLE DISTRIBUTION (ELECTRONICS)
    HOLES (ELECTRON DEFICIENCIES)
    INDIUM ANTIMONIDES
    INDIUM TELLURIDES
    INTERMETALLICS
    MAJORITY CARRIERS
    ∞MATERIALS
    MELTS (CRYSTAL GROWTH)
    METALLOIDS
    MINORITY CARRIERS
    ORGANIC CHARGE TRANSFER SALTS
    POLYACETYLENE
    PSEUDOPOTENTIALS
    RESISTORS
    SCREEN EFFECT
    SEMICONDUCTOR DEVICES
    SEMICONDUCTOR PLASMAS
    ∞SOLID STATE PHYSICS
    THERMOELECTRIC MATERIALS

SEMIEMPIRICAL EQUATIONS
RT   ALGEBRA
    ∞EQUATIONS
    PARAMETERIZATION

SEMIMETALS
USE   METALLOIDS

SEMISOLIDS
RT   PLASTIC PROPERTIES
    THIXOTROPY
    VISCOUS FLUIDS

SEMISPAN MODELS
GS   MODELS
    . **SEMISPAN MODELS**
RT   AERODYNAMIC CONFIGURATIONS
    AIRCRAFT MODELS
    SCALE MODELS

SEMISPAN MODELS-*(CONT.)*
    WIND TUNNEL MODELS

SENARMONT POLARISCOPES
GS   MEASURING INSTRUMENTS
    . POLARISCOPES
    . . **SENARMONT POLARISCOPES**
    OPTICAL EQUIPMENT
    . POLARISCOPES
    . . **SENARMONT POLARISCOPES**
RT   LASERS
    OPTICAL MEASURING INSTRUMENTS

SENDERS
USE   TRANSMITTERS

SENECA AIRCRAFT
USE   PA-34 SENECA AIRCRAFT

SENEGAL
GS   NATIONS
    . **SENEGAL**
RT   AFRICA

SENSE ORGANS
GS   ANATOMY
    . **SENSE ORGANS**
    . . BARORECEPTORS
    . . CHEMORECEPTORS
    . . EAR
    . . . CORTI ORGAN
    . . . EARDRUMS
    . . . LABYRINTH
    . . . . COCHLEA
    . . . . VESTIBULES
    . . . MASTOIDS
    . . . MIDDLE EAR
    . . . SEMICIRCULAR CANALS
    . . EUSTACHIAN TUBES
    . . EYE (ANATOMY)
    . . . CHOROID MEMBRANES
    . . . CONJUNCTIVA
    . . . CORNEA
    . . . OCULOMOTOR NERVES
    . . . PUPILS
    . . . RETINA
    . . . . FOVEA
    . . GRAVIRECEPTORS
    . . . OTOLITH ORGANS
    . . MECHANORECEPTORS
    . . PHOTORECEPTORS
    . . PROPRIOCEPTORS
    . . THERMORECEPTORS
RT   FINGERS
    HEAD (ANATOMY)
    NERVOUS SYSTEM
    OLFACTORY PERCEPTION
    PERCEPTUAL TIME CONSTANT
    RECEPTORS (PHYSIOLOGY)
    SENSITOMETRY
    SKIN (ANATOMY)

SENSES
USE   SENSORY PERCEPTION

SENSIBILITY
USE   SENSITIVITY

SENSING
USE   DETECTION

SENSITIVITY
UF   INSENSITIVITY
    SENSIBILITY
GS   **SENSITIVITY**
    . ANAPHYLAXIS
    . IMPACT RESISTANCE
    . NOTCH SENSITIVITY
    . PAIN SENSITIVITY
    . PHOTOSENSITIVITY
    . . LIGHT ADAPTATION
    . . PHOTOTROPISM
    . PROPELLANT SENSITIVITY
    . RADIATION TOLERANCE
    . SENSITOMETRY
    . SPECTRAL SENSITIVITY
RT   ACUITY
    ADAPTATION
    AMPLIFICATION
    AUDITORY PERCEPTION
    DYNAMIC CHARACTERISTICS
    DYNAMIC RESPONSE
    ∞FREQUENCY RESPONSE
    ITCHING

SENSITIVITY-*(CONT.)*
    PERCEPTION
    PRECISION
    RANGE (EXTREMES)
    REACTION TIME
    ∞RESISTANCE
    RESOLUTION
    SENSITIZING
    SHOCK RESISTANCE
    ∞THRESHOLDS
    THRESHOLDS (PERCEPTION)
    TOLERANCES (MECHANICS)
    TRANSFER FUNCTIONS
    TRANSIENT RESPONSE
    VISIBILITY
    VULNERABILITY

SENSITIZING
RT   ACTIVATION
    ACTUATION
    ANAPHYLAXIS
    CORROSION PREVENTION
    SENSITIVITY

SENSITOMETRY
GS   SENSITIVITY
    . **SENSITOMETRY**
RT   GRAVIRECEPTORS
    MECHANORECEPTORS
    PHOTORECEPTORS
    PHOTOSENSITIVITY
    PROPRIOCEPTORS
    RADIATION MEASUREMENT
    RECEPTORS (PHYSIOLOGY)
    SENSE ORGANS
    THERMORECEPTORS

SENSORIMOTOR PERFORMANCE
GS   **SENSORIMOTOR PERFORMANCE**
    . PSYCHOMOTOR PERFORMANCE
    . . PSYCHOSOMATICS
RT   AFFERENT NERVOUS SYSTEMS
    EFFERENT NERVOUS SYSTEMS
    HUMAN PERFORMANCE
    HUMAN REACTIONS
    PERCEPTUAL TIME CONSTANT
    PHYSIOLOGICAL TESTS
    PILOT PERFORMANCE
    REACTION TIME
    SENSORY FEEDBACK

∞ **SENSORS**
SN   *(USE OF A MORE SPECIFIC TERM IS*
    RECOMMENDED--CONSULT THE TERMS
    LISTED BELOW)
UF   PICKOFFS
    PICKUPS
RT   BIOINSTRUMENTATION
    CHARACTER RECOGNITION
    CHARGE FLOW DEVICES
    CONTOUR SENSORS
    CROP IDENTIFICATION
    DATA ACQUISITION
    ELECTRONIC TRANSDUCERS
    FLIR DETECTORS
    GAS DETECTORS
    GUIDANCE SENSORS
    IMAGE VELOCITY SENSORS
    LASER GYROSCOPES
    MEASURING INSTRUMENTS
    MICROWAVE SENSORS
    MULTISPECTRAL LINEAR ARRAYS
    REMOTE SENSORS
    SERVOMOTORS
    TRANSDUCERS

SENSORY DEPRIVATION
GS   DEPRIVATION
    . **SENSORY DEPRIVATION**
RT   CONFINEMENT
    CONFINING
    MONOTONY
    PERCEPTION

SENSORY DISCRIMINATION
GS   DISCRIMINATION
    . **SENSORY DISCRIMINATION**
    . . TACTILE DISCRIMINATION
    . . VISUAL DISCRIMINATION
RT   SPEECH RECOGNITION
    TIME DISCRIMINATION

SENSORY FEEDBACK
UF   FEELINGS
GS   FEEDBACK

SENSORY FEEDBACK-*(CONT.)*
. BIOFEEDBACK
. . **SENSORY FEEDBACK**
RT   BIOFEEDBACK
     EMOTIONAL FACTORS
     EMOTIONS
     MOODS
     MOON ILLUSION
     NONLINEAR FEEDBACK
     PERCEPTION
     SENSORIMOTOR PERFORMANCE

SENSORY PERCEPTION
UF   SENSES
GS   PERCEPTION
     . **SENSORY PERCEPTION**
     . . CONSCIOUSNESS
     . . EXTRASENSORY PERCEPTION
     . . KINESTHESIA
     . . OLFACTORY PERCEPTION
     . . PAIN
     . . PAIN SENSITIVITY
     . . PROPRIOCEPTION
     . . . AUTOKINESIS
     . . TASTE
     . . TOUCH
     . . . TACTILE DISCRIMINATION
     . . VERTICAL PERCEPTION
     . . VIBRATION PERCEPTION
     . . VISUAL PERCEPTION
     . . . CRITICAL FLICKER FUSION
     . . . SPACE PERCEPTION
     . . . . AUTOKINESIS
     . . . VISUAL DISCRIMINATION
RT   AFTERIMAGES
     ANESTHESIA
     ELECTROCUTANEOUS COMMUNICATION
     ITCHING

SENSORY STIMULATION
GS   STIMULATION
     . **SENSORY STIMULATION**
RT   CHRONAXY
     EMOTIONAL FACTORS
     SUBLIMINAL STIMULI

SENTENCES
GS   LANGUAGES
     . **SENTENCES**
     . . WORDS (LANGUAGE)
     . . . SYLLABLES
     LINGUISTICS
     . SYNTAX
     . . **SENTENCES**
     . . . WORDS (LANGUAGE)
     . . . . SYLLABLES
RT   COMMUNICATION THEORY
     MESSAGES
     SEMANTICS
     SIGNAL RECEPTION
     SIGNAL TRANSMISSION
     SPEECH
     TALKING

SENTINEL SYSTEM
GS   WEAPON SYSTEMS
     . **SENTINEL SYSTEM**
RT   ANTIMISSILE DEFENSE
     ANTIMISSILE MISSILES
     CIVIL DEFENSE
     NIKE MISSILES
     SAFEGUARD SYSTEM
     SPARTAN MISSILE
     SPRINT MISSILE
     SURFACE TO AIR MISSILES
     ∞SYSTEMS

SEO (INDIAN SPACECRAFT)
USE   INDIAN SPACECRAFT

SEOCS (SATELLITE)
GS   SATELLITES
     . ARTIFICIAL SATELLITES
     . . METEOROLOGICAL SATELLITES
     . . . **SEOCS (SATELLITE)**

SEOS
USE   SYNCHRONOUS EARTH OBSERVATORY
      SATELLITE

SEPAC (PAYLOAD)
UF   SPACE EXPER WITH PARTICLE
     ACCELERATORS
GS   PAYLOADS
     . **SEPAC (PAYLOAD)**

SEPAC (PAYLOAD)-*(CONT.)*
RT   ∞ACCELERATORS
     PARTICLE ACCELERATORS
     SPACELAB

SEPARATED FLOW
UF   FLOW SEPARATION
GS   FLUID FLOW
     . VISCOUS FLOW
     . . BOUNDARY LAYER FLOW
     . . . **SEPARATED FLOW**
     . . . . BOUNDARY LAYER SEPARATION
RT   CAVITATION FLOW
     CONICAL FLOW
     CROCCO-LEE THEORY
     FLOW CHARACTERISTICS
     FLOW DISTRIBUTION
     REATTACHED FLOW
     REVERSED FLOW
     ∞SEPARATION
     SURFACE ROUGHNESS EFFECTS
     TURBULENCE EFFECTS
     VORTEX FLAPS

∞ **SEPARATION**
SN   *(USE OF A MORE SPECIFIC TERM IS*
     RECOMMENDED--CONSULT THE TERMS
     LISTED BELOW)
UF   SEGREGATION
RT   ADSORPTION
     AERATION
     AGGLOMERATION
     AGITATION
     BENEFICIATION
     BOUNDARY LAYER SEPARATION
     BREAKING
     CENTRIFUGING
     CHEMICAL FRACTIONATION
     CHIPPING
     CLASSIFIERS
     CLEANING
     COAGULATION
     COALESCING
     COANDA EFFECT
     COLLOIDS
     CONCENTRATING
     CONDENSING
     CRYSTALLIZATION
     CUTTING
     DECONTAMINATION
     DEGASSING
     DEHUMIDIFICATION
     DEHYDRATION
     DEIONIZATION
     DELAMINATING
     DEMINERALIZING
     DEOXYGENATION
     DEPOSITION
     DESCALING
     DESORPTION
     DIALYSIS
     DIFFUSION
     DISPERSING
     DISSOLVING
     DISTILLATION
     DIVERTERS
     ∞DIVISION
     DRYING
     ELECTRODIALYSIS
     ELECTROSTATIC PRECIPITATORS
     ELIMINATION
     ELUTION
     EVAPORATION
     EXCHANGING
     EXCLUSION
     EXTERNAL STORE SEPARATION
     EXTRACTION
     FILTRATION
     FLAKING
     FLASHING (VAPORIZING)
     FLOTATION
     FLUSHING
     FOAMING
     FRACTIONATION
     FRACTURING
     HOMOGENIZING
     ION EXCHANGING
     ION EXTRACTION
     ION STRIPPING
     ISOLATION
     LEACHING
     MATERIALS RECOVERY
     MELTING
     MIXING
     OSMOSIS

SEPARATION-*(CONT.)*
     PERCOLATION
     POLARIZATION (CHARGE SEPARATION)
     PRECIPITATION (CHEMISTRY)
     PURGING
     PURIFICATION
     RADIOCHEMICAL SEPARATION
     RECRYSTALLIZATION
     REFINING
     REMOVAL
     SCRAPERS
     SEPARATED FLOW
     SEPARATORS
     SETTLING
     SHAKING
     SHEARING
     SIZE SEPARATION
     SLICING
     SOLVENT EXTRACTION
     SORPTION
     SPACING
     SPLITTING
     SPREADING
     STAGE SEPARATION
     STRIPPING (DISTILLATION)
     SUBLIMATION
     SWIRLING
     THERMAL DIFFUSION
     TUMBLING MOTION
     VAPORIZING
     VENTING
     WASHING
     ZONE MELTING

SEPARATORS
UF   BATTERY SEPARATORS
GS   **SEPARATORS**
     . CLASSIFIERS
     . . SIZING SCREENS
     . . THICKENERS (EQUIPMENT)
     . DIVIDERS
     . DRYING APPARATUS
     . . DESICCATORS
     . DUST COLLECTORS
     . EVAPORATORS
     . FLUID FILTERS
       AIR FILTERS
     . PRECIPITATORS
     . . ELECTROSTATIC PRECIPITATORS
     . SIEVES
     . SPIRALS (CONCENTRATORS)
     . STILLS
RT   CENTRIFUGES
     CLEANERS
     COLUMNS (PROCESS ENGINEERING)
     CONCENTRATING
     CONCENTRATORS
     CONDENSERS (LIQUEFIERS)
     CURTAINS
     ∞DIFFUSERS
     DIVERTERS
     ∞FILTERS
     FLOATS
     FLUIDIZED BED PROCESSORS
     FURNACES
     ION EXCHANGE MEMBRANE
       ELECTROLYTES
     MIXERS
     ∞SEPARATION
     SHAKERS
     SPACERS
     TRAPS
     VAPORIZERS
     WASHERS (CLEANERS)
     WASHERS (SPACERS)
     WINDOWS (APERTURES)

SEPTUM
RT   MEDIASTINUM
     MEMBRANES
     ∞PARTITIONS

SEQUENCING
RT   CONSECUTIVE EVENTS
     COORDINATION
     CRITICAL PATH METHOD
     INTERRUPTION
     ∞OPERATIONS
     OPERATIONS RESEARCH
     PACKET SWITCHING
     PETRI NETS
     PLANNING
     PRIORITIES
     RANKING
     SCHEDULING

SEQUENCING-*(CONT.)*
    SEQUENTIAL CONTROL
    SWITCHING
    SWITCHING THEORY
    TURNAROUND (STS)

SEQUENTIAL ANALYSIS
GS   STATISTICAL ANALYSIS
    . **SEQUENTIAL ANALYSIS**
RT   QUALITY CONTROL
    SAMPLING

SEQUENTIAL COMPUTERS
GS   DATA PROCESSING EQUIPMENT
    . COMPUTERS
    . . DIGITAL COMPUTERS
    . . . **SEQUENTIAL COMPUTERS**

SEQUENTIAL CONTROL
GS   AUTOMATIC CONTROL
    . **SEQUENTIAL CONTROL**
RT   ACCURACY
    COMPUTER PROGRAMMING
    CONSECUTIVE EVENTS
    ∞CONTROL
    DATA FLOW ANALYSIS
    NUMERICAL CONTROL
    SEQUENCING

SERGEANT MISSILES
GS   MISSILES
    . SURFACE TO SURFACE MISSILES
    . . **SERGEANT MISSILES**
RT   JUNO 1 LAUNCH VEHICLE
    JUNO 2 LAUNCH VEHICLE
    JUPITER C ROCKET VEHICLE
    LITTLE JOE 2 LAUNCH VEHICLE
    SOLID PROPELLANT ROCKET ENGINES

SERGENIUM
GS   CHEMICAL ELEMENTS
    . ACTINIDE SERIES
    . . TRANSURANIUM ELEMENTS
    . . . **SERGENIUM**
    . NUCLIDES
    . . ISOTOPES
    . . . RADIOACTIVE ISOTOPES
    . . . . TRANSURANIUM ELEMENTS
    . . . . . **SERGENIUM**
    HEAVY ELEMENTS
    . TRANSURANIUM ELEMENTS
    . . **SERGENIUM**
    METALS
    . ACTINIDE SERIES
    . . TRANSURANIUM ELEMENTS
    . . . **SERGENIUM**

SERIES (MATHEMATICS)
GS   ANALYSIS (MATHEMATICS)
    . CALCULUS
    . . **SERIES (MATHEMATICS)**
    . . . ASYMPTOTIC SERIES
    . . . CAMPBELL-HAUSDORFF SERIES
    . . . COSINE SERIES
    . . . FOURIER SERIES
    . . . PADE APPROXIMATION
    . . . POWER SERIES
    . . . . TAYLOR SERIES
    . . . . . MACLAURIN SERIES
    . . . PROGRESSIONS
    . . . PRONY SERIES
    . . . SINE SERIES
    . REAL VARIABLES
    . . **SERIES (MATHEMATICS)**
    . . . ASYMPTOTIC SERIES
    . . . CAMPBELL-HAUSDORFF SERIES
    . . . COSINE SERIES
    . . . FOURIER SERIES
    . . . PADE APPROXIMATION
    . . . POWER SERIES
    . . . . TAYLOR SERIES
    . . . . . MACLAURIN SERIES
    . . . PROGRESSIONS
    . . . PRONY SERIES
    . . . SINE SERIES
RT   ABEL FUNCTION
    CHEBYSHEV APPROXIMATION
    DIVERGENCE
    FORM FACTORS
    FOURIER-BESSEL TRANSFORMATIONS
    FUNCTION SPACE
    FUNCTIONAL ANALYSIS
    GIBBS PHENOMENON
    INFINITY
    SERIES EXPANSION

SERIES (MATHEMATICS)-*(CONT.)*
    SUMS

SERIES EXPANSION
GS   EXPANSION
    . **SERIES EXPANSION**
RT   ASYMPTOTIC SERIES
    DIVERGENCE
    ∞MATHEMATICS
    SERIES (MATHEMATICS)

SEROTONIN
GS   AMINES
    . TRYPTAMINES
    . . **SEROTONIN**
    ORGANIC COMPOUNDS
    . **SEROTONIN**

SERPENTINE
GS   MINERALS
    . **SERPENTINE**
RT   ASBESTOS
    CHROMITES
    ROCKS
    SOILS

SERRATIA
GS   MICROORGANISMS
    . BACTERIA
    . . **SERRATIA**

SERT (ROCKET TESTS)
USE   SPACE ELECTRIC ROCKET TESTS

SERT 1 SPACECRAFT
RT   ELECTRIC PROPULSION
    ELECTRIC ROCKET ENGINES
    ENGINE TESTS
    SPACE ELECTRIC ROCKET TESTS
    ∞SPACECRAFT

SERT 2 SPACECRAFT
RT   ELECTRIC PROPULSION
    ELECTRIC ROCKET ENGINES
    ENGINE TESTS
    SPACE ELECTRIC ROCKET TESTS
    ∞SPACECRAFT

SERUMS
GS   **SERUMS**
    . INOCULUM
RT   ANTISERUMS
    ∞FLUIDS
    PROTEINS

SERVICE LIFE
UF   MACHINE LIFE
GS   LIFE (DURABILITY)
    . **SERVICE LIFE**
RT   ACCELERATED LIFE TESTS
    ∞EQUIPMENT
    FATIGUE LIFE
    MAINTENANCE
    RETIREMENT FOR CAUSE

SERVICE MODULES
GS   MODULES
    . **SERVICE MODULES**
    SPACECRAFT COMPONENTS
    . **SERVICE MODULES**
RT   APOLLO SPACECRAFT
    COMMAND MODULES
    SPACECRAFT DOCKING MODULES
    SPACECRAFT MODULES

SERVICES
GS   **SERVICES**
    . MEDICAL SERVICES
    . METEOROLOGICAL SERVICES
RT   ∞FOOD
    GOVERNMENT PROCUREMENT
    INVENTORY MANAGEMENT
    LOGISTICS
    LOGISTICS MANAGEMENT
    MATERIALS HANDLING
    PERSONNEL
    PROCUREMENT
    PROCUREMENT MANAGEMENT
    PRODUCTS
    SITE SELECTION
    SUPPORT SYSTEMS
    TRANSPORTATION
    UTILITIES

SERVOAMPLIFIERS
GS   AMPLIFIERS
    . **SERVOAMPLIFIERS**
    CONTROL EQUIPMENT
    . **SERVOAMPLIFIERS**
    CONTROLLERS
    . SERVOMECHANISMS
    . . **SERVOAMPLIFIERS**
RT   FEEDBACK AMPLIFIERS
    FLY BY TUBE CONTROL
    SERVOCONTROL

SERVOCONTROL
UF   SERVOSTABILITY CONTROL
RT   AIRCRAFT HYDRAULIC SYSTEMS
    AUTOMATIC CONTROL
    ∞CONTROL
    CONTROL MOMENT GYROSCOPES
    CONTROL THEORY
    DIGITAL COMMAND SYSTEMS
    FEEDBACK CONTROL
    HYDRAULIC EQUIPMENT
    MANIPULATORS
    MANUAL CONTROL
    OFF-ON CONTROL
    PNEUMATIC EQUIPMENT
    PROPORTIONAL CONTROL
    REMOTE CONTROL
    ROCKET ENGINE CONTROL
    SERVOAMPLIFIERS
    SERVOMECHANISMS
    SERVOMOTORS
    STEPPING MOTORS
    TURBOJET ENGINE CONTROL
    VISUAL CONTROL

SERVOMECHANISMS
GS   CONTROLLERS
    . **SERVOMECHANISMS**
    . . SERVOAMPLIFIERS
    . . SERVOMOTORS
RT   ACTIVE CONTROL
    ACTUATORS
    AIRCRAFT HYDRAULIC SYSTEMS
    AUTOMATIC CONTROL
    AUTOMATIC CONTROL VALVES
    ∞AUTOMATION
    ∞CONTROL
    CONTROL MOMENT GYROSCOPES
    ELECTRIC MOTORS
    FEEDBACK CONTROL
    HYDRAULIC EQUIPMENT
    PNEUMATIC EQUIPMENT
    RADAR EQUIPMENT
    REMOTE CONTROL
    ROBOTS
    SELF ALIGNMENT
    SERVOCONTROL
    STEPPING MOTORS

SERVOMOTORS
UF   MAGNESYN (TRADEMARK)
    SELSYNS (TRADEMARK)
    SERVOS
GS   CONTROLLERS
    . SERVOMECHANISMS
    . . **SERVOMOTORS**
    MOTORS
    . **SERVOMOTORS**
RT   ACTUATORS
    AMPLIDYNES
    AUTOMATIC CONTROL
    ELECTRIC MOTORS
    HELIOSTATS
    ∞ROTATING ELECTRICAL MACHINES
    ∞SENSORS
    SERVOCONTROL
    SLEWING
    SYNCHRONIZERS
    TORQUE MOTORS

SERVOS
USE   SERVOMOTORS

SERVOSTABILITY CONTROL
USE   SERVOCONTROL

SES
USE   SURFACE EFFECT SHIPS

SET
SN   (EXCLUDES SET THEORY)
GS   MECHANICAL PROPERTIES
    . **SET**
RT   DEFORMATION

SET-*(CONT.)*
    SHEAR PROPERTIES

SET THEORY
UF   SUBSETS (MATHEMATICS)
GS   MATHEMATICAL LOGIC
    . **SET THEORY**
    . . BOREL SETS
    . . EQUIVALENCE
    . . THRESHOLD LOGIC
RT   BOOLEAN ALGEBRA
    BRANCHING (MATHEMATICS)
    COMBINATORIAL ANALYSIS
    ∞CONJUNCTION
    FIBONACCI NUMBERS
    FRACTALS
    FUZZY SETS
    FUZZY SYSTEMS
    GRAPH THEORY
    HOMOTROPY
    HYPERPLANES
    LATTICES (MATHEMATICS)
    LEBESGUE THEOREM
    ORLICZ SPACE
    PERMUTATIONS
    ∞SPACE
    SUBDIVISIONS
    SUBGROUPS
    ∞THEORIES

SETI
USE   PROJECT SETI

∞ **SETTING**
SN   *(USE OF A MORE SPECIFIC TERM IS RECOMMENDED--CONSULT THE TERMS LISTED BELOW)*
RT   ADJUSTING
    COAGULATION
    CURING
    HARDENING (MATERIALS)
    POLYMERIZATION
    POSITIONING
    SOLIDIFICATION

SETTLING
RT   ACCUMULATIONS
    AGGLOMERATION
    AGITATION
    BENEFICIATION
    COAGULATION
    COALESCING
    CONCENTRATING
    CRYSTALLIZATION
    DEPOSITION
    EFFLUENTS
    FLOCCULATING
    FLOTATION
    PARTICLE MOTION
    PRECIPITATION (CHEMISTRY)
    ∞PROCESSING
    ∞SEPARATION
    SIZE SEPARATION
    STOKES LAW (FLUID MECHANICS)
    SUBSIDENCE
    WATER TREATMENT

SETUPS
RT   MACHINING
    TOOLING

SEVERE STORMS OBSERVING SATELLITE
USE   STORMSAT SATELLITE

SEWAGE
GS   WASTES
    . **SEWAGE**
    . . SOLID WASTES
RT   ACTIVATED SLUDGE
    EFFLUENTS
    ENVIRONMENT EFFECTS
    HUMAN WASTES
    LIQUID WASTES
    METABOLIC WASTES
    ORGANIC WASTES (FUEL CONVERSION)
    SEWERS
    WASTE DISPOSAL
    WATER TREATMENT

SEWAGE TREATMENT
GS   WASTE TREATMENT
    . **SEWAGE TREATMENT**
RT   AEROBES
    ANAEROBES
    CHEMICAL STERILIZATION

SEWAGE TREATMENT-*(CONT.)*
    FILTRATION
    MODULAR INTEGRATED UTILITY SYSTEM
    PURIFICATION
    SEDIMENTS
    SLUDGE
    ∞TREATMENT
    WASTE DISPOSAL

SEWERS
GS   PIPELINES
    . **SEWERS**
RT   DRAINAGE
    EFFLUENTS
    GARBAGE
    HUMAN WASTES
    METABOLIC WASTES
    SANITATION
    SEWAGE
    WASTE DISPOSAL
    WASTES

SEWING
RT   BINDING
    ∞JOINING
    NEEDLES
    WEAVING

SEX
RT   ∞DRIVES
    FEMALES
    MALES
    SEX FACTOR

SEX FACTOR
RT   FEMALES
    MALES
    PHYSIOLOGICAL FACTORS
    PHYSIOLOGY
    PSYCHOLOGICAL FACTORS
    SEX
    SEX GLANDS

SEX GLANDS
GS   ANATOMY
    . GLANDS (ANATOMY)
    . . **SEX GLANDS**
    . . . GONADS
    . . . OVARIES
    . . . PROSTATE GLAND
    . . . TESTES
    VISCERA
    . **SEX GLANDS**
    . . GONADS
    . . OVARIES
    . . PROSTATE GLAND
    . . TESTES
RT   ESTROGENS
    ∞REPRODUCTION
    SEX FACTOR

SEXTANTS
GS   MEASURING INSTRUMENTS
    . OPTICAL MEASURING INSTRUMENTS
    . . **SEXTANTS**
    OPTICAL EQUIPMENT
    . OPTICAL MEASURING INSTRUMENTS
    . . **SEXTANTS**
RT   NAVIGATION AIDS
    POSITION INDICATORS
    STADIMETERS
    THEODOLITES
    TRANSITS

SEYFERT GALAXIES
GS   CELESTIAL BODIES
    . GALAXIES
    . . **SEYFERT GALAXIES**
RT   GALACTIC NUCLEI
    INFRARED RADIATION
    LINE SPECTRA
    LUMINOUS INTENSITY
    STELLAR SPECTRA
    ULTRAVIOLET RADIATION

SFAR
USE   SOUND FIXING AND RANGING

SFERICS
USE   ATMOSPHERICS

SGEMP
USE   SYSTEM GENERATED ELECTROMAGNETIC PULSES

SGR (NUCLEAR REACTORS)
USE   SODIUM GRAPHITE REACTORS

SH-3 HELICOPTER
UF   HSS-2 HELICOPTER
    SEA KING HELICOPTER
    SIKORSKY HSS-2 HELICOPTER
GS   ANTISUBMARINE WARFARE AIRCRAFT
    . **SH-3 HELICOPTER**
    SIKORSKY AIRCRAFT
    . **SH-3 HELICOPTER**
    TRANSPORT AIRCRAFT
    . **SH-3 HELICOPTER**
    V/STOL AIRCRAFT
    . ROTARY WING AIRCRAFT
    . . HELICOPTERS
    . . . **SH-3 HELICOPTER**
RT   S-61 HELICOPTER
    SH-4 HELICOPTER

SH-4 HELICOPTER
GS   ANTISUBMARINE WARFARE AIRCRAFT
    . **SH-4 HELICOPTER**
    SIKORSKY AIRCRAFT
    . **SH-4 HELICOPTER**
    TRANSPORT AIRCRAFT
    . **SH-4 HELICOPTER**
    V/STOL AIRCRAFT
    . ROTARY WING AIRCRAFT
    . . HELICOPTERS
    . . . **SH-4 HELICOPTER**
RT   S-61 HELICOPTER
    SH-3 HELICOPTER

SHACKLETON BOMBER
GS   ATTACK AIRCRAFT
    . BOMBER AIRCRAFT
    . . **SHACKLETON BOMBER**
    HAWKER SIDDELEY AIRCRAFT
    . **SHACKLETON BOMBER**
    MONOPLANES
    . **SHACKLETON BOMBER**

SHADES
RT   LOUVERS
    SHIELDING
    ∞SHUTTERS

SHADOWGRAPH PHOTOGRAPHY
UF   SHADOWGRAPHS
    SPARK SHADOWGRAPH PHOTOGRAPHY
GS   IMAGERY
    . **SHADOWGRAPH PHOTOGRAPHY**
    . . SCHLIEREN PHOTOGRAPHY
    PHOTOGRAPHY
    . **SHADOWGRAPH PHOTOGRAPHY**
    . . SCHLIEREN PHOTOGRAPHY
RT   BLACK AND WHITE PHOTOGRAPHY
    COLOR PHOTOGRAPHY
    FLOW VISUALIZATION
    WIND TUNNEL MODELS

SHADOWGRAPHS
USE   SHADOWGRAPH PHOTOGRAPHY

SHADOWS
GS   **SHADOWS**
    . LUNAR SHADOW
    . PENUMBRAS
RT   CLOUD COVER
    CLOUDS (METEOROLOGY)
    DARKNESS
    ILLUMINATING
    LIGHT (VISIBLE RADIATION)
    NIGHT
    SELF SHADOWING
    UMBRAS

SHAFTS (MACHINE ELEMENTS)
UF   AXLES
    JOURNALS (SHAFTS)
    TRUNNIONS
GS   ROTATING SHAFTS
    . **SHAFTS (MACHINE ELEMENTS)**
    . . TURBOSHAFTS
RT   AXES OF ROTATION
    BEARINGS
    BUSHINGS
    ∞JOURNALS
    ∞LOADING
    LOADS (FORCES)
    MANDRELS
    MECHANICAL DRIVES
    PACKINGS (SEALS)
    PINTLES

SHAFTS (MACHINE ELEMENTS)-*(CONT.)*
    PIVOTS
    ROTATING CYLINDERS
    SPINDLES
    SUPPORTS
    TORQUE
    TRANSMISSIONS (MACHINE ELEMENTS)
    VEHICLE WHEELS
    WHEELS

SHAKERS
RT CLASSIFIERS
    MIXERS
    SEPARATORS
    SHAKING
    SIEVES
    SIZING SCREENS
    VERTICAL MOTION SIMULATORS
    VIBRATION SIMULATORS

SHAKING
GS **SHAKING**
    . DITHERS
RT AGITATION
    BUFFETING
    DISPERSING
    EPILEPSY
    FLAPPING
    FLUTTER
    MIXING
    ∞SEPARATION
    SHAKERS
    STRUCTURAL VIBRATION
    SUSPENDING (MIXING)
    SWIRLING
    VIBRATION

SHALE OIL
GS OILS
    . SHALE OIL
RT FUEL OILS
    FUELS
    GASOLINE
    HYDROCARBON FUELS
    KEROGEN
    KEROSENE
    LUBRICATING OILS
    PARAFFINS
    RETORT PROCESSING

SHALES
GS ROCKS
    . SHALES
    SEDIMENTARY ROCKS
    . SHALES
RT BOREHOLES
    CLAYS
    EARTH RESOURCES
    MINERALS
    SOILS

SHALLOW SHELL EQUATIONS
RT END PLATES
    ∞EQUATIONS
    PRESSURE VESSELS
    STRESS ANALYSIS

SHALLOW SHELLS
GS SHELLS (STRUCTURAL FORMS)
    . SHALLOW SHELLS
RT CRITICAL LOADING
    SHELL STABILITY
    SHELL THEORY

SHALLOW WATER
GS WATER
    . SHALLOW WATER
RT CNOIDAL WAVES
    FISHERIES
    OCEANOGRAPHY
    OCEANS
    REEFS
    SEAS
    SHORELINES
    TOPOGRAPHY
    WATER DEPTH

SHANKS
USE JOINTS (JUNCTIONS)

SHANNON INFORMATION THEORY
USE INFORMATION THEORY

SHANNON-WIENER MEASURE
RT ENTROPY
    INFORMATION THEORY
    RANDOM VARIABLES

SHAPE CONTROL
RT ACTUATORS
    ∞CONTROL
    CONTROL THEORY
    FLEXIBLE SPACECRAFT
    LARGE SPACE STRUCTURES
    SPACE PLATFORMS
    SPACECRAFT CONTROL

SHAPE MEMORY ALLOYS
GS ALLOYS
    . SHAPE MEMORY ALLOYS
    . . NITINOL ALLOYS
RT MICROSTRUCTURE
    NICKEL ALLOYS
    PHASE TRANSFORMATIONS
    PLASTIC MEMORY
    STRESS-STRAIN DIAGRAMS
    TEMPERATURE EFFECTS
    TITANIUM ALLOYS
    TRANSITION METALS

SHAPED CHARGES
GS EXPLOSIVE DEVICES
    . SHAPED CHARGES
RT AMMUNITION
    BOMBS (ORDNANCE)
    EXPLOSIVE FORMING
    EXPLOSIVES
    PROJECTILES
    TORPEDOES
    WARHEADS
    WEAPONS

SHAPERS
GS TOOLS
    . MACHINE TOOLS
    . . SHAPERS
RT GRINDING MACHINES
    MILLING MACHINES

SHAPES
UF CURVED SURFACES
    FORM
GS **SHAPES**
    . CONVEXITY
    . ELLIPTICITY
    . FLATNESS
    . LINE SHAPE
    . OGEE SHAPE
    . ROSETTE SHAPES
    . T SHAPE
RT ASYMMETRY
    CONCAVITY
    CONTOUR SENSORS
    CONTOURS
    CORNERS
    ∞CROSS SECTIONS
    CURVATURE
    CURVED PANELS
    GEOIDS
    GEOMETRY
    MORPHOLOGY
    OBLATE SPHEROIDS
    PLANFORMS
    ∞PROFILES
    PROFILOMETERS
    PROLATENESS
    ∞SURFACE GEOMETRY
    SYMMETRY
    TOPOLOGY

SHARKS
GS ANIMALS
    . VERTEBRATES
    . . FISHES
    . . . SHARKS

SHARP LEADING EDGES
GS EDGES
    . LEADING EDGES
    . . SHARP LEADING EDGES

∞ **SHARPNESS**
SN *(USE OF A MORE SPECIFIC TERM IS
    RECOMMENDED--CONSULT THE TERMS
    LISTED BELOW)*
RT CLARITY
    CONTRAST
    PRECISION

SHATTER CONES
GS CONES
    . SHATTER CONES
    ROCKS
    . SHATTER CONES
RT CARBONACEOUS ROCKS
    CRUSTAL FRACTURES
    FORMATIONS
    GEOLOGY
    GEOMORPHOLOGY
    METEORITE COLLISIONS
    METEORITE CRATERS
    SEDIMENTARY ROCKS
    SHOCK LOADS
    STRIATION
    STRUCTURAL PROPERTIES (GEOLOGY)

SHATTERING
USE FRAGMENTATION

SHAWNEE HELICOPTER
USE CH-21 HELICOPTER

∞ **SHEAR**
SN *(USE OF A MORE SPECIFIC TERM IS
    RECOMMENDED--CONSULT THE TERMS
    LISTED BELOW)*
RT DILATATIONAL WAVES
    SHEARING
    SHEARS

SHEAR CREEP
GS MECHANICAL PROPERTIES
    . CREEP PROPERTIES
    . . SHEAR CREEP
RT PLASTIC DEFORMATION
    TENSILE CREEP

SHEAR DISTURBANCES
USE S WAVES

SHEAR FATIGUE
USE SHEAR STRESS

SHEAR FLOW
GS FLUID FLOW
    . SHEAR FLOW
RT COAXIAL FLOW
    CORE FLOW
    CREEP PROPERTIES
    ∞FLOW
    GRAZING FLOW
    KOLMOGOROFF THEORY
    KROOK EQUATION
    MIXING LENGTH FLOW THEORY
    PLASTIC FLOW
    RICHARDSON NUMBER
    STRATIFIED FLOW

SHEAR LAYERS
UF CHAPMAN SHEAR LAYER
RT BOUNDARY LAYERS
    IONOSPHERE
    ∞LAYERS
    SHOCK LAYERS
    SHOCK WAVE CONTROL
    ∞TRANSITION LAYERS

SHEAR PROPERTIES
GS MECHANICAL PROPERTIES
    . SHEAR PROPERTIES
    . . SHEAR STRENGTH
RT CREEP PROPERTIES
    DUCTILITY
    FATIGUE (MATERIALS)
    HOOKES LAW
    HYSTERESIS
    IMPACT STRENGTH
    MODULUS OF ELASTICITY
    ∞PROPERTIES
    RESILIENCE
    SET
    STRESS RELAXATION
    STRESS-STRAIN DIAGRAMS
    STRESSES
    TEMPERATURE INVERSIONS
    TOUGHNESS

SHEAR STRAIN
RT MECHANICAL PROPERTIES
    STRUCTURAL STRAIN
    TORSIONAL VIBRATION

SHEAR STRENGTH
GS   MECHANICAL PROPERTIES
     . SHEAR PROPERTIES
     . . **SHEAR STRENGTH**
RT   COMPRESSIVE STRENGTH
     FIBER STRENGTH
     HIGH STRENGTH
     INTERFACIAL ENERGY
     ∞STRENGTH
     TENSILE STRENGTH

SHEAR STRESS
UF   SHEAR FATIGUE
     SHEARING STRESS
GS   STRESSES
     . **SHEAR STRESS**
     . . TORSIONAL STRESS
RT   MECHANICAL PROPERTIES

SHEAR WAVES
USE   S WAVES

SHEARING
GS   CUTTING
     . **SHEARING**
RT   BLANKING (CUTTING)
     COLD WORKING
     FAILURE
     FAILURE MODES
     HOT WORKING
     LOADS (FORCES)
     METAL CUTTING
     METAL WORKING
     ∞SEPARATION
     ∞SHEAR
     SHEARS
     STAMPING
     STRUCTURAL STRAIN

SHEARING STRESS
USE   SHEAR STRESS

SHEARS
GS   CUTTERS
     SHEARS
     TOOLS
     . **SHEARS**
RT   MACHINE TOOLS
     SAWS
     ∞SHEAR
     SHEARING

SHEATHS
GS   **SHEATHS**
     . ION SHEATHS
     . PLASMA SHEATHS
RT   ∞CASING
     ENCAPSULATING
     FAIRINGS
     JACKETS
     LININGS
     PROTECTORS
     ROOFS
     WALLS

SHEDDING
RT   EJECTION
     MOLTING
     PEELING
     VORTEX SHEDDING

SHEDS
RT   SHELTERS

SHEEP
GS   ANIMALS
     . VERTEBRATES
     . . MAMMALS
     . . . **SHEEP**
RT   LIVESTOCK
     WOOL

SHEET METAL
USE   METAL SHEETS

∞ **SHEETS**
SN   *(USE OF A MORE SPECIFIC TERM IS*
     RECOMMENDED--CONSULT THE TERMS
     LISTED BELOW)
RT   COATINGS
     CURRENT SHEETS
     ELASTIC SHEETS
     FABRICS
     FLAT PLATES

SHEETS-*(CONT.)*
     LAMINATES
     MEMBRANE STRUCTURES
     MEMBRANES
     METAL FOILS
     METAL SHEETS
     MULTILAYER INSULATION
     NEUTRAL SHEETS
     PANELS
     PAPERS
     POLYMERIC FILMS
     THICK PLATES
     THIN PLATES
     VORTEX SHEETS
     VORTEX STREETS
     WEBS (SHEETS)

SHELL ANODES
GS   ELECTRODES
     . ANODES
     . . **SHELL ANODES**
RT   HEAT MEASUREMENT

SHELL STABILITY
GS   MECHANICAL PROPERTIES
     . DIMENSIONAL STABILITY
     . . STRUCTURAL STABILITY
     . . . **SHELL STABILITY**
     STABILITY
     . STATIC STABILITY
     . . DIMENSIONAL STABILITY
     . . . STRUCTURAL STABILITY
     . . . . **SHELL STABILITY**
RT   BUCKLING
     FLUID FILLED SHELLS
     LIQUID FILLED SHELLS
     ORTHOTROPIC SHELLS
     REINFORCED SHELLS
     SHALLOW SHELLS

SHELL THEORY
RT   PERFORATED SHELLS
     SELF CONSISTENT FIELDS
     SHALLOW SHELLS
     ∞THEORIES

SHELLFISH
RT   COASTAL WATER
     MARINE BIOLOGY
     MARINE ENVIRONMENTS
     MARINE RESOURCES
     MOLLUSKS

SHELLS (STRUCTURAL FORMS)
GS   **SHELLS (STRUCTURAL FORMS)**
     . ANISOTROPIC SHELLS
     . CIRCULAR SHELLS
     . CONICAL SHELLS
     . CORRUGATED SHELLS
     . CYLINDRICAL SHELLS
     . DOMES (STRUCTURAL FORMS)
     . . RADOMES
     . ELASTIC SHELLS
     . FLUID FILLED SHELLS
     . . LIQUID FILLED SHELLS
     . HEMISPHERICAL SHELLS
     . METAL SHELLS
     . ORTHOTROPIC SHELLS
     . PERFORATED SHELLS
     . REINFORCED SHELLS
     . SHALLOW SHELLS
     . SPHERICAL SHELLS
     . . SPHERICAL CAPS
     . THIN WALLED SHELLS
     . TOROIDAL SHELLS
RT   AIRCRAFT STRUCTURES
     ARCHES
     BAYS (STRUCTURAL UNITS)
     ∞CAPSULES
     COVERINGS
     COWLINGS
     ENCLOSURES
     FAIRINGS
     HOUSINGS
     HULLS (STRUCTURES)
     ISOTENSOID STRUCTURES
     MEMBRANE STRUCTURES
     MEMBRANES
     MONOCOQUE STRUCTURES
     NACELLES
     PRESSURE VESSEL DESIGN
     PROTUBERANCES
     ROCKET ENGINE CASES
     SKIN (STRUCTURAL MEMBER)
     WALLS

SHELTERS
GS   **SHELTERS**
     . LUNAR SHELTERS
RT   ∞BUILDINGS
     CIVIL DEFENSE
     ENVIRONMENTAL ENGINEERING
     HABITABILITY
     SHEDS
     STARSITE PROGRAM
     SURVIVAL

∞ **SHELVES**
SN   *(USE OF A MORE SPECIFIC TERM IS*
     RECOMMENDED--CONSULT THE TERMS
     LISTED BELOW)
RT   BEDROCK
     CASES (CONTAINERS)
     CLIFFS
     CONTINENTAL SHELVES
     RACKS (FRAMES)
     REEFS

SHENANDOAH VALLEY (VA)
GS   VALLEYS
     . **SHENANDOAH VALLEY (VA)**
RT   RIVER BASINS
     VIRGINIA

SHIELDING
GS   **SHIELDING**
     . ELECTROMAGNETIC SHIELDING
     . . RADIO FREQUENCY SHIELDING
     . ELECTROSTATIC SHIELDING
     . HEAT SHIELDING
     . . REENTRY SHIELDING
     . . REUSABLE HEAT SHIELDING
     . MAGNETIC SHIELDING
     . RADIATION SHIELDING
     . . SOLAR RADIATION SHIELDING
     . SPACECRAFT SHIELDING
RT   ABLATIVE NOSE CONES
     ABSORBERS (MATERIALS)
     ARMOR
     ATTENUATION
     ATTENUATORS
     BAFFLES
     ∞BARRIERS
     BLAST DEFLECTORS
     BLINDS
     DEFLECTORS
     DIVERTERS
     ENCLOSURES
     FLAME DEFLECTORS
     GUARDS (SHIELDS)
     HOUSINGS
     LININGS
     LOUVERS
     MANIPULATORS
     PANELS
     PROTECTION
     PROTECTORS
     SAFETY DEVICES
     ∞SCREENS
     SHADES
     SUPPRESSORS
     WINDOWS (APERTURES)
     WINDSHIELDS

SHIELDS (GEOLOGY)
USE   BEDROCK

∞ **SHIFT**
SN   *(USE OF A MORE SPECIFIC TERM IS*
     RECOMMENDED--CONSULT THE TERMS
     LISTED BELOW)
RT   EXCHANGING
     FREQUENCY SHIFT
     PHASE SHIFT
     SHIFT REGISTERS
     TRANSFERRING

SHIFT REGISTERS
RT   COMPUTER COMPONENTS
     COMPUTER STORAGE DEVICES
     DELAY LINES (COMPUTER STORAGE)
     DIGITAL TECHNIQUES
     REGISTERS (COMPUTERS)
     ∞SHIFT

SHIFTING EQUILIBRIUM FLOW
GS   FLUID FLOW
     . GAS FLOW
     . . EQUILIBRIUM FLOW
     . . . **SHIFTING EQUILIBRIUM FLOW**
RT   FROZEN EQUILIBRIUM FLOW

SHILLELAGH MISSILES
  GS   MISSILES
        . SURFACE TO SURFACE MISSILES
        .. ANTITANK MISSILES
        ... SHILLELAGH MISSILES

SHIP HULLS
  GS   HULLS (STRUCTURES)
        . SHIP HULLS
  RT   HYDRODYNAMIC COEFFICIENTS
        HYDRODYNAMICS
        SHIPS
        STRUCTURAL DESIGN
        SUBMARINES

SHIP TERMINALS
  GS   TERMINAL FACILITIES
        . SHIP TERMINALS
  RT   ARTIFICIAL HARBORS
        DEEPWATER TERMINALS
        HARBORS
        MAROTS (ESA)
        OFFSHORE DOCKING
        TANKER TERMINALS
        ∞TERMINALS
        WHARVES

SHIP TO SHORE COMMUNICATION
  GS   TELECOMMUNICATION
        . COMMUNICATION
        .. SHIP TO SHORE COMMUNICATION
  RT   DATA TRANSMISSION
        RADIO COMMUNICATION
        SHIPS
        TELEMETRY

SHIPS
  GS   WATER VEHICLES
        . SHIPS
        .. ADVANCED RANGE
              INSTRUMENTATION SHIP
        .. AIRCRAFT CARRIERS
        .. CARGO SHIPS
        ... SAVANNAH NUCLEAR SHIP
        ... TANKER SHIPS
        .. NUCLEAR POWERED SHIPS
        ... SAVANNAH NUCLEAR SHIP
        .. SATELLITE COMMUNICATIONS SHIPS
        .. SUBMARINES
        ... BALLISTIC MISSILE SUBMARINES
        ... GUIDED MISSILE SUBMARINES
        ... TRIDENT SUBMARINE
        .. SURFACE EFFECT SHIPS
        .. SWATH (SHIP)
  RT   AMPHIBIOUS VEHICLES
        ANTISHIP MISSILES
        ANTISHIP WARFARE
        BOATS
        HARBORS
        HYDROFOIL CRAFT
        HYDROFOILS
        KEELS
        MARINE TRANSPORTATION
        ∞MILITARY VEHICLES
        NAVY
        OCEAN DATA ACQUISITIONS SYSTEMS
        PROPELLERS
        RESEARCH VEHICLES
        SHIP HULLS
        SHIP TO SHORE COMMUNICATION
        SHIPYARDS
        SURFACE NAVIGATION
        SURFACE VEHICLES
        ∞TRANSPORT VEHICLES
        TRANSPORTATION ENERGY
        UNDERWATER VEHICLES
        ∞VESSELS

SHIPYARDS
  RT   CARGO SHIPS
        CONSTRUCTION
        ENCLOSURES
        INDUSTRIAL AREAS
        INDUSTRIES
        LOGISTICS
        MAINTENANCE
        OCEANOGRAPHY
        ∞PORTS
        SHIPS
        TANKER SHIPS
        WATER VEHICLES

SHIVA LASER SYSTEM
  GS   STIMULATED EMISSION DEVICES
        . LASERS

SHIVA LASER SYSTEM-(CONT.)
        .. HIGH POWER LASERS
        ... SHIVA LASER SYSTEM
  RT   COHERENT LIGHT
        LASER FUSION
        LASER OUTPUTS
        NOVA LASER SYSTEM
        ∞SYSTEMS

SHIVERING
  GS   SHIVERING
        . DITHERS
  RT   BODY TEMPERATURE

SHOALS
  GS   WATER
        . SHOALS
  RT   BEACHES
        LAKES
        OCEANOGRAPHY
        OCEANS
        REEFS
        RIVERS
        SEAS
        WATER DEPTH

∞ SHOCK
  SN   (USE OF A MORE SPECIFIC TERM IS
        RECOMMENDED--CONSULT THE TERMS
        LISTED BELOW)
  RT   CONVULSIONS
        MECHANICAL SHOCK
        SHOCK (PHYSIOLOGY)
        SHOCK RESISTANCE
        THERMAL SHOCK

SHOCK (PHYSIOLOGY)
  RT   HUMAN REACTIONS
        HUMAN TOLERANCES
        PHYSIOLOGICAL EFFECTS
        ∞SHOCK

SHOCK ABSORBERS
  RT   ∞ABSORBERS
        ABSORBERS (EQUIPMENT)
        CUSHIONS
        DAMPING
        ENERGY ABSORPTION
        HYDRAULIC EQUIPMENT
        IMPACT
        IMPACT ACCELERATION
        ISOLATORS
        LANDING GEAR
        MECHANICAL SHOCK
        PNEUMATIC EQUIPMENT
        SILENCERS
        SPRINGS (ELASTIC)
        SUSPENSION SYSTEMS (VEHICLES)
        VIBRATION DAMPING
        VIBRATION ISOLATORS

SHOCK DIFFUSERS
  USE   DIFFUSERS
        SHOCK WAVE ATTENUATION

SHOCK DISCONTINUITY
  GS   DISCONTINUITY
        . SHOCK DISCONTINUITY
  RT   DENSITY DISTRIBUTION
        WAVE FRONTS

SHOCK FRONTS
  GS   WAVE FRONTS
        . SHOCK FRONTS
  RT   ∞FRONTS
        WAVE PROPAGATION
        WAVE SCATTERING

SHOCK HEATING
  GS   HEATING
        . AERODYNAMIC HEATING
        .. SHOCK HEATING
        . KINETIC HEATING
        .. SHOCK HEATING
        . TRANSIENT HEATING
        .. SHOCK HEATING
  RT   MAGNETOHYDRODYNAMIC SHEAR
              HEATING
        PLASMA HEATING

SHOCK LAYERS
  RT   ∞LAYERS
        NORMAL SHOCK WAVES
        OBLIQUE SHOCK WAVES

SHOCK LAYERS-(CONT.)
        SHEAR LAYERS
        STRESS WAVES
        ∞TRANSITION LAYERS

SHOCK LOADS
  GS   LOADS (FORCES)
        . DYNAMIC LOADS
        .. TRANSIENT LOADS
        ... SHOCK LOADS
        .... BLAST LOADS
  RT   AERODYNAMIC LOADS
        AXIAL COMPRESSION LOADS
        COMPRESSION LOADS
        CRUSTAL FRACTURES
        IMPACT LOADS
        LANDING LOADS
        SHATTER CONES
        STRUCTURAL DESIGN CRITERIA

SHOCK MEASURING INSTRUMENTS
  GS   MEASURING INSTRUMENTS
        . SHOCK MEASURING INSTRUMENTS
  RT   ACCELEROMETERS
        PRESSURE GAGES
        SEISMOGRAPHS
        STRAIN GAGES

SHOCK RESISTANCE
  GS   SHOCK RESISTANCE
        . IMPACT RESISTANCE
  RT   EARTHQUAKE RESISTANCE
        HIGH ACCELERATION
        IMPACT
        MECHANICAL PROPERTIES
        MECHANICAL SHOCK
        PROPELLANT SENSITIVITY
        ∞RESISTANCE
        SENSITIVITY
        ∞SHOCK
        THERMAL SHOCK
        VIBRATION

SHOCK SIMULATORS
  GS   SIMULATORS
        . SHOCK SIMULATORS
  RT   VERTICAL MOTION SIMULATORS
        VIBRATION SIMULATORS

SHOCK SPECTRA
  GS   SPECTRA
        . SHOCK SPECTRA
  RT   DYNAMIC STRUCTURAL ANALYSIS
        ENERGY SPECTRA
        MECHANICAL SHOCK
        NOISE SPECTRA
        STROKING TESTS
        STRUCTURAL DESIGN
        STRUCTURAL VIBRATION
        VIBRATIONAL SPECTRA

SHOCK TESTS
  RT   DROP TESTS
        IMPACT TESTS
        LOAD TESTS
        RAILROAD HUMPING TESTS
        ∞TESTS
        VIBRATION TESTS

SHOCK TUBES
  GS   SHOCK WAVE GENERATORS
        . SHOCK TUBES
        .. MAGNETIC ANNULAR SHOCK TUBES
        .. SHOCK TUNNELS
  RT   GAS TEMPERATURE
        HOTSHOT WIND TUNNELS
        HYPERSONIC FLOW
        HYPERSONIC WIND TUNNELS
        HYPERVELOCITY WIND TUNNELS
        LOW DENSITY RESEARCH
        LOW DENSITY WIND TUNNELS
        MAGNETIC PISTONS
        TEST FACILITIES
        TUBE LASERS
        ∞TUBES

SHOCK TUNNELS
  GS   SHOCK WAVE GENERATORS
        . SHOCK TUBES
        .. SHOCK TUNNELS
        TEST FACILITIES
        . WIND TUNNELS
        .. HYPERSONIC WIND TUNNELS
        ... SHOCK TUNNELS
        .. HYPERVELOCITY WIND TUNNELS

SHOCK TUNNELS-*(CONT.)*
. . . **SHOCK TUNNELS**
RT   CASCADE WIND TUNNELS
     HOTSHOT WIND TUNNELS
     HYPERSONIC FLOW
     LOW DENSITY RESEARCH
     LOW DENSITY WIND TUNNELS
     SUPERSONIC WIND TUNNELS

SHOCK WAVE ATTENUATION
UF   SHOCK DIFFUSERS
GS   ATTENUATION
    . ACOUSTIC ATTENUATION
    . . **SHOCK WAVE ATTENUATION**
    . WAVE ATTENUATION
    . . **SHOCK WAVE ATTENUATION**
RT   ATMOSPHERIC ATTENUATION
     NOISE REDUCTION
     WAVE PROPAGATION

SHOCK WAVE CONTROL
RT   ∞CONTROL
     SECONDARY INJECTION
     SHEAR LAYERS

SHOCK WAVE GENERATORS
GS   **SHOCK WAVE GENERATORS**
    . SHOCK TUBES
    . . MAGNETIC ANNULAR SHOCK TUBES
    . . SHOCK TUNNELS
RT   ∞GENERATORS
     MAGNETIC PISTONS
     PRESSURE SENSORS
     PULSE GENERATORS
     WAVE GENERATION

SHOCK WAVE INTERACTION
GS   WAVE INTERACTION
    . **SHOCK WAVE INTERACTION**
RT   ∞INTERACTIONS
     PROPAGATION MODES
     SCATTERING
     WAVE DEGRADATION

SHOCK WAVE LUMINESCENCE
GS   DECAY
    . EMISSION
    . . LIGHT EMISSION
    . . . LUMINESCENCE
    . . . . **SHOCK WAVE LUMINESCENCE**
RT   LOW DENSITY RESEARCH
     WAVE INTERACTION

SHOCK WAVE PROFILES
RT   KROOK EQUATION
     PRESSURE DISTRIBUTION
    ∞PROFILES
     VELOCITY DISTRIBUTION
     WAVE INTERACTION

SHOCK WAVE PROPAGATION
GS   TRANSMISSION
    . WAVE PROPAGATION
    . . **SHOCK WAVE PROPAGATION**
RT   ATMOSPHERIC ATTENUATION
     BURGER EQUATION
     CROCCO METHOD
     HIGH TEMPERATURE GASES
     NONEQUILIBRIUM RADIATION
     RANKINE-HUGONIOT RELATION
     SECONDARY INJECTION
     SOUND PROPAGATION
     TWO FLUID MODELS
     WAVE ATTENUATION
     WAVE INTERACTION

SHOCK WAVES
UF   BOW SHOCK WAVES
GS   ELASTIC WAVES
    . **SHOCK WAVES**
    . . DETONATION WAVES
    . . MACH CONES
    . . NORMAL SHOCK WAVES
    . . OBLIQUE SHOCK WAVES
    . . RIEMANN WAVES
    . . SONIC BOOMS
RT   ADIABATIC EQUATIONS
     AERODYNAMIC NOISE
     BLAST LOADS
    ∞BLASTS
     BOW WAVES
     CAUSTIC LINES
     CRUSTAL FRACTURES
     DETONATION
     EARTHQUAKE DAMAGE

SHOCK WAVES-*(CONT.)*
     EARTHQUAKE RESISTANCE
     EARTHQUAKE RESISTANT STRUCTURES
     EARTHQUAKES
     ELECTROSTATIC WAVES
     EXPLODING WIRES
     EXPLOSIONS
     GAS TEMPERATURE
     GEODYNAMICS
     HUGONIOT EQUATION OF STATE
     HYPERSONIC FLOW
     HYPERSONIC SHOCK
     HYPERSONIC WAKES
     IMPACT
     IMPLOSIONS
     LONGITUDINAL WAVES
     MACH NUMBER
     MACH REFLECTION
     MAGNETOHYDRODYNAMIC WAVES
     MECHANICAL SHOCK
     MOLECULAR RELAXATION
     NOISE (SOUND)
     NOVAE
     PLANE WAVES
     PLANETARY QUAKES
     PLASMA WAVES
     PLUMES
     PRESSURE PULSES
     SEISMIC WAVES
     SOUND PRESSURE
     SOUND WAVES
     STRESS WAVES
     SUPERSONIC FLOW
    ∞TRANSITION LAYERS
     TRANSONIC FLOW
     TSUNAMI WAVES
     UNDERWATER ACOUSTICS
     UNDERWATER COMMUNICATION
    ∞WAVES
     WEDGE FLOW
     WHITHAM RULE

SHOES
GS   CLOTHING
    . **SHOES**
RT   BOOTS (FOOTWEAR)
     LEATHER
     PROTECTIVE CLOTHING
     SOCKS

SHOOTING STAR AIRCRAFT
USE   T-33 AIRCRAFT

SHOPS
RT   MAINTENANCE

SHORAN
UF   SHORT RANGE NAVIGATION
GS   NAVIGATION
    . RADIO NAVIGATION
    . . HYPERBOLIC NAVIGATION
    . . . **SHORAN**
RT   AIR NAVIGATION
     DECCA NAVIGATION
     DISTANCE MEASURING EQUIPMENT
     NAVIGATION AIDS
     SOLAR COMPASSES

SHORELINES
RT   BEACHES
     COASTAL WATER
     COASTS
     LAKES
     OCEANOGRAPHY
     OCEANS
     RIVERS
     SHALLOW WATER
     TIDAL FLATS
     WETLANDS

SHORT BELFAST C MK-1 AIRCRAFT
USE   SC-5 AIRCRAFT

SHORT CIRCUIT CURRENTS
GS   ELECTRIC CURRENT
    . **SHORT CIRCUIT CURRENTS**
RT   OPEN CIRCUIT VOLTAGE
     PHOTOVOLTAIC CELLS
     SHORT CIRCUITS
     SOLAR CELLS
     VOLT-AMPERE CHARACTERISTICS

SHORT CIRCUITS
GS   ELECTRICAL FAULTS
    . **SHORT CIRCUITS**

SHORT CIRCUITS-*(CONT.)*
RT   CIRCUITS
     ELECTRIC ARCS
     FAILURE
     JUMPERS
     SHORT CIRCUIT CURRENTS
     SNEAK CIRCUIT ANALYSIS
     SYSTEM FAILURES

SHORT HAUL AIRCRAFT
GS   TRANSPORT AIRCRAFT
    . **SHORT HAUL AIRCRAFT**
    . . EUROPEAN AIRBUS
    . . . A-310 AIRCRAFT
    . . . A-320 AIRCRAFT
    . . MERCURE AIRCRAFT
RT   AIR TRANSPORTATION
    ∞AIRCRAFT
     AIRCRAFT DESIGN
     AIRLINE OPERATIONS
     PASSENGER AIRCRAFT
     V/STOL AIRCRAFT

SHORT RANGE BALLISTIC MISSILES
GS   MISSILES
    . BALLISTIC MISSILES
    . . **SHORT RANGE BALLISTIC MISSILES**
    . SURFACE TO SURFACE MISSILES
    . . **SHORT RANGE BALLISTIC MISSILES**
RT   FIELD ARMY BALLISTIC MISSILES
     INTERMEDIATE RANGE BALLISTIC
      MISSILES

SHORT RANGE NAVIGATION
USE   SHORAN

SHORT SC-1 AIRCRAFT
USE   SC-1 AIRCRAFT

SHORT SC-5 AIRCRAFT
USE   SC-5 AIRCRAFT

SHORT SC-7 AIRCRAFT
USE   SC-7 AIRCRAFT

SHORT TAKEOFF AIRCRAFT
UF   STOL AIRCRAFT
GS   V/STOL AIRCRAFT
    . **SHORT TAKEOFF AIRCRAFT**
    . . BREGUET 940 AIRCRAFT
    . . BREGUET 941 AIRCRAFT
    . . C-15 AIRCRAFT
    . . C-123 AIRCRAFT
    . . DHC 4 AIRCRAFT
    . . DHC 5 AIRCRAFT
    . . QUESTOL
    . . U-10 AIRCRAFT
RT   ∞AIRCRAFT
     CIRCULATION CONTROL AIRFOILS
     COMPOUND HELICOPTERS
     EXTERNALLY BLOWN FLAPS
     FAN IN WING AIRCRAFT
     HELICOPTERS
     JATO ENGINES
     JET AIRCRAFT
     JET FLAPS
     LIFT FANS
     LIFTING ROTORS
    ∞MILITARY AIRCRAFT
     POWERED LIFT AIRCRAFT
     ROTARY WING AIRCRAFT
    ∞SUBSONIC AIRCRAFT
     TAKEOFF RUNS
     TILT WING AIRCRAFT
     VERTICAL TAKEOFF AIRCRAFT
    ∞WINGED VEHICLES

SHORT WAVE RADIATION
SN   (RADIO WAVES)
GS   ELECTROMAGNETIC RADIATION
    . RADIO WAVES
    . . **SHORT WAVE RADIATION**
    . . . MICROWAVES
    . . . . CENTIMETER WAVES
    . . . . DECIMETER WAVES
    . . . . MICROWAVE EMISSION
    . . . . MILLIMETER WAVES
    . . . SUBMILLIMETER WAVES
RT   FAR INFRARED RADIATION
     HIGH FREQUENCIES
     LONG WAVE RADIATION
     MONOCHROMATIC RADIATION
    ∞RADIATION

SHORT WAVE RADIO TRANSMISSION
GS   TRANSMISSION
. ELECTROMAGNETIC WAVE
TRANSMISSION
.. RADIO TRANSMISSION
... **SHORT WAVE RADIO
TRANSMISSION**
. SIGNAL TRANSMISSION
.. RADIO TRANSMISSION
... **SHORT WAVE RADIO
TRANSMISSION**
RT   HIGH FREQUENCIES
WAVE PROPAGATION

SHORTENING
USE   REDUCTION

∞ **SHOT**
SN   *(USE OF A MORE SPECIFIC TERM IS
RECOMMENDED--CONSULT THE TERMS
LISTED BELOW)*
RT   AMMUNITION
LAUNCHING
ORBITAL SHOTS
PELLETS
SHOT NOISE
SHOT PEENING

SHOT NOISE
GS   ELECTROMAGNETIC INTERFERENCE
. RADIO FREQUENCY INTERFERENCE
.. BLACKOUT (PROPAGATION)
... ELECTROMAGNETIC NOISE
.... **SHOT NOISE**
RT   BARRITT DIODES
∞ SHOT
THERMAL NOISE

SHOT PEENING
GS   HARDENING (MATERIALS)
. **SHOT PEENING**
METAL FINISHING
. PEENING
.. **SHOT PEENING**
RT   COLD WORKING
DESCALING
FATIGUE (MATERIALS)
METAL WORKING
∞ SHOT
STRAIN HARDENING
SURFACE FINISHING
WORK HARDENING

SHOULDERS
RT   JOINTS (ANATOMY)
SCAPULA

∞ **SHOWERS**
SN   *(USE OF A MORE SPECIFIC TERM IS
RECOMMENDED--CONSULT THE TERMS
LISTED BELOW)*
RT   COSMIC RAY SHOWERS
FLOOD PREDICTIONS
METEOROID SHOWERS
RAIN
RAIN FORESTS
RAINSTORMS

SHRAPNEL
RT   FRAGMENTATION
FRAGMENTS
PROJECTILES
WEAPONS

SHREDDING
GS   COMMINUTION
. **SHREDDING**
RT   COMPOSTING
CUTTING
TEARING

SHREWS
GS   ANIMALS
. VERTEBRATES
.. MAMMALS
... **SHREWS**

SHRIKE MISSILE
GS   MISSILES
. AIR TO SURFACE MISSILES
.. **SHRIKE MISSILE**
RT   SOLID PROPELLANT ROCKET ENGINES

SHRINKAGE
RT   CASTING
CONTRACTION
GROWTH
∞ REDUCTION
SINTERING
TEMPERATURE INVERSIONS
WARPAGE

SHROUDED BODIES
USE   SHROUDS

SHROUDED NOZZLES
RT   ANNULAR NOZZLES
NOZZLE GEOMETRY
NOZZLE WALLS
∞ NOZZLES

SHROUDED PROPELLERS
UF   DUCTED PROPELLERS
GS   PROPELLERS
. **SHROUDED PROPELLERS**
RT   DUCTED FANS
RING WINGS
THRUST AUGMENTATION
XV-11A AIRCRAFT

SHROUDED TURBINES
GS   TURBOMACHINERY
. TURBINES
.. **SHROUDED TURBINES**

SHROUDS
UF   SHROUDED BODIES
RT   COVERINGS
DUCTED BODIES
RIGGING

SHUNTS
USE   BYPASSES
CIRCUITS

SHUTDOWNS
RT   DEACTIVATION
ENGINES
∞ SCRAM

∞ **SHUTTERS**
SN   *(USE OF A MORE SPECIFIC TERM IS
RECOMMENDED--CONSULT THE TERMS
LISTED BELOW)*
RT   BLINDS
CAMERA SHUTTERS
LOUVERS
SHADES

SHUTTLE AVIONICS INTEGRATION LABORATORY
USE   SAIL PROJECT

SHUTTLE BOOSTERS
USE   SPACE SHUTTLE BOOSTERS

SHUTTLE DERIVED VEHICLES
UF   SDV
GS   MANNED SPACECRAFT
. **SHUTTLE DERIVED VEHICLES**
RT   SPACE SHUTTLE ORBITERS
SPACE SHUTTLES
∞ SPACECRAFT
SPACECRAFT DESIGN

SHUTTLE ENGINEERING SIMULATOR
GS   SIMULATORS
. **SHUTTLE ENGINEERING SIMULATOR**
RT   SPACE SHUTTLES

SHUTTLE IMAGING RADAR
UF   SIR-A
SIR-B
GS   PAYLOADS
. **SHUTTLE IMAGING RADAR**
RT   RADAR GEOLOGY
RADAR IMAGERY
REMOTE SENSING
SPACE SHUTTLE PAYLOADS
SYNTHETIC APERTURE RADAR

SHUTTLE MISSION SIMULATOR
GS   SIMULATORS
. **SHUTTLE MISSION SIMULATOR**

SHUTTLE ORBITERS
USE   SPACE SHUTTLE ORBITERS

SHUTTLE PALLET SATELLITES
UF   SPAS (ESA PLATFORMS)
GS   SATELLITES
. ARTIFICIAL SATELLITES
.. **SHUTTLE PALLET SATELLITES**
RT   SPACE SHUTTLES

SI
USE   INTERNATIONAL SYSTEM OF UNITS

SIALON
GS   MIXTURES
. **SIALON**
RT   ALUMINUM
CERAMICS
HIGH TEMPERATURE
NITROGEN
OXYGEN
REACTION BONDING
REFRACTORY MATERIALS
SILICON NITRIDES
SINTERING

SIAM MISSILES
UF   SELF INITIATED ANTIAIRCRAFT MISSILES
GS   MISSILES
. ANTIAIRCRAFT MISSILES
.. **SIAM MISSILES**
RT   AIR TO AIR MISSILES
ANTIMISSILE MISSILES

SIBERIA
RT   ARCTIC REGIONS
ASIA
U.S.S.R.

SIC (COEFFICIENT)
USE   STRUCTURAL INFLUENCE COEFFICIENTS

SICILY
GS   LANDFORMS
. ISLANDS
.. **SICILY**
RT   ITALY
MEDITERRANEAN SEA

SICKNESSES
GS   **SICKNESSES**
. ALTITUDE SICKNESS
. DECOMPRESSION SICKNESS

SID (IONOSPHERIC DISTURBANCES)
USE   SUDDEN IONOSPHERIC DISTURBANCES

SIDE INLETS
GS   INTAKE SYSTEMS
. **SIDE INLETS**
RT   AIR INTAKES
BYPASS RATIO
HYPERSONIC INLETS
INLET AIRFRAME CONFIGURATIONS
NOSE INLETS
SCOOPS
SUPERSONIC INLETS
∞ WATER INTAKES

SIDE-LOOKING RADAR
GS   RADAR
. SYNTHETIC APERTURE RADAR
.. **SIDE-LOOKING RADAR**
RT   CHANGE DETECTION
IMAGING RADAR
RADAR SCANNING
SEARCH RADAR

SIDEBANDS
RT   ∞ BANDS
DOUBLE SIDEBAND TRANSMISSION
SELECTIVE FADING
SINGLE SIDEBAND TRANSMISSION

SIDELOBE REDUCTION
GS   ATTENUATION
. **SIDELOBE REDUCTION**
RT   HORN ANTENNAS
RADAR ANTENNAS
RADAR ATTENUATION
RADAR RECEPTION
RADAR REFLECTORS
RADAR RESOLUTION
∞ REDUCTION
SIDELOBES

SIDELOBES
GS   DISTRIBUTION (PROPERTY)
    . RADIATION DISTRIBUTION
    . . ANTENNA RADIATION PATTERNS
    . . . **SIDELOBES**
RT   ANTENNA DESIGN
    ∞LOBES
    NEAR FIELDS
    SIDELOBE REDUCTION

SIDEREAL TIME
GS   TIME
    . **SIDEREAL TIME**
RT   ASTRONOMY
    EARTH ROTATION
    STELLAR MOTIONS
    TIME MEASUREMENT
    UNITS OF MEASUREMENT

SIDERITE METEORITES
USE   IRON METEORITES

SIDERITES
GS   CARBON COMPOUNDS
    . CARBONATES
    . . **SIDERITES**
    IRON COMPOUNDS
    . **SIDERITES**
    MINERALS
    . **SIDERITES**

SIDES
RT   EDGES
    GEOMETRY
    RIMS
    WALLS

SIDESLIP
GS   MANEUVERS
    . **SIDESLIP**
RT   ROLL
    SKIDDING
    ∞SLIP
    SPACECRAFT MOTION
    YAW

SIDEWASH
USE   BACKWASH

SIDEWINDER MISSILES
GS   MISSILES
    . AIR TO AIR MISSILES
    . . **SIDEWINDER MISSILES**
    . ANTIAIRCRAFT MISSILES
    . . **SIDEWINDER MISSILES**

SIEBEL AIRCRAFT
RT   ∞AIRCRAFT

SIEMENS 2002 COMPUTER
GS   DATA PROCESSING EQUIPMENT
    . COMPUTERS
    . . **SIEMENS 2002 COMPUTER**

SIERRA LEONE
GS   NATIONS
    . **SIERRA LEONE**
RT   AFRICA

SIERRA NEVADA MOUNTAINS (CA)
GS   LANDFORMS
    . MOUNTAINS
    . . **SIERRA NEVADA MOUNTAINS (CA)**
RT   CALIFORNIA

SIEVES
GS   SEPARATORS
    . **SIEVES**
RT   FLUID FILTERS
    SHAKERS
    SIZING SCREENS
    WIRE CLOTH

SIGHT
USE   VISUAL PERCEPTION

SIGMA COMPUTERS
GS   DATA PROCESSING EQUIPMENT
    . COMPUTERS
    . . DIGITAL COMPUTERS
    . . . **SIGMA COMPUTERS**
    . . . . SIGMA 9 COMPUTER

SIGMA ORIONIS
GS   CELESTIAL BODIES
    . STARS
    . . **SIGMA ORIONIS**
RT   ORION CONSTELLATION

SIGMA 5 COMPUTER
GS   DATA PROCESSING EQUIPMENT
    . COMPUTERS
    . . ANALOG COMPUTERS
    . . . **SIGMA 5 COMPUTER**
    . . DIGITAL COMPUTERS
    . . . **SIGMA 5 COMPUTER**

SIGMA 7
GS   MANNED SPACECRAFT
    . MERCURY SPACECRAFT
    . . **SIGMA 7**
    REENTRY VEHICLES
    . RECOVERABLE SPACECRAFT
    . . MERCURY SPACECRAFT
    . . . **SIGMA 7**
    SOFT LANDING SPACECRAFT
    . MERCURY SPACECRAFT
    . . **SIGMA 7**
    SPACE CAPSULES
    . MERCURY SPACECRAFT
    . . **SIGMA 7**
RT   MERCURY MA-8 FLIGHT

SIGMA 9 COMPUTER
GS   DATA PROCESSING EQUIPMENT
    . COMPUTERS
    . . DIGITAL COMPUTERS
    . . . SIGMA COMPUTERS
    . . . . **SIGMA 9 COMPUTER**

SIGMA-MESONS
GS   PARTICLES
    . ELEMENTARY PARTICLES
    . . BOSONS
    . . . MESONS
    . . . . VECTOR MESONS
    . . . . . **SIGMA-MESONS**
    . . FERMIONS
    . . . BARYONS
    . . . . **SIGMA-MESONS**
    . . HADRONS
    . . . BARYONS
    . . . . **SIGMA-MESONS**
    . . . MESONS
    . . . . VECTOR MESONS
    . . . . . **SIGMA-MESONS**
    . NUCLEAR PARTICLES
    . . BOSONS
    . . . MESONS
    . . . . VECTOR MESONS
    . . . . . **SIGMA-MESONS**
RT   CHARGED PARTICLES
    ETA-MESONS

SIGNAL ANALYSIS
GS   CORRELATION
    . DATA CORRELATION
    . . **SIGNAL ANALYSIS**
    DATA PROCESSING
    . DATA CORRELATION
    . . **SIGNAL ANALYSIS**
RT   ∞ANALYZING
    DIGITAL RADAR SYSTEMS
    FREQUENCY ANALYZERS
    SPECTRUM ANALYSIS

SIGNAL ANALYZERS
GS   MEASURING INSTRUMENTS
    . ANALYZERS
    . . **SIGNAL ANALYZERS**
RT   ANALOG COMPUTERS
    AUTODYNES

SIGNAL DETECTION
GS   DETECTION
    . **SIGNAL DETECTION**
    . . CORRELATION DETECTION
RT   AUTODYNES
    ∞DETECTORS
    DISCRIMINATION
    PHASE DETECTORS
    PREAMPLIFIERS
    RADAR DETECTION
    SOUND TRANSDUCERS
    TELECOMMUNICATION

SIGNAL DETECTORS
UF   SIGNAL DISCRIMINATORS

SIGNAL DETECTORS-(CONT.)
RT   AUTODYNES
    ∞DETECTORS
    DISCRIMINATION
    MICROWAVE SENSORS
    PREAMPLIFIERS
    SOUND TRANSDUCERS
    TELECOMMUNICATION

SIGNAL DISCRIMINATORS
USE   SIGNAL DETECTORS

SIGNAL DISTORTION
GS   DISTORTION
    . **SIGNAL DISTORTION**
RT   INTERSYMBOLIC INTERFERENCE
    RADIO SIGNALS
    SCRAMBLING (COMMUNICATION)

SIGNAL ENCODING
GS   CODING
    . **SIGNAL ENCODING**
RT   CONCATENATED CODES
    DIGITAL TO ANALOG CONVERTERS
    REDUNDANCY ENCODING
    SCRAMBLING (COMMUNICATION)
    TELECOMMUNICATION
    TRANSMITTERS
    VOICE DATA PROCESSING

SIGNAL FADEOUT
USE   SIGNAL FADING

SIGNAL FADING
UF   SIGNAL FADEOUT
GS   FADING
    . **SIGNAL FADING**
    . . SELECTIVE FADING
RT   ACOUSTIC INSTABILITY
    ATMOSPHERIC SCATTERING
    ATTENUATION
    DIFFRACTION PATTERNS
    ELECTROMAGNETIC ABSORPTION
    GROUND EFFECT (COMMUNICATIONS)
    RADIO FREQUENCY INTERFERENCE
    RADIO SCATTERING
    RECEPTION DIVERSITY
    SMEAR
    SOUND INTENSITY

SIGNAL FADING RATE
GS   RATES (PER TIME)
    . **SIGNAL FADING RATE**
RT   FADING
    SELECTIVE FADING
    SOUND INTENSITY

SIGNAL FLOW GRAPHS
RT   DUALITY PRINCIPLE
    ∞FLOW GRAPHS
    NETWORK ANALYSIS
    ∞NETWORKS
    RICHARDS THEOREM
    SNEAK CIRCUIT ANALYSIS

SIGNAL GENERATORS
GS   **SIGNAL GENERATORS**
    . FREQUENCY SYNTHESIZERS
    . FUNCTION GENERATORS
RT   CIRCUITS
    ∞GENERATORS
    HALL GENERATORS
    OSCILLATORS
    SIRENS
    SOLID STATE DEVICES
    SOUND GENERATORS
    SUBHARMONIC GENERATORS
    VOLTAGE GENERATORS

SIGNAL MEASUREMENT
UF   ELECTRONIC SIGNAL MEASUREMENT
RT   ∞MEASUREMENT

SIGNAL MIXING
GS   MIXING
    . **SIGNAL MIXING**
RT   AUDITORY SIGNALS
    ERROR SIGNALS
    MAGNETIC SIGNALS
    RADIO SIGNALS

SIGNAL PROCESSING
GS   DATA PROCESSING
    . **SIGNAL PROCESSING**

SIGNAL PROCESSING-(CONT.)
RT   AUDIO SIGNALS
      COMPANDING
      DIRECTION FINDING
      EQUALIZERS (CIRCUITS)
      INTERFERENCE IMMUNITY
      MAXIMUM ENTROPY METHOD
      MESSAGE PROCESSING
      ONBOARD DATA PROCESSING
      ∞ PROCESSING
      SMOKE DETECTORS
      SURFACE ACOUSTIC WAVE DEVICES
      TELEMETRY
      VHSIC (CIRCUITS)
      VIDEO SIGNALS

SIGNAL RECEPTION
GS   **SIGNAL RECEPTION**
    . SYLLABLES
    . SYMBOLS
    . TELEVISION RECEPTION
RT   HOMODYNE RECEPTION
      PREAMPLIFIERS
      ∞ RECEIVING
      SENTENCES
      VOCODERS

SIGNAL REFLECTION
GS   ECHOES
    . **SIGNAL REFLECTION**
    REFLECTION
    . **SIGNAL REFLECTION**
RT   CEPSTRAL ANALYSIS
      SPREAD REFLECTION
      TRANSMISSION
      WAVE REFLECTION

SIGNAL STABILIZATION
GS   STABILIZATION
    . **SIGNAL STABILIZATION**
RT   FREQUENCY CONTROL
      TRANSMISSION CIRCUITS

SIGNAL TO NOISE RATIOS
GS   RATIOS
    . **SIGNAL TO NOISE RATIOS**
RT   AMPLITUDE DISTRIBUTION ANALYSIS
      ATTENUATION
      BACKGROUND NOISE
      BIT ERROR RATE
      CARRIER TO NOISE RATIOS
      CHANNEL NOISE
      COMMUNICATION THEORY
      COMPANDING
      CORRELATION DETECTION
      ELECTROMAGNETIC INTERFERENCE
      ELECTROMAGNETIC NOISE
      IMAGE CONTRAST
      IMAGE ENHANCEMENT
      INTERFERENCE IMMUNITY
      LOW NOISE
      MATCHED FILTERS
      MAXIMUM ENTROPY METHOD
      ∞ NOISE
      NOISE PROPAGATION
      NOISE SPECTRA
      NOISE THRESHOLD
      RANDOM NOISE
      RANDOM SIGNALS
      ∞ SIGNALS
      WHITE NOISE

SIGNAL TRANSMISSION
GS   TRANSMISSION
    . **SIGNAL TRANSMISSION**
    .. DATA TRANSMISSION
    ... AUTOMATIC PICTURE
        TRANSMISSION
    ... MULTIPLE ACCESS
    .... CODE DIVISION MULTIPLE ACCESS
    .... FREQUENCY DIVISION MULTIPLE
        ACCESS
    .. SINGLE CHANNEL PER CARRIER
        TRANSMISSION
    .. PCM TELEMETRY
    .. RADAR TRANSMISSION
    .. RADIO TRANSMISSION
    ... DOUBLE SIDEBAND TRANSMISSION
    ... IONOSPHERIC PROPAGATION
    .... IONOSPHERIC F-SCATTER
        PROPAGATION
    ... MICROWAVE ATTENUATION
    ... MICROWAVE TRANSMISSION
    ... MULTIPATH TRANSMISSION

SIGNAL TRANSMISSION-(CONT.)
    ... SHORT WAVE RADIO
        TRANSMISSION
    ... SINGLE SIDEBAND TRANSMISSION
    ... TRANSEQUATORIAL PROPAGATION
    ... TRANSHORIZON RADIO
        PROPAGATION
    .. SATELLITE TRANSMISSION
    .. TELEVISION TRANSMISSION
RT   AUDIO SIGNALS
      CODE DIVISION MULTIPLEXING
      MESSAGE PROCESSING
      MESSAGES
      MULTIPLEXING
      ORTHOGONAL MULTIPLEXING THEORY
      PACKET SWITCHING
      PULSE COMMUNICATION
      RADAR ATTENUATION
      RADIO ATTENUATION
      RADIO SCATTERING
      SENTENCES
      SOUND TRANSMISSION
      SYLLABLES
      TALKING
      TELECOMMUNICATION
      TRANSMISSION EFFICIENCY
      VIDEO SIGNALS
      WIRELESS COMMUNICATION

∞ SIGNALS
SN   *(USE OF A MORE SPECIFIC TERM IS*
      RECOMMENDED--CONSULT THE TERMS
      BELOW)
RT   AUDIO SIGNALS
      AUDITORY SIGNALS
      BEACONS
      BELLS
      CHIRP SIGNALS
      ELECTRIC PULSES
      ERROR SIGNALS
      HORNS
      MAGNETIC SIGNALS
      MESSAGES
      PSEUDORANDOM SEQUENCES
      PYROTECHNICS
      RADIO SIGNALS
      RANDOM SIGNALS
      SIGNAL TO NOISE RATIOS
      SIRENS
      SOUND GENERATORS
      TELECOMMUNICATION
      TIME SIGNALS
      VIDEO SIGNALS
      VISUAL SIGNALS
      VISUAL STIMULI

SIGNATURE ANALYSIS
RT   ∞ ANALYZING
      CEPSTRAL ANALYSIS
      DETECTION
      IMAGERY
      INFRARED SIGNATURES
      MISSILE SIGNATURES
      RADAR SIGNATURES
      SIGNATURES
      TARGET RECOGNITION

SIGNATURES
GS   **SIGNATURES**
    . INFRARED SIGNATURES
    . MAGNETIC SIGNATURES
    . MISSILE SIGNATURES
    . RADAR SIGNATURES
    . SPECTRAL SIGNATURES
RT   AMPLITUDE DISTRIBUTION ANALYSIS
      DETECTION
      REPRESENTATIONS
      SIGNATURE ANALYSIS
      TARGET RECOGNITION
      VIDEO LANDMARK ACQUISITION AND
        TRACKING

SIGNIFICANCE
RT   CONFIDENCE LIMITS
      CORRELATION
      COVARIANCE
      DEGREES OF FREEDOM
      FINITE DIFFERENCE THEORY
      NULL HYPOTHESIS
      NUMERICAL ANALYSIS
      REGRESSION ANALYSIS
      STATISTICAL TESTS

SIGNS (SYMBOLS)
USE   SYMBOLS

SIGNS AND SYMPTOMS
UF   SYMPTOMS
      SYNDROMES
GS   **SIGNS AND SYMPTOMS**
    . BRADYCARDIA
    . COUGH
    . DYSPNEA
    . HEADACHE
    . HEMATURIA
    . LEUKOPENIA
    . VERTIGO
RT   ASPHYXIA
      DISEASES
      HALLUCINATIONS
      ∞ INDICATION
      SYMPTOMOLOGY

SIKHOTE-ALIN METEORITE
GS   CELESTIAL BODIES
    . METEORITES
    .. IRON METEORITES
    ... **SIKHOTE-ALIN METEORITE**

SIKKIM
GS   NATIONS
    . **SIKKIM**
RT   ASIA
      BHUTAAN
      HIMALAYAS
      INDIA

SIKORSKY AIRCRAFT
GS   **SIKORSKY AIRCRAFT**
    . CH-3 HELICOPTER
    . CH-34 HELICOPTER
    . CH-54 HELICOPTER
    . H-19 HELICOPTER
    . H-53 HELICOPTER
    . H-56 HELICOPTER
    . H-60 HELICOPTER
    . S-58 HELICOPTER
    . S-61 HELICOPTER
    . S-67 HELICOPTER
    . SH-3 HELICOPTER
    . SH-4 HELICOPTER
    . SIKORSKY WHIRLWIND HELICOPTER
    . UH-34 HELICOPTER
    . UH-60A HELICOPTER
    . UH-61A HELICOPTER
RT   ∞ AIRCRAFT

SIKORSKY HSS-2 HELICOPTER
USE   SH-3 HELICOPTER

SIKORSKY S-58 HELICOPTER
USE   S-58 HELICOPTER

SIKORSKY S-61 HELICOPTER
USE   S-61 HELICOPTER

SIKORSKY S-64 HELICOPTER
USE   CH-54 HELICOPTER

SIKORSKY S-65 HELICOPTER
USE   H-53 HELICOPTER

SIKORSKY S-67 HELICOPTER
USE   S-67 HELICOPTER

SIKORSKY WHIRLWIND HELICOPTER
GS   SIKORSKY AIRCRAFT
    . **SIKORSKY WHIRLWIND HELICOPTER**
    V/STOL AIRCRAFT
    . ROTARY WING AIRCRAFT
    .. HELICOPTERS
    ... **SIKORSKY WHIRLWIND**
        HELICOPTER
RT   ∞ AIRCRAFT

SILANES
GS   HYDROGEN COMPOUNDS
    . HYDRIDES
    .. **SILANES**
    ... CHLOROSILANES
    .... METHYL CHLOROSILANES
    SILICON COMPOUNDS
    . **SILANES**
    .. CHLOROSILANES
    .. METHYL CHLOROSILANES
RT   DISILICIDES

SILENCE
RT   NOISE REDUCTION
      TRANSMISSION LOSS

SILENCERS
RT ATTENUATORS
DAMPING
INHIBITORS
MUFFLERS
SHOCK ABSORBERS
SQUELCH CIRCUITS
SUPPRESSORS
ZERO SOUND

SILICA
USE SILICON DIOXIDE

SILICA GEL
GS GELS
. SILICA GEL
RT DEHUMIDIFICATION
DEHYDRATION
DRYING
SILICON DIOXIDE

SILICA GLASS
GS GLASS
. SILICA GLASS
RT GLASS COATINGS
GLASS ELECTRODES
GLASS FIBERS
GLASSWARE
SANDS
SILICON DIOXIDE

SILICATES
GS SILICON COMPOUNDS
. SILICATES
. . ANDESITE
. . ARAGONITE
. . BERYL
. . CALCIUM SILICATES
. . . GEHLENITE
. . CORDIERITE
. . FAYALITE
. . FELDSPARS
. . FLUOROSILICATES
. . FORSTERITE
. . GARNETS
. . . YTTRIUM-ALUMINUM GARNET
. . . YTTRIUM-IRON GARNET
. . KAOLINITE
. . MERWINITE
. . MONTICELLITE
. . MONTMORILLONITE
. . NEPHELINE
. . POTASSIUM SILICATES
. . PYROPHYLLITE
. . PYROXENES
. . . ENSTATITE
. . SODIUM SILICATES
. . . SPODUMENE
. . . TALC
. . . TOURMALINE
. . ZEOLITES
RT AKERMANITE
DISILICIDES
MINERALS
SILICIDES
SILICON DIOXIDE
TETRAETHYL ORTHOSILICATE

SILICIDES
GS SILICON COMPOUNDS
. SILICIDES
. . DISILICIDES
RT INTERMETALLICS
SILICATES

SILICON
GS CHEMICAL ELEMENTS
. METALLOIDS
. . SILICON
. . . SILICON ISOTOPES
RT FLOAT ZONES
REACTION BONDING
SCHOTTKY DIODES

SILICON ALLOYS
GS ALLOYS
. SILICON ALLOYS

SILICON CARBIDES
GS CARBON COMPOUNDS
. CARBIDES
. . SILICON CARBIDES
SILICON COMPOUNDS
. SILICON CARBIDES
RT ABRASIVES

SILICON CARBIDES-(CONT.)
CARBORUNDUM (TRADEMARK)

SILICON COMPOUNDS
GS SILICON COMPOUNDS
. ALUMINUM SILICATES
. . ANDESITE
. . KAOLINITE
MONTMORILLONITE
. . PYROPHYLLITE
. FLINT
. ORGANIC SILICON COMPOUNDS
. TRIPHENYL SILICON
. SILANES
. . CHLOROSILANES
. . METHYL CHLOROSILANES
. SILICATES
. . ANDESITE
. . ARAGONITE
. . BERYL
. . CALCIUM SILICATES
. . . GEHLENITE
. . CORDIERITE
. . FAYALITE
. . FELDSPARS
. . FLUOROSILICATES
. . FORSTERITE
. . GARNETS
. . . YTTRIUM-ALUMINUM GARNET
. . . YTTRIUM-IRON GARNET
. . KAOLINITE
. . MERWINITE
. . MONTICELLITE
. . MONTMORILLONITE
. . NEPHELINE
. . POTASSIUM SILICATES
. . PYROPHYLLITE
. . PYROXENES
. . . ENSTATITE
. . SODIUM SILICATES
. . . SPODUMENE
. . . TALC
. . . TOURMALINE
. . ZEOLITES
. SILICIDES
. . DISILICIDES
. SILICON CARBIDES
. SILICON NITRIDES
. SILICON OXIDES
. . MUSCOVITE
. . NEPHELITE
. . SILICON DIOXIDE
. . . QUARTZ
. . . . COESITE
. . SPODUMENE
. SILICON TETRACHLORIDE
RT AKERMANITE
∞CHEMICAL COMPOUNDS
∞GROUP 4A COMPOUNDS
METHYL POLYSILOXANE
SILICONES
SILOXANES

SILICON CONTROLLED RECTIFIERS
UF SCR (RECTIFIERS)
GS ELECTRONIC EQUIPMENT
. SOLID STATE DEVICES
. . SEMICONDUCTOR DEVICES
. . . THYRISTORS
. . . . SILICON CONTROLLED
RECTIFIERS
RECTIFIERS
. THYRISTORS
. . SILICON CONTROLLED RECTIFIERS
RT CURRENT CONVERTERS (AC TO DC)
THYRATRONS

SILICON DIOXIDE
UF REFRASIL (TRADEMARK)
SILICA
GS CHALCOGENIDES
. OXIDES
. . DIOXIDES
. . . SILICON DIOXIDE
. . . . QUARTZ
. . . . . COESITE
. . SILICON OXIDES
. . . SILICON DIOXIDE
. . . . QUARTZ
. . . . . COESITE
SILICON COMPOUNDS
. SILICON OXIDES
. . SILICON DIOXIDE
. . . QUARTZ
. . . . COESITE

SILICON DIOXIDE-(CONT.)
RT BOROSILICATE GLASS
CERAMICS
E GLASS
GLASS
METALLIC GLASSES
OBSIDIAN
PORCELAIN
QUARTZ CRYSTALS
S GLASS
SANDS
SILICA GEL
SILICA GLASS
SILICATES
VYCOR

SILICON FILMS
RT ∞FILMS
SEMICONDUCTOR DEVICES
THIN FILMS

SILICON ISOTOPES
GS CHEMICAL ELEMENTS
. METALLOIDS
. . SILICON
. . . SILICON ISOTOPES

SILICON JUNCTIONS
GS SEMICONDUCTOR JUNCTIONS
. SILICON JUNCTIONS
RT HETEROJUNCTIONS
HOMOJUNCTIONS
SIS (SEMICONDUCTORS)

SILICON NITRIDES
GS NITROGEN COMPOUNDS
. NITRIDES
. . SILICON NITRIDES
SILICON COMPOUNDS
. SILICON NITRIDES
RT CERAMIC MATRIX COMPOSITES
REACTION BONDING
SIALON

SILICON OXIDES
GS CHALCOGENIDES
. OXIDES
. . SILICON OXIDES
. . . MUSCOVITE
. . . NEPHELITE
. . . SILICON DIOXIDE
. . . . QUARTZ
. . . . . COESITE
. . . SPODUMENE
SILICON COMPOUNDS
. SILICON OXIDES
. . MUSCOVITE
. . NEPHELITE
. . SILICON DIOXIDE
. . . QUARTZ
. . . . COESITE
. . SPODUMENE
RT AKERMANITE

SILICON POLYMERS
GS SILICON POLYMERS
. METHYL POLYSILOXANE
. SILICONE RESINS
. SILICONES
. SILOXANES
RT ∞POLYMERS

SILICON RADIATION DETECTORS
GS MEASURING INSTRUMENTS
. RADIATION MEASURING INSTRUMENTS
. . RADIATION DETECTORS
. . . SILICON RADIATION DETECTORS
RT ∞RADIATION

SILICON RECTIFIERS
USE CRYSTAL RECTIFIERS

SILICON SOLAR CELLS
USE SOLAR CELLS

SILICON TETRACHLORIDE
GS HALOGEN COMPOUNDS
. CHLORINE COMPOUNDS
. . CHLORIDES
. . . SILICON TETRACHLORIDE
. HALIDES
. . CHLORIDES
. . . SILICON TETRACHLORIDE
SILICON COMPOUNDS

SILICON TETRACHLORIDE-*(CONT.)*
. SILICON TETRACHLORIDE

SILICON TRANSISTORS
GS   ELECTRONIC EQUIPMENT
. SOLID STATE DEVICES
. . SEMICONDUCTOR DEVICES
. . . TRANSISTORS
. . . . **SILICON TRANSISTORS**
†   . . . . . SOS (SEMICONDUCTORS)

SILICON-ON-SAPPHIRE JUNCTIONS
USE   SOS (SEMICONDUCTORS)

SILICON-ON-SAPPHIRE SEMICONDUCTORS
USE   SOS (SEMICONDUCTORS)

SILICON-ON-SAPPHIRE TRANSISTORS
USE   SOS (SEMICONDUCTORS)

SILICONE RESINS
GS   RESINS
. **SILICONE RESINS**
SILICON POLYMERS
. **SILICONE RESINS**
RT   THERMOSETTING RESINS

SILICONE RUBBER
GS   RUBBER
. **SILICONE RUBBER**
. . RTV-40 RUBBER (TRADEMARK)
. . RTV-60 RUBBER (TRADEMARK)
RT   ELASTOMERS
SYNTHETIC RUBBERS

SILICONES
GS   SILICON POLYMERS
. **SILICONES**
. . SILOXANES
RT   ∞POLYMERS
SILICON COMPOUNDS

SILICONIZING
GS   HARDENING (MATERIALS)
. **SILICONIZING**
RT   COATING
COATINGS
CORROSION PREVENTION
CORROSION RESISTANCE
OXIDATION RESISTANCE
PASSIVITY

SILK
GS   FABRICS
. **SILK**
FIBERS
. **SILK**
RT   CREPE
ORGANIC MATERIALS

SILKWORMS
GS   ANIMALS
. INVERTEBRATES
. . ARTHROPODS
. . . INSECTS
. . . . LARVAE
. . . . . **SILKWORMS**
. . . . MOTHS
. . . . . **SILKWORMS**
RT   INFESTATION

SILOS (MISSILE STORAGE)
USE   MISSILE SILOS

SILOXANES
GS   SILICON POLYMERS
. SILICONES
. . **SILOXANES**
RT   ∞POLYMERS
SILICON COMPOUNDS

SILTS
USE   SEDIMENTS

SILVER
GS   CHEMICAL ELEMENTS
. **SILVER**
. . SILVER ISOTOPES
METALS
. NOBLE METALS
. . **SILVER**
. . . SILVER ISOTOPES
. TRANSITION METALS
. . **SILVER**

SILVER-*(CONT.)*
. . . SILVER ISOTOPES

SILVER ALLOYS
GS   ALLOYS
. **SILVER ALLOYS**
RT   BEARING ALLOYS

SILVER BROMIDES
GS   HALOGEN COMPOUNDS
. BROMINE COMPOUNDS
. . BROMIDES
. . . **SILVER BROMIDES**
. HALIDES
. . BROMIDES
. . . **SILVER BROMIDES**
. . METAL HALIDES
. . . SILVER HALIDES
. . . . **SILVER BROMIDES**
SILVER COMPOUNDS
. SILVER HALIDES
. . **SILVER BROMIDES**

SILVER CADMIUM BATTERIES
UF   CADMIUM SILVER BATTERIES
GS   ELECTROCHEMICAL CELLS
. ELECTRIC BATTERIES
. . STORAGE BATTERIES
. . . **SILVER CADMIUM BATTERIES**
RT   NICKEL CADMIUM BATTERIES

SILVER CHLORIDES
GS   HALOGEN COMPOUNDS
. CHLORINE COMPOUNDS
. . CHLORIDES
. . . **SILVER CHLORIDES**
. HALIDES
. . CHLORIDES
. . . **SILVER CHLORIDES**
. . METAL HALIDES
. . . SILVER HALIDES
. . . . **SILVER CHLORIDES**
SILVER COMPOUNDS
. SILVER HALIDES
. . **SILVER CHLORIDES**

SILVER COMPOUNDS
GS   **SILVER COMPOUNDS**
. SILVER HALIDES
. . SILVER BROMIDES
. . SILVER CHLORIDES
. . SILVER IODIDES
. SILVER NITRATES
. SILVER OXIDES
RT   ∞CHEMICAL COMPOUNDS
∞GROUP 1B COMPOUNDS
∞METAL COMPOUNDS

SILVER HALIDES
GS   HALOGEN COMPOUNDS
. HALIDES
. . METAL HALIDES
. . . **SILVER HALIDES**
. . . . SILVER BROMIDES
. . . . SILVER CHLORIDES
. . . . SILVER IODIDES
SILVER COMPOUNDS
. **SILVER HALIDES**
. . SILVER BROMIDES
. . SILVER CHLORIDES
. . SILVER IODIDES

SILVER HYDROGEN BATTERIES
GS   ELECTROCHEMICAL CELLS
. ELECTRIC BATTERIES
. . STORAGE BATTERIES
. . . **SILVER HYDROGEN BATTERIES**

SILVER IODIDES
GS   HALOGEN COMPOUNDS
. HALIDES
. . METAL HALIDES
. . . SILVER HALIDES
. . . . **SILVER IODIDES**
. IODINE COMPOUNDS
. . IODIDES
. . . **SILVER IODIDES**
SILVER COMPOUNDS
. SILVER HALIDES
. . **SILVER IODIDES**

SILVER ISOTOPES
GS   CHEMICAL ELEMENTS
. NUCLIDES

SILVER ISOTOPES-*(CONT.)*
. . ISOTOPES
. . . **SILVER ISOTOPES**
. SILVER
. . **SILVER ISOTOPES**
METALS
. NOBLE METALS
. SILVER
. . . **SILVER ISOTOPES**
. TRANSITION METALS
. SILVER
. . . **SILVER ISOTOPES**

SILVER NITRATES
GS   NITROGEN COMPOUNDS
. NITRATES
. . INORGANIC NITRATES
. . . **SILVER NITRATES**
SILVER COMPOUNDS
. **SILVER NITRATES**

SILVER OXIDE ZINC BATTERIES
USE   SILVER ZINC BATTERIES

SILVER OXIDES
GS   CHALCOGENIDES
. OXIDES
. . METAL OXIDES
. . . **SILVER OXIDES**
SILVER COMPOUNDS
. **SILVER OXIDES**

SILVER ZINC BATTERIES
UF   SILVER OXIDE ZINC BATTERIES
ZINC SILVER BATTERIES
ZINC SILVER OXIDE BATTERIES
GS   ELECTROCHEMICAL CELLS
. ELECTRIC BATTERIES
. . STORAGE BATTERIES
. . . **SILVER ZINC BATTERIES**

SILVICULTURE
RT   AGRICULTURE
BIOMASS
BOTANY
CULTIVATION
FORESTS
ORCHARDS
PLANTING
TREES (PLANTS)

SIM
UF   SCIENTIFIC INSTRUMENT MODULES
GS   MODULES
. SPACECRAFT MODULES
. . **SIM**
RT   APOLLO PROJECT
APOLLO 15 FLIGHT
CAMERAS
INSTRUMENT PACKAGES
∞INSTRUMENTS

SIMICOR (IMAGE CORRELATOR)
USE   IMAGE CORRELATORS

SIMILARITIES
USE   ANALOGIES

SIMILARITY NUMBERS
GS   RATIOS
. DIMENSIONLESS NUMBERS
. . **SIMILARITY NUMBERS**
RT   DIMENSIONAL ANALYSIS
SCALING LAWS

SIMILARITY THEOREM
GS   THEOREMS
. **SIMILARITY THEOREM**
. . LAGRANGE SIMILARITY HYPOTHESIS
RT   DYNAMIC MODELS
MATHEMATICAL MODELS
SCALE MODELS

SIMILITUDE LAW
GS   LAWS
. **SIMILITUDE LAW**
RT   GRAVITATION
INERTIA
SCALE MODELS
VISCOSITY

SIMPLE HARMONIC MOTION
GS   HARMONIC MOTION
. **SIMPLE HARMONIC MOTION**

SIMPLE HARMONIC MOTION-*(CONT.)*
HARMONICS
. **SIMPLE HARMONIC MOTION**
RT   ACOUSTICS
FOURIER ANALYSIS
HARMONIC EXCITATION

SIMPLEX METHOD
GS   MATHEMATICAL LOGIC
. ALGORITHMS
. . **SIMPLEX METHOD**
RT   LINEAR PROGRAMMING
MATHEMATICAL PROGRAMMING
MATRICES (MATHEMATICS)
∞METHODOLOGY
OPTIMIZATION
PROBLEM SOLVING

SIMPLIFICATION
RT   ASSUMPTIONS
LINEARIZATION

SIMULATED ALTITUDE
USE   ALTITUDE SIMULATION

SIMULATION
GS   **SIMULATION**
. COMPUTER SYSTEMS SIMULATION
. COMPUTERIZED SIMULATION
. . ANALOG SIMULATION
. . DIGITAL SIMULATION
. CONTROL SIMULATION
. ENVIRONMENT SIMULATION
. . ACOUSTIC SIMULATION
. . ALTITUDE SIMULATION
. . SPACE ENVIRONMENT SIMULATION
. . . WEIGHTLESSNESS SIMULATION
. . THERMAL SIMULATION
. EXHAUST FLOW SIMULATION
. . ATMOSPHERIC ENTRY SIMULATION
. FLIGHT SIMULATION
. LANDING SIMULATION
. MOTION SIMULATION
. RHEOELECTRICAL SIMULATION
. SOLAR SIMULATION
. SYSTEMS SIMULATION
RT   ANALOGIES
BIONICS
BOND GRAPHS
DATA PROCESSING EQUIPMENT
DATA SIMULATION
DECEPTION
GAME THEORY
HEURISTIC METHODS
HYPERVELOCITY PROJECTILES
MATHEMATICAL MODELS
MONTE CARLO METHOD
OPERATIONS RESEARCH
SIMULATORS
SPACECRAFT CABIN SIMULATORS
SYSTEMS ANALYSIS
VALIDITY
WAR GAMES

SIMULATOR TRAINING
USE   TRAINING SIMULATORS

SIMULATORS
GS   **SIMULATORS**
. CONTROL SIMULATION
. ENVIRONMENT SIMULATORS
. . LUNAR GRAVITY SIMULATOR
. . SOLAR SIMULATORS
. . SPACE SIMULATORS
. . . HIGH VACUUM ORBITAL SIMULATOR
. . . LANGLEY COMPLEX COORDINATOR
. LUNAR ORBIT AND LANDING
   SIMULATORS
. MOTION SIMULATORS
. SHOCK SIMULATORS
. SHUTTLE ENGINEERING SIMULATOR
. SHUTTLE MISSION SIMULATOR
. TARGET SIMULATORS
. TRAINING SIMULATORS
. . FLIGHT SIMULATORS
. . . COCKPIT SIMULATORS
. . SPACECRAFT CABIN SIMULATORS
. VIBRATION SIMULATORS
. . VERTICAL MOTION SIMULATORS
RT   ANALOGS
COMPUTER SYSTEMS SIMULATION
DUMMIES
∞MISSILE SIMULATORS
MODELS
SIMULATION

SIMULATORS-*(CONT.)*
∞TEST EQUIPMENT
TEST FACILITIES
TRAINING DEVICES

SIMULTANEOUS EQUATIONS
RT   ∞EQUATIONS
LEAST SQUARES METHOD
MATRICES (MATHEMATICS)

SIMULTANEOUS IMAGE CORRELATOR
USE   IMAGE CORRELATORS

SINE SERIES
GS   ANALYSIS (MATHEMATICS)
. CALCULUS
. . SERIES (MATHEMATICS)
. . . **SINE SERIES**
. REAL VARIABLES
. . PERIODIC FUNCTIONS
. . . TRIGONOMETRIC FUNCTIONS
. . . . **SINE SERIES**
. . SERIES (MATHEMATICS)
. . . **SINE SERIES**
FUNCTIONS (MATHEMATICS)
. TRANSCENDENTAL FUNCTIONS
. PERIODIC FUNCTIONS
. . . TRIGONOMETRIC FUNCTIONS
. . . . **SINE SERIES**

SINE WAVES
UF   SINUSOIDS
RT   ELASTIC WAVES
ELECTROMAGNETIC RADIATION
TRIGONOMETRIC FUNCTIONS
∞WAVES

SINGAPORE
GS   NATIONS
. **SINGAPORE**
RT   ASIA

SINGLE CHANNEL PER CARRIER TRANSMISSION
UF   SCPC TRANSMISSION
GS   TELECOMMUNICATION
. **SINGLE CHANNEL PER CARRIER
   TRANSMISSION**
TRANSMISSION
. SIGNAL TRANSMISSION
. . DATA TRANSMISSION
. . . **SINGLE CHANNEL PER CARRIER
   TRANSMISSION**
RT   CARRIER FREQUENCIES
CHANNELS (DATA TRANSMISSION)
SATELLITE TRANSMISSION
SPACECRAFT COMMUNICATION
TELEGRAPH SYSTEMS
TELEMETRY
TELEPHONY
VOICE COMMUNICATION
VOICE DATA PROCESSING

SINGLE CRYSTALS
UF   MONOCRYSTALS
GS   CRYSTALS
. **SINGLE CRYSTALS**
RT   BICRYSTALS
BOULES
BRAVAIS CRYSTALS
BRIDGMAN METHOD
CRYSTAL LATTICES
DIAMONDS
GRAPHITE
NEEDLES
PIEZOELECTRIC CRYSTALS
POLYCRYSTALS
SPACE PROCESSING
ULTRAPURE METALS

SINGLE EVENT UPSETS
GS   RADIATION EFFECTS
. **SINGLE EVENT UPSETS**
RT   ASTRIONICS
AVIONICS
CHARGED PARTICLES
COSMIC RAYS
ELECTRON-HOLE DROPS
INNER RADIATION BELT
IONIZATION
MICROELECTRONICS
RADIATION DAMAGE
RADIATION DOSAGE
SATELLITE-BORNE INSTRUMENTS
SECONDARY COSMIC RAYS
SPACECRAFT CHARGING

SINGLE EVENT UPSETS-*(CONT.)*
SPACECRAFT ELECTRONIC EQUIPMENT

SINGLE SIDEBAND MODULATION
USE   SINGLE SIDEBAND TRANSMISSION

SINGLE SIDEBAND TRANSMISSION
UF   SINGLE SIDEBAND MODULATION
GS   TRANSMISSION
. ELECTROMAGNETIC WAVE
   TRANSMISSION
. . RADIO TRANSMISSION
. . . **SINGLE SIDEBAND TRANSMISSION**
. SIGNAL TRANSMISSION
. . RADIO TRANSMISSION
. . . **SINGLE SIDEBAND TRANSMISSION**
RT   AMPLITUDE MODULATION
DOUBLE SIDEBAND TRANSMISSION
SIDEBANDS
TELEVISION TRANSMISSION
VOICE COMMUNICATION
WAVE PROPAGATION

SINGLE STAGE ROCKET VEHICLES
GS   ROCKET VEHICLES
. **SINGLE STAGE ROCKET VEHICLES**
. . AGENA ROCKET VEHICLES
. . . AGENA A ROCKET VEHICLE
. . . AGENA B ROCKET VEHICLE
. . . AGENA C ROCKET VEHICLE
. . . AGENA D ROCKET VEHICLE
. . ARCAS ROCKET VEHICLES
. . BLACK BRANT SOUNDING ROCKETS
. . . BLACK BRANT 1 SOUNDING
   ROCKET
. . . BLACK BRANT 2 SOUNDING
   ROCKET
. . . BLACK BRANT 3 SOUNDING
   ROCKET
. . . BLACK BRANT 4 SOUNDING
   ROCKET
. . . BLACK BRANT 5 SOUNDING
   ROCKET
. . BLACK KNIGHT ROCKET VEHICLE
. . DORNIER PARAGLIDER ROCKET
   VEHICLE
. . GENIE ROCKET VEHICLE
. . HONEST JOHN ROCKET VEHICLE
. . HYLA-STAR ROCKET VEHICLE
. . LITTLE JOHN ROCKET VEHICLE
. . LOKI ROCKET VEHICLE
. . NOMAD LAUNCH VEHICLE
. . VERONIQUE ROCKET VEHICLES
. . VIKING ROCKET VEHICLE
. . ZUNI ROCKET VEHICLE
RT   MAULER MISSILE
ROCKET ENGINES
∞VEHICLES

SINGLE STAGE TO ORBIT VEHICLES
GS   LAUNCH VEHICLES
. REUSABLE LAUNCH VEHICLES
. . **SINGLE STAGE TO ORBIT VEHICLES**
RT   NASA PROGRAMS
SPACE SHUTTLES
SPACE TRANSPORTATION
∞VEHICLES

SINGLE-PHASE FLOW
UF   ONE-PHASE FLOW
UNIPHASE FLOW
GS   FLUID FLOW
. **SINGLE-PHASE FLOW**
RT   CRITICAL FLOW
GAS FLOW
LAMINAR FLOW
LIQUID FLOW
MASS FLOW
MULTIPHASE FLOW
ORIFICE FLOW
PIPE FLOW
STEADY FLOW
STEAM FLOW
SUBCRITICAL FLOW
SUPERCRITICAL FLOW
TURBULENT FLOW
TWO PHASE FLOW
UNIFORM FLOW
UNSTEADY FLOW

SINGULAR INTEGRAL EQUATIONS
GS   ANALYSIS (MATHEMATICS)
. FUNCTIONAL ANALYSIS
. . INTEGRAL EQUATIONS
. . . **SINGULAR INTEGRAL EQUATIONS**

SINGULAR INTEGRAL EQUATIONS-*(CONT.)*
RT   ∞EQUATIONS

SINGULARITY (MATHEMATICS)
GS   ANALYSIS (MATHEMATICS)
. COMPLEX VARIABLES
. . **SINGULARITY (MATHEMATICS)**
. . . NAKED SINGULARITIES
RT   POINTS (MATHEMATICS)
UNIQUENESS

SINKHOLES
GS   LANDFORMS
. STRUCTURAL BASINS
. . KARST
. . . **SINKHOLES**
RT   KETTLES (GEOLOGY)
LANDFORMS
STRUCTURAL PROPERTIES (GEOLOGY)

SINKING
RT   FALLING
REFRACTION
SUBMERGING
WATER IMMERSION

SINKS
SN   (EXCLUDES PLUMBING
FIXTURES--LIMITED TO AREAS FOR
ABSORPTIVE DISPOSAL OF HEAT OR
FLUIDS)
GS   **SINKS**
. HEAT SINKS
RT   ABSORBERS (MATERIALS)
DISPOSAL
∞SOURCES

SINKS (GEOLOGY)
USE   STRUCTURAL BASINS

SINTERED ALUMINUM POWDER
GS   CHEMICAL ELEMENTS
. ALUMINUM
. . **SINTERED ALUMINUM POWDER**
METALS
. ALUMINUM
. . **SINTERED ALUMINUM POWDER**
RT   POWDER METALLURGY
POWDERED ALUMINUM

SINTERING
UF   PRESINTERING
RT   AGGLOMERATION
FURNACES
GROWTH
HEATING
HOT PRESSING
METAL POWDER
MIXED CRYSTALS
POROSITY
POWDER METALLURGY
PYROMETALLURGY
REACTION BONDING
ROASTING
SHRINKAGE
SIALON

SINUSES
GS   **SINUSES**
. PARANASAL SINUSES
RT   CAROTID SINUS BODY
CAROTID SINUS REFLEX
NOSE (ANATOMY)
RESPIRATION

SINUSOIDS
USE   SINE WAVES

SIOUX HELICOPTER
USE   OH-13 HELICOPTER

SIPHONING
RT   ∞FLUIDS
SIPHONS
THERMOSIPHONS

SIPHONS
RT   MATERIALS HANDLING
PIPELINES
PIPES (TUBES)
PUMPS
SIPHONING
∞TUBES

SIR-A
USE   SHUTTLE IMAGING RADAR

SIR-B
USE   SHUTTLE IMAGING RADAR

SIRENS
RT   HORNS
NOISE INTENSITY
SIGNAL GENERATORS
∞SIGNALS
SOUND GENERATORS
SOUND INTENSITY
SOUND TRANSMISSION
WARNING SYSTEMS

SIRIO SATELLITE
GS   SATELLITES
. ARTIFICIAL SATELLITES
. . SYNCHRONOUS SATELLITES
. . . **SIRIO SATELLITE**
RT   ITALIAN SPACE PROGRAM
ITALY

SIRS B SATELLITE
GS   SATELLITES
. ARTIFICIAL SATELLITES
. . METEOROLOGICAL SATELLITES
. . . **SIRS B SATELLITE**
RT   METEOROLOGICAL FLIGHT
METEOROLOGICAL INSTRUMENTS
SATELLITE OBSERVATION
UNMANNED SPACECRAFT

SIS (SEMICONDUCTORS)
UF   SEMICONDUCTOR INSULATOR
SEMICONDUCTORS
GS   ELECTRONIC EQUIPMENT
. SOLID STATE DEVICES
. . **SIS (SEMICONDUCTORS)**
RT   BARRIER LAYERS
MIM (SEMICONDUCTORS)
MIS (SEMICONDUCTORS)
P-N JUNCTIONS
PHOTODIODES
PHOTOVOLTAIC CELLS
SCHOTTKY DIODES
SEMICONDUCTOR DIODES
SEMICONDUCTOR JUNCTIONS
SILICON JUNCTIONS
SOLAR CELLS
SOS (SEMICONDUCTORS)
TIN OXIDES
TRANSISTORS

SITE DATA PROCESSORS
UF   SDP (COMPUTERS)
GS   DATA PROCESSING EQUIPMENT
. COMPUTERS
. . **SITE DATA PROCESSORS**
RT   APOLLO PROJECT
∞DATA
DATA LINKS
DATA PROCESSING

SITE SELECTION
GS   SELECTION
. **SITE SELECTION**
RT   AIRPORTS
CERTIFICATION
∞FACILITIES
INDUSTRIAL AREAS
LAND USE
LEASING
LOGISTICS
OPTIONS
RESOURCES
ROADS
ROUTES
SERVICES
SITES
TERMINAL FACILITIES
TRANSPORTATION
UTILITIES

SITES
UF   TRACTS
GS   **SITES**
. ARIZONA REGIONAL ECOLOGICAL
TEST SITE
. CENTRAL ATLANTIC REGIONAL ECOL
TEST SITE
. LANDING SITES
. . LUNAR LANDING SITES
. LAUNCHING SITES

SITES-*(CONT.)*
. . LAUNCHING PADS
. OFFSHORE REACTOR SITES
RT   AIRPORT PLANNING
BARREN LAND
∞FACILITIES
LAND
∞PLOTS
POSITION (LOCATION)
REGIONS
RURAL LAND USE
SITE SELECTION

SITTING POSITION
RT   PRONE POSITION
REST
SEATS
SUPINE POSITION

SIZE (DIMENSIONS)
GS   **SIZE (DIMENSIONS)**
. GRAIN SIZE
RT   FINENESS

SIZE DETERMINATION
GS   **SIZE DETERMINATION**
. PRECIPITATION PARTICLE
MEASUREMENT
RT   BODY MEASUREMENT (BIOLOGY)
CLASSIFIERS
DIMENSIONAL MEASUREMENT
∞MEASUREMENT
PARTICLE SIZE DISTRIBUTION
∞SIZING

SIZE DISTRIBUTION
GS   **SIZE DISTRIBUTION**
. PARTICLE SIZE DISTRIBUTION
RT   DROP SIZE
MASS DISTRIBUTION
STATISTICAL DISTRIBUTIONS

SIZE SEPARATION
UF   SIZING (SEPARATION)
RT   BENEFICIATION
CLASSIFIERS
∞CLASSIFYING
CONCENTRATORS
FILTRATION
FLOTATION
METAL POWDER
PARTICLE SIZE DISTRIBUTION
POWDER (PARTICLES)
∞SEPARATION
SETTLING
∞SIZING

∞ SIZING
SN   *(USE OF A MORE SPECIFIC TERM IS
RECOMMENDED--CONSULT THE TERMS
LISTED BELOW)*
RT   BODY MEASUREMENT (BIOLOGY)
SIZE DETERMINATION
SIZE SEPARATION
SIZING (SHAPING)
SIZING (SURFACE TREATMENT)
SIZING MATERIALS

SIZING (SEPARATION)
USE   SIZE SEPARATION

SIZING (SHAPING)
GS   METAL WORKING
. **SIZING (SHAPING)**
RT   COINING
PRESSING (FORMING)
∞SIZING

SIZING (SURFACE TREATMENT)
SN   (EXCLUDES MECHANICAL SHAPING OR
REMOVAL OF SURFACE MATERIALS)
RT   FINISHES
SIZING MATERIALS
∞SURFACES

SIZING MATERIALS
SN   (MATERIALS USED FOR SURFACE
TREATMENT)
RT   BINDERS (MATERIALS)
CLAYS
FILLERS
GLUES
∞MATERIALS
SIZING (SURFACE TREATMENT)

SIZING MATERIALS-*(CONT.)*
  STARCHES

SIZING SCREENS
  GS   SEPARATORS
        . CLASSIFIERS
        . . **SIZING SCREENS**
  RT   AGITATION
        CONCENTRATORS
        FLUID FILTERS
        ∞ SCREENS
        SHAKERS
        SIEVES

SKELETON
  USE   MUSCULOSKELETAL SYSTEM

SKEWNESS
  RT   ASYMMETRY
        DEFORMATION
        DISPLACEMENT
        DISTORTION
        DISTRIBUTION MOMENTS
        ECCENTRICITY
        MOMENTS

SKID LANDINGS
  GS   LANDING
        . AIRCRAFT LANDING
        . . **SKID LANDINGS**
  RT   AIR CUSHION LANDING SYSTEMS
        CRASH LANDING
        HYDROPLANING
        SKIDDING

SKIDDING
  RT   HYDROPLANING
        LANDING GEAR
        SIDESLIP
        SKID LANDINGS
        SLEDS
        YAW

SKILLS
  USE   ABILITIES

SKIN (ANATOMY)
  GS   **SKIN (ANATOMY)**
        . EPIDERMIS
        . EPITHELIUM
        . LEATHER
  RT   ALBINISM
        ∞ BLISTERS
        CAROTENE
        CHLOROPHYLLS
        COLLAGENS
        CONTACT DERMATITIS
        CYTOCHROMES
        DERMATITIS
        DERMATOLOGY
        EVAPORATION
        HOMEOSTASIS
        MELANIN
        MEMBRANES
        PERSPIRATION
        PETECHIA
        PIGMENTS
        SECRETIONS
        SENSE ORGANS
        THERMORECEPTORS
        TOUCH

SKIN (STRUCTURAL MEMBER)
  GS   MEMBRANES
        . MEMBRANE STRUCTURES
        . . **SKIN (STRUCTURAL MEMBER)**
        STRUCTURAL MEMBERS
        . MEMBRANE STRUCTURES
        . . **SKIN (STRUCTURAL MEMBER)**
  RT   AIRCRAFT CONSTRUCTION MATERIALS
        ∞ CONSTRUCTION MATERIALS
        HULLS (STRUCTURES)
        METAL SHELLS
        SHELLS (STRUCTURAL FORMS)
        STRESSED-SKIN STRUCTURES
        THIN WALLED SHELLS
        THIN WALLS
        TOROIDAL SHELLS
        WEBS (SUPPORTS)

SKIN FRICTION
  UF   FRICTION PRESSURE DROP
  GS   FRICTION
        . **SKIN FRICTION**

SKIN FRICTION-*(CONT.)*
        . . FRICTION DRAG
        . . . SUPERSONIC DRAG
        . . . VISCOUS DRAG
  RT   AERODYNAMIC HEATING
        DRAG
        DRAG DEVICES
        FLOW RESISTANCE
        FLUID FLOW
        FRICTION FACTOR
        STREAMLINING

SKIN GRAFTS
  RT   SURGERY
        THERAPY

SKIN RESISTANCE
  GS   IMPEDANCE
        . ELECTRICAL IMPEDANCE
        . . ELECTRICAL RESISTANCE
        . . . **SKIN RESISTANCE**
  RT   ∞ RESISTANCE

SKIN TEMPERATURE (BIOLOGY)
  GS   TEMPERATURE
        . **SKIN TEMPERATURE (BIOLOGY)**
  RT   ∞ BIOLOGY
        FEVER
        HYPERTHERMIA
        HYPOTHERMIA

SKIN TEMPERATURE (NON-BIOLOGICAL)
  GS   SURFACE PROPERTIES
        . SURFACE TEMPERATURE
        . . **SKIN TEMPERATURE
             (NON-BIOLOGICAL)**
        TEMPERATURE
        . SURFACE TEMPERATURE
        . . **SKIN TEMPERATURE
             (NON-BIOLOGICAL)**
  RT   AERODYNAMIC HEATING
        AEROTHERMODYNAMICS

SKINNER BOXES
  RT   BEHAVIOR
        PSYCHOLOGICAL TESTS
        PSYCHOMETRICS

SKIRTS
  RT   AFTERBODIES
        BOATTAILS
        CONICAL NOZZLES
        EXHAUST NOZZLES
        FOUNDATIONS
        ∞ JET NOZZLES
        ROCKET NOZZLES

SKIS
  RT   HYDROFOILS
        HYDROPLANES (SURFACES)
        LANDING GEAR

SKUA ROCKET VEHICLES
  GS   ROCKET VEHICLES
        . SOUNDING ROCKETS
        . . **SKUA ROCKET VEHICLES**
  RT   SOLID PROPELLANT ROCKET ENGINES
        ∞ VEHICLES

SKULL
  GS   ANATOMY
        . HEAD (ANATOMY)
        . . **SKULL**
        . . . CRANIUM
        . . . . INTRACRANIAL CAVITY
        . MUSCULOSKELETAL SYSTEM
        . . BONES
        . . . **SKULL**
        . . . . CRANIUM
        . . . . . INTRACRANIAL CAVITY
  RT   FOREHEAD
        INTERCRANIAL CIRCULATION

SKY
  GS   **SKY**
        . NIGHT SKY
  RT   CLOUD COVER
        CLOUDS (METEOROLOGY)
        DAYGLOW
        RAYLEIGH SCATTERING
        SUNLIGHT

SKY BRIGHTNESS
  GS   ELECTROMAGNETIC PROPERTIES

SKY BRIGHTNESS-*(CONT.)*
        . OPTICAL PROPERTIES
        . . **SKY BRIGHTNESS**
  RT   AIRGLOW
        AURORAS
        BRIGHTNESS
        CLOUD COVER
        DAYTIME
        GEGENSCHEIN
        GLARE
        LIGHT (VISIBLE RADIATION)
        LIGHT EMISSION
        LUMINANCE
        NIGHT
        NIGHT SKY
        NIGHTGLOW
        SOLAR RADIATION
        SUNLIGHT
        ZODIACAL LIGHT

SKY RADIATION
  GS   ATMOSPHERIC RADIATION
        . **SKY RADIATION**
        . . AIRGLOW
        . . . GEOCORONAL EMISSIONS
        . . . NIGHTGLOW
        . . . TWILIGHT GLOW
        . . DAYGLOW
        ELECTROMAGNETIC RADIATION
        . LIGHT (VISIBLE RADIATION)
        . . **SKY RADIATION**
        . . . AIRGLOW
        . . . . GEOCORONAL EMISSIONS
        . . . . NIGHTGLOW
        . . . . TWILIGHT GLOW
        . . . DAYGLOW
  RT   BACKGROUND RADIATION
        PYRANOMETERS
        ∞ RADIATION
        STRATOSPHERE RADIATION
        SUNLIGHT
        THERMAL RADIATION
        TROPOSPHERIC RADIATION

SKY WAVES
  GS   ELECTROMAGNETIC RADIATION
        . RADIO WAVES
        . . **SKY WAVES**
  RT   GROUND WAVE PROPAGATION
        IONOSPHERIC NOISE

SKYBOLT MISSILE
  GS   MISSILES
        . BALLISTIC MISSILES
        . . **SKYBOLT MISSILE**
  RT   SOLID PROPELLANT ROCKET ENGINES

SKYCRANE HELICOPTER
  USE   CH-54 HELICOPTER

SKYDROL (TRADEMARK)
  GS   LIQUIDS
        . HYDRAULIC FLUIDS
        . . **SKYDROL (TRADEMARK)**
  RT   ESTERS
        PHOSPHATES
        PLASTICIZERS

SKYHAWK AIRCRAFT
  USE   A-4 AIRCRAFT

SKYHOOK BALLOONS
  GS   EXPANDABLE STRUCTURES
        . INFLATABLE STRUCTURES
        . . BALLOONS
        . . . HIGH ALTITUDE BALLOONS
        . . . . **SKYHOOK BALLOONS**
  RT   HIGH ALTITUDE
        METEOROLOGICAL BALLOONS
        ROBIN BALLOONS
        ROCKOONS

SKYLAB PROGRAM
  GS   PROGRAMS
        . NASA PROGRAMS
        . . **SKYLAB PROGRAM**
        . NASA SPACE PROGRAMS
        . . **SKYLAB PROGRAM**
  RT   AAP 1 MISSION
        AAP 2 MISSION
        AAP 3 MISSION
        AAP 4 MISSION
        AIRLOCK MODULES
        APOLLO APPLICATIONS PROGRAM
        APOLLO FLIGHTS

SKYLAB PROGRAM-*(CONT.)*
    APOLLO PROJECT
    APOLLO SPACECRAFT
    APOLLO TELESCOPE MOUNT
    EARTH RESOURCES INFORMATION
      SYSTEM
    EARTH RESOURCES PROGRAM
    EARTH RESOURCES SURVEY PROGRAM
    ORBITAL WORKSHOPS
    SATURN WORKSHOPS
    SATURN 1 WORKSHOP
    SATURN 5 WORKSHOP
    SPACELAB

SKYLAB SPACE STATION (UNMANNED)
USE   SKYLAB 1

SKYLAB 1
UF   SKYLAB SPACE STATION (UNMANNED)
      SL 1
GS   MANNED SPACECRAFT
    . SPACE STATIONS
    . . ORBITAL SPACE STATIONS
    . . . **SKYLAB 1**
RT   AIRLOCK MODULES
    ARTIFICIAL SATELLITES
    COMMAND SERVICE MODULES
    EREP
    MULTIPLE DOCKING ADAPTERS
    ORBITAL WORKSHOPS
    SPACE MISSIONS

SKYLAB 2
UF   SL 2
GS   MANNED SPACECRAFT
    . SPACE STATIONS
    . . ORBITAL SPACE STATIONS
    . . . **SKYLAB 2**
RT   AIRLOCK MODULES
    ARTIFICIAL SATELLITES
    COMMAND SERVICE MODULES
    EREP
    MULTIPLE DOCKING ADAPTERS
    ORBITAL WORKSHOPS
    SATURN 1B LAUNCH VEHICLES
    SATURN 5 LAUNCH VEHICLES
    SPACE MISSIONS

SKYLAB 3
UF   SL 3
GS   MANNED SPACECRAFT
    . SPACE STATIONS
    . . ORBITAL SPACE STATIONS
    . . . **SKYLAB 3**
RT   AIRLOCK MODULES
    ARTIFICIAL SATELLITES
    COMMAND SERVICE MODULES
    EREP
    MULTIPLE DOCKING ADAPTERS
    ORBITAL WORKSHOPS
    SATURN 1B LAUNCH VEHICLES
    SATURN 5 LAUNCH VEHICLES
    SPACE MISSIONS

SKYLAB 4
UF   SL 4
GS   MANNED SPACECRAFT
    . SPACE STATIONS
    . . ORBITAL SPACE STATIONS
    . . . **SKYLAB 4**
RT   AIRLOCK MODULES
    ARTIFICIAL SATELLITES
    COMMAND SERVICE MODULES
    EREP
    MULTIPLE DOCKING ADAPTERS
    ORBITAL WORKSHOPS
    SATURN 1B LAUNCH VEHICLES
    SATURN 5 LAUNCH VEHICLES
    SPACE MISSIONS

SKYLARK
USE   SKYLARK ROCKET VEHICLE

SKYLARK ROCKET VEHICLE
UF   SKYLARK
GS   ROCKET VEHICLES
    . MULTISTAGE ROCKET VEHICLES
    . . **SKYLARK ROCKET VEHICLE**
    . SOUNDING ROCKETS
    . . **SKYLARK ROCKET VEHICLE**
RT   SOLID PROPELLANT ROCKET ENGINES

SKYMASTER AIRCRAFT
USE   C-54 AIRCRAFT

SKYNET SATELLITES
GS   SATELLITES
    . ARTIFICIAL SATELLITES
    . . **SKYNET SATELLITES**
RT   COMMUNICATION SATELLITES
    SATELLITE NETWORKS
    UK SATELLITES

SKYRAIDER AIRCRAFT
USE   A-1 AIRCRAFT

SKYROCKET AIRCRAFT
USE   D-558 AIRCRAFT

SKYSTREAK AIRCRAFT
USE   D-558 AIRCRAFT

SKYVAN AIRCRAFT
USE   SC-7 AIRCRAFT

SKYWARRIOR AIRCRAFT
USE   A-3 AIRCRAFT

SL 1
USE   SKYLAB 1

SL 2
USE   SKYLAB 2

SL 3
USE   SKYLAB 3

SL 4
USE   SKYLAB 4

SL-3 ROCKET ENGINE
GS   ENGINES
    . ROCKET ENGINES
    . . SOLID PROPELLANT ROCKET
      ENGINES
    . . . **SL-3 ROCKET ENGINE**

SLABS
RT   BILLETS
    BLOCKS
    FLAT PLATES
    METAL PLATES
    PLATES (STRUCTURAL MEMBERS)
    ∞PLATFORMS
    STRUCTURAL MEMBERS

SLAGS
RT   AGGREGATES
    REACTION PRODUCTS
    WASTES

SLAM
USE   SUPERSONIC LOW ALTITUDE MISSILE

SLAMMING
RT   FLUID DYNAMICS

SLANT
USE   SLOPES

SLANT PERCEPTION
USE   SPACE PERCEPTION

SLASHES
USE   CLEARINGS (OPENINGS)

SLATER ORBITALS
GS   ORBITALS
    . **SLATER ORBITALS**
RT   HARTREE-FOCK-SLATER METHOD
    ORBITAL ELEMENTS

SLEDS
GS   SURFACE VEHICLES
    . **SLEDS**
    . . ROCKET PROPELLED SLEDS
RT   DOLLIES
    SKIDDING
    TOWED BODIES
    TRACTORS
    TRAILERS

SLEEP
UF   DROWSINESS
GS   **SLEEP**
    . HYPERSOMNIA
    . HYPNOSIS
    . INSOMNIA

SLEEP-*(CONT.)*
RT   DREAMS
    ∞DRIVES
    RAPID EYE MOVEMENT STATE
    REST

SLEEP DEPRIVATION
GS   DEPRIVATION
    . **SLEEP DEPRIVATION**
RT   CONSCIOUSNESS
    INSOMNIA
    WAKEFULNESS

SLEEVES
SN   (EXCLUDES CLOTHING)
RT   CONNECTORS
    COUPLINGS
    FASTENERS
    FITTINGS
    JOINTS (JUNCTIONS)

SLENDER BODIES
GS   **SLENDER BODIES**
    . SLENDER CONES
RT   AERODYNAMIC CONFIGURATIONS
    AERODYNAMICS
    AXISYMMETRIC BODIES
    ∞BODIES
    DUCTED BODIES
    FINENESS RATIO
    MISSILE BODIES
    STREAMLINED BODIES
    SYMMETRICAL BODIES
    ∞THIN BODIES

SLENDER CONES
GS   CONES
    . CONICAL BODIES
    . . **SLENDER CONES**
    SLENDER BODIES
    . **SLENDER CONES**
    SYMMETRICAL BODIES
    . BODIES OF REVOLUTION
    . . CONICAL BODIES
    . . . **SLENDER CONES**
RT   AERODYNAMIC CONFIGURATIONS
    AXISYMMETRIC BODIES

SLENDER WINGS
UF   HIGH ASPECT RATIO WINGS
GS   AIRFOILS
    . WINGS
    . . **SLENDER WINGS**
    . . . INFINITE SPAN WINGS
RT   FIXED WINGS
    WING PLANFORMS

SLEUTH (PROGRAMMING LANGUAGE)
GS   LANGUAGES
    . PROGRAMMING LANGUAGES
    . . MACHINE ORIENTED LANGUAGES
    . . . **SLEUTH (PROGRAMMING
      LANGUAGE)**
RT   COMPUTER PROGRAMMING

SLEWING
RT   ANTENNAS
    ERROR SIGNALS
    POSITIONING DEVICES (MACHINERY)
    RADAR TRACKING
    SERVOMOTORS
    SPINNERS

SLICING
GS   CUTTING
    . **SLICING**
RT   METAL CUTTING
    PLANING
    PLANNING
    SCARFING
    ∞SEPARATION
    SPLITTING

SLICKS
USE   OIL SLICKS

SLIDES
USE   CHUTES

SLIDES (MICROSCOPY)
RT   MICROSCOPY

SLIDING
RT   INTERFACIAL TENSION

SLIDING-*(CONT.)*
LUBRICATION
MASS FLOW
∞ SLIP
SLUMPING
STATIC FRICTION

∞ SLIDING CONTACT
SN   *(USE OF A MORE SPECIFIC TERM IS
      RECOMMENDED--CONSULT THE TERMS
      LISTED BELOW)*
RT   ELECTRIC CONTACTS
      SLIDING FRICTION

SLIDING FRICTION
GS   FRICTION
      . KINETIC FRICTION
      . . **SLIDING FRICTION**
RT   COEFFICIENT OF FRICTION
      DRY FRICTION
      ELECTRIC CONTACTS
      ∞ SLIDING CONTACT
      STATIC FRICTION
      WEAR

∞ SLIP
SN   *(USE OF A MORE SPECIFIC TERM IS
      RECOMMENDED--CONSULT THE TERMS
      LISTED BELOW)*
RT   PLASTIC DEFORMATION
      POLYSLIPS
      SIDESLIP
      SLIDING

SLIP BANDS
USE   EDGE DISLOCATIONS

SLIP CASTING
GS   FORMING TECHNIQUES
      . CASTING
      . . **SLIP CASTING**

SLIP FLOW
SN   (RAREFIED GAS FLOW IN THE REGION
      BETWEEN KNUDSEN NUMBERS 0.01
      AND 0.1 ONLY, EXCLUDES TRANSITION
      FLOW, FREE MOLECULE FLOW CREEP,
      SHEAR FLOW, AND PLASTIC FLOW)
GS   FLUID FLOW
      . GAS FLOW
      . . MOLECULAR FLOW
      . . . **SLIP FLOW**
RT   CONTINUUM FLOW
      FREE MOLECULAR FLOW
      LOW DENSITY WIND TUNNELS
      RAREFIED GAS DYNAMICS
      TRANSITION FLOW

SLIPSTREAMS
GS   WAKES
      . AIRCRAFT WAKES
      . . **SLIPSTREAMS**
      . . . PROPELLER SLIPSTREAMS
      . TURBULENT WAKES
      . . **SLIPSTREAMS**
      . . . PROPELLER SLIPSTREAMS
RT   BACKWASH
      STROUHAL NUMBER
      TURBULENCE

SLITS
GS   OPENINGS
      . **SLITS**
RT   APERTURES
      FRESNEL REFLECTORS
      SLOTS

SLIVERS
RT   FIBERS
      WOOD

SLOPES
UF   CANT
      SLANT
      STEEPNESS
GS   **SLOPES**
      . GLIDE PATHS
RT   ANGLES (GEOMETRY)
      CLIFFS
      ESCARPMENTS
      ∞ GRADE
      GRADIENTS
      HEIGHT
      ∞ INCLINATION

SLOPES-*(CONT.)*
LANDFORMS
LANDSLIDES
LEVEL (HORIZONTAL)
PITCH (INCLINATION)
∞ PROFILES
∞ RAKES
RAMP FUNCTIONS
RAMPS (STRUCTURES)
TOPOGRAPHY

SLOSHING
USE   LIQUID SLOSHING

SLOT ANTENNAS
UF   SLOTTED ANTENNAS
GS   ANTENNAS
      . DIRECTIONAL ANTENNAS
      . . **SLOT ANTENNAS**
      MICROWAVE EQUIPMENT
      . MICROWAVE ANTENNAS
      . . **SLOT ANTENNAS**
RT   ANTENNA DESIGN
      HORN ANTENNAS
      RADANT
      RADAR ANTENNAS
      WAVEGUIDE ANTENNAS

SLOTS
GS   **SLOTS**
      . WING SLOTS
RT   LIFT DEVICES
      LOUVERS
      OPENINGS
      SLITS

SLOTTED ANTENNAS
USE   SLOT ANTENNAS

SLOTTED WIND TUNNELS
GS   TEST FACILITIES
      . WIND TUNNELS
      . . **SLOTTED WIND TUNNELS**
RT   SUPERSONIC WIND TUNNELS
      TRANSONIC WIND TUNNELS
      TRISONIC WIND TUNNELS
      VENTS

SLOW NEUTRONS
USE   THERMAL NEUTRONS

SLUDGE
GS   **SLUDGE**
      . ACTIVATED SLUDGE
RT   DEPOSITS
      LIQUID WASTES
      MUD
      OCEAN BOTTOM
      ORGANIC WASTES (FUEL CONVERSION)
      REACTION PRODUCTS
      SEDIMENTS
      SEWAGE TREATMENT
      SOLID WASTES
      WASTE TREATMENT
      WASTES

SLUMPING
RT   GEOMORPHOLOGY
      MASS FLOW
      SLIDING

SLURRIES
GS   MIXTURES
      . **SLURRIES**
RT   DISPERSIONS
      EMULSIONS
      GELS
      RHEOCASTING
      SLURRY PROPELLANTS

SLURRY PROPELLANTS
GS   LIQUIDS
      . LIQUID FUELS
      . . LIQUID ROCKET PROPELLANTS
      . . . **SLURRY PROPELLANTS**
      PROPELLANTS
      . ROCKET PROPELLANTS
      . . LIQUID ROCKET PROPELLANTS
      . . . **SLURRY PROPELLANTS**
RT   AIRCRAFT FUELS
      COLLOIDAL PROPELLANTS
      DISPERSIONS
      GELLED ROCKET PROPELLANTS
      METAL FUELS

SLURRY PROPELLANTS-*(CONT.)*
METAL PROPELLANTS
MONOPROPELLANTS
SLURRIES
SOLID ROCKET PROPELLANTS

SLUSH
RT   BAY ICE
      CRYOGENIC ROCKET PROPELLANTS
      ICE
      RUNWAY CONDITIONS
      SNOW
      WATER

SLV
USE   STANDARD LAUNCH VEHICLES

SLV (SOFT LANDING VEHICLES)
USE   SOFT LANDING SPACECRAFT

SM-65 MISSILE
USE   ATLAS LAUNCH VEHICLES

SM-68 MISSILE
USE   TITAN 1 ICBM

SM-68B MISSILE
USE   TITAN 2 ICBM

SMALL ASTRONOMY SATELLITE 1
USE   SAS-1

SMALL ASTRONOMY SATELLITE 2
USE   SAS-2

SMALL ASTRONOMY SATELLITE 3
USE   SAS-3

SMALL ASTRONOMY SATELLITES
USE   SAS

SMALL PERTURBATION FLOW
GS   FLUID FLOW
      . **SMALL PERTURBATION FLOW**
RT   FLOW DISTORTION
      OSCILLATING FLOW

SMALL SCIENTIFIC SATELLITES
GS   SATELLITES
      . ARTIFICIAL SATELLITES
      . . SCIENTIFIC SATELLITES
      . . . **SMALL SCIENTIFIC SATELLITES**

SMALL WATER PLANE AREA TWIN HULL
USE   SWATH (SHIP)

SMALLPOX
GS   DISEASES
      . INFECTIOUS DISEASES
      . . **SMALLPOX**

SMEAR
RT   ∞ FREQUENCY RESPONSE
      IMAGE CONTRAST
      SIGNAL FADING
      TELEVISION TRANSMISSION
      VIDEO DATA

SMELL
USE   OLFACTORY PERCEPTION

SMELTING
RT   MELTING
      ∞ METALLURGY
      REDUCTION (CHEMISTRY)
      REFINING

SMITH CHART
RT   ELECTRICAL IMPEDANCE
      IMPEDANCE
      POLAR COORDINATES
      REACTANCE
      STANDING WAVE RATIOS
      TRANSMISSION LINES
      WAVEGUIDES

SMM-A
USE   SOLAR MAXIMUM MISSION-A

SMOG
RT   AIR POLLUTION
      AIR SAMPLING
      CARBON MONOXIDE

SMOG-*(CONT.)*
    COMBUSTION PRODUCTS
    ENVIRONMENTAL CHEMISTRY
    EXHAUST GASES
    FLAMES
    FOG
    HYDROCARBON COMBUSTION
    HYDROCARBON POISONING
    LEAD POISONING
    SMOKE

SMOKE
GS   MIXTURES
    . DISPERSIONS
    . . PLASTISOLS
    . . . **SMOKE**
RT   AEROSOLS
    AIR POLLUTION
    COMBUSTION PRODUCTS
    DUST
    EXHAUST GASES
    FIRE DAMAGE
    FOG
    FOREST FIRES
    FUMES
    HAZE DETECTION
    ∞MARKERS
    PARTICLES
    SMOG
    SOOT
    VAPORS
    VISIBILITY

SMOKE ABATEMENT
RT   AEROSOLS
    AIR POLLUTION
    CARBON DIOXIDE REMOVAL
    EXHAUST GASES
    POLLUTION
    SOOT

SMOKE DETECTORS
GS   MEASURING INSTRUMENTS
    . INDICATING INSTRUMENTS
    . . **SMOKE DETECTORS**
RT   FIRE PREVENTION
    FUMES
    GAS DETECTORS
    SAFETY DEVICES
    SIGNAL PROCESSING

SMOKE TRAILS
RT   ∞TRACKS
    WIND DIRECTION
    WIND MEASUREMENT
    WIND PROFILES

SMOOTHING
GS   **SMOOTHING**
    . DATA SMOOTHING
RT   ADJUSTING
    FLATTENING
    HONING
    LEVELING
    PLANING
    POLISHING
    ROUGHNESS

SMS
USE SYNCHRONOUS METEOROLOGICAL
      SATELLITE

SMS 1
GS   SATELLITES
    . ARTIFICIAL SATELLITES
    . . METEOROLOGICAL SATELLITES
    . . . SYNCHRONOUS EARTH
         OBSERVATORY SATELLITE
    . . . . **SMS 1**
    . . . SYNCHRONOUS METEOROLOGICAL
         SATELLITE
    . . . . **SMS 1**
    . . SYNCHRONOUS SATELLITES
    . . . SYNCHRONOUS EARTH
         OBSERVATORY SATELLITE
    . . . . **SMS 1**
    . . . SYNCHRONOUS METEOROLOGICAL
         SATELLITE
    . . . . **SMS 1**
RT   GOES 2
    NOAA SATELLITES

SMS 2
GS   SATELLITES
    . ARTIFICIAL SATELLITES

SMS 2-*(CONT.)*
    . . METEOROLOGICAL SATELLITES
    . . . SYNCHRONOUS EARTH
         OBSERVATORY SATELLITE
    . . . . **SMS 2**
    . . . SYNCHRONOUS METEOROLOGICAL
         SATELLITE
    . . . . **SMS 2**
    . . SYNCHRONOUS SATELLITES
    . . . SYNCHRONOUS EARTH
         OBSERVATORY SATELLITE
    . . . . **SMS 2**
    . . . SYNCHRONOUS METEOROLOGICAL
         SATELLITE
    . . . . **SMS 2**
RT   GOES 2
    NOAA SATELLITES

SMU (MANEUVERING UNITS)
USE SELF MANEUVERING UNITS

SNAILS
GS   ANIMALS
    . INVERTEBRATES
    . . ARTHROPODS
    . . . CEPHALOPODS
    . . . . **SNAILS**

SNAKES
GS   ANIMALS
    . POIKILOTHERMIA
    . . REPTILES
    . . . **SNAKES**
    . VERTEBRATES
    . . REPTILES
    . . . **SNAKES**

SNAKING
USE LATERAL OSCILLATION

SNAP
UF   SYSTEMS FOR NUCLEAR AUXILIARY
     POWER
GS   AUXILIARY POWER SOURCES
    . NUCLEAR AUXILIARY POWER UNITS
    . . **SNAP**
    . . . FISSION ELECTRIC CELLS
    . . . . SNAP 2
    . . . . SNAP 4
    . . . . SNAP 8
    . . . . SNAP 10A
    . . . SNAP 1
    . . . SNAP 3
    . . . SNAP 7
    . . . SNAP 9A
    . . . SNAP 11
    . . . SNAP 13
    . . . SNAP 15
    . . . SNAP 17
    . . . SNAP 19
    . . . SNAP 21
    . . . SNAP 23
    . . . SNAP 27
    . . . SNAP 29
    . . . SNAP 50
    NUCLEAR ELECTRIC POWER
     GENERATION
    . NUCLEAR AUXILIARY POWER UNITS
    . . **SNAP**
    . . . FISSION ELECTRIC CELLS
    . . . . SNAP 2
    . . . . SNAP 4
    . . . . SNAP 8
    . . . . SNAP 10A
    . . . SNAP 1
    . . . SNAP 3
    . . . SNAP 7
    . . . SNAP 9A
    . . . SNAP 11
    . . . SNAP 13
    . . . SNAP 15
    . . . SNAP 17
    . . . SNAP 19
    . . . SNAP 21
    . . . SNAP 23
    . . . SNAP 27
    . . . SNAP 29
    . . . SNAP 50
RT   ELECTRIC GENERATORS
    HEAT EXCHANGERS
    NUCLEAR POWER REACTORS
    SNAPTRAN REACTOR
    SPACE POWER UNIT REACTORS
    ∞SYSTEMS
    THERMIONIC CONVERTERS

SNAP-*(CONT.)*
    THERMIONIC POWER GENERATION
    THERMOELECTRIC GENERATORS
    THERMOELECTRIC POWER GENERATION
    TRANSIENT REACTOR TEST FACILITY
    TURBOGENERATORS

SNAP 1
GS   AUXILIARY POWER SOURCES
    . NUCLEAR AUXILIARY POWER UNITS
    . . SNAP
    . . . **SNAP 1**
    NUCLEAR ELECTRIC POWER
     GENERATION
    . NUCLEAR AUXILIARY POWER UNITS
    . . SNAP
    . . . **SNAP 1**
RT   HEAT EXCHANGERS
    TURBOGENERATORS

SNAP 2
GS   AUXILIARY POWER SOURCES
    . NUCLEAR AUXILIARY POWER UNITS
    . . SNAP
    . . . FISSION ELECTRIC CELLS
    . . . . **SNAP 2**
    . . SPACE POWER REACTORS
    . . . FISSION ELECTRIC CELLS
    . . . . **SNAP 2**
    NUCLEAR ELECTRIC POWER
     GENERATION
    . NUCLEAR AUXILIARY POWER UNITS
    . . SNAP
    . . . FISSION ELECTRIC CELLS
    . . . . **SNAP 2**
    . . SPACE POWER REACTORS
    . . . FISSION ELECTRIC CELLS
    . . . . **SNAP 2**
    . NUCLEAR POWER REACTORS
    . . SPACE POWER REACTORS
    . . . FISSION ELECTRIC CELLS
    . . . . **SNAP 2**
    NUCLEAR REACTORS
    . NUCLEAR POWER REACTORS
    . . SPACE POWER REACTORS
    . . . FISSION ELECTRIC CELLS
    . . . . **SNAP 2**
RT   HEAT EXCHANGERS
    SPACE POWER UNIT REACTORS
    TURBOGENERATORS

SNAP 3
GS   AUXILIARY POWER SOURCES
    . NUCLEAR AUXILIARY POWER UNITS
    . . SNAP
    . . . **SNAP 3**
    ELECTRIC GENERATORS
    . DIRECT POWER GENERATORS
    . . THERMOELECTRIC GENERATORS
    . . . **SNAP 3**
    NUCLEAR ELECTRIC POWER
     GENERATION
    . NUCLEAR AUXILIARY POWER UNITS
    . . SNAP
    . . . **SNAP 3**

SNAP 4
GS   AUXILIARY POWER SOURCES
    . NUCLEAR AUXILIARY POWER UNITS
    . . SNAP
    . . . FISSION ELECTRIC CELLS
    . . . . **SNAP 4**
    . . SPACE POWER REACTORS
    . . . FISSION ELECTRIC CELLS
    . . . . **SNAP 4**
    NUCLEAR ELECTRIC POWER
     GENERATION
    . NUCLEAR AUXILIARY POWER UNITS
    . . SNAP
    . . . FISSION ELECTRIC CELLS
    . . . . **SNAP 4**
    . . SPACE POWER REACTORS
    . . . FISSION ELECTRIC CELLS
    . . . . **SNAP 4**
    . NUCLEAR POWER REACTORS
    . . SPACE POWER REACTORS
    . . . FISSION ELECTRIC CELLS
    . . . . **SNAP 4**
    NUCLEAR REACTORS
    . NUCLEAR POWER REACTORS
    . . SPACE POWER REACTORS
    . . . FISSION ELECTRIC CELLS
    . . . . **SNAP 4**
RT   SPACE POWER UNIT REACTORS

SNAP 7
GS   AUXILIARY POWER SOURCES
        . NUCLEAR AUXILIARY POWER UNITS
        . . SNAP
        . . . **SNAP 7**
        ELECTRIC GENERATORS
        . DIRECT POWER GENERATORS
        . . RADIOISOTOPE BATTERIES
        . . . **SNAP 7**
        . . THERMOELECTRIC GENERATORS
        . . . **SNAP 7**
        NUCLEAR ELECTRIC POWER
          GENERATION
        . NUCLEAR AUXILIARY POWER UNITS
        . . SNAP
        . . . **SNAP 7**

SNAP 8
GS   AUXILIARY POWER SOURCES
        . NUCLEAR AUXILIARY POWER UNITS
        . . SNAP
        . . . FISSION ELECTRIC CELLS
        . . . . **SNAP 8**
        . SPACE POWER REACTORS
        . . . FISSION ELECTRIC CELLS
        . . . . **SNAP 8**
        NUCLEAR ELECTRIC POWER
          GENERATION
        . NUCLEAR AUXILIARY POWER UNITS
        . . SNAP
        . . . FISSION ELECTRIC CELLS
        . . . . **SNAP 8**
        . SPACE POWER REACTORS
        . . . FISSION ELECTRIC CELLS
        . . . . **SNAP 8**
        . NUCLEAR POWER REACTORS
        . . SPACE POWER REACTORS
        . . . FISSION ELECTRIC CELLS
        . . . . **SNAP 8**
        NUCLEAR REACTORS
        . NUCLEAR POWER REACTORS
        . . SPACE POWER REACTORS
        . . . FISSION ELECTRIC CELLS
        . . . . **SNAP 8**
RT   HEAT EXCHANGERS
        SPACE POWER UNIT REACTORS
        TURBOGENERATORS

SNAP 9A
GS   AUXILIARY POWER SOURCES
        . NUCLEAR AUXILIARY POWER UNITS
        . . SNAP
        . . . **SNAP 9A**
        ELECTRIC GENERATORS
        . DIRECT POWER GENERATORS
        . . RADIOISOTOPE BATTERIES
        . . . **SNAP 9A**
        . . THERMOELECTRIC GENERATORS
        . . . **SNAP 9A**
        NUCLEAR ELECTRIC POWER
          GENERATION
        . NUCLEAR AUXILIARY POWER UNITS
        . . SNAP
        . . . **SNAP 9A**

SNAP 10A
GS   AUXILIARY POWER SOURCES
        . NUCLEAR AUXILIARY POWER UNITS
        . . SNAP
        . . . FISSION ELECTRIC CELLS
        . . . . **SNAP 10A**
        . SPACE POWER REACTORS
        . . . FISSION ELECTRIC CELLS
        . . . . **SNAP 10A**
        ELECTRIC GENERATORS
        . DIRECT POWER GENERATORS
        . . THERMOELECTRIC GENERATORS
        . . . **SNAP 10A**
        NUCLEAR ELECTRIC POWER
          GENERATION
        . NUCLEAR AUXILIARY POWER UNITS
        . . SNAP
        . . . FISSION ELECTRIC CELLS
        . . . . **SNAP 10A**
        . SPACE POWER REACTORS
        . . . FISSION ELECTRIC CELLS
        . . . . **SNAP 10A**
        . NUCLEAR POWER REACTORS
        . . SPACE POWER REACTORS
        . . . FISSION ELECTRIC CELLS
        . . . . **SNAP 10A**
        NUCLEAR REACTORS
        . NUCLEAR POWER REACTORS
        . . SPACE POWER REACTORS
        . . . FISSION ELECTRIC CELLS

SNAP 10A-*(CONT.)*
        . . . . **SNAP 10A**
RT   HEAT EXCHANGERS
        SNAPSHOT SATELLITE

SNAP 11
GS   AUXILIARY POWER SOURCES
        . NUCLEAR AUXILIARY POWER UNITS
        . . SNAP
        . . . **SNAP 11**
        ELECTRIC GENERATORS
        . DIRECT POWER GENERATORS
        . . RADIOISOTOPE BATTERIES
        . . . **SNAP 11**
        . . THERMOELECTRIC GENERATORS
        . . . **SNAP 11**
        NUCLEAR ELECTRIC POWER
          GENERATION
        . NUCLEAR AUXILIARY POWER UNITS
        . . SNAP
        . . . **SNAP 11**

SNAP 13
GS   AUXILIARY POWER SOURCES
        . NUCLEAR AUXILIARY POWER UNITS
        . . SNAP
        . . . **SNAP 13**
        ELECTRIC GENERATORS
        . DIRECT POWER GENERATORS
        . . RADIOISOTOPE BATTERIES
        . . . **SNAP 13**
        . . THERMIONIC CONVERTERS
        . . . **SNAP 13**
        NUCLEAR ELECTRIC POWER
          GENERATION
        . NUCLEAR AUXILIARY POWER UNITS
        . . SNAP
        . . . **SNAP 13**
RT   THERMIONIC POWER GENERATION

SNAP 15
GS   AUXILIARY POWER SOURCES
        . NUCLEAR AUXILIARY POWER UNITS
        . . SNAP
        . . . **SNAP 15**
        ELECTRIC GENERATORS
        . DIRECT POWER GENERATORS
        . . RADIOISOTOPE BATTERIES
        . . . **SNAP 15**
        . . THERMOELECTRIC GENERATORS
        . . . **SNAP 15**
        NUCLEAR ELECTRIC POWER
          GENERATION
        . NUCLEAR AUXILIARY POWER UNITS
        . . SNAP
        . . . **SNAP 15**

SNAP 17
GS   AUXILIARY POWER SOURCES
        . NUCLEAR AUXILIARY POWER UNITS
        . . SNAP
        . . . **SNAP 17**
        ELECTRIC GENERATORS
        . DIRECT POWER GENERATORS
        . . RADIOISOTOPE BATTERIES
        . . . **SNAP 17**
        . . THERMOELECTRIC GENERATORS
        . . . **SNAP 17**
        NUCLEAR ELECTRIC POWER
          GENERATION
        . NUCLEAR AUXILIARY POWER UNITS
        . . SNAP
        . . . **SNAP 17**

SNAP 19
GS   AUXILIARY POWER SOURCES
        . NUCLEAR AUXILIARY POWER UNITS
        . . SNAP
        . . . **SNAP 19**
        ELECTRIC GENERATORS
        . DIRECT POWER GENERATORS
        . . RADIOISOTOPE BATTERIES
        . . . **SNAP 19**
        . . THERMOELECTRIC GENERATORS
        . . . **SNAP 19**
        NUCLEAR ELECTRIC POWER
          GENERATION
        . NUCLEAR AUXILIARY POWER UNITS
        . . SNAP
        . . . **SNAP 19**

SNAP 21
GS   AUXILIARY POWER SOURCES
        . NUCLEAR AUXILIARY POWER UNITS
        . . SNAP

SNAP 21-*(CONT.)*
        . . . **SNAP 21**
        ELECTRIC GENERATORS
        . DIRECT POWER GENERATORS
        . . RADIOISOTOPE BATTERIES
        . . . **SNAP 21**
        . . THERMOELECTRIC GENERATORS
        . . . **SNAP 21**
        NUCLEAR ELECTRIC POWER
          GENERATION
        . NUCLEAR AUXILIARY POWER UNITS
        . . SNAP
        . . . **SNAP 21**

SNAP 23
GS   AUXILIARY POWER SOURCES
        . NUCLEAR AUXILIARY POWER UNITS
        . . SNAP
        . . . **SNAP 23**
        ELECTRIC GENERATORS
        . DIRECT POWER GENERATORS
        . . RADIOISOTOPE BATTERIES
        . . . **SNAP 23**
        . . THERMOELECTRIC GENERATORS
        . . . **SNAP 23**
        NUCLEAR ELECTRIC POWER
          GENERATION
        . NUCLEAR AUXILIARY POWER UNITS
        . . SNAP
        . . . **SNAP 23**

SNAP 27
GS   AUXILIARY POWER SOURCES
        . NUCLEAR AUXILIARY POWER UNITS
        . . SNAP
        . . . **SNAP 27**
        ELECTRIC GENERATORS
        . DIRECT POWER GENERATORS
        . . RADIOISOTOPE BATTERIES
        . . . **SNAP 27**
        . . THERMOELECTRIC GENERATORS
        . . . **SNAP 27**
        NUCLEAR ELECTRIC POWER
          GENERATION
        . NUCLEAR AUXILIARY POWER UNITS
        . . SNAP
        . . . **SNAP 27**

SNAP 29
GS   AUXILIARY POWER SOURCES
        . NUCLEAR AUXILIARY POWER UNITS
        . . SNAP
        . . . **SNAP 29**
        ELECTRIC GENERATORS
        . DIRECT POWER GENERATORS
        . . RADIOISOTOPE BATTERIES
        . . . **SNAP 29**
        . . THERMOELECTRIC GENERATORS
        . . . **SNAP 29**
        NUCLEAR ELECTRIC POWER
          GENERATION
        . NUCLEAR AUXILIARY POWER UNITS
        . . SNAP
        . . . **SNAP 29**

SNAP 50
GS   AUXILIARY POWER SOURCES
        . NUCLEAR AUXILIARY POWER UNITS
        . . SNAP
        . . . **SNAP 50**
        . . SPACE POWER REACTORS
        . . . **SNAP 50**
        NUCLEAR ELECTRIC POWER
          GENERATION
        . NUCLEAR AUXILIARY POWER UNITS
        . . SNAP
        . . . **SNAP 50**
        . . SPACE POWER REACTORS
        . . . **SNAP 50**
RT   SPACE POWER UNIT REACTORS

SNAPSHOT SATELLITE
GS   SATELLITES
        . ARTIFICIAL SATELLITES
        . . **SNAPSHOT SATELLITE**
RT   SNAP 10A

SNAPTRAN REACTOR
RT   ∞REACTORS
        SNAP

SNATCHING
USE   SPACECRAFT RECOVERY

SNEAK CIRCUIT ANALYSIS

```
SNEAK CIRCUIT ANALYSIS
  GS    NETWORK ANALYSIS
        . SNEAK CIRCUIT ANALYSIS
  RT    AUTOMATIC TEST EQUIPMENT
        CIRCUIT PROTECTION
        CIRCUIT RELIABILITY
        CRITICAL PATH METHOD
        ELECTRIC NETWORKS
        ELECTRICAL FAULTS
        RELIABILITY ENGINEERING
        SHORT CIRCUITS
        SIGNAL FLOW GRAPHS
        TREES (MATHEMATICS)

SNEEZING
  GS    REFLEXES
        . RESPIRATORY REFLEXES
        . . SNEEZING
  RT    INVOLUNTARY ACTIONS
        VASOCONSTRICTION

SNELLEN TESTS
  RT    ∞TESTS
        VISUAL ACUITY

SNELLS LAW
  GS    LAWS
        . SNELLS LAW
  RT    GEOMETRICAL OPTICS
        ∞OPTICS
        REFRACTION
        REFRACTIVITY

SNOW
  GS    PRECIPITATION (METEOROLOGY)
        . SNOW
  RT    ACID RAIN
        CIRQUES (LANDFORMS)
        CLOUD GLACIATION
        ICE FORMATION
        SLUSH
        STORMS (METEOROLOGY)

SNOW AERIAL APPLICATOR AIRCRAFT S-2B
  USE   S-2 AIRCRAFT

SNOW AIRCRAFT
  GS    SNOW AIRCRAFT
        . S-2 AIRCRAFT
  RT    ∞AIRCRAFT
        UTILITY AIRCRAFT

SNOW COVER
  GS    PRECIPITATION (METEOROLOGY)
        . SNOW COVER
  RT    CLOUD GLACIATION
        COLD WEATHER
        STORM DAMAGE
        STORMS
        STORMS (METEOROLOGY)

SNOW S-2 AIRCRAFT
  USE   S-2 AIRCRAFT

SNOWPLOW EFFECT
  USE   PLASMA DYNAMICS

SNOWSTORMS
  GS    STORMS
        . STORMS (METEOROLOGY)
        . . SNOWSTORMS
  RT    CLIMATOLOGY
        PRECIPITATION (METEOROLOGY)
        STORM ENHANCEMENT
        STORM SUPPRESSION
        WEATHER FORECASTING
        WEATHER MODIFICATION

∞ SOAKING
  SN    (USE OF A MORE SPECIFIC TERM IS
        RECOMMENDED--CONSULT THE TERMS
        LISTED BELOW)
  RT    BATHS
        HEAT TREATMENT
        HEATING
        SUBMERGING
        WATER IMMERSION
        WETTING

SOAPS
  GS    CONSUMABLES (SPACECRAFT)
        . CONSUMABLES (SPACECREW
          SUPPLIES)
        . . SOAPS
```

```
SOAPS-(CONT.)
  RT    DETERGENTS
        STEARATES
        SURFACTANTS
        WETTING

SOARING
  RT    CLIMBING FLIGHT
        COASTING FLIGHT
        ∞FLIGHT
        GLIDERS
        GLIDING
        HANG GLIDERS
        HORIZONTAL FLIGHT
        MAN POWERED AIRCRAFT
        VERTICAL AIR CURRENTS

SOBOLEV SPACE
  GS    ALGEBRA
        . VECTOR SPACES
        . . HILBERT SPACE
        . . . BANACH SPACE
        . . . . SOBOLEV SPACE
        ANALYSIS (MATHEMATICS)
        . FUNCTION SPACE
        . HILBERT SPACE
        . . BANACH SPACE
        . . . SOBOLEV SPACE
        . FUNCTIONAL ANALYSIS
        . . HILBERT SPACE
        . . . BANACH SPACE
        . . . . SOBOLEV SPACE
  RT    BOUNDARY VALUE PROBLEMS
        EUCLIDEAN GEOMETRY

SOCIAL FACTORS
  GS    SOCIOLOGY
        . SOCIAL FACTORS
        . . ETHNIC FACTORS
  RT    ANTHROPOLOGY
        CRIME
        CULTURE (SOCIAL SCIENCES)
        POLICE
        RACE FACTORS
        SPACE PSYCHOLOGY
        STARSITE PROGRAM
        URBAN PLANNING
        URBAN RESEARCH
        ∞VARIABLE

SOCIAL ISOLATION
  GS    ISOLATION
        . SOCIAL ISOLATION
  RT    SOCIOLOGY

SOCIAL PSYCHIATRY
  GS    MEDICAL SCIENCE
        . PSYCHIATRY
        . . SOCIAL PSYCHIATRY
  RT    SOCIOLOGY

SOCIOLOGY
  GS    SOCIOLOGY
        . SOCIAL FACTORS
        . . ETHNIC FACTORS
  RT    ANTHROPOLOGY
        CASE HISTORIES
        CITIES
        COMMUNITIES
        CULTURE (SOCIAL SCIENCES)
        DEMOGRAPHY
        DEPENDENCE
        ETHNIC FACTORS
        GROUP DYNAMICS
        HUMAN RELATIONS
        MINORITIES
        POLITICS
        RACE FACTORS
        RETIREMENT
        SOCIAL ISOLATION
        SOCIAL PSYCHIATRY
        SYSTEMS ANALYSIS
        URBAN PLANNING

SOCKS
  GS    CLOTHING
        . SOCKS
  RT    FABRICS
        SHOES

SOD
  RT    CANOPIES (VEGETATION)
        FARMLANDS
        GRASSES
        GRASSLANDS
```

```
SOD-(CONT.)
        LAND
        PLOWING

SODALITE
  RT    CHEMICAL REACTIONS
        PHOTOCHEMICAL REACTIONS
        PHOTOCHROMISM

SODAR
  RT    ACOUSTIC SCATTERING
        ATMOSPHERIC TEMPERATURE
        ∞INSTRUMENTS
        MEASURING INSTRUMENTS
        METEOROLOGICAL INSTRUMENTS
        METEOROLOGY
        SOUND DETECTING AND RANGING
        TEMPERATURE MEASUREMENT

SODIUM
  GS    CHEMICAL ELEMENTS
        . ALKALI METALS
        . . SODIUM
        . . . LIQUID SODIUM
        . . . SODIUM ISOTOPES
        . . . . SODIUM 22
        . . . . SODIUM 24
        . . . SODIUM VAPOR
        METALS
        . ALKALI METALS
        . . SODIUM
        . . . LIQUID SODIUM
        . . . SODIUM ISOTOPES
        . . . . SODIUM 22
        . . . . SODIUM 24
        . . . SODIUM VAPOR
  RT    DAWSONITE
        LIQUID METAL COOLED REACTORS

SODIUM ALLOYS
  GS    ALLOYS
        . SODIUM ALLOYS

SODIUM AZIDES
  GS    NITROGEN COMPOUNDS
        . AZIDES (INORGANIC)
        . . SODIUM AZIDES
        . AZIDES (ORGANIC)
        . SODIUM AZIDES
        SODIUM COMPOUNDS
        . SODIUM AZIDES
  RT    DETONATORS
        EXPLOSIVES

SODIUM BROMIDES
  GS    HALOGEN COMPOUNDS
        . BROMINE COMPOUNDS
        . . BROMIDES
        . . . SODIUM BROMIDES
        . HALIDES
        . . BROMIDES
        . . . SODIUM BROMIDES
        . METAL HALIDES
        . . . ALKALI HALIDES
        . . . . SODIUM BROMIDES
        SODIUM COMPOUNDS
        . SODIUM BROMIDES

SODIUM CARBONATES
  GS    CARBON COMPOUNDS
        . CARBONATES
        . . SODIUM CARBONATES
        SODIUM COMPOUNDS
        . SODIUM CARBONATES

SODIUM CHLORIDES
  GS    HALOGEN COMPOUNDS
        . CHLORINE COMPOUNDS
        . . CHLORIDES
        . . . SODIUM CHLORIDES
        . HALIDES
        . . CHLORIDES
        . . . SODIUM CHLORIDES
        . METAL HALIDES
        . . . ALKALI HALIDES
        . . . . SODIUM CHLORIDES
        SODIUM COMPOUNDS
        . SODIUM CHLORIDES
  RT    MOLTEN SALTS
        SALT BEDS
        ∞SALTS

SODIUM CHLORODIFLUOROACETATES
  GS    ACETATES
```

SODIUM CHLORODIFLUOROACETATES-*(CONT.)*
. **SODIUM CHLORODIFLUOROACETATES**
ALIPHATIC COMPOUNDS
. **SODIUM CHLORODIFLUOROACETATES**
ESTERS
. **SODIUM CHLORODIFLUOROACETATES**

SODIUM CHROMITES
GS   SODIUM COMPOUNDS
. **SODIUM CHROMITES**

SODIUM COMPOUNDS
GS   **SODIUM COMPOUNDS**
. CRYOLITE
. NEMBUTAL (TRADEMARK)
. NEPHELINE
. SODIUM AZIDES
. SODIUM BROMIDES
. SODIUM CARBONATES
. SODIUM CHLORIDES
. SODIUM CHROMITES
. SODIUM FLUORIDES
. SODIUM GALLATES
. SODIUM HYDRIDES
. SODIUM HYDROXIDES
. SODIUM IODIDES
. SODIUM NITRATES
. SODIUM PEROXIDES
. SODIUM SALICYLATES
. SODIUM SILICATES
. . SPODUMENE
. . TALC
. . TOURMALINE
. SODIUM SULFATES
. SODIUM SULFITES
RT   ∞ALKALI METAL COMPOUNDS
BLOEDITE
∞CHEMICAL COMPOUNDS
∞METAL COMPOUNDS

SODIUM COOLING
SN   (COOLING WITH SODIUM)
GS   COOLING
. **SODIUM COOLING**
RT   COOLANTS
LIQUID COOLED REACTORS
LIQUID COOLING

SODIUM FLUORIDES
GS   HALOGEN COMPOUNDS
. FLUORINE COMPOUNDS
. . FLUORIDES
. . . METAL FLUORIDES
. . . . **SODIUM FLUORIDES**
. HALIDES
. . METAL HALIDES
. . . ALKALI HALIDES
. . . . **SODIUM FLUORIDES**
SODIUM COMPOUNDS
. **SODIUM FLUORIDES**

SODIUM GALLATES
GS   GALLIUM COMPOUNDS
. GALLATES
. . **SODIUM GALLATES**
SODIUM COMPOUNDS
. **SODIUM GALLATES**

SODIUM GRAPHITE REACTORS
UF   SGR (NUCLEAR REACTORS)
GS   NUCLEAR REACTORS
. LIQUID COOLED REACTORS
. . LIQUID METAL COOLED REACTORS
. . . **SODIUM GRAPHITE REACTORS**
RT   HALLAM NUCLEAR POWER FACILITY
NUCLEAR POWER REACTORS

SODIUM HYDRIDES
GS   HYDROGEN COMPOUNDS
. HYDRIDES
. . METAL HYDRIDES
. . . **SODIUM HYDRIDES**
SODIUM COMPOUNDS
. **SODIUM HYDRIDES**

SODIUM HYDROXIDES
GS   ALKALIES
. **SODIUM HYDROXIDES**
HYDROXIDES
. **SODIUM HYDROXIDES**
SODIUM COMPOUNDS
. **SODIUM HYDROXIDES**

SODIUM IODIDES
GS   HALOGEN COMPOUNDS
. HALIDES
. . METAL HALIDES
. . . ALKALI HALIDES
. . . . **SODIUM IODIDES**
. IODINE COMPOUNDS
. . IODIDES
. . . **SODIUM IODIDES**
SODIUM COMPOUNDS
. **SODIUM IODIDES**

SODIUM ISOTOPES
GS   CHEMICAL ELEMENTS
. ALKALI METALS
. . SODIUM
. . . **SODIUM ISOTOPES**
. . . . SODIUM 22
. . . . SODIUM 24
. NUCLIDES
. ISOTOPES
. . **SODIUM ISOTOPES**
. . . . SODIUM 22
. . . . SODIUM 24
METALS
. ALKALI METALS
. . SODIUM
. . . **SODIUM ISOTOPES**
. . . . SODIUM 22
. . . . SODIUM 24

SODIUM NITRATES
GS   NITROGEN COMPOUNDS
. NITRATES
. . INORGANIC NITRATES
. . . **SODIUM NITRATES**
SODIUM COMPOUNDS
. **SODIUM NITRATES**

SODIUM PEROXIDES
GS   CHALCOGENIDES
. OXIDES
. . ANHYDRIDES
. . . PEROXIDES
. . . . **SODIUM PEROXIDES**
. . METAL OXIDES
. . . **SODIUM PEROXIDES**
SODIUM COMPOUNDS
. **SODIUM PEROXIDES**

SODIUM REACTOR EXPERIMENT
UF   SRE REACTOR
GS   NUCLEAR REACTORS
. LIQUID COOLED REACTORS
. . LIQUID METAL COOLED REACTORS
. . . **SODIUM REACTOR EXPERIMENT**
. NUCLEAR RESEARCH AND TEST
REACTORS
. . **SODIUM REACTOR EXPERIMENT**
RT   LIQUID METAL COOLED REACTORS

SODIUM SALICYLATES
GS   ESTERS
. **SODIUM SALICYLATES**
SALICYLATES
. **SODIUM SALICYLATES**
SODIUM COMPOUNDS
. **SODIUM SALICYLATES**

SODIUM SILICATES
GS   SILICON COMPOUNDS
. SILICATES
. . **SODIUM SILICATES**
. . . SPODUMENE
. . . TALC
. . . TOURMALINE
SODIUM COMPOUNDS
. **SODIUM SILICATES**
. . SPODUMENE
. . TALC
. . TOURMALINE
RT   MINERALS

SODIUM SULFATES
GS   SODIUM COMPOUNDS
. **SODIUM SULFATES**
SULFUR COMPOUNDS
. SULFATES
. . **SODIUM SULFATES**

SODIUM SULFITES
GS   SODIUM COMPOUNDS
. **SODIUM SULFITES**
SULFUR COMPOUNDS
. SULFITES

SODIUM SULFITES-*(CONT.)*
. . **SODIUM SULFITES**

SODIUM SULFUR BATTERIES
GS   ELECTROCHEMICAL CELLS
. ELECTRIC BATTERIES
. . PRIMARY BATTERIES
. . . **SODIUM SULFUR BATTERIES**
RT   ∞ELECTRIC CELLS

SODIUM VAPOR
GS   CHEMICAL ELEMENTS
. ALKALI METALS
. . SODIUM
. . . **SODIUM VAPOR**
METALS
. ALKALI METALS
. . SODIUM
. . . **SODIUM VAPOR**
. METAL VAPORS
. . **SODIUM VAPOR**
VAPORS
. METAL VAPORS
. . **SODIUM VAPOR**
RT   MERCURY VAPOR

SODIUM 22
GS   CHEMICAL ELEMENTS
. ALKALI METALS
. . SODIUM
. . . SODIUM ISOTOPES
. . . . **SODIUM 22**
. NUCLIDES
. . ISOTOPES
. . . RADIOACTIVE ISOTOPES
. . . . **SODIUM 22**
. . . SODIUM ISOTOPES
. . . . **SODIUM 22**
METALS
. ALKALI METALS
. . SODIUM
. . . SODIUM ISOTOPES
. . . . **SODIUM 22**

SODIUM 24
GS   CHEMICAL ELEMENTS
. ALKALI METALS
. . SODIUM
. . . SODIUM ISOTOPES
. . . . **SODIUM 24**
. NUCLIDES
. . ISOTOPES
. . . RADIOACTIVE ISOTOPES
. . . . **SODIUM 24**
. . . SODIUM ISOTOPES
. . . . **SODIUM 24**
METALS
. ALKALI METALS
. . SODIUM
. . . SODIUM ISOTOPES
. . . . **SODIUM 24**

SOFAR
USE   SOUND FIXING AND RANGING

SOFT LANDING
SN   (SPACECRAFT OR AIRCRAFT)
UF   SOFT RECOVERY
GS   LANDING
. **SOFT LANDING**
RT   AIRCRAFT LANDING
∞ASTRONAUTICS
CRASH LANDING
GLIDE LANDINGS
HARD LANDING
HORIZONTAL SPACECRAFT LANDING
LUNAR LANDING
MARS LANDING
PLANETARY LANDING
SPACECRAFT LANDING
SURVEYOR PROJECT
VIKING 75 ENTRY VEHICLE
WATER LANDING

SOFT LANDING SPACECRAFT
UF   SLV (SOFT LANDING VEHICLES)
GS   **SOFT LANDING SPACECRAFT**
. AEROSPACEPLANES
. . ASTROPLANE
. APOLLO SPACECRAFT
. . APOLLO LUNAR EXPERIMENT
MODULE
. ASTRO VEHICLE
. GEMINI B SPACECRAFT
. GEMINI SPACECRAFT

SOFT LANDING SPACECRAFT-*(CONT.)*
- . . GEMINI (GT-1) SPACECRAFT
- . . GEMINI 2 SPACECRAFT
- . JANUS SPACECRAFT
- . LANDING MODULES
- . . LUNAR LANDING MODULES
- . . . LUNAR MODULE
- . . . . APOLLO LUNAR EXPERIMENT
-             MODULE
- . . . . LSSM
- . . MARS EXCURSION MODULE
- . MERCURY SPACECRAFT
- . . AURORA 7
- . . FAITH 7
- . . FRIENDSHIP 7
- . . SIGMA 7
- . SURVEYOR LUNAR PROBES
- . . SURVEYOR 1 LUNAR PROBE
- . . SURVEYOR 2 LUNAR PROBE
- . . SURVEYOR 3 LUNAR PROBE
- . . SURVEYOR 4 LUNAR PROBE
- . . SURVEYOR 5 LUNAR PROBE
- . . SURVEYOR 6 LUNAR PROBE
- . . SURVEYOR 7 LUNAR PROBE
- . VOSKHOD MANNED SPACECRAFT
- . . VOSKHOD 1 SPACECRAFT
- . . VOSKHOD 2 SPACECRAFT
- . VOSTOK SPACECRAFT
- . . VOSTOK 1 SPACECRAFT
- . . VOSTOK 2 SPACECRAFT
- . . VOSTOK 3 SPACECRAFT
- . . VOSTOK 4 SPACECRAFT
- . . VOSTOK 5 SPACECRAFT
- . . VOSTOK 6 SPACECRAFT

RT   APOLLO PROJECT
     FERRY SPACECRAFT
     HOVERING ROCKET VEHICLES
     LUNAR PROBES
     REUSABLE SPACECRAFT
     SPACE CAPSULES
    ∞SPACECRAFT
     SPACECRAFT LANDING
     SURVEYOR PROJECT
     X-20 AIRCRAFT

SOFT RECOVERY
USE   SOFT LANDING

SOFTENING
SN   (EXCLUDES WATER SOFTENING)
GS   **SOFTENING**
    . WORK SOFTENING
RT   ANNEALING
    DEIONIZATION
    DEMINERALIZING
    DIGESTING
    DISSOLVING
    HARDENING (MATERIALS)
    ION EXCHANGING

SOFTNESS
RT   DUCTILITY
    ELASTIC PROPERTIES
    FLEXIBILITY
    HARDNESS
    STIFFNESS

SOFTWARE (COMPUTERS)
USE   COMPUTER PROGRAMS
    COMPUTER SYSTEMS PROGRAMS

SOFTWARE ENGINEERING
GS   **SOFTWARE ENGINEERING**
    . COMPUTER PROGRAMMING
    . . ASSEMBLER ROUTINES
    . . LANGUAGE PROGRAMMING
    . . LOGIC PROGRAMMING
    . . MICROPROGRAMMING
    . . MULTIPROGRAMMING
    . . ON-LINE PROGRAMMING
    . . PARALLEL PROGRAMMING
    . . SYMBOLIC PROGRAMMING
RT   COMPUTER PROGRAMS
    COMPUTER SYSTEMS DESIGN
    COMPUTER SYSTEMS PROGRAMS
    DATA BASES
    SOFTWARE TOOLS
    SYSTEMS ENGINEERING

SOFTWARE TOOLS
RT   ARCHITECTURE (COMPUTERS)
    COMPUTER PROGRAMMING
    COMPUTER PROGRAMS
    COMPUTER SYSTEMS DESIGN
    COMPUTER SYSTEMS PROGRAMS

SOFTWARE TOOLS-*(CONT.)*
    DATA BASE MANAGEMENT SYSTEMS
    PROGRAM VERIFICATION (COMPUTERS)
    SOFTWARE ENGINEERING

† **SOIL EROSION**
GS   EROSION
    . **SOIL EROSION**
RT   ABRASION
    ATMOSPHERIC EFFECTS
    DETERIORATION
    ENVIRONMENT EFFECTS
    HYDROGEOLOGY
    LANDSLIDES
    OUTLIERS (LANDFORMS)
    PENEPLAINS
    RAIN EROSION
    RAIN IMPACT DAMAGE
    SOILS
    WATER EROSION
    WEATHERING
    WIND EFFECTS

SOIL MAPPING
GS   MAPPING
    . **SOIL MAPPING**
RT   GEOGRAPHIC APPLICATIONS PROGRAM
    MAPS
    PHOTOMAPPING
    SOILS
    SPOT (FRENCH SATELLITE)
    SURVEYS
    TERRAIN ANALYSIS

SOIL MECHANICS
RT   CRUSTAL FRACTURES
    FRACTURE MECHANICS
    GEOTECHNICAL ENGINEERING
    GEOTECHNICAL FABRICS
    ROCK MECHANICS

SOIL MOISTURE
GS   MOISTURE
    . **SOIL MOISTURE**
RT   LYSIMETERS
    MOISTURE CONTENT
    PLANT STRESS
    SOILS
    VEGETATION GROWTH

SOIL SCIENCE
UF   PEDOLOGY
RT   AGRICULTURE
    CONSERVATION
    EROSION
    SOILS
    VEGETATION GROWTH

SOILS
GS   **SOILS**
    . ALLUVIUM
    . DIRT
    . GRAVELS
    . LATERITES
    . LUNAR SOIL
    . . LUNAR DUST
    . MUD
    . PERMAFROST
    . SANDS
    . . MONAZITE SANDS
    . . TAR SANDS
RT   ANDESITE
    ANORTHOSITE
    ATAXITE
    BARREN LAND
    BASALT
    BEDROCK
    BENTONITE
    BOREHOLES
    BRECCIA
    CARBONACEOUS ROCKS
    CLAYS
    COAL
    CONSERVATION
    CULTIVATION
    DELTAS
    DIORITE
    DUNITE
    EARTH RESOURCES
    ECLOGITE
    ENSTATITE
    FORMATIONS
    GEOLOGY
    GNEISS
    GRANITE

SOILS-*(CONT.)*
    IGNEOUS ROCKS
    ILLITE
    KAOLINITE
    LAND
    LANDSLIDES
    LAVA
    LIMESTONE
    LYSIMETERS
    MAGMA
    MINERALS
    MOLDAVITE
    MUSKEGS
    OBSIDIAN
    OLIVINE
    PERIDOTITE
    PLANTING
    POLYURETHANE FOAM
    POROUS MATERIALS
    PUMICE
    PYROXENES
    QUARTZ
    ROCKS
    SANDSTONES
    SEDIMENTARY ROCKS
    SERPENTINE
    SHALES
    SOIL EROSION
    SOIL MAPPING
    SOIL MOISTURE
    SOIL SCIENCE
    STRIP MINING
    SYENITE
    TRACHYTE
    TUNNELING (EXCAVATION)
    VADOSE WATER
    VEGETATION GROWTH

SOL-GEL PROCESSES
RT   CERAMIC NUCLEAR FUELS
    NUCLEAR FUELS
    ∞PROCESSES

SOLAR ACTIVITY
GS   STELLAR ACTIVITY
    . **SOLAR ACTIVITY**
    . . FACULAE
    . . SOLAR PROMINENCES
    . . SOLAR STORMS
    . . SPICULES
    . . STELLAR FLARES
    . . . SOLAR FLARES
    . . SUNSPOTS
RT   ∞ACTIVITY
    AURORAS
    ∞DISTURBANCES
    INTERNATIONAL QUIET SUN YEAR
    IRIS SATELLITES
    MAGNETIC DISTURBANCES
    PROMINENCES
    RADIO AURORAS
    SOLAR PLANETARY INTERACTIONS
    STARSPOTS
    SUN
    SUNSPOT CYCLE

SOLAR ACTIVITY EFFECTS
RT   BLACKOUT (PROPAGATION)
    ∞EFFECTS
    GALACTIC COSMIC RAYS
    HELIOSPHERE
    MAGNETIC DISTURBANCES
    SECULAR VARIATIONS
    SOLAR OSCILLATIONS
    SOLAR PLANETARY INTERACTIONS
    SUDDEN IONOSPHERIC DISTURBANCES
    SUDDEN STORM COMMENCEMENTS
    SUN

SOLAR ARRAYS
UF   ROLLUP SOLAR ARRAYS
GS   ARRAYS
    . **SOLAR ARRAYS**
    . . SOLAR BLANKETS
RT   ELECTROSTATIC BONDING
    PAYLOAD DELIVERY (STS)
    POWER MODULES (STS)
    SATELLITE POWER TRANSMISSION (TO
     EARTH)
    SELF SHADOWING
    SOLAR ATRIUMS
    SUN

SOLAR ATMOSPHERE
GS   ENVIRONMENTS

SOLAR ATMOSPHERE-*(CONT.)*
. EXTRATERRESTRIAL ENVIRONMENTS
. . STELLAR ATMOSPHERES
. . . **SOLAR ATMOSPHERE**
RT ∞ATMOSPHERES
CHROMOSPHERE
M REGION
PHOTOSPHERE
SOLAR OSCILLATIONS
SPICULES
STELLAR STRUCTURE
SUN

SOLAR ATRIUMS
RT SOLAR ARRAYS
SOLAR HEATING
SOLAR REFLECTORS
SPACE HEATING (BUILDINGS)
SUN

SOLAR AUXILIARY POWER UNITS
GS AUXILIARY POWER SOURCES
. **SOLAR AUXILIARY POWER UNITS**
. . SUNFLOWER POWER SYSTEM
ELECTRIC GENERATORS
. SOLAR GENERATORS
. . **SOLAR AUXILIARY POWER UNITS**
. . . ASTEC SOLAR TURBOELECTRIC
GENERATOR
. . . SUNFLOWER POWER SYSTEM
RT SUN

SOLAR AZIMUTH
USE AZIMUTH
SOLAR POSITION

SOLAR BACKSCATTER UV SPECTROMETER
GS MEASURING INSTRUMENTS
. SPECTROMETERS
. . **SOLAR BACKSCATTER UV
SPECTROMETER**
RT IRRADIANCE
SATELLITE-BORNE INSTRUMENTS

SOLAR BLANKETS
GS ARRAYS
. SOLAR ARRAYS
. . **SOLAR BLANKETS**
ELECTRIC GENERATORS
. DIRECT POWER GENERATORS
. . THERMIONIC CONVERTERS
. . . **SOLAR BLANKETS**
RT ∞BLANKETS
∞CONVERTERS
SUN

SOLAR CELL CALIBRATION FACILITY
UF SCCF
GS PAYLOADS
. SPACELAB PAYLOADS
. . **SOLAR CELL CALIBRATION FACILITY**
RT CALIBRATING
∞FACILITIES
SPACE TRANSPORTATION SYSTEM
FLIGHTS

SOLAR CELLS
UF SILICON SOLAR CELLS
WRAPAROUND CONTACT SOLAR CELLS
GS ELECTRIC GENERATORS
. DIRECT POWER GENERATORS
. . **SOLAR CELLS**
. . . HOMOJUNCTIONS
. . . VERTICAL JUNCTION SOLAR CELLS
. SOLAR GENERATORS
. . **SOLAR CELLS**
. . . HOMOJUNCTIONS
. . . VERTICAL JUNCTION SOLAR CELLS
RT ANTIREFLECTION COATINGS
CARRIER LIFETIME
CARRIER TRANSPORT (SOLID STATE)
∞CELLS
∞ELECTRIC CELLS
ELECTROSTATIC BONDING
FLOAT ZONES
FUEL CELLS
HETEROJUNCTIONS
OPEN CIRCUIT VOLTAGE
PHOTODIODES
PHOTOELECTRIC CELLS
PHOTOVOLTAIC CELLS
QUANTUM EFFICIENCY
SATELLITE POWER TRANSMISSION (TO
EARTH)

SOLAR CELLS-*(CONT.)*
SATELLITE SOLAR ENERGY
CONVERSION
SATELLITE SOLAR POWER STATIONS
SEMICONDUCTOR DEVICES
SHORT CIRCUIT CURRENTS
SIS (SEMICONDUCTORS)
SOLAR POWERED AIRCRAFT
SPECTROPHOTOVOLTAICS
SUN
THERMIONIC CONVERTERS
THERMOELECTRIC GENERATORS

SOLAR COLLECTORS
UF SOLAR RECEIVERS
GS ACCUMULATORS
. **SOLAR COLLECTORS**
. . SOLAR REFLECTORS
REFLECTORS
. **SOLAR COLLECTORS**
. . SOLAR REFLECTORS
RT CONCENTRATORS
MIRRORS
SELECTIVE SURFACES
SPACE COOLING (BUILDINGS)
SPECTROPHOTOVOLTAICS
SUN
SUNFLOWER POWER SYSTEM

SOLAR COMPASSES
GS AIRCRAFT INSTRUMENTS
. COMPASSES
. . **SOLAR COMPASSES**
NAVIGATION AIDS
. NAVIGATION INSTRUMENTS
. . COMPASSES
. . . **SOLAR COMPASSES**
RT AIR NAVIGATION
AIR TRAFFIC CONTROL
AIRCRAFT SAFETY
AIRPORT BEACONS
ALL-WEATHER AIR NAVIGATION
APPROACH INDICATORS
AUTOMATIC FLIGHT CONTROL
AUTOMATIC PILOTS
BEACONS
DECCA NAVIGATION
DISPLAY DEVICES
DISTANCE MEASURING EQUIPMENT
FLIGHT CONTROL
FLIGHT INSTRUMENTS
FLIGHT PATHS
GYROCOMPASSES
HELIPORTS
HOMING DEVICES
LANDING AIDS
LORAN
MAGNETIC COMPASSES
OMNIDIRECTIONAL RADIO RANGES
POSITION INDICATORS
RADAR BEACONS
RADIO BEACONS
RADIO NAVIGATION
SELF CALIBRATING OMNIRANGE
SHORAN
SUN
TACAN
VHF OMNIRANGE NAVIGATION
WEATHER

SOLAR CONSTANT
GS CONSTANTS
. **SOLAR CONSTANT**
RATES (PER TIME)
. FLUX DENSITY
. . RADIANT FLUX DENSITY
. . . IRRADIANCE
. . . . **SOLAR CONSTANT**
. . . SOLAR FLUX DENSITY
. . . . **SOLAR CONSTANT**
RT ILLUMINANCE
PARTICLE FLUX DENSITY
SUN

SOLAR CONVERTERS
USE SOLAR GENERATORS

SOLAR COOLING
GS COOLING
. **SOLAR COOLING**
RT COOLING SYSTEMS
DOMESTIC ENERGY
ENERGY TECHNOLOGY
RESIDENTIAL ENERGY
SPACE COOLING (BUILDINGS)

SOLAR COOLING-*(CONT.)*
SUN

SOLAR CORONA
UF SOLAR NEBULA
GS CORONAS
. STELLAR CORONAS
. . **SOLAR CORONA**
. . . CORONAL LOOPS
RT CHROMOSPHERE
ELECTRIC CORONA
NEBULAE
STELLAR STRUCTURE
SUN

SOLAR CORPUSCULAR RADIATION
UF SOLAR STREAMS
GS EXTRATERRESTRIAL RADIATION
. SOLAR RADIATION
. . **SOLAR CORPUSCULAR RADIATION**
. . . SOLAR ELECTRONS
. . . SOLAR NEUTRINOS
. . . SOLAR PROTONS
PARTICLES
. CORPUSCULAR RADIATION
. . **SOLAR CORPUSCULAR RADIATION**
. . . SOLAR ELECTRONS
. . . SOLAR PROTONS
RT M REGION
∞RADIATION
SOLAR PLANETARY INTERACTIONS
SUDDEN STORM COMMENCEMENTS
SUN

SOLAR COSMIC RAYS
GS EXTRATERRESTRIAL RADIATION
. PRIMARY COSMIC RAYS
. . **SOLAR COSMIC RAYS**
. SOLAR RADIATION
. . **SOLAR COSMIC RAYS**
IONIZING RADIATION
. COSMIC RAYS
. . PRIMARY COSMIC RAYS
. . . **SOLAR COSMIC RAYS**
PARTICLES
. CORPUSCULAR RADIATION
. . PRIMARY COSMIC RAYS
. . . **SOLAR COSMIC RAYS**
RT ELECTRON ACCELERATION
ENERGETIC PARTICLES
GRIST (TELESCOPE)
SUN

SOLAR CYCLES
GS CYCLES
. **SOLAR CYCLES**
. . SUNSPOT CYCLE
RT INTERNATIONAL QUIET SUN YEAR
IRIS SATELLITES
SECULAR VARIATIONS
SUN
SUNSPOTS
TWENTY-SEVEN DAY VARIATION

SOLAR DIAMETER
RT ASTROMETRY
∞SCIENCE
SOLAR ECLIPSES

SOLAR DISK
USE SUN

SOLAR DYNAMICS
USE HELIOSEISMOLOGY

SOLAR ECLIPSES
GS ECLIPSES
. **SOLAR ECLIPSES**
OCCULTATION
. LUNAR OCCULTATION
. . **SOLAR ECLIPSES**
RT LUNAR SHADOW
SOLAR DIAMETER
SUN

SOLAR ELECTRIC PROPULSION
GS PROPULSION
. ELECTRIC PROPULSION
. . **SOLAR ELECTRIC PROPULSION**
. LOW THRUST PROPULSION
. . SOLAR PROPULSION
. . . **SOLAR ELECTRIC PROPULSION**
. SPACECRAFT PROPULSION
. . SOLAR PROPULSION

SOLAR ELECTRIC PROPULSION-*(CONT.)*
... **SOLAR ELECTRIC PROPULSION**
RT   SOLAR POWERED AIRCRAFT
     SOLAR THERMAL PROPULSION
     SUN

SOLAR ELECTRONS
GS   EXTRATERRESTRIAL RADIATION
     . SOLAR RADIATION
     .. SOLAR CORPUSCULAR RADIATION
     ... **SOLAR ELECTRONS**
     PARTICLES
     . CORPUSCULAR RADIATION
     .. SOLAR CORPUSCULAR RADIATION
     ... **SOLAR ELECTRONS**
RT   SUN

SOLAR ENERGY
RT   CIRCUMSOLAR TELESCOPES
     CLEAN ENERGY
     ∞ENERGY
     ENERGY ABSORPTION FILMS
     IRIS SATELLITES
     PHASE CHANGE MATERIALS
     SOLAR THERMAL ELECTRIC POWER
       PLANTS
     SOLETTAS
     SUN

SOLAR ENERGY ABSORBERS
RT   ABSORBERS (MATERIALS)
     ELECTROMAGNETIC ABSORPTION
     PHOTOTHERMAL CONVERSION
     SELECTIVE SURFACES
     SUN
     TROMBE WALLS

SOLAR ENERGY CONVERSION
GS   ENERGY CONVERSION
     . **SOLAR ENERGY CONVERSION**
     .. SOLAR TOTAL ENERGY SYSTEMS
RT   COGENERATION
     ∞CONVERSION
     ENERGY TECHNOLOGY
     HETEROJUNCTION DEVICES
     HYDROGEN PRODUCTION
     PHASE CHANGE MATERIALS
     PHOTOELECTRIC GENERATORS
     PHOTOELECTROCHEMICAL DEVICES
     PHOTOTHERMAL CONVERSION
     SOLAR-PUMPED LASERS
     SPACE COOLING (BUILDINGS)
     SUN

SOLAR FACULAE
USE   FACULAE

SOLAR FLARES
GS   STELLAR ACTIVITY
     . SOLAR ACTIVITY
     .. STELLAR FLARES
     ... **SOLAR FLARES**
RT   CORONAL LOOPS
     FLARE STARS
     ∞FLARES
     ∞FLASH
     FORBUSH DECREASES
     IRIS SATELLITES
     MAGNETIC DISTURBANCES
     SOLAR MAXIMUM MISSION
     SUDDEN STORM COMMENCEMENTS
     SUN
     SUNSPOTS

SOLAR FLUX
SN   (ENERGY OR PARTICLES EMITTED
     FROM THE SUN PER UNIT TIME--DO
     NOT CONFUSE WITH SOLAR FLUX
     DENSITY THE ENERGY OR PARTICLE
     EMISSION OR DETECTION RATE PER
     UNIT AREA)
GS   RATES (PER TIME)
     . FLUX (RATE)
     .. **SOLAR FLUX**
RT   HEAT FLUX
     LIMB BRIGHTENING
     SUN

SOLAR FLUX DENSITY
GS   RATES (PER TIME)
     . FLUX DENSITY
     .. RADIANT FLUX DENSITY
     ... **SOLAR FLUX DENSITY**
     .... SOLAR CONSTANT
RT   ELECTRON FLUX DENSITY

SOLAR FLUX DENSITY-*(CONT.)*
     HELIOS SATELLITES
     ILLUMINANCE
     IRRADIANCE
     LIMB BRIGHTENING
     LUMINANCE
     LUMINOUS INTENSITY
     PARTICLE FLUX DENSITY
     PROTON FLUX DENSITY
     RADIANCE
     RADIANCY
     RADIATION PRESSURE
     SUN

SOLAR FURNACES
GS   HEATING EQUIPMENT
     . FURNACES
     .. **SOLAR FURNACES**
RT   FORBUSH DECREASES
     MELTING
     SUN
     VACUUM FURNACES

SOLAR GENERATORS
UF   SOLAR CONVERTERS
     SOLAR POWER GENERATION
     SOLAR POWER SOURCES
GS   ELECTRIC GENERATORS
     . **SOLAR GENERATORS**
     .. SOLAR AUXILIARY POWER UNITS
     ... ASTEC SOLAR TURBOELECTRIC
         GENERATOR
     ... SUNFLOWER POWER SYSTEM
     .. SOLAR CELLS
     ... HOMOJUNCTIONS
     ... VERTICAL JUNCTION SOLAR CELLS
RT   DIRECT POWER GENERATORS
     FUEL CELLS
     PADDLES
     PHOTOELECTRIC CELLS
     PHOTOELECTRIC GENERATORS
     PHOTOVOLTAIC CELLS
     ∞POWER SUPPLIES
     RANKINE CYCLE
     SOLAR SEA POWER PLANTS
     SUN
     THERMOELECTRIC GENERATORS
     TURBOGENERATORS

SOLAR GRANULATION
GS   ELECTROMAGNETIC PROPERTIES
     . OPTICAL PROPERTIES
     .. BRIGHTNESS
     ... **SOLAR GRANULATION**
     PHOTOSPHERE
     . **SOLAR GRANULATION**
RT   BENARD CELLS
     BRIGHTNESS DISTRIBUTION
     CONVECTION CURRENTS
     LIMB BRIGHTENING
     SUN
     SURFACE LAYERS
     TEMPERATURE EFFECTS

SOLAR GRAVITATION
UF   EVECTION
GS   GRAVITATION
     . **SOLAR GRAVITATION**
RT   SUN

SOLAR HEATING
GS   HEATING
     . **SOLAR HEATING**
RT   BIOCONVERSION
     HYDROTHERMAL SYSTEMS
     INSOLATION
     PHASE CHANGE MATERIALS
     RADIANT HEATING
     RESIDENTIAL ENERGY
     SOLAR ATRIUMS
     SPACE HEATING (BUILDINGS)
     SUN
     SUNLIGHT
     TROMBE WALLS

SOLAR HOUSES
RT   ∞BUILDINGS
     DOMESTIC ENERGY
     ENERGY TECHNOLOGY
     HEAT STORAGE
     RESIDENTIAL ENERGY
     SPACE HEATING (BUILDINGS)
     SUN
     TROMBE WALLS

SOLAR INSTRUMENTS
GS   **SOLAR INSTRUMENTS**
     . SPECTROHELIOGRAPHS
RT   CELESCOPES
     OPTICAL MEASURING INSTRUMENTS
     RADIATION MEASURING INSTRUMENTS
     SOLAR OPTICAL TELESCOPE
     SPECTROMETERS
     SUN
     TELESCOPES

SOLAR LASERS
USE   SOLAR-PUMPED LASERS

SOLAR LIMB
RT   CORONAL LOOPS
     LIMB BRIGHTENING
     LIMB DARKENING
     ∞LIMBS
     PLANETARY LIMB
     SUN

SOLAR LONGITUDE
GS   LONGITUDE
     . **SOLAR LONGITUDE**
RT   ASTRONOMICAL COORDINATES
     CELESTIAL REFERENCE SYSTEMS
     SUN

SOLAR MAGNETIC FIELD
UF   HELIOMAGNETISM
GS   MAGNETIC FIELDS
     . STELLAR MAGNETIC FIELDS
     .. **SOLAR MAGNETIC FIELD**
RT   ELECTROMAGNETIC FIELDS
     INTERPLANETARY MAGNETIC FIELDS
     SUN

SOLAR MAXIMUM MISSION
GS   SPACE MISSIONS
     . **SOLAR MAXIMUM MISSION**
     .. SOLAR MAXIMUM MISSION-A
RT   ∞FLARES
     FLUX DENSITY
     GAMMA RAY SPECTROMETERS
     ∞MISSIONS
     MULTIMISSION MODULAR SPACECRAFT
     POLARIMETERS
     PROGRAMS
     SOLAR FLARES
     SPACE PROGRAMS
     SUN
     ULTRAVIOLET SPECTROMETERS
     ULYSSES MISSION

SOLAR MAXIMUM MISSION-A
UF   SMM-A
GS   SPACE MISSIONS
     . SOLAR MAXIMUM MISSION
     .. **SOLAR MAXIMUM MISSION-A**
RT   ∞MISSIONS
     SPACE EXPLORATION
     ∞SPACECRAFT
     SUN

SOLAR MESOSPHERE EXPLORER
GS   SATELLITES
     . ARTIFICIAL SATELLITES
     .. EXPLORER SATELLITES
     ... **SOLAR MESOSPHERE EXPLORER**
RT   ATMOSPHERIC COMPOSITION
     MESOSPHERE
     OZONE
     SUN

SOLAR NEBULA
USE   SOLAR CORONA

SOLAR NEUTRINOS
GS   EXTRATERRESTRIAL RADIATION
     . SOLAR RADIATION
     .. SOLAR CORPUSCULAR RADIATION
     ... **SOLAR NEUTRINOS**
     PARTICLES
     . ELEMENTARY PARTICLES
     .. FERMIONS
     ... LEPTONS
     .... NEUTRINOS
     ..... **SOLAR NEUTRINOS**
RT   ASTRONOMICAL MODELS
     ASTROPHYSICS
     NUCLEAR REACTIONS
     STELLAR MODELS
     SUN

SOLAR NOISE
USE   SOLAR RADIO EMISSION

SOLAR OBLATENESS
RT   OBLATE SPHEROIDS
       SUN

SOLAR OBSERVATORIES
GS   OBSERVATORIES
       . **SOLAR OBSERVATORIES**
       . . OSO
       . . . OSO-C
       . . . OSO-1
       . . . OSO-2
       . . . OSO-3
       . . . OSO-4
       . . . OSO-5
       . . . OSO-6
       . . . OSO-7
       . . . OSO-8
       UNMANNED SPACECRAFT
       . **SOLAR OBSERVATORIES**
       . . OSO
       . . . AOSO
       . . . OSO-C
       . . . OSO-1
       . . . OSO-2
       . . . OSO-3
       . . . OSO-4
       . . . OSO-5
       . . . OSO-6
       . . . OSO-7
       . . . OSO-8
RT   CORONAGRAPHS
       SUN

SOLAR OPTICAL TELESCOPE
UF   SOT
GS   TELESCOPES
       . SPACEBORNE TELESCOPES
       . . **SOLAR OPTICAL TELESCOPE**
RT   ASTRONOMICAL TELESCOPES
       SOLAR INSTRUMENTS
       SOLAR PHYSICS

SOLAR ORBITS
SN   (RESTRICTED TO ORBITS AROUND THE
       SUN)
UF   HELIOCENTRIC ORBITS
       PLANETARY MOTION
GS   ORBITS
       . **SOLAR ORBITS**
       . . PERIHELIONS
RT   CIRCULAR ORBITS
       ∞EARTH MOTION
       ECLIPTIC
       ELLIPTICAL ORBITS
       HEOS SATELLITES
       INTERPLANETARY TRAJECTORIES
       ∞MOTION
       PROTOPLANETS
       SPACECRAFT ORBITS
       SUN
       TRANSFER ORBITS

SOLAR OSCILLATIONS
GS   OSCILLATIONS
       . STELLAR OSCILLATIONS
       . . **SOLAR OSCILLATIONS**
RT   ASTRONOMICAL MODELS
       ATMOSPHERIC MODELS
       CATACLYSMIC VARIABLES
       SOLAR ACTIVITY EFFECTS
       SOLAR ATMOSPHERE
       STELLAR MODELS
       SUN
       VARIABLE STARS

SOLAR PARALLAX
GS   PARALLAX
       . **SOLAR PARALLAX**
RT   ASTRONOMY
       STELLAR PARALLAX
       SUN

SOLAR PHYSICS
GS   ASTROPHYSICS
       . STELLAR PHYSICS
       . . **SOLAR PHYSICS**
RT   HELIOSEISMOLOGY
       INTERNATIONAL QUIET SUN YEAR
       PHOTOSPHERE
       ∞PHYSICS
       PLASMAS (PHYSICS)
       ∞SCIENCE

SOLAR PHYSICS-*(CONT.)*
       SOLAR OPTICAL TELESCOPE
       SPARTAN SATELLITES
       SUN

SOLAR PLANETARY INTERACTIONS
GS   **SOLAR PLANETARY INTERACTIONS**
       . SOLAR TERRESTRIAL INTERACTIONS
RT   MAGNETIC DISTURBANCES
       MAGNETOSPHERE
       PLANETARY ATMOSPHERES
       PLANETARY MAGNETIC FIELDS
       PLASMA INTERACTIONS
       SOLAR ACTIVITY
       SOLAR ACTIVITY EFFECTS
       SOLAR CORPUSCULAR RADIATION
       SOLAR WIND
       SOLAR WIND VELOCITY

SOLAR PLASMA (RADIATION)
USE   SOLAR WIND

SOLAR PONDS (HEAT STORAGE)
RT   ELECTRIC GENERATORS
       ENERGY CONVERSION
       PONDS
       RESERVOIRS
       SUN

SOLAR POSITION
UF   SOLAR AZIMUTH
GS   POSITION (LOCATION)
       . **SOLAR POSITION**
RT   ASTROLABES
       CELESTIAL NAVIGATION
       EQUINOXES
       SEASONS
       SOLSTICES
       SUN
       ZENITH

SOLAR POWER GENERATION
USE   SOLAR GENERATORS

SOLAR POWER SATELLITES
GS   SATELLITES
       . ARTIFICIAL SATELLITES
       . . **SOLAR POWER SATELLITES**
RT   LARGE SPACE STRUCTURES
       POWER TRANSMISSION (LASERS)
       SATELLITE POWER TRANSMISSION (TO
       EARTH)
       SUN

SOLAR POWER SOURCES
USE   SOLAR GENERATORS

SOLAR POWERED AIRCRAFT
RT   ∞AIRCRAFT
       SOLAR CELLS
       SOLAR ELECTRIC PROPULSION
       SOLAR PROPULSION
       SUN

SOLAR PROBES
GS   UNMANNED SPACECRAFT
       . SPACE PROBES
       . . **SOLAR PROBES**
       . . . HELIOS A
       . . . HELIOS B
       . . . HELIOS 1
       . . . HELIOS 2
       . . . SUNBLAZER SPACE PROBE
RT   HELIOS PROJECT
       PIONEER SPACE PROBES
       SUN
       ULYSSES MISSION

SOLAR PROMINENCES
UF   FILAMENTS (SOLAR PHYSICS)
GS   PROMINENCES
       . **SOLAR PROMINENCES**
       STELLAR ACTIVITY
       . SOLAR ACTIVITY
       . . **SOLAR PROMINENCES**
RT   CHROMOSPHERE
       SUN

SOLAR PROPULSION
GS   PROPULSION
       . LOW THRUST PROPULSION
       . . **SOLAR PROPULSION**
       . . . SOLAR ELECTRIC PROPULSION
       . . . SOLAR THERMAL PROPULSION

SOLAR PROPULSION-*(CONT.)*
       . SPACECRAFT PROPULSION
       . . **SOLAR PROPULSION**
       . . . SOLAR ELECTRIC PROPULSION
       . . . SOLAR THERMAL PROPULSION
RT   SOLAR POWERED AIRCRAFT
       SUN

SOLAR PROTONS
GS   EXTRATERRESTRIAL RADIATION
       . SOLAR RADIATION
       . . SOLAR CORPUSCULAR RADIATION
       . . . **SOLAR PROTONS**
       IONS
       . PROTONS
       . **SOLAR PROTONS**
       PARTICLES
       . CHARGED PARTICLES
       . . PROTONS
       . . . **SOLAR PROTONS**
       CORPUSCULAR RADIATION
       . . SOLAR CORPUSCULAR RADIATION
       . . . **SOLAR PROTONS**
       . ELEMENTARY PARTICLES
       . . FERMIONS
       . . . PROTONS
       . . . . **SOLAR PROTONS**
RT   BARYONS
       SUN

SOLAR RADAR ECHOES
GS   ECHOES
       . RADAR ECHOES
       . . **SOLAR RADAR ECHOES**
RT   SUN

SOLAR RADIATION
GS   EXTRATERRESTRIAL RADIATION
       . **SOLAR RADIATION**
       . . CIRCUMSOLAR RADIATION
       . . SOLAR CORPUSCULAR RADIATION
       . . . SOLAR ELECTRONS
       . . . SOLAR NEUTRINOS
       . . . SOLAR PROTONS
       . . SOLAR COSMIC RAYS
       . . SOLAR RADIO EMISSION
       . . . SOLAR RADIO BURSTS
       . . . . TYPE 2 BURSTS
       . . . . TYPE 3 BURSTS
       . . . . TYPE 4 BURSTS
       . . . . TYPE 5 BURSTS
       . . SOLAR WIND
       . . SOLAR X-RAYS
       . . SUNLIGHT
RT   AEROSPACE ENVIRONMENTS
       ALBEDO
       ATMOSPHERIC REFRACTION
       CIRCUMSOLAR TELESCOPES
       CLIMATOLOGY
       CLOUD COVER
       CORPUSCULAR RADIATION
       COSMIC NOISE
       COSMIC RAYS
       DAYGLOW
       ELECTROMAGNETIC RADIATION
       EXTREME ULTRAVIOLET RADIATION
       GEGENSCHEIN
       INFRARED RADIATION
       INSOLATION
       IONIZING RADIATION
       IRIS SATELLITES
       LIGHT (VISIBLE RADIATION)
       LONG WAVE RADIATION
       LONGITUDINAL WAVES
       ∞RADIATION
       RADIATION BELTS
       RADIATION PRESSURE
       RADIATIVE TRANSFER
       RADIO WAVES
       RECTENNAS
       SKY BRIGHTNESS
       SOLAR-PUMPED LASERS
       STELLAR RADIATION
       SUN
       THERMAL RADIATION
       ULTRAVIOLET RADIATION
       ZODIACAL LIGHT

SOLAR RADIATION SHIELDING
GS   PROTECTION
       . RADIATION PROTECTION
       . . RADIATION SHIELDING
       . . . **SOLAR RADIATION SHIELDING**
       SHIELDING
       . RADIATION SHIELDING

SOLAR RADIATION SHIELDING-*(CONT.)*
  . . **SOLAR RADIATION SHIELDING**
RT ∞RADIATION
    SATELLITE TEMPERATURE
    SPACECRAFT SHIELDING
    SUN

SOLAR RADIATION 1 SATELLITE
GS SATELLITES
    . ARTIFICIAL SATELLITES
    . . **SOLAR RADIATION 1 SATELLITE**
RT GALACTIC RADIATION
    ∞RADIATION
    SUN

SOLAR RADIATION 3 SATELLITE
GS SATELLITES
    . ARTIFICIAL SATELLITES
    . . **SOLAR RADIATION 3 SATELLITE**
RT GALACTIC RADIATION
    ∞RADIATION
    SUN

SOLAR RADIO BURSTS
GS BURSTS
    . RADIO BURSTS
    . . **SOLAR RADIO BURSTS**
    . . . TYPE 2 BURSTS
    . . . TYPE 3 BURSTS
    . . . TYPE 4 BURSTS
    . . . TYPE 5 BURSTS
    DECAY
    . EMISSION
    . . RADIO EMISSION
    . . . SOLAR RADIO EMISSION
    . . . . **SOLAR RADIO BURSTS**
    . . . . . TYPE 2 BURSTS
    . . . . . TYPE 3 BURSTS
    . . . . . TYPE 4 BURSTS
    . . . . . TYPE 5 BURSTS
    ELECTROMAGNETIC RADIATION
    . RADIO WAVES
    . . EXTRATERRESTRIAL RADIO WAVES
    . . . RADIO BURSTS
    . . . . **SOLAR RADIO BURSTS**
    . . . . . TYPE 2 BURSTS
    . . . . . TYPE 3 BURSTS
    . . . . . TYPE 4 BURSTS
    . . . . . TYPE 5 BURSTS
    . . SOLAR RADIO EMISSION
    . . . . **SOLAR RADIO BURSTS**
    . . . . . TYPE 2 BURSTS
    . . . . . TYPE 3 BURSTS
    . . . . . TYPE 4 BURSTS
    . . . . . TYPE 5 BURSTS
    . . RADIO EMISSION
    . . . RADIO BURSTS
    . . . . **SOLAR RADIO BURSTS**
    . . . . . TYPE 2 BURSTS
    . . . . . TYPE 3 BURSTS
    . . . . . TYPE 4 BURSTS
    . . . . . TYPE 5 BURSTS
    . . SOLAR RADIO EMISSION
    . . . . **SOLAR RADIO BURSTS**
    . . . . . TYPE 2 BURSTS
    . . . . . TYPE 3 BURSTS
    . . . . . TYPE 4 BURSTS
    . . . . . TYPE 5 BURSTS
    EXTRATERRESTRIAL RADIATION
    . EXTRATERRESTRIAL RADIO WAVES
    . . RADIO BURSTS
    . . . **SOLAR RADIO BURSTS**
    . . . . TYPE 2 BURSTS
    . . . . TYPE 3 BURSTS
    . . . . TYPE 4 BURSTS
    . . . . TYPE 5 BURSTS
    . . SOLAR RADIO EMISSION
    . . . **SOLAR RADIO BURSTS**
    . . . . TYPE 2 BURSTS
    . . . . TYPE 3 BURSTS
    . . . . TYPE 4 BURSTS
    . . . . TYPE 5 BURSTS
    . SOLAR RADIATION
    . . SOLAR RADIO EMISSION
    . . . **SOLAR RADIO BURSTS**
    . . . . TYPE 2 BURSTS
    . . . . TYPE 3 BURSTS
    . . . . TYPE 4 BURSTS
    . . . . TYPE 5 BURSTS
RT SUN

SOLAR RADIO EMISSION
UF SOLAR NOISE
    SOLAR RADIO WAVES
GS DECAY

SOLAR RADIO EMISSION-*(CONT.)*
  . EMISSION
  . . RADIO EMISSION
  . . . **SOLAR RADIO EMISSION**
  . . . . SOLAR RADIO BURSTS
  . . . . . TYPE 2 BURSTS
  . . . . . TYPE 3 BURSTS
  . . . . . TYPE 4 BURSTS
  . . . . . TYPE 5 BURSTS
  ELECTROMAGNETIC RADIATION
  . RADIO WAVES
  . . EXTRATERRESTRIAL RADIO WAVES
  . . . **SOLAR RADIO EMISSION**
  . . . . SOLAR RADIO BURSTS
  . . . . . TYPE 2 BURSTS
  . . . . . TYPE 3 BURSTS
  . . . . . TYPE 4 BURSTS
  . . . . . TYPE 5 BURSTS
  . . RADIO EMISSION
  . . . **SOLAR RADIO EMISSION**
  . . . . SOLAR RADIO BURSTS
  . . . . . TYPE 2 BURSTS
  . . . . . TYPE 3 BURSTS
  . . . . . TYPE 4 BURSTS
  . . . . . TYPE 5 BURSTS
  EXTRATERRESTRIAL RADIATION
  . EXTRATERRESTRIAL RADIO WAVES
  . . **SOLAR RADIO EMISSION**
  . . . SOLAR RADIO BURSTS
  . . . . TYPE 2 BURSTS
  . . . . TYPE 3 BURSTS
  . . . . TYPE 4 BURSTS
  . . . . TYPE 5 BURSTS
  . SOLAR RADIATION
  . . **SOLAR RADIO EMISSION**
  . . . SOLAR RADIO BURSTS
  . . . . TYPE 2 BURSTS
  . . . . TYPE 3 BURSTS
  . . . . TYPE 4 BURSTS
  . . . . TYPE 5 BURSTS
RT CORONAL HOLES
    COSMIC NOISE
    DECIMETER WAVES
    ELECTROMAGNETIC NOISE
    MILLIMETER WAVES
    RADIO BURSTS

SOLAR RADIO WAVES
USE SOLAR RADIO EMISSION

SOLAR RECEIVERS
USE SOLAR COLLECTORS

SOLAR REFLECTORS
GS ACCUMULATORS
    . SOLAR COLLECTORS
    . . **SOLAR REFLECTORS**
    REFLECTORS
    . SOLAR COLLECTORS
    . . **SOLAR REFLECTORS**
RT FOCUSING
    HEAT SHIELDING
    HELIOSTATS
    MIRRORS
    PARABOLIC REFLECTORS
    PARABOLOID MIRRORS
    RADIANT FLUX DENSITY
    SOLAR ATRIUMS
    SPACECRAFT RADIATORS
    SUN

SOLAR ROTATION
UF CARRINGTON ROTATION
GS GYRATION
    . ROTATION
    . . STELLAR ROTATION
    . . . **SOLAR ROTATION**
RT SUN
    TWENTY-SEVEN DAY VARIATION

SOLAR SAILS
GS SAILS
    . **SOLAR SAILS**
RT PROPULSION
    SPACE FLIGHT
    SPACECRAFT PROPULSION
    SUN

SOLAR SEA POWER PLANTS
GS ELECTRIC GENERATORS
    . DIRECT POWER GENERATORS
    . . THERMOELECTRIC GENERATORS
    . . . **SOLAR SEA POWER PLANTS**
RT ELECTRIC POWER PLANTS
    ENERGY CONVERSION

SOLAR SEA POWER PLANTS-*(CONT.)*
  ∞GENERATORS
    OCEAN TEMPERATURE
    OCEAN THERMAL ENERGY CONVERSION
  ∞POWER PLANTS
    SOLAR GENERATORS
    SUN

SOLAR SEISMOLOGY
USE HELIOSEISMOLOGY

SOLAR SELECTIVE COATINGS
USE SELECTIVE SURFACES

SOLAR SENSORS
UF SUN SENSORS
RT ATTITUDE CONTROL
    GUIDANCE SENSORS
    IRIS SATELLITES
    NAVIGATION AIDS
    NAVIGATION INSTRUMENTS
    STAR TRACKERS
    SUN
    TRACKING (POSITION)

SOLAR SIMULATION
GS SIMULATION
    . **SOLAR SIMULATION**
RT SPACE ENVIRONMENT SIMULATION
    SUN
    THERMAL SIMULATION

SOLAR SIMULATORS
GS SIMULATORS
    . ENVIRONMENT SIMULATORS
    . . **SOLAR SIMULATORS**
RT SPACE SIMULATORS
    SUN
    TEST FACILITIES

SOLAR SPECTRA
GS SPECTRA
    . RADIATION SPECTRA
    . . ELECTROMAGNETIC SPECTRA
    . . . STELLAR SPECTRA
    . . . . **SOLAR SPECTRA**
RT ABSORPTION SPECTRA
    ASTRONOMICAL SPECTROSCOPY
    CONTINUOUS SPECTRA
    CORONAS
    D LINES
    EMISSION SPECTRA
    FRAUNHOFER LINES
    H ALPHA LINE
    H BETA LINE
    H GAMMA LINE
    H LINES
    INFRARED SPECTRA
    LINE SPECTRA
    LYMAN SPECTRA
    MOLECULAR SPECTRA
    OXYGEN SPECTRA
    SUN
    ULTRAVIOLET SPECTRA
    VISIBLE SPECTRUM
    X RAY SPECTRA

SOLAR SPECTROMETERS
GS MEASURING INSTRUMENTS
    . RADIATION MEASURING INSTRUMENTS
    . . ACTINOMETERS
    . . . **SOLAR SPECTROMETERS**
    . SPECTROMETERS
    . . **SOLAR SPECTROMETERS**
RT ABSORPTION SPECTRA
    EMISSION SPECTRA
    FILTER WHEEL INFRARED
      SPECTROMETERS
    INFRARED SPECTROMETERS
    SPECTROHELIOGRAPHS
    SUN
    ULTRAVIOLET SPECTROMETERS

SOLAR STORMS
GS STELLAR ACTIVITY
    . SOLAR ACTIVITY
    . . **SOLAR STORMS**
    STORMS
    . **SOLAR STORMS**
RT FORBUSH DECREASES
    IONOSPHERIC STORMS
    MAGNETIC STORMS
    NOISE STORMS
    SUN

SOLAR STREAMS
USE   SOLAR CORPUSCULAR RADIATION

SOLAR SYSTEM
GS   CELESTIAL BODIES
        . **SOLAR SYSTEM**
        . . HALLEY'S COMET
RT   AMALTHEA
        AMOR ASTEROID
        APOLLO ASTEROIDS
        AREND-ROLAND COMET
        ASTEROID BELTS
        ASTEROID CAPTURE
        ASTEROIDS
        CELESTIAL MECHANICS
        CHARON
        CHIRON
        COMET HEADS
        COMET NUCLEI
        COMET TAILS
        COMETS
        EARTH-MOON SYSTEM
        GAS GIANT PLANETS
        GRIGG-SKJELLERUP COMET
        IRAS-ARAKI-ALCOCK COMET
        JUPITER SATELLITES
        KOHOUTEK COMET
        METEOROIDS
        NATURAL SATELLITES
        PLANETS
        PROTOPLANETS
        RHEA (ASTRONOMY)
        SATURN RINGS
        SUN
        ∞SYSTEMS
        TEMPEL 2 COMET
        TERRESTRIAL PLANETS
        TORO ASTEROID
        VENUS SURFACE
        VESTA ASTEROID
        VOYAGER 1977 MISSION
        WEST COMET

SOLAR TEMPERATURE
GS   TEMPERATURE
        . **SOLAR TEMPERATURE**
RT   SUN

SOLAR TERRESTRIAL INTERACTIONS
GS   SOLAR PLANETARY INTERACTIONS
        . **SOLAR TERRESTRIAL INTERACTIONS**
RT   CORPUSCULAR RADIATION
        ∞FLARES
        ∞INTERACTIONS
        INTERNATIONAL
          GEOSPHERE-BIOSPHERE PROGRAM
        MAGNETIC DISTURBANCES
        MAGNETIC STORMS
        MAGNETOSPHERE
        STORMS
        SUN
        SUNSPOTS
        WEATHER

SOLAR THERMAL ELECTRIC POWER PLANTS
GS   ELECTRIC POWER PLANTS
        . **SOLAR THERMAL ELECTRIC POWER
          PLANTS**
RT   ∞POWER PLANTS
        SOLAR ENERGY
        THERMAL ENERGY

SOLAR THERMAL PROPULSION
GS   PROPULSION
        . LOW THRUST PROPULSION
        . . SOLAR PROPULSION
        . . . **SOLAR THERMAL PROPULSION**
        . SPACECRAFT PROPULSION
        . . SOLAR PROPULSION
        . . . **SOLAR THERMAL PROPULSION**
RT   PROPULSION SYSTEM PERFORMANCE
        SOLAR ELECTRIC PROPULSION
        SUN

SOLAR TOTAL ENERGY SYSTEMS
GS   ENERGY CONVERSION
        . SOLAR ENERGY CONVERSION
        . . **SOLAR TOTAL ENERGY SYSTEMS**
        TOTAL ENERGY SYSTEMS
        . **SOLAR TOTAL ENERGY SYSTEMS**
RT   ∞CONVERSION
        DIRECT POWER GENERATORS
        ∞ENERGY
        SUN
        ∞SYSTEMS

SOLAR VELOCITY
GS   RATES (PER TIME)
        . **SOLAR VELOCITY**
        VELOCITY
        . **SOLAR VELOCITY**
RT   SUN

SOLAR WIND
UF   SOLAR PLASMA (RADIATION)
GS   EXTRATERRESTRIAL RADIATION
        . SOLAR RADIATION
        . . **SOLAR WIND**
        GASES
        . IONIZED GASES
        . . CHARGED PARTICLES
        . . . **SOLAR WIND**
        PARTICLES
        . CHARGED PARTICLES
        . . ENERGETIC PARTICLES
        . . . PLASMAS (PHYSICS)
        . . . . **SOLAR WIND**
RT   AMPTE (SATELLITES)
        CHAPMAN-FERRARO PROBLEM
        COMET TAILS
        CORONAL HOLES
        COSMIC PLASMA
        GALACTIC COSMIC RAYS
        GRIGG-SKJELLERUP COMET
        HELIOSPHERE
        HYDROGEN PLASMA
        INTERPLANETARY GAS
        INTERPLANETARY MEDIUM
        M REGION
        MAGNETOPAUSE
        PLASMAPAUSE
        PLASMAS (PHYSICS)
        ∞RADIATION
        RADIATION PRESSURE
        SOLAR PLANETARY INTERACTIONS
        SPACE PLASMAS
        STELLAR WINDS
        SUN

SOLAR WIND VELOCITY
GS   RATES (PER TIME)
        . FLOW VELOCITY
        . . **SOLAR WIND VELOCITY**
        WIND VELOCITY
        . . **SOLAR WIND VELOCITY**
        VELOCITY
        . FLOW VELOCITY
        . . **SOLAR WIND VELOCITY**
        . WIND VELOCITY
        . . **SOLAR WIND VELOCITY**
RT   ALPHA PARTICLES
        MAGNETIC DISTURBANCES
        MAGNETOHYDRODYNAMIC FLOW
        MAGNETOSPHERE
        SOLAR PLANETARY INTERACTIONS
        STELLAR WINDS
        SUN
        VELOCITY MEASUREMENT

SOLAR X-RAYS
GS   ELECTROMAGNETIC RADIATION
        . X RAYS
        . . **SOLAR X-RAYS**
        EXTRATERRESTRIAL RADIATION
        . SOLAR RADIATION
        . . **SOLAR X-RAYS**
        IONIZING RADIATION
        . X RAYS
        . . **SOLAR X-RAYS**
RT   CORONAL HOLES
        SUN

SOLAR-PUMPED LASERS
UF   SOLAR LASERS
GS   STIMULATED EMISSION DEVICES
        . LASERS
        . . **SOLAR-PUMPED LASERS**
RT   LASER PUMPING
        OPTICAL PUMPING
        SOLAR ENERGY CONVERSION
        SOLAR RADIATION

SOLDERED JOINTS
GS   JOINTS (JUNCTIONS)
        . METAL JOINTS
        . . **SOLDERED JOINTS**
RT   BEAM LEADS
        BUTT JOINTS
        LAP JOINTS
        SOLDERING

SOLDERING
GS   **SOLDERING**
        . ULTRASONIC SOLDERING
RT   BONDING
        BRAZING
        FLUXES
        ∞JOINING
        LASER WELDING
        LOW TEMPERATURE BRAZING
        METAL BONDING
        METAL-METAL BONDING
        SEALING
        SOLDERED JOINTS
        SOLDERS
        WELDING

SOLDERS
GS   ALLOYS
        . **SOLDERS**
RT   LEAD ALLOYS
        SEALERS
        SOLDERING
        TIN ALLOYS
        ZINC ALLOYS

SOLENOID VALVES
GS   VALVES
        . **SOLENOID VALVES**
RT   AUTOMATIC CONTROL VALVES
        ELECTRIC CONTROL
        ∞ELECTRIC EQUIPMENT
        ELECTRIC RELAYS
        ELECTRIC SWITCHES
        HYDRAULIC CONTROL
        OFF-ON CONTROL
        SOLENOIDS

SOLENOIDS
SN   (EXCLUDES METEOROLOGICAL
        SOLENOIDS)
RT   ACTUATORS
        ELECTRIC RELAYS
        ELECTROMAGNETS
        MAGNET COILS
        SOLENOID VALVES
        TOROIDAL PLASMAS

SOLETTAS
GS   MIRRORS
        . **SOLETTAS**
RT   ∞PLATFORMS
        SATELLITES
        SOLAR ENERGY
        ∞SPACECRAFT

SOLID ARGON
USE   SOLIDIFIED GASES

SOLID CRYOGEN COOLING
GS   COOLING
        . **SOLID CRYOGEN COOLING**
RT   CRYOGENIC FLUIDS
        CRYOGENICS
        LIQUEFIED GASES

SOLID CRYOGENS
GS   GASES
        . SOLIDIFIED GASES
        . . **SOLID CRYOGENS**
        SOLIDS
        . SOLIDIFIED GASES
        . . **SOLID CRYOGENS**
RT   COOLING SYSTEMS
        CRYOGENIC EQUIPMENT
        CRYOGENICS
        LIQUID NITROGEN
        SOLID CRYOGENS
        SOLID NITROGEN

SOLID ELECTRODES
GS   ELECTRODES
        . **SOLID ELECTRODES**

SOLID ELECTROLYTES
GS   CONDUCTORS
        . ELECTROLYTES
        . . **SOLID ELECTROLYTES**

SOLID LUBRICANTS
SN   (EXCLUDES SEMISOLIDS SUCH AS
        GREASES)
GS   LUBRICANTS
        . **SOLID LUBRICANTS**
RT   BINDERS (MATERIALS)

SOLID LUBRICANTS-*(CONT.)*
    GAS LUBRICANTS
    GRAPHITE
    SELF LUBRICATING MATERIALS

SOLID MECHANICS
RT   CONTINUUM MECHANICS
    FINITE ELEMENT METHOD
    MECHANICAL PROPERTIES
  ∞MECHANICS (PHYSICS)
  ∞SCIENCE
    SOLIDS
    STRUCTURAL ANALYSIS

SOLID NITROGEN
GS   CHEMICAL ELEMENTS
  . NITROGEN
  . . **SOLID NITROGEN**
RT   CRYOGENICS
    REFRIGERANTS
    SOLID CRYOGENS

SOLID PHASES
RT   EUTECTICS
    GAS-METAL INTERACTIONS
    GAS-SOLID INTERFACES
    LIQUID PHASES
    LIQUID-SOLID INTERFACES
    LIQUIDUS
    METALLIC HYDROGEN
    PHASE DIAGRAMS
  ∞PHASES
    SOLIDIFIED GASES
    SOLIDUS
    SYNTECTIC ALLOYS

SOLID PROPELLANT COMBUSTION
GS   COMBUSTION
  . PROPELLANT COMBUSTION
  . . **SOLID PROPELLANT COMBUSTION**
  . . . SOLID PROPELLANT IGNITION
RT   BURNING RATE
    COMBUSTION STABILITY
    EROSIVE BURNING
    FUEL COMBUSTION
    HEAT GENERATION
    METAL COMBUSTION
    PROPELLANT CHEMISTRY

SOLID PROPELLANT IGNITION
GS   COMBUSTION
  . PROPELLANT COMBUSTION
  . . SOLID PROPELLANT COMBUSTION
  . . . **SOLID PROPELLANT IGNITION**
    IGNITION
  . **SOLID PROPELLANT IGNITION**
RT   HYBRID PROPELLANTS
    HYPERGOLIC ROCKET PROPELLANTS
    IGNITERS
    IGNITION TEMPERATURE
    INHIBITORS
    METAL COMBUSTION
    PYROPHORIC MATERIALS
    SQUIBS

SOLID PROPELLANT ROCKET ENGINES
GS   ENGINES
  . ROCKET ENGINES
  . . **SOLID PROPELLANT ROCKET ENGINES**
  . . . ALGOL ENGINE
  . . . APOGEE BOOST MOTORS
  . . . ASROC ENGINE
  . . . HERCULES ENGINE
  . . . M-46 ENGINE
  . . . M-55 ENGINE
  . . . M-56 ENGINE
  . . . M-57 ENGINE
  . . . NIKE BOOSTER ROCKET ENGINES
  . . . P-1 ENGINE
  . . . SL-3 ROCKET ENGINE
  . . . SYNCOM APOGEE ENGINES
  . . . TU-121 ENGINE
  . . . TX-77 ENGINE
  . . . TX-354 ENGINE
  . . . X-248 ENGINE
  . . . X-254 ENGINE
  . . . X-258 ENGINES
  . . . . X-258-B1 ENGINE
  . . . X-259 ENGINE
  . . . XM-33 ENGINE
RT   AIR SLEW MISSILES
    ANTARES ROCKET VEHICLE
    ARCAS ROCKET VEHICLES
    ARGO ROCKET VEHICLES

SOLID PROPELLANT ROCKET ENGINES-*(CONT.)*
    ASTROBEE ROCKET VEHICLES
    ASTROBEE 1500 ROCKET VEHICLE
    ATHENA ROCKET VEHICLE
    BE-3 ENGINE
    BERENICE ROCKET VEHICLE
    BLACK BRANT SOUNDING ROCKETS
    BLACK BRANT 1 SOUNDING ROCKET
    BLACK BRANT 2 SOUNDING ROCKET
    BLACK BRANT 3 SOUNDING ROCKET
    BLACK BRANT 4 SOUNDING ROCKET
    BLACK BRANT 5 SOUNDING ROCKET
    BLUE GOOSE MISSILE
    BLUE SCOUT ROCKET VEHICLE
    BOMARC A MISSILE
    BOMARC B MISSILE
    BOOSTER ROCKET ENGINES
    BURNING RATE
    BURNOUT
    CAJUN ROCKET VEHICLE
    DIAMANT LAUNCH VEHICLE
    DUCTED ROCKET ENGINES
    EXOS SOUNDING ROCKET
    FALCON MISSILE
    FOLDING FIN AIRCRAFT ROCKET VEHICLE
    GENIE ROCKET VEHICLE
    HAWK MISSILE
    HONEST JOHN ROCKET VEHICLE
    HYBRID PROPELLANT ROCKET ENGINES
    INTEGRAL ROCKET RAMJETS
    INTERNAL COMBUSTION ENGINES
    JAGUAR ROCKET VEHICLE
    JATO ENGINES
    JAVELIN ROCKET VEHICLE
    JUNO LAUNCH VEHICLES
    JUNO 1 LAUNCH VEHICLE
    JUNO 2 LAUNCH VEHICLE
    JUPITER C ROCKET VEHICLE
    KAPPA ROCKET VEHICLES
    KAPPA 8 ROCKET VEHICLE
    KAPPA 9 ROCKET VEHICLE
    LAMBDA ROCKET VEHICLES
    LIQUID PROPELLANT ROCKET ENGINES
    LITTLE JOE 2 LAUNCH VEHICLE
    LITTLE JOHN ROCKET VEHICLE
    LOKI ROCKET VEHICLE
    MACE MISSILES
    MATRA MISSILE
    MAULER MISSILE
    METEOR 1 ROCKET VEHICLE
    MINUTEMAN ICBM
    NIKE-AJAX MISSILE
    NIKE-APACHE ROCKET VEHICLE
    NIKE-CAJUN ROCKET VEHICLE
    NIKE-HERCULES MISSILE
    NIKE-JAVELIN ROCKET VEHICLE
    NIKE-TOMAHAWK ROCKET VEHICLE
    NIKE-ZEUS MISSILE
    PERSHING MISSILE
    PHOENIX SOUNDING ROCKET
    POLARIS MISSILES
    RAM B LAUNCH VEHICLE
    REDEYE MISSILE
    REGULUS MISSILE
    RESTARTABLE ROCKET ENGINES
    RETROROCKET ENGINES
    RUBIS ROCKET VEHICLE
    SCOUT LAUNCH VEHICLE
    SERGEANT MISSILES
    SHRIKE MISSILE
    SKUA ROCKET VEHICLES
    SKYBOLT MISSILE
    SKYLARK ROCKET VEHICLE
    SPACE SHUTTLE UPPER STAGE D
    SPARROW MISSILES
    SPARROW 2 MISSILE
    SPRINT MISSILE
    SS-11 MISSILE
    SUNBLAZER SPACE PROBE
    SUSTAINER ROCKET ENGINES
    TALOS MISSILE
    TARTAR MISSILE
    TERRIER MISSILE
    THOR ABLE ROCKET VEHICLE
    THOR DELTA LAUNCH VEHICLE
    THOR LAUNCH VEHICLES
    TITAN LAUNCH VEHICLES
    TRAILBLAZER 1 REENTRY VEHICLE
    TRAILBLAZER 2 REENTRY VEHICLE
    ULLAGE ROCKET ENGINES
    VANGUARD 2 LAUNCH VEHICLE
    VERNIER ENGINES
    WASP SOUNDING ROCKET
    X-17 REENTRY VEHICLE
    ZUNI ROCKET VEHICLE

SOLID PROPELLANTS
GS   PROPELLANTS
  . **SOLID PROPELLANTS**
  . . CASE BONDED PROPELLANTS
  . . COMPOSITE PROPELLANTS
  . . NITRAMINE PROPELLANTS
  . . PLASTIC PROPELLANTS
  . . SOLID ROCKET PROPELLANTS
  . . . DOUBLE BASE ROCKET PROPELLANTS
  . . . HTPB PROPELLANTS
  . . . METAL PROPELLANTS
RT   AIRCRAFT FUELS
    CHEMICAL FUELS
    COLLOIDAL PROPELLANTS
    GELLED PROPELLANTS
    HIGH TEMPERATURE PROPELLANTS
    HYBRID PROPELLANTS
    INHIBITORS
    METAL FUELS
    NITROGUANIDINE
    PLASTICIZERS
    PROPELLANT BINDERS
    PROPELLANT GRAINS
    RDX
    ROCKET PROPELLANTS
    STORABLE PROPELLANTS

SOLID ROCKET BINDERS
GS   ADDITIVES
  . PROPELLANT ADDITIVES
  . . PROPELLANT BINDERS
  . . . **SOLID ROCKET BINDERS**
    BINDERS (MATERIALS)
  . PROPELLANT BINDERS
  . . **SOLID ROCKET BINDERS**
RT   PLASTICIZERS

SOLID ROCKET PROPELLANTS
GS   FUELS
  . **SOLID ROCKET PROPELLANTS**
  . . DOUBLE BASE ROCKET PROPELLANTS
  . . METAL PROPELLANTS
    PROPELLANTS
  . ROCKET PROPELLANTS
  . . **SOLID ROCKET PROPELLANTS**
  . . . DOUBLE BASE ROCKET PROPELLANTS
  . . . METAL PROPELLANTS
  . SOLID PROPELLANTS
  . . **SOLID ROCKET PROPELLANTS**
  . . . DOUBLE BASE ROCKET PROPELLANTS
  . . . HTPB PROPELLANTS
  . . . METAL PROPELLANTS
RT   AMMONIUM PERCHLORATES
    BURNING RATE
    CASE BONDED PROPELLANTS
    COMPOSITE PROPELLANTS
    DOMINO PROPELLANTS
    GELLED ROCKET PROPELLANTS
    HYBRID PROPELLANTS
    LIQUID ROCKET PROPELLANTS
    MONOPROPELLANTS
    POTASSIUM PERCHLORATES
    PROPELLANT GRAINS
    RDX
    ROCKET ENGINES
    SLURRY PROPELLANTS

SOLID ROTATION
USE   ROTATING BODIES

SOLID SOLUTIONS
GS   MIXTURES
  . SOLUTIONS
  . . **SOLID SOLUTIONS**
RT   AGING (METALLURGY)
    ALLOYS
    LIQUID PHASES
    LIQUIDUS
    MELTING POINTS
    ORDER-DISORDER TRANSFORMATIONS
    PHASE DIAGRAMS
    PRECIPITATION HARDENING
    SOLIDS
    SUPERSATURATION
    TERNARY SYSTEMS

SOLID STATE
RT   CRYSTALLIZATION
    ENERGY GAPS (SOLID STATE)
    MELTING POINTS
    METALLIC HYDROGEN

SOLID STATE-*(CONT.)*
   SOLIDS

SOLID STATE DEVICES
GS   ELECTRONIC EQUIPMENT
   . **SOLID STATE DEVICES**
   . . CRYOTRONS
   . . CRYSTAL RECTIFIERS
   . . METAL-NITRIDE-OXIDE-SEMICONDUCT
      ORS
   . . MULTISPECTRAL LINEAR ARRAYS
   . . SEMICONDUCTOR DEVICES
   . . . AVALANCHE DIODES
   . . . . CRYOSAR
   . . . BARRITT DIODES
   . . . GERMANIUM DIODES
   . . . HETEROJUNCTION DEVICES
   . . . . HIGH ELECTRON MOBILITY
        TRANSISTORS
   . . . JUNCTION DIODES
   . . . . MIM DIODES
   . . . . STEP RECOVERY DIODES
   . . . LIGHT EMITTING DIODES
   . . METAL OXIDE SEMICONDUCTORS
   . . . CHARGE TRANSFER DEVICES
   . . . . BUCKET BRIGADE DEVICES
   . . . . . CHARGE COUPLED DEVICES
   . . . . CHARGE INJECTION DEVICES
   . . . CMOS
   . . . SOS (SEMICONDUCTORS)
   . . MIM (SEMICONDUCTORS)
   . . . MIS (SEMICONDUCTORS)
   . . . NDM SEMICONDUCTOR DEVICES
   . . NEURISTORS
   . . PARAMETRIC DIODES
   . . PHOTODIODES
   . . PHOTOVOLTAIC CELLS
   . . SCHOTTKY DIODES
   . . SEMICONDUCTOR LASERS
   . . . GALLIUM ARSENIDE LASERS
   . . . THERMISTORS
   . . . THYRISTORS
   . . . . SILICON CONTROLLED
        RECTIFIERS
   . . . TRANSFERRED ELECTRON DEVICES
   . . . TRANSISTOR AMPLIFIERS
   . . . TRANSISTORS
   . . . . BIPOLAR TRANSISTORS
   . . . . FIELD EFFECT TRANSISTORS
   . . . . . CHARGE FLOW DEVICES
   . . . . . JFET
   . . . . HIGH ELECTRON MOBILITY
        TRANSISTORS
   . . . . JUNCTION TRANSISTORS
   . . . . . JFET
   . . . . PHOTOTRANSISTORS
   . . . . SILICON TRANSISTORS
   . . . . . SOS (SEMICONDUCTORS)
   . . . TRAPATT DEVICES
   . . . VARACTOR DIODES
   . . . VARISTORS
   . . SIS (SEMICONDUCTORS)
   . . SOLID STATE LASERS
   . . . GALLIUM ARSENIDE LASERS
   . . . RUBY LASERS
   . . . YAG LASERS
RT   AMPLIFIERS
   BUBBLE TECHNIQUE
   CAPACITORS
   CIRCUITS
   ∞ DEVICES
   ELECTRIC BRIDGES
   LASER CAVITIES
   LASERS
   MINIATURE ELECTRONIC EQUIPMENT
   OSCILLATORS
   PHOTOMASKS
   RECTIFIERS
   RESISTORS
   SIGNAL GENERATORS
   SUPERCONDUCTORS
   THIN FILMS
   TRANSFORMERS
   WAFERS

SOLID STATE LASERS
GS   ELECTRONIC EQUIPMENT
   . SOLID STATE DEVICES
   . . **SOLID STATE LASERS**
   . . . GALLIUM ARSENIDE LASERS
   . . . RUBY LASERS
   . . . YAG LASERS
   STIMULATED EMISSION DEVICES
   . LASERS
   . . **SOLID STATE LASERS**

SOLID STATE LASERS-*(CONT.)*
   . . . GALLIUM ARSENIDE LASERS
   . . . RUBY LASERS
   . . . YAG LASERS
RT   CONTINUOUS WAVE LASERS
   DISTRIBUTED FEEDBACK LASERS
   INFRARED LASERS
   LASER CAVITIES
   Q SWITCHED LASERS
   SEMICONDUCTOR LASERS

∞ SOLID STATE PHYSICS
SN   *(USE OF A MORE SPECIFIC TERM IS*
   RECOMMENDED--CONSULT THE TERMS
   LISTED BELOW
RT   CRYSTALLOGRAPHY
   ELECTRICAL PROPERTIES
   ELECTRON MOBILITY
   ENERGY GAPS (SOLID STATE)
   FORBIDDEN TRANSITIONS
   HOLE MOBILITY
   MAGNETIC PROPERTIES
   OPTICAL PROPERTIES
   ∞ PHYSICS
   ∞ SCIENCE
   SEMICONDUCTORS (MATERIALS)
   SUPERCONDUCTIVITY
   THEORETICAL PHYSICS
   THIN FILMS
   TRANSPORT PROPERTIES

SOLID SURFACES
GS   **SOLID SURFACES**
   . CRYSTAL SURFACES
RT   LIQUID SURFACES
   METAL SURFACES
   SURFACE CRACKS
   SURFACE FINISHING
   SURFACE PROPERTIES
   ∞ SURFACES

SOLID SUSPENSIONS
GS   MIXTURES
   . **SOLID SUSPENSIONS**
RT   COLLOIDAL PROPELLANTS
   COMPOSITE MATERIALS
   METALLOGRAPHY
   PHASE DIAGRAMS
   ∞ SUSPENSIONS

SOLID WASTES
GS   WASTES
   . SEWAGE
   . . **SOLID WASTES**
RT   COMPOSTING
   GARBAGE
   HUMAN WASTES
   INDUSTRIAL WASTES
   LANDFILLS
   LIQUID WASTES
   METABOLIC WASTES
   POLLUTION
   RADIOACTIVE WASTES
   RESIDUES
   SLUDGE
   WASTE DISPOSAL
   WASTE ENERGY UTILIZATION
   WASTE UTILIZATION

SOLID-SOLID INTERFACES
GS   INTERFACES
   . **SOLID-SOLID INTERFACES**
RT   GAS-SOLID INTERFACES
   LIQUID-SOLID INTERFACES
   SURFACE PROPERTIES

SOLIDIFICATION
GS   **SOLIDIFICATION**
   . MELT SPINNING
RT   CASTING
   CASTINGS
   COAGULATION
   CRYSTALLIZATION
   FREEZING
   GELATION
   INGOTS
   MELTING POINTS
   MUSHY ZONES
   OCCLUSION
   PHASE TRANSFORMATIONS
   RHEOCASTING
   ∞ SETTING
   SOLIDIFIED GASES
   TRANSITION TEMPERATURE
   VITRIFICATION

SOLIDIFIED GASES
UF   SOLID ARGON
GS   GASES
   . **SOLIDIFIED GASES**
   . . SOLID CRYOGENS
   SOLIDS
   . **SOLIDIFIED GASES**
   . . SOLID CRYOGENS
RT   CRYOGENICS
   FREEZING
   LOW TEMPERATURE PHYSICS
   MELTING POINTS
   METALLIC HYDROGEN
   SOLID PHASES
   SOLIDIFICATION
   ULTRALOW TEMPERATURES

SOLIDS
GS   **SOLIDS**
   . SOLIDIFIED GASES
   . . SOLID CRYOGENS
   . THREE DIMENSIONAL BODIES
RT   ∞ BODIES
   ∞ FLUIDS
   ∞ MATERIALS
   METALLIC HYDROGEN
   ORGANIC SOLIDS
   PHASE TRANSFORMATIONS
   SOLID MECHANICS
   SOLID SOLUTIONS
   SOLID STATE
   THERMOCHROMATIC MATERIALS
   VAPOR PHASES

SOLIDS FLOW
GS   FLUID FLOW
   . **SOLIDS FLOW**
RT   ∞ FLOW
   FLOW MEASUREMENT
   FLOW THEORY
   MASS FLOW
   MULTIPHASE FLOW
   PARTICLE SIZE DISTRIBUTION
   STEADY FLOW
   TWO PHASE FLOW
   UNIFORM FLOW
   UNSTEADY FLOW

SOLIDUS
RT   BINARY SYSTEMS (MATERIALS)
   LIQUID PHASES
   LIQUIDUS
   SOLID PHASES

SOLIONS
RT   CIRCUITS
   DIODES
   INTEGRATORS
   ION CURRENTS

SOLITARY WAVES
UF   SOLITONS
GS   TRAVELING WAVES
   . **SOLITARY WAVES**
RT   BACKWARD WAVES
   CNOIDAL WAVES
   ELASTIC WAVES
   ELECTROMAGNETIC RADIATION
   PLANE WAVES
   PULSES
   RADIO WAVES
   RATES (PER TIME)
   VELOCITY

SOLITHANES
RT   ELASTOMERS
   ∞ POLYMERS
   SYNTHETIC RUBBERS

SOLITONS
USE   SOLITARY WAVES

SOLOMON COMPUTERS
GS   DATA PROCESSING EQUIPMENT
   . COMPUTERS
   . . DIGITAL COMPUTERS
   . . . **SOLOMON COMPUTERS**

SOLRAD 10 SATELLITE
USE   EXPLORER 44 SATELLITE

SOLSTICES
RT   EQUINOXES
   SEASONS

SOLSTICES-(CONT.)
    SOLAR POSITION
    SUMMER
    WINTER

SOLUBILITY
UF   IMMISCIBILITY
    MISCIBILITY
RT   CLARITY
    CONCENTRATION (COMPOSITION)
    DIFFUSIVITY
    DISSOLVED GASES
    DISSOLVING
    GAS-SOLID INTERFACES
    HENRY LAW
    HYGROSCOPICITY
    INCOMPATIBILITY
    LIQUID PHASES
    LIQUID-GAS MIXTURES
    LIQUID-LIQUID INTERFACES
    LIQUID-VAPOR INTERFACES
    MIXTURES
    PHASE DIAGRAMS
    PRECIPITATION (CHEMISTRY)
    SOLUTIONS
    SUPERCRITICAL FLUIDS
    THERMODYNAMIC PROPERTIES
    THIXOTROPY
    TURBIDITY
    VISCOSITY

SOLUTES
RT   DISSOLVING
    SOLUTIONS

∞ SOLUTION
SN   (USE OF A MORE SPECIFIC TERM IS
    RECOMMENDED--CONSULT THE TERMS
    LISTED BELOW)
RT   DISSOLVING
    PROBLEM SOLVING
    SOLUTIONS

SOLUTIONS
GS   MIXTURES
   . **SOLUTIONS**
   . . AQUEOUS SOLUTIONS
   . . GAS MIXTURES
   . . . DETONABLE GAS MIXTURES
   . . PHOTOGRAPHIC EMULSIONS
   . . . NUCLEAR EMULSIONS
   . . SOLID SOLUTIONS
RT   AZEOTROPES
    COMPOSITION (PROPERTY)
    DISSOLVED GASES
    EMULSIONS
    EUTECTICS
    HENRY LAW
    RAOULT LAW
    SOLUBILITY
    SOLUTES
  ∞ SOLUTION
    SOLVENTS
    TITRATION

SOLVATION
RT   AQUEOUS SOLUTIONS
    CHEMICAL REACTIONS
    REACTION KINETICS
    SOLVENTS

SOLVENT EXTRACTION
GS   EXTRACTION
   . **SOLVENT EXTRACTION**
RT   ION EXTRACTION
    PURIFICATION
  ∞ SEPARATION

SOLVENT REFINED COAL
GS   RESOURCES
   . EARTH RESOURCES
   . . FOSSIL FUELS
   . . . COAL
   . . . . **SOLVENT REFINED COAL**
   ROCKS
   . COAL
   . . **SOLVENT REFINED COAL**
RT   BENZENE
    BITUMENS
    CARBONACEOUS MATERIALS
    COAL LIQUEFACTION
    COAL UTILIZATION
    FRACTIONATION
    FUEL OILS
    GASOLINE

SOLVENT REFINED COAL-(CONT.)
    HYDROCARBON FUEL PRODUCTION
    METHANE
    PRODUCT DEVELOPMENT

SOLVENT RETENTION
RT   DISSOLVING
    SOLVENTS

SOLVENTS
UF   THINNERS
GS   **SOLVENTS**
   . TETRAHYDROFURAN
   . TURPENTINE
RT   ADDITIVES
    COATINGS
    DILUENTS
    DISSOLVING
    EXTRACTION
    FURANS
    SOLUTIONS
    SOLVATION
    SOLVENT RETENTION
    SOLVOLYSIS
    TOLUENE
    TRIACETIN

SOLVOLYSIS
GS   RECLAMATION
   . MATERIALS RECOVERY
   . . **SOLVOLYSIS**
RT   RECYCLING
    SOLVENTS

SOMALIA
GS   NATIONS
   . **SOMALIA**
RT   AFRICA

SOMMERFELD APPROXIMATION
GS   ANALYSIS (MATHEMATICS)
   . NUMERICAL ANALYSIS
   . . APPROXIMATION
   . . . **SOMMERFELD APPROXIMATION**
RT   ANTENNA RADIATION PATTERNS
    DIRECTIONAL ANTENNAS
    ELECTROMAGNETIC FIELDS
    RADIO TRANSMISSION

SOMMERFELD WAVES
GS   ELECTROMAGNETIC RADIATION
   . **SOMMERFELD WAVES**
   SURFACE WAVES
   . **SOMMERFELD WAVES**
RT   DIELECTRIC PROPERTIES
    ELECTRIC CONDUCTORS

SONAR
GS   **SONAR**
   . SONOBUOYS
RT   DISTANCE MEASURING EQUIPMENT
    ECHO SOUNDING
    ECHO SUPPRESSORS
    HYDROPHONES
    LOFAR
    NAVIGATION AIDS
    SOUND LOCALIZATION
    SOUND RANGING
    ULTRASONIC WAVE TRANSDUCERS
    UNDERWATER ACOUSTICS
    UNDERWATER COMMUNICATION

SONDES
UF   METEOROLOGICAL PROBES
GS   MEASURING INSTRUMENTS
   . **SONDES**
   . . DROPSONDES
   . . JUDI-DART ROCKET
   . . RADIOSONDES
   . . . ENDORADIOSONDES
   . . . IONOSONDES
   . . . RAWINSONDES
RT   APACHE ROCKET VEHICLE
    ASP ROCKET VEHICLE
    CAJUN ROCKET VEHICLE
  ∞ PROBES
    SOUNDING
    SOUNDING ROCKETS

SONIC ANEMOMETERS
GS   AIRCRAFT INSTRUMENTS
   . SPEED INDICATORS
   . . ANEMOMETERS
   . . . **SONIC ANEMOMETERS**

SONIC ANEMOMETERS-(CONT.)
    DISPLAY DEVICES
   . SPEED INDICATORS
   . . ANEMOMETERS
   . . . **SONIC ANEMOMETERS**
   MEASURING INSTRUMENTS
   . INDICATING INSTRUMENTS
   . SPEED INDICATORS
   . . . ANEMOMETERS
   . . . . **SONIC ANEMOMETERS**
RT   ACOUSTICS
    FLOWMETERS
    HOT-FILM ANEMOMETERS

SONIC BOOMS
GS   ELASTIC WAVES
   . SHOCK WAVES
   . . **SONIC BOOMS**
   . SOUND WAVES
   . . NOISE (SOUND)
   . . . AIRCRAFT NOISE
   . . . . **SONIC BOOMS**
RT   ACOUSTIC VELOCITY
    AERODYNAMIC NOISE
  ∞ BOOM
    CAUSTIC LINES
    JET AIRCRAFT NOISE
    SUPERSONIC FLIGHT
    TRANSONIC FLIGHT

SONIC FATIGUE
USE ACOUSTIC FATIGUE

SONIC FLOW
USE TRANSONIC FLOW

SONIC NOZZLES
RT   ACOUSTIC NOZZLES
    CONICAL NOZZLES
  ∞ NOZZLES
    SUPERSONIC NOZZLES
    TRANSONIC FLOW
    TRANSONIC NOZZLES

SONIC SOLDERING
USE ULTRASONIC SOLDERING

SONIC SPEED
USE ACOUSTIC VELOCITY

SONIC WAVEGUIDES
USE ACOUSTIC DELAY LINES

SONOBUOYS
GS   RADIO EQUIPMENT
   . RADIO TRANSMITTERS
   . . **SONOBUOYS**
   SONAR
   . **SONOBUOYS**
   TRANSMITTERS
   . RADIO TRANSMITTERS
   . . **SONOBUOYS**
RT   ANTISUBMARINE WARFARE
    HYDROPHONES
    UNDERWATER ACOUSTICS
    UNDERWATER COMMUNICATION

SONOGRAMS
RT   RECORDING INSTRUMENTS
    SOUND WAVES
    WHISTLER RECORDERS
    WHISTLERS

SONOHOLOGRAPHY
USE ACOUSTICAL HOLOGRAPHY

SONOLUMINESCENCE
GS   DECAY
   . EMISSION
   . . LIGHT EMISSION
   . . . LUMINESCENCE
   . . . . **SONOLUMINESCENCE**

SOOT
GS   PARTICLES
   . **SOOT**
RT   AIR POLLUTION
    CARBON
    COMBUSTION PRODUCTS
    FIRE DAMAGE
    SMOKE
    SMOKE ABATEMENT

SORBATES
RT   SORBENTS
      SORPTION

SORBENTS
GS   **SORBENTS**
      . ABSORBENTS
      . ADSORBENTS
RT   SORBATES
      SORPTION

SORET COEFFICIENT
GS   COEFFICIENTS
      . DIFFUSION COEFFICIENT
      . . **SORET COEFFICIENT**
      TRANSPORT PROPERTIES
      . DIFFUSION COEFFICIENT
      . . **SORET COEFFICIENT**
RT   COEFFICIENTS
      LIQUID FLOW
      THERMAL DIFFUSION

SORGHUM
GS   FARM CROPS
      . GRAINS (FOOD)
      . . **SORGHUM**
      PLANTS (BOTANY)
      . GRASSES
      . . **SORGHUM**
RT   AGRICULTURE
      CROP IDENTIFICATION
      ∞ CROPS
      EARTH RESOURCES

SORPTION
UF   CRYOSORPTION
GS   **SORPTION**
      . ADSORPTION
      . . CHEMISORPTION
RT   ∞ ABSORPTION
      CHROMATOGRAPHY
      CONCENTRATING
      EXTRACTION
      GAS CHROMATOGRAPHY
      LIQUID CHROMATOGRAPHY
      MATERIAL ABSORPTION
      PERMEATING
      ∞ SEPARATION
      SORBATES
      SORBENTS
      SURFACE PROPERTIES

SORTIE CAN
USE   SORTIE SYSTEMS

SORTIE LAB
USE   SORTIE SYSTEMS

SORTIE SYSTEMS
UF   SORTIE CAN
      SORTIE LAB
GS   PAYLOADS
      . **SORTIE SYSTEMS**
RT   SPACE LABORATORIES
      SPACE SHUTTLE PAYLOADS
      SPACE SHUTTLES
      SPACE STATIONS
      SPACELAB PAYLOADS

SORTING
USE   CLASSIFYING

SOS (SEMICONDUCTORS)
UF   SILICON-ON-SAPPHIRE JUNCTIONS
      SILICON-ON-SAPPHIRE
        SEMICONDUCTORS
      SILICON-ON-SAPPHIRE TRANSISTORS
GS   ELECTRONIC EQUIPMENT
      . SOLID STATE DEVICES
      . . SEMICONDUCTOR DEVICES
      . . . METAL OXIDE SEMICONDUCTORS
      . . . . **SOS (SEMICONDUCTORS)**
      . . . . TRANSISTORS
      . . . . SILICON TRANSISTORS
      . . . . . **SOS (SEMICONDUCTORS)**
      SEMICONDUCTORS (MATERIALS)
      . METAL OXIDE SEMICONDUCTORS
      . . **SOS (SEMICONDUCTORS)**
RT   SIS (SEMICONDUCTORS)

SOT
USE   SOLAR OPTICAL TELESCOPE

SOUND
USE   ACOUSTICS

SOUND ABSORPTION
USE   SOUND TRANSMISSION

SOUND AMPLIFICATION
GS   AMPLIFICATION
      . **SOUND AMPLIFICATION**
RT   ACOUSTIC ATTENUATION
      ACOUSTIC EXCITATION
      ACOUSTICS

SOUND BARRIER
USE   ACOUSTIC VELOCITY

SOUND DETECTING AND RANGING
RT   ACOUSTIC SCATTERING
      ATMOSPHERIC TEMPERATURE
      ∞ INSTRUMENTS
      MEASURING INSTRUMENTS
      METEOROLOGICAL INSTRUMENTS
      METEOROLOGY
      SODAR
      TEMPERATURE MEASUREMENT

SOUND DETECTORS
USE   SOUND TRANSDUCERS

SOUND FIELDS
RT   ACOUSTICS
      FIELD THEORY (PHYSICS)
      MICROSONICS

SOUND FIXING AND RANGING
UF   SFAR
      SOFAR
RT   SOUND RANGING
      SOUND TRANSMISSION
†   UNDERWATER ACOUSTICS

SOUND GENERATORS
UF   ACOUSTIC GENERATORS
RT   ACOUSTIC NOZZLES
      AUDIO FREQUENCIES
      AUDITORY STIMULI
      BELLS
      CONTINUOUS NOISE
      ∞ GENERATORS
      HORNS
      LOUDSPEAKERS
      NOISE GENERATORS
      RADIATION SOURCES
      SIGNAL GENERATORS
      ∞ SIGNALS
      SIRENS
      WARNING SYSTEMS

SOUND HOLOGRAPHY
USE   ACOUSTICAL HOLOGRAPHY

SOUND INTENSITY
GS   ACOUSTIC PROPERTIES
      . **SOUND INTENSITY**
      . . ZERO SOUND
      RATES (PER TIME)
      . FLUX DENSITY
      . . **SOUND INTENSITY**
      . . . ZERO SOUND
RT   AUDITORY STIMULI
      BIOACOUSTICS
      EFFECTIVE PERCEIVED NOISE LEVELS
      LOUDNESS
      NOISE INTENSITY
      NOISE MEASUREMENT
      RADIANT FLUX DENSITY
      SIGNAL FADING
      SIGNAL FADING RATE
      SIRENS

SOUND LOCALIZATION
GS   PERCEPTION
      . **SOUND LOCALIZATION**
RT   AUDITORY PERCEPTION
      BEARING (DIRECTION)
      BINAURAL HEARING
      DETECTION
      ECHO SOUNDING
      ∞ ORIENTATION
      POSITION INDICATORS
      RANGE FINDERS
      SONAR
      SPACE PERCEPTION
      TRACKING (POSITION)

SOUND MEASUREMENT
USE   ACOUSTIC MEASUREMENT

SOUND PERCEPTION
USE   AUDITORY PERCEPTION

SOUND PRESSURE
GS   PRESSURE
      . RADIATION PRESSURE
      . . **SOUND PRESSURE**
RT   ACOUSTIC MEASUREMENT
      ACOUSTIC VELOCITY
      EXPLOSIONS
      FLUX DENSITY
      LOUDNESS
      NOISE (SOUND)
      SHOCK WAVES
      STATIC PRESSURE

SOUND PROPAGATION
GS   **SOUND PROPAGATION**
      . VOICE
RT   ACOUSTIC PROPAGATION
      ACOUSTICS
      ATTENUATION
      ∞ CONDUCTION
      DIFFUSION
      NOISE GENERATORS
      NOISE PROPAGATION
      SHOCK WAVE PROPAGATION

SOUND RANGING
GS   RANGEFINDING
      . **SOUND RANGING**
RT   DETECTION
      DISTANCE MEASURING EQUIPMENT
      ECHO SOUNDING
      POSITION (LOCATION)
      SONAR
      SOUND FIXING AND RANGING
      TARGET ACQUISITION
      TRACKING (POSITION)

SOUND TRANSDUCERS
UF   SOUND DETECTORS
GS   TRANSDUCERS
      . **SOUND TRANSDUCERS**
      . . ELECTROACOUSTIC TRANSDUCERS
      . . . HYDROPHONES
      . . . LOUDSPEAKERS
      . . . MICROPHONES
RT   ∞ RADIATORS
      SIGNAL DETECTION
      SIGNAL DETECTORS
      UNDERWATER ACOUSTICS
      UNDERWATER COMMUNICATION

SOUND TRANSMISSION
UF   SOUND ABSORPTION
GS   TRANSMISSION
      . **SOUND TRANSMISSION**
RT   ∞ ABSORPTION
      ACOUSTICS
      ATTENUATION
      AUDIO FREQUENCIES
      ∞ CONDUCTION
      EARPHONES
      ELASTIC WAVES
      ENERGY ABSORPTION
      MONAURAL SIGNALS
      MULTIPATH TRANSMISSION
      ∞ PATHS
      SIGNAL TRANSMISSION
      SIRENS
      SOUND FIXING AND RANGING
      TELEPHONY
      THERMOCLINES
      WAVE PROPAGATION

SOUND VELOCITY
USE   ACOUSTIC VELOCITY

SOUND WAVES
SN   (ELASTIC WAVES IN THE AUDIBLE
      RANGE)
UF   ACOUSTIC RADIATION
      ACOUSTIC VIBRATIONS
GS   ELASTIC WAVES
      . **SOUND WAVES**
      . . ELECTROACOUSTIC WAVES
      . . ION ACOUSTIC WAVES
      . . LAMB WAVES
      . . NOISE (SOUND)
      . . . AERODYNAMIC NOISE
      . . . . AIRCRAFT NOISE

SOUND WAVES-*(CONT.)*
```
. . . . JET AIRCRAFT NOISE
. . . . SONIC BOOMS
. . . ENGINE NOISE
. . . . ROCKET ENGINE NOISE
. . . THERMAL NOISE
```
RT   ACOUSTIC MEASUREMENT
```
      ACOUSTIC PROPERTIES
      ACOUSTIC STREAMING
      ACOUSTICAL HOLOGRAPHY
      ACOUSTICS
      AEOLIAN TONES
      AUDIO FREQUENCIES
      AUDITORY PERCEPTION
   ∞ BLASTS
      DEEP SCATTERING LAYERS
      DETONATION WAVES
      DIFFUSION
      LONGITUDINAL WAVES
      LOUDNESS
      MACH CONES
      MAGNETOELASTIC WAVES
      MICROSONICS
      NOISE POLLUTION
      NOISE PREDICTION (AIRCRAFT)
      PHONONS
      PLANE WAVES
      POLARIZED ELASTIC WAVES
   ∞ RADIATION
      REVERBERATION
      SHOCK WAVES
      SONOGRAMS
      SURFACE ACOUSTIC WAVE DEVICES
      ULTRASONIC RADIATION
   ∞ WAVES
```

SOUND-SOUND INTERACTIONS
RT   HARMONICS
```
   ∞ INTERACTIONS
      INTERMODULATION
      WAVE DISPERSION
```

SOUNDERS
USE   SOUNDING

SOUNDING
UF   SOUNDERS
GS   **SOUNDING**
```
   . ACOUSTIC SOUNDING
   . ATMOSPHERIC SOUNDING
   . BALLOON SOUNDING
   . ECHO SOUNDING
   . IONOSPHERIC SOUNDING
   . MICROWAVE SOUNDING
   . ROCKET SOUNDING
   . SATELLITE SOUNDING
```
RT   BATHYMETERS
```
      DEPTH MEASUREMENT
   ∞ MEASUREMENT
      METEOROLOGICAL BALLOONS
      METEOROLOGICAL FLIGHT
      ROBIN BALLOONS
      SONDES
```

SOUNDING ROCKETS
UF   METEOROLOGICAL ROCKETS
```
      ROCKET SONDES
```
GS   ROCKET VEHICLES
```
   . SOUNDING ROCKETS
   . . AEROBEE ROCKET VEHICLE
   . . ANTARES ROCKET VEHICLE
   . . APACHE ROCKET VEHICLE
   . . ARCAS ROCKET VEHICLES
   . . ARIES SOUNDING ROCKET
   . . ASP ROCKET VEHICLE
   . . ASTROBEE ROCKET VEHICLES
   . . . ASTROBEE 1500 ROCKET VEHICLE
   . . BLACK BRANT SOUNDING ROCKETS
   . . . BLACK BRANT 1 SOUNDING
           ROCKET
   . . . BLACK BRANT 2 SOUNDING
           ROCKET
   . . . BLACK BRANT 3 SOUNDING
           ROCKET
   . . . BLACK BRANT 4 SOUNDING
           ROCKET
   . . . BLACK BRANT 5 SOUNDING
           ROCKET
   . . CAJUN ROCKET VEHICLE
   . . DORNIER PARAGLIDER ROCKET
         VEHICLE
   . . EXOS SOUNDING ROCKET
   . . JAGUAR ROCKET VEHICLE
   . . JUDI-DART ROCKET
   . . KAPPA ROCKET VEHICLES
```

SOUNDING ROCKETS-*(CONT.)*
```
   . . . KAPPA 8 ROCKET VEHICLE
   . . . KAPPA 9 ROCKET VEHICLE
   . . LAMBDA ROCKET VEHICLES
   . . LOKI ROCKET VEHICLE
   . . PETREL SOUNDING ROCKET
   . . PHOENIX SOUNDING ROCKET
   . . SKUA ROCKET VEHICLES
   . . SKYLARK ROCKET VEHICLE
   . . VENUS FLY TRAP ROCKET VEHICLE
   . . VERONIQUE ROCKET VEHICLES
   . . VERTICAL 8 ROCKET
   . . WASP SOUNDING ROCKET
```
RT   ACOUSTIC SOUNDING
```
      ARGO ROCKET VEHICLES
      IONOSONDES
      JAVELIN ROCKET VEHICLE
      METEOROLOGICAL INSTRUMENTS
      METEOROLOGICAL SATELLITES
      NIKE-JAVELIN ROCKET VEHICLE
      PAYLOAD CONTROL
      RADIOSONDES
      ROCKET SOUNDING
      SONDES
      VIKING ROCKET VEHICLE
```

SOUNDS (TOPOGRAPHIC FEATURES)
GS   **SOUNDS (TOPOGRAPHIC FEATURES)**
```
   . BLOCK ISLAND SOUND (RI)
   . PRINCE WILLIAM SOUND (AK)
```
RT   CHESAPEAKE BAY (US)
```
      INLETS (TOPOGRAPHY)
      OCEANS
      RIVERS
      WATER
```

SOURCE PROGRAMS
GS   COMPUTER PROGRAMS
```
   . SOURCE PROGRAMS
      PROGRAMS
   . SOURCE PROGRAMS
```

∞ **SOURCES**
SN   *(USE OF A MORE SPECIFIC TERM IS*
```
      RECOMMENDED--CONSULT THE TERMS
      LISTED BELOW)
```
RT   CAUSES
```
      DERIVATION
      ELECTRON SOURCES
      EXTRAGALACTIC RADIO SOURCES
      ION SOURCES
      NONPOINT SOURCES
      RADIATION SOURCES
      RADIO SOURCES (ASTRONOMY)
      SINKS
```

SOUTH AFRICA
USE   REPUBLIC OF SOUTH AFRICA

SOUTH AMERICA
GS   CONTINENTS
```
   . SOUTH AMERICA
```
RT   ANDES MOUNTAINS (SOUTH AMERICA)
```
      ARGENTINA
      BOLIVIA
      BRAZIL
      CENTRAL AMERICA
      CHILE
      COLOMBIA
      ECUADOR
      FRENCH GUIANA
      GUYANA
      MAGDALENA-CAUCA VALLEY (COLOMBIA)
      PARAGUAY
      PERU
      SURINAM
      TRINIDAD AND TOBAGO
      URUGUAY
      VENEZUELA
```

SOUTH CAROLINA
GS   NATIONS
```
   . UNITED STATES
   . . SOUTH CAROLINA
```
RT   SAND HILLS REGION (GA-NC-SC)

SOUTH DAKOTA
GS   NATIONS
```
   . UNITED STATES
   . . SOUTH DAKOTA
```
RT   BLACK HILLS (SD-WY)
```
      MISSOURI RIVER (US)
```

SOUTH KOREA
UF   REPUBLIC OF KOREA

SOUTH KOREA-*(CONT.)*
GS   NATIONS
```
   . SOUTH KOREA
```
RT   ASIA
```
   ∞ KOREA
      NORTH KOREA
```

SOUTH VIETNAM
USE   VIETNAM

SOUTH WEST AFRICA
USE   NAMIBIA

SOUTHEAST ASIA
GS   REGIONS
```
   . SOUTHEAST ASIA
```
RT   ASIA
```
      VIETNAM
```

SOUTHERN CALIFORNIA
GS   REGIONS
```
   . SOUTHERN CALIFORNIA
```
RT   CALIFORNIA
```
      MEXICO
      NEVADA
      PACIFIC OCEAN
      UNITED STATES
```

SOUTHERN HEMISPHERE
GS   **SOUTHERN HEMISPHERE**
```
   . ANTARCTIC REGIONS
   . . MCMURDO SOUND
   . . ROSS ICE SHELF
```
RT   ∞ HEMISPHERES
```
      NORTHERN HEMISPHERE
      SOUTHERN SKY
```

SOUTHERN SKY
RT   ASTRONOMICAL CATALOGS
```
      ASTRONOMICAL OBSERVATORIES
      ASTRONOMICAL PHOTOGRAPHY
      ASTRONOMICAL SPECTROSCOPY
      ASTRONOMY
      NORTHERN SKY
      SOUTHERN HEMISPHERE
```

SOUTHERN YEMEN
UF   ADEN
GS   NATIONS
```
   . SOUTHERN YEMEN
```
RT   ASIA

SOVEREIGNTY
RT   INTERNATIONAL COOPERATION
```
      INTERNATIONAL LAW
      POLITICS
      VOTING
```

SOVIET SATELLITES
GS   SATELLITES
```
   . ARTIFICIAL SATELLITES
   . . SOVIET SATELLITES
   . . . COSMOS SATELLITES
   . . . . COSMOS 2 SATELLITE
   . . . . COSMOS 3 SATELLITE
   . . . . COSMOS 5 SATELLITE
   . . . . COSMOS 6 SATELLITE
   . . . . COSMOS 14 SATELLITE
   . . . . COSMOS 44 SATELLITE
   . . . . COSMOS 54 SATELLITE
   . . . . COSMOS 71 SATELLITE
   . . . . COSMOS 110 SATELLITE
   . . . . COSMOS 137 SATELLITE
   . . . . COSMOS 144 SATELLITE
   . . . . COSMOS 149 SATELLITE
   . . . . COSMOS 166 SATELLITE
   . . . . COSMOS 186 SATELLITE
   . . . . COSMOS 188 SATELLITE
   . . . . COSMOS 206 SATELLITE
   . . . . COSMOS 213 SATELLITE
   . . . . COSMOS 224 SATELLITE
   . . . . COSMOS 225 SATELLITE
   . . . . COSMOS 381 SATELLITE
   . . . . COSMOS 954 SATELLITE
   . . . . COSMOS 1129 SATELLITE
   . . . . INTERCOSMOS SATELLITES
   . . . COSMOS 782 SATELLITE
   . . . COSMOS 936 SATELLITE
   . . . MOLNIYA SATELLITES
   . . . PROGNOZ SATELLITES
   . . . PROTON SATELLITES
   . . . . PROTON 1 SATELLITE
   . . . . PROTON 2 SATELLITE
```

SOVIET SATELLITES-*(CONT.)*
```
. . . . PROTON 3 SATELLITE
. . . . PROTON 4 SATELLITE
. . . RADUGA SATELLITE
. . . SPUTNIK SATELLITES
. . . . SPUTNIK 1 SATELLITE
. . . . SPUTNIK 2 SATELLITE
. . . . SPUTNIK 3 SATELLITE
. . . . SPUTNIK 4 SATELLITE
. . . . SPUTNIK 5 SATELLITE
. . . VENERA SATELLITES
. . . . VENERA 9 SATELLITE
. . . . VENERA 10 SATELLITE
. . . VENERA 11 SATELLITE
. . . . VENERA 12 SATELLITE
```

SOVIET SPACECRAFT
```
GS   SOVIET SPACECRAFT
. LUNIK LUNAR PROBES
. . LUNIK 2 LUNAR PROBE
. . LUNIK 3 LUNAR PROBE
. . LUNIK 9 LUNAR PROBE
. . LUNIK 10 LUNAR PROBE
. . LUNIK 11 LUNAR PROBE
. . LUNIK 12 LUNAR PROBE
. . LUNIK 13 LUNAR PROBE
. . LUNIK 14 LUNAR PROBE
. . LUNIK 16 LUNAR PROBE
. . LUNIK 17 LUNAR PROBE
. . LUNIK 19 LUNAR PROBE
. . LUNIK 20 LUNAR PROBE
. . LUNIK 22 LUNAR PROBE
. MARS 1 SPACECRAFT
. MARS 2 SPACECRAFT
. MARS 3 SPACECRAFT
. MARS 4 SPACECRAFT
. MARS 5 SPACECRAFT
. MARS 6 SPACECRAFT
. MARS 7 SPACECRAFT
. SALYUT SPACE STATION
. SOYUZ SPACECRAFT
. ZOND SPACE PROBES
. . ZOND 1 SPACE PROBE
. . ZOND 2 SPACE PROBE
. . ZOND 3 SPACE PROBE
. . ZOND 4 SPACE PROBE
. . ZOND 5 SPACE PROBE
. . ZOND 6 SPACE PROBE
. . ZOND 7 SPACE PROBE
. . ZOND 8 SPACE PROBE
RT   ∞SPACECRAFT
```

SOVIET UNION
```
USE   U.S.S.R.
```

SOYBEANS
```
RT   ∞FOOD
```

SOYUZ SPACECRAFT
```
GS   MANNED SPACECRAFT
. SOYUZ SPACECRAFT
SOVIET SPACECRAFT
. SOYUZ SPACECRAFT
RT   APOLLO SOYUZ TEST PROJECT
SALYUT SPACE STATION
U.S.S.R. SPACE PROGRAM
```

∞ SPACE
```
SN   (USE OF A MORE SPECIFIC TERM IS
     RECOMMENDED--CONSULT THE TERMS
     LISTED BELOW)
RT   ALGEBRA
     ANALYSIS (MATHEMATICS)
     CARTAN SPACE
     CISLUNAR SPACE
     DEEP SPACE
     FRACTALS
     FUNCTION SPACE
     HYPERSPACES
     SET THEORY
```

SPACE ADAPTATION SYNDROME
```
RT   AEROSPACE MEDICINE
     BIOASTRONAUTICS
     BIOLOGICAL EFFECTS
     LONG DURATION SPACE FLIGHT
     MANNED SPACE FLIGHT
     MOTION SICKNESS
     PHYSIOLOGICAL EFFECTS
```

SPACE ADAPTATION SYNDROME-*(CONT.)*
```
PSYCHOLOGICAL EFFECTS
SPACE FLIGHT STRESS
SPACE PSYCHOLOGY
WEIGHTLESSNESS
```

SPACE ARROW SATELLITE
```
USE   COSMOS 149 SATELLITE
```

SPACE BASE COMMAND CENTER
```
GS   MANNED SPACECRAFT
. SPACE STATIONS
. . SPACE BASE COMMAND CENTER
STATIONS
. SPACE STATIONS
. . SPACE BASE COMMAND CENTER
```

SPACE BASED RADAR
```
GS   RADAR
. SPACE BASED RADAR
RT   ANTENNA ARRAYS
     RADIATION EFFECTS
     RADIATION SHIELDING
```

SPACE BASES
```
GS   SPACE BASES
. SPACE COLONIES
RT   ∞BASES
     ORBITAL SPACE STATIONS
     SALYUT SPACE STATION
     STATIONS
```

SPACE BIOLOGY
```
USE   EXOBIOLOGY
```

SPACE BUSES
```
USE   FERRY SPACECRAFT
```

SPACE CAPSULES
```
UF   CAPSULES (SPACECRAFT)
GS   SPACE CAPSULES
. DISCOVERER RECOVERY CAPSULES
. ESCAPE CAPSULES
. MERCURY SPACECRAFT
. . AURORA 7
. . FAITH 7
. . FRIENDSHIP 7
. . SIGMA 7
RT   ARTIFICIAL SATELLITES
     BIOSATELLITES
     CABIN ATMOSPHERES
     ∞CAPSULES
     COCKPITS
     GEMINI SPACECRAFT
     INTERPLANETARY SPACECRAFT
     LANDING MODULES
     LUNAR SPACECRAFT
     MANNED SPACECRAFT
     MERCURY FLIGHTS
     RECOVERABLE SPACECRAFT
     REENTRY VEHICLES
     RENDEZVOUS SPACECRAFT
     SATELLITES
     SOFT LANDING SPACECRAFT
     ∞SPACECRAFT
     SPACECRAFT CABINS
     SPACECRAFT MODULES
     UNMANNED SPACECRAFT
     VOSKHOD MANNED SPACECRAFT
     VOSTOK SPACECRAFT
```

SPACE CHARGE
```
GS   ELECTRIC CHARGE
. SPACE CHARGE
RT   BUNCHING
     CHILD-LANGMUIR LAW
     ELECTRIC DISCHARGES
     ELECTRON CLOUDS
     LANDAU DAMPING
     MAGNETOHYDRODYNAMICS
     NONOHMIC EFFECT
     ORBITRONS
     PERVEANCE
     PLASMAS (PHYSICS)
     STATIC ELECTRICITY
```

SPACE COLONIES
```
GS   SPACE BASES
. SPACE COLONIES
RT   LUNAR BASES
     LUNAR SHELTERS
     ORBITAL SPACE STATIONS
     SPACE HABITATS
```

SPACE COMMERCIALIZATION
```
RT   AEROSPACE INDUSTRY
     COMMERCE LAB
     COMMERCIAL SPACECRAFT
     COMMUNICATION SATELLITES
     ∞MICROGRAVITY APPLICATIONS
     SPACE INDUSTRIALIZATION
     SPACE MANUFACTURING
     SPACE PROCESSING
     SPACECRAFT LAUNCHING
     TECHNOLOGY TRANSFER
```

SPACE COMMUNICATION
```
GS   TELECOMMUNICATION
. SPACE COMMUNICATION
. . EXTRATERRESTRIAL
      COMMUNICATION
. . INTERPLANETARY COMMUNICATION
. . LUNAR COMMUNICATION
. . . CIRCUMLUNAR COMMUNICATION
. . SPACECRAFT COMMUNICATION
. . . REENTRY COMMUNICATION
RT   COMMUNICATION SATELLITES
     DEFENSE COMMUNICATIONS SATELLITE
       SYSTEM
     EXTRATERRESTRIAL INTELLIGENCE
     FURLABLE ANTENNAS
     INTERSTELLAR COMMUNICATION
     LASERS
     LINE OF SIGHT COMMUNICATION
     MANNED SPACE FLIGHT
     OPTICAL COMMUNICATION
     PULSE COMMUNICATION
     RADIO COMMUNICATION
     RADIO TELEMETRY
     TELEVISION SYSTEMS
     WIRELESS COMMUNICATION
```

SPACE COOLING (BUILDINGS)
```
GS   COOLING
. SPACE COOLING (BUILDINGS)
RT   COOLING SYSTEMS
     ENERGY TECHNOLOGY
     HEAT EXCHANGERS
     HEAT PUMPS
     LIQUID COOLING
     RESIDENTIAL ENERGY
     SOLAR COLLECTORS
     SOLAR COOLING
     SOLAR ENERGY CONVERSION
     TEMPERATURE CONTROL
```

SPACE DEBRIS
```
GS   DEBRIS
. SPACE DEBRIS
RT   ASTEROID BELTS
     ASTEROIDS
     CHIRON
     COSMIC DUST
     DUST
     METEOROIDS
     MICROMETEOROIDS
     ∞SPACECRAFT
     SPACECRAFT DESIGN
     TORO ASTEROID
     VESTA ASTEROID
```

SPACE DENSITY
```
GS   DENSITY (MASS/VOLUME)
. SPACE DENSITY
DENSITY (NUMBER/VOLUME)
. SPACE DENSITY
RT   ATMOSPHERIC DENSITY
     ELECTRON DENSITY (CONCENTRATION)
     ION DENSITY (CONCENTRATION)
     PARTICLE DENSITY (CONCENTRATION)
     PLASMA DENSITY
     PLASMA INTERACTION EXPERIMENT
     PROTON DENSITY (CONCENTRATION)
```

SPACE DETECTION AND TRACKING SYSTEM
```
UF   SPADATS (TRACKING SYSTEM)
GS   STATIONS
. GROUND STATIONS
. . SPACE DETECTION AND TRACKING
      SYSTEM
. TRACKING STATIONS
. . SPACE DETECTION AND TRACKING
      SYSTEM
TRACKING (POSITION)
. SPACE DETECTION AND TRACKING
    SYSTEM
TRACKING NETWORKS
. SPACE DETECTION AND TRACKING
    SYSTEM
```

SPACE DETECTION AND TRACKING-*(CONT.)*
RT   MINITRACK SYSTEM
      MISSILE TRACKING
      OPTICAL TRACKING
      PHOTOGRAPHIC TRACKING
      POLYSTATION DOPPLER TRACKING
        SYSTEM
      SPACECRAFT TRACKING
      STDN (NETWORK)
      ∞SYSTEMS

SPACE DIVERSITY
USE   RECEPTION DIVERSITY

SPACE ELECTRIC ROCKET TESTS
UF   SERT (ROCKET TESTS)
GS   ENGINE TESTS
      . **SPACE ELECTRIC ROCKET TESTS**
RT   ELECTRIC ROCKET ENGINES
      FLIGHT TESTS
      GROUND TESTS
      SERT 1 SPACECRAFT
      SERT 2 SPACECRAFT
      ∞TESTS

SPACE ENVIRONMENT
USE   AEROSPACE ENVIRONMENTS

SPACE ENVIRONMENT SIMULATION
GS   SIMULATION
      . ENVIRONMENT SIMULATION
      . . **SPACE ENVIRONMENT SIMULATION**
      . . . WEIGHTLESSNESS SIMULATION
RT   ALTITUDE SIMULATION
      ATMOSPHERIC ENTRY SIMULATION
      FLIGHT SIMULATION
      FLIGHT SIMULATORS
      HIGH VACUUM ORBITAL SIMULATOR
      LANGLEY COMPLEX COORDINATOR
      MOTION SIMULATORS
      SOLAR SIMULATION
      THERMAL SIMULATION
      VACUUM CHAMBERS

SPACE ENVIRONMENTAL LUBRICATION
USE   SPACECRAFT LUBRICATION

SPACE ERECTABLE STRUCTURES
GS   **SPACE ERECTABLE STRUCTURES**
      . BEACON EXPLORER A
      . INFLATABLE SPACECRAFT
      . . BEACON SATELLITES
      . . . EXPLORER 22 SATELLITE
RT   EXPANDABLE STRUCTURES
      FOLDING STRUCTURES
      INFLATABLE STRUCTURES
      LARGE SPACE STRUCTURES
      MAYPOLE ANTENNAS
      ORBITAL ASSEMBLY
      RIGID STRUCTURES
      SELF ERECTING DEVICES
      SPACE TECHNOLOGY EXPERIMENTS
      SPACECRAFT MODULES
      SPACECRAFT STRUCTURES
      ∞STRUCTURES

SPACE EXPER WITH PARTICLE ACCELERATORS
USE   SEPAC (PAYLOAD)

SPACE EXPLORATION
UF   PLANETARY EXPLORATION
GS   EXPLORATION
      . **SPACE EXPLORATION**
RT   AEROSPACE ENVIRONMENTS
      ASTEROID MISSIONS
      ASTRODYNAMICS
      ∞ASTRONAUTICS
      BIOASTRONAUTICS
      EXTRATERRESTRIAL ENVIRONMENTS
      EXTRATERRESTRIAL RESOURCES
      FRENCH SPACE PROGRAMS
      GULLIVER PROGRAM
      INTERPLANETARY FLIGHT
      INTERPLANETARY SPACECRAFT
      INTERSTELLAR SPACECRAFT
      JUPITER RINGS
      LUNAR EXPLORATION
      MANNED SPACE FLIGHT
      MARS 69 PROJECT
      MARS 71 PROJECT
      PLANETARY BASES
      PLANETARY COMPOSITION
      SOLAR MAXIMUM MISSION-A
      TOPS (SPACECRAFT)
      VIKING LANDER SPACECRAFT

SPACE EXPLORATION-*(CONT.)*
      VIKING LANDER 1
      VIKING LANDER 2
      VIKING ORBITER SPACECRAFT
      VIKING ORBITER 1
      VIKING ORBITER 2
      VIKING 1 SPACECRAFT
      VIKING 2 SPACECRAFT

SPACE FLIGHT
GS   **SPACE FLIGHT**
      . HYPERBOLIC REENTRY
      . HYPERSONIC REENTRY
      . INTERPLANETARY FLIGHT
      . . LONG DURATION SPACE FLIGHT
      . INTERSTELLAR TRAVEL
      . LUNAR FLIGHT
      . MANNED SPACE FLIGHT
      . . APOLLO FLIGHTS
      . . . APOLLO 5 FLIGHT
      . . . APOLLO 6 FLIGHT
      . . . APOLLO 7 FLIGHT
      . . . APOLLO 8 FLIGHT
      . . . APOLLO 9 FLIGHT
      . . . APOLLO 10 FLIGHT
      . . . APOLLO 11 FLIGHT
      . . . APOLLO 12 FLIGHT
      . . . APOLLO 13 FLIGHT
      . . . APOLLO 14 FLIGHT
      . . . APOLLO 15 FLIGHT
      . . . APOLLO 16 FLIGHT
      . . . APOLLO 17 FLIGHT
      . . GEMINI FLIGHTS
      . . . GEMINI 3 FLIGHT
      . . . GEMINI 4 FLIGHT
      . . . GEMINI 5 FLIGHT
      . . . GEMINI 6 FLIGHT
      . . . GEMINI 7 FLIGHT
      . . . GEMINI 8 FLIGHT
      . . . GEMINI 9 FLIGHT
      . . . GEMINI 10 FLIGHT
      . . . GEMINI 11 FLIGHT
      . . . GEMINI 12 FLIGHT
      . . MANNED REENTRY
      . . MERCURY FLIGHTS
      . . . MERCURY MA-1 FLIGHT
      . . . MERCURY MA-2 FLIGHT
      . . . MERCURY MA-3 FLIGHT
      . . . MERCURY MA-4 FLIGHT
      . . . MERCURY MA-5 FLIGHT
      . . . MERCURY MA-6 FLIGHT
      . . . MERCURY MA-7 FLIGHT
      . . . MERCURY MA-8 FLIGHT
      . . . MERCURY MA-9 FLIGHT
      . . . MERCURY MR-1 FLIGHT
      . . . MERCURY MR-2 FLIGHT
      . . . MERCURY MR-3 FLIGHT
      . . . MERCURY MR-4 FLIGHT
      . . SPACE SHUTTLE MISSION 31-A
      . . SPACE SHUTTLE MISSION 31-B
      . . SPACE SHUTTLE MISSION 31-C
      . . SPACE SHUTTLE MISSION 31-D
      . . SPACE SHUTTLE MISSION 41-A
      . . SPACE SHUTTLE MISSION 41-B
      . . SPACE SHUTTLE MISSION 41-C
      . . SPACE SHUTTLE MISSION 41-D
      . . SPACE SHUTTLE MISSION 41-G
      . . SPACE SHUTTLE MISSION 51-A
      . . SPACE SHUTTLE MISSION 51-B
      . . SPACE SHUTTLE MISSION 51-C
      . . SPACE SHUTTLE MISSION 51-D
      . . SPACE SHUTTLE MISSION 51-E
      . . SPACE SHUTTLE MISSION 51-F
      . . SPACE SHUTTLE MISSION 51-G
      . . SPACE SHUTTLE MISSION 51-H
      . . SPACE SHUTTLE MISSION 51-I
      . . SPACE SHUTTLE MISSION 51-J
      . . SPACE SHUTTLE MISSION 51-L
      . . SPACE SHUTTLE MISSION 61-A
      . . SPACE SHUTTLE MISSION 61-B
      . . SPACE SHUTTLE MISSION 61-C
      . . SPACE SHUTTLE MISSION 61-E
      . RETURN TO EARTH SPACE FLIGHT
      . SPACE SHUTTLE MISSIONS
      . . SPACE SHUTTLE MISSION 31-A
      . . SPACE SHUTTLE MISSION 31-B
      . . SPACE SHUTTLE MISSION 31-C
      . . SPACE SHUTTLE MISSION 31-D
      . . SPACE SHUTTLE MISSION 41-A
      . . SPACE SHUTTLE MISSION 41-B
      . . SPACE SHUTTLE MISSION 41-C
      . . SPACE SHUTTLE MISSION 41-D
      . . SPACE SHUTTLE MISSION 41-G
      . . SPACE SHUTTLE MISSION 51-A
      . . SPACE SHUTTLE MISSION 51-B
      . . SPACE SHUTTLE MISSION 51-C

SPACE FLIGHT-*(CONT.)*
      . . SPACE SHUTTLE MISSION 51-D
      . . SPACE SHUTTLE MISSION 51-E
      . . SPACE SHUTTLE MISSION 51-F
      . . SPACE SHUTTLE MISSION 51-G
      . . SPACE SHUTTLE MISSION 51-H
      . . SPACE SHUTTLE MISSION 51-I
      . . SPACE SHUTTLE MISSION 51-J
      . . SPACE SHUTTLE MISSION 51-L
      . . SPACE SHUTTLE MISSION 61-A
      . . SPACE SHUTTLE MISSION 61-B
      . . SPACE SHUTTLE MISSION 61-C
      . . SPACE SHUTTLE MISSION 61-E
      . SPACECRAFT REENTRY
      . VIKING MARS PROGRAM
RT   AEROSPACE ENVIRONMENTS
      APOLLO SOYUZ TEST PROJECT
      ASCENT PROPULSION SYSTEMS
      ASTRODYNAMICS
      ∞ASTRONAUTICS
      ATMOSPHERIC ENTRY
      AUXILIARY PROPULSION
      BIOASTRONAUTICS
      CELESTIAL BODIES
      EXPEDITIONS
      EXPLORATION
      EXTRAVEHICULAR ACTIVITY
      ∞FLIGHT
      FLIGHT MECHANICS
      FLIGHT OPTIMIZATION
      FLIGHT SIMULATION
      FLYBY MISSIONS
      GRAND TOURS
      MARINER JUPITER-SATURN FLYBY
      MARINER JUPITER-URANUS FLYBY
      METEOROLOGICAL FLIGHT
      ∞MISSIONS
      ORBITS
      PHYSICS AND CHEMISTRY EXPERIMENT
        IN SPACE
      POINTING CONTROL SYSTEMS
      PROPULSION
      REENTRY
      ROCKET FLIGHT
      ∞ROCKETS
      SOLAR SAILS
      SPACE TRANSPORTATION SYSTEM
        FLIGHTS
      SPACECRAFT GUIDANCE
      SPACECRAFT MANEUVERS
      SPACECRAFT PROPULSION
      SUBORBITAL FLIGHT
      TRAJECTORIES
      VIKING LANDER SPACECRAFT
      VIKING LANDER 1
      VIKING LANDER 2
      VIKING ORBITER SPACECRAFT
      VIKING ORBITER 1
      VIKING ORBITER 2
      VIKING 1 SPACECRAFT
      VIKING 2 SPACECRAFT

SPACE FLIGHT FEEDING
RT   CONSUMABLES (SPACECREW SUPPLIES)
      DEHYDRATED FOOD
      DIETS
      EATING
      ∞FOOD
      FOOD INTAKE
      LIFE SUPPORT SYSTEMS
      NUTRITION
      NUTRITIONAL REQUIREMENTS
      WASTE DISPOSAL

SPACE FLIGHT STRESS
GS   FLIGHT STRESS (BIOLOGY)
      . **SPACE FLIGHT STRESS**
RT   BOREDOM
      ∞FLIGHT STRESS
      GRAVITATIONAL PHYSIOLOGY
      LOWER BODY NEGATIVE PRESSURE
      MANNED SPACE FLIGHT
      SPACE ADAPTATION SYNDROME
      SPACE PSYCHOLOGY
      STRESS (PHYSIOLOGY)
      STRESS (PSYCHOLOGY)
      WEIGHTLESSNESS

SPACE FLIGHT TRACKING AND DATA NETWORK
GS   TRACKING NETWORKS
      . **SPACE FLIGHT TRACKING AND DATA
        NETWORK**
RT   ∞DATA
      DATA ACQUISITION
      GLOBAL TRACKING NETWORK

SPACE FLIGHT TRACKING AND DATA-*(CONT.)*
  GROUND STATIONS
  SATELLITE TRACKING
  STATIONS
  STDN (NETWORK)
  TRACKING STATIONS

SPACE FLIGHT TRAINING
GS EDUCATION
  . FLIGHT TRAINING
  . . **SPACE FLIGHT TRAINING**
RT ASTRONAUT TRAINING
  PILOT TRAINING
  SPACECRAFT CABIN SIMULATORS
  TRAINING SIMULATORS

SPACE GLIDERS
USE LIFTING REENTRY VEHICLES

SPACE GLOSSARIES
RT BIBLIOGRAPHIES
  DICTIONARIES
  DOCUMENTATION
  INDEXES (DOCUMENTATION)
  INFORMATION RETRIEVAL
  THESAURI

SPACE HABITATS
RT AEROSPACE ENVIRONMENTS
  CLOSED ECOLOGICAL SYSTEMS
  LIFE SUPPORT SYSTEMS
  SPACE COLONIES
  SPACE STATIONS
  SPACECREWS

SPACE HEATING (BUILDINGS)
GS HEATING
  . **SPACE HEATING (BUILDINGS)**
RT AIR CONDITIONING
  ENVIRONMENTAL ENGINEERING
  HEATING EQUIPMENT
  RESIDENTIAL ENERGY
  SOLAR ATRIUMS
  SOLAR HEATING
  SOLAR HOUSES
  TEMPERATURE CONTROL
  WASTE ENERGY UTILIZATION

SPACE INDUSTRIALIZATION
GS **SPACE INDUSTRIALIZATION**
  . SPACE MANUFACTURING
  . SPACE PROCESSING
RT COMMERCIAL SPACECRAFT
  ECONOMIC DEVELOPMENT
  ENERGY CONVERSION
  INDUSTRIES
  MANUFACTURING
  ∞PROCESSES
  PRODUCT DEVELOPMENT
  PRODUCTS
  RESEARCH FACILITIES
  SPACE COMMERCIALIZATION

SPACE INFRARED TELESCOPE FACILITY
GS OBSERVATORIES
  . ASTRONOMICAL OBSERVATORIES
  . ASTRONOMICAL SATELLITES
  . . . **SPACE INFRARED TELESCOPE FACILITY**
  TELESCOPES
  . ASTRONOMICAL TELESCOPES
  . . INFRARED TELESCOPES
  . . . **SPACE INFRARED TELESCOPE FACILITY**
  . SPACEBORNE TELESCOPES
  . . **SPACE INFRARED TELESCOPE FACILITY**
RT INFRARED ASTRONOMY
  INFRARED TELESCOPES

SPACE LABORATORIES
GS LABORATORIES
  . **SPACE LABORATORIES**
  . . ATMOSPHERIC CLOUD PHYSICS LAB (SPACELAB)
  . . EARTH VIEWING APPLICATIONS LABORATORY
  . . MANNED ORBITAL LABORATORIES
  . . . MANNED ORBITAL RESEARCH LABORATORIES
RT ∞AEROSPACE SCIENCES
  ARTIFICIAL SATELLITES
  GEOPHYSICAL SATELLITES
  LONG DURATION EXPOSURE FACILITY
  RESEARCH FACILITIES

SPACE LABORATORIES-*(CONT.)*
  RESEARCH VEHICLES
  SAIL PROJECT
  SALYUT SPACE STATION
  SORTIE SYSTEMS
  SPACEBORNE EXPERIMENTS
  ∞SPACECRAFT

SPACE LAW
GS LAW (JURISPRUDENCE)
  . INTERNATIONAL LAW
  . . **SPACE LAW**
RT AIR LAW
  OUTER SPACE TREATY
  SABOTAGE

SPACE LOGISTICS
GS CONSUMABLES (SPACECRAFT)
  . **SPACE LOGISTICS**
  LOGISTICS
  . **SPACE LOGISTICS**
RT CONSUMABLES (SPACECREW SUPPLIES)
  EXTRATERRESTRIAL RESOURCES
  MANNED SPACE FLIGHT
  SPACECRAFT CABIN SIMULATORS
  SPACECREW TRANSFER
  STOWAGE (ONBOARD EQUIPMENT)

SPACE MAINTENANCE
GS MAINTENANCE
  . **SPACE MAINTENANCE**
RT ASTRONAUT TRAINING
  ∞ASTRONAUTICS
  EXTRAVEHICULAR ACTIVITY
  ORBITAL WORKERS
  PAYLOAD TRANSFER
  REMOTE MANIPULATOR SYSTEM

SPACE MANUFACTURING
GS FABRICATION
  . **SPACE MANUFACTURING**
  MANUFACTURING
  . **SPACE MANUFACTURING**
  SPACE INDUSTRIALIZATION
  . **SPACE MANUFACTURING**
RT AEROSPACE ENVIRONMENTS
  ASSEMBLING
  COMMERCIAL SPACECRAFT
  CONSTRUCTION
  HIGH VACUUM
  INDUSTRIES
  LEVITATION MELTING
  LOW GRAVITY MANUFACTURING
  ∞MICROGRAVITY APPLICATIONS
  SPACE COMMERCIALIZATION
  SPACE PROCESSING
  SPACEBORNE EXPERIMENTS
  TECHNOLOGIES
  VACUUM EFFECTS
  WEIGHTLESSNESS

SPACE MECHANICS
GS CLASSICAL MECHANICS
  . **SPACE MECHANICS**
  . . ASTRODYNAMICS
  . . CELESTIAL MECHANICS
  . . ORBITAL MECHANICS
  . . . KEPLER LAWS
  . . . MINIMUM VARIANCE ORBIT DETERMINATION
RT FLIGHT MECHANICS
  MAGNETOHYDRODYNAMICS
  ORBITAL SPACE STATIONS
  ORBITAL SPACE TESTS
  QUADRATURES

SPACE MISSIONS
GS **SPACE MISSIONS**
  . ASTEROID MISSIONS
  . GIOTTO MISSION
  . SOLAR MAXIMUM MISSION
  . . SOLAR MAXIMUM MISSION-A
  . ULYSSES MISSION
RT APOLLO SOYUZ TEST PROJECT
  CHINESE SPACE PROGRAM
  EARTH-VENUS TRAJECTORIES
  FLYBY MISSIONS
  FRENCH SPACE PROGRAMS
  GRAND TOURS
  INDIAN SPACE PROGRAM
  JAPANESE SPACE PROGRAM
  MARINER JUPITER-SATURN FLYBY
  MARINER JUPITER-URANUS FLYBY
  ∞MISSIONS
  SKYLAB 1

SPACE MISSIONS-*(CONT.)*
  SKYLAB 2
  SKYLAB 3
  SKYLAB 4
  ∞SPACECRAFT
  TOPS (SPACECRAFT)

SPACE NAVIGATION
GS NAVIGATION
  . **SPACE NAVIGATION**
  . . INTERPLANETARY NAVIGATION
RT AIR NAVIGATION
  ∞ASTRONAUTICS
  ASTRONAVIGATION
  AUTONOMOUS NAVIGATION
  CELESTIAL NAVIGATION
  DIGITAL NAVIGATION
  EARTH-VENUS TRAJECTORIES
  GLOBAL POSITIONING SYSTEM
  INERTIAL NAVIGATION
  INTERPLANETARY FLIGHT
  INTERPLANETARY TRAJECTORIES
  MANNED SPACECRAFT
  ORBITAL MANEUVERS
  ORBITAL MECHANICS
  ORBITS
  RADAR NAVIGATION
  RADIO NAVIGATION
  REFERENCE STARS
  SATELLITE GUIDANCE
  SATELLITE NAVIGATION SYSTEMS
  SPACECRAFT GUIDANCE
  SPACECRAFT POSITION INDICATORS
  STANDARDIZED SPACE GUIDANCE

SPACE OBSERVATIONS (FROM EARTH)
GS OBSERVATION
  . **SPACE OBSERVATIONS (FROM EARTH)**
RT DETECTION
  RADIO OBSERVATION
  RECONNAISSANCE
  SPACE SURVEILLANCE (GROUND BASED)
  VISUAL OBSERVATION

SPACE OPERATIONS CENTER (NASA)
GS MANNED SPACECRAFT
  . SPACE STATIONS
  . . ORBITAL SPACE STATIONS
  . . . **SPACE OPERATIONS CENTER (NASA)**
  STATIONS
  . SPACE STATIONS
  . . ORBITAL SPACE STATIONS
  . . . **SPACE OPERATIONS CENTER (NASA)**
RT LARGE SPACE STRUCTURES
  ORBITAL ASSEMBLY
  ORBITAL SERVICING

∞ **SPACE ORIENTATION**
SN *(USE OF A MORE SPECIFIC TERM IS RECOMMENDED--CONSULT THE TERMS LISTED BELOW)*
RT ATTITUDE (INCLINATION)
  BEARING (DIRECTION)
  VERTICAL PERCEPTION
  VISUAL PERCEPTION

SPACE PERCEPTION
UF DEPTH PERCEPTION
  DISTANCE PERCEPTION
  FORM PERCEPTION
  SLANT PERCEPTION
GS PERCEPTION
  . SENSORY PERCEPTION
  . . VISUAL PERCEPTION
  . . . **SPACE PERCEPTION**
  . . . . AUTOKINESIS
RT BINOCULAR VISION
  MONOCULAR VISION
  PERIPHERAL VISION
  RANGE FINDERS
  SOUND LOCALIZATION
  VISUAL FIELDS

SPACE PHOTOGRAPHY
USE SPACEBORNE PHOTOGRAPHY

SPACE PLASMA H/V INTERACTION EXPERIMENTS
USE SPHINX

SPACE PLASMAS
GS   PARTICLES
    . CHARGED PARTICLES
    . . ENERGETIC PARTICLES
    . . . PLASMAS (PHYSICS)
    . . . . **SPACE PLASMAS**
RT   AMPTE (SATELLITES)
    GEOMAGNETISM
    IONOPAUSE
    MAGNETOHYDRODYNAMIC STABILITY
    MAGNETOHYDRODYNAMICS
    MAGNETOSPHERE
    OPEN PROJECT
    PLASMA DENSITY
    PLASMA DIAGNOSTICS
    PLASMA INTERACTIONS
    PLASMA LAYERS
    PLASMA PHYSICS
    PLASMA WAVES
    PLASMA-ELECTROMAGNETIC
      INTERACTION
    POLAR CUSPS
    SOLAR WIND

SPACE PLATFORMS
GS   **SPACE PLATFORMS**
    . EURECA (ESA)
    . SYNCHRONOUS PLATFORMS
RT   INTRAORBIT TRANSFER VEHICLES
    ORBITAL SERVICING
    SHAPE CONTROL

SPACE POWER REACTORS
GS   AUXILIARY POWER SOURCES
    . NUCLEAR AUXILIARY POWER UNITS
    . . **SPACE POWER REACTORS**
    . . . FISSION ELECTRIC CELLS
    . . . . SNAP 2
    . . . . SNAP 4
    . . . . SNAP 8
    . . . . SNAP 10A
    . . . SNAP 50
    . . . SPACE POWER UNIT REACTORS
    NUCLEAR ELECTRIC POWER
      GENERATION
    . NUCLEAR AUXILIARY POWER UNITS
    . . **SPACE POWER REACTORS**
    . . . FISSION ELECTRIC CELLS
    . . . . SNAP 2
    . . . . SNAP 4
    . . . . SNAP 8
    . . . . SNAP 10A
    . . . SNAP 50
    . . . SPACE POWER UNIT REACTORS
    . NUCLEAR POWER REACTORS
    . . **SPACE POWER REACTORS**
    . . . FISSION ELECTRIC CELLS
    . . . . SNAP 2
    . . . . SNAP 4
    . . . . SNAP 8
    . . . . SNAP 10A
    . . . SPACE POWER UNIT REACTORS
    NUCLEAR REACTORS
    . NUCLEAR POWER REACTORS
    . . **SPACE POWER REACTORS**
    . . . FISSION ELECTRIC CELLS
    . . . . SNAP 2
    . . . . SNAP 4
    . . . . SNAP 8
    . . . . SNAP 10A
    . . . SPACE POWER UNIT REACTORS
RT   HEAT EXCHANGERS
    TURBOGENERATORS

SPACE POWER UNIT REACTORS
UF   SPUR (REACTORS)
GS   AUXILIARY POWER SOURCES
    . NUCLEAR AUXILIARY POWER UNITS
    . . SPACE POWER REACTORS
    . . . **SPACE POWER UNIT REACTORS**
    NUCLEAR ELECTRIC POWER
      GENERATION
    . NUCLEAR AUXILIARY POWER UNITS
    . . SPACE POWER REACTORS
    . . . **SPACE POWER UNIT REACTORS**
    . NUCLEAR POWER REACTORS
    . . SPACE POWER REACTORS
    . . . **SPACE POWER UNIT REACTORS**
    NUCLEAR REACTORS
    . NUCLEAR POWER REACTORS
    . . SPACE POWER REACTORS
    . . . **SPACE POWER UNIT REACTORS**
RT   FISSION ELECTRIC CELLS
    HEAT EXCHANGERS
    SNAP

SPACE POWER UNIT REACTORS-*(CONT.)*
    SNAP 2
    SNAP 4
    SNAP 8
    SNAP 50
    TURBOGENERATORS

SPACE PROBES
GS   UNMANNED SPACECRAFT
    . **SPACE PROBES**
    . . EXPLORER 18 SATELLITE
    . . GIOTTO MISSION
    . . JUPITER PROBES
    . . . GALILEO PROBE
    . . . GALILEO SPACECRAFT
    . . LUNAR PROBES
    . . . LUNIK LUNAR PROBES
    . . . . LUNIK 2 LUNAR PROBE
    . . . . LUNIK 3 LUNAR PROBE
    . . . . LUNIK 9 LUNAR PROBE
    . . . . LUNIK 10 LUNAR PROBE
    . . . . LUNIK 11 LUNAR PROBE
    . . . . LUNIK 12 LUNAR PROBE
    . . . . LUNIK 13 LUNAR PROBE
    . . . . LUNIK 14 LUNAR PROBE
    . . . . LUNIK 16 LUNAR PROBE
    . . . . LUNIK 17 LUNAR PROBE
    . . . . LUNIK 19 LUNAR PROBE
    . . . . LUNIK 20 LUNAR PROBE
    . . . . LUNIK 22 LUNAR PROBE
    . . . RANGER LUNAR PROBES
    . . . . RANGER LUNAR LANDING
        VEHICLES
    . . . . RANGER 1 LUNAR PROBE
    . . . . RANGER 2 LUNAR PROBE
    . . . . RANGER 3 LUNAR PROBE
    . . . . RANGER 4 LUNAR PROBE
    . . . . RANGER 5 LUNAR PROBE
    . . . . RANGER 6 LUNAR PROBE
    . . . . RANGER 7 LUNAR PROBE
    . . . . RANGER 8 LUNAR PROBE
    . . . . RANGER 9 LUNAR PROBE
    . . . SURVEYOR LUNAR PROBES
    . . . . SURVEYOR 1 LUNAR PROBE
    . . . . SURVEYOR 2 LUNAR PROBE
    . . . . SURVEYOR 3 LUNAR PROBE
    . . . . SURVEYOR 4 LUNAR PROBE
    . . . . SURVEYOR 5 LUNAR PROBE
    . . . . SURVEYOR 6 LUNAR PROBE
    . . . . SURVEYOR 7 LUNAR PROBE
    . . MARINER SPACE PROBES
    . . . MARINER R 2 SPACE PROBE
    . . . MARINER VENUS-MERCURY 1973
    . . . MARINER 1 SPACE PROBE
    . . . MARINER 2 SPACE PROBE
    . . . MARINER 3 SPACE PROBE
    . . . MARINER 4 SPACE PROBE
    . . . MARINER 5 SPACE PROBE
    . . . MARINER 6 SPACE PROBE
    . . . MARINER 7 SPACE PROBE
    . . . MARINER 8 SPACE PROBE
    . . . MARINER 9 SPACE PROBE
    . . . MARINER 10 SPACE PROBE
    . . . MARINER 11 SPACE PROBE
    . . . MARINER-MERCURY 1973
    . . . MARINER SPACECRAFT
    . . . MARINER C SPACECRAFT
    . . . MARINER VENUS 67 SPACECRAFT
    . . MARS PROBES
    . . . ADVANCED RECONN ELECTRIC
        SPACECRAFT
    . . . MARINER 3 SPACE PROBE
    . . . MARINER 4 SPACE PROBE
    . . . MARINER 6 SPACE PROBE
    . . . MARINER 7 SPACE PROBE
    . . . MARINER 8 SPACE PROBE
    . . . MARINER 10 SPACE PROBE
    . . . MARS 1 SPACECRAFT
    . . . MARS 2 SPACECRAFT
    . . . MARS 3 SPACECRAFT
    . . . MARS 4 SPACECRAFT
    . . . MARS 5 SPACECRAFT
    . . . MARS 6 SPACECRAFT
    . . . MARS 7 SPACECRAFT
    . . . VIKING ORBITER 1975
    . . . ZOND 2 SPACE PROBE
    . . PIONEER SPACE PROBES
    . . . PIONEER VENUS 2 ENTRY PROBES
    . . . . PIONEER VENUS 2 NIGHT PROBE
    . . . PIONEER 1 SPACE PROBE
    . . . PIONEER 2 SPACE PROBE
    . . . PIONEER 3 SPACE PROBE
    . . . PIONEER 4 SPACE PROBE
    . . . PIONEER 5 SPACE PROBE
    . . . PIONEER 6 SPACE PROBE
    . . . PIONEER 7 SPACE PROBE

SPACE PROBES-*(CONT.)*
    . . . PIONEER 8 SPACE PROBE
    . . . PIONEER 9 SPACE PROBE
    . . . PIONEER 10 SPACE PROBE
    . . . PIONEER 11 SPACE PROBE
    . . SOLAR PROBES
    . . . HELIOS A
    . . . HELIOS B
    . . . HELIOS 1
    . . . HELIOS 2
    . . . SUNBLAZER SPACE PROBE
    . . VENUS PROBES
    . . . MARINER 1 SPACE PROBE
    . . . MARINER 2 SPACE PROBE
    . . . MARINER 5 SPACE PROBE
    . . . VENERA SATELLITES
    . . . . VENERA 2 SATELLITE
    . . . . VENERA 3 SATELLITE
    . . . . VENERA 4 SATELLITE
    . . . . VENERA 5 SATELLITE
    . . . . VENERA 6 SATELLITE
    . . . . VENERA 7 SATELLITE
    . . . . VENERA 8 SATELLITE
    . . . . VENERA 9 SATELLITE
    . . . . VENERA 10 SATELLITE
    . . . . VENERA 11 SATELLITE
    . . . . VENERA 12 SATELLITE
    . . . ZOND 1 SPACE PROBE
    . . . ZOND 3 SPACE PROBE
    . . . ZOND 4 SPACE PROBE
    . . . ZOND 5 SPACE PROBE
    . . . ZOND 6 SPACE PROBE
    . . . ZOND 7 SPACE PROBE
    . . . ZOND 8 SPACE PROBE
RT   ATLAS ABLE 5 LAUNCH VEHICLE
    INTERPLANETARY SPACECRAFT
    MAGNETIC PROBES
    MANEUVERABLE SPACECRAFT
    MARINER PROGRAM
    METEOROLOGICAL SATELLITES
    PIONEER PROJECT
    PIONEER VENUS SPACECRAFT
    PIONEER VENUS 1 SPACECRAFT
∞   PROBES
    RADIO OCCULTATION
    SATELLITE TELEVISION
    SATELLITES
    VIKING LANDER SPACECRAFT
    VIKING LANDER 1
    VIKING LANDER 2
    VIKING ORBITER SPACECRAFT
    VIKING ORBITER 1
    VIKING ORBITER 2
    VIKING 1 SPACECRAFT
    VIKING 2 SPACECRAFT
    VOYAGER PROJECT
    VOYAGER 1 SPACECRAFT
    VOYAGER 2 SPACECRAFT
    VOYAGER 1977 MISSION

SPACE PROCESSING
GS   SPACE INDUSTRIALIZATION
    . **SPACE PROCESSING**
RT   ACOUSTIC LEVITATION
    BIOPROCESSING
    COMMERCIAL SPACECRAFT
    CONTAINERLESS MELTS
    CRYSTAL GROWTH
    ELECTRIC FURNACES
    FLOAT ZONES
    LEVITATION MELTING
    LOW GRAVITY MANUFACTURING
    MARANGONI CONVECTION
∞   MICROGRAVITY APPLICATIONS
    ORBITAL WORKSHOPS
    SINGLE CRYSTALS
    SPACE COMMERCIALIZATION
    SPACE MANUFACTURING
    SPACEBORNE EXPERIMENTS
    ULTRAPURE METALS

SPACE PROCESSING APPLICATIONS ROCKET
UF   SPAR (ROCKET)
RT   LAUNCH VEHICLES
    METAL FOAMS
    PAYLOADS
    ROCKET VEHICLES
    WEIGHTLESSNESS

SPACE PROGRAMS
GS   PROGRAMS
    . **SPACE PROGRAMS**
    . . BRAZILIAN SPACE PROGRAM
    . . CANADIAN SPACE PROGRAM
    . . CHINESE SPACE PROGRAM

SPACE PROGRAMS-(CONT.)

```
          .. EUROPEAN SPACE PROGRAMS
          .. FRENCH SPACE PROGRAMS
          .. GEOGRAPHIC APPLICATIONS
              PROGRAM
          .. INDIAN SPACE PROGRAM
          .. INDONESIAN SPACE PROGRAM
          .. ITALIAN SPACE PROGRAM
          .. JAPANESE SPACE PROGRAM
          .. SAUDI ARABIAN SPACE PROGRAM
          .. SWEDISH SPACE PROGRAM
          .. SWISS SPACE PROGRAM
          .. U.S.S.R. SPACE PROGRAM
          .. UK SPACE PROGRAM
RT     APOLLO SOYUZ TEST PROJECT
       EUROPEAN SPACE AGENCY
       ISRO
       MANNED SPACE FLIGHT
       NASA PROGRAMS
       NATIONAL LAUNCH VEHICLE PROGRAM
       ∞ RESEARCH PROJECTS
       SOLAR MAXIMUM MISSION
```

SPACE PSYCHOLOGY

```
GS     AEROSPACE MEDICINE
       . SPACE PSYCHOLOGY
       PSYCHOLOGY
       . SPACE PSYCHOLOGY
RT     ASTRONAUT PERFORMANCE
       ASTRONAUT TRAINING
       AVIATION PSYCHOLOGY
       MANNED SPACE FLIGHT
       MILITARY PSYCHOLOGY
       PSYCHOLOGICAL EFFECTS
       PSYCHOLOGICAL FACTORS
       SOCIAL FACTORS
       SPACE ADAPTATION SYNDROME
       SPACE FLIGHT STRESS
       STRESS (PSYCHOLOGY)
```

SPACE RADIATION
```
USE    EXTRATERRESTRIAL RADIATION
```

SPACE RADIATORS
```
USE    SPACECRAFT RADIATORS
```

SPACE RATIONS
```
GS     CONSUMABLES (SPACECRAFT)
       . CONSUMABLES (SPACECREW
           SUPPLIES)
       .. SPACE RATIONS
       RATIONS
       . SPACE RATIONS
RT     ∞ FOOD
       PROVISIONING
       STOWAGE (ONBOARD EQUIPMENT)
```

SPACE RENDEZVOUS
```
UF     SPACECRAFT RENDEZVOUS
GS     RENDEZVOUS
       . SPACE RENDEZVOUS
       .. ORBITAL RENDEZVOUS
       ... LUNAR ORBITAL RENDEZVOUS
RT     APOLLO SOYUZ TEST PROJECT
       RENDEZVOUS TRAJECTORIES
       SPACECRAFT DOCKING
       TRANSFER ORBITS
```

SPACE SCIENCES
```
USE    AEROSPACE SCIENCES
```

SPACE SELF MANEUVERING UNITS
```
USE    SELF MANEUVERING UNITS
```

SPACE SHUTTLE ASCENT STAGE
```
RT     ASCENT
       ASCENT PROPULSION SYSTEMS
       ASCENT TRAJECTORIES
       SPACE SHUTTLE BOOSTERS
       SPACE SHUTTLE UPPER STAGES
       SPACE SHUTTLES
       SPACE TRANSPORTATION SYSTEM
           FLIGHTS
       SPACECRAFT LAUNCHING
       STAGE SEPARATION
       UPPER STAGE ROCKET ENGINES
```

SPACE SHUTTLE BOOSTERS
```
UF     SHUTTLE BOOSTERS
GS     TRANSPORTATION
       . SPACE TRANSPORTATION
       .. SPACE TRANSPORTATION SYSTEM
       ... SPACE SHUTTLE BOOSTERS
```

SPACE SHUTTLE BOOSTERS-(CONT.)
```
       .... AEROMANEUVERING ORBIT TO
           ORBIT SHUTTLE
RT     ∞ BOOSTERS
       MANNED SPACECRAFT
       REUSABLE SPACECRAFT
       SPACE SHUTTLE ASCENT STAGE
```

SPACE SHUTTLE MAIN ENGINE
```
GS     ENGINES
       . ROCKET ENGINES
       .. LIQUID PROPELLANT ROCKET
           ENGINES
       ... SPACE SHUTTLE MAIN ENGINE
RT     PROPULSION
       SPACE TRANSPORTATION SYSTEM
       SPACE TRANSPORTATION SYSTEM
           FLIGHTS
```

SPACE SHUTTLE MISSION 31-A
```
UF     STS-5
GS     SPACE FLIGHT
       . MANNED SPACE FLIGHT
       .. SPACE SHUTTLE MISSION 31-A
       . SPACE SHUTTLE MISSIONS
       .. SPACE SHUTTLE MISSION 31-A
       TRANSPORTATION
       . SPACE TRANSPORTATION
       .. SPACE TRANSPORTATION SYSTEM
       ... SPACE SHUTTLE MISSIONS
       .... SPACE SHUTTLE MISSION 31-A
RT     COLUMBIA (ORBITER)
```

SPACE SHUTTLE MISSION 31-B
```
UF     STS-6
GS     SPACE FLIGHT
       . MANNED SPACE FLIGHT
       .. SPACE SHUTTLE MISSION 31-B
       . SPACE SHUTTLE MISSIONS
       .. SPACE SHUTTLE MISSION 31-B
       TRANSPORTATION
       . SPACE TRANSPORTATION
       .. SPACE TRANSPORTATION SYSTEM
       ... SPACE SHUTTLE MISSIONS
       .... SPACE SHUTTLE MISSION 31-B
RT     CHALLENGER (ORBITER)
```

SPACE SHUTTLE MISSION 31-C
```
UF     SPACE SHUTTLE ORBITAL FLIGHT 7
       STS-7
GS     SPACE FLIGHT
       . MANNED SPACE FLIGHT
       .. SPACE SHUTTLE MISSION 31-C
       . SPACE SHUTTLE MISSIONS
       .. SPACE SHUTTLE MISSION 31-C
       TRANSPORTATION
       . SPACE TRANSPORTATION
       .. SPACE TRANSPORTATION SYSTEM
       ... SPACE SHUTTLE MISSIONS
       .... SPACE SHUTTLE MISSION 31-C
RT     CHALLENGER (ORBITER)
```

SPACE SHUTTLE MISSION 31-D
```
UF     SPACE SHUTTLE ORBITAL FLIGHT 8
       STS-8
GS     SPACE FLIGHT
       . MANNED SPACE FLIGHT
       .. SPACE SHUTTLE MISSION 31-D
       . SPACE SHUTTLE MISSIONS
       .. SPACE SHUTTLE MISSION 31-D
       TRANSPORTATION
       . SPACE TRANSPORTATION
       .. SPACE TRANSPORTATION SYSTEM
       ... SPACE SHUTTLE MISSIONS
       .... SPACE SHUTTLE MISSION 31-D
RT     CHALLENGER (ORBITER)
```

SPACE SHUTTLE MISSION 41-A
```
UF     SPACE SHUTTLE ORBITAL FLIGHT 9
       STS-9
GS     SPACE FLIGHT
       . MANNED SPACE FLIGHT
       .. SPACE SHUTTLE MISSION 41-A
       . SPACE SHUTTLE MISSIONS
       .. SPACE SHUTTLE MISSION 41-A
       TRANSPORTATION
       . SPACE TRANSPORTATION
       .. SPACE TRANSPORTATION SYSTEM
       ... SPACE SHUTTLE MISSIONS
       .... SPACE SHUTTLE MISSION 41-A
RT     COLUMBIA (ORBITER)
```

SPACE SHUTTLE MISSION 41-B
```
UF     STS-11
GS     SPACE FLIGHT
```

SPACE SHUTTLE MISSION 41-B-(CONT.)
```
       . MANNED SPACE FLIGHT
       .. SPACE SHUTTLE MISSION 41-B
       . SPACE SHUTTLE MISSIONS
       .. SPACE SHUTTLE MISSION 41-B
       TRANSPORTATION
       . SPACE TRANSPORTATION
       .. SPACE TRANSPORTATION SYSTEM
       ... SPACE SHUTTLE MISSIONS
       .... SPACE SHUTTLE MISSION 41-B
RT     CHALLENGER (ORBITER)
```

SPACE SHUTTLE MISSION 41-C
```
UF     STS-13
GS     SPACE FLIGHT
       . MANNED SPACE FLIGHT
       .. SPACE SHUTTLE MISSION 41-C
       . SPACE SHUTTLE MISSIONS
       .. SPACE SHUTTLE MISSION 41-C
       TRANSPORTATION
       . SPACE TRANSPORTATION
       .. SPACE TRANSPORTATION SYSTEM
       ... SPACE SHUTTLE MISSIONS
       .... SPACE SHUTTLE MISSION 41-C
RT     CHALLENGER (ORBITER)
```

SPACE SHUTTLE MISSION 41-D
```
UF     STS-14
GS     SPACE FLIGHT
       . MANNED SPACE FLIGHT
       .. SPACE SHUTTLE MISSION 41-D
       . SPACE SHUTTLE MISSIONS
       .. SPACE SHUTTLE MISSION 41-D
       TRANSPORTATION
       . SPACE TRANSPORTATION
       .. SPACE TRANSPORTATION SYSTEM
       ... SPACE SHUTTLE MISSIONS
       .. SPACE SHUTTLE MISSION 41-D
RT     DISCOVERY (ORBITER)
```

SPACE SHUTTLE MISSION 41-G
```
UF     STS-17
GS     SPACE FLIGHT
       . MANNED SPACE FLIGHT
       .. SPACE SHUTTLE MISSION 41-G
       . SPACE SHUTTLE MISSIONS
       .. SPACE SHUTTLE MISSION 41-G
       TRANSPORTATION
       . SPACE TRANSPORTATION
       .. SPACE TRANSPORTATION SYSTEM
       ... SPACE SHUTTLE MISSIONS
       .... SPACE SHUTTLE MISSION 41-G
RT     CHALLENGER (ORBITER)
```

SPACE SHUTTLE MISSION 51-A
```
UF     STS-19
GS     SPACE FLIGHT
       . MANNED SPACE FLIGHT
       .. SPACE SHUTTLE MISSION 51-A
       . SPACE SHUTTLE MISSIONS
       .. SPACE SHUTTLE MISSION 51-A
       TRANSPORTATION
       . SPACE TRANSPORTATION
       .. SPACE TRANSPORTATION SYSTEM
       ... SPACE SHUTTLE MISSIONS
       .... SPACE SHUTTLE MISSION 51-A
RT     DISCOVERY (ORBITER)
```

SPACE SHUTTLE MISSION 51-B
```
UF     STS-24
GS     SPACE FLIGHT
       . MANNED SPACE FLIGHT
       .. SPACE SHUTTLE MISSION 51-B
       . SPACE SHUTTLE MISSIONS
       .. SPACE SHUTTLE MISSION 51-B
       TRANSPORTATION
       . SPACE TRANSPORTATION
       .. SPACE TRANSPORTATION SYSTEM
       ... SPACE SHUTTLE MISSIONS
       .... SPACE SHUTTLE MISSION 51-B
RT     CHALLENGER (ORBITER)
```

SPACE SHUTTLE MISSION 51-C
```
UF     STS-20
GS     SPACE FLIGHT
       . MANNED SPACE FLIGHT
       .. SPACE SHUTTLE MISSION 51-C
       . SPACE SHUTTLE MISSIONS
       .. SPACE SHUTTLE MISSION 51-C
       TRANSPORTATION
       . SPACE TRANSPORTATION
       .. SPACE TRANSPORTATION SYSTEM
       ... SPACE SHUTTLE MISSIONS
       .... SPACE SHUTTLE MISSION 51-C
RT     DISCOVERY (ORBITER)
```

SPACE SHUTTLE ORBITERS-*(CONT.)*
. SPACE TRANSPORTATION
. . SPACE TRANSPORTATION SYSTEM
. . . **SPACE SHUTTLE ORBITERS**
. . . . ATLANTIS (ORBITER)
. . . . CHALLENGER (ORBITER)
. . . . COLUMBIA (ORBITER)
. . . . DISCOVERY (ORBITER)
. . . . ENTERPRISE (ORBITER)
RT   INERTIAL UPPER STAGE
        MANNED SPACE FLIGHT
        MICROWAVE SCANNING BEAM LANDING
          SYSTEM
        PAYLOAD INTEGRATION PLAN
        RECOVERABLE SPACECRAFT
        REUSABLE SPACECRAFT
        SHUTTLE DERIVED VEHICLES
        SPACECRAFT RECOVERY
        TERMINAL AREA ENERGY MANAGEMENT

SPACE SHUTTLE PAYLOADS
GS   PAYLOADS
        . **SPACE SHUTTLE PAYLOADS**
        . . ELECTROMAGNETIC ENVIRONMENT
            EXPERIMENT
        . . HALOGEN OCCULTATION
            EXPERIMENT
        . . OSS-1 PAYLOAD
        . . OSTA-1 PAYLOAD
        . . SPACEBORNE EXPERIMENTS
        . . . ATMOSPHERIC GENERAL
              CIRCULATION EXPERIMENT
        . . . EARTH RADIATION BUDGET
              EXPERIMENT
        . . . PHYSICS AND CHEMISTRY
              EXPERIMENT IN SPACE
        . . . PLASMA INTERACTION EXPERIMENT
        . . SPACELAB
        . . X RAY ASTROPHYSICS FACILITY
RT   COMMERCE LAB
        EXTRAVEHICULAR ACTIVITY
        FEATURE IDENTIFICATION AND
          LOCATION EXPER
        ORBITAL SERVICING
        PAYLOAD ASSIST MODULE
        PAYLOAD INTEGRATION
        PAYLOAD INTEGRATION PLAN
        SHUTTLE IMAGING RADAR
        SORTIE SYSTEMS
        SPACE TECHNOLOGY EXPERIMENTS
        SPACE TRANSPORTATION SYSTEM

SPACE SHUTTLE UPPER STAGE A
UF   SSUS-A
GS   SPACE SHUTTLE UPPER STAGES
        . **SPACE SHUTTLE UPPER STAGE A**
RT   ATLAS CENTAUR LAUNCH VEHICLE

SPACE SHUTTLE UPPER STAGE D
UF   SSUS-D
GS   SPACE SHUTTLE UPPER STAGES
        . **SPACE SHUTTLE UPPER STAGE D**
RT   DELTA LAUNCH VEHICLE
        SOLID PROPELLANT ROCKET ENGINES
        SPIN STABILIZATION

SPACE SHUTTLE UPPER STAGES
GS   **SPACE SHUTTLE UPPER STAGES**
        . SPACE SHUTTLE UPPER STAGE A
        . SPACE SHUTTLE UPPER STAGE D
        . SPINNING SOLID UPPER STAGE
RT   SPACE SHUTTLE ASCENT STAGE

SPACE SHUTTLES
GS   MANNED SPACECRAFT
        . **SPACE SHUTTLES**
        . . HERMES MANNED SPACEPLANE
RT   ADVANCED TECHNOLOGY LABORATORY
        AEPS
        AEROMANEUVERING ORBIT TO ORBIT
          SHUTTLE
        ANNULAR SUSPENSION AND POINTING
          SYSTEM
        APPROACH AND LANDING TESTS (STS)
        ASSESS PROGRAM
        ATMOSPHERIC CLOUD PHYSICS LAB
          (SPACELAB)
        AUXILIARY PROPULSION
        BESS (SATELLITE)
        EARTH VIEWING APPLICATIONS
          LABORATORY
        ENTRY GUIDANCE (STS)
        EURECA (ESA)
        EXPENDABLE STAGES (SPACECRAFT)
        HUBBLE SPACE TELESCOPE

SPACE SHUTTLES-*(CONT.)*
        INERTIAL UPPER STAGE
        INTERIM STAGES (SPACECRAFT)
        INTRAORBIT TRANSFER VEHICLES
        LIRTS (TELESCOPE)
        MANNED SPACE FLIGHT
        ORBIT MANEUVERING ENGINE (SPACE
          SHUTTLE)
        ORBIT TRANSFER VEHICLES
        ORBITAL MANEUVERS
        PAYLOAD CONTROL
        PAYLOAD DEPLOYMENT & RETRIEVAL
          SYSTEM
        PAYLOAD RETRIEVAL (STS)
        SAIL PROJECT
        SHUTTLE DERIVED VEHICLES
        SHUTTLE ENGINEERING SIMULATOR
        SHUTTLE PALLET SATELLITES
        SINGLE STAGE TO ORBIT VEHICLES
        SORTIE SYSTEMS
        SPACE SHUTTLE ASCENT STAGE
        SPACE TRANSPORTATION SYSTEM
          FLIGHTS
∞ SPACECRAFT
        SPACECRAFT RECOVERY
        SPACELAB

SPACE SIMULATORS
UF   ORBITAL SIMULATORS
GS   SIMULATORS
        . ENVIRONMENT SIMULATORS
        . **SPACE SIMULATORS**
        . . . HIGH VACUUM ORBITAL SIMULATOR
        . . . LANGLEY COMPLEX COORDINATOR
RT   CENTRIFUGES
        FLIGHT SIMULATORS
        SOLAR SIMULATORS
        VACUUM CHAMBERS

SPACE STATIONS
UF   SELF DEPLOYING SPACE STATIONS
GS   MANNED SPACECRAFT
        . **SPACE STATIONS**
        . . ORBITAL SPACE STATIONS
        . . . LONG DURATION EXPOSURE
              FACILITY
        . . . ORBITAL WORKSHOPS
        . . . SALYUT SPACE STATION
        . . . SKYLAB 1
        . . . SKYLAB 2
        . . . SKYLAB 3
        . . . SKYLAB 4
        . . . SPACE OPERATIONS CENTER
              (NASA)
        . . SPACE BASE COMMAND CENTER
        STATIONS
        . **SPACE STATIONS**
        . . ORBITAL SPACE STATIONS
        . . . HALO ORBIT SPACE STATION
        . . . LONG DURATION EXPOSURE
              FACILITY
        . . . ORBITAL WORKSHOPS
        . . . ORBITING LUNAR STATIONS
        . . . SALYUT SPACE STATION
        . . . SPACE OPERATIONS CENTER
              (NASA)
        . . SPACE BASE COMMAND CENTER
RT   AEPS
        ARTIFICIAL SATELLITES
        BIOASTRONAUTICS
        FERRY SPACECRAFT
        HUBBLE SPACE TELESCOPE
        INFLATABLE STRUCTURES
        MILITARY SPACECRAFT
        ORBITAL SPACE STATIONS
        ORBITAL SPACE TESTS
∞ PLATFORMS
        RENDEZVOUS SPACECRAFT
        SATELLITES
        SATURN WORKSHOPS
        SATURN 1 WORKSHOP
        SATURN 5 WORKSHOP
        SORTIE SYSTEMS
        SPACE HABITATS
        SPIN STABILIZATION

SPACE STORAGE
RT   CRYOGENIC FLUID STORAGE
        CRYOGENIC ROCKET PROPELLANTS
        PROPELLANT STORAGE
        STORABLE PROPELLANTS
∞ STORAGE
        STORAGE TANKS

SPACE SUITS
GS   CLOTHING
        . PROTECTIVE CLOTHING
        . . PRESSURE SUITS
        . . . **SPACE SUITS**
        . SUITS
        . . PRESSURE SUITS
        . . . **SPACE SUITS**
        SAFETY DEVICES
        . **SPACE SUITS**
RT   MANNED MANEUVERING UNITS

∞ **SPACE SURVEILLANCE**
SN   *(USE OF A MORE SPECIFIC TERM IS
        RECOMMENDED--CONSULT THE TERMS
        LISTED BELOW)*
RT   SPACE SURVEILLANCE (GROUND
          BASED)
        SPACE SURVEILLANCE (SPACEBORNE)

SPACE SURVEILLANCE (GROUND BASED)
GS   SURVEILLANCE
        . **SPACE SURVEILLANCE (GROUND
          BASED)**
RT   AIR DEFENSE
        ANTIMISSILE DEFENSE
        MINITRACK SYSTEM
        SPACE OBSERVATIONS (FROM EARTH)
∞ SPACE SURVEILLANCE
        SPACECRAFT TRACKING

SPACE SURVEILLANCE (SPACEBORNE)
GS   SURVEILLANCE
        . **SPACE SURVEILLANCE (SPACEBORNE)**
RT   AIR DEFENSE
        ANTIMISSILE DEFENSE
        HIGH ALTITUDE NUCLEAR DETECTION
        ICE MAPPING
        ICE REPORTING
        MILITARY SPACECRAFT
        OPTICAL COUNTERMEASURES
        RECONNAISSANCE
        SATELLITE-BORNE PHOTOGRAPHY
        SATELLITE-TO-SATELLITE TRACKING
∞ SPACE SURVEILLANCE
        SPACECRAFT TRACKING

SPACE SYSTEMS ENGINEERING
USE   AEROSPACE ENGINEERING

SPACE TECHNOLOGY EXPERIMENTS
RT   ANTENNA DESIGN
        ANTENNAS
        LARGE SPACE STRUCTURES
        SPACE ERECTABLE STRUCTURES
        SPACE SHUTTLE PAYLOADS
        SPACEBORNE EXPERIMENTS

SPACE TELESCOPE
USE   HUBBLE SPACE TELESCOPE

SPACE TEMPERATURE
GS   TEMPERATURE
        . **SPACE TEMPERATURE**
RT   ELECTRON ENERGY
        ION TEMPERATURE
        ULTRALOW TEMPERATURES

SPACE TOOLS
GS   TOOLS
        . **SPACE TOOLS**
RT   LOW GRAVITY MANUFACTURING
        ORBITAL WORKERS

SPACE TRANSPORTATION
GS   TRANSPORTATION
        . **SPACE TRANSPORTATION**
        . . SPACE TRANSPORTATION SYSTEM
        . . . SPACE SHUTTLE BOOSTERS
        . . . . AEROMANEUVERING ORBIT TO
              ORBIT SHUTTLE
        . . . SPACE SHUTTLE MISSIONS
        . . . . SPACE SHUTTLE MISSION 31-A
        . . . . SPACE SHUTTLE MISSION 31-B
        . . . . SPACE SHUTTLE MISSION 31-C
        . . . . SPACE SHUTTLE MISSION 31-D
        . . . . SPACE SHUTTLE MISSION 41-A
        . . . . SPACE SHUTTLE MISSION 41-B
        . . . . SPACE SHUTTLE MISSION 41-C
        . . . . SPACE SHUTTLE MISSION 41-D
        . . . . SPACE SHUTTLE MISSION 41-G
        . . . . SPACE SHUTTLE MISSION 51-A
        . . . . SPACE SHUTTLE MISSION 51-B
        . . . . SPACE SHUTTLE MISSION 51-C

SPACE TRANSPORTATION-*(CONT.)*
```
.... SPACE SHUTTLE MISSION 51-D
.... SPACE SHUTTLE MISSION 51-E
.... SPACE SHUTTLE MISSION 51-F
.... SPACE SHUTTLE MISSION 51-G
.... SPACE SHUTTLE MISSION 51-H
.... SPACE SHUTTLE MISSION 51-I
.... SPACE SHUTTLE MISSION 51-J
.... SPACE SHUTTLE MISSION 51-L
.... SPACE SHUTTLE MISSION 61-A
.... SPACE SHUTTLE MISSION 61-B
.... SPACE SHUTTLE MISSION 61-C
.... SPACE SHUTTLE MISSION 61-E
... SPACE SHUTTLE ORBITERS
.... ATLANTIS (ORBITER)
.... CHALLENGER (ORBITER)
.... COLUMBIA (ORBITER)
.... DISCOVERY (ORBITER)
.... ENTERPRISE (ORBITER)
```
RT   INERTIAL UPPER STAGE
     JAPANESE SPACE PROGRAM
     ORBIT TRANSFER VEHICLES
     PAYLOAD STATIONS
     PAYLOADS
     SINGLE STAGE TO ORBIT VEHICLES
     TERMINAL AREA ENERGY MANAGEMENT

SPACE TRANSPORTATION SYSTEM
UF   STS
GS   TRANSPORTATION
```
. SPACE TRANSPORTATION
.. SPACE TRANSPORTATION SYSTEM
... SPACE SHUTTLE BOOSTERS
.... AEROMANEUVERING ORBIT TO
        ORBIT SHUTTLE
... SPACE SHUTTLE MISSIONS
.... SPACE SHUTTLE MISSION 31-A
.... SPACE SHUTTLE MISSION 31-B
.... SPACE SHUTTLE MISSION 31-C
.... SPACE SHUTTLE MISSION 31-D
.... SPACE SHUTTLE MISSION 41-A
.... SPACE SHUTTLE MISSION 41-B
.... SPACE SHUTTLE MISSION 41-C
.... SPACE SHUTTLE MISSION 41-D
.... SPACE SHUTTLE MISSION 41-G
.... SPACE SHUTTLE MISSION 51-A
.... SPACE SHUTTLE MISSION 51-B
.... SPACE SHUTTLE MISSION 51-C
.... SPACE SHUTTLE MISSION 51-D
.... SPACE SHUTTLE MISSION 51-E
.... SPACE SHUTTLE MISSION 51-F
.... SPACE SHUTTLE MISSION 51-G
.... SPACE SHUTTLE MISSION 51-H
.... SPACE SHUTTLE MISSION 51-I
.... SPACE SHUTTLE MISSION 51-J
.... SPACE SHUTTLE MISSION 51-L
.... SPACE SHUTTLE MISSION 61-A
.... SPACE SHUTTLE MISSION 61-B
.... SPACE SHUTTLE MISSION 61-C
.... SPACE SHUTTLE MISSION 61-E
... SPACE SHUTTLE ORBITERS
.... ATLANTIS (ORBITER)
.... CHALLENGER (ORBITER)
.... COLUMBIA (ORBITER)
.... DISCOVERY (ORBITER)
.... ENTERPRISE (ORBITER)
```
RT   ANNULAR SUSPENSION AND POINTING
       SYSTEM
     APPROACH AND LANDING TESTS (STS)
     ATMOSPHERIC GENERAL CIRCULATION
       EXPERIMENT
     DEFENSE PROGRAM
     HERMES MANNED SPACEPLANE
     INERTIAL UPPER STAGE
     NASA PROGRAMS
     ORBITAL SERVICING
     OSS-1 PAYLOAD
     OSTA-1 PAYLOAD
     OSTA-2 PAYLOAD
     PAYLOAD ASSIST MODULE
     PAYLOAD DELIVERY (STS)
     PAYLOAD INTEGRATION PLAN
     PAYLOAD RETRIEVAL (STS)
     POWER MODULES (STS)
     REMOTE MANIPULATOR SYSTEM
     SPACE SHUTTLE MAIN ENGINE
     SPACE SHUTTLE MISSION 61-A
     SPACE SHUTTLE PAYLOADS
     SPACE TRANSPORTATION SYSTEM
       FLIGHTS
     SPACE TUGS
     SPACELAB
     SYNCOM 4 SATELLITE
∞ SYSTEMS
     TRANSPORTATION

SPACE TRANSPORTATION SYSTEM FLIGHTS
UF   OFT
     ORBITAL FLIGHT TESTS (SHUTTLE)
     SPACE SHUTTLE ORBITAL FLIGHT
       TESTS
     SPACE SHUTTLE ORBITAL FLIGHTS
GS   FLIGHT TESTS
```
. SPACE TRANSPORTATION SYSTEM
   FLIGHTS
.. SPACE TRANSPORTATION SYSTEM 1
      FLIGHT
.. SPACE TRANSPORTATION SYSTEM 2
      FLIGHT
.. SPACE TRANSPORTATION SYSTEM 3
      FLIGHT
.. SPACE TRANSPORTATION SYSTEM 4
      FLIGHT
```
RT   ENTRY GUIDANCE (STS)
     GEOPHYSICAL FLUID FLOW CELLS
     SOLAR CELL CALIBRATION FACILITY
     SPACE FLIGHT
     SPACE SHUTTLE ASCENT STAGE
     SPACE SHUTTLE MAIN ENGINE
     SPACE SHUTTLES
     SPACE TRANSPORTATION SYSTEM
∞ SYSTEMS
∞ TESTS

SPACE TRANSPORTATION SYSTEM 1 FLIGHT
UF   OFT 1
     ORBITAL FLIGHT TEST 1 (SHUTTLE)
     SPACE SHUTTLE ORBITAL FLIGHT TEST
       1
     STS-1
GS   FLIGHT TESTS
```
. SPACE TRANSPORTATION SYSTEM
   FLIGHTS
.. SPACE TRANSPORTATION SYSTEM 1
      FLIGHT
```

SPACE TRANSPORTATION SYSTEM 2 FLIGHT
UF   OFT 2
     ORBITAL FLIGHT TEST 2 (SHUTTLE)
     SPACE SHUTTLE ORBITAL FLIGHT TEST
       2
     STS-2
GS   FLIGHT TESTS
```
. SPACE TRANSPORTATION SYSTEM
   FLIGHTS
.. SPACE TRANSPORTATION SYSTEM 2
      FLIGHT
```

SPACE TRANSPORTATION SYSTEM 3 FLIGHT
UF   OFT 3
     ORBITAL FLIGHT TEST 3 (SHUTTLE)
     SPACE SHUTTLE ORBITAL FLIGHT TEST
       3
     STS-3
GS   FLIGHT TESTS
```
. SPACE TRANSPORTATION SYSTEM
   FLIGHTS
.. SPACE TRANSPORTATION SYSTEM 3
      FLIGHT
```

SPACE TRANSPORTATION SYSTEM 4 FLIGHT
UF   OFT 4
     ORBITAL FLIGHT TEST 4 (SHUTTLE)
     SPACE SHUTTLE ORBITAL FLIGHT TEST
       4
     STS-4
GS   FLIGHT TESTS
```
. SPACE TRANSPORTATION SYSTEM
   FLIGHTS
.. SPACE TRANSPORTATION SYSTEM 4
      FLIGHT
```

SPACE TUGS
RT   INERTIAL UPPER STAGE
     MODULES
     ORBIT TRANSFER VEHICLES
     ORBITAL SERVICING
     PAYLOADS
     PROPULSION
     SPACE TRANSPORTATION SYSTEM
     SPACECRAFT PROPULSION

SPACE VEHICLE CHECKOUT PROGRAM
UF   SPACECRAFT PRELAUNCH TESTS
RT   CHECKOUT
     COUNTDOWN
     PERFORMANCE TESTS
     PREFIRING TESTS
     SPACECRAFT MAINTENANCE
∞ TESTS

SPACE VEHICLE CONTROL
USE   SPACECRAFT CONTROL

SPACE VEHICLES
USE   SPACECRAFT

SPACE WEAPONS
GS   WEAPONS
```
. SPACE WEAPONS
.. LASER WEAPONS
```
RT   AIR TO AIR MISSILES
     ANTIMISSILE MISSILES
     CHAPARRAL MISSILE
     MINUTEMAN ICBM
     NUCLEAR WEAPONS
     SURFACE TO AIR MISSILES
     WEAPON SYSTEMS
     WEAPONS DELIVERY

SPACE-TIME CONTINUUM
USE   RELATIVITY

SPACE-TIME FUNCTIONS
UF   SPACE-TIME METRIC
GS   FUNCTIONS (MATHEMATICS)
```
. SPACE-TIME FUNCTIONS
```
RT   MINKOWSKI SPACE
     NAKED SINGULARITIES
     RELATIVITY
     YANG-MILLS THEORY

SPACE-TIME METRIC
USE   SPACE-TIME FUNCTIONS

SPACEBORNE ASTRONOMY
GS   ASTRONOMY
```
. SPACEBORNE ASTRONOMY
```
RT   ASTRONOMICAL SATELLITES
     COSMIC BACKGROUND EXPLORER
       SATELLITE
     FAINT OBJECT CAMERA
     HIPPARCOS SATELLITE
     HUBBLE SPACE TELESCOPE
     IUE
     MAGELLAN MISSION
     QUASAT
     ROSAT MISSION
     SAS-2
     SAS-3
     STARSAT TELESCOPE
     TELESCOPES
     ULTRAVIOLET TELESCOPES

SPACEBORNE EXPERIMENTS
GS   PAYLOADS
```
. SPACE SHUTTLE PAYLOADS
.. SPACEBORNE EXPERIMENTS
... ATMOSPHERIC GENERAL
        CIRCULATION EXPERIMENT
... EARTH RADIATION BUDGET
        EXPERIMENT
... PHYSICS AND CHEMISTRY
        EXPERIMENT IN SPACE
... PLASMA INTERACTION EXPERIMENT
```
RT   AEROSPACE ENVIRONMENTS
     AMPTE (SATELLITES)
     BIOPROCESSING
     EXPERIMENTATION
     GEOPHYSICAL FLUID FLOW CELLS
     MOLECULAR SHIELDS
     OSS-1 PAYLOAD
     OSTA-1 PAYLOAD
     PAYLOAD ASSIST MODULE
     PAYLOAD INTEGRATION PLAN
     PAYLOADS
     SPACE LABORATORIES
     SPACE MANUFACTURING
     SPACE PROCESSING
     SPACE TECHNOLOGY EXPERIMENTS
     SPACELAB
     WEIGHTLESSNESS
     X RAY ASTROPHYSICS FACILITY

SPACEBORNE LASERS
GS   STIMULATED EMISSION DEVICES
```
. LASERS
.. SPACEBORNE LASERS
```
RT   AIRBORNE LASERS
     LASER APPLICATIONS
     REMOTE SENSORS

SPACEBORNE PHOTOGRAPHY
UF   SPACE PHOTOGRAPHY
GS   IMAGERY

SPACEBORNE PHOTOGRAPHY-(CONT.)
. **SPACEBORNE PHOTOGRAPHY**
.. SATELLITE-BORNE PHOTOGRAPHY
PHOTOGRAPHY
. **SPACEBORNE PHOTOGRAPHY**
.. SATELLITE-BORNE PHOTOGRAPHY
RT   AERIAL PHOTOGRAPHY
ASTRONOMICAL PHOTOGRAPHY
BLACK AND WHITE PHOTOGRAPHY
CLOUD PHOTOGRAPHS
CLOUD PHOTOGRAPHY
DIFFRACTION LIMITED CAMERAS
EARTH RESOURCES
LUNAR PHOTOGRAPHS
LUNAR PHOTOGRAPHY
MARS PHOTOGRAPHS
MULTISPECTRAL BAND SCANNERS
PHOTOMAPPING
PHOTOMAPS
ROCKET-BORNE PHOTOGRAPHY
SATELLITE OBSERVATION

SPACEBORNE TELESCOPES
GS   TELESCOPES
. **SPACEBORNE TELESCOPES**
.. HUBBLE SPACE TELESCOPE
.. LIRTS (TELESCOPE)
.. SOLAR OPTICAL TELESCOPE
.. SPACE INFRARED TELESCOPE
   FACILITY
.. STARSAT TELESCOPE
RT   ASTRONOMICAL OBSERVATORIES
ASTRONOMICAL PHOTOGRAPHY
ASTRONOMICAL TELESCOPES
ASTRONOMY
DIFFRACTION LIMITED CAMERAS
FAINT OBJECT CAMERA
MULTI-ANODE MICROCHANNEL ARRAYS
OPTICAL TRANSFER FUNCTION
ROSAT MISSION

∞ SPACECRAFT
SN   (USE OF A MORE SPECIFIC TERM IS
      RECOMMENDED--CONSULT THE TERMS
      LISTED BELOW)
UF   SPACE VEHICLES
RT   ADVANCED RECONN ELECTRIC
         SPACECRAFT
AEROSPACE VEHICLES
ASTRO VEHICLE
ASTRODYNAMICS
ATLANTIS (ORBITER)
AUXILIARY PROPULSION
BIOSATELLITES
CANADIAN SPACECRAFT
CARGO SPACECRAFT
CHALLENGER (ORBITER)
CHINESE SPACECRAFT
COLUMBIA (ORBITER)
CZECHOSLOVAKIAN SPACECRAFT
DISCOVERY (ORBITER)
DUAL SPIN SPACECRAFT
ENTERPRISE (ORBITER)
ESA SPACECRAFT
ESCAPE ROCKETS
EXPANDABLE STRUCTURES
FLEXIBLE SPACECRAFT
FLIGHT TEST VEHICLES
FRENCH SPACE PROGRAMS
GALILEO PROBE
GALILEO SPACECRAFT
GROUND SUPPORT EQUIPMENT
HYPERSONIC VEHICLES
INDIAN SPACE PROGRAM
INDIAN SPACECRAFT
INFLATABLE SPACECRAFT
INTERPLANETARY SPACECRAFT
JAPANESE SPACE PROGRAM
JAPANESE SPACECRAFT
LAUNCH VEHICLES
LUNAR SPACECRAFT
MANEUVERABLE SPACECRAFT
MANNED ORBITAL RESEARCH
   LABORATORIES
MANNED SPACECRAFT
MARINER MARK 2 SPACECRAFT
MARK 1 SPACECRAFT
MILITARY SPACECRAFT
ORBIT TRANSFER VEHICLES
ORBITAL MANEUVERING VEHICLES
ORBITING LUNAR STATIONS
OUTER PLANETS EXPLORERS
PHOTO RECONNAISSANCE SPACECRAFT
PIONEER VENUS 2 SPACECRAFT
POWER LIMITED SPACECRAFT

SPACECRAFT-(CONT.)
RADIATION METEOROID SPACECRAFT
RECOVERABLE SPACECRAFT
REENTRY VEHICLES
RESEARCH VEHICLES
SATELLITES
SERT 1 SPACECRAFT
SERT 2 SPACECRAFT
SHUTTLE DERIVED VEHICLES
SOFT LANDING SPACECRAFT
SOLAR MAXIMUM MISSION-A
SOLETTAS
SOVIET SPACECRAFT
SPACE CAPSULES
SPACE DEBRIS
SPACE LABORATORIES
SPACE MISSIONS
SPACE SHUTTLES
SPACECRAFT CABIN SIMULATORS
SPACECRAFT MODULES
TECHNOLOGY FEASIBILITY SPACECRAFT
TEST VEHICLES
TOPS (SPACECRAFT)
TRANSATMOSPHERIC VEHICLES
UNIDENTIFIED FLYING OBJECTS
UNMANNED SPACECRAFT
∞VEHICLES
VIKING ORBITER SPACECRAFT
VIKING ORBITER 1
VIKING ORBITER 2
VOYAGER 1 SPACECRAFT
VOYAGER 2 SPACECRAFT

SPACECRAFT ANTENNAS
GS   ANTENNAS
. **SPACECRAFT ANTENNAS**
RADIO EQUIPMENT
. **SPACECRAFT ANTENNAS**
TELECOMMUNICATION
. **SPACECRAFT ANTENNAS**
RT   FURLABLE ANTENNAS

SPACECRAFT CABIN ATMOSPHERES
GS   CONTROLLED ATMOSPHERES
. CABIN ATMOSPHERES
.. **SPACECRAFT CABIN ATMOSPHERES**
RT   CARBON DIOXIDE CONCENTRATION
CLOSED ECOLOGICAL SYSTEMS
COCKPITS
ENVIRONMENTAL CONTROL
HIGH PRESSURE OXYGEN
PRESSURIZED CABINS
REBREATHING

SPACECRAFT CABIN SIMULATORS
GS   SIMULATORS
. TRAINING SIMULATORS
.. **SPACECRAFT CABIN SIMULATORS**
RT   AEROSPACE ENVIRONMENTS
COCKPIT SIMULATORS
MANNED SPACECRAFT
SIMULATION
SPACE FLIGHT TRAINING
SPACE LOGISTICS
∞SPACECRAFT
TEST FACILITIES

SPACECRAFT CABINS
GS   COMPARTMENTS
. **SPACECRAFT CABINS**
SPACECRAFT COMPONENTS
. **SPACECRAFT CABINS**
RT   ∞CABINS
COCKPITS
CREW EXPERIMENT STATIONS
CREW OBSERVATION STATIONS
CREW STATIONS
CREW WORKSTATIONS
PRESSURIZED CABINS
SPACE CAPSULES

SPACECRAFT CHARGING
RT   ELECTRIC FIELDS
EXTERNAL SURFACE CURRENTS
SINGLE EVENT UPSETS
SYSTEM GENERATED
   ELECTROMAGNETIC PULSES

SPACECRAFT COMMUNICATION
SN   (COMMUNICATION OF SPACECRAFT
      WITH GROUND OR OTHER
      SPACECRAFT)
UF   SATELLITE COMMUNICATION
GS   TELECOMMUNICATION
. SPACE COMMUNICATION

SPACECRAFT COMMUNICATION-(CONT.)
.. **SPACECRAFT COMMUNICATION**
... REENTRY COMMUNICATION
RT   ARPA COMPUTER NETWORK
ASTRIONICS
CIRCUMLUNAR COMMUNICATION
EARTH TERMINALS
FACSIMILE COMMUNICATION
GROUND-AIR-GROUND COMMUNICATION
HOOP COLUMN ANTENNAS
INTERPLANETARY COMMUNICATION
LUNAR COMMUNICATION
OPTICAL COMMUNICATION
PACKET TRANSMISSION
PLASMA ANTENNAS
RADIO COMMUNICATION
SATELLITE COMMUNICATIONS SHIPS
SATELLITE GROUND SUPPORT
SINGLE CHANNEL PER CARRIER
   TRANSMISSION
SYSTEM GENERATED
   ELECTROMAGNETIC PULSES
UNIFIED S BAND
WIRELESS COMMUNICATION

SPACECRAFT COMPONENTS
GS   **SPACECRAFT COMPONENTS**
. APOLLO LUNAR EXPERIMENT MODULE
. SERVICE MODULES
. SPACECRAFT CABINS
. SPACECRAFT DOCKING MODULES
. SPACECRAFT MODULES
.. COMMAND MODULES
RT   AIRBORNE/SPACEBORNE COMPUTERS
BORON-EPOXY COMPOUNDS
COMMONALITY
∞COMPONENTS
NOSE CONES
ROCKET ENGINES

SPACECRAFT CONFIGURATIONS
GS   **SPACECRAFT CONFIGURATIONS**
. APOLLO TELESCOPE MOUNT
. SATELLITE CONFIGURATIONS
RT   AERODYNAMIC CONFIGURATIONS
AIRCRAFT CONFIGURATIONS
APOLLO SHORT STACK
∞CONFIGURATIONS
FLARED BODIES
LAUNCH VEHICLE CONFIGURATIONS
PROPULSION SYSTEM CONFIGURATIONS
RADIATION METEOROID SPACECRAFT
REENTRY VEHICLES
UPPER STAGE ROCKET ENGINES

SPACECRAFT CONSTRUCTION MATERIALS
RT   ∞CONSTRUCTION MATERIALS
LUDOX (TRADEMARK)
∞MATERIALS

SPACECRAFT CONTAMINATION
GS   CONTAMINATION
. **SPACECRAFT CONTAMINATION**
RT   DECONTAMINATION
EXOBIOLOGY

SPACECRAFT CONTROL
UF   SPACE VEHICLE CONTROL
GS   **SPACECRAFT CONTROL**
. SATELLITE CONTROL
.. SATELLITE ATTITUDE CONTROL
RT   ATTITUDE CONTROL
AUTOMATIC CONTROL
∞CONTROL
CONTROL SIMULATION
CREW PROCEDURES (PREFLIGHT)
ENGINE CONTROL
FLEXIBLE SPACECRAFT
FLIGHT CONTROL
FLY BY WIRE CONTROL
GROUND BASED CONTROL
MANUAL CONTROL
MARQUARDT R4D ENGINE
MISSILE CONTROL
POINTING CONTROL SYSTEMS
RADIO CONTROL
REMOTE CONTROL
ROCKET ENGINE CONTROL
SHAPE CONTROL
STATIONKEEPING
THRUST VECTOR CONTROL
VISUAL CONTROL

SPACECRAFT DEFENSE
UF   SATELLITE DEFENSE

SPACECRAFT DEFENSE-*(CONT.)*
RT   SPACECRAFT SURVIVABILITY
        VULNERABILITY

SPACECRAFT DESIGN
GS   **SPACECRAFT DESIGN**
      . IPAD
      . MARS 3 SPACECRAFT
      . SATELLITE DESIGN
RT   COMPUTER AIDED DESIGN
      ∞DESIGN
      ENGINE DESIGN
      HERMES MANNED SPACEPLANE
      INDIAN SPACE PROGRAM
      JAPANESE SPACE PROGRAM
      LOFTING
      MARS 3 SPACECRAFT
      PRODUCT DEVELOPMENT
      SATELLITE TEMPERATURE
      SHUTTLE DERIVED VEHICLES
      SPACE DEBRIS
      SPACECRAFT TEMPERATURE
      STRUCTURAL DESIGN
      SYSTEMS ENGINEERING
      TRANSATMOSPHERIC VEHICLES
      WEIGHT REDUCTION

SPACECRAFT DOCKING
UF   DOCKING
GS   MANEUVERS
      . **SPACECRAFT DOCKING**
RT   ∞ASTRONAUTICS
      INTERCEPTION
      MOORING
      MULTIPLE DOCKING ADAPTERS
      ORBITAL RENDEZVOUS
      ORBITAL SPACE STATIONS
      RENDEZVOUS TRAJECTORIES
      SALYUT SPACE STATION
      SPACE RENDEZVOUS
      TRANSFER ORBITS

SPACECRAFT DOCKING MODULES
GS   MODULES
      . **SPACECRAFT DOCKING MODULES**
      SPACECRAFT COMPONENTS
      . **SPACECRAFT DOCKING MODULES**
RT   AIRLOCK MODULES
      COMMAND MODULES
      COMMAND SERVICE MODULES
      LANDING MODULES
      SERVICE MODULES

SPACECRAFT ELECTRONIC EQUIPMENT
GS   ELECTRONIC EQUIPMENT
      . **SPACECRAFT ELECTRONIC
         EQUIPMENT**
      ONBOARD EQUIPMENT
      . SPACECRAFT EQUIPMENT
      . . **SPACECRAFT ELECTRONIC
         EQUIPMENT**
RT   AIRBORNE/SPACEBORNE COMPUTERS
      ASTRIONICS
      SINGLE EVENT UPSETS

SPACECRAFT ENVIRONMENTS
GS   ENVIRONMENTS
      . **SPACECRAFT ENVIRONMENTS**
RT   AEROSPACE MEDICINE
      ASTRONAUTS
      BIOASTRONAUTICS
      CLOSED ECOLOGICAL SYSTEMS
      CONTROLLED ATMOSPHERES
      COSMONAUTS
      COUCHES
      ENVIRONMENTAL CONTROL
      EXOBIOLOGY
      EXTRATERRESTRIAL ENVIRONMENTS
      GRAVITATION
      INTRAVEHICULAR ACTIVITY
      LANGLEY COMPLEX COORDINATOR
      LIFE SUPPORT SYSTEMS
      ROTATING ENVIRONMENTS
      SATELLITE TEMPERATURE
      THERMAL ENVIRONMENTS
      WEIGHTLESSNESS

SPACECRAFT EQUIPMENT
GS   ONBOARD EQUIPMENT
      . **SPACECRAFT EQUIPMENT**
      . . SPACECRAFT ELECTRONIC
         EQUIPMENT
RT   ∞EQUIPMENT
      SPACECRAFT INSTRUMENTS

SPACECRAFT GUIDANCE
GS   GUIDANCE (MOTION)
      . **SPACECRAFT GUIDANCE**
      . . SATELLITE GUIDANCE
RT   AUTOMATIC CONTROL
      AUTONOMOUS NAVIGATION
      CCD STAR TRACKER
      CELESTIAL NAVIGATION
      COMMAND GUIDANCE
      GROUND BASED CONTROL
      INERTIAL GUIDANCE
      INJECTION GUIDANCE
      INTERPLANETARY FLIGHT
      INTERPLANETARY TRAJECTORIES
      LASER GYROSCOPES
      MANUAL CONTROL
      MIDCOURSE GUIDANCE
      ORBITS
      REENTRY GUIDANCE
      RENDEZVOUS GUIDANCE
      SPACE FLIGHT
      SPACE NAVIGATION
      STAR TRACKERS
      TERMINAL GUIDANCE

SPACECRAFT INSTRUMENTS
UF   SPACECRAFT SENSORS
GS   **SPACECRAFT INSTRUMENTS**
      . LASER ALTIMETERS
      . SATELLITE INSTRUMENTS
      . . MULTISPECTRAL LINEAR ARRAYS
      . SPACECRAFT POSITION INDICATORS
RT   ASTRIONICS
      ATMOSPHERIC CLOUD PHYSICS LAB
        (SPACELAB)
      AUTONOMOUS SPACECRAFT CLOCKS
      BUBBLE TECHNIQUE
      FLIGHT INSTRUMENTS
      FLIGHT TEST INSTRUMENTS
      GUIDANCE SENSORS
      INSTRUMENT PACKAGES
      ∞INSTRUMENTS
      I2S CAMERAS
      MEASURING INSTRUMENTS
      ONBOARD EQUIPMENT
      SPACECRAFT EQUIPMENT

SPACECRAFT LANDING
GS   LANDING
      . **SPACECRAFT LANDING**
      . . HORIZONTAL SPACECRAFT LANDING
      . . LUNAR LANDING
      . . MARS LANDING
      . . PLANETARY LANDING
RT   AIRCRAFT LANDING
      APPROACH AND LANDING TESTS (STS)
      CRASH LANDING
      GLIDE LANDINGS
      HARD LANDING
      LANDING SIMULATION
      SOFT LANDING
      SOFT LANDING SPACECRAFT
      TERMINAL AREA ENERGY MANAGEMENT
      TOUCHDOWN
      VERTICAL LANDING
      WATER LANDING

SPACECRAFT LAUNCHING
UF   SATELLITE LAUNCHING
GS   LAUNCHING
      . **SPACECRAFT LAUNCHING**
RT   COUNTDOWN
      HEAVY LIFT LAUNCH VEHICLES
      LAUNCH DATES
      LAUNCH VEHICLES
      LAUNCH WINDOWS
      LAUNCHING PADS
      MISSILES
      ORBITAL LAUNCHING
      ORBITAL SHOTS
      POSTLAUNCH REPORTS
      PRELAUNCH SUMMARIES
      RAILGUN ACCELERATORS
      REUSABLE LAUNCH VEHICLES
      ROCKET LAUNCHING
      SPACE COMMERCIALIZATION
      SPACE SHUTTLE ASCENT STAGE

SPACECRAFT LUBRICATION
UF   SPACE ENVIRONMENTAL LUBRICATION
GS   LUBRICATION
      . **SPACECRAFT LUBRICATION**
RT   SELF LUBRICATING MATERIALS

SPACECRAFT MAINTENANCE
GS   MAINTENANCE
      . **SPACECRAFT MAINTENANCE**
RT   CHECKOUT
      PRELAUNCH TESTS
      SPACE VEHICLE CHECKOUT PROGRAM
      SPACECRAFT RELIABILITY
      TURNAROUND (STS)

SPACECRAFT MANEUVERS
UF   SATELLITE MANEUVERS
GS   MANEUVERS
      . **SPACECRAFT MANEUVERS**
      . . ORBITAL MANEUVERS
RT   CONTROL SIMULATION
      MANEUVERABILITY
      MANEUVERABLE SPACECRAFT
      SPACE FLIGHT

SPACECRAFT MODELS
GS   MODELS
      . **SPACECRAFT MODELS**
RT   AIRCRAFT MODELS
      DYNAMIC MODELS
      MATHEMATICAL MODELS
      SCALE MODELS

SPACECRAFT MODULES
GS   MODULES
      . **SPACECRAFT MODULES**
      . . COMMAND MODULES
      . . COMMAND SERVICE MODULES
      . . LANDING MODULES
      . . . LUNAR LANDING MODULES
      . . . . LUNAR MODULE
      . . . . . LSSM
      . . . MARS EXCURSION MODULE
      . . SIM
      SPACECRAFT COMPONENTS
      . **SPACECRAFT MODULES**
      . . COMMAND MODULES
RT   COMMAND MODULES
      COMPARTMENTS
      ORBITAL ASSEMBLY
      SERVICE MODULES
      SPACE CAPSULES
      SPACE ERECTABLE STRUCTURES

SPACECRAFT MOTION
SN   (NONTRAJECTORY MOTION)
RT   AERODYNAMIC BALANCE
      AERODYNAMIC STABILITY
      ATTITUDE STABILITY
      BUFFETING
      CONTROL STABILITY
      DYNAMIC STABILITY
      FLEXIBLE SPACECRAFT
      FLUTTER
      ∞MOTION
      MOTION STABILITY
      OSCILLATIONS
      SIDESLIP
      STABILITY
      TUMBLING MOTION
      VIBRATION

SPACECRAFT ORBITAL ASSEMBLY
USE   ORBITAL ASSEMBLY

SPACECRAFT ORBITS
GS   ORBITS
      . **SPACECRAFT ORBITS**
      . . SATELLITE ORBITS
      . . . GEOSYNCHRONOUS ORBITS
      . . . PARKING ORBITS
      . . . POLAR ORBITS
      . . . STATIONARY ORBITS
      . . . TWENTY-FOUR HOUR ORBITS
      . . TRANSFER ORBITS
      . . . INTERPLANETARY TRANSFER
         ORBITS
      . . TROJAN ORBITS
RT   CIRCULAR ORBITS
      EARTH ORBITS
      ELLIPTICAL ORBITS
      EQUATORIAL ORBITS
      LUNAR ORBITS
      ORBITAL MECHANICS
      ORBITAL POSITION ESTIMATION
      PLANETARY ORBITS
      SOLAR ORBITS

SPACECRAFT PERFORMANCE
RT   ASTRONAUT PERFORMANCE
      ∞PERFORMANCE

SPACECRAFT PERFORMANCE-*(CONT.)*
    POSTLAUNCH REPORTS

SPACECRAFT POSITION INDICATORS
GS   AIRCRAFT INSTRUMENTS
    . POSITION INDICATORS
    . . **SPACECRAFT POSITION INDICATORS**
    DISPLAY DEVICES
    . POSITION INDICATORS
    . . **SPACECRAFT POSITION INDICATORS**
    MEASURING INSTRUMENTS
    . INDICATING INSTRUMENTS
    . . POSITION INDICATORS
    . . . **SPACECRAFT POSITION INDICATORS**
    SPACECRAFT INSTRUMENTS
    . **SPACECRAFT POSITION INDICATORS**
RT   FLIGHT INSTRUMENTS
    HEAD-UP DISPLAYS
    ORBITAL POSITION ESTIMATION
    SPACE NAVIGATION

SPACECRAFT POWER SUPPLIES
GS   ELECTRIC POWER SUPPLIES
    . **SPACECRAFT POWER SUPPLIES**
RT   AUXILIARY POWER SOURCES
    CRYOCYCLE PRINCIPLE
    DIRECT POWER GENERATORS
    ELECTRIC BATTERIES
    ∞ENERGY SOURCES
    NICKEL HYDROGEN BATTERIES
    NUCLEAR AUXILIARY POWER UNITS
    POWER MODULES (STS)
    ∞POWER SUPPLIES
    PROPELLANTS

SPACECRAFT PRELAUNCH TESTS
USE   SPACE VEHICLE CHECKOUT PROGRAM

SPACECRAFT PROPULSION
GS   PROPULSION
    . **SPACECRAFT PROPULSION**
    . . ELECTROMAGNETIC PROPULSION
    . . . MASS DRIVERS (PAYLOAD DELIVERY)
    . . ELECTROSTATIC PROPULSION
    . . . ION PROPULSION
    . . PHOTONIC PROPULSION
    . . PLASMA PROPULSION
    . . SOLAR PROPULSION
    . . . SOLAR ELECTRIC PROPULSION
    . . . SOLAR THERMAL PROPULSION
RT   CHEMICAL PROPULSION
    DESCENT PROPULSION SYSTEMS
    ELECTRIC PROPULSION
    LASER PROPULSION
    LOW THRUST PROPULSION
    MAGNETOPLASMADYNAMICS
    NUCLEAR ELECTRIC PROPULSION
    NUCLEAR PROPULSION
    POST BOOST PROPULSION SYSTEM
    PROPELLANTS
    ROCKET ENGINES
    ROVER PROJECT
    SOLAR SAILS
    SPACE FLIGHT
    SPACE TUGS

SPACECRAFT RADIATORS
UF   SPACE RADIATORS
GS   HEAT RADIATORS
    . **SPACECRAFT RADIATORS**
RT   CONDENSERS (LIQUEFIERS)
    COOLING
    COOLING SYSTEMS
    RADIATIVE HEAT TRANSFER
    SOLAR REFLECTORS

SPACECRAFT RECOVERY
UF   SATELLITE CAPTURE
    SNATCHING
RT   BOOSTER RECOVERY
    DISCOVERER RECOVERY CAPSULES
    RECOVERABLE SPACECRAFT
    ∞RECOVERY
    RECOVERY PARACHUTES
    RECOVERY ZONES
    RESCUE OPERATIONS
    REUSABLE LAUNCH VEHICLES
    SPACE SHUTTLE ORBITERS
    SPACE SHUTTLES
    WATER LANDING

SPACECRAFT REENTRY
GS   ATMOSPHERIC ENTRY

SPACECRAFT REENTRY-*(CONT.)*
    . REENTRY
    . . **SPACECRAFT REENTRY**
    . . . UNCONTROLLED REENTRY (SPACECRAFT)
    SPACE FLIGHT
    . **SPACECRAFT REENTRY**
RT   EARTH-VENUS TRAJECTORIES
    ENTRY GUIDANCE (STS)
    FLIGHT MECHANICS
    HYPERSONIC REENTRY
    LIFTING REENTRY VEHICLES
    MANNED REENTRY
    RETURN TO EARTH SPACE FLIGHT
    SATELLITE LIFETIME

SPACECRAFT RELIABILITY
GS   RELIABILITY
    . **SPACECRAFT RELIABILITY**
RT   CIRCUIT RELIABILITY
    COMPONENT RELIABILITY
    CONTROLLABILITY
    PRELAUNCH PROBLEMS
    QUALITY CONTROL
    SPACECRAFT MAINTENANCE

SPACECRAFT RENDEZVOUS
USE   SPACE RENDEZVOUS

SPACECRAFT SENSORS
USE   SPACECRAFT INSTRUMENTS

SPACECRAFT SHIELDING
GS   SHIELDING
    . **SPACECRAFT SHIELDING**
RT   HEAT SHIELDING
    ∞INSULATED STRUCTURES
    METEOROID PROTECTION
    NOSE CONES
    RADIATION SHIELDING
    RADIO FREQUENCY SHIELDING
    REENTRY SHIELDING
    REUSABLE HEAT SHIELDING
    SAFETY DEVICES
    SOLAR RADIATION SHIELDING

SPACECRAFT STABILITY
UF   SATELLITE ATTITUDE DISTURBANCE
GS   DYNAMIC CHARACTERISTICS
    . DYNAMIC STABILITY
    . . MOTION STABILITY
    . . . **SPACECRAFT STABILITY**
    STABILITY
    . DYNAMIC STABILITY
    . . MOTION STABILITY
    . . . **SPACECRAFT STABILITY**
RT   AERODYNAMIC BALANCE
    AERODYNAMIC STABILITY
    ATTITUDE STABILITY
    BUFFETING
    CONTROL STABILITY
    COUNTERBALANCES
    DIRECTIONAL STABILITY
    DISCOS (SATELLITE ATTITUDE CONTROL)
    DUAL SPIN SPACECRAFT
    LATERAL STABILITY
    LIQUID SLOSHING
    LONGITUDINAL STABILITY
    LOW SPEED STABILITY
    NUTATION DAMPERS
    SATELLITE PERTURBATION
    TUMBLING MOTION
    WIND TUNNEL STABILITY TESTS

SPACECRAFT STERILIZATION
GS   STERILIZATION
    . **SPACECRAFT STERILIZATION**
RT   CHEMICAL STERILIZATION
    EXOBIOLOGY
    PLANETARY QUARANTINE
    PURIFICATION
    STERILIZATION EFFECTS

SPACECRAFT STRUCTURES
GS   **SPACECRAFT STRUCTURES**
    . MARS 3 SPACECRAFT
RT   AIRCRAFT STRUCTURES
    FOLDING STRUCTURES
    FUEL TANKS
    MARS 3 SPACECRAFT
    METEOROID PROTECTION
    ORBITAL ASSEMBLY
    ROCKET ENGINES
    SATELLITE DESIGN

SPACECRAFT STRUCTURES-*(CONT.)*
    SPACE ERECTABLE STRUCTURES
    STRUCTURAL DESIGN
    ∞STRUCTURES

SPACECRAFT SURVIVABILITY
RT   AIRCRAFT SURVIVABILITY
    SPACECRAFT DEFENSE
    SURVIVAL
    UNCONTROLLED REENTRY (SPACECRAFT)
    VULNERABILITY

SPACECRAFT TELEVISION
GS   COMMUNICATION EQUIPMENT
    . **SPACECRAFT TELEVISION**
    . . DIGITAL SPACECRAFT TELEVISION
    . . RANGER BLOCK 3 TELEVISION SYSTEM
    . . SATELLITE TELEVISION
    TELECOMMUNICATION
    . **SPACECRAFT TELEVISION**
    . . DIGITAL SPACECRAFT TELEVISION
    . . SATELLITE TELEVISION
    TELEVISION SYSTEMS
    . **SPACECRAFT TELEVISION**
    . . DIGITAL SPACECRAFT TELEVISION
    . . RANGER BLOCK 3 TELEVISION SYSTEM
    . . SATELLITE TELEVISION
RT   COLOR TELEVISION
    SATELLITE TRANSMISSION
    STEREOTELEVISION
    TELEVISION TRANSMISSION

SPACECRAFT TEMPERATURE
RT   HEAT PIPES
    SATELLITE TEMPERATURE
    SPACECRAFT DESIGN
    TEMPERATURE CONTROL

SPACECRAFT TRACKING
GS   TRACKING (POSITION)
    . **SPACECRAFT TRACKING**
    . . SATELLITE TRACKING
    . . . SATELLITE-TO-SATELLITE TRACKING
RT   ADVANCED RANGE INSTRUMENTATION SHIP
    DEEP SPACE NETWORK
    MINITRACK SYSTEM
    MISSILE TRACKING
    OPTICAL TRACKING
    POLYSTATION DOPPLER TRACKING SYSTEM
    RADAR TRACKING
    RADIO TRACKING
    SPACE DETECTION AND TRACKING SYSTEM
    SPACE SURVEILLANCE (GROUND BASED)
    SPACE SURVEILLANCE (SPACEBORNE)
    TRACKING NETWORKS
    TRACKING STATIONS
    TRANSPONDER CONTROL GROUP
    UNIFIED S BAND

SPACECRAFT TRACKING AND DATA NETWORK
USE   STDN (NETWORK)

SPACECRAFT TRAJECTORIES
GS   TRAJECTORIES
    . **SPACECRAFT TRAJECTORIES**
    . . EARTH-VENUS TRAJECTORIES
    . . INTERPLANETARY TRAJECTORIES
    . . . EARTH-MARS TRAJECTORIES
    . . . EARTH-MERCURY TRAJECTORIES
    . . LUNAR TRAJECTORIES
    . . . CIRCUMLUNAR TRAJECTORIES
    . . . EARTH-MOON TRAJECTORIES
    . . . MOON-EARTH TRAJECTORIES
RT   ASCENT TRAJECTORIES
    DESCENT TRAJECTORIES
    EARTH ORBITAL RENDEZVOUS
    FLIGHT MECHANICS
    GODDARD TRAJECTORY DETERMINATION SYSTEM
    HYPERBOLIC TRAJECTORIES
    INTERORBITAL TRAJECTORIES
    LUNAR ORBITAL RENDEZVOUS
    ∞MOTION
    ORBITAL RENDEZVOUS
    RADIO OCCULTATION
    REENTRY TRAJECTORIES
    RENDEZVOUS TRAJECTORIES
    ROUND TRIP TRAJECTORIES

SPACECREW TRANSFER
UF   INTERVEHICLE SPACECREW TRANSFER
RT   APOLLO SOYUZ TEST PROJECT
      COMMAND MODULES
      MANNED SPACE FLIGHT
      RENDEZVOUS SPACECRAFT
      SPACE LOGISTICS

SPACECREWS
GS   PERSONNEL
      . CREWS
      . . FLIGHT CREWS
      . . . **SPACECREWS**
      . FLYING PERSONNEL
      . . FLIGHT CREWS
      . . . **SPACECREWS**
RT   ASTRONAUTS
      COSMONAUTS
      CREW EXPERIMENT STATIONS
      CREW OBSERVATION STATIONS
      CREW PROCEDURES (INFLIGHT)
      CREW PROCEDURES (PREFLIGHT)
      CREW STATIONS
      CREW WORKSTATIONS
      SPACE HABITATS

SPACELAB
GS   PAYLOADS
      . SPACE SHUTTLE PAYLOADS
      . . **SPACELAB**
RT   ADVANCED TECHNOLOGY LABORATORY
      ANNULAR SUSPENSION AND POINTING
        SYSTEM
      EXPOS (SPACELAB PAYLOAD)
      GEOPHYSICAL FLUID FLOW CELLS
      GRIST (TELESCOPE)
      LIRTS (TELESCOPE)
      NASA PROGRAMS
      OSTA-2 PAYLOAD
      SEPAC (PAYLOAD)
      SKYLAB PROGRAM
      SPACE SHUTTLES
      SPACE TRANSPORTATION SYSTEM
      SPACEBORNE EXPERIMENTS
      STARLAB

SPACELAB PAYLOADS
GS   PAYLOADS
      . **SPACELAB PAYLOADS**
      . . ATMOSPHERIC GENERAL
            CIRCULATION EXPERIMENT
      . . GEOPHYSICAL FLUID FLOW CELLS
      . . SOLAR CELL CALIBRATION FACILITY
RT   ANNULAR SUSPENSION AND POINTING
        SYSTEM
      SORTIE SYSTEMS

SPACELAB SIMULATION FLIGHTS
USE   ASSESS PROGRAM

SPACELAB UV-OPTICAL TELESCOPE FACILITY
USE   STARLAB

SPACERS
RT   BUSHINGS
      DIVIDERS
      FASTENERS
      INSERTS
      ISOLATORS
      SEPARATORS
      SPACING
      WASHERS (SPACERS)

SPACETENNAS
GS   ANTENNAS
      . RADIO ANTENNAS
      . . MICROWAVE ANTENNAS
      . . . **SPACETENNAS**
      MICROWAVE EQUIPMENT
      . MICROWAVE ANTENNAS
      . . **SPACETENNAS**
RT   MICROWAVE TRANSMISSION
      RECTENNAS

SPACING
GS   **SPACING**
      . AIRCRAFT APPROACH SPACING
RT   ALTITUDE CONTROL
      ATTITUDE CONTROL
      CLEARANCES
      INTERVALS
      ISOLATION
      POSITIONING
      ∞ SEPARATION
      SPACERS

SPACING-(CONT.)
      THICKNESS

SPADATS (TRACKING SYSTEM)
USE   SPACE DETECTION AND TRACKING
        SYSTEM

SPAIN
GS   NATIONS
      . **SPAIN**
RT   ANDORRA
      EUROPE
      PYRENEES MOUNTAINS (EUROPE)
      SPANISH SAHARA

SPALLATION
GS   NUCLEAR RADIATION
      . **SPALLATION**
      NUCLEAR REACTIONS
      . **SPALLATION**
RT   PARTICLE PRODUCTION

SPALLING
RT   CHIPPING
      FLAKING
      FRACTURING
      FRAGMENTATION
      NUCLEAR REACTIONS
      WEAR
      WEAR TESTS

∞ **SPAN**
SN   (USE OF A MORE SPECIFIC TERM IS
      RECOMMENDED--CONSULT THE TERMS
      LISTED BELOW)
RT   ASPECT RATIO
      DIMENSIONS
      LIFE SPAN
      WIDTH
      WING SPAN

SPANISH SAHARA
RT   AFRICA
      NATIONS
      SPAIN

SPANLOADER AIRCRAFT
GS   TRANSPORT AIRCRAFT
      . CARGO AIRCRAFT
      . . **SPANLOADER AIRCRAFT**
RT   ∞ AIRCRAFT
      SUPERCRITICAL WINGS
      SWEPT WINGS

SPANWISE BLOWING
GS   **SPANWISE BLOWING**
      . BLOWING
RT   CROSS FLOW
      EXTERNALLY BLOWN FLAPS
      JET FLOW
      LIFT AUGMENTATION
      PRESSURE DISTRIBUTION
      WING SPAN

SPAR (ROCKET)
USE   SPACE PROCESSING APPLICATIONS
        ROCKET

SPARE PARTS
RT   ∞ COMPONENTS
      DAMAGE ASSESSMENT
      DOWNTIME
      ENGINE PARTS
      INVENTORY MANAGEMENT
      LOGISTICS MANAGEMENT
      MAINTENANCE
      MODULES
      REDUNDANT COMPONENTS
      RETIREMENT FOR CAUSE

SPARK CHAMBERS
GS   IONIZATION CHAMBERS
      . **SPARK CHAMBERS**
      MEASURING INSTRUMENTS
      . COUNTERS
      . . RADIATION COUNTERS
      . . . **SPARK CHAMBERS**
      . RADIATION MEASURING INSTRUMENTS
      . . RADIATION COUNTERS
      . . . **SPARK CHAMBERS**
RT   BUBBLE CHAMBERS
      ∞ CHAMBERS
      CLOUD CHAMBERS
      ELECTRIC SPARKS

SPARK CHAMBERS-(CONT.)
      NEUTRON COUNTERS

SPARK DISCHARGES
USE   ELECTRIC SPARKS

SPARK GAPS
GS   GAPS
      . **SPARK GAPS**
RT   ARC GENERATORS
      DIELECTRICS
      ELECTRIC FIELDS
      ELECTRIC SPARKS
      ELECTRICAL FAULTS
      MULTIPACTOR DISCHARGES
      POTENTIAL GRADIENTS
      TRIGATRONS

SPARK IGNITION
GS   IGNITION
      . **SPARK IGNITION**
RT   COMBUSTION
      ELECTRIC IGNITION
      ELECTRIC SPARKS
      EUDIOMETERS

SPARK MACHINING
UF   ELECTROEROSION
      ELECTROSTATIC EROSION
GS   CUTTING
      . **SPARK MACHINING**
      MACHINING
      . **SPARK MACHINING**
RT   ELECTROFORMING
      EROSION
      METAL CUTTING
      PIERCING

SPARK PLUGS
RT   ARC GENERATORS
      COMBUSTION CHAMBERS
      ELECTRIC SPARKS
      IGNITERS
      IGNITION SYSTEMS
      INTERNAL COMBUSTION ENGINES

SPARK SHADOWGRAPH PHOTOGRAPHY
USE   SHADOWGRAPH PHOTOGRAPHY

SPARKS
GS   **SPARKS**
      . ELECTRIC SPARKS
RT   IGNITION

SPARROW MISSILES
GS   MISSILES
      . AIR TO AIR MISSILES
      . . **SPARROW MISSILES**
      . . . SPARROW 2 MISSILE
      . . . SPARROW 3 MISSILE
RT   SOLID PROPELLANT ROCKET ENGINES

SPARROW 2 MISSILE
GS   MISSILES
      . AIR TO AIR MISSILES
      . . SPARROW MISSILES
      . . . **SPARROW 2 MISSILE**
RT   SOLID PROPELLANT ROCKET ENGINES

SPARROW 3 MISSILE
GS   MISSILES
      . AIR TO AIR MISSILES
      . . SPARROW MISSILES
      . . . **SPARROW 3 MISSILE**
RT   LIQUID PROPELLANT ROCKET ENGINES

SPARTAN MISSILE
GS   MISSILES
      . ANTIMISSILE MISSILES
      . . **SPARTAN MISSILE**
RT   NIKE-ZEUS MISSILE
      SENTINEL SYSTEM
      SPRINT MISSILE
      SURFACE TO AIR MISSILES

SPARTAN SATELLITES
GS   OBSERVATORIES
      . ASTRONOMICAL OBSERVATORIES
      . . ASTRONOMICAL SATELLITES
      . . . **SPARTAN SATELLITES**
RT   ASTROPHYSICS
      SOLAR PHYSICS
      ULTRAVIOLET ASTRONOMY

SPAS (ESA PLATFORMS)
USE   SHUTTLE PALLET SATELLITES

SPASMS
GS   MUSCULAR FUNCTION
    . **SPASMS**
RT   CONTRACTION
    INVOLUNTARY ACTIONS
    MUSCLES

SPATIAL DEPENDENCIES
GS   DEPENDENCE
    . **SPATIAL DEPENDENCIES**

SPATIAL DISTRIBUTION
UF   MOLIERE FORMULA
    SPATIAL ISOTROPY
GS   DISTRIBUTION (PROPERTY)
    . **SPATIAL DISTRIBUTION**
    .. STAR DISTRIBUTION
RT   ANISOTROPY
    ION DISTRIBUTION
    METEOROID CONCENTRATION
    PARTICLE DENSITY (CONCENTRATION)
    POSITION (LOCATION)
    STEREOCHEMISTRY
    TEMPORAL DISTRIBUTION
    VERTICAL DISTRIBUTION

SPATIAL FILTERING
GS   FILTRATION
    . **SPATIAL FILTERING**
RT   ABERRATION
    ATTENUATION
    AUGMENTATION
    BLURRING
    ∞FILTERS
    HOLOGRAPHY
    IMAGES
    ∞NOISE
    PHOTOGRAPHS
    PHOTOINTERPRETATION
    PLANE WAVES
    RESOLUTION

SPATIAL ISOTROPY
USE   ISOTROPY
    SPATIAL DISTRIBUTION

SPATIAL MARCHING
RT   ACOUSTIC DUCTS
    DUCT GEOMETRY
    NUMERICAL ANALYSIS
    TIME MARCHING

SPATIAL ORIENTATION
USE   ATTITUDE (INCLINATION)

SPATIAL RESOLUTION
GS   RESOLUTION
    . **SPATIAL RESOLUTION**
RT   ATMOSPHERIC CORRECTION
    HIGH RESOLUTION
    IMAGE PROCESSING
    IMAGE RESOLUTION
    IMAGING TECHNIQUES
    SPECTRAL RESOLUTION
    TEMPORAL RESOLUTION

SPECIES DIFFUSION
GS   DIFFUSION
    . **SPECIES DIFFUSION**
RT   EVOLUTION (DEVELOPMENT)
    GENETICS

SPECIFIC GRAVITY
USE   DENSITY (MASS/VOLUME)

SPECIFIC HEAT
UF   DEBYE TEMPERATURE
    HEAT CAPACITY
GS   THERMODYNAMIC PROPERTIES
    . THERMOPHYSICAL PROPERTIES
    .. **SPECIFIC HEAT**
    ... HEAT OF SOLUTION
RT   ENTHALPY
    EQUIPARTITION THEOREM
    GRUNEISEN CONSTANT
    HEAT BUDGET
    HEAT OF FUSION
    ION TEMPERATURE
    LEWIS NUMBERS
    MELTING POINTS
    NEEL TEMPERATURE

SPECIFIC HEAT-(CONT.)
    THERMAL CONDUCTIVITY
    THERMAL RESISTANCE

SPECIFIC IMPULSE
RT   IMPULSES
    MASS FLOW RATE
    PROPELLANT MASS RATIO
    PROPELLANTS
    PROPULSION SYSTEM PERFORMANCE
    PROPULSIVE EFFICIENCY
    ROCKET PROPELLANTS
    ROCKET THRUST
    THERMODYNAMIC EFFICIENCY
    THRUST

SPECIFICATIONS
GS   **SPECIFICATIONS**
    . AIRCRAFT SPECIFICATIONS
    . EQUIPMENT SPECIFICATIONS
    . FUNCTIONAL DESIGN SPECIFICATIONS
RT   AIRCRAFT PERFORMANCE
    COMMONALITY
    DRAWINGS
    INSPECTION
    MAINTENANCE
    ∞MATERIALS TESTS
    MECHANICAL PROPERTIES
    NAMING
    PERFORMANCE TESTS
    PROCESS CONTROL (INDUSTRY)
    PROCUREMENT
    QUALITY
    QUALITY CONTROL
    RELIABILITY
    REQUIREMENTS
    STANDARDIZATION
    STANDARDS
    TECHNICAL WRITING
    TOLERANCES (MECHANICS)
    USER REQUIREMENTS

SPECIMEN GEOMETRY
GS   GEOMETRY
    . **SPECIMEN GEOMETRY**
RT   FATIGUE TESTS
    LOAD TESTS
    MECHANICAL PROPERTIES
    TENSILE TESTS

SPECIMENS
RT   MARS SURFACE SAMPLES
    SAMPLES
    SAMPLING

SPECKLE PATTERNS
RT   DIFFRACTION PATTERNS
    HOLOGRAPHY
    LASER OUTPUTS
    LIGHT SCATTERING
    ∞PATTERNS
    SURFACE ROUGHNESS EFFECTS

SPECTRA
UF   OPTICAL SPECTRUM
GS   **SPECTRA**
    . ABSORPTION SPECTRA
    .. FRAUNHOFER LINES
    .. HERZBERG BANDS
    .. TELLURIC LINES
    . ATOMIC SPECTRA
    . CONTINUOUS SPECTRA
    . ENERGY SPECTRA
    .. ELECTRONIC SPECTRA
    . NEUTRON SPECTRA
    . MASS SPECTRA
    . MOLECULAR SPECTRA
    . ELECTRONIC SPECTRA
    .. RAMAN SPECTRA
    . VIBRATIONAL SPECTRA
    . NOISE SPECTRA
    . OXYGEN SPECTRA
    . PLASMA SPECTRA
    . POWER SPECTRA
    .. CEPSTRA
    . RADIATION SPECTRA
    .. ELECTROMAGNETIC SPECTRA
    ... GAMMA RAY SPECTRA
    ... INFRARED SPECTRA
    ... LINE SPECTRA
    .... BALMER SERIES
    .... D LINES
    .... ELECTRONIC SPECTRA
    .... FRAUNHOFER LINES
    .... H LINES

SPECTRA-(CONT.)
    ..... H ALPHA LINE
    ..... H BETA LINE
    ..... H GAMMA LINE
    ... K LINES
    ... LYMAN SPECTRA
    .... PASCHEN SERIES
    ... RYDBERG SERIES
    .... TELLURIC LINES
    ... RADIO SPECTRA
    ... MICROWAVE SPECTRA
    .. RAMAN SPECTRA
    . STELLAR SPECTRA
    ... SOLAR SPECTRA
    .. UBV SPECTRA
    .. ULTRAVIOLET SPECTRA
    . VIBRATIONAL SPECTRA
    .. VISIBLE SPECTRUM
    . X RAY SPECTRA
    .. EMISSION SPECTRA
    . HERZBERG BANDS
    . SHOCK SPECTRA
    . SPECTRAL BANDS
    . FRAUNHOFER LINES
    .. HERZBERG BANDS
    .. PHOTOLUMINESCENT BANDS
    . SCHUMANN-RUNGE BANDS
    .. SWAN BANDS
    . TELLURIC LINES
    .. VEGARD-KAPLAN BANDS
RT   ASTRONOMICAL SPECTROSCOPY
    COLOR
    EXCITONS
    FLUX DENSITY
    GAMMA RAY SPECTRA
    GAMMA RAY SPECTROMETERS
    ISOELECTRONIC SEQUENCE
    SPECTRAL SENSITIVITY
    SPECTRAL SHIFT CONTROL
    SPECTRAL THEORY
    SPECTROGRAMS
    SPECTROGRAPHS
    SPECTROMETERS
    SPECTROSCOPY
    SPECTRUM ANALYSIS
    TRANSITION PROBABILITIES

SPECTRAL ABSORPTION
USE   ABSORPTION SPECTRA

SPECTRAL ANALYSIS
USE   SPECTRUM ANALYSIS

SPECTRAL BANDS
GS   SPECTRA
    . **SPECTRAL BANDS**
    .. FRAUNHOFER LINES
    .. HERZBERG BANDS
    .. PHOTOLUMINESCENT BANDS
    .. SCHUMANN-RUNGE BANDS
    .. SWAN BANDS
    .. TELLURIC LINES
    .. VEGARD-KAPLAN BANDS
RT   BAND RATIOING
    ∞BANDS
    ELECTRONIC SPECTRA
    ENERGY BANDS
    FREQUENCIES
    LINE SPECTRA
    VISIBLE SPECTRUM
    WHITE NOISE

SPECTRAL CORRELATION
GS   CORRELATION
    . **SPECTRAL CORRELATION**
RT   ELECTROMAGNETIC SPECTRA
    SPECTROPHOTOGRAPHY

SPECTRAL EMISSION
GS   DECAY
    . EMISSION
    .. **SPECTRAL EMISSION**
RT   CONTINUOUS SPECTRA
    ELECTROMAGNETIC RADIATION
    EMITTANCE
    INCANDESCENCE
    LIGHT EMISSION
    LINE SPECTRA
    NONGRAY GAS
    ∞RADIATION
    SPECTROGRAMS
    SPECTROSCOPY
    SPECTRUM ANALYSIS
    SPONTANEOUS EMISSION
    WAVELENGTHS

SPECTRAL ENERGY DISTRIBUTION
GS   DISTRIBUTION (PROPERTY)
.   ENERGY DISTRIBUTION
. . **SPECTRAL ENERGY DISTRIBUTION**
RT   ∞ DISTRIBUTION
ELECTROMAGNETIC RADIATION
ENERGY SPECTRA
FINE STRUCTURE
LINE SPECTRA

SPECTRAL LINE WIDTH
GS   BANDWIDTH
.   **SPECTRAL LINE WIDTH**
RT   LINE SPECTRA
OSCILLATOR STRENGTHS

SPECTRAL LINES
USE   LINE SPECTRA

SPECTRAL METHODS
RT   COMPUTATIONAL FLUID DYNAMICS
DIFFERENTIAL EQUATIONS
SPECTRUM ANALYSIS

SPECTRAL NOISE
USE   WHITE NOISE

SPECTRAL RECONNAISSANCE
GS   RECONNAISSANCE
.   **SPECTRAL RECONNAISSANCE**
RT   EARTH RESOURCES
ELECTROMAGNETIC SPECTRA
MULTISPECTRAL BAND SCANNERS
MULTISPECTRAL PHOTOGRAPHY
MULTISPECTRAL RADAR
PHOTORECONNAISSANCE
RADAR PHOTOGRAPHY
SATELLITE-BORNE PHOTOGRAPHY
SPECTROPHOTOGRAPHY

SPECTRAL REFLECTANCE
GS   SURFACE PROPERTIES
.   **SPECTRAL REFLECTANCE**
RT   ELECTROMAGNETIC PROPERTIES
PLANT STRESS
SPECTROMETERS
SPECTROSCOPY
SPECTRUM ANALYSIS
VEGETATIVE INDEX

SPECTRAL RESOLUTION
GS   RESOLUTION
.   **SPECTRAL RESOLUTION**
RT   ANALOG COMPUTERS
LINE SPECTRA
Q FACTORS
RADAR RESOLUTION
RADIOMETRIC RESOLUTION
SPATIAL RESOLUTION
SPECTRUM ANALYSIS

SPECTRAL SENSITIVITY
GS   SENSITIVITY
.   **SPECTRAL SENSITIVITY**
RT   ∞ FREQUENCY RESPONSE
INSTRUMENT ERRORS
PHOTOTHERMAL CONVERSION
SPECTRA

SPECTRAL SHIFT CONTROL
RT   ∞ CONTROL
SPECTRA

SPECTRAL SHIFT CONTROL REACTOR
GS   NUCLEAR REACTORS
.   LIQUID COOLED REACTORS
. . WATER COOLED REACTORS
. . . PRESSURIZED WATER REACTORS
. . . . **SPECTRAL SHIFT CONTROL
REACTOR**
RT   ∞ CONTROL

SPECTRAL SIGNATURES
GS   SIGNATURES
.   **SPECTRAL SIGNATURES**
RT   CEPSTRAL ANALYSIS
CHEMICAL ANALYSIS
CHEMICAL COMPOSITION
CROP IDENTIFICATION
EMISSION SPECTRA
IDENTIFYING
OPTICAL MEASUREMENT
SPECTRUM ANALYSIS

SPECTRAL THEORY
RT   LYMAN SPECTRA
SPECTRA
∞ THEORIES

SPECTROGRAMS
RT   LINE SPECTRA
SPECTRA
SPECTRAL EMISSION
SPECTROGRAPHS
SPECTROPHOTOGRAPHY
SPECTROSCOPY
SPECTRUM ANALYSIS

SPECTROGRAPHS
GS   **SPECTROGRAPHS**
.   HIGH DISPERSION SPECTROGRAPHS
RT   SPECTRA
SPECTROGRAMS
SPECTROMETERS
SPECTROSCOPIC ANALYSIS
SPECTROSCOPY

SPECTROHELIOGRAPHS
UF   HELIOGRAPHS
HELIOGRAPHY
SPECTROHELIOSCOPES
GS   IMAGERY
.   **SPECTROHELIOGRAPHS**
MEASURING INSTRUMENTS
.   RADIATION MEASURING INSTRUMENTS
. . ACTINOMETERS
. . . **SPECTROHELIOGRAPHS**
.   SPECTROMETERS
. . **SPECTROHELIOGRAPHS**
OPTICAL EQUIPMENT
.   **SPECTROHELIOGRAPHS**
SOLAR INSTRUMENTS
.   **SPECTROHELIOGRAPHS**
RT   BLACK AND WHITE PHOTOGRAPHY
CORONAGRAPHS
SOLAR SPECTROMETERS
STARSAT TELESCOPE

SPECTROHELIOSCOPES
USE   SPECTROHELIOGRAPHS

SPECTROMETERS
UF   SPECTROMETRY
SPECTROSCOPES
GS   MEASURING INSTRUMENTS
.   **SPECTROMETERS**
. . EBERT SPECTROMETERS
. . FABRY-PEROT SPECTROMETERS
. . GAMMA RAY SPECTROMETERS
. . INFRARED SPECTROMETERS
. . . FILTER WHEEL INFRARED
SPECTROMETERS
. . LASER SPECTROMETERS
. . MASS SPECTROMETERS
. . MICROWAVE SPECTROMETERS
. . NEUTRON SPECTROMETERS
. . SOLAR BACKSCATTER UV
SPECTROMETER
. . SOLAR SPECTROMETERS
. . SPECTROHELIOGRAPHS
. . TIME OF FLIGHT SPECTROMETERS
. . ULTRAVIOLET SPECTROMETERS
. . . HIGH DISPERSION
SPECTROGRAPHS
RT   ACTINOMETERS
CHEMICAL ANALYSIS
DIFFRACTOMETERS
ELECTRON PROBES
GONIOMETERS
INFRARED SPECTROSCOPY
MICHELSON INTERFEROMETERS
OPTICAL EQUIPMENT
OPTICAL MEASUREMENT
PHOTOGONIOMETERS
PHOTOGRAPHIC MEASUREMENT
PHOTOMETERS
RADIATION COUNTERS
SOLAR INSTRUMENTS
SPECTRA
SPECTRAL REFLECTANCE
SPECTROGRAPHS
SPECTRORADIOMETERS
SPECTROSCOPIC ANALYSIS
SPECTROSCOPY
SPECTRUM ANALYSIS

SPECTROMETRY
USE   SPECTROMETERS

SPECTROPHOTOGRAPHY
GS   IMAGERY
.   **SPECTROPHOTOGRAPHY**
PHOTOGRAPHY
.   **SPECTROPHOTOGRAPHY**
SPECTROSCOPY
.   **SPECTROPHOTOGRAPHY**
RT   BLACK AND WHITE PHOTOGRAPHY
GROUND TRUTH
SPECTRAL CORRELATION
SPECTRAL RECONNAISSANCE
SPECTROGRAMS

SPECTROPHOTOMETERS
GS   MEASURING INSTRUMENTS
.   OPTICAL MEASURING INSTRUMENTS
. . **SPECTROPHOTOMETERS**
. . . INFRARED SPECTROPHOTOMETERS
. . . ULTRAVIOLET
SPECTROPHOTOMETERS
.   RADIATION MEASURING INSTRUMENTS
. . ACTINOMETERS
. . . **SPECTROPHOTOMETERS**
. . . . INFRARED
SPECTROPHOTOMETERS
. . . . ULTRAVIOLET
SPECTROPHOTOMETERS
OPTICAL EQUIPMENT
.   OPTICAL MEASURING INSTRUMENTS
. . **SPECTROPHOTOMETERS**
. . . INFRARED SPECTROPHOTOMETERS
. . . ULTRAVIOLET
SPECTROPHOTOMETERS
RT   CHEMICAL ANALYSIS
DUOCHROMATORS
MONOCHROMATORS
OPTICAL MEASUREMENT
PHOTOMETERS
RADIOMETERS
SPECTRORADIOMETERS
SPECTROSCOPIC ANALYSIS
SPECTROSCOPY

SPECTROPHOTOMETRY
GS   OPTICAL MEASUREMENT
.   PHOTOMETRY
. . **SPECTROPHOTOMETRY**
. . . STELLAR SPECTROPHOTOMETRY
SPECTROSCOPY
.   **SPECTROPHOTOMETRY**
. . STELLAR SPECTROPHOTOMETRY
RT   ASTRONOMICAL PHOTOMETRY
COLORIMETRY
SPECTROSCOPIC ANALYSIS

SPECTROPHOTOVOLTAICS
RT   ENERGY CONVERSION EFFICIENCY
ENERGY SPECTRA
SOLAR CELLS
SOLAR COLLECTORS

SPECTROPOLARIMETERS
USE   POLARIMETERS

SPECTRORADIOMETERS
GS   MEASURING INSTRUMENTS
.   RADIATION MEASURING INSTRUMENTS
. . ACTINOMETERS
. . . RADIOMETERS
. . . . **SPECTRORADIOMETERS**
RT   SPECTROMETERS
SPECTROPHOTOMETERS

SPECTROSCOPES
USE   SPECTROMETERS

SPECTROSCOPIC ANALYSIS
SN   (FOR SPECTROSCOPIC TOOLS IN
CHEMICAL ANALYSIS)
GS   CHEMICAL TESTS
.   CHEMICAL ANALYSIS
. . **SPECTROSCOPIC ANALYSIS**
SPECTROSCOPY
.   **SPECTROSCOPIC ANALYSIS**
. . FLAME SPECTROSCOPY
RT   AUGER SPECTROSCOPY
AURORAL SPECTROSCOPY
ELECTROPHOTOMETRY
FRAUNHOFER LINE DISCRIMINATORS
GAS SPECTROSCOPY
INFRARED SPECTROSCOPY
LASER SPECTROSCOPY
MAGNETIC SPECTROSCOPY
MASS SPECTROSCOPY
METALLICITY

SPECTROSCOPIC ANALYSIS-*(CONT.)*
MICROANALYSIS
MOLECULAR SPECTROSCOPY
NEUTRON ACTIVATION ANALYSIS
NUCLEAR RADIATION SPECTROSCOPY
PHOTOELECTRON SPECTROSCOPY
QUALITATIVE ANALYSIS
QUANTITATIVE ANALYSIS
RAMAN SPECTROSCOPY
SPECTROGRAPHS
SPECTROMETERS
SPECTROPHOTOMETERS
SPECTROPHOTOMETRY
ULTRAVIOLET SPECTROSCOPY
VACUUM SPECTROSCOPY
X RAY SPECTROSCOPY

SPECTROSCOPIC TELESCOPES
UF   DIFFRACTION TELESCOPES
GS   OPTICAL EQUIPMENT
. ASTRONOMICAL TELESCOPES
. . **SPECTROSCOPIC TELESCOPES**
. . . MULTISPECTRAL TRACKING
TELESCOPES
. . . STRATOSCOPE TELESCOPES
TELESCOPES
. ASTRONOMICAL TELESCOPES
. **SPECTROSCOPIC TELESCOPES**
. . . MULTISPECTRAL TRACKING
TELESCOPES
. . . STRATOSCOPE TELESCOPES
RT   ASTRONOMICAL SPECTROSCOPY
REFLECTING TELESCOPES
REFRACTING TELESCOPES
STELLAR SPECTROPHOTOMETRY

SPECTROSCOPY
GS   **SPECTROSCOPY**
. ABSORPTION SPECTROSCOPY
. . OPTOGALVANIC SPECTROSCOPY
. ASTRONOMICAL SPECTROSCOPY
. AUGER SPECTROSCOPY
. AURORAL SPECTROSCOPY
. ELECTRON SPECTROSCOPY
. GAS SPECTROSCOPY
. . FLAME SPECTROSCOPY
. HOLOGRAPHIC SPECTROSCOPY
. INFRARED SPECTROSCOPY
. MAGNETIC SPECTROSCOPY
. MASS SPECTROSCOPY
. MOLECULAR SPECTROSCOPY
. . RAMAN SPECTROSCOPY
. NUCLEAR RADIATION SPECTROSCOPY
. OPTICAL EMISSION SPECTROSCOPY
. . LASER SPECTROSCOPY
. PHOTOACOUSTIC SPECTROSCOPY
. PHOTOELECTRON SPECTROSCOPY
. RADIO SPECTROSCOPY
. SPECTROPHOTOGRAPHY
. SPECTROPHOTOMETRY
. . STELLAR SPECTROPHOTOMETRY
. SPECTROSCOPIC ANALYSIS
. . FLAME SPECTROSCOPY
. ULTRASONIC SPECTROSCOPY
. ULTRAVIOLET SPECTROSCOPY
. VACUUM SPECTROSCOPY
. X RAY SPECTROSCOPY
RT   CHEMICAL ANALYSIS
CINESPECTROGRAPHS
COLORIMETRY
ELECTROPHOTOMETRY
FRAUNHOFER LINE DISCRIMINATORS
ISOELECTRONIC SEQUENCE
LALLEMAND CAMERAS
∞OPTICS
PHOTOMETRY
PRESSURE BROADENING
SPECTRA
SPECTRAL EMISSION
SPECTRAL REFLECTANCE
SPECTROGRAMS
SPECTROGRAPHS
SPECTROMETERS
SPECTROPHOTOMETERS
SPECTRUM ANALYSIS
TIME OF FLIGHT SPECTROMETERS
VISIBLE SPECTRUM
ZEEMAN EFFECT

SPECTRUM ANALYSIS
UF   SPECTRAL ANALYSIS
GS   **SPECTRUM ANALYSIS**
. CEPSTRAL ANALYSIS
. FLAME SPECTROSCOPY
. MAXIMUM ENTROPY METHOD

SPECTRUM ANALYSIS-*(CONT.)*
RT   ABSORPTION SPECTRA
∞ANALYZING
EMISSION SPECTRA
FREQUENCY ANALYZERS
FREQUENCY SCANNING
GAMMA RAY SPECTROMETERS
HOLOGRAPHIC SPECTROSCOPY
HYPERFINE STRUCTURE
KRAMERS-KRONIG FORMULA
LASER SPECTROSCOPY
LINE SPECTRA
MAGNETIC RESONANCE
OPTICAL RESONANCE
SIGNAL ANALYSIS
SPECTRA
SPECTRAL EMISSION
SPECTRAL METHODS
SPECTRAL REFLECTANCE
SPECTRAL RESOLUTION
SPECTRAL SIGNATURES
SPECTROGRAMS
SPECTROMETERS
SPECTROSCOPY
STARK EFFECT
TOROIDAL DISCHARGE
ULTRASONIC SPECTROSCOPY
ULTRAVIOLET SPECTROSCOPY
ZEEMAN EFFECT

SPECULAR REFLECTION
GS   REFLECTION
. **SPECULAR REFLECTION**
RT   DIFFUSE RADIATION
GLARE
MIRRORS

SPEECH
GS   **SPEECH**
. ARTICULATION
. CONVERSATION
. PHONEMES
. PHONETICS
. TALKING
. . WORDS (LANGUAGE)
. . . SYLLABLES
RT   ACOUSTICS
AUDITORY PERCEPTION
CONSONANTS (SPEECH)
ENGLISH LANGUAGE
LANGUAGES
LECTURES
LINGUISTICS
PHONEMICS
PUBLIC SPEAKING
SEMANTICS
SENTENCES
SYNTAX
VOICE
VOICE COMMUNICATION

SPEECH BASEBAND COMPRESSION
GS   COMPRESSING
. **SPEECH BASEBAND COMPRESSION**
RT   BANDWIDTH
VERBAL COMMUNICATION
VOCODERS
VOICE COMMUNICATION
WAVEFORMS

SPEECH DEFECTS
GS   DEFECTS
. **SPEECH DEFECTS**
RT   ARTICULATION
PHONEMICS
PHONETICS

SPEECH DISCRIMINATION
USE   SPEECH RECOGNITION

SPEECH RECOGNITION
UF   SPEECH DISCRIMINATION
GS   INTELLIGIBILITY
. **SPEECH RECOGNITION**
RECOGNITION
. **SPEECH RECOGNITION**
RT   CEPSTRAL ANALYSIS
PHONEMES
PHONEMICS
PHONETICS
SENSORY DISCRIMINATION
VOICE CONTROL

SPEECHES
USE   LECTURES

SPEED
USE   VELOCITY

SPEED CONTROL
UF   SPEED REGULATION
RT   AUTOMATIC CONTROL
∞CONTROL
CONTROL EQUIPMENT
CONTROLLERS
ENGINE CONTROL
HELICOPTER CONTROL
MANUAL CONTROL
REGULATORS

SPEED INDICATORS
UF   PRESTON TUBES
SPEEDOMETERS
GS   AIRCRAFT INSTRUMENTS
. **SPEED INDICATORS**
. . ANEMOMETERS
. . . DRAG FORCE ANEMOMETERS
. . . HOT-WIRE ANEMOMETERS
. . . SONIC ANEMOMETERS
. . TACHOMETERS
DISPLAY DEVICES
. **SPEED INDICATORS**
. . ANEMOMETERS
. . . DRAG FORCE ANEMOMETERS
. . . HOT-FILM ANEMOMETERS
. . . HOT-WIRE ANEMOMETERS
. . . SONIC ANEMOMETERS
. . TACHOMETERS
MEASURING INSTRUMENTS
. INDICATING INSTRUMENTS
. . **SPEED INDICATORS**
. . . ANEMOMETERS
. . . . DRAG FORCE ANEMOMETERS
. . . . HOT-FILM ANEMOMETERS
. . . . HOT-WIRE ANEMOMETERS
. . . . LASER ANEMOMETERS
. . . . SONIC ANEMOMETERS
. . . TACHOMETERS
RT   ACCELEROMETERS
APPROACH INDICATORS
FLIGHT INSTRUMENTS
FLOWMETERS
LANDING INSTRUMENTS
PITOT TUBES
VELOCITY MEASUREMENT

SPEED REGULATION
USE   SPEED CONTROL

SPEED REGULATORS
UF   GOVERNORS
GS   CONTROL EQUIPMENT
. **SPEED REGULATORS**
REGULATORS
. **SPEED REGULATORS**
RT   CONTROLLERS
ENGINES

SPEEDOMETERS
USE   SPEED INDICATORS

SPENT FUELS
GS   FUELS
. NUCLEAR FUELS
. . **SPENT FUELS**
RT   FUEL CAPSULES
NEUTRON SOURCES
∞NUCLEAR ENERGY
NUCLEAR FUEL REPROCESSING
REACTOR MATERIALS
RECYCLING

SPERMATOCYTES
USE   GAMETOCYTES

SPERMATOGENESIS
RT   ABIOGENESIS
GAMETOCYTES
SPERMATOZOA

SPERMATOZOA
RT   FERTILIZATION
∞REPRODUCTION
SPERMATOGENESIS

SPERT REACTORS
GS   NUCLEAR REACTORS
. LIQUID COOLED REACTORS
. . WATER COOLED REACTORS
. . . BOILING WATER REACTORS

SPERT REACTORS-*(CONT.)*
    . . . . **SPERT REACTORS**
    . NUCLEAR RESEARCH AND TEST
        REACTORS
    . . **SPERT REACTORS**

SPHALERITE
USE   ZINCBLENDE

SPHERES
GS   SYMMETRICAL BODIES
    . BODIES OF REVOLUTION
    . . **SPHERES**
    . . . CELESTIAL SPHERE
    . . . CONCENTRIC SPHERES
    . . . FALLING SPHERES
    . . . POINCARE SPHERES
    . . . ROTATING SPHERES
RT   AERODYNAMIC CONFIGURATIONS
    ASPHERICITY ·
    BALLS
    CIRCLES (GEOMETRY)
    EUCLIDEAN GEOMETRY
    GEOMETRY
    ∞GLOBES
    GLOBULES
    ∞HEMISPHERES
    HEMISPHERICAL SHELLS
    MICROBALLOONS
    NODULES
    OGIVES
    SPHERICAL SHELLS
    SPHEROIDS
    SPHERULES

SPHERICAL ANTENNAS
GS   ANTENNAS
    . **SPHERICAL ANTENNAS**
RT   COMMUNICATION EQUIPMENT
    ELECTRONIC EQUIPMENT

SPHERICAL CAPS
GS   SHELLS (STRUCTURAL FORMS)
    . SPHERICAL SHELLS
    . . **SPHERICAL CAPS**
RT   ∞CAPS
    COVERINGS
    NOSE CONES
    SEALS (STOPPERS)

SPHERICAL COORDINATES
UF   CURVILINEAR COORDINATES
GS   COORDINATES
    . **SPHERICAL COORDINATES**
RT   ASTRONOMICAL COORDINATES
    CELESTIAL REFERENCE SYSTEMS
    GEOCENTRIC COORDINATES
    PLANETOCENTRIC COORDINATES
    POLAR COORDINATES
    POSITION (LOCATION)
    ∞REFERENCE SYSTEMS

SPHERICAL HARMONICS
GS   ANALYSIS (MATHEMATICS)
    . COMPLEX VARIABLES
    . . **SPHERICAL HARMONICS**
    FUNCTIONS (MATHEMATICS)
    . **SPHERICAL HARMONICS**
    HARMONICS
    . **SPHERICAL HARMONICS**
RT   LEGENDRE FUNCTIONS

SPHERICAL PLASMAS
GS   PARTICLES
    . CHARGED PARTICLES
    . . ENERGETIC PARTICLES
    . . . PLASMAS (PHYSICS)
    . . . . **SPHERICAL PLASMAS**

SPHERICAL SHELLS
GS   SHELLS (STRUCTURAL FORMS)
    . **SPHERICAL SHELLS**
    . . SPHERICAL CAPS
RT   BODIES OF REVOLUTION
    CIRCULAR SHELLS
    HEMISPHERICAL SHELLS
    METAL SHELLS
    REINFORCED SHELLS
    ROTATING SPHERES
    SPHERES
    STRESSED-SKIN STRUCTURES
    THIN WALLED SHELLS

SPHERICAL TANKS
GS   TANKS (CONTAINERS)
    . **SPHERICAL TANKS**
RT   FUEL TANKS
    PRESSURE VESSELS
    PROPELLANT TANKS
    STORAGE TANKS

SPHERICAL WAVES
RT   CYLINDRICAL WAVES
    DIFFRACTION PATHS
    DIFFRACTION PROPAGATION
    ELASTIC WAVES
    ELECTROMAGNETIC RADIATION
    HUYGENS PRINCIPLE
    PLANE WAVES
    POINT SOURCES
    THREE DIMENSIONAL FLOW
    ∞WAVES

SPHEROIDS
GS   GEOMETRY
    . EUCLIDEAN GEOMETRY
    . . ANALYTIC GEOMETRY
    . . . **SPHEROIDS**
    . . . . OBLATE SPHEROIDS
    . . . . PROLATE SPHEROIDS
RT   FALLING SPHERES
    GEOIDS
    SPHERES

SPHEROMAKS
GS   NUCLEAR REACTORS
    . FUSION REACTORS
    . . **SPHEROMAKS**
RT   DENSE PLASMAS
    MAGNETIC FIELD CONFIGURATIONS
    MAGNETIC MIRRORS
    PLASMA CURRENTS
    ∞REACTORS
    TOKAMAK DEVICES
    TOROIDAL PLASMAS

SPHERULES
GS   **SPHERULES**
    . SPHERULITES
RT   CRYSTALS
    SPHERES

SPHERULITES
GS   CRYSTALS
    . CRYSTALLITES
    . . **SPHERULITES**
    SPHERULES
    . **SPHERULITES**
RT   CRYSTAL STRUCTURE
    MICROCRYSTALS
    MICROSTRUCTURE
    NODULES
    ROSETTE SHAPES

SPHINX
SN   (SPACE PLASMA HIGH VOLTAGE
    INTERACTION EXPERIMENTS)RAFT)
UF   SPACE PLASMA H/V INTERACTION
    EXPERIMENTS
GS   EXPERIMENTATION
    . **SPHINX**
RT   HIGH ENERGY INTERACTIONS
    PARTICLE INTERACTIONS
    PLASMA INTERACTION EXPERIMENT
    PLASMA-PARTICLE INTERACTIONS
    PLASMAS (PHYSICS)
    ∞RESEARCH PROJECTS

SPHYGMOGRAPHY
RT   ARTERIES
    BIOINSTRUMENTATION
    BLOOD PRESSURE
    HEART RATE
    ∞MEASUREMENT
    RECORDING INSTRUMENTS

SPICULES
GS   STELLAR ACTIVITY
    . SOLAR ACTIVITY
    . . **SPICULES**
RT   CHROMOSPHERE
    PHOTOSPHERE
    SOLAR ATMOSPHERE

SPIDERS
GS   ANIMALS
    . INVERTEBRATES

SPIDERS-*(CONT.)*
    . . ARTHROPODS
    . . . **SPIDERS**

SPIKE ANTENNAS
USE   MONOPOLE ANTENNAS

SPIKE NOZZLES
GS   EXHAUST NOZZLES
    . **SPIKE NOZZLES**
RT   CONICAL NOZZLES
    NOZZLE GEOMETRY
    ∞NOZZLES
    PLUG NOZZLES
    ROCKET NOZZLES
    ∞SPIKES

SPIKE POTENTIALS
GS   POTENTIAL ENERGY
    . ELECTRIC POTENTIAL
    . . **SPIKE POTENTIALS**
RT   BIOELECTRICITY
    DEPOLARIZATION
    ∞SPIKES

∞ **SPIKES**
SN   *(USE OF A MORE SPECIFIC TERM IS*
    RECOMMENDED--CONSULT THE TERMS
    LISTED BELOW)
RT   FASTENERS
    HOLDERS
    MONOPOLE ANTENNAS
    PINS
    SPIKE NOZZLES
    SPIKE POTENTIALS
    SPIKES (AERODYNAMIC
      CONFIGURATIONS)

SPIKES (AERODYNAMIC CONFIGURATIONS)
RT   ∞SPIKES
    WIND TUNNELS

SPIKING
RT   ELECTRON BEAM WELDING
    MELTING
    METAL CUTTING

SPILLING
RT   DUMPING
    EMPTYING
    JETTISONING
    OIL SLICKS
    RELEASING
    SPREADING

SPIN
GS   **SPIN**
    . METAL SPINNING
    . . HYDROSPINNING
    . PARTICLE SPIN
    . . ELECTRON SPIN
    . . ISOTOPIC SPIN
    . . NUCLEAR SPIN
    . SPIN-ORBIT INTERACTIONS
    . . ELECTRON CAPTURE
    . SPIN-SPIN COUPLING
RT   ANGULAR MOMENTUM
    NUCLEAR CAPTURE
    SPINNERS
    YO-YO DEVICES

SPIN DECOUPLING
GS   DECOUPLING
    . **SPIN DECOUPLING**
RT   PHOTOMAGNETIC EFFECTS

SPIN DYNAMICS
RT   AIRCRAFT SPIN
    ARTIFICIAL GRAVITY
    DYNAMIC TESTS
    ∞DYNAMICS
    ELECTRON SPIN
    GYRATION
    LUNAR ROTATION
    ROTATING MATTER

SPIN EXCHANGE
GS   EXCHANGING
    . **SPIN EXCHANGE**
RT   RESONANCE CHARGE EXCHANGE

SPIN FORGING
USE   METAL SPINNING

SPIN GLASS
GS   GLASS
      . **SPIN GLASS**
RT   AMORPHOUS MATERIALS
      MAGNETIC PROPERTIES
      METALLIC GLASSES
      SPIN-LATTICE RELAXATION
      SUPERCONDUCTIVITY

SPIN REDUCTION
UF   DESPINNING
      JET DAMPING
GS   RATES (PER TIME)
      . ACCELERATION (PHYSICS)
      . . DECELERATION
      . . . **SPIN REDUCTION**
RT   ANGULAR ACCELERATION
      DESTABILIZATION
      GRAVITY GRADIENT SATELLITES
      ∞REDUCTION
      SATELLITE ROTATION
      YO-YO DEVICES

SPIN RESONANCE
GS   RESONANCE
      . **SPIN RESONANCE**
RT   NUCLEAR MAGNETIC RESONANCE
      PARTICLE SPIN

SPIN STABILIZATION
GS   STABILIZATION
      . **SPIN STABILIZATION**
RT   ATTITUDE CONTROL
      DUAL SPIN SPACECRAFT
      MISSILES
      OV-1 SATELLITES
      OV-2 SATELLITES
      OV-3 SATELLITES
      OV-4 SATELLITES
      OV-5 SATELLITES
      SATELLITE ORIENTATION
      SATELLITE ROTATION
      SATELLITES
      SPACE SHUTTLE UPPER STAGE D
      SPACE STATIONS

SPIN TEMPERATURE
SN   (LIMITED TO ASTROPHYSICS)
GS   TEMPERATURE
      . **SPIN TEMPERATURE**
RT   ABSORPTION SPECTRA
      ASTROPHYSICS
      HYDROGEN CLOUDS
      INTERSTELLAR GAS
      INTERSTELLAR MATTER

SPIN TESTS
UF   WHIRLING TESTS
RT   ANGULAR MOMENTUM
      DYNAMIC TESTS
      ENVIRONMENTAL TESTS
      LOAD TESTS
      POLARIZATION (SPIN ALIGNMENT)
      ∞TESTS
      WHIRL TOWERS

SPIN WAVES
USE   MAGNONS

SPIN-LATTICE RELAXATION
GS   MAGNETIC PROPERTIES
      . MAGNETIC RELAXATION
      . . **SPIN-LATTICE RELAXATION**
      RELAXATION (MECHANICS)
      . **SPIN-LATTICE RELAXATION**
RT   LATTICE VIBRATIONS
      NUCLEAR MAGNETIC RESONANCE
      RELAXATION TIME
      SPIN GLASS

SPIN-ORBIT INTERACTIONS
GS   NUCLEAR REACTIONS
      . NUCLEAR INTERACTIONS
      . . **SPIN-ORBIT INTERACTIONS**
      . . . ELECTRON CAPTURE
      PARTICLE INTERACTIONS
      . NUCLEAR INTERACTIONS
      . . **SPIN-ORBIT INTERACTIONS**
      . . . ELECTRON CAPTURE
      SPIN
      . **SPIN-ORBIT INTERACTIONS**
      . . ELECTRON CAPTURE
RT   ∞INTERACTIONS

SPIN-SPIN COUPLING
GS   COUPLING
      . **SPIN-SPIN COUPLING**
      SPIN
      . **SPIN-SPIN COUPLING**
RT   COUPLES
      CROSS RELAXATION

SPINACH
GS   FARM CROPS
      . **SPINACH**
      VEGETABLES
      . **SPINACH**
RT   ∞FOOD
      PLANTS (BOTANY)

SPINAL CORD
GS   NERVOUS SYSTEM
      . CENTRAL NERVOUS SYSTEM
      . . **SPINAL CORD**
      . . . SPINE
RT   BONES
      BRAIN

SPINDLES
RT   SHAFTS (MACHINE ELEMENTS)
      SPOOLS
      WINDING

SPINE
GS   NERVOUS SYSTEM
      . CENTRAL NERVOUS SYSTEM
      . . SPINAL CORD
      . . . **SPINE**
RT   MUSCULOSKELETAL SYSTEM
      VERTEBRAE

SPINEL
GS   MINERALS
      . **SPINEL**
RT   ALUMINATES
      FERRITES
      IGNEOUS ROCKS

SPINNERS
RT   DIRECTIONAL ANTENNAS
      RADAR ANTENNAS
      SLEWING
      SPIN

SPINNING (METALLURGY)
USE   METAL SPINNING

SPINNING SOLID UPPER STAGE
GS   SPACE SHUTTLE UPPER STAGES
      . **SPINNING SOLID UPPER STAGE**
RT   BOOSTER ROCKET ENGINES
      UPPER STAGE ROCKET ENGINES

SPINNING UNGUIDED ROCKET TRAJECTORY
UF   SPURT (TRAJECTORIES)
GS   TRAJECTORIES
      . **SPINNING UNGUIDED ROCKET
         TRAJECTORY**
RT   EQUATIONS OF MOTION
      MISSILE TRAJECTORIES
      ROTATING BODIES
      SYMMETRICAL BODIES

SPINOR GROUPS
GS   ALGEBRA
      . LIE GROUPS
      . . **SPINOR GROUPS**
      GEOMETRY
      . DIFFERENTIAL GEOMETRY
      . . LIE GROUPS
      . . . **SPINOR GROUPS**

SPIRAL ANTENNAS
GS   ANTENNAS
      . **SPIRAL ANTENNAS**
      . . LOG SPIRAL ANTENNAS
RT   ANTENNA DESIGN
      BROADBAND
      TELEMETRY

SPIRAL GALAXIES
GS   CELESTIAL BODIES
      . GALAXIES
      . . **SPIRAL GALAXIES**
      . . . BARRED GALAXIES
      . . . MILKY WAY GALAXY
RT   ANDROMEDA GALAXIES
      COROTATION

SPIRAL GALAXIES-*(CONT.)*
      DENSITY WAVE MODEL
      DISK GALAXIES
      ELLIPTICAL GALAXIES
      LOCAL GROUP (ASTRONOMY)
      MAFFEI GALAXIES
      VIRGO GALACTIC CLUSTER

SPIRAL WRAPPING
RT   COMPOSITE MATERIALS
      COMPOSITE WRAPPING
      ISOTENSOID STRUCTURES
      PACKAGING
      ∞SPIRALS
      WINDING

∞ **SPIRALS**
SN   *(USE OF A MORE SPECIFIC TERM IS
      RECOMMENDED--CONSULT THE TERMS
      LISTED BELOW)*
RT   CURVES (GEOMETRY)
      SPIRAL WRAPPING
      SPIRALS (CONCENTRATORS)

SPIRALS (CONCENTRATORS)
GS   CONCENTRATORS
      . **SPIRALS (CONCENTRATORS)**
      SEPARATORS
      . **SPIRALS (CONCENTRATORS)**
RT   CLASSIFIERS
      ∞SPIRALS

SPIROMETERS
RT   HEART MINUTE VOLUME
      LUNGS
      RESPIRATORY RATE

SPITSBERGEN (NORWAY)
RT   ARCHIPELAGOES
      ARCTIC OCEAN
      OCEAN CURRENTS

SPLASHING
UF   SWASH
RT   AGITATION
      SURFACE WAVES
      ULLAGE
      WATER LANDING

SPLEEN
GS   ANATOMY
      . ORGANS
      . . **SPLEEN**
      VISCERA
      . ORGANS
      . . **SPLEEN**
RT   POLYCYTHEMIA

SPLICING
RT   FASTENERS
      ∞JOINING
      ∞TAPES
      WIRING

SPLINE FUNCTIONS
GS   FUNCTIONS (MATHEMATICS)
      . **SPLINE FUNCTIONS**
RT   APPROXIMATION
      MATRIX METHODS

SPLINES
RT   COUPLINGS
      FASTENERS
      HOLDERS

SPLINTS
RT   BONES
      CASTS
      FIRST AID

SPLIT FLAPS
GS   AIRFOILS
      . FLAPS (CONTROL SURFACES)
      . . **SPLIT FLAPS**
      BRAKES (FOR ARRESTING MOTION)
      . AERODYNAMIC BRAKES
      . . **SPLIT FLAPS**
      . AIRCRAFT BRAKES
      . . **SPLIT FLAPS**
      CONTROL SURFACES
      . FLAPS (CONTROL SURFACES)
      . . **SPLIT FLAPS**
      DRAG DEVICES
      . AERODYNAMIC BRAKES

SPLIT FLAPS-*(CONT.)*
　　. . **SPLIT FLAPS**
RT　JET FLAPS
　　LEADING EDGE SLATS
　　TRAILING EDGE FLAPS
　　WING FLAPS

SPLITS (GEOLOGY)
USE　GEOLOGICAL FAULTS

SPLITTING
RT　CHIPPING
　　CUTTING
　　FISSION
　　FLAKING
　　FRACTURING
　　LASER CUTTING
　　∞SEPARATION
　　SLICING

SPODUMENE
GS　ALUMINUM COMPOUNDS
　　. **SPODUMENE**
　　CHALCOGENIDES
　　. OXIDES
　　. . SILICON OXIDES
　　. . . **SPODUMENE**
　　LITHIUM COMPOUNDS
　　. **SPODUMENE**
　　MINERALS
　　. **SPODUMENE**
　　SILICON COMPOUNDS
　　. SILICATES
　　. . SODIUM SILICATES
　　. . . **SPODUMENE**
　　. SILICON OXIDES
　　. . **SPODUMENE**
　　SODIUM COMPOUNDS
　　. SODIUM SILICATES
　　. . **SPODUMENE**

SPOILER SLOT AILERONS
GS　AIRFOILS
　　. AILERONS
　　. . **SPOILER SLOT AILERONS**
　　CONTROL SURFACES
　　. AILERONS
　　. . **SPOILER SLOT AILERONS**
RT　SPOILERS

SPOILERS
GS　AIRFOILS
　　. **SPOILERS**
　　CONTROL SURFACES
　　. **SPOILERS**
　　DRAG DEVICES
　　. **SPOILERS**
RT　AERODYNAMIC BRAKES
　　BOUNDARY LAYER CONTROL
　　DEFLECTORS
　　FLAPS (CONTROL SURFACES)
　　GUST ALLEVIATORS
　　LEADING EDGE SLATS
　　SPOILER SLOT AILERONS
　　VORTEX ALLEVIATION
　　WINGS

SPOKES
RT　HUBS
　　WHEELS

SPONGES (MATERIALS)
SN　(ORGANIC OPEN-CELL STRUCTURES)
RT　ELASTOMERS
　　∞MATERIALS
　　POLYURETHANE FOAM
　　POROUS MATERIALS

SPONTANEOUS COMBUSTION
GS　COMBUSTION
　　. **SPONTANEOUS COMBUSTION**
RT　COMBUSTION TEMPERATURE
　　EXPLOSIONS
　　FIRE POINT
　　FIRE PREVENTION
　　FLAMMABILITY
　　FLASH POINT
　　FUEL COMBUSTION
　　HAZARDS
　　HYPERGOLIC ROCKET PROPELLANTS
　　IGNITION
　　IGNITION TEMPERATURE
　　PROPELLANT SENSITIVITY
　　PYROPHORIC MATERIALS

SPONTANEOUS EMISSION
GS　DECAY
　　. EMISSION
　　. . **SPONTANEOUS EMISSION**
RT　ATOMIC ENERGY LEVELS
　　ELECTROMAGNETIC RADIATION
　　EMISSION SPECTRA
　　SPECTRAL EMISSION

SPOOLS
RT　∞CONTAINERS
　　INSERTS
　　MAGAZINES (SUPPLY CHAMBERS)
　　REELS
　　SPINDLES

SPORADIC E LAYER
GS　EARTH ATMOSPHERE
　　. UPPER ATMOSPHERE
　　. . IONOSPHERE
　　. . . E REGION
　　. . . . **SPORADIC E LAYER**
RT　E-1 LAYER
　　E-2 LAYER
　　MIDLATITUDE ATMOSPHERE

SPORADIC METEOROIDS
SN　(METEOROIDS NOT ASSOCIATED WITH A
　　METEOROID SHOWER OR STREAM)
GS　CELESTIAL BODIES
　　. METEOROIDS
　　. . **SPORADIC METEOROIDS**
RT　METEOR TRAILS
　　METEOROID CONCENTRATION

SPORES
GS　ANIMALS
　　. INVERTEBRATES
　　. . **SPORES**
　　. . . MICROSPORES
　　FUNGI
　　. **SPORES**
　　. . MICROSPORES
　　MICROORGANISMS
　　. **SPORES**
　　. . MICROSPORES
RT　PLANTS (BOTANY)
　　∞REPRODUCTION
　　TETRAD THEORY

SPORTS MEDICINE
RT　AEROSPACE MEDICINE
　　ATHLETES
　　CLINICAL MEDICINE
　　EXERCISE PHYSIOLOGY
　　MEDICAL SCIENCE
　　PHYSICAL FITNESS
　　PHYSIOLOGICAL EFFECTS

SPOT (FRENCH SATELLITE)
GS　OBSERVATION
　　. EARTH OBSERVATIONS (FROM SPACE)
　　. . **SPOT (FRENCH SATELLITE)**
　　SATELLITES
　　. ARTIFICIAL SATELLITES
　　. . FRENCH SATELLITES
　　. . . **SPOT (FRENCH SATELLITE)**
RT　CROP IDENTIFICATION
　　EARTH RESOURCES
　　LAND USE
　　MAPPING
　　SOIL MAPPING
　　STEREOPHOTOGRAPHY

SPOT WELDS
GS　JOINTS (JUNCTIONS)
　　. METAL JOINTS
　　. . WELDED JOINTS
　　. . . **SPOT WELDS**
RT　ARC WELDING
　　BEADS
　　ELECTRIC WELDING
　　FUSION WELDING
　　PRESSURE WELDING
　　ULTRASONIC WELDING

SPRAY CHARACTERISTICS
RT　∞CHARACTERISTICS
　　SPRAYERS
　　SPRAYING

SPRAY CONDENSERS
RT　JET CONDENSERS
　　SPRAYERS

SPRAY INGESTION
RT　GAS TURBINES
　　LANDING GEAR
　　SALT SPRAY TESTS

SPRAY NOZZLES
RT　ANNULAR NOZZLES
　　CONICAL NOZZLES
　　FUEL INJECTION
　　FUEL SYSTEMS
　　INJECTORS
　　∞NOZZLES
　　ORIFICES
　　SPRAYERS

SPRAYED COATINGS
UF　SPRAYED PROTECTIVE COATINGS
GS　COATINGS
　　. **SPRAYED COATINGS**
RT　CERAMIC COATINGS
　　FINISHES
　　LACQUERS
　　METAL COATINGS
　　PAINTS
　　PLASMA SPRAYING
　　PLASTIC COATINGS
　　PRIMERS (COATINGS)
　　PROTECTIVE COATINGS
　　VARNISHES

SPRAYED PROTECTIVE COATINGS
USE　PROTECTIVE COATINGS
　　SPRAYED COATINGS

SPRAYERS
UF　SPRAYING APPARATUS
　　SPRAYS
RT　ATOMIZERS
　　BLOWERS
　　COLLOIDAL GENERATORS
　　CONTACTORS
　　∞CONTAINERS
　　∞DIFFUSERS
　　DISPENSERS
　　DISTRIBUTORS
　　DROPS (LIQUIDS)
　　EJECTORS
　　FUEL SPRAYS
　　∞JETS
　　MATERIALS HANDLING
　　MIXERS
　　∞NOZZLES
　　PROPELLANT SPRAYS
　　SPRAY CHARACTERISTICS
　　SPRAY CONDENSERS
　　SPRAY NOZZLES
　　SPRAYING
　　VAPORIZERS

SPRAYING
GS　**SPRAYING**
　　. ARC SPRAYING
　　. CROP DUSTING
　　. FLAME SPRAYING
　　. METAL SPRAYING
　　. PLASMA SPRAYING
RT　AERATION
　　AEROSOLS
　　ATOMIZING
　　BLOWING
　　COATING
　　COATINGS
　　DIFFUSION
　　DISPERSING
　　ENTRAINMENT
　　FORMING TECHNIQUES
　　FUMIGATION
　　LIQUID ATOMIZATION
　　METALLIZING
　　MIXING
　　PREMIXING
　　SEALING
　　SPRAY CHARACTERISTICS
　　SPRAYERS
　　SPRINKLING
　　VAPORIZING
　　WETTING

SPRAYING APPARATUS
USE　SPRAYERS

SPRAYS
USE　SPRAYERS

SPREAD F
RT   F 2 REGION
        IONOSPHERIC STORMS
        MAGNETIC STORMS

SPREAD REFLECTION
GS   REFLECTION
      . **SPREAD REFLECTION**
RT   GLARE
        INFRARED REFLECTION
        OPTICAL REFLECTION
        SCATTERING
        SIGNAL REFLECTION
        ULTRAVIOLET REFLECTION
        WAVE REFLECTION

SPREAD SPECTRUM TRANSMISSION
GS   TRANSMISSION
      . ELECTROMAGNETIC WAVE
        TRANSMISSION
      . . RADIO TRANSMISSION
      . . . **SPREAD SPECTRUM TRANSMISSION**
RT   COMMUNICATION
        FREQUENCY HOPPING

SPREADING
RT   ADHESION
        COHESION
        DIFFUSION
        DISPERSING
        DISPOSAL
        DUMPING
        EMPTYING
        INTERFACIAL TENSION
        INTERNAL PRESSURE
        MATERIALS HANDLING
        SCATTERING
      ∞SEPARATION
        SPILLING
        SWELLING
        THROWING
        UNLOADING

SPRING (SEASON)
GS   SEASONS
      . **SPRING (SEASON)**
RT   AUTUMN
        SUMMER
        WINTER

SPRINGS (ELASTIC)
RT   ∞COILS
        ENERGY STORAGE
        FRAMES
        OSCILLATION DAMPERS
        OSCILLATIONS
        RESILIENCE
        SHOCK ABSORBERS
        SUSPENSION SYSTEMS (VEHICLES)
        VIBRATION ISOLATORS

SPRINGS (WATER)
GS   RESOURCES
      . EARTH RESOURCES
      . . **SPRINGS (WATER)**
        WATER
      . **SPRINGS (WATER)**
RT   AQUIFERS
        FRESH WATER
        GROUND WATER
        INLAND WATERS
        LAKES
        OASES
        POTABLE WATER
        WATER TABLES
        WELLS

SPRINKLING
RT   SCATTERING
        SPRAYING
        WETTING

SPRINT MISSILE
GS   MISSILES
      . ANTIMISSILE MISSILES
      . . **SPRINT MISSILE**
      . SURFACE TO AIR MISSILES
      . . **SPRINT MISSILE**
RT   NIKE-ZEUS MISSILE
        SENTINEL SYSTEM
        SOLID PROPELLANT ROCKET ENGINES
        SPARTAN MISSILE

SPUR (REACTORS)
USE   SPACE POWER UNIT REACTORS

SPURT (TRAJECTORIES)
USE   SPINNING UNGUIDED ROCKET
        TRAJECTORY

SPUTNIK SATELLITES
GS   SATELLITES
      . ARTIFICIAL SATELLITES
      . . SOVIET SATELLITES
      . . . **SPUTNIK SATELLITES**
      . . . . SPUTNIK 1 SATELLITE
      . . . . SPUTNIK 2 SATELLITE
      . . . . SPUTNIK 3 SATELLITE
      . . . . SPUTNIK 4 SATELLITE
      . . . . SPUTNIK 5 SATELLITE

SPUTNIK 1 SATELLITE
GS   SATELLITES
      . ARTIFICIAL SATELLITES
      . . METEOROLOGICAL SATELLITES
      . . . **SPUTNIK 1 SATELLITE**
      . SOVIET SATELLITES
      . . SPUTNIK SATELLITES
      . . . . **SPUTNIK 1 SATELLITE**

SPUTNIK 2 SATELLITE
GS   SATELLITES
      . ARTIFICIAL SATELLITES
      . . BIOSATELLITES
      . . **SPUTNIK 2 SATELLITE**
      . . METEOROLOGICAL SATELLITES
      . . . **SPUTNIK 2 SATELLITE**
      . . SOVIET SATELLITES
      . . . SPUTNIK SATELLITES
      . . . . **SPUTNIK 2 SATELLITE**
      UNMANNED SPACECRAFT
      . BIOSATELLITES
      . . **SPUTNIK 2 SATELLITE**

SPUTNIK 3 SATELLITE
GS   SATELLITES
      . ARTIFICIAL SATELLITES
      . . GEOPHYSICAL SATELLITES
      . . . **SPUTNIK 3 SATELLITE**
      . . METEOROLOGICAL SATELLITES
      . . . **SPUTNIK 3 SATELLITE**
      . . SOVIET SATELLITES
      . . . SPUTNIK SATELLITES
      . . . . **SPUTNIK 3 SATELLITE**

SPUTNIK 4 SATELLITE
GS   SATELLITES
      . ARTIFICIAL SATELLITES
      . . SOVIET SATELLITES
      . . . SPUTNIK SATELLITES
      . . . . **SPUTNIK 4 SATELLITE**

SPUTNIK 5 SATELLITE
GS   SATELLITES
      . ARTIFICIAL SATELLITES
      . . SOVIET SATELLITES
      . . . SPUTNIK SATELLITES
      . . . . **SPUTNIK 5 SATELLITE**
RT   VENUS PROBES

SPUTTERING
GS   **SPUTTERING**
      . MAGNETRON SPUTTERING
RT   ARC WELDING
      ∞BOMBARDMENT
      DEPOSITION

SPUTTERING-*(CONT.)*
      DUOPLASMATRONS
      ELECTRON BOMBARDMENT
      EMISSION
      ION PLATING
      ION SOURCES
      METAL PARTICLES
      PLASMA ETCHING
      SURFACE FINISHING
      THERMAL INSTABILITY

SPUTTERING GAGES
GS   MEASURING INSTRUMENTS
      . **SPUTTERING GAGES**
RT   METAL FILMS
        THIN FILMS

SQUALLS
GS   WIND (METEOROLOGY)
      . **SQUALLS**
RT   GROUND WIND
        STORMS (METEOROLOGY)

SQUAMA
RT   FISHES

SQUARE WAVES
GS   WAVEFORMS
      . **SQUARE WAVES**
RT   FORM FACTORS
        PULSE AMPLITUDE
        PULSE DURATION
        SAWTOOTH WAVEFORMS
        TIME FUNCTIONS
        WAVE FUNCTIONS
        WAVE PROPAGATION
      ∞WAVES

SQUARE WELLS
RT   ELECTRON MOBILITY
        MAGNETIC FIELDS
        PHOTOCONDUCTIVITY
        VACANCIES (CRYSTAL DEFECTS)
        WELLS

SQUARES (MATHEMATICS)
GS   GEOMETRY
      . EUCLIDEAN GEOMETRY
      . . POLYGONS
      . . . TETRAGONS
      . . . . **SQUARES (MATHEMATICS)**

SQUEEZE FILMS
GS   FLUID FILMS
      . **SQUEEZE FILMS**
RT   BOUNDARY LUBRICATION
        ELASTOHYDRODYNAMICS
      ∞FILMS
        GAS BEARINGS
        GAS LUBRICANTS
        LIQUID-SOLID INTERFACES
        LUBRICANTS
        THIN FILMS
        VISCOELASTICITY
        VISCOUS FLUIDS

†

SQUEEZING
USE   COMPRESSING

SQUELCH CIRCUITS
GS   CIRCUITS
      . **SQUELCH CIRCUITS**
RT   BACKGROUND NOISE
        ELECTROMAGNETIC NOISE
        NOISE REDUCTION
        SILENCERS
        SUPPRESSORS
        SWITCHING CIRCUITS

SQUIBS
UF   XM-6 SQUIB
      XM-8 SQUIB
GS   IGNITERS
      . **SQUIBS**
RT   ELECTRIC IGNITION
        IGNITION SYSTEMS
        PRIMERS (EXPLOSIVES)
        SOLID PROPELLANT IGNITION
        STARTERS

SQUID (DETECTORS)
UF   SUPERCONDUCTING QUANTUM
        INTERFEROMETERS
RT   ∞DETECTORS

SQUID (DETECTORS)-*(CONT.)*
    JOSEPHSON JUNCTIONS
    MAGNETIC MEASUREMENT
    QUANTUM COUNTERS
    SUPERCONDUCTORS

SQUID PROJECT
GS   PROGRAMS
    . PROJECTS
    . . **SQUID PROJECT**
RT   JET PROPULSION

SQUIRRELS
GS   ANIMALS
    . VERTEBRATES
    . . MAMMALS
    . . . RODENTS
    . . . . **SQUIRRELS**
    . . . . . GROUND SQUIRRELS

SR (REACTORS)
USE   SATURABLE REACTORS

SR-N2 GROUND EFFECT MACHINE
USE   WESTLAND GROUND EFFECT MACHINES

SR-N3 GROUND EFFECT MACHINE
USE   WESTLAND GROUND EFFECT MACHINES

SR-N5 GROUND EFFECT MACHINE
USE   WESTLAND GROUND EFFECT MACHINES

SRE REACTOR
USE   SODIUM REACTOR EXPERIMENT

SRET SATELLITES
GS   SATELLITES
    . ARTIFICIAL SATELLITES
    . . FRENCH SATELLITES
    . . . **SRET SATELLITES**
    . . . . SRET 1 SATELLITE
    . . . . SRET 2 SATELLITE
    . . METEOROLOGICAL SATELLITES
    . . . **SRET SATELLITES**
    . . . . SRET 1 SATELLITE
    . . . . SRET 2 SATELLITE
RT   FRENCH SPACE PROGRAMS

SRET 1 SATELLITE
GS   SATELLITES
    . ARTIFICIAL SATELLITES
    . . FRENCH SATELLITES
    . . . SRET SATELLITES
    . . . . **SRET 1 SATELLITE**
    . . METEOROLOGICAL SATELLITES
    . . . SRET SATELLITES
    . . . . **SRET 1 SATELLITE**
RT   FRENCH SPACE PROGRAMS

SRET 2 SATELLITE
GS   SATELLITES
    . ARTIFICIAL SATELLITES
    . . FRENCH SATELLITES
    . . . SRET SATELLITES
    . . . . **SRET 2 SATELLITE**
    . . METEOROLOGICAL SATELLITES
    . . . SRET SATELLITES
    . . . . **SRET 2 SATELLITE**

† **SRI LANKA**
UF   CEYLON
GS   NATIONS
    . **SRI LANKA**

SS-11 MISSILE
GS   MISSILES
    . **SS-11 MISSILE**
RT   MULTISTAGE ROCKET VEHICLES
    SOLID PROPELLANT ROCKET ENGINES

SSGS (STANDARDIZED SPACE GUIDANCE)
USE   STANDARDIZED SPACE GUIDANCE

SSUS-A
USE   SPACE SHUTTLE UPPER STAGE A

SSUS-D
USE   SPACE SHUTTLE UPPER STAGE D

ST LAWRENCE VALLEY (NORTH AMERICA)
GS   LANDFORMS
    . **ST LAWRENCE VALLEY (NORTH AMERICA)**
    VALLEYS

ST LAWRENCE VALLEY (NORTH-*(CONT.)*
    . **ST LAWRENCE VALLEY (NORTH AMERICA)**
RT   CANADA
    MAINE
    NEW HAMPSHIRE
    NEW YORK
    VERMONT

ST LOUIS-KANSAS CITY CORRIDOR (MO)
GS   CORRIDORS
    . **ST LOUIS-KANSAS CITY CORRIDOR (MO)**
RT   MISSOURI
    REGIONAL PLANNING

ST VENANT FLEXURE PROBLEM
USE   SAINT VENANT PRINCIPLE

STABILITY
UF   INSTABILITY
GS   **STABILITY**
    . ACOUSTIC INSTABILITY
    . BAROCLINIC INSTABILITY
    . DYNAMIC STABILITY
    . . COMBUSTION STABILITY
    . . . FLAME STABILITY
    . . CONTROL STABILITY
    . . FREQUENCY STABILITY
    . . MOTION STABILITY
    . . AERODYNAMIC STABILITY
    . . . AIRCRAFT STABILITY
    . . . . HOVERING STABILITY
    . . . . ATTITUDE STABILITY
    . . . . DIRECTIONAL STABILITY
    . . . . . GYROSCOPIC STABILITY
    . . . . LATERAL STABILITY
    . . . . LONGITUDINAL STABILITY
    . . . FLOW STABILITY
    . . . . BOUNDARY LAYER STABILITY
    . . . . FLAME STABILITY
    . . . . MAGNETOHYDRODYNAMIC STABILITY
    . . . LOW SPEED STABILITY
    . . . ROTARY STABILITY
    . . . . GYROSCOPIC STABILITY
    . . . SPACECRAFT STABILITY
    . GOERTLER INSTABILITY
    . MAGNETOSPHERIC INSTABILITY
    . STATIC STABILITY
    . . DIMENSIONAL STABILITY
    . . . STRUCTURAL STABILITY
    . . . . SHELL STABILITY
    . STORAGE STABILITY
    . SURFACE STABILITY
    . SYSTEMS STABILITY
    . THERMAL STABILITY
RT   AMPLIFICATION
    BALLAST (MASS)
    COMPATIBILITY
    CONTROLLABILITY
    ∞ DRIFT
    DRIFT RATE
    DURABILITY
    DYNAMIC CHARACTERISTICS
    EQUATIONS OF MOTION
    ∞ EQUILIBRIUM
    METASTABLE STATE
    QUALITY
    RELIABILITY
    ∞ RESISTANCE
    SAFETY FACTORS
    SPACECRAFT MOTION
    STABILIZERS (AGENTS)
    STEADY STATE
    TOLERANCES (MECHANICS)
    TRESCA FLOW
    UNITY
    UNSTEADY STATE
    VARIABILITY
    VLASOV EQUATIONS
    VULNERABILITY

STABILITY AUGMENTATION
GS   AUGMENTATION
    . **STABILITY AUGMENTATION**
RT   AERODYNAMIC STABILITY
    AIRCRAFT CONTROL
    ATTITUDE (INCLINATION)
    AUTOMATIC CONTROL
    CONTROL STABILITY
    DIRECTIONAL STABILITY
    FEEDBACK CONTROL
    FLIGHT CONTROL
    LATERAL OSCILLATION

STABILITY AUGMENTATION-*(CONT.)*
    PITCH (INCLINATION)

STABILITY DERIVATIVES
UF   AERODYNAMIC MOMENTS
GS   MOMENTS
    . **STABILITY DERIVATIVES**
    . . PITCHING MOMENTS
    . . ROLLING MOMENTS
    . . YAWING MOMENTS
RT   COMPLEX VARIABLES
    DAMPING
    DIFFERENTIAL EQUATIONS
    MOMENTS OF INERTIA
    REAL VARIABLES
    VECTOR ANALYSIS

STABILITY TESTS
GS   **STABILITY TESTS**
    . FLIGHT STABILITY TESTS
    . WIND TUNNEL STABILITY TESTS
RT   CORROSION TESTS
    DAMPING TESTS
    ELECTRONIC EQUIPMENT TESTS
    FLIGHT TESTS
    FUEL TESTS
    GROUND TESTS
    MISSILE TESTS
    PROPELLANT TESTS
    RESONANCE TESTING
    ∞ TESTS
    VIBRATION TESTS

STABILIZATION
UF   MISSILE STABILIZATION
GS   **STABILIZATION**
    . SIGNAL STABILIZATION
    . SPIN STABILIZATION
    . THREE AXIS STABILIZATION
RT   ACID BASE EQUILIBRIUM
    BALANCING
    CONSOLIDATION
    ∞ CONTROL
    ∞ EQUILIBRIUM
    HEAT TREATMENT
    HORIZONTAL ORIENTATION
    LASER GYROSCOPES
    STABILIZERS (AGENTS)
    STRESS RELIEVING
    VERTICAL ORIENTATION

STABILIZED PLATFORMS
RT   GIMBALS
    GYROSCOPIC STABILITY
    GYROSTABILIZERS
    INERTIAL GUIDANCE
    ∞ PLATFORMS
    THREE AXIS STABILIZATION

∞ **STABILIZERS**
SN   *(USE OF A MORE SPECIFIC TERM IS RECOMMENDED--CONSULT THE TERMS LISTED BELOW)*
RT   GYROSCOPES
    SEA KEEPING
    STABILIZERS (AGENTS)
    STABILIZERS (FLUID DYNAMICS)

STABILIZERS (AGENTS)
RT   ADDITIVES
    ∞ AGENTS
    ANTICOAGULANTS
    ANTIOXIDANTS
    NEUTRALIZERS
    PRESERVATIVES
    RETARDANTS
    STABILITY
    STABILIZATION
    ∞ STABILIZERS

STABILIZERS (FLUID DYNAMICS)
UF   HORIZONTAL STABILIZERS
    VERTICAL STABILIZERS
    VERTICAL TAILS
GS   **STABILIZERS (FLUID DYNAMICS)**
    . HORIZONTAL TAIL SURFACES
RT   AERIAL RUDDERS
    AIRFOILS
    CONTROL SURFACES
    ∞ DYNAMICS
    ELEVATORS (CONTROL SURFACES)
    FINS
    KEELS
    RUDDERS
    ∞ STABILIZERS

STABILIZERS (FLUID DYNAMICS)-(CONT.)
        SWEPTBACK TAIL SURFACES
        T TAIL SURFACES
        TABS (CONTROL SURFACES)
        TAIL ASSEMBLIES
        TAIL SURFACES
        TRAPEZOIDAL TAIL SURFACES

STABLE OSCILLATIONS
  GS   OSCILLATIONS
        . **STABLE OSCILLATIONS**
  RT   DYNAMIC STABILITY
        FREQUENCY STABILITY
        GYROSCOPIC STABILITY
        MOTION STABILITY
        NONSTABILIZED OSCILLATION
        PILOT INDUCED OSCILLATION
        RESONANT VIBRATION
        TRANSVERSE OSCILLATION
        UNDAMPED OSCILLATIONS
        WING OSCILLATIONS

STACKING FAULT ENERGY
  RT   CRYSTAL DEFECTS
        ∞ENERGY
        TWINNING

STACKING FAULTS
  USE   CRYSTAL DEFECTS

STACKS
  RT   CHIMNEYS
        CRYSTAL DEFECTS
        MATERIALS HANDLING

STADAN (SATELLITE TRACKING NETWORK)
  USE   STDN (NETWORK)

STADIMETERS
  GS   MEASURING INSTRUMENTS
        . DISTANCE MEASURING EQUIPMENT
        . . **STADIMETERS**
  RT   RANGE FINDERS
        SEXTANTS

STAGE SEPARATION
  UF   STAGING (ROCKETS)
  RT   BOOSTER ROCKET ENGINES
        EXPENDABLE STAGES (SPACECRAFT)
        INTERIM STAGES (SPACECRAFT)
        LUNAR MODULE ASCENT STAGE
        MISSILES
        MULTISTAGE ROCKET VEHICLES
        PROPELLANT MASS RATIO
        ROCKET VEHICLES
        ∞SEPARATION
        SPACE SHUTTLE ASCENT STAGE
        SUSTAINER ROCKET ENGINES
        THRUST TERMINATION
        UPPER STAGE ROCKET ENGINES

STAGGERING
  RT   ∞CONFIGURATIONS
        DISORIENTATION

STAGING (ROCKETS)
  USE   STAGE SEPARATION

STAGNATION FLOW
  GS   FLUID FLOW
        . INVISCID FLOW
        . . **STAGNATION FLOW**
  RT   BOUNDARY LAYER FLOW
        BOUNDARY LAYER SEPARATION
        COMPRESSIBLE FLOW

STAGNATION POINT
  UF   STAGNATION REGION
  RT   FLUID DYNAMICS

STAGNATION PRESSURE
  GS   PRESSURE
        . **STAGNATION PRESSURE**
  RT   COMPRESSIBLE FLOW
        INLET PRESSURE

STAGNATION REGION
  USE   STAGNATION POINT

STAGNATION TEMPERATURE
  GS   TEMPERATURE
        . **STAGNATION TEMPERATURE**
  RT   ADIABATIC FLOW
        COMPRESSIBLE FLOW

STAGNATION TEMPERATURE-(CONT.)
        INVISCID FLOW

STAINING
  RT   CHEMICAL TESTS
        DISCOLORATION
        MARKING
        METHYLENE
        METHYLENE BLUE

STAINLESS STEELS
  GS   ALLOYS
        . IRON ALLOYS
        . . STEELS
        . . . **STAINLESS STEELS**
        . . . . AUSTENITIC STAINLESS STEELS
        . . . . FERRITIC STAINLESS STEELS
        . . . . MARTENSITIC STAINLESS STEELS
  RT   CHROMIUM ALLOYS
        MARAGING STEELS
        MOLYBDENUM ALLOYS
        NICKEL ALLOYS
        NICKEL STEELS

STAIRCASES
  USE   STAIRWAYS

STAIRSTEPS
  RT   BACKWARD FACING STEPS
        FORMATIONS
        ∞STEPS
        TOPOGRAPHY

STAIRWAYS
  UF   STAIRCASES
  RT   ∞BUILDINGS
        ESCALATORS
        LADDERS
        TREADS

∞ **STALLING**
  SN   (USE OF A MORE SPECIFIC TERM IS
        RECOMMENDED--CONSULT THE TERMS
        LISTED BELOW)
  RT   AERODYNAMIC STALLING
        BOUNDARY LAYER SEPARATION
        ENGINE FAILURE

STAMPING
  SN   (EXCLUDES IDENTIFICATION MARKING)
  GS   FORMING TECHNIQUES
        . PRESSING (FORMING)
        . . **STAMPING**
  RT   ∞BLANKING
        BLANKING (CUTTING)
        COINING
        COLD WORKING
        DIES
        DIMPLING
        FORGING
        HOT PRESSING
        METAL WORKING
        PUNCHES
        SHEARING
        SWAGING
        UPSETTING

STANDARD ATMOSPHERES
  USE   REFERENCE ATMOSPHERES

STANDARD DEVIATION
  GS   MOMENTS
        . DISTRIBUTION MOMENTS
        . . **STANDARD DEVIATION**
        STATISTICAL ANALYSIS
        . **STANDARD DEVIATION**
  RT   CONFIDENCE LIMITS
        ESTIMATING
        HETEROGENEITY
        QUALITY CONTROL
        RANGE (EXTREMES)
        VARIABILITY
        VARIANCE (STATISTICS)

STANDARD LAUNCH VEHICLE 3
  USE   ATLAS SLV-3 LAUNCH VEHICLE

STANDARD LAUNCH VEHICLE 5
  GS   LAUNCH VEHICLES
        . STANDARD LAUNCH VEHICLES
        . . **STANDARD LAUNCH VEHICLE 5**
        ROCKET VEHICLES
        . STANDARD LAUNCH VEHICLES
        . . **STANDARD LAUNCH VEHICLE 5**

STANDARD LAUNCH VEHICLES
  UF   SLV
  GS   LAUNCH VEHICLES
        . **STANDARD LAUNCH VEHICLES**
        . . ATLAS SLV-3 LAUNCH VEHICLE
        . . STANDARD LAUNCH VEHICLE 5
        ROCKET VEHICLES
        . **STANDARD LAUNCH VEHICLES**
        . . ATLAS SLV-3 LAUNCH VEHICLE
        . . STANDARD LAUNCH VEHICLE 5
  RT   ATLAS D ICBM
        ∞VEHICLES

STANDARDIZATION
  GS   **STANDARDIZATION**
        . COMMONALITY
  RT   CALIBRATING
        METRICATION
        NAMING
        NUMERICAL CONTROL
        PRODUCT DEVELOPMENT
        PRODUCTION ENGINEERING
        QUALITY CONTROL
        SPECIFICATIONS
        STANDARDS
        VARIABILITY

STANDARDIZED SPACE GUIDANCE
  UF   SSGS (STANDARDIZED SPACE
        GUIDANCE)
  GS   GUIDANCE (MOTION)
        . **STANDARDIZED SPACE GUIDANCE**
  RT   SPACE NAVIGATION

STANDARDS
  UF   REFERENCES (STANDARDS)
  GS   **STANDARDS**
        . FREQUENCY STANDARDS
        . REFERENCE ATMOSPHERES
  RT   ACCEPTABILITY
        ACCURACY
        CALIBRATING
        ∞CODES
        CONVENTIONS
        CRITERIA
        INSPECTION
        ∞MEASUREMENT
        ∞MEASURES
        METROLOGY
        ∞PERFORMANCE
        PERFORMANCE TESTS
        QUALITY CONTROL
        RELIABILITY
        SAMPLING
        SPECIFICATIONS
        STANDARDIZATION
        TEMPERATURE SCALES
        TOLERANCES (MECHANICS)
        VALIDITY
        VALUE ENGINEERING

STANDING WAVE RATIOS
  GS   RATIOS
        . **STANDING WAVE RATIOS**
  RT   AMPLITUDES
        ELECTRICAL PROPERTIES
        SMITH CHART
        TRANSMISSION LINES

STANDING WAVES
  RT   ANTINODES
        BEAT FREQUENCIES
        FREQUENCIES
        HARMONICS
        NODES (STANDING WAVES)
        ∞RADIATION
        RESONANT FREQUENCIES
        VIBRATION
        WAVELENGTHS
        ∞WAVES

STANDS
  USE   SUPPORTS

STANNATES
  GS   TIN COMPOUNDS
        . **STANNATES**
  RT   ∞OXYGEN COMPOUNDS

STANNIDES
  GS   TIN COMPOUNDS
        . **STANNIDES**
        . . NIOBIUM STANNIDES
  RT   TIN ALLOYS

STANTON NUMBER
GS   RATIOS
      . DIMENSIONLESS NUMBERS
      . . **STANTON NUMBER**
RT   FORCED CONVECTION
      HEAT TRANSFER

STAPHYLOCOCCUS
GS   MICROORGANISMS
      . BACTERIA
      . . **STAPHYLOCOCCUS**
RT   PLEUROTIN

STAR CLUSTERS
GS   CELESTIAL BODIES
      . **STAR CLUSTERS**
      . . PRAESEPE STAR CLUSTERS
      . . VIRGO GALACTIC CLUSTER
RT   BARRED GALAXIES
      BINARY STARS
      COLOR-MAGNITUDE DIAGRAM
      DISK GALAXIES
      ELLIPTICAL GALAXIES
      GALACTIC CLUSTERS
      GALAXIES
      MAGELLANIC CLOUDS
      METALLICITY
      STARS
      VIRGO GALACTIC CLUSTER

STAR DISTRIBUTION
UF   STAR FIELDS
      STELLAR FIELDS
GS   DISTRIBUTION (PROPERTY)
      . SPATIAL DISTRIBUTION
      . . **STAR DISTRIBUTION**
      . VERTICAL DISTRIBUTION
      . . **STAR DISTRIBUTION**
RT   ANGULAR DISTRIBUTION
      ASTROLABES
      BARRED GALAXIES
      COSMOLOGY
      GALACTIC CLUSTERS
      GALACTIC EVOLUTION
      GLOBULAR CLUSTERS
      MASS DISTRIBUTION
      RADIAL DISTRIBUTION
      VIRGO GALACTIC CLUSTER

STAR FIELDS
USE   STAR DISTRIBUTION

STAR TRACKERS
UF   STAR TRACKING
GS   TRACKING (POSITION)
      . **STAR TRACKERS**
      . . CCD STAR TRACKER
RT   ASTROGUIDE NAVIGATION SYSTEM
      ASTROLABES
      ATTITUDE CONTROL
      CELESTIAL NAVIGATION
      CHARGE INJECTION DEVICES
      FLIGHT INSTRUMENTS
      GUIDANCE SENSORS
      INERTIAL NAVIGATION
      MISSILE CONTROL
      NAVIGATION
      NAVIGATION AIDS
      NAVIGATION INSTRUMENTS
      SOLAR SENSORS
      SPACECRAFT GUIDANCE

STAR TRACKING
USE   STAR TRACKERS

STARCHES
GS   ALIPHATIC COMPOUNDS
      . POLYSACCHARIDES
      . . **STARCHES**
      CARBOHYDRATES
      . POLYSACCHARIDES
      . . **STARCHES**
RT   CHITIN
      ∞FOOD
      SIZING MATERIALS

STARFIGHTER AIRCRAFT
USE   F-104 AIRCRAFT

STARK EFFECT
RT   ∞EFFECTS
      ELECTRIC FIELDS
      ELECTRO-OPTICS
      HYDROGEN PLASMA

STARK EFFECT-(CONT.)
      LINE SPECTRA
      RESOLUTION
      SPECTRUM ANALYSIS
      ZEEMAN EFFECT

STARLAB
UF   SPACELAB UV-OPTICAL TELESCOPE
      FACILITY
GS   PAYLOADS
      . **STARLAB**
RT   ∞OPTICS
      SPACELAB

STARLIFTER AIRCRAFT
USE   C-141 AIRCRAFT

STARS
GS   CELESTIAL BODIES
      . **STARS**
      . . COOL STARS
      . . DOUBLE STARS
      . . . BINARY STARS
      . . . . COMPANION STARS
      . . . . ECLIPSING BINARY STARS
      . . . . . DWARF NOVAE
      . . DWARF STARS
      . . . DWARF NOVAE
      . . . FLARE STARS
      . . . RED DWARF STARS
      . . . SUBDWARF STARS
      . . . WHITE DWARF STARS
      . . EARLY STARS
      . . . PROTOSTARS
      . . . . T TAURI STARS
      . . EXTARS
      . . HERBIG-HARO OBJECTS
      . . HOT STARS
      . . . B STARS
      . . . BLUE STARS
      . . . . SYMBIOTIC STARS
      . . . O STARS
      . . . WHITE DWARF STARS
      . . . WOLF-RAYET STARS
      . . INFRARED STARS
      . . LAMBDA TAURI STARS
      . . LATE STARS
      . . M STARS
      . . . FLARE STARS
      . . MAGNETIC STARS
      . . MAIN SEQUENCE STARS
      . . . PRE-MAIN SEQUENCE STARS
      . . METALLIC STARS
      . . NEUTRON STARS
      . . PULSARS
      . . OMICRON CETI STAR
      . . PECULIAR STARS
      . . SYMBIOTIC STARS
      . . PRAESEPE STAR CLUSTERS
      . . RADIO STARS
      . . . PULSARS
      . . REFERENCE STARS
      . . S STARS
      . . GIANT STARS
      . . . RED GIANT STARS
      . . . . CARBON STARS
      . . SIGMA ORIONIS
      . . SUBGIANT STARS
      . . SUN
      . . SUPERGIANT STARS
      . . SUPERMASSIVE STARS
      . . VAN BIESBROECK STAR
      . . VARIABLE STARS
      . . . CATACLYSMIC VARIABLES
      . . . CEPHEID VARIABLES
      . . . NOVAE
      . . . . DWARF NOVAE
      . . . . HERCULES NOVA
      . . . SUPERNOVAE
      . . SYMBIOTIC STARS
      . . . T TAURI STARS
      . . WHITE HOLES (ASTRONOMY)
      . . ZETA AURIGAE STAR
RT   ARIES CONSTELLATION
      ASTROLABES
      BARRED GALAXIES
      CELESTIAL MECHANICS
      CENTAURUS CONSTELLATION
      CONSTELLATIONS
      CORONA BOREALIS CONSTELLATION
      CYGNUS CONSTELLATION
      GALAXIES
      LYRA CONSTELLATION
      MAGELLANIC CLOUDS
      METALLICITY

STARS-(CONT.)
      MILKY WAY GALAXY
      QUASARS
      STAR CLUSTERS
      STARSPOTS
      STELLAR ACTIVITY
      STELLAR COMPOSITION
      STELLAR CORES
      STELLAR MAGNITUDE
      STELLAR OSCILLATIONS
      VIRGO GALACTIC CLUSTER

STARS (MATHEMATICS)
RT   ∞MATHEMATICS

STARSAT TELESCOPE
GS   TELESCOPES
      . ASTRONOMICAL TELESCOPES
      . . **STARSAT TELESCOPE**
      . REFLECTING TELESCOPES
      . . **STARSAT TELESCOPE**
      . SPACEBORNE TELESCOPES
      . . **STARSAT TELESCOPE**
RT   CORONAGRAPHS
      SPACEBORNE ASTRONOMY
      SPECTROHELIOGRAPHS
      ULTRAVIOLET ASTRONOMY

STARSITE PROGRAM
GS   PROGRAMS
      . **STARSITE PROGRAM**
RT   ARCHITECTURE
      ∞BUILDINGS
      COMMUNITIES
      CONFERENCES
      CONSTRUCTION
      DECISION MAKING
      ∞DEVELOPMENT
      ENVIRONMENTAL ENGINEERING
      INFORMATION RETRIEVAL
      LAND USE
      MANAGEMENT METHODS
      NASA PROGRAMS
      PLANNING
      RECREATION
      SHELTERS
      SOCIAL FACTORS
      TECHNOLOGY TRANSFER
      URBAN DEVELOPMENT
      URBAN PLANNING

STARSPOTS
GS   STELLAR ACTIVITY
      . **STARSPOTS**
RT   FACULAE
      MAGNETIC DISTURBANCES
      PHOTOSPHERE
      SOLAR ACTIVITY
      STARS
      STELLAR ATMOSPHERES
      STELLAR LUMINOSITY
      STELLAR MAGNETIC FIELDS
      STELLAR RADIATION
      SUNSPOT CYCLE
      TWENTY-SEVEN DAY VARIATION

STARTERS
GS   **STARTERS**
      . ENGINE STARTERS
RT   ACTUATORS
      IGNITION SYSTEMS
      SQUIBS
      STARTING

STARTING
GS   **STARTING**
      . AIR START
RT   ACTIVATION
      ACTUATION
      CYCLES
      ELECTRIC IGNITION
      ENGINE PRIMERS
      EXCITATION
      FIRING (IGNITING)
      IGNITION
      INITIATION
      LAUNCHING
      ∞PRIMING
      REACTOR STARTUP TESTS
      STARTERS
      STIMULATION

STATE EQUATIONS
USE   EQUATIONS OF STATE

STATE ESTIMATION
RT   ALGORITHMS
       KALMAN FILTERS
       LINEAR SYSTEMS
       ORBITAL POSITION ESTIMATION
       STATE VECTORS
       STOCHASTIC PROCESSES

STATE VECTORS
GS   ALGEBRA
       . VECTOR SPACES
       . . VECTORS (MATHEMATICS)
       . . . **STATE VECTORS**
RT   OBSERVABILITY (SYSTEMS)
       PHASE-SPACE INTEGRAL
       STATE ESTIMATION
       STEADY STATE
       STRANGE ATTRACTORS

STATIC AERODYNAMIC CHARACTERISTICS
GS   AERODYNAMIC CHARACTERISTICS
       . **STATIC AERODYNAMIC
         CHARACTERISTICS**
       STATIC CHARACTERISTICS
       . **STATIC AERODYNAMIC
         CHARACTERISTICS**
RT   AERODYNAMIC BALANCE
       AERODYNAMIC STABILITY
      ∞CHARACTERISTICS

STATIC ALTERNATORS
GS   ELECTRIC GENERATORS
       . AC GENERATORS
       . . **STATIC ALTERNATORS**
       . ROTATING GENERATORS
       . . **STATIC ALTERNATORS**

STATIC CHARACTERISTICS
SN   (EXCLUDES STATICS)
GS   **STATIC CHARACTERISTICS**
       . STATIC AERODYNAMIC
         CHARACTERISTICS
RT   STATIC LOADS
       STATIC STABILITY
       STATIC TESTS

STATIC DEFORMATION
GS   DEFORMATION
       . **STATIC DEFORMATION**
RT   CREEP PROPERTIES
       MAXWELL-MOHR METHOD
       SAINT VENANT PRINCIPLE

STATIC DISCHARGERS
UF   ANTISTATIC DEVICES
GS   DISCHARGERS
       . **STATIC DISCHARGERS**

STATIC ELECTRICITY
GS   ELECTRICITY
       . **STATIC ELECTRICITY**
RT   ATMOSPHERIC ELECTRICITY
       ATMOSPHERICS
       ELECTRIC CORONA
       ELECTRIC FIELDS
       ELECTRIC POTENTIAL
       ELECTRIC SPARKS
       ELECTROSTATIC CHARGE
       ELECTROSTATICS
       LIGHTNING
       OPEN CIRCUIT VOLTAGE
       SPACE CHARGE

STATIC FIRING
GS   CAPTIVE TESTS
       . STATIC TESTS
       . . **STATIC FIRING**
       ENGINE TESTS
       . **STATIC FIRING**
       FIRING (IGNITING)
       . TEST FIRING
       . . **STATIC FIRING**
       GROUND TESTS
       . PRELAUNCH TESTS
       . . **STATIC FIRING**
       NONDESTRUCTIVE TESTS
       . PRELAUNCH TESTS
       . . **STATIC FIRING**
RT   ROCKET FIRING

STATIC FRICTION
GS   FRICTION
       . **STATIC FRICTION**
RT   COEFFICIENT OF FRICTION

STATIC FRICTION-(CONT.)
       DRY FRICTION
       FRICTION MEASUREMENT
       KINETIC FRICTION
       SLIDING
       SLIDING FRICTION

STATIC INVERTERS
GS   INVERTERS
       . **STATIC INVERTERS**
RT   ELECTRIC GENERATORS

STATIC LOADS
UF   DEADWEIGHT
GS   LOADS (FORCES)
       . **STATIC LOADS**
RT   AERODYNAMIC LOADS
       AXIAL COMPRESSION LOADS
       AXIAL LOADS
       BALLAST (MASS)
       BENDING MOMENTS
       COMPRESSION LOADS
       CRITICAL LOADING
       DYNAMIC LOADS
       EDGE LOADING
       LOADING MOMENTS
       MASS DISTRIBUTION
       MOMENT DISTRIBUTION
       PRESSURE DISTRIBUTION
       RANDOM LOADS
       SAINT VENANT PRINCIPLE
       STATIC CHARACTERISTICS
       STRUCTURAL DESIGN CRITERIA
       WING LOADING

STATIC MODELS
GS   MODELS
       . **STATIC MODELS**
RT   APPROXIMATION
       DYNAMIC MODELS
       OPTIMIZATION

STATIC PRESSURE
GS   PRESSURE
       . **STATIC PRESSURE**
       . . HYDROSTATIC PRESSURE
RT   ISOSTATIC PRESSURE
       PITOT TUBES
       SOUND PRESSURE

STATIC STABILITY
GS   STABILITY
       . **STATIC STABILITY**
       . . DIMENSIONAL STABILITY
       . . . STRUCTURAL STABILITY
       . . . . SHELL STABILITY
RT   AIRCRAFT STABILITY
       COUNTERBALANCES
       DRIFT (INSTRUMENTATION)
       DYNAMIC STABILITY
       MAGNETOHYDROSTATICS
       STATIC CHARACTERISTICS
       STORAGE STABILITY
       STRATIFICATION
       SURFACE STABILITY

STATIC TESTS
SN   (ENCOMPASSES MATERIALS, ENGINE,
       AND VEHICLE TESTS)
GS   CAPTIVE TESTS
       . **STATIC TESTS**
       . . STATIC FIRING
RT   COLD FLOW TESTS
       COMPRESSION TESTS
       CREEP TESTS
       DYNAMIC TESTS
       ENGINE TESTS
       FATIGUE TESTS
       GROUND TESTS
       HARDNESS TESTS
       INSPECTION
       LOAD TESTS
      ∞MATERIALS TESTS
       MISSILE TESTS
       NONDESTRUCTIVE TESTS
       PREFIRING TESTS
       PRELAUNCH TESTS
       QUALITY CONTROL
       RESONANCE TESTING
       STATIC CHARACTERISTICS
       TENSILE TESTS
       TEST FIRING
      ∞TESTS
       VIBRATION TESTS
       WEAR TESTS

STATIC THRUST
GS   THRUST
       . **STATIC THRUST**
RT   JET THRUST
       ROCKET THRUST

STATICS
GS   **STATICS**
       . AEROSTATICS
       . ELECTROSTATICS
       . HYDROSTATICS
       . MAGNETOHYDROSTATICS
RT   ∞DYNAMICS
       ELASTOSTATICS
      ∞EQUILIBRIUM
       FLUID MECHANICS
      ∞MECHANICS (PHYSICS)

STATIONARY ORBITS
GS   ORBITS
       . CIRCULAR ORBITS
       . . **STATIONARY ORBITS**
       . EQUATORIAL ORBITS
       . . **STATIONARY ORBITS**
       . SPACECRAFT ORBITS
       . SATELLITE ORBITS
       . . . **STATIONARY ORBITS**
RT   EARTH ORBITS
       GEOSYNCHRONOUS ORBITS
       SYNCHRONOUS SATELLITES
       TWENTY-FOUR HOUR ORBITS

STATIONKEEPING
RT   GUIDANCE (MOTION)
       NAVIGATION
       ORBITAL MECHANICS
       ORBITS
       PAYLOAD RETRIEVAL (STS)
       POSITIONING
       SPACECRAFT CONTROL

STATIONS
GS   **STATIONS**
       . AUTOMATIC WEATHER STATIONS
       . CREW STATIONS
       . . CREW EXPERIMENT STATIONS
       . . CREW OBSERVATION STATIONS
       . . CREW WORKSTATIONS
       . GROUND STATIONS
       . . DEEP SPACE INSTRUMENTATION
         FACILITY
       . . EARTH TERMINALS
       . . INTEGRATED MISSION CONTROL
         CENTER
       . . POLYSTATION DOPPLER TRACKING
         SYSTEM
       . . SPACE DETECTION AND TRACKING
         SYSTEM
       . . STDN (NETWORK)
       . HYDROELECTRIC POWER STATIONS
       . PAYLOAD STATIONS
       . SPACE STATIONS
       . . ORBITAL SPACE STATIONS
       . . . HALO ORBIT SPACE STATION
       . . . LONG DURATION EXPOSURE
         FACILITY
       . . . ORBITAL WORKSHOPS
       . . . ORBITING LUNAR STATIONS
       . . . SALYUT SPACE STATION
       . . . SPACE OPERATIONS CENTER
         (NASA)
       . . SPACE BASE COMMAND CENTER
       . TRACKING STATIONS
       . . DEEP SPACE INSTRUMENTATION
         FACILITY
       . . GLOBAL TRACKING NETWORK
       . . POLYSTATION DOPPLER TRACKING
         SYSTEM
       . . SPACE DETECTION AND TRACKING
         SYSTEM
       . . STDN (NETWORK)
       . WEATHER STATIONS
       . WORKSTATIONS
RT   ∞BASES
      ∞FACILITIES
       LUNAR BASES
       MILITARY AIR FACILITIES
       PLANETARY BASES
       POSITION (LOCATION)
       SPACE BASES
       SPACE FLIGHT TRACKING AND DATA
         NETWORK

STATISTICAL ANALYSIS
GS   **STATISTICAL ANALYSIS**

STATISTICAL ANALYSIS-*(CONT.)*

- . AMPLITUDE DISTRIBUTION ANALYSIS
- . CORRELATION COEFFICIENTS
- . DISCRIMINANT ANALYSIS (STATISTICS)
- . FACTOR ANALYSIS
- . GOODNESS OF FIT
- . LIKELIHOOD RATIO
- . MAXWELL-BOLTZMANN DENSITY
    FUNCTION
- . NONPARAMETRIC STATISTICS
- . POISSON DENSITY FUNCTIONS
- . PROBABILITY DENSITY FUNCTIONS
- . . NORMAL DENSITY FUNCTIONS
- . . PEARSON DISTRIBUTIONS
- . . RAYLEIGH DISTRIBUTION
- . . WEIBULL DENSITY FUNCTIONS
- . PROBABILITY DISTRIBUTION
    FUNCTIONS
- . QUANTILES
- . SEQUENTIAL ANALYSIS
- . STANDARD DEVIATION
- . STATISTICAL CORRELATION
- . STATISTICAL DECISION THEORY
- . STATISTICAL TESTS
- . . KOLMOGOROFF-SMIRNOFF TEST
- . . MANN-WHITNEY-WILCOXON U TEST
- . . RANK TESTS
- . VARIANCE (STATISTICS)
- . . ANALYSIS OF VARIANCE
- . . MULTIVARIATE STATISTICAL
    ANALYSIS
- . . . BIVARIATE ANALYSIS
- . . . COVARIANCE
- . . . REGRESSION ANALYSIS

RT   ∞ANALYZING
- ∞APPLICATIONS OF MATHEMATICS
- APPROXIMATION
- AUTOREGRESSIVE PROCESSES
- BINOMIAL THEOREM
- BIOMETRICS
- CENSORED DATA (MATHEMATICS)
- CHARTS
- CHEBYSHEV APPROXIMATION
- CLUSTER ANALYSIS
- COEFFICIENTS
- CONFIDENCE
- CONFIDENCE LIMITS
- CONTINUITY (MATHEMATICS)
- CORRELATION
- ∞DATA
- DATA CORRELATION
- DECISION THEORY
- DISCRETE FUNCTIONS
- ∞DISPERSION
- ECONOMICS
- ESTIMATES
- ESTIMATING
- EVENTS
- EXPECTANCY HYPOTHESIS
- EXPERIMENT DESIGN
- EXPONENTIAL FUNCTIONS
- EXTRAPOLATION
- FACTORIAL DESIGN
- FAILURE ANALYSIS
- FORECASTING
- GAME THEORY
- GAUSS-MARKOV THEOREM
- GRAPHS (CHARTS)
- INFORMATION THEORY
- INSPECTION
- INTERPOLATION
- LINEAR PREDICTION
- MANAGEMENT
- ∞MATHEMATICS
- MAXIMUM ENTROPY METHOD
- MEAN
- MEDIAN (STATISTICS)
- MILLS RATIO
- MINIMUM VARIANCE ORBIT
    DETERMINATION
- MONTE CARLO METHOD
- MTBF
- OPERATIONS RESEARCH
- OUTLIERS (STATISTICS)
- PARAMETER IDENTIFICATION
- PROBABILITY THEORY
- QUALITY CONTROL
- QUARTILES
- QUEUEING THEORY
- RANDOM PROCESSES
- RELIABILITY
- ROOT-MEAN-SQUARE ERRORS
- SAMPLING
- ∞STATISTICS
- STOCHASTIC PROCESSES
- SYSTEM IDENTIFICATION

STATISTICAL ANALYSIS-*(CONT.)*

- SYSTEMS ANALYSIS
- SYSTEMS ENGINEERING
- TABLES (DATA)
- TRAVELING SALESMAN PROBLEM
- WIENER FILTERING
- YANG-MILLS THEORY

STATISTICAL COMMUNICATION THEORY
USE   COMMUNICATION THEORY

STATISTICAL CORRELATION
GS   CORRELATION
- . **STATISTICAL CORRELATION**
- STATISTICAL ANALYSIS
- . **STATISTICAL CORRELATION**

RT   CORRELATION COEFFICIENTS
- DATA CORRELATION
- ECONOMETRICS
- EVALUATION
- QUALITY CONTROL
- ∞STATISTICS

STATISTICAL DECISION THEORY
GS   DECISION THEORY
- . **STATISTICAL DECISION THEORY**
- STATISTICAL ANALYSIS
- . **STATISTICAL DECISION THEORY**

RT   GAME THEORY
- ∞THEORIES

STATISTICAL DISTRIBUTIONS
UF   RANDOM DISTRIBUTIONS
GS   **STATISTICAL DISTRIBUTIONS**
- . BRIGHTNESS DISTRIBUTION
- . PEARSON DISTRIBUTIONS
- . PROBABILITY DISTRIBUTION
    FUNCTIONS
- . RAYLEIGH DISTRIBUTION

RT   BINOMIAL THEOREM
- CENSORED DATA (MATHEMATICS)
- COMPLEXITY
- CURVE FITTING
- ∞DISTRIBUTION
- DISTRIBUTION (PROPERTY)
- DISTRIBUTION FUNCTIONS
- DISTRIBUTION MOMENTS
- ERROR FUNCTIONS
- EVENTS
- EXPECTANCY HYPOTHESIS
- FORECASTING
- GAMMA FUNCTION
- GOODNESS OF FIT
- KURTOSIS
- MATHEMATICAL MODELS
- OUTLIERS (STATISTICS)
- PROBABILITY THEORY
- QUALITY CONTROL
- QUANTILES
- QUANTUM THEORY
- QUARTILES
- RELIABILITY
- SCATTERING
- SIZE DISTRIBUTION

STATISTICAL MECHANICS
RT   BOLTZMANN DISTRIBUTION
- BOLTZMANN TRANSPORT EQUATION
- CLASSICAL MECHANICS
- CLOSURE LAW
- CONTINUUM MECHANICS
- ENERGY DISTRIBUTION
- FLUCTUATION THEORY
- FUNCTION SPACE
- LIOUVILLE EQUATIONS
- MACROSCOPIC EQUATIONS
- MALKUS THEORY
- MANY BODY PROBLEM
- MAXWELL-BOLTZMANN DENSITY
    FUNCTION
- ∞MECHANICS (PHYSICS)
- ONSAGER PHENOMENOLOGICAL
    COEFFICIENT
- QUANTUM MECHANICS
- QUANTUM THEORY
- THERMODYNAMIC EQUILIBRIUM
- WEIGHTING FUNCTIONS

STATISTICAL MOMENTS
USE   DISTRIBUTION MOMENTS

STATISTICAL PROBABILITY
USE   PROBABILITY THEORY

STATISTICAL TESTS
UF   BRUCETON TEST
GS   STATISTICAL ANALYSIS
- . **STATISTICAL TESTS**
- . . KOLMOGOROFF-SMIRNOFF TEST
- . . MANN-WHITNEY-WILCOXON U TEST
- . . RANK TESTS

RT   CHARTS
- CONFIDENCE LIMITS
- CURVE FITTING
- ∞DATA
- ESTIMATES
- ESTIMATING
- FACTOR ANALYSIS
- GOODNESS OF FIT
- HETEROGENEITY
- HOMOGENEITY
- LIKELIHOOD RATIO
- NORMALITY
- NULL HYPOTHESIS
- OUTLIERS (STATISTICS)
- QUALITY CONTROL
- RANGE (EXTREMES)
- REGRESSION ANALYSIS
- RELIABILITY
- SIGNIFICANCE
- ∞TESTS
- VALIDITY

STATISTICAL WEATHER FORECASTING
GS   FORECASTING
- . WEATHER FORECASTING
- . . **STATISTICAL WEATHER
    FORECASTING**
- METEOROLOGY
- . WEATHER FORECASTING
- . . **STATISTICAL WEATHER
    FORECASTING**

RT   LONG RANGE WEATHER FORECASTING
- NUMERICAL WEATHER FORECASTING

∞ STATISTICS
SN   *(USE OF A MORE SPECIFIC TERM IS
      RECOMMENDED--CONSULT THE TERMS
      LISTED BELOW)*

RT   ARRAYS
- BIOMETRICS
- CENSUS
- ∞DATA
- DEMOGRAPHY
- ENTROPY (STATISTICS)
- ESTIMATES
- ESTIMATING
- FERMI-DIRAC STATISTICS
- INFORMATION THEORY
- NONPARAMETRIC STATISTICS
- POPULATIONS
- PROBABILITY THEORY
- QUANTILES
- QUANTUM STATISTICS
- RANDOM VARIABLES
- RELIABILITY
- SAMPLING
- STATISTICAL ANALYSIS
- STATISTICAL CORRELATION
- STOCHASTIC PROCESSES
- SURVEYS
- SYSTEMS ENGINEERING
- TABLES (DATA)
- TIME SERIES ANALYSIS

STATOR BLADES
GS   TURBOMACHINE BLADES
- . **STATOR BLADES**

RT   ∞BLADES
- COMPRESSOR BLADES
- ROTOR BLADES (TURBOMACHINERY)
- STATORS
- TURBINE BLADES
- VANES

STATORS
RT   COMPRESSORS
- ELECTRIC MOTORS
- ∞GENERATORS
- IMPELLERS
- MOTORS
- PUMPS
- ∞ROTATING ELECTRICAL MACHINES
- ROTORS
- STATOR BLADES
- TURBINES

STAYS
USE   GUY WIRES

STDN (NETWORK)
- UF   SATELLITE TRACKING AND DATA ACQ
  - NETWORK
  - SPACECRAFT TRACKING AND DATA
  - NETWORK
  - STADAN (SATELLITE TRACKING
  - NETWORK)
- GS   STATIONS
  - . GROUND STATIONS
  - . . **STDN (NETWORK)**
  - . TRACKING STATIONS
  - . . **STDN (NETWORK)**
  - TRACKING NETWORKS
  - . **STDN (NETWORK)**
- RT   DATA ACQUISITION
  - GLOBAL TRACKING NETWORK
  - MINITRACK SYSTEM
  - OPTICAL TRACKING
  - RANGE AND RANGE RATE TRACKING
  - SATELLITE TRACKING
  - SPACE DETECTION AND TRACKING
  - SYSTEM
  - SPACE FLIGHT TRACKING AND DATA
  - NETWORK

STEADY FLOW
- GS   FLUID FLOW
  - . **STEADY FLOW**
  - . . COUETTE FLOW
  - . . HARTMANN FLOW
- RT   BELTRAMI FLOW
  - CONTINUITY EQUATION
  - CRITICAL FLOW
  - CROCCO METHOD
  - EQUILIBRIUM FLOW
  - ∞FLOW
  - FLOW CHARACTERISTICS
  - FLOW GEOMETRY
  - FLOW STABILITY
  - GAS FLOW
  - HEAT TRANSMISSION
  - HYDRODYNAMIC COEFFICIENTS
  - LAMINAR FLOW
  - LIQUID FLOW
  - LOW TURBULENCE
  - MASS FLOW
  - METHOD OF CHARACTERISTICS
  - MULTIPHASE FLOW
  - NONNEWTONIAN FLOW
  - ORIFICE FLOW
  - PARALLEL FLOW
  - PIPE FLOW
  - PRESSURE GRADIENTS
  - QUASI-STEADY STATES
  - SINGLE-PHASE FLOW
  - SOLIDS FLOW
  - STEAM FLOW
  - STOKES FLOW
  - SUBCRITICAL FLOW
  - SUPERCRITICAL FLOW
  - TURBULENCE
  - TURBULENT FLOW
  - TWO DIMENSIONAL FLOW
  - UNIFORM FLOW
  - UNSTEADY FLOW

STEADY STATE
- RT   ∞EQUILIBRIUM
  - FLUID DYNAMICS
  - METASTABLE STATE
  - STABILITY
  - STATE VECTORS
  - UNSTEADY FLOW
  - UNSTEADY STATE

STEADY STATE CREEP
- GS   MECHANICAL PROPERTIES
  - . CREEP PROPERTIES
  - . . **STEADY STATE CREEP**
- RT   PLASTIC FLOW
  - QUASI-STEADY STATES

STEADY STATE FLOW
- USE   EQUILIBRIUM FLOW

STEAM
- RT   BOILERS
  - FOG
  - SUPERHEATING
  - THERMODYNAMICS
  - WATER
  - WATER VAPOR

STEAM FLOW
- GS   FLUID FLOW

STEAM FLOW-(CONT.)
- . **STEAM FLOW**
- RT   CRITICAL FLOW
  - GAS FLOW
  - LAMINAR FLOW
  - MASS FLOW
  - MULTIPHASE FLOW
  - ORIFICE FLOW
  - PIPE FLOW
  - PIPELINES
  - PRESSURE GRADIENTS
  - SINGLE-PHASE FLOW
  - STEADY FLOW
  - SUBCRITICAL FLOW
  - SUPERCRITICAL FLOW
  - TURBULENT FLOW
  - UNIFORM FLOW
  - UNSTEADY FLOW

STEAM GENERATORS
- USE   BOILERS

STEAM TURBINES
- GS   TURBOMACHINERY
  - . TURBINES
  - . . **STEAM TURBINES**
- RT   AXIAL FLOW TURBINES
  - COMBINED CYCLE POWER GENERATION
  - GAS TURBINE ENGINES
  - GAS TURBINES
  - TURBOGENERATORS
  - TWO STAGE TURBINES

STEARATES
- GS   ALIPHATIC COMPOUNDS
  - . **STEARATES**
  - ESTERS
  - . **STEARATES**
- RT   SOAPS

STEAROTHERMOPHILUS
- GS   MICROORGANISMS
  - . BACTERIA
  - . . **STEAROTHERMOPHILUS**

STEATITE
- USE   TALC

STEEL STRUCTURES
- GS   WELDED STRUCTURES
  - . **STEEL STRUCTURES**
- RT   COMPOSITE STRUCTURES
  - CONSTRUCTION
  - RIGID STRUCTURES
  - ∞STRUCTURES

STEELS
- GS   ALLOYS
  - . IRON ALLOYS
  - . . **STEELS**
  - . . . BAINITIC STEEL
  - . . . CARBON STEELS
  - . . . . LOW CARBON STEELS
  - . . . CHROMIUM STEELS
  - . . . CROLOY
  - . . . HIGH STRENGTH STEELS
  - . . . . MARAGING STEELS
  - . . . NICKEL STEELS
  - . . . STAINLESS STEELS
  - . . . . AUSTENITIC STAINLESS STEELS
  - . . . . FERRITIC STAINLESS STEELS
  - . . . . MARTENSITIC STAINLESS STEELS
- RT   AUSTENITE
  - BAINITE
  - CEMENTITE
  - FERRITES
  - HYDROGEN EMBRITTLEMENT
  - MARTENSITE
  - PEARLITE

STEEP GRADIENT AIRCRAFT
- USE   V/STOL AIRCRAFT

STEEPEST ASCENT METHOD
- USE   STEEPEST DESCENT METHOD

STEEPEST DESCENT METHOD
- UF   STEEPEST ASCENT METHOD
- RT   CALCULUS OF VARIATIONS
  - DYNAMIC PROGRAMMING
  - ∞METHODOLOGY
  - MINIMA
  - OPTIMIZATION
  - PARAMETER IDENTIFICATION

STEEPEST DESCENT METHOD-(CONT.)
- SYSTEM IDENTIFICATION

STEEPNESS
- USE   SLOPES

STEERABLE ANTENNAS
- SN   (ANTENNAS DESIGNED OR ARRANGED
  - TO PERMIT CHANGES IN THE
  - DIRECTION OF AIM BY ALTERATIONS
  - OF PHASE RELATIONS)
- GS   ANTENNAS
  - . DIRECTIONAL ANTENNAS
  - . . **STEERABLE ANTENNAS**
  - . . . INERTIALESS STEERABLE
  - ANTENNAS
  - ARRAYS
  - . ANTENNA ARRAYS
  - . . **STEERABLE ANTENNAS**
  - . . . INERTIALESS STEERABLE
  - ANTENNAS
- RT   PHASED ARRAYS
  - RADAR ANTENNAS

STEERING
- RT   ∞CONTROL
  - CONTROL ROCKETS
  - CONTROLLABILITY
  - ELECTRON OPTICS
  - ∞FLIGHT
  - FOCUSING
  - SUSPENSION SYSTEMS (VEHICLES)

STEERING ROCKETS
- USE   CONTROL ROCKETS

STEFAN-BOLTZMANN LAW
- GS   LAWS
  - . RADIATION LAWS
  - . . **STEFAN-BOLTZMANN LAW**
- RT   ELECTROMAGNETIC RADIATION
  - EMISSIVITY
  - FLUX (RATE)
  - HEAT RADIATORS
  - KIRCHHOFF LAW OF RADIATION
  - RADIATIVE HEAT TRANSFER

STELLAR (STAR TRACKER)
- USE   CCD STAR TRACKER

STELLAR ACTIVITY
- GS   **STELLAR ACTIVITY**
  - . SOLAR ACTIVITY
  - . . FACULAE
  - . . SOLAR PROMINENCES
  - . . SOLAR STORMS
  - . . SPICULES
  - . . STELLAR FLARES
  - . . . SOLAR FLARES
  - . . SUNSPOTS
  - . STARSPOTS
- RT   FLARE STARS
  - ∞FLARES
  - MAGNETIC DISTURBANCES
  - MAGNETOHYDRODYNAMICS
  - PHOTOSPHERE
  - STARS
  - STELLAR LUMINOSITY
  - STELLAR MAGNETIC FIELDS
  - STELLAR MASS EJECTION
  - STELLAR OSCILLATIONS
  - STELLAR PHYSICS
  - STELLAR RADIATION
  - SUNSPOT CYCLE

STELLAR ATMOSPHERES
- GS   ENVIRONMENTS
  - . EXTRATERRESTRIAL ENVIRONMENTS
  - . . **STELLAR ATMOSPHERES**
  - . . . CHROMOSPHERE
  - . . . SOLAR ATMOSPHERE
- RT   ∞ATMOSPHERES
  - COOL STARS
  - LIMB BRIGHTENING
  - LIMB DARKENING
  - METALLIC STARS
  - RADIATIVE TRANSFER
  - SATELLITE ATMOSPHERES
  - STARSPOTS
  - STELLAR CORONAS

STELLAR COLOR
- RT   COLOR-MAGNITUDE DIAGRAM
  - STELLAR LUMINOSITY

STELLAR COLOR-(CONT.)
  STELLAR MAGNITUDE
  STELLAR SPECTRA
  STELLAR SPECTROPHOTOMETRY

STELLAR COMPOSITION
 GS COMPOSITION (PROPERTY)
  . CHEMICAL COMPOSITION
  . . **STELLAR COMPOSITION**
 RT ABUNDANCE
  B STARS
  CARBON STARS
  STARS
  STELLAR MODELS
  STELLAR PHYSICS
  STELLAR STRUCTURE

STELLAR CORES
 GS CORES
  . **STELLAR CORES**
 RT ASTROPHYSICS
  GRAVITATIONAL COLLAPSE
  LUNAR CORE
  PLANETARY CORES
  STARS
  STELLAR CORONAS
  STELLAR STRUCTURE

STELLAR CORONAS
 GS CORONAS
  . **STELLAR CORONAS**
  . . SOLAR CORONA
  . . . CORONAL LOOPS
 RT IONIZATION
  ORION NEBULA
  STELLAR ATMOSPHERES
  STELLAR CORES

STELLAR DOPPLER SHIFT
 USE DOPPLER EFFECT
  EXTRATERRESTRIAL RADIATION

STELLAR ENVELOPES
 UF CIRCUMSTELLAR MATTER
 RT ASTROPHYSICS
  COOL STARS
  SYMBIOTIC STARS
  WOLF-RAYET STARS

STELLAR EVOLUTION
 GS EVOLUTION (DEVELOPMENT)
  . **STELLAR EVOLUTION**
  . . STELLAR MASS ACCRETION
 RT ASTROPHYSICS
  COLOR-MAGNITUDE DIAGRAM
  COSMOLOGY
  GALACTIC EVOLUTION
  HERTZSPRUNG-RUSSELL DIAGRAM
  HORIZONTAL BRANCH STARS
  INTERSTELLAR EXTINCTION
  LATE STARS
  MAIN SEQUENCE STARS
  NEUTRAL CURRENTS
  PLANETARY EVOLUTION
  PRE-MAIN SEQUENCE STARS
  PROTOPLANETS
  PROTOSTARS
  RED GIANT STARS
  STELLAR PHYSICS
  SUBGIANT STARS

STELLAR FIELDS
 USE STAR DISTRIBUTION

STELLAR FLARES
 GS STELLAR ACTIVITY
  . SOLAR ACTIVITY
  . . **STELLAR FLARES**
  . . . SOLAR FLARES
 RT CATACLYSMIC VARIABLES
  FLARE STARS
  ∞FLARES
  STELLAR LUMINOSITY
  STELLAR PHYSICS
  STELLAR RADIATION

STELLAR GRAVITATION
 GS GRAVITATIONAL EFFECTS
  . **STELLAR GRAVITATION**
  GRAVITATIONAL FIELDS
  . **STELLAR GRAVITATION**
 RT GRAVITATIONAL LENSES

STELLAR LUMINOSITY
 GS ELECTROMAGNETIC PROPERTIES
  . OPTICAL PROPERTIES
  . . LUMINOSITY
  . . . **STELLAR LUMINOSITY**
 RT BRIGHTNESS
  BRIGHTNESS DISTRIBUTION
  HERTZSPRUNG-RUSSELL DIAGRAM
  HORIZONTAL BRANCH STARS
  LIMB BRIGHTENING
  LIMB DARKENING
  LUMINESCENCE
  MASS TO LIGHT RATIOS
  RED DWARF STARS
  RED GIANT STARS
  STARSPOTS
  STELLAR ACTIVITY
  STELLAR COLOR
  STELLAR FLARES
  STELLAR PARALLAX
  STELLAR PHYSICS
  WOLF-RAYET STARS

STELLAR MAGNETIC FIELDS
 GS MAGNETIC FIELDS
  . **STELLAR MAGNETIC FIELDS**
  . . SOLAR MAGNETIC FIELD
 RT ELECTROMAGNETIC FIELDS
  INTERSTELLAR MAGNETIC FIELDS
  MAGNETIC FIELD CONFIGURATIONS
  PLASMAS (PHYSICS)
  STARSPOTS
  STELLAR ACTIVITY

STELLAR MAGNITUDE
 GS MAGNITUDE
  . **STELLAR MAGNITUDE**
 RT ASTRONOMY
  COLOR-MAGNITUDE DIAGRAM
  ∞INTENSITY
  LUMINANCE
  LUMINOUS INTENSITY
  RED DWARF STARS
  STARS
  STELLAR COLOR
  STELLAR PARALLAX

STELLAR MASS
 GS MASS
  . **STELLAR MASS**
 RT MAIN SEQUENCE STARS
  MASS TO LIGHT RATIOS
  NOVAE
  STELLAR TEMPERATURE
  SUPERNOVAE
  VARIABLE STARS

STELLAR MASS ACCRETION
 GS EVOLUTION (DEVELOPMENT)
  . STELLAR EVOLUTION
  . . **STELLAR MASS ACCRETION**
 RT ACCRETION DISKS
  COSMOLOGY
  DWARF NOVAE
  GALACTIC EVOLUTION
  GRAVITATIONAL EFFECTS
  INTERSTELLAR GAS
  INTERSTELLAR MATTER
  PROTOSTARS
  STELLAR PHYSICS
  SYMBIOTIC STARS
  X RAY BINARIES

STELLAR MASS EJECTION
 GS EJECTION
  . **STELLAR MASS EJECTION**
 RT CATACLYSMIC VARIABLES
  DWARF NOVAE
  NOVAE
  STELLAR ACTIVITY
  SUPERNOVAE
  VARIABLE STARS
  WOLF-RAYET STARS

STELLAR MODELS
 GS MODELS
  . ASTRONOMICAL MODELS
  . . **STELLAR MODELS**
 RT ASTRONOMY
  SOLAR NEUTRINOS
  SOLAR OSCILLATIONS
  STELLAR COMPOSITION
  SUPERMASSIVE STARS

STELLAR MOTIONS
 GS **STELLAR MOTIONS**
  . DOUBLE STARS
  . STELLAR ORBITS
  . STELLAR OSCILLATIONS
  . STELLAR ROTATION
 RT COMPANION STARS
  COROTATION
  DOPPLER EFFECT
  DOPPLER-FIZEAU EFFECT
  GALACTIC ROTATION
  HIPPARCOS SATELLITE
  ∞MOTION
  SIDEREAL TIME
  STELLAR PARALLAX

STELLAR OCCULTATION
 GS OCCULTATION
  . **STELLAR OCCULTATION**
 RT ECLIPSING BINARY STARS
  LUNAR OCCULTATION

STELLAR ORBITS
 SN (EXCLUDES PLANETARY ORBITS)
 GS ORBITS
  . **STELLAR ORBITS**
  STELLAR MOTIONS
  . **STELLAR ORBITS**
 RT CELESTIAL MECHANICS

STELLAR OSCILLATIONS
 GS OSCILLATIONS
  . **STELLAR OSCILLATIONS**
  . . SOLAR OSCILLATIONS
  STELLAR MOTIONS
  . **STELLAR OSCILLATIONS**
 RT ASTRONOMICAL MODELS
  ASTRONOMY
  ASTROPHYSICS
  ATMOSPHERIC MODELS
  CATACLYSMIC VARIABLES
  STARS
  STELLAR ACTIVITY
  SYMBIOTIC STARS
  VARIABLE STARS

STELLAR PARALLAX
 GS PARALLAX
  . **STELLAR PARALLAX**
 RT ASTROMETRY
  BINARY STARS
  HIPPARCOS SATELLITE
  SOLAR PARALLAX
  STELLAR LUMINOSITY
  STELLAR MAGNITUDE
  STELLAR MOTIONS

STELLAR PHYSICS
 GS ASTROPHYSICS
  . **STELLAR PHYSICS**
  . . SOLAR PHYSICS
 RT NUCLEAR FUSION
  ∞SCIENCE
  STELLAR ACTIVITY
  STELLAR COMPOSITION
  STELLAR EVOLUTION
  STELLAR FLARES
  STELLAR LUMINOSITY
  STELLAR MASS ACCRETION
  STELLAR RADIATION
  STELLAR ROTATION
  STELLAR STRUCTURE
  SUPERNOVAE

STELLAR RADIATION
 GS EXTRATERRESTRIAL RADIATION
  . **STELLAR RADIATION**
  . . STELLAR WINDS
 RT COSMIC RAYS
  ELECTROMAGNETIC RADIATION
  EXTARS
  GALACTIC RADIATION
  GAMMA RAY BURSTS
  HERBIG-HARO OBJECTS
  INTERSTELLAR EXTINCTION
  INTERSTELLAR RADIATION
  LIGHT CURVE
  MICROWAVE EMISSION
  POLARIZED ELECTROMAGNETIC
   RADIATION
  ∞RADIATION
  RADIATIVE TRANSFER
  RADIO BURSTS
  RADIO STARS
  SOLAR RADIATION

STELLAR RADIATION-(CONT.)
STARSPOTS
STELLAR ACTIVITY
STELLAR FLARES
STELLAR PHYSICS

STELLAR ROTATION
GS   GYRATION
. ROTATION
. . **STELLAR ROTATION**
. . . SOLAR ROTATION
STELLAR MOTIONS
. **STELLAR ROTATION**
RT   ANGULAR MOMENTUM
COROTATION
PLANETARY ROTATION
STELLAR PHYSICS

STELLAR SPECTRA
GS   SPECTRA
. RADIATION SPECTRA
. . ELECTROMAGNETIC SPECTRA
. . . **STELLAR SPECTRA**
. . . . SOLAR SPECTRA
RT   ABSORPTION SPECTRA
ASTRONOMICAL SPECTROSCOPY
CONTINUOUS SPECTRA
COOL STARS
EMISSION SPECTRA
HERBIG-HARO OBJECTS
HERTZSPRUNG-RUSSELL DIAGRAM
INFRARED SPECTRA
LINE SPECTRA
MOLECULAR SPECTRA
PECULIAR STARS
SEYFERT GALAXIES
STELLAR COLOR
SYMBIOTIC STARS
ULTRAVIOLET SPECTRA
VISIBLE SPECTRUM
X RAY SPECTRA

STELLAR SPECTROPHOTOMETRY
GS   OPTICAL MEASUREMENT
. ASTRONOMICAL PHOTOMETRY
. . **STELLAR SPECTROPHOTOMETRY**
. PHOTOMETRY
. . SPECTROPHOTOMETRY
. . . **STELLAR SPECTROPHOTOMETRY**
SPECTROSCOPY
. SPECTROPHOTOMETRY
. . **STELLAR SPECTROPHOTOMETRY**
RT   HORIZONTAL BRANCH STARS
INFRARED PHOTOMETRY
PECULIAR STARS
SPECTROSCOPIC TELESCOPES
STELLAR COLOR

STELLAR STRUCTURE
RT   CHROMOSPHERE
CORONAL HOLES
DENSE PLASMAS
METALLIC STARS
PECULIAR STARS
SOLAR ATMOSPHERE
SOLAR CORONA
STELLAR COMPOSITION
STELLAR CORES
STELLAR PHYSICS
∞STRUCTURES
SUPERMASSIVE STARS

STELLAR TEMPERATURE
GS   TEMPERATURE
. **STELLAR TEMPERATURE**
RT   COOL STARS
STELLAR MASS
SYMBIOTIC STARS

STELLAR WINDS
GS   EXTRATERRESTRIAL RADIATION
. STELLAR RADIATION
. . **STELLAR WINDS**
GASES
. IONIZED GASES
. . CHARGED PARTICLES
. . . **STELLAR WINDS**
PARTICLES
. CHARGED PARTICLES
. . ENERGETIC PARTICLES
. . . PLASMAS (PHYSICS)
. . . . **STELLAR WINDS**
RT   CHROMOSPHERE
COSMIC PLASMA

STELLAR WINDS-(CONT.)
INTERGALACTIC MEDIA
INTERSTELLAR GAS
RADIATION PRESSURE
SOLAR WIND
SOLAR WIND VELOCITY

STELLARATORS
RT   MAGNETOHYDRODYNAMICS
NUCLEAR REACTORS
PINCH EFFECT
THERMAL INSTABILITY
THERMONUCLEAR POWER GENERATION
THERMONUCLEAR REACTIONS
TOROIDAL PLASMAS

STELLITE (TRADEMARK)
UF   HAYNES STELLITE
RT   CHROMIUM ALLOYS
COBALT ALLOYS
TUNGSTEN ALLOYS

STEMS
RT   PLANTS (BOTANY)

STENCIL PROCESSES
RT   PRINTING
REPRODUCTION (COPYING)

STEP FAULTS
USE   GEOLOGICAL FAULTS

STEP FUNCTIONS
GS   FUNCTIONS (MATHEMATICS)
. **STEP FUNCTIONS**
RT   DYNAMIC RESPONSE
∞FREQUENCY RESPONSE
INTERVALS
RAMP FUNCTIONS
REACTION TIME
∞STEPS

STEP RECOVERY DIODES
GS   ELECTRONIC EQUIPMENT
. SOLID STATE DEVICES
. . SEMICONDUCTOR DEVICES
. . . JUNCTION DIODES
. . . . **STEP RECOVERY DIODES**
RT   SWITCHING

STEPPES
GS   LANDFORMS
. **STEPPES**
RT   ARID LANDS
DESERTIFICATION
GRASSLANDS
PLAINS

STEPPING MOTORS
GS   MOTORS
. ELECTRIC MOTORS
. . **STEPPING MOTORS**
RT   ACTUATORS
SERVOCONTROL
SERVOMECHANISMS

STEPPING SWITCHES
GS   SWITCHES
. ELECTRIC SWITCHES
. . **STEPPING SWITCHES**

∞ STEPS
SN   *(USE OF A MORE SPECIFIC TERM IS RECOMMENDED--CONSULT THE TERMS LISTED BELOW)*
RT   BACKWARD FACING STEPS
PROGRAMMING (SCHEDULING)
STAIRSTEPS
STEP FUNCTIONS

STEREOCHEMISTRY
RT   CARBOHYDRATES
∞CHEMISTRY
ISOMERS
OPTICAL ACTIVITY
SPATIAL DISTRIBUTION
X RAY ANALYSIS

STEREOGRAPHY
USE   STEREOPHOTOGRAPHY

STEREOPHONICS
RT   ACOUSTICS
HEARING

STEREOPHOTOGRAPHY
UF   STEREOGRAPHY
STEREOSCOPIC PHOTOGRAPHY
GS   IMAGERY
. STEREOSCOPY
. . **STEREOPHOTOGRAPHY**
PHOTOGRAPHY
. STEREOSCOPY
. . **STEREOPHOTOGRAPHY**
RT   AERIAL PHOTOGRAPHY
BLACK AND WHITE PHOTOGRAPHY
CINEMATOGRAPHY
COLOR PHOTOGRAPHY
MAPSAT
PHOTOGRAMMETRY
SPOT (FRENCH SATELLITE)

STEREOSCOPIC PHOTOGRAPHY
USE   STEREOPHOTOGRAPHY

STEREOSCOPIC VISION
GS   VISION
. **STEREOSCOPIC VISION**
RT   BINOCULAR VISION
STEREOSCOPY

STEREOSCOPY
GS   IMAGERY
. **STEREOSCOPY**
. . STEREOPHOTOGRAPHY
PHOTOGRAPHY
. **STEREOSCOPY**
. . STEREOPHOTOGRAPHY
RT   STEREOSCOPIC VISION

STEREOTELEVISION
GS   COMMUNICATION EQUIPMENT
. **STEREOTELEVISION**
TELECOMMUNICATION
. **STEREOTELEVISION**
TELEVISION SYSTEMS
. **STEREOTELEVISION**
RT   CLOSED CIRCUIT TELEVISION
COLOR TELEVISION
COMMUNICATING
EDUCATIONAL TELEVISION
SATELLITE TELEVISION
SPACECRAFT TELEVISION

STERILIZATION
GS   **STERILIZATION**
. CHEMICAL STERILIZATION
. HOUSEKEEPING (SPACECRAFT)
. SPACECRAFT STERILIZATION
RT   AIR PURIFICATION
ANTIFOULING
ANTISEPTICS
BACTERICIDES
BAKING
CLEANING
DECONTAMINATION
FUMIGATION
GNOTOBIOTICS
IONIZING RADIATION
MERCURY LAMPS
PASTEURIZING
PURIFICATION
ULTRAVIOLET RADIATION

STERILIZATION EFFECTS
GS   **STERILIZATION EFFECTS**
. CHEMICAL EFFECTS
. DECONTAMINATION
. THERMAL DEGRADATION
RT   CORROSION
DEGRADATION
∞DEOXIFICATION
∞EFFECTS
SPACECRAFT STERILIZATION
TEMPERATURE EFFECTS
WAVE DEGRADATION

STERNS
USE   AFTERBODIES

STERNUM
GS   ANATOMY
. MUSCULOSKELETAL SYSTEM
. . BONES
. . . . **STERNUM**
RT   THORAX

STEROIDS
GS   **STEROIDS**

STEROIDS-(CONT.)
. ACTINOMYCIN
. CHOLESTEROL
. CORTICOSTEROIDS
. . ALDOSTERONE
. . HYDROXYCORTICOSTEROID
. . . CORTISONE
. PENICILLIN
. STREPTOMYCIN
. TETRACYCLINES
RT   ANTIBIOTICS

STETHOSCOPES
GS   MEDICAL EQUIPMENT
      . **STETHOSCOPES**
RT   PHYSICIANS

STIELTJES INTEGRAL
GS   ANALYSIS (MATHEMATICS)
      . REAL VARIABLES
      . . MEASURE AND INTEGRATION
      . . . **STIELTJES INTEGRAL**
RT   PROBABILITY THEORY

STIFF STRUCTURES
USE   RIGID STRUCTURES

STIFFENING
RT   REINFORCEMENT (STRUCTURES)
      RIBS (SUPPORTS)
      WEBS (SUPPORTS)

STIFFNESS
GS   MECHANICAL PROPERTIES
      . **STIFFNESS**
RT   BENDING
      DEFORMATION
      FLEXIBILITY
      MODULUS OF ELASTICITY
      ∞RIGIDITY
      SOFTNESS
      STRUCTURAL STABILITY

STIFFNESS MATRIX
GS   ALGEBRA
      . VECTOR SPACES
      . . MATRICES (MATHEMATICS)
      . . . **STIFFNESS MATRIX**
RT   STRUCTURAL ANALYSIS
      STRUCTURAL MEMBERS

STIGMATISM
GS   ELECTROMAGNETIC PROPERTIES
      . OPTICAL PROPERTIES
      . . **STIGMATISM**
RT   ASTIGMATISM
      FOCUSING
      LENS DESIGN
      LENSES

STILBENE
GS   HYDROCARBONS
      . **STILBENE**
RT   DYES
      HEXANITROSTILBENE

STILLS
GS   SEPARATORS
      . **STILLS**
RT   CONCENTRATORS
      DISTILLATION EQUIPMENT

STIMULANTS
GS   DRUGS
      . **STIMULANTS**
      . . ATROPINE
      . . CAFFEINE
      . CENTRAL NERVOUS SYSTEM
         STIMULANTS
      . . NORADRENALINE
      . . NOREPINEPHRINE
RT   AMINOPHYLLINE
      EPINEPHRINE
      STRYCHNINE

STIMULATED EMISSION
GS   DECAY
      . EMISSION
      . . **STIMULATED EMISSION**
      . . . WATER MASERS
RT   ARGON LASERS
      CARBON DIOXIDE LASERS
      CARBON LASERS
      CARBON MONOXIDE LASERS

STIMULATED EMISSION-(CONT.)
      COHERENT ELECTROMAGNETIC
         RADIATION
      COHERENT LIGHT
      ELECTRON EMISSION
      ELECTRON PUMPING
      GALLIUM ARSENIDE LASERS
      GAS LASERS
      GAS MASERS
      HCN LASERS
      INTERSTELLAR MASERS
      LASERS
      LIGHT EMISSION
      MASERS
      NUCLEAR PUMPING
      OPTICAL PUMPING
      PARTICLE EMISSION
      PHOTOELECTRIC EMISSION
      POPULATION INVERSION
      RAPID BALLISTICS IDENTIFICATION
      RARE GAS-HALIDE LASERS
      SELF SUSTAINED EMISSION
      TEA LASERS
      TWO-WAVELENGTH LASERS
      ULTRASHORT PULSED LASERS
      ULTRAVIOLET LASERS

STIMULATED EMISSION DEVICES
UF   QUANTUM GENERATORS
GS   **STIMULATED EMISSION DEVICES**
      . LASERS
      . . AIRBORNE LASERS
      . . ARGON LASERS
      . . ATMOSPHERIC LASERS
      . . CARBON LASERS
      . . CHEMICAL LASERS
      . . . HCL LASERS
      . . CONTINUOUS WAVE LASERS
      . . DISTRIBUTED FEEDBACK LASERS
      . FREE ELECTRON LASERS
      . GAMMA RAY LASERS
      . GAS LASERS
      . . CARBON DIOXIDE LASERS
      . . CARBON MONOXIDE LASERS
      . . DF LASERS
      . . EXCIMER LASERS
      . . HCL LASERS
      . . . HCL ARGON LASERS
      . . HCN LASERS
      . . HELIUM-NEON LASERS
      . . HF LASERS
      . . KRYPTON FLUORIDE LASERS
      . . NITROGEN LASERS
      . . TEA LASERS
      . . ULTRAVIOLET LASERS
      . . XENON CHLORIDE LASERS
      . . XENON FLUORIDE LASERS
      . . GASDYNAMIC LASERS
      . GLASS LASERS
      . HIGH POWER LASERS
      . . NOVA LASER SYSTEM
      . . SHIVA LASER SYSTEM
      . . INFRARED LASERS
      . . INJECTION LASERS
      . . IODINE LASERS
      . . LIQUID LASERS
      . . METAL VAPOR LASERS
      . . NEODYMIUM LASERS
      . . NUCLEAR PUMPED LASERS
      . . ORGANIC LASERS
      . . . DYE LASERS
      . . PLASMADYNAMIC LASERS
      . . PULSED LASERS
      . . . Q SWITCHED LASERS
      . . . ULTRASHORT PULSED LASERS
      . . . ULTRAVIOLET LASERS
      . . RAMAN LASERS
      . . RARE GAS-HALIDE LASERS
      . . . KRYPTON FLUORIDE LASERS
      . . . XENON CHLORIDE LASERS
      . . . XENON FLUORIDE LASERS
      . . RING LASERS
      . SEMICONDUCTOR LASERS
      . . GALLIUM ARSENIDE LASERS
      . SOLAR-PUMPED LASERS
      . SOLID STATE LASERS
      . . GALLIUM ARSENIDE LASERS
      . . RUBY LASERS
      . . YAG LASERS
      . SPACEBORNE LASERS
      . TUNABLE LASERS
      . TWO-WAVELENGTH LASERS
      . WAVEGUIDE LASERS
      . X RAY LASERS
      . MASERS
      . . GAS MASERS

STIMULATED EMISSION DEVICES-(CONT.)
      . . . HYDROGEN MASERS
      . . INTERSTELLAR MASERS
      . . PROTON MASERS
      . . TRAVELING WAVE MASERS
      . . WATER MASERS
RT   AMPLIFIERS
      COHERENT ELECTROMAGNETIC
         RADIATION
      ELECTRON PUMPING
      ∞GENERATORS
      LASER CAVITIES
      LASER PUMPING
      LASER WEAPONS
      LASING
      LIGHT TRANSMISSION
      NUCLEAR PUMPING
      OPTICAL PUMPING
      RAPID BALLISTICS IDENTIFICATION
      SUBHARMONIC GENERATORS
      TRANSIENT OSCILLATIONS

STIMULATION
GS   **STIMULATION**
      . AUDITORY STIMULI
      . SENSORY STIMULATION
RT   ACTIVATION
      ACTIVATION (BIOLOGY)
      ACTUATION
      CLOUD SEEDING
      GAS INJECTION
      INITIATION
      PRESSURIZING
      STARTING

∞ STIMULI
SN   *(USE OF A MORE SPECIFIC TERM IS*
      RECOMMENDED--CONSULT THE TERMS
      LISTED BELOW)
RT   AROUSAL
      AUDITORY STIMULI
      CALORIC STIMULI
      ELECTRIC STIMULI
      MOTIVATION
      PSYCHOLOGICAL FACTORS
      SUBLIMINAL STIMULI
      VISUAL STIMULI

STIRLING CYCLE
GS   CYCLES
      . THERMODYNAMIC CYCLES
      . . **STIRLING CYCLE**
RT   CARNOT CYCLE

STIRRING
RT   AERATION
      DISPERSING
      MIXERS
      SUSPENDING (MIXING)
      SWIRLING

STISHOVITE
GS   MINERALS
      . QUARTZ
      . . **STISHOVITE**
RT   COESITE
      EARTH CRUST
      EARTH MANTLE
      RUTILE

STOCHASTIC PROCESSES
UF   POISSON PROCESS
GS   **STOCHASTIC PROCESSES**
      . MARKOV CHAINS
      . MARKOV PROCESSES
      . RANDOM PROCESSES
      . . RANDOM WALK
RT   ∞APPLICATIONS OF MATHEMATICS
      CHAOS
      COHERENCE COEFFICIENT
      DECISION THEORY
      ERGODIC PROCESS
      EVENTS
      FOKKER-PLANCK EQUATION
      GAME THEORY
      INFORMATION THEORY
      KAKUTANI THEOREM
      KALMAN-SCHMIDT FILTERING
      MARTINGALES
      MATHEMATICAL MODELS
      MONTE CARLO METHOD
      OPERATIONS RESEARCH
      PROBABILITY THEORY
      QUEUEING THEORY
      RANDOM ERRORS

STOCHASTIC PROCESSES-(CONT.)
        RANDOM NOISE
        RANDOM SIGNALS
        STATE ESTIMATION
        STATISTICAL ANALYSIS
    ∞ STATISTICS
        TIME DEPENDENCE
        TIME FUNCTIONS
        TIME SERIES ANALYSIS

STOCKPILING
    RT   ACCUMULATIONS
          COLLECTION
          INVENTORY MANAGEMENT
          LOGISTICS
          RESERVES
     ∞ STORAGE
          STRATEGIC MATERIALS

STOICHIOMETRY
    RT   CHEMICAL REACTIONS
      ∞ CHEMISTRY
      ∞ COMPOSITION
          COMPOSITION (PROPERTY)
          FORMULATIONS
          MATERIAL BALANCE
          PHASE DIAGRAMS

STOKES FLOW
    GS   FLUID FLOW
          . INCOMPRESSIBLE FLOW
          . . STOKES FLOW
          . VISCOUS FLOW
          . . STOKES FLOW
    RT   OSEEN APPROXIMATION
          STEADY FLOW

∞ STOKES LAW
    SN   (USE OF A MORE SPECIFIC TERM IS
          RECOMMENDED--CONSULT THE TERMS
          LISTED BELOW)
    RT   LAWS
          MAXWELL EQUATION
          STOKES LAW (FLUID MECHANICS)
          STOKES THEOREM (VECTOR CALCULUS)

STOKES LAW (FLUID MECHANICS)
    RT   SETTLING
      ∞ STOKES LAW
          VISCOSITY

STOKES LAW OF RADIATION
    GS   LAWS
          . RADIATION LAWS
          . . STOKES LAW OF RADIATION
    RT   INCIDENT RADIATION
          LUMINESCENCE
      ∞ RADIATION
          WAVELENGTHS

STOKES THEOREM (VECTOR CALCULUS)
    GS   ALGEBRA
          . VECTOR SPACES
          . . STOKES THEOREM (VECTOR
                CALCULUS)
          THEOREMS
          . STOKES THEOREM (VECTOR
                CALCULUS)
    RT   ∞ STOKES LAW

STOKES-BELTRAMI EQUATION
    RT   ∞ EQUATIONS
          LAPLACE EQUATION
          STREAM FUNCTIONS (FLUIDS)

STOL AIRCRAFT
    USE   SHORT TAKEOFF AIRCRAFT

STOMACH
    GS   ANATOMY
          . DIGESTIVE SYSTEM
          . . GASTROINTESTINAL SYSTEM
          . . . STOMACH
          . ORGANS
          . . STOMACH
          VISCERA
          . ORGANS
          . . STOMACH
    RT   ABDOMEN

STONES (ROCKS)
    USE   ROCKS

STONY METEORITES
    GS   CELESTIAL BODIES
          . METEORITES
          . . STONY METEORITES
          . . . ACHONDRITES
          . . . . BONDOC METEORITE
          . . . . KAPOETA ACHONDRITE
          . . . . NORTON COUNTY ACHONDRITE
          . . . CHONDRITES
          . . . . BRUDERHEIM METEORITE
          . . . . CARBONACEOUS CHONDRITES
          . . . . . ALLENDE METEORITE
          . . . . . MURCHISON METEORITE
          . . . . CARBONACEOUS METEORITES
          . . . . . ALAIS METEORITE
          . . . . . COLD BOKKEVELD METEORITE
          . . . . . IVUNA METEORITE
          . . . . . MURRAY METEORITE
          . . . . . ORGUEIL METEORITE
          . . . . . TONK METEORITE
          . . . . HVITTIS CHONDRITE
          . . . . PANTAR CHONDRITES
          . . . . PRIBRAM METEORITE
          . . . TEKTITES
          . . . . AUSTRALITES
          . . . . BEDIASITES
          . . . . TUNGUSK METEORITE
    RT   CARBONACEOUS CHONDRITES
          COESITE
          HARLETON METEORITE
          IRON METEORITES
          LAZAREV METEORITE
          METEORITIC COMPOSITION
          METEORITIC MICROSTRUCTURES
          OKHANSK METEORITE
          SCHREIBERSITE

STOPCOCKS
    USE   COCKS

STOPPING
    UF   TERMINATING
    GS   STOPPING
          . THRUST TERMINATION
    RT   BLOCKING
          CANCELLATION
          CLOSING
          CONSTRICTIONS
          CONTAINMENT
          DAMPING
          DECELERATION
          DELAY
          ELIMINATION
      ∞ HOLDING
      ∞ INHIBITION
          OPTIMIZATION
          PLUGS
          PREVENTION
      ∞ REDUCTION
          RETARDING
          SEALING

STOPPING POWER
    RT   ABSORBERS (MATERIALS)
          ABSORPTION CROSS SECTIONS
      ∞ CROSS SECTIONS
          DENSITY (MASS/VOLUME)
          NEUTRON CROSS SECTIONS
          RADIATION ABSORPTION
          RADIATION SHIELDING
          SCATTERING CROSS SECTIONS

STORABLE PROPELLANTS
    GS   CONSUMABLES (SPACECRAFT)
          . STORABLE PROPELLANTS
          PROPELLANTS
          . STORABLE PROPELLANTS
          . . RP-1 ROCKET PROPELLANTS
    RT   CRYOGENIC ROCKET PROPELLANTS
          GASEOUS ROCKET PROPELLANTS
          GELLED ROCKET PROPELLANTS
          GROUND SUPPORT EQUIPMENT
          HIGH TEMPERATURE PROPELLANTS
          HYDROCARBON FUELS
          HYPERGOLIC ROCKET PROPELLANTS
          LIQUID ROCKET PROPELLANTS
          PROPELLANT ADDITIVES
          PROPELLANT DECOMPOSITION
          PROPELLANT EVAPORATION
          PROPELLANT SENSITIVITY
          PROPELLANT STORABILITY
          PROPELLANT STORAGE
          ROCKET PROPELLANTS
          SOLID PROPELLANTS
          SPACE STORAGE

∞ STORAGE
    SN   (USE OF A MORE SPECIFIC TERM IS
          RECOMMENDED--CONSULT THE TERMS
          LISTED BELOW)
    RT   BUFFER STORAGE
          COMPUTER STORAGE DEVICES
          CORE STORAGE
          CRYOGENIC FLUID STORAGE
          DATA STORAGE
          DISPOSAL
          DOCUMENT STORAGE
          ENERGY STORAGE
          EXTERNAL STORE SEPARATION
          EXTERNAL STORES
          FLUID FILLED SHELLS
          HANDLING EQUIPMENT
          INVENTORIES
          INVENTORY CONTROLS
          INVENTORY MANAGEMENT
          ION STORAGE
          LIQUID FILLED SHELLS
          LOGISTICS
          LOGISTICS MANAGEMENT
          MAGNETIC STORAGE
          MATERIALS HANDLING
          MISSILE SILOS
          MISSILE STORAGE
          PACKAGING
          PIPELINES
          PRESERVING
          PROPELLANT STORAGE
          RACKS (FRAMES)
          RECORDING
          RESERVES
          RETAINING
          SAFETY
          SPACE STORAGE
          STOCKPILING
          STORAGE TANKS
          STOWAGE (ONBOARD EQUIPMENT)
          UNDERGROUND STORAGE
          WASTE DISPOSAL
          WING-FUSELAGE STORES

STORAGE BATTERIES
    SN   (RECHARGEABLE BATTERIES)
    UF   SECONDARY BATTERIES
    GS   ELECTROCHEMICAL CELLS
          . ELECTRIC BATTERIES
          . . STORAGE BATTERIES
          . . . LEAD ACID BATTERIES
          . . . NICKEL CADMIUM BATTERIES
          . . . NICKEL HYDROGEN BATTERIES
          . . . NICKEL ZINC BATTERIES
          . . . SILVER CADMIUM BATTERIES
          . . . SILVER HYDROGEN BATTERIES
          . . . SILVER ZINC BATTERIES
          . . . ZINC-BROMIDE BATTERIES
          . . . ZINC-CHLORINE BATTERIES
    RT   ALKALINE BATTERIES
          BATTERY CHARGERS
          CHARGE EFFICIENCY
          DRY CELLS
          ELECTROLYTES
          METAL AIR BATTERIES
          NICKEL IRON BATTERIES
          NONAQUEOUS ELECTROLYTES
          PRIMARY BATTERIES
          PULSE CHARGING
          REGENERATIVE FUEL CELLS

STORAGE RINGS (PARTICLE ACCELERATORS)
    UF   ELECTRON RING ACCELERATORS
    GS   PARTICLE ACCELERATORS
          . CYCLIC ACCELERATORS
          . . SYNCHROTRONS
          . . . STORAGE RINGS (PARTICLE
                ACCELERATORS)
    RT   ∞ ACCELERATORS
      ∞ RINGS

STORAGE STABILITY
    GS   LIFE (DURABILITY)
          . STORAGE STABILITY
          STABILITY
          . STORAGE STABILITY
    RT   DECOMPOSITION
          LIQUID SLOSHING
          LONG TERM EFFECTS
          STATIC STABILITY
          SURFACE STABILITY
          THERMAL STABILITY

STORAGE TANKS
    GS   TANKS (CONTAINERS)

STORAGE TANKS-(CONT.)
. **STORAGE TANKS**
RT   CRYOGENIC FLUID STORAGE
    CYLINDRICAL TANKS
    EXPULSION BLADDERS
    EXTERNAL TANKS
    FUEL TANKS
    PIPELINES
    PRESSURE VESSELS
    PROPELLANT TANKS
    SPACE STORAGE
    SPHERICAL TANKS
  ∞STORAGE
    TANK GEOMETRY
    UNDERGROUND STORAGE
    WING-FUSELAGE STORES

STORE RELEASE
USE   EXTERNAL STORE SEPARATION

STORM DAMAGE
GS   DAMAGE
  . **STORM DAMAGE**
RT   CYCLONES
    FLOOD CONTROL
    FLOODS
    GUSTS
    HAILSTORMS
    HURRICANES
    LANDSLIDES
    PRECIPITATION (METEOROLOGY)
    RAINSTORMS
    SNOW COVER
    STORM SURGES
    STORMS
    THUNDERSTORMS
    TORNADOES
    TROPICAL STORMS
    TYPHOONS
    WIND (METEOROLOGY)

STORM ENHANCEMENT
GS   WEATHER MODIFICATION
  . **STORM ENHANCEMENT**
RT   CLIMATOLOGY
    HAILSTORMS
    PRECIPITATION (METEOROLOGY)
    RAINSTORMS
    SNOWSTORMS
    STORMS
    STORMS (METEOROLOGY)

STORM SUPPRESSION
GS   WEATHER MODIFICATION
  . **STORM SUPPRESSION**
RT   CLIMATOLOGY
    HAILSTORMS
    ICE PREVENTION
    PRECIPITATION (METEOROLOGY)
    RAINSTORMS
    SNOWSTORMS
    STORMS

STORM SURGES
RT   COASTS
    HURRICANES
    OCEAN SURFACE
    OCEANOGRAPHY
    STORM DAMAGE
    STORMS (METEOROLOGY)
    SURGES

STORMS
GS   **STORMS**
  . CYCLONES
  . . CYCLOGENESIS
  . . HURRICANES
  . . . ANNA HURRICANE
  . . TYPHOONS
  . IONOSPHERIC STORMS
  . . SUDDEN IONOSPHERIC
      DISTURBANCES
  . MAGNETIC STORMS
  . NOISE STORMS
  . SOLAR STORMS
  . STORMS (METEOROLOGY)
  . . DUST STORMS
  . . HAILSTORMS
  . . POLAR SUBSTORMS
  . . RAINSTORMS
  . . . THUNDERSTORMS
  . . SNOWSTORMS
  . . TORNADOES
  . . TROPICAL STORMS
  . . . HURRICANES

STORMS-(CONT.)
  . . . . ANNA HURRICANE
  . . . TYPHOONS
RT   CLIMATOLOGY
    COLD FRONTS
  ∞DISTURBANCES
    FLOOD DAMAGE
    FLOODS
    FRONTS (METEOROLOGY)
    GUSTS
    PRECIPITATION (METEOROLOGY)
    SNOW COVER
    SOLAR TERRESTRIAL INTERACTIONS
    STORM DAMAGE
    STORM ENHANCEMENT
    STORM SUPPRESSION
    SUDDEN STORM COMMENCEMENTS
    WARM FRONTS
    WEATHER FORECASTING
    WIND (METEOROLOGY)

STORMS (METEOROLOGY)
GS   STORMS
  . . **STORMS (METEOROLOGY)**
  . . DUST STORMS
  . . HAILSTORMS
  . . POLAR SUBSTORMS
  . . RAINSTORMS
  . . . THUNDERSTORMS
  . . SNOWSTORMS
  . . TORNADOES
  . . TROPICAL STORMS
  . . . HURRICANES
  . . . . ANNA HURRICANE
  . . . TYPHOONS
RT   ALPINE METEOROLOGY
    CLIMATOLOGY
    FLIGHT CONDITIONS
    FLOOD CONTROL
    FLOOD DAMAGE
    FLOOD PREDICTIONS
    FLOODS
    GROUND WIND
    GUSTS
    HAIL
    ICE
    METEOROLOGICAL PARAMETERS
    METEOROLOGY
    PRECIPITATION (METEOROLOGY)
    SNOW
    SNOW COVER
    SQUALLS
    STORM ENHANCEMENT
    STORM SURGES
    WATERSHEDS
    WEATHER FORECASTING
    WIND (METEOROLOGY)

STORMSAT SATELLITE
UF   SEVERE STORMS OBSERVING
     SATELLITE
GS   SATELLITES
  . ARTIFICIAL SATELLITES
  . . SYNCHRONOUS SATELLITES
  . . . **STORMSAT SATELLITE**
RT   NASA PROGRAMS

STOSS-AND-LEE TOPOGRAPHY
USE   GLACIAL DRIFT

STOWAGE (ONBOARD EQUIPMENT)
RT   LOGISTICS
    ONBOARD EQUIPMENT
    PORTABLE EQUIPMENT
    PROVISIONING
    SPACE LOGISTICS
    SPACE RATIONS
  ∞STORAGE

STRAIGHT WINGS
USE   RECTANGULAR WINGS

STRAIN AGING
USE   PRECIPITATION HARDENING

STRAIN DISTRIBUTION
USE   STRESS CONCENTRATION

STRAIN ENERGY METHODS
GS   STRUCTURAL ANALYSIS
  . ENERGY METHODS
  . . **STRAIN ENERGY METHODS**
RT   ∞ENERGY
  ∞METHODOLOGY
    SEISMIC ENERGY

STRAIN FATIGUE
USE   FATIGUE (MATERIALS)

STRAIN GAGE ACCELEROMETERS
GS   MEASURING INSTRUMENTS
  . ACCELEROMETERS
  . . **STRAIN GAGE ACCELEROMETERS**
RT   PRESSURE GAGES

STRAIN GAGE BALANCES
GS   MEASURING INSTRUMENTS
  . INDICATING INSTRUMENTS
  . . WEIGHT INDICATORS
  . . . **STRAIN GAGE BALANCES**
RT   PRESSURE GAGES

STRAIN GAGES
GS   MEASURING INSTRUMENTS
  . **STRAIN GAGES**
RT   CABLE FORCE RECORDERS
    DEFORMETERS
    ELASTOMETERS
    EXTENSOMETERS
    FLIGHT LOAD RECORDERS
    MECHANICAL MEASUREMENT
    PIEZOELECTRIC GAGES
    PRESSURE GAGES
    ROSETTE SHAPES
    SHOCK MEASURING INSTRUMENTS
    STRAIN MEASUREMENT
    STRESS MEASUREMENT
    TEMPERATURE INVERSIONS
    TENSOMETERS
    TRANSDUCERS
    WEIGHT INDICATORS

STRAIN HARDENING
GS   HARDENING (MATERIALS)
  . WORK HARDENING
  . . **STRAIN HARDENING**
RT   AGING (MATERIALS)
    AGING (METALLURGY)
    PRECIPITATION HARDENING
    RESIDUAL STRESS
    SHOT PEENING
    STRESS RELIEVING
    TEMPERATURE INVERSIONS

STRAIN MEASUREMENT
RT   ∞MEASUREMENT
    STRAIN GAGES
    STRAIN RATE
    STRESS-STRAIN DIAGRAMS
    STRESS-STRAIN RELATIONSHIPS
    STRUCTURAL STRAIN

STRAIN RATE
GS   RATES (PER TIME)
  . **STRAIN RATE**
RT   IMPACT TESTS
    LOADING RATE
    MECHANICAL PROPERTIES
    STRAIN MEASUREMENT
    TEMPERATURE INVERSIONS

STRAIN SOFTENING
USE   PLASTIC DEFORMATION

STRAITS
GS   PASSAGEWAYS
  . **STRAITS**
  . . TORRES STRAIT
RT   CANALS
    LAKES
    SEAS
    WATER
    WATERWAYS

STRAKES
GS   STRUCTURAL MEMBERS
  . **STRAKES**
RT   AERODYNAMIC CONFIGURATIONS
    HULLS (STRUCTURES)
    LONGERONS
    METAL STRIPS
    RECTANGULAR PANELS
    REINFORCEMENT (STRUCTURES)
    WATER TUNNEL TESTS

STRANDS
RT   CABLES (ROPES)
    CORDAGE
    FIBERS
  ∞FILAMENTS

STRANDS-*(CONT.)*
  MESH
  YARNS

STRANGE ATTRACTORS
RT CHAOS
  FRACTALS
  IMBEDDINGS (MATHEMATICS)
  ITERATIVE SOLUTION
  NONLINEAR SYSTEMS
  NUMERICAL STABILITY
  PERTURBATION THEORY
  ∞PHYSICS
  RECURSIVE FUNCTIONS
  STATE VECTORS
  THEORETICAL PHYSICS
  TURBULENCE

STRANGENESS
GS DECAY
  . **STRANGENESS**
RT HYPERONS
  MESONS
  PARITY
  QUANTUM MECHANICS

STRAPDOWN INERTIAL GUIDANCE
GS GUIDANCE (MOTION)
  . INERTIAL GUIDANCE
  . . **STRAPDOWN INERTIAL GUIDANCE**
RT INERTIAL NAVIGATION

STRAPS
RT ANCHORS (FASTENERS)
  ∞BANDS
  CLAMPS
  FASTENERS
  HOLDERS

STRATA
UF STRATIFIED LAYERS
GS **STRATA**
  . SUBSTRATES
RT ANTICLINES
  BEDROCK
  BEDS (GEOLOGY)
  CROSSBEDDING (GEOLOGY)
  FLAT LAYERS
  FOLDS (GEOLOGY)
  GEOSYNCLINES
  ∞LAYERS
  STRATIFICATION
  SYNCLINES
  UNDERGROUND ACOUSTICS

STRATEGIC MATERIALS
RT CHROMIUM
  COBALT
  MANGANESE
  ∞MATERIALS
  METALS
  STOCKPILING
  TECHNOLOGY ASSESSMENT

STRATEGY
RT DECISION THEORY
  DEPLOYMENT
  ELECTRONIC WARFARE
  GAME THEORY
  ∞OPERATIONS
  OPERATIONS RESEARCH
  RISK
  WARFARE

STRATIFICATION
GS **STRATIFICATION**
  . ATMOSPHERIC STRATIFICATION
  . INTERCALATION
RT ANTICLINES
  BEDROCK
  CROSSBEDDING (GEOLOGY)
  FLAT LAYERS
  FOLDS (GEOLOGY)
  GEOSYNCLINES
  ∞LAYERS
  STATIC STABILITY
  STRATA
  . STRATIFIED FLOW
  STRATIGRAPHY
  SYNCLINES
  TEMPERATURE GRADIENTS
  THERMOCLINES

STRATIFIED FLOW
GS FLUID FLOW
  . LAMINAR FLOW
  . . **STRATIFIED FLOW**
RT BAROCLINIC WAVES
  BAROCLINITY
  COAXIAL FLOW
  FLOW GEOMETRY
  SHEAR FLOW
  STRATIFICATION

STRATIFIED LAYERS
USE STRATA

STRATIGRAPHY
RT ANTICLINES
  BEDROCK
  BEDS (GEOLOGY)
  CROSSBEDDING (GEOLOGY)
  ∞FORMATION
  FORMATIONS
  GEOCHRONOLOGY
  GEOLOGY
  GEOPHYSICS
  GEOSYNCLINES
  HYDROGEOLOGY
  MINES (EXCAVATIONS)
  PALEONTOLOGY
  PARTICLE TRACKS
  PETROLOGY
  PLATEAUS
  REGOLITH
  ROCKS
  SEDIMENTARY ROCKS
  STRATIFICATION
  SYNCLINES
  WELLS

STRATOCUMULUS CLOUDS
GS CLOUDS
  . CLOUDS (METEOROLOGY)
  . . CONVECTION CLOUDS
  . . . **STRATOCUMULUS CLOUDS**
RT CUMULUS CLOUDS
  STRATUS CLOUDS

STRATOFORTRESS AIRCRAFT
USE B-52 AIRCRAFT

STRATOJET AIRCRAFT
USE B-47 AIRCRAFT

STRATOPAUSE
GS EARTH ATMOSPHERE
  . **STRATOPAUSE**
RT MESOPAUSE
  MESOSPHERE
  MIDDLE ATMOSPHERE
  STRATOSPHERE

STRATOSCOPE TELESCOPES
UF STRATOSCOPE 1 TELESCOPE
  STRATOSCOPE 2 TELESCOPE
GS OPTICAL EQUIPMENT
  . ASTRONOMICAL TELESCOPES
  . . SPECTROSCOPIC TELESCOPES
  . . . **STRATOSCOPE TELESCOPES**
  TELESCOPES
  . ASTRONOMICAL TELESCOPES
  . . SPECTROSCOPIC TELESCOPES
  . . . **STRATOSCOPE TELESCOPES**
RT BALLOONS
  REFLECTING TELESCOPES
  REFRACTING TELESCOPES

STRATOSCOPE 1 TELESCOPE
USE STRATOSCOPE TELESCOPES

STRATOSCOPE 2 TELESCOPE
USE STRATOSCOPE TELESCOPES

STRATOSPHERE
GS EARTH ATMOSPHERE
  . HOMOSPHERE
  . . MIDDLE ATMOSPHERE
  . . . **STRATOSPHERE**
RT CHEMOSPHERE
  HOMOSPHERE
  ISOTHERMAL LAYERS
  STRATOPAUSE

STRATOSPHERE RADIATION
GS ATMOSPHERIC RADIATION
  . **STRATOSPHERE RADIATION**

STRATOSPHERE RADIATION-*(CONT.)*
RT CORPUSCULAR RADIATION
  ELECTROMAGNETIC RADIATION
  ∞RADIATION
  SKY RADIATION
  TROPOSPHERIC RADIATION

STRATOSPHERIC AEROSOL & GAS EXPERIMENT
USE SAGE SATELLITE

STRATOTANKER AIRCRAFT
USE C-135 AIRCRAFT

STRATUS CLOUDS
GS CLOUDS
  . CLOUDS (METEOROLOGY)
  . . CONVECTION CLOUDS
  . . . **STRATUS CLOUDS**
RT FOG
  NIMBOSTRATUS CLOUDS
  STRATOCUMULUS CLOUDS

STREAK CAMERAS
GS OPTICAL EQUIPMENT
  . CAMERAS
  . . **STREAK CAMERAS**
  PHOTOGRAPHIC EQUIPMENT
  . CAMERAS
  . . **STREAK CAMERAS**
RT CAMERA SHUTTERS
  CINEMATOGRAPHY
  LENSES

STREAK PHOTOGRAPHY
GS PHOTOGRAPHY
  . **STREAK PHOTOGRAPHY**
RT CAMERAS
  ELECTRO-OPTICAL PHOTOGRAPHY
  HIGH SPEED CAMERAS
  IMAGING TECHNIQUES

STREAM FUNCTIONS (FLUIDS)
RT INCOMPRESSIBLE FLOW
  POTENTIAL THEORY
  STOKES-BELTRAMI EQUATION
  STREAMS
  TWO DIMENSIONAL FLOW

STREAMLINE FLOW
USE LAMINAR FLOW

STREAMLINED BODIES
GS SYMMETRICAL BODIES
  . **STREAMLINED BODIES**
  . . FAIRINGS
RT AERODYNAMIC CONFIGURATIONS
  AIRFOILS
  AXISYMMETRIC BODIES
  ∞BODIES
  BODIES OF REVOLUTION
  MISSILE BODIES
  OGIVES
  SLENDER BODIES
  STREAMLINING
  TOWED BODIES

STREAMLINING
RT ACOUSTIC STREAMING
  AIR FLOW
  AIRCRAFT DESIGN
  AIRCRAFT STRUCTURES
  AIRFOIL PROFILES
  AIRFOILS
  FAIRINGS
  FLUID DYNAMICS
  FRICTION REDUCTION
  HELICOPTER DESIGN
  HYDROFOILS
  ∞PROFILES
  SKIN FRICTION
  STREAMLINED BODIES

STREAMS
GS **STREAMS**
  . GAS STREAMS
RT AIR FLOW
  ALLUVIUM
  AQUIFERS
  DELAWARE RIVER BASIN (US)
  FLUID FLOW
  GAS FLOW
  HYDROLOGY
  HYDROLOGY MODELS

STREAMS-*(CONT.)*
    INTERNATIONAL HYDROLOGICAL
       DECADE
    LAKE ERIE
    LAKE HURON
    LAKE MICHIGAN
    LAKE ONTARIO
    LAKE SUPERIOR
    LIMNOLOGY
    MEANDERS
    RAPIDS
    RESERVOIRS
    RIVERS
    STREAM FUNCTIONS (FLUIDS)
    SURFACE WATER
    SUSQUEHANNA RIVER BASIN
      (MD-NY-PA)
    WADIS

STREETS
RT   HIGHWAYS
     INTERSECTIONS
     PAVEMENTS
     ROADS
     ∞TUNNELS
     URBAN PLANNING
     URBAN RESEARCH

∞ **STRENGTH**
SN   *(USE OF A MORE SPECIFIC TERM IS*
     RECOMMENDED--CONSULT THE TERMS
     LISTED BELOW)
RT   COLD STRENGTH
     COMPRESSIVE STRENGTH
     CREEP RUPTURE STRENGTH
     CREEP STRENGTH
     ELECTRIC FIELD STRENGTH
     FIBER STRENGTH
     FIELD STRENGTH
     FRACTURE STRENGTH
     HIGH STRENGTH
     IMPACT STRENGTH
     MECHANICAL PROPERTIES
     MICROYIELD STRENGTH
     MUSCULAR STRENGTH
     NOTCH STRENGTH
     RESIDUAL STRENGTH
     SHEAR STRENGTH
     TENSILE STRENGTH
     WELD STRENGTH
     YIELD STRENGTH

STRENGTH OF MATERIALS
USE MECHANICAL PROPERTIES

STREPTOCOCCUS
GS   MICROORGANISMS
    . BACTERIA
    . . **STREPTOCOCCUS**

STREPTOMYCETES
GS   FUNGI
    . **STREPTOMYCETES**
    MICROORGANISMS
    . BACTERIA
    . . **STREPTOMYCETES**

STREPTOMYCIN
GS   DRUGS
    . ANTIBIOTICS
    . . **STREPTOMYCIN**
    STEROIDS
    . **STREPTOMYCIN**

∞ **STRESS (BIOLOGY)**
SN   *(USE OF A MORE SPECIFIC TERM IS*
     RECOMMENDED--CONSULT THE TERMS
     LISTED BELOW)
RT   ∞BIOLOGY
     PATHOLOGICAL EFFECTS
     STRESS (PHYSIOLOGY)
     STRESS (PSYCHOLOGY)

STRESS (PHYSIOLOGY)
GS   **STRESS (PHYSIOLOGY)**
    . ACCELERATION STRESSES
      (PHYSIOLOGY)
    . . CENTRIFUGING STRESS
RT   ACCELERATION (PHYSICS)
     ACCLIMATIZATION
     AEROEMBOLISM
     ANGINA PECTORIS
     ANOXIA
     BIODYNAMICS
     DEPRIVATION

STRESS (PHYSIOLOGY)-*(CONT.)*
    EXERCISE PHYSIOLOGY
    FATIGUE (BIOLOGY)
    FLIGHT STRESS (BIOLOGY)
    GRAVITATIONAL PHYSIOLOGY
    HOMEOSTASIS
    HYPERKINESIA
    HYPOXIA
    LOWER BODY NEGATIVE PRESSURE
    MUSCULAR FATIGUE
    PALMAR SWEAT INDEX
    PHYSIOLOGY
    PRESSURE BREATHING
    SPACE FLIGHT STRESS
   ∞STRESS (BIOLOGY)
    STRESS (PSYCHOLOGY)
    UNDERWATER PHYSIOLOGY

STRESS (PSYCHOLOGY)
UF   MENTAL STRESS
RT   FATIGUE (BIOLOGY)
     FLIGHT STRESS (BIOLOGY)
     MENTAL PERFORMANCE
     PALMAR SWEAT INDEX
     PSYCHOLOGICAL EFFECTS
     PSYCHOLOGICAL FACTORS
     PSYCHOLOGY
     SPACE FLIGHT STRESS
     SPACE PSYCHOLOGY
     ∞STRESS (BIOLOGY)
     STRESS (PHYSIOLOGY)
     WORKLOADS (PSYCHOPHYSIOLOGY)

STRESS ANALYSIS
UF   STRESS CALCULATIONS
GS   **STRESS ANALYSIS**
    . BOUNDARY ELEMENT METHOD
    . SCHWARTZ METHOD
    . X RAY STRESS ANALYSIS
RT   AIRY FUNCTION
     ∞ANALYZING
     BENDING MOMENTS
     BENDING THEORY
     CASTIGLIANO VARIATIONAL THEOREM
     COMBINED STRESS
     CONSTRUCTION
     CREEP ANALYSIS
     DONNELL EQUATIONS
     ENERGY METHODS
     EULER BUCKLING
     ∞FLIGHT STRESS
     FRINGE MULTIPLICATION
     INELASTIC STRESS
     INFLUENCE COEFFICIENT
     ISOPARAMETRIC FINITE ELEMENTS
     MECHANICAL ENGINEERING
     MICHELL THEOREM
     MOIRE FRINGES
     MOMENTS OF INERTIA
     NASTRAN
     PHOTOELASTIC ANALYSIS
     PHOTOELASTICITY
     REISSNER THEORY
     S-N DIAGRAMS
     SAINT VENANT PRINCIPLE
     SHALLOW SHELL EQUATIONS
     STRESSES
     STRUCTURAL ANALYSIS
     STRUCTURAL DESIGN
     STRUCTURAL ENGINEERING
     TEMPERATURE INVERSIONS
     X RAY ANALYSIS

STRESS CALCULATIONS
USE STRESS ANALYSIS

STRESS CONCENTRATION
UF   STRAIN DISTRIBUTION
     STRESS DISTRIBUTION
     STRESS-STRAIN DISTRIBUTION
GS   DISTRIBUTION (PROPERTY)
    . **STRESS CONCENTRATION**
RT   COMBINED STRESS
     CONCENTRATING
     CRACK INITIATION
     CRACKING (FRACTURING)
     ELBER EQUATION
     FATIGUE (MATERIALS)
     FATIGUE TESTS
     FORCE DISTRIBUTION
     FRINGE MULTIPLICATION
     HOLE DISTRIBUTION (MECHANICS)
     HOLE GEOMETRY (MECHANICS)
     IMPACT STRENGTH
     IMPACT TESTS

STRESS CONCENTRATION-*(CONT.)*
    LOADS (FORCES)
    MECHANICAL PROPERTIES
    MICROMECHANICS
    MOIRE FRINGES
    MOMENT DISTRIBUTION
    NOTCH STRENGTH
    NOTCH TESTS
    PERFORATED PLATES
    PERFORATED SHELLS
    SAINT VENANT PRINCIPLE
    STRESS-STRAIN RELATIONSHIPS
    STRESSES
    STRUCTURAL STRAIN

STRESS CORROSION
GS   CORROSION
    . **STRESS CORROSION**
RT   CRACKING (FRACTURING)
     FRETTING CORROSION
     INTERGRANULAR CORROSION
     METAL FATIGUE
     SALT SPRAY TESTS
     TRANSGRANULAR CORROSION

STRESS CORROSION CRACKING
GS   CRACKING (FRACTURING)
    . **STRESS CORROSION CRACKING**
RT   CORROSION TESTS
     CRACK CLOSURE
     CRACK INITIATION
     CRACK PROPAGATION
     METAL FATIGUE

STRESS CYCLES
GS   CYCLES
    . **STRESS CYCLES**
    MECHANICAL PROPERTIES
    . **STRESS CYCLES**
RT   CYCLIC LOADS
     ELBER EQUATION
     FATIGUE (MATERIALS)
     FATIGUE LIFE
     FATIGUE TESTS
     S-N DIAGRAMS
     STRESSES

STRESS DISTRIBUTION
USE STRESS CONCENTRATION

STRESS FUNCTIONS
UF   VON MISES THEORY
GS   FUNCTIONS (MATHEMATICS)
    . **STRESS FUNCTIONS**
RT   FRACTURING

STRESS INTENSITY FACTORS
RT   BENDING THEORY
     COMBINED STRESS
     CRACK INITIATION
     CRACK PROPAGATION
     CRACKING (FRACTURING)
     FORCE DISTRIBUTION
     FRACTURE MECHANICS
     HOLE GEOMETRY (MECHANICS)
     LOADS (FORCES)
     NOTCH STRENGTH
     PLANE STRAIN
     TENSILE STRESS

STRESS MEASUREMENT
GS   MECHANICAL MEASUREMENT
    . **STRESS MEASUREMENT**
    . . X RAY STRESS MEASUREMENT
RT   DEFORMETERS
     EXTENSOMETERS
     PHOTOELASTIC ANALYSIS
     S-N DIAGRAMS
     STRAIN GAGES
     TENSOMETERS
     VIBRATION MEASUREMENT

STRESS PROPAGATION
GS   TRANSMISSION
    . **STRESS PROPAGATION**
RT   ELASTIC WAVES
     PLASTIC DEFORMATION
     ∞PROPAGATION

STRESS RATIO
GS   MECHANICAL PROPERTIES
    . **STRESS RATIO**
    RATIOS
    . **STRESS RATIO**

STRESS RATIO-*(CONT.)*
RT   FATIGUE (MATERIALS)
       FATIGUE TESTS
       MODULAR RATIOS
       PRESSURE RATIO
       S-N DIAGRAMS

STRESS RELAXATION
GS   MECHANICAL PROPERTIES
      . **STRESS RELAXATION**
       RELAXATION (MECHANICS)
      . **STRESS RELAXATION**
RT   ANELASTICITY
       BORDONI PEAKS
       CREEP ANALYSIS
       CREEP DIAGRAMS
       CREEP PROPERTIES
       DUCTILITY
       FATIGUE (MATERIALS)
       PLASTIC DEFORMATION
       PLASTIC FLOW
       PLASTIC MEMORY
       PLASTIC PROPERTIES
      ∞RECOVERY
       RESIDUAL STRESS
       SHEAR PROPERTIES
       STRESSES
       TEMPERATURE INVERSIONS

STRESS RELIEVING
GS   HEAT TREATMENT
      . **STRESS RELIEVING**
       RELIEVING
      . **STRESS RELIEVING**
RT   ALLOYS
       ANNEALING
       FATIGUE (MATERIALS)
      ∞RECOVERY
       RESIDUAL STRESS
       STABILIZATION
       STRAIN HARDENING
       TEMPERING

STRESS RUPTURE STRENGTH
USE   CREEP RUPTURE STRENGTH

STRESS TENSORS
GS   ALGEBRA
      . TENSORS
      . . **STRESS TENSORS**
RT   CONTINUUM MECHANICS
       ELASTIC PROPERTIES
       FRACTURE MECHANICS
       PLASTIC PROPERTIES
       STRUCTURAL DESIGN

STRESS WAVES
GS   ELASTIC WAVES
      . **STRESS WAVES**
RT   ACOUSTIC EMISSION
       SHOCK LAYERS
       SHOCK WAVES
       STRESSES
       TEMPERATURE INVERSIONS
       WAVE PROPAGATION
      ∞WAVES

STRESS-STRAIN DIAGRAMS
GS   DIAGRAMS
      . **STRESS-STRAIN DIAGRAMS**
RT   AXIAL STRAIN
       HOOKES LAW
       INELASTIC STRESS
       MODULUS OF ELASTICITY
       POISSON RATIO
       PROPORTIONAL LIMIT
       SHAPE MEMORY ALLOYS
       SHEAR PROPERTIES
       STRAIN MEASUREMENT
       STRUCTURAL STRAIN
       YIELD STRENGTH

STRESS-STRAIN DISTRIBUTION
USE   STRESS CONCENTRATION

STRESS-STRAIN RELATIONSHIPS
RT   ELASTIC DEFORMATION
       PLANE STRAIN
       PLASTIC DEFORMATION
      ∞RELATIONSHIPS
       RESIDUAL STRESS
       STRAIN MEASUREMENT
       STRESS CONCENTRATION
       STRUCTURAL STRAIN
       YIELD STRENGTH

STRESS-STRAIN-TIME RELATIONS
RT   CREEP DIAGRAMS
       NEWTONIAN FLUIDS
       THERMOVISCOELASTICITY

STRESSED-SKIN STRUCTURES
RT   MONOCOQUE STRUCTURES
       SKIN (STRUCTURAL MEMBER)
       SPHERICAL SHELLS
      ∞STRUCTURES
       THIN WALLED SHELLS

STRESSES
GS   **STRESSES**
      . AXIAL STRESS
      . COMBINED STRESS
      . CRITICAL LOADING
      . PHOTOSTRESSES
      . RESIDUAL STRESS
      . REYNOLDS STRESS
      . SHEAR STRESS
      . . TORSIONAL STRESS
      . TENSILE STRESS
      . THERMAL STRESSES
      . TRIAXIAL STRESSES
      . VIBRATIONAL STRESS
RT   BUCKLING
       CRACKS
       CREEP PROPERTIES
       DESTRUCTION
       FATIGUE (MATERIALS)
      ∞FLIGHT STRESS
       IMPACT
       LOADS (FORCES)
       MECHANICAL PROPERTIES
       MICROYIELD STRENGTH
       PRESTRESSING
       ROLLING CONTACT LOADS
       SHEAR PROPERTIES
       STRESS ANALYSIS
       STRESS CONCENTRATION
       STRESS CYCLES
       STRESS RELAXATION
       STRESS WAVES
       STRUCTURAL STRAIN
       TEMPERATURE INVERSIONS
       TRIBOLUMINESCENCE
       X RAY STRESS ANALYSIS
       YIELD STRENGTH

STRETCH FORMING
RT   BULGING
       COLD WORKING
      ∞DRAWING
       METAL DRAWING
       METAL WORKING
       STRETCHING

STRETCHERS
GS   MEDICAL EQUIPMENT
      . **STRETCHERS**
RT   FIRST AID

STRETCHING
UF   DILATATION
RT   COLD WORKING
       DEEP DRAWING
       DILATATIONAL WAVES
       DISTORTION
      ∞DRAWING
       DUCTILITY
       ELASTIC DEFORMATION
       ELONGATION
       METAL WORKING
       PLASTIC DEFORMATION
       STRETCH FORMING
       TEMPERING
      ∞TENSION
       WINDING

STRIATION
RT   GROOVING
       MUSCULOSKELETAL SYSTEM
       SHATTER CONES

STRINGERS
GS   STRUCTURAL MEMBERS
      . **STRINGERS**
RT   LONGERONS
       REINFORCEMENT (STRUCTURES)
       STRUCTURAL STABILITY

STRINGS
RT   ASSEMBLIES
       CORDAGE

∞ **STRIP**
SN   *(USE OF A MORE SPECIFIC TERM IS*
       RECOMMENDED--CONSULT THE TERMS
       LISTED BELOW)
RT   AIRPORTS
       CIRCUITS
       DISPLAY DEVICES
       METAL STRIPS
       RIBBONS
       RUNWAYS

STRIP MINING
GS   MINING
      . **STRIP MINING**
RT   CLAYS
       COAL
       EARTH RESOURCES
       EXCAVATION
       EXPLOITATION
       MINERAL DEPOSITS
       MINES (EXCAVATIONS)
       SOILS

STRIP TRANSMISSION LINES
GS   TRANSMISSION LINES
      . **STRIP TRANSMISSION LINES**
      . . MICROSTRIP TRANSMISSION LINES
RT   ANTENNA FEEDS
       TRANSMISSION CIRCUITS

∞ **STRIPPING**
SN   *(USE OF A MORE SPECIFIC TERM IS*
       RECOMMENDED--CONSULT THE TERMS
       LISTED BELOW)
RT   ANODIC STRIPPING
       ION STRIPPING
       PEELING
       STRIPPING (DISTILLATION)

STRIPPING (DISTILLATION)
GS   DISTILLATION
      . **STRIPPING (DISTILLATION)**
RT   ∞SEPARATION
      ∞STRIPPING
       VAPORIZING

STROBOSCOPES
GS   OPTICAL EQUIPMENT
      . **STROBOSCOPES**
RT   BALLISTIC CAMERAS
       HIGH SPEED CAMERAS
       OPTICAL MEASUREMENT
       SYNCHRONISM
       TIME MEASUREMENT
       VELOCITY MEASUREMENT

∞ **STROKES**
SN   *(USE OF A MORE SPECIFIC TERM IS*
       RECOMMENDED--CONSULT THE TERMS
       LISTED BELOW)
RT   CEREBRAL VASCULAR ACCIDENTS
       THERMODYNAMIC CYCLES

STROKING TESTS
GS   VIBRATION TESTS
      . DAMPING TESTS
      . . **STROKING TESTS**
RT   DYNAMIC RESPONSE
      ∞FREQUENCY RESPONSE
       MODAL RESPONSE
       SHOCK SPECTRA
      ∞TESTS
       TRANSIENT RESPONSE
       WAVE EXCITATION

STRONG INTERACTIONS (FIELD THEORY)
GS   FIELD THEORY (PHYSICS)
      . **STRONG INTERACTIONS (FIELD**
        THEORY)
       PARTICLE INTERACTIONS
      . NUCLEAR INTERACTIONS
      . . **STRONG INTERACTIONS (FIELD**
        THEORY)
RT   ∞INTERACTIONS
      ∞THEORIES
       WEAK INTERACTIONS (FIELD THEORY)

STRONGLY COUPLED PLASMAS
GS   GASES
      . IONIZED GASES
      . . CHARGED PARTICLES
      . . . COLLISIONAL PLASMAS
      . . . . **STRONGLY COUPLED PLASMAS**
      . . . DENSE PLASMAS

STRONGLY COUPLED PLASMAS-*(CONT.)*	**STRONTIUM ISOTOPES**-*(CONT.)*

STRONGLY COUPLED PLASMAS-*(CONT.)*
```
        . . . . STRONGLY COUPLED PLASMAS
        PARTICLES
        . CHARGED PARTICLES
        . . ENERGETIC PARTICLES
        . . . PLASMAS (PHYSICS)
        . . . . COLLISIONAL PLASMAS
        . . . . . STRONGLY COUPLED PLASMAS
        . . . DENSE PLASMAS
        . . . . . STRONGLY COUPLED PLASMAS
   RT   CONTROLLED FUSION
        COSMIC PLASMA
        COUPLED MODES
        HIGH TEMPERATURE PLASMAS
        INERTIAL CONFINEMENT FUSION
        MAGNETOHYDRODYNAMIC STABILITY
        PLASMA COMPRESSION
        PLASMA CONDUCTIVITY
        PLASMA DENSITY
        PLASMA EQUILIBRIUM
        PLASMA FOCUS
```

STRONTIUM
```
   GS   CHEMICAL ELEMENTS
        . STRONTIUM
        . . STRONTIUM ISOTOPES
        . . . STRONTIUM 85
        . . . STRONTIUM 87
        . . . STRONTIUM 89
        . . . STRONTIUM 90
        METALS
        . STRONTIUM
        . . STRONTIUM ISOTOPES
        . . . STRONTIUM 85
        . . . STRONTIUM 87
        . . . STRONTIUM 89
        . . . STRONTIUM 90
```

STRONTIUM BROMIDES
```
   GS   HALOGEN COMPOUNDS
        . BROMINE COMPOUNDS
        . . BROMIDES
        . . . STRONTIUM BROMIDES
        . HALIDES
        . . BROMIDES
        . . . STRONTIUM BROMIDES
        . . METAL HALIDES
        . . . STRONTIUM BROMIDES
        STRONTIUM COMPOUNDS
        . STRONTIUM BROMIDES
```

STRONTIUM COMPOUNDS
```
   GS   STRONTIUM COMPOUNDS
        . STRONTIUM BROMIDES
        . STRONTIUM FLUORIDES
        . STRONTIUM SULFIDES
        . STRONTIUM TITANATES
        . STRONTIUM ZIRCONATES
   RT   ∞ALKALINE EARTH COMPOUNDS
        ∞CHEMICAL COMPOUNDS
        ∞METAL COMPOUNDS
```

STRONTIUM FLUORIDES
```
   GS   HALOGEN COMPOUNDS
        . FLUORINE COMPOUNDS
        . . FLUORIDES
        . . . METAL FLUORIDES
        . . . . STRONTIUM FLUORIDES
        . HALIDES
        . . METAL HALIDES
        . . . STRONTIUM FLUORIDES
        STRONTIUM COMPOUNDS
        . STRONTIUM FLUORIDES
```

STRONTIUM ISOTOPES
```
   GS   CHEMICAL ELEMENTS
        . NUCLIDES
        . . ISOTOPES
        . . . STRONTIUM ISOTOPES
        . . . . STRONTIUM 85
        . . . . STRONTIUM 87
        . . . . STRONTIUM 89
        . . . . STRONTIUM 90
        . STRONTIUM
        . . STRONTIUM ISOTOPES
        . . . STRONTIUM 85
        . . . STRONTIUM 87
        . . . STRONTIUM 89
        . . . STRONTIUM 90
        METALS
        . STRONTIUM
        . . STRONTIUM ISOTOPES
        . . . STRONTIUM 85
        . . . STRONTIUM 87
        . . . STRONTIUM 89
```

STRONTIUM ISOTOPES-*(CONT.)*
```
        . . STRONTIUM 90
```

STRONTIUM SULFIDES
```
   GS   CHALCOGENIDES
        . SULFIDES
        . . INORGANIC SULFIDES
        . . . STRONTIUM SULFIDES
        STRONTIUM COMPOUNDS
        . STRONTIUM SULFIDES
        SULFUR COMPOUNDS
        . SULFIDES
        . . INORGANIC SULFIDES
        . . . STRONTIUM SULFIDES
```

STRONTIUM TITANATES
```
   GS   STRONTIUM COMPOUNDS
        . STRONTIUM TITANATES
        TITANIUM COMPOUNDS
        . TITANATES
        . . STRONTIUM TITANATES
```

STRONTIUM ZIRCONATES
```
   GS   STRONTIUM COMPOUNDS
        . STRONTIUM ZIRCONATES
        ZIRCONIUM COMPOUNDS
        . ZIRCONATES
        . . STRONTIUM ZIRCONATES
```

STRONTIUM 85
```
   GS   CHEMICAL ELEMENTS
        . NUCLIDES
        . . ISOTOPES
        . . . RADIOACTIVE ISOTOPES
        . . . . STRONTIUM 85
        . . . STRONTIUM ISOTOPES
        . . . . STRONTIUM 85
        . STRONTIUM
        . . STRONTIUM ISOTOPES
        . . . STRONTIUM 85
        METALS
        . STRONTIUM
        . . STRONTIUM ISOTOPES
        . . . STRONTIUM 85
```

STRONTIUM 87
```
   GS   CHEMICAL ELEMENTS
        . NUCLIDES
        . . ISOTOPES
        . . . STRONTIUM ISOTOPES
        . . . . STRONTIUM 87
        . STRONTIUM
        . . STRONTIUM ISOTOPES
        . . . STRONTIUM 87
        METALS
        . STRONTIUM
        . . STRONTIUM ISOTOPES
        . . . STRONTIUM 87
```

STRONTIUM 88
```
   GS   CHEMICAL ELEMENTS
        . NUCLIDES
        . . ISOTOPES
        . . . RADIOACTIVE ISOTOPES
        . . . . STRONTIUM 88
```

STRONTIUM 89
```
   GS   CHEMICAL ELEMENTS
        . NUCLIDES
        . . ISOTOPES
        . . . RADIOACTIVE ISOTOPES
        . . . . STRONTIUM 89
        . . . STRONTIUM ISOTOPES
        . . . . STRONTIUM 89
        . STRONTIUM
        . . STRONTIUM ISOTOPES
        . . . STRONTIUM 89
        METALS
        . STRONTIUM
        . . STRONTIUM ISOTOPES
        . . . STRONTIUM 89
```

STRONTIUM 90
```
   GS   CHEMICAL ELEMENTS
        . NUCLIDES
        . . ISOTOPES
        . . . RADIOACTIVE ISOTOPES
        . . . . STRONTIUM 90
        . . . STRONTIUM ISOTOPES
        . . . . STRONTIUM 90
        . STRONTIUM
        . . STRONTIUM ISOTOPES
        . . . STRONTIUM 90
        METALS
        . STRONTIUM
```

STRONTIUM 90-*(CONT.)*
```
        . . STRONTIUM ISOTOPES
        . . . STRONTIUM 90
```

STROUHAL NUMBER
```
   GS   RATIOS
        . DIMENSIONLESS NUMBERS
        . . STROUHAL NUMBER
   RT   BACKWASH
        BUFFETING
        FLOW CHARACTERISTICS
        FLOW DISTRIBUTION
        FLOW STABILITY
        FROUDE NUMBER
        OSCILLATING FLOW
        SLIPSTREAMS
        TURBULENCE
        UNSTEADY FLOW
        VORTICES
        WAKES
```

STRUCTURAL ANALYSIS
```
   UF   MEMBRANE ANALOGY
        MEMBRANE THEORY
   GS   STRUCTURAL ANALYSIS
        . DYNAMIC STRUCTURAL ANALYSIS
        . ENERGY METHODS
        . . BERNSTEIN ENERGY PRINCIPLE
        . . STRAIN ENERGY METHODS
        . EQUILIBRIUM METHODS
        . FLUTTER ANALYSIS
        . MATRIX METHODS
   RT   ∞ANALYZING
        CASTIGLIANO VARIATIONAL THEOREM
        CONSTRUCTION
        CONTINUUM MODELING
        CREEP ANALYSIS
        HOLE GEOMETRY (MECHANICS)
        INFLUENCE COEFFICIENT
        J INTEGRAL
        LOADING MOMENTS
        MEGAMECHANICS
        MICHELL THEOREM
        MODULAR RATIOS
        MOMENT DISTRIBUTION
        NASTRAN
        ORBITAL SPACE STATIONS
        ORBITAL SPACE TESTS
        PLATE THEORY
        SOLID MECHANICS
        STIFFNESS MATRIX
        STRESS ANALYSIS
```

STRUCTURAL BASINS
```
   UF   BASINS
        CLOSED BASINS
        DEPRESSIONS (TOPOGRAPHY)
        SINKS (GEOLOGY)
   GS   LANDFORMS
        . STRUCTURAL BASINS
        . . CIRQUES (LANDFORMS)
        . . GREAT BASIN (US)
        . . KALIHARI BASIN (AFRICA)
        . . KARST
        . . SINKHOLES
        . . KETTLES (GEOLOGY)
        . . LAKE CHAMPLAIN BASIN (NY-VT)
        . . RIVER BASINS
        . . . ATCHAFALAYA RIVER BASIN (LA)
        . . . CHENA RIVER BASIN (AK)
        . . . COLUMBIA RIVER BASIN (ID-OR-WA)
        . . . DELAWARE RIVER BASIN (US)
        . . . FEATHER RIVER BASIN (CA)
        . . . MISSOURI RIVER BASIN (US)
        . . . SUSQUEHANNA RIVER BASIN
                (MD-NY-PA)
        . . . WABASH RIVER BASIN (IL-IN-OH)
        . . . WADIS
        . . WATERSHEDS
        . . WILLISTON BASIN (NORTH AMERICA)
   RT   GEOLOGY
        SEAMOUNTS
        VALLEYS
```

STRUCTURAL BEAMS
```
   USE  BEAMS (SUPPORTS)
```

STRUCTURAL DESIGN
```
   GS   STRUCTURAL DESIGN
        . PRESSURE VESSEL DESIGN
   RT   AEROELASTIC RESEARCH WINGS
        AIRCRAFT DESIGN
        AIRFRAME MATERIALS
        ARCHITECTURE
        BREAKWATERS
```

STRUCTURAL DESIGN-*(CONT.)*
    COMPUTER AIDED DESIGN
    CONSTRUCTION
  ∞DESIGN
    HELICOPTER DESIGN
    LOFTING
    MISSILE DESIGN
    PLANT DESIGN
    PRODUCT DEVELOPMENT
    SATELLITE DESIGN
    SHIP HULLS
    SHOCK SPECTRA
    SPACECRAFT DESIGN
    SPACECRAFT STRUCTURES
    STRESS ANALYSIS
    STRESS TENSORS
    SUBSTRUCTURES
    UNDERWATER STRUCTURES
    WEIGHT REDUCTION

STRUCTURAL DESIGN CRITERIA
GS   CRITERIA
    . **STRUCTURAL DESIGN CRITERIA**
RT   AERODYNAMIC LOADS
    AXIAL COMPRESSION LOADS
    AXIAL LOADS
    BENDING MOMENTS
    COMPRESSION LOADS
    CYCLIC LOADS
  ∞DESIGN
    DYNAMIC LOADS
    GEOTECHNICAL ENGINEERING
    GUST LOADS
    IMPACT LOADS
    LANDING LOADS
    LOADS (FORCES)
    MASS DISTRIBUTION
    MOMENT DISTRIBUTION
    PRESSURE DISTRIBUTION
    RANDOM LOADS
    ROLLING CONTACT LOADS
    SHOCK LOADS
    STATIC LOADS
    THRUST LOADS
    TRANSIENT LOADS
    VIBRATORY LOADS

STRUCTURAL DYNAMICS
USE   DYNAMIC STRUCTURAL ANALYSIS

STRUCTURAL ENGINEERING
RT   AERONAUTICAL ENGINEERING
    AEROSPACE ENGINEERING
    CONSTRUCTION
  ∞ENGINEERING
    GEOTECHNICAL ENGINEERING
    MEGAMECHANICS
    MODULAR RATIOS
    STRESS ANALYSIS

STRUCTURAL FAILURE
GS   FAILURE
    . **STRUCTURAL FAILURE**
RT   BENDING
    BUCKLING
    COLLAPSE
    CRACKING (FRACTURING)
    CREEP PROPERTIES
    DEFORMATION
    FATIGUE (MATERIALS)
    FRACTURING
    MECHANICAL PROPERTIES
    SYSTEM FAILURES

STRUCTURAL FATIGUE
USE   FATIGUE (MATERIALS)

STRUCTURAL FOUNDATIONS
USE   FOUNDATIONS

STRUCTURAL INFLUENCE COEFFICIENTS ·
UF   SIC (COEFFICIENT)
GS   COEFFICIENTS
    . INFLUENCE COEFFICIENT
    . . **STRUCTURAL INFLUENCE
       COEFFICIENTS**

STRUCTURAL MATERIALS
USE   CONSTRUCTION MATERIALS

STRUCTURAL MEMBERS
GS   **STRUCTURAL MEMBERS**
    . BEAMS (SUPPORTS)
    . . BOX BEAMS

STRUCTURAL MEMBERS-*(CONT.)*
    . . CANTILEVER BEAMS
    . . CURVED BEAMS
    . . I BEAMS
    . . RECTANGULAR BEAMS
    . . TIMOSHENKO BEAMS
    . COLUMNS (SUPPORTS)
    . . TAPERED COLUMNS
    . FLAT PLATES
    . GIRDERS
    . LONGERONS
    . MEMBRANE STRUCTURES
    . . SKIN (STRUCTURAL MEMBER)
    . PLATES (STRUCTURAL MEMBERS)
    . . ANISOTROPIC PLATES
    . . ANNULAR PLATES
    . . CANTILEVER PLATES
    . . CIRCULAR PLATES
    . . CORRUGATED PLATES
    . . ELASTIC PLATES
    . . END PLATES
    . . GIRDER WEBS
    . . METAL PLATES
    . . . BOILER PLATE
    . . ORTHOTROPIC PLATES
    . . PERFORATED PLATES
    . . POROUS PLATES
    . . REINFORCED PLATES
    . SADDLES (SUPPORTS)
    . STRAKES
    . STRINGERS
    . STRUTS
    . TRUSSES
    . WING PANELS
RT   AIRCRAFT CONSTRUCTION MATERIALS
    AIRFRAME MATERIALS
    BARS
  ∞CHANNELS
  ∞COMPONENTS
    CONCRETES
    CONSTRUCTION
  ∞CONSTRUCTION MATERIALS
    FASTENERS
    FOUNDATIONS
    GUY WIRES
    JOINTS (JUNCTIONS)
    MASONRY
    PYLON MOUNTING
    PYLONS
    RECTANGULAR PANELS
    REINFORCEMENT (STRUCTURES)
    RODS
    SLABS
    STIFFNESS MATRIX
  ∞STRUCTURES
    SUBSTRUCTURES
    THICK WALLS

STRUCTURAL PROPERTIES (GEOLOGY)
UF   LINEAMENT
GS   GEOLOGY
    . **STRUCTURAL PROPERTIES (GEOLOGY)**
RT   EARTH CORE
    EARTH CRUST
    EARTH MANTLE
    EARTH PLANETARY STRUCTURE
    EARTH SURFACE
    FISSURES (GEOLOGY)
    GEOPHYSICS
    GREAT BASIN (US)
    HYDROLOGY
    INLIERS (LANDFORMS)
    LANDFORMS
    PLANETARY COMPOSITION
    PLATES (TECTONICS)
  ∞PROPERTIES
    ROCK MECHANICS
    SHATTER CONES
    SINKHOLES
    SUBDUCTION (GEOLOGY)

STRUCTURAL RELIABILITY
GS   RELIABILITY
    . **STRUCTURAL RELIABILITY**
RT   AIRCRAFT RELIABILITY
    COMPONENT RELIABILITY
    CUMULATIVE DAMAGE
    QUALITY CONTROL

STRUCTURAL RIGIDITY
USE   STRUCTURAL STABILITY

STRUCTURAL STABILITY
UF   STRUCTURAL RIGIDITY
GS   MECHANICAL PROPERTIES

STRUCTURAL STABILITY-*(CONT.)*
    . DIMENSIONAL STABILITY
    . . **STRUCTURAL STABILITY**
    . . . SHELL STABILITY
    STABILITY
    . STATIC STABILITY
    . DIMENSIONAL STABILITY
    . . . **STRUCTURAL STABILITY**
    . . . . SHELL STABILITY
RT   AIRCRAFT STABILITY
    COMBUSTION VIBRATION
    HYBRID STRUCTURES
    LONGERONS
    PLASTIC PROPERTIES
    REINFORCEMENT (STRUCTURES)
    RESONANCE TESTING
  ∞RIGIDITY
    STIFFNESS
    STRINGERS
    WAVE RESISTANCE

STRUCTURAL STRAIN
GS   FATIGUE (MATERIALS)
    . **STRUCTURAL STRAIN**
RT   AXIAL STRAIN
    BENDING
    BUCKLING
    CRACKING (FRACTURING)
    DEFLECTION
    DEFORMATION
    ELASTIC DEFORMATION
    FAILURE
    MOMENTS OF INERTIA
    PLASTIC DEFORMATION
    PRESTRESSING
    REINFORCEMENT (STRUCTURES)
    RUPTURING
    SHEAR STRAIN
    SHEARING
    STRAIN MEASUREMENT
    STRESS CONCENTRATION
    STRESS-STRAIN DIAGRAMS
    STRESS-STRAIN RELATIONSHIPS
    STRESSES
    SYSTEM FAILURES
    TEMPERATURE INVERSIONS
    TWISTING
    VOLUMETRIC STRAIN
    WARPAGE

STRUCTURAL VIBRATION
GS   VIBRATION
    . **STRUCTURAL VIBRATION**
    . . BENDING VIBRATION
    . . BREATHING VIBRATION
    . . FLUTTER
    . . . PANEL FLUTTER
    . . . SUBSONIC FLUTTER
    . . . SUPERSONIC FLUTTER
    . . . TRANSONIC FLUTTER
    . . LINEAR VIBRATION
    . . MISSILE VIBRATION
    . . SELF INDUCED VIBRATION
    . . . PANEL FLUTTER
    . . . SUBSONIC FLUTTER
    . . . SUPERSONIC FLUTTER
    . . . TRANSONIC FLUTTER
    . . TORSIONAL VIBRATION
RT   EARTHQUAKE RESISTANT STRUCTURES
    FLEXIBLE SPACECRAFT
    FLUTTER ANALYSIS
    GYRODAMPERS
    RANDOM VIBRATION
    RESONANT VIBRATION
    SHAKING
    SHOCK SPECTRA
    VIBRATION TESTS

STRUCTURAL WEIGHT
GS   WEIGHT (MASS)
    . **STRUCTURAL WEIGHT**
RT   MASS RATIOS
    NEW MOONS PROJECT
    WEIGHT ANALYSIS
    WEIGHT REDUCTION

∞ **STRUCTURES**
SN   *(USE OF A MORE SPECIFIC TERM IS
    RECOMMENDED--CONSULT THE TERMS
    LISTED BELOW)*
RT   AIRCRAFT STRUCTURES
    ARCHITECTURE
    ATOMIC STRUCTURE
    BREAKWATERS
    BRIDGES (STRUCTURES)

STRUCTURES-*(CONT.)*
    COMPOSITE STRUCTURES
    CONCRETE STRUCTURES
    CONFIGURATION INTERACTION
    CRYSTAL STRUCTURE
    EARTH PLANETARY STRUCTURE
    EARTHQUAKE RESISTANT STRUCTURES
    EXPANDABLE STRUCTURES
    FINE STRUCTURE
    FOLDING STRUCTURES
    FOUNDATIONS
    FRAMES
    GALACTIC STRUCTURE
    HONEYCOMB STRUCTURES
    HYBRID STRUCTURES
    HYPERFINE STRUCTURE
    INFLATABLE STRUCTURES
    INTRAMOLECULAR STRUCTURES
    ISOTENSOID STRUCTURES
    LARGE SPACE STRUCTURES
    MEMBRANE STRUCTURES
    MICROSTRUCTURE
    MISSILE STRUCTURES
    MOLECULAR STRUCTURE
    MONOCOQUE STRUCTURES
    PLANAR STRUCTURES
    REDUNDANT COMPONENTS
    RIGID STRUCTURES
    RING STRUCTURES
    SANDWICH STRUCTURES
    SPACE ERECTABLE STRUCTURES
    SPACECRAFT STRUCTURES
    STEEL STRUCTURES
    STELLAR STRUCTURE
    STRESSED-SKIN STRUCTURES
    STRUCTURAL MEMBERS
    SUBSTRUCTURES
    TANKS (CONTAINERS)
    TOWERS
    TRUSSES
    UNIMOLECULAR STRUCTURES
    VARIABLE GEOMETRY STRUCTURES
    WELDED STRUCTURES
    WOODEN STRUCTURES

STRUTS
GS STRUCTURAL MEMBERS
   . **STRUTS**
RT CHASSIS
    COLUMNS (SUPPORTS)
    FRAMES
    PYLONS
    SUPPORTS
    TRUSSES

STRYCHNINE
GS POISONS
   . **STRYCHNINE**
RT ALKALOIDS
    STIMULANTS

STS
USE SPACE TRANSPORTATION SYSTEM

STS-1
USE SPACE TRANSPORTATION SYSTEM 1
     FLIGHT

STS-2
USE SPACE TRANSPORTATION SYSTEM 2
     FLIGHT

STS-3
USE SPACE TRANSPORTATION SYSTEM 3
     FLIGHT

STS-4
USE SPACE TRANSPORTATION SYSTEM 4
     FLIGHT

STS-5
USE SPACE SHUTTLE MISSION 31-A

STS-6
USE SPACE SHUTTLE MISSION 31-B

STS-7
USE SPACE SHUTTLE MISSION 31-C

STS-8
USE SPACE SHUTTLE MISSION 31-D

STS-9
USE SPACE SHUTTLE MISSION 41-A

STS-11
USE SPACE SHUTTLE MISSION 41-B

STS-13
USE SPACE SHUTTLE MISSION 41-C

STS-14
USE SPACE SHUTTLE MISSION 41-D

STS-17
USE SPACE SHUTTLE MISSION 41-G

STS-19
USE SPACE SHUTTLE MISSION 51-A

STS-20
USE SPACE SHUTTLE MISSION 51-C

STS-22
USE SPACE SHUTTLE MISSION 51-E

STS-23
USE SPACE SHUTTLE MISSION 51-D

STS-24
USE SPACE SHUTTLE MISSION 51-B

STS-25
USE SPACE SHUTTLE MISSION 51-G

STS-26
USE SPACE SHUTTLE MISSION 51-F

STS-27
USE SPACE SHUTTLE MISSION 51-I

STS-28
USE SPACE SHUTTLE MISSION 51-J

STS-29
USE SPACE SHUTTLE MISSION 61-A

STS-30
USE SPACE SHUTTLE MISSION 61-A

STS-31
USE SPACE SHUTTLE MISSION 51-H
     SPACE SHUTTLE MISSION 61-B

STS-32
USE SPACE SHUTTLE MISSION 61-C

STS-33
USE SPACE SHUTTLE MISSION 51-L

STS-34
USE SPACE SHUTTLE MISSION 61-E

STUDENTS
UF TRAINEES
RT EDUCATION
    INSTRUCTORS
    LEARNING
    TRAINING EVALUATION
    UNIVERSITIES

STUDIES
USE INVESTIGATION

STUDS (STRUCTURAL MEMBERS)
RT ANCHORS (FASTENERS)
    BOLTS
    COLUMNS (SUPPORTS)
    FASTENERS
    HOLDERS
    LUGS
    PINS
    SCREWS
    WALLS

STURM-LIOUVILLE OPERATOR
USE STURM-LIOUVILLE THEORY

STURM-LIOUVILLE THEORY
UF STURM-LIOUVILLE OPERATOR
GS ANALYSIS (MATHEMATICS)
   . REAL VARIABLES
   . . **STURM-LIOUVILLE THEORY**
RT DIFFERENTIAL EQUATIONS
    LAMB WAVES
    LAME WAVE EQUATIONS
   ∞ THEORIES

STYLUSES
USE PENS

STYPHNATES
GS EXPLOSIVES
   . **STYPHNATES**
RT ∞ CHEMICAL COMPOUNDS
    ∞ INITIATORS
    INITIATORS (EXPLOSIVES)

STYRENES
GS **STYRENES**
   . POLYSTYRENE
   . . STYROFOAM (TRADEMARK)
RT BUNA (TRADEMARK)

STYROFOAM (TRADEMARK)
GS STYRENES
   . POLYSTYRENE
   . . **STYROFOAM (TRADEMARK)**
RT FOAMS
    PLASTICS
   ∞ POLYMERS

SUBARCTIC REGIONS
GS REGIONS
   . **SUBARCTIC REGIONS**
RT ARCTIC REGIONS

SUBASSEMBLIES
UF SUBCIRCUITS
GS ASSEMBLIES
   . **SUBASSEMBLIES**
RT ACCESSORIES
   ∞ COMPONENTS

SUBAUDIBLE FREQUENCIES
GS FREQUENCIES
   . RADIO FREQUENCIES
   . . LOW FREQUENCIES
   . . . **SUBAUDIBLE FREQUENCIES**
   RANGE (EXTREMES)
   . FREQUENCY RANGES
   . . **SUBAUDIBLE FREQUENCIES**
RT FREQUENCY DISTRIBUTION
    HARMONICS
    ZERO SOUND

SUBCARRIER WAVES
USE CARRIER WAVES

SUBCIRCUITS
USE CIRCUITS
    SUBASSEMBLIES

SUBCONTRACTS
GS CONTRACTS
   . **SUBCONTRACTS**
RT AGREEMENTS
    CONTRACT MANAGEMENT
    CONTRACT NEGOTIATION
    CONTRACTORS
    ESTIMATES
    GRANTS
    OPTIONS
    PROCUREMENT

SUBCRITICAL FLOW
GS FLUID FLOW
   . **SUBCRITICAL FLOW**
RT CRITICAL FLOW
    FLOW CHARACTERISTICS
    GAS FLOW
    LIQUID FLOW
    MULTIPHASE FLOW
    ORIFICE FLOW
    PIPE FLOW
    PRESSURE GRADIENTS
    SINGLE-PHASE FLOW
    STEADY FLOW
    STEAM FLOW
    SUPERCRITICAL FLOW
    TURBULENT FLOW
    UNIFORM FLOW
    UNSTEADY FLOW

SUBCRITICAL MASS
GS MASS
   . **SUBCRITICAL MASS**
RT CRITICAL MASS
    NUCLEAR FISSION
    NUCLEAR REACTIONS

SUBDIVISIONS
RT ∞DIVISION
    ∞GROUPS
    ∞SECTIONS
    SET THEORY
    SUBGROUPS
    SUBSIDIARIES

SUBDUCTION (GEOLOGY)
GS GEOLOGY
    . **SUBDUCTION (GEOLOGY)**
RT EARTH MANTLE
    EARTHQUAKES
    LITHOSPHERE
    PLATES (TECTONICS)
    SEISMOLOGY
    STRUCTURAL PROPERTIES (GEOLOGY)
    TECTONICS

SUBDWARF STARS
GS CELESTIAL BODIES
    . STARS
    . . DWARF STARS
    . . . **SUBDWARF STARS**
RT MAIN SEQUENCE STARS
    RED DWARF STARS
    WHITE DWARF STARS

SUBGIANT STARS
GS CELESTIAL BODIES
    . STARS
    . . **SUBGIANT STARS**
RT CARBON STARS
    DWARF STARS
    GIANT STARS
    LATE STARS
    M STARS
    MAIN SEQUENCE STARS
    STELLAR EVOLUTION
    SUPERGIANT STARS

SUBGRAVITY
USE REDUCED GRAVITY

SUBGROUPS
UF SUBLATTICES
GS ALGEBRA
    . GROUP THEORY
    . . HOMOMORPHISMS
    . . . **SUBGROUPS**
RT MATRICES (MATHEMATICS)
    NUMBER THEORY
    PROBABILITY THEORY
    SET THEORY
    SUBDIVISIONS
    SUBSIDIARIES

SUBHARMONIC GENERATORS
RT DAMPING
    ∞GENERATORS
    HARMONIC GENERATORS
    HARMONIC OSCILLATORS
    HARMONICS
    OSCILLATORS
    SIGNAL GENERATORS
    STIMULATED EMISSION DEVICES

SUBIC PROJECT
USE SUBMARINE INTEGRATED CONTROL
     PROJECT

SUBJECTS
GS CLASSIFICATIONS
    . **SUBJECTS**
RT HANDBOOKS
    INFORMATION RETRIEVAL
    TEXTBOOKS

SUBLATTICES
USE LATTICES (MATHEMATICS)
     SUBGROUPS

SUBLAYERS
USE SUBSTRATES

SUBLETHAL DOSAGE
GS DOSAGE
    . **SUBLETHAL DOSAGE**
RT DRUGS

SUBLIMATION
GS PHASE TRANSFORMATIONS
    . VAPORIZING
    . . **SUBLIMATION**

SUBLIMATION-(CONT.)
RT ABLATION
    BENEFICIATION
    CONDENSING
    CRYSTALLIZATION
    DESORPTION
    DIFFUSION
    EVAPORATION
    GAS-METAL INTERACTIONS
    GAS-SOLID INTERFACES
    PHASE CHANGE MATERIALS
    PURIFICATION
    PYROMETALLURGY
    REFINING
    ∞SEPARATION
    VAPOR PRESSURE

SUBLIMINAL STIMULI
RT PSYCHOLOGY
    SENSORY STIMULATION
    ∞STIMULI

SUBMARINE CABLES
GS TRANSMISSION LINES
    . **SUBMARINE CABLES**
RT ∞CABLES
    COAXIAL CABLES
    COMMUNICATION CABLES
    POWER LINES

SUBMARINE INTEGRATED CONTROL PROJECT
UF SUBIC PROJECT
GS PROGRAMS
    . PROJECTS
    . . **SUBMARINE INTEGRATED CONTROL
         PROJECT**
RT ∞CONTROL

SUBMARINE PROPULSION
GS PROPULSION
    . MARINE PROPULSION
    . . UNDERWATER PROPULSION
    . . . **SUBMARINE PROPULSION**

SUBMARINES
GS WATER VEHICLES
    . SHIPS
    . . **SUBMARINES**
    . . . BALLISTIC MISSILE SUBMARINES
    . . . GUIDED MISSILE SUBMARINES
    . . . TRIDENT SUBMARINE
    . UNDERWATER VEHICLES
    . . **SUBMARINES**
    . . . BALLISTIC MISSILE SUBMARINES
    . . . GUIDED MISSILE SUBMARINES
    . . . TRIDENT SUBMARINE
RT ANTISHIP MISSILES
    ANTISHIP WARFARE
    ANTISUBMARINE WARFARE
    ∞MILITARY VEHICLES
    NAVY
    NUCLEAR POWERED SHIPS
    SEAFARER PROJECT
    SHIP HULLS
    SUBMERGED BODIES

SUBMERGED BODIES
GS **SUBMERGED BODIES**
    . DIVING (UNDERWATER)
    . UNDERWATER RESEARCH
      LABORATORIES
RT SUBMARINES
    TORPEDOES
    TOWED BODIES
    UNDERWATER ENGINEERING
    UNDERWATER PHOTOGRAPHY
    UNDERWATER STRUCTURES
    UNDERWATER VEHICLES
    WATER IMMERSION

SUBMERGING
UF IMMERSION
RT BATHS
    DIPPING
    QUENCHING (COOLING)
    SINKING
    ∞SOAKING
    WATER IMMERSION
    WEIGHTLESSNESS SIMULATION
    WETTING

SUBMERSIBLE AIRCRAFT
RT ∞AIRCRAFT
    ANTISUBMARINE WARFARE AIRCRAFT
    LIGHT AIRCRAFT

SUBMERSIBLE AIRCRAFT-(CONT.)
    ∞MILITARY AIRCRAFT
    RECONNAISSANCE AIRCRAFT
    RESEARCH AIRCRAFT
    WATER TAKEOFF AND LANDING
      AIRCRAFT

SUBMILLIMETER WAVES
SN (BELOW 1 MILLIMETER)
GS ELECTROMAGNETIC RADIATION
    . RADIO WAVES
    . . SHORT WAVE RADIATION
    . . . **SUBMILLIMETER WAVES**
RT BEAMS (RADIATION)
    ELECTROMAGNETIC NOISE
    FAR INFRARED RADIATION
    FREQUENCIES
    MICROWAVES
    MILLIMETER WAVES
    WAVELENGTHS

SUBMINIATURIZATION
GS MINIATURIZATION
    . **SUBMINIATURIZATION**
RT ELECTRONIC MODULES
    MICROMINIATURIZATION
    MINIATURE ELECTRONIC EQUIPMENT
    PRINTED CIRCUITS

SUBORBITAL FLIGHT
RT ∞FLIGHT
    MANNED SPACE FLIGHT
    ORBITS
    PARABOLIC FLIGHT
    ROCKET FLIGHT
    SPACE FLIGHT
    WEIGHTLESSNESS

SUBREFLECTORS
RT CASSEGRAIN ANTENNAS
    CONDUCTORS
    REFLECTORS
    SCANNERS

SUBROC MISSILE
GS MISSILES
    . BALLISTIC MISSILES
    . . FIELD ARMY BALLISTIC MISSILES
    . . . **SUBROC MISSILE**
    . SURFACE TO SURFACE MISSILES
    . . FLEET BALLISTIC MISSILES
    . . . **SUBROC MISSILE**
    . UNDERWATER TO SURFACE MISSILES
    . . **SUBROC MISSILE**
RT UNDERWATER TRAJECTORIES

SUBROUTINE LIBRARIES (COMPUTERS)
GS COMPUTER PROGRAMS
    . COMPUTER SYSTEMS PROGRAMS
    . . **SUBROUTINE LIBRARIES
         (COMPUTERS)**
RT SUBROUTINES

SUBROUTINES
GS COMPUTER PROGRAMS
    . **SUBROUTINES**
    DATA CONVERSION ROUTINES
    . **SUBROUTINES**
RT COMPILERS
    PARSING ALGORITHMS
    SUBROUTINE LIBRARIES (COMPUTERS)
    USER MANUALS (COMPUTER
      PROGRAMS)

SUBSETS (MATHEMATICS)
USE SET THEORY

SUBSIDENCE
RT ISOSTASY
    MINES (EXCAVATIONS)
    SETTLING

SUBSIDIARIES
RT ∞DIVISION
    ∞SECTIONS
    SUBDIVISIONS
    SUBGROUPS

∞ **SUBSONIC AIRCRAFT**
SN (USE OF A MORE SPECIFIC TERM IS
    RECOMMENDED--CONSULT THE TERMS
    LISTED BELOW)
RT ∞AIRCRAFT
    FLYING PLATFORMS

SUBSONIC AIRCRAFT-*(CONT.)*
 GENERAL AVIATION AIRCRAFT
 GETOL AIRCRAFT
 GLIDERS
 GROUND EFFECT MACHINES
 HELICOPTERS
 JET AIRCRAFT
 LIGHT AIRCRAFT
 PARAGLIDERS
 PASSENGER AIRCRAFT
 ROTARY WING AIRCRAFT
 SHORT TAKEOFF AIRCRAFT
 SUPERSONIC AIRCRAFT
 TANDEM WING AIRCRAFT
 TERRAIN FOLLOWING AIRCRAFT
 TRAINING AIRCRAFT
 TRANSPORT AIRCRAFT
 TURBOPROP AIRCRAFT
 UTILITY AIRCRAFT
 VERTICAL TAKEOFF AIRCRAFT
 WATER TAKEOFF AND LANDING
  AIRCRAFT

SUBSONIC FLOW
GS FLUID FLOW
 . **SUBSONIC FLOW**
RT AERODYNAMICS
 COMPRESSIBLE FLOW
 FLOW VELOCITY
 GAS FLOW
 INCOMPRESSIBLE FLOW
 KARMAN VORTEX STREET
 TRANSONIC FLOW

SUBSONIC FLUTTER
GS VIBRATION
 . STRUCTURAL VIBRATION
 . . FLUTTER
 . . . **SUBSONIC FLUTTER**
 . . SELF INDUCED VIBRATION
 . . . **SUBSONIC FLUTTER**
RT TRANSONIC FLUTTER

SUBSONIC SPEED
GS RATES (PER TIME)
 . **SUBSONIC SPEED**
 VELOCITY
 . **SUBSONIC SPEED**
RT ACOUSTIC VELOCITY
 LOW SPEED
 TRANSONIC SPEED

SUBSONIC WIND TUNNELS
GS TEST FACILITIES
 . WIND TUNNELS
 . . LOW SPEED WIND TUNNELS
 . . . **SUBSONIC WIND TUNNELS**
RT BLOWDOWN WIND TUNNELS
 HYPERSONIC WIND TUNNELS
 RECTANGULAR WIND TUNNELS
 SUPERSONIC WIND TUNNELS
 TRANSONIC WIND TUNNELS

SUBSTANCES
USE MATERIALS

SUBSTITUTES
UF SUBSTITUTION
RT ALTERNATIVES
 REPLACING
 VARIATIONS

SUBSTITUTION
USE SUBSTITUTES

SUBSTRATES
UF SUBLAYERS
GS STRATA
 . **SUBSTRATES**
RT COATINGS
 LAMINATES
 ∞ LAYERS
 METALLIZING
 PHOTOMASKS
 PLATING
 PLY ORIENTATION
 PRIMERS (COATINGS)

SUBSTRUCTURES
RT FLOORS
 FOUNDATIONS
 STRUCTURAL DESIGN
 STRUCTURAL MEMBERS
 ∞ STRUCTURES

SUBSTRUCTURES-*(CONT.)*
 SUPPORTS
 UNDERCARRIAGES
 WALLS

SUBTRACTION
GS NUMBER THEORY
 . **SUBTRACTION**
RT ARITHMETIC
 COMPUTATION

SUBTROPICAL REGIONS
USE TEMPERATE REGIONS
 TROPICAL REGIONS

SUBURBAN AREAS
RT CITIES
 LAND USE
 MEGALOPOLISES
 REGIONAL PLANNING
 RESIDENTIAL AREAS
 RURAL AREAS

SUBZERO TEMPERATURE
GS TEMPERATURE
 . **SUBZERO TEMPERATURE**
RT ABSOLUTE ZERO
 ATMOSPHERIC TEMPERATURE
 COLD ACCLIMATIZATION
 COLD TOLERANCE
 COLD WEATHER

SUCCESS PROJECT
GS PROGRAMS
 . PROJECTS
 . . **SUCCESS PROJECT**
 WEAPON SYSTEMS
 . **SUCCESS PROJECT**

SUCCINIMIDES
GS NITROGEN COMPOUNDS
 . AMIDES
 . . **SUCCINIMIDES**
 . IMIDES
 . . **SUCCINIMIDES**

SUCROSE
GS ALIPHATIC COMPOUNDS
 . SUGARS
 . . **SUCROSE**
 CARBOHYDRATES
 . SUGARS
 . . **SUCROSE**

SUCTION
RT EVACUATING (VACUUM)
 PRESSURE EFFECTS
 PRESSURE GRADIENTS
 VACUUM
 VACUUM APPARATUS
 VACUUM PUMPS

SUD AVIATION AIRCRAFT
GS **SUD AVIATION AIRCRAFT**
 . ALOUETTE HELICOPTERS
 . . SA-330 HELICOPTER
 . . SE-3160 HELICOPTER
 . CONCORDE AIRCRAFT
 . SA-321 HELICOPTER
 . SE-210 AIRCRAFT
 . VJ-101 AIRCRAFT
RT ∞ AIRCRAFT

SUD AVIATION SA-321 HELICOPTER
USE SA-321 HELICOPTER

SUD AVIATION SA-330 HELICOPTER
USE SA-330 HELICOPTER

SUD AVIATION SE-210 AIRCRAFT
USE SE-210 AIRCRAFT

SUD AVIATION SE-3160 HELICOPTER
USE SE-3160 HELICOPTER

SUD VJ-101 AIRCRAFT
USE VJ-101 AIRCRAFT

SUDAN
GS NATIONS
 . **SUDAN**
RT AFRICA

SUDDEN ENHANCEMENT OF ATMOSPHERICS
GS ELECTROMAGNETIC INTERFERENCE
 . RADIO FREQUENCY INTERFERENCE
 . . BLACKOUT (PROPAGATION)
 . . . ELECTROMAGNETIC NOISE
 . . . . ATMOSPHERICS
 . . . . . **SUDDEN ENHANCEMENT OF
   ATMOSPHERICS**

SUDDEN IONOSPHERIC DISTURBANCES
UF GEOMAGNETIC CROTCHETS
 SID (IONOSPHERIC DISTURBANCES)
GS IONOSPHERIC DISTURBANCES
 . IONOSPHERIC STORMS
 . . **SUDDEN IONOSPHERIC
   DISTURBANCES**
 STORMS
 . IONOSPHERIC STORMS
 . . **SUDDEN IONOSPHERIC
   DISTURBANCES**
RT ∞ DISTURBANCES
 MAGNETIC DISTURBANCES
 MAGNETIC STORMS
 SOLAR ACTIVITY EFFECTS
 TRAVELING IONOSPHERIC
  DISTURBANCES

SUDDEN STORM COMMENCEMENTS
RT MAGNETIC DISTURBANCES
 MAGNETIC STORMS
 SOLAR ACTIVITY EFFECTS
 SOLAR CORPUSCULAR RADIATION
 SOLAR FLARES
 STORMS

SUGAR BEETS
GS FARM CROPS
 . **SUGAR BEETS**
RT AGRICULTURE
 BOTANY
 CROP GROWTH
 CROP VIGOR
 EARTH RESOURCES
 FARMLANDS
 IRRIGATION
 SEEDS
 SUGARS

SUGAR CANE
GS FARM CROPS
 . **SUGAR CANE**
 PLANTS (BOTANY)
 . **SUGAR CANE**
RT AGRICULTURE
 BOTANY
 CROP GROWTH
 CROP VIGOR
 EARTH RESOURCES
 FARMLANDS
 ∞ FOOD
 IRRIGATION
 SEEDS

SUGARS
GS ALIPHATIC COMPOUNDS
 . **SUGARS**
 . . DEXTRANS
 . . GALACTOSE
 . . GLUCOSE
 . . HEXOSES
 . . INOSITOLS
 . . LACTOSE
 . . MANNITOL
 . . MONOSACCHARIDES
 . . . RIBOSE
 . . . XYLOSE
 . . PENTOSE
 . . . RIBOSE
 . . . XYLOSE
 . . SUCROSE
 CARBOHYDRATES
 . **SUGARS**
 . . DEXTRANS
 . . GALACTOSE
 . . GLUCOSE
 . . HEXOSES
 . . INOSITOLS
 . . LACTOSE
 . . MANNITOL
 . . MONOSACCHARIDES
 . . . RIBOSE
 . . . XYLOSE
 . . PENTOSE
 . . . RIBOSE
 . . . XYLOSE

SUGARS-*(CONT.)*
. . SUCROSE
RT   ∞FOOD
      SUGAR BEETS

SUGGESTION
GS   RECOMMENDATIONS
      . SUGGESTION
RT   HYPNOSIS

SUHL EFFECT
RT   CARRIER INJECTION
      ∞EFFECTS
      ELECTRONS
      EXCITONS
      HOLES (ELECTRON DEFICIENCIES)
      MAGNETIC FIELDS
      N-TYPE SEMICONDUCTORS
      RECOMBINATION REACTIONS

SUITABILITY
RT   ACCEPTABILITY
      COMPATIBILITY

SUITS
GS   CLOTHING
      . SUITS
      . . PRESSURE SUITS
      . . . SPACE SUITS
RT   GARMENTS

SULFATES
GS   SULFUR COMPOUNDS
      . SULFATES
      . . AMMONIUM SULFATES
      . . BARITE
      . . HYDROXYLAMINE SULFATE
      . . LITHIUM SULFATES
      . . MAGNESIUM SULFATES
      . . . HEXAHEDRITE
      . . SODIUM SULFATES
RT   GYPSUM
      SULFURIC ACID

SULFATION
GS   CHEMICAL REACTIONS
      . SULFATION
RT   HYDROMETALLURGY
      SULFIDATION

SULFIDATION
GS   CHEMICAL REACTIONS
      . SULFIDATION
RT   CORROSION RESISTANCE
      GAS-METAL INTERACTIONS
      HEAT RESISTANT ALLOYS
      NICKEL ALLOYS
      SULFATION
      SULFIDES

SULFIDES
GS   CHALCOGENIDES
      . SULFIDES
      . . DISULFIDES
      . . . CARBON DISULFIDE
      . . INORGANIC SULFIDES
      . . . BARIUM SULFIDES
      . . . BISMUTH SULFIDES
      . . . CADMIUM SULFIDES
      . . . CALCIUM SULFIDES
      . . . COPPER SULFIDES
      . . . . ENARGITE
      . . . HYDROGEN SULFIDE
      . . . INDIUM SULFIDES
      . . . LEAD SULFIDES
      . . . MOLYBDENUM SULFIDES
      . . . . MOLYBDENUM DISULFIDES
      . . POLYSULFIDES
      . . . STRONTIUM SULFIDES
      . . . ZINC SULFIDES
      . . . . WURTZITE
      . . . . ZINCBLENDE
      . . PYRITES
      . . PYRRHOTITE
      . . . TROILITE
      SULFUR COMPOUNDS
      . SULFIDES
      . . DISULFIDES
      . . . CARBON DISULFIDE
      . . INORGANIC SULFIDES
      . . . BARIUM SULFIDES
      . . . BISMUTH SULFIDES
      . . . CADMIUM SULFIDES
      . . . CALCIUM SULFIDES
      . . . COPPER SULFIDES

SULFIDES-*(CONT.)*
      . . . . ENARGITE
      . . . HYDROGEN SULFIDE
      . . . INDIUM SULFIDES
      . . . LEAD SULFIDES
      . . . MOLYBDENUM SULFIDES
      . . . . MOLYBDENUM DISULFIDES
      . . POLYSULFIDES
      . . . STRONTIUM SULFIDES
      . . . ZINC SULFIDES
      . . . . WURTZITE
      . . . . ZINCBLENDE
      . . PYRITES
      . . PYRRHOTITE
      . . . TROILITE
RT   SULFIDATION
      THIOPLASTICS

SULFITES
GS   SULFUR COMPOUNDS
      . SULFITES
      . . HYDROSULFITES
      . . SODIUM SULFITES

SULFONATES
GS   ESTERS
      . SULFONATES
      SULFUR COMPOUNDS
      . SULFONATES
RT   ∞SALTS

SULFONES
GS   SULFUR COMPOUNDS
      . SULFONES
RT   SULFONIC ACID

SULFONIC ACID
GS   ACIDS
      . SULFONIC ACID
RT   SULFONES
      SULFUR COMPOUNDS

SULFUR
GS   CHEMICAL ELEMENTS
      . SULFUR
      . . SULFUR ISOTOPES

SULFUR CHLORIDES
GS   HALOGEN COMPOUNDS
      . CHLORINE COMPOUNDS
      . . CHLORIDES
      . . . SULFUR CHLORIDES
      . HALIDES
      . . CHLORIDES
      . . . SULFUR CHLORIDES
      SULFUR COMPOUNDS
      . SULFUR CHLORIDES

SULFUR COMPOUNDS
GS   SULFUR COMPOUNDS
      . ALUM
      . ORGANIC SULFUR COMPOUNDS
      . SULFATES
      . . AMMONIUM SULFATES
      . . BARITE
      . . HYDROXYLAMINE SULFATE
      . . LITHIUM SULFATES
      . . MAGNESIUM SULFATES
      . . . HEXAHEDRITE
      . . SODIUM SULFATES
      . SULFIDES
      . . DISULFIDES
      . . . CARBON DISULFIDE
      . . INORGANIC SULFIDES
      . . . BARIUM SULFIDES
      . . . BISMUTH SULFIDES
      . . . CADMIUM SULFIDES
      . . . CALCIUM SULFIDES
      . . . COPPER SULFIDES
      . . . . ENARGITE
      . . . HYDROGEN SULFIDE
      . . . INDIUM SULFIDES
      . . . LEAD SULFIDES
      . . . MOLYBDENUM SULFIDES
      . . . . MOLYBDENUM DISULFIDES
      . . POLYSULFIDES
      . . . STRONTIUM SULFIDES
      . . . ZINC SULFIDES
      . . . . WURTZITE
      . . . . ZINCBLENDE
      . . PYRITES
      . . PYRRHOTITE
      . . . TROILITE
      . SULFITES
      . . HYDROSULFITES

SULFUR COMPOUNDS-*(CONT.)*
      . . SODIUM SULFITES
      . SULFONATES
      . SULFONES
      . SULFUR CHLORIDES
      . SULFUR FLUORIDES
      . SULFUR OXIDES
      . . SULFUR DIOXIDES
      . SULFURIC ACID
      . THIAZINE (TRADEMARK)
      . THIOLS
      . . CYSTEINE
      . . DIMERCAPROL
RT   BLOEDITE
      ∞CHEMICAL COMPOUNDS
      ∞GROUP 6A COMPOUNDS
      SULFONIC ACID

SULFUR DIOXIDES
GS   CHALCOGENIDES
      . OXIDES
      . . DIOXIDES
      . . . SULFUR DIOXIDES
      . SULFUR OXIDES
      . . . SULFUR DIOXIDES
      SULFUR COMPOUNDS
      . SULFUR OXIDES
      . . SULFUR DIOXIDES

SULFUR FLUORIDES
GS   HALOGEN COMPOUNDS
      . FLUORINE COMPOUNDS
      . . FLUORIDES
      . . . SULFUR FLUORIDES
      . HALIDES
      . FLUORIDES
      . . . SULFUR FLUORIDES
      SULFUR COMPOUNDS
      . SULFUR FLUORIDES

SULFUR ISOTOPES
GS   CHEMICAL ELEMENTS
      . SULFUR
      . . SULFUR ISOTOPES

SULFUR OXIDES
GS   CHALCOGENIDES
      . OXIDES
      . SULFUR OXIDES
      . . . SULFUR DIOXIDES
      SULFUR COMPOUNDS
      . SULFUR OXIDES
      . . SULFUR DIOXIDES
RT   ACID RAIN
      DIOXIDES

SULFURIC ACID
GS   ACIDS
      . SULFURIC ACID
      SULFUR COMPOUNDS
      . SULFURIC ACID
RT   SULFATES

SUM RULES
GS   RULES
      . SUM RULES
RT   SUMS

SUMMARIES
GS   SUMMARIES
      . ABSTRACTS
      . PRELAUNCH SUMMARIES
RT   ANNOTATIONS
      BIBLIOGRAPHIES
      DOCUMENTATION
      INDEXES (DOCUMENTATION)
      INFORMATION DISSEMINATION
      POSTLAUNCH REPORTS
      REPORTS

SUMMER
GS   SEASONS
      . SUMMER
RT   AUTUMN
      HOT WEATHER
      SOLSTICES
      SPRING (SEASON)
      WINTER

SUMPS
RT   DRAINAGE
      PITS (EXCAVATIONS)
      WASTE DISPOSAL

SUMS
RT ALGEBRA
   AMOUNT
   ARITHMETIC
   COMPUTATION
   SERIES (MATHEMATICS)
   SUM RULES

SUN
UF SOLAR DISK
GS CELESTIAL BODIES
   . STARS
   . . SUN
RT AOSO
   ASTEC SOLAR TURBOELECTRIC
      GENERATOR
   CELESTIAL MECHANICS
   GRIST (TELESCOPE)
   LIGHT SOURCES
   OSO
   PHOTOSPHERE
   PLANETS
   SATELLITE SOLAR ENERGY
      CONVERSION
   SATELLITE SOLAR POWER STATIONS
   SOLAR ACTIVITY
   SOLAR ACTIVITY EFFECTS
   SOLAR ARRAYS
   SOLAR ATMOSPHERE
   SOLAR ATRIUMS
   SOLAR AUXILIARY POWER UNITS
   SOLAR BLANKETS
   SOLAR CELLS
   SOLAR COLLECTORS
   SOLAR COMPASSES
   SOLAR CONSTANT
   SOLAR COOLING
   SOLAR CORONA
   SOLAR CORPUSCULAR RADIATION
   SOLAR COSMIC RAYS
   SOLAR CYCLES
   SOLAR ECLIPSES
   SOLAR ELECTRIC PROPULSION
   SOLAR ELECTRONS
   SOLAR ENERGY
   SOLAR ENERGY ABSORBERS
   SOLAR ENERGY CONVERSION
   SOLAR FLARES
   SOLAR FLUX
   SOLAR FLUX DENSITY
   SOLAR FURNACES
   SOLAR GENERATORS
   SOLAR GRANULATION
   SOLAR GRAVITATION
   SOLAR HEATING
   SOLAR HOUSES
   SOLAR INSTRUMENTS
   SOLAR LIMB
   SOLAR LONGITUDE
   SOLAR MAGNETIC FIELD
   SOLAR MAXIMUM MISSION
   SOLAR MAXIMUM MISSION-A
   SOLAR MESOSPHERE EXPLORER
   SOLAR NEUTRINOS
   SOLAR OBLATENESS
   SOLAR OBSERVATORIES
   SOLAR ORBITS
   SOLAR OSCILLATIONS
   SOLAR PARALLAX
   SOLAR PHYSICS
   SOLAR PONDS (HEAT STORAGE)
   SOLAR POSITION
   SOLAR POWER SATELLITES
   SOLAR POWERED AIRCRAFT
   SOLAR PROBES
   SOLAR PROMINENCES
   SOLAR PROPULSION
   SOLAR PROTONS
   SOLAR RADAR ECHOES
   SOLAR RADIATION
   SOLAR RADIATION SHIELDING
   SOLAR RADIATION 1 SATELLITE
   SOLAR RADIATION 3 SATELLITE
   SOLAR RADIO BURSTS
   SOLAR REFLECTORS
   SOLAR ROTATION
   SOLAR SAILS
   SOLAR SEA POWER PLANTS
   SOLAR SENSORS
   SOLAR SIMULATION
   SOLAR SIMULATORS
   SOLAR SPECTRA
   SOLAR SPECTROMETERS
   SOLAR STORMS
   SOLAR SYSTEM
   SOLAR TEMPERATURE

SUN-(CONT.)
   SOLAR TERRESTRIAL INTERACTIONS
   SOLAR THERMAL PROPULSION
   SOLAR TOTAL ENERGY SYSTEMS
   SOLAR VELOCITY
   SOLAR WIND
   SOLAR WIND VELOCITY
   SOLAR X-RAYS
   SUNLIGHT
   ULYSSES MISSION

SUN SENSORS
USE SOLAR SENSORS

SUNBLAZER SPACE PROBE
GS UNMANNED SPACECRAFT
   . SPACE PROBES
   . . SOLAR PROBES
   . . . SUNBLAZER SPACE PROBE
RT MULTISTAGE ROCKET VEHICLES
   SOLID PROPELLANT ROCKET ENGINES

SUNFLOWER POWER SYSTEM
GS AUXILIARY POWER SOURCES
   . SOLAR AUXILIARY POWER UNITS
   . . SUNFLOWER POWER SYSTEM
   ELECTRIC GENERATORS
   . SOLAR GENERATORS
   . . SOLAR AUXILIARY POWER UNITS
   . . . SUNFLOWER POWER SYSTEM
RT SOLAR COLLECTORS
   ∞SYSTEMS
   TURBOGENERATORS

SUNFLOWERS
GS FARM CROPS
   . SUNFLOWERS
   PLANTS (BOTANY)
   . SUNFLOWERS
RT AGRICULTURE
   CROP IDENTIFICATION
   ∞CROPS
   EARTH RESOURCES

SUNGLASSES
RT EYE PROTECTION
   EYEPIECES
   GOGGLES
   OPTICAL FILTERS
   VISORS

SUNLIGHT
GS ELECTROMAGNETIC RADIATION
   . LIGHT (VISIBLE RADIATION)
   . . SUNLIGHT
   EXTRATERRESTRIAL RADIATION
   . SOLAR RADIATION
   . . SUNLIGHT
RT BLACK BODY RADIATION
   CIRCUMSOLAR RADIATION
   CLIMATOLOGY
   CLOUD COVER
   INFRARED RADIATION
   INSOLATION
   SKY
   SKY BRIGHTNESS
   SKY RADIATION
   SOLAR HEATING
   SUN
   THERMAL RADIATION
   ULTRAVIOLET RADIATION
   UMKEHR EFFECT
   ZODIACAL LIGHT

SUNRISE
RT MORNING
   ∞SCIENCE
   SUNSET
   TERMINATOR LINES

SUNSET
RT EVENING
   ∞SCIENCE
   SUNRISE
   TERMINATOR LINES

SUNSPOT CYCLE
GS CYCLES
   . SOLAR CYCLES
   . . SUNSPOT CYCLE
RT SOLAR ACTIVITY
   STARSPOTS
   STELLAR ACTIVITY

SUNSPOTS
GS STELLAR ACTIVITY
   . SOLAR ACTIVITY
   . . SUNSPOTS
RT FACULAE
   MAGNETIC DISTURBANCES
   PHOTOSPHERE
   SOLAR CYCLES
   SOLAR FLARES
   SOLAR TERRESTRIAL INTERACTIONS
   TWENTY-SEVEN DAY VARIATION

SUPER FORTRESS AIRCRAFT
USE RB-50 AIRCRAFT

SUPER SABRE AIRCRAFT
USE F-100 AIRCRAFT

SUPERALLOYS
USE HEAT RESISTANT ALLOYS

SUPERCAVITATING FLOW
UF SUPERCAVITATION
GS FLUID FLOW
   . TURBULENT FLOW
   . . SUPERCAVITATING FLOW
RT CAVITATION FLOW
   HYDROFOIL OSCILLATIONS

SUPERCAVITATION
USE SUPERCAVITATING FLOW

SUPERCHARGERS
UF SUPERCHARGING
   TURBOCHARGERS
GS COMPRESSORS
   . SUPERCHARGERS
RT AIR INTAKES
   BLOWERS
   CENTRIFUGAL COMPRESSORS
   COMPRESSING
   INTERNAL COMBUSTION ENGINES
   TURBOCOMPRESSORS
   TURBOMACHINERY

SUPERCHARGING
USE SUPERCHARGERS

SUPERCOMPUTERS
GS SUPERCOMPUTERS
   . CRAY COMPUTERS

SUPERCONDUCTING MAGNETS
GS MAGNETS
   . ELECTROMAGNETS
   . . SUPERCONDUCTING MAGNETS
RT CRYOGENIC MAGNETS
   FLUX PUMPS
   HIGH FIELD MAGNETS
   MAGNET COILS
   MAGNETIC ENERGY STORAGE

SUPERCONDUCTING POWER TRANSMISSION
RT ∞CONDUCTIVITY
   CRYOGENICS
   ELECTRIC POWER TRANSMISSION
   LOW TEMPERATURE PHYSICS
   POWER LINES
   TRANSITION TEMPERATURE
   TRANSMISSION LINES

SUPERCONDUCTING QUANTUM
INTERFEROMETERS
USE SQUID (DETECTORS)

SUPERCONDUCTIVITY
UF MEISSNER EFFECT
GS TRANSPORT PROPERTIES
   . ELECTRICAL RESISTIVITY
   . . SUPERCONDUCTIVITY
   . . . KONDO EFFECT
RT ABRIKOSOV THEORY
   BCS THEORY
   BLOCH BAND
   ∞CONDUCTIVITY
   CRYOGENICS
   CRYOTRONS
   ELECTRON PHONON INTERACTIONS
   ELECTRON TUNNELING
   FLUX PINNING
   FLUX PUMPS
   JOSEPHSON JUNCTIONS
   LANDAU FACTOR
   LANDAU-GINZBURG EQUATIONS

SUPERCONDUCTIVITY-(CONT.)
- LOW TEMPERATURE PHYSICS
- ∞ SOLID STATE PHYSICS
- SPIN GLASS
- TRANSITION TEMPERATURE
- TRAPPED MAGNETIC FIELDS
- VECTOR CURRENTS
- VORTICES

SUPERCONDUCTORS
- GS   CONDUCTORS
- . **SUPERCONDUCTORS**
- RT   ABRIKOSOV THEORY
- CARRIER MOBILITY
- CRYOGENIC COMPUTER STORAGE
- CRYOTRONS
- ELECTRON GAS
- ENERGY STORAGE
- PINNING
- PROXIMITY EFFECT (ELECTRICITY)
- SOLID STATE DEVICES
- SQUID (DETECTORS)
- THERMODYNAMIC COUPLING

SUPERCOOLING
- GS   COOLING
- . **SUPERCOOLING**
- . . CRYOGENIC COOLING
- RT   AGING (METALLURGY)
- AITKEN NUCLEI
- CONDENSING
- CONVECTION CLOUDS
- CRYSTALLIZATION
- HEAT TREATMENT
- MECHANICAL PROPERTIES
- NUCLEATION
- QUENCHING (COOLING)
- SUPERSATURATION

SUPERCRITICAL AIRFOILS
- GS   AIRFOILS
- . **SUPERCRITICAL AIRFOILS**
- . . SUPERCRITICAL WINGS
- RT   AIRFOIL PROFILES

SUPERCRITICAL FLOW
- GS   FLUID FLOW
- . **SUPERCRITICAL FLOW**
- RT   CRITICAL FLOW
- FLOW CHARACTERISTICS
- GAS FLOW
- LIQUID FLOW
- MULTIPHASE FLOW
- ORIFICE FLOW
- PIPE FLOW
- PRESSURE GRADIENTS
- SINGLE-PHASE FLOW
- STEADY FLOW
- STEAM FLOW
- SUBCRITICAL FLOW
- TURBULENT FLOW
- UNSTEADY FLOW

SUPERCRITICAL FLUIDS
- RT   FLUID MECHANICS
- ∞ FLUIDS
- SOLUBILITY
- SUPERCRITICAL PRESSURES

SUPERCRITICAL PRESSURES
- GS   PRESSURE
- . **SUPERCRITICAL PRESSURES**
- THERMODYNAMIC PROPERTIES
- . THERMOPHYSICAL PROPERTIES
- . . **SUPERCRITICAL PRESSURES**
- RT   CRITICAL PRESSURE
- HIGH PRESSURE
- LIQUID PHASES
- SUPERCRITICAL FLUIDS
- VAPOR PHASES
- VAPOR PRESSURE

SUPERCRITICAL WINGS
- GS   AIRFOILS
- . SUPERCRITICAL AIRFOILS
- . . **SUPERCRITICAL WINGS**
- . WINGS
- . . **SUPERCRITICAL WINGS**
- RT   CL-600 CHALLENGER AIRCRAFT
- SPANLOADER AIRCRAFT
- WING PROFILES
- ∞ WINGED VEHICLES

SUPERFLUID FLOW
- USE   SUPERFLUIDITY

SUPERFLUIDITY
- UF   SUPERFLUID FLOW
- RT   COMPRESSIBLE FLUIDS
- ∞ FLUIDS
- INCOMPRESSIBLE FLUIDS
- KELVIN-HELMHOLTZ INSTABILITY
- LIQUID HELIUM
- LIQUID HELIUM 2
- MANY BODY PROBLEM
- QUANTUM STATISTICS
- TWO FLUID MODELS
- VISCOSITY
- VORTICES

SUPERGIANT STARS
- GS   CELESTIAL BODIES
- . STARS
- . . **SUPERGIANT STARS**
- RT   GIANT STARS
- M STARS
- SUBGIANT STARS

SUPERHARMONICS
- GS   HARMONICS
- . **SUPERHARMONICS**
- RT   CYCLES
- FREQUENCIES
- MACH NUMBER
- SUPERSONIC WIND TUNNELS
- SUPERSONICS

SUPERHEATING
- GS   HEATING
- . **SUPERHEATING**
- RT   STEAM

SUPERHETERODYNE RECEIVERS
- GS   COMMUNICATION EQUIPMENT
- . RADIO RECEIVERS
- . . **SUPERHETERODYNE RECEIVERS**
- RADIO EQUIPMENT
- . RADIO RECEIVERS
- . . **SUPERHETERODYNE RECEIVERS**
- RECEIVERS
- . RADIO RECEIVERS
- . . **SUPERHETERODYNE RECEIVERS**
- RT   BEAT FREQUENCIES
- HETERODYNING

SUPERHIGH FREQUENCIES
- SN   (3 TO 30 GHZ)
- UF   KU BAND
- S BAND
- X BAND
- GS   FREQUENCIES
- . RADIO FREQUENCIES
- . . MICROWAVE FREQUENCIES
- . . . **SUPERHIGH FREQUENCIES**
- RT   C BAND
- CENTIMETER WAVES
- UNIFIED S BAND

SUPERHYBRID MATERIALS
- GS   COMPOSITE MATERIALS
- . **SUPERHYBRID MATERIALS**
- . . GRAPHITE-EPOXY COMPOSITES
- RT   BORON-EPOXY COMPOUNDS
- CARBON FIBER REINFORCED PLASTICS
- FIBER COMPOSITES
- ∞ MATERIALS
- REINFORCING FIBERS

SUPERIMPOSITION (MATHEMATICS)
- USE   SUPERPOSITION (MATHEMATICS)

SUPERLATTICES
- GS   CRYSTAL LATTICES
- . **SUPERLATTICES**
- SEMICONDUCTORS (MATERIALS)
- . **SUPERLATTICES**
- RT   CRYSTAL DISLOCATIONS
- CRYSTAL STRUCTURE
- LATTICE PARAMETERS

SUPERMAGNETS
- USE   HIGH FIELD MAGNETS

SUPERMASSIVE STARS
- GS   CELESTIAL BODIES
- . STARS
- . . **SUPERMASSIVE STARS**
- RT   STELLAR MODELS
- STELLAR STRUCTURE

SUPERNOVA REMNANTS
- RT   BLACK HOLES (ASTRONOMY)
- NEUTRON STARS
- NORTH POLAR SPUR (ASTRONOMY)
- PULSARS
- RED DWARF STARS
- SUPERNOVAE
- WHITE DWARF STARS
- WHITE HOLES (ASTRONOMY)

SUPERNOVAE
- GS   CELESTIAL BODIES
- . STARS
- . . VARIABLE STARS
- . . . **SUPERNOVAE**
- RT   CRAB NEBULA
- GRAVITATIONAL COLLAPSE
- NEBULAE
- NOVAE
- OPIK THEORY
- ORION NEBULA
- STELLAR MASS
- STELLAR MASS EJECTION
- STELLAR PHYSICS
- SUPERNOVA REMNANTS

SUPEROXIDES
- USE   INORGANIC PEROXIDES

SUPERPLASTICITY
- GS   MECHANICAL PROPERTIES
- . PLASTIC PROPERTIES
- . . **SUPERPLASTICITY**
- RT   CREEP PROPERTIES
- CRYSTAL DISLOCATIONS
- ELONGATION
- EUTECTIC ALLOYS
- HEAT RESISTANT ALLOYS
- PLASTIC DEFORMATION
- PLASTIC FLOW

SUPERPOSITION (MATHEMATICS)
- UF   SUPERIMPOSITION (MATHEMATICS)
- RT   EQUIVALENT CIRCUITS
- LINEAR CIRCUITS
- ∞ MATHEMATICS
- NETWORK ANALYSIS
- NETWORK SYNTHESIS

SUPERPRESSURE BALLOONS
- UF   CONSTANT VOLUME BALLOONS
- TETROONS
- GS   EXPANDABLE STRUCTURES
- . INFLATABLE STRUCTURES
- . . BALLOONS
- . . . HIGH ALTITUDE BALLOONS
- . . . . **SUPERPRESSURE BALLOONS**
- RT   BALLOON SOUNDING
- METEOROLOGICAL BALLOONS

SUPERROTATION
- RT   ATMOSPHERIC CIRCULATION
- EARTH ATMOSPHERE
- EARTH ROTATION
- ROTATING FLUIDS

SUPERSATURATION
- RT   CONDENSING
- CRYSTALLIZATION
- HEAT TREATMENT
- MAYER PROBLEM
- PRECIPITATION (CHEMISTRY)
- PRECIPITATION HARDENING
- QUENCHING (COOLING)
- SOLID SOLUTIONS
- SUPERCOOLING

SUPERSONIC AIRCRAFT
- SN   (AIRCRAFT DESIGNED TO FLY AT SPEEDS ABOVE MACH 1 AND BELOW MACH 5)
- UF   TRANSONIC AIRCRAFT
- GS   **SUPERSONIC AIRCRAFT**
- . A-5 AIRCRAFT
- . B-58 AIRCRAFT
- . B-70 AIRCRAFT
- . BOEING 733 AIRCRAFT
- . D-558 AIRCRAFT
- . F-5 AIRCRAFT
- . F-8 AIRCRAFT
- . F-14 AIRCRAFT
- . F-15 AIRCRAFT
- . F-16 AIRCRAFT
- . F-17 AIRCRAFT
- . F-100 AIRCRAFT

SUPERSONIC AIRCRAFT-*(CONT.)*
- . F-101 AIRCRAFT
- . F-102 AIRCRAFT
- . F-104 AIRCRAFT
- . F-106 AIRCRAFT
- . F-111 AIRCRAFT
- . FIREBEE 2 TARGET DRONE AIRCRAFT
- . G-95/4 AIRCRAFT
- . JAGUAR AIRCRAFT
- . MIG AIRCRAFT
- . MIRAGE AIRCRAFT
- . . MIRAGE 3 AIRCRAFT
- . NORD 1500 AIRCRAFT
- . P-1154 AIRCRAFT
- . PHANTOM AIRCRAFT
- . . F-4 AIRCRAFT
- . . RF-4 AIRCRAFT
- . SAAB 37 AIRCRAFT
- . SUPERSONIC TRANSPORTS
- . . CL-823 AIRCRAFT
- . . CONCORDE AIRCRAFT
- . . L-2000 AIRCRAFT
- . . SUPERSONIC COMMERCIAL AIR TRANSPORT
- . . . BOEING 2707 AIRCRAFT
- . T-38 AIRCRAFT
- . TSR-2 AIRCRAFT
- . VJ-101 AIRCRAFT
- . X-1 AIRCRAFT
- . X-2 AIRCRAFT
- . X-3 AIRCRAFT
- . X-15 AIRCRAFT

RT   ∞AIRCRAFT
     ATTACK AIRCRAFT
     FIGHTER AIRCRAFT
     HYPERSONIC AIRCRAFT
     JET AIRCRAFT
     PASSENGER AIRCRAFT
     RECONNAISSANCE AIRCRAFT
     RESEARCH AIRCRAFT
     ∞SUBSONIC AIRCRAFT
     SUPERSONIC CRUISE AIRCRAFT RESEARCH
     SUPERSONICS
     SWEPTBACK WINGS
     TRANSPORT AIRCRAFT
     TRAPEZOIDAL TAIL SURFACES
     VARIABLE CYCLE ENGINES
     VARIABLE STREAM CONTROL ENGINES

SUPERSONIC AIRFOILS
GS   AIRFOILS
- . **SUPERSONIC AIRFOILS**

RT   SWEEPBACK
     SWEPTBACK TAIL SURFACES
     SWEPTBACK WINGS

SUPERSONIC BOUNDARY LAYERS
GS   BOUNDARY LAYERS
- . **SUPERSONIC BOUNDARY LAYERS**

RT   FLUID FLOW
     LAMINAR BOUNDARY LAYER
     SUPERSONICS
     TURBULENT BOUNDARY LAYER
     TWO DIMENSIONAL BOUNDARY LAYER

SUPERSONIC COMBUSTION
GS   COMBUSTION
- . **SUPERSONIC COMBUSTION**

RT   ENGINES
     FUEL COMBUSTION

SUPERSONIC COMBUSTION RAMJET ENGINES
UF   SCRAMJET ENGINES
     SCRAMJETS
GS   ENGINES
- . AIR BREATHING ENGINES
- . . GAS TURBINE ENGINES
- . . . JET ENGINES
- . . . . RAMJET ENGINES
- . . . . . **SUPERSONIC COMBUSTION RAMJET ENGINES**
- . INTERNAL COMBUSTION ENGINES
- . . GAS TURBINE ENGINES
- . . JET ENGINES
- . . . RAMJET ENGINES
- . . . . **SUPERSONIC COMBUSTION RAMJET ENGINES**
- . TURBINE ENGINES
- . . GAS TURBINE ENGINES
- . . . JET ENGINES
- . . . . RAMJET ENGINES
- . . . . . **SUPERSONIC COMBUSTION RAMJET ENGINES**

RT   COMBUSTION

SUPERSONIC COMBUSTION RAMJET-*(CONT.)*
     MISSILES
     RAMJET MISSILES
     ∞SCRAM

SUPERSONIC COMMERCIAL AIR TRANSPORT
UF   SCAT
GS   COMMERCIAL AIRCRAFT
- . **SUPERSONIC COMMERCIAL AIR TRANSPORT**
- . . BOEING 2707 AIRCRAFT
- . . TU-144 AIRCRAFT
     SUPERSONIC AIRCRAFT
- . SUPERSONIC TRANSPORTS
- . . **SUPERSONIC COMMERCIAL AIR TRANSPORT**
- . . . BOEING 2707 AIRCRAFT

SUPERSONIC COMPRESSORS
GS   COMPRESSORS
- . **SUPERSONIC COMPRESSORS**

RT   OBLIQUE SHOCK WAVES
     TRANSONIC COMPRESSORS
     TURBOCOMPRESSORS

SUPERSONIC CRUISE AIRCRAFT RESEARCH
UF   SCAR PROGRAM
GS   PROGRAMS
- . NASA PROGRAMS
- . . **SUPERSONIC CRUISE AIRCRAFT RESEARCH**

RT   ∞AIRCRAFT
     SUPERSONIC AIRCRAFT
     SUPERSONIC TRANSPORTS

SUPERSONIC DIFFUSERS
RT   AIR INTAKES
     ∞DIFFUSERS
     EXHAUST DIFFUSERS
     FLOW STABILITY
     VANELESS DIFFUSERS

SUPERSONIC DRAG
GS   AERODYNAMIC CHARACTERISTICS
- . AERODYNAMIC DRAG
- . . **SUPERSONIC DRAG**
     AERODYNAMIC FORCES
- . AERODYNAMIC DRAG
- . . **SUPERSONIC DRAG**
     DYNAMIC CHARACTERISTICS
- . DRAG
- . . FRICTION DRAG
- . . . AERODYNAMIC DRAG
- . . . . **SUPERSONIC DRAG**
- . PRESSURE DRAG
- . . **SUPERSONIC DRAG**
     FRICTION
- . AERODYNAMIC DRAG
- . . **SUPERSONIC DRAG**
- . FLOW RESISTANCE
- . . FRICTION DRAG
- . . . **SUPERSONIC DRAG**
- . SKIN FRICTION
- . . FRICTION DRAG
- . . . **SUPERSONIC DRAG**

RT   INTERFERENCE DRAG
     WAVE DRAG

SUPERSONIC FLIGHT
RT   CAUSTIC LINES
     ∞FLIGHT
     HYPERSONIC FLIGHT
     JET LAG
     MACH CONES
     MISSILES
     ROCKET FLIGHT
     SONIC BOOMS
     SUPERSONICS
     TRANSONIC FLIGHT

SUPERSONIC FLOW
GS   FLUID FLOW
- . **SUPERSONIC FLOW**

RT   AERODYNAMICS
     COMPRESSIBILITY EFFECTS
     COMPRESSIBLE FLOW
     FLOW VELOCITY
     GAS FLOW
     HYPERSONIC FLOW
     HYPERVELOCITY FLOW
     MACH CONES
     PRANDTL-MEYER EXPANSION
     SECONDARY INJECTION
     SHOCK WAVES
     TRANSONIC FLOW

SUPERSONIC FLOW-*(CONT.)*
     WEDGE FLOW
     WIND TUNNELS

SUPERSONIC FLOW INLETS
USE   SUPERSONIC INLETS

SUPERSONIC FLUTTER
GS   VIBRATION
- . STRUCTURAL VIBRATION
- . . FLUTTER
- . . . **SUPERSONIC FLUTTER**
- . SELF INDUCED VIBRATION
- . . . **SUPERSONIC FLUTTER**

RT   MISSILE VIBRATION
     TRANSONIC FLUTTER

SUPERSONIC HEAT TRANSFER
GS   TRANSMISSION
- . HEAT TRANSMISSION
- . . HEAT TRANSFER
- . . . AERODYNAMIC HEAT TRANSFER
- . . . . **SUPERSONIC HEAT TRANSFER**

RT   HYPERSONIC HEAT TRANSFER
     SUPERSONICS

SUPERSONIC INLETS
UF   SUPERSONIC FLOW INLETS
     TRANSONIC INLETS
GS   INTAKE SYSTEMS
- . AIR INTAKES
- . . **SUPERSONIC INLETS**

RT   BYPASS RATIO
     HYPERSONIC INLETS
     INLET AIRFRAME CONFIGURATIONS
     INLET FLOW
     INTERNAL COMPRESSION INLETS
     NOSE INLETS
     SIDE INLETS

SUPERSONIC JET FLOW
GS   FLUID FLOW
- . JET FLOW
- . . **SUPERSONIC JET FLOW**

RT   GAS FLOW
     NOZZLE FLOW

SUPERSONIC LOW ALTITUDE MISSILE
UF   SLAM
GS   MISSILES
- . RAMJET MISSILES
- . . **SUPERSONIC LOW ALTITUDE MISSILE**
- . SURFACE TO SURFACE MISSILES
- . . **SUPERSONIC LOW ALTITUDE MISSILE**

RT   NUCLEAR RAMJET ENGINES
     PLUTO REACTORS
     RAMJET ENGINES

SUPERSONIC NOZZLES
RT   COAXIAL NOZZLES
     CONICAL NOZZLES
     CONVERGENT-DIVERGENT NOZZLES
     HYPERSONIC NOZZLES
     ∞NOZZLES
     ROCKET NOZZLES
     SONIC NOZZLES
     TRANSONIC NOZZLES
     VARIABLE STREAM CONTROL ENGINES
     WIND TUNNEL NOZZLES

SUPERSONIC SPEEDS
GS   RATES (PER TIME)
- . **SUPERSONIC SPEEDS**
     VELOCITY
- . **SUPERSONIC SPEEDS**

RT   ACOUSTIC VELOCITY
     HIGH SPEED
     HYPERSONIC SPEED
     HYPERSONICS
     SUPERSONICS
     TRANSONIC SPEED

SUPERSONIC TEST APPARATUS
RT   HYPERSONIC TEST APPARATUS
     SUPERSONICS
     ∞TEST EQUIPMENT
     WIND TUNNEL APPARATUS

SUPERSONIC TRANSPORTS
GS   SUPERSONIC AIRCRAFT
- . **SUPERSONIC TRANSPORTS**
- . . CL-823 AIRCRAFT

SUPERSONIC TRANSPORTS-*(CONT.)*
- .. CONCORDE AIRCRAFT
- .. L-2000 AIRCRAFT
- .. SUPERSONIC COMMERCIAL AIR
    TRANSPORT
- ... BOEING 2707 AIRCRAFT
RT CARGO AIRCRAFT
     COMMERCIAL AIRCRAFT
     PASSENGER AIRCRAFT
     SUPERSONIC CRUISE AIRCRAFT
       RESEARCH

SUPERSONIC TURBINES
UF TRANSONIC TURBINES
GS TURBOMACHINERY
- . TURBINES
- .. **SUPERSONIC TURBINES**
RT GAS TURBINE ENGINES
     GAS TURBINES

SUPERSONIC WAKES
GS WAKES
- . **SUPERSONIC WAKES**
RT AIRCRAFT WAKES
     HYPERSONIC WAKES

SUPERSONIC WIND TUNNELS
SN (MACH 1 TO 5)
GS TEST FACILITIES
- . WIND TUNNELS
- .. **SUPERSONIC WIND TUNNELS**
RT BLOWDOWN WIND TUNNELS
     HYPERSONIC WIND TUNNELS
     HYPERVELOCITY WIND TUNNELS
     LOW DENSITY WIND TUNNELS
     SHOCK TUNNELS
     SLOTTED WIND TUNNELS
     SUBSONIC WIND TUNNELS
     SUPERHARMONICS
     TRANSONIC WIND TUNNELS

SUPERSONICS
GS FLUID MECHANICS
- . FLUID DYNAMICS
- .. GAS DYNAMICS
- ... AERODYNAMICS
- .... **SUPERSONICS**
RT AEROTHERMODYNAMICS
     HYPERSONICS
     MACH CONES
     SUPERHARMONICS
     SUPERSONIC AIRCRAFT
     SUPERSONIC BOUNDARY LAYERS
     SUPERSONIC FLIGHT
     SUPERSONIC HEAT TRANSFER
     SUPERSONIC SPEEDS
     SUPERSONIC TEST APPARATUS

SUPINE POSITION
RT ACCELERATION PROTECTION
     PRONE POSITION
     REST
     SITTING POSITION

SUPPLEMENTS
GS DOCUMENTS
- . **SUPPLEMENTS**
RT CONTRACTS
     EXTENSIONS
     INDEXES (DOCUMENTATION)
     MOTION PICTURES
     RECORDS
     REPORTS

SUPPLYING
RT COMMERCE
     CONSUMPTION
     DEMAND (ECONOMICS)
     FILLING
     INJECTION
     INPUT
     MARKETING
     OUTPUT

SUPPORT INTERFERENCE
RT ANTENNA RADIATION PATTERNS
     ∞INTERFERENCE
     SUPPORTS
     VIBRATION EFFECTS

SUPPORT SYSTEMS
GS **SUPPORT SYSTEMS**
- . GROUND OPERATIONAL SUPPORT
    SYSTEM

SUPPORT SYSTEMS-*(CONT.)*
- . GROUND SUPPORT SYSTEMS
- . LIFE SUPPORT SYSTEMS
- .. BIOPAKS
- .. CLOSED ECOLOGICAL SYSTEMS
- .. EMERGENCY LIFE SUSTAINING
    SYSTEMS
- ... AEPS
- .. PORTABLE LIFE SUPPORT SYSTEMS
- ... AEPS
RT SELF SEALING
     SERVICES
     ∞SYSTEMS

SUPPORTS
UF MOUNTS
     STANDS
GS **SUPPORTS**
- . PYLONS
- . SADDLES (SUPPORTS)
- . TRIPODS
RT BEARINGS
     CARRIAGES
     CHASSIS
     FOUNDATIONS
     FRAMES
     GIMBALS
     ∞HEADERS
     LUGS
     PIVOTS
     ∞PLATFORMS
     PYLON MOUNTING
     RACKS (FRAMES)
     REINFORCEMENT (STRUCTURES)
     SHAFTS (MACHINE ELEMENTS)
     STRUTS
     SUBSTRUCTURES
     SUPPORT INTERFERENCE
     ∞SUSTAINING
     TRUSSES

SUPPRESSION
USE RETARDING

SUPPRESSORS
GS **SUPPRESSORS**
- . ECHO SUPPRESSORS
RT ABSORBERS (MATERIALS)
     ADDITIVES
     ATTENUATORS
     BAFFLES
     CIRCUIT PROTECTION
     DAMPING
     INFRARED SUPPRESSION
     INHIBITORS
     INSULATION
     ISOLATORS
     MUFFLERS
     NEUTRALIZERS
     NOISE REDUCTION
     RETARDANTS
     SHIELDING
     SILENCERS
     SQUELCH CIRCUITS

SURFACE ACOUSTIC WAVE DEVICES
UF S-A-W DEVICES
RT ACOUSTIC DELAY LINES
     BULK ACOUSTIC WAVE DEVICES
     ∞DEVICES
     ELECTROACOUSTIC TRANSDUCERS
     INTERDIGITAL TRANSDUCERS
     SIGNAL PROCESSING
     SOUND WAVES
     ULTRASONIC WAVE TRANSDUCERS

SURFACE COOLING
GS COOLING
- . **SURFACE COOLING**
RT CONVECTIVE HEAT TRANSFER
     EVAPORATIVE COOLING
     FILM COOLING
     RADIANT COOLING
     RADIATIVE HEAT TRANSFER
     ∞SURFACES
     SWEAT COOLING
     TEMPERATURE

SURFACE CRACKS
UF CRAZING
GS CRACKS
- . **SURFACE CRACKS**
    SURFACE PROPERTIES
- . **SURFACE CRACKS**
RT CRACK CLOSURE

SURFACE CRACKS-*(CONT.)*
     CRACK GEOMETRY
     CRACK INITIATION
     CRACK PROPAGATION
     MICROCRACKS
     SOLID SURFACES
     ∞SURFACES

SURFACE DEFECTS
GS DEFECTS
- . **SURFACE DEFECTS**
    SURFACE PROPERTIES
- . **SURFACE DEFECTS**
RT CAUSTICS (OPTICS)
     CRACK INITIATION
     CRYSTAL DEFECTS
     CRYSTAL DISLOCATIONS
     FATIGUE (MATERIALS)
     MECHANICAL PROPERTIES
     POINT DEFECTS
     ∞SURFACES

SURFACE DIFFUSION
GS DIFFUSION
- . **SURFACE DIFFUSION**
RT MOLECULAR DIFFUSION
     ∞SURFACES
     THERMAL DIFFUSION

SURFACE DISTORTION
GS DISTORTION
- . **SURFACE DISTORTION**
RT LAMBERT SURFACE
     ∞SURFACE GEOMETRY
     ∞SURFACES
     WARPAGE

SURFACE EFFECT SHIPS
UF SES
GS SURFACE VEHICLES
- . **SURFACE EFFECT SHIPS**
    WATER VEHICLES
- . SHIPS
- .. **SURFACE EFFECT SHIPS**
RT CAPTURED AIR BUBBLE VEHICLES
     ∞EFFECTS
     RESEARCH VEHICLES
     ∞SURFACES
     SWATH (SHIP)
     ∞VEHICLES

SURFACE ENERGY
GS SURFACE PROPERTIES
- . **SURFACE ENERGY**
    THERMODYNAMIC PROPERTIES
- . **SURFACE ENERGY**
RT ACTIVATION ENERGY
     ELECTRON ENERGY
     ∞ENERGY
     INTERFACIAL ENERGY
     INTERFACIAL TENSION
     PROTON ENERGY
     ∞SURFACES
     THERMOPHYSICAL PROPERTIES

SURFACE FINISHING
UF SURFACE TREATMENT
RT CLEANING
     COATING
     COATINGS
     CORROSION PREVENTION
     CORROSION RESISTANCE
     ELECTROPLATING
     ELECTROPOLISHING
     FINISHES
     MACHINING
     METAL FINISHING
     METAL GRINDING
     METAL POLISHING
     METAL SPRAYING
     METAL SURFACES
     POLISHING
     PROTECTIVE COATINGS
     SHOT PEENING
     SOLID SURFACES
     SPUTTERING
     ∞SURFACES
     WEAR

∞ SURFACE GEOMETRY
SN *(USE OF A MORE SPECIFIC TERM IS*
    RECOMMENDED--CONSULT THE TERMS
    LISTED BELOW)
RT CONCAVITY
     CONVEXITY

SURFACE GEOMETRY-*(CONT.)*
      COSSERAT SURFACES
      FLAT SURFACES
      FLATNESS
      GEOMETRY
      LAMBERT SURFACE
      LOFTING
      PLANFORMS
      SHAPES
      SURFACE DISTORTION
      SURFACE LAYERS
      SURFACE PROPERTIES
      SURFACE REACTIONS
      SURFACE ROUGHNESS
      SURFACE STABILITY
    ∞SURFACES

SURFACE INTERACTIONS
  USE   SURFACE REACTIONS

SURFACE IONIZATION
  GS   IONIZATION
         . **SURFACE IONIZATION**
  RT   IONIZERS
         ∞SURFACES

SURFACE LAYERS
  GS   **SURFACE LAYERS**
         . MONOMOLECULAR FILMS
  RT   ATMOSPHERIC STRATIFICATION
         BARRIER LAYERS
         BOUNDARY LAYERS
         CRYSTAL SURFACES
         ∞LAYERS
         LUNAR SURFACE
         OXIDE FILMS
         SOLAR GRANULATION
         ∞SURFACE GEOMETRY
         ∞SURFACES
         THERMOCLINES
         ∞TRANSITION LAYERS

SURFACE NAVIGATION
  UF   MARINE NAVIGATION
  GS   NAVIGATION
         . **SURFACE NAVIGATION**
  RT   CELESTIAL NAVIGATION
         DEAD RECKONING
         DECCA NAVIGATION
         DIGITAL NAVIGATION
         HYPERBOLIC NAVIGATION
         INERTIAL NAVIGATION
         LORAC NAVIGATION SYSTEM
         LORAN
         NAUTICAL CHARTS
         NAVIGATION AIDS
         RADAR NAVIGATION
         RADIO NAVIGATION
         SHIPS
         ∞SURFACES

SURFACE NOISE INTERACTIONS
  RT   ACOUSTIC EXCITATION
         ACOUSTIC SCATTERING
         AEROACOUSTICS
         AERODYNAMIC NOISE
         TURBULENCE

SURFACE PRESSURE
  USE   PRESSURE

SURFACE PROPERTIES
  UF   BARDEEN APPROXIMATION
  GS   **SURFACE PROPERTIES**
         . ADHESION
         . ADSORPTIVITY
         . COEFFICIENT OF FRICTION
         . INTERFACIAL TENSION
         . SPECTRAL REFLECTANCE
         . SURFACE CRACKS
         . SURFACE DEFECTS
         . SURFACE ENERGY
         . SURFACE ROUGHNESS
         . SURFACE STABILITY
         . SURFACE TEMPERATURE
         . . SKIN TEMPERATURE
               (NON-BIOLOGICAL)
         . . WALL TEMPERATURE
  RT   ABSORPTANCE
         ALBEDO
         COARSENESS
         COATING
         COATINGS
         COLOR
         CONTACT POTENTIALS

SURFACE PROPERTIES-*(CONT.)*
      CONTACT RESISTANCE
      CORROSION
      COSSERAT SURFACES
      DIFFUSION
      EFFERVESCENCE
      EMISSIVITY
      EVANESCENCE
      FINISHES
      FLAT SURFACES
      FOAMING
      FRICTION
      HARDNESS
      HOT CORROSION
      INTERFACES
      JUPITER RED SPOT
      LUNAR ALBEDO
      LUNAR SURFACE
      LUNAR TOPOGRAPHY
      MECHANICAL PROPERTIES
      METAL SURFACES
      OPTICAL PROPERTIES
      PERMEABILITY
    ∞PHYSICAL PROPERTIES
      PLANAR STRUCTURES
      PLANETARY SURFACES
      PROFILOMETERS
    ∞PROPERTIES
      REFLECTANCE
      ROUGHNESS
      SELENOGRAPHY
      SOLID SURFACES
      SOLID-SOLID INTERFACES
      SORPTION
    ∞SURFACE GEOMETRY
    ∞SURFACES
      TEXTURES
      VISCOSITY
      VOID RATIO
      WETTABILITY

SURFACE REACTIONS
  UF   SURFACE INTERACTIONS
  RT   CHEMICAL REACTIONS
         EROSION
         FLUID-SOLID INTERACTIONS
         GAS-LIQUID INTERACTIONS
         INTERFACES
         METAL SURFACES
         METAL-WATER REACTIONS
         ∞SURFACE GEOMETRY
         ∞SURFACES
         SURFACTANTS
         VAPORIZING

SURFACE ROUGHNESS
  GS   ROUGHNESS
         . **SURFACE ROUGHNESS**
         SURFACE PROPERTIES
         . **SURFACE ROUGHNESS**
  RT   COARSENESS
         FRICTION
         LUNAR TOPOGRAPHY
         MACHINING
         MECHANICAL PROPERTIES
         PROFILOMETERS
         RUNWAY CONDITIONS
         ∞SURFACE GEOMETRY
         ∞SURFACES
         TOPOGRAPHY

SURFACE ROUGHNESS EFFECTS
  RT   ∞EFFECTS
         FRICTION DRAG
         REFLECTANCE
         SEPARATED FLOW
         SPECKLE PATTERNS
         ∞SURFACES

SURFACE STABILITY
  GS   STABILITY
         . **SURFACE STABILITY**
         SURFACE PROPERTIES
         . **SURFACE STABILITY**
  RT   COARSENESS
         DYNAMIC STABILITY
         INTERFACIAL TENSION
         MOTION STABILITY
         STATIC STABILITY
         STORAGE STABILITY
         ∞SURFACE GEOMETRY
         ∞SURFACES
         THERMAL STABILITY

SURFACE TEMPERATURE
  GS   SURFACE PROPERTIES
         . **SURFACE TEMPERATURE**
         . . SKIN TEMPERATURE
               (NON-BIOLOGICAL)
         . . WALL TEMPERATURE
         TEMPERATURE
         . **SURFACE TEMPERATURE**
         . . SKIN TEMPERATURE
               (NON-BIOLOGICAL)
         . . WALL TEMPERATURE
  RT   COARSENESS
         GEOTHERMAL ANOMALIES
         OCEAN TEMPERATURE
         SEA SURFACE TEMPERATURE
         ∞SURFACES
         THERMOCLINES
         WATER TEMPERATURE

SURFACE TENSION
  USE   INTERFACIAL TENSION

SURFACE TO AIR MISSILES
  UF   GROUND-TO-AIR MISSILES
  GS   MISSILES
         . **SURFACE TO AIR MISSILES**
         . . BLUE GOOSE MISSILE
         . . BOMARC MISSILES
         . . . BOMARC A MISSILE
         . . . BOMARC B MISSILE
         . . CHAPARRAL MISSILE
         . . HAWK MISSILE
         . . MAULER MISSILE
         . . NIKE MISSILES
         . . . NIKE-AJAX MISSILE
         . . . NIKE-HERCULES MISSILE
         . . . NIKE-ZEUS MISSILE
         . . PATRIOT MISSILE
         . . REDEYE MISSILE
         . . SPRINT MISSILE
         . . TALOS MISSILE
         . . TARTAR MISSILE
         . . TERRIER MISSILE
  RT   AIR TO AIR MISSILES
         AIR TO SURFACE MISSILES
         ANTIAIRCRAFT MISSILES
         ANTIMISSILE MISSILES
         NIKE X SYSTEMS
         RAMJET MISSILES
         ∞ROCKETS
         SENTINEL SYSTEM
         SPACE WEAPONS
         SPARTAN MISSILE
         ∞SURFACES

SURFACE TO SURFACE MISSILES
  GS   MISSILES
         . **SURFACE TO SURFACE MISSILES**
         . . ANTITANK MISSILES
         . . . SHILLELAGH MISSILES
         . . . TOW MISSILES
         . . CORPORAL MISSILE
         . . CRUISE MISSILES
         . . . NAVAHO MISSILE
         . . TOMAHAWK MISSILES
         . . FLEET BALLISTIC MISSILES
         . . . POLARIS A1 MISSILE
         . . . POLARIS A2 MISSILE
         . . . POLARIS A3 MISSILE
         . . POSEIDON MISSILES
         . . SUBROC MISSILE
         . . INTERCONTINENTAL BALLISTIC
               MISSILES
         . . . ATLAS ICBM
         . . . . ATLAS D ICBM
         . . . . ATLAS E ICBM
         . . . . ATLAS F ICBM
         . . . MINUTEMAN ICBM
         . . . MX MISSILE
         . . . TITAN ICBM
         . . . . TITAN 1 ICBM
         . . . . TITAN 2 ICBM
         . . INTERMEDIATE RANGE BALLISTIC
               MISSILES
         . . . BLUE STREAK MISSILE
         . . . JUPITER MISSILE
         . . POLARIS MISSILES
         . . . . POLARIS A1 MISSILE
         . . . . POLARIS A2 MISSILE
         . . . . POLARIS A3 MISSILE
         . . LANCE MISSILE
         . . MACE MISSILES
         . . PERSHING MISSILE
         . . REGULUS MISSILE
         . . SERGEANT MISSILES

SURFACE TO SURFACE MISSILES-_(CONT.)_
    . . SHORT RANGE BALLISTIC MISSILES
    . . SUPERSONIC LOW ALTITUDE MISSILE
    . . V-1 MISSILE
RT   AIR TO SURFACE MISSILES
    BALLISTIC MISSILES
    HARPOON MISSILE
    RAMJET MISSILES
    ∞ROCKETS
    ∞SURFACES

SURFACE TO SURFACE ROCKETS
GS   ROCKET VEHICLES
    . **SURFACE TO SURFACE ROCKETS**
    . . HONEST JOHN ROCKET VEHICLE
    . . LITTLE JOHN ROCKET VEHICLE
RT   ∞ROCKETS
    ∞SURFACES

SURFACE TREATMENT
USE   SURFACE FINISHING

SURFACE VEHICLES
GS   **SURFACE VEHICLES**
    . AIRCRAFT CARRIERS
    . AUTOMATED TRANSIT VEHICLES
    . . AUTOMATED GUIDEWAY TRANSIT
        VEHICLES
    . BOATS
    . . LIFEBOATS
    . CAPTURED AIR BUBBLE VEHICLES
    . CARGO SHIPS
    . . SAVANNAH NUCLEAR SHIP
    . . TANKER SHIPS
    . DOLLIES
    . ELECTRIC HYBRID VEHICLES
    . LUNAR SURFACE VEHICLES
    . . LUNAR MOBILE LABORATORIES
    . . LUNAR ROVING VEHICLES
    . . . LUNOKHOD LUNAR ROVING
        VEHICLES
    . . . MANNED LUNAR SURFACE
        VEHICLES
    . MAGNETIC LEVITATION VEHICLES
    . MOTOR VEHICLES
    . . AUTOMATED MIXED TRAFFIC
        VEHICLES
    . . AUTOMOBILES
    . . . ELECTRIC AUTOMOBILES
    . . ELECTRIC MOTOR VEHICLES
    . . TRACTORS
    . . . CRAWLER TRACTORS
    . . TRACKED VEHICLES
    . . TRUCKS
    . . . TANK TRUCKS
    . NUCLEAR POWERED SHIPS
    . . SAVANNAH NUCLEAR SHIP
    . ROADWAY POWERED VEHICLES
    . ROVING VEHICLES
    . . LUNAR ROVING VEHICLES
    . . . LUNOKHOD LUNAR ROVING
        VEHICLES
    . SATELLITE COMMUNICATIONS SHIPS
    . SLEDS
    . . ROCKET PROPELLED SLEDS
    . SURFACE EFFECT SHIPS
    . SWATH (SHIP)
    . TANKS (COMBAT VEHICLES)
    . TRANSPORTER
    . WALKING MACHINES
RT   AMPHIBIOUS VEHICLES
    ∞BICYCLE
    GROUND EFFECT MACHINES
    RAIL TRANSPORTATION
    RAILS
    SHIPS
    ∞SURFACES
    UNDERWATER VEHICLES
    URBAN TRANSPORTATION
    ∞VEHICLES
    VEHICULAR TRACKS
    WATER VEHICLES

SURFACE WATER
GS   WATER
    . **SURFACE WATER**
RT   EARTH RESOURCES
    GROUND WATER
    LAKES
    PONDS
    RIVERS
    STREAMS
    ∞SURFACES

SURFACE WAVES
SN   (EXCLUDES SURFACE RADIO WAVES)
GS   **SURFACE WAVES**
    . CAPILLARY WAVES
    . . GRAVITY WAVES
    . . . BAROCLINIC WAVES
    . . . RIPPLES
    . ELECTROMAGNETIC SURFACE WAVES
    . SOMMERFELD WAVES
RT   BOW WAVES
    CNOIDAL WAVES
    CRUSTAL FRACTURES
    ELASTIC WAVES
    INTERNAL WAVES
    LEE WAVES
    LIQUID SURFACES
    LOVE WAVES
    MICROSONICS
    P WAVES
    S WAVES
    SEA ROUGHNESS
    SEISMIC WAVES
    SPLASHING
    ∞SURFACES
    TROPOSPHERIC WAVES
    TSUNAMI WAVES
    WATER CURRENTS
    WATER WAVES
    ∞WAVES

∞ **SURFACES**
SN   _(USE OF A MORE SPECIFIC TERM IS_
    RECOMMENDED--CONSULT THE TERMS
    LISTED BELOW)
UF   CURVED SURFACES
    LIFTING SURFACES
RT   AIR TO SURFACE MISSILES
    AIRFIELD SURFACE MOVEMENTS
    AIRPORT SURFACE DETECTION
        EQUIPMENT
    APOLLO LUNAR SURFACE EXPERIMENTS
        PACKAGE
    AREA
    COLD SURFACES
    CONTROL SURFACES
    COSSERAT SURFACES
    CRYSTAL SURFACES
    EARTH SURFACE
    EASEP
    ELECTROMAGNETIC SURFACE WAVES
    ELEVATORS (CONTROL SURFACES)
    EXTERNAL SURFACE CURRENTS
    FERMI SURFACES
    FLAPS (CONTROL SURFACES)
    FLAT SURFACES
    HORIZONTAL TAIL SURFACES
    HOT SURFACES
    INTERFACES
    INTERFACIAL TENSION
    LAMBERT SURFACE
    LIQUID SURFACES
    LSSM
    LUNAR SURFACE
    LUNAR SURFACE VEHICLES
    MANNED LUNAR SURFACE VEHICLES
    MARS SURFACE
    MARS SURFACE SAMPLES
    MENISCI
    METAL SURFACES
    MINIMAL SURFACES
    OCEAN SURFACE
    PLANETARY SURFACES
    SATELLITE SURFACES
    SIZING (SURFACE TREATMENT)
    SOLID SURFACES
    SURFACE COOLING
    SURFACE CRACKS
    SURFACE DEFECTS
    SURFACE DIFFUSION
    SURFACE DISTORTION
    SURFACE EFFECT SHIPS
    SURFACE ENERGY
    SURFACE FINISHING
    ∞SURFACE GEOMETRY
    SURFACE IONIZATION
    SURFACE LAYERS
    SURFACE NAVIGATION
    SURFACE PROPERTIES
    SURFACE REACTIONS
    SURFACE ROUGHNESS
    SURFACE ROUGHNESS EFFECTS
    SURFACE STABILITY
    SURFACE TEMPERATURE
    SURFACE TO AIR MISSILES
    SURFACE TO SURFACE MISSILES
    SURFACE TO SURFACE ROCKETS

SURFACES-_(CONT.)_
    SURFACE VEHICLES
    SURFACE WATER
    SURFACE WAVES
    SWEPTBACK TAIL SURFACES
    T TAIL SURFACES
    TABS (CONTROL SURFACES)
    TAIL SURFACES
    TOWNSEND AVALANCHE
    TRAPEZOIDAL TAIL SURFACES
    TWO DIMENSIONAL BODIES
    UNDER SURFACE BLOWING
    UNDERWATER TO SURFACE MISSILES
    UPPER SURFACE BLOWING
    UPPER SURFACE BLOWN FLAPS
    VENUS SURFACE
    WEAR

SURFACTANTS
RT   ADMIXTURES
    ∞AGENTS
    DETERGENTS
    MONOMOLECULAR FILMS
    PLASTICIZERS
    RETARDANTS
    SOAPS
    SURFACE REACTIONS

SURGEONS
GS   PERSONNEL
    . MEDICAL PERSONNEL
    . . **SURGEONS**
    . . . FLIGHT SURGEONS

SURGERY
GS   **SURGERY**
    . LABYRINTHECTOMY
RT   CLINICAL MEDICINE
    HEART IMPLANTATION
    ∞OPERATIONS
    SKIN GRAFTS
    TRANSPLANTATION
    VETERINARY MEDICINE

SURGES
UF   TRANSIENTS (SURGES)
RT   CIRCUIT PROTECTION
    FLUID FLOW
    OVERVOLTAGE
    STORM SURGES
    VARIATIONS
    WATER HAMMER
    ∞WAVES

SURGICAL INSTRUMENTS
GS   MEDICAL EQUIPMENT
    . **SURGICAL INSTRUMENTS**
RT   ∞INSTRUMENTS
    NEEDLES

SURINAM
GS   NATIONS
    . **SURINAM**
RT   CARIBBEAN REGION
    NETHERLANDS
    SOUTH AMERICA

SURVEILLANCE
GS   **SURVEILLANCE**
    . SPACE SURVEILLANCE (GROUND
        BASED)
    . SPACE SURVEILLANCE (SPACEBORNE)
RT   COMMAND AND CONTROL
    CONICAL SCANNING
    CRIME
    DETECTION
    EARTH RESOURCES
    FOREST FIRE DETECTION
    ICE MAPPING
    ICE REPORTING
    INSPECTION
    INTELLIGENCE
    PANORAMIC SCANNING
    RADAR SCANNING
    RECONNAISSANCE
    SCANNING
    TARGET ACQUISITION
    TARGET RECOGNITION
    TARGETS

SURVEILLANCE RADAR
GS   RADAR
    . **SURVEILLANCE RADAR**
    . . AIRBORNE SURVEILLANCE RADAR
    . . COBRA DANE (RADAR)

SURVEILLANCE RADAR-*(CONT.)*
```
        . . MULTISTATIC RADAR
RT      AIR TRAFFIC CONTROL
        AIRPORT SURFACE DETECTION
           EQUIPMENT
        COHERENT RADAR
        CONTINUOUS WAVE RADAR
        DIGITAL RADAR SYSTEMS
        DOPPLER RADAR
        METEOROLOGICAL RADAR
        PULSE RADAR
        RADAR APPROACH CONTROL
        RADAR TRACKING
        RADARSCOPES
        SATELLITE-BORNE RADAR
        SEARCH RADAR
        SYNTHETIC APERTURE RADAR
        TRACKING RADAR
```

SURVEYING
```
USE     SURVEYS
```

SURVEYOR LUNAR PROBES
```
GS      LUNAR SPACECRAFT
        . LUNAR PROBES
        . . SURVEYOR LUNAR PROBES
        . . . SURVEYOR 1 LUNAR PROBE
        . . . SURVEYOR 2 LUNAR PROBE
        . . . SURVEYOR 3 LUNAR PROBE
        . . . SURVEYOR 4 LUNAR PROBE
        . . . SURVEYOR 5 LUNAR PROBE
        . . . SURVEYOR 6 LUNAR PROBE
        . . . SURVEYOR 7 LUNAR PROBE
        SOFT LANDING SPACECRAFT
        . SURVEYOR LUNAR PROBES
        . . SURVEYOR 1 LUNAR PROBE
        . . SURVEYOR 2 LUNAR PROBE
        . . SURVEYOR 3 LUNAR PROBE
        . . SURVEYOR 4 LUNAR PROBE
        . . SURVEYOR 5 LUNAR PROBE
        . . SURVEYOR 6 LUNAR PROBE
        . . SURVEYOR 7 LUNAR PROBE
        UNMANNED SPACECRAFT
        . SPACE PROBES
        . . LUNAR PROBES
        . . . SURVEYOR LUNAR PROBES
        . . . . SURVEYOR 1 LUNAR PROBE
        . . . . SURVEYOR 2 LUNAR PROBE
        . . . . SURVEYOR 3 LUNAR PROBE
        . . . . SURVEYOR 4 LUNAR PROBE
        . . . . SURVEYOR 5 LUNAR PROBE
        . . . . SURVEYOR 6 LUNAR PROBE
        . . . . SURVEYOR 7 LUNAR PROBE
```

SURVEYOR PROJECT
```
GS      PROGRAMS
        . LUNAR PROGRAMS
        . . SURVEYOR PROJECT
        . NASA PROGRAMS
        . . SURVEYOR PROJECT
        . NASA SPACE PROGRAMS
        . . SURVEYOR PROJECT
        . PROJECTS
        . . SURVEYOR PROJECT
RT      ATLAS CENTAUR LAUNCH VEHICLE
        CENTAUR PROJECT
        LUNAR LANDING
        LUNAR PROBES
        LUNAR SPACECRAFT
        SOFT LANDING
        SOFT LANDING SPACECRAFT
```

SURVEYOR 1 LUNAR PROBE
```
GS      LUNAR SPACECRAFT
        . LUNAR PROBES
        . . SURVEYOR LUNAR PROBES
        . . . SURVEYOR 1 LUNAR PROBE
        SOFT LANDING SPACECRAFT
        . SURVEYOR LUNAR PROBES
        . . SURVEYOR 1 LUNAR PROBE
        UNMANNED SPACECRAFT
        . SPACE PROBES
        . . LUNAR PROBES
        . . . SURVEYOR LUNAR PROBES
        . . . . SURVEYOR 1 LUNAR PROBE
RT      ATLAS CENTAUR LAUNCH VEHICLE
```

SURVEYOR 2 LUNAR PROBE
```
GS      LUNAR SPACECRAFT
        . LUNAR PROBES
        . . SURVEYOR LUNAR PROBES
        . . . SURVEYOR 2 LUNAR PROBE
        SOFT LANDING SPACECRAFT
        . SURVEYOR LUNAR PROBES
        . . SURVEYOR 2 LUNAR PROBE
```

SURVEYOR 2 LUNAR PROBE-*(CONT.)*
```
        UNMANNED SPACECRAFT
        . SPACE PROBES
        . . LUNAR PROBES
        . . . SURVEYOR LUNAR PROBES
        . . . . SURVEYOR 2 LUNAR PROBE
RT      ATLAS CENTAUR LAUNCH VEHICLE
```

SURVEYOR 3 LUNAR PROBE
```
GS      LUNAR SPACECRAFT
        . LUNAR PROBES
        . . SURVEYOR LUNAR PROBES
        . . . SURVEYOR 3 LUNAR PROBE
        SOFT LANDING SPACECRAFT
        . SURVEYOR LUNAR PROBES
        . . SURVEYOR 3 LUNAR PROBE
        UNMANNED SPACECRAFT
        . SPACE PROBES
        . . LUNAR PROBES
        . . . SURVEYOR LUNAR PROBES
        . . . . SURVEYOR 3 LUNAR PROBE
RT      ATLAS CENTAUR LAUNCH VEHICLE
```

SURVEYOR 4 LUNAR PROBE
```
GS      LUNAR SPACECRAFT
        . LUNAR PROBES
        . . SURVEYOR LUNAR PROBES
        . . . SURVEYOR 4 LUNAR PROBE
        SOFT LANDING SPACECRAFT
        . SURVEYOR LUNAR PROBES
        . . SURVEYOR 4 LUNAR PROBE
        UNMANNED SPACECRAFT
        . SPACE PROBES
        . . LUNAR PROBES
        . . . SURVEYOR LUNAR PROBES
        . . . . SURVEYOR 4 LUNAR PROBE
RT      ATLAS CENTAUR LAUNCH VEHICLE
```

SURVEYOR 5 LUNAR PROBE
```
GS      LUNAR SPACECRAFT
        . LUNAR PROBES
        . . SURVEYOR LUNAR PROBES
        . . . SURVEYOR 5 LUNAR PROBE
        SOFT LANDING SPACECRAFT
        . SURVEYOR LUNAR PROBES
        . . SURVEYOR 5 LUNAR PROBE
        UNMANNED SPACECRAFT
        . SPACE PROBES
        . . LUNAR PROBES
        . . . SURVEYOR LUNAR PROBES
        . . . . SURVEYOR 5 LUNAR PROBE
RT      ATLAS CENTAUR LAUNCH VEHICLE
```

SURVEYOR 6 LUNAR PROBE
```
GS      LUNAR SPACECRAFT
        . LUNAR PROBES
        . . SURVEYOR LUNAR PROBES
        . . . SURVEYOR 6 LUNAR PROBE
        SOFT LANDING SPACECRAFT
        . SURVEYOR LUNAR PROBES
        . . SURVEYOR 6 LUNAR PROBE
        UNMANNED SPACECRAFT
        . SPACE PROBES
        . . LUNAR PROBES
        . . . SURVEYOR LUNAR PROBES
        . . . . SURVEYOR 6 LUNAR PROBE
RT      ATLAS CENTAUR LAUNCH VEHICLE
```

SURVEYOR 7 LUNAR PROBE
```
GS      LUNAR SPACECRAFT
        . LUNAR PROBES
        . . SURVEYOR LUNAR PROBES
        . . . SURVEYOR 7 LUNAR PROBE
        SOFT LANDING SPACECRAFT
        . SURVEYOR LUNAR PROBES
        . . SURVEYOR 7 LUNAR PROBE
        UNMANNED SPACECRAFT
        . SPACE PROBES
        . . LUNAR PROBES
        . . . SURVEYOR LUNAR PROBES
        . . . . SURVEYOR 7 LUNAR PROBE
RT      ATLAS CENTAUR LAUNCH VEHICLE
```

SURVEYS
```
UF      SURVEYING
GS      SURVEYS
        . GEODETIC SURVEYS
        . GEOLOGICAL SURVEYS
        . WAGE SURVEYS
RT      ACCURACY
        CONSTRUCTION
        ∞ CROSS SECTIONS
        DATA ACQUISITION
        DATA MANAGEMENT
        DATUM (ELEVATION)
```

SURVEYS-*(CONT.)*
```
        EXPLORATION
        GEOMETRY
        LAYOUTS
        LORAN
        MAPPING
        MAPS
        PHOTOGRAMMETRY
        POSITION (LOCATION)
        RECONNAISSANCE
        SOIL MAPPING
        ∞ STATISTICS
```

SURVIVAL
```
RT      AIRCRAFT SURVIVABILITY
        CIVIL DEFENSE
        CLOSED ECOLOGICAL SYSTEMS
        DESERT ADAPTATION
        KITS
        LIFE SUPPORT SYSTEMS
        LUNAR SHELTERS
        SHELTERS
        SPACECRAFT SURVIVABILITY
```

SURVIVAL EQUIPMENT
```
RT      AEPS
        AIRCRAFT SURVIVABILITY
        CONSUMABLES (SPACECREW SUPPLIES)
        EMERGENCY LIFE SUSTAINING
           SYSTEMS
        ∞ EQUIPMENT
        LIFEBOATS
        ONBOARD EQUIPMENT
        OXYGEN SUPPLY EQUIPMENT
        RAFTS
```

SUSCEPTIBILITY (MAGNETISM)
```
USE     MAGNETIC PERMEABILITY
```

SUSPENDING (HANGING)
```
GS      SUSPENDING (HANGING)
        . MAGNETIC SUSPENSION
RT      GYROSCOPE FLUIDS
        MOUNTING
        SUSPENSION SYSTEMS (VEHICLES)
        ∞ SUSPENSIONS
```

SUSPENDING (MIXING)
```
GS      MIXING
        . SUSPENDING (MIXING)
RT      AERATION
        AGITATION
   .    COLLOIDING
        DISPERSING
        DISPERSIONS
        ENTRAINMENT
        FERROFLUIDS
        HOMOGENIZING
        SHAKING
        STIRRING
        ∞ SUSPENSIONS
```

SUSPENSION SYSTEMS (VEHICLES)
```
RT      BEARINGS
        FLOTATION
        LEVITATION
        MAGNETIC LEVITATION VEHICLES
        RIDING QUALITY
        SHOCK ABSORBERS
        SPRINGS (ELASTIC)
        STEERING
        SUSPENDING (HANGING)
        ∞ SUSPENSIONS
        ∞ SYSTEMS
        TOROIDAL WHEELS
        UNDERCARRIAGES
        VEHICLE WHEELS
        VEHICULAR TRACKS
        VIBRATION ISOLATORS
```

∞ SUSPENSIONS
```
SN      (USE OF A MORE SPECIFIC TERM IS
        RECOMMENDED--CONSULT THE TERMS
        LISTED BELOW)
RT      BROWNIAN MOVEMENTS
        DISPERSIONS
        FERROFLUIDS
        SOLID SUSPENSIONS
        SUSPENDING (HANGING)
        SUSPENDING (MIXING)
        SUSPENSION SYSTEMS (VEHICLES)
```

SUSQUEHANNA RIVER BASIN (MD-NY-PA)
```
GS      LANDFORMS
        . STRUCTURAL BASINS
```

SUSQUEHANNA RIVER BASIN-*(CONT.)*
  . . RIVER BASINS
  . . . **SUSQUEHANNA RIVER BASIN**
      (MD-NY-PA)
  RT   MARYLAND
      NEW YORK
      PENNSYLVANIA
      RIVERS
      STREAMS
      VALLEYS

SUSTAINER ROCKET ENGINES
  GS   ENGINES
     . ROCKET ENGINES
     . . **SUSTAINER ROCKET ENGINES**
  RT   BOOSTER ROCKET ENGINES
      DUCTED ROCKET ENGINES
      ELECTRIC ROCKET ENGINES
      ELECTROSTATIC ENGINES
      ELECTROTHERMAL ENGINES
      HYBRID PROPELLANT ROCKET ENGINES
      INTERNAL COMBUSTION ENGINES
      ION ENGINES
      LAUNCH VEHICLES
      LIQUID AIR CYCLE ENGINES
      LIQUID PROPELLANT ROCKET ENGINES
      NUCLEAR ENGINE FOR ROCKET
        VEHICLES
      NUCLEAR ROCKET ENGINES
      RESTARTABLE ROCKET ENGINES
      SOLID PROPELLANT ROCKET ENGINES
      STAGE SEPARATION
      ∞SUSTAINING
      TURBOROCKET ENGINES
      TX-354 ENGINE

∞ **SUSTAINING**
  SN   *(USE OF A MORE SPECIFIC TERM IS*
      RECOMMENDED--CONSULT THE TERMS
      LISTED BELOW)
  RT   LIFE SUPPORT SYSTEMS
      SUPPORTS
      SUSTAINER ROCKET ENGINES

SWAGING
  RT   COLD WORKING
      METAL WORKING
      STAMPING

SWALLOWING
  RT   DRINKING
      EATING
      INGESTION (BIOLOGY)

SWAMPS
  USE MARSHLANDS

SWAN BANDS
  GS   SPECTRA
     . SPECTRAL BANDS
     . . **SWAN BANDS**
  RT   ∞BANDS
      CARBON COMPOUNDS
      CHEMICAL BONDS
      EMISSION SPECTRA
      MOLECULAR SPECTRA

SWARMING
  RT   BEES
      ∞MOTION

SWASH
  USE SPLASHING

SWATH (SHIP)
  UF   SMALL WATER PLANE AREA TWIN HULL
  GS   SURFACE VEHICLES
     . **SWATH (SHIP)**
     WATER VEHICLES
     . SHIPS
     . . **SWATH (SHIP)**
  RT   CAPTURED AIR BUBBLE VEHICLES
      HULLS (STRUCTURES)
      SURFACE EFFECT SHIPS
      ∞VEHICLES

SWATH WIDTH
  RT   AGRICULTURAL AIRCRAFT
      FLIGHT PATHS
      REMOTE SENSING
      SATELLITE OBSERVATION

SWAZILAND
  GS   NATIONS

SWAZILAND-*(CONT.)*
     . **SWAZILAND**
  RT   AFRICA
      REPUBLIC OF SOUTH AFRICA

SWEAT
  GS   BODY FLUIDS
     . **SWEAT**
     SECRETIONS
     . **SWEAT**
  RT   MILIARIA
      PERSPIRATION

SWEAT COOLING
  UF   TRANSPIRATION COOLING
  GS   COOLING
     . EVAPORATIVE COOLING
     . . **SWEAT COOLING**
  RT   FILM COOLING
      LIQUID COOLING
      SURFACE COOLING

SWEATING
  USE PERSPIRATION

SWEDEN
  GS   NATIONS
     . **SWEDEN**
  RT   EUROPE
      SCANDINAVIA
      SWEDISH SPACE PROGRAM

SWEDISH SPACE PROGRAM
  GS   PROGRAMS
     . SPACE PROGRAMS
     . . **SWEDISH SPACE PROGRAM**
  RT   EUROPEAN SPACE PROGRAMS
      SWEDEN

SWEEP ANGLE
  GS   GEOMETRY
     . EUCLIDEAN GEOMETRY
     . . ANGLES (GEOMETRY)
     . . . **SWEEP ANGLE**
     . . . . SWEEPBACK
     . . . . . LEADING EDGE SWEEP
  RT   AERODYNAMIC STALLING
      ANGLE OF ATTACK
      BOUNDARY LAYER SEPARATION
      MACH NUMBER

SWEEP CIRCUITS
  GS   CIRCUITS
     . **SWEEP CIRCUITS**
  RT   FREQUENCY SCANNING
      OSCILLOSCOPES
      SAMPLING

SWEEP EFFECT
  RT   ∞EFFECTS
      FORCE DISTRIBUTION
      LIFT
      ∞LOADING
      WING LOADING

SWEEP FREQUENCY
  UF   ELECTRON SWEEPING
  GS   FREQUENCIES
     . **SWEEP FREQUENCY**
  RT   CARRIER FREQUENCIES
      FREQUENCY ANALYZERS
      FREQUENCY SCANNING
      FREQUENCY SYNCHRONIZATION
      OSCILLOSCOPES
      TELEVISION TRANSMISSION

SWEEPBACK
  UF   SWEEPBACK ANGLES
  GS   GEOMETRY
     . EUCLIDEAN GEOMETRY
     . . ANGLES (GEOMETRY)
     . . . SWEEP ANGLE
     . . . . **SWEEPBACK**
     . . . . . LEADING EDGE SWEEP
  RT   SUPERSONIC AIRFOILS

SWEEPBACK ANGLES
  USE SWEEPBACK

SWELLING
  RT   DISTORTION
      EXPANSION
      GROWTH
      INCREASING

SWELLING-*(CONT.)*
      INFLATING
      SPREADING

SWEPT FORWARD WINGS
  GS   AIRFOILS
     . WINGS
     . . SWEPT WINGS
     . . . **SWEPT FORWARD WINGS**
     . . . . TRAPEZOIDAL WINGS
     PLANFORMS
     . WING PLANFORMS
     . . **SWEPT FORWARD WINGS**
     . . . TRAPEZOIDAL WINGS
  RT   SWEPTBACK WINGS
      VARIABLE SWEEP WINGS
      X-29 AIRCRAFT

SWEPT WINGS
  UF   CRANKED WINGS
      DIAMOND WINGS
      TAPERED WINGS
  GS   AIRFOILS
     . WINGS
     . . **SWEPT WINGS**
     . . . SWEPT FORWARD WINGS
     . . . . TRAPEZOIDAL WINGS
     . . . SWEPTBACK WINGS
     . . . . ARROW WINGS
     . . . . DELTA WINGS
     . . . . TRAPEZOIDAL WINGS
  RT   A-300 AIRCRAFT
      A-310 AIRCRAFT
      A-320 AIRCRAFT
      FIXED WINGS
      SPANLOADER AIRCRAFT
      UNSWEPT WINGS
      WING PLANFORMS

SWEPTBACK TAIL SURFACES
  GS   PLANFORMS
     . **SWEPTBACK TAIL SURFACES**
     TAIL SURFACES
     . **SWEPTBACK TAIL SURFACES**
  RT   CONTROL SURFACES
      HYPERSONIC AIRCRAFT
      RUDDERS
      STABILIZERS (FLUID DYNAMICS)
      SUPERSONIC AIRFOILS
      ∞SURFACES
      T TAIL SURFACES
      TRAPEZOIDAL TAIL SURFACES

SWEPTBACK WINGS
  GS   AIRFOILS
     . WINGS
     . . SWEPT WINGS
     . . . **SWEPTBACK WINGS**
     . . . . ARROW WINGS
     . . . . DELTA WINGS
     . . . . TRAPEZOIDAL WINGS
     PLANFORMS
     . WING PLANFORMS
     . . **SWEPTBACK WINGS**
     . . . ARROW WINGS
     . . . DELTA WINGS
     . . . TRAPEZOIDAL WINGS
  RT   HYPERSONIC AIRCRAFT
      SUPERSONIC AIRCRAFT
      SUPERSONIC AIRFOILS
      SWEPT FORWARD WINGS
      VARIABLE SWEEP WINGS

SWIMMING
  RT   PHYSICAL EXERCISE
      PHYSICAL FITNESS

SWIMMING POOL REACTORS
  GS   NUCLEAR REACTORS
     . LIQUID COOLED REACTORS
     . . WATER COOLED REACTORS
     . . . **SWIMMING POOL REACTORS**
  RT   ∞REACTORS

SWINE
  SN   (EXCLUDES GUINEA PIGS)
  UF   PIGS (SWINE)
  GS   ANIMALS
     . VERTEBRATES
     . . MAMMALS
     . . . **SWINE**
  RT   GRAZING
      LIVESTOCK

SWING TAIL ASSEMBLIES
GS   ASSEMBLIES
     . TAIL ASSEMBLIES
     . . **SWING TAIL ASSEMBLIES**
RT   AFTERBODIES
     AIRCRAFT PARTS
     AIRCRAFT STRUCTURES

SWING WINGS
GS   AIRFOILS
     . WINGS
     . . **SWING WINGS**
RT   AIRCRAFT PARTS
     AIRCRAFT STRUCTURES
     WING PLANFORMS
     WING PROFILES

SWINGBY TECHNIQUE
RT   FLYBY MISSIONS
     GRAVITATIONAL EFFECTS
     INTERPLANETARY TRANSFER ORBITS
     ORBITAL MECHANICS
     PLANETARY ORBITS
     ROUND TRIP TRAJECTORIES

SWIRLING
RT   AGITATION
     CENTRIFUGING
     DISPERSING
     FOAMING
     MIXING
     ∞SEPARATION
     SHAKING
     STIRRING

SWIRLING WAKES
USE   TURBULENT WAKES

SWISS SPACE PROGRAM
GS   PROGRAMS
     . SPACE PROGRAMS
     . . **SWISS SPACE PROGRAM**
RT   EUROPEAN SPACE PROGRAMS
     SWITZERLAND

SWITCHES
GS   **SWITCHES**
     . CAPACITANCE SWITCHES
     . ELECTRIC RELAYS
     . ELECTRIC SWITCHES
     . . CRYOTRONS
     . . STEPPING SWITCHES
     . . THERMOSTATS
     . . VACUUM ARC SWITCHES
     . PRESSURE SWITCHES
     . SWITCHING CIRCUITS
     . . FLUID SWITCHING ELEMENTS
     . TRIGATRONS
RT   CIRCUIT BREAKERS
     DROPOUTS
     ECHO SUPPRESSORS
     ELECTRIC CONNECTORS
     ELECTRIC CONTACTS
     INTERRUPTION
     SELECTORS
     SWITCHING

SWITCHING
GS   **SWITCHING**
     . BEAM SWITCHING
     . MAGNETIC SWITCHING
     . MICROWAVE SWITCHING
     . PACKET TRANSMISSION
     . . PACKET SWITCHING
RT   CODE DIVISION MULTIPLE ACCESS
     INTERRUPTION
     SEQUENCING
     STEP RECOVERY DIODES
     SWITCHES
     TIME DIVISION MULTIPLE ACCESS

SWITCHING CIRCUITS
UF   ELECTRONIC SWITCHES
     SWITCHING ELEMENTS
GS   CIRCUITS
     . **SWITCHING CIRCUITS**
     . . FLUID SWITCHING ELEMENTS
     SWITCHES
     . **SWITCHING CIRCUITS**
     . . FLUID SWITCHING ELEMENTS
RT   ARPA COMPUTER NETWORK
     CAPACITANCE SWITCHES
     CIRCUIT BREAKERS
     CURRENT REGULATORS
     DUPLEX OPERATION

SWITCHING CIRCUITS-(CONT.)
     DUPLEXERS
     ELECTRIC RELAYS
     ELECTRIC SWITCHES
     GATES (CIRCUITS)
     LATCH-UP
     LOGIC CIRCUITS
     MATRICES (CIRCUITS)
     MICROWAVE SWITCHING
     MULTIVIBRATORS
     OPTICAL BISTABILITY
     PACKET SWITCHING
     SELECTORS
     SQUELCH CIRCUITS
     VACUUM ARC SWITCHES
     VOLTAGE REGULATORS

SWITCHING ELEMENTS
USE   SWITCHING CIRCUITS

SWITCHING THEORY
RT   BOOLEAN ALGEBRA
     BRANCHING (MATHEMATICS)
     COMMUNICATION THEORY
     COMMUTATION
     LOGIC DESIGN
     NETWORK SYNTHESIS
     PACKET SWITCHING
     SEQUENCING
     ∞THEORIES
     TOPOLOGY

SWITZERLAND
GS   NATIONS
     . **SWITZERLAND**
RT   ALPS MOUNTAINS (EUROPE)
     EUROPE
     SWISS SPACE PROGRAM

SWIVELS
RT   BEARINGS
     GIMBALS
     HINGES
     HOOKS
     JOINTS (JUNCTIONS)
     PIVOTS

SYENITE
GS   ROCKS
     . IGNEOUS ROCKS
     . . **SYENITE**
RT   SOILS
     TRACHYTE

SYLLABLES
GS   COMMUNICATION THEORY
     . WORDS (LANGUAGE)
     . . **SYLLABLES**
     LANGUAGES
     . SENTENCES
     . . WORDS (LANGUAGE)
     . . . **SYLLABLES**
     LINGUISTICS
     . SYNTAX
     . . SENTENCES
     . . . WORDS (LANGUAGE)
     . . . . **SYLLABLES**
     SIGNAL RECEPTION
     . **SYLLABLES**
     SPEECH
     . TALKING
     . . WORDS (LANGUAGE)
     . . . **SYLLABLES**
RT   MESSAGES
     PSYCHOLINGUISTICS
     SEMANTICS
     SIGNAL TRANSMISSION

SYMBIOSIS
RT   ECOLOGY
     LICHENS

SYMBIOTIC STARS
GS   CELESTIAL BODIES
     . STARS
     . . HOT STARS
     . . . BLUE STARS
     . . . . **SYMBIOTIC STARS**
     . . PECULIAR STARS
     . . . **SYMBIOTIC STARS**
     . . VARIABLE STARS
     . . . **SYMBIOTIC STARS**
RT   ABSORPTION SPECTRA
     ECLIPSING BINARY STARS
     EMISSION SPECTRA

SYMBIOTIC STARS-(CONT.)
     FLARE STARS
     M STARS
     NOVAE
     STELLAR ENVELOPES
     STELLAR MASS ACCRETION
     STELLAR OSCILLATIONS
     STELLAR SPECTRA
     STELLAR TEMPERATURE

SYMBOLIC PROGRAMMING
GS   SOFTWARE ENGINEERING
     . COMPUTER PROGRAMMING
     . . **SYMBOLIC PROGRAMMING**
RT   CODING
     COMPUTER ASSISTED INSTRUCTION
     CONTEXT FREE LANGUAGES
     LANGUAGE PROGRAMMING
     MNEMONICS

SYMBOLS
UF   CHARACTERS
     LETTERS (SYMBOLS)
     SIGNS (SYMBOLS)
GS   SIGNAL RECEPTION
     . **SYMBOLS**
RT   ALPHABETS
     ALPHANUMERIC CHARACTERS
     CHARACTER RECOGNITION
     ∞CODES
     CODING
     COLOR
     DATA PROCESSING
     DIGITS
     HIGH LEVEL LANGUAGES
     LANGUAGES
     LEGIBILITY
     ∞MATHEMATICS
     MESSAGE PROCESSING
     MESSAGES
     MNEMONICS
     NOMENCLATURES
     PERCEPTION
     READING
     SEMANTICS
     UNITS OF MEASUREMENT
     VISIBILITY

SYMMETRICAL BODIES
GS   **SYMMETRICAL BODIES**
     . AXISYMMETRIC BODIES
     . BODIES OF REVOLUTION
     . . CONICAL BODIES
     . . . SLENDER CONES
     . . CYLINDRICAL BODIES
     . . . ROTATING CYLINDERS
     . . PARABOLIC BODIES
     . . SPHERES
     . . . CELESTIAL SPHERE
     . . . CONCENTRIC SPHERES
     . . . FALLING SPHERES
     . . . POINCARE SPHERES
     . . . ROTATING SPHERES
     . . TORUSES
     . ELLIPSOIDS
     . LENTICULAR BODIES
     . STREAMLINED BODIES
     . . FAIRINGS
RT   AXES OF ROTATION
     BLUNT BODIES
     ∞BODIES
     CONES
     FINNED BODIES
     FLARED BODIES
     GEOIDS
     OGIVES
     SLENDER BODIES
     SPINNING UNGUIDED ROCKET
       TRAJECTORY

SYMMETRY
UF   AXISYMMETRY
GS   **SYMMETRY**
     . BROKEN SYMMETRY
RT   ANTISYMMETRY
     ASYMMETRY
     CONGRUENCES
     CONTINUITY (MATHEMATICS)
     ECCENTRICITY
     GEOMETRY
     ISOTROPISM
     QUANTILES
     SHAPES

SYMMETRY BREAKING
USE   BROKEN SYMMETRY

SYMPATHETIC NERVOUS SYSTEM
GS   NERVOUS SYSTEM
. AUTONOMIC NERVOUS SYSTEM
. . **SYMPATHETIC NERVOUS SYSTEM**
RT   ∞SYSTEMS

SYMPATHOMIMETICS
USE   ADRENERGICS

SYMPHONIE SATELLITES
GS   SATELLITES
. ARTIFICIAL SATELLITES
. . COMMUNICATION SATELLITES
. . . **SYMPHONIE SATELLITES**
RT   ARCOMSAT
BROADCASTING
EUROPEAN SPACE PROGRAMS
FRENCH SATELLITES
INTERNATIONAL COOPERATION
RADIO TRANSMISSION
SATELLITE TELEVISION
SYNCHRONOUS SATELLITES
TELEPHONY

SYMPTOMOLOGY
GS   MEDICAL SCIENCE
. **SYMPTOMOLOGY**
RT   DISEASES
SIGNS AND SYMPTOMS

SYMPTOMS
USE   SIGNS AND SYMPTOMS

SYNAPSES
GS   NERVOUS SYSTEM
. **SYNAPSES**
RT   NEUROMUSCULAR TRANSMISSION
SYNCODERS

SYNCHROCYCLOTRONS
GS   PARTICLE ACCELERATORS
. CYCLIC ACCELERATORS
. . **SYNCHROCYCLOTRONS**
. CYCLOTRONS
. . **SYNCHROCYCLOTRONS**
RT   BEVATRON
SYNCHROTRONS

SYNCHRONISM
UF   BEAT
SYNCHRONIZATION
GS   **SYNCHRONISM**
. BIT SYNCHRONIZATION
. FREQUENCY SYNCHRONIZATION
RT   COINCIDENCE CIRCUITS
DINING PHILOSOPHERS PROBLEM
PHASE DETECTORS
STROBOSCOPES
SYNCHRONIZERS
SYNCHROPHASING
TIME
TIME MEASUREMENT

SYNCHRONIZATION
USE   SYNCHRONISM

SYNCHRONIZED OSCILLATORS
GS   OSCILLATORS
. **SYNCHRONIZED OSCILLATORS**
RT   FREQUENCY SYNCHRONIZATION
PHASE LOCKED SYSTEMS
SYNCHROSCOPES

SYNCHRONIZERS
RT   HELIOSTATS
PULSE RADAR
SERVOMOTORS
SYNCHRONISM

SYNCHRONOUS COMMUNICATION SATELLITES
USE   SYNCOM SATELLITES

SYNCHRONOUS COMMUNICATIONS SATELLITE PROJ
SN   IONS SATELLITE PROJECT)
GS   PROGRAMS
. NASA PROGRAMS
. . **SYNCHRONOUS COMMUNICATIONS SATELLITE PROJ**
. NASA SPACE PROGRAMS

SYNCHRONOUS COMMUNICATIONS-*(CONT.)*
. . **SYNCHRONOUS COMMUNICATIONS SATELLITE PROJ**
. PROJECTS
. . **SYNCHRONOUS COMMUNICATIONS SATELLITE PROJ**
RT   COMMUNICATION SATELLITES
SATELLITES
TWENTY-FOUR HOUR ORBITS

SYNCHRONOUS DETECTORS
USE   CORRELATORS

SYNCHRONOUS EARTH OBSERVATORY SATELLITE
UF   SEOS
GS   SATELLITES
. ARTIFICIAL SATELLITES
. . METEOROLOGICAL SATELLITES
. . . **SYNCHRONOUS EARTH OBSERVATORY SATELLITE**
. . . . SMS 1
. . . . SMS 2
. . SYNCHRONOUS SATELLITES
. . . **SYNCHRONOUS EARTH OBSERVATORY SATELLITE**
. . . . SMS 1
. . . . SMS 2
RT   EARLY WARNING SYSTEMS
LANDSAT SATELLITES
NASA PROGRAMS
PROGRAMS
SATELLITE OBSERVATION
SEASAT SATELLITES
SYNCHRONOUS METEOROLOGICAL SATELLITE
TECHNOLOGY UTILIZATION

SYNCHRONOUS METEOROLOGICAL SATELLITE
UF   SMS
GS   SATELLITES
. ARTIFICIAL SATELLITES
. . METEOROLOGICAL SATELLITES
. . . **SYNCHRONOUS METEOROLOGICAL SATELLITE**
. . . . SMS 1
. . . . SMS 2
. . SYNCHRONOUS SATELLITES
. . . **SYNCHRONOUS METEOROLOGICAL SATELLITE**
. . . . SMS 1
. . . . SMS 2
RT   COMMUNICATION SATELLITES
SYNCHRONOUS EARTH OBSERVATORY SATELLITE

SYNCHRONOUS MOTORS
GS   MOTORS
. ELECTRIC MOTORS
. . **SYNCHRONOUS MOTORS**
RT   ASYNCHRONOUS MOTORS
INDUCTION MOTORS

SYNCHRONOUS PLATFORMS
UF   GEOSTATIONARY PLATFORMS
GS   SPACE PLATFORMS
. **SYNCHRONOUS PLATFORMS**
RT   COMMUNICATION SATELLITES
GEOSYNCHRONOUS ORBITS
∞PLATFORMS

SYNCHRONOUS SATELLITES
UF   GEOSTATIONARY SATELLITES
GS   SATELLITES
. ARTIFICIAL SATELLITES
. . **SYNCHRONOUS SATELLITES**
. . . AEROS SATELLITE
. . . AEROSAT SATELLITES
. . . ANIK SATELLITES
. . . . ANIK 1
. . . . ANIK 2
. . . . ANIK 3
. . . GOES SATELLITES
. . . . GOES 1
. . . . GOES 2
. . . . GOES 3
. . . . GOES 4
. . . . GOES 5
. . . MIRANDA SATELLITE
. . . SIRIO SATELLITE
. . . STORMSAT SATELLITE
. . . SYNCHRONOUS EARTH OBSERVATORY SATELLITE
. . . . SMS 1
. . . . SMS 2

SYNCHRONOUS SATELLITES-*(CONT.)*
. . . SYNCHRONOUS METEOROLOGICAL SATELLITE
. . . . SMS 1
. . . . SMS 2
. . . SYNCOM SATELLITES
. . . . EARLY BIRD SATELLITES
. . . . SYNCOM 1 SATELLITE
. . . . SYNCOM 2 SATELLITE
. . . . SYNCOM 3 SATELLITE
. . . TD SATELLITES
. . . . TD-1 SATELLITE
RT   ACTIVE SATELLITES
ARCOMSAT
CANADIAN SPACE PROGRAM
COMMUNICATION SATELLITES
COMMUNICATIONS TECHNOLOGY SATELLITE
MILITARY SPACECRAFT
NAVIGATION SATELLITES
PASSIVE SATELLITES
REFSAT
STATIONARY ORBITS
SYMPHONIE SATELLITES
TWENTY-FOUR HOUR ORBITS

SYNCHROPHASING
RT   AIRCRAFT NOISE
NOISE REDUCTION
PROPELLER BLADES
SYNCHRONISM

SYNCHROPHASOTRONS
GS   PARTICLE ACCELERATORS
. **SYNCHROPHASOTRONS**
RT   ∞ACCELERATORS
SYNCHROTRONS

SYNCHROSCOPES
GS   CIRCUITS
. PHASE DETECTORS
. . **SYNCHROSCOPES**
RT   CORRELATORS
MEASURING INSTRUMENTS
OSCILLOSCOPES
SYNCHRONIZED OSCILLATORS

SYNCHROTRON RADIATION
GS   ELECTROMAGNETIC RADIATION
. POLARIZED ELECTROMAGNETIC RADIATION
. . **SYNCHROTRON RADIATION**
POLARIZED RADIATION
. POLARIZED ELECTROMAGNETIC RADIATION
. . **SYNCHROTRON RADIATION**
RT   BREMSSTRAHLUNG
EXTRATERRESTRIAL RADIATION
∞RADIATION
RADIATION PROTECTION
SYNCHROTRONS
X RAYS

SYNCHROTRONS
GS   PARTICLE ACCELERATORS
. CYCLIC ACCELERATORS
. . **SYNCHROTRONS**
. . . BEVATRON
. . . STORAGE RINGS (PARTICLE ACCELERATORS)
RT   BETATRONS
CYCLOTRONS
ELECTRON ACCELERATORS
ION ACCELERATORS
MICROTRONS
SYNCHROCYCLOTRONS
SYNCHROPHASOTRONS
SYNCHROTRON RADIATION

SYNCLINES
UF   SYNCLINORIA
RT   ANTICLINES
DOMES (GEOLOGY)
GEOLOGICAL FAULTS
GEOSYNCLINES
∞LAYERS
STRATA
STRATIFICATION
STRATIGRAPHY

SYNCLINORIA
USE   SYNCLINES

SYNCODERS
RT   BIONICS

SYNCODERS-*(CONT.)*
    NEURONS
    SYNAPSES

SYNCOM APOGEE ENGINES
  GS   ENGINES
      . ROCKET ENGINES
      . . SOLID PROPELLANT ROCKET
         ENGINES
      . . . **SYNCOM APOGEE ENGINES**
      . . VERNIER ENGINES
      . . . **SYNCOM APOGEE ENGINES**
      . TORPEDO ENGINES
      . . VERNIER ENGINES
      . . . **SYNCOM APOGEE ENGINES**

SYNCOM SATELLITES
  UF   SYNCHRONOUS COMMUNICATION
       SATELLITES
  GS   SATELLITES
      . ACTIVE SATELLITES
      . . **SYNCOM SATELLITES**
      . . . EARLY BIRD SATELLITES
      . . . SYNCOM 1 SATELLITE
      . . . SYNCOM 2 SATELLITE
      . . . SYNCOM 3 SATELLITE
      . ARTIFICIAL SATELLITES
      . . COMMUNICATION SATELLITES
      . . . **SYNCOM SATELLITES**
      . . . . EARLY BIRD SATELLITES
      . . . . SYNCOM 1 SATELLITE
      . . . . SYNCOM 2 SATELLITE
      . . . . SYNCOM 3 SATELLITE
      . . . . SYNCOM 4 SATELLITE
      . . SYNCHRONOUS SATELLITES
      . . . **SYNCOM SATELLITES**
      . . . . EARLY BIRD SATELLITES
      . . . . SYNCOM 1 SATELLITE
      . . . . SYNCOM 2 SATELLITE
      . . . . SYNCOM 3 SATELLITE
  RT   THOR DELTA LAUNCH VEHICLE

SYNCOM 1 SATELLITE
  GS   SATELLITES
      . ACTIVE SATELLITES
      . . SYNCOM SATELLITES
      . . . **SYNCOM 1 SATELLITE**
      . ARTIFICIAL SATELLITES
      . . COMMUNICATION SATELLITES
      . . . SYNCOM SATELLITES
      . . . . **SYNCOM 1 SATELLITE**
      . . SYNCHRONOUS SATELLITES
      . . . SYNCOM SATELLITES
      . . . . **SYNCOM 1 SATELLITE**
  RT   DELTA LAUNCH VEHICLE

SYNCOM 2 SATELLITE
  GS   SATELLITES
      . ACTIVE SATELLITES
      . . SYNCOM SATELLITES
      . . . **SYNCOM 2 SATELLITE**
      . ARTIFICIAL SATELLITES
      . . COMMUNICATION SATELLITES
      . . . SYNCOM SATELLITES
      . . . . **SYNCOM 2 SATELLITE**
      . . SYNCHRONOUS SATELLITES
      . . . SYNCOM SATELLITES
      . . . . **SYNCOM 2 SATELLITE**
  RT   DELTA LAUNCH VEHICLE

SYNCOM 3 SATELLITE
  GS   SATELLITES
      . ACTIVE SATELLITES
      . . SYNCOM SATELLITES
      . . . **SYNCOM 3 SATELLITE**
      . ARTIFICIAL SATELLITES
      . . COMMUNICATION SATELLITES
      . . . SYNCOM SATELLITES
      . . . . **SYNCOM 3 SATELLITE**
      . . SYNCHRONOUS SATELLITES
      . . . SYNCOM SATELLITES
      . . . . **SYNCOM 3 SATELLITE**
  RT   DELTA LAUNCH VEHICLE

SYNCOM 4 SATELLITE
  GS   SATELLITES
      . ARTIFICIAL SATELLITES
      . . COMMUNICATION SATELLITES
      . . . SYNCOM SATELLITES
      . . . . **SYNCOM 4 SATELLITE**
  RT   SPACE TRANSPORTATION SYSTEM

SYNCOPE
  UF   FAINTING
  GS   **SYNCOPE**

SYNCOPE-*(CONT.)*
    . BLACKOUT (PHYSIOLOGY)
    . . BLACKOUT PREVENTION
  RT   UNCONSCIOUSNESS

SYNDROMES
  USE SIGNS AND SYMPTOMS

SYNOPTIC MEASUREMENT
  RT   ∞MEASUREMENT
       NEPHANALYSIS

SYNOPTIC METEOROLOGY
  GS   METEOROLOGY
      . **SYNOPTIC METEOROLOGY**
  RT   AIR MASSES
       ANTICYCLONES
       COLD FRONTS
       CYCLONES
       FRONTS (METEOROLOGY)
       METEOROLOGICAL CHARTS
       NEPHANALYSIS
       WARM FRONTS
       WEATHER FORECASTING

SYNTAX
  GS   LINGUISTICS
      . **SYNTAX**
      . . SENTENCES
      . . . WORDS (LANGUAGE)
      . . . . SYLLABLES
  RT   FORMAT
       GRAMMARS
       ∞INTERPRETATION
       LANGUAGES
       ORTHOGRAPHY
       PARSING ALGORITHMS
       PSYCHOLINGUISTICS
       SEMANTICS
       SPEECH

SYNTECTIC ALLOYS
  GS   ALLOYS
      . **SYNTECTIC ALLOYS**
  RT   EUTECTICS
       LIQUID PHASES
       METALS
       MIXTURES
       PHASE TRANSFORMATIONS
       SOLID PHASES

SYNTHANE
  UF   SYNTHETIC METHANE
  GS   FUELS
      . CHEMICAL FUELS
      . . HYDROCARBON FUELS
      . . . **SYNTHANE**
      . . SYNTHETIC FUELS
      . . . **SYNTHANE**
  RT   AUTOMOBILE FUELS
       CARBON DIOXIDE
       CARBON MONOXIDE
       COAL
       COAL GASIFICATION
       GASIFICATION
       HYDROGEN
       LIGNITE
       METHANE

∞ **SYNTHESIS**
  SN   *(USE OF A MORE SPECIFIC TERM IS*
       RECOMMENDED--CONSULT THE TERMS
       LISTED BELOW)
  RT   BIOSYNTHESIS
       CHEMICAL REACTIONS
       DECISION THEORY
       ∞DESIGN
       NETWORK SYNTHESIS
       NUCLEAR FUSION
       OPERATIONS RESEARCH
       PLASMA JET SYNTHESIS
       SYNTHESIS (CHEMISTRY)
       SYNTHETIC FUELS
       SYSTEMS ENGINEERING

SYNTHESIS (CHEMISTRY)
  RT   ADDITION RESINS
       CHEMICAL REACTIONS
       ∞CHEMISTRY
       FISCHER-TROPSCH PROCESS
       OPERATIONS RESEARCH
       ∞SYNTHESIS
       SYNTHETIC FIBERS
       SYNTHETIC FUELS
       SYNTHETIC RESINS

SYNTHESIS (CHEMISTRY)-*(CONT.)*
    SYNTHETIC RUBBERS
    SYSTEMS ENGINEERING

SYNTHESIZERS
  RT   CHEMICAL REACTORS
       FREQUENCY SYNTHESIZERS

SYNTHETIC APERTURE RADAR
  GS   RADAR
      . **SYNTHETIC APERTURE RADAR**
      . . SIDE-LOOKING RADAR
  RT   EARTHNET
       IMAGING RADAR
       MICROWAVE IMAGERY
       MICROWAVE SENSORS
       RADAR EQUIPMENT
       RADARSAT
       SATELLITE-BORNE RADAR
       SHUTTLE IMAGING RADAR
       SURVEILLANCE RADAR
       SYNTHETIC APERTURES
       VENUS ORBITING IMAGING RADAR
         (SPACECRAFT)

SYNTHETIC APERTURES
  GS   OPENINGS
      . APERTURES
      . . **SYNTHETIC APERTURES**
  RT   IMAGING TECHNIQUES
       SYNTHETIC APERTURE RADAR

SYNTHETIC ARRAYS
  GS   ARRAYS
      . **SYNTHETIC ARRAYS**
  RT   ANTENNA RADIATION PATTERNS
       APERTURES
       DISTRIBUTION (PROPERTY)
       EARTH RESOURCES SHUTTLE IMAGING
         RADAR
       ∞PATTERNS

SYNTHETIC FIBERS
  GS   FIBERS
      . **SYNTHETIC FIBERS**
      . . DACRON (TRADEMARK)
      . . FORTISAN (TRADEMARK)
      . . GLASS FIBERS
      . . NYLON (TRADEMARK)
      . . RAYON
      . . VYCOR
  RT   ADDITION RESINS
       FLAME RETARDANTS
       KEVLAR (TRADEMARK)
       POLYBENZIMIDAZOLE
       POLYESTERS
       REINFORCING FIBERS
       SYNTHESIS (CHEMISTRY)
       WET SPINNING

SYNTHETIC FOOD
  RT   AMINO ACIDS
       BIOSYNTHESIS
       CARBOHYDRATES
       CELLULOSE
       EATING
       FATS
       ∞FOOD
       FOOD INTAKE
       NUTRITIONAL REQUIREMENTS
       PROTEIN METABOLISM
       PROTEINS
       TASTE

SYNTHETIC FUELS
  GS   FUELS
      . CHEMICAL FUELS
      . . **SYNTHETIC FUELS**
      . . . GASOHOL (FUEL)
      . . . . SYNTHANE
  RT   CHEMICAL REACTIONS
       CLEAN FUELS
       FISCHER-TROPSCH PROCESS
       HYDROCARBON FUELS
       LIQUID FUELS
       ∞SYNTHESIS
       SYNTHESIS (CHEMISTRY)

SYNTHETIC METALS
  RT   CRYSTAL LATTICES
       GRAPHITE
       ORGANOMETALLIC COMPOUNDS

SYNTHETIC METHANE
USE   SYNTHANE

SYNTHETIC RESINS
GS   PLASTICS
. **SYNTHETIC RESINS**
. . ADDITION RESINS
. . . ACRYLIC RESINS
. . . VINYL COPOLYMERS
. . POLYESTER RESINS
. . POLYETHER RESINS
. . . POLYMETHYL METHACRYLATE
. . THERMOPLASTIC RESINS
. . . QUINOXALINES
. . . THERMOPLASTIC FILMS
. . THERMOSETTING RESINS
. . EPOXY RESINS
. . . FURAN RESINS
. . . . POLYAMIDE RESINS
. . . . . KEVLAR (TRADEMARK)
. . . PHENOLIC RESINS
. . . . MICARTA
RESINS
. **SYNTHETIC RESINS**
. . ADDITION RESINS
. . . VINYL COPOLYMERS
. . POLYESTER RESINS
. . POLYETHER RESINS
. . . POLYMETHYL METHACRYLATE
. . THERMOPLASTIC RESINS
. . QUINOXALINES
. . . THERMOPLASTIC FILMS
. . THERMOSETTING RESINS
. . . EPOXY RESINS
. . FURAN RESINS
. . . POLYAMIDE RESINS
. . . . . KEVLAR (TRADEMARK)
. . PHENOLIC RESINS
. . . MICARTA
. . . . PHENOLIC EPOXY RESINS
RT   POLYETHYLENES
∞POLYMERS
POLYPROPYLENE
POLYSTYRENE
POLYTETRAFLUOROETHYLENE
POLYVINYL ALCOHOL
POLYVINYL CHLORIDE
SYNTHESIS (CHEMISTRY)
TEFLON (TRADEMARK)

SYNTHETIC RUBBERS
GS   RUBBER
. **SYNTHETIC RUBBERS**
. . BUNA (TRADEMARK)
. . ELASTOMERS
. . . CHLOROPRENE RESINS
. . . THIOPLASTICS
. . . VITON RUBBER (TRADEMARK)
. . . VULCANIZED ELASTOMERS
. . RTV-40 RUBBER (TRADEMARK)
. . RTV-60 RUBBER (TRADEMARK)
RT   LATEX
POLYBUTADIENE
POLYISOBUTYLENE
POLYISOPRENES
SILICONE RUBBER
SOLITHANES
SYNTHESIS (CHEMISTRY)

SYNTONY
RT   FREQUENCY SYNCHRONIZATION
OSCILLATIONS
RESONANCE

SYPHILIS
GS   DISEASES
. INFECTIOUS DISEASES
. . **SYPHILIS**

SYRIA
GS   NATIONS
. **SYRIA**
RT   ASIA

SYRINGES
GS   LABORATORY EQUIPMENT
. **SYRINGES**
MEDICAL EQUIPMENT
. **SYRINGES**
RT   BULBS
∞EQUIPMENT
FLUID FLOW
PIPES (TUBES)
TRANSFUSION

SYSTEM EFFECTIVENESS
GS   EFFECTIVENESS
. **SYSTEM EFFECTIVENESS**
RT   MODULATION TRANSFER FUNCTION
OPTICAL TRANSFER FUNCTION
RELIABILITY
RELIABILITY ENGINEERING
∞SYSTEMS
SYSTEMS ENGINEERING
SYSTEMS INTEGRATION

SYSTEM FAILURES
GS   FAILURE
. **SYSTEM FAILURES**
RT   ∞BREAKDOWN
DETERIORATION
DOWNTIME
FATIGUE (MATERIALS)
MALFUNCTIONS
SHORT CIRCUITS
STRUCTURAL FAILURE
STRUCTURAL STRAIN
∞SYSTEMS
WEAR

**SYSTEM GENERATED ELECTROMAGNETIC
PULSES**
UF   SGEMP
GS   ELECTROMAGNETIC FIELDS
. **SYSTEM GENERATED
ELECTROMAGNETIC PULSES**
ELECTROMAGNETIC RADIATION
. ELECTROMAGNETIC PULSES
. . **SYSTEM GENERATED
ELECTROMAGNETIC PULSES**
PULSED RADIATION
. ELECTROMAGNETIC PULSES
. . **SYSTEM GENERATED
ELECTROMAGNETIC PULSES**
PULSES
. ELECTROMAGNETIC PULSES
. . **SYSTEM GENERATED
ELECTROMAGNETIC PULSES**
RT   ELECTRIC CURRENT
ELECTRIC PULSES
ELECTROMAGNETIC INTERFERENCE
ELECTRONIC EQUIPMENT
EXTERNAL SURFACE CURRENTS
EXTRATERRESTRIAL RADIATION
IONIZING RADIATION
PLASMA SHEATHS
SPACECRAFT CHARGING
SPACECRAFT COMMUNICATION
X RAYS

SYSTEM IDENTIFICATION
GS   ESTIMATING
. **SYSTEM IDENTIFICATION**
IDENTIFYING
. **SYSTEM IDENTIFICATION**
SYSTEMS ANALYSIS
. **SYSTEM IDENTIFICATION**
RT   COMPLEX SYSTEMS
CONTROL SYSTEMS DESIGN
DYNAMIC RESPONSE
ESTIMATES
FUZZY SYSTEMS
MATHEMATICAL MODELS
MAXIMUM LIKELIHOOD ESTIMATES
OBSERVABILITY (SYSTEMS)
OPTIMIZATION
PARAMETER IDENTIFICATION
PARAMETERIZATION
PREDICTION ANALYSIS TECHNIQUES
PROBABILITY THEORY
RELIABILITY ENGINEERING
STATISTICAL ANALYSIS
STEEPEST DESCENT METHOD
∞SYSTEMS
SYSTEMS ENGINEERING

SYSTEM 10 COMPUTER
USE   PDP 10 COMPUTER

∞ **SYSTEMS**
SN   *(USE OF A MORE SPECIFIC TERM IS
RECOMMENDED--CONSULT THE TERMS
LISTED BELOW)*
RT   ADVANCED VIDICON CAMERA SYSTEM
(AVCS)
AEROSPACE SYSTEMS
AFFERENT NERVOUS SYSTEMS
AFRICAN RIFT SYSTEM
AIR CUSHION LANDING SYSTEMS

SYSTEMS-*(CONT.)*
AIRBORNE INTEGRATED
RECONNAISSANCE SYSTEM
AIRCRAFT FUEL SYSTEMS
AIRCRAFT HYDRAULIC SYSTEMS
ALL-WEATHER LANDING SYSTEMS
ALOHA SYSTEM
ANNULAR SUSPENSION AND POINTING
SYSTEM
APOLLO EXTENSION SYSTEM
ASCENT PROPULSION SYSTEMS
ASTROGUIDE NAVIGATION SYSTEM
ATMOSPHERIC & OCEANOGRAPHIC
INFORM SYS
AUTOMATED PILOT ADVISORY SYSTEM
AUTOMATED RADAR TERMINAL SYSTEM
AUTOMATIC TRAFFIC ADVISORY AND
RESOLUTION
AUTONOMIC NERVOUS SYSTEM
BALLISTIC MISSILE EARLY WARNING
SYSTEM
BEACON COLLISION AVOIDANCE
SYSTEM
BINARY SYSTEMS (MATERIALS)
BIOASTRONAUTICAL ORBITAL SPACE
SYSTEM
BIOCONTROL SYSTEMS
CARDIOVASCULAR SYSTEM
CELESTIAL REFERENCE SYSTEMS
CENTRAL ELECTRONIC MANAGEMENT
SYSTEM
CENTRAL NERVOUS SYSTEM
CENTRAL NERVOUS SYSTEM
DEPRESSANTS
CENTRAL NERVOUS SYSTEM
STIMULANTS
CHOKES (FUEL SYSTEMS)
CLOSED ECOLOGICAL SYSTEMS
COMPLEX SYSTEMS
COMPUTER SYSTEMS DESIGN
COMPUTER SYSTEMS PERFORMANCE
COMPUTER SYSTEMS PROGRAMS
COMPUTER SYSTEMS SIMULATION
COOLING SYSTEMS
CYBERNETICS
DATA BASE MANAGEMENT SYSTEMS
DATA SYSTEMS
DEFENSE COMMUNICATIONS SATELLITE
SYSTEM
DEFENSE COMMUNICATIONS SYSTEM
(DCS)
DESCENT PROPULSION SYSTEMS
DIGESTIVE SYSTEM
DIGITAL COMMAND SYSTEMS
DIGITAL SYSTEMS
DISCRETE ADDRESS BEACON SYSTEM
DISPLAY DEVICES
DISTRIBUTED PARAMETER SYSTEMS
DOMESTIC SATELLITE COMMUNICATIONS
SYSTEMS
EARLY WARNING SYSTEMS
EARTH RESOURCES INFORMATION
SYSTEM
EARTH TERMINAL MEASUREMENT
SYSTEM
EARTH-MOON SYSTEM
ECOSYSTEMS
EFFERENT NERVOUS SYSTEMS
EISCAT RADAR SYSTEM (EUROPE)
∞ELASTIC SYSTEMS
ELECTRONIC RECORDING SYSTEMS
EMERGENCY LIFE SUSTAINING
SYSTEMS
END-TO-END DATA SYSTEMS
ENDOCRINE SYSTEMS
ESCAPE SYSTEMS
EXHAUST SYSTEMS
FAIL-SAFE SYSTEMS
FEED SYSTEMS
FEEDBACK
FLEET SATELLITE COMMUNICATION
SYSTEM
FUEL SYSTEMS
FUZZY SYSTEMS
GASTROINTESTINAL SYSTEM
GENITOURINARY SYSTEM
GLOBAL POSITIONING SYSTEM
GODDARD TRAJECTORY
DETERMINATION SYSTEM
GROUND OPERATIONAL SUPPORT
SYSTEM
GROUND SUPPORT SYSTEMS
GUIDANCE (MOTION)
HARDENING (SYSTEMS)
HEMATOPOIETIC SYSTEM
HYBRID NAVIGATION SYSTEMS

SYSTEMS-*(CONT.)*
HYDRAULIC EQUIPMENT
HYDROPLANES (SURFACES)
HYDROTHERMAL SYSTEMS
∞HYPERBOLIC SYSTEMS
IFF SYSTEMS (IDENTIFICATION)
IGNITION SYSTEMS
IMLSS
INERTIAL REFERENCE SYSTEMS
INFORMATION ADAPTIVE SYSTEM
INFORMATION SYSTEMS
INSTRUMENT LANDING SYSTEMS
INTAKE SYSTEMS
INTEGRATED ENERGY SYSTEMS
INTEGRATED GLOBAL OCEAN STATION
SYSTEMS
INTERNATIONAL SYSTEM OF UNITS
INTRAVASCULAR SYSTEM
JETTISON SYSTEMS
LAUNCH ESCAPE SYSTEMS
LIFE SUPPORT SYSTEMS
LIGHT AIRBORNE MULTIPURPOSE
SYSTEM
LINEAR SYSTEMS
LOCATES SYSTEM
LORAC NAVIGATION SYSTEM
LUBRICATION SYSTEMS
LUMPED PARAMETER SYSTEMS
LUNAR EXPLORATION SYSTEM FOR
APOLLO
MAN MACHINE SYSTEMS
MAN OPERATED PROPULSION SYSTEMS
MANAGEMENT INFORMATION SYSTEMS
MANAGEMENT SYSTEMS
METAL-GAS SYSTEMS
METHOXY SYSTEMS
MICROWAVE LANDING SYSTEMS
MICROWAVE SCANNING BEAM LANDING
SYSTEM
MINITRACK SYSTEM
MIROS SYSTEM
MISSILE SYSTEMS
MODULAR INTEGRATED UTILITY SYSTEM
MUSCULOSKELETAL SYSTEM
NASA INTERACTIVE PLANNING SYSTEM
NATIONAL AIRSPACE UTILIZATION
SYSTEM
NATIONAL AVIATION SYSTEM
NATIONAL OCEANIC SATELLITE SYSTEM
NAVIGATION
NEEDS (DATA SYSTEM)
NERVOUS SYSTEM
NIKE X SYSTEMS
NOESS
NONLINEAR SYSTEMS
NOVA LASER SYSTEM
OBSERVABILITY (SYSTEMS)
OMEGA NAVIGATION SYSTEM
ON-LINE SYSTEMS
OPERATING SYSTEMS (COMPUTERS)
OPTICAL RELAY SYSTEMS
PAYLOAD DEPLOYMENT & RETRIEVAL
SYSTEM
PERIPHERAL NERVOUS SYSTEM
PHASE LOCKED SYSTEMS
PIGGYBACK SYSTEMS
PLAT SYSTEM
PNEUMATIC EQUIPMENT
POINTING CONTROL SYSTEMS
POLYSTATION DOPPLER TRACKING
SYSTEM
PORTABLE LIFE SUPPORT SYSTEMS
POST BOOST PROPULSION SYSTEM
PROPULSION SYSTEM CONFIGURATIONS
PROPULSION SYSTEM PERFORMANCE
PUBLIC ADDRESS SYSTEMS
QUALITY CONTROL
RADIO RELAY SYSTEMS
RANGER BLOCK 3 TELEVISION SYSTEM
RAPID TRANSIT SYSTEMS
∞REFERENCE SYSTEMS
REMOTE MANIPULATOR SYSTEM
REPRODUCTIVE SYSTEM
RESPIRATORY SYSTEM
ROTOR SYSTEMS RESEARCH AIRCRAFT
SAFEGUARD SYSTEM
SAGE AIR DEFENSE SYSTEM
SATELLITE NAVIGATION SYSTEMS
SELF ADAPTIVE CONTROL SYSTEMS
SELF ORGANIZING SYSTEMS
SENTINEL SYSTEM
SHIVA LASER SYSTEM
SNAP
SOLAR SYSTEM
SOLAR TOTAL ENERGY SYSTEMS

SYSTEMS-*(CONT.)*
SPACE DETECTION AND TRACKING
SYSTEM
SPACE TRANSPORTATION SYSTEM
SPACE TRANSPORTATION SYSTEM
FLIGHTS
SUNFLOWER POWER SYSTEM
SUPPORT SYSTEMS
SUSPENSION SYSTEMS (VEHICLES)
SYMPATHETIC NERVOUS SYSTEM
SYSTEM EFFECTIVENESS
SYSTEM FAILURES
SYSTEM IDENTIFICATION
SYSTEMS ANALYSIS
SYSTEMS ENGINEERING
SYSTEMS INTEGRATION
SYSTEMS MANAGEMENT
SYSTEMS SIMULATION
SYSTEMS STABILITY
TELECOMMUNICATION
TELEGRAPH SYSTEMS
TELETYPEWRITER SYSTEMS
TELEVISION SYSTEMS
TERCOM
TERNARY SYSTEMS
TIROS OPERATIONAL SATELLITE
SYSTEM
TOTAL ENERGY SYSTEMS
TRADEX RADAR SYSTEM
TRANSCONTINENTAL SYSTEMS
TRANSFER FUNCTIONS
TRANSOCEANIC SYSTEMS
TYPHON WEAPON SYSTEM
VACUUM SYSTEMS
VARIABLE MASS SYSTEMS
VASCULAR SYSTEM
VORTEX ADVISORY SYSTEM
WARNING SYSTEMS
WEAPON SYSTEM MANAGEMENT
WEAPON SYSTEMS

SYSTEMS ANALYSIS
GS   **SYSTEMS ANALYSIS**
.   SYSTEM IDENTIFICATION
RT   ∞ANALYZING
BLOCK DIAGRAMS
BOND GRAPHS
COMPLEX SYSTEMS
COMPUTER PROGRAMMING
COMPUTER SYSTEMS PROGRAMS
COMPUTER SYSTEMS SIMULATION
CONTROL SYSTEMS DESIGN
FEASIBILITY ANALYSIS
FUZZY SYSTEMS
MAN MACHINE SYSTEMS
MATHEMATICAL MODELS
MODULATION TRANSFER FUNCTION
OBSERVABILITY (SYSTEMS)
OPERATING COSTS
OPERATIONS RESEARCH
OPTICAL TRANSFER FUNCTION
PARAMETER IDENTIFICATION
PREFLIGHT ANALYSIS
PROCEDURES
PROGRAM VERIFICATION (COMPUTERS)
SIMULATION
SOCIOLOGY
STATISTICAL ANALYSIS
∞SYSTEMS
TRAJECTORY ANALYSIS
WEIGHT ANALYSIS
WEIGHT REDUCTION

SYSTEMS COMPATIBILITY
GS   COMPATIBILITY
.   **SYSTEMS COMPATIBILITY**
RT   RELIABILITY
RELIABILITY ENGINEERING

SYSTEMS DESIGN
USE   SYSTEMS ENGINEERING

SYSTEMS ENGINEERING
UF   SYSTEMS DESIGN
GS   **SYSTEMS ENGINEERING**
.   COMPUTER SYSTEMS DESIGN
.   CONTROL SYSTEMS DESIGN
RT   AEROSPACE SYSTEMS
AIRCRAFT DESIGN
∞AUTOMATION
BIONICS
BOND GRAPHS
COMMUNICATING
CONTRACT MANAGEMENT
∞CONTROL

SYSTEMS ENGINEERING-*(CONT.)*
CRITICAL PATH METHOD
CYBERNETICS
DATA PROCESSING
DECISION MAKING
DECISION THEORY
∞DESIGN
ELECTRICAL ENGINEERING
∞ENGINEERING
EXPERIMENT DESIGN
FLIGHT MANAGEMENT SYSTEMS
FORECASTING
FUNCTIONAL DESIGN SPECIFICATIONS
HUMAN FACTORS ENGINEERING
INFORMATION THEORY
LIFE CYCLE COSTS
MAN MACHINE SYSTEMS
MANAGEMENT
MANAGEMENT PLANNING
MATHEMATICAL MODELS
MECHANIZATION
MISSILE DESIGN
MODULARITY
OBSERVABILITY (SYSTEMS)
∞OPERATIONS
OPERATIONS RESEARCH
OPTICAL TRANSFER FUNCTION
ORBIT SPECTRUM UTILIZATION
PARAMETER IDENTIFICATION
RELIABILITY
RELIABILITY ENGINEERING
RESEARCH AND DEVELOPMENT
SATELLITE DESIGN
SOFTWARE ENGINEERING
SPACECRAFT DESIGN
STATISTICAL ANALYSIS
∞STATISTICS
∞SYNTHESIS
SYNTHESIS (CHEMISTRY)
SYSTEM EFFECTIVENESS
SYSTEM IDENTIFICATION
SYSTEMS INTEGRATION

SYSTEMS FOR NUCLEAR AUXILIARY POWER
USE   SNAP

SYSTEMS INTEGRATION
RT   AIRBORNE/SPACEBORNE COMPUTERS
AVIONICS
CONTROL SYSTEMS DESIGN
DIGITAL SYSTEMS
SYSTEM EFFECTIVENESS
∞SYSTEMS
SYSTEMS ENGINEERING
SYSTEMS SIMULATION

SYSTEMS MANAGEMENT
GS   MANAGEMENT
.   **SYSTEMS MANAGEMENT**
RT   INDUSTRIAL MANAGEMENT
INFORMATION SYSTEMS
MAN MACHINE SYSTEMS
MANAGEMENT METHODS
OPERATIONS RESEARCH

SYSTEMS SIMULATION
GS   SIMULATION
.   **SYSTEMS SIMULATION**
RT   ANALOG SIMULATION
COMPUTERIZED SIMULATION
DYNAMIC MODELS
DYNAMICAL SYSTEMS
FLIGHT SIMULATION
MATHEMATICAL MODELS
OPERATIONS RESEARCH
∞SYSTEMS
SYSTEMS INTEGRATION

SYSTEMS STABILITY
GS   STABILITY
.   **SYSTEMS STABILITY**
RT   CONTROL STABILITY
DYNAMIC STABILITY
EQUATIONS OF MOTION
∞EQUILIBRIUM
FLOW STABILITY
∞SYSTEMS
UNSTEADY STATE

SYSTOLE
GS   ANATOMY
.   CARDIOVASCULAR SYSTEM
. .   **SYSTOLE**
RATES (PER TIME)
.   HEART RATE

SYSTOLE-*(CONT.)*
　. . **SYSTOLE**
RT　BLOOD FLOW
　　BLOOD PRESSURE
　　CARDIAC VENTRICLES
　　DIASTOLE
　　SYSTOLIC PRESSURE

SYSTOLIC PRESSURE
GS　PRESSURE
　. BLOOD PRESSURE
　. . **SYSTOLIC PRESSURE**
RT　SYSTOLE

T

T SHAPE
UF　TEE
GS　SHAPES
　. **T SHAPE**
RT　BEAMS (SUPPORTS)

T TAIL SURFACES
GS　TAIL SURFACES
　. **T TAIL SURFACES**
RT　CONTROL SURFACES
　　STABILIZERS (FLUID DYNAMICS)
　　∞SURFACES
　　SWEPTBACK TAIL SURFACES
　　TAIL ASSEMBLIES

T TAURI STARS
GS　CELESTIAL BODIES
　. STARS
　. . EARLY STARS
　. . . PROTOSTARS
　. . . . **T TAURI STARS**
　. . VARIABLE STARS
　. . . **T TAURI STARS**
RT　HERBIG-HARO OBJECTS
　　TAURUS CONSTELLATION

T-2 AIRCRAFT
UF　BUCKEYE AIRCRAFT
　　T2J AIRCRAFT
　　YT-2 AIRCRAFT
GS　ATTACK AIRCRAFT
　. **T-2 AIRCRAFT**
　　JET AIRCRAFT
　. **T-2 AIRCRAFT**
　　MONOPLANES
　. **T-2 AIRCRAFT**
　　NORTH AMERICAN AIRCRAFT
　. **T-2 AIRCRAFT**
　　TRAINING AIRCRAFT
　. **T-2 AIRCRAFT**
RT　∞AIRCRAFT

T-28 AIRCRAFT
UF　TROJAN AIRCRAFT
GS　MONOPLANES
　. **T-28 AIRCRAFT**
　　NORTH AMERICAN AIRCRAFT
　. **T-28 AIRCRAFT**
　　TRAINING AIRCRAFT
　. **T-28 AIRCRAFT**
RT　∞AIRCRAFT

T-33 AIRCRAFT
UF　F-80 AIRCRAFT
　　SHOOTING STAR AIRCRAFT
GS　JET AIRCRAFT
　. **T-33 AIRCRAFT**
　　LOCKHEED AIRCRAFT
　. **T-33 AIRCRAFT**
　　MONOPLANES
　. **T-33 AIRCRAFT**
　　TRAINING AIRCRAFT
　. **T-33 AIRCRAFT**
RT　∞AIRCRAFT

T-34 ENGINE
GS　AIRCRAFT ENGINES
　. **T-34 ENGINE**
　　ENGINES
　. INTERNAL COMBUSTION ENGINES
　. . GAS TURBINE ENGINES
　. . . JET ENGINES
　. . . . TURBOJET ENGINES
　. . . . . TURBOPROP ENGINES
　. . . . . . **T-34 ENGINE**

T-34 ENGINE-*(CONT.)*
　. . TURBINE ENGINES
　. . GAS TURBINE ENGINES
　. . . JET ENGINES
　. . . . TURBOJET ENGINES
　. . . . . TURBOPROP ENGINES
　. . . . . . **T-34 ENGINE**
RT　C-133 AIRCRAFT

T-37 AIRCRAFT
GS　CESSNA AIRCRAFT
　. **T-37 AIRCRAFT**
　　JET AIRCRAFT
　. **T-37 AIRCRAFT**
　　MONOPLANES
　. **T-37 AIRCRAFT**
　　TRAINING AIRCRAFT
　. **T-37 AIRCRAFT**
RT　A-37 AIRCRAFT
　　∞AIRCRAFT

T-38 AIRCRAFT
UF　TALON AIRCRAFT
GS　JET AIRCRAFT
　. **T-38 AIRCRAFT**
　　MONOPLANES
　. **T-38 AIRCRAFT**
　　NORTHROP AIRCRAFT
　. **T-38 AIRCRAFT**
　　SUPERSONIC AIRCRAFT
　. **T-38 AIRCRAFT**
　　TRAINING AIRCRAFT
　. **T-38 AIRCRAFT**
RT　∞AIRCRAFT

T-38 ENGINE
GS　AIRCRAFT ENGINES
　. **T-38 ENGINE**
　　ENGINES
　. INTERNAL COMBUSTION ENGINES
　. . GAS TURBINE ENGINES
　. . . JET ENGINES
　. . . . TURBOJET ENGINES
　. . . . . TURBOPROP ENGINES
　. . . . . . **T-38 ENGINE**
　. TURBINE ENGINES
　. . GAS TURBINE ENGINES
　. . . JET ENGINES
　. . . . TURBOJET ENGINES
　. . . . . TURBOPROP ENGINES
　. . . . . . **T-38 ENGINE**

T-39 AIRCRAFT
UF　SABRELINER AIRCRAFT
　　T3J AIRCRAFT
GS　JET AIRCRAFT
　. **T-39 AIRCRAFT**
　　MONOPLANES
　. **T-39 AIRCRAFT**
　　NORTH AMERICAN AIRCRAFT
　. **T-39 AIRCRAFT**
　　PASSENGER AIRCRAFT
　. **T-39 AIRCRAFT**
　　TRAINING AIRCRAFT
　. **T-39 AIRCRAFT**
　　UTILITY AIRCRAFT
　. **T-39 AIRCRAFT**
RT　∞AIRCRAFT
　　CARGO AIRCRAFT

T-53 ENGINE
GS　ENGINES
　. AIR BREATHING ENGINES
　. . GAS TURBINE ENGINES
　. . . JET ENGINES
　. . . . TURBOJET ENGINES
　. . . . . TURBOPROP ENGINES
　. . . . . . **T-53 ENGINE**
　. INTERNAL COMBUSTION ENGINES
　. . GAS TURBINE ENGINES
　. . . JET ENGINES
　. . . . TURBOJET ENGINES
　. . . . . TURBOPROP ENGINES
　. . . . . . **T-53 ENGINE**
　. TURBINE ENGINES
　. . GAS TURBINE ENGINES
　. . . JET ENGINES
　. . . . TURBOJET ENGINES
　. . . . . TURBOPROP ENGINES
　. . . . . . **T-53 ENGINE**
RT　HELICOPTER ENGINES

T-55 ENGINE
GS　AIRCRAFT ENGINES
　. **T-55 ENGINE**

T-55 ENGINE-*(CONT.)*
RT　HELICOPTER ENGINES

T-56 ENGINE
GS　ENGINES
　. AIR BREATHING ENGINES
　. . GAS TURBINE ENGINES
　. . . JET ENGINES
　. . . . TURBOJET ENGINES
　. . . . . TURBOPROP ENGINES
　. . . . . . **T-56 ENGINE**
　. INTERNAL COMBUSTION ENGINES
　. . GAS TURBINE ENGINES
　. . . JET ENGINES
　. . . . TURBOJET ENGINES
　. . . . . TURBOPROP ENGINES
　. . . . . . **T-56 ENGINE**
　. TURBINE ENGINES
　. . GAS TURBINE ENGINES
　. . . JET ENGINES
　. . . . TURBOJET ENGINES
　. . . . . TURBOPROP ENGINES
　. . . . . . **T-56 ENGINE**
RT　C-130 AIRCRAFT

T-58 ENGINE
GS　ENGINES
　. INTERNAL COMBUSTION ENGINES
　. . GAS TURBINE ENGINES
　. . . **T-58 ENGINE**
　. TURBINE ENGINES
　. . GAS TURBINE ENGINES
　. . . **T-58 ENGINE**
RT　AIRCRAFT ENGINES
　　HELICOPTER ENGINES
　　VERTICAL TAKEOFF AIRCRAFT

T-58-GE-8B ENGINE
GS　ENGINES
　. AIR BREATHING ENGINES
　. . GAS TURBINE ENGINES
　. . . **T-58-GE-8B ENGINE**
　. INTERNAL COMBUSTION ENGINES
　. . GAS TURBINE ENGINES
　. . . **T-58-GE-8B ENGINE**
　. TURBINE ENGINES
　. . GAS TURBINE ENGINES
　. . . **T-58-GE-8B ENGINE**
RT　AIRCRAFT ENGINES
　　HELICOPTER ENGINES
　　VERTICAL TAKEOFF AIRCRAFT

T-63 ENGINE
GS　AIRCRAFT ENGINES
　. **T-63 ENGINE**
　　ENGINES
　. INTERNAL COMBUSTION ENGINES
　. . GAS TURBINE ENGINES
　. . . JET ENGINES
　. . . . **T-63 ENGINE**
　. TURBINE ENGINES
　. . GAS TURBINE ENGINES
　. . . JET ENGINES
　. . . . **T-63 ENGINE**
RT　HELICOPTER ENGINES

T-64 ENGINE
GS　ENGINES
　. AIR BREATHING ENGINES
　. . GAS TURBINE ENGINES
　. . . JET ENGINES
　. . . . TURBOJET ENGINES
　. . . . . TURBOPROP ENGINES
　. . . . . . **T-64 ENGINE**
　. INTERNAL COMBUSTION ENGINES
　. . GAS TURBINE ENGINES
　. . . JET ENGINES
　. . . . TURBOJET ENGINES
　. . . . . TURBOPROP ENGINES
　. . . . . . **T-64 ENGINE**
　. TURBINE ENGINES
　. . GAS TURBINE ENGINES
　. . . JET ENGINES
　. . . . TURBOJET ENGINES
　. . . . . TURBOPROP ENGINES
　. . . . . . **T-64 ENGINE**
RT　HELICOPTER ENGINES

T-74 ENGINE
GS　ENGINES
　. AIR BREATHING ENGINES
　. . GAS TURBINE ENGINES
　. . . JET ENGINES
　. . . . TURBOJET ENGINES
　. . . . . TURBOPROP ENGINES

T-74 ENGINE-*(CONT.)*
```
      . . . . . T-74 ENGINE
      . INTERNAL COMBUSTION ENGINES
      . . GAS TURBINE ENGINES
      . . . JET ENGINES
      . . . . TURBOJET ENGINES
      . . . . . TURBOPROP ENGINES
      . . . . . . T-74 ENGINE
      . TURBINE ENGINES
      . . GAS TURBINE ENGINES
      . . . JET ENGINES
      . . . . TURBOJET ENGINES
      . . . . . TURBOPROP ENGINES
      . . . . . . T-74 ENGINE
  RT  HELICOPTER ENGINES
```

T-76 ENGINE
```
  GS  AIRCRAFT ENGINES
      . T-76 ENGINE
      ENGINES
      . INTERNAL COMBUSTION ENGINES
      . . GAS TURBINE ENGINES
      . . . JET ENGINES
      . . . . T-76 ENGINE
      . TURBINE ENGINES
      . . GAS TURBINE ENGINES
      . . . JET ENGINES
      . . . . T-76 ENGINE
  RT  HELICOPTER ENGINES
```

T-78 ENGINE
```
  GS  AIRCRAFT ENGINES
      . T-78 ENGINE
      ENGINES
      . INTERNAL COMBUSTION ENGINES
      . . GAS TURBINE ENGINES
      . . . JET ENGINES
      . . . . TURBOJET ENGINES
      . . . . . TURBOPROP ENGINES
      . . . . . . T-78 ENGINE
      . TURBINE ENGINES
      . . GAS TURBINE ENGINES
      . . . JET ENGINES
      . . . . TURBOJET ENGINES
      . . . . . TURBOPROP ENGINES
      . . . . . . T-78 ENGINE
```

TABLASER

TABLES (DATA)
```
  GS  TABLES (DATA)
      . CONVERSION TABLES
      . INTERFERENCE FACTOR TABLE
      . MATHEMATICAL TABLES
  RT  ASTRONOMICAL CATALOGS
      ∞DATA
      DATA ACQUISITION
      DATA MANAGEMENT
      DÁTA PROCESSING
      DATA RECORDING
      DATA REDUCTION
      DATA RETRIEVAL
      PRINTOUTS
      STATISTICAL ANALYSIS
      ∞STATISTICS
      ∞TABULATION
      TABULATION PROCESSES
```

TABLETS
```
  RT  BRIQUETS
      ∞CAPSULES
      MOLDS
```

TABS (CONTROL SURFACES)
```
  GS  AIRFOILS
      . TABS (CONTROL SURFACES)
      CONTROL SURFACES
      . TABS (CONTROL SURFACES)
  RT  AERIAL RUDDERS
      AILERONS
      ∞CONTROL
      ELEVATORS (CONTROL SURFACES)
      ELEVONS
      RUDDERS
      STABILIZERS (FLUID DYNAMICS)
      ∞SURFACES
```

TABULATING
```
  USE  TABULATION PROCESSES
```

∞ TABULATION
```
  SN  (USE OF A MORE SPECIFIC TERM IS
      RECOMMENDED--CONSULT THE TERMS
      LISTED BELOW)
  RT  TABLES (DATA)
```

TABULATION-*(CONT.)*
```
      TABULATION PROCESSES
```

TABULATION PROCESSES
```
  UF  TABULATING
  RT  DATA PROCESSING
      DATA RECORDING
      TABLES (DATA)
      ∞TABULATION
```

TACAN
```
  UF  TACTICAL AIR NAVIGATION
  GS  NAVIGATION
      . RADIO NAVIGATION
      . . TACAN
  RT  AIR NAVIGATION
      ALL-WEATHER AIR NAVIGATION
      FLIGHT PATHS
      NAVIGATION AIDS
      RADAR NAVIGATION
      SOLAR COMPASSES
```

TACHISTOSCOPES
```
  RT  VISUAL PERCEPTION
```

TACHOMETERS
```
  GS  AIRCRAFT INSTRUMENTS
      . SPEED INDICATORS
      . . TACHOMETERS
      DISPLAY DEVICES
      . SPEED INDICATORS
      . . TACHOMETERS
      MEASURING INSTRUMENTS
      . INDICATING INSTRUMENTS
      . . SPEED INDICATORS
      . . . TACHOMETERS
  RT  ANGULAR VELOCITY
      TIMING DEVICES
      VELOCITY MEASUREMENT
```

TACHYCARDIA
```
  GS  DISEASES
      . TACHYCARDIA
      RATES (PER TIME)
      . HEART RATE
      . . TACHYCARDIA
```

TACHYONS
```
  GS  PARTICLES
      . ELEMENTARY PARTICLES
      . . TACHYONS
```

TACHYPNEA
```
  GS  RATES (PER TIME)
      . RESPIRATORY RATE
      . . TACHYPNEA
```

TACKINESS
```
  RT  ADHESION
```

TACT PROGRAM
```
  UF  TRANSONIC AIRCRAFT TECHNOLOGY
          PROGRAM
  GS  PROGRAMS
      . NASA PROGRAMS
      . . TACT PROGRAM
  RT  ∞AERONAUTICS
      ∞AIRCRAFT
```

TACTICAL AIR NAVIGATION
```
  USE  TACAN
```

TACTICS
```
  RT  ATTACKING (ASSAULTING)
      DEPLOYMENT
      EVASIVE ACTIONS
      MILITARY OPERATIONS
      MILITARY TECHNOLOGY
      OBSTACLE AVOIDANCE
```

TACTILE DISCRIMINATION
```
  GS  DISCRIMINATION
      . SENSORY DISCRIMINATION
      . . TACTILE DISCRIMINATION
      PERCEPTION
      . SENSORY PERCEPTION
      . . TOUCH
      . . . TACTILE DISCRIMINATION
```

TACTILE SENSATION
```
  USE  TOUCH
```

TAFEL LAW
```
  GS  LAWS
```

TAFEL LAW-*(CONT.)*
```
      . TAFEL LAW
  RT  ELECTRODES
      ELECTROLYSIS
      FICKS EQUATION
      POLARIZATION (CHARGE SEPARATION)
```

TAGGING
```
  USE  MARKING
```

TAGN
```
  UF  TRIAMINOGUANIDINENITRATE
  GS  OXIDIZERS
      . ROCKET OXIDIZERS
      . . TAGN
      PROPELLANTS
      . ROCKET PROPELLANTS
      . . TAGN
  RT  EXPLOSIVES
```

TAIL ASSEMBLIES
```
  UF  EMPENNAGE
      TAIL MOUNTINGS
      TAILS (ASSEMBLIES)
      VERTICAL TAILS
  GS  ASSEMBLIES
      . TAIL ASSEMBLIES
      . . SWING TAIL ASSEMBLIES
  RT  AERIAL RUDDERS
      AFTERBODIES
      AIRCRAFT PARTS
      AIRCRAFT STRUCTURES
      AIRFOILS
      AIRFRAMES
      BOATTAILS
      BODY-WING AND TAIL CONFIGURATIONS
      ∞BOOM
      CONTROL SURFACES
      ELEVATORS (CONTROL SURFACES)
      FINS
      HORIZONTAL TAIL SURFACES
      HYDROFOILS
      MARINE RUDDERS
      MISSILE STRUCTURES
      RUDDERS
      SAILS
      STABILIZERS (FLUID DYNAMICS)
      T TAIL SURFACES
      VANES
```

TAIL MOUNTINGS
```
  USE  TAIL ASSEMBLIES
```

TAIL PLANES
```
  USE  HORIZONTAL TAIL SURFACES
```

TAIL ROTORS
```
  GS  ROTATING BODIES
      . ROTORS
      . . TAIL ROTORS
      . . . HELICOPTER TAIL ROTORS
  RT  HELICOPTER CONTROL
      ROTARY WINGS
      ∞ROTOR BLADES
```

TAIL SURFACES
```
  GS  TAIL SURFACES
      . HORIZONTAL TAIL SURFACES
      . SWEPTBACK TAIL SURFACES
      . T TAIL SURFACES
      . TRAPEZOIDAL TAIL SURFACES
  RT  CONTROL SURFACES
      ELEVATORS (CONTROL SURFACES)
      RUDDERS
      STABILIZERS (FLUID DYNAMICS)
      ∞SURFACES
```

TAILLESS AIRCRAFT
```
  UF  FLYING WING AIRCRAFT
  GS  TAILLESS AIRCRAFT
      . AVRO 707 AIRCRAFT
      . B-58 AIRCRAFT
      . F-102 AIRCRAFT
      . F-106 AIRCRAFT
      . FD 2 AIRCRAFT
      . HP-115 AIRCRAFT
      . MIRAGE 3 AIRCRAFT
      . SC-1 AIRCRAFT
      . VULCAN AIRCRAFT
  RT  ∞AIRCRAFT
      JET AIRCRAFT
      ∞LOW WING AIRCRAFT
      ∞MILITARY AIRCRAFT
      MONOPLANES
      RESEARCH AIRCRAFT
```

TAILORING
USE   DESIGN

TAILS (ASSEMBLIES)
USE   TAIL ASSEMBLIES

TAIWAN
UF   REPUBLIC OF CHINA
GS   NATIONS
       . **TAIWAN**
RT   ASIA
       CHINA
       CHINESE SPACE PROGRAM
       CHINESE SPACECRAFT
       HONG KONG

TAKEOFF
GS   **TAKEOFF**
       . VERTICAL TAKEOFF
RT   AIR TRAFFIC CONTROL
       AIRCRAFT LANDING
       ASCENT
       CLIMBING FLIGHT
       JATO ENGINES
       LANDING
       MANEUVERS
       RUNWAYS

TAKEOFF RUNS
RT   AIRCRAFT PERFORMANCE
       DISTANCE
       RUNWAY ALIGNMENT
       SHORT TAKEOFF AIRCRAFT

TAKEOFF SYSTEMS
USE   AIRCRAFT LAUNCHING DEVICES

TALC
UF   STEATITE
GS   MAGNESIUM COMPOUNDS
       . **TALC**
       MINERALS
       . **TALC**
       SILICON COMPOUNDS
       . SILICATES
       . . SODIUM SILICATES
       . . . **TALC**
       SODIUM COMPOUNDS
       . SODIUM SILICATES
       . . **TALC**

TALKING
GS   SPEECH
       . **TALKING**
       . . WORDS (LANGUAGE)
       . . . SYLLABLES
RT   SENTENCES
       SIGNAL TRANSMISSION

TALON AIRCRAFT
USE   T-38 AIRCRAFT

TALOS MISSILE
GS   MISSILES
       . SURFACE TO AIR MISSILES
       . . **TALOS MISSILE**
RT   BUMBLEBEE PROJECT
       LIQUID PROPELLANT ROCKET ENGINES
       MULTISTAGE ROCKET VEHICLES
       SOLID PROPELLANT ROCKET ENGINES

TANDEM ROTOR HELICOPTERS
GS   V/STOL AIRCRAFT
       . ROTARY WING AIRCRAFT
       . . HELICOPTERS
       . . . **TANDEM ROTOR HELICOPTERS**
       . . . . CH-46 HELICOPTER
       . . . . CH-47 HELICOPTER
       . . . . H-25 HELICOPTER
RT   ∞AIRCRAFT

TANDEM WING AIRCRAFT
GS   **TANDEM WING AIRCRAFT**
       . X-19 AIRCRAFT
       . X-22A AIRCRAFT
RT   ∞AIRCRAFT
       BIPLANES
       CANARD CONFIGURATIONS
       DUAL WING CONFIGURATIONS
       JET AIRCRAFT
       RESEARCH AIRCRAFT
       ∞SUBSONIC AIRCRAFT
       X-22 AIRCRAFT

TANGENTS
GS   ANALYSIS (MATHEMATICS)
       . REAL VARIABLES
       . . PERIODIC FUNCTIONS
       . . . TRIGONOMETRIC FUNCTIONS
       . . . . **TANGENTS**
       FUNCTIONS (MATHEMATICS)
       . TRANSCENDENTAL FUNCTIONS
       . . PERIODIC FUNCTIONS
       . . . TRIGONOMETRIC FUNCTIONS
       . . . . **TANGENTS**
       GEOMETRY
       . EUCLIDEAN GEOMETRY
       . . ANALYTIC GEOMETRY
       . . . **TANGENTS**
RT   CHORDS (GEOMETRY)

TANGLING
RT   CONFUSION
       ENTRAPMENT
       MIXING

TANK GEOMETRY
GS   GEOMETRY
       . **TANK GEOMETRY**
RT   LIQUID SLOSHING
       PROPELLANT TANKS
       STORAGE TANKS
       TANKS (CONTAINERS)
       ULLAGE

TANK TRUCKS
GS   SURFACE VEHICLES
       . MOTOR VEHICLES
       . . TRUCKS
       . . . **TANK TRUCKS**
RT   ∞TANKERS
       TRAILERS

TANKER AIRCRAFT
GS   TRANSPORT AIRCRAFT
       . **TANKER AIRCRAFT**
RT   AIR TO AIR REFUELING
       ∞AIRCRAFT
       AIRCRAFT FUELS
       BOMBER AIRCRAFT
       FUEL TANKS
       ∞MILITARY AIRCRAFT
       ∞TANKERS
       VALIANT AIRCRAFT

TANKER SHIPS
GS   SURFACE VEHICLES
       . CARGO SHIPS
       . . **TANKER SHIPS**
       WATER VEHICLES
       . SHIPS
       . . CARGO SHIPS
       . . . **TANKER SHIPS**
RT   ARTIFICIAL HARBORS
       DEEPWATER TERMINALS
       HARBORS
       MARINE TRANSPORTATION
       OFFSHORE DOCKING
       OFFSHORE PLATFORMS
       SHIPYARDS
       ∞TANKERS
       WHARVES

TANKER TERMINALS
RT   ARTIFICIAL HARBORS
       CARGO SHIPS
       DEEPWATER TERMINALS
       MARINE TECHNOLOGY
       OCEANOGRAPHY
       OFFSHORE DOCKING
       OFFSHORE PLATFORMS
       SHIP TERMINALS
       ∞TANKERS
       TERMINAL FACILITIES
       TRANSPORTATION

∞ **TANKERS**
SN   *(USE OF A MORE SPECIFIC TERM IS*
       RECOMMENDED--CONSULT THE TERMS
       LISTED BELOW)
RT   ARTIFICIAL HARBORS
       DEEPWATER TERMINALS
       OFFSHORE DOCKING
       OFFSHORE PLATFORMS
       TANK TRUCKS
       TANKER AIRCRAFT
       TANKER SHIPS
       TANKER TERMINALS
       TRANSPORTATION ENERGY

TANKS (COMBAT VEHICLES)
GS   SURFACE VEHICLES
       . **TANKS (COMBAT VEHICLES)**
RT   ARMED FORCES
       MILITARY OPERATIONS
       ∞MILITARY VEHICLES
       ORDNANCE
       ∞VEHICLES
       WEAPONS

TANKS (CONTAINERS)
GS   **TANKS (CONTAINERS)**
       . BUNKERS (FUEL)
       . CYLINDRICAL TANKS
       . EXTERNAL TANKS
       . FUEL TANKS
       . . WING TANKS
       . PROPELLANT TANKS
       . SPHERICAL TANKS
       . STORAGE TANKS
RT   BASINS (CONTAINERS)
       BOTTLES
       CHEMICAL REACTORS
       ∞CONTAINERS
       DRUMS (CONTAINERS)
       FLUID FILLED SHELLS
       LIQUID FILLED SHELLS
       MATERIALS HANDLING
       PIPE NOZZLES
       PRESSURE VESSELS
       RECEIVERS
       ∞STRUCTURES
       TANK GEOMETRY
       TOWERS
       WING-FUSELAGE STORES

TANTALUM
GS   CHEMICAL ELEMENTS
       . REFRACTORY METALS
       . . **TANTALUM**
       . . . TANTALUM ISOTOPES
       METALS
       . TRANSITION METALS
       . . REFRACTORY METALS
       . . . **TANTALUM**
       . . . . TANTALUM ISOTOPES
       REFRACTORY MATERIALS
       . REFRACTORY METALS
       . . **TANTALUM**
       . . . TANTALUM ISOTOPES

TANTALUM ALLOYS
GS   ALLOYS
       . HEAT RESISTANT ALLOYS
       . . REFRACTORY METAL ALLOYS
       . . . **TANTALUM ALLOYS**
       REFRACTORY MATERIALS
       . REFRACTORY METAL ALLOYS
       . . **TANTALUM ALLOYS**

TANTALUM CARBIDES
GS   CARBON COMPOUNDS
       . CARBIDES
       . . **TANTALUM CARBIDES**
       TANTALUM COMPOUNDS
       . **TANTALUM CARBIDES**

TANTALUM COMPOUNDS
GS   **TANTALUM COMPOUNDS**
       . TANTALUM CARBIDES
       . TANTALUM NITRIDES
       . TANTALUM OXIDES
RT   ∞CHEMICAL COMPOUNDS
       ∞GROUP 5B COMPOUNDS
       ∞METAL COMPOUNDS

TANTALUM ISOTOPES
GS   CHEMICAL ELEMENTS
       . NUCLIDES
       . . ISOTOPES
       . . . **TANTALUM ISOTOPES**
       . REFRACTORY METALS
       . . TANTALUM
       . . . **TANTALUM ISOTOPES**
       METALS
       . TRANSITION METALS
       . . REFRACTORY METALS
       . . . TANTALUM
       . . . . **TANTALUM ISOTOPES**
       REFRACTORY MATERIALS
       . REFRACTORY METALS
       . . TANTALUM
       . . . **TANTALUM ISOTOPES**

TANTALUM NITRIDES
GS   NITROGEN COMPOUNDS
      . NITRIDES
      . . **TANTALUM NITRIDES**
      TANTALUM COMPOUNDS
      . **TANTALUM NITRIDES**
RT   METAL NITRIDES

TANTALUM OXIDES
GS   CHALCOGENIDES
      . OXIDES
      . . METAL OXIDES
      . . . **TANTALUM OXIDES**
      TANTALUM COMPOUNDS
      . **TANTALUM OXIDES**

TANZANIA
GS   NATIONS
      . **TANZANIA**
RT   AFRICA

TAPE RECORDERS
UF   MAGNETIC TAPE RECORDERS
RT   DATA RECORDERS
      ELECTRONIC RECORDING SYSTEMS
      MAGNETIC TAPE TRANSPORTS
      MAGNETIC TAPES
      ∞RECORDERS
      RECORDING HEADS
      RECORDING INSTRUMENTS

TAPER
USE   TAPERING

TAPERED COLUMNS
GS   STRUCTURAL MEMBERS
      . COLUMNS (SUPPORTS)
      . . **TAPERED COLUMNS**

TAPERED WINGS
USE   SWEPT WINGS

TAPERING
UF   TAPER
RT   CONVERGENCE
      DECELERATION
      ∞REDUCTION

∞ **TAPES**
SN   (USE OF A MORE SPECIFIC TERM IS
      RECOMMENDED--CONSULT THE TERMS
      LISTED BELOW)
RT   ADHESIVES
      COMPUTER COMPATIBLE TAPES
      FASTENERS
      HEAT TAPES
      MAGNETIC TAPES
      PLASTIC TAPES
      PLAYBACKS
      PUNCHED TAPES
      RIBBONS
      SEALS (STOPPERS)
      SPLICING

TAPS
RT   CUTTERS
      DRILLS
      MACHINE TOOLS
      TOOLS

TAR SANDS
GS   RESOURCES
      . EARTH RESOURCES
      . . **TAR SANDS**
      SEDIMENTS
      . SANDS
      . . **TAR SANDS**
      SOILS
      . SANDS
      . . **TAR SANDS**
RT   DISTILLATION
      OIL EXPLORATION
      OIL FIELDS
      OILS
      TARS

TARE (DATA REDUCTION)
USE   DATA REDUCTION

TARGET ACQUISITION
GS   ACQUISITION
      . **TARGET ACQUISITION**
RT   DETECTION

TARGET ACQUISITION-(CONT.)
      HIGH ALT TARGET AND BACKGROUND
          MEASUREMENT
      MATTS (SYSTEMS)
      MISSILE DETECTION
      MOVING TARGET INDICATORS
      SOUND RANGING
      SURVEILLANCE
      TARGETS

TARGET DRONE AIRCRAFT
GS   DRONE VEHICLES
      . DRONE AIRCRAFT
      . . **TARGET DRONE AIRCRAFT**
      . . . **FIREBEE 2 TARGET DRONE
              AIRCRAFT**
      . . . JINDIVIK TARGET AIRCRAFT
      PILOTLESS AIRCRAFT
      . DRONE AIRCRAFT
      . . **TARGET DRONE AIRCRAFT**
      . . . **FIREBEE 2 TARGET DRONE
              AIRCRAFT**
      . . . JINDIVIK TARGET AIRCRAFT
RT   ∞AIRCRAFT
      ∞MILITARY AIRCRAFT
      REMOTELY PILOTED VEHICLES
      TARGETS

TARGET MASKING
GS   MASKING
      . **TARGET MASKING**
RT   COUNTERMEASURES
      TARGETS

TARGET PENETRATION
USE   TERMINAL BALLISTICS

TARGET RECOGNITION
GS   DETECTION
      . **TARGET RECOGNITION**
      RECOGNITION
      . **TARGET RECOGNITION**
RT   DISCRIMINATION
      LASER TARGET DESIGNATORS
      MISSILE DETECTION
      MISSILE SIGNATURES
      MULTISTATIC RADAR
      NAP-OF-THE-EARTH NAVIGATION
      RADAR HOMING MISSILES
      RADAR SIGNATURES
      SIGNATURE ANALYSIS
      SIGNATURES
      SURVEILLANCE
      TARGETS
      TRADEX RADAR SYSTEM

TARGET SIMULATORS
GS   SIMULATORS
      . **TARGET SIMULATORS**
RT   COMPUTERIZED SIMULATION
      DISPLAY DEVICES

TARGET THICKNESS
GS   DIMENSIONS
      . **TARGET THICKNESS**
RT   PARTICLE ACCELERATOR TARGETS
      TARGETS
      THICKNESS

TARGETS
UF   TOWED TARGETS
GS   **TARGETS**
      . JINDIVIK TARGET AIRCRAFT
      . LASER TARGETS
      . PARTICLE ACCELERATOR TARGETS
      . RADAR TARGETS
RT   AIRBORNE INTEGRATED
          RECONNAISSANCE SYSTEM
      COMMAND AND CONTROL
      DETECTION
      FIREBEE 2 TARGET DRONE AIRCRAFT
      IRRADIATION
      LASER TARGET DESIGNATORS
      LASER TARGET INTERACTIONS
      LINE OF SIGHT
      MICROBALLOONS
      ∞MISSIONS
      RADAR ECHOES
      RADAR TARGET SCATTER SITE
          PROGRAM
      SANDPIPER TARGET MISSILE
      SURVEILLANCE
      TARGET ACQUISITION
      TARGET DRONE AIRCRAFT
      TARGET MASKING

TARGETS-(CONT.)
      TARGET RECOGNITION
      TARGET THICKNESS

TARS
GS   PRODUCTS
      . PETROLEUM PRODUCTS
      . . **TARS**
RT   ASPHALT
      GUMS (SUBSTANCES)
      PITCH (MATERIAL)
      TAR SANDS

TARTAR MISSILE
GS   MISSILES
      . ANTIAIRCRAFT MISSILES
      . . **TARTAR MISSILE**
      . SURFACE TO AIR MISSILES
      . . **TARTAR MISSILE**
RT   AJ-10 ENGINE
      BUMBLEBEE PROJECT
      SOLID PROPELLANT ROCKET ENGINES

TASK COMPLEXITY
GS   COMPLEXITY
      . **TASK COMPLEXITY**
RT   COSTS
      ∞PERFORMANCE
      QUALITY CONTROL
      SCHEDULING

TASKS
UF   JOBS
GS   **TASKS**
      . AUDITORY TASKS
      . VISUAL TASKS
RT   COSTS
      CREW PROCEDURES (INFLIGHT)
      CREW PROCEDURES (PREFLIGHT)
      ∞ELEMENTS
      MATRIX MANAGEMENT
      PHYSICAL WORK
      PROJECTS
      QUALITY CONTROL
      RETRAINING
      SCHEDULING
      ∞TESTS

TASMANIA
GS   LANDFORMS
      . ISLANDS
      . . **TASMANIA**
RT   AUSTRALIA

TASTE
UF   GUSTATORY PERCEPTION
GS   PERCEPTION
      . SENSORY PERCEPTION
      . . **TASTE**
RT   CHEMORECEPTORS
      SYNTHETIC FOOD

TATB
UF   TRIAMINOTRINITROBENZENE
GS   EXPLOSIVES
      . **TATB**
      PROPELLANTS
      . ROCKET PROPELLANTS
      . . **TATB**
RT   ROCKET OXIDIZERS

TAURID METEOROIDS
GS   CELESTIAL BODIES
      . METEOROID SHOWERS
      . . **TAURID METEOROIDS**
      . METEOROIDS
      . . **TAURID METEOROIDS**

TAURUS CONSTELLATION
GS   CONSTELLATIONS
      . **TAURUS CONSTELLATION**
RT   CRAB NEBULA
      T TAURI STARS

TAUTOMERS
RT   ISOMERS

TAXIING
RT   AIR TRAFFIC CONTROL
      AIRFIELD SURFACE MOVEMENTS
      RUNWAYS

TAXONOMY
RT   CLASSIFICATIONS

TAXONOMY-*(CONT.)*
- ∞ CLASSIFYING
- ∞ SCIENCE
- ∞ ZOOLOGY

TAYLOR INSTABILITY
- RT   DENSITY DISTRIBUTION
-     GOERTLER INSTABILITY
-     INTERFACE STABILITY
-     PERTURBATION THEORY
-     ROTATING FLUIDS
-     TWO DIMENSIONAL FLOW

TAYLOR MANIFEST ANXIETY SCALE
- RT   ANXIETY
-     PHYSIOLOGICAL TESTS
-     PSYCHOLOGICAL EFFECTS
-     PSYCHOLOGICAL TESTS

TAYLOR SERIES
- UF   TAYLOR THEOREM
- GS   ANALYSIS (MATHEMATICS)
-     . CALCULUS
-     .. SERIES (MATHEMATICS)
-     ... POWER SERIES
-     .... **TAYLOR SERIES**
-     ..... MACLAURIN SERIES
-     . REAL VARIABLES
-     .. SERIES (MATHEMATICS)
-     ... POWER SERIES
-     .... **TAYLOR SERIES**
-     ..... MACLAURIN SERIES
- RT   THEOREMS

TAYLOR THEOREM
- USE   TAYLOR SERIES

TAYLOR-GOERTLER INSTABILITY
- USE   GOERTLER INSTABILITY

TCG (TRACKING)
- USE   TRANSPONDER CONTROL GROUP

TCV PROGRAM
- USE   TERMINAL CONFIGURED VEHICLE PROGRAM

TD SATELLITES
- GS   ESA SPACECRAFT
-     . ESA SATELLITES
-     .. **TD SATELLITES**
-     ... TD-1 SATELLITE
-     SATELLITES
-     . ARTIFICIAL SATELLITES
-     .. ESA SATELLITES
-     ... **TD SATELLITES**
-     .... TD-1 SATELLITE
-     .. SYNCHRONOUS SATELLITES
-     ... **TD SATELLITES**
-     .... TD-1 SATELLITE

TD-1 SATELLITE
- GS   ESA SPACECRAFT
-     . ESA SATELLITES
-     .. TD SATELLITES
-     ... **TD-1 SATELLITE**
-     SATELLITES
-     . ARTIFICIAL SATELLITES
-     .. ESA SATELLITES
-     ... TD SATELLITES
-     .... **TD-1 SATELLITE**
-     .. SYNCHRONOUS SATELLITES
-     ... TD SATELLITES
-     .... **TD-1 SATELLITE**

TDMA
- USE   TIME DIVISION MULTIPLE ACCESS

TDR SATELLITES
- UF   TRACKING AND DATA RELAY SATELLITES
- GS   SATELLITES
-     . **TDR SATELLITES**
- RT   AUTONOMOUS SPACECRAFT CLOCKS
-     COMMUNICATING
-     DATA TRANSMISSION
-     RADIO RELAY SYSTEMS
-     SATELLITE NETWORKS
-     SATELLITE TRANSMISSION
-     TELECOMMUNICATION
-     TELEMETRY

TEA LASERS
- UF   TRANSVERSELY EXCITED ATMOSPHERIC LASERS
- GS   STIMULATED EMISSION DEVICES
-     . LASERS
-     .. GAS LASERS
-     ... **TEA LASERS**
- RT   ATMOSPHERIC LASERS
-     CARBON DIOXIDE LASERS
-     CARBON MONOXIDE LASERS
-     CHEMICAL LASERS
-     GAS MASERS
-     HF LASERS
-     LASER MODES
-     PULSED LASERS
-     STIMULATED EMISSION

TEACHING
- USE   EDUCATION

TEACHING MACHINES
- GS   TRAINING DEVICES
-     . **TEACHING MACHINES**
- RT   LEARNING
-     LEARNING MACHINES
-     ∞ MACHINERY

TEAMS
- RT   BUREAUS (ORGANIZATIONS)
-     FEDERATIONS
-     INSTITUTIONS
-     ORGANIZATIONS
-     PROJECTS
-     UNIVERSITY PROGRAM

TEARING
- RT   MECHANICAL PROPERTIES
-     RUPTURING
-     SHREDDING

TEARING MODES (PLASMAS)
- RT   BALLOONING MODES
-     MODES
-     PLASMAS (PHYSICS)

TECHNETIUM
- GS   CHEMICAL ELEMENTS
-     . **TECHNETIUM**
-     METALS
-     . TRANSITION METALS
-     .. **TECHNETIUM**

TECHNETIUM COMPOUNDS
- GS   **TECHNETIUM COMPOUNDS**
-     . TECHNETIUM FLUORIDES
- RT   ∞ CHEMICAL COMPOUNDS
-     ∞ GROUP 7B COMPOUNDS
-     ∞ METAL COMPOUNDS

TECHNETIUM FLUORIDES
- GS   HALOGEN COMPOUNDS
-     . FLUORINE COMPOUNDS
-     .. FLUORIDES
-     ... **TECHNETIUM FLUORIDES**
-     . HALIDES
-     .. FLUORIDES
-     ... **TECHNETIUM FLUORIDES**
-     .. METAL HALIDES
-     ... **TECHNETIUM FLUORIDES**
-     TECHNETIUM COMPOUNDS
-     . **TECHNETIUM FLUORIDES**

TECHNETIUM ISOTOPES
- GS   METALS
-     . TRANSITION METALS
-     .. **TECHNETIUM ISOTOPES**

TECHNICAL WRITING
- RT   ABSTRACTS
-     DOCUMENTATION
-     EDITING
-     RECORDS
-     SPECIFICATIONS
-     TRANSLATING

TECHNIQUES
- USE   METHODOLOGY

TECHNOLOGICAL FORECASTING
- GS   FORECASTING
-     . **TECHNOLOGICAL FORECASTING**
-     .. PATTERN METHOD (FORECASTING)
-     .. PROBE METHOD (FORECASTING)
-     .. PROFILE METHOD (FORECASTING)

TECHNOLOGICAL FORECASTING-*(CONT.)*
- RT   AEROSPACE TECHNOLOGY TRANSFER
-     ESTIMATING
-     PREDICTIONS
-     TECHNOLOGY TRANSFER

TECHNOLOGIES
- GS   **TECHNOLOGIES**
-     . BIOTECHNOLOGY
-     . BUBBLE TECHNIQUE
-     . ENERGY TECHNOLOGY
-     .. GEOTHERMAL TECHNOLOGY
-     . MARINE TECHNOLOGY
-     . MILITARY TECHNOLOGY
-     . REACTOR TECHNOLOGY
- RT   INDUSTRIES
-     LOW GRAVITY MANUFACTURING
-     MANUFACTURING
-     NUCLEONICS
-     SPACE MANUFACTURING
-     TECHNOLOGY ASSESSMENT
-     TECHNOLOGY UTILIZATION
-     URBAN DEVELOPMENT

TECHNOLOGY ASSESSMENT
- GS   ASSESSMENTS
-     . **TECHNOLOGY ASSESSMENT**
- RT   CANADIAN SPACE PROGRAM
-     COMMUNICATIONS TECHNOLOGY SATELLITE
-     DELPHI METHOD (FORECASTING)
-     EVALUATION
-     FEASIBILITY ANALYSIS
-     INDUSTRIES
-     MANUFACTURING
-     PATTERN METHOD (FORECASTING)
-     PROBE METHOD (FORECASTING)
-     PROFILE METHOD (FORECASTING)
-     STRATEGIC MATERIALS
-     TECHNOLOGIES
-     VALUE

TECHNOLOGY FEASIBILITY SPACECRAFT
- GS   UNMANNED SPACECRAFT
-     . **TECHNOLOGY FEASIBILITY SPACECRAFT**
- RT   SCIENTIFIC SATELLITES
-     ∞ SPACECRAFT

TECHNOLOGY TRANSFER
- GS   **TECHNOLOGY TRANSFER**
-     . AEROSPACE TECHNOLOGY TRANSFER
- RT   COMMUNICATING
-     COMMUNICATION
-     DOCUMENTATION
-     DOCUMENTS
-     INFORMATION FLOW
-     INFORMATION MANAGEMENT
-     INFORMATION TRANSFER
-     REPORTS
-     SELECTIVE DISSEMINATION OF INFORMATION
-     SPACE COMMERCIALIZATION
-     STARSITE PROGRAM
-     TECHNOLOGICAL FORECASTING
-     TRANSFERRING

TECHNOLOGY UTILIZATION
- RT   AEROSPACE TECHNOLOGY TRANSFER
-     CANADIAN SPACE PROGRAM
-     COMMUNICATIONS TECHNOLOGY SATELLITE
-     CONTROL CONFIGURED VEHICLES
-     INDIAN SPACE PROGRAM
-     INDUSTRIES
-     INFORMATION TRANSFER
-     LASER APPLICATIONS
-     MANUFACTURING
-     NASA PROGRAMS
-     PATENT APPLICATIONS
-     RESEARCH AND DEVELOPMENT
-     SYNCHRONOUS EARTH OBSERVATORY SATELLITE
-     TECHNOLOGIES
-     UTILIZATION

TECTONIC MOVEMENT
- USE   TECTONICS

TECTONICS
- UF   TECTONIC MOVEMENT
- GS   GEOLOGY
-     . **TECTONICS**
- RT   ∞ DEPRESSION
-     EARTH MOVEMENTS

TECTONICS-*(CONT.)*
        EARTH PLANETARY STRUCTURE
        FISSURES (GEOLOGY)
        GEOPHYSICS
        PLATES (TECTONICS)
        SUBDUCTION (GEOLOGY)

TED
  USE   TRANSFERRED ELECTRON DEVICES

TEE
  USE   T SHAPE

TEETERING
  RT   ∞MOTION

TEETH
  SN   (EXCLUDES GEAR TEETH AND OTHER
         MECHANICAL DEVICES)
  GS   ANATOMY
      . DIGESTIVE SYSTEM
      . . **TEETH**
  RT   DENTAL CALCULI
        DENTISTRY
        MASTICATION
        MOUTH
        ORAL HYGIENE
        TOOTH DISEASES

TEFLON (TRADEMARK)
  GS   PLASTICS
      . **TEFLON (TRADEMARK)**
  RT   ∞POLYMERS
        POLYTETRAFLUOROETHYLENE
        RESINS
        SYNTHETIC RESINS

TEKTITE PROJECT
  GS   PROGRAMS
      . NASA PROGRAMS
      . . **TEKTITE PROJECT**
      . NASA SPACE PROGRAMS
      . . **TEKTITE PROJECT**

TEKTITES
  GS   CELESTIAL BODIES
      . METEORITES
      . . STONY METEORITES
      . . . **TEKTITES**
      . . . . AUSTRALITES
      . . . . BEDIASITES
  RT   CHONDRITES
        COESITE
        CYRILLID METEOROIDS
        METEORITIC COMPOSITION
        METEORITIC MICROSTRUCTURES
        MICROMETEORITES
        NATURAL SATELLITES

TELECHIRICS
  USE   REMOTE HANDLING

TELECOMMUNICATION
  UF   COMMUNICATION SYSTEMS
  GS   **TELECOMMUNICATION**
      . AIRCRAFT COMMUNICATION
      . BROADCASTING
      . CLOSED CIRCUIT TELEVISION
      . COLOR TELEVISION
      . COMMUNICATION
      . . FACSIMILE COMMUNICATION
      . . LINE OF SIGHT COMMUNICATION
      . . OPTICAL COMMUNICATION
      . . SHIP TO SHORE COMMUNICATION
      . . UNDERWATER COMMUNICATION
      . DATA LINKS
      . DEFENSE COMMUNICATIONS
        SATELLITE SYSTEM
      . . FLEET SATELLITE COMMUNICATION
        SYSTEM
      . DEFENSE COMMUNICATIONS SYSTEM
        (DCS)
      . EDUCATIONAL TELEVISION
      . ELECTRONIC MAIL
      . GROUND-AIR-GROUND
        COMMUNICATION
      . HET EXPERIMENT
      . MULTICHANNEL COMMUNICATION
      . MULTIPLE ACCESS
      . . ALOHA SYSTEM
      . . CODE DIVISION MULTIPLE ACCESS
      . . DEMAND ASSIGNMENT MULTIPLE
        ACCESS

TELECOMMUNICATION-*(CONT.)*
      . . FREQUENCY DIVISION MULTIPLE
        ACCESS
      . . TIME DIVISION MULTIPLE ACCESS
      . PLAT SYSTEM
      . PULSE COMMUNICATION
      . DIGITAL SPACECRAFT TELEVISION
      . RADIO COMMUNICATION
      . . RADIO RELAY SYSTEMS
      . . . CODE DIVISION MULTIPLE ACCESS
      . . . TIME DIVISION MULTIPLE ACCESS
      . . RADIO TELEGRAPHY
      . . RADIO TELEMETRY
      . . . PULSE FREQUENCY MODULATION
        TELEMETRY
      . . TELEPHONY
      . RADIOTELEPHONES
      . SELECTIVE DISSEMINATION OF
        INFORMATION
      . SINGLE CHANNEL PER CARRIER
        TRANSMISSION
      . SPACE COMMUNICATION
      . . EXTRATERRESTRIAL
        COMMUNICATION
      . . INTERPLANETARY COMMUNICATION
      . . LUNAR COMMUNICATION
      . . . CIRCUMLUNAR COMMUNICATION
      . . SPACECRAFT COMMUNICATION
      . . . REENTRY COMMUNICATION
      . SPACECRAFT ANTENNAS
      . SPACECRAFT TELEVISION
      . DIGITAL SPACECRAFT TELEVISION
      . . SATELLITE TELEVISION
      . STEREOTELEVISION
      . TELEMETRY
      . . BIOTELEMETRY
      . . P.A.C.M. TELEMETRY
      . . PCM TELEMETRY
      . . RADIO TELEMETRY
      . . . PULSE FREQUENCY MODULATION
        TELEMETRY
      . TRANSOCEANIC COMMUNICATION
      . VIDEO COMMUNICATION
      . VOICE COMMUNICATION
      . TELEPHONY
      . . VOICE DATA PROCESSING
      . WIDEBAND COMMUNICATION
      . WIRELESS COMMUNICATION
  RT   ACCESS CONTROL
        ANTENNAS
        ARPA COMPUTER NETWORK
      ∞CHANNELS
        CODE DIVISION MULTIPLEXING
        COMMUNICATING
        COMMUNICATION NETWORKS
        COMMUNICATION SATELLITES
        COMPUTERS
        DATA COMPRESSION
        DATA PROCESSING
        DATA SAMPLING
        DATA TRANSMISSION
        DEMODULATION
      ∞DETECTORS
        DIGITAL SYSTEMS
        ELECTROMAGNETIC RADIATION
        FLEET SATELLITE COMMUNICATION
        SYSTEM
        FREQUENCY DIVISION MULTIPLE
        ACCESS
        FREQUENCY DIVISION MULTIPLEXING
        INFORMATION THEORY
        INTERFACES
        INTERPHONES
        MODULATION
        MOLNIYA SATELLITES
        MORSE CODE
        NASCOM NETWORK
        ONBOARD EQUIPMENT
        PACKET SWITCHING
        POINT TO POINT COMMUNICATION
        SATELLITE ANTENNAS
        SEAFARER PROJECT
        SIGNAL DETECTION
        SIGNAL DETECTORS
        SIGNAL ENCODING
        SIGNAL TRANSMISSION
      ∞SIGNALS
      ∞SYSTEMS
        TDR SATELLITES
        TELECONFERENCING
        TELEGRAPH SYSTEMS
        TELETYPEWRITER SYSTEMS
        TELEVISION SYSTEMS
        TRANSCONTINENTAL SYSTEMS
        TRANSMISSION
        TRANSMISSION CIRCUITS

TELECOMMUNICATION-*(CONT.)*
        TRANSMISSION LINES
        TRANSMITTERS
        TRANSOCEANIC SYSTEMS
        VIDEO DATA
        WESTAR SATELLITES

TELECONFERENCING
  GS   **TELECONFERENCING**
      . HET EXPERIMENT
  RT   COMMUNICATION SATELLITES
        CONFERENCES
        MULTICHANNEL COMMUNICATION
        SATELLITE NETWORKS
        SATELLITES
†   TELECOMMUNICATION

TELEGRAPH SYSTEMS
  UF   TELEGRAPHY
  RT   PULSE COMMUNICATION
        RADIO COMMUNICATION
        SINGLE CHANNEL PER CARRIER
        TRANSMISSION
      ∞SYSTEMS
        TELECOMMUNICATION
        TELEPRINTERS
        TELETYPEWRITERS
        WESTAR SATELLITES

TELEGRAPHY
  USE   TELEGRAPH SYSTEMS

TELEMETERS
  USE   TELEMETRY

TELEMETRY
  UF   TELEMETERS
  GS   TELECOMMUNICATION
      . **TELEMETRY**
      . . BIOTELEMETRY
      . . P.A.C.M. TELEMETRY
      . . PCM TELEMETRY
      . . RADIO TELEMETRY
      . . . PULSE FREQUENCY MODULATION
        TELEMETRY
  RT   ADVANCED RANGE INSTRUMENTATION
        AIRCRAFT
        COMMUNICATION EQUIPMENT
        DATA COMPRESSION
        DATA LINKS
        DATA RETRIEVAL
        DATA TRANSMISSION
        DECOMMUTATORS
        DIFFERENTIAL PULSE CODE
        MODULATION
        IN-FLIGHT MONITORING
        MEASURING INSTRUMENTS
        PULSE COMMUNICATION
        RADIO COMMUNICATION
        SHIP TO SHORE COMMUNICATION
        SIGNAL PROCESSING
        SINGLE CHANNEL PER CARRIER
        TRANSMISSION
        SPIRAL ANTENNAS
        TDR SATELLITES
        TIME DIVISION MULTIPLEXING
        TRAJECTORY MEASUREMENT
        TRANSPONDER CONTROL GROUP
        WEATHER DATA RECORDERS
        WIRELESS COMMUNICATION

TELEOPERATOR MANEUVERING SYSTEM
  USE   TELEOPERATORS

TELEOPERATORS
  UF   TELEOPERATOR MANEUVERING SYSTEM
  GS   CONTROL EQUIPMENT
      . **TELEOPERATORS**
  RT   HUMAN FACTORS ENGINEERING
        MAN MACHINE SYSTEMS
        MANIPULATORS
        REMOTE CONTROL
        REMOTE HANDLING
        ROBOTICS

TELEPHONES
  GS   **TELEPHONES**
      . RADIOTELEPHONES
  RT   EARPHONES
        TELEPHONY
        UTILITIES

TELEPHONY
  GS   TELECOMMUNICATION

TELEPHONY-(CONT.)

. RADIO COMMUNICATION
. . **TELEPHONY**
. VOICE COMMUNICATION
. . **TELEPHONY**
TRANSMISSION
. **TELEPHONY**
RT COMMUNICATION EQUIPMENT
CROSSTALK
ECHO SUPPRESSORS
RADIOTELEPHONES
SINGLE CHANNEL PER CARRIER
    TRANSMISSION
SOUND TRANSMISSION
SYMPHONIE SATELLITES
TELEPHONES
VERBAL COMMUNICATION

TELEPHOTOMETERS
USE TELEPHOTOMETRY

TELEPHOTOMETRY
UF TELEPHOTOMETERS
GS OPTICAL MEASUREMENT
. PHOTOMETRY
. . **TELEPHOTOMETRY**
RT ASTRONOMICAL PHOTOMETRY
OPTICAL MEASURING INSTRUMENTS
PHOTOMETERS
TRANSMISSOMETERS

TELEPRINTERS
GS PRINTERS
. **TELEPRINTERS**
TYPEWRITERS
. TELETYPEWRITERS
. . **TELEPRINTERS**
RT KEYING
PRINTERS (DATA PROCESSING)
RECEIVERS
TELEGRAPH SYSTEMS

TELESAT CANADA A
USE ANIK 1

TELESAT CANADA B
USE ANIK 2

TELESAT CANADA C
USE ANIK 3

TELESAT CANADA 3
USE ANIK 3

TELESCOPES
GS **TELESCOPES**
. APOLLO TELESCOPE MOUNT
. ASTRONOMICAL TELESCOPES
. . HELIOMETERS
. . . PYROHELIOMETERS
. . INFRARED TELESCOPES
. . . SPACE INFRARED TELESCOPE
        FACILITY
. . KILOMETER WAVE ORBITING
        TELESCOPE
. . SPECTROSCOPIC TELESCOPES
. . . MULTISPECTRAL TRACKING
        TELESCOPES
. . . STRATOSCOPE TELESCOPES
. . STARSAT TELESCOPE
. . ULTRAVIOLET TELESCOPES
. . X RAY TELESCOPES
. CELESCOPES
. CIRCUMSOLAR TELESCOPES
. GAMMA RAY TELESCOPES
. GRIST (TELESCOPE)
. MANNED ORBITAL TELESCOPES
. PARTICLE TELESCOPES
. RADIO TELESCOPES
. . KILOMETER WAVE ORBITING
        TELESCOPE
. REFLECTING TELESCOPES
. . STARSAT TELESCOPE
. REFRACTING TELESCOPES
. SCHMIDT CAMERAS
. SCHMIDT TELESCOPES
. SPACEBORNE TELESCOPES
. . HUBBLE SPACE TELESCOPE
. . LIRTS (TELESCOPE)
. . SOLAR OPTICAL TELESCOPE
. . SPACE INFRARED TELESCOPE
        FACILITY
. STARSAT TELESCOPE
RT ANTENNAS
ASTRONOMICAL OBSERVATORIES

TELESCOPES-(CONT.)
ASTRONOMY
BINOCULARS
EYEPIECES
GAMMA RAY OBSERVATORY
LENSES
MIRRORS
OPTICAL EQUIPMENT
OPTICAL MEASURING INSTRUMENTS
OPTICAL TRANSFER FUNCTION
PERISCOPES
REFLECTORS
SOLAR INSTRUMENTS
SPACEBORNE ASTRONOMY
ULTRAVIOLET ASTRONOMY
X RAY ASTROPHYSICS FACILITY

TELESCOPING STRUCTURES
USE FOLDING STRUCTURES

TELETYPEWRITER SYSTEMS
RT FACSIMILE COMMUNICATION
MICROWAVE TRANSMISSION
∞SYSTEMS
TELECOMMUNICATION
TELETYPEWRITERS

TELETYPEWRITERS
GS TYPEWRITERS
. **TELETYPEWRITERS**
. . TELEPRINTERS
RT KEYING
RECEIVERS
TELEGRAPH SYSTEMS
TELETYPEWRITER SYSTEMS

TELEVISION CAMERAS
GS OPTICAL EQUIPMENT
. CAMERAS
. . **TELEVISION CAMERAS**
PHOTOGRAPHIC EQUIPMENT
. CAMERAS
. . **TELEVISION CAMERAS**
TELEVISION EQUIPMENT
. **TELEVISION CAMERAS**
RT CAMERA TUBES
CLOSED CIRCUIT TELEVISION
LALLEMAND CAMERAS
OPTICAL SCANNERS
ORTHICONS
RETURN BEAM VIDICONS
SATELLITE TELEVISION

TELEVISION EQUIPMENT
GS **TELEVISION EQUIPMENT**
. IMAGE DISSECTOR TUBES
. MONOSCOPES
. TELEVISION CAMERAS
. TELEVISION RECEIVERS
RT CATHODE RAY TUBES
DIPLEXERS
∞EQUIPMENT
FLYING SPOT SCANNERS
ORTHICONS
PICTURE TUBES
VIDEO EQUIPMENT

TELEVISION RECEIVERS
GS RECEIVERS
. **TELEVISION RECEIVERS**
TELEVISION EQUIPMENT
. **TELEVISION RECEIVERS**
RT CLOSED CIRCUIT TELEVISION
TUNERS

TELEVISION RECEPTION
GS SIGNAL RECEPTION
. **TELEVISION RECEPTION**
RT COLOR TELEVISION
RADIO RECEIVERS
RADIO RECEPTION
∞RECEIVING

TELEVISION SYSTEMS
GS **TELEVISION SYSTEMS**
. ADVANCED VIDICON CAMERA SYSTEM
    (AVCS)
. CLOSED CIRCUIT TELEVISION
. COLOR TELEVISION
. EDUCATIONAL TELEVISION
. PLAT SYSTEM
. SPACECRAFT TELEVISION
. . DIGITAL SPACECRAFT TELEVISION
. . RANGER BLOCK 3 TELEVISION
        SYSTEM

TELEVISION SYSTEMS-(CONT.)
. . SATELLITE TELEVISION
. STEREOTELEVISION
RT COMMUNICATION EQUIPMENT
DIGITAL TELEVISION
EARTH TERMINALS
FACSIMILE COMMUNICATION
IMAGING TECHNIQUES
ORBIT SPECTRUM UTILIZATION
RADIO COMMUNICATION
SPACE COMMUNICATION
∞SYSTEMS
TELECOMMUNICATION
VIDEO COMMUNICATION
VIDEO DATA

TELEVISION TRANSMISSION
GS TRANSMISSION
. ELECTROMAGNETIC WAVE
    TRANSMISSION
. . **TELEVISION TRANSMISSION**
. SIGNAL TRANSMISSION
. . **TELEVISION TRANSMISSION**
RT AUTOMATIC PICTURE TRANSMISSION
CLOSED CIRCUIT TELEVISION
COLOR TELEVISION
DIGITAL TELEVISION
DOUBLE SIDEBAND TRANSMISSION
LINE OF SIGHT COMMUNICATION
MOLNIYA SATELLITES
RADIO TRANSMITTERS
SATELLITE TELEVISION
SATELLITE TRANSMISSION
SINGLE SIDEBAND TRANSMISSION
SMEAR
SPACECRAFT TELEVISION
SWEEP FREQUENCY
TIME DIVISION MULTIPLEXING
TRANSMITTERS
WAVE PROPAGATION

TELLEGEN THEORY
USE GYRATORS
NETWORK ANALYSIS
NETWORK SYNTHESIS

TELLURIC CURRENTS
UF EARTH CURRENTS
GS ELECTRIC CURRENT
. **TELLURIC CURRENTS**
ELECTRICITY
. GEOELECTRICITY
. . **TELLURIC CURRENTS**
RT ATMOSPHERIC ELECTRICITY
AURORAL ELECTROJETS
DYNAMO THEORY
GEOMAGNETIC MICROPULSATIONS

TELLURIC LINES
GS SPECTRA
. ABSORPTION SPECTRA
. . **TELLURIC LINES**
. RADIATION SPECTRA
. . ELECTROMAGNETIC SPECTRA
. . . LINE SPECTRA
. . . . **TELLURIC LINES**
. SPECTRAL BANDS
. . **TELLURIC LINES**
RT H LINES

TELLURIDES
GS CHALCOGENIDES
. **TELLURIDES**
. . BISMUTH TELLURIDES
. . CADMIUM TELLURIDES
. . INDIUM TELLURIDES
. . LANTHANUM TELLURIDES
. . LEAD TELLURIDES
. . MERCURY TELLURIDES
. . TIN TELLURIDES
. . ZINC TELLURIDES
TELLURIUM COMPOUNDS
. **TELLURIDES**
. . BISMUTH TELLURIDES
. . CADMIUM TELLURIDES
. . INDIUM TELLURIDES
. . LANTHANUM TELLURIDES
. . LEAD TELLURIDES
. . MERCURY TELLURIDES
. . TIN TELLURIDES
. . ZINC TELLURIDES
RT INTERMETALLICS

TELLURIUM
GS CHEMICAL ELEMENTS

TELLURIUM-(CONT.)
  . METALLOIDS
  . . **TELLURIUM**
  . . . TELLURIUM ISOTOPES
  . NUCLIDES
  . . ISOTOPES
  . . . **TELLURIUM**
  . . . . TELLURIUM ISOTOPES

TELLURIUM ALLOYS
GS ALLOYS
  . **TELLURIUM ALLOYS**

TELLURIUM COMPOUNDS
GS **TELLURIUM COMPOUNDS**
  . TELLURIDES
  . . BISMUTH TELLURIDES
  . . CADMIUM TELLURIDES
  . . INDIUM TELLURIDES
  . . LANTHANUM TELLURIDES
  . . LEAD TELLURIDES
  . . MERCURY TELLURIDES
  . . TIN TELLURIDES
  . . ZINC TELLURIDES
RT ∞CHEMICAL COMPOUNDS
  ∞GROUP 6A COMPOUNDS

TELLURIUM ISOTOPES
UF TELLURIUM 119
GS CHEMICAL ELEMENTS
  . METALLOIDS
  . . TELLURIUM
  . . . **TELLURIUM ISOTOPES**
  . NUCLIDES
  . . ISOTOPES
  . . . TELLURIUM
  . . . . **TELLURIUM ISOTOPES**

TELLURIUM 119
USE TELLURIUM ISOTOPES

TELLUROMETERS
GS MEASURING INSTRUMENTS
  . DISTANCE MEASURING EQUIPMENT
  . . **TELLUROMETERS**
RT GEODIMETERS
  RANGE FINDERS

TELSTAR PROJECT
GS PROGRAMS
  . PROJECTS
  . . **TELSTAR PROJECT**
RT ARTIFICIAL SATELLITES
  COMMUNICATION SATELLITES
  COMSAT PROGRAM

TELSTAR SATELLITES
GS SATELLITES
  . ARTIFICIAL SATELLITES
  . . **TELSTAR SATELLITES**
  . . . TELSTAR 1 SATELLITE
  . . . TELSTAR 2 SATELLITE
RT COMSAT PROGRAM
  THOR DELTA LAUNCH VEHICLE

TELSTAR 1 SATELLITE
GS SATELLITES
  . ARTIFICIAL SATELLITES
  . . TELSTAR SATELLITES
  . . . **TELSTAR 1 SATELLITE**

TELSTAR 2 SATELLITE
GS SATELLITES
  . ARTIFICIAL SATELLITES
  . . TELSTAR SATELLITES
  . . . **TELSTAR 2 SATELLITE**

TEMPEL 2 COMET
GS CELESTIAL BODIES
  . COMETS
  . . **TEMPEL 2 COMET**
RT ∞COMA
  METEOROIDS
  SOLAR SYSTEM

TEMPER (METALLURGY)
RT COLD WORKING
  DUCTILITY
  HARDNESS
  HEAT TREATMENT

TEMPERATE REGIONS
UF MIDLATITUDES
  SUBTROPICAL REGIONS

TEMPERATE REGIONS-(CONT.)
GS REGIONS
  . **TEMPERATE REGIONS**
RT CLIMATOLOGY
  GEOGRAPHY
  POLAR REGIONS
  TROPICAL REGIONS

TEMPERATURE
UF BODY TEMPERATURE (NON-BIOLOGICAL)
GS **TEMPERATURE**
  . ABSOLUTE ZERO
  . AMBIENT TEMPERATURE
  . ATMOSPHERIC TEMPERATURE
  . . AURORAL TEMPERATURE
  . . IONOSPHERIC TEMPERATURE
  . BODY TEMPERATURE
  . BRIGHTNESS TEMPERATURE
  . COMBUSTION TEMPERATURE
  . CRITICAL TEMPERATURE
  . CURIE TEMPERATURE
  . FLAME TEMPERATURE
  . GAS TEMPERATURE
  . HIGH TEMPERATURE
  . IGNITION TEMPERATURE
  . . FLASH POINT
  . INLET TEMPERATURE
  . ION TEMPERATURE
  . LOW TEMPERATURE
  . . ULTRALOW TEMPERATURES
  . LUNAR TEMPERATURE
  . NEEL TEMPERATURE
  . NOISE TEMPERATURE
  . OPERATING TEMPERATURE
  . PLANETARY TEMPERATURE
  . PLASMA TEMPERATURE
  . ROOM TEMPERATURE
  . SATELLITE TEMPERATURE
  . SKIN TEMPERATURE (BIOLOGY)
  . SOLAR TEMPERATURE
  . SPACE TEMPERATURE
  . SPIN TEMPERATURE
  . STAGNATION TEMPERATURE
  . STELLAR TEMPERATURE
  . SUBZERO TEMPERATURE
  . SURFACE TEMPERATURE
  . . SKIN TEMPERATURE
      (NON-BIOLOGICAL)
  . . WALL TEMPERATURE
  . TRANSITION TEMPERATURE
  . WATER TEMPERATURE
  . . OCEAN TEMPERATURE
  . . . SEA SURFACE TEMPERATURE
RT ABLATIVE MATERIALS
  ADIABATIC CONDITIONS
  AIR CONDITIONING
  BIOLOGICAL EFFECTS
  CLIMATOLOGY
  COMFORT
  CONVECTIVE FLOW
  ELECTRON ENERGY
  EMISSIVITY
  ENVIRONMENTS
  FREE CONVECTION
  GEOTEMPERATURE
  GIBBS-HELMHOLTZ EQUATIONS
  HEAT
  HEAT SHIELDING
  HEAT STORAGE
  HEATING
  HUMIDITY
  ISOTHERMS
  LAPSE RATE
  MELTING POINTS
  METEOROLOGY
  OCEAN THERMAL ENERGY CONVERSION
  REFRIGERATING
  SAHA EQUATIONS
  SURFACE COOLING
  TEMPERATURE COMPENSATION
  TEMPERATURE CONTROL
  TEMPERATURE DEPENDENCE
  TEMPERATURE DISTRIBUTION
  TEMPERATURE EFFECTS
  TEMPERATURE GRADIENTS
  TEMPERATURE INVERSIONS
  TEMPERATURE MEASUREMENT
  TEMPERATURE MEASURING
    INSTRUMENTS
  TEMPERATURE PROBES
  TEMPERATURE PROFILES
  TEMPERATURE RATIO
  TEMPERATURE SCALES
  TEMPERATURE SENSORS
  TEPHIGRAMS
  THERMAL ABSORPTION

TEMPERATURE-(CONT.)
  THERMAL ANALYSIS
  THERMAL BLOOMING
  THERMAL BOUNDARY LAYER
  THERMAL BUCKLING
  THERMAL COMFORT
  THERMAL CONDUCTIVITY
  THERMAL CONDUCTIVITY GAGES
  THERMAL CONDUCTORS
  THERMAL CONTROL COATINGS
  THERMAL CYCLING TESTS
  THERMAL DECOMPOSITION
  THERMAL DEGRADATION
  THERMAL DIFFUSION
  THERMAL DIFFUSIVITY
  THERMAL DISSOCIATION
  THERMAL EMISSION
  THERMAL ENERGY
  THERMAL ENVIRONMENTS
  THERMAL EXPANSION
  THERMAL FATIGUE
  THERMAL INSTABILITY
  THERMAL INSULATION
  THERMAL MAPPING
  THERMAL NEUTRONS
  THERMAL NOISE
  THERMAL PLASMAS
  THERMAL POLLUTION
  THERMAL PROTECTION
  THERMAL RADIATION
  THERMAL REACTORS
  THERMAL RESISTANCE
  THERMAL RESOURCES
  THERMAL SHOCK
  THERMAL SIMULATION
  THERMAL STABILITY
  THERMAL STRESSES
  THERMAL VACUUM TESTS
  THERMODYNAMIC EFFICIENCY
  THERMODYNAMIC PROPERTIES
  VENTILATION

TEMPERATURE COMPENSATION
GS INSTRUMENT COMPENSATION
  . **TEMPERATURE COMPENSATION**
RT ∞COMPENSATION
  TEMPERATURE

TEMPERATURE CONTROL
UF HEAT REGULATION
RT AIR CONDITIONING
  AUTOMATIC CONTROL
  AUTOMATIC CONTROL VALVES
  CHEMICAL REACTION CONTROL
  COMBUSTION CONTROL
  ∞CONTROL
  CONTROLLERS
  COOLING
  COOLING SYSTEMS
  CRYOSTATS
  ENGINE CONTROL
  ENVIRONMENTAL CONTROL
  ENVIRONMENTAL ENGINEERING
  EXHAUST SYSTEMS
  HEAT SHIELDING
  HEATING
  HEATING EQUIPMENT
  HIGH TEMPERATURE TESTS
  INFRARED SUPPRESSION
  LOW TEMPERATURE TESTS
  MANUAL CONTROL
  PLASMA COOLING
  REFRIGERATING
  REFRIGERATING MACHINERY
  REMOTE CONTROL
  REUSABLE HEAT SHIELDING
  SPACE COOLING (BUILDINGS)
  SPACE HEATING (BUILDINGS)
  SPACECRAFT TEMPERATURE
  TEMPERATURE
  THERMAL CONTROL COATINGS
  THERMAL CYCLING TESTS
  THERMAL INSULATION
  THERMOMETERS
  THERMOREGULATION
  THERMOSTATS
  TRANSPIRATION
  VENTILATION
  WATER HEATING

TEMPERATURE DEPENDENCE
GS DEPENDENCE
  . **TEMPERATURE DEPENDENCE**
RT TEMPERATURE

TEMPERATURE DIFFERENCES
USE   TEMPERATURE GRADIENTS

TEMPERATURE DISTRIBUTION
UF   TEMPERATURE FIELDS
GS   DISTRIBUTION (PROPERTY)
 .   TEMPERATURE DISTRIBUTION
RT   AIR CONDITIONING
      COOLING
      COOLING SYSTEMS
      ENVIRONMENTAL ENGINEERING
      FIELD THEORY (PHYSICS)
      HEAT TREATMENT
      HEATING
      ISOTHERMAL FLOW
      ISOTHERMAL LAYERS
      ISOTHERMS
      OCEAN TEMPERATURE
      REFRIGERATING
      SATELLITE TEMPERATURE
      TEMPERATURE
      THERMAL MAPPING
      THERMAL RESOURCES
      THERMAL SHOCK
      THERMAL STRESSES
      THERMOGRAPHY
      VENTILATION
      VERTICAL DISTRIBUTION
      WATER TEMPERATURE

TEMPERATURE EFFECTS
UF   HEAT EFFECTS
      PHOTOTHERMOTROPISM
      RICHARDSON-DUSHMAN EQUATION
      THERMAL EFFECTS
      THERMOTROPISM
RT   ABLATION
      ABSOLUTE ZERO
      CHEMICAL EFFECTS
      ∞EFFECTS
      ETTINGSHAUSEN EFFECT
      JET BLAST EFFECTS
      MAGNETIC EFFECTS
      NERNST-ETTINGSHAUSEN EFFECT
      PELTIER EFFECTS
      PRESSURE EFFECTS
      ∞RADIATION
      REENTRY EFFECTS
      SEEBECK EFFECT
      SHAPE MEMORY ALLOYS
      SOLAR GRANULATION
      STERILIZATION EFFECTS
      TEMPERATURE
      THERMAL BUCKLING
      THERMAL DEGRADATION
      THERMAL DISSOCIATION
      THERMAL RESISTANCE
      THERMAL STRESSES
      THERMOGRAVIMETRY
      THERMOLUMINESCENCE
      THERMOPLASTICITY
      TIME TEMPERATURE PARAMETER

TEMPERATURE FIELDS
USE   TEMPERATURE DISTRIBUTION

TEMPERATURE GRADIENTS
UF   TEMPERATURE DIFFERENCES
GS   GRADIENTS
 .   TEMPERATURE GRADIENTS
 . . THERMOCLINES
RT   ATMOSPHERIC TEMPERATURE
      BATHYTHERMOGRAPHS
      CHAPMAN-ENSKOG THEORY
      CONVECTIVE HEAT TRANSFER
      ISOTHERMAL LAYERS
      ISOTHERMS
      NONISOTHERMAL PROCESSES
      OCEAN TEMPERATURE
      POTENTIAL GRADIENTS
      STRATIFICATION
      TEMPERATURE
      THERMAL ANALYSIS
      THERMAL MAPPING
      THERMOMIGRATION

TEMPERATURE INDICATORS
USE   INDICATING INSTRUMENTS
       TEMPERATURE MEASURING
        INSTRUMENTS

TEMPERATURE INSTRUMENTS
USE   TEMPERATURE MEASURING
        INSTRUMENTS

TEMPERATURE INVERSIONS
GS   INVERSIONS
 .   TEMPERATURE INVERSIONS
RT   AIR POLLUTION
      ATMOSPHERIC TEMPERATURE
      BENDING
      BIREFRINGENCE
      BUCKLING
      CRACKING (FRACTURING)
      CRACKS
      CREEP PROPERTIES
      DEFLECTION
      DEFORMATION
      DISPLACEMENT
      DISTORTION
      FAILURE
      FATIGUE (MATERIALS)
      INTERNAL PRESSURE
      LAPSE RATE
      MECHANICAL PROPERTIES
      METEOROLOGICAL PARAMETERS
      METEOROLOGY
      PHOTOELASTIC ANALYSIS
      PLASTIC DEFORMATION
      PRESSURE
      PRESSURE EFFECTS
      RESIDUAL STRESS
      SAINT VENANT PRINCIPLE
      SHEAR PROPERTIES
      SHRINKAGE
      STRAIN GAGES
      STRAIN HARDENING
      STRAIN RATE
      STRESS ANALYSIS
      STRESS RELAXATION
      STRESS WAVES
      STRESSES
      STRUCTURAL STRAIN
      TEMPERATURE
      TENSILE DEFORMATION
      ∞TENSION
      TEPHIGRAMS
      TORSION
      VOLUMETRIC STRAIN
      X RAY STRESS ANALYSIS
      YIELD STRENGTH

TEMPERATURE MEASUREMENT
UF   PYROMETRY
      THERMOMETRY
RT   ANOMALOUS TEMPERATURE ZONES
      BOLOMETERS
      BRIGHTNESS TEMPERATURE
      CRAYONS
      GAS TEMPERATURE
      HIGH TEMPERATURE
      ∞MEASUREMENT
      NOISE TEMPERATURE
      PYROMETERS
      RADIATION PYROMETERS
      RESISTANCE THERMOMETERS
      SATELLITE TEMPERATURE
      SODAR
      SOUND DETECTING AND RANGING
      TEMPERATURE
      THERMOCOUPLE PYROMETERS
      THERMOCOUPLES
      THERMOGRAPHY
      THERMOMETERS
      WIND TUNNEL CALIBRATION

TEMPERATURE MEASURING INSTRUMENTS
UF   TEMPERATURE INDICATORS
      TEMPERATURE INSTRUMENTS
      THERMOGRAMS
GS   MEASURING INSTRUMENTS
 .   TEMPERATURE MEASURING
       INSTRUMENTS
 . . BATHYTHERMOGRAPHS
 . . OPTICAL PYROMETERS
 . . PYROMETERS
 . . . RADIATION PYROMETERS
 . . . THERMOCOUPLE PYROMETERS
 . . TEMPERATURE PROBES
 . . . PNEUMATIC PROBES
 . . THERMOMETERS
 . . . RESISTANCE THERMOMETERS
RT   ANOMALOUS TEMPERATURE ZONES
      BOLOMETERS
      BOMB CALORIMETERS
      CALORIMETERS
      DROP CALORIMETERS
      FLAME CALORIMETERS
      FLAME PROBES
      TEMPERATURE

TEMPERATURE MEASURING-*(CONT.)*
      THERMISTORS
      THERMOCOUPLES
      THERMOPILES
      THERMOSTATS
      TRANSDUCERS

TEMPERATURE PROBES
GS   MEASURING INSTRUMENTS
 .   TEMPERATURE MEASURING
       INSTRUMENTS
 . . TEMPERATURE PROBES
 . . . PNEUMATIC PROBES
RT   TEMPERATURE
      THERMOCOUPLES

TEMPERATURE PROFILES
RT   HEAT TRANSFER
      TEMPERATURE
      THERMAL ANALYSIS

TEMPERATURE RATIO
RT   DATA CORRELATION
      HEAT TRANSFER
      RATIOS
      TEMPERATURE

TEMPERATURE SCALES
UF   FAHRENHEIT TEMPERATURE SCALE
      INTERNATIONAL PRACTICAL
       TEMPERATURE
RT   ABSOLUTE ZERO
      ANOMALOUS TEMPERATURE ZONES
      CALIBRATING
      ∞SCALE
      STANDARDS
      TEMPERATURE
      THERMOMETERS

TEMPERATURE SENSORS
GS   TEMPERATURE SENSORS
 .   THERMISTORS
RT   ANOMALOUS TEMPERATURE ZONES
      TEMPERATURE

TEMPERING
GS   HEAT TREATMENT
 .   TEMPERING
RT   ANNEALING
      ∞DRAWING
      HARDENING (MATERIALS)
      LASER ANNEALING
      METAL WORKING
      NORMALIZING (HEAT TREATMENT)
      STRESS RELIEVING
      STRETCHING

TEMPLATES
RT   LOFTING
      MOLDS
      ∞PATTERNS

TEMPORAL DISTRIBUTION
RT   ANNUAL VARIATIONS
      SPATIAL DISTRIBUTION
      TIME DEPENDENCE
      ∞TIME RESPONSE

TEMPORAL RESOLUTION
UF   MULTITEMPORAL ANALYSIS
GS   RESOLUTION
 .   TEMPORAL RESOLUTION
RT   SPATIAL RESOLUTION

TENDENCIES
RT   ∞INCLINATION

TENDONS
RT   CONNECTIVE TISSUE
      FIBROBLASTS

TENITE
RT   CELLULOSE
      MOLDING MATERIALS

TENNESSEE
GS   NATIONS
 .   UNITED STATES
 . . TENNESSEE
RT   GREAT SMOKY MOUNTAINS (NC-TN)
      TENNESSEE VALLEY (AL-KY-TN)

TENNESSEE VALLEY (AL-KY-TN)
GS   VALLEYS

TENNESSEE VALLEY (AL-KY-TN)-*(CONT.)*
. **TENNESSEE VALLEY (AL-KY-TN)**
RT   ALABAMA
     KENTUCKY
     TENNESSEE

TENSILE CREEP
GS   MECHANICAL PROPERTIES
     . CREEP PROPERTIES
     . . **TENSILE CREEP**
RT   PLASTIC DEFORMATION
     SHEAR CREEP

TENSILE DEFORMATION
GS   DEFORMATION
     . **TENSILE DEFORMATION**
RT   ELASTIC DEFORMATION
     ELONGATION
     PLASTIC DEFORMATION
     TEMPERATURE INVERSIONS

TENSILE PROPERTIES
GS   MECHANICAL PROPERTIES
     . **TENSILE PROPERTIES**
RT   ELASTIC PROPERTIES
     HIGH STRENGTH ALLOYS
     ∞PROPERTIES

TENSILE STRENGTH
GS   MECHANICAL PROPERTIES
     . **TENSILE STRENGTH**
RT   DUCTILITY
     ELASTIC PROPERTIES
     ELONGATION
     FIBER STRENGTH
     HIGH STRENGTH
     HYSTERESIS
     POISSON RATIO
     RESIDUAL STRENGTH
     RESILIENCE
     SHEAR STRENGTH
     ∞STRENGTH
     TOUGHNESS

TENSILE STRESS
GS   STRESSES
     . **TENSILE STRESS**
RT   AXIAL STRESS
     HIGH STRENGTH
     HOOPS
     STRESS INTENSITY FACTORS
     ∞TENSION
     TRIAXIAL STRESSES

TENSILE TESTS
RT   DESTRUCTIVE TESTS
     FATIGUE TESTS
     LOAD TESTS
     SPECIMEN GEOMETRY
     STATIC TESTS
     ∞TESTS

TENSIOMETERS
GS   MEASURING INSTRUMENTS
     . **TENSIOMETERS**
RT   CABLE FORCE RECORDERS
     MECHANICAL MEASUREMENT

∞ TENSION
SN   *(USE OF A MORE SPECIFIC TERM IS
     RECOMMENDED--CONSULT THE TERMS
     LISTED BELOW)*
RT   BLOOD PRESSURE
     INTERFACIAL TENSION
     PARTIAL PRESSURE
     STRETCHING
     TEMPERATURE INVERSIONS
     TENSILE STRESS

TENSOMETERS
GS   MEASURING INSTRUMENTS
     . **TENSOMETERS**
RT   DEFORMETERS
     EXTENSOMETERS
     STRAIN GAGES
     STRESS MEASUREMENT
     WEIGHT INDICATORS

TENSOR ANALYSIS
GS   GEOMETRY
     . DIFFERENTIAL GEOMETRY
     . . **TENSOR ANALYSIS**
RT   RELATIVITY
     SCALARS

TENSOR ANALYSIS-*(CONT.)*
     SCALERS
     TENSORS

TENSOR FIELDS
USE   TENSORS

TENSORS
UF   TENSOR FIELDS
     TRANSFORMATION TENSORS
GS   ALGEBRA
     . **TENSORS**
     . . STRESS TENSORS
RT   FIELD THEORY (PHYSICS)
     JORDAN FORM
     SCALARS
     TENSOR ANALYSIS

TEPHIGRAMS
GS   DIAGRAMS
     . **TEPHIGRAMS**
RT   ATMOSPHERIC TURBULENCE
     ENTROPY
     LAPSE RATE
     TEMPERATURE
     TEMPERATURE INVERSIONS
     THERMODYNAMIC PROPERTIES

TERBIUM
GS   CHEMICAL ELEMENTS
     . RARE EARTH ELEMENTS
     . . **TERBIUM**
     . . . TERBIUM ISOTOPES
     METALS
     . RARE EARTH ELEMENTS
     . . **TERBIUM**
     . . . TERBIUM ISOTOPES

TERBIUM ISOTOPES
UF   TERBIUM 155
     TERBIUM 161
GS   CHEMICAL ELEMENTS
     . NUCLIDES
     . . ISOTOPES
     . . . **TERBIUM ISOTOPES**
     . RARE EARTH ELEMENTS
     . . TERBIUM
     . . . **TERBIUM ISOTOPES**
     METALS
     . RARE EARTH ELEMENTS
     . . TERBIUM
     . . . **TERBIUM ISOTOPES**

TERBIUM 155
USE   TERBIUM ISOTOPES

TERBIUM 161
USE   TERBIUM ISOTOPES

TERCOM
UF   TERRAIN CONTOUR MATCHING
     NAVIGATION SYSTEM
GS   NAVIGATION AIDS
     . **TERCOM**
     ONBOARD EQUIPMENT
     . AIRBORNE EQUIPMENT
     . . **TERCOM**
     . AIRCRAFT EQUIPMENT
     . . **TERCOM**
RT   DISPLAY DEVICES
     FLIGHT INSTRUMENTS
     MAP MATCHING GUIDANCE
     NAVIGATION INSTRUMENTS
     ∞SYSTEMS
     VIDEO LANDMARK ACQUISITION AND
       TRACKING

TEREPHTHALATE
GS   ACIDS
     . FATTY ACIDS
     . . CARBOXYLIC ACIDS
     . . . DICARBOXYLIC ACIDS
     . . . . **TEREPHTHALATE**

TERMINAL AREA ENERGY MANAGEMENT
GS   MANAGEMENT
     . **TERMINAL AREA ENERGY
       MANAGEMENT**
RT   DIGITAL TECHNIQUES
     SPACE SHUTTLE ORBITERS
     SPACE TRANSPORTATION
     SPACECRAFT LANDING

TERMINAL BALLISTICS
UF   PENETRATION BALLISTICS
     PROJECTILE PENETRATION
     TARGET PENETRATION
GS   BALLISTICS
     . **TERMINAL BALLISTICS**
RT   ENERGY TRANSFER
     FRAGMENTATION
     MISSILES
     PENETRATION
     PROJECTILES

TERMINAL CONFIGURED VEHICLE PROGRAM
UF   TCV PROGRAM
GS   PROGRAMS
     . NASA PROGRAMS
     . . **TERMINAL CONFIGURED VEHICLE
       PROGRAM**
RT   AIRCRAFT DESIGN
     AUTOMATIC CONTROL
     AUTOMATIC FLIGHT CONTROL
     AUTOMATIC LANDING CONTROL
     ELECTRONIC CONTROL
     FEEDBACK CONTROL
     ∞VEHICLES

TERMINAL FACILITIES
GS   **TERMINAL FACILITIES**
     . SHIP TERMINALS
RT   ARTIFICIAL HARBORS
     DEEPWATER TERMINALS
     ∞FACILITIES
     HARBORS
     OFFSHORE DOCKING
     OFFSHORE PLATFORMS
     SITE SELECTION
     TANKER TERMINALS
     ∞TERMINALS
     TRANSPORTATION
     WHARVES

TERMINAL GUIDANCE
GS   GUIDANCE (MOTION)
     . **TERMINAL GUIDANCE**
     . . LASER GUIDANCE
RT   COMMAND GUIDANCE
     DESCENT TRAJECTORIES
     ENTRY GUIDANCE (STS)
     GLIDE PATHS
     HOMING
     INERTIAL GUIDANCE
     MIDCOURSE GUIDANCE
     REENTRY
     REENTRY GUIDANCE
     REENTRY TRAJECTORIES
     REENTRY VEHICLES
     RENDEZVOUS GUIDANCE
     SPACECRAFT GUIDANCE

TERMINAL VELOCITY
GS   RATES (PER TIME)
     . **TERMINAL VELOCITY**
     VELOCITY
     . **TERMINAL VELOCITY**
RT   GRAVITATION

∞ TERMINALS
SN   *(USE OF A MORE SPECIFIC TERM IS
     RECOMMENDED--CONSULT THE TERMS
     LISTED BELOW)*
RT   CONNECTORS
     DATA PROCESSING TERMINALS
     ELECTRIC TERMINALS
     ∞HEADERS
     JUMPERS
     OUTLETS
     SHIP TERMINALS
     TERMINAL FACILITIES

TERMINATING
USE   STOPPING

TERMINATOR LINES
RT   ∞LINES
     LUNAR PHASES
     ∞PHASES
     SUNRISE
     SUNSET

TERMINOLOGY
RT   DICTIONARIES
     NOMENCLATURES
     THESAURI
     WORDS (LANGUAGE)

TERMS
RT   INFORMATION THEORY
         THESAURI

TERNARY ALLOYS
GS   ALLOYS
      . **TERNARY ALLOYS**
      . . ASTROLOY (TRADEMARK)

TERNARY SYSTEMS
RT   ALLOYS
         BINARY SYSTEMS (MATERIALS)
         SOLID SOLUTIONS
         ∞SYSTEMS

TERNARY SYSTEMS (DIGITAL)
USE   DIGITAL SYSTEMS

TERPENES
GS   **TERPENES**
      . AZULENE
      . CAMPHOR
      . MECAMYLAMINE
      . MENTHOL
      . TURPENTINE
RT   ALKENES

TERPHENYLS
GS   PHENYLS
      . **TERPHENYLS**

TERRACES (LANDFORMS)
GS   LANDFORMS
      . **TERRACES (LANDFORMS)**
      . . PLATEAUS
      . . . ALLEGHENY PLATEAU (US)
      . . . COLORADO PLATEAU (US)
      . . . MESAS
      . . . . BUTTES
      . . . PIEDMONTS
      . . . . CENTRAL PIEDMONT (US)
RT   FORMATIONS
         MOUNTAINS

TERRADYNAMICS
RT   ∞DYNAMICS
         EARTH SURFACE
         GEODYNAMICS
         PROJECTILES

TERRAIN
UF   LANDSCAPE
GS   TOPOGRAPHY
      . **TERRAIN**
RT   GEOMORPHOLOGY
         LANDFORMS
         LANDMARKS

TERRAIN ANALYSIS
UF   SATAN (SENSOR)
RT   ∞ANALYZING
         CHANGE DETECTION
         EARTH RESOURCES
         EROS (SATELLITES)
         GEOGRAPHIC APPLICATIONS PROGRAM
         HOLOGRAMMETRY
         MAPPING
         NAP-OF-THE-EARTH NAVIGATION
         PHOTOGRAMMETRY
         RECONNAISSANCE
         SATELLITE SURFACES
         SOIL MAPPING
         VIDEO LANDMARK ACQUISITION AND
            TRACKING

**TERRAIN CONTOUR MATCHING NAVIGATION
SYSTEM**
USE   TERCOM

TERRAIN FOLLOWING AIRCRAFT
GS   **TERRAIN FOLLOWING AIRCRAFT**
      . TSR-2 AIRCRAFT
RT   AH-1G HELICOPTER
         AH-63 HELICOPTER
         AH-64 HELICOPTER
         ∞AIRCRAFT
         ATTACK AIRCRAFT
         JET AIRCRAFT
         LIGHT AIRCRAFT
         ∞MILITARY AIRCRAFT
         NAP-OF-THE-EARTH NAVIGATION
         OBSERVATION AIRCRAFT
         ∞SUBSONIC AIRCRAFT
         UTILITY AIRCRAFT

TERRESTRIAL DUST BELT
GS   DUST
      . **TERRESTRIAL DUST BELT**
RT   ∞BELTS
         COSMIC DUST
         GEGENSCHEIN
         METEOROID DUST CLOUDS
         MICROMETEOROIDS
         ZODIACAL DUST

TERRESTRIAL MAGNETISM
USE   GEOMAGNETISM

TERRESTRIAL PLANETS
GS   CELESTIAL BODIES
      . PLANETS
      . . **TERRESTRIAL PLANETS**
      . . . EARTH (PLANET)
      . . . MARS (PLANET)
      . . . MERCURY (PLANET)
      . . . VENUS (PLANET)
RT   CELESTIAL MECHANICS
         PLANETARY ENVIRONMENTS
         PLANETOLOGY
         SOLAR SYSTEM

TERRESTRIAL RADIATION
SN   (EXCLUDES ATMOSPHERIC RADIATION
         AND REFLECTED VISIBLE LIGHT)
UF   EARTH RADIATION
GS   ELECTROMAGNETIC RADIATION
      . **TERRESTRIAL RADIATION**
RT   ATMOSPHERIC RADIATION
         EARTH (PLANET)
         EARTH ALBEDO
         EARTH RADIATION BUDGET
            EXPERIMENT
         EXTRATERRESTRIAL RADIATION
         FAR INFRARED RADIATION
         GREENHOUSE EFFECT
         INFRARED RADIATION
         NEAR INFRARED RADIATION
         PLANETARY RADIATION
         ∞RADIATION
         TROPOSPHERIC RADIATION

TERRIER MISSILE
GS   MISSILES
      . ANTIAIRCRAFT MISSILES
      . . **TERRIER MISSILE**
      . SURFACE TO AIR MISSILES
      . . **TERRIER MISSILE**
RT   BUMBLEBEE PROJECT
         MULTISTAGE ROCKET VEHICLES
         SOLID PROPELLANT ROCKET ENGINES

TESSERAL HARMONICS
GS   ANALYSIS (MATHEMATICS)
      . FUNCTIONAL ANALYSIS
      . . HARMONIC ANALYSIS
      . . . **TESSERAL HARMONICS**
      HARMONICS
      . **TESSERAL HARMONICS**
RT   SATELLITE PERTURBATION

TEST BEDS
USE   TEST EQUIPMENT

TEST CHAMBERS
UF   ENVIRONMENTAL CHAMBERS
GS   COMPARTMENTS
      . **TEST CHAMBERS**
      . . ANECHOIC CHAMBERS
      . . PRESSURE CHAMBERS
      . . . HYPERBARIC CHAMBERS
      . . . VACUUM CHAMBERS
RT   ∞CAPSULES
         ∞CHAMBERS
         CRYOGENIC WIND TUNNELS
         ENVIRONMENT MODELS
         ENVIRONMENT SIMULATORS
         ENVIRONMENTAL CONTROL
         ENVIRONMENTAL LABORATORIES
         ENVIRONMENTAL TESTS
         THERMAL VACUUM TESTS
         VACUUM TESTS
         WIND TUNNELS

∞ **TEST EQUIPMENT**
SN   *(USE OF A MORE SPECIFIC TERM IS
         RECOMMENDED--CONSULT THE TERMS
         LISTED BELOW)*

TEST EQUIPMENT-(CONT.)
UF   CHECKOUT EQUIPMENT
         TEST BEDS
         TESTERS
         TESTING MACHINES
RT   ANALYZERS
         ASTRIONICS
         AUTOMATIC TEST EQUIPMENT
         AVIONICS
         ∞CAPSULES
         CEFOAM CHECKOUT EQUIPMENT
         CENTRIFUGES
         CHECKOUT
         DYNAMOMETERS
         EARTH TERMINAL MEASUREMENT
            SYSTEM
         ELECTRONIC EQUIPMENT TESTS
         ∞EQUIPMENT
         FATIGUE TESTING MACHINES
         FREE FLIGHT TEST APPARATUS
         FREQUENCY ANALYZERS
         GEOPHYSICAL FLUID FLOW CELLS
         GROUND SUPPORT EQUIPMENT
         HYPERSONIC TEST APPARATUS
         IMPACT TESTING MACHINES
         LOAD TESTING MACHINES
         MEASURING INSTRUMENTS
         MONOSCOPES
         ONBOARD EQUIPMENT
         ROCKET PROPELLED SLEDS
         SAMPLERS
         SIMULATORS
         SUPERSONIC TEST APPARATUS
         TEST FACILITIES
         TEST PATTERN GENERATORS
         TEST STANDS
         WIND TUNNEL MODELS
         WIND TUNNELS

TEST FACILITIES
GS   **TEST FACILITIES**
      . ANECHOIC CHAMBERS
      . CENTRAL ATLANTIC REGIONAL ECOL
         TEST SITE
      . ENGINE TESTING LABORATORIES
      . ENVIRONMENTAL LABORATORIES
      . HYDRAULIC TEST TUNNELS
      . ROCKET TEST FACILITIES
      . TEST RANGES
      . . BALLISTIC RANGES
      . . MISSILE RANGES
      . TEST STANDS
      . TRANSIENT REACTOR TEST FACILITY
      . WIND TUNNELS
      . . BLOWDOWN WIND TUNNELS
      . . COMBUSTION WIND TUNNELS
      . . CRYOGENIC WIND TUNNELS
      . . HYPERSONIC WIND TUNNELS
      . . . CASCADE WIND TUNNELS
      . . . HOTSHOT WIND TUNNELS
      . . . PLASMA JET WIND TUNNELS
      . . . SHOCK TUNNELS
      . . HYPERVELOCITY WIND TUNNELS
      . . . CASCADE WIND TUNNELS
      . . . HOTSHOT WIND TUNNELS
      . . . PLASMA JET WIND TUNNELS
      . . . SHOCK TUNNELS
      . . LOW DENSITY WIND TUNNELS
      . . LOW SPEED WIND TUNNELS
      . . . SUBSONIC WIND TUNNELS
      . . RECTANGULAR WIND TUNNELS
      . . SLOTTED WIND TUNNELS
      . . SUPERSONIC WIND TUNNELS
      . . TRANSONIC WIND TUNNELS
      . . TRISONIC WIND TUNNELS
RT   ARIZONA REGIONAL ECOLOGICAL TEST
            SITE
         ∞FACILITIES
         FLIGHT SIMULATORS
         LABORATORIES
         MODELS
         MOTION SIMULATORS
         RESEARCH FACILITIES
         SHOCK TUBES
         SIMULATORS
         SOLAR SIMULATORS
         SPACECRAFT CABIN SIMULATORS
         ∞TEST EQUIPMENT
         ∞TESTS

TEST FIRING
GS   FIRING (IGNITING)
      . **TEST FIRING**
      . . STATIC FIRING
RT   ENGINE TESTS

TEST FIRING-*(CONT.)*
    FUEL TESTS
    GROUND TESTS
    MISSILE TESTS
    PREFIRING TESTS
    PRELAUNCH TESTS
    ROCKET FIRING
    ROCKET TEST FACILITIES
    STATIC TESTS
    ∞ TESTS

TEST PATTERN GENERATORS
RT   ∞ FAULTS
     ∞ GENERATORS
     ∞ PATTERNS
     ∞ TEST EQUIPMENT

TEST PILOTS
GS   PERSONNEL
     . FLYING PERSONNEL
     . . PILOTS (PERSONNEL)
     . . . AIRCRAFT PILOTS
     . . . . **TEST PILOTS**
     . OPERATORS (PERSONNEL)
     . . PILOTS (PERSONNEL)
     . . . AIRCRAFT PILOTS
     . . . . **TEST PILOTS**
RT   ∞ PILOTS

TEST RANGES
GS   RANGES (FACILITIES)
     . **TEST RANGES**
     . . BALLISTIC RANGES
     . . MISSILE RANGES
     TEST FACILITIES
     . **TEST RANGES**
     . . BALLISTIC RANGES
     . . MISSILE RANGES
RT   DOWNRANGE
     DOWNRANGE MEASUREMENT
     RANGE SAFETY
     ROCKET TEST FACILITIES

TEST STANDS
GS   TEST FACILITIES
     . **TEST STANDS**
RT   ENGINE TESTS
     FLAME DEFLECTORS
     PREFIRING TESTS
     PRELAUNCH TESTS
     ROCKET TEST FACILITIES
     ∞ TEST EQUIPMENT

TEST VEHICLES
GS   **TEST VEHICLES**
     . FLIGHT TEST VEHICLES
RT   ∞ AIRCRAFT
     ALTITUDE TESTS
     ∞ BALLISTIC VEHICLES
     ∞ CAPSULES
     ELECTRIC MOTOR VEHICLES
     HIGH ALTITUDE TESTS
     HYPERSONIC VEHICLES
     LAUNCH VEHICLES
     MISSILE TESTS
     MISSILES
     REENTRY VEHICLES
     RESEARCH AIRCRAFT
     ROCKET VEHICLES
     ∞ SPACECRAFT
     ∞ TESTS
     TOWED BODIES
     ∞ VEHICLES

TESTERS
USE   TEST EQUIPMENT

TESTES
GS   ANATOMY
     . GENITOURINARY SYSTEM
     . . **TESTES**
     . GLANDS (ANATOMY)
     . . SEX GLANDS
     . . . **TESTES**
     . ORGANS
     . . **TESTES**
     REPRODUCTIVE SYSTEMS
     . **TESTES**
     VISCERA
     . ORGANS
     . . **TESTES**
     . SEX GLANDS
     . . **TESTES**
RT   ∞ REPRODUCTION

TESTING
USE   TESTS

TESTING MACHINES
USE   TEST EQUIPMENT

TESTING TIME
GS   TIME
     . **TESTING TIME**
RT   BURNING TIME
     ENGINE TESTS
     FATIGUE TESTS
     FLIGHT TIME
     ∞ TESTS
     TURNAROUND (STS)
     WINDOWS (INTERVALS)

∞ **TESTS**
SN   *(USE OF A MORE SPECIFIC TERM IS*
     RECOMMENDED--CONSULT THE TERMS
     LISTED BELOW)
UF   PRETESTS
     TESTING
RT   ACCELERATED LIFE TESTS
     ACCEPTABILITY
     ACCURACY
     ADHESION TESTS
     ALTITUDE TESTS
     APPROACH AND LANDING TESTS (STS)
     BEND TESTS
     CAPTIVE TESTS
     CHECKOUT
     CHEMICAL ANALYSIS
     CHEMICAL TESTS
     COLD FLOW TESTS
     COLD WEATHER TESTS
     COMPRESSION TESTS
     COMPUTATIONAL CHEMISTRY
     CONFIDENCE LIMITS
     CORROSION TEST LOOPS
     CORROSION TESTS
     CREEP TESTS
     CREW PROCEDURES (INFLIGHT)
     CREW PROCEDURES (PREFLIGHT)
     DAMPING TESTS
     DESTRUCTIVE TESTS
     DROP TESTS
     DYNAMIC TESTS
     EDUCATION
     ELECTRIC EQUIPMENT TESTS
     ELECTRONIC EQUIPMENT TESTS
     EMPLOYMENT
     ENGINE TESTS
     ENVIRONMENTAL TESTS
     ERRORS
     EVALUATION
     EXAMINATION
     EXTRAPOLATION
     FATIGUE TESTS
     FLIGHT STABILITY TESTS
     FLIGHT TESTS
     FUEL TESTS
     FULL SCALE TESTS
     GROUND TESTS
     HARDNESS TESTS
     HIGH ALTITUDE TESTS
     HIGH TEMPERATURE TESTS
     IMPACT TESTS
     LABORATORIES
     LOAD TESTS
     LOW TEMPERATURE TESTS
     LUBRICANT TESTS
∞ MATERIALS TESTS
     MEDIAN (STATISTICS)
     MISSILE TESTS
     NONDESTRUCTIVE TESTS
     NOTCH TESTS
     ORBITAL SPACE STATIONS
     ORBITAL SPACE TESTS
     PATCH TESTS
     PERFORMANCE TESTS
     PERSONALITY TESTS
     PHYSIOLOGICAL TESTS
     PREFIRING TESTS
     PREFLIGHT ANALYSIS
     PRELAUNCH TESTS
     PROGRAM VERIFICATION (COMPUTERS)
     PROPELLANT TESTS
     PROVING
     PSYCHOLOGICAL TESTS
     QUALIFICATIONS
     QUALITY
     QUALITY CONTROL
     RAILROAD HUMPING TESTS
     RANK TESTS

TESTS-*(CONT.)*
    REACTOR STARTUP TESTS
    RECORDS
    RELIABILITY
    RESONANCE TESTING
    RORSCHACH TESTS
    SALT SPRAY TESTS
    SAMPLING
    SELECTION
    SHOCK TESTS
    SNELLEN TESTS
    SPACE ELECTRIC ROCKET TESTS
    SPACE TRANSPORTATION SYSTEM
      FLIGHTS
    SPACE VEHICLE CHECKOUT PROGRAM
    SPIN TESTS
    STABILITY TESTS
    STATIC TESTS
    STATISTICAL TESTS
    STROKING TESTS
    TASKS
    TENSILE TESTS
    TEST FACILITIES
    TEST FIRING
    TEST VEHICLES
    TESTING TIME
    THERMAL CYCLING TESTS
    THERMAL VACUUM TESTS
    ULTRASONIC TESTS
    VACUUM TESTS
    VIBRATION TESTS
    WATER TUNNEL TESTS
    WEAR TESTS
    WELD TESTS
    WIND TUNNEL STABILITY TESTS
    WIND TUNNEL TESTS
    WING FLOW METHOD TESTS
    X RAY INSPECTION

TETHERED BALLOONS
UF   KITE BALLOONS
GS   EXPANDABLE STRUCTURES
     . INFLATABLE STRUCTURES
     . . BALLOONS
     . . . **TETHERED BALLOONS**
RT   METEOROLOGICAL BALLOONS
     REELS

TETHERED SATELLITES
GS   SATELLITES
     . ARTIFICIAL SATELLITES
     . . **TETHERED SATELLITES**
RT   REELS

TETHERING
RT   ORBITAL RENDEZVOUS
     REELS
     TETHERLINES

TETHERLINES
RT   ANCHORS (FASTENERS)
     ∞ CABLES
     ∞ LINES
     TETHERING
     UMBILICAL CONNECTORS

TETHYS
GS   SATELLITES
     . NATURAL SATELLITES
     . . SATURN SATELLITES
     . . . **TETHYS**
RT   SATURN (PLANET)

TETRABUTYLS
GS   ALIPHATIC COMPOUNDS
     . **TETRABUTYLS**

TETRACHLORIDES
GS   HALOGEN COMPOUNDS
     . CHLORINE COMPOUNDS
     . . CHLORIDES
     . . . **TETRACHLORIDES**
     . HALIDES
     . . CHLORIDES
     . . . **TETRACHLORIDES**

TETRACHLOROMETHANE
USE   CARBON TETRACHLORIDE

TETRACYCLINES
GS   DRUGS
     . ANTIBIOTICS
     . . **TETRACYCLINES**
     HETEROCYCLIC COMPOUNDS

TETRACYCLINES-(CONT.)
. **TETRACYCLINES**
STEROIDS
. **TETRACYCLINES**

TETRAD THEORY
RT   CHROMOSOMES
MIOSIS
SPORES
∞THEORIES

TETRAETHYL ORTHOCARBONATES
GS   CARBON COMPOUNDS
. CARBONATES
. . **TETRAETHYL ORTHOCARBONATES**

TETRAETHYL ORTHOSILICATE
GS   ADHESIVES
. **TETRAETHYL ORTHOSILICATE**
RT   ETHYL COMPOUNDS
GLUES
SILICATES

TETRAFLUOROHYDRAZINE
GS   ALIPHATIC COMPOUNDS
. HYDRAZINES
. . **TETRAFLUOROHYDRAZINE**
AMINES
. **TETRAFLUOROHYDRAZINE**

TETRAGONS
GS   GEOMETRY
. EUCLIDEAN GEOMETRY
. . POLYGONS
. . . **TETRAGONS**
. . . . PARALLELOGRAMS
. . . . . RHOMBOIDS
. . . . RECTANGLES
. . . . SQUARES (MATHEMATICS)
. . . . TRAPEZOIDS

TETRAHEDRONS
GS   GEOMETRY
. EUCLIDEAN GEOMETRY
. . POLYHEDRONS
. . . **TETRAHEDRONS**
RT   TRIANGLES

TETRAHYDROFURAN
UF   BUTYLENE OXIDES
GS   HETEROCYCLIC COMPOUNDS
. FURANS
. . **TETRAHYDROFURAN**
ORGANIC COMPOUNDS
. FURANS
. . **TETRAHYDROFURAN**
SOLVENTS
. **TETRAHYDROFURAN**
RT   ADDITIVES
∞CHEMICAL COMPOUNDS
PLASTICS
POLYVINYL CHLORIDE

TETRANITROTETRAZACYCLOOCTANE
USE   HMX

TETRAPHENYLS
GS   PHENYLS
. POLYPHENYLS
. . **TETRAPHENYLS**

TETRAZOLES
GS   HETEROCYCLIC COMPOUNDS
. **TETRAZOLES**

TETRODES
RT   ELECTRON TUBES
PENTODES
SEMICONDUCTOR DEVICES
TRANSISTORS
TRIODES

TETROONS
USE   SUPERPRESSURE BALLOONS

TETRYL
GS   AMINES
. **TETRYL**
EXPLOSIVES
. **TETRYL**
NITROGEN COMPOUNDS
. NITRO COMPOUNDS
. . **TETRYL**
PROPELLANTS

TETRYL-(CONT.)
. **TETRYL**

TEXAS
GS   NATIONS
. UNITED STATES
. . **TEXAS**
RT   GULF OF MEXICO
HOUSTON (TX)
LAKE TEXOMA (OK-TX)
RIO GRANDE (NORTH AMERICA)

TEXTBOOKS
GS   DOCUMENTS
. **TEXTBOOKS**
RT   EDUCATION
HANDBOOKS
KNOWLEDGE
LEARNING
LIBRARIES
MANUALS
SUBJECTS

TEXTILES
GS   **TEXTILES**
. COTTON FIBERS
. LINEN
. RAYON
RT   CLOTHING
COTTON
FABRICS
FIBERS
VAPOR BARRIER CLOTHING
WET SPINNING

TEXTS
GS   DOCUMENTS
. **TEXTS**
RT   FORMAT
RECORDS

TEXTURES
RT   CURL (MATERIALS)
FINENESS
MECHANICAL PROPERTIES
POROSITY
SURFACE PROPERTIES

TF-30 ENGINE
GS   AIRCRAFT ENGINES
. **TF-30 ENGINE**
ENGINES
. INTERNAL COMBUSTION ENGINES
. . GAS TURBINE ENGINES
. . . JET ENGINES
. . . . TURBOJET ENGINES
. . . . . TURBOFAN ENGINES
. . . . . . **TF-30 ENGINE**
. TURBINE ENGINES
. . GAS TURBINE ENGINES
. . . JET ENGINES
. . . . TURBOJET ENGINES
. . . . . TURBOFAN ENGINES
. . . . . . **TF-30 ENGINE**

TF-34 ENGINE
GS   AIRCRAFT ENGINES
. **TF-34 ENGINE**

TF-41 ENGINE
GS   AIRCRAFT ENGINES
. **TF-41 ENGINE**
ENGINES
. AIR BREATHING ENGINES
. . GAS TURBINE ENGINES
. . . JET ENGINES
. . . . TURBOJET ENGINES
. . . . . TURBOFAN ENGINES
. . . . . . **TF-41 ENGINE**
. INTERNAL COMBUSTION ENGINES
. . GAS TURBINE ENGINES
. . . JET ENGINES
. . . . TURBOJET ENGINES
. . . . . TURBOFAN ENGINES
. . . . . . **TF-41 ENGINE**
. TURBINE ENGINES
. . GAS TURBINE ENGINES
. . . JET ENGINES
. . . . TURBOJET ENGINES
. . . . . TURBOFAN ENGINES
. . . . . . **TF-41 ENGINE**

TFX AIRCRAFT
USE   F-111 AIRCRAFT

TH-55 HELICOPTER
GS   HUGHES AIRCRAFT
. **TH-55 HELICOPTER**
V/STOL AIRCRAFT
. ROTARY WING AIRCRAFT
. . HELICOPTERS
. . . **TH-55 HELICOPTER**

THAILAND
GS   NATIONS
. **THAILAND**
RT   ASIA

THALAMUS
GS   NERVOUS SYSTEM
. CENTRAL NERVOUS SYSTEM
. . **THALAMUS**
RT   BRAIN

THALLIUM
GS   CHEMICAL ELEMENTS
. **THALLIUM**
. . THALLIUM ISOTOPES
METALS
. **THALLIUM**
. . THALLIUM ISOTOPES
RT   THALLIUM COMPOUNDS

THALLIUM ALLOYS
GS   ALLOYS
. **THALLIUM ALLOYS**

THALLIUM COMPOUNDS
RT   ∞METAL COMPOUNDS
THALLIUM

THALLIUM ISOTOPES
GS   CHEMICAL ELEMENTS
. THALLIUM
. . **THALLIUM ISOTOPES**
METALS
. THALLIUM
. . **THALLIUM ISOTOPES**

THAWING
USE   MELTING

THEMATIC MAPPING
GS   MAPPING
. **THEMATIC MAPPING**
RT   CADASTRAL MAPPING
MAPS
PHOTOGEOLOGY
PHOTOMAPPING
PHOTOMAPS

THEMIS PROJECT
GS   PROGRAMS
. PROJECTS
. . **THEMIS PROJECT**

THEODOLITES
GS   MEASURING INSTRUMENTS
. OPTICAL MEASURING INSTRUMENTS
. . TRANSITS
. . . **THEODOLITES**
. . . . CINETHEODOLITES
OPTICAL EQUIPMENT
. OPTICAL MEASURING INSTRUMENTS
. . TRANSITS
. . . **THEODOLITES**
. . . . CINETHEODOLITES
RT   SEXTANTS

THEODORSEN TRANSFORMATION
RT   AIRFOIL PROFILES
COMPLEX VARIABLES
CONFORMAL MAPPING
COORDINATE TRANSFORMATIONS
JOUKOWSKI TRANSFORMATION
PRESSURE DISTRIBUTION

THEOREM PROVING
GS   PROBLEM SOLVING
. **THEOREM PROVING**
PROVING
. **THEOREM PROVING**
RT   ARTIFICIAL INTELLIGENCE
COMPUTER PROGRAMMING
THEOREMS

THEOREMS
UF   LEMMAS
GS   **THEOREMS**

THEOREMS-(CONT.)

. ADDITION THEOREM
. BAYES THEOREM
. BERNOULLI THEOREM
. BINOMIAL THEOREM
. CASTIGLIANO VARIATIONAL THEOREM
. DUALITY THEOREM
. EQUIPARTITION THEOREM
. EXISTENCE THEOREMS
. FLOQUET THEOREM
. GAUSS-MARKOV THEOREM
. HELLMANN-FEYNMAN THEOREM
. KAKUTANI THEOREM
. LEBESGUE THEOREM
. LIOUVILLE THEOREM
. MICHELL THEOREM
. POMERANCHUK THEOREM
. POYNTING THEOREM
. RECIPROCAL THEOREMS
. RECIPROCITY THEOREM
. RICHARDS THEOREM
. RIESZ THEOREM
. SCHAUDER FIXPOINT THEOREM
. SIMILARITY THEOREM
. . LAGRANGE SIMILARITY HYPOTHESIS
. STOKES THEOREM (VECTOR
   CALCULUS)
. UNIQUENESS THEOREM
. VIRIAL THEOREM
RT HYPOTHESES
    MATHEMATICAL LOGIC
  ∞MATHEMATICS
    TAYLOR SERIES
    THEOREM PROVING

THEORETICAL PHYSICS

GS **THEORETICAL PHYSICS**
    . NEWTON THEORY
    . QUANTUM THEORY
    . . BOHR THEORY
RT ASTROPHYSICS
    BROKEN SYMMETRY
    CHARM (PARTICLE PHYSICS)
    ELECTROPHYSICS
    FLAVOR (PARTICLE PHYSICS)
    GEOPHYSICS
    NAKED SINGULARITIES
    NUCLEAR PHYSICS
  ∞PHYSICS
    PLASMA PHYSICS
    RADIO PHYSICS
  ∞SCIENCE
  ∞SOLID STATE PHYSICS
    STRANGE ATTRACTORS
    UNIFIED FIELD THEORY
    YANG-MILLS THEORY

∞ THEORIES

SN *(USE OF A MORE SPECIFIC TERM IS
    RECOMMENDED--CONSULT THE TERMS
    LISTED BELOW)*
RT ABRIKOSOV THEORY
    ASSUMPTIONS
    ATOMIC THEORY
    AUTOMATA THEORY
    BCS THEORY
    BELLMAN THEORY
    BENDING THEORY
    BESSEL-BREDICHIN THEORY
    BIMETRIC THEORIES
    BOGOLIUBOV THEORY
    BOHR THEORY
    BORN-INFELD THEORY
    CATASTROPHE THEORY
    CHAPMAN-ENSKOG THEORY
    COMMUNICATION THEORY
    CONTROL THEORY
    CROCCO-LEE THEORY
    DEBYE-HUCKEL THEORY
    DECISION THEORY
    DIFFUSION THEORY
    DYNAMO THEORY
    DYSON THEORY
    EYRING THEORY
    FIELD MODE THEORY
    FIELD THEORY (ALGEBRA)
    FIELD THEORY (PHYSICS)
    FINITE DIFFERENCE THEORY
    FLOW THEORY
    FLUCTUATION THEORY
    FOSTER THEORY
    GAME THEORY
    GAUGE THEORY
    GEOMETRICAL THEORY OF
      DIFFRACTION

THEORIES-(CONT.)

GESTALT THEORY
GLAUBER THEORY
GOAL THEORY
GRAPH THEORY
GRAVITATION THEORY
GRIFFITH CRACK
GROUP THEORY
HANSEN LUNAR THEORY
HEISENBERG THEORY
HILL LUNAR THEORY
HOMOTOPY THEORY
HUECKEL THEORY
HYPOTHESES
INFORMATION THEORY
JEANS THEORY
KINETIC THEORY
KOLMOGOROFF THEORY
LEARNING THEORY
MALKUS THEORY
MANNING THEORY
MATRIX THEORY
MICHAELIS THEORY
MIXING LENGTH FLOW THEORY
MOLECULAR THEORY
MOMENTUM THEORY
NEWTON THEORY
NONADIABATIC THEORY
NUMBER THEORY
NUMERICAL DIFFERENTIATION
OPIK THEORY
ORTHOGONAL MULTIPLEXING THEORY
PARTICLE THEORY
PERTURBATION THEORY
PHYSICAL OPTICS
PISTON THEORY
PLATE THEORY
POPULATION THEORY
POTENTIAL THEORY
PROBABILITY THEORY
QUANTUM CHROMODYNAMICS
QUANTUM THEORY
QUEUEING THEORY
REISSNER THEORY
RELATIVISTIC THEORY
S MATRIX THEORY
SADDLE POINTS (GAME THEORY)
SET THEORY
SHELL THEORY
SPECTRAL THEORY
STATISTICAL DECISION THEORY
STRONG INTERACTIONS (FIELD
  THEORY)
STURM-LIOUVILLE THEORY
SWITCHING THEORY
TETRAD THEORY
TRANSPORT THEORY
VINTI THEORY
WEAK INTERACTIONS (FIELD THEORY)
YANG-MILLS THEORY
YOUNG-HELMHOLTZ THEORY

THERAPY

GS **THERAPY**
    . CHEMOTHERAPY
    . MASSAGING
    . PSYCHOTHERAPY
    . RADIATION THERAPY
RT CURES
    DISEASES
    HEALING
    MEDICAL EQUIPMENT
    PATIENTS
    RESPIRATORS
    SKIN GRAFTS

THERMAL ABSORPTION

GS ENERGY ABSORPTION
    . **THERMAL ABSORPTION**
    . . POLAR CAP ABSORPTION
RT ABLATION
  ∞ABSORPTION
    ATMOSPHERIC ATTENUATION
    CHARRING
    GRAY GAS
    HEAT SINKS
    PYROLYSIS
    TEMPERATURE

THERMAL ACCOMMODATION COEFFICIENTS

USE ACCOMMODATION COEFFICIENT

THERMAL AGITATION

USE THERMAL ENERGY

THERMAL ANALYSIS

UF DIFFERENTIAL THERMAL ANALYSIS
    DTA (ANALYSIS)
RT ∞ANALYZING
    HEAT TRANSMISSION
    TEMPERATURE
    TEMPERATURE GRADIENTS
    TEMPERATURE PROFILES

THERMAL BATTERIES

GS ELECTRIC GENERATORS
    . DIRECT POWER GENERATORS
    . . PRIMARY BATTERIES
    . . . **THERMAL BATTERIES**
    ELECTROCHEMICAL CELLS
    . ELECTRIC BATTERIES
    . . PRIMARY BATTERIES
    . . . **THERMAL BATTERIES**
RT ALKALINE BATTERIES
    DRY CELLS

THERMAL BLOOMING

UF LASER BEAM DEFOCUSING
    THERMAL DEFOCUSING
RT LASER CUTTING
    LASER HEATING
    LASER OUTPUTS
    LASERS
    PHOTON BEAMS
    TEMPERATURE

THERMAL BOUNDARY LAYER

GS BOUNDARY LAYERS
    . **THERMAL BOUNDARY LAYER**
RT HYPERSONIC BOUNDARY LAYER
    LAMINAR BOUNDARY LAYER
    RAYLEIGH-BENARD CONVECTION
    TEMPERATURE
    TURBULENT BOUNDARY LAYER

THERMAL BUCKLING

GS BUCKLING
    . **THERMAL BUCKLING**
    THERMODYNAMIC PROPERTIES
    . THERMAL EXPANSION
    . . **THERMAL BUCKLING**
RT EXPANSION
    TEMPERATURE
    TEMPERATURE EFFECTS

THERMAL COMFORT

RT HEAT STROKE
    TEMPERATURE
    THERMAL ENVIRONMENTS

THERMAL CONDUCTIVITY

GS THERMODYNAMIC PROPERTIES
    . THERMOPHYSICAL PROPERTIES
    . . **THERMAL CONDUCTIVITY**
    TRANSPORT PROPERTIES
    . **THERMAL CONDUCTIVITY**
RT AIR CONDUCTIVITY
    ATMOSPHERIC CONDUCTIVITY
    CONDUCTIVE HEAT TRANSFER
  ∞CONDUCTIVITY
    FOURIER LAW
    HOT-WIRE FLOWMETERS
    LEWIS NUMBERS
    SPECIFIC HEAT
    TEMPERATURE
    THERMOHYDRAULICS

THERMAL CONDUCTIVITY GAGES

SN (GAGES FOR MEASURING THERMAL
    CONDUCTIVITY--EXCLUDES GAGES
    USING THERMAL CONDUCTIVITY TO
    MEASURE OTHER PROPERTIES OR
    VARIABLES)
GS MEASURING INSTRUMENTS
    . **THERMAL CONDUCTIVITY GAGES**
RT TEMPERATURE

THERMAL CONDUCTORS

GS CONDUCTORS
    . **THERMAL CONDUCTORS**
RT ∞CONDUCTION
    CONDUCTIVE HEAT TRANSFER
    ELECTRIC CONDUCTORS
    TEMPERATURE

THERMAL CONTROL COATINGS

GS COATINGS
    . **THERMAL CONTROL COATINGS**
RT ABLATIVE MATERIALS

THERMAL CONTROL COATINGS-*(CONT.)*
　∞CONTROL
　　HEAT SHIELDING
　　REENTRY SHIELDING
　　REUSABLE HEAT SHIELDING
　　TEMPERATURE
　　TEMPERATURE CONTROL

THERMAL CONVECTION
　USE　FREE CONVECTION

THERMAL CURRENTS
　USE　CONVECTIVE FLOW

THERMAL CYCLING TESTS
　RT　CLOSED CYCLES
　　COOLING
　　FATIGUE TESTS
　　HEATING
　　TEMPERATURE
　　TEMPERATURE CONTROL
　　∞TESTS
　　THERMODYNAMIC PROPERTIES

THERMAL DECOMPOSITION
　GS　CHEMICAL REACTIONS
　　. THERMAL DECOMPOSITION
　　. . PYROLYSIS
　　DECOMPOSITION
　　. THERMAL DECOMPOSITION
　RT　ABLATION
　　ENDOTHERMIC REACTIONS
　　EXOTHERMIC REACTIONS
　　TEMPERATURE
　　THERMOCHEMISTRY
　　THERMOGRAVIMETRY

THERMAL DEFOCUSING
　USE　THERMAL BLOOMING

THERMAL DEGRADATION
　GS　DEGRADATION
　　. THERMAL DEGRADATION
　　STERILIZATION EFFECTS
　　. THERMAL DEGRADATION
　RT　PYROLYSIS
　　TEMPERATURE
　　TEMPERATURE EFFECTS

THERMAL DIFFUSION
　GS　DIFFUSION
　　. THERMAL DIFFUSION
　　THERMODYNAMIC PROPERTIES
　　. THERMOPHYSICAL PROPERTIES
　　. . THERMAL DIFFUSION
　RT　CHAPMAN-ENSKOG THEORY
　　∞CONDUCTION
　　CONVECTIVE FLOW
　　ELECTRON DIFFUSION
　　GAS HEATING
　　GASEOUS DIFFUSION
　　HEAT TRANSFER
　　KIRKENDALL EFFECT
　　PECLET NUMBER
　　∞SEPARATION
　　SORET COEFFICIENT
　　SURFACE DIFFUSION
　　TEMPERATURE
　　THERMOCHEMISTRY
　　THERMOHYDRAULICS
　　VISCOSITY

THERMAL DIFFUSIVITY
　GS　THERMODYNAMIC PROPERTIES
　　. THERMOPHYSICAL PROPERTIES
　　. . THERMAL DIFFUSIVITY
　　TRANSPORT PROPERTIES
　　. THERMAL DIFFUSIVITY
　RT　TEMPERATURE
　　VISCOSITY

THERMAL DISSOCIATION
　GS　CHEMICAL REACTIONS
　　. THERMAL DISSOCIATION
　　DISSOCIATION
　　. THERMAL DISSOCIATION
　RT　CRACKING (CHEMICAL ENGINEERING)
　　DECOMPOSITION
　　DEGRADATION
　　GAS DISSOCIATION
　　HEAT OF DISSOCIATION
　　HYDROGEN PRODUCTION
　　IONIZATION
　　PLASMAS (PHYSICS)

THERMAL DISSOCIATION-*(CONT.)*
　　TEMPERATURE
　　TEMPERATURE EFFECTS

THERMAL EFFECTS
　USE　TEMPERATURE EFFECTS

THERMAL EFFICIENCY
　USE　THERMODYNAMIC EFFICIENCY

THERMAL EMISSION
　GS　DECAY
　　. EMISSION
　　. . THERMAL EMISSION
　　. . . THERMIONIC EMISSION
　RT　ELECTRON EMISSION
　　EMISSIVITY
　　EXHAUST EMISSION
　　INCANDESCENCE
　　INFRARED ABSORPTION
　　TEMPERATURE

THERMAL ENERGY
　UF　THERMAL AGITATION
　RT　COGENERATION
　　∞ENERGY
　　FREE ENERGY
　　GEOTHERMAL ENERGY CONVERSION
　　GEOTHERMAL RESOURCES
　　HEAT
　　HEAT OF FUSION
　　HEAT OF SOLUTION
　　INTERNAL ENERGY
　　KINETIC ENERGY
　　LATTICE VIBRATIONS
　　PHOTOTHERMAL CONVERSION
　　SOLAR THERMAL ELECTRIC POWER
　　　PLANTS
　　TEMPERATURE

THERMAL ENERGY STORAGE
　USE　HEAT STORAGE

THERMAL ENVIRONMENTS
　GS　ENVIRONMENTS
　　. THERMAL ENVIRONMENTS
　RT　ADIABATIC CONDITIONS
　　AEROSPACE ENVIRONMENTS
　　HEAT STROKE
　　HIGH TEMPERATURE ENVIRONMENTS
　　LIFE SUPPORT SYSTEMS
　　LOW TEMPERATURE ENVIRONMENTS
　　LUNAR ENVIRONMENT
　　PLANETARY ENVIRONMENTS
　　SATELLITE TEMPERATURE
　　SPACECRAFT ENVIRONMENTS
　　TEMPERATURE
　　THERMAL COMFORT

THERMAL EXPANSION
　GS　EXPANSION
　　. THERMAL EXPANSION
　　THERMODYNAMIC PROPERTIES
　　. THERMAL EXPANSION
　　. . THERMAL BUCKLING
　RT　BOUSSINESQ APPROXIMATION
　　DILATOMETRY
　　EXTENSOMETERS
　　GRUNEISEN CONSTANT
　　HEAT TRANSFER
　　HIGH TEMPERATURE TESTS
　　LOW TEMPERATURE TESTS
　　NEEL TEMPERATURE
　　∞PHYSICAL PROPERTIES
　　TEMPERATURE
　　THERMOPHYSICAL PROPERTIES
　　WARPAGE

THERMAL FATIGUE
　GS　FATIGUE (MATERIALS)
　　. THERMAL FATIGUE
　RT　HIGH TEMPERATURE ENVIRONMENTS
　　METAL FATIGUE
　　TEMPERATURE

THERMAL GRAVIMETRY
　USE　THERMOGRAVIMETRY

THERMAL INSTABILITY
　GS　THERMODYNAMIC PROPERTIES
　　. THERMAL INSTABILITY
　RT　CLEAR AIR TURBULENCE
　　COMBUSTION STABILITY
　　MAGNETOHYDRODYNAMIC STABILITY

THERMAL INSTABILITY-*(CONT.)*
　　PYROLYSIS
　　SPUTTERING
　　STELLARATORS
　　TEMPERATURE

THERMAL INSULATION
　GS　INSULATION
　　. THERMAL INSULATION
　RT　AIR CONDITIONING
　　AMBERLITE (TRADEMARK)
　　ASBESTOS
　　CORK (MATERIALS)
　　CRYOGENIC FLUID STORAGE
　　HEAT
　　HEAT SHIELDING
　　HEAT SINKS
　　HEAT TRANSFER
　　HEAT TRANSMISSION
　　HEATING EQUIPMENT
　　REENTRY SHIELDING
　　REFRACTORIES
　　REFRACTORY COATINGS
　　TEMPERATURE
　　TEMPERATURE CONTROL
　　TROMBE WALLS

THERMAL MAPPING
　GS　MAPPING
　　. THERMAL MAPPING
　RT　AERIAL RECONNAISSANCE
　　EARTH RESOURCES
　　GEOTHERMAL ANOMALIES
　　GEOTHERMAL RESOURCES
　　HEAT CAPACITY MAPPING MISSION
　　INFRARED RADIOMETERS
　　INFRARED SCANNERS
　　ISOTHERMAL LAYERS
　　ISOTHERMS
　　PHOTOMAPPING
　　PLANETARY MAPPING
　　TEMPERATURE
　　TEMPERATURE DISTRIBUTION
　　TEMPERATURE GRADIENTS
　　THERMOGRAPHY

THERMAL NEUTRONS
　UF　SLOW NEUTRONS
　GS　NUCLEAR RADIATION
　　. THERMAL NEUTRONS
　　PARTICLES
　　. ELEMENTARY PARTICLES
　　. . FERMIONS
　　. . . NEUTRONS
　　. . . . THERMAL NEUTRONS
　　. NEUTRAL PARTICLES
　　. . NEUTRONS
　　. . . THERMAL NEUTRONS
　RT　BARYONS
　　FAST NEUTRONS
　　NUCLEAR REACTORS
　　TEMPERATURE
　　THERMALIZATION (ENERGY
　　　ABSORPTION)

THERMAL NOISE
　GS　ELASTIC WAVES
　　. SOUND WAVES
　　. . NOISE (SOUND)
　　. . . THERMAL NOISE
　　ELECTROMAGNETIC INTERFERENCE
　　. RADIO FREQUENCY INTERFERENCE
　　. . BLACKOUT (PROPAGATION)
　　. . . ELECTROMAGNETIC NOISE
　　. . . . WHITE NOISE
　　. . . . . THERMAL NOISE
　RT　CHANNEL NOISE
　　ELECTROMAGNETIC NOISE
　　MEASUREMENT
　　NOISE TEMPERATURE
　　SHOT NOISE
　　TEMPERATURE

THERMAL PLASMAS
　GS　GASES
　　. IONIZED GASES
　　. . CHARGED PARTICLES
　　. . . THERMAL PLASMAS
　　PARTICLES
　　. CHARGED PARTICLES
　　. . ENERGETIC PARTICLES
　　. . . PLASMAS (PHYSICS)
　　. . . . THERMAL PLASMAS
　RT　ELECTRON PLASMA
　　HIGH TEMPERATURE PLASMAS

THERMAL PLASMAS-*(CONT.)*
        PLASMA GENERATORS
        PLASMA TEMPERATURE
        TEMPERATURE

THERMAL POLLUTION
GS   POLLUTION
      . **THERMAL POLLUTION**
RT   BIOLOGICAL EFFECTS
      COASTAL ECOLOGY
      ENVIRONMENT EFFECTS
      ENVIRONMENT POLLUTION
      ENVIRONMENTAL QUALITY
      ENVIRONMENTAL SURVEYS
      ENVIRONMENTS
      HEAT TRANSFER
      LAKES
      LIQUID COOLING
      MARINE BIOLOGY
      NUCLEAR REACTORS
      OCEAN TEMPERATURE
      OCEANS
      PLANKTON
      POLLUTION TRANSPORT
      SEAS
      TEMPERATURE
      WATER POLLUTION
      WATER TEMPERATURE

THERMAL POWER
USE   TURBOGENERATORS

THERMAL PROPERTIES
USE   THERMODYNAMIC PROPERTIES

THERMAL PROTECTION
GS   PROTECTION
      . **THERMAL PROTECTION**
RT   ABLATIVE MATERIALS
      CARBON-CARBON COMPOSITES
      HEAT SHIELDING
      LUDOX (TRADEMARK)
      RADIATION PROTECTION
      REENTRY SHIELDING
      REUSABLE HEAT SHIELDING
      TEMPERATURE

THERMAL RADIATION
SN   (EMITTED AS THE RESULT OF THERMAL
      EXCITATION OF MOLECULES)
GS   ELECTROMAGNETIC RADIATION
      . **THERMAL RADIATION**
      . . PHONON BEAMS
RT   BLACK BODY RADIATION
      CONCENTRATORS
      GREENHOUSE EFFECT
      HEAT
      INFRARED RADIATION
      LIGHT (VISIBLE RADIATION)
      NEAR INFRARED RADIATION
      NONGRAY GAS
      PLANCKS CONSTANT
      PLANETARY RADIATION
      ∞RADIATION
      RADIO WAVES
      SKY RADIATION
      SOLAR RADIATION
      SUNLIGHT
      TEMPERATURE
      THERMODYNAMIC PROPERTIES
      ULTRAVIOLET RADIATION

THERMAL REACTORS
GS   NUCLEAR REACTORS
      . **THERMAL REACTORS**
RT   ∞REACTORS
      TEMPERATURE

THERMAL RESISTANCE
UF   HEAT RESISTANCE
GS   MECHANICAL PROPERTIES
      . **THERMAL RESISTANCE**
RT   CARBON-CARBON COMPOSITES
      ∞HIGH RESISTANCE
      HIGH TEMPERATURE LUBRICANTS
      HIGH TEMPERATURE TESTS
      ∞LOW RESISTANCE
      OXIDATION
      OXIDATION RESISTANCE
      ∞RESISTANCE
      SPECIFIC HEAT
      TEMPERATURE
      TEMPERATURE EFFECTS
      THERMODYNAMIC PROPERTIES

THERMAL RESOURCES
GS   HEAT SOURCES
      . **THERMAL RESOURCES**
      . . GEOTHERMAL RESOURCES
      . . . GEYSERS
      RESOURCES
      . EARTH RESOURCES
      . . **THERMAL RESOURCES**
      . . . GEOTHERMAL RESOURCES
      . . . . GEYSERS
RT   AGROMETEOROLOGY
      ATMOSPHERIC TEMPERATURE
      CROP GROWTH
      CROP VIGOR
      GEOTHERMAL TECHNOLOGY
      RESOURCES MANAGEMENT
      TEMPERATURE
      TEMPERATURE DISTRIBUTION

THERMAL SHIELDING
USE   HEAT SHIELDING

THERMAL SHOCK
RT   COOLING
      HEATING
      HIGH TEMPERATURE TESTS
      ∞SHOCK
      SHOCK RESISTANCE
      TEMPERATURE
      TEMPERATURE DISTRIBUTION
      THERMODYNAMIC PROPERTIES

THERMAL SIMULATION
GS   SIMULATION
      . ENVIRONMENT SIMULATION
      . . **THERMAL SIMULATION**
RT   ALTITUDE SIMULATION
      SOLAR SIMULATION
      SPACE ENVIRONMENT SIMULATION
      TEMPERATURE

THERMAL STABILITY
UF   THERMOSTABILITY
GS   STABILITY
      . **THERMAL STABILITY**
      THERMODYNAMIC PROPERTIES
      . THERMOPHYSICAL PROPERTIES
      . . **THERMAL STABILITY**
RT   DIMENSIONAL STABILITY
      HIGH TEMPERATURE TESTS
      LOW TEMPERATURE TESTS
      STORAGE STABILITY
      SURFACE STABILITY
      TEMPERATURE

THERMAL STRESSES
SN   (EXCLUDES BIOLOGICAL STRESSES)
GS   STRESSES
      . **THERMAL STRESSES**
RT   COOLING
      FATIGUE (MATERIALS)
      HEATING
      TEMPERATURE
      TEMPERATURE DISTRIBUTION
      TEMPERATURE EFFECTS

THERMAL VACUUM TESTS
GS   VACUUM TESTS
      . **THERMAL VACUUM TESTS**
RT   ENVIRONMENTAL TESTS
      HIGH ALTITUDE ENVIRONMENTS
      TEMPERATURE
      TEST CHAMBERS
      ∞TESTS
      VACUUM CHAMBERS

THERMALIZATION (ENERGY ABSORPTION)
GS   ENERGY ABSORPTION
      . MODERATION (ENERGY ABSORPTION)
      . . **THERMALIZATION (ENERGY
          ABSORPTION)**
      . . . NEUTRON THERMALIZATION
RT   THERMAL NEUTRONS

THERMICONS
GS   ELECTRON TUBES
      . CAMERA TUBES
      . . VIDICONS
      . . . RETURN BEAM VIDICONS
      . . . . **THERMICONS**
      . VACUUM TUBES
      . . VACUUM TUBE OSCILLATORS
      . . . MICROWAVE TUBES
      . . . . PLANOTRONS
      . . . . . **THERMICONS**

THERMICONS-*(CONT.)*
        MICROWAVE EQUIPMENT
        . MICROWAVE TUBES
        . . IMAGE TUBES
        . . . **THERMICONS**
        . . PLANOTRONS
        . . . **THERMICONS**
        . VIDICONS
        . . **THERMICONS**
        OPTICAL EQUIPMENT
        . IMAGE CONVERTERS
        . . IMAGE TUBES
        . . . **THERMICONS**

THERMIONIC CATHODES
GS   ELECTRODES
      . CATHODES
      . . TUBE CATHODES
      . . . **THERMIONIC CATHODES**
      EMITTERS
      . **THERMIONIC CATHODES**
RT   HOT CATHODES

THERMIONIC CONVERSION SYSTEMS
USE   THERMIONIC POWER GENERATION

THERMIONIC CONVERTERS
GS   ELECTRIC GENERATORS
      . DIRECT POWER GENERATORS
      . . **THERMIONIC CONVERTERS**
      . . . SNAP 13
      . . . SOLAR BLANKETS
RT   CESIUM DIODES
      CESIUM PLASMA
      ∞CONVERTERS
      FUEL CELLS
      ION PRODUCTION RATES
      MAGNETOHYDRODYNAMIC GENERATORS
      PLASMA POWER SOURCES
      RADIOISOTOPE BATTERIES
      SNAP
      SOLAR CELLS
      THERMOELECTRIC GENERATORS

THERMIONIC DIODES
GS   ELECTRON TUBES
      . VACUUM TUBES
      . . VACUUM TUBE OSCILLATORS
      . . . MICROWAVE TUBES
      . . . . PLANOTRONS
      . . . . . **THERMIONIC DIODES**
      MICROWAVE EQUIPMENT
      . MICROWAVE TUBES
      . . PLANOTRONS
      . . . **THERMIONIC DIODES**
RT   CHILD-LANGMUIR LAW
      PERVEANCE

THERMIONIC EMISSION
UF   RICHARDSON-DUSHMAN EQUATION
GS   DECAY
      . EMISSION
      . . PARTICLE EMISSION
      . . . **THERMIONIC EMISSION**
      . . THERMAL EMISSION
      . . . **THERMIONIC EMISSION**
RT   ELECTRON EMISSION
      ION EMISSION
      THERMOELECTRICITY
      WORK FUNCTIONS

THERMIONIC EMITTERS
GS   EMITTERS
      . **THERMIONIC EMITTERS**

THERMIONIC POWER GENERATION
UF   THERMIONIC CONVERSION SYSTEMS
RT   ∞CONVERSION
      SNAP
      SNAP 13

THERMIONIC REACTORS
USE   ION ENGINES
      NUCLEAR ROCKET ENGINES

THERMIONICS
RT   CATHODES
      ELECTRON EMISSION
      ∞ELECTRONICS
      ION EMISSION

THERMISTORS
GS   ATTENUATORS
      . RESISTORS

THERMISTORS-(CONT.)
```
    . . THERMISTORS
    ELECTRONIC EQUIPMENT
    . SOLID STATE DEVICES
    . . SEMICONDUCTOR DEVICES
    . . . THERMISTORS
    TEMPERATURE SENSORS
    . THERMISTORS
RT  RADIOMETERS
    TEMPERATURE MEASURING
        INSTRUMENTS
    VARISTORS
```

THERMITES
```
RT  ALUMINUM OXIDES
    AUGER SPECTROSCOPY
    BARIUM ION CLOUDS
    COPPER OXIDES
    IGNITION TEMPERATURE
    PYROTECHNICS
```

THERMOBALANCES
```
GS  MEASURING INSTRUMENTS
    . INDICATING INSTRUMENTS
    . . WEIGHT INDICATORS
    . . . THERMOBALANCES
RT  THERMOGRAVIMETRY
```

THERMOCHEMICAL PROPERTIES
```
GS  CHEMICAL PROPERTIES
    . THERMOCHEMICAL PROPERTIES
    . . HEAT OF COMBUSTION
    . . HEAT OF FORMATION
    . . HEAT OF FUSION
    . . HEAT OF VAPORIZATION
    THERMODYNAMIC PROPERTIES
    . THERMOCHEMICAL PROPERTIES
    . . HEAT OF COMBUSTION
    . . HEAT OF FORMATION
    . . HEAT OF FUSION
    . . HEAT OF VAPORIZATION
RT  HEAT BALANCE
    ∞PROPERTIES
```

THERMOCHEMISTRY
```
GS  THERMOCHEMISTRY
    . AEROTHERMOCHEMISTRY
RT  CHEMICAL ENGINEERING
    CHEMICAL REACTIONS
    ∞CHEMISTRY
    COMBUSTION PHYSICS
    ENTHALPY
    ENTROPY
    HEAT
    HEAT BALANCE
    HEAT OF DISSOCIATION
    HEAT OF FUSION
    HEAT OF SOLUTION
    HEAT TREATMENT
    PHYSICAL CHEMISTRY
    PROPELLANT CHEMISTRY
    PYROMETALLURGY
    THERMAL DECOMPOSITION
    THERMAL DIFFUSION
    THERMODYNAMIC PROPERTIES
    THERMODYNAMICS
    THERMOGRAVIMETRY
    THERMOPHYSICAL PROPERTIES
```

THERMOCHROMATIC MATERIALS
```
RT  COLOR
    COLORIMETRY
    ∞INORGANIC MATERIALS
    ∞MATERIALS
    OPTICAL PROPERTIES
    ORGANIC MATERIALS
    SOLIDS
```

THERMOCLINES
```
GS  GRADIENTS
    . TEMPERATURE GRADIENTS
    . . THERMOCLINES
RT  OCEANOGRAPHY
    SEA WATER
    SOUND TRANSMISSION
    STRATIFICATION
    SURFACE LAYERS
    SURFACE TEMPERATURE
    UNDERWATER ACOUSTICS
```

THERMOCOUPLE PYROMETERS
```
GS  MEASURING INSTRUMENTS
    . TEMPERATURE MEASURING
        INSTRUMENTS
    . . PYROMETERS
```

THERMOCOUPLE PYROMETERS-(CONT.)
```
    . . . THERMOCOUPLE PYROMETERS
RT  GALVANOMETERS
    POTENTIOMETERS (INSTRUMENTS)
    RADIATION PYROMETERS
    RESISTANCE THERMOMETERS
    TEMPERATURE MEASUREMENT
    THERMOCOUPLES
    THERMOELEMENT AMMETERS
```

THERMOCOUPLES
```
GS  THERMOCOUPLES
    . THERMOPILES
RT  CONSTANTAN
    INDICATING INSTRUMENTS
    MANGANIN (TRADEMARK)
    PELTIER EFFECTS
    POTENTIOMETERS (INSTRUMENTS)
    SEEBECK EFFECT
    TEMPERATURE MEASUREMENT
    TEMPERATURE MEASURING
        INSTRUMENTS
    TEMPERATURE PROBES
    THERMOCOUPLE PYROMETERS
    THERMOELECTRIC GENERATORS
    THERMOELECTRICITY
```

THERMODYNAMIC COUPLING
```
GS  COUPLING
    . THERMODYNAMIC COUPLING
RT  BCS THEORY
    ELECTRON PHONON INTERACTIONS
    SUPERCONDUCTORS
```

THERMODYNAMIC CYCLES
```
GS  CYCLES
    . THERMODYNAMIC CYCLES
    . . BRAYTON CYCLE
    . . CARNOT CYCLE
    . . OTTO CYCLE
    . . RANKINE CYCLE
    . . STIRLING CYCLE
RT  ADIABATIC CONDITIONS
    CLOSED CYCLES
    INTERNAL COMBUSTION ENGINES
    LASER PROPULSION
    ∞STROKES
    THERMODYNAMICS
```

THERMODYNAMIC EFFICIENCY
```
UF  THERMAL EFFICIENCY
GS  EFFICIENCY
    . THERMODYNAMIC EFFICIENCY
RT  COMBUSTION EFFICIENCY
    COMPRESSOR EFFICIENCY
    ENGINES
    HEAT SOURCES
    INTERNAL COMBUSTION ENGINES
    NOZZLE EFFICIENCY
    POWER EFFICIENCY
    PROPULSION SYSTEM PERFORMANCE
    PROPULSIVE EFFICIENCY
    SPECIFIC IMPULSE
    TEMPERATURE
    THERMODYNAMICS
```

THERMODYNAMIC EQUILIBRIUM
```
RT  ACID BASE EQUILIBRIUM
    ADIABATIC CONDITIONS
    CHEMICAL EQUILIBRIUM
    ∞EQUILIBRIUM
    HEAT OF DISSOCIATION
    ISENTROPIC PROCESSES
    ISOCHORIC PROCESSES
    ISOENERGETIC PROCESSES
    ISOTHERMAL PROCESSES
    LIQUID-VAPOR EQUILIBRIUM
    STATISTICAL MECHANICS
```

THERMODYNAMIC PROPERTIES
```
UF  THERMAL PROPERTIES
GS  THERMODYNAMIC PROPERTIES
    . ENTHALPY
    . ENTROPY
    . FREE ENERGY
    . . GIBBS FREE ENERGY
    . SURFACE ENERGY
    . THERMAL EXPANSION
    . . THERMAL BUCKLING
    . THERMAL INSTABILITY
    . THERMOCHEMICAL PROPERTIES
    . . HEAT OF COMBUSTION
    . . HEAT OF FORMATION
    . . HEAT OF FUSION
    . . HEAT OF VAPORIZATION
```

THERMODYNAMIC PROPERTIES-(CONT.)
```
    . THERMOPHYSICAL PROPERTIES
    . . CRITICAL POINT
    . . CRITICAL PRESSURE
    . . CRITICAL TEMPERATURE
    . . EMISSIVITY
    . . FUSIBILITY
    . . HEAT OF FUSION
    . . MELTING POINTS
    . . PYROELECTRICITY
    . . SPECIFIC HEAT
    . . . HEAT OF SOLUTION
    . . SUPERCRITICAL PRESSURES
    . . THERMAL CONDUCTIVITY
    . . THERMAL DIFFUSION
    . . THERMAL DIFFUSIVITY
    . . THERMAL STABILITY
    . . VAPOR PRESSURE
    . . VOLATILITY
RT  CHEMICAL PROPERTIES
    DIFFUSIVITY
    EMITTANCE
    ∞EQUILIBRIUM
    HEAT
    HEAT BALANCE
    HIGH TEMPERATURE TESTS
    JOULE-THOMSON EFFECT
    OPTICAL PROPERTIES
    ∞PHYSICAL PROPERTIES
    PRANDTL NUMBER
    PROPELLANT PROPERTIES
    ∞PROPERTIES
    SEEBECK EFFECT
    SOLUBILITY
    TEMPERATURE
    TEPHIGRAMS
    THERMAL CYCLING TESTS
    THERMAL RADIATION
    THERMAL RESISTANCE
    THERMAL SHOCK
    THERMOCHEMISTRY
    THERMODYNAMICS
    THERMOLUMINESCENCE
    ZERO POINT ENERGY
```

THERMODYNAMICS
```
UF  HEAT EQUATIONS
    THERMOMECHANICS
    THERMOPHYSICS
GS  THERMODYNAMICS
    . AEROTHERMODYNAMICS
    . COMBUSTION PHYSICS
    . NONEQUILIBRIUM THERMODYNAMICS
RT  AERODYNAMICS
    ∞DYNAMICS
    ENGINES
    ENTHALPY
    ENTROPY
    ∞EQUATIONS
    EQUATIONS OF STATE
    ∞EQUILIBRIUM
    ERGODIC PROCESS
    FLUID MECHANICS
    FREE ENERGY
    GAS DYNAMICS
    HEAT
    HEAT OF FUSION
    HEAT OF SOLUTION
    HEAT TRANSFER
    INTERNAL ENERGY
    IRREVERSIBLE PROCESSES
    ISOTHERMS
    JOULE-THOMSON EFFECT
    KIRCHHOFF LAW OF RADIATION
    MECHANICAL ENGINEERING
    MOLECULAR RELAXATION
    MOLLIER DIAGRAM
    NONADIABATIC CONDITIONS
    NONGRAY GAS
    NONISOTHERMAL PROCESSES
    ONSAGER RELATIONSHIP
    ∞PATHS
    PFAFF EQUATION
    PHOTOTHERMAL CONVERSION
    PHYSICAL CHEMISTRY
    PLASMA PHYSICS
    PLASMAS (PHYSICS)
    POLYTROPIC PROCESSES
    RANKINE CYCLE
    RAYLEIGH EQUATIONS
    STEAM
    THERMOCHEMISTRY
    THERMODYNAMIC CYCLES
    THERMODYNAMIC EFFICIENCY
    THERMODYNAMIC PROPERTIES
    THERMOELECTRIC COOLING
```

THERMODYNAMICS-*(CONT.)*
    UNSTEADY STATE

THERMOELASTICITY
  GS   MECHANICAL PROPERTIES
     . ELASTIC PROPERTIES
     . . **THERMOELASTICITY**
     . . . AEROTHERMOELASTICITY
  RT   AEROELASTICITY
     AEROTHERMODYNAMICS
     HYDROELASTICITY
     THERMOELECTRIC GENERATORS
     THERMOELECTRIC MATERIALS

THERMOELECTRIC CONVERSION SYSTEMS
  USE   THERMOELECTRIC POWER GENERATION

THERMOELECTRIC COOLING
  UF   ETTINGSHAUSEN COOLERS
  GS   COOLING
     . **THERMOELECTRIC COOLING**
  RT   CRYOGENICS
     ETTINGSHAUSEN EFFECT
     HEAT PUMPS
     PELTIER EFFECTS
     REFRIGERATING
     REFRIGERATING MACHINERY
     THERMODYNAMICS
     THERMOELECTRICITY
     THERMOMAGNETIC COOLING

THERMOELECTRIC GENERATORS
  GS   ELECTRIC GENERATORS
     . DIRECT POWER GENERATORS
     . . **THERMOELECTRIC GENERATORS**
     . . . SNAP 3
     . . . SNAP 7
     . . . SNAP 9A
     . . . SNAP 10A
     . . . SNAP 11
     . . . SNAP 15
     . . . SNAP 17
     . . . SNAP 19
     . . . SNAP 21
     . . . SNAP 23
     . . . SNAP 27
     . . . SNAP 29
     . . . SOLAR SEA POWER PLANTS
  RT   ASTEC SOLAR TURBOELECTRIC
       GENERATOR
     FUEL CELLS
     ∞GENERATORS
     MAGNETOHYDRODYNAMIC GENERATORS
     NUCLEAR AUXILIARY POWER UNITS
     PHOTOELECTRIC GENERATORS
     RADIOISOTOPE BATTERIES
     SNAP
     SOLAR CELLS
     SOLAR GENERATORS
     THERMIONIC CONVERTERS
     THERMOCOUPLES
     THERMOELASTICITY
     THERMOELECTRICITY

THERMOELECTRIC MATERIALS
  RT   ∞MATERIALS
     SEMICONDUCTORS (MATERIALS)
     THERMOELASTICITY
     THERMOELECTRICITY

THERMOELECTRIC OUTER PLANET SPACECRAFT
  USE   TOPS (SPACECRAFT)

THERMOELECTRIC POWER GENERATION
  UF   THERMOELECTRIC CONVERSION
       SYSTEMS
  RT   ∞CONVERSION
     NUCLEAR AUXILIARY POWER UNITS
     SNAP
     THERMOELECTRICITY

THERMOELECTRIC SPACECRAFT
  USE   TOPS (SPACECRAFT)

THERMOELECTRICITY
  UF   THOMSON EFFECT
  RT   ETTINGSHAUSEN EFFECT
     PELTIER EFFECTS
     SEEBECK EFFECT
     THERMIONIC EMISSION
     THERMOCOUPLES
     THERMOELECTRIC COOLING
     THERMOELECTRIC GENERATORS
     THERMOELECTRIC MATERIALS

THERMOELECTRICITY-*(CONT.)*
    THERMOELECTRIC POWER GENERATION
    THERMOPILES
    TRANSPORT PROPERTIES

THERMOELEMENT AMMETERS
  GS   MEASURING INSTRUMENTS
     . AMMETERS
     . . **THERMOELEMENT AMMETERS**
  RT   THERMOCOUPLE PYROMETERS

THERMOGRAMS
  USE   RECORDING INSTRUMENTS
     TEMPERATURE MEASURING
      INSTRUMENTS

THERMOGRAPHY
  RT   INFRARED IMAGERY
     NONDESTRUCTIVE TESTS
     TEMPERATURE DISTRIBUTION
     TEMPERATURE MEASUREMENT
     THERMAL MAPPING

THERMOGRAVIMETRY
  UF   THERMAL GRAVIMETRY
  RT   CHEMICAL ANALYSIS
     DEHYDRATION
     PYROLYSIS
     TEMPERATURE EFFECTS
     THERMAL DECOMPOSITION
     THERMOBALANCES
     THERMOCHEMISTRY

THERMOHYDRAULICS
  RT   CONVECTIVE HEAT TRANSFER
     FLUID DYNAMICS
     FLUID FLOW
     HEAT TRANSMISSION
     ∞HYDRAULICS
     HYDRODYNAMICS
     LAMINAR HEAT TRANSFER
     RADIATIVE HEAT TRANSFER
     THERMAL CONDUCTIVITY
     THERMAL DIFFUSION
     TURBULENT HEAT TRANSFER

THERMOLUMINESCENCE
  GS   DECAY
     . EMISSION
     . . LIGHT EMISSION
     . . . LUMINESCENCE
     . . . . **THERMOLUMINESCENCE**
  RT   TEMPERATURE EFFECTS
     THERMODYNAMIC PROPERTIES

THERMOMAGNADYNAMICS
  USE   THERMOMAGNETIC EFFECTS

THERMOMAGNETIC COOLING
  UF   NERNST GENERATORS
  GS   COOLING
     . **THERMOMAGNETIC COOLING**
  RT   CRYOGENICS
     ETTINGSHAUSEN EFFECT
     THERMOELECTRIC COOLING

THERMOMAGNETIC EFFECTS
  UF   THERMOMAGNADYNAMICS
     THERMOMAGNETISM
  GS   MAGNETIC PROPERTIES
     . **THERMOMAGNETIC EFFECTS**
  RT   ∞EFFECTS
     ETTINGSHAUSEN EFFECT
     NERNST-ETTINGSHAUSEN EFFECT

THERMOMAGNETISM
  USE   THERMOMAGNETIC EFFECTS

THERMOMECHANICAL TREATMENT
  RT   HEAT TREATMENT
     ∞METALLURGY
     MICROSTRUCTURE
     PLASTIC DEFORMATION
     QUENCHING (COOLING)
     ∞TREATMENT

THERMOMECHANICS
  USE   THERMODYNAMICS

THERMOMETERS
  GS   MEASURING INSTRUMENTS
     . TEMPERATURE MEASURING
      INSTRUMENTS
     . . **THERMOMETERS**

THERMOMETERS-*(CONT.)*
    . . . RESISTANCE THERMOMETERS
  RT   TEMPERATURE CONTROL
     TEMPERATURE MEASUREMENT
     TEMPERATURE SCALES

THERMOMETRY
  USE   TEMPERATURE MEASUREMENT

THERMOMIGRATION
  RT   ELECTROMIGRATION
     HEAT TRANSFER
     TEMPERATURE GRADIENTS

THERMONUCLEAR ENERGY
  USE   THERMONUCLEAR POWER GENERATION

THERMONUCLEAR EXPLOSIONS
  GS   EXPLOSIONS
     . NUCLEAR EXPLOSIONS
     . . **THERMONUCLEAR EXPLOSIONS**
  RT   AERIAL EXPLOSIONS
     ARGUS PROJECT
     FISSION WEAPONS
     NUCLEAR DEVICES
     NUCLEAR VULNERABILITY
     UNDERGROUND EXPLOSIONS
     UNDERWATER EXPLOSIONS

THERMONUCLEAR POWER GENERATION
  UF   THERMONUCLEAR ENERGY
  GS   NUCLEAR ELECTRIC POWER
      GENERATION
     . **THERMONUCLEAR POWER**
      GENERATION
  RT   ASTRON THERMONUCLEAR REACTOR
     CONTROLLED FUSION
     ELECTRIC GENERATORS
     ∞ENERGY
     PINCH EFFECT
     PLASMA GENERATORS
     STELLARATORS
     ZETA THERMONUCLEAR REACTOR

THERMONUCLEAR PROPULSION
  USE   NUCLEAR PROPULSION

THERMONUCLEAR REACTIONS
  GS   NUCLEAR REACTIONS
     . **THERMONUCLEAR REACTIONS**
     . . NUCLEAR FUSION
     . . . CONTROLLED FUSION
  RT   ASTRON THERMONUCLEAR REACTOR
     HIGH ENERGY INTERACTIONS
     MAGNETOHYDRODYNAMICS
     PINCH EFFECT
     PLASMAS (PHYSICS)
     PROTON-PROTON REACTIONS
     RADIOACTIVE DECAY
     SCYLLA
     STELLARATORS
     ZETA THERMONUCLEAR REACTOR

THERMOPHILES
  RT   ALGAE
     FUNGI
     MESOPHILES
     PSYCHROPHILES

THERMOPHILIC PLANTS
  GS   PLANTS (BOTANY)
     . **THERMOPHILIC PLANTS**
  RT   ALGAE

†

THERMOPHYSICAL PROPERTIES
  GS   THERMODYNAMIC PROPERTIES
     . **THERMOPHYSICAL PROPERTIES**
     . . CRITICAL POINT
     . . CRITICAL PRESSURE
     . . CRITICAL TEMPERATURE
     . . EMISSIVITY
     . . FUSIBILITY
     . . HEAT OF FUSION
     . . MELTING POINTS
     . . PYROELECTRICITY
     . . SPECIFIC HEAT
     . . . HEAT OF SOLUTION
     . . SUPERCRITICAL PRESSURES
     . . THERMAL CONDUCTIVITY
     . . THERMAL DIFFUSION
     . . THERMAL DIFFUSIVITY
     . . THERMAL STABILITY
     . . VAPOR PRESSURE
     . . VOLATILITY

THERMOPHYSICAL PROPERTIES-(CONT.)
RT   PELTIER EFFECTS
     ∞PROPERTIES
     SEEBECK EFFECT
     SURFACE ENERGY
     THERMAL EXPANSION
     THERMOCHEMISTRY

THERMOPHYSICS
USE   THERMODYNAMICS

THERMOPILES
GS   THERMOCOUPLES
     . THERMOPILES
     TRANSDUCERS
     . THERMOPILES
RT   DICKE RADIOMETERS
     INDICATING INSTRUMENTS
     TEMPERATURE MEASURING
      INSTRUMENTS
     THERMOELECTRICITY

THERMOPLASTIC FILMS
GS   PLASTICS
     . SYNTHETIC RESINS
     . . THERMOPLASTIC RESINS
     . . . THERMOPLASTIC FILMS
     RESINS
     . SYNTHETIC RESINS
     . . THERMOPLASTIC RESINS
     . . . THERMOPLASTIC FILMS
RT   ∞FILMS

THERMOPLASTIC RESINS
GS   PLASTICS
     . SYNTHETIC RESINS
     . . THERMOPLASTIC RESINS
     . . . QUINOXALINES
     . . . THERMOPLASTIC FILMS
     RESINS
     . SYNTHETIC RESINS
     . . THERMOPLASTIC RESINS
     . . . QUINOXALINES
     . . . THERMOPLASTIC FILMS
RT   ACRYLIC RESINS
     GLASS FIBER REINFORCED PLASTICS
     POLYETHYLENES
     POLYSTYRENE
     THERMOPLASTICITY
     THERMOSETTING RESINS
     VULCANIZED ELASTOMERS

THERMOPLASTICITY
GS   MECHANICAL PROPERTIES
     . PLASTIC PROPERTIES
     . . THERMOPLASTICITY
RT   BOUGUER LAW
     TEMPERATURE EFFECTS
     THERMOPLASTIC RESINS

THERMORECEPTORS
GS   ANATOMY
     . SENSE ORGANS
     . . THERMORECEPTORS
     RECEPTORS (PHYSIOLOGY)
     . THERMORECEPTORS
RT   BODY TEMPERATURE
     SENSITOMETRY
     SKIN (ANATOMY)
     THERMOREGULATION

THERMOREGULATION
UF   BODY TEMPERATURE REGULATION
RT   BODY TEMPERATURE
     COLD TOLERANCE
     HIBERNATION
     HOMEOSTASIS
     HYPERTHERMIA
     HYPOTHERMIA
     METABOLISM
     TEMPERATURE CONTROL
     THERMORECEPTORS

THERMOSETTING RESINS
GS   PLASTICS
     . SYNTHETIC RESINS
     . . THERMOSETTING RESINS
     . . . EPOXY RESINS
     . . . FURAN RESINS
     . . . . POLYAMIDE RESINS
     . . . . . KEVLAR (TRADEMARK)
     . . . PHENOLIC RESINS
     . . . . MICARTA
     RESINS
     . SYNTHETIC RESINS

THERMOSETTING RESINS-(CONT.)
     . . THERMOSETTING RESINS
     . . . EPOXY RESINS
     . . . FURAN RESINS
     . . . . POLYAMIDE RESINS
     . . . . . KEVLAR (TRADEMARK)
     . . . PHENOLIC RESINS
     . . . . MICARTA
     . . . . PHENOLIC EPOXY RESINS
RT   BAKELITE (TRADEMARK)
     COMPOSITE MATERIALS
     FORMICA
     GLASS FIBER REINFORCED PLASTICS
     LAMINATES
     POLYESTER RESINS
     REINFORCED PLASTICS
     SILICONE RESINS
     THERMOPLASTIC RESINS

THERMOSIPHONS
GS   REGENERATORS
     . THERMOSIPHONS
RT   CONVECTIVE HEAT TRANSFER
     FREE CONVECTION
     ∞RADIATORS
     SIPHONING

THERMOSPHERE
GS   EARTH ATMOSPHERE
     UPPER ATMOSPHERE
     . . THERMOSPHERE
     . . . TURBOPAUSE
RT   CHEMOSPHERE
     EXOSPHERE
     HETEROSPHERE
     HOMOSPHERE
     IONOSPHERE
     MAGNETOSPHERE

THERMOSTABILITY
USE   THERMAL STABILITY

THERMOSTATS
GS   CONTROL EQUIPMENT
     . THERMOSTATS
     REGULATORS
     . THERMOSTATS
     SWITCHES
     . ELECTRIC SWITCHES
     . . THERMOSTATS
RT   AUTOMATIC CONTROL
     CONTROLLERS
     CRYOSTATS
     TEMPERATURE CONTROL
     TEMPERATURE MEASURING
      INSTRUMENTS

THERMOTROPISM
USE   ANISOTROPY
     TEMPERATURE EFFECTS

THERMOVISCOELASTICITY
GS   MECHANICAL PROPERTIES
     . ELASTIC PROPERTIES
     . . VISCOELASTICITY
     . . . THERMOVISCOELASTICITY
RT   IRREVERSIBLE PROCESSES
     STRESS-STRAIN-TIME RELATIONS

THESAURI
RT   INDEXES (DOCUMENTATION)
     INFORMATION RETRIEVAL
     KWIC INDEXES
     NOMENCLATURES
     SPACE GLOSSARIES
     TERMINOLOGY
     TERMS
     WORDS (LANGUAGE)

THESES
GS   DOCUMENTS
     . THESES
RT   HYPOTHESES

THETA PINCH
GS   PINCH EFFECT
     . PLASMA PINCH
     . . THETA PINCH
RT   LASER PLASMA INTERACTIONS
     PLASMA COMPRESSION
     ROTATING PLASMAS
     SCREW PINCH
     ZETA PINCH

THIAMINE
UF   VITAMIN B
GS   HETEROCYCLIC COMPOUNDS
     . THIAMINE
     VITAMINS
     . THIAMINE

THIAZINE (TRADEMARK)
GS   DYES
     . THIAZITE (TRADEMARK)
     HETEROCYCLIC COMPOUNDS
     . THIAZINE (TRADEMARK)
     NITROGEN COMPOUNDS
     . THIAZINE (TRADEMARK)
     SULFUR COMPOUNDS
     . THIAZINE (TRADEMARK)

THICK FILMS
RT   ELECTRONIC PACKAGING
     ∞FILMS
     INTEGRATED CIRCUITS
     MICROMINIATURIZATION
     PRINTED CIRCUITS
     SEMICONDUCTING FILMS
     THIN FILMS

THICK PLATES
RT   FLAT PLATES
     METAL PLATES
     ∞PLATES
     PLATES (STRUCTURAL MEMBERS)
     ∞SHEETS
     THICKNESS
     THIN PLATES

THICK WALLS
GS   WALLS
     . THICK WALLS
RT   BOILER PLATE
     BULKHEADS
     REINFORCEMENT (STRUCTURES)
     STRUCTURAL MEMBERS
     THIN WALLS
     WALL PRESSURE
     WALL TEMPERATURE

∞ THICKENERS
SN   (USE OF A MORE SPECIFIC TERM IS
     RECOMMENDED--CONSULT THE TERMS
     LISTED BELOW)
RT   THICKENERS (EQUIPMENT)
     THICKENERS (MATERIALS)

THICKENERS (EQUIPMENT)
GS   SEPARATORS
     . CLASSIFIERS
     . . THICKENERS (EQUIPMENT)
RT   COALESCING
     PRECIPITATORS
     ∞THICKENERS

THICKENERS (MATERIALS)
RT   ADDITIVES
     GELS
     GREASES
     ∞MATERIALS
     ∞THICKENERS

THICKNESS
RT   AIRFOIL PROFILES
     DEPTH
     DIAMETERS
     DIMENSIONS
     FILM THICKNESS
     LENGTH
     OPTICAL THICKNESS
     SPACING
     TARGET THICKNESS
     THICK PLATES
     THICKNESS RATIO
     VOLUME

THICKNESS RATIO
GS   RATIOS
     . THICKNESS RATIO
RT   AIRFOIL PROFILES
     AIRFOILS
     FINENESS RATIO
     THICKNESS
     THIN AIRFOILS
     THIN WINGS

THIGH
GS   ANATOMY

THIGH-(CONT.)
    . THIGH
  RT   LEG (ANATOMY)

THIN AIRFOILS
  GS   AIRFOILS
      . **THIN AIRFOILS**
      . . THIN WINGS
      . . . INFINITE SPAN WINGS
  RT   AIRFOIL PROFILES
      THICKNESS RATIO

∞ **THIN BODIES**
  SN   *(USE OF A MORE SPECIFIC TERM IS*
      RECOMMENDED--CONSULT THE TERMS
      LISTED BELOW)
  RT   SLENDER BODIES
      THIN PLATES
      THIN WALLS
      THIN WINGS

THIN FILMS
  SN   (SOLID STATE PHYSICS AND
      ELECTRONICS)
  GS   **THIN FILMS**
      . ENERGY ABSORPTION FILMS
      . FERROMAGNETIC FILMS
      . MONOMOLECULAR FILMS
  RT   COATINGS
      COMPUTER STORAGE DEVICES
      ELECTROCHROMISM
      ELECTRODE FILM BARRIERS
    ∞FILMS
      HETEROJUNCTIONS
      INTEGRATED CIRCUITS
      INTEGRATED OPTICS
      ION PLATING
      METAL FILMS
      MICROCHANNEL PLATES
      MICROMINIATURIZATION
      MINIATURE ELECTRONIC EQUIPMENT
      MOLECULAR ELECTRONICS
      OXIDE FILMS
      PARAMETRONS
      PELLICLE
      PHOTOTHERMAL CONVERSION
      PLATING
      PRAETERSONIC DEVICES
      RECTIFIERS
      SEMICONDUCTING FILMS
      SILICON FILMS
      SOLID STATE DEVICES
    ∞SOLID STATE PHYSICS
      SPUTTERING GAGES
      SQUEEZE FILMS
      THICK FILMS
      WAFERS

THIN LAYER CHROMATOGRAPHY
  GS   CHROMATOGRAPHY
      . **THIN LAYER CHROMATOGRAPHY**
  RT   GAS CHROMATOGRAPHY
      MONOMOLECULAR FILMS

THIN PLATES
  SN   (EXCLUDES THIN SURFACE COATINGS
      AND FILMS)
  RT   DIAPHRAGMS (MECHANICS)
      FLAT PLATES
      FOILS (MATERIALS)
      METAL PLATES
      PANELS
      PARALLEL PLATES
    ∞PLATES
      PLATES (STRUCTURAL MEMBERS)
    ∞SHEETS
      THICK PLATES
    ∞THIN BODIES

THIN WALLED SHELLS
  GS   SHELLS (STRUCTURAL FORMS)
      . **THIN WALLED SHELLS**
  RT   CYLINDRICAL SHELLS
      MEMBRANE STRUCTURES
      METAL SHELLS
      ORTHOTROPIC SHELLS
      REINFORCED SHELLS
      SKIN (STRUCTURAL MEMBER)
      SPHERICAL SHELLS
      STRESSED-SKIN STRUCTURES
      TOROIDAL SHELLS

THIN WALLS
  GS   WALLS
      . **THIN WALLS**

THIN WALLS-(CONT.)
  RT   BULKHEADS
      DIAPHRAGMS (MECHANICS)
      PARTITIONS (STRUCTURES)
      SKIN (STRUCTURAL MEMBER)
      THICK WALLS
    ∞THIN BODIES

THIN WINGS
  GS   AIRFOILS
      . THIN AIRFOILS
      . . **THIN WINGS**
      . . . INFINITE SPAN WINGS
      WINGS
      . . **THIN WINGS**
      . . . INFINITE SPAN WINGS
  RT   AIRFOIL PROFILES
      FIXED WINGS
      FLEXIBLE WINGS
      THICKNESS RATIO
    ∞THIN BODIES
      UNCAMBERED WINGS

THINNERS
  USE   SOLVENTS

THIOLS
  UF   DITHIOLS
      MERCAPTAN
      MERCAPTO COMPOUNDS
  GS   ALIPHATIC COMPOUNDS
      . **THIOLS**
      . . CYSTEINE
      . . DIMERCAPROL
      SULFUR COMPOUNDS
      . **THIOLS**
      . . CYSTEINE
      . . DIMERCAPROL
  RT   ALCOHOLS
    ∞CHEMICAL COMPOUNDS
      PHENOLS

THIOPLASTICS
  GS   RUBBER
      . SYNTHETIC RUBBERS
      . . ELASTOMERS
      . . . **THIOPLASTICS**
  RT   PLASTICS
      SULFIDES

THIOUREAS
  GS   NITROGEN COMPOUNDS
      . AMIDES
      . . UREAS
      . . . **THIOUREAS**

THIURONIUM
  GS   AMINES
      . **THIURONIUM**
      NITROGEN COMPOUNDS
      . AMIDES
      . . UREAS
      . . . **THIURONIUM**

THIXOTROPIC PROPELLANTS
  USE   GELLED ROCKET PROPELLANTS

THIXOTROPY
  RT   GELATION
      GELS
      LIQUEFACTION
      NONNEWTONIAN FLOW
    ∞PHYSICAL PROPERTIES
      SEMISOLIDS
      SOLUBILITY
      VISCOSITY

THOMAS-FERMI MODEL
  UF   THOMAS-FERMI THEORY
  GS   MODELS
      . MATHEMATICAL MODELS
      . . **THOMAS-FERMI MODEL**
  RT   ATOMIC STRUCTURE
      ELECTRON DISTRIBUTION
      PLASMA COMPOSITION
      QUANTUM STATISTICS

THOMAS-FERMI THEORY
  USE   THOMAS-FERMI MODEL

THOMSON EFFECT
  USE   THERMOELECTRICITY

THOMSON SCATTERING
  GS   SCATTERING
      . WAVE SCATTERING
      . . ELECTROMAGNETIC SCATTERING
      . . . **THOMSON SCATTERING**
  RT   ELECTROMAGNETIC RADIATION

THOR ABLE ROCKET VEHICLE
  GS   LAUNCH VEHICLES
      . THOR LAUNCH VEHICLES
      . . **THOR ABLE ROCKET VEHICLE**
      . THORAD LAUNCH VEHICLES
      . . **THOR ABLE ROCKET VEHICLE**
      ROCKET VEHICLES
      . MULTISTAGE ROCKET VEHICLES
      . . THOR LAUNCH VEHICLES
      . . . **THOR ABLE ROCKET VEHICLE**
      . THORAD LAUNCH VEHICLES
      . . **THOR ABLE ROCKET VEHICLE**
  RT   EXPLORER 6 SATELLITE
      LIQUID PROPELLANT ROCKET ENGINES
      PIONEER 1 SPACE PROBE
      PIONEER 5 SPACE PROBE
      SOLID PROPELLANT ROCKET ENGINES
      TIROS 1 SATELLITE

THOR AGENA LAUNCH VEHICLE
  GS   LAUNCH VEHICLES
      . THOR LAUNCH VEHICLES
      . . **THOR AGENA LAUNCH VEHICLE**
      . THORAD LAUNCH VEHICLES
      . . **THOR AGENA LAUNCH VEHICLE**
      ROCKET VEHICLES
      . MULTISTAGE ROCKET VEHICLES
      . . THOR LAUNCH VEHICLES
      . . . **THOR AGENA LAUNCH VEHICLE**
      . THORAD LAUNCH VEHICLES
      . . **THOR AGENA LAUNCH VEHICLE**
  RT   AGENA A ROCKET VEHICLE
      AGENA B RANGER PROGRAM
      AGENA ROCKET VEHICLES
      DISCOVERER SATELLITES
      EXPLORER 31 SATELLITE
      EXPLORER 34 SATELLITE
      EXPLORER 35 SATELLITE
      EXPLORER 36 SATELLITE
      LIQUID PROPELLANT ROCKET ENGINES
      NIMBUS SATELLITES
      NIMBUS 1 SATELLITE
      NIMBUS 2 SATELLITE
      OGO-3

THOR DELTA LAUNCH VEHICLE
  UF   ECHO 1 CARRIER ROCKET
  GS   LAUNCH VEHICLES
      . THOR LAUNCH VEHICLES
      . . **THOR DELTA LAUNCH VEHICLE**
      . THORAD LAUNCH VEHICLES
      . . **THOR DELTA LAUNCH VEHICLE**
      ROCKET VEHICLES
      . MULTISTAGE ROCKET VEHICLES
      . . THOR LAUNCH VEHICLES
      . . . **THOR DELTA LAUNCH VEHICLE**
      . THORAD LAUNCH VEHICLES
      . . **THOR DELTA LAUNCH VEHICLE**
  RT   ARIEL SATELLITES
      ECHO 1 SATELLITE
      EXPLORER SATELLITES
      LIQUID PROPELLANT ROCKET ENGINES
      OSO
      RELAY SATELLITES
      SOLID PROPELLANT ROCKET ENGINES
      SYNCOM SATELLITES
      TELSTAR SATELLITES

THOR LAUNCH VEHICLES
  GS   LAUNCH VEHICLES
      . **THOR LAUNCH VEHICLES**
      . . THOR ABLE ROCKET VEHICLE
      . . THOR AGENA LAUNCH VEHICLE
      . . THOR DELTA LAUNCH VEHICLE
      ROCKET VEHICLES
      . MULTISTAGE ROCKET VEHICLES
      . . **THOR LAUNCH VEHICLES**
      . . . THOR ABLE ROCKET VEHICLE
      . . . THOR AGENA LAUNCH VEHICLE
      . . . THOR DELTA LAUNCH VEHICLE
  RT   LIQUID PROPELLANT ROCKET ENGINES
      SOLID PROPELLANT ROCKET ENGINES
      THORAD LAUNCH VEHICLES
    ∞VEHICLES

THORAD LAUNCH VEHICLES
  GS   LAUNCH VEHICLES
      . **THORAD LAUNCH VEHICLES**

THORAD LAUNCH VEHICLES-*(CONT.)*
.. THOR ABLE ROCKET VEHICLE
.. THOR AGENA LAUNCH VEHICLE
.. THOR DELTA LAUNCH VEHICLE
ROCKET VEHICLES
. **THORAD LAUNCH VEHICLES**
.. THOR ABLE ROCKET VEHICLE
.. THOR AGENA LAUNCH VEHICLE
.. THOR DELTA LAUNCH VEHICLE
RT   LIQUID PROPELLANT ROCKET ENGINES
THOR LAUNCH VEHICLES
∞ VEHICLES

THORAX
RT   CHEST
DIAPHRAGM (ANATOMY)
STERNUM

THORIUM
GS   CHEMICAL ELEMENTS
. ACTINIDE SERIES
.. **THORIUM**
... THORIUM ISOTOPES
METALS
. ACTINIDE SERIES
.. **THORIUM**
... THORIUM ISOTOPES
RT   NUCLEAR FUELS

THORIUM ALLOYS
GS   ALLOYS
. **THORIUM ALLOYS**
RT   NUCLEAR FUELS

THORIUM COMPOUNDS
GS   ACTINIDE SERIES COMPOUNDS
. **THORIUM COMPOUNDS**
.. THORIUM FLUORIDES
.. THORIUM OXIDES
RT   CERAMIC NUCLEAR FUELS
∞ CHEMICAL COMPOUNDS
∞ METAL COMPOUNDS
NUCLEAR FUELS

THORIUM FLUORIDES
GS   ACTINIDE SERIES COMPOUNDS
. THORIUM COMPOUNDS
.. **THORIUM FLUORIDES**
HALOGEN COMPOUNDS
. FLUORINE COMPOUNDS
.. FLUORIDES
... METAL FLUORIDES
.... **THORIUM FLUORIDES**
. HALIDES
.. METAL HALIDES
... **THORIUM FLUORIDES**

THORIUM ISOTOPES
UF   THORIUM 228
THORIUM 230
THORIUM 234
GS   CHEMICAL ELEMENTS
. ACTINIDE SERIES
.. THORIUM
... **THORIUM ISOTOPES**
NUCLIDES
. ISOTOPES
... **THORIUM ISOTOPES**
METALS
. ACTINIDE SERIES
.. THORIUM
... **THORIUM ISOTOPES**

THORIUM OXIDES
GS   ACTINIDE SERIES COMPOUNDS
. THORIUM COMPOUNDS
.. **THORIUM OXIDES**
CHALCOGENIDES
. OXIDES
.. METAL OXIDES
... **THORIUM OXIDES**
RT   DIOXIDES

THORIUM 228
USE   THORIUM ISOTOPES

THORIUM 230
USE   THORIUM ISOTOPES

THORIUM 234
USE   THORIUM ISOTOPES

THORON
USE   RADON ISOTOPES

THREADS
SN   (EXCLUDES TEXTILES AND
FILAMENTARY FORMS)
RT   BOLTS
NUTS (FASTENERS)
SCREWS

THREAT EVALUATION
RT   AIRCRAFT HAZARDS
AIRCRAFT SAFETY
COLLISION AVOIDANCE
MIDAIR COLLISIONS
RADAR TRACKING
WARNING SYSTEMS

THREE AXIS STABILIZATION
GS   STABILIZATION
. **THREE AXIS STABILIZATION**
RT   INERTIAL PLATFORMS
SATELLITE ATTITUDE CONTROL
SATELLITE ORIENTATION
STABILIZED PLATFORMS

THREE BODY PROBLEM
RT   CELESTIAL MECHANICS
FOUR BODY PROBLEM
MANY BODY PROBLEM
ORBITS
PERTURBATION
∞ PROBLEMS
TROJAN ORBITS
TWO BODY PROBLEM

THREE DIMENSIONAL BODIES
GS   SOLIDS
. **THREE DIMENSIONAL BODIES**
RT   AERODYNAMIC CONFIGURATIONS
∞ BODIES
BOUNDARY VALUE PROBLEMS
FLOW DISTRIBUTION

THREE DIMENSIONAL BOUNDARY LAYER
GS   BOUNDARY LAYERS
. **THREE DIMENSIONAL BOUNDARY
LAYER**
RT   AXISYMMETRIC FLOW
BOUNDARY LAYER TRANSITION
COMPRESSIBLE BOUNDARY LAYER
LAMINAR BOUNDARY LAYER
∞ LAYERS
SECONDARY FLOW
TURBULENT BOUNDARY LAYER
VELOCITY DISTRIBUTION

THREE DIMENSIONAL COMPOSITES
GS   COMPOSITE MATERIALS
. **THREE DIMENSIONAL COMPOSITES**
RT   FIBER COMPOSITES
∞ MATERIALS

THREE DIMENSIONAL FLOW
GS   FLUID FLOW
. PARALLEL FLOW
.. **THREE DIMENSIONAL FLOW**
TRANSLATIONAL MOTION
. THREE DIMENSIONAL MOTION
.. **THREE DIMENSIONAL FLOW**
... SECONDARY FLOW
RT   AXIAL FLOW
CONICAL FLOW
FLOW GEOMETRY
HELICAL FLOW
ONE DIMENSIONAL FLOW
ROSHKO PREDICTION
SPHERICAL WAVES
TWO DIMENSIONAL FLOW
WEDGE FLOW

THREE DIMENSIONAL MOTION
GS   TRANSLATIONAL MOTION
. **THREE DIMENSIONAL MOTION**
.. THREE DIMENSIONAL FLOW
... SECONDARY FLOW
RT   DEGREES OF FREEDOM

THRESHOLD CURRENTS
GS   ELECTRIC CURRENT
. **THRESHOLD CURRENTS**
RT   LASERS
∞ THRESHOLDS

THRESHOLD DETECTORS (DOSIMETERS)
GS   MEASURING INSTRUMENTS
. RADIATION MEASURING INSTRUMENTS

THRESHOLD DETECTORS-*(CONT.)*
.. RADIATION DETECTORS
... DOSIMETERS
.... **THRESHOLD DETECTORS
(DOSIMETERS)**
RT   IONIZATION CHAMBERS
∞ THRESHOLDS

THRESHOLD GATES
GS   CIRCUITS
. GATES (CIRCUITS)
.. **THRESHOLD GATES**
. LOGIC CIRCUITS
.. **THRESHOLD GATES**
RT   ∞ THRESHOLDS
TRIGGER CIRCUITS

THRESHOLD LOGIC
GS   MATHEMATICAL LOGIC
. SET THEORY
.. **THRESHOLD LOGIC**
RT   GATES (CIRCUITS)
∞ LOGIC
LOGIC CIRCUITS
∞ THRESHOLDS
TRANSISTOR LOGIC
TRIGGER CIRCUITS

THRESHOLD SHIFT
USE   THRESHOLDS

†

∞ **THRESHOLDS**
SN   (USE OF A MORE SPECIFIC TERM IS
RECOMMENDED--CONSULT THE TERMS
LISTED BELOW)
UF   THRESHOLD SHIFT
RT   DOORS
ENTRANCES
NOISE THRESHOLD
RESOLUTION
RESPONSES
SENSITIVITY
THRESHOLD CURRENTS
THRESHOLD DETECTORS (DOSIMETERS)
THRESHOLD GATES
THRESHOLD LOGIC
THRESHOLDS (PERCEPTION)

THRESHOLDS (PERCEPTION)
RT   ACUITY
ADAPTATION
AUDIOMETRY
AUDITORY PERCEPTION
AUDITORY SENSATION AREAS
AUDITORY STIMULI
CHRONAXY
∞ FREQUENCY RESPONSE
HEARING
LIGHT ADAPTATION
LIMEN
NEUROLOGY
PERCEPTION
PHOTOSENSITIVITY
SENSITIVITY
∞ THRESHOLDS
VISION
VISUAL PERCEPTION

THROATS
SN   (NON BIOLOGICAL)
RT   CARBURETORS
∞ CHANNELS
CHOKES (RESTRICTIONS)
DUCTS
NOZZLE GEOMETRY
NOZZLE INSERTS
NOZZLE WALLS
ORIFICES

THROMBIN
GS   ANATOMY
. CARDIOVASCULAR SYSTEM
.. **THROMBIN**
BODY FLUIDS
. BLOOD
.. **THROMBIN**
ENZYMES
. **THROMBIN**
PROTEINS
. **THROMBIN**
RT   FIBRINOGEN
HEMOSTATICS
PROTHROMBIN
THROMBOPLASTIN

THROMBOCYTES
- RT   BLOOD COAGULATION
-      CLOTTING

THROMBOPENIA
- GS   DISEASES
-      . **THROMBOPENIA**
- RT   COAGULATION

THROMBOPLASTIN
- GS   ANATOMY
-      . CARDIOVASCULAR SYSTEM
-      . . **THROMBOPLASTIN**
-      BODY FLUIDS
-      . BLOOD
-      . . **THROMBOPLASTIN**
- RT   CLOTTING
-      HEMOSTATICS
-      HOMEOSTASIS
-      PLATELETS
-      THROMBIN

THROMBOSIS
- GS   DISEASES
-      . **THROMBOSIS**
- RT   BLOOD COAGULATION
-      INFARCTION
-      MYOCARDIAL INFARCTION

THROTTLING
- RT   JOULE-THOMSON EFFECT
-      VARIABLE THRUST

THROWING
- RT   EJECTION
-      SPREADING

THRUST
- UF   THRUST POWER
- GS   **THRUST**
-      . HIGH THRUST
-      . JET THRUST
-      . LEADING EDGE THRUST
-      . LOW THRUST
-      . . MICROTHRUST
-      . ROCKET THRUST
-      . . RETROTHRUST
-      . STATIC THRUST
-      . VARIABLE THRUST
- RT   ACCELERATION (PHYSICS)
-      AUXILIARY PROPULSION
-      BURNING TIME
-      DUAL THRUST NOZZLES
-      ∞FORCE
-      JET ENGINES
-      NOZZLE THRUST COEFFICIENTS
-      ∞POWER
-      PROPULSION
-      ∞REACTION
-      ROCKET ENGINES
-      ROCKET PROPELLANTS
-      SPECIFIC IMPULSE

THRUST AUGMENTATION
- GS   AUGMENTATION
-      . **THRUST AUGMENTATION**
- RT   AFTERBURNING
-      COANDA EFFECT
-      HIGH THRUST
-      SECONDARY INJECTION
-      SHROUDED PROPELLERS
-      VARIABLE THRUST
-      WATER INJECTION

THRUST BEARINGS
- GS   BEARINGS
-      . **THRUST BEARINGS**
- RT   ANTIFRICTION BEARINGS
-      BALL BEARINGS
-      GAS BEARINGS
-      ROLLER BEARINGS

THRUST CHAMBER PRESSURE
- GS   PRESSURE
-      . **THRUST CHAMBER PRESSURE**

THRUST CHAMBERS
- UF   ROCKET CHAMBERS
- RT   ARC CHAMBERS
-      ∞CHAMBERS
-      COMBUSTION CHAMBERS
-      DIVERGENT NOZZLES
-      ROCKET ENGINE CASES

THRUST CONTROL
- GS   **THRUST CONTROL**
-      . THRUST VECTOR CONTROL
- RT   ATTITUDE CONTROL
-      ∞CONTROL
-      CONTROL ROCKETS
-      ENGINE CONTROL
-      JET CONTROL
-      ∞REACTION CONTROL
-      ROCKET ENGINE CONTROL
-      SATELLITE CONTROL
-      TURBOJET ENGINE CONTROL
-      VARIABLE THRUST

THRUST DISTRIBUTION
- RT   AERODYNAMIC FORCES
-      ∞DISTRIBUTION
-      FORCE DISTRIBUTION
-      LEADING EDGES
-      PRESSURE DISTRIBUTION
-      VORTICES
-      WING PLANFORMS

THRUST FAULTS
- USE   GEOLOGICAL FAULTS

THRUST LOADS
- GS   LOADS (FORCES)
-      . DYNAMIC LOADS
-      . . THRUST LOADS
- RT   AERODYNAMIC LOADS
-      AXIAL COMPRESSION LOADS
-      AXIAL LOADS
-      COMPRESSION LOADS
-      JET THRUST
-      ROCKET THRUST
-      STRUCTURAL DESIGN CRITERIA

THRUST MEASUREMENT
- GS   MECHANICAL MEASUREMENT
-      . **THRUST MEASUREMENT**
- RT   ACCELEROMETERS
-      DYNAMOMETERS
-      ∞FORCE
-      ∞MEASUREMENT

THRUST POWER
- USE   THRUST

THRUST PROGRAMMING
- UF   OPTIMUM THRUST PROGRAMMING
- GS   PROGRAMMING (SCHEDULING)
-      . **THRUST PROGRAMMING**
- RT   FLIGHT MECHANICS
-      FLIGHT OPTIMIZATION
-      FLIGHT PLANS
-      ORBITAL MECHANICS
-      PARKING ORBITS
-      PROPULSIVE EFFICIENCY
-      TRAJECTORY CONTROL

THRUST REVERSAL
- RT   AIRCRAFT BRAKES
-      BRAKES (FOR ARRESTING MOTION)
-      BRAKING
-      DECELERATION

THRUST TERMINATION
- GS   STOPPING
-      . **THRUST TERMINATION**
- RT   BURNOUT
-      ROCKET THRUST
-      STAGE SEPARATION
-      VARIABLE THRUST

THRUST VECTOR CONTROL
- UF   TVC (CONTROL)
- GS   ATTITUDE CONTROL
-      . DIRECTIONAL CONTROL
-      . . **THRUST VECTOR CONTROL**
-      FLIGHT CONTROL
-      . **THRUST VECTOR CONTROL**
-      THRUST CONTROL
-      . **THRUST VECTOR CONTROL**
- RT   AIR SLEW MISSILES
-      AUTOMATIC CONTROL
-      AUTOMATIC FLIGHT CONTROL
-      ∞CONTROL
-      GUIDE VANES
-      GYROSTABILIZERS
-      JET VANES
-      LIQUID INJECTION
-      MANEUVERABLE SPACECRAFT
-      MISSILE CONTROL

THRUST VECTOR CONTROL-(CONT.)
-      NOZZLE THRUST COEFFICIENTS
-      ROCKET ENGINES
-      SECONDARY INJECTION
-      SPACECRAFT CONTROL
-      VARIABLE THRUST
-      VERNIER ENGINES

THRUST-WEIGHT RATIO
- GS   RATIOS
-      . **THRUST-WEIGHT RATIO**
- RT   ACCELERATION (PHYSICS)
-      MASS RATIOS
-      PRESSURE RATIO
-      ROCKET ENGINES

∞ THRUSTORS
- SN   *(USE OF A MORE SPECIFIC TERM IS RECOMMENDED--CONSULT THE TERMS LISTED BELOW)*
- RT   ION ENGINES
-      ROCKET ENGINES

THULIUM
- GS   CHEMICAL ELEMENTS
-      . RARE EARTH ELEMENTS
-      . . **THULIUM**
-      . . . THULIUM ISOTOPES
-      METALS
-      . RARE EARTH ELEMENTS
-      . . **THULIUM**
-      . . . THULIUM ISOTOPES

THULIUM COMPOUNDS
- GS   RARE EARTH COMPOUNDS
-      . **THULIUM COMPOUNDS**
- RT   ∞CHEMICAL COMPOUNDS
-      ∞METAL COMPOUNDS

THULIUM ISOTOPES
- UF   THULIUM 171
- GS   CHEMICAL ELEMENTS
-      . NUCLIDES
-      . . ISOTOPES
-      . . . **THULIUM ISOTOPES**
-      . RARE EARTH ELEMENTS
-      . . THULIUM
-      . . . **THULIUM ISOTOPES**
-      METALS
-      . RARE EARTH ELEMENTS
-      . . THULIUM
-      . . . **THULIUM ISOTOPES**

THULIUM 171
- USE   THULIUM ISOTOPES

THUNDERCHIEF AIRCRAFT
- USE   F-105 AIRCRAFT

THUNDERSTORMS
- GS   STORMS
-      . STORMS (METEOROLOGY)
-      . . RAINSTORMS
-      . . . **THUNDERSTORMS**
- RT   ANVIL CLOUDS
-      ATMOSPHERICS
-      CIRROCUMULUS CLOUDS
-      CIRROSTRATUS CLOUDS
-      CLOUDS (METEOROLOGY)
-      COLD FRONTS
-      CUMULONIMBUS CLOUDS
-      FRONTS (METEOROLOGY)
-      HAIL
-      HAILSTORMS
-      LIGHTNING
-      LIGHTNING SUPPRESSION
-      RAIN
-      STORM DAMAGE
-      WARM FRONTS
-      WIND (METEOROLOGY)

THYMIDINE
- GS   ACIDS
-      . **THYMIDINE**
-      HETEROCYCLIC COMPOUNDS
-      . **THYMIDINE**
-      PYRIMIDINES
-      . **THYMIDINE**
- RT   ALLOXAN
-      DEOXYRIBONUCLEIC ACID
-      NUCLEOSIDES

THYMINE
- GS   ACIDS

THYMINE-*(CONT.)*
. **THYMINE**
HETEROCYCLIC COMPOUNDS
. **THYMINE**
NITROGEN COMPOUNDS
. **THYMINE**
PYRIMIDINES
. **THYMINE**
RT   ALLOXAN
DEOXYRIBONUCLEIC ACID

THYMOL
GS   HYDROXYL COMPOUNDS
. ALCOHOLS
. . PHENOLS
. . . **THYMOL**

THYMUS GLAND
GS   ANATOMY
. GLANDS (ANATOMY)
. . ENDOCRINE GLANDS
. . . **THYMUS GLAND**
VISCERA
. ENDOCRINE GLANDS
. . **THYMUS GLAND**

THYRATRONS
GS   ELECTRON TUBES
. VACUUM TUBES
. . VACUUM TUBE OSCILLATORS
. . . GAS DISCHARGE TUBES
. . . . **THYRATRONS**
. . . MICROWAVE TUBES
. . . . PLANOTRONS
. . . . . **THYRATRONS**
MICROWAVE EQUIPMENT
. MICROWAVE TUBES
. . PLANOTRONS
. . . **THYRATRONS**
RECTIFIERS
. **THYRATRONS**
RT   CURRENT CONVERTERS (AC TO DC)
SILICON CONTROLLED RECTIFIERS
THYRISTORS

THYRISTORS
GS   ELECTRONIC EQUIPMENT
. SOLID STATE DEVICES
. . SEMICONDUCTOR DEVICES
. . . **THYRISTORS**
. . . . SILICON CONTROLLED
RECTIFIERS
. **THYRISTORS**
. . SILICON CONTROLLED RECTIFIERS
RT   JUNCTION TRANSISTORS
P-N-P-N JUNCTIONS
THYRATRONS
TRIGGER CIRCUITS
TRIODES

THYROID GLAND
GS   ANATOMY
. GLANDS (ANATOMY)
. . ENDOCRINE GLANDS
. . . **THYROID GLAND**
VISCERA
. ENDOCRINE GLANDS
. . **THYROID GLAND**
RT   CALCIUM METABOLISM
HYPOMETABOLISM

THYROXINE
GS   ACIDS
. AMINO ACIDS
. . **THYROXINE**
ORGANIC COMPOUNDS
. AMINO ACIDS
. . **THYROXINE**

TIBET
GS   NATIONS
. **TIBET**
RT   ASIA
BHUTAN
HIMALAYAS

TIBIA
GS   ANATOMY
. MUSCULOSKELETAL SYSTEM
. . BONES
. . . **TIBIA**
RT   LEG (ANATOMY)

TID
USE   TRAVELING IONOSPHERIC
DISTURBANCES

TIDAL FLATS
GS   LANDFORMS
. FLATS (LANDFORMS)
. . **TIDAL FLATS**
RT   AQUICULTURE
COASTS
ESTUARIES
FISHERIES
MARSHLANDS
MUD
OCEANS
SHORELINES
TIDES

TIDAL OSCILLATION
USE   TIDES

TIDAL WAVES
GS   WATER WAVES
. **TIDAL WAVES**
RT   OCEAN CURRENTS
OCEAN SURFACE
OCEANOGRAPHY
SEA BREEZE
SEA ROUGHNESS
SEISMOLOGY
TSUNAMI WAVES
∞WAVES
WIND (METEOROLOGY)

TIDE POWERED GENERATORS
RT   ELECTRIC GENERATORS
ENERGY CONVERSION EFFICIENCY
∞GENERATORS
OCEAN CURRENTS
OCEAN SURFACE
OCEANOGRAPHY
OCEANS
SEA ROUGHNESS
TIDEPOWER
TIDES
WATERWAVE ENERGY CONVERSION
WATERWAVE POWERED MACHINES

TIDE POWERED MACHINES
RT   ∞MACHINERY
OCEAN CURRENTS
OCEAN SURFACE
SEA ROUGHNESS
TIDEPOWER
TIDES
WATERWAVE ENERGY CONVERSION
WATERWAVE POWERED MACHINES

TIDEPOWER
RT   CLEAN ENERGY
EARTH RESOURCES
∞ENERGY SOURCES
OCEAN CURRENTS
OCEAN SURFACE
OCEANOGRAPHY
SEA ROUGHNESS
TIDE POWERED GENERATORS
TIDE POWERED MACHINES
TIDES
WATERWAVE ENERGY
WATERWAVE ENERGY CONVERSION
WATERWAVE POWERED MACHINES

TIDES
UF   TIDAL OSCILLATION
GS   **TIDES**
. ATMOSPHERIC TIDES
. EARTH TIDES
. LUNAR TIDES
RT   COASTAL CURRENTS
ESTUARIES
FLOOD DAMAGE
FLOODS
OCEAN CURRENTS
OCEAN SURFACE
OCEANOGRAPHY
PRESSURE ICE
SEA ROUGHNESS
TIDAL FLATS
TIDE POWERED GENERATORS
TIDE POWERED MACHINES
TIDEPOWER
WATER CURRENTS
WATERWAVE ENERGY CONVERSION
WATERWAVE POWERED MACHINES

TIDES-*(CONT.)*
WETLANDS

TIEBOLTS
GS   FASTENERS
. BOLTS
. . **TIEBOLTS**

TIG WELDING
USE   GAS TUNGSTEN ARC WELDING

TIGHTNESS
RT   CLEARANCES
CLOSURES
PROXIMITY

TILES
RT   CERAMICS
FLOORS
GROUT
LUDOX (TRADEMARK)
MASONRY
WALLS

TILT
USE   ATTITUDE (INCLINATION)

TILT ROTOR AIRCRAFT
GS   V/STOL AIRCRAFT
. ROTARY WING AIRCRAFT
. . **TILT ROTOR AIRCRAFT**
. . . XV-15 AIRCRAFT
RT   ∞AIRCRAFT
HELICOPTERS
TILTING ROTORS

TILT ROTOR RESEARCH AIRCRAFT PROGRAM
GS   PROGRAMS
. NASA PROGRAMS
. . **TILT ROTOR RESEARCH AIRCRAFT
PROGRAM**
RT   ∞AIRCRAFT
HELICOPTERS
ROTARY WINGS
XV-15 AIRCRAFT

TILT WING AIRCRAFT
UF   PIVOTED WING AIRCRAFT
GS   **TILT WING AIRCRAFT**
. CL-84 AIRCRAFT
. L-29 JET TRAINER
. VZ-2 AIRCRAFT
. XC-142 AIRCRAFT
RT   ∞AIRCRAFT
FAN IN WING AIRCRAFT
RESEARCH AIRCRAFT
SHORT TAKEOFF AIRCRAFT
V/STOL AIRCRAFT
VERTICAL TAKEOFF AIRCRAFT
X-22 AIRCRAFT

TILTED PROPELLERS
GS   PROPELLERS
. **TILTED PROPELLERS**
RT   HELICOPTER PROPELLER DRIVE

TILTING
USE   ATTITUDE (INCLINATION)

TILTING ROTORS
GS   AIRFOILS
. WINGS
. . ROTARY WINGS
. . . **TILTING ROTORS**
ROTATING BODIES
. ROTORS
. . ROTARY WINGS
. . . **TILTING ROTORS**
RT   TILT ROTOR AIRCRAFT
XV-3 AIRCRAFT

TILTMETERS
GS   MEASURING INSTRUMENTS
. **TILTMETERS**
RT   ATTITUDE (INCLINATION)
GEOPHYSICS
SEISMOGRAPHS

TIMBER IDENTIFICATION
GS   IDENTIFYING
. **TIMBER IDENTIFICATION**
RECOGNITION
. **TIMBER IDENTIFICATION**
RT   CONIFERS

TIMBER IDENTIFICATION-*(CONT.)*
    CROP IDENTIFICATION
    DECIDUOUS TREES
    EARTH RESOURCES
    EVALUATION
    FORESTS
    TREES (PLANTS)

TIMBER INVENTORY
  GS   INVENTORIES
      . **TIMBER INVENTORY**
  RT   AERIAL PHOTOGRAPHY
      EARTH RESOURCES
      FOREST MANAGEMENT
      FORESTS
      INFRARED PHOTOGRAPHY
      PHOTOGRAPHY
      REFORESTATION
      SATELLITE-BORNE PHOTOGRAPHY
      TREES (PLANTS)

TIMBER VIGOR
  RT   FOLIAGE
      FORESTS
      GROWTH
      TIMBERLINE
      TREES (PLANTS)

TIMBERLINE
  RT   DENDROCHRONOLOGY
      FORESTS
      GROWTH
      HIGH ALTITUDE ENVIRONMENTS
      POLAR REGIONS
      TIMBER VIGOR
      TREES (PLANTS)

TIME
  UF   DURATION
  GS   **TIME**
      . ACCESS TIME
      . BURNING TIME
      . DOWNTIME
      . EPHEMERIS TIME
      . FLIGHT TIME
      . MTBF
      . REACTION TIME
      . . CHRONAXY
      . RELAXATION TIME
      . RESPONSE TIME (COMPUTERS)
      . SIDEREAL TIME
      . TESTING TIME
      . TRANSIT TIME
      . UNIVERSAL TIME
  RT   CALENDARS
      CELESTIAL GEODESY
      CHRONOLOGY
      EXPOSURE
      INTERVALS
      LAUNCH DATES
      MONTH
      PROLONGATION
      RELATIVISTIC EFFECTS
      SCHEDULES
      SYNCHRONISM
      TIME MEASUREMENT
      UNITS OF MEASUREMENT

TIME CONSTANT
  GS   CONSTANTS
      . **TIME CONSTANT**
      . . PERCEPTUAL TIME CONSTANT
  RT   ACCESS TIME
      ∞CONSTANT
      DAMPING
      DYNAMIC CHARACTERISTICS
      DYNAMIC RESPONSE
      IMPEDANCE
      LC CIRCUITS
      RC CIRCUITS
      REACTION TIME
      RELAXATION TIME
      RL CIRCUITS
      RLC CIRCUITS
      ∞TIME RESPONSE
      TRANSFER FUNCTIONS
      TRANSIENT RESPONSE

TIME DELAY
  USE   TIME LAG

TIME DEPENDENCE
  GS   DEPENDENCE
      . **TIME DEPENDENCE**
  RT   ∞HELMHOLTZ EQUATIONS

TIME DEPENDENCE-*(CONT.)*
    STOCHASTIC PROCESSES
    TEMPORAL DISTRIBUTION
    ∞TIME RESPONSE

TIME DISCRIMINATION
  RT   COMPARATOR CIRCUITS
      SENSORY DISCRIMINATION

TIME DIVISION MULTIPLE ACCESS
  UF   TDMA
  GS   TELECOMMUNICATION
      . MULTIPLE ACCESS
      . . **TIME DIVISION MULTIPLE ACCESS**
      . RADIO COMMUNICATION
      . . RADIO RELAY SYSTEMS
      . . . **TIME DIVISION MULTIPLE ACCESS**
  RT   ALOHA SYSTEM
      CHANNEL NOISE
      FREQUENCY DIVISION MULTIPLE
        ACCESS
      MULTICHANNEL COMMUNICATION
      PACKET SWITCHING
      PULSE COMMUNICATION
      SATELLITE NETWORKS
      SWITCHING
      WIDEBAND COMMUNICATION

TIME DIVISION MULTIPLEXING
  GS   TRANSMISSION
      . MULTIPLEXING
      . . **TIME DIVISION MULTIPLEXING**
  RT   DEMULTIPLEXING
      FREQUENCY DIVISION MULTIPLEXING
      PULSE MODULATION
      TELEMETRY
      TELEVISION TRANSMISSION
      WAVELENGTH DIVISION MULTIPLEXING

TIME FUNCTIONS
  GS   FUNCTIONS (MATHEMATICS)
      . **TIME FUNCTIONS**
  RT   RATES (PER TIME)
      SQUARE WAVES
      STOCHASTIC PROCESSES
      WAVE FUNCTIONS
      WAVEFORMS

TIME LAG
  UF   CHRONOTRONS
      LAG (DELAY)
      TIME DELAY
  RT   CEPSTRAL ANALYSIS
      DELAY
      DELAY LINES
      ELECTRIC RELAYS
      HYSTERESIS
      INVENTORY CONTROLS
      REACTION TIME
      REFRACTORY PERIOD
      RESPONSES
      SCHEDULES
      ∞TIME RESPONSE

TIME LAPSE PHOTOGRAPHY
  USE   CHRONOPHOTOGRAPHY

TIME MARCHING
  RT   FINITE DIFFERENCE THEORY
      NUMERICAL ANALYSIS
      SPATIAL MARCHING

TIME MEASUREMENT
  UF   DATING
      EPOCHS
      TIMING
  GS   **TIME MEASUREMENT**
      . CLOCK PARADOX
  RT   ATOMIC CLOCKS
      CHRONOMETERS
      CLOCKS
      CONSECUTIVE EVENTS
      FREQUENCY MEASUREMENT
      ∞MEASUREMENT
      OSCILLOGRAPHS
      RADIOACTIVE AGE DETERMINATION
      RATES (PER TIME)
      SIDEREAL TIME
      STROBOSCOPES
      SYNCHRONISM
      ∞TIME RESPONSE
      TIMING DEVICES
      VELOCITY
      VELOCITY MEASUREMENT
      WINDOWS (INTERVALS)

TIME MEASURING INSTRUMENTS
  GS   MEASURING INSTRUMENTS
      . **TIME MEASURING INSTRUMENTS**
      . . CLOCKS
      . . . ATOMIC CLOCKS
      . . . AUTONOMOUS SPACECRAFT
         CLOCKS
      . . . CHRONOMETERS
      . . TIMING DEVICES

TIME OF FLIGHT SPECTROMETERS
  GS   MEASURING INSTRUMENTS
      . SPECTROMETERS
      . . **TIME OF FLIGHT SPECTROMETERS**
  RT   SPECTROSCOPY

TIME OPTIMAL CONTROL
  GS   AUTOMATIC CONTROL
      . OPTIMAL CONTROL
      . . **TIME OPTIMAL CONTROL**
      OPTIMIZATION
      . OPTIMAL CONTROL
      . . **TIME OPTIMAL CONTROL**
  RT   ∞CONTROL

∞ **TIME RESPONSE**
  SN   *(USE OF A MORE SPECIFIC TERM IS*
      RECOMMENDED--CONSULT THE TERMS
      LISTED BELOW)
  RT   ACCESS TIME
      DELAY
      RESPONSE BIAS
      RESPONSES
      TEMPORAL DISTRIBUTION
      TIME CONSTANT
      TIME DEPENDENCE
      TIME LAG
      TIME MEASUREMENT

TIME SERIES ANALYSIS
  RT   ∞APPLICATIONS OF MATHEMATICS
      AUTOCORRELATION
      CORRELATION
      CURVE FITTING
      DATA SAMPLING
      EXTRAPOLATION
      FORECASTING
      FOURIER ANALYSIS
      KALMAN-SCHMIDT FILTERING
      MAXIMUM ENTROPY METHOD
      SCHEDULING
      ∞STATISTICS
      STOCHASTIC PROCESSES
      TRENDS

TIME SHARING
  RT   COMPUTER PROGRAMMING
      COORDINATION
      MULTIPLE OUTPUT PROGRAMS
      MULTIPROCESSING (COMPUTERS)
      MULTIPROGRAMMING
      PIPELINING (COMPUTERS)
      RUN TIME (COMPUTERS)

TIME SIGNALS
  RT   CLOCK PARADOX
      FREQUENCY STANDARDS
      PICOSECOND PULSES
      PULSE DURATION
      ∞SIGNALS
      TIMING DEVICES

TIME TEMPERATURE PARAMETER
  RT   AGING (METALLURGY)
      AUSTENITIC STAINLESS STEELS
      EMBRITTLEMENT
      FRACTURE MECHANICS
      LONG TERM EFFECTS
      METALLOGRAPHY
      PRECIPITATION HARDENING
      TEMPERATURE EFFECTS

TIMERS
  USE   TIMING DEVICES

TIMING
  USE   TIME MEASUREMENT

TIMING DEVICES
  UF   TIMERS
  GS   MEASURING INSTRUMENTS
      . TIME MEASURING INSTRUMENTS
      . . **TIMING DEVICES**
  RT   CHRONOMETERS

TIMING DEVICES-*(CONT.)*
    CLOCK PARADOX
    CLOCKS
    DWELL
    PENDULUMS
    TACHOMETERS
    TIME MEASUREMENT
    TIME SIGNALS

TIMOSHENKO BEAMS
GS   STRUCTURAL MEMBERS
    . BEAMS (SUPPORTS)
    . . **TIMOSHENKO BEAMS**
RT   COLUMNS (SUPPORTS)
    TRUSSES

TIN
GS   CHEMICAL ELEMENTS
    . **TIN**
    . . TIN ISOTOPES
    METALS
    . **TIN**
    . . TIN ISOTOPES

TIN ALLOYS
GS   ALLOYS
    . **TIN ALLOYS**
    . . BABBITT METAL
RT   BEARING ALLOYS
    SOLDERS
    STANNIDES
    ZIRCALOYS (TRADEMARK)

TIN COMPOUNDS
GS   **TIN COMPOUNDS**
    . ORGANIC TIN COMPOUNDS
    . STANNATES
    . STANNIDES
    . . NIOBIUM STANNIDES
    . TIN OXIDES
    . TIN TELLURIDES
RT   ∞CHEMICAL COMPOUNDS
    ∞GROUP 4A COMPOUNDS
    ∞METAL COMPOUNDS

TIN ISOTOPES
GS   CHEMICAL ELEMENTS
    . NUCLIDES
    . . ISOTOPES
    . . . **TIN ISOTOPES**
    . TIN
    . . **TIN ISOTOPES**
    METALS
    . TIN
    . . **TIN ISOTOPES**

TIN OXIDES
GS   CHALCOGENIDES
    . OXIDES
    . . METAL OXIDES
    . . . **TIN OXIDES**
    TIN COMPOUNDS
    . **TIN OXIDES**
RT   SIS (SEMICONDUCTORS)

TIN TELLURIDES
GS   CHALCOGENIDES
    . TELLURIDES
    . . **TIN TELLURIDES**
    TELLURIUM COMPOUNDS
    . TELLURIDES
    . . **TIN TELLURIDES**
    TIN COMPOUNDS
    . **TIN TELLURIDES**

TIP DRIVEN ROTORS
UF   HOT CYCLE PROPULSION SYSTEM
GS   AIRFOILS
    . WINGS
    . . ROTARY WINGS
    . . . **TIP DRIVEN ROTORS**
    ROTATING BODIES
    . ROTORS
    . . ROTARY WINGS
    . . . **TIP DRIVEN ROTORS**
RT   XV-9A AIRCRAFT

TIP SPEED
GS   RATES (PER TIME)
    . **TIP SPEED**
    VELOCITY
    . **TIP SPEED**
RT   ANGULAR VELOCITY
    CRITICAL VELOCITY

TIP SPEED-*(CONT.)*
    ROTOR SPEED

TIP VANES
GS   ROTATING BODIES
    . ROTORS
    . . **TIP VANES**
    TURBOMACHINERY
    . TURBINES
    . . WIND TURBINES
    . . . **TIP VANES**

TIPS
GS   **TIPS**
    . BLADE TIPS
    . CRACK TIPS
    . NOSE TIPS
    . WING TIPS
RT   AIRFOIL PROFILES
    EDGES

TIRES
GS   **TIRES**
    . AIRCRAFT TIRES
RT   BLOWOUTS
    INFLATABLE STRUCTURES
    LANDING GEAR
    ROLLERS
    TOROIDAL WHEELS
    TREADS
    VEHICLE WHEELS
    WHEEL BRAKES
    WHEELS

TIROS D SATELLITE
USE   TIROS 4 SATELLITE

TIROS E SATELLITE
USE   TIROS 5 SATELLITE

TIROS F SATELLITE
USE   TIROS 6 SATELLITE

TIROS G SATELLITE
USE   TIROS 7 SATELLITE

TIROS H SATELLITE
USE   TIROS 8 SATELLITE

TIROS M
GS   SATELLITES
    . ARTIFICIAL SATELLITES
    . . METEOROLOGICAL SATELLITES
    . . . TIROS SATELLITES
    . . . . **TIROS M**
RT   IMPROVED TIROS OPERATIONAL
      SATELLITES
    ITOS SATELLITES
    ITOS 1
    ITOS 2
    ITOS 3
    ITOS 4

TIROS N SERIES SATELLITES
GS   SATELLITES
    . ARTIFICIAL SATELLITES
    . . METEOROLOGICAL SATELLITES
    . . . TIROS SATELLITES
    . . . . **TIROS N SERIES SATELLITES**
RT   IMPROVED TIROS OPERATIONAL
      SATELLITES
    ITOS 1
    ITOS 2
    ITOS 3
    ITOS 4
    NOAA 6 SATELLITE
    NOAA 7 SATELLITE
    TIROS OPERATIONAL SATELLITE
      SYSTEM

TIROS OPERATIONAL SATELLITE SYSTEM
RT   CLOUD PHOTOGRAPHY
    ITOS 1
    ITOS 2
    ITOS 3
    ITOS 4
    SATELLITE OBSERVATION
    ∞SYSTEMS
    TIROS N SERIES SATELLITES

TIROS PROJECT
GS   PROGRAMS
    . NASA PROGRAMS
    . . **TIROS PROJECT**

TIROS PROJECT-*(CONT.)*
    . NASA SPACE PROGRAMS
    . . **TIROS PROJECT**
    . PROJECTS
    . . **TIROS PROJECT**
RT   CLOUD PHOTOGRAPHS
    CLOUD PHOTOGRAPHY
    METEOROLOGICAL SATELLITES

TIROS SATELLITES
GS   SATELLITES
    . ARTIFICIAL SATELLITES
    . . METEOROLOGICAL SATELLITES
    . . . **TIROS SATELLITES**
    . . . . IMPROVED TIROS OPERATIONAL
         SATELLITES
    . . . . . ITOS 1
    . . . . . ITOS 2
    . . . . . ITOS 3
    . . . . . ITOS 4
    . . . . ITOS SATELLITES
    . . . . . ITOS 1
    . . . . . ITOS 2
    . . . . . ITOS 3
    . . . . . ITOS 4
    . . . TIROS M
    . . . TIROS N SERIES SATELLITES
    . . . TIROS 1 SATELLITE
    . . . TIROS 2 SATELLITE
    . . . TIROS 3 SATELLITE
    . . . TIROS 4 SATELLITE
    . . . TIROS 5 SATELLITE
    . . . TIROS 6 SATELLITE
    . . . TIROS 7 SATELLITE
    . . . TIROS 8 SATELLITE
    . . . TIROS 9 SATELLITE
    . . . TIROS 10 SATELLITE
RT   CLOUD PHOTOGRAPHY
    ESSA SATELLITES
    POLAR ORBITS
    SATELLITE OBSERVATION

TIROS WHEEL SATELLITE
USE   TIROS 9 SATELLITE

TIROS 1 SATELLITE
GS   SATELLITES
    . ARTIFICIAL SATELLITES
    . . METEOROLOGICAL SATELLITES
    . . . TIROS SATELLITES
    . . . . **TIROS 1 SATELLITE**
RT   THOR ABLE ROCKET VEHICLE

TIROS 2 SATELLITE
GS   SATELLITES
    . ARTIFICIAL SATELLITES
    . . METEOROLOGICAL SATELLITES
    . . . TIROS SATELLITES
    . . . . **TIROS 2 SATELLITE**
RT   DELTA LAUNCH VEHICLE

TIROS 3 SATELLITE
GS   SATELLITES
    . ARTIFICIAL SATELLITES
    . . METEOROLOGICAL SATELLITES
    . . . TIROS SATELLITES
    . . . . **TIROS 3 SATELLITE**
RT   DELTA LAUNCH VEHICLE

TIROS 4 SATELLITE
UF   TIROS D SATELLITE
GS   SATELLITES
    . ARTIFICIAL SATELLITES
    . . METEOROLOGICAL SATELLITES
    . . . TIROS SATELLITES
    . . . . **TIROS 4 SATELLITE**
RT   DELTA LAUNCH VEHICLE

TIROS 5 SATELLITE
UF   TIROS E SATELLITE
GS   SATELLITES
    . ARTIFICIAL SATELLITES
    . . METEOROLOGICAL SATELLITES
    . . . TIROS SATELLITES
    . . . . **TIROS 5 SATELLITE**
RT   DELTA LAUNCH VEHICLE

TIROS 6 SATELLITE
UF   TIROS F SATELLITE
GS   SATELLITES
    . ARTIFICIAL SATELLITES
    . . METEOROLOGICAL SATELLITES
    . . . TIROS SATELLITES
    . . . . **TIROS 6 SATELLITE**
RT   DELTA LAUNCH VEHICLE

TIROS 7 SATELLITE

TIROS 7 SATELLITE
UF   TIROS G SATELLITE
GS   SATELLITES
     . ARTIFICIAL SATELLITES
     . . METEOROLOGICAL SATELLITES
     . . . TIROS SATELLITES
     . . . . **TIROS 7 SATELLITE**
RT   DELTA LAUNCH VEHICLE

TIROS 8 SATELLITE
UF   TIROS H SATELLITE
GS   SATELLITES
     . ARTIFICIAL SATELLITES
     . . METEOROLOGICAL SATELLITES
     . . . TIROS SATELLITES
     . . . . **TIROS 8 SATELLITE**
RT   DELTA LAUNCH VEHICLE

TIROS 9 SATELLITE
UF   TIROS WHEEL SATELLITE
GS   SATELLITES
     . ARTIFICIAL SATELLITES
     . . METEOROLOGICAL SATELLITES
     . . . TIROS SATELLITES
     . . . . **TIROS 9 SATELLITE**
RT   DELTA LAUNCH VEHICLE

TIROS 10 SATELLITE
GS   SATELLITES
     . ARTIFICIAL SATELLITES
     . . METEOROLOGICAL SATELLITES
     . . . TIROS SATELLITES
     . . . . **TIROS 10 SATELLITE**
RT   DELTA LAUNCH VEHICLE

TISSUES (BIOLOGY)
GS   **TISSUES (BIOLOGY)**
     . ADIPOSE TISSUES
     . ENDOTHELIUM
     . EPICARDIUM
     . EPITHELIUM
     . HYPODERMIS
     . NEUROGLIA
     . PERITONEUM
     . PLANTAR TISSUES
     . SCARS
RT   ATROPHY
     ∞BIOLOGY
     CANCER
     ∞CELLS
     CELLS (BIOLOGY)
     CULTIVATION
     CYSTIC FIBROSIS
     CYSTS
     FIBROBLASTS
     FIBROSIS
     HISTOCHEMICAL ANALYSIS
     INFARCTION
     MACROPHAGES
     MEDIASTINUM

TITAN
GS   CELESTIAL BODIES
     . NATURAL SATELLITES
     . . **TITAN**
     SATELLITES
     . NATURAL SATELLITES
     . SATURN SATELLITES
     . . . **TITAN**
RT   ATMOSPHERIC COMPOSITION
     CHARON
     SATELLITE ATMOSPHERES
     SATURN (PLANET)
     TRITON

TITAN CENTAUR LAUNCH VEHICLE
GS   LAUNCH VEHICLES
     . **TITAN CENTAUR LAUNCH VEHICLE**
     ROCKET VEHICLES
     . **TITAN CENTAUR LAUNCH VEHICLE**
RT   CENTAUR LAUNCH VEHICLE

TITAN ICBM
GS   MISSILES
     . BALLISTIC MISSILES
     . . INTERCONTINENTAL BALLISTIC
        MISSILES
     . . . **TITAN ICBM**
     . . . . TITAN 1 ICBM
     . . . . TITAN 2 ICBM
     . SURFACE TO SURFACE MISSILES
     . . INTERCONTINENTAL BALLISTIC
        MISSILES
     . . . **TITAN ICBM**
     . . . . TITAN 1 ICBM

TITAN ICBM-(CONT.)

TITAN ICBM-*(CONT.)*
     . . . . TITAN 2 ICBM
RT   LIQUID PROPELLANT ROCKET ENGINES
     LR-91-AJ-5 ENGINE
     MULTISTAGE ROCKET VEHICLES
     YLR-91-AJ-1 ENGINE

TITAN LAUNCH VEHICLES
GS   LAUNCH VEHICLES
     . **TITAN LAUNCH VEHICLES**
     . . TITAN 3 LAUNCH VEHICLE
     ROCKET VEHICLES
     . MULTISTAGE ROCKET VEHICLES
     . . **TITAN LAUNCH VEHICLES**
     . . . TITAN 3 LAUNCH VEHICLE
RT   GEMINI 3 FLIGHT
     GEMINI 7 FLIGHT
     GEMINI 8 FLIGHT
     GEMINI 9 FLIGHT
     GEMINI 10 FLIGHT
     GEMINI 11 FLIGHT
     GEMINI 12 FLIGHT
     LIQUID PROPELLANT ROCKET ENGINES
     SOLID PROPELLANT ROCKET ENGINES
     ∞VEHICLES

TITAN PROJECT
GS   PROGRAMS
     . NASA PROGRAMS
     . . **TITAN PROJECT**
     . NASA SPACE PROGRAMS
     . . **TITAN PROJECT**
     . PROJECTS
     . . **TITAN PROJECT**
RT   ∞BOOSTERS
     GEMINI PROJECT
     GEMINI SPACECRAFT
     LAUNCH VEHICLES
     LAUNCHERS
     LAUNCHING

TITAN 1 ICBM
UF   SM-68 MISSILE
GS   MISSILES
     . BALLISTIC MISSILES
     . . INTERCONTINENTAL BALLISTIC
        MISSILES
     . . . TITAN ICBM
     . . . . **TITAN 1 ICBM**
     . SURFACE TO SURFACE MISSILES
     . . INTERCONTINENTAL BALLISTIC
        MISSILES
     . . . TITAN ICBM
     . . . . **TITAN 1 ICBM**
RT   LR-87-AJ-5 ENGINE

TITAN 2 ICBM
UF   SM-68B MISSILE
GS   MISSILES
     . BALLISTIC MISSILES
     . . INTERCONTINENTAL BALLISTIC
        MISSILES
     . . . TITAN ICBM
     . . . . **TITAN 2 ICBM**
     . SURFACE TO SURFACE MISSILES
     . . INTERCONTINENTAL BALLISTIC
        MISSILES
     . . . TITAN ICBM
     . . . . **TITAN 2 ICBM**
RT   HYLA-STAR ROCKET VEHICLE

TITAN 3 LAUNCH VEHICLE
GS   LAUNCH VEHICLES
     . TITAN LAUNCH VEHICLES
     . . **TITAN 3 LAUNCH VEHICLE**
     ROCKET VEHICLES
     . MULTISTAGE ROCKET VEHICLES
     . . TITAN LAUNCH VEHICLES
     . . . **TITAN 3 LAUNCH VEHICLE**
RT   MANNED ORBITAL LABORATORIES

TITANATES
GS   TITANIUM COMPOUNDS
     . **TITANATES**
     . . BARIUM TITANATES
     . . ILMENITE
     . . LEAD TITANATES
     . . MAGNESIUM TITANATES
     . . PEROVSKITES
     . . PIEZOELECTRIC CERAMICS
     . . . LEAD ZIRCONATE TITANATES
     . . STRONTIUM TITANATES
     . . ZIRCONIUM TITANATES
RT   EUXENITE

TITANIUM

TITANIUM
GS   CHEMICAL ELEMENTS
     . **TITANIUM**
     . . TITANIUM ISOTOPES
     METALS
     . TRANSITION METALS
     . . **TITANIUM**
     . . . TITANIUM ISOTOPES

TITANIUM ALLOYS
GS   ALLOYS
     . **TITANIUM ALLOYS**
     . . NITINOL ALLOYS
RT   SHAPE MEMORY ALLOYS

TITANIUM BORIDES
GS   BORON COMPOUNDS
     . BORIDES
     . . **TITANIUM BORIDES**
     TITANIUM COMPOUNDS
     . **TITANIUM BORIDES**

TITANIUM CARBIDES
GS   CARBON COMPOUNDS
     . CARBIDES
     . . **TITANIUM CARBIDES**
     TITANIUM COMPOUNDS
     . **TITANIUM CARBIDES**
RT   CERAMIC MATRIX COMPOSITES

TITANIUM CHLORIDES
GS   HALOGEN COMPOUNDS
     . CHLORINE COMPOUNDS
     . . CHLORIDES
     . . . **TITANIUM CHLORIDES**
     . HALIDES
     . . CHLORIDES
     . . . **TITANIUM CHLORIDES**
     . METAL HALIDES
     . . . **TITANIUM CHLORIDES**
     TITANIUM COMPOUNDS
     . **TITANIUM CHLORIDES**

TITANIUM COMPOUNDS
GS   **TITANIUM COMPOUNDS**
     . TITANATES
     . . BARIUM TITANATES
     . . ILMENITE
     . . LEAD TITANATES
     . . MAGNESIUM TITANATES
     . . PEROVSKITES
     . . PIEZOELECTRIC CERAMICS
     . . . LEAD ZIRCONATE TITANATES
     . . STRONTIUM TITANATES
     . . ZIRCONIUM TITANATES
     . TITANIUM BORIDES
     . TITANIUM CARBIDES
     . TITANIUM CHLORIDES
     . TITANIUM NITRIDES
     . TITANIUM OXIDES
     . . ANATASE
     . . ILMENITE
     . . RUTILE
RT   ∞CHEMICAL COMPOUNDS
     ∞GROUP 4B COMPOUNDS
     ∞METAL COMPOUNDS

TITANIUM DIOXIDE
USE   TITANIUM OXIDES

TITANIUM ISOTOPES
GS   CHEMICAL ELEMENTS
     . NUCLIDES
     . . ISOTOPES
     . . . **TITANIUM ISOTOPES**
     . TITANIUM
     . . **TITANIUM ISOTOPES**
     METALS
     . TRANSITION METALS
     . . TITANIUM
     . . . **TITANIUM ISOTOPES**

TITANIUM NITRIDES
GS   NITROGEN COMPOUNDS
     . NITRIDES
     . . **TITANIUM NITRIDES**
     TITANIUM COMPOUNDS
     . **TITANIUM NITRIDES**
RT   CERAMIC MATRIX COMPOSITES
     METAL NITRIDES

TITANIUM OXIDES
UF   TITANIUM DIOXIDE
GS   CHALCOGENIDES

TITANIUM OXIDES-*(CONT.)*
```
      . OXIDES
      .. METAL OXIDES
      ... TITANIUM OXIDES
      .... ANATASE
      .... ILMENITE
      .... RUTILE
      TITANIUM COMPOUNDS
      . TITANIUM OXIDES
      .. ANATASE
      .. ILMENITE
      .. RUTILE
RT    DIOXIDES
```

TITRATION
```
GS    CHEMICAL REACTIONS
      . TITRATION
RT    ACIDITY
      COULOMETERS
      IODIMETRY
      ION CONCENTRATION
      KJELDAHL METHOD
      SOLUTIONS
```

TITRIMETERS
```
GS    MEASURING INSTRUMENTS
      . TITRIMETERS
RT    CHEMICAL ANALYSIS
```

TNT (TRINITROTOLUENE)
```
USE   TRINITROTOLUENE
```

TOBACCO
```
GS    PLANTS (BOTANY)
      . TOBACCO
RT    NICOTINE
```

TOCOPHEROL
```
UF    VITAMIN E
GS    HETEROCYCLIC COMPOUNDS
      . TOCOPHEROL
      VITAMINS
      . TOCOPHEROL
```

TOGO
```
GS    NATIONS
      . TOGO
RT    AFRICA
```

† **TOKAMAK DEVICES**
```
GS    NUCLEAR REACTORS
      . TOKAMAK DEVICES
      .. JOINT EUROPEAN TORUS
      PLASMA GENERATORS
      . TOKAMAK DEVICES
      .. JOINT EUROPEAN TORUS
RT    BEAM INJECTION
      BETA FACTOR
      BUMPY TORUSES
      ∞ELECTRIC POWER
      LIMITERS (FUSION REACTORS)
      NUCLEAR FUSION
      PLASMA COMPRESSION
      PLASMA CONTROL
      PLASMA PHYSICS
      POLOIDAL FLUX
      ∞REACTORS
      SPHEROMAKS
```

TOLERANCES (MECHANICS)
```
GS    TOLERANCES (MECHANICS)
      . IMPACT TOLERANCES
RT    ACCEPTABILITY
      ACCURACY
      ALLOWANCES
      CLEARANCES
      CONSISTENCY
      DIMENSIONAL STABILITY
      DRIFT (INSTRUMENTATION)
      ERRORS
      HYSTERESIS
      INSPECTION
      LINEARITY
      MECHANICAL PROPERTIES
      NONDESTRUCTIVE TESTS
      PRECISION
      QUALITY CONTROL
      RADIATION TOLERANCE
      RANGE (EXTREMES)
      RELIABILITY
      RESOLUTION
      SENSITIVITY
      SPECIFICATIONS
      STABILITY
      STANDARDS
```

TOLERANCES (PHYSIOLOGY)
```
GS    TOLERANCES (PHYSIOLOGY)
      . ACCELERATION TOLERANCE
      . ALTITUDE TOLERANCE
      . COLD TOLERANCE
      . HEAT TOLERANCE
      . HUMAN TOLERANCES
      . RADIATION TOLERANCE
RT    ACCLIMATIZATION
      BARANY CHAIR
      BIOCONTROL SYSTEMS
      IMPACT RESISTANCE
      NOISE TOLERANCE
      ORTHOSTATIC TOLERANCE
      RANGE (EXTREMES)
      ∞RESISTANCE
```

TOLLMEIN-SCHLICHTING WAVES
```
GS    ELASTIC WAVES
      . TOLLMEIN-SCHLICHTING WAVES
RT    BLASIUS FLOW
      BOUNDARY LAYER FLOW
      BOUNDARY LAYER TRANSITION
      LAMINAR FLOW
      TURBULENT FLOW
```

TOLUENE
```
GS    HYDROCARBONS
      . TOLUENE
RT    SOLVENTS
      XYLENE
```

TOMAHAWK MISSILES
```
GS    MISSILES
      . SURFACE TO SURFACE MISSILES
      .. CRUISE MISSILES
      ... TOMAHAWK MISSILES
RT    WEAPONS
```

TOMBOLOS
```
USE   BARS (LANDFORMS)
```

TOMOGRAPHY
```
UF    PLANIGRAPHY
GS    IMAGERY
      . RADIOGRAPHY
      .. TOMOGRAPHY
      ... COMPUTER AIDED TOMOGRAPHY
RT    COMPUTER GRAPHICS
      IMAGE ENHANCEMENT
      OPTICAL DATA PROCESSING
      X RAY ANALYSIS
```

TONE
```
USE   PITCH
```

TONGUE
```
GS    ANATOMY
      . DIGESTIVE SYSTEM
      .. TONGUE
RT    MOUTH
      VOICE
```

TONK METEORITE
```
GS    CELESTIAL BODIES
      . METEORITES
      .. STONY METEORITES
      ... CHONDRITES
      .... CARBONACEOUS METEORITES
      ..... TONK METEORITE
```

TONOMETRY
```
USE   INTRAOCULAR PRESSURE
      PRESSURE MEASUREMENT
```

TONUS
```
USE   MUSCULAR TONUS
```

TOOLING
```
RT    ∞AUTOMATION
      MACHINING
      MECHANIZATION
      SETUPS
      TOOLS
```

TOOLS
```
GS    TOOLS
      . BORING MACHINES
      . DRILL BITS
      . FILES (TOOLS)
      . MACHINE TOOLS
      .. GRINDING MACHINES
      .. LATHES
      ... TURRET LATHES
```

TOOLS-*(CONT.)*
```
      .. MILLING MACHINES
      .. SHAPERS
      . SAWS
      . SHEARS
      . SPACE TOOLS
      . WRENCHES
RT    ANTIQUITIES
      ANVILS
      CUTTERS
      DRILLS
      FIXTURES
      HAMMERS
      HARDWARE
      JIGS
      KITS
      ∞MACHINERY
      MECHANICAL DEVICES
      MECHANIZATION
      PLATENS
      PRESSES
      ∞PRODUCTION
      TAPS
      TOOLING
      ULTRASONIC CLEANING
```

TOOTH DISEASES
```
UF    AERODONTALGIA
GS    DISEASES
      . TOOTH DISEASES
      MEDICAL SCIENCE
      . TOOTH DISEASES
RT    CAVITIES
      DENTAL CALCULI
      DENTISTRY
      ORAL HYGIENE
      TEETH
```

TOPEX
```
RT    GULF STREAM
      MARITIME SATELLITES
      OCEAN CURRENTS
      OCEAN SURFACE
      OCEANOGRAPHY
      SATELLITE OBSERVATION
      SEA STATES
      TOPOGRAPHY
```

TOPOGRAPHY
```
UF    LANDSCAPE
GS    TOPOGRAPHY
      . LUNAR TOPOGRAPHY
      . TERRAIN
RT    BADLANDS
      BARREN LAND
      BEACHES
      CLIFFS
      CONTOUR SENSORS
      CONTOURS
      CUSPS (LANDFORMS)
      ∞DEPRESSION
      DESERTLINE
      DESERTS
      DUNES
      EARTH SURFACE
      ELEVATION
      ELEVATION ANGLE
      ESCARPMENTS
      GEODESY
      GEODETIC SURVEYS
      GEOMORPHOLOGY
      GEOPHYSICS
      GULFS
      HIGHLANDS
      HYPSOGRAPHY
      ISTHMUSES
      JUPITER RED SPOT
      LAGOONS
      LAND
      LANDFORMS
      LANDMARKS
      LEDGES
      MAPPING
      MARIA
      MARS SURFACE
      MEANDERS
      MUSKEGS
      OCEANOGRAPHY
      PEAKS (LANDFORMS)
      PENEPLAINS
      PHOTOMAPPING
      PLAINS
      PLANETARY SURFACES
      ∞PROFILES
      RAVINES
```

TOPOGRAPHY-(CONT.)
RELIEF MAPS
SHALLOW WATER
SLOPES
STAIRSTEPS
SURFACE ROUGHNESS
TOPEX
VALLEYS
VENUS SURFACE
WADIS

TOPOLOGY
GS   GEOMETRY
. **TOPOLOGY**
. . FIXED POINTS (MATHEMATICS)
. . HOMOTOPY THEORY
. . IMBEDDINGS (MATHEMATICS)
. . . INVARIANT IMBEDDINGS
. . LINKS (MATHEMATICS)
. . METRIC SPACE
RT   CATASTROPHE THEORY
∞CELLS
CONTINUITY
CONTINUITY (MATHEMATICS)
CONTINUUMS
DEFORMATION
DIMENSIONS
FAULT TREES
FIBERS (MATHEMATICS)
GRAPH THEORY
HOMOLOGY
HOMOTROPY
INTERVALS
ISOPERIMETRIC PROBLEM
MANIFOLDS (MATHEMATICS)
MAPPING
NETWORK SYNTHESIS
∞NETWORKS
SHAPES
SWITCHING THEORY
TORUSES
TREES (MATHEMATICS)

TOPPING CYCLE ENGINES
RT   AIRCRAFT ENGINES
LIQUID HYDROGEN
PROPULSION SYSTEM CONFIGURATIONS

TOPS (SPACECRAFT)
UF   THERMOELECTRIC OUTER PLANET
SPACECRAFT
THERMOELECTRIC SPACECRAFT
GS   INTERPLANETARY SPACECRAFT
. **TOPS (SPACECRAFT)**
RT   FLYBY MISSIONS
INTERPLANETARY FLIGHT
OUTER PLANETS EXPLORERS
SPACE EXPLORATION
SPACE MISSIONS
∞SPACECRAFT

TORCHES
RT   CUTTING
PYROGEN
WELDING
WELDING MACHINES

TORNADO AIRCRAFT
USE   MRCA AIRCRAFT

TORNADOES
GS   STORMS
. STORMS (METEOROLOGY)
. . **TORNADOES**
RT   ATMOSPHERIC CIRCULATION
COLD FRONTS
CUMULONIMBUS CLOUDS
CYCLONES
FRONTS (METEOROLOGY)
GROUND WIND
HURRICANES
NATIONAL SEVERE STORMS PROJECT
RAINSTORMS
STORM DAMAGE
TROPICAL STORMS
TYPHOONS
WARM FRONTS
WIND (METEOROLOGY)

TORO ASTEROID
GS   CELESTIAL BODIES
. ASTEROID BELTS
. . ASTEROIDS
. . . **TORO ASTEROID**
RT   METEOROIDS

TORO ASTEROID-(CONT.)
SOLAR SYSTEM
SPACE DEBRIS

TOROIDAL DISCHARGE
GS   ELECTRIC CURRENT
. ELECTRIC DISCHARGES
. . TOWNSEND DISCHARGE
. . . GAS DISCHARGES
. . . . **TOROIDAL DISCHARGE**
. . . . . RING DISCHARGE
RT   ELECTRODELESS DISCHARGES
HIGH FREQUENCIES
PLASMA JETS
SPECTRUM ANALYSIS

TOROIDAL PLASMAS
UF   PLASMA RINGS
GS   PARTICLES
. CHARGED PARTICLES
. . ENERGETIC PARTICLES
. . . PLASMAS (PHYSICS)
. . . . **TOROIDAL PLASMAS**
RT   BEAM INJECTION
BETA FACTOR
BUMPY TORUSES
ELLIPTICAL PLASMAS
LIMITERS (FUSION REACTORS)
PLASMA CONTROL
PLASMA CURRENTS
POLOIDAL FLUX
REVERSE FIELD PINCH
∞RINGS
ROTATING PLASMAS
SOLENOIDS
SPHEROMAKS
STELLARATORS

TOROIDAL SHELLS
GS   SHELLS (STRUCTURAL FORMS)
. **TOROIDAL SHELLS**
RT   METAL SHELLS
REINFORCED SHELLS
SKIN (STRUCTURAL MEMBER)
THIN WALLED SHELLS
TOROIDS

TOROIDAL WHEELS
UF   DOUGHNUT SHAPE WHEELS
GS   WHEELS
. **TOROIDAL WHEELS**
RT   ROVING VEHICLES
SUSPENSION SYSTEMS (VEHICLES)
TIRES
VEHICLE WHEELS

TOROIDS
RT   ∞COILS
∞CURVES
GEOMETRY
INDUCTORS
ION IMPACT
MAGNET COILS
MAGNETIC CORES
TOROIDAL SHELLS
TRANSFORMERS

TORPEDO ENGINES
GS   ENGINES
. **TORPEDO ENGINES**
. . TURBOROCKET ENGINES
. . ULLAGE ROCKET ENGINES
. . VERNIER ENGINES
. . . CONTROL ROCKETS
. . . SYNCOM APOGEE ENGINES
RT   INTERNAL COMBUSTION ENGINES
ROCKET PROPELLANTS
TURBINE ENGINES
UNDERWATER PROPULSION

TORPEDOES
UF   RETORC (TORPEDOES)
GS   EXPLOSIVE DEVICES
. **TORPEDOES**
RT   AMMUNITION
ANTISUBMARINE WARFARE
ASROC ENGINE
BOMBS (ORDNANCE)
∞CONFIGURATIONS
COUNTERMEASURES
EXPLOSIVES
HYDROBALLISTICS
MISSILES
NUCLEAR WEAPONS
PROPELLANTS

TORPEDOES-(CONT.)
∞ROCKETS
SEA LAUNCHING
SHAPED CHARGES
SUBMERGED BODIES
UNDERWATER TRAJECTORIES
WARHEADS
WEAPONS

TORQUE
UF   HINGE MOMENTS
GS   MOMENTS
. **TORQUE**
RT   BENDING MOMENTS
∞FORCE
LOADING MOMENTS
MOMENTS OF INERTIA
PITCHING MOMENTS
ROLLING MOMENTS
ROTATION
SHAFTS (MACHINE ELEMENTS)
TORQUEMETERS
TORSION
TORSIONAL STRESS
TORSIONAL VIBRATION
TWISTING
YAWING MOMENTS

TORQUE CONVERTERS
RT   ∞CONVERTERS
POWER CONVERTERS
TRANSMISSIONS (MACHINE ELEMENTS)

TORQUE MEASURING APPARATUS
USE   TORQUEMETERS

TORQUE MOTORS
GS   MOTORS
. ELECTRIC MOTORS
. . **TORQUE MOTORS**
RT   ACTUATORS
SERVOMOTORS
TRANSMISSIONS (MACHINE ELEMENTS)

TORQUEMETERS
UF   TORQUE MEASURING APPARATUS
GS   MEASURING INSTRUMENTS
. **TORQUEMETERS**
RT   DYNAMOMETERS
MECHANICAL MEASUREMENT
TORQUE

TORQUERS
GS   TRANSDUCERS
. **TORQUERS**
RT   DEGREES OF FREEDOM
GYROSCOPES
SEA KEEPING

TORRES STRAIT
GS   PASSAGEWAYS
. STRAITS
. . **TORRES STRAIT**
RT   AUSTRALIA
NEW GUINEA (ISLAND)

TORSION
RT   BUCKLING
DEFLECTION
DEFORMATION
∞FORCE
MOMENTS
TEMPERATURE INVERSIONS
TORQUE
TORSIONAL STRESS
TORSIONAL VIBRATION
TWISTING

TORSIONAL STRESS
GS   STRESSES
. SHEAR STRESS
. . **TORSIONAL STRESS**
RT   TORQUE
TORSION

TORSIONAL VIBRATION
GS   VIBRATION
. STRUCTURAL VIBRATION
. . **TORSIONAL VIBRATION**
RT   MISSILE VIBRATION
RANDOM VIBRATION
SELF INDUCED VIBRATION
SHEAR STRAIN
TORQUE

TORSIONAL VIBRATION-(CONT.)
      TORSION
      TWISTING

TORSO
  GS   ANATOMY
       . **TORSO**
  RT   CHEST

TORUSES
  GS   GEOMETRY
       . EUCLIDEAN GEOMETRY
       . . ANALYTIC GEOMETRY
       . . . **TORUSES**
       SYMMETRICAL BODIES
       . BODIES OF REVOLUTION
       . . **TORUSES**
  RT   DESCRIPTIVE GEOMETRY
       LOOPS
       ∞RINGS
       TOPOLOGY

TORY 2 REACTOR
  GS   NUCLEAR ELECTRIC POWER
         GENERATION
       . NUCLEAR POWER REACTORS
       . . **TORY 2 REACTOR**
       NUCLEAR REACTORS
       . GAS COOLED REACTORS
       . . **TORY 2 REACTOR**
       . NUCLEAR POWER REACTORS
       . . **TORY 2 REACTOR**
       . NUCLEAR RESEARCH AND TEST
         REACTORS
       . . **TORY 2 REACTOR**

TORY 2-A REACTOR
  GS   NUCLEAR ELECTRIC POWER
         GENERATION
       . NUCLEAR POWER REACTORS
       . . **TORY 2-A REACTOR**
       NUCLEAR REACTORS
       . GAS COOLED REACTORS
       . . **TORY 2-A REACTOR**
       . NUCLEAR POWER REACTORS
       . . **TORY 2-A REACTOR**
       . NUCLEAR RESEARCH AND TEST
         REACTORS
       . . **TORY 2-A REACTOR**

TORY 2-C REACTOR
  GS   NUCLEAR ELECTRIC POWER
         GENERATION
       . NUCLEAR POWER REACTORS
       . . **TORY 2-C REACTOR**
       NUCLEAR REACTORS
       . GAS COOLED REACTORS
       . . **TORY 2-C REACTOR**
       . NUCLEAR POWER REACTORS
       . . **TORY 2-C REACTOR**
       . NUCLEAR RESEARCH AND TEST
         REACTORS
       . . **TORY 2-C REACTOR**

TOS-A
  USE ESSA 3 SATELLITE

TOTAL ENERGY SYSTEMS
  GS   **TOTAL ENERGY SYSTEMS**
       . SOLAR TOTAL ENERGY SYSTEMS
  RT   INTEGRATED ENERGY SYSTEMS
       PHOSPHORIC ACID FUEL CELLS
       ∞SYSTEMS

TOUCH
  UF   CUTANEOUS PERCEPTION
       TACTILE SENSATION
  GS   PERCEPTION
       . SENSORY PERCEPTION
       . . **TOUCH**
       . . . TACTILE DISCRIMINATION
  RT   ELECTROCUTANEOUS COMMUNICATION
       SKIN (ANATOMY)

TOUCHDOWN
  GS   LANDING
       . **TOUCHDOWN**
  RT   AIRCRAFT LANDING
       APPROACH
       APPROACH AND LANDING TESTS (STS)
       DOWNRANGE
       SPACECRAFT LANDING
       VERTICAL LANDING
       VERTICAL MOTION

TOUCHDOWN-(CONT.)
      WATER LANDING

TOUGHNESS
  GS   MECHANICAL PROPERTIES
       . **TOUGHNESS**
       . . NOTCH SENSITIVITY
  RT   ABRASION RESISTANCE
       BRITTLENESS
       COMPRESSIVE STRENGTH
       CRACK INITIATION
       DUCTILITY
       FRACTURE STRENGTH
       HARDNESS
       IMPACT TESTS
       J INTEGRAL
       SHEAR PROPERTIES
       TENSILE STRENGTH

TOURMALINE
  GS   ALUMINUM COMPOUNDS
       . **TOURMALINE**
       BORON COMPOUNDS
       . **TOURMALINE**
       MINERALS
       . **TOURMALINE**
       SILICON COMPOUNDS
       . SILICATES
       . . SODIUM SILICATES
       . . . **TOURMALINE**
       SODIUM COMPOUNDS
       . SODIUM SILICATES
       . . **TOURMALINE**
  RT   IGNEOUS ROCKS

TOURNESOLE SATELLITE
  USE D-2 SATELLITES

TOURNIQUETS
  GS   MEDICAL EQUIPMENT
       . **TOURNIQUETS**
  RT   BLOOD CIRCULATION
       BLOOD FLOW
       FIRST AID

TOW MISSILES
  GS   MISSILES
       . SURFACE TO SURFACE MISSILES
       . . ANTITANK MISSILES
       . . . **TOW MISSILES**

TOWED BODIES
  UF   DROGUES
       TOWED TARGETS
  RT   AIRCRAFT BRAKES
       ∞BODIES
       BRAKES (FOR ARRESTING MOTION)
       DRAG CHUTES
       GLIDERS
       LIFTING BODIES
       PARACHUTES
       SLEDS
       STREAMLINED BODIES
       SUBMERGED BODIES
       TEST VEHICLES
       TOWING
       TRAILERS

TOWED TARGETS
  USE TARGETS
       TOWED BODIES

TOWER SHIELDING REACTOR 2
  GS   NUCLEAR REACTORS
       . NUCLEAR RESEARCH AND TEST
         REACTORS
       . . **TOWER SHIELDING REACTOR 2**
  RT   RADIATION SHIELDING

TOWERS
  GS   **TOWERS**
       . AIRPORT TOWERS
       . UMBILICAL TOWERS
  RT   AIR TRAFFIC CONTROL
       ANTENNAS
       BRIDGES (STRUCTURES)
       COLUMNS (SUPPORTS)
       CONCRETE STRUCTURES
       CONSTRUCTION INDUSTRY
       CRANES
       PYLONS
       ∞STRUCTURES
       TANKS (CONTAINERS)

TOWING
  RT   CABLES (ROPES)
       TOWED BODIES
       TRACTORS
       TRAILERS

TOWNSEND AVALANCHE
  UF   TOWNSEND SURFACES
  GS   AVALANCHES
       . **TOWNSEND AVALANCHE**
  RT   ELECTROMAGNETIC ABSORPTION
       ELECTRON AVALANCHE
       ION IMPACT
       SECONDARY EMISSION
       ∞SURFACES

TOWNSEND DISCHARGE
  GS   ELECTRIC CURRENT
       . ELECTRIC DISCHARGES
       . . **TOWNSEND DISCHARGE**
       . . . GAS DISCHARGES
       . . . . TOROIDAL DISCHARGE
       . . . . . RING DISCHARGE
  RT   ELECTRODELESS DISCHARGES
       ION IMPACT

TOWNSEND SURFACES
  USE TOWNSEND AVALANCHE

TOXIC DISEASES
  UF   POISONING (TOXICOLOGY)
  GS   DISEASES
       . **TOXIC DISEASES**
       . . CARBON MONOXIDE POISONING
       . . LEAD POISONING
  RT   CLOSTRIDIUM BOTULINUM
       DIPHTHERIA
       HYPEROXIA
       ∞POISONING
       TOXICITY
       TOXICOLOGY

TOXIC HAZARDS
  GS   HAZARDS
       . **TOXIC HAZARDS**
  RT   AIRCRAFT HAZARDS
       FLIGHT HAZARDS
       ∞POISONING
       TOXICITY

TOXICITY
  GS   **TOXICITY**
       . CARBON MONOXIDE POISONING
       . LEAD POISONING
  RT   ACIDOSIS
       ALKALOSIS
       CHEMICAL PROPERTIES
       ENDANGERED SPECIES
       HERBICIDES
       HYPEROXIA
       POISONS
       TOXIC DISEASES
       TOXIC HAZARDS
       TOXICOLOGY
       TOXINS AND ANTITOXINS
       VIRULENCE

TOXICITY AND SAFETY HAZARD
  GS   IRRITATION
       . **TOXICITY AND SAFETY HAZARD**
  RT   ACROLEINS
       BENZENE POISONING
       BERYLLIUM POISONING
       CARBON TETRACHLORIDE POISONING
       CHEMICAL PROPERTIES
       HAZARDOUS MATERIAL DISPOSAL (IN
         SPACE)
       HYDROCARBON POISONING
       INTOXICATION

TOXICOLOGY
  RT   BENZENE POISONING
       BERYLLIUM POISONING
       CARBON TETRACHLORIDE POISONING
       CURARE
       ENDOTOXINS
       FUNGICIDES
       HAZARDS
       HEMOPERFUSION
       HYDROCARBON POISONING
       INSECTICIDES
       INTOXICATION
       NONPOINT SOURCES
       PESTICIDES
       POLLUTION

TOXICOLOGY-*(CONT.)*
    RED TIDE
    TOXIC DISEASES
    TOXICITY
    VACCINES

TOXINS AND ANTITOXINS
GS   **TOXINS AND ANTITOXINS**
    . ENDOTOXINS
    . POLYBROMINATED BIPHENYLS
RT   IMMUNITY
    TOXICITY
    VACCINES

TRAAC SATELLITE
USE   TRANSIT ATTITUDE CONTROL
      SATELLITE

TRACE CONTAMINANTS
GS   CONTAMINANTS
    . **TRACE CONTAMINANTS**
RT   CHEMICAL ELEMENTS
    IMPURITIES
    POLLUTION TRANSPORT
    PURITY
    ∞ TRACING

TRACE ELEMENTS
GS   CHEMICAL ELEMENTS
    . **TRACE ELEMENTS**
RT   ISOTOPIC LABELING
    ∞ NUTRIENTS
    PARTICLE TRACKS
    ∞ TRACERS
    ∞ TRACING

∞ **TRACERS**
SN   *(USE OF A MORE SPECIFIC TERM IS*
    RECOMMENDED--CONSULT THE TERMS
    LISTED BELOW)
RT   AMMUNITION
    ISOTOPIC LABELING
    MARKING
    TRACE ELEMENTS

TRACHEA
GS   ANATOMY
    . RESPIRATORY SYSTEM
    . . BRONCHIAL TUBE
    . . . **TRACHEA**
RT   BRONCHI
    ∞ TUBES

TRACHYTE
GS   ROCKS
    . IGNEOUS ROCKS
    . . **TRACHYTE**
RT   SOILS
    SYENITE

∞ **TRACING**
SN   *(USE OF A MORE SPECIFIC TERM IS*
    RECOMMENDED--CONSULT THE TERMS
    LISTED BELOW)
RT   DRAWINGS
    TRACE CONTAMINANTS
    TRACE ELEMENTS

TRACKED VEHICLES
GS   SURFACE VEHICLES
    . MOTOR VEHICLES
    . . TRACTORS
    . . . **TRACKED VEHICLES**
RT   CRAWLER TRACTORS
    ∞ VEHICLES
    VEHICULAR TRACKS

TRACKING (POSITION)
UF   TRACKING STUDIES
GS   **TRACKING (POSITION)**
    . COMPENSATORY TRACKING
    . INFRARED TRACKING
    . MISSILE TRACKING
    . OPTICAL TRACKING
    . PHOTOGRAPHIC TRACKING
    . POLYSTATION DOPPLER TRACKING
      SYSTEM
    . PURSUIT TRACKING
    . RADAR TRACKING
    . RADIO TRACKING
    . . WILDLIFE RADIOLOCATION
    . RANGE AND RANGE RATE TRACKING
    . SPACE DETECTION AND TRACKING
      SYSTEM

TRACKING (POSITION)-*(CONT.)*
    . SPACECRAFT TRACKING
    . . SATELLITE TRACKING
    . . . SATELLITE-TO-SATELLITE TRACKING
    . STAR TRACKERS
    . . CCD STAR TRACKER
    . VIDEO LANDMARK ACQUISITION AND
      TRACKING
RT   AIR TRAFFIC CONTROL
    AIRCRAFT DETECTION
    APPROACH CONTROL
    DETECTION
    IDENTIFYING
    INSTRUMENT LANDING SYSTEMS
    LASER RANGER/TRACKER
    MULTISPECTRAL TRACKING
      TELESCOPES
    POSITION (LOCATION)
    RANGEFINDING
    RAY TRACING
    SATELLITE DOPPLER POSITIONING
    SOLAR SENSORS
    SOUND LOCALIZATION
    SOUND RANGING
    TRACKING PROBLEM
    ∞ TRACKS

TRACKING AND DATA RELAY SATELLITES
USE   TDR SATELLITES

TRACKING ANTENNAS
USE   DIRECTIONAL ANTENNAS

TRACKING FILTERS
GS   ELECTROMAGNETIC WAVE FILTERS
    . BANDPASS FILTERS
    . . **TRACKING FILTERS**
    . ELECTRIC FILTERS
    . . **TRACKING FILTERS**
RT   ADAPTIVE FILTERS
    BANDSTOP FILTERS
    BANDWIDTH
    PHASE LOCKED SYSTEMS
    VIDEO LANDMARK ACQUISITION AND
      TRACKING

TRACKING NETWORKS
GS   **TRACKING NETWORKS**
    . DEEP SPACE NETWORK
    . GLOBAL TRACKING NETWORK
    . MANNED SPACE FLIGHT NETWORK
    . MATTS (SYSTEMS)
    . POLYSTATION DOPPLER TRACKING
      SYSTEM
    . RADAR NETWORKS
    . SPACE DETECTION AND TRACKING
      SYSTEM
    . SPACE FLIGHT TRACKING AND DATA
      NETWORK
    . STDN (NETWORK)
RT   ADVANCED RANGE INSTRUMENTATION
      SHIP
    DATA ACQUISITION
    GROUND SUPPORT EQUIPMENT
    MINITRACK SYSTEM
    MISSILE TRACKING
    ∞ NETWORKS
    ORION (RADIO INTERFEROMETRY
      NETWORK)
    SATELLITE TRACKING
    SATELLITE-TO-SATELLITE TRACKING
    SPACECRAFT TRACKING

TRACKING PROBLEM
RT   AUTOMATIC CONTROL
    CONTROL THEORY
    FEEDBACK CONTROL
    LINEAR SYSTEMS
    NONLINEAR SYSTEMS
    OPTIMAL CONTROL
    OUTPUT
    ∞ PROBLEMS
    TRACKING (POSITION)
    TRAJECTORY CONTROL
    TRAJECTORY OPTIMIZATION

TRACKING RADAR
GS   RADAR
    . **TRACKING RADAR**
    . . COBRA DANE (RADAR)
RT   COHERENT RADAR
    CONTINUOUS WAVE RADAR
    DIGITAL RADAR SYSTEMS
    MONOPULSE RADAR
    PULSE RADAR

TRACKING RADAR-*(CONT.)*
    RADAR TRACKING
    SATELLITE-BORNE RADAR
    SEARCH RADAR
    SURVEILLANCE RADAR
    TRADEX RADAR SYSTEM
    TRAJECTORY MEASUREMENT

TRACKING STATIONS
GS   STATIONS
    . **TRACKING STATIONS**
    . . DEEP SPACE INSTRUMENTATION
      FACILITY
    . . GLOBAL TRACKING NETWORK
    . . POLYSTATION DOPPLER TRACKING
      SYSTEM
    . . SPACE DETECTION AND TRACKING
      SYSTEM
    . . STDN (NETWORK)
RT   ∞ FENCES
    GROUND STATIONS
    GROUND SUPPORT EQUIPMENT
    JODRELL BANK OBSERVATORY
    ∞ MARS
    MINITRACK SYSTEM
    MISSILE TRACKING
    RADAR NETWORKS
    SATELLITE TRACKING
    SPACE FLIGHT TRACKING AND DATA
      NETWORK
    SPACECRAFT TRACKING

TRACKING STUDIES
USE   TRACKING (POSITION)

∞ **TRACKS**
SN   *(USE OF A MORE SPECIFIC TERM IS*
    RECOMMENDED--CONSULT THE TERMS
    LISTED BELOW)
UF   TRAILS
RT   CONVEYORS
    GROUND TRACKS
    METEOR TRAILS
    MINITRACK SYSTEM
    PARTICLE TRACKS
    PARTICLE TRAJECTORIES
    SMOKE TRAILS
    TRACKING (POSITION)
    VEHICULAR TRACKS

TRACTION
RT   ADHESION
    FRICTION
    PULLING

TRACTORS
GS   SURFACE VEHICLES
    . MOTOR VEHICLES
    . . **TRACTORS**
    . . . CRAWLER TRACTORS
    . . . TRACKED VEHICLES
RT   AGRICULTURE
    ELECTRIC MOTOR VEHICLES
    GROUND HANDLING
    HANDLING EQUIPMENT
    MATERIALS HANDLING
    PLANTING
    PLOWING
    SLEDS
    TOWING
    TRANSPORTATION
    TRUCKS
    ∞ VEHICLES

TRACTS
USE   SITES

TRADEOFFS
RT   DECISION MAKING
    MANAGEMENT ANALYSIS
    MANAGEMENT PLANNING

TRADER AIRCRAFT
USE   C-1A AIRCRAFT

TRADESCANTIA
RT   PLANTS (BOTANY)

TRADEX RADAR SYSTEM
GS   RADAR
    . **TRADEX RADAR SYSTEM**
RT   SEARCH RADAR
    ∞ SYSTEMS
    TARGET RECOGNITION

TRADEX RADAR SYSTEM-*(CONT.)*
    TRACKING RADAR

TRAFFIC
  GS   **TRAFFIC**
      . AIR TRAFFIC
  RT   ACCIDENTS
      AVOIDANCE
      HARBORS
      TRANSPORTATION

TRAFFIC CONTROL
  GS   **TRAFFIC CONTROL**
      . AIR TRAFFIC CONTROL
      .. AUTOMATED EN ROUTE ATC
      .. RADAR APPROACH CONTROL
  RT   AIR TRAFFIC CONTROLLERS
      (PERSONNEL)
      AIRPORT TOWERS
      APPROACH CONTROL
      AVOIDANCE
      COLLISION AVOIDANCE
      ∞CONTROL
      GROUND BASED CONTROL
      NATIONAL AVIATION SYSTEM

TRAGACANTH
  GS   PLANTS (BOTANY)
      . **TRAGACANTH**

TRAILBLAZER 1 REENTRY VEHICLE
  UF   TRAILBLAZER 1 ROCKET VEHICLE
  GS   REENTRY VEHICLES
      . **TRAILBLAZER 1 REENTRY VEHICLE**
  RT   HONEST JOHN ROCKET VEHICLE
      LANCE MISSILE
      MULTISTAGE ROCKET VEHICLES
      NIKE-AJAX MISSILE
      ROCKET VEHICLES
      SOLID PROPELLANT ROCKET ENGINES

TRAILBLAZER 1 ROCKET VEHICLE
  USE   TRAILBLAZER 1 REENTRY VEHICLE

TRAILBLAZER 2 REENTRY VEHICLE
  UF   TRAILBLAZER 2 ROCKET VEHICLE
  GS   REENTRY VEHICLES
      . **TRAILBLAZER 2 REENTRY VEHICLE**
  RT   MULTISTAGE ROCKET VEHICLES
      ROCKET VEHICLES
      SOLID PROPELLANT ROCKET ENGINES
      TX-354 ENGINE

TRAILBLAZER 2 ROCKET VEHICLE
  USE   TRAILBLAZER 2 REENTRY VEHICLE

TRAILERS
  RT   AUTOMOBILES
      COUPLINGS
      SLEDS
      TANK TRUCKS
      TOWED BODIES
      TOWING
      TRUCKS

TRAILING EDGE FLAPS
  UF   VARIABLE AREA WINGS
  GS   AIRFOILS
      . FLAPS (CONTROL SURFACES)
      .. WING FLAPS
      ... **TRAILING EDGE FLAPS**
      BRAKES (FOR ARRESTING MOTION)
      . AERODYNAMIC BRAKES
      .. WING FLAPS
      ... **TRAILING EDGE FLAPS**
      . AIRCRAFT BRAKES
      .. WING FLAPS
      ... **TRAILING EDGE FLAPS**
      CONTROL SURFACES
      . FLAPS (CONTROL SURFACES)
      .. WING FLAPS
      ... **TRAILING EDGE FLAPS**
      DRAG DEVICES
      . AERODYNAMIC BRAKES
      .. WING FLAPS
      ... **TRAILING EDGE FLAPS**
  RT   JET FLAPS
      LEADING EDGE SLATS
      SPLIT FLAPS
      VORTEX FLAPS

TRAILING EDGES
  GS   EDGES
      TRAILING EDGES

TRAILING EDGES-*(CONT.)*
      .. BLUNT TRAILING EDGES
  RT   AIRFOILS
      BLUNT LEADING EDGES
      LEADING EDGES
      VORTEX FLAPS

TRAILS
  USE   TRACKS

TRAINEES
  USE   STUDENTS

TRAINERS
  USE   TRAINING DEVICES

TRAINING
  USE   EDUCATION

TRAINING AIRCRAFT
  GS   **TRAINING AIRCRAFT**
      . ALPHA JET AIRCRAFT
      . CL-41 AIRCRAFT
      . DH 115 AIRCRAFT
      . G-91 AIRCRAFT
      . JAGUAR AIRCRAFT
      . JET PROVOST AIRCRAFT
      . L-29 JET TRAINER
      . T-2 AIRCRAFT
      . T-28 AIRCRAFT
      . T-33 AIRCRAFT
      . T-37 AIRCRAFT
      . T-38 AIRCRAFT
      . T-39 AIRCRAFT
      . TS-11 AIRCRAFT
  RT   ∞AIRCRAFT
      BOMBER AIRCRAFT
      FIGHTER AIRCRAFT
      GENERAL AVIATION AIRCRAFT
      JET AIRCRAFT
      LIGHT AIRCRAFT
      ∞MILITARY AIRCRAFT
      ∞SUBSONIC AIRCRAFT

TRAINING ANALYSIS
  RT   ∞ANALYZING
      ∞DEVELOPMENT
      EDUCATION
      HANDBOOKS
      LEARNING
      PERSONNEL DEVELOPMENT
      PLANNING
      RETRAINING

TRAINING DEVICES
  UF   AUDIO VISUAL EQUIPMENT
      TRAINERS
  GS   **TRAINING DEVICES**
      . TEACHING MACHINES
  RT   ALTITUDE SIMULATION
      CHILD DEVICE
      COCKPIT SIMULATORS
      ∞DEVICES
      EDUCATION
      EDUCATIONAL TELEVISION
      FLIGHT SIMULATORS
      ONBOARD EQUIPMENT
      SIMULATORS
      VISUAL AIDS

TRAINING EVALUATION
  GS   EVALUATION
      . **TRAINING EVALUATION**
  RT   CERTIFICATION
      EXAMINATION
      INSTRUCTORS
      KNOWLEDGE
      LEARNING
      ∞PERFORMANCE
      REVIEWING
      SCHOOLS
      STUDENTS

TRAINING SIMULATORS
  UF   SIMULATOR TRAINING
  GS   SIMULATORS
      . **TRAINING SIMULATORS**
      .. FLIGHT SIMULATORS
      ... COCKPIT SIMULATORS
      .. SPACECRAFT CABIN SIMULATORS
  RT   ASTRONAUT TRAINING
      CENTRIFUGES
      CONTROL SIMULATION
      FLIGHT SIMULATION

TRAINING SIMULATORS-*(CONT.)*
      FLIGHT TRAINING
      LANDING SIMULATION
      LUNAR ORBIT AND LANDING
      SIMULATORS
      ∞MISSILE SIMULATORS
      PILOT TRAINING
      SPACE FLIGHT TRAINING

TRAJECTORIES
  GS   **TRAJECTORIES**
      . ABORT TRAJECTORIES
      . ASCENT TRAJECTORIES
      . BALLISTIC TRAJECTORIES
      . DESCENT TRAJECTORIES
      . REENTRY TRAJECTORIES
      . HYPERBOLIC TRAJECTORIES
      . INTERORBITAL TRAJECTORIES
      . MIDCOURSE TRAJECTORIES
      . MISSILE TRAJECTORIES
      . MOLECULAR TRAJECTORIES
      . PARTICLE TRAJECTORIES
      .. ELECTRON TRAJECTORIES
      . RENDEZVOUS TRAJECTORIES
      . ROUND TRIP TRAJECTORIES
      .. CIRCUMLUNAR TRAJECTORIES
      . SPACECRAFT TRAJECTORIES
      .. EARTH-VENUS TRAJECTORIES
      .. INTERPLANETARY TRAJECTORIES
      ... EARTH-MARS TRAJECTORIES
      ... EARTH-MERCURY TRAJECTORIES
      .. LUNAR TRAJECTORIES
      ... CIRCUMLUNAR TRAJECTORIES
      ... EARTH-MOON TRAJECTORIES
      ... MOON-EARTH TRAJECTORIES
      . SPINNING UNGUIDED ROCKET
      TRAJECTORY
      . UNDERWATER TRAJECTORIES
  RT   APEXES
      BALLISTICS
      ∞CURVES
      DOWNRANGE
      EQUATIONS OF MOTION
      ∞FLIGHT
      FLIGHT MECHANICS
      FLIGHT OPTIMIZATION
      FLIGHT PATHS
      FLIGHT TIME
      GREAT CIRCLES
      MISSILES
      ORBITS
      ORDNANCE
      PARABOLIC FLIGHT
      ∞PATHS
      ROCKET FLIGHT
      SPACE FLIGHT
      TRANSFER ORBITS

TRAJECTORY ANALYSIS
  RT   ∞ANALYZING
      ASTRODYNAMICS
      BALLISTICS
      CAPTURE EFFECT
      CELESTIAL MECHANICS
      EQUATIONS OF MOTION
      GODDARD TRAJECTORY
      DETERMINATION SYSTEM
      IMPACT PREDICTION
      MATHEMATICAL MODELS
      NUMERICAL ANALYSIS
      ORBITAL MECHANICS
      PREFLIGHT ANALYSIS
      SYSTEMS ANALYSIS

TRAJECTORY CONTROL
  UF   RANGE CONTROL
  GS   **TRAJECTORY CONTROL**
      . TRAJECTORY OPTIMIZATION
  RT   ATTITUDE CONTROL
      ∞CONTROL
      DRIFT RATE
      GUIDANCE (MOTION)
      HOMING DEVICES
      LANDING SITES
      OPTIMAL CONTROL
      OPTIMIZATION
      POST BOOST PROPULSION SYSTEM
      RANGE SAFETY
      THRUST PROGRAMMING
      TRACKING PROBLEM

TRAJECTORY MEASUREMENT
  RT   BALLISTIC CAMERAS
      BALLISTICS
      FLIGHT MECHANICS

TRAJECTORY MEASUREMENT-*(CONT.)*
  ∞ MEASUREMENT
  TELEMETRY
  TRACKING RADAR

TRAJECTORY OPTIMIZATION
  GS OPTIMIZATION
    . **TRAJECTORY OPTIMIZATION**
    TRAJECTORY CONTROL
    . **TRAJECTORY OPTIMIZATION**
  RT AIRCRAFT MANEUVERS
    FLIGHT MECHANICS
    FLIGHT OPTIMIZATION
    GODDARD TRAJECTORY
      DETERMINATION SYSTEM
    TRACKING PROBLEM

TRANQUILIZERS
  GS DRUGS
    . **TRANQUILIZERS**
  RT CENTRAL NERVOUS SYSTEM
    DEPRESSANTS
    HYPERTENSION
    SEDATIVES

TRANSALL C-160 AIRCRAFT
  USE C-160 AIRCRAFT

TRANSATMOSPHERIC VEHICLES
  RT AEROSPACE VEHICLES
    AEROSPACEPLANES
    ∞ AIRCRAFT
    AIRCRAFT DESIGN
    ∞ SPACECRAFT
    SPACECRAFT DESIGN

TRANSCEIVERS
  USE TRANSMITTER RECEIVERS

TRANSCENDENTAL FUNCTIONS
  GS FUNCTIONS (MATHEMATICS)
    . **TRANSCENDENTAL FUNCTIONS**
    . . EXPONENTIAL FUNCTIONS
    . . . LOGARITHMS
    . . PERIODIC FUNCTIONS
    . . . TRIGONOMETRIC FUNCTIONS
    . . . . COSINE SERIES
    . . . . SINE SERIES
† . . . . TANGENTS

TRANSCONTINENTAL SYSTEMS
  RT CONTINENTS
    ∞ SYSTEMS
    TELECOMMUNICATION
    TRANSPORTATION

TRANSDUCERS
  GS **TRANSDUCERS**
    . DIGITAL TRANSDUCERS
    . ELECTRONIC TRANSDUCERS
    . IMAGE TRANSDUCERS
    . INTERDIGITAL TRANSDUCERS
    . MAGNETIC TRANSDUCERS
    . MODE TRANSFORMERS
    . PIEZOELECTRIC TRANSDUCERS
    . . PIEZOELECTRIC GAGES
    . PIEZORESISTIVE TRANSDUCERS
    . . PIEZOELECTRIC GAGES
    . PRESSURE SENSORS
    . . BOURDON TUBES
    . QUARTZ TRANSDUCERS
    . SOUND TRANSDUCERS
    . . ELECTROACOUSTIC TRANSDUCERS
    . . . HYDROPHONES
    . . . LOUDSPEAKERS
    . . . MICROPHONES
    . THERMOPILES
    . TORQUERS
    . ULTRASONIC WAVE TRANSDUCERS
  RT BULK ACOUSTIC WAVE DEVICES
    CONTROL EQUIPMENT
    ∞ CONVERTERS
    DATA CONVERTERS
    ∞ DETECTORS
    ENERGY CONVERSION EFFICIENCY
    EXTENSOMETERS
    FORM FACTORS
    INSTRUMENT RECEIVERS
    INSTRUMENT TRANSMITTERS
    ∞ INSTRUMENTS
    MEASURING INSTRUMENTS
    METEOROLOGICAL INSTRUMENTS
    PHOTOELECTRIC CELLS
    ∞ PROBES
    RECORDING INSTRUMENTS

TRANSDUCERS-*(CONT.)*
  REMOTE SENSORS
  ∞ SENSORS
  STRAIN GAGES
  TEMPERATURE MEASURING
    INSTRUMENTS
  ULTRASONIC CLEANING
  VIBRATION METERS

TRANSEARTH INJECTION
  GS INJECTION
    . **TRANSEARTH INJECTION**
  RT INJECTION GUIDANCE
    MIDCOURSE GUIDANCE
    ORBITAL MECHANICS
    TRANSFER ORBITS

TRANSEQUATORIAL PROPAGATION
  GS TRANSMISSION
    . ELECTROMAGNETIC WAVE
      TRANSMISSION
    . . RADIO TRANSMISSION
    . . . **TRANSEQUATORIAL PROPAGATION**
    . SIGNAL TRANSMISSION
    . . RADIO TRANSMISSION
    . . . **TRANSEQUATORIAL PROPAGATION**
    . WAVE PROPAGATION
    . . **TRANSEQUATORIAL PROPAGATION**
  RT EQUATORS
    F 2 REGION
    ∞ PROPAGATION

TRANSFER
  USE TRANSFERRING

TRANSFER FUNCTIONS
  GS FUNCTIONS (MATHEMATICS)
    . **TRANSFER FUNCTIONS**
    . . MODULATION TRANSFER FUNCTION
    . . OPTICAL TRANSFER FUNCTION
  RT AMPLIFICATION
    AUTOMATIC CONTROL
    BANDWIDTH
    COUPLING COEFFICIENTS
    DAMPING
    DYNAMIC CHARACTERISTICS
    DYNAMIC RESPONSE
    FEEDBACK
    FEEDBACK CIRCUITS
    HIGH GAIN
    IMPEDANCE MATCHING
    LOGARITHMIC RECEIVERS
    NEGATIVE FEEDBACK
    NONLINEAR FEEDBACK
    NYQUIST DIAGRAM
    OUTPUT
    POSITIVE FEEDBACK
    SELF OSCILLATION
    SENSITIVITY
    ∞ SYSTEMS
    TIME CONSTANT
    TRANSIENT RESPONSE

TRANSFER OF TRAINING
  GS LEARNING
    . **TRANSFER OF TRAINING**
  RT ABILITIES
    EDUCATION
    GENERALIZATION (PSYCHOLOGY)

TRANSFER ORBITS
  UF HOHMANN TRAJECTORIES
    HOHMANN TRANSFER ORBITS
    ORBITAL TRANSFER
  GS ORBITS
    . ELLIPTICAL ORBITS
    . . **TRANSFER ORBITS**
    . . . INTERPLANETARY TRANSFER
        ORBITS
    . SPACECRAFT ORBITS
    . . **TRANSFER ORBITS**
    . . . INTERPLANETARY TRANSFER
        ORBITS
  RT AEROASSIST
    AEROBRAKING
    AEROCAPTURE
    AEROMANEUVERING
    CIRCUMLUNAR TRAJECTORIES
    EARTH ORBITAL RENDEZVOUS
    EARTH ORBITS
    EARTH-MARS TRAJECTORIES
    EARTH-MERCURY TRAJECTORIES
    EARTH-MOON TRAJECTORIES
    EARTH-VENUS TRAJECTORIES
    INTERPLANETARY TRAJECTORIES

TRANSFER ORBITS-*(CONT.)*
  LUNAR ORBITS
  LUNAR TRAJECTORIES
  MOON-EARTH TRAJECTORIES
  ORBITAL LAUNCHING
  ORBITAL MECHANICS
  PARKING ORBITS
  PLANETARY ORBITS
  SATELLITE ORBITS
  SOLAR ORBITS
  SPACE RENDEZVOUS
  SPACECRAFT DOCKING
  TRAJECTORIES
  TRANSEARTH INJECTION
  TRANSFERRING
  TRANSLUNAR INJECTION

TRANSFER TUNNELS
  GS PASSAGEWAYS
    . **TRANSFER TUNNELS**
  RT ENTRANCES
    ∞ TUNNELS

TRANSFERRED ELECTRON DEVICES
  UF TED
  GS ELECTRONIC EQUIPMENT
    . SOLID STATE DEVICES
    . . SEMICONDUCTOR DEVICES
    . . . **TRANSFERRED ELECTRON DEVICES**
  RT ELECTRON TRANSFER
    GALLIUM ARSENIDES
    INDIUM PHOSPHIDES
    MICROWAVE AMPLIFIERS
    MICROWAVE OSCILLATORS

TRANSFERRING
  UF TRANSFER
  GS **TRANSFERRING**
    . DROP TRANSFER
  RT CHARGE TRANSFER
    ELECTRON TRANSFER
    ENERGY TRANSFER
    EXCHANGING
    HEAT TRANSFER
    MASS TRANSFER
    MATERIALS HANDLING
    MOMENTUM TRANSFER
    ∞ SHIFT
    TECHNOLOGY TRANSFER
    TRANSFER ORBITS
    TRANSPORTATION

TRANSFORM INTEGRALS
  USE INTEGRAL TRANSFORMATIONS

TRANSFORMATION TENSORS
  USE TENSORS

∞ **TRANSFORMATIONS**
  SN *(USE OF A MORE SPECIFIC TERM IS*
     RECOMMENDED--CONSULT THE TERMS
     LISTED BELOW)
  RT FUJITA METHOD
    FUNCTIONS (MATHEMATICS)
    ORDER-DISORDER TRANSFORMATIONS
    PHASE TRANSFORMATIONS
    TRANSFORMATIONS (MATHEMATICS)

TRANSFORMATIONS (MATHEMATICS)
  UF TRANSFORMS
  GS **TRANSFORMATIONS (MATHEMATICS)**
    . COORDINATE TRANSFORMATIONS
    . FOURIER-BESSEL TRANSFORMATIONS
    . HOUSEHOLDER TRANSFORMATIONS
    . INTEGRAL TRANSFORMATIONS
    . . FAST FOURIER TRANSFORMATIONS
    . LAPLACE TRANSFORMATION
  RT FUJITA METHOD
    FUNCTIONS (MATHEMATICS)
    GAUGE INVARIANCE
    RACAH COEFFICIENT
    ∞ TRANSFORMATIONS

TRANSFORMERS
  GS **TRANSFORMERS**
    . INSTRUMENT TRANSFORMERS
    . MODE TRANSFORMERS
    . VOLTAGE CONVERTERS (AC TO AC)
  RT AMPLIFIERS
    BALLASTS (IMPEDANCES)
    CIRCUIT PROTECTION
    ∞ CONVERTERS
    COUPLING CIRCUITS
    DIPLEXERS
    ELECTRIC COILS

TRANSFORMERS-(CONT.)
ELECTRIC FILTERS
ELECTRIC MOTORS
ELECTRIC REACTORS
ELECTRICAL GROUNDING
INDUCTANCE
MAGNET COILS
MAGNETIC CIRCUITS
MAGNETIC CORES
OSCILLATORS
PHASE CONTROL
PLASMA CONTROL
POWER SUPPLY CIRCUITS
RESOLVERS
SATURABLE REACTORS
SOLID STATE DEVICES
TOROIDS
UP-CONVERTERS
VOLTAGE REGULATORS

TRANSFORMS
USE   TRANSFORMATIONS (MATHEMATICS)

TRANSFUSION
RT   BLOOD
FIRST AID
MEDICAL SCIENCE
SYRINGES
VEINS

TRANSGRANULAR CORROSION
GS   CHEMICAL ATTACK
. TRANSGRANULAR CORROSION
CORROSION
. TRANSGRANULAR CORROSION
RT   CORROSION TESTS
INTERGRANULAR CORROSION
METAL FATIGUE
PROTECTIVE COATINGS
STRESS CORROSION

TRANSHORIZON RADIO PROPAGATION
GS   TRANSMISSION
. ELECTROMAGNETIC WAVE
TRANSMISSION
.. RADIO TRANSMISSION
... TRANSHORIZON RADIO
PROPAGATION
. SIGNAL TRANSMISSION
.. RADIO TRANSMISSION
... TRANSHORIZON RADIO
PROPAGATION
RT   RADIO ATTENUATION
RADIO SIGNALS

TRANSIENT HEATING
GS   HEATING
. TRANSIENT HEATING
.. PULSE HEATING
.. SHOCK HEATING
RT   AERODYNAMIC HEATING

TRANSIENT LOADS
SN   (LIMITED TO FORCE LOADS)
GS   LOADS (FORCES)
. DYNAMIC LOADS
.. TRANSIENT LOADS
... GUST LOADS
... IMPACT LOADS
... LANDING LOADS
... SHOCK LOADS
.... BLAST LOADS
RT   AERODYNAMIC LOADS
CYCLIC LOADS
RANDOM LOADS
STRUCTURAL DESIGN CRITERIA

TRANSIENT OSCILLATIONS
GS   OSCILLATIONS
. TRANSIENT OSCILLATIONS
RT   DAMPING
ELECTRON OSCILLATIONS
LASERS
MASERS
PILOT INDUCED OSCILLATION
STIMULATED EMISSION DEVICES
TRANSVERSE OSCILLATION
VIBRATIONAL SPECTRA

TRANSIENT PRESSURES
GS   PRESSURE
. TRANSIENT PRESSURES
RT   MASS FLOW RATE
PRESSURE
PRESSURE SENSORS

TRANSIENT REACTOR TEST FACILITY
UF   TREAT (TEST FACILITY)
GS   TEST FACILITIES
. TRANSIENT REACTOR TEST FACILITY
RT   NUCLEAR RESEARCH AND TEST
REACTORS
REACTOR SAFETY
SNAP

TRANSIENT RESPONSE
GS   DYNAMIC CHARACTERISTICS
. TRANSIENT RESPONSE
RESPONSES
. DYNAMIC RESPONSE
.. TRANSIENT RESPONSE
RT   AMPLIFICATION
∞COMPENSATION
DAMPING
DYNAMIC STABILITY
IMPEDANCE
PRESSURE SENSORS
RESONANT FREQUENCIES
RESPONSE BIAS
SENSITIVITY
STROKING TESTS
TIME CONSTANT
TRANSFER FUNCTIONS

TRANSIENTS (SURGES)
USE   SURGES

TRANSISTOR AMPLIFIERS
GS   AMPLIFIERS
. TRANSISTOR AMPLIFIERS
ELECTRONIC EQUIPMENT
. SOLID STATE DEVICES
.. SEMICONDUCTOR DEVICES
... TRANSISTOR AMPLIFIERS
RT   CURRENT AMPLIFIERS
DIFFERENTIAL AMPLIFIERS
FEEDBACK AMPLIFIERS
INTERMEDIATE FREQUENCY AMPLIFIERS
OPERATIONAL AMPLIFIERS
POWER AMPLIFIERS
PREAMPLIFIERS
TRANSISTORS

TRANSISTOR CIRCUITS
GS   CIRCUITS
. TRANSISTOR CIRCUITS
RT   DTL INTEGRATED CIRCUITS
∞ELECTRONICS
HYBRID CIRCUITS
INTEGRATED CIRCUITS
LINEAR INTEGRATED CIRCUITS
LOGIC CIRCUITS
MICROELECTRONICS
PRINTED CIRCUITS
TTL INTEGRATED CIRCUITS

TRANSISTOR LOGIC
RT   BOOLEAN ALGEBRA
∞LOGIC
LOGIC CIRCUITS
LOGIC DESIGN
THRESHOLD LOGIC

TRANSISTOR-TRANSISTOR-LOGIC INTEG CIRCUITS
USE   TTL INTEGRATED CIRCUITS

TRANSISTORS
GS   ELECTRONIC EQUIPMENT
. SOLID STATE DEVICES
.. SEMICONDUCTOR DEVICES
... TRANSISTORS
.... BIPOLAR TRANSISTORS
.... FIELD EFFECT TRANSISTORS
..... CHARGE FLOW DEVICES
..... JFET
.... HIGH ELECTRON MOBILITY
TRANSISTORS
..... JUNCTION TRANSISTORS
..... JFET
.... PHOTOTRANSISTORS
.... SILICON TRANSISTORS
..... SOS (SEMICONDUCTORS)
RT   GERMANIUM DIODES
ION IMPLANTATION
MINIATURIZATION
PENTODES
SIS (SEMICONDUCTORS)
TETRODES
TRANSISTOR AMPLIFIERS
TRAPATT DEVICES

TRANSISTORS-(CONT.)
TRIODES

∞ TRANSIT
SN   (USE OF A MORE SPECIFIC TERM IS
RECOMMENDED--CONSULT THE TERMS
LISTED BELOW)
RT   OCCULTATION
TRANSIT SATELLITES
TRANSITS

TRANSIT ATTITUDE CONTROL SATELLITE
UF   TRAAC SATELLITE
GS   SATELLITES
. ARTIFICIAL SATELLITES
.. NAVIGATION SATELLITES
... TRANSIT ATTITUDE CONTROL
SATELLITE
UNMANNED SPACECRAFT
. NAVIGATION SATELLITES
.. TRANSIT ATTITUDE CONTROL
SATELLITE
RT   ∞CONTROL
SATELLITE ATTITUDE CONTROL

TRANSIT NAVIGATION SYSTEM
GS   SATELLITE NAVIGATION SYSTEMS
. TRANSIT NAVIGATION SYSTEM
RT   NASA PROGRAMS
NAVIGATION SATELLITES
NOVA SATELLITES
TRANSIT SATELLITES

TRANSIT SATELLITES
GS   SATELLITES
. ARTIFICIAL SATELLITES
.. NAVIGATION SATELLITES
... TRANSIT SATELLITES
UNMANNED SPACECRAFT
. NAVIGATION SATELLITES
.. TRANSIT SATELLITES
RT   DISCOS (SATELLITE ATTITUDE
CONTROL)
∞TRANSIT
TRANSIT NAVIGATION SYSTEM

TRANSIT TIME
SN   (NOT LIMITED TO ASTRONOMICAL
TIMES OF TRANSIT)
GS   TIME
. TRANSIT TIME
RT   BARRITT DIODES
CATT DEVICES
FLIGHT TIME
∞MOTION

∞ TRANSITION
SN   (USE OF A MORE SPECIFIC TERM IS
RECOMMENDED--CONSULT THE TERMS
LISTED BELOW)
RT   BOUNDARY LAYER TRANSITION
ELECTRON TRANSITIONS
FORBIDDEN TRANSITIONS
PHASE TRANSFORMATIONS

TRANSITION FLOW
GS   FLUID FLOW
. GAS FLOW
.. MOLECULAR FLOW
... TRANSITION FLOW
RT   BOUNDARY LAYER TRANSITION
FREE MOLECULAR FLOW
RAREFIED GAS DYNAMICS
SLIP FLOW

∞ TRANSITION LAYERS
SN   (USE OF A MORE SPECIFIC TERM IS
RECOMMENDED--CONSULT THE TERMS
LISTED BELOW)
RT   BOUNDARY LAYER TRANSITION
INTERLAYERS
LAMINAR FLOW
PLASMA LAYERS
SHEAR LAYERS
SHOCK LAYERS
SHOCK WAVES
SURFACE LAYERS
TURBULENT FLOW

TRANSITION METALS
GS   METALS
. TRANSITION METALS
.. CADMIUM
... CADMIUM ISOTOPES

TRANSITION METALS-(CONT.)
```
.. COBALT
... COBALT ISOTOPES
.... COBALT 58
.... COBALT 60
.. GOLD
... GOLD ISOTOPES
.... GOLD 198
.. HAFNIUM
... HAFNIUM ISOTOPES
.. IRON
... IRON ISOTOPES
.... IRON 57
.... IRON 58
.... IRON 59
.. MANGANESE
... MANGANESE ISOTOPES
.. NICKEL
... NICKEL ISOTOPES
.. PALLADIUM
.. PLATINUM
... PLATINUM BLACK
... PLATINUM ISOTOPES
.. REFRACTORY METALS
... CHROMIUM
.... CHROMIUM ISOTOPES
... IRIDIUM
.... IRIDIUM ISOTOPES
.. MOLYBDENUM
.. NIOBIUM
... NIOBIUM ISOTOPES
.... NIOBIUM 95
.. OSMIUM
... OSMIUM ISOTOPES
.. RHENIUM
... RHENIUM ISOTOPES
.. TANTALUM
... TANTALUM ISOTOPES
.. TUNGSTEN
... TUNGSTEN ISOTOPES
.. RHODIUM
... RHODIUM ISOTOPES
.. RUTHENIUM
... RUTHENIUM ISOTOPES
.. SCANDIUM
... SCANDIUM ISOTOPES
.. SILVER
... SILVER ISOTOPES
.. TECHNETIUM
... TECHNETIUM ISOTOPES
.. TITANIUM
... TITANIUM ISOTOPES
.. VANADIUM
... VANADIUM ISOTOPES
.. YTTRIUM
... YTTRIUM ISOTOPES
.. ZINC
... ZINC ISOTOPES
.. ZIRCONIUM
... ZIRCONIUM ISOTOPES
.... ZIRCONIUM 95
```
RT   ACTINIDE SERIES
     COMPLEX COMPOUNDS
     METAL NITRIDES
     PALLADIUM COMPOUNDS
     RARE EARTH ELEMENTS
     RUTHENIUM COMPOUNDS
     SHAPE MEMORY ALLOYS
     TRANSURANIUM ELEMENTS

TRANSITION POINTS
RT   BOUNDARY LAYER TRANSITION
∞EQUILIBRIUM
     KNUDSEN FLOW
     PHASE DIAGRAMS
     REYNOLDS NUMBER

TRANSITION PRESSURE
GS   PRESSURE
     . TRANSITION PRESSURE
RT   HIGH PRESSURE
     HYDROSTATIC PRESSURE
     PHASE TRANSFORMATIONS
     PRESSURE EFFECTS

TRANSITION PROBABILITIES
RT   ELECTRON TRANSITIONS
     EXCITATION
     FERMI SURFACES
     NUCLEAR CAPTURE
     PROBABILITY THEORY
     SPECTRA

TRANSITION TEMPERATURE
GS   TEMPERATURE

TRANSITION TEMPERATURE-(CONT.)
     . TRANSITION TEMPERATURE
RT   HEAT OF FUSION
     KONDO EFFECT
     LIQUID PHASES
     MELTING POINTS
     PHASE DIAGRAMS
     PHASE TRANSFORMATIONS
     SOLIDIFICATION
     SUPERCONDUCTING POWER
       TRANSMISSION
     SUPERCONDUCTIVITY

TRANSITS
SN   (EXCLUDES PARTIAL OR TOTAL
       OCCULTATION OF ONE BODY BY
       ANOTHER)
GS   MEASURING INSTRUMENTS
     . OPTICAL MEASURING INSTRUMENTS
     .. TRANSITS
     ... THEODOLITES
     .... CINETHEODOLITES
     OPTICAL EQUIPMENT
     . OPTICAL MEASURING INSTRUMENTS
     .. TRANSITS
     ... THEODOLITES
     .... CINETHEODOLITES
RT   COMPASSES
     SEXTANTS
∞TRANSIT

TRANSLATING
GS   TRANSLATING
     . MACHINE TRANSLATION
RT   DECODING
     DOCUMENTATION
∞INTERPRETATION
     LANGUAGES
     TECHNICAL WRITING
∞TRANSLATORS

TRANSLATIONAL MOTION
GS   TRANSLATIONAL MOTION
     . KARMAN-BODEWADT FLOW
     . THREE DIMENSIONAL MOTION
     .. THREE DIMENSIONAL FLOW
     ... SECONDARY FLOW
RT   ∞MOTION
     RACKS (GEARS)
     RIGID STRUCTURES

∞ TRANSLATORS
SN   (USE OF A MORE SPECIFIC TERM IS
       RECOMMENDED--CONSULT THE TERMS
       LISTED BELOW)
RT   COMPUTER PROGRAMS
     DECODERS
     DIGITAL TO VOICE TRANSLATORS
     LANGUAGE PROGRAMMING
     REPEATERS
     TRANSLATING

TRANSLUCENCE
GS   ELECTROMAGNETIC PROPERTIES
     . OPTICAL PROPERTIES
     .. TRANSLUCENCE
RT   LIGHT TRANSMISSION
     OPACITY
     OPTICAL DENSITY
     TRANSMISSIVITY
     TRANSPARENCE

TRANSLUNAR INJECTION
GS   INJECTION
     . TRANSLUNAR INJECTION
RT   INJECTION GUIDANCE
     MIDCOURSE GUIDANCE
     ORBITAL MECHANICS
     TRANSFER ORBITS

TRANSLUNAR SPACE
USE INTERPLANETARY SPACE

TRANSMISSION
UF   COAXIAL TRANSMISSION
GS   TRANSMISSION
     . ACOUSTIC PROPAGATION
     . DEMULTIPLEXING
     . ELECTRIC POWER TRANSMISSION
     . ELECTROMAGNETIC WAVE
       TRANSMISSION
     .. LIGHT TRANSMISSION
     ... LIGHT SCATTERING
     .... HALOS
     .. RADAR TRANSMISSION

TRANSMISSION-(CONT.)
```
.. RADIO TRANSMISSION
... DOUBLE SIDEBAND TRANSMISSION
... IONOSPHERIC PROPAGATION
.... IONOSPHERIC F-SCATTER
       PROPAGATION
... MICROWAVE ATTENUATION
... MICROWAVE TRANSMISSION
... MULTIPATH TRANSMISSION
... SHORT WAVE RADIO
       TRANSMISSION
... SINGLE SIDEBAND TRANSMISSION
... SPREAD SPECTRUM TRANSMISSION
... TRANSEQUATORIAL PROPAGATION
... TRANSHORIZON RADIO
       PROPAGATION
.. SCATTER PROPAGATION
... IONOSPHERIC F-SCATTER
       PROPAGATION
.. TELEVISION TRANSMISSION
. HEAT TRANSMISSION
.. HEAT TRANSFER
... AERODYNAMIC HEAT TRANSFER
.... HYPERSONIC HEAT TRANSFER
.... SUPERSONIC HEAT TRANSFER
... CONDUCTIVE HEAT TRANSFER
... CONVECTIVE HEAT TRANSFER
... LAMINAR HEAT TRANSFER
... RADIATIVE HEAT TRANSFER
.. TURBULENT HEAT TRANSFER
. MULTIPLEXING
.. CODE DIVISION MULTIPLEXING
.. FREQUENCY DIVISION MULTIPLEXING
.. TIME DIVISION MULTIPLEXING
.. WAVELENGTH DIVISION
       MULTIPLEXING
. SELF PROPAGATION
. SIGNAL TRANSMISSION
.. DATA TRANSMISSION
... AUTOMATIC PICTURE
       TRANSMISSION
... MULTIPLE ACCESS
.... CODE DIVISION MULTIPLE ACCESS
.... FREQUENCY DIVISION MULTIPLE
       ACCESS
... SINGLE CHANNEL PER CARRIER
       TRANSMISSION
.. PCM TELEMETRY
. RADAR TRANSMISSION
.. RADIO TRANSMISSION
... DOUBLE SIDEBAND TRANSMISSION
... IONOSPHERIC PROPAGATION
.... IONOSPHERIC F-SCATTER
       PROPAGATION
... MICROWAVE ATTENUATION
... MICROWAVE TRANSMISSION
... MULTIPATH TRANSMISSION
... SHORT WAVE RADIO
       TRANSMISSION
... SINGLE SIDEBAND TRANSMISSION
... TRANSEQUATORIAL PROPAGATION
... TRANSHORIZON RADIO
       PROPAGATION
.. SATELLITE TRANSMISSION
. TELEVISION TRANSMISSION
. SOUND TRANSMISSION
. STRESS PROPAGATION
. TELEPHONY
. WAVE PROPAGATION
.. DIFFRACTION PROPAGATION
.. GROUND WAVE PROPAGATION
.. IONOSPHERIC PROPAGATION
... IONOSPHERIC F-SCATTER
       PROPAGATION
.. LIGHT SCATTERING
... HALOS
.. SCATTER PROPAGATION
... IONOSPHERIC F-SCATTER
       PROPAGATION
.. SHOCK WAVE PROPAGATION
.. TRANSEQUATORIAL PROPAGATION
```
RT   ABSORPTANCE
     ALOHA SYSTEM
     ATMOSPHERIC ATTENUATION
     ATTENUATION
     BROADCASTING
∞CONDUCTION
     DIFFRACTION
     ELECTROMAGNETIC ABSORPTION
     ELECTROMAGNETIC RADIATION
     OPTICAL FILTERS
     OUTPUT
∞PROPAGATION
     RADAR ATTENUATION
     RADIO ATTENUATION
     REFLECTION

TRANSMISSION-*(CONT.)*
    REFRACTION
    SIGNAL REFLECTION
    TELECOMMUNICATION
    TRANSMISSIVITY
    TRANSMITTANCE
    WAVE DISPERSION

TRANSMISSION CIRCUITS
GS   CIRCUITS
    . **TRANSMISSION CIRCUITS**
RT   CIRCUIT PROTECTION
    ELECTRIC POWER TRANSMISSION
    SIGNAL STABILIZATION
    STRIP TRANSMISSION LINES
    TELECOMMUNICATION

TRANSMISSION EFFICIENCY
GS   EFFICIENCY
    . **TRANSMISSION EFFICIENCY**
RT   ALOHA SYSTEM
    ATTENUATION COEFFICIENTS
    BIT ERROR RATE
    CARRIER TO NOISE RATIOS
    DATA TRANSMISSION
    DOWNLINKING
    ELECTROMAGNETIC WAVE
      TRANSMISSION
    FREQUENCY HOPPING
    INTERSYMBOLIC INTERFERENCE
    NETWORK CONTROL
    OPACITY
    PACKET TRANSMISSION
    PACKETS (COMMUNICATION)
    POWER EFFICIENCY
    SIGNAL TRANSMISSION
    TRANSMITTANCE
    UPLINKING

TRANSMISSION FLUIDS
RT   FLUID TRANSMISSION LINES
   ∞FLUIDS
    HYDRAULIC FLUIDS
    WORKING FLUIDS

TRANSMISSION LINES
UF   TRUNKS (LINES)
GS   **TRANSMISSION LINES**
    . COMMUNICATION CABLES
    . . COAXIAL CABLES
    . . WAVEGUIDES
    . . . BEAM WAVEGUIDES
    . . . CIRCULAR WAVEGUIDES
    . . . OPTICAL WAVEGUIDES
    . . . PLASMAGUIDES
    . . . RECTANGULAR WAVEGUIDES
    . FLUID TRANSMISSION LINES
    . POWER LINES
    . STRIP TRANSMISSION LINES
    . . MICROSTRIP TRANSMISSION LINES
    . SUBMARINE CABLES
    . UNDERGROUND TRANSMISSION LINES
RT   ACOUSTIC DELAY LINES
    ANTENNA COUPLERS
    ANTENNA FEEDS
    BACKWARD WAVES
   ∞CABLES
    CIRCUIT PROTECTION
    CIRCUITS
    DELTA ANTENNAS
    DIRECTIONAL COUPLERS
    DISTRIBUTED AMPLIFIERS
    ELECTRIC CONDUCTORS
    ELECTRIC CURRENT
    ELECTRIC POWER TRANSMISSION
    ELECTRIC WIRE
    ELECTRICAL ENGINEERING
    ELECTRIFICATION
    HARNESSES
    IMPEDANCE MATCHING
    INSULATORS
   ∞LINES
    MODE TRANSFORMERS
   ∞NETWORKS
    NONRESONANCE
    SMITH CHART
    STANDING WAVE RATIOS
    SUPERCONDUCTING POWER
      TRANSMISSION
    TELECOMMUNICATION
    WIRING

TRANSMISSION LOSS
RT   ATTENUATION
    CURRENT REGULATORS

TRANSMISSION LOSS-*(CONT.)*
    ELECTRIC POWER TRANSMISSION
    INSERTION
    INSERTION LOSS
    LOSSES
    LOSSY MEDIA
    SILENCE
    VOLTAGE REGULATORS
    WAVE DISPERSION

TRANSMISSIONS (MACHINE ELEMENTS)
GS   MECHANICAL DRIVES
    . **TRANSMISSIONS (MACHINE**
      ELEMENTS)
RT   GEARS
    SHAFTS (MACHINE ELEMENTS)
    TORQUE CONVERTERS
    TORQUE MOTORS
    VEHICLE WHEELS

TRANSMISSIVITY
GS   ELECTROMAGNETIC PROPERTIES
    . OPTICAL PROPERTIES
    . . **TRANSMISSIVITY**
RT   ABSORPTANCE
    ABSORPTIVITY
    DENSITY (MASS/VOLUME)
    LIGHT SCATTERING
    OPACITY
   ∞PHYSICAL PROPERTIES
    TRANSLUCENCE
    TRANSMISSION
    TRANSMITTANCE
    TRANSPARENCE
    TRANSPONDERS
    VISIBILITY

TRANSMISSOMETERS
GS   MEASURING INSTRUMENTS
    . OPTICAL MEASURING INSTRUMENTS
    . . **TRANSMISSOMETERS**
    OPTICAL EQUIPMENT
    . OPTICAL MEASURING INSTRUMENTS
    . . **TRANSMISSOMETERS**
RT   DENSITOMETERS
    PHOTOMETERS
    RADIANCE
    TELEPHOTOMETRY
    TRANSMITTANCE

TRANSMITTANCE
GS   ELECTROMAGNETIC PROPERTIES
    . OPTICAL PROPERTIES
    . . **TRANSMITTANCE**
RT   ABSORPTANCE
    ATTENUATION COEFFICIENTS
    DENSITY (MASS/VOLUME)
    ELECTROMAGNETIC ABSORPTION
    INFRARED ABSORPTION
    LIGHT (VISIBLE RADIATION)
    OPTICAL DENSITY
    PHOTOMETRY
    RAY TRACING
    REFLECTANCE
    SCATTERING
    TRANSMISSION
    TRANSMISSION EFFICIENCY
    TRANSMISSIVITY
    TRANSMISSOMETERS
    TRANSPARENCE

TRANSMITTER RECEIVERS
UF   TRANSCEIVERS
GS   COMMUNICATION EQUIPMENT
    . RADIO RECEIVERS
    . . **TRANSMITTER RECEIVERS**
    RADIO EQUIPMENT
    . RADIO RECEIVERS
    . . **TRANSMITTER RECEIVERS**
    . RADIO TRANSMITTERS
    . . **TRANSMITTER RECEIVERS**
    TRANSMITTERS
    . RADIO TRANSMITTERS
    . . **TRANSMITTER RECEIVERS**
RT   INTERROGATION
    TRANSPONDERS

TRANSMITTERS
UF   SENDERS
GS   **TRANSMITTERS**
    . EMERGENCY LOCATOR
      TRANSMITTERS
    . INSTRUMENT TRANSMITTERS
    . RADAR TRANSMITTERS
    . RADIO TRANSMITTERS

TRANSMITTERS-*(CONT.)*
    . . RADIO BEACONS
    . . . OMNIDIRECTIONAL RADIO RANGES
    . . . . SELF CALIBRATING OMNIRANGE
    . . RADIOMETEOROGRAPHS
    . . RADIOSONDES
    . . . ENDORADIOSONDES
    . . . IONOSONDES
    . . . RAWINSONDES
    . . RADIOTELEPHONES
    . . SONOBUOYS
    . . TRANSMITTER RECEIVERS
    . REPEATERS
RT   ANTENNAS
    ATTENUATION
    DUPLEXERS
   ∞INSTRUMENTS
    MICROPHONES
    RECEIVERS
    SIGNAL ENCODING
    TELECOMMUNICATION
    TELEVISION TRANSMISSION
    TRANSPONDERS

TRANSMUTATION
RT   NEUTRON IRRADIATION
    NUCLEAR REACTIONS
    RADIOGENIC MATERIALS

TRANSOCEANIC COMMUNICATION
GS   TELECOMMUNICATION
    . **TRANSOCEANIC COMMUNICATION**
    TRANSOCEANIC SYSTEMS
    . **TRANSOCEANIC COMMUNICATION**
RT   FACSIMILE COMMUNICATION
    RADIO COMMUNICATION
    RELAY SATELLITES

TRANSOCEANIC FLIGHT
RT   ∞FLIGHT

TRANSOCEANIC SYSTEMS
GS   **TRANSOCEANIC SYSTEMS**
    . TRANSOCEANIC COMMUNICATION
RT   INTERCONTINENTAL BALLISTIC MISSILES
    OCEAN DATA ACQUISITIONS SYSTEMS
   ∞SYSTEMS
    TELECOMMUNICATION
    TRANSPORTATION
    WORLD DATA CENTERS

TRANSONIC AIRCRAFT
USE   SUPERSONIC AIRCRAFT

TRANSONIC AIRCRAFT TECHNOLOGY PROGRAM
USE   TACT PROGRAM

TRANSONIC COMPRESSORS
GS   COMPRESSORS
    . **TRANSONIC COMPRESSORS**
RT   SUPERSONIC COMPRESSORS
    TURBOCOMPRESSORS

TRANSONIC FLIGHT
RT   ∞FLIGHT
    ROCKET FLIGHT
    SONIC BOOMS
    SUPERSONIC FLIGHT

TRANSONIC FLOW
UF   SONIC FLOW
    TRANSONICS
GS   FLUID FLOW
    . COMPRESSIBLE FLOW
    . . **TRANSONIC FLOW**
RT   AERODYNAMICS
    COMPRESSIBILITY EFFECTS
   ∞FLOW
    FLOW VELOCITY
    GAS FLOW
    NOZZLE FLOW
    SHOCK WAVES
    SONIC NOZZLES
    SUBSONIC FLOW
    SUPERSONIC FLOW
    TRISONIC WIND TUNNELS
    WIND TUNNELS

TRANSONIC FLUTTER
GS   VIBRATION
    . STRUCTURAL VIBRATION
    . . FLUTTER
    . . . **TRANSONIC FLUTTER**
    . . SELF INDUCED VIBRATION

TRANSONIC FLUTTER-(CONT.)
```
      . . . TRANSONIC FLUTTER
RT    MISSILE VIBRATION
      SUBSONIC FLUTTER
      SUPERSONIC FLUTTER
```

TRANSONIC INLETS
```
USE   SUPERSONIC INLETS
```

TRANSONIC NOZZLES
```
RT    CONICAL NOZZLES
      CONVERGENT-DIVERGENT NOZZLES
      HYPERSONIC NOZZLES
      ∞ NOZZLES
      SONIC NOZZLES
      SUPERSONIC NOZZLES
      WIND TUNNEL NOZZLES
```

TRANSONIC SPEED
```
GS    RATES (PER TIME)
      . TRANSONIC SPEED
      VELOCITY
      . TRANSONIC SPEED
RT    ACOUSTIC VELOCITY
      SUBSONIC SPEED
      SUPERSONIC SPEEDS
```

TRANSONIC TURBINES
```
USE   SUPERSONIC TURBINES
```

TRANSONIC WIND TUNNELS
```
GS    TEST FACILITIES
      . WIND TUNNELS
      . . TRANSONIC WIND TUNNELS
RT    BLOWDOWN WIND TUNNELS
      HYPERSONIC WIND TUNNELS
      SLOTTED WIND TUNNELS
      SUBSONIC WIND TUNNELS
      SUPERSONIC WIND TUNNELS
      WING FLOW METHOD TESTS
```

TRANSONICS
```
USE   TRANSONIC FLOW
```

TRANSPARENCE
```
UF    TRANSPARENT MATERIALS
GS    ELECTROMAGNETIC PROPERTIES
      . OPTICAL PROPERTIES
      . . TRANSPARENCE
RT    ABSORPTANCE
      ABSORPTIVITY
      ATMOSPHERIC OPTICS
      CLARITY
      DENSITY (MASS/VOLUME)
      ELECTROMAGNETIC ABSORPTION
      HAZE
      LIGHT TRANSMISSION
      OPACITY
      OPTICAL DENSITY
      RADOME MATERIALS
      TRANSLUCENCE
      TRANSMISSIVITY
      TRANSMITTANCE
      TURBIDITY
```

TRANSPARENT MATERIALS
```
USE   TRANSPARENCE
```

TRANSPIRATION
```
UF    FLUID TRANSPIRATION
GS    PHASE TRANSFORMATIONS
      . VAPORIZING
      . . EVAPORATION
      . . . TRANSPIRATION
RT    COOLING
      COOLING SYSTEMS
      EVANESCENCE
      EVAPOTRANSPIRATION
      EVOLUTION (LIBERATION)
      GAS EVOLUTION
      MASS TRANSFER
      MOLECULAR FLOW
      OUTGASSING
      PERMEATING
      PERSPIRATION
      TEMPERATURE CONTROL
```

TRANSPIRATION COOLING
```
USE   SWEAT COOLING
```

TRANSPLANTATION
```
RT    CLINICAL MEDICINE
      HEART IMPLANTATION
      SURGERY
```

TRANSPONDER CONTROL GROUP
```
UF    TCG (TRACKING)
RT    ∞ CONTROL
      RADAR TRACKING
      SATELLITE TRACKING
      SPACECRAFT TRACKING
      TELEMETRY
      TRANSPONDERS
```

TRANSPONDERS
```
UF    RESPONDERS
GS    RADIO EQUIPMENT
      . TRANSPONDERS
RT    AIR TRAFFIC CONTROL
      BEACON COLLISION AVOIDANCE
        SYSTEM
      INTERROGATION
      RADAR BEACONS
      RADAR EQUIPMENT
      RADIO RECEIVERS
      RADIO TRANSMITTERS
      TRANSMISSIVITY
      TRANSMITTER RECEIVERS
      TRANSMITTERS
      TRANSPONDER CONTROL GROUP
```

TRANSPORT AIRCRAFT
```
GS    TRANSPORT AIRCRAFT
      . A-300 AIRCRAFT
      . ALADIN 2 AIRCRAFT
      . AN-2 AIRCRAFT
      . AN-22 AIRCRAFT
      . AN-24 AIRCRAFT
      . ARGOSY MK-1 AIRCRAFT
      . BAC 111 AIRCRAFT
      . BOEING 707 AIRCRAFT
      . BOEING 720 AIRCRAFT
      . BOEING 727 AIRCRAFT
      . BOEING 733 AIRCRAFT
      . BOEING 737 AIRCRAFT
      . BOEING 747 AIRCRAFT
      . BOEING 757 AIRCRAFT
      . BOEING 767 AIRCRAFT
      . BOEING 2707 AIRCRAFT
      . CARGO AIRCRAFT
      . . BREGUET 941 AIRCRAFT
      . . C-1A AIRCRAFT
      . . C-2 AIRCRAFT
      . . C-5 AIRCRAFT
      . . C-9 AIRCRAFT
      . . C-46 AIRCRAFT
      . . C-47 AIRCRAFT
      . . C-54 AIRCRAFT
      . . C-118 AIRCRAFT
      . . C-119 AIRCRAFT
      . . C-121 AIRCRAFT
      . . C-123 AIRCRAFT
      . . C-124 AIRCRAFT
      . . C-130 AIRCRAFT
      . . C-131 AIRCRAFT
      . . C-133 AIRCRAFT
      . . C-135 AIRCRAFT
      . . C-140 AIRCRAFT
      . . C-141 AIRCRAFT
      . . C-160 AIRCRAFT
      . . CL-44 AIRCRAFT
      . . DC 3 AIRCRAFT
      . . DC 7 AIRCRAFT
      . . F-27 AIRCRAFT
      . . P-166 AIRCRAFT
      . . SPANLOADER AIRCRAFT
      . . YC-14 AIRCRAFT
      . CH-3 HELICOPTER
      . CH-34 HELICOPTER
      . CH-46 HELICOPTER
      . CH-47 HELICOPTER
      . CH-54 HELICOPTER
      . CL-84 AIRCRAFT
      . CL-823 AIRCRAFT
      . CONCORDE AIRCRAFT
      . CV-880 AIRCRAFT
      . DC 8 AIRCRAFT
      . DC 9 AIRCRAFT
      . DC 10 AIRCRAFT
      . DH 121 AIRCRAFT
      . DH 125 AIRCRAFT
      . DHC 2 AIRCRAFT
      . DHC 4 AIRCRAFT
      . DHC 5 AIRCRAFT
      . DO-31 AIRCRAFT
      . EC-121 AIRCRAFT
      . ELECTRA AIRCRAFT
      . F-28 TRANSPORT AIRCRAFT
      . G-1 AIRCRAFT
      . G-222 AIRCRAFT
```

TRANSPORT AIRCRAFT-(CONT.)
```
      . H-19 HELICOPTER
      . H-53 HELICOPTER
      . H-56 HELICOPTER
      . HC-3 HELICOPTER
      . HFB-320 AIRCRAFT
      . IL-14 AIRCRAFT
      . L-1011 AIRCRAFT
      . L-2000 AIRCRAFT
      . LIGHT INTRATHEATER TRANSPORT
      . LIGHT TRANSPORT AIRCRAFT
      . LOCKHEED MODEL 18 AIRCRAFT
      . MH-262 AIRCRAFT
      . MYSTERE 20 AIRCRAFT
      . S-58 HELICOPTER
      . S-61 HELICOPTER
      . SA-330 HELICOPTER
      . SC-5 AIRCRAFT
      . SC-7 AIRCRAFT
      . SH-3 HELICOPTER
      . SH-4 HELICOPTER
      . SHORT HAUL AIRCRAFT
      . . EUROPEAN AIRBUS
      . . . A-310 AIRCRAFT
      . . . A-320 AIRCRAFT
      . . MERCURE AIRCRAFT
      . TANKER AIRCRAFT
      . TU-124 AIRCRAFT
      . TU-144 AIRCRAFT
      . TU-154 AIRCRAFT
      . UH-34 HELICOPTER
      . UH-60A HELICOPTER
      . UH-61A HELICOPTER
      . VC-10 AIRCRAFT
      . VISCOUNT AIRCRAFT
      . XC-142 AIRCRAFT
      . YS-11 AIRCRAFT
RT    AIR TRANSPORTATION
      ∞ AIRCRAFT
      COMMERCIAL AIRCRAFT
      GENERAL AVIATION AIRCRAFT
      JET AIRCRAFT
      LIGHT AIRCRAFT
      ∞ LOW WING AIRCRAFT
      ∞ MILITARY AIRCRAFT
      MYSTERE 50 AIRCRAFT
      PASSENGER AIRCRAFT
      ROTARY WING AIRCRAFT
      ∞ SUBSONIC AIRCRAFT
      SUPERSONIC AIRCRAFT
      ∞ TRANSPORT VEHICLES
      TURBOFAN AIRCRAFT
      TURBOPROP AIRCRAFT
      UTILITY AIRCRAFT
      V/STOL AIRCRAFT
      WATER TAKEOFF AND LANDING
        AIRCRAFT
```

TRANSPORT COEFFICIENTS
```
USE   TRANSPORT PROPERTIES
```

TRANSPORT PROPERTIES
```
UF    TRANSPORT COEFFICIENTS
GS    TRANSPORT PROPERTIES
      . ATMOSPHERIC CONDUCTIVITY
      . . IONOSPHERIC CONDUCTIVITY
      . CARRIER MOBILITY
      . . ELECTRON MOBILITY
      . . HOLE MOBILITY
      . DIFFUSION COEFFICIENT
      . . SORET COEFFICIENT
      . ELECTRICAL RESISTIVITY
      . IONOSPHERIC CONDUCTIVITY
      . MAGNETORESISTIVITY
      . PHOTOCONDUCTIVITY
      . PLASMA CONDUCTIVITY
      . SUPERCONDUCTIVITY
      . . KONDO EFFECT
      . GASEOUS DIFFUSION
      . . GASEOUS SELF-DIFFUSION
      . IONIC MOBILITY
      . THERMAL CONDUCTIVITY
      . THERMAL DIFFUSIVITY
      . VISCOSITY
      . . EDDY VISCOSITY
      . . GAS VISCOSITY
RT    BINARY FLUIDS
      BOLTZMANN TRANSPORT EQUATION
      ∞ CONDUCTIVITY
      DIFFUSION
      FLOW COEFFICIENTS
      HALL EFFECT
      HEAT TRANSFER
      HIGH TEMPERATURE TESTS
      KINETIC THEORY
```

TRANSPORT PROPERTIES-(CONT.)
LIGHTHILL GAS MODEL
MOBILITY
∞ PHYSICAL PROPERTIES
POLLUTION TRANSPORT
∞ PROPERTIES
RADIATION TRANSPORT
SEEBECK EFFECT
∞ SOLID STATE PHYSICS
THERMOELECTRICITY

TRANSPORT THEORY
GS   KINETIC THEORY
   . **TRANSPORT THEORY**
   . . EYRING THEORY
   . . MIXING LENGTH FLOW THEORY
RT   BOLTZMANN TRANSPORT EQUATION
   DIFFUSION THEORY
   GAS TRANSPORT
   INTEGRAL EQUATIONS
   MOLECULAR INTERACTIONS
   MONTE CARLO METHOD
   POLLUTION TRANSPORT
   ∞ THEORIES

∞ **TRANSPORT VEHICLES**
SN   *(USE OF A MORE SPECIFIC TERM IS*
   RECOMMENDED--CONSULT THE TERMS
   LISTED BELOW)
RT   CRAWLER TRACTORS
   GROUND EFFECT MACHINES
   RAPID TRANSIT SYSTEMS
   SHIPS
   TRANSPORT AIRCRAFT
   ∞ VEHICLES

TRANSPORTATION
GS   **TRANSPORTATION**
   . AIR TRANSPORTATION
   . MARINE TRANSPORTATION
   . RAIL TRANSPORTATION
   . RAPID TRANSIT SYSTEMS
   . SPACE TRANSPORTATION
   . . SPACE TRANSPORTATION SYSTEM
   . . . SPACE SHUTTLE BOOSTERS
   . . . . AEROMANEUVERING ORBIT TO
       ORBIT SHUTTLE
   . . . SPACE SHUTTLE MISSIONS
   . . . . SPACE SHUTTLE MISSION 31-A
   . . . . SPACE SHUTTLE MISSION 31-B
   . . . . SPACE SHUTTLE MISSION 31-C
   . . . . SPACE SHUTTLE MISSION 31-D
   . . . . SPACE SHUTTLE MISSION 41-A
   . . . . SPACE SHUTTLE MISSION 41-B
   . . . . SPACE SHUTTLE MISSION 41-C
   . . . . SPACE SHUTTLE MISSION 41-D
   . . . . SPACE SHUTTLE MISSION 41-G
   . . . . SPACE SHUTTLE MISSION 51-A
   . . . . SPACE SHUTTLE MISSION 51-B
   . . . . SPACE SHUTTLE MISSION 51-C
   . . . . SPACE SHUTTLE MISSION 51-D
   . . . . SPACE SHUTTLE MISSION 51-E
   . . . . SPACE SHUTTLE MISSION 51-F
   . . . . SPACE SHUTTLE MISSION 51-G
   . . . . SPACE SHUTTLE MISSION 51-H
   . . . . SPACE SHUTTLE MISSION 51-I
   . . . . SPACE SHUTTLE MISSION 51-J
   . . . . SPACE SHUTTLE MISSION 51-L
   . . . . SPACE SHUTTLE MISSION 61-A
   . . . . SPACE SHUTTLE MISSION 61-B
   . . . . SPACE SHUTTLE MISSION 61-C
   . . . . SPACE SHUTTLE MISSION 61-E
   . . . SPACE SHUTTLE ORBITERS
   . . . . ATLANTIS (ORBITER)
   . . . . CHALLENGER (ORBITER)
   . . . . COLUMBIA (ORBITER)
   . . . . DISCOVERY (ORBITER)
   . . . . ENTERPRISE (ORBITER)
   . URBAN TRANSPORTATION
RT   ARTIFICIAL HARBORS
   AUTOMATED GUIDEWAY TRANSIT
      VEHICLES
   AUTOMATED TRANSIT VEHICLES
   CARGO
   CONTRACTORS
   CONVEYORS
   DEEPWATER TERMINALS
   DELIVERY
   DISTRIBUTING
   ∞ DISTRIBUTION
   ELECTRIC AUTOMOBILES
   EVACUATING (TRANSPORTATION)
   FREIGHT COSTS
   FREIGHTERS
   HANDLING EQUIPMENT

TRANSPORTATION-(CONT.)
HAULING
HIGHWAYS
LOGISTICS
MATERIALS HANDLING
MISSILES
MOTOR VEHICLES
OFFSHORE DOCKING
OFFSHORE PLATFORMS
PACKAGING
PASSENGERS
PIPELINES
RIDING QUALITY
ROADS
ROUTES
SERVICES
SITE SELECTION
SPACE SHUTTLE MISSION 61-A
SPACE TRANSPORTATION SYSTEM
TANKER TERMINALS
TERMINAL FACILITIES
TRACTORS
TRAFFIC
TRANSCONTINENTAL SYSTEMS
TRANSFERRING
TRANSOCEANIC SYSTEMS
TRANSPORTATION NETWORKS
∞ TRAVEL
TRUCKS

TRANSPORTATION ENERGY
RT   ALLOCATIONS
   CARGO
   COMMERCIAL ENERGY
   DISTRIBUTING
   DOMESTIC ENERGY
   ECONOMIC FACTORS
   ∞ ENERGY
   ENERGY CONVERSION
   ENGINES
   FUELS
   HAULING
   INDUSTRIAL ENERGY
   SHIPS
   ∞ TANKERS
   TRUCKS

TRANSPORTATION NETWORKS
RT   HIGHWAYS
   INTERSECTIONS
   ∞ NETWORKS
   RAPID TRANSIT SYSTEMS
   ROADS
   TRANSPORTATION

TRANSPORTER
GS   SURFACE VEHICLES
   . **TRANSPORTER**
RT   ∞ CONTAINERS
   ∞ VEHICLES

TRANSURANIUM ELEMENTS
GS   CHEMICAL ELEMENTS
   . ACTINIDE SERIES
   . . **TRANSURANIUM ELEMENTS**
   . . . AMERICIUM
   . . . . AMERICIUM ISOTOPES
   . . . . . AMERICIUM 241
   . . . BERKELIUM
   . . . CALIFORNIUM
   . . . . CALIFORNIUM ISOTOPES
   . . . CURIUM
   . . . . CURIUM ISOTOPES
   . . . . . CURIUM 242
   . . . . . CURIUM 244
   . . . EINSTEINIUM
   . . . FERMIUM
   . . . LAWRENCIUM
   . . . MENDELEVIUM
   . . . NEPTUNIUM
   . . . . NEPTUNIUM ISOTOPES
   . . . NOBELIUM
   . . . PLUTONIUM
   . . . . PLUTONIUM ISOTOPES
   . . . . . PLUTONIUM 238
   . . . . . PLUTONIUM 239
   . . . . . PLUTONIUM 240
   . . . . . PLUTONIUM 241
   . . . . . PLUTONIUM 244
   . . . SERGENIUM
   . NUCLIDES
   . . ISOTOPES
   . . . RADIOACTIVE ISOTOPES
   . . . . **TRANSURANIUM ELEMENTS**
   . . . . . AMERICIUM

TRANSURANIUM ELEMENTS-(CONT.)
. . . . . . AMERICIUM ISOTOPES
. . . . . BERKELIUM
. . . . . CALIFORNIUM
. . . . . . CALIFORNIUM ISOTOPES
. . . . . CURIUM
. . . . . . CURIUM ISOTOPES
. . . . . EINSTEINIUM
. . . . . FERMIUM
. . . . . LAWRENCIUM
. . . . . MENDELEVIUM
. . . . . NEPTUNIUM
. . . . . . NEPTUNIUM ISOTOPES
. . . . . NOBELIUM
. . . . . PLUTONIUM
. . . . . . PLUTONIUM ISOTOPES
. . . . . SERGENIUM
HEAVY ELEMENTS
. **TRANSURANIUM ELEMENTS**
. . AMERICIUM ISOTOPES
. . . AMERICIUM 241
. . BERKELIUM
. . CALIFORNIUM
. . . CALIFORNIUM ISOTOPES
. . CURIUM
. . . CURIUM ISOTOPES
. . . . CURIUM 242
. . . . CURIUM 244
. . EINSTEINIUM
. . FERMIUM
. . LAWRENCIUM
. . MENDELEVIUM
. . NEPTUNIUM
. . . NEPTUNIUM ISOTOPES
. . NOBELIUM
. . PLUTONIUM
. . . PLUTONIUM ISOTOPES
. . . . PLUTONIUM 238
. . . . PLUTONIUM 239
. . . . PLUTONIUM 240
. . . . PLUTONIUM 241
. . . . PLUTONIUM 244
. . SERGENIUM
METALS
. ACTINIDE SERIES
. . **TRANSURANIUM ELEMENTS**
. . . AMERICIUM
. . . . AMERICIUM ISOTOPES
. . . . . AMERICIUM 241
. . . BERKELIUM
. . . CALIFORNIUM
. . . . CALIFORNIUM ISOTOPES
. . . CURIUM
. . . . CURIUM ISOTOPES
. . . . . CURIUM 242
. . . . . CURIUM 244
. . . EINSTEINIUM
. . . FERMIUM
. . . LAWRENCIUM
. . . MENDELEVIUM
. . . NEPTUNIUM
. . . . NEPTUNIUM ISOTOPES
. . . NOBELIUM
. . . PLUTONIUM
. . . . PLUTONIUM ISOTOPES
. . . . . PLUTONIUM 238
. . . . . PLUTONIUM 239
. . . . . PLUTONIUM 240
. . . . . PLUTONIUM 241
. . . . . PLUTONIUM 244
. . . SERGENIUM
RT   TRANSITION METALS

TRANSVERSE ACCELERATION
GS   RATES (PER TIME)
   . ACCELERATION (PHYSICS)
   . . **TRANSVERSE ACCELERATION**
RT   ∞ ACCELERATION
   ACCELERATION STRESSES
      (PHYSIOLOGY)
   ANGULAR ACCELERATION

TRANSVERSE OSCILLATION
UF   TRANSVERSE VIBRATION
GS   OSCILLATIONS
   . **TRANSVERSE OSCILLATION**
   . . H WAVES
RT   GAMMA RAYS
   HARMONIC OSCILLATION
   LATERAL OSCILLATION
   STABLE OSCILLATIONS
   TRANSIENT OSCILLATIONS

TRANSVERSE VIBRATION
USE   TRANSVERSE OSCILLATION

TRANSVERSE WAVES
GS   **TRANSVERSE WAVES**
     . H WAVES
RT   ELASTIC WAVES
     ELECTROMAGNETIC RADIATION
     GAMMA RAYS
     LONGITUDINAL WAVES
     MAGNETOHYDRODYNAMIC FLOW
     PLANE WAVES
     RADIO WAVES
     S WAVES
     VIBRATION MODE
     WAVE PACKETS
     ∞ WAVES

TRANSVERSELY EXCITED ATMOSPHERIC LASERS
USE   TEA LASERS

TRAP PROGRAM
GS   PROGRAMS
     . **TRAP PROGRAM**
RT   PLASMA CONTROL
     ∞ RADIATION

TRAPATT DEVICES
UF   TRAPPED PLASMA AVALANCHE
     TRIGGERED TRANSIT
GS   ELECTRONIC EQUIPMENT
     . SOLID STATE DEVICES
     . . SEMICONDUCTOR DEVICES
     . . . **TRAPATT DEVICES**
RT   AVALANCHE DIODES
     ∞ DEVICES
     DIODES
     TRANSISTORS

TRAPATT DIODES
USE   AVALANCHE DIODES

TRAPEZOIDAL TAIL SURFACES
GS   PLANFORMS
     . **TRAPEZOIDAL TAIL SURFACES**
     TAIL SURFACES
     . **TRAPEZOIDAL TAIL SURFACES**
RT   CONTROL SURFACES
     HORIZONTAL TAIL SURFACES
     HYPERSONIC AIRCRAFT
     RUDDERS
     STABILIZERS (FLUID DYNAMICS)
     SUPERSONIC AIRCRAFT
     ∞ SURFACES
     SWEPTBACK TAIL SURFACES

TRAPEZOIDAL WINGS
GS   AIRFOILS
     . WINGS
     . . LOW ASPECT RATIO WINGS
     . . . **TRAPEZOIDAL WINGS**
     . . SWEPT WINGS
     . . . SWEPT FORWARD WINGS
     . . . . **TRAPEZOIDAL WINGS**
     . . . SWEPTBACK WINGS
     . . . . **TRAPEZOIDAL WINGS**
     PLANFORMS
     . WING PLANFORMS
     . . SWEPT FORWARD WINGS
     . . . **TRAPEZOIDAL WINGS**
     . . SWEPTBACK WINGS
     . . . **TRAPEZOIDAL WINGS**

TRAPEZOIDS
GS   GEOMETRY
     . EUCLIDEAN GEOMETRY
     . . POLYGONS
     . . . TETRAGONS
     . . . . **TRAPEZOIDS**

TRAPPED MAGNETIC FIELDS
GS   MAGNETIC FIELDS
     . **TRAPPED MAGNETIC FIELDS**
RT   FLUX PINNING
     MAGNETICALLY TRAPPED PARTICLES
     PLASMA CONTROL
     SUPERCONDUCTIVITY
     TRAPPING

TRAPPED PARTICLES
GS   PARTICLES
     . **TRAPPED PARTICLES**
     . . MAGNETICALLY TRAPPED PARTICLES
     . . . RADIATION BELTS
     . . . . ARTIFICIAL RADIATION BELTS
     . . . . INNER RADIATION BELT
     . . . . OUTER RADIATION BELT

TRAPPED PARTICLES-*(CONT.)*
     . . . . PROTON BELTS
RT   CHARGED PARTICLES
     ELECTRON PRECIPITATION
     PROTON PRECIPITATION
     TRAPPING

TRAPPED PLASMA AVALANCHE TRIGGERED TRANSIT
USE   TRAPATT DEVICES

TRAPPED VORTEXES
UF   VORTEX TRAPS
GS   VORTICES
     . **TRAPPED VORTEXES**
RT   COUNTERFLOW
     FLOW DISTRIBUTION
     MIXING
     ROTATING FLUIDS
     ROTATING LIQUIDS
     TURBULENT MIXING
     TURBULENT WAKES
     VORTEX RINGS
     VORTICITY

TRAPPING
GS   **TRAPPING**
     . CRYOTRAPPING
RT   CONDUCTION BANDS
     CRYSTAL DEFECTS
     FLUX PINNING
     ION STORAGE
     PHOSPHORESCENCE
     RADIATION BELTS
     TRAPPED MAGNETIC FIELDS
     TRAPPED PARTICLES

TRAPS
GS   **TRAPS**
     . COLD TRAPS
     . ION TRAPS (INSTRUMENTATION)
     . VAPOR TRAPS
RT   CONCENTRATORS
     ENTRAPMENT
     SEPARATORS
     VALVES

∞ **TRAVEL**
SN   *(USE OF A MORE SPECIFIC TERM IS RECOMMENDED--CONSULT THE TERMS LISTED BELOW)*
RT   DISTANCE
     HARBORS
     LOGISTICS
     RANGE (EXTREMES)
     TRANSPORTATION

TRAVELING CHARGE
GS   ELECTRIC CHARGE
     . **TRAVELING CHARGE**
RT   ELECTRODYNAMICS
     ENERGY DISSIPATION
     FIELD THEORY (PHYSICS)

TRAVELING IONOSPHERIC DISTURBANCES
UF   TID
GS   IONOSPHERIC DISTURBANCES
     . **TRAVELING IONOSPHERIC DISTURBANCES**
RT   IONOSPHERIC CURRENTS
     IONOSPHERIC PROPAGATION
     IONOSPHERIC STORMS
     IONOSPHERIC TILTS
     MAGNETIC VARIATIONS
     SUDDEN IONOSPHERIC DISTURBANCES

TRAVELING SALESMAN PROBLEM
RT   OPERATIONS RESEARCH
     PROBABILITY THEORY
     ∞ PROBLEMS
     STATISTICAL ANALYSIS

TRAVELING SOLVENT METHOD
SN   (LIMITED TO CRYSTAL GROWTH TECHNIQUES)
GS   GROWTH
     . CRYSTAL GROWTH
     . . **TRAVELING SOLVENT METHOD**
RT   ADDITIVES
     CARRIER INJECTION
     ELECTROEPITAXY
     ∞ METHODOLOGY

TRAVELING WAVE AMPLIFIERS
GS   AMPLIFIERS
     . **TRAVELING WAVE AMPLIFIERS**

TRAVELING WAVE MASERS
GS   STIMULATED EMISSION DEVICES
     . MASERS
     . . **TRAVELING WAVE MASERS**
RT   AMPLIFIERS
     CAVITY RESONATORS
     COHERENT ELECTROMAGNETIC RADIATION

TRAVELING WAVE MODULATION
GS   MODULATION
     . **TRAVELING WAVE MODULATION**
RT   LASERS
     LIGHT MODULATION
     WAVE DIFFRACTION

TRAVELING WAVE TUBES
UF   CRESTATRONS
     HELIX TUBES
GS   ELECTRON TUBES
     . VACUUM TUBES
     . . VACUUM TUBE OSCILLATORS
     . . . MICROWAVE TUBES
     . . . . **TRAVELING WAVE TUBES**
     . . . . . CARCINOTRONS
     . . . . . HELITRONS
     MICROWAVE EQUIPMENT
     . MICROWAVE TUBES
     . . **TRAVELING WAVE TUBES**
     . . . CARCINOTRONS
     . . . HELITRONS
RT   BACKWARD WAVES
     BRILLOUIN FLOW
     CROSSED FIELD AMPLIFIERS
     CYCLOTRON RESONANCE DEVICES
     ELECTRON BUNCHING
     MAGNETOSTATIC AMPLIFIERS
     MAGNETRONS
     OSCILLATIONS
     SCALLOPING

TRAVELING WAVES
GS   **TRAVELING WAVES**
     . SOLITARY WAVES
RT   BACKWARD WAVES
     ELASTIC WAVES
     ELECTROMAGNETIC RADIATION
     NONRESONANCE
     PHASE VELOCITY
     PLANE WAVES
     RADIO WAVES

TRAYS
RT   ∞ BUCKETS
     ∞ CONTAINERS
     ∞ PLATES

TREADMILLS
RT   PHYSICAL EXERCISE
     PHYSICAL FITNESS
     PHYSICAL WORK
     PHYSIOLOGICAL TESTS

TREADS
RT   STAIRWAYS
     TIRES
     VEHICULAR TRACKS

TREAT (TEST FACILITY)
USE   TRANSIENT REACTOR TEST FACILITY

∞ **TREATMENT**
SN   *(USE OF A MORE SPECIFIC TERM IS RECOMMENDED--CONSULT THE TERMS LISTED BELOW)*
UF   CONDITIONING (TREATING)
RT   AIR CONDITIONING
     CLINICAL MEDICINE
     HEAT TREATMENT
     PREWHITENING
     SEWAGE TREATMENT
     THERMOMECHANICAL TREATMENT
     WASTE TREATMENT
     WATER TREATMENT

TREE RING DATING
USE   DENDROCHRONOLOGY

∞ TREES
SN   (USE OF A MORE SPECIFIC TERM IS
        RECOMMENDED--CONSULT THE TERMS
        LISTED BELOW)
RT   CONIFERS
        TREES (MATHEMATICS)
        TREES (PLANTS)

TREES (MATHEMATICS)
GS   TREES (MATHEMATICS)
        . FAULT TREES
RT   ANALYSIS (MATHEMATICS)
        CIRCUITS
        GRAPH THEORY
        GRAPHS (CHARTS)
        PETRI NETS
        SNEAK CIRCUIT ANALYSIS
        TOPOLOGY
        ∞ TREES

TREES (PLANTS)
GS   PLANTS (BOTANY)
        . TREES (PLANTS)
        . . CITRUS TREES
        . . CONIFERS
        . . DECIDUOUS TREES
RT   BALSA
        CANOPIES (VEGETATION)
        CHAPARRAL
        CLEARINGS (OPENINGS)
        DEFOLIANTS
        DEFOLIATION
        DENDROCHRONOLOGY
        FORESTS
        GEOBOTANY
        HERBICIDES
        LOGGING (INDUSTRY)
        MASONITE (TRADEMARK)
        ORCHARDS
        PHREATOPHYTES
        PLYWOOD
        SILVICULTURE
        TIMBER IDENTIFICATION
        TIMBER INVENTORY
        TIMBER VIGOR
        TIMBERLINE
        ∞ TREES
        VEGETATION
        WOOD

TREMORS
RT   EARTHQUAKE RESISTANCE
        EARTHQUAKES
        PARALYSIS
        PARKINSON DISEASE
        ROUSE BELTS

TRENDS
RT   EXTRAPOLATION
        FORECASTING
        GROWTH
        PERIODIC VARIATIONS
        ∞ PROJECTION
        TIME SERIES ANALYSIS

TRESCA FLOW
GS   FLUID FLOW
        . PLASTIC FLOW
        . . TRESCA FLOW
RT   DUCTILITY
        STABILITY
        YIELD POINT

TRIACETIN
GS   ACETATES
        . TRIACETIN
        ESTERS
        . TRIACETIN
RT   ACETIC ACID
        GLYCEROLS
        PLASTICIZERS
        SOLVENTS

TRIAMINOGUANIDINENITRATE
USE   TAGN

TRIAMINOGUANIDINIUM AZIDE
GS   ALIPHATIC COMPOUNDS
        . GUANIDINES
        . . TRIAMINOGUANIDINIUM AZIDE
        AMINES
        . DIAMINES
        . . GUANIDINES
        . . . TRIAMINOGUANIDINIUM AZIDE
        NITROGEN COMPOUNDS

TRIAMINOGUANIDINIUM AZIDE-(CONT.)
        . AZIDES (ORGANIC)
        . . TRIAMINOGUANIDINIUM AZIDE

TRIAMINOTRINITROBENZENE
USE   TATB

TRIANGLES
GS   GEOMETRY
        . EUCLIDEAN GEOMETRY
        . . POLYGONS
        . . . TRIANGLES
RT   TETRAHEDRONS
        TRIGONOMETRY

TRIANGULAR WINGS
USE   DELTA WINGS

TRIANGULATION
RT   ANGLES (GEOMETRY)
        MAPPING
        NAVIGATION
        TRIGONOMETRY
        WILDLIFE RADIOLOCATION

TRIATOMIC MOLECULES
GS   MOLECULES
        . POLYATOMIC MOLECULES
        . . TRIATOMIC MOLECULES
RT   DIATOMIC MOLECULES

TRIAXIAL STRESSES
UF   TRIAXIALITY
GS   STRESSES
        . TRIAXIAL STRESSES
RT   MECHANICAL PROPERTIES
        TENSILE STRESS

TRIAXIALITY
USE   TRIAXIAL STRESSES

TRIBOLIA
GS   ANIMALS
        . INVERTEBRATES
        . . ARTHROPODS
        . . . INSECTS
        . . . . CRICKETS
        . . . . . BEETLES
        . . . . . . TRIBOLIA

TRIBOLOGY
RT   ABRASION
        CORROSION
        EROSION
        EROSIVE BURNING
        FRETTING
        FRICTION
        INTERFACIAL TENSION
        LUBRICATION
        TRIBOLUMINESCENCE
        WEAR

TRIBOLUMINESCENCE
GS   DECAY
        . EMISSION
        . . LIGHT EMISSION
        . . . LUMINESCENCE
        . . . . PHOTOLUMINESCENCE
        . . . . . TRIBOLUMINESCENCE
RT   FLUORESCENCE
        FRICTION
        MECHANICAL PROPERTIES
        PHOTOLUMINESCENT BANDS
        STRESSES
        TRIBOLOGY

TRIBUTARIES
RT   DRAINAGE PATTERNS
        EARTH RESOURCES
        ESTUARIES
        RIVERS

TRICHLORIDES
USE   CHLORIDES

TRIDENT AIRCRAFT
USE   DH 121 AIRCRAFT

TRIDENT SUBMARINE
GS   WATER VEHICLES
        . SHIPS
        . . SUBMARINES
        . . . TRIDENT SUBMARINE
        . UNDERWATER VEHICLES

TRIDENT SUBMARINE-(CONT.)
        . . SUBMARINES
        . . . TRIDENT SUBMARINE
RT   NAVY
        NUCLEAR PROPULSION

TRIENES
GS   ALIPHATIC COMPOUNDS
        . ALKENES
        . . TRIENES
        HYDROCARBONS
        . ALKENES
        . . TRIENES

TRIETHYL COMPOUNDS
GS   ALIPHATIC COMPOUNDS
        . ALKYL COMPOUNDS
        . . TRIETHYL COMPOUNDS
RT   ∞ CHEMICAL COMPOUNDS
        ETHYL COMPOUNDS

TRIFLUOROAMINE OXIDE
GS   ALIPHATIC COMPOUNDS
        . FLUOROAMINES
        . . TRIFLUOROAMINE OXIDE
        AMINES
        . FLUOROAMINES
        . . TRIFLUOROAMINE OXIDE
        HALOGEN COMPOUNDS
        . FLUORINE COMPOUNDS
        . . FLUORO COMPOUNDS
        . . . FLUORINE ORGANIC COMPOUNDS
        . . . . FLUOROAMINES
        . . . . . TRIFLUOROAMINE OXIDE
        ORGANIC COMPOUNDS
        . FLUORINE ORGANIC COMPOUNDS
        . . FLUOROAMINES
        . . . TRIFLUOROAMINE OXIDE

TRIGATRONS
GS   SWITCHES
        . TRIGATRONS
RT   ∞ GAS TUBES
        PULSE MODULATION
        SPARK GAPS
        TRIGGER CIRCUITS

TRIGGER CIRCUITS
GS   CIRCUITS
        . TRIGGER CIRCUITS
RT   BISTABLE CIRCUITS
        GATES (CIRCUITS)
        MULTIVIBRATORS
        THRESHOLD GATES
        THRESHOLD LOGIC
        THYRISTORS
        TRIGATRONS

TRIGGERS
USE   ACTUATORS

TRIGONOMETRIC FUNCTIONS
GS   ANALYSIS (MATHEMATICS)
        . REAL VARIABLES
        . . PERIODIC FUNCTIONS
        . . . TRIGONOMETRIC FUNCTIONS
        . . . . COSINE SERIES
        . . . . SINE SERIES
        . . . . TANGENTS
        FUNCTIONS (MATHEMATICS)
        . TRANSCENDENTAL FUNCTIONS
        . . PERIODIC FUNCTIONS
        . . . TRIGONOMETRIC FUNCTIONS
        . . . . COSINE SERIES
        . . . . SINE SERIES
        . . . . TANGENTS
RT   FRESNEL INTEGRALS
        SINE WAVES
        TRIGONOMETRY

TRIGONOMETRY
GS   GEOMETRY
        . EUCLIDEAN GEOMETRY
        . . ANALYTIC GEOMETRY
        . . . TRIGONOMETRY
RT   ANGLES (GEOMETRY)
        ∞ SCIENCE
        TRIANGLES
        TRIANGULATION
        TRIGONOMETRIC FUNCTIONS

TRIM (BALANCE)
USE   AERODYNAMIC BALANCE

TRIMERS
- GS PREPOLYMERS
- . **TRIMERS**
- RT DIMERS
-     MONOMERS

TRIMETHADIONE
- GS ALIPHATIC COMPOUNDS
- . KETONES
- . . **TRIMETHADIONE**
-     DRUGS
- . **TRIMETHADIONE**
-     HETEROCYCLIC COMPOUNDS
- . **TRIMETHADIONE**

TRIMETHYL COMPOUNDS
- GS ALIPHATIC COMPOUNDS
- . ALKYL COMPOUNDS
- . . **TRIMETHYL COMPOUNDS**
- RT ∞CHEMICAL COMPOUNDS
-     METHYL COMPOUNDS

TRINIDAD AND TOBAGO
- GS LANDFORMS
- . ISLANDS
- . . WEST INDIES
- . . . **TRINIDAD AND TOBAGO**
-     NATIONS
- . **TRINIDAD AND TOBAGO**
- RT CARIBBEAN REGION
-     SOUTH AMERICA

TRINITRAMINE
- GS ALIPHATIC COMPOUNDS
- . **TRINITRAMINE**
-     AMINES
- . **TRINITRAMINE**
-     NITROGEN COMPOUNDS
- . **TRINITRAMINE**

TRINITRO COMPOUNDS
- GS NITROGEN COMPOUNDS
- . NITRO COMPOUNDS
- . . **TRINITRO COMPOUNDS**
- RT ∞CHEMICAL COMPOUNDS

TRINITROTOLUENE
- UF TNT (TRINITROTOLUENE)
- GS EXPLOSIVES
- . **TRINITROTOLUENE**
-     NITROGEN COMPOUNDS
- . NITRO COMPOUNDS
- . . NITROBENZENES
- . . . **TRINITROTOLUENE**
- RT EXPOSURE

TRINITROTRIAZOCYCLOHEXANE
- USE RDX

TRIODES
- RT CATT DEVICES
-     DIODES
-     ELECTRON TUBES
-     MICROWAVE TUBES
-     SEMICONDUCTOR DEVICES
-     TETRODES
-     THYRISTORS
-     TRANSISTORS

TRIOLS
- GS ALIPHATIC COMPOUNDS
- . **TRIOLS**
- . . CYANURIC ACID
-     HYDROXYL COMPOUNDS
- . ALCOHOLS
- . . **TRIOLS**
- . . . CYANURIC ACID

TRIPHENYL SILICON
- GS SILICON COMPOUNDS
- . ORGANIC SILICON COMPOUNDS
- . . **TRIPHENYL SILICON**

TRIPHENYLS
- GS HYDROCARBONS
- . **TRIPHENYLS**

TRIPLE AXIS SPECTROMETERS
- USE NEUTRON SPECTROMETERS

TRIPLET EXCITATION
- USE ATOMIC ENERGY LEVELS

TRIPLET STATE
- USE ATOMIC ENERGY LEVELS

TRIPODS
- GS SUPPORTS
- . **TRIPODS**
- RT OPTICAL EQUIPMENT

TRIPROPELLANTS
- USE LIQUID ROCKET PROPELLANTS

TRISONIC WIND TUNNELS
- GS TEST FACILITIES
- . WIND TUNNELS
- . . **TRISONIC WIND TUNNELS**
- RT SLOTTED WIND TUNNELS
-     TRANSONIC FLOW
-     WIND TUNNEL TESTS

TRITIUM
- UF HYDROGEN 3
- GS CHEMICAL ELEMENTS
- . HYDROGEN
- . . HYDROGEN ISOTOPES
- . . . **TRITIUM**
- . NUCLIDES
- . . ISOTOPES
- . . . HYDROGEN ISOTOPES
- . . . . **TRITIUM**
- . . . RADIOACTIVE ISOTOPES
- . . . . **TRITIUM**
-     GASES
- . HYDROGEN
- . . HYDROGEN ISOTOPES
- . . . **TRITIUM**
- RT HEAVY WATER
-     NUCLEAR FUELS

TRITON
- GS CELESTIAL BODIES
- . NATURAL SATELLITES
- . . **TRITON**
-     SATELLITES
- . NATURAL SATELLITES
- . . **TRITON**
- RT GALILEAN SATELLITES
-     NEPTUNE (PLANET)
-     NEPTUNE ATMOSPHERE
-     SATELLITE ATMOSPHERES
-     TITAN

TRITONS
- GS IONS
- . **TRITONS**
- RT ALPHA PARTICLES
-     PROTONS

TRIVALENT IONS
- GS IONS
- . **TRIVALENT IONS**
- RT FREE RADICALS
-     POSITIVE IONS
-     VALENCE

TROCHOIDS
- USE PIVOTS

TROILITE
- GS CHALCOGENIDES
- . SULFIDES
- . . PYRRHOTITE
- . . . **TROILITE**
-     IRON COMPOUNDS
- . PYRRHOTITE
- . . **TROILITE**
-     MINERALS
- . PYRRHOTITE
- . . **TROILITE**
-     SULFUR COMPOUNDS
- . SULFIDES
- . . PYRRHOTITE
- . . . **TROILITE**
- RT IRON METEORITES
-     METEORITIC COMPOSITION

TROJAN AIRCRAFT
- USE T-28 AIRCRAFT

TROJAN ORBITS
- GS ORBITS
- . SPACECRAFT ORBITS
- . . **TROJAN ORBITS**
- RT CELESTIAL MECHANICS
-     MANY BODY PROBLEM

TROJAN ORBITS-(CONT.)
-     THREE BODY PROBLEM

TROMBE WALLS
- GS WALLS
- . **TROMBE WALLS**
- RT ENERGY TECHNOLOGY
-     HEAT STORAGE
-     PHASE CHANGE MATERIALS
-     RADIATIVE HEAT TRANSFER
-     SOLAR ENERGY ABSORBERS
-     SOLAR HEATING
-     SOLAR HOUSES
-     THERMAL INSULATION

TROPICAL METEOROLOGY
- GS METEOROLOGY
- . **TROPICAL METEOROLOGY**
- RT AGROMETEOROLOGY
-     EL NINO
-     EQUATORIAL ATMOSPHERE
-     GARP ATLANTIC TROPICAL EXPERIMENT
-     INTERTROPICAL CONVERGENT ZONES
-     METEOROLOGICAL PARAMETERS

TROPICAL REGIONS
- UF JUNGLES
-     LOW LATITUDES
-     SUBTROPICAL REGIONS
-     TROPICS
- GS REGIONS
- . **TROPICAL REGIONS**
- . . AMAZON REGION (SOUTH AMERICA)
- RT CLIMATOLOGY
-     EQUATORIAL ATMOSPHERE
-     EQUATORIAL REGIONS
-     GARP ATLANTIC TROPICAL EXPERIMENT
-     GEOGRAPHY
-     HOT WEATHER
-     INTERTROPICAL CONVERGENT ZONES
-     LATERITES
-     LOMONOSOV CURRENT
-     METEOROLOGY
-     RAIN FORESTS
-     TEMPERATE REGIONS
-     VIRGIN ISLANDS

TROPICAL STORMS
- GS STORMS
- . STORMS (METEOROLOGY)
- . . **TROPICAL STORMS**
- . . . HURRICANES
- . . . . ANNA HURRICANE
- . . . TYPHOONS
- RT ATMOSPHERIC CIRCULATION
-     CYCLONES
-     METEOROLOGY
-     STORM DAMAGE
-     TORNADOES

TROPICS
- USE TROPICAL REGIONS

TROPISM
- GS **TROPISM**
- . AEOLOTROPISM
- . GEOTROPISM
- . GRAVITROPISM
- . GYROTROPISM
- . NEUROTROPISM

TROPOPAUSE
- SN (ALTITUDE APPROXIMATELY 15 TO 20 KM)
- GS EARTH ATMOSPHERE
- . **TROPOPAUSE**
- RT DIURNAL VARIATIONS
-     ISOTHERMAL LAYERS
-     LOWER ATMOSPHERE
-     MIDDLE ATMOSPHERE
-     TROPOSPHERE

TROPOSPHERE
- GS EARTH ATMOSPHERE
- . LOWER ATMOSPHERE
- . . **TROPOSPHERE**
- RT CHEMOSPHERE
-     HOMOSPHERE
-     INTASAT SATELLITE
-     TROPOPAUSE

TROPOSPHERIC RADIATION
- SN (EXCLUDES TERRESTRIAL RADIATION)
- GS ATMOSPHERIC RADIATION

TROPOSPHERIC RADIATION-(CONT.)
. **TROPOSPHERIC RADIATION**
ELECTROMAGNETIC RADIATION
. **TROPOSPHERIC RADIATION**
RT ∞RADIATION
SKY RADIATION
STRATOSPHERE RADIATION
TERRESTRIAL RADIATION

TROPOSPHERIC SCATTERING
GS SCATTERING
. WAVE SCATTERING
. . ATMOSPHERIC SCATTERING
. . . **TROPOSPHERIC SCATTERING**
RT LIGHT SCATTERING
WIDEBAND COMMUNICATION

TROPOSPHERIC WAVES
SN (EXCLUDES RADIO WAVES)
GS **TROPOSPHERIC WAVES**
. PLANETARY WAVES
RT ELASTIC WAVES
LEE WAVES
RADIO WAVES
SURFACE WAVES
∞WAVES

TROPYL COMPOUNDS
GS HETEROCYCLIC COMPOUNDS
. ALKALOIDS
. . **TROPYL COMPOUNDS**
NITROGEN COMPOUNDS
. ALKALOIDS
. . **TROPYL COMPOUNDS**
RT ∞CHEMICAL COMPOUNDS

TROUBLESHOOTING
USE MAINTENANCE

TROUGHS
RT CANALS
DITCHES
IRRIGATION
LOW PRESSURE

TRUCKS
SN (EXCLUDES UNDERCARRIAGES)
UF VANS
GS SURFACE VEHICLES
. MOTOR VEHICLES
. . **TRUCKS**
. . . TANK TRUCKS
RT ANTISKID DEVICES
AUTOMOBILES
CARGO
DELIVERY
DOLLIES
ELECTRIC MOTOR VEHICLES
GROUND HANDLING
HAULING
MATERIALS HANDLING
∞MILITARY VEHICLES
RECOVERY VEHICLES
TRACTORS
TRAILERS
TRANSPORTATION
TRANSPORTATION ENERGY

TRUNCATION (MATHEMATICS)
USE APPROXIMATION

TRUNCATION ERRORS
GS ERRORS
. **TRUNCATION ERRORS**
RT PRECISION

TRUNKS (LINES)
USE TRANSMISSION LINES

TRUNNIONS
USE SHAFTS (MACHINE ELEMENTS)

TRUSSES
GS STRUCTURAL MEMBERS
. **TRUSSES**
RT ARCHES
BEAMS (SUPPORTS)
CONSTRUCTION INDUSTRY
FRAMES
GIRDERS
I BEAMS
LOOPS
MAXWELL-MOHR METHOD
MEGAMECHANICS

TRUSSES-(CONT.)
∞STRUCTURES
STRUTS
SUPPORTS
TIMOSHENKO BEAMS

TRYPANOSOME
GS ANIMALS
. INVERTEBRATES
. . PROTOZOA
. . . FLAGELLATA
. . . . **TRYPANOSOME**
MICROORGANISMS
. PROTOZOA
. . FLAGELLATA
. . . **TRYPANOSOME**
RT PARASITIC DISEASES

TRYPSIN
GS ENZYMES
. **TRYPSIN**
RT PANCREAS

TRYPTAMINES
GS AMINES
. **TRYPTAMINES**
. . SEROTONIN

TRYPTOPHAN
GS ACIDS
. AMINO ACIDS
. . **TRYPTOPHAN**
HETEROCYCLIC COMPOUNDS
. AZOLES
. . PYRROLES
. . . INDOLES
. . . . **TRYPTOPHAN**
NITROGEN COMPOUNDS
. **TRYPTOPHAN**
ORGANIC COMPOUNDS
. AMINO ACIDS
. . **TRYPTOPHAN**

TS-11 AIRCRAFT
UF ISKRA AIRCRAFT
POLISH TS-11 AIRCRAFT
GS JET AIRCRAFT
. **TS-11 AIRCRAFT**
MONOPLANES
. **TS-11 AIRCRAFT**
TRAINING AIRCRAFT
. **TS-11 AIRCRAFT**
RT ∞AIRCRAFT

TSR-2 AIRCRAFT
UF BAC TSR 2 AIRCRAFT
GS ATTACK AIRCRAFT
. **TSR-2 AIRCRAFT**
BAC AIRCRAFT
. **TSR-2 AIRCRAFT**
JET AIRCRAFT
. **TSR-2 AIRCRAFT**
MONOPLANES
. **TSR-2 AIRCRAFT**
OBSERVATION AIRCRAFT
. **TSR-2 AIRCRAFT**
RECONNAISSANCE AIRCRAFT
. **TSR-2 AIRCRAFT**
SUPERSONIC AIRCRAFT
. **TSR-2 AIRCRAFT**
TERRAIN FOLLOWING AIRCRAFT
. **TSR-2 AIRCRAFT**
RT ∞AIRCRAFT

TSUNAMI WAVES
RT EARTH MOVEMENTS
EARTHQUAKE DAMAGE
EARTHQUAKES
FRONTAL WAVES
SEISMIC WAVES
SHOCK WAVES
SURFACE WAVES
TIDAL WAVES
WATER WAVES

TTL INTEGRATED CIRCUITS
SN (TRANSISTOR-TRANSISTOR-LOGIC
INTEGRATED CIRCUITS)
UF TRANSISTOR-TRANSISTOR-LOGIC INTEG
CIRCUITS
GS CIRCUITS
. INTEGRATED CIRCUITS
. . **TTL INTEGRATED CIRCUITS**
RT ELECTRONIC PACKAGING
LARGE SCALE INTEGRATION

TTL INTEGRATED CIRCUITS-(CONT.)
MICROMINIATURIZATION
MOLECULAR ELECTRONICS
TRANSISTOR CIRCUITS

TU-104 AIRCRAFT
UF CAMEL AIRCRAFT
GS COMMERCIAL AIRCRAFT
. **TU-104 AIRCRAFT**
JET AIRCRAFT
. **TU-104 AIRCRAFT**
MONOPLANES
. **TU-104 AIRCRAFT**
PASSENGER AIRCRAFT
. **TU-104 AIRCRAFT**
TUPOLEV AIRCRAFT
. **TU-104 AIRCRAFT**
RT ∞AIRCRAFT
TU-154 AIRCRAFT

TU-121 ENGINE
GS ENGINES
. ROCKET ENGINES
. . SOLID PROPELLANT ROCKET
ENGINES
. . . **TU-121 ENGINE**

TU-124 AIRCRAFT
UF COOKPOT AIRCRAFT
GS COMMERCIAL AIRCRAFT
. **TU-124 AIRCRAFT**
JET AIRCRAFT
. **TU-124 AIRCRAFT**
MONOPLANES
. **TU-124 AIRCRAFT**
PASSENGER AIRCRAFT
. **TU-124 AIRCRAFT**
TRANSPORT AIRCRAFT
. **TU-124 AIRCRAFT**
TUPOLEV AIRCRAFT
. **TU-124 AIRCRAFT**
RT ∞AIRCRAFT
TURBOFAN ENGINES

TU-134 AIRCRAFT
GS COMMERCIAL AIRCRAFT
. **TU-134 AIRCRAFT**
JET AIRCRAFT
. TURBOFAN AIRCRAFT
. . **TU-134 AIRCRAFT**
MONOPLANES
. **TU-134 AIRCRAFT**
PASSENGER AIRCRAFT
. **TU-134 AIRCRAFT**
TUPOLEV AIRCRAFT
. **TU-134 AIRCRAFT**
RT ∞AIRCRAFT

TU-144 AIRCRAFT
GS COMMERCIAL AIRCRAFT
. SUPERSONIC COMMERCIAL AIR
TRANSPORT
. . **TU-144 AIRCRAFT**
JET AIRCRAFT
. TURBOFAN AIRCRAFT
. . **TU-144 AIRCRAFT**
PASSENGER AIRCRAFT
. **TU-144 AIRCRAFT**
TRANSPORT AIRCRAFT
. **TU-144 AIRCRAFT**
RT ∞AIRCRAFT

TU-154 AIRCRAFT
GS COMMERCIAL AIRCRAFT
. **TU-154 AIRCRAFT**
TRANSPORT AIRCRAFT
. **TU-154 AIRCRAFT**
TUPOLEV AIRCRAFT
. **TU-154 AIRCRAFT**
RT ∞AIRCRAFT
AIRCRAFT DESIGN
CARGO AIRCRAFT
JET AIRCRAFT
PASSENGER AIRCRAFT
TU-104 AIRCRAFT

TUBE ANODES
GS ELECTRODES
. ANODES
. . **TUBE ANODES**
RT CATHODES
ELECTRODE MATERIALS
ELECTRON GUNS

TUBE CATHODES

TUBE CATHODES
- GS ELECTRODES
  - . CATHODES
    - . . **TUBE CATHODES**
      - . . . COLD CATHODES
      - . . . HOT CATHODES
      - . . . PHOTOCATHODES
      - . . . THERMIONIC CATHODES
      - . . . TUNNEL CATHODES
- RT COLD CATHODE TUBES
  - ELECTRON GUNS
  - HOLLOW CATHODES

TUBE GRIDS
- GS ELECTRODES
  - . **TUBE GRIDS**
- RT BIAS
  - ELECTRON GUNS
  - ELECTRON TUBES
  - ∞ GRIDS
  - IONIZERS

TUBE HEAT EXCHANGERS
- GS HEAT EXCHANGERS
  - . **TUBE HEAT EXCHANGERS**
- RT REGENERATORS

TUBE LASERS
- RT CHEMICAL LASERS
  - GASDYNAMIC LASERS
  - LASER OUTPUTS
  - PULSED LASERS
  - SHOCK TUBES
  - WAVEGUIDE LASERS

TUBERCULOSIS
- GS DISEASES
  - . INFECTIOUS DISEASES
    - . . **TUBERCULOSIS**
  - . RESPIRATORY DISEASES
    - . . **TUBERCULOSIS**
- RT PATCH TESTS

∞ TUBES
- SN *(USE OF A MORE SPECIFIC TERM IS RECOMMENDED--CONSULT THE TERMS LISTED BELOW)*
- RT BOURDON TUBES
  - BRONCHIAL TUBE
  - BURETTES
  - CANNULAE
  - CAPILLARY TUBES
  - CIRCULAR TUBES
  - DUCTS
  - ELECTRON TUBES
  - EUSTACHIAN TUBES
  - HILSCH TUBES
  - HOSES
  - LININGS
  - MANIFOLDS
  - MICROWAVE TUBES
  - PIPES (TUBES)
  - PITOT TUBES
  - SHOCK TUBES
  - SIPHONS
  - TRACHEA
  - VENTURI TUBES

TUBING
- USE PIPES (TUBES)

TUMBLING MOTION
- RT ATTITUDE STABILITY
  - DESTABILIZATION
  - MIXERS
  - ∞ MOTION
  - ROTATING ENVIRONMENTS
  - SATELLITE ROTATION
  - ∞ SEPARATION
  - SPACECRAFT MOTION
  - SPACECRAFT STABILITY

TUMORS
- GS DISEASES
  - . **TUMORS**
    - . . NEOPLASMS
      - . . . CANCER
        - . . . . LEUKEMIAS
- RT CYSTS

TUNABLE LASERS
- GS STIMULATED EMISSION DEVICES
  - . LASERS
    - . . **TUNABLE LASERS**

TUNABLE LASERS-*(CONT.)*
- RT LIGHT MODULATION
  - OPTICAL COMMUNICATION
  - TUNING
  - WIGGLER MAGNETS

TUNDRA
- GS LAND
  - . PLAINS
    - . . **TUNDRA**
  - LANDFORMS
  - . **TUNDRA**
- RT ARCTIC REGIONS
  - ASIA
  - GEOGRAPHY
  - NORTH AMERICA

TUNERS
- GS **TUNERS**
  - . WAVEGUIDE TUNERS
- RT RADIO RECEIVERS
  - RESONANCE PROBES
  - RESONANT FREQUENCIES
  - TELEVISION RECEIVERS

TUNGSTATES
- GS TUNGSTEN COMPOUNDS
  - . **TUNGSTATES**
    - . . LEAD TUNGSTATES
    - . . ZINC TUNGSTATES

TUNGSTEN
- UF WOLFRAM
- GS CHEMICAL ELEMENTS
  - . REFRACTORY METALS
    - . . **TUNGSTEN**
  - METALS
  - . TRANSITION METALS
    - . . REFRACTORY METALS
      - . . . **TUNGSTEN**
  - REFRACTORY MATERIALS
  - . REFRACTORY METALS
    - . . **TUNGSTEN**

TUNGSTEN ALLOYS
- GS ALLOYS
  - . HEAT RESISTANT ALLOYS
    - . . REFRACTORY METAL ALLOYS
      - . . . **TUNGSTEN ALLOYS**
  - REFRACTORY MATERIALS
  - . REFRACTORY METAL ALLOYS
    - . . **TUNGSTEN ALLOYS**
- RT STELLITE (TRADEMARK)

TUNGSTEN CARBIDES
- GS CARBON COMPOUNDS
  - . CARBIDES
    - . . **TUNGSTEN CARBIDES**
  - TUNGSTEN COMPOUNDS
  - . **TUNGSTEN CARBIDES**

TUNGSTEN CHLORIDES
- GS HALOGEN COMPOUNDS
  - . CHLORINE COMPOUNDS
    - . . CHLORIDES
      - . . . **TUNGSTEN CHLORIDES**
  - . HALIDES
    - . . CHLORIDES
      - . . . **TUNGSTEN CHLORIDES**
    - . . METAL HALIDES
      - . . . TUNGSTEN HALIDES
        - . . . . **TUNGSTEN CHLORIDES**
  - TUNGSTEN COMPOUNDS
  - . TUNGSTEN HALIDES
    - . . **TUNGSTEN CHLORIDES**

TUNGSTEN COMPOUNDS
- GS **TUNGSTEN COMPOUNDS**
  - . CALCIUM TUNGSTATES
  - . TUNGSTATES
    - . . LEAD TUNGSTATES
    - . . ZINC TUNGSTATES
  - . TUNGSTEN CARBIDES
  - . TUNGSTEN HALIDES
    - . . TUNGSTEN CHLORIDES
    - . . TUNGSTEN FLUORIDES
  - . TUNGSTEN OXIDES
    - . . SCHEELITE
- RT ∞ CHEMICAL COMPOUNDS
  - ∞ GROUP 6B COMPOUNDS
  - ∞ METAL COMPOUNDS

TUNGSTEN FLUORIDES
- GS HALOGEN COMPOUNDS

TUNGSTEN FLUORIDES-*(CONT.)*
- . FLUORINE COMPOUNDS
  - . . FLUORIDES
    - . . . METAL FLUORIDES
      - . . . . **TUNGSTEN FLUORIDES**
- . HALIDES
  - . METAL HALIDES
    - . . TUNGSTEN HALIDES
      - . . . . **TUNGSTEN FLUORIDES**
- TUNGSTEN COMPOUNDS
- . TUNGSTEN HALIDES
  - . . **TUNGSTEN FLUORIDES**

TUNGSTEN HALIDES
- GS HALOGEN COMPOUNDS
  - . HALIDES
    - . . METAL HALIDES
      - . . . **TUNGSTEN HALIDES**
        - . . . . TUNGSTEN CHLORIDES
        - . . . . TUNGSTEN FLUORIDES
  - TUNGSTEN COMPOUNDS
  - . **TUNGSTEN HALIDES**
    - . . TUNGSTEN CHLORIDES
    - . . TUNGSTEN FLUORIDES

TUNGSTEN INERT GAS WELDING
- USE GAS TUNGSTEN ARC WELDING

TUNGSTEN ISOTOPES
- GS CHEMICAL ELEMENTS
  - . REFRACTORY METALS
    - . . **TUNGSTEN ISOTOPES**
  - METALS
  - . TRANSITION METALS
    - . . REFRACTORY METALS
      - . . . **TUNGSTEN ISOTOPES**
  - REFRACTORY MATERIALS
  - . REFRACTORY METALS
    - . . **TUNGSTEN ISOTOPES**

TUNGSTEN OXIDES
- GS CHALCOGENIDES
  - . OXIDES
    - . . METAL OXIDES
      - . . . **TUNGSTEN OXIDES**
        - . . . . SCHEELITE
  - TUNGSTEN COMPOUNDS
  - . **TUNGSTEN OXIDES**
    - . . SCHEELITE

TUNGUSK METEORITE
- GS CELESTIAL BODIES
  - . METEORITES
    - . . STONY METEORITES
      - . . . **TUNGUSK METEORITE**
- RT METEORITE CRATERS

TUNING
- GS **TUNING**
  - . SCHULER TUNING
- RT AUTOMATIC FREQUENCY CONTROL
  - AUTOMATIC GAIN CONTROL
  - DYE LASERS
  - Q FACTORS
  - RESONANCE
  - RESONANT FREQUENCIES
  - RESONATORS
  - TUNABLE LASERS

TUNING FORK GYROSCOPES
- GS GYROSCOPES
  - . **TUNING FORK GYROSCOPES**
- RT RESONATORS

TUNISIA
- GS NATIONS
  - . **TUNISIA**
- RT AFRICA

TUNNEL CATHODES
- GS ELECTRODES
  - . CATHODES
    - . . TUBE CATHODES
      - . . . **TUNNEL CATHODES**
- RT COLD CATHODE TUBES
  - COLD CATHODES
  - ELECTRON TUBES
  - HOLLOW CATHODES
  - ∞ TUNNELS

TUNNEL DIODES
- UF ESAKI DIODES
- GS ELECTRONIC EQUIPMENT
  - . DIODES

TUNNEL DIODES-(CONT.)
```
  . . SEMICONDUCTOR DIODES
  . . . TUNNEL DIODES
RT    ELECTRON TUNNELING
      JUNCTION DIODES
      MIM DIODES
      NEGATIVE CONDUCTANCE
      NEGATIVE RESISTANCE CIRCUITS
   ∞ TUNNELS
```

TUNNEL RESISTORS
```
USE   ELECTRON TUNNELING
      RESISTORS
```

∞ TUNNELING
```
SN    (USE OF A MORE SPECIFIC TERM IS
      RECOMMENDED--CONSULT THE TERMS
      LISTED BELOW)
RT    ELECTRON TUNNELING
      TUNNELING (EXCAVATION)
```

TUNNELING (EXCAVATION)
```
GS    EXCAVATION
      . TUNNELING (EXCAVATION)
RT    BEDROCK
      CONSTRUCTION
      DRAINAGE
      DRILLING
      JACKS (LIFTS)
      LINING PROCESSES
      ROCKS
      SOILS
   ∞ TUNNELING
      UNDERGROUND STRUCTURES
```

∞ TUNNELS
```
SN    (USE OF A MORE SPECIFIC TERM IS
      RECOMMENDED--CONSULT THE TERMS
      LISTED BELOW)
RT    GAPS
      HYDRAULIC TEST TUNNELS
      LUNAR SHELTERS
      PASSAGEWAYS
      STREETS
      TRANSFER TUNNELS
      TUNNEL CATHODES
      TUNNEL DIODES
      WIND TUNNELS
```

TUPOLEV AIRCRAFT
```
GS    TUPOLEV AIRCRAFT
      . TU-104 AIRCRAFT
      . TU-124 AIRCRAFT
      . TU-134 AIRCRAFT
      . TU-154 AIRCRAFT
RT    ∞ AIRCRAFT
```

TURBIDITY
```
GS    ELECTROMAGNETIC PROPERTIES
      . OPTICAL PROPERTIES
      . . TURBIDITY
RT    ABSORPTANCE
      CLARITY
      HAZE
      LIGHT TRANSMISSION
      OPACITY
      OPTICAL DENSITY
   ∞ PROPERTIES
      SOLUBILITY
      TRANSPARENCE
```

TURBINE BLADES
```
GS    TURBOMACHINE BLADES
      . TURBINE BLADES
RT    ∞ BLADES
      COMPRESSOR BLADES
      ENGINE PARTS
      FAN BLADES
      ROTOR BLADES (TURBOMACHINERY)
      STATOR BLADES
      TURBINES
```

TURBINE ENGINES
```
GS    ENGINES
      . TURBINE ENGINES
      . . GAS TURBINE ENGINES
      . . . JET ENGINES
      . . . . RAMJET ENGINES
      . . . . . LOW VOLUME RAMJET ENGINES
      . . . . . PULSEJET ENGINES
      . . . . . SUPERSONIC COMBUSTION
                RAMJET ENGINES
      . . . . . TURBORAMJET ENGINES
      . . . . T-63 ENGINE
      . . . . T-76 ENGINE
```

TURBINE ENGINES-(CONT.)
```
      . . . . TURBOJET ENGINES
      . . . . . BRISTOL-SIDDELEY OLYMPUS
                593 ENGINE
      . . . . . BRISTOL-SIDDELEY VIPER
                ENGINE
      . . . . DUCTED FAN ENGINES
      . . . . J-33 ENGINE
      . . . . J-34 ENGINE
      . . . . J-47 ENGINE
      . . . . J-52 ENGINE
      . . . . J-57 ENGINE
      . . . . J-57-P-20 ENGINE
      . . . . J-65 ENGINE
      . . . . J-69-T-25 ENGINE
      . . . . J-71 ENGINE
      . . . . J-73 ENGINE
      . . . . J-75 ENGINE
      . . . . J-79 ENGINE
      . . . . J-85 ENGINE
      . . . . J-93 ENGINE
      . . . . RA-28 ENGINE
      . . . . TURBOFAN ENGINES
      . . . . . BRISTOL-SIDDELEY BS 53
                ENGINE
      . . . . . CF-700 ENGINE
      . . . . . J-97 ENGINE
      . . . . . TF-30 ENGINE
      . . . . . TF-41 ENGINE
      . . . . TURBOPROP ENGINES
      . . . . . T-34 ENGINE
      . . . . . T-38 ENGINE
      . . . . . T-53 ENGINE
      . . . . . T-56 ENGINE
      . . . . . T-64 ENGINE
      . . . . . T-74 ENGINE
      . . . . . T-78 ENGINE
      . . . . TURBORAMJET ENGINES
      . . . T-58 ENGINE
      . . . T-58-GE-8B ENGINE
RT    AIRCRAFT ENGINES
      AUTOMOBILE ENGINES
      CONVERGENT NOZZLES
      GAS BEARINGS
      INTEGRAL ROCKET RAMJETS
      TORPEDO ENGINES
```

TURBINE EXHAUST NOZZLES
```
GS    EXHAUST NOZZLES
      . TURBINE EXHAUST NOZZLES
RT    CONICAL NOZZLES
      CONVERGENT-DIVERGENT NOZZLES
```

TURBINE INSTRUMENTS
```
RT    FLOWMETERS
   ∞ INSTRUMENTS
      TURBOMACHINERY
```

TURBINE PUMPS
```
UF    TURBOPUMPS
GS    PUMPS
      . AXIAL FLOW PUMPS
      . . TURBINE PUMPS
      TURBOMACHINERY
      . TURBINE PUMPS
RT    CENTRIFUGAL PUMPS
      FUEL PUMPS
      JET PUMPS
      PREBURNERS
      TURBINES
      TURBOCOMPRESSORS
```

TURBINE WHEELS
```
UF    ROTOR DISKS
      TURBOROTORS
GS    ROTATING BODIES
      . ROTORS
      . . TURBINE WHEELS
      WHEELS
      . TURBINE WHEELS
RT    COMPRESSOR ROTORS
      ENGINE PARTS
      HYDRAULIC EQUIPMENT
      IMPELLERS
      TURBINES
      TURBOMACHINE BLADES
      WATER WHEELS
```

TURBINES
```
GS    TURBOMACHINERY
      . TURBINES
      . . AXIAL FLOW TURBINES
      . . GAS TURBINES
      . . SHROUDED TURBINES
      . . STEAM TURBINES
```

TURBINES-(CONT.)
```
      . . SUPERSONIC TURBINES
      . . TWO STAGE TURBINES
      . . WIND TURBINES
      . . . TIP VANES
RT    ENGINES
      GEOTHERMAL ENERGY CONVERSION
      GEOTHERMAL ENERGY EXTRACTION
      IMPELLERS
      IMPULSE GENERATORS
      JET ENGINE FUELS
      JET PROPULSION
   ∞ NOZZLES
      REFRACTORIES
      ROTATING GENERATORS
      ROTORS
      STATORS
      TURBINE BLADES
      TURBINE PUMPS
      TURBINE WHEELS
      TURBOGENERATORS
      TURBOSHAFTS
```

TURBO-SKYVAN AIRCRAFT
```
USE   SC-7 AIRCRAFT
```

TURBOCHARGERS
```
USE   SUPERCHARGERS
      TURBOCOMPRESSORS
```

TURBOCOMPRESSORS
```
UF    AXIAL COMPRESSORS
      AXIAL FLOW COMPRESSORS
      MULTISTAGE COMPRESSORS
      TURBOCHARGERS
GS    COMPRESSORS
      . TURBOCOMPRESSORS
      TURBOMACHINERY
      . TURBOCOMPRESSORS
RT    CENTRIFUGAL COMPRESSORS
      CENTRIFUGAL PUMPS
      COMPRESSOR BLADES
      COMPRESSOR ROTORS
      ROTATING STALLS
      ROTORS
      SUPERCHARGERS
      SUPERSONIC COMPRESSORS
      TRANSONIC COMPRESSORS
      TURBINE PUMPS
      TURBOFANS
```

TURBOCONVERTERS
```
USE   TURBOGENERATORS
```

TURBOELECTRIC CONVERSION
```
USE   TURBOGENERATORS
```

TURBOFAN AIRCRAFT
```
GS    JET AIRCRAFT
      . TURBOFAN AIRCRAFT
      . . A-7 AIRCRAFT
      . . BAC 111 AIRCRAFT
      . . BOEING 707 AIRCRAFT
      . . BOEING 720 AIRCRAFT
      . . BOEING 727 AIRCRAFT
      . . BOEING 733 AIRCRAFT
      . . BOEING 737 AIRCRAFT
      . . BOEING 757 AIRCRAFT
      . . BOEING 767 AIRCRAFT
      . . C-141 AIRCRAFT
      . . CONCORDE AIRCRAFT
      . . CV-990 AIRCRAFT
      . . DC 8 AIRCRAFT
      . . DH 121 AIRCRAFT
      . . DO-31 AIRCRAFT
      . . F-28 TRANSPORT AIRCRAFT
      . . F-111 AIRCRAFT
      . . IL-62 AIRCRAFT
      . . MYSTERE 20 AIRCRAFT
      . . P-1127 AIRCRAFT
      . . P-1154 AIRCRAFT
      . . SAAB 37 AIRCRAFT
      . . SAAB 105 AIRCRAFT
      . . SE-210 AIRCRAFT
      . . TU-134 AIRCRAFT
      . . TU-144 AIRCRAFT
RT    ∞ AIRCRAFT
      C-135 AIRCRAFT
   ∞ LOW WING AIRCRAFT
      MYSTERE 50 AIRCRAFT
      PASSENGER AIRCRAFT
      TRANSPORT AIRCRAFT
      TURBOPROP AIRCRAFT
```

TURBOFAN ENGINES
GS ENGINES
  . AIR BREATHING ENGINES
  . . GAS TURBINE ENGINES
  . . . JET ENGINES
  . . . . TURBOJET ENGINES
  . . . . . **TURBOFAN ENGINES**
  . . . . . . BRISTOL-SIDDELEY BS 53
      ENGINE
  . . . . . . CF-700 ENGINE
  . . . . . . J-97 ENGINE
  . . . . . . TF-41 ENGINE
  . INTERNAL COMBUSTION ENGINES
  . . GAS TURBINE ENGINES
  . . . JET ENGINES
  . . . . TURBOJET ENGINES
  . . . . . **TURBOFAN ENGINES**
  . . . . . . BRISTOL-SIDDELEY BS 53
      ENGINE
  . . . . . . CF-700 ENGINE
  . . . . . . J-97 ENGINE
  . . . . . . TF-30 ENGINE
  . . . . . . TF-41 ENGINE
  . TURBINE ENGINES
  . . GAS TURBINE ENGINES
  . . . JET ENGINES
  . . . . TURBOJET ENGINES
  . . . . . **TURBOFAN ENGINES**
  . . . . . . BRISTOL-SIDDELEY BS 53
      ENGINE
  . . . . . . CF-700 ENGINE
  . . . . . . J-97 ENGINE
  . . . . . . TF-30 ENGINE
  . . . . . . TF-41 ENGINE
RT B-52 AIRCRAFT
  BOEING 747 AIRCRAFT
  BOEING 767 AIRCRAFT
  C-5 AIRCRAFT
  C-141 AIRCRAFT
  DC 10 AIRCRAFT
  DUCTED FAN ENGINES
  L-1011 AIRCRAFT
  P-1127 AIRCRAFT
  P-1154 AIRCRAFT
  TU-124 AIRCRAFT
  TURBOFANS
  TURBOPROP ENGINES

TURBOFANS
GS TURBOMACHINERY
  . **TURBOFANS**
RT DUCTED FANS
 ∞FANS
  LIFT FANS
  TURBOCOMPRESSORS
  TURBOFAN ENGINES

TURBOGENERATORS
UF THERMAL POWER
  TURBOCONVERTERS
  TURBOELECTRIC CONVERSION
GS ELECTRIC GENERATORS
  . ROTATING GENERATORS
  . . **TURBOGENERATORS**
  . . . ASTEC SOLAR TURBOELECTRIC
    GENERATOR
  TURBOMACHINERY
  . **TURBOGENERATORS**
  . . ASTEC SOLAR TURBOELECTRIC
    GENERATOR
RT AC GENERATORS
 ∞CONVERSION
 ∞ELECTRIC POWER
  ELECTRICAL ENGINEERING
  GAS TURBINE ENGINES
  GAS TURBINES
 ∞GENERATORS
  GEOTHERMAL ENERGY CONVERSION
  GEOTHERMAL ENERGY EXTRACTION
  GEOTHERMAL ENERGY UTILIZATION
  HYDROELECTRIC POWER STATIONS
  HYDROELECTRICITY
  SNAP
  SNAP 1
  SNAP 2
  SNAP 8
  SOLAR GENERATORS
  SPACE POWER REACTORS
  SPACE POWER UNIT REACTORS
  STEAM TURBINES
  SUNFLOWER POWER SYSTEM
  TURBINES
  WIND TURBINES

TURBOJET AIRCRAFT
USE JET AIRCRAFT

TURBOJET ENGINE CONTROL
GS ENGINE CONTROL
  . **TURBOJET ENGINE CONTROL**
RT AIRCRAFT CONTROL
  AUTOMATIC CONTROL
 ∞CONTROL
  FLIGHT CONTROL
  FUEL CONTROL
  REMOTE CONTROL
  SERVOCONTROL
  THRUST CONTROL

TURBOJET ENGINES
GS ENGINES
  . AIR BREATHING ENGINES
  . . GAS TURBINE ENGINES
  . . . JET ENGINES
  . . . . **TURBOJET ENGINES**
  . . . . . BRISTOL-SIDDELEY OLYMPUS
     593 ENGINE
  . . . . . BRISTOL-SIDDELEY VIPER
     ENGINE
  . . . . . DUCTED FAN ENGINES
  . . . . . J-33 ENGINE
  . . . . . J-34 ENGINE
  . . . . . J-47 ENGINE
  . . . . . J-57 ENGINE
  . . . . . J-57-P-20 ENGINE
  . . . . . J-65 ENGINE
  . . . . . J-69-T-25 ENGINE
  . . . . . J-71 ENGINE
  . . . . . J-73 ENGINE
  . . . . . J-75 ENGINE
  . . . . . J-79 ENGINE
  . . . . . J-85 ENGINE
  . . . . . J-93 ENGINE
  . . . . . RA-28 ENGINE
  . . . . . TURBOFAN ENGINES
  . . . . . . BRISTOL-SIDDELEY BS 53
      ENGINE
  . . . . . . CF-700 ENGINE
  . . . . . . J-97 ENGINE
  . . . . . . TF-41 ENGINE
  . . . . . TURBOPROP ENGINES
  . . . . . . T-53 ENGINE
  . . . . . . T-56 ENGINE
  . . . . . . T-64 ENGINE
  . . . . . . T-74 ENGINE
  . . . . . TURBORAMJET ENGINES
  . INTERNAL COMBUSTION ENGINES
  . . GAS TURBINE ENGINES
  . . . JET ENGINES
  . . . . **TURBOJET ENGINES**
  . . . . . BRISTOL-SIDDELEY OLYMPUS
     593 ENGINE
  . . . . . BRISTOL-SIDDELEY VIPER
     ENGINE
  . . . . . DUCTED FAN ENGINES
  . . . . . J-33 ENGINE
  . . . . . J-34 ENGINE
  . . . . . J-47 ENGINE
  . . . . . J-52 ENGINE
  . . . . . J-57 ENGINE
  . . . . . J-57-P-20 ENGINE
  . . . . . J-65 ENGINE
  . . . . . J-69-T-25 ENGINE
  . . . . . J-71 ENGINE
  . . . . . J-73 ENGINE
  . . . . . J-75 ENGINE
  . . . . . J-79 ENGINE
  . . . . . J-85 ENGINE
  . . . . . J-93 ENGINE
  . . . . . RA-28 ENGINE
  . . . . . TURBOFAN ENGINES
  . . . . . . BRISTOL-SIDDELEY BS 53
      ENGINE
  . . . . . . CF-700 ENGINE
  . . . . . . J-97 ENGINE
  . . . . . . TF-30 ENGINE
  . . . . . . TF-41 ENGINE
  . . . . . TURBOPROP ENGINES
  . . . . . . T-34 ENGINE
  . . . . . . T-38 ENGINE
  . . . . . . T-53 ENGINE
  . . . . . . T-56 ENGINE
  . . . . . . T-64 ENGINE
  . . . . . . T-74 ENGINE
  . . . . . . T-78 ENGINE
  . . . . . TURBORAMJET ENGINES
  . TURBINE ENGINES
  . . GAS TURBINE ENGINES
  . . . JET ENGINES

TURBOJET ENGINES-(CONT.)
  . . . . **TURBOJET ENGINES**
  . . . . . BRISTOL-SIDDELEY OLYMPUS
     593 ENGINE
  . . . . . BRISTOL-SIDDELEY VIPER
     ENGINE
  . . . . . DUCTED FAN ENGINES
  . . . . . J-33 ENGINE
  . . . . . J-34 ENGINE
  . . . . . J-47 ENGINE
  . . . . . J-52 ENGINE
  . . . . . J-57 ENGINE
  . . . . . J-57-P-20 ENGINE
  . . . . . J-65 ENGINE
  . . . . . J-69-T-25 ENGINE
  . . . . . J-71 ENGINE
  . . . . . J-73 ENGINE
  . . . . . J-75 ENGINE
  . . . . . J-79 ENGINE
  . . . . . J-85 ENGINE
  . . . . . J-93 ENGINE
  . . . . . RA-28 ENGINE
  . . . . . TURBOFAN ENGINES
  . . . . . . BRISTOL-SIDDELEY BS 53
      ENGINE
  . . . . . . CF-700 ENGINE
  . . . . . . J-97 ENGINE
  . . . . . . TF-30 ENGINE
  . . . . . . TF-41 ENGINE
  . . . . . TURBOPROP ENGINES
  . . . . . . T-34 ENGINE
  . . . . . . T-38 ENGINE
  . . . . . . T-53 ENGINE
  . . . . . . T-56 ENGINE
  . . . . . . T-64 ENGINE
  . . . . . . T-74 ENGINE
  . . . . . . T-78 ENGINE
  . . . . . TURBORAMJET ENGINES
RT CONVERGENT NOZZLES
  HOUND DOG MISSILE
  JET AIRCRAFT
  MACE MISSILES
  QUAIL MISSILE
  RAMJET ENGINES
  REGULUS MISSILE

TURBOMACHINE BLADES
GS **TURBOMACHINE BLADES**
  . COMPRESSOR BLADES
  . ROTOR·BLADES (TURBOMACHINERY)
  . STATOR BLADES
  . TURBINE BLADES
RT AIRFOILS
 ∞BLADES
 ∞BUCKETS
  CASCADE FLOW
  FAN BLADES
  IMPELLERS
  PADDLES
  ROTORS
  TURBINE WHEELS
  VANES

TURBOMACHINERY
GS **TURBOMACHINERY**
  . CENTRIFUGAL COMPRESSORS
  . CENTRIFUGAL PUMPS
  . J-33 ENGINE
  . TURBINE PUMPS
  . TURBINES
  . AXIAL FLOW TURBINES
  . . GAS TURBINES
  . . SHROUDED TURBINES
  . . STEAM TURBINES
  . . SUPERSONIC TURBINES
  . . TWO STAGE TURBINES
  . . WIND TURBINES
  . . . TIP VANES
  . TURBOCOMPRESSORS
  . TURBOFANS
  . TURBOGENERATORS
  . ASTEC SOLAR TURBOELECTRIC
    GENERATOR
RT BLOWERS
  COMPRESSORS
 ∞MACHINERY
  PUMPS
  ROTATING GENERATORS
  SUPERCHARGERS
  TURBINE INSTRUMENTS

TURBOPAUSE
GS EARTH ATMOSPHERE
  . UPPER ATMOSPHERE
  . . THERMOSPHERE

TURBOPAUSE-*(CONT.)*
. . . TURBOPAUSE
RT   ATMOSPHERIC CIRCULATION
      ATMOSPHERIC PHYSICS
      ATMOSPHERIC TURBULENCE

TURBOPROP AIRCRAFT
GS   JET AIRCRAFT
      . TURBOPROP AIRCRAFT
      . . AN-22 AIRCRAFT
      . . AN-24 AIRCRAFT
      . . ARGOSY MK-1 AIRCRAFT
      . . BREGUET 941 AIRCRAFT
      . . BREGUET 1150 AIRCRAFT
      . . C-2 AIRCRAFT
      . . C-133 AIRCRAFT
      . . C-160 AIRCRAFT
      . . CL-44 AIRCRAFT
      . . CL-84 AIRCRAFT
      . . DHC 5 AIRCRAFT
      . . ELECTRA AIRCRAFT
      . . F-27 AIRCRAFT
      . . G-222 AIRCRAFT
      . . HS-748 AIRCRAFT
      . . MH-262 AIRCRAFT
      . . OV-1 AIRCRAFT
      . . OV-10 AIRCRAFT
      . . SC-5 AIRCRAFT
      . . VISCOUNT AIRCRAFT
      . . YS-11 AIRCRAFT
RT   ∞AIRCRAFT
      GENERAL AVIATION AIRCRAFT
      ∞LOW WING AIRCRAFT
      PASSENGER AIRCRAFT
      ∞SUBSONIC AIRCRAFT
      TRANSPORT AIRCRAFT
      TURBOFAN AIRCRAFT

TURBOPROP ENGINES
UF   DART TURBOPROP ENGINES
GS   ENGINES
      . AIR BREATHING ENGINES
      . . GAS TURBINE ENGINES
      . . . JET ENGINES
      . . . . TURBOJET ENGINES
      . . . . . TURBOPROP ENGINES
      . . . . . . T-53 ENGINE
      . . . . . . T-56 ENGINE
      . . . . . . T-64 ENGINE
      . . . . . . T-74 ENGINE
      . INTERNAL COMBUSTION ENGINES
      . . GAS TURBINE ENGINES
      . . . JET ENGINES
      . . . . TURBOJET ENGINES
      . . . . . TURBOPROP ENGINES
      . . . . . . T-34 ENGINE
      . . . . . . T-38 ENGINE
      . . . . . . T-53 ENGINE
      . . . . . . T-56 ENGINE
      . . . . . . T-64 ENGINE
      . . . . . . T-74 ENGINE
      . . . . . . T-78 ENGINE
      . TURBINE ENGINES
      . . GAS TURBINE ENGINES
      . . . JET ENGINES
      . . . . TURBOJET ENGINES
      . . . . . TURBOPROP ENGINES
      . . . . . . T-34 ENGINE
      . . . . . . T-38 ENGINE
      . . . . . . T-53 ENGINE
      . . . . . . T-56 ENGINE
      . . . . . . T-64 ENGINE
      . . . . . . T-74 ENGINE
      . . . . . . T-78 ENGINE
RT   C-160 AIRCRAFT
      CONTRAROTATING PROPELLERS
      E-2 AIRCRAFT
      P-3 AIRCRAFT
      PROP-FAN TECHNOLOGY
      TURBOFAN ENGINES
      XC-142 AIRCRAFT

TURBOPUMPS
USE   TURBINE PUMPS

TURBORAMJET ENGINES
GS   ENGINES
      . AIR BREATHING ENGINES
      . . GAS TURBINE ENGINES
      . . . JET ENGINES
      . . . . RAMJET ENGINES
      . . . . . TURBORAMJET ENGINES
      . . . . TURBOJET ENGINES
      . . . . . TURBORAMJET ENGINES
      . INTERNAL COMBUSTION ENGINES

TURBORAMJET ENGINES-*(CONT.)*
. . GAS TURBINE ENGINES
. . . JET ENGINES
. . . . RAMJET ENGINES
. . . . . TURBORAMJET ENGINES
. . . . TURBOJET ENGINES
. . . . . TURBORAMJET ENGINES
. TURBINE ENGINES
. . GAS TURBINE ENGINES
. . . JET ENGINES
. . . . RAMJET ENGINES
. . . . . TURBORAMJET ENGINES
. . . . TURBOJET ENGINES
. . . . . TURBORAMJET ENGINES

TURBOROCKET ENGINES
GS   ENGINES
      . ROCKET ENGINES
      . . TURBOROCKET ENGINES
      . TORPEDO ENGINES
      . . TURBOROCKET ENGINES
RT   BOOSTER ROCKET ENGINES
      HYDRAZINE ENGINES
      HYDROGEN OXYGEN ENGINES
      LIQUID AIR CYCLE ENGINES
      RESTARTABLE ROCKET ENGINES
      SUSTAINER ROCKET ENGINES

TURBOROTORS
USE   TURBINE WHEELS

TURBOSHAFTS
GS   ROTATING SHAFTS
      . SHAFTS (MACHINE ELEMENTS)
      . . TURBOSHAFTS
RT   ROTORS
      TURBINES

TURBULENCE
GS   TURBULENCE
      . ATMOSPHERIC TURBULENCE
      . . CLEAR AIR TURBULENCE
      . . GUSTS
      . LOW LEVEL TURBULENCE
      . HOMOGENEOUS TURBULENCE
      . ISOTROPIC TURBULENCE
      . LOW TURBULENCE
      . MAGNETOHYDRODYNAMIC
        TURBULENCE
      . . PLASMA TURBULENCE
RT   AERODYNAMIC DRAG
      ATMOSPHERIC EFFECTS
      BACKWASH
      BOUNDARY LAYER CONTROL
      BOUNDARY LAYER TRANSITION
      FLOW CHARACTERISTICS
      FLUID DYNAMICS
      GAS STREAMS
      MICROMETEOROLOGY
      MIXING
      ∞MOTION
      NONUNIFORMITY
      PANEL METHOD (FLUID DYNAMICS)
      SEA ROUGHNESS
      SLIPSTREAMS
      STEADY FLOW
      STRANGE ATTRACTORS
      STROUHAL NUMBER
      SURFACE NOISE INTERACTIONS
      TURBULENT BOUNDARY LAYER
      TURBULENT FLOW
      UNSTEADY FLOW
      VERTICAL AIR CURRENTS
      VORTEX FILAMENTS
      VORTICES
      VORTICITY
      WAKES
      WIND EFFECTS

TURBULENCE EFFECTS
RT   AERODYNAMIC STABILITY
      BUFFETING
      ∞EFFECTS
      FLUTTER
      SEPARATED FLOW

TURBULENCE METERS
UF   HOT-WIRE TURBULENCE METERS
GS   MEASURING INSTRUMENTS
      . TURBULENCE METERS
RT   HOT-WIRE FLOWMETERS

TURBULENT BOUNDARY LAYER
GS   BOUNDARY LAYERS
      TURBULENT BOUNDARY LAYER

TURBULENT BOUNDARY LAYER-*(CONT.)*
RT   BOUNDARY LAYER TRANSITION
      COMPRESSIBLE BOUNDARY LAYER
      EKMAN LAYER
      HYPERSONIC BOUNDARY LAYER
      INCOMPRESSIBLE BOUNDARY LAYER
      LAMINAR BOUNDARY LAYER
      ∞LAYERS
      REYNOLDS STRESS
      SUPERSONIC BOUNDARY LAYERS
      THERMAL BOUNDARY LAYER
      THREE DIMENSIONAL BOUNDARY LAYER
      TURBULENCE
      TWO DIMENSIONAL BOUNDARY LAYER

TURBULENT DIFFUSION
UF   EDDY DIFFUSION
GS   DIFFUSION
      . TURBULENT DIFFUSION
RT   ATMOSPHERIC DIFFUSION
      ATMOSPHERIC TURBULENCE
      CLEAR AIR TURBULENCE
      COUNTERFLOW

TURBULENT FLOW
GS   FLUID FLOW
      . TURBULENT FLOW
      . . CAVITATION FLOW
      . . SUPERCAVITATING FLOW
RT   AERODYNAMIC INTERFERENCE
      AERODYNAMICS
      ANNULAR FLOW
      ATMOSPHERIC TURBULENCE
      BLASIUS FLOW
      BOUNDARY LAYER TRANSITION
      CLOSURE LAW
      COMBUSTIBLE FLOW
      COUNTERFLOW
      CRITICAL FLOW
      EDDY VISCOSITY
      FLOW CHARACTERISTICS
      FLOW STABILITY
      FLUID AMPLIFIERS
      FLUID DYNAMICS
      FREE CONVECTION
      GAS FLOW
      GUST ALLEVIATORS
      INVISCID FLOW
      ISOTROPIC TURBULENCE
      KOLMOGOROFF THEORY
      LAGRANGE SIMILARITY HYPOTHESIS
      LAMINAR FLOW
      LIQUID FLOW
      MASS FLOW
      MIXING LENGTH FLOW THEORY
      MULTIPHASE FLOW
      NONUNIFORM FLOW
      OPEN CHANNEL FLOW
      ORIFICE FLOW
      PARTICLE LADEN JETS
      PIPE FLOW
      PRESSURE OSCILLATIONS
      RECIRCULATIVE FLUID FLOW
      REYNOLDS NUMBER
      REYNOLDS STRESS
      ROTATING FLUIDS
      SINGLE-PHASE FLOW
      STEADY FLOW
      STEAM FLOW
      SUBCRITICAL FLOW
      SUPERCRITICAL FLOW
      TOLLMEIN-SCHLICHTING WAVES
      ∞TRANSITION LAYERS
      TURBULENCE
      TWO PHASE FLOW
      UNIFORM FLOW
      VISCOUS DRAG
      VISCOUS FLOW
      VORTEX AVOIDANCE
      VORTEX BREAKDOWN
      VORTICES
      VORTICITY TRANSPORT HYPOTHESIS

TURBULENT HEAT TRANSFER
GS   TRANSMISSION
      . HEAT TRANSMISSION
      . . HEAT TRANSFER
      . . . TURBULENT HEAT TRANSFER
RT   AERODYNAMIC HEAT TRANSFER
      CONVECTIVE HEAT TRANSFER
      LAMINAR HEAT TRANSFER
      THERMOHYDRAULICS

TURBULENT JETS
RT   FLUID AMPLIFIERS

TURBULENT JETS-*(CONT.)*
    JET STREAMS (METEOROLOGY)
    ∞JETS

TURBULENT MIXING
GS   MIXING
    . **TURBULENT MIXING**
RT   AGITATION
    LAMINAR MIXING
    MIXING LENGTH FLOW THEORY
    RECIRCULATIVE FLUID FLOW
    TRAPPED VORTEXES
    VORTICES

TURBULENT WAKES
UF   SWIRLING WAKES
GS   WAKES
    . **TURBULENT WAKES**
    . . SLIPSTREAMS
    . . . PROPELLER SLIPSTREAMS
RT   AIRCRAFT WAKES
    LAMINAR WAKES
    TRAPPED VORTEXES
    VORTEX ADVISORY SYSTEM
    VORTEX SHEETS
    VORTEX STREETS

TURING MACHINES
UF   FINITE-STATE MACHINES
RT   AUTOMATA THEORY
    DIGITAL COMPUTERS
    ∞MACHINERY
    MATHEMATICAL LOGIC
    SELF ORGANIZING SYSTEMS

TURKEY
GS   NATIONS
    . **TURKEY**
RT   BLACK SEA
    EUROPE

TURKEYS
GS   ANIMALS
    . VERTEBRATES
    . . BIRDS
    . . . **TURKEYS**
RT   LIVESTOCK

TURNAROUND (STS)
RT   DOWNTIME
    FLIGHT TIME
    LAUNCH DATES
    SCHEDULES
    SEQUENCING
    SPACECRAFT MAINTENANCE
    TESTING TIME

TURNING FLIGHT
UF   BANKING FLIGHT
GS   **TURNING FLIGHT**
    . MINOR CIRCLE TURNING FLIGHT
RT   AERODYNAMIC BALANCE
    AIRCRAFT MANEUVERS
    AIRCRAFT STABILITY
    CLIMBING FLIGHT
    ∞FLIGHT
    FLIGHT PATHS
    HORIZONTAL FLIGHT
    LATERAL OSCILLATION
    LATERAL STABILITY
    MANEUVERS
    MOMENTUM
    ROLL
    YAW

TURNSTILE ANTENNAS
GS   ANTENNAS
    . OMNIDIRECTIONAL ANTENNAS
    . . **TURNSTILE ANTENNAS**
    ARRAYS
    . ANTENNA ARRAYS
    . . **TURNSTILE ANTENNAS**
RT   DIPOLE ANTENNAS
    ∞GRIDS
    WIRE GRID LENSES

TURPENTINE
GS   SOLVENTS
    . **TURPENTINE**
    TERPENES
    . **TURPENTINE**
RT   PAINTS

∞ **TURRET**
SN   *(USE OF A MORE SPECIFIC TERM IS*
    RECOMMENDED--CONSULT THE TERMS
    LISTED BELOW)
RT   GUN TURRETS
    TURRET LATHES

TURRET LATHES
GS   TOOLS
    . MACHINE TOOLS
    . . LATHES
    . . . **TURRET LATHES**
RT   ∞TURRET

TURTLES
GS   ANIMALS
    . POIKILOTHERMIA
    . REPTILES
    . . . **TURTLES**
    . VERTEBRATES
    . REPTILES
    . . . **TURTLES**

TUTOR AIRCRAFT
USE   CL-41 AIRCRAFT

TVC (CONTROL)
USE   THRUST VECTOR CONTROL

TWENTY-FOUR HOUR ORBITS
GS   ORBITS
    . SPACECRAFT ORBITS
    . . SATELLITE ORBITS
    . . . **TWENTY-FOUR HOUR ORBITS**
RT   CIRCULAR ORBITS
    EARTH ORBITS
    EQUATORIAL ORBITS
    GEOSYNCHRONOUS ORBITS
    ORBITAL MECHANICS
    PAS
    PLANETARY ORBITS
    POLAR ORBITS
    STATIONARY ORBITS
    SYNCHRONOUS COMMUNICATIONS
      SATELLITE PROJ
    SYNCHRONOUS SATELLITES

TWENTY-SEVEN DAY VARIATION
GS   VARIATIONS
    . **TWENTY-SEVEN DAY VARIATION**
RT   SOLAR CYCLES
    SOLAR ROTATION
    STARSPOTS
    SUNSPOTS

TWILIGHT GLOW
GS   ATMOSPHERIC RADIATION
    . SKY RADIATION
    . . AIRGLOW
    . . . **TWILIGHT GLOW**
    ELECTROMAGNETIC RADIATION
    . LIGHT (VISIBLE RADIATION)
    . . SKY RADIATION
    . . . AIRGLOW
    . . . . **TWILIGHT GLOW**
RT   DAYGLOW
    NIGHT
    NIGHT SKY

TWINNING
GS   **TWINNING**
    . MECHANICAL TWINNING
RT   CRYSTAL DEFECTS
    CRYSTAL GROWTH
    CRYSTAL STRUCTURE
    GRAIN BOUNDARIES
    STACKING FAULT ENERGY

TWISTED WINGS
GS   AIRFOILS
    . WINGS
    . . **TWISTED WINGS**
RT   CAMBERED WINGS
    FIXED WINGS
    FLEXIBLE WINGS
    RING WINGS
    UNCAMBERED WINGS

TWISTING
UF   PRETWISTING
RT   BENDING
    BUCKLING
    DEFORMATION
    DISTORTION

TWISTING-*(CONT.)*
    STRUCTURAL STRAIN
    TORQUE
    TORSION
    TORSIONAL VIBRATION
    WARPAGE
    WINDING

TWITCHING
RT   INVOLUNTARY ACTIONS
    MUSCLES
    MUSCULAR FUNCTION

TWO BODY ORBITS
USE   TWO BODY PROBLEM

TWO BODY PROBLEM
UF   TWO BODY ORBITS
RT   BINARY STARS
    CELESTIAL MECHANICS
    EARTH-MOON SYSTEM
    HYLLERAAS COORDINATES
    MANY BODY PROBLEM
    ORBITAL MECHANICS
    ORBITS
    PERTURBATION
    ∞PROBLEMS
    ROCHE LIMIT
    THREE BODY PROBLEM

TWO DIMENSIONAL BODIES
RT   ∞BODIES
    ∞CROSS SECTIONS
    DUCTED BODIES
    MATHEMATICAL MODELS
    ∞SURFACES

TWO DIMENSIONAL BOUNDARY LAYER
GS   BOUNDARY LAYERS
    . **TWO DIMENSIONAL BOUNDARY
      LAYER**
RT   LAMINAR BOUNDARY LAYER
    SUPERSONIC BOUNDARY LAYERS
    TURBULENT BOUNDARY LAYER

TWO DIMENSIONAL FLOW
GS   FLUID FLOW
    . **TWO DIMENSIONAL FLOW**
    . . COUETTE FLOW
RT   AXIAL FLOW
    BLASIUS FLOW
    CAPILLARY WAVES
    COAXIAL FLOW
    FLOW GEOMETRY
    HARTMANN FLOW
    ONE DIMENSIONAL FLOW
    PRANDTL-MEYER EXPANSION
    RADIAL FLOW
    RAYLEIGH WAVES
    STEADY FLOW
    STREAM FUNCTIONS (FLUIDS)
    TAYLOR INSTABILITY
    THREE DIMENSIONAL FLOW
    WALL FLOW
    WEDGE FLOW

TWO DIMENSIONAL JETS
RT   JET FLOW
    JET MIXING FLOW
    ∞JETS
    WALL FLOW

TWO FLUID MODELS
RT   BOLTZMANN DISTRIBUTION
    LIQUID HELIUM
    MAGNETOHYDRODYNAMIC FLOW
    ROTATING PLASMAS
    SHOCK WAVE PROPAGATION
    SUPERFLUIDITY

TWO PHASE FLOW
GS   FLUID FLOW
    . MULTIPHASE FLOW
    . . **TWO PHASE FLOW**
RT   GAS FLOW
    LAMINAR FLOW
    LIQUID FLOW
    ∞PRESSURE DROP
    SINGLE-PHASE FLOW
    SOLIDS FLOW
    TURBULENT FLOW

TWO PHASE SYSTEMS
USE   BINARY SYSTEMS (MATERIALS)

† **TWO REFLECTOR ANTENNAS**
GS   ANTENNAS
. DIRECTIONAL ANTENNAS
.. **TWO REFLECTOR ANTENNAS**
RT   CASSEGRAIN ANTENNAS
RADIO ANTENNAS
REFLECTOMETERS
REFLECTORS

TWO STAGE PLASMA ENGINES
GS   PLASMA POWER SOURCES
. PLASMA ENGINES
.. **TWO STAGE PLASMA ENGINES**
RT   ELECTRIC PROPULSION
PLASMAS (PHYSICS)

TWO STAGE TURBINES
GS   TURBOMACHINERY
. TURBINES
.. **TWO STAGE TURBINES**
RT   GAS TURBINE ENGINES
GAS TURBINES
STEAM TURBINES

TWO-WAVELENGTH LASERS
GS   STIMULATED EMISSION DEVICES
. LASERS
.. **TWO-WAVELENGTH LASERS**
RT   COHERENT LIGHT
DYE LASERS
LASER OUTPUTS
MASERS
MOLECULAR OSCILLATORS
QUANTUM AMPLIFIERS
STIMULATED EMISSION

TX-33-39 ENGINE
USE   XM-33 ENGINE

TX-77 ENGINE
GS   ENGINES
. ROCKET ENGINES
.. SOLID PROPELLANT ROCKET
ENGINES
... **TX-77 ENGINE**
RT   LANCE MISSILE

TX-354 ENGINE
UF   CASTOR 2 ENGINE
GS   ENGINES
. ROCKET ENGINES
.. SOLID PROPELLANT ROCKET
ENGINES
... **TX-354 ENGINE**
RT   BOOSTER ROCKET ENGINES
LITTLE JOE 2 LAUNCH VEHICLE
RAM B LAUNCH VEHICLE
SCOUT LAUNCH VEHICLE
SUSTAINER ROCKET ENGINES
TRAILBLAZER 2 REENTRY VEHICLE
XM-33 ENGINE

TYCHO CRATER
GS   CRATERS
. LUNAR CRATERS
.. **TYCHO CRATER**
RT   METEORITE CRATERS

TYPE 2 BURSTS
GS   BURSTS
. RADIO BURSTS
.. SOLAR RADIO BURSTS
... **TYPE 2 BURSTS**
DECAY
. EMISSION
.. RADIO EMISSION
... SOLAR RADIO EMISSION
.... SOLAR RADIO BURSTS
..... **TYPE 2 BURSTS**
ELECTROMAGNETIC RADIATION
. RADIO WAVES
.. EXTRATERRESTRIAL RADIO WAVES
... RADIO BURSTS
.... SOLAR RADIO BURSTS
..... **TYPE 2 BURSTS**
... SOLAR RADIO EMISSION
.... SOLAR RADIO BURSTS
..... **TYPE 2 BURSTS**
.. RADIO EMISSION
... RADIO BURSTS
.... SOLAR RADIO BURSTS
..... **TYPE 2 BURSTS**
... SOLAR RADIO EMISSION
.... SOLAR RADIO BURSTS
..... **TYPE 2 BURSTS**

TYPE 2 BURSTS-(CONT.)
ELECTROMAGNETIC RADIATION
. EXTRATERRESTRIAL RADIO WAVES
.. RADIO BURSTS
... SOLAR RADIO BURSTS
.... **TYPE 2 BURSTS**
.. SOLAR RADIO EMISSION
... SOLAR RADIO BURSTS
.... **TYPE 2 BURSTS**
. SOLAR RADIATION
.. SOLAR RADIO EMISSION
... SOLAR RADIO BURSTS
.... **TYPE 2 BURSTS**

TYPE 3 BURSTS
GS   BURSTS
. RADIO BURSTS
.. SOLAR RADIO BURSTS
... **TYPE 3 BURSTS**
DECAY
. EMISSION
.. RADIO EMISSION
... SOLAR RADIO EMISSION
.... SOLAR RADIO BURSTS
..... **TYPE 3 BURSTS**
ELECTROMAGNETIC RADIATION
. RADIO WAVES
.. EXTRATERRESTRIAL RADIO WAVES
... RADIO BURSTS
.... SOLAR RADIO BURSTS
..... **TYPE 3 BURSTS**
... SOLAR RADIO EMISSION
.... SOLAR RADIO BURSTS
..... **TYPE 3 BURSTS**
.. RADIO EMISSION
... RADIO BURSTS
.... SOLAR RADIO BURSTS
..... **TYPE 3 BURSTS**
.. SOLAR RADIO EMISSION
... SOLAR RADIO BURSTS
..... **TYPE 3 BURSTS**
EXTRATERRESTRIAL RADIATION
. EXTRATERRESTRIAL RADIO WAVES
.. RADIO BURSTS
... SOLAR RADIO BURSTS
.... **TYPE 3 BURSTS**
.. SOLAR RADIO EMISSION
... SOLAR RADIO BURSTS
.... **TYPE 3 BURSTS**
. SOLAR RADIATION
.. SOLAR RADIO EMISSION
... SOLAR RADIO BURSTS
.... **TYPE 3 BURSTS**

TYPE 4 BURSTS
GS   BURSTS
. RADIO BURSTS
.. SOLAR RADIO BURSTS
... **TYPE 4 BURSTS**
DECAY
. EMISSION
.. RADIO EMISSION
... SOLAR RADIO EMISSION
.... SOLAR RADIO BURSTS
..... **TYPE 4 BURSTS**
ELECTROMAGNETIC RADIATION
. RADIO WAVES
.. EXTRATERRESTRIAL RADIO WAVES
... RADIO BURSTS
.... SOLAR RADIO BURSTS
..... **TYPE 4 BURSTS**
... SOLAR RADIO EMISSION
.... SOLAR RADIO BURSTS
..... **TYPE 4 BURSTS**
.. RADIO EMISSION
... RADIO BURSTS
.... SOLAR RADIO BURSTS
..... **TYPE 4 BURSTS**
... SOLAR RADIO EMISSION
.... SOLAR RADIO BURSTS
..... **TYPE 4 BURSTS**
EXTRATERRESTRIAL RADIATION
. EXTRATERRESTRIAL RADIO WAVES
.. RADIO BURSTS
... SOLAR RADIO BURSTS
.... **TYPE 4 BURSTS**
.. SOLAR RADIO EMISSION
... SOLAR RADIO BURSTS
.... **TYPE 4 BURSTS**
. SOLAR RADIATION
.. SOLAR RADIO EMISSION
... SOLAR RADIO BURSTS
.... **TYPE 4 BURSTS**

TYPE 5 BURSTS
GS   BURSTS
. RADIO BURSTS
.. SOLAR RADIO BURSTS
... **TYPE 5 BURSTS**
DECAY
. EMISSION
.. RADIO EMISSION
... SOLAR RADIO EMISSION
.... SOLAR RADIO BURSTS
..... **TYPE 5 BURSTS**
ELECTROMAGNETIC RADIATION
. RADIO WAVES
.. EXTRATERRESTRIAL RADIO WAVES
... RADIO BURSTS
.... SOLAR RADIO BURSTS
..... **TYPE 5 BURSTS**
... SOLAR RADIO EMISSION
.... SOLAR RADIO BURSTS
..... **TYPE 5 BURSTS**
.. RADIO EMISSION
... RADIO BURSTS
.... SOLAR RADIO BURSTS
..... **TYPE 5 BURSTS**
... SOLAR RADIO EMISSION
.... SOLAR RADIO BURSTS
..... **TYPE 5 BURSTS**
EXTRATERRESTRIAL RADIATION
. EXTRATERRESTRIAL RADIO WAVES
.. RADIO BURSTS
... SOLAR RADIO BURSTS
.... **TYPE 5 BURSTS**
.. SOLAR RADIO EMISSION
... SOLAR RADIO BURSTS
.... **TYPE 5 BURSTS**
. SOLAR RADIATION
.. SOLAR RADIO EMISSION
... SOLAR RADIO BURSTS
.... **TYPE 5 BURSTS**

TYPEWRITERS
GS   **TYPEWRITERS**
. AUTOMATIC TYPEWRITERS
. TELETYPEWRITERS
.. TELEPRINTERS
RT   PRINTERS

TYPHOID
GS   DISEASES
. INFECTIOUS DISEASES
.. **TYPHOID**

TYPHON WEAPON SYSTEM
GS   WEAPON SYSTEMS
. **TYPHON WEAPON SYSTEM**
RT   BUMBLEBEE PROJECT
∞SYSTEMS

TYPHOONS
GS   STORMS
. CYCLONES
.. **TYPHOONS**
. STORMS (METEOROLOGY)
.. TROPICAL STORMS
... **TYPHOONS**
RT   ATMOSPHERIC CIRCULATION
HURRICANES
MARINE METEOROLOGY
METEOROLOGY
STORM DAMAGE
TORNADOES

TYPHUS
GS   DISEASES
. INFECTIOUS DISEASES
.. **TYPHUS**

TYROSINE
GS   ACIDS
. AMINO ACIDS
.. **TYROSINE**
ORGANIC COMPOUNDS
. AMINO ACIDS
.. **TYROSINE**
RT   ENZYME ACTIVITY
LIVER

T2J AIRCRAFT
USE   T-2 AIRCRAFT

T3J AIRCRAFT
USE   T-39 AIRCRAFT

U

U BENDS
GS   PIPES (TUBES)
. **U BENDS**
RT   FITTINGS

U SPIN SPACE
GS   ALGEBRA
. VECTOR SPACES
. . **U SPIN SPACE**
RT   MATRICES (MATHEMATICS)
QUANTUM MECHANICS

U TUBES
USE   MANOMETERS

U.S.S.R.
UF   SOVIET UNION
GS   NATIONS
. **U.S.S.R.**
RT   ASIA
BARENTS SEA
BLACK SEA
CAUCASUS MOUNTAAINS (U.S.S.R.)
EUROPE
KURILE ISLANDS
MOSCOW
SEA OF OKHOTSK
SIBERIA

U.S.S.R. SPACE PROGRAM
GS   PROGRAMS
. SPACE PROGRAMS
. . **U.S.S.R. SPACE PROGRAM**
RT   APOLLO SOYUZ TEST PROJECT
EUROPEAN SPACE PROGRAMS
INTERNATIONAL COOPERATION
INTERNATIONAL RELATIONS
INTERNATIONAL SATELLITE GEODESY
EXPERIMENT
LUNAR RETROREFLECTORS
LUNIK LUNAR PROBES
LUNIK 19 LUNAR PROBE
LUNIK 22 LUNAR PROBE
LUNOKHOD LUNAR ROVING VEHICLES
MARS 2 SPACECRAFT
MARS 3 SPACECRAFT
MARS 4 SPACECRAFT
MARS 5 SPACECRAFT
MARS 6 SPACECRAFT
MARS 7 SPACECRAFT
MOLNIYA SATELLITES
PROTON SATELLITES
SALYUT SPACE STATION
SOYUZ SPACECRAFT
VEGA PROJECT
VENERA SATELLITES
VENERA 8 SATELLITE
VENERA 10 SATELLITE
VENERA 11 SATELLITE
VENERA 12 SATELLITE

U-2 AIRCRAFT
UF   LOCKHEED U-2 AIRCRAFT
WU-2 AIRCRAFT
GS   JET AIRCRAFT
. **U-2 AIRCRAFT**
LOCKHEED AIRCRAFT
. **U-2 AIRCRAFT**
MONOPLANES
. **U-2 AIRCRAFT**
OBSERVATION AIRCRAFT
. **U-2 AIRCRAFT**
RECONNAISSANCE AIRCRAFT
. **U-2 AIRCRAFT**
RESEARCH AIRCRAFT
. **U-2 AIRCRAFT**
UTILITY AIRCRAFT
. **U-2 AIRCRAFT**
RT   ∞ AIRCRAFT

U-10 AIRCRAFT
UF   COURIER AIRCRAFT
L-28 AIRCRAFT
GS   HELIO AIRCRAFT
. **U-10 AIRCRAFT**
LIGHT AIRCRAFT
. **U-10 AIRCRAFT**
MONOPLANES
. **U-10 AIRCRAFT**
PASSENGER AIRCRAFT
. **U-10 AIRCRAFT**
UTILITY AIRCRAFT
. **U-10 AIRCRAFT**

U-10 AIRCRAFT-*(CONT.)*
V/STOL AIRCRAFT
. SHORT TAKEOFF AIRCRAFT
. . **U-10 AIRCRAFT**
RT   ∞ AIRCRAFT

UBV SPECTRA
GS   SPECTRA
. RADIATION SPECTRA
. . ELECTROMAGNETIC SPECTRA
. . . **UBV SPECTRA**

UDIMET ALLOYS
GS   ALLOYS
. HEAT RESISTANT ALLOYS
. . **UDIMET ALLOYS**
. NICKEL ALLOYS
. . **UDIMET ALLOYS**

UFO
USE   UNIDENTIFIED FLYING OBJECTS

UGANDA
GS   NATIONS
. **UGANDA**
RT   AFRICA

UH-1 HELICOPTER
UF   HU-1 HELICOPTER
IROQUOIS HELICOPTER
RH-2 HELICOPTER
YHU-1 HELICOPTER
YUH-1 HELICOPTER
GS   BELL AIRCRAFT
. **UH-1 HELICOPTER**
UTILITY AIRCRAFT
. **UH-1 HELICOPTER**
V/STOL AIRCRAFT
. ROTARY WING AIRCRAFT
. . HELICOPTERS
. . . MILITARY HELICOPTERS
. . . . **UH-1 HELICOPTER**
RT   UH-60A HELICOPTER
UH-61A HELICOPTER

UH-2 HELICOPTER
UF   HU2K-1 HELICOPTER
KAMAN UH-2A HELICOPTER
SEASPRITE HELICOPTER
GS   KAMAN AIRCRAFT
. **UH-2 HELICOPTER**
UTILITY AIRCRAFT
. **UH-2 HELICOPTER**
V/STOL AIRCRAFT
. ROTARY WING AIRCRAFT
. . HELICOPTERS
. . . MILITARY HELICOPTERS
. . . . **UH-2 HELICOPTER**

UH-12 HELICOPTER
USE   OH-23 HELICOPTER

UH-13 HELICOPTER
USE   OH-13 HELICOPTER

UH-34 HELICOPTER
UF   HUS-1 HELICOPTER
SEAHORSE HELICOPTER
GS   SIKORSKY AIRCRAFT
. **UH-34 HELICOPTER**
TRANSPORT AIRCRAFT
. **UH-34 HELICOPTER**
UTILITY AIRCRAFT
. **UH-34 HELICOPTER**
V/STOL AIRCRAFT
. ROTARY WING AIRCRAFT
. . HELICOPTERS
. . . MILITARY HELICOPTERS
. . . . **UH-34 HELICOPTER**
RT   S-58 HELICOPTER

UH-60A HELICOPTER
UF   YUH-60A HELICOPTER
GS   SIKORSKY AIRCRAFT
. **UH-60A HELICOPTER**
TRANSPORT AIRCRAFT
. **UH-60A HELICOPTER**
UTILITY AIRCRAFT
. **UH-60A HELICOPTER**
V/STOL AIRCRAFT
. ROTARY WING AIRCRAFT
. . HELICOPTERS
. . . MILITARY HELICOPTERS
. . . . **UH-60A HELICOPTER**

UH-60A HELICOPTER-*(CONT.)*
RT   HELICOPTER DESIGN
UH-1 HELICOPTER

UH-61A HELICOPTER
UF   YUH-61A HELICOPTER
GS   SIKORSKY AIRCRAFT
. **UH-61A HELICOPTER**
TRANSPORT AIRCRAFT
. **UH-61A HELICOPTER**
UTILITY AIRCRAFT
. **UH-61A HELICOPTER**
V/STOL AIRCRAFT
. ROTARY WING AIRCRAFT
. . HELICOPTERS
. . . MILITARY HELICOPTERS
. . . . **UH-61A HELICOPTER**
RT   HELICOPTER DESIGN
UH-1 HELICOPTER

UHTREX (NUCLEAR REACTORS)
USE   HIGH TEMPERATURE NUCLEAR
REACTORS

UHURU SATELLITE
UF   EXPLORER 42 SATELLITE
GS   SATELLITES
. ARTIFICIAL SATELLITES
. . **UHURU SATELLITE**
RT   EXTARS
GALACTIC RADIATION
SAS
SATELLITE OBSERVATION
X RAY ASTRONOMY

UK SATELLITES
UF   UNITED KINGDOM SATELLITES
GS   SATELLITES
. ARTIFICIAL SATELLITES
. . SCIENTIFIC SATELLITES
. . . **UK SATELLITES**
. . . . UK 4 SATELLITE
RT   SKYNET SATELLITES
UK SPACE PROGRAM

UK SPACE PROGRAM
GS   PROGRAMS
. SPACE PROGRAMS
. . **UK SPACE PROGRAM**
RT   UK SATELLITES
UNITED KINGDOM

UK 4 SATELLITE
GS   SATELLITES
. ARTIFICIAL SATELLITES
. . SCIENTIFIC SATELLITES
. . . UK SATELLITES
. . . . **UK 4 SATELLITE**

ULCERS
GS   DISEASES
. **ULCERS**
RT   CANCER

ULLAGE
RT   FUEL TANK PRESSURIZATION
FUEL TANKS
INTERFACE STABILITY
LIQUID SLOSHING
PROPELLANT TANKS
SPLASHING
TANK GEOMETRY
ULLAGE ROCKET ENGINES

ULLAGE ROCKET ENGINES
GS   ENGINES
. ROCKET ENGINES
. . **ULLAGE ROCKET ENGINES**
. TORPEDO ENGINES
. . **ULLAGE ROCKET ENGINES**
RT   SOLID PROPELLANT ROCKET ENGINES
ULLAGE

ULM (LIGHT MODULATION)
USE   ULTRASONIC LIGHT MODULATION

ULNA
GS   ANATOMY
. MUSCULOSKELETAL SYSTEM
. . BONES
. . . **ULNA**
RT   ARM (ANATOMY)
ELBOW (ANATOMY)

ULTRA SHORT WAVE RADIO EQUIPMENT
USE   VERY HIGH FREQUENCY RADIO
      EQUIPMENT

ULTRAHIGH FREQUENCIES
UF   L BAND
     S BAND
GS   FREQUENCIES
   . RADIO FREQUENCIES
   .. **ULTRAHIGH FREQUENCIES**
   ... P BAND
RT   DECIMETER WAVES
     EISCAT RADAR SYSTEM (EUROPE)
     FLEET SATELLITE COMMUNICATION
      SYSTEM
     LOW FREQUENCY BANDS
     PASSIVE L-BAND RADIOMETERS
     PRAETERSONIC DEVICES
     UNIFIED S BAND
     VERY HIGH FREQUENCY RADIO
      EQUIPMENT

ULTRAHIGH VACUUM
GS   PRESSURE
   . VACUUM
   .. **ULTRAHIGH VACUUM**
RT   HIGH VACUUM
     LOW DENSITY RESEARCH
     RESIDUAL GAS
     VACUUM APPARATUS
     VACUUM TESTS

ULTRALIGHT AIRCRAFT
RT   ∞AIRCRAFT
     HANG GLIDERS
     LIGHT AIRCRAFT
     MAN POWERED AIRCRAFT
     ∞WINGED VEHICLES

ULTRALOW FREQUENCIES
USE   EXTREMELY LOW RADIO FREQUENCIES

ULTRALOW TEMPERATURES
GS   TEMPERATURE
   . LOW TEMPERATURE
   .. **ULTRALOW TEMPERATURES**
RT   ABSOLUTE ZERO
     COLD TRAPS
     CRITICAL TEMPERATURE
     CRYOGENIC FLUIDS
     CRYOGENICS
     CURIE TEMPERATURE
     SOLIDIFIED GASES
     SPACE TEMPERATURE

ULTRAPURE METALS
GS   METALS
   . **ULTRAPURE METALS**
RT   CRYSTAL LATTICES
     IMPURITIES
     PURIFICATION
     PURITY
     SINGLE CRYSTALS
     SPACE PROCESSING
     VAPOR DEPOSITION
     ZONE MELTING

ULTRASHORT PULSED LASERS
GS   STIMULATED EMISSION DEVICES
   . LASERS
   .. PULSED LASERS
   ... **ULTRASHORT PULSED LASERS**
RT   GLASS LASERS
     LASER APPLICATIONS
     LIGHT AMPLIFIERS
     PULSE DURATION
     QUANTUM AMPLIFIERS
     STIMULATED EMISSION

ULTRASONIC AGITATION
GS   AGITATION
   . **ULTRASONIC AGITATION**
RT   ULTRASONICS

ULTRASONIC CLEANING
GS   CLEANING
   . **ULTRASONIC CLEANING**
RT   ACOUSTICS
     CAVITATION FLOW
     CLEANERS
     ETCHING
     FLUID FLOW
     GRINDING MACHINES
     MACHINE TOOLS

ULTRASONIC CLEANING-*(CONT.)*
     PIEZOELECTRIC TRANSDUCERS
     POLISHING
     TOOLS
     TRANSDUCERS
     ULTRASONICS

ULTRASONIC DENSIMETERS
GS   MEASURING INSTRUMENTS
   . DENSIMETERS
   .. **ULTRASONIC DENSIMETERS**
RT   DENSITY (MASS/VOLUME)
     DENSITY MEASUREMENT
   ∞INSTRUMENTS
   ∞MEASUREMENT

ULTRASONIC FLAW DETECTION
GS   DETECTION
   . **ULTRASONIC FLAW DETECTION**
RT   ∞DETECTORS
     EXAMINATION
     IDENTIFYING
     INSPECTION
     NONDESTRUCTIVE TESTS
     QUALITY CONTROL
     ULTRASONIC SCANNERS

ULTRASONIC GRINDING MACHINES
USE   ULTRASONIC MACHINING

ULTRASONIC LIGHT MODULATION
UF   ULM (LIGHT MODULATION)
GS   MODULATION
   . LIGHT MODULATION
   .. **ULTRASONIC LIGHT MODULATION**
RT   ULTRASONICS

ULTRASONIC MACHINING
UF   ULTRASONIC GRINDING MACHINES
GS   MACHINING
   . **ULTRASONIC MACHINING**
RT   ULTRASONICS

ULTRASONIC RADIATION
UF   ULTRASONIC WAVES
GS   ELASTIC WAVES
   . **ULTRASONIC RADIATION**
RT   COHERENT ACOUSTIC RADIATION
     MAGNETOELASTIC WAVES
   ∞RADIATION
     SOUND WAVES
     ULTRASONICS
     UNDERWATER ACOUSTICS

ULTRASONIC SCANNERS
GS   SCANNERS
   . **ULTRASONIC SCANNERS**
RT   ACOUSTICS
     IMAGING TECHNIQUES
     MEASURING INSTRUMENTS
     SCANNING
     ULTRASONIC FLAW DETECTION
     ULTRASONICS

ULTRASONIC SOLDERING
UF   SONIC SOLDERING
GS   SOLDERING
   . **ULTRASONIC SOLDERING**
RT   ∞JOINING
     ULTRASONICS

ULTRASONIC SPECTROSCOPY
GS   SPECTROSCOPY
   . **ULTRASONIC SPECTROSCOPY**
RT   CRACKS
     NONDESTRUCTIVE TESTS
     SPECTRUM ANALYSIS

ULTRASONIC TESTS
RT   ACOUSTIC MEASUREMENT
     ACOUSTIC SOUNDING
     DYNAMIC MODULUS OF ELASTICITY
     LAMB WAVES
   ∞MATERIALS TESTS
     NONDESTRUCTIVE TESTS
   ∞TESTS
     ULTRASONICS

ULTRASONIC WAVE TRANSDUCERS
GS   TRANSDUCERS
   . **ULTRASONIC WAVE TRANSDUCERS**
RT   ELECTRONIC TRANSDUCERS
     MICROPHONES
     PRESSURE SENSORS

ULTRASONIC WAVE TRANSDUCERS-*(CONT.)*
     SONAR
     SURFACE ACOUSTIC WAVE DEVICES
     ULTRASONICS
     UNDERWATER ACOUSTICS

ULTRASONIC WAVES
USE   ULTRASONIC RADIATION

ULTRASONIC WELDING
GS   WELDING
   . PRESSURE WELDING
   .. **ULTRASONIC WELDING**
RT   SPOT WELDS
     ULTRASONICS

ULTRASONICS
RT   ACOUSTICS
     ULTRASONIC AGITATION
     ULTRASONIC CLEANING
     ULTRASONIC LIGHT MODULATION
     ULTRASONIC MACHINING
     ULTRASONIC RADIATION
     ULTRASONIC SCANNERS
     ULTRASONIC SOLDERING
     ULTRASONIC TESTS
     ULTRASONIC WAVE TRANSDUCERS
     ULTRASONIC WELDING

ULTRAVIOLET ABSORPTION
GS   ENERGY ABSORPTION
   . ELECTROMAGNETIC ABSORPTION
   .. **ULTRAVIOLET ABSORPTION**
   RADIATION ABSORPTION
   . ELECTROMAGNETIC ABSORPTION
   .. **ULTRAVIOLET ABSORPTION**
RT   ∞ABSORPTION

ULTRAVIOLET ASTRONOMY
GS   ASTRONOMY
   . **ULTRAVIOLET ASTRONOMY**
RT   ELECTROMAGNETIC RADIATION
     EXTREME ULTRAVIOLET EXPLORER
      SATELLITE
     LYMAN ALPHA RADIATION
     LYMAN BETA RADIATION
     SPARTAN SATELLITES
     STARSAT TELESCOPE
     TELESCOPES
     ULTRAVIOLET TELESCOPES

ULTRAVIOLET FILTERS
GS   ELECTROMAGNETIC WAVE FILTERS
   . OPTICAL FILTERS
   .. **ULTRAVIOLET FILTERS**
RT   BANDPASS FILTERS
     ELECTRIC FILTERS
     INFRARED FILTERS

ULTRAVIOLET LASERS
UF   UV LASERS
GS   STIMULATED EMISSION DEVICES
   . LASERS
   .. GAS LASERS
   ... **ULTRAVIOLET LASERS**
   .. PULSED LASERS
   ... **ULTRAVIOLET LASERS**
RT   COHERENT LIGHT
     LASER OUTPUTS
     LIGHT AMPLIFIERS
     LIGHT TRANSMISSION
     MASERS
     MOLECULAR OSCILLATORS
     NITROGEN LASERS
     QUANTUM AMPLIFIERS
     STIMULATED EMISSION
     XENON CHLORIDE LASERS

ULTRAVIOLET LIGHT
USE   ULTRAVIOLET RADIATION

ULTRAVIOLET MICROSCOPY
GS   MICROSCOPY
   . **ULTRAVIOLET MICROSCOPY**
RT   MICROSCOPES

ULTRAVIOLET PHOTOGRAPHY
GS   PHOTOGRAPHY
   . **ULTRAVIOLET PHOTOGRAPHY**
RT   AERIAL PHOTOGRAPHY
     CAMERAS
     COLOR PHOTOGRAPHY
     FAINT OBJECT CAMERA
     INFRARED PHOTOGRAPHY

ULTRAVIOLET PHOTOGRAPHY-(CONT.)
    RADAR PHOTOGRAPHY

ULTRAVIOLET PHOTOMETRY
GS   IMAGERY
    . **ULTRAVIOLET PHOTOMETRY**
    OPTICAL MEASUREMENT
    . PHOTOMETRY
    . . **ULTRAVIOLET PHOTOMETRY**
    PHOTOGRAPHY
    . **ULTRAVIOLET PHOTOMETRY**
RT   BLACK AND WHITE PHOTOGRAPHY

ULTRAVIOLET RADIATION
UF   ULTRAVIOLET LIGHT
GS   ELECTROMAGNETIC RADIATION
    . **ULTRAVIOLET RADIATION**
    . . EXTREME ULTRAVIOLET RADIATION
    . . FAR ULTRAVIOLET RADIATION
    . . . LYMAN ALPHA RADIATION
    . . . LYMAN BETA RADIATION
    . . NEAR ULTRAVIOLET RADIATION
    IONIZING RADIATION
    . **ULTRAVIOLET RADIATION**
    . . EXTREME ULTRAVIOLET RADIATION
    . . FAR ULTRAVIOLET RADIATION
    . . . LYMAN ALPHA RADIATION
    . . . LYMAN BETA RADIATION
    . . NEAR ULTRAVIOLET RADIATION
RT   BEAMS (RADIATION)
    BLACK BODY RADIATION
    CERENKOV RADIATION
    COHERENT ELECTROMAGNETIC
      RADIATION
    CORONAL HOLES
    DAYGLOW
    IUE
    MICROCHANNELS
    MONOCHROMATIC RADIATION
    POLARIZED ELECTROMAGNETIC
      RADIATION
    ∞RADIATION
    SEYFERT GALAXIES
    SOLAR RADIATION
    STERILIZATION
    SUNLIGHT
    THERMAL RADIATION
    UMKEHR EFFECT

ULTRAVIOLET REFLECTION
GS   REFLECTION
    . **ULTRAVIOLET REFLECTION**
RT   INFRARED REFLECTION
    RADIO ECHOES
    REFLECTOMETERS
    SPREAD REFLECTION

ULTRAVIOLET SPECTRA
GS   SPECTRA
    . RADIATION SPECTRA
    . . ELECTROMAGNETIC SPECTRA
    . . . **ULTRAVIOLET SPECTRA**
RT   ABSORPTION SPECTRA
    EMISSION SPECTRA
    HERZBERG BANDS
    LIGHT (VISIBLE RADIATION)
    LINE SPECTRA
    LYMAN SPECTRA
    MOLECULAR SPECTRA
    RADIO SPECTROSCOPY
    SOLAR SPECTRA
    STELLAR SPECTRA

ULTRAVIOLET SPECTROGRAPHS
USE   ULTRAVIOLET SPECTROMETERS

ULTRAVIOLET SPECTROMETERS
UF   ULTRAVIOLET SPECTROGRAPHS
GS   MEASURING INSTRUMENTS
    . OPTICAL MEASURING INSTRUMENTS
    . . PHOTOMETERS
    . . . **ULTRAVIOLET SPECTROMETERS**
    . RADIATION MEASURING INSTRUMENTS
    . . ACTINOMETERS
    . . . **ULTRAVIOLET SPECTROMETERS**
    . . PHOTOMETERS
    . . . **ULTRAVIOLET SPECTROMETERS**
    . SPECTROMETERS
    . . **ULTRAVIOLET SPECTROMETERS**
    . . . HIGH DISPERSION
        SPECTROGRAPHS
    OPTICAL EQUIPMENT
    . OPTICAL MEASURING INSTRUMENTS
    . . PHOTOMETERS
    . . . **ULTRAVIOLET SPECTROMETERS**

ULTRAVIOLET SPECTROMETERS-(CONT.)
RT   EBERT SPECTROMETERS
    SOLAR MAXIMUM MISSION
    SOLAR SPECTROMETERS

ULTRAVIOLET SPECTROPHOTOMETERS
GS   MEASURING INSTRUMENTS
    . OPTICAL MEASURING INSTRUMENTS
    . . PHOTOMETERS
    . . . **ULTRAVIOLET**
        SPECTROPHOTOMETERS
    . . SPECTROPHOTOMETERS
    . . . **ULTRAVIOLET**
        SPECTROPHOTOMETERS
    . RADIATION MEASURING INSTRUMENTS
    . . ACTINOMETERS
    . . . SPECTROPHOTOMETERS
    . . . . **ULTRAVIOLET**
        SPECTROPHOTOMETERS
    . . PHOTOMETERS
    . . . **ULTRAVIOLET**
        SPECTROPHOTOMETERS
    OPTICAL EQUIPMENT
    . OPTICAL MEASURING INSTRUMENTS
    . . PHOTOMETERS
    . . . **ULTRAVIOLET**
        SPECTROPHOTOMETERS
    . . SPECTROPHOTOMETERS
    . . . **ULTRAVIOLET**
        SPECTROPHOTOMETERS

ULTRAVIOLET SPECTROSCOPY
GS   SPECTROSCOPY
    . **ULTRAVIOLET SPECTROSCOPY**
RT   ABSORPTION SPECTROSCOPY
    ASTRONOMICAL SPECTROSCOPY
    MOLECULAR SPECTROSCOPY
    OPTOGALVANIC SPECTROSCOPY
    RADIO SPECTROSCOPY
    SPECTROSCOPIC ANALYSIS
    SPECTRUM ANALYSIS
    VACUUM SPECTROSCOPY
    X RAY SPECTROSCOPY

ULTRAVIOLET TELESCOPES
GS   TELESCOPES
    . ASTRONOMICAL TELESCOPES
    . . **ULTRAVIOLET TELESCOPES**
RT   FAR ULTRAVIOLET RADIATION
    SPACEBORNE ASTRONOMY
    ULTRAVIOLET ASTRONOMY
    X RAY ASTRONOMY

ULYSSES MISSION
UF   INTERNATIONAL SOLAR POLAR MISSION
GS   SPACE MISSIONS
    . **ULYSSES MISSION**
RT   INERTIAL UPPER STAGE
    MISSION PLANNING
    ∞MISSIONS
    SOLAR MAXIMUM MISSION
    SOLAR PROBES
    SUN

UMBILICAL CONNECTORS
GS   CONNECTORS
    . **UMBILICAL CONNECTORS**
RT   BUNDLES
    EXTRAVEHICULAR ACTIVITY
    TETHERLINES

UMBILICAL TOWERS
GS   TOWERS
    . **UMBILICAL TOWERS**
RT   GANTRY CRANES
    LAUNCHING PADS

UMBRAS
RT   ECLIPSES
    PENUMBRAS
    SHADOWS

†

UMKEHR EFFECT
RT   ∞EFFECTS
    LIGHT SCATTERING
    OZONOSPHERE
    SUNLIGHT
    ULTRAVIOLET RADIATION

UMKLAPP PROCESS
RT   ELECTRON SCATTERING
    PHONON BEAMS
    PHONONS
    PHOTON-ELECTRON INTERACTION

UMKLAPP PROCESS-(CONT.)
    ∞PROCESSES

UNCAMBERED WINGS
GS   AIRFOILS
    . WINGS
    . . **UNCAMBERED WINGS**
    . . . RING WINGS
RT   CAMBERED WINGS
    FIXED WINGS
    THIN WINGS
    TWISTED WINGS

UNCONSCIOUSNESS
GS   **UNCONSCIOUSNESS**
    . BLACKOUT (PHYSIOLOGY)
    . . BLACKOUT PREVENTION
    . NARCOSIS
RT   ANESTHESIA
    ∞COMA
    SYNCOPE

UNCONTROLLED REENTRY (SPACECRAFT)
GS   ATMOSPHERIC ENTRY
    . REENTRY
    . . HYPERSONIC REENTRY
    . . . **UNCONTROLLED REENTRY**
        (SPACECRAFT)
    . . SPACECRAFT REENTRY
    . . . **UNCONTROLLED REENTRY**
        (SPACECRAFT)
RT   AERODYNAMIC HEATING
    COSMOS 954 SATELLITE
    DESCENT
    ∞ENTRY
    FLIGHT PATHS
    PLASMA SHEATHS
    SPACECRAFT SURVIVABILITY

UNCOUPLED MODES
GS   MODES
    . VIBRATION MODE
    . . **UNCOUPLED MODES**
RT   COUPLED MODES
    COUPLES
    MODES (STANDING WAVES)

UNDAMPED OSCILLATIONS
GS   OSCILLATIONS
    . **UNDAMPED OSCILLATIONS**
RT   FLAPPING
    FLUTTER
    RESONANT VIBRATION
    STABLE OSCILLATIONS
    WING OSCILLATIONS

UNDER SURFACE BLOWING
RT   AERODYNAMIC CHARACTERISTICS
    AIRCRAFT CONFIGURATIONS
    CIRCULATION CONTROL AIRFOILS
    LIFT
    ∞SURFACES
    UPPER SURFACE BLOWING

UNDERCARRIAGES
GS   FRAMES
    . **UNDERCARRIAGES**
RT   CARRIAGES
    CARTS
    CHASSIS
    DOLLIES
    LANDING GEAR
    SUBSTRUCTURES
    SUSPENSION SYSTEMS (VEHICLES)

UNDERGROUND ACOUSTICS
RT   ACOUSTIC SOUNDING
    EXPLORATION
    MINERALS
    STRATA

UNDERGROUND COMMUNICATION
GS   COMMUNICATING
    . **UNDERGROUND COMMUNICATION**
RT   RADIO COMMUNICATION

UNDERGROUND EXPLOSIONS
GS   EXPLOSIONS
    . **UNDERGROUND EXPLOSIONS**
RT   CHEMICAL EXPLOSIONS
    GAS EXPLOSIONS
    MINES (EXCAVATIONS)
    NUCLEAR EXPLOSIONS
    SEISMIC WAVES

UNDERGROUND EXPLOSIONS-*(CONT.)*
    THERMONUCLEAR EXPLOSIONS

UNDERGROUND RADIO ANTENNA GRID (NAVY)
USE   SEAFARER PROJECT

UNDERGROUND STORAGE
RT   DECOMMISSIONING
    MINES (EXCAVATIONS)
    MISSILE STORAGE
    PROPELLANT STORAGE
    ∞STORAGE
    STORAGE TANKS

UNDERGROUND STRUCTURES
RT   CAVES
    EXCAVATION
    FOUNDATIONS
    MINES (EXCAVATIONS)
    MINING
    PASSAGEWAYS
    TUNNELING (EXCAVATION)

UNDERGROUND TRANSMISSION LINES
GS   TRANSMISSION LINES
    . **UNDERGROUND TRANSMISSION LINES**
RT   CIRCUITS
    ELECTRIC POWER TRANSMISSION
    ∞LINES
    ∞NETWORKS
    POWER LINES

UNDERWATER ACOUSTICS
UF   HYDROACOUSTICS
    UNDERWATER SOUND
GS   ACOUSTICS
    . **UNDERWATER ACOUSTICS**
RT   ACOUSTIC SCATTERING
    COHERENT ACOUSTIC RADIATION
    DEEP SCATTERING LAYERS
    ECHO SOUNDING
    ELASTIC WAVES
    LOFAR
    NOISE (SOUND)
    SHOCK WAVES
    SONAR
    SONOBUOYS
    SOUND FIXING AND RANGING
    SOUND TRANSDUCERS
    THERMOCLINES
    ULTRASONIC RADIATION
    ULTRASONIC WAVE TRANSDUCERS

UNDERWATER BREATHING APPARATUS
GS   BREATHING APPARATUS
    . **UNDERWATER BREATHING
      APPARATUS**
RT   ARGON-OXYGEN ATMOSPHERES
    BIOENGINEERING
    HELIUM-OXYGEN ATMOSPHERES
    LIFE SUPPORT SYSTEMS

UNDERWATER COMMUNICATION
GS   TELECOMMUNICATION
    . COMMUNICATION
    . . **UNDERWATER COMMUNICATION**
RT   SEAFARER PROJECT
    SHOCK WAVES
    SONAR
    SONOBUOYS
    SOUND TRANSDUCERS

UNDERWATER ENGINEERING
RT   BREAKWATERS
    ∞ENGINEERING
    SUBMERGED BODIES

UNDERWATER EXPLOSIONS
GS   EXPLOSIONS
    . **UNDERWATER EXPLOSIONS**
RT   ANTISUBMARINE WARFARE
    CHEMICAL EXPLOSIONS
    HYDROBALLISTICS
    NUCLEAR EXPLOSIONS
    THERMONUCLEAR EXPLOSIONS

UNDERWATER OPTICS
RT   DIFFRACTION PATTERNS
    DIFFRACTION PROPAGATION
    GEOMETRICAL OPTICS
    OPACITY
    OPTICAL DENSITY
    OPTICAL PATHS
    ∞OPTICS

UNDERWATER OPTICS-*(CONT.)*
    REFRACTIVITY

UNDERWATER PHOTOGRAPHY
GS   PHOTOGRAPHY
    . **UNDERWATER PHOTOGRAPHY**
RT   CAMERAS
    COLOR PHOTOGRAPHY
    SEA WATER
    SEAS
    SUBMERGED BODIES

UNDERWATER PHYSIOLOGY
GS   PHYSIOLOGY
    . **UNDERWATER PHYSIOLOGY**
RT   DIVING (UNDERWATER)
    ∞SCIENCE
    STRESS (PHYSIOLOGY)

UNDERWATER PROPULSION
GS   PROPULSION
    . MARINE PROPULSION
    . . **UNDERWATER PROPULSION**
    . . . SUBMARINE PROPULSION
RT   AEROQUATIC VEHICLES
    CHEMICAL PROPULSION
    ELECTRIC PROPULSION
    NUCLEAR PROPULSION
    PROPELLER DRIVE
    TORPEDO ENGINES

UNDERWATER RESEARCH LABORATORIES
GS   LABORATORIES
    . **UNDERWATER RESEARCH
      LABORATORIES**
    RESEARCH VEHICLES
    . **UNDERWATER RESEARCH
      LABORATORIES**
    SUBMERGED BODIES
    . **UNDERWATER RESEARCH
      LABORATORIES**
    WATER VEHICLES
    . UNDERWATER VEHICLES
    . . **UNDERWATER RESEARCH
      LABORATORIES**
RT   BATHYMETERS
    OCEAN DATA ACQUISITIONS SYSTEMS
    OCEANOGRAPHY

UNDERWATER RESOURCES
GS   RESOURCES
    . EARTH RESOURCES
    . . **UNDERWATER RESOURCES**
RT   CRUDE OIL
    DREDGING
    FOSSIL FUELS
    GEOTHERMAL RESOURCES
    MARINE RESOURCES
    MINERAL DEPOSITS
    OCEAN BOTTOM
    OCEANOGRAPHY
    OIL EXPLORATION
    SEA WATER
    WATER RESOURCES

UNDERWATER SOUND
USE   UNDERWATER ACOUSTICS

UNDERWATER STRUCTURES
RT   BREAKWATERS
    STRUCTURAL DESIGN
    SUBMERGED BODIES

UNDERWATER TESTS
GS   ENVIRONMENTAL TESTS
    . **UNDERWATER TESTS**
RT   CORROSION TESTS
    DIVING (UNDERWATER)
    WATER IMMERSION

UNDERWATER TO SURFACE MISSILES
GS   MISSILES
    . **UNDERWATER TO SURFACE MISSILES**
    . . SUBROC MISSILE
RT   ∞SURFACES

UNDERWATER TRAJECTORIES
GS   TRAJECTORIES
    . **UNDERWATER TRAJECTORIES**
RT   ANTISUBMARINE WARFARE
    HYDROBALLISTICS
    MISSILE TRAJECTORIES
    SUBROC MISSILE
    TORPEDOES

UNDERWATER VEHICLES
GS   WATER VEHICLES
    . **UNDERWATER VEHICLES**
    . . SUBMARINES
    . . . BALLISTIC MISSILE SUBMARINES
    . . . . GUIDED MISSILE SUBMARINES
    . . . TRIDENT SUBMARINE
    . . UNDERWATER RESEARCH
      LABORATORIES
RT   AEROQUATIC VEHICLES
    BOATS
    ∞MILITARY VEHICLES
    RESEARCH VEHICLES
    SHIPS
    SUBMERGED BODIES
    SURFACE VEHICLES
    ∞VEHICLES

UNIAXIAL STRAIN
USE   AXIAL STRAIN

UNIDENTIFIED FLYING OBJECTS
UF   UFO
RT   ∞AIRCRAFT
    EXTRATERRESTRIAL INTELLIGENCE
    SATELLITES
    ∞SPACECRAFT
    ∞VEHICLES

UNIFIED FIELD THEORY
GS   FIELD THEORY (PHYSICS)
    . **UNIFIED FIELD THEORY**
RT   EINSTEIN EQUATIONS
    ELECTROMAGNETIC FIELDS
    ELECTROMAGNETIC INTERACTIONS
    ELECTROMAGNETISM
    GRAVITATION THEORY
    GRAVITATIONAL FIELDS
    PARTICLE THEORY
    PLASMA PHYSICS
    RELATIVITY
    THEORETICAL PHYSICS

UNIFIED S BAND
RT   APOLLO SPACECRAFT
    CARRIER FREQUENCIES
    CIRCUMLUNAR COMMUNICATION
    COMMUNICATION EQUIPMENT
    DIFFERENTIAL PULSE CODE
      MODULATION
    MANNED SPACE FLIGHT NETWORK
    PULSE CODE MODULATION
    SPACECRAFT COMMUNICATION
    SPACECRAFT TRACKING
    SUPERHIGH FREQUENCIES
    ULTRAHIGH FREQUENCIES

UNIFORM FLOW
GS   FLUID FLOW
    . **UNIFORM FLOW**
    . . BLASIUS FLOW
RT   AERODYNAMICS
    FLUID DYNAMICS
    GAS FLOW
    HEAT TRANSMISSION
    LAMINAR FLOW
    LIQUID FLOW
    MASS FLOW
    MULTIPHASE FLOW
    NONUNIFORM FLOW
    PIPE FLOW
    PRESSURE GRADIENTS
    QUASI-STEADY STATES
    SINGLE-PHASE FLOW
    SOLIDS FLOW
    STEADY FLOW
    STEAM FLOW
    SUBCRITICAL FLOW
    TURBULENT FLOW
    UNSTEADY FLOW

UNIMOLECULAR STRUCTURES
RT   MOLECULAR STRUCTURE
    ∞STRUCTURES

UNIONIZATION
RT   FEDERATIONS
    ORGANIZING
    PERSONNEL
    ∞UNIONS

∞ UNIONS
SN   *(USE OF A MORE SPECIFIC TERM IS
    RECOMMENDED--CONSULT THE TERMS
    LISTED BELOW)*

UNIONS-*(CONT.)*
RT   BOOLEAN ALGEBRA
       UNIONIZATION
       UNIONS (CONNECTORS)

UNIONS (CONNECTORS)
GS   CONNECTORS
       . **UNIONS (CONNECTORS)**
RT   COUPLINGS
       FASTENERS
       FITTINGS
       JOINTS (JUNCTIONS)
       LINKAGES
       ∞ UNIONS

UNIPHASE FLOW
USE   SINGLE-PHASE FLOW

UNIPOLAR TRANSISTORS
USE   FIELD EFFECT TRANSISTORS

UNIQUENESS
RT   ABNORMALITIES
       SINGULARITY (MATHEMATICS)

UNIQUENESS THEOREM
GS   THEOREMS
       . **UNIQUENESS THEOREM**
RT   ALGEBRA
       COMPLEX VARIABLES
       GEOMETRY
       NUMBER THEORY
       PROBABILITY THEORY
       REAL VARIABLES

UNITED ARAB EMIRATES
GS   NATIONS
       . **UNITED ARAB EMIRATES**

UNITED KINGDOM
UF   GREAT BRITAIN
GS   NATIONS
       . **UNITED KINGDOM**
       . . ENGLAND
       . . SCOTLAND
RT   ENGLISH CHANNEL
       EUROPE
       UK SPACE PROGRAM
       UNITED KINGDOM
       UNITED KINGDOM

UNITED KINGDOM SATELLITES
USE   UK SATELLITES

UNITED NATIONS
RT   COMMUNITIES
       DEVELOPING NATIONS
       FEDERATIONS
       INTERNATIONAL COOPERATION
       INTERNATIONAL LAW
       NATIONS
       ORGANIZATIONS
       POLITICS
       SEA LAW
       WORLD METEOROLOGICAL
          ORGANIZATION

UNITED STATES
UF   USA (UNITED STATES)
GS   NATIONS
       . **UNITED STATES**
       . . ALABAMA
       . . ALASKA
       . . ARIZONA
       . . ARKANSAS
       . . CALIFORNIA
       . . COLORADO
       . . CONNECTICUT
       . . DELAWARE
       . . FLORIDA
       . . GEORGIA
       . . HAWAII
       . . IDAHO
       . . ILLINOIS
       . . INDIANA
       . . IOWA
       . . KANSAS
       . . KENTUCKY
       . . LOUISIANA
       . . MAINE
       . . MARYLAND
       . . MASSACHUSETTS
       . . MICHIGAN
       . . MINNESOTA

UNITED STATES-*(CONT.)*
       . . MISSISSIPPI
       . . MISSOURI
       . . MONTANA
       . . NEBRASKA
       . . NEVADA
       . . NEW HAMPSHIRE
       . . NEW JERSEY
       . . NEW MEXICO
       . . NEW YORK
       . . NORTH CAROLINA
       . . NORTH DAKOTA
       . . OHIO
       . . OKLAHOMA
       . . OREGON
       . . PENNSYLVANIA
       . . RHODE ISLAND
       . . SOUTH CAROLINA
       . . SOUTH DAKOTA
       . . TENNESSEE
       . . TEXAS
       . . UTAH
       . . VERMONT
       . . VIRGINIA
       . . WASHINGTON
       . . WEST VIRGINIA
       . . WISCONSIN
       . . WYOMING
RT   ALEUTIAN ISLANDS (US)
       CASCADE RANGE (CA-OR-WA)
       CENTRAL ATLANTIC REGION (US)
       DISTRICT OF COLUMBIA
       GREAT LAKES (NORTH AMERICA)
       GREAT PLAINS CORRIDOR (NORTH
          AMERICA)
       GUAM
       INTERNATIONAL FIELD YEAR FOR
          GREAT LAKES
       INTERNATIONAL HYDROLOGICAL
          DECADE
       MISSOURI RIVER (US)
       NEW ENGLAND (US)
       NORTH AMERICA
       PACIFIC NORTHWEST (US)
       PANAMA CANAL ZONE
       PUERTO RICO
       ROCKY MOUNTAINS (NORTH AMERICA)
       SOUTHERN CALIFORNIA
       VIRGIN ISLANDS

UNITS OF MEASUREMENT
GS   **UNITS OF MEASUREMENT**
       . INTERNATIONAL SYSTEM OF UNITS
RT   CONVERSION TABLES
       DIMENSIONAL ANALYSIS
       DIMENSIONS
       ∞ MEASUREMENT
       METRICATION
       METROLOGY
       MONTH
       PARAMETERIZATION
       SIDEREAL TIME
       SYMBOLS
       TIME

UNITY
RT   HOMOGENEITY
       STABILITY

UNIVAC COMPUTERS
GS   DATA PROCESSING EQUIPMENT
       . COMPUTERS
       . . **UNIVAC COMPUTERS**
       . . . UNIVAC LARC COMPUTER
       . . . UNIVAC 80 COMPUTER
       . . . UNIVAC 418 COMPUTER
       . . . UNIVAC 490 COMPUTER
       . . . UNIVAC 494 COMPUTER
       . . . UNIVAC 1100 SERIES COMPUTERS
       . . . . UNIVAC 1105 COMPUTER
       . . . . UNIVAC 1106 COMPUTER
       . . . . UNIVAC 1107 COMPUTER
       . . . . UNIVAC 1108 COMPUTER
       . . . . UNIVAC 1110 COMPUTER
       . . . UNIVAC 1230 COMPUTER
RT   DIGITAL COMPUTERS

UNIVAC LARC COMPUTER
GS   DATA PROCESSING EQUIPMENT
       . COMPUTERS
       . . DIGITAL COMPUTERS
       . . . **UNIVAC LARC COMPUTER**
       . . UNIVAC COMPUTERS
       . . . **UNIVAC LARC COMPUTER**

UNIVAC 80 COMPUTER
GS   DATA PROCESSING EQUIPMENT
       . COMPUTERS
       . . DIGITAL COMPUTERS
       . . . **UNIVAC 80 COMPUTER**
       . . UNIVAC COMPUTERS
       . . . **UNIVAC 80 COMPUTER**

UNIVAC 418 COMPUTER
GS   DATA PROCESSING EQUIPMENT
       . COMPUTERS
       . . DIGITAL COMPUTERS
       . . . **UNIVAC 418 COMPUTER**
       . . UNIVAC COMPUTERS
       . . . **UNIVAC 418 COMPUTER**

UNIVAC 490 COMPUTER
GS   DATA PROCESSING EQUIPMENT
       . COMPUTERS
       . . DIGITAL COMPUTERS
       . . . **UNIVAC 490 COMPUTER**
       . . UNIVAC COMPUTERS
       . . . **UNIVAC 490 COMPUTER**

UNIVAC 494 COMPUTER
GS   DATA PROCESSING EQUIPMENT
       . COMPUTERS
       . . DIGITAL COMPUTERS
       . . . **UNIVAC 494 COMPUTER**
       . . UNIVAC COMPUTERS
       . . . **UNIVAC 494 COMPUTER**

UNIVAC 1100 SERIES COMPUTERS
GS   DATA PROCESSING EQUIPMENT
       . COMPUTERS
       . . ANALOG COMPUTERS
       . . . **UNIVAC 1100 SERIES COMPUTERS**
       . . DIGITAL COMPUTERS
       . . . **UNIVAC 1100 SERIES COMPUTERS**
       . . . . UNIVAC 1105 COMPUTER
       . . . . UNIVAC 1106 COMPUTER
       . . . . UNIVAC 1107 COMPUTER
       . . . . UNIVAC 1108 COMPUTER
       . . . . UNIVAC 1110 COMPUTER
       . . UNIVAC COMPUTERS
       . . . **UNIVAC 1100 SERIES COMPUTERS**
       . . . . UNIVAC 1105 COMPUTER
       . . . . UNIVAC 1106 COMPUTER
       . . . . UNIVAC 1107 COMPUTER
       . . . . UNIVAC 1108 COMPUTER
       . . . . UNIVAC 1110 COMPUTER

UNIVAC 1105 COMPUTER
GS   DATA PROCESSING EQUIPMENT
       . COMPUTERS
       . . DIGITAL COMPUTERS
       . . . UNIVAC 1100 SERIES COMPUTERS
       . . . . **UNIVAC 1105 COMPUTER**
       . . UNIVAC COMPUTERS
       . . . UNIVAC 1100 SERIES COMPUTERS
       . . . . **UNIVAC 1105 COMPUTER**

UNIVAC 1106 COMPUTER
GS   DATA PROCESSING EQUIPMENT
       . COMPUTERS
       . . DIGITAL COMPUTERS
       . . . UNIVAC 1100 SERIES COMPUTERS
       . . . . **UNIVAC 1106 COMPUTER**
       . . UNIVAC COMPUTERS
       . . . UNIVAC 1100 SERIES COMPUTERS
       . . . . **UNIVAC 1106 COMPUTER**

UNIVAC 1107 COMPUTER
GS   DATA PROCESSING EQUIPMENT
       . COMPUTERS
       . . DIGITAL COMPUTERS
       . . . UNIVAC 1100 SERIES COMPUTERS
       . . . . **UNIVAC 1107 COMPUTER**
       . . UNIVAC COMPUTERS
       . . . UNIVAC 1100 SERIES COMPUTERS
       . . . . **UNIVAC 1107 COMPUTER**

UNIVAC 1108 COMPUTER
GS   DATA PROCESSING EQUIPMENT
       . COMPUTERS
       . . DIGITAL COMPUTERS
       . . . UNIVAC 1100 SERIES COMPUTERS
       . . . . **UNIVAC 1108 COMPUTER**
       . . UNIVAC COMPUTERS
       . . . UNIVAC 1100 SERIES COMPUTERS
       . . . . **UNIVAC 1108 COMPUTER**

UNIVAC 1110 COMPUTER
GS   DATA PROCESSING EQUIPMENT

UNIVAC 1110 COMPUTER-*(CONT.)*
. COMPUTERS
. . DIGITAL COMPUTERS
. . . UNIVAC 1100 SERIES COMPUTERS
. . . . **UNIVAC 1110 COMPUTER**
. . UNIVAC COMPUTERS
. . . UNIVAC 1100 SERIES COMPUTERS
. . . . **UNIVAC 1110 COMPUTER**

UNIVAC 1230 COMPUTER
GS   DATA PROCESSING EQUIPMENT
. COMPUTERS
. . DIGITAL COMPUTERS
. . . **UNIVAC 1230 COMPUTER**
. . UNIVAC COMPUTERS
. . . **UNIVAC 1230 COMPUTER**

UNIVERSAL TIME
GS   TIME
. **UNIVERSAL TIME**
RT   EPHEMERIS TIME

UNIVERSE
UF   METAGALAXY
RT   BIG BANG COSMOLOGY
        CELESTIAL BODIES
        COSMOLOGY
        ∞COSMOS
        RELIC RADIATION

UNIVERSITIES
UF   COLLEGES
RT   EDUCATION
        INSTRUCTORS
        LEARNING
        SCHOOLS
        STUDENTS

UNIVERSITY PROGRAM
GS   PROGRAMS
. **UNIVERSITY PROGRAM**
RT   BUREAUS (ORGANIZATIONS)
        INVESTIGATION
        NASA PROGRAMS
        TEAMS

UNLOADING
RT   ∞DISCHARGE
        DISPOSAL
        DUMPING
        EJECTION
        EMPTYING
        EVACUATING (TRANSPORTATION)
        EXPULSION
        LOADING OPERATIONS
        MATERIALS HANDLING
        RELEASING
        REMOVAL
        SPREADING

UNLOADING WAVES
GS   ELASTIC WAVES
. **UNLOADING WAVES**

UNMANNED SPACECRAFT
GS   **UNMANNED SPACECRAFT**
. BIOSATELLITES
. . BIOSATELLITE 1
. . BIOSATELLITE 2
. . BIOSATELLITE 3
. . SPUTNIK 2 SATELLITE
. EXOSAT SATELLITE
. EXPLORER 55 SATELLITE
. GEODETIC SATELLITES
. . ANNA SATELLITES
. . EXPLORER 29 SATELLITE
. . EXPLORER 36 SATELLITE
. . GEOS 1 SATELLITE
. . GEOS 2 SATELLITE
. . GEOS 3 SATELLITE
. . LARGOS SATELLITE
. . PAGEOS SATELLITE
. . VANGUARD 1 SATELLITE
. HEAO 1
. HEAO 2
. HEAO 3
. NAVIGATION SATELLITES
. . EXPLORER 22 SATELLITE
. . NAVSTAR SATELLITES
. . NOVA SATELLITES
. . REFSAT
. . TRANSIT ATTITUDE CONTROL
        SATELLITE
. . TRANSIT SATELLITES
. PASSIVE SATELLITES

UNMANNED SPACECRAFT-*(CONT.)*
. . BEACON SATELLITES
. . . BEACON EXPLORER A
. . . EXPLORER 22 SATELLITE
. . ECHO SATELLITES
. . . ECHO 1 SATELLITE
. . . ECHO 2 SATELLITE
. . LAGEOS (SATELLITE)
. . PAGEOS SATELLITE
. PIONEER VENUS SPACECRAFT
. . PIONEER VENUS 1 SPACECRAFT
. . PIONEER VENUS 2 SPACECRAFT
. . . PIONEER VENUS 2 TRANSPORTER
        BUS
. POLYOT SATELLITES
. SOLAR OBSERVATORIES
. . OSO
. . . AOSO
. . . OSO-C
. . . OSO-1
. . . OSO-2
. . . OSO-3
. . . OSO-4
. . . OSO-5
. . . OSO-6
. . . OSO-7
. . . OSO-8
. SPACE PROBES
. . EXPLORER 18 SATELLITE
. . GIOTTO MISSION
. . JUPITER PROBES
. . . GALILEO PROBE
. . . GALILEO SPACECRAFT
. . LUNAR PROBES
. . . LUNIK LUNAR PROBES
. . . . LUNIK 2 LUNAR PROBE
. . . . LUNIK 3 LUNAR PROBE
. . . . LUNIK 9 LUNAR PROBE
. . . . LUNIK 10 LUNAR PROBE
. . . . LUNIK 11 LUNAR PROBE
. . . . LUNIK 12 LUNAR PROBE
. . . . LUNIK 13 LUNAR PROBE
. . . . LUNIK 14 LUNAR PROBE
. . . . LUNIK 16 LUNAR PROBE
. . . . LUNIK 17 LUNAR PROBE
. . . . LUNIK 19 LUNAR PROBE
. . . . LUNIK 20 LUNAR PROBE
. . . . LUNIK 22 LUNAR PROBE
. . . RANGER LUNAR PROBES
. . . . RANGER LUNAR LANDING
          VEHICLES
. . . . RANGER 1 LUNAR PROBE
. . . . RANGER 2 LUNAR PROBE
. . . . RANGER 3 LUNAR PROBE
. . . . RANGER 4 LUNAR PROBE
. . . . RANGER 5 LUNAR PROBE
. . . . RANGER 6 LUNAR PROBE
. . . . RANGER 7 LUNAR PROBE
. . . . RANGER 8 LUNAR PROBE
. . . . RANGER 9 LUNAR PROBE
. . . SURVEYOR LUNAR PROBES
. . . . SURVEYOR 1 LUNAR PROBE
. . . . SURVEYOR 2 LUNAR PROBE
. . . . SURVEYOR 3 LUNAR PROBE
. . . . SURVEYOR 4 LUNAR PROBE
. . . . SURVEYOR 5 LUNAR PROBE
. . . . SURVEYOR 6 LUNAR PROBE
. . . . SURVEYOR 7 LUNAR PROBE
. . MARINER SPACE PROBES
. . . MARINER R 2 SPACE PROBE
. . . MARINER VENUS-MERCURY 1973
. . . MARINER 1 SPACE PROBE
. . . MARINER 2 SPACE PROBE
. . . MARINER 3 SPACE PROBE
. . . MARINER 4 SPACE PROBE
. . . MARINER 5 SPACE PROBE
. . . MARINER 6 SPACE PROBE
. . . MARINER 7 SPACE PROBE
. . . MARINER 8 SPACE PROBE
. . . MARINER 9 SPACE PROBE
. . . MARINER 10 SPACE PROBE
. . . MARINER 11 SPACE PROBE
. . . MARINER-MERCURY 1973
. . MARINER SPACECRAFT
. . . MARINER C SPACECRAFT
. . . MARINER VENUS 67 SPACECRAFT
. . MARS PROBES
. . . ADVANCED RECONN ELECTRIC
        SPACECRAFT
. . . MARINER 3 SPACE PROBE
. . . MARINER 4 SPACE PROBE
. . . MARINER 6 SPACE PROBE
. . . MARINER 7 SPACE PROBE
. . . MARINER 8 SPACE PROBE
. . . MARINER 10 SPACE PROBE
. . . MARS 1 SPACECRAFT

UNMANNED SPACECRAFT-*(CONT.)*
. . . MARS 2 SPACECRAFT
. . . MARS 3 SPACECRAFT
. . . MARS 4 SPACECRAFT
. . . MARS 5 SPACECRAFT
. . . MARS 6 SPACECRAFT
. . . MARS 7 SPACECRAFT
. . . VIKING ORBITER 1975
. . . ZOND 2 SPACE PROBE
. . PIONEER SPACE PROBES
. . . PIONEER VENUS 2 ENTRY PROBES
. . . . PIONEER VENUS 2 NIGHT PROBE
. . . PIONEER 1 SPACE PROBE
. . . PIONEER 2 SPACE PROBE
. . . PIONEER 3 SPACE PROBE
. . . PIONEER 4 SPACE PROBE
. . . PIONEER 5 SPACE PROBE
. . . PIONEER 6 SPACE PROBE
. . . PIONEER 7 SPACE PROBE
. . . PIONEER 8 SPACE PROBE
. . . PIONEER 9 SPACE PROBE
. . . PIONEER 10 SPACE PROBE
. . . PIONEER 11 SPACE PROBE
. . SOLAR PROBES
. . . HELIOS A
. . . HELIOS B
. . . HELIOS 1
. . . HELIOS 2
. . . SUNBLAZER SPACE PROBE
. . VENUS PROBES
. . . MARINER 1 SPACE PROBE
. . . MARINER 2 SPACE PROBE
. . . MARINER 5 SPACE PROBE
. . . VENERA SATELLITES
. . . . VENERA 2 SATELLITE
. . . . VENERA 3 SATELLITE
. . . . VENERA 4 SATELLITE
. . . . VENERA 5 SATELLITE
. . . . VENERA 6 SATELLITE
. . . . VENERA 7 SATELLITE
. . . . VENERA 8 SATELLITE
. . . . VENERA 9 SATELLITE
. . . . VENERA 10 SATELLITE
. . . . VENERA 11 SATELLITE
. . . . VENERA 12 SATELLITE
. . . ZOND 1 SPACE PROBE
. . . ZOND 3 SPACE PROBE
. . . ZOND 4 SPACE PROBE
. . . ZOND 5 SPACE PROBE
. . . ZOND 6 SPACE PROBE
. . . ZOND 7 SPACE PROBE
. . . ZOND 8 SPACE PROBE
. TECHNOLOGY FEASIBILITY
        SPACECRAFT
. VOYAGER 1 SPACECRAFT
. VOYAGER 2 SPACECRAFT
. ZOND SPACE PROBES
. . ZOND 1 SPACE PROBE
. . ZOND 2 SPACE PROBE
. . ZOND 3 SPACE PROBE
. . ZOND 4 SPACE PROBE
. . ZOND 5 SPACE PROBE
. . ZOND 6 SPACE PROBE
. . ZOND 7 SPACE PROBE
. . ZOND 8 SPACE PROBE
RT   ARTIFICIAL SATELLITES
        COMMUNICATION SATELLITES
        GEOPHYSICAL SATELLITES
        GRAVITY GRADIENT SATELLITES
        INFLATABLE SPACECRAFT
        INTERPLANETARY SPACECRAFT
        LUNAR LANDING MODULES
        LUNAR SATELLITES
        LUNAR SPACECRAFT
        MANNED SPACECRAFT
        MARINER PROGRAM
        METEOROLOGICAL SATELLITES
        MILITARY SPACECRAFT
        RECONNAISSANCE SPACECRAFT
        RECOVERABLE SPACECRAFT
        RENDEZVOUS SPACECRAFT
        REUSABLE SPACECRAFT
        SATELLITES
        SIRS B SATELLITE
        SPACE CAPSULES
        ∞SPACECRAFT
        VIKING LANDER SPACECRAFT
        VIKING LANDER 1
        VIKING LANDER 2
        VIKING ORBITER SPACECRAFT
        VIKING ORBITER 1
        VIKING ORBITER 2
        VIKING 1 SPACECRAFT
        VIKING 2 SPACECRAFT
        VOYAGER PROJECT

UNSATURATION (CHEMISTRY)
RT   CHEMICAL BONDS
    ∞CHEMISTRY
    PRECIPITATION (CHEMISTRY)
    ∞SATURATION
    SATURATION (CHEMISTRY)

UNSTEADY FLOW
UF   PULSATING FLOW
GS   FLUID FLOW
   . **UNSTEADY FLOW**
   . . OSCILLATING FLOW
RT   AERODYNAMICS
    CRITICAL FLOW
    ∞FLOW
    FLOW STABILITY
    FLOW VELOCITY
    FLUID DYNAMICS
    GAS FLOW
    HEAT TRANSMISSION
    HYDRODYNAMIC COEFFICIENTS
    LAMINAR FLOW
    LIQUID FLOW
    MASS FLOW
    METHOD OF CHARACTERISTICS
    MULTIPHASE FLOW
    NONEQUILIBRIUM FLOW
    NONNEWTONIAN FLOW
    NONUNIFORM FLOW
    ORIFICE FLOW
    PIPE FLOW
    PRESSURE GRADIENTS
    SINGLE-PHASE FLOW
    SOLIDS FLOW
    STEADY FLOW
    STEADY STATE
    STEAM FLOW
    STROUHAL NUMBER
    SUBCRITICAL FLOW
    SUPERCRITICAL FLOW
    TURBULENCE
    UNIFORM FLOW

UNSTEADY STATE
RT   ∞EQUILIBRIUM
    FLUID DYNAMICS
    METASTABLE STATE
    NONEQUILIBRIUM CONDITIONS
    STABILITY
    STEADY STATE
    SYSTEMS STABILITY
    THERMODYNAMICS

UNSWEPT WINGS
GS   AIRFOILS
   . WINGS
   . . **UNSWEPT WINGS**
   . . . INFINITE SPAN WINGS
   . . . RECTANGULAR WINGS
   . . . RING WINGS
RT   FIXED WINGS
    SWEPT WINGS
    WING PLANFORMS

UP-CONVERTERS
GS   FREQUENCY CONVERTERS
   . **UP-CONVERTERS**
RT   ∞CONVERTERS
    PARAMETRIC FREQUENCY CONVERTERS
    TRANSFORMERS

UPDRAFTS
USE   VERTICAL AIR CURRENTS

UPGRADING
RT   BENEFICIATION
    CONCENTRATING
    ENRICHMENT
    EXPERIENCE
    IMPROVEMENT
    PROMOTION
    PUBLIC RELATIONS
    PURIFICATION
    QUALITY
    REFINING

UPLINKING
RT   CARRIER TO NOISE RATIOS
    COMMUNICATION SATELLITES
    DOWNLINKING
    FREQUENCY REUSE
    MICROWAVE TRANSMISSION
    SATELLITE TRANSMISSION
    TRANSMISSION EFFICIENCY

UPPER AIR
USE   UPPER ATMOSPHERE

UPPER ATMOSPHERE
UF   UPPER AIR
GS   EARTH ATMOSPHERE
   . **UPPER ATMOSPHERE**
   . . EXOSPHERE
   . . IONOSPHERE
   . . . E REGION
   . . . . E-1 LAYER
   . . . . E-2 LAYER
   . . . . SPORADIC E LAYER
   . . . LOWER IONOSPHERE
   . . . . D REGION
   . . . UPPER IONOSPHERE
   . . . . F REGION
   . . . . . F 1 REGION
   . . . . . F 2 REGION
   . . MAGNETOSPHERE
   . . . GEOMAGNETIC TAIL
   . . . MAGNETOPAUSE
   . . MESOPAUSE
   . . MESOSPHERE
   . . THERMOSPHERE
   . . . TURBOPAUSE
RT   ACOUSTIC SOUNDING
    AERONOMY
    CHEMOSPHERE
    HETEROSPHERE
    HIGH ALTITUDE
    HOMOSPHERE
    METEOR TRAILS
    METEOROLOGICAL BALLOONS
    MIDDLE ATMOSPHERE
    OZONOSPHERE
    PLASMASPHERE
    PROTON PRECIPITATION
    RADIATION BELTS
    SATELLITE ATMOSPHERES

UPPER IONOSPHERE
GS   EARTH ATMOSPHERE
   . UPPER ATMOSPHERE
   . . IONOSPHERE
   . . . **UPPER IONOSPHERE**
   . . . . F REGION
   . . . . . F 1 REGION
   . . . . . F 2 REGION
RT   E REGION

UPPER STAGE ROCKET ENGINES
GS   ENGINES
   . ROCKET ENGINES
   . . **UPPER STAGE ROCKET ENGINES**
RT   INERTIAL UPPER STAGE
    MULTISTAGE ROCKET VEHICLES
    SPACE SHUTTLE ASCENT STAGE
    SPACECRAFT CONFIGURATIONS
    SPINNING SOLID UPPER STAGE
    STAGE SEPARATION

UPPER SURFACE BLOWING
RT   AERODYNAMIC CHARACTERISTICS
    AIRCRAFT CONFIGURATIONS
    CIRCULATION CONTROL AIRFOILS
    LIFT
    ∞SURFACES
    UNDER SURFACE BLOWING

UPPER SURFACE BLOWN FLAPS
GS   AIRFOILS
   . FLAPS (CONTROL SURFACES)
   . . EXTERNALLY BLOWN FLAPS
   . . . **UPPER SURFACE BLOWN FLAPS**
   CONTROL SURFACES
   . FLAPS (CONTROL SURFACES)
   . . EXTERNALLY BLOWN FLAPS
   . . . **UPPER SURFACE BLOWN FLAPS**
RT   AIRCRAFT STABILITY
    BOUNDARY LAYER CONTROL
    CONTROL SURFACES
    LIFT AUGMENTATION
    LIFT DEVICES
    ∞SURFACES

UPPER VOLTA
USE   BURKINA

UPSETTING
RT   COLD PRESSING
    COLD WORKING
    FORMING TECHNIQUES
    HOT PRESSING
    HOT WORKING

UPSETTING-*(CONT.)*
    PRESSING (FORMING)
    STAMPING

UPSTREAM
RT   AIR CURRENTS
    WATER CURRENTS
    WIND DIRECTION

UPWASH
RT   DOWNWASH
    ∞DRAFT
    INTERFERENCE DRAG
    INTERFERENCE LIFT

UPWELLING
USE   UPWELLING WATER

UPWELLING WATER
UF   UPWELLING
RT   ATMOSPHERIC CIRCULATION
    COASTS
    OCEAN CURRENTS
    WIND (METEOROLOGY)
    WIND DIRECTION

URACIL
GS   HETEROCYCLIC COMPOUNDS
   . **URACIL**
   NITROGEN COMPOUNDS
   . **URACIL**
   PYRIMIDINES
   . **URACIL**
RT   ALLOXAN
    URIDYLIC ACID

URANIUM
GS   CHEMICAL ELEMENTS
   . ACTINIDE SERIES
   . . **URANIUM**
   . . . URANIUM ISOTOPES
   . . . . URANIUM 232
   . . . . URANIUM 233
   . . . . URANIUM 234
   . . . . URANIUM 235
   . . . . URANIUM 238
   . . . URANIUM PLASMAS
   METALS
   . ACTINIDE SERIES
   . . **URANIUM**
   . . . URANIUM ISOTOPES
   . . . . URANIUM 232
   . . . . URANIUM 233
   . . . . URANIUM 234
   . . . . URANIUM 235
   . . . . URANIUM 238
   . . . URANIUM PLASMAS
RT   FISSIONABLE MATERIALS
    JET MEMBRANE PROCESS
    NUCLEAR FUELS

URANIUM ALLOYS
GS   ALLOYS
   . **URANIUM ALLOYS**
RT   NUCLEAR FUEL ELEMENTS
    NUCLEAR FUELS

URANIUM CARBIDES
GS   ACTINIDE SERIES COMPOUNDS
   . URANIUM COMPOUNDS
   . . **URANIUM CARBIDES**
   CARBON COMPOUNDS
   . CARBIDES
   . . **URANIUM CARBIDES**
RT   CERAMIC NUCLEAR FUELS
    NUCLEAR FUEL ELEMENTS
    NUCLEAR FUELS

URANIUM COMPOUNDS
GS   ACTINIDE SERIES COMPOUNDS
   . **URANIUM COMPOUNDS**
   . . URANIUM CARBIDES
   . . URANIUM FLUORIDES
   . . URANIUM OXIDES
RT   CERAMIC NUCLEAR FUELS
    ∞CHEMICAL COMPOUNDS
    ∞METAL COMPOUNDS
    NUCLEAR FUELS

URANIUM FLUORIDES
GS   ACTINIDE SERIES COMPOUNDS
   . URANIUM COMPOUNDS
   . . **URANIUM FLUORIDES**
   HALOGEN COMPOUNDS

URANIUM FLUORIDES-*(CONT.)*
```
          .  FLUORINE COMPOUNDS
          .  .  FLUORIDES
          .  .  METAL FLUORIDES
          .  .  .  .  URANIUM FLUORIDES
          .  HALIDES
          .  .  METAL HALIDES
          .  .  .  URANIUM FLUORIDES
```

URANIUM ISOTOPES
```
GS        CHEMICAL ELEMENTS
          .  ACTINIDE SERIES
          .  .  URANIUM
          .  .  .  URANIUM ISOTOPES
          .  .  .  .  URANIUM 232
          .  .  .  .  URANIUM 233
          .  .  .  .  URANIUM 234
          .  .  .  .  URANIUM 235
          .  .  .  .  URANIUM 238
          .  NUCLIDES
          .  .  ISOTOPES
          .  .  .  URANIUM ISOTOPES
          .  .  .  .  URANIUM 232
          .  .  .  .  URANIUM 233
          .  .  .  .  URANIUM 234
          .  .  .  .  URANIUM 235
          .  .  .  .  URANIUM 238
          METALS
          .  ACTINIDE SERIES
          .  .  URANIUM
          .  .  .  URANIUM ISOTOPES
          .  .  .  .  URANIUM 232
          .  .  .  .  URANIUM 233
          .  .  .  .  URANIUM 234
          .  .  .  .  URANIUM 235
          .  .  .  .  URANIUM 238
```

URANIUM OXIDES
```
GS        ACTINIDE SERIES COMPOUNDS
          .  URANIUM COMPOUNDS
          .  .  URANIUM OXIDES
          CHALCOGENIDES
          .  OXIDES
          .  .  METAL OXIDES
          .  .  .  URANIUM OXIDES
RT        CERAMIC NUCLEAR FUELS
          MIXED OXIDES
          NUCLEAR FUELS
```

URANIUM PLASMAS
```
GS        CHEMICAL ELEMENTS
          .  ACTINIDE SERIES
          .  .  URANIUM
          .  .  .  URANIUM PLASMAS
          METALS
          .  ACTINIDE SERIES
          .  .  URANIUM
          .  .  .  URANIUM PLASMAS
RT        MAGNETOHYDRODYNAMICS
          PLASMA COMPOSITION
          PLASMA PHYSICS
          RADIOACTIVE MATERIALS
```

URANIUM 232
```
GS        CHEMICAL ELEMENTS
          .  ACTINIDE SERIES
          .  .  URANIUM
          .  .  .  URANIUM ISOTOPES
          .  .  .  .  URANIUM 232
          .  NUCLIDES
          .  .  ISOTOPES
          .  .  .  RADIOACTIVE ISOTOPES
          .  .  .  .  URANIUM 232
          .  .  .  URANIUM ISOTOPES
          .  .  .  .  URANIUM 232
          METALS
          .  ACTINIDE SERIES
          .  .  URANIUM
          .  .  .  URANIUM ISOTOPES
          .  .  .  .  URANIUM 232
```

URANIUM 233
```
GS        CHEMICAL ELEMENTS
          .  ACTINIDE SERIES
          .  .  URANIUM
          .  .  .  URANIUM ISOTOPES
          .  .  .  .  URANIUM 233
          .  NUCLIDES
          .  .  ISOTOPES
          .  .  .  RADIOACTIVE ISOTOPES
          .  .  .  .  URANIUM 233
          .  .  .  URANIUM ISOTOPES
          .  .  .  .  URANIUM 233
          METALS
          .  ACTINIDE SERIES
```

URANIUM 233-*(CONT.)*
```
          .  .  URANIUM
          .  .  .  URANIUM ISOTOPES
          .  .  .  .  URANIUM 233
RT        NUCLEAR FUELS
```

URANIUM 234
```
GS        CHEMICAL ELEMENTS
          .  ACTINIDE SERIES
          .  .  URANIUM
          .  .  .  URANIUM ISOTOPES
          .  .  .  .  URANIUM 234
          .  NUCLIDES
          .  .  ISOTOPES
          .  .  .  URANIUM ISOTOPES
          .  .  .  .  URANIUM 234
          METALS
          .  ACTINIDE SERIES
          .  .  URANIUM
          .  .  .  URANIUM ISOTOPES
          .  .  .  .  URANIUM 234
```

URANIUM 235
```
GS        CHEMICAL ELEMENTS
          .  ACTINIDE SERIES
          .  .  URANIUM
          .  .  .  URANIUM ISOTOPES
          .  .  .  .  URANIUM 235
          .  NUCLIDES
          .  .  ISOTOPES
          .  .  .  URANIUM ISOTOPES
          .  .  .  .  URANIUM 235
          METALS
          .  ACTINIDE SERIES
          .  .  URANIUM
          .  .  .  URANIUM ISOTOPES
          .  .  .  .  URANIUM 235
RT        NUCLEAR FUELS
```

URANIUM 238
```
GS        CHEMICAL ELEMENTS
          .  ACTINIDE SERIES
          .  .  URANIUM
          .  .  .  URANIUM ISOTOPES
          .  .  .  .  URANIUM 238
          .  NUCLIDES
          .  .  ISOTOPES
          .  .  .  RADIOACTIVE ISOTOPES
          .  .  .  .  URANIUM 238
          .  .  .  URANIUM ISOTOPES
          .  .  .  .  URANIUM 238
          METALS
          .  ACTINIDE SERIES
          .  .  URANIUM
          .  .  .  URANIUM ISOTOPES
          .  .  .  .  URANIUM 238
RT        NUCLEAR FUELS
```

URANUS (PLANET)
```
GS        CELESTIAL BODIES
          .  PLANETS
          .  .  GAS GIANT PLANETS
          .  .  .  URANUS (PLANET)
          .  .  .  .  URANUS RINGS
          .  .  .  .  URANUS SATELLITES
RT        URANUS ATMOSPHERE
```

URANUS ATMOSPHERE
```
GS        ENVIRONMENTS
          .  EXTRATERRESTRIAL ENVIRONMENTS
          .  .  PLANETARY ENVIRONMENTS
          .  .  .  PLANETARY ATMOSPHERES
          .  .  .  .  URANUS ATMOSPHERE
RT        AEROSPACE ENVIRONMENTS
          ∞ATMOSPHERES
          GAS GIANT PLANETS
          HYDROGEN
          METHANE
          URANUS (PLANET)
```

URANUS RINGS
```
GS        CELESTIAL BODIES
          .  PLANETARY RINGS
          .  .  URANUS RINGS
          .  PLANETS
          .  .  GAS GIANT PLANETS
          .  .  .  URANUS (PLANET)
          .  .  .  .  URANUS RINGS
RT        JUPITER RINGS
          NATURAL SATELLITES
          PLANETARY STRUCTURE
          ∞RINGS
          SATURN RINGS
```

URANUS SATELLITES
```
GS        CELESTIAL BODIES
          .  NATURAL SATELLITES
          .  .  URANUS SATELLITES
          .  PLANETS
          .  .  GAS GIANT PLANETS
          .  .  .  URANUS (PLANET)
          .  .  .  .  URANUS SATELLITES
```

URBAN AREAS
```
USE       CITIES
```

URBAN DEVELOPMENT
```
RT        CITIES
          COMMUNITIES
          ∞DEVELOPMENT
          ECONOMIC DEVELOPMENT
          INDUSTRIAL AREAS
          LAND USE
          MEGALOPOLISES
          OPERATIONS RESEARCH
          PARKS
          PLANNING
          ∞PLANS
          REGIONAL PLANNING
          RESIDENTIAL AREAS
          RESOURCES
          STARSITE PROGRAM
          TECHNOLOGIES
```

URBAN PLANNING
```
GS        PLANNING
          .  REGIONAL PLANNING
          .  .  URBAN PLANNING
RT        CENSUS
          CITIES
          COMMUNITIES
          HEAT ISLANDS
          HIGHWAYS
          LAND MANAGEMENT
          LAND USE
          PARKS
          PUBLIC HEALTH
          RECREATION
          SOCIAL FACTORS
          SOCIOLOGY
          STARSITE PROGRAM
          STREETS
```

URBAN RESEARCH
```
RT        CITIES
          COMMUNITIES
          LAND USE
          RECREATION
          SOCIAL FACTORS
          STREETS
```

URBAN TRANSPORTATION
```
GS        TRANSPORTATION
          .  URBAN TRANSPORTATION
RT        AUTOMATED GUIDEWAY TRANSIT
             VEHICLES
          AUTOMATED MIXED TRAFFIC VEHICLES
          AUTOMATED TRANSIT VEHICLES
          INDUSTRIAL AREAS
          MEGALOPOLISES
          RAIL TRANSPORTATION
          RAPID TRANSIT SYSTEMS
          REGIONAL PLANNING
          SURFACE VEHICLES
```

UREAS
```
GS        NITROGEN COMPOUNDS
          .  AMIDES
          .  .  UREAS
          .  .  .  DIFLUOROUREA
          .  .  .  THIOUREAS
          .  .  .  THIURONIUM
RT        DIURETICS
          FERTILIZERS
          URINE
```

URETHANES
```
GS        ALIPHATIC COMPOUNDS
          .  CARBAMATES (TRADENAME)
          .  .  URETHANES
          ESTERS
          .  CARBAMATES (TRADENAME)
          .  .  URETHANES
          POISONS
          .  CARBAMATES (TRADENAME)
          .  .  URETHANES
RT        CYANATES
```

URIC ACID
GS   ACIDS
      . **URIC ACID**
      FUNGICIDES
      . XANTHINES
      . . **URIC ACID**
      HETEROCYCLIC COMPOUNDS
      . XANTHINES
      . . **URIC ACID**
      NITROGEN COMPOUNDS
      . XANTHINES
      . . **URIC ACID**
      PURINES
      . XANTHINES
      . . **URIC ACID**
RT   ALLOXAN

URIDYLIC ACID
GS   ACIDS
      . AMINO ACIDS
      . . **URIDYLIC ACID**
      . NUCLEIC ACIDS
      . . **URIDYLIC ACID**
      ORGANIC COMPOUNDS
      . AMINO ACIDS
      . . **URIDYLIC ACID**
      . NUCLEOTIDES
      . . **URIDYLIC ACID**
      PHOSPHORUS COMPOUNDS
      . ORGANIC PHOSPHORUS COMPOUNDS
      . . **URIDYLIC ACID**
      . PHOSPHATES
      . . **URIDYLIC ACID**
      PROTEINS
      . NUCLEOTIDES
      . . **URIDYLIC ACID**
RT   URACIL

URINALYSIS
GS   CHEMICAL TESTS
      . CHEMICAL ANALYSIS
      . . **URINALYSIS**
RT   DIABETES MELLITUS
      PHYSIOLOGICAL TESTS
      URINE

URINATION
UF   MICTURITION
RT   DIURESIS
      URINE
      WATER BALANCE

URINE
GS   BODY FLUIDS
      . **URINE**
      WASTES
      . LIQUID WASTES
      . . **URINE**
      . METABOLIC WASTES
      . . HUMAN WASTES
      . . . **URINE**
RT   ANTIDIURETICS
      CREATININE
      EXCRETION
      FECES
      HEMATURIA
      KIDNEYS
      UREAS
      URINALYSIS
      URINATION

UROGRAPHY
GS   IMAGERY
      . RADIOGRAPHY
      . . **UROGRAPHY**
      PHOTOGRAPHY
      . **UROGRAPHY**
RT   BLACK AND WHITE PHOTOGRAPHY

UROLITHIASIS
GS   DISEASES
      . **UROLITHIASIS**
RT   CALCULI
      KIDNEYS
      UROLOGY

UROLOGY
GS   MEDICAL SCIENCE
      . **UROLOGY**
RT   BLADDER
      GENITOURINARY SYSTEM
      KIDNEYS
      UROLITHIASIS

URUGUAY
GS   NATIONS
      . **URUGUAY**
RT   SOUTH AMERICA

US-2A AIRCRAFT
USE   S-2 AIRCRAFT

USA (UNITED STATES)
USE   UNITED STATES

USER MANUALS (COMPUTER PROGRAMS)
GS   DOCUMENTS
      . HANDBOOKS
      . . **USER MANUALS (COMPUTER
          PROGRAMS)**
      . MANUALS
      . . **USER MANUALS (COMPUTER
          PROGRAMS)**
RT   COMPUTER PROGRAMS
      PROGRAMS
      REPORT GENERATORS
      ∞ROUTINES
      SUBROUTINES

USER REQUIREMENTS
RT   COMMERCE LAB
      INTERNATIONAL COOPERATION
      REQUIREMENTS
      SPECIFICATIONS

USNS KINGSPORT
USE   SATELLITE COMMUNICATIONS SHIPS

UTAH
GS   NATIONS
      . UNITED STATES
      . . **UTAH**
RT   COLORADO PLATEAU (US)
      COLORADO RIVER (NORTH AMERICA)
      GREAT BASIN (US)
      GREAT SALT LAKE (UT)

UTERUS
GS   ANATOMY
      . GENITOURINARY SYSTEM
      . . **UTERUS**
      REPRODUCTIVE SYSTEMS
      . **UTERUS**
RT   ∞REPRODUCTION

UTILITIES
RT   ∞ELECTRIC EQUIPMENT
      ∞ELECTRIC POWER
      GARBAGE
      INDUSTRIES
      INTEGRATED ENERGY SYSTEMS
      LOGISTICS
      MODULAR INTEGRATED UTILITY SYSTEM
      SERVICES
      SITE SELECTION
      TELEPHONES
      WASTE DISPOSAL
      WATER

UTILITY AIRCRAFT
GS   **UTILITY AIRCRAFT**
      . BO-105 HELICOPTER
      . C-140 AIRCRAFT
      . DHC 4 AIRCRAFT
      . DHC 5 AIRCRAFT
      . DO-27 AIRCRAFT
      . DO-28 AIRCRAFT
      . HC-3 HELICOPTER
      . HH-43 HELICOPTER
      . OH-13 HELICOPTER
      . OH-23 HELICOPTER
      . P-531 HELICOPTER
      . PD-808 AIRCRAFT
      . S-2 AIRCRAFT
      . SAAB 105 AIRCRAFT
      . T-39 AIRCRAFT
      . U-2 AIRCRAFT
      . U-10 AIRCRAFT
      . UH-1 HELICOPTER
      . UH-2 HELICOPTER
      . UH-34 HELICOPTER
      . UH-60A HELICOPTER
      . UH-61A HELICOPTER
      . WESTLAND WHIRLWIND HELICOPTER
      . XV-8A AIRCRAFT
      . Z-37 AIRCRAFT
RT   ∞AIRCRAFT
      BIPLANES

UTILITY AIRCRAFT-(CONT.)
      CARGO AIRCRAFT
      COMMERCIAL AIRCRAFT
      GENERAL AVIATION AIRCRAFT
      HELICOPTERS
      LIGHT AIRCRAFT
      ∞MILITARY AIRCRAFT
      OBSERVATION AIRCRAFT
      RECONNAISSANCE AIRCRAFT
      SNOW AIRCRAFT
      ∞SUBSONIC AIRCRAFT
      TERRAIN FOLLOWING AIRCRAFT
      TRANSPORT AIRCRAFT
      V/STOL AIRCRAFT
      WATER TAKEOFF AND LANDING
        AIRCRAFT

UTILIZATION
UF   APPLICATION
GS   **UTILIZATION**
      . COAL UTILIZATION
      . GEOTHERMAL ENERGY UTILIZATION
      . LASER APPLICATIONS
      . . LASER CUTTING
      . . LASER FUSION
      . REUSE
      . WASTE ENERGY UTILIZATION
      . WASTE UTILIZATION
      . WINDPOWER UTILIZATION
RT   CONSUMPTION
      DEPLETION
      EFFICIENCY
      TECHNOLOGY UTILIZATION

UTRICLE
RT   PLANTS (BOTANY)
      SEEDS

UV CETI STARS
USE   FLARE STARS

UV LASERS
USE   ULTRAVIOLET LASERS

V

V BAND
USE   EXTREMELY HIGH FREQUENCIES

V GROOVES
GS   GROOVES
      . **V GROOVES**
RT   MACHINING
      NOTCHES

V-1 MISSILE
GS   MISSILES
      . SURFACE TO SURFACE MISSILES
      . . **V-1 MISSILE**
RT   LIQUID PROPELLANT ROCKET ENGINES
      PULSEJET ENGINES

V-2 MISSILE
GS   MISSILES
      . BALLISTIC MISSILES
      . . **V-2 MISSILE**
RT   LIQUID PROPELLANT ROCKET ENGINES

V-3 AIRCRAFT
USE   XV-3 AIRCRAFT

V-4 AIRCRAFT
USE   XV-4 AIRCRAFT

V-5 AIRCRAFT
USE   XV-5 AIRCRAFT

V-9 AIRCRAFT
USE   XV-9A AIRCRAFT

V/STOL AIRCRAFT
UF   CONVERTAPLANES
      STEEP GRADIENT AIRCRAFT
GS   **V/STOL AIRCRAFT**
      . CL-84 AIRCRAFT
      . DO-31 AIRCRAFT
      . FV-12A AIRCRAFT
      . G-95/4 AIRCRAFT
      . G-222 AIRCRAFT
      . L-29 JET TRAINER
      . P-1127 AIRCRAFT

V/STOL AIRCRAFT-*(CONT.)*
. P-1154 AIRCRAFT
. ROTARY WING AIRCRAFT
. . AUTOGYROS
. . . AVIAN 2/180 AUTOGIRO
. . HELICOPTERS
. . . ALOUETTE HELICOPTERS
. . . . SA-330 HELICOPTER
. . . . SE-3160 HELICOPTER
. . . BELL 214A HELICOPTER
. . . BO-105 HELICOPTER
. . . CH-21 HELICOPTER
. . . COMPOUND HELICOPTERS
. . . H-17 HELICOPTER
. . . H-54 HELICOPTER
. . . HC-3 HELICOPTER
. . MILITARY HELICOPTERS
. . . . AH-1G HELICOPTER
. . . . AH-64 HELICOPTER
. . . . CH-3 HELICOPTER
. . . . CH-34 HELICOPTER
. . . . CH-46 HELICOPTER
. . . . CH-47 HELICOPTER
. . . . CH-54 HELICOPTER
. . . . H-43 HELICOPTER
. . . . H-53 HELICOPTER
. . . . H-56 HELICOPTER
. . . . H-60 HELICOPTER
. . . . HEAVY LIFT HELICOPTERS
. . . . . CH-62 HELICOPTER
. . . . HH-43 HELICOPTER
. . . . OH-4 HELICOPTER
. . . . OH-5 HELICOPTER
. . . . OH-6 HELICOPTER
. . . . OH-13 HELICOPTER
. . . . OH-23 HELICOPTER
. . . . OH-58 HELICOPTER
. . . . QH-50 HELICOPTER
. . . . S-58 HELICOPTER
. . . . S-61 HELICOPTER
. . . . SA-330 HELICOPTER
. . . . UH-1 HELICOPTER
. . . . UH-2 HELICOPTER
. . . . UH-34 HELICOPTER
. . . . UH-60A HELICOPTER
. . . . UH-61A HELICOPTER
. . . P-531 HELICOPTER
. . . RIGID ROTOR HELICOPTERS
. . . . CH-3 HELICOPTER
. . . . F-28 HELICOPTER
. . . . XH-51 HELICOPTER
. . . SH-3 HELICOPTER
. . . SH-4 HELICOPTER
. . . SIKORSKY WHIRLWIND HELICOPTER
. . . TANDEM ROTOR HELICOPTERS
. . . . CH-46 HELICOPTER
. . . . CH-47 HELICOPTER
. . . . H-25 HELICOPTER
. . . . TH-55 HELICOPTER
. . . WESTLAND WHIRLWIND
        HELICOPTER
. . . XV-9A AIRCRAFT
. . TILT ROTOR AIRCRAFT
. . . XV-15 AIRCRAFT
. SHORT TAKEOFF AIRCRAFT
. . BREGUET 940 AIRCRAFT
. . BREGUET 941 AIRCRAFT
. . C-15 AIRCRAFT
. . C-123 AIRCRAFT
. . DHC 4 AIRCRAFT
. . DHC 5 AIRCRAFT
. . QUESTOL
. . U-10 AIRCRAFT
. VERTICAL TAKEOFF AIRCRAFT
. . FLYING PLATFORMS
. . SC-1 AIRCRAFT
. . VJ-101 AIRCRAFT
. . VZ-8 AIRCRAFT
. . X-13 AIRCRAFT
. . X-14 AIRCRAFT
. . X-19 AIRCRAFT
. . X-22 AIRCRAFT
. . X-22A AIRCRAFT
. . XV-4 AIRCRAFT
. . XV-11A AIRCRAFT
. VZ-2 AIRCRAFT
. XV-3 AIRCRAFT
. XV-5 AIRCRAFT
. XV-8A AIRCRAFT
RT ∞AIRCRAFT
     ANTISUBMARINE WARFARE AIRCRAFT
     ATTACK AIRCRAFT
     COMMERCIAL AIRCRAFT
     DRONE AIRCRAFT
     FAN IN WING AIRCRAFT
     FIGHTER AIRCRAFT

V/STOL AIRCRAFT-*(CONT.)*
GROUND EFFECT MACHINES
H-19 HELICOPTER
HELIPORTS
HOVERING
JET AIRCRAFT
∞MILITARY AIRCRAFT
PASSENGER AIRCRAFT
RECONNAISSANCE AIRCRAFT
RESEARCH AIRCRAFT
SHORT HAUL AIRCRAFT
TILT WING AIRCRAFT
TRANSPORT AIRCRAFT
UTILITY AIRCRAFT
VERTICAL FLIGHT
WESER AIRCRAFT
WESTLAND AIRCRAFT

VACANCIES (CRYSTAL DEFECTS)
GS   DEFECTS
. CRYSTAL DEFECTS
. . POINT DEFECTS
. . . **VACANCIES (CRYSTAL DEFECTS)**
. . . . FRENKEL DEFECTS
RT   HOLES (ELECTRON DEFICIENCIES)
       SQUARE WELLS

VACCINES
GS   **VACCINES**
. INOCULUM
RT   ANTIBODIES
       ANTIGENS
       ANTISERUMS
       BACTERIOLOGY
       BIOCOMPATIBILITY
       DISEASES
       DRUGS
       EPIDEMIOLOGY
       INOCULATION
       TOXICOLOGY
       TOXINS AND ANTITOXINS

VACILLATION
RT   DITHERS
       HUMAN REACTIONS

VACUUM
UF   ASPIRATION
GS   PRESSURE
. **VACUUM**
. . HIGH VACUUM
. . LOW VACUUM
. . ULTRAHIGH VACUUM
RT   AEROSPACE ENVIRONMENTS
       BOUNDARY LAYER CONTROL
       EVACUATING (VACUUM)
       GETTERS
       HIGH PRESSURE
       KNUDSEN FLOW
       LOW PRESSURE
       MEAN FREE PATH
       OFFGASSING
       OUTGASSING
       PRESSURE MEASUREMENT
       RAREFACTION
       SUCTION

VACUUM APPARATUS
GS   **VACUUM APPARATUS**
. VACUUM CHAMBERS
. VACUUM FURNACES
. VACUUM GAGES
. . IONIZATION GAGES
. . . ALPHATRONS
. . . BAYARD-ALPERT IONIZATION
         GAGES
. . . PENNING GAGES
. . . PHILIPS IONIZATION GAGES
. . KNUDSEN GAGES
. . MCLEOD GAGES
. PIRANI GAGES
. VACUUM PUMPS
. . CONDENSATION PUMPS
. . ION PUMPS
. . MOLECULAR PUMPS
RT   COLD TRAPS
       DIFFUSION PUMPS
       HIGH VACUUM
       LOW DENSITY RESEARCH
       RESIDUAL GAS
       SUCTION
       ULTRAHIGH VACUUM
       VACUUM ARC SWITCHES

VACUUM ARC SWITCHES
GS   SWITCHES
. ELECTRIC SWITCHES
. . **VACUUM ARC SWITCHES**
RT   AIRBORNE EQUIPMENT
       SWITCHING CIRCUITS
       VACUUM APPARATUS
       VACUUM EFFECTS

VACUUM CHAMBERS
UF   LOW PRESSURE CHAMBERS
GS   COMPARTMENTS
. TEST CHAMBERS
. . PRESSURE CHAMBERS
. . . **VACUUM CHAMBERS**
       VACUUM APPARATUS
. **VACUUM CHAMBERS**
RT   ALTITUDE SIMULATION
       ∞CHAMBERS
       HIGH ALTITUDE ENVIRONMENTS
       HIGH ALTITUDE PRESSURE
       HYPERBARIC CHAMBERS
       PRESSURE
       SPACE ENVIRONMENT SIMULATION
       SPACE SIMULATORS
       THERMAL VACUUM TESTS
       WIND TUNNEL DRIVES

VACUUM DEPOSITION
GS   DEPOSITION
. VAPOR DEPOSITION
. . **VACUUM DEPOSITION**
RT   CERAMIC COATINGS
       ELECTROLESS DEPOSITION
       ION PLATING

VACUUM EFFECTS
RT   COLD WELDING
       ∞EFFECTS
       ENVIRONMENTS
       OFFGASSING
       PRESSURE EFFECTS
       SPACE MANUFACTURING
       VACUUM ARC SWITCHES

VACUUM FURNACES
GS   HEATING EQUIPMENT
. FURNACES
. . **VACUUM FURNACES**
       VACUUM APPARATUS
. **VACUUM FURNACES**
RT   SOLAR FURNACES

VACUUM GAGES
GS   MEASURING INSTRUMENTS
. PRESSURE GAGES
. . **VACUUM GAGES**
. . . IONIZATION GAGES
. . . . ALPHATRONS
. . . . BAYARD-ALPERT IONIZATION
           GAGES
. . . . PENNING GAGES
. . . . PHILIPS IONIZATION GAGES
. . . KNUDSEN GAGES
. . . MCLEOD GAGES
. . . PIRANI GAGES
       VACUUM APPARATUS
. **VACUUM GAGES**
. . IONIZATION GAGES
. . . ALPHATRONS
. . . BAYARD-ALPERT IONIZATION
         GAGES
. . . PENNING GAGES
. . . PHILIPS IONIZATION GAGES
. . KNUDSEN GAGES
. . MCLEOD GAGES
. . PIRANI GAGES
RT   BAROMETERS
       MANOMETERS
       ORBITRONS
       PRESSURE MEASUREMENT

VACUUM MELTING
GS   PHASE TRANSFORMATIONS
. MELTING
. . **VACUUM MELTING**
RT   ARC MELTING
       INDUCTION HEATING
       LEVITATION
       POWDER METALLURGY
       ZONE MELTING

VACUUM PUMPS
GS   PUMPS
. **VACUUM PUMPS**

VACUUM PUMPS-(CONT.)
.. CONDENSATION PUMPS
. ION PUMPS
.. MOLECULAR PUMPS
VACUUM APPARATUS
. **VACUUM PUMPS**
.. CONDENSATION PUMPS
.. ION PUMPS
.. MOLECULAR PUMPS
RT COMPRESSORS
CRYOPUMPING
DIFFUSION PUMPS
EJECTORS
EVACUATING (VACUUM)
JET PUMPS
MATERIALS HANDLING
OUTGASSING
SUCTION

VACUUM SPECTROSCOPY
GS SPECTROSCOPY
. **VACUUM SPECTROSCOPY**
RT GAS SPECTROSCOPY
INFRARED SPECTROSCOPY
MAGNETIC SPECTROSCOPY
MASS SPECTROSCOPY
MOLECULAR SPECTROSCOPY
NUCLEAR RADIATION SPECTROSCOPY
SPECTROSCOPIC ANALYSIS
ULTRAVIOLET SPECTROSCOPY
X RAY SPECTROSCOPY

VACUUM SYSTEMS
RT AMPOULES
∞SYSTEMS

VACUUM TESTS
GS **VACUUM TESTS**
. THERMAL VACUUM TESTS
RT HIGH VACUUM
HYPOBARIC ATMOSPHERES
TEST CHAMBERS
∞TESTS
ULTRAHIGH VACUUM

VACUUM TUBE OSCILLATORS
GS ELECTRON TUBES
. VACUUM TUBES
.. **VACUUM TUBE OSCILLATORS**
... CATHODE RAY TUBES
.... PICTURE TUBES
.. GAS DISCHARGE TUBES
.... IGNITRONS
.... THYRATRONS
.. MICROWAVE TUBES
.... KLYSTRONS
.... MAGNETRONS
..... NIGOTRONS
.... MICROWAVE OSCILLATORS
.... PLANOTRONS
..... CARCINOTRONS
..... CELESCOPES
..... HELITRONS
..... IGNITRONS
.... THERMICONS
..... THERMIONIC DIODES
..... THYRATRONS
.... TRAVELING WAVE TUBES
..... CARCINOTRONS
..... HELITRONS
OSCILLATORS
. **VACUUM TUBE OSCILLATORS**
.. CATHODE RAY TUBES
. PICTURE TUBES
.. GAS DISCHARGE TUBES
.. MICROWAVE TUBES
.. MAGNETRONS
.... NIGOTRONS
.. MICROWAVE OSCILLATORS
.. PLANOTRONS
.... CARCINOTRONS
.... HELITRONS
RT AUTODYNES

VACUUM TUBES
GS ELECTRON TUBES
. **VACUUM TUBES**
.. VACUUM TUBE OSCILLATORS
... CATHODE RAY TUBES
.... PICTURE TUBES
... GAS DISCHARGE TUBES
.... IGNITRONS
.... THYRATRONS
... MICROWAVE TUBES
.... KLYSTRONS

VACUUM TUBES-(CONT.)
.... MAGNETRONS
..... NIGOTRONS
.... MICROWAVE OSCILLATORS
.... PLANOTRONS
..... CARCINOTRONS
..... CELESCOPES
..... HELITRONS
..... IGNITRONS
.... THERMICONS
..... THERMIONIC DIODES
..... THYRATRONS
.... TRAVELING WAVE TUBES
..... CARCINOTRONS
..... HELITRONS
RT PENTODES
PERVEANCE
RESIDUAL GAS

VACUUM ULTRAVIOLET RADIATION
USE FAR ULTRAVIOLET RADIATION

VADOSE WATER
GS WATER
. **VADOSE WATER**
RT COASTAL WATER
EVAPOTRANSPIRATION
LAKE TEXOMA (OK-TX)
NEARSHORE WATER
RIVER BASINS
SOILS
WATER TABLES

VALENCE
GS **VALENCE**
. OCTETS
RT CHEMICAL BONDS
CONDUCTION ELECTRONS
ION CHARGE
IONS
POSITIVE IONS
QUANTUM WELLS
TRIVALENT IONS

VALERIC ACID
GS ACIDS
. FATTY ACIDS
.. CARBOXYLIC ACIDS
... **VALERIC ACID**
ALIPHATIC COMPOUNDS
. FATTY ACIDS
.. **VALERIC ACID**
LIPIDS
. **VALERIC ACID**
ORGANIC COMPOUNDS
. FATTY ACIDS
.. **VALERIC ACID**

VALIANT AIRCRAFT
UF VICKERS VALIANT AIRCRAFT
GS ATTACK AIRCRAFT
. BOMBER AIRCRAFT
.. **VALIANT AIRCRAFT**
BAC AIRCRAFT
. **VALIANT AIRCRAFT**
JET AIRCRAFT
. **VALIANT AIRCRAFT**
MONOPLANES
. **VALIANT AIRCRAFT**
RT ∞AIRCRAFT
RECONNAISSANCE AIRCRAFT
TANKER AIRCRAFT

VALIDATION
USE PROVING

VALIDITY
RT ACCEPTABILITY
ACCURACY
ADEQUACY
CONSISTENCY
CORRELATION
EXISTENCE
MATHEMATICAL MODELS
PRECISION
QUALITY
RELIABILITY
SIMULATION
STANDARDS
STATISTICAL TESTS
VARIABILITY

VALKYRIE AIRCRAFT
USE B-70 AIRCRAFT

VALLEYS
UF INTERMONTANE FLOORS
RIFT VALLEYS
RILLS
GS **VALLEYS**
. COACHELLA VALLEY (CA)
. DEATH VALLEY (CA)
. IMPERIAL VALLEY (CA)
. MAGDALENA-CAUCA VALLEY
(COLOMBIA)
. PALO VERDE VALLEY (CA)
. POTOMAC RIVER VALLEY (MD-VA-WV)
. SACRAMENTO VALLEY (CA)
. SAN JOAQUIN VALLEY (CA)
. SHENANDOAH VALLEY (VA)
. ST LAWRENCE VALLEY (NORTH
AMERICA)
. TENNESSEE VALLEY (AL-KY-TN)
RT CANYONS
DELAWARE RIVER BASIN (US)
EROSION
MEANDERS
MISSOURI RIVER (US)
RAVINES
RIVERS
STRUCTURAL BASINS
SUSQUEHANNA RIVER BASIN
(MD-NY-PA)
TOPOGRAPHY
WADIS
WATERSHEDS

VALSALVA EXERCISE
UF VALSALVA MANEUVER
RT RESPIRATION

VALSALVA MANEUVER
USE VALSALVA EXERCISE

VALUE
GS **VALUE**
. Q VALUES
RT AMOUNT
ASSESSMENTS
COSTS
DAMAGE ASSESSMENT
ESTIMATES
ESTIMATING
EVALUATION
FIGURE OF MERIT
LEVEL (QUANTITY)
NORMS
RANKING
TECHNOLOGY ASSESSMENT

VALUE ENGINEERING
RT COST ANALYSIS
COST ESTIMATES
COST INCENTIVES
COST REDUCTION
DESIGN ANALYSIS
ECONOMIC ANALYSIS
∞ENGINEERING
INCENTIVE TECHNIQUES
LIFE CYCLE COSTS
MANAGEMENT PLANNING
QUALITY CONTROL
RELIABILITY ENGINEERING
STANDARDS

VALVES
UF HYDRAULIC VALVES
GS **VALVES**
. ARTIFICIAL HEART VALVES
. AUTOMATIC CONTROL VALVES
.. PRESSURE REGULATORS
.. RELIEF VALVES
. BUTTERFLY VALVES
.. DAMPERS (VALVES)
. COCKS
. CONTROL VALVES
. FUEL VALVES
. GAS VALVES
. HEART VALVES
. SOLENOID VALVES
RT BALLS
CHOKES (RESTRICTIONS)
CLOSURES
DIVERTERS
ENGINE PARTS
HYDRAULIC EQUIPMENT
PACKINGS (SEALS)
PNEUMATIC CIRCUITS
PNEUMATIC EQUIPMENT
SEALS (STOPPERS)

VALVES-*(CONT.)*
TRAPS
WATER HAMMER

VAMPIRE AIRCRAFT
USE   DH 115 AIRCRAFT

VAMPIRE MK 35 AIRCRAFT
GS   ATTACK AIRCRAFT
. FIGHTER AIRCRAFT
. . **VAMPIRE MK 35 AIRCRAFT**
HAWKER SIDDELEY AIRCRAFT
. **VAMPIRE MK 35 AIRCRAFT**
JET AIRCRAFT
. **VAMPIRE MK 35 AIRCRAFT**
RT   ∞AIRCRAFT
BOMBER AIRCRAFT
HARRIER AIRCRAFT

VAN ALLEN RADIATION BELTS
USE   RADIATION BELTS

VAN BIESBROECK STAR
GS   CELESTIAL BODIES
. STARS
. . **VAN BIESBROECK STAR**

VAN DE GRAAFF ACCELERATORS
GS   PARTICLE ACCELERATORS
. **VAN DE GRAAFF ACCELERATORS**
RT   ∞ACCELERATORS
ELECTRON ACCELERATORS

VAN DER WAAL FORCES
RT   DIPOLE MOMENTS
∞FORCE
INTERATOMIC FORCES
INTERMOLECULAR FORCES

VAN SLYKE METHOD
GS   CHEMICAL TESTS
. CHEMICAL ANALYSIS
. . GAS ANALYSIS
. . . **VAN SLYKE METHOD**
. . QUANTITATIVE ANALYSIS
. . . **VAN SLYKE METHOD**
RT   ∞METHODOLOGY

VANADATES
GS   VANADIUM COMPOUNDS
. **VANADATES**
RT   METAL OXIDES
VANADIUM OXIDES

VANADIUM
GS   CHEMICAL ELEMENTS
. **VANADIUM**
. . VANADIUM ISOTOPES
METALS
. TRANSITION METALS
. . **VANADIUM**
. . . VANADIUM ISOTOPES

VANADIUM ALLOYS
GS   ALLOYS
. **VANADIUM ALLOYS**

VANADIUM CARBIDES
GS   CARBON COMPOUNDS
. CARBIDES
. . **VANADIUM CARBIDES**
VANADIUM COMPOUNDS
. **VANADIUM CARBIDES**

VANADIUM COMPOUNDS
GS   **VANADIUM COMPOUNDS**
. CALCIUM VANADATES
. VANADATES
. VANADIUM CARBIDES
. VANADIUM OXIDES
. VANADYL COMPOUNDS
RT   ∞CHEMICAL COMPOUNDS
∞GROUP 5B COMPOUNDS
∞METAL COMPOUNDS

VANADIUM ISOTOPES
GS   CHEMICAL ELEMENTS
. NUCLIDES
. . ISOTOPES
. . . **VANADIUM ISOTOPES**
. VANADIUM
. . **VANADIUM ISOTOPES**
METALS
. TRANSITION METALS

VANADIUM ISOTOPES-*(CONT.)*
. . VANADIUM
. . . **VANADIUM ISOTOPES**

VANADIUM OXIDES
GS   CHALCOGENIDES
. OXIDES
. . METAL OXIDES
. . . **VANADIUM OXIDES**
VANADIUM COMPOUNDS
. **VANADIUM OXIDES**
RT   VANADATES

VANADYL COMPOUNDS
GS   VANADIUM COMPOUNDS
. **VANADYL COMPOUNDS**
RT   ∞CHEMICAL COMPOUNDS
∞METAL COMPOUNDS

VANADYL RADICAL
GS   IONS
. CATIONS
. . METAL IONS
. . . **VANADYL RADICAL**
RADICALS
. **VANADYL RADICAL**

VANELESS DIFFUSERS
RT   COMPRESSORS
∞DIFFUSERS
EXHAUST DIFFUSERS
PUMPS
SUPERSONIC DIFFUSERS

VANES
GS   **VANES**
. GUIDE VANES
. . JET VANES
. WIND VANES
RT   AIRFOILS
∞BLADES
COMPRESSOR BLADES
CONTROL SURFACES
FINS
IMPELLERS
NOSE FINS
STATOR BLADES
TAIL ASSEMBLIES
TURBOMACHINE BLADES
WINDPOWER UTILIZATION
WINDPOWERED GENERATORS
WINDPOWERED PUMPS

VANGUARD PROJECT
GS   PROGRAMS
. NASA PROGRAMS
. . **VANGUARD PROJECT**
. NASA SPACE PROGRAMS
. . **VANGUARD PROJECT**
. PROJECTS
. . **VANGUARD PROJECT**
RT   X-405 ENGINE

VANGUARD SATELLITES
GS   SATELLITES
. ARTIFICIAL SATELLITES
. . **VANGUARD SATELLITES**
. . . VANGUARD 1 SATELLITE
. . . VANGUARD 2 SATELLITE
. . . VANGUARD 3 SATELLITE
RT   GEODETIC SATELLITES
GEOPHYSICAL SATELLITES
INTERNATIONAL GEOPHYSICAL YEAR
METEOROLOGICAL SATELLITES

VANGUARD 1 SATELLITE
GS   SATELLITES
. ARTIFICIAL SATELLITES
. . GEODETIC SATELLITES
. . . **VANGUARD 1 SATELLITE**
. VANGUARD SATELLITES
. . **VANGUARD 1 SATELLITE**
UNMANNED SPACECRAFT
. GEODETIC SATELLITES
. . **VANGUARD 1 SATELLITE**

VANGUARD 2 LAUNCH VEHICLE
GS   LAUNCH VEHICLES
. **VANGUARD 2 LAUNCH VEHICLE**
ROCKET VEHICLES
. MULTISTAGE ROCKET VEHICLES
. . **VANGUARD 2 LAUNCH VEHICLE**
RT   LIQUID PROPELLANT ROCKET ENGINES
SOLID PROPELLANT ROCKET ENGINES

VANGUARD 2 LAUNCH VEHICLE-*(CONT.)*
VIKING ROCKET VEHICLE
X-248 ENGINE

VANGUARD 2 SATELLITE
GS   SATELLITES
. ARTIFICIAL SATELLITES
. . METEOROLOGICAL SATELLITES
. . . **VANGUARD 2 SATELLITE**
. VANGUARD SATELLITES
. . . **VANGUARD 2 SATELLITE**

VANGUARD 3 SATELLITE
GS   SATELLITES
. ARTIFICIAL SATELLITES
. . GEOPHYSICAL SATELLITES
. . . **VANGUARD 3 SATELLITE**
. . VANGUARD SATELLITES
. . . **VANGUARD 3 SATELLITE**

VANS
USE   TRUCKS

VAPOR BARRIER CLOTHING
GS   CLOTHING
. PROTECTIVE CLOTHING
. . **VAPOR BARRIER CLOTHING**
RT   ∞BARRIERS
LIFE SUPPORT SYSTEMS
TEXTILES

VAPOR DEPOSITION
GS   DEPOSITION
. **VAPOR DEPOSITION**
. . VACUUM DEPOSITION
RT   COATING
COATINGS
CRYSTAL GROWTH
ELECTROLESS DEPOSITION
METAL VAPORS
METALLIZING
ULTRAPURE METALS
VAPORIZERS
VAPORIZING

VAPOR GENERATORS
USE   VAPORIZERS

VAPOR JETS
GS   FLUID JETS
. **VAPOR JETS**
RT   AIR JETS
GAS FLOW
JET FLOW
PLASMA JETS

VAPOR LIQUID EQUILIBRIUM
USE   LIQUID-VAPOR EQUILIBRIUM

VAPOR PHASE EPITAXY
GS   GROWTH
. CRYSTAL GROWTH
. . EPITAXY
. . . **VAPOR PHASE EPITAXY**
RT   CRYSTAL STRUCTURE
LIQUID PHASE EPITAXY
LIQUID PHASES

VAPOR PHASES
UF   GAS PHASES
RT   ASSOCIATION REACTIONS
CRITICAL PRESSURE
GAS-METAL INTERACTIONS
GAS-SOLID INTERFACES
GASES
HYDROGEN CLOUDS
LIQUID PHASES
LIQUID-GAS MIXTURES
LIQUID-VAPOR INTERFACES
LIQUIDS
METAL-GAS SYSTEMS
PHASE DIAGRAMS
∞PHASES
PREVAPORIZATION
SOLIDS
SUPERCRITICAL PRESSURES
VAPORS
VOLATILITY

VAPOR PRESSURE
GS   PRESSURE
. **VAPOR PRESSURE**
THERMODYNAMIC PROPERTIES
. THERMOPHYSICAL PROPERTIES

VAPOR PRESSURE-*(CONT.)*
. . **VAPOR PRESSURE**
RT DALTON LAW
    FLASH POINT
    FUEL TANK PRESSURIZATION
    HENRY LAW
    HUMIDITY
    INTERFACIAL TENSION
    LIQUID-GAS MIXTURES
    LIQUID-VAPOR INTERFACES
    PARTIAL PRESSURE
    RAOULT LAW
    SUBLIMATION
    SUPERCRITICAL PRESSURES
    VOLATILITY

VAPOR TRAILS
USE CONTRAILS

VAPOR TRAPS
GS TRAPS
    . **VAPOR TRAPS**
RT COLD TRAPS
    GETTERS
    ION TRAPS (INSTRUMENTATION)

VAPORIZATION HEAT
USE HEAT OF VAPORIZATION

VAPORIZERS
UF VAPOR GENERATORS
GS HEATING EQUIPMENT
    . **VAPORIZERS**
    . . EVAPORATORS
RT BOILERS
    CAVITY VAPOR GENERATORS
    COLLOIDAL GENERATORS
    COLUMNS (PROCESS ENGINEERING)
    CONDENSERS (LIQUEFIERS)
    GAS GENERATORS
    ∞GENERATORS
    ∞HEATERS
    SEPARATORS
    SPRAYERS
    VAPOR DEPOSITION
    VAPORIZING
    VAPORS

VAPORIZING
UF VOLATILIZATION
GS PHASE TRANSFORMATIONS
    . **VAPORIZING**
    . . BOILING
    . . . FILM BOILING
    . . . NUCLEATE BOILING
    . . . . LEIDENFROST PHENOMENON
    . . EVAPORATION
    . . . EVAPOTRANSPIRATION
    . . . PROPELLANT EVAPORATION
    . . . TRANSPIRATION
    . . FLASHING (VAPORIZING)
    . . PREVAPORIZATION
    . . SUBLIMATION
RT ABLATION
    CONCENTRATING
    DESALINIZATION
    DISTILLATION
    EVOLUTION (LIBERATION)
    HEAT OF VAPORIZATION
    HEATING
    ∞SEPARATION
    SPRAYING
    STRIPPING (DISTILLATION)
    SURFACE REACTIONS
    VAPOR DEPOSITION
    VAPORIZERS
    VAPORS
    VOLATILITY

VAPORS
GS **VAPORS**
    . CESIUM VAPOR
    . METAL VAPORS
    . . MERCURY VAPOR
    . . SODIUM VAPOR
    . WATER VAPOR
RT CAVITY VAPOR GENERATORS
    COMBUSTION PRODUCTS
    CONDENSATES
    EXHAUST GASES
    FUMES
    GASES
    HAZE DETECTION
    HYDROGEN CLOUDS
    LIQUID-VAPOR EQUILIBRIUM

VAPORS-*(CONT.)*
    PREVAPORIZATION
    SMOKE
    VAPOR PHASES
    VAPORIZERS
    VAPORIZING

VARACTOR DIODE CIRCUITS
GS CIRCUITS
    . **VARACTOR DIODE CIRCUITS**
RT DIODES

VARACTOR DIODES
UF VARACTORS
GS ELECTRONIC EQUIPMENT
    . DIODES
    . . SEMICONDUCTOR DIODES
    . . . **VARACTOR DIODES**
    . SOLID STATE DEVICES
    . . SEMICONDUCTOR DEVICES
    . . . **VARACTOR DIODES**
RT JUNCTION DIODES
    PARAMETRIC DIODES
    VARISTORS

VARACTORS
USE VARACTOR DIODES

VARIABILITY
RT CONSISTENCY
    CONTINUITY
    CONVERGENCE
    CORRELATION
    COVARIANCE
    ∞DISPERSION
    ECCENTRICITY
    ∞EQUILIBRIUM
    FACTOR ANALYSIS
    HETEROGENEITY
    LINEARITY
    NONLINEARITY
    PERIODIC VARIATIONS
    QUALITY
    QUALITY CONTROL
    RANGE (EXTREMES)
    REGRESSION ANALYSIS
    RELIABILITY
    SAMPLING
    STABILITY
    STANDARD DEVIATION
    STANDARDIZATION
    VALIDITY
    ∞VARIABLE
    VARIANCE (STATISTICS)

∞ **VARIABLE**
SN *(USE OF A MORE SPECIFIC TERM IS
    RECOMMENDED--CONSULT THE TERMS
    LISTED BELOW)*
UF FACTORS
RT COMPLEX VARIABLES
    DEPENDENT VARIABLES
    FORM FACTORS
    INDEPENDENT VARIABLES
    LATIN SQUARE METHOD
    RACE FACTORS
    RANDOM VARIABLES
    REAL VARIABLES
    SOCIAL FACTORS
    VARIABILITY

VARIABLE AREA WINGS
USE TRAILING EDGE FLAPS

VARIABLE CYCLE ENGINES
UF VCE
GS AIRCRAFT ENGINES
    . **VARIABLE CYCLE ENGINES**
    ENGINES
    . **VARIABLE CYCLE ENGINES**
RT COAXIAL NOZZLES
    SUPERSONIC AIRCRAFT
    VARIABLE STREAM CONTROL ENGINES

VARIABLE GEOMETRY STRUCTURES
RT EXPANDABLE STRUCTURES
    FOLDING STRUCTURES
    INFLATABLE STRUCTURES
    ∞STRUCTURES

VARIABLE LIFT
USE LIFT

VARIABLE MASS SYSTEMS
GS KINETICS
    . **VARIABLE MASS SYSTEMS**
RT EQUATIONS OF MOTION
    ∞MASS BALANCE
    MASS DISTRIBUTION
    ∞SYSTEMS

VARIABLE PITCH PROPELLERS
UF CONSTANT SPEED PROPELLERS
GS PROPELLERS
    . **VARIABLE PITCH PROPELLERS**
RT HELICOPTER PROPELLER DRIVE
    PITCH (INCLINATION)

VARIABLE STARS
GS CELESTIAL BODIES
    . STARS
    . . **VARIABLE STARS**
    . . . CATACLYSMIC VARIABLES
    . . . CEPHEID VARIABLES
    . . . NOVAE
    . . . . DWARF NOVAE
    . . . . HERCULES NOVA
    . . . SUPERNOVAE
    . . . SYMBIOTIC STARS
    . . . T TAURI STARS
RT BINARY STARS
    COMPANION STARS
    ECLIPSING BINARY STARS
    FLARE STARS
    LAMBDA TAURI STARS
    OMICRON CETI STAR
    PERIODIC VARIATIONS
    SOLAR OSCILLATIONS
    STELLAR MASS
    STELLAR MASS EJECTION
    STELLAR OSCILLATIONS

VARIABLE STREAM CONTROL ENGINES
GS AIRCRAFT ENGINES
    . **VARIABLE STREAM CONTROL
      ENGINES**
    ENGINES
    . **VARIABLE STREAM CONTROL
      ENGINES**
RT ∞CONTROL
    ENGINE CONTROL
    SUPERSONIC AIRCRAFT
    SUPERSONIC NOZZLES
    VARIABLE CYCLE ENGINES

VARIABLE SWEEP WINGS
UF M WINGS
    OGEE WINGS
    W WINGS
GS AIRFOILS
    . WINGS
    . . **VARIABLE SWEEP WINGS**
    PLANFORMS
    . WING PLANFORMS
    . . **VARIABLE SWEEP WINGS**
RT ARROW WINGS
    BOEING 733 AIRCRAFT
    DELTA WINGS
    F-111 AIRCRAFT
    FOLDING STRUCTURES
    OGEE SHAPE
    PANAVIA MILITARY AIRCRAFT
    SWEPT FORWARD WINGS
    SWEPTBACK WINGS

VARIABLE THRUST
GS THRUST
    . **VARIABLE THRUST**
RT CONTROL ROCKETS
    HIGH THRUST
    JET CONTROL
    JET THRUST
    LOW THRUST
    LOW THRUST PROPULSION
    MICROTHRUST
    ROCKET THRUST
    THROTTLING
    THRUST AUGMENTATION
    THRUST CONTROL
    THRUST TERMINATION
    THRUST VECTOR CONTROL

∞ **VARIANCE**
SN *(USE OF A MORE SPECIFIC TERM IS
    RECOMMENDED--CONSULT THE TERMS
    LISTED BELOW)*
RT ANALYSIS OF VARIANCE
    DEGREES OF FREEDOM

VARIANCE-*(CONT.)*
    MULTIVARIATE STATISTICAL ANALYSIS
    VARIANCE (STATISTICS)

VARIANCE (STATISTICS)
GS   STATISTICAL ANALYSIS
    . **VARIANCE (STATISTICS)**
    . . ANALYSIS OF VARIANCE
    . . MULTIVARIATE STATISTICAL
        ANALYSIS
    . . . BIVARIATE ANALYSIS
    . . . COVARIANCE
    . . . REGRESSION ANALYSIS
RT   CONFIDENCE LIMITS
    CORRELATION
    DISTRIBUTION MOMENTS
    EXPERIMENT DESIGN
    FACTOR ANALYSIS
    GAUSS-MARKOV THEOREM
    GOODNESS OF FIT
    HETEROGENEITY
    HOMOGENEITY
    KRIGING
    MEAN
    MOMENTS
    QUALITY CONTROL
    RANGE (EXTREMES)
    STANDARD DEVIATION
    VARIABILITY
    ∞VARIANCE

VARIATION METHOD
USE   CALCULUS OF VARIATIONS

VARIATIONAL PRINCIPLES
RT   CALCULUS OF VARIATIONS
    ∞DYNAMICS
    EQUILIBRIUM METHODS
    IRREVERSIBLE PROCESSES
    ONSAGER PHENOMENOLOGICAL
        COEFFICIENT
    RAYLEIGH-RITZ METHOD

VARIATIONS
UF   FLUCTUATION
GS   **VARIATIONS**
    . ALTERNATIONS
    . MAGNETIC VARIATIONS
    . . GEOMAGNETIC PULSATIONS
    . . . GEOMAGNETIC MICROPULSATIONS
    . . NOCTURNAL VARIATIONS
    . PERIODIC VARIATIONS
    . . ANNUAL VARIATIONS
    . . DIURNAL VARIATIONS
    . . NOCTURNAL VARIATIONS
    . . SECULAR VARIATIONS
    . TWENTY-SEVEN DAY VARIATION
    . WIND VARIATIONS
RT   ALTERNATIVES
    ASYMMETRY
    DEFLECTION
    DEVIATION
    DIFFERENCES
    DISPLACEMENT
    DISTORTION
    DIVERGENCE
    ECCENTRICITY
    GRADIENTS
    MICROPULSATIONS
    PERTURBATION
    REVISIONS
    SUBSTITUTES
    SURGES

VARIOMETERS
UF   MAGNETOVARIOGRAPHS
GS   MEASURING INSTRUMENTS
    . MAGNETOMETERS
    . . **VARIOMETERS**
RT   GEOMAGNETISM

VARISTORS
GS   ELECTRONIC EQUIPMENT
    . SOLID STATE DEVICES
    . . SEMICONDUCTOR DEVICES
    . . . **VARISTORS**
RT   RESISTORS
    THERMISTORS
    VARACTOR DIODES

VARNISHES
RT   FILLERS
    FINISHES
    PAINTS
    PRIMERS (COATINGS)

VARNISHES-*(CONT.)*
    PROTECTIVE COATINGS
    SEALERS
    SPRAYED COATINGS

VASCULAR SYSTEM
GS   ANATOMY
    . CIRCULATORY SYSTEM
    . . **VASCULAR SYSTEM**
    . . . BLOOD VESSELS
    . . . . ARTERIES
    . . . . . AORTA
    . . . . CAPILLARIES (ANATOMY)
    . . . . GLOMERULUS
    . . . . VEINS
RT   BLOOD CIRCULATION
    CARDIOVASCULAR SYSTEM
    CAROTID SINUS BODY
    CAROTID SINUS REFLEX
    ∞SYSTEMS

VASOCONSTRICTION
RT   BLOOD VESSELS
    BODY TEMPERATURE
    COLD TOLERANCE
    ISCHEMIA
    REFLEXES
    SNEEZING

VASOCONSTRICTOR DRUGS
UF   PRESSORS
GS   DRUGS
    . **VASOCONSTRICTOR DRUGS**
    . . HYPERTENSIN
RT   PHARMACOLOGY

VASODILATION
RT   BLOOD VESSELS
    BODY TEMPERATURE
    CONGESTION
    REFLEXES

VASOMOTOR NERVOUS SYSTEM
USE   NERVOUS SYSTEM

VATICAN CITY
GS   CITIES
    . **VATICAN CITY**
RT   EUROPE
    ITALY
    NATIONS

VATOL AIRCRAFT
SN   (VERTICAL ATTITUDE TAKEOFF AND
        LANDING AIRCRAFT)
UF   VERTICAL ATTITUDE TAKEOFF-LANDING
        AIRCRAFT
    XBQM-180A AIRCRAFT
RT   ∞AIRCRAFT
    DELTA WINGS
    REMOTELY PILOTED VEHICLES
    VERTICAL LANDING
    VERTICAL TAKEOFF AIRCRAFT

VAX COMPUTERS
GS   DATA PROCESSING EQUIPMENT
    . COMPUTERS
    . . DIGITAL COMPUTERS
    . . . **VAX COMPUTERS**
    . . . . VAX-11 SERIES COMPUTERS
    . . . . . VAX-11/780 COMPUTER

VAX-11 SERIES COMPUTERS
GS   DATA PROCESSING EQUIPMENT
    . COMPUTERS
    . . DIGITAL COMPUTERS
    . . . VAX COMPUTERS
    . . . . **VAX-11 SERIES COMPUTERS**
    . . . . . VAX-11/780 COMPUTER

VAX-11/780 COMPUTER
GS   DATA PROCESSING EQUIPMENT
    . COMPUTERS
    . . DIGITAL COMPUTERS
    . . . VAX COMPUTERS
    . . . . VAX-11 SERIES COMPUTERS
    . . . . . **VAX-11/780 COMPUTER**

VC-10 AIRCRAFT
UF   VICKERS VC-10 AIRCRAFT
    VICKERS 1100 AIRCRAFT
GS   BAC AIRCRAFT
    . **VC-10 AIRCRAFT**
    COMMERCIAL AIRCRAFT

VC-10 AIRCRAFT-*(CONT.)*
    . **VC-10 AIRCRAFT**
    JET AIRCRAFT
    . **VC-10 AIRCRAFT**
    MONOPLANES
    . **VC-10 AIRCRAFT**
    PASSENGER AIRCRAFT
    . **VC-10 AIRCRAFT**
    TRANSPORT AIRCRAFT
    . **VC-10 AIRCRAFT**
RT   ∞AIRCRAFT
    CARGO AIRCRAFT

VCE
USE   VARIABLE CYCLE ENGINES

VCO
USE   VOLTAGE CONTROLLED OSCILLATORS

VECTOR ANALYSIS
GS   ANALYSIS (MATHEMATICS)
    . CALCULUS
    . . **VECTOR ANALYSIS**
    . . . COLLINEARITY
    . . . COPLANARITY
    . . . CURL (VECTORS)
    . . . . VORTICITY
    . REAL VARIABLES
    . . **VECTOR ANALYSIS**
    . . . COLLINEARITY
    . . . COPLANARITY
    . . . CURL (VECTORS)
    . . . . VORTICITY
    GEOMETRY
    . **VECTOR ANALYSIS**
    . . COLLINEARITY
    . . COPLANARITY
    . . CURL (VECTORS)
    . . . VORTICITY
RT   DIFFERENTIAL EQUATIONS
    EULER-CAUCHY EQUATIONS
    GRADIENTS
    POYNTING THEOREM
    RESULTANTS
    STABILITY DERIVATIVES

VECTOR CALCULUS
USE   VECTOR SPACES

VECTOR CONTROL
USE   DIRECTIONAL CONTROL

VECTOR CURRENTS
RT   CURRENT ALGEBRA
    PARITY
    RADIOACTIVE DECAY
    SUPERCONDUCTIVITY

VECTOR DOMINANCE MODEL
GS   MODELS
    . **VECTOR DOMINANCE MODEL**
RT   HADRONS
    HIGH ENERGY INTERACTIONS
    NUCLEONS
    PHOTONEUTRONS
    PHOTOPRODUCTION

VECTOR MESONS
GS   PARTICLES
    . ELEMENTARY PARTICLES
    . . BOSONS
    . . . MESONS
    . . . . **VECTOR MESONS**
    . . . . . RHO-MESONS
    . . . . . SIGMA-MESONS
    . . HADRONS
    . . MESONS
    . . . . **VECTOR MESONS**
    . . . . . RHO-MESONS
    . . . . . SIGMA-MESONS
    . NUCLEAR PARTICLES
    . . BOSONS
    . . . MESONS
    . . . . **VECTOR MESONS**
    . . . . . RHO-MESONS
    . . . . . SIGMA-MESONS

VECTOR SPACES
UF   GRASSMANN ALGEBRA
    VECTOR CALCULUS
GS   ALGEBRA
    . **VECTOR SPACES**
    . . HILBERT SPACE
    . . . BANACH SPACE

VECTOR SPACES-*(CONT.)*
```
         . . . . SOBOLEV SPACE
         . . MATRICES (MATHEMATICS)
         . . . ADJOINTS
         . . . CANONICAL FORMS
         . . . EIGENVALUES
         . . . EIGENVECTORS
         . . . JORDAN FORM
         . . . STIFFNESS MATRIX
         . . STOKES THEOREM (VECTOR
                CALCULUS)
         . . U SPIN SPACE
         . . VECTORS (MATHEMATICS)
         . . . EIGENVECTORS
         . . . STATE VECTORS
         . . . VORTICITY
RT       ANALYSIS (MATHEMATICS)
         CHAPLYGIN EQUATION
         HERMITIAN POLYNOMIAL
         HODOGRAPHS
         KAKUTANI THEOREM
         LINEAR TRANSFORMATIONS
```

VECTORCARDIOGRAPHY
```
GS       BIOENGINEERING
         . BIOMETRICS
         . . CARDIOGRAPHY
         . . . VECTORCARDIOGRAPHY
RT       ELECTROCARDIOGRAPHY
         PHONOCARDIOGRAPHY
```

VECTORS (MATHEMATICS)
```
GS       ALGEBRA
         . VECTOR SPACES
         . . VECTORS (MATHEMATICS)
         . . . EIGENVECTORS
         . . . STATE VECTORS
         . . . VORTICITY
RT       DYADICS
         FUNCTION SPACE
         SCHWARTZ INEQUALITY
```

VEGA LAUNCH VEHICLE
```
UF       VEGA ROCKET VEHICLE
GS       LAUNCH VEHICLES
         . VEGA LAUNCH VEHICLE
         ROCKET VEHICLES
         . MULTISTAGE ROCKET VEHICLES
         . . VEGA LAUNCH VEHICLE
RT       ATLAS D ICBM
         LIQUID PROPELLANT ROCKET ENGINES
```

VEGA PROJECT
```
RT       FLYBY MISSIONS
         HALLEY'S COMET
         INTERNATIONAL COOPERATION
         U.S.S.R. SPACE PROGRAM
         VENERA SATELLITES
         VENUS (PLANET)
```

VEGA ROCKET VEHICLE
```
USE      VEGA LAUNCH VEHICLE
```

VEGARD-KAPLAN BANDS
```
GS       SPECTRA
         . SPECTRAL BANDS
         . . VEGARD-KAPLAN BANDS
RT       ∞ BANDS
         EMISSION SPECTRA
         MOLECULAR SPECTRA
         NITROGEN
```

VEGETABLES
```
GS       VEGETABLES
         . POTATOES
         . SPINACH
RT       ANGIOSPERMS
         ∞ FOOD
         LEGUMINOUS PLANTS
         PLANTING
         SEEDS
```

VEGETATION
```
GS       VEGETATION
         . CANOPIES (VEGETATION)
RT       BIOCONVERSION
         BIOMASS ENERGY PRODUCTION
         EARTH RESOURCES
         LOCUSTS
         OASES
         PLANTS (BOTANY)
         RAIN FORESTS
         RESOURCES
         SEA GRASSES
         TREES (PLANTS)
```

VEGETATION GROWTH
```
GS       GROWTH
         . VEGETATION GROWTH
RT       AGRICULTURE
         BIOCHEMISTRY
         BOTANY
         CROP VIGOR
         ECOLOGY
         FERTILIZERS
         GRAVITROPISM
         HYDROPONICS
         IRRIGATION
         PLANT ROOTS
         PLANT STRESS
         SOIL MOISTURE
         SOIL SCIENCE
         SOILS
         VEGETATIVE INDEX
```

VEGETATIVE INDEX
```
RT       AGRISTARS PROJECT
         ATMOSPHERIC ATTENUATION
         ATMOSPHERIC EFFECTS
         ATMOSPHERIC OPTICS
         ATMOSPHERIC SCATTERING
         CANOPIES (VEGETATION)
         COLOR
         CORRECTION
         CROP IDENTIFICATION
         CROP INVENTORIES
         IMAGE ENHANCEMENT
         IMAGING TECHNIQUES
         MULTISPECTRAL BAND SCANNERS
         RADIOMETRIC CORRECTION
         REFLECTANCE
         REMOTE SENSING
         SATELLITE IMAGERY
         SATELLITE OBSERVATION
         SPECTRAL REFLECTANCE
         VEGETATION GROWTH
```

VEHICLE WHEELS
```
GS       WHEELS
         . VEHICLE WHEELS
         . . NOSE WHEELS
RT       AIRCRAFT TIRES
         BRAKES (FOR ARRESTING MOTION)
         LANDING GEAR
         MECHANICAL DRIVES
         ROLLERS
         SHAFTS (MACHINE ELEMENTS)
         SUSPENSION SYSTEMS (VEHICLES)
         TIRES
         TOROIDAL WHEELS
         TRANSMISSIONS (MACHINE ELEMENTS)
         WHEEL BRAKES
```

∞ VEHICLES
```
SN       (USE OF A MORE SPECIFIC TERM IS
         RECOMMENDED--CONSULT THE TERMS
         LISTED BELOW)
UF       CRAFT
         ROTATING VEHICLES
RT       AMPHIBIOUS VEHICLES
         ARCAS ROCKET VEHICLES
         ARCON ROCKET VEHICLE
         ARGO ROCKET VEHICLES
         ASTROBEE ROCKET VEHICLES
         ATLAS AGENA LAUNCH VEHICLES
         ATLAS LAUNCH VEHICLES
         AUTOMATED GUIDEWAY TRANSIT
            VEHICLES
         AUTOMATED MIXED TRAFFIC VEHICLES
         AUTOMATED TRANSIT VEHICLES
         AUTOMOBILES
         ∞ BALLISTIC VEHICLES
         BOOSTGLIDE VEHICLES
         CAPTURED AIR BUBBLE VEHICLES
         CONTROL CONFIGURED VEHICLES
         CRAWLER TRACTORS
         DRONE VEHICLES
         ELECTRIC HYBRID VEHICLES
         ELECTRIC MOTOR VEHICLES
         ENGINES
         EUROPA LAUNCH VEHICLES
         FLIGHT TEST VEHICLES
         ∞ FLIGHT VEHICLES
         GROUND EFFECT MACHINES
         HEAVY LIFT LAUNCH VEHICLES
         HOVERING ROCKET VEHICLES
         HYDROPLANES (VEHICLES)
         HYPERSONIC VEHICLES
         INTRAORBIT TRANSFER VEHICLES
         JUNO LAUNCH VEHICLES
         KAPPA ROCKET VEHICLES
```

VEHICLES-*(CONT.)*
```
         LAMBDA ROCKET VEHICLES
         LAUNCH VEHICLES
         LOW OBSERVABLE REENTRY VEHICLES
         LUNAR FLYING VEHICLES
         LUNAR ROVING VEHICLES
         LUNAR SURFACE VEHICLES
         LUNOKHOD LUNAR ROVING VEHICLES
         MAGNETIC LEVITATION VEHICLES
         MANNED LUNAR SURFACE VEHICLES
         ∞ MILITARY VEHICLES
         MISSILES
         MOTOR VEHICLES
         MULTIENGINE VEHICLES
         MULTISTAGE ROCKET VEHICLES
         NIKE ROCKET VEHICLES
         NIKE-HYDAC ROCKET VEHICLE
         NIKE-IROQUOIS ROCKET VEHICLE
         NOVA LAUNCH VEHICLES
         NUCLEAR ENGINE FOR ROCKET
            VEHICLES
         ORBIT TRANSFER VEHICLES
         RANGER LUNAR LANDING VEHICLES
         RECOVERABLE LAUNCH VEHICLES
         RECOVERY VEHICLES
         REENTRY VEHICLES
         REMOTELY PILOTED VEHICLES
         RESEARCH VEHICLES
         REUSABLE LAUNCH VEHICLES
         ROCKET VEHICLES
         ROVING VEHICLES
         SATURN LAUNCH VEHICLES
         SINGLE STAGE ROCKET VEHICLES
         SINGLE STAGE TO ORBIT VEHICLES
         SKUA ROCKET VEHICLES
         ∞ SPACECRAFT
         STANDARD LAUNCH VEHICLES
         SURFACE EFFECT SHIPS
         SURFACE VEHICLES
         SWATH (SHIP)
         TANKS (COMBAT VEHICLES)
         TERMINAL CONFIGURED VEHICLE
            PROGRAM
         TEST VEHICLES
         THOR LAUNCH VEHICLES
         THORAD LAUNCH VEHICLES
         TITAN LAUNCH VEHICLES
         TRACKED VEHICLES
         TRACTORS
         ∞ TRANSPORT VEHICLES
         TRANSPORTER
         UNDERWATER VEHICLES
         UNIDENTIFIED FLYING OBJECTS
         VERONIQUE ROCKET VEHICLES
         WATER VEHICLES
         ∞ WINGED VEHICLES
```

VEHICULAR TRACKS
```
RT       IDLERS
         SURFACE VEHICLES
         SUSPENSION SYSTEMS (VEHICLES)
         TRACKED VEHICLES
         ∞ TRACKS
         TREADS
```

VEINS
```
GS       ANATOMY
         . CARDIOVASCULAR SYSTEM
         . . BLOOD VESSELS
         . . . VEINS
         . CIRCULATORY SYSTEM
         . . VASCULAR SYSTEM
         . . . BLOOD VESSELS
         . . . . VEINS
RT       ARTERIES
         BIFURCATION (BIOLOGY)
         TRANSFUSION
```

VELA SATELLITES
```
GS       MILITARY SPACECRAFT
         . VELA SATELLITES
         SATELLITES
         . ARTIFICIAL SATELLITES
         . . VELA SATELLITES
RT       FISHBOWL OPERATION
         HIGH ALTITUDE NUCLEAR DETECTION
         HIGH ALTITUDE TESTS
         NUCLEAR EXPLOSIONS
         NUCLEAR RADIATION
         POST-BLAST NUCLEAR RADIATION
         RADIATION DETECTORS
         RADIATION MEASURING INSTRUMENTS
         SATELLITE OBSERVATION
```

VELOCITY
UF   SPEED
GS   **VELOCITY**
. ACOUSTIC VELOCITY
. AIRSPEED
. ANGULAR VELOCITY
. CRITICAL VELOCITY
. ESCAPE VELOCITY
. EXHAUST VELOCITY
. FLOW VELOCITY
. . SOLAR WIND VELOCITY
. GROUND SPEED
. GROUP VELOCITY
. HIGH SPEED
. HYPERSONIC SPEED
. LANDING SPEED
. LIGHT SPEED
. LOW SPEED
. ORBITAL VELOCITY
. PHASE VELOCITY
. PROPAGATION VELOCITY
. RADIAL VELOCITY
. RELATIVISTIC VELOCITY
. ROTOR SPEED
. SOLAR VELOCITY
. SUBSONIC SPEED
. SUPERSONIC SPEEDS
. TERMINAL VELOCITY
. TIP SPEED
. TRANSONIC SPEED
. WIND VELOCITY
. . SOLAR WIND VELOCITY
RT   ACCELERATION (PHYSICS)
BODY KINEMATICS
DE BROGLIE WAVELENGTHS
∞DYNAMICS
FERMAT PRINCIPLE
KINEMATICS
KINETICS
LOADING RATE
∞MOTION
PERCEPTUAL TIME CONSTANT
PRESSURE MEASUREMENT
RELATIVISTIC EFFECTS
SOLITARY WAVES
TIME MEASUREMENT

VELOCITY COUPLING
RT   BURNING RATE
COMBUSTION STABILITY
COUPLING
PROPELLANT COMBUSTION

VELOCITY DISTRIBUTION
UF   VELOCITY FIELDS
VELOCITY PROFILES
GS   DISTRIBUTION (PROPERTY)
. **VELOCITY DISTRIBUTION**
RT   CIRCULATION DISTRIBUTION
FLOW DISTRIBUTION
FLOW VELOCITY
GALACTIC ROTATION
ORR-SOMMERFELD EQUATIONS
POHLHAUSEN METHOD
PRESSURE DISTRIBUTION
SHOCK WAVE PROFILES
THREE DIMENSIONAL BOUNDARY LAYER

VELOCITY ERRORS
GS   ERRORS
. **VELOCITY ERRORS**
RT   ESCAPE VELOCITY
ORBITAL VELOCITY
POSITION ERRORS

VELOCITY FIELDS
USE   VELOCITY DISTRIBUTION

VELOCITY MEASUREMENT
UF   ANEMOMETRY
GS   MECHANICAL MEASUREMENT
. **VELOCITY MEASUREMENT**
. . WIND VELOCITY MEASUREMENT
RT   ACCELEROMETERS
ANEMOMETERS
DRAG FORCE ANEMOMETERS
FLOW MEASUREMENT
FLOW VELOCITY
FLOWMETERS
HOT-FILM ANEMOMETERS
HOT-WIRE ANEMOMETERS
HUBBLE CONSTANT
HUBBLE DIAGRAM
LASER DOPPLER VELOCIMETERS
∞MEASUREMENT

VELOCITY MEASUREMENT-(CONT.)
PITOT TUBES
PRESSURE MEASUREMENT
RADIAL VELOCITY
SOLAR WIND VELOCITY
SPEED INDICATORS
STROBOSCOPES
TACHOMETERS
TIME MEASUREMENT
VENTURI TUBES
VORTEX PRECESSION

VELOCITY MODULATION
GS   MODULATION
. **VELOCITY MODULATION**
RT   BUNCHING
CAVITY RESONATORS
ELECTRON BUNCHING
ELECTRON TUBES

VELOCITY PROFILES
USE   VELOCITY DISTRIBUTION

VENEERS
RT   COATINGS
FINISHES
LAMINATES
MASONRY

VENERA SATELLITES
GS   INTERPLANETARY SPACECRAFT
. VENUS PROBES
. . **VENERA SATELLITES**
. . . VENERA 2 SATELLITE
. . . VENERA 3 SATELLITE
. . . VENERA 8 SATELLITE
. . . VENERA 9 SATELLITE
. . . VENERA 10 SATELLITE
. . . VENERA 11 SATELLITE
. . . VENERA 12 SATELLITE
SATELLITES
. ARTIFICIAL SATELLITES
. . SOVIET SATELLITES
. . . **VENERA SATELLITES**
. . . . VENERA 9 SATELLITE
. . . . VENERA 10 SATELLITE
. . . . VENERA 11 SATELLITE
. . . . VENERA 12 SATELLITE
UNMANNED SPACECRAFT
. SPACE PROBES
. . VENUS PROBES
. . . **VENERA SATELLITES**
. . . . VENERA 2 SATELLITE
. . . . VENERA 3 SATELLITE
. . . . VENERA 4 SATELLITE
. . . . VENERA 5 SATELLITE
. . . . VENERA 6 SATELLITE
. . . . VENERA 7 SATELLITE
. . . . VENERA 8 SATELLITE
. . . . VENERA 9 SATELLITE
. . . . VENERA 10 SATELLITE
. . . . VENERA 11 SATELLITE
. . . . VENERA 12 SATELLITE
RT   U.S.S.R. SPACE PROGRAM
VEGA PROJECT

VENERA 2 SATELLITE
GS   INTERPLANETARY SPACECRAFT
. VENUS PROBES
. . VENERA SATELLITES
. . . **VENERA 2 SATELLITE**
UNMANNED SPACECRAFT
. SPACE PROBES
. . VENUS PROBES
. . . VENERA SATELLITES
. . . . **VENERA 2 SATELLITE**

VENERA 3 SATELLITE
GS   INTERPLANETARY SPACECRAFT
. VENUS PROBES
. . VENERA SATELLITES
. . . **VENERA 3 SATELLITE**
UNMANNED SPACECRAFT
. SPACE PROBES
. . VENUS PROBES
. . . VENERA SATELLITES
. . . . **VENERA 3 SATELLITE**

VENERA 4 SATELLITE
GS   UNMANNED SPACECRAFT
. SPACE PROBES
. . VENUS PROBES
. . . VENERA SATELLITES
. . . . **VENERA 4 SATELLITE**

VENERA 5 SATELLITE
GS   UNMANNED SPACECRAFT
. SPACE PROBES
. . VENUS PROBES
. . . VENERA SATELLITES
. . . . **VENERA 5 SATELLITE**

VENERA 6 SATELLITE
GS   UNMANNED SPACECRAFT
. SPACE PROBES
. . VENUS PROBES
. . . VENERA SATELLITES
. . . . **VENERA 6 SATELLITE**

VENERA 7 SATELLITE
GS   UNMANNED SPACECRAFT
. SPACE PROBES
. . VENUS PROBES
. . . VENERA SATELLITES
. . . . **VENERA 7 SATELLITE**

VENERA 8 SATELLITE
GS   INTERPLANETARY SPACECRAFT
. VENUS PROBES
. . VENERA SATELLITES
. . . **VENERA 8 SATELLITE**
UNMANNED SPACECRAFT
. SPACE PROBES
. . VENUS PROBES
. . . VENERA SATELLITES
. . . . **VENERA 8 SATELLITE**
RT   U.S.S.R. SPACE PROGRAM
VENUS (PLANET)

VENERA 9 SATELLITE
GS   INTERPLANETARY SPACECRAFT
. VENUS PROBES
. . VENERA SATELLITES
. . . **VENERA 9 SATELLITE**
SATELLITES
. ARTIFICIAL SATELLITES
. . SOVIET SATELLITES
. . . VENERA SATELLITES
. . . . **VENERA 9 SATELLITE**
UNMANNED SPACECRAFT
. SPACE PROBES
. . VENUS PROBES
. . . VENERA SATELLITES
. . . . **VENERA 9 SATELLITE**

VENERA 10 SATELLITE
GS   INTERPLANETARY SPACECRAFT
. VENUS PROBES
. . VENERA SATELLITES
. . . **VENERA 10 SATELLITE**
SATELLITES
. ARTIFICIAL SATELLITES
. . SOVIET SATELLITES
. . . VENERA SATELLITES
. . . . **VENERA 10 SATELLITE**
UNMANNED SPACECRAFT
. SPACE PROBES
. . VENUS PROBES
. . . VENERA SATELLITES
. . . . **VENERA 10 SATELLITE**
RT   U.S.S.R. SPACE PROGRAM
VENUS (PLANET)

VENERA 11 SATELLITE
GS   INTERPLANETARY SPACECRAFT
. VENUS PROBES
. . VENERA SATELLITES
. . . **VENERA 11 SATELLITE**
SATELLITES
. ARTIFICIAL SATELLITES
. . SOVIET SATELLITES
. . . VENERA SATELLITES
. . . . **VENERA 11 SATELLITE**
UNMANNED SPACECRAFT
. SPACE PROBES
. . VENUS PROBES
. . . VENERA SATELLITES
. . . . **VENERA 11 SATELLITE**
RT   U.S.S.R. SPACE PROGRAM
VENUS (PLANET)
VENUS ATMOSPHERE
VENUS SURFACE

VENERA 12 SATELLITE
GS   INTERPLANETARY SPACECRAFT
. VENUS PROBES
. . VENERA SATELLITES
. . . **VENERA 12 SATELLITE**
SATELLITES
. ARTIFICIAL SATELLITES

VENERA 12 SATELLITE-*(CONT.)*
.. SOVIET SATELLITES
... VENERA SATELLITES
.... **VENERA 12 SATELLITE**
UNMANNED SPACECRAFT
. SPACE PROBES
.. VENUS PROBES
... VENERA SATELLITES
.... **VENERA 12 SATELLITE**
RT   U.S.S.R. SPACE PROGRAM
VENUS (PLANET)
VENUS ATMOSPHERE
VENUS SURFACE

VENEZIANO MODEL
GS   MODELS
. MATHEMATICAL MODELS
.. **VENEZIANO MODEL**
RT   ELEMENTARY PARTICLE INTERACTIONS

VENEZUELA
GS   NATIONS
. **VENEZUELA**
RT   SOUTH AMERICA

VENN DIAGRAMS
GS   DIAGRAMS
. **VENN DIAGRAMS**
RT   ANALYSIS (MATHEMATICS)
GEOMETRY
MATHEMATICAL LOGIC

VENOM AIRCRAFT
USE   DH 112 AIRCRAFT

VENTILATION
RT   AIR CONDITIONING
AIR COOLING
AIR FILTERS
AIR FLOW
AIR INTAKES
AIR PURIFICATION
BLOWERS
COMFORT
COOLING
COOLING SYSTEMS
DRAFT (GAS FLOW)
DUCTS
ENVIRONMENTAL ENGINEERING
EXHAUST SYSTEMS
EXHAUSTING
LIFE SUPPORT SYSTEMS
REFRIGERATING
TEMPERATURE
TEMPERATURE CONTROL
TEMPERATURE DISTRIBUTION
VENTILATORS
VENTING
VENTS

VENTILATION FANS
RT   BLOWERS
COOLING
COOLING SYSTEMS
DUCTED FANS
FAN BLADES
∞FANS
VENTILATORS

VENTILATORS
RT   AIR DUCTS
AIR INTAKES
BLOWERS
∞DIFFUSERS
EXHAUST SYSTEMS
VENTILATION
VENTILATION FANS
VENTS

VENTING
RT   BREATHING VIBRATION
COOLING
∞DISCHARGE
EVACUATING (VACUUM)
EXHAUSTING
FLUSHING
PURGING
RELEASING
RELIEF VALVES
∞SEPARATION
VENTILATION
VENTS

VENTRAL SECTIONS
RT   ABDOMEN

VENTS
GS   OUTLETS
. **VENTS**
RT   ANNULAR DUCTS
APERTURES
CAVITIES
CHIMNEYS
COOLING SYSTEMS
DUCTS
EVACUATING (VACUUM)
EXHAUST SYSTEMS
FLUES
GATES (OPENINGS)
LOUVERS
∞NOZZLES
OPENINGS
PORTS (OPENINGS)
RELIEF VALVES
SLOTTED WIND TUNNELS
VENTILATION
VENTILATORS
VENTING
WINDOWS (APERTURES)

VENTURI TUBES
RT   ∞DETECTORS
FLOW MEASUREMENT
FLOWMETERS
GAS METERS
MEASURING INSTRUMENTS
ORIFICES
PITOT TUBES
PRESSURE GRADIENTS
PRESSURE MEASUREMENT
∞TUBES
VELOCITY MEASUREMENT

VENUS (PLANET)
GS   CELESTIAL BODIES
. PLANETS
.. TERRESTRIAL PLANETS
... **VENUS (PLANET)**
RT   PLANETARY CRATERS
VEGA PROJECT
VENERA 8 SATELLITE
VENERA 10 SATELLITE
VENERA 11 SATELLITE
VENERA 12 SATELLITE

VENUS ATMOSPHERE
GS   ENVIRONMENTS
. EXTRATERRESTRIAL ENVIRONMENTS
.. PLANETARY ENVIRONMENTS
... PLANETARY ATMOSPHERES
.... **VENUS ATMOSPHERE**
RT   AEROSPACE ENVIRONMENTS
IONOPAUSE
VENERA 11 SATELLITE
VENERA 12 SATELLITE
VENUS ORBITING IMAGING RADAR
(SPACECRAFT)

VENUS CLOUDS
GS   CLOUDS
. **VENUS CLOUDS**
RT   ATMOSPHERIC MODELS
CLOUD COVER
CLOUD PHYSICS
GREENHOUSE EFFECT

VENUS FLY TRAP ROCKET VEHICLE
GS   ROCKET VEHICLES
. SOUNDING ROCKETS
.. **VENUS FLY TRAP ROCKET VEHICLE**
RT   COSMIC DUST
EXTRATERRESTRIAL MATTER

**VENUS ORBITING IMAGING RADAR
(SPACECRAFT)**
GS   RADAR
. **VENUS ORBITING IMAGING RADAR
(SPACECRAFT)**
RT   SYNTHETIC APERTURE RADAR
VENUS ATMOSPHERE
VENUS PROBES
VENUS SURFACE

VENUS PROBES
GS   INTERPLANETARY SPACECRAFT
. **VENUS PROBES**
.. MARINER 1 SPACE PROBE
.. MARINER 2 SPACE PROBE

VENUS PROBES-*(CONT.)*
.. MARINER 5 SPACE PROBE
.. PIONEER VENUS 2 SPACECRAFT
... PIONEER VENUS 2 TRANSPORTER
BUS
.. VENERA SATELLITES
... VENERA 2 SATELLITE
... VENERA 3 SATELLITE
... VENERA 8 SATELLITE
... VENERA 9 SATELLITE
... VENERA 10 SATELLITE
... VENERA 11 SATELLITE
... VENERA 12 SATELLITE
.. ZOND 1 SPACE PROBE
.. ZOND 3 SPACE PROBE
.. ZOND 4 SPACE PROBE
.. ZOND 5 SPACE PROBE
.. ZOND 6 SPACE PROBE
.. ZOND 7 SPACE PROBE
.. ZOND 8 SPACE PROBE
UNMANNED SPACECRAFT
. SPACE PROBES
.. **VENUS PROBES**
... MARINER 1 SPACE PROBE
... MARINER 2 SPACE PROBE
... MARINER 5 SPACE PROBE
... VENERA SATELLITES
.... VENERA 2 SATELLITE
.... VENERA 3 SATELLITE
.... VENERA 4 SATELLITE
.... VENERA 5 SATELLITE
.... VENERA 6 SATELLITE
.... VENERA 7 SATELLITE
.... VENERA 8 SATELLITE
.... VENERA 9 SATELLITE
.... VENERA 10 SATELLITE
.... VENERA 11 SATELLITE
.... VENERA 12 SATELLITE
... ZOND 1 SPACE PROBE
... ZOND 3 SPACE PROBE
... ZOND 4 SPACE PROBE
... ZOND 5 SPACE PROBE
... ZOND 6 SPACE PROBE
... ZOND 7 SPACE PROBE
... ZOND 8 SPACE PROBE
RT   MARINER PROGRAM
MARINER VENUS 67 SPACECRAFT
MARS PROBES
OUTER PLANETS EXPLORERS
SPUTNIK 5 SATELLITE
VENUS ORBITING IMAGING RADAR
(SPACECRAFT)
VOYAGER PROJECT

VENUS RADAR ECHOES
GS   ECHOES
. RADAR ECHOES
.. **VENUS RADAR ECHOES**

VENUS SURFACE
GS   PLANETARY SURFACES
. **VENUS SURFACE**
RT   CLOUD COVER
EXTRATERRESTRIAL ENVIRONMENTS
PLANETARY CRATERS
SOLAR SYSTEM
∞SURFACES
TOPOGRAPHY
VENERA 11 SATELLITE
VENERA 12 SATELLITE
VENUS ORBITING IMAGING RADAR
(SPACECRAFT)

VERBAL COMMUNICATION
GS   COMMUNICATING
. **VERBAL COMMUNICATION**
RT   ACOUSTICS
CONVERSATION
LANGUAGES
LECTURES
PHONETICS
SPEECH BASEBAND COMPRESSION
TELEPHONY
VOICE COMMUNICATION
VOICE DATA PROCESSING
WORDS (LANGUAGE)

VERIFICATION (PROVING)
USE   PROVING

VERMICULITE
GS   CLAYS
. **VERMICULITE**
MINERALS
. **VERMICULITE**

VERMICULITE-*(CONT.)*
RT   INSULATION
      MICA
      PACKAGING

VERMONT
GS   NATIONS
      . UNITED STATES
      . . **VERMONT**
RT   LAKE CHAMPLAIN BASIN (NY-VT)
      ST LAWRENCE VALLEY (NORTH
          AMERICA)

VERNEUIL PROCESS
GS   GROWTH
      . CRYSTAL GROWTH
      . . **VERNEUIL PROCESS**
RT   CZOCHRALSKI METHOD
      RUBY LASERS

VERNIER ENGINES
GS   ENGINES
      . ROCKET ENGINES
      . . **VERNIER ENGINES**
      . . . SYNCOM APOGEE ENGINES
      . TORPEDO ENGINES
      . . **VERNIER ENGINES**
      . . . CONTROL ROCKETS
      . . . SYNCOM APOGEE ENGINES
RT   ELECTRIC ROCKET ENGINES
      ELECTROSTATIC ENGINES
      HYBRID PROPELLANT ROCKET ENGINES
      INTERNAL COMBUSTION ENGINES
      LAUNCH VEHICLES
      LIQUID PROPELLANT ROCKET ENGINES
      MA-2 ENGINE
      MA-3 ENGINE
      MA-5 ENGINE
      MICROROCKET ENGINES
      RESTARTABLE ROCKET ENGINES
      SOLID PROPELLANT ROCKET ENGINES
      THRUST VECTOR CONTROL

VERNINE
USE   GUANOSINES

VERONIQUE ROCKET VEHICLES
GS   ROCKET VEHICLES
      . SINGLE STAGE ROCKET VEHICLES
      . . **VERONIQUE ROCKET VEHICLES**
      . SOUNDING ROCKETS
      . . **VERONIQUE ROCKET VEHICLES**
RT   LIQUID PROPELLANT ROCKET ENGINES
      ∞ VEHICLES

VERSATILITY
RT   COMPATIBILITY
      FLEXIBILITY

VERTEBRAE
GS   ANATOMY
      . MUSCULOSKELETAL SYSTEM
      . . BONES
      . . . **VERTEBRAE**
RT   INTERVERTEBRAL DISKS
      NECK (ANATOMY)
      SPINE

VERTEBRAL COLUMN
GS   ANATOMY
      . MUSCULOSKELETAL SYSTEM
      . . **VERTEBRAL COLUMN**
RT   BONES

VERTEBRATES
GS   ANIMALS
      . **VERTEBRATES**
      . . AMPHIBIA
      . . . FROGS
      . . BIRDS
      . . . CHICKENS
      . . . PIGEONS
      . . . TURKEYS
      . . . WATERFOWL
      . . FISHES
      . . . SCHOOLS (FISH)
      . . . SHARKS
      . . MAMMALS
      . . . BATS
      . . . BEARS
      . . . CATS
      . . . CATTLE
      . . . . CALVES
      . . . DEER

VERTEBRATES-*(CONT.)*
      . . . CARIBOUS
      . . . DOGS
      . . . DOLPHINS
      . . GOATS
      . . HORSES
      . . . MANATEES
      . . . MARINE MAMMALS
      . . . PORPOISES
      . . . PRIMATES
      . . . . APES
      . . . . . CHIMPANZEES
      . . . . BABOONS
      . . . . HUMAN BEINGS
      . . . . MONKEYS
      . . . RODENTS
      . . . . GUINEA PIGS
      . . . . HAMSTERS
      . . . . MICE
      . . . . . JERBOAS
      . . . . . POCKET MICE
      . . . . RABBITS
      . . . . RATS
      . . . . SQUIRRELS
      . . . . . GROUND SQUIRRELS
      . . . SEALS (ANIMALS)
      . . . SHEEP
      . . . SHREWS
      . . . SWINE
      . . . WHALES
      . . . WOLVES
      . . REPTILES
      . . . LIZARDS
      . . . SNAKES
      . . . TURTLES
RT   HOMEOTHERMS

VERTICAL AIR CURRENTS
UF   UPDRAFTS
GS   FLUID FLOW
      . GAS FLOW
      . . AIR FLOW
      . . . AIR CURRENTS
      . . . . **VERTICAL AIR CURRENTS**
RT   ATMOSPHERIC CIRCULATION
      CONVECTION CLOUDS
      CONVECTION CURRENTS
      LEE WAVES
      MIXING HEIGHT
      SOARING
      TURBULENCE
      WIND (METEOROLOGY)
      WINDS ALOFT

VERTICAL ATTITUDE TAKEOFF-LANDING
AIRCRAFT
USE   VATOL AIRCRAFT

VERTICAL DISTRIBUTION
GS   DISTRIBUTION (PROPERTY)
      . **VERTICAL DISTRIBUTION**
      . . STAR DISTRIBUTION
RT   ELECTRON DISTRIBUTION
      ION DISTRIBUTION
      PRESSURE DISTRIBUTION
      RADIATION DISTRIBUTION
      SPATIAL DISTRIBUTION
      TEMPERATURE DISTRIBUTION
      WIND PROFILES

VERTICAL FINS
USE   FINS

VERTICAL FLIGHT
RT   BALLOON FLIGHT
      CLIMBING FLIGHT
      ∞ FLIGHT
      FLIGHT PATHS
      HOVERING
      ROCKET FLIGHT
      V/STOL AIRCRAFT

VERTICAL JUNCTION SOLAR CELLS
GS   ELECTRIC GENERATORS
      . DIRECT POWER GENERATORS
      . . SOLAR CELLS
      . . . **VERTICAL JUNCTION SOLAR CELLS**
      . SOLAR GENERATORS
      . SOLAR CELLS
      . . . **VERTICAL JUNCTION SOLAR CELLS**
RT   WAFERS

VERTICAL LANDING
UF   VERTICAL TAKEOFF AND LANDING
      VTOL

VERTICAL LANDING-*(CONT.)*
GS   LANDING
      . **VERTICAL LANDING**
RT   AIRCRAFT LANDING
      SPACECRAFT LANDING
      TOUCHDOWN
      VATOL AIRCRAFT

VERTICAL MOTION
RT   FALLING
      ∞ MOTION
      TOUCHDOWN

VERTICAL MOTION SIMULATORS
GS   SIMULATORS
      . VIBRATION SIMULATORS
      . . **VERTICAL MOTION SIMULATORS**
RT   ∞ MOTION
      SHAKERS
      SHOCK SIMULATORS
      VIBRATORY LOADS

VERTICAL ORIENTATION
RT   ALIGNMENT
      ATTITUDE (INCLINATION)
      DIRECTIONAL STABILITY
      DYNAMIC STABILITY
      HORIZONTAL ORIENTATION
      LATERAL STABILITY
      ∞ ORIENTATION
      STABILIZATION

VERTICAL PERCEPTION
GS   PERCEPTION
      . SENSORY PERCEPTION
      . . **VERTICAL PERCEPTION**
RT   BODY SWAY TEST
      GRAVIRECEPTORS
      OCULOGRAVIC ILLUSIONS
      ∞ ORIENTATION
      OTOLITH ORGANS
      ∞ SPACE ORIENTATION
      VESTIBULAR TESTS

VERTICAL STABILIZERS
USE   STABILIZERS (FLUID DYNAMICS)

VERTICAL TAILS
USE   STABILIZERS (FLUID DYNAMICS)
      TAIL ASSEMBLIES

VERTICAL TAKEOFF
UF   VERTICAL TAKEOFF AND LANDING
      VTOL
GS   TAKEOFF
      . **VERTICAL TAKEOFF**

VERTICAL TAKEOFF AIRCRAFT
UF   VTOL AIRCRAFT
GS   V/STOL AIRCRAFT
      . **VERTICAL TAKEOFF AIRCRAFT**
      . . FLYING PLATFORMS
      . . SC-1 AIRCRAFT
      . . VJ-101 AIRCRAFT
      . . VZ-8 AIRCRAFT
      . . X-13 AIRCRAFT
      . . X-14 AIRCRAFT
      . . X-19 AIRCRAFT
      . . X-22 AIRCRAFT
      . . X-22A AIRCRAFT
      . . XV-4 AIRCRAFT
      . . XV-11A AIRCRAFT
RT   ∞ AIRCRAFT
      BELL 214A HELICOPTER
      CF-700 ENGINE
      CIRCULATION CONTROL ROTORS
      COMPOUND HELICOPTERS
      CUSHIONCRAFT GROUND EFFECT
          MACHINE
      FAN IN WING AIRCRAFT
      GETOL AIRCRAFT
      HELICOPTERS
      LIFT FANS
      LIFTING ROTORS
      ∞ MILITARY AIRCRAFT
      POWERED LIFT AIRCRAFT
      RESEARCH AIRCRAFT
      ROTARY WING AIRCRAFT
      SHORT TAKEOFF AIRCRAFT
      ∞ SUBSONIC AIRCRAFT
      T-58 ENGINE
      T-58-GE-8B ENGINE
      TILT WING AIRCRAFT
      VATOL AIRCRAFT
      ∞ WINGED VEHICLES

VERTICAL TAKEOFF AND LANDING
USE   VERTICAL LANDING
          VERTICAL TAKEOFF

VERTICAL 8 ROCKET
GS   ROCKET VEHICLES
      . SOUNDING ROCKETS
      . . **VERTICAL 8 ROCKET**
RT   PAYLOADS
      ∞ROCKETS

VERTICES
USE   APEXES

VERTIGO
GS   SIGNS AND SYMPTOMS
      . **VERTIGO**
RT   BARANY CHAIR
      EAR PRESSURE TEST
      VESTIBULAR TESTS

VERTOL MILITARY HELICOPTERS
USE   BOEING AIRCRAFT

VERY HIGH FREQUENCIES
SN   (30 TO 300 MHZ)
GS   FREQUENCIES
      . RADIO FREQUENCIES
      . . **VERY HIGH FREQUENCIES**
      . . . P BAND
RT   DECAMETRIC WAVES
      LOW FREQUENCY BANDS
      MAXIMUM USABLE FREQUENCY

VERY HIGH FREQUENCY RADIO EQUIPMENT
UF   ULTRA SHORT WAVE RADIO EQUIPMENT
GS   RADIO EQUIPMENT
      . **VERY HIGH FREQUENCY RADIO
        EQUIPMENT**
RT   RADIO ASTRONOMY
      ULTRAHIGH FREQUENCIES

VERY HIGH SPEED INTEGRATED CIRCUITS
USE   VHSIC (CIRCUITS)

VERY LARGE SCALE INTEGRATION
UF   VLSI
GS   CIRCUITS
      . INTEGRATED CIRCUITS
      . . **VERY LARGE SCALE INTEGRATION**
RT   ARCHITECTURE (COMPUTERS)
      CHIPS (ELECTRONICS)
      LARGE SCALE INTEGRATION

VERY LONG BASE INTERFEROMETRY
UF   VLBI
GS   INTERFEROMETRY
      . **VERY LONG BASE INTERFEROMETRY**
RT   DIFFRACTION PATTERNS
      INTERFEROMETERS
      NULL ZONES
      QUASAT
      RADIO ASTRONOMY
      RADIO INTERFEROMETERS

VERY LOW FREQUENCIES
SN   (3 TO 30 KH2)
GS   FREQUENCIES
      . RADIO FREQUENCIES
      . . LOW FREQUENCIES
      . . . **VERY LOW FREQUENCIES**
      . . LOW FREQUENCY BANDS
      . . . **VERY LOW FREQUENCIES**
RT   AUDIO FREQUENCIES

∞ **VESSELS**
SN   *(USE OF A MORE SPECIFIC TERM IS
      RECOMMENDED--CONSULT THE TERMS
      LISTED BELOW)*
RT   BLOOD VESSELS
      ∞CAPSULES
      FLUID FILLED SHELLS
      LIQUID FILLED SHELLS
      NAVY
      PRESSURE VESSELS
      SHIPS

VESTA ASTEROID
GS   CELESTIAL BODIES
      . ASTEROID BELTS
      . . ASTEROIDS
      . . . **VESTA ASTEROID**
RT   METEOROIDS
      SOLAR SYSTEM

VESTA ASTEROID-*(CONT.)*
      SPACE DEBRIS

VESTIBULAR NYSTAGMUS
GS   REFLEXES
      . **VESTIBULAR NYSTAGMUS**
RT   ANATOMY
      EYE (ANATOMY)
      OPHTHALMOLOGY

VESTIBULAR TESTS
GS   PHYSIOLOGICAL TESTS
      . **VESTIBULAR TESTS**
RT   BODY SWAY TEST
      CORIOLIS EFFECT
      EAR PRESSURE TEST
      VERTICAL PERCEPTION
      VERTIGO

VESTIBULES
GS   ANATOMY
      . SENSE ORGANS
      . . EAR
      . . . LABYRINTH
      . . . . **VESTIBULES**
RT   MORPHOLOGY
      PASSAGEWAYS
      SEMICIRCULAR CANALS

VESTS
RT   CLOTHING
      GARMENTS

VETERINARY MEDICINE
RT   ∞BIOLOGY
      DIAGNOSIS
      DISEASES
      EPIDEMIOLOGY
      IMMUNOLOGY
      INJURIES
      ∞MEDICINE
      PATHOLOGY
      PHARMACOLOGY
      SURGERY

VFR (RULES)
USE   VISUAL FLIGHT RULES

VHF OMNIRANGE NAVIGATION
UF   OMNIRANGE NAVIGATION
      VOR SYSTEMS
GS   NAVIGATION
      . RADIO NAVIGATION
      . . **VHF OMNIRANGE NAVIGATION**
RT   AIR NAVIGATION
      NAVIGATION AIDS
      RADIO DIRECTION FINDERS
      SOLAR COMPASSES

VHSIC (CIRCUITS)
UF   VERY HIGH SPEED INTEGRATED
        CIRCUITS
GS   CIRCUITS
      . INTEGRATED CIRCUITS
      . . **VHSIC (CIRCUITS)**
RT   CHIPS (ELECTRONICS)
      LARGE SCALE INTEGRATION
      SIGNAL PROCESSING

VIABILITY
RT   ANIMALS
      CARBON CYCLE
      CROP VIGOR
      GERMINATION
      GROWTH
      PLANTS (BOTANY)
      SEEDS

VIBRATION
UF   JITTER
GS   **VIBRATION**
      . COMBUSTION VIBRATION
      . FORCED VIBRATION
      . FREE VIBRATION
      . LATTICE VIBRATIONS
      . POGO EFFECTS
      . RANDOM VIBRATION
      . RESONANT VIBRATION
      . STRUCTURAL VIBRATION
      . . BENDING VIBRATION
      . . BREATHING VIBRATION
      . . FLUTTER
      . . . PANEL FLUTTER
      . . . SUBSONIC FLUTTER

VIBRATION-*(CONT.)*
      . . . SUPERSONIC FLUTTER
      . . . TRANSONIC FLUTTER
      . . LINEAR VIBRATION
      . . MISSILE VIBRATION
      . . SELF INDUCED VIBRATION
      . . . PANEL FLUTTER
      . . . SUBSONIC FLUTTER
      . . . SUPERSONIC FLUTTER
      . . . TRANSONIC FLUTTER
      . . TORSIONAL VIBRATION
RT   ACOUSTICS
      AMPLITUDES
      ANTINODES
      COMPACTING
      CYCLIC LOADS
      DISPLACEMENT
      ∞DYNAMICS
      ELASTIC WAVES
      FATIGUE (MATERIALS)
      FLAPPING
      HARMONICS
      ISOLATORS
      MECHANICAL OSCILLATORS
      MECHANICAL SHOCK
      MODES (STANDING WAVES)
      ∞MOTION
      NODES (STANDING WAVES)
      NUTATION
      OSCILLATING CYLINDERS
      OSCILLATIONS
      OSCILLATORS
      RESONANCE
      SHAKING
      SHOCK RESISTANCE
      SPACECRAFT MOTION
      STANDING WAVES
      VIBRATIONAL STRESS
      VIBRATORY LOADS
      VIBRATORY POLISHING
      ∞WAVES
      WING OSCILLATIONS

VIBRATION DAMPERS
USE   VIBRATION ISOLATORS

VIBRATION DAMPING
GS   DAMPING
      . **VIBRATION DAMPING**
RT   ACOUSTICS
      ATTENUATION
      DAST PROGRAM
      ELASTIC DAMPING
      FLEXIBLE SPACECRAFT
      GYRODAMPERS
      HARMONIC CONTROL
      MOLECULAR RELAXATION
      NONOSCILLATORY ACTION
      NONSTABILIZED OSCILLATION
      SHOCK ABSORBERS

VIBRATION EFFECTS
GS   **VIBRATION EFFECTS**
      . POGO EFFECTS
RT   ∞EFFECTS
      SUPPORT INTERFERENCE
      VIBRATIONAL STRESS

VIBRATION ISOLATORS
UF   VIBRATION DAMPERS
      VIBRATION PROTECTION
GS   ISOLATORS
      . **VIBRATION ISOLATORS**
RT   ∞ABSORBERS
      ACOUSTIC RETROFITTING
      CUSHIONS
      ∞DAMPERS
      DAMPERS (VALVES)
      DAMPING
      ENERGY ABSORPTION
      NOISE REDUCTION
      OSCILLATION DAMPERS
      SHOCK ABSORBERS
      SPRINGS (ELASTIC)
      SUSPENSION SYSTEMS (VEHICLES)

VIBRATION MEASUREMENT
GS   MECHANICAL MEASUREMENT
      . **VIBRATION MEASUREMENT**
RT   CEPSTRAL ANALYSIS
      DAMPING TESTS
      FREQUENCY ANALYZERS
      FREQUENCY MEASUREMENT
      ∞MEASUREMENT
      STRESS MEASUREMENT

VIBRATION MEASUREMENT-*(CONT.)*
    VIBRATIONAL SPECTRA

VIBRATION METERS
UF   VIBROMETERS
GS   MEASURING INSTRUMENTS
    . **VIBRATION METERS**
    . . SEISMOGRAPHS
    . . . LUNAR SEISMOGRAPHS
RT   ACCELEROMETERS
    TRANSDUCERS

VIBRATION MODE
UF   MODE OF VIBRATION
GS   MODES
    . **VIBRATION MODE**
    . . UNCOUPLED MODES
RT   FREE VIBRATION
    LINEAR VIBRATION
    MODE TRANSFORMERS
    TRANSVERSE WAVES

VIBRATION PERCEPTION
GS   PERCEPTION
    . SENSORY PERCEPTION
    . . **VIBRATION PERCEPTION**

VIBRATION PROTECTION
USE   VIBRATION ISOLATORS

VIBRATION SIMULATORS
UF   VIBRATION TESTING MACHINES
GS   SIMULATORS
    . **VIBRATION SIMULATORS**
    . . VERTICAL MOTION SIMULATORS
RT   FLUTTER
    ∞MACHINERY
    SHAKERS
    SHOCK SIMULATORS
    VIBRATORY LOADS

VIBRATION TESTING MACHINES
USE   VIBRATION SIMULATORS

VIBRATION TESTS
GS   **VIBRATION TESTS**
    . DAMPING TESTS
    . . STROKING TESTS
RT   DESTRUCTIVE TESTS
    DYNAMIC TESTS
    ELECTRONIC EQUIPMENT TESTS
    ENGINE TESTS
    ENVIRONMENTAL TESTS
    FLIGHT TESTS
    FLUTTER
    MECHANICAL ENGINEERING
    OSCILLATIONS
    RESONANCE TESTING
    SHOCK TESTS
    STABILITY TESTS
    STATIC TESTS
    STRUCTURAL VIBRATION
    ∞TESTS

VIBRATIONAL FREEZING
GS   PHASE TRANSFORMATIONS
    . FREEZING
    . . **VIBRATIONAL FREEZING**

VIBRATIONAL FREQUENCIES
USE   VIBRATIONAL SPECTRA

VIBRATIONAL RELAXATION
USE   MOLECULAR RELAXATION

VIBRATIONAL SPECTRA
UF   VIBRATIONAL FREQUENCIES
GS   SPECTRA
    . MOLECULAR SPECTRA
    . . **VIBRATIONAL SPECTRA**
    . RADIATION SPECTRA
    . . ELECTROMAGNETIC SPECTRA
    . . . **VIBRATIONAL SPECTRA**
RT   ELECTRONIC SPECTRA
    ENERGY SPECTRA
    MOLECULAR RELAXATION
    RAMAN SPECTRA
    SHOCK SPECTRA
    TRANSIENT OSCILLATIONS
    VIBRATION MEASUREMENT

VIBRATIONAL STRESS
GS   STRESSES
    VIBRATIONAL STRESS

VIBRATIONAL STRESS-*(CONT.)*
RT   FLUTTER
    VIBRATION
    VIBRATION EFFECTS
    VIBRATORY LOADS

VIBRATORY LOADS
GS   LOADS (FORCES)
    . DYNAMIC LOADS
    . . **VIBRATORY LOADS**
RT   AERODYNAMIC LOADS
    CYCLIC LOADS
    STRUCTURAL DESIGN CRITERIA
    VERTICAL MOTION SIMULATORS
    VIBRATION
    VIBRATION SIMULATORS
    VIBRATIONAL STRESS

VIBRATORY POLISHING
GS   POLISHING
    . **VIBRATORY POLISHING**
RT   METALLOGRAPHY
    VIBRATION

VIBROCARDIOGRAPHY
USE   PHONOCARDIOGRAPHY

VIBROMETERS
USE   VIBRATION METERS

VICKERS SCIMITAR AIRCRAFT
USE   SCIMITAR AIRCRAFT

VICKERS VALIANT AIRCRAFT
USE   VALIANT AIRCRAFT

VICKERS VC-10 AIRCRAFT
USE   VC-10 AIRCRAFT

VICKERS 1100 AIRCRAFT
USE   VC-10 AIRCRAFT

VICTOR MK-1 AIRCRAFT
GS   ATTACK AIRCRAFT
    . BOMBER AIRCRAFT
    . . **VICTOR MK-1 AIRCRAFT**
    HANDLEY PAGE AIRCRAFT
    . **VICTOR MK-1 AIRCRAFT**
    JET AIRCRAFT
    . **VICTOR MK-1 AIRCRAFT**
    MONOPLANES
    . **VICTOR MK-1 AIRCRAFT**
    RECONNAISSANCE AIRCRAFT
    . **VICTOR MK-1 AIRCRAFT**
RT   ∞AIRCRAFT

VIDEO COMMUNICATION
GS   TELECOMMUNICATION
    . **VIDEO COMMUNICATION**
RT   TELEVISION SYSTEMS
    VIDEO SIGNALS

VIDEO DATA
RT   ANALOG DATA
    ∞DATA
    DATA CONVERTERS
    DATA TRANSMISSION
    DIGITAL DATA
    DISPLAY DEVICES
    RADAR DATA
    SMEAR
    TELECOMMUNICATION
    TELEVISION SYSTEMS
    VIDEO DISKS
    VIDEO SIGNALS

VIDEO DISKS
GS   DOCUMENTS
    . RECORDS
    . . **VIDEO DISKS**
RT   DATA RECORDERS
    DATA RECORDING
    DATA STORAGE
    DISKS (SHAPES)
    MAGNETIC DISKS
    MEMORY (COMPUTERS)
    OPTICAL DATA STORAGE MATERIALS
    OPTICAL DISKS
    OPTICAL MEMORY (DATA STORAGE)
    PLAYBACKS
    VIDEO DATA
    VIDEO EQUIPMENT

VIDEO EQUIPMENT
GS   **VIDEO EQUIPMENT**
    . PICTURE TUBES
RT   ADVANCED VIDICON CAMERA SYSTEM
    (AVCS)
    CAMERA TUBES
    CATHODE RAY TUBES
    COMPENSATORS
    DISPLAY DEVICES
    FLYING SPOT SCANNERS
    MOTION PICTURES
    OPTICAL EQUIPMENT
    OSCILLOSCOPES
    RECORDING HEADS
    TELEVISION EQUIPMENT
    VIDEO DISKS
    VIDICONS

VIDEO LANDMARK ACQUISITION AND TRACKING
GS   TRACKING (POSITION)
    . **VIDEO LANDMARK ACQUISITION AND TRACKING**
RT   AVIONICS
    IMAGE CORRELATORS
    MAP MATCHING GUIDANCE
    SCENE ANALYSIS
    SIGNATURES
    TERCOM
    TERRAIN ANALYSIS
    TRACKING FILTERS

VIDEO SIGNALS
RT   SIGNAL PROCESSING
    SIGNAL TRANSMISSION
    ∞SIGNALS
    VIDEO COMMUNICATION
    VIDEO DATA

VIDICONS
GS   ELECTRON TUBES
    . CAMERA TUBES
    . . **VIDICONS**
    . . . RETURN BEAM VIDICONS
    . . . . THERMICONS
    MICROWAVE EQUIPMENT
    . **VIDICONS**
    . . THERMICONS
RT   ADVANCED VIDICON CAMERA SYSTEM
    (AVCS)
    FIBER OPTICS
    VIDEO EQUIPMENT

VIETNAM
UF   NORTH VIETNAM
    REPUBLIC OF VIETNAM
    SOUTH VIETNAM
GS   NATIONS
    . **VIETNAM**
RT   ASIA
    SOUTHEAST ASIA

VIEW EFFECTS
SN   (EFFECTS OF CHANGE IN ANGULAR SIZE OF FIELD OF VIEW UPON RECEPTORS OF RADIATION)
RT   ANGULAR CORRELATION
    ∞EFFECTS
    RADIANT FLUX DENSITY
    RADIATION MEASURING INSTRUMENTS
    RADIATIVE HEAT TRANSFER
    VIEWING

VIEWING
GS   **VIEWING**
    . FIELD OF VIEW
RT   DISPLAY DEVICES
    PERISCOPES
    VIEW EFFECTS
    VISIBILITY
    VISION

VIGILANTE AIRCRAFT
USE   A-5 AIRCRAFT

VIGNETTING
RT   DEFECTS
    FOCUSING
    LENSES

VIKING LANDER SPACECRAFT
GS   INTERPLANETARY SPACECRAFT
    . MARS PROBES
    . . VIKING SPACECRAFT
    . . . VIKING 1 SPACECRAFT

VIKING LANDER SPACECRAFT-(CONT.)
```
        . . . . VIKING LANDER SPACECRAFT
        . . . . . VIKING LANDER 1
        . . . VIKING 2 SPACECRAFT
        . . . . VIKING LANDER SPACECRAFT
        . . . . . VIKING LANDER 2
  RT    INTERPLANETARY TRAJECTORIES
        SPACE EXPLORATION
        SPACE FLIGHT
        SPACE PROBES
        UNMANNED SPACECRAFT
```

VIKING LANDER 1
```
  GS    INTERPLANETARY SPACECRAFT
        . MARS PROBES
        . . VIKING SPACECRAFT
        . . . VIKING 1 SPACECRAFT
        . . . . VIKING LANDER SPACECRAFT
        . . . . . VIKING LANDER 1
  RT    INTERPLANETARY TRAJECTORIES
        MARS SURFACE SAMPLES
        SPACE EXPLORATION
        SPACE FLIGHT
        SPACE PROBES
        UNMANNED SPACECRAFT
```

VIKING LANDER 2
```
  GS    INTERPLANETARY SPACECRAFT
        . MARS PROBES
        . . VIKING SPACECRAFT
        . . . VIKING 2 SPACECRAFT
        . . . . VIKING LANDER SPACECRAFT
        . . . . . VIKING LANDER 2
  RT    INTERPLANETARY TRAJECTORIES
        MARS SURFACE SAMPLES
        SPACE EXPLORATION
        SPACE FLIGHT
        SPACE PROBES
        UNMANNED SPACECRAFT
```

VIKING MARS PROGRAM
```
  GS    SPACE FLIGHT
        . VIKING MARS PROGRAM
  RT    NASA PROGRAMS
```

VIKING ORBITER SPACECRAFT
```
  GS    INTERPLANETARY SPACECRAFT
        . MARS PROBES
        . . VIKING SPACECRAFT
        . . . VIKING 1 SPACECRAFT
        . . . . VIKING ORBITER SPACECRAFT
        . . . . . VIKING ORBITER 1
        . . . VIKING 2 SPACECRAFT
        . . . . VIKING ORBITER SPACECRAFT
        . . . . . VIKING ORBITER 2
  RT    INTERPLANETARY TRAJECTORIES
        PLANETARY ORBITS
        SATELLITES
        SPACE EXPLORATION
        SPACE FLIGHT
        SPACE PROBES
        ∞ SPACECRAFT
        UNMANNED SPACECRAFT
```

VIKING ORBITER 1
```
  GS    INTERPLANETARY SPACECRAFT
        . MARS PROBES
        . . VIKING SPACECRAFT
        . . . VIKING 1 SPACECRAFT
        . . . . VIKING ORBITER SPACECRAFT
        . . . . . VIKING ORBITER 1
  RT    INTERPLANETARY TRAJECTORIES
        SPACE EXPLORATION
        SPACE FLIGHT
        SPACE PROBES
        ∞ SPACECRAFT
        UNMANNED SPACECRAFT
```

VIKING ORBITER 2
```
  GS    INTERPLANETARY SPACECRAFT
        . MARS PROBES
        . . VIKING SPACECRAFT
        . . . VIKING 2 SPACECRAFT
        . . . . VIKING ORBITER SPACECRAFT
        . . . . . VIKING ORBITER 2
  RT    INTERPLANETARY TRAJECTORIES
        SPACE EXPLORATION
        SPACE FLIGHT
        SPACE PROBES
        ∞ SPACECRAFT
        UNMANNED SPACECRAFT
```

VIKING ORBITER 1975
```
  GS    UNMANNED SPACECRAFT
        . SPACE PROBES
```

VIKING ORBITER 1975-(CONT.)
```
        . . MARS PROBES
        . . . VIKING ORBITER 1975
```

VIKING ROCKET VEHICLE
```
  GS    ROCKET VEHICLES
        . SINGLE STAGE ROCKET VEHICLES
        . . VIKING ROCKET VEHICLE
  RT    LIQUID PROPELLANT ROCKET ENGINES
        SOUNDING ROCKETS
        VANGUARD 2 LAUNCH VEHICLE
```

VIKING SPACECRAFT
```
  GS    INTERPLANETARY SPACECRAFT
        . MARS PROBES
        . . VIKING SPACECRAFT
        . . . VIKING 1 SPACECRAFT
        . . . . VIKING LANDER SPACECRAFT
        . . . . . VIKING LANDER 1
        . . . . VIKING ORBITER SPACECRAFT
        . . . . . VIKING ORBITER 1
        . . . VIKING 2 SPACECRAFT
        . . . . VIKING LANDER SPACECRAFT
        . . . . . VIKING LANDER 2
        . . . . VIKING ORBITER SPACECRAFT
        . . . . . VIKING ORBITER 2
```

VIKING 1 SPACECRAFT
```
  GS    INTERPLANETARY SPACECRAFT
        . MARS PROBES
        . . VIKING SPACECRAFT
        . . . VIKING 1 SPACECRAFT
        . . . . VIKING LANDER SPACECRAFT
        . . . . . VIKING LANDER 1
        . . . . VIKING ORBITER SPACECRAFT
        . . . . . VIKING ORBITER 1
  RT    INTERPLANETARY TRAJECTORIES
        SPACE EXPLORATION
        SPACE FLIGHT
        SPACE PROBES
        UNMANNED SPACECRAFT
```

VIKING 2 SPACECRAFT
```
  GS    INTERPLANETARY SPACECRAFT
        . MARS PROBES
        . . VIKING SPACECRAFT
        . . . VIKING 2 SPACECRAFT
        . . . . VIKING LANDER SPACECRAFT
        . . . . . VIKING LANDER 2
        . . . . VIKING ORBITER SPACECRAFT
        . . . . . VIKING ORBITER 2
  RT    INTERPLANETARY TRAJECTORIES
        SPACE EXPLORATION
        SPACE FLIGHT
        SPACE PROBES
        UNMANNED SPACECRAFT
```

VIKING 75 ENTRY VEHICLE
```
  GS    INTERPLANETARY SPACECRAFT
        . MARS PROBES
        . . VIKING 75 ENTRY VEHICLE
  RT    MARS LANDING
        SOFT LANDING
```

VINEYARDS
```
  RT    AGRICULTURE
        BLIGHT
        BOTANY
        CROP GROWTH
        CROP VIGOR
        ∞ CROPS
        EARTH RESOURCES
        FARM CROPS
        ∞ FOOD
        IRRIGATION
        PLANTS (BOTANY)
        WINES
```

VINTI THEORY
```
  GS    PERTURBATION THEORY
        . VINTI THEORY
  RT    GEODESY
        ORBIT PERTURBATION
        ∞ THEORIES
```

VINYL COPOLYMERS
```
  GS    PLASTICS
        . SYNTHETIC RESINS
        . . ADDITION RESINS
        . . . VINYL COPOLYMERS
        POLYMERIZATION
        . VINYL COPOLYMERS
        RESINS
        . SYNTHETIC RESINS
        . . ADDITION RESINS
```

VINYL COPOLYMERS-(CONT.)
```
        . . . VINYL COPOLYMERS
  RT    ADDITIVES
        COPOLYMERS
        ∞ POLYMERS
```

VINYL CYANIDE
```
  USE   ACRYLONITRILES
```

VINYL ETHYLENE
```
  USE   BUTADIENE
```

VINYL POLYMERS
```
  GS    VINYL POLYMERS
        . POLYVINYL FLUORIDE
  RT    ∞ POLYMERS
        POLYVINYL ALCOHOL
        POLYVINYL CHLORIDE
```

VINYL RADICAL
```
  GS    RADICALS
        . VINYL RADICAL
  RT    FREE RADICALS
```

VINYLIDENE
```
  GS    ALIPHATIC COMPOUNDS
        . ALKENES
        . . ETHYLENE
        . . . VINYLIDENE
        HYDROCARBONS
        . ALKENES
        . . ETHYLENE
        . . . VINYLIDENE
```

VIOLENCE
```
  GS    VIOLENCE
        . ATTACKING (ASSAULTING)
  RT    CRIME
        DISORDERS
        POLICE
        WARFARE
```

VIRGIN ISLANDS
```
  GS    LANDFORMS
        . ISLANDS
        . . WEST INDIES
        . . . VIRGIN ISLANDS
  RT    ARCHIPELAGOES
        CARIBBEAN REGION
        CARIBBEAN SEA
        TROPICAL REGIONS
        UNITED STATES
```

VIRGINIA
```
  GS    NATIONS
        . UNITED STATES
        . . VIRGINIA
  RT    ALLEGHENY PLATEAU (US)
        ASSATEAGUE ISLAND (MD-VA)
        CHESAPEAKE BAY (US)
        DELMARVA PENINSULA (DE-MD-VA)
        POTOMAC RIVER VALLEY (MD-VA-WV)
        SHENANDOAH VALLEY (VA)
        WALLOPS ISLAND
```

VIRGO GALACTIC CLUSTER
```
  UF    VIRGO STAR CLUSTER
  GS    CELESTIAL BODIES
        . GALAXIES
        . . GALACTIC CLUSTERS
        . . . VIRGO GALACTIC CLUSTER
        . STAR CLUSTERS
        . . VIRGO GALACTIC CLUSTER
  RT    AGGLOMERATION
        BARRED GALAXIES
        DISK GALAXIES
        ELLIPTICAL GALAXIES
        LOCAL GROUP (ASTRONOMY)
        SPIRAL GALAXIES
        STAR CLUSTERS
        STAR DISTRIBUTION
        STARS
```

VIRGO STAR CLUSTER
```
  USE   VIRGO GALACTIC CLUSTER
```

VIRIAL COEFFICIENTS
```
  GS    COEFFICIENTS
        . VIRIAL COEFFICIENTS
  RT    EQUATIONS OF STATE
        INTERMOLECULAR FORCES
        VIRIAL THEOREM
```

VIRIAL THEOREM
GS   THEOREMS
      . **VIRIAL THEOREM**
RT   KINETIC ENERGY
      KINETIC EQUATIONS
      ∞MECHANICS (PHYSICS)
      MISSING MASS (ASTROPHYSICS)
      VIRIAL COEFFICIENTS

VIRTUAL MEMORY SYSTEMS
RT   COMPUTER SYSTEMS DESIGN
      DATA MANAGEMENT
      DATA STORAGE
      MAGNETIC STORAGE

VIRTUAL PROPERTIES
RT   ACCURACY
      ∞PHYSICAL PROPERTIES
      ∞PROPERTIES

VIRULENCE
RT   MICROORGANISMS
      TOXICITY
      VIRUSES

VIRUSES
GS   MICROORGANISMS
      . **VIRUSES**
      . . ADENOVIRUSES
      . . BACTERIOPHAGES
RT   ∞BLISTERS
      INTERFERON
      PROTOBIOLOGY
      VIRULENCE

VISCERA
GS   **VISCERA**
      . ABDOMEN
      . APPENDIX (ANATOMY)
      . ENDOCRINE GLANDS
      . . ADRENAL GLAND
      . . GONADS
      . . OVARIES
      . . PANCREAS
      . . PARATHYROID GLAND
      . . PINEAL GLAND
      . . PITUITARY GLAND
      . . PROSTATE GLAND
      . . THYMUS GLAND
      . . THYROID GLAND
      . INTESTINES
      . . RECTUM
      . ORGANS
      . . BLADDER
      . . ESOPHAGUS
      . . KIDNEYS
      . . LIVER
      . . LUNGS
      . . OVARIES
      . . PITUITARY GLAND
      . . SPLEEN
      . . STOMACH
      . . TESTES
      . SEX GLANDS
      . . GONADS
      . . OVARIES
      . . PROSTATE GLAND
      . . TESTES

VISCOELASTIC CYLINDERS
RT   ∞CYLINDERS
      CYLINDRICAL BODIES
      CYLINDRICAL SHELLS

VISCOELASTIC DAMPING
GS   DAMPING
      . ELASTIC DAMPING
      . . **VISCOELASTIC DAMPING**
      . VISCOUS DAMPING
      . . **VISCOELASTIC DAMPING**

VISCOELASTIC FLOW
USE   VISCOELASTICITY

VISCOELASTICITY
UF   VISCOELASTIC FLOW
GS   MECHANICAL PROPERTIES
      . ELASTIC PROPERTIES
      . . **VISCOELASTICITY**
      . . . PHOTOVISCOELASTICITY
      . . . THERMOVISCOELASTICITY
RT   HYDROELASTICITY
      HYSTERESIS
      MAXWELL FLUIDS

VISCOELASTICITY-*(CONT.)*
      NONNEWTONIAN FLOW
      NONNEWTONIAN FLUIDS
      PLASTIC FLOW
      RELAXATION (MECHANICS)
      SQUEEZE FILMS
      VISCOPLASTICITY
      VISCOUS DAMPING

VISCOMETERS
GS   MEASURING INSTRUMENTS
      . **VISCOMETERS**
RT   ROTATING CYLINDERS
      VISCOMETRY
      VISCOSITY

VISCOMETRY
RT   ROTATING CYLINDERS
      VISCOMETERS
      VISCOSITY
      VISCOUS DRAG
      VISCOUS FLOW

VISCOPLASTIC FLOW
USE   VISCOPLASTICITY

VISCOPLASTICITY
UF   VISCOPLASTIC FLOW
GS   MECHANICAL PROPERTIES
      . PLASTIC PROPERTIES
      . . **VISCOPLASTICITY**
RT   HYSTERESIS
      NONNEWTONIAN FLOW
      NONNEWTONIAN FLUIDS
      PLASTIC ANISOTROPY
      PLASTIC FLOW
      RELAXATION (MECHANICS)
      VISCOELASTICITY
      VISCOUS DAMPING

VISCOPUMPS
GS   PUMPS
      . **VISCOPUMPS**
RT   VISCOUS FLOW

VISCOSITY
GS   TRANSPORT PROPERTIES
      . **VISCOSITY**
      . . EDDY VISCOSITY
      . . GAS VISCOSITY
RT   DENSITY (MASS/VOLUME)
      FLOW CHARACTERISTICS
      FLOW RESISTANCE
      INTERNAL FRICTION
      ∞MOTION
      ∞PHYSICAL PROPERTIES
      RHEOLOGY
      SIMILITUDE LAW
      SOLUBILITY
      STOKES LAW (FLUID MECHANICS)
      SUPERFLUIDITY
      SURFACE PROPERTIES
      THERMAL DIFFUSION
      THERMAL DIFFUSIVITY
      THIXOTROPY
      VISCOMETERS
      VISCOMETRY
      VISCOUS FLOW

VISCOUNT AIRCRAFT
GS   BAC AIRCRAFT
      . **VISCOUNT AIRCRAFT**
      JET AIRCRAFT
      . TURBOPROP AIRCRAFT
      . . **VISCOUNT AIRCRAFT**
      MONOPLANES
      . **VISCOUNT AIRCRAFT**
      PASSENGER AIRCRAFT
      . **VISCOUNT AIRCRAFT**
      TRANSPORT AIRCRAFT
      . **VISCOUNT AIRCRAFT**
RT   ∞AIRCRAFT

VISCOUS DAMPING
GS   DAMPING
      . **VISCOUS DAMPING**
      . . VISCOELASTIC DAMPING
RT   ELASTIC DAMPING
      RESONANCE TESTING
      VISCOELASTICITY
      VISCOPLASTICITY

VISCOUS DRAG
GS   DYNAMIC CHARACTERISTICS

VISCOUS DRAG-*(CONT.)*
      . DRAG
      . . FRICTION DRAG
      . . . **VISCOUS DRAG**
      FRICTION
      . FLOW RESISTANCE
      . . FRICTION DRAG
      . . . **VISCOUS DRAG**
      . SKIN FRICTION
      . . FRICTION DRAG
      . . . **VISCOUS DRAG**
RT   EDDY VISCOSITY
      HARTMANN NUMBER
      LAMINAR FLOW
      TURBULENT FLOW
      VISCOMETRY

VISCOUS FLOW
GS   FLUID FLOW
      . **VISCOUS FLOW**
      . . BOUNDARY LAYER FLOW
      . . . REATTACHED FLOW
      . . . SECONDARY FLOW
      . . . SEPARATED FLOW
      . . . . BOUNDARY LAYER SEPARATION
      . . COUETTE FLOW
      . . KARMAN-BODEWADT FLOW
      . . STOKES FLOW
RT   AERODYNAMICS
      BAROTROPIC FLOW
      EDDY VISCOSITY
      ∞FLOW
      FLOW CHARACTERISTICS
      GAS FLOW
      INVISCID FLOW
      KNUDSEN FLOW
      LAMINAR FLOW
      MAGNETOHYDRODYNAMIC SHEAR
        HEATING
      MAXWELL FLUIDS
      MILNE-THOMSON METHOD
      NAVIER-STOKES EQUATION
      POHLHAUSEN METHOD
      PRANDTL NUMBER
      REYNOLDS NUMBER
      TURBULENT FLOW
      VISCOMETRY
      VISCOPUMPS
      VISCOSITY
      WEDGE FLOW

VISCOUS FLUIDS
RT   FLOW STABILITY
      ∞FLUIDS
      MAXWELL FLUIDS
      NAVIER-STOKES EQUATION
      NEWTONIAN FLUIDS
      NONNEWTONIAN FLUIDS
      OSEEN APPROXIMATION
      SEMISOLIDS
      SQUEEZE FILMS
      WEIGHTLESS FLUIDS

VISIBILITY
UF   INVISIBILITY
GS   **VISIBILITY**
      . LOW VISIBILITY
RT   APPEARANCE
      BRIGHTNESS
      CEILINGS (METEOROLOGY)
      CHARACTER RECOGNITION
      COLOR
      CONTRAST
      DARKENING
      FOG
      GLARE
      HAZE
      HUMAN FACTORS ENGINEERING
      ILLUMINANCE
      IMAGE CONTRAST
      LEGIBILITY
      LIGHT (VISIBLE RADIATION)
      LIGHT TRANSMISSION
      LUMINESCENCE
      LUMINOSITY
      NIGHT FLIGHTS (AIRCRAFT)
      OPACITY
      OPTICAL PROPERTIES
      PERCEPTION
      RADIANCE
      READING
      RESOLUTION
      RETINAL ADAPTATION
      SENSITIVITY
      SMOKE

VISIBILITY-*(CONT.)*
    SYMBOLS
    TRANSMISSIVITY
    VIEWING
    VISION
    VISUAL CONTROL
    WHITEOUT

VISIBLE INFRARED SPIN SCAN RADIOMETER
RT   ATMOSPHERIC SOUNDING
    INFRARED RADIOMETERS
    SATELLITE SOUNDING
    SATELLITE-BORNE INSTRUMENTS

VISIBLE RADIATION
USE   LIGHT (VISIBLE RADIATION)

VISIBLE SPECTRUM
GS   SPECTRA
    . RADIATION SPECTRA
    . . ELECTROMAGNETIC SPECTRA
    . . . **VISIBLE SPECTRUM**
RT   ∞ABSORPTION
    ABSORPTION SPECTRA
    ASTRONOMICAL SPECTROSCOPY
    AURORAL SPECTROSCOPY
    CATHODOLUMINESCENCE
    EMISSION SPECTRA
    GAS SPECTROSCOPY
    LIGHT (VISIBLE RADIATION)
    LINE SPECTRA
    MOLECULAR SPECTRA
    SOLAR SPECTRA
    SPECTRAL BANDS
    SPECTROSCOPY
    STELLAR SPECTRA

VISION
UF   MACULAR VISION
GS   **VISION**
    . BINOCULAR VISION
    . COLOR VISION
    . MONOCULAR VISION
    . NIGHT VISION
    . PERIPHERAL VISION
    . STEREOSCOPIC VISION
RT   ADAPTATION
    ANASTIGMATISM
    BLINDNESS
    BRIGHTNESS
    CHOROID MEMBRANES
    COLOR
    CONJUNCTIVA
    CONTRAST
    CORNEA
    DARK ADAPTATION
    EYE (ANATOMY)
    EYE DOMINANCE
    FLASH BLINDNESS
    GLARE
    HETEROPHORIA
    HUMAN FACTORS ENGINEERING
    HYPEROPIA
    ILLUSIONS
    IMAGES
    LEGIBILITY
    LIGHT ADAPTATION
    MIOSIS
    MYOPIA
    OCULOMOTOR NERVES
    OPHTHALMODYNAMOMETRY
    OPTOMETRY
    PERCEPTION
    PHOSPHENE
    PRESBYOPIA
    PUPILS
    RESOLUTION
    RETINA
    RETINAL ADAPTATION
    THRESHOLDS (PERCEPTION)
    VIEWING
    VISIBILITY
    VISUAL ACUITY

VISORS
RT   EYE PROTECTION
    RADIATION PROTECTION
    SUNGLASSES

VISUAL ACCOMMODATION
RT   ACCOMMODATION

VISUAL ACUITY
GS   ACUITY
    . **VISUAL ACUITY**

VISUAL ACUITY-*(CONT.)*
    . . HYPEROPIA
RT   PERIPHERAL VISION
    SNELLEN TESTS
    VISION

VISUAL AIDS
UF   AUDIO VISUAL EQUIPMENT
RT   ∞AIDS
    CHARTS
    DIAGRAMS
    DISPLAY DEVICES
    DRAWINGS
    PHOTOGRAPHS
    TRAINING DEVICES

VISUAL CONTROL
GS   MANUAL CONTROL
    . **VISUAL CONTROL**
RT   AIRCRAFT CONTROL
    APPROACH CONTROL
    ATTITUDE CONTROL
    ∞CONTROL
    DISPLAY DEVICES
    GUIDANCE (MOTION)
    MISSILE CONTROL
    REMOTE CONTROL
    RUNWAY LIGHTS
    SERVOCONTROL
    SPACECRAFT CONTROL
    VISIBILITY

VISUAL DISCRIMINATION
GS   DISCRIMINATION
    . SENSORY DISCRIMINATION
    . . **VISUAL DISCRIMINATION**
    PERCEPTION
    . SENSORY PERCEPTION
    . . VISUAL PERCEPTION
    . . . **VISUAL DISCRIMINATION**
RT   ∞RECOVERY

VISUAL DISPLAYS
USE   DISPLAY DEVICES

VISUAL FIELDS
RT   FIELD OF VIEW
    ∞FIELDS
    PERIPHERAL VISION
    RETINA
    SACCADIC EYE MOVEMENTS
    SPACE PERCEPTION

VISUAL FLIGHT
RT   AIR NAVIGATION
    COLLISION AVOIDANCE
    ∞FLIGHT
    FLIGHT CONDITIONS
    FLIGHT PATHS
    FLIGHT SAFETY
    LANDING
    WHITEOUT

VISUAL FLIGHT RULES
UF   VFR (RULES)
GS   RULES
    . FLIGHT RULES
    . . **VISUAL FLIGHT RULES**

VISUAL OBSERVATION
GS   OBSERVATION
    . **VISUAL OBSERVATION**
RT   COMPANION STARS
    SPACE OBSERVATIONS (FROM EARTH)

VISUAL PERCEPTION
UF   SIGHT
GS   PERCEPTION
    . SENSORY PERCEPTION
    . . **VISUAL PERCEPTION**
    . . . CRITICAL FLICKER FUSION
    . . . SPACE PERCEPTION
    . . . . AUTOKINESIS
    . . . VISUAL DISCRIMINATION
RT   AFTERIMAGES
    BLINKING
    BRIGHTNESS DISCRIMINATION
    ELEVATOR ILLUSION
    MOTION PERCEPTION
    ∞ORIENTATION
    PERCEPTUAL ERRORS
    ∞SPACE ORIENTATION
    TACHISTOSCOPES
    THRESHOLDS (PERCEPTION)

VISUAL PHOTOMETRY
GS   OPTICAL MEASUREMENT
    . PHOTOMETRY
    . . **VISUAL PHOTOMETRY**

VISUAL PIGMENTS
GS   PIGMENTS
    . **VISUAL PIGMENTS**
RT   DARK ADAPTATION
    PHOTORECEPTORS
    PHOTOSENSITIVITY

VISUAL SIGNALS
RT   BEACONS
    CUES
    LUMINAIRES
    OPTICAL COMMUNICATION
    ∞SIGNALS

VISUAL STIMULI
RT   PERCEPTUAL ERRORS
    ∞SIGNALS
    ∞STIMULI

VISUAL TASKS
GS   TASKS
    VISUAL TASKS

VISUAL TRACKING
USE   OPTICAL TRACKING

VISUALIZATION OF FLOW
USE   FLOW VISUALIZATION

VITAMIN A
USE   RETINENE

VITAMIN B
USE   THIAMINE

VITAMIN B COMPLEX
USE   BIOTIN

VITAMIN B 2
USE   RIBOFLAVIN

VITAMIN B 6
USE   PYRIDOXINE

VITAMIN B 12
USE   CYANOCOBALAMIN

VITAMIN C
USE   ASCORBIC ACID

VITAMIN D
USE   CALCIFEROL

VITAMIN E
USE   TOCOPHEROL

VITAMIN G
USE   RIBOFLAVIN

VITAMIN K
USE   PHYLLOQUINONE

VITAMIN M
USE   FOLIC ACID

VITAMIN P
USE   BIOFLAVONOIDS

VITAMINS
GS   **VITAMINS**
    . ASCORBIC ACID
    . BIOFLAVONOIDS
    . BIOTIN
    . CALCIFEROL
    . CARNITINE
    . CYANOCOBALAMIN
    . FOLIC ACID
    . NICOTINAMIDE
    . NICOTINIC ACID
    . PHYLLOQUINONE
    . PYRIDOXINE
    . RETINENE
    . RIBOFLAVIN
    . THIAMINE
    . TOCOPHEROL
RT   ASCORBIC ACID METABOLISM
    DRUGS
    ∞FOOD

VITAMINS-(CONT.)
∞ NUTRIENTS

VITON
GS   COPOLYMERS
. **VITON**
PLASTICS
. **VITON**
RT   FLUOROHYDROCARBONS

VITON RUBBER (TRADEMARK)
GS   RUBBER
. SYNTHETIC RUBBERS
. . ELASTOMERS
. . . **VITON RUBBER (TRADEMARK)**

VITREOUS MATERIALS
RT   FRIT
GLASS
∞ INORGANIC MATERIALS
∞ MATERIALS
METALLIC GLASSES
PORCELAIN
VITRIFICATION

VITRIFICATION
RT   CERAMICS
GLASS
PORCELAIN
SOLIDIFICATION
VITREOUS MATERIALS

VJ-101 AIRCRAFT
UF   SUD VJ-101 AIRCRAFT
GS   ATTACK AIRCRAFT
. FIGHTER AIRCRAFT
. . **VJ-101 AIRCRAFT**
JET AIRCRAFT
. **VJ-101 AIRCRAFT**
MONOPLANES
. **VJ-101 AIRCRAFT**
SUD AVIATION AIRCRAFT
. **VJ-101 AIRCRAFT**
SUPERSONIC AIRCRAFT
. **VJ-101 AIRCRAFT**
V/STOL AIRCRAFT
. VERTICAL TAKEOFF AIRCRAFT
. . **VJ-101 AIRCRAFT**
RT   ∞ AIRCRAFT

VLASOV EQUATIONS
GS   ANALYSIS (MATHEMATICS)
. REAL VARIABLES
. . DIFFERENTIAL EQUATIONS
. . . PARTIAL DIFFERENTIAL EQUATIONS
. . . . **VLASOV EQUATIONS**
RT   ∞ EQUATIONS
STABILITY

VLBI
USE   VERY LONG BASE INTERFEROMETRY

VLF EMISSION RECORDERS
RT   ATMOSPHERIC RADIATION
ATMOSPHERICS
COSMIC RAYS
ELECTROMAGNETIC RADIATION
PLANETARY RADIATION
∞ RECORDERS
RECORDING INSTRUMENTS

VLSI
USE   VERY LARGE SCALE INTEGRATION

VOCAL CORDS
RT   GLOTTIS
LARYNX

VOCODERS
RT   BANDPASS FILTERS
COMPUTERS
DIGITAL TO VOICE TRANSLATORS
FREQUENCY MODULATION
MESSAGES
RADIO COMMUNICATION
SCRAMBLING (COMMUNICATION)
SIGNAL RECEPTION
SPEECH BASEBAND COMPRESSION
VOICE COMMUNICATION
VOICE DATA PROCESSING

VOICE
GS   SOUND PROPAGATION
. **VOICE**

VOICE-(CONT.)
RT   AUDIO FREQUENCIES
SPEECH
TONGUE

VOICE COMMUNICATION
GS   TELECOMMUNICATION
. **VOICE COMMUNICATION**
. . TELEPHONY
. . VOICE DATA PROCESSING
RT   ACOUSTICS
CONVERSATION
ECHO SUPPRESSORS
GROUND-AIR-GROUND COMMUNICATION
RADIO COMMUNICATION
RADIOTELEPHONES
REENTRY COMMUNICATION
SCRAMBLING (COMMUNICATION)
SINGLE CHANNEL PER CARRIER
TRANSMISSION
SINGLE SIDEBAND TRANSMISSION
SPEECH
SPEECH BASEBAND COMPRESSION
VERBAL COMMUNICATION
VOCODERS
VOICE CONTROL
WIRELESS COMMUNICATION
WORDS (LANGUAGE)

VOICE CONTROL
SN   (DEVICE OPERATION BY VOICE)
RT   BIOENGINEERING
∞ CONTROL
ROBOTICS
ROBOTS
SPEECH RECOGNITION
VOICE COMMUNICATION
VOICE DATA PROCESSING

VOICE DATA PROCESSING
GS   DATA PROCESSING
. **VOICE DATA PROCESSING**
TELECOMMUNICATION
. VOICE COMMUNICATION
. . **VOICE DATA PROCESSING**
RT   ARTIFICIAL INTELLIGENCE
∞ DATA
DIGITAL TO VOICE TRANSLATORS
SIGNAL ENCODING
SINGLE CHANNEL PER CARRIER
TRANSMISSION
VERBAL COMMUNICATION
VOCODERS
VOICE CONTROL

VOICE OF AMERICA
RT   BROADCASTING
RADIO TRANSMISSION

VOID RATIO
UF   COMPACTNESS
GS   RATIOS
. **VOID RATIO**
RT   ∞ CONDUCTIVITY
DENSITY (MASS/VOLUME)
FREE FLOW
HOLE DISTRIBUTION (MECHANICS)
PACKING DENSITY
PERMEABILITY
POROSITY
REACTOR CORES
SURFACE PROPERTIES
VOIDS

VOIDS
RT   BUOYANCY
CAVITIES
CRACK GEOMETRY
DEFECTS
INCLUSIONS
INFILTRATION
INTERSTICES
PERCOLATION
PERMEABILITY
POROSITY
VOID RATIO

VOIGT EFFECT
RT   BIREFRINGENCE
∞ EFFECTS
OPTICAL PATHS
REFRACTION
ZEEMAN EFFECT

VOLATILITY
GS   THERMODYNAMIC PROPERTIES
. THERMOPHYSICAL PROPERTIES
. . **VOLATILITY**
RT   COAL GASIFICATION
EVAPORATION
FLASH POINT
PREVAPORIZATION
VAPOR PHASES
VAPOR PRESSURE
VAPORIZING

VOLATILIZATION
USE   VAPORIZING

VOLCANICS
USE   VOLCANOLOGY

VOLCANOES
UF   ACTIVE VOLCANOES
GS   GEOLOGY
. **VOLCANOES**
. . MARS VOLCANOES
LANDFORMS
. **VOLCANOES**
. . MARS VOLCANOES
RT   BASALT
CALDERAS
CONES (VOLCANOES)
EFFUSIVES
GEOMORPHOLOGY
LAVA
MOUNTAINS
OROGRAPHY
PALEOMAGNETISM
PETROLOGY
ROUSE BELTS
VOLCANOLOGY

VOLCANOLOGY
UF   VOLCANICS
GS   GEOLOGY
. **VOLCANOLOGY**
RT   BASALT
CALDERAS
CONES (VOLCANOES)
EFFUSIVES
GEOMORPHOLOGY
LAVA
MARS VOLCANOES
MOUNTAINS
OROGRAPHY
PALEOMAGNETISM
PETROLOGY
ROUSE BELTS
VOLCANOES

VOLT-AMPERE CHARACTERISTICS
RT   CAPACITANCE-VOLTAGE
CHARACTERISTICS
∞ CHARACTERISTICS
ELECTRIC CURRENT
ELECTRIC POTENTIAL
∞ ELECTRONICS
LINEAR CIRCUITS
OHMS LAW
OPEN CIRCUIT VOLTAGE
OPTOGALVANIC SPECTROSCOPY
PROXIMITY EFFECT (ELECTRICITY)
QUANTUM EFFICIENCY
SHORT CIRCUIT CURRENTS
VOLTAGE AMPLIFIERS

VOLTAGE
USE   ELECTRIC POTENTIAL

VOLTAGE AMPLIFIERS
GS   AMPLIFIERS
. **VOLTAGE AMPLIFIERS**
RT   CURRENT AMPLIFIERS
FEEDBACK AMPLIFIERS
MAGNETIC AMPLIFIERS
PREAMPLIFIERS
VOLT-AMPERE CHARACTERISTICS

VOLTAGE BREAKDOWN
USE   ELECTRICAL FAULTS

VOLTAGE CONTROLLED OSCILLATORS
UF   VCO
GS   OSCILLATORS
. **VOLTAGE CONTROLLED
OSCILLATORS**
RT   CIRCUITS

VOLTAGE CONTROLLED OSCILLATORS-_(CONT.)_
  ELECTRIC CONTROL
  ELECTRIC NETWORKS
  FREQUENCY MODULATION
  FREQUENCY STABILITY
  MICROWAVE OSCILLATORS
  VOLTAGE REGULATORS

VOLTAGE CONVERTERS (AC TO AC)
GS TRANSFORMERS
  . **VOLTAGE CONVERTERS (AC TO AC)**
RT ALTERNATING CURRENT
  AUXILIARY POWER SOURCES
  ∞ CONVERTERS
  ∞ ELECTRIC EQUIPMENT
  ∞ ELECTRIC POWER
  ∞ POWER SUPPLIES

VOLTAGE CONVERTERS (DC TO DC)
RT AUXILIARY POWER SOURCES
  ∞ CONVERTERS
  DIRECT CURRENT
  ELECTRIC BATTERIES
  ∞ ELECTRIC EQUIPMENT
  ∞ ELECTRIC POWER
  ∞ POWER SUPPLIES
  POWER SUPPLY CIRCUITS

VOLTAGE GENERATORS
GS **VOLTAGE GENERATORS**
  . PHOTOVOLTAIC CELLS
RT ARC GENERATORS
  ELECTROSTATIC GENERATORS
  FUNCTION GENERATORS
  ∞ GENERATORS
  SIGNAL GENERATORS

VOLTAGE MEASUREMENT
USE ELECTRICAL MEASUREMENT

VOLTAGE REGULATORS
GS REGULATORS
  . **VOLTAGE REGULATORS**
RT AVALANCHE DIODES
  CIRCUIT PROTECTION
  CONTROLLERS
  CURRENT REGULATORS
  ELECTRIC SWITCHES
  ELECTRONIC CONTROL
  POWER FACTOR CONTROLLERS
  POWER SUPPLY CIRCUITS
  SWITCHING CIRCUITS
  TRANSFORMERS
  TRANSMISSION LOSS
  VOLTAGE CONTROLLED OSCILLATORS

VOLTAGE VARIATION INDICATORS
USE VOLTMETERS

VOLTERRA EQUATIONS
GS ANALYSIS (MATHEMATICS)
  . FUNCTIONAL ANALYSIS
  . . INTEGRAL EQUATIONS
  . . . **VOLTERRA EQUATIONS**
RT ∞ EQUATIONS
  NONLINEARITY
  ∞ RADIATION

VOLTMETERS
UF VOLTAGE VARIATION INDICATORS
GS MEASURING INSTRUMENTS
  . **VOLTMETERS**
  . . MILLIVOLTMETERS
RT AMMETERS
  COULOMETERS
  ELECTROMETERS
  POTENTIOMETERS (INSTRUMENTS)

VOLUME
GS **VOLUME**
  . BODY VOLUME (BIOLOGY)
RT AREA
  ∞ CAPACITY
  DIMENSIONS
  FRUSTUMS
  GEOMETRY
  ISOCHORIC PROCESSES
  RATES (PER TIME)
  THICKNESS
  VOLUMETRIC ANALYSIS
  WEIGHT (MASS)

VOLUMETRIC ANALYSIS
GS CHEMICAL TESTS

VOLUMETRIC ANALYSIS-_(CONT.)_
  . CHEMICAL ANALYSIS
  . . **VOLUMETRIC ANALYSIS**
RT ANALYTICAL CHEMISTRY
  GAS ANALYSIS
  QUANTITATIVE ANALYSIS
  VOLUME

VOLUMETRIC EFFICIENCY
RT ENERGY CONVERSION EFFICIENCY
  ENGINE DESIGN
  FUEL-AIR RATIO
  LASER OUTPUTS

VOLUMETRIC STRAIN
GS FATIGUE (MATERIALS)
  . **VOLUMETRIC STRAIN**
RT DEFORMATION
  STRUCTURAL STRAIN
  TEMPERATURE INVERSIONS

VOMITING
RT MOTION SICKNESS
  NAUSEA

VON KARMAN EQUATION
GS FLOW EQUATIONS
  . **VON KARMAN EQUATION**
RT ∞ EQUATIONS
  FLOW STABILITY
  KARMAN VORTEX STREET
  VORTEX BREAKDOWN
  VORTEX STREETS
  VORTICITY EQUATIONS

VON MISES THEORY
USE STRESS FUNCTIONS

VON ZEIPEL METHOD
RT EQUATIONS OF MOTION
  HAMILTONIAN FUNCTIONS
  ∞ METHODOLOGY
  PERTURBATION THEORY

VOODOO AIRCRAFT
USE F-101 AIRCRAFT

VOR SYSTEMS
USE VHF OMNIRANGE NAVIGATION

VORTEX ADVISORY SYSTEM
RT AIR TRAFFIC CONTROL
  AIRCRAFT APPROACH SPACING
  AIRCRAFT WAKES
  ∞ SYSTEMS
  TURBULENT WAKES
  VORTEX ALLEVIATION

VORTEX ALLEVIATION
RT AIRCRAFT WAKES
  DRAG DEVICES
  GUST ALLEVIATORS
  PROTUBERANCES
  SPOILERS
  VORTEX ADVISORY SYSTEM
  VORTICES
  WAKES
  WINGLETS

VORTEX AVOIDANCE
GS AVOIDANCE
  . **VORTEX AVOIDANCE**
RT AERODYNAMIC STABILITY
  AIR TRAFFIC CONTROL
  AIRCRAFT APPROACH SPACING
  AIRCRAFT LANDING
  BUFFETING
  GUSTS
  ROTATION
  SAFETY
  TURBULENT FLOW
  VORTICES
  WINGLETS

VORTEX BREAKDOWN
RT FLOW STABILITY
  TURBULENT FLOW
  VON KARMAN EQUATION
  VORTEX FLAPS

VORTEX COLUMNS
USE VORTICES

VORTEX DISTURBANCES
USE VORTICES

VORTEX FILAMENTS
RT ∞ FILAMENTS
  FLOW STABILITY
  FLUID DYNAMICS
  TURBULENCE
  VORTICES

VORTEX FLAPS
GS AIRFOILS
  . FLAPS (CONTROL SURFACES)
  . . WING FLAPS
  . . . **VORTEX FLAPS**
RT AERODYNAMIC DRAG
  CONTROL SURFACES
  JET FLAPS
  LEADING EDGE FLAPS
  LEADING EDGES
  LIFT AUGMENTATION
  SEPARATED FLOW
  TRAILING EDGE FLAPS
  TRAILING EDGES
  VORTEX BREAKDOWN
  VORTICES
  WING LOADING

VORTEX FLOW
USE VORTICES

VORTEX GENERATION
USE VORTEX GENERATORS

VORTEX GENERATORS
UF VORTEX GENERATION
RT AIRFOIL FENCES
  BOUNDARY LAYER CONTROL
  BOUNDARY LAYER SEPARATION
  ∞ GENERATORS
  HILSCH TUBES
  INLET FLOW
  VORTICES
  WING SLOTS

VORTEX INJECTORS
GS INJECTORS
  . **VORTEX INJECTORS**

VORTEX PRECESSION
RT FLOW VELOCITY
  FLOWMETERS
  PRECESSION
  VELOCITY MEASUREMENT
  VORTICES

VORTEX RINGS
RT ∞ RINGS
  TRAPPED VORTEXES
  VORTICES

VORTEX SHEDDING
GS FLUID MECHANICS
  . FLUID DYNAMICS
  . . **VORTEX SHEDDING**
RT SHEDDING
  VORTICES

VORTEX SHEETS
RT AIRCRAFT DESIGN
  FLOW DISTRIBUTION
  ROTATING FLUIDS
  ∞ SHEETS
  TURBULENT WAKES
  VORTICES
  VORTICITY

VORTEX STREETS
GS **VORTEX STREETS**
  . KARMAN VORTEX STREET
RT DISCONTINUITY
  ∞ SHEETS
  TURBULENT WAKES
  VON KARMAN EQUATION
  VORTICES

VORTEX TRAPS
USE TRAPPED VORTEXES

VORTEX TUBES
USE HILSCH TUBES
  VORTICES

VORTICES
UF   EDDIES
      ROTATIONAL FLOW
      VORTEX COLUMNS
      VORTEX DISTURBANCES
      VORTEX FLOW
      VORTEX TUBES
GS   **VORTICES**
     . TRAPPED VORTEXES
     . WING TIP VORTICES
RT   ABRIKOSOV THEORY
      AGITATION
      CAVITATION FLOW
      COUNTERFLOW
      ∞DISTURBANCES
      DIVERGENCE
      FLOW DISTORTION
      FLOW STABILITY
      FLUID FLOW
      GOERTLER INSTABILITY
      HILSCH TUBES
      KOLMOGOROFF THEORY
      METEOROLOGICAL SOLENOIDS
      MIXING
      PLANETARY WAVES
      RECIRCULATIVE FLUID FLOW
      ROTATING FLUIDS
      ROTATING LIQUIDS
      ROTATION
      SECONDARY FLOW
      STROUHAL NUMBER
      SUPERCONDUCTIVITY
      SUPERFLUIDITY
      THRUST DISTRIBUTION
      TURBULENCE
      TURBULENT FLOW
      TURBULENT MIXING
      VORTEX ALLEVIATION
      VORTEX AVOIDANCE
      VORTEX FILAMENTS
      VORTEX FLAPS
      VORTEX GENERATORS
      VORTEX PRECESSION
      VORTEX RINGS
      VORTEX SHEDDING
      VORTEX SHEETS
      VORTEX STREETS
      VORTICITY
      WAKES

VORTICITY
UF   ENSTROPHY
GS   ALGEBRA
     . VECTOR SPACES
     . . VECTORS (MATHEMATICS)
     . . . **VORTICITY**
     ANALYSIS (MATHEMATICS)
     . CALCULUS
     . . VECTOR ANALYSIS
     . . . CURL (VECTORS)
     . . . . **VORTICITY**
     . REAL VARIABLES
     . . VECTOR ANALYSIS
     . . . CURL (VECTORS)
     . . . . **VORTICITY**
     GEOMETRY
     . VECTOR ANALYSIS
     . . CURL (VECTORS)
     . . . **VORTICITY**
RT   ATMOSPHERIC CIRCULATION
      BELTRAMI FLOW
      CROCCO METHOD
      FLOW STABILITY
      HELMHOLTZ VORTICITY EQUATION
      POTENTIAL FLOW
      SECONDARY FLOW
      TRAPPED VORTEXES
      TURBULENCE
      VORTEX SHEETS
      VORTICES

VORTICITY EQUATIONS
GS   FLOW EQUATIONS
     . **VORTICITY EQUATIONS**
RT   ANALYSIS (MATHEMATICS)
      ∞EQUATIONS
      KARMAN VORTEX STREET
      VON KARMAN EQUATION

VORTICITY TRANSPORT HYPOTHESIS
GS   HYPOTHESES
     . **VORTICITY TRANSPORT HYPOTHESIS**
RT   CONSERVATION EQUATIONS
      EDDY CURRENTS
      MIXING LENGTH FLOW THEORY

VORTICITY TRANSPORT HYPOTHESIS-*(CONT.)*
      TURBULENT FLOW

VOSKHOD MANNED SPACECRAFT
GS   MANNED SPACECRAFT
     . **VOSKHOD MANNED SPACECRAFT**
     . . VOSKHOD 1 SPACECRAFT
     . . VOSKHOD 2 SPACECRAFT
     REENTRY VEHICLES
     . RECOVERABLE SPACECRAFT
     . . **VOSKHOD MANNED SPACECRAFT**
     . . . VOSKHOD 1 SPACECRAFT
     . . . VOSKHOD 2 SPACECRAFT
     SOFT LANDING SPACECRAFT
     . **VOSKHOD MANNED SPACECRAFT**
     . . VOSKHOD 1 SPACECRAFT
     . . VOSKHOD 2 SPACECRAFT
RT   SPACE CAPSULES

VOSKHOD 1 SPACECRAFT
GS   MANNED SPACECRAFT
     . VOSKHOD MANNED SPACECRAFT
     . . **VOSKHOD 1 SPACECRAFT**
     REENTRY VEHICLES
     . RECOVERABLE SPACECRAFT
     . . VOSKHOD MANNED SPACECRAFT
     . . . **VOSKHOD 1 SPACECRAFT**
     SOFT LANDING SPACECRAFT
     . VOSKHOD MANNED SPACECRAFT
     . . **VOSKHOD 1 SPACECRAFT**

VOSKHOD 2 SPACECRAFT
GS   MANNED SPACECRAFT
     . VOSKHOD MANNED SPACECRAFT
     . . **VOSKHOD 2 SPACECRAFT**
     REENTRY VEHICLES
     . RECOVERABLE SPACECRAFT
     . . VOSKHOD MANNED SPACECRAFT
     . . . **VOSKHOD 2 SPACECRAFT**
     SOFT LANDING SPACECRAFT
     . VOSKHOD MANNED SPACECRAFT
     . . **VOSKHOD 2 SPACECRAFT**

VOSTOK SPACECRAFT
GS   MANNED SPACECRAFT
     . **VOSTOK SPACECRAFT**
     . . VOSTOK 1 SPACECRAFT
     . . VOSTOK 2 SPACECRAFT
     . . VOSTOK 3 SPACECRAFT
     . . VOSTOK 4 SPACECRAFT
     . . VOSTOK 5 SPACECRAFT
     . . VOSTOK 6 SPACECRAFT
     REENTRY VEHICLES
     . RECOVERABLE SPACECRAFT
     . . **VOSTOK SPACECRAFT**
     . . . VOSTOK 1 SPACECRAFT
     . . . VOSTOK 2 SPACECRAFT
     . . . VOSTOK 3 SPACECRAFT
     . . . VOSTOK 4 SPACECRAFT
     . . . VOSTOK 5 SPACECRAFT
     . . . VOSTOK 6 SPACECRAFT
     SOFT LANDING SPACECRAFT
     . **VOSTOK SPACECRAFT**
     . . VOSTOK 1 SPACECRAFT
     . . VOSTOK 2 SPACECRAFT
     . . VOSTOK 3 SPACECRAFT
     . . VOSTOK 4 SPACECRAFT
     . . VOSTOK 5 SPACECRAFT
     . . VOSTOK 6 SPACECRAFT
RT   SPACE CAPSULES

VOSTOK 1 SPACECRAFT
GS   MANNED SPACECRAFT
     . VOSTOK SPACECRAFT
     . . **VOSTOK 1 SPACECRAFT**
     REENTRY VEHICLES
     . RECOVERABLE SPACECRAFT
     . . VOSTOK SPACECRAFT
     . . . **VOSTOK 1 SPACECRAFT**
     SOFT LANDING SPACECRAFT
     . VOSTOK SPACECRAFT
     . . **VOSTOK 1 SPACECRAFT**

VOSTOK 2 SPACECRAFT
GS   MANNED SPACECRAFT
     . VOSTOK SPACECRAFT
     . . **VOSTOK 2 SPACECRAFT**
     REENTRY VEHICLES
     . RECOVERABLE SPACECRAFT
     . . VOSTOK SPACECRAFT
     . . . **VOSTOK 2 SPACECRAFT**
     SOFT LANDING SPACECRAFT
     . VOSTOK SPACECRAFT
     . . **VOSTOK 2 SPACECRAFT**

VOSTOK 3 SPACECRAFT
GS   MANNED SPACECRAFT
     . VOSTOK SPACECRAFT
     . . **VOSTOK 3 SPACECRAFT**
     REENTRY VEHICLES
     . RECOVERABLE SPACECRAFT
     . . VOSTOK SPACECRAFT
     . . . **VOSTOK 3 SPACECRAFT**
     SOFT LANDING SPACECRAFT
     . VOSTOK SPACECRAFT
     . . **VOSTOK 3 SPACECRAFT**

VOSTOK 4 SPACECRAFT
GS   MANNED SPACECRAFT
     . VOSTOK SPACECRAFT
     . . **VOSTOK 4 SPACECRAFT**
     REENTRY VEHICLES
     . RECOVERABLE SPACECRAFT
     . . VOSTOK SPACECRAFT
     . . . **VOSTOK 4 SPACECRAFT**
     SOFT LANDING SPACECRAFT
     . VOSTOK SPACECRAFT
     . . **VOSTOK 4 SPACECRAFT**

VOSTOK 5 SPACECRAFT
GS   MANNED SPACECRAFT
     . VOSTOK SPACECRAFT
     . . **VOSTOK 5 SPACECRAFT**
     REENTRY VEHICLES
     . RECOVERABLE SPACECRAFT
     . . VOSTOK SPACECRAFT
     . . . **VOSTOK 5 SPACECRAFT**
     SOFT LANDING SPACECRAFT
     . VOSTOK SPACECRAFT
     . . **VOSTOK 5 SPACECRAFT**

VOSTOK 6 SPACECRAFT
GS   MANNED SPACECRAFT
     . VOSTOK SPACECRAFT
     . . **VOSTOK 6 SPACECRAFT**
     REENTRY VEHICLES
     . RECOVERABLE SPACECRAFT
     . . VOSTOK SPACECRAFT
     . . . **VOSTOK 6 SPACECRAFT**
     SOFT LANDING SPACECRAFT
     . VOSTOK SPACECRAFT
     . . **VOSTOK 6 SPACECRAFT**

VOTING
RT   GOVERNMENTS
      LAW (JURISPRUDENCE)
      MINORITIES
      POLITICS
      SOVEREIGNTY

VOWELS
RT   CONSONANTS (SPEECH)
      GRAMMARS
      LANGUAGES
      WORDS (LANGUAGE)

VOYAGER PROJECT
GS   PROGRAMS
     . NASA PROGRAMS
     . . **VOYAGER PROJECT**
     . NASA SPACE PROGRAMS
     . . **VOYAGER PROJECT**
     . PROJECTS
     . . **VOYAGER PROJECT**
RT   MARS PROBES
      SATURN PROJECT
      SPACE PROBES
      UNMANNED SPACECRAFT
      VENUS PROBES

VOYAGER 1 SPACECRAFT
GS   INTERPLANETARY SPACECRAFT
     . **VOYAGER 1 SPACECRAFT**
     UNMANNED SPACECRAFT
     . **VOYAGER 1 SPACECRAFT**
RT   FLYBY MISSIONS
      GRAND TOURS
      JUPITER (PLANET)
      JUPITER PROBES
      JUPITER RINGS
      SPACE PROBES
      ∞SPACECRAFT

VOYAGER 2 SPACECRAFT
GS   INTERPLANETARY SPACECRAFT
     . **VOYAGER 2 SPACECRAFT**
     UNMANNED SPACECRAFT
     . **VOYAGER 2 SPACECRAFT**
RT   FLYBY MISSIONS
      GRAND TOURS

VOYAGER 2 SPACECRAFT-*(CONT.)*
- JUPITER (PLANET)
- JUPITER PROBES
- SPACE PROBES
- ∞ SPACECRAFT

VOYAGER 1977 MISSION
GS   FLYBY MISSIONS
- . GRAND TOURS
- . . **VOYAGER 1977 MISSION**
RT   INTERPLANETARY SPACECRAFT
- JUPITER (PLANET)
- JUPITER PROBES
- ∞ MISSIONS
- SOLAR SYSTEM
- SPACE PROBES

VOYAGEUR HELICOPTER
USE   CH-46 HELICOPTER

VTOL
USE   VERTICAL LANDING
      VERTICAL TAKEOFF

VTOL AIRCRAFT
USE   VERTICAL TAKEOFF AIRCRAFT

VULCAN AIRCRAFT
UF   AVRO 698 AIRCRAFT
GS   ATTACK AIRCRAFT
- . BOMBER AIRCRAFT
- . . **VULCAN AIRCRAFT**
- HAWKER SIDDELEY AIRCRAFT
- . **VULCAN AIRCRAFT**
- JET AIRCRAFT
- . **VULCAN AIRCRAFT**
- TAILLESS AIRCRAFT
- . **VULCAN AIRCRAFT**
RT   ∞ AIRCRAFT
- AVRO 707 AIRCRAFT
- HARRIER AIRCRAFT

VULCANIZATES
USE   VULCANIZED ELASTOMERS

VULCANIZED ELASTOMERS
UF   GUM VULCANIZATES
      VULCANIZATES
GS   RUBBER
- . SYNTHETIC RUBBERS
- . . ELASTOMERS
- . . . **VULCANIZED ELASTOMERS**
RT   ADDITION RESINS
- POLYETHER RESINS
- THERMOPLASTIC RESINS
- VULCANIZING

VULCANIZING
GS   CROSSLINKING
- . **VULCANIZING**
RT   CURING
- VULCANIZED ELASTOMERS

VULNERABILITY
GS   **VULNERABILITY**
- . NUCLEAR VULNERABILITY
RT   AIRCRAFT RELIABILITY
- AIRCRAFT SURVIVABILITY
- AIRPORT SECURITY
- DURABILITY
- INTEGRITY
- LIFE (DURABILITY)
- OBSTACLE AVOIDANCE
- PENETRATION
- PIERCING
- RELIABILITY
- ∞ RESISTANCE
- SECURITY
- SENSITIVITY
- SPACECRAFT DEFENSE
- SPACECRAFT SURVIVABILITY
- STABILITY

VYCOR
GS   FIBERS
- . SYNTHETIC FIBERS
- . . **VYCOR**
- GLASS
- . **VYCOR**
- SEMICONDUCTORS (MATERIALS)
- . **VYCOR**
RT   GLASS FIBERS
- ∞ MATERIALS
- SILICON DIOXIDE

VZ-2 AIRCRAFT
GS   BOEING AIRCRAFT
- . **VZ-2 AIRCRAFT**
- RESEARCH AIRCRAFT
- . **VZ-2 AIRCRAFT**
- TILT WING AIRCRAFT
- . **VZ-2 AIRCRAFT**
- V/STOL AIRCRAFT
- . **VZ-2 AIRCRAFT**
RT   ∞ AIRCRAFT

VZ-8 AIRCRAFT
UF   AIRGEEP AIRCRAFT
GS   LIGHT AIRCRAFT
- . **VZ-8 AIRCRAFT**
- PIASECKI AIRCRAFT
- . **VZ-8 AIRCRAFT**
- RESEARCH AIRCRAFT
- . **VZ-8 AIRCRAFT**
- V/STOL AIRCRAFT
- . VERTICAL TAKEOFF AIRCRAFT
- . . **VZ-8 AIRCRAFT**
RT   ∞ AIRCRAFT
- FLYING PLATFORMS

VZ-10 AIRCRAFT
USE   XV-4 AIRCRAFT

VZ-11 AIRCRAFT
USE   XV-5 AIRCRAFT

VZ-12 AIRCRAFT
USE   P-1127 AIRCRAFT

W

W WINGS
USE   VARIABLE SWEEP WINGS

W-R STARS
USE   WOLF-RAYET STARS

WABASH RIVER BASIN (IL-IN-OH)
GS   LANDFORMS
- . STRUCTURAL BASINS
- . . RIVER BASINS
- . . . **WABASH RIVER BASIN (IL-IN-OH)**
RT   ILLINOIS
- INDIANA
- OHIO
- RIVERS

WADIS
GS   LANDFORMS
- . STRUCTURAL BASINS
- . . RIVER BASINS
- . . . **WADIS**
RT   ARID LANDS
- DESERTIFICATION
- RIVERS
- STREAMS
- TOPOGRAPHY
- VALLEYS
- WATER RUNOFF

WAFERS
RT   MICROELECTRONICS
- MICROMINIATURIZATION
- MINIATURIZATION
- PHOTOMASKS
- SEMICONDUCTOR DEVICES
- SOLID STATE DEVICES
- THIN FILMS
- VERTICAL JUNCTION SOLAR CELLS

WAGE SURVEYS
GS   REPORTS
- . **WAGE SURVEYS**
- SURVEYS
- . **WAGE SURVEYS**
RT   COST ANALYSIS
- COST ESTIMATES
- COST REDUCTION
- EMPLOYEE RELATIONS
- FINANCE
- PERSONNEL

WAKEFULNESS
RT   ALERTNESS
- SLEEP DEPRIVATION

WAKES
GS   **WAKES**
- . AIRCRAFT WAKES
- . . HELICOPTER WAKES
- . . SLIPSTREAMS
- . . . PROPELLER SLIPSTREAMS
- . HYPERSONIC WAKES
- . LAMINAR WAKES
- . NEAR WAKES
- . SUPERSONIC WAKES
- . TURBULENT WAKES
- . . SLIPSTREAMS
- . . . PROPELLER SLIPSTREAMS
RT   BACKWASH
- BASE FLOW
- BUBBLES
- CAVITATION FLOW
- CONTRAILS
- DOWNWASH
- ∞ DRAFT
- DRAG
- GROUND EFFECT (AERODYNAMICS)
- STROUHAL NUMBER
- TURBULENCE
- VORTEX ALLEVIATION
- VORTICES

WALKING
GS   LOCOMOTION
- . **WALKING**
RT   PHYSICAL EXERCISE

WALKING MACHINES
GS   SURFACE VEHICLES
- . **WALKING MACHINES**
RT   ASTRONAUT MANEUVERING EQUIPMENT
- LUNAR SURFACE VEHICLES
- ∞ MACHINERY
- MANNED LUNAR SURFACE VEHICLES
- PROSTHETIC DEVICES

WALL FLOW
GS   FLUID FLOW
- . **WALL FLOW**
RT   BOUNDARY LAYER FLOW
- CHANNEL FLOW
- CONICAL FLOW
- DISCHARGE COEFFICIENT
- DUCTED FLOW
- GOERTLER INSTABILITY
- HEAT TRANSMISSION
- MANNING THEORY
- TWO DIMENSIONAL FLOW
- TWO DIMENSIONAL JETS

WALL JETS
RT   FLUID AMPLIFIERS
- JET BOUNDARIES
- JET FLOW
- JET VANES
- ∞ JETS

WALL PRESSURE
GS   PRESSURE
- . **WALL PRESSURE**
RT   BOUNDARY LAYERS
- PRESSURE DISTRIBUTION
- PRESSURE VESSELS
- THICK WALLS

WALL TEMPERATURE
GS   SURFACE PROPERTIES
- . SURFACE TEMPERATURE
- . . **WALL TEMPERATURE**
- TEMPERATURE
- . SURFACE TEMPERATURE
- . . **WALL TEMPERATURE**
RT   OPERATING TEMPERATURE
- THICK WALLS

WALLOPS ISLAND
GS   LANDFORMS
- . ISLANDS
- . . **WALLOPS ISLAND**
RT   ATLANTIC OCEAN
- VIRGINIA

WALLS
UF   COLD WALLS
GS   **WALLS**
- . BULKHEADS
- . NOZZLE WALLS
- . POROUS WALLS
- . THICK WALLS
- . THIN WALLS

WALLS-*(CONT.)*
    . TROMBE WALLS
    . WIND TUNNEL WALLS
RT ∞BARRIERS
    ∞BUILDINGS
    CURTAINS
    ENCLOSURES
    FLOORS
    GATES (OPENINGS)
    HOUSINGS
    LIMITERS (FUSION REACTORS)
    PANELS
    PARTITIONS (STRUCTURES)
    SANDWICH STRUCTURES
    SHEATHS
    SHELLS (STRUCTURAL FORMS)
    SIDES
    STUDS (STRUCTURAL MEMBERS)
    SUBSTRUCTURES
    TILES

WALSH FUNCTION
GS FUNCTIONS (MATHEMATICS)
    . ORTHOGONAL FUNCTIONS
    . . **WALSH FUNCTION**
RT FAST FOURIER TRANSFORMATIONS
    FOURIER TRANSFORMATION
    FUNCTIONAL ANALYSIS
    MATRICES (MATHEMATICS)

WANKEL ENGINES
GS ENGINES
    . INTERNAL COMBUSTION ENGINES
    . . ROTARY ENGINES
    . . . **WANKEL ENGINES**
RT AIRCRAFT ENGINES
    AUTOMOBILE ENGINES
    PISTON ENGINES

WAR GAMES
RT DIGITAL SIMULATION
    GAME THEORY
    MATHEMATICAL MODELS
    OPERATIONS RESEARCH
    SIMULATION

WARFARE
GS **WARFARE**
    . ANTISHIP WARFARE
    . ANTISUBMARINE WARFARE
    . CHEMICAL WARFARE
    . COMBAT
    . ELECTRONIC WARFARE
    . NUCLEAR WARFARE
RT ATTACKING (ASSAULTING)
    B-1 AIRCRAFT
    CHEMICAL DEFENSE
    EVASIVE ACTIONS
    INFILTRATION
    INTERNATIONAL LAW
    ORDNANCE
    PEACETIME
    POLITICS
    STRATEGY
    VIOLENCE

WARHEADS
GS WEAPONS
    . **WARHEADS**
    . . NUCLEAR WARHEADS
    . . PRECISION GUIDED PROJECTILES
RT AMMUNITION
    ANTISHIP WARFARE
    BOMBS (ORDNANCE)
    EXPLOSIVE DEVICES
    EXPLOSIVES
    ∞FUSES
    FUSES (ORDNANCE)
    MISSILE COMPONENTS
    MISSILES
    NOSE CONES
    NUCLEAR DEVICES
    NUCLEAR WEAPONS
    PAYLOADS
    PROJECTILES
    ∞ROCKETS
    SHAPED CHARGES
    TORPEDOES

WARM FRONTS
GS FRONTS (METEOROLOGY)
    . **WARM FRONTS**
RT AIR MASSES
    COLD FRONTS
    ∞FRONTS

WARM FRONTS-*(CONT.)*
    METEOROLOGICAL PARAMETERS
    METEOROLOGY
    STORMS
    SYNOPTIC METEOROLOGY
    THUNDERSTORMS
    TORNADOES
    WEATHER FORECASTING

WARMING
USE HEATING

WARNING
RT ACCIDENT PREVENTION
    AUDITORY SIGNALS
    BELLS
    CIVIL DEFENSE
    COLLISION AVOIDANCE
    DETECTION
    ∞DETECTORS
    EARLY WARNING SYSTEMS
    FIRE PREVENTION
    HORNS
    MINE DETECTORS
    MONITORS
    PROTECTION
    SAFETY
    SAFETY DEVICES

WARNING DEVICES
USE WARNING SYSTEMS

WARNING SIGNALS
USE WARNING SYSTEMS

WARNING STAR AIRCRAFT
USE EC-121 AIRCRAFT

WARNING SYSTEMS
UF ALARMS
    COLLISION WARNING DEVICES
    WARNING DEVICES
    WARNING SIGNALS
GS **WARNING SYSTEMS**
    . EARLY WARNING SYSTEMS
    . . BALLISTIC MISSILE EARLY WARNING
        SYSTEM
    . MINE DETECTORS
RT ACCIDENT PREVENTION
    AUDITORY SIGNALS
    AVOIDANCE
    BELLS
    CIVIL DEFENSE
    COLLISION AVOIDANCE
    DETECTION
    ∞DETECTORS
    DISPLAY DEVICES
    EXPLOSIONS
    FIRE PREVENTION
    FIRES
    GAS DETECTORS
    HAZARDS
    HEAD-UP DISPLAYS
    HORNS
    MONITORS
    NATIONAL SEVERE STORMS PROJECT
    POLLUTION MONITORING
    PROTECTION
    PUBLIC ADDRESS SYSTEMS
    SAFETY
    SAFETY DEVICES
    SAFETY MANAGEMENT
    SANITATION
    SIRENS
    SOUND GENERATORS
    ∞SYSTEMS
    THREAT EVALUATION

WARPAGE
RT BENDING
    BUCKLING
    CAMBER
    DAMAGE
    DEFORMATION
    DISTORTION
    GROWTH
    HEAVING
    PLASTIC DEFORMATION
    SHRINKAGE
    STRUCTURAL STRAIN
    SURFACE DISTORTION
    THERMAL EXPANSION
    TWISTING

∞ WASHERS
SN *(USE OF A MORE SPECIFIC TERM IS RECOMMENDED--CONSULT THE TERMS LISTED BELOW)*
RT WASHERS (CLEANERS)
    WASHERS (SPACERS)

WASHERS (CLEANERS)
RT CLEANERS
    CONCENTRATORS
    EXTRACTION
    SEPARATORS
    ∞WASHERS
    WASHING

WASHERS (SPACERS)
GS FASTENERS
    . **WASHERS (SPACERS)**
RT INSERTS
    SEPARATORS
    SPACERS
    ∞WASHERS

WASHING
UF SCRUBBING
GS **WASHING**
    . HOUSEKEEPING (SPACECRAFT)
RT BATHING
    BENEFICIATION
    CLEANING
    DECONTAMINATION
    DISSOLVING
    DISTILLATION
    ELUTION
    FLUSHING
    PURIFICATION
    SCRUBBERS
    ∞SEPARATION
    WASHERS (CLEANERS)
    WASTE WATER

WASHINGTON
GS NATIONS
    . UNITED STATES
    . . **WASHINGTON**
RT CASCADE RANGE (CA-OR-WA)
    COLUMBIA RIVER BASIN (ID-OR-WA)

WASHOUT (RADIOACTIVITY)
USE FALLOUT

WASP SOUNDING ROCKET
UF HIGH ALTITUDE SOUNDING PROJECTILE
    WINDOW ATMOSPHERE SOUNDING
       PROJECTILE
GS ROCKET VEHICLES
    . MULTISTAGE ROCKET VEHICLES
    . . **WASP SOUNDING ROCKET**
    . SOUNDING ROCKETS
    . . **WASP SOUNDING ROCKET**
RT LOKI ROCKET VEHICLE
    SOLID PROPELLANT ROCKET ENGINES

WASPALOY
GS ALLOYS
    . HEAT RESISTANT ALLOYS
    . . **WASPALOY**
    . NICKEL ALLOYS
    . . **WASPALOY**
RT CHROMIUM ALLOYS
    COBALT ALLOYS
    WROUGHT ALLOYS

WASTE DISPOSAL
GS DISPOSAL
    . **WASTE DISPOSAL**
    . . COMPOSTING
    . . HAZARDOUS MATERIAL DISPOSAL (IN
        SPACE)
RT AIR POLLUTION
    DEEP WELL INJECTION (WASTES)
    DEWATERING
    DILUTION
    DISSIPATION
    DRAINAGE
    EFFLUENTS
    ELIMINATION
    ENVIRONMENT EFFECTS
    ENVIRONMENT POLLUTION
    ENVIRONMENT PROTECTION
    ENVIRONMENTAL CHEMISTRY
    ENVIRONMENTAL ENGINEERING
    ENVIRONMENTAL SURVEYS
    EXHAUST GASES
    EXHAUST SYSTEMS

WASTE DISPOSAL-(CONT.)
    GARBAGE
    HUMAN WASTES
    INCINERATORS
    INDUSTRIAL WASTES
    LANDFILLS
    MANURES
    MATERIALS HANDLING
    METABOLIC WASTES
    MINES (EXCAVATIONS)
    MODULAR INTEGRATED UTILITY SYSTEM
    PIPELINES
    PLASMA CORE REACTORS
    POLLUTION
    PONDS
    RADIOACTIVE WASTES
    SANITATION
    SEWAGE
    SEWAGE TREATMENT
    SEWERS
    SOLID WASTES
    SPACE FLIGHT FEEDING
   ∞STORAGE
    SUMPS
    UTILITIES
    WASTES
    WATER POLLUTION

WASTE ENERGY UTILIZATION
GS   UTILIZATION
    . **WASTE ENERGY UTILIZATION**
RT   BOILERS
    BURNERS
    CHIMNEYS
    COGENERATION
    ENERGY CONVERSION
    EXHAUST GASES
    FURNACES
    HEAT TRANSFER
    HEATING
    INCINERATORS
    LIGHTING EQUIPMENT
    OVENS
    SOLID WASTES
    SPACE HEATING (BUILDINGS)
    WASTE HEAT
    WASTES

WASTE HEAT
RT   ENERGY TECHNOLOGY
    HEAT EXCHANGERS
    HEAT PUMPS
    WASTE ENERGY UTILIZATION

WASTE TREATMENT
GS   **WASTE TREATMENT**
    . SEWAGE TREATMENT
RT   BACTERIA
    COMPOSTING
    GARBAGE
    RESIDUES
    SLUDGE
   ∞TREATMENT
    WASTES

WASTE UTILIZATION
GS   UTILIZATION
    . **WASTE UTILIZATION**
RT   BIOMASS ENERGY PRODUCTION
    COMPOSTING
    HYDROCARBON FUEL PRODUCTION
    INDUSTRIAL WASTES
    LANDFILLS
    MANURES
    SOLID WASTES
    WASTES

WASTE WATER
GS   WASTES
    . LIQUID WASTES
    . . **WASTE WATER**
    WATER
    . **WASTE WATER**
RT   BATHING
    CLEANING
    FLUSHING
    INDUSTRIAL WASTES
    RESIDUES
    WASHING
    WATER COOLED REACTORS

WASTES
GS   **WASTES**
    . FECES
    . GARBAGE

WASTES-(CONT.)
    . INDUSTRIAL WASTES
    . LIQUID WASTES
    . . URINE
    . . WASTE WATER
    . MANURES
    . METABOLIC WASTES
    . . HUMAN WASTES
    . . . URINE
    . RADIOACTIVE WASTES
    . SEWAGE
    . SOLID WASTES
RT   ACTIVATED SLUDGE
    AIR POLLUTION
    BENEFICIATION
    BY-PRODUCTS
    COMBUSTION PRODUCTS
    CONTAMINANTS
    DEBRIS
    EFFLUENTS
    ENVIRONMENT EFFECTS
    EXHAUST GASES
    FOREST FIRES
    FUMES
    GAS RECOVERY
    IMPURITIES
    LEAKAGE
    LOSSES
    NONPOINT SOURCES
    ORGANIC WASTES (FUEL CONVERSION)
    POLLUTION
    RESIDUES
    SCRAP
    SEWERS
    SLAGS
    SLUDGE
    WASTE DISPOSAL
    WASTE ENERGY UTILIZATION
    WASTE TREATMENT
    WASTE UTILIZATION

WATCHES
USE   CLOCKS

WATER
GS   **WATER**
    . COLD WATER
    . FRESH WATER
    . GREAT SALT LAKE (UT)
    . HEAVY WATER
    . INLAND WATERS
    . . GROUND WATER
    . LIGHT WATER
    . NEARSHORE WATER
    . COASTAL WATER
    . POLYWATER
    . POTABLE WATER
    . SEA WATER
    . SHALLOW WATER
    . SHOALS
    . SPRINGS (WATER)
    . SURFACE WATER
    . VADOSE WATER
    . WASTE WATER
RT   AQUIFERS
    ARROYOS
    BAY ICE
    BODY FLUIDS
    CAVITATION FLOW
    FIORDS
    HUMIDITY
    HYDRATES
   ∞HYDRAULICS
    HYDRODYNAMICS
    HYDROGEN COMPOUNDS
    HYDROLOGY
    HYDROMECHANICS
    HYDROSTATICS
    ICE
    ISTHMUSES
    LAKE ERIE
    LAKE HURON
    LAKE ICE
    LAKE MICHIGAN
    LAKE ONTARIO
    LAKE SUPERIOR
    LATERITES
    LIFE SUPPORT SYSTEMS
    LIMNOLOGY
    LIQUIDS
    MODERATORS
    MOISTURE
    MOISTURE CONTENT
    MUSKEGS
    OXIDES

WATER-(CONT.)
    PENINSULAS
    POLLUTION
    PRECIPITATION (METEOROLOGY)
    RUNWAY CONDITIONS
    SLUSH
    SOUNDS (TOPOGRAPHIC FEATURES)
    STEAM
    STRAITS
    UTILITIES
    WATERSHEDS
    WHARVES
    WINDPOWERED PUMPS

WATER BALANCE
GS   MATERIAL BALANCE
    . **WATER BALANCE**
RT   BODY FLUIDS
    EDEMA
   ∞EQUILIBRIUM
    HOMEOSTASIS
    HYDROMETEOROLOGY
    LYSIMETERS
    OSMOSIS
    URINATION

WATER CIRCULATION
GS   CIRCULATION
    . **WATER CIRCULATION**
    . . WATER CURRENTS
    . . . OCEAN CURRENTS
    . . . . COASTAL CURRENTS
    . . . . EL NINO
    . . . . GULF STREAM
    . . . . LOMONOSOV CURRENT
RT   LAKES
    OCEANOGRAPHY
    POLLUTION TRANSPORT
    WIND EFFECTS

WATER COLOR
GS   ELECTROMAGNETIC PROPERTIES
    . OPTICAL PROPERTIES
    . . COLOR
    . . . **WATER COLOR**
RT   LAKES
    OCEAN COLOR SCANNER
    OCEANS
    RIVERS

WATER CONSUMPTION
GS   CONSUMPTION
    . **WATER CONSUMPTION**
RT   DROUGHT
    IRRIGATION
    SEEPAGE

WATER CONTENT
USE   MOISTURE CONTENT

WATER COOLED REACTORS
UF   PHYSICAL CONSTANTS TESTING
      REACTOR
GS   NUCLEAR REACTORS
    . LIQUID COOLED REACTORS
    . . **WATER COOLED REACTORS**
    . . . BOILING WATER REACTORS
    . . . . EXPERIMENTAL BOILING WATER
        REACTORS
    . . . . HALDEN BOILING WATER
        REACTOR
    . . . . LOS ALAMOS WATER BOILER
        REACTOR
    . . . . PATHFINDER NUCLEAR REACTOR
    . . . . SPERT REACTORS
    . . . HEAVY WATER REACTORS
    . . . . HEAVY WATER COMPONENTS
        TEST REACTORS
    . . . . PLUTONIUM RECYCLE TEST
        REACTOR
    . . . . ZERO POWER REACTOR 2
    . . . LIGHT WATER REACTORS
    . . . NRX REACTORS
    . . . PLUM BROOK REACTOR
    . . . PRESSURIZED WATER REACTORS
    . . . . SPECTRAL SHIFT CONTROL
        REACTOR
    . . . SWIMMING POOL REACTORS
    . . . ZERO POWER REACTORS
    . . . . ZERO POWER REACTOR 2
    . . . . ZERO POWER REACTOR 3
    . . . . ZERO POWER REACTOR 6
    . . . . ZERO POWER REACTOR 9
RT   CHEMICAL REACTORS

WATER COOLED REACTORS-*(CONT.)*
    NUCLEAR ENGINE FOR ROCKET
      VEHICLES
    ∞REACTORS
    WASTE WATER

WATER COOLING
USE   LIQUID COOLING

WATER CURRENTS
UF   CURRENTS (OCEANOGRAPHY)
GS   CIRCULATION
    . WATER CIRCULATION
    . . **WATER CURRENTS**
    . . . OCEAN CURRENTS
    . . . . COASTAL CURRENTS
    . . . . EL NINO
    . . . . GULF STREAM
    . . . . LOMONOSOV CURRENT
RT   ARROYOS
    ∞CURRENTS
    OCEANOGRAPHY
    RAPIDS
    SEA ROUGHNESS
    SEA STATES
    SURFACE WAVES
    TIDES
    UPSTREAM

WATER DEPRIVATION
GS   DEPRIVATION
    . **WATER DEPRIVATION**

WATER DEPTH
RT   CNOIDAL WAVES
    COASTAL WATER
    LAKES
    NEARSHORE WATER
    OCEANS
    SHALLOW WATER
    SHOALS

WATER EROSION
GS   EROSION
    . **WATER EROSION**
RT   ARROYOS
    CANYONS
    DRAINAGE PATTERNS
    FLOOD DAMAGE
    RAIN IMPACT DAMAGE
    RAVINES
    SOIL EROSION
    WIND EROSION

WATER FLOW
GS   FLUID FLOW
    . LIQUID FLOW
    . . **WATER FLOW**
RT   ALLUVIUM
    CANALS
    DRAINAGE
    DRAINAGE PATTERNS
    FLOOD DAMAGE
    FLOODS
    FLOW MEASUREMENT
    GREAT LAKES (NORTH AMERICA)
    GROUND WATER
    ∞HYDRAULICS
    HYDRODYNAMICS
    HYDROLOGY MODELS
    OPEN CHANNEL FLOW
    PIPE FLOW
    RAPIDS
    WATERSHEDS

WATER HAMMER
RT   HYDRAULIC EQUIPMENT
    HYDRODYNAMICS
    PIPE FLOW
    PIPELINES
    RAMS (PUMPS)
    SURGES
    VALVES

WATER HEATING
GS   HEATING
    . **WATER HEATING**
RT   DOMESTIC ENERGY
    GEOTHERMAL ENERGY EXTRACTION
    HEAT EXCHANGERS
    ∞HEATERS
    HEATING EQUIPMENT
    RESIDENTIAL ENERGY
    TEMPERATURE CONTROL

WATER IMMERSION
RT   BATHS
    LIQUID COOLING
    QUENCHING (COOLING)
    SINKING
    ∞SOAKING
    SUBMERGED BODIES
    SUBMERGING
    UNDERWATER TESTS

WATER INJECTION
GS   INJECTION
    . FLUID INJECTION
    . . LIQUID INJECTION
    . . . **WATER INJECTION**
RT   GAS INJECTION
    PERFORATING
    THRUST AUGMENTATION

∞ **WATER INTAKES**
SN   *(USE OF A MORE SPECIFIC TERM IS*
    RECOMMENDED--CONSULT THE TERMS
    LISTED BELOW)
RT   AIR INTAKES
    INTAKE SYSTEMS
    MANIFOLDS
    NOSE INLETS
    PLENUM CHAMBERS
    SCOOPS
    SIDE INLETS

WATER JETS
USE   HYDRAULIC JETS

WATER LANDING
GS   LANDING
    . **WATER LANDING**
    . . DITCHING (LANDING)
RT   AIRCRAFT LANDING
    CRASH LANDING
    GLIDE LANDINGS
    HARD LANDING
    HORIZONTAL SPACECRAFT LANDING
    HYDROPLANING
    PLANETARY LANDING
    SOFT LANDING
    SPACECRAFT LANDING
    SPACECRAFT RECOVERY
    SPLASHING
    TOUCHDOWN

WATER LOSS
RT   DEHYDRATION
    DRYING
    EVAPORATION
    LOSSES

WATER MANAGEMENT
GS   MANAGEMENT
    . **WATER MANAGEMENT**
RT   CONSERVATION
    DROUGHT
    ENVIRONMENT MANAGEMENT
    FLOODS
    HYDROLOGY
    LAKE ERIE
    LAKE HURON
    LAKE MICHIGAN
    LAKE ONTARIO
    LAKE SUPERIOR
    LIMNOLOGY
    POTABLE WATER
    PYRAMID LAKE (NV)
    WATERSHEDS

WATER MASERS
GS   DECAY
    . EMISSION
    . . STIMULATED EMISSION
    . . . **WATER MASERS**
    STIMULATED EMISSION DEVICES
    . MASERS
    . . **WATER MASERS**
RT   GAS LASERS
    GAS MASERS
    INTERSTELLAR MASERS
    MASER OUTPUTS

WATER MODERATED REACTORS
GS   NUCLEAR REACTORS
    . **WATER MODERATED REACTORS**
    . . EXPERIMENTAL BOILING WATER
      REACTORS
    . . HEAVY WATER COMPONENTS TEST
      REACTORS

WATER MODERATED REACTORS-*(CONT.)*
    . . PLUTONIUM RECYCLE TEST
      REACTOR
RT   LIGHT WATER REACTORS

WATER POLLUTION
GS   POLLUTION
    . ENVIRONMENT POLLUTION
    . . **WATER POLLUTION**
    . . . OIL POLLUTION
RT   ALGAE
    ALKALINITY
    BIOCHEMICAL OXYGEN DEMAND
    CLEAN ENERGY
    CONTAMINATION
    DROUGHT
    ENVIRONMENT EFFECTS
    ENVIRONMENT PROTECTION
    ENVIRONMENTAL CHEMISTRY
    ENVIRONMENTAL QUALITY
    ENVIRONMENTAL SURVEYS
    INLAND WATERS
    LANDFILLS
    LIMNOLOGY
    LYSIMETERS
    MARINE RESOURCES
    OIL SLICKS
    POLLUTION MONITORING
    POLLUTION TRANSPORT
    PURITY
    THERMAL POLLUTION
    WASTE DISPOSAL

WATER PRESSURE
GS   PRESSURE
    . FLUID PRESSURE
    . . **WATER PRESSURE**
RT   ∞HYDRAULICS
    HYDRODYNAMICS
    HYDROSTATIC PRESSURE
    HYDROSTATICS
    INLET PRESSURE
    PIPE FLOW

WATER PURIFICATION
USE   WATER TREATMENT

WATER QUALITY
GS   QUALITY
    . ENVIRONMENTAL QUALITY
    . . **WATER QUALITY**
RT   ALKALINITY
    ENVIRONMENT EFFECTS

WATER RECLAMATION
UF   WATER RECOVERY
GS   RECLAMATION
    . MATERIALS RECOVERY
    . . **WATER RECLAMATION**
RT   CONSERVATION
    DEWATERING
    DROUGHT
    POLLUTION

WATER RECOVERY
USE   WATER RECLAMATION

WATER RESOURCES
GS   RESOURCES
    . EARTH RESOURCES
    . . **WATER RESOURCES**
RT   ENVIRONMENT EFFECTS
    ENVIRONMENT MANAGEMENT
    GREAT LAKES (NORTH AMERICA)
    GROUND WATER
    HYDROLOGY
    INLAND WATERS
    INTERNATIONAL HYDROLOGICAL
      DECADE
    LAKES
    LIMNOLOGY
    OCEANS
    PONDS
    POTABLE WATER
    PRECIPITATION (METEOROLOGY)
    PYRAMID LAKE (NV)
    RAINMAKING
    RESERVOIRS
    SEA WATER
    UNDERWATER RESOURCES
    WETLANDS
    WINDPOWERED PUMPS

WATER RUNOFF
RT   DRAINAGE

WATER RUNOFF-(CONT.)
GROUND WATER
INLAND WATERS
RESOURCES MANAGEMENT
RIVERS
WADIS

WATER TABLES
RT   AQUIFERS
DRAINAGE
GROUND WATER
POTABLE WATER
SPRINGS (WATER)
VADOSE WATER
WATERSHEDS

WATER TAKEOFF AND LANDING AIRCRAFT
GS   WATER TAKEOFF AND LANDING
AIRCRAFT
. SEAPLANES
RT   ∞AIRCRAFT
AMPHIBIOUS AIRCRAFT
ANTISUBMARINE WARFARE AIRCRAFT
COMMERCIAL AIRCRAFT
GROUND EFFECT MACHINES
HOVERCRAFT GROUND EFFECT
MACHINES
LIGHT AIRCRAFT
MONOPLANES
PASSENGER AIRCRAFT
RECONNAISSANCE AIRCRAFT
S-61 HELICOPTER
SEA LAUNCHING
SUBMERSIBLE AIRCRAFT
∞SUBSONIC AIRCRAFT
TRANSPORT AIRCRAFT
UTILITY AIRCRAFT

WATER TEMPERATURE
GS   TEMPERATURE
. WATER TEMPERATURE
. . OCEAN TEMPERATURE
. . . SEA SURFACE TEMPERATURE
RT   SURFACE TEMPERATURE
TEMPERATURE DISTRIBUTION
THERMAL POLLUTION

WATER TREATMENT
UF   WATER PURIFICATION
RT   ACTIVATED CARBON
ADSORPTION
AERATION
AGITATION
BENTONITE
BIOCHEMICAL OXYGEN DEMAND
CHLORINATION
COAGULATION
CONTAMINANTS
CORROSION PREVENTION
DEMINERALIZING
DESALINIZATION
FILTRATION
FLOCCULATING
FLOTATION
ION EXCHANGING
MATERIAL ABSORPTION
POLLUTION
POTABLE WATER
PURIFICATION
∞SCREENING
SETTLING
SEWAGE
∞TREATMENT

WATER TUNNEL TESTS
RT   AIR WATER INTERACTIONS
CROSS FLOW
FLOW DISTRIBUTION
FLOW VISUALIZATION
STRAKES
∞TESTS
WATER WAVES
WIND TUNNEL TESTS

WATER TUNNELS
USE   HYDRAULIC TEST TUNNELS

WATER VAPOR
GS   VAPORS
. WATER VAPOR
RT   ATMOSPHERIC MOISTURE
DEW
HUMIDITY
MOISTURE
MOISTURE CONTENT

WATER VAPOR-(CONT.)
GS   STEAM

WATER VEHICLES
GS   WATER VEHICLES
. BOATS
. . LIFEBOATS
. CAPTURED AIR BUBBLE VEHICLES
. SHIPS
. . ADVANCED RANGE
INSTRUMENTATION SHIP
. . AIRCRAFT CARRIERS
. . CARGO SHIPS
. . . SAVANNAH NUCLEAR SHIP
. . . TANKER SHIPS
. . NUCLEAR POWERED SHIPS
. . . SAVANNAH NUCLEAR SHIP
. . SATELLITE COMMUNICATIONS SHIPS
. . SUBMARINES
. . . BALLISTIC MISSILE SUBMARINES
. . . GUIDED MISSILE SUBMARINES
. . . TRIDENT SUBMARINE
. . SURFACE EFFECT SHIPS
. . SWATH (SHIP)
. UNDERWATER VEHICLES
. . SUBMARINES
. . . BALLISTIC MISSILE SUBMARINES
. . . GUIDED MISSILE SUBMARINES
. . . TRIDENT SUBMARINE
. . UNDERWATER RESEARCH
LABORATORIES
RT   AMPHIBIOUS VEHICLES
HARBORS
MARINE TRANSPORTATION
∞MILITARY VEHICLES
RESEARCH VEHICLES
SHIPYARDS
SURFACE VEHICLES
∞VEHICLES

WATER WAVES
GS   WATER WAVES
. TIDAL WAVES
RT   BREAKWATERS
CAPILLARY WAVES
CNOIDAL WAVES
ELASTOHYDRODYNAMICS
FRONTAL WAVES
GRAVITY WAVES
HYDRODYNAMIC COEFFICIENTS
LITTORAL TRANSPORT
OCEAN DYNAMICS
RIPPLES
SEA ROUGHNESS
SEA STATES
SURFACE WAVES
TSUNAMI WAVES
WATER TUNNEL TESTS
WATERWAVE ENERGY
WATERWAVE ENERGY CONVERSION
WATERWAVE POWERED MACHINES
∞WAVES

WATER WHEELS
GS   WHEELS
. WATER WHEELS
RT   HYDROELECTRIC POWER STATIONS
TURBINE WHEELS

WATERFOWL
GS   ANIMALS
. VERTEBRATES
. . BIRDS
. . . WATERFOWL
RT   BEACHES
COASTAL ECOLOGY
MARINE BIOLOGY
MARINE ENVIRONMENTS
MARSHLANDS
MIGRATION
OCEANOGRAPHY
WETLANDS

WATERPROOFING
RT   BARRIER LAYERS
CAULKING
COATINGS
INSULATION
MOISTURE RESISTANCE
PROTECTIVE COATINGS
SEALING
WEATHERPROOFING

WATERSHEDS
UF   CATCHMENT AREAS

WATERSHEDS-(CONT.)
GS   LANDFORMS
. STRUCTURAL BASINS
. . WATERSHEDS
RT   DIVIDES (LANDFORMS)
DRAINAGE PATTERNS
FLOOD CONTROL
FLOODS
HYDROGEOLOGY
HYDROLOGY
INTERNATIONAL HYDROLOGICAL
DECADE
MISSOURI RIVER BASIN (US)
MOUNTAINS
PONDS
PRECIPITATION (METEOROLOGY)
RAIN
RIVER BASINS
RIVERS
STORMS (METEOROLOGY)
VALLEYS
WATER
WATER FLOW
WATER MANAGEMENT
WATER TABLES

WATERWAVE ENERGY
RT   CLEAN ENERGY
EARTH RESOURCES
∞ENERGY
OCEANOGRAPHY
TIDEPOWER
WATER WAVES
∞WAVES

WATERWAVE ENERGY CONVERSION
GS   ENERGY CONVERSION
. WATERWAVE ENERGY CONVERSION
RT   ∞CONVERSION
EARTH RESOURCES
ENERGY CONVERSION EFFICIENCY
∞ENERGY SOURCES
OCEAN CURRENTS
OCEAN SURFACE
OCEANOGRAPHY
OCEANS
SEA ROUGHNESS
TIDE POWERED GENERATORS
TIDE POWERED MACHINES
TIDEPOWER
TIDES
WATER WAVES

WATERWAVE POWERED MACHINES
RT   ∞MACHINERY
OCEAN CURRENTS
OCEAN SURFACE
SEA ROUGHNESS
TIDE POWERED GENERATORS
TIDE POWERED MACHINES
TIDEPOWER
TIDES
WATER WAVES

WATERWAYS
GS   WATERWAYS
. CANALS
. HARBORS
. . ARTIFICIAL HARBORS
RT   LAKES
RIVERS
STRAITS

WATTMETERS
GS   MEASURING INSTRUMENTS
. WATTMETERS
RT   ELECTRICAL MEASUREMENT
ELECTROMETERS

WAVE AMPLIFICATION
GS   AMPLIFICATION
. WAVE AMPLIFICATION
RT   BAROCLINIC WAVES
ELECTROMAGNETIC RADIATION
∞WAVES

WAVE ATTENUATION
GS   ATTENUATION
. WAVE ATTENUATION
. . RADAR ATTENUATION
. . RADIO ATTENUATION
. . . MANDELSTAM REPRESENTATION
. . SHOCK WAVE ATTENUATION
RT   ATMOSPHERIC ATTENUATION
ELECTROMAGNETIC ABSORPTION

WAVE ATTENUATION-*(CONT.)*
    INFRARED ABSORPTION
    RADAR TRANSMISSION
    RADIO TRANSMISSION
    SHOCK WAVE PROPAGATION

WAVE DEGRADATION
GS   DEGRADATION
    . **WAVE DEGRADATION**
RT   ATTENUATION
    SCATTERING
    SHOCK WAVE INTERACTION
    STERILIZATION EFFECTS

WAVE DIFFRACTION
GS   DIFFRACTION
    . **WAVE DIFFRACTION**
RT   ATTENUATION
    CROSSTALK
    FRESNEL INTEGRALS
    GEOMETRICAL THEORY OF
      DIFFRACTION
    ∞INTERFERENCE
    TRAVELING WAVE MODULATION

WAVE DISPERSION
RT   ACOUSTIC PROPERTIES
    ATMOSPHERIC REFRACTION
    ATTENUATION
    ∞COHERENCE
    COLOR
    DEFLECTION
    DIFFRACTION
    ∞DISPERSION
    ELASTIC WAVES
    ELECTROMAGNETIC RADIATION
    FADING
    LIGHT TRANSMISSION
    OPTICAL PATHS
    OPTICAL PROPERTIES
    RADIATION DISTRIBUTION
    RADIO WAVE REFRACTION
    REFRACTION
    SCATTERING
    SOUND-SOUND INTERACTIONS
    TRANSMISSION
    TRANSMISSION LOSS

WAVE DRAG
GS   DYNAMIC CHARACTERISTICS
    . DRAG
    . . PRESSURE DRAG
    . . . **WAVE DRAG**
    . . . . INTERFERENCE DRAG
RT   FRICTION DRAG
    SUPERSONIC DRAG

WAVE EQUATIONS
SN   (NOT EQUATIONS OF MOTION)
GS   **WAVE EQUATIONS**
    . DIRAC EQUATION
    . EIKONAL EQUATION
    . KLEIN-GORDON EQUATION
    . KORTEWEG-DEVRIES EQUATION
    . LAME WAVE EQUATIONS
    . SCHROEDINGER EQUATION
RT   BOLTZMANN-VLASOV EQUATION
    DENSITY WAVE MODEL
    ∞EQUATIONS
    FORBIDDEN BANDS
    ∞HELMHOLTZ EQUATIONS
    HYPERBOLIC DIFFERENTIAL EQUATIONS
    PARTIAL DIFFERENTIAL EQUATIONS
    QUANTUM THEORY

WAVE EXCITATION
GS   EXCITATION
    . **WAVE EXCITATION**
    . . ACOUSTIC EXCITATION
    . . HARMONIC EXCITATION
RT   STROKING TESTS
    ∞WAVES

WAVE FRONT DEFORMATION
GS   DEFORMATION
    . **WAVE FRONT DEFORMATION**
RT   ∞INTERFERENCE

WAVE FRONT RECONSTRUCTION
GS   RECONSTRUCTION
    . **WAVE FRONT RECONSTRUCTION**
RT   ACOUSTICAL HOLOGRAPHY
    DIFFRACTOMETERS
    HOLOGRAPHIC INTERFEROMETRY
    HOLOGRAPHIC SPECTROSCOPY

WAVE FRONT RECONSTRUCTION-*(CONT.)*
    HOLOGRAPHY
    MICROWAVE HOLOGRAPHY
    PHOTOGRAPHY
    WHITE LIGHT HOLOGRAPHY

WAVE FRONTS
GS   **WAVE FRONTS**
    . SHOCK FRONTS
RT   CAUSTIC LINES
    EIKONAL EQUATION
    ∞FRONTS
    HUYGENS PRINCIPLE
    PHASE COHERENCE
    PHASE VELOCITY
    SHOCK DISCONTINUITY
    ∞WAVES

WAVE FUNCTIONS
GS   **WAVE FUNCTIONS**
    . MOLECULAR ORBITALS
    . PAULI EXCLUSION PRINCIPLE
RT   FORBIDDEN TRANSITIONS
    HARTREE APPROXIMATION
    PERTURBATION THEORY
    SQUARE WAVES
    TIME FUNCTIONS

WAVE GENERATION
RT   ELECTROMAGNETIC RADIATION
    FUNCTION GENERATORS
    ∞GENERATORS
    HARMONIC GENERATIONS
    SHOCK WAVE GENERATORS

WAVE INCIDENCE CONTROL
RT   ∞CONTROL
    INCIDENT RADIATION

WAVE INTERACTION
GS   **WAVE INTERACTION**
    . SHOCK WAVE INTERACTION
RT   COUPLING
    DAMPING
    ELECTROACOUSTIC WAVES
    ELECTROMAGNETIC INTERACTIONS
    ∞INTERACTIONS
    INTERMODULATION
    MODULATION
    ORTHOGONAL MULTIPLEXING THEORY
    PLASMA INTERACTIONS
    PROPAGATION MODES
    SCATTERING
    SHOCK WAVE LUMINESCENCE
    SHOCK WAVE PROFILES
    SHOCK WAVE PROPAGATION

WAVE MOTION
USE WAVES

WAVE OSCILLATORS
USE OSCILLATORS

WAVE PACKETS
RT   LONGITUDINAL WAVES
    PACKETS (COMMUNICATION)
    PLASMA WAVES
    QUANTUM MECHANICS
    TRANSVERSE WAVES

WAVE PROPAGATION
UF   KIRCHHOFF-HUYGENS PRINCIPLE
GS   TRANSMISSION
    . **WAVE PROPAGATION**
    . . DIFFRACTION PROPAGATION
    . . GROUND WAVE PROPAGATION
    . . IONOSPHERIC PROPAGATION
    . . . IONOSPHERIC F-SCATTER
        PROPAGATION
    . . LIGHT SCATTERING
    . . . HALOS
    . . SCATTER PROPAGATION
    . . . IONOSPHERIC F-SCATTER
        PROPAGATION
    . . SHOCK WAVE PROPAGATION
    . . TRANSEQUATORIAL PROPAGATION
RT   ACOUSTIC ATTENUATION
    ACOUSTIC MICROSCOPES
    ATMOSPHERIC ATTENUATION
    ATTENUATION
    AUTOMATIC PICTURE TRANSMISSION
    BEAM WAVEGUIDES
    ∞COHERENCE
    COHERENT RADIATION

WAVE PROPAGATION-*(CONT.)*
    ∞CONDUCTION
    DIFFRACTION
    DOUBLE SIDEBAND TRANSMISSION
    ELECTROMAGNETIC ABSORPTION
    ELECTROMAGNETIC WAVE
      TRANSMISSION
    GEOMETRICAL ACOUSTICS
    GROUP VELOCITY
    HUYGENS PRINCIPLE
    HYDRAULIC ANALOGIES
    ION ACOUSTIC WAVES
    LAME WAVE EQUATIONS
    LIGHT TRANSMISSION
    LOSSY MEDIA
    MICROWAVE ATTENUATION
    MICROWAVE TRANSMISSION
    MULTIPATH TRANSMISSION
    NONADIABATIC THEORY
    PHASE VELOCITY
    PLASMA ACCELERATION
    PLASMAGUIDES
    ∞PROPAGATION
    PROPAGATION MODES
    PROPAGATION VELOCITY
    PULSE DIFFRACTION
    RADAR ATTENUATION
    RADAR TRANSMISSION
    RADIO ATTENUATION
    RADIO TRANSMISSION
    REFRACTION
    SAGNAC EFFECT
    SCREEN EFFECT
    SHOCK FRONTS
    SHOCK WAVE ATTENUATION
    SHORT WAVE RADIO TRANSMISSION
    SINGLE SIDEBAND TRANSMISSION
    SOUND TRANSMISSION
    SQUARE WAVES
    STRESS WAVES
    TELEVISION TRANSMISSION
    WAVEFORMS
    WHITHAM RULE

WAVE RADIATION
USE ELECTROMAGNETIC RADIATION

WAVE REFLECTION
GS   REFLECTION
    . **WAVE REFLECTION**
    . . MACH REFLECTION
RT   GROUND EFFECT (COMMUNICATIONS)
    REFLECTED WAVES
    SIGNAL REFLECTION
    SPREAD REFLECTION

WAVE RESISTANCE
RT   BLAST LOADS
    EROSION
    IMPACT STRENGTH
    ∞RESISTANCE
    STRUCTURAL STABILITY

WAVE SCATTERING
GS   SCATTERING
    . **WAVE SCATTERING**
    . . ACOUSTIC SCATTERING
    . . . REVERBERATION
    . . ATMOSPHERIC SCATTERING
    . . . TROPOSPHERIC SCATTERING
    . . ELECTROMAGNETIC SCATTERING
    . . . IONOSPHERIC F-SCATTER
        PROPAGATION
    . . . LIGHT SCATTERING
    . . . . HALOS
    . . . MICROWAVE SCATTERING
    . . . MIE SCATTERING
    . . . . RAYLEIGH SCATTERING
    . . . RAMAN SPECTRA
    . . . THOMSON SCATTERING
    . . . X RAY SCATTERING
RT   FADDEEV EQUATIONS
    MAGNETIC DISPERSION
    RECIPROCITY THEOREM
    SCATTERING AMPLITUDE
    SCATTEROMETERS
    SHOCK FRONTS

WAVEFORMS
GS   **WAVEFORMS**
    . PULSE AMPLITUDE
    . PULSE DURATION
    . SAWTOOTH WAVEFORMS
    . SQUARE WAVES
RT   FORM FACTORS

WAVEFORMS-*(CONT.)*
        SPEECH BASEBAND COMPRESSION
        TIME FUNCTIONS
        WAVE PROPAGATION

WAVEGUIDE ANTENNAS
GS   ANTENNAS
        . **WAVEGUIDE ANTENNAS**
        . . HORN ANTENNAS
RT   LENS ANTENNAS
        MICROWAVE ANTENNAS
        MONOPULSE ANTENNAS
        RADANT
        SLOT ANTENNAS
        YAGI ANTENNAS

WAVEGUIDE FILTERS
GS   ELECTROMAGNETIC WAVE FILTERS
        . ELECTRIC FILTERS
        . . **WAVEGUIDE FILTERS**
RT   BANDSTOP FILTERS
        MICROWAVE FILTERS
        RADAR FILTERS
        WAVEGUIDES

WAVEGUIDE LASERS
GS   STIMULATED EMISSION DEVICES
        . LASERS
        . . **WAVEGUIDE LASERS**
RT   CARBON DIOXIDE LASERS
        GALLIUM ARSENIDE LASERS
        HETEROJUNCTION DEVICES
        INFRARED LASERS
        LASER MODES
        LASER OUTPUTS
        OPTICAL WAVEGUIDES
        PULSED LASERS
        SEMICONDUCTOR LASERS
        TUBE LASERS

WAVEGUIDE TUNERS
GS   TUNERS
        . **WAVEGUIDE TUNERS**
RT   IMPEDANCE MATCHING
        MODE TRANSFORMERS
        YTTRIUM-IRON GARNET

WAVEGUIDE WINDOWS
RT   IMPEDANCE MATCHING
        IRISES (MECHANICAL APERTURES)
        ∞WINDOWS

WAVEGUIDES
GS   TRANSMISSION LINES
        . COMMUNICATION CABLES
        . . **WAVEGUIDES**
        . . . BEAM WAVEGUIDES
        . . . CIRCULAR WAVEGUIDES
        . . . OPTICAL WAVEGUIDES
        . . . PLASMAGUIDES
        . . . RECTANGULAR WAVEGUIDES
RT   ANTENNA FEEDS
        COAXIAL CABLES
        CROSSED FIELDS
        ELECTROMAGNETIC SURFACE WAVES
        GYRATORS
        IRISES (MECHANICAL APERTURES)
        MICROWAVE SWITCHING
        PARALLEL PLATES
        PROPAGATION MODES
        SMITH CHART
        WAVEGUIDE FILTERS

WAVELENGTH DIVISION MULTIPLEXING
GS   TRANSMISSION
        . MULTIPLEXING
        . . **WAVELENGTH DIVISION
            MULTIPLEXING**
RT   CODE DIVISION MULTIPLEXING
        DEMULTIPLEXING
        FREQUENCY DIVISION MULTIPLEXING
        ORTHOGONAL MULTIPLEXING THEORY
        TIME DIVISION MULTIPLEXING

WAVELENGTHS
GS   **WAVELENGTHS**
        . DE BROGLIE WAVELENGTHS
RT   ANTINODES
        HARMONICS
        INFRARED RADIATION
        LASER MODES
        LASER OUTPUTS
        LONGITUDINAL WAVES
        MASER OUTPUTS
        MILLIMETER WAVES

WAVELENGTHS-*(CONT.)*
        NODES (STANDING WAVES)
        SPECTRAL EMISSION
        STANDING WAVES
        STOKES LAW OF RADIATION
        SUBMILLIMETER WAVES

∞ **WAVES**
SN   *(USE OF A MORE SPECIFIC TERM IS
        RECOMMENDED--CONSULT THE TERMS
        LISTED BELOW)*
UF   CRESTS
        WAVE MOTION
RT   BAROCLINIC WAVES
        BREAKWATERS
        CNOIDAL WAVES
        CORRUGATING
        CYLINDRICAL WAVES
        DETONATION WAVES
        DILATATIONAL WAVES
        EIKONAL EQUATION
        ELASTIC WAVES
        ELECTROACOUSTIC WAVES
        ELECTROMAGNETIC RADIATION
        ELECTROMAGNETIC SURFACE WAVES
        FRONTAL WAVES
        GRAVITATIONAL WAVES
        INTERNAL WAVES
        IONIC WAVES
        KILOMETRIC WAVES
        LITTORAL TRANSPORT
        LONGITUDINAL WAVES
        NODES (STANDING WAVES)
        PLANE WAVES
        PLANETARY WAVES
        REFRACTED WAVES
        SEISMIC WAVES
        SHOCK WAVES
        SINE WAVES
        SOUND WAVES
        SPHERICAL WAVES
        SQUARE WAVES
        STANDING WAVES
        STRESS WAVES
        SURFACE WAVES
        SURGES
        TIDAL WAVES
        TRANSVERSE WAVES
        TROPOSPHERIC WAVES
        VIBRATION
        WATER WAVES
        WATERWAVE ENERGY
        WAVE AMPLIFICATION
        WAVE EXCITATION
        WAVE FRONTS

WAXES
GS   **WAXES**
        . CERESIN
RT   ALKANES
        COATINGS
        CRUDE OIL
        FINISHES
        PHASE CHANGE MATERIALS

WEAK ENERGY INTERACTIONS
GS   DECAY
        . **WEAK ENERGY INTERACTIONS**
        P⌐   ⌐E INTERACTIONS
        .   ⌐ **ENERGY INTERACTIONS**
        . . WEAK INTERACTIONS (FIELD
            THEORY)
RT   BETA PARTICLES
        GRAVITINOS
        ∞INTERACTIONS
        PARTICLE THEORY

WEAK INTERACTIONS (FIELD THEORY)
UF   BETA INTERACTIONS
GS   FIELD THEORY (PHYSICS)
        . **WEAK INTERACTIONS (FIELD THEORY)**
        PARTICLE INTERACTIONS
        . NUCLEAR INTERACTIONS
        . . **WEAK INTERACTIONS (FIELD
              THEORY)**
        . WEAK ENERGY INTERACTIONS
        . . **WEAK INTERACTIONS (FIELD
              THEORY)**
RT   ∞INTERACTIONS
        STRONG INTERACTIONS (FIELD
          THEORY)
        ∞THEORIES

WEAPON SYSTEM MANAGEMENT
GS   MANAGEMENT

WEAPON SYSTEM MANAGEMENT-*(CONT.)*
        . **WEAPON SYSTEM MANAGEMENT**
RT   PROJECT MANAGEMENT
        ∞SYSTEMS

WEAPON SYSTEM 107A-1
GS   WEAPON SYSTEMS
        . **WEAPON SYSTEM 107A-1**

WEAPON SYSTEM 107A-2
GS   WEAPON SYSTEMS
        . **WEAPON SYSTEM 107A-2**

WEAPON SYSTEM 133A
GS   WEAPON SYSTEMS
        . **WEAPON SYSTEM 133A**

WEAPON SYSTEM 133B
GS   WEAPON SYSTEMS
        . **WEAPON SYSTEM 133B**

WEAPON SYSTEM 315A
GS   WEAPON SYSTEMS
        . **WEAPON SYSTEM 315A**

WEAPON SYSTEMS
GS   **WEAPON SYSTEMS**
        . GROUND OPERATIONAL SUPPORT
          SYSTEM
        . LASER WEAPONS
        . MISSILE SYSTEMS
        . . NIKE X SYSTEMS
        . . SAFEGUARD SYSTEM
        . SENTINEL SYSTEM
        . SUCCESS PROJECT
        . TYPHON WEAPON SYSTEM
        . WEAPON SYSTEM 107A-1
        . WEAPON SYSTEM 107A-2
        . WEAPON SYSTEM 133A
        . WEAPON SYSTEM 133B
        . WEAPON SYSTEM 315A
RT   AIR TO SURFACE MISSILES
        ANTISHIP MISSILES
        FIRE CONTROL
        HARPOON MISSILE
        ∞MILITARY AIRCRAFT
        MILITARY SPACECRAFT
        MISSILE LAUNCHERS
        MISSILES
        MOBILE MISSILE LAUNCHERS
        NUCLEAR WEAPONS
        ORDNANCE
        PANAVIA MILITARY AIRCRAFT
        ∞ROCKETS
        SPACE WEAPONS
        ∞SYSTEMS
        WEAPONS
        WEAPONS DEVELOPMENT

WEAPONS
GS   **WEAPONS**
        . ARTILLERY
        . . HOWITZERS
        . . PRECISION GUIDED PROJECTILES
        . FUSION WEAPONS
        . GUNS (ORDNANCE)
        . . HOWITZERS
        . . PRECISION GUIDED PROJECTILES
        . . RIFLES
        . MINES (ORDNANCE)
        . NUCLEAR WEAPONS
        . . FISSION WEAPONS
        . SPACE WEAPONS
        . . LASER WEAPONS
        . WARHEADS
        . . NUCLEAR WARHEADS
        . . PRECISION GUIDED PROJECTILES
RT   AMMUNITION
        ANTIQUITIES
        ANTISHIP WARFARE
        ARMED FORCES (FOREIGN)
        ARMED FORCES (UNITED STATES)
        ∞BALLISTIC VEHICLES
        DISARMAMENT
        FIRE CONTROL
        GUNNERY TRAINING
        MILITARY TECHNOLOGY
        MISSILES
        ORDNANCE
        PATRIOT MISSILE
        PROJECTILES
        SHAPED CHARGES
        SHRAPNEL
        TANKS (COMBAT VEHICLES)
        TOMAHAWK MISSILES

WEAPONS-(CONT.)
    TORPEDOES
    WEAPON SYSTEMS
    WEAPONS DELIVERY
    WING-FUSELAGE STORES

WEAPONS DELIVERY
GS   DELIVERY
    . **WEAPONS DELIVERY**
RT   AIR DEFENSE
    ∞AIRCRAFT
    DEFENSE PROGRAM
    MILITARY TECHNOLOGY
    MISSILE DEFENSE
    NUCLEAR WEAPONS
    ∞ROCKETS
    SPACE WEAPONS
    WEAPONS

WEAPONS DEVELOPMENT
GS   PRODUCT DEVELOPMENT
    . **WEAPONS DEVELOPMENT**
RT   RESEARCH AND DEVELOPMENT
    WEAPON SYSTEMS

WEAPONS INDUSTRY
GS   INDUSTRIES
    . DEFENSE INDUSTRY
    . . **WEAPONS INDUSTRY**
RT   ARMED FORCES (UNITED STATES)
    MILITARY TECHNOLOGY

WEAR
RT   ABRASION
    CHIPPING
    CORROSION
    DAMAGE
    DEPRECIATION
    DETERIORATION
    DURABILITY
    EROSION
    FAILURE
    FLAKING
    FRETTING CORROSION
    FRICTION
    GRINDING (MATERIAL REMOVAL)
    HARDNESS
    REMOVAL
    SCORING
    SLIDING FRICTION
    SPALLING
    SURFACE FINISHING
    ∞SURFACES
    SYSTEM FAILURES
    TRIBOLOGY

WEAR INHIBITORS
GS   INHIBITORS
    . **WEAR INHIBITORS**
† RT   RETARDANTS

WEAR TESTS
RT   CUMULATIVE DAMAGE
    DESTRUCTIVE TESTS
    EROSION
    FERROGRAPHY
    FRETTING
    FRICTION
    HARDNESS TESTS
    ∞MATERIALS TESTS
    PATCH TESTS
    QUALITY CONTROL
    SPALLING
    STATIC TESTS
    ∞TESTS

WEATHER
UF   WEATHER CONDITIONS
GS   **WEATHER**
    . COLD WEATHER
    . HOT WEATHER
RT   AIRCRAFT ACCIDENTS
    AIRCRAFT HAZARDS
    AIRCRAFT SAFETY
    ALPINE METEOROLOGY
    ANNUAL VARIATIONS
    ANVIL CLOUDS
    ATMOSPHERIC & OCEANOGRAPHIC
      INFORM SYS
    ATMOSPHERIC PRESSURE
    ATMOSPHERIC TEMPERATURE
    CAP CLOUDS
    CIRROCUMULUS CLOUDS
    CIRROSTRATUS CLOUDS
    CLIMATE

WEATHER-(CONT.)
    CLIMATOLOGY
    CLOUDS (METEOROLOGY)
    FLIGHT HAZARDS
    FLIGHT PLANS
    GLOBAL ATMOSPHERIC RESEARCH
      PROGRAM
    LONG TERM EFFECTS
    METEOROLOGICAL PARAMETERS
    METEOROLOGY
    METEOSAT SATELLITE
    NAVIGATION AIDS
    PRECIPITATION (METEOROLOGY)
    RUNWAY CONDITIONS
    SEASONS
    SOLAR COMPASSES
    SOLAR TERRESTRIAL INTERACTIONS
    WIND (METEOROLOGY)

WEATHER CHARTS
USE   METEOROLOGICAL CHARTS

WEATHER CONDITIONS
USE   WEATHER

WEATHER CONTROL
USE   WEATHER MODIFICATION

WEATHER DATA RECORDERS
GS   MEASURING INSTRUMENTS
    . METEOROLOGICAL INSTRUMENTS
    . . **WEATHER DATA RECORDERS**
    RECORDING INSTRUMENTS
    . **WEATHER DATA RECORDERS**
RT   AUTOMATIC WEATHER STATIONS
    ∞DATA
    TELEMETRY

WEATHER FORECASTING
GS   FORECASTING
    . **WEATHER FORECASTING**
    . . LONG RANGE WEATHER
        FORECASTING
    . . NOWCASTING
    . . NUMERICAL WEATHER FORECASTING
    . . STATISTICAL WEATHER
        FORECASTING
    METEOROLOGY
    . **WEATHER FORECASTING**
    . . LONG RANGE WEATHER
        FORECASTING
    . . NOWCASTING
    . . NUMERICAL WEATHER FORECASTING
    . . STATISTICAL WEATHER
        FORECASTING
RT   AIR MASSES
    ATMOSPHERIC MODELS
    CIRRUS SHIELDS
    CLOUD COVER
    COLD FRONTS
    ENVIRONMENTAL MONITORING
    FLIGHT CONDITIONS
    FLOOD PREDICTIONS
    GARP ATLANTIC TROPICAL EXPERIMENT
    HUMIDITY
    METEOROLOGICAL BALLOONS
    METEOROLOGICAL FLIGHT
    METEOROLOGICAL RADAR
    METEOROLOGICAL SATELLITES
    METEOROLOGICAL SERVICES
    NEPHANALYSIS
    PRECIPITATION (METEOROLOGY)
    SNOWSTORMS
    STORMS
    STORMS (METEOROLOGY)
    SYNOPTIC METEOROLOGY
    WARM FRONTS
    WIND (METEOROLOGY)

WEATHER FRONTS
USE   FRONTS (METEOROLOGY)

WEATHER MAPS
USE   METEOROLOGICAL CHARTS

WEATHER MODIFICATION
UF   WEATHER CONTROL
GS   **WEATHER MODIFICATION**
    . CLOUD DISPERSAL
    . CLOUD SEEDING
    . FOG DISPERSAL
    . LIGHTNING SUPPRESSION
    . RAINMAKING
    . STORM ENHANCEMENT
    . STORM SUPPRESSION

WEATHER MODIFICATION-(CONT.)
RT   ARTIFICIAL CLOUDS
    CLOUD PHYSICS
    ∞CONTROL
    ENVIRONMENTAL CONTROL
    HEAT ISLANDS
    SNOWSTORMS

WEATHER RADAR
USE   METEOROLOGICAL RADAR

WEATHER RECONNAISSANCE AIRCRAFT
GS   RECONNAISSANCE AIRCRAFT
    . **WEATHER RECONNAISSANCE**
        AIRCRAFT
RT   ∞AIRCRAFT
    GLOBAL ATMOSPHERIC RESEARCH
      PROGRAM
    METEOROLOGICAL INSTRUMENTS
    OBSERVATION AIRCRAFT
    RB-50 AIRCRAFT

WEATHER STATIONS
UF   METEOROLOGICAL STATIONS
GS   STATIONS
    . **WEATHER STATIONS**
RT   GROUND STATIONS
    INSTRUMENT PACKAGES
    INTEGRATED GLOBAL OCEAN STATION
      SYSTEMS
    METEOROLOGICAL INSTRUMENTS
    METEOROLOGICAL SATELLITES
    METEOROLOGICAL SERVICES
    OCEAN DATA ACQUISITIONS SYSTEMS

WEATHERING
RT   CORROSION
    CORROSION TESTS
    CURING
    DAMAGE
    DEGRADATION
    DETERIORATION
    EARTH ATMOSPHERE
    EROSION
    EXPOSURE
    MECHANICAL PROPERTIES
    RUSTING
    SOIL EROSION

WEATHERPROOFING
RT   COATINGS
    COLD WEATHER
    CORROSION PREVENTION
    MOISTURE RESISTANCE
    PACKAGING
    PRESERVING
    WATERPROOFING

WEAVING
RT   FABRICS
    SEWING

WEBBING
RT   FABRICS
    MESH
    ∞WEBS
    WEBS (SHEETS)

WEBER TEST
GS   PHYSIOLOGICAL TESTS
    . **WEBER TEST**
RT   AUDITORY PERCEPTION
    BINAURAL HEARING

WEBER-FECHNER LAW
GS   LAWS
    . **WEBER-FECHNER LAW**

∞ **WEBS**
SN   *(USE OF A MORE SPECIFIC TERM IS*
      RECOMMENDED-CONSULT THE TERMS
      LISTED BELOW)
RT   MEMBRANES
    MESH
    WEBBING
    WEBS (SHEETS)
    WEBS (SUPPORTS)

WEBS (MEMBRANES)
USE   MEMBRANES

WEBS (SHEETS)
SN   *(EXCLUDES POLYMERIC FILMS AND*
      STRUCTURAL REINFORCEMENTS)

WEBS (SHEETS)-*(CONT.)*
- RT DIAPHRAGMS (MECHANICS)
  - ELASTIC SHEETS
  - FABRICS
  - ∞FILMS
  - MEMBRANES
  - PAPER (MATERIAL)
  - PAPERS
  - ∞ROVINGS
  - ∞SHEETS
  - WEBBING
  - WEBS (SUPPORTS)

WEBS (SUPPORTS)
- GS **WEBS (SUPPORTS)**
  - . GIRDER WEBS
- RT DIAPHRAGMS (MECHANICS)
  - ELASTIC SHEETS
  - MEMBRANE STRUCTURES
  - MEMBRANES
  - RIBS (SUPPORTS)
  - SKIN (STRUCTURAL MEMBER)
  - STIFFENING
  - WEBS (SHEETS)

WEDGE FLOW
- GS FLUID FLOW
  - . **WEDGE FLOW**
- RT BLASIUS FLOW
  - CONICAL FLOW
  - FALKNER-SKAN EQUATION
  - FLOW GEOMETRY
  - LAMINAR FLOW
  - SHOCK WAVES
  - SUPERSONIC FLOW
  - THREE DIMENSIONAL FLOW
  - TWO DIMENSIONAL FLOW
  - VISCOUS FLOW

WEDGES
- RT AERODYNAMIC CONFIGURATIONS
  - AIRFOIL PROFILES
  - AIRFOILS

WEIBEL INSTABILITY
- GS DYNAMIC CHARACTERISTICS
  - . DYNAMIC STABILITY
  - . . MOTION STABILITY
  - . . . FLOW STABILITY
  - . . . . MAGNETOHYDRODYNAMIC
      STABILITY
  - . . . . . **WEIBEL INSTABILITY**
  - . FLOW CHARACTERISTICS
  - . . FLOW STABILITY
  - . . . MAGNETOHYDRODYNAMIC
      STABILITY
  - . . . . **WEIBEL INSTABILITY**
- RT PLASMA INTERACTIONS

WEIBULL DENSITY FUNCTIONS
- GS FUNCTIONS (MATHEMATICS)
  - . PROBABILITY DENSITY FUNCTIONS
  - . . **WEIBULL DENSITY FUNCTIONS**
  - STATISTICAL ANALYSIS
  - . PROBABILITY DENSITY FUNCTIONS
  - . . **WEIBULL DENSITY FUNCTIONS**
- RT EXPONENTIAL FUNCTIONS
  - FATIGUE TESTS
  - SAMPLING

WEIERSTRASS FUNCTIONS
- GS ANALYSIS (MATHEMATICS)
  - . REAL VARIABLES
  - . . **WEIERSTRASS FUNCTIONS**
- RT ELLIPTIC FUNCTIONS
  - JACOBI INTEGRAL

∞ **WEIGHT**
- SN *(USE OF A MORE SPECIFIC TERM IS*
  RECOMMENDED--CONSULT THE TERMS
  LISTED BELOW)
- RT ATOMIC WEIGHTS
  - BIOMASS
  - COEFFICIENTS
  - PAYLOADS
  - WEIGHT (MASS)

WEIGHT (MASS)
- UF WEIGHT FACTORS
- GS **WEIGHT (MASS)**
  - . ATOMIC WEIGHTS
  - . BIOMASS
  - . BODY WEIGHT
  - . ORGAN WEIGHT
  - . STRUCTURAL WEIGHT

WEIGHT (MASS)-*(CONT.)*
- RT CENTER OF MASS
  - ∞FORCE
  - GRAVITATION
  - LOADS (FORCES)
  - LOW MOLECULAR WEIGHTS
  - MASCONS
  - MASS
  - MOLECULAR WEIGHT
  - PAYLOADS
  - PRESSURE
  - VOLUME
  - ∞WEIGHT

WEIGHT ANALYSIS
- RT ∞ANALYZING
  - NEW MOONS PROJECT
  - PREFLIGHT ANALYSIS
  - STRUCTURAL WEIGHT
  - SYSTEMS ANALYSIS

WEIGHT FACTORS
- USE WEIGHT (MASS)

WEIGHT INDICATORS
- UF WIND TUNNEL BALANCES
- GS MEASURING INSTRUMENTS
  - . INDICATING INSTRUMENTS
  - . . **WEIGHT INDICATORS**
  - . . . MICROBALANCES
  - . . . STRAIN GAGE BALANCES
  - . . . THERMOBALANCES
- RT BALANCE
  - MECHANICAL MEASUREMENT
  - PRESSURE GAGES
  - PRESSURE MEASUREMENT
  - ∞SCALE
  - STRAIN GAGES
  - TENSOMETERS

WEIGHT MEASUREMENT
- UF MICROWEIGHING
- RT DENSITY (MASS/VOLUME)
  - HYDROMETERS
  - ∞MEASUREMENT

WEIGHT REDUCTION
- RT AIRCRAFT DESIGN
  - SPACECRAFT DESIGN
  - STRUCTURAL DESIGN
  - STRUCTURAL WEIGHT
  - SYSTEMS ANALYSIS

WEIGHTING FUNCTIONS
- GS ANALYSIS (MATHEMATICS)
  - . REAL VARIABLES
  - . . MEASURE AND INTEGRATION
  - . . . **WEIGHTING FUNCTIONS**
  - FUNCTIONS (MATHEMATICS)
  - . **WEIGHTING FUNCTIONS**
- RT STATISTICAL MECHANICS

WEIGHTLESS FLUIDS
- RT ∞FLUIDS
  - VISCOUS FLUIDS

WEIGHTLESSNESS
- UF ZERO GRAVITY
- RT AEROSPACE MEDICINE
  - ARTIFICIAL GRAVITY
  - ASTRONAUT PERFORMANCE
  - ∞ASTRONAUTICS
  - BIOPROCESSING
  - BLACKOUT PREVENTION
  - BODY WEIGHT
  - BONE DEMINERALIZATION
  - CONTAINERLESS MELTS
  - DISORIENTATION
  - DROP TOWERS
  - ENVIRONMENTS
  - EXTRAVEHICULAR ACTIVITY
  - FLIGHT STRESS (BIOLOGY)
  - FREE FALL
  - GRAVITATION
  - GRAVITATIONAL EFFECTS
  - INTRAVEHICULAR ACTIVITY
  - LIFE SUPPORT SYSTEMS
  - LOW WEIGHT
  - LOWER BODY NEGATIVE PRESSURE
  - PARABOLIC FLIGHT
  - SPACE ADAPTATION SYNDROME
  - SPACE FLIGHT STRESS
  - SPACE MANUFACTURING
  - SPACE PROCESSING APPLICATIONS
    ROCKET

WEIGHTLESSNESS-*(CONT.)*
- SPACEBORNE EXPERIMENTS
- SPACECRAFT ENVIRONMENTS
- SUBORBITAL FLIGHT

WEIGHTLESSNESS SIMULATION
- GS SIMULATION
  - . ENVIRONMENT SIMULATION
  - . . SPACE ENVIRONMENT SIMULATION
  - . . . **WEIGHTLESSNESS SIMULATION**
- RT FLIGHT SIMULATION
  - LANGLEY COMPLEX COORDINATOR
  - PARABOLIC FLIGHT
  - SUBMERGING

WELD STRENGTH
- GS MECHANICAL PROPERTIES
  - . **WELD STRENGTH**
- RT PATCH TESTS
  - ∞STRENGTH
  - WELDABILITY
  - WELDED JOINTS

WELD TESTS
- RT FATIGUE TESTS
  - ∞TESTS
  - WELDED JOINTS

WELDABILITY
- RT BRITTLENESS
  - DUCTILITY
  - WELD STRENGTH

WELDED JOINTS
- GS JOINTS (JUNCTIONS)
  - . METAL JOINTS
  - . . **WELDED JOINTS**
  - . . . SPOT WELDS
- RT BEADS
  - BUTT JOINTS
  - LAP JOINTS
  - RIVETED JOINTS
  - WELD STRENGTH
  - WELD TESTS
  - WELDING

WELDED STRUCTURES
- GS **WELDED STRUCTURES**
  - . STEEL STRUCTURES
- RT RIGID STRUCTURES
  - ∞STRUCTURES

WELDING
- GS **WELDING**
  - . COLD WELDING
  - . ELECTRIC WELDING
  - . . ARC WELDING
  - . . . GAS TUNGSTEN ARC WELDING
  - . . . PLASMA ARC WELDING
  - . . ELECTRON BEAM WELDING
  - . . ELECTROSLAG WELDING
  - . FLASH WELDING
  - . FRICTION WELDING
  - . LASER WELDING
  - . . FUSION WELDING
  - . . . GAS WELDING
  - . . . . BRAZING
  - . . . . . LOW TEMPERATURE BRAZING
  - . PRESSURE WELDING
  - . . DIFFUSION WELDING
  - . . EXPLOSIVE WELDING
  - . . ULTRASONIC WELDING
- RT BACKUPS
  - BEADS
  - BONDING
  - CONSTRUCTION
  - FILLETS
  - FLAME PLATING
  - FLUXES
  - FUSIBILITY
  - ∞JOINING
  - METAL BONDING
  - METAL-METAL BONDING
  - SEALING
  - SOLDERING
  - TORCHES
  - WELDED JOINTS

WELDING MACHINES
- RT ∞ELECTRIC EQUIPMENT
  - ELECTRIC WELDING
  - ∞MACHINERY
  - TORCHES

WELLS
RT   AQUIFERS
     DRILLING
     GROUND WATER
     LIMNOLOGY
     OASES
     SPRINGS (WATER)
     SQUARE WELLS
     STRATIGRAPHY

WENTZEL-KRAMER-BRILLOUIN METHOD
UF   WKB APPROXIMATION
RT   DE BROGLIE WAVELENGTHS
     ∞METHODOLOGY
     PERTURBATION THEORY
     PLANCKS CONSTANT
     SCHROEDINGER EQUATION

WESER AIRCRAFT
RT   ∞AIRCRAFT
     HELICOPTERS
     ROTARY WING AIRCRAFT
     V/STOL AIRCRAFT

WEST COMET
GS   CELESTIAL BODIES
     . COMETS
     . . WEST COMET
RT   SOLAR SYSTEM

WEST FORD PROJECT
GS   PROGRAMS
     . PROJECTS
     . . WEST FORD PROJECT

WEST GERMANY
UF   FEDERAL REPUBLIC OF GERMANY
GS   NATIONS
     . WEST GERMANY
RT   ALPS MOUNTAINS (EUROPE)
     AZUR SATELLITE
     CENTRAL EUROPE
     EAST GERMANY
     EUROPE
     ∞GERMANY

WEST INDIES
GS   LANDFORMS
     . ISLANDS
     . . WEST INDIES
     . . . BAHAMAS
     . . . BARBADOS
     . . . CUBA
     . . . DOMINICA
     . . . GUADELOUPE
     . . . HAITI
     . . . JAMAICA
     . . . LESSER ANTILLES
     . . . MARTINIQUE
     . . . PUERTO RICO
     . . . TRINIDAD AND TOBAGO
     . . . VIRGIN ISLANDS
RT   ATLANTIC OCEAN
     CARIBBEAN REGION

WEST PAKISTAN
USE BANGLADESH

WEST VIRGINIA
GS   NATIONS
     . UNITED STATES
     . . WEST VIRGINIA
RT   ALLEGHENY PLATEAU (US)
     OHIO RIVER (US)
     POTOMAC RIVER VALLEY (MD-VA-WV)

WESTAR SATELLITES
GS   SATELLITES
     . ARTIFICIAL SATELLITES
     . . COMMUNICATION SATELLITES
     . . . WESTAR SATELLITES
RT   POINT TO POINT COMMUNICATION
     TELECOMMUNICATION
     TELEGRAPH SYSTEMS

WESTERN HEMISPHERE
RT   EARTH (PLANET)
     EASTERN HEMISPHERE
     GEOGRAPHY

WESTLAND AIRCRAFT
GS   WESTLAND AIRCRAFT
     . P-531 HELICOPTER
     . WESTLAND WHIRLWIND HELICOPTER

WESTLAND AIRCRAFT-(CONT.)
RT   ∞AIRCRAFT
     HELICOPTERS
     ROTARY WING AIRCRAFT
     V/STOL AIRCRAFT

WESTLAND GROUND EFFECT MACHINES
UF   SR-N2 GROUND EFFECT MACHINE
     SR-N3 GROUND EFFECT MACHINE
     SR-N5 GROUND EFFECT MACHINE
     WESTLAND SR-N2 GROUND EFFECT
        MACHINE
     WESTLAND SR-N2 HOVERCRAFT
     WESTLAND SR-N3 GROUND EFFECT
        MACHINE
     WESTLAND SR-N3 HOVERCRAFT
     WESTLAND SR-N5 GROUND EFFECT
        MACHINE
GS   GROUND EFFECT MACHINES
     . WESTLAND GROUND EFFECT
        MACHINES
RT   ∞AIRCRAFT

WESTLAND MK-10 HELICOPTER
USE WESTLAND WHIRLWIND HELICOPTER

WESTLAND P-531 HELICOPTER
USE P-531 HELICOPTER

WESTLAND SR-N2 GROUND EFFECT MACHINE
USE WESTLAND GROUND EFFECT MACHINES

WESTLAND SR-N2 HOVERCRAFT
USE WESTLAND GROUND EFFECT MACHINES

WESTLAND SR-N3 GROUND EFFECT MACHINE
USE WESTLAND GROUND EFFECT MACHINES

WESTLAND SR-N3 HOVERCRAFT
USE WESTLAND GROUND EFFECT MACHINES

WESTLAND SR-N5 GROUND EFFECT MACHINE
USE WESTLAND GROUND EFFECT MACHINES

WESTLAND WHIRLWIND HELICOPTER
UF   WESTLAND MK-10 HELICOPTER
     WHIRLWIND MK-10 HELICOPTER
GS   UTILITY AIRCRAFT
     . WESTLAND WHIRLWIND HELICOPTER
     V/STOL AIRCRAFT
     . ROTARY WING AIRCRAFT
     . . HELICOPTERS
     . . . WESTLAND WHIRLWIND
        HELICOPTER
     WESTLAND AIRCRAFT
     . WESTLAND WHIRLWIND HELICOPTER
RT   ∞AIRCRAFT

WET CELLS
GS   ELECTROCHEMICAL CELLS
     . ELECTRIC BATTERIES
     . . WET CELLS
RT   ∞ELECTRIC CELLS
     ELECTROLYTES
     FUEL CELLS
     NONAQUEOUS ELECTROLYTES
     PRIMARY BATTERIES

WET SPINNING
RT   EXTRUDING
     FIBERS
     ∞FILAMENTS
     ∞PROCESSING
     SYNTHETIC FIBERS
     TEXTILES
     WETTING

WETLANDS
GS   LAND
     . WETLANDS
RT   COASTAL CURRENTS
     COASTAL ECOLOGY
     COASTAL PLAINS
     COASTAL WATER
     ENVIRONMENT EFFECTS
     FISHERIES
     MARINE BIOLOGY
     MARINE ENVIRONMENTS
     MARINE RESOURCES
     MARSHLANDS
     NEARSHORE WATER
     OCEANOGRAPHY
     OIL POLLUTION
     SEA GRASSES

WETLANDS-(CONT.)
     SHORELINES
     TIDES
     WATER RESOURCES
     WATERFOWL
     WILDLIFE

WETNESS
USE MOISTURE CONTENT

WETTABILITY
RT   ADHESION
     ADHESION TESTS
     FORMATIONS
     HYGROSCOPICITY
     PERMEABILITY
     POROSITY
     SURFACE PROPERTIES
     WETTING

WETTING
RT   COOLING
     DIPPING
     FOAMING
     INTERFACIAL TENSION
     ∞SATURATION
     ∞SOAKING
     SOAPS
     SPRAYING
     SPRINKLING
     SUBMERGING
     WET SPINNING
     WETTABILITY

WHALES
GS   ANIMALS
     . VERTEBRATES
     . . MAMMALS
     . . . WHALES
RT   MARINE MAMMALS

WHARVES
UF   PIERS
RT   CARGO SHIPS
     DAMS
     EARTH RESOURCES
     FREIGHTERS
     HARBORS
     MARINE TECHNOLOGY
     MATERIALS HANDLING
     ∞PORTS
     RIVERS
     SHIP TERMINALS
     TANKER SHIPS
     TERMINAL FACILITIES
     WATER

WHEAT
GS   FARM CROPS
     . GRAINS (FOOD)
     . . WHEAT
RT   CROP GROWTH
     CROP VIGOR
     ∞CROPS
     RICE

WHEATSTONE BRIDGES
GS   CIRCUITS
     . ELECTRIC BRIDGES
     . . WIRE BRIDGE CIRCUITS
     . . . WHEATSTONE BRIDGES
RT   MEASURING INSTRUMENTS
     OHMMETERS

WHEEL BRAKES
GS   BRAKES (FOR ARRESTING MOTION)
     . WHEEL BRAKES
RT   AIRCRAFT BRAKES
     AIRCRAFT SAFETY
     ANTISKID DEVICES
     CONTROLLABILITY
     FRICTION
     HYDRAULIC EQUIPMENT
     LANDING GEAR
     TIRES
     VEHICLE WHEELS

WHEELCHAIRS
RT   HANDICAPS
     HUMAN FACTORS ENGINEERING
     LOCOMOTION
     RAMPS (STRUCTURES)

WHEELS
- GS   **WHEELS**
  - . COUNTER-ROTATING WHEELS
  - . FLYWHEELS
  - . REACTION WHEELS
  - . TOROIDAL WHEELS
  - . TURBINE WHEELS
  - . VEHICLE WHEELS
  - . . NOSE WHEELS
  - . WATER WHEELS
- RT   BEARINGS
  - BRAKES (FOR ARRESTING MOTION)
  - GEARS
  - HUBS
  - LANDING GEAR
  - PULLEYS
  - ROLLERS
  - ROTORS
  - SHAFTS (MACHINE ELEMENTS)
  - SPOKES
  - TIRES

WHIP ANTENNAS
- GS   ANTENNAS
  - . OMNIDIRECTIONAL ANTENNAS
  - . . MONOPOLE ANTENNAS
  - . . . **WHIP ANTENNAS**
- RT   RADIO ANTENNAS

WHIPLASH INJURIES
- GS   INJURIES
  - . **WHIPLASH INJURIES**
- RT   BACK INJURIES
  - CRASH INJURIES

WHIRL
- USE   ROTATION

WHIRL INSTABILITY
- USE   ROTARY STABILITY

WHIRL TOWERS
- RT   HELICOPTER DESIGN
  - HOVERING
  - HOVERING STABILITY
  - PARACHUTES
  - ROTARY WINGS
  - ROTOR AERODYNAMICS
  - SPIN TESTS

WHIRLING
- USE   ROTATION

WHIRLING TESTS
- USE   SPIN TESTS

WHIRLWIND MK-10 HELICOPTER
- USE   WESTLAND WHIRLWIND HELICOPTER

WHISKER COMPOSITES
- UF   METAL WHISKER REINFORCEMENT
- GS   COMPOSITE MATERIALS
  - . **WHISKER COMPOSITES**
- RT   EUTECTIC ALLOYS
  - METAL MATRIX COMPOSITES
  - REINFORCING FIBERS

WHISKERS (CRYSTALS)
- GS   CRYSTALS
  - . **WHISKERS (CRYSTALS)**
- RT   DENDRITIC CRYSTALS
  - FIBERS
  - ∞FILAMENTS

WHISTLER RECORDERS
- GS   COMMUNICATION EQUIPMENT
  - . RADIO RECEIVERS
  - . . **WHISTLER RECORDERS**
  - RADIO EQUIPMENT
  - . RADIO RECEIVERS
  - . . **WHISTLER RECORDERS**
  - RECORDING INSTRUMENTS
  - . **WHISTLER RECORDERS**
- RT   SONOGRAMS

WHISTLERS
- GS   ATMOSPHERIC RADIATION
  - . IONOSPHERIC NOISE
  - . . **WHISTLERS**
  - ELECTROMAGNETIC INTERFERENCE
  - . RADIO FREQUENCY INTERFERENCE
  - . . BLACKOUT (PROPAGATION)
  - . . . ELECTROMAGNETIC NOISE
  - . . . . ATMOSPHERICS

WHISTLERS-*(CONT.)*
  - . . . . . **WHISTLERS**
  - . . . . IONOSPHERIC NOISE
  - . . . . . **WHISTLERS**
- RT   DAWN CHORUS
  - ELECTROMAGNETIC FIELDS
  - LIGHTNING
  - MICROWAVES
  - RADIO SIGNALS
  - SONOGRAMS

WHITE BLOOD CELLS
- GS   BODY FLUIDS
  - . BLOOD
  - . . **WHITE BLOOD CELLS**
- RT   LEUKOCYTES

WHITE DWARF STARS
- GS   CELESTIAL BODIES
  - . STARS
  - . . DWARF STARS
  - . . . **WHITE DWARF STARS**
  - . . HOT STARS
  - . . . **WHITE DWARF STARS**
- RT   CATACLYSMIC VARIABLES
  - DWARF NOVAE
  - RED DWARF STARS
  - SUBDWARF STARS
  - SUPERNOVA REMNANTS
  - WOLF-RAYET STARS

WHITE HOLES (ASTRONOMY)
- GS   CELESTIAL BODIES
  - . STARS
  - . . **WHITE HOLES (ASTRONOMY)**
  - GRAVITATIONAL COLLAPSE
  - . **WHITE HOLES (ASTRONOMY)**
- RT   BLACK HOLES (ASTRONOMY)
  - COSMOLOGY
  - ELECTROMAGNETIC RADIATION
  - GRAVITATIONAL LENSES
  - LIGHT EMISSION
  - NAKED SINGULARITIES
  - SUPERNOVA REMNANTS

WHITE LIGHT HOLOGRAPHY
- GS   IMAGERY
  - . HOLOGRAPHY
  - . . **WHITE LIGHT HOLOGRAPHY**
  - PHOTOGRAPHY
  - . HOLOGRAPHY
  - . . **WHITE LIGHT HOLOGRAPHY**
- RT   DATA STORAGE
  - WAVE FRONT RECONSTRUCTION

WHITE NOISE
- UF   SPECTRAL NOISE
- GS   ELECTROMAGNETIC INTERFERENCE
  - . RADIO FREQUENCY INTERFERENCE
  - . . BLACKOUT (PROPAGATION)
  - . . . ELECTROMAGNETIC NOISE
  - . . . . **WHITE NOISE**
  - . . . . . THERMAL NOISE
- RT   ELECTROMAGNETIC NOISE
  - MEASUREMENT
  - JAMMING
  - ∞NOISE
  - NOISE (SOUND)
  - NOISE SPECTRA
  - RANDOM NOISE
  - SIGNAL TO NOISE RATIOS
  - SPECTRAL BANDS

WHITEOUT
- RT   VISIBILITY
  - VISUAL FLIGHT

WHITHAM RULE
- GS   RULES
  - . **WHITHAM RULE**
- RT   SHOCK WAVES
  - WAVE PROPAGATION

WHITTAKER FUNCTIONS
- GS   ANALYSIS (MATHEMATICS)
  - . REAL VARIABLES
  - . . **WHITTAKER FUNCTIONS**
  - FUNCTIONS (MATHEMATICS)
  - . **WHITTAKER FUNCTIONS**
- RT   DIFFERENTIAL EQUATIONS

WICKS
- RT   FUSES (ORDNANCE)

WIDE ANGLE LENSES
- GS   LENSES
  - . **WIDE ANGLE LENSES**
  - OPTICAL EQUIPMENT
  - . **WIDE ANGLE LENSES**
- RT   ALL SKY PHOTOGRAPHY
  - CAMERAS
  - PANORAMIC CAMERAS

WIDEBAND
- USE   BROADBAND

WIDEBAND COMMUNICATION
- GS   TELECOMMUNICATION
  - . **WIDEBAND COMMUNICATION**
- RT   BROADBAND AMPLIFIERS
  - CODE DIVISION MULTIPLE ACCESS
  - MULTIPLE ACCESS
  - POINT TO POINT COMMUNICATION
  - TIME DIVISION MULTIPLE ACCESS
  - TROPOSPHERIC SCATTERING

WIDMANSTATTEN STRUCTURE
- GS   CRYSTAL STRUCTURE
  - . **WIDMANSTATTEN STRUCTURE**
  - MICROSTRUCTURE
  - . **WIDMANSTATTEN STRUCTURE**
- RT   IRON METEORITES
  - METALLOGRAPHY
  - METEORITIC MICROSTRUCTURES
  - ∞PATTERNS

WIDTH
- GS   DIMENSIONS
  - . **WIDTH**
- RT   BANDWIDTH
  - ∞SPAN

WIENER FILTERING
- RT   ELECTRIC FILTERS
  - OPTIMIZATION
  - STATISTICAL ANALYSIS

WIENER HOPF EQUATIONS
- GS   ANALYSIS (MATHEMATICS)
  - . FUNCTIONAL ANALYSIS
  - . . INTEGRAL EQUATIONS
  - . . . **WIENER HOPF EQUATIONS**
- RT   ∞EQUATIONS

WIGGLER MAGNETS
- GS   MAGNETS
  - . **WIGGLER MAGNETS**
  - REFLECTORS
  - . **WIGGLER MAGNETS**
- RT   FREE ELECTRON LASERS
  - LASER PUMPING
  - TUNABLE LASERS

WIGHTMAN THEORY
- USE   FIELD THEORY (PHYSICS)
  - QUANTUM THEORY
  - RELATIVISTIC THEORY

WIGNER COEFFICIENT
- GS   COEFFICIENTS
  - . **WIGNER COEFFICIENT**
- RT   ANGULAR MOMENTUM
  - ∞MECHANICS (PHYSICS)

WILDERNESS
- RT   DESERTS
  - FORESTS
  - LAND MANAGEMENT
  - PLAINS
  - REMOTE REGIONS
  - RURAL AREAS

WILDLIFE
- GS   ANIMALS
  - . **WILDLIFE**
  - . . BABOONS
  - . . BATS
  - . . BIRDS
  - . . CHIMPANZEES
  - . . MONKEYS
- RT   ENDANGERED SPECIES
  - ENVIRONMENT EFFECTS
  - HABITATS
  - WETLANDS

WILDLIFE RADIOLOCATION
- GS   TRACKING (POSITION)
  - . RADIO TRACKING

WILDLIFE RADIOLOCATION-_(CONT.)_
.. **WILDLIFE RADIOLOCATION**
RT ANIMALS
BIOINSTRUMENTATION
BIOTELEMETRY
RADIO TRANSMITTERS
REMOTE SENSORS
SATELLITE INSTRUMENTS
SATELLITE OBSERVATION
TRIANGULATION

WILLISTON BASIN (NORTH AMERICA)
GS LANDFORMS
. STRUCTURAL BASINS
.. **WILLISTON BASIN (NORTH AMERICA)**
RT CANADA
MONTANA
NORTH AMERICA
NORTH DAKOTA

WINCHES
RT CRANES
ELEVATORS (LIFTS)
∞LIFTS
PULLEYS

WIND (METEOROLOGY)
GS **WIND (METEOROLOGY)**
. CIRCUMPOLAR WESTERLIES
. CYCLOGENESIS
. GROUND WIND
. GUSTS
. MONSOONS
. SEA BREEZE
. SQUALLS
. WINDS ALOFT
.. GEOSTROPHIC WIND
.. JET STREAMS (METEOROLOGY)
RT AEOLIAN TONES
AEROLOGY
AIR CURRENTS
AIR POLLUTION
ALPINE METEOROLOGY
ANEMOMETERS
ATMOSPHERIC CIRCULATION
BAROTROPIC FLOW
∞BARRIERS
BLOWING
CLIMATOLOGY
CYCLONES
GRAVITY WAVES
HOT-FILM ANEMOMETERS
JIMSPHERE BALLOONS
MARINE METEOROLOGY
MERIDIONAL FLOW
MESOSCALE PHENOMENA
METEOROLOGY
MIXING HEIGHT
PRESSURE ICE
RIPPLES
STORM DAMAGE
STORMS
STORMS (METEOROLOGY)
THUNDERSTORMS
TIDAL WAVES
TORNADOES
UPWELLING WATER
VERTICAL AIR CURRENTS
WEATHER
WEATHER FORECASTING
WINDMILLS (WINDPOWERED MACHINES)
WINDPOWER UTILIZATION
WINDPOWERED GENERATORS

WIND CIRCULATION
USE ATMOSPHERIC CIRCULATION

WIND DIRECTION
RT ATMOSPHERIC CIRCULATION
GROUND WIND
MERIDIONAL FLOW
SEA BREEZE
SMOKE TRAILS
UPSTREAM
UPWELLING WATER
WINDMILLS (WINDPOWERED MACHINES)
WINDPOWERED GENERATORS

WIND EFFECTS
RT ATMOSPHERIC EFFECTS
DUNES
DUST STORMS
∞EFFECTS
EROSION
GROUND WIND

WIND EFFECTS-_(CONT.)_
PRESSURE EFFECTS
SEA BREEZE
SEA ROUGHNESS
SEA STATES
SOIL EROSION
TURBULENCE
WATER CIRCULATION

WIND ENERGY
USE WINDPOWER UTILIZATION

WIND EROSION
GS EROSION
. **WIND EROSION**
RT ATMOSPHERIC EFFECTS
GROUND WIND
PENEPLAINS
SEA BREEZE
WATER EROSION

WIND MEASUREMENT
GS MECHANICAL MEASUREMENT
. **WIND MEASUREMENT**
.. WIND VELOCITY MEASUREMENT
RT AERODYNAMICS
ANEMOMETERS
HOT-FILM ANEMOMETERS
∞MEASUREMENT
METEOROLOGICAL PARAMETERS
METEOROLOGY
RAWINSONDES
SEA BREEZE
SMOKE TRAILS

WIND PRESSURE
GS PRESSURE
. **WIND PRESSURE**
RT DYNAMIC LOADS
GROUND WIND
GUST LOADS
LOADS (FORCES)
WINDPOWER UTILIZATION
WINDPOWERED GENERATORS

WIND PROFILES
RT ATMOSPHERIC CIRCULATION
GROUND WIND
∞PROFILES
RADIAL DISTRIBUTION
SMOKE TRAILS
VERTICAL DISTRIBUTION

WIND RIVER RANGE (WY)
GS LANDFORMS
. MOUNTAINS
.. **WIND RIVER RANGE (WY)**
RT WYOMING

WIND SHEAR
UF DUNGEYS WIND SHEAR MECHANISM
RT BAROTROPIC FLOW
CLEAR AIR TURBULENCE
GEOSTROPHIC WIND
GROUND WIND

WIND TUNNEL APPARATUS
UF WIND TUNNEL BALANCES
GS **WIND TUNNEL APPARATUS**
. WIND TUNNEL DRIVES
. WIND TUNNEL NOZZLES
RT ∞EQUIPMENT
SUPERSONIC TEST APPARATUS

WIND TUNNEL BALANCES
USE WEIGHT INDICATORS
WIND TUNNEL APPARATUS

WIND TUNNEL CALIBRATION
GS CALIBRATING
. **WIND TUNNEL CALIBRATION**
RT MEASURING INSTRUMENTS
PRESSURE MEASUREMENT
SCALING LAWS
TEMPERATURE MEASUREMENT

WIND TUNNEL DRIVES
GS WIND TUNNEL APPARATUS
. **WIND TUNNEL DRIVES**
RT ∞DRIVES
∞FANS
MECHANICAL DRIVES
PLASMA GENERATORS
PRESSURE CHAMBERS

WIND TUNNEL DRIVES-_(CONT.)_
VACUUM CHAMBERS

WIND TUNNEL MODELS
GS MODELS
. **WIND TUNNEL MODELS**
.. POWERED MODELS
RT AERODYNAMIC CONFIGURATIONS
AIRCRAFT MODELS
DYNAMIC MODELS
FLOW VISUALIZATION
∞MISSILE SIMULATORS
PYLON MOUNTING
SCALE MODELS
SEMISPAN MODELS
SHADOWGRAPH PHOTOGRAPHY
∞TEST EQUIPMENT

WIND TUNNEL NOZZLES
GS WIND TUNNEL APPARATUS
. **WIND TUNNEL NOZZLES**
RT CONICAL NOZZLES
CONVERGENT-DIVERGENT NOZZLES
DIVERGENT NOZZLES
HYPERSONIC NOZZLES
∞NOZZLES
SUPERSONIC NOZZLES
TRANSONIC NOZZLES

WIND TUNNEL STABILITY TESTS
GS STABILITY TESTS
. **WIND TUNNEL STABILITY TESTS**
RT AERODYNAMIC STABILITY
AIRCRAFT STABILITY
MISSILE TESTS
SPACECRAFT STABILITY
∞TESTS

WIND TUNNEL TESTS
RT AERODYNAMIC CHARACTERISTICS
DENSITY MEASUREMENT
FLOW DISTRIBUTION
PRESSURE MEASUREMENT
∞TESTS
TRISONIC WIND TUNNELS
WATER TUNNEL TESTS

WIND TUNNEL WALLS
GS WALLS
. **WIND TUNNEL WALLS**
RT PRESSURE VESSELS
REINFORCED SHELLS

WIND TUNNELS
GS TEST FACILITIES
. **WIND TUNNELS**
.. BLOWDOWN WIND TUNNELS
.. COMBUSTION WIND TUNNELS
.. CRYOGENIC WIND TUNNELS
.. HYPERSONIC WIND TUNNELS
... CASCADE WIND TUNNELS
... HOTSHOT WIND TUNNELS
... PLASMA JET WIND TUNNELS
... SHOCK TUNNELS
.. HYPERVELOCITY WIND TUNNELS
... CASCADE WIND TUNNELS
... HOTSHOT WIND TUNNELS
... PLASMA JET WIND TUNNELS
... SHOCK TUNNELS
.. LOW DENSITY WIND TUNNELS
.. LOW SPEED WIND TUNNELS
.. SUBSONIC WIND TUNNELS
.. RECTANGULAR WIND TUNNELS
.. SLOTTED WIND TUNNELS
.. SUPERSONIC WIND TUNNELS
.. TRANSONIC WIND TUNNELS
.. TRISONIC WIND TUNNELS
RT AERODYNAMICS
EXHAUST FLOW SIMULATION
FLIGHT SIMULATORS
GAS GUNS
GAS STREAMS
HYPERSONIC FLOW
SPIKES (AERODYNAMIC
CONFIGURATIONS)
SUPERSONIC FLOW
TEST CHAMBERS
∞TEST EQUIPMENT
TRANSONIC FLOW
∞TUNNELS

WIND TURBINES
GS TURBOMACHINERY
. TURBINES
.. **WIND TURBINES**

WIND TURBINES-*(CONT.)*
```
        . . . TIP VANES
RT      TURBOGENERATORS
        WIND VELOCITY
        WINDMILLS (WINDPOWERED MACHINES)
        WINDPOWER UTILIZATION
        WINDPOWERED GENERATORS
```

WIND VANES
```
GS      DISPLAY DEVICES
        . FLOW DIRECTION INDICATORS
        . . WIND VANES
        MEASURING INSTRUMENTS
        . INDICATING INSTRUMENTS
        . . FLOW DIRECTION INDICATORS
        . . . WIND VANES
        . METEOROLOGICAL INSTRUMENTS
        . . WIND VANES
        VANES
        . WIND VANES
RT      ANEMOMETERS
        HOT-FILM ANEMOMETERS
```

WIND VARIATIONS
```
GS      VARIATIONS
        . WIND VARIATIONS
RT      ANNUAL VARIATIONS
        ATMOSPHERIC TURBULENCE
        DIURNAL VARIATIONS
        SEASONS
```

WIND VELOCITY
```
GS      RATES (PER TIME)
        . WIND VELOCITY
        . . SOLAR WIND VELOCITY
        VELOCITY
        . WIND VELOCITY
        . . SOLAR WIND VELOCITY
RT      AIRSPEED
        ANEMOMETERS
        FLOW MEASUREMENT
        GROUND WIND
        HOT-FILM ANEMOMETERS
        SEA ROUGHNESS
        WIND TURBINES
        WINDMILLS (WINDPOWERED MACHINES)
        WINDPOWER UTILIZATION
        WINDPOWERED GENERATORS
```

WIND VELOCITY MEASUREMENT
```
GS      MECHANICAL MEASUREMENT
        . VELOCITY MEASUREMENT
        . . WIND VELOCITY MEASUREMENT
        . WIND MEASUREMENT
        . . WIND VELOCITY MEASUREMENT
RT      ANEMOMETERS
        HOT-FILM ANEMOMETERS
```

WINDING
```
GS      WINDING
        . FILAMENT WINDING
        . HELICAL WINDINGS
        . WIRE WINDING
RT      COLD WORKING
        LEVELING
        METAL WORKING
        SPINDLES
        SPIRAL WRAPPING
        STRETCHING
        TWISTING
```

WINDMILLING
```
USE     AUTOROTATION
```

WINDMILLS (WINDPOWERED MACHINES)
```
RT      ELECTRIC GENERATORS
        GEARS
        GROUND WIND
        ∞MACHINERY
        MECHANICAL DRIVES
        ∞POWER TRANSMISSION
        ∞PUMPING
        WIND (METEOROLOGY)
        WIND DIRECTION
        WIND TURBINES
        WIND VELOCITY
        WINDPOWER UTILIZATION
        WINDPOWERED GENERATORS
        WINDPOWERED PUMPS
```

WINDOW ATMOSPHERE SOUNDING PROJECTILE
```
USE     WASP SOUNDING ROCKET
```

∞ WINDOWS
```
SN      (USE OF A MORE SPECIFIC TERM IS
        RECOMMENDED--CONSULT THE TERMS
        LISTED BELOW)
RT      INFRARED WINDOWS
        PORTS (OPENINGS)
        WAVEGUIDE WINDOWS
        WINDOWS (APERTURES)
        WINDOWS (INTERVALS)
```

WINDOWS (APERTURES)
```
SN      (EXCLUDES INTERVALS IN TIME,
        FREQUENCY, ENERGY ETC)
RT      APERTURES
        ∞BARRIERS
        CURTAINS
        DOORS
        DUCTS
        OPENINGS
        PORTS (OPENINGS)
        SEPARATORS
        SHIELDING
        VENTS
        ∞WINDOWS
        WINDSHIELDS
```

WINDOWS (INTERVALS)
```
SN      (EXCLUDES INTERVALS IN SPACE
        CONTINUUM)
GS      WINDOWS (INTERVALS)
        . LASER WINDOWS
        . LAUNCH WINDOWS
RT      BANDWIDTH
        BURNING TIME
        COUNTDOWN
        ENERGY BANDS
        FLIGHT TIME
        TESTING TIME
        TIME MEASUREMENT
        ∞WINDOWS
```

WINDPOWER UTILIZATION
```
UF      WIND ENERGY
GS      UTILIZATION
        . WINDPOWER UTILIZATION
RT      AIR CURRENTS
        AIR MASSES
        ATMOSPHERIC CIRCULATION
        CLEAN ENERGY
        EARTH RESOURCES
        GROUND WIND
        SEA BREEZE
        VANES
        WIND (METEOROLOGY)
        WIND PRESSURE
        WIND TURBINES
        WIND VELOCITY
        WINDMILLS (WINDPOWERED MACHINES)
        WINDPOWERED GENERATORS
        WINDPOWERED PUMPS
```

WINDPOWERED GENERATORS
```
RT      ELECTRIC GENERATORS
        ∞GENERATORS
        GROUND WIND
        VANES
        WIND (METEOROLOGY)
        WIND DIRECTION
        WIND PRESSURE
        WIND TURBINES
        WIND VELOCITY
        WINDMILLS (WINDPOWERED MACHINES)
        WINDPOWER UTILIZATION
```

WINDPOWERED PUMPS
```
GS      PUMPS
        . WINDPOWERED PUMPS
RT      PONDS
        ∞PUMPING
        RESERVOIRS
        VANES
        WATER
        WATER RESOURCES
        WINDMILLS (WINDPOWERED MACHINES)
        WINDPOWER UTILIZATION
```

WINDS ALOFT
```
GS      WIND (METEOROLOGY)
        . WINDS ALOFT
        . . GEOSTROPHIC WIND
        . . JET STREAMS (METEOROLOGY)
RT      CIRCUMPOLAR WESTERLIES
        SEA BREEZE
        VERTICAL AIR CURRENTS
```

WINDSCREENS
```
USE     WINDSHIELDS
```

WINDSHIELDS
```
UF      WINDSCREENS
RT      AIRCRAFT COMPARTMENTS
        CANOPIES
        COCKPITS
        ENVIRONMENTAL CONTROL
        LOCOMOTIVES
        SHIELDING
        WINDOWS (APERTURES)
```

WINES
```
GS      LIQUIDS
        . POTABLE LIQUIDS
        . . BEVERAGES
        . . . WINES
RT      VINEYARDS
```

WING CAMBER
```
GS      CAMBER
        . WING CAMBER
RT      CAMBERED WINGS
        CONICAL CAMBER
```

WING FLAPS
```
UF      JET AUGMENTED WING FLAPS
GS      AIRFOILS
        . FLAPS (CONTROL SURFACES)
        . . WING FLAPS
        . . . LEADING EDGE SLATS
        . . . TRAILING EDGE FLAPS
        . . . VORTEX FLAPS
        BRAKES (FOR ARRESTING MOTION)
        . AERODYNAMIC BRAKES
        . . WING FLAPS
        . . . LEADING EDGE SLATS
        . . . TRAILING EDGE FLAPS
        . AIRCRAFT BRAKES
        . . WING FLAPS
        . . . LEADING EDGE SLATS
        . . . TRAILING EDGE FLAPS
        CONTROL SURFACES
        . FLAPS (CONTROL SURFACES)
        . . WING FLAPS
        . . . LEADING EDGE FLAPS
        . . . LEADING EDGE SLATS
        . . . TRAILING EDGE FLAPS
        DRAG DEVICES
        . AERODYNAMIC BRAKES
        . . WING FLAPS
        . . . LEADING EDGE SLATS
        . . . TRAILING EDGE FLAPS
RT      EXTERNALLY BLOWN FLAPS
        JET FLAPS
        SPLIT FLAPS
```

WING FLOW METHOD TESTS
```
RT      FLIGHT TESTS
        FLUID FLOW
        GROUND TESTS
        ∞METHODOLOGY
        ∞TESTS
        TRANSONIC WIND TUNNELS
```

WING LOADING
```
GS      AERODYNAMIC FORCES
        . WING LOADING
        LOADS (FORCES)
        . DYNAMIC LOADS
        . . WING LOADING
RT      AERODYNAMIC LOADS
        AEROELASTICITY
        EDGE LOADING
        FORCE DISTRIBUTION
        GUST LOADS
        LEADING EDGE THRUST
        STATIC LOADS
        SWEEP EFFECT
        VORTEX FLAPS
```

WING NACELLE CONFIGURATIONS
```
GS      AERODYNAMIC CONFIGURATIONS
        . WING NACELLE CONFIGURATIONS
RT      ∞AIRCRAFT
        AIRFRAMES
        EXTERNALLY BLOWN FLAPS
```

WING OSCILLATIONS
```
GS      OSCILLATIONS
        . WING OSCILLATIONS
RT      AERODYNAMIC STABILITY
        FLAPPING
        FLUTTER
```

WING OSCILLATIONS-*(CONT.)*
    STABLE OSCILLATIONS
    UNDAMPED OSCILLATIONS
    VIBRATION

WING PANELS
GS   PANELS
    . **WING PANELS**
    STRUCTURAL MEMBERS
    . **WING PANELS**
RT   CURVED PANELS
    RECTANGULAR PANELS
    WINGS

WING PLANFORMS
GS   PLANFORMS
    . **WING PLANFORMS**
    . . CHANNEL WINGS
    . . INFINITE SPAN WINGS
    . . SWEPT FORWARD WINGS
    . . . TRAPEZOIDAL WINGS
    . . SWEPTBACK WINGS
    . . . ARROW WINGS
    . . . DELTA WINGS
    . . . TRAPEZOIDAL WINGS
    . . VARIABLE SWEEP WINGS
RT   HP-115 AIRCRAFT
    LOW ASPECT RATIO WINGS
    MONOPLANES
    OBLIQUE WINGS
    RECTANGULAR PLANFORMS
    SLENDER WINGS
    SWEPT WINGS
    SWING WINGS
    THRUST DISTRIBUTION
    UNSWEPT WINGS

WING PROFILES
GS   AIRFOIL PROFILES
    . **WING PROFILES**
    . . WING SPAN
RT   AERODYNAMIC INTERFERENCE
    GAW-1 AIRFOIL
    GAW-2 AIRFOIL
    MONOPLANES
    SUPERCRITICAL WINGS
    SWING WINGS
    WINGS

WING ROOTS
RT   AERODYNAMIC CONFIGURATIONS
    AIRCRAFT CONFIGURATIONS
    DROOPED AIRFOILS
    FAIRINGS
    ∞ROOTS

WING SLATS
USE   LEADING EDGE SLATS

WING SLOTS
GS   SLOTS
    . **WING SLOTS**
RT   BOUNDARY LAYER CONTROL
    LEADING EDGE SLATS
    VORTEX GENERATORS

WING SPAN
GS   AIRFOIL PROFILES
    . WING PROFILES
    . . **WING SPAN**
RT   ∞SPAN
    SPANWISE BLOWING
    WINGS

WING TANKS
GS   TANKS (CONTAINERS)
    . FUEL TANKS
    . . **WING TANKS**
RT   ∞CONTAINERS
    EXTERNAL STORE SEPARATION
    EXTERNAL STORES
    EXTERNAL TANKS
    JETTISON SYSTEMS
    WING-FUSELAGE STORES

WING TIP VORTICES
GS   VORTICES
    . **WING TIP VORTICES**
RT   FLOW DISTORTION
    ROTATING FLUIDS

WING TIPS
GS   TIPS
    . **WING TIPS**

WING TIPS-*(CONT.)*
RT   AIRFOIL PROFILES
    BLADE TIPS
    WINGS

WING-FUSELAGE STORES
RT   EXTERNAL STORE SEPARATION
    EXTERNAL STORES
    FUSELAGES
    NACELLES
    PODS (EXTERNAL STORES)
    PROTUBERANCES
    ∞STORAGE
    STORAGE TANKS
    TANKS (CONTAINERS)
    WEAPONS
    WING TANKS

∞ **WINGED VEHICLES**
SN   *(USE OF A MORE SPECIFIC TERM IS*
    RECOMMENDED--CONSULT THE TERMS
    LISTED BELOW)
RT   ∞AIRCRAFT
    B-1 AIRCRAFT
    DRONE VEHICLES
    FIREBEE 2 TARGET DRONE AIRCRAFT
    GLIDERS
    HANG GLIDERS
    HYPERSONIC VEHICLES
    JET AIRCRAFT
    KA-6 SAILPLANES
    LAUNCH VEHICLES
    LEADING EDGE FLAPS
    MAN POWERED AIRCRAFT
    MISSILES
    MONOPLANES
    RECOVERABLE LAUNCH VEHICLES
    RECOVERABLE SPACECRAFT
    REENTRY VEHICLES
    RESEARCH AIRCRAFT
    ROCKET VEHICLES
    SHORT TAKEOFF AIRCRAFT
    SUPERCRITICAL WINGS
    ULTRALIGHT AIRCRAFT
    ∞VEHICLES
    VERTICAL TAKEOFF AIRCRAFT
    WINGS

WINGLETS
RT   DRAG REDUCTION
    FINS
    PROTUBERANCES
    VORTEX ALLEVIATION
    VORTEX AVOIDANCE
    WINGS

WINGS
UF   CANTILEVER WINGS
GS   AIRFOILS
    . **WINGS**
    . . AEROELASTIC RESEARCH WINGS
    . . CAMBERED WINGS
    . . CARET WINGS
    . . CHANNEL WINGS
    . . CRUCIFORM WINGS
    . . FIXED WINGS
    . . FLEXIBLE WINGS
    . . . PARAWINGS
    . . GAW-1 AIRFOIL
    . . GAW-2 AIRFOIL
    . . LOW ASPECT RATIO WINGS
    . . . DELTA WINGS
    . . . TRAPEZOIDAL WINGS
    . . OBLIQUE WINGS
    . . RIGID WINGS
    . . ROTARY WINGS
    . . . LIFTING ROTORS
    . . . . BEARINGLESS ROTORS
    . . . RIGID ROTORS
    . . . TILTING ROTORS
    . . . TIP DRIVEN ROTORS
    . . SLENDER WINGS
    . . . INFINITE SPAN WINGS
    . . SUPERCRITICAL WINGS
    . . SWEPT WINGS
    . . . SWEPT FORWARD WINGS
    . . . . TRAPEZOIDAL WINGS
    . . . SWEPTBACK WINGS
    . . . . ARROW WINGS
    . . . . DELTA WINGS
    . . . . TRAPEZOIDAL WINGS
    . . SWING WINGS
    . . THIN WINGS
    . . . INFINITE SPAN WINGS
    . . TWISTED WINGS

WINGS-*(CONT.)*
    . . UNCAMBERED WINGS
    . . . RING WINGS
    . . UNSWEPT WINGS
    . . . INFINITE SPAN WINGS
    . . . RECTANGULAR WINGS
    . . . RING WINGS
    . . VARIABLE SWEEP WINGS
RT   AIRCRAFT CONSTRUCTION MATERIALS
    AIRCRAFT PARTS
    AIRCRAFT STRUCTURES
    AIRFOIL FENCES
    AIRFRAMES
    ASPECT RATIO
    BLUNT TRAILING EDGES
    BODY-WING AND TAIL CONFIGURATIONS
    BODY-WING CONFIGURATIONS
    COATINGS
    CONTROL SURFACES
    DROOPED AIRFOILS
    DUAL WING CONFIGURATIONS
    LEADING EDGE FLAPS
    MISSILE COMPONENTS
    ROTORS
    SPOILERS
    WING PANELS
    WING PROFILES
    WING SPAN
    WING TIPS
    ∞WINGED VEHICLES
    WINGLETS
    X WING ROTORS

WINTER
GS   SEASONS
    . **WINTER**
RT   AUTUMN
    COLD WEATHER
    EQUINOXES
    PRESSURE ICE
    SOLSTICES
    SPRING (SEASON)
    SUMMER

WIRE
GS   **WIRE**
    . ELECTRIC WIRE
    . EXPLODING WIRES
    . GUY WIRES
RT   BILLETS
    CABLES (ROPES)
    ∞COILS
    CORDAGE
    FASTENERS
    ∞FILAMENTS
    FLAT CONDUCTORS
    JUMPERS
    REINFORCEMENT (STRUCTURES)
    RODS
    WIRING

WIRE BRIDGE CIRCUITS
GS   CIRCUITS
    . ELECTRIC BRIDGES
    . . **WIRE BRIDGE CIRCUITS**
    . . . WHEATSTONE BRIDGES
RT   ELECTRIC WIRE
    EXPLODING WIRES

WIRE CLOTH
UF   WIRE MESH
RT   FABRICS
    REINFORCEMENT (STRUCTURES)
    ∞SCREENS
    SIEVES

WIRE GRID LENSES
GS   LENSES
    . **WIRE GRID LENSES**
    RETICLES
    . **WIRE GRID LENSES**
RT   ∞GRIDS
    LENS ANTENNAS
    MAGNETIC LENSES
    TURNSTILE ANTENNAS

WIRE MESH
USE   WIRE CLOTH

WIRE WINDING
GS   WINDING
    . **WIRE WINDING**
RT   MAGNET COILS
    WIRING

WIRELESS COMMUNICATION
- UF   CARRIER SYSTEMS
- GS   TELECOMMUNICATION
  - . **WIRELESS COMMUNICATION**
- RT   AIRCRAFT COMMUNICATION
  - CLOSED CIRCUIT TELEVISION
  - COMMUNICATION SATELLITES
  - DATA LINKS
  - DATA TRANSMISSION
  - DIGITAL SPACECRAFT TELEVISION
  - FACSIMILE COMMUNICATION
  - OPTICAL COMMUNICATION
  - RADIO TELEMETRY
  - SIGNAL TRANSMISSION
  - SPACE COMMUNICATION
  - SPACECRAFT COMMUNICATION
  - TELEMETRY
  - VOICE COMMUNICATION

WIRING
- SN   (PROCESS--AS DISTINGUISHED FROM MATERIAL)
- UF   ELECTRIC WIRING
  - WIRING SYSTEMS
- GS   **WIRING**
  - . BEAM LEADS
- RT   BUNDLES
  - CIRCUITS
  - ELECTRICAL INSULATION
  - FLAT CONDUCTORS
  - SPLICING
  - TRANSMISSION LINES
  - WIRE
  - WIRE WINDING

WIRING SYSTEMS
- USE   WIRING

WISCONSIN
- GS   NATIONS
  - . UNITED STATES
  - . . **WISCONSIN**

WISWESSER NOTATIONS
- GS   CODING
  - . **WISWESSER NOTATIONS**
- RT   ∞CHEMICAL COMPOUNDS
  - ∞CHEMISTRY
  - CLASSIFICATIONS
  - IDENTIFYING
  - MOLECULAR STRUCTURE
  - ∞REFERENCE SYSTEMS

WKB APPROXIMATION
- USE   WENTZEL-KRAMER-BRILLOUIN METHOD

WOLF-RAYET STARS
- UF   W-R STARS
- GS   CELESTIAL BODIES
  - . STARS
  - . . HOT STARS
  - . . . **WOLF-RAYET STARS**
- RT   A STARS
  - ASTROPHYSICS
  - B STARS
  - CARBON STARS
  - CELESTIAL MECHANICS
  - EJECTA
  - HELIUM
  - NITROGEN
  - O STARS
  - STELLAR ENVELOPES
  - STELLAR LUMINOSITY
  - STELLAR MASS EJECTION
  - WHITE DWARF STARS

WOLFRAM
- USE   TUNGSTEN

WOLVES
- GS   ANIMALS
  - . VERTEBRATES
  - . . MAMMALS
  - . . . **WOLVES**

WOMEN
- USE   FEMALES

WOOD
- GS   **WOOD**
  - . CORK (MATERIALS)
  - . PLYWOOD
- RT   BALSA
  - CELLULOSE

WOOD-*(CONT.)*
-     MASONITE (TRADEMARK)
-     ORGANIC MATERIALS
-     PAPER (MATERIAL)
-     PLANTS (BOTANY)
-     SLIVERS
-     TREES (PLANTS)
-     WOODEN STRUCTURES

WOODEN STRUCTURES
- RT   PLYWOOD
  - ∞STRUCTURES
  - WOOD

WOOL
- SN   (LIMITED TO ANIMAL FIBERS)
- GS   FABRICS
  - . **WOOL**
  - FIBERS
  - . **WOOL**
- RT   FELTS
  - HAIR
  - KERATINS
  - ORGANIC MATERIALS
  - SHEEP
  - YARNS

WORD PROCESSING
- RT   COMPUTER TECHNIQUES
  - DATA PROCESSING
  - WORDS (LANGUAGE)

WORDS (LANGUAGE)
- GS   COMMUNICATION THEORY
  - . **WORDS (LANGUAGE)**
  - . . SYLLABLES
  - LANGUAGES
  - . SENTENCES
  - . . **WORDS (LANGUAGE)**
  - . . . SYLLABLES
  - LINGUISTICS
  - . SYNTAX
  - . . SENTENCES
  - . . . **WORDS (LANGUAGE)**
  - . . . . SYLLABLES
  - SPEECH
  - . TALKING
  - . . **WORDS (LANGUAGE)**
  - . . . SYLLABLES
- RT   CONSONANTS (SPEECH)
  - CONVERSATION
  - ENGLISH LANGUAGE
  - GRAMMARS
  - MESSAGES
  - ORTHOGRAPHY
  - PHONEMES
  - PHONEMICS
  - PHONETICS
  - SEMANTICS
  - TERMINOLOGY
  - THESAURI
  - VERBAL COMMUNICATION
  - VOICE COMMUNICATION
  - VOWELS
  - WORD PROCESSING

WORK
- GS   **WORK**
  - . PHYSICAL WORK
- RT   ∞ENERGY
  - HEAT
  - HORSEPOWER
  - KINETIC ENERGY
  - OCCUPATION
  - PHYSICAL FACTORS

WORK CAPACITY
- RT   HYPERKINESIA
  - ORBITAL WORKERS
  - PHYSICAL FITNESS
  - PHYSICAL WORK
  - WORKLOADS (PSYCHOPHYSIOLOGY)

WORK FUNCTIONS
- UF   SCHOTTKY EFFECT
- RT   ELECTRON EMISSION
  - ∞FUNCTIONS
  - IONIZATION POTENTIALS
  - PERVEANCE
  - PHOTOELECTRIC EMISSION
  - SCHOTTKY DIODES
  - THERMIONIC EMISSION

WORK HARDENING
- GS   HARDENING (MATERIALS)

WORK HARDENING-*(CONT.)*
-     . **WORK HARDENING**
-     . . STRAIN HARDENING
- RT   COLD HARDENING
  - MECHANICAL TWINNING
  - METAL WORKING
  - PEENING
  - SHOT PEENING
  - WORK SOFTENING

WORK SOFTENING
- GS   SOFTENING
  - . **WORK SOFTENING**
- RT   MICROSTRUCTURE
  - PLASTIC DEFORMATION
  - WORK HARDENING

WORK-REST CYCLE
- GS   CYCLES
  - . **WORK-REST CYCLE**
- RT   FATIGUE (BIOLOGY)
  - RELAXATION (PHYSIOLOGY)

WORKHORSE HELICOPTER
- USE   CH-21 HELICOPTER

WORKING FLUIDS
- GS   CONSUMABLES (SPACECRAFT)
  - . **WORKING FLUIDS**
- RT   FLUID POWER
  - FLUID TRANSMISSION LINES
  - ∞FLUIDS
  - HIGH TEMPERATURE FLUIDS
  - HYDRAULIC FLUIDS
  - JET CONDENSERS
  - PHASE CHANGE MATERIALS
  - TRANSMISSION FLUIDS

WORKLOADS (PSYCHOPHYSIOLOGY)
- GS   PHYSIOLOGY
  - . PSYCHOPHYSIOLOGY
  - . . **WORKLOADS (PSYCHOPHYSIOLOGY)**
- RT   FATIGUE (BIOLOGY)
  - HUMAN PERFORMANCE
  - MENTAL PERFORMANCE
  - PHYSICAL WORK
  - PSYCHOMOTOR PERFORMANCE
  - STRESS (PSYCHOLOGY)
  - WORK CAPACITY

WORKSTATIONS
- GS   STATIONS
  - . **WORKSTATIONS**
- RT   HUMAN FACTORS ENGINEERING
  - MAN MACHINE SYSTEMS
  - PRODUCTIVITY

WORLD
- USE   EARTH (PLANET)

WORLD DATA CENTERS
- RT   ∞CENTERS
  - ∞DATA
  - DATA RETRIEVAL
  - DATA STORAGE
  - INTERNATIONAL GEOPHYSICAL YEAR
  - LIBRARIES
  - TRANSOCEANIC SYSTEMS

WORLD METEOROLOGICAL ORGANIZATION
- GS   ORGANIZATIONS
  - . **WORLD METEOROLOGICAL ORGANIZATION**
- RT   INTERNATIONAL COOPERATION
  - METEOROLOGY
  - UNITED NATIONS

WORMS
- GS   ANIMALS
  - . INVERTEBRATES
  - . . **WORMS**
  - . . . FLATWORMS
- RT   INFESTATION
  - LARVAE

WOUND HEALING
- GS   HEALING
  - . **WOUND HEALING**
- RT   INJURIES

WRANGELL MOUNTAINS (AK)
- GS   LANDFORMS
  - . MOUNTAINS
  - . . **WRANGELL MOUNTAINS (AK)**

WRANGELL MOUNTAINS (AK)-(CONT.)
RT   ALASKA

∞ **WRAP**
SN   *(USE OF A MORE SPECIFIC TERM IS*
RECOMMENDED--CONSULT THE TERMS
LISTED BELOW)
RT   COMPOSITE WRAPPING
PACKAGING

WRAPAROUND CONTACT SOLAR CELLS
USE   SOLAR CELLS

WRECKAGE
RT   ACCIDENT INVESTIGATION
ACCIDENTS
CRASHES
SABOTAGE

WRENCHES
GS   TOOLS
. **WRENCHES**

WRINKLING
GS   **WRINKLING**
. FLANGE WRINKLING
RT   BUCKLING
DEFORMATION
DISTORTION
∞ RIDGES

WRIST
GS   ANATOMY
. MUSCULOSKELETAL SYSTEM
. . JOINTS (ANATOMY)
. . . **WRIST**
RT   ARM (ANATOMY)
HAND (ANATOMY)

WROUGHT ALLOYS
GS   ALLOYS
. **WROUGHT ALLOYS**
RT   RENE 41
RENE 63
RENE 77
WASPALOY

WU-2 AIRCRAFT
USE   U-2 AIRCRAFT

WURTZITE
GS   CHALCOGENIDES
. SULFIDES
. . INORGANIC SULFIDES
. . . ZINC SULFIDES
. . . . **WURTZITE**
MINERALS
. **WURTZITE**
SULFUR COMPOUNDS
. SULFIDES
. . INORGANIC SULFIDES
. . . ZINC SULFIDES
. . . . **WURTZITE**
ZINC COMPOUNDS
. **WURTZITE**

WYOMING
GS   NATIONS
. UNITED STATES
. . **WYOMING**
RT   BIGHORN MOUNTAINS (MT-WY)
BLACK HILLS (SD-WY)
WIND RIVER RANGE (WY)
YELLOWSTONE NATIONAL PARK
(ID-MT-WY)

W2F AIRCRAFT
USE   E-2 AIRCRAFT

X

X BAND
USE   SUPERHIGH FREQUENCIES

X MESONS
GS   PARTICLES
. ELEMENTARY PARTICLES
. . BOSONS
. . . MESONS
. . . . MESON RESONANCE
. . . . . **X MESONS**

X MESONS-(CONT.)
. NUCLEAR PARTICLES
. . BOSONS
. . . MESONS
. . . . MESON RESONANCE
. . . . . **X MESONS**
RESONANCE
. MESON RESONANCE
. . **X MESONS**

X RAY ABSORPTION
GS   ENERGY ABSORPTION
. ELECTROMAGNETIC ABSORPTION
. . **X RAY ABSORPTION**
RADIATION ABSORPTION
. ELECTROMAGNETIC ABSORPTION
. . **X RAY ABSORPTION**
RT   ∞ ABSORPTION
ELECTRON SPECTROSCOPY

X RAY ANALYSIS
GS   **X RAY ANALYSIS**
. LAUE METHOD
RT   ∞ ANALYZING
CHEMICAL ANALYSIS
CRYSTALLOGRAPHY
DEFECTS
FLUOROSCOPY
LATTICE PARAMETERS
∞ MATERIALS TESTS
MICROANALYSIS
MICROBEAMS
RADIOGRAPHY
RADIOLOGY
STEREOCHEMISTRY
STRESS ANALYSIS
TOMOGRAPHY

X RAY APPARATUS
GS   MEDICAL EQUIPMENT
. **X RAY APPARATUS**
. . LIXISCOPES
. . X RAY TUBES
RT   ∞ EQUIPMENT
RADIOGRAPHY

X RAY ASTRONOMY
GS   ASTRONOMY
. **X RAY ASTRONOMY**
. . X RAY SOURCES
. . . X RAY BINARIES
RT   COSMIC X RAYS
EXOSAT SATELLITE
EXTARS
GAMMA RAY ASTRONOMY
GAMMA RAY BURSTS
LIXISCOPES
RADIOGRAPHY
ROSAT MISSION
SAS-3
UHURU SATELLITE
ULTRAVIOLET TELESCOPES

X RAY ASTROPHYSICS FACILITY
UF   ADVANCED X RAY ASTROPHYSICAL
FACILITY
ADVANCED X RAY ASTROPHYSICS
FACILITY
AXAF
GS   PAYLOADS
. SPACE SHUTTLE PAYLOADS
. . **X RAY ASTROPHYSICS FACILITY**
RT   ASTROPHYSICS
∞ FACILITIES
SPACEBORNE EXPERIMENTS
TELESCOPES

X RAY BINARIES
GS   ASTRONOMY
. X RAY ASTRONOMY
. . X RAY SOURCES
. . . **X RAY BINARIES**
RT   ACCRETION DISKS
ASTROPHYSICS
BLACK HOLES (ASTRONOMY)
COMPANION STARS
COSMIC X RAYS
ECLIPSING BINARY STARS
EXTARS
NEUTRON STARS
STELLAR MASS ACCRETION
X RAYS

X RAY DENSITY MEASUREMENT
GS   DENSITY MEASUREMENT

X RAY DENSITY MEASUREMENT-(CONT.)
. **X RAY DENSITY MEASUREMENT**
RT   FLUX DENSITY

X RAY DIFFRACTION
GS   DIFFRACTION
. **X RAY DIFFRACTION**
RT   CRYSTALLOGRAPHY
ELECTRON DIFFRACTION
LAUE METHOD
METALLOGRAPHY
RADIOGRAPHY

X RAY FLUORESCENCE
GS   DECAY
. EMISSION
. . LIGHT EMISSION
. . . LUMINESCENCE
. . . . FLUORESCENCE
. . . . . **X RAY FLUORESCENCE**
. . . . PHOTOLUMINESCENCE
. . . . . **X RAY FLUORESCENCE**
RT   RADIOGRAPHY

X RAY IMAGERY
GS   IMAGERY
. **X RAY IMAGERY**
RT   IMAGING TECHNIQUES
INFRARED IMAGERY
LIXISCOPES
MICROWAVE IMAGERY
RADAR IMAGERY
RADIOGRAPHY

X RAY INSPECTION
GS   INSPECTION
. **X RAY INSPECTION**
RT   NONDESTRUCTIVE TESTS
RADIOGRAPHY
∞ TESTS

X RAY IRRADIATION
GS   IRRADIATION
. **X RAY IRRADIATION**

X RAY LASERS
GS   STIMULATED EMISSION DEVICES
. LASERS
. . **X RAY LASERS**
RT   ELECTRON TRANSITIONS
LASER OUTPUTS

X RAY SCATTERING
GS   SCATTERING
. WAVE SCATTERING
. . ELECTROMAGNETIC SCATTERING
. . . **X RAY SCATTERING**
RT   FORM FACTORS

X RAY SOURCES
GS   ASTRONOMY
. X RAY ASTRONOMY
. . **X RAY SOURCES**
. . . X RAY BINARIES
RT   EXOSAT SATELLITE
EXTARS
∞ RADIATION
ROSAT MISSION

X RAY SPECTRA
GS   SPECTRA
. RADIATION SPECTRA
. . ELECTROMAGNETIC SPECTRA
. . . **X RAY SPECTRA**
RT   NORTH POLAR SPUR (ASTRONOMY)
QUASARS
SOLAR SPECTRA
STELLAR SPECTRA

X RAY SPECTROGRAPHY
USE   X RAY SPECTROSCOPY

X RAY SPECTROMETRY
USE   X RAY SPECTROSCOPY

X RAY SPECTROPOLARIMETRY PAYLOAD
USE   EXPOS (SPACELAB PAYLOAD)

X RAY SPECTROSCOPY
UF   X RAY SPECTROGRAPHY
X RAY SPECTROMETRY
GS   SPECTROSCOPY
. **X RAY SPECTROSCOPY**
RT   ASTRONOMICAL SPECTROSCOPY

X RAY SPECTROSCOPY-*(CONT.)*
∞MATERIALS TESTS
  MOLECULAR SPECTROSCOPY
  RADIO SPECTROSCOPY
  RADIOGRAPHY
  SPECTROSCOPIC ANALYSIS
  ULTRAVIOLET SPECTROSCOPY
  VACUUM SPECTROSCOPY

X RAY STRESS ANALYSIS
GS STRESS ANALYSIS
  . **X RAY STRESS ANALYSIS**
RT STRESSES
  TEMPERATURE INVERSIONS

X RAY STRESS MEASUREMENT
GS MECHANICAL MEASUREMENT
  . STRESS MEASUREMENT
  . . **X RAY STRESS MEASUREMENT**

X RAY TELESCOPES
GS OPTICAL EQUIPMENT
  . ASTRONOMICAL TELESCOPES
  . . **X RAY TELESCOPES**
  TELESCOPES
  . ASTRONOMICAL TELESCOPES
  . . **X RAY TELESCOPES**
RT EXTARS
  RADIOGRAPHY
  ROSAT MISSION

X RAY TIMING EXPLORER
GS SATELLITES
  . ARTIFICIAL SATELLITES
  . . EXPLORER SATELLITES
  . . . **X RAY TIMING EXPLORER**

X RAY TUBES
GS MEDICAL EQUIPMENT
  . X RAY APPARATUS
  . . **X RAY TUBES**
RT ELECTRON TUBES
  RADIOGRAPHY

X RAYS
GS ELECTROMAGNETIC RADIATION
  . **X RAYS**
  . . COSMIC X RAYS
  . . SOLAR X-RAYS
  IONIZING RADIATION
  . **X RAYS**
  . . COSMIC X RAYS
  . . SOLAR X-RAYS
RT AURORAS
  BLACKOUT (PROPAGATION)
  BREMSSTRAHLUNG
  COSMIC RAYS
  EMISSION SPECTRA
  EXTARS
  EXTRATERRESTRIAL RADIATION
  FAR ULTRAVIOLET RADIATION
  GAMMA RAYS
  MONOCHROMATIC RADIATION
  RADIOGRAPHY
  RADIOLOGY
  SYNCHROTRON RADIATION
  SYSTEM GENERATED
    ELECTROMAGNETIC PULSES
  X RAY BINARIES

X WING ROTORS
GS ROTATING BODIES
  . ROTORS
  . . ROTARY WINGS
  . . . **X WING ROTORS**
RT CIRCULATION CONTROL ROTORS
  ∞ROTOR BLADES
  WINGS

X-Y PLOTTERS
GS RECORDING INSTRUMENTS
  . PLOTTERS
  . . **X-Y PLOTTERS**
RT DIGITAL TO ANALOG CONVERTERS

X-1 AIRCRAFT
GS BELL AIRCRAFT
  . **X-1 AIRCRAFT**
  MONOPLANES
  . **X-1 AIRCRAFT**
  RESEARCH AIRCRAFT
  . **X-1 AIRCRAFT**
  SUPERSONIC AIRCRAFT
  . **X-1 AIRCRAFT**

X-1 AIRCRAFT-*(CONT.)*
RT ∞AIRCRAFT
  ROCKET VEHICLES

X-2 AIRCRAFT
GS BELL AIRCRAFT
  . **X-2 AIRCRAFT**
  MONOPLANES
  . **X-2 AIRCRAFT**
  RESEARCH AIRCRAFT
  . **X-2 AIRCRAFT**
  SUPERSONIC AIRCRAFT
  . **X-2 AIRCRAFT**
RT ∞AIRCRAFT
  ROCKET VEHICLES

X-3 AIRCRAFT
GS JET AIRCRAFT
  . **X-3 AIRCRAFT**
  MCDONNELL DOUGLAS AIRCRAFT
  . DOUGLAS AIRCRAFT
  . . **X-3 AIRCRAFT**
  MONOPLANES
  . **X-3 AIRCRAFT**
  RESEARCH AIRCRAFT
  . **X-3 AIRCRAFT**
  SUPERSONIC AIRCRAFT
  . **X-3 AIRCRAFT**
RT ∞AIRCRAFT

X-5 AIRCRAFT
GS BELL AIRCRAFT
  . **X-5 AIRCRAFT**
  JET AIRCRAFT
  . **X-5 AIRCRAFT**
  MONOPLANES
  . **X-5 AIRCRAFT**
  RESEARCH AIRCRAFT
  . **X-5 AIRCRAFT**
RT ∞AIRCRAFT

X-13 AIRCRAFT
GS JET AIRCRAFT
  . **X-13 AIRCRAFT**
  MONOPLANES
  . **X-13 AIRCRAFT**
  RESEARCH AIRCRAFT
  . **X-13 AIRCRAFT**
  RYAN AIRCRAFT
  . **X-13 AIRCRAFT**
  V/STOL AIRCRAFT
  . VERTICAL TAKEOFF AIRCRAFT
  . . **X-13 AIRCRAFT**
RT ∞AIRCRAFT

X-14 AIRCRAFT
GS BELL AIRCRAFT
  . **X-14 AIRCRAFT**
  JET AIRCRAFT
  . **X-14 AIRCRAFT**
  MONOPLANES
  . **X-14 AIRCRAFT**
  RESEARCH AIRCRAFT
  . **X-14 AIRCRAFT**
  V/STOL AIRCRAFT
  . VERTICAL TAKEOFF AIRCRAFT
  . . **X-14 AIRCRAFT**
RT ∞AIRCRAFT

X-15 AIRCRAFT
GS NORTH AMERICAN AIRCRAFT
  . **X-15 AIRCRAFT**
  RESEARCH AIRCRAFT
  . **X-15 AIRCRAFT**
  SUPERSONIC AIRCRAFT
  . **X-15 AIRCRAFT**
RT ∞AIRCRAFT
  LR-99 ENGINE
  ROCKET VEHICLES
  XLR-99 ENGINE

X-17 REENTRY VEHICLE
GS REENTRY VEHICLES
  . **X-17 REENTRY VEHICLE**
RT ROCKET VEHICLES
  SOLID PROPELLANT ROCKET ENGINES

X-19 AIRCRAFT
GS CURTISS-WRIGHT AIRCRAFT
  . **X-19 AIRCRAFT**
  RESEARCH AIRCRAFT
  . **X-19 AIRCRAFT**
  TANDEM WING AIRCRAFT
  . **X-19 AIRCRAFT**
  V/STOL AIRCRAFT

X-19 AIRCRAFT-*(CONT.)*
  . VERTICAL TAKEOFF AIRCRAFT
  . . **X-19 AIRCRAFT**
RT ∞AIRCRAFT

X-20 AIRCRAFT
UF DYNA-SOAR SPACE GLIDER
GS BOEING AIRCRAFT
  . **X-20 AIRCRAFT**
  GLIDERS
  . BOOSTGLIDE VEHICLES
  . . **X-20 AIRCRAFT**
  . HYPERSONIC GLIDERS
  . . **X-20 AIRCRAFT**
  HYPERSONIC VEHICLES
  . HYPERSONIC AIRCRAFT
  . . HYPERSONIC GLIDERS
  . . . **X-20 AIRCRAFT**
  . LIFTING REENTRY VEHICLES
  . . **X-20 AIRCRAFT**
  LIFTING BODIES
  . LIFTING REENTRY VEHICLES
  . . **X-20 AIRCRAFT**
  MANEUVERABLE SPACECRAFT
  . LIFTING REENTRY VEHICLES
  . . **X-20 AIRCRAFT**
  REENTRY VEHICLES
  . BOOSTGLIDE VEHICLES
  . . **X-20 AIRCRAFT**
  . LIFTING REENTRY VEHICLES
  . . **X-20 AIRCRAFT**
  RESEARCH AIRCRAFT
  . **X-20 AIRCRAFT**
RT AEROSPACEPLANES
  ∞AIRCRAFT
  ASTROPLANE
  MANNED SPACECRAFT
  SOFT LANDING SPACECRAFT

X-21 AIRCRAFT
GS JET AIRCRAFT
  . **X-21 AIRCRAFT**
  MONOPLANES
  . **X-21 AIRCRAFT**
  NORTHROP AIRCRAFT
  . **X-21 AIRCRAFT**
  RESEARCH AIRCRAFT
  . **X-21 AIRCRAFT**
RT ∞AIRCRAFT
  BOUNDARY LAYER CONTROL
  LAMINAR BOUNDARY LAYER

X-21A AIRCRAFT
GS JET AIRCRAFT
  . **X-21A AIRCRAFT**
  MONOPLANES
  . **X-21A AIRCRAFT**
  NORTHROP AIRCRAFT
  . **X-21A AIRCRAFT**
  RESEARCH AIRCRAFT
  . **X-21A AIRCRAFT**
RT ∞AIRCRAFT
  LAMINAR FLOW

X-22 AIRCRAFT
GS BELL AIRCRAFT
  . **X-22 AIRCRAFT**
  RESEARCH AIRCRAFT
  . **X-22 AIRCRAFT**
  V/STOL AIRCRAFT
  . VERTICAL TAKEOFF AIRCRAFT
  . . **X-22 AIRCRAFT**
RT ∞AIRCRAFT
  TANDEM WING AIRCRAFT
  TILT WING AIRCRAFT

X-22A AIRCRAFT
GS RESEARCH AIRCRAFT
  . **X-22A AIRCRAFT**
  TANDEM WING AIRCRAFT
  . **X-22A AIRCRAFT**
  V/STOL AIRCRAFT
  . VERTICAL TAKEOFF AIRCRAFT
  . . **X-22A AIRCRAFT**
RT ∞AIRCRAFT

X-24 AIRCRAFT
GS LIFTING BODIES
  . LIFTING REENTRY VEHICLES
  . . **X-24 AIRCRAFT**
  RESEARCH AIRCRAFT
  . **X-24 AIRCRAFT**
RT ∞AIRCRAFT

X-29 AIRCRAFT
RT　∞AIRCRAFT
　　SWEPT FORWARD WINGS

X-248 ENGINE
UF　ALTAIR ENGINE
GS　ENGINES
　　. ROCKET ENGINES
　　. . SOLID PROPELLANT ROCKET
　　　　ENGINES
　　. . . X-248 **ENGINE**
RT　BLUE SCOUT ROCKET VEHICLE
　　SCOUT LAUNCH VEHICLE
　　VANGUARD 2 LAUNCH VEHICLE

X-254 ENGINE
GS　ENGINES
　　. ROCKET ENGINES
　　. . SOLID PROPELLANT ROCKET
　　　　ENGINES
　　. . . X-254 **ENGINE**
RT　ANTARES ROCKET VEHICLE
　　BLUE SCOUT ROCKET VEHICLE
　　SCOUT LAUNCH VEHICLE

X-258 ENGINES
GS　ENGINES
　　. ROCKET ENGINES
　　. . SOLID PROPELLANT ROCKET
　　　　ENGINES
　　. . . X-258 **ENGINES**
　　. . . . X-258-B1 ENGINE
RT　SCOUT LAUNCH VEHICLE

X-258-B1 ENGINE
GS　ENGINES
　　. ROCKET ENGINES
　　. . SOLID PROPELLANT ROCKET
　　　　ENGINES
　　. . . X-258 ENGINES
　　. . . . **X-258-B1 ENGINE**

X-259 ENGINE
GS　ENGINES
　　. ROCKET ENGINES
　　. . SOLID PROPELLANT ROCKET
　　　　ENGINES
　　. . . **X-259 ENGINE**
RT　SCOUT LAUNCH VEHICLE

X-405 ENGINE
GS　ENGINES
　　. ROCKET ENGINES
　　. . BOOSTER ROCKET ENGINES
　　. . . **X-405 ENGINE**
　　. . LIQUID PROPELLANT ROCKET
　　　　ENGINES
　　. . . **X-405 ENGINE**
RT　VANGUARD PROJECT

XANTHIC ACIDS
GS　ACIDS
　　. **XANTHIC ACIDS**
RT　ORGANIC LIQUIDS

XANTHINES
GS　FUNGICIDES
　　. **XANTHINES**
　　. . GUANINES
　　. . URIC ACID
　　HETEROCYCLIC COMPOUNDS
　　. **XANTHINES**
　　. . CAFFEINE
　　. . GUANINES
　　. . URIC ACID
　　NITROGEN COMPOUNDS
　　. **XANTHINES**
　　. . CAFFEINE
　　. . GUANINES
　　. . URIC ACID
　　PURINES
　　. **XANTHINES**
　　. . CAFFEINE
　　. . GUANINES
　　. . URIC ACID

XB-47 AIRCRAFT
USE　B-47 AIRCRAFT

XB-70 AIRCRAFT
USE　B-70 AIRCRAFT

XBQM-180A AIRCRAFT
USE　VATOL AIRCRAFT

XC-142 AIRCRAFT
UF　C-142 AIRCRAFT
GS　FAIRCHILD-HILLER AIRCRAFT
　　. **XC-142 AIRCRAFT**
　　JET AIRCRAFT
　　. **XC-142 AIRCRAFT**
　　LING-TEMCO-VOUGHT AIRCRAFT
　　. **XC-142 AIRCRAFT**
　　MONOPLANES
　　. **XC-142 AIRCRAFT**
　　RYAN AIRCRAFT
　　. **XC-142 AIRCRAFT**
　　TILT WING AIRCRAFT
　　. **XC-142 AIRCRAFT**
　　TRANSPORT AIRCRAFT
　　. **XC-142 AIRCRAFT**
RT　∞AIRCRAFT
　　TURBOPROP ENGINES

XENON
GS　CHEMICAL ELEMENTS
　　. RARE GASES
　　. . **XENON**
　　. . . XENON ISOTOPES
　　. . . . XENON 129
　　. . . . XENON 133
　　. . . . XENON 135
　　GASES
　　. RARE GASES
　　. . **XENON**
　　. . . XENON ISOTOPES
　　. . . . XENON 129
　　. . . . XENON 133
　　. . . . XENON 135

XENON CHLORIDE LASERS
GS　STIMULATED EMISSION DEVICES
　　. LASERS
　　. . GAS LASERS
　　. . . **XENON CHLORIDE LASERS**
　　. . RARE GAS-HALIDE LASERS
　　. . . **XENON CHLORIDE LASERS**
RT　ELECTRON TRANSITIONS
　　EXCIMER LASERS
　　LASER MATERIALS
　　LASER OUTPUTS
　　ULTRAVIOLET LASERS

XENON COMPOUNDS
RT　∞CHEMICAL COMPOUNDS
　　∞RARE GAS COMPOUNDS

XENON FLUORIDE LASERS
GS　STIMULATED EMISSION DEVICES
　　. LASERS
　　. . GAS LASERS
　　. . . **XENON FLUORIDE LASERS**
　　. . RARE GAS-HALIDE LASERS
　　. . . **XENON FLUORIDE LASERS**
RT　ELECTRON TRANSITIONS
　　EXCIMER LASERS
　　LASER MATERIALS
　　LASER OUTPUTS

XENON ISOTOPES
GS　CHEMICAL ELEMENTS
　　. NUCLIDES
　　. . ISOTOPES
　　. . . **XENON ISOTOPES**
　　. . . . XENON 129
　　. . . . XENON 133
　　. . . . XENON 135
　　. RARE GASES
　　. . XENON
　　. . . **XENON ISOTOPES**
　　. . . . XENON 129
　　. . . . XENON 133
　　. . . . XENON 135
　　GASES
　　. RARE GASES
　　. . XENON
　　. . . **XENON ISOTOPES**
　　. . . . XENON 129
　　. . . . XENON 133
　　. . . . XENON 135

XENON LAMPS
GS　LIGHTING EQUIPMENT
　　. LUMINAIRES
　　. . **XENON LAMPS**
RT　ARC LAMPS
　　FLASH LAMPS
　　INFRARED RADIATION
　　MERCURY LAMPS

XENON 129
GS　CHEMICAL ELEMENTS
　　. NUCLIDES
　　. . ISOTOPES
　　. . . XENON ISOTOPES
　　. . . . **XENON 129**
　　. RARE GASES
　　. . XENON
　　. . . XENON ISOTOPES
　　. . . . **XENON 129**
　　GASES
　　. RARE GASES
　　. . XENON
　　. . . XENON ISOTOPES
　　. . . . **XENON 129**

XENON 133
GS　CHEMICAL ELEMENTS
　　. NUCLIDES
　　. . ISOTOPES
　　. . . RADIOACTIVE ISOTOPES
　　. . . . **XENON 133**
　　. . . XENON ISOTOPES
　　. . . . **XENON 133**
　　. RARE GASES
　　. . XENON
　　. . . XENON ISOTOPES
　　. . . . **XENON 133**
　　GASES
　　. RARE GASES
　　. . XENON
　　. . . XENON ISOTOPES
　　. . . . **XENON 133**

XENON 135
GS　CHEMICAL ELEMENTS
　　. NUCLIDES
　　. . ISOTOPES
　　. . . RADIOACTIVE ISOTOPES
　　. . . . **XENON 135**
　　. . . XENON ISOTOPES
　　. . . . **XENON 135**
　　. RARE GASES
　　. . XENON
　　. . . XENON ISOTOPES
　　. . . . **XENON 135**
　　GASES
　　. RARE GASES
　　. . XENON
　　. . . XENON ISOTOPES
　　. . . . **XENON 135**

XEROGRAPHY
GS　IMAGERY
　　. REPRODUCTION (COPYING)
　　. . **XEROGRAPHY**
RT　ELECTROSTATIC CHARGE
　　PHOTOGRAPHS
　　PHOTOGRAPHY

XH-51 HELICOPTER
UF　AEROGYRO HELICOPTERS
　　CL-595 HELICOPTER
　　H-51 HELICOPTER
　　LOCKHEED CL-595 HELICOPTER
　　LOCKHEED 186 HELICOPTER
GS　LOCKHEED AIRCRAFT
　　. **XH-51 HELICOPTER**
　　RESEARCH AIRCRAFT
　　. **XH-51 HELICOPTER**
　　V/STOL AIRCRAFT
　　. ROTARY WING AIRCRAFT
　　. . HELICOPTERS
　　. . . RIGID ROTOR HELICOPTERS
　　. . . . **XH-51 HELICOPTER**

XI HYPERONS
GS　PARTICLES
　　. ELEMENTARY PARTICLES
　　. . BOSONS
　　. . . **XI HYPERONS**
　　. FERMIONS
　　. . BARYONS
　　. . . HYPERONS
　　. . . . **XI HYPERONS**
　　. NUCLEAR PARTICLES
　　. . BOSONS
　　. . . **XI HYPERONS**

XJ-34-WE-32 ENGINE
USE　J-34 ENGINE

XJ-79-GE-1 ENGINE
USE　J-79 ENGINE

XLR-91-AJ-5 ENGINE
USE   LR-91-AJ-5 ENGINE

XLR-99 ENGINE
GS   ENGINES
. ROCKET ENGINES
. . LIQUID PROPELLANT ROCKET
ENGINES
. . . **XLR-99 ENGINE**
RT   X-15 AIRCRAFT

XM-6 SQUIB
USE   SQUIBS

XM-8 SQUIB
USE   SQUIBS

XM-33 ENGINE
UF   TX-33-39 ENGINE
GS   ENGINES
. ROCKET ENGINES
. . SOLID PROPELLANT ROCKET
ENGINES
. . . **XM-33 ENGINE**
RT   BLUE SCOUT ROCKET VEHICLE
EXOS SOUNDING ROCKET
LITTLE JOE 2 LAUNCH VEHICLE
POLARIS MISSILES
SCOUT LAUNCH VEHICLE
TX-354 ENGINE

XV-3 AIRCRAFT
UF   V-3 AIRCRAFT
GS   BELL AIRCRAFT
. **XV-3 AIRCRAFT**
V/STOL AIRCRAFT
. **XV-3 AIRCRAFT**
RT   ∞AIRCRAFT
TILTING ROTORS

XV-4 AIRCRAFT
UF   HUMMINGBIRD AIRCRAFT
LOCKHEED XV-4A AIRCRAFT
V-4 AIRCRAFT
VZ-10 AIRCRAFT
GS   JET AIRCRAFT
. **XV-4 AIRCRAFT**
LOCKHEED AIRCRAFT
. **XV-4 AIRCRAFT**
MONOPLANES
. **XV-4 AIRCRAFT**
RESEARCH AIRCRAFT
. **XV-4 AIRCRAFT**
V/STOL AIRCRAFT
. VERTICAL TAKEOFF AIRCRAFT
. . **XV-4 AIRCRAFT**
RT   ∞AIRCRAFT

XV-5 AIRCRAFT
UF   V-5 AIRCRAFT
VZ-11 AIRCRAFT
XV-5A AIRCRAFT
GS   FAN IN WING AIRCRAFT
. **XV-5 AIRCRAFT**
JET AIRCRAFT
. **XV-5 AIRCRAFT**
MONOPLANES
. **XV-5 AIRCRAFT**
RESEARCH AIRCRAFT
. **XV-5 AIRCRAFT**
RYAN AIRCRAFT
. **XV-5 AIRCRAFT**
V/STOL AIRCRAFT
. **XV-5 AIRCRAFT**
RT   ∞AIRCRAFT

XV-5A AIRCRAFT
USE   XV-5 AIRCRAFT

XV-6A AIRCRAFT
USE   P-1127 AIRCRAFT

XV-8A AIRCRAFT
GS   RESEARCH AIRCRAFT
. **XV-8A AIRCRAFT**
RYAN AIRCRAFT
. **XV-8A AIRCRAFT**
UTILITY AIRCRAFT
. **XV-8A AIRCRAFT**
V/STOL AIRCRAFT
. **XV-8A AIRCRAFT**
RT   ∞AIRCRAFT
FLEXIBLE WINGS

XV-9A AIRCRAFT
UF   V-9 AIRCRAFT
GS   HUGHES AIRCRAFT
. **XV-9A AIRCRAFT**
JET AIRCRAFT
. **XV-9A AIRCRAFT**
RESEARCH AIRCRAFT
. **XV-9A AIRCRAFT**
V/STOL AIRCRAFT
. ROTARY WING AIRCRAFT
. . HELICOPTERS
. . . **XV-9A AIRCRAFT**
RT   ∞AIRCRAFT
TIP DRIVEN ROTORS

XV-11A AIRCRAFT
GS   RESEARCH AIRCRAFT
. **XV-11A AIRCRAFT**
V/STOL AIRCRAFT
. VERTICAL TAKEOFF AIRCRAFT
. . **XV-11A AIRCRAFT**
RT   ∞AIRCRAFT
LIFT FANS
SHROUDED PROPELLERS

XV-15 AIRCRAFT
GS   BELL AIRCRAFT
. **XV-15 AIRCRAFT**
V/STOL AIRCRAFT
. ROTARY WING AIRCRAFT
. . TILT ROTOR AIRCRAFT
. . . **XV-15 AIRCRAFT**
RT   ∞AIRCRAFT
HELICOPTERS
TILT ROTOR RESEARCH AIRCRAFT
PROGRAM

XYLENE
GS   HYDROCARBONS
. **XYLENE**
RT   TOLUENE

XYLOSE
GS   ALIPHATIC COMPOUNDS
. SUGARS
. . MONOSACCHARIDES
. . . **XYLOSE**
. . PENTOSE
. . . **XYLOSE**
CARBOHYDRATES
. SUGARS
. . MONOSACCHARIDES
. . . **XYLOSE**
. . PENTOSE
. . . **XYLOSE**

Y

YAG (GARNET)
USE   YTTRIUM-ALUMINUM GARNET

YAG LASERS
GS   ELECTRONIC EQUIPMENT
. SOLID STATE DEVICES
. . SOLID STATE LASERS
. . . **YAG LASERS**
STIMULATED EMISSION DEVICES
. LASERS
. . SOLID STATE LASERS
. . . **YAG LASERS**
RT   LASER HEATING
LASER MATERIALS
LASER OUTPUTS

YAGI ANTENNAS
GS   ANTENNAS
. DIRECTIONAL ANTENNAS
. . **YAGI ANTENNAS**
ARRAYS
. ANTENNA ARRAYS
. . LINEAR ARRAYS
. . . ENDFIRE ARRAYS
. . . . **YAGI ANTENNAS**
RT   ANTENNA DESIGN
DIPOLE ANTENNAS
DIRECTORS (ANTENNA ELEMENTS)
WAVEGUIDE ANTENNAS

YAK 40 AIRCRAFT
GS   GENERAL AVIATION AIRCRAFT
. **YAK 40 AIRCRAFT**

YAK 40 AIRCRAFT-*(CONT.)*
JET AIRCRAFT
. **YAK 40 AIRCRAFT**
LIGHT AIRCRAFT
. **YAK 40 AIRCRAFT**
PASSENGER AIRCRAFT
. **YAK 40 AIRCRAFT**
RT   ∞AIRCRAFT

YANG-MILLS FIELDS
RT   ELECTROMAGNETIC FIELDS
FIELD THEORY (PHYSICS)
GAUGE THEORY
GRAVITATIONAL FIELDS
PERTURBATION THEORY
YANG-MILLS THEORY

YANG-MILLS THEORY
RT   FIELD THEORY (PHYSICS)
GAUGE THEORY
PERTURBATION THEORY
SPACE-TIME FUNCTIONS
STATISTICAL ANALYSIS
THEORETICAL PHYSICS
∞THEORIES
YANG-MILLS FIELDS

YARNS
RT   CORDAGE
COTTON
FIBERS
∞ROVINGS
STRANDS
WOOL

YAW
UF   DAMPING IN YAW
FISHTAILING
YAWMETERS
GS   ATTITUDE (INCLINATION)
. **YAW**
RT   AERODYNAMIC STABILITY
DIRECTIONAL CONTROL
DIRECTIONAL STABILITY
LATERAL OSCILLATION
∞MOTION
PITCH (INCLINATION)
ROLL
ROTATION
SIDESLIP
SKIDDING
TURNING FLIGHT
YAWING MOMENTS

YAWING MOMENTS
GS   MOMENTS
. STABILITY DERIVATIVES
. . **YAWING MOMENTS**
RT   AERODYNAMIC COEFFICIENTS
LATERAL OSCILLATION
MOMENTS OF INERTIA
PITCHING MOMENTS
ROLLING MOMENTS
TORQUE
YAW

YAWMETERS
USE   ATTITUDE INDICATORS
YAW

YC-14 AIRCRAFT
GS   TRANSPORT AIRCRAFT
. CARGO AIRCRAFT
. . **YC-14 AIRCRAFT**
RT   ∞AIRCRAFT
BOEING AIRCRAFT
∞MILITARY AIRCRAFT

YC-15 AIRCRAFT
USE   C-15 AIRCRAFT

YC-123 AIRCRAFT
USE   C-123 AIRCRAFT

YEAST
GS   FUNGI
. **YEAST**
RT   ∞FOOD

YELLOWSTONE NATIONAL PARK (ID-MT-WY)
GS   LAND
. PARKS
. . NATIONAL PARKS

YELLOWSTONE NATIONAL PARK-*(CONT.)*
   . . . YELLOWSTONE NATIONAL PARK
      (ID-MT-WY)
RT   IDAHO
       MONTANA
       WYOMING

YEMEN
GS   NATIONS
      . **YEMEN**
RT   ASIA

YF-12 AIRCRAFT
GS   ATTACK AIRCRAFT
      . FIGHTER AIRCRAFT
      . . **YF-12 AIRCRAFT**
RT   ∞AIRCRAFT
       AIRCRAFT DESIGN
       ∞INTERCEPTORS
       JET AIRCRAFT
       ∞MILITARY AIRCRAFT
       RECONNAISSANCE AIRCRAFT
       RESEARCH AIRCRAFT

YF-16 AIRCRAFT
GS   ATTACK AIRCRAFT
      . FIGHTER AIRCRAFT
      . . **YF-16 AIRCRAFT**
RT   ∞AIRCRAFT
       ∞MILITARY AIRCRAFT

YF-17 AIRCRAFT
USE   F-17 AIRCRAFT

YF-102 AIRCRAFT
USE   F-102 AIRCRAFT

YHU-1 HELICOPTER
USE   UH-1 HELICOPTER

YIELD
RT   LOSSES
       OUTPUT

YIELD POINT
UF   DAMAGE THRESHOLD
       LUDER BANDS
GS   MECHANICAL PROPERTIES
      . PLASTIC PROPERTIES
      . . **YIELD POINT**
RT   MICROYIELD STRENGTH
       TRESCA FLOW

YIELD STRENGTH
GS   MECHANICAL PROPERTIES
      . **YIELD STRENGTH**
      . . MICROYIELD STRENGTH
RT   ELASTIC PROPERTIES
       FRACTURE STRENGTH
       HIGH STRENGTH
       J INTEGRAL
       PLASTIC DEFORMATION
       ∞STRENGTH
       STRESS-STRAIN DIAGRAMS
       STRESS-STRAIN RELATIONSHIPS
       STRESSES
       TEMPERATURE INVERSIONS

YIG (GARNET)
USE   YTTRIUM-IRON GARNET

YJ-73-GE-3 ENGINE
USE   J-73 ENGINE

YJ-79 ENGINE
USE   J-79 ENGINE

YJ-85 ENGINE
USE   J-85 ENGINE

YJ-93 ENGINE
USE   J-93 ENGINE

YJ-93-GE-3 ENGINE
USE   J-93 ENGINE

YJ73 TURBOJET ENGINE
USE   J-73 ENGINE

YLR-91-AJ-1 ENGINE
GS   ENGINES
      . ROCKET ENGINES
      . . LIQUID PROPELLANT ROCKET
         ENGINES

YLR-91-AJ-1 ENGINE-*(CONT.)*
      . . . **YLR-91-AJ-1 ENGINE**
RT   TITAN ICBM

YLR-99-RM-1 ENGINE
USE   LR-99 ENGINE

YO-YO DEVICES
RT   ANGULAR ACCELERATION
       GYROSCOPIC STABILITY
       SATELLITE ROTATION
       SPIN
       SPIN REDUCTION

YOKES
RT   BEAM WAVEGUIDES
       CONNECTORS
       COUPLERS
       COUPLES
       DEFLECTION
       DIRECTIONAL COUPLERS
       FERROMAGNETIC MATERIALS
       ∞JOINING
       LINKAGES
       MAGNET COILS

YOUNG MODULUS
USE   MODULUS OF ELASTICITY

YOUNG-HELMHOLTZ THEORY
RT   COLOR VISION
       PHOTORECEPTORS
       ∞THEORIES

YOUTH
RT   GROWTH
       HUMAN BEINGS

YS-11 AIRCRAFT
UF   NIHON YS-11 AIRCRAFT
GS   JET AIRCRAFT
      . TURBOPROP AIRCRAFT
      . . **YS-11 AIRCRAFT**
      MONOPLANES
      . **YS-11 AIRCRAFT**
      NIHON AIRCRAFT
      . **YS-11 AIRCRAFT**
      PASSENGER AIRCRAFT
      . **YS-11 AIRCRAFT**
      TRANSPORT AIRCRAFT
      . **YS-11 AIRCRAFT**

YT-2 AIRCRAFT
USE   T-2 AIRCRAFT

YTTERBIUM
GS   CHEMICAL ELEMENTS
      . RARE EARTH ELEMENTS
      . . **YTTERBIUM**
      . . . YTTERBIUM ISOTOPES
      METALS
      . RARE EARTH ELEMENTS
      . . **YTTERBIUM**
      . . . YTTERBIUM ISOTOPES

YTTERBIUM COMPOUNDS
GS   RARE EARTH COMPOUNDS
      . **YTTERBIUM COMPOUNDS**
RT   ∞CHEMICAL COMPOUNDS
       ∞METAL COMPOUNDS

YTTERBIUM ISOTOPES
GS   CHEMICAL ELEMENTS
      . RARE EARTH ELEMENTS
      . . YTTERBIUM
      . . . **YTTERBIUM ISOTOPES**
      METALS
      . RARE EARTH ELEMENTS
      . . YTTERBIUM
      . . . **YTTERBIUM ISOTOPES**

YTTRIUM
GS   CHEMICAL ELEMENTS
      . RARE EARTH ELEMENTS
      . . **YTTRIUM**
      . . . YTTRIUM ISOTOPES
      METALS
      . RARE EARTH ELEMENTS
      . . **YTTRIUM**
      . . . YTTRIUM ISOTOPES
      . TRANSITION METALS
      . . **YTTRIUM**
      . . . YTTRIUM ISOTOPES

YTTRIUM ALLOYS
GS   ALLOYS
      . **YTTRIUM ALLOYS**
RT   RARE EARTH ALLOYS

YTTRIUM COMPOUNDS
GS   **YTTRIUM COMPOUNDS**
      . YTTRIUM OXIDES
      . YTTRIUM-ALUMINUM GARNET
      . YTTRIUM-IRON GARNET
RT   ∞CHEMICAL COMPOUNDS
       ∞GROUP 3B COMPOUNDS
       ∞METAL COMPOUNDS

YTTRIUM ISOTOPES
GS   CHEMICAL ELEMENTS
      . NUCLIDES
      . . ISOTOPES
      . . . **YTTRIUM ISOTOPES**
      . RARE EARTH ELEMENTS
      . . YTTRIUM
      . . . **YTTRIUM ISOTOPES**
      METALS
      . RARE EARTH ELEMENTS
      . . YTTRIUM
      . . . **YTTRIUM ISOTOPES**
      . TRANSITION METALS
      . . YTTRIUM
      . . . **YTTRIUM ISOTOPES**

YTTRIUM OXIDES
GS   CHALCOGENIDES
      . OXIDES
      . . METAL OXIDES
      . . . **YTTRIUM OXIDES**
      YTTRIUM COMPOUNDS
      . **YTTRIUM OXIDES**

YTTRIUM-ALUMINUM GARNET
UF   YAG (GARNET)
GS   SILICON COMPOUNDS
      . SILICATES
      . . GARNETS
      . . . **YTTRIUM-ALUMINUM GARNET**
      YTTRIUM COMPOUNDS
      . **YTTRIUM-ALUMINUM GARNET**
RT   FERRITES
       MAGNETOSTATIC AMPLIFIERS
       MINERALS

YTTRIUM-IRON GARNET
UF   YIG (GARNET)
GS   SILICON COMPOUNDS
      . SILICATES
      . . GARNETS
      . . . **YTTRIUM-IRON GARNET**
      YTTRIUM COMPOUNDS
      . **YTTRIUM-IRON GARNET**
RT   FERRITES
       MAGNETOSTATIC AMPLIFIERS
       MINERALS
       WAVEGUIDE TUNERS

YUGOSLAVIA
GS   NATIONS
      . **YUGOSLAVIA**
RT   ADRIATIC SEA
       EUROPE

YUH-1 HELICOPTER
USE   UH-1 HELICOPTER

YUH-60A HELICOPTER
USE   UH-60A HELICOPTER

YUH-61A HELICOPTER
USE   UH-61A HELICOPTER

YUKAWA POTENTIAL
RT   MESON-NUCLEON INTERACTIONS
       ∞POTENTIAL

YUKON AIRCRAFT
USE   CL-44 AIRCRAFT

YUKON TERRITORY
GS   NATIONS
      . CANADA
      . . **YUKON TERRITORY**

Z

Z-37 AIRCRAFT
UF   OMNIPOL Z-37 AIRCRAFT
GS   MONOPLANES
.   **Z-37 AIRCRAFT**
UTILITY AIRCRAFT
.   **Z-37 AIRCRAFT**

ZAIRE
UF   BELGIAN CONGO
CONGO (KINSHASA)
GS   NATIONS
.   **ZAIRE**
RT   AFRICA

ZAMBIA
GS   NATIONS
.   **ZAMBIA**
RT   AFRICA

ZEEMAN EFFECT
RT   ∞EFFECTS
MAGNETIC FIELDS
SPECTROSCOPY
SPECTRUM ANALYSIS
STARK EFFECT
VOIGT EFFECT

ZENER DIODES
USE   AVALANCHE DIODES

ZENER EFFECT
RT   BARRIER LAYERS
CARRIER DENSITY (SOLID STATE)
∞CARRIERS
∞EFFECTS
ELECTRIC DISCHARGES
FIELD EMISSION

ZENITH
RT   ANTIPODES
APEXES
CELESTIAL SPHERE
MAXIMA
NOON
SOLAR POSITION

ZEOLITES
GS   SILICON COMPOUNDS
.   SILICATES
. .   **ZEOLITES**
RT   ION EXCHANGE RESINS
MINERALS

ZERO ANGLE OF ATTACK
GS   GEOMETRY
.   EUCLIDEAN GEOMETRY
. .   ANGLES (GEOMETRY)
. . .   ANGLE OF ATTACK
. . . .   **ZERO ANGLE OF ATTACK**

ZERO CROSSINGS
USE   ROOTS OF EQUATIONS

ZERO FORCE CURVES
RT   CURVATURE
∞CURVES
∞FORCE

ZERO GRAVITY
USE   WEIGHTLESSNESS

ZERO LIFT
GS   AERODYNAMIC CHARACTERISTICS
.   LIFT
. .   **ZERO LIFT**
AERODYNAMIC FORCES
.   LIFT
. .   **ZERO LIFT**
DYNAMIC CHARACTERISTICS
.   LIFT
. .   **ZERO LIFT**
RT   AERODYNAMIC STALLING
BOUNDARY LAYER SEPARATION
DISTRIBUTION (PROPERTY)

ZERO POINT ENERGY
RT   ABSOLUTE ZERO
FIELD THEORY (PHYSICS)
KINETIC ENERGY
THERMODYNAMIC PROPERTIES

ZERO POWER REACTOR 2
GS   NUCLEAR REACTORS
.   LIQUID COOLED REACTORS
. .   WATER COOLED REACTORS
. . .   HEAVY WATER REACTORS
. . . .   **ZERO POWER REACTOR 2**
. . .   ZERO POWER REACTORS
. . . .   **ZERO POWER REACTOR 2**
.   NUCLEAR RESEARCH AND TEST
REACTORS
. .   **ZERO POWER REACTOR 2**

ZERO POWER REACTOR 3
GS   NUCLEAR REACTORS
.   LIQUID COOLED REACTORS
. .   WATER COOLED REACTORS
. . .   ZERO POWER REACTORS
. . . .   **ZERO POWER REACTOR 3**
.   NUCLEAR RESEARCH AND TEST
REACTORS
. .   **ZERO POWER REACTOR 3**

ZERO POWER REACTOR 6
GS   NUCLEAR REACTORS
.   LIQUID COOLED REACTORS
. .   WATER COOLED REACTORS
. . .   ZERO POWER REACTORS
. . . .   **ZERO POWER REACTOR 6**
.   NUCLEAR RESEARCH AND TEST
REACTORS
. .   **ZERO POWER REACTOR 6**

ZERO POWER REACTOR 9
GS   NUCLEAR REACTORS
.   LIQUID COOLED REACTORS
. .   WATER COOLED REACTORS
. . .   ZERO POWER REACTORS
. . . .   **ZERO POWER REACTOR 9**
.   NUCLEAR RESEARCH AND TEST
REACTORS
. .   **ZERO POWER REACTOR 9**

ZERO POWER REACTORS
UF   ZPR REACTORS
GS   NUCLEAR REACTORS
.   LIQUID COOLED REACTORS
. .   WATER COOLED REACTORS
. . .   **ZERO POWER REACTORS**
. . . .   ZERO POWER REACTOR 2
. . . .   ZERO POWER REACTOR 3
. . . .   ZERO POWER REACTOR 6
. . . .   ZERO POWER REACTOR 9

ZERO SOUND
GS   ACOUSTIC PROPERTIES
.   SOUND INTENSITY
. .   **ZERO SOUND**
RATES (PER TIME)
.   FLUX DENSITY
. .   SOUND INTENSITY
. . .   **ZERO SOUND**
RT   ACOUSTIC ATTENUATION
ACOUSTICS
ANECHOIC CHAMBERS
REFLECTION
SILENCERS
SUBAUDIBLE FREQUENCIES

ZERO-G ACPL (SPACELAB)
USE   ATMOSPHERIC CLOUD PHYSICS LAB
(SPACELAB)

ZETA AURIGAE STAR
GS   CELESTIAL BODIES
.   STARS
. .   **ZETA AURIGAE STAR**
RT   AURIGA CONSTELLATION

ZETA PINCH
GS   PINCH EFFECT
.   PLASMA PINCH
. .   **ZETA PINCH**
RT   CONTROLLED FUSION
MAGNETOHYDRODYNAMIC STABILITY
PLASMA COMPRESSION
PLASMA CONTROL
PLASMA ELECTRODES
PLASMA FOCUS
Q DEVICES
ROTATING PLASMAS
SCREW PINCH
THETA PINCH

ZETA THERMONUCLEAR REACTOR
RT   PINCH EFFECT
THERMONUCLEAR POWER GENERATION
THERMONUCLEAR REACTIONS

ZEUS MISSILE
USE   NIKE-ZEUS MISSILE

ZIEGLER CATALYST
GS   CATALYSTS
.   **ZIEGLER CATALYST**
RT   POLYMERIZATION

ZIMBABWE
UF   RHODESIA
GS   NATIONS
.   **ZIMBABWE**
RT   AFRICA

ZINC
GS   CHEMICAL ELEMENTS
.   **ZINC**
. .   ZINC ISOTOPES
METALS
.   TRANSITION METALS
. .   **ZINC**
. . .   ZINC ISOTOPES

ZINC ALLOYS
GS   ALLOYS
.   **ZINC ALLOYS**
RT   BEARING ALLOYS
SOLDERS

ZINC ANTIMONIDES
GS   ANTIMONY COMPOUNDS
.   ANTIMONIDES
. .   **ZINC ANTIMONIDES**
ZINC COMPOUNDS
.   **ZINC ANTIMONIDES**

ZINC CHLORIDES
GS   HALOGEN COMPOUNDS
.   CHLORINE COMPOUNDS
. .   CHLORIDES
. . .   **ZINC CHLORIDES**
.   HALIDES
. .   CHLORIDES
. . .   **ZINC CHLORIDES**
. .   METAL HALIDES
. . .   **ZINC CHLORIDES**
ZINC COMPOUNDS
.   **ZINC CHLORIDES**

ZINC COATINGS
UF   GALVANIZING
GS   COATINGS
.   METAL COATINGS
. .   **ZINC COATINGS**
METALS
.   METAL COATINGS
. .   **ZINC COATINGS**
RT   PROTECTIVE COATINGS

ZINC COMPOUNDS
GS   **ZINC COMPOUNDS**
.   WURTZITE
.   ZINC ANTIMONIDES
.   ZINC CHLORIDES
.   ZINC FLUORIDES
.   ZINC OXIDES
.   ZINC SELENIDES
.   ZINC SULFIDES
.   ZINCBLENDE
.   ZINC TELLURIDES
.   ZINC TUNGSTATES
RT   ∞CHEMICAL COMPOUNDS
∞GROUP 2B COMPOUNDS
∞METAL COMPOUNDS

ZINC FLUORIDES
GS   HALOGEN COMPOUNDS
.   FLUORINE COMPOUNDS
. .   FLUORIDES
. . .   METAL FLUORIDES
. . . .   **ZINC FLUORIDES**
.   HALIDES
. .   METAL HALIDES
. . .   **ZINC FLUORIDES**
ZINC COMPOUNDS
.   **ZINC FLUORIDES**

ZINC ISOTOPES
GS   CHEMICAL ELEMENTS

ZINC ISOTOPES-*(CONT.)*
. NUCLIDES
. . ISOTOPES
. . . **ZINC ISOTOPES**
. ZINC
. . **ZINC ISOTOPES**
METALS
. TRANSITION METALS
. . ZINC
. . . **ZINC ISOTOPES**

ZINC NICKEL BATTERIES
USE   NICKEL ZINC BATTERIES

ZINC OXIDES
GS   CHALCOGENIDES
. OXIDES
. . METAL OXIDES
. . . **ZINC OXIDES**
ZINC COMPOUNDS
. **ZINC OXIDES**

ZINC SELENIDES
GS   CHALCOGENIDES
. SELENIDES
. . **ZINC SELENIDES**
SELENIUM COMPOUNDS
. SELENIDES
. . **ZINC SELENIDES**
ZINC COMPOUNDS
. **ZINC SELENIDES**
RT   SCHOTTKY DIODES

ZINC SILVER BATTERIES
USE   SILVER ZINC BATTERIES

ZINC SILVER OXIDE BATTERIES
USE   SILVER ZINC BATTERIES

ZINC SULFIDES
GS   CHALCOGENIDES
. SULFIDES
. . INORGANIC SULFIDES
. . . **ZINC SULFIDES**
. . . . WURTZITE
. . . . ZINCBLENDE
SULFUR COMPOUNDS
. SULFIDES
. . INORGANIC SULFIDES
. . . **ZINC SULFIDES**
. . . . WURTZITE
. . . . ZINCBLENDE
ZINC COMPOUNDS
. **ZINC SULFIDES**
. . ZINCBLENDE

ZINC TELLURIDES
GS   CHALCOGENIDES
. TELLURIDES
. . **ZINC TELLURIDES**
TELLURIUM COMPOUNDS
. TELLURIDES
. . **ZINC TELLURIDES**
ZINC COMPOUNDS
. **ZINC TELLURIDES**

ZINC TUNGSTATES
GS   TUNGSTEN COMPOUNDS
. TUNGSTATES
. . **ZINC TUNGSTATES**
ZINC COMPOUNDS
. **ZINC TUNGSTATES**

ZINC-BROMIDE BATTERIES
GS   ELECTROCHEMICAL CELLS
. ELECTRIC BATTERIES
. . STORAGE BATTERIES
. . . **ZINC-BROMIDE BATTERIES**
RT   ZINC-CHLORINE BATTERIES

ZINC-CHLORINE BATTERIES
GS   ELECTROCHEMICAL CELLS
. ELECTRIC BATTERIES
. . STORAGE BATTERIES
. . . **ZINC-CHLORINE BATTERIES**
RT   ZINC-BROMIDE BATTERIES

ZINC-OXYGEN BATTERIES
GS   ELECTRIC GENERATORS
. DIRECT POWER GENERATORS
. . PRIMARY BATTERIES
. . . METAL AIR BATTERIES
. . . . **ZINC-OXYGEN BATTERIES**
ELECTROCHEMICAL CELLS

ZINC-OXYGEN BATTERIES-*(CONT.)*
. ELECTRIC BATTERIES
. . PRIMARY BATTERIES
. . . METAL AIR BATTERIES
. . . . **ZINC-OXYGEN BATTERIES**

ZINCBLENDE
UF   SPHALERITE
GS   CHALCOGENIDES
. SULFIDES
. . INORGANIC SULFIDES
. . . ZINC SULFIDES
. . . . **ZINCBLENDE**
MINERALS
. **ZINCBLENDE**
SULFUR COMPOUNDS
. SULFIDES
. . INORGANIC SULFIDES
. . . ZINC SULFIDES
. . . . **ZINCBLENDE**
ZINC COMPOUNDS
. ZINC SULFIDES
. . **ZINCBLENDE**

ZIPPERS
GS   FASTENERS
. **ZIPPERS**
RT   HOLDERS

ZIRCALOY 2 (TRADEMARK)
GS   ALLOYS
. ZIRCONIUM ALLOYS
. . ZIRCALOYS (TRADEMARK)
. . . **ZIRCALOY 2 (TRADEMARK)**

ZIRCALOYS (TRADEMARK)
GS   ALLOYS
. ZIRCONIUM ALLOYS
. . **ZIRCALOYS (TRADEMARK)**
. . . ZIRCALOY 2 (TRADEMARK)
RT   IRON ALLOYS
TIN ALLOYS

ZIRCONATES
GS   ZIRCONIUM COMPOUNDS
. **ZIRCONATES**
. . BARIUM ZIRCONATES
. . STRONTIUM ZIRCONATES

ZIRCONIUM
GS   CHEMICAL ELEMENTS
. **ZIRCONIUM**
. . ZIRCONIUM ISOTOPES
. . . ZIRCONIUM 95
METALS
. TRANSITION METALS
. . **ZIRCONIUM**
. . . ZIRCONIUM ISOTOPES
. . . . ZIRCONIUM 95

ZIRCONIUM ALLOYS
GS   ALLOYS
. **ZIRCONIUM ALLOYS**
. . ZIRCALOYS (TRADEMARK)
. . . ZIRCALOY 2 (TRADEMARK)

ZIRCONIUM CARBIDES
GS   CARBON COMPOUNDS
. CARBIDES
. . **ZIRCONIUM CARBIDES**
ZIRCONIUM COMPOUNDS
. **ZIRCONIUM CARBIDES**

ZIRCONIUM COMPOUNDS
GS   **ZIRCONIUM COMPOUNDS**
. ZIRCONATES
. . BARIUM ZIRCONATES
. . STRONTIUM ZIRCONATES
. ZIRCONIUM CARBIDES
. ZIRCONIUM HYDRIDES
. ZIRCONIUM IODIDES
. ZIRCONIUM NITRIDES
. ZIRCONIUM OXIDES
. ZIRCONIUM TITANATES
RT   ∞CHEMICAL COMPOUNDS
∞GROUP 4B COMPOUNDS
∞METAL COMPOUNDS

ZIRCONIUM HYDRIDES
GS   HYDROGEN COMPOUNDS
. HYDRIDES
. . **ZIRCONIUM HYDRIDES**
ZIRCONIUM COMPOUNDS
. **ZIRCONIUM HYDRIDES**

ZIRCONIUM IODIDES
GS   HALOGEN COMPOUNDS
. HALIDES
. . METAL HALIDES
. . . **ZIRCONIUM IODIDES**
. IODINE COMPOUNDS
. . IODIDES
. . . **ZIRCONIUM IODIDES**
ZIRCONIUM COMPOUNDS
. **ZIRCONIUM IODIDES**

ZIRCONIUM ISOTOPES
GS   CHEMICAL ELEMENTS
. NUCLIDES
. ISOTOPES
. . . **ZIRCONIUM ISOTOPES**
. . . . ZIRCONIUM 95
. ZIRCONIUM
. **ZIRCONIUM ISOTOPES**
. . . ZIRCONIUM 95
METALS
. TRANSITION METALS
. . ZIRCONIUM
. . . **ZIRCONIUM ISOTOPES**
. . . . ZIRCONIUM 95

ZIRCONIUM NITRIDES
GS   NITROGEN COMPOUNDS
. NITRIDES
. . **ZIRCONIUM NITRIDES**
ZIRCONIUM COMPOUNDS
. **ZIRCONIUM NITRIDES**
RT   METAL NITRIDES

ZIRCONIUM OXIDES
GS   CHALCOGENIDES
. OXIDES
. . METAL OXIDES
. . . **ZIRCONIUM OXIDES**
ZIRCONIUM COMPOUNDS
. **ZIRCONIUM OXIDES**

ZIRCONIUM TITANATES
GS   TITANIUM COMPOUNDS
. TITANATES
. . **ZIRCONIUM TITANATES**
ZIRCONIUM COMPOUNDS
. **ZIRCONIUM TITANATES**

ZIRCONIUM 95
GS   CHEMICAL ELEMENTS
. NUCLIDES
. . ISOTOPES
. . . RADIOACTIVE ISOTOPES
. . . . **ZIRCONIUM 95**
. . . ZIRCONIUM ISOTOPES
. . . . **ZIRCONIUM 95**
. ZIRCONIUM
. . ZIRCONIUM ISOTOPES
. . . **ZIRCONIUM 95**
METALS
. TRANSITION METALS
. . ZIRCONIUM
. . . ZIRCONIUM ISOTOPES
. . . . **ZIRCONIUM 95**

ZODIAC
RT   CONSTELLATIONS
ECLIPTIC
SCORPIUS CONSTELLATION
SCUTUM CONSTELLATION

ZODIACAL DUST
GS   CELESTIAL BODIES
. METEOROIDS
. . MICROMETEOROIDS
. . . METEOROID DUST CLOUDS
. . . . **ZODIACAL DUST**
DUST
. COSMIC DUST
. . INTERPLANETARY DUST
. . . METEOROID DUST CLOUDS
. . . . **ZODIACAL DUST**
MEDIA
. INTERPLANETARY MEDIUM
. . INTERPLANETARY DUST
. . . METEOROID DUST CLOUDS
. . . . **ZODIACAL DUST**
RT   EXPLORER SATELLITES
MICROMETEORITES
POYNTING-ROBERTSON EFFECT
TERRESTRIAL DUST BELT

ZODIACAL LIGHT
GS   ELECTROMAGNETIC RADIATION

ZODIACAL LIGHT-*(CONT.)*
  . LIGHT (VISIBLE RADIATION)
  . . **ZODIACAL LIGHT**
  EXTRATERRESTRIAL RADIATION
  . **ZODIACAL LIGHT**
RT GEGENSCHEIN
  HELIOS PROJECT
  MICROMETEOROIDS
  NIGHT SKY
  POLARIZED LIGHT
  POYNTING-ROBERTSON EFFECT
  SKY BRIGHTNESS
  SOLAR RADIATION
  SUNLIGHT

ZONAL EARTH ENERGY BUDGET EXPERIMENT
USE LZEEBE SATELLITE

ZONAL HARMONICS
GS ANALYSIS (MATHEMATICS)
  . FUNCTIONAL ANALYSIS
  . . HARMONIC ANALYSIS
  . . . **ZONAL HARMONICS**
  HARMONICS
  . **ZONAL HARMONICS**

ZOND SPACE PROBES
GS INTERPLANETARY SPACECRAFT
  . **ZOND SPACE PROBES**
  . . ZOND 1 SPACE PROBE
  . . ZOND 2 SPACE PROBE
  . . ZOND 3 SPACE PROBE
  . . ZOND 4 SPACE PROBE
  . . ZOND 5 SPACE PROBE
  . . ZOND 6 SPACE PROBE
  . . ZOND 7 SPACE PROBE
  . . ZOND 8 SPACE PROBE
  SOVIET SPACECRAFT
  . **ZOND SPACE PROBES**
  . . ZOND 1 SPACE PROBE
  . . ZOND 2 SPACE PROBE
  . . ZOND 3 SPACE PROBE
  . . ZOND 4 SPACE PROBE
  . . ZOND 5 SPACE PROBE
  . . ZOND 6 SPACE PROBE
  . . ZOND 7 SPACE PROBE
  . . ZOND 8 SPACE PROBE
  UNMANNED SPACECRAFT
  . **ZOND SPACE PROBES**
  . . ZOND 1 SPACE PROBE
  . . ZOND 2 SPACE PROBE
  . . ZOND 3 SPACE PROBE
  . . ZOND 4 SPACE PROBE
  . . ZOND 5 SPACE PROBE
  . . ZOND 6 SPACE PROBE
  . . ZOND 7 SPACE PROBE
  . . ZOND 8 SPACE PROBE
RT MARS PROBES

ZOND 1 SPACE PROBE
GS INTERPLANETARY SPACECRAFT
  . VENUS PROBES
  . . **ZOND 1 SPACE PROBE**
  . ZOND SPACE PROBES
  . . **ZOND 1 SPACE PROBE**
  SOVIET SPACECRAFT
  . ZOND SPACE PROBES
  . . **ZOND 1 SPACE PROBE**
  UNMANNED SPACECRAFT
  . SPACE PROBES
  . . VENUS PROBES
  . . . **ZOND 1 SPACE PROBE**
  . ZOND SPACE PROBES
  . . **ZOND 1 SPACE PROBE**

ZOND 2 SPACE PROBE
GS INTERPLANETARY SPACECRAFT
  . MARS PROBES
  . . **ZOND 2 SPACE PROBE**
  . ZOND SPACE PROBES
  . . **ZOND 2 SPACE PROBE**
  SOVIET SPACECRAFT
  . ZOND SPACE PROBES
  . . **ZOND 2 SPACE PROBE**
  UNMANNED SPACECRAFT
  . SPACE PROBES
  . . MARS PROBES
  . . . **ZOND 2 SPACE PROBE**
  . ZOND SPACE PROBES
  . . **ZOND 2 SPACE PROBE**

ZOND 3 SPACE PROBE
GS INTERPLANETARY SPACECRAFT
  . VENUS PROBES
  . . **ZOND 3 SPACE PROBE**

ZOND 3 SPACE PROBE-*(CONT.)*
  . ZOND SPACE PROBES
  . . **ZOND 3 SPACE PROBE**
  SOVIET SPACECRAFT
  . ZOND SPACE PROBES
  . . **ZOND 3 SPACE PROBE**
  UNMANNED SPACECRAFT
  . SPACE PROBES
  . . VENUS PROBES
  . . . **ZOND 3 SPACE PROBE**
  . ZOND SPACE PROBES
  . . **ZOND 3 SPACE PROBE**

ZOND 4 SPACE PROBE
GS INTERPLANETARY SPACECRAFT
  . VENUS PROBES
  . . **ZOND 4 SPACE PROBE**
  . ZOND SPACE PROBES
  . . **ZOND 4 SPACE PROBE**
  SOVIET SPACECRAFT
  . ZOND SPACE PROBES
  . . **ZOND 4 SPACE PROBE**
  UNMANNED SPACECRAFT
  . SPACE PROBES
  . . VENUS PROBES
  . . . **ZOND 4 SPACE PROBE**
  . ZOND SPACE PROBES
  . . **ZOND 4 SPACE PROBE**

ZOND 5 SPACE PROBE
GS INTERPLANETARY SPACECRAFT
  . VENUS PROBES
  . . **ZOND 5 SPACE PROBE**
  . ZOND SPACE PROBES
  . . **ZOND 5 SPACE PROBE**
  SOVIET SPACECRAFT
  . ZOND SPACE PROBES
  . . **ZOND 5 SPACE PROBE**
  UNMANNED SPACECRAFT
  . SPACE PROBES
  . . VENUS PROBES
  . . . **ZOND 5 SPACE PROBE**
  . ZOND SPACE PROBES
  . . **ZOND 5 SPACE PROBE**

ZOND 6 SPACE PROBE
GS INTERPLANETARY SPACECRAFT
  . VENUS PROBES
  . . **ZOND 6 SPACE PROBE**
  . ZOND SPACE PROBES
  . . **ZOND 6 SPACE PROBE**
  SOVIET SPACECRAFT
  . ZOND SPACE PROBES
  . . **ZOND 6 SPACE PROBE**
  UNMANNED SPACECRAFT
  . SPACE PROBES
  . . VENUS PROBES
  . . . **ZOND 6 SPACE PROBE**
  . ZOND SPACE PROBES
  . . **ZOND 6 SPACE PROBE**

ZOND 7 SPACE PROBE
GS INTERPLANETARY SPACECRAFT
  . VENUS PROBES
  . . **ZOND 7 SPACE PROBE**
  . ZOND SPACE PROBES
  . . **ZOND 7 SPACE PROBE**
  SOVIET SPACECRAFT
  . ZOND SPACE PROBES
  . . **ZOND 7 SPACE PROBE**
  UNMANNED SPACECRAFT
  . SPACE PROBES
  . . VENUS PROBES
  . . . **ZOND 7 SPACE PROBE**
  . ZOND SPACE PROBES
  . . **ZOND 7 SPACE PROBE**

ZOND 8 SPACE PROBE
GS INTERPLANETARY SPACECRAFT
  . VENUS PROBES
  . . **ZOND 8 SPACE PROBE**
  . ZOND SPACE PROBES
  . . **ZOND 8 SPACE PROBE**
  SOVIET SPACECRAFT
  . ZOND SPACE PROBES
  . . **ZOND 8 SPACE PROBE**
  UNMANNED SPACECRAFT
  . SPACE PROBES
  . . VENUS PROBES
  . . . **ZOND 8 SPACE PROBE**
  . ZOND SPACE PROBES
  . . **ZOND 8 SPACE PROBE**

ZONE MELTING
UF ZONE REFINING

ZONE MELTING-*(CONT.)*
GS PHASE TRANSFORMATIONS
  . FREEZING
  . . **ZONE MELTING**
RT ARC MELTING
  CRYSTALLIZATION
  FLOAT ZONES
  MELTING
  PURIFICATION
  REFINING
  ∞SEPARATION
  ULTRAPURE METALS
  VACUUM MELTING

ZONE REFINING
USE ZONE MELTING

ZONES
USE REGIONS

∞ **ZOOLOGY**
SN *(USE OF A MORE SPECIFIC TERM IS*
  RECOMMENDED--CONSULT THE TERMS
  LISTED BELOW)
RT ANIMALS
  BOTANY
  ENTOMOLOGY
  ∞SCIENCE
  TAXONOMY

ZOOM LENSES
GS LENSES
  . **ZOOM LENSES**
RT LENS DESIGN

ZPR REACTORS
USE ZERO POWER REACTORS

ZUNI ROCKET VEHICLE
GS ROCKET VEHICLES
  . SINGLE STAGE ROCKET VEHICLES
  . . **ZUNI ROCKET VEHICLE**
RT SOLID PROPELLANT ROCKET ENGINES

A

ACOUSTIC FREQUENCIES
UF	SOUND FREQUENCIES
GS	FREQUENCIES
	. **ACOUSTIC FREQUENCIES**
	. . AUDIO FREQUENCIES
RT	ACOUSTIC MEASUREMENT
	ACOUSTIC PROPERTIES
	ACOUSTICS
	FREQUENCY RANGES
	NOISE SPECTRA
	PRESSURE OSCILLATIONS
	RESONANT FREQUENCIES
	SOUND WAVES
	SUBAUDIBLE FREQUENCIES
	ULTRASONIC RADIATION

AIR DATA SYSTEMS
SN	(LIMITED TO FLIGHT DATA SYSTEMS)
GS	DATA SYSTEMS
	. **AIR DATA SYSTEMS**
RT	AIR
	∝ AIRCRAFT
	∝ SPACECRAFT
	TABLES (DATA)
	WIND TUNNEL TESTS

ARIEL
GS	CELESTIAL BODIES
	. NATURAL SATELLITES
	. . URANUS SATELLITES
	. . . **ARIEL**
RT	URANUS (PLANET)

C

CERAMIC FIBERS
GS	FIBERS
	. SYNTHETIC FIBERS
	. . **CERAMIC FIBERS**
RT	BORON CARBIDES
	CERAMIC MATRIX COMPOSITES
	CERAMICS
	COMPOSITE WRAPPING
	CORDAGE
	FIBER COMPOSITES
	FIBER ORIENTATION
	FIBER STRENGTH
	FILAMENT WINDING
	∝ FILAMENTS
	REINFORCING FIBERS
	SILICON CARBIDES
	STRANDS
	TITANIUM CARBIDES

CEYLON
USE	SRI LANKA

CHRONOBIOLOGY
USE	RHYTHM (BIOLOGY)

COMBUSTION CHEMISTRY
GS	THERMOCHEMISTRY
	. **COMBUSTION CHEMISTRY**
RT	CHEMICAL ENGINEERING
	CHEMICAL REACTIONS
	COMBUSTION
	COMBUSTION PHYSICS
	COMBUSTION PRODUCTS
	COMBUSTION STABILITY
	EXOTHERMIC REACTIONS
	FLAME TEMPERATURE
	OXIDATION

COMBUSTION CHEMISTRY-(CONT.)
	REACTION KINETICS

CONTINUOUS FLOW ELECTROPHORESIS
USE	ELECTROPHORESIS

CYLINDRICAL COORDINATES
GS	COORDINATES
	. **CYLINDRICAL COORDINATES**
RT	ASTRONOMICAL COORDINATES
	CARTESIAN COORDINATES
	CYLINDRICAL BODIES

E

ELECTROACOUSTICS
GS	ACOUSTICS
	. **ELECTROACOUSTICS**
RT	ELECTROACOUSTIC TRANSDUCERS
	ELECTROACOUSTIC WAVES
	SOUND TRANSDUCERS
	SURFACE ACOUSTIC WAVE DEVICES
	ULTRASONIC WAVE TRANSDUCERS
	ULTRASONICS

END EFFECTORS
UF	MECHANICAL FINGERS
	MECHANICAL HANDS
RT	MANIPULATORS
	ROBOTICS
	ROBOTS

F

FOOD PRODUCTION (IN SPACE)
RT	CLOSED ECOLOGICAL SYSTEMS
	CONSUMABLES (SPACECREW SUPPLIES)
	∝ FOOD
	∝ PRODUCTION
	SPACE FLIGHT FEEDING
	SPACE RATIONS

G

GEO ENVIRONMENTS
USE	EARTH ORBITAL ENVIRONMENTS

GEOSYNCHRONOUS EARTH ORBITAL ENVIRONMENTS
USE	EARTH ORBITAL ENVIRONMENTS

∝ GRADIOMETERS
SN	*(USE OF A MORE SPECIFIC TERM IS RECOMMENDED-CONSULT THE TERMS LISTED BELOW)*
RT	GRAVITY GRADIOMETERS
	MAGNETOMETERS

GRAND UNIFIED THEORY
UF	GUT
GS	FIELD THEORY (PHYSICS)
	. **GRAND UNIFIED THEORY**
	. . UNIFIED FIELD THEORY
RT	ASTROPHYSICS
	BIG BANG COSMOLOGY

GRAND UNIFIED THEORY-(CONT.)
	BROKEN SYMMETRY
	COSMOLOGY
	EINSTEIN EQUATIONS
	ELECTROMAGNETIC FIELDS
	ELECTROMAGNETIC INTERACTIONS
	ELECTROMAGNETISM
	GRAVITATION THEORY
	GRAVITATIONAL FIELDS
	PARTICLE THEORY
	PLASMA PHYSICS
	RELATIVITY
	STRONG INTERACTIONS (FIELD THEORY)
	SYMMETRY
	THEORETICAL PHYSICS
	WEAK ENERGY INTERACTIONS
	WEAK INTERACTIONS (FIELD THEORY)

GUT
USE	GRAND UNIFIED THEORY

I

INDIUM-TIN-OXIDE SEMICONDUCTORS
USE	ITO (SEMICONDUCTORS)

INTERFERENCE FIT
GS	JOINTS (JUNCTIONS)
	. **INTERFERENCE FIT**
RT	AIRCRAFT STRUCTURES
	FASTENERS
	FATIGUE LIFE
	FITTING
	MECHANICAL PROPERTIES
	STRESS ANALYSIS

INTERNATIONAL COMETARY EXPLORER
USE	INTERNATIONAL SUN EARTH EXPLORER 3

IRREGULAR GALAXIES
GS	CELESTIAL BODIES
	. GALAXIES
	. . **IRREGULAR GALAXIES**
RT	BL LACERTAE OBJECTS
	GALACTIC RADIATION
	GALACTIC ROTATION
	GALACTIC STRUCTURE
	GUM NEBULA
	HUBBLE CONSTANT
	HUBBLE DIAGRAM
	NEBULAE
	ORION NEBULA
	QUASARS
	RADIO SOURCES (ASTRONOMY)
	RED SHIFT
	STAR CLUSTERS
	STARS

L

LASER INDUCED FLUORESCENCE
UF	LIF (FLUORESCENCE)
GS	DECAY
	. EMISSION
	. . LIGHT EMISSION
	. . . LUMINESCENCE
	. . . . FLUORESCENCE

LASER INDUCED FLUORESCENCE-*(CONT.)*
```
. . . . . LASER INDUCED
              FLUORESCENCE
```

RT   ELECTROMAGNETIC ABSORPTION
      EXCITATION
      EXTINCTION
      IRRADIATION
      LASER OUTPUTS
      LASER SPECTROSCOPY
      MOSSBAUER EFFECT
      PHOSPHORS
      PHOTOIONIZATION
      PHOTOLUMINESCENCE
      PLASMA RADIATION

LATENT HEAT

GS   CHEMICAL PROPERTIES
      . THERMOCHEMICAL PROPERTIES
      . . **LATENT HEAT**
      . . . HEAT OF FUSION
      . . . HEAT OF VAPORIZATION
      HEAT
      . **LATENT HEAT**
      . . HEAT OF FUSION
      . . HEAT OF VAPORIZATION
      THERMODYNAMIC PROPERTIES
      . THERMOCHEMICAL PROPERTIES
      . . **LATENT HEAT**
      . . . HEAT OF FUSION
      . . . HEAT OF VAPORIZATION
      . THERMOPHYSICAL PROPERTIES
      . . **LATENT HEAT**
      . . . HEAT OF FUSION
      . . . HEAT OF VAPORIZATION

LEO ENVIRONMENTS

USE   EARTH ORBITAL ENVIRONMENTS

LIF (FLUORESCENCE)

USE   LASER INDUCED FLUORESCENCE

LOW EARTH ORBITAL ENVIRONMENTS

USE   EARTH ORBITAL ENVIRONMENTS

M

MAGNETIC FIELD RECONNECTION

GS   MAGNETIC PROPERTIES
      . MAGNETOACTIVITY
      . . **MAGNETIC FIELD RECONNECTION**
RT   INTERPLANETARY MAGNETIC FIELDS
      MAGNETIC FIELD CONFIGURATIONS
      MAGNETIC FIELDS
      MAGNETIC FLUX
      SOLAR MAGNETIC FIELD
      SPACE PLASMAS

MALAYA

USE   MALAYSIA

MALAYSIA

UF   MALAYA
GS   NATIONS
      . **MALAYSIA**
RT   ASIA

MECHANICAL FINGERS

USE   END EFFECTORS

MECHANICAL HANDS

USE   END EFFECTORS

METAL-SEMICONDUCTOR-METAL SEMICONDUCTORS

USE   MSM (SEMICONDUCTORS)

MIRANDA

GS   CELESTIAL BODIES
      . NATURAL SATELLITES
      . . URANUS SATELLITES
      . . . **MIRANDA**
RT   URANUS (PLANET)

MSM (SEMICONDUCTORS)

UF   METAL-SEMICONDUCTOR-METAL
      SEMICONDUCTORS
GS   ELECTRONIC EQUIPMENT
      . SOLID STATE DEVICES
      . . SEMICONDUCTOR DEVICES
      . . . **MSM (SEMICONDUCTORS)**
RT   MIM DIODES
      PHOTODIODES
      SCHOTTKY DIODES
      SEMICONDUCTOR JUNCTIONS
      SIS (SEMICONDUCTORS)

O

OBERON

GS   CELESTIAL BODIES
      . NATURAL SATELLITES
      . . URANUS SATELLITES
      . . . **OBERON**
RT   URANUS (PLANET)

P

PAPUA NEW GUINEA

GS   NATIONS
      . **PAPUA NEW GUINEA**
RT   ASIA
      AUSTRALIA
      NEW GUINEA (ISLAND)

PARTICULATES

GS   PARTICLES
      . **PARTICULATES**
      . . SOOT
RT   AEROSOLS
      AIR POLLUTION
      AIR QUALITY
      AIR SAMPLING
      ATMOSPHERIC COMPOSITION
      COMBUSTION PRODUCTS
      CONTAMINANTS
      DISPERSIONS
      EXHAUST GASES
      FLY ASH
      PARTICLE SIZE DISTRIBUTION
      PARTICULATE SAMPLING
      POLLUTION CONTROL
      POLLUTION MONITORING
      SMOG
      SMOKE
      SOLID SUSPENSIONS

R

RADIO JETS (ASTRONOMY)

GS   CELESTIAL BODIES
      . RADIO SOURCES (ASTRONOMY)
      . . EXTRAGALACTIC RADIO SOURCES
      . . . **RADIO JETS (ASTRONOMY)**
RT   ASTROPHYSICS
      ENERGETIC PARTICLES
      EXTRATERRESTRIAL RADIATION
      EXTRATERRESTRIAL RADIO WAVES
      GALACTIC NUCLEI
      GALACTIC RADIO WAVES
      PLASMA JETS
      QUASARS
      RADIO ASTRONOMY
      RADIO EMISSION

REFLECTOR ANTENNAS

GS   ANTENNAS
      . DIRECTIONAL ANTENNAS
      . . **REFLECTOR ANTENNAS**
      . . . PARABOLIC ANTENNAS

REFLECTOR ANTENNAS-*(CONT.)*
      . . . TWO REFLECTOR ANTENNAS
RT   ANTENNA FEEDS
      ANTENNA RADIATION PATTERNS
      CASSEGRAIN ANTENNAS
      MICROWAVE ANTENNAS
      PARABOLIC REFLECTORS
      RADAR ANTENNAS
      RADAR CORNER REFLECTORS
      RADAR REFLECTORS
      RADIO ANTENNAS
      REFLECTOMETERS
      REFLECTORS
      SUBREFLECTORS

S

SILICON-ON-INSULATOR SEMICONDUCTORS

USE   SOI (SEMICONDUCTORS)

SOI (SEMICONDUCTORS)

UF   SILICON-ON-INSULATOR
      SEMICONDUCTORS
GS   ELECTRONIC EQUIPMENT
      . SOLID STATE DEVICES
      . . SEMICONDUCTOR DEVICES
      . . . **SOI (SEMICONDUCTORS)**
RT   FIELD EFFECT TRANSISTORS
      METAL OXIDE SEMICONDUCTORS
      SILICON FILMS
      SILICON JUNCTIONS
      SILICON TRANSISTORS
      SIS (SEMICONDUCTORS)
      SOS (SEMICONDUCTORS)

SOUND FREQUENCIES

USE   ACOUSTIC FREQUENCIES

SQUEEZED STATES (QUANTUM THEORY)

UF   TWO PHOTON COHERENT STATES
RT   COHERENT ELECTROMAGNETIC
      RADIATION
      COHERENT LIGHT
      ELECTROMAGNETIC FIELDS
      FLUCTUATION THEORY
      LIGHT TRANSMISSION
    ∞ OPTICS
      PHOTON DENSITY
      QUANTUM MECHANICS
      QUANTUM THEORY

SRI LANKA

UF   CEYLON
GS   NATIONS
      . **SRI LANKA**
RT   ASIA

T

TELECONNECTIONS (METEOROLOGY)

RT   CLIMATOLOGY
      CORRELATION
      DATA CORRELATION
      EARTH ATMOSPHERE
      METEOROLOGICAL PARAMETERS
      METEOROLOGY
      SECULAR VARIATIONS
      SIGNIFICANCE
      SPATIAL DISTRIBUTION
      STATISTICAL ANALYSIS
      STATISTICAL CORRELATION
      SYNOPTIC METEOROLOGY
      TEMPORAL DISTRIBUTION

THERMOPHORESIS

RT   AEROSOLS
      DEPOSITION
      DIFFUSION
      PARTICLE DIFFUSION
      PARTICLE MOTION
      PARTICLE SIZE DISTRIBUTION
    ∞ SEPARATION
      TEMPERATURE EFFECTS
      TEMPERATURE GRADIENTS

THRESHOLD VOLTAGE

GS   POTENTIAL ENERGY
      . ELECTRIC POTENTIAL
      . . **THRESHOLD VOLTAGE**
RT   PHOTOVOLTAGES
      PHOTOVOLTAIC EFFECT
      SEMICONDUCTOR JUNCTIONS
      SILICON JUNCTIONS
      SOLID STATE DEVICES
      THRESHOLD CURRENTS
      VOLT-AMPERE CHARACTERISTICS

TITANIA

GS   CELESTIAL BODIES
      . NATURAL SATELLITES
      . . URANUS SATELLITES
      . . . **TITANIA**
RT   URANUS (PLANET)

TOILETS

RT   HUMAN WASTES
      SANITATION
      SPACECREWS
      WASTE DISPOSAL

TRANSCONDUCTANCE

GS   IMPEDANCE
      . ELECTRICAL IMPEDANCE
      . . ELECTRICAL RESISTANCE
      . . . **TRANSCONDUCTANCE**
RT   ∝CONDUCTIVITY
      ELECTRIC POTENTIAL
      ELECTRODES
      ELECTRON TUBES
      LINEAR CIRCUITS
      LOW CONDUCTIVITY
      ∝LOW RESISTANCE
      OHMMETERS
      OHMS LAW
      RC CIRCUITS
      REACTANCE
      ∝RESISTANCE
      RL CIRCUITS
      RLC CIRCUITS
      SOLID ELECTRODES
      VOLT-AMPERE CHARACTERISTICS

TWO PHOTON COHERENT STATES

USE   SQUEEZED STATES (QUANTUM THEORY)

U

ULTRAVIOLET DETECTORS

GS   MEASURING INSTRUMENTS
      . OPTICAL MEASURING INSTRUMENTS
      . . **ULTRAVIOLET DETECTORS**
      . . . ULTRAVIOLET SPECTROMETERS
      . . . ULTRAVIOLET
           SPECTROPHOTOMETERS
      . RADIATION MEASURING INSTRUMENTS
      . . ACTINOMETERS
      . . . **ULTRAVIOLET DETECTORS**
      . . . . ULTRAVIOLET SPECTROMETERS
      . . . . ULTRAVIOLET
           SPECTROPHOTOMETERS
RT   ∝DETECTORS
      PHOTOMETERS
      RADIOMETERS
      ULTRAVIOLET ABSORPTION
      ULTRAVIOLET RADIATION
      ULTRAVIOLET SPECTRA

UMBRIEL

GS   CELESTIAL BODIES
      . NATURAL SATELLITES
      . . URANUS SATELLITES
      . . . **UMBRIEL**
RT   URANUS (PLANET)

W

WEAR RESISTANCE

GS   MECHANICAL PROPERTIES
      . **WEAR RESISTANCE**
      . . ABRASION RESISTANCE
RT   ABRASION

WEAR RESISTANCE-*(CONT.)*

      BOUNDARY LUBRICATION
      COEFFICIENT OF FRICTION
      DETERIORATION
      HARDNESS
      LUBRICANT TESTS
      ∝RESISTANCE
      SLIDING FRICTION
      TOUGHNESS
      WEAR
      WEAR INHIBITORS
      WEAR TESTS

VOLUME 2
ACCESS VOCABULARY

A

A, Air Density Explorer
  USE   EXPLORER 19 SATELLITE

A, Anik
  USE   ANIK 1

A, Atmosphere Explorer
  USE   EXPLORER 17 SATELLITE

A, BE
  USE   BEACON EXPLORER A

A, Beacon Explorer
  USE   BEACON EXPLORER A

A, Cassiopeia
  USE   CASSIOPEIA A

A, Compound
  USE   COMPOUND A

A Computer, CDC 160-
  USE   CDC 160-A COMPUTER

A, Energetic Particle Explorer
  USE   EXPLORER 12 SATELLITE

A, EOS-
  USE   LANDSAT E

A, EPE-
  USE   EXPLORER 12 SATELLITE

A, ERTS-
  USE   LANDSAT 1

A, HEAO
  USE   HEAO 1

A, Helios
  USE   HELIOS A

A, High Energy Astronomy Observatory
  USE   HEAO 1

A, IMP-
  USE   EXPLORER 18 SATELLITE

A, Ionosphere Explorer
  USE   EXPLORER 20 SATELLITE

A, ISIS-
  USE   ISIS-A

A, Lunar Orbiter
  USE   LUNAR ORBITER 1

A Missile, Bomarc
  USE   BOMARC A MISSILE

A, OAO-
  USE   OAO 1

A, OGO-
  USE   OGO-A

A, OSO-
  USE   OSO-1

A Reactor, Tory 2-
  USE   TORY 2-A REACTOR

A Rocket Vehicle, Agena
  USE   AGENA A ROCKET VEHICLE

A Satellite, AD-
  USE   EXPLORER 19 SATELLITE

A Satellite, AE-
  USE   EXPLORER 17 SATELLITE

A Satellite, DME-
  USE   EXPLORER 31 SATELLITE

A Satellite, HEOS
  USE   HEOS A SATELLITE

A Satellite, Magsat
  USE   MAGSAT A SATELLITE

A, SE-
  USE   EXPLORER 30 SATELLITE

A, Sir-
  USE   SHUTTLE IMAGING RADAR

A, SMM-
  USE   SOLAR MAXIMUM MISSION-A

A, Solar Maximum Mission-
  USE   SOLAR MAXIMUM MISSION-A

A, Space Shuttle Mission 31-
  USE   SPACE SHUTTLE MISSION 31-A

A, Space Shuttle Mission 41-
  USE   SPACE SHUTTLE MISSION 41-A

A, Space Shuttle Mission 51-
  USE   SPACE SHUTTLE MISSION 51-A

A, Space Shuttle Mission 61-
  USE   SPACE SHUTTLE MISSION 61-A

A, Space Shuttle Upper Stage
  USE   SPACE SHUTTLE UPPER STAGE A

A, SSUS-
  USE   SPACE SHUTTLE UPPER STAGE A

A STARS

A, TELESAT Canada
  USE   ANIK 1

A, TOS-
  USE   ESSA 3 SATELLITE

A, Vitamin
  USE   RETINENE

A-W Devices, B-
  USE   BULK ACOUSTIC WAVE DEVICES

A-W Devices, S-
  USE   SURFACE ACOUSTIC WAVE DEVICES

A-1 AIRCRAFT

A-1 Engine, RL-10-
  USE   RL-10-A-1 ENGINE

A-2 AIRCRAFT

A-3 AIRCRAFT

A-3 Engine, RL-10-
  USE   RL-10-A-3 ENGINE

A-4 AIRCRAFT

A-5 AIRCRAFT

A-6 AIRCRAFT

A-7 AIRCRAFT

A-9 AIRCRAFT

A-10 AIRCRAFT

A-11 Satellite
  USE   ECHO 1 SATELLITE

A-12 Satellite
  USE   ECHO 2 SATELLITE

A-37 AIRCRAFT

A-300 AIRCRAFT

A-310 AIRCRAFT

A-320 AIRCRAFT

AAP 1 MISSION

AAP 2 MISSION

AAP 3 MISSION

AAP 4 MISSION

(Abandonment), Escape
  USE   ESCAPE (ABANDONMENT)

Abatement, Smoke
  USE   SMOKE ABATEMENT

ABDOMEN

ABEL FUNCTION

ABERRATION

ABILITIES

ABIOGENESIS

Ablated Nosetips
  USE   PANT PROGRAM

ABLATION

ABLATIVE MATERIALS

ABLATIVE NOSE CONES

Able Rocket Vehicle, Thor
  USE   THOR ABLE ROCKET VEHICLE

Able 5 Launch Vehicle, Atlas
  USE   ATLAS ABLE 5 LAUNCH VEHICLE

ABLESTAR LAUNCH VEHICLE

ABM
  USE   APOGEE BOOST MOTORS

ABNORMALITIES

ABORIGINES

ABORT APPARATUS

ABORT TRAJECTORIES

ABORTED MISSIONS

ABRASION

ABRASION RESISTANCE

ABRASIVES

ABRIKOSOV THEORY

ABSOLUTE ZERO

ABSORBENTS

ABSORBERS

ABSORBERS (EQUIPMENT)

ABSORBERS (MATERIALS)

Absorbers, Neutron
USE   NEUTRON ABSORBERS

Absorbers, Radar
USE   RADAR ABSORBERS

Absorbers, Shock
USE   SHOCK ABSORBERS

Absorbers, Solar Energy
USE   SOLAR ENERGY ABSORBERS

Absorbing Materials, Radar
USE   ANTIRADAR COATINGS

ABSORPTANCE

Absorptiometry, Gamma Ray
USE   GAMMA RAY ABSORPTIOMETRY

Absorptiometry, Photon
USE   PHOTON ABSORPTIOMETRY

ABSORPTION

Absorption, Atmospheric
USE   ATMOSPHERIC ATTENUATION

Absorption, Auroral
USE   AURORAL ABSORPTION

Absorption Bands
USE   ABSORPTION SPECTRA

Absorption Coefficient
USE   ABSORPTIVITY

ABSORPTION COOLING

ABSORPTION CROSS SECTIONS

Absorption, Electromagnetic
USE   ELECTROMAGNETIC ABSORPTION

Absorption, Energy
USE   ENERGY ABSORPTION

Absorption Films, Energy
USE   ENERGY ABSORPTION FILMS

Absorption, Gamma Ray
USE   GAMMA RAY ABSORPTION

Absorption, Infrared
USE   INFRARED ABSORPTION

Absorption, Ionospheric
USE   IONOSPHERIC PROPAGATION
      ELECTROMAGNETIC ABSORPTION

Absorption, Light
USE   ELECTROMAGNETIC ABSORPTION

Absorption, Magnetic
USE   ELECTROMAGNETIC ABSORPTION

Absorption, Material
USE   MATERIAL ABSORPTION

Absorption), Moderation (Energy
USE   MODERATION (ENERGY ABSORPTION)

Absorption, Molecular
USE   MOLECULAR ABSORPTION

Absorption, Multiphoton
USE   MULTIPHOTON ABSORPTION

Absorption, Optical
USE   LIGHT TRANSMISSION
      ELECTROMAGNETIC ABSORPTION

Absorption, Photo
USE   PHOTOABSORPTION

Absorption, Polar Cap
USE   POLAR CAP ABSORPTION

Absorption, Radiation
USE   RADIATION ABSORPTION

Absorption, Self
USE   SELF ABSORPTION

Absorption, Sound
USE   SOUND TRANSMISSION

ABSORPTION SPECTRA

Absorption, Spectral
USE   ABSORPTION SPECTRA

ABSORPTION SPECTROSCOPY

Absorption, Thermal
USE   THERMAL ABSORPTION

Absorption), Thermalization (Energy
USE   THERMALIZATION (ENERGY ABSORPTION)

Absorption, Ultraviolet
USE   ULTRAVIOLET ABSORPTION

Absorption, X Ray
USE   X RAY ABSORPTION

Absorptive Index
USE   ABSORPTIVITY

ABSORPTIVITY

ABSTRACTS

ABUNDANCE

Abundance, Element
USE   ABUNDANCE

Ac
USE   ACTINIUM

AC (Current)
USE   ALTERNATING CURRENT

AC GENERATORS

AC), Inverted Converters (DC To
USE   INVERTED CONVERTERS (DC TO AC)

(AC To AC), Voltage Converters
USE   VOLTAGE CONVERTERS (AC TO AC)

(AC To DC), Current Converters
USE   CURRENT CONVERTERS (AC TO DC)

AC), Voltage Converters (AC To
USE   VOLTAGE CONVERTERS (AC TO AC)

AC-1 Aircraft
USE   DHC 4 AIRCRAFT

ACCELERATED LIFE TESTS

ACCELERATING AGENTS

ACCELERATION

Acceleration, Angular
USE   ANGULAR ACCELERATION

Acceleration, Electromagnetic
USE   ELECTROMAGNETIC ACCELERATION

Acceleration, Electron
USE   ELECTRON ACCELERATION

Acceleration, High
USE   HIGH ACCELERATION

(Acceleration), High Gravity
USE   HIGH GRAVITY ENVIRONMENTS

Acceleration, Impact
USE   IMPACT ACCELERATION

Acceleration, Magnetohydrodynamic
USE   PLASMA ACCELERATION

Acceleration, Particle
USE   PARTICLE ACCELERATION

ACCELERATION (PHYSICS)

Acceleration, Physiological
USE   PHYSIOLOGICAL ACCELERATION

Acceleration, Plasma
USE   PLASMA ACCELERATION

ACCELERATION PROTECTION

ACCELERATION STRESSES (PHYSIOLOGY)

ACCELERATION TOLERANCE

Acceleration, Transverse
USE   TRANSVERSE ACCELERATION

Accelerator, Cyclops Plasma
USE   CYCLOPS PLASMA ACCELERATOR

Accelerator, Nimrod
USE   NIMROD ACCELERATOR

Accelerator Targets, Particle
USE   PARTICLE ACCELERATOR TARGETS

ACCELERATORS

Accelerators, Coaxial Plasma
USE   COAXIAL PLASMA ACCELERATORS

Accelerators, Cyclic
USE   CYCLIC ACCELERATORS

Accelerators, Electron
USE   ELECTRON ACCELERATORS

Accelerators, Electron Ring
USE   STORAGE RINGS (PARTICLE
      ACCELERATORS)

Accelerators, Hall
USE   HALL ACCELERATORS

Accelerators, Hypervelocity
USE   HYPERVELOCITY GUNS

Accelerators, Ion
USE   ION ACCELERATORS

Accelerators, Linear
USE   LINEAR ACCELERATORS

Accelerators, Particle
USE   PARTICLE ACCELERATORS

Accelerators, Plasma
USE   PLASMA ACCELERATORS

Accelerators), Racetracks (Particle
USE   RACETRACKS (PARTICLE ACCELERATORS)

Accelerators, Railgun
USE   RAILGUN ACCELERATORS

Accelerators, Space Exper With Particle
USE   SEPAC (PAYLOAD)

Accelerators), Storage Rings (Particle
USE   STORAGE RINGS (PARTICLE
      ACCELERATORS)

Accelerators, Van De Graaff
USE   VAN DE GRAAFF ACCELERATORS

ACCELEROMETERS

Accelerometers, Strain Gage
USE   STRAIN GAGE ACCELEROMETERS

ACCEPTABILITY

Acceptance
USE   ACCEPTABILITY

ACCEPTOR MATERIALS

Access, Code Division Multiple
USE   CODE DIVISION MULTIPLE ACCESS

ACCESS CONTROL

Access, Demand Assignment Multiple
USE   DEMAND ASSIGNMENT MULTIPLE ACCESS

Access, Frequency Division Multiple
USE   FREQUENCY DIVISION MULTIPLE ACCESS

Access Memory, Random
USE   RANDOM ACCESS MEMORY

Access, Multiple
USE   MULTIPLE ACCESS

Access, Random
USE   RANDOM ACCESS

ACCESS TIME

Access, Time Division Multiple
USE   TIME DIVISION MULTIPLE ACCESS

ACCESSORIES

ACCIDENT INVESTIGATION

Accident Investigation, Aircraft
USE   AIRCRAFT ACCIDENT INVESTIGATION

ACCIDENT PREVENTION

ACCIDENT PRONENESS

ACCIDENTS

Accidents, Aircraft
USE   AIRCRAFT ACCIDENTS

Accidents, Automobile
USE   AUTOMOBILE ACCIDENTS

Accidents, Cerebral Vascular
USE   CEREBRAL VASCULAR ACCIDENTS

ACCLIMATIZATION

Acclimatization, Altitude
USE   ALTITUDE ACCLIMATIZATION

Acclimatization, Cold
USE   COLD ACCLIMATIZATION

Acclimatization, Heat
USE   HEAT ACCLIMATIZATION

ACCOMMODATION

ACCOMMODATION COEFFICIENT

Accommodation Coefficients, Thermal
USE   ACCOMMODATION COEFFICIENT

Accommodation, Visual
USE   VISUAL ACCOMMODATION

ACCOUNTING

Accretion
USE   DEPOSITION

ACCRETION DISKS

Accretion, Stellar Mass
USE   STELLAR MASS ACCRETION

ACCUMULATIONS

ACCUMULATORS

ACCUMULATORS (COMPUTERS)

ACCURACY

Accuracy, Geodetic
USE   GEODETIC ACCURACY

Accuracy, Geometric
USE   GEOMETRIC ACCURACY

ACEE PROGRAM

ACETALDEHYDE

ACETALS

ACETANILIDE

ACETATES

Acetates, Cobalt
USE   COBALT ACETATES

Acetates, Lead
USE   LEAD ACETATES

Acetation
USE   ACETYLATION

ACETAZOLAMIDE

ACETIC ACID

ACETONE

Acetone, Acetyl
USE   ACETYLACETONE

ACETYL COMPOUNDS

ACETYLACETONE

ACETYLATION

ACETYLENE

Acetylene, Oxy
USE   OXYACETYLENE

ACETYLSALICYLIC ACID

ACHIEVEMENT

Achondrite, Kapoeta
USE   KAPOETA ACHONDRITE

Achondrite, Norton County
USE   NORTON COUNTY ACHONDRITE

ACHONDRITES

Acid, Acetic
USE   ACETIC ACID

Acid, Acetylsalicylic
USE   ACETYLSALICYLIC ACID

Acid, Acrylic
USE   ACRYLIC ACID

Acid, Ascorbic
USE   ASCORBIC ACID

Acid, Aspartic
USE   ASPARTIC ACID

ACID BASE EQUILIBRIUM

Acid Batteries, Lead
USE   LEAD ACID BATTERIES

Acid, Benzilic
USE   BENZILIC ACID

Acid, Benzoic
USE   BENZOIC ACID

Acid, Butyric
USE   BUTYRIC ACID

Acid, Carbonic
USE   CARBONIC ACID

Acid, Chromic
USE   CHROMIC ACID

Acid, Citric
USE   CITRIC ACID

Acid, Cyanuric
USE   CYANURIC ACID

Acid, Cytidylic
USE   CYTIDYLIC ACID

Acid, Deoxyribonucleic
USE   DEOXYRIBONUCLEIC ACID

Acid, Folic
USE   FOLIC ACID

Acid, Formhydroxamic
USE   FORMHYDROXAMIC ACID

Acid, Formic
USE   FORMIC ACID

Acid Fuel Cells, Phosphoric
USE   PHOSPHORIC ACID FUEL CELLS

Acid, Glutamic
USE   GLUTAMIC ACID

Acid, Hippuric
USE   HIPPURIC ACID

Acid, Hydrazoic
USE   HYDRAZOIC ACID

Acid, Hydrobromic
USE   HYDROBROMIC ACID

Acid, Hydrochloric
USE   HYDROCHLORIC ACID

Acid, Hydrocyanic
USE   HYDROCYANIC ACID

Acid, Hydrofluoric
USE   HYDROFLUORIC ACID

Acid, Iodoacetic
USE   IODOACETIC ACID

Acid, Lactic
USE   LACTIC ACID

Acid, Lipoic
USE   LIPOIC ACID

Acid Metabolism, Ascorbic
USE   ASCORBIC ACID METABOLISM

Acid, Nicotinic
USE   NICOTINIC ACID

Acid, Nitric
USE   NITRIC ACID

Acid, Nitrous
USE   NITROUS ACID

Acid, Oleic
USE   OLEIC ACID

Acid, Oxalic
USE   OXALIC ACID

Acid, Palmitic
USE   PALMITIC ACID

Acid, Perchloric
USE   PERCHLORIC ACID

Acid, Phosphoric
USE   PHOSPHORIC ACID

Acid, Propionic
USE   PROPIONIC ACID

Acid, Prussic
USE   HYDROCYANIC ACID

ACID RAIN

Acid, Sebacic
USE   SEBACIC ACID

Acid, Sulfonic
USE   SULFONIC ACID

Acid, Sulfuric
USE   SULFURIC ACID

Acid, Uric
USE   URIC ACID

Acid, Uridylic
USE   URIDYLIC ACID

Acid, Valeric
USE   VALERIC ACID

ACIDITY

ACIDOSIS

ACIDS

Acids, Amino
USE   AMINO ACIDS

Acids, Boric
USE   BORIC ACIDS

Acids, Carboxylic
USE   CARBOXYLIC ACIDS

Acids, Dicarboxylic
USE   DICARBOXYLIC ACIDS

Acids, Ethylenediaminetetraacetic
USE   ETHYLENEDIAMINETETRAACETIC ACIDS

Acids, Fatty
USE   FATTY ACIDS

Acids, Nucleic
USE   NUCLEIC ACIDS

Acids, Oxamic
USE   OXAMIC ACIDS

Acids, Ribonucleic
USE   RIBONUCLEIC ACIDS

Acids, Xanthic
USE   XANTHIC ACIDS

ACOUSTIC ATTENUATION

Acoustic Combustion
USE   COMBUSTION STABILITY

ACOUSTIC DELAY LINES

ACOUSTIC DUCTS

ACOUSTIC EMISSION

ACOUSTIC EXCITATION

ACOUSTIC FATIGUE

Acoustic Generators
USE   SOUND GENERATORS

ACOUSTIC IMPEDANCE

ACOUSTIC INSTABILITY

ACOUSTIC LEVITATION

ACOUSTIC MEASUREMENT

Acoustic Microscope (Slam), Scanning Laser
USE   ACOUSTIC MICROSCOPES

ACOUSTIC MICROSCOPES

ACOUSTIC NOZZLES

ACOUSTIC PROPAGATION

ACOUSTIC PROPERTIES

Acoustic Radiation
USE   SOUND WAVES

Acoustic Radiation, Coherent
USE   COHERENT ACOUSTIC RADIATION

ACOUSTIC RETROFITTING

ACOUSTIC SCATTERING

ACOUSTIC SIMULATION

ACOUSTIC SOUNDING

Acoustic Stability
USE   FREQUENCY STABILITY

ACOUSTIC STREAMING

ACOUSTIC VELOCITY

Acoustic Vibrations
USE   SOUND WAVES

Acoustic Wave Devices, Bulk
USE   BULK ACOUSTIC WAVE DEVICES

Acoustic Wave Devices, Surface
USE   SURFACE ACOUSTIC WAVE DEVICES

Acoustic Waves, Ion
USE   ION ACOUSTIC WAVES

ACOUSTICAL HOLOGRAPHY

ACOUSTICS

Acoustics, Aero
USE   AEROACOUSTICS

Acoustics, Bio
USE   BIOACOUSTICS

Acoustics, Geometrical
USE   GEOMETRICAL ACOUSTICS

Acoustics, Magneto
USE   MAGNETOACOUSTICS

Acoustics, Psycho
USE   PSYCHOACOUSTICS

Acoustics, Ray
USE   GEOMETRICAL ACOUSTICS

Acoustics, Underground
USE   UNDERGROUND ACOUSTICS

Acoustics, Underwater
USE   UNDERWATER ACOUSTICS

ACOUSTO-OPTICS

ACPL (Spacelab)
USE   ATMOSPHERIC CLOUD PHYSICS LAB
            (SPACELAB)

ACPL (Spacelab), Zero-G
USE   ATMOSPHERIC CLOUD PHYSICS LAB
            (SPACELAB)

Acq Network, Satellite Tracking And Data
USE   STDN (NETWORK)

ACQUISITION

Acquisition And Tracking, Video Landmark
USE   VIDEO LANDMARK ACQUISITION AND
            TRACKING

Acquisition, Data
USE   DATA ACQUISITION

Acquisition, Target
USE   TARGET ACQUISITION

Acquisitions Systems, Ocean Data
USE   OCEAN DATA ACQUISITIONS SYSTEMS

ACRIFLAVINE

ACROBATICS

ACROLEINS

ACRYLATES

ACRYLIC ACID

ACRYLIC RESINS

ACRYLONITRILES

ACTH
USE   ADRENOCORTICOTROPIN (ACTH)

(ACTH), Adrenocorticotropin
USE   ADRENOCORTICOTROPIN (ACTH)

ACTINIDE SERIES

ACTINIDE SERIES COMPOUNDS

ACTINIUM

Actinographs
USE   ACTINOMETERS

ACTINOMETERS

ACTINOMYCETES

ACTINOMYCIN

Action, Nonoscillatory
USE   NONOSCILLATORY ACTION

Actions, Evasive
USE   EVASIVE ACTIONS

Actions, Involuntary
USE   INVOLUNTARY ACTIONS

ACTIVATED CARBON

ACTIVATED SLUDGE

ACTIVATION

ACTIVATION ANALYSIS

Activation Analysis, Neutron
USE   NEUTRON ACTIVATION ANALYSIS

ACTIVATION (BIOLOGY)

ACTIVATION ENERGY

ACTIVE CONTROL

Active Glaciers
USE   GLACIERS

Active Magneto Particle Tracer Explorers
USE   AMPTE (SATELLITES)

ACTIVE SATELLITES

Active Volcanoes
USE   VOLCANOES

ACTIVITY

Activity, Auroral
USE   AURORAS

Activity, Biological
USE   ACTIVITY (BIOLOGY)

ACTIVITY (BIOLOGY)

Activity, Catalytic
USE   CATALYTIC ACTIVITY

ACTIVITY CYCLES (BIOLOGY)

Activity Effects, Solar
USE   SOLAR ACTIVITY EFFECTS

Activity, Enzyme
USE   ENZYME ACTIVITY

Activity, Extravehicular
USE   EXTRAVEHICULAR ACTIVITY

Activity, Intravehicular
USE   INTRAVEHICULAR ACTIVITY

Activity, Magneto
USE   MAGNETOACTIVITY

Activity, Optical
USE   OPTICAL ACTIVITY

Activity, Plasma Renin
USE   IMMUNOASSAY

Activity, Radio
USE   RADIOACTIVITY

Activity, Solar
USE   SOLAR ACTIVITY

Activity, Stellar
USE   STELLAR ACTIVITY

Actuated Devices, Cartridge
USE   ACTUATORS
       EXPLOSIVE DEVICES

Actuated Devices, Propellant
USE   PROPELLANT ACTUATED DEVICES

Actuated Instruments, Propellant
USE   PROPELLANT ACTUATED INSTRUMENTS

ACTUATION

ACTUATOR DISKS

ACTUATORS

Actuators, Hydraulic
USE   ACTUATORS
       HYDRAULIC EQUIPMENT

ACUITY

Acuity, Visual
USE   VISUAL ACUITY

ACYLATION

AD-A Satellite
USE   EXPLORER 19 SATELLITE

AD/I B
USE   EXPLORER 25 SATELLITE

AD/I Satellite
USE   EXPLORER 24 SATELLITE

ADA (PROGRAMMING LANGUAGE)

ADAPTATION

Adaptation, Dark
USE   DARK ADAPTATION

Adaptation, Desert
USE   DESERT ADAPTATION

Adaptation, Light
USE   LIGHT ADAPTATION

Adaptation, Retinal
USE   RETINAL ADAPTATION

Adaptation Syndrome, Space
USE   SPACE ADAPTATION SYNDROME

ADAPTERS

Adapters, Multiple Docking
USE   MULTIPLE DOCKING ADAPTERS

ADAPTIVE CONTROL

Adaptive Control Systems
USE   ADAPTIVE CONTROL

Adaptive Control Systems, Self
USE   SELF ADAPTIVE CONTROL SYSTEMS

Adaptive Evaluator/monitor, Data
USE   DATA PROCESSING
       DATA REDUCTION
       DATA TRANSMISSION

ADAPTIVE FILTERS

ADAPTIVE OPTICS

Adaptive System, Information
USE   INFORMATION ADAPTIVE SYSTEM

Adders (Circuits)
USE   ADDING CIRCUITS

ADDING CIRCUITS

ADDITION

ADDITION RESINS

ADDITION THEOREM

ADDITIVES

Additives, Antiicing
USE   ANTIICING ADDITIVES

Additives, Antiknock
USE   ANTIKNOCK ADDITIVES

(Additives), Doping
USE   ADDITIVES

Additives, Oil
USE   OIL ADDITIVES

Additives, Propellant
USE   PROPELLANT ADDITIVES

Address Beacon System, Discrete
USE   DISCRETE ADDRESS BEACON SYSTEM

Address Systems, Public
USE   PUBLIC ADDRESS SYSTEMS

ADDRESSING

ADDUCTS

Aden
USE   SOUTHERN YEMEN

ADENINES

ADENOSINE DIPHOSPHATE

Adenosine Monophosphate, Cyclic
USE   CYCLIC AMP

ADENOSINE TRIPHOSPHATE

ADENOSINES

ADENOVIRUSES

Adept Computer, Honeywell
USE   HONEYWELL ADEPT COMPUTER

ADEQUACY

Adherometers
USE   ADHESION TESTS

ADHESION

ADHESION TESTS

ADHESIVE BONDING

ADHESIVES

(Adhesives), Binders
USE   ADHESIVES

Adiabat, Hugoniot
USE   HUGONIOT EQUATION OF STATE

ADIABATIC CONDITIONS

ADIABATIC DEMAGNETIZATION COOLING

ADIABATIC EQUATIONS

ADIABATIC FLOW

ADIPOSE TISSUES

ADIPRENE (TRADEMARK)

ADIRONDACK MOUNTAINS (NY)

ADJOINTS

ADJUSTING

Adjustment
USE   ADJUSTING

Administration
USE   MANAGEMENT

Admittance
USE   ELECTRICAL IMPEDANCE

ADMIXTURES

Adobe Flats
USE   FLATS (LANDFORMS)

ADP
USE   ADENOSINE DIPHOSPHATE

ADRENAL GLAND

ADRENAL METABOLISM

Adrenaline
USE   EPINEPHRINE

ADRENERGICS

Adrenergics, Anti
USE   ANTIADRENERGICS

ADRENOCORTICOTROPIN (ACTH)

ADRIATIC SEA

ADSORBENTS

ADSORPTION

Adsorption Equation, Gibbs
USE   GIBBS ADSORPTION EQUATION

ADSORPTIVITY

Advanced Airborne Command Post
USE   E-4A AIRCRAFT

Advanced EVA Protection Systems
USE   AEPS

Advanced Orbiting Solar Observatory
USE   AOSO

ADVANCED RANGE INSTRUMENTATION AIRCRAFT

ADVANCED RANGE INSTRUMENTATION SHIP

ADVANCED RECONN ELECTRIC SPACECRAFT

ADVANCED SODIUM COOLED REACTOR

ADVANCED TECHNOLOGY LABORATORY

Advanced Technology Light Twin Aircraft
USE   ATLIT PROJECT

ADVANCED TEST REACTORS

ADVANCED VIDICON CAMERA SYSTEM (AVCS)

Advanced X Ray Astrophysical Facility
USE   X RAY ASTROPHYSICS FACILITY

Advanced X Ray Astrophysics Facility
USE   X RAY ASTROPHYSICS FACILITY

Advancing Glaciers
USE   GLACIERS

Advancing Shorelines
USE   BEACHES

ADVECTION

ADVENT PROJECT

Advisory And Resolution, Automatic Traffic
USE   AUTOMATIC TRAFFIC ADVISORY AND
       RESOLUTION

Advisory System, Automated Pilot
USE   AUTOMATED PILOT ADVISORY SYSTEM

Advisory System, Vortex
USE   VORTEX ADVISORY SYSTEM

AE-A Satellite
USE   EXPLORER 17 SATELLITE

AE-B Satellite
USE   EXPLORER 32 SATELLITE

AE-C Satellite
USE   EXPLORER 51 SATELLITE

AE-D Satellite
USE   EXPLORER 54 SATELLITE

AE-E Satellite
USE   EXPLORER 55 SATELLITE

AEOLIAN TONES

AEOLOTROPISM

AEPS

AERATION

Aerial Applicator Aircraft S-2b, Snow
USE   S-2 AIRCRAFT

AERIAL EXPLOSIONS

Aerial Imagery
USE   AERIAL PHOTOGRAPHY

AERIAL PHOTOGRAPHY

AERIAL RECONNAISSANCE

AERIAL RUDDERS

AEROACOUSTICS

AEROASSIST

AEROBEE ROCKET VEHICLE

AEROBES

Aerobes, An
USE   ANAEROBES

AEROBIOLOGY

AEROBRAKING

AEROCAPTURE

Aerodontalgia
USE   TOOTH DISEASES

Aerodynamic And Struct Test, Drones For
USE   DAST PROGRAM

Aerodynamic Axis
USE   AERODYNAMIC BALANCE

AERODYNAMIC BALANCE

AERODYNAMIC BRAKES

Aerodynamic Buzz
USE   FLUTTER

Aerodynamic Center
USE   AERODYNAMIC BALANCE

AERODYNAMIC CHARACTERISTICS

Aerodynamic Characteristics, Static
USE   STATIC AERODYNAMIC CHARACTERISTICS

Aerodynamic Chords
USE   AIRFOIL PROFILES
       CHORDS (GEOMETRY)

AERODYNAMIC COEFFICIENTS

AERODYNAMIC CONFIGURATIONS

(Aerodynamic Configurations), Spikes
USE   SPIKES (AERODYNAMIC CONFIGURATIONS)

AERODYNAMIC DRAG

AERODYNAMIC FORCES

AERODYNAMIC HEAT TRANSFER

AERODYNAMIC HEATING

AERODYNAMIC INTERFERENCE

Aerodynamic Lift
USE   LIFT

AERODYNAMIC LOADS

Aerodynamic Moments
USE   STABILITY DERIVATIVES

AERODYNAMIC NOISE

Aerodynamic Reusable Spaceship, Manned
USE   MARS (MANNED REUSABLE SPACECRAFT)

AERODYNAMIC STABILITY

AERODYNAMIC STALLING

Aerodynamic Vehicles
USE   AIRCRAFT

AERODYNAMICS

(Aerodynamics), Ground Effect
USE   GROUND EFFECT (AERODYNAMICS)

Aerodynamics, Interactional
USE   INTERACTIONAL AERODYNAMICS

Aerodynamics, Rotor
USE   ROTOR AERODYNAMICS

AEROELASTIC RESEARCH WINGS

AEROELASTICITY

AEROEMBOLISM

Aerogyro Helicopters
USE   XH-51 HELICOPTER

AEROLOGY

AEROMAGNETISM

Aeromagneto Flutter
USE   FLUTTER

AEROMANEUVERING

AEROMANEUVERING ORBIT TO ORBIT SHUTTLE

AERONAUTICAL ENGINEERING

AERONAUTICAL SATELLITES

AERONAUTICS

AERONOMY

Aerophysics
USE   ATMOSPHERIC PHYSICS

AEROQUATIC VEHICLES

AEROS SATELLITE

AEROSAT SATELLITES

AEROSINUSITIS

Aerosol & Gas Experiment, Stratospheric
USE   SAGE SATELLITE

AEROSOLS

AEROSPACE ENGINEERING

AEROSPACE ENVIRONMENTS

AEROSPACE INDUSTRY

AEROSPACE MEDICINE

AEROSPACE SAFETY

AEROSPACE SCIENCES

AEROSPACE SYSTEMS

AEROSPACE TECHNOLOGY TRANSFER

Aerospace Veh Design, Integ Program For
USE   IPAD

AEROSPACE VEHICLES

AEROSPACEPLANES

AEROSTATICS

Aerostats
USE   AIRSHIPS

AEROTHERMOCHEMISTRY

AEROTHERMODYNAMICS

AEROTHERMOELASTICITY

AEROZINE

AFC (Control)
USE   AUTOMATIC FREQUENCY CONTROL

AFCS (Control System)
USE   AUTOMATIC FLIGHT CONTROL

Affects
USE   EFFECTS

AFFERENT NERVOUS SYSTEMS

AFFINITY

AFGHANISTAN

AFRICA

(Africa), Kalihari Basin
USE   KALIHARI BASIN (AFRICA)

Africa, Republic Of South
USE   REPUBLIC OF SOUTH AFRICA

(Africa), Sahara Desert
USE   SAHARA DESERT (AFRICA)

Africa, South
USE   REPUBLIC OF SOUTH AFRICA

Africa, South West
USE   NAMIBIA

African Republic, Central
USE   CENTRAL AFRICAN REPUBLIC

AFRICAN RIFT SYSTEM

AFTERBODIES

Afterbodies, Cylindrical
USE   AFTERBODIES
       CYLINDRICAL BODIES

Afterburners
USE   AFTERBURNING

AFTERBURNING

Aftereffects, Motion
USE   MOTION AFTEREFFECTS

Afterglow, Helium
USE   HELIUM AFTERGLOW

Afterglow, Oxygen
USE   OXYGEN AFTERGLOW

AFTERGLOWS

AFTERIMAGES

Ag
USE   SILVER

AGC (Control)
USE   AUTOMATIC GAIN CONTROL

Age Determination
USE   CHRONOLOGY

Age Determination, Radioactive
USE   RADIOACTIVE AGE DETERMINATION

AGE FACTOR

Age Hardening
USE   PRECIPITATION HARDENING

AGENA A ROCKET VEHICLE

Agena B Launch Vehicle, Atlas
USE   ATLAS AGENA B LAUNCH VEHICLE

AGENA B RANGER PROGRAM

AGENA B ROCKET VEHICLE

AGENA C ROCKET VEHICLE

AGENA D ROCKET VEHICLE

Agena Launch Vehicle, Thor
USE   THOR AGENA LAUNCH VEHICLE

Agena Launch Vehicles, Atlas
USE   ATLAS AGENA LAUNCH VEHICLES

AGENA ROCKET VEHICLES

Agency, European Space
USE   EUROPEAN SPACE AGENCY

AGENTS

Agents, Accelerating
USE   ACCELERATING AGENTS

Agents, Antihypertensive
USE   ANTIHYPERTENSIVE AGENTS

Agents, Cholinergic Blocking
USE   ANTICHOLINERGICS

Agents, Radioprotective
USE   ANTIRADIATION DRUGS

(Agents), Stabilizers
USE   STABILIZERS (AGENTS)

AGGLOMERATION

AGGLUTINATION

AGGREGATES

AGING

AGING (BIOLOGY)

AGING (MATERIALS)

AGING (METALLURGY)

Aging, Strain
USE   PRECIPITATION HARDENING

AGITATION

Agitation, Thermal
USE   THERMAL ENERGY

Agitation, Ultrasonic
USE   ULTRASONIC AGITATION

AGREEMENTS

AGRICULTURAL AIRCRAFT

AGRICULTURE

AGRISTARS PROJECT

AGROCLIMATOLOGY

AGROMETEOROLOGY

AGROPHYSICAL UNITS

AGT
USE   AUTOMATED GUIDEWAY TRANSIT
       VEHICLES

AH-1G HELICOPTER

AH-63 HELICOPTER

AH-64 HELICOPTER

Aid, First
USE   FIRST AID

Aid, Microvision Landing
USE   MICROVISION LANDING AID

Aid Television System, Pilot Landing
USE   PLAT SYSTEM

Aided Design, Computer
USE   COMPUTER AIDED DESIGN

Aided Manufacturing, Computer
USE   COMPUTER AIDED MANUFACTURING

Aided Mapping, Computer
USE   COMPUTER AIDED MAPPING

Aided Tomography, Computer
USE   COMPUTER AIDED TOMOGRAPHY

AIDS

Aids, Landing
USE   LANDING AIDS

Aids, Navigation
USE   NAVIGATION AIDS

Aids, Visual
USE   VISUAL AIDS

AILERONS

Ailerons, Spoiler Slot
USE   SPOILER SLOT AILERONS

AIMP-D
USE   EXPLORER 33 SATELLITE

AIMP-E
USE   EXPLORER 35 SATELLITE

AIMP-1
USE   EXPLORER 33 SATELLITE

AIMP-2
USE   EXPLORER 35 SATELLITE

AIR

Air, Alveolar
USE   ALVEOLAR AIR

AIR BAG RESTRAINT DEVICES

Air Batteries, Metal
USE   METAL AIR BATTERIES

Air Bearings
USE   GAS BEARINGS

Air Blasts
USE   AERIAL EXPLOSIONS

AIR BREATHING BOOSTERS

AIR BREATHING ENGINES

Air Bubble Vehicles, Captured
USE   CAPTURED AIR BUBBLE VEHICLES

AIR CARGO

(Air Circulation), Registers
USE   REGISTERS (AIR CIRCULATION)

Air, Compressed
USE   COMPRESSED AIR

AIR CONDITIONING

AIR CONDITIONING EQUIPMENT

AIR CONDUCTIVITY

AIR COOLING

AIR CURRENTS

Air Currents, Vertical
USE   VERTICAL AIR CURRENTS

AIR CUSHION LANDING SYSTEMS

Air Cushion Vehicles
USE   GROUND EFFECT MACHINES

Air Cycle Engines, Liquid
USE   LIQUID AIR CYCLE ENGINES

AIR DEFENSE

Air Defense System, Sage
USE   SAGE AIR DEFENSE SYSTEM

Air Density Explorer A
USE   EXPLORER 19 SATELLITE

Air Density Explorer, Dual
USE   DUAL AIR DENSITY EXPLORER

Air Density/injun Explorer B
USE   EXPLORER 25 SATELLITE

AIR DROP OPERATIONS

AIR DUCTS

Air, Expired
USE   EXPIRED AIR

Air Facilities, Military
USE   MILITARY AIR FACILITIES

AIR FILTERS

AIR FLOW

Air Freight
USE   AIR CARGO

Air Fuel Cells, Hydrogen
USE   HYDROGEN OXYGEN FUEL CELLS

Air, High Temperature
USE   HIGH TEMPERATURE AIR

Air, Hot
USE   HIGH TEMPERATURE AIR

Air Inlets
USE   AIR INTAKES

AIR INTAKES

AIR JETS

AIR LAND INTERACTIONS

AIR LAUNCHING

AIR LAW

Air, Liquid
USE   LIQUID AIR

AIR LOCKS

AIR MAIL

AIR MASSES

Air Missiles, Air To
USE   AIR TO AIR MISSILES

Air Missiles, Ground-To-
USE   SURFACE TO AIR MISSILES

Air Missiles, Surface To
USE   SURFACE TO AIR MISSILES

AIR NAVIGATION

Air Navigation, All-Weather
USE   ALL-WEATHER AIR NAVIGATION

Air Navigation, Tactical
USE   TACAN

AIR PIRACY

AIR POLLUTION

Air Pollution, Global
USE   GLOBAL AIR POLLUTION

Air Pollution, Indoor
USE   INDOOR AIR POLLUTION

AIR PURIFICATION

AIR QUALITY

Air Ratio, Fuel-
USE   FUEL-AIR RATIO

Air Refueling, Air To
USE   AIR TO AIR REFUELING

Air Rockets, Air To
USE   AIR TO AIR MISSILES

AIR SAMPLING

Air Sampling Program, Global
USE   GLOBAL AIR SAMPLING PROGRAM

AIR SEA ICE INTERACTIONS

Air Sea Interactions
USE   AIR WATER INTERACTIONS

Air Sickness
USE   MOTION SICKNESS

AIR SLEW MISSILES

AIR START

AIR TO AIR MISSILES

AIR TO AIR REFUELING

Air To Air Rockets
USE   AIR TO AIR MISSILES

AIR TO SURFACE MISSILES

AIR TRAFFIC

AIR TRAFFIC CONTROL

AIR TRAFFIC CONTROLLERS (PERSONNEL)

Air Traffic Satellites, Location Of
USE   LOCATES SYSTEM

Air Transport, Supersonic Commercial
USE   SUPERSONIC COMMERCIAL AIR
        TRANSPORT

AIR TRANSPORTATION

Air Turbulence, Clear
USE   CLEAR AIR TURBULENCE

Air, Upper
USE   UPPER ATMOSPHERE

AIR WATER INTERACTIONS

Air-Ground Communication, Ground-
USE   GROUND-AIR-GROUND COMMUNICATION

Airborne Command Post, Advanced
USE   E-4A AIRCRAFT

AIRBORNE EQUIPMENT

AIRBORNE INFECTION

AIRBORNE INTEGRATED RECONNAISSANCE
SYSTEM

AIRBORNE LASERS

Airborne Multipurpose System, Light
USE   LIGHT AIRBORNE MULTIPURPOSE SYSTEM

Airborne Observatory, Kuiper
USE   C-141 AIRCRAFT

AIRBORNE RADAR APPROACH

AIRBORNE RANGE AND ORBIT DETERMINATION

AIRBORNE SURVEILLANCE RADAR

Airborne Warning And Control System
USE   AWACS AIRCRAFT

AIRBORNE/SPACEBORNE COMPUTERS

Airbus
USE   EUROPEAN AIRBUS

Airbus, European
USE   EUROPEAN AIRBUS

AIRCRAFT

Aircraft, A-1
USE   A-1 AIRCRAFT

Aircraft, A-2
USE   A-2 AIRCRAFT

Aircraft, A-3
USE   A-3 AIRCRAFT

Aircraft, A-4
USE   A-4 AIRCRAFT

Aircraft, A-5
USE   A-5 AIRCRAFT

Aircraft, A-6
USE   A-6 AIRCRAFT

Aircraft, A-7
USE   A-7 AIRCRAFT

Aircraft, A-9
USE   A-9 AIRCRAFT

Aircraft, A-10
USE   A-10 AIRCRAFT

Aircraft, A-37
USE   A-37 AIRCRAFT

Aircraft, A-300
USE   A-300 AIRCRAFT

Aircraft, A-310
USE   A-310 AIRCRAFT

Aircraft, A-320
USE   A-320 AIRCRAFT

Aircraft, AC-1
USE   DHC 4 AIRCRAFT

AIRCRAFT ACCIDENT INVESTIGATION

AIRCRAFT ACCIDENTS

Aircraft, Advanced Range Instrumentation
USE   ADVANCED RANGE INSTRUMENTATION
        AIRCRAFT

Aircraft, Advanced Technology Light Twin
USE   ATLIT PROJECT

Aircraft, Agricultural
USE   AGRICULTURAL AIRCRAFT

Aircraft, Airgeep
USE   VZ-8 AIRCRAFT

Aircraft, Aladin 2
USE   ALADIN 2 AIRCRAFT

Aircraft, Alpha Jet
USE   ALPHA JET AIRCRAFT

Aircraft, Amphibious
USE   AMPHIBIOUS AIRCRAFT

Aircraft, AN-2
USE   AN-2 AIRCRAFT

Aircraft, AN-22
USE   AN-22 AIRCRAFT

Aircraft, AN-24
USE   AN-24 AIRCRAFT

AIRCRAFT ANTENNAS

Aircraft, Antheus
USE   AN-22 AIRCRAFT

Aircraft, Antisubmarine Warfare
USE   ANTISUBMARINE WARFARE AIRCRAFT

Aircraft, Antonov
USE   ANTONOV AIRCRAFT

Aircraft, Antonov AN-22
USE   AN-22 AIRCRAFT

Aircraft, Antonov AN-24
USE   AN-24 AIRCRAFT

Aircraft, AO-1
USE   OV-1 AIRCRAFT

AIRCRAFT APPROACH SPACING

Aircraft, Argosy MK-1
USE   ARGOSY MK-1 AIRCRAFT

Aircraft, Atlantic
USE   BREGUET 1150 AIRCRAFT

Aircraft, Attack
USE   ATTACK AIRCRAFT

Aircraft, AV-8A
USE   HARRIER AIRCRAFT

Aircraft, AV-8B
USE   HARRIER AIRCRAFT

Aircraft, AVRO Whitworth HS-748
USE   HS-748 AIRCRAFT

Aircraft, AVRO 698
USE   VULCAN AIRCRAFT

Aircraft, AVRO 707
USE   AVRO 707 AIRCRAFT

Aircraft, Awacs
USE   AWACS AIRCRAFT

Aircraft, A2F
USE   A-6 AIRCRAFT

Aircraft, A3D
USE   A-3 AIRCRAFT

Aircraft, A3J
USE   A-5 AIRCRAFT

Aircraft, A4D
USE   A-4 AIRCRAFT

Aircraft, B-1
USE   B-1 AIRCRAFT

Aircraft, B-26
USE   B-26 AIRCRAFT

Aircraft, B-47
USE   B-47 AIRCRAFT

Aircraft, B-50
USE   B-50 AIRCRAFT

Aircraft, B-52
USE   B-52 AIRCRAFT

Aircraft, B-57
USE   B-57 AIRCRAFT

Aircraft, B-58
USE   B-58 AIRCRAFT

Aircraft, B-66
USE   B-66 AIRCRAFT

Aircraft, B-70
USE   B-70 AIRCRAFT

Aircraft, B-103
USE   BUCCANEER AIRCRAFT

Aircraft, BAC
USE   BAC AIRCRAFT

Aircraft, BAC TSR 2
USE   TSR-2 AIRCRAFT

Aircraft, BAC 111
USE   BAC 111 AIRCRAFT

Aircraft Bases
USE   MILITARY AIR FACILITIES

Aircraft, Beagle
USE   BEAGLE AIRCRAFT

Aircraft, Beech
USE   BEECHCRAFT AIRCRAFT

Aircraft, Beech C-33
USE   C-33 AIRCRAFT

Aircraft, Beech S-35
USE   C-35 AIRCRAFT

Aircraft, Beech 99
USE   BEECH 99 AIRCRAFT

Aircraft, Beechcraft
USE   BEECHCRAFT AIRCRAFT

Aircraft, Beechcraft 18
USE   BEECHCRAFT 18 AIRCRAFT

Aircraft, Belfast
USE   SC-5 AIRCRAFT

Aircraft, Bell
USE   BELL AIRCRAFT

Aircraft, Blackburn B-103
USE   BUCCANEER AIRCRAFT

Aircraft, Boeing
USE   BOEING AIRCRAFT

Aircraft, Boeing Military
USE   MILITARY AIRCRAFT

Aircraft, Boeing 707
USE   BOEING 707 AIRCRAFT

Aircraft, Boeing 720
USE   BOEING 720 AIRCRAFT

Aircraft, Boeing 727
USE   BOEING 727 AIRCRAFT

Aircraft, Boeing 733
USE   BOEING 733 AIRCRAFT

Aircraft, Boeing 737
USE   BOEING 737 AIRCRAFT

Aircraft, Boeing 747
USE   BOEING 747 AIRCRAFT

Aircraft, Boeing 747B
USE   E-4A AIRCRAFT

Aircraft, Boeing 757
USE   BOEING 757 AIRCRAFT

Aircraft, Boeing 767
USE   BOEING 767 AIRCRAFT

Aircraft, Boeing 2707
USE   BOEING 2707 AIRCRAFT

Aircraft, Bolkow
USE   BOLKOW AIRCRAFT

Aircraft, Bomber
USE   BOMBER AIRCRAFT

Aircraft, Bonanza
USE   C-35 AIRCRAFT

AIRCRAFT BRAKES

Aircraft, Breguet
USE   BREGUET AIRCRAFT

Aircraft, Breguet 940
USE   BREGUET 940 AIRCRAFT

Aircraft, Breguet 941
USE   BREGUET 941 AIRCRAFT

Aircraft, Breguet 1150
USE   BREGUET 1150 AIRCRAFT

Aircraft, British Aircraft Corp
USE   BAC AIRCRAFT

Aircraft, Buccaneer
USE   BUCCANEER AIRCRAFT

Aircraft, Buckeye
USE   T-2 AIRCRAFT

Aircraft, Buffalo
USE   DHC 5 AIRCRAFT

Aircraft, C-1A
USE   C-1A AIRCRAFT

Aircraft, C-2
USE   C-2 AIRCRAFT

Aircraft, C-5
USE   C-5 AIRCRAFT

Aircraft, C-8A Augmentor Wing
USE   C-8A AUGMENTOR WING AIRCRAFT

Aircraft, C-9
USE   C-9 AIRCRAFT

Aircraft, C-15
USE   C-15 AIRCRAFT

Aircraft, C-33
USE   C-33 AIRCRAFT

Aircraft, C-35
USE   C-35 AIRCRAFT

Aircraft, C-46
USE   C-46 AIRCRAFT

Aircraft, C-47
USE   C-47 AIRCRAFT

Aircraft, C-54
USE   C-54 AIRCRAFT

Aircraft, C-118
USE   C-118 AIRCRAFT

Aircraft, C-119
USE   C-119 AIRCRAFT

Aircraft, C-121
USE   C-121 AIRCRAFT

Aircraft, C-123
USE   C-123 AIRCRAFT

Aircraft, C-124
USE   C-124 AIRCRAFT

Aircraft, C-130
USE   C-130 AIRCRAFT

Aircraft, C-131
USE   C-131 AIRCRAFT

Aircraft, C-133
USE   C-133 AIRCRAFT

Aircraft, C-135
USE   C-135 AIRCRAFT

Aircraft, C-140
USE   C-140 AIRCRAFT

Aircraft, C-141
USE   C-141 AIRCRAFT

Aircraft, C-142
USE   XC-142 AIRCRAFT

Aircraft, C-160
USE   C-160 AIRCRAFT

Aircraft Cabins
USE   AIRCRAFT COMPARTMENTS

Aircraft, Camel
USE   TU-104 AIRCRAFT

Aircraft, Canadair
USE   CANADAIR AIRCRAFT

Aircraft, Canadair CF-104
USE   CANADAIR AIRCRAFT
        F-104 AIRCRAFT

Aircraft, Canadair CL-41
USE   CL-41 AIRCRAFT

Aircraft, Canadair CL-44
USE   CL-44 AIRCRAFT

Aircraft, Canadair CL-84
USE   CL-84 AIRCRAFT

Aircraft, Canberra
USE   CANBERRA AIRCRAFT

(Aircraft Capability), Ceiling
USE   CEILING (AIRCRAFT CAPABILITY)

Aircraft, Caravelle
USE   SE-210 AIRCRAFT

Aircraft, Cargo
USE   CARGO AIRCRAFT

Aircraft, Cargomaster
USE   C-133 AIRCRAFT

Aircraft, Caribou
   USE   DHC 4 AIRCRAFT

AIRCRAFT CARRIERS

Aircraft, CC-106
   USE   CL-44 AIRCRAFT

Aircraft, Centurion
   USE   CESSNA 210 AIRCRAFT

Aircraft, Cessna
   USE   CESSNA AIRCRAFT

Aircraft, Cessna L-19
   USE   CESSNA L-19 AIRCRAFT

Aircraft, Cessna Military
   USE   MILITARY AIRCRAFT

Aircraft, Cessna 172
   USE   CESSNA 172 AIRCRAFT

Aircraft, Cessna 205
   USE   CESSNA 205 AIRCRAFT

Aircraft, Cessna 210
   USE   CESSNA 210 AIRCRAFT

Aircraft, Cessna 402B
   USE   CESSNA 402B AIRCRAFT

Aircraft, CF-104
   USE   CANADAIR AIRCRAFT
          F-104 AIRCRAFT

Aircraft, Chance-Vought
   USE   CHANCE-VOUGHT AIRCRAFT

Aircraft, Chance-Vought Military
   USE   CHANCE-VOUGHT AIRCRAFT
          MILITARY AIRCRAFT

Aircraft, Chinese
   USE   CHINESE AIRCRAFT

Aircraft, CL-41
   USE   CL-41 AIRCRAFT

Aircraft, CL-44
   USE   CL-44 AIRCRAFT

Aircraft, CL-84
   USE   CL-84 AIRCRAFT

Aircraft, CL-600 Challenger
   USE   CL-600 CHALLENGER AIRCRAFT

Aircraft, CL-823
   USE   CL-823 AIRCRAFT

Aircraft, Classic
   USE   IL-62 AIRCRAFT

Aircraft, Cock
   USE   AN-22 AIRCRAFT

Aircraft, COD
   USE   C-2 AIRCRAFT

Aircraft, COIN
   USE   COIN AIRCRAFT

Aircraft, Coke
   USE   AN-24 AIRCRAFT

Aircraft Collisions, Bird-
   USE   BIRD-AIRCRAFT COLLISIONS

Aircraft, Comet 4
   USE   COMET 4 AIRCRAFT

Aircraft, Commando
   USE   C-46 AIRCRAFT

Aircraft, Commercial
   USE   COMMERCIAL AIRCRAFT

AIRCRAFT COMMUNICATION

AIRCRAFT COMPARTMENTS

Aircraft, Concorde
   USE   CONCORDE AIRCRAFT

AIRCRAFT CONFIGURATIONS

Aircraft Construction
   USE   AIRCRAFT STRUCTURES

AIRCRAFT CONSTRUCTION MATERIALS

AIRCRAFT CONTROL

Aircraft, Convair Military
   USE   MILITARY AIRCRAFT
          GENERAL DYNAMICS AIRCRAFT

Aircraft, Convair 340
   USE   CV-340 AIRCRAFT

Aircraft, Convair 440
   USE   CV-440 AIRCRAFT

Aircraft, Convair 880
   USE   CV-880 AIRCRAFT

Aircraft, Convair 990
   USE   CV-990 AIRCRAFT

Aircraft, Cookpot
   USE   TU-124 AIRCRAFT

Aircraft Corp Aircraft, British
   USE   BAC AIRCRAFT

Aircraft, Corsair
   USE   A-7 AIRCRAFT

Aircraft, Cougar
   USE   F-9 AIRCRAFT

Aircraft, Courier
   USE   U-10 AIRCRAFT

Aircraft, Crusader
   USE   F-8 AIRCRAFT

Aircraft, CT-114
   USE   CL-41 AIRCRAFT

Aircraft, Curtiss C-46
   USE   C-46 AIRCRAFT

Aircraft, Curtiss-Wright
   USE   CURTISS-WRIGHT AIRCRAFT

Aircraft, Curtiss-Wright Military
   USE   CURTISS-WRIGHT AIRCRAFT
          MILITARY AIRCRAFT

Aircraft, CV-2
   USE   DHC 4 AIRCRAFT

Aircraft, CV-7
   USE   DHC 5 AIRCRAFT

Aircraft, CV-340
   USE   CV-340 AIRCRAFT

Aircraft, CV-440
   USE   CV-440 AIRCRAFT

Aircraft, CV-880
   USE   CV-880 AIRCRAFT

Aircraft, CV-990
   USE   CV-990 AIRCRAFT

Aircraft, D-558
   USE   D-558 AIRCRAFT

Aircraft, Dakota
   USE   C-47 AIRCRAFT

Aircraft, Dassault
   USE   DASSAULT AIRCRAFT

Aircraft, Dassault Mirage 3
   USE   MIRAGE 3 AIRCRAFT

Aircraft, Dassault Mystere 20
   USE   MYSTERE 20 AIRCRAFT

Aircraft, Dassault Mystere 50
   USE   MYSTERE 50 AIRCRAFT

Aircraft, DC 3
   USE   DC 3 AIRCRAFT

Aircraft, DC 7
   USE   DC 7 AIRCRAFT

Aircraft, DC 8
   USE   DC 8 AIRCRAFT

Aircraft, DC 9
   USE   DC 9 AIRCRAFT

Aircraft, DC 10
   USE   DC 10 AIRCRAFT

Aircraft, De Havilland
   USE   DE HAVILLAND AIRCRAFT

Aircraft, De Havilland DH 106
   USE   COMET 4 AIRCRAFT

Aircraft, De Havilland DH 112
   USE   DH 112 AIRCRAFT

Aircraft, De Havilland DH 115
   USE   DH 115 AIRCRAFT

Aircraft, De Havilland DH 121
   USE   DH 121 AIRCRAFT

Aircraft, De Havilland DH 125
   USE   DH 125 AIRCRAFT

Aircraft, De Havilland DHC 4
   USE   DHC 4 AIRCRAFT

Aircraft, De Havilland DHC 5
   USE   DHC 5 AIRCRAFT

Aircraft, De Havilland Venom
   USE   DH 112 AIRCRAFT

Aircraft, Debonair
   USE   C-33 AIRCRAFT

Aircraft, Delfin
   USE   L-29 JET TRAINER

Aircraft, Delta Dagger
   USE   F-102 AIRCRAFT

Aircraft, Delta Dart
   USE   F-106 AIRCRAFT

AIRCRAFT DESIGN

Aircraft, Destroyer
   USE   B-66 AIRCRAFT

AIRCRAFT DETECTION

Aircraft, DH 106
   USE   COMET 4 AIRCRAFT

Aircraft, DH 112
   USE   DH 112 AIRCRAFT

Aircraft, DH 115
   USE   DH 115 AIRCRAFT

Aircraft, DH 121
   USE   DH 121 AIRCRAFT

Aircraft, DH 125
   USE   DH 125 AIRCRAFT

Aircraft, DHC Beaver
   USE   DHC 2 AIRCRAFT

Aircraft, DHC 2
   USE   DHC 2 AIRCRAFT

Aircraft, DHC 4
   USE   DHC 4 AIRCRAFT

Aircraft, DHC 5
   USE   DHC 5 AIRCRAFT

Aircraft, DO-27
USE   DO-27 AIRCRAFT

Aircraft, DO-28
USE   DO-28 AIRCRAFT

Aircraft, DO-31
USE   DO-31 AIRCRAFT

Aircraft, Dornier
USE   DORNIER AIRCRAFT

Aircraft, Dornier DO-27
USE   DO-27 AIRCRAFT

Aircraft, Dornier DO-28
USE   DO-28 AIRCRAFT

Aircraft, Dornier DO-31
USE   DO-31 AIRCRAFT

Aircraft, Douglas
USE   DOUGLAS AIRCRAFT

Aircraft, Douglas D-558
USE   D-558 AIRCRAFT

Aircraft, Douglas DC-3
USE   DC 3 AIRCRAFT

Aircraft, Douglas DC-7
USE   DC 7 AIRCRAFT

Aircraft, Douglas DC-8
USE   DC 8 AIRCRAFT

Aircraft, Douglas DC-9
USE   DC 9 AIRCRAFT

Aircraft, Douglas Military
USE   DOUGLAS AIRCRAFT
MILITARY AIRCRAFT

Aircraft, Douglas PD-808
USE   PD-808 AIRCRAFT

Aircraft, Drone
USE   DRONE AIRCRAFT

Aircraft, E-2
USE   E-2 AIRCRAFT

Aircraft, E-3A
USE   E-3A AIRCRAFT

Aircraft, E-4A
USE   E-4A AIRCRAFT

Aircraft, Earth Resources Survey
USE   EARTH RESOURCES SURVEY AIRCRAFT

Aircraft, EC-121
USE   EC-121 AIRCRAFT

Aircraft, Electra
USE   ELECTRA AIRCRAFT

Aircraft, Electric
USE   FLY BY WIRE CONTROL

Aircraft, Electronic
USE   ELECTRONIC AIRCRAFT

Aircraft Energy Efficiency Program
USE   ACEE PROGRAM

AIRCRAFT ENGINES

Aircraft, English Electric Canberra
USE   CANBERRA AIRCRAFT

AIRCRAFT EQUIPMENT

Aircraft, Executive
USE   GENERAL AVIATION AIRCRAFT
PASSENGER AIRCRAFT

Aircraft, F-2
USE   F-2 AIRCRAFT

Aircraft, F-4
USE   F-4 AIRCRAFT

Aircraft, F-5
USE   F-5 AIRCRAFT

Aircraft, F-8
USE   F-8 AIRCRAFT

Aircraft, F-9
USE   F-9 AIRCRAFT

Aircraft, F-14
USE   F-14 AIRCRAFT

Aircraft, F-15
USE   F-15 AIRCRAFT

Aircraft, F-16
USE   F-16 AIRCRAFT

Aircraft, F-17
USE   F-17 AIRCRAFT

Aircraft, F-18
USE   F-18 AIRCRAFT

Aircraft, F-20
USE   F-20 AIRCRAFT

Aircraft, F-27
USE   F-27 AIRCRAFT

Aircraft, F-28 Transport
USE   F-28 TRANSPORT AIRCRAFT

Aircraft, F-80
USE   T-33 AIRCRAFT

Aircraft, F-84
USE   F-84 AIRCRAFT

Aircraft, F-86
USE   F-86 AIRCRAFT

Aircraft, F-89
USE   F-89 AIRCRAFT

Aircraft, F-94
USE   F-94 AIRCRAFT

Aircraft, F-100
USE   F-100 AIRCRAFT

Aircraft, F-101
USE   F-101 AIRCRAFT

Aircraft, F-102
USE   F-102 AIRCRAFT

Aircraft, F-104
USE   F-104 AIRCRAFT

Aircraft, F-105
USE   F-105 AIRCRAFT

Aircraft, F-106
USE   F-106 AIRCRAFT

Aircraft, F-110
USE   F-4 AIRCRAFT

Aircraft, F-111
USE   F-111 AIRCRAFT

Aircraft, Fairchild Military
USE   FAIRCHILD-HILLER AIRCRAFT
MILITARY AIRCRAFT

Aircraft, Fairchild-Hiller
USE   FAIRCHILD-HILLER AIRCRAFT

Aircraft, Fairey
USE   FAIREY AIRCRAFT

Aircraft, Fairey Delta 2
USE   FD 2 AIRCRAFT

Aircraft, Fan In Wing
USE   FAN IN WING AIRCRAFT

Aircraft, FD 2
USE   FD 2 AIRCRAFT

Aircraft, Fellowship
USE   F-28 TRANSPORT AIRCRAFT

Aircraft, Fiat
USE   FIAT AIRCRAFT

Aircraft, Fiat G-91
USE   G-91 AIRCRAFT

Aircraft, Fiat G-95/4
USE   G-95/4 AIRCRAFT

Aircraft, Fiat G-222
USE   G-222 AIRCRAFT

Aircraft, Fighter
USE   FIGHTER AIRCRAFT

Aircraft, Firebee 2 Target Drone
USE   FIREBEE 2 TARGET DRONE AIRCRAFT

Aircraft, Fixed-Wing
USE   FIXED WINGS
AIRCRAFT CONFIGURATIONS

Aircraft, Flying Bedstead
USE   FLYING PLATFORMS

Aircraft, Flying Wing
USE   TAILLESS AIRCRAFT

Aircraft, Fokker
USE   FOKKER AIRCRAFT

Aircraft, Fokker F 27
USE   F-27 AIRCRAFT

Aircraft, Fokker F 28
USE   F-28 TRANSPORT AIRCRAFT

Aircraft, Fokker Friendship
USE   F-27 AIRCRAFT

Aircraft, Free Wing
USE   FREE WING AIRCRAFT

Aircraft, Freedom Fighter
USE   F-5 AIRCRAFT

AIRCRAFT FUEL SYSTEMS

AIRCRAFT FUELS

Aircraft, FV-12A
USE   FV-12A AIRCRAFT

Aircraft, F4H
USE   F-4 AIRCRAFT

Aircraft, F8U
USE   F-8 AIRCRAFT

Aircraft, F9F
USE   F-9 AIRCRAFT

Aircraft, G-1
USE   G-1 AIRCRAFT

Aircraft, G-91
USE   G-91 AIRCRAFT

Aircraft, G-95/4
USE   G-95/4 AIRCRAFT

Aircraft, G-222
USE   G-222 AIRCRAFT

Aircraft, GA-5
USE   GA-5 AIRCRAFT

Aircraft, Galaxy
USE   C-5 AIRCRAFT

Aircraft, GC-130
USE   C-130 AIRCRAFT

Aircraft, General Aviation
USE   GENERAL AVIATION AIRCRAFT

Aircraft, General Dynamics
USE   GENERAL DYNAMICS AIRCRAFT

Aircraft, General Dynamics Military
USE   GENERAL DYNAMICS AIRCRAFT
       MILITARY AIRCRAFT

Aircraft, GETOL
USE   GETOL AIRCRAFT

Aircraft, Gloster GA-5
USE   GA-5 AIRCRAFT

Aircraft, Griffon
USE   NORD 1500 AIRCRAFT

Aircraft, Grumman
USE   GRUMMAN AIRCRAFT

Aircraft, Grumman OV-1C
USE   OV-1 AIRCRAFT

AIRCRAFT GUIDANCE

Aircraft, Gyrodyne
USE   GYRODYNE AIRCRAFT

Aircraft, Gyrodyne Military
USE   QH-50 HELICOPTER

Aircraft, H-126
USE   H-126 AIRCRAFT

Aircraft, Hamburger
USE   HAMBURGER AIRCRAFT

Aircraft, Hamburger HFB-320
USE   HFB-320 AIRCRAFT

Aircraft, Handley Page
USE   HANDLEY PAGE AIRCRAFT

Aircraft, Handley Page HP-115
USE   HP-115 AIRCRAFT

Aircraft, Harrier
USE   HARRIER AIRCRAFT

Aircraft, Hawker Hunter
USE   F-2 AIRCRAFT

Aircraft, Hawker P-1052
USE   P-1052 AIRCRAFT

Aircraft, Hawker P-1127
USE   P-1127 AIRCRAFT

Aircraft, Hawker P-1154
USE   P-1154 AIRCRAFT

Aircraft, Hawker Siddeley
USE   HAWKER SIDDELEY AIRCRAFT

Aircraft, Hawkeye
USE   E-2 AIRCRAFT

AIRCRAFT HAZARDS

Aircraft, Heinkel
USE   HEINKEL AIRCRAFT

Aircraft, Helio
USE   HELIO AIRCRAFT

Aircraft, Helio Military
USE   HELIO AIRCRAFT

Aircraft, Hercules
USE   C-130 AIRCRAFT

Aircraft, HFB-320
USE   HFB-320 AIRCRAFT

Aircraft, Highly Maneuverable
USE   HIGHLY MANEUVERABLE AIRCRAFT

Aircraft, Hiller
USE   HILLER AIRCRAFT

Aircraft, Hiller Military
USE   HILLER AIRCRAFT
       MILITARY AIRCRAFT

Aircraft, HP-115
USE   HP-115 AIRCRAFT

Aircraft, HS-125
USE   DH 125 AIRCRAFT

Aircraft, HS-748
USE   HS-748 AIRCRAFT

Aircraft, HS-801
USE   HS-801 AIRCRAFT

Aircraft, Hughes
USE   HUGHES AIRCRAFT

Aircraft, Hughes Military
USE   HUGHES AIRCRAFT
       MILITARY AIRCRAFT

Aircraft, Hummingbird
USE   XV-4 AIRCRAFT

Aircraft, Hunter F-2
USE   F-2 AIRCRAFT

Aircraft, Hunting H-126
USE   H-126 AIRCRAFT

Aircraft, Hunting P-84
USE   JET PROVOST AIRCRAFT

Aircraft, Hustler
USE   B-58 AIRCRAFT

AIRCRAFT HYDRAULIC SYSTEMS

Aircraft, Hypersonic
USE   HYPERSONIC AIRCRAFT

Aircraft, IL-14
USE   IL-14 AIRCRAFT

Aircraft, IL-62
USE   IL-62 AIRCRAFT

Aircraft, Ilyushin
USE   ILYUSHIN AIRCRAFT

Aircraft, Ilyushin IL-14
USE   IL-14 AIRCRAFT

Aircraft, Ilyushin IL-62
USE   IL-62 AIRCRAFT

AIRCRAFT INDUSTRY

AIRCRAFT INSTRUMENTS

Aircraft, Interceptor
USE   FIGHTER AIRCRAFT

Aircraft, Intruder
USE   A-6 AIRCRAFT

Aircraft, Invader
USE   B-26 AIRCRAFT

Aircraft, Iskra
USE   TS-11 AIRCRAFT

Aircraft, Jaguar
USE   JAGUAR AIRCRAFT

Aircraft, Javelin
USE   GA-5 AIRCRAFT

Aircraft, JC-130
USE   C-130 AIRCRAFT

Aircraft, Jet
USE   JET AIRCRAFT

Aircraft, Jet Dragon
USE   DH 125 AIRCRAFT

Aircraft, Jet Provost
USE   JET PROVOST AIRCRAFT

Aircraft, Jet Star
USE   C-140 AIRCRAFT

Aircraft, Jetstream
USE   JETSTREAM AIRCRAFT

Aircraft, JF 101
USE   F-101 AIRCRAFT

Aircraft, Jindivik Target
USE   JINDIVIK TARGET AIRCRAFT

Aircraft, Kaman
USE   KAMAN AIRCRAFT

Aircraft, Kawasaki
USE   KAWASAKI AIRCRAFT

Aircraft, KC-130
USE   C-130 AIRCRAFT

Aircraft, KC-135
USE   C-135 AIRCRAFT

Aircraft, Kestrel
USE   P-1127 AIRCRAFT

Aircraft, L-28
USE   U-10 AIRCRAFT

Aircraft, L-29
USE   L-29 JET TRAINER

Aircraft, L-1011
USE   L-1011 AIRCRAFT

Aircraft, L-2000
USE   L-2000 AIRCRAFT

AIRCRAFT LANDING

Aircraft, Lara
USE   COIN AIRCRAFT

AIRCRAFT LAUNCHING DEVICES

Aircraft, Lear Jet
USE   LEAR JET AIRCRAFT

Aircraft, Light
USE   LIGHT AIRCRAFT

Aircraft, Light Armed Reconnaissance
USE   COIN AIRCRAFT

Aircraft, Light Transport
USE   LIGHT TRANSPORT AIRCRAFT

AIRCRAFT LIGHTS

Aircraft, Ling-Temco-Vought
USE   LING-TEMCO-VOUGHT AIRCRAFT

Aircraft, Lockheed
USE   LOCKHEED AIRCRAFT

Aircraft, Lockheed C-5
USE   C-5 AIRCRAFT

Aircraft, Lockheed CL-823
USE   CL-823 AIRCRAFT

Aircraft, Lockheed Constellation
USE   C-121 AIRCRAFT

Aircraft, Lockheed L-2000
USE   L-2000 AIRCRAFT

Aircraft, Lockheed Model 18
USE   LOCKHEED MODEL 18 AIRCRAFT

Aircraft, Lockheed U-2
USE   U-2 AIRCRAFT

Aircraft, Lockheed XV-4A
USE   XV-4 AIRCRAFT

Aircraft, Low Wing
USE   LOW WING AIRCRAFT

Aircraft, LTV
USE   LING-TEMCO-VOUGHT AIRCRAFT

AIRCRAFT MAINTENANCE

Aircraft, Man Powered
USE   MAN POWERED AIRCRAFT

AIRCRAFT MANEUVERS

Aircraft, Martin
USE   MARTIN AIRCRAFT

Aircraft, Max Holste MH-262
USE   MH-262 AIRCRAFT

Aircraft, Mcdonnell
USE   MCDONNELL AIRCRAFT

Aircraft, Mcdonnell Douglas
USE   MCDONNELL DOUGLAS AIRCRAFT

Aircraft, ME P-160
USE   P-160 AIRCRAFT

Aircraft, ME P-308
USE   P-308 AIRCRAFT

Aircraft, Mercure
USE   MERCURE AIRCRAFT

Aircraft, Messerschmitt ME P-160
USE   P-160 AIRCRAFT

Aircraft, Messerschmitt ME P-308
USE   P-308 AIRCRAFT

Aircraft, Meteorological Research
USE   METEOROLOGICAL RESEARCH AIRCRAFT

Aircraft, Metropolitan
USE   CV-440 AIRCRAFT

Aircraft, MH-262
USE   MH-262 AIRCRAFT

Aircraft, MIG
USE   MIG AIRCRAFT

Aircraft, Mil
USE   MIL AIRCRAFT

Aircraft, Military
USE   MILITARY AIRCRAFT

Aircraft, Mirage
USE   MIRAGE AIRCRAFT

Aircraft, Mirage 3
USE   MIRAGE 3 AIRCRAFT

AIRCRAFT MODELS

Aircraft, Mohawk
USE   OV-1 AIRCRAFT

Aircraft, MRCA
USE   MRCA AIRCRAFT

Aircraft, Multi-Role Combat
USE   MRCA AIRCRAFT

Aircraft, Mustang
USE   P-51 AIRCRAFT

Aircraft, Mystere 20
USE   MYSTERE 20 AIRCRAFT

Aircraft, Mystere 50
USE   MYSTERE 50 AIRCRAFT

Aircraft, N-156
USE   F-5 AIRCRAFT

Aircraft, NA-300
USE   OV-10 AIRCRAFT

Aircraft, NAMC
USE   NIHON AIRCRAFT

Aircraft, Navion
USE   NAVION AIRCRAFT

Aircraft, Navion G-1
USE   G-1 AIRCRAFT

Aircraft, Navion Rangemaster
USE   G-1 AIRCRAFT

Aircraft, NC-130
USE   C-130 AIRCRAFT

(Aircraft), Night Flights
USE   NIGHT FLIGHTS (AIRCRAFT)

Aircraft, Nihon
USE   NIHON AIRCRAFT

Aircraft, Nihon YS-11
USE   YS-11 AIRCRAFT

AIRCRAFT NOISE

Aircraft Noise, Jet
USE   JET AIRCRAFT NOISE

Aircraft Noise Prediction
USE   NOISE PREDICTION (AIRCRAFT)

(Aircraft), Noise Prediction
USE   NOISE PREDICTION (AIRCRAFT)

Aircraft, Nord
USE   NORD AIRCRAFT

Aircraft, Nord 262
USE   MH-262 AIRCRAFT

Aircraft, Nord 1500
USE   NORD 1500 AIRCRAFT

Aircraft, North American
USE   NORTH AMERICAN AIRCRAFT

Aircraft, Northrop
USE   NORTHROP AIRCRAFT

Aircraft, Nuclear Propelled
USE   NUCLEAR PROPELLED AIRCRAFT

Aircraft, Observation
USE   OBSERVATION AIRCRAFT

Aircraft, Omnipol L-29
USE   L-29 JET TRAINER

Aircraft, Omnipol Z-37
USE   Z-37 AIRCRAFT

Aircraft, Orion
USE   P-3 AIRCRAFT

Aircraft, Ornithopter
USE   RESEARCH AIRCRAFT

Aircraft, OV-1
USE   OV-1 AIRCRAFT

Aircraft, OV-10
USE   OV-10 AIRCRAFT

Aircraft, P-3
USE   P-3 AIRCRAFT

Aircraft, P-51
USE   P-51 AIRCRAFT

Aircraft, P-84
USE   JET PROVOST AIRCRAFT

Aircraft, P-160
USE   P-160 AIRCRAFT

Aircraft, P-166
USE   P-166 AIRCRAFT

Aircraft, P-308
USE   P-308 AIRCRAFT

Aircraft, P-1052
USE   P-1052 AIRCRAFT

Aircraft, P-1127
USE   P-1127 AIRCRAFT

Aircraft, P-1154
USE   P-1154 AIRCRAFT

Aircraft, Pa-34 Seneca
USE   PA-34 SENECA AIRCRAFT

Aircraft, Panavia Military
USE   PANAVIA MILITARY AIRCRAFT

Aircraft, Panther
USE   F-9 AIRCRAFT

AIRCRAFT PARTS

Aircraft, Passenger
USE   PASSENGER AIRCRAFT

Aircraft, PD-808
USE   PD-808 AIRCRAFT

AIRCRAFT PERFORMANCE

Aircraft, Phantom
USE   PHANTOM AIRCRAFT

Aircraft, Piaggio
USE   PIAGGIO AIRCRAFT

Aircraft, Piaggio P-166
USE   P-166 AIRCRAFT

Aircraft, Piaggio-Douglas PD-808
USE   PD-808 AIRCRAFT

Aircraft, Piasecki
USE   PIASECKI AIRCRAFT

Aircraft, Pilotless
USE   PILOTLESS AIRCRAFT

AIRCRAFT PILOTS

Aircraft, Piper
USE   PIPER AIRCRAFT

Aircraft, Pivoted Wing
USE   TILT WING AIRCRAFT

Aircraft, Polish TS-11
USE   TS-11 AIRCRAFT

Aircraft, Potez
USE   POTEZ AIRCRAFT

Aircraft Power Sources
USE   AIRCRAFT ENGINES

AIRCRAFT POWER SUPPLIES

Aircraft, Powered Lift
USE   POWERED LIFT AIRCRAFT

Aircraft, Private
USE   GENERAL AVIATION AIRCRAFT

AIRCRAFT PRODUCTION

AIRCRAFT PRODUCTION COSTS

Aircraft Program, Tilt Rotor Research
USE   TILT ROTOR RESEARCH AIRCRAFT
       PROGRAM

Aircraft, Provider
USE   C-123 AIRCRAFT

Aircraft, P3V
USE   P-3 AIRCRAFT

Aircraft, Rangemaster
USE   G-1 AIRCRAFT

Aircraft, RB-47
USE   B-47 AIRCRAFT

Aircraft, RB-50
USE   RB-50 AIRCRAFT

Aircraft, RB-57
　　USE　B-57 AIRCRAFT

Aircraft, RB-66
　　USE　B-66 AIRCRAFT

Aircraft Readiness Monitor, Automatic Light
　　USE　ALARM PROJECT

Aircraft, Reconnaissance
　　USE　RECONNAISSANCE AIRCRAFT

AIRCRAFT RELIABILITY

Aircraft, Republic
　　USE　REPUBLIC AIRCRAFT

Aircraft, Republic Military
　　USE　MILITARY AIRCRAFT

Aircraft, Research
　　USE　RESEARCH AIRCRAFT

Aircraft Research, Supersonic Cruise
　　USE　SUPERSONIC CRUISE AIRCRAFT
　　　　　RESEARCH

Aircraft, RF-4
　　USE　RF-4 AIRCRAFT

Aircraft, RF-8
　　USE　F-8 AIRCRAFT

Aircraft Rocket Vehicle, Folding Fin
　　USE　FOLDING FIN AIRCRAFT ROCKET VEHICLE

Aircraft, Rotary Wing
　　USE　ROTARY WING AIRCRAFT

Aircraft, Rotor Systems Research
　　USE　ROTOR SYSTEMS RESEARCH AIRCRAFT

Aircraft, Rotorcraft
　　USE　ROTORCRAFT AIRCRAFT

AIRCRAFT RUNUP

Aircraft, Ryan
　　USE　RYAN AIRCRAFT

Aircraft, Ryan Military
　　USE　RYAN AIRCRAFT

Aircraft, R5D
　　USE　C-54 AIRCRAFT

Aircraft, R7V
　　USE　C-121 AIRCRAFT
　　　　　EC-121 AIRCRAFT

Aircraft, S-2
　　USE　S-2 AIRCRAFT

Aircraft S-2b, Snow Aerial Applicator
　　USE　S-2 AIRCRAFT

Aircraft, S-3
　　USE　S-3 AIRCRAFT

Aircraft, Saab
　　USE　SAAB AIRCRAFT

Aircraft, Saab 37
　　USE　SAAB 37 AIRCRAFT

Aircraft, Saab 105
　　USE　SAAB 105 AIRCRAFT

Aircraft, Sabre
　　USE　F-86 AIRCRAFT

Aircraft, Sabreliner
　　USE　T-39 AIRCRAFT

AIRCRAFT SAFETY

Aircraft, Samaritan
　　USE　C-131 AIRCRAFT

Aircraft, Savage
　　USE　A-2 AIRCRAFT

Aircraft, SC-1
　　USE　SC-1 AIRCRAFT

Aircraft, SC-5
　　USE　SC-5 AIRCRAFT

Aircraft, SC-7
　　USE　SC-7 AIRCRAFT

Aircraft, Schleicher
　　USE　SCHLEICHER AIRCRAFT

Aircraft, Scimitar
　　USE　SCIMITAR AIRCRAFT

Aircraft, SE-210
　　USE　SE-210 AIRCRAFT

Aircraft, Seneca
　　USE　PA-34 SENECA AIRCRAFT

Aircraft, Shooting Star
　　USE　T-33 AIRCRAFT

Aircraft, Short Belfast C MK-1
　　USE　SC-5 AIRCRAFT

Aircraft, Short Haul
　　USE　SHORT HAUL AIRCRAFT

Aircraft, Short SC-1
　　USE　SC-1 AIRCRAFT

Aircraft, Short SC-5
　　USE　SC-5 AIRCRAFT

Aircraft, Short SC-7
　　USE　SC-7 AIRCRAFT

Aircraft, Short Takeoff
　　USE　SHORT TAKEOFF AIRCRAFT

Aircraft, Siebel
　　USE　SIEBEL AIRCRAFT

Aircraft, Sikorsky
　　USE　SIKORSKY AIRCRAFT

Aircraft, Skyhawk
　　USE　A-4 AIRCRAFT

Aircraft, Skymaster
　　USE　C-54 AIRCRAFT

Aircraft, Skyraider
　　USE　A-1 AIRCRAFT

Aircraft, Skyrocket
　　USE　D-558 AIRCRAFT

Aircraft, Skystreak
　　USE　D-558 AIRCRAFT

Aircraft, Skyvan
　　USE　SC-7 AIRCRAFT

Aircraft, Skywarrior
　　USE　A-3 AIRCRAFT

Aircraft, Snow
　　USE　SNOW AIRCRAFT

Aircraft, Snow S-2
　　USE　S-2 AIRCRAFT

Aircraft, Solar Powered
　　USE　SOLAR POWERED AIRCRAFT

Aircraft, Spanloader
　　USE　SPANLOADER AIRCRAFT

AIRCRAFT SPECIFICATIONS

AIRCRAFT SPIN

AIRCRAFT STABILITY

Aircraft, Starfighter
　　USE　F-104 AIRCRAFT

Aircraft, Starlifter
　　USE　C-141 AIRCRAFT

Aircraft, Steep Gradient
　　USE　V/STOL AIRCRAFT

Aircraft, STOL
　　USE　SHORT TAKEOFF AIRCRAFT

Aircraft, Stratofortress
　　USE　B-52 AIRCRAFT

Aircraft, Stratojet
　　USE　B-47 AIRCRAFT

Aircraft, Stratotanker
　　USE　C-135 AIRCRAFT

AIRCRAFT STRUCTURES

Aircraft Structures, Plastic
　　USE　PLASTIC AIRCRAFT STRUCTURES

Aircraft, Submersible
　　USE　SUBMERSIBLE AIRCRAFT

Aircraft, Subsonic
　　USE　SUBSONIC AIRCRAFT

Aircraft, Sud Aviation
　　USE　SUD AVIATION AIRCRAFT

Aircraft, Sud Aviation SE-210
　　USE　SE-210 AIRCRAFT

Aircraft, Sud VJ-101
　　USE　VJ-101 AIRCRAFT

Aircraft, Super Fortress
　　USE　RB-50 AIRCRAFT

Aircraft, Super Sabre
　　USE　F-100 AIRCRAFT

Aircraft, Supersonic
　　USE　SUPERSONIC AIRCRAFT

AIRCRAFT SURVIVABILITY

Aircraft, T-2
　　USE　T-2 AIRCRAFT

Aircraft, T-28
　　USE　T-28 AIRCRAFT

Aircraft, T-33
　　USE　T-33 AIRCRAFT

Aircraft, T-37
　　USE　T-37 AIRCRAFT

Aircraft, T-38
　　USE　T-38 AIRCRAFT

Aircraft, T-39
　　USE　T-39 AIRCRAFT

Aircraft, Tailless
　　USE　TAILLESS AIRCRAFT

Aircraft, Talon
　　USE　T-38 AIRCRAFT

Aircraft, Tandem Wing
　　USE　TANDEM WING AIRCRAFT

Aircraft, Tanker
　　USE　TANKER AIRCRAFT

Aircraft, Target Drone
　　USE　TARGET DRONE AIRCRAFT

Aircraft Technology Program, Transonic
　　USE　TACT PROGRAM

Aircraft, Terrain Following
　　USE　TERRAIN FOLLOWING AIRCRAFT

Aircraft, TFX
　　USE　F-111 AIRCRAFT

Aircraft, Thunderchief
USE   F-105 AIRCRAFT

Aircraft, Tilt Rotor
USE   TILT ROTOR AIRCRAFT

Aircraft, Tilt Wing
USE   TILT WING AIRCRAFT

AIRCRAFT TIRES

Aircraft, Tornado
USE   MRCA AIRCRAFT

Aircraft, Trader
USE   C-1A AIRCRAFT

Aircraft, Training
USE   TRAINING AIRCRAFT

Aircraft, Transall C-160
USE   C-160 AIRCRAFT

Aircraft, Transonic
USE   SUPERSONIC AIRCRAFT

Aircraft, Transport
USE   TRANSPORT AIRCRAFT

Aircraft, Trident
USE   DH 121 AIRCRAFT

Aircraft, Trojan
USE   T-28 AIRCRAFT

Aircraft, TS-11
USE   TS-11 AIRCRAFT

Aircraft, TSR-2
USE   TSR-2 AIRCRAFT

Aircraft, TU-104
USE   TU-104 AIRCRAFT

Aircraft, TU-124
USE   TU-124 AIRCRAFT

Aircraft, TU-134
USE   TU-134 AIRCRAFT

Aircraft, TU-144
USE   TU-144 AIRCRAFT

Aircraft, TU-154
USE   TU-154 AIRCRAFT

Aircraft, Tupolev
USE   TUPOLEV AIRCRAFT

Aircraft, Turbo-Skyvan
USE   SC-7 AIRCRAFT

Aircraft, Turbofan
USE   TURBOFAN AIRCRAFT

Aircraft, Turbojet
USE   JET AIRCRAFT

Aircraft, Turboprop
USE   TURBOPROP AIRCRAFT

Aircraft, Tutor
USE   CL-41 AIRCRAFT

Aircraft, T2J
USE   T-2 AIRCRAFT

Aircraft, T3J
USE   T-39 AIRCRAFT

Aircraft, U-2
USE   U-2 AIRCRAFT

Aircraft, U-10
USE   U-10 AIRCRAFT

Aircraft, Ultralight
USE   ULTRALIGHT AIRCRAFT

Aircraft, US-2A
USE   S-2 AIRCRAFT

Aircraft, Utility
USE   UTILITY AIRCRAFT

Aircraft, V-3
USE   XV-3 AIRCRAFT

Aircraft, V-4
USE   XV-4 AIRCRAFT

Aircraft, V-5
USE   XV-5 AIRCRAFT

Aircraft, V-9
USE   XV-9A AIRCRAFT

Aircraft, V/STOL
USE   V/STOL AIRCRAFT

Aircraft, Valiant
USE   VALIANT AIRCRAFT

Aircraft, Valkyrie
USE   B-70 AIRCRAFT

Aircraft, Vampire
USE   DH 115 AIRCRAFT

Aircraft, Vampire MK 35
USE   VAMPIRE MK 35 AIRCRAFT

Aircraft, Vatol
USE   VATOL AIRCRAFT

Aircraft, VC-10
USE   VC-10 AIRCRAFT

Aircraft, Venom
USE   DH 112 AIRCRAFT

Aircraft, Vertical Attitude Takeoff-Landing
USE   VATOL AIRCRAFT

Aircraft, Vertical Takeoff
USE   VERTICAL TAKEOFF AIRCRAFT

Aircraft, Vickers Scimitar
USE   SCIMITAR AIRCRAFT

Aircraft, Vickers Valiant
USE   VALIANT AIRCRAFT

Aircraft, Vickers VC-10
USE   VC-10 AIRCRAFT

Aircraft, Vickers 1100
USE   VC-10 AIRCRAFT

Aircraft, Victor MK-1
USE   VICTOR MK-1 AIRCRAFT

Aircraft, Vigilante
USE   A-5 AIRCRAFT

Aircraft, Viscount
USE   VISCOUNT AIRCRAFT

Aircraft, VJ-101
USE   VJ-101 AIRCRAFT

Aircraft, Voodoo
USE   F-101 AIRCRAFT

Aircraft, VTOL
USE   VERTICAL TAKEOFF AIRCRAFT

Aircraft, Vulcan
USE   VULCAN AIRCRAFT

Aircraft, VZ-2
USE   VZ-2 AIRCRAFT

Aircraft, VZ-8
USE   VZ-8 AIRCRAFT

Aircraft, VZ-10
USE   XV-4 AIRCRAFT

Aircraft, VZ-11
USE   XV-5 AIRCRAFT

Aircraft, VZ-12
USE   P-1127 AIRCRAFT

AIRCRAFT WAKES

Aircraft, Warning Star
USE   EC-121 AIRCRAFT

Aircraft, Water Takeoff And Landing
USE   WATER TAKEOFF AND LANDING AIRCRAFT

Aircraft, Weather Reconnaissance
USE   WEATHER RECONNAISSANCE AIRCRAFT

Aircraft, Weser
USE   WESER AIRCRAFT

Aircraft, Westland
USE   WESTLAND AIRCRAFT

Aircraft, WU-2
USE   U-2 AIRCRAFT

Aircraft, W2F
USE   E-2 AIRCRAFT

Aircraft, X-1
USE   X-1 AIRCRAFT

Aircraft, X-2
USE   X-2 AIRCRAFT

Aircraft, X-3
USE   X-3 AIRCRAFT

Aircraft, X-5
USE   X-5 AIRCRAFT

Aircraft, X-13
USE   X-13 AIRCRAFT

Aircraft, X-14
USE   X-14 AIRCRAFT

Aircraft, X-15
USE   X-15 AIRCRAFT

Aircraft, X-19
USE   X-19 AIRCRAFT

Aircraft, X-20
USE   X-20 AIRCRAFT

Aircraft, X-21
USE   X-21 AIRCRAFT

Aircraft, X-21A
USE   X-21A AIRCRAFT

Aircraft, X-22
USE   X-22 AIRCRAFT

Aircraft, X-22A
USE   X-22A AIRCRAFT

Aircraft, X-24
USE   X-24 AIRCRAFT

Aircraft, X-29
USE   X-29 AIRCRAFT

Aircraft, XB-47
USE   B-47 AIRCRAFT

Aircraft, XB-70
USE   B-70 AIRCRAFT

Aircraft, Xbqm-180a
USE   VATOL AIRCRAFT

Aircraft, XC-142
USE   XC-142 AIRCRAFT

Aircraft, XV-3
USE   XV-3 AIRCRAFT

Aircraft, XV-4
USE   XV-4 AIRCRAFT

Aircraft, XV-5
USE   XV-5 AIRCRAFT

Aircraft, XV-5A
USE   XV-5 AIRCRAFT

Aircraft, XV-6A
USE   P-1127 AIRCRAFT

Aircraft, XV-8A
USE   XV-8A AIRCRAFT

Aircraft, XV-9A
USE   XV-9A AIRCRAFT

Aircraft, XV-11A
USE   XV-11A AIRCRAFT

Aircraft, XV-15
USE   XV-15 AIRCRAFT

Aircraft, Yak 40
USE   YAK 40 AIRCRAFT

Aircraft, YC-14
USE   YC-14 AIRCRAFT

Aircraft, YC-15
USE   C-15 AIRCRAFT

Aircraft, YC-123
USE   C-123 AIRCRAFT

Aircraft, YF-12
USE   YF-12 AIRCRAFT

Aircraft, YF-16
USE   YF-16 AIRCRAFT

Aircraft, YF-17
USE   F-17 AIRCRAFT

Aircraft, YF-102
USE   F-102 AIRCRAFT

Aircraft, YS-11
USE   YS-11 AIRCRAFT

Aircraft, YT-2
USE   T-2 AIRCRAFT

Aircraft, Yukon
USE   CL-44 AIRCRAFT

Aircraft, Z-37
USE   Z-37 AIRCRAFT

Aircrews
USE   FLIGHT CREWS

AIRDROPS

AIRFIELD SURFACE MOVEMENTS

Airfields
USE   AIRPORTS

Airfoil Characteristics
USE   AIRFOILS

Airfoil, Clark Y
USE   AIRFOIL PROFILES

AIRFOIL FENCES

Airfoil, Gaw-1
USE   GAW-1 AIRFOIL

Airfoil, Gaw-2
USE   GAW-2 AIRFOIL

Airfoil, General Aviation Whitcomb
USE   GAW-1 AIRFOIL
       GAW-2 AIRFOIL

AIRFOIL PROFILES

Airfoil Sections
USE   AIRFOIL PROFILES

Airfoil Thickness
USE   AIRFOIL PROFILES

AIRFOILS

Airfoils, Circulation Control
USE   CIRCULATION CONTROL AIRFOILS

Airfoils, Drooped
USE   DROOPED AIRFOILS

Airfoils, Laminar Flow
USE   LAMINAR FLOW AIRFOILS

Airfoils, Supercritical
USE   SUPERCRITICAL AIRFOILS

Airfoils, Supersonic
USE   SUPERSONIC AIRFOILS

Airfoils, Thin
USE   THIN AIRFOILS

Airframe Configurations, Inlet
USE   INLET AIRFRAME CONFIGURATIONS

Airframe Integration, Engine
USE   ENGINE AIRFRAME INTEGRATION

AIRFRAME MATERIALS

AIRFRAMES

Airgeep Aircraft
USE   VZ-8 AIRCRAFT

AIRGLOW

Airglow, Night
USE   NIGHTGLOW

AIRLINE OPERATIONS

AIRLOCK MODULES

Airplane, Experimental STOL Transport Rsch
USE   QUESTOL

AIRPORT BEACONS

AIRPORT LIGHTS

AIRPORT PLANNING

AIRPORT SECURITY

AIRPORT SURFACE DETECTION EQUIPMENT

AIRPORT TOWERS

AIRPORTS

Airs (Reconnaissance Sys)
USE   AIRBORNE INTEGRATED
       RECONNAISSANCE SYSTEM

AIRSHIPS

Airships, Heavy Lift
USE   HEAVY LIFT AIRSHIPS

AIRSPACE

Airspace System, National
USE   NATIONAL AIRSPACE SYSTEM

Airspace Utilization System, National
USE   NATIONAL AIRSPACE UTILIZATION SYSTEM

AIRSPEED

Airstreams, Jet
USE   JET STREAMS (METEOROLOGY)

Airworthiness
USE   AIRCRAFT RELIABILITY

Airworthiness Requirements
USE   AIRCRAFT RELIABILITY

AIRY FUNCTION

AITKEN NUCLEI

AJ-1 Engine, YLR-91-
USE   YLR-91-AJ-1 ENGINE

AJ-5 Engine, LR-87-
USE   LR-87-AJ-5 ENGINE

AJ-5 Engine, LR-91-
USE   LR-91-AJ-5 ENGINE

AJ-5 Engine, XLR-91-
USE   LR-91-AJ-5 ENGINE

AJ-10 ENGINE

AJ-1000 Engine
USE   M-1 ENGINE

Ajax Missile, Nike-
USE   NIKE-AJAX MISSILE

AK
USE   ALASKA

(AK), Chena River Basin
USE   CHENA RIVER BASIN (AK)

(AK), Cook Inlet
USE   COOK INLET (AK)

(AK), Prince William Sound
USE   PRINCE WILLIAM SOUND (AK)

(AK), Wrangell Mountains
USE   WRANGELL MOUNTAINS (AK)

AKERMANITE

Al
USE   ALUMINUM

AL
USE   ALABAMA

(AL-KY-TN), Tennessee Valley
USE   TENNESSEE VALLEY (AL-KY-TN)

ALABAMA

ALADIN 2 AIRCRAFT

ALAIS METEORITE

Alamos Molten Plutonium Reactor, Los
USE   LOS ALAMOS MOLTEN PLUTONIUM
       REACTOR

Alamos Turret Reactor, Los
USE   HIGH TEMPERATURE NUCLEAR REACTORS

Alamos Water Boiler Reactor, Los
USE   LOS ALAMOS WATER BOILER REACTOR

ALANINE

Alanine, Phenyl
USE   PHENYLALANINE

ALARM PROJECT

Alarms
USE   WARNING SYSTEMS

ALASKA

Alaska, Gulf Of
USE   GULF OF ALASKA

ALBANIA

ALBEDO

Albedo, Cosmic Ray
USE   COSMIC RAY ALBEDO

Albedo, Earth
USE   EARTH ALBEDO

Albedo, Lunar
USE   LUNAR ALBEDO

ALBERTA

ALBINISM

ALBUMINS

Alcock Comet, Iras-Araki-
   USE   IRAS-ARAKI-ALCOCK COMET

Alcohol, Ethyl
   USE   ETHYL ALCOHOL

Alcohol, Furfuryl
   USE   FURFURYL ALCOHOL

Alcohol, Isopropyl
   USE   ISOPROPYL ALCOHOL

Alcohol, Polyvinyl
   USE   POLYVINYL ALCOHOL

ALCOHOLS

Alcohols, Methyl
   USE   METHYL ALCOHOLS

Aldehyde, Acet
   USE   ACETALDEHYDE

Aldehyde, Form
   USE   FORMALDEHYDE

ALDEHYDES

Alder Reactions, Diels-
   USE   DIELS-ALDER REACTIONS

ALDOLASE

ALDOSTERONE

ALERTNESS

ALEUTIAN ISLANDS (US)

ALFALFA

Alfven Waves
   USE   MAGNETOHYDRODYNAMIC WAVES

Algaas
   USE   ALUMINUM GALLIUM ARSENIDES

ALGAE

Algae, Blue Green
   USE   BLUE GREEN ALGAE

Algal Bloom
   USE   ALGAE

ALGEBRA

Algebra, Boolean
   USE   BOOLEAN ALGEBRA

Algebra, Current
   USE   CURRENT ALGEBRA

Algebra, Differential
   USE   DIFFERENTIAL CALCULUS
          MATRICES (MATHEMATICS)

(Algebra), Field Theory
   USE   FIELD THEORY (ALGEBRA)

Algebra, Grassmann
   USE   VECTOR SPACES

ALGERIA

ALGOL

ALGOL ENGINE

ALGORITHMS

Algorithms, Parsing
   USE   PARSING ALGORITHMS

ALIGNMENT

Alignment, Mis
   USE   MISALIGNMENT

Alignment), Polarization (Spin
   USE   POLARIZATION (SPIN ALIGNMENT)

Alignment, Runway
   USE   RUNWAY ALIGNMENT

Alignment, Self
   USE   SELF ALIGNMENT

Alin Meteorite, Sikhote-
   USE   SIKHOTE-ALIN METEORITE

ALIPHATIC COMPOUNDS

ALKALI HALIDES

ALKALI METAL COMPOUNDS

ALKALI METALS

ALKALI VAPOR LAMPS

ALKALIES

ALKALINE BATTERIES

ALKALINE EARTH COMPOUNDS

ALKALINE EARTH METALS

ALKALINE EARTH OXIDES

ALKALINITY

ALKALOIDS

ALKALOSIS

Alkane, Perfluoro
   USE   PERFLUOROALKANE

ALKANES

ALKENES

ALKYD RESINS

ALKYL COMPOUNDS

ALKYLATES

ALKYLATION

ALKYLFERROCENE

ALKYLIDENE

ALKYNES

ALL SKY PHOTOGRAPHY

ALL-WEATHER AIR NAVIGATION

ALL-WEATHER LANDING SYSTEMS

ALLEGHENY PLATEAU (US)

Allen Radiation Belts, Van
   USE   RADIATION BELTS

ALLENDE METEORITE

ALLERGIC DISEASES

Alleviation, Vortex
   USE   VORTEX ALLEVIATION

Alleviators, Gust
   USE   GUST ALLEVIATORS

Allocation, Resource
   USE   RESOURCE ALLOCATION

ALLOCATIONS

ALLOTROPY

ALLOWANCES

ALLOXAN

(Alloy), Mulberry
   USE   MULBERRY (ALLOY)

Alloy Steels, Low
   USE   HIGH STRENGTH STEELS

ALLOYS

Alloys, Aluminum
   USE   ALUMINUM ALLOYS

Alloys, Antimony
   USE   ANTIMONY ALLOYS

Alloys, Arsenic
   USE   ARSENIC ALLOYS

Alloys, Barium
   USE   BARIUM ALLOYS

Alloys, Bearing
   USE   BEARING ALLOYS

Alloys, Beryllium
   USE   BERYLLIUM ALLOYS

Alloys, Binary
   USE   BINARY ALLOYS

Alloys, Bismuth
   USE   BISMUTH ALLOYS

Alloys, Boron
   USE   BORON ALLOYS

Alloys, Cadmium
   USE   CADMIUM ALLOYS

Alloys, Cast
   USE   CAST ALLOYS

Alloys, Cesium
   USE   CESIUM ALLOYS

Alloys, Chromium
   USE   CHROMIUM ALLOYS

Alloys, Cobalt
   USE   COBALT ALLOYS

Alloys, Copper
   USE   COPPER ALLOYS

Alloys, Erbium
   USE   ERBIUM ALLOYS

Alloys, Eutectic
   USE   EUTECTIC ALLOYS

Alloys, Gadolinium
   USE   GADOLINIUM ALLOYS

Alloys, Gallium
   USE   GALLIUM ALLOYS

Alloys, Germanium
   USE   GERMANIUM ALLOYS

Alloys, Gold
   USE   GOLD ALLOYS

Alloys, Hafnium
   USE   HAFNIUM ALLOYS

Alloys, Heat Resistant
   USE   HEAT RESISTANT ALLOYS

Alloys, High Strength
   USE   HIGH STRENGTH ALLOYS

Alloys, High Temperature
   USE   HEAT RESISTANT ALLOYS

Alloys, Indium
   USE   INDIUM ALLOYS

Alloys, Iron
USE   IRON ALLOYS

Alloys, Lanthanum
USE   LANTHANUM ALLOYS

Alloys, Lead
USE   LEAD ALLOYS

Alloys, Light
USE   LIGHT ALLOYS

Alloys, Liquid
USE   LIQUID ALLOYS

Alloys, Lithium
USE   LITHIUM ALLOYS

Alloys, Magnesium
USE   MAGNESIUM ALLOYS

Alloys, Manganese
USE   MANGANESE ALLOYS

Alloys, Mercury
USE   MERCURY ALLOYS

Alloys, Molybdenum
USE   MOLYBDENUM ALLOYS

Alloys, Monotectic
USE   MONOTECTIC ALLOYS

Alloys, Neodymium
USE   NEODYMIUM ALLOYS

Alloys, Nickel
USE   NICKEL ALLOYS

Alloys, Nimonic
USE   NIMONIC ALLOYS

Alloys, Niobium
USE   NIOBIUM ALLOYS

Alloys, Nitinol
USE   NITINOL ALLOYS

Alloys, Osmium
USE   OSMIUM ALLOYS

Alloys, Palladium
USE   PALLADIUM ALLOYS

Alloys, Platinum
USE   PLATINUM ALLOYS

Alloys, Plutonium
USE   PLUTONIUM ALLOYS

Alloys, Potassium
USE   POTASSIUM ALLOYS

Alloys, Quaternary
USE   QUATERNARY ALLOYS

Alloys, Rare Earth
USE   RARE EARTH ALLOYS

Alloys, Refractory Metal
USE   REFRACTORY METAL ALLOYS

Alloys, Rhenium
USE   RHENIUM ALLOYS

Alloys, Rhodium
USE   RHODIUM ALLOYS

Alloys, Ruthenium
USE   RUTHENIUM ALLOYS

Alloys, Selenium
USE   SELENIUM ALLOYS

Alloys, Shape Memory
USE   SHAPE MEMORY ALLOYS

Alloys, Silicon
USE   SILICON ALLOYS

Alloys, Silver
USE   SILVER ALLOYS

Alloys, Sodium
USE   SODIUM ALLOYS

Alloys, Syntectic
USE   SYNTECTIC ALLOYS

Alloys, Tantalum
USE   TANTALUM ALLOYS

Alloys, Tellurium
USE   TELLURIUM ALLOYS

Alloys, Ternary
USE   TERNARY ALLOYS

Alloys, Thallium
USE   THALLIUM ALLOYS

Alloys, Thorium
USE   THORIUM ALLOYS

Alloys, Tin
USE   TIN ALLOYS

Alloys, Titanium
USE   TITANIUM ALLOYS

Alloys, Tungsten
USE   TUNGSTEN ALLOYS

Alloys, Udimet
USE   UDIMET ALLOYS

Alloys, Uranium
USE   URANIUM ALLOYS

Alloys, Vanadium
USE   VANADIUM ALLOYS

Alloys, Wrought
USE   WROUGHT ALLOYS

Alloys, Yttrium
USE   YTTRIUM ALLOYS

Alloys, Zinc
USE   ZINC ALLOYS

Alloys, Zirconium
USE   ZIRCONIUM ALLOYS

ALLUVIUM

ALLYL COMPOUNDS

Almucantar
USE   ELEVATION ANGLE

Aloft, Winds
USE   WINDS ALOFT

ALOHA SYSTEM

ALOUETTE B SATELLITE

ALOUETTE HELICOPTERS

ALOUETTE PROJECT

ALOUETTE SATELLITES

ALOUETTE 1 SATELLITE

ALOUETTE 2 SATELLITE

Alouette 3 Helicopter
USE   SE-3160 HELICOPTER

Alpert Ionization Gages, Bayard-
USE   BAYARD-ALPERT IONIZATION GAGES

ALPHA DECAY

ALPHA JET AIRCRAFT

Alpha Line, H
USE   H ALPHA LINE

ALPHA PARTICLES

ALPHA PLASMA DEVICES

Alpha Radiation
USE   ALPHA PARTICLES

Alpha Radiation, Lyman
USE   LYMAN ALPHA RADIATION

ALPHABETS

ALPHANUMERIC CHARACTERS

ALPHATRONS

ALPINE METEOROLOGY

ALPS MOUNTAINS (EUROPE)

ALSEP
USE   APOLLO LUNAR SURFACE EXPERIMENTS
          PACKAGE

Alt Target And Background Measurement, High
USE   HIGH ALT TARGET AND BACKGROUND
          MEASUREMENT

Altair Engine
USE   X-248 ENGINE

Alteration
USE   REVISIONS

ALTERNATING CURRENT

Alternating Current Generators
USE   AC GENERATORS

ALTERNATIONS

ALTERNATIVES

Alternators (Generators)
USE   AC GENERATORS

Alternators, Static
USE   STATIC ALTERNATORS

ALTIMETERS

Altimeters, Laser
USE   LASER ALTIMETERS

Altimeters, Radar
USE   RADIO ALTIMETERS

Altimeters, Radio
USE   RADIO ALTIMETERS

ALTITUDE

ALTITUDE ACCLIMATIZATION

Altitude Balloons, High
USE   HIGH ALTITUDE BALLOONS

Altitude Breathing, High
USE   HIGH ALTITUDE BREATHING

ALTITUDE CONTROL

Altitude Environments, High
USE   HIGH ALTITUDE ENVIRONMENTS

Altitude, Flight
USE   FLIGHT ALTITUDE

Altitude Flight, High
USE   FLIGHT
          HIGH ALTITUDE

Altitude, High
USE   HIGH ALTITUDE

Altitude, Low
USE   LOW ALTITUDE

Altitude Missile, Supersonic Low
USE   SUPERSONIC LOW ALTITUDE MISSILE

Altitude Nuclear Detection, High
USE   HIGH ALTITUDE NUCLEAR DETECTION

Altitude Pressure, High
USE   HIGH ALTITUDE PRESSURE

ALTITUDE SICKNESS

Altitude, Simulated
USE   ALTITUDE SIMULATION

ALTITUDE SIMULATION

Altitude Sounding Projectile, High
USE   WASP SOUNDING ROCKET

ALTITUDE TESTS

Altitude Tests, High
USE   HIGH ALTITUDE TESTS

ALTITUDE TOLERANCE

ALU (Computer Components)
USE   ARITHMETIC AND LOGIC UNITS

ALUM

Alumina
USE   ALUMINUM OXIDES

ALUMINATES

Aluminizing
USE   ALUMINUM COATINGS

ALUMINUM

ALUMINUM ALLOYS

ALUMINUM ANTIMONIDES

ALUMINUM ARSENIDES

ALUMINUM BOROHYDRIDES

ALUMINUM BORON COMPOSITES

ALUMINUM CARBIDES

ALUMINUM CHLORIDES

ALUMINUM COATINGS

ALUMINUM COMPOUNDS

Aluminum Compounds, Organic
USE   ORGANIC ALUMINUM COMPOUNDS

ALUMINUM FLUORIDES

ALUMINUM GALLIUM ARSENIDES

Aluminum Garnet, Yttrium-
USE   YTTRIUM-ALUMINUM GARNET

ALUMINUM GRAPHITE COMPOSITES

ALUMINUM HYDRIDES

Aluminum Hydrides, Lithium
USE   LITHIUM ALUMINUM HYDRIDES

ALUMINUM ISOTOPES

ALUMINUM NITRIDES

ALUMINUM OXIDES

ALUMINUM PERCHLORATES

Aluminum Powder, Sintered
USE   SINTERED ALUMINUM POWDER

Aluminum, Powdered
USE   POWDERED ALUMINUM

ALUMINUM SILICATES

ALUMINUM 26

ALUMINUM 27

ALVEOLAR AIR

ALVEOLI

Am
USE   AMERICIUM

Amalgams
USE   MERCURY AMALGAMS

Amalgams, Mercury
USE   MERCURY AMALGAMS

AMALTHEA

AMAZON REGION (SOUTH AMERICA)

AMBERLITE (TRADEMARK)

AMBIENCE

AMBIENT TEMPERATURE

AMBIGUITY

AMBIPOLAR DIFFUSION

Ambit
USE   FIELD THEORY (PHYSICS)

AMBULANCES

America), Amazon Region (South
USE   AMAZON REGION (SOUTH AMERICA)

America), Andes Mountains (South
USE   ANDES MOUNTAINS (SOUTH AMERICA)

America), Appalachian Mountains (North
USE   APPALACHIAN MOUNTAINS (NORTH
        AMERICA)

America), Beaufort Sea (North
USE   BEAUFORT SEA (NORTH AMERICA)

America, Central
USE   CENTRAL AMERICA

America), Colorado River (North
USE   COLORADO RIVER (NORTH AMERICA)

America), Great Lakes (North
USE   GREAT LAKES (NORTH AMERICA)

America), Great Plains Corridor (North
USE   GREAT PLAINS CORRIDOR (NORTH
        AMERICA)

America, North
USE   NORTH AMERICA

America), Rio Grande (North
USE   RIO GRANDE (NORTH AMERICA)

America), Rocky Mountains (North
USE   ROCKY MOUNTAINS (NORTH AMERICA)

America, South
USE   SOUTH AMERICA

America), St Lawrence Valley (North
USE   ST LAWRENCE VALLEY (NORTH AMERICA)

America, Voice Of
USE   VOICE OF AMERICA

America), Williston Basin (North
USE   WILLISTON BASIN (NORTH AMERICA)

American Aircraft, North
USE   NORTH AMERICAN AIRCRAFT

AMERICAN INDIANS

American Search And Ranging Radar, North
USE   NORTH AMERICAN SEARCH AND RANGING
        RADAR

AMERICIUM

AMERICIUM ISOTOPES

AMERICIUM 241

AMIDASE

Amide, Acetazol
USE   ACETAZOLAMIDE

Amide, Lyserg
USE   LYSERGAMIDE

AMIDES

Amides, Carb
USE   CARBAMIDES

Amine, Catechol
USE   CATECHOLAMINE

Amine, Ergot
USE   ERGOTAMINE

Amine, Ethylenedi
USE   ETHYLENEDIAMINE

Amine, Hexamethylenetetr
USE   HEXAMETHYLENETETRAMINE

Amine, Mecamyl
USE   MECAMYLAMINE

Amine, Mel
USE   MELAMINE

Amine, Methamphet
USE   METHAMPHETAMINE

Amine, Nitros
USE   NITROSAMINE

Amine, Trinitr
USE   TRINITRAMINE

AMINES

Amines, Amphet
USE   AMPHETAMINES*

Amines, Di
USE   DIAMINES

Amines, Fluoro
USE   FLUOROAMINES

Amines, Hist
USE   HISTAMINES

Amines, Nitro
USE   NITROAMINES

Amines, Trypt
USE   TRYPTAMINES

AMINO ACIDS

AMINOPHYLLINE

AMMETERS

Ammeters, Micromilli
USE   MICROMILLIAMMETERS

Ammeters, Thermoelement
USE   THERMOELEMENT AMMETERS

AMMINES

AMMONIA

Ammonia, Liquid
USE   LIQUID AMMONIA

AMMONIUM BROMIDES

AMMONIUM CHLORIDES

AMMONIUM COMPOUNDS

AMMONIUM NITRATES

AMMONIUM PERCHLORATES

AMMONIUM PHOSPHATES

AMMONIUM PICRATES

AMMONIUM SULFATES

AMMONOLYSIS

AMMUNITION

Ammunition, Incendiary
　　USE　INCENDIARY AMMUNITION

AMOBARBITAL

AMOEBA

AMOOS
　　USE　AEROMANEUVERING ORBIT TO ORBIT
　　　　　SHUTTLE

AMOR ASTEROID

AMORPHOUS MATERIALS

AMORPHOUS SEMICONDUCTORS

AMOUNT

Amp, Cyclic
　　USE　CYCLIC AMP

Amperage
　　USE　ELECTRIC CURRENT

Ampere Characteristics, Volt-
　　USE　VOLT-AMPERE CHARACTERISTICS

Ampere Equation, Monge-
　　USE　MONGE-AMPERE EQUATION

Amphetamine, Meth
　　USE　METHAMPHETAMINE

AMPHETAMINES

AMPHIBIA

AMPHIBIOUS AIRCRAFT

AMPHIBIOUS VEHICLES

AMPHIBOLES

AMPHITRITE ASTEROID

AMPLIDYNES

AMPLIFICATION

Amplification Factor
　　USE　AMPLIFICATION

Amplification, Fluid
　　USE　FLUID AMPLIFIERS

(Amplification), Gain
　　USE　AMPLIFICATION

Amplification, Sound
　　USE　SOUND AMPLIFICATION

Amplification, Wave
　　USE　WAVE AMPLIFICATION

AMPLIFIER DESIGN

AMPLIFIERS

Amplifiers, Balanced
　　USE　PUSH-PULL AMPLIFIERS

Amplifiers, Beam Plasma
　　USE　BEAM PLASMA AMPLIFIERS

Amplifiers, Bistable
　　USE　FLIP-FLOPS

Amplifiers, Broadband
　　USE　BROADBAND AMPLIFIERS

Amplifiers, Crossed Field
　　USE　CROSSED FIELD AMPLIFIERS

Amplifiers, Current
　　USE　CURRENT AMPLIFIERS

Amplifiers, Differential
　　USE　DIFFERENTIAL AMPLIFIERS

Amplifiers, Distributed
　　USE　DISTRIBUTED AMPLIFIERS

Amplifiers, Electronic
　　USE　AMPLIFIERS

Amplifiers, Feedback
　　USE　FEEDBACK AMPLIFIERS

Amplifiers, Fluid
　　USE　FLUID AMPLIFIERS

Amplifiers, Fluid Jet
　　USE　FLUID AMPLIFIERS
　　　　　JET AMPLIFIERS

Amplifiers, Intermediate Frequency
　　USE　INTERMEDIATE FREQUENCY AMPLIFIERS

Amplifiers, Jet
　　USE　JET AMPLIFIERS

Amplifiers, Light
　　USE　LIGHT AMPLIFIERS

Amplifiers, Limiter
　　USE　LIMITER AMPLIFIERS

Amplifiers, Linear
　　USE　LINEAR AMPLIFIERS

Amplifiers, Magnetic
　　USE　MAGNETIC AMPLIFIERS

Amplifiers, Magnetostatic
　　USE　MAGNETOSTATIC AMPLIFIERS

Amplifiers, Microwave
　　USE　MICROWAVE AMPLIFIERS

Amplifiers, Operational
　　USE　OPERATIONAL AMPLIFIERS

Amplifiers, Optical
　　USE　LIGHT AMPLIFIERS

Amplifiers, Paramagnetic
　　USE　MASERS

Amplifiers, Parametric
　　USE　PARAMETRIC AMPLIFIERS

Amplifiers, Power
　　USE　POWER AMPLIFIERS

Amplifiers, Push-Pull
　　USE　PUSH-PULL AMPLIFIERS

Amplifiers, Quantum
　　USE　QUANTUM AMPLIFIERS

Amplifiers, Servo
　　USE　SERVOAMPLIFIERS

Amplifiers, Transistor
　　USE　TRANSISTOR AMPLIFIERS

Amplifiers, Traveling Wave
　　USE　TRAVELING WAVE AMPLIFIERS

Amplifiers, Voltage
　　USE　VOLTAGE AMPLIFIERS

Amplitrons (Trademark)
　　USE　PLANOTRONS

Amplitude Converters, Pulse Width
　　USE　PULSE WIDTH AMPLITUDE CONVERTERS

AMPLITUDE DISTRIBUTION ANALYSIS

AMPLITUDE MODULATION

Amplitude Modulation, Pulse
　　USE　PULSE AMPLITUDE MODULATION

Amplitude Probability Analysis
　　USE　AMPLITUDE DISTRIBUTION ANALYSIS

Amplitude, Pulse
　　USE　PULSE AMPLITUDE

Amplitude, Scattering
　　USE　SCATTERING AMPLITUDE

AMPLITUDES

AMPOULES

AMPS (SATELLITE PAYLOAD)

AMPTE (SATELLITES)

AMTV
　　USE　AUTOMATED MIXED TRAFFIC VEHICLES

AN-2 AIRCRAFT

AN-22 AIRCRAFT

AN-22 Aircraft, Antonov
　　USE　AN-22 AIRCRAFT

AN-24 AIRCRAFT

AN-24 Aircraft, Antonov
　　USE　AN-24 AIRCRAFT

ANABAENA

ANAEROBES

ANALGESIA

ANALOG CIRCUITS

ANALOG COMPUTERS

Analog Converters, Digital To
　　USE　DIGITAL TO ANALOG CONVERTERS

ANALOG DATA

ANALOG SIMULATION

ANALOG TO DIGITAL CONVERTERS

ANALOGIES

Analogies, Hydraulic
　　USE　HYDRAULIC ANALOGIES

ANALOGS

Analogy, Membrane
　　USE　MEMBRANE STRUCTURES
　　　　　STRUCTURAL ANALYSIS

Analysis
　　USE　ANALYZING

Analysis, Activation
　　USE　ACTIVATION ANALYSIS

Analysis, Amplitude Distribution
　　USE　AMPLITUDE DISTRIBUTION ANALYSIS

Analysis, Amplitude Probability
　　USE　AMPLITUDE DISTRIBUTION ANALYSIS

Analysis, Biological
　　USE　BIOASSAY

Analysis, Bivariate
　　USE　BIVARIATE ANALYSIS

Analysis, Cepstral
　　USE　CEPSTRAL ANALYSIS

Analysis, Chemical
USE   CHEMICAL ANALYSIS

Analysis, Cluster
USE   CLUSTER ANALYSIS

Analysis, Combinatorial
USE   COMBINATORIAL ANALYSIS

Analysis, Cost
USE   COST ANALYSIS

Analysis, Creep
USE   CREEP ANALYSIS

Analysis), DAEMO (Data
USE   DATA TRANSMISSION
      DATA REDUCTION
      DATA PROCESSING

Analysis, Data
USE   DATA PROCESSING
      DATA REDUCTION

Analysis, Data Flow
USE   DATA FLOW ANALYSIS

Analysis, Design
USE   DESIGN ANALYSIS

Analysis, Differential Thermal
USE   THERMAL ANALYSIS

Analysis, Dimensional
USE   DIMENSIONAL ANALYSIS

(Analysis), DTA
USE   THERMAL ANALYSIS

Analysis, Dynamic Structural
USE   DYNAMIC STRUCTURAL ANALYSIS

Analysis, Economic
USE   ECONOMIC ANALYSIS

Analysis, Error
USE   ERROR ANALYSIS

Analysis, Factor
USE   FACTOR ANALYSIS

Analysis, Failure
USE   FAILURE ANALYSIS

Analysis, Feasibility
USE   FEASIBILITY ANALYSIS

Analysis, Flutter
USE   FLUTTER ANALYSIS

Analysis, Fourier
USE   FOURIER ANALYSIS

Analysis, Functional
USE   FUNCTIONAL ANALYSIS

Analysis, Gas
USE   GAS ANALYSIS

Analysis, Gas Path
USE   GAS PATH ANALYSIS

Analysis, Harmonic
USE   HARMONIC ANALYSIS

Analysis, Histochemical
USE   HISTOCHEMICAL ANALYSIS

Analysis, Hydrothermal Stress
USE   HYDROTHERMAL STRESS ANALYSIS

Analysis, Image
USE   IMAGE ANALYSIS

Analysis, Instrumental
USE   ANALYZING
      AUTOMATION

Analysis, Management
USE   MANAGEMENT ANALYSIS

Analysis, Mathematical
USE   APPLICATIONS OF MATHEMATICS

ANALYSIS (MATHEMATICS)

Analysis, Matrix
USE   MATRICES (MATHEMATICS)

Analysis, Micro
USE   MICROANALYSIS

Analysis, Multitemporal
USE   TEMPORAL RESOLUTION

Analysis, Multivariate Statistical
USE   MULTIVARIATE STATISTICAL ANALYSIS

Analysis, Neph
USE   NEPHANALYSIS

Analysis, Network
USE   NETWORK ANALYSIS

Analysis, Neutron Activation
USE   NEUTRON ACTIVATION ANALYSIS

Analysis, Numerical
USE   NUMERICAL ANALYSIS

ANALYSIS OF VARIANCE

Analysis, Photoelastic
USE   PHOTOELASTIC ANALYSIS

Analysis, Postflight
USE   POSTFLIGHT ANALYSIS

Analysis, Potentiometric
USE   POTENTIOMETRIC ANALYSIS

Analysis, Preflight
USE   PREFLIGHT ANALYSIS

Analysis, Principal Components
USE   PRINCIPAL COMPONENTS ANALYSIS

Analysis Program, NASA Structural
USE   NASTRAN

Analysis, Program Trend Line
USE   PROGRAM TREND LINE ANALYSIS

Analysis, Qualitative
USE   QUALITATIVE ANALYSIS

Analysis, Quantitative
USE   QUANTITATIVE ANALYSIS

Analysis, Regression
USE   REGRESSION ANALYSIS

Analysis, Reliability
USE   RELIABILITY ANALYSIS

Analysis, Scene
USE   SCENE ANALYSIS

Analysis, Sequential
USE   SEQUENTIAL ANALYSIS

Analysis, Signal
USE   SIGNAL ANALYSIS

Analysis, Signature
USE   SIGNATURE ANALYSIS

Analysis, Sneak Circuit
USE   SNEAK CIRCUIT ANALYSIS

Analysis (Spacecraft), Postmission
USE   POSTMISSION ANALYSIS (SPACECRAFT)

Analysis, Spectral
USE   SPECTRUM ANALYSIS

Analysis, Spectroscopic
USE   SPECTROSCOPIC ANALYSIS

Analysis, Spectrum
USE   SPECTRUM ANALYSIS

Analysis, Statistical
USE   STATISTICAL ANALYSIS

Analysis (Statistics), Discriminant
USE   DISCRIMINANT ANALYSIS (STATISTICS)

Analysis, Stress
USE   STRESS ANALYSIS

Analysis, Structural
USE   STRUCTURAL ANALYSIS

Analysis, Systems
USE   SYSTEMS ANALYSIS

Analysis Techniques, Prediction
USE   PREDICTION ANALYSIS TECHNIQUES

Analysis, Tensor
USE   TENSOR ANALYSIS

Analysis, Terrain
USE   TERRAIN ANALYSIS

Analysis, Thermal
USE   THERMAL ANALYSIS

Analysis, Time Series
USE   TIME SERIES ANALYSIS

Analysis, Training
USE   TRAINING ANALYSIS

Analysis, Trajectory
USE   TRAJECTORY ANALYSIS

Analysis, Vector
USE   VECTOR ANALYSIS

Analysis, Volumetric
USE   VOLUMETRIC ANALYSIS

Analysis, Weight
USE   WEIGHT ANALYSIS

Analysis, X Ray
USE   X RAY ANALYSIS

Analysis, X Ray Stress
USE   X RAY STRESS ANALYSIS

ANALYTIC FUNCTIONS

ANALYTIC GEOMETRY

ANALYTICAL CHEMISTRY

ANALYZERS

Analyzers, Differential
USE   DIFFERENTIAL ANALYZERS

Analyzers, Engine
USE   ENGINE ANALYZERS

Analyzers, Frequency
USE   FREQUENCY ANALYZERS

Analyzers, Oxygen
USE   OXYGEN ANALYZERS

Analyzers, Signal
USE   SIGNAL ANALYZERS

ANALYZING

ANAPHYLAXIS

ANASTIGMATISM

ANATASE

ANATOMY

(Anatomy), Appendix
USE   APPENDIX (ANATOMY)

(Anatomy), Arm
USE   ARM (ANATOMY)

(Anatomy), Capillaries
   USE   CAPILLARIES (ANATOMY)

(Anatomy), Diaphragm
   USE   DIAPHRAGM (ANATOMY)

(Anatomy), Elbow
   USE   ELBOW (ANATOMY)

(Anatomy), Eye
   USE   EYE (ANATOMY)

(Anatomy), Face
   USE   FACE (ANATOMY)

(Anatomy), Feet
   USE   FEET (ANATOMY)

(Anatomy), Glands
   USE   GLANDS (ANATOMY)

(Anatomy), Hand
   USE   HAND (ANATOMY)

(Anatomy), Head
   USE   HEAD (ANATOMY)

(Anatomy), Joints
   USE   JOINTS (ANATOMY)

(Anatomy), Knee
   USE   KNEE (ANATOMY)

(Anatomy), Leg
   USE   LEG (ANATOMY)

(Anatomy), Limbs
   USE   LIMBS (ANATOMY)

(Anatomy), Lips
   USE   LIPS (ANATOMY)

(Anatomy), Neck
   USE   NECK (ANATOMY)

(Anatomy), Nose
   USE   NOSE (ANATOMY)

(Anatomy), Skin
   USE   SKIN (ANATOMY)

ANCHORS (FASTENERS)

ANDES MOUNTAINS (SOUTH AMERICA)

ANDESITE

ANDORRA

Andreas Fault Experiment, San
   USE   SAN ANDREAS FAULT EXPERIMENT

Andreas Fault, San
   USE   SAN ANDREAS FAULT

ANDROMEDA

ANDROMEDA CONSTELLATION

ANDROMEDA GALAXIES

ANECHOIC CHAMBERS

ANELASTICITY

ANEMIAS

ANEMOMETERS

Anemometers, Drag Force
   USE   DRAG FORCE ANEMOMETERS

Anemometers, Hot-Film
   USE   HOT-FILM ANEMOMETERS

Anemometers, Hot-Wire
   USE   HOT-WIRE ANEMOMETERS

Anemometers, Laser
   USE   LASER ANEMOMETERS

Anemometers, Sonic
   USE   SONIC ANEMOMETERS

Anemometry
   USE   VELOCITY MEASUREMENT

ANESTHESIA

Anesthesia, Electro
   USE   ELECTROANESTHESIA

ANESTHESIOLOGY

ANESTHETICS

ANGELS

ANGINA PECTORIS

ANGIOGRAPHY

ANGIOSPERMS

Angle, Bragg
   USE   BRAGG ANGLE

Angle, Brewster
   USE   BREWSTER ANGLE

Angle, Dihedral
   USE   DIHEDRAL ANGLE

Angle, Elevation
   USE   ELEVATION ANGLE

Angle Lenses, Wide
   USE   WIDE ANGLE LENSES

ANGLE OF ATTACK

Angle Of Attack, Zero
   USE   ZERO ANGLE OF ATTACK

Angle, Phase
   USE   PHASE SHIFT

Angle, Sweep
   USE   SWEEP ANGLE

Angles, Apsidal
   USE   APSIDES

Angles (Electronics), Look
   USE   LOOK ANGLES (ELECTRONICS)

ANGLES (GEOMETRY)

Angles, Glide
   USE   GLIDE PATHS

Angles, Pitch
   USE   PITCH (INCLINATION)

Angles, Sweepback
   USE   SWEEPBACK

Angles (Tracking), Look
   USE   LOOK ANGLES (TRACKING)

ANGOLA

ANGULAR ACCELERATION

ANGULAR CORRELATION

ANGULAR DISTRIBUTION

ANGULAR MOMENTUM

Angular Motion
   USE   ANGULAR VELOCITY

ANGULAR RESOLUTION

ANGULAR VELOCITY

Anhydrase, Carbonic
   USE   CARBONIC ANHYDRASE

ANHYDRIDES

Anik A
   USE   ANIK 1

Anik B
   USE   ANIK 2

Anik C
   USE   ANIK 3

ANIK SATELLITES

ANIK 1

ANIK 2

ANIK 3

ANILINE

ANIMALS

(Animals), Seals
   USE   SEALS (ANIMALS)

Animation
   USE   MOTION

ANIONS

ANISOLE

ANISOTROPIC FLUIDS

ANISOTROPIC MEDIA

ANISOTROPIC PLATES

ANISOTROPIC SHELLS

ANISOTROPY

Anisotropy, Elastic
   USE   ELASTIC ANISOTROPY

Anisotropy, Plastic
   USE   PLASTIC ANISOTROPY

ANNA HURRICANE

ANNA SATELLITES

ANNEALING

Annealing, Laser
   USE   LASER ANNEALING

Annihilation, Positron
   USE   POSITRON ANNIHILATION

ANNIHILATION REACTIONS

ANNOTATIONS

ANNUAL VARIATIONS

Annular Arc, Magnetic
   USE   MAGNETIC ANNULAR ARC

ANNULAR CORE PULSE REACTORS

ANNULAR DUCTS

ANNULAR FLOW

ANNULAR NOZZLES

ANNULAR PLATES

Annular Shock Tubes, Magnetic
   USE   MAGNETIC ANNULAR SHOCK TUBES

ANNULAR SUSPENSION AND POINTING SYSTEM

ANNULI

Anode Microchannel Arrays, Multi-
   USE   MULTI-ANODE MICROCHANNEL ARRAYS

ANODES

Anodes, Cell
USE   CELL ANODES

Anodes, Shell
USE   SHELL ANODES

Anodes, Tube
USE   TUBE ANODES

ANODIC COATINGS

ANODIC STRIPPING

ANODIZING

ANOLYTES

ANOMALIES

Anomalies, Congenital
USE   CONGENITAL ANOMALIES

Anomalies, Geomagnetic
USE   MAGNETIC ANOMALIES

Anomalies, Geothermal
USE   GEOTHERMAL ANOMALIES

Anomalies, Gravity
USE   GRAVITY ANOMALIES

Anomalies, Magnetic
USE   MAGNETIC ANOMALIES

ANOMALOUS TEMPERATURE ZONES

ANORTHOSITE

ANOXIA

ANS
USE   ASTRONOMICAL NETHERLANDS SATELLITE

Antarctic Environment
USE   ICE ENVIRONMENTS

ANTARCTIC REGIONS

Antarctica
USE   ANTARCTIC REGIONS

ANTARES ROCKET VEHICLE

ANTELOPE MISSILE

ANTENNA ARRAYS

ANTENNA COMPONENTS

ANTENNA COUPLERS

ANTENNA DESIGN

(Antenna Elements), Directors
USE   DIRECTORS (ANTENNA ELEMENTS)

ANTENNA FEEDS

Antenna Fields
USE   ANTENNA RADIATION PATTERNS

Antenna Grid (Navy), Global Communications
USE   SEAFARER PROJECT

Antenna Grid (Navy), Underground Radio
USE   SEAFARER PROJECT

ANTENNA RADIATION PATTERNS

ANTENNAS

Antennas, Aircraft
USE   AIRCRAFT ANTENNAS

Antennas, Cassegrain
USE   CASSEGRAIN ANTENNAS

Antennas, Cylindrical
USE   CYLINDRICAL ANTENNAS

Antennas, Delta
USE   DELTA ANTENNAS

Antennas, Dipole
USE   DIPOLE ANTENNAS

Antennas, Directional
USE   DIRECTIONAL ANTENNAS

Antennas, Furlable
USE   FURLABLE ANTENNAS

Antennas, Gravitational Wave
USE   GRAVITATIONAL WAVE ANTENNAS

Antennas, Gregorian
USE   GREGORIAN ANTENNAS

Antennas, Helical
USE   HELICAL ANTENNAS

Antennas, High Resolution Coverage
USE   HIGH RESOLUTION COVERAGE ANTENNAS

Antennas, Hoop Column
USE   HOOP COLUMN ANTENNAS

Antennas, Horn
USE   HORN ANTENNAS

Antennas, Inertialess Steerable
USE   INERTIALESS STEERABLE ANTENNAS

Antennas, Lens
USE   LENS ANTENNAS

Antennas, Log Periodic
USE   LOG PERIODIC ANTENNAS

Antennas, Log Spiral
USE   LOG SPIRAL ANTENNAS

Antennas, Loop
USE   LOOP ANTENNAS

Antennas, Maypole
USE   MAYPOLE ANTENNAS

Antennas, Microwave
USE   MICROWAVE ANTENNAS

Antennas, Missile
USE   MISSILE ANTENNAS

Antennas, Monopole
USE   MONOPOLE ANTENNAS

Antennas, Monopulse
USE   MONOPULSE ANTENNAS

Antennas, Multibeam
USE   MULTIBEAM ANTENNAS

Antennas, Omnidirectional
USE   OMNIDIRECTIONAL ANTENNAS

Antennas, Parabolic
USE   PARABOLIC ANTENNAS

Antennas, Plasma
USE   PLASMA ANTENNAS

Antennas, Radar
USE   RADAR ANTENNAS

Antennas, Radio
USE   RADIO ANTENNAS

Antennas, Rectifier
USE   RECTENNAS

Antennas, Rhombic
USE   RHOMBIC ANTENNAS

Antennas, Satellite
USE   SATELLITE ANTENNAS

Antennas, Schwarzschild
USE   SCHWARZSCHILD ANTENNAS

Antennas, Slot
USE   SLOT ANTENNAS

Antennas, Slotted
USE   SLOT ANTENNAS

Antennas, Spacecraft
USE   SPACECRAFT ANTENNAS

Antennas, Spherical
USE   SPHERICAL ANTENNAS

Antennas, Spike
USE   MONOPOLE ANTENNAS

Antennas, Spiral
USE   SPIRAL ANTENNAS

Antennas, Steerable
USE   STEERABLE ANTENNAS

Antennas, Tracking
USE   DIRECTIONAL ANTENNAS

Antennas, Turnstile
USE   TURNSTILE ANTENNAS

Antennas, Two Reflector
USE   TWO REFLECTOR ANTENNAS

Antennas, Waveguide
USE   WAVEGUIDE ANTENNAS

Antennas, Whip
USE   WHIP ANTENNAS

Antennas, Yagi
USE   YAGI ANTENNAS

Antheus Aircraft
USE   AN-22 AIRCRAFT

ANTHRACENE

ANTHRAQUINONES

ANTHROPOLOGY

ANTHROPOMETRY

Anti-Stokes Raman Spectroscopy, Coherent
USE   RAMAN SPECTROSCOPY

ANTIADRENERGICS

ANTIAIRCRAFT MISSILES

Antiaircraft Missiles, Self Initiated
USE   SIAM MISSILES

Antibacterials, Antiinfectives And
USE   ANTIINFECTIVES AND ANTIBACTERIALS

ANTIBIOTICS

ANTIBODIES

ANTICHOLINERGICS

ANTICLINES

Anticlinoria
USE   ANTICLINES

ANTICOAGULANTS

ANTICONVULSANTS

ANTICYCLONES

ANTIDIURETICS

ANTIDOTES

ANTIEMETICS AND ANTINAUSEANTS

ANTIFERROELECTRICITY

ANTIFERROMAGNETISM

ANTIFOULING

ANTIFREEZES

ANTIFRICTION BEARINGS

ANTIGENS

ANTIGRAVITY

ANTIHISTAMINICS

ANTIHYPERTENSIVE AGENTS

ANTIICING ADDITIVES

ANTIINFECTIVES AND ANTIBACTERIALS

ANTIKNOCK ADDITIVES

Antilles, Lesser
    USE   LESSER ANTILLES

ANTIMATTER

ANTIMISSILE DEFENSE

Antimissile Measurement Program, Downrange
    USE   DOWNRANGE ANTIMISSILE MEASUREMENT
            PROGRAM

ANTIMISSILE MISSILES

ANTIMISTING FUELS

ANTIMONIDES

Antimonides, Aluminum
    USE   ALUMINUM ANTIMONIDES

Antimonides, Cadmium
    USE   CADMIUM ANTIMONIDES

Antimonides, Cesium
    USE   CESIUM ANTIMONIDES

Antimonides, Gallium
    USE   GALLIUM ANTIMONIDES

Antimonides, Germanium
    USE   GERMANIUM ANTIMONIDES

Antimonides, Indium
    USE   INDIUM ANTIMONIDES

Antimonides, Zinc
    USE   ZINC ANTIMONIDES

ANTIMONY

ANTIMONY ALLOYS

ANTIMONY COMPOUNDS

ANTIMONY FLUORIDES

ANTIMONY ISOTOPES

Antinauseants, Antiemetics And
    USE   ANTIEMETICS AND ANTINAUSEANTS

ANTINEUTRINOS

ANTINODES

ANTINUCLEONS

ANTIOXIDANTS

ANTIPARTICLES

ANTIPODES

ANTIPROTONS

ANTIQUITIES

ANTIRADAR COATINGS

ANTIRADIATION DRUGS

ANTIRADIATION MISSILES

ANTIREFLECTION COATINGS

ANTISEPTICS

ANTISERUMS

ANTISHIP MISSILES

ANTISHIP WARFARE

ANTISKID DEVICES

Antistatic Devices
    USE   STATIC DISCHARGERS

ANTISUBMARINE WARFARE

ANTISUBMARINE WARFARE AIRCRAFT

ANTISYMMETRY

ANTITANK MISSILES

Antitoxins, Toxins And
    USE   TOXINS AND ANTITOXINS

ANTONOV AIRCRAFT

Antonov AN-22 Aircraft
    USE   AN-22 AIRCRAFT

Antonov AN-24 Aircraft
    USE   AN-24 AIRCRAFT

ANVIL CLOUDS

ANVILS

ANXIETY

Anxiety Scale, Taylor Manifest
    USE   TAYLOR MANIFEST ANXIETY SCALE

AO-1 Aircraft
    USE   OV-1 AIRCRAFT

AOIPS
    USE   ATMOSPHERIC & OCEANOGRAPHIC
            INFORM SYS

AORTA

AOSO

APACHE ROCKET VEHICLE

Apache Rocket Vehicle, Nike-
    USE   NIKE-APACHE ROCKET VEHICLE

Apatites
    USE   MINERALS
            CALCIUM PHOSPHATES

APERIODIC FUNCTIONS

Aperture Radar, Synthetic
    USE   SYNTHETIC APERTURE RADAR

Aperture Seismic Array, Large
    USE   LARGE APERTURE SEISMIC ARRAY

APERTURES

Apertures), Irises (Mechanical
    USE   IRISES (MECHANICAL APERTURES)

Apertures, Synthetic
    USE   SYNTHETIC APERTURES

(Apertures), Windows
    USE   WINDOWS (APERTURES)

APES

APEXES

APHELIONS

APL (PROGRAMMING LANGUAGE)

Apnea
    USE   RESPIRATION

APOGEE BOOST MOTORS

Apogee Engines, SYNCOM
    USE   SYNCOM APOGEE ENGINES

Apogee Satellites, Perigee-
    USE   PAS

APOGEES

APOLLO APPLICATIONS PROGRAM

APOLLO ASTEROIDS

APOLLO EXTENSION SYSTEM

APOLLO FLIGHTS

APOLLO LUNAR EXPERIMENT MODULE

Apollo, Lunar Exploration System For
    USE   LUNAR EXPLORATION SYSTEM FOR
            APOLLO

APOLLO LUNAR SURFACE EXPERIMENTS PACKAGE

APOLLO PROJECT

APOLLO SHORT STACK

APOLLO SOYUZ TEST PROJECT

APOLLO SPACECRAFT

Apollo Surface Experiments Package, Early
    USE   EASEP

APOLLO TELESCOPE MOUNT

APOLLO 5 FLIGHT

APOLLO 6 FLIGHT

APOLLO 7 FLIGHT

APOLLO 8 FLIGHT

APOLLO 9 FLIGHT

APOLLO 10 FLIGHT

APOLLO 11 FLIGHT

APOLLO 12 FLIGHT

APOLLO 13 FLIGHT

APOLLO 14 FLIGHT

APOLLO 15 FLIGHT

APOLLO 16 FLIGHT

APOLLO 17 FLIGHT

APPALACHIAN MOUNTAINS (NORTH AMERICA)

Apparatus
    USE   EQUIPMENT

Apparatus, Abort
    USE   ABORT APPARATUS

Apparatus, Breathing
    USE   BREATHING APPARATUS

Apparatus, Drying
    USE   DRYING APPARATUS

Apparatus, Free Flight Test
    USE   FREE FLIGHT TEST APPARATUS

Apparatus, Hypersonic Test
    USE   HYPERSONIC TEST APPARATUS

Apparatus, Spraying
    USE   SPRAYERS

Apparatus, Supersonic Test
USE   SUPERSONIC TEST APPARATUS

Apparatus, Torque Measuring
USE   TORQUEMETERS

Apparatus, Underwater Breathing
USE   UNDERWATER BREATHING APPARATUS

Apparatus, Vacuum
USE   VACUUM APPARATUS

Apparatus, Wind Tunnel
USE   WIND TUNNEL APPARATUS

Apparatus, X Ray
USE   X RAY APPARATUS

APPEARANCE

APPENDAGES

APPENDIX (ANATOMY)

Appleton Approximation, Hartree-
USE   HARTREE APPROXIMATION

Appliances, Electric
USE   ELECTRIC EQUIPMENT

Applic Payloads, Office Of Space & Terrestr
USE   OSTA-2 PAYLOAD
       OSTA-1 PAYLOAD

Application
USE   UTILIZATION

APPLICATIONS EXPLORER SATELLITES

Applications Laboratory, Earth Viewing
USE   EARTH VIEWING APPLICATIONS
       LABORATORY

Applications, Laser
USE   LASER APPLICATIONS

Applications, Microgravity
USE   MICROGRAVITY APPLICATIONS

Applications, Multisensor
USE   MULTISENSOR APPLICATIONS

APPLICATIONS OF MATHEMATICS

Applications, Patent
USE   PATENT APPLICATIONS

Applications Program, Apollo
USE   APOLLO APPLICATIONS PROGRAM

Applications Program, Earth & Ocean Physics
USE   EARTH & OCEAN PHYSICS APPLICATIONS
       PROGRAM

Applications Program, Geographic
USE   GEOGRAPHIC APPLICATIONS PROGRAM

APPLICATIONS PROGRAMS (COMPUTERS)

Applications Rocket, Space Processing
USE   SPACE PROCESSING APPLICATIONS
       ROCKET

Applications Technology Satellites
USE   ATS

Applicator Aircraft S-2b, Snow Aerial
USE   S-2 AIRCRAFT

APPROACH

Approach, Airborne Radar
USE   AIRBORNE RADAR APPROACH

APPROACH AND LANDING TESTS (STS)

APPROACH CONTROL

Approach Control, Radar
USE   RADAR APPROACH CONTROL

Approach, Delayed Flap
USE   DELAYED FLAP APPROACH

APPROACH INDICATORS

Approach, Instrument
USE   INSTRUMENT APPROACH

Approach Spacing, Aircraft
USE   AIRCRAFT APPROACH SPACING

APPROPRIATIONS

APPROXIMATION

Approximation, Bardeen
USE   BARRIER LAYERS
       ELECTRICAL PROPERTIES
       SURFACE PROPERTIES

Approximation, Born
USE   BORN APPROXIMATION

Approximation, Born-Oppenheimer
USE   BORN-OPPENHEIMER APPROXIMATION

Approximation, Boussinesq
USE   BOUSSINESQ APPROXIMATION

Approximation, Chebyshev
USE   CHEBYSHEV APPROXIMATION

Approximation, Eddington
USE   EDDINGTON APPROXIMATION

Approximation, Hartree
USE   HARTREE APPROXIMATION

Approximation, Hartree-Appleton
USE   HARTREE APPROXIMATION

Approximation, Hartree-Fock
USE   HARTREE APPROXIMATION

Approximation Methods
USE   APPROXIMATION

Approximation, Oseen
USE   OSEEN APPROXIMATION

Approximation, Pade
USE   PADE APPROXIMATION

Approximation, Quadrature
USE   QUADRATURES

Approximation, Sommerfeld
USE   SOMMERFELD APPROXIMATION

Approximation, WKB
USE   WENTZEL-KRAMER-BRILLOUIN METHOD

Apsidal Angles
USE   APSIDES

APSIDES

APT (Picture Transmission)
USE   AUTOMATIC PICTURE TRANSMISSION

APTITUDE

AQUARID METEOROIDS

AQUATIC PLANTS

AQUEOUS SOLUTIONS

AQUICULTURE

AQUIFERS

Ar
USE   ARGON

AR
USE   ARKANSAS

Arab Emirates, United
USE   UNITED ARAB EMIRATES

Arabia, Saudi
USE   SAUDI ARABIA

Arabian Commercial Satellite
USE   ARCOMSAT

ARABIAN SEA

Arabian Space Program, Saudi
USE   SAUDI ARABIAN SPACE PROGRAM

ARABSAT

ARAGONITE

Araki-Alcock Comet, Iras-
USE   IRAS-ARAKI-ALCOCK COMET

ARC CHAMBERS

ARC CLOUDS

Arc Cutting, Plasma
USE   PLASMA ARC CUTTING

ARC DISCHARGES

ARC GENERATORS

Arc Heaters, Gerdien
USE   ARC HEATING
       HEATING EQUIPMENT

ARC HEATING

ARC JET ENGINES

ARC LAMPS

Arc, Magnetic Annular
USE   MAGNETIC ANNULAR ARC

ARC MELTING

ARC SPRAYING

Arc Spraying, Plasma
USE   ARC SPRAYING

Arc Switches, Vacuum
USE   VACUUM ARC SWITCHES

ARC WELDING

Arc Welding, Gas Tungsten
USE   GAS TUNGSTEN ARC WELDING

Arc Welding, Plasma
USE   PLASMA ARC WELDING

ARCAS ROCKET VEHICLES

ARCHAEOLOGY

ARCHES

ARCHIPELAGOES

ARCHITECTURE

(Architecture), Ceilings
USE   CEILINGS (ARCHITECTURE)

ARCHITECTURE (COMPUTERS)

ARCOMSAT

ARCON ROCKET VEHICLE

ARCS

Arcs, Auroral
USE   AURORAL ARCS

Arcs, Carbon
USE   CARBON ARCS

Arcs, Electric
USE   ELECTRIC ARCS

Arcs, Island
USE   ISLAND ARCS

Arcs, Mercury
USE   MERCURY ARCS

Arcs, Plasma
USE   PLASMA JETS

Arcs, Red
USE   RED ARCS

Arctic Environments
USE   ICE ENVIRONMENTS

ARCTIC OCEAN

ARCTIC REGIONS

AREA

Area Crop Inventory Experiment, Large
USE   LARGE AREA CROP INVENTORY
         EXPERIMENT

Area Energy Management, Terminal
USE   TERMINAL AREA ENERGY MANAGEMENT

Area), Flux (Rate Per Unit
USE   FLUX DENSITY

Area (Mexico), Leon-Queretaro
USE   LEON-QUERETARO AREA (MEXICO)

AREA NAVIGATION

Area Twin Hull, Small Water Plane
USE   SWATH (SHIP)

Area Wings, Variable
USE   TRAILING EDGE FLAPS

Areas, Auditory Sensation
USE   AUDITORY SENSATION AREAS

Areas, Catchment
USE   WATERSHEDS

Areas, Industrial
USE   INDUSTRIAL AREAS

Areas, Lumbering
USE   FORESTS

Areas (Meteorology), Frontal
USE   FRONTS (METEOROLOGY)

Areas, Metropolitan
USE   CITIES

Areas, Residential
USE   RESIDENTIAL AREAS

Areas, Rural
USE   RURAL AREAS

Areas, Suburban
USE   SUBURBAN AREAS

Areas, Urban
USE   CITIES

AREND-ROLAND COMET

ARES (Spacecraft)
USE   ADVANCED RECONN ELECTRIC
         SPACECRAFT

ARETS
USE   ARIZONA REGIONAL ECOLOGICAL TEST
         SITE

ARGENTINA

ARGO ROCKET VEHICLES

ARGON

ARGON ISOTOPES

ARGON LASERS

Argon Lasers, HCL
USE   HCL ARGON LASERS

ARGON PLASMA

Argon, Solid
USE   SOLIDIFIED GASES

ARGON-OXYGEN ATMOSPHERES

ARGOSY MK-1 AIRCRAFT

Arguments (Mathematics)
USE   INDEPENDENT VARIABLES

ARGUS PROJECT

ARIANE LAUNCH VEHICLE

† ARID LANDS

ARIEL SATELLITES

ARIEL 1 SATELLITE

ARIEL 2 SATELLITE

ARIEL 3 SATELLITE

ARIEL 4 SATELLITE

ARIEL 5 SATELLITE

ARIES CONSTELLATION

ARIES SOUNDING ROCKET

ARIETID METEOROIDS

ARIP (Impact Prediction)
USE   IMPACT PREDICTION
         COMPUTERIZED SIMULATION

ARIS Instrumentation Ship
USE   ADVANCED RANGE INSTRUMENTATION
         SHIP

ARITHMETIC

ARITHMETIC AND LOGIC UNITS

Arithmetic, Double Precision
USE   DOUBLE PRECISION ARITHMETIC

Arithmetic, Fixed Point
USE   FIXED POINT ARITHMETIC

Arithmetic, Floating Point
USE   FLOATING POINT ARITHMETIC

ARIZONA

ARIZONA REGIONAL ECOLOGICAL TEST SITE

ARKANSAS

ARM (ANATOMY)

ARMATURES

ARMED FORCES

ARMED FORCES (FOREIGN)

ARMED FORCES (UNITED STATES)

Armed Reconnaissance Aircraft, Light
USE   COIN AIRCRAFT

ARMOR

Army Ballistic Missiles, Field
USE   FIELD ARMY BALLISTIC MISSILES

ARMY-NAVY INSTRUMENTATION PROGRAM

AROD (Range-Orbit Determination)
USE   AIRBORNE RANGE AND ORBIT
         DETERMINATION

AROMATIC COMPOUNDS

Aromatics, Chloro
USE   CHLOROAROMATICS

AROOS METEORITE

AROUSAL

ARPA COMPUTER NETWORK

Array, Large Aperture Seismic
USE   LARGE APERTURE SEISMIC ARRAY

ARRAYS

Arrays, Antenna
USE   ANTENNA ARRAYS

Arrays, Endfire
USE   ENDFIRE ARRAYS

Arrays, Linear
USE   LINEAR ARRAYS

Arrays, Multi-Anode Microchannel
USE   MULTI-ANODE MICROCHANNEL ARRAYS

Arrays, Multispectral Linear
USE   MULTISPECTRAL LINEAR ARRAYS

Arrays, Phased
USE   PHASED ARRAYS

Arrays, Rollup Solar
USE   SOLAR ARRAYS

Arrays, Solar
USE   SOLAR ARRAYS

Arrays, Synthetic
USE   SYNTHETIC ARRAYS

Arrest, Crack
USE   CRACK ARREST

ARRESTERS

ARRESTING GEAR

Arresting Motion), Brakes (For
USE   BRAKES (FOR ARRESTING MOTION)

ARRHYTHMIA

ARRIVALS

Arrow Launch Vehicle, Black
USE   BLACK KNIGHT ROCKET VEHICLE

Arrow Satellite, Space
USE   COSMOS 149 SATELLITE

ARROW WINGS

ARROYOS

ARSENATES

ARSENIC

ARSENIC ALLOYS

ARSENIC COMPOUNDS

ARSENIC ISOTOPES

Arsenide Lasers, Gallium
USE   GALLIUM ARSENIDE LASERS

ARSENIDES

Arsenides, Aluminum
USE   ALUMINUM ARSENIDES

Arsenides, Aluminum Gallium
USE   ALUMINUM GALLIUM ARSENIDES

Arsenides, Gallium
USE   GALLIUM ARSENIDES

Arsenides, Indium
USE   INDIUM ARSENIDES

ARTEMIA

ARTERIES

ARTERIOSCLEROSIS

Artery Disease, Coronary
USE   CORONARY ARTERY DISEASE

ARTHRITIS

ARTHROPODS

ARTICULATION

ARTIFACTS

ARTIFICIAL CARDIAC PACEMAKER

ARTIFICIAL CLOUDS

ARTIFICIAL EARS

ARTIFICIAL GRAVITY

ARTIFICIAL HARBORS

ARTIFICIAL HEART VALVES

ARTIFICIAL INTELLIGENCE

ARTIFICIAL RADIATION BELTS

Artificial Respiration
USE   RESUSCITATION

ARTIFICIAL SATELLITES

ARTILLERY

ARTILLERY FIRE

ARTS

Arts, Graphic
USE   GRAPHIC ARTS

Aryabhata
USE   INDIAN SPACECRAFT

Aryl Compounds
USE   AROMATIC COMPOUNDS

As
USE   ARSENIC

ASA
USE   ACETYLSALICYLIC ACID

ASBESTOS

ASCENT

Ascent Method, Steepest
USE   STEEPEST DESCENT METHOD

ASCENT PROPULSION SYSTEMS

Ascent Stage, Lunar Module
USE   LUNAR MODULE ASCENT STAGE

Ascent Stage, Space Shuttle
USE   SPACE SHUTTLE ASCENT STAGE

ASCENT TRAJECTORIES

ASCORBIC ACID

ASCORBIC ACID METABOLISM

ASCR Reactor
USE   ADVANCED SODIUM COOLED REACTOR

ASDE
USE   AIRPORT SURFACE DETECTION
         EQUIPMENT

Ash, Fly
USE   FLY ASH

ASHES

ASIA

Asia, Southeast
USE   SOUTHEAST ASIA

Asp Rocket, Nike-
USE   ASP ROCKET VEHICLE

ASP ROCKET VEHICLE

ASPARTATES

ASPARTIC ACID

ASPECT RATIO

Aspect Ratio, High
USE   HIGH ASPECT RATIO

Aspect Ratio, Low
USE   LOW ASPECT RATIO

Aspect Ratio Wings, High
USE   SLENDER WINGS

Aspect Ratio Wings, Low
USE   LOW ASPECT RATIO WINGS

ASPERGILLUS

ASPHALT

ASPHALTENES

ASPHERICITY

ASPHYXIA

Aspiration
USE   VACUUM

ASROC ENGINE

ASSATEAGUE ISLAND (MD-VA)

Assault Helicopter, Black Hawk
USE   H-60 HELICOPTER

Assaulting
USE   ATTACKING (ASSAULTING)

(Assaulting), Attacking
USE   ATTACKING (ASSAULTING)

Assay, Immuno
USE   IMMUNOASSAY

Assay, Radioimmuno
USE   RADIOIMMUNOASSAY

ASSAYING

ASSEMBLER ROUTINES

ASSEMBLIES

Assemblies, Sub
USE   SUBASSEMBLIES

Assemblies, Swing Tail
USE   SWING TAIL ASSEMBLIES

Assemblies, Tail
USE   TAIL ASSEMBLIES

(Assemblies), Tails
USE   TAIL ASSEMBLIES

ASSEMBLING

ASSEMBLY

ASSEMBLY LANGUAGE

Assembly, Orbital
USE   ORBITAL ASSEMBLY

Assembly, Spacecraft Orbital
USE   ORBITAL ASSEMBLY

ASSESS PROGRAM

Assessment, Damage
USE   DAMAGE ASSESSMENT

Assessment, Technology
USE   TECHNOLOGY ASSESSMENT

ASSESSMENTS

ASSET GLIDERS

ASSET PROJECT

Assignment
USE   ALLOCATIONS

Assignment, Frequency
USE   FREQUENCY ASSIGNMENT

Assignment Multiple Access, Demand
USE   DEMAND ASSIGNMENT MULTIPLE ACCESS

ASSIMILATION

Assist Module, Payload
USE   PAYLOAD ASSIST MODULE

Assisted Instruction, Computer
USE   COMPUTER ASSISTED INSTRUCTION

Assisted Takeoff, Jet
USE   JATO ENGINES

ASSOCIATION REACTIONS

Associations
USE   ORGANIZATIONS

ASSOCIATIVE PROCESSING (COMPUTERS)

ASSUMPTIONS

ASSURANCE

ASTATINE

ASTATINE ISOTOPES

ASTEC SOLAR TURBOELECTRIC GENERATOR

Asteroid, Amor
USE   AMOR ASTEROID

Asteroid, Amphitrite
USE   AMPHITRITE ASTEROID

ASTEROID BELTS

ASTEROID CAPTURE

Asteroid, Ceres
USE   CERES ASTEROID

Asteroid, Icarus
USE   ICARUS ASTEROID

ASTEROID MISSIONS

Asteroid, Toro
USE   TORO ASTEROID

Asteroid, Vesta
USE   VESTA ASTEROID

ASTEROIDS

Asteroids, Apollo
USE   APOLLO ASTEROIDS

ASTHENOPIA

ASTHMA

ASTIGMATISM

Astigmatism, An
USE   ANASTIGMATISM

ASTP
USE APOLLO SOYUZ TEST PROJECT

ASTRIONICS

ASTRO MISSIONS (STS)

ASTRO VEHICLE

ASTROBEE ROCKET VEHICLES

ASTROBEE 1500 ROCKET VEHICLE

Astrobiology
USE EXOBIOLOGY

ASTRODYNAMICS

ASTROGRAPHY

ASTROGUIDE NAVIGATION SYSTEM

ASTROLABES

ASTROLOY (TRADEMARK)

Astromasts
USE LONGERONS

ASTROMETRY

ASTRON THERMONUCLEAR REACTOR

ASTRONAUT LOCOMOTION

ASTRONAUT MANEUVERING EQUIPMENT

ASTRONAUT PERFORMANCE

ASTRONAUT TRAINING

ASTRONAUTICS

ASTRONAUTS

ASTRONAVIGATION

ASTRONOMICAL CATALOGS

ASTRONOMICAL COORDINATES

ASTRONOMICAL MAPS

ASTRONOMICAL MODELS

ASTRONOMICAL NETHERLANDS SATELLITE

ASTRONOMICAL OBSERVATORIES

Astronomical Observatory, Orbiting
USE OAO

ASTRONOMICAL PHOTOGRAPHY

ASTRONOMICAL PHOTOMETRY

ASTRONOMICAL SATELLITES

ASTRONOMICAL SPECTROSCOPY

ASTRONOMICAL TELESCOPES

ASTRONOMY

(Astronomy), Black Holes
USE BLACK HOLES (ASTRONOMY)

Astronomy Explorer B, Radio
USE EXPLORER 49 SATELLITE

Astronomy Explorer, Gamma Ray
USE EXPLORER 11 SATELLITE

Astronomy Explorer Satellite, Radio
USE RADIO ASTRONOMY EXPLORER SATELLITE

Astronomy Explorer 2, Radio
USE EXPLORER 49 SATELLITE

Astronomy, Gamma Ray
USE GAMMA RAY ASTRONOMY

Astronomy, Infrared
USE INFRARED ASTRONOMY

(Astronomy), Local Group
USE LOCAL GROUP (ASTRONOMY)

(Astronomy), North Polar SPUR
USE NORTH POLAR SPUR (ASTRONOMY)

Astronomy Observatories, High Energy
USE HEAO

Astronomy Observatory A, High Energy
USE HEAO 1

Astronomy Observatory B, High Energy
USE HEAO 2

Astronomy Observatory C, High Energy
USE HEAO 3

Astronomy Observatory 1, High Energy
USE HEAO 1

Astronomy Observatory 2, High Energy
USE HEAO 2

Astronomy Observatory 3, High Energy
USE HEAO 3

Astronomy, Radar
USE RADAR ASTRONOMY

Astronomy, Radio
USE RADIO ASTRONOMY

(Astronomy), Radio Sources
USE RADIO SOURCES (ASTRONOMY)

(Astronomy), Rhea
USE RHEA (ASTRONOMY)

Astronomy Satellite, Infrared
USE INFRARED ASTRONOMY SATELLITE

Astronomy Satellite 1, Small
USE SAS-1

Astronomy Satellite 2, Small
USE SAS-2

Astronomy Satellite 3, Small
USE SAS-3

Astronomy Satellites, Small
USE SAS

Astronomy, Spaceborne
USE SPACEBORNE ASTRONOMY

Astronomy, Ultraviolet
USE ULTRAVIOLET ASTRONOMY

(Astronomy), White Holes
USE WHITE HOLES (ASTRONOMY)

Astronomy, X Ray
USE X RAY ASTRONOMY

Astrophysical Facility, Advanced X Ray
USE X RAY ASTROPHYSICS FACILITY

ASTROPHYSICS

Astrophysics, Computational
USE COMPUTATIONAL ASTROPHYSICS

Astrophysics Facility, Advanced X Ray
USE X RAY ASTROPHYSICS FACILITY

Astrophysics Facility, X Ray
USE X RAY ASTROPHYSICS FACILITY

(Astrophysics), Missing Mass
USE MISSING MASS (ASTROPHYSICS)

ASTROPLANE

ASYMMETRY

ASYMPTOTES

ASYMPTOTIC METHODS

ASYMPTOTIC PROPERTIES

ASYMPTOTIC SERIES

ASYNCHRONOUS MOTORS

At
USE ASTATINE

Atars
USE AUTOMATIC TRAFFIC ADVISORY AND
RESOLUTION

ATAXIA

ATAXITE

ATC, Automated En Route
USE AUTOMATED EN ROUTE ATC

ATCHAFALAYA RIVER BASIN (LA)

ATELECTASIS

ATHENA ROCKET VEHICLE

Atherosclerosis
USE ARTERIOSCLEROSIS

ATHLETES

Athodyds
USE RAMJET ENGINES

ATLANTA (GA)

Atlantic Aircraft
USE BREGUET 1150 AIRCRAFT

ATLANTIC OCEAN

Atlantic Region (US), Central
USE CENTRAL ATLANTIC REGION (US)

Atlantic Regional Ecol Test Site, Central
USE CENTRAL ATLANTIC REGIONAL ECOL
TEST SITE

Atlantic Treaty Organization (NATO), North
USE NORTH ATLANTIC TREATY ORGANIZATION
(NATO)

Atlantic Tropical Experiment, GARP
USE GARP ATLANTIC TROPICAL EXPERIMENT

ATLANTIS (ORBITER)

ATLAS ABLE 5 LAUNCH VEHICLE

ATLAS AGENA B LAUNCH VEHICLE

ATLAS AGENA LAUNCH VEHICLES

ATLAS CENTAUR LAUNCH VEHICLE

ATLAS D ICBM

ATLAS E ICBM

ATLAS F ICBM

ATLAS ICBM

ATLAS LAUNCH VEHICLES

ATLAS SLV-3 LAUNCH VEHICLE

ATLIT PROJECT

Atmosphere, Earth
USE EARTH ATMOSPHERE

Atmosphere, Equatorial
USE EQUATORIAL ATMOSPHERE

Atmosphere Explorer A
USE   EXPLORER 17 SATELLITE

Atmosphere Explorer B
USE   EXPLORER 32 SATELLITE

Atmosphere Explorer C
USE   EXPLORER 51 SATELLITE

Atmosphere Explorer D
USE   EXPLORER 54 SATELLITE

Atmosphere Explorer E
USE   EXPLORER 55 SATELLITE

Atmosphere, Free
USE   FREE ATMOSPHERE

Atmosphere, Inert
USE   INERT ATMOSPHERE

Atmosphere, Jupiter
USE   JUPITER ATMOSPHERE

Atmosphere, Lower
USE   LOWER ATMOSPHERE

Atmosphere, Lunar
USE   LUNAR ATMOSPHERE

Atmosphere, Mars
USE   MARS ATMOSPHERE

Atmosphere, Middle
USE   MIDDLE ATMOSPHERE

Atmosphere, Midlatitude
USE   MIDLATITUDE ATMOSPHERE

Atmosphere, Neptune
USE   NEPTUNE ATMOSPHERE

Atmosphere, Primitive Earth
USE   PRIMITIVE EARTH ATMOSPHERE

Atmosphere, Saturn
USE   SATURN ATMOSPHERE

Atmosphere, Solar
USE   SOLAR ATMOSPHERE

Atmosphere Sounding Projectile, Window
USE   WASP SOUNDING ROCKET

Atmosphere, Upper
USE   UPPER ATMOSPHERE

Atmosphere, Uranus
USE   URANUS ATMOSPHERE

Atmosphere, Venus
USE   VENUS ATMOSPHERE

ATMOSPHERES

Atmospheres, Argon-Oxygen
USE   ARGON-OXYGEN ATMOSPHERES

Atmospheres, Cabin
USE   CABIN ATMOSPHERES

Atmospheres, Cometary
USE   COMETARY ATMOSPHERES

Atmospheres, Controlled
USE   CONTROLLED ATMOSPHERES

Atmospheres, Helium Hydrogen
USE   HELIUM HYDROGEN ATMOSPHERES

Atmospheres, Helium-Oxygen
USE   HELIUM-OXYGEN ATMOSPHERES

Atmospheres, Hypobaric
USE   HYPOBARIC ATMOSPHERES

Atmospheres, Neutral
USE   NEUTRAL ATMOSPHERES

Atmospheres, Nongray
USE   NONGRAY ATMOSPHERES

Atmospheres, Planetary
USE   PLANETARY ATMOSPHERES

Atmospheres, Reference
USE   REFERENCE ATMOSPHERES

Atmospheres, Satellite
USE   SATELLITE ATMOSPHERES

Atmospheres, Spacecraft Cabin
USE   SPACECRAFT CABIN ATMOSPHERES

Atmospheres, Standard
USE   REFERENCE ATMOSPHERES

Atmospheres, Stellar
USE   STELLAR ATMOSPHERES

ATMOSPHERIC & OCEANOGRAPHIC INFORM SYS

Atmospheric Absorption
USE   ATMOSPHERIC ATTENUATION

Atmospheric And Magnetospheric Payload
USE   AMPS (SATELLITE PAYLOAD)

ATMOSPHERIC ATTENUATION

ATMOSPHERIC BOUNDARY LAYER

ATMOSPHERIC CHEMISTRY

ATMOSPHERIC CIRCULATION

ATMOSPHERIC CLOUD PHYSICS LAB (SPACELAB)

ATMOSPHERIC COMPOSITION

Atmospheric Composition Experiment, Lower
USE   LACATE (EXPERIMENT)

Atmospheric Conditions
USE   METEOROLOGY

ATMOSPHERIC CONDUCTIVITY

ATMOSPHERIC CORRECTION

ATMOSPHERIC DENSITY

ATMOSPHERIC DIFFUSION

ATMOSPHERIC EFFECTS

ATMOSPHERIC ELECTRICITY

Atmospheric Emission
USE   AIRGLOW

ATMOSPHERIC ENERGY SOURCES

ATMOSPHERIC ENTRY

ATMOSPHERIC ENTRY SIMULATION

ATMOSPHERIC GENERAL CIRCULATION EXPERIMENT

ATMOSPHERIC HEAT BUDGET

ATMOSPHERIC HEATING

Atmospheric Impurities
USE   AIR POLLUTION

ATMOSPHERIC IONIZATION

ATMOSPHERIC LASERS

Atmospheric Lasers, Transversely Excited
USE   TEA LASERS

Atmospheric Loading
USE   POLLUTION TRANSPORT

ATMOSPHERIC MODELS

ATMOSPHERIC MOISTURE

Atmospheric Noise
USE   ATMOSPHERICS

ATMOSPHERIC OPTICS

ATMOSPHERIC PHYSICS

ATMOSPHERIC PRESSURE

ATMOSPHERIC RADIATION

ATMOSPHERIC REFRACTION

Atmospheric Research Program, Global
USE   GLOBAL ATMOSPHERIC RESEARCH
         PROGRAM

ATMOSPHERIC SCATTERING

Atmospheric Shells
USE   ATMOSPHERIC STRATIFICATION

ATMOSPHERIC SOUNDING

ATMOSPHERIC STRATIFICATION

ATMOSPHERIC TEMPERATURE

ATMOSPHERIC TIDES

ATMOSPHERIC TURBULENCE

ATMOSPHERIC WINDOWS

ATMOSPHERICS

Atmospherics, Sudden Enhancement Of
USE   SUDDEN ENHANCEMENT OF
         ATMOSPHERICS

Atoll Reefs
USE   CORAL REEFS

ATOLLS

ATOM CONCENTRATION

Atom Interactions, Ion
USE   ION ATOM INTERACTIONS

Atomic Batteries
USE   RADIOISOTOPE BATTERIES

ATOMIC BEAMS

Atomic Bombs
USE   FISSION WEAPONS

ATOMIC CLOCKS

ATOMIC COLLISIONS

Atomic Energy
USE   NUCLEAR ENERGY

ATOMIC ENERGY LEVELS

ATOMIC EXCITATIONS

Atomic Explosions
USE   NUCLEAR EXPLOSIONS

Atomic Gases
USE   MONATOMIC GASES

ATOMIC INTERACTIONS

Atomic Mass
USE   ATOMIC WEIGHTS

ATOMIC MOBILITIES

ATOMIC PHYSICS

(Atomic Physics), Quenching
USE   QUENCHING (ATOMIC PHYSICS)

Atomic Power Plant, Enrico Fermi
USE   ENRICO FERMI ATOMIC POWER PLANT

ATOMIC RECOMBINATION

ATOMIC SPECTRA

ATOMIC STRUCTURE

ATOMIC THEORY

ATOMIC WEIGHTS

Atomization
USE   ATOMIZING

Atomization, Gas
USE   GAS ATOMIZATION

Atomization, Liquid
USE   LIQUID ATOMIZATION

ATOMIZERS

ATOMIZING

ATOMS

Atoms, Helium
USE   HELIUM ATOMS

Atoms, Hot
USE   HOT ATOMS

Atoms, Hydrogen
USE   HYDROGEN ATOMS

Atoms, Metastable
USE   METASTABLE ATOMS

Atoms, Neutral
USE   NEUTRAL ATOMS

Atoms, Nitrogen
USE   NITROGEN ATOMS

Atoms, Oxygen
USE   OXYGEN ATOMS

Atoms, Recoil
USE   RECOIL ATOMS

ATP
USE   ADENOSINE TRIPHOSPHATE

ATR Reactor
USE   ADVANCED TEST REACTORS

Atriums, Solar
USE   SOLAR ATRIUMS

ATROPHY

ATROPINE

ATS

ATS 1

ATS 2

ATS 3

ATS 4

ATS 5

ATS 6

ATS 7

ATS 8

ATTACHMENT

Attachment, Electron
USE   ELECTRON ATTACHMENT

Attachments
USE   ACCESSORIES

ATTACK

ATTACK AIRCRAFT

Attack, Angle Of
USE   ANGLE OF ATTACK

Attack, Chemical
USE   CHEMICAL ATTACK

Attack, Zero Angle Of
USE   ZERO ANGLE OF ATTACK

ATTACKING (ASSAULTING)

ATTENTION

ATTENUATION

Attenuation, Acoustic
USE   ACOUSTIC ATTENUATION

Attenuation, Atmospheric
USE   ATMOSPHERIC ATTENUATION

ATTENUATION COEFFICIENTS

Attenuation Measurement Project, Radio
USE   RADIO ATTENUATION MEASUREMENT
           PROJECT

Attenuation, Microwave
USE   MICROWAVE ATTENUATION

Attenuation, Noise
USE   NOISE REDUCTION

Attenuation, Radar
USE   RADAR ATTENUATION

Attenuation, Radio
USE   RADIO ATTENUATION

Attenuation, Radio Signal
USE   RADIO ATTENUATION

Attenuation, Shock Wave
USE   SHOCK WAVE ATTENUATION

Attenuation, Wave
USE   WAVE ATTENUATION

ATTENUATORS

ATTITUDE CONTROL

Attitude Control), DISCOS (Satellite
USE   DISCOS (SATELLITE ATTITUDE CONTROL)

Attitude Control, Pitch
USE   LONGITUDINAL CONTROL

Attitude Control, Satellite
USE   SATELLITE ATTITUDE CONTROL

Attitude Control Satellite, Transit
USE   TRANSIT ATTITUDE CONTROL SATELLITE

Attitude Disturbance, Satellite
USE   SPACECRAFT STABILITY
           ATTITUDE STABILITY

ATTITUDE GYROS

ATTITUDE (INCLINATION)

ATTITUDE INDICATORS

Attitude Indicators, Helicopter
USE   HELICOPTERS
           ATTITUDE INDICATORS

ATTITUDE STABILITY

Attitude Takeoff-Landing Aircraft, Vertical
USE   VATOL AIRCRAFT

ATTRACTION

Attractors, Strange
USE   STRANGE ATTRACTORS

Attributes
USE   PROPERTIES

Attrition (Materials)
USE   COMMINUTION

Au
USE   GOLD

AUDIO DATA

AUDIO EQUIPMENT

AUDIO FREQUENCIES

AUDIO SIGNALS

Audio Visual Equipment
USE   VISUAL AIDS
           TRAINING DEVICES

AUDIOLOGY

AUDIOMETRY

AUDITORY DEFECTS

AUDITORY FATIGUE

AUDITORY PERCEPTION

AUDITORY SENSATION AREAS

AUDITORY SIGNALS

AUDITORY STIMULI

AUDITORY TASKS

AUFEIS (ICE)

AUGER EFFECT

AUGER SPECTROSCOPY

AUGMENTATION

Augmentation, Lift
USE   LIFT AUGMENTATION

Augmentation, Stability
USE   STABILITY AUGMENTATION

Augmentation, Thrust
USE   THRUST AUGMENTATION

Augmented Wing Flaps, Jet
USE   WING FLAPS
           JET FLAPS

Augmentor Wing Aircraft, C-8A
USE   C-8A AUGMENTOR WING AIRCRAFT

Auricles, Cardiac
USE   CARDIAC AURICLES

AURIGA CONSTELLATION

Aurigae Star, Zeta
USE   ZETA AURIGAE STAR

AURORA 7

AURORAL ABSORPTION

Auroral Activity
USE   AURORAS

AURORAL ARCS

AURORAL ECHOES

AURORAL ELECTROJETS

AURORAL IONIZATION

AURORAL IRRADIATION

AURORAL SPECTROSCOPY

AURORAL TEMPERATURE

AURORAL ZONES

AURORAS

Auroras, Polar
USE   AURORAS

Auroras, Radio
USE   RADIO AURORAS

AUSFORMING

AUSTENITE

AUSTENITIC STAINLESS STEELS

AUSTRALIA

AUSTRALITES

AUSTRIA

AUTOCATALYSIS

AUTOCLAVES

AUTOCLAVING

AUTOCODERS

Autocollimators
USE   COLLIMATORS

AUTOCORRELATION

AUTODYNES

Autogiro, Avian 2/180
USE   AVIAN 2/180 AUTOGIRO

AUTOGYROS

AUTOIONIZATION

AUTOKINESIS

AUTOMATA THEORY

AUTOMATED EN ROUTE ATC

AUTOMATED GUIDEWAY TRANSIT VEHICLES

AUTOMATED MIXED TRAFFIC VEHICLES

AUTOMATED PILOT ADVISORY SYSTEM

AUTOMATED RADAR TERMINAL SYSTEM

AUTOMATED TRANSIT VEHICLES

AUTOMATIC CONTROL

AUTOMATIC CONTROL VALVES

Automatic Data Processing
USE   DATA PROCESSING

AUTOMATIC FLIGHT CONTROL

AUTOMATIC FREQUENCY CONTROL

AUTOMATIC GAIN CONTROL

AUTOMATIC LANDING CONTROL

Automatic Light Aircraft Readiness Monitor
USE   ALARM PROJECT

Automatic Pattern Recognition
USE   PATTERN RECOGNITION

AUTOMATIC PICTURE TRANSMISSION

AUTOMATIC PILOTS

Automatic Rocket Impact Predictors
USE   COMPUTERIZED SIMULATION
       IMPACT PREDICTION

AUTOMATIC TEST EQUIPMENT

AUTOMATIC TRAFFIC ADVISORY AND RESOLUTION

AUTOMATIC TYPEWRITERS

AUTOMATIC WEATHER STATIONS

AUTOMATION

AUTOMOBILE ACCIDENTS

AUTOMOBILE ENGINES

AUTOMOBILE FUELS

AUTOMOBILES

Automobiles, Electric
USE   ELECTRIC AUTOMOBILES

AUTOMORPHISMS

AUTONOMIC NERVOUS SYSTEM

AUTONOMOUS NAVIGATION

AUTONOMOUS SPACECRAFT CLOCKS

AUTONOMY

Autopilots
USE   AUTOMATIC PILOTS

AUTOPSIES

AUTORADIOGRAPHY

AUTOREGRESSIVE PROCESSES

AUTOROTATION

AUTOTROPHS

AUTUMN

AUXILIARY EQUIPMENT (COMPUTERS)

AUXILIARY POWER SOURCES

Auxiliary Power, Systems For Nuclear
USE   SNAP

Auxiliary Power Units, Chemical
USE   CHEMICAL AUXILIARY POWER UNITS

Auxiliary Power Units, Nuclear
USE   NUCLEAR AUXILIARY POWER UNITS

Auxiliary Power Units, Solar
USE   SOLAR AUXILIARY POWER UNITS

AUXILIARY PROPULSION

AV-8A Aircraft
USE   HARRIER AIRCRAFT

AV-8B Aircraft
USE   HARRIER AIRCRAFT

AVAILABILITY

AVALANCHE DIODES

Avalanche, Electron
USE   ELECTRON AVALANCHE

Avalanche, Townsend
USE   TOWNSEND AVALANCHE

Avalanche Transit Time Devices, Controlled
USE   CATT DEVICES

Avalanche Triggered Transit, Trapped Plasma
USE   TRAPATT DEVICES

AVALANCHES

AVCS
USE   ADVANCED VIDICON CAMERA SYSTEM
       (AVCS)

(AVCS), Advanced Vidicon Camera System
USE   ADVANCED VIDICON CAMERA SYSTEM
       (AVCS)

AVERAGE

Averaging Method, Ritz
USE   RITZ AVERAGING METHOD

AVIAN 2/180 AUTOGIRO

Aviation
USE   AERONAUTICS

Aviation Aircraft, General
USE   GENERAL AVIATION AIRCRAFT

Aviation Aircraft, Sud
USE   SUD AVIATION AIRCRAFT

Aviation, Civil
USE   CIVIL AVIATION

Aviation, Commercial
USE   CIVIL AVIATION
       COMMERCIAL AIRCRAFT

Aviation, Military
USE   MILITARY AVIATION

AVIATION PSYCHOLOGY

Aviation SA-321 Helicopter, Sud
USE   SA-321 HELICOPTER

Aviation SA-330 Helicopter, Sud
USE   SA-330 HELICOPTER

Aviation SE-210 Aircraft, Sud
USE   SE-210 AIRCRAFT

Aviation SE-3160 Helicopter, Sud
USE   SE-3160 HELICOPTER

Aviation System, National
USE   NATIONAL AVIATION SYSTEM

Aviation Whitcomb Airfoil, General
USE   GAW-2 AIRFOIL
       GAW-1 AIRFOIL

Aviators
USE   AIRCRAFT PILOTS

AVIONICS

Avionics Integration Laboratory, Shuttle
USE   SAIL PROJECT

AVOIDANCE

Avoidance, Collision
USE   COLLISION AVOIDANCE

Avoidance, Obstacle
USE   OBSTACLE AVOIDANCE

Avoidance System, Beacon Collision
USE   BEACON COLLISION AVOIDANCE SYSTEM

Avoidance, Vortex
USE   VORTEX AVOIDANCE

AVRO Whitworth HS-748 Aircraft
USE   HS-748 AIRCRAFT

AVRO 698 Aircraft
USE   VULCAN AIRCRAFT

AVRO 707 AIRCRAFT

AWACS AIRCRAFT

AWARDS

AXAF
USE   X RAY ASTROPHYSICS FACILITY

Axes (Coordinates)
 USE   COORDINATES

AXES OF ROTATION

AXES (REFERENCE LINES)

AXIAL COMPRESSION LOADS

Axial Compressors
 USE   TURBOCOMPRESSORS

AXIAL FLOW

Axial Flow Compressors
 USE   TURBOCOMPRESSORS

AXIAL FLOW PUMPS

AXIAL FLOW TURBINES

AXIAL LOADS

AXIAL MODES

AXIAL STRAIN

AXIAL STRESS

AXIOMS

Axis, Aerodynamic
 USE   AERODYNAMIC BALANCE

Axis, Earth
 USE   EARTH AXIS

Axis Spectrometers, Triple
 USE   NEUTRON SPECTROMETERS

Axis Stabilization, Three
 USE   THREE AXIS STABILIZATION

AXISYMMETRIC BODIES

Axisymmetric Deformation
 USE   AXIAL STRAIN

AXISYMMETRIC FLOW

Axisymmetry
 USE   SYMMETRY

Axles
 USE   SHAFTS (MACHINE ELEMENTS)

AXONS

AZ
 USE   ARIZONA

(AZ), Grand Canyon
 USE   GRAND CANYON (AZ)

(AZ), Phoenix
 USE   PHOENIX (AZ)

(AZ), Phoenix Quadrangle
 USE   PHOENIX QUADRANGLE (AZ)

AZEOTROPES

Azide, Triaminoguanidinium
 USE   TRIAMINOGUANIDINIUM AZIDE

Azides, Hydrogen
 USE   HYDROGEN AZIDES

AZIDES (INORGANIC)

AZIDES (ORGANIC)

Azides, Sodium
 USE   SODIUM AZIDES

AZIMUTH

Azimuth, Solar
 USE   AZIMUTH
        SOLAR POSITION

AZINES

AZO COMPOUNDS

AZOLES

Azoles, Carb
 USE   CARBAZOLES

Azoles, Tetr
 USE   TETRAZOLES

AZOTOBACTER

AZULENE

AZUR SATELLITE

A1 Missile, Polaris
 USE   POLARIS A1 MISSILE

A2 Missile, Polaris
 USE   POLARIS A2 MISSILE

A2, OAO-
 USE   OAO 2

A2F Aircraft
 USE   A-6 AIRCRAFT

A3 Missile, Polaris
 USE   POLARIS A3 MISSILE

A3D Aircraft
 USE   A-3 AIRCRAFT

A3J Aircraft
 USE   A-5 AIRCRAFT

A4D Aircraft
 USE   A-4 AIRCRAFT

B

B, AD/I
 USE   EXPLORER 25 SATELLITE

B, Air Density/injun Explorer
 USE   EXPLORER 25 SATELLITE

B, Anik
 USE   ANIK 2

B, Atmosphere Explorer
 USE   EXPLORER 32 SATELLITE

B, BE
 USE   EXPLORER 22 SATELLITE

B, Beacon Explorer
 USE   EXPLORER 22 SATELLITE

B Complex, Vitamin
 USE   BIOTIN

B, Earth Resources Technology Satellite
 USE   LANDSAT 2

B, Energetic Particle Explorer
 USE   EXPLORER 14 SATELLITE

B, EOS-
 USE   LANDSAT F

B, EPE-
 USE   EXPLORER 14 SATELLITE

B, ERTS-
 USE   LANDSAT 2

B, Geostationary Operatl Environ Satellite
 USE   GOES 2

B, Gravity Probe
 USE   GRAVITY PROBE B

B, HEAO
 USE   HEAO 2

B, Helios
 USE   HELIOS B

B, High Energy Astronomy Observatory
 USE   HEAO 2

B, IMP-
 USE   EXPLORER 21 SATELLITE

B, ISIS-
 USE   ISIS-B

B Launch Vehicle, Atlas Agena
 USE   ATLAS AGENA B LAUNCH VEHICLE

B Launch Vehicle, RAM
 USE   RAM B LAUNCH VEHICLE

B, Lunar Orbiter
 USE   LUNAR ORBITER 2

B Missile, Bomarc
 USE   BOMARC B MISSILE

B Missile, Bullpup
 USE   BULLPUP B MISSILE

B, OGO-
 USE   OGO-3

B, OSO-
 USE   OSO-2

B, Radio Astronomy Explorer
 USE   EXPLORER 49 SATELLITE

B, RAE
 USE   EXPLORER 49 SATELLITE

B Ranger Program, Agena
 USE   AGENA B RANGER PROGRAM

B Reactors, KIWI
 USE   KIWI B REACTORS

B Rocket Vehicle, Agena
 USE   AGENA B ROCKET VEHICLE

B Satellite, AE-
 USE   EXPLORER 32 SATELLITE

B Satellite, Alouette
 USE   ALOUETTE B SATELLITE

B Satellite, COS-
 USE   COS-B SATELLITE

B Satellite, GEOS-
 USE   GEOS 2 SATELLITE

B Satellite, HEOS
 USE   HEOS B SATELLITE

B Satellite, Magsat
 USE   MAGSAT B SATELLITE

B Satellite, Palapa
 USE   PALAPA 2 SATELLITE

B Satellite, SEASAT-
 USE   SEASAT-B SATELLITE

B Satellite, SIRS
 USE   SIRS B SATELLITE

B, Sir-
 USE   SHUTTLE IMAGING RADAR

B, Space Shuttle Mission 31-
 USE   SPACE SHUTTLE MISSION 31-B

B, Space Shuttle Mission 41-
 USE   SPACE SHUTTLE MISSION 41-B

B, Space Shuttle Mission 51-
 USE   SPACE SHUTTLE MISSION 51-B

B, Space Shuttle Mission 61-
 USE   SPACE SHUTTLE MISSION 61-B

B Spacecraft, Gemini
  USE   GEMINI B SPACECRAFT

B STARS

B, TELESAT Canada
  USE   ANIK 2

B, Vitamin
  USE   THIAMINE

B 2, Vitamin
  USE   RIBOFLAVIN

B 6, Vitamin
  USE   PYRIDOXINE

B 12, Vitamin
  USE   CYANOCOBALAMIN

B-A-W Devices
  USE   BULK ACOUSTIC WAVE DEVICES

B-1 AIRCRAFT

B-1 Reactor, KIWI
  USE   KIWI B-1 REACTOR

B-4 Reactor, KIWI
  USE   KIWI B-4 REACTOR

B-26 AIRCRAFT

B-47 AIRCRAFT

B-50 AIRCRAFT

B-52 AIRCRAFT

B-57 AIRCRAFT

B-58 AIRCRAFT

B-66 AIRCRAFT

B-70 AIRCRAFT

B-103 Aircraft
  USE   BUCCANEER AIRCRAFT

B-103 Aircraft, Blackburn
  USE   BUCCANEER AIRCRAFT

Ba
  USE   BARIUM

BABBITT METAL

BABOONS

BAC AIRCRAFT

BAC TSR 2 Aircraft
  USE   TSR-2 AIRCRAFT

BAC 111 AIRCRAFT

BACILLUS

BACK INJURIES

BACKFIRE

Background Explorer Satellite, Cosmic
  USE   COSMIC BACKGROUND EXPLORER
         SATELLITE

Background Measurement, High Alt Target And
  USE   HIGH ALT TARGET AND BACKGROUND
         MEASUREMENT

BACKGROUND NOISE

BACKGROUND RADIATION

Background Sats, Galactic Radiation Exp
  USE   GREB SATELLITES

Backings
  USE   BACKUPS

BACKLOBES

Backpacks, Reaction Jet
  USE   SELF MANEUVERING UNITS

Backscatter UV Spectrometer, Solar
  USE   SOLAR BACKSCATTER UV
         SPECTROMETER

BACKSCATTERING

Backshores
  USE   BEACHES

BACKUPS

BACKWARD DIFFERENCING

BACKWARD FACING STEPS

BACKWARD WAVE TUBES

BACKWARD WAVES

BACKWASH

BACTERIA

BACTERICIDES

BACTERIOLOGY

BACTERIOPHAGES

BADLANDS

BAFFLES

Bag Restraint Devices, Air
  USE   AIR BAG RESTRAINT DEVICES

BAGGAGE

BAGS

Bags, Gas
  USE   GAS BAGS

BAHAMAS

BAHRAIN

BAILOUT

BAINITE

BAINITIC STEEL

Baja California
  USE   LOWER CALIFORNIA (MEXICO)

Bajadas
  USE   FANS (LANDFORMS)

BAKELITE (TRADEMARK)

Bakeout
  USE   DEGASSING

BAKER-NUNN CAMERA

BAKING

BALANCE

Balance, Aerodynamic
  USE   AERODYNAMIC BALANCE

Balance, Drag
  USE   AERODYNAMIC BALANCE
         LIFT DRAG RATIO

Balance Equations
  USE   EQUATIONS

Balance, Heat
  USE   HEAT BALANCE

Balance, Mass
  USE   MASS BALANCE

Balance, Material
  USE   MATERIAL BALANCE

(Balance), Trim
  USE   AERODYNAMIC BALANCE

Balance, Water
  USE   WATER BALANCE

Balanced Amplifiers
  USE   PUSH-PULL AMPLIFIERS

Balances, Counter
  USE   COUNTERBALANCES

Balances, Micro
  USE   MICROBALANCES

Balances, Strain Gage
  USE   STRAIN GAGE BALANCES

Balances, Thermo
  USE   THERMOBALANCES

Balances, Wind Tunnel
  USE   WEIGHT INDICATORS
         WIND TUNNEL APPARATUS

BALANCING

BALL BEARINGS

BALL LIGHTNING

BALLAST

BALLAST (MASS)

BALLASTS (IMPEDANCES)

BALLISTIC CAMERAS

BALLISTIC MISSILE DECOYS

BALLISTIC MISSILE EARLY WARNING SYSTEM

BALLISTIC MISSILE SUBMARINES

BALLISTIC MISSILES

Ballistic Missiles, Field Army
  USE   FIELD ARMY BALLISTIC MISSILES

Ballistic Missiles, Fleet
  USE   FLEET BALLISTIC MISSILES

Ballistic Missiles, Intercontinental
  USE   INTERCONTINENTAL BALLISTIC MISSILES

Ballistic Missiles, Intermediate Range
  USE   INTERMEDIATE RANGE BALLISTIC
         MISSILES

Ballistic Missiles, Short Range
  USE   SHORT RANGE BALLISTIC MISSILES

BALLISTIC RANGES

BALLISTIC TRAJECTORIES

BALLISTIC VEHICLES

BALLISTICS

Ballistics, Hydro
  USE   HYDROBALLISTICS

Ballistics Identification, Rapid
  USE   RAPID BALLISTICS IDENTIFICATION

Ballistics, Interior
  USE   INTERIOR BALLISTICS

Ballistics, Penetration
  USE   TERMINAL BALLISTICS

Ballistics, Terminal
  USE   TERMINAL BALLISTICS

BALLISTOCARDIOGRAPHY

BALLOON FLIGHT

BALLOON SOUNDING

BALLOON-BORNE INSTRUMENTS

BALLOONING MODES

BALLOONS

Balloons, Constant Volume
USE   SUPERPRESSURE BALLOONS

Balloons, High Altitude
USE   HIGH ALTITUDE BALLOONS

Balloons, Jimsphere
USE   JIMSPHERE BALLOONS

Balloons, Kite
USE   TETHERED BALLOONS

Balloons, Meteorological
USE   METEOROLOGICAL BALLOONS

Balloons, Robin
USE   ROBIN BALLOONS

Balloons, Skyhook
USE   SKYHOOK BALLOONS

Balloons, Superpressure
USE   SUPERPRESSURE BALLOONS

Balloons, Tethered
USE   TETHERED BALLOONS

BALLS

Balls, Fire
USE   FIREBALLS

BALLUTES

BALMER SERIES

BALSA

BALTIC SEA

BALTIC SHIELD (EUROPE)

BANACH SPACE

Band, Bloch
USE   BLOCH BAND

Band, Broad
USE   BROADBAND

Band, C
USE   C BAND

Band Cameras, Multispectral
USE   MULTISPECTRAL BAND CAMERAS

Band, Error
USE   ACCURACY

Band, K
USE   EXTREMELY HIGH FREQUENCIES

Band, KA
USE   EXTREMELY HIGH FREQUENCIES

Band, KU
USE   SUPERHIGH FREQUENCIES

Band, L
USE   ULTRAHIGH FREQUENCIES

Band, P
USE   P BAND

Band Radiometers, Passive L-
USE   PASSIVE L-BAND RADIOMETERS

BAND RATIOING

Band, S
USE   ULTRAHIGH FREQUENCIES
        SUPERHIGH FREQUENCIES

Band Scanners, Multispectral
USE   MULTISPECTRAL BAND SCANNERS

BAND STRUCTURE OF SOLIDS

Band, Unified S
USE   UNIFIED S BAND

Band, V
USE   EXTREMELY HIGH FREQUENCIES

Band, X
USE   SUPERHIGH FREQUENCIES

Bandgap
USE   ENERGY GAPS (SOLID STATE)

BANDPASS FILTERS

BANDS

Bands, Absorption
USE   ABSORPTION SPECTRA

Bands, Conduction
USE   CONDUCTION BANDS

Bands, Energy
USE   ENERGY BANDS

Bands, Forbidden
USE   FORBIDDEN BANDS

Bands, Frequency
USE   FREQUENCIES

Bands, Herzberg
USE   HERZBERG BANDS

Bands, Low Frequency
USE   LOW FREQUENCY BANDS

Bands, Luder
USE   YIELD POINT
        PLASTIC DEFORMATION

Bands, Photoluminescent
USE   PHOTOLUMINESCENT BANDS

Bands, Schumann-Runge
USE   SCHUMANN-RUNGE BANDS

Bands, Side
USE   SIDEBANDS

Bands, Slip
USE   EDGE DISLOCATIONS

Bands, Spectral
USE   SPECTRAL BANDS

Bands, Swan
USE   SWAN BANDS

Bands, Vegard-Kaplan
USE   VEGARD-KAPLAN BANDS

BANDSTOP FILTERS

BANDWIDTH

Bang Control, Bang-
USE   OFF-ON CONTROL

Bang Cosmology, Big
USE   BIG BANG COSMOLOGY

Bang-Bang Control
USE   OFF-ON CONTROL

BANGLADESH

Bank Observatory, Jodrell
USE   JODRELL BANK OBSERVATORY

Banking Flight
USE   TURNING FLIGHT

Banks (NC), Outer
USE   OUTER BANKS (NC)

BARANY CHAIR

BARBADOS

Barchans
USE   DUNES

Bardeen Approximation
USE   ELECTRICAL PROPERTIES
        BARRIER LAYERS
        SURFACE PROPERTIES

Bardeen-Cooper-Schrieffer Theory
USE   BCS THEORY

BARENTS SEA

BARITE

BARIUM

BARIUM ALLOYS

BARIUM COMPOUNDS

BARIUM FERRATES

BARIUM FLUORIDES

BARIUM ION CLOUDS

BARIUM ISOTOPES

BARIUM OXIDES

BARIUM SULFIDES

BARIUM TITANATES

BARIUM ZIRCONATES

BARKHAUSEN EFFECT

BARLEY

BAROCLINIC INSTABILITY

BAROCLINIC WAVES

BAROCLINITY

BAROMETERS

Barometric Pressure
USE   ATMOSPHERIC PRESSURE

BARORECEPTORS

BAROTRAUMA

BAROTROPIC FLOW

BAROTROPISM

BARRAGES

BARRED GALAXIES

BARRELS

BARRELS (CONTAINERS)

BARREN LAND

Barrens
USE   BARREN LAND

Barricades
USE   BARRIERS

Barrier, Blood-Brain
USE   BLOOD-BRAIN BARRIER

Barrier Clothing, Vapor
USE   VAPOR BARRIER CLOTHING

Barrier Diodes, Schottky
USE   SCHOTTKY DIODES

Barrier Injection Transit Time Diodes
USE   BARRITT DIODES

BARRIER LAYERS

Barrier, Sound
USE   ACOUSTIC VELOCITY

Barrier-Metal Junctions, Metal-
USE   MBM JUNCTIONS

BARRIERS

Barriers, Electrode Film
USE   ELECTRODE FILM BARRIERS

(Barriers), Fences
USE   FENCES (BARRIERS)

BARRIERS (LANDFORMS)

BARRITT DIODES

BARS

Bars, Elastic
USE   ELASTIC BARS

BARS (LANDFORMS)

Bars, Prismatic
USE   PRISMATIC BARS

Barycenter
USE   CENTER OF GRAVITY

BARYON RESONANCE

BARYONS

BASALT

Base Command Center, Space
USE   SPACE BASE COMMAND CENTER

Base Equilibrium, Acid
USE   ACID BASE EQUILIBRIUM

BASE FLOW

BASE HEATING

Base Interferometry, Very Long
USE   VERY LONG BASE INTERFEROMETRY

Base, Lewis
USE   LEWIS BASE

Base Management Systems, Data
USE   DATA BASE MANAGEMENT SYSTEMS

BASE PRESSURE

Base Propellants, Double
USE   DOUBLE BASE PROPELLANTS

Base Rocket Propellants, Double
USE   DOUBLE BASE ROCKET PROPELLANTS

Baseband Compression, Speech
USE   SPEECH BASEBAND COMPRESSION

Based Control, Ground
USE   GROUND BASED CONTROL

Based Energy, Hydrogen-
USE   HYDROGEN-BASED ENERGY

Based Radar, Space
USE   SPACE BASED RADAR

Based), Space Surveillance (Ground
USE   SPACE SURVEILLANCE (GROUND BASED)

BASEMENTS

BASES

Bases, Aircraft
USE   MILITARY AIR FACILITIES

BASES (CHEMICAL)

Bases, Data
USE   DATA BASES

Bases (Foundations)
USE   FOUNDATIONS

Bases, Launching
USE   LAUNCHING BASES

Bases, Lunar
USE   LUNAR BASES

Bases, Numerical Data
USE   NUMERICAL DATA BASES

Bases, Planetary
USE   PLANETARY BASES

Bases, Schiff
USE   IMINES

Bases, Space
USE   SPACE BASES

BASIC (PROGRAMMING LANGUAGE)

Basin (Africa), Kalihari
USE   KALIHARI BASIN (AFRICA)

Basin (AK), Chena River
USE   CHENA RIVER BASIN (AK)

Basin (CA), Feather River
USE   FEATHER RIVER BASIN (CA)

Basin (ID-OR-WA), Columbia River
USE   COLUMBIA RIVER BASIN (ID-OR-WA)

Basin (IL-IN-OH), Wabash River
USE   WABASH RIVER BASIN (IL-IN-OH)

Basin (LA), Atchafalaya River
USE   ATCHAFALAYA RIVER BASIN (LA)

Basin (MD-NY-PA), Susquehanna River
USE   SUSQUEHANNA RIVER BASIN (MD-NY-PA)

Basin (North America), Williston
USE   WILLISTON BASIN (NORTH AMERICA)

Basin (NY-VT), Lake Champlain
USE   LAKE CHAMPLAIN BASIN (NY-VT)

Basin (US), Delaware River
USE   DELAWARE RIVER BASIN (US)

Basin (US), Great
USE   GREAT BASIN (US)

Basin (US), Missouri River
USE   MISSOURI RIVER BASIN (US)

Basins
USE   STRUCTURAL BASINS

Basins, Closed
USE   STRUCTURAL BASINS

BASINS (CONTAINERS)

Basins, River
USE   RIVER BASINS

Basins, Structural
USE   STRUCTURAL BASINS

BASKETS

BASTNASITE

BATCH PROCESSING

BATHING

BATHOLITHS

BATHS

Baths, Salt
USE   SALT BATHS

BATHYMETERS

Bathymetry
USE   BATHYMETERS

BATHYTHERMOGRAPHS

BATS

Batteries
USE   ELECTRIC BATTERIES

Batteries, Alkaline
USE   ALKALINE BATTERIES

Batteries, Atomic
USE   RADIOISOTOPE BATTERIES

Batteries, Cadmium Nickel
USE   NICKEL CADMIUM BATTERIES

Batteries, Cadmium Silver
USE   SILVER CADMIUM BATTERIES

Batteries, Electric
USE   ELECTRIC BATTERIES

Batteries, Lead Acid
USE   LEAD ACID BATTERIES

Batteries, Lithium Sulfur
USE   LITHIUM SULFUR BATTERIES

Batteries, Metal Air
USE   METAL AIR BATTERIES

Batteries, Nickel Cadmium
USE   NICKEL CADMIUM BATTERIES

Batteries, Nickel Hydrogen
USE   NICKEL HYDROGEN BATTERIES

Batteries, Nickel Iron
USE   NICKEL IRON BATTERIES

Batteries, Nickel Zinc
USE   NICKEL ZINC BATTERIES

Batteries, Primary
USE   PRIMARY BATTERIES

Batteries, Radioisotope
USE   RADIOISOTOPE BATTERIES

Batteries, Secondary
USE   STORAGE BATTERIES

Batteries, Silver Cadmium
USE   SILVER CADMIUM BATTERIES

Batteries, Silver Hydrogen
USE   SILVER HYDROGEN BATTERIES

Batteries, Silver Oxide Zinc
USE   SILVER ZINC BATTERIES

Batteries, Silver Zinc
USE   SILVER ZINC BATTERIES

Batteries, Sodium Sulfur
USE   SODIUM SULFUR BATTERIES

Batteries, Storage
USE   STORAGE BATTERIES

Batteries, Thermal
USE   THERMAL BATTERIES

Batteries, Zinc Nickel
USE   NICKEL ZINC BATTERIES

Batteries, Zinc Silver
USE   SILVER ZINC BATTERIES

Batteries, Zinc Silver Oxide
USE   SILVER ZINC BATTERIES

Batteries, Zinc-Bromide
USE   ZINC-BROMIDE BATTERIES

Batteries, Zinc-Chlorine
USE   ZINC-CHLORINE BATTERIES

Batteries, Zinc-Oxygen
USE   ZINC-OXYGEN BATTERIES

BATTERY CHARGERS

Battery Separators
USE   SEPARATORS

BAUSCHINGER EFFECT

BAUXITE

Bay (CA), Monterey
USE   MONTEREY BAY (CA)

Bay (CA), San Francisco
USE   SAN FRANCISCO BAY (CA)

Bay (CA), San Pablo
USE   SAN PABLO BAY (CA)

BAY ICE

Bay (MI), Saginaw
USE   SAGINAW BAY (MI)

Bay (US), Chesapeake
USE   CHESAPEAKE BAY (US)

Bay (US), Delaware
USE   DELAWARE BAY (US)

BAYARD-ALPERT IONIZATION GAGES

BAYES THEOREM

Bayesian Statistics
USE   BAYES THEOREM

BAYOUS

BAYS

BAYS (STRUCTURAL UNITS)

BAYS (TOPOGRAPHIC FEATURES)

BBGKY HIERARCHY

BCAS
USE   BEACON COLLISION AVOIDANCE SYSTEM

BCC Lattices
USE   BODY CENTERED CUBIC LATTICES

BCH CODES

BCS THEORY

Be
USE   BERYLLIUM

BE A
USE   BEACON EXPLORER A

BE B
USE   EXPLORER 22 SATELLITE

BE C
USE   EXPLORER 27 SATELLITE

BE-3 ENGINE

BEACHES

BEACON COLLISION AVOIDANCE SYSTEM

BEACON EXPLORER A

Beacon Explorer B
USE   EXPLORER 22 SATELLITE

Beacon Explorer C
USE   EXPLORER 27 SATELLITE

Beacon Ionospheric Sounder, Orbiting Radio
USE   ORBIS

Beacon, Polar Ionosphere
USE   BEACON SATELLITES

BEACON SATELLITES

Beacon System, Discrete Address
USE   DISCRETE ADDRESS BEACON SYSTEM

BEACONS

Beacons, Airport
USE   AIRPORT BEACONS

Beacons, RACON
USE   RADAR BEACONS

Beacons, Radar
USE   RADAR BEACONS

Beacons, Radio
USE   RADIO BEACONS

BEADS

BEAGLE AIRCRAFT

BEAM CURRENTS

Beam Defocusing, Laser
USE   THERMAL BLOOMING

Beam Epitaxy, Molecular
USE   MOLECULAR BEAM EPITAXY

BEAM INJECTION

BEAM INTERACTIONS

Beam Interval Scanners, Multiple
USE   MULTIPLE BEAM INTERVAL SCANNERS

Beam Landing System, Microwave Scanning
USE   MICROWAVE SCANNING BEAM LANDING
        SYSTEM

BEAM LEADS

BEAM NEUTRALIZATION

BEAM PLASMA AMPLIFIERS

Beam Reactors, High Flux
USE   HIGH FLUX BEAM REACTORS

BEAM RIDER GUIDANCE

BEAM SPLITTERS

BEAM SWITCHING

Beam Vidicons, Return
USE   RETURN BEAM VIDICONS

BEAM WAVEGUIDES

Beam Welding, Electron
USE   ELECTRON BEAM WELDING

BEAMS

Beams, Atomic
USE   ATOMIC BEAMS

Beams, Box
USE   BOX BEAMS

Beams, Cantilever
USE   CANTILEVER BEAMS

Beams, Curved
USE   CURVED BEAMS

Beams, Electron
USE   ELECTRON BEAMS

Beams, Gamma Ray
USE   GAMMA RAY BEAMS

Beams, I
USE   I BEAMS

Beams, Ion
USE   ION BEAMS

Beams, Light
USE   LIGHT BEAMS

Beams, Micro
USE   MICROBEAMS

Beams, Molecular
USE   MOLECULAR BEAMS

Beams, Neutral
USE   NEUTRAL BEAMS

Beams, Neutrino
USE   NEUTRINO BEAMS

Beams, Neutron
USE   NEUTRON BEAMS

Beams, Particle
USE   PARTICLE BEAMS

Beams, Phonon
USE   PHONON BEAMS

Beams, Photon
USE   PHOTON BEAMS

Beams, Pion
USE   PION BEAMS

Beams, Proton
USE   PROTON BEAMS

Beams, Radar
USE   RADAR BEAMS

BEAMS (RADIATION)

Beams, Rectangular
USE   RECTANGULAR BEAMS

Beams, Relativistic Electron
USE   RELATIVISTIC ELECTRON BEAMS

Beams, Structural
USE   BEAMS (SUPPORTS)

BEAMS (SUPPORTS)

Beams, Timoshenko
USE   TIMOSHENKO BEAMS

Beamshaping
USE   COLLIMATION

BEARING

BEARING ALLOYS

BEARING (DIRECTION)

BEARINGLESS ROTORS

BEARINGS

Bearings, Air
USE   GAS BEARINGS

Bearings, Antifriction
USE   ANTIFRICTION BEARINGS

Bearings, Ball
USE   BALL BEARINGS

Bearings, Foil
USE   FOIL BEARINGS

Bearings, Gas
USE   GAS BEARINGS

Bearings, Gas Lubricated
USE   GAS BEARINGS

Bearings, Journal
USE   JOURNAL BEARINGS

Bearings, Liquid
USE   LIQUID BEARINGS

Bearings, Magnetic
USE   MAGNETIC BEARINGS

Bearings, Needle
USE   NEEDLE BEARINGS

Bearings, Roller
USE   ROLLER BEARINGS

Bearings, Thrust
USE   THRUST BEARINGS

BEARS

Beat
USE   SYNCHRONISM

BEAT FREQUENCIES

BEAUFORT SEA (NORTH AMERICA)

Beaver Aircraft, DHC
USE   DHC 2 AIRCRAFT

Bed Processors, Fluidized
USE   FLUIDIZED BED PROCESSORS

Bed Reactors, Pebble
USE   PEBBLE BED REACTORS

BED REST

BEDDING EQUIPMENT

BEDIASITES

BEDROCK

BEDS

BEDS (GEOLOGY)

Beds, Lake
USE   BEDS (GEOLOGY)

BEDS (PROCESS ENGINEERING)

Beds, Salt
USE   SALT BEDS

Beds, Test
USE   TEST EQUIPMENT

Bedstead Aircraft, Flying
USE   FLYING PLATFORMS

Beech Aircraft
USE   BEECHCRAFT AIRCRAFT

Beech C-33 Aircraft
USE   C-33 AIRCRAFT

Beech S-35 Aircraft
USE   C-35 AIRCRAFT

BEECH 99 AIRCRAFT

BEECHCRAFT AIRCRAFT

BEECHCRAFT 18 AIRCRAFT

BEER LAW

BEES

BEETLES

Beets, Sugar
USE   SUGAR BEETS

BEHAVIOR

Behavior, Group
USE   GROUP DYNAMICS

Behavior, Human
USE   HUMAN BEHAVIOR

Behavioral Lab Measur System, Integ Med And
USE   IMBLMS

Beings, Human
USE   HUMAN BEINGS

Belfast Aircraft
USE   SC-5 AIRCRAFT

Belfast C MK-1 Aircraft, Short
USE   SC-5 AIRCRAFT

Belgian Congo
USE   ZAIRE

BELGIUM

BELIZE

BELL AIRCRAFT

BELL 214A HELICOPTER

BELLMAN THEORY

BELLOWS

BELLS

Belt, Inner Radiation
USE   INNER RADIATION BELT

Belt, Outer Radiation
USE   OUTER RADIATION BELT

Belt, Terrestrial Dust
USE   TERRESTRIAL DUST BELT

Beltrami Equation, Stokes-
USE   STOKES-BELTRAMI EQUATION

BELTRAMI FLOW

BELTS

Belts, Artificial Radiation
USE   ARTIFICIAL RADIATION BELTS

Belts, Asteroid
USE   ASTEROID BELTS

Belts, Proton
USE   PROTON BELTS

Belts, Radiation
USE   RADIATION BELTS

Belts, Rouse
USE   ROUSE BELTS

Belts, Seat
USE   SEAT BELTS

Belts, Van Allen Radiation
USE   RADIATION BELTS

BENARD CELLS

Benard Convection, Rayleigh-
USE   RAYLEIGH-BENARD CONVECTION

Benches
USE   SEATS

BEND TESTS

BENDING

Bending), Brakes (Forming OR
USE   BRAKES (FORMING OR BENDING)

BENDING DIAGRAMS

Bending, Elastic
USE   ELASTIC BENDING

BENDING FATIGUE

BENDING MOMENTS

BENDING THEORY

BENDING VIBRATION

Bends (Physiology)
USE   DECOMPRESSION SICKNESS

Bends, U
USE   U BENDS

BENEFICIATION

BENIN

BENTONITE

BENZENE

BENZENE POISONING

Benzenes, Chloro
USE   CHLOROBENZENES

Benzenes, Nitro
USE   NITROBENZENES

BENZILIC ACID

BENZOIC ACID

BERENICE ROCKET VEHICLE

BERGMAN OPERATOR

BERING SEA

BERKELIUM

BERMUDA

Bernoulli Equation
USE   BERNOULLI THEOREM

BERNOULLI THEOREM

BERNSTEIN ENERGY PRINCIPLE

BERYL

BERYLLIUM

BERYLLIUM ALLOYS

BERYLLIUM BOROHYDRIDES

BERYLLIUM CHLORIDES

BERYLLIUM COMPOUNDS

BERYLLIUM FLUORIDES

BERYLLIUM HYDRIDES

BERYLLIUM ISOTOPES

BERYLLIUM NITRIDES

BERYLLIUM OXIDES

BERYLLIUM POISONING

BERYLLIUM 7

BERYLLIUM 9

BERYLLIUM 10

BESS (SATELLITE)

BESSEL FUNCTIONS

Bessel Transformations, Fourier-
USE   FOURIER-BESSEL TRANSFORMATIONS

BESSEL-BREDICHIN THEORY

BETA FACTOR

Beta Interactions
USE   WEAK INTERACTIONS (FIELD THEORY)

Beta Line, H
　　USE　H BETA LINE

BETA PARTICLES

Beta Radiation, Lyman
　　USE　LYMAN BETA RADIATION

BETAINES

BETATRONS

BETHE-HEITLER FORMULA

BETHE-SALPETER EQUATION

Between Failures, Mean Time
　　USE　MTBF

BEVATRON

BEVERAGES

BHUTAN

Bi
　　USE　BISMUTH

Bibs
　　USE　BIBLIOGRAPHIES

BIAS

Bias, Response
　　USE　RESPONSE BIAS

BIBLIOGRAPHIES

Bicarbonates
　　USE　CARBONATES

BICRYSTALS

BICYCLE

Biesbroeck Star, Van
　　USE　VAN BIESBROECK STAR

BIFURCATION (BIOLOGY)

Bifurcation (Mathematics)
　　USE　BRANCHING (MATHEMATICS)

BIG BANG COSMOLOGY

BIG SHOT PROJECT

BIGHORN MOUNTAINS (MT-WY)

Bights
　　USE　BAYS (TOPOGRAPHIC FEATURES)

BIHARMONIC EQUATIONS

BILLETS

BIMETALS

BIMETRIC THEORIES

Binaries, X Ray
　　USE　X RAY BINARIES

BINARY ALLOYS

BINARY CODES

Binary Converters, Decimal To
　　USE　DECIMAL TO BINARY CONVERTERS

BINARY DATA

BINARY DIGITS

BINARY FLUIDS

BINARY INTEGRATION

BINARY MIXTURES

BINARY STARS

Binary Stars, Eclipsing
　　USE　ECLIPSING BINARY STARS

Binary Summators
　　USE　ADDING CIRCUITS

Binary Systems (Digital)
　　USE　DIGITAL SYSTEMS

BINARY SYSTEMS (MATERIALS)

BINARY TO DECIMAL CONVERTERS

BINAURAL HEARING

Binders (Adhesives)
　　USE　ADHESIVES

BINDERS (MATERIALS)

Binders, Propellant
　　USE　PROPELLANT BINDERS

Binders, Solid Rocket
　　USE　SOLID ROCKET BINDERS

BINDING

Binding Energy, Nuclear
　　USE　NUCLEAR BINDING ENERGY

BINOCULAR VISION

BINOCULARS

BINOMIAL COEFFICIENTS

BINOMIAL THEOREM

BINOMIALS

BIOACOUSTICS

BIOASSAY

BIOASTRONAUTICAL ORBITAL SPACE SYSTEM

BIOASTRONAUTICS

BIOCHEMICAL FUEL CELLS

BIOCHEMICAL OXYGEN DEMAND

BIOCHEMISTRY

Bioclimatology
　　USE　BIOMETEOROLOGY

BIOCOMPATIBILITY

BIOCONTROL SYSTEMS

BIOCONVERSION

BIODEGRADABILITY

BIODEGRADATION

BIODYNAMICS

BIOELECTRIC POTENTIAL

BIOELECTRICITY

BIOENGINEERING

BIOFEEDBACK

BIOFLAVONOIDS

Biogenesis
　　USE　BIOLOGICAL EVOLUTION

BIOGENY

BIOGEOCHEMISTRY

BIOGRAPHY

BIOINSTRUMENTATION

Biological Activity
　　USE　ACTIVITY (BIOLOGY)

Biological Analysis
　　USE　BIOASSAY

Biological), Body Temperature (Non-
　　USE　TEMPERATURE

Biological Cells
　　USE　CELLS (BIOLOGY)

Biological), Cellular Materials (Non
　　USE　FOAMS

Biological Clocks
　　USE　RHYTHM (BIOLOGY)

Biological Effectiveness (RBE), Relative
　　USE　RELATIVE BIOLOGICAL EFFECTIVENESS
　　　　　(RBE)

BIOLOGICAL EFFECTS

BIOLOGICAL EVOLUTION

Biological Models
　　USE　BIONICS

BIOLOGICAL MODELS (MATHEMATICS)

Biological Rhythm
　　USE　RHYTHM (BIOLOGY)

Biological), Skin Temperature (Non-
　　USE　SKIN TEMPERATURE (NON-BIOLOGICAL)

BIOLOGY

(Biology), Activation
　　USE　ACTIVATION (BIOLOGY)

(Biology), Activity
　　USE　ACTIVITY (BIOLOGY)

(Biology), Activity Cycles
　　USE　ACTIVITY CYCLES (BIOLOGY)

Biology, Aero
　　USE　AEROBIOLOGY

(Biology), Aging
　　USE　AGING (BIOLOGY)

(Biology), Bifurcation
　　USE　BIFURCATION (BIOLOGY)

(Biology), Body Composition
　　USE　BODY COMPOSITION (BIOLOGY)

(Biology), Body Measurement
　　USE　BODY MEASUREMENT (BIOLOGY)

(Biology), Body Size
　　USE　BODY SIZE (BIOLOGY)

(Biology), Body Volume
　　USE　BODY VOLUME (BIOLOGY)

(Biology), Cells
　　USE　CELLS (BIOLOGY)

(Biology), Complement
　　USE　COMPLEMENT (BIOLOGY)

(Biology), Desynchronization
　　USE　DESYNCHRONIZATION (BIOLOGY)

(Biology), Differentiation
　　USE　DIFFERENTIATION (BIOLOGY)

Biology, Exo
　　USE　EXOBIOLOGY

(Biology), Fatigue
　　USE　FATIGUE (BIOLOGY)

(Biology), Flight Stress
　　USE　FLIGHT STRESS (BIOLOGY)

(Biology), Hybrids
USE GENETIC ENGINEERING

(Biology), Implanted Electrodes
USE IMPLANTED ELECTRODES (BIOLOGY)

(Biology), Ingestion
USE INGESTION (BIOLOGY)

(Biology), Life
USE LIFE SCIENCES

Biology, Marine
USE MARINE BIOLOGY

Biology, Micro
USE MICROBIOLOGY

Biology, Molecular
USE MOLECULAR BIOLOGY

(Biology), Motor Systems
USE EFFERENT NERVOUS SYSTEMS

(Biology), Periodicity
USE RHYTHM (BIOLOGY)

Biology, Proto
USE PROTOBIOLOGY

Biology, Radio
USE RADIOBIOLOGY

(Biology), Reproduction
USE REPRODUCTION (BIOLOGY)

(Biology), Rhythm
USE RHYTHM (BIOLOGY)

(Biology), Skin Temperature
USE SKIN TEMPERATURE (BIOLOGY)

Biology, Space
USE EXOBIOLOGY

(Biology), Stress
USE STRESS (BIOLOGY)

(Biology), Tissues
USE TISSUES (BIOLOGY)

BIOLUMINESCENCE

BIOMAGNETISM

BIOMASS

BIOMASS ENERGY PRODUCTION

Biomechanics
USE BIODYNAMICS

BIOMEDICAL DATA

Biomedical Experiment Scientific Satellite
USE BESS (SATELLITE)

BIOMETEOROLOGY

BIOMETRICS

BIONICS

BIOPAKS

BIOPHYSICS

BIOPROCESSING

BIOREACTORS

Bioregeneration
USE REGENERATION (PHYSIOLOGY)

Bioregenerative Life Support Systems
USE CLOSED ECOLOGICAL SYSTEMS

BIOS PROJECT

BIOSATELLITE 1

BIOSATELLITE 2

BIOSATELLITE 3

BIOSATELLITES

Biosensors
USE BIOINSTRUMENTATION

Biosimulation
USE BIONICS

BIOSPHERE

Biosphere Program, International Geosphere-
USE INTERNATIONAL GEOSPHERE-BIOSPHERE
PROGRAM

BIOSYNTHESIS

BIOT METHOD

BIOT NUMBER

BIOTECHNOLOGY

BIOTELEMETRY

BIOTIN

BIOTITE

Biphenyls, Polybrominated
USE POLYBROMINATED BIPHENYLS

Biphenyls, Polychlorinated
USE POLYCHLORINATED BIPHENYLS

BIPLANES

BIPOLAR TRANSISTORS

BIPOLARITY

Bipropellants
USE LIQUID ROCKET PROPELLANTS

Bird Satellites, Early
USE EARLY BIRD SATELLITES

BIRD-AIRCRAFT COLLISIONS

BIRDS

BIREFRINGENCE

BIREFRINGENT COATINGS

BIREFRINGENT FILTERS

BIRTH

BISMUTH

BISMUTH ALLOYS

BISMUTH COMPOUNDS

BISMUTH ISOTOPES

BISMUTH OXIDES

BISMUTH SULFIDES

BISMUTH TELLURIDES

Bismuth 205
USE BISMUTH ISOTOPES

BISPHENOLS

Bistability, Optical
USE OPTICAL BISTABILITY

Bistable Amplifiers
USE FLIP-FLOPS

BISTABLE CIRCUITS

Bistatic Radar
USE MULTISTATIC RADAR

BISTATIC REFLECTIVITY

BIT ERROR RATE

BIT SYNCHRONIZATION

BITERNARY CODE

BITS

Bits, Drill
USE DRILL BITS

BITUMENS

BIVARIATE ANALYSIS

Bk
USE BERKELIUM

BL LACERTAE OBJECTS

BLACK AND WHITE PHOTOGRAPHY

Black Arrow Launch Vehicle
USE BLACK KNIGHT ROCKET VEHICLE

BLACK BODY RADIATION

BLACK BRANT SOUNDING ROCKETS

BLACK BRANT 1 SOUNDING ROCKET

BLACK BRANT 2 SOUNDING ROCKET

BLACK BRANT 3 SOUNDING ROCKET

BLACK BRANT 4 SOUNDING ROCKET

BLACK BRANT 5 SOUNDING ROCKET

Black Hawk Assault Helicopter
USE H-60 HELICOPTER

BLACK HILLS (SD-WY)

BLACK HOLES (ASTRONOMY)

BLACK KNIGHT ROCKET VEHICLE

Black, Platinum
USE PLATINUM BLACK

BLACK SEA

Blackburn B-103 Aircraft
USE BUCCANEER AIRCRAFT

BLACKOUT

Blackout, Ionospheric
USE BLACKOUT (PROPAGATION)

BLACKOUT (PHYSIOLOGY)

Blackout, Polar Radio
USE POLAR RADIO BLACKOUT

BLACKOUT PREVENTION

BLACKOUT (PROPAGATION)

BLADDER

Bladders, Expulsion
USE EXPULSION BLADDERS

Bladders (Mechanics)
USE DIAPHRAGMS (MECHANICS)

BLADE SLAP NOISE

BLADE TIPS

BLADES

Blades, Compressor
   USE   COMPRESSOR BLADES

BLADES (CUTTERS)

Blades, Fan
   USE   FAN BLADES

Blades, Hinged Rotor
   USE   ROTARY WINGS
           HINGES

Blades, Impeller
   USE   ROTOR BLADES (TURBOMACHINERY)

Blades, Propeller
   USE   PROPELLER BLADES

Blades, Razor
   USE   RAZOR BLADES

Blades, Rotor
   USE   ROTOR BLADES

Blades, Stator
   USE   STATOR BLADES

Blades, Turbine
   USE   TURBINE BLADES

Blades, Turbomachine
   USE   TURBOMACHINE BLADES

Blades (Turbomachinery), Rotor
   USE   ROTOR BLADES (TURBOMACHINERY)

BLANKETS

BLANKETS (FISSION REACTORS)

BLANKETS (FUSION REACTORS)

Blankets, Solar
   USE   SOLAR BLANKETS

BLANKING

BLANKING (CUTTING)

BLANKS

BLASIUS EQUATION

BLASIUS FLOW

BLAST DEFLECTORS

Blast Effects, Jet
   USE   JET BLAST EFFECTS

BLAST LOADS

Blast Nuclear Radiation, Post-
   USE   POST-BLAST NUCLEAR RADIATION

Blastoff
   USE   ROCKET LAUNCHING

BLASTS

Blasts, Air
   USE   AERIAL EXPLOSIONS

Blattidae
   USE   COCKROACHES

BLEACHING

Bleed-Off
   USE   PRESSURE REDUCTION

BLEEDING

Blends
   USE   MIXTURES

BLIGHT

BLIND LANDING

BLINDNESS

Blindness, Flash
   USE   FLASH BLINDNESS

BLINDS

BLINKING

BLISTERS

BLOCH BAND

BLOCK DIAGRAMS

BLOCK ISLAND SOUND (RI)

Block 3 Television System, Ranger
   USE   RANGER BLOCK 3 TELEVISION SYSTEM

BLOCKING

Blocking Agents, Cholinergic
   USE   ANTICHOLINERGICS

BLOCKS

BLOEDITE

BLOOD

Blood Cells, Red
   USE   ERYTHROCYTES

Blood Cells, White
   USE   WHITE BLOOD CELLS

BLOOD CIRCULATION

BLOOD COAGULATION

BLOOD FLOW

BLOOD GROUPS

BLOOD PLASMA

BLOOD PRESSURE

BLOOD PUMPS

BLOOD VESSELS

BLOOD VOLUME

BLOOD-BRAIN BARRIER

Bloom, Algal
   USE   ALGAE

Bloom, Plankton
   USE   PLANKTON

Blooming, Thermal
   USE   THERMAL BLOOMING

BLOWDOWN WIND TUNNELS

BLOWERS

BLOWING

Blowing, Spanwise
   USE   SPANWISE BLOWING

Blowing, Under Surface
   USE   UNDER SURFACE BLOWING

Blowing, Upper Surface
   USE   UPPER SURFACE BLOWING

Blown Flaps
   USE   EXTERNALLY BLOWN FLAPS

Blown Flaps, Externally
   USE   EXTERNALLY BLOWN FLAPS

Blown Flaps, Upper Surface
   USE   UPPER SURFACE BLOWN FLAPS

BLOWOUTS

BLUE GOOSE MISSILE

BLUE GREEN ALGAE

Blue, Methylene
   USE   METHYLENE BLUE

BLUE SCOUT ROCKET VEHICLE

BLUE STARS

BLUE STEEL MISSILE

BLUE STREAK LAUNCH VEHICLE

BLUE STREAK MISSILE

BLUEPRINTS

BLUFF BODIES

Bluffs (Landforms)
   USE   CLIFFS

BLUNT BODIES

BLUNT LEADING EDGES

BLUNT TRAILING EDGES

BLURRING

BMC
   USE   BONE MINERAL CONTENT

BMEWS
   USE   BALLISTIC MISSILE EARLY WARNING
           SYSTEM

BO-105 HELICOPTER

Boards, Circuit
   USE   CIRCUIT BOARDS

Boards, Control
   USE   CONTROL BOARDS

BOARDS (PAPER)

BOATS

Boats, Hydrofoil
   USE   HYDROFOIL CRAFT

BOATTAILS

BOD
   USE   BIOCHEMICAL OXYGEN DEMAND

Bodewadt Flow, Karman-
   USE   KARMAN-BODEWADT FLOW

BODIES

Bodies, After
   USE   AFTERBODIES

Bodies, Anti
   USE   ANTIBODIES

Bodies, Axisymmetric
   USE   AXISYMMETRIC BODIES

Bodies, Bluff
   USE   BLUFF BODIES

Bodies, Blunt
   USE   BLUNT BODIES

Bodies, Celestial
   USE   CELESTIAL BODIES

Bodies, Center
   USE   CENTERBODIES

Bodies, Conical
   USE   CONICAL BODIES

Bodies, Cylindrical
   USE   CYLINDRICAL BODIES

Bodies, Ducted
USE   DUCTED BODIES

Bodies, Elastic
USE   ELASTIC BODIES

Bodies, Finned
USE   FINNED BODIES

Bodies, Flared
USE   FLARED BODIES

Bodies, Flexible
USE   FLEXIBLE BODIES

Bodies, Fore
USE   FOREBODIES

Bodies, Foreign
USE   FOREIGN BODIES

Bodies, Hemisphere Cylinder
USE   HEMISPHERE CYLINDER BODIES

Bodies, Inelastic
USE   RIGID STRUCTURES

Bodies, Lenticular
USE   LENTICULAR BODIES

Bodies, Lifting
USE   LIFTING BODIES

Bodies, Maneuverable Reentry
USE   MANEUVERABLE REENTRY BODIES

Bodies, Maxwell
USE   MAXWELL BODIES

Bodies, Missile
USE   MISSILE BODIES

BODIES OF REVOLUTION

Bodies, Parabolic
USE   PARABOLIC BODIES

Bodies, Pyramidal
USE   PYRAMIDAL BODIES

Bodies, Reentry
USE   REENTRY VEHICLES

Bodies, Rigid
USE   RIGID STRUCTURES

Bodies, Rotating
USE   ROTATING BODIES

Bodies, Shrouded
USE   SHROUDS

Bodies, Slender
USE   SLENDER BODIES

Bodies, Streamlined
USE   STREAMLINED BODIES

Bodies, Submerged
USE   SUBMERGED BODIES

Bodies, Symmetrical
USE   SYMMETRICAL BODIES

Bodies, Thin
USE   THIN BODIES

Bodies, Three Dimensional
USE   THREE DIMENSIONAL BODIES

Bodies, Towed
USE   TOWED BODIES

Bodies, Two Dimensional
USE   TWO DIMENSIONAL BODIES

Body, Carotid Sinus
USE   CAROTID SINUS BODY

BODY CENTERED CUBIC LATTICES

BODY COMPOSITION (BIOLOGY)

BODY FLUIDS

Body, Human
USE   HUMAN BODY

Body Interactions, Rotor
USE   ROTOR BODY INTERACTIONS

BODY KINEMATICS

Body, M-2 Lifting
USE   M-2 LIFTING BODY

Body, M-2F2 Lifting
USE   M-2F2 LIFTING BODY

Body, M-2F3 Lifting
USE   M-2F3 LIFTING BODY

Body, Mark 1 Reentry
USE   MARK 1 REENTRY BODY

Body, Mark 2 Reentry
USE   MARK 2 REENTRY BODY

Body, Mark 3 Reentry
USE   MARK 3 REENTRY BODY

Body, Mark 4 Reentry
USE   MARK 4 REENTRY BODY

Body, Mark 5 Reentry
USE   MARK 5 REENTRY BODY

Body, Mark 6 Reentry
USE   MARK 6 REENTRY BODY

Body, Mark 11 Reentry
USE   MARK 11 REENTRY BODY

Body, Mark 12 Reentry
USE   MARK 12 REENTRY BODY

Body, Mark 17 Reentry
USE   MARK 17 REENTRY BODY

BODY MEASUREMENT (BIOLOGY)

Body Negative Pressure, Lower
USE   LOWER BODY NEGATIVE PRESSURE

Body Orbits, Two
USE   TWO BODY PROBLEM

Body Problem, Four
USE   FOUR BODY PROBLEM

Body Problem, Many
USE   MANY BODY PROBLEM

Body Problem, Three
USE   THREE BODY PROBLEM

Body Problem, Two
USE   TWO BODY PROBLEM

Body Radiation, Black
USE   BLACK BODY RADIATION

BODY SIZE (BIOLOGY)

BODY SWAY TEST

BODY TEMPERATURE

Body Temperature (Non-Biological)
USE   TEMPERATURE

Body Temperature Regulation
USE   THERMOREGULATION

BODY VOLUME (BIOLOGY)

BODY WEIGHT

BODY-WING AND TAIL CONFIGURATIONS

BODY-WING CONFIGURATIONS

BOEING AIRCRAFT

Boeing Military Aircraft
USE   MILITARY AIRCRAFT

BOEING 707 AIRCRAFT

BOEING 720 AIRCRAFT

BOEING 727 AIRCRAFT

BOEING 733 AIRCRAFT

BOEING 737 AIRCRAFT

BOEING 747 AIRCRAFT

Boeing 747B Aircraft
USE   E-4A AIRCRAFT

BOEING 757 AIRCRAFT

BOEING 767 AIRCRAFT

BOEING 2707 AIRCRAFT

BOGOLIUBOV THEORY

Bogs
USE   MARSHLANDS

BOHR MAGNETON

BOHR THEORY

BOILER PLATE

Boiler Reactor, Los Alamos Water
USE   LOS ALAMOS WATER BOILER REACTOR

BOILERS

BOILING

Boiling, Film
USE   FILM BOILING

Boiling, Nucleate
USE   NUCLEATE BOILING

Boiling Water Reactor, Halden
USE   HALDEN BOILING WATER REACTOR

BOILING WATER REACTORS

Boiling Water Reactors, Experimental
USE   EXPERIMENTAL BOILING WATER
             REACTORS

Bokkeveld Meteorite, Cold
USE   COLD BOKKEVELD METEORITE

BOLIDES

BOLIVIA

BOLKOW AIRCRAFT

BOLL WEEVILS

BOLLWORMS

Bolograms
USE   BOLOMETERS

BOLOMETERS

BOLTS

Bolts, Rock
USE   ROCK BOLTS

Boltzmann Density Function, Maxwell-
USE   MAXWELL-BOLTZMANN DENSITY
             FUNCTION

BOLTZMANN DISTRIBUTION

Boltzmann Law, Stefan-
USE   STEFAN-BOLTZMANN LAW

BOLTZMANN TRANSPORT EQUATION

BOLTZMANN-VLASOV EQUATION

BOLZA PROBLEMS

BOMARC A MISSILE

BOMARC B MISSILE

BOMARC MISSILES

BOMB CALORIMETERS

BOMBARDMENT

Bombardment, Electron
USE   ELECTRON BOMBARDMENT

BOMBER AIRCRAFT

Bomber, Canberra
USE   B-57 AIRCRAFT

Bomber, Shackleton
USE   SHACKLETON BOMBER

BOMBING EQUIPMENT

BOMBS

Bombs, Atomic
USE   FISSION WEAPONS

Bombs, Hydrogen
USE   FUSION WEAPONS

BOMBS (ORDNANCE)

Bombs (Pressure Gages)
USE   PRESSURE GAGES

Bombs (Samplers)
USE   SAMPLERS

Bonanza Aircraft
USE   C-35 AIRCRAFT

BOND GRAPHS

Bond Testers, Fokker
USE   ADHESION TESTS

Bonded Propellants, Case
USE   CASE BONDED PROPELLANTS

BONDING

Bonding, Adhesive
USE   ADHESIVE BONDING

Bonding, Ceramic
USE   CERAMIC BONDING

Bonding, Diffusion
USE   DIFFUSION WELDING

Bonding, Electrostatic
USE   ELECTROSTATIC BONDING

Bonding, Inertia
USE   INERTIA BONDING

Bonding, Metal
USE   METAL BONDING

Bonding, Metal-Metal
USE   METAL-METAL BONDING

Bonding, Reaction
USE   REACTION BONDING

Bonding, Resin
USE   RESIN BONDING

BONDOC METEORITE

Bonds, Chemical
USE   CHEMICAL BONDS

Bonds, Covalent
USE   COVALENT BONDS

Bonds, Hydrogen
USE   HYDROGEN BONDS

Bonds, Molecular
USE   CHEMICAL BONDS

BONE DEMINERALIZATION

BONE MARROW

BONE MINERAL CONTENT

BONES

BONNE PROJECTION

Books, Hand
USE   HANDBOOKS

Books, Text
USE   TEXTBOOKS

BOOLEAN ALGEBRA

BOOLEAN FUNCTIONS

BOOM

BOOMS (EQUIPMENT)

Booms, Sonic
USE   SONIC BOOMS

Boost
USE   ACCELERATION (PHYSICS)

Boost Motors, Apogee
USE   APOGEE BOOST MOTORS

Boost Propulsion System, Post
USE   POST BOOST PROPULSION SYSTEM

BOOSTER RECOVERY

BOOSTER ROCKET ENGINES

Booster Rocket Engines, Nike
USE   NIKE BOOSTER ROCKET ENGINES

BOOSTER ROCKETS

BOOSTERS

Boosters, Air Breathing
USE   AIR BREATHING BOOSTERS

BOOSTERS (EXPLOSIVES)

Boosters, Rocket
USE   BOOSTER ROCKET ENGINES

Boosters, Shuttle
USE   SPACE SHUTTLE BOOSTERS

Boosters, Space Shuttle
USE   SPACE SHUTTLE BOOSTERS

BOOSTGLIDE VEHICLES

BOOTS (FOOTWEAR)

BORAL

Borane, Di
USE   DIBORANE

Borane, Hydrazine
USE   HYDRAZINE BORANE

BORANES

BORATES

Borates, Lithium
USE   LITHIUM BORATES

Borazon (Trademark)
USE   BORON NITRIDES

BORDERS

BORDONI PEAKS

Borealis Constellation, Corona
USE   CORONA BOREALIS CONSTELLATION

BOREDOM

BOREHOLES

BOREL SETS

Bores
USE   CAVITIES

Borescopes
USE   ENDOSCOPES

BORESIGHT ERROR

BORESIGHTS

BORIC ACIDS

BORIDES

Borides, Chromium
USE   CHROMIUM BORIDES

Borides, Titanium
USE   TITANIUM BORIDES

BORING MACHINES

BORN APPROXIMATION

BORN-INFELD THEORY

Born-Mayer Equation
USE   BORN APPROXIMATION

BORN-OPPENHEIMER APPROXIMATION

Borne Instruments, Balloon-
USE   BALLOON-BORNE INSTRUMENTS

Borne Instruments, Rocket-
USE   ROCKET-BORNE INSTRUMENTS

Borne Instruments, Satellite-
USE   SATELLITE-BORNE INSTRUMENTS

Borne Photography, Rocket-
USE   ROCKET-BORNE PHOTOGRAPHY

Borne Photography, Satellite-
USE   SATELLITE-BORNE PHOTOGRAPHY

Borne Radar, Satellite-
USE   SATELLITE-BORNE RADAR

BOROHYDRIDES

Borohydrides, Aluminum
USE   ALUMINUM BOROHYDRIDES

Borohydrides, Beryllium
USE   BERYLLIUM BOROHYDRIDES

BORON

BORON ALLOYS

BORON CARBIDES

BORON CHLORIDES

Boron Composites, Aluminum
USE   ALUMINUM BORON COMPOSITES

BORON COMPOUNDS

Boron Compounds, Organic
USE   ORGANIC BORON COMPOUNDS

BORON FIBERS

BORON FLUORIDES

BORON HYDRIDES

BORON ISOTOPES

BORON NITRIDES

BORON OXIDES

BORON PHOSPHIDES

BORON REINFORCED MATERIALS

Boron Trifluoride
  USE   BORON FLUORIDES

BORON 10

BORON-EPOXY COMPOUNDS

BOROSILICATE GLASS

BORSIC (TRADENAME)

BOSE GEOMETRY

Bose-Chaudhuri-Hocquenghem Codes
  USE   BCH CODES

Bose-Einstein Statistics
  USE   QUANTUM STATISTICS

BOSON FIELDS

BOSONS

BOTANY

(Botany), Brush
  USE   BRUSH (BOTANY)

(Botany), Cortexes
  USE   CORTEXES (BOTANY)

Botany, Geo
  USE   GEOBOTANY

(Botany), Plants
  USE   PLANTS (BOTANY)

(Botany), Rusts
  USE   RUST FUNGI

(Botany), Scrubs
  USE   BRUSH (BOTANY)

BOTSWANA

BOTTLES

Bottom, Ocean
  USE   OCEAN BOTTOM

Botulinum, Clostridium
  USE   CLOSTRIDIUM BOTULINUM

BOUGUER LAW

BOULES

BOUNDARIES

Boundaries, Fluid
  USE   FLUID BOUNDARIES

Boundaries, Free
  USE   FREE BOUNDARIES

Boundaries, Grain
  USE   GRAIN BOUNDARIES

Boundaries, Jet
  USE   JET BOUNDARIES

BOUNDARY ELEMENT METHOD

BOUNDARY INTEGRAL METHOD

Boundary Layer, Atmospheric
  USE   ATMOSPHERIC BOUNDARY LAYER

BOUNDARY LAYER COMBUSTION

Boundary Layer, Compressible
  USE   COMPRESSIBLE BOUNDARY LAYER

BOUNDARY LAYER CONTROL

Boundary Layer Control, Porous
  USE   POROUS BOUNDARY LAYER CONTROL

BOUNDARY LAYER EQUATIONS

BOUNDARY LAYER FLOW

Boundary Layer, Hypersonic
  USE   HYPERSONIC BOUNDARY LAYER

Boundary Layer, Incompressible
  USE   INCOMPRESSIBLE BOUNDARY LAYER

Boundary Layer, Laminar
  USE   LAMINAR BOUNDARY LAYER

Boundary Layer Noise
  USE   BOUNDARY LAYERS
         AERODYNAMIC NOISE

Boundary Layer, Planetary
  USE   PLANETARY BOUNDARY LAYER

BOUNDARY LAYER PLASMAS

BOUNDARY LAYER SEPARATION

Boundary Layer Separation, Laminar
  USE   LAMINAR BOUNDARY LAYER

BOUNDARY LAYER STABILITY

Boundary Layer, Thermal
  USE   THERMAL BOUNDARY LAYER

Boundary Layer, Three Dimensional
  USE   THREE DIMENSIONAL BOUNDARY LAYER

BOUNDARY LAYER TRANSITION

Boundary Layer, Turbulent
  USE   TURBULENT BOUNDARY LAYER

Boundary Layer, Two Dimensional
  USE   TWO DIMENSIONAL BOUNDARY LAYER

BOUNDARY LAYERS

Boundary Layers, Supersonic
  USE   SUPERSONIC BOUNDARY LAYERS

BOUNDARY LUBRICATION

BOUNDARY VALUE PROBLEMS

BOURDON TUBES

BOUSSINESQ APPROXIMATION

Bow Shock Waves
  USE   BOW WAVES
         SHOCK WAVES

BOW WAVES

BOWS

Bows, Rain
  USE   RAINBOWS

BOX BEAMS

BOXES

BOXES (CONTAINERS)

Boxes, Skinner
  USE   SKINNER BOXES

Br
  USE   BROMINE

BRACKETS

BRADYCARDIA

BRAGG ANGLE

BRAGG CURVE

BRAILLE

BRAIN

Brain Barrier, Blood-
  USE   BLOOD-BRAIN BARRIER

BRAIN CIRCULATION

BRAIN DAMAGE

BRAIN STEM

BRAKES

Brakes, Aerodynamic
  USE   AERODYNAMIC BRAKES

Brakes, Aircraft
  USE   AIRCRAFT BRAKES

BRAKES (FOR ARRESTING MOTION)

BRAKES (FORMING OR BENDING)

Brakes, Wheel
  USE   WHEEL BRAKES

BRAKING

Braking, Aero
  USE   AEROBRAKING

Branch Stars, Horizontal
  USE   HORIZONTAL BRANCH STARS

BRANCHING (MATHEMATICS)

BRANCHING (PHYSICS)

Brant Sounding Rockets, Black
  USE   BLACK BRANT SOUNDING ROCKETS

Brant 1 Sounding Rocket, Black
  USE   BLACK BRANT 1 SOUNDING ROCKET

Brant 2 Sounding Rocket, Black
  USE   BLACK BRANT 2 SOUNDING ROCKET

Brant 3 Sounding Rocket, Black
  USE   BLACK BRANT 3 SOUNDING ROCKET

Brant 4 Sounding Rocket, Black
  USE   BLACK BRANT 4 SOUNDING ROCKET

Brant 5 Sounding Rocket, Black
  USE   BLACK BRANT 5 SOUNDING ROCKET

BRASSES

BRAVAIS CRYSTALS

BRAYTON CYCLE

BRAZIL

BRAZILIAN SPACE PROGRAM

BRAZING

Brazing, Low Temperature
  USE   LOW TEMPERATURE BRAZING

Brazzaville
  USE   CONGO (BRAZZAVILLE)

(Brazzaville), Congo
  USE   CONGO (BRAZZAVILLE)

BREADBOARD MODELS

Breakaway
  USE   BOUNDARY LAYER SEPARATION

BREAKDOWN

Breakdown, Electrical
USE   ELECTRICAL FAULTS

Breakdown, Voltage
USE   ELECTRICAL FAULTS

Breakdown, Vortex
USE   VORTEX BREAKDOWN

Breakers, Circuit
USE   CIRCUIT BREAKERS

Breakers (Electric)
USE   CIRCUIT BREAKERS

BREAKING

Breaking, Symmetry
USE   BROKEN SYMMETRY

BREAKWATERS

BREATHING

BREATHING APPARATUS

Breathing Apparatus, Underwater
USE   UNDERWATER BREATHING APPARATUS

Breathing Boosters, Air
USE   AIR BREATHING BOOSTERS

Breathing Engines, Air
USE   AIR BREATHING ENGINES

Breathing, High Altitude
USE   HIGH ALTITUDE BREATHING

Breathing, Liquid
USE   LIQUID BREATHING

Breathing, Oxygen
USE   OXYGEN BREATHING

Breathing, Pressure
USE   PRESSURE BREATHING

Breathing, Re
USE   REBREATHING

Breathing Techniques, Emergency
USE   EMERGENCY BREATHING TECHNIQUES

BREATHING VIBRATION

BRECCIA

Bredichin Theory, Bessel-
USE   BESSEL-BREDICHIN THEORY

Breeder Reactor 1, Experimental
USE   EXPERIMENTAL BREEDER REACTOR 1

Breeder Reactor 2, Experimental
USE   EXPERIMENTAL BREEDER REACTOR 2

BREEDER REACTORS

Breeder Reactors, Light Water
USE   LIGHT WATER BREEDER REACTORS

Breeder Reactors, Liquid Metal Fast
USE   LIQUID METAL FAST BREEDER REACTORS

BREEDING (REPRODUCTION)

Breeze, Sea
USE   SEA BREEZE

BREGUET AIRCRAFT

BREGUET 940 AIRCRAFT

BREGUET 941 AIRCRAFT

BREGUET 1150 AIRCRAFT

BREMSSTRAHLUNG

Brever Reflex, Hering-
USE   HERING-BREVER REFLEX

BREWSTER ANGLE

BRICKS

Bridge Circuits, Wire
USE   WIRE BRIDGE CIRCUITS

BRIDGES

Bridges, Electric
USE   ELECTRIC BRIDGES

BRIDGES (LANDFORMS)

BRIDGES (STRUCTURES)

Bridges, Wheatstone
USE   WHEATSTONE BRIDGES

BRIDGMAN METHOD

Brigade Devices, Bucket
USE   BUCKET BRIGADE DEVICES

Brightening, Limb
USE   LIMB BRIGHTENING

BRIGHTNESS

BRIGHTNESS DISCRIMINATION

BRIGHTNESS DISTRIBUTION

Brightness, Sky
USE   SKY BRIGHTNESS

BRIGHTNESS TEMPERATURE

BRILLOUIN EFFECT

BRILLOUIN FLOW

Brillouin Method, Wentzel-Kramer-
USE   WENTZEL-KRAMER-BRILLOUIN METHOD

BRILLOUIN ZONES

BRILLOUIN-WIGNER EQUATION

BRINES

BRIQUETS

BRISTOL-SIDDELEY BS 53 ENGINE

BRISTOL-SIDDELEY OLYMPUS 593 ENGINE

BRISTOL-SIDDELEY VIPER ENGINE

Britain, Great
USE   UNITED KINGDOM

British Aircraft Corp Aircraft
USE   BAC AIRCRAFT

BRITISH COLUMBIA

British Guinea
USE   GUYANA

British Honduras
USE   BELIZE

BRITTLE MATERIALS

BRITTLENESS

BROADBAND

BROADBAND AMPLIFIERS

BROADCASTING

Broadcasting, Radio
USE   BROADCASTING

Broadening, Pressure
USE   PRESSURE BROADENING

Broglie Wavelengths, De
USE   DE BROGLIE WAVELENGTHS

BROKEN SYMMETRY

BROMATES

Bromide Batteries, Zinc-
USE   ZINC-BROMIDE BATTERIES

BROMIDES

Bromides, Ammonium
USE   AMMONIUM BROMIDES

Bromides, Cesium
USE   CESIUM BROMIDES

Bromides, Chromium
USE   CHROMIUM BROMIDES

Bromides, Di
USE   DIBROMIDES

Bromides, Hydro
USE   HYDROBROMIDES

Bromides, Magnesium
USE   MAGNESIUM BROMIDES

Bromides, Potassium
USE   POTASSIUM BROMIDES

Bromides, Silver
USE   SILVER BROMIDES

Bromides, Sodium
USE   SODIUM BROMIDES

Bromides, Strontium
USE   STRONTIUM BROMIDES

BROMINATION

BROMINE

BROMINE COMPOUNDS

BROMINE ISOTOPES

Bromine 82
USE   BROMINE ISOTOPES

Bromine 87
USE   BROMINE ISOTOPES

BRONCHI

BRONCHIAL TUBE

BRONZES

Brook Reactor, Plum
USE   PLUM BROOK REACTOR

BROTHS

BROWN WAVE EFFECT

BROWNIAN MOVEMENTS

Bruceton Test
USE   STATISTICAL TESTS

BRUCITE

BRUDERHEIM METEORITE

BRUNEI

Brunswick, New
USE   NEW BRUNSWICK

BRUNT-VAISALA FREQUENCY

BRUSH (BOTANY)

BRUSHES

BRUSHES (ELECTRICAL CONTACTS)

BRYOPHYTES

BS 53 Engine, Bristol-Siddeley
    USE   BRISTOL-SIDDELEY BS 53 ENGINE

BSX

BUBBLE CHAMBERS

BUBBLE MEMORY DEVICES

BUBBLE TECHNIQUE

Bubble Vehicles, Captured Air
    USE   CAPTURED AIR BUBBLE VEHICLES

BUBBLES

Bubbles, Plasma
    USE   PLASMA BUBBLES

BUCCANEER AIRCRAFT

BUCKET BRIGADE DEVICES

BUCKETS

Buckeye Aircraft
    USE   T-2 AIRCRAFT

BUCKLING

Buckling, Creep
    USE   CREEP BUCKLING

Buckling, Elastic
    USE   ELASTIC BUCKLING

Buckling, Euler
    USE   EULER BUCKLING

Buckling, Thermal
    USE   THERMAL BUCKLING

Budget, Atmospheric Heat
    USE   ATMOSPHERIC HEAT BUDGET

Budget Experiment, Earth Energy
    USE   LZEEBE SATELLITE

Budget Experiment, Earth Radiation
    USE   EARTH RADIATION BUDGET EXPERIMENT

Budget Experiment, Zonal Earth Energy
    USE   LZEEBE SATELLITE

Budget, Heat
    USE   HEAT BUDGET

BUDGETING

BUDGETS

Budgets, Energy
    USE   ENERGY BUDGETS

Budgets, Federal
    USE   FEDERAL BUDGETS

Buffalo Aircraft
    USE   DHC 5 AIRCRAFT

BUFFER STORAGE

BUFFERS

BUFFERS (CHEMISTRY)

BUFFETING

Building Materials
    USE   CONSTRUCTION MATERIALS

Building Structures
    USE   BUILDINGS

BUILDINGS

(Buildings), Space Cooling
    USE   SPACE COOLING (BUILDINGS)

(Buildings), Space Heating
    USE   SPACE HEATING (BUILDINGS)

BULBS

Bulbs, Light
    USE   LUMINAIRES

BULGARIA

BULGING

BULK ACOUSTIC WAVE DEVICES

BULK MODULUS

BULKHEADS

BULLPUP B MISSILE

BULLPUP MISSILES

BUMBLEBEE PROJECT

BUMPERS

BUMPY TORUSES

BUNA (TRADEMARK)

BUNCHING

Bunching, Electron
    USE   ELECTRON BUNCHING

BUNDLE DRAWING

Bundle, His
    USE   HIS BUNDLE

BUNDLES

BUNKERS (FUEL)

BUOYANCY

BUOYS

Buoys, Sono
    USE   SONOBUOYS

BUREAUS (ORGANIZATIONS)

BURETTES

BURGER EQUATION

BURKINA

BURMA

BURN-IN

BURNERS

Burners, Pre
    USE   PREBURNERS

Burning
    USE   COMBUSTION

Burning, After
    USE   AFTERBURNING

Burning, Erosive
    USE   EROSIVE BURNING

Burning, Hole
    USE   HOLE BURNING

Burning Process
    USE   COMBUSTION

BURNING RATE

BURNING TIME

BURNOUT

BURNS (INJURIES)

BURNTHROUGH (FAILURE)

Burnup, Nuclear Fuel
    USE   NUCLEAR FUEL BURNUP

BURST TESTS

BURSTS

Bursts, Cosmic Gamma Ray
    USE   GAMMA RAY BURSTS

Bursts, Gamma Ray
    USE   GAMMA RAY BURSTS

Bursts, Meteor
    USE   METEOROID SHOWERS

Bursts, Radio
    USE   RADIO BURSTS

Bursts, Solar Radio
    USE   SOLAR RADIO BURSTS

Bursts, Type 2
    USE   TYPE 2 BURSTS

Bursts, Type 3
    USE   TYPE 3 BURSTS

Bursts, Type 4
    USE   TYPE 4 BURSTS

Bursts, Type 5
    USE   TYPE 5 BURSTS

BURUNDI

BUS CONDUCTORS

Bus, Pioneer Venus 2 Transporter
    USE   PIONEER VENUS 2 TRANSPORTER BUS

Busemann Law, Newton-
    USE   NEWTON-BUSEMANN LAW

Buses, Space
    USE   FERRY SPACECRAFT

BUSHINGS

Business Management
    USE   INDUSTRIAL MANAGEMENT

Busses, Data
    USE   CHANNELS (DATA TRANSMISSION)

BUTADIENE

Butadiene, Poly
    USE   POLYBUTADIENE

Butane, Cyclo
    USE   CYCLOBUTANE

BUTANES

BUTENES

BUTT JOINTS

BUTTERFLY VALVES

BUTTES

BUTTONS

Butylene
    USE   BUTENES

Butylene Oxides
    USE   TETRAHYDROFURAN

Butyls, Tetra
    USE   TETRABUTYLS

BUTYRIC ACID

Buzz, Aerodynamic
  USE   FLUTTER

BY-PRODUCTS

BYPASS RATIO

BYPASSES

B1 Engine, X-258-
  USE   X-258-B1 ENGINE

C

C, Anik
  USE   ANIK 3

C, Atmosphere Explorer
  USE   EXPLORER 51 SATELLITE

C BAND

C, BE
  USE   EXPLORER 27 SATELLITE

C, Beacon Explorer
  USE   EXPLORER 27 SATELLITE

C, Comstar
  USE   COMSTAR C

C, Earth Resources Technology Satellite
  USE   LANDSAT 3

C, Energetic Particle Explorer
  USE   EXPLORER 15 SATELLITE

C, EPE-
  USE   EXPLORER 15 SATELLITE

C, ERTS-
  USE   LANDSAT 3

C, HEAO
  USE   HEAO 3

C, High Energy Astronomy Observatory
  USE   HEAO 3

C, IMP-
  USE   EXPLORER 28 SATELLITE

C, LORAN
  USE   LORAN C

C, Lunar Orbiter
  USE   LUNAR ORBITER 3

C MK-1 Aircraft, Short Belfast
  USE   SC-5 AIRCRAFT

C, OAO-
  USE   OAO 3

C, OGO-
  USE   OGO-C

C, OSO-
  USE   OSO-C

C Reactor, Tory 2-
  USE   TORY 2-C REACTOR

C Rocket Vehicle, Agena
  USE   AGENA C ROCKET VEHICLE

C Rocket Vehicle, Jupiter
  USE   JUPITER C ROCKET VEHICLE

C Satellite, AE-
  USE   EXPLORER 51 SATELLITE

C Satellite, GEOS-
  USE   GEOS 3 SATELLITE

C, Space Shuttle Mission 31-
  USE   SPACE SHUTTLE MISSION 31-C

C, Space Shuttle Mission 41-
  USE   SPACE SHUTTLE MISSION 41-C

C, Space Shuttle Mission 51-
  USE   SPACE SHUTTLE MISSION 51-C

C, Space Shuttle Mission 61-
  USE   SPACE SHUTTLE MISSION 61-C

C Spacecraft, Mariner
  USE   MARINER C SPACECRAFT

C, TELESAT Canada
  USE   ANIK 3

C, Vitamin
  USE   ASCORBIC ACID

C-M Diagram
  USE   COLOR-MAGNITUDE DIAGRAM

C-1A AIRCRAFT

C-2 AIRCRAFT

C-5 AIRCRAFT

C-5 Aircraft, Lockheed
  USE   C-5 AIRCRAFT

C-8A AUGMENTOR WING AIRCRAFT

C-9 AIRCRAFT

C-15 AIRCRAFT

C-33 AIRCRAFT

C-33 Aircraft, Beech
  USE   C-33 AIRCRAFT

C-35 AIRCRAFT

C-46 AIRCRAFT

C-46 Aircraft, Curtiss
  USE   C-46 AIRCRAFT

C-47 AIRCRAFT

C-54 AIRCRAFT

C-118 AIRCRAFT

C-119 AIRCRAFT

C-121 AIRCRAFT

C-123 AIRCRAFT

C-124 AIRCRAFT

C-130 AIRCRAFT

C-131 AIRCRAFT

C-133 AIRCRAFT

C-135 AIRCRAFT

C-140 AIRCRAFT

C-141 AIRCRAFT

C-142 Aircraft
  USE   XC-142 AIRCRAFT

C-160 AIRCRAFT

C-160 Aircraft, Transall
  USE   C-160 AIRCRAFT

Ca
  USE   CALCIUM

CA
  USE   CALIFORNIA

(CA), Coachella Valley
  USE   COACHELLA VALLEY (CA)

(CA), Coastal Ranges
  USE   COASTAL RANGES (CA)

(CA), Death Valley
  USE   DEATH VALLEY (CA)

(CA), Feather River Basin
  USE   FEATHER RIVER BASIN (CA)

(CA), Imperial Valley
  USE   IMPERIAL VALLEY (CA)

(CA), Mojave Desert
  USE   MOJAVE DESERT (CA)

(CA), Monterey Bay
  USE   MONTEREY BAY (CA)

(CA), Palo Verde Valley
  USE   PALO VERDE VALLEY (CA)

(CA), Peninsular Ranges
  USE   PENINSULAR RANGES (CA)

(CA), Sacramento Valley
  USE   SACRAMENTO VALLEY (CA)

(CA), Salton Sea
  USE   SALTON SEA (CA)

(CA), San Francisco
  USE   SAN FRANCISCO (CA)

(CA), San Francisco Bay
  USE   SAN FRANCISCO BAY (CA)

(CA), San Joaquin Valley
  USE   SAN JOAQUIN VALLEY (CA)

(CA), San Pablo Bay
  USE   SAN PABLO BAY (CA)

(CA), Sierra Nevada Mountains
  USE   SIERRA NEVADA MOUNTAINS (CA)

(CA-NV), Lake Tahoe
  USE   LAKE TAHOE (CA-NV)

(CA-OR-WA), Cascade Range
  USE   CASCADE RANGE (CA-OR-WA)

CABIN ATMOSPHERES

Cabin Atmospheres, Spacecraft
  USE   SPACECRAFT CABIN ATMOSPHERES

Cabin Simulators, Spacecraft
  USE   SPACECRAFT CABIN SIMULATORS

CABINS

Cabins, Aircraft
  USE   AIRCRAFT COMPARTMENTS

Cabins, Pressure
  USE   PRESSURIZED CABINS

Cabins, Pressurized
  USE   PRESSURIZED CABINS

Cabins, Spacecraft
  USE   SPACECRAFT CABINS

CABLE FORCE RECORDERS

CABLES

Cables, Coaxial
  USE   COAXIAL CABLES

Cables, Communication
  USE   COMMUNICATION CABLES

CABLES (ROPES)

Cables, Submarine
  USE   SUBMARINE CABLES

CAD (Design)
  USE   COMPUTER AIDED DESIGN

CADASTRAL MAPPING

CADMIUM

CADMIUM ALLOYS

CADMIUM ANTIMONIDES

Cadmium Batteries, Nickel
USE   NICKEL CADMIUM BATTERIES

Cadmium Batteries, Silver
USE   SILVER CADMIUM BATTERIES

CADMIUM CHLORIDES

CADMIUM COMPOUNDS

CADMIUM FLUORIDES

CADMIUM ISOTOPES

Cadmium Mercury Tellurides
USE   MERCURY CADMIUM TELLURIDES

Cadmium Nickel Batteries
USE   NICKEL CADMIUM BATTERIES

CADMIUM SELENIDES

Cadmium Silver Batteries
USE   SILVER CADMIUM BATTERIES

CADMIUM SULFIDES

CADMIUM TELLURIDES

Cadmium Tellurides, Mercury
USE   MERCURY CADMIUM TELLURIDES

Cadmium 114
USE   CADMIUM ISOTOPES

CAFFEINE

CAI
USE   COMPUTER ASSISTED INSTRUCTION

CAISSONS

CAJUN ROCKET VEHICLE

Cajun Rocket Vehicle, Nike-
USE   NIKE-CAJUN ROCKET VEHICLE

Cal Satellite, ORBIS
USE   ORBIS CAL SATELLITE

CALCIFEROL

CALCIFICATION

Calcination
USE   ROASTING

CALCITE

CALCIUM

CALCIUM CARBONATES

CALCIUM CHLORIDES

CALCIUM COMPOUNDS

CALCIUM FLUORIDES

CALCIUM ISOTOPES

CALCIUM METABOLISM

CALCIUM OXIDES

CALCIUM PHOSPHATES

CALCIUM SILICATES

CALCIUM SULFIDES

CALCIUM TUNGSTATES

CALCIUM VANADATES

Calcium 45
USE   CALCIUM ISOTOPES

Calculation
USE   COMPUTATION

Calculation, Matrix Stress
USE   MATRIX METHODS

Calculation, Orbit
USE   ORBIT CALCULATION

Calculation, Satellite Orbit
USE   ORBIT CALCULATION

Calculations, Stress
USE   STRESS ANALYSIS

CALCULATORS

CALCULI

Calculi, Dental
USE   DENTAL CALCULI

Calculi, Renal
USE   CALCULI

CALCULUS

Calculus, Derivation
USE   DIFFERENTIAL CALCULUS

Calculus, Differential
USE   DIFFERENTIAL CALCULUS

Calculus, Graeff
USE   GRAEFF CALCULUS

Calculus, Integral
USE   INTEGRAL CALCULUS

CALCULUS OF VARIATIONS

Calculus, Operational
USE   OPERATIONAL CALCULUS

Calculus), Stokes Theorem (Vector
USE   STOKES THEOREM (VECTOR CALCULUS)

Calculus, Vector
USE   VECTOR SPACES

CALDERAS

CALENDARS

Calendars, Crop
USE   CROP CALENDARS

CALIBRATING

Calibrating Omnirange, Self
USE   SELF CALIBRATING OMNIRANGE

Calibration Facility, Solar Cell
USE   SOLAR CELL CALIBRATION FACILITY

Calibration, Wind Tunnel
USE   WIND TUNNEL CALIBRATION

CALIFORNIA

California, Baja
USE   LOWER CALIFORNIA (MEXICO)

California (Mexico), Gulf Of
USE   GULF OF CALIFORNIA (MEXICO)

California (Mexico), Lower
USE   LOWER CALIFORNIA (MEXICO)

California, Southern
USE   SOUTHERN CALIFORNIA

CALIFORNIUM

CALIFORNIUM ISOTOPES

Californium 252
USE   CALIFORNIUM ISOTOPES

CALLISTO

CALORIC REQUIREMENTS

CALORIC STIMULI

CALORIMETERS

Calorimeters, Bomb
USE   BOMB CALORIMETERS

Calorimeters, Drop
USE   DROP CALORIMETERS

Calorimeters, Flame
USE   FLAME CALORIMETERS

Calorimetry
USE   HEAT MEASUREMENT

Calutrons
USE   CYCLOTRONS

CALVES

CAM (Manufacturing)
USE   COMPUTER AIDED MANUFACTURING

CAMBER

Camber, Conical
USE   CONICAL CAMBER

Camber, Wing
USE   WING CAMBER

CAMBERED WINGS

CAMBODIA

Camel Aircraft
USE   TU-104 AIRCRAFT

Camera, Baker-Nunn
USE   BAKER-NUNN CAMERA

Camera, Delft
USE   DELFT CAMERA

Camera, Faint Object
USE   FAINT OBJECT CAMERA

CAMERA SHUTTERS

Camera System (AVCS), Advanced Vidicon
USE   ADVANCED VIDICON CAMERA SYSTEM
       (AVCS)

CAMERA TUBES

CAMERAS

Cameras, Ballistic
USE   BALLISTIC CAMERAS

Cameras, Diffraction Limited
USE   DIFFRACTION LIMITED CAMERAS

Cameras, Framing
USE   FRAMING CAMERAS

Cameras, High Speed
USE   HIGH SPEED CAMERAS

Cameras, I2S
USE   I2S CAMERAS

Cameras, Lallemand
USE   LALLEMAND CAMERAS

Cameras, Multispectral Band
USE   MULTISPECTRAL BAND CAMERAS

Cameras, Panoramic
USE   PANORAMIC CAMERAS

Cameras, Pinhole
USE   PINHOLE CAMERAS

Cameras, Schmidt
   USE   SCHMIDT CAMERAS

Cameras, Streak
   USE   STREAK CAMERAS

Cameras, Television
   USE   TELEVISION CAMERAS

CAMEROON

CAMOUFLAGE

CAMPBELL-HAUSDORFF SERIES

CAMPHOR

CAMS

Can, Sortie
   USE   SORTIE SYSTEMS

CANADA

Canada A, TELESAT
   USE   ANIK 1

Canada B, TELESAT
   USE   ANIK 2

Canada C, TELESAT
   USE   ANIK 3

Canada 3, TELESAT
   USE   ANIK 3

CANADAIR AIRCRAFT

Canadair CF-104 Aircraft
   USE   CANADAIR AIRCRAFT
          F-104 AIRCRAFT

Canadair CL-41 Aircraft
   USE   CL-41 AIRCRAFT

Canadair CL-44 Aircraft
   USE   CL-44 AIRCRAFT

Canadair CL-84 Aircraft
   USE   CL-84 AIRCRAFT

CANADIAN SHIELD

CANADIAN SPACE PROGRAM

CANADIAN SPACECRAFT

Canal Zone, Panama
   USE   PANAMA CANAL ZONE

CANALS

Canals, Semicircular
   USE   SEMICIRCULAR CANALS

CANARD CONFIGURATIONS

CANBERRA AIRCRAFT

Canberra Aircraft, English Electric
   USE   CANBERRA AIRCRAFT

Canberra Bomber
   USE   B-57 AIRCRAFT

CANCELLATION

CANCELLATION CIRCUITS

CANCER

Cane, Sugar
   USE   SUGAR CANE

Canisters
   USE   CANS

CANNING

CANNONBALL 2 SATELLITE

Cannons
   USE   GUNS (ORDNANCE)

CANNULAE

CANONICAL FORMS

CANOPIES

CANOPIES (VEGETATION)

CANS

Cant
   USE   SLOPES

CANTILEVER BEAMS

CANTILEVER MEMBERS

CANTILEVER PLATES

Cantilever Wings
   USE   WINGS

Canyon (AZ), Grand
   USE   GRAND CANYON (AZ)

CANYONS

Cap Absorption, Polar
   USE   POLAR CAP ABSORPTION

CAP CLOUDS

Capability), Ceiling (Aircraft
   USE   CEILING (AIRCRAFT CAPABILITY)

CAPACITANCE

CAPACITANCE SWITCHES

CAPACITANCE-VOLTAGE CHARACTERISTICS

CAPACITIVE FUEL GAGES

CAPACITORS

CAPACITY

Capacity, Channel
   USE   CHANNEL CAPACITY

Capacity, Heat
   USE   SPECIFIC HEAT

Capacity Mapping Mission, Heat
   USE   HEAT CAPACITY MAPPING MISSION

Capacity, Work
   USE   WORK CAPACITY

CAPE HATTERAS (NC)

CAPE KENNEDY LAUNCH COMPLEX

CAPE VERDE

CAPES (LANDFORMS)

CAPILLARIES

CAPILLARIES (ANATOMY)

Capillary Circulation
   USE   CAPILLARY FLOW

CAPILLARY FLOW

CAPILLARY TUBES

CAPILLARY WAVES

CAPS

CAPS (EXPLOSIVES)

Caps, Nose
   USE   NOSE CONES

Caps, Polar
   USE   POLAR CAPS

Caps, Spherical
   USE   SPHERICAL CAPS

(Capsule), DRC
   USE   DISCOVERER RECOVERY CAPSULES

CAPSULES

Capsules, Discoverer Recovery
   USE   DISCOVERER RECOVERY CAPSULES

Capsules, Escape
   USE   ESCAPE CAPSULES

Capsules, Fuel
   USE   FUEL CAPSULES

Capsules, Space
   USE   SPACE CAPSULES

Capsules (Spacecraft)
   USE   SPACE CAPSULES

CAPTIVE TESTS

Capture, Aero
   USE   AEROCAPTURE

Capture, Asteroid
   USE   ASTEROID CAPTURE

Capture Cross Sections
   USE   ABSORPTION CROSS SECTIONS

CAPTURE EFFECT

Capture, Electron
   USE   ELECTRON CAPTURE

Capture, Nuclear
   USE   NUCLEAR CAPTURE

Capture, Satellite
   USE   SPACECRAFT RECOVERY

CAPTURED AIR BUBBLE VEHICLES

Caravelle Aircraft
   USE   SE-210 AIRCRAFT

CARBAMATES (TRADENAME)

CARBAMIDES

CARBAZOLES

CARBENES

CARBIDES

Carbides, Aluminum
   USE   ALUMINUM CARBIDES

Carbides, Boron
   USE   BORON CARBIDES

Carbides, Chromium
   USE   CHROMIUM CARBIDES

Carbides, Hafnium
   USE   HAFNIUM CARBIDES

Carbides, Molybdenum
   USE   MOLYBDENUM CARBIDES

Carbides, Niobium
   USE   NIOBIUM CARBIDES

Carbides, Plutonium
   USE   PLUTONIUM COMPOUNDS

Carbides, Silicon
   USE   SILICON CARBIDES

Carbides, Tantalum
   USE   TANTALUM CARBIDES

Carbides, Titanium
USE   TITANIUM CARBIDES

Carbides, Tungsten
USE   TUNGSTEN CARBIDES

Carbides, Uranium
USE   URANIUM CARBIDES

Carbides, Vanadium
USE   VANADIUM CARBIDES

Carbides, Zirconium
USE   ZIRCONIUM CARBIDES

CARBOHYDRATE METABOLISM

CARBOHYDRATES

CARBON

Carbon, Activated
USE   ACTIVATED CARBON

CARBON ARCS

Carbon Composites, Carbon-
USE   CARBON-CARBON COMPOSITES

CARBON COMPOUNDS

CARBON CYCLE

CARBON DIOXIDE

CARBON DIOXIDE CONCENTRATION

CARBON DIOXIDE LASERS

CARBON DIOXIDE REMOVAL

CARBON DIOXIDE TENSION

CARBON DISULFIDE

CARBON FIBER REINFORCED PLASTICS

CARBON FIBERS

Carbon, Glassy
USE   GLASSY CARBON

CARBON ISOTOPES

CARBON LASERS

CARBON MONOXIDE

CARBON MONOXIDE LASERS

CARBON MONOXIDE POISONING

CARBON STARS

CARBON STEELS

Carbon Steels, Low
USE   LOW CARBON STEELS

CARBON SUBOXIDES

CARBON TETRACHLORIDE

CARBON TETRACHLORIDE POISONING

CARBON TETRAFLUORIDE

CARBON 12

CARBON 13

CARBON 14

CARBON-CARBON COMPOSITES

CARBONACEOUS CHONDRITES

CARBONACEOUS MATERIALS

CARBONACEOUS METEORITES

CARBONACEOUS ROCKS

CARBONATES

Carbonates, Calcium
USE   CALCIUM CARBONATES

Carbonates, Poly
USE   POLYCARBONATES

Carbonates, Sodium
USE   SODIUM CARBONATES

CARBONIC ACID

CARBONIC ANHYDRASE

CARBONIZATION

Carbons, Chloro
USE   CHLOROCARBONS

Carbons, Fluoro
USE   FLUOROCARBONS

Carbons, Fluorohydro
USE   FLUOROHYDROCARBONS

Carbons, Hydro
USE   HYDROCARBONS

CARBONYL COMPOUNDS

CARBORANE

CARBORUNDUM (TRADEMARK)

CARBOXYHEMOGLOBIN

CARBOXYHEMOGLOBIN TEST

CARBOXYL GROUP

CARBOXYLATES

CARBOXYLATION

CARBOXYLIC ACIDS

CARBURETORS

Carburetors, Injection
USE   CARBURETORS
       FUEL INJECTION

CARBURIZING

CARCINOGENS

Carcinoma
USE   CANCER

CARCINOTRONS

CARDIAC AURICLES

Cardiac Pacemaker, Artificial
USE   ARTIFICIAL CARDIAC PACEMAKER

CARDIAC VENTRICLES

CARDIOGRAMS

CARDIOGRAPHY

Cardiography, Echo
USE   ECOSYSTEMS

Cardiography, Electro
USE   ELECTROCARDIOGRAPHY

Cardiography, Magneto
USE   MAGNETOCARDIOGRAPHY

Cardiography, Phono
USE   PHONOCARDIOGRAPHY

Cardiography, Radio
USE   RADIOCARDIOGRAPHY

Cardiography, Vector
USE   VECTORCARDIOGRAPHY

CARDIOLOGY

CARDIOTACHOMETERS

CARDIOVASCULAR SYSTEM

CARDS

Cards, Punched
USE   PUNCHED CARDS

CARET WINGS

CARETS (Test Site)
USE   CENTRAL ATLANTIC REGIONAL ECOL
       TEST SITE

CARGO

Cargo, Air
USE   AIR CARGO

CARGO AIRCRAFT

CARGO SHIPS

Cargo Ships, LOTS
USE   CARGO SHIPS

CARGO SPACECRAFT

Cargomaster Aircraft
USE   C-133 AIRCRAFT

CARIBBEAN REGION

CARIBBEAN SEA

Caribou Aircraft
USE   DHC 4 AIRCRAFT

CARIBOUS

Carlo Method, Monte
USE   MONTE CARLO METHOD

CARNITINE

CARNOT CYCLE

Carolina, North
USE   NORTH CAROLINA

Carolina, South
USE   SOUTH CAROLINA

CAROTENE

CAROTID SINUS BODY

CAROTID SINUS REFLEX

CARPATHIAN MOUNTAINS (EUROPE)

CARRIAGES

Carriages, Under
USE   UNDERCARRIAGES

CARRIER DENSITY (SOLID STATE)

Carrier, European Retrievable
USE   EURECA (ESA)

CARRIER FREQUENCIES

CARRIER INJECTION

CARRIER LIFETIME

Carrier, Logistics Over The Shore (LOTS)
USE   LOGISTICS OVER THE SHORE (LOTS)
       CARRIER

CARRIER MOBILITY

Carrier Modulation
    USE   MODULATION

Carrier Rocket, Echo 1
    USE   THOR DELTA LAUNCH VEHICLE

Carrier Rockets
    USE   LAUNCH VEHICLES

Carrier Systems
    USE   WIRELESS COMMUNICATION

CARRIER TO NOISE RATIOS

Carrier Transmission, Single Channel Per
    USE   SINGLE CHANNEL PER CARRIER
                    TRANSMISSION

CARRIER TRANSPORT (SOLID STATE)

CARRIER WAVES

CARRIERS

Carriers, Aircraft
    USE   AIRCRAFT CARRIERS

Carriers, Charge
    USE   CHARGE CARRIERS

Carriers, Majority
    USE   MAJORITY CARRIERS

Carriers, Minority
    USE   MINORITY CARRIERS

Carrington Rotation
    USE   SOLAR ROTATION

CARTAN SPACE

CARTESIAN COORDINATES

CARTILAGE

Cartography
    USE   MAPPING

Cartridge Actuated Devices
    USE   ACTUATORS
                    EXPLOSIVE DEVICES

CARTRIDGES

CARTS

CASCADE CONTROL

CASCADE FLOW

CASCADE RANGE (CA-OR-WA)

CASCADE WIND TUNNELS

CASCADES

Cascades, Electron Photon
    USE   ELECTRON PHOTON CASCADES

Cascades (Fluid Dynamics)
    USE   FLUID DYNAMICS

Cascode MOSFET
    USE   FIELD EFFECT TRANSISTORS

CASE BONDED PROPELLANTS

CASE HISTORIES

CASES (CONTAINERS)

Cases, Missile
    USE   MISSILE BODIES

Cases, Missile Engine
    USE   ROCKET ENGINE CASES

Cases, Rocket Engine
    USE   ROCKET ENGINE CASES

Cases, Rocket Motor
    USE   ROCKET ENGINE CASES

CASING

Casks
    USE   BARRELS (CONTAINERS)

CASPIAN SEA

CASSEGRAIN ANTENNAS

CASSEGRAIN OPTICS

CASSIOPEIA A

CASSIOPEIA CONSTELLATION

CAST ALLOYS

CASTIGLIANO VARIATIONAL THEOREM

CASTING

Casting, Centrifugal
    USE   CENTRIFUGAL CASTING

Casting, Fore
    USE   FORECASTING

Casting, Investment
    USE   INVESTMENT CASTING

Casting, Propellant
    USE   PROPELLANT CASTING

Casting, Sand
    USE   SAND CASTING

Casting, Slip
    USE   SLIP CASTING

Casting Solvents
    USE   PLASTICIZERS

CASTINGS

CASTOR OIL

Castor 2 Engine
    USE   TX-354 ENGINE

CASTS

CASUALTIES

Cat Scanner
    USE   COMPUTER AIDED TOMOGRAPHY

CATABOLISM

CATACLYSMIC VARIABLES

CATALASE

CATALOGS

Catalogs, Astronomical
    USE   ASTRONOMICAL CATALOGS

CATALOGS (PUBLICATIONS)

CATALYSIS

Catalysis, Auto
    USE   AUTOCATALYSIS

Catalyst, Ziegler
    USE   ZIEGLER CATALYST

CATALYSTS

Catalysts, Electro
    USE   ELECTROCATALYSTS

Catalysts, Fuel Cell
    USE   ELECTROCATALYSTS

CATALYTIC ACTIVITY

CATAPULTS

Catapults, Rocket
    USE   ROCKET CATAPULTS

CATARACTS

CATASTROPHE THEORY

CATCHERS

Catchment Areas
    USE   WATERSHEDS

CATECHOLAMINE

CATEGORIES

CATENARIES

CATHETERIZATION

CATHETOMETERS

CATHODE GLOW

CATHODE RAY TUBES

Cathode Tubes, Cold
    USE   COLD CATHODE TUBES

CATHODES

Cathodes, Cell
    USE   CELL CATHODES

Cathodes, Cold
    USE   COLD CATHODES

Cathodes, Hollow
    USE   HOLLOW CATHODES

Cathodes, Hot
    USE   HOT CATHODES

Cathodes, Photo
    USE   PHOTOCATHODES

Cathodes, Thermionic
    USE   THERMIONIC CATHODES

Cathodes, Tube
    USE   TUBE CATHODES

Cathodes, Tunnel
    USE   TUNNEL CATHODES

CATHODIC COATINGS

CATHODOLUMINESCENCE

CATHOLYTES

CATIONS

CATS

CATT DEVICES

CATTLE

Cauca Valley (Colombia), Magdalena-
    USE   MAGDALENA-CAUCA VALLEY (COLOMBIA)

CAUCASUS MOUNTAINS (U.S.S.R.)

Cauchy Equations, Euler-
    USE   EULER-CAUCHY EQUATIONS

CAUCHY INTEGRAL FORMULA

CAUCHY PROBLEM

CAUCHY-RIEMANN EQUATIONS

CAULKING

Cause, Retirement For
    USE   RETIREMENT FOR CAUSE

CAUSES

CAUSTIC LINES

Caustics
  USE   ALKALIES

CAUSTICS (OPTICS)

CAVES

Cavitation
  USE   CAVITATION FLOW

CAVITATION CORROSION

CAVITATION FLOW

Cavitation, Gaseous
  USE   CAVITATION FLOW
          GAS FLOW

CAVITIES

Cavities, Laser
  USE   LASER CAVITIES

Cavities, Resonant
  USE   CAVITY RESONATORS

CAVITONS

Cavity, Intracranial
  USE   INTRACRANIAL CAVITY

CAVITY RESONATORS

CAVITY VAPOR GENERATORS

Cays
  USE   KEYS (ISLANDS)

CC-106 Aircraft
  USE   CL-44 AIRCRAFT

CCD
  USE   CHARGE COUPLED DEVICES

CCD STAR TRACKER

Cd
  USE   CADMIUM

CDC COMPUTERS

CDC CYBER 74 COMPUTER

CDC CYBER 170 SERIES COMPUTERS

CDC CYBER 174 COMPUTER

CDC CYBER 175 COMPUTER

CDC CYBER 203 COMPUTER

CDC CYBER 205 COMPUTER

CDC STAR 100 COMPUTER

CDC 160-A COMPUTER

CDC 1604 COMPUTER

CDC 3100 COMPUTER

CDC 3200 COMPUTER

CDC 3600 COMPUTER

CDC 3800 COMPUTER

CDC 6000 SERIES COMPUTERS

CDC 6400 COMPUTER

CDC 6600 COMPUTER

CDC 6700 COMPUTER

CDC 7000 SERIES COMPUTERS

CDC 7600 COMPUTER

CDC 8090 COMPUTER

CDMA
  USE   CODE DIVISION MULTIPLE ACCESS

Ce
  USE   CERIUM

CEDAR RAPIDS (IA)

CEFOAM CHECKOUT EQUIPMENT

CEILING (AIRCRAFT CAPABILITY)

CEILINGS

CEILINGS (ARCHITECTURE)

CEILINGS (METEOROLOGY)

Ceilometers
  USE   CLOUD HEIGHT INDICATORS

CELESCOPES

CELESTIAL BODIES

CELESTIAL GEODESY

CELESTIAL MECHANICS

CELESTIAL NAVIGATION

Celestial Observation
  USE   ASTRONOMY

CELESTIAL REFERENCE SYSTEMS

CELESTIAL SPHERE

CELL ANODES

Cell Calibration Facility, Solar
  USE   SOLAR CELL CALIBRATION FACILITY

Cell Catalysts, Fuel
  USE   ELECTROCATALYSTS

CELL CATHODES

CELL DIVISION

Cell Power Plants, Fuel
  USE   FUEL CELL POWER PLANTS

Cell, Resolution
  USE   RESOLUTION CELL

Cell Technique, Particle In
  USE   PARTICLE IN CELL TECHNIQUE

CELLOPHANE

CELLS

Cells, Benard
  USE   BENARD CELLS

Cells, Biochemical Fuel
  USE   BIOCHEMICAL FUEL CELLS

Cells, Biological
  USE   CELLS (BIOLOGY)

CELLS (BIOLOGY)

Cells, Dry
  USE   DRY CELLS

Cells, Electric
  USE   ELECTRIC CELLS

Cells, Electrochemical
  USE   ELECTROCHEMICAL CELLS

Cells, Electrolytic
  USE   ELECTROLYTIC CELLS

Cells, Fission Electric
  USE   FISSION ELECTRIC CELLS

Cells, Fuel
  USE   FUEL CELLS

Cells, Galvanic
  USE   ELECTROLYTIC CELLS

Cells, Geophysical Fluid Flow
  USE   GEOPHYSICAL FLUID FLOW CELLS

Cells, Golay Detector
  USE   GOLAY DETECTOR CELLS

Cells, Hexagonal
  USE   HEXAGONAL CELLS

Cells, Hydrogen Air Fuel
  USE   HYDROGEN OXYGEN FUEL CELLS

Cells, Hydrogen Oxygen Fuel
  USE   HYDROGEN OXYGEN FUEL CELLS

Cells, Kerr
  USE   KERR CELLS

Cells, Knudsen
  USE   KNUDSEN GAGES

Cells, Magnesium
  USE   MAGNESIUM CELLS

Cells, Phosphoric Acid Fuel
  USE   PHOSPHORIC ACID FUEL CELLS

Cells, Photoconductive
  USE   PHOTOCONDUCTIVE CELLS

Cells, Photoelectric
  USE   PHOTOELECTRIC CELLS

Cells, Photovoltaic
  USE   PHOTOVOLTAIC CELLS

Cells, Red Blood
  USE   ERYTHROCYTES

Cells, Redox
  USE   REDOX CELLS

Cells, Regenerative Fuel
  USE   REGENERATIVE FUEL CELLS

Cells, Silicon Solar
  USE   SOLAR CELLS

Cells, Solar
  USE   SOLAR CELLS

Cells, Vertical Junction Solar
  USE   VERTICAL JUNCTION SOLAR CELLS

Cells, Wet
  USE   WET CELLS

Cells, White Blood
  USE   WHITE BLOOD CELLS

Cells, Wraparound Contact Solar
  USE   SOLAR CELLS

Cellular Materials (Non Biological)
  USE   FOAMS

CELLULOSE

CELLULOSE NITRATE

CEMENTATION

CEMENTITE

CEMENTS

CEMS System
  USE   CENTRAL ELECTRONIC MANAGEMENT
          SYSTEM

CENSORED DATA (MATHEMATICS)

CENSUS

CENTAUR LAUNCH VEHICLE

Centaur Launch Vehicle, Atlas
    USE   ATLAS CENTAUR LAUNCH VEHICLE

Centaur Launch Vehicle, Titan
    USE   TITAN CENTAUR LAUNCH VEHICLE

CENTAUR PROJECT

Centaur Vehicle
    USE   CENTAUR LAUNCH VEHICLE

CENTAURUS CONSTELLATION

Center, Aerodynamic
    USE   AERODYNAMIC BALANCE

Center), IMCC (Control
    USE   INTEGRATED MISSION CONTROL CENTER

Center, Integrated Mission Control
    USE   INTEGRATED MISSION CONTROL CENTER

Center (NASA), Space Operations
    USE   SPACE OPERATIONS CENTER (NASA)

CENTER OF GRAVITY

CENTER OF MASS

CENTER OF PRESSURE

Center, Space Base Command
    USE   SPACE BASE COMMAND CENTER

CENTERBODIES

Centered Cubic Lattices, Body
    USE   BODY CENTERED CUBIC LATTICES

Centered Cubic Lattices, Face
    USE   FACE CENTERED CUBIC LATTICES

CENTERS

Centers, Color
    USE   COLOR CENTERS

Centers, F
    USE   COLOR CENTERS

Centers, World Data
    USE   WORLD DATA CENTERS

CENTIMETER WAVES

CENTRAL AFRICAN REPUBLIC

CENTRAL AMERICA

CENTRAL ATLANTIC REGION (US)

CENTRAL ATLANTIC REGIONAL ECOL TEST SITE

CENTRAL ELECTRONIC MANAGEMENT SYSTEM

CENTRAL EUROPE

CENTRAL NERVOUS SYSTEM

CENTRAL NERVOUS SYSTEM DEPRESSANTS

CENTRAL NERVOUS SYSTEM STIMULANTS

CENTRAL PIEDMONT (US)

CENTRAL PROCESSING UNITS

CENTRIFUGAL CASTING

CENTRIFUGAL COMPRESSORS

CENTRIFUGAL FORCE

CENTRIFUGAL PUMPS

CENTRIFUGES

Centrifuges, Human
    USE   HUMAN CENTRIFUGES

Centrifuges, Piloted
    USE   HUMAN CENTRIFUGES

CENTRIFUGING

CENTRIFUGING STRESS

CENTRIPETAL FORCE

CENTROIDS

Centurion Aircraft
    USE   CESSNA 210 AIRCRAFT

Cephalagia
    USE   HEADACHE

CEPHALOPODS

CEPHEID VARIABLES

CEPHEUS CONSTELLATION

CEPSTRA

CEPSTRAL ANALYSIS

Ceramal Protective Coatings
    USE   CERMETS
           PROTECTIVE COATINGS

Ceramals
    USE   CERMETS

CERAMIC BONDING

CERAMIC COATINGS

CERAMIC HONEYCOMBS

CERAMIC MATRIX COMPOSITES

CERAMIC NUCLEAR FUELS

CERAMICS

Ceramics, Piezoelectric
    USE   PIEZOELECTRIC CERAMICS

CEREBELLUM

CEREBRAL CORTEX

CEREBRAL VASCULAR ACCIDENTS

CEREBRAL VENTRICLES

CEREBROSPINAL FLUID

CEREBRUM

CERENKOV COUNTERS

Cerenkov Effect
    USE   CERENKOV RADIATION

CERENKOV RADIATION

CERES ASTEROID

CERESIN

CERIUM

CERIUM COMPOUNDS

CERIUM ISOTOPES

CERIUM OXIDES

CERIUM 137

CERIUM 144

CERMETS

CERTIFICATION

CESIUM

CESIUM ALLOYS

CESIUM ANTIMONIDES

CESIUM BROMIDES

CESIUM COMPOUNDS

CESIUM DIODES

CESIUM ENGINES

CESIUM FLUORIDES

CESIUM HALIDES

CESIUM HYDRIDES

CESIUM IODIDES

CESIUM IONS

CESIUM ISOTOPES

CESIUM OXIDES

CESIUM PLASMA

CESIUM VAPOR

CESIUM 133

CESIUM 134

CESIUM 137

CESIUM 144

CESSNA AIRCRAFT

CESSNA L-19 AIRCRAFT

Cessna Military Aircraft
    USE   MILITARY AIRCRAFT

CESSNA 172 AIRCRAFT

CESSNA 205 AIRCRAFT

CESSNA 210 AIRCRAFT

CESSNA 402B AIRCRAFT

CETANE

Ceti Star, Omicron
    USE   OMICRON CETI STAR

Ceti Stars, UV
    USE   FLARE STARS

CETYL COMPOUNDS

CEYLON
    USE   SRI LANKA

Cf
    USE   CALIFORNIUM

CF-104 Aircraft
    USE   CANADAIR AIRCRAFT
           F-104 AIRCRAFT

CF-104 Aircraft, Canadair
    USE   F-104 AIRCRAFT
           CANADAIR AIRCRAFT

CF-700 ENGINE

CFD
    USE   CHARGE FLOW DEVICES

CFRP
    USE   CARBON FIBER REINFORCED PLASTICS

CH-3 HELICOPTER

CH-21 HELICOPTER

CH-34 HELICOPTER

CH-46 HELICOPTER

CH-47 HELICOPTER

CH-53 Helicopter
  USE   H-53 HELICOPTER

CH-54 HELICOPTER

CH-62 HELICOPTER

CH-113 Helicopter
  USE   CH-46 HELICOPTER

CHAD

CHAFF

Chain, Food
  USE   FOOD CHAIN

CHAINS

Chains, Markov
  USE   MARKOV CHAINS

Chains, Molecular
  USE   MOLECULAR CHAINS

Chair, Barany
  USE   BARANY CHAIR

Chairs
  USE   SEATS

CHALCOGENIDES

CHALK

Challenger Aircraft, CL-600
  USE   CL-600 CHALLENGER AIRCRAFT

CHALLENGER (ORBITER)

Chamber Pressure, Thrust
  USE   THRUST CHAMBER PRESSURE

CHAMBERS

Chambers, Anechoic
  USE   ANECHOIC CHAMBERS

Chambers, Arc
  USE   ARC CHAMBERS

Chambers, Bubble
  USE   BUBBLE CHAMBERS

Chambers, Cloud
  USE   CLOUD CHAMBERS

Chambers, Combustion
  USE   COMBUSTION CHAMBERS

Chambers, Cylindrical
  USE   CYLINDRICAL CHAMBERS

Chambers, Environmental
  USE   TEST CHAMBERS

Chambers, Flow
  USE   FLOW CHAMBERS

Chambers, Growth
  USE   PHYTOTRONS

Chambers, Hyperbaric
  USE   HYPERBARIC CHAMBERS

Chambers, Ion
  USE   IONIZATION CHAMBERS

Chambers, Ionization
  USE   IONIZATION CHAMBERS

Chambers, Low Pressure
  USE   VACUUM CHAMBERS

Chambers), Magazines (Supply
  USE   MAGAZINES (SUPPLY CHAMBERS)

Chambers, Plenum
  USE   PLENUM CHAMBERS

Chambers, Pressure
  USE   PRESSURE CHAMBERS

Chambers, Rocket
  USE   THRUST CHAMBERS

Chambers, Spark
  USE   SPARK CHAMBERS

Chambers, Test
  USE   TEST CHAMBERS

Chambers, Thrust
  USE   THRUST CHAMBERS

Chambers, Vacuum
  USE   VACUUM CHAMBERS

Champlain Basin (NY-VT), Lake
  USE   LAKE CHAMPLAIN BASIN (NY-VT)

CHANCE-VOUGHT AIRCRAFT

Chance-Vought Military Aircraft
  USE   CHANCE-VOUGHT AIRCRAFT
         MILITARY AIRCRAFT

Chandler Motion
  USE   POLAR WANDERING (GEOLOGY)

CHANDRASEKHAR EQUATION

CHANGE DETECTION

Change Materials, Phase
  USE   PHASE CHANGE MATERIALS

CHANNEL CAPACITY

Channel, English
  USE   ENGLISH CHANNEL

CHANNEL FLOW

Channel Flow, Open
  USE   OPEN CHANNEL FLOW

CHANNEL MULTIPLIERS

CHANNEL NOISE

Channel Per Carrier Transmission, Single
  USE   SINGLE CHANNEL PER CARRIER
         TRANSMISSION

CHANNEL WINGS

CHANNELS

CHANNELS (DATA TRANSMISSION)

Channels, Micro
  USE   MICROCHANNELS

Channeltrons
  USE   CHANNEL MULTIPLIERS

CHAOS

Chaotic Cloud Patterns
  USE   CLOUDS (METEOROLOGY)

CHAPARRAL

CHAPARRAL MISSILE

CHAPLYGIN EQUATION

Chapman Shear Layer
  USE   SHEAR LAYERS

Chapman Theory, Enskog-
  USE   CHAPMAN-ENSKOG THEORY

CHAPMAN-ENSKOG THEORY

CHAPMAN-FERRARO PROBLEM

Chapman-Jouget Flame
  USE   CHEMICAL EQUILIBRIUM
         DETONATION
         FLAME PROPAGATION

CHARACTER RECOGNITION

Characteristic Equations
  USE   EIGENVALUES
         EIGENVECTORS

Characteristic Functions
  USE   EIGENVECTORS
         EIGENVALUES

Characteristic Method
  USE   METHOD OF CHARACTERISTICS

Characteristic, Segre
  USE   SEGRE CHARACTERISTIC

CHARACTERISTICS

Characteristics, Aerodynamic
  USE   AERODYNAMIC CHARACTERISTICS

Characteristics, Airfoil
  USE   AIRFOILS

Characteristics, Capacitance-Voltage
  USE   CAPACITANCE-VOLTAGE
         CHARACTERISTICS

Characteristics, Dynamic
  USE   DYNAMIC CHARACTERISTICS

Characteristics, Flight
  USE   FLIGHT CHARACTERISTICS

Characteristics, Flow
  USE   FLOW CHARACTERISTICS

Characteristics, Method Of
  USE   METHOD OF CHARACTERISTICS

Characteristics, Polarization
  USE   POLARIZATION CHARACTERISTICS

Characteristics, Spray
  USE   SPRAY CHARACTERISTICS

Characteristics, Static
  USE   STATIC CHARACTERISTICS

Characteristics, Static Aerodynamic
  USE   STATIC AERODYNAMIC CHARACTERISTICS

Characteristics, Volt-Ampere
  USE   VOLT-AMPERE CHARACTERISTICS

CHARACTERIZATION

Characters
  USE   SYMBOLS

Characters, Alphanumeric
  USE   ALPHANUMERIC CHARACTERS

CHARCOAL

CHARGE CARRIERS

CHARGE COUPLED DEVICES

Charge Density, Magnetic
  USE   MAGNETIC CHARGE DENSITY

CHARGE DISTRIBUTION

CHARGE EFFICIENCY

Charge, Electric
  USE   ELECTRIC CHARGE

Charge, Electrostatic
  USE   ELECTROSTATIC CHARGE

CHARGE EXCHANGE

Charge Exchange, Resonance
   USE   RESONANCE CHARGE EXCHANGE

CHARGE FLOW DEVICES

CHARGE INJECTION DEVICES

Charge, Ion
   USE   ION CHARGE

Charge, Scalar Magnetic
   USE   MAGNETIC CHARGE DENSITY

Charge Separation
   USE   POLARIZATION (CHARGE SEPARATION)

(Charge Separation), Polarization
   USE   POLARIZATION (CHARGE SEPARATION)

Charge, Space
   USE   SPACE CHARGE

CHARGE TRANSFER

CHARGE TRANSFER DEVICES

Charge Transfer Salts, Organic
   USE   ORGANIC CHARGE TRANSFER SALTS

Charge, Traveling
   USE   TRAVELING CHARGE

CHARGED PARTICLES

Chargers, Battery
   USE   BATTERY CHARGERS

Charges, Shaped
   USE   SHAPED CHARGES

CHARGING

Charging, Particle
   USE   PARTICLE CHARGING

Charging, Pulse
   USE   PULSE CHARGING

Charging, Spacecraft
   USE   SPACECRAFT CHARGING

CHARM (PARTICLE PHYSICS)

CHARON

CHARPY IMPACT TEST

CHARRING

Chart, Smith
   USE   SMITH CHART

CHARTS

Charts, Flow
   USE   FLOW CHARTS

(Charts), Graphs
   USE   GRAPHS (CHARTS)

Charts, Meteorological
   USE   METEOROLOGICAL CHARTS

Charts, Nautical
   USE   NAUTICAL CHARTS

Charts, Polarization
   USE   POLARIZATION (WAVES)
            GRAPHS (CHARTS)

Charts, Weather
   USE   METEOROLOGICAL CHARTS

CHASSIS

Chaudhuri-Hocquenghem Codes, Bose-
   USE   BCH CODES

CHEBYSHEV APPROXIMATION

CHECKOUT

Checkout Equipment
   USE   TEST EQUIPMENT

Checkout Equipment, Cefoam
   USE   CEFOAM CHECKOUT EQUIPMENT

Checkout Program, Space Vehicle
   USE   SPACE VEHICLE CHECKOUT PROGRAM

Chelate Compounds
   USE   CHELATES

CHELATES

CHELATION

CHEMICAL ANALYSIS

CHEMICAL ATTACK

CHEMICAL AUXILIARY POWER UNITS

(Chemical), Bases
   USE   BASES (CHEMICAL)

CHEMICAL BONDS

CHEMICAL CLEANING

CHEMICAL CLOUDS

CHEMICAL COMPOSITION

CHEMICAL COMPOUNDS

CHEMICAL DEFENSE

CHEMICAL EFFECTS

CHEMICAL ELEMENTS

CHEMICAL ENERGY

CHEMICAL ENGINEERING

(Chemical Engineering), Cracking
   USE   CRACKING (CHEMICAL ENGINEERING)

CHEMICAL EQUILIBRIUM

CHEMICAL EVOLUTION

CHEMICAL EXPLOSIONS

Chemical Extinguishers
   USE   FIRE EXTINGUISHERS

CHEMICAL FRACTIONATION

CHEMICAL FUELS

CHEMICAL INDICATORS

Chemical Kinetics
   USE   REACTION KINETICS

CHEMICAL LASERS

CHEMICAL MACHINING

Chemical Milling
   USE   CHEMICAL MACHINING

CHEMICAL PROPERTIES

CHEMICAL PROPULSION

CHEMICAL REACTION CONTROL

CHEMICAL REACTIONS

CHEMICAL REACTORS

Chemical Relaxation
   USE   MOLECULAR RELAXATION

CHEMICAL RELEASE MODULES

Chemical Shift
   USE   CHEMICAL EQUILIBRIUM

CHEMICAL STERILIZATION

CHEMICAL TESTS

CHEMICAL WARFARE

CHEMICALS

CHEMILUMINESCENCE

CHEMISORPTION

CHEMISTRY

Chemistry, Aerothermo
   USE   AEROTHERMOCHEMISTRY

Chemistry, Analytical
   USE   ANALYTICAL CHEMISTRY

Chemistry, Atmospheric
   USE   ATMOSPHERIC CHEMISTRY

Chemistry, Bio
   USE   BIOCHEMISTRY

Chemistry, Biogeo
   USE   BIOGEOCHEMISTRY

(Chemistry), Buffers
   USE   BUFFERS (CHEMISTRY)

Chemistry, Computational
   USE   COMPUTATIONAL CHEMISTRY

Chemistry, Cryo
   USE   CRYOCHEMISTRY

Chemistry, Electro
   USE   ELECTROCHEMISTRY

Chemistry, Environmental
   USE   ENVIRONMENTAL CHEMISTRY

Chemistry Experiment In Space, Physics And
   USE   PHYSICS AND CHEMISTRY EXPERIMENT IN
           SPACE

Chemistry, Geo
   USE   GEOCHEMISTRY

Chemistry, Inorganic
   USE   INORGANIC CHEMISTRY

Chemistry, Interstellar
   USE   INTERSTELLAR CHEMISTRY

Chemistry, Marine
   USE   MARINE CHEMISTRY

Chemistry, Nuclear
   USE   NUCLEAR CHEMISTRY

Chemistry, Organic
   USE   ORGANIC CHEMISTRY

Chemistry, Photoelectro
   USE   PHOTOELECTROCHEMISTRY

Chemistry, Physical
   USE   PHYSICAL CHEMISTRY

Chemistry, Physio
   USE   PHYSIOCHEMISTRY

Chemistry, Plasma
   USE   PLASMA CHEMISTRY

Chemistry, Polymer
   USE   POLYMER CHEMISTRY

(Chemistry), Precipitation
   USE   PRECIPITATION (CHEMISTRY)

Chemistry, Propellant
   USE   PROPELLANT CHEMISTRY

Chemistry, Quantum
   USE   QUANTUM CHEMISTRY

Chemistry, Radiation
USE   RADIATION CHEMISTRY

Chemistry, Radio
USE   RADIOCHEMISTRY

Chemistry, Reactor
USE   RADIOCHEMISTRY

(Chemistry), Reduction
USE   REDUCTION (CHEMISTRY)

(Chemistry), Saturation
USE   SATURATION (CHEMISTRY)

Chemistry, Stereo
USE   STEREOCHEMISTRY

(Chemistry), Synthesis
USE   SYNTHESIS (CHEMISTRY)

Chemistry, Thermo
USE   THERMOCHEMISTRY

(Chemistry), Unsaturation
USE   UNSATURATION (CHEMISTRY)

Chemonuclear Propulsion
USE   CHEMICAL PROPULSION
       NUCLEAR PROPULSION

CHEMORECEPTORS

CHEMOSPHERE

CHEMOTHERAPY

CHENA RIVER BASIN (AK)

CHESAPEAKE BAY (US)

CHEST

Chewing
USE   MASTICATION

CHIAPAS (MEXICO)

CHIASMS

CHICKENS

CHILD DEVICE

CHILD-LANGMUIR LAW

CHILDREN

CHILE

Chilling
USE   COOLING

Chilling, Heat Dissipation
USE   COOLING

Chimes
USE   AUDITORY SIGNALS

CHIMNEYS

CHIMPANZEES

CHIN

CHINA

China (Communist) Mainland
USE   CHINA

China, Republic Of
USE   TAIWAN

CHINESE AIRCRAFT

Chinese Peoples Republic
USE   CHINA

CHINESE SPACE PROGRAM

CHINESE SPACECRAFT

Chinook Helicopter
USE   CH-47 HELICOPTER

CHIPPING

CHIPS

CHIPS (ELECTRONICS)

CHIPS (MEMORY DEVICES)

CHIRAL DYNAMICS

CHIRON

CHIRONOMUS FLIES

CHIRP

CHIRP SIGNALS

CHITIN

CHLORAL

CHLORATES

Chlorates, Per
USE   PERCHLORATES

CHLORELLA

Chloride Lasers, Xenon
USE   XENON CHLORIDE LASERS

Chloride, Methyl
USE   METHYL CHLORIDE

Chloride, Polyvinyl
USE   POLYVINYL CHLORIDE

CHLORIDES

Chlorides, Aluminum
USE   ALUMINUM CHLORIDES

Chlorides, Ammonium
USE   AMMONIUM CHLORIDES

Chlorides, Beryllium
USE   BERYLLIUM CHLORIDES

Chlorides, Boron
USE   BORON CHLORIDES

Chlorides, Cadmium
USE   CADMIUM CHLORIDES

Chlorides, Calcium
USE   CALCIUM CHLORIDES

Chlorides, Copper
USE   COPPER CHLORIDES

Chlorides, Di
USE   DICHLORIDES

Chlorides, Germanium
USE   GERMANIUM CHLORIDES

Chlorides, Hydro
USE   HYDROCHLORIDES

Chlorides, Hydrogen
USE   HYDROGEN CHLORIDES

Chlorides, Iron
USE   IRON CHLORIDES

Chlorides, Lanthanum
USE   LANTHANUM CHLORIDES

Chlorides, Lead
USE   LEAD CHLORIDES

Chlorides, Lithium
USE   LITHIUM CHLORIDES

Chlorides, Magnesium
USE   MAGNESIUM CHLORIDES

Chlorides, Nitrosyl
USE   NITROSYL CHLORIDES

Chlorides, Nitroxy
USE   NITROXYCHLORIDES

Chlorides, Nitryl
USE   NITRYL CHLORIDES

Chlorides, Potassium
USE   POTASSIUM CHLORIDES

Chlorides, Silver
USE   SILVER CHLORIDES

Chlorides, Sodium
USE   SODIUM CHLORIDES

Chlorides, Sulfur
USE   SULFUR CHLORIDES

Chlorides, Tetra
USE   TETRACHLORIDES

Chlorides, Titanium
USE   TITANIUM CHLORIDES

Chlorides, Tungsten
USE   TUNGSTEN CHLORIDES

Chlorides, Zinc
USE   ZINC CHLORIDES

CHLORINATION

CHLORINE

Chlorine Batteries, Zinc-
USE   ZINC-CHLORINE BATTERIES

CHLORINE COMPOUNDS

CHLORINE FLUORIDES

CHLORINE OXIDES

CHLOROAROMATICS

CHLOROBENZENES

CHLOROCARBONS

Chlorodifluoroacetates, Sodium
USE   SODIUM CHLORODIFLUOROACETATES

CHLOROETHYLENE

CHLOROFORM

CHLOROFORMATE

CHLOROPHYLLS

CHLOROPLASTS

CHLOROPRENE RESINS

CHLOROSILANES

Chlorosilanes, Methyl
USE   METHYL CHLOROSILANES

CHLORPROMAZINE

Choctaw Helicopter
USE   CH-34 HELICOPTER

Choice
USE   SELECTION

CHOKES

CHOKES (FUEL SYSTEMS)

CHOKES (RESTRICTIONS)

CHOLERA

CHOLESKY FACTORIZATION

CHOLESTEROL

CHOLINE

Cholinergic Blocking Agents
USE   ANTICHOLINERGICS

CHOLINERGICS

CHOLINESTERASE

Chondrite, Hvittis
USE   HVITTIS CHONDRITE

CHONDRITES

Chondrites, Carbonaceous
USE   CARBONACEOUS CHONDRITES

Chondrites, Pantar
USE   PANTAR CHONDRITES

CHONDRULE

Choppers (Electric)
USE   ELECTRIC CHOPPERS

Choppers, Electric
USE   ELECTRIC CHOPPERS

Chords, Aerodynamic
USE   AIRFOIL PROFILES
       CHORDS (GEOMETRY)

CHORDS (GEOMETRY)

CHOROID MEMBRANES

Chorus, Dawn
USE   DAWN CHORUS

Chorus (Dawn Phenomenon)
USE   DAWN CHORUS

Chorus Phenomenon
USE   DAWN CHORUS

Christoffel Transformation, Schwarz-
USE   SCHWARZ-CHRISTOFFEL
       TRANSFORMATION

CHROMATES

Chromates, Potassium
USE   POTASSIUM CHROMATES

CHROMATOGRAPHY

Chromatography, Gas
USE   GAS CHROMATOGRAPHY

Chromatography, Gel Permeation
USE   LIQUID CHROMATOGRAPHY

Chromatography, Liquid
USE   LIQUID CHROMATOGRAPHY

Chromatography, Paper
USE   PAPER CHROMATOGRAPHY

Chromatography, Thin Layer
USE   THIN LAYER CHROMATOGRAPHY

Chrome
USE   CHROMIUM

CHROMIC ACID

CHROMITES

Chromites, Sodium
USE   SODIUM CHROMITES

CHROMIUM

CHROMIUM ALLOYS

CHROMIUM BORIDES

CHROMIUM BROMIDES

CHROMIUM CARBIDES

CHROMIUM COMPOUNDS

CHROMIUM FLUORIDES

CHROMIUM ISOTOPES

CHROMIUM OXIDES

CHROMIUM STEELS

Chromodynamics, Quantum
USE   QUANTUM CHROMODYNAMICS

CHROMOSOMES

CHROMOSPHERE

CHRONAXY

CHRONIC CONDITIONS

Chronographs
USE   CHRONOMETERS

CHRONOLOGY

Chronology, Geo
USE   GEOCHRONOLOGY

CHRONOMETERS

CHRONOPHOTOGRAPHY

Chronotrons
USE   TIME LAG
       PULSE RATE

CHUCKCHI SEA

Chugging
USE   COMBUSTION STABILITY

CHUTES

Chutes, Drag
USE   DRAG CHUTES

CID
USE   CHARGE INJECTION DEVICES

Cinder Cones
USE   CONES (VOLCANOES)

Cinefluorography
USE   MOTION PICTURES
       RADIOGRAPHY

CINEMATOGRAPHY

Cinematography, Lunar
USE   LUNAR PHOTOGRAPHY

Cineradiography
USE   RADIOGRAPHY
       MOTION PICTURES

CINESPECTROGRAPHS

CINETHEODOLITES

CIRCADIAN RHYTHMS

Circle Turning Flight, Minor
USE   MINOR CIRCLE TURNING FLIGHT

CIRCLES (GEOMETRY)

Circles, Great
USE   GREAT CIRCLES

Circles, Mohr
USE   FRACTURE MECHANICS

Circles, Rowland
USE   ROWLAND CIRCLES

Circuit Analysis, Sneak
USE   SNEAK CIRCUIT ANALYSIS

CIRCUIT BOARDS

CIRCUIT BREAKERS

Circuit Currents, Short
USE   SHORT CIRCUIT CURRENTS

CIRCUIT DIAGRAMS

CIRCUIT PROTECTION

CIRCUIT RELIABILITY

Circuit Television, Closed
USE   CLOSED CIRCUIT TELEVISION

Circuit Voltage, Open
USE   OPEN CIRCUIT VOLTAGE

CIRCUITS

(Circuits), Adders
USE   ADDING CIRCUITS

Circuits, Adding
USE   ADDING CIRCUITS

Circuits, Analog
USE   ANALOG CIRCUITS

Circuits, Bistable
USE   BISTABLE CIRCUITS

Circuits, Cancellation
USE   CANCELLATION CIRCUITS

Circuits), Circulators (Phase Shift
USE   CIRCULATORS (PHASE SHIFT CIRCUITS)

Circuits, Clamping
USE   CLAMPING CIRCUITS

Circuits, Clipper
USE   CLIPPER CIRCUITS

Circuits, Coincidence
USE   COINCIDENCE CIRCUITS

Circuits, Comparator
USE   COMPARATOR CIRCUITS

Circuits, Conjugated
USE   CONJUGATED CIRCUITS

Circuits, Counting
USE   COUNTING CIRCUITS

Circuits, Coupling
USE   COUPLING CIRCUITS

Circuits, Delay
USE   DELAY CIRCUITS

Circuits, Diode-Transistor-Logic Integ
USE   DTL INTEGRATED CIRCUITS

Circuits, DTL Integrated
USE   DTL INTEGRATED CIRCUITS

Circuits, Electric
USE   CIRCUITS

(Circuits), Equalizers
USE   EQUALIZERS (CIRCUITS)

Circuits, Equivalent
USE   EQUIVALENT CIRCUITS

Circuits, Exploding Conductor
USE   EXPLODING WIRES
       CIRCUITS

Circuits, Feedback
USE   FEEDBACK CIRCUITS

Circuits, Fire Control
USE   FIRE CONTROL CIRCUITS

Circuits, Fluidic
USE   FLUIDIC CIRCUITS

(Circuits), Gates
    USE   GATES (CIRCUITS)

Circuits, Hybrid
    USE   HYBRID CIRCUITS

Circuits, Integrated
    USE   INTEGRATED CIRCUITS

Circuits, LC
    USE   LC CIRCUITS

Circuits, Limiter
    USE   LIMITER CIRCUITS

Circuits, Linear
    USE   LINEAR CIRCUITS

Circuits, Linear Integrated
    USE   LINEAR INTEGRATED CIRCUITS

Circuits, Logic
    USE   LOGIC CIRCUITS

Circuits, LR
    USE   RL CIRCUITS

Circuits, LRC
    USE   RLC CIRCUITS

Circuits, Magnetic
    USE   MAGNETIC CIRCUITS

(Circuits), Matrices
    USE   MATRICES (CIRCUITS)

Circuits, Microwave
    USE   MICROWAVE CIRCUITS

Circuits, Mixing
    USE   MIXING CIRCUITS

Circuits, Monolithic
    USE   INTEGRATED CIRCUITS

Circuits, Negative Resistance
    USE   NEGATIVE RESISTANCE CIRCUITS

Circuits, Phase Shift
    USE   PHASE SHIFT CIRCUITS

Circuits, Pneumatic
    USE   PNEUMATIC CIRCUITS

Circuits, Power Supply
    USE   POWER SUPPLY CIRCUITS

Circuits, Printed
    USE   PRINTED CIRCUITS

Circuits, RC
    USE   RC CIRCUITS

Circuits, RL
    USE   RL CIRCUITS

Circuits, RLC
    USE   RLC CIRCUITS

Circuits, Short
    USE   SHORT CIRCUITS

Circuits, Squelch
    USE   SQUELCH CIRCUITS

Circuits, Sweep
    USE   SWEEP CIRCUITS

Circuits, Switching
    USE   SWITCHING CIRCUITS

Circuits, Transistor
    USE   TRANSISTOR CIRCUITS

Circuits, Transistor-Transistor-Logic Integ
    USE   TTL INTEGRATED CIRCUITS

Circuits, Transmission
    USE   TRANSMISSION CIRCUITS

Circuits, Trigger
    USE   TRIGGER CIRCUITS

Circuits, TTL Integrated
    USE   TTL INTEGRATED CIRCUITS

Circuits, Varactor Diode
    USE   VARACTOR DIODE CIRCUITS

Circuits, Very High Speed Integrated
    USE   VHSIC (CIRCUITS)

(Circuits), Vhsic
    USE   VHSIC (CIRCUITS)

Circuits, Wire Bridge
    USE   WIRE BRIDGE CIRCUITS

CIRCULAR CONES

CIRCULAR CYLINDERS

CIRCULAR ORBITS

CIRCULAR PLATES

CIRCULAR POLARIZATION

CIRCULAR SHELLS

CIRCULAR TUBES

CIRCULAR WAVEGUIDES

CIRCULATION

Circulation, Atmospheric
    USE   ATMOSPHERIC CIRCULATION

Circulation, Blood
    USE   BLOOD CIRCULATION

Circulation, Brain
    USE   BRAIN CIRCULATION

Circulation, Capillary
    USE   CAPILLARY FLOW

CIRCULATION CONTROL AIRFOILS

CIRCULATION CONTROL ROTORS

Circulation, Coronary
    USE   CORONARY CIRCULATION

CIRCULATION DISTRIBUTION

Circulation Experiment, Atmospheric General
    USE   ATMOSPHERIC GENERAL CIRCULATION
           EXPERIMENT

Circulation, Intercranial
    USE   INTERCRANIAL CIRCULATION

Circulation, Ocular
    USE   OCULAR CIRCULATION

Circulation, Peripheral
    USE   PERIPHERAL CIRCULATION

Circulation, Pulmonary
    USE   PULMONARY CIRCULATION

Circulation), Registers (Air
    USE   REGISTERS (AIR CIRCULATION)

Circulation, Water
    USE   WATER CIRCULATION

Circulation, Wind
    USE   ATMOSPHERIC CIRCULATION

CIRCULATORS (PHASE SHIFT CIRCUITS)

CIRCULATORY SYSTEM

CIRCUMFERENCES

CIRCUMLUNAR COMMUNICATION

CIRCUMLUNAR TRAJECTORIES

CIRCUMPOLAR WESTERLIES

CIRCUMSOLAR RADIATION

CIRCUMSOLAR TELESCOPES

Circumstellar Matter
    USE   STELLAR ENVELOPES

CIRQUES (LANDFORMS)

CIRROCUMULUS CLOUDS

CIRROSTRATUS CLOUDS

CIRRUS CLOUDS

CIRRUS SHIELDS

CISLUNAR SPACE

CITIES

CITRATES

CITRIC ACID

CITRUS TREES

City Corridor (MO), St Louis-Kansas
    USE   ST LOUIS-KANSAS CITY CORRIDOR (MO)

City (NY), New York
    USE   NEW YORK CITY (NY)

City, Vatican
    USE   VATICAN CITY

CIVIL AVIATION

CIVIL DEFENSE

CI
    USE   CHLORINE

CL-41 AIRCRAFT

CL-41 Aircraft, Canadair
    USE   CL-41 AIRCRAFT

CL-44 AIRCRAFT

CL-44 Aircraft, Canadair
    USE   CL-44 AIRCRAFT

CL-84 AIRCRAFT

CL-84 Aircraft, Canadair
    USE   CL-84 AIRCRAFT

CL-595 Helicopter
    USE   XH-51 HELICOPTER

CL-595 Helicopter, Lockheed
    USE   XH-51 HELICOPTER

CL-600 CHALLENGER AIRCRAFT

CL-823 AIRCRAFT

CL-823 Aircraft, Lockheed
    USE   CL-823 AIRCRAFT

CLADDING

CLAIMING

CLAMPING CIRCUITS

CLAMPS

CLARITY

Clark Y Airfoil
    USE   AIRFOIL PROFILES

CLASSES

Classic Aircraft
    USE   IL-62 AIRCRAFT

CLASSICAL MECHANICS

CLASSIFICATIONS

CLASSIFIERS

CLASSIFYING

CLATHRATES

CLAYS

CLEAN ENERGY

CLEAN FUELS

CLEAN ROOMS

CLEANERS

(Cleaners), Washers
USE   WASHERS (CLEANERS)

CLEANING

Cleaning, Chemical
USE   CHEMICAL CLEANING

Cleaning, Ultrasonic
USE   ULTRASONIC CLEANING

CLEANLINESS

CLEAR AIR TURBULENCE

CLEARANCES

CLEARING

CLEARINGS (OPENINGS)

CLEAVAGE

CLEBSCH-GORDAN COEFFICIENTS

CLIFFS

CLIMATE

CLIMATOLOGY

Climatology, Agro
USE   AGROCLIMATOLOGY

Climatology, Micro
USE   MICROCLIMATOLOGY

Climb Indicators, Rate Of
USE   RATE OF CLIMB INDICATORS

CLIMBING FLIGHT

CLINICAL MEDICINE

CLIPPER CIRCUITS

CLIPS

CLOCK PARADOX

CLOCKS

Clocks, Atomic
USE   ATOMIC CLOCKS

Clocks, Autonomous Spacecraft
USE   AUTONOMOUS SPACECRAFT CLOCKS

Clocks, Biological
USE   RHYTHM (BIOLOGY)

Clogging
USE   PLUGGING

CLOSE PACKED LATTICES

Closed Basins
USE   STRUCTURAL BASINS

CLOSED CIRCUIT TELEVISION

CLOSED CYCLES

CLOSED ECOLOGICAL SYSTEMS

Closed Faults
USE   GEOLOGICAL FAULTS

Closed Loop Systems
USE   FEEDBACK CONTROL

CLOSING

CLOSTRIDIUM BOTULINUM

Closure, Crack
USE   CRACK CLOSURE

CLOSURE LAW

CLOSURES

Cloth
USE   FABRICS

Cloth, Wire
USE   WIRE CLOTH

CLOTHING

Clothing, Flight
USE   FLIGHT CLOTHING

Clothing, Protective
USE   PROTECTIVE CLOTHING

Clothing, Vapor Barrier
USE   VAPOR BARRIER CLOTHING

CLOTTING

CLOUD CHAMBERS

CLOUD COVER

CLOUD DISPERSAL

CLOUD GLACIATION

CLOUD HEIGHT INDICATORS

Cloud Patterns, Chaotic
USE   CLOUDS (METEOROLOGY)

CLOUD PHOTOGRAPHS

CLOUD PHOTOGRAPHY

CLOUD PHYSICS

Cloud Physics Lab (Spacelab), Atmospheric
USE   ATMOSPHERIC CLOUD PHYSICS LAB
      (SPACELAB)

CLOUD SEEDING

CLOUDS

Clouds, Anvil
USE   ANVIL CLOUDS

Clouds, Arc
USE   ARC CLOUDS

Clouds, Artificial
USE   ARTIFICIAL CLOUDS

Clouds, Barium Ion
USE   BARIUM ION CLOUDS

Clouds, Cap
USE   CAP CLOUDS

Clouds, Chemical
USE   CHEMICAL CLOUDS

Clouds, Cirrocumulus
USE   CIRROCUMULUS CLOUDS

Clouds, Cirrostratus
USE   CIRROSTRATUS CLOUDS

Clouds, Cirrus
USE   CIRRUS CLOUDS

Clouds, Convection
USE   CONVECTION CLOUDS

Clouds, Cumulonimbus
USE   CUMULONIMBUS CLOUDS

Clouds, Cumulus
USE   CUMULUS CLOUDS

Clouds, Electron
USE   ELECTRON CLOUDS

Clouds, Hydrogen
USE   HYDROGEN CLOUDS

Clouds, Magellanic
USE   MAGELLANIC CLOUDS

Clouds, Meteoroid Dust
USE   METEOROID DUST CLOUDS

CLOUDS (METEOROLOGY)

Clouds, Molecular
USE   MOLECULAR CLOUDS

Clouds, Nimbostratus
USE   NIMBOSTRATUS CLOUDS

Clouds, Nimbus
USE   NIMBOSTRATUS CLOUDS

Clouds, Noctilucent
USE   NOCTILUCENT CLOUDS

Clouds, Ophiuchi
USE   OPHIUCHI CLOUDS

Clouds, Orographic
USE   CAP CLOUDS

Clouds, Plasma
USE   PLASMA CLOUDS

Clouds, Stratocumulus
USE   STRATOCUMULUS CLOUDS

Clouds, Stratus
USE   STRATUS CLOUDS

Clouds, Venus
USE   VENUS CLOUDS

CLUMPS

CLUSTER ANALYSIS

Cluster, Virgo Galactic
USE   VIRGO GALACTIC CLUSTER

Cluster, Virgo Star
USE   VIRGO GALACTIC CLUSTER

Clusters
USE   CLUMPS

Clusters, Galactic
USE   GALACTIC CLUSTERS

Clusters, Globular
USE   GLOBULAR CLUSTERS

Clusters, Praesepe Star
USE   PRAESEPE STAR CLUSTERS

Clusters, Star
USE   STAR CLUSTERS

CLUTCHES

CLUTTER

Clutter Maps, Radar
USE   RADAR CLUTTER MAPS

Cm
USE   CURIUM

CMOS

CN EMISSION

CNOIDAL WAVES

Co
    USE   COBALT

CO
    USE   COLORADO

(CO), Manitou
    USE   MANITOU (CO)

(CO), Pike's Peak
    USE   PIKE'S PEAK (CO)

(CO), San Juan Mountains
    USE   SAN JUAN MOUNTAINS (CO)

COACHELLA VALLEY (CA)

COAGULATION

Coagulation, Blood
    USE   BLOOD COAGULATION

COAL

Coal, Char
    USE   CHARCOAL

COAL DERIVED GASES

COAL DERIVED LIQUIDS

COAL GASIFICATION

COAL LIQUEFACTION

Coal, Solvent Refined
    USE   SOLVENT REFINED COAL

COAL UTILIZATION

Coalescence
    USE   COALESCING

COALESCING

COANDA EFFECT

COARSENESS

Coast, Ivory
    USE   IVORY COAST

COASTAL CURRENTS

Coastal Dunes
    USE   DUNES

COASTAL ECOLOGY

Coastal Marshlands
    USE   MARSHLANDS

COASTAL PLAINS

COASTAL RANGES (CA)

COASTAL WATER

COASTAL ZONE COLOR SCANNER

COASTING FLIGHT

COASTS

COATING

COATINGS

Coatings, Aluminum
    USE   ALUMINUM COATINGS

Coatings, Anodic
    USE   ANODIC COATINGS

Coatings, Antiradar
    USE   ANTIRADAR COATINGS

Coatings, Antireflection
    USE   ANTIREFLECTION COATINGS

Coatings, Birefringent
    USE   BIREFRINGENT COATINGS

Coatings, Cathodic
    USE   CATHODIC COATINGS

Coatings, Ceramal Protective
    USE   PROTECTIVE COATINGS
           CERMETS

Coatings, Ceramic
    USE   CERAMIC COATINGS

Coatings, Glass
    USE   GLASS COATINGS

Coatings, Gold
    USE   GOLD COATINGS

Coatings, Inorganic
    USE   INORGANIC COATINGS

Coatings, Metal
    USE   METAL COATINGS

Coatings, Nickel
    USE   NICKEL COATINGS

Coatings, Plastic
    USE   PLASTIC COATINGS

(Coatings), Primers
    USE   PRIMERS (COATINGS)

Coatings, Protective
    USE   PROTECTIVE COATINGS

Coatings, Refractory
    USE   REFRACTORY COATINGS

Coatings, Rubber
    USE   RUBBER COATINGS

Coatings, Solar Selective
    USE   SELECTIVE SURFACES

Coatings, Sprayed
    USE   SPRAYED COATINGS

Coatings, Sprayed Protective
    USE   SPRAYED COATINGS
           PROTECTIVE COATINGS

Coatings, Thermal Control
    USE   THERMAL CONTROL COATINGS

Coatings, Zinc
    USE   ZINC COATINGS

COAXIAL CABLES

COAXIAL FLOW

COAXIAL NOZZLES

COAXIAL PLASMA ACCELERATORS

Coaxial Transmission
    USE   COAXIAL CABLES
           TRANSMISSION

Coaxial Transmission Lines, Flat
    USE   MICROSTRIP TRANSMISSION LINES

COBALT

COBALT ACETATES

COBALT ALLOYS

COBALT COMPOUNDS

COBALT FLUORIDES

COBALT ISOTOPES

COBALT OXALATES

COBALT OXIDES

COBALT 58

COBALT 60

COBE
    USE   COSMIC BACKGROUND EXPLORER
           SATELLITE

COBOL

COBRA DANE (RADAR)

COCCOMYCES

COCHLEA

Cock Aircraft
    USE   AN-22 AIRCRAFT

COCKPIT SIMULATORS

COCKPITS

COCKROACHES

COCKS

COD Aircraft
    USE   C-2 AIRCRAFT

Code, Biternary
    USE   BITERNARY CODE

CODE DIVISION MULTIPLE ACCESS

CODE DIVISION MULTIPLEXING

Code, Genetic
    USE   GENETIC CODE

Code, Legendre
    USE   COMPUTER PROGRAMMING
           NEUTRON SCATTERING

Code Modulation, Differential Pulse
    USE   DIFFERENTIAL PULSE CODE MODULATION

Code Modulation, Pulse
    USE   PULSE CODE MODULATION

Code, Morse
    USE   MORSE CODE

CODERS

Coders, Auto
    USE   AUTOCODERS

Coders, De
    USE   DECODERS

Coders, Vo
    USE   VOCODERS

CODES

Codes, BCH
    USE   BCH CODES

Codes, Binary
    USE   BINARY CODES

Codes, Bose-Chaudhuri-Hocquenghem
    USE   BCH CODES

Codes, Concatenated
    USE   CONCATENATED CODES

Codes, Error Correcting
    USE   ERROR CORRECTING CODES

Codes, Error Detection
    USE   ERROR DETECTION CODES

CODING

Coding, Color
USE   COLOR CODING

Coding, De
USE   DECODING

Coefficient, Absorption
USE   ABSORPTIVITY

Coefficient, Accommodation
USE   ACCOMMODATION COEFFICIENT

Coefficient, Coherence
USE   COHERENCE COEFFICIENT

Coefficient, Diffusion
USE   DIFFUSION COEFFICIENT

Coefficient, Discharge
USE   DISCHARGE COEFFICIENT

Coefficient, Friction
USE   COEFFICIENT OF FRICTION

Coefficient, Friction Loss
USE   FRICTION FACTOR

Coefficient, Glauert
USE   MACH NUMBER
       AERODYNAMIC FORCES

Coefficient, Hall
USE   HALL EFFECT

Coefficient, Influence
USE   INFLUENCE COEFFICIENT

Coefficient, Nozzle
USE   NOZZLE FLOW

COEFFICIENT OF FRICTION

Coefficient, Onsager Phenomenological
USE   ONSAGER PHENOMENOLOGICAL
       COEFFICIENT

Coefficient, Racah
USE   RACAH COEFFICIENT

Coefficient, Recombination
USE   RECOMBINATION COEFFICIENT

Coefficient, Reflection
USE   REFLECTANCE

Coefficient, Seebeck
USE   SEEBECK EFFECT

(Coefficient), SIC
USE   STRUCTURAL INFLUENCE COEFFICIENTS

Coefficient, Soret
USE   SORET COEFFICIENT

Coefficient, Wigner
USE   WIGNER COEFFICIENT

COEFFICIENTS

Coefficients, Aerodynamic
USE   AERODYNAMIC COEFFICIENTS

Coefficients, Attenuation
USE   ATTENUATION COEFFICIENTS

Coefficients, Binomial
USE   BINOMIAL COEFFICIENTS

Coefficients, Clebsch-Gordan
USE   CLEBSCH-GORDAN COEFFICIENTS

Coefficients, Correlation
USE   CORRELATION COEFFICIENTS

Coefficients, Coupling
USE   COUPLING COEFFICIENTS

Coefficients, Drag
USE   DRAG COEFFICIENTS

Coefficients, Flow
USE   FLOW COEFFICIENTS

Coefficients, Heat Transfer
USE   HEAT TRANSFER COEFFICIENTS

Coefficients, Hydrodynamic
USE   HYDRODYNAMIC COEFFICIENTS

Coefficients, Ionization
USE   IONIZATION COEFFICIENTS

Coefficients, Lift
USE   LIFT
       AERODYNAMIC COEFFICIENTS

Coefficients, Nozzle Thrust
USE   NOZZLE THRUST COEFFICIENTS

Coefficients, Regression
USE   REGRESSION COEFFICIENTS

Coefficients, Resistance
USE   RESISTANCE

Coefficients, Scattering
USE   SCATTERING COEFFICIENTS

Coefficients, Structural Influence
USE   STRUCTURAL INFLUENCE COEFFICIENTS

Coefficients, Thermal Accommodation
USE   ACCOMMODATION COEFFICIENT

Coefficients, Transport
USE   TRANSPORT PROPERTIES

Coefficients, Virial
USE   VIRIAL COEFFICIENTS

COENZYMES

COERCIVITY

COESITE

COFFEE

COFFIN-MANSON LAW

COGENERATION

COGNITION

COGNITIVE PSYCHOLOGY

COGO (PROGRAMMING LANGUAGE)

COHENITE

COHERENCE

COHERENCE COEFFICIENT

Coherence, In
USE   INCOHERENCE

Coherence, Phase
USE   PHASE COHERENCE

COHERENT ACOUSTIC RADIATION

Coherent Anti-Stokes Raman Spectroscopy
USE   RAMAN SPECTROSCOPY

COHERENT ELECTROMAGNETIC RADIATION

COHERENT LIGHT

COHERENT RADAR

COHERENT RADIATION

COHERENT SCATTERING

Coherent Sources
USE   COHERENT RADIATION
       RADIATION SOURCES

Coherent Transmission
USE   COHERENT RADIATION

COHESION

Cohomology
USE   HOMOLOGY

COILS

Coils, Electric
USE   ELECTRIC COILS

Coils, Field
USE   FIELD COILS

Coils, Magnet
USE   MAGNET COILS

Coils, Magnetic
USE   MAGNETIC COILS

COIN AIRCRAFT

COINCIDENCE CIRCUITS

COINING

COKE

Coke Aircraft
USE   AN-24 AIRCRAFT

COLCHICINE

COLD ACCLIMATIZATION

COLD BOKKEVELD METEORITE

COLD CATHODE TUBES

COLD CATHODES

COLD DRAWING

COLD FLOW TESTS

Cold Forming
USE   COLD WORKING

COLD FRONTS

COLD GAS

COLD HARDENING

COLD NEUTRONS

COLD PLASMAS

COLD PRESSING

COLD ROLLING

COLD STRENGTH

COLD SURFACES

COLD TOLERANCE

COLD TRAPS

Cold Walls
USE   COLD SURFACES
       WALLS

COLD WATER

COLD WEATHER

COLD WEATHER TESTS

COLD WELDING

COLD WORKING

COLEOPTERA

COLIC

COLLAGENS

COLLAPSE

Collapse, Gravitational
    USE   GRAVITATIONAL COLLAPSE

COLLATING

COLLECTION

Collection Platforms, Data
    USE   DATA COLLECTION PLATFORMS

Collectors
    USE   ACCUMULATORS

Collectors, Dust
    USE   DUST COLLECTORS

Collectors, Solar
    USE   SOLAR COLLECTORS

Colleges
    USE   UNIVERSITIES

COLLIMATION

COLLIMATORS

COLLINEARITY

COLLISION AVOIDANCE

Collision Avoidance System, Beacon
    USE   BEACON COLLISION AVOIDANCE SYSTEM

COLLISION PARAMETERS

COLLISION RATES

Collision Warning Devices
    USE   COLLISION AVOIDANCE
            WARNING SYSTEMS

COLLISIONAL PLASMAS

COLLISIONLESS PLASMAS

COLLISIONS

Collisions, Atomic
    USE   ATOMIC COLLISIONS

Collisions, Bird-Aircraft
    USE   BIRD-AIRCRAFT COLLISIONS

Collisions, Coulomb
    USE   COULOMB COLLISIONS

Collisions, Elastic
    USE   ELASTIC SCATTERING

Collisions, Electron
    USE   ELECTRON SCATTERING

Collisions, Inelastic
    USE   INELASTIC COLLISIONS

Collisions, Ionic
    USE   IONIC COLLISIONS

Collisions, Meteorite
    USE   METEORITE COLLISIONS

Collisions, Midair
    USE   MIDAIR COLLISIONS

Collisions, Molecular
    USE   MOLECULAR COLLISIONS

Collisions, Particle
    USE   PARTICLE COLLISIONS

COLLOCATION

COLLOIDAL GENERATORS

COLLOIDAL PROPELLANTS

COLLOIDING

COLLOIDS

COLOMBIA

(Colombia), Llanos Orientales
    USE   LLANOS ORIENTALES (COLOMBIA)

(Colombia), Magdalena-Cauca Valley
    USE   MAGDALENA-CAUCA VALLEY (COLOMBIA)

COLONIES

Colonies, Space
    USE   SPACE COLONIES

COLOR

COLOR CENTERS

COLOR CODING

COLOR INFRARED PHOTOGRAPHY

Color (Particle Physics)
    USE   QUANTUM CHROMODYNAMICS

Color Perception
    USE   COLOR VISION

COLOR PHOTOGRAPHY

Color Scanner, Coastal Zone
    USE   COASTAL ZONE COLOR SCANNER

Color Scanner, Ocean
    USE   OCEAN COLOR SCANNER

Color, Stellar
    USE   STELLAR COLOR

COLOR TELEVISION

COLOR VISION

Color, Water
    USE   WATER COLOR

COLOR-MAGNITUDE DIAGRAM

COLORADO

COLORADO PLATEAU (US)

COLORADO RIVER (NORTH AMERICA)

Coloration
    USE   COLOR

COLORIMETRY

Cols
    USE   GAPS (GEOLOGY)

Columbia, British
    USE   BRITISH COLUMBIA

Columbia, District Of
    USE   DISTRICT OF COLUMBIA

COLUMBIA (ORBITER)

COLUMBIA RIVER BASIN (ID-OR-WA)

Columbium
    USE   NIOBIUM

Column Antennas, Hoop
    USE   HOOP COLUMN ANTENNAS

Column, Vertebral
    USE   VERTEBRAL COLUMN

COLUMNS

COLUMNS (PROCESS ENGINEERING)

COLUMNS (SUPPORTS)

Columns, Tapered
    USE   TAPERED COLUMNS

Columns, Vortex
    USE   VORTICES

COMA

COMBAT

Combat Aircraft, Multi-Role
    USE   MRCA AIRCRAFT

(Combat Vehicles), Tanks
    USE   TANKS (COMBAT VEHICLES)

COMBINATION

COMBINATIONS (MATHEMATICS)

COMBINATORIAL ANALYSIS

COMBINED CYCLE POWER GENERATION

COMBINED STRESS

Combustibility
    USE   FLAMMABILITY

COMBUSTIBLE FLOW

COMBUSTION

Combustion, Acoustic
    USE   COMBUSTION STABILITY

Combustion, Boundary Layer
    USE   BOUNDARY LAYER COMBUSTION

COMBUSTION CHAMBERS

COMBUSTION CONTROL

COMBUSTION EFFICIENCY

Combustion Engines, External
    USE   EXTERNAL COMBUSTION ENGINES

Combustion Engines, Internal
    USE   INTERNAL COMBUSTION ENGINES

Combustion, Fuel
    USE   FUEL COMBUSTION

Combustion Heat
    USE   HEAT OF COMBUSTION

Combustion, Heat Of
    USE   HEAT OF COMBUSTION

Combustion, Hybrid
    USE   HYBRID PROPELLANT ROCKET ENGINES

Combustion, Hydrocarbon
    USE   HYDROCARBON COMBUSTION

Combustion, Hypersonic
    USE   HYPERSONIC COMBUSTION

Combustion Instability
    USE   COMBUSTION STABILITY

Combustion, Metal
    USE   METAL COMBUSTION

COMBUSTION PHYSICS

COMBUSTION PRODUCTS

Combustion, Propellant
    USE   PROPELLANT COMBUSTION

Combustion Ramjet Engines, Supersonic
    USE   SUPERSONIC COMBUSTION RAMJET
            ENGINES

Combustion, Solid Propellant
    USE   SOLID PROPELLANT COMBUSTION

Combustion, Spontaneous
    USE   SPONTANEOUS COMBUSTION

COMBUSTION STABILITY

Combustion, Supersonic
    USE   SUPERSONIC COMBUSTION

COMBUSTION TEMPERATURE

COMBUSTION VIBRATION

Combustion Waves
  USE   FLAME PROPAGATION

COMBUSTION WIND TUNNELS

Combustors
  USE   COMBUSTION CHAMBERS

Comet, Arend-Roland
  USE   AREND-ROLAND COMET

Comet, Encke
  USE   ENCKE COMET

Comet, Giacobini-Zinner
  USE   GIACOBINI-ZINNER COMET

Comet, Grigg-Skjellerup
  USE   GRIGG-SKJELLERUP COMET

Comet, Halley's
  USE   HALLEY'S COMET

COMET HEADS

Comet, Humason
  USE   HUMASON COMET

Comet, Iras-Araki-Alcock
  USE   IRAS-ARAKI-ALCOCK COMET

Comet, Kohoutek
  USE   KOHOUTEK COMET

Comet, Morehouse
  USE   MOREHOUSE COMET

Comet, Mrkos
  USE   MRKOS COMET

COMET NUCLEI

Comet, Schwassmann-Wachmann
  USE   SCHWASSMANN-WACHMANN COMET

COMET TAILS

Comet, Tempel 2
  USE   TEMPEL 2 COMET

Comet, West
  USE   WEST COMET

COMET 4 AIRCRAFT

COMETARY ATMOSPHERES

COMETS

COMFORT

Comfort, Thermal
  USE   THERMAL COMFORT

COMMAND AND CONTROL

Command Center, Space Base
  USE   SPACE BASE COMMAND CENTER

COMMAND GUIDANCE

COMMAND LANGUAGES

COMMAND MODULES

Command Post, Advanced Airborne
  USE   E-4A AIRCRAFT

COMMAND SERVICE MODULES

Command Systems
  USE   COMMAND GUIDANCE

Command Systems, Digital
  USE   DIGITAL COMMAND SYSTEMS

Command-Control
  USE   COMMAND AND CONTROL

Commando Aircraft
  USE   C-46 AIRCRAFT

COMMANDS

Commencements, Sudden Storm
  USE   SUDDEN STORM COMMENCEMENTS

COMMERCE

COMMERCE LAB

Commercial Air Transport, Supersonic
  USE   SUPERSONIC COMMERCIAL AIR
            TRANSPORT

COMMERCIAL AIRCRAFT

Commercial Aviation
  USE   COMMERCIAL AIRCRAFT
            CIVIL AVIATION

COMMERCIAL ENERGY

Commercial Satellite, Arabian
  USE   ARCOMSAT

COMMERCIAL SPACECRAFT

Commercialization, Space
  USE   SPACE COMMERCIALIZATION

COMMINUTION

(Comminution), Grinding
  USE   GRINDING (COMMINUTION)

(Committee), COSPAR
  USE   COMMITTEE ON SPACE RESEARCH

COMMITTEE ON SPACE RESEARCH

COMMODITIES

COMMONALITY

COMMUNICATING

COMMUNICATION

Communication, Aircraft
  USE   AIRCRAFT COMMUNICATION

COMMUNICATION CABLES

Communication, Circumlunar
  USE   CIRCUMLUNAR COMMUNICATION

Communication, Digital
  USE   PULSE COMMUNICATION

Communication, Electrocutaneous
  USE   ELECTROCUTANEOUS COMMUNICATION

COMMUNICATION EQUIPMENT

Communication, Extraterrestrial
  USE   EXTRATERRESTRIAL COMMUNICATION

Communication, Facsimile
  USE   FACSIMILE COMMUNICATION

Communication, Ground-Air-Ground
  USE   GROUND-AIR-GROUND COMMUNICATION

Communication, Interplanetary
  USE   INTERPLANETARY COMMUNICATION

Communication, Interprocessor
  USE   INTERPROCESSOR COMMUNICATION

Communication, Interstellar
  USE   INTERSTELLAR COMMUNICATION

Communication, Laser
  USE   OPTICAL COMMUNICATION

Communication, Light
  USE   OPTICAL COMMUNICATION

Communication, Line Of Sight
  USE   LINE OF SIGHT COMMUNICATION

Communication, Lunar
  USE   LUNAR COMMUNICATION

Communication, Multichannel
  USE   MULTICHANNEL COMMUNICATION

Communication Network, NASA
  USE   NASCOM NETWORK

COMMUNICATION NETWORKS

Communication, Optical
  USE   OPTICAL COMMUNICATION

(Communication), Packets
  USE   PACKETS (COMMUNICATION)

Communication, Point To Point
  USE   POINT TO POINT COMMUNICATION

Communication, Pulse
  USE   PULSE COMMUNICATION

Communication, Radio
  USE   RADIO COMMUNICATION

Communication, Reentry
  USE   REENTRY COMMUNICATION

Communication, Satellite
  USE   SPACECRAFT COMMUNICATION

Communication Satellite (Esa), Maritime
  USE   MAROTS (ESA)

COMMUNICATION SATELLITES

Communication Satellites, Synchronous
  USE   SYNCOM SATELLITES

(Communication), Scrambling
  USE   SCRAMBLING (COMMUNICATION)

Communication, Ship To Shore
  USE   SHIP TO SHORE COMMUNICATION

Communication, Space
  USE   SPACE COMMUNICATION

Communication, Spacecraft
  USE   SPACECRAFT COMMUNICATION

Communication System, Fleet Satellite
  USE   FLEET SATELLITE COMMUNICATION
            SYSTEM

Communication Systems
  USE   TELECOMMUNICATION

Communication Systems, Mobile
  USE   MOBILE COMMUNICATION SYSTEMS

Communication, Tele
  USE   TELECOMMUNICATION

COMMUNICATION THEORY

Communication Theory, Statistical
  USE   COMMUNICATION THEORY

Communication, Transoceanic
  USE   TRANSOCEANIC COMMUNICATION

Communication, Underground
  USE   UNDERGROUND COMMUNICATION

Communication, Underwater
  USE   UNDERWATER COMMUNICATION

Communication, Verbal
  USE   VERBAL COMMUNICATION

Communication, Video
  USE   VIDEO COMMUNICATION

Communication, Voice
  USE   VOICE COMMUNICATION

Communication, Wideband
USE   WIDEBAND COMMUNICATION

Communication, Wireless
USE   WIRELESS COMMUNICATION

Communications Antenna Grid (Navy), Global
USE   SEAFARER PROJECT

(Communications), Ground Effect
USE   GROUND EFFECT (COMMUNICATIONS)

Communications Satellite, European
USE   EUROPEAN COMMUNICATIONS SATELLITE

Communications Satellite Proj, Synchronous
USE   SYNCHRONOUS COMMUNICATIONS
         SATELLITE PROJ

Communications Satellite System, Defense
USE   DEFENSE COMMUNICATIONS SATELLITE
         SYSTEM

Communications Ships, Satellite
USE   SATELLITE COMMUNICATIONS SHIPS

Communications System (DCS), Defense
USE   DEFENSE COMMUNICATIONS SYSTEM
         (DCS)

Communications Systems, Domestic Satellite
USE   DOMESTIC SATELLITE COMMUNICATIONS
         SYSTEMS

COMMUNICATIONS TECHNOLOGY SATELLITE

(Communist) Mainland, China
USE   CHINA

COMMUNITIES

COMMUTATION

COMMUTATORS

Commutators, De
USE   DECOMMUTATORS

Compact Reactors, Military
USE   MILITARY COMPACT REACTORS

COMPACTING

Compaction, Data
USE   DATA COMPRESSION

Compactness
USE   VOID RATIO

COMPANDING

COMPANION STARS

COMPARATOR CIRCUITS

COMPARATORS

COMPARISON

Compartmentation
USE   COMPARTMENTS

COMPARTMENTS

Compartments, Aircraft
USE   AIRCRAFT COMPARTMENTS

COMPASS (PROGRAMMING LANGUAGE)

COMPASSES

Compasses, Gyro
USE   GYROCOMPASSES

Compasses, Magnetic
USE   MAGNETIC COMPASSES

Compasses, Solar
USE   SOLAR COMPASSES

COMPATIBILITY

Compatibility, Electromagnetic
USE   ELECTROMAGNETIC COMPATIBILITY

Compatibility, In
USE   INCOMPATIBILITY

Compatibility, Systems
USE   SYSTEMS COMPATIBILITY

Compatible Tapes, Computer
USE   COMPUTER COMPATIBLE TAPES

COMPENSATION

Compensation, Image Motion
USE   IMAGE MOTION COMPENSATION

Compensation, Instrument
USE   INSTRUMENT COMPENSATION

Compensation, Temperature
USE   TEMPERATURE COMPENSATION

COMPENSATORS

COMPENSATORY TRACKING

COMPETITION

Compilation (Computers)
USE   COMPILERS

Compiler Programs
USE   COMPILERS

COMPILERS

COMPLEMENT

COMPLEMENT (BIOLOGY)

Complementary Metal Oxide Semiconductors
USE   CMOS

COMPLEMENTS (MATHEMATICS)

COMPLETENESS

Complex, Cape Kennedy Launch
USE   CAPE KENNEDY LAUNCH COMPLEX

COMPLEX COMPOUNDS

Complex Coordinator, Langley
USE   LANGLEY COMPLEX COORDINATOR

COMPLEX NUMBERS

COMPLEX SYSTEMS

COMPLEX VARIABLES

Complex, Vitamin B
USE   BIOTIN

Complexes, Launch
USE   LAUNCHING BASES

COMPLEXITY

Complexity, Task
USE   TASK COMPLEXITY

Compliance (Elasticity)
USE   MODULUS OF ELASTICITY

Complication
USE   COMPLEXITY

COMPONENT RELIABILITY

COMPONENTS

Components), ALU (Computer
USE   ARITHMETIC AND LOGIC UNITS

Components Analysis, Principal
USE   PRINCIPAL COMPONENTS ANALYSIS

Components, Antenna
USE   ANTENNA COMPONENTS

Components, Computer
USE   COMPUTER COMPONENTS

Components, Missile
USE   MISSILE COMPONENTS

Components, Redundant
USE   REDUNDANT COMPONENTS

Components, Spacecraft
USE   SPACECRAFT COMPONENTS

Components Test Reactors, Heavy Water
USE   HEAVY WATER COMPONENTS TEST
         REACTORS

COMPOSITE FUNCTIONS

COMPOSITE MATERIALS

COMPOSITE PROPELLANTS

COMPOSITE STRUCTURES

COMPOSITE WRAPPING

Composites
USE   COMPOSITE MATERIALS

Composites, Aluminum Boron
USE   ALUMINUM BORON COMPOSITES

Composites, Aluminum Graphite
USE   ALUMINUM GRAPHITE COMPOSITES

Composites, Carbon-Carbon
USE   CARBON-CARBON COMPOSITES

Composites, Ceramic Matrix
USE   CERAMIC MATRIX COMPOSITES

Composites, Epoxy Matrix
USE   EPOXY MATRIX COMPOSITES

Composites, Eutectic
USE   EUTECTIC COMPOSITES

Composites, Fiber
USE   FIBER COMPOSITES

Composites, Fiber Reinforced
USE   FIBER REINFORCED COMPOSITES

Composites, Graphite-Epoxy
USE   GRAPHITE-EPOXY COMPOSITES

Composites, Graphite-Polyimide
USE   GRAPHITE-POLYIMIDE COMPOSITES

Composites, Metal Matrix
USE   METAL MATRIX COMPOSITES

Composites, Polymer Matrix
USE   POLYMER MATRIX COMPOSITES

Composites, Resin Matrix
USE   RESIN MATRIX COMPOSITES

Composites, Three Dimensional
USE   THREE DIMENSIONAL COMPOSITES

Composites, Whisker
USE   WHISKER COMPOSITES

COMPOSITION

Composition, Atmospheric
USE   ATMOSPHERIC COMPOSITION

Composition (Biology), Body
USE   BODY COMPOSITION (BIOLOGY)

Composition, Chemical
USE   CHEMICAL COMPOSITION

(Composition), Concentration
USE   CONCENTRATION (COMPOSITION)

Composition, De
USE   DECOMPOSITION

Composition Experiment, Lower Atmospheric
USE   LACATE (EXPERIMENT)

Composition, Gas
USE   GAS COMPOSITION

Composition, Ionospheric
USE   IONOSPHERIC COMPOSITION

Composition, Lunar
USE   LUNAR COMPOSITION

Composition, Meteoritic
USE   METEORITIC COMPOSITION

Composition, Photode
USE   PHOTODECOMPOSITION

Composition, Planetary
USE   PLANETARY COMPOSITION

Composition, Plasma
USE   PLASMA COMPOSITION

COMPOSITION (PROPERTY)

Composition, Stellar
USE   STELLAR COMPOSITION

COMPOSTING

COMPOUND A

COMPOUND HELICOPTERS

COMPOUNDING

COMPOUNDS

Compounds, Acetyl
USE   ACETYL COMPOUNDS

Compounds, Actinide Series
USE   ACTINIDE SERIES COMPOUNDS

Compounds, Aliphatic
USE   ALIPHATIC COMPOUNDS

Compounds, Alkali Metal
USE   ALKALI METAL COMPOUNDS

Compounds, Alkaline Earth
USE   ALKALINE EARTH COMPOUNDS

Compounds, Alkyl
USE   ALKYL COMPOUNDS

Compounds, Allyl
USE   ALLYL COMPOUNDS

Compounds, Aluminum
USE   ALUMINUM COMPOUNDS

Compounds, Ammonium
USE   AMMONIUM COMPOUNDS

Compounds, Antimony
USE   ANTIMONY COMPOUNDS

Compounds, Aromatic
USE   AROMATIC COMPOUNDS

Compounds, Arsenic
USE   ARSENIC COMPOUNDS

Compounds, Aryl
USE   AROMATIC COMPOUNDS

Compounds, Azo
USE   AZO COMPOUNDS

Compounds, Barium
USE   BARIUM COMPOUNDS

Compounds, Beryllium
USE   BERYLLIUM COMPOUNDS

Compounds, Bismuth
USE   BISMUTH COMPOUNDS

Compounds, Boron
USE   BORON COMPOUNDS

Compounds, Boron-Epoxy
USE   BORON-EPOXY COMPOUNDS

Compounds, Bromine
USE   BROMINE COMPOUNDS

Compounds, Cadmium
~ USE   CADMIUM COMPOUNDS

Compounds, Calcium
USE   CALCIUM COMPOUNDS

Compounds, Carbon
USE   CARBON COMPOUNDS

Compounds, Carbonyl
USE   CARBONYL COMPOUNDS

Compounds, Cerium
USE   CERIUM COMPOUNDS

Compounds, Cesium
USE   CESIUM COMPOUNDS

Compounds, Cetyl
USE   CETYL COMPOUNDS

Compounds, Chelate
USE   CHELATES

Compounds, Chemical
USE   CHEMICAL COMPOUNDS

Compounds, Chlorine
USE   CHLORINE COMPOUNDS

Compounds, Chromium
USE   CHROMIUM COMPOUNDS

Compounds, Cobalt
USE   COBALT COMPOUNDS

Compounds, Complex
USE   COMPLEX COMPOUNDS

Compounds, Copper
USE   COPPER COMPOUNDS

Compounds, Curium
USE   CURIUM COMPOUNDS

Compounds, Cyano
USE   CYANO COMPOUNDS

Compounds, Cyclic
USE   CYCLIC COMPOUNDS

Compounds, Deuterium
USE   DEUTERIUM COMPOUNDS

Compounds, Diallyl
USE   DIALLYL COMPOUNDS

Compounds, Dibasic
USE   DIBASIC COMPOUNDS

Compounds, Dibutyl
USE   DIBUTYL COMPOUNDS

Compounds, Difluoro
USE   DIFLUORO COMPOUNDS

Compounds, Diphenyl
USE   DIPHENYL COMPOUNDS

Compounds, Dysprosium
USE   DYSPROSIUM COMPOUNDS

Compounds, Electron
USE   INTERMETALLICS

Compounds, Epoxy
USE   EPOXY COMPOUNDS

Compounds, Erbium
USE   ERBIUM COMPOUNDS

Compounds, Ethyl
USE   ETHYL COMPOUNDS

Compounds, Ethylene
USE   ETHYLENE COMPOUNDS

Compounds, Europium
USE   EUROPIUM COMPOUNDS

Compounds, Fluorine
USE   FLUORINE COMPOUNDS

Compounds, Fluorine Organic
USE   FLUORINE ORGANIC COMPOUNDS

Compounds, Fluoro
USE   FLUORO COMPOUNDS

Compounds, Gallium
USE   GALLIUM COMPOUNDS

Compounds, Germanium
USE   GERMANIUM COMPOUNDS

Compounds, Group 1A
USE   ALKALI METAL COMPOUNDS

Compounds, Group 1B
USE   GROUP 1B COMPOUNDS

Compounds, Group 2A
USE   ALKALINE EARTH COMPOUNDS

Compounds, Group 2B
USE   GROUP 2B COMPOUNDS

Compounds, Group 3A
USE   GROUP 3A COMPOUNDS

Compounds, Group 3B
USE   GROUP 3B COMPOUNDS

Compounds, Group 4A
USE   GROUP 4A COMPOUNDS

Compounds, Group 4B
USE   GROUP 4B COMPOUNDS

Compounds, Group 5A
USE   GROUP 5A COMPOUNDS

Compounds, Group 5B
USE   GROUP 5B COMPOUNDS

Compounds, Group 6A
USE   GROUP 6A COMPOUNDS

Compounds, Group 6B
USE   GROUP 6B COMPOUNDS

Compounds, Group 7A
USE   HALOGEN COMPOUNDS

Compounds, Group 7B
USE   GROUP 7B COMPOUNDS

Compounds, Group 8
USE   GROUP 8 COMPOUNDS

Compounds, Hafnium
USE   HAFNIUM COMPOUNDS

Compounds, Halogen
USE   HALOGEN COMPOUNDS

Compounds, Helium
USE   HELIUM COMPOUNDS

Compounds, Heterocyclic
USE   HETEROCYCLIC COMPOUNDS

Compounds, Hexyl
USE   HEXYL COMPOUNDS

Compounds, High Melting
USE   REFRACTORY MATERIALS

Compounds, Hydrazinium
USE   HYDRAZINIUM COMPOUNDS

Compounds, Hydrazonium
    USE   HYDRAZONIUM COMPOUNDS

Compounds, Hydrogen
    USE   HYDROGEN COMPOUNDS

Compounds, Hydroxyl
    USE   HYDROXYL COMPOUNDS

Compounds, Indium
    USE   INDIUM COMPOUNDS

Compounds, Inorganic
    USE   INORGANIC COMPOUNDS

Compounds, Iodine
    USE   IODINE COMPOUNDS

Compounds, Iron
    USE   IRON COMPOUNDS

Compounds, Isopropyl
    USE   ISOPROPYL COMPOUNDS

Compounds, Lanthanum
    USE   LANTHANUM COMPOUNDS

Compounds, Lead
    USE   LEAD COMPOUNDS

Compounds, Lead Organic
    USE   LEAD ORGANIC COMPOUNDS

Compounds, Lithium
    USE   LITHIUM COMPOUNDS

Compounds, Lutetium
    USE   LUTETIUM COMPOUNDS

Compounds, Magnesium
    USE   MAGNESIUM COMPOUNDS

Compounds, Manganese
    USE   MANGANESE COMPOUNDS

Compounds, Mercapto
    USE   THIOLS

Compounds, Mercury
    USE   MERCURY COMPOUNDS

Compounds, Metal
    USE   METAL COMPOUNDS

Compounds, Metallorganic
    USE   ORGANOMETALLIC COMPOUNDS

Compounds, Methyl
    USE   METHYL COMPOUNDS

Compounds, Molybdenum
    USE   MOLYBDENUM COMPOUNDS

Compounds, Neodymium
    USE   NEODYMIUM COMPOUNDS

Compounds, Neptunium
    USE   NEPTUNIUM COMPOUNDS

Compounds, Nickel
    USE   NICKEL COMPOUNDS

Compounds, Niobium
    USE   NIOBIUM COMPOUNDS

Compounds, Nitro
    USE   NITRO COMPOUNDS

Compounds, Nitrogen
    USE   NITROGEN COMPOUNDS

Compounds, Nitronium
    USE   NITRONIUM COMPOUNDS

Compounds, Nitroso
    USE   NITROSO COMPOUNDS

Compounds, Organic
    USE   ORGANIC COMPOUNDS

Compounds, Organic Aluminum
    USE   ORGANIC ALUMINUM COMPOUNDS

Compounds, Organic Boron
    USE   ORGANIC BORON COMPOUNDS

Compounds, Organic Fluorine
    USE   FLUORINE ORGANIC COMPOUNDS

Compounds, Organic Germanium
    USE   ORGANIC GERMANIUM COMPOUNDS

Compounds, Organic Lithium
    USE   ORGANIC LITHIUM COMPOUNDS

Compounds, Organic Phosphorus
    USE   ORGANIC PHOSPHORUS COMPOUNDS

Compounds, Organic Silicon
    USE   ORGANIC SILICON COMPOUNDS

Compounds, Organic Sulfur
    USE   ORGANIC SULFUR COMPOUNDS

Compounds, Organic Tin
    USE   ORGANIC TIN COMPOUNDS

Compounds, Organometallic
    USE   ORGANOMETALLIC COMPOUNDS

Compounds, Osmium
    USE   OSMIUM COMPOUNDS

Compounds, Oxygen
    USE   OXYGEN COMPOUNDS

Compounds, Palladium
    USE   PALLADIUM COMPOUNDS

Compounds, Perfluoro
    USE   PERFLUORO COMPOUNDS

Compounds, Phosphonium
    USE   PHOSPHONIUM COMPOUNDS

Compounds, Phosphorus
    USE   PHOSPHORUS COMPOUNDS

Compounds, Platinum
    USE   PLATINUM COMPOUNDS

Compounds, Plutonium
    USE   PLUTONIUM COMPOUNDS

Compounds, Polonium
    USE   POLONIUM COMPOUNDS

Compounds, Polynuclear Organic
    USE   POLYNUCLEAR ORGANIC COMPOUNDS

Compounds, Potassium
    USE   POTASSIUM COMPOUNDS

Compounds, Potting
    USE   POTTING COMPOUNDS

Compounds, Propyl
    USE   PROPYL COMPOUNDS

Compounds, Protactinium
    USE   PROTACTINIUM COMPOUNDS

Compounds, Rare Earth
    USE   RARE EARTH COMPOUNDS

Compounds, Rare Gas
    USE   RARE GAS COMPOUNDS

Compounds, Rhenium
    USE   RHENIUM COMPOUNDS

Compounds, Rhodium
    USE   RHODIUM COMPOUNDS

Compounds, Rubidium
    USE   RUBIDIUM COMPOUNDS

Compounds, Ruthenium
    USE   RUTHENIUM COMPOUNDS

Compounds, Samarium
    USE   SAMARIUM COMPOUNDS

Compounds, Scandium
    USE   SCANDIUM COMPOUNDS

Compounds, Selenium
    USE   SELENIUM COMPOUNDS

Compounds, Silicon
    USE   SILICON COMPOUNDS

Compounds, Silver
    USE   SILVER COMPOUNDS

Compounds, Sodium
    USE   SODIUM COMPOUNDS

Compounds, Strontium
    USE   STRONTIUM COMPOUNDS

Compounds, Sulfur
    USE   SULFUR COMPOUNDS

Compounds, Tantalum
    USE   TANTALUM COMPOUNDS

Compounds, Technetium
    USE   TECHNETIUM COMPOUNDS

Compounds, Tellurium
    USE   TELLURIUM COMPOUNDS

Compounds, Thallium
    USE   THALLIUM COMPOUNDS

Compounds, Thorium
    USE   THORIUM COMPOUNDS

Compounds, Thulium
    USE   THULIUM COMPOUNDS

Compounds, Tin
    USE   TIN COMPOUNDS

Compounds, Titanium
    USE   TITANIUM COMPOUNDS

Compounds, Triethyl
    USE   TRIETHYL COMPOUNDS

Compounds, Trimethyl
    USE   TRIMETHYL COMPOUNDS

Compounds, Trinitro
    USE   TRINITRO COMPOUNDS

Compounds, Tropyl
    USE   TROPYL COMPOUNDS

Compounds, Tungsten
    USE   TUNGSTEN COMPOUNDS

Compounds, Uranium
    USE   URANIUM COMPOUNDS

Compounds, Vanadium
    USE   VANADIUM COMPOUNDS

Compounds, Vanadyl
    USE   VANADYL COMPOUNDS

Compounds, Xenon
    USE   XENON COMPOUNDS

Compounds, Ytterbium
    USE   YTTERBIUM COMPOUNDS

Compounds, Yttrium
    USE   YTTRIUM COMPOUNDS

Compounds, Zinc
    USE   ZINC COMPOUNDS

Compounds, Zirconium
    USE   ZIRCONIUM COMPOUNDS

COMPRESSED AIR

COMPRESSED GAS

COMPRESSIBILITY

COMPRESSIBILITY EFFECTS

COMPRESSIBLE BOUNDARY LAYER

COMPRESSIBLE FLOW

COMPRESSIBLE FLUIDS

COMPRESSING

Compression, Data
    USE   DATA COMPRESSION

Compression Demodulators, Frequency
    USE   FREQUENCY COMPRESSION
              DEMODULATORS

Compression Inlets, Internal
    USE   INTERNAL COMPRESSION INLETS

COMPRESSION LOADS

Compression Loads, Axial
    USE   AXIAL COMPRESSION LOADS

Compression, Magnetic
    USE   MAGNETIC COMPRESSION

Compression, Plasma
    USE   PLASMA COMPRESSION

Compression, Pulse
    USE   PULSE COMPRESSION

COMPRESSION RATIO

Compression, Speech Baseband
    USE   SPEECH BASEBAND COMPRESSION

Compression Testers
    USE   COMPRESSION TESTS

COMPRESSION TESTS

Compression Tests, Meteorite
    USE   COMPRESSION TESTS
              METEORITES
              MECHANICAL PROPERTIES

COMPRESSION WAVES

COMPRESSIVE STRENGTH

COMPRESSOR BLADES

COMPRESSOR EFFICIENCY

COMPRESSOR ROTORS

COMPRESSORS

Compressors, Axial
    USE   TURBOCOMPRESSORS

Compressors, Axial Flow
    USE   TURBOCOMPRESSORS

Compressors, Centrifugal
    USE   CENTRIFUGAL COMPRESSORS

Compressors, Multistage
    USE   TURBOCOMPRESSORS

Compressors, Supersonic
    USE   SUPERSONIC COMPRESSORS

Compressors, Transonic
    USE   TRANSONIC COMPRESSORS

Compressors, Turbo
    USE   TURBOCOMPRESSORS

COMPTON EFFECT

COMPULSATORS

COMPUTATION

COMPUTATIONAL ASTROPHYSICS

COMPUTATIONAL CHEMISTRY

COMPUTATIONAL FLUID DYNAMICS

COMPUTATIONAL GRIDS

COMPUTER AIDED DESIGN

COMPUTER AIDED MANUFACTURING

COMPUTER AIDED MAPPING

COMPUTER AIDED TOMOGRAPHY

COMPUTER ASSISTED INSTRUCTION

Computer, CDC Cyber 74
    USE   CDC CYBER 74 COMPUTER

Computer, CDC Cyber 174
    USE   CDC CYBER 174 COMPUTER

Computer, CDC Cyber 175
    USE   CDC CYBER 175 COMPUTER

Computer, CDC Cyber 203
    USE   CDC CYBER 203 COMPUTER

Computer, CDC Cyber 205
    USE   CDC CYBER 205 COMPUTER

Computer, CDC Star 100
    USE   CDC STAR 100 COMPUTER

Computer, CDC 160-A
    USE   CDC 160-A COMPUTER

Computer, CDC 1604
    USE   CDC 1604 COMPUTER

Computer, CDC 3100
    USE   CDC 3100 COMPUTER

Computer, CDC 3200
    USE   CDC 3200 COMPUTER

Computer, CDC 3600
    USE   CDC 3600 COMPUTER

Computer, CDC 3800
    USE   CDC 3800 COMPUTER

Computer, CDC 6400
    USE   CDC 6400 COMPUTER

Computer, CDC 6600
    USE   CDC 6600 COMPUTER

Computer, CDC 6700
    USE   CDC 6700 COMPUTER

Computer, CDC 7600
    USE   CDC 7600 COMPUTER

Computer, CDC 8090
    USE   CDC 8090 COMPUTER

COMPUTER COMPATIBLE TAPES

COMPUTER COMPONENTS

(Computer Components), ALU
    USE   ARITHMETIC AND LOGIC UNITS

Computer, Cyber 74
    USE   CDC CYBER 74 COMPUTER

Computer, DDP 516
    USE   DDP 516 COMPUTER

COMPUTER DESIGN

Computer, EAI 680
    USE   EAI 680 COMPUTER

Computer, EAI 8400
    USE   EAI 8400 COMPUTER

Computer, EAI 8900
    USE   EAI 8900 COMPUTER

Computer, EMR 6050
    USE   EMR 6050 COMPUTER

Computer, Ferranti Mercury
    USE   FERRANTI MERCURY COMPUTER

Computer, GE 625
    USE   GE 625 COMPUTER

Computer, GE 635
    USE   GE 635 COMPUTER

COMPUTER GRAPHICS

Computer, Honeywell Adept
    USE   HONEYWELL ADEPT COMPUTER

Computer, Honeywell DDP 116
    USE   HONEYWELL DDP 116 COMPUTER

Computer, Honeywell 600/6000
    USE   HONEYWELL 600/6000 COMPUTER

Computer, IBM 360
    USE   IBM 360 COMPUTER

Computer, IBM 370
    USE   IBM 370 COMPUTER

Computer, IBM 650
    USE   IBM 650 COMPUTER

Computer, IBM 704
    USE   IBM 704 COMPUTER

Computer, IBM 709
    USE   IBM 709 COMPUTER

Computer, IBM 1130
    USE   IBM 1130 COMPUTER

Computer, IBM 1401
    USE   IBM 1401 COMPUTER

Computer, IBM 1410
    USE   IBM 1410 COMPUTER

Computer, IBM 1620
    USE   IBM 1620 COMPUTER

Computer, IBM 2250
    USE   IBM 2250 COMPUTER

Computer, IBM 7030
    USE   IBM 7030 COMPUTER

Computer, IBM 7040
    USE   IBM 7040 COMPUTER

Computer, IBM 7044
    USE   IBM 7044 COMPUTER

Computer, IBM 7070
    USE   IBM 7070 COMPUTER

Computer, IBM 7074
    USE   IBM 7074 COMPUTER

Computer, IBM 7090
    USE   IBM 7090 COMPUTER

Computer, IBM 7094
    USE   IBM 7094 COMPUTER

Computer, Illiac 3
    USE   ILLIAC 3 COMPUTER

Computer, Illiac 4
    USE   ILLIAC 4 COMPUTER

COMPUTER INFORMATION SECURITY

Computer Methods
    USE   COMPUTER PROGRAMS

Computer, Minos
    USE   MINOS COMPUTER

Computer, Modcomp II
    USE   MODCOMP II COMPUTER

Computer, Modcomp IV
 USE   MODCOMP IV COMPUTER

Computer Network, Arpa
 USE   ARPA COMPUTER NETWORK

COMPUTER NETWORKS

Computer, PDP 7
 USE   PDP 7 COMPUTER

Computer, PDP 8
 USE   PDP 8 COMPUTER

Computer, PDP 9
 USE   PDP 9 COMPUTER

Computer, PDP 10
 USE   PDP 10 COMPUTER

Computer, PDP 11
 USE   PDP 11 COMPUTER

Computer, PDP 11/20
 USE   PDP 11/20 COMPUTER

Computer, PDP 11/40
 USE   PDP 11/40 COMPUTER

Computer, PDP 11/45
 USE   PDP 11/45 COMPUTER

Computer, PDP 11/50
 USE   PDP 11/50 COMPUTER

Computer, PDP 11/70
 USE   PDP 11/70 COMPUTER

Computer, PDP 12
 USE   PDP 12 COMPUTER

Computer, PDP 15
 USE   PDP 15 COMPUTER

Computer, Pegasus
 USE   PEGASUS COMPUTER

Computer, Philco 2000
 USE   PHILCO 2000 COMPUTER

COMPUTER PROGRAM INTEGRITY

COMPUTER PROGRAMMING

COMPUTER PROGRAMS

(Computer Programs), User Manuals
 USE   USER MANUALS (COMPUTER PROGRAMS)

Computer, RCA Spectra 70
 USE   RCA SPECTRA 70 COMPUTER

Computer, SDS 930
 USE   SDS 930 COMPUTER

Computer, SDS 9300
 USE   SDS 9300 COMPUTER

Computer, Siemens 2002
 USE   SIEMENS 2002 COMPUTER

Computer, Sigma 5
 USE   SIGMA 5 COMPUTER

Computer, Sigma 9
 USE   SIGMA 9 COMPUTER

Computer Simulation
 USE   COMPUTERIZED SIMULATION

Computer Storage, Cryogenic
 USE   CRYOGENIC COMPUTER STORAGE

(Computer Storage), Delay Lines
 USE   DELAY LINES (COMPUTER STORAGE)

COMPUTER STORAGE DEVICES

Computer, System 10
 USE   PDP 10 COMPUTER

COMPUTER SYSTEMS DESIGN

Computer Systems, Embedded
 USE   EMBEDDED COMPUTER SYSTEMS

COMPUTER SYSTEMS PERFORMANCE

COMPUTER SYSTEMS PROGRAMS

COMPUTER SYSTEMS SIMULATION

COMPUTER TECHNIQUES

Computer, Univac Larc
 USE   UNIVAC LARC COMPUTER

Computer, Univac 80
 USE   UNIVAC 80 COMPUTER

Computer, Univac 418
 USE   UNIVAC 418 COMPUTER

Computer, Univac 490
 USE   UNIVAC 490 COMPUTER

Computer, Univac 494
 USE   UNIVAC 494 COMPUTER

Computer, Univac 1105
 USE   UNIVAC 1105 COMPUTER

Computer, Univac 1106
 USE   UNIVAC 1106 COMPUTER

Computer, Univac 1107
 USE   UNIVAC 1107 COMPUTER

Computer, Univac 1108
 USE   UNIVAC 1108 COMPUTER

Computer, Univac 1110
 USE   UNIVAC 1110 COMPUTER

Computer, Univac 1230
 USE   UNIVAC 1230 COMPUTER

Computer, Vax-11/780
 USE   VAX-11/780 COMPUTER

COMPUTER VISION

Computerized Control
 USE   NUMERICAL CONTROL

Computerized Design
 USE   COMPUTER AIDED DESIGN

COMPUTERIZED SIMULATION

COMPUTERS

(Computers), Accumulators
 USE   ACCUMULATORS (COMPUTERS)

Computers, Airborne/spaceborne
 USE   AIRBORNE/SPACEBORNE COMPUTERS

Computers, Analog
 USE   ANALOG COMPUTERS

(Computers), Applications Programs
 USE   APPLICATIONS PROGRAMS (COMPUTERS)

(Computers), Architecture
 USE   ARCHITECTURE (COMPUTERS)

(Computers), Associative Processing
 USE   ASSOCIATIVE PROCESSING (COMPUTERS)

(Computers), Auxiliary Equipment
 USE   AUXILIARY EQUIPMENT (COMPUTERS)

Computers, CDC
 USE   CDC COMPUTERS

Computers, CDC Cyber 170 Series
 USE   CDC CYBER 170 SERIES COMPUTERS

Computers, CDC 6000 Series
 USE   CDC 6000 SERIES COMPUTERS

Computers, CDC 7000 Series
 USE   CDC 7000 SERIES COMPUTERS

(Computers), Compilation
 USE   COMPILERS

(Computers), Control Data
 USE   CONTROL DATA (COMPUTERS)

(Computers), Control Units
 USE   CONTROL UNITS (COMPUTERS)

Computers, Counting Rate
 USE   COUNTING RATE COMPUTERS

Computers, Cray
 USE   CRAY COMPUTERS

Computers, DDP
 USE   DDP COMPUTERS

Computers, Digital
 USE   DIGITAL COMPUTERS

(Computers), Editing Routines
 USE   EDITING ROUTINES (COMPUTERS)

(Computers), File Maintenance
 USE   FILE MAINTENANCE (COMPUTERS)

Computers, Flight
 USE   AIRBORNE/SPACEBORNE COMPUTERS

Computers, GE
 USE   GE COMPUTERS

Computers, General Electric
 USE   GE COMPUTERS

Computers, Hewlett-Packard
 USE   HEWLETT-PACKARD COMPUTERS

Computers, Honeywell
 USE   HONEYWELL COMPUTERS

Computers, Hybrid
 USE   HYBRID COMPUTERS

Computers, IBM
 USE   IBM COMPUTERS

Computers, IBM 7000 Series
 USE   IBM 7000 SERIES COMPUTERS

Computers, Icl
 USE   ICL COMPUTERS

Computers, Illiac
 USE   ILLIAC COMPUTERS

(Computers), Instruction Sets
 USE   INSTRUCTION SETS (COMPUTERS)

Computers Limited, International
 USE   ICL COMPUTERS

(Computers), Memory
 USE   MEMORY (COMPUTERS)

Computers, Micro
 USE   MICROCOMPUTERS

Computers, Mini
 USE   MINICOMPUTERS

(Computers), Multiprocessing
 USE   MULTIPROCESSING (COMPUTERS)

(Computers), Natural Language
 USE   NATURAL LANGUAGE (COMPUTERS)

Computers, Nova
 USE   NOVA COMPUTERS

Computers, Onboard
 USE   AIRBORNE/SPACEBORNE COMPUTERS

(Computers), Operating Systems
 USE   OPERATING SYSTEMS (COMPUTERS)

Computers, Optical
    USE   OPTICAL COMPUTERS

Computers, Parallel
    USE   PARALLEL COMPUTERS

(Computers), Parallel Processing
    USE   PARALLEL PROCESSING (COMPUTERS)

Computers, PDP
    USE   PDP COMPUTERS

(Computers), Peripheral Equipment
    USE   PERIPHERAL EQUIPMENT (COMPUTERS)

Computers, Personal
    USE   PERSONAL COMPUTERS

(Computers), Pipelining
    USE   PIPELINING (COMPUTERS)

(Computers), Processors
    USE   CENTRAL PROCESSING UNITS

(Computers), Program Verification
    USE   PROGRAM VERIFICATION (COMPUTERS)

(Computers), Protocol
    USE   PROTOCOL (COMPUTERS)

Computers, Raytheon
    USE   RAYTHEON COMPUTERS

Computers, RCA
    USE   RCA COMPUTERS

Computers, RCA-110
    USE   RCA-110 COMPUTERS

(Computers), Registers
    USE   REGISTERS (COMPUTERS)

(Computers), Response Time
    USE   RESPONSE TIME (COMPUTERS)

(Computers), Run Time
    USE   RUN TIME (COMPUTERS)

(Computers), SDP
    USE   SITE DATA PROCESSORS

Computers, SDS 900 Series
    USE   SDS 900 SERIES COMPUTERS

Computers, SEL
    USE   SEL COMPUTERS

Computers, Sequential
    USE   SEQUENTIAL COMPUTERS

Computers, Sigma
    USE   SIGMA COMPUTERS

(Computers), Software
    USE   COMPUTER SYSTEMS PROGRAMS
            COMPUTER PROGRAMS

Computers, Solomon
    USE   SOLOMON COMPUTERS

(Computers), Subroutine Libraries
    USE   SUBROUTINE LIBRARIES (COMPUTERS)

Computers, Super
    USE   SUPERCOMPUTERS

Computers, Univac
    USE   UNIVAC COMPUTERS

Computers, Univac 1100 Series
    USE   UNIVAC 1100 SERIES COMPUTERS

Computers, VAX
    USE   VAX COMPUTERS

Computers, Vax-11 Series
    USE   VAX-11 SERIES COMPUTERS

COMSAT PROGRAM

COMSTAR C

COMSTAR SATELLITES

CONCATENATED CODES

CONCAVITY

CONCENTRATING

CONCENTRATION

Concentration, Atom
    USE   ATOM CONCENTRATION

Concentration, Carbon Dioxide
    USE   CARBON DIOXIDE CONCENTRATION

CONCENTRATION (COMPOSITION)

(Concentration), Electron Density
    USE   ELECTRON DENSITY (CONCENTRATION)

Concentration, Ion
    USE   ION CONCENTRATION

(Concentration), Ion Density
    USE   ION DENSITY (CONCENTRATION)

Concentration, Meteoroid
    USE   METEOROID CONCENTRATION

(Concentration), Particle Density
    USE   PARTICLE DENSITY (CONCENTRATION)

(Concentration), Proton Density
    USE   PROTON DENSITY (CONCENTRATION)

Concentration, Stress
    USE   STRESS CONCENTRATION

Concentrations, Low
    USE   LOW CONCENTRATIONS

CONCENTRATORS

(Concentrators), Spirals
    USE   SPIRALS (CONCENTRATORS)

CONCENTRIC CYLINDERS

CONCENTRIC SPHERES

CONCENTRICITY

CONCORDE AIRCRAFT

CONCRETE STRUCTURES

CONCRETES

CONCURRENT PROCESSING

CONDENSATES

CONDENSATION

Condensation, Film
    USE   FILM CONDENSATION

CONDENSATION NUCLEI

CONDENSATION PUMPS

Condensation Trails
    USE   CONTRAILS

Condenser Radiators
    USE   CONDENSERS (LIQUEFIERS)
            HEAT RADIATORS

CONDENSERS

Condensers, Gerdien
    USE   GERDIEN CONDENSERS

Condensers, Jet
    USE   JET CONDENSERS

CONDENSERS (LIQUEFIERS)

Condensers, Spray
    USE   SPRAY CONDENSERS

CONDENSING

Condition, Kutta-Joukowski
    USE   KUTTA-JOUKOWSKI CONDITION

Condition, Lipschitz
    USE   LIPSCHITZ CONDITION

CONDITIONED REFLEXES

Conditioned Responses
    USE   CONDITIONING (LEARNING)

CONDITIONING

Conditioning, Air
    USE   AIR CONDITIONING

Conditioning, De
    USE   DECONDITIONING

Conditioning Equipment, Air
    USE   AIR CONDITIONING EQUIPMENT

CONDITIONING (LEARNING)

Conditioning, Power
    USE   POWER CONDITIONING

Conditioning (Treating)
    USE   TREATMENT

CONDITIONS

Conditions, Adiabatic
    USE   ADIABATIC CONDITIONS

Conditions, Atmospheric
    USE   METEOROLOGY

Conditions, Chronic
    USE   CHRONIC CONDITIONS

Conditions, Congenital
    USE   CONGENITAL ANOMALIES

Conditions, Drought
    USE   DROUGHT

Conditions, Flight
    USE   FLIGHT CONDITIONS

Conditions, Nonadiabatic
    USE   NONADIABATIC CONDITIONS

Conditions, Nonequilibrium
    USE   NONEQUILIBRIUM CONDITIONS

Conditions, Runway
    USE   RUNWAY CONDITIONS

Conditions, Weather
    USE   WEATHER

Condon Principle, Franck-
    USE   FRANCK-CONDON PRINCIPLE

CONDOR MISSILE

Conductance
    USE   RESISTANCE

Conductance, Negative
    USE   NEGATIVE CONDUCTANCE

Conducting
    USE   CONDUCTION

CONDUCTING FLUIDS

Conducting Media
    USE   CONDUCTORS

CONDUCTION

CONDUCTION BANDS

CONDUCTION ELECTRONS

Conduction, Heat
    USE   CONDUCTIVE HEAT TRANSFER**

CONDUCTIVE HEAT TRANSFER

CONDUCTIVITY

Conductivity, Air
USE   AIR CONDUCTIVITY

Conductivity, Atmospheric
USE   ATMOSPHERIC CONDUCTIVITY

Conductivity, Electrical
USE   ELECTRICAL RESISTIVITY

Conductivity Gages, Thermal
USE   THERMAL CONDUCTIVITY GAGES

Conductivity, Ionic
USE   ION CURRENTS

Conductivity, Ionospheric
USE   IONOSPHERIC CONDUCTIVITY

Conductivity, Low
USE   LOW CONDUCTIVITY

CONDUCTIVITY METERS

Conductivity Meters, Electrical
USE   ELECTRICAL CONDUCTIVITY METERS

Conductivity, Photo
USE   PHOTOCONDUCTIVITY

Conductivity, Plasma
USE   PLASMA CONDUCTIVITY

Conductivity, Super
USE   SUPERCONDUCTIVITY

Conductivity, Thermal
USE   THERMAL CONDUCTIVITY

Conductor Circuits, Exploding
USE   EXPLODING WIRES
          CIRCUITS

CONDUCTORS

Conductors, Bus
USE   BUS CONDUCTORS

Conductors, Electric
USE   ELECTRIC CONDUCTORS

Conductors, Exploding
USE   EXPLODING WIRES

Conductors, Flat
USE   FLAT CONDUCTORS

Conductors, Photo
USE   PHOTOCONDUCTORS

Conductors, Super
USE   SUPERCONDUCTORS

Conductors, Thermal
USE   THERMAL CONDUCTORS

Cone Expansion, Light-
USE   LIGHT-CONE EXPANSION

CONES

Cones, Ablative Nose
USE   ABLATIVE NOSE CONES

Cones, Cinder
USE   CONES (VOLCANOES)

Cones, Circular
USE   CIRCULAR CONES

Cones, Half
USE   HALF CONES

Cones, Mach
USE   MACH CONES

Cones, Nose
USE   NOSE CONES

Cones, Rocket Nose
USE   ROCKET NOSE CONES

Cones, Shatter
USE   SHATTER CONES

Cones, Slender
USE   SLENDER CONES

CONES (VOLCANOES)

CONFERENCES

CONFIDENCE

CONFIDENCE LIMITS

Configuration, Hammerhead
USE   HAMMERHEAD CONFIGURATION

CONFIGURATION INTERACTION

CONFIGURATION MANAGEMENT

CONFIGURATIONS

Configurations, Aerodynamic
USE   AERODYNAMIC CONFIGURATIONS

Configurations, Aircraft
USE   AIRCRAFT CONFIGURATIONS

Configurations, Body-Wing
USE   BODY-WING CONFIGURATIONS

Configurations, Body-Wing And Tail
USE   BODY-WING AND TAIL CONFIGURATIONS

Configurations, Canard
USE   CANARD CONFIGURATIONS

Configurations, Dual Wing
USE   DUAL WING CONFIGURATIONS

Configurations, Inlet Airframe
USE   INLET AIRFRAME CONFIGURATIONS

Configurations, Launch Vehicle
USE   LAUNCH VEHICLE CONFIGURATIONS

Configurations, Magnetic Field
USE   MAGNETIC FIELD CONFIGURATIONS

Configurations, Missile
USE   MISSILE CONFIGURATIONS

Configurations, Propulsion System
USE   PROPULSION SYSTEM CONFIGURATIONS

Configurations, Satellite
USE   SATELLITE CONFIGURATIONS

Configurations, Spacecraft
USE   SPACECRAFT CONFIGURATIONS

Configurations), Spikes (Aerodynamic
USE   SPIKES (AERODYNAMIC CONFIGURATIONS)

Configurations, Wing Nacelle
USE   WING NACELLE CONFIGURATIONS

Configured Vehicle Program, Terminal
USE   TERMINAL CONFIGURED VEHICLE
          PROGRAM

Configured Vehicles, Control
USE   CONTROL CONFIGURED VEHICLES

CONFINEMENT

Confinement Fusion, Inertial
USE   INERTIAL CONFINEMENT FUSION

Confinement, Plasma
USE   PLASMA CONTROL

CONFINING

Confirmation
USE   PROVING

Confluence
USE   CONVERGENCE

CONFORMAL MAPPING

Conformal Transformations
USE   CONFORMAL MAPPING

CONFUSION

CONGENERS

CONGENITAL ANOMALIES

Congenital Conditions
USE   CONGENITAL ANOMALIES

Congestants, De
USE   DECONGESTANTS

CONGESTION

Congo, Belgian
USE   ZAIRE

CONGO (BRAZZAVILLE)

Congo, French Equatorial
USE   CONGO (BRAZZAVILLE)

Congo (Kinshasa)
USE   ZAIRE

CONGRESSIONAL REPORTS

CONGRUENCES

CONICAL BODIES

CONICAL CAMBER

Conical Flare
USE   CONES

CONICAL FLOW

CONICAL INLETS

CONICAL NOZZLES

CONICAL SCANNING

CONICAL SHELLS

CONICS

CONIFERS

CONJUGATE GRADIENT METHOD

CONJUGATE POINTS

CONJUGATED CIRCUITS

CONJUGATES

CONJUGATION

Conjugation, Phase
USE   PHASE CONJUGATION

CONJUNCTION

CONJUNCTIVA

CONJUNCTIVITIS

CONNECTICUT

Connections
USE   JOINTS (JUNCTIONS)

CONNECTIVE TISSUE

CONNECTORS

Connectors (Electric)
USE   ELECTRIC CONNECTORS

Connectors, Electric
USE   ELECTRIC CONNECTORS

Connectors, Umbilical
USE   UMBILICAL CONNECTORS

(Connectors), Unions
USE   UNIONS (CONNECTORS)

Conoids
USE   CONICAL BODIES

CONSCIOUSNESS

Consciousness, Un
USE   UNCONSCIOUSNESS

CONSECUTIVE EVENTS

CONSERVATION

Conservation, Energy
USE   ENERGY CONSERVATION

CONSERVATION EQUATIONS

CONSERVATION LAWS

CONSISTENCY

(Consistency), Paste
USE   PASTE (CONSISTENCY)

Consistent Fields, Self
USE   SELF CONSISTENT FIELDS

CONSOLES

Consoles, Remote
USE   REMOTE CONSOLES

CONSOLIDATION

Consolidation, Over
USE   OVERCONSOLIDATION

CONSONANTS (SPEECH)

CONSTANT

Constant, Dielectric
USE   PERMITTIVITY

Constant, Gravitational
USE   GRAVITATIONAL CONSTANT

Constant, Gruneisen
USE   GRUNEISEN CONSTANT

Constant, Hubble
USE   HUBBLE CONSTANT

Constant, Perceptual Time
USE   PERCEPTUAL TIME CONSTANT

Constant, Plancks
USE   PLANCKS CONSTANT

Constant, Solar
USE   SOLAR CONSTANT

Constant Speed Propellers
USE   VARIABLE PITCH PROPELLERS

Constant, Time
USE   TIME CONSTANT

Constant Volume Balloons
USE   SUPERPRESSURE BALLOONS

CONSTANTAN

CONSTANTS

Constants, Elastic
USE   ELASTIC PROPERTIES

Constants Testing Reactor, Physical
USE   WATER COOLED REACTORS
      NUCLEAR RESEARCH AND TEST
      REACTORS

Constellation Aircraft, Lockheed
USE   C-121 AIRCRAFT

Constellation, Andromeda
USE   ANDROMEDA CONSTELLATION

Constellation, Aries
USE   ARIES CONSTELLATION

Constellation, Auriga
USE   AURIGA CONSTELLATION

Constellation, Cassiopeia
USE   CASSIOPEIA CONSTELLATION

Constellation, Centaurus
USE   CENTAURUS CONSTELLATION

Constellation, Cepheus
USE   CEPHEUS CONSTELLATION

Constellation, Corona Borealis
USE   CORONA BOREALIS CONSTELLATION

Constellation, Cygnus
USE   CYGNUS CONSTELLATION

Constellation, Lyra
USE   LYRA CONSTELLATION

Constellation, Orion
USE   ORION CONSTELLATION

Constellation, Sagittarius
USE   SAGITTARIUS CONSTELLATION

Constellation, Scorpio
USE   SCORPIUS CONSTELLATION

Constellation, Scorpius
USE   SCORPIUS CONSTELLATION

Constellation, Scutum
USE   SCUTUM CONSTELLATION

Constellation, Taurus
USE   TAURUS CONSTELLATION

CONSTELLATIONS

CONSTITUTION

Constitutional Diagrams
USE   PHASE DIAGRAMS

CONSTITUTIVE EQUATIONS

CONSTRAINTS

Constriction, Vaso
USE   VASOCONSTRICTION

CONSTRICTIONS

CONSTRICTORS

CONSTRUCTION

Construction, Aircraft
USE   AIRCRAFT STRUCTURES

Construction, Filament Wound
USE   FILAMENT WINDING

Construction In Space
USE   ORBITAL ASSEMBLY

CONSTRUCTION INDUSTRY

CONSTRUCTION MATERIALS

Construction Materials, Aircraft
USE   AIRCRAFT CONSTRUCTION MATERIALS

Construction Materials, Spacecraft
USE   SPACECRAFT CONSTRUCTION MATERIALS

Construction, Missile
USE   MISSILE STRUCTURES

Construction, Sandwich
USE   SANDWICH STRUCTURES

CONSULTING

CONSUMABLES (SPACECRAFT)

CONSUMABLES (SPACECREW SUPPLIES)

CONSUMERS

CONSUMPTION

Consumption, Energy
USE   ENERGY CONSUMPTION

Consumption, Fuel
USE   FUEL CONSUMPTION

Consumption, Oxygen
USE   OXYGEN CONSUMPTION

Consumption, Water
USE   WATER CONSUMPTION

CONTACT DERMATITIS

CONTACT LENSES

Contact Loads, Rolling
USE   ROLLING CONTACT LOADS

CONTACT POTENTIALS

CONTACT RESISTANCE

Contact, Sliding
USE   SLIDING CONTACT

Contact Solar Cells, Wraparound
USE   SOLAR CELLS

CONTACTORS

Contacts), Brushes (Electrical
USE   BRUSHES (ELECTRICAL CONTACTS)

Contacts (Electric)
USE   ELECTRIC CONTACTS

Contacts, Electric
USE   ELECTRIC CONTACTS

CONTACTS (GEOLOGY)

CONTAINERLESS MELTS

CONTAINERS

(Containers), Barrels
USE   BARRELS (CONTAINERS)

(Containers), Basins
USE   BASINS (CONTAINERS)

(Containers), Boxes
USE   BOXES (CONTAINERS)

(Containers), Cases
USE   CASES (CONTAINERS)

(Containers), Drums
USE   DRUMS (CONTAINERS)

(Containers), Receptacles
USE   CONTAINERS

(Containers), Tanks
USE   TANKS (CONTAINERS)

CONTAINMENT

CONTAMINANTS

Contaminants, Radioactive
USE   RADIOACTIVE CONTAMINANTS

Contaminants, Trace
USE   TRACE CONTAMINANTS

CONTAMINATION

Contamination, De
USE   DECONTAMINATION

Contamination, Fuel
USE   FUEL CONTAMINATION

Contamination, Spacecraft
USE   SPACECRAFT CONTAMINATION

CONTENT

Content, Bone Mineral
USE   BONE MINERAL CONTENT

Content, Heat
USE   ENTHALPY

Content, Moisture
USE   MOISTURE CONTENT

Content, Water
USE   MOISTURE CONTENT

CONTEXT

CONTEXT FREE LANGUAGES

CONTINENTAL DRIFT

Continental Margins
USE   CONTINENTAL SHELVES

CONTINENTAL SHELVES

CONTINENTS

CONTINGENCY

CONTINUITY

Continuity, Dis
USE   DISCONTINUITY

CONTINUITY EQUATION

CONTINUITY (MATHEMATICS)

CONTINUOUS NOISE

CONTINUOUS RADIATION

Continuous Radiation, Modulated
USE   MODULATED CONTINUOUS RADIATION

CONTINUOUS SPECTRA

CONTINUOUS WAVE LASERS

CONTINUOUS WAVE RADAR

Continuous Waves
USE   CONTINUOUS RADIATION

CONTINUUM FLOW

CONTINUUM MECHANICS

CONTINUUM MODELING

Continuum, Space-Time
USE   RELATIVITY

CONTINUUMS

Contour Matching Navigation System, Terrain
USE   TERCOM

CONTOUR SENSORS

CONTOURS

CONTRACT INCENTIVES

CONTRACT MANAGEMENT

CONTRACT NEGOTIATION

CONTRACTION

Contraction, Fitzgerald-Lorentz
USE   LORENTZ CONTRACTION

Contraction, Lorentz
USE   LORENTZ CONTRACTION

CONTRACTORS

CONTRACTS

CONTRAILS

CONTRALATERAL FUNCTIONS

CONTRAROTATING PROPELLERS

CONTRAST

Contrast, Image
USE   IMAGE CONTRAST

Contrast, Phase
USE   PHASE CONTRAST

CONTROL

Control, Access
USE   ACCESS CONTROL

Control, Active
USE   ACTIVE CONTROL

Control, Adaptive
USE   ADAPTIVE CONTROL

(Control), AFC
USE   AUTOMATIC FREQUENCY CONTROL

(Control), AGC
USE   AUTOMATIC GAIN CONTROL

Control, Air Traffic
USE   AIR TRAFFIC CONTROL

Control, Aircraft
USE   AIRCRAFT CONTROL

Control Airfoils, Circulation
USE   CIRCULATION CONTROL AIRFOILS

Control, Altitude
USE   ALTITUDE CONTROL

Control, Approach
USE   APPROACH CONTROL

Control, Attitude
USE   ATTITUDE CONTROL

Control, Automatic
USE   AUTOMATIC CONTROL

Control, Automatic Flight
USE   AUTOMATIC FLIGHT CONTROL

Control, Automatic Frequency
USE   AUTOMATIC FREQUENCY CONTROL

Control, Automatic Gain
USE   AUTOMATIC GAIN CONTROL

Control, Automatic Landing
USE   AUTOMATIC LANDING CONTROL

Control, Bang-Bang
USE   OFF-ON CONTROL

CONTROL BOARDS

Control, Boundary Layer
USE   BOUNDARY LAYER CONTROL

Control, Cascade
USE   CASCADE CONTROL

(Control Center), IMCC
USE   INTEGRATED MISSION CONTROL CENTER

Control Center, Integrated Mission
USE   INTEGRATED MISSION CONTROL CENTER

Control, Chemical Reaction
USE   CHEMICAL REACTION CONTROL

Control Circuits, Fire
USE   FIRE CONTROL CIRCUITS

Control Coatings, Thermal
USE   THERMAL CONTROL COATINGS

Control, Combustion
USE   COMBUSTION CONTROL

Control, Command And
USE   COMMAND AND CONTROL

Control, Command-
USE   COMMAND AND CONTROL

Control, Computerized
USE   NUMERICAL CONTROL

CONTROL CONFIGURED VEHICLES

CONTROL DATA (COMPUTERS)

Control Devices
USE   CONTROL EQUIPMENT

Control, Directional
USE   DIRECTIONAL CONTROL

Control), DISCOS (Satellite Attitude
USE   DISCOS (SATELLITE ATTITUDE CONTROL)

Control, Dynamic
USE   DYNAMIC CONTROL

Control, Electric
USE   ELECTRIC CONTROL

Control, Electrohydraulic
USE   ELECTRIC CONTROL
       HYDRAULIC CONTROL

Control, Electromagnetic
USE   REMOTE CONTROL
       ELECTROMAGNETS

Control, Electronic
USE   ELECTRONIC CONTROL

Control, Engine
USE   ENGINE CONTROL

Control Engines, Variable Stream
USE   VARIABLE STREAM CONTROL ENGINES

Control, Environmental
USE   ENVIRONMENTAL CONTROL

CONTROL EQUIPMENT

Control, Feedback
USE   FEEDBACK CONTROL

Control, Feedforward
USE   FEEDFORWARD CONTROL

Control, Fire
USE   FIRE CONTROL

Control, Flap
USE   FLAPS (CONTROL SURFACES)
       AIRCRAFT CONTROL

Control, Flight
USE   FLIGHT CONTROL

Control, Flood
USE   FLOOD CONTROL

Control, Fly By Tube
USE   FLY BY TUBE CONTROL

Control, Fly By Wire
USE   FLY BY WIRE CONTROL

Control, Frequency
USE   FREQUENCY CONTROL

Control, Fuel
  USE   FUEL CONTROL

Control, Ground Based
  USE   GROUND BASED CONTROL

Control Group, Transponder
  USE   TRANSPONDER CONTROL GROUP

Control, Harmonic
  USE   HARMONIC CONTROL

Control, Helicopter
  USE   HELICOPTER CONTROL

Control, Hydraulic
  USE   HYDRAULIC CONTROL

Control (Industry), Process
  USE   PROCESS CONTROL (INDUSTRY)

Control, Interactive
  USE   INTERACTIVE CONTROL

Control, Jet
  USE   JET CONTROL

Control, Laminar Flow
  USE   LAMINAR BOUNDARY LAYER
         BOUNDARY LAYER CONTROL

Control, Lateral
  USE   LATERAL CONTROL

Control, Longitudinal
  USE   LONGITUDINAL CONTROL

Control, Magnetic
  USE   MAGNETIC CONTROL

Control, Manual
  USE   MANUAL CONTROL

Control, Missile
  USE   MISSILE CONTROL

CONTROL MOMENT GYROSCOPES

Control, Network
  USE   NETWORK CONTROL

Control, Nuclear Reactor
  USE   NUCLEAR REACTOR CONTROL

Control, Numerical
  USE   NUMERICAL CONTROL

Control, Off-On
  USE   OFF-ON CONTROL

Control, Optimal
  USE   OPTIMAL CONTROL

Control, Optimum
  USE   OPTIMAL CONTROL

Control Panels
  USE   CONTROL BOARDS

Control, Payload
  USE   PAYLOAD CONTROL

Control, Phase
  USE   PHASE CONTROL

Control, Pitch Attitude
  USE   LONGITUDINAL CONTROL

Control, Plasma
  USE   PLASMA CONTROL

Control, Pneumatic
  USE   PNEUMATIC CONTROL

Control, Pollution
  USE   POLLUTION CONTROL

Control, Porous Boundary Layer
  USE   POROUS BOUNDARY LAYER CONTROL

Control Project, Submarine Integrated
  USE   SUBMARINE INTEGRATED CONTROL
         PROJECT

Control, Proportional
  USE   PROPORTIONAL CONTROL

Control, Quality
  USE   QUALITY CONTROL

Control, Radar Approach
  USE   RADAR APPROACH CONTROL

Control, Radio
  USE   RADIO CONTROL

Control, Range
  USE   TRAJECTORY CONTROL

(Control), RAPCON
  USE   RADAR APPROACH CONTROL

Control, Reaction
  USE   REACTION CONTROL

Control Reactor, Spectral Shift
  USE   SPECTRAL SHIFT CONTROL REACTOR

Control, Reliability
  USE   RELIABILITY ENGINEERING
         QUALITY CONTROL

Control, Remote
  USE   REMOTE CONTROL

Control, Rocket Engine
  USE   ROCKET ENGINE CONTROL

CONTROL ROCKETS

CONTROL RODS

Control, Roll
  USE   LATERAL CONTROL

Control Rotors, Circulation
  USE   CIRCULATION CONTROL ROTORS

Control, Satellite
  USE   SATELLITE CONTROL

Control, Satellite Attitude
  USE   SATELLITE ATTITUDE CONTROL

Control Satellite, Transit Attitude
  USE   TRANSIT ATTITUDE CONTROL SATELLITE

Control, Sequential
  USE   SEQUENTIAL CONTROL

Control, Servo
  USE   SERVOCONTROL

Control, Servostability
  USE   SERVOCONTROL

Control, Shape
  USE   SHAPE CONTROL

Control, Shock Wave
  USE   SHOCK WAVE CONTROL

CONTROL SIMULATION

Control, Space Vehicle
  USE   SPACECRAFT CONTROL

Control, Spacecraft
  USE   SPACECRAFT CONTROL

Control, Spectral Shift
  USE   SPECTRAL SHIFT CONTROL

Control, Speed
  USE   SPEED CONTROL

CONTROL STABILITY

CONTROL STICKS

CONTROL SURFACES

(Control Surfaces), Elevators
  USE   ELEVATORS (CONTROL SURFACES)

(Control Surfaces), Flaps
  USE   FLAPS (CONTROL SURFACES)

(Control Surfaces), Tabs
  USE   TABS (CONTROL SURFACES)

(Control System), AFCS
  USE   AUTOMATIC FLIGHT CONTROL

Control System, Airborne Warning And
  USE   AWACS AIRCRAFT

Control Systems
  USE   CONTROL

Control Systems, Adaptive
  USE   ADAPTIVE CONTROL

CONTROL SYSTEMS DESIGN

Control Systems, Pointing
  USE   POINTING CONTROL SYSTEMS

Control Systems, Self Adaptive
  USE   SELF ADAPTIVE CONTROL SYSTEMS

Control, Temperature
  USE   TEMPERATURE CONTROL

CONTROL THEORY

Control, Thrust
  USE   THRUST CONTROL

Control, Thrust Vector
  USE   THRUST VECTOR CONTROL

Control, Time Optimal
  USE   TIME OPTIMAL CONTROL

Control, Traffic
  USE   TRAFFIC CONTROL

Control, Trajectory
  USE   TRAJECTORY CONTROL

Control, Turbojet Engine
  USE   TURBOJET ENGINE CONTROL

(Control), TVC
  USE   THRUST VECTOR CONTROL

CONTROL UNITS (COMPUTERS)

CONTROL VALVES

Control Valves, Automatic
  USE   AUTOMATIC CONTROL VALVES

Control, Vector
  USE   DIRECTIONAL CONTROL

Control, Visual
  USE   VISUAL CONTROL

Control, Voice
  USE   VOICE CONTROL

Control, Wave Incidence
  USE   WAVE INCIDENCE CONTROL

Control, Weather
  USE   WEATHER MODIFICATION

CONTROLLABILITY

CONTROLLED ATMOSPHERES

Controlled Avalanche Transit Time Devices
  USE   CATT DEVICES

CONTROLLED FUSION

Controlled Oscillators, Voltage
  USE   VOLTAGE CONTROLLED OSCILLATORS

Controlled Rectifiers, Silicon
  USE   SILICON CONTROLLED RECTIFIERS

Controlled Stability
  USE   CONTROL

CONTROLLERS

Controllers (Personnel), Air Traffic
  USE   AIR TRAFFIC CONTROLLERS (PERSONNEL)

Controllers, Power Factor
  USE   POWER FACTOR CONTROLLERS

Controls, Direct Lift
  USE   DIRECT LIFT CONTROLS

Controls, Inventory
  USE   INVENTORY CONTROLS

Convair Military Aircraft
  USE   MILITARY AIRCRAFT
         GENERAL DYNAMICS AIRCRAFT

Convair 340 Aircraft
  USE   CV-340 AIRCRAFT

Convair 440 Aircraft
  USE   CV-440 AIRCRAFT

Convair 880 Aircraft
  USE   CV-880 AIRCRAFT

Convair 990 Aircraft
  USE   CV-990 AIRCRAFT

CONVECTION

CONVECTION CLOUDS

CONVECTION CURRENTS

Convection, Forced
  USE   FORCED CONVECTION

Convection, Free
  USE   FREE CONVECTION

Convection, Marangoni
  USE   MARANGONI CONVECTION

Convection, Rayleigh-Benard
  USE   RAYLEIGH-BENARD CONVECTION

Convection, Thermal
  USE   FREE CONVECTION

CONVECTIVE FLOW

CONVECTIVE HEAT TRANSFER

CONVENTIONS

CONVERGENCE

CONVERGENT NOZZLES

Convergent Zones, Intertropical
  USE   INTERTROPICAL CONVERGENT ZONES

CONVERGENT-DIVERGENT NOZZLES

CONVERSATION

CONVERSION

Conversion, Bio
  USE   BIOCONVERSION

Conversion Efficiency, Energy
  USE   ENERGY CONVERSION EFFICIENCY

Conversion, Electric Power
  USE   ELECTRIC GENERATORS

Conversion, Energy
  USE   ENERGY CONVERSION

Conversion, Frequency
  USE   FREQUENCY CONVERTERS

Conversion, Geothermal Energy
  USE   GEOTHERMAL ENERGY CONVERSION

Conversion, Internal
  USE   INTERNAL CONVERSION

Conversion, Metric
  USE   METRICATION

Conversion, Ocean Thermal Energy
  USE   OCEAN THERMAL ENERGY CONVERSION

Conversion), Organic Wastes (Fuel
  USE   ORGANIC WASTES (FUEL CONVERSION)

Conversion, Ortho Para
  USE   ORTHO PARA CONVERSION

Conversion, Photothermal
  USE   PHOTOTHERMAL CONVERSION

Conversion, Photovoltaic
  USE   PHOTOVOLTAIC CONVERSION

Conversion Routines, Data
  USE   DATA CONVERSION ROUTINES

Conversion, Satellite Solar Energy
  USE   SATELLITE SOLAR ENERGY CONVERSION

Conversion, Solar Energy
  USE   SOLAR ENERGY CONVERSION

Conversion Systems, Thermionic
  USE   THERMIONIC POWER GENERATION

Conversion Systems, Thermoelectric
  USE   THERMOELECTRIC POWER GENERATION

CONVERSION TABLES

Conversion, Turboelectric
  USE   TURBOGENERATORS

Conversion, Waterwave Energy
  USE   WATERWAVE ENERGY CONVERSION

Convertaplanes
  USE   V/STOL AIRCRAFT

CONVERTERS

Converters (AC To AC), Voltage
  USE   VOLTAGE CONVERTERS (AC TO AC)

Converters (AC To DC), Current
  USE   CURRENT CONVERTERS (AC TO DC)

Converters, Analog To Digital
  USE   ANALOG TO DIGITAL CONVERTERS

Converters, Binary To Decimal
  USE   BINARY TO DECIMAL CONVERTERS

Converters, Data
  USE   DATA CONVERTERS

Converters (DC To AC), Inverted
  USE   INVERTED CONVERTERS (DC TO AC)

Converters (DC To DC), Voltage
  USE   VOLTAGE CONVERTERS (DC TO DC)

Converters, Decimal To Binary
  USE   DECIMAL TO BINARY CONVERTERS

Converters, Digital To Analog
  USE   DIGITAL TO ANALOG CONVERTERS

Converters, Down-
  USE   DOWN-CONVERTERS

Converters, Energy
  USE   DIRECT POWER GENERATORS

Converters, Frequency
  USE   FREQUENCY CONVERTERS

Converters, Image
  USE   IMAGE CONVERTERS

Converters, Parametric Frequency
  USE   PARAMETRIC FREQUENCY CONVERTERS

Converters, Power
  USE   POWER CONVERTERS

Converters, Pulse Width Amplitude
  USE   PULSE WIDTH AMPLITUDE CONVERTERS

Converters, Solar
  USE   SOLAR GENERATORS

Converters, Thermionic
  USE   THERMIONIC CONVERTERS

Converters, Torque
  USE   TORQUE CONVERTERS

Converters, Up-
  USE   UP-CONVERTERS

CONVEXITY

CONVEYORS

CONVOLUTION INTEGRALS

Convolutions (Mathematics)
  USE   CONVOLUTION INTEGRALS

Convulsants, Anti
  USE   ANTICONVULSANTS

CONVULSIONS

COOK INLET (AK)

Cookpot Aircraft
  USE   TU-124 AIRCRAFT

COOL STARS

Coolant Loss
  USE   LOSS OF COOLANT

Coolant, Loss Of
  USE   LOSS OF COOLANT

COOLANTS

Coolants, Engine
  USE   ENGINE COOLANTS

Coolants, Organic
  USE   ORGANIC COOLANTS

Cooled Fast Reactors, Gas
  USE   GAS COOLED FAST REACTORS

Cooled Reactor, Advanced Sodium
  USE   ADVANCED SODIUM COOLED REACTOR

Cooled Reactor Experiment, Lithium
  USE   LITHIUM COOLED REACTOR EXPERIMENT

Cooled Reactors, Experimental Gas
  USE   EXPERIMENTAL GAS COOLED REACTORS

Cooled Reactors, Experimental Organic
  USE   EXPERIMENTAL ORGANIC COOLED
         REACTORS

Cooled Reactors, Gas
  USE   GAS COOLED REACTORS

Cooled Reactors, High Temperature Gas
  USE   HIGH TEMPERATURE GAS COOLED
         REACTORS

Cooled Reactors, Liquid
  USE   LIQUID COOLED REACTORS

Cooled Reactors, Liquid Metal
  USE   LIQUID METAL COOLED REACTORS

Cooled Reactors, Organic
  USE   ORGANIC COOLED REACTORS

Cooled Reactors, Water
  USE   WATER COOLED REACTORS

COOLERS

Coolers, Ettingshausen
USE   ETTINGSHAUSEN EFFECT
      THERMOELECTRIC COOLING

COOLING

Cooling, Absorption
USE   ABSORPTION COOLING

Cooling, Adiabatic Demagnetization
USE   ADIABATIC DEMAGNETIZATION COOLING

Cooling, Air
USE   AIR COOLING

Cooling (Buildings), Space
USE   SPACE COOLING (BUILDINGS)

Cooling, Cryogenic
USE   CRYOGENIC COOLING

Cooling, Evaporative
USE   EVAPORATIVE COOLING

Cooling, Film
USE   FILM COOLING

COOLING FINS

Cooling, Gas
USE   GAS COOLING

Cooling, Liquid
USE   LIQUID COOLING

Cooling, Magnetic
USE   MAGNETIC COOLING

Cooling, Plasma
USE   PLASMA COOLING

(Cooling), Quenching
USE   QUENCHING (COOLING)

Cooling, Radiant
USE   RADIANT COOLING

Cooling, Regenerative
USE   REGENERATIVE COOLING

Cooling, Sodium
USE   SODIUM COOLING

Cooling, Solar
USE   SOLAR COOLING

Cooling, Solid Cryogen
USE   SOLID CRYOGEN COOLING

Cooling, Super
USE   SUPERCOOLING

Cooling, Surface
USE   SURFACE COOLING

Cooling, Sweat
USE   SWEAT COOLING

COOLING SYSTEMS

Cooling, Thermoelectric
USE   THERMOELECTRIC COOLING

Cooling, Thermomagnetic
USE   THERMOMAGNETIC COOLING

Cooling, Transpiration
USE   SWEAT COOLING

Cooling, Water
USE   LIQUID COOLING

Cooper-Schrieffer Theory, Bardeen-
USE   BCS THEORY

COOPERATION

Cooperation, International
USE   INTERNATIONAL COOPERATION

Coordinate Geometry Language
USE   COGO (PROGRAMMING LANGUAGE)

Coordinate Systems
USE   COORDINATES

COORDINATE TRANSFORMATIONS

COORDINATES

Coordinates, Astronomical
USE   ASTRONOMICAL COORDINATES

(Coordinates), Axes
USE   COORDINATES

Coordinates, Cartesian
USE   CARTESIAN COORDINATES

Coordinates, Curvilinear
USE   SPHERICAL COORDINATES

Coordinates, Cylindrical
USE   CARTESIAN COORDINATES

Coordinates, Geocentric
USE   GEOCENTRIC COORDINATES

Coordinates, Geodetic
USE   GEODETIC COORDINATES

Coordinates, Hylleraas
USE   HYLLERAAS COORDINATES

Coordinates, Hyperbolic
USE   HYPERBOLIC COORDINATES

Coordinates, Inertial
USE   INERTIAL COORDINATES

Coordinates, Lagrange
USE   LAGRANGE COORDINATES

Coordinates, Oblique
USE   OBLIQUE COORDINATES

Coordinates, Planetocentric
USE   PLANETOCENTRIC COORDINATES

Coordinates, Polar
USE   POLAR COORDINATES

Coordinates, Rectangular
USE   CARTESIAN COORDINATES

Coordinates, Spherical
USE   SPHERICAL COORDINATES

COORDINATION

COORDINATION POLYMERS

Coordinator, Langley Complex
USE   LANGLEY COMPLEX COORDINATOR

Copernicus Spacecraft
USE   OAO 3

Copilots
USE   AIRCRAFT PILOTS

COPLANARITY

COPOLYMERIZATION

COPOLYMERS

Copolymers, Vinyl
USE   VINYL COPOLYMERS

COPPER

COPPER ALLOYS

COPPER CHLORIDES

COPPER COMPOUNDS

COPPER FLUORIDES

COPPER ISOTOPES

COPPER OXIDES

COPPER SELENIDES

COPPER SULFIDES

(Copying), Reproduction
USE   REPRODUCTION (COPYING)

COPYRIGHTS

Coral Heads
USE   CORAL REEFS

CORAL REEFS

Cord, Spinal
USE   SPINAL CORD

CORDAGE

CORDIERITE

Cordite
USE   COLLOIDAL PROPELLANTS
      DOUBLE BASE PROPELLANTS

Cords, Vocal
USE   VOCAL CORDS

Core, Earth
USE   EARTH CORE

CORE FLOW

Core, Lunar
USE   LUNAR CORE

Core Pulse Reactors, Annular
USE   ANNULAR CORE PULSE REACTORS

Core Reactors, Plasma
USE   PLASMA CORE REACTORS

CORE SAMPLING

CORE STORAGE

CORES

Cores, Honeycomb
USE   HONEYCOMB CORES

Cores, Magnetic
USE   MAGNETIC CORES

Cores, Planetary
USE   PLANETARY CORES

Cores, Reactor
USE   REACTOR CORES

Cores, Stellar
USE   STELLAR CORES

CORIOLIS EFFECT

CORK (MATERIALS)

CORN

CORNEA

CORNER FLOW

Corner Reflectors, Radar
USE   RADAR CORNER REFLECTORS

CORNERS

CORONA BOREALIS CONSTELLATION

Corona Discharges
USE   ELECTRIC CORONA

Corona, Electric
USE   ELECTRIC CORONA

Corona, Solar
USE   SOLAR CORONA

CORONAGRAPHS

CORONAL HOLES

CORONAL LOOPS

CORONARY ARTERY DISEASE

CORONARY CIRCULATION

CORONAS

Coronas, Stellar
    USE   STELLAR CORONAS

COROTATION

Corp Aircraft, British Aircraft
    USE   BAC AIRCRAFT

CORPORAL MISSILE

CORPUSCLES

CORPUSCULAR RADIATION

Corpuscular Radiation, Solar
    USE   SOLAR CORPUSCULAR RADIATION

Correcting Codes, Error
    USE   ERROR CORRECTING CODES

Correcting Devices, Error
    USE   ERROR CORRECTING DEVICES

CORRECTION

Correction, Atmospheric
    USE   ATMOSPHERIC CORRECTION

Correction Procedure, Optical
    USE   OPTICAL CORRECTION PROCEDURE

Correction, Radiometric
    USE   RADIOMETRIC CORRECTION

CORRELATION

Correlation, Angular
    USE   ANGULAR CORRELATION

Correlation, Auto
    USE   AUTOCORRELATION

CORRELATION COEFFICIENTS

Correlation, Cross
    USE   CROSS CORRELATION

Correlation, Data
    USE   DATA CORRELATION

CORRELATION DETECTION

Correlation Functions
    USE   CORRELATION

Correlation, Spectral
    USE   SPECTRAL CORRELATION

Correlation, Statistical
    USE   STATISTICAL CORRELATION

Correlator), SIMICOR (Image
    USE   IMAGE CORRELATORS

Correlator, Simultaneous Image
    USE   IMAGE CORRELATORS

CORRELATORS

Correlators, Image
    USE   IMAGE CORRELATORS

Corridor (MO), St Louis-Kansas City
    USE   ST LOUIS-KANSAS CITY CORRIDOR (MO)

Corridor (North America), Great Plains
    USE   GREAT PLAINS CORRIDOR (NORTH
           AMERICA)

CORRIDORS

CORROSION

Corrosion, Cavitation
    USE   CAVITATION CORROSION

Corrosion Cracking, Stress
    USE   STRESS CORROSION CRACKING

Corrosion, Electrochemical
    USE   ELECTROCHEMICAL CORROSION

Corrosion, Fretting
    USE   FRETTING CORROSION

Corrosion, Fuel
    USE   FUEL CORROSION

Corrosion, Hot
    USE   HOT CORROSION

Corrosion, Intergranular
    USE   INTERGRANULAR CORROSION

Corrosion, Metal
    USE   CORROSION

CORROSION PREVENTION

CORROSION RESISTANCE

(Corrosion), Scale
    USE   SCALE (CORROSION)

Corrosion, Stress
    USE   STRESS CORROSION

CORROSION TEST LOOPS

CORROSION TESTS

Corrosion, Transgranular
    USE   TRANSGRANULAR CORROSION

CORRUGATED PLATES

CORRUGATED SHELLS

CORRUGATING

Corsair Aircraft
    USE   A-7 AIRCRAFT

Cortex, Cerebral
    USE   CEREBRAL CORTEX

CORTEXES

CORTEXES (BOTANY)

CORTI ORGAN

Corticosteroid, Hydroxy
    USE   HYDROXYCORTICOSTEROID

CORTICOSTEROIDS

CORTISONE

Corundum
    USE   ALUMINUM OXIDES

CORVUS MISSILE

COS-B SATELLITE

COSINE SERIES

COSMIC BACKGROUND EXPLORER SATELLITE

COSMIC DUST

Cosmic Gamma Ray Bursts
    USE   GAMMA RAY BURSTS

COSMIC GASES

COSMIC NOISE

COSMIC PLASMA

Cosmic Radiation
    USE   COSMIC RAYS

Cosmic Radio Waves
    USE   EXTRATERRESTRIAL RADIO WAVES

COSMIC RAY ALBEDO

Cosmic Ray Primaries, Heavy
    USE   HEAVY NUCLEI
           PRIMARY COSMIC RAYS

COSMIC RAY SHOWERS

COSMIC RAYS

Cosmic Rays, Galactic
    USE   GALACTIC COSMIC RAYS

Cosmic Rays, Primary
    USE   PRIMARY COSMIC RAYS

Cosmic Rays, Secondary
    USE   SECONDARY COSMIC RAYS

Cosmic Rays, Solar
    USE   SOLAR COSMIC RAYS

COSMIC X RAYS

COSMOCHEMISTRY

Cosmogony
    USE   COSMOLOGY

COSMOLOGY

Cosmology, Big Bang
    USE   BIG BANG COSMOLOGY

COSMONAUTS

COSMOS

COSMOS SATELLITES

COSMOS 2 SATELLITE

COSMOS 3 SATELLITE

COSMOS 5 SATELLITE

COSMOS 6 SATELLITE

COSMOS 14 SATELLITE

COSMOS 44 SATELLITE

COSMOS 54 SATELLITE

COSMOS 71 SATELLITE

COSMOS 110 SATELLITE

COSMOS 137 SATELLITE

COSMOS 144 SATELLITE

COSMOS 149 SATELLITE

COSMOS 166 SATELLITE

COSMOS 186 SATELLITE

COSMOS 188 SATELLITE

COSMOS 206 SATELLITE

COSMOS 213 SATELLITE

COSMOS 224 SATELLITE

COSMOS 225 SATELLITE

COSMOS 381 SATELLITE

COSMOS 782 SATELLITE

COSMOS 936 SATELLITE

COSMOS 954 SATELLITE

COSMOS 1129 SATELLITE

COSPAR (Committee)
    USE   COMMITTEE ON SPACE RESEARCH

COSPAS

COSSERAT SURFACES

COST ANALYSIS

Cost, Design To
    USE   DESIGN TO COST

COST EFFECTIVENESS

COST ESTIMATES

COST INCENTIVES

Cost, Low
    USE   LOW COST

COST REDUCTION

COSTA RICA

COSTS

Costs, Aircraft Production
    USE   AIRCRAFT PRODUCTION COSTS

Costs, Freight
    USE   FREIGHT COSTS

Costs, Life Cycle
    USE   LIFE CYCLE COSTS

Costs, Operating
    USE   OPERATING COSTS

Costs, Production
    USE   PRODUCTION COSTS

COTTON

COTTON FIBERS

COUCHES

COUETTE FLOW

Cougar Aircraft
    USE   F-9 AIRCRAFT

COUGH

Coulees
    USE   CANYONS

COULOMB COLLISIONS

COULOMB POTENTIAL

COULOMETERS

COULOMETRY

COUNTDOWN

COUNTER ROTATION

COUNTER-ROTATING WHEELS

COUNTERBALANCES

COUNTERFLOW

COUNTERMEASURES

Countermeasures, Electronic
    USE   ELECTRONIC COUNTERMEASURES

Countermeasures, Optical
    USE   OPTICAL COUNTERMEASURES

COUNTERS

Counters, Cerenkov
    USE   CERENKOV COUNTERS

Counters, Electron
    USE   ELECTRON COUNTERS

Counters, Gas Discharge
    USE   GAS DISCHARGE TUBES
           COUNTERS

Counters, Geiger
    USE   GEIGER COUNTERS

Counters, Ionization
    USE   RADIATION COUNTERS
           IONIZATION CHAMBERS

Counters, Neutron
    USE   NEUTRON COUNTERS

Counters, Particle
    USE   RADIATION COUNTERS

Counters, Proportional
    USE   PROPORTIONAL COUNTERS

Counters, Quantum
    USE   QUANTUM COUNTERS

Counters, Radiation
    USE   RADIATION COUNTERS

Counters, Scintillation
    USE   SCINTILLATION COUNTERS

COUNTERSINKING

COUNTING

COUNTING CIRCUITS

COUNTING RATE COMPUTERS

County Achondrite, Norton
    USE   NORTON COUNTY ACHONDRITE

Coupled Devices, Charge
    USE   CHARGE COUPLED DEVICES

COUPLED MODES

Coupled Plasmas, Strongly
    USE   STRONGLY COUPLED PLASMAS

COUPLERS

Couplers, Antenna
    USE   ANTENNA COUPLERS

Couplers, Directional
    USE   DIRECTIONAL COUPLERS

COUPLES

COUPLING

COUPLING CIRCUITS

COUPLING COEFFICIENTS

Coupling, Cross
    USE   CROSS COUPLING

Coupling, De
    USE   DECOUPLING

Coupling, Gyroscopic
    USE   GYROSCOPIC COUPLING

Coupling, Microwave
    USE   MICROWAVE COUPLING

Coupling, Mode
    USE   COUPLED MODES

Coupling, Optical
    USE   OPTICAL COUPLING

Coupling, Spin-Spin
    USE   SPIN-SPIN COUPLING

Coupling, Thermodynamic
    USE   THERMODYNAMIC COUPLING

Coupling, Velocity
    USE   VELOCITY COUPLING

COUPLINGS

Courier Aircraft
    USE   U-10 AIRCRAFT

COURIER SATELLITE

Courses
    USE   PATHS

COVALENCE

COVALENT BONDS

COVARIANCE

Cover, Cloud
    USE   CLOUD COVER

Cover, Snow
    USE   SNOW COVER

Coverage Antennas, High Resolution
    USE   HIGH RESOLUTION COVERAGE ANTENNAS

COVERALLS

COVERINGS

Coves
    USE   BAYS (TOPOGRAPHIC FEATURES)

Cowell Method
    USE   NUMERICAL INTEGRATION

COWLINGS

Cr
    USE   CHROMIUM

CRAB NEBULA

CRABS

CRACK ARREST

CRACK CLOSURE

Crack Formation
    USE   CRACK INITIATION

CRACK GEOMETRY

Crack, Griffith
    USE   GRIFFITH CRACK

CRACK INITIATION

CRACK PROPAGATION

CRACK TIPS

CRACKING (CHEMICAL ENGINEERING)

CRACKING (FRACTURING)

Cracking, Stress Corrosion
    USE   STRESS CORROSION CRACKING

CRACKS

Cracks, Micro
    USE   MICROCRACKS

Cracks, Surface
    USE   SURFACE CRACKS

Craft
    USE   VEHICLES

Craft, Hydrofoil
    USE   HYDROFOIL CRAFT

Craft Reaction, Friedel-
    USE   FRIEDEL-CRAFT REACTION

CRAMPS

Crane Helicopter, Flying
    USE   H-17 HELICOPTER

CRANES

Cranes, Gantry
    USE   GANTRY CRANES

CRANIUM

CRANK-NICHOLSON METHOD

Cranked Wings
    USE   SWEPT WINGS

Cranks
    USE   ECCENTRICS

CRASH INJURIES

CRASH LANDING

CRASHES

CRASHWORTHINESS

Crater, Ptolemaeus
    USE   PTOLEMAEUS CRATER

Crater, Tycho
    USE   TYCHO CRATER

CRATERING

Cratering, Hypervelocity
    USE   PROJECTILE CRATERING
           HYPERVELOCITY PROJECTILES

Cratering, Projectile
    USE   PROJECTILE CRATERING

CRATERS

Craters, Fossil Meteorite
    USE   FOSSILS
           METEORITE CRATERS

Craters, Lunar
    USE   LUNAR CRATERS

Craters, Mars
    USE   MARS CRATERS

Craters, Meteor
    USE   CRATERS

Craters, Meteorite
    USE   METEORITE CRATERS

Craters, Meteoroid
    USE   METEORITE CRATERS

Craters, Planetary
    USE   PLANETARY CRATERS

CRATONS

CRAWLER TRACTORS

CRAY COMPUTERS

CRAYONS

Crazing
    USE   SURFACE CRACKS

CREATINE

CREATININE

Creation
    USE   CREATIVITY

CREATIVITY

CREEP ANALYSIS

CREEP BUCKLING

CREEP DIAGRAMS

CREEP PROPERTIES

Creep Resistance
    USE   CREEP STRENGTH

CREEP RUPTURE STRENGTH

Creep, Shear
    USE   SHEAR CREEP

Creep, Steady State
    USE   STEADY STATE CREEP

CREEP STRENGTH

Creep, Tensile
    USE   TENSILE CREEP

CREEP TESTS

CREPE

CRESOLS

Crestatrons
    USE   TRAVELING WAVE TUBES

Crests
    USE   WAVES

CREVASSES

Crevices
    USE   CRACKS

CREW EXPERIMENT STATIONS

CREW OBSERVATION STATIONS

CREW PROCEDURES (INFLIGHT)

CREW PROCEDURES (PREFLIGHT)

CREW SIZE

CREW STATIONS

CREW WORKSTATIONS

CREWS

Crews, Flight
    USE   FLIGHT CREWS

Crews, Ground
    USE   GROUND CREWS

Crews, Space
    USE   SPACECREWS

CRICKETS

CRIME

Crimping
    USE   FOLDING

CRITERIA

Criteria, Structural Design
    USE   STRUCTURAL DESIGN CRITERIA

CRITICAL EXPERIMENTS

CRITICAL FLICKER FUSION

CRITICAL FLOW

CRITICAL FREQUENCIES

CRITICAL LOADING

Critical Mach Number
    USE   CRITICAL VELOCITY
           MACH NUMBER

CRITICAL MASS

CRITICAL PATH METHOD

CRITICAL POINT

CRITICAL PRESSURE

Critical Reynolds Number
    USE   CRITICAL VELOCITY
           REYNOLDS NUMBER

Critical Speed
    USE   CRITICAL VELOCITY

Critical Stress
    USE   CRITICAL LOADING

CRITICAL TEMPERATURE

CRITICAL VELOCITY

CROCCO METHOD

CROCCO-LEE THEORY

CROLOY

CROP CALENDARS

CROP DUSTING

CROP GROWTH

CROP IDENTIFICATION

CROP INVENTORIES

Crop Inventories By Remote Sensing
    USE   AGRISTARS PROJECT

Crop Inventory Experiment, Large Area
    USE   LARGE AREA CROP INVENTORY
           EXPERIMENT

CROP VIGOR

Croplands
    USE   FARMLANDS

CROPS

Crops, Farm
    USE   FARM CROPS

CROSS CORRELATION

CROSS COUPLING

Cross Faults
    USE   GEOLOGICAL FAULTS

CROSS FLOW

Cross Modulation, Ionospheric
    USE   IONOSPHERIC CROSS MODULATION

CROSS POLARIZATION

CROSS RELAXATION

CROSS SECTIONS

Cross Sections, Absorption
    USE   ABSORPTION CROSS SECTIONS

Cross Sections, Capture
    USE   ABSORPTION CROSS SECTIONS

Cross Sections, Ionization
    USE   IONIZATION CROSS SECTIONS

Cross Sections, Neutron
    USE   NEUTRON CROSS SECTIONS

Cross Sections, Radar
    USE   RADAR CROSS SECTIONS

Cross Sections, Scattering
    USE   SCATTERING CROSS SECTIONS

CROSSBEDDING (GEOLOGY)

CROSSED FIELD AMPLIFIERS

CROSSED FIELD GUNS

CROSSED FIELDS

CROSSINGS

Crossings, Zero
    USE   ROOTS OF EQUATIONS

CROSSLINKING

CROSSOVERS

CROSSTALK

Crotchets, Geomagnetic
    USE   SUDDEN IONOSPHERIC DISTURBANCES

CROWDING

CRUCIBLES

CRUCIFORM WINGS

CRUDE OIL

Cruise Aircraft Research, Supersonic
    USE   SUPERSONIC CRUISE AIRCRAFT
            RESEARCH

CRUISE MISSILES

CRUISING FLIGHT

Crusader Aircraft
    USE   F-8 AIRCRAFT

CRUSHERS

CRUSHING

Crust, Earth
    USE   EARTH CRUST

Crust, Lunar
    USE   LUNAR CRUST

CRUSTAL FRACTURES

CRUSTS

CRYOCHEMISTRY

CRYOCYCLE PRINCIPLE

CRYODEPOSITS

Cryogen Cooling, Solid
    USE   SOLID CRYOGEN COOLING

CRYOGENIC COMPUTER STORAGE

CRYOGENIC COOLING

CRYOGENIC EQUIPMENT

CRYOGENIC FLUID STORAGE

CRYOGENIC FLUIDS

CRYOGENIC GYROSCOPES

CRYOGENIC MAGNETS

CRYOGENIC ROCKET PROPELLANTS

CRYOGENIC STORAGE

CRYOGENIC WIND TUNNELS

CRYOGENICS

Cryogens, Solid
    USE   SOLID CRYOGENS

CRYOLITE

CRYOPUMPING

CRYOSAR

Cryosorption
    USE   SORPTION

CRYOSTATS

CRYOTRAPPING

CRYOTRONS

CRYPTOGRAPHY

CRYSTAL DEFECTS

(Crystal Defects), Vacancies
    USE   VACANCIES (CRYSTAL DEFECTS)

CRYSTAL DISLOCATIONS

CRYSTAL FILTERS

CRYSTAL GROWTH

Crystal Growth, Hydrothermal
    USE   HYDROTHERMAL CRYSTAL GROWTH

(Crystal Growth), Melts
    USE   MELTS (CRYSTAL GROWTH)

CRYSTAL LATTICES

CRYSTAL OPTICS

CRYSTAL OSCILLATORS

CRYSTAL RECTIFIERS

CRYSTAL STRUCTURE

CRYSTAL SURFACES

CRYSTALLINITY

CRYSTALLITES

CRYSTALLIZATION

CRYSTALLOGRAPHY

CRYSTALS

Crystals, Bravais
    USE   BRAVAIS CRYSTALS

Crystals, Dendritic
    USE   DENDRITIC CRYSTALS

(Crystals), Directional Solidification
    USE   DIRECTIONAL SOLIDIFICATION (CRYSTALS)

Crystals, Doped
    USE   DOPED CRYSTALS

Crystals, Ionic
    USE   IONIC CRYSTALS

Crystals, Liquid
    USE   LIQUID CRYSTALS

Crystals, Metal
    USE   METAL CRYSTALS

Crystals, Micro
    USE   MICROCRYSTALS

Crystals, Mixed
    USE   MIXED CRYSTALS

Crystals, Piezoelectric
    USE   PIEZOELECTRIC CRYSTALS

Crystals, Poly
    USE   POLYCRYSTALS

Crystals, Quartz
    USE   QUARTZ CRYSTALS

Crystals, Single
    USE   SINGLE CRYSTALS

(Crystals), Whiskers
    USE   WHISKERS (CRYSTALS)

Cs
    USE   CESIUM

CSM
    USE   COMMAND SERVICE MODULES

CT
    USE   CONNECTICUT

(CT), New Haven
    USE   NEW HAVEN (CT)

CT-114 Aircraft
    USE   CL-41 AIRCRAFT

CTD
    USE   CHARGE TRANSFER DEVICES

Cu
    USE   COPPER

CUBA

CUBANE

CUBES (MATHEMATICS)

CUBIC EQUATIONS

CUBIC LATTICES

Cubic Lattices, Body Centered
    USE   BODY CENTERED CUBIC LATTICES

Cubic Lattices, Face Centered
    USE   FACE CENTERED CUBIC LATTICES

CUES

Cuestas
    USE   RIDGES

CUFFS

CULTIVATION

CULTURAL RESOURCES

CULTURE (SOCIAL SCIENCES)

CULTURE TECHNIQUES

CUMULATIVE DAMAGE

CUMULONIMBUS CLOUDS

CUMULUS CLOUDS

CUPOLAS

CURARE

CURES

CURIE TEMPERATURE

CURIE-WEISS LAW

CURING

CURIUM

CURIUM COMPOUNDS

CURIUM ISOTOPES

CURIUM 242

CURIUM 244

CURL

CURL (MATERIALS)

CURL (VECTORS)

(Current), AC
USE ALTERNATING CURRENT

CURRENT ALGEBRA

Current, Alternating
USE ALTERNATING CURRENT

CURRENT AMPLIFIERS

CURRENT CONVERTERS (AC TO DC)

(Current), DC
USE DIRECT CURRENT

CURRENT DENSITY

Current, Direct
USE DIRECT CURRENT

CURRENT DISTRIBUTION

Current, Electric
USE ELECTRIC CURRENT

Current Generators, Alternating
USE AC GENERATORS

Current, High
USE HIGH CURRENT

Current, Line
USE LINE CURRENT

Current, Lomonosov
USE LOMONOSOV CURRENT

CURRENT REGULATORS

CURRENT SHEETS

Current Stabilizers
USE CURRENT REGULATORS

CURRENTS

Currents, Air
USE AIR CURRENTS

Currents, Beam
USE BEAM CURRENTS

Currents, Coastal
USE COASTAL CURRENTS

Currents, Convection
USE CONVECTION CURRENTS

Currents, Earth
USE TELLURIC CURRENTS

Currents, Eddy
USE EDDY CURRENTS

Currents, External Surface
USE EXTERNAL SURFACE CURRENTS

Currents, Hall
USE HALL EFFECT
   ELECTRIC CURRENT

Currents, Ion
USE ION CURRENTS

Currents, Ionospheric
USE IONOSPHERIC CURRENTS

Currents, Littoral
USE COASTAL CURRENTS

Currents, Longshore
USE COASTAL CURRENTS

Currents, Low
USE LOW CURRENTS

Currents, Neutral
USE NEUTRAL CURRENTS

Currents, Ocean
USE OCEAN CURRENTS

Currents (Oceanography)
USE WATER CURRENTS

Currents, Plasma
USE PLASMA CURRENTS

Currents, Ring
USE RING CURRENTS

Currents, Short Circuit
USE SHORT CIRCUIT CURRENTS

Currents, Telluric
USE TELLURIC CURRENTS

Currents, Thermal
USE CONVECTIVE FLOW

Currents, Threshold
USE THRESHOLD CURRENTS

Currents, Vector
USE VECTOR CURRENTS

Currents, Vertical Air
USE VERTICAL AIR CURRENTS

Currents, Water
USE WATER CURRENTS

CURTAINS

Curtiss C-46 Aircraft
USE C-46 AIRCRAFT

CURTISS-WRIGHT AIRCRAFT

Curtiss-Wright Military Aircraft
USE CURTISS-WRIGHT AIRCRAFT
   MILITARY AIRCRAFT

CURVATURE

Curve, Bragg
USE BRAGG CURVE

CURVE FITTING

Curve, Light
USE LIGHT CURVE

CURVED BEAMS

CURVED PANELS

Curved Surfaces
USE CONTOURS
   SHAPES
   SURFACES

CURVES

CURVES (GEOMETRY)

Curves, Gompertz
USE GOMPERTZ CURVES

Curves, Hill
USE HILL METHOD

Curves, Learning
USE LEARNING CURVES

Curves, S
USE S CURVES

Curves, Zero Force
USE ZERO FORCE CURVES

Curvilinear Coordinates
USE SPHERICAL COORDINATES

Cushion Landing Systems, Air
USE AIR CUSHION LANDING SYSTEMS

Cushion Vehicles, Air
USE GROUND EFFECT MACHINES

CUSHIONCRAFT GROUND EFFECT MACHINE

CUSHIONS

CUSPS

Cusps, Double
USE DOUBLE CUSPS

CUSPS (LANDFORMS)

CUSPS (MATHEMATICS)

Cusps, Polar
USE POLAR CUSPS

CUT-OFF

Cut-Outs
USE OPENINGS

Cutaneous Perception
USE TOUCH

CUTTERS

(Cutters), Blades
USE BLADES (CUTTERS)

CUTTING

(Cutting), Blanking
USE BLANKING (CUTTING)

Cutting, Laser
USE LASER CUTTING

Cutting, Metal
USE METAL CUTTING

Cutting, Plasma Arc
USE PLASMA ARC CUTTING

CV-2 Aircraft
USE DHC 4 AIRCRAFT

CV-7 Aircraft
USE DHC 5 AIRCRAFT

CV-340 AIRCRAFT

CV-440 AIRCRAFT

CV-880 AIRCRAFT

CV-990 AIRCRAFT

CW Radar
USE CONTINUOUS WAVE RADAR

CYANAMIDES

CYANATES

Cyanates, Diiso
USE DIISOCYANATES

Cyanates, Iso
USE ISOCYANATES

Cyanide Emission
USE CN EMISSION

Cyanide, Vinyl
USE ACRYLONITRILES

CYANIDES

Cyanides, Hydrogen
USE HYDROCYANIC ACID

Cyanides, Iron
USE IRON CYANIDES

CYANO COMPOUNDS

CYANOCOBALAMIN

CYANOGEN

Cyanophyta
USE BLUE GREEN ALGAE

CYANOSIS

CYANURATES

CYANURIC ACID *

Cyber 74 Computer
USE CDC CYBER 74 COMPUTER

Cyber 74 Computer, CDC
USE CDC CYBER 74 COMPUTER

Cyber 170 Series Computers, CDC
USE CDC CYBER 170 SERIES COMPUTERS

Cyber 174 Computer, CDC
USE CDC CYBER 174 COMPUTER

Cyber 175 Computer, CDC
USE CDC CYBER 175 COMPUTER

Cyber 203 Computer, CDC
USE CDC CYBER 203 COMPUTER

Cyber 205 Computer, CDC
USE CDC CYBER 205 COMPUTER

CYBERNETICS

Cycle, Brayton
USE BRAYTON CYCLE

Cycle, Carbon
USE CARBON CYCLE

Cycle, Carnot
USE CARNOT CYCLE

Cycle Costs, Life
USE LIFE CYCLE COSTS

Cycle Engines, Liquid Air
USE LIQUID AIR CYCLE ENGINES

Cycle Engines, Topping
USE TOPPING CYCLE ENGINES

Cycle Engines, Variable
USE VARIABLE CYCLE ENGINES

Cycle, Krebs
USE KREBS CYCLE

Cycle, Otto
USE OTTO CYCLE

Cycle Power Generation, Combined
USE COMBINED CYCLE POWER GENERATION

Cycle Propulsion System, Hot
USE TIP DRIVEN ROTORS

Cycle, Rankine
USE RANKINE CYCLE

Cycle, Stirling
USE STIRLING CYCLE

Cycle, Sunspot
USE SUNSPOT CYCLE

Cycle, Work-Rest
USE WORK-REST CYCLE

CYCLES

Cycles (Biology), Activity
USE ACTIVITY CYCLES (BIOLOGY)

Cycles, Closed
USE CLOSED CYCLES

Cycles, Regenerative
USE REGENERATION (ENGINEERING)

Cycles, Solar
USE SOLAR CYCLES

Cycles, Stress
USE STRESS CYCLES

Cycles, Thermodynamic
USE THERMODYNAMIC CYCLES

CYCLIC ACCELERATORS

Cyclic Adenosine Monophosphate
USE CYCLIC AMP

CYCLIC AMP

CYCLIC COMPOUNDS

CYCLIC HYDROCARBONS

CYCLIC LOADS

Cycling
USE CYCLES

Cycling Tests, Thermal
USE THERMAL CYCLING TESTS

CYCLOBUTANE

CYCLOGENESIS

CYCLOHEXANE

CYCLOIDS

Cycloids, Epi
USE EPICYCLOIDS

CYCLONES

Cyclones, Anti
USE ANTICYCLONES

Cyclones (Equipment)
USE CENTRIFUGES

CYCLOPROPANE

CYCLOPS PLASMA ACCELERATOR

Cyclotetramethylene Tetranitramine
USE HMX

Cyclotrimethylene Trinitramine
USE RDX

CYCLOTRON FREQUENCY

Cyclotron Heating, Electron
USE ELECTRON CYCLOTRON HEATING

Cyclotron, Oak Ridge Isochronous
USE OAK RIDGE ISOCHRONOUS CYCLOTRON

Cyclotron, ORIC
USE OAK RIDGE ISOCHRONOUS CYCLOTRON

CYCLOTRON RADIATION

Cyclotron Radiation, Ion
USE ION CYCLOTRON RADIATION

CYCLOTRON RESONANCE

CYCLOTRON RESONANCE DEVICES

CYCLOTRONS

Cyclotrons, Geo
USE GEOCYCLOTRONS

Cyclotrons, Synchro
USE SYNCHROCYCLOTRONS

CYGNUS CONSTELLATION

Cylinder Bodies, Hemisphere
USE HEMISPHERE CYLINDER BODIES

CYLINDERS

Cylinders, Circular
USE CIRCULAR CYLINDERS

Cylinders, Concentric
USE CONCENTRIC CYLINDERS

Cylinders, Elastic
USE ELASTIC CYLINDERS

Cylinders, Elliptical
USE ELLIPTICAL CYLINDERS

Cylinders, Orthotropic
USE ORTHOTROPIC CYLINDERS

Cylinders, Oscillating
USE OSCILLATING CYLINDERS

Cylinders, Plasma
USE PLASMA CYLINDERS

Cylinders, Rotating
USE ROTATING CYLINDERS

Cylinders, Viscoelastic
USE VISCOELASTIC CYLINDERS

Cylindrical Afterbodies
USE CYLINDRICAL BODIES
AFTERBODIES

CYLINDRICAL ANTENNAS

CYLINDRICAL BODIES

CYLINDRICAL CHAMBERS

Cylindrical Coordinates
USE CARTESIAN COORDINATES

CYLINDRICAL PLASMAS

CYLINDRICAL SHELLS

CYLINDRICAL TANKS

CYLINDRICAL WAVES

Cylindroids
USE CYLINDRICAL BODIES

CYPRUS

CYRILLID METEOROIDS

CYSTEAMINE

CYSTEINE

CYSTIC FIBROSIS

CYSTS

CYTIDYLIC ACID

CYTOCHROMES

CYTOGENESIS

CYTOLOGY

CYTOPLASM

CZECHOSLOVAKIA

CZECHOSLOVAKIAN SPACECRAFT

CZOCHRALSKI METHOD

D

D, AIMP-
USE EXPLORER 33 SATELLITE

D, Atmosphere Explorer
USE EXPLORER 54 SATELLITE

D, Earth Resources Technology Satellite
USE LANDSAT 4

D, Energetic Particle Explorer
USE   EXPLORER 26 SATELLITE

D, EPE-
USE   EXPLORER 26 SATELLITE

D, ERTS-
USE   LANDSAT 4

D ICBM, Atlas
USE   ATLAS D ICBM

D, IMP-
USE   EXPLORER 33 SATELLITE

D Launch Vehicle, Saturn
USE   SATURN D LAUNCH VEHICLE

D Layer
USE   D REGION

D LINES

D, LORAN
USE   LORAN D

D, Lunar Orbiter
USE   LUNAR ORBITER 4

D, OGO-
USE   OGO-4

D, OSO-
USE   OSO-4

D REGION

D Rocket Vehicle, Agena
USE   AGENA D ROCKET VEHICLE

D, SAS-
USE   IUE

D Satellite, AE-
USE   EXPLORER 54 SATELLITE

D Satellite, GEOS-
USE   GEOS-D SATELLITE

D Satellite, TIROS
USE   TIROS 4 SATELLITE

D, Space Shuttle Mission 31-
USE   SPACE SHUTTLE MISSION 31-D

D, Space Shuttle Mission 41-
USE   SPACE SHUTTLE MISSION 41-D

D, Space Shuttle Mission 51-
USE   SPACE SHUTTLE MISSION 51-D

D, Space Shuttle Upper Stage
USE   SPACE SHUTTLE UPPER STAGE D

D, SSUS-
USE   SPACE SHUTTLE UPPER STAGE D

D, Vitamin
USE   CALCIFEROL

D-1 SATELLITE

D-2 SATELLITES

D-2B Satellite
USE   D-2 SATELLITES

D-558 AIRCRAFT

D-558 Aircraft, Douglas
USE   D-558 AIRCRAFT

DACRON (TRADEMARK)

DAD Explorer
USE   DUAL AIR DENSITY EXPLORER

DAEMO (Data Analysis)
USE   DATA PROCESSING
       DATA REDUCTION
       DATA TRANSMISSION

Dagger Aircraft, Delta
USE   F-102 AIRCRAFT

Dahomey
USE   BENIN

Dakota Aircraft
USE   C-47 AIRCRAFT

Dakota, North
USE   NORTH DAKOTA

Dakota, South
USE   SOUTH DAKOTA

DALTON LAW

DAMA
USE   DEMAND ASSIGNMENT MULTIPLE ACCESS

DAMAGE

DAMAGE ASSESSMENT

Damage, Brain
USE   BRAIN DAMAGE

Damage, Cumulative
USE   CUMULATIVE DAMAGE

Damage, Earthquake
USE   EARTHQUAKE DAMAGE

Damage, Fire
USE   FIRE DAMAGE

Damage, Flood
USE   FLOOD DAMAGE

Damage, Frost
USE   FROST DAMAGE

Damage, Impact
USE   IMPACT DAMAGE

Damage, Insect
USE   INFESTATION

Damage, Laser
USE   LASER DAMAGE

Damage, Meteoritic
USE   METEORITIC DAMAGE

Damage, Proton
USE   PROTON DAMAGE

Damage, Radiation
USE   RADIATION DAMAGE

Damage, Rain Impact
USE   RAIN IMPACT DAMAGE

Damage, Storm
USE   STORM DAMAGE

Damage Threshold
USE   YIELD POINT

DAMKOHLER NUMBER

DAMP Program
USE   DOWNRANGE ANTIMISSILE MEASUREMENT
       PROGRAM

DAMPERS

Dampers, Gyro
USE   GYRODAMPERS

Dampers, Nutation
USE   NUTATION DAMPERS

Dampers, Oscillation
USE   OSCILLATION DAMPERS

DAMPERS (VALVES)

Dampers, Vibration
USE   VIBRATION ISOLATORS

DAMPING

Damping, Elastic
USE   ELASTIC DAMPING

Damping Factor
USE   DAMPING

Damping In Pitch
USE   DAMPING
       PITCH (INCLINATION)

Damping In Roll
USE   ROLL
       DAMPING

Damping In Yaw
USE   DAMPING
       YAW

Damping, Jet
USE   DAMPING
       SPIN REDUCTION

Damping, Landau
USE   LANDAU DAMPING

DAMPING TESTS

Damping, Vibration
USE   VIBRATION DAMPING

Damping, Viscoelastic
USE   VISCOELASTIC DAMPING

Damping, Viscous
USE   VISCOUS DAMPING

Dampness
USE   MOISTURE CONTENT

DAMS

Dane (Radar), Cobra
USE   COBRA DANE (RADAR)

Danger
USE   HAZARDS

DARK ADAPTATION

Dark Space, Faraday
USE   FARADAY DARK SPACE

DARKENING

Darkening, Limb
USE   LIMB DARKENING

DARKNESS

DARKROOMS

Dart Aircraft, Delta
USE   F-106 AIRCRAFT

Dart Rocket, Judi-
USE   JUDI-DART ROCKET

Dart Turboprop Engines
USE   TURBOPROP ENGINES

Dash Helicopter
USE   QH-50 HELICOPTER

DASSAULT AIRCRAFT

Dassault Mirage 3 Aircraft
USE   MIRAGE 3 AIRCRAFT

Dassault Mystere 20 Aircraft
USE   MYSTERE 20 AIRCRAFT

Dassault Mystere 50 Aircraft
USE   MYSTERE 50 AIRCRAFT

DAST PROGRAM

DATA

Data Acq Network, Satellite Tracking And
USE   STDN (NETWORK)

DATA ACQUISITION

Data Acquisitions Systems, Ocean
USE   OCEAN DATA ACQUISITIONS SYSTEMS

Data Adaptive Evaluator/monitor
USE   DATA PROCESSING
       DATA TRANSMISSION
       DATA REDUCTION

Data, Analog
USE   ANALOG DATA

Data Analysis
USE   DATA PROCESSING
       DATA REDUCTION

(Data Analysis), DAEMO
USE   DATA PROCESSING
       DATA REDUCTION
       DATA TRANSMISSION

Data, Audio
USE   AUDIO DATA

DATA BASE MANAGEMENT SYSTEMS

DATA BASES

Data Bases, Numerical
USE   NUMERICAL DATA BASES

Data, Binary
USE   BINARY DATA

Data, Biomedical
USE   BIOMEDICAL DATA

Data Busses
USE   CHANNELS (DATA TRANSMISSION)

Data Centers, World
USE   WORLD DATA CENTERS

DATA COLLECTION PLATFORMS

Data Compaction
USE   DATA COMPRESSION

DATA COMPRESSION

Data (Computers), Control
USE   CONTROL DATA (COMPUTERS)

DATA CONVERSION ROUTINES

DATA CONVERTERS

DATA CORRELATION

Data, Digital
USE   DIGITAL DATA

(Data Exchange), IDEP
USE   INTERSERVICE DATA EXCHANGE
       PROGRAM

Data Exchange Program, Interservice
USE   INTERSERVICE DATA EXCHANGE
       PROGRAM

DATA FLOW ANALYSIS

Data Handling Systems
USE   DATA SYSTEMS

DATA INTEGRATION

DATA LINKS

DATA MANAGEMENT

Data (Mathematics), Censored
USE   CENSORED DATA (MATHEMATICS)

Data Network, Space Flight Tracking And
USE   SPACE FLIGHT TRACKING AND DATA
       NETWORK

Data Network, Spacecraft Tracking And
USE   STDN (NETWORK)

Data Platforms, Ocean
USE   OCEAN DATA ACQUISITIONS SYSTEMS

DATA PROCESSING

Data Processing, Automatic
USE   DATA PROCESSING

DATA PROCESSING EQUIPMENT

(Data Processing), Frames
USE   FRAMES (DATA PROCESSING)

Data Processing, Onboard
USE   ONBOARD DATA PROCESSING

Data Processing, Optical
USE   OPTICAL DATA PROCESSING

(Data Processing), Printers
USE   PRINTERS (DATA PROCESSING)

DATA PROCESSING TERMINALS

Data Processing, Voice
USE   VOICE DATA PROCESSING

Data Processors
USE   DATA PROCESSING EQUIPMENT

Data Processors, Site
USE   SITE DATA PROCESSORS

Data, Radar
USE   RADAR DATA

Data Readout Systems
USE   DATA SYSTEMS
       DISPLAY DEVICES

DATA RECORDERS

Data Recorders, Weather
USE   WEATHER DATA RECORDERS

DATA RECORDING

DATA REDUCTION

(Data Reduction), TARE
USE   DATA REDUCTION

Data Relay Satellites, Tracking And
USE   TDR SATELLITES

DATA RETRIEVAL

Data, Sampled
USE   DATA SAMPLING

DATA SAMPLING

DATA SIMULATION

DATA SMOOTHING

Data Stations, Ocean
USE   OCEAN DATA ACQUISITIONS SYSTEMS

DATA STORAGE

Data Storage Materials, Optical
USE   OPTICAL DATA STORAGE MATERIALS

(Data Storage), Optical Memory
USE   OPTICAL MEMORY (DATA STORAGE)

DATA STRUCTURES

Data System, NASA End-To-End
USE   NEEDS (DATA SYSTEM)

(Data System), Needs
USE   NEEDS (DATA SYSTEM)

DATA SYSTEMS

Data Systems, End-To-End
USE   END-TO-END DATA SYSTEMS

Data Systems, Sampled
USE   DATA SAMPLING

(Data), Tables
USE   TABLES (DATA)

DATA TRANSMISSION

(Data Transmission), Channels
USE   CHANNELS (DATA TRANSMISSION)

Data, Video
USE   VIDEO DATA

Dates, Launch
USE   LAUNCH DATES

Dating
USE   CHRONOLOGY
       TIME MEASUREMENT

Dating, Radioactive
USE   RADIOACTIVE AGE DETERMINATION

Dating, Tree Ring
USE   DENDROCHRONOLOGY

DATUM (ELEVATION)

DAWN CHORUS

(Dawn Phenomenon), Chorus
USE   DAWN CHORUS

DAWSONITE

Day Variation, Twenty-Seven
USE   TWENTY-SEVEN DAY VARIATION

DAYGLOW

DAYTIME

DC
USE   DIRECT CURRENT

DC (Current)
USE   DIRECT CURRENT

DC), Current Converters (AC To
USE   CURRENT CONVERTERS (AC TO DC)

(DC To AC), Inverted Converters
USE   INVERTED CONVERTERS (DC TO AC)

(DC To DC), Voltage Converters
USE   VOLTAGE CONVERTERS (DC TO DC)

DC), Voltage Converters (DC To
USE   VOLTAGE CONVERTERS (DC TO DC)

DC 3 AIRCRAFT

DC 7 AIRCRAFT

DC 8 AIRCRAFT

DC 9 AIRCRAFT

DC 10 AIRCRAFT

DC-3 Aircraft, Douglas
USE   DC 3 AIRCRAFT

DC-7 Aircraft, Douglas
USE   DC 7 AIRCRAFT

DC-8 Aircraft, Douglas
USE   DC 8 AIRCRAFT

DC-9 Aircraft, Douglas
USE   DC 9 AIRCRAFT

(DCS), Defense Communications System
USE   DEFENSE COMMUNICATIONS SYSTEM
       (DCS)

DDP COMPUTERS

DDP 116 Computer, Honeywell
USE   HONEYWELL DDP 116 COMPUTER

DDP 516 COMPUTER

DDT

DE
USE   DELAWARE

DE BROGLIE WAVELENGTHS

De Graaff Accelerators, Van
USE   VAN DE GRAAFF ACCELERATORS

DE HAVILLAND AIRCRAFT

De Havilland DH 106 Aircraft
USE   COMET 4 AIRCRAFT

De Havilland DH 112 Aircraft
USE   DH 112 AIRCRAFT

De Havilland DH 115 Aircraft
USE   DH 115 AIRCRAFT

De Havilland DH 121 Aircraft
USE   DH 121 AIRCRAFT

De Havilland DH 125 Aircraft
USE   DH 125 AIRCRAFT

De Havilland DHC 4 Aircraft
USE   DHC 4 AIRCRAFT

De Havilland DHC 5 Aircraft
USE   DHC 5 AIRCRAFT

De Havilland Venom Aircraft
USE   DH 112 AIRCRAFT

De Laval Nozzles
USE   CONVERGENT-DIVERGENT NOZZLES

(De-MD-VA), Delmarva Peninsula
USE   DELMARVA PENINSULA (DE-MD-VA)

Deacclimatization
USE   ACCLIMATIZATION

DEACTIVATION

DEAD RECKONING

Deadweight
USE   STATIC LOADS

Deafness
USE   AUDITORY DEFECTS

DEATH

DEATH VALLEY (CA)

Debonair Aircraft
USE   C-33 AIRCRAFT

DEBRIS

Debris, Radioactive
USE   RADIOACTIVE DEBRIS

Debris, Space
USE   SPACE DEBRIS

Debugging
USE   CHECKOUT

DEBYE LENGTH

Debye Temperature
USE   SPECIFIC HEAT

DEBYE-HUCKEL THEORY

DEBYE-SCHERRER METHOD

Decade, International Hydrological
USE   INTERNATIONAL HYDROLOGICAL DECADE

DECAMETRIC WAVES

DECARBONATION

DECARBOXYLATION

DECARBURIZATION

DECAY

Decay, Alpha
USE   ALPHA DECAY

Decay, Neutron
USE   NEUTRON DECAY

Decay, Orbit
USE   ORBIT DECAY

Decay, Particle
USE   RADIOACTIVE DECAY

Decay, Plasma
USE   PLASMA DECAY

Decay, Radioactive
USE   RADIOACTIVE DECAY

Decay Rate, Electron
USE   ELECTRON DECAY RATE

DECAY RATES

DECCA NAVIGATION

DECELERATION

Deceleration, Impact
USE   DECELERATION
       IMPACT ACCELERATION

Decelerators
USE   BRAKES (FOR ARRESTING MOTION)

DECEPTION

DECIDUOUS TREES

Decimal Converters, Binary To
USE   BINARY TO DECIMAL CONVERTERS

DECIMAL TO BINARY CONVERTERS

DECIMALS

DECIMETER WAVES

Decision Elements
USE   LOGICAL ELEMENTS

DECISION MAKING

DECISION THEORY

Decision Theory, Statistical
USE   STATISTICAL DECISION THEORY

DECISIONS

Decks (Floors)
USE   FLOORS

DECLINATION

DECODERS

DECODING

DECOMMISSIONING

DECOMMUTATORS

DECOMPOSITION

Decomposition, Photo
USE   PHOTODECOMPOSITION

Decomposition, Propellant
USE   PROPELLANT DECOMPOSITION

Decomposition, Thermal
USE   THERMAL DECOMPOSITION

Decompression
USE   PRESSURE REDUCTION

Decompression, Explosive
USE   EXPLOSIVE DECOMPRESSION

DECOMPRESSION SICKNESS

DECONDITIONING

DECONGESTANTS

DECONTAMINATION

DECOUPLING

Decoupling, Spin
USE   SPIN DECOUPLING

DECOYS

Decoys, Ballistic Missile
USE   BALLISTIC MISSILE DECOYS

Decoys, Reentry
USE   REENTRY DECOYS

Decreases, Forbush
USE   FORBUSH DECREASES

Decrementing
USE   REDUCTION

DEDUCTION

Deduction, Electromagnetic
USE   MAGNETIC INDUCTION

DEEP DRAWING

DEEP SCATTERING LAYERS

DEEP SPACE

DEEP SPACE INSTRUMENTATION FACILITY

DEEP SPACE NETWORK

DEEP WELL INJECTION (WASTES)

DEEPWATER TERMINALS

DEER

DEFECTS

Defects, Auditory
USE   AUDITORY DEFECTS

Defects, Crystal
USE   CRYSTAL DEFECTS

Defects, Frenkel
USE   FRENKEL DEFECTS

Defects, Point
USE   POINT DEFECTS

Defects, Speech
USE   SPEECH DEFECTS

Defects, Surface
USE   SURFACE DEFECTS

Defects), Vacancies (Crystal
USE   VACANCIES (CRYSTAL DEFECTS)

DEFENDER PROJECT

DEFENSE

Defense, Air
USE   AIR DEFENSE

Defense, Antimissile
USE   ANTIMISSILE DEFENSE

Defense, Chemical
USE   CHEMICAL DEFENSE

Defense, Civil
USE   CIVIL DEFENSE

DEFENSE COMMUNICATIONS SATELLITE SYSTEM

DEFENSE COMMUNICATIONS SYSTEM (DCS)

DEFENSE INDUSTRY

Defense Meteorological Satellite Program
USE   DMSP SATELLITES

Defense, Missile
USE   MISSILE DEFENSE

DEFENSE PROGRAM

Defense, Satellite
USE   SPACECRAFT DEFENSE

Defense, Spacecraft
USE   SPACECRAFT DEFENSE

Defense System, Sage Air
USE   SAGE AIR DEFENSE SYSTEM

Defenses, Physiological
USE   PHYSIOLOGICAL DEFENSES

Deficiencies), Holes (Electron
USE   HOLES (ELECTRON DEFICIENCIES)

Deficiency, Oxygen
USE   HYPOXIA

DEFINITION

DEFLAGRATION

Deflating
USE   PRESSURE REDUCTION
           INFLATABLE STRUCTURES

DEFLECTION

Deflection, Flow
USE   FLOW DEFLECTION

DEFLECTORS

Deflectors, Blast
USE   BLAST DEFLECTORS

Deflectors, Flame
USE   FLAME DEFLECTORS

DEFLUORINATION

DEFOCUSING

Defocusing, Laser Beam
USE   THERMAL BLOOMING

Defocusing, Thermal
USE   THERMAL BLOOMING

DEFOLIANTS

DEFOLIATION

DEFORESTATION

DEFORMATION

Deformation, Axisymmetric
USE   AXIAL STRAIN

Deformation, Elastic
USE   ELASTIC DEFORMATION

Deformation, Nuclear
USE   NUCLEAR DEFORMATION

Deformation, Plastic
USE   PLASTIC DEFORMATION

Deformation, Static
USE   STATIC DEFORMATION

Deformation, Tensile
USE   TENSILE DEFORMATION

Deformation, Wave Front
USE   WAVE FRONT DEFORMATION

DEFORMETERS

DEFROSTING

DEGASSING

DEGENERATION

Degenerative Feedback
USE   NEGATIVE FEEDBACK

DEGRADATION

Degradation, Thermal
USE   THERMAL DEGRADATION

Degradation, Wave
USE   WAVE DEGRADATION

DEGREES OF FREEDOM

DEHP
USE   DIETHYL HYDROGEN PHOSPHITE (DEHP)

(DEHP), Diethyl Hydrogen Phosphite
USE   DIETHYL HYDROGEN PHOSPHITE (DEHP)

DEHUMIDIFICATION

DEHYDRATED FOOD

DEHYDRATION

DEHYDROGENATION

DEICERS

DEICING

Deicing Systems
USE   DEICERS

DEIMOS

DEIONIZATION

Dekatrons
USE   COUNTERS

DELAMINATING

DELAWARE

DELAWARE BAY (US)

DELAWARE RIVER BASIN (US)

DELAY

DELAY CIRCUITS

(Delay), Lag
USE   TIME LAG

DELAY LINES

Delay Lines, Acoustic
USE   ACOUSTIC DELAY LINES

DELAY LINES (COMPUTER STORAGE)

Delay, Time
USE   TIME LAG

DELAYED FLAP APPROACH

DELETION

Delfin Aircraft
USE   L-29 JET TRAINER

DELFT CAMERA

DELINEATION

DELIVERY

Delivery), Mass Drivers (Payload
USE   MASS DRIVERS (PAYLOAD DELIVERY)

Delivery (STS), Payload
USE   PAYLOAD DELIVERY (STS)

Delivery, Weapons
USE   WEAPONS DELIVERY

DELMARVA PENINSULA (DE-MD-VA)

DELPHI METHOD (FORECASTING)

DELRIN (TRADEMARK)

DELTA ANTENNAS

Delta Dagger Aircraft
USE   F-102 AIRCRAFT

Delta Dart Aircraft
USE   F-106 AIRCRAFT

Delta (France), Rhone
USE   RHONE DELTA (FRANCE)

DELTA FUNCTION

Delta (LA), Mississippi
USE   MISSISSIPPI DELTA (LA)

DELTA LAUNCH VEHICLE

Delta Launch Vehicle, Thor
USE   THOR DELTA LAUNCH VEHICLE

DELTA MODULATION

DELTA WINGS

Delta 2 Aircraft, Fairey
USE   FD 2 AIRCRAFT

DELTAS

DEMAGNETIZATION

Demagnetization Cooling, Adiabatic
USE   ADIABATIC DEMAGNETIZATION COOLING

DEMAND ASSIGNMENT MULTIPLE ACCESS

Demand, Biochemical Oxygen
USE   BIOCHEMICAL OXYGEN DEMAND

DEMAND (ECONOMICS)

Demineralization, Bone
USE   BONE DEMINERALIZATION

DEMINERALIZING

Democratic Peoples Republic Of Korea
USE   NORTH KOREA

Democratic Republic, German
USE   EAST GERMANY

Democratic Republic Of Germany, Peoples
USE   EAST GERMANY

DEMODULATION

DEMODULATORS

Demodulators, Frequency Compression
USE   FREQUENCY COMPRESSION
           DEMODULATORS

Demodulators, Modulators-
USE   MODEMS

Demodulators, Phase
USE   PHASE DEMODULATORS

Demodulators, Phase Lock
USE   PHASE LOCK DEMODULATORS

DEMOGRAPHY

Demonstration
USE   PROVING

DEMULTIPLEXING

DENDRITIC CRYSTALS

Dendritic Drainage
USE   DRAINAGE PATTERNS

DENDROCHRONOLOGY

DENITROGENATION

DENMARK

DENSE PLASMAS

DENSIFICATION

DENSIMETERS

Densimeters, Ultrasonic
USE   ULTRASONIC DENSIMETERS

DENSITOMETERS

Densitometers, Micro
USE   MICRODENSITOMETERS

DENSITY

Density, Atmospheric
USE   ATMOSPHERIC DENSITY

Density (Concentration), Electron
USE   ELECTRON DENSITY (CONCENTRATION)

Density (Concentration), Ion
USE   ION DENSITY (CONCENTRATION)

Density (Concentration), Particle
USE   PARTICLE DENSITY (CONCENTRATION)

Density (Concentration), Proton
USE   PROTON DENSITY (CONCENTRATION)

Density, Current
USE   CURRENT DENSITY

DENSITY DISTRIBUTION

Density (Electromagnetic), Power
USE   RADIANT FLUX DENSITY

Density, Electron Flux
USE   ELECTRON FLUX DENSITY

Density, Energy
USE   FLUX DENSITY

Density Explorer A, Air
USE   EXPLORER 19 SATELLITE

Density Explorer, Dual Air
USE   DUAL AIR DENSITY EXPLORER

Density Flow, Low
USE   LOW DENSITY FLOW

Density, Flux
USE   FLUX DENSITY

Density Function, Maxwell-Boltzmann
USE   MAXWELL-BOLTZMANN DENSITY
          FUNCTION

Density Functions, Normal
USE   NORMAL DENSITY FUNCTIONS

Density Functions, Poisson
USE   POISSON DENSITY FUNCTIONS

Density Functions, Probability
USE   PROBABILITY DENSITY FUNCTIONS

Density Functions, Weibull
USE   WEIBULL DENSITY FUNCTIONS

Density, Gas
USE   GAS DENSITY

Density Gases, Low
USE   RAREFIED GASES

Density, Ionospheric Electron
USE   IONOSPHERIC ELECTRON DENSITY

Density, Ionospheric Ion
USE   IONOSPHERIC ION DENSITY

Density, Luminous Flux
USE   LUMINOUS INTENSITY

Density, Magnetic Charge
USE   MAGNETIC CHARGE DENSITY

Density, Magnetospheric Electron
USE   MAGNETOSPHERIC ELECTRON DENSITY

Density, Magnetospheric Ion
USE   MAGNETOSPHERIC ION DENSITY

Density, Magnetospheric Proton
USE   MAGNETOSPHERIC PROTON DENSITY

DENSITY (MASS/VOLUME)

Density Materials, Low
USE   LOW DENSITY MATERIALS

(Density), Maxwellian Distribution
USE   MAXWELL-BOLTZMANN DENSITY
          FUNCTION

DENSITY MEASUREMENT

Density Measurement, X Ray
USE   X RAY DENSITY MEASUREMENT

Density, Neutron Flux
USE   NEUTRON FLUX DENSITY

DENSITY (NUMBER/VOLUME)

Density, Optical
USE   OPTICAL DENSITY

Density, Packing
USE   PACKING DENSITY

Density, Particle Flux
USE   PARTICLE FLUX DENSITY

Density, Photon
USE   PHOTON DENSITY

Density, Plasma
USE   PLASMA DENSITY

Density Profiles, Electron
USE   ELECTRON DENSITY PROFILES

Density, Proton Flux
USE   PROTON FLUX DENSITY

Density, Radiant Flux
USE   RADIANT FLUX DENSITY

Density (Rate/area)
USE   FLUX DENSITY

Density Research, Low
USE   LOW DENSITY RESEARCH

Density, Solar Flux
USE   SOLAR FLUX DENSITY

Density (Solid State), Carrier
USE   CARRIER DENSITY (SOLID STATE)

Density, Space
USE   SPACE DENSITY

DENSITY WAVE MODEL

Density Wind Tunnels, Low
USE   LOW DENSITY WIND TUNNELS

Density/injun Explorer B, Air
USE   EXPLORER 25 SATELLITE

DENTAL CALCULI

DENTISTRY

DEOXIDIZING

DEOXIFICATION

DEOXYGENATION

DEOXYRIBONUCLEIC ACID

DEPENDENCE

Dependence, Pressure
USE   PRESSURE DEPENDENCE

Dependence, Temperature
USE   TEMPERATURE DEPENDENCE

Dependence, Time
USE   TIME DEPENDENCE

Dependencies, Spatial
USE   SPATIAL DEPENDENCIES

Dependency
USE   DEPENDENCE

DEPENDENT VARIABLES

DEPERSONALIZATION

DEPLETION

Deploying Space Stations, Self
USE   SPACE STATIONS
          SELF ERECTING DEVICES

DEPLOYMENT

Deployment & Retrieval System, Payload
USE   PAYLOAD DEPLOYMENT & RETRIEVAL
          SYSTEM

DEPOLARIZATION

Depolarization, Optical
USE   OPTICAL DEPOLARIZATION

Depolarizers
USE   DEPOLARIZATION

DEPOLYMERIZATION

DEPOSITION

Deposition, Electro
USE   ELECTRODEPOSITION

Deposition, Electroless
USE   ELECTROLESS DEPOSITION

Deposition, Vacuum
USE   VACUUM DEPOSITION

Deposition, Vapor
USE   VAPOR DEPOSITION

DEPOSITS

Deposits, Cryo
USE   CRYODEPOSITS

Deposits, Glaciofluvial
USE   GLACIAL DRIFT

Deposits, Gravel
USE   GRAVELS

Deposits, Mineral
USE   MINERAL DEPOSITS

DEPRECIATION

DEPRESSANTS

Depressants, Central Nervous System
USE   CENTRAL NERVOUS SYSTEM
          DEPRESSANTS

DEPRESSION

Depression, Neurotic
USE   NEUROTIC DEPRESSION

Depression, Psychotic
USE   PSYCHOTIC DEPRESSION

Depressions (Topography)
USE   STRUCTURAL BASINS

Depressurization
USE   PRESSURE REDUCTION

DEPRIVATION

Deprivation, Sensory
USE   SENSORY DEPRIVATION

Deprivation, Sleep
USE   SLEEP DEPRIVATION

Deprivation, Water
USE   WATER DEPRIVATION

DEPTH

DEPTH MEASUREMENT

Depth, Mixing
USE   MIXING HEIGHT

Depth Perception
USE   SPACE PERCEPTION

Depth, Water
USE   WATER DEPTH

Der Waal Forces, Van
USE   VAN DER WAAL FORCES

DERIVATION

Derivation Calculus
USE   DIFFERENTIAL CALCULUS

Derivatives, Stability
USE   STABILITY DERIVATIVES

Derived Gases, Coal
USE   COAL DERIVED GASES

Derived Liquids, Coal
USE   COAL DERIVED LIQUIDS

Derived Vehicles, Shuttle
USE   SHUTTLE DERIVED VEHICLES

DERMATITIS

Dermatitis, Contact
USE   CONTACT DERMATITIS

DERMATOLOGY

DESALINIZATION

DESATURATION

DESCALING

DESCENT

Descent Method, Steepest
USE   STEEPEST DESCENT METHOD

Descent, Parachute
USE   PARACHUTE DESCENT

DESCENT PROPULSION SYSTEMS

DESCENT TRAJECTORIES

DESCRIPTIONS

DESCRIPTIVE GEOMETRY

DESENSITIZING

DESERT ADAPTATION

Desert (Africa), Sahara
USE   SAHARA DESERT (AFRICA)

Desert (CA), Mojave
USE   MOJAVE DESERT (CA)

Desert, Gobi
USE   GOBI DESERT

Desert, Libyan
USE   LIBYAN DESERT

DESERTIFICATION

DESERTLINE

DESERTS

DESICCANTS

Desiccation
USE   DRYING

DESICCATORS

DESIGN

Design, Aircraft
USE   AIRCRAFT DESIGN

Design, Amplifier
USE   AMPLIFIER DESIGN

DESIGN ANALYSIS

Design, Antenna
USE   ANTENNA DESIGN

(Design), CAD
USE   COMPUTER AIDED DESIGN

Design, Computer
USE   COMPUTER DESIGN

Design, Computer Aided
USE   COMPUTER AIDED DESIGN

Design, Computer Systems
USE   COMPUTER SYSTEMS DESIGN

Design, Computerized
USE   COMPUTER AIDED DESIGN

Design, Control Systems
USE   CONTROL SYSTEMS DESIGN

Design Criteria, Structural
USE   STRUCTURAL DESIGN CRITERIA

Design, Engine
USE   ENGINE DESIGN

Design, Experiment
USE   EXPERIMENT DESIGN

Design, Factorial
USE   FACTORIAL DESIGN

Design, Helicopter
USE   HELICOPTER DESIGN

Design, Integ Program For Aerospace Veh
USE   IPAD

Design, Lens
USE   LENS DESIGN

Design, Logic
USE   LOGIC DESIGN

Design, Missile
USE   MISSILE DESIGN

Design, Nozzle
USE   NOZZLE DESIGN

Design Of Experiments
USE   EXPERIMENT DESIGN

Design, Plant
USE   PLANT DESIGN

Design, Pressure Vessel
USE   PRESSURE VESSEL DESIGN

Design, Reactor
USE   REACTOR DESIGN

Design, Rocket Engine
USE   ROCKET ENGINE DESIGN

Design, Satellite
USE   SATELLITE DESIGN

Design, Spacecraft
USE   SPACECRAFT DESIGN

Design Specifications, Functional
USE   FUNCTIONAL DESIGN SPECIFICATIONS

Design, Structural
USE   STRUCTURAL DESIGN

Design, Systems
USE   SYSTEMS ENGINEERING

DESIGN TO COST

Designators, Laser Target
USE   LASER TARGET DESIGNATORS

DESORPTION

Despinning
USE   SPIN REDUCTION

DESTABILIZATION

Destroyer Aircraft
USE   B-66 AIRCRAFT

DESTRUCTION

DESTRUCTIVE TESTS

DESULFURIZING

DESYNCHRONIZATION (BIOLOGY)

Desynchronized Sleep
USE   RAPID EYE MOVEMENT STATE

DETACHMENT

Detachment, Photo
USE   PHOTODETACHMENT

Detecting And Ranging, Sound
USE   SOUND DETECTING AND RANGING

DETECTION

Detection, Aircraft
USE   AIRCRAFT DETECTION

Detection And Tracking System, Space
USE   SPACE DETECTION AND TRACKING
        SYSTEM

Detection, Change
USE   CHANGE DETECTION

Detection Codes, Error
USE   ERROR DETECTION CODES

Detection, Correlation
USE   CORRELATION DETECTION

Detection Equipment, Airport Surface
USE   AIRPORT SURFACE DETECTION
        EQUIPMENT

Detection, Flaw
USE   NONDESTRUCTIVE TESTS

Detection, Forest Fire
USE   FOREST FIRE DETECTION

Detection, Haze
USE   HAZE DETECTION

Detection, High Altitude Nuclear
USE   HIGH ALTITUDE NUCLEAR DETECTION

Detection, Missile
USE   MISSILE DETECTION

Detection, Radar
USE   RADAR DETECTION

Detection, Signal
USE   SIGNAL DETECTION

Detection, Ultrasonic Flaw
USE   ULTRASONIC FLAW DETECTION

Detector Cells, Golay
USE   GOLAY DETECTOR CELLS

DETECTORS

Detectors (Dosimeters), Threshold
USE   THRESHOLD DETECTORS (DOSIMETERS)

Detectors, Electron
USE   ELECTRON COUNTERS

Detectors, Flir
USE   FLIR DETECTORS

Detectors, Forward Looking Infrared
USE   FLIR DETECTORS

Detectors, Gas
USE   GAS DETECTORS

Detectors, Infrared
USE   INFRARED DETECTORS

Detectors, Life
USE   LIFE DETECTORS

Detectors, Mine
USE   MINE DETECTORS

Detectors, Moisture
USE   MOISTURE METERS

Detectors, Neutron
USE   NEUTRON COUNTERS

Detectors, Oxygen
USE   OXYGEN ANALYZERS

Detectors, Particle
USE   RADIATION COUNTERS

Detectors, Phase
USE   PHASE DETECTORS

Detectors, Photoelectromagnetic
USE   RADIATION MEASURING INSTRUMENTS
        PHOTOELECTROMAGNETIC EFFECTS

Detectors, Radiation
USE   RADIATION DETECTORS

Detectors, Signal
USE   SIGNAL DETECTORS

Detectors, Silicon Radiation
USE   SILICON RADIATION DETECTORS

Detectors, Smoke
USE   SMOKE DETECTORS

Detectors, Sound
USE   SOUND TRANSDUCERS

(Detectors), Squid
USE   SQUID (DETECTORS)

Detectors, Synchronous
USE   CORRELATORS

DETERGENTS

DETERIORATION

Determinant, Hill
USE   HILL DETERMINANT

DETERMINANTS

Determination
USE   MEASUREMENT

Determination, Age
USE   CHRONOLOGY

Determination, Airborne Range And Orbit
USE   AIRBORNE RANGE AND ORBIT
        DETERMINATION

Determination), AROD (Range-Orbit
USE   AIRBORNE RANGE AND ORBIT
        DETERMINATION

Determination, Minimum Variance Orbit
USE   MINIMUM VARIANCE ORBIT
        DETERMINATION

Determination, MINIVAR Orbit
USE   MINIMUM VARIANCE ORBIT
        DETERMINATION

Determination, Radioactive Age
USE   RADIOACTIVE AGE DETERMINATION

Determination, Size
USE   SIZE DETERMINATION

Determination System, Goddard Trajectory
USE   GODDARD TRAJECTORY DETERMINATION
        SYSTEM

DETONABLE GAS MIXTURES

DETONATION

DETONATION WAVES

DETONATORS

DEUTERIDES

DEUTERIUM

DEUTERIUM COMPOUNDS

Deuterium Fluoride Lasers
USE   DF LASERS

DEUTERIUM FLUORIDES

Deuterium Oxide, Hydrogen
USE   HEAVY WATER

Deuterium Oxides
USE   HEAVY WATER

DEUTERIUM PLASMA

DEUTERON IRRADIATION

DEUTERONS

Developers, Photographic
USE   PHOTOGRAPHIC DEVELOPERS

Developers (Photography)
USE   PHOTOGRAPHIC DEVELOPERS

DEVELOPING NATIONS

DEVELOPMENT

Development, Economic
USE   ECONOMIC DEVELOPMENT

Development, Engineering
USE   PRODUCT DEVELOPMENT

(Development), Evolution
USE   EVOLUTION (DEVELOPMENT)

Development, Personnel
USE   PERSONNEL DEVELOPMENT

Development, Product
USE   PRODUCT DEVELOPMENT

Development, Research And
USE   RESEARCH AND DEVELOPMENT

Development, Urban
USE   URBAN DEVELOPMENT

Development, Weapons
USE   WEAPONS DEVELOPMENT

DEVIATION

Deviation, Phase
USE   PHASE DEVIATION

Deviation, Standard
USE   STANDARD DEVIATION

Device, Child
USE   CHILD DEVICE

DEVICES

Devices, Air Bag Restraint
USE   AIR BAG RESTRAINT DEVICES

Devices, Aircraft Launching
USE   AIRCRAFT LAUNCHING DEVICES

Devices, Alpha Plasma
USE   ALPHA PLASMA DEVICES

Devices, Antiskid
USE   ANTISKID DEVICES

Devices, Antistatic
USE   STATIC DISCHARGERS

Devices, B-A-W
USE   BULK ACOUSTIC WAVE DEVICES

Devices, Bubble Memory
USE   BUBBLE MEMORY DEVICES

Devices, Bucket Brigade
USE   BUCKET BRIGADE DEVICES

Devices, Bulk Acoustic Wave
USE   BULK ACOUSTIC WAVE DEVICES

Devices, Cartridge Actuated
USE   ACTUATORS
        EXPLOSIVE DEVICES

Devices, CATT
USE   CATT DEVICES

Devices, Charge Coupled
USE   CHARGE COUPLED DEVICES

Devices, Charge Flow
USE   CHARGE FLOW DEVICES

Devices, Charge Injection
USE   CHARGE INJECTION DEVICES

Devices, Charge Transfer
USE   CHARGE TRANSFER DEVICES

Devices), Chips (Memory
USE   CHIPS (MEMORY DEVICES)

Devices, Collision Warning
USE   COLLISION AVOIDANCE
        WARNING SYSTEMS

Devices, Computer Storage
USE   COMPUTER STORAGE DEVICES

Devices, Control
USE   CONTROL EQUIPMENT

Devices, Controlled Avalanche Transit Time
USE   CATT DEVICES

Devices, Cyclotron Resonance
USE   CYCLOTRON RESONANCE DEVICES

Devices, Disconnect
USE   DISCONNECT DEVICES

Devices, Display
USE   DISPLAY DEVICES

Devices, Drag
USE   DRAG DEVICES

Devices, Electroexplosive
USE   INITIATORS (EXPLOSIVES)

Devices, Electromechanical
USE   ELECTROMECHANICAL DEVICES

Devices, Energy Storage
USE   ENERGY STORAGE

Devices, Error Correcting
USE   ERROR CORRECTING DEVICES

Devices, Explosive
USE   EXPLOSIVE DEVICES

Devices, Fanlift
USE   LIFT FANS

Devices, Heat Rejection
USE   HEAT RADIATORS

Devices, Heterojunction
USE   HETEROJUNCTION DEVICES

Devices, Homing
USE   HOMING DEVICES

Devices, Inflatable
USE   INFLATABLE STRUCTURES

(Devices), Inlets
USE   INTAKE SYSTEMS

Devices, Launching
USE   LAUNCHERS

Devices, Lift
USE   LIFT DEVICES

Devices, Lunar Escape
USE   LUNAR ESCAPE DEVICES

Devices (Machinery), Positioning
USE   POSITIONING DEVICES (MACHINERY)

Devices, Mechanical
USE   MECHANICAL DEVICES

Devices, Microminiaturized Electronic
USE   MICROMINIATURIZED ELECTRONIC
          DEVICES

Devices, NDM Semiconductor
USE   NDM SEMICONDUCTOR DEVICES

Devices, Negative Resistance
USE   NEGATIVE RESISTANCE DEVICES

Devices, Nuclear
USE   NUCLEAR DEVICES

Devices, Photoelectrochemical
USE   PHOTOELECTROCHEMICAL DEVICES

Devices, Plasma Display
USE   PLASMA DISPLAY DEVICES

Devices, Praetersonic
USE   PRAETERSONIC DEVICES

Devices, Propellant Actuated
USE   PROPELLANT ACTUATED DEVICES

Devices, Prosthetic
USE   PROSTHETIC DEVICES

Devices, Q
USE   Q DEVICES

Devices, Read-Only Memory
USE   READ-ONLY MEMORY DEVICES

(Devices), Retarders
USE   RETARDERS (DEVICES)

Devices, S-A-W
USE   SURFACE ACOUSTIC WAVE DEVICES

Devices, Safety
USE   SAFETY DEVICES

Devices, Sampling
USE   SAMPLERS

Devices, Scanning
USE   SCANNERS

Devices, Self Erecting
USE   SELF ERECTING DEVICES

Devices, Self Repairing
USE   SELF REPAIRING DEVICES

Devices, Semiconductor
USE   SEMICONDUCTOR DEVICES

Devices, Solid State
USE   SOLID STATE DEVICES

Devices, Stimulated Emission
USE   STIMULATED EMISSION DEVICES

Devices, Surface Acoustic Wave
USE   SURFACE ACOUSTIC WAVE DEVICES

Devices, Timing
USE   TIMING DEVICES

Devices, Tokamak
USE   TOKAMAK DEVICES

Devices, Training
USE   TRAINING DEVICES

Devices, Transferred Electron
USE   TRANSFERRED ELECTRON DEVICES

Devices, TRAPATT
USE   TRAPATT DEVICES

Devices, Warning
USE   WARNING SYSTEMS

Devices, Yo-Yo
USE   YO-YO DEVICES

Devitrification
USE   CRYSTALLIZATION

Devries Equation, Korteweg-
USE   KORTEWEG-DEVRIES EQUATION

DEW

DEW POINT

Dewar Systems
USE   CRYOGENIC EQUIPMENT

DEWATERING

DEWAXING

Dewetting
USE   DRYING

DEXTRANS

DF
USE   DEUTERIUM FLUORIDES

DF LASERS

Dfa
USE   DELAYED FLAP APPROACH

DH 106 Aircraft
USE   COMET 4 AIRCRAFT

DH 106 Aircraft, De Havilland
USE   COMET 4 AIRCRAFT

DH 112 AIRCRAFT

DH 112 Aircraft, De Havilland
USE   DH 112 AIRCRAFT

DH 115 AIRCRAFT

DH 115 Aircraft, De Havilland
USE   DH 115 AIRCRAFT

DH 121 AIRCRAFT

DH 121 Aircraft, De Havilland
USE   DH 121 AIRCRAFT

DH 125 AIRCRAFT

DH 125 Aircraft, De Havilland
USE   DH 125 AIRCRAFT

DHC Beaver Aircraft
USE   DHC 2 AIRCRAFT

DHC 2 AIRCRAFT

DHC 4 AIRCRAFT

DHC 4 Aircraft, De Havilland
USE   DHC 4 AIRCRAFT

DHC 5 AIRCRAFT

DHC 5 Aircraft, De Havilland
USE   DHC 5 AIRCRAFT

DIABETES MELLITUS

DIADEME SATELLITES

DIAGNOSIS

Diagnostics, Plasma
USE   PLASMA DIAGNOSTICS

Diagram, C-M
USE   COLOR-MAGNITUDE DIAGRAM

Diagram, Color-Magnitude
USE   COLOR-MAGNITUDE DIAGRAM

Diagram, Hertzsprung-Russell
USE   HERTZSPRUNG-RUSSELL DIAGRAM

Diagram, HR
USE   HERTZSPRUNG-RUSSELL DIAGRAM

Diagram, Hubble
USE   HUBBLE DIAGRAM

Diagram, Mollier
USE   MOLLIER DIAGRAM

Diagram, Nyquist
USE   NYQUIST DIAGRAM

DIAGRAMS

Diagrams, Bending
USE   BENDING DIAGRAMS

Diagrams, Block
USE   BLOCK DIAGRAMS

Diagrams, Circuit
USE   CIRCUIT DIAGRAMS

Diagrams, Constitutional
USE   PHASE DIAGRAMS

Diagrams, Creep
USE   CREEP DIAGRAMS

Diagrams, Enthalpy-Entropy
USE   MOLLIER DIAGRAM

Diagrams, Equilibrium
USE   PHASE DIAGRAMS

Diagrams, Eutectic
USE   PHASE DIAGRAMS

Diagrams, Fatigue
USE   S-N DIAGRAMS

Diagrams, Feynman
USE   FEYNMAN DIAGRAMS

Diagrams, Phase
USE   PHASE DIAGRAMS

Diagrams, S-N
  USE   S-N DIAGRAMS

Diagrams, Stress-Strain
  USE   STRESS-STRAIN DIAGRAMS

Diagrams, Venn
  USE   VENN DIAGRAMS

DIAL SATELLITE

DIALLYL COMPOUNDS

DIALS

DIALYSIS

Dialysis, Electro
  USE   ELECTRODIALYSIS

DIAMAGNETISM

DIAMANT LAUNCH VEHICLE

Diameter, Solar
  USE   SOLAR DIAMETER

DIAMETERS

Diamine, Ethylene
  USE   ETHYLENEDIAMINE

Diamine, Methylene
  USE   METHYLENE DIAMINE

DIAMINES

Diamond Wings
  USE   LOW ASPECT RATIO WINGS
        SWEPT WINGS

DIAMONDS

Diamonds, Meteoritic
  USE   METEORITIC DIAMONDS

DIAPHRAGM (ANATOMY)

DIAPHRAGMS

DIAPHRAGMS (MECHANICS)

DIASTOLE

DIASTOLIC PRESSURE

DIATOMIC GASES

DIATOMIC MOLECULES

DIBASIC COMPOUNDS

DIBORANE

DIBROMIDES

DIBUTYL COMPOUNDS

DICARBOXYLIC ACIDS

DICHLORIDES

Dichlorodiphenyltrichloroethane
  USE   DDT

DICHOTOMIES

DICHROISM

Dichromates
  USE   CHROMATES

DICKE RADIOMETERS

Dicke Type Radiometers
  USE   DICKE RADIOMETERS

DICTIONARIES

DIDYMIUM

DIELDRIN

Dielectric Constant
  USE   PERMITTIVITY

Dielectric Materials
  USE   DIELECTRICS

DIELECTRIC PERMEABILITY

DIELECTRIC POLARIZATION

DIELECTRIC PROPERTIES

DIELECTRICS

Dielectronic Satellite Lines
  USE   RESONANCE LINES

DIELS-ALDER REACTIONS

DIENCEPHALON

DIENES

DIES

DIESEL ENGINES

DIESEL FUELS

DIETHYL ETHER

DIETHYL HYDROGEN PHOSPHITE (DEHP)

DIETS

Diff Mobility Semiconductors, Negative
  USE   NDM SEMICONDUCTOR DEVICES

DIFFERENCE EQUATIONS

Difference Theory, Finite
  USE   FINITE DIFFERENCE THEORY

DIFFERENCES

Differences, Temperature
  USE   TEMPERATURE GRADIENTS

Differencing, Backward
  USE   BACKWARD DIFFERENCING

Differential Algebra
  USE   DIFFERENTIAL CALCULUS
        MATRICES (MATHEMATICS)

DIFFERENTIAL AMPLIFIERS

DIFFERENTIAL ANALYZERS

DIFFERENTIAL CALCULUS

Differential Equation, Duffing
  USE   DUFFING DIFFERENTIAL EQUATION

DIFFERENTIAL EQUATIONS

Differential Equations, Elliptic
  USE   ELLIPTIC DIFFERENTIAL EQUATIONS

Differential Equations, Hyperbolic
  USE   HYPERBOLIC DIFFERENTIAL EQUATIONS

Differential Equations, Parabolic
  USE   PARABOLIC DIFFERENTIAL EQUATIONS

Differential Equations, Partial
  USE   PARTIAL DIFFERENTIAL EQUATIONS

DIFFERENTIAL GEOMETRY

DIFFERENTIAL INTERFEROMETRY

Differential Operators
  USE   DIFFERENTIAL EQUATIONS
        OPERATORS (MATHEMATICS)

DIFFERENTIAL PRESSURE

DIFFERENTIAL PULSE CODE MODULATION

Differential Thermal Analysis
  USE   THERMAL ANALYSIS

DIFFERENTIATION

DIFFERENTIATION (BIOLOGY)

Differentiation, Numerical
  USE   NUMERICAL DIFFERENTIATION

DIFFERENTIATORS

DIFFRACTION

Diffraction, Electron
  USE   ELECTRON DIFFRACTION

Diffraction, Fresnel
  USE   FRESNEL DIFFRACTION

Diffraction, Geometrical Theory Of
  USE   GEOMETRICAL THEORY OF DIFFRACTION

Diffraction Gratings
  USE   GRATINGS (SPECTRA)

DIFFRACTION LIMITED CAMERAS

Diffraction, Neutron
  USE   NEUTRON DIFFRACTION

DIFFRACTION PATHS

DIFFRACTION PATTERNS

DIFFRACTION PROPAGATION

Diffraction, Pulse
  USE   PULSE DIFFRACTION

Diffraction Telescopes
  USE   SPECTROSCOPIC TELESCOPES

Diffraction, Wave
  USE   WAVE DIFFRACTION

Diffraction, X Ray
  USE   X RAY DIFFRACTION

DIFFRACTOMETERS

DIFFUSE RADIATION

DIFFUSERS

Diffusers, Exhaust
  USE   EXHAUST DIFFUSERS

Diffusers, Shock
  USE   DIFFUSERS
        SHOCK WAVE ATTENUATION

Diffusers, Supersonic
  USE   SUPERSONIC DIFFUSERS

Diffusers, Vaneless
  USE   VANELESS DIFFUSERS

DIFFUSION

Diffusion, Ambipolar
  USE   AMBIPOLAR DIFFUSION

Diffusion, Atmospheric
  USE   ATMOSPHERIC DIFFUSION

Diffusion Bonding
  USE   DIFFUSION WELDING

DIFFUSION COEFFICIENT

Diffusion, Eddy
  USE   TURBULENT DIFFUSION

Diffusion Effect
  USE   DIFFUSION

DIFFUSION ELECTRODES

Diffusion, Electron
USE   ELECTRON DIFFUSION

DIFFUSION FLAMES

Diffusion, Gas
USE   GASEOUS DIFFUSION

Diffusion, Gaseous
USE   GASEOUS DIFFUSION

Diffusion, Gaseous Self-
USE   GASEOUS SELF-DIFFUSION

Diffusion, Ionic
USE   IONIC DIFFUSION

Diffusion, Magnetic
USE   MAGNETIC DIFFUSION

Diffusion, Molecular
USE   MOLECULAR DIFFUSION

Diffusion, Particle
USE   PARTICLE DIFFUSION

Diffusion, Plasma
USE   PLASMA DIFFUSION

DIFFUSION PUMPS

Diffusion (Solid State), Self
USE   SELF DIFFUSION (SOLID STATE)

Diffusion, Species
USE   SPECIES DIFFUSION

Diffusion, Surface
USE   SURFACE DIFFUSION

DIFFUSION THEORY

Diffusion, Thermal
USE   THERMAL DIFFUSION

Diffusion, Turbulent
USE   TURBULENT DIFFUSION

DIFFUSION WAVES

DIFFUSION WELDING

DIFFUSIVITY

Diffusivity, Thermal
USE   THERMAL DIFFUSIVITY

DIFLUORIDES

DIFLUORO COMPOUNDS

DIFLUOROUREA

DIGESTING

DIGESTIVE SYSTEM

(Digital), Binary Systems
USE   DIGITAL SYSTEMS

DIGITAL COMMAND SYSTEMS

Digital Communication
USE   PULSE COMMUNICATION

DIGITAL COMPUTERS

Digital Converters, Analog To
USE   ANALOG TO DIGITAL CONVERTERS

DIGITAL DATA

DIGITAL FILTERS

DIGITAL INTEGRATORS

DIGITAL NAVIGATION

DIGITAL RADAR SYSTEMS

DIGITAL SIMULATION

DIGITAL SPACECRAFT TELEVISION

DIGITAL SYSTEMS

DIGITAL TECHNIQUES

DIGITAL TELEVISION

(Digital), Ternary Systems
USE   DIGITAL SYSTEMS

DIGITAL TO ANALOG CONVERTERS

DIGITAL TO VOICE TRANSLATORS

DIGITAL TRANSDUCERS

DIGITALIS

Digitizers
USE   ANALOG TO DIGITAL CONVERTERS

DIGITS

Digits, Binary
USE   BINARY DIGITS

DIHEDRAL ANGLE

Dihedral Effect
USE   LATERAL STABILITY

DIHYDRAZINE

Dihydrazine, Ethylene
USE   ETHYLENE DIHYDRAZINE

DIHYDRIDES

Dihydroxyphenylalanine
USE   DOPA

DIISOCYANATES

Dikes (Geology)
USE   ROCK INTRUSIONS

Dilatation
USE   STRETCHING

DILATATIONAL WAVES

Dilation, Vaso
USE   VASODILATION

Dilatometers
USE   EXTENSOMETERS

DILATOMETRY

DILUENTS

DILUTION

Dilution Of Precision, Geometric
USE   GEOMETRIC DILUTION OF PRECISION

DIMENHYDRINATE

DIMENSIONAL ANALYSIS

Dimensional Bodies, Three
USE   THREE DIMENSIONAL BODIES

Dimensional Bodies, Two
USE   TWO DIMENSIONAL BODIES

Dimensional Boundary Layer, Three
USE   THREE DIMENSIONAL BOUNDARY LAYER

Dimensional Boundary Layer, Two
USE   TWO DIMENSIONAL BOUNDARY LAYER

Dimensional Composites, Three
USE   THREE DIMENSIONAL COMPOSITES

Dimensional Flow, One
USE   ONE DIMENSIONAL FLOW

Dimensional Flow, Three
USE   THREE DIMENSIONAL FLOW

Dimensional Flow, Two
USE   TWO DIMENSIONAL FLOW

Dimensional Jets, Two
USE   TWO DIMENSIONAL JETS

DIMENSIONAL MEASUREMENT

Dimensional Motion, Three
USE   THREE DIMENSIONAL MOTION

DIMENSIONAL STABILITY

DIMENSIONLESS NUMBERS

DIMENSIONS

(Dimensions), Size
USE   SIZE (DIMENSIONS)

DIMERCAPROL

DIMERIZATION

DIMERS

DIMETHYLHYDRAZINES

Diminution
USE   REDUCTION

DIMMING

DIMPLING

DINING PHILOSOPHERS PROBLEM

DINITRATES

Diode Circuits, Varactor
USE   VARACTOR DIODE CIRCUITS

Diode-Transistor-Logic Integ Circuits
USE   DTL INTEGRATED CIRCUITS

DIODES

Diodes, Avalanche
USE   AVALANCHE DIODES

Diodes, Barrier Injection Transit Time
USE   BARRITT DIODES

Diodes, Barritt
USE   BARRITT DIODES

Diodes, Cesium
USE   CESIUM DIODES

Diodes, Esaki
USE   TUNNEL DIODES

Diodes, Germanium
USE   GERMANIUM DIODES

Diodes, Gunn
USE   GUNN DIODES

Diodes, IMPATT
USE   AVALANCHE DIODES

Diodes, Junction
USE   JUNCTION DIODES

(Diodes), LED
USE   LIGHT EMITTING DIODES

Diodes, Light Emitting
USE   LIGHT EMITTING DIODES

Diodes, Metal-Insulator-Metal
USE   MIM DIODES

Diodes, MIM
USE   MIM DIODES

Diodes, P-I-N
USE   P-I-N JUNCTIONS
DIODES

Diodes, Parametric
  USE   PARAMETRIC DIODES

Diodes, Photo
  USE   PHOTODIODES

Diodes, Plasma
  USE   PLASMA DIODES

Diodes, Schottky
  USE   SCHOTTKY DIODES

Diodes, Schottky Barrier
  USE   SCHOTTKY DIODES

Diodes, Semiconductor
  USE   SEMICONDUCTOR DIODES

Diodes, Step Recovery
  USE   STEP RECOVERY DIODES

Diodes, Thermionic
  USE   THERMIONIC DIODES

Diodes, TRAPATT
  USE   AVALANCHE DIODES

Diodes, Tunnel
  USE   TUNNEL DIODES

Diodes, Varactor
  USE   VARACTOR DIODES

Diodes, Zener
  USE   AVALANCHE DIODES

DIONE

DIOPHANTINE EQUATION

DIORITE

Dioxide, Carbon
  USE   CARBON DIOXIDE

Dioxide Concentration, Carbon
  USE   CARBON DIOXIDE CONCENTRATION

Dioxide Lasers, Carbon
  USE   CARBON DIOXIDE LASERS

Dioxide, Nitrogen
  USE   NITROGEN DIOXIDE

Dioxide Removal, Carbon
  USE   CARBON DIOXIDE REMOVAL

Dioxide, Silicon
  USE   SILICON DIOXIDE

Dioxide Tension, Carbon
  USE   CARBON DIOXIDE TENSION

Dioxide, Titanium
  USE   TITANIUM OXIDES

DIOXIDES

Dioxides, Sulfur
  USE   SULFUR DIOXIDES

DIPHENYL COMPOUNDS

DIPHENYL HYDANTOIN

Diphosphate, Adenosine
  USE   ADENOSINE DIPHOSPHATE

DIPHOSPHATES

DIPHTHERIA

DIPLEXERS

DIPOLE ANTENNAS

DIPOLE MOMENTS

DIPOLES

Dipoles, Electric
  USE   ELECTRIC DIPOLES

Dipoles, Magnetic
  USE   MAGNETIC DIPOLES

Dipoles, Orbiting
  USE   ORBITING DIPOLES

DIPPING

DIRAC EQUATION

Dirac Statistics, Fermi-
  USE   FERMI-DIRAC STATISTICS

DIRECT CURRENT

DIRECT LIFT CONTROLS

DIRECT POWER GENERATORS

DIRECTION

(Direction), Bearing
  USE   BEARING (DIRECTION)

Direction Finders, Radar
  USE   RADIO DIRECTION FINDERS

Direction Finders, Radio
  USE   RADIO DIRECTION FINDERS

Direction Finders (Radio)
  USE   RADIO DIRECTION FINDERS

DIRECTION FINDING

Direction Indicators, Flow
  USE   FLOW DIRECTION INDICATORS

Direction, Wind
  USE   WIND DIRECTION

DIRECTIONAL ANTENNAS

DIRECTIONAL CONTROL

DIRECTIONAL COUPLERS

DIRECTIONAL SOLIDIFICATION (CRYSTALS)

DIRECTIONAL STABILITY

DIRECTIVITY

DIRECTORIES

DIRECTORS (ANTENNA ELEMENTS)

DIRICHLET PROBLEM

Dirigibles
  USE   AIRSHIPS

DIRT

DISARMAMENT

DISASTERS

DISCHARGE

DISCHARGE COEFFICIENT

Discharge Counters, Gas
  USE   COUNTERS
          GAS DISCHARGE TUBES

Discharge, Penning
  USE   PENNING DISCHARGE

Discharge, Radio Frequency
  USE   RADIO FREQUENCY DISCHARGE

Discharge, Ring
  USE   RING DISCHARGE

Discharge, Toroidal
  USE   TOROIDAL DISCHARGE

Discharge, Townsend
  USE   TOWNSEND DISCHARGE

Discharge Tubes
  USE   GAS DISCHARGE TUBES

Discharge Tubes, Gas
  USE   GAS DISCHARGE TUBES

DISCHARGERS

Dischargers, Static
  USE   STATIC DISCHARGERS

Discharges, Arc
  USE   ARC DISCHARGES

Discharges, Corona
  USE   ELECTRIC CORONA

Discharges, Electric
  USE   ELECTRIC DISCHARGES

Discharges, Electrodeless
  USE   ELECTRODELESS DISCHARGES

Discharges, Gas
  USE   GAS DISCHARGES

Discharges, Glow
  USE   GLOW DISCHARGES

Discharges, Multipactor
  USE   MULTIPACTOR DISCHARGES

Discharges, Plasma
  USE   PLASMA JETS

Discharges, Spark
  USE   ELECTRIC SPARKS

DISCIPLINING

DISCOLORATION

DISCONNECT DEVICES

Disconnectors
  USE   DISCONNECT DEVICES

DISCONTINUITY

Discontinuity, Shock
  USE   SHOCK DISCONTINUITY

DISCOS (SATELLITE ATTITUDE CONTROL)

DISCOVERER RECOVERY CAPSULES

DISCOVERER SATELLITES

Discovering
  USE   EXPLORATION

DISCOVERY (ORBITER)

DISCRETE ADDRESS BEACON SYSTEM

DISCRETE FUNCTIONS

DISCRIMINANT ANALYSIS (STATISTICS)

Discriminant Functions
  USE   DISCRIMINANT ANALYSIS (STATISTICS)

DISCRIMINATION

Discrimination, Brightness
  USE   BRIGHTNESS DISCRIMINATION

Discrimination, Sensory
  USE   SENSORY DISCRIMINATION

Discrimination, Speech
  USE   SPEECH RECOGNITION

Discrimination, Tactile
  USE   TACTILE DISCRIMINATION

Discrimination, Time
  USE   TIME DISCRIMINATION

Discrimination, Visual
USE   VISUAL DISCRIMINATION

DISCRIMINATORS

Discriminators, Fraunhofer Line
USE   FRAUNHOFER LINE DISCRIMINATORS

Discriminators, Frequency
USE   FREQUENCY DISCRIMINATORS

Discriminators, Signal
USE   SIGNAL DETECTORS

DISCUSSION

Disease, Coronary Artery
USE   CORONARY ARTERY DISEASE

Disease, Parkinson
USE   PARKINSON DISEASE

Diseased Vegetation
USE   BLIGHT

DISEASES

Diseases, Allergic
USE   ALLERGIC DISEASES

Diseases, Eye
USE   EYE DISEASES

Diseases, Heart
USE   HEART DISEASES

Diseases, Infectious
USE   INFECTIOUS DISEASES

Diseases, Kidney
USE   KIDNEY DISEASES

Diseases, Metabolic
USE   METABOLIC DISEASES

Diseases, Parasitic
USE   PARASITIC DISEASES

Diseases, Respiratory
USE   RESPIRATORY DISEASES

Diseases, Rheumatic
USE   RHEUMATIC DISEASES

Diseases, Tooth
USE   TOOTH DISEASES

Diseases, Toxic
USE   TOXIC DISEASES

Dishes
USE   PARABOLIC REFLECTORS

DISILICIDES

Disinfectants
USE   ANTISEPTICS

DISINTEGRATION

DISK GALAXIES

Disk, Solar
USE   SUN

DISKS

Disks, Accretion
USE   ACCRETION DISKS

Disks, Actuator
USE   ACTUATOR DISKS

Disks, Intervertebral
USE   INTERVERTEBRAL DISKS

Disks, Magnetic
USE   MAGNETIC DISKS

Disks, Optical
USE   OPTICAL DISKS

Disks, Rotating
USE   ROTATING DISKS

Disks, Rotor
USE   TURBINE WHEELS

DISKS (SHAPES)

Disks, Video
USE   VIDEO DISKS

Dislocations, Crystal
USE   CRYSTAL DISLOCATIONS

Dislocations, Edge
USE   EDGE DISLOCATIONS

DISLOCATIONS (MATERIALS)

Dislocations, Screw
USE   SCREW DISLOCATIONS

Disorder Transformations, Order-
USE   ORDER-DISORDER TRANSFORMATIONS

DISORDERS

DISORIENTATION

Dispatching
USE   DISTRIBUTING

DISPENSERS

Dispersal, Cloud
USE   CLOUD DISPERSAL

Dispersal, Fog
USE   FOG DISPERSAL

DISPERSING

DISPERSION

Dispersion, Magnetic
USE   MAGNETIC DISPERSION

Dispersion, Plasma
USE   PLASMA DIFFUSION

Dispersion Precipitation Hardening
USE   PRECIPITATION HARDENING

Dispersion Spectrographs, High
USE   HIGH DISPERSION SPECTROGRAPHS

Dispersion, Wave
USE   WAVE DISPERSION

DISPERSIONS

DISPLACEMENT

DISPLACEMENT MEASUREMENT

DISPLAY DEVICES

Display Devices, Plasma
USE   PLASMA DISPLAY DEVICES

Display Systems
USE   DISPLAY DEVICES

Displays, F
USE   F REGION

Displays, Head-Up
USE   HEAD-UP DISPLAYS

Displays, Helmet Mounted
USE   HELMET MOUNTED DISPLAYS

Displays, Radar
USE   RADARSCOPES

Displays, Visual
USE   DISPLAY DEVICES

DISPOSAL

Disposal (In Space), Hazardous Material
USE   HAZARDOUS MATERIAL DISPOSAL (IN
SPACE)

Disposal, Waste
USE   WASTE DISPOSAL

DISRUPTING

DISSECTION

Dissector Tubes, Image
USE   IMAGE DISSECTOR TUBES

Dissemination, Information
USE   INFORMATION DISSEMINATION

Dissemination Of Information, Selective
USE   SELECTIVE DISSEMINATION OF
INFORMATION

DISSIPATION

Dissipation Chilling, Heat
USE   COOLING

Dissipation, Energy
USE   ENERGY DISSIPATION

Dissipation, Heat
USE   COOLING

Dissipation, Ohmic
USE   OHMIC DISSIPATION

Dissipators
USE   DISSIPATION

DISSOCIATION

Dissociation, Gas
USE   GAS DISSOCIATION

Dissociation, Heat Of
USE   HEAT OF DISSOCIATION

Dissociation, Molecular
USE   DISSOCIATION

Dissociation, Photo
USE   PHOTODISSOCIATION

Dissociation, Thermal
USE   THERMAL DISSOCIATION

Dissolution
USE   DISSOLVING

DISSOLVED GASES

DISSOLVING

Dissymmetry
USE   ASYMMETRY

DISTANCE

DISTANCE MEASURING EQUIPMENT

Distance, Miss
USE   MISS DISTANCE

Distance Perception
USE   SPACE PERCEPTION

DISTILLATION

DISTILLATION EQUIPMENT

(Distillation), Stripping
USE   STRIPPING (DISTILLATION)

DISTORTION

Distortion, Flow
USE   FLOW DISTORTION

Distortion, Signal
USE   SIGNAL DISTORTION

Distortion, Surface
 USE SURFACE DISTORTION

DISTRIBUTED AMPLIFIERS

DISTRIBUTED FEEDBACK LASERS

DISTRIBUTED PARAMETER SYSTEMS

DISTRIBUTED PROCESSING

DISTRIBUTING

DISTRIBUTION

Distribution Analysis, Amplitude
 USE AMPLITUDE DISTRIBUTION ANALYSIS

Distribution, Angular
 USE ANGULAR DISTRIBUTION

Distribution, Boltzmann
 USE BOLTZMANN DISTRIBUTION

Distribution, Brightness
 USE BRIGHTNESS DISTRIBUTION

Distribution, Charge
 USE CHARGE DISTRIBUTION

Distribution, Circulation
 USE CIRCULATION DISTRIBUTION

Distribution, Current
 USE CURRENT DISTRIBUTION

Distribution, Density
 USE DENSITY DISTRIBUTION

Distribution (Density), Maxwellian
 USE MAXWELL-BOLTZMANN DENSITY
    FUNCTION

Distribution, Electron
 USE ELECTRON DISTRIBUTION

Distribution (Electronics), Hole
 USE HOLE DISTRIBUTION (ELECTRONICS)

Distribution, Energy
 USE ENERGY DISTRIBUTION

Distribution, Flow
 USE FLOW DISTRIBUTION

Distribution, Force
 USE FORCE DISTRIBUTION

Distribution (Forces), Load
 USE LOAD DISTRIBUTION (FORCES)

Distribution, Frequency
 USE FREQUENCY DISTRIBUTION

DISTRIBUTION FUNCTIONS

Distribution Functions, Probability
 USE PROBABILITY DISTRIBUTION FUNCTIONS

Distribution, Hole
 USE HOLE DISTRIBUTION

Distribution, Ion
 USE ION DISTRIBUTION

Distribution, Lift
 USE LIFT
    FORCE DISTRIBUTION

Distribution, Mass
 USE MASS DISTRIBUTION

Distribution (Mechanics), Hole
 USE HOLE DISTRIBUTION (MECHANICS)

Distribution, Moment
 USE MOMENT DISTRIBUTION

DISTRIBUTION MOMENTS

Distribution, Neutron
 USE NEUTRON DISTRIBUTION

Distribution, Normal Force
 USE FORCE DISTRIBUTION

Distribution, Particle Size
 USE PARTICLE SIZE DISTRIBUTION

Distribution, Pattern
 USE DISTRIBUTION (PROPERTY)

Distribution, Pressure
 USE PRESSURE DISTRIBUTION

DISTRIBUTION (PROPERTY)

Distribution, Radial
 USE RADIAL DISTRIBUTION

Distribution, Radiation
 USE RADIATION DISTRIBUTION

Distribution, Rayleigh
 USE RAYLEIGH DISTRIBUTION

Distribution, Size
 USE SIZE DISTRIBUTION

Distribution, Spatial
 USE SPATIAL DISTRIBUTION

Distribution, Spectral Energy
 USE SPECTRAL ENERGY DISTRIBUTION

Distribution, Star
 USE STAR DISTRIBUTION

Distribution, Strain
 USE STRESS CONCENTRATION

Distribution, Stress
 USE STRESS CONCENTRATION

Distribution, Stress-Strain
 USE STRESS CONCENTRATION

Distribution, Temperature
 USE TEMPERATURE DISTRIBUTION

Distribution, Temporal
 USE TEMPORAL DISTRIBUTION

Distribution, Thrust
 USE THRUST DISTRIBUTION

Distribution, Velocity
 USE VELOCITY DISTRIBUTION

Distribution, Vertical
 USE VERTICAL DISTRIBUTION

Distributions, Gaussian
 USE NORMAL DENSITY FUNCTIONS

Distributions, Normal
 USE NORMAL DENSITY FUNCTIONS

Distributions, Pearson
 USE PEARSON DISTRIBUTIONS

Distributions, Random
 USE STATISTICAL DISTRIBUTIONS

Distributions, Statistical
 USE STATISTICAL DISTRIBUTIONS

DISTRIBUTORS

DISTRICT OF COLUMBIA

Disturbance, Satellite Attitude
 USE SPACECRAFT STABILITY
    ATTITUDE STABILITY

Disturbance Theory
 USE PERTURBATION THEORY

DISTURBANCES

Disturbances, Ionospheric
 USE IONOSPHERIC DISTURBANCES

Disturbances, Magnetic
 USE MAGNETIC DISTURBANCES

Disturbances, Shear
 USE S WAVES

Disturbances) SID (Ionospheric
 USE SUDDEN IONOSPHERIC DISTURBANCES

Disturbances, Sudden Ionospheric
 USE SUDDEN IONOSPHERIC DISTURBANCES

Disturbances, Traveling Ionospheric
 USE TRAVELING IONOSPHERIC DISTURBANCES

Disturbances, Vortex
 USE VORTICES

DISTURBING FUNCTIONS

Disulfide, Carbon
 USE CARBON DISULFIDE

DISULFIDES

Disulfides, Molybdenum
 USE MOLYBDENUM DISULFIDES

DITCHES

DITCHING

Ditching (Excavation)
 USE EXCAVATION

DITCHING (LANDING)

DITHERS

Dithiols
 USE THIOLS

DIURESIS

DIURETICS

Diuretics, Anti
 USE ANTIDIURETICS

Diurnal Rhythms
 USE CIRCADIAN RHYTHMS

DIURNAL VARIATIONS

DIVERGENCE

DIVERGENT NOZZLES

Divergent Nozzles, Convergent-
 USE CONVERGENT-DIVERGENT NOZZLES

Diversity, Reception
 USE RECEPTION DIVERSITY

Diversity, Space
 USE RECEPTION DIVERSITY

DIVERTERS

DIVIDERS

Dividers, Frequency
 USE FREQUENCY DIVIDERS

DIVIDES (LANDFORMS)

DIVIDING (MATHEMATICS)

DIVING (UNDERWATER)

DIVISION

Division, Cell
 USE CELL DIVISION

Division Multiple Access, Code
 USE CODE DIVISION MULTIPLE ACCESS

Division Multiple Access, Frequency
  USE   FREQUENCY DIVISION MULTIPLE ACCESS

Division Multiple Access, Time
  USE   TIME DIVISION MULTIPLE ACCESS

Division Multiplexing, Code
  USE   CODE DIVISION MULTIPLEXING

Division Multiplexing, Frequency
  USE   FREQUENCY DIVISION MULTIPLEXING

Division Multiplexing, Time
  USE   TIME DIVISION MULTIPLEXING

Division Multiplexing, Wavelength
  USE   WAVELENGTH DIVISION MULTIPLEXING

Divisions, Sub
  USE   SUBDIVISIONS

DIVOT (Voice Translators)
  USE   DIGITAL TO VOICE TRANSLATORS

DME-A Satellite
  USE   EXPLORER 31 SATELLITE

DMSP SATELLITES

DNA
  USE   DEOXYRIBONUCLEIC ACID

DO-27 AIRCRAFT

DO-27 Aircraft, Dornier
  USE   DO-27 AIRCRAFT

DO-28 AIRCRAFT

DO-28 Aircraft, Dornier
  USE   DO-28 AIRCRAFT

DO-31 AIRCRAFT

DO-31 Aircraft, Dornier
  USE   DO-31 AIRCRAFT

Docking
  USE   SPACECRAFT DOCKING

Docking Adapters, Multiple
  USE   MULTIPLE DOCKING ADAPTERS

Docking Modules, Spacecraft
  USE   SPACECRAFT DOCKING MODULES

Docking, Offshore
  USE   OFFSHORE DOCKING

Docking, Spacecraft
  USE   SPACECRAFT DOCKING

DOCUMENT STORAGE

DOCUMENTATION

(Documentation), Indexes
  USE   INDEXES (DOCUMENTATION)

DOCUMENTS

(Documents), Journals
  USE   PERIODICALS

DODGE SATELLITE

Dog Missile, Hound
  USE   HOUND DOG MISSILE

DOGHOUSES (ELECTRONICS)

DOGS

DOLLIES

DOLOMITE (MINERAL)

DOLPHINS

DOMAIN WALL

DOMAINS

Domains, Magnetic
  USE   MAGNETIC DOMAINS

DOMES

DOMES (GEOLOGY)

DOMES (STRUCTURAL FORMS)

DOMESTIC ENERGY

DOMESTIC SATELLITE COMMUNICATIONS SYSTEMS

DOMINANCE

Dominance, Eye
  USE   EYE DOMINANCE

Dominance Model, Vector
  USE   VECTOR DOMINANCE MODEL

DOMINICA

DOMINICAN REPUBLIC

DOMINO PROPELLANTS

DONNELL EQUATIONS

DONOR MATERIALS

DOORS

(Doors), Exits
  USE   DOORS

DOPA

DOPED CRYSTALS

DOPES

Doping (Additives)
  USE   ADDITIVES

DOPPLER EFFECT

DOPPLER NAVIGATION

Doppler Positioning, Satellite
  USE   SATELLITE DOPPLER POSITIONING

DOPPLER RADAR

Doppler Radar, Pulse
  USE   PULSE DOPPLER RADAR

Doppler Shift, Stellar
  USE   DOPPLER EFFECT
        EXTRATERRESTRIAL RADIATION

Doppler Tracking System, Polystation
  USE   POLYSTATION DOPPLER TRACKING
        SYSTEM

Doppler Velocimeters, Laser
  USE   LASER DOPPLER VELOCIMETERS

DOPPLER-FIZEAU EFFECT

DORNIER AIRCRAFT

Dornier DO-27 Aircraft
  USE   DO-27 AIRCRAFT

Dornier DO-28 Aircraft
  USE   DO-28 AIRCRAFT

Dornier DO-31 Aircraft
  USE   DO-31 AIRCRAFT

DORNIER PARAGLIDER ROCKET VEHICLE

DORSAL SECTIONS

DOSAGE

Dosage, Radiation
  USE   RADIATION DOSAGE

Dosage, Sublethal
  USE   SUBLETHAL DOSAGE

Dose
  USE   DOSAGE

DOSIMETERS

(Dosimeters), Threshold Detectors
  USE   THRESHOLD DETECTORS (DOSIMETERS)

Dosimetry
  USE   DOSIMETERS

DOUBLE BASE PROPELLANTS

DOUBLE BASE ROCKET PROPELLANTS

DOUBLE CUSPS

DOUBLE PRECISION ARITHMETIC

DOUBLE SIDEBAND TRANSMISSION

DOUBLE STARS

Doughnut Shape Wheels
  USE   TOROIDAL WHEELS

DOUGLAS AIRCRAFT

Douglas Aircraft, Mcdonnell
  USE   MCDONNELL DOUGLAS AIRCRAFT

Douglas D-558 Aircraft
  USE   D-558 AIRCRAFT

Douglas DC-3 Aircraft
  USE   DC 3 AIRCRAFT

Douglas DC-7 Aircraft
  USE   DC 7 AIRCRAFT

Douglas DC-8 Aircraft
  USE   DC 8 AIRCRAFT

Douglas DC-9 Aircraft
  USE   DC 9 AIRCRAFT

Douglas Military Aircraft
  USE   DOUGLAS AIRCRAFT
        MILITARY AIRCRAFT

Douglas PD-808 Aircraft
  USE   PD-808 AIRCRAFT

Douglas PD-808 Aircraft, Piaggio-
  USE   PD-808 AIRCRAFT

DOVAP
  USE   DOPPLER EFFECT

DOWN-CONVERTERS

DOWNLINKING

DOWNRANGE

DOWNRANGE ANTIMISSILE MEASUREMENT PROGRAM

DOWNRANGE MEASUREMENT

DOWNTIME

DOWNWASH

DPCM (Modulation)
  USE   DIFFERENTIAL PULSE CODE MODULATION

DRACONID METEOROIDS

DRAFT

DRAFT (GAS FLOW)

DRAFTING (DRAWING)

DRAFTING MACHINES

DRAG

Drag, Aerodynamic
USE   AERODYNAMIC DRAG

Drag Balance
USE   AERODYNAMIC BALANCE
       LIFT DRAG RATIO

DRAG CHUTES

DRAG COEFFICIENTS

DRAG DEVICES

Drag Effect
USE   DRAG

Drag, Electrostatic
USE   ELECTROSTATIC DRAG

DRAG FORCE ANEMOMETERS

Drag, Friction
USE   FRICTION DRAG

Drag, Interference
USE   INTERFERENCE DRAG

DRAG MEASUREMENT

Drag, Minimum
USE   MINIMUM DRAG

Drag, Nonequilibrium
USE   FRICTION DRAG

Drag, Pressure
USE   PRESSURE DRAG

Drag Ratio, Lift
USE   LIFT DRAG RATIO

DRAG REDUCTION

Drag, Satellite
USE   SATELLITE DRAG

Drag, Supersonic
USE   SUPERSONIC DRAG

Drag, Viscous
USE   VISCOUS DRAG

Drag, Wave
USE   WAVE DRAG

Dragon Aircraft, Jet
USE   DH 125 AIRCRAFT

Dragulators
USE   DRAG DEVICES
       BRAKES (FOR ARRESTING MOTION)

DRAINAGE

Drainage, Dendritic
USE   DRAINAGE PATTERNS

Drainage, Interlacing
USE   DRAINAGE PATTERNS

DRAINAGE PATTERNS

Drainage Patterns, Radial
USE   DRAINAGE PATTERNS

Drainage, Rectangular
USE   DRAINAGE PATTERNS

Draining
USE   DRAINAGE

DRAWING

Drawing, Bundle
USE   BUNDLE DRAWING

Drawing, Cold
USE   COLD DRAWING

Drawing, Deep
USE   DEEP DRAWING

(Drawing), Drafting
USE   DRAFTING (DRAWING)

Drawing, Metal
USE   METAL DRAWING

DRAWINGS

(Drawings), Elevations
USE   DRAWINGS

Drawings, Engineering
USE   ENGINEERING DRAWINGS

Drawings, Mechanical
USE   ENGINEERING DRAWINGS

DRC (Capsule)
USE   DISCOVERER RECOVERY CAPSULES

DREAMS

DREDGED MATERIALS

DREDGING

DRIFT

Drift, Continental
USE   CONTINENTAL DRIFT

Drift, Glacial
USE   GLACIAL DRIFT

Drift, Gyroscopic
USE   GYROSCOPES
       GYROSCOPIC STABILITY

Drift, Instrument
USE   DRIFT (INSTRUMENTATION)

DRIFT (INSTRUMENTATION)

Drift, Ionospheric
USE   IONOSPHERIC DRIFT

Drift, Littoral
USE   LITTORAL DRIFT

Drift, Plasma
USE   PLASMA DRIFT

DRIFT RATE

DRILL BITS

DRILLING

Drilling, Laser
USE   LASER DRILLING

DRILLS

DRINKING

Drive, Helicopter Propeller
USE   HELICOPTER PROPELLER DRIVE

Drive, Jet
USE   JET PROPULSION

Drive, Propeller
USE   PROPELLER DRIVE

Driven Rotors, Tip
USE   TIP DRIVEN ROTORS

Drivers (Payload Delivery), Mass
USE   MASS DRIVERS (PAYLOAD DELIVERY)

DRIVES

Drives, Mechanical
USE   MECHANICAL DRIVES

Drives, Rotary
USE   MECHANICAL DRIVES

Drives, Wind Tunnel
USE   WIND TUNNEL DRIVES

Drogue Parachutes
USE   DRAG CHUTES

Drogues
USE   TOWED BODIES

DRONE AIRCRAFT

Drone Aircraft, Firebee 2 Target
USE   FIREBEE 2 TARGET DRONE AIRCRAFT

Drone Aircraft, Target
USE   TARGET DRONE AIRCRAFT

Drone Helicopters
USE   HELICOPTERS
       DRONE AIRCRAFT

DRONE VEHICLES

Drones For Aerodynamic And Struct Test
USE   DAST PROGRAM

DROOPED AIRFOILS

DROP

DROP CALORIMETERS

Drop, Friction Pressure
USE   SKIN FRICTION

Drop Operations, Air
USE   AIR DROP OPERATIONS

Drop, Pressure
USE   PRESSURE DROP

DROP SIZE

DROP TESTS

DROP TOWERS

DROP TRANSFER

Drop Tubes
USE   DROP TOWERS

Drop Weight Tests
USE   DROP TESTS

DROPOUTS

Drops, Electron-Hole
USE   ELECTRON-HOLE DROPS

Drops, Liquid
USE   DROPS (LIQUIDS)

DROPS (LIQUIDS)

Drops, Rain
USE   RAINDROPS

DROPSONDES

DROSOPHILA

DROUGHT

Drought Conditions
USE   DROUGHT

Drowsiness
USE   SLEEP

Drug Therapy
USE   CHEMOTHERAPY

DRUGS

Drugs, Antiradiation
USE   ANTIRADIATION DRUGS

Drugs, Motion Sickness
   USE   MOTION SICKNESS DRUGS

Drugs, Psychotropic
   USE   PSYCHOTROPIC DRUGS

Drugs, Vasoconstrictor
   USE   VASOCONSTRICTOR DRUGS

Drumlins
   USE   GLACIAL DRIFT

DRUMS

DRUMS (CONTAINERS)

Drums, Magnetic
   USE   MAGNETIC DRUMS

DRY CELLS

DRY FRICTION

DRY HEAT

DRYDOCKS

Dryers (Equipment)
   USE   DRYING APPARATUS

DRYING

DRYING APPARATUS

Drying, Freeze
   USE   FREEZE DRYING

DSIF (Instrumentation Facility)
   USE   DEEP SPACE INSTRUMENTATION FACILITY

DSN Helicopter
   USE   QH-50 HELICOPTER

DSN-3 Helicopter, Gyrodyne
   USE   QH-50 HELICOPTER

DTA (Analysis)
   USE   THERMAL ANALYSIS

DTL INTEGRATED CIRCUITS

DTMB-111 Ground Effect Machine
   USE   GROUND EFFECT MACHINES

DTMB-430 Ground Effect Machine
   USE   GROUND EFFECT MACHINES

DUAL AIR DENSITY EXPLORER

Dual Mode Propulsion
   USE   HYBRID PROPULSION

DUAL SPIN SPACECRAFT

DUAL THRUST NOZZLES

DUAL WING CONFIGURATIONS

DUALITY PRINCIPLE

DUALITY THEOREM

DUCT GEOMETRY

DUCTED BODIES

DUCTED FAN ENGINES

DUCTED FANS

DUCTED FLOW

Ducted Propellers
   USE   SHROUDED PROPELLERS

DUCTED ROCKET ENGINES

DUCTILITY

DUCTS

Ducts, Acoustic
   USE   ACOUSTIC DUCTS

Ducts, Air
   USE   AIR DUCTS

Ducts, Annular
   USE   ANNULAR DUCTS

DUFFING DIFFERENTIAL EQUATION

Dullness
   USE   LUSTER

DUMMIES

Dummy Loads
   USE   OUTPUT
           LOADING
           IMPEDANCE

DUMPING

DUNALIELLA

DUNES

Dunes, Coastal
   USE   DUNES

Dunes, Sand
   USE   DUNES

Dungeys Wind Shear Mechanism
   USE   WIND SHEAR

Dunham Potential, Klein-
   USE   KLEIN-DUNHAM POTENTIAL

DUNITE

DUOCHROMATORS

DUOPLASMATRONS

DUPLEX OPERATION

DUPLEXERS

Duplicating
   USE   REPRODUCTION (COPYING)

DURABILITY

(Durability), Life
   USE   LIFE (DURABILITY)

(Durability), Lifetime
   USE   LIFE (DURABILITY)

Duration
   USE   TIME

Duration Exposure Facility, Long
   USE   LONG DURATION EXPOSURE FACILITY

Duration, Light
   USE   PULSE DURATION
           FLASH

Duration Modulation, Pulse
   USE   PULSE DURATION MODULATION

Duration, Pulse
   USE   PULSE DURATION

Duration Space Flight, Extended
   USE   LONG DURATION SPACE FLIGHT

Duration Space Flight, Long
   USE   LONG DURATION SPACE FLIGHT

DURENE

Dushman Equation, Richardson-
   USE   THERMIONIC EMISSION
           TEMPERATURE EFFECTS

DUST

Dust Belt, Terrestrial
   USE   TERRESTRIAL DUST BELT

Dust Clouds, Meteoroid
   USE   METEOROID DUST CLOUDS

DUST COLLECTORS

Dust, Cosmic
   USE   COSMIC DUST

Dust, Interplanetary
   USE   INTERPLANETARY DUST

Dust, Lunar
   USE   LUNAR DUST

Dust, Meteoritic
   USE   MICROMETEOROIDS

DUST STORMS

Dust, Zodiacal
   USE   ZODIACAL DUST

Dusting, Crop
   USE   CROP DUSTING

DWARF GALAXIES

DWARF NOVAE

DWARF STARS

Dwarf Stars, Red
   USE   RED DWARF STARS

Dwarf Stars, White
   USE   WHITE DWARF STARS

DWELL

Dy
   USE   DYSPROSIUM

DYADICS

DYE LASERS

DYES

Dyna-Soar Space Glider
   USE   X-20 AIRCRAFT

DYNAMIC CHARACTERISTICS

DYNAMIC CONTROL

DYNAMIC LOADS

DYNAMIC MODELS

DYNAMIC MODULUS OF ELASTICITY

DYNAMIC PRESSURE

DYNAMIC PROGRAMMING

Dynamic Properties
   USE   DYNAMIC CHARACTERISTICS

DYNAMIC RESPONSE

DYNAMIC STABILITY

DYNAMIC STRUCTURAL ANALYSIS

DYNAMIC TESTS

DYNAMICAL SYSTEMS

DYNAMICS

Dynamics, Aero
   USE   AERODYNAMICS

Dynamics, Aerothermo
   USE   AEROTHERMODYNAMICS**

Dynamics Aircraft, General
USE   GENERAL DYNAMICS AIRCRAFT

Dynamics, Astro
USE   ASTRODYNAMICS

Dynamics, Bio
USE   BIODYNAMICS

Dynamics), Cascades (Fluid
USE   FLUID DYNAMICS

Dynamics, Chiral
USE   CHIRAL DYNAMICS

Dynamics, Computational Fluid
USE   COMPUTATIONAL FLUID DYNAMICS

Dynamics, Elasto
USE   ELASTODYNAMICS

Dynamics, Electro
USE   ELECTRODYNAMICS

DYNAMICS EXPLORER SATELLITES

DYNAMICS EXPLORER 1 SATELLITE

DYNAMICS EXPLORER 2 SATELLITE

Dynamics, Fluid
USE   FLUID DYNAMICS

Dynamics, Gas
USE   GAS DYNAMICS

Dynamics, Geo
USE   GEODYNAMICS

Dynamics, Group
USE   GROUP DYNAMICS

Dynamics, Hemo
USE   HEMODYNAMICS

Dynamics, Hydro
USE   HYDRODYNAMICS

Dynamics, Magnetohydro
USE   MAGNETOHYDRODYNAMICS

Dynamics Military Aircraft, General
USE   GENERAL DYNAMICS AIRCRAFT
       MILITARY AIRCRAFT

Dynamics, Ocean
USE   OCEAN DYNAMICS

Dynamics), Panel Method (Fluid
USE   PANEL METHOD (FLUID DYNAMICS)

Dynamics, Plasma
USE   PLASMA DYNAMICS

Dynamics, Rarefied Gas
USE   RAREFIED GAS DYNAMICS

Dynamics, Solar
USE   HELIOSEISMOLOGY

Dynamics, Spin
USE   SPIN DYNAMICS

Dynamics), Stabilizers (Fluid
USE   STABILIZERS (FLUID DYNAMICS)

Dynamics, Structural
USE   DYNAMIC STRUCTURAL ANALYSIS

Dynamics, Terra
USE   TERRADYNAMICS

Dynamics, Thermo
USE   THERMODYNAMICS

DYNAMITE

DYNAMO THEORY

DYNAMOMETERS

Dynamometry, Ophthalmo
USE   OPHTHALMODYNAMOMETRY

Dynamos
USE   ROTATING GENERATORS

Dynes, Auto
USE   AUTODYNES

DYNODES

DYSON THEORY

DYSPNEA

DYSPROSIUM

DYSPROSIUM COMPOUNDS

DYSPROSIUM ISOTOPES

Dysprosium 161
USE   DYSPROSIUM ISOTOPES

E

E, AIMP-
USE   EXPLORER 35 SATELLITE

E, Atmosphere Explorer
USE   EXPLORER 55 SATELLITE

E, Earth Resources Technology Satellite
USE   LANDSAT E

E, ERTS-
USE   LANDSAT E

E GLASS

E ICBM, Atlas
USE   ATLAS E ICBM

E, IMP-
USE   EXPLORER 35 SATELLITE

E, LANDSAT
USE   LANDSAT E

E Layer, Night
USE   E REGION
       NIGHT SKY

E Layer, Sporadic
USE   SPORADIC E LAYER

E Layers
USE   E REGION

E, Lunar Orbiter
USE   LUNAR ORBITER 5

E, NOAA
USE   NOAA 8 SATELLITE

E, OGO-
USE   OGO-5

E, OSO-
USE   OSO-3

E REGION

E Satellite, AE-
USE   EXPLORER 55 SATELLITE

E Satellite, TIROS
USE   TIROS 5 SATELLITE

E, Space Shuttle Mission 51-
USE   SPACE SHUTTLE MISSION 51-E

E, Space Shuttle Mission 61-
USE   SPACE SHUTTLE MISSION 61-E

E, Vitamin
USE   TOCOPHEROL

E-1 LAYER

E-2 AIRCRAFT

E-2 LAYER

E-3A AIRCRAFT

E-4A AIRCRAFT

EAI 680 COMPUTER

EAI 8400 COMPUTER

EAI 8900 COMPUTER

EAR

Ear, Middle
USE   MIDDLE EAR

Ear Pressure, Middle
USE   MIDDLE EAR PRESSURE

EAR PRESSURE TEST

EAR PROTECTORS

EARDRUMS

Early Apollo Surface Experiments Package
USE   EASEP

EARLY BIRD SATELLITES

EARLY STARS

Early Warning System, Ballistic Missile
USE   BALLISTIC MISSILE EARLY WARNING
       SYSTEM

EARLY WARNING SYSTEMS

EARPHONES

Ears, Artificial
USE   ARTIFICIAL EARS

**EARTH & OCEAN PHYSICS APPLICATIONS
PROGRAM**

EARTH ALBEDO

Earth Alloys, Rare
USE   RARE EARTH ALLOYS

EARTH ATMOSPHERE

Earth Atmosphere, Primitive
USE   PRIMITIVE EARTH ATMOSPHERE

EARTH AXIS

Earth Compounds, Alkaline
USE   ALKALINE EARTH COMPOUNDS

Earth Compounds, Rare
USE   RARE EARTH COMPOUNDS

EARTH CORE

EARTH CRUST

Earth Currents
USE   TELLURIC CURRENTS

Earth Elements, Rare
USE   RARE EARTH ELEMENTS

Earth Energy Budget Experiment
USE   LZEEBE SATELLITE

Earth Energy Budget Experiment, Zonal
USE   LZEEBE SATELLITE

Earth Energy Experiment, Long Term Zonal
USE   LZEEBE SATELLITE

EARTH ENVIRONMENT

Earth Explorer 1, International Sun
USE   INTERNATIONAL SUN EARTH EXPLORER 1

Earth Explorer 2, International Sun
USE   INTERNATIONAL SUN EARTH EXPLORER 2

Earth Explorer 3, International Sun
USE   INTERNATIONAL SUN EARTH EXPLORER 3

Earth Explorers, International Sun
USE   INTERNATIONAL SUN EARTH EXPLORERS

Earth Figure
USE   GEODESY

EARTH HYDROSPHERE

(Earth), Hydrosphere
USE   EARTH HYDROSPHERE

EARTH LIMB

EARTH MANTLE

Earth Metals, Alkaline
USE   ALKALINE EARTH METALS

EARTH MOTION

EARTH MOVEMENTS

Earth Navigation, Nap-Of-The-
USE   NAP-OF-THE-EARTH NAVIGATION

Earth Neighborhood, Origin Of Plasmas In
USE   OPEN PROJECT

EARTH OBSERVATIONS (FROM SPACE)

Earth Observatory Satellite, Synchronous
USE   SYNCHRONOUS EARTH OBSERVATORY
         SATELLITE

EARTH ORBITAL RENDEZVOUS

Earth Orbiting Space Stations
USE   EOSS

EARTH ORBITS

Earth Oxides, Alkaline
USE   ALKALINE EARTH OXIDES

EARTH (PLANET)

EARTH PLANETARY STRUCTURE

Earth Radiation
USE   TERRESTRIAL RADIATION

EARTH RADIATION BUDGET EXPERIMENT

EARTH RESOURCES

Earth Resources Experiment Package
USE   EREP

EARTH RESOURCES INFORMATION SYSTEM

Earth Resources Observation Satellites
USE   EROS (SATELLITES)

EARTH RESOURCES PROGRAM

EARTH RESOURCES SHUTTLE IMAGING RADAR

EARTH RESOURCES SURVEY AIRCRAFT

EARTH RESOURCES SURVEY PROGRAM

Earth Resources Technology Satellite B
USE   LANDSAT 2

Earth Resources Technology Satellite C
USE   LANDSAT 3

Earth Resources Technology Satellite D
USE   LANDSAT 4

Earth Resources Technology Satellite E
USE   LANDSAT E

Earth Resources Technology Satellite F
USE   LANDSAT F

Earth Resources Technology Satellite 1
USE   LANDSAT 1

Earth Resources Technology Satellites
USE   LANDSAT SATELLITES

EARTH ROTATION

Earth), Satellite Power Transmission (To
USE   SATELLITE POWER TRANSMISSION (TO
         EARTH)

Earth Shape
USE   GEODESY

Earth Space Flight, Return To
USE   RETURN TO EARTH SPACE FLIGHT

Earth), Space Observations (From
USE   SPACE OBSERVATIONS (FROM EARTH)

(Earth Structure), Mantle
USE   EARTH MANTLE

EARTH SURFACE

EARTH TERMINAL MEASUREMENT SYSTEM

EARTH TERMINALS

EARTH TIDES

Earth Trajectories, Moon-
USE   MOON-EARTH TRAJECTORIES

EARTH VIEWING APPLICATIONS LABORATORY

EARTH-MARS TRAJECTORIES

EARTH-MERCURY TRAJECTORIES

EARTH-MOON SYSTEM

EARTH-MOON TRAJECTORIES

EARTH-VENUS TRAJECTORIES

EARTHNET

EARTHQUAKE DAMAGE

EARTHQUAKE RESISTANCE

EARTHQUAKE RESISTANT STRUCTURES

EARTHQUAKES

EASEP

EAST GERMANY

EASTERN HEMISPHERE

EATING

EBERT SPECTROMETERS

EBF
USE   EXTERNALLY BLOWN FLAPS

EBR-1 Reactor
USE   EXPERIMENTAL BREEDER REACTOR 1

EBR-2 Reactor
USE   EXPERIMENTAL BREEDER REACTOR 2

Ebullition
USE   BOILING

EBWR (Reactor)
USE   EXPERIMENTAL BOILING WATER
         REACTORS

EC-121 AIRCRAFT

Eccentric Geophysical Observatory
USE   EGO

Eccentric Lunar Occultation Satellite, High
USE   EXOSAT SATELLITE

Eccentric Orbit Geophysical Observatory
USE   EGO

Eccentric Orbit Satellites, Highly
USE   HEOS SATELLITES

ECCENTRIC ORBITS

ECCENTRICITY

ECCENTRICS

ECHELETTE GRATINGS

Echelon Faults
USE   GEOLOGICAL FAULTS

ECHO PROJECT

ECHO SATELLITES

ECHO SOUNDING

ECHO SUPPRESSORS

Echo 1 Carrier Rocket
USE   THOR DELTA LAUNCH VEHICLE

ECHO 1 SATELLITE

ECHO 2 SATELLITE

ECHOCARDIOGRAPHY

ECHOENCEPHALOGRAPHY

ECHOES

Echoes, Auroral
USE   AURORAL ECHOES

Echoes, Lunar
USE   LUNAR ECHOES

Echoes, Lunar Radar
USE   LUNAR RADAR ECHOES

Echoes, Radar
USE   RADAR ECHOES

Echoes, Radio
USE   RADIO ECHOES

Echoes, Solar Radar
USE   SOLAR RADAR ECHOES

Echoes, Venus Radar
USE   VENUS RADAR ECHOES

ECLIPSE PROJECT

ECLIPSES

Eclipses, Lunar
USE   LUNAR ECLIPSES

Eclipses, Solar
USE   SOLAR ECLIPSES

ECLIPSING BINARY STARS

ECLIPTIC

ECLOGITE

Ecol Test Site, Central Atlantic Regional
USE   CENTRAL ATLANTIC REGIONAL ECOL
         TEST SITE

Ecological Systems
USE   ECOLOGY

Ecological Systems, Closed
USE   CLOSED ECOLOGICAL SYSTEMS

Ecological Test Site, Arizona Regional
USE   ARIZONA REGIONAL ECOLOGICAL TEST
         SITE

ECOLOGY

Ecology, Coastal
USE   COASTAL ECOLOGY

ECONOMETRICS

ECONOMIC ANALYSIS

ECONOMIC DEVELOPMENT

ECONOMIC FACTORS

ECONOMIC IMPACT

ECONOMICS

(Economics), Demand
USE   DEMAND (ECONOMICS)

ECONOMY

ECOSYSTEMS

ECS
USE   EUROPEAN COMMUNICATIONS SATELLITE

ECUADOR

Eddies
USE   VORTICES

EDDINGTON APPROXIMATION

EDDY CURRENTS

Eddy Diffusion
USE   TURBULENT DIFFUSION

EDDY VISCOSITY

EDEMA

EDGE DISLOCATIONS

Edge Flaps, Leading
USE   LEADING EDGE FLAPS

Edge Flaps, Trailing
USE   TRAILING EDGE FLAPS

EDGE LOADING

Edge Slats, Leading
USE   LEADING EDGE SLATS

Edge Sweep, Leading
USE   LEADING EDGE SWEEP

Edge Thrust, Leading
USE   LEADING EDGE THRUST

EDGES

Edges, Blunt Leading
USE   BLUNT LEADING EDGES

Edges, Blunt Trailing
USE   BLUNT TRAILING EDGES

Edges, Leading
USE   LEADING EDGES

Edges, Sharp Leading
USE   SHARP LEADING EDGES

Edges, Trailing
USE   TRAILING EDGES

EDITING

EDITING ROUTINES (COMPUTERS)

EDTA
USE   ETHYLENEDIAMINETETRAACETIC ACIDS

EDUCATION

Education Telecommunications Exp, Health-
USE   HET EXPERIMENT

EDUCATIONAL TELEVISION

Edward Island, Prince
USE   PRINCE EDWARD ISLAND

EEG (Electroencephalograms)
USE   ELECTROENCEPHALOGRAPHY

Effect (Aerodynamics), Ground
USE   GROUND EFFECT (AERODYNAMICS)

Effect, Auger
USE   AUGER EFFECT

Effect, Barkhausen
USE   BARKHAUSEN EFFECT

Effect, Bauschinger
USE   BAUSCHINGER EFFECT

Effect, Brillouin
USE   BRILLOUIN EFFECT

Effect, Brown Wave
USE   BROWN WAVE EFFECT

Effect, Capture
USE   CAPTURE EFFECT

Effect, Cerenkov
USE   CERENKOV RADIATION

Effect, Coanda
USE   COANDA EFFECT

Effect (Communications), Ground
USE   GROUND EFFECT (COMMUNICATIONS)

Effect, Compton
USE   COMPTON EFFECT

Effect, Coriolis
USE   CORIOLIS EFFECT

Effect, Diffusion
USE   DIFFUSION

Effect, Dihedral
USE   LATERAL STABILITY

Effect, Doppler
USE   DOPPLER EFFECT

Effect, Doppler-Fizeau
USE   DOPPLER-FIZEAU EFFECT

Effect, Drag
USE   DRAG

Effect (Electricity), Proximity
USE   PROXIMITY EFFECT (ELECTRICITY)

Effect, Electro-Optical
USE   ELECTRO-OPTICAL EFFECT

Effect, Electroseismic
USE   ELECTRIC CURRENT
         SEISMIC WAVES

Effect, Ettingshausen
USE   ETTINGSHAUSEN EFFECT

Effect, Faraday
USE   FARADAY EFFECT

Effect, Fizeau
USE   FIZEAU EFFECT

Effect, Forbush
USE   FORBUSH DECREASES

Effect, Green Wave
USE   GREEN WAVE EFFECT

Effect, Greenhouse
USE   GREENHOUSE EFFECT

Effect, Gunn
USE   GUNN EFFECT

Effect, Hall
USE   HALL EFFECT

Effect, Hydrodynamic RAM
USE   HYDRODYNAMIC RAM EFFECT

Effect, Isotope
USE   ISOTOPE EFFECT

Effect, Jahn-Teller
USE   JAHN-TELLER EFFECT

Effect, Joule-Thomson
USE   JOULE-THOMSON EFFECT

Effect, Kerr Electrooptical
USE   KERR ELECTROOPTICAL EFFECT

Effect, Kerr Magnetooptical
USE   KERR MAGNETOOPTICAL EFFECT

Effect, Kirkendall
USE   KIRKENDALL EFFECT

Effect, Kondo
USE   KONDO EFFECT

Effect, Luxembourg
USE   LUXEMBOURG EFFECT

Effect Machine, Cushioncraft Ground
USE   CUSHIONCRAFT GROUND EFFECT
         MACHINE

Effect Machine, DTMB-111 Ground
USE   GROUND EFFECT MACHINES

Effect Machine, DTMB-430 Ground
USE   GROUND EFFECT MACHINES

Effect Machine, SR-N2 Ground
USE   WESTLAND GROUND EFFECT MACHINES

Effect Machine, SR-N3 Ground
USE   WESTLAND GROUND EFFECT MACHINES

Effect Machine, SR-N5 Ground
USE   WESTLAND GROUND EFFECT MACHINES

Effect Machine, Westland SR-N2 Ground
USE   WESTLAND GROUND EFFECT MACHINES

Effect Machine, Westland SR-N3 Ground
USE   WESTLAND GROUND EFFECT MACHINES

Effect Machine, Westland SR-N5 Ground
USE   WESTLAND GROUND EFFECT MACHINES

Effect Machines, Ground
USE   GROUND EFFECT MACHINES

Effect Machines, HD-1 Ground
USE   HOVERCRAFT GROUND EFFECT
         MACHINES

Effect Machines, Hovercraft Ground
USE   HOVERCRAFT GROUND EFFECT
         MACHINES

Effect Machines, Westland Ground
USE   WESTLAND GROUND EFFECT MACHINES

Effect, Magnus
USE   MAGNUS EFFECT

Effect, Meissner
USE   SUPERCONDUCTIVITY
         DIAMAGNETISM

Effect, Mossbauer
USE   MOSSBAUER EFFECT

Effect, Nernst-Ettingshausen
USE   NERNST-ETTINGSHAUSEN EFFECT

Effect, Nonohmic
USE   NONOHMIC EFFECT

Effect, Nuclear Explosion
USE   NUCLEAR EXPLOSION EFFECT

Effect, Overhauser
USE   OVERHAUSER EFFECT

Effect, Penning
USE   PENNING EFFECT

Effect, Photoelectric
USE   PHOTOELECTRIC EFFECT

Effect, Photomechanical
USE   PHOTOMECHANICAL EFFECT

Effect, Photovoltaic
USE   PHOTOVOLTAIC EFFECT

Effect, Pinch
USE   PINCH EFFECT

Effect, Pockels
USE   BIREFRINGENCE

Effect, Poynting-Robertson
USE   POYNTING-ROBERTSON EFFECT

Effect, Raman
USE   RAMAN SPECTRA

Effect, Ramsauer
USE   RAMSAUER EFFECT

Effect, Sagnac
USE   SAGNAC EFFECT

Effect, Scale
USE   SCALE EFFECT

Effect, Schach
USE   SCHACH EFFECT

Effect, Schottky
USE   WORK FUNCTIONS

Effect, Screen
USE   SCREEN EFFECT

Effect, Seebeck
USE   SEEBECK EFFECT

Effect Ships, Surface
USE   SURFACE EFFECT SHIPS

Effect, Snowplow
USE   PLASMA DYNAMICS

Effect, Stark
USE   STARK EFFECT

Effect, Suhl
USE   SUHL EFFECT

Effect, Sweep
USE   SWEEP EFFECT

Effect, Thomson
USE   THERMOELECTRICITY

Effect Transistors, Field
USE   FIELD EFFECT TRANSISTORS

Effect Transistors, Junction Field
USE   JFET

Effect, Umkehr
USE   UMKEHR EFFECT

Effect, Voigt
USE   VOIGT EFFECT

Effect, Zeeman
USE   ZEEMAN EFFECT

Effect, Zener
USE   ZENER EFFECT

EFFECTIVE PERCEIVED NOISE LEVELS

EFFECTIVENESS

Effectiveness, Cost
USE   COST EFFECTIVENESS

Effectiveness (RBE), Relative Biological
USE   RELATIVE BIOLOGICAL EFFECTIVENESS
(RBE)

Effectiveness, System
USE   SYSTEM EFFECTIVENESS

Effectors
USE   CONTROL EQUIPMENT

EFFECTS

Effects, Atmospheric
USE   ATMOSPHERIC EFFECTS

Effects, Biological
USE   BIOLOGICAL EFFECTS

Effects, Chemical
USE   CHEMICAL EFFECTS

Effects, Compressibility
USE   COMPRESSIBILITY EFFECTS

Effects, Environment
USE   ENVIRONMENT EFFECTS

Effects, Free Stream
USE   FREE FLOW

Effects, Galvanomagnetic
USE   GALVANOMAGNETIC EFFECTS

Effects, Geomagnetic
USE   MAGNETIC EFFECTS

Effects, Gravitational
USE   GRAVITATIONAL EFFECTS

Effects, Heat
USE   TEMPERATURE EFFECTS

Effects, Jet Blast
USE   JET BLAST EFFECTS

Effects, Kerr
USE   KERR EFFECTS

Effects, Long Term
USE   LONG TERM EFFECTS

Effects, Lunar
USE   LUNAR EFFECTS

Effects, Lunar Gravitational
USE   LUNAR GRAVITATIONAL EFFECTS

Effects, Magnetic
USE   MAGNETIC EFFECTS

Effects, Many Electron
USE   MANY ELECTRON EFFECTS

Effects, Moire
USE   MOIRE EFFECTS

Effects, Pathological
USE   PATHOLOGICAL EFFECTS

Effects, Peltier
USE   PELTIER EFFECTS

Effects, Photoelectromagnetic
USE   PHOTOELECTROMAGNETIC EFFECTS

Effects, Photomagnetic
USE   PHOTOMAGNETIC EFFECTS

Effects, Physiological
USE   PHYSIOLOGICAL EFFECTS

Effects, Pogo
USE   POGO EFFECTS

Effects, Pressure
USE   PRESSURE EFFECTS

Effects, Psychological
USE   PSYCHOLOGICAL EFFECTS

Effects, Radiation
USE   RADIATION EFFECTS

Effects, Reentry
USE   REENTRY EFFECTS

Effects, Relativistic
USE   RELATIVISTIC EFFECTS

Effects, Solar Activity
USE   SOLAR ACTIVITY EFFECTS

Effects, Sterilization
USE   STERILIZATION EFFECTS

Effects, Surface Roughness
USE   SURFACE ROUGHNESS EFFECTS

Effects, Temperature
USE   TEMPERATURE EFFECTS

Effects, Thermal
USE   TEMPERATURE EFFECTS

Effects, Thermomagnetic
USE   THERMOMAGNETIC EFFECTS

Effects, Turbulence
USE   TURBULENCE EFFECTS

Effects, Vacuum
USE   VACUUM EFFECTS

Effects, Vibration
USE   VIBRATION EFFECTS

Effects, View
USE   VIEW EFFECTS

Effects, Wind
USE   WIND EFFECTS

EFFERENT NERVOUS SYSTEMS

EFFERVESCENCE

EFFICIENCY

Efficiency, Charge
USE   CHARGE EFFICIENCY

Efficiency, Combustion
USE   COMBUSTION EFFICIENCY

Efficiency, Compressor
USE   COMPRESSOR EFFICIENCY

Efficiency, Energy Conversion
USE   ENERGY CONVERSION EFFICIENCY

Efficiency, Nozzle
USE   NOZZLE EFFICIENCY

Efficiency, Power
USE   POWER EFFICIENCY

Efficiency Program, Aircraft Energy
USE   ACEE PROGRAM

Efficiency, Propeller
USE   PROPELLER EFFICIENCY

Efficiency, Propulsive
USE   PROPULSIVE EFFICIENCY

Efficiency, Quantum
USE   QUANTUM EFFICIENCY

Efficiency, Thermal
USE   THERMODYNAMIC EFFICIENCY

Efficiency, Thermodynamic
USE   THERMODYNAMIC EFFICIENCY

Efficiency, Transmission
USE   TRANSMISSION EFFICIENCY

Efficiency Transport Program, Energy
USE   ACEE PROGRAM

Efficiency, Volumetric
USE   VOLUMETRIC EFFICIENCY

EFFLUENTS

EFFLUX

EFFORT

EFFUSIVES

EGCR (Reactor)
USE   EXPERIMENTAL GAS COOLED REACTORS

EGGS

EGO

EGRESS

EGYPT

Eigenfunctions
USE   EIGENVECTORS

Eigenstates
USE   EIGENVECTORS

EIGENVALUES

EIGENVECTORS

EIKONAL EQUATION

EINSTEIN EQUATIONS

Einstein Observatory
USE   HEAO 2

Einstein Statistics, Bose-
USE   QUANTUM STATISTICS

EINSTEINIUM

EISCAT RADAR SYSTEM (EUROPE)

EJECTA

EJECTION

EJECTION INJURIES

EJECTION SEATS

Ejection Seats, Flying
USE   FLYING EJECTION SEATS

Ejection, Stellar Mass
USE   STELLAR MASS EJECTION

EJECTION TRAINING

EJECTORS

EKMAN LAYER

EL NINO

EL SALVADOR

ELASTIC ANISOTROPY

ELASTIC BARS

ELASTIC BENDING

ELASTIC BODIES

ELASTIC BUCKLING

Elastic Collisions
USE   ELASTIC SCATTERING

Elastic Constants
USE   ELASTIC PROPERTIES

ELASTIC CYLINDERS

ELASTIC DAMPING

ELASTIC DEFORMATION

ELASTIC MEDIA

Elastic Modulus
USE   MODULUS OF ELASTICITY

ELASTIC PLATES

ELASTIC PROPERTIES

ELASTIC SCATTERING

ELASTIC SHEETS

ELASTIC SHELLS

(Elastic), Springs
USE   SPRINGS (ELASTIC)

Elastic Stability
USE   DAMPING

Elastic Strength
USE   PROPORTIONAL LIMIT

ELASTIC SYSTEMS

ELASTIC WAVES

Elastic Waves, Polarized
USE   POLARIZED ELASTIC WAVES

Elasticity
USE   ELASTIC PROPERTIES

Elasticity, Aero
USE   AEROELASTICITY

Elasticity, Aerothermo
USE   AEROTHERMOELASTICITY

Elasticity, An
USE   ANELASTICITY

(Elasticity), Compliance
USE   MODULUS OF ELASTICITY

Elasticity, Dynamic Modulus Of
USE   DYNAMIC MODULUS OF ELASTICITY

Elasticity, Hydro
USE   HYDROELASTICITY

Elasticity, Hypo
USE   HYPOELASTICITY

Elasticity, Modulus Of
USE   MODULUS OF ELASTICITY

Elasticity, Photo
USE   PHOTOELASTICITY

Elasticity, Photovisco
USE   PHOTOVISCOELASTICITY

Elasticity, Thermo
USE   THERMOELASTICITY

Elasticity, Thermovisco
USE   THERMOVISCOELASTICITY

Elasticity, Visco
USE   VISCOELASTICITY

Elasticizers
USE   PLASTICIZERS

ELASTIN

ELASTODYNAMICS

ELASTOHYDRODYNAMICS

ELASTOMERS

Elastomers, Vulcanized
USE   VULCANIZED ELASTOMERS

ELASTOMETERS

ELASTOPLASTICITY

ELASTOSTATICS

ELBER EQUATION

ELBOW (ANATOMY)

ELDO LAUNCH VEHICLE

ELECTRA AIRCRAFT

ELECTRETS

Electric Aircraft
USE   FLY BY WIRE CONTROL

Electric Appliances
USE   ELECTRIC EQUIPMENT

ELECTRIC ARCS

ELECTRIC AUTOMOBILES

ELECTRIC BATTERIES

(Electric), Breakers
USE   CIRCUIT BREAKERS

ELECTRIC BRIDGES

Electric Canberra Aircraft, English
USE   CANBERRA AIRCRAFT

ELECTRIC CELLS

Electric Cells, Fission
USE   FISSION ELECTRIC CELLS

ELECTRIC CHARGE

ELECTRIC CHOPPERS

(Electric), Choppers
USE   ELECTRIC CHOPPERS

Electric Circuits
USE   CIRCUITS

ELECTRIC COILS

Electric Computers, General
USE   GE COMPUTERS

ELECTRIC CONDUCTORS

ELECTRIC CONNECTORS

(Electric), Connectors
USE   ELECTRIC CONNECTORS

(Electric), Contacts
USE   ELECTRIC CONTACTS

ELECTRIC CONTACTS

ELECTRIC CONTROL

ELECTRIC CORONA

ELECTRIC CURRENT

ELECTRIC DIPOLES

ELECTRIC DISCHARGES

ELECTRIC ENERGY STORAGE

ELECTRIC EQUIPMENT

ELECTRIC EQUIPMENT TESTS

ELECTRIC FIELD STRENGTH

ELECTRIC FIELDS

ELECTRIC FILTERS

ELECTRIC FURNACES

ELECTRIC FUSES

ELECTRIC GENERATORS

ELECTRIC HYBRID VEHICLES

ELECTRIC IGNITION

Electric Impulses
USE   ELECTRIC PULSES

ELECTRIC MOMENTS

ELECTRIC MOTOR VEHICLES

ELECTRIC MOTORS

ELECTRIC NETWORKS

ELECTRIC OUTLETS

ELECTRIC POTENTIAL

ELECTRIC POWER

Electric Power Conversion
USE   ELECTRIC GENERATORS

Electric Power Generation, Nuclear
USE   NUCLEAR ELECTRIC POWER GENERATION

ELECTRIC POWER PLANTS

Electric Power Plants, Solar Thermal
USE   SOLAR THERMAL ELECTRIC POWER
PLANTS

ELECTRIC POWER SUPPLIES

ELECTRIC POWER TRANSMISSION

ELECTRIC PROPULSION

Electric Propulsion, Nuclear
USE   NUCLEAR ELECTRIC PROPULSION

Electric Propulsion, Solar
USE   SOLAR ELECTRIC PROPULSION

ELECTRIC PULSES

ELECTRIC REACTORS

ELECTRIC RELAYS

ELECTRIC ROCKET ENGINES

Electric Rocket Tests, Space
USE   SPACE ELECTRIC ROCKET TESTS

Electric Spacecraft, Advanced Reconn
USE   ADVANCED RECONN ELECTRIC
SPACECRAFT

ELECTRIC SPARKS

ELECTRIC STIMULI

ELECTRIC SWITCHES

ELECTRIC TERMINALS

ELECTRIC WELDING

ELECTRIC WIRE

Electric Wiring
USE   ELECTRIC WIRE
WIRING

Electrical Breakdown
USE   ELECTRICAL FAULTS

Electrical Conductivity
USE   ELECTRICAL RESISTIVITY

ELECTRICAL CONDUCTIVITY METERS

(Electrical Contacts), Brushes
USE   BRUSHES (ELECTRICAL CONTACTS)

Electrical Energy
USE   ELECTRIC POWER

ELECTRICAL ENGINEERING

ELECTRICAL FAULTS

ELECTRICAL GROUNDING

ELECTRICAL IMPEDANCE

ELECTRICAL INSULATION

(Electrical), Jacks
USE   ELECTRIC CONNECTORS

Electrical Leads
USE   ELECTRIC CONDUCTORS

Electrical Machines, Rotating
USE   ROTATING ELECTRICAL MACHINES

ELECTRICAL MEASUREMENT

(Electrical), Mismatch
USE   MISMATCH (ELECTRICAL)

ELECTRICAL PROPERTIES

ELECTRICAL RESISTANCE

ELECTRICAL RESISTIVITY

Electrically Suspended Gyroscopes
USE   ELECTROSTATIC GYROSCOPES

ELECTRICITY

Electricity, Antiferro
USE   ANTIFERROELECTRICITY

Electricity, Atmospheric
USE   ATMOSPHERIC ELECTRICITY

Electricity, Bio
USE   BIOELECTRICITY

Electricity, Ferro
USE   FERROELECTRICITY

Electricity, Geo
USE   GEOELECTRICITY

Electricity, Myo
USE   MYOELECTRICITY

Electricity, Photo
USE   PHOTOELECTRICITY

Electricity, Piezo
USE   PIEZOELECTRICITY

(Electricity), Proximity Effect
USE   PROXIMITY EFFECT (ELECTRICITY)

Electricity, Pyro
USE   PYROELECTRICITY

Electricity, Static
USE   STATIC ELECTRICITY

Electricity, Thermo
USE   THERMOELECTRICITY

ELECTRIFICATION

ELECTRO-OPTICAL EFFECT

ELECTRO-OPTICAL PHOTOGRAPHY

ELECTRO-OPTICS

ELECTROACOUSTIC TRANSDUCERS

ELECTROACOUSTIC WAVES

ELECTROANESTHESIA

Electrocardiograms
USE   ELECTROCARDIOGRAPHY

ELECTROCARDIOGRAPHY

ELECTROCATALYSTS

ELECTROCHEMICAL CELLS

ELECTROCHEMICAL CORROSION

ELECTROCHEMICAL MACHINING

ELECTROCHEMICAL OXIDATION

ELECTROCHEMISTRY

Electrochemistry, Photo
USE   PHOTOELECTROCHEMISTRY

ELECTROCHROMISM

Electroconductivity
USE   ELECTRICAL RESISTIVITY

ELECTROCUTANEOUS COMMUNICATION

ELECTRODE FILM BARRIERS

ELECTRODE MATERIALS

ELECTRODELESS DISCHARGES

ELECTRODEPOSITION

Electrodermal Response
USE   GALVANIC SKIN RESPONSE

ELECTRODES

Electrodes (Biology), Implanted
USE   IMPLANTED ELECTRODES (BIOLOGY)

Electrodes, Diffusion
USE   DIFFUSION ELECTRODES

Electrodes, Glass
USE   GLASS ELECTRODES

Electrodes, Ion Selective
USE   ION SELECTIVE ELECTRODES

Electrodes, Plasma
USE   PLASMA ELECTRODES

Electrodes, Solid
USE   SOLID ELECTRODES

ELECTRODIALYSIS

ELECTRODISSOLUTION

ELECTRODYNAMICS

Electrodynamics, Quantum
USE   QUANTUM ELECTRODYNAMICS

Electrodynamometers
USE   DYNAMOMETERS

Electroencephalogram
USE   ELECTROENCEPHALOGRAPHY

(Electroencephalograms), EEG
USE   ELECTROENCEPHALOGRAPHY

ELECTROENCEPHALOGRAPHY

ELECTROEPITAXY

Electroerosion
USE   SPARK MACHINING

Electroexplosive Devices
USE   INITIATORS (EXPLOSIVES)

ELECTROFORMING

Electrogenerators
USE   ELECTRIC GENERATORS

Electrohydraulic Control
USE   ELECTRIC CONTROL
HYDRAULIC CONTROL

ELECTROHYDRAULIC FORMING

ELECTROHYDRODYNAMICS

Electrojet, Equatorial
USE   EQUATORIAL ELECTROJET

ELECTROJETS

Electrojets, Auroral
USE   AURORAL ELECTROJETS

ELECTROKINETICS

ELECTROLESS DEPOSITION

ELECTROLUMINESCENCE

Electroluminescent Lamps
USE   ELECTROLUMINESCENCE
      LUMINAIRES

ELECTROLYSIS

ELECTROLYTE METABOLISM

ELECTROLYTES

Electrolytes, Ion Exchange Membrane
USE   ION EXCHANGE MEMBRANE
      ELECTROLYTES

Electrolytes, Molten Salt
USE   MOLTEN SALT ELECTROLYTES

Electrolytes, Non
USE   NONELECTROLYTES

Electrolytes, Nonaqueous
USE   NONAQUEOUS ELECTROLYTES

Electrolytes, Solid
USE   SOLID ELECTROLYTES

ELECTROLYTIC CELLS

Electrolytic Grinding
USE   ELECTROCHEMICAL MACHINING

ELECTROLYTIC POLARIZATION

Electrolytic Polishing
USE   ELECTROPOLISHING

ELECTROMAGNETIC ABSORPTION

ELECTROMAGNETIC ACCELERATION

ELECTROMAGNETIC COMPATIBILITY

Electromagnetic Control
USE   ELECTROMAGNETS
      REMOTE CONTROL

Electromagnetic Deduction
USE   MAGNETIC INDUCTION

ELECTROMAGNETIC ENVIRONMENT EXPERIMENT

ELECTROMAGNETIC FIELDS

ELECTROMAGNETIC HAMMERS

Electromagnetic Interaction, Plasma-
USE   PLASMA-ELECTROMAGNETIC INTERACTION

ELECTROMAGNETIC INTERACTIONS

ELECTROMAGNETIC INTERFERENCE

ELECTROMAGNETIC MEASUREMENT

ELECTROMAGNETIC NOISE

ELECTROMAGNETIC NOISE MEASUREMENT

(Electromagnetic), Power Density
USE   RADIANT FLUX DENSITY

Electromagnetic Propagation
USE   ELECTROMAGNETIC WAVE TRANSMISSION

ELECTROMAGNETIC PROPERTIES

ELECTROMAGNETIC PROPULSION

ELECTROMAGNETIC PULSES

Electromagnetic Pulses, System Generated
USE   SYSTEM GENERATED ELECTROMAGNETIC
      PULSES

ELECTROMAGNETIC PUMPS

ELECTROMAGNETIC RADIATION

Electromagnetic Radiation, Coherent
USE   COHERENT ELECTROMAGNETIC
      RADIATION

Electromagnetic Radiation, Polarized
USE   POLARIZED ELECTROMAGNETIC
      RADIATION

ELECTROMAGNETIC SCATTERING

ELECTROMAGNETIC SHIELDING

ELECTROMAGNETIC SPECTRA

ELECTROMAGNETIC SURFACE WAVES

ELECTROMAGNETIC WAVE FILTERS

ELECTROMAGNETIC WAVE TRANSMISSION

Electromagnetic Waves
USE   ELECTROMAGNETIC RADIATION

Electromagnetics
USE   ELECTROMAGNETISM

ELECTROMAGNETISM

ELECTROMAGNETS

ELECTROMECHANICAL DEVICES

ELECTROMECHANICS

ELECTROMETERS

ELECTROMIGRATION

ELECTROMOTIVE FORCES

Electromyograms
USE   ELECTROMYOGRAPHY

Electromyographs
USE   ELECTROMYOGRAPHY

ELECTROMYOGRAPHY

ELECTRON ACCELERATION

ELECTRON ACCELERATORS

ELECTRON ATTACHMENT

ELECTRON AVALANCHE

ELECTRON BEAM WELDING

ELECTRON BEAMS

Electron Beams, Relativistic
USE   RELATIVISTIC ELECTRON BEAMS

ELECTRON BOMBARDMENT

ELECTRON BUNCHING

ELECTRON CAPTURE

ELECTRON CLOUDS

Electron Collisions
USE   ELECTRON SCATTERING

Electron Compounds
USE   INTERMETALLICS

ELECTRON COUNTERS

ELECTRON CYCLOTRON HEATING

ELECTRON DECAY RATE

(Electron Deficiencies), Holes
USE   HOLES (ELECTRON DEFICIENCIES)

ELECTRON DENSITY (CONCENTRATION)

Electron Density, Ionospheric
USE   IONOSPHERIC ELECTRON DENSITY

Electron Density, Magnetospheric
USE   MAGNETOSPHERIC ELECTRON DENSITY

ELECTRON DENSITY PROFILES

Electron Detectors
USE   ELECTRON COUNTERS

Electron Devices, Transferred
USE   TRANSFERRED ELECTRON DEVICES

ELECTRON DIFFRACTION

ELECTRON DIFFUSION

ELECTRON DISTRIBUTION

Electron Effects, Many
USE   MANY ELECTRON EFFECTS

ELECTRON EMISSION

ELECTRON ENERGY

Electron Flux
USE   ELECTRONS
      FLUX (RATE)

ELECTRON FLUX DENSITY

ELECTRON GAS

ELECTRON GUNS

ELECTRON IMPACT

Electron Intensity
USE   ELECTRON FLUX DENSITY

Electron Interaction, Photon-
USE   PHOTON-ELECTRON INTERACTION

Electron Interactions
USE   ELECTRON SCATTERING

Electron Ionization
USE   IONIZATION

ELECTRON IRRADIATION

Electron Lasers, Free
USE   FREE ELECTRON LASERS

ELECTRON MASS

ELECTRON MICROSCOPES

ELECTRON MICROSCOPY

ELECTRON MOBILITY

Electron Mobility Transistors, High
USE   HIGH ELECTRON MOBILITY TRANSISTORS

Electron Multipliers
USE   PHOTOMULTIPLIER TUBES

ELECTRON OPTICS

ELECTRON ORBITALS

ELECTRON OSCILLATIONS

ELECTRON PARAMAGNETIC RESONANCE

Electron Paths
    USE   ELECTRON TRAJECTORIES

ELECTRON PHONON INTERACTIONS

ELECTRON PHOTOGRAPHY

ELECTRON PHOTON CASCADES

ELECTRON PLASMA

ELECTRON PRECIPITATION

ELECTRON PRESSURE

ELECTRON PROBES

ELECTRON PUMPING

ELECTRON RADIATION

ELECTRON RECOMBINATION

Electron Ring Accelerators
    USE   STORAGE RINGS (PARTICLE
               ACCELERATORS)

ELECTRON RUNAWAY (PLASMA PHYSICS)

ELECTRON SCATTERING

ELECTRON SOURCES

ELECTRON SPECTROSCOPY

ELECTRON SPIN

Electron Spin Resonance
    USE   ELECTRON PARAMAGNETIC RESONANCE

ELECTRON STATES

Electron Sweeping
    USE   SWEEP FREQUENCY

Electron Telescopes
    USE   PARTICLE TELESCOPES

Electron Temperature
    USE   ELECTRON ENERGY

ELECTRON TRAJECTORIES

ELECTRON TRANSFER

ELECTRON TRANSITIONS

ELECTRON TUBES

ELECTRON TUNNELING

ELECTRON-HOLE DROPS

ELECTRON-ION RECOMBINATION

ELECTRONARCOSIS

ELECTRONIC AIRCRAFT

Electronic Amplifiers
    USE   AMPLIFIERS

ELECTRONIC CONTROL

ELECTRONIC COUNTERMEASURES

Electronic Devices, Microminiaturized
    USE   MICROMINIATURIZED ELECTRONIC
               DEVICES

ELECTRONIC EQUIPMENT

Electronic Equipment, Miniature
    USE   MINIATURE ELECTRONIC EQUIPMENT

Electronic Equipment, Spacecraft
    USE   SPACECRAFT ELECTRONIC EQUIPMENT

ELECTRONIC EQUIPMENT TESTS

ELECTRONIC FILTERS

Electronic Levels
    USE   ELECTRON ENERGY
               ENERGY LEVELS

ELECTRONIC MAIL

Electronic Management System, Central
    USE   CENTRAL ELECTRONIC MANAGEMENT
               SYSTEM

ELECTRONIC MODULES

ELECTRONIC PACKAGING

Electronic Photography
    USE   ELECTRO-OPTICAL PHOTOGRAPHY

ELECTRONIC RECORDING SYSTEMS

Electronic Signal Measurement
    USE   SIGNAL MEASUREMENT

ELECTRONIC SPECTRA

Electronic Structure
    USE   ATOMIC STRUCTURE

Electronic Switches
    USE   SWITCHING CIRCUITS

ELECTRONIC TRANSDUCERS

ELECTRONIC WARFARE

ELECTRONICS

(Electronics), Chips
    USE   CHIPS (ELECTRONICS)

(Electronics), Doghouses
    USE   DOGHOUSES (ELECTRONICS)

(Electronics), Hole Distribution
    USE   HOLE DISTRIBUTION (ELECTRONICS)

(Electronics), Look Angles
    USE   LOOK ANGLES (ELECTRONICS)

Electronics, Medical
    USE   MEDICAL ELECTRONICS

Electronics, Micro
    USE   MICROELECTRONICS

Electronics, Molecular
    USE   MOLECULAR ELECTRONICS

Electronics, Quantum
    USE   QUANTUM ELECTRONICS

Electronics, Radio
    USE   RADIO ELECTRONICS

ELECTRONOGRAPHY

ELECTRONS

Electrons, Conduction
    USE   CONDUCTION ELECTRONS

Electrons, Free
    USE   FREE ELECTRONS

Electrons, High Energy
    USE   HIGH ENERGY ELECTRONS

Electrons, Hot
    USE   HOT ELECTRONS

Electrons, N
    USE   N ELECTRONS

Electrons, Nonrelativistic
    USE   ELECTRONS

Electrons, Photo
    USE   PHOTOELECTRONS

Electrons, Pi-
    USE   PI-ELECTRONS

Electrons, Solar
    USE   SOLAR ELECTRONS

ELECTRONYSTAGMOGRAPHY

Electrooptical Effect, Kerr
    USE   KERR ELECTROOPTICAL EFFECT

ELECTROPHORESIS

ELECTROPHOTOMETERS

ELECTROPHOTOMETRY

ELECTROPHYSICS

ELECTROPHYSIOLOGY

ELECTROPLATING

ELECTROPLETHYSMOGRAPHY

ELECTROPOLISHING

ELECTROREFINING

ELECTRORETINOGRAPHY

Electroseismic Effect
    USE   ELECTRIC CURRENT
               SEISMIC WAVES

ELECTROSLAG PROCESS

ELECTROSLAG REFINING

ELECTROSLAG WELDING

ELECTROSTATIC BONDING

ELECTROSTATIC CHARGE

ELECTROSTATIC DRAG

ELECTROSTATIC ENGINES

Electrostatic Erosion
    USE   SPARK MACHINING

Electrostatic Fields
    USE   ELECTRIC FIELDS

ELECTROSTATIC GENERATORS

ELECTROSTATIC GYROSCOPES

Electrostatic Plasma
    USE   PLASMAS (PHYSICS)

ELECTROSTATIC PRECIPITATORS

ELECTROSTATIC PROBES

ELECTROSTATIC PROPULSION

ELECTROSTATIC SHIELDING

ELECTROSTATIC WAVES

ELECTROSTATICS

ELECTROSTRICTION

ELECTROTHERMAL ENGINES

ELECTROWINNING

ELEKTRON SATELLITES

ELEKTRON 1 SATELLITE

ELEKTRON 2 SATELLITE

ELEKTRON 4 SATELLITE

Element Abundance
    USE   ABUNDANCE

Element Method, Boundary
USE   BOUNDARY ELEMENT METHOD

Element Method, Finite
USE   FINITE ELEMENT METHOD

ELEMENT 104

ELEMENT 105

ELEMENTARY EXCITATIONS

ELEMENTARY PARTICLE INTERACTIONS

ELEMENTARY PARTICLES

ELEMENTS

Elements, Chemical
USE   CHEMICAL ELEMENTS

Elements, Decision
USE   LOGICAL ELEMENTS

Elements), Directors (Antenna
USE   DIRECTORS (ANTENNA ELEMENTS)

Elements, Fluid Switching
USE   FLUID SWITCHING ELEMENTS

Elements, Heavy
USE   HEAVY ELEMENTS

Elements, Isoparametric Finite
USE   ISOPARAMETRIC FINITE ELEMENTS

Elements, Light
USE   LIGHT ELEMENTS

Elements, Logical
USE   LOGICAL ELEMENTS

Elements, Nuclear Fuel
USE   NUCLEAR FUEL ELEMENTS

Elements (Nuclear Reactors), Fuel
USE   NUCLEAR FUEL ELEMENTS

Elements, Orbital
USE   ORBITAL ELEMENTS

Elements, Radioactive
USE   RADIOACTIVE ISOTOPES

Elements, Rare Earth
USE   RARE EARTH ELEMENTS

Elements), Shafts (Machine
USE   SHAFTS (MACHINE ELEMENTS)

Elements, Switching
USE   SWITCHING CIRCUITS

Elements, Trace
USE   TRACE ELEMENTS

Elements), Transmissions (Machine
USE   TRANSMISSIONS (MACHINE ELEMENTS)

Elements, Transuranium
USE   TRANSURANIUM ELEMENTS

ELEVATION

ELEVATION ANGLE

(Elevation), Datum
USE   DATUM (ELEVATION)

Elevations (Drawings)
USE   DRAWINGS

ELEVATOR ILLUSION

ELEVATORS (CONTROL SURFACES)

ELEVATORS (LIFTS)

ELEVONS

ELIMINATION

Elimination, Noise
USE   NOISE REDUCTION

ELLIPSES

Ellipsoid, Izsak
USE   ELLIPSOIDS
       GEODESY

ELLIPSOIDS

ELLIPSOMETERS

ELLIPTIC DIFFERENTIAL EQUATIONS

ELLIPTIC FUNCTIONS

Elliptic Integrals
USE   ELLIPTIC FUNCTIONS

ELLIPTICAL CYLINDERS

ELLIPTICAL GALAXIES

ELLIPTICAL ORBITS

ELLIPTICAL PLASMAS

ELLIPTICAL POLARIZATION

ELLIPTICITY

Elmo Fire, Saint
USE   SAINT ELMO FIRE

ELONGATION

ELUTION

Elutriation
USE   ELUTION

Emanation
USE   EMISSION

EMBEDDED COMPUTER SYSTEMS

EMBEDDING

Embolism, Aero
USE   AEROEMBOLISM

EMBOLISMS

Embolisms, Fat
USE   FAT EMBOLISMS

EMBOSSING

EMBRITTLEMENT

Embrittlement, Hydrogen
USE   HYDROGEN EMBRITTLEMENT

EMBRYOLOGY

EMBRYOS

Emerald
USE   BERYL

EMERGENCIES

EMERGENCY BREATHING TECHNIQUES

EMERGENCY LIFE SUSTAINING SYSTEMS

EMERGENCY LOCATOR TRANSMITTERS

EMERGING

Emirates, United Arab
USE   UNITED ARAB EMIRATES

EMISSION

Emission, Acoustic
USE   ACOUSTIC EMISSION

Emission, Atmospheric
USE   AIRGLOW

Emission, Cn
USE   CN EMISSION

Emission, Cyanide
USE   CN EMISSION

Emission Devices, Stimulated
USE   STIMULATED EMISSION DEVICES

Emission, Electron
USE   ELECTRON EMISSION

Emission, Exhaust
USE   EXHAUST EMISSION

Emission, Field
USE   FIELD EMISSION

Emission, Fluorescent
USE   FLUORESCENCE

Emission, Hydroxyl
USE   HYDROXYL EMISSION

Emission, Ion
USE   ION EMISSION

Emission, Light
USE   LIGHT EMISSION

Emission, Microwave
USE   MICROWAVE EMISSION

Emission, Neutron
USE   NEUTRON EMISSION

Emission, Optical
USE   LIGHT EMISSION

Emission, Particle
USE   PARTICLE EMISSION

Emission, Photoelectric
USE   PHOTOELECTRIC EMISSION

Emission, Radiation
USE   RADIATION

Emission, Radio
USE   RADIO EMISSION

Emission Recorders, VLF
USE   VLF EMISSION RECORDERS

Emission, Secondary
USE   SECONDARY EMISSION

Emission, Self Sustained
USE   SELF SUSTAINED EMISSION

Emission, Solar Radio
USE   SOLAR RADIO EMISSION

EMISSION SPECTRA

Emission, Spectral
USE   SPECTRAL EMISSION

Emission Spectroscopy, Optical
USE   OPTICAL EMISSION SPECTROSCOPY

Emission, Spontaneous
USE   SPONTANEOUS EMISSION

Emission, Stimulated
USE   STIMULATED EMISSION

Emission, Thermal
USE   THERMAL EMISSION

Emission, Thermionic
USE   THERMIONIC EMISSION

Emissions, Geocoronal
USE   GEOCORONAL EMISSIONS

EMISSIVITY

Emissographs
USE   ACTINOMETERS
           RECORDING INSTRUMENTS

EMITTANCE

EMITTERS

Emitters, Thermionic
USE   THERMIONIC EMITTERS

Emitting Diodes, Light
USE   LIGHT EMITTING DIODES

EMOTIONAL FACTORS

EMOTIONS

Empennage
USE   TAIL ASSEMBLIES

EMPHYSEMA

EMPLOYEE RELATIONS

EMPLOYMENT

EMPTYING

EMR 6050 COMPUTER

EMULSIONS

Emulsions, Nuclear
USE   NUCLEAR EMULSIONS

Emulsions, Photographic
USE   PHOTOGRAPHIC EMULSIONS

En Route ATC, Automated
USE   AUTOMATED EN ROUTE ATC

ENAMELS

ENARGITE

ENCAPSULATED MICROCIRCUITS

ENCAPSULATING

ENCELADUS

ENCEPHALITIS

Encephalography, Echo
USE   ECOSYSTEMS

Encephalography, Electro
USE   ELECTROENCEPHALOGRAPHY

Encephalography, Rheo
USE   RHEOENCEPHALOGRAPHY

ENCKE COMET

ENCKE METHOD

ENCLOSURE

ENCLOSURES

Encoders
USE   CODERS

Encoding
USE   CODING

Encoding, Redundancy
USE   REDUNDANCY ENCODING

Encoding, Signal
USE   SIGNAL ENCODING

ENCOUNTERS

End Data System, NASA End-To-
USE   NEEDS (DATA SYSTEM)

End Data Systems, End-To-
USE   END-TO-END DATA SYSTEMS

End Moraines
USE   GLACIAL DRIFT

END PLATES

End-To-End Data System, NASA
USE   NEEDS (DATA SYSTEM)

END-TO-END DATA SYSTEMS

ENDANGERED SPECIES

ENDFIRE ARRAYS

ENDOCRINE GLANDS

ENDOCRINE SECRETIONS

ENDOCRINE SYSTEMS

ENDOCRINOLOGY

ENDOLYMPH

ENDORADIOSONDES

ENDOSCOPES

ENDOTHELIUM

ENDOTHERMIC FUELS

ENDOTHERMIC REACTIONS

ENDOTOXINS

ENDRIN

ENDURANCE

Endurance, Physical
USE   PHYSICAL FITNESS

ENEMY PERSONNEL

Energetic Particle Explorer A
USE   EXPLORER 12 SATELLITE

Energetic Particle Explorer B
USE   EXPLORER 14 SATELLITE

Energetic Particle Explorer C
USE   EXPLORER 15 SATELLITE

Energetic Particle Explorer D
USE   EXPLORER 26 SATELLITE

ENERGETIC PARTICLES

ENERGY

Energy Absorbers, Solar
USE   SOLAR ENERGY ABSORBERS

ENERGY ABSORPTION

ENERGY ABSORPTION FILMS

(Energy Absorption), Moderation
USE   MODERATION (ENERGY ABSORPTION)

(Energy Absorption), Thermalization
USE   THERMALIZATION (ENERGY ABSORPTION)

Energy, Activation
USE   ACTIVATION ENERGY

Energy Astronomy Observatories, High
USE   HEAO

Energy Astronomy Observatory A, High
USE   HEAO 1

Energy Astronomy Observatory B, High
USE   HEAO 2

Energy Astronomy Observatory C, High
USE   HEAO 3

Energy Astronomy Observatory 1, High
USE   HEAO 1

Energy Astronomy Observatory 2, High
USE   HEAO 2

Energy Astronomy Observatory 3, High
USE   HEAO 3

Energy, Atomic
USE   NUCLEAR ENERGY

ENERGY BANDS

Energy Budget Experiment, Earth
USE   LZEEBE SATELLITE

Energy Budget Experiment, Zonal Earth
USE   LZEEBE SATELLITE

ENERGY BUDGETS

Energy, Chemical
USE   CHEMICAL ENERGY

Energy, Clean
USE   CLEAN ENERGY

Energy, Commercial
USE   COMMERCIAL ENERGY

ENERGY CONSERVATION

ENERGY CONSUMPTION

ENERGY CONVERSION

ENERGY CONVERSION EFFICIENCY

Energy Conversion, Geothermal
USE   GEOTHERMAL ENERGY CONVERSION

Energy Conversion, Ocean Thermal
USE   OCEAN THERMAL ENERGY CONVERSION

Energy Conversion, Satellite Solar
USE   SATELLITE SOLAR ENERGY CONVERSION

Energy Conversion, Solar
USE   SOLAR ENERGY CONVERSION

Energy Conversion, Waterwave
USE   WATERWAVE ENERGY CONVERSION

Energy Converters
USE   DIRECT POWER GENERATORS

Energy Density
USE   FLUX DENSITY

ENERGY DISSIPATION

ENERGY DISTRIBUTION

Energy Distribution, Spectral
USE   SPECTRAL ENERGY DISTRIBUTION

Energy, Domestic
USE   DOMESTIC ENERGY

Energy Efficiency Program, Aircraft
USE   ACEE PROGRAM

Energy Efficiency Transport Program
USE   ACEE PROGRAM

Energy, Electrical
USE   ELECTRIC POWER

Energy, Electron
USE   ELECTRON ENERGY

Energy Electrons, High
USE   HIGH ENERGY ELECTRONS

Energy Equipartition
USE   EQUIPARTITION THEOREM

Energy Exchange
USE   ENERGY TRANSFER

Energy Experiment, Long Term Zonal Earth
USE   LZEEBE SATELLITE

Energy Extraction, Geothermal
USE   GEOTHERMAL ENERGY EXTRACTION

Energy, Free
USE   FREE ENERGY

Energy Fuels), HEF (High
USE   HIGH ENERGY FUELS

Energy Fuels, High
USE   HIGH ENERGY FUELS

ENERGY GAPS (SOLID STATE)

Energy, Gibbs Free
USE   GIBBS FREE ENERGY

Energy, Hydrogen-Based
USE   HYDROGEN-BASED ENERGY

Energy, Industrial
USE   INDUSTRIAL ENERGY

Energy Interactions, High
USE   HIGH ENERGY INTERACTIONS

Energy Interactions, Weak
USE   WEAK ENERGY INTERACTIONS

Energy, Interfacial
USE   INTERFACIAL ENERGY

Energy, Internal
USE   INTERNAL ENERGY

Energy, Kinetic
USE   KINETIC ENERGY

ENERGY LEVELS

Energy Levels, Atomic
USE   ATOMIC ENERGY LEVELS

Energy Levels, Molecular
USE   MOLECULAR ENERGY LEVELS

Energy Losses
USE   ENERGY DISSIPATION

Energy Management, Terminal Area
USE   TERMINAL AREA ENERGY MANAGEMENT

ENERGY METHODS

Energy Methods, Strain
USE   STRAIN ENERGY METHODS

Energy, Momentum
USE   KINETIC ENERGY

Energy, Nuclear
USE   NUCLEAR ENERGY

Energy, Nuclear Binding
USE   NUCLEAR BINDING ENERGY

ENERGY OF FORMATION

Energy Oxidizers, High
USE   HIGH ENERGY OXIDIZERS

Energy, Particle
USE   PARTICLE ENERGY

ENERGY POLICY

Energy, Potential
USE   POTENTIAL ENERGY

Energy Principle, Bernstein
USE   BERNSTEIN ENERGY PRINCIPLE

Energy Production, Biomass
USE   BIOMASS ENERGY PRODUCTION

Energy Propellants, High
USE   HIGH ENERGY PROPELLANTS

Energy, Proton
USE   PROTON ENERGY

Energy, Radiant
USE   RADIATION

ENERGY REQUIREMENTS

Energy, Residential
USE   RESIDENTIAL ENERGY

Energy, Seismic
USE   SEISMIC ENERGY

Energy, Solar
USE   SOLAR ENERGY

ENERGY SOURCES

Energy Sources, Atmospheric
USE   ATMOSPHERIC ENERGY SOURCES

Energy Sources, Offshore
USE   OFFSHORE ENERGY SOURCES

ENERGY SPECTRA

Energy, Stacking Fault
USE   STACKING FAULT ENERGY

ENERGY STORAGE

Energy Storage Devices
USE   ENERGY STORAGE

Energy Storage, Electric
USE   ELECTRIC ENERGY STORAGE

Energy Storage, Magnetic
USE   MAGNETIC ENERGY STORAGE

Energy Storage, Thermal
USE   HEAT STORAGE

Energy, Surface
USE   SURFACE ENERGY

Energy Systems, Integrated
USE   INTEGRATED ENERGY SYSTEMS

Energy Systems, Solar Total
USE   SOLAR TOTAL ENERGY SYSTEMS

Energy Systems, Total
USE   TOTAL ENERGY SYSTEMS

ENERGY TECHNOLOGY

Energy, Thermal
USE   THERMAL ENERGY

Energy, Thermonuclear
USE   THERMONUCLEAR POWER GENERATION

ENERGY TRANSFER

Energy Transfer (LET), Linear
USE   LINEAR ENERGY TRANSFER (LET)

Energy, Transportation
USE   TRANSPORTATION ENERGY

Energy Utilization, Geothermal
USE   GEOTHERMAL ENERGY UTILIZATION

Energy Utilization, Waste
USE   WASTE ENERGY UTILIZATION

Energy, Waterwave
USE   WATERWAVE ENERGY

Energy, Wind
USE   WINDPOWER UTILIZATION

Energy, Zero Point
USE   ZERO POINT ENERGY

ENGINE AIRFRAME INTEGRATION

Engine, AJ-10
USE   AJ-10 ENGINE

Engine, AJ-1000
USE   M-1 ENGINE

Engine, ALGOL
USE   ALGOL ENGINE

Engine, Altair
USE   X-248 ENGINE

ENGINE ANALYZERS

Engine, ASROC
USE   ASROC ENGINE

Engine, BE-3
USE   BE-3 ENGINE

Engine, Bristol-Siddeley BS 53
USE   BRISTOL-SIDDELEY BS 53 ENGINE

Engine, Bristol-Siddeley Olympus 593
USE   BRISTOL-SIDDELEY OLYMPUS 593 ENGINE

Engine, Bristol-Siddeley Viper
USE   BRISTOL-SIDDELEY VIPER ENGINE

Engine Cases, Missile
USE   ROCKET ENGINE CASES

Engine Cases, Rocket
USE   ROCKET ENGINE CASES

Engine, Castor 2
USE   TX-354 ENGINE

Engine, CF-700
USE   CF-700 ENGINE

ENGINE CONTROL

Engine Control, Rocket
USE   ROCKET ENGINE CONTROL

Engine Control, Turbojet
USE   TURBOJET ENGINE CONTROL

ENGINE COOLANTS

ENGINE DESIGN

Engine Design, Rocket
USE   ROCKET ENGINE DESIGN

Engine, F-1 Rocket
USE   F-1 ROCKET ENGINE

ENGINE FAILURE

Engine For Rocket Vehicles, Nuclear
USE   NUCLEAR ENGINE FOR ROCKET VEHICLES

Engine Fuels, Jet
USE   JET ENGINE FUELS

Engine, H-1
USE   H-1 ENGINE

Engine, Hercules
USE   HERCULES ENGINE

ENGINE INLETS

Engine, J-2
USE   J-2 ENGINE

Engine, J-33
USE   J-33 ENGINE

Engine, J-34
USE   J-34 ENGINE

Engine, J-47
USE   J-47 ENGINE

Engine, J-52
USE   J-52 ENGINE

Engine, J-57
USE   J-57 ENGINE

Engine, J-57-P-20
　USE　J-57-P-20 ENGINE

Engine, J-58
　USE　J-58 ENGINE

Engine, J-65
　USE　J-65 ENGINE

Engine, J-69-T-25
　USE　J-69-T-25 ENGINE

Engine, J-71
　USE　J-71 ENGINE

Engine, J-73
　USE　J-73 ENGINE

Engine, J-75
　USE　J-75 ENGINE

Engine, J-79
　USE　J-79 ENGINE

Engine, J-85
　USE　J-85 ENGINE

Engine, J-93
　USE　J-93 ENGINE

Engine, J-97
　USE　J-97 ENGINE

Engine, J93-MJ252H
　USE　J-93 ENGINE

Engine, J93-MJ280G
　USE　J-93 ENGINE

(Engine), LACE
　USE　LIQUID AIR CYCLE ENGINES

Engine, LR-62-RM-2
　USE　LR-62-RM-2 ENGINE

Engine, LR-87-AJ-5
　USE　LR-87-AJ-5 ENGINE

Engine, LR-91-AJ-5
　USE　LR-91-AJ-5 ENGINE

Engine, LR-99
　USE　LR-99 ENGINE

Engine, M-1
　USE　M-1 ENGINE

Engine, M-46
　USE　M-46 ENGINE

Engine, M-55
　USE　M-55 ENGINE

Engine, M-56
　USE　M-56 ENGINE

Engine, M-57
　USE　M-57 ENGINE

Engine, M-100
　USE　M-100 ENGINE

Engine, MA-2
　USE　MA-2 ENGINE

Engine, MA-3
　USE　MA-3 ENGINE

Engine, MA-5
　USE　MA-5 ENGINE

Engine, Marbore 2
　USE　J-69-T-25 ENGINE

Engine, Marquardt R4D
　USE　MARQUARDT R4D ENGINE

ENGINE MONITORING INSTRUMENTS

(Engine), NERVA
　USE　NUCLEAR ENGINE FOR ROCKET VEHICLES

(Engine), NIMPHE
　USE　HYDRAZINE ENGINES

ENGINE NOISE

Engine Noise, Rocket
　USE　ROCKET ENGINE NOISE

Engine, P-1
　USE　P-1 ENGINE

ENGINE PARTS

Engine, Pegasus
　USE　BRISTOL-SIDDELEY BS 53 ENGINE

ENGINE PRIMERS

Engine Program, Quiet
　USE　QUIET ENGINE PROGRAM

Engine, RA-28
　USE　RA-28 ENGINE

Engine, RL-10-A-1
　USE　RL-10-A-1 ENGINE

Engine, RL-10-A-3
　USE　RL-10-A-3 ENGINE

Engine, SL-3 Rocket
　USE　SL-3 ROCKET ENGINE

Engine, Space Shuttle Main
　USE　SPACE SHUTTLE MAIN ENGINE

Engine (Space Shuttle), Orbit Maneuvering
　USE　ORBIT MANEUVERING ENGINE (SPACE SHUTTLE)

ENGINE STARTERS

Engine, T-34
　USE　T-34 ENGINE

Engine, T-38
　USE　T-38 ENGINE

Engine, T-53
　USE　T-53 ENGINE

Engine, T-55
　USE　T-55 ENGINE

Engine, T-56
　USE　T-56 ENGINE

Engine, T-58
　USE　T-58 ENGINE

Engine, T-58-GE-8B
　USE　T-58-GE-8B ENGINE

Engine, T-63
　USE　T-63 ENGINE

Engine, T-64
　USE　T-64 ENGINE

Engine, T-74
　USE　T-74 ENGINE

Engine, T-76
　USE　T-76 ENGINE

Engine, T-78
　USE　T-78 ENGINE

ENGINE TESTING LABORATORIES

ENGINE TESTS

Engine, TF-30
　USE　TF-30 ENGINE

Engine, TF-34
　USE　TF-34 ENGINE

Engine, TF-41
　USE　TF-41 ENGINE

Engine, TU-121
　USE　TU-121 ENGINE

Engine, TX-33-39
　USE　XM-33 ENGINE

Engine, TX-77
　USE　TX-77 ENGINE

Engine, TX-354
　USE　TX-354 ENGINE

Engine, X-248
　USE　X-248 ENGINE

Engine, X-254
　USE　X-254 ENGINE

Engine, X-258-B1
　USE　X-258-B1 ENGINE

Engine, X-259
　USE　X-259 ENGINE

Engine, X-405
　USE　X-405 ENGINE

Engine, XJ-34-WE-32
　USE　J-34 ENGINE

Engine, XJ-79-GE-1
　USE　J-79 ENGINE

Engine, XLR-91-AJ-5
　USE　LR-91-AJ-5 ENGINE

Engine, XLR-99
　USE　XLR-99 ENGINE

Engine, XM-33
　USE　XM-33 ENGINE

Engine, YJ-73-GE-3
　USE　J-73 ENGINE

Engine, YJ-79
　USE　J-79 ENGINE

Engine, YJ-85
　USE　J-85 ENGINE

Engine, YJ-93
　USE　J-93 ENGINE

Engine, YJ-93-GE-3
　USE　J-93 ENGINE

Engine, YJ73 Turbojet
　USE　J-73 ENGINE

Engine, YLR-91-AJ-1
　USE　YLR-91-AJ-1 ENGINE

Engine, YLR-99-RM-1
　USE　LR-99 ENGINE

Engine 9KS-11000, Rocket
　USE　ROCKET ENGINE 9KS-11000

ENGINEERING

Engineering, Aeronautical
　USE　AERONAUTICAL ENGINEERING

Engineering, Aerospace
　USE　AEROSPACE ENGINEERING

Engineering), Beds (Process
　USE　BEDS (PROCESS ENGINEERING)

Engineering, Bio
　USE　BIOENGINEERING

Engineering, Chemical
　USE　CHEMICAL ENGINEERING

Engineering), Columns (Process
　USE　COLUMNS (PROCESS ENGINEERING)

Engineering), Cracking (Chemical
　USE　CRACKING (CHEMICAL ENGINEERING)

Engineering Development
  USE   PRODUCT DEVELOPMENT

ENGINEERING DRAWINGS

Engineering, Electrical
  USE   ELECTRICAL ENGINEERING

Engineering, Environmental
  USE   ENVIRONMENTAL ENGINEERING

Engineering, Genetic
  USE   GENETIC ENGINEERING

Engineering, Geotechnical
  USE   GEOTECHNICAL ENGINEERING

Engineering, Human
  USE   HUMAN FACTORS ENGINEERING

Engineering, Human Factors
  USE   HUMAN FACTORS ENGINEERING

Engineering, Knowledge
  USE   EXPERT SYSTEMS

ENGINEERING MANAGEMENT

Engineering, Mechanical
  USE   MECHANICAL ENGINEERING

Engineering, Production
  USE   PRODUCTION ENGINEERING

(Engineering), Regeneration
  USE   REGENERATION (ENGINEERING)

Engineering, Reliability
  USE   RELIABILITY ENGINEERING

Engineering Simulator, Shuttle
  USE   SHUTTLE ENGINEERING SIMULATOR

Engineering, Software
  USE   SOFTWARE ENGINEERING

Engineering, Space Systems
  USE   AEROSPACE ENGINEERING

Engineering, Structural
  USE   STRUCTURAL ENGINEERING

Engineering, Systems
  USE   SYSTEMS ENGINEERING

ENGINEERING TEST REACTORS

Engineering, Underwater
  USE   UNDERWATER ENGINEERING

Engineering, Value
  USE   VALUE ENGINEERING

ENGINES

Engines, Air Breathing
  USE   AIR BREATHING ENGINES

Engines, Aircraft
  USE   AIRCRAFT ENGINES

Engines, Arc Jet
  USE   ARC JET ENGINES

Engines, Automobile
  USE   AUTOMOBILE ENGINES

Engines, Booster Rocket
  USE   BOOSTER ROCKET ENGINES

Engines, Cesium
  USE   CESIUM ENGINES

Engines, Dart Turboprop
  USE   TURBOPROP ENGINES

Engines, Diesel
  USE   DIESEL ENGINES

Engines, Ducted Fan
  USE   DUCTED FAN ENGINES

Engines, Ducted Rocket
  USE   DUCTED ROCKET ENGINES

Engines, Electric Rocket
  USE   ELECTRIC ROCKET ENGINES

Engines, Electrostatic
  USE   ELECTROSTATIC ENGINES

Engines, Electrothermal
  USE   ELECTROTHERMAL ENGINES

Engines, External Combustion
  USE   EXTERNAL COMBUSTION ENGINES

Engines, Gas Generator
  USE   GAS GENERATORS
        ENGINES

Engines, Gas Turbine
  USE   GAS TURBINE ENGINES

Engines, Helicopter
  USE   HELICOPTER ENGINES

Engines, Heus Rocket
  USE   HEUS ROCKET ENGINES

Engines, Hot Water Rocket
  USE   HOT WATER ROCKET ENGINES

Engines, Hybrid Propellant Rocket
  USE   HYBRID PROPELLANT ROCKET ENGINES

Engines, Hybrid Rocket
  USE   HYBRID ROCKET ENGINES

Engines, Hydrazine
  USE   HYDRAZINE ENGINES

Engines, Hydrogen
  USE   HYDROGEN ENGINES

Engines, Hydrogen Oxygen
  USE   HYDROGEN OXYGEN ENGINES

Engines, Hydrox
  USE   HYDROGEN OXYGEN ENGINES

(Engines), Ingestion
  USE   INGESTION (ENGINES)

Engines, Internal Combustion
  USE   INTERNAL COMBUSTION ENGINES

Engines, Ion
  USE   ION ENGINES

Engines, JATO
  USE   JATO ENGINES

Engines, Jet
  USE   JET ENGINES

Engines, Liquid Air Cycle
  USE   LIQUID AIR CYCLE ENGINES

Engines, Liquid Propellant Rocket
  USE   LIQUID PROPELLANT ROCKET ENGINES

Engines, Lithergol Rocket
  USE   LITHERGOL ROCKET ENGINES

Engines, Low Volume Ramjet
  USE   LOW VOLUME RAMJET ENGINES

Engines, LOX-Hydrogen
  USE   HYDROGEN OXYGEN ENGINES

Engines, Mercury Ion
  USE   MERCURY ION ENGINES

Engines, Microrocket
  USE   MICROROCKET ENGINES

Engines, Nike Booster Rocket
  USE   NIKE BOOSTER ROCKET ENGINES

Engines, Nozzleless Rocket
  USE   NOZZLELESS ROCKET ENGINES

Engines, Nuclear Lightbulb
  USE   NUCLEAR LIGHTBULB ENGINES

Engines, Nuclear Ramjet
  USE   NUCLEAR RAMJET ENGINES

Engines, Nuclear Rocket
  USE   NUCLEAR ROCKET ENGINES

Engines, Piston
  USE   PISTON ENGINES

Engines, Plasma
  USE   PLASMA ENGINES

Engines, Pulsed Jet
  USE   PULSED JET ENGINES

Engines, Pulsejet
  USE   PULSEJET ENGINES

Engines, Radio Frequency Ion Thrustor
  USE   RIT ENGINES

Engines, Ramjet
  USE   RAMJET ENGINES

Engines, Reciprocating
  USE   PISTON ENGINES

Engines, Resistojet
  USE   RESISTOJET ENGINES

Engines, Restartable Rocket
  USE   RESTARTABLE ROCKET ENGINES

Engines, Retrorocket
  USE   RETROROCKET ENGINES

Engines, Reusable Rocket
  USE   REUSABLE ROCKET ENGINES

Engines, Rit
  USE   RIT ENGINES

Engines, RL-10
  USE   RL-10 ENGINES

Engines, Rocket
  USE   ROCKET ENGINES

Engines, Rotary
  USE   ROTARY ENGINES

Engines, Scramjet
  USE   SUPERSONIC COMBUSTION RAMJET
        ENGINES

Engines, Solid Propellant Rocket
  USE   SOLID PROPELLANT ROCKET ENGINES

Engines, Supersonic Combustion Ramjet
  USE   SUPERSONIC COMBUSTION RAMJET
        ENGINES

Engines, Sustainer Rocket
  USE   SUSTAINER ROCKET ENGINES

Engines, SYNCOM Apogee
  USE   SYNCOM APOGEE ENGINES

Engines, Topping Cycle
  USE   TOPPING CYCLE ENGINES

Engines, Torpedo
  USE   TORPEDO ENGINES

Engines, Turbine
  USE   TURBINE ENGINES

Engines, Turbofan
  USE   TURBOFAN ENGINES

Engines, Turbojet
  USE   TURBOJET ENGINES

Engines, Turboprop
  USE   TURBOPROP ENGINES

Engines, Turboramjet
  USE   TURBORAMJET ENGINES

Engines, Turborocket
USE   TURBOROCKET ENGINES

Engines, Two Stage Plasma
USE   TWO STAGE PLASMA ENGINES

Engines, Ullage Rocket
USE   ULLAGE ROCKET ENGINES

Engines, Upper Stage Rocket
USE   UPPER STAGE ROCKET ENGINES

Engines, Variable Cycle
USE   VARIABLE CYCLE ENGINES

Engines, Variable Stream Control
USE   VARIABLE STREAM CONTROL ENGINES

Engines, Vernier
USE   VERNIER ENGINES

Engines, Wankel
USE   WANKEL ENGINES

Engines, X-258
USE   X-258 ENGINES

ENGLAND

England (US), New
USE   NEW ENGLAND (US)

ENGLISH CHANNEL

English Electric Canberra Aircraft
USE   CANBERRA AIRCRAFT

ENGLISH LANGUAGE

ENGRAVING

Engraving, Photo
USE   PHOTOENGRAVING

Enhancement
USE   AUGMENTATION

Enhancement, Image
USE   IMAGE ENHANCEMENT

Enhancement Of Atmospherics, Sudden
USE   SUDDEN ENHANCEMENT OF
         ATMOSPHERICS

Enhancement, Storm
USE   STORM ENHANCEMENT

Enlarging
USE   EXPANSION

ENRICHMENT

Enrichment, Isotopic
USE   ISOTOPIC ENRICHMENT

ENRICO FERMI ATOMIC POWER PLANT

Enskog Theory, Chapman-
USE   CHAPMAN-ENSKOG THEORY

Enskog-Chapman Theory
USE   CHAPMAN-ENSKOG THEORY

ENSTATITE

Enstrophy
USE   VORTICITY

ENTERPRISE (ORBITER)

ENTHALPY

Enthalpy-Entropy Diagrams
USE   MOLLIER DIAGRAM

ENTIRE FUNCTIONS

ENTOMOLOGY

ENTRAINMENT

ENTRANCES

ENTRAPMENT

ENTROPY

Entropy Diagrams, Enthalpy-
USE   MOLLIER DIAGRAM

Entropy Method, Maximum
USE   MAXIMUM ENTROPY METHOD

Entropy Method, Minimum
USE   MINIMUM ENTROPY METHOD

ENTROPY (STATISTICS)

ENTRY

Entry, Atmospheric
USE   ATMOSPHERIC ENTRY

ENTRY GUIDANCE (STS)

Entry, Planetary
USE   ATMOSPHERIC ENTRY

Entry Probes, Pioneer Venus 2
USE   PIONEER VENUS 2 ENTRY PROBES

Entry Simulation, Atmospheric
USE   ATMOSPHERIC ENTRY SIMULATION

Entry Vehicle, Viking 75
USE   VIKING 75 ENTRY VEHICLE

ENUMERATION

ENVELOPES

Envelopes, Stellar
USE   STELLAR ENVELOPES

Environ Satellite B, Geostationary Operatl
USE   GOES 2

Environ Sats, Geostationary Operational
USE   GOES SATELLITES

Environment, Antarctic
USE   ICE ENVIRONMENTS

Environment, Earth
USE   EARTH ENVIRONMENT

ENVIRONMENT EFFECTS

Environment Experiment, Electromagnetic
USE   ELECTROMAGNETIC ENVIRONMENT
         EXPERIMENT

Environment Interactions, Man
USE   MAN ENVIRONMENT INTERACTIONS

Environment, Lunar
USE   LUNAR ENVIRONMENT

ENVIRONMENT MANAGEMENT

Environment, Mars
USE   MARS ENVIRONMENT

ENVIRONMENT MODELS

ENVIRONMENT POLLUTION

ENVIRONMENT PROTECTION

ENVIRONMENT SIMULATION

Environment Simulation, Space
USE   SPACE ENVIRONMENT SIMULATION

ENVIRONMENT SIMULATORS

Environment, Space
USE   AEROSPACE ENVIRONMENTS

Environmental Chambers
USE   TEST CHAMBERS

ENVIRONMENTAL CHEMISTRY

ENVIRONMENTAL CONTROL

ENVIRONMENTAL ENGINEERING

ENVIRONMENTAL INDEX

ENVIRONMENTAL LABORATORIES

Environmental Lubrication, Space
USE   SPACECRAFT LUBRICATION

ENVIRONMENTAL MONITORING

ENVIRONMENTAL QUALITY

ENVIRONMENTAL RESEARCH SATELLITES

Environmental Sat Sys, National Operational
USE   NOESS

ENVIRONMENTAL SURVEYS

Environmental Temperature
USE   AMBIENT TEMPERATURE

ENVIRONMENTAL TESTS

ENVIRONMENTS

Environments, Aerospace
USE   AEROSPACE ENVIRONMENTS

Environments, Arctic
USE   ICE ENVIRONMENTS

Environments, Extraterrestrial
USE   EXTRATERRESTRIAL ENVIRONMENTS

Environments, Frictionless
USE   FRICTIONLESS ENVIRONMENTS

Environments, High Altitude
USE   HIGH ALTITUDE ENVIRONMENTS

Environments, High Gravity
USE   HIGH GRAVITY ENVIRONMENTS

Environments, High Temperature
USE   HIGH TEMPERATURE ENVIRONMENTS

Environments, Ice
USE   ICE ENVIRONMENTS

Environments, Low Temperature
USE   LOW TEMPERATURE ENVIRONMENTS

Environments, Marine
USE   MARINE ENVIRONMENTS

Environments, Planetary
USE   PLANETARY ENVIRONMENTS

Environments, Rotating
USE   ROTATING ENVIRONMENTS

Environments, Spacecraft
USE   SPACECRAFT ENVIRONMENTS

Environments, Thermal
USE   THERMAL ENVIRONMENTS

ENZYME ACTIVITY

ENZYMES

Enzymes, Co
USE   COENZYMES

ENZYMOLOGY

EOCR (Reactor)
USE   EXPERIMENTAL ORGANIC COOLED
         REACTORS

EOGO
USE   EGO

EOLE SATELLITES

EOPAP
USE   EARTH & OCEAN PHYSICS APPLICATIONS
      PROGRAM

EOR (Rendezvous)
USE   EARTH ORBITAL RENDEZVOUS

EOS
USE   LANDSAT SATELLITES

EOS-A
USE   LANDSAT E

EOS-B
USE   LANDSAT F

EOSINOPHILS

EOSS

EPE-A
USE   EXPLORER 12 SATELLITE

EPE-B
USE   EXPLORER 14 SATELLITE

EPE-C
USE   EXPLORER 15 SATELLITE

EPE-D
USE   EXPLORER 26 SATELLITE

EPHEMERIDES

Ephemerides, Planet
USE   PLANET EPHEMERIDES

EPHEMERIS TIME

EPICARDIUM

EPICYCLOIDS

EPIDEMIOLOGY

EPIDERMIS

EPILEPSY

EPINEPHRINE

EPITAXY

Epitaxy, Grapho
USE   GRAPHOEPITAXY

Epitaxy, Liquid Phase
USE   LIQUID PHASE EPITAXY

Epitaxy, Molecular Beam
USE   MOLECULAR BEAM EPITAXY

Epitaxy, Vapor Phase
USE   VAPOR PHASE EPITAXY

EPITHELIUM

EPNL
USE   EFFECTIVE PERCEIVED NOISE LEVELS

Epochs
USE   TIME MEASUREMENT

EPOXIDATION

Epoxides
USE   EPOXY COMPOUNDS

Epoxy Composites, Graphite-
USE   GRAPHITE-EPOXY COMPOSITES

EPOXY COMPOUNDS

Epoxy Compounds, Boron-
USE   BORON-EPOXY COMPOUNDS

EPOXY MATRIX COMPOSITES

EPOXY RESINS

Epoxy Resins, Phenolic
USE   PHENOLIC EPOXY RESINS

EQUALIZERS (CIRCUITS)

Equation, Bernoulli
USE   BERNOULLI THEOREM

Equation, Bethe-Salpeter
USE   BETHE-SALPETER EQUATION

Equation, Blasius
USE   BLASIUS EQUATION

Equation, Boltzmann Transport
USE   BOLTZMANN TRANSPORT EQUATION

Equation, Boltzmann-Vlasov
USE   BOLTZMANN-VLASOV EQUATION

Equation, Born-Mayer
USE   BORN APPROXIMATION

Equation, Brillouin-Wigner
USE   BRILLOUIN-WIGNER EQUATION

Equation, Burger
USE   BURGER EQUATION

Equation, Chandrasekhar
USE   CHANDRASEKHAR EQUATION

Equation, Chaplygin
USE   CHAPLYGIN EQUATION

Equation, Continuity
USE   CONTINUITY EQUATION

Equation, Diophantine
USE   DIOPHANTINE EQUATION

Equation, Dirac
USE   DIRAC EQUATION

Equation, Duffing Differential
USE   DUFFING DIFFERENTIAL EQUATION

Equation, Eikonal
USE   EIKONAL EQUATION

Equation, Elber
USE   ELBER EQUATION

Equation, Euler-Lagrange
USE   EULER-LAGRANGE EQUATION

Equation, Euler-Lambert
USE   EULER-LAMBERT EQUATION

Equation, Falkner-Skan
USE   FALKNER-SKAN EQUATION

Equation, Ficks
USE   FICKS EQUATION

Equation, Fokker-Planck
USE   FOKKER-PLANCK EQUATION

Equation, Gauss
USE   GAUSS EQUATION

Equation, Gibbs Adsorption
USE   GIBBS ADSORPTION EQUATION

Equation, Hamilton-Jacobi
USE   HAMILTON-JACOBI EQUATION

Equation, Helmholtz Vorticity
USE   HELMHOLTZ VORTICITY EQUATION

Equation, Inhour
USE   INHOUR EQUATION

Equation, Klein-Gordon
USE   KLEIN-GORDON EQUATION

Equation, Korteweg-Devries
USE   KORTEWEG-DEVRIES EQUATION

Equation, Krook
USE   KROOK EQUATION

Equation, Laplace
USE   LAPLACE EQUATION

Equation, Mathieu
USE   MATHIEU FUNCTION

Equation, Maxwell
USE   MAXWELL EQUATION

Equation, Monge-Ampere
USE   MONGE-AMPERE EQUATION

Equation, Navier-Stokes
USE   NAVIER-STOKES EQUATION

Equation Of State, Hugoniot
USE   HUGONIOT EQUATION OF STATE

Equation, Pfaff
USE   PFAFF EQUATION

Equation, Poisson
USE   POISSON EQUATION

Equation, Reynolds
USE   REYNOLDS EQUATION

Equation, Riccati
USE   RICCATI EQUATION

Equation, Richardson-Dushman
USE   THERMIONIC EMISSION
      TEMPERATURE EFFECTS

Equation, Schroedinger
USE   SCHROEDINGER EQUATION

Equation, Stokes-Beltrami
USE   STOKES-BELTRAMI EQUATION

Equation, Von Karman
USE   VON KARMAN EQUATION

EQUATIONS

Equations, Adiabatic
USE   ADIABATIC EQUATIONS

Equations, Balance
USE   EQUATIONS

Equations, Biharmonic
USE   BIHARMONIC EQUATIONS

Equations, Boundary Layer
USE   BOUNDARY LAYER EQUATIONS

Equations, Cauchy-Riemann
USE   CAUCHY-RIEMANN EQUATIONS

Equations, Characteristic
USE   EIGENVECTORS
      EIGENVALUES

Equations, Conservation
USE   CONSERVATION EQUATIONS

Equations, Constitutive
USE   CONSTITUTIVE EQUATIONS

Equations, Cubic
USE   CUBIC EQUATIONS

Equations, Difference
USE   DIFFERENCE EQUATIONS

Equations, Differential
USE   DIFFERENTIAL EQUATIONS

Equations, Donnell
USE   DONNELL EQUATIONS

Equations, Einstein
USE   EINSTEIN EQUATIONS

Equations, Elliptic Differential
USE   ELLIPTIC DIFFERENTIAL EQUATIONS

Equations, Equilibrium
USE   EQUILIBRIUM EQUATIONS

Equations, Euler-Cauchy
USE EULER-CAUCHY EQUATIONS

Equations, Faddeev
USE FADDEEV EQUATIONS

Equations, Flow
USE FLOW EQUATIONS

Equations, Forced Vibratory Motion
USE FORCED VIBRATION
EQUATIONS

Equations, Fredholm
USE FREDHOLM EQUATIONS

Equations, Gibbs
USE GIBBS EQUATIONS

Equations, Gibbs-Helmholtz
USE GIBBS-HELMHOLTZ EQUATIONS

Equations, Heat
USE THERMODYNAMICS

Equations, Helmholtz
USE HELMHOLTZ EQUATIONS

Equations, Hydrodynamic
USE HYDRODYNAMIC EQUATIONS

Equations, Hyperbolic Differential
USE HYPERBOLIC DIFFERENTIAL EQUATIONS

Equations, Integral
USE INTEGRAL EQUATIONS

Equations, Integrodifferential
USE INTEGRAL EQUATIONS
DIFFERENTIAL EQUATIONS

Equations, Kinematic
USE KINEMATIC EQUATIONS

Equations, Kinetic
USE KINETIC EQUATIONS

Equations, Lame Wave
USE LAME WAVE EQUATIONS

Equations, Landau-Ginzburg
USE LANDAU-GINZBURG EQUATIONS

Equations, Linear
USE LINEAR EQUATIONS

Equations, Linear Evolution
USE LINEAR EVOLUTION EQUATIONS

Equations, Liouville
USE LIOUVILLE EQUATIONS

Equations, Macroscopic
USE MACROSCOPIC EQUATIONS

Equations, Motion
USE EQUATIONS OF MOTION

Equations, Nonholonomic
USE NONHOLONOMIC EQUATIONS

Equations, Nonlinear
USE NONLINEAR EQUATIONS

Equations, Nonlinear Evolution
USE NONLINEAR EVOLUTION EQUATIONS

EQUATIONS OF MOTION

Equations Of Motion, Euler
USE EULER EQUATIONS OF MOTION

Equations Of Motion, Lagrange
USE EULER-LAGRANGE EQUATION

EQUATIONS OF STATE

Equations, Orbit
USE ORBITAL MECHANICS

Equations, Orr-Sommerfeld
USE ORR-SOMMERFELD EQUATIONS

Equations, Parabolic Differential
USE PARABOLIC DIFFERENTIAL EQUATIONS

Equations, Partial Differential
USE PARTIAL DIFFERENTIAL EQUATIONS

Equations, Period
USE PERIODIC FUNCTIONS

Equations, Primitive
USE PRIMITIVE EQUATIONS

Equations, Quadratic
USE QUADRATIC EQUATIONS

Equations, Quartic
USE QUARTIC EQUATIONS

Equations, Rayleigh
USE RAYLEIGH EQUATIONS

Equations, Roots Of
USE ROOTS OF EQUATIONS

Equations, Saha
USE SAHA EQUATIONS

Equations, Semiempirical
USE SEMIEMPIRICAL EQUATIONS

Equations, Shallow Shell
USE SHALLOW SHELL EQUATIONS

Equations, Simultaneous
USE SIMULTANEOUS EQUATIONS

Equations, Singular Integral
USE SINGULAR INTEGRAL EQUATIONS

Equations, State
USE EQUATIONS OF STATE

Equations, Vlasov
USE VLASOV EQUATIONS

Equations, Volterra
USE VOLTERRA EQUATIONS

Equations, Vorticity
USE VORTICITY EQUATIONS

Equations, Wave
USE WAVE EQUATIONS

Equations, Wiener Hopf
USE WIENER HOPF EQUATIONS

Equator, Geomagnetic
USE MAGNETIC EQUATOR

Equator, Lunar
USE LUNAR EQUATOR

Equator, Magnetic
USE MAGNETIC EQUATOR

EQUATORIAL ATMOSPHERE

Equatorial Congo, French
USE CONGO (BRAZZAVILLE)

EQUATORIAL ELECTROJET

EQUATORIAL ORBITS

EQUATORIAL REGIONS

EQUATORS

EQUILIBRIUM

Equilibrium, Acid Base
USE ACID BASE EQUILIBRIUM

Equilibrium, Chemical
USE CHEMICAL EQUILIBRIUM

Equilibrium Diagrams
USE PHASE DIAGRAMS

EQUILIBRIUM EQUATIONS

EQUILIBRIUM FLOW

Equilibrium Flow, Frozen
USE FROZEN EQUILIBRIUM FLOW

Equilibrium Flow, Shifting
USE SHIFTING EQUILIBRIUM FLOW

Equilibrium, Liquid-Vapor
USE LIQUID-VAPOR EQUILIBRIUM

EQUILIBRIUM METHODS

Equilibrium, Plasma
USE PLASMA EQUILIBRIUM

Equilibrium Points, Lagrangian
USE LAGRANGIAN EQUILIBRIUM POINTS

Equilibrium, Thermodynamic
USE THERMODYNAMIC EQUILIBRIUM

Equilibrium, Vapor Liquid
USE LIQUID-VAPOR EQUILIBRIUM

EQUINOXES

Equipartition, Energy
USE EQUIPARTITION THEOREM

EQUIPARTITION THEOREM

EQUIPMENT

(Equipment), Absorbers
USE ABSORBERS (EQUIPMENT)

Equipment, Air Conditioning
USE AIR CONDITIONING EQUIPMENT

Equipment, Airborne
USE AIRBORNE EQUIPMENT

Equipment, Aircraft
USE AIRCRAFT EQUIPMENT

Equipment, Airport Surface Detection
USE AIRPORT SURFACE DETECTION
EQUIPMENT

Equipment, Astronaut Maneuvering
USE ASTRONAUT MANEUVERING EQUIPMENT

Equipment, Audio
USE AUDIO EQUIPMENT

Equipment, Audio Visual
USE VISUAL AIDS
TRAINING DEVICES

Equipment, Automatic Test
USE AUTOMATIC TEST EQUIPMENT

Equipment, Bedding
USE BEDDING EQUIPMENT

Equipment, Bombing
USE BOMBING EQUIPMENT

(Equipment), Booms
USE BOOMS (EQUIPMENT)

Equipment, Cefoam Checkout
USE CEFOAM CHECKOUT EQUIPMENT

Equipment, Checkout
USE TEST EQUIPMENT

Equipment, Communication
USE COMMUNICATION EQUIPMENT

Equipment (Computers), Auxiliary
USE AUXILIARY EQUIPMENT (COMPUTERS)

Equipment (Computers), Peripheral
USE PERIPHERAL EQUIPMENT (COMPUTERS)

Equipment, Control
USE   CONTROL EQUIPMENT

Equipment, Cryogenic
USE   CRYOGENIC EQUIPMENT

(Equipment), Cyclones
USE   CENTRIFUGES

Equipment, Data Processing
USE   DATA PROCESSING EQUIPMENT

Equipment, Distance Measuring
USE   DISTANCE MEASURING EQUIPMENT

Equipment, Distillation
USE   DISTILLATION EQUIPMENT

(Equipment), Dryers
USE   DRYING APPARATUS

Equipment, Electric
USE   ELECTRIC EQUIPMENT

Equipment, Electronic
USE   ELECTRONIC EQUIPMENT

Equipment, Ground Support
USE   GROUND SUPPORT EQUIPMENT

Equipment, Handling
USE   HANDLING EQUIPMENT

Equipment, Heating
USE   HEATING EQUIPMENT

Equipment, Hydraulic
USE   HYDRAULIC EQUIPMENT

Equipment, Jacking
USE   JACKS (LIFTS)

Equipment, Laboratory
USE   LABORATORY EQUIPMENT

Equipment, Lighting
USE   LIGHTING EQUIPMENT

Equipment, Lossless
USE   LOSSLESS EQUIPMENT

Equipment, Medical
USE   MEDICAL EQUIPMENT

Equipment, Microwave
USE   MICROWAVE EQUIPMENT

Equipment, Miniature Electronic
USE   MINIATURE ELECTRONIC EQUIPMENT

Equipment, Onboard
USE   ONBOARD EQUIPMENT

Equipment, Optical
USE   OPTICAL EQUIPMENT

Equipment, Oxygen Supply
USE   OXYGEN SUPPLY EQUIPMENT

Equipment, Photographic
USE   PHOTOGRAPHIC EQUIPMENT

Equipment, Photographic Processing
USE   PHOTOGRAPHIC PROCESSING EQUIPMENT

Equipment, Pneumatic
USE   PNEUMATIC EQUIPMENT

Equipment, Portable
USE   PORTABLE EQUIPMENT

Equipment, Radar
USE   RADAR EQUIPMENT

Equipment, Radio
USE   RADIO EQUIPMENT

Equipment, Retractable
USE   RETRACTABLE EQUIPMENT

Equipment, Spacecraft
USE   SPACECRAFT EQUIPMENT

Equipment, Spacecraft Electronic
USE   SPACECRAFT ELECTRONIC EQUIPMENT

EQUIPMENT SPECIFICATIONS

Equipment), Stowage (Onboard
USE   STOWAGE (ONBOARD EQUIPMENT)

Equipment, Survival
USE   SURVIVAL EQUIPMENT

Equipment, Television
USE   TELEVISION EQUIPMENT

Equipment, Test
USE   TEST EQUIPMENT

Equipment Tests, Electric
USE   ELECTRIC EQUIPMENT TESTS

Equipment Tests, Electronic
USE   ELECTRONIC EQUIPMENT TESTS

(Equipment), Thickeners
USE   THICKENERS (EQUIPMENT)

Equipment, Ultra Short Wave Radio
USE   VERY HIGH FREQUENCY RADIO
          EQUIPMENT

Equipment, Very High Frequency Radio
USE   VERY HIGH FREQUENCY RADIO
          EQUIPMENT

Equipment, Video
USE   VIDEO EQUIPMENT

EQUIPOTENTIALS

EQUIVALENCE

EQUIVALENT CIRCUITS

Er
USE   ERBIUM

ERBE
USE   EARTH RADIATION BUDGET EXPERIMENT

ERBIUM

ERBIUM ALLOYS

ERBIUM COMPOUNDS

ERBIUM ISOTOPES

Erbium 169
USE   ERBIUM ISOTOPES

Erbium 171
USE   ERBIUM ISOTOPES

Erectable Structures, Space
USE   SPACE ERECTABLE STRUCTURES

Erecting Devices, Self
USE   SELF ERECTING DEVICES

Erection
USE   CONSTRUCTION

EREP

ERGODIC PROCESS

ERGOMETERS

Ergonomics
USE   HUMAN FACTORS ENGINEERING

ERGOTAMINE

Erie, Lake
USE   LAKE ERIE

EROS Project
USE   EXPERIMENTAL REFLECTOR ORBITAL
          SHOT PROJ

EROS (SATELLITES)

EROSION

Erosion, Electrostatic
USE   SPARK MACHINING

Erosion, Rain
USE   RAIN EROSION

Erosion, Soil
USE   SOIL EROSION

Erosion, Water
USE   WATER EROSION

Erosion, Wind
USE   WIND EROSION

EROSIVE BURNING

ERROR ANALYSIS

Error Band
USE   ACCURACY

Error, Boresight
USE   BORESIGHT ERROR

ERROR CORRECTING CODES

ERROR CORRECTING DEVICES

ERROR DETECTION CODES

Error, Flight Technical
USE   PILOT ERROR

ERROR FUNCTIONS

Error, Phase
USE   PHASE ERROR

Error, Pilot
USE   PILOT ERROR

Error Rate, Bit
USE   BIT ERROR RATE

ERROR SIGNALS

ERRORS

Errors, Instrument
USE   INSTRUMENT ERRORS

Errors, Perceptual
USE   PERCEPTUAL ERRORS

Errors, Position
USE   POSITION ERRORS

Errors, Random
USE   RANDOM ERRORS

Errors, Range
USE   RANGE ERRORS

Errors, Root-Mean-Square
USE   ROOT-MEAN-SQUARE ERRORS

Errors, Truncation
USE   TRUNCATION ERRORS

Errors, Velocity
USE   VELOCITY ERRORS

ERS 17

ERS 18

ERS-1 (ESA SATELLITE)

ERTS
USE   LANDSAT SATELLITES

ERTS-A
USE   LANDSAT 1

ERTS-B
USE   LANDSAT 2

ERTS-C
USE   LANDSAT 3

ERTS-D
USE   LANDSAT 4

ERTS-E
USE   LANDSAT E

ERTS-F
USE   LANDSAT F

ERYTHROCYTES

Es
USE   EINSTEINIUM

ESA
USE   EUROPEAN SPACE AGENCY

(ESA), EURECA
USE   EURECA (ESA)

(ESA), GEOS Satellites
USE   GEOS SATELLITES (ESA)

(Esa), Maritime Communication Satellite
USE   MAROTS (ESA)

(ESA), Marots
USE   MAROTS (ESA)

(Esa), Orbital Test Satellite
USE   OTS (ESA)

(ESA), Ots
USE   OTS (ESA)

(ESA Platforms), SPAS
USE   SHUTTLE PALLET SATELLITES

(ESA Satellite), ERS-1
USE   ERS-1 (ESA SATELLITE)

ESA SATELLITES

ESA SPACECRAFT

Esaki Diodes
USE   TUNNEL DIODES

ESCALATORS

ESCAPE

ESCAPE (ABANDONMENT)

ESCAPE CAPSULES

Escape Devices, Lunar
USE   LUNAR ESCAPE DEVICES

ESCAPE ROCKETS

ESCAPE SYSTEMS

Escape Systems, Launch
USE   LAUNCH ESCAPE SYSTEMS

(Escape Systems), LES
USE   LAUNCH ESCAPE SYSTEMS

ESCAPE VELOCITY

ESCARPMENTS

ESCHERICHIA

ESG (Gyroscopes)
USE   ELECTROSTATIC GYROSCOPES

Eskers
USE   GLACIAL DRIFT

ESKIMOS

ESOPHAGUS

ESRO
USE   EUROPEAN SPACE AGENCY

(Esro), GEOS Satellites
USE   GEOS SATELLITES (ESA)

ESRO Satellites
USE   ESA SATELLITES

ESRO 1 SATELLITE

ESRO 2 SATELLITE

ESRO 4 SATELLITE

ESSA SATELLITES

ESSA 1 SATELLITE

ESSA 2 SATELLITE

ESSA 3 SATELLITE

ESSA 4 SATELLITE

ESSA 5 SATELLITE

ESSA 6 SATELLITE

ESSA 7 SATELLITE

ESSA 8 SATELLITE

ESSA 9 SATELLITE

ESTERS

Esters, Nitrate
USE   NITRATE ESTERS

Esters, Poly
USE   POLYESTERS

ESTIMATES

Estimates, Cost
USE   COST ESTIMATES

Estimates, Maximum Likelihood
USE   MAXIMUM LIKELIHOOD ESTIMATES

ESTIMATING

Estimation, Orbital Position
USE   ORBITAL POSITION ESTIMATION

Estimation, State
USE   STATE ESTIMATION

ESTIMATORS

ESTONIA

ESTROGENS

ESTUARIES

ETA-MESONS

ETCHANTS

ETCHING

Etching, Plasma
USE   PLASMA ETCHING

ETHANE

Ether, Diethyl
USE   DIETHYL ETHER

Ether, Polyphenyl
USE   POLYPHENYL ETHER

ETHERS

ETHICS

ETHIOPIA

ETHNIC FACTORS

ETHOXY ETHYLENE

ETHYL ALCOHOL

ETHYL COMPOUNDS

ETHYLENE

Ethylene, Chloro
USE   CHLOROETHYLENE

ETHYLENE COMPOUNDS

ETHYLENE DIHYDRAZINE

Ethylene, Ethoxy
USE   ETHOXY ETHYLENE

ETHYLENE OXIDE

Ethylene, Polytetrafluoro
USE   POLYTETRAFLUOROETHYLENE

Ethylene, Vinyl
USE   BUTADIENE

ETHYLENEDIAMINE

ETHYLENEDIAMINETETRAACETIC ACIDS

Ethylenes, Poly
USE   POLYETHYLENES

ETIOLOGY

ETR (Reactors)
USE   ENGINEERING TEST REACTORS

Ettingshausen Coolers
USE   ETTINGSHAUSEN EFFECT
       THERMOELECTRIC COOLING

ETTINGSHAUSEN EFFECT

Ettingshausen Effect, Nernst-
USE   NERNST-ETTINGSHAUSEN EFFECT

Eu
USE   EUROPIUM

EUCLIDEAN GEOMETRY

Euclidean Space
USE   EUCLIDEAN GEOMETRY

EUDIOMETERS

EUGLENA

EULER BUCKLING

EULER EQUATIONS OF MOTION

EULER-CAUCHY EQUATIONS

EULER-LAGRANGE EQUATION

EULER-LAMBERT EQUATION

EURECA (ESA)

EUROPA

EUROPA LAUNCH VEHICLES

EUROPA 1 LAUNCH VEHICLE

EUROPA 2 LAUNCH VEHICLE

EUROPA 3 LAUNCH VEHICLE

EUROPA 4 LAUNCH VEHICLE

EUROPE

(Europe), Alps Mountains
   USE   ALPS MOUNTAINS (EUROPE)

(Europe), Baltic Shield
   USE   BALTIC SHIELD (EUROPE)

(Europe), Carpathian Mountains
   USE   CARPATHIAN MOUNTAINS (EUROPE)

Europe, Central
   USE   CENTRAL EUROPE

(Europe), Eiscat Radar System
   USE   EISCAT RADAR SYSTEM (EUROPE)

(Europe), Pyrenees Mountains
   USE   PYRENEES MOUNTAINS (EUROPE)

EUROPEAN AIRBUS

EUROPEAN COMMUNICATIONS SATELLITE

European Incoherent Scatter Radar
   USE   EISCAT RADAR SYSTEM (EUROPE)

European Large Telecomm Satellite
   USE   L-SAT

European Retrievable Carrier
   USE   EURECA (ESA)

EUROPEAN SPACE AGENCY

EUROPEAN SPACE PROGRAMS

European Space Research Organization
   USE   EUROPEAN SPACE AGENCY

European Space Research Organization Sat
   USE   ESA SATELLITES

European Torus, Joint
   USE   JOINT EUROPEAN TORUS

EUROPEAN 1 SPACECRAFT

EUROPIUM

EUROPIUM COMPOUNDS

EUROPIUM ISOTOPES

EUSTACHIAN TUBES

EUTECTIC ALLOYS

EUTECTIC COMPOSITES

Eutectic Diagrams
   USE   PHASE DIAGRAMS

EUTECTICS

EUTROPHICATION

EUVE
   USE   EXTREME ULTRAVIOLET EXPLORER
       SATELLITE

EUXENITE

EVA Protection Systems, Advanced
   USE   AEPS

EVACUATING

Evacuating, Gas
   USE   EVACUATING (VACUUM)

EVACUATING (TRANSPORTATION)

EVACUATING (VACUUM)

EVAL
   USE   EARTH VIEWING APPLICATIONS
       LABORATORY

EVALUATION

Evaluation And Review Techniques, Graphic
   USE   GERT

Evaluation, Threat
   USE   THREAT EVALUATION

Evaluation, Training
   USE   TRAINING EVALUATION

Evaluator/monitor, Data Adaptive
   USE   DATA TRANSMISSION
       DATA REDUCTION
       DATA PROCESSING

EVANESCENCE

EVAPORATION

Evaporation, Propellant
   USE   PROPELLANT EVAPORATION

EVAPORATION RATE

EVAPORATIVE COOLING

EVAPORATORS

EVAPOROGRAPHY

EVAPOTRANSPIRATION

EVASIVE ACTIONS

EVASIVE SATELLITES

Evection
   USE   ORBIT PERTURBATION
       LUNAR ORBITS
       SOLAR GRAVITATION

Even Nuclei, Even-
   USE   EVEN-EVEN NUCLEI

Even Nuclei, Odd-
   USE   ODD-EVEN NUCLEI

EVEN-EVEN NUCLEI

EVENING

Event Upsets, Single
   USE   SINGLE EVENT UPSETS

EVENTS

Events, Consecutive
   USE   CONSECUTIVE EVENTS

EVERGLADES (FL)

EVOKED RESPONSE (PSYCHOPHYSIOLOGY)

EVOLUTION

Evolution, Biological
   USE   BIOLOGICAL EVOLUTION

Evolution, Chemical
   USE   CHEMICAL EVOLUTION

EVOLUTION (DEVELOPMENT)

Evolution Equations, Linear
   USE   LINEAR EVOLUTION EQUATIONS

Evolution Equations, Nonlinear
   USE   NONLINEAR EVOLUTION EQUATIONS

Evolution, Galactic
   USE   GALACTIC EVOLUTION

Evolution, Gas
   USE   GAS EVOLUTION

EVOLUTION (LIBERATION)

Evolution, Lunar
   USE   LUNAR EVOLUTION

Evolution, Planetary
   USE   PLANETARY EVOLUTION

Evolution, Stellar
   USE   STELLAR EVOLUTION

Exactness
   USE   PRECISION

EXAMINATION

Examinations, Eye
   USE   EYE EXAMINATIONS

Examinations, Physical
   USE   PHYSICAL EXAMINATIONS

EXCAVATION

(Excavation), Ditching
   USE   EXCAVATION

(Excavation), Tunneling
   USE   TUNNELING (EXCAVATION)

(Excavations), Mines
   USE   MINES (EXCAVATIONS)

(Excavations), Pits
   USE   PITS (EXCAVATIONS)

Exchange, Charge
   USE   CHARGE EXCHANGE

Exchange, Energy
   USE   ENERGY TRANSFER

Exchange, Gas
   USE   GAS EXCHANGE

Exchange), IDEP (Data
   USE   INTERSERVICE DATA EXCHANGE
       PROGRAM

Exchange Membrane Electrolytes, Ion
   USE   ION EXCHANGE MEMBRANE
       ELECTROLYTES

Exchange Program, Interservice Data
   USE   INTERSERVICE DATA EXCHANGE
       PROGRAM

Exchange Resins, Ion
   USE   ION EXCHANGE RESINS

Exchange, Resonance Charge
   USE   RESONANCE CHARGE EXCHANGE

Exchange, Spin
   USE   SPIN EXCHANGE

EXCHANGERS

Exchangers, Heat
   USE   HEAT EXCHANGERS

Exchangers, Tube Heat
   USE   TUBE HEAT EXCHANGERS

EXCHANGING

Exchanging, Ion
   USE   ION EXCHANGING

EXCIMER LASERS

EXCIMERS

EXCITATION

Excitation, Acoustic
   USE   ACOUSTIC EXCITATION

Excitation, Harmonic
   USE   HARMONIC EXCITATION

Excitation, Molecular
   USE   MOLECULAR EXCITATION

Excitation, Self
   USE   SELF EXCITATION

Excitation, Triplet
   USE   ATOMIC ENERGY LEVELS

Excitation, Wave
   USE   WAVE EXCITATION

Excitations, Atomic
USE   ATOMIC EXCITATIONS

Excitations, Elementary
USE   ELEMENTARY EXCITATIONS

Excited Atmospheric Lasers, Transversely
USE   TEA LASERS

Excited States
USE   EXCITATION

EXCITONS

EXCLUSION

Exclusion Principle, Pauli
USE   PAULI EXCLUSION PRINCIPLE

EXCRETION

Excursion Module, Mars
USE   MARS EXCURSION MODULE

(Excursion Module), MEM
USE   MARS EXCURSION MODULE

Executive Aircraft
USE   GENERAL AVIATION AIRCRAFT
     PASSENGER AIRCRAFT

Exercise
USE   PHYSICAL EXERCISE

Exercise, Physical
USE   PHYSICAL EXERCISE

EXERCISE PHYSIOLOGY

Exercise, Valsalva
USE   VALSALVA EXERCISE

Exertion
USE   PHYSICAL WORK

EXHALATION

EXHAUST DIFFUSERS

EXHAUST EMISSION

EXHAUST FLOW SIMULATION

EXHAUST GASES

Exhaust, Hot Jet
USE   JET EXHAUST
     HIGH TEMPERATURE GASES

Exhaust, Jet
USE   JET EXHAUST

Exhaust Jets
USE   EXHAUST GASES

EXHAUST NOZZLES

Exhaust Nozzles, Turbine
USE   TURBINE EXHAUST NOZZLES

Exhaust, Rocket
USE   ROCKET EXHAUST

EXHAUST SYSTEMS

EXHAUST VELOCITY

EXHAUSTING

EXHAUSTION

EXISTENCE

EXISTENCE THEOREMS

Exits (Doors)
USE   DOORS

EXOBIOLOGY

Exophoria
USE   HETEROPHORIA

EXOS SOUNDING ROCKET

EXOSAT SATELLITE

EXOSKELETONS

EXOSPHERE

EXOTHERMIC REACTIONS

Exp Background Sats, Galactic Radiation
USE   GREB SATELLITES

Exp, Health-Education Telecommunications
USE   HET EXPERIMENT

EXPANDABLE STRUCTURES

EXPANSION

Expansion, Gas
USE   GAS EXPANSION

Expansion, Karhunen-Loeve
USE   KARHUNEN-LOEVE EXPANSION

Expansion, Light-Cone
USE   LIGHT-CONE EXPANSION

Expansion, Prandtl-Meyer
USE   PRANDTL-MEYER EXPANSION

Expansion, Series
USE   SERIES EXPANSION

Expansion, Thermal
USE   THERMAL EXPANSION

Expansion Waves
USE   ELASTIC WAVES

EXPECTANCY HYPOTHESIS

EXPECTATION

EXPEDITIONS

EXPELLANTS

EXPENDABLE STAGES (SPACECRAFT)

Exper, Feature Identification And Location
USE   FEATURE IDENTIFICATION AND LOCATION
     EXPER

Exper With Particle Accelerators, Space
USE   SEPAC (PAYLOAD)

EXPERIENCE

Experiment, Atmospheric General Circulation
USE   ATMOSPHERIC GENERAL CIRCULATION
     EXPERIMENT

EXPERIMENT DESIGN

Experiment, Earth Energy Budget
USE   LZEEBE SATELLITE

Experiment, Earth Radiation Budget
USE   EARTH RADIATION BUDGET EXPERIMENT

Experiment, Electromagnetic Environment
USE   ELECTROMAGNETIC ENVIRONMENT
     EXPERIMENT

Experiment, GARP Atlantic Tropical
USE   GARP ATLANTIC TROPICAL EXPERIMENT

(Experiment), GATE
USE   GARP ATLANTIC TROPICAL EXPERIMENT

Experiment, Halogen Occultation
USE   HALOGEN OCCULTATION EXPERIMENT

Experiment, HET
USE   HET EXPERIMENT

Experiment In Space, Physics And Chemistry
USE   PHYSICS AND CHEMISTRY EXPERIMENT IN
     SPACE

Experiment, International Satellite Geodesy
USE   INTERNATIONAL SATELLITE GEODESY
     EXPERIMENT

(Experiment), LACATE
USE   LACATE (EXPERIMENT)

Experiment, Large Area Crop Inventory
USE   LARGE AREA CROP INVENTORY
     EXPERIMENT

Experiment, Lithium Cooled Reactor
USE   LITHIUM COOLED REACTOR EXPERIMENT

Experiment, Long Term Zonal Earth Energy
USE   LZEEBE SATELLITE

Experiment, Lower Atmospheric Composition
USE   LACATE (EXPERIMENT)

Experiment Module, Apollo Lunar
USE   APOLLO LUNAR EXPERIMENT MODULE

Experiment Package, Earth Resources
USE   EREP

Experiment Package Telescope, Goddard
USE   PARTICLE TELESCOPES

Experiment, Plasma Interaction
USE   PLASMA INTERACTION EXPERIMENT

Experiment, San Andreas Fault
USE   SAN ANDREAS FAULT EXPERIMENT

Experiment Scientific Satellite, Biomedical
USE   BESS (SATELLITE)

Experiment, Sodium Reactor
USE   SODIUM REACTOR EXPERIMENT

Experiment Stations, Crew
USE   CREW EXPERIMENT STATIONS

Experiment, Stratospheric Aerosol & Gas
USE   SAGE SATELLITE

Experiment, Zonal Earth Energy Budget
USE   LZEEBE SATELLITE

EXPERIMENTAL BOILING WATER REACTORS

EXPERIMENTAL BREEDER REACTOR 1

EXPERIMENTAL BREEDER REACTOR 2

EXPERIMENTAL GAS COOLED REACTORS

Experimental Ocean Satellite, Geodynamic
USE   GEOS-D SATELLITE

EXPERIMENTAL ORGANIC COOLED REACTORS

EXPERIMENTAL REFLECTOR ORBITAL SHOT PROJ

Experimental Satellites, Lincoln
USE   LINCOLN EXPERIMENTAL SATELLITES

Experimental STOL Transport Rsch Airplane
USE   QUESTOL

EXPERIMENTATION

Experiments, Critical
USE   CRITICAL EXPERIMENTS

Experiments, Design Of
USE   EXPERIMENT DESIGN

Experiments Package, Apollo Lunar Surface
USE   APOLLO LUNAR SURFACE EXPERIMENTS
     PACKAGE

Experiments Package, Early Apollo Surface
USE   EASEP

Experiments, Space Plasma H/v Interaction
　　USE　　SPHINX

Experiments, Space Technology
　　USE　　SPACE TECHNOLOGY EXPERIMENTS

Experiments, Spaceborne
　　USE　　SPACEBORNE EXPERIMENTS

EXPERT SYSTEMS

EXPIRATION

EXPIRED AIR

Exploding Conductor Circuits
　　USE　　EXPLODING WIRES
　　　　　　CIRCUITS

Exploding Conductors
　　USE　　EXPLODING WIRES

EXPLODING WIRES

EXPLOITATION

EXPLORATION

Exploration, Lunar
　　USE　　LUNAR EXPLORATION

Exploration, Mineral
　　USE　　MINERAL EXPLORATION

Exploration, Natural Gas
　　USE　　NATURAL GAS EXPLORATION

Exploration, Oil
　　USE　　OIL EXPLORATION

Exploration, Planetary
　　USE　　SPACE EXPLORATION

Exploration, Space
　　USE　　SPACE EXPLORATION

Exploration System For Apollo, Lunar
　　USE　　LUNAR EXPLORATION SYSTEM FOR
　　　　　　APOLLO

Exploration System), LESA (Lunar
　　USE　　LUNAR EXPLORATION SYSTEM FOR
　　　　　　APOLLO

Explorer A, Air Density
　　USE　　EXPLORER 19 SATELLITE

Explorer A, Atmosphere
　　USE　　EXPLORER 17 SATELLITE

Explorer A, Beacon
　　USE　　BEACON EXPLORER A

Explorer A, Energetic Particle
　　USE　　EXPLORER 12 SATELLITE

Explorer A, Ionosphere
　　USE　　EXPLORER 20 SATELLITE

Explorer B, Air Density/injun
　　USE　　EXPLORER 25 SATELLITE

Explorer B, Atmosphere
　　USE　　EXPLORER 32 SATELLITE

Explorer B, Beacon
　　USE　　EXPLORER 22 SATELLITE

Explorer B, Energetic Particle
　　USE　　EXPLORER 14 SATELLITE

Explorer B, Radio Astronomy
　　USE　　EXPLORER 49 SATELLITE

Explorer C, Atmosphere
　　USE　　EXPLORER 51 SATELLITE

Explorer C, Beacon
　　USE　　EXPLORER 27 SATELLITE

Explorer C, Energetic Particle
　　USE　　EXPLORER 15 SATELLITE

Explorer D, Atmosphere
　　USE　　EXPLORER 54 SATELLITE

Explorer D, Energetic Particle
　　USE　　EXPLORER 26 SATELLITE

Explorer, DAD
　　USE　　DUAL AIR DENSITY EXPLORER

Explorer, Dual Air Density
　　USE　　DUAL AIR DENSITY EXPLORER

Explorer E, Atmosphere
　　USE　　EXPLORER 55 SATELLITE

Explorer, Far UV Spectroscopic
　　USE　　FAR UV SPECTROSCOPIC EXPLORER

Explorer, Gamma Ray Astronomy
　　USE　　EXPLORER 11 SATELLITE

Explorer, Injun
　　USE　　EXPLORER 25 SATELLITE

Explorer, International Magnetospheric
　　USE　　INTERNATIONAL MAGNETOSPHERIC
　　　　　　EXPLORER

Explorer, International Ultraviolet
　　USE　　IUE

Explorer, Interplanetary
　　USE　　EXPLORER 18 SATELLITE

Explorer, Planetary
　　USE　　OUTER PLANETS EXPLORERS

Explorer Satellite, Cosmic Background
　　USE　　COSMIC BACKGROUND EXPLORER
　　　　　　SATELLITE

Explorer Satellite, Extreme Ultraviolet
　　USE　　EXTREME ULTRAVIOLET EXPLORER
　　　　　　SATELLITE

Explorer Satellite, Radio Astronomy
　　USE　　RADIO ASTRONOMY EXPLORER SATELLITE

EXPLORER SATELLITES

Explorer Satellites, Applications
　　USE　　APPLICATIONS EXPLORER SATELLITES

Explorer Satellites, Dynamics
　　USE　　DYNAMICS EXPLORER SATELLITES

Explorer Satellites, Micrometeoroid
　　USE　　MICROMETEOROID EXPLORER SATELLITES

Explorer, Solar Mesosphere
　　USE　　SOLAR MESOSPHERE EXPLORER

Explorer, X Ray Timing
　　USE　　X RAY TIMING EXPLORER

Explorer 1, International Sun Earth
　　USE　　INTERNATIONAL SUN EARTH EXPLORER 1

EXPLORER 1 SATELLITE

Explorer 1 Satellite, Dynamics
　　USE　　DYNAMICS EXPLORER 1 SATELLITE

Explorer 2, International Sun Earth
　　USE　　INTERNATIONAL SUN EARTH EXPLORER 2

Explorer 2, Radio Astronomy
　　USE　　EXPLORER 49 SATELLITE

EXPLORER 2 SATELLITE

Explorer 2 Satellite, Dynamics
　　USE　　DYNAMICS EXPLORER 2 SATELLITE

Explorer 3, International Sun Earth
　　USE　　INTERNATIONAL SUN EARTH EXPLORER 3

EXPLORER 3 SATELLITE

EXPLORER 4 SATELLITE

EXPLORER 5 SATELLITE

EXPLORER 6 SATELLITE

EXPLORER 7 SATELLITE

EXPLORER 8 SATELLITE

EXPLORER 9 SATELLITE

EXPLORER 10 SATELLITE

EXPLORER 11 SATELLITE

EXPLORER 12 SATELLITE

EXPLORER 14 SATELLITE

EXPLORER 15 SATELLITE

EXPLORER 16 SATELLITE

EXPLORER 17 SATELLITE

EXPLORER 18 SATELLITE

EXPLORER 19 SATELLITE

EXPLORER 20 SATELLITE

EXPLORER 21 SATELLITE

EXPLORER 22 SATELLITE

EXPLORER 23 SATELLITE

EXPLORER 24 SATELLITE

EXPLORER 25 SATELLITE

EXPLORER 26 SATELLITE

EXPLORER 27 SATELLITE

EXPLORER 28 SATELLITE

EXPLORER 29 SATELLITE

EXPLORER 30 SATELLITE

EXPLORER 31 SATELLITE

EXPLORER 32 SATELLITE

EXPLORER 33 SATELLITE

EXPLORER 34 SATELLITE

EXPLORER 35 SATELLITE

EXPLORER 36 SATELLITE

EXPLORER 37 SATELLITE

EXPLORER 38 SATELLITE

EXPLORER 39 SATELLITE

EXPLORER 40 SATELLITE

EXPLORER 41 SATELLITE

Explorer 42 Satellite
　　USE　　UHURU SATELLITE

EXPLORER 43 SATELLITE

EXPLORER 44 SATELLITE

EXPLORER 45 SATELLITE

EXPLORER 46 SATELLITE

EXPLORER 47 SATELLITE

EXPLORER 48 SATELLITE

EXPLORER 49 SATELLITE

EXPLORER 50 SATELLITE

EXPLORER 51 SATELLITE

EXPLORER 52 SATELLITE

EXPLORER 53 SATELLITE

EXPLORER 54 SATELLITE

EXPLORER 55 SATELLITE

Explorers, Active Magneto Particle Tracer
    USE   AMPTE (SATELLITES)

Explorers, International Sun Earth
    USE   INTERNATIONAL SUN EARTH EXPLORERS

Explorers, Outer Planets
    USE   OUTER PLANETS EXPLORERS

Explosion Effect, Nuclear
    USE   NUCLEAR EXPLOSION EFFECT

EXPLOSION SUPPRESSION

EXPLOSIONS

Explosions, Aerial
    USE   AERIAL EXPLOSIONS

Explosions, Atomic
    USE   NUCLEAR EXPLOSIONS

Explosions, Chemical
    USE   CHEMICAL EXPLOSIONS

Explosions, Gas
    USE   GAS EXPLOSIONS

Explosions, Nuclear
    USE   NUCLEAR EXPLOSIONS

Explosions, Propellant
    USE   PROPELLANT EXPLOSIONS

Explosions, Thermonuclear
    USE   THERMONUCLEAR EXPLOSIONS

Explosions, Underground
    USE   UNDERGROUND EXPLOSIONS

Explosions, Underwater
    USE   UNDERWATER EXPLOSIONS

EXPLOSIVE DECOMPRESSION

EXPLOSIVE DEVICES

EXPLOSIVE FORMING

Explosive Gases
    USE   FLAMMABLE GASES

(Explosive), Octol
    USE   OCTOL (EXPLOSIVE)

EXPLOSIVE WELDING

EXPLOSIVES

(Explosives), Boosters
    USE   BOOSTERS (EXPLOSIVES)

(Explosives), Caps
    USE   CAPS (EXPLOSIVES)

(Explosives), Initiators
    USE   INITIATORS (EXPLOSIVES)

Explosives, Nitrasol
    USE   NITRASOL EXPLOSIVES

(Explosives), Primers
    USE   PRIMERS (EXPLOSIVES)

EXPONENTIAL FUNCTIONS

EXPONENTS

Exports
    USE   INTERNATIONAL TRADE

EXPOS (SPACELAB PAYLOAD)

EXPOSURE

Exposure Facility, Long Duration
    USE   LONG DURATION EXPOSURE FACILITY

Exposure, Radiation
    USE   RADIATION DOSAGE

Expressions (Mathematics)
    USE   FORMULAS (MATHEMATICS)

EXPULSION

EXPULSION BLADDERS

EXTARS

Extended Duration Space Flight
    USE   LONG DURATION SPACE FLIGHT

(Extension), Propagation
    USE   PROPAGATION (EXTENSION)

Extension System, Apollo
    USE   APOLLO EXTENSION SYSTEM

EXTENSIONS

EXTENSOMETERS

EXTERNAL COMBUSTION ENGINES

EXTERNAL STORE SEPARATION

EXTERNAL STORES

(External Stores), Pods
    USE   PODS (EXTERNAL STORES)

EXTERNAL SURFACE CURRENTS

EXTERNAL TANKS

EXTERNALLY BLOWN FLAPS

EXTINCTION

Extinction, Interstellar
    USE   INTERSTELLAR EXTINCTION

Extinguishers
    USE   FIRE EXTINGUISHERS

Extinguishers, Chemical
    USE   FIRE EXTINGUISHERS

Extinguishers, Fire
    USE   FIRE EXTINGUISHERS

EXTINGUISHING

EXTRACTION

Extraction, Feature
    USE   PATTERN RECOGNITION

Extraction, Geothermal Energy
    USE   GEOTHERMAL ENERGY EXTRACTION

Extraction, Ion
    USE   ION EXTRACTION

Extraction, Solvent
    USE   SOLVENT EXTRACTION

Extragalactic Light
    USE   LIGHT (VISIBLE RADIATION)
           EXTRATERRESTRIAL RADIATION

Extragalactic Media
    USE   INTERGALACTIC MEDIA

EXTRAGALACTIC RADIO SOURCES

EXTRAPOLATION

EXTRASENSORY PERCEPTION

EXTRASOLAR PLANETS

EXTRATERRESTRIAL COMMUNICATION

EXTRATERRESTRIAL ENVIRONMENTS

EXTRATERRESTRIAL INTELLIGENCE

Extraterrestrial Intelligence, Search For
    USE   PROJECT SETI

EXTRATERRESTRIAL LIFE

EXTRATERRESTRIAL MATTER

EXTRATERRESTRIAL RADIATION

EXTRATERRESTRIAL RADIO WAVES

EXTRATERRESTRIAL RESOURCES

Extraterrestrial Roving Vehicles
    USE   ROVING VEHICLES

EXTRAVEHICULAR ACTIVITY

EXTRAVEHICULAR MOBILITY UNITS

Extrema
    USE   RANGE (EXTREMES)

EXTREME ULTRAVIOLET EXPLORER SATELLITE

EXTREME ULTRAVIOLET RADIATION

EXTREMELY HIGH FREQUENCIES

EXTREMELY LOW FREQUENCIES

EXTREMELY LOW RADIO FREQUENCIES

(Extremes), Range
    USE   RANGE (EXTREMES)

EXTREMUM VALUES

EXTROVERSION

EXTRUDING

Extruding, Hot
    USE   EXTRUDING

EYE (ANATOMY)

EYE DISEASES

EYE DOMINANCE

EYE EXAMINATIONS

Eye Movement State, Rapid
    USE   RAPID EYE MOVEMENT STATE

EYE MOVEMENTS

Eye Movements, Saccadic
    USE   SACCADIC EYE MOVEMENTS

EYE PROTECTION

EYEPIECES

EYRING THEORY

F

F Centers
    USE   COLOR CENTERS

F Displays
    USE   F REGION

F, Earth Resources Technology Satellite
    USE   LANDSAT F

F, ERTS-
USE   LANDSAT F

F ICBM, Atlas
USE   ATLAS F ICBM

F, IMP-
USE   EXPLORER 34 SATELLITE

F, KEL-
USE   KEL-F

F, LANDSAT
USE   LANDSAT F

F Layer
USE   F REGION

F Layer, Night
USE   F REGION
          NIGHT SKY

F, OGO-
USE   OGO-6

F, OSO-
USE   OSO-5

F REGION

F Satellite, TIROS
USE   TIROS 6 SATELLITE

F Space Probe, Pioneer
USE   PIONEER 10 SPACE PROBE

F, Space Shuttle Mission 51-
USE   SPACE SHUTTLE MISSION 51-F

F, Spread
USE   SPREAD F

F 1 REGION

F 2 REGION

F 27 Aircraft, Fokker
USE   F-27 AIRCRAFT

F 28 Aircraft, Fokker
USE   F-28 TRANSPORT AIRCRAFT

F-Scatter Propagation, Ionospheric
USE   IONOSPHERIC F-SCATTER PROPAGATION

F-1 ROCKET ENGINE

F-2 AIRCRAFT

F-2 Aircraft, Hunter
USE   F-2 AIRCRAFT

F-4 AIRCRAFT

F-5 AIRCRAFT

F-8 AIRCRAFT

F-9 AIRCRAFT

F-14 AIRCRAFT

F-15 AIRCRAFT

F-16 AIRCRAFT

F-17 AIRCRAFT

F-18 AIRCRAFT

F-20 AIRCRAFT

F-27 AIRCRAFT

F-28 HELICOPTER

F-28 TRANSPORT AIRCRAFT

F-80 Aircraft
USE   T-33 AIRCRAFT

F-84 AIRCRAFT

F-86 AIRCRAFT

F-89 AIRCRAFT

F-94 AIRCRAFT

F-100 AIRCRAFT

F-101 AIRCRAFT

F-102 AIRCRAFT

F-104 AIRCRAFT

F-105 AIRCRAFT

F-106 AIRCRAFT

F-110 Aircraft
USE   F-4 AIRCRAFT

F-111 AIRCRAFT

FAB (Programming Language)
USE   FORTRAN

FABRICATION

FABRICS

Fabrics, Geotechnical
USE   GEOTECHNICAL FABRICS

Fabrics, Parachute
USE   PARACHUTE FABRICS

FABRY-PEROT INTERFEROMETERS

Fabry-Perot Lasers
USE   LASERS

FABRY-PEROT SPECTROMETERS

FACE (ANATOMY)

FACE CENTERED CUBIC LATTICES

Faces, Inter
USE   INTERFACES

Facets
USE   FLAT SURFACES

FACILITIES

Facilities, Military Air
USE   MILITARY AIR FACILITIES

(Facilities), Ranges
USE   RANGES (FACILITIES)

Facilities, Research
USE   RESEARCH FACILITIES

Facilities, Rocket Test
USE   ROCKET TEST FACILITIES

Facilities, Terminal
USE   TERMINAL FACILITIES

Facilities, Test
USE   TEST FACILITIES

Facility, Advanced X Ray Astrophysical
USE   X RAY ASTROPHYSICS FACILITY

Facility, Advanced X Ray Astrophysics
USE   X RAY ASTROPHYSICS FACILITY

Facility, Deep Space Instrumentation
USE   DEEP SPACE INSTRUMENTATION FACILITY

Facility), DSIF (Instrumentation
USE   DEEP SPACE INSTRUMENTATION FACILITY

Facility, Hallam Nuclear Power
USE   HALLAM NUCLEAR POWER FACILITY

Facility), HNPF (Hallam Nuclear Power
USE   HALLAM NUCLEAR POWER FACILITY

Facility, Long Duration Exposure
USE   LONG DURATION EXPOSURE FACILITY

Facility, Mobile Quarantine
USE   MOBILE QUARANTINE FACILITY

Facility, Solar Cell Calibration
USE   SOLAR CELL CALIBRATION FACILITY

Facility, Space Infrared Telescope
USE   SPACE INFRARED TELESCOPE FACILITY

Facility, Spacelab UV-Optical Telescope
USE   STARLAB

Facility, Transient Reactor Test
USE   TRANSIENT REACTOR TEST FACILITY

Facility), TREAT (Test
USE   TRANSIENT REACTOR TEST FACILITY

Facility, X Ray Astrophysics
USE   X RAY ASTROPHYSICS FACILITY

Facing Steps, Backward
USE   BACKWARD FACING STEPS

Facing Steps, Rearward
USE   BACKWARD FACING STEPS

FACSIMILE COMMUNICATION

Facsimile Transmission
USE   FACSIMILE COMMUNICATION

Factor, Age
USE   AGE FACTOR

Factor, Amplification
USE   AMPLIFICATION

FACTOR ANALYSIS

Factor, Beta
USE   BETA FACTOR

Factor Controllers, Power
USE   POWER FACTOR CONTROLLERS

Factor, Damping
USE   DAMPING

Factor, Friction
USE   FRICTION FACTOR

Factor, Landau
USE   LANDAU FACTOR

Factor, Nu
USE   NU FACTOR

Factor, Ph
USE   PH FACTOR

Factor, Rhesus
USE   RHESUS FACTOR

Factor, Sex
USE   SEX FACTOR

Factor Table, Interference
USE   INTERFERENCE FACTOR TABLE

FACTORIAL DESIGN

FACTORIALS

Factories
USE   INDUSTRIAL PLANTS

FACTORIZATION

Factorization, Cholesky
USE   CHOLESKY FACTORIZATION

Factors
USE   VARIABLE

Factors, Economic
USE ECONOMIC FACTORS

Factors, Emotional
USE EMOTIONAL FACTORS

Factors Engineering, Human
USE HUMAN FACTORS ENGINEERING

Factors, Ethnic
USE ETHNIC FACTORS

Factors, Form
USE FORM FACTORS

Factors Laboratories, Human
USE HUMAN FACTORS LABORATORIES

Factors, Load
USE LOADS (FORCES)

Factors, Mass Flow
USE MASS FLOW FACTORS

Factors, Physical
USE PHYSICAL FACTORS

Factors, Physiological
USE PHYSIOLOGICAL FACTORS

Factors, Psychological
USE PSYCHOLOGICAL FACTORS

Factors, Q
USE Q FACTORS

Factors, Quality
USE Q FACTORS

Factors, Race
USE RACE FACTORS

Factors, Safety
USE SAFETY FACTORS

Factors, Social
USE SOCIAL FACTORS

Factors, Stress Intensity
USE STRESS INTENSITY FACTORS

Factors, Weight
USE WEIGHT (MASS)

FACULAE

(Faculae), Plages
USE FACULAE

Faculae, Solar
USE FACULAE

FADDEEV EQUATIONS

Fadeout, Signal
USE SIGNAL FADING

FADING

Fading Rate, Signal
USE SIGNAL FADING RATE

Fading, Selective
USE SELECTIVE FADING

Fading, Signal
USE SIGNAL FADING

Fahrenheit Temperature Scale
USE TEMPERATURE SCALES

FAIL-SAFE SYSTEMS

FAILURE

FAILURE ANALYSIS

(Failure), Burnthrough
USE BURNTHROUGH (FAILURE)

Failure, Engine
USE ENGINE FAILURE

FAILURE MODES

Failure, Structural
USE STRUCTURAL FAILURE

Failures, Mean Time Between
USE MTBF

Failures, System
USE SYSTEM FAILURES

FAINT OBJECT CAMERA

Fainting
USE SYNCOPE

Fairchild Military Aircraft
USE FAIRCHILD-HILLER AIRCRAFT
MILITARY AIRCRAFT

FAIRCHILD-HILLER AIRCRAFT

FAIREY AIRCRAFT

Fairey Delta 2 Aircraft
USE FD 2 AIRCRAFT

FAIRINGS

FAITH 7

FALCON MISSILE

FALKNER-SKAN EQUATION

Fall, Free
USE FREE FALL

FALLING

FALLING SPHERES

FALLOUT

FAN BLADES

Fan Engines, Ducted
USE DUCTED FAN ENGINES

FAN IN WING AIRCRAFT

Fan Technology, Prop-
USE PROP-FAN TECHNOLOGY

Fanlift Devices
USE LIFT FANS

FANS

Fans, Ducted
USE DUCTED FANS

FANS (LANDFORMS)

Fans, Lift
USE LIFT FANS

Fans, Propeller
USE PROPELLER FANS

Fans, Turbo
USE TURBOFANS

Fans, Ventilation
USE VENTILATION FANS

FAR FIELDS

FAR INFRARED RADIATION

Far Side, Lunar
USE LUNAR FAR SIDE

FAR ULTRAVIOLET RADIATION

FAR UV SPECTROSCOPIC EXPLORER

FARADAY DARK SPACE

FARADAY EFFECT

Faraday Rotation
USE FARADAY EFFECT

FARM CROPS

FARMLANDS

Fast Breeder Reactors, Liquid Metal
USE LIQUID METAL FAST BREEDER REACTORS

FAST FOURIER TRANSFORMATIONS

FAST NEUTRONS

FAST NUCLEAR REACTORS

FAST OXIDE REACTORS

Fast Reactors, Gas Cooled
USE GAS COOLED FAST REACTORS

FAST TEST REACTORS

FASTENERS

(Fasteners), Anchors
USE ANCHORS (FASTENERS)

(Fasteners), Locks
USE LOCKS (FASTENERS)

(Fasteners), Nuts
USE NUTS (FASTENERS)

FASTING

FAT EMBOLISMS

Fatigue, Acoustic
USE ACOUSTIC FATIGUE

Fatigue, Auditory
USE AUDITORY FATIGUE

Fatigue, Bending
USE BENDING FATIGUE

FATIGUE (BIOLOGY)

Fatigue Diagrams
USE S-N DIAGRAMS

Fatigue, Flight
USE FLIGHT FATIGUE

FATIGUE LIFE

FATIGUE (MATERIALS)

Fatigue, Metal
USE METAL FATIGUE

Fatigue, Muscular
USE MUSCULAR FATIGUE

Fatigue, Shear
USE SHEAR STRESS

Fatigue, Sonic
USE ACOUSTIC FATIGUE

Fatigue, Strain
USE FATIGUE (MATERIALS)

Fatigue, Structural
USE FATIGUE (MATERIALS)

FATIGUE TESTING MACHINES

FATIGUE TESTS

Fatigue, Thermal
USE THERMAL FATIGUE

FATS

FATTY ACIDS

Fault Energy, Stacking
 USE   STACKING FAULT ENERGY

Fault Experiment, San Andreas
 USE   SAN ANDREAS FAULT EXPERIMENT

Fault Mechanics
 USE   FRACTURE MECHANICS

Fault, San Andreas
 USE   SAN ANDREAS FAULT

FAULT TOLERANCE

FAULT TREES

FAULTS

Faults, Closed
 USE   GEOLOGICAL FAULTS

Faults, Cross
 USE   GEOLOGICAL FAULTS

Faults, Echelon
 USE   GEOLOGICAL FAULTS

Faults, Electrical
 USE   ELECTRICAL FAULTS

Faults, Geological
 USE   GEOLOGICAL FAULTS

Faults, Stacking
 USE   CRYSTAL DEFECTS

Faults, Step
 USE   GEOLOGICAL FAULTS

Faults, Thrust
 USE   GEOLOGICAL FAULTS

Fauna
 USE   ANIMALS

FAYALITE

FBFM (Modulation)
 USE   FEEDBACK FREQUENCY MODULATION

FBM (Missiles)
 USE   FLEET BALLISTIC MISSILES

FCC Lattices
 USE   FACE CENTERED CUBIC LATTICES

FD 2 AIRCRAFT

FDL-5 REENTRY VEHICLE

FDMA
 USE   FREQUENCY DIVISION MULTIPLE ACCESS

Fe
 USE   IRON

FEAR

FEAR OF FLYING

FEASIBILITY

FEASIBILITY ANALYSIS

Feasibility Spacecraft, Technology
 USE   TECHNOLOGY FEASIBILITY SPACECRAFT

FEATHER RIVER BASIN (CA)

FEATHERING

Feature Extraction
 USE   PATTERN RECOGNITION

FEATURE IDENTIFICATION AND LOCATION EXPER

Features), Bays (Topographic
 USE   BAYS (TOPOGRAPHIC FEATURES)

Features), Sounds (Topographic
 USE   SOUNDS (TOPOGRAPHIC FEATURES)

FECES

Fechner Law, Weber-
 USE   WEBER-FECHNER LAW

FEDERAL BUDGETS

Federal Republic Of Germany
 USE   WEST GERMANY

FEDERATIONS

FEED SYSTEMS

FEEDBACK

FEEDBACK AMPLIFIERS

Feedback, Bio
 USE   BIOFEEDBACK

FEEDBACK CIRCUITS

FEEDBACK CONTROL

Feedback, Degenerative
 USE   NEGATIVE FEEDBACK

FEEDBACK FREQUENCY MODULATION

Feedback Lasers, Distributed
 USE   DISTRIBUTED FEEDBACK LASERS

Feedback, Negative
 USE   NEGATIVE FEEDBACK

Feedback, Nonlinear
 USE   NONLINEAR FEEDBACK

Feedback, Positive
 USE   POSITIVE FEEDBACK

Feedback, Regenerative
 USE   POSITIVE FEEDBACK

Feedback, Sensory
 USE   SENSORY FEEDBACK

FEEDERS

FEEDFORWARD CONTROL

Feeding, Space Flight
 USE   SPACE FLIGHT FEEDING

FEEDING (SUPPLYING)

Feeds, Antenna
 USE   ANTENNA FEEDS

Feelings
 USE   SENSORY FEEDBACK

FEET (ANATOMY)

FELDSPARS

Fellowship Aircraft
 USE   F-28 TRANSPORT AIRCRAFT

FELSITE

FELTS

FEMALES

FEMUR

FENCES

Fences, Airfoil
 USE   AIRFOIL FENCES

FENCES (BARRIERS)

FERMAT PRINCIPLE

FERMENTATION

Fermi Atomic Power Plant, Enrico
 USE   ENRICO FERMI ATOMIC POWER PLANT

FERMI LIQUIDS

Fermi Model, Thomas-
 USE   THOMAS-FERMI MODEL

FERMI SURFACES

Fermi Theory, Thomas-
 USE   THOMAS-FERMI MODEL

FERMI-DIRAC STATISTICS

FERMIONS

FERMIUM

FERRANTI MERCURY COMPUTER

Ferraro Problem, Chapman-
 USE   CHAPMAN-FERRARO PROBLEM

FERRATES

Ferrates, Barium
 USE   BARIUM FERRATES

FERRIC IONS

FERRIMAGNETIC MATERIALS

FERRIMAGNETISM

FERRIMAGNETS

FERRITES

FERRITIC STAINLESS STEELS

Ferroalloys
 USE   IRON ALLOYS

Ferrocene, Alkyl
 USE   ALKYLFERROCENE

FERROCENES

FERROELECTRICITY

Ferroelectricity, Anti
 USE   ANTIFERROELECTRICITY

FERROFLUIDS

FERROGRAPHY

FERROMAGNETIC FILMS

FERROMAGNETIC MATERIALS

FERROMAGNETIC RESONANCE

FERROMAGNETISM

Ferromagnetism, Anti
 USE   ANTIFERROMAGNETISM

FERROUS METALS

FERRY SPACECRAFT

FERTILITY

FERTILIZATION

FERTILIZERS

FET (Transistors)
 USE   FIELD EFFECT TRANSISTORS

FETUSES

FEVER

FEYNMAN DIAGRAMS

Feynman Theorem, Hellmann-
 USE   HELLMANN-FEYNMAN THEOREM

FFAR Rocket Vehicle
USE   FOLDING FIN AIRCRAFT ROCKET VEHICLE

FFT
USE   FAST FOURIER TRANSFORMATIONS

FH-1100 Helicopter
USE   OH-5 HELICOPTER

FIAT AIRCRAFT

Fiat G-91 Aircraft
USE   G-91 AIRCRAFT

Fiat G-95/4 Aircraft
USE   G-95/4 AIRCRAFT

Fiat G-222 Aircraft
USE   G-222 AIRCRAFT

FIBER COMPOSITES

FIBER OPTICS

FIBER ORIENTATION

FIBER REINFORCED COMPOSITES

Fiber Reinforced Plastics, Carbon
USE   CARBON FIBER REINFORCED PLASTICS

Fiber Reinforced Plastics, Glass
USE   GLASS FIBER REINFORCED PLASTICS

FIBER RELEASE

FIBER STRENGTH

Fiberboard
USE   BOARDS (PAPER)

Fiberglass
USE   GLASS FIBERS

FIBERS

Fibers, Boron
USE   BORON FIBERS

Fibers, Carbon
USE   CARBON FIBERS

Fibers, Cotton
USE   COTTON FIBERS

Fibers, Glass
USE   GLASS FIBERS

FIBERS (MATHEMATICS)

Fibers, Metal
USE   METAL FIBERS

Fibers, Micro
USE   MICROFIBERS

Fibers, Reinforcing
USE   REINFORCING FIBERS

Fibers, Synthetic
USE   SYNTHETIC FIBERS

FIBONACCI NUMBERS

FIBRILLATION

FIBRIN

FIBRINOGEN

FIBROBLASTS

FIBROSIS

Fibrosis, Cystic
USE   CYSTIC FIBROSIS

Fibrous Materials
USE   FIBERS

FICKS EQUATION

Fidelity
USE   ACCURACY

FIDUCIARIES

Field Amplifiers, Crossed
USE   CROSSED FIELD AMPLIFIERS

FIELD ARMY BALLISTIC MISSILES

FIELD COILS

Field Configurations, Magnetic
USE   MAGNETIC FIELD CONFIGURATIONS

FIELD EFFECT TRANSISTORS

Field Effect Transistors, Junction
USE   JFET

FIELD EMISSION

Field, Geomagnetic
USE   GEOMAGNETISM

Field Guns, Crossed
USE   CROSSED FIELD GUNS

Field Intensity, Magnetic
USE   MAGNETIC FLUX

FIELD INTENSITY METERS

Field Inversions, Magnetic
USE   MAGNETIC FIELD INVERSIONS

Field Magnets, High
USE   HIGH FIELD MAGNETS

FIELD MODE THEORY

FIELD OF VIEW

Field Pinch, Reverse
USE   REVERSE FIELD PINCH

Field, Solar Magnetic
USE   SOLAR MAGNETIC FIELD

FIELD STRENGTH

Field Strength, Electric
USE   ELECTRIC FIELD STRENGTH

FIELD THEORY (ALGEBRA)

FIELD THEORY (PHYSICS)

(Field Theory), Strong Interactions
USE   STRONG INTERACTIONS (FIELD THEORY)

Field Theory, Unified
USE   UNIFIED FIELD THEORY

(Field Theory), Weak Interactions
USE   WEAK INTERACTIONS (FIELD THEORY)

Field Year For Great Lakes, International
USE   INTERNATIONAL FIELD YEAR FOR GREAT
            LAKES

FIELDS

Fields, Antenna
USE   ANTENNA RADIATION PATTERNS

Fields, Boson
USE   BOSON FIELDS

Fields, Crossed
USE   CROSSED FIELDS

Fields, Electric
USE   ELECTRIC FIELDS

Fields, Electromagnetic
USE   ELECTROMAGNETIC FIELDS

Fields, Electrostatic
USE   ELECTRIC FIELDS

Fields, Far
USE   FAR FIELDS

Fields, Flow
USE   FLOW DISTRIBUTION

Fields, Force
USE   FIELD THEORY (PHYSICS)

Fields, Force-Free Magnetic
USE   FORCE-FREE MAGNETIC FIELDS

Fields, Galactic Magnetic
USE   INTERSTELLAR MAGNETIC FIELDS

Fields, Gravitational
USE   GRAVITATIONAL FIELDS

Fields, Interplanetary Magnetic
USE   INTERPLANETARY MAGNETIC FIELDS

Fields, Interstellar Magnetic
USE   INTERSTELLAR MAGNETIC FIELDS

Fields, Lunar Magnetic
USE   LUNAR MAGNETIC FIELDS

Fields, Magnetic
USE   MAGNETIC FIELDS

Fields, Magnetostatic
USE   MAGNETOSTATIC FIELDS

Fields, Multipolar
USE   MULTIPOLAR FIELDS

Fields, Near
USE   NEAR FIELDS

Fields, Nonuniform Magnetic
USE   NONUNIFORM MAGNETIC FIELDS

Fields, Oil
USE   OIL FIELDS

Fields, Planetary Magnetic
USE   PLANETARY MAGNETIC FIELDS

Fields, Plowed
USE   FARMLANDS

Fields, Potential
USE   POTENTIAL FIELDS

Fields, Pressure
USE   PRESSURE DISTRIBUTION

Fields, Radiation
USE   RADIATION DISTRIBUTION

Fields, Self Consistent
USE   SELF CONSISTENT FIELDS

Fields, Sound
USE   SOUND FIELDS

Fields, Star
USE   STAR DISTRIBUTION

Fields, Stellar .
USE   STAR DISTRIBUTION

Fields, Stellar Magnetic
USE   STELLAR MAGNETIC FIELDS

Fields, Temperature
USE   TEMPERATURE DISTRIBUTION

Fields, Tensor
USE   TENSORS

Fields, Trapped Magnetic
USE   TRAPPED MAGNETIC FIELDS

Fields, Velocity
USE   VELOCITY DISTRIBUTION

Fields, Visual
USE   VISUAL FIELDS

Fields, Yang-Mills
USE   YANG-MILLS FIELDS

FIGHTER AIRCRAFT

Fighter Aircraft, Freedom
USE   F-5 AIRCRAFT

Fighting, Fire
USE   FIRE FIGHTING

Figure, Earth
USE   GEODESY

Figure, Lunar
USE   LUNAR FIGURE

FIGURE OF MERIT

Figures, Lissajous
USE   LISSAJOUS FIGURES

FILAMENT WINDING

Filament Wound Construction
USE   FILAMENT WINDING

FILAMENTS

Filaments (Solar Physics)
USE   SOLAR PROMINENCES

Filaments, Vortex
USE   VORTEX FILAMENTS

FILE MAINTENANCE (COMPUTERS)

FILES

FILES (TOOLS)

Filled Shells, Fluid
USE   FLUID FILLED SHELLS

Filled Shells, Liquid
USE   LIQUID FILLED SHELLS

FILLERS

FILLETS

FILLING

Film Anemometers, Hot-
USE   HOT-FILM ANEMOMETERS

Film Barriers, Electrode
USE   ELECTRODE FILM BARRIERS

FILM BOILING

FILM CONDENSATION

FILM COOLING

Film, Helium
USE   HELIUM FILM

Film, Photographic
USE   PHOTOGRAPHIC FILM

FILM THICKNESS

FILMS

Films, Energy Absorption
USE   ENERGY ABSORPTION FILMS

Films, Ferromagnetic
USE   FERROMAGNETIC FILMS

Films, Fluid
USE   FLUID FILMS

Films, Magnetic
USE   MAGNETIC FILMS

Films, Metal
USE   METAL FILMS

Films, Micro
USE   MICROFILMS

Films, Monomolecular
USE   MONOMOLECULAR FILMS

Films, Oxide
USE   OXIDE FILMS

Films, Plastic
USE   POLYMERIC FILMS

Films, Polymeric
USE   POLYMERIC FILMS

Films, Semiconducting
USE   SEMICONDUCTING FILMS

Films, Silicon
USE   SILICON FILMS

Films, Squeeze
USE   SQUEEZE FILMS

Films, Thermoplastic
USE   THERMOPLASTIC FILMS

Films, Thick
USE   THICK FILMS

Films, Thin
USE   THIN FILMS

FILTER WHEEL INFRARED SPECTROMETERS

Filtering
USE   FILTRATION

Filtering, Kalman-Schmidt
USE   KALMAN-SCHMIDT FILTERING

Filtering, Spatial
USE   SPATIAL FILTERING

Filtering, Wiener
USE   WIENER FILTERING

FILTERS

Filters, Adaptive
USE   ADAPTIVE FILTERS

Filters, Air
USE   AIR FILTERS

Filters, Bandpass
USE   BANDPASS FILTERS

Filters, Bandstop
USE   BANDSTOP FILTERS

Filters, Birefringent
USE   BIREFRINGENT FILTERS

Filters, Crystal
USE   CRYSTAL FILTERS

Filters, Digital
USE   DIGITAL FILTERS

Filters, Electric
USE   ELECTRIC FILTERS

Filters, Electromagnetic Wave
USE   ELECTROMAGNETIC WAVE FILTERS

Filters, Electronic
USE   ELECTRONIC FILTERS

Filters, Finite Impulse Response
USE   FIR FILTERS

Filters, Fir
USE   FIR FILTERS

Filters, Fluid
USE   FLUID FILTERS

Filters, High Pass
USE   HIGH PASS FILTERS

Filters, Image
USE   IMAGE FILTERS

Filters, Infrared
USE   INFRARED FILTERS

Filters, Kalman
USE   KALMAN FILTERS

Filters, Linear
USE   LINEAR FILTERS

Filters, Low Pass
USE   LOW PASS FILTERS

Filters, Mass
USE   FLUID FILTERS

Filters, Matched
USE   MATCHED FILTERS

Filters, Microwave
USE   MICROWAVE FILTERS

Filters, Nonlinear
USE   NONLINEAR FILTERS

Filters, Optical
USE   OPTICAL FILTERS

Filters, Particulate
USE   FLUID FILTERS

Filters, Radar
USE   RADAR FILTERS

Filters, Radio
USE   RADIO FILTERS

Filters, Reduced Order
USE   REDUCED ORDER FILTERS

Filters, Tracking
USE   TRACKING FILTERS

Filters, Ultraviolet
USE   ULTRAVIOLET FILTERS

Filters, Waveguide
USE   WAVEGUIDE FILTERS

FILTRATION

Filtration, In
USE   INFILTRATION

Fin Aircraft Rocket Vehicle, Folding
USE   FOLDING FIN AIRCRAFT ROCKET VEHICLE

FINANCE

FINANCIAL MANAGEMENT

Finders, Laser Range
USE   LASER RANGE FINDERS

Finders, Optical Range
USE   OPTICAL RANGE FINDERS

Finders, Radar Direction
USE   RADIO DIRECTION FINDERS

Finders, Radio Direction
USE   RADIO DIRECTION FINDERS

Finders (Radio), Direction
USE   RADIO DIRECTION FINDERS

Finders, Range
USE   RANGE FINDERS

Finding, Direction
USE   DIRECTION FINDING

FINE

FINE STRUCTURE

FINENESS

FINENESS RATIO

FINES

FINGERS

FINISHES

Finishing, Metal
USE   METAL FINISHING

Finishing, Surface
USE   SURFACE FINISHING

FINITE DIFFERENCE THEORY

FINITE ELEMENT METHOD

Finite Elements, Isoparametric
USE   ISOPARAMETRIC FINITE ELEMENTS

Finite Impulse Response Filters
USE   FIR FILTERS

FINITE VOLUME METHOD

Finite-State Machines
USE   TURING MACHINES

FINLAND

FINNED BODIES

FINS

Fins, Cooling
USE   COOLING FINS

Fins, Nose
USE   NOSE FINS

Fins, Vertical
USE   FINS

FIORDS

FIR FILTERS

Fire, Artillery
USE   ARTILLERY FIRE

FIRE CONTROL

FIRE CONTROL CIRCUITS

FIRE DAMAGE

Fire Detection, Forest
USE   FOREST FIRE DETECTION

FIRE EXTINGUISHERS

FIRE FIGHTING

FIRE POINT

FIRE PREVENTION

Fire Resistance
USE   FLAMMABILITY

Fire Retardants
USE   FLAME RETARDANTS

Fire, Saint Elmo
USE   SAINT ELMO FIRE

FIREBALLS

FIREBEE 2 TARGET DRONE AIRCRAFT

FIREBREAKS

FIREFLIES

FIREPROOFING

FIRES

Fires, Forest
USE   FOREST FIRES

Fireworks
USE   PYROTECHNICS

FIRING (IGNITING)

Firing, Retro
USE   RETROFIRING

Firing, Rocket
USE   ROCKET FIRING

Firing, Static
USE   STATIC FIRING

Firing, Test
USE   TEST FIRING

Firing Time
USE   BURNING TIME

FIRMWARE

FIRST AID

Fischer Reagent, Karl
USE   KARL FISCHER REAGENT

FISCHER-TROPSCH PROCESS

Fish
USE   FISHES

(Fish), Schools
USE   SCHOOLS (FISH)

Fish, Shell
USE   SHELLFISH

FISHBOWL OPERATION

FISHERIES

FISHES

Fishtailing
USE   YAW

FISSILE FUELS

Fissile Materials
USE   FISSIONABLE MATERIALS

FISSION

FISSION ELECTRIC CELLS

Fission Hybrid Reactors, Fusion-
USE   FUSION-FISSION HYBRID REACTORS

Fission, Nuclear
USE   NUCLEAR FISSION

FISSION PRODUCTS

(Fission Reactors), Blankets
USE   BLANKETS (FISSION REACTORS)

Fission Reactors, Gaseous
USE   GASEOUS FISSION REACTORS

FISSION WEAPONS

FISSIONABLE MATERIALS

FISSIUM

FISSURES (GEOLOGY)

Fit, Goodness Of
USE   GOODNESS OF FIT

FITNESS

Fitness, Flight
USE   FLIGHT FITNESS

Fitness, Physical
USE   PHYSICAL FITNESS

FITTING

Fitting, Curve
USE   CURVE FITTING

FITTINGS

Fitzgerald-Lorentz Contraction
USE   LORENTZ CONTRACTION

Fix
USE   FIXING

Fixation, Nitrogen
USE   NITROGENATION

FIXED POINT ARITHMETIC

FIXED POINTS (MATHEMATICS)

FIXED WINGS

Fixed-Wing Aircraft
USE   FIXED WINGS
          AIRCRAFT CONFIGURATIONS

FIXING

Fixing And Ranging, Sound
USE   SOUND FIXING AND RANGING

Fixpoint Theorem, Schauder
USE   SCHAUDER FIXPOINT THEOREM

FIXTURES

FIZEAU EFFECT

Fizeau Effect, Doppler-
USE   DOPPLER-FIZEAU EFFECT

FL
USE   FLORIDA

(FL), Everglades
USE   EVERGLADES (FL)

(FL), Merritt Island
USE   MERRITT ISLAND (FL)

FLAGELLATA

FLAKES

FLAKING

FLAME CALORIMETERS

Flame, Chapman-Jouget
USE   FLAME PROPAGATION
          CHEMICAL EQUILIBRIUM
          DETONATION

FLAME DEFLECTORS

Flame Fronts
USE   FLAME PROPAGATION

FLAME HOLDERS

Flame Interaction
USE   FLAME PROPAGATION
          CHEMICAL REACTIONS

FLAME IONIZATION

FLAME PLATING

FLAME PROBES

FLAME PROPAGATION

Flame Quenching
USE   EXTINGUISHING
          QUENCHING (COOLING)

FLAME RETARDANTS

FLAME SPECTROSCOPY

FLAME SPRAYING

FLAME STABILITY

FLAME TEMPERATURE

FLAMEOUT

FLAMES

Flames, Diffusion
USE   DIFFUSION FLAMES

Flames, Jet
USE   FLAMES
JET FLOW

Flames, Laminar
USE   LAMINAR FLOW
FLAMES

Flames, Premixed
USE   PREMIXED FLAMES

FLAMMABILITY

FLAMMABLE GASES

FLANGE WRINKLING

FLANGES

Flap Approach, Delayed
USE   DELAYED FLAP APPROACH

Flap Control
USE   FLAPS (CONTROL SURFACES)
AIRCRAFT CONTROL

FLAPERONS

FLAPPING

FLAPPING HINGES

Flaps, Blown
USE   EXTERNALLY BLOWN FLAPS

FLAPS (CONTROL SURFACES)

Flaps, Externally Blown
USE   EXTERNALLY BLOWN FLAPS

Flaps, Jet
USE   JET FLAPS

Flaps, Jet Augmented Wing
USE   JET FLAPS
WING FLAPS

Flaps, Leading Edge
USE   LEADING EDGE FLAPS

Flaps, Split
USE   SPLIT FLAPS

Flaps, Trailing Edge
USE   TRAILING EDGE FLAPS

Flaps, Upper Surface Blown
USE   UPPER SURFACE BLOWN FLAPS

Flaps, Vortex
USE   VORTEX FLAPS

Flaps, Wing
USE   WING FLAPS

Flare, Conical
USE   CONES

FLARE STARS

FLARED BODIES

FLARES

Flares, Solar
USE   SOLAR FLARES

Flares, Stellar
USE   STELLAR FLARES

FLASH

FLASH BLINDNESS

FLASH LAMPS

FLASH POINT

Flash Tubes
USE   FLASH LAMPS

FLASH WELDING

FLASHBACK

FLASHING (VAPORIZING)

FLASHOVER

FLASKS

Flat Coaxial Transmission Lines
USE   MICROSTRIP TRANSMISSION LINES

FLAT CONDUCTORS

FLAT LAYERS

FLAT PATTERNS

FLAT PLATES

FLAT SURFACES

FLATNESS

Flats, Adobe
USE   FLATS (LANDFORMS)

FLATS (LANDFORMS)

Flats, Salt
USE   FLATS (LANDFORMS)

Flats, Tidal
USE   TIDAL FLATS

FLATTENING

FLATWORMS

FLAVOR (PARTICLE PHYSICS)

Flaw Detection
USE   NONDESTRUCTIVE TESTS

Flaw Detection, Ultrasonic
USE   ULTRASONIC FLAW DETECTION

Flaws
USE   DEFECTS

FLEET BALLISTIC MISSILES

FLEET SATELLITE COMMUNICATION SYSTEM

FLEETSATCOM
USE   FLEET SATELLITE COMMUNICATION
SYSTEM

FLEXIBILITY

FLEXIBLE BODIES

FLEXIBLE SPACECRAFT

FLEXIBLE WINGS

FLEXING

FLEXORS

Flexowriters (Trademark)
USE   AUTOMATIC TYPEWRITERS

Flexure
USE   FLEXING

Flexure Problem, Saint Venant
USE   SAINT VENANT PRINCIPLE

Flexure Problem, St Venant
USE   SAINT VENANT PRINCIPLE

FLICKER

Flicker Fusion, Critical
USE   CRITICAL FLICKER FUSION

Flicker Fusion Frequency
USE   CRITICAL FLICKER FUSION

Flies, Chironomus
USE   CHIRONOMUS FLIES

FLIGHT

FLIGHT ALTITUDE

Flight, Apollo 5
USE   APOLLO 5 FLIGHT

Flight, Apollo 6
USE   APOLLO 6 FLIGHT

Flight, Apollo 7
USE   APOLLO 7 FLIGHT

Flight, Apollo 8
USE   APOLLO 8 FLIGHT

Flight, Apollo 9
USE   APOLLO 9 FLIGHT

Flight, Apollo 10
USE   APOLLO 10 FLIGHT

Flight, Apollo 11
USE   APOLLO 11 FLIGHT

Flight, Apollo 12
USE   APOLLO 12 FLIGHT

Flight, Apollo 13
USE   APOLLO 13 FLIGHT

Flight, Apollo 14
USE   APOLLO 14 FLIGHT

Flight, Apollo 15
USE   APOLLO 15 FLIGHT

Flight, Apollo 16
USE   APOLLO 16 FLIGHT

Flight, Apollo 17
USE   APOLLO 17 FLIGHT

Flight, Balloon
USE   BALLOON FLIGHT

Flight, Banking
USE   TURNING FLIGHT

FLIGHT CHARACTERISTICS

Flight, Climbing
USE   CLIMBING FLIGHT

FLIGHT CLOTHING

Flight, Coasting
USE   COASTING FLIGHT

Flight Computers
USE   AIRBORNE/SPACEBORNE COMPUTERS

FLIGHT CONDITIONS

FLIGHT CONTROL

Flight Control, Automatic
USE   AUTOMATIC FLIGHT CONTROL

FLIGHT CREWS

Flight, Cruising
USE   CRUISING FLIGHT

Flight, Extended Duration Space
USE   LONG DURATION SPACE FLIGHT

FLIGHT FATIGUE

Flight Feeding, Space
USE   SPACE FLIGHT FEEDING

FLIGHT FITNESS

Flight, Free
USE   FREE FLIGHT

Flight, Gemini 3
USE   GEMINI 3 FLIGHT

Flight, Gemini 4
USE   GEMINI 4 FLIGHT

Flight, Gemini 5
USE   GEMINI 5 FLIGHT

Flight, Gemini 6
USE   GEMINI 6 FLIGHT

Flight, Gemini 7
USE   GEMINI 7 FLIGHT

Flight, Gemini 8
USE   GEMINI 8 FLIGHT

Flight, Gemini 9
USE   GEMINI 9 FLIGHT

Flight, Gemini 10
USE   GEMINI 10 FLIGHT

Flight, Gemini 11
USE   GEMINI 11 FLIGHT

Flight, Gemini 12
USE   GEMINI 12 FLIGHT

FLIGHT HAZARDS

Flight, High Altitude
USE   FLIGHT
        HIGH ALTITUDE

Flight, High Speed
USE   HIGH SPEED
        FLIGHT

Flight, Horizontal
USE   HORIZONTAL FLIGHT

Flight, Hypersonic
USE   HYPERSONIC FLIGHT

FLIGHT INSTRUMENTS

Flight, Interplanetary
USE   INTERPLANETARY FLIGHT

Flight, Jet
USE   JET AIRCRAFT

FLIGHT LOAD RECORDERS

Flight, Long Duration Space
USE   LONG DURATION SPACE FLIGHT

Flight, Lunar
USE   LUNAR FLIGHT

Flight, MA-3
USE   MERCURY MA-3 FLIGHT

Flight, MA-4
USE   MERCURY MA-4 FLIGHT

Flight, MA-5
USE   MERCURY MA-5 FLIGHT

Flight, MA-8
USE   MERCURY MA-8 FLIGHT

Flight, MA-9
USE   MERCURY MA-9 FLIGHT

FLIGHT MANAGEMENT SYSTEMS

Flight, Manned Space
USE   MANNED SPACE FLIGHT

FLIGHT MECHANICS

Flight, Mercury MA-1
USE   MERCURY MA-1 FLIGHT

Flight, Mercury MA-2
USE   MERCURY MA-2 FLIGHT

Flight, Mercury MA-3
USE   MERCURY MA-3 FLIGHT

Flight, Mercury MA-4
USE   MERCURY MA-4 FLIGHT

Flight, Mercury MA-5
USE   MERCURY MA-5 FLIGHT

Flight, Mercury MA-6
USE   MERCURY MA-6 FLIGHT

Flight, Mercury MA-7
USE   MERCURY MA-7 FLIGHT

Flight, Mercury MA-8
USE   MERCURY MA-8 FLIGHT

Flight, Mercury MA-9
USE   MERCURY MA-9 FLIGHT

Flight, Mercury MR-1
USE   MERCURY MR-1 FLIGHT

Flight, Mercury MR-2
USE   MERCURY MR-2 FLIGHT

Flight, Mercury MR-3
USE   MERCURY MR-3 FLIGHT

Flight, Mercury MR-4
USE   MERCURY MR-4 FLIGHT

Flight, Meteorological
USE   METEOROLOGICAL FLIGHT

Flight, Minor Circle Turning
USE   MINOR CIRCLE TURNING FLIGHT

Flight Monitoring, IN-
USE   IN-FLIGHT MONITORING

Flight, MR-3
USE   MERCURY MR-3 FLIGHT

Flight Network, Manned Space
USE   MANNED SPACE FLIGHT NETWORK

FLIGHT NURSES

FLIGHT OPERATIONS

FLIGHT OPTIMIZATION

Flight, Parabolic
USE   PARABOLIC FLIGHT

FLIGHT PATHS

Flight Performance
USE   FLIGHT CHARACTERISTICS

Flight, Planetary Space
USE   INTERPLANETARY FLIGHT

FLIGHT PLANS

FLIGHT RECORDERS

Flight, Return To Earth Space
USE   RETURN TO EARTH SPACE FLIGHT

Flight, Rocket
USE   ROCKET FLIGHT

FLIGHT RULES

Flight Rules, Instrument
USE   INSTRUMENT FLIGHT RULES

Flight Rules, Visual
USE   VISUAL FLIGHT RULES

FLIGHT SAFETY

FLIGHT SIMULATION

FLIGHT SIMULATORS

Flight, Space
USE   SPACE FLIGHT

Flight, Space Transportation System 1
USE   SPACE TRANSPORTATION SYSTEM 1
        FLIGHT

Flight, Space Transportation System 2
USE   SPACE TRANSPORTATION SYSTEM 2
        FLIGHT

Flight, Space Transportation System 3
USE   SPACE TRANSPORTATION SYSTEM 3
        FLIGHT

Flight, Space Transportation System 4
USE   SPACE TRANSPORTATION SYSTEM 4
        FLIGHT

Flight Spectrometers, Time Of
USE   TIME OF FLIGHT SPECTROMETERS

FLIGHT STABILITY TESTS

FLIGHT STRESS

FLIGHT STRESS (BIOLOGY)

Flight Stress, Space
USE   SPACE FLIGHT STRESS

Flight, Suborbital
USE   SUBORBITAL FLIGHT

Flight, Supersonic
USE   SUPERSONIC FLIGHT

FLIGHT SURGEONS

Flight Technical Error
USE   PILOT ERROR

Flight Test Apparatus, Free
USE   FREE FLIGHT TEST APPARATUS

FLIGHT TEST INSTRUMENTS

Flight Test Program, Reactor In
USE   RIFT (REACTOR IN FLIGHT TEST)

Flight Test), Rift (Reactor In
USE   RIFT (REACTOR IN FLIGHT TEST)

FLIGHT TEST VEHICLES

Flight Test 1 (Shuttle), Orbital
USE   SPACE TRANSPORTATION SYSTEM 1
        FLIGHT

Flight Test 1, Space Shuttle Orbital
USE   SPACE TRANSPORTATION SYSTEM 1
        FLIGHT

Flight Test 2 (Shuttle), Orbital
USE   SPACE TRANSPORTATION SYSTEM 2
        FLIGHT

Flight Test 2, Space Shuttle Orbital
USE   SPACE TRANSPORTATION SYSTEM 2
        FLIGHT

Flight Test 3 (Shuttle), Orbital
USE   SPACE TRANSPORTATION SYSTEM 3
        FLIGHT

Flight Test 3, Space Shuttle Orbital
USE   SPACE TRANSPORTATION SYSTEM 3
        FLIGHT

Flight Test 4 (Shuttle), Orbital
USE   SPACE TRANSPORTATION SYSTEM 4
        FLIGHT

Flight Test 4, Space Shuttle Orbital
USE   SPACE TRANSPORTATION SYSTEM 4
        FLIGHT

FLIGHT TESTS

Flight Tests (Shuttle), Orbital
USE   SPACE TRANSPORTATION SYSTEM
          FLIGHTS

Flight Tests, Space Shuttle Orbital
USE   SPACE TRANSPORTATION SYSTEM
          FLIGHTS

FLIGHT TIME

Flight Tracking And Data Network, Space
USE   SPACE FLIGHT TRACKING AND DATA
          NETWORK

FLIGHT TRAINING

Flight Training, Space
USE   SPACE FLIGHT TRAINING

Flight, Transoceanic
USE   TRANSOCEANIC FLIGHT

Flight, Transonic
USE   TRANSONIC FLIGHT

Flight, Turning
USE   TURNING FLIGHT

FLIGHT VEHICLES

Flight, Vertical
USE   VERTICAL FLIGHT

Flight, Visual
USE   VISUAL FLIGHT

Flight 7, Space Shuttle Orbital
USE   SPACE SHUTTLE MISSION 31-C

Flight 8, Space Shuttle Orbital
USE   SPACE SHUTTLE MISSION 31-D

Flight 9, Space Shuttle Orbital
USE   SPACE SHUTTLE MISSION 41-A

Flights (Aircraft), Night
USE   NIGHT FLIGHTS (AIRCRAFT)

Flights, Apollo
USE   APOLLO FLIGHTS

Flights, Gemini
USE   GEMINI FLIGHTS

Flights, Mercury
USE   MERCURY FLIGHTS

Flights, Space Shuttle Orbital
USE   SPACE TRANSPORTATION SYSTEM
          FLIGHTS

Flights, Space Transportation System
USE   SPACE TRANSPORTATION SYSTEM
          FLIGHTS

Flights, Spacelab Simulation
USE   ASSESS PROGRAM

FLINT

FLIP-FLOPS

FLIR DETECTORS

FLOAT ZONES

FLOATING

FLOATING POINT ARITHMETIC

FLOATS

FLOCCULATING

Floes, Ice
USE   ICE FLOES

FLOOD CONTROL

FLOOD DAMAGE

FLOOD PLAINS

FLOOD PREDICTIONS

FLOODS

FLOORS

(Floors), Decks
USE   FLOORS

Floors, Intermontane
USE   VALLEYS

Flops, Flip-
USE   FLIP-FLOPS

FLOQUET THEOREM

Flora
USE   PLANTS (BOTANY)

FLORIDA

FLOTATION

Flotation Systems
USE   FLOATS

FLOUR

FLOUR (FOOD)

FLOW

Flow, Adiabatic
USE   ADIABATIC FLOW

Flow, Air
USE   AIR FLOW

Flow Airfoils, Laminar
USE   LAMINAR FLOW AIRFOILS

Flow Analysis, Data
USE   DATA FLOW ANALYSIS

Flow, Annular
USE   ANNULAR FLOW

Flow, Axial
USE   AXIAL FLOW

Flow, Axisymmetric
USE   AXISYMMETRIC FLOW

Flow, Barotropic
USE   BAROTROPIC FLOW

Flow, Base
USE   BASE FLOW

Flow, Beltrami
USE   BELTRAMI FLOW

Flow, Blasius
USE   BLASIUS FLOW

Flow, Blood
USE   BLOOD FLOW

Flow, Boundary Layer
USE   BOUNDARY LAYER FLOW

Flow, Brillouin
USE   BRILLOUIN FLOW

Flow, Capillary
USE   CAPILLARY FLOW

Flow, Cascade
USE   CASCADE FLOW

Flow, Cavitation
USE   CAVITATION FLOW

Flow Cells, Geophysical Fluid
USE   GEOPHYSICAL FLUID FLOW CELLS

FLOW CHAMBERS

Flow, Channel
USE   CHANNEL FLOW

FLOW CHARACTERISTICS

FLOW CHARTS

Flow, Coaxial
USE   COAXIAL FLOW

FLOW COEFFICIENTS

Flow, Combustible
USE   COMBUSTIBLE FLOW

Flow, Compressible
USE   COMPRESSIBLE FLOW

Flow Compressors, Axial
USE   TURBOCOMPRESSORS

Flow, Conical
USE   CONICAL FLOW

Flow, Continuum
USE   CONTINUUM FLOW

Flow Control, Laminar
USE   BOUNDARY LAYER CONTROL
          LAMINAR BOUNDARY LAYER

Flow, Convective
USE   CONVECTIVE FLOW

Flow, Core
USE   CORE FLOW

Flow, Corner
USE   CORNER FLOW

Flow, Couette
USE   COUETTE FLOW

Flow, Counter
USE   COUNTERFLOW

Flow, Critical
USE   CRITICAL FLOW

Flow, Cross
USE   CROSS FLOW

FLOW DEFLECTION

Flow Devices, Charge
USE   CHARGE FLOW DEVICES

FLOW DIRECTION INDICATORS

FLOW DISTORTION

FLOW DISTRIBUTION

Flow), Draft (Gas
USE   DRAFT (GAS FLOW)

Flow, Ducted
USE   DUCTED FLOW

FLOW EQUATIONS

Flow, Equilibrium
USE   EQUILIBRIUM FLOW

Flow Factors, Mass
USE   MASS FLOW FACTORS

Flow Fields
USE   FLOW DISTRIBUTION

Flow, Fluid
USE   FLUID FLOW

Flow, Free
USE   FREE FLOW

Flow, Free Molecular
USE   FREE MOLECULAR FLOW

Flow, Frozen Equilibrium
USE   FROZEN EQUILIBRIUM FLOW

Flow, Fuel
USE   FUEL FLOW

Flow, Gas
USE   GAS FLOW

FLOW GEOMETRY

FLOW GRAPHS

Flow Graphs, Signal
USE   SIGNAL FLOW GRAPHS

Flow, Grazing
USE   GRAZING FLOW

Flow, Hartmann
USE   HARTMANN FLOW

Flow, Head
USE   HEAD FLOW

Flow, Heat
USE   HEAT TRANSMISSION

Flow, Helical
USE   HELICAL FLOW

Flow, Hydromagnetic
USE   MAGNETOHYDRODYNAMIC FLOW

Flow, Hypersonic
USE   HYPERSONIC FLOW

Flow, Hypervelocity
USE   HYPERVELOCITY FLOW

Flow, Incompressible
USE   INCOMPRESSIBLE FLOW

Flow, Induced Fluid
USE   FLUID FLOW

Flow, Information
USE   INFORMATION FLOW

Flow, Inlet
USE   INLET FLOW

Flow Inlets, Supersonic
USE   SUPERSONIC INLETS

Flow, Inviscid
USE   INVISCID FLOW

Flow, Irrotational
USE   POTENTIAL FLOW

Flow, Isothermal
USE   ISOTHERMAL FLOW

Flow, Jet
USE   JET FLOW

Flow, Jet Mixing
USE   JET MIXING FLOW

Flow, Karman-Bodewadt
USE   KARMAN-BODEWADT FLOW

Flow, Kirchhoff-Helmholtz
USE   PIPE FLOW

Flow, Knudsen
USE   KNUDSEN FLOW

Flow, Laminar
USE   LAMINAR FLOW

Flow, Liquid
USE   LIQUID FLOW

Flow, Low Density
USE   LOW DENSITY FLOW

Flow, Magnetohydrodynamic
USE   MAGNETOHYDRODYNAMIC FLOW

Flow, Mass
USE   MASS FLOW

FLOW MEASUREMENT

Flow, Meridional
USE   MERIDIONAL FLOW

Flow Method Tests, Wing
USE   WING FLOW METHOD TESTS

Flow, Mixed
USE   MULTIPHASE FLOW

Flow, Molecular
USE   MOLECULAR FLOW

Flow, Multiphase
USE   MULTIPHASE FLOW

FLOW NETS

Flow, Nonequilibrium
USE   NONEQUILIBRIUM FLOW

Flow, Nonnewtonian
USE   NONNEWTONIAN FLOW

Flow, Nonuniform
USE   NONUNIFORM FLOW

Flow, Nonviscous
USE   INVISCID FLOW

Flow, Nozzle
USE   NOZZLE FLOW

Flow, One Dimensional
USE   ONE DIMENSIONAL FLOW

Flow, One-Phase
USE   SINGLE-PHASE FLOW

Flow, Open Channel
USE   OPEN CHANNEL FLOW

Flow, Orifice
USE   ORIFICE FLOW

Flow, Oscillating
USE   OSCILLATING FLOW

Flow, Outlet
USE   OUTLET FLOW

Flow, Parallel
USE   PARALLEL FLOW

Flow Patterns
USE   FLOW DISTRIBUTION

Flow, Peripheral Jet
USE   PERIPHERAL JET FLOW

Flow, Pipe
USE   PIPE FLOW

Flow, Plasma
USE   MAGNETOHYDRODYNAMIC FLOW

Flow, Plastic
USE   PLASTIC FLOW

Flow, Poiseuille
USE   LAMINAR FLOW

Flow, Potential
USE   POTENTIAL FLOW

Flow, Pulsating
USE   UNSTEADY FLOW

Flow Pumps, Axial
USE   AXIAL FLOW PUMPS

Flow, Radial
USE   RADIAL FLOW

Flow Rate
USE   FLOW VELOCITY

Flow Rate, Mass
USE   MASS FLOW RATE

Flow, Reattached
USE   REATTACHED FLOW

Flow, Recirculative Fluid
USE   RECIRCULATIVE FLUID FLOW

FLOW REGULATORS

Flow Regulators, Fuel
USE   FUEL FLOW REGULATORS

FLOW RESISTANCE

Flow, Reversed
USE   REVERSED FLOW

Flow, Rotational
USE   VORTICES
        FLUID FLOW

Flow, Secondary
USE   SECONDARY FLOW

Flow, Separated
USE   SEPARATED FLOW

Flow Separation
USE   SEPARATED FLOW
        BOUNDARY LAYER SEPARATION

Flow, Shear
USE   SHEAR FLOW

Flow, Shifting Equilibrium
USE   SHIFTING EQUILIBRIUM FLOW

Flow Simulation, Exhaust
USE   EXHAUST FLOW SIMULATION

Flow, Single-Phase
USE   SINGLE-PHASE FLOW

Flow, Slip
USE   SLIP FLOW

Flow, Small Perturbation
USE   SMALL PERTURBATION FLOW

Flow, Solids
USE   SOLIDS FLOW

Flow, Sonic
USE   TRANSONIC FLOW

FLOW STABILITY

Flow, Stagnation
USE   STAGNATION FLOW

Flow, Steady
USE   STEADY FLOW

Flow, Steady State
USE   EQUILIBRIUM FLOW

Flow, Steam
USE   STEAM FLOW

Flow, Stokes
USE   STOKES FLOW

Flow, Stratified
USE   STRATIFIED FLOW

Flow, Streamline
USE   LAMINAR FLOW

Flow, Subcritical
USE   SUBCRITICAL FLOW

Flow, Subsonic
USE   SUBSONIC FLOW

Flow, Supercavitating
USE   SUPERCAVITATING FLOW

Flow, Supercritical
USE   SUPERCRITICAL FLOW

Flow, Superfluid
USE   SUPERFLUIDITY

Flow, Supersonic
USE   SUPERSONIC FLOW

Flow, Supersonic Jet
USE   SUPERSONIC JET FLOW

Flow Tests, Cold
USE   COLD FLOW TESTS

FLOW THEORY

Flow Theory, Mixing Length
USE   MIXING LENGTH FLOW THEORY

Flow, Three Dimensional
USE   THREE DIMENSIONAL FLOW

Flow, Transition
USE   TRANSITION FLOW

Flow, Transonic
USE   TRANSONIC FLOW

Flow, Tresca
USE   TRESCA FLOW

Flow Turbines, Axial
USE   AXIAL FLOW TURBINES

Flow, Turbulent
USE   TURBULENT FLOW

Flow, Two Dimensional
USE   TWO DIMENSIONAL FLOW

Flow, Two Phase
USE   TWO PHASE FLOW

Flow, Uniform
USE   UNIFORM FLOW

Flow, Uniphase
USE   SINGLE-PHASE FLOW

Flow, Unsteady
USE   UNSTEADY FLOW

FLOW VELOCITY

Flow, Viscoelastic
USE   VISCOELASTICITY

Flow, Viscoplastic
USE   VISCOPLASTICITY

Flow, Viscous
USE   VISCOUS FLOW

FLOW VISUALIZATION

Flow Visualization, Numerical
USE   NUMERICAL FLOW VISUALIZATION

Flow, Visualization Of
USE   FLOW VISUALIZATION

Flow, Vortex
USE   VORTICES

Flow, Wall
USE   WALL FLOW

Flow, Water
USE   WATER FLOW

Flow, Wedge
USE   WEDGE FLOW

Flowers, Sun
USE   SUNFLOWERS

FLOWMETERS

Flowmeters, Hot-Wire
USE   HOT-WIRE FLOWMETERS

FLOX

Fltsatcom
USE   FLEET SATELLITE COMMUNICATION
       SYSTEM

Fluctuation
USE   VARIATIONS

FLUCTUATION THEORY

FLUE GASES

FLUENCE

FLUERICS

FLUES

Fluid Amplification
USE   FLUID AMPLIFIERS

FLUID AMPLIFIERS

FLUID BOUNDARIES

Fluid, Cerebrospinal
USE   CEREBROSPINAL FLUID

FLUID DYNAMICS

(Fluid Dynamics), Cascades
USE   FLUID DYNAMICS

Fluid Dynamics, Computational
USE   COMPUTATIONAL FLUID DYNAMICS

(Fluid Dynamics), Panel Method
USE   PANEL METHOD (FLUID DYNAMICS)

(Fluid Dynamics), Stabilizers
USE   STABILIZERS (FLUID DYNAMICS)

FLUID FILLED SHELLS

FLUID FILMS

FLUID FILTERS

FLUID FLOW

Fluid Flow Cells, Geophysical
USE   GEOPHYSICAL FLUID FLOW CELLS

Fluid Flow, Induced
USE   FLUID FLOW

Fluid Flow, Recirculative
USE   RECIRCULATIVE FLUID FLOW

FLUID INJECTION

Fluid Jet Amplifiers
USE   FLUID AMPLIFIERS
       JET AMPLIFIERS

FLUID JETS

FLUID LOGIC

FLUID MANAGEMENT

FLUID MECHANICS

(Fluid Mechanics), Head
USE   HEAD (FLUID MECHANICS)

(Fluid Mechanics), Stokes Law
USE   STOKES LAW (FLUID MECHANICS)

Fluid Models, Two
USE   TWO FLUID MODELS

FLUID POWER

FLUID PRESSURE

FLUID ROTOR GYROSCOPES

Fluid Storage, Cryogenic
USE   CRYOGENIC FLUID STORAGE

FLUID SWITCHING ELEMENTS

FLUID TRANSMISSION LINES

Fluid Transpiration
USE   TRANSPIRATION

FLUID-SOLID INTERACTIONS

FLUIDIC CIRCUITS

FLUIDICS

FLUIDIZED BED PROCESSORS

FLUIDS

Fluids, Anisotropic
USE   ANISOTROPIC FLUIDS

Fluids, Binary
USE   BINARY FLUIDS

Fluids, Body
USE   BODY FLUIDS

Fluids, Compressible
USE   COMPRESSIBLE FLUIDS

Fluids, Conducting
USE   CONDUCTING FLUIDS

Fluids, Cryogenic
USE   CRYOGENIC FLUIDS

Fluids, Ferro
USE   FERROFLUIDS

Fluids, Geophysical
USE   GEOPHYSICAL FLUIDS

Fluids, Gyroscope
USE   GYROSCOPE FLUIDS

Fluids, High Temperature
USE   HIGH TEMPERATURE FLUIDS

Fluids, Hydraulic
USE   HYDRAULIC FLUIDS

Fluids, Ideal
USE   IDEAL FLUIDS

Fluids, Incompressible
USE   INCOMPRESSIBLE FLUIDS

Fluids, Maxwell
USE   MAXWELL FLUIDS

Fluids, Micropolar
USE   MICROPOLAR FLUIDS

Fluids, Newtonian
USE   NEWTONIAN FLUIDS

Fluids, Nonnewtonian
USE   NONNEWTONIAN FLUIDS

Fluids, Rotating
USE   ROTATING FLUIDS

(Fluids), Stream Functions
USE   STREAM FUNCTIONS (FLUIDS)

Fluids, Supercritical
USE   SUPERCRITICAL FLUIDS

Fluids, Transmission
USE   TRANSMISSION FLUIDS

Fluids, Viscous
USE   VISCOUS FLUIDS

Fluids, Weightless
USE   WEIGHTLESS FLUIDS

Fluids, Working
USE   WORKING FLUIDS

FLUORESCENCE

Fluorescence, Resonance
USE   RESONANCE FLUORESCENCE

Fluorescence, X Ray
  USE   X RAY FLUORESCENCE

Fluorescent Emission
  USE   FLUORESCENCE

Fluoride Lasers, Deuterium
  USE   DF LASERS

Fluoride Lasers, Krypton
  USE   KRYPTON FLUORIDE LASERS

Fluoride Lasers, Xenon
  USE   XENON FLUORIDE LASERS

Fluoride, Ozone
  USE   OZONE FLUORIDE

Fluoride, Polyvinyl
  USE   POLYVINYL FLUORIDE

FLUORIDES

Fluorides, Aluminum
  USE   ALUMINUM FLUORIDES

Fluorides, Antimony
  USE   ANTIMONY FLUORIDES

Fluorides, Barium
  USE   BARIUM FLUORIDES

Fluorides, Beryllium
  USE   BERYLLIUM FLUORIDES

Fluorides, Boron
  USE   BORON FLUORIDES

Fluorides, Cadmium
  USE   CADMIUM FLUORIDES

Fluorides, Calcium
  USE   CALCIUM FLUORIDES

Fluorides, Cesium
  USE   CESIUM FLUORIDES

Fluorides, Chlorine
  USE   CHLORINE FLUORIDES

Fluorides, Chromium
  USE   CHROMIUM FLUORIDES

Fluorides, Cobalt
  USE   COBALT FLUORIDES

Fluorides, Copper
  USE   COPPER FLUORIDES

Fluorides, Deuterium
  USE   DEUTERIUM FLUORIDES

Fluorides, Di
  USE   DIFLUORIDES

Fluorides, Hydrogen
  USE   HYDROFLUORIC ACID

Fluorides, Lanthanum
  USE   LANTHANUM FLUORIDES

Fluorides, Lithium
  USE   LITHIUM FLUORIDES

Fluorides, Magnesium
  USE   MAGNESIUM FLUORIDES

Fluorides, Metal
  USE   METAL FLUORIDES

Fluorides, Nickel
  USE   NICKEL FLUORIDES

Fluorides, Nitrogen
  USE   NITROGEN FLUORIDES

Fluorides, Nitryl
  USE   NITRYL FLUORIDES

Fluorides, Oxy
  USE   OXYFLUORIDES

Fluorides, Oxygen
  USE   OXYGEN FLUORIDES

Fluorides, Perchloryl
  USE   PERCHLORYL FLUORIDES

Fluorides, Plutonium
  USE   PLUTONIUM FLUORIDES

Fluorides, Protactinium
  USE   PROTACTINIUM FLUORIDES

Fluorides, Sodium
  USE   SODIUM FLUORIDES

Fluorides, Strontium
  USE   STRONTIUM FLUORIDES

Fluorides, Sulfur
  USE   SULFUR FLUORIDES

Fluorides, Technetium
  USE   TECHNETIUM FLUORIDES

Fluorides, Thorium
  USE   THORIUM FLUORIDES

Fluorides, Tungsten
  USE   TUNGSTEN FLUORIDES

Fluorides, Uranium
  USE   URANIUM FLUORIDES

Fluorides, Zinc
  USE   ZINC FLUORIDES

FLUORINATION

Fluorination, De
  USE   DEFLUORINATION

FLUORINE

FLUORINE COMPOUNDS

Fluorine Compounds, Organic
  USE   FLUORINE ORGANIC COMPOUNDS

FLUORINE ISOTOPES

Fluorine, Liquid
  USE   LIQUID FLUORINE

FLUORINE ORGANIC COMPOUNDS

Fluorine-Liquid Oxygen
  USE   FLOX

FLUORITE

FLUORO COMPOUNDS

FLUOROAMINES

FLUOROCARBONS

FLUOROHYDROCARBONS

Fluoromica
  USE   FLUOROSILICATES
         MICA

FLUOROPHLOGOPITE

Fluoroplastics
  USE   FLUOROPOLYMERS

FLUOROPOLYMERS

FLUOROSCOPY

FLUOROSILICATES

FLUORSPAR

FLUSHING

Fluting
  USE   GROOVING

FLUTTER

Flutter, Aeromagneto
  USE   FLUTTER

FLUTTER ANALYSIS

Flutter, Panel
  USE   PANEL FLUTTER

Flutter, Subsonic
  USE   SUBSONIC FLUTTER

Flutter, Supersonic
  USE   SUPERSONIC FLUTTER

Flutter, Transonic
  USE   TRANSONIC FLUTTER

FLUX

Flux Beam Reactors, High
  USE   HIGH FLUX BEAM REACTORS

FLUX DENSITY

Flux Density, Electron
  USE   ELECTRON FLUX DENSITY

Flux Density, Luminous
  USE   LUMINOUS INTENSITY

Flux Density, Neutron
  USE   NEUTRON FLUX DENSITY

Flux Density, Particle
  USE   PARTICLE FLUX DENSITY

Flux Density, Proton
  USE   PROTON FLUX DENSITY

Flux Density, Radiant
  USE   RADIANT FLUX DENSITY

Flux Density, Solar
  USE   SOLAR FLUX DENSITY

Flux, Electron
  USE   ELECTRONS
         FLUX (RATE)

Flux, Heat
  USE   HEAT FLUX

Flux Isotope Reactors, High
  USE   HIGH FLUX ISOTOPE REACTORS

Flux, Magnetic
  USE   MAGNETIC FLUX

Flux Mapping
  USE   MAPPING
         FLUX DENSITY

Flux Measurement, Plasma
  USE   PLASMA FLUX MEASUREMENT

Flux, Neutron
  USE   FLUX (RATE)

Flux, Particle
  USE   FLUX (RATE)

FLUX PINNING

Flux, Poloidal
  USE   POLOIDAL FLUX

FLUX PUMPS

FLUX QUANTIZATION

FLUX (RATE)

Flux (Rate Per Unit Area)
  USE   FLUX DENSITY

Flux, Solar
  USE   SOLAR FLUX

FLUXES

Fluxmeters
USE   MAGNETIC MEASUREMENT
       MEASURING INSTRUMENTS

FLY ASH

FLY BY TUBE CONTROL

FLY BY WIRE CONTROL

Fly TRAP Rocket Vehicle, Venus
USE   VENUS FLY TRAP ROCKET VEHICLE

Flyby, Mariner Jupiter-Saturn
USE   MARINER JUPITER-SATURN FLYBY

Flyby, Mariner Jupiter-Uranus
USE   MARINER JUPITER-URANUS FLYBY

FLYBY MISSIONS

Flying
USE   FLIGHT

Flying Bedstead Aircraft
USE   FLYING PLATFORMS

Flying Crane Helicopter
USE   H-17 HELICOPTER

FLYING EJECTION SEATS

Flying, Fear Of
USE   FEAR OF FLYING

Flying Objects, Unidentified
USE   UNIDENTIFIED FLYING OBJECTS

FLYING PERSONNEL

Flying Platform Stability
USE   FLYING PLATFORMS
       AERODYNAMIC STABILITY

FLYING PLATFORMS

Flying Qualities
USE   FLIGHT CHARACTERISTICS

FLYING SPOT SCANNERS

Flying Vehicles, Lunar
USE   LUNAR FLYING VEHICLES

Flying Wing Aircraft
USE   TAILLESS AIRCRAFT

FLYWHEELS

Fm
USE   FERMIUM

FM
USE   FREQUENCY MODULATION

FM/PM (MODULATION)

Foam, Polyurethane
USE   POLYURETHANE FOAM

FOAMING

FOAMS

Foams, Metal
USE   METAL FOAMS

FOCI

Fock Approximation, Hartree-
USE   HARTREE APPROXIMATION

Fock-Slater Method, Hartree-
USE   HARTREE-FOCK-SLATER METHOD

Focus, Plasma
USE   PLASMA FOCUS

FOCUSING

Focusing, De
USE   DEFOCUSING

Focusing, Self
USE   SELF FOCUSING

Foe, Identify Friend OR
USE   IFF SYSTEMS (IDENTIFICATION)

Foetuses
USE   FETUSES

FOG

FOG DISPERSAL

FOIL BEARINGS

FOILS

Foils, Air
USE   AIRFOILS

Foils, Hydro
USE   HYDROFOILS

FOILS (MATERIALS)

Foils, Metal
USE   METAL FOILS

FOKKER AIRCRAFT

Fokker Bond Testers
USE   ADHESION TESTS

Fokker F 27 Aircraft
USE   F-27 AIRCRAFT

Fokker F 28 Aircraft
USE   F-28 TRANSPORT AIRCRAFT

Fokker Friendship Aircraft
USE   F-27 AIRCRAFT

FOKKER-PLANCK EQUATION

FOLDING

FOLDING FIN AIRCRAFT ROCKET VEHICLE

FOLDING STRUCTURES

FOLDS (GEOLOGY)

FOLIAGE

FOLIC ACID

Follow-On Missions, LANDSAT
USE   LANDSAT FOLLOW-ON MISSIONS

Following Aircraft, Terrain
USE   TERRAIN FOLLOWING AIRCRAFT

FOOD

FOOD CHAIN

Food, Dehydrated
USE   DEHYDRATED FOOD

(Food), Flour
USE   FLOUR (FOOD)

(Food), Grains
USE   GRAINS (FOOD)

FOOD INTAKE

FOOD PROCESSING

Food, Synthetic
USE   SYNTHETIC FOOD

Foods, Frozen
USE   FROZEN FOODS

FOOTPRINTS

(Footwear), Boots
USE   BOOTS (FOOTWEAR)

FORBIDDEN BANDS

FORBIDDEN TRANSITIONS

FORBUSH DECREASES

Forbush Effect
USE   FORBUSH DECREASES

FORCE

Force Anemometers, Drag
USE   DRAG FORCE ANEMOMETERS

Force, Centrifugal
USE   CENTRIFUGAL FORCE

Force, Centripetal
USE   CENTRIPETAL FORCE

Force Curves, Zero
USE   ZERO FORCE CURVES

FORCE DISTRIBUTION

Force Distribution, Normal
USE   FORCE DISTRIBUTION

Force Fields
USE   FIELD THEORY (PHYSICS)

Force, G
USE   ACCELERATION (PHYSICS)

Force, Lines Of
USE   LINES OF FORCE

Force, Lorentz
USE   LORENTZ FORCE

Force Recorders, Cable
USE   CABLE FORCE RECORDERS

FORCE VECTOR RECORDERS

FORCE-FREE MAGNETIC FIELDS

FORCED CONVECTION

Forced Oscillation
USE   FORCED VIBRATION

FORCED VIBRATION

Forced Vibratory Motion Equations
USE   FORCED VIBRATION
       EQUATIONS

Forces, Aerodynamic
USE   AERODYNAMIC FORCES

Forces, Armed
USE   ARMED FORCES

Forces, Electromotive
USE   ELECTROMOTIVE FORCES

Forces (Foreign), Armed
USE   ARMED FORCES (FOREIGN)

Forces, Hypersonic
USE   HYPERSONIC FORCES

Forces, Inertial
USE   INERTIA

Forces, Interatomic
USE   INTERATOMIC FORCES

Forces, Intermolecular
USE   INTERMOLECULAR FORCES

Forces, Lift
USE   LIFT

(Forces), Load Distribution
USE   LOAD DISTRIBUTION (FORCES)

Forces, Loading
USE   LOADS (FORCES)

(Forces), Loads
USE   LOADS (FORCES)

Forces, Nonconservative
USE   NONCONSERVATIVE FORCES

Forces, Ponderomotive
USE   PONDEROMOTIVE FORCES

Forces (United States), Armed
USE   ARMED FORCES (UNITED STATES)

Forces, Van Der Waal
USE   VAN DER WAAL FORCES

Ford Project, West
USE   WEST FORD PROJECT

FOREARM

FOREBODIES

(Forebodies), Noses
USE   NOSES (FOREBODIES)

FORECASTING

(Forecasting), Delphi Method
USE   DELPHI METHOD (FORECASTING)

Forecasting, Long Range Weather
USE   LONG RANGE WEATHER FORECASTING

Forecasting, Numerical Weather
USE   NUMERICAL WEATHER FORECASTING

(Forecasting), Pattern Method
USE   PATTERN METHOD (FORECASTING)

(Forecasting), Probe Method
USE   PROBE METHOD (FORECASTING)

(Forecasting), Profile Method
USE   PROFILE METHOD (FORECASTING)

Forecasting, Statistical Weather
USE   STATISTICAL WEATHER FORECASTING

Forecasting, Technological
USE   TECHNOLOGICAL FORECASTING

Forecasting, Weather
USE   WEATHER FORECASTING

Forecasts
USE   FORECASTING

FOREHEAD

(Foreign), Armed Forces
USE   ARMED FORCES (FOREIGN)

FOREIGN BODIES

FOREIGN POLICY

FOREIGN TRADE

Forensic Sciences
USE   LAW (JURISPRUDENCE)

FOREST FIRE DETECTION

FOREST FIRES

FOREST MANAGEMENT

FORESTS

Forests, Rain
USE   RAIN FORESTS

FORGING

Forging, Metal
USE   FORGING

Forging, Spin
USE   METAL SPINNING

Fork Gyroscopes, Tuning
USE   TUNING FORK GYROSCOPES

FORKS

Form
USE   SHAPES

FORM FACTORS

Form, Jordan
USE   JORDAN FORM

Form Perception
USE   SPACE PERCEPTION

FORMALDEHYDE

Formaldehyde, Phenol
USE   PHENOL FORMALDEHYDE

FORMALISM

FORMAT

Formate, Chloro
USE   CHLOROFORMATE

FORMATES

Formates, Nitro
USE   NITROFORMATES

FORMATION

Formation, Crack
USE   CRACK INITIATION

Formation, Energy Of
USE   ENERGY OF FORMATION

Formation Heat
USE   HEAT OF FORMATION

Formation, Heat Of
USE   HEAT OF FORMATION

Formation, Ice
USE   ICE FORMATION

FORMATIONS

FORMHYDROXAMIC ACID

FORMIC ACID

FORMICA

Forming, Aus
USE   AUSFORMING

Forming, Cold
USE   COLD WORKING

Forming, Electro
USE   ELECTROFORMING

Forming, Electrohydraulic
USE   ELECTROHYDRAULIC FORMING

Forming, Explosive
USE   EXPLOSIVE FORMING

Forming, Hot
USE   HOT WORKING

Forming, Hydro
USE   HYDROFORMING

Forming, Magnetic
USE   MAGNETIC FORMING

Forming, Metal
USE   METAL WORKING
        FORMING TECHNIQUES

(Forming OR Bending), Brakes
USE   BRAKES (FORMING OR BENDING)

(Forming), Pressing
USE   PRESSING (FORMING)

Forming, Roll
USE   ROLL FORMING

Forming, Stretch
USE   STRETCH FORMING

FORMING TECHNIQUES

Forms, Canonical
USE   CANONICAL FORMS

Forms), Domes (Structural
USE   DOMES (STRUCTURAL FORMS)

Forms, Land
USE   LANDFORMS

Forms, Nitro
USE   NITROFORMS

FORMS (PAPER)

Forms, Plan
USE   PLANFORMS

Forms), Shells (Structural
USE   SHELLS (STRUCTURAL FORMS)

Forms, Wave
USE   WAVEFORMS

Formula, Bethe-Heitler
USE   BETHE-HEITLER FORMULA

Formula, Cauchy Integral
USE   CAUCHY INTEGRAL FORMULA

Formula, Kramers-Kronig
USE   KRAMERS-KRONIG FORMULA

Formula, Langevin
USE   LANGEVIN FORMULA

Formula, Moliere
USE   SPATIAL DISTRIBUTION
        SECONDARY COSMIC RAYS
        COSMIC RAY SHOWERS

FORMULAS

FORMULAS (MATHEMATICS)

Formulas, Recursion
USE   RECURSIVE FUNCTIONS

FORMULATIONS

FORMYL IONS

FORSTERITE

FORTISAN (TRADEMARK)

FORTRAN

Fortress Aircraft, Super
USE   RB-50 AIRCRAFT

Forward Looking Infrared Detectors
USE   FLIR DETECTORS

FORWARD SCATTERING

Forward Wings, Swept
USE   SWEPT FORWARD WINGS

FOSSIL FUELS

Fossil Meteorite Craters
USE   FOSSILS
        METEORITE CRATERS

FOSSILS

FOSTER THEORY

FOULING

Fouling, Anti
　USE　ANTIFOULING

FOUNDATIONS

(Foundations), Bases
　USE　FOUNDATIONS

Foundations, Pile
　USE　PILE FOUNDATIONS

Foundations, Structural
　USE　FOUNDATIONS

FOUNDRIES

FOUR BODY PROBLEM

Four Hour Orbits, Twenty-
　USE　TWENTY-FOUR HOUR ORBITS

FOURIER ANALYSIS

FOURIER LAW

FOURIER SERIES

FOURIER TRANSFORMATION

Fourier Transformations, Fast
　USE　FAST FOURIER TRANSFORMATIONS

FOURIER-BESSEL TRANSFORMATIONS

FOVEA

Fr
　USE　FRANCIUM

FR-1 SATELLITE

FRACTALS

FRACTIONATION

Fractionation, Chemical
　USE　CHEMICAL FRACTIONATION

FRACTIONS

FRACTOGRAPHY

FRACTURE MECHANICS

Fracture Resistance
　USE　FRACTURE STRENGTH

FRACTURE STRENGTH

Fracture Toughness
　USE　FRACTURE STRENGTH

Fractures, Crustal
　USE　CRUSTAL FRACTURES

FRACTURES (MATERIALS)

FRACTURING

(Fracturing), Cracking
　USE　CRACKING (FRACTURING)

FRAGMENTATION

FRAGMENTS

FRAME PHOTOGRAPHY

FRAMES

Frames, Air
　USE　AIRFRAMES

FRAMES (DATA PROCESSING)

(Frames), Racks
　USE　RACKS (FRAMES)

FRAMING CAMERAS

FRANCE

(France), Rhone Delta
　USE　RHONE DELTA (FRANCE)

Francisco Bay (CA), San
　USE　SAN FRANCISCO BAY (CA)

Francisco (CA), San
　USE　SAN FRANCISCO (CA)

FRANCIUM

FRANCK-CONDON PRINCIPLE

FRAUNHOFER LINE DISCRIMINATORS

FRAUNHOFER LINES

Fraunhofer Region
　USE　FAR FIELDS

FREDHOLM EQUATIONS

Fredholm Operators
　USE　FREDHOLM EQUATIONS
　　　　OPERATORS (MATHEMATICS)

FREE ATMOSPHERE

FREE BOUNDARIES

FREE CONVECTION

FREE ELECTRON LASERS

FREE ELECTRONS

FREE ENERGY

Free Energy, Gibbs
　USE　GIBBS FREE ENERGY

FREE FALL

FREE FLIGHT

FREE FLIGHT TEST APPARATUS

FREE FLOW

FREE JETS

Free Languages, Context
　USE　CONTEXT FREE LANGUAGES

Free Magnetic Fields, Force-
　USE　FORCE-FREE MAGNETIC FIELDS

FREE MOLECULAR FLOW

Free Oscillations
　USE　FREE VIBRATION

Free Path, Mean
　USE　MEAN FREE PATH

FREE RADICALS

Free Stream Effects
　USE　FREE FLOW

Free Streams
　USE　FREE FLOW

FREE VIBRATION

FREE WING AIRCRAFT

Freedom, Degrees Of
　USE　DEGREES OF FREEDOM

Freedom Fighter Aircraft
　USE　F-5 AIRCRAFT

FREEZE DRYING

FREEZING

Freezing Points
　USE　MELTING POINTS

Freezing, Vibrational
　USE　VIBRATIONAL FREEZING

Freight
　USE　CARGO

Freight, Air
　USE　AIR CARGO

FREIGHT COSTS

FREIGHTERS

French Equatorial Congo
　USE　CONGO (BRAZZAVILLE)

FRENCH GUIANA

(French Satellite), Spot
　USE　SPOT (FRENCH SATELLITE)

FRENCH SATELLITES

FRENCH SPACE PROGRAMS

FRENKEL DEFECTS

FREON

FREQUENCIES

Frequencies, Audio
　USE　AUDIO FREQUENCIES

Frequencies, Beat
　USE　BEAT FREQUENCIES

Frequencies, Carrier
　USE　CARRIER FREQUENCIES

Frequencies, Critical
　USE　CRITICAL FREQUENCIES

Frequencies, Extremely High
　USE　EXTREMELY HIGH FREQUENCIES

Frequencies, Extremely Low
　USE　EXTREMELY LOW FREQUENCIES

Frequencies, Extremely Low Radio
　USE　EXTREMELY LOW RADIO FREQUENCIES

Frequencies, High
　USE　HIGH FREQUENCIES

Frequencies, Infrasonic
　USE　INFRASONIC FREQUENCIES

Frequencies, Intermediate
　USE　INTERMEDIATE FREQUENCIES

Frequencies, Ionization
　USE　IONIZATION FREQUENCIES

Frequencies, Low
　USE　LOW FREQUENCIES

Frequencies, Microwave
　USE　MICROWAVE FREQUENCIES

Frequencies, Natural
　USE　RESONANT FREQUENCIES

Frequencies, Nyquist
　USE　NYQUIST FREQUENCIES

Frequencies, Plasma
　USE　PLASMA FREQUENCIES

Frequencies, Radio
　USE　RADIO FREQUENCIES

Frequencies, Resonant
　USE　RESONANT FREQUENCIES

Frequencies, Subaudible
　USE　SUBAUDIBLE FREQUENCIES

Frequencies, Superhigh
USE   SUPERHIGH FREQUENCIES

Frequencies, Ultrahigh
USE   ULTRAHIGH FREQUENCIES

Frequencies, Ultralow
USE   EXTREMELY LOW RADIO FREQUENCIES

Frequencies, Very High
USE   VERY HIGH FREQUENCIES

Frequencies, Very Low
USE   VERY LOW FREQUENCIES

Frequencies, Vibrational
USE   VIBRATIONAL SPECTRA

Frequency Amplifiers, Intermediate
USE   INTERMEDIATE FREQUENCY AMPLIFIERS

FREQUENCY ANALYZERS

FREQUENCY ASSIGNMENT

Frequency Bands
USE   FREQUENCIES

Frequency Bands, Low
USE   LOW FREQUENCY BANDS

Frequency, Brunt-Vaisala
USE   BRUNT-VAISALA FREQUENCY

FREQUENCY COMPRESSION DEMODULATORS

FREQUENCY CONTROL

Frequency Control, Automatic
USE   AUTOMATIC FREQUENCY CONTROL

Frequency Conversion
USE   FREQUENCY CONVERTERS

FREQUENCY CONVERTERS

Frequency Converters, Parametric
USE   PARAMETRIC FREQUENCY CONVERTERS

Frequency, Cyclotron
USE   CYCLOTRON FREQUENCY

Frequency Discharge, Radio
USE   RADIO FREQUENCY DISCHARGE

FREQUENCY DISCRIMINATORS

FREQUENCY DISTRIBUTION

FREQUENCY DIVIDERS

FREQUENCY DIVISION MULTIPLE ACCESS

FREQUENCY DIVISION MULTIPLEXING

Frequency, Flicker Fusion
USE   CRITICAL FLICKER FUSION

Frequency, Gyro
USE   GYROFREQUENCY

Frequency Heating, Radio
USE   RADIO FREQUENCY HEATING

FREQUENCY HOPPING

Frequency Impedance Probes, Radio
USE   RADIO FREQUENCY IMPEDANCE PROBES

Frequency Interference, Radio
USE   RADIO FREQUENCY INTERFERENCE

Frequency Ion Thrustor Engines, Radio
USE   RIT ENGINES

Frequency, Maximum Usable
USE   MAXIMUM USABLE FREQUENCY

FREQUENCY MEASUREMENT

FREQUENCY MODULATION

Frequency Modulation, Feedback
USE   FEEDBACK FREQUENCY MODULATION

FREQUENCY MODULATION PHOTOMULTIPLIERS

Frequency Modulation, Pulse
USE   PULSE FREQUENCY MODULATION

Frequency Modulation Telemetry, Pulse
USE   PULSE FREQUENCY MODULATION
        TELEMETRY

FREQUENCY MULTIPLIERS

Frequency Noise, Radio
USE   ELECTROMAGNETIC NOISE

Frequency Radiation, Radio
USE   RADIO WAVES

Frequency Radio Equipment, Very High
USE   VERY HIGH FREQUENCY RADIO
        EQUIPMENT

FREQUENCY RANGES

Frequency Regulation
USE   FREQUENCY CONTROL

FREQUENCY RESPONSE

FREQUENCY REUSE

FREQUENCY SCANNING

Frequency Shielding, Radio
USE   RADIO FREQUENCY SHIELDING

FREQUENCY SHIFT

FREQUENCY SHIFT KEYING

FREQUENCY STABILITY

FREQUENCY STANDARDS

Frequency, Sweep
USE   SWEEP FREQUENCY

FREQUENCY SYNCHRONIZATION

FREQUENCY SYNTHESIZERS

Frequency Transionospheric Satellites, Low
USE   LOW FREQUENCY TRANSIONOSPHERIC
        SATELLITES

Frequency Translation
USE   FREQUENCY CONVERTERS

FRESH WATER

FRESNEL DIFFRACTION

FRESNEL INTEGRALS

FRESNEL LENSES

FRESNEL REFLECTORS

FRESNEL REGION

Fresnel-Kirchhoff Integrals
USE   FRESNEL INTEGRALS

FRETTING

FRETTING CORROSION

FRICTION

Friction Coefficient
USE   COEFFICIENT OF FRICTION

Friction, Coefficient Of
USE   COEFFICIENT OF FRICTION

FRICTION DRAG

Friction, Dry
USE   DRY FRICTION

FRICTION FACTOR

Friction, Internal
USE   INTERNAL FRICTION

Friction, Kinetic
USE   KINETIC FRICTION

Friction Loss Coefficient
USE   FRICTION FACTOR

FRICTION MEASUREMENT

Friction Pressure Drop
USE   SKIN FRICTION

FRICTION REDUCTION

Friction, Skin
USE   SKIN FRICTION

Friction, Sliding
USE   SLIDING FRICTION

Friction, Static
USE   STATIC FRICTION

FRICTION WELDING

FRICTIONLESS ENVIRONMENTS

FRIEDEL-CRAFT REACTION

Friend OR Foe, Identify
USE   IFF SYSTEMS (IDENTIFICATION)

Friendship Aircraft, Fokker
USE   F-27 AIRCRAFT

FRIENDSHIP 7

FRINGE MULTIPLICATION

Fringe Patterns
USE   DIFFRACTION PATTERNS

Fringes, Moire
USE   MOIRE FRINGES

FRIT

Frog Otolith, Orbiting
USE   ORBITING FROG OTOLITH

FROGS

(From Earth), Space Observations
USE   SPACE OBSERVATIONS (FROM EARTH)

(From Space), Earth Observations
USE   EARTH OBSERVATIONS (FROM SPACE)

Front Deformation, Wave
USE   WAVE FRONT DEFORMATION

Front Reconstruction, Wave
USE   WAVE FRONT RECONSTRUCTION

Frontal Areas (Meteorology)
USE   FRONTS (METEOROLOGY)

FRONTAL WAVES

FRONTS

Fronts, Cold
USE   COLD FRONTS

Fronts, Flame
USE   FLAME PROPAGATION

FRONTS (METEOROLOGY)

Fronts, Shock
USE   SHOCK FRONTS

Fronts, Warm
USE   WARM FRONTS

Fronts, Wave
USE   WAVE FRONTS

Fronts, Weather
USE   FRONTS (METEOROLOGY)

FROST

FROST DAMAGE

Frost, Perma
USE   PERMAFROST

FROSTBITE

FROUDE NUMBER

FROZEN EQUILIBRIUM FLOW

FROZEN FOODS

Frozen Soils
USE   PERMAFROST

FRUITS

(Fruits), Nuts
USE   NUTS (FRUITS)

FRUSTRATION

FRUSTUMS

(Fuel), Bunkers
USE   BUNKERS (FUEL)

Fuel Burnup, Nuclear
USE   NUCLEAR FUEL BURNUP

FUEL CAPSULES

Fuel Cell Catalysts
USE   ELECTROCATALYSTS

FUEL CELL POWER PLANTS

FUEL CELLS

Fuel Cells, Biochemical
USE   BIOCHEMICAL FUEL CELLS

Fuel Cells, Hydrogen Air
USE   HYDROGEN OXYGEN FUEL CELLS

Fuel Cells, Hydrogen Oxygen
USE   HYDROGEN OXYGEN FUEL CELLS

Fuel Cells, Phosphoric Acid
USE   PHOSPHORIC ACID FUEL CELLS

Fuel Cells, Regenerative
USE   REGENERATIVE FUEL CELLS

FUEL COMBUSTION

FUEL CONSUMPTION

FUEL CONTAMINATION

FUEL CONTROL

(Fuel Conversion), Organic Wastes
USE   ORGANIC WASTES (FUEL CONVERSION)

FUEL CORROSION

Fuel Elements, Nuclear
USE   NUCLEAR FUEL ELEMENTS

Fuel Elements (Nuclear Reactors)
USE   NUCLEAR FUEL ELEMENTS

FUEL FLOW

FUEL FLOW REGULATORS

FUEL GAGES

Fuel Gages, Capacitive
USE   CAPACITIVE FUEL GAGES

(Fuel), Gasohol
USE   GASOHOL (FUEL)

FUEL INJECTION

Fuel, JP-4 Jet
USE   JP-4 JET FUEL

Fuel, JP-5 Jet
USE   JP-5 JET FUEL

Fuel, JP-6 Jet
USE   JP-6 JET FUEL

Fuel, JP-8 Jet
USE   JP-8 JET FUEL

FUEL OILS

FUEL PRODUCTION

Fuel Production, Hydrocarbon
USE   HYDROCARBON FUEL PRODUCTION

FUEL PUMPS

Fuel Reprocessing, Nuclear
USE   NUCLEAR FUEL REPROCESSING

FUEL SPRAYS

FUEL SYSTEMS

Fuel Systems, Aircraft
USE   AIRCRAFT FUEL SYSTEMS

(Fuel Systems), Chokes
USE   CHOKES (FUEL SYSTEMS)

FUEL TANK PRESSURIZATION

FUEL TANKS

FUEL TESTS

FUEL VALVES

FUEL-AIR RATIO

Fueling
USE   REFUELING

FUELS

Fuels, Aircraft
USE   AIRCRAFT FUELS

Fuels, Antimisting
USE   ANTIMISTING FUELS

Fuels, Automobile
USE   AUTOMOBILE FUELS

Fuels, Ceramic Nuclear
USE   CERAMIC NUCLEAR FUELS

Fuels, Chemical
USE   CHEMICAL FUELS

Fuels, Clean
USE   CLEAN FUELS

Fuels, Diesel
USE   DIESEL FUELS

Fuels, Endothermic
USE   ENDOTHERMIC FUELS

Fuels, Fissile
USE   FISSILE FUELS

Fuels, Fossil
USE   FOSSIL FUELS

Fuels, Gaseous
USE   GASEOUS FUELS

Fuels), HEF (High Energy
USE   HIGH ENERGY FUELS

Fuels, High Energy
USE   HIGH ENERGY FUELS

Fuels, Hydrocarbon
USE   HYDROCARBON FUELS

Fuels, Hydrogen
USE   HYDROGEN FUELS

Fuels, Jet
USE   JET ENGINE FUELS

Fuels, Jet Engine
USE   JET ENGINE FUELS

Fuels, Liquid
USE   LIQUID FUELS

Fuels, Metal
USE   METAL FUELS

Fuels, Nuclear
USE   NUCLEAR FUELS

Fuels, Reactor
USE   NUCLEAR FUELS

Fuels, Spent
USE   SPENT FUELS

Fuels, Synthetic
USE   SYNTHETIC FUELS

FUJITA METHOD

FULL SCALE TESTS

FULMINATES

FUMES

FUMIGATION

Function, Abel
USE   ABEL FUNCTION

Function, Airy
USE   AIRY FUNCTION

Function, Delta
USE   DELTA FUNCTION

Function, Gamma
USE   GAMMA FUNCTION

Function, Gauss
USE   GAUSS EQUATION

FUNCTION GENERATORS

Function, Heart
USE   HEART FUNCTION

Function, Mathieu
USE   MATHIEU FUNCTION

Function, Maxwell-Boltzmann Density
USE   MAXWELL-BOLTZMANN DENSITY
           FUNCTION

Function, Modulation Transfer
USE   MODULATION TRANSFER FUNCTION

Function, Muscular
USE   MUSCULAR FUNCTION

Function, Optical Transfer
USE   OPTICAL TRANSFER FUNCTION

Function, Penalty
USE   PENALTY FUNCTION

Function, Renal
USE   RENAL FUNCTION

FUNCTION SPACE

Function, Walsh
USE   WALSH FUNCTION

FUNCTIONAL ANALYSIS

FUNCTIONAL DESIGN SPECIFICATIONS

FUNCTIONAL INTEGRATION

FUNCTIONALS

FUNCTIONS

Functions, Analytic
USE   ANALYTIC FUNCTIONS

Functions, Aperiodic
USE   APERIODIC FUNCTIONS

Functions, Bessel
USE   BESSEL FUNCTIONS

Functions, Boolean
USE   BOOLEAN FUNCTIONS

Functions, Characteristic
USE   EIGENVECTORS
           EIGENVALUES

Functions, Composite
USE   COMPOSITE FUNCTIONS

Functions, Contralateral
USE   CONTRALATERAL FUNCTIONS

Functions, Correlation
USE   CORRELATION

Functions, Discrete
USE   DISCRETE FUNCTIONS

Functions, Discriminant
USE   DISCRIMINANT ANALYSIS (STATISTICS)

Functions, Distribution
USE   DISTRIBUTION FUNCTIONS

Functions, Disturbing
USE   DISTURBING FUNCTIONS

Functions, Elliptic
USE   ELLIPTIC FUNCTIONS

Functions, Entire
USE   ENTIRE FUNCTIONS

Functions, Error
USE   ERROR FUNCTIONS

Functions, Exponential
USE   EXPONENTIAL FUNCTIONS

Functions (Fluids), Stream
USE   STREAM FUNCTIONS (FLUIDS)

Functions, Green's
USE   GREEN'S FUNCTIONS

Functions, Hamiltonian
USE   HAMILTONIAN FUNCTIONS

Functions, Hankel
USE   HANKEL FUNCTIONS

Functions, Harmonic
USE   HARMONIC FUNCTIONS

Functions, Hyperbolic
USE   HYPERBOLIC FUNCTIONS

Functions, Hypergeometric
USE   HYPERGEOMETRIC FUNCTIONS

Functions, Integral
USE   ENTIRE FUNCTIONS

Functions, Kernel
USE   KERNEL FUNCTIONS

Functions, Laguerre
USE   LAGUERRE FUNCTIONS

Functions, Lame
USE   LAME FUNCTIONS

Functions, Legendre
USE   LEGENDRE FUNCTIONS

Functions, Liapunov
USE   LIAPUNOV FUNCTIONS

Functions, Lyapunov
USE   LIAPUNOV FUNCTIONS

Functions, Mal
USE   MALFUNCTIONS

FUNCTIONS (MATHEMATICS)

Functions, Meromorphic
USE   MEROMORPHIC FUNCTIONS

Functions, Monotone
USE   MONOTONE FUNCTIONS

Functions, Normal Density
USE   NORMAL DENSITY FUNCTIONS

Functions, Orthogonal
USE   ORTHOGONAL FUNCTIONS

Functions, Orthonormal
USE   ORTHONORMAL FUNCTIONS

Functions, Parenteral
USE   PARENTERAL FUNCTIONS

Functions, Periodic
USE   PERIODIC FUNCTIONS

Functions, Point Spread
USE   POINT SPREAD FUNCTIONS

Functions, Poisson Density
USE   POISSON DENSITY FUNCTIONS

Functions, Probability Density
USE   PROBABILITY DENSITY FUNCTIONS

Functions, Probability Distribution
USE   PROBABILITY DISTRIBUTION FUNCTIONS

Functions, Pulmonary
USE   PULMONARY FUNCTIONS

Functions, Ramp
USE   RAMP FUNCTIONS

Functions, Rational
USE   RATIONAL FUNCTIONS

Functions, Recursive
USE   RECURSIVE FUNCTIONS

Functions, Scattering
USE   SCATTERING FUNCTIONS

Functions, Space-Time
USE   SPACE-TIME FUNCTIONS

Functions, Spline
USE   SPLINE FUNCTIONS

Functions, Step
USE   STEP FUNCTIONS

Functions, Stress
USE   STRESS FUNCTIONS

Functions, Time
USE   TIME FUNCTIONS

Functions, Transcendental
USE   TRANSCENDENTAL FUNCTIONS

Functions, Transfer
USE   TRANSFER FUNCTIONS

Functions, Trigonometric
USE   TRIGONOMETRIC FUNCTIONS

Functions, Wave
USE   WAVE FUNCTIONS

Functions, Weibull Density
USE   WEIBULL DENSITY FUNCTIONS

Functions, Weierstrass
USE   WEIERSTRASS FUNCTIONS

Functions, Weighting
USE   WEIGHTING FUNCTIONS

Functions, Whittaker
USE   WHITTAKER FUNCTIONS

Functions, Work
USE   WORK FUNCTIONS

FUNGI

Fungi, Rust
USE   RUST FUNGI

FUNGICIDES

FUNNELS

FURAN RESINS

FURANS

FURFURYL ALCOHOL

FURLABLE ANTENNAS

FURNACES

Furnaces, Electric
USE   ELECTRIC FURNACES

Furnaces, Image
USE   IMAGE FURNACES

Furnaces, Solar
USE   SOLAR FURNACES

Furnaces, Vacuum
USE   VACUUM FURNACES

Fuselage Mounting
USE   AIRCRAFT PRODUCTION

Fuselage Stores, Wing-
USE   WING-FUSELAGE STORES

FUSELAGES

FUSES

Fuses, Electric
USE   ELECTRIC FUSES

FUSES (ORDNANCE)

FUSIBILITY

Fusiform Shapes
USE   CONES

FUSION

Fusion, Controlled
USE   CONTROLLED FUSION

Fusion, Critical Flicker
USE   CRITICAL FLICKER FUSION

Fusion Frequency, Flicker
USE   CRITICAL FLICKER FUSION

Fusion, Heat Of
USE   HEAT OF FUSION

Fusion, Impact
USE   IMPACT FUSION

Fusion, Inertial Confinement
USE   INERTIAL CONFINEMENT FUSION

Fusion, Laser
USE   LASER FUSION

Fusion, Latent Heat Of
USE   HEAT OF FUSION

FUSION (MELTING)

Fusion, Mirror
USE   MIRROR FUSION

Fusion, Nuclear
USE   NUCLEAR FUSION

Fusion (Reactor), Inertial
USE   INERTIAL FUSION (REACTOR)

FUSION REACTORS

(Fusion Reactors), Blankets
USE   BLANKETS (FUSION REACTORS)

(Fusion Reactors), Limiters
USE   LIMITERS (FUSION REACTORS)

FUSION WEAPONS

FUSION WELDING

FUSION-FISSION HYBRID REACTORS

FUZZY SETS

FUZZY SYSTEMS

FV-12A AIRCRAFT

F4H Aircraft
USE   F-4 AIRCRAFT

F8U Aircraft
USE   F-8 AIRCRAFT

F9F Aircraft
USE   F-9 AIRCRAFT

G

G ACPL (Spacelab), Zero-
USE   ATMOSPHERIC CLOUD PHYSICS LAB
            (SPACELAB)

G Force
USE   ACCELERATION (PHYSICS)

G, IMP-
USE   EXPLORER 41 SATELLITE

G, OSO-
USE   OSO-6

G Satellite, TIROS
USE   TIROS 7 SATELLITE

G Space Probe, Pioneer
USE   PIONEER 11 SPACE PROBE

G, Space Shuttle Mission 41-
USE   SPACE SHUTTLE MISSION 41-G

G, Space Shuttle Mission 51-
USE   SPACE SHUTTLE MISSION 51-G

G, Vitamin
USE   RIBOFLAVIN

G-1 AIRCRAFT

G-1 Aircraft, Navion
USE   G-1 AIRCRAFT

G-91 AIRCRAFT

G-91 Aircraft, Fiat
USE   G-91 AIRCRAFT

G-95/4 AIRCRAFT

G-95/4 Aircraft, Fiat
USE   G-95/4 AIRCRAFT

G-222 AIRCRAFT

G-222 Aircraft, Fiat
USE   G-222 AIRCRAFT

Ga
USE   GALLIUM

GA
USE   GEORGIA

(GA), Atlanta
USE   ATLANTA (GA)

(GA-NC-SC), Sand Hills Region
USE   SAND HILLS REGION (GA-NC-SC)

GA-5 AIRCRAFT

GA-5 Aircraft, Gloster
USE   GA-5 AIRCRAFT

GABON

GADOLINIUM

GADOLINIUM ALLOYS

GADOLINIUM ISOTOPES

Gage Accelerometers, Strain
USE   STRAIN GAGE ACCELEROMETERS

Gage Balances, Strain
USE   STRAIN GAGE BALANCES

Gages
USE   MEASURING INSTRUMENTS

Gages, Bayard-Alpert Ionization
USE   BAYARD-ALPERT IONIZATION GAGES

Gages), Bombs (Pressure
USE   PRESSURE GAGES

Gages, Capacitive Fuel
USE   CAPACITIVE FUEL GAGES

Gages, Fuel
USE   FUEL GAGES

Gages, Ion
USE   IONIZATION GAGES

Gages, Ionization
USE   IONIZATION GAGES

Gages, Knudsen
USE   KNUDSEN GAGES

Gages, Mcleod
USE   MCLEOD GAGES

Gages, Penning
USE   PENNING GAGES

Gages, Philips Ionization
USE   PHILIPS IONIZATION GAGES

Gages, Piezoelectric
USE   PIEZOELECTRIC GAGES

Gages, Pirani
USE   PIRANI GAGES

Gages, Pressure
USE   PRESSURE GAGES

Gages, Rain
USE   RAIN GAGES

Gages, Sputtering
USE   SPUTTERING GAGES

Gages, Strain
USE   STRAIN GAGES

Gages, Thermal Conductivity
USE   THERMAL CONDUCTIVITY GAGES

Gages, Vacuum
USE   VACUUM GAGES

Gain (Amplification)
USE   AMPLIFICATION

Gain Control, Automatic
USE   AUTOMATIC GAIN CONTROL

Gain, Heat
USE   HEATING

Gain, High
USE   HIGH GAIN

Gain, Power
USE   POWER GAIN

Galactic Cluster, Virgo
USE   VIRGO GALACTIC CLUSTER

GALACTIC CLUSTERS

GALACTIC COSMIC RAYS

GALACTIC EVOLUTION

Galactic Magnetic Fields
USE   INTERSTELLAR MAGNETIC FIELDS

GALACTIC NUCLEI

GALACTIC RADIATION

Galactic Radiation Exp Background Sats
USE   GREB SATELLITES

GALACTIC RADIO WAVES

GALACTIC ROTATION

GALACTIC STRUCTURE

GALACTOSE

GALAXIES

Galaxies, Andromeda
USE   ANDROMEDA GALAXIES

Galaxies, Barred
USE   BARRED GALAXIES

Galaxies, Disk
USE   DISK GALAXIES

Galaxies, Dwarf
USE   DWARF GALAXIES

Galaxies, Elliptical
USE   ELLIPTICAL GALAXIES

Galaxies, Maffei
USE   MAFFEI GALAXIES

Galaxies, Radio
USE   RADIO GALAXIES

Galaxies, Seyfert
USE   SEYFERT GALAXIES

Galaxies, Spiral
USE   SPIRAL GALAXIES

Galaxy Aircraft
USE   C-5 AIRCRAFT

Galaxy, Milky Way
USE   MILKY WAY GALAXY

GALERKIN METHOD

GALILEAN SATELLITES

Galileo Mission
USE   GALILEO PROJECT

GALILEO PROBE

GALILEO PROJECT

GALILEO SPACECRAFT

GALL

GALLAMINE TRIETHIODIDE

GALLATES

Gallates, Sodium
USE   SODIUM GALLATES

GALLIUM

GALLIUM ALLOYS

GALLIUM ANTIMONIDES

GALLIUM ARSENIDE LASERS

GALLIUM ARSENIDES

Gallium Arsenides, Aluminum
USE   ALUMINUM GALLIUM ARSENIDES

GALLIUM COMPOUNDS

GALLIUM ISOTOPES

GALLIUM NITRIDES

GALLIUM OXIDES

GALLIUM PHOSPHIDES

GALLIUM SELENIDES

Galvanic Cells
USE   ELECTROLYTIC CELLS

GALVANIC SKIN RESPONSE

Galvanizing
USE   ZINC COATINGS

GALVANOMAGNETIC EFFECTS

Galvanomagnetism
USE   GALVANOMAGNETIC EFFECTS

GALVANOMETERS

GAMBIA

GAME THEORY

(Game Theory), Saddle Points
USE   SADDLE POINTS (GAME THEORY)

Games, War
USE   WAR GAMES

GAMETOCYTES

GAMMA FUNCTION

GAMMA GLOBULIN

Gamma Line, H
USE   H GAMMA LINE

Gamma Radiation
USE   GAMMA RAYS

GAMMA RAY ABSORPTIOMETRY

GAMMA RAY ABSORPTION

GAMMA RAY ASTRONOMY

Gamma Ray Astronomy Explorer
USE   EXPLORER 11 SATELLITE

GAMMA RAY BEAMS

GAMMA RAY BURSTS

Gamma Ray Bursts, Cosmic
USE   GAMMA RAY BURSTS

GAMMA RAY LASERS

GAMMA RAY OBSERVATORY

GAMMA RAY SPECTRA

GAMMA RAY SPECTROMETERS

GAMMA RAY TELESCOPES

GAMMA RAYS

GANGLIA

Gantries
USE   GANTRY CRANES

GANTRY CRANES

GANYMEDE

GAPS

GAPS (GEOLOGY)

Gaps (Solid State), Energy
USE   ENERGY GAPS (SOLID STATE)

Gaps, Spark
USE   SPARK GAPS

GARBAGE

GARMENTS

(Garnet), YAG
USE   YTTRIUM-ALUMINUM GARNET

(Garnet), YIG
USE   YTTRIUM-IRON GARNET

Garnet, Yttrium-Aluminum
USE   YTTRIUM-ALUMINUM GARNET

Garnet, Yttrium-Iron
USE   YTTRIUM-IRON GARNET

GARNETS

GARP
USE   GLOBAL ATMOSPHERIC RESEARCH
PROGRAM

GARP ATLANTIC TROPICAL EXPERIMENT

GAS ANALYSIS

GAS ATOMIZATION

GAS BAGS

GAS BEARINGS

GAS CHROMATOGRAPHY

Gas, Cold
USE   COLD GAS

GAS COMPOSITION

Gas Compounds, Rare
USE   RARE GAS COMPOUNDS

Gas, Compressed
USE   COMPRESSED GAS

GAS COOLED FAST REACTORS

GAS COOLED REACTORS

Gas Cooled Reactors, Experimental
USE   EXPERIMENTAL GAS COOLED REACTORS

Gas Cooled Reactors, High Temperature
USE   HIGH TEMPERATURE GAS COOLED
REACTORS

GAS COOLING

GAS DENSITY

GAS DETECTORS

Gas Diffusion
USE   GASEOUS DIFFUSION

Gas Discharge Counters
USE   GAS DISCHARGE TUBES
COUNTERS

GAS DISCHARGE TUBES

GAS DISCHARGES

GAS DISSOCIATION

GAS DYNAMICS

Gas Dynamics, Rarefied
USE   RAREFIED GAS DYNAMICS

Gas, Electron
USE   ELECTRON GAS

Gas Evacuating
USE   EVACUATING (VACUUM)

GAS EVOLUTION

GAS EXCHANGE

GAS EXPANSION

Gas Experiment, Stratospheric Aerosol &
USE   SAGE SATELLITE

Gas Exploration, Natural
USE   NATURAL GAS EXPLORATION

GAS EXPLOSIONS

GAS FLOW

(Gas Flow), Draft
USE   DRAFT (GAS FLOW)

Gas Generator Engines
USE   ENGINES
GAS GENERATORS

GAS GENERATORS

GAS GIANT PLANETS

Gas, Gray
USE   GRAY GAS

GAS GUNS

Gas Guns, Light
USE   LIGHT GAS GUNS

GAS HEATING

Gas, Ideal
USE   IDEAL GAS

GAS INJECTION

Gas Interactions, Gas-
USE   GAS-GAS INTERACTIONS

Gas Interactions, Ion-
USE   GAS-ION INTERACTIONS

Gas, Interplanetary
USE   INTERPLANETARY GAS

Gas, Interstellar
USE   INTERSTELLAR GAS

GAS IONIZATION

GAS JETS

GAS LASERS

Gas, Lennard-Jones
USE   LENNARD-JONES GAS

Gas Liquefaction
USE   CONDENSING

Gas, Liquefied Natural
USE   LIQUEFIED NATURAL GAS

Gas, Lorentz
USE   LORENTZ GAS

GAS LUBRICANTS

Gas Lubricated Bearings
USE   GAS BEARINGS

GAS MASERS

GAS METERS

GAS MIXTURES

Gas Mixtures, Detonable
    USE   DETONABLE GAS MIXTURES

Gas Mixtures, Liquid-
    USE   LIQUID-GAS MIXTURES

Gas Model, Lighthill
    USE   LIGHTHILL GAS MODEL

Gas, Natural
    USE   NATURAL GAS

Gas, Nongray
    USE   NONGRAY GAS

GAS PATH ANALYSIS

Gas, Perfect
    USE   IDEAL GAS

Gas Phases
    USE   VAPOR PHASES

GAS PIPES

GAS POCKETS

GAS PRESSURE

GAS REACTORS

GAS RECOVERY

Gas, Residual
    USE   RESIDUAL GAS

GAS SPECTROSCOPY

GAS STREAMS

Gas Systems, Hot
    USE   HIGH TEMPERATURE GASES

Gas Systems, Metal-
    USE   METAL-GAS SYSTEMS

GAS TEMPERATURE

GAS TRANSPORT

GAS TUBES

GAS TUNGSTEN ARC WELDING

GAS TURBINE ENGINES

GAS TURBINES

GAS VALVES

GAS VISCOSITY

GAS WELDING

Gas Welding, Tungsten Inert
    USE   GAS TUNGSTEN ARC WELDING

GAS-GAS INTERACTIONS

Gas-Halide Lasers, Rare
    USE   RARE GAS-HALIDE LASERS

GAS-ION INTERACTIONS

GAS-LIQUID INTERACTIONS

GAS-METAL INTERACTIONS

GAS-SOLID INTERACTIONS

GAS-SOLID INTERFACES

GASDYNAMIC LASERS

Gaseous Cavitation
    USE   GAS FLOW
           CAVITATION FLOW

GASEOUS DIFFUSION

GASEOUS FISSION REACTORS

GASEOUS FUELS

GASEOUS ROCKET PROPELLANTS

GASEOUS SELF-DIFFUSION

GASES

Gases, Atomic
    USE   MONATOMIC GASES

Gases, Coal Derived
    USE   COAL DERIVED GASES

Gases, Cosmic
    USE   COSMIC GASES

Gases, Diatomic
    USE   DIATOMIC GASES

Gases, Dissolved
    USE   DISSOLVED GASES

Gases, Exhaust
    USE   EXHAUST GASES

Gases, Explosive
    USE   FLAMMABLE GASES

Gases, Flammable
    USE   FLAMMABLE GASES

Gases, Flue
    USE   FLUE GASES

Gases, High Temperature
    USE   HIGH TEMPERATURE GASES

Gases, Hot
    USE   HIGH TEMPERATURE GASES

Gases, Inert
    USE   RARE GASES

Gases, Ionized
    USE   IONIZED GASES

Gases, Liquefied
    USE   LIQUEFIED GASES

Gases, Low Density
    USE   RAREFIED GASES

Gases, Molecular
    USE   MOLECULAR GASES

Gases, Monatomic
    USE   MONATOMIC GASES

Gases, Neutral
    USE   NEUTRAL GASES

Gases, Noble
    USE   RARE GASES

Gases, Noncondensable
    USE   NONCONDENSABLE GASES

Gases, Nonpolar
    USE   NONPOLAR GASES

Gases, Polar
    USE   POLAR GASES

Gases, Polyatomic
    USE   POLYATOMIC GASES

Gases, Rare
    USE   RARE GASES

Gases, Rarefied
    USE   RAREFIED GASES

Gases, Real
    USE   REAL GASES

Gases, Solidified
    USE   SOLIDIFIED GASES

GASIFICATION

Gasification, Coal
    USE   COAL GASIFICATION

GASKETS

GASOHOL (FUEL)

GASOLINE

GASP
    USE   GLOBAL AIR SAMPLING PROGRAM

Gassing, De
    USE   DEGASSING

Gassing, Off
    USE   OFFGASSING

Gassing, Out
    USE   OUTGASSING

GASTROINTESTINAL SYSTEM

GATE (Experiment)
    USE   GARP ATLANTIC TROPICAL EXPERIMENT

GATES

GATES (CIRCUITS)

GATES (OPENINGS)

Gates, OR-
    USE   GATES (CIRCUITS)

Gates, Threshold
    USE   THRESHOLD GATES

GAUGE INVARIANCE

GAUGE THEORY

GAUSS EQUATION

Gauss Function
    USE   GAUSS EQUATION

GAUSS-MARKOV THEOREM

Gaussian Distributions
    USE   NORMAL DENSITY FUNCTIONS

Gaussian Noise
    USE   RANDOM NOISE

Gaussmeters
    USE   MAGNETOMETERS

GAUZE

GAW-1 AIRFOIL

GAW-2 AIRFOIL

GC-130 Aircraft
    USE   C-130 AIRCRAFT

GCR (Reactors)
    USE   GAS COOLED REACTORS

Gd
    USE   GADOLINIUM

GDOP
    USE   GEOMETRIC DILUTION OF PRECISION

Ge
    USE   GERMANIUM

GE COMPUTERS

GE 625 COMPUTER

GE 635 COMPUTER

GE-1 Engine, XJ-79-
USE   J-79 ENGINE

GE-3 Engine, YJ-73-
USE   J-73 ENGINE

GE-3 Engine, YJ-93-
USE   J-93 ENGINE

GE-8B Engine, T-58-
USE   T-58-GE-8B ENGINE

GEAR

Gear, Arresting
USE   ARRESTING GEAR

Gear, Landing
USE   LANDING GEAR

Gear, Retractable Landing
USE   LANDING GEAR
RETRACTABLE EQUIPMENT

GEAR TEETH

GEARS

(Gears), Racks
USE   RACKS (GEARS)

GEGENSCHEIN

GEHLENITE

GEIGER COUNTERS

Geiger-Mueller Tubes
USE   GEIGER COUNTERS

Gel Permeation Chromatography
USE   LIQUID CHROMATOGRAPHY

Gel Processes, Sol-
USE   SOL-GEL PROCESSES

Gel, Silica
USE   SILICA GEL

GELATINS

GELATION

GELLED PROPELLANTS

GELLED ROCKET PROPELLANTS

GELS

GEMINI B SPACECRAFT

GEMINI FLIGHTS

GEMINI (GT-1) SPACECRAFT

GEMINI PROJECT

GEMINI SPACECRAFT

GEMINI 2 SPACECRAFT

GEMINI 3 FLIGHT

GEMINI 4 FLIGHT

GEMINI 5 FLIGHT

GEMINI 6 FLIGHT

GEMINI 7 FLIGHT

GEMINI 8 FLIGHT

GEMINI 9 FLIGHT

GEMINI 10 FLIGHT

GEMINI 11 FLIGHT

GEMINI 12 FLIGHT

GEMINID METEOROIDS

GENERAL AVIATION AIRCRAFT

General Aviation Whitcomb Airfoil
USE   GAW-2 AIRFOIL
GAW-1 AIRFOIL

General Circulation Experiment, Atmospheric
USE   ATMOSPHERIC GENERAL CIRCULATION
EXPERIMENT

GENERAL DYNAMICS AIRCRAFT

General Dynamics Military Aircraft
USE   GENERAL DYNAMICS AIRCRAFT
MILITARY AIRCRAFT

General Electric Computers
USE   GE COMPUTERS

GENERALIZATION (PSYCHOLOGY)

Generated Electromagnetic Pulses, System
USE   SYSTEM GENERATED ELECTROMAGNETIC
PULSES

GENERATION

Generation, Combined Cycle Power
USE   COMBINED CYCLE POWER GENERATION

Generation, Heat
USE   HEAT GENERATION

Generation, Nuclear Electric Power
USE   NUCLEAR ELECTRIC POWER GENERATION

Generation, Nuclear Power
USE   NUCLEAR ELECTRIC POWER GENERATION

Generation, Plasma
USE   PLASMA GENERATORS

Generation, Solar Power
USE   SOLAR GENERATORS

Generation, Thermionic Power
USE   THERMIONIC POWER GENERATION

Generation, Thermoelectric Power
USE   THERMOELECTRIC POWER GENERATION

Generation, Thermonuclear Power
USE   THERMONUCLEAR POWER GENERATION

Generation, Vortex
USE   VORTEX GENERATORS

Generation, Wave
USE   WAVE GENERATION

Generations, Harmonic
USE   HARMONIC GENERATIONS

Generator, ASTEC Solar Turboelectric
USE   ASTEC SOLAR TURBOELECTRIC
GENERATOR

Generator Engines, Gas
USE   GAS GENERATORS
ENGINES

GENERATORS

Generators, AC
USE   AC GENERATORS

Generators, Acoustic
USE   SOUND GENERATORS

Generators, Alternating Current
USE   AC GENERATORS

(Generators), Alternators
USE   AC GENERATORS

Generators, Arc
USE   ARC GENERATORS

Generators, Cavity Vapor
USE   CAVITY VAPOR GENERATORS

Generators, Colloidal
USE   COLLOIDAL GENERATORS

Generators, Direct Power
USE   DIRECT POWER GENERATORS

Generators, Electric
USE   ELECTRIC GENERATORS

Generators, Electrostatic
USE   ELECTROSTATIC GENERATORS

Generators, Function
USE   FUNCTION GENERATORS

Generators, Gas
USE   GAS GENERATORS

Generators, Hall
USE   HALL GENERATORS

Generators, Harmonic
USE   HARMONIC GENERATORS

Generators, Homopolar
USE   HOMOPOLAR GENERATORS

Generators, Impulse
USE   IMPULSE GENERATORS

Generators, Magnetohydrodynamic
USE   MAGNETOHYDRODYNAMIC GENERATORS

Generators, Nernst
USE   THERMOMAGNETIC COOLING

Generators, Noise
USE   NOISE GENERATORS

Generators, Optical
USE   LASER CAVITIES

Generators, Photoelectric
USE   PHOTOELECTRIC GENERATORS

Generators, Plasma
USE   PLASMA GENERATORS

Generators, Power
USE   ELECTRIC GENERATORS

Generators, Pulse
USE   PULSE GENERATORS

Generators, Quantum
USE   STIMULATED EMISSION DEVICES

Generators, Report
USE   REPORT GENERATORS

Generators, Rotating
USE   ROTATING GENERATORS

Generators, Shock Wave
USE   SHOCK WAVE GENERATORS

Generators, Signal
USE   SIGNAL GENERATORS

Generators, Solar
USE   SOLAR GENERATORS

Generators, Sound
USE   SOUND GENERATORS

Generators, Steam
USE   BOILERS

Generators, Subharmonic
USE   SUBHARMONIC GENERATORS

Generators, Test Pattern
USE   TEST PATTERN GENERATORS

Generators, Thermoelectric
USE   THERMOELECTRIC GENERATORS

Generators, Tide Powered
USE TIDE POWERED GENERATORS

Generators, Turbo
USE TURBOGENERATORS

Generators, Vapor
USE VAPORIZERS

Generators, Voltage
USE VOLTAGE GENERATORS

Generators, Vortex
USE VORTEX GENERATORS

Generators, Windpowered
USE WINDPOWERED GENERATORS

Genesis, Ablo
USE ABIOGENESIS

Genesis, Cyclo
USE CYCLOGENESIS

Genesis, Cyto
USE CYTOGENESIS

Genesis, Lyso
USE LYSOGENESIS

Genesis, Spermato
USE SPERMATOGENESIS

GENETIC CODE

GENETIC ENGINEERING

GENETICS

GENIE ROCKET VEHICLE

GENITOURINARY SYSTEM

Geoastrophysics
USE GEOPHYSICS
     ASTROPHYSICS

GEOBOTANY

GEOCENTRIC COORDINATES

GEOCHEMISTRY

Geochemistry, Bio
USE BIOGEOCHEMISTRY

GEOCHRONOLOGY

GEOCORONAL EMISSIONS

GEOCYCLOTRONS

GEODESIC LINES

GEODESY

Geodesy, Celestial
USE CELESTIAL GEODESY

Geodesy Experiment, International Satellite
USE INTERNATIONAL SATELLITE GEODESY
     EXPERIMENT

GEODETIC ACCURACY

GEODETIC COORDINATES

GEODETIC SATELLITES

GEODETIC SURVEYS

GEODIMETERS

Geodynamic Experimental Ocean Satellite
USE GEOS-D SATELLITE

Geodynamic Satellite, Laser
USE LAGEOS (SATELLITE)

GEODYNAMICS

GEOELECTRICITY

Geofabrics
USE GEOTECHNICAL FABRICS

Geofractures
USE GEOLOGICAL FAULTS

GEOGRAPHIC APPLICATIONS PROGRAM

GEOGRAPHIC INFORMATION SYSTEMS

GEOGRAPHY

GEOIDS

GEOLE SATELLITES

GEOLOGICAL FAULTS

GEOLOGICAL SURVEYS

GEOLOGY

(Geology), Beds
USE BEDS (GEOLOGY)

(Geology), Contacts
USE CONTACTS (GEOLOGY)

(Geology), Crossbedding
USE CROSSBEDDING (GEOLOGY)

(Geology), Dikes
USE ROCK INTRUSIONS

(Geology), Domes
USE DOMES (GEOLOGY)

(Geology), Fissures
USE FISSURES (GEOLOGY)

(Geology), Folds
USE FOLDS (GEOLOGY)

(Geology), Gaps
USE GAPS (GEOLOGY)

Geology, Hydro
USE HYDROGEOLOGY

(Geology), Kettles
USE KETTLES (GEOLOGY)

Geology, Lunar
USE LUNAR GEOLOGY

Geology, Marine
USE HYDROGEOLOGY

(Geology), Metamorphism
USE METAMORPHISM (GEOLOGY)

(Geology), Outlets
USE ESTUARIES

Geology, Photo
USE PHOTOGEOLOGY

Geology, Planetary
USE PLANETARY GEOLOGY

(Geology), Polar Wandering
USE POLAR WANDERING (GEOLOGY)

Geology, Radar
USE RADAR GEOLOGY

(Geology), Scars
USE EROSION

(Geology), Shields
USE BEDROCK

(Geology), Sinks
USE STRUCTURAL BASINS

(Geology), Splits
USE GEOLOGICAL FAULTS

(Geology), Structural Properties
USE STRUCTURAL PROPERTIES (GEOLOGY)

(Geology), Subduction
USE SUBDUCTION (GEOLOGY)

Geomagnetic Anomalies
USE MAGNETIC ANOMALIES

Geomagnetic Crotchets
USE SUDDEN IONOSPHERIC DISTURBANCES

Geomagnetic Effects
USE MAGNETIC EFFECTS

Geomagnetic Equator
USE MAGNETIC EQUATOR

Geomagnetic Field
USE GEOMAGNETISM

GEOMAGNETIC HOLLOW

GEOMAGNETIC LATITUDE

GEOMAGNETIC MICROPULSATIONS

GEOMAGNETIC PULSATIONS

Geomagnetic Storms
USE MAGNETIC STORMS

GEOMAGNETIC TAIL

Geomagnetically Trapped Particles
USE RADIATION BELTS

GEOMAGNETISM

GEOMETRIC ACCURACY

GEOMETRIC DILUTION OF PRECISION

GEOMETRIC RECTIFICATION (IMAGERY)

GEOMETRICAL ACOUSTICS

Geometrical Hydromagnetics
USE MAGNETOHYDRODYNAMICS

GEOMETRICAL OPTICS

GEOMETRICAL THEORY OF DIFFRACTION

Geometrodynamics
USE RELATIVITY

GEOMETRY

Geometry, Analytic
USE ANALYTIC GEOMETRY

(Geometry), Angles
USE ANGLES (GEOMETRY)

Geometry, Bose
USE BOSE GEOMETRY

(Geometry), Chords
USE CHORDS (GEOMETRY)

(Geometry), Circles
USE CIRCLES (GEOMETRY)

Geometry, Crack
USE CRACK GEOMETRY

(Geometry), Curves
USE CURVES (GEOMETRY)

Geometry, Descriptive
USE DESCRIPTIVE GEOMETRY

Geometry, Differential
USE DIFFERENTIAL GEOMETRY

Geometry, Duct
USE DUCT GEOMETRY

Geometry, Euclidean
USE EUCLIDEAN GEOMETRY

Geometry, Flow
USE   FLOW GEOMETRY

Geometry Language, Coordinate
USE   COGO (PROGRAMMING LANGUAGE)

(Geometry), Lines
USE   LINES (GEOMETRY)

Geometry (Mechanics), Hole
USE   HOLE GEOMETRY (MECHANICS)

Geometry, Noneuclidian
USE   DIFFERENTIAL GEOMETRY

Geometry, Nozzle
USE   NOZZLE GEOMETRY

Geometry, Projective
USE   PROJECTIVE GEOMETRY

Geometry, Specimen
USE   SPECIMEN GEOMETRY

Geometry Structures, Variable
USE   VARIABLE GEOMETRY STRUCTURES

Geometry, Surface
USE   SURFACE GEOMETRY

Geometry, Tank
USE   TANK GEOMETRY

GEOMORPHOLOGY

Geon (Trademark)
USE   POLYVINYL CHLORIDE

GEOPHYSICAL FLUID FLOW CELLS

GEOPHYSICAL FLUIDS

GEOPHYSICAL OBSERVATORIES

Geophysical Observatory, Eccentric
USE   EGO

Geophysical Observatory, Eccentric Orbit
USE   EGO

Geophysical Observatory, Orbiting
USE   OGO

Geophysical Observatory, Polar Orbit
USE   POGO

GEOPHYSICAL SATELLITES

(Geophysical Year), IGY
USE   INTERNATIONAL GEOPHYSICAL YEAR

Geophysical Year, International
USE   INTERNATIONAL GEOPHYSICAL YEAR

GEOPHYSICS

GEOPOTENTIAL

GEOPOTENTIAL HEIGHT

GEOPRESSURE

GEORGIA

GEOS SATELLITES (ESA)

GEOS Satellites (ESRO)
USE   GEOS SATELLITES (ESA)

GEOS 1 SATELLITE

GEOS 2 SATELLITE

GEOS 3 SATELLITE

GEOS-B Satellite
USE   GEOS 2 SATELLITE

GEOS-C Satellite
USE   GEOS 3 SATELLITE

GEOS-D SATELLITE

GEOSARI PROJECT

Geosphere-Biosphere Program, International
USE   INTERNATIONAL GEOSPHERE-BIOSPHERE
        PROGRAM

Geostationary Operational Environ Sats
USE   GOES SATELLITES

Geostationary Operatl Environ Satellite B
USE   GOES 2

Geostationary Platforms
USE   SYNCHRONOUS PLATFORMS

Geostationary Satellites
USE   SYNCHRONOUS SATELLITES

GEOSTROPHIC WIND

GEOSYNCHRONOUS ORBITS

GEOSYNCLINES

GEOTECHNICAL ENGINEERING

GEOTECHNICAL FABRICS

GEOTEMPERATURE

Geotextiles
USE   GEOTECHNICAL FABRICS

GEOTHERMAL ANOMALIES

GEOTHERMAL ENERGY CONVERSION

GEOTHERMAL ENERGY EXTRACTION

GEOTHERMAL ENERGY UTILIZATION

GEOTHERMAL RESOURCES

GEOTHERMAL TECHNOLOGY

Geothermometry
USE   GEOTEMPERATURE

GEOTROPISM

GEP Telescopes
USE   PARTICLE TELESCOPES

Gerdien Arc Heaters
USE   HEATING EQUIPMENT
        ARC HEATING

GERDIEN CONDENSERS

GERIATRICS

German Democratic Republic
USE   EAST GERMANY

GERMANATES

Germanates, Magnesium
USE   MAGNESIUM GERMANATES

GERMANIDES

Germanides, Magnesium
USE   MAGNESIUM GERMANIDES

GERMANIUM

GERMANIUM ALLOYS

GERMANIUM ANTIMONIDES

GERMANIUM CHLORIDES

GERMANIUM COMPOUNDS

Germanium Compounds, Organic
USE   ORGANIC GERMANIUM COMPOUNDS

GERMANIUM DIODES

GERMANIUM ISOTOPES

GERMANIUM OXIDES

Germanium Rectifiers
USE   GERMANIUM DIODES

GERMANY

Germany, East
USE   EAST GERMANY

Germany, Federal Republic Of
USE   WEST GERMANY

Germany, Peoples Democratic Republic Of
USE   EAST GERMANY

Germany, West
USE   WEST GERMANY

Germicides
USE   BACTERICIDES

GERMINATION

Germinators
USE   PHYTOTRONS

GERONTOLOGY

GERT

GESTALT THEORY

GETOL AIRCRAFT

GETTERS

GEYSERS

GHANA

GHOSTS

GIACOBINI-ZINNER COMET

Giant Planets, Gas
USE   GAS GIANT PLANETS

GIANT STARS

Giant Stars, Red
USE   RED GIANT STARS

GIBBERELLINS

GIBBS ADSORPTION EQUATION

GIBBS EQUATIONS

GIBBS FREE ENERGY

GIBBS PHENOMENON

GIBBS-HELMHOLTZ EQUATIONS

GIMBALLESS INERTIAL NAVIGATION

GIMBALS

Ginzburg Equations, Landau-
USE   LANDAU-GINZBURG EQUATIONS

GIOTTO MISSION

GIRDER WEBS

GIRDERS

GIRDLES

GLACIAL DRIFT

Glaciation, Cloud
USE   CLOUD GLACIATION

GLACIERS

Glaciers, Active
USE GLACIERS

Glaciers, Advancing
USE GLACIERS

Glaciofluvial Deposits
USE GLACIAL DRIFT

GLACIOLOGY

Gland, Adrenal
USE ADRENAL GLAND

Gland, Parathyroid
USE PARATHYROID GLAND

Gland, Parotid
USE SALIVARY GLANDS

Gland, Pineal
USE PINEAL GLAND

Gland, Pituitary
USE PITUITARY GLAND

Gland, Prostate
USE PROSTATE GLAND

Gland, Thymus
USE THYMUS GLAND

Gland, Thyroid
USE THYROID GLAND

GLANDS

GLANDS (ANATOMY)

Glands, Endocrine
USE ENDOCRINE GLANDS

Glands, Mammary
USE MAMMARY GLANDS

Glands, Salivary
USE SALIVARY GLANDS

GLANDS (SEALS)

Glands, Sebaceous
USE SEBACEOUS GLANDS

Glands, Sex
USE SEX GLANDS

GLARE

GLASS

Glass, Borosilicate
USE BOROSILICATE GLASS

GLASS COATINGS

Glass, E
USE E GLASS

GLASS ELECTRODES

GLASS FIBER REINFORCED PLASTICS

GLASS FIBERS

GLASS LASERS

Glass, Obsidian
USE OBSIDIAN GLASS

Glass, S
USE S GLASS

Glass, Silica
USE SILICA GLASS

Glass, Spin
USE SPIN GLASS

Glasses, Metallic
USE METALLIC GLASSES

Glasses, Sun
USE SUNGLASSES

GLASSWARE

GLASSY CARBON

GLAUBER THEORY

GLAUCOMA

Glauert Coefficient
USE MACH NUMBER
AERODYNAMIC FORCES

GLAZES

Glide Angles
USE GLIDE PATHS

GLIDE LANDINGS

GLIDE PATHS

Glide Slopes
USE GLIDE PATHS

Glider, Dyna-Soar Space
USE X-20 AIRCRAFT

GLIDERS

Gliders, ASSET
USE ASSET GLIDERS

Gliders, Hang
USE HANG GLIDERS

Gliders, Hypersonic
USE HYPERSONIC GLIDERS

Gliders, Inflatable
USE INFLATABLE GLIDERS

Gliders, Para
USE PARAMAGNETISM

Gliders, Reentry
USE LIFTING REENTRY VEHICLES

Gliders, Space
USE LIFTING REENTRY VEHICLES

GLIDING

GLIMM METHOD

GLINT

GLOBAL AIR POLLUTION

GLOBAL AIR SAMPLING PROGRAM

GLOBAL ATMOSPHERIC RESEARCH PROGRAM

Global Communications Antenna Grid (Navy)
USE SEAFARER PROJECT

Global Ocean Station Systems, Integrated
USE INTEGRATED GLOBAL OCEAN STATION
SYSTEMS

GLOBAL POSITIONING SYSTEM

GLOBAL TRACKING NETWORK

GLOBES

GLOBULAR CLUSTERS

GLOBULES

Globulin, Gamma
USE GAMMA GLOBULIN

GLOBULINS

GLOMERULUS

Glossaries
USE DICTIONARIES

Glossaries, Space
USE SPACE GLOSSARIES

Gloster GA-5 Aircraft
USE GA-5 AIRCRAFT

GLOTRAC (Tracking Network)
USE GLOBAL TRACKING NETWORK

GLOTTIS

GLOVES

Glow
USE LUMINESCENCE

Glow, Air
USE AIRGLOW

Glow, Cathode
USE CATHODE GLOW

Glow, Day
USE DAYGLOW

GLOW DISCHARGES

Glow, Twilight
USE TWILIGHT GLOW

Glows, After
USE AFTERGLOWS

GLUCOSE

GLUCOSIDES

GLUES

GLUONS

GLUTAMATES

GLUTAMIC ACID

GLUTAMINE

GLUTATHIONE

GLYCERIDES

Glycerin, Nitro
USE NITROGLYCERIN

Glycerins
USE GLYCEROLS

GLYCEROLS

GLYCINE

GLYCOGENS

GLYCOLS

GLYCOLYSIS

Glycosides
USE GLUCOSIDES

GNEISS

GNOMONIC PROJECTION

GNOTOBIOTICS

GNP
USE GROSS NATIONAL PRODUCT

GOAL THEORY

GOALS

GOATS

GOBI DESERT

Goddard Experiment Package Telescope
USE PARTICLE TELESCOPES

GODDARD TRAJECTORY DETERMINATION SYSTEM

GOERTLER INSTABILITY

Goertler Instability, Taylor-
    USE   GOERTLER INSTABILITY

GOES SATELLITES

GOES 1

GOES 2

GOES 3

GOES 4

GOES 5

GOGGLES

GOLAY DETECTOR CELLS

GOLD

GOLD ALLOYS

GOLD COATINGS

GOLD ISOTOPES

Gold Plate
    USE   GOLD COATINGS

GOLD 198

GOMPERTZ CURVES

GONADS

GONDOLAS

GONIOMETERS

Goniometers, Photo
    USE   PHOTOGONIOMETERS

Goniometers, Radio
    USE   RADIOGONIOMETERS

GOODNESS OF FIT

Goose Missile, Blue
    USE   BLUE GOOSE MISSILE

Gordan Coefficients, Clebsch-
    USE   CLEBSCH-GORDAN COEFFICIENTS

Gordon Equation, Klein-
    USE   KLEIN-GORDON EQUATION

GORES

Gorges
    USE   CANYONS

GOSS (Support System)
    USE   GROUND OPERATIONAL SUPPORT
           SYSTEM

GOVERNMENT PROCUREMENT

GOVERNMENT/INDUSTRY RELATIONS

GOVERNMENTS

Governors
    USE   SPEED REGULATORS

Graaff Accelerators, Van De
    USE   VAN DE GRAAFF ACCELERATORS

Grabens
    USE   GEOLOGICAL FAULTS

GRADE

Gradient Aircraft, Steep
    USE   V/STOL AIRCRAFT

GRADIENT INDEX OPTICS

Gradient Method, Conjugate
    USE   CONJUGATE GRADIENT METHOD

Gradient Satellites, Gravity
    USE   GRAVITY GRADIENT SATELLITES

GRADIENTS

Gradients, Potential
    USE   POTENTIAL GRADIENTS

Gradients, Pressure
    USE   PRESSURE GRADIENTS

Gradients, Temperature
    USE   TEMPERATURE GRADIENTS

Gradiometers
    USE   MAGNETOMETERS

Gradiometers, Gravity
    USE   GRAVITY GRADIOMETERS

Graduation
    USE   CALIBRATING

GRAEFF CALCULUS

GRAFTING

Grafts, Skin
    USE   SKIN GRAFTS

GRAIN BOUNDARIES

GRAIN SIZE

GRAINS

GRAINS (FOOD)

Grains, Propellant
    USE   PROPELLANT GRAINS

GRAMMARS

GRAND CANYON (AZ)

GRAND TOURS

Grande (North America), Rio
    USE   RIO GRANDE (NORTH AMERICA)

GRANITE

GRANTS

GRANULAR MATERIALS

Granulation, Solar
    USE   SOLAR GRANULATION

GRAPH THEORY

GRAPHIC ARTS

Graphic Evaluation And Review Techniques
    USE   GERT

Graphics, Computer
    USE   COMPUTER GRAPHICS

Graphics, Interactive
    USE   COMPUTER GRAPHICS

GRAPHITE

Graphite Composites, Aluminum
    USE   ALUMINUM GRAPHITE COMPOSITES

Graphite, Pyrolytic
    USE   PYROLYTIC GRAPHITE

Graphite Reactors, Sodium
    USE   SODIUM GRAPHITE REACTORS

GRAPHITE-EPOXY COMPOSITES

GRAPHITE-POLYIMIDE COMPOSITES

GRAPHITIZATION

GRAPHOEPITAXY

GRAPHOLOGY

Graphs, Bond
    USE   BOND GRAPHS

GRAPHS (CHARTS)

Graphs, Flow
    USE   FLOW GRAPHS

Graphs, Signal Flow
    USE   SIGNAL FLOW GRAPHS

GRASHOF NUMBER

GRASSES

Grasses, Sea
    USE   SEA GRASSES

GRASSHOPPERS

GRASSLANDS

Grassmann Algebra
    USE   VECTOR SPACES

Grating, Interference
    USE   INTERFERENCE GRATING

GRATINGS

Gratings, Diffraction
    USE   GRATINGS (SPECTRA)

Gratings, Echelette
    USE   ECHELETTE GRATINGS

GRATINGS (SPECTRA)

Gravel Deposits
    USE   GRAVELS

GRAVELS

GRAVIMETERS

GRAVIMETRY

Gravimetry, Thermal
    USE   THERMOGRAVIMETRY

Gravimetry, Thermo
    USE   THERMOGRAVIMETRY

GRAVIRECEPTORS

GRAVITATION

Gravitation, Lunar
    USE   LUNAR GRAVITATION

Gravitation, Planetary
    USE   PLANETARY GRAVITATION

Gravitation, Solar
    USE   SOLAR GRAVITATION

Gravitation, Stellar
    USE   STELLAR GRAVITATION

GRAVITATION THEORY

GRAVITATIONAL COLLAPSE

GRAVITATIONAL CONSTANT

GRAVITATIONAL EFFECTS

Gravitational Effects, Lunar
    USE   LUNAR GRAVITATIONAL EFFECTS

GRAVITATIONAL FIELDS

GRAVITATIONAL LENSES

GRAVITATIONAL PHYSIOLOGY

Gravitational Potential
    USE   GRAVITATIONAL FIELDS

Gravitational Radiation
    USE   GRAVITATIONAL WAVES

GRAVITATIONAL WAVE ANTENNAS

GRAVITATIONAL WAVES

GRAVITINOS

GRAVITONS

GRAVITROPISM

Gravity
    USE   GRAVITATION

Gravity (Acceleration), High
    USE   HIGH GRAVITY ENVIRONMENTS

GRAVITY ANOMALIES

Gravity, Anti
    USE   ANTIGRAVITY

Gravity, Artificial
    USE   ARTIFICIAL GRAVITY

Gravity, Center Of
    USE   CENTER OF GRAVITY

Gravity Environments, High
    USE   HIGH GRAVITY ENVIRONMENTS

GRAVITY GRADIENT SATELLITES

GRAVITY GRADIOMETERS

Gravity, Low
    USE   REDUCED GRAVITY

Gravity Manufacturing, Low
    USE   LOW GRAVITY MANUFACTURING

GRAVITY PROBE B

Gravity, Reduced
    USE   REDUCED GRAVITY

Gravity Simulator, Lunar
    USE   LUNAR GRAVITY SIMULATOR

Gravity, Specific
    USE   DENSITY (MASS/VOLUME)

GRAVITY WAVES

Gravity, Zero
    USE   WEIGHTLESSNESS

GRAVSAT SATELLITE

GRAY GAS

GRAY SCALE

GRAZING

GRAZING FLOW

GRAZING INCIDENCE

Grazing Incidence Solar Telescope
    USE   GRIST (TELESCOPE)

Grazing Lands
    USE   GRASSLANDS

GREASES

GREAT BASIN (US)

Great Britain
    USE   UNITED KINGDOM

GREAT CIRCLES

Great Lakes, International Field Year For
    USE   INTERNATIONAL FIELD YEAR FOR GREAT
           LAKES

GREAT LAKES (NORTH AMERICA)

GREAT PLAINS CORRIDOR (NORTH AMERICA)

GREAT SALT LAKE (UT)

GREAT SMOKY MOUNTAINS (NC-TN)

GREB SATELLITES

GREECE

Green Algae, Blue
    USE   BLUE GREEN ALGAE

Green Theorem
    USE   GREEN'S FUNCTIONS

GREEN WAVE EFFECT

GREEN'S FUNCTIONS

GREENHOUSE EFFECT

GREENHOUSES

GREENLAND

GREGORIAN ANTENNAS

GRENADES

Grid Lenses, Wire
    USE   WIRE GRID LENSES

Grid (Navy), Global Communications Antenna
    USE   SEAFARER PROJECT

Grid (Navy), Underground Radio Antenna
    USE   SEAFARER PROJECT

GRIDS

Grids, Computational
    USE   COMPUTATIONAL GRIDS

Grids (Mathematics)
    USE   COMPUTATIONAL GRIDS

Grids, Tube
    USE   TUBE GRIDS

GRIFFITH CRACK

Griffon Aircraft
    USE   NORD 1500 AIRCRAFT

GRIGG-SKJELLERUP COMET

GRIGNARD REACTIONS

GRINDING

GRINDING (COMMINUTION)

Grinding, Electrolytic
    USE   ELECTROCHEMICAL MACHINING

GRINDING MACHINES

Grinding Machines, Ultrasonic
    USE   ULTRASONIC MACHINING

GRINDING (MATERIAL REMOVAL)

Grinding, Metal
    USE   METAL GRINDING

GRINDING MILLS

GRIST (TELESCOPE)

GRIT

GROOVES

Grooves, V
    USE   V GROOVES

GROOVING

GROSS NATIONAL PRODUCT

GROUND BASED CONTROL

(Ground Based), Space Surveillance
    USE   SPACE SURVEILLANCE (GROUND BASED)

Ground Communication, Ground-Air-
    USE   GROUND-AIR-GROUND COMMUNICATION

GROUND CREWS

GROUND EFFECT (AERODYNAMICS)

GROUND EFFECT (COMMUNICATIONS)

Ground Effect Machine, Cushioncraft
    USE   CUSHIONCRAFT GROUND EFFECT
           MACHINE

Ground Effect Machine, DTMB-111
    USE   GROUND EFFECT MACHINES

Ground Effect Machine, DTMB-430
    USE   GROUND EFFECT MACHINES

Ground Effect Machine, SR-N2
    USE   WESTLAND GROUND EFFECT MACHINES

Ground Effect Machine, SR-N3
    USE   WESTLAND GROUND EFFECT MACHINES

Ground Effect Machine, SR-N5
    USE   WESTLAND GROUND EFFECT MACHINES

Ground Effect Machine, Westland SR-N2
    USE   WESTLAND GROUND EFFECT MACHINES

Ground Effect Machine, Westland SR-N3
    USE   WESTLAND GROUND EFFECT MACHINES

Ground Effect Machine, Westland SR-N5
    USE   WESTLAND GROUND EFFECT MACHINES

GROUND EFFECT MACHINES

Ground Effect Machines, HD-1
    USE   HOVERCRAFT GROUND EFFECT
           MACHINES

Ground Effect Machines, Hovercraft
    USE   HOVERCRAFT GROUND EFFECT
           MACHINES

Ground Effect Machines, Westland
    USE   WESTLAND GROUND EFFECT MACHINES

GROUND HANDLING

GROUND OPERATIONAL SUPPORT SYSTEM

GROUND RESONANCE

GROUND SPEED

GROUND SQUIRRELS

GROUND STATE

GROUND STATIONS

GROUND SUPPORT EQUIPMENT

Ground Support, Satellite
    USE   SATELLITE GROUND SUPPORT

GROUND SUPPORT SYSTEMS

GROUND TESTS

GROUND TRACKS

Ground Tracks, Satellite
    USE   SATELLITE GROUND TRACKS

GROUND TRUTH

GROUND WATER

GROUND WAVE PROPAGATION

GROUND WIND

GROUND-AIR-GROUND COMMUNICATION

Ground-To-Air Missiles
USE   SURFACE TO AIR MISSILES

Grounding, Electrical
USE   ELECTRICAL GROUNDING

Group (Astronomy), Local
USE   LOCAL GROUP (ASTRONOMY)

Group Behavior
USE   GROUP DYNAMICS

Group, Carboxyl
USE   CARBOXYL GROUP

GROUP DYNAMICS

GROUP THEORY

Group, Transponder Control
USE   TRANSPONDER CONTROL GROUP

GROUP VELOCITY

Group 1A Compounds
USE   ALKALI METAL COMPOUNDS

GROUP 1B COMPOUNDS

Group 2A Compounds
USE   ALKALINE EARTH COMPOUNDS

GROUP 2B COMPOUNDS

GROUP 3A COMPOUNDS

GROUP 3B COMPOUNDS

GROUP 4A COMPOUNDS

GROUP 4B COMPOUNDS

GROUP 5A COMPOUNDS

GROUP 5B COMPOUNDS

GROUP 6A COMPOUNDS

GROUP 6B COMPOUNDS

Group 7A Compounds
USE   HALOGEN COMPOUNDS

GROUP 7B COMPOUNDS

GROUP 8 COMPOUNDS

GROUPS

Groups, Blood
USE   BLOOD GROUPS

Groups, Lie
USE   LIE GROUPS

Groups, Propargyl
USE   PROPARGYL GROUPS

Groups, Spinor
USE   SPINOR GROUPS

Groups, Sub
USE   SUBGROUPS

GROUT

GROWTH

Growth Chambers
USE   PHYTOTRONS

Growth, Crop
USE   CROP GROWTH

Growth, Crystal
USE   CRYSTAL GROWTH

Growth, Hydrothermal Crystal
USE   HYDROTHERMAL CRYSTAL GROWTH

Growth), Melts (Crystal
USE   MELTS (CRYSTAL GROWTH)

Growth, Vegetation
USE   VEGETATION GROWTH

GRUMMAN AIRCRAFT

Grumman OV-1C Aircraft
USE   OV-1 AIRCRAFT

GRUNEISEN CONSTANT

(GT-1) Spacecraft, Gemini
USE   GEMINI (GT-1) SPACECRAFT

GTDS
USE   GODDARD TRAJECTORY DETERMINATION
        SYSTEM

GUADELOUPE

GUAM

GUANETHIDINE

Guanidine, Nitro
USE   NITROGUANIDINE

Guanidine, Perfluoro
USE   PERFLUOROGUANIDINE

GUANIDINES

GUANINES

GUANOSINES

GUARDS (SHIELDS)

GUATEMALA

GUAYULE

Guiana, French
USE   FRENCH GUIANA

Guidance, Aircraft
USE   AIRCRAFT GUIDANCE

Guidance, Beam Rider
USE   BEAM RIDER GUIDANCE

Guidance, Command
USE   COMMAND GUIDANCE

Guidance, Inertial
USE   INERTIAL GUIDANCE

Guidance, Injection
USE   INJECTION GUIDANCE

Guidance, Laser
USE   LASER GUIDANCE

Guidance, Map Matching
USE   MAP MATCHING GUIDANCE

Guidance, Midcourse
USE   MIDCOURSE GUIDANCE

Guidance, Missile
USE   MISSILE CONTROL

GUIDANCE (MOTION)

Guidance, Reentry
USE   REENTRY GUIDANCE

Guidance, Rendezvous
USE   RENDEZVOUS GUIDANCE

Guidance, Satellite
USE   SATELLITE GUIDANCE

GUIDANCE SENSORS

Guidance, Spacecraft
USE   SPACECRAFT GUIDANCE

Guidance), SSGS (Standardized Space
USE   STANDARDIZED SPACE GUIDANCE

Guidance, Standardized Space
USE   STANDARDIZED SPACE GUIDANCE

Guidance, Strapdown Inertial
USE   STRAPDOWN INERTIAL GUIDANCE

Guidance (STS), Entry
USE   ENTRY GUIDANCE (STS)

Guidance, Terminal
USE   TERMINAL GUIDANCE

GUIDE VANES

GUIDED MISSILE SUBMARINES

Guided Projectiles, Precision
USE   PRECISION GUIDED PROJECTILES

Guides, Wave
USE   WAVEGUIDES

Guideway Transit Vehicles, Automated
USE   AUTOMATED GUIDEWAY TRANSIT
        VEHICLES

GUINEA

Guinea, British
USE   GUYANA

Guinea (Island), New
USE   NEW GUINEA (ISLAND)

GUINEA PIGS

GULF OF ALASKA

GULF OF CALIFORNIA (MEXICO)

GULF OF MEXICO

Gulf, Persian
USE   PERSIAN GULF

GULF STREAM

GULFS

GULLIVER PROGRAM

GUM NEBULA

Gum Vulcanizates
USE   VULCANIZED ELASTOMERS

Gumbel Theory
USE   RANGE (EXTREMES)

GUMS (SUBSTANCES)

GUN LAUNCHERS

GUN PROPELLANTS

GUN TURRETS

GUNFIRE

GUNN DIODES

GUNN EFFECT

GUNNERY TRAINING

Gunpowder
USE   GUN PROPELLANTS

GUNS

Guns, Crossed Field
USE   CROSSED FIELD GUNS

Guns, Electron
USE   ELECTRON GUNS

Guns, Gas
USE   GAS GUNS

Guns, Hypervelocity
USE   HYPERVELOCITY GUNS

Guns, Light Gas
USE   LIGHT GAS GUNS

GUNS (ORDNANCE)

Guns, Plasma
USE   PLASMA GUNS

GUST ALLEVIATORS

GUST LOADS

Gustatory Perception
USE   TASTE

GUSTS

GUTENBERG ZONE

GUY WIRES

GUYANA

Gymnastics
USE   PHYSICAL EXERCISE

GYNECOLOGY

GYPSUM

GYRATION

GYRATORS

GYRES

GYRO HORIZONS

GYROCOMPASSES

GYRODAMPERS

GYRODYNE AIRCRAFT

Gyrodyne DSN-3 Helicopter
USE   QH-50 HELICOPTER

Gyrodyne Military Aircraft
USE   QH-50 HELICOPTER

GYROFREQUENCY

Gyrointeraction
USE   MAGNETIC RIGIDITY

GYROMAGNETISM

Gyroplanes
USE   HELICOPTERS

Gyros
USE   GYROSCOPES

Gyros, Attitude
USE   ATTITUDE GYROS

GYROSCOPE FLUIDS

GYROSCOPES

Gyroscopes, Control Moment
USE   CONTROL MOMENT GYROSCOPES

Gyroscopes, Cryogenic
USE   CRYOGENIC GYROSCOPES

Gyroscopes, Electrically Suspended
USE   ELECTROSTATIC GYROSCOPES

Gyroscopes, Electrostatic
USE   ELECTROSTATIC GYROSCOPES

(Gyroscopes), ESG
USE   ELECTROSTATIC GYROSCOPES

Gyroscopes, Fluid Rotor
USE   FLUID ROTOR GYROSCOPES

Gyroscopes, Laser
USE   LASER GYROSCOPES

Gyroscopes, Nuclear
USE   NUCLEAR GYROSCOPES

Gyroscopes, Optical
USE   OPTICAL GYROSCOPES

Gyroscopes, Pendulous
USE   GYROSCOPIC PENDULUMS

Gyroscopes, Rotary
USE   ROTARY GYROSCOPES

Gyroscopes, Tuning Fork
USE   TUNING FORK GYROSCOPES

GYROSCOPIC COUPLING

Gyroscopic Drift
USE   GYROSCOPES
        GYROSCOPIC STABILITY

GYROSCOPIC PENDULUMS

GYROSCOPIC STABILITY

GYROSTABILIZERS

Gyrostats
USE   GYROSCOPES

Gyrotrons
USE   CYCLOTRON RESONANCE DEVICES

GYROTROPISM

H

H ALPHA LINE

H BETA LINE

H GAMMA LINE

H, IMP-
USE   EXPLORER 47 SATELLITE

H LINES

H, OSO-
USE   OSO-7

H Satellite, TIROS
USE   TIROS 8 SATELLITE

H, Space Shuttle Mission 51-
USE   SPACE SHUTTLE MISSION 51-H

H WAVES

H-1 ENGINE

H-13 Helicopter
USE   OH-13 HELICOPTER

H-17 HELICOPTER

H-19 HELICOPTER

H-21 Helicopter
USE   CH-21 HELICOPTER

H-23 Helicopter
USE   OH-23 HELICOPTER

H-25 HELICOPTER

H-34 Helicopter
USE   CH-34 HELICOPTER

H-43 HELICOPTER

H-51 Helicopter
USE   XH-51 HELICOPTER

H-53 HELICOPTER

H-54 HELICOPTER

H-56 HELICOPTER

H-60 HELICOPTER

H-126 AIRCRAFT

H-126 Aircraft, Hunting
USE   H-126 AIRCRAFT

H/v Interaction Experiments, Space Plasma
USE   SPHINX

HABITABILITY

HABITATS

Habitats, Space
USE   SPACE HABITATS

HABITS

HABITUATION (LEARNING)

HADRONS

HAFNIUM

HAFNIUM ALLOYS

HAFNIUM CARBIDES

HAFNIUM COMPOUNDS

HAFNIUM IODIDES

HAFNIUM ISOTOPES

HAFNIUM OXIDES

HAIL

Hailstones
USE   HAIL

HAILSTORMS

HAIR

HAITI

HAL/S (LANGUAGE)

HALDEN BOILING WATER REACTOR

Halden Reactor
USE   HALDEN BOILING WATER REACTOR

HALF CONES

HALF LIFE

HALF PLANES

HALF SPACES

Halide Lasers, Rare Gas-
USE   RARE GAS-HALIDE LASERS

HALIDES

Halides, Alkali
USE   ALKALI HALIDES

Halides, Cesium
USE   CESIUM HALIDES

Halides, Metal
USE   METAL HALIDES

Halides, Oxy
USE   OXYHALIDES

Halides, Silver
USE   SILVER HALIDES

Halides, Tungsten
USE   TUNGSTEN HALIDES

HALITES

HALL ACCELERATORS

Hall Coefficient
USE   HALL EFFECT

Hall Currents
USE   HALL EFFECT
          ELECTRIC CURRENT

HALL EFFECT

HALL GENERATORS

HALLAM NUCLEAR POWER FACILITY

(Hallam Nuclear Power Facility), HNPF
USE   HALLAM NUCLEAR POWER FACILITY

HALLEY'S COMET

HALLUCINATIONS

HALO ORBIT SPACE STATION

HALOCARBONS

Haloe
USE   HALOGEN OCCULTATION EXPERIMENT

HALOGEN COMPOUNDS

HALOGEN OCCULTATION EXPERIMENT

HALOGENATION

HALOGENS

HALOPHILES

HALOS

HALPHEN METHOD

HAMBURGER AIRCRAFT

Hamburger HFB-320 Aircraft
USE   HFB-320 AIRCRAFT

HAMILTON-JACOBI EQUATION

HAMILTONIAN FUNCTIONS

Hammer, Water
USE   WATER HAMMER

HAMMERHEAD CONFIGURATION

HAMMERS

Hammers, Electromagnetic
USE   ELECTROMAGNETIC HAMMERS

Hampshire, New
USE   NEW HAMPSHIRE

HAMSTERS

HAND (ANATOMY)

HANDBOOKS

HANDEDNESS

HANDICAPS

HANDLES

HANDLEY PAGE AIRCRAFT

Handley Page HP-115 Aircraft
USE   HP-115 AIRCRAFT

HANDLING EQUIPMENT

Handling, Ground
USE   GROUND HANDLING

Handling, Materials
USE   MATERIALS HANDLING

Handling Qualities
USE   CONTROLLABILITY

Handling, Remote
USE   REMOTE HANDLING

Handling Systems, Data
USE   DATA SYSTEMS

HANDWRITING

HANFORD REACTORS

HANG GLIDERS

HANGARS

(Hanging), Suspending
USE   SUSPENDING (HANGING)

HANKEL FUNCTIONS

HANSEN LUNAR THEORY

HAPLOSCOPES

HARBORS

Harbors, Artificial
USE   ARTIFICIAL HARBORS

HARD LANDING

HARDENERS

HARDENING

Hardening, Age
USE   PRECIPITATION HARDENING

Hardening, Cold
USE   COLD HARDENING

Hardening, Dispersion Precipitation
USE   PRECIPITATION HARDENING

HARDENING (MATERIALS)

Hardening, Metal
USE   HARDENING (MATERIALS)

Hardening, Precipitation
USE   PRECIPITATION HARDENING

Hardening, Radiation
USE   RADIATION HARDENING

Hardening, Strain
USE   STRAIN HARDENING

HARDENING (SYSTEMS)

Hardening, Work
USE   WORK HARDENING

HARDNESS

Hardness, Knoop
USE   KNOOP HARDNESS

Hardness, Micro
USE   MICROHARDNESS

Hardness, Rockwell
USE   ROCKWELL HARDNESS

HARDNESS TESTS

HARDWARE

HARDWARE UTILIZATION LISTS

HARLETON METEORITE

HARMONIC ANALYSIS

HARMONIC CONTROL

HARMONIC EXCITATION

HARMONIC FUNCTIONS

HARMONIC GENERATIONS

HARMONIC GENERATORS

HARMONIC MOTION

Harmonic Motion, Simple
USE   SIMPLE HARMONIC MOTION

HARMONIC OSCILLATION

HARMONIC OSCILLATORS

HARMONIC RADIATION

HARMONICS

Harmonics, Spherical
USE   SPHERICAL HARMONICS

Harmonics, Super
USE   SUPERHARMONICS

Harmonics, Tesseral
USE   TESSERAL HARMONICS

Harmonics, Zonal
USE   ZONAL HARMONICS

HARNESSES

Haro Objects, Herbig-
USE   HERBIG-HARO OBJECTS

HARPOON MISSILE

HARRIER AIRCRAFT

HARTMANN FLOW

HARTMANN NUMBER

HARTREE APPROXIMATION

Hartree-Appleton Approximation
USE   HARTREE APPROXIMATION

Hartree-Fock Approximation
USE   HARTREE APPROXIMATION

HARTREE-FOCK-SLATER METHOD

HARVARD RADIO METEOR PROJECT

HASTELLOY (TRADEMARK)

HATCHES

Hatteras (NC), Cape
USE   CAPE HATTERAS (NC)

Haul Aircraft, Short
USE   SHORT HAUL AIRCRAFT

HAULING

Hausdorff Series, Campbell-
USE   CAMPBELL-HAUSDORFF SERIES

Haven (CT), New
USE   NEW HAVEN (CT)

Havilland Aircraft, De
USE   DE HAVILLAND AIRCRAFT

Havilland DH 106 Aircraft, De
USE   COMET 4 AIRCRAFT

Havilland DH 112 Aircraft, De
USE   DH 112 AIRCRAFT

Havilland DH 115 Aircraft, De
USE   DH 115 AIRCRAFT

Havilland DH 121 Aircraft, De
USE   DH 121 AIRCRAFT

Havilland DH 125 Aircraft, De
USE   DH 125 AIRCRAFT

Havilland DHC 4 Aircraft, De
USE   DHC 4 AIRCRAFT

Havilland DHC 5 Aircraft, De
USE   DHC 5 AIRCRAFT

Havilland Venom Aircraft, De
USE   DH 112 AIRCRAFT

HAWAII

Hawk Assault Helicopter, Black
USE   H-60 HELICOPTER

HAWK MISSILE

Hawker Hunter Aircraft
USE   F-2 AIRCRAFT

Hawker P-1052 Aircraft
USE   P-1052 AIRCRAFT

Hawker P-1127 Aircraft
USE   P-1127 AIRCRAFT

Hawker P-1154 Aircraft
USE   P-1154 AIRCRAFT

HAWKER SIDDELEY AIRCRAFT

Hawkeye Aircraft
USE   E-2 AIRCRAFT

HAWKEYE SATELLITES

Hawkeye 1 Satellite
USE   EXPLORER 52 SATELLITE

HAY

Haynes Stellite
USE   STELLITE (TRADEMARK)

Hazard, Toxicity And Safety
USE   TOXICITY AND SAFETY HAZARD

HAZARDOUS MATERIAL DISPOSAL (IN SPACE)

HAZARDS

Hazards, Aircraft
USE   AIRCRAFT HAZARDS

Hazards, Flight
USE   FLIGHT HAZARDS

Hazards, Meteor
USE   METEOROID HAZARDS

Hazards, Meteoroid
USE   METEOROID HAZARDS

Hazards, Noise
USE   HAZARDS
       NOISE (SOUND)

Hazards, Operational
USE   OPERATIONAL HAZARDS

Hazards, Radiation
USE   RADIATION HAZARDS

Hazards, Toxic
USE   TOXIC HAZARDS

HAZE

HAZE DETECTION

HBNQ
USE   NITROGUANIDINE

HBr
USE   HYDROBROMIC ACID

HBWR Reactor
USE   HALDEN BOILING WATER REACTOR

HC-1 Helicopter
USE   CH-47 HELICOPTER

HC-3 HELICOPTER

HC-3 Helicopter, Omnipol
USE   HC-3 HELICOPTER

HCl
USE   HYDROCHLORIC ACID

HCL ARGON LASERS

HCL LASERS

HCMM
USE   HEAT CAPACITY MAPPING MISSION

HCN
USE   HYDROCYANIC ACID

HCN LASERS

HD-1 Ground Effect Machines
USE   HOVERCRAFT GROUND EFFECT
       MACHINES

He
USE   HELIUM

HEAD (ANATOMY)

HEAD FLOW

HEAD (FLUID MECHANICS)

Head, Fore
USE   FOREHEAD

HEAD MOVEMENT

Head (Pressure)
USE   PRESSURE HEADS

HEAD-UP DISPLAYS

HEADACHE

HEADERS

Heads, Comet
USE   COMET HEADS

Heads, Coral
USE   CORAL REEFS

Heads, Pressure
USE   PRESSURE HEADS

Heads, Recording
USE   RECORDING HEADS

Heads, War
USE   WARHEADS

Headsets
USE   EARPHONES

HEALING

Healing, Wound
USE   WOUND HEALING

HEALTH

Health, Mental
USE   MENTAL HEALTH

HEALTH PHYSICS

HEALTH PHYSICS RESEARCH REACTOR

Health, Public
USE   PUBLIC HEALTH

Health-Education Telecommunications Exp
USE   HET EXPERIMENT

HEAO

HEAO A
USE   HEAO 1

HEAO B
USE   HEAO 2

HEAO C
USE   HEAO 3

HEAO 1

HEAO 2

HEAO 3

HEARING

Hearing, Binaural
USE   BINAURAL HEARING

Hearing Loss
USE   AUDITORY DEFECTS

HEART

HEART DISEASES

HEART FUNCTION

HEART IMPLANTATION

HEART MINUTE VOLUME

HEART RATE

HEART VALVES

Heart Valves, Artificial
USE   ARTIFICIAL HEART VALVES

HEARTHS

HEAT

HEAT ACCLIMATIZATION

HEAT BALANCE

HEAT BUDGET

Heat Budget, Atmospheric
USE   ATMOSPHERIC HEAT BUDGET

Heat Capacity
USE   SPECIFIC HEAT

HEAT CAPACITY MAPPING MISSION

Heat, Combustion
USE   HEAT OF COMBUSTION

Heat Conduction
USE   CONDUCTIVE HEAT TRANSFER

Heat Content
USE   ENTHALPY

Heat Dissipation
USE   COOLING

Heat Dissipation Chilling
USE   COOLING

Heat, Dry
USE   DRY HEAT

Heat Effects
USE   TEMPERATURE EFFECTS

Heat Equations
USE   THERMODYNAMICS

HEAT EXCHANGERS

Heat Exchangers, Tube
USE   TUBE HEAT EXCHANGERS

Heat Flow
USE   HEAT TRANSMISSION

HEAT FLUX

Heat, Formation
USE   HEAT OF FORMATION

Heat Gain
USE   HEATING

HEAT GENERATION

HEAT ISLANDS

HEAT MEASUREMENT

Heat, Nuclear
USE   NUCLEAR HEAT

HEAT OF COMBUSTION

HEAT OF DISSOCIATION

HEAT OF FORMATION

HEAT OF FUSION

Heat Of Fusion, Latent
USE   HEAT OF FUSION

HEAT OF SOLUTION

HEAT OF VAPORIZATION

HEAT PIPES

Heat, Process
USE   PROCESS HEAT

HEAT PUMPS

HEAT RADIATORS

Heat Regulation
USE   TEMPERATURE CONTROL

Heat Rejection Devices
USE   HEAT RADIATORS

Heat Resistance
USE   THERMAL RESISTANCE

HEAT RESISTANT ALLOYS

HEAT SHIELDING

Heat Shielding, Reusable
USE   REUSABLE HEAT SHIELDING

HEAT SINKS

HEAT SOURCES

Heat, Specific
USE   SPECIFIC HEAT

HEAT STORAGE

(Heat Storage), Solar Ponds
USE   SOLAR PONDS (HEAT STORAGE)

HEAT STROKE

HEAT TAPES

Heat Tests
USE   HIGH TEMPERATURE TESTS

Heat Theorem, Nernst
USE   NERNST-ETTINGSHAUSEN EFFECT

HEAT TOLERANCE

HEAT TRANSFER

Heat Transfer, Aerodynamic
USE   AERODYNAMIC HEAT TRANSFER

HEAT TRANSFER COEFFICIENTS

Heat Transfer, Conductive
USE   CONDUCTIVE HEAT TRANSFER

Heat Transfer, Convective
USE   CONVECTIVE HEAT TRANSFER

Heat Transfer, Hypersonic
USE   HYPERSONIC HEAT TRANSFER

Heat Transfer, Laminar
USE   LAMINAR HEAT TRANSFER

Heat Transfer, Radiative
USE   RADIATIVE HEAT TRANSFER

Heat Transfer, Supersonic
USE   SUPERSONIC HEAT TRANSFER

Heat Transfer, Turbulent
USE   TURBULENT HEAT TRANSFER

HEAT TRANSMISSION

HEAT TREATMENT

(Heat Treatment), Normalizing
USE   NORMALIZING (HEAT TREATMENT)

Heat, Vaporization
USE   HEAT OF VAPORIZATION

Heat, Waste
USE   WASTE HEAT

HEATERS

Heaters, Gerdien Arc
USE   HEATING EQUIPMENT
       ARC HEATING

HEATING

Heating, Aerodynamic
USE   AERODYNAMIC HEATING

Heating, Arc
USE   ARC HEATING

Heating, Atmospheric
USE   ATMOSPHERIC HEATING

Heating, Base
USE   BASE HEATING

Heating (Buildings), Space
USE   SPACE HEATING (BUILDINGS)

Heating, Electron Cyclotron
USE   ELECTRON CYCLOTRON HEATING

HEATING EQUIPMENT

Heating, Gas
USE   GAS HEATING

Heating, Induction
USE   INDUCTION HEATING

Heating, Ionospheric
USE   IONOSPHERIC HEATING

Heating, Joule
USE   RESISTANCE HEATING
       OHMIC DISSIPATION

Heating, Kinetic
USE   KINETIC HEATING

Heating, Laser
USE   LASER HEATING

Heating, Magnetohydrodynamic Shear
USE   MAGNETOHYDRODYNAMIC SHEAR
       HEATING

Heating, Plasma
USE   PLASMA HEATING

Heating, Pulse
USE   PULSE HEATING

Heating, Radiant
USE   RADIANT HEATING

Heating, Radiation
USE   RADIANT HEATING

Heating, Radio Frequency
USE   RADIO FREQUENCY HEATING

Heating, Resistance
USE   RESISTANCE HEATING

Heating, Shock
USE   SHOCK HEATING

Heating, Solar
USE   SOLAR HEATING

Heating Sources, Hydraulic
USE   HYDRAULIC EQUIPMENT
       HEAT SOURCES

Heating, Super
USE   SUPERHEATING

Heating, Transient
USE   TRANSIENT HEATING

Heating, Water
USE   WATER HEATING

HEAVING

Heavy Cosmic Ray Primaries
USE   HEAVY NUCLEI
       PRIMARY COSMIC RAYS

HEAVY ELEMENTS

HEAVY IONS

HEAVY LIFT AIRSHIPS

HEAVY LIFT HELICOPTERS

HEAVY LIFT LAUNCH VEHICLES

HEAVY NUCLEI

HEAVY WATER

HEAVY WATER COMPONENTS TEST REACTORS

HEAVY WATER REACTORS

HEF (High Energy Fuels)
USE   HIGH ENERGY FUELS

HEIGHT

Height, Geopotential
USE   GEOPOTENTIAL HEIGHT

Height Indicators, Cloud
USE   CLOUD HEIGHT INDICATORS

Height, Mixing
USE   MIXING HEIGHT

Height, Pulse
USE   PULSE AMPLITUDE

Height, Scale
USE   SCALE HEIGHT

HEINKEL AIRCRAFT

HEISENBERG THEORY

Heitler Formula, Bethe-
USE   BETHE-HEITLER FORMULA

HELICAL ANTENNAS

HELICAL FLOW

HELICAL INDUCERS

HELICAL WINDINGS

Helicopter, Ah-1g
USE   AH-1G HELICOPTER

Helicopter, Ah-63
USE   AH-63 HELICOPTER

Helicopter, Ah-64
USE   AH-64 HELICOPTER

Helicopter, Alouette 3
USE   SE-3160 HELICOPTER

Helicopter Attitude Indicators
USE   ATTITUDE INDICATORS
       HELICOPTERS

Helicopter, Bell 214a
USE   BELL 214A HELICOPTER

Helicopter, Black Hawk Assault
USE   H-60 HELICOPTER

Helicopter, BO-105
USE   BO-105 HELICOPTER

Helicopter, CH-3
USE   CH-3 HELICOPTER

Helicopter, CH-21
USE   CH-21 HELICOPTER

Helicopter, CH-34
USE   CH-34 HELICOPTER

Helicopter, CH-46
USE   CH-46 HELICOPTER

Helicopter, CH-47
USE   CH-47 HELICOPTER

Helicopter, CH-53
USE   H-53 HELICOPTER

Helicopter, CH-54
USE   CH-54 HELICOPTER

Helicopter, CH-62
USE   CH-62 HELICOPTER

Helicopter, CH-113
USE   CH-46 HELICOPTER

Helicopter, Chinook
USE   CH-47 HELICOPTER

Helicopter, Choctaw
USE   CH-34 HELICOPTER

Helicopter, CL-595
USE   XH-51 HELICOPTER

HELICOPTER CONTROL

Helicopter, Dash
USE   QH-50 HELICOPTER

HELICOPTER DESIGN

Helicopter, DSN
USE   QH-50 HELICOPTER

HELICOPTER ENGINES

Helicopter, F-28
USE   F-28 HELICOPTER

Helicopter, FH-1100
USE   OH-5 HELICOPTER

Helicopter, Flying Crane
USE   H-17 HELICOPTER

Helicopter, Gyrodyne DSN-3
USE   QH-50 HELICOPTER

Helicopter, H-13
USE   OH-13 HELICOPTER

Helicopter, H-17
USE   H-17 HELICOPTER

Helicopter, H-19
USE   H-19 HELICOPTER

Helicopter, H-21
USE   CH-21 HELICOPTER

Helicopter, H-23
USE   OH-23 HELICOPTER

Helicopter, H-25
USE   H-25 HELICOPTER

Helicopter, H-34
USE   CH-34 HELICOPTER

Helicopter, H-43
USE   H 43 HELICOPTER

Helicopter, H-51
USE   XH-51 HELICOPTER

Helicopter, H-53
USE   H-53 HELICOPTER

Helicopter, H-54
USE   H-54 HELICOPTER

Helicopter, H-56
USE   H-56 HELICOPTER

Helicopter, H-60
USE   H-60 HELICOPTER

Helicopter, HC-1
USE   CH-47 HELICOPTER

Helicopter, HC-3
USE   HC-3 HELICOPTER

Helicopter, HH-43
USE   HH-43 HELICOPTER

Helicopter, HH-43B
USE   HH-43 HELICOPTER

Helicopter, HHX
USE   H-53 HELICOPTER

Helicopter, HO-4
USE   OH-4 HELICOPTER

Helicopter, HO-5
USE   OH-5 HELICOPTER

Helicopter, HO-6
USE   OH-6 HELICOPTER

Helicopter, HRB-1
USE   CH-46 HELICOPTER

Helicopter, HSS-2
USE   SH-3 HELICOPTER

Helicopter, HU-1
USE   UH-1 HELICOPTER

Helicopter, HUS-1
USE   UH-34 HELICOPTER

Helicopter, Huskie
USE   HH-43 HELICOPTER

Helicopter, HU2K-1
USE   UH-2 HELICOPTER

Helicopter, Iroquois
USE   UH-1 HELICOPTER

Helicopter, Kaman UH-2A
USE   UH-2 HELICOPTER

Helicopter, Lockheed CL-595
USE   XH-51 HELICOPTER

Helicopter, Lockheed 186
USE   XH-51 HELICOPTER

Helicopter, LOH
USE   OH-6 HELICOPTER

Helicopter, OH-4
USE   OH-4 HELICOPTER

Helicopter, OH-5
USE   OH-5 HELICOPTER

Helicopter, OH-6
USE   OH-6 HELICOPTER

Helicopter, OH-13
USE   OH-13 HELICOPTER

Helicopter, OH-23
USE   OH-23 HELICOPTER

Helicopter, OH-58
USE   OH-58 HELICOPTER

Helicopter, Omnipol HC-3
USE   HC-3 HELICOPTER

Helicopter, P-531
USE   P-531 HELICOPTER

HELICOPTER PERFORMANCE

HELICOPTER PROPELLER DRIVE

Helicopter, QH-50
USE   QH-50 HELICOPTER

Helicopter, Raven
USE   OH-23 HELICOPTER

Helicopter, RH-2
USE   UH-1 HELICOPTER

Helicopter Rotors
USE   ROTARY WINGS

Helicopter, S-58
USE   S-58 HELICOPTER

Helicopter, S-61
USE   S-61 HELICOPTER

Helicopter, S-64
USE   CH-54 HELICOPTER

Helicopter, S-67
USE   S-67 HELICOPTER

Helicopter, SA-321
USE   SA-321 HELICOPTER

Helicopter, SA-330
USE   SA-330 HELICOPTER

Helicopter, Scout
USE   P-531 HELICOPTER

Helicopter, SE-3160
USE   SE-3160 HELICOPTER

Helicopter, Sea King
USE   SH-3 HELICOPTER

Helicopter, Sea Knight
USE   CH-46 HELICOPTER

Helicopter, Seahorse
USE   UH-34 HELICOPTER

Helicopter, Seasprite
USE   UH-2 HELICOPTER

Helicopter, SH-3
USE   SH-3 HELICOPTER

Helicopter, SH-4
USE   SH-4 HELICOPTER

Helicopter, Shawnee
USE   CH-21 HELICOPTER

Helicopter, Sikorsky HSS-2
USE   SH-3 HELICOPTER

Helicopter, Sikorsky S-58
USE   S-58 HELICOPTER

Helicopter, Sikorsky S-61
USE   S-61 HELICOPTER

Helicopter, Sikorsky S-64
USE   CH-54 HELICOPTER

Helicopter, Sikorsky S-65
USE   H-53 HELICOPTER

Helicopter, Sikorsky S-67
USE   S-67 HELICOPTER

Helicopter, Sikorsky Whirlwind
USE   SIKORSKY WHIRLWIND HELICOPTER

Helicopter, Sioux
USE   OH-13 HELICOPTER

Helicopter, Skycrane
USE   CH-54 HELICOPTER

Helicopter, Sud Aviation SA-321
USE   SA-321 HELICOPTER

Helicopter, Sud Aviation SA-330
USE   SA-330 HELICOPTER

Helicopter, Sud Aviation SE-3160
USE   SE-3160 HELICOPTER

HELICOPTER TAIL ROTORS

Helicopter, TH-55
USE   TH-55 HELICOPTER

Helicopter, UH-1
USE   UH-1 HELICOPTER

Helicopter, UH-2
USE   UH-2 HELICOPTER

Helicopter, UH-12
USE   OH-23 HELICOPTER

Helicopter, UH-13
USE   OH-13 HELICOPTER

Helicopter, UH-34
USE   UH-34 HELICOPTER

Helicopter, UH-60a
USE   UH-60A HELICOPTER

Helicopter, UH-61a
USE   UH-61A HELICOPTER

Helicopter, Voyageur
USE   CH-46 HELICOPTER

HELICOPTER WAKES

Helicopter, Westland MK-10
USE   WESTLAND WHIRLWIND HELICOPTER

Helicopter, Westland P-531
USE   P-531 HELICOPTER

Helicopter, Westland Whirlwind
USE   WESTLAND WHIRLWIND HELICOPTER

Helicopter, Whirlwind MK-10
USE   WESTLAND WHIRLWIND HELICOPTER

Helicopter, Workhorse
USE   CH-21 HELICOPTER

Helicopter, XH-51
USE   XH-51 HELICOPTER

Helicopter, YHU-1
USE   UH-1 HELICOPTER

Helicopter, YUH-1
USE   UH-1 HELICOPTER

Helicopter, YUH-60a
USE   UH-60A HELICOPTER

Helicopter, YUH-61a
USE   UH-61A HELICOPTER

HELICOPTERS

Helicopters, Aerogyro
USE   XH-51 HELICOPTER

Helicopters, Alouette
USE   ALOUETTE HELICOPTERS

Helicopters, Compound
USE   COMPOUND HELICOPTERS

Helicopters, Drone
USE   DRONE AIRCRAFT
       HELICOPTERS

Helicopters, Heavy Lift
USE   HEAVY LIFT HELICOPTERS

Helicopters, Military
USE   MILITARY HELICOPTERS

Helicopters, Rigid Rotor
USE   RIGID ROTOR HELICOPTERS

Helicopters, Tandem Rotor
USE   TANDEM ROTOR HELICOPTERS

Helicopters, Vertol Military
USE   BOEING AIRCRAFT

HELIO AIRCRAFT

Helio Military Aircraft
USE   HELIO AIRCRAFT

Heliocentric Orbits
USE   SOLAR ORBITS

Heliographs
USE   SPECTROHELIOGRAPHS

Heliographs, Spectro
USE   SPECTROHELIOGRAPHS

Heliography
USE   SPECTROHELIOGRAPHS

Heliomagnetism
USE   SOLAR MAGNETIC FIELD

HELIOMETERS

Heliometry
USE   HELIOMETERS
       PYROHELIOMETERS

HELIOS A

HELIOS B

HELIOS PROJECT

HELIOS SATELLITES

HELIOS 1

HELIOS 2

HELIOSEISMOLOGY

HELIOSPHERE

HELIOSTATS

HELIPORTS

HELITRONS

HELIUM

HELIUM AFTERGLOW

HELIUM ATOMS

HELIUM COMPOUNDS

HELIUM FILM

HELIUM HYDROGEN ATMOSPHERES

HELIUM IONS

HELIUM ISOTOPES

Helium, Liquid
USE   LIQUID HELIUM

HELIUM PLASMA

Helium Stars
USE   B STARS

Helium 2
USE   HELIUM ISOTOPES
       LIQUID HELIUM

Helium 2, Liquid
USE   LIQUID HELIUM 2

Helium 3
USE   HELIUM ISOTOPES

Helium 4
USE   HELIUM ISOTOPES

HELIUM-NEON LASERS

HELIUM-OXYGEN ATMOSPHERES

Helix Tubes
USE   TRAVELING WAVE TUBES

Helixes
USE   CURVES (GEOMETRY)

HELLMANN-FEYNMAN THEOREM

HELMET MOUNTED DISPLAYS

HELMETS

HELMHOLTZ EQUATIONS

Helmholtz Equations, Gibbs-
USE   GIBBS-HELMHOLTZ EQUATIONS

Helmholtz Flow, Kirchhoff-
USE   PIPE FLOW

Helmholtz Instability, Kelvin-
USE   KELVIN-HELMHOLTZ INSTABILITY

HELMHOLTZ RESONATORS

Helmholtz Theory, Young-
USE   YOUNG-HELMHOLTZ THEORY

HELMHOLTZ VORTICITY EQUATION

HELOS (Satellite)
USE   EXOSAT SATELLITE

HEMATITE

HEMATOCRIT

HEMATOCRIT RATIO

HEMATOLOGY

HEMATOPOIESIS

HEMATOPOIETIC SYSTEM

HEMATURIA

HEMISPHERE CYLINDER BODIES

Hemisphere, Eastern
USE   EASTERN HEMISPHERE

Hemisphere, Northern
USE   NORTHERN HEMISPHERE

Hemisphere, Southern
USE   SOUTHERN HEMISPHERE

Hemisphere, Western
USE   WESTERN HEMISPHERE

HEMISPHERES

HEMISPHERICAL SHELLS

HEMOCYTES

HEMODYNAMIC RESPONSES

HEMODYNAMICS

HEMOGLOBIN

Hemoglobin, Carboxy
USE   CARBOXYHEMOGLOBIN

Hemoglobin, Oxy
USE   OXYHEMOGLOBIN

HEMOLYSIS

HEMOPERFUSION

HEMORRHAGES

Hemostasis
USE   HEMOSTATICS

HEMOSTATICS

HENRY LAW

HEOS A SATELLITE

HEOS B SATELLITE

HEOS SATELLITES

HEPARINS

HEPATITIS

HEPTADIENE

HEPTANES

HERBICIDES

HERBIG-HARO OBJECTS

Hercules Aircraft
USE   C-130 AIRCRAFT

HERCULES ENGINE

Hercules Missile, Nike-
USE   NIKE-HERCULES MISSILE

HERCULES NOVA

HEREDITY

HERING-BREVER REFLEX

HERMES MANNED SPACEPLANE

Hermes Satellite
USE   COMMUNICATIONS TECHNOLOGY
       SATELLITE

HERMETIC SEALS

HERMITIAN POLYNOMIAL

HERO REACTOR

HERTZSPRUNG-RUSSELL DIAGRAM

HERZBERG BANDS

HESSIAN MATRICES

HET EXPERIMENT

HETEROCYCLIC COMPOUNDS

HETERODYNING

Heterodyning, Optical
USE   OPTICAL HETERODYNING

HETEROGENEITY

HETEROJUNCTION DEVICES

HETEROJUNCTIONS

HETEROPHORIA

HETEROSPHERE

HETEROTROPHS

HEURISTIC METHODS

HEUS ROCKET ENGINES

HEWLETT-PACKARD COMPUTERS

HEXADIENE

HEXAGONAL CELLS

HEXAGONS

HEXAHEDRITE

HEXAMETHONIUM

HEXAMETHYLENETETRAMINE

HEXANITROSTILBENE

HEXENES

HEXOGENES (TRADEMARK)

HEXOKINASE

HEXOSES

HEXYL COMPOUNDS

Hf
USE   HAFNIUM

HF
USE   HYDROFLUORIC ACID

HF LASERS

HFB-320 AIRCRAFT

HFB-320 Aircraft, Hamburger
USE   HFB-320 AIRCRAFT

HFIR
USE   HIGH FLUX ISOTOPE REACTORS

HFIR (Reactor)
USE   HIGH FLUX ISOTOPE REACTORS

Hg
USE   MERCURY (METAL)

HH-43 HELICOPTER

HH-43B Helicopter
USE   HH-43 HELICOPTER

HHX Helicopter
USE   H-53 HELICOPTER

HI
USE   HAWAII

HIBERNATION

HICAT Project
USE   HIGH RESOLUTION COVERAGE ANTENNAS

HICAT (Radar Technique)
USE   HIGH RESOLUTION COVERAGE ANTENNAS

HIERARCHIES

Hierarchy, BBGKY
USE   BBGKY HIERARCHY

HIGH ACCELERATION

HIGH ALT TARGET AND BACKGROUND
MEASUREMENT

HIGH ALTITUDE

HIGH ALTITUDE BALLOONS

HIGH ALTITUDE BREATHING

HIGH ALTITUDE ENVIRONMENTS

High Altitude Flight
USE   HIGH ALTITUDE
       FLIGHT

HIGH ALTITUDE NUCLEAR DETECTION

HIGH ALTITUDE PRESSURE

High Altitude Sounding Projectile
USE   WASP SOUNDING ROCKET

HIGH ALTITUDE TESTS

HIGH ASPECT RATIO

High Aspect Ratio Wings
USE   SLENDER WINGS

HIGH CURRENT

HIGH DISPERSION SPECTROGRAPHS

High Eccentric Lunar Occultation Satellite
USE   EXOSAT SATELLITE

HIGH ELECTRON MOBILITY TRANSISTORS

High Energy Astronomy Observatories
USE   HEAO

High Energy Astronomy Observatory A
USE   HEAO 1

High Energy Astronomy Observatory B
USE   HEAO 2

High Energy Astronomy Observatory C
USE   HEAO 3

High Energy Astronomy Observatory 1
USE   HEAO 1

High Energy Astronomy Observatory 2
USE   HEAO 2

High Energy Astronomy Observatory 3
USE   HEAO 3

HIGH ENERGY ELECTRONS

HIGH ENERGY FUELS

(High Energy Fuels), HEF
USE   HIGH ENERGY FUELS

HIGH ENERGY INTERACTIONS

HIGH ENERGY OXIDIZERS

HIGH ENERGY PROPELLANTS

HIGH FIELD MAGNETS

HIGH FLUX BEAM REACTORS

HIGH FLUX ISOTOPE REACTORS

HIGH FREQUENCIES

High Frequencies, Extremely
USE   EXTREMELY HIGH FREQUENCIES

High Frequencies, Very
USE   VERY HIGH FREQUENCIES

High Frequency Radio Equipment, Very
USE   VERY HIGH FREQUENCY RADIO
       EQUIPMENT

HIGH GAIN

High Gravity (Acceleration)
USE   HIGH GRAVITY ENVIRONMENTS

HIGH GRAVITY ENVIRONMENTS

HIGH IMPULSE

High Intensity Lasers
USE   HIGH POWER LASERS

High Latitudes
USE   POLAR REGIONS

HIGH LEVEL LANGUAGES

High Melting Compounds
USE   REFRACTORY MATERIALS

HIGH PASS FILTERS

HIGH POLYMERS

HIGH POWER LASERS

HIGH PRESSURE

HIGH PRESSURE OXYGEN

High Q
USE   Q FACTORS

HIGH RESISTANCE

HIGH RESOLUTION

HIGH RESOLUTION COVERAGE ANTENNAS

HIGH REYNOLDS NUMBER

HIGH SPEED

HIGH SPEED CAMERAS

High Speed Flight
USE   HIGH SPEED
        FLIGHT

High Speed Integrated Circuits, Very
USE   VHSIC (CIRCUITS)

HIGH SPEED PHOTOGRAPHY

High Speed Transportation
USE   RAPID TRANSIT SYSTEMS

HIGH STRENGTH

HIGH STRENGTH ALLOYS

HIGH STRENGTH STEELS

HIGH TEMPERATURE

HIGH TEMPERATURE AIR

High Temperature Alloys
USE   HEAT RESISTANT ALLOYS

HIGH TEMPERATURE ENVIRONMENTS

HIGH TEMPERATURE FLUIDS

HIGH TEMPERATURE GAS COOLED REACTORS

HIGH TEMPERATURE GASES

HIGH TEMPERATURE LUBRICANTS

High Temperature Materials
USE   REFRACTORY MATERIALS

HIGH TEMPERATURE NUCLEAR REACTORS

HIGH TEMPERATURE PLASMAS

HIGH TEMPERATURE PROPELLANTS

HIGH TEMPERATURE RESEARCH

HIGH TEMPERATURE TESTS

HIGH THRUST

HIGH VACUUM

HIGH VACUUM ORBITAL SIMULATOR

HIGH VOLTAGES

HIGHLANDS

Highly Eccentric Orbit Satellites
USE   HEOS SATELLITES

HIGHLY MANEUVERABLE AIRCRAFT

HIGHWAYS

Hijacking
USE   AIR PIRACY

HILBERT SPACE

HILBERT TRANSFORMATION

Hill Curves
USE   HILL METHOD

HILL DETERMINANT

HILL LUNAR THEORY

HILL METHOD

HILLER AIRCRAFT

Hiller Aircraft, Fairchild-
USE   FAIRCHILD-HILLER AIRCRAFT

Hiller Military Aircraft
USE   HILLER AIRCRAFT
        MILITARY AIRCRAFT

Hills Region (GA-NC-SC), Sand
USE   SAND HILLS REGION (GA-NC-SC)

Hills Region (NE), Sand
USE   SAND HILLS REGION (NE)

Hills (SD-WY), Black
USE   BLACK HILLS (SD-WY)

HILSCH TUBES

HIMALAYAS

HIMAT
USE   HIGHLY MANEUVERABLE AIRCRAFT

Hindrance
USE   CONSTRAINTS

Hinge Moments
USE   TORQUE

Hinged Rotor Blades
USE   ROTARY WINGS
        HINGES

Hingeless Rotors
USE   RIGID ROTORS

HINGES

Hinges, Flapping
USE   FLAPPING HINGES

HIPPARCOS SATELLITE

HIPPOCAMPUS

HIPPURIC ACID

HIS BUNDLE

HISS

HISTAMINES

HISTIDINE

HISTOCHEMICAL ANALYSIS

HISTOGRAMS

HISTOLOGY

HISTORIES

Histories, Case
USE   CASE HISTORIES

HITAB Program
USE   HIGH ALT TARGET AND BACKGROUND
        MEASUREMENT

HIVOS (Simulator)
USE   HIGH VACUUM ORBITAL SIMULATOR

HL-10 REENTRY VEHICLE

HLD-35 REENTRY VEHICLE

Hllv
USE   HEAVY LIFT LAUNCH VEHICLES

HMX

HNPF (Hallam Nuclear Power Facility)
USE   HALLAM NUCLEAR POWER FACILITY

HNST
USE   HEXANITROSTILBENE

Ho
USE   HOLMIUM

HO-4 Helicopter
USE   OH-4 HELICOPTER

HO-5 Helicopter
USE   OH-5 HELICOPTER

HO-6 Helicopter
USE   OH-6 HELICOPTER

Hocquenghem Codes, Bose-Chaudhuri-
USE   BCH CODES

HODOGRAPHS

HODOSCOPES

Hogbacks
USE   RIDGES

HOHLRAUMS

Hohmann Trajectories
USE   ELLIPTICAL ORBITS
        TRANSFER ORBITS

Hohmann Transfer Orbits
USE   TRANSFER ORBITS
        ELLIPTICAL ORBITS

HOLDERS

Holders, Flame
USE   FLAME HOLDERS

HOLDING

HOLE BURNING

HOLE DISTRIBUTION

HOLE DISTRIBUTION (ELECTRONICS)

HOLE DISTRIBUTION (MECHANICS)

Hole Drops, Electron-
USE   ELECTRON-HOLE DROPS

HOLE GEOMETRY (MECHANICS)

HOLE MOBILITY

HOLES

Holes (Astronomy), Black
USE   BLACK HOLES (ASTRONOMY)**

Holes (Astronomy), White
  USE   WHITE HOLES (ASTRONOMY)

Holes, Coronal
  USE   CORONAL HOLES

HOLES (ELECTRON DEFICIENCIES)

Holes, Sink
  USE   SLIPSTREAMS

Holland
  USE   NETHERLANDS

HOLLOW

HOLLOW CATHODES

Hollow, Geomagnetic
  USE   GEOMAGNETIC HOLLOW

HOLMIUM

HOLMIUM ISOTOPES

HOLOGRAMMETRY

HOLOGRAPHIC INTERFEROMETRY

HOLOGRAPHIC SPECTROSCOPY

HOLOGRAPHIC SUBTRACTION

HOLOGRAPHY

Holography, Acoustical
  USE   ACOUSTICAL HOLOGRAPHY

Holography, Microwave
  USE   MICROWAVE HOLOGRAPHY

Holography, Self Subtraction
  USE   HOLOGRAPHIC SUBTRACTION

Holography, Sound
  USE   ACOUSTICAL HOLOGRAPHY

Holography, White Light
  USE   WHITE LIGHT HOLOGRAPHY

Holomorphism
  USE   ANALYTIC FUNCTIONS

Holste MH-262 Aircraft, Max
  USE   MH-262 AIRCRAFT

HOMEOSTASIS

HOMEOTHERMS

HOMING

HOMING DEVICES

Homing Missiles, Radar
  USE   RADAR HOMING MISSILES

HOMODYNE RECEPTION

HOMOGENEITY

Homogeneity, In
  USE   INHOMOGENEITY

HOMOGENEOUS TURBULENCE

Homogenization
  USE   HOMOGENIZING

HOMOGENIZING

HOMOJUNCTIONS

HOMOLOGY

HOMOMORPHISMS

HOMOPOLAR GENERATORS

HOMOSPHERE

HOMOTOPY THEORY

HOMOTROPY

HONDURAS

Honduras, British
  USE   BELIZE

HONEST JOHN ROCKET VEHICLE

HONEYCOMB CORES

HONEYCOMB STRUCTURES

Honeycombs, Ceramic
  USE   CERAMIC HONEYCOMBS

HONEYWELL ADEPT COMPUTER

HONEYWELL COMPUTERS

HONEYWELL DDP 116 COMPUTER

HONEYWELL 600/6000 COMPUTER

HONG KONG

HONING

HOOKES LAW

HOOKS

HOOP COLUMN ANTENNAS

HOOPS

HOPCALITE (TRADEMARK)

Hopf Equations, Wiener
  USE   WIENER HOPF EQUATIONS

HOPPERS

Hopping, Frequency
  USE   FREQUENCY HOPPING

HORIZON

Horizon Radar, Over-The-
  USE   OVER-THE-HORIZON RADAR

HORIZON SCANNERS

Horizon Scanners, Infrared
  USE   HORIZON SCANNERS
         INFRARED SCANNERS

Horizon Sensing
  USE   HORIZON SCANNERS

Horizons, Gyro
  USE   GYRO HORIZONS

Horizons, Radio
  USE   RADIO HORIZONS

HORIZONTAL BRANCH STARS

HORIZONTAL FLIGHT

(Horizontal), Level
  USE   LEVEL (HORIZONTAL)

HORIZONTAL ORIENTATION

HORIZONTAL SPACECRAFT LANDING

Horizontal Stabilizers
  USE   STABILIZERS (FLUID DYNAMICS)

HORIZONTAL TAIL SURFACES

HORMONE METABOLISMS

HORMONES

Hormones, Pituitary
  USE   PITUITARY HORMONES

HORN ANTENNAS

HORNS

HORSEPOWER

HORSES

HOSES

HOSPITALS

Hot Air
  USE   HIGH TEMPERATURE AIR

HOT ATOMS

HOT CATHODES

HOT CORROSION

Hot Cycle Propulsion System
  USE   TIP DRIVEN ROTORS

HOT ELECTRONS

Hot Extruding
  USE   EXTRUDING

Hot Forming
  USE   HOT WORKING

Hot Gas Systems
  USE   HIGH TEMPERATURE GASES

Hot Gases
  USE   HIGH TEMPERATURE GASES

Hot Jet Exhaust
  USE   HIGH TEMPERATURE GASES
         JET EXHAUST

Hot Jets
  USE   JET FLOW

HOT MACHINING

Hot Plasmas
  USE   HIGH TEMPERATURE PLASMAS

HOT PRESSING

HOT STARS

HOT SURFACES

HOT WATER ROCKET ENGINES

HOT WEATHER

HOT WORKING

HOT-FILM ANEMOMETERS

HOT-WIRE ANEMOMETERS

HOT-WIRE FLOWMETERS

Hot-Wire Turbulence Meters
  USE   HOT-WIRE FLOWMETERS
         TURBULENCE METERS

HOTSHOT WIND TUNNELS

HOUND DOG MISSILE

Hour Orbits, Twenty-Four
  USE   TWENTY-FOUR HOUR ORBITS

HOUSEHOLDER TRANSFORMATIONS

HOUSEKEEPING (SPACECRAFT)

Houses, Green
  USE   GREENHOUSES

Houses, Solar
  USE   SOLAR HOUSES

HOUSINGS

HOUSTON (TX)

Hovercraft
USE   GROUND EFFECT MACHINES

HOVERCRAFT GROUND EFFECT MACHINES

Hovercraft, Westland SR-N2
USE   WESTLAND GROUND EFFECT MACHINES

Hovercraft, Westland SR-N3
USE   WESTLAND GROUND EFFECT MACHINES

HOVERING

HOVERING ROCKET VEHICLES

HOVERING STABILITY

HOWITZERS

HP-115 AIRCRAFT

HP-115 Aircraft, Handley Page
USE   HP-115 AIRCRAFT

HPRR
USE   HEALTH PHYSICS RESEARCH REACTOR

HR Diagram
USE   HERTZSPRUNG-RUSSELL DIAGRAM

HRB-1 Helicopter
USE   CH-46 HELICOPTER

HS-125 Aircraft
USE   DH 125 AIRCRAFT

HS-748 AIRCRAFT

HS-748 Aircraft, AVRO Whitworth
USE   HS-748 AIRCRAFT

HS-801 AIRCRAFT

HSS-2 Helicopter
USE   SH-3 HELICOPTER

HSS-2 Helicopter, Sikorsky
USE   SH-3 HELICOPTER

HTGR
USE   HIGH TEMPERATURE GAS COOLED
           REACTORS

HTPB PROPELLANTS

HU-1 Helicopter
USE   UH-1 HELICOPTER

HUBBLE CONSTANT

HUBBLE DIAGRAM

HUBBLE SPACE TELESCOPE

HUBS

Hubs, Rotor
USE   HUBS
           ROTORS

Huckel Theory, Debye-
USE   DEBYE-HUCKEL THEORY

HUDSON RIVER (NY-NJ)

HUECKEL THEORY

HUGHES AIRCRAFT

Hughes Military Aircraft
USE   HUGHES AIRCRAFT
           MILITARY AIRCRAFT

Hugoniot Adiabat
USE   HUGONIOT EQUATION OF STATE

HUGONIOT EQUATION OF STATE

Hugoniot Relation, Rankine-
USE   RANKINE-HUGONIOT RELATION

HUL
USE   HARDWARE UTILIZATION LISTS

Hull, Small Water Plane Area Twin
USE   SWATH (SHIP)

Hulls, Ship
USE   SHIP HULLS

HULLS (STRUCTURES)

HUM

HUMAN BEHAVIOR

HUMAN BEINGS

HUMAN BODY

HUMAN CENTRIFUGES

Human Engineering
USE   HUMAN FACTORS ENGINEERING

HUMAN FACTORS ENGINEERING

HUMAN FACTORS LABORATORIES

HUMAN PATHOLOGY

HUMAN PERFORMANCE

HUMAN REACTIONS

HUMAN RELATIONS

HUMAN RESOURCES

HUMAN TOLERANCES

HUMAN WASTES

HUMASON COMET

HUMERUS

HUMIDITY

HUMIDITY MEASUREMENT

Hummingbird Aircraft
USE   XV-4 AIRCRAFT

Humping Tests, Railroad
USE   RAILROAD HUMPING TESTS

HUNGARY

Hunter Aircraft, Hawker
USE   F-2 AIRCRAFT

Hunter F-2 Aircraft
USE   F-2 AIRCRAFT

Hunting H-126 Aircraft
USE   H-126 AIRCRAFT

Hunting P-84 Aircraft
USE   JET PROVOST AIRCRAFT

Huron, Lake
USE   LAKE HURON

Hurricane, ANNA
USE   ANNA HURRICANE

HURRICANES

HUS-1 Helicopter
USE   UH-34 HELICOPTER

Huskie Helicopter
USE   HH-43 HELICOPTER

Hustler Aircraft
USE   B-58 AIRCRAFT

HUYGENS PRINCIPLE

Huygens Principle, Kirchhoff-
USE   WAVE PROPAGATION
           DIFFRACTION

HU2K-1 Helicopter
USE   UH-2 HELICOPTER

HVITTIS CHONDRITE

HYBRID CIRCUITS

Hybrid Combustion
USE   HYBRID PROPELLANT ROCKET ENGINES

HYBRID COMPUTERS

HYBRID NAVIGATION SYSTEMS

HYBRID PROPELLANT ROCKET ENGINES

HYBRID PROPELLANTS

HYBRID PROPULSION

Hybrid Reactors, Fusion-Fission
USE   FUSION-FISSION HYBRID REACTORS

HYBRID ROCKET ENGINES

HYBRID STRUCTURES

Hybrid Vehicles, Electric
USE   ELECTRIC HYBRID VEHICLES

Hybrids (Biology)
USE   GENETIC ENGINEERING

Hydac Rocket Vehicle, Nike-
USE   NIKE-HYDAC ROCKET VEHICLE

Hydantoin, Diphenyl
USE   DIPHENYL HYDANTOIN

HYDRATES

Hydrates, Carbo
USE   CARBOHYDRATES

HYDRATION

Hydration, De
USE   DEHYDRATION

Hydraulic Actuators
USE   HYDRAULIC EQUIPMENT
           ACTUATORS

HYDRAULIC ANALOGIES

HYDRAULIC CONTROL

HYDRAULIC EQUIPMENT

HYDRAULIC FLUIDS

Hydraulic Heating Sources
USE   HYDRAULIC EQUIPMENT
           HEAT SOURCES

HYDRAULIC JETS

Hydraulic Pumps
USE   HYDRAULIC EQUIPMENT
           PUMPS

HYDRAULIC SHOCK

Hydraulic Systems
USE   HYDRAULIC EQUIPMENT

Hydraulic Systems, Aircraft
USE   AIRCRAFT HYDRAULIC SYSTEMS

HYDRAULIC TEST TUNNELS

Hydraulic Valves
USE   HYDRAULIC EQUIPMENT
           VALVES

HYDRAULICS

Hydraulics, Thermo
    USE   THERMOHYDRAULICS

HYDRAZIDES

HYDRAZINE BORANE

Hydrazine, Di
    USE   DIHYDRAZINE

HYDRAZINE ENGINES

Hydrazine, Methyl
    USE   METHYLHYDRAZINE

HYDRAZINE NITRATE

HYDRAZINE NITROFORM

HYDRAZINE PERCHLORATES

HYDRAZINES

Hydrazines, Dimethyl
    USE   DIMETHYLHYDRAZINES

HYDRAZINIUM COMPOUNDS

HYDRAZOIC ACID

HYDRAZONES

HYDRAZONIUM COMPOUNDS

HYDRIDES

Hydrides, Aluminum
    USE   ALUMINUM HYDRIDES

Hydrides, An
    USE   ANHYDRIDES

Hydrides, Beryllium
    USE   BERYLLIUM HYDRIDES

Hydrides, Boro
    USE   BOROHYDRIDES

Hydrides, Boron
    USE   BORON HYDRIDES

Hydrides, Cesium
    USE   CESIUM HYDRIDES

Hydrides, Di
    USE   DIHYDRIDES

Hydrides, Lithium
    USE   LITHIUM HYDRIDES

Hydrides, Lithium Aluminum
    USE   LITHIUM ALUMINUM HYDRIDES

Hydrides, Metal
    USE   METAL HYDRIDES

Hydrides, Nitrogen
    USE   NITROGEN HYDRIDES

Hydrides, Potassium
    USE   POTASSIUM HYDRIDES

Hydrides, Sodium
    USE   SODIUM HYDRIDES

Hydrides, Zirconium
    USE   ZIRCONIUM HYDRIDES

Hydroacoustics
    USE   UNDERWATER ACOUSTICS

Hydroaeromechanics
    USE   AERODYNAMICS

HYDROBALLISTICS

Hydrobarophones
    USE   HYDROPHONES

HYDROBORATION

HYDROBROMIC ACID

HYDROBROMIDES

HYDROCARBON COMBUSTION

HYDROCARBON FUEL PRODUCTION

HYDROCARBON FUELS

HYDROCARBON POISONING

HYDROCARBONS

Hydrocarbons, Cyclic
    USE   CYCLIC HYDROCARBONS

Hydrocarbons, Fluoro
    USE   FLUOROHYDROCARBONS

Hydrocarbons, Saturated
    USE   ALKANES

HYDROCHLORIC ACID

HYDROCHLORIDES

HYDROCLIMATOLOGY

HYDROCRACKING

HYDROCYANIC ACID

HYDRODYNAMIC COEFFICIENTS

HYDRODYNAMIC EQUATIONS

HYDRODYNAMIC RAM EFFECT

Hydrodynamic Stability
    USE   FLOW STABILITY

Hydrodynamic Tunnels
    USE   PLASMA JET WIND TUNNELS

HYDRODYNAMICS

Hydrodynamics, Magneto
    USE   MAGNETOHYDRODYNAMICS

HYDROELASTICITY

HYDROELECTRIC POWER STATIONS

HYDROELECTRICITY

HYDROFLUORIC ACID

Hydrofoil Boats
    USE   HYDROFOIL CRAFT

HYDROFOIL CRAFT

HYDROFOIL OSCILLATIONS

HYDROFOILS

HYDROFORMING

HYDROGEN

Hydrogen Air Fuel Cells
    USE   HYDROGEN OXYGEN FUEL CELLS

Hydrogen Atmospheres, Helium
    USE   HELIUM HYDROGEN ATMOSPHERES

HYDROGEN ATOMS

HYDROGEN AZIDES

Hydrogen Batteries, Nickel
    USE   NICKEL HYDROGEN BATTERIES

Hydrogen Batteries, Silver
    USE   SILVER HYDROGEN BATTERIES

Hydrogen Bombs
    USE   FUSION WEAPONS

HYDROGEN BONDS

HYDROGEN CHLORIDES

HYDROGEN CLOUDS

HYDROGEN COMPOUNDS

Hydrogen Cyanides
    USE   HYDROCYANIC ACID

Hydrogen Deuterium Oxide
    USE   HEAVY WATER

HYDROGEN EMBRITTLEMENT

HYDROGEN ENGINES

Hydrogen Engines, LOX-
    USE   HYDROGEN OXYGEN ENGINES

Hydrogen Fluorides
    USE   HYDROFLUORIC ACID

HYDROGEN FUELS

HYDROGEN IONS

HYDROGEN ISOTOPES

Hydrogen, Liquid
    USE   LIQUID HYDROGEN

HYDROGEN MASERS

HYDROGEN METABOLISM

Hydrogen, Metallic
    USE   METALLIC HYDROGEN

Hydrogen, Ortho
    USE   ORTHO HYDROGEN

HYDROGEN OXYGEN ENGINES

HYDROGEN OXYGEN FUEL CELLS

Hydrogen, Para
    USE   PARA HYDROGEN

HYDROGEN PERCHLORATE

HYDROGEN PEROXIDE

Hydrogen Phosphite (DEHP), Diethyl
    USE   DIETHYL HYDROGEN PHOSPHITE (DEHP)

HYDROGEN PLASMA

HYDROGEN PRODUCTION

HYDROGEN RECOMBINATIONS

HYDROGEN SULFIDE

Hydrogen 2
    USE   DEUTERIUM

Hydrogen 3
    USE   TRITIUM

HYDROGEN 4

HYDROGEN-BASED ENERGY

HYDROGENATION

Hydrogenation, De
    USE   DEHYDROGENATION

HYDROGENOLYSIS

HYDROGENOMONAS

HYDROGEOLOGY

HYDROGRAPHY

Hydrokinetics
USE   HYDROMECHANICS

Hydrological Decade, International
USE   INTERNATIONAL HYDROLOGICAL DECADE

HYDROLOGY

HYDROLOGY MODELS

HYDROLYSIS

Hydrolysis, Pyro
USE   PYROHYDROLYSIS

Hydromagnetic Flow
USE   MAGNETOHYDRODYNAMIC FLOW

Hydromagnetic Stability
USE   MAGNETOHYDRODYNAMIC STABILITY

Hydromagnetic Waves
USE   MAGNETOHYDRODYNAMIC WAVES

Hydromagnetics
USE   MAGNETOHYDRODYNAMICS

Hydromagnetics, Geometrical
USE   MAGNETOHYDRODYNAMICS

Hydromagnetism
USE   MAGNETOHYDRODYNAMICS

HYDROMECHANICS

HYDROMETALLURGY

HYDROMETEOROLOGY

HYDROMETERS

HYDRONIUM IONS

HYDROPHONES

HYDROPLANES (SURFACES)

HYDROPLANES (VEHICLES)

HYDROPLANING

HYDROPONICS

Hydropower Stations
USE   HYDROELECTRIC POWER STATIONS

HYDROPYROLYSIS

Hydroskis
USE   HYDROPLANES (SURFACES)

Hydrosphere (Earth)
USE   EARTH HYDROSPHERE

Hydrosphere, Earth
USE   EARTH HYDROSPHERE

HYDROSPINNING

HYDROSTATIC PRESSURE

HYDROSTATICS

Hydrostatics, Magneto
USE   MAGNETOHYDROSTATICS

HYDROSULFITES

HYDROTHERMAL CRYSTAL GROWTH

HYDROTHERMAL STRESS ANALYSIS

HYDROTHERMAL SYSTEMS

Hydrox Engines
USE   HYDROGEN OXYGEN ENGINES

HYDROXIDES

Hydroxides, Lithium
USE   LITHIUM HYDROXIDES

Hydroxides, Potassium
USE   POTASSIUM HYDROXIDES

Hydroxides, Sodium
USE   SODIUM HYDROXIDES

HYDROXYCORTICOSTEROID

HYDROXYL COMPOUNDS

HYDROXYL EMISSION

HYDROXYL RADICALS

HYDROXYLAMINE SULFATE

HYDROXYLAMMONIUM PERCHLORATES

HYGIENE

Hygiene, Oral
USE   ORAL HYGIENE

HYGRAL PROPERTIES

HYGROMETERS

HYGROSCOPICITY

HYLA-STAR ROCKET VEHICLE

HYLLERAAS COORDINATES

HYOSCINE

HYPERBARIC CHAMBERS

HYPERBOLAS

HYPERBOLIC COORDINATES

HYPERBOLIC DIFFERENTIAL EQUATIONS

HYPERBOLIC FUNCTIONS

HYPERBOLIC NAVIGATION

HYPERBOLIC REENTRY

Hyperbolic Space
USE   HYPERBOLIC COORDINATES

HYPERBOLIC SYSTEMS

HYPERBOLIC TRAJECTORIES

HYPERCAPNIA

HYPERFINE STRUCTURE

HYPERGEOMETRIC FUNCTIONS

Hypergeometry
USE   HYPERSPACES

HYPERGLYCEMIA

HYPERGOLIC ROCKET PROPELLANTS

HYPERION

HYPERKINESIA

HYPERNEA

HYPERNUCLEI

HYPERONS

Hyperons, Xi
USE   XI HYPERONS

HYPEROPIA

HYPEROXIA

HYPERPLANES

HYPERPNEA

HYPERSOMNIA

HYPERSONIC AIRCRAFT

HYPERSONIC BOUNDARY LAYER

HYPERSONIC COMBUSTION

HYPERSONIC FLIGHT

HYPERSONIC FLOW

HYPERSONIC FORCES

HYPERSONIC GLIDERS

HYPERSONIC HEAT TRANSFER

HYPERSONIC INLETS

HYPERSONIC NOZZLES

HYPERSONIC REENTRY

HYPERSONIC SHOCK

HYPERSONIC SPEED

HYPERSONIC TEST APPARATUS

HYPERSONIC VEHICLES

HYPERSONIC WAKES

HYPERSONIC WIND TUNNELS

HYPERSONICS

HYPERSPACES

HYPERSPHERES

HYPERTENSIN

HYPERTENSION

HYPERTHERMIA

Hypertonia
USE   OSMOSIS

Hypertrophy
USE   GROWTH

HYPERVELOCITY

Hypervelocity Accelerators
USE   HYPERVELOCITY GUNS

Hypervelocity Cratering
USE   HYPERVELOCITY PROJECTILES
PROJECTILE CRATERING

HYPERVELOCITY FLOW

HYPERVELOCITY GUNS

HYPERVELOCITY IMPACT

HYPERVELOCITY LAUNCHERS

HYPERVELOCITY PROJECTILES

HYPERVELOCITY WIND TUNNELS

HYPERVENTILATION

HYPERVOLEMIA

HYPNOSIS

HYPOBARIC ATMOSPHERES

HYPOCAPNIA

HYPODERMIS

HYPODYNAMIA

HYPOELASTICITY

HYPOGLYCEMIA

HYPOKINESIA

HYPOMETABOLISM

HYPOTENSION

HYPOTHALAMUS

HYPOTHERMIA

HYPOTHESES

Hypothesis, Expectancy
  USE   EXPECTANCY HYPOTHESIS

Hypothesis, Intermittency
  USE   INTERMITTENCY HYPOTHESIS

Hypothesis, Lagrange Similarity
  USE   LAGRANGE SIMILARITY HYPOTHESIS

Hypothesis, Null
  USE   NULL HYPOTHESIS

Hypothesis, Vorticity Transport
  USE   VORTICITY TRANSPORT HYPOTHESIS

HYPOTONIA

HYPOVENTILATION

HYPOVOLEMIA

HYPOXEMIA

HYPOXIA

HYPSOGRAPHY

HYPSOMETERS

HYSTERESIS

I

I BEAMS

I, IMP-
  USE   EXPLORER 43 SATELLITE

I, Space Shuttle Mission 51-
  USE   SPACE SHUTTLE MISSION 51-I

I-N Diodes, P-
  USE   DIODES
        P-I-N JUNCTIONS

I-N Junctions, P-
  USE   P-I-N JUNCTIONS

IA
  USE   IOWA

(IA), Cedar Rapids
  USE   CEDAR RAPIDS (IA)

IAPETUS

IBM COMPUTERS

IBM 360 COMPUTER

IBM 370 COMPUTER

IBM 650 COMPUTER

IBM 704 COMPUTER

IBM 709 COMPUTER

IBM 1130 COMPUTER

IBM 1401 COMPUTER

IBM 1410 COMPUTER

IBM 1620 COMPUTER

IBM 2250 COMPUTER

IBM 7000 SERIES COMPUTERS

IBM 7030 COMPUTER

IBM 7040 COMPUTER

IBM 7044 COMPUTER

IBM 7070 COMPUTER

IBM 7074 COMPUTER

IBM 7090 COMPUTER

IBM 7094 COMPUTER

ICARUS ASTEROID

ICBM, Atlas
  USE   ATLAS ICBM

ICBM, Atlas D
  USE   ATLAS D ICBM

ICBM, Atlas E
  USE   ATLAS E ICBM

ICBM, Atlas F
  USE   ATLAS F ICBM

ICBM, Minuteman
  USE   MINUTEMAN ICBM

ICBM (Missiles)
  USE   INTERCONTINENTAL BALLISTIC MISSILES

ICBM, Titan
  USE   TITAN ICBM

ICBM, Titan 1
  USE   TITAN 1 ICBM

ICBM, Titan 2
  USE   TITAN 2 ICBM

ICE

(Ice), Aufeis
  USE   AUFEIS (ICE)

Ice, Bay
  USE   BAY ICE

ICE ENVIRONMENTS

ICE FLOES

ICE FORMATION

Ice Interactions, Air Sea
  USE   AIR SEA ICE INTERACTIONS

Ice, Lake
  USE   LAKE ICE

Ice, Land
  USE   LAND ICE

ICE MAPPING

ICE NUCLEI

Ice Observation
  USE   ICE REPORTING

Ice Packs
  USE   SEA ICE

Ice, Pressure
  USE   PRESSURE ICE

ICE PREVENTION

ICE REPORTING

Ice, Sea
  USE   SEA ICE

Ice Shelf, Ross
  USE   ROSS ICE SHELF

Ice Shelves
  USE   LAND ICE

ICEBERGS

ICELAND

ICHTHYOLOGY

Icing
  USE   ICE FORMATION

ICL COMPUTERS

ICOSAHEDRONS

ID
  USE   IDAHO

(ID-MT-WY), Yellowstone National Park
  USE   YELLOWSTONE NATIONAL PARK
        (ID-MT-WY)

(ID-OR-WA), Columbia River Basin
  USE   COLUMBIA RIVER BASIN (ID-OR-WA)

IDAHO

IDEAL FLUIDS

IDEAL GAS

Identification And Location Exper, Feature
  USE   FEATURE IDENTIFICATION AND LOCATION
        EXPER

Identification, Crop
  USE   CROP IDENTIFICATION

(Identification), IFF Systems
  USE   IFF SYSTEMS (IDENTIFICATION)

Identification, Parameter
  USE   PARAMETER IDENTIFICATION

Identification, Rapid Ballistics
  USE   RAPID BALLISTICS IDENTIFICATION

Identification, System
  USE   SYSTEM IDENTIFICATION

Identification, Timber
  USE   TIMBER IDENTIFICATION

Identify Friend OR Foe
  USE   IFF SYSTEMS (IDENTIFICATION)

IDENTIFYING

IDENTITIES

IDEP (Data Exchange)
  USE   INTERSERVICE DATA EXCHANGE
        PROGRAM

IDLERS

IFF SYSTEMS (IDENTIFICATION)

IFR (Rules)
  USE   INSTRUMENT FLIGHT RULES

IGFET
  USE   FIELD EFFECT TRANSISTORS

IGNEOUS ROCKS

Ignimbrite
  USE   IGNEOUS ROCKS

IGNITERS

(Igniting), Firing
  USE   FIRING (IGNITING)

IGNITION

Ignition, Electric
USE   ELECTRIC IGNITION

IGNITION LIMITS

Ignition, Solid Propellant
USE   SOLID PROPELLANT IGNITION

Ignition, Spark
USE   SPARK IGNITION

IGNITION SYSTEMS

IGNITION TEMPERATURE

IGNITRONS

IGOSS
USE   INTEGRATED GLOBAL OCEAN STATION
SYSTEMS

IGY (Geophysical Year)
USE   INTERNATIONAL GEOPHYSICAL YEAR

II Computer, Modcomp
USE   MODCOMP II COMPUTER

IL
USE   ILLINOIS

(IL-IN-OH), Wabash River Basin
USE   WABASH RIVER BASIN (IL-IN-OH)

IL-14 AIRCRAFT

IL-14 Aircraft, Ilyushin
USE   IL-14 AIRCRAFT

IL-62 AIRCRAFT

IL-62 Aircraft, Ilyushin
USE   IL-62 AIRCRAFT

ILLIAC COMPUTERS

ILLIAC 3 COMPUTER

ILLIAC 4 COMPUTER

ILLINOIS

ILLITE

ILLUMINANCE

ILLUMINATING

ILLUMINATION

ILLUMINATORS

Illusion, Elevator
USE   ELEVATOR ILLUSION

Illusion, Moon
USE   MOON ILLUSION

Illusion, Optical
USE   OPTICAL ILLUSION

ILLUSIONS

Illusions, Oculogravic
USE   OCULOGRAVIC ILLUSIONS

ILMENITE

ILS (Landing Systems)
USE   INSTRUMENT LANDING SYSTEMS

ILYUSHIN AIRCRAFT

Ilyushin IL-14 Aircraft
USE   IL-14 AIRCRAFT

Ilyushin IL-62 Aircraft
USE   IL-62 AIRCRAFT

IMAGE ANALYSIS

IMAGE CONTRAST

IMAGE CONVERTERS

(Image Correlator), SIMICOR
USE   IMAGE CORRELATORS

Image Correlator, Simultaneous
USE   IMAGE CORRELATORS

IMAGE CORRELATORS

IMAGE DISSECTOR TUBES

IMAGE ENHANCEMENT

IMAGE FILTERS

IMAGE FURNACES

IMAGE INTENSIFIERS

IMAGE MOTION COMPENSATION

IMAGE ORTHICONS

IMAGE PROCESSING

IMAGE RECONSTRUCTION

IMAGE RESOLUTION

IMAGE ROTATION

IMAGE TRANSDUCERS

IMAGE TUBES

IMAGE VELOCITY SENSORS

IMAGERY

Imagery, Aerial
USE   AERIAL PHOTOGRAPHY

(Imagery), Geometric Rectification
USE   GEOMETRIC RECTIFICATION (IMAGERY)

Imagery, Infrared
USE   INFRARED IMAGERY

Imagery, Microwave
USE   MICROWAVE IMAGERY

Imagery, Radar
USE   RADAR IMAGERY

Imagery, Satellite
USE   SATELLITE IMAGERY

Imagery, X Ray
USE   X RAY IMAGERY

IMAGES

Images, After
USE   AFTERIMAGES

Images, Optical
USE   IMAGES

Images, Retinal
USE   RETINAL IMAGES

IMAGING RADAR

Imaging Radar, Earth Resources Shuttle
USE   EARTH RESOURCES SHUTTLE IMAGING
RADAR

Imaging Radar, Shuttle
USE   SHUTTLE IMAGING RADAR

Imaging Radar (Spacecraft), Venus Orbiting
USE   VENUS ORBITING IMAGING RADAR
(SPACECRAFT)

Imaging Scope, Low Intensity X Ray
USE   LIXISCOPES

IMAGING TECHNIQUES

IMBEDDINGS

Imbeddings, Invariant
USE   INVARIANT IMBEDDINGS

IMBEDDINGS (MATHEMATICS)

IMBLMS

Imbrian Period, Pre-
USE   PRE-IMBRIAN PERIOD

IMCC (Control Center)
USE   INTEGRATED MISSION CONTROL CENTER

IME Satellite
USE   INTERNATIONAL MAGNETOSPHERIC
EXPLORER

IMIDES

Imides, Poly
USE   POLYIMIDES

IMINES

IMLSS

Immersion
USE   SUBMERGING

Immersion, Water
USE   WATER IMMERSION

Immiscibility
USE   SOLUBILITY

Immittance
USE   ELECTRICAL IMPEDANCE

IMMOBILIZATION

IMMUNITY

Immunity, Interference
USE   INTERFERENCE IMMUNITY

IMMUNOASSAY

Immunoassay, Radio
USE   RADIOIMMUNOASSAY

IMMUNOLOGY

IMP

IMP-A
USE   EXPLORER 18 SATELLITE

IMP-B
USE   EXPLORER 21 SATELLITE

IMP-C
USE   EXPLORER 28 SATELLITE

IMP-D
USE   EXPLORER 33 SATELLITE

IMP-E
USE   EXPLORER 35 SATELLITE

IMP-F
USE   EXPLORER 34 SATELLITE

IMP-G
USE   EXPLORER 41 SATELLITE

IMP-H
USE   EXPLORER 47 SATELLITE

IMP-I
USE   EXPLORER 43 SATELLITE

IMP-J
USE   EXPLORER 50 SATELLITE

IMP-1
USE   EXPLORER 18 SATELLITE

IMP-2
USE   EXPLORER 21 SATELLITE

IMP-3
USE   EXPLORER 28 SATELLITE

IMP-4
USE   EXPLORER 34 SATELLITE

IMP-5
USE   EXPLORER 41 SATELLITE

IMP-6
USE   EXPLORER 43 SATELLITE

IMP-7
USE   EXPLORER 47 SATELLITE

IMP-8
USE   EXPLORER 50 SATELLITE

IMPACT

IMPACT ACCELERATION

IMPACT DAMAGE

Impact Damage, Rain
USE   RAIN IMPACT DAMAGE

Impact Deceleration
USE   IMPACT ACCELERATION
       DECELERATION

Impact, Economic
USE   ECONOMIC IMPACT

Impact, Electron
USE   ELECTRON IMPACT

IMPACT FUSION

Impact, Hypervelocity
USE   HYPERVELOCITY IMPACT

Impact, Ion
USE   ION IMPACT

IMPACT LOADS

IMPACT MELTS

Impact, Point
USE   POINT IMPACT

IMPACT PREDICTION

(Impact Prediction), ARIP
USE   IMPACT PREDICTION
       COMPUTERIZED SIMULATION

(Impact Prediction), IP
USE   COMPUTERIZED SIMULATION

Impact Predictors, Automatic Rocket
USE   COMPUTERIZED SIMULATION
       IMPACT PREDICTION

Impact Pressures
USE   IMPACT LOADS

Impact, Proton
USE   PROTON IMPACT

IMPACT RESISTANCE

Impact Sensitivity
USE   IMPACT RESISTANCE

IMPACT STRENGTH

Impact Test, Charpy
USE   CHARPY IMPACT TEST

IMPACT TESTING MACHINES

IMPACT TESTS

IMPACT TOLERANCES

IMPACTORS

IMPAIRMENT

IMPATT Diodes
USE   AVALANCHE DIODES

IMPEDANCE

Impedance, Acoustic
USE   ACOUSTIC IMPEDANCE

Impedance, Electrical
USE   ELECTRICAL IMPEDANCE

IMPEDANCE MATCHING

IMPEDANCE MEASUREMENT

Impedance, Mechanical
USE   MECHANICAL IMPEDANCE

IMPEDANCE PROBES

Impedance Probes, Radio Frequency
USE   RADIO FREQUENCY IMPEDANCE PROBES

Impedance, Respiratory
USE   RESPIRATORY IMPEDANCE

(Impedances), Ballasts
USE   BALLASTS (IMPEDANCES)

Impeller Blades
USE   ROTOR BLADES (TURBOMACHINERY)

IMPELLERS

Impellers, Pump
USE   PUMP IMPELLERS

Imperfections
USE   DEFECTS

Imperfections, Lattice
USE   CRYSTAL DEFECTS

IMPERIAL VALLEY (CA)

IMPINGEMENT

Impingement, Jet
USE   JET IMPINGEMENT

IMPLANTATION

Implantation, Heart
USE   HEART IMPLANTATION

Implantation, Ion
USE   ION IMPLANTATION

IMPLANTED ELECTRODES (BIOLOGY)

IMPLICATION

IMPLOSIONS

IMPREGNATING

IMPROVED TIROS OPERATIONAL SATELLITES

IMPROVEMENT

IMPULSE GENERATORS

Impulse, High
USE   HIGH IMPULSE

Impulse Response Filters, Finite
USE   FIR FILTERS

Impulse, Specific
USE   SPECIFIC IMPULSE

IMPULSES

Impulses, Electric
USE   ELECTRIC PULSES

IMPURITIES

Impurities, Atmospheric
USE   AIR POLLUTION

IMS
USE   INTERNATIONAL MAGNETOSPHERIC
       STUDY

In
USE   INDIUM

IN
USE   INDIANA

In, Burn-
USE   BURN-IN

In Earth Neighborhood, Origin Of Plasmas
USE   OPEN PROJECT

(In Space), Hazardous Material Disposal
USE   HAZARDOUS MATERIAL DISPOSAL (IN
       SPACE)

IN-FLIGHT MONITORING

IN-OH), Wabash River Basin (IL-
USE   WABASH RIVER BASIN (IL-IN-OH)

Inactivation
USE   DEACTIVATION

INCANDESCENCE

INCENDIARY AMMUNITION

INCENTIVE TECHNIQUES

INCENTIVES

Incentives, Contract
USE   CONTRACT INCENTIVES

Incentives, Cost
USE   COST INCENTIVES

INCIDENCE

Incidence Control, Wave
USE   WAVE INCIDENCE CONTROL

Incidence, Grazing
USE   GRAZING INCIDENCE

Incidence Solar Telescope, Grazing
USE   GRIST (TELESCOPE)

INCIDENT RADIATION

Incineration
USE   INCINERATORS

INCINERATORS

INCLINATION

(Inclination), Attitude
USE   ATTITUDE (INCLINATION)

(Inclination), Pitch
USE   PITCH (INCLINATION)

INCLUSIONS

INCOHERENCE

INCOHERENT SCATTER RADAR

Incoherent Scatter Radar, European
USE   EISCAT RADAR SYSTEM (EUROPE)

INCOHERENT SCATTERING

INCOME

INCOMPATIBILITY

INCOMPRESSIBILITY

INCOMPRESSIBLE BOUNDARY LAYER

INCOMPRESSIBLE FLOW

INCOMPRESSIBLE FLUIDS

INCONEL (TRADEMARK)

INCREASING

INDENE

INDENTATION

Independent Programs, Machine-
USE   MACHINE-INDEPENDENT PROGRAMS

INDEPENDENT VARIABLES

Index, Absorptive
USE   ABSORPTIVITY

Index, Environmental
USE   ENVIRONMENTAL INDEX

Index, KP
USE   KP INDEX

Index Optics, Gradient
USE   GRADIENT INDEX OPTICS

Index, Palmar Sweat
USE   PALMAR SWEAT INDEX

Index, Refractive
USE   REFRACTIVITY

Index, Vegetative
USE   VEGETATIVE INDEX

INDEXES

INDEXES (DOCUMENTATION)

Indexes, Kwic
USE   KWIC INDEXES

Indexes, Morphological
USE   MORPHOLOGICAL INDEXES

Indexes, Psychological
USE   PSYCHOLOGICAL TESTS

INDEXES (RATIOS)

INDIA

INDIAN OCEAN

INDIAN SPACE PROGRAM

Indian Space Research Organization
USE   ISRO

INDIAN SPACECRAFT

(Indian Spacecraft), IRS
USE   INDIAN SPACECRAFT

(Indian Spacecraft), SEO
USE   INDIAN SPACECRAFT

INDIANA

Indians, American
USE   AMERICAN INDIANS

INDICATING INSTRUMENTS

INDICATION

INDICATORS

Indicators, Approach
USE   APPROACH INDICATORS

Indicators, Attitude
USE   ATTITUDE INDICATORS

Indicators, Chemical
USE   CHEMICAL INDICATORS

Indicators, Cloud Height
USE   CLOUD HEIGHT INDICATORS

Indicators, Flow Direction
USE   FLOW DIRECTION INDICATORS

Indicators, Helicopter Attitude
USE   HELICOPTERS
        ATTITUDE INDICATORS

Indicators, Moving Target
USE   MOVING TARGET INDICATORS

Indicators, Plan Position
USE   PLAN POSITION INDICATORS

Indicators, Position
USE   POSITION INDICATORS

Indicators), PPI (Position
USE   PLAN POSITION INDICATORS

Indicators, Range
USE   RANGE FINDERS

Indicators, Rate Of Climb
USE   RATE OF CLIMB INDICATORS

Indicators, Spacecraft Position
USE   SPACECRAFT POSITION INDICATORS

Indicators, Speed
USE   SPEED INDICATORS

Indicators, Temperature
USE   TEMPERATURE MEASURING INSTRUMENTS
        INDICATING INSTRUMENTS

Indicators, Voltage Variation
USE   VOLTMETERS

Indicators, Weight
USE   WEIGHT INDICATORS

Indies, West
USE   WEST INDIES

INDIUM

INDIUM ALLOYS

INDIUM ANTIMONIDES

INDIUM ARSENIDES

INDIUM COMPOUNDS

INDIUM ISOTOPES

INDIUM PHOSPHATES

INDIUM PHOSPHIDES

INDIUM SULFIDES

INDIUM TELLURIDES

INDOLES

INDONESIA

INDONESIAN SPACE PROGRAM

INDOOR AIR POLLUTION

Induced Fluid Flow
USE   FLUID FLOW

Induced Oscillation, Pilot
USE   PILOT INDUCED OSCILLATION

Induced Vibration, Self
USE   SELF INDUCED VIBRATION

Inducers, Helical
USE   HELICAL INDUCERS

INDUCTANCE

INDUCTION

INDUCTION HEATING

Induction, Magnetic
USE   MAGNETIC INDUCTION

INDUCTION (MATHEMATICS)

INDUCTION MOTORS

Induction Probes, Magnetic
USE   MAGNETIC PROBES

Induction Systems
USE   INTAKE SYSTEMS

INDUCTORS

INDUSTRIAL AREAS

INDUSTRIAL ENERGY

INDUSTRIAL MANAGEMENT

INDUSTRIAL PLANTS

INDUSTRIAL SAFETY

INDUSTRIAL WASTES

Industrialization, Space
USE   SPACE INDUSTRIALIZATION

INDUSTRIES

(Industries), Plants
USE   INDUSTRIAL PLANTS

Industry, Aerospace
USE   AEROSPACE INDUSTRY

Industry, Aircraft
USE   AIRCRAFT INDUSTRY

Industry, Construction
USE   CONSTRUCTION INDUSTRY

Industry, Defense
USE   DEFENSE INDUSTRY

(Industry), Logging
USE   LOGGING (INDUSTRY)

(Industry), Process Control
USE   PROCESS CONTROL (INDUSTRY)

Industry, Weapons
USE   WEAPONS INDUSTRY

Inelastic Bodies
USE   RIGID STRUCTURES

INELASTIC COLLISIONS

INELASTIC SCATTERING

INELASTIC STRESS

INEQUALITIES

Inequality, Schwartz
USE   SCHWARTZ INEQUALITY

INERT ATMOSPHERE

Inert Gas Welding, Tungsten
USE   GAS TUNGSTEN ARC WELDING

Inert Gases
USE   RARE GASES

INERTIA

INERTIA BONDING

Inertia Moments
USE   MOMENTS OF INERTIA

Inertia, Moments Of
USE   MOMENTS OF INERTIA

INERTIA PRINCIPLE

Inertia Principle, Mach
USE   MACH INERTIA PRINCIPLE

Inertia Wheels
USE   REACTION WHEELS
        COUNTER-ROTATING WHEELS

INERTIAL CONFINEMENT FUSION

INERTIAL COORDINATES

Inertial Forces
USE   INERTIA

INERTIAL FUSION (REACTOR)

INERTIAL GUIDANCE

Inertial Guidance, Strapdown
USE   STRAPDOWN INERTIAL GUIDANCE

Inertial Measuring Units
USE   INERTIAL PLATFORMS

INERTIAL NAVIGATION

Inertial Navigation, Gimballess
USE   GIMBALLESS INERTIAL NAVIGATION

INERTIAL PLATFORMS

INERTIAL REFERENCE SYSTEMS

INERTIAL UPPER STAGE

INERTIALESS STEERABLE ANTENNAS

INFARCTION

Infarction, Myocardial
USE   MYOCARDIAL INFARCTION

Infection, Airborne
USE   AIRBORNE INFECTION

Infections
USE   INFECTIOUS DISEASES

INFECTIOUS DISEASES

Infeld Theory, Born-
USE   BORN-INFELD THEORY

INFERENCE

INFESTATION

INFILTRATION

INFINITE SPAN WINGS

INFINITY

Inflatable Devices
USE   INFLATABLE STRUCTURES

INFLATABLE GLIDERS

INFLATABLE SPACECRAFT

INFLATABLE STRUCTURES

INFLATING

INFLECTION POINTS

(Inflight), Crew Procedures
USE   CREW PROCEDURES (INFLIGHT)

INFLUENCE COEFFICIENT

Influence Coefficients, Structural
USE   STRUCTURAL INFLUENCE COEFFICIENTS

INFLUENZA

Inform Sys, Atmospheric & Oceanographic
USE   ATMOSPHERIC & OCEANOGRAPHIC
       INFORM SYS

INFORMATION

INFORMATION ADAPTIVE SYSTEM

INFORMATION DISSEMINATION

INFORMATION FLOW

INFORMATION MANAGEMENT

INFORMATION RETRIEVAL

Information Security, Computer
USE   COMPUTER INFORMATION SECURITY

Information, Selective Dissemination Of
USE   SELECTIVE DISSEMINATION OF
       INFORMATION

Information System, Earth Resources
USE   EARTH RESOURCES INFORMATION
       SYSTEM

INFORMATION SYSTEMS

Information Systems, Geographic
USE   GEOGRAPHIC INFORMATION SYSTEMS

Information Systems, Management
USE   MANAGEMENT INFORMATION SYSTEMS

INFORMATION THEORY

Information Theory, Shannon
USE   INFORMATION THEORY

INFORMATION TRANSFER

Information Transmission
USE   DATA TRANSMISSION

INFRARED ABSORPTION

INFRARED ASTRONOMY

INFRARED ASTRONOMY SATELLITE

INFRARED DETECTORS

Infrared Detectors, Forward Looking
USE   FLIR DETECTORS

INFRARED FILTERS

Infrared Horizon Scanners
USE   INFRARED SCANNERS
       HORIZON SCANNERS

INFRARED IMAGERY

INFRARED INSPECTION

INFRARED INSTRUMENTS

INFRARED INTERFEROMETERS

INFRARED LASERS

Infrared Masers
USE   INFRARED LASERS

INFRARED PHOTOGRAPHY

Infrared Photography, Color
USE   COLOR INFRARED PHOTOGRAPHY

INFRARED PHOTOMETRY

INFRARED RADAR

INFRARED RADIATION

Infrared Radiation, Far
USE   FAR INFRARED RADIATION

Infrared Radiation, Near
USE   NEAR INFRARED RADIATION

INFRARED RADIOMETERS

INFRARED REFLECTION

INFRARED SCANNERS

INFRARED SIGNATURES

INFRARED SPECTRA

INFRARED SPECTROMETERS

Infrared Spectrometers, Filter Wheel
USE   FILTER WHEEL INFRARED
       SPECTROMETERS

INFRARED SPECTROPHOTOMETERS

INFRARED SPECTROSCOPY

Infrared Spin Scan Radiometer, Visible
USE   VISIBLE INFRARED SPIN SCAN
       RADIOMETER

INFRARED STARS

INFRARED SUPPRESSION

Infrared Telescope Facility, Space
USE   SPACE INFRARED TELESCOPE FACILITY

Infrared Telescope On Spacelab, Large
USE   LIRTS (TELESCOPE)

INFRARED TELESCOPES

INFRARED TRACKING

INFRARED WINDOWS

INFRASONIC FREQUENCIES

INGESTION

INGESTION (BIOLOGY)

INGESTION (ENGINES)

Ingestion, Spray
USE   SPRAY INGESTION

INGOTS

INGREDIENTS

INGRESS (SPACECRAFT PASSAGEWAY)

INHABITANTS

Inhabitants, Mountain
USE   MOUNTAIN INHABITANTS

Inhalation
USE   RESPIRATION

INHIBITION

Inhibition), Poisoning (Reaction
USE   POISONING (REACTION INHIBITION)

INHIBITION (PSYCHOLOGY)

INHIBITORS

Inhibitors, Wear
USE   WEAR INHIBITORS

INHOMOGENEITY

INHOUR EQUATION

Initial Value Problems
USE   BOUNDARY VALUE PROBLEMS

Initiated Antiaircraft Missiles, Self
USE   SIAM MISSILES

INITIATION

Initiation, Crack
USE   CRACK INITIATION

INITIATORS

INITIATORS (EXPLOSIVES)

INJECTION

Injection, Beam
USE   BEAM INJECTION

Injection Carburetors
USE   CARBURETORS
          FUEL INJECTION

Injection, Carrier
USE   CARRIER INJECTION

Injection Devices, Charge
USE   CHARGE INJECTION DEVICES

Injection, Fluid
USE   FLUID INJECTION

Injection, Fuel
USE   FUEL INJECTION

Injection, Gas
USE   GAS INJECTION

INJECTION GUIDANCE

Injection, Ion
USE   ION INJECTION

INJECTION LASERS

Injection, Liquid
USE   LIQUID INJECTION

INJECTION LOCKING

INJECTION MOLDING

Injection, Secondary
USE   SECONDARY INJECTION

Injection, Transearth
USE   TRANSEARTH INJECTION

Injection Transit Time Diodes, Barrier
USE   BARRITT DIODES

Injection, Translunar
USE   TRANSLUNAR INJECTION

Injection (Wastes), Deep Well
USE   DEEP WELL INJECTION (WASTES)

Injection, Water
USE   WATER INJECTION

INJECTORS

Injectors, Vortex
USE   VORTEX INJECTORS

Injun Explorer
USE   EXPLORER 25 SATELLITE

INJUN SATELLITES

INJUN 1 SATELLITE

INJUN 3 SATELLITE

INJUN 4 SATELLITE

Injun 5 Satellite
USE   EXPLORER 40 SATELLITE

INJURIES

Injuries, Back
USE   BACK INJURIES

(Injuries), Burns
USE   BURNS (INJURIES)

Injuries, Crash
USE   CRASH INJURIES

Injuries, Ejection
USE   EJECTION INJURIES

Injuries, Noise
USE   NOISE INJURIES

Injuries, Radiation
USE   RADIATION INJURIES

Injuries, Whiplash
USE   WHIPLASH INJURIES

Injury, Parachuting
USE   PARACHUTING INJURY

INKS

INLAND WATERS

INLET AIRFRAME CONFIGURATIONS

Inlet (AK), Cook
USE   COOK INLET (AK)

INLET FLOW

INLET NOZZLES

INLET PRESSURE

INLET TEMPERATURE

Inlets, Air
USE   AIR INTAKES

Inlets, Conical
USE   CONICAL INLETS

Inlets (Devices)
USE   INTAKE SYSTEMS

Inlets, Engine
USE   ENGINE INLETS

Inlets, Hypersonic
USE   HYPERSONIC INLETS

Inlets, Internal Compression
USE   INTERNAL COMPRESSION INLETS

Inlets, Nose
USE   NOSE INLETS

Inlets, Side
USE   SIDE INLETS

Inlets, Supersonic
USE   SUPERSONIC INLETS

Inlets, Supersonic Flow
USE   SUPERSONIC INLETS

INLETS (TOPOGRAPHY)

Inlets, Transonic
USE   SUPERSONIC INLETS

INLIERS (LANDFORMS)

INNER RADIATION BELT

INOCULATION

(Inoculation), Seeding
USE   INOCULATION

INOCULUM

(Inorganic), Azides
USE   AZIDES (INORGANIC)

INORGANIC CHEMISTRY

INORGANIC COATINGS

INORGANIC COMPOUNDS

INORGANIC MATERIALS

INORGANIC NITRATES

INORGANIC PEROXIDES

INORGANIC SULFIDES

INOSITOLS

INPUT

INPUT/OUTPUT ROUTINES

INSAT Satellites
USE   INDIAN SPACECRAFT

Insect Damage
USE   INFESTATION

INSECTICIDES

INSECTS

Insensitivity
USE   SENSITIVITY

INSERTION

INSERTION LOSS

INSERTS

Inserts, Nozzle
USE   NOZZLE INSERTS

Inshore Zones
USE   BEACHES

INSOLATION

INSOMNIA

INSPECTION

Inspection, Infrared
USE   INFRARED INSPECTION

Inspection, X Ray
USE   X RAY INSPECTION

INSPECTOR SATELLITE

INSPIRATION

Instability
USE   STABILITY

Instability, Acoustic
USE   ACOUSTIC INSTABILITY

Instability, Baroclinic
USE   BAROCLINIC INSTABILITY

Instability, Combustion
USE   COMBUSTION STABILITY

Instability, Goertler
USE   GOERTLER INSTABILITY

Instability, Kelvin-Helmholtz
USE   KELVIN-HELMHOLTZ INSTABILITY

Instability, Magnetospheric
USE   MAGNETOSPHERIC INSTABILITY

Instability, Plasma
USE   MAGNETOHYDRODYNAMIC STABILITY

Instability, Taylor
USE   TAYLOR INSTABILITY

Instability, Taylor-Goertler
USE   GOERTLER INSTABILITY

Instability, Thermal
USE   THERMAL INSTABILITY

Instability, Weibel
USE   WEIBEL INSTABILITY

Instability, Whirl
USE   ROTARY STABILITY

Installation
USE   INSTALLING

INSTALLATION MANUALS

INSTALLING

INSTANTONS

INSTITUTIONS

Instruction, Computer Assisted
USE　COMPUTER ASSISTED INSTRUCTION

Instruction, Programmed
USE　PROGRAMMED INSTRUCTION

INSTRUCTION SETS (COMPUTERS)

Instructions
USE　EDUCATION

INSTRUCTORS

INSTRUMENT APPROACH

INSTRUMENT COMPENSATION

Instrument Drift
USE　DRIFT (INSTRUMENTATION)

INSTRUMENT ERRORS

INSTRUMENT FLIGHT RULES

INSTRUMENT LANDING SYSTEMS

Instrument Modules, Scientific
USE　SIM

INSTRUMENT ORIENTATION

INSTRUMENT PACKAGES

INSTRUMENT RECEIVERS

INSTRUMENT TRANSFORMERS

INSTRUMENT TRANSMITTERS

Instrumental Analysis
USE　ANALYZING
　　　AUTOMATION

Instrumentation
USE　INSTRUMENTS

Instrumentation Aircraft, Advanced Range
USE　ADVANCED RANGE INSTRUMENTATION
　　　AIRCRAFT

Instrumentation, Bio
USE　BIOINSTRUMENTATION

(Instrumentation), Drift
USE　DRIFT (INSTRUMENTATION)

Instrumentation Facility, Deep Space
USE　DEEP SPACE INSTRUMENTATION FACILITY

(Instrumentation Facility), DSIF
USE　DEEP SPACE INSTRUMENTATION FACILITY

(Instrumentation), Ion Traps
USE　ION TRAPS (INSTRUMENTATION)

Instrumentation, Micro
USE　MICROINSTRUMENTATION

Instrumentation Program, Army-Navy
USE　ARMY-NAVY INSTRUMENTATION PROGRAM

Instrumentation Ship, Advanced Range
USE　ADVANCED RANGE INSTRUMENTATION
　　　SHIP

Instrumentation Ship, ARIS
USE　ADVANCED RANGE INSTRUMENTATION
　　　SHIP

INSTRUMENTS

Instruments, Aircraft
USE　AIRCRAFT INSTRUMENTS

Instruments, Balloon-Borne
USE　BALLOON-BORNE INSTRUMENTS

Instruments, Engine Monitoring
USE　ENGINE MONITORING INSTRUMENTS

Instruments, Flight
USE　FLIGHT INSTRUMENTS

Instruments, Flight Test
USE　FLIGHT TEST INSTRUMENTS

Instruments, Indicating
USE　INDICATING INSTRUMENTS

Instruments, Infrared
USE　INFRARED INSTRUMENTS

Instruments, Landing
USE　LANDING INSTRUMENTS

Instruments, Measuring
USE　MEASURING INSTRUMENTS

Instruments, Meteorological
USE　METEOROLOGICAL INSTRUMENTS

Instruments, Navigation
USE　NAVIGATION INSTRUMENTS

Instruments, Optical Measuring
USE　OPTICAL MEASURING INSTRUMENTS

Instruments, Plotting
USE　PLOTTERS

(Instruments), Potentiometers
USE　POTENTIOMETERS (INSTRUMENTS)

Instruments, Propellant Actuated
USE　PROPELLANT ACTUATED INSTRUMENTS

Instruments, Radiation Measuring
USE　RADIATION MEASURING INSTRUMENTS

Instruments, Recording
USE　RECORDING INSTRUMENTS

Instruments, Rocket-Borne
USE　ROCKET-BORNE INSTRUMENTS

Instruments, Satellite
USE　SATELLITE INSTRUMENTS

Instruments, Satellite-Borne
USE　SATELLITE-BORNE INSTRUMENTS

Instruments, Shock Measuring
USE　SHOCK MEASURING INSTRUMENTS

Instruments, Solar
USE　SOLAR INSTRUMENTS

Instruments, Spacecraft
USE　SPACECRAFT INSTRUMENTS

Instruments, Surgical
USE　SURGICAL INSTRUMENTS

Instruments, Temperature
USE　TEMPERATURE MEASURING INSTRUMENTS

Instruments, Temperature Measuring
USE　TEMPERATURE MEASURING INSTRUMENTS

Instruments, Time Measuring
USE　TIME MEASURING INSTRUMENTS

Instruments, Turbine
USE　TURBINE INSTRUMENTS

INSULATED STRUCTURES

Insulating Materials
USE　INSULATION

INSULATION

Insulation, Electrical
USE　ELECTRICAL INSULATION

Insulation, Multilayer
USE　MULTILAYER INSULATION

Insulation, Thermal
USE　THERMAL INSULATION

Insulator Semiconductors, Metal
USE　MIS (SEMICONDUCTORS)

Insulator Semiconductors, Semiconductor
USE　SIS (SEMICONDUCTORS)

Insulator-Metal Diodes, Metal-
USE　MIM DIODES

Insulator-Metal Semiconductors, Metal-
USE　MIM (SEMICONDUCTORS)

INSULATORS

INSULIN

Intake, Food
USE　FOOD INTAKE

INTAKE SYSTEMS

Intakes, Air
USE　AIR INTAKES

Intakes, Water
USE　WATER INTAKES

INTASAT SATELLITE

Integ Circuits, Diode-Transistor-Logic
USE　DTL INTEGRATED CIRCUITS

Integ Circuits, Transistor-Transistor-Logic
USE　TTL INTEGRATED CIRCUITS

Integ Med And Behavioral Lab Measur System
USE　IMBLMS

Integ Program For Aerospace Veh Design
USE　IPAD

INTEGERS

INTEGRAL CALCULUS

INTEGRAL EQUATIONS

Integral Equations, Singular
USE　SINGULAR INTEGRAL EQUATIONS

Integral Formula, Cauchy
USE　CAUCHY INTEGRAL FORMULA

Integral Functions
USE　ENTIRE FUNCTIONS

Integral, J
USE　J INTEGRAL

Integral, Jacobi
USE　JACOBI INTEGRAL

Integral Method, Boundary
USE　BOUNDARY INTEGRAL METHOD

Integral, Phase-Space
USE　PHASE-SPACE INTEGRAL

Integral, Riemann
USE　MEASURE AND INTEGRATION

INTEGRAL ROCKET RAMJETS

Integral, Stieltjes
USE　STIELTJES INTEGRAL

INTEGRAL TRANSFORMATIONS

INTEGRALS

Integrals, Convolution
USE　CONVOLUTION INTEGRALS

Integrals, Elliptic
USE　ELLIPTIC FUNCTIONS

Integrals, Fresnel
USE　FRESNEL INTEGRALS

Integrals, Fresnel-Kirchhoff
USE　FRESNEL INTEGRALS

Integrals, Transform
USE   INTEGRAL TRANSFORMATIONS

INTEGRATED CIRCUITS

Integrated Circuits, DTL
USE   DTL INTEGRATED CIRCUITS

Integrated Circuits, Linear
USE   LINEAR INTEGRATED CIRCUITS

Integrated Circuits, TTL
USE   TTL INTEGRATED CIRCUITS

Integrated Circuits, Very High Speed
USE   VHSIC (CIRCUITS)

Integrated Control Project, Submarine
USE   SUBMARINE INTEGRATED CONTROL
        PROJECT

INTEGRATED ENERGY SYSTEMS

INTEGRATED GLOBAL OCEAN STATION SYSTEMS

INTEGRATED LIBRARY SYSTEMS

Integrated Maneuvering Life Support Sys
USE   IMLSS

INTEGRATED MISSION CONTROL CENTER

INTEGRATED OPTICS

Integrated Reconnaissance System, Airborne
USE   AIRBORNE INTEGRATED
        RECONNAISSANCE SYSTEM

Integrated Utility System, Modular
USE   MODULAR INTEGRATED UTILITY SYSTEM

Integration, Binary
USE   BINARY INTEGRATION

Integration, Data
USE   DATA INTEGRATION

Integration, Engine Airframe
USE   ENGINE AIRFRAME INTEGRATION

Integration, Functional
USE   FUNCTIONAL INTEGRATION

Integration Laboratory, Shuttle Avionics
USE   SAIL PROJECT

Integration, Large Scale
USE   LARGE SCALE INTEGRATION

Integration, Measure And
USE   MEASURE AND INTEGRATION

Integration, Medium Scale
USE   MEDIUM SCALE INTEGRATION

Integration, Numerical
USE   NUMERICAL INTEGRATION

Integration, Payload
USE   PAYLOAD INTEGRATION

Integration Plan, Payload
USE   PAYLOAD INTEGRATION PLAN

Integration (Real Variables)
USE   MEASURE AND INTEGRATION

Integration, Systems
USE   SYSTEMS INTEGRATION

Integration, Very Large Scale
USE   VERY LARGE SCALE INTEGRATION

INTEGRATORS

Integrators, Digital
USE   DIGITAL INTEGRATORS

INTEGRITY

Integrity, Computer Program
USE   COMPUTER PROGRAM INTEGRITY

Integrodifferential Equations
USE   DIFFERENTIAL EQUATIONS
        INTEGRAL EQUATIONS

INTEL 8080 MICROPROCESSOR

INTELLECT

INTELLIGENCE

Intelligence, Artificial
USE   ARTIFICIAL INTELLIGENCE

Intelligence, Extraterrestrial
USE   EXTRATERRESTRIAL INTELLIGENCE

Intelligence, Search For Extraterrestrial
USE   PROJECT SETI

INTELLIGIBILITY

INTELSAT SATELLITES

Intensification
USE   AMPLIFICATION

Intensifier Tubes
USE   IMAGE INTENSIFIERS

INTENSIFIERS

Intensifiers, Image
USE   IMAGE INTENSIFIERS

INTENSITY

Intensity, Electron
USE   ELECTRON FLUX DENSITY

Intensity Factors, Stress
USE   STRESS INTENSITY FACTORS

Intensity Lasers, High
USE   HIGH POWER LASERS

Intensity, Light
USE   LUMINOUS INTENSITY

Intensity, Luminescent
USE   LUMINOUS INTENSITY

Intensity, Luminous
USE   LUMINOUS INTENSITY

Intensity, Magnetic Field
USE   MAGNETIC FLUX

Intensity Meters, Field
USE   FIELD INTENSITY METERS

Intensity, Noise
USE   NOISE INTENSITY

Intensity, Particle
USE   PARTICLE INTENSITY

Intensity, Radiant
USE   RADIANT FLUX DENSITY

Intensity, Radiation
USE   RADIANT FLUX DENSITY

Intensity, Sound
USE   SOUND INTENSITY

Intensity X Ray Imaging Scope, Low
USE   LIXISCOPES

Interaction, Configuration
USE   CONFIGURATION INTERACTION

Interaction Experiment, Plasma
USE   PLASMA INTERACTION EXPERIMENT

Interaction Experiments, Space Plasma H/v
USE   SPHINX

Interaction, Flame
USE   FLAME PROPAGATION
        CHEMICAL REACTIONS

Interaction, Photon-Electron
USE   PHOTON-ELECTRON INTERACTION

Interaction, Plasma-Electromagnetic
USE   PLASMA-ELECTROMAGNETIC INTERACTION

Interaction, Shock Wave
USE   SHOCK WAVE INTERACTION

Interaction, Wave
USE   WAVE INTERACTION

INTERACTIONAL AERODYNAMICS

INTERACTIONS

Interactions, Air Land
USE   AIR LAND INTERACTIONS

Interactions, Air Sea
USE   AIR WATER INTERACTIONS

Interactions, Air Sea Ice
USE   AIR SEA ICE INTERACTIONS

Interactions, Air Water
USE   AIR WATER INTERACTIONS

Interactions, Atomic
USE   ATOMIC INTERACTIONS

Interactions, Beam
USE   BEAM INTERACTIONS

Interactions, Beta
USE   WEAK INTERACTIONS (FIELD THEORY)

Interactions, Electromagnetic
USE   ELECTROMAGNETIC INTERACTIONS

Interactions, Electron
USE   ELECTRON SCATTERING

Interactions, Electron Phonon
USE   ELECTRON PHONON INTERACTIONS

Interactions, Elementary Particle
USE   ELEMENTARY PARTICLE INTERACTIONS

Interactions (Field Theory), Strong
USE   STRONG INTERACTIONS (FIELD THEORY)

Interactions (Field Theory), Weak
USE   WEAK INTERACTIONS (FIELD THEORY)

Interactions, Fluid-Solid
USE   FLUID-SOLID INTERACTIONS

Interactions, Gas-Gas
USE   GAS-GAS INTERACTIONS

Interactions, Gas-Ion
USE   GAS-ION INTERACTIONS

Interactions, Gas-Liquid
USE   GAS-LIQUID INTERACTIONS

Interactions, Gas-Metal
USE   GAS-METAL INTERACTIONS

Interactions, Gas-Solid
USE   GAS-SOLID INTERACTIONS

Interactions, High Energy
USE   HIGH ENERGY INTERACTIONS

Interactions, Ion Atom
USE   ION ATOM INTERACTIONS

Interactions, Ion-Gas
USE   GAS-ION INTERACTIONS

Interactions, Laser Plasma
USE   LASER PLASMA INTERACTIONS

Interactions, Laser Target
USE   LASER TARGET INTERACTIONS

Interactions, Man Environment
USE   MAN ENVIRONMENT INTERACTIONS

Interactions, Meson-Meson
USE   MESON-MESON INTERACTIONS

Interactions, Meson-Nucleon
USE   MESON-NUCLEON INTERACTIONS

Interactions, Molecular
USE   MOLECULAR INTERACTIONS

Interactions, Nuclear
USE   NUCLEAR INTERACTIONS

Interactions, Nucleon-Nucleon
USE   NUCLEON-NUCLEON INTERACTIONS

Interactions, Particle
USE   PARTICLE INTERACTIONS

Interactions, Plasma
USE   PLASMA INTERACTIONS

Interactions, Plasma-Particle
USE   PLASMA-PARTICLE INTERACTIONS

Interactions, Rotor Body
USE   ROTOR BODY INTERACTIONS

Interactions, Solar Planetary
USE   SOLAR PLANETARY INTERACTIONS

Interactions, Solar Terrestrial
USE   SOLAR TERRESTRIAL INTERACTIONS

Interactions, Sound-Sound
USE   SOUND-SOUND INTERACTIONS

Interactions, Spin-Orbit
USE   SPIN-ORBIT INTERACTIONS

Interactions, Surface
USE   SURFACE REACTIONS

Interactions, Surface Noise
USE   SURFACE NOISE INTERACTIONS

Interactions, Weak Energy
USE   WEAK ENERGY INTERACTIONS

INTERACTIVE CONTROL

Interactive Graphics
USE   COMPUTER GRAPHICS

Interactive Planning System, NASA
USE   NASA INTERACTIVE PLANNING SYSTEM

INTERATOMIC FORCES

INTERCALATION

INTERCEPTION

Interceptor Aircraft
USE   FIGHTER AIRCRAFT

INTERCEPTORS

Interceptors, Satellite
USE   SATELLITE INTERCEPTORS

Interconnection
USE   JOINING

INTERCONTINENTAL BALLISTIC MISSILES

INTERCOSMOS SATELLITES

INTERCRANIAL CIRCULATION

INTERDIGITAL TRANSDUCERS

INTERFACE STABILITY

INTERFACES

Interfaces, Gas-Solid
USE   GAS-SOLID INTERFACES

Interfaces, Liquid-Liquid
USE   LIQUID-LIQUID INTERFACES

Interfaces, Liquid-Solid
USE   LIQUID-SOLID INTERFACES

Interfaces, Liquid-Vapor
USE   LIQUID-VAPOR INTERFACES

Interfaces, Solid-Solid
USE   SOLID-SOLID INTERFACES

INTERFACIAL ENERGY

Interfacial Strain
USE   INTERFACIAL TENSION

INTERFACIAL TENSION

INTERFERENCE

Interference, Aerodynamic
USE   AERODYNAMIC INTERFERENCE

INTERFERENCE DRAG

Interference, Electromagnetic
USE   ELECTROMAGNETIC INTERFERENCE

INTERFERENCE FACTOR TABLE

INTERFERENCE GRATING

INTERFERENCE IMMUNITY

Interference, Intersymbolic
USE   INTERSYMBOLIC INTERFERENCE

INTERFERENCE LIFT

Interference Monochromatization
USE   MONOCHROMATIZATION
       DIFFRACTION

Interference, Radio
USE   RADIO FREQUENCY INTERFERENCE

Interference, Radio Frequency
USE   RADIO FREQUENCY INTERFERENCE

Interference, Support
USE   SUPPORT INTERFERENCE

Interferograms
USE   INTERFEROMETRY

INTERFEROMETERS

Interferometers, Fabry-Perot
USE   FABRY-PEROT INTERFEROMETERS

Interferometers, Infrared
USE   INFRARED INTERFEROMETERS

Interferometers, Mach-Zehnder
USE   MACH-ZEHNDER INTERFEROMETERS

Interferometers, Michelson
USE   MICHELSON INTERFEROMETERS

Interferometers, Microwave
USE   MICROWAVE INTERFEROMETERS

Interferometers, Phase Switching
USE   PHASE SWITCHING INTERFEROMETERS

Interferometers, Radio
USE   RADIO INTERFEROMETERS

Interferometers, Superconducting Quantum
USE   SQUID (DETECTORS)

INTERFEROMETRY

Interferometry, Differential
USE   DIFFERENTIAL INTERFEROMETRY

Interferometry, Holographic
USE   HOLOGRAPHIC INTERFEROMETRY

Interferometry, Laser
USE   LASER INTERFEROMETRY

Interferometry, Moire
USE   MOIRE INTERFEROMETRY

Interferometry Network), Orion (Radio
USE   ORION (RADIO INTERFEROMETRY
       NETWORK)

Interferometry, Very Long Base
USE   VERY LONG BASE INTERFEROMETRY

INTERFERON

INTERGALACTIC MEDIA

INTERGRANULAR CORROSION

INTERIM STAGES (SPACECRAFT)

Interim Upper Stage (STS)
USE   INERTIAL UPPER STAGE

INTERIOR BALLISTICS

Interlacing Drainage
USE   DRAINAGE PATTERNS

INTERLAYERS

Interlocking
USE   LOCKING

INTERMEDIATE FREQUENCIES

INTERMEDIATE FREQUENCY AMPLIFIERS

INTERMEDIATE RANGE BALLISTIC MISSILES

INTERMETALLICS

INTERMITTENCY

INTERMITTENCY HYPOTHESIS

INTERMODULATION

INTERMOLECULAR FORCES

Intermontane Floors
USE   VALLEYS

INTERNAL COMBUSTION ENGINES

INTERNAL COMPRESSION INLETS

INTERNAL CONVERSION

INTERNAL ENERGY

INTERNAL FRICTION

INTERNAL PRESSURE

Internal Stress
USE   RESIDUAL STRESS

INTERNAL WAVES

International Computers Limited
USE   ICL COMPUTERS

INTERNATIONAL COOPERATION

INTERNATIONAL FIELD YEAR FOR GREAT LAKES

INTERNATIONAL GEOPHYSICAL YEAR

INTERNATIONAL GEOSPHERE-BIOSPHERE
PROGRAM

INTERNATIONAL HYDROLOGICAL DECADE

INTERNATIONAL LAW

INTERNATIONAL MAGNETOSPHERIC EXPLORER

INTERNATIONAL MAGNETOSPHERIC STUDY

International Practical Temperature
    USE   TEMPERATURE SCALES

INTERNATIONAL QUIET SUN YEAR

INTERNATIONAL RELATIONS

INTERNATIONAL SATELLITE GEODESY EXPERIMENT

International Sats For Ionospheric Study
    USE   ISIS SATELLITES

International Solar Polar Mission
    USE   ULYSSES MISSION

INTERNATIONAL SUN EARTH EXPLORER 1

INTERNATIONAL SUN EARTH EXPLORER 2

INTERNATIONAL SUN EARTH EXPLORER 3

INTERNATIONAL SUN EARTH EXPLORERS

INTERNATIONAL SYSTEM OF UNITS

INTERNATIONAL TRADE

International Ultraviolet Explorer
    USE   IUE

(International Year), IQSY
    USE   INTERNATIONAL QUIET SUN YEAR

INTERNUCLEAR PROPERTIES

INTERORBITAL TRAJECTORIES

Interpersonal Relations
    USE   HUMAN RELATIONS

INTERPHONES

INTERPLANETARY COMMUNICATION

INTERPLANETARY DUST

Interplanetary Explorer
    USE   EXPLORER 18 SATELLITE

INTERPLANETARY FLIGHT

INTERPLANETARY GAS

INTERPLANETARY MAGNETIC FIELDS

INTERPLANETARY MEDIUM

Interplanetary Monitoring Platform
    USE   IMP

INTERPLANETARY NAVIGATION

Interplanetary Propulsion
    USE   INTERPLANETARY SPACECRAFT
          ROCKET ENGINES

INTERPLANETARY SPACE

INTERPLANETARY SPACECRAFT

INTERPLANETARY TRAJECTORIES

INTERPLANETARY TRANSFER ORBITS

INTERPOLATION

Interpolators
    USE   REPEATERS

INTERPRETATION

Interpretation, Photo
    USE   PHOTOINTERPRETATION

Interpretation, Photograph
    USE   PHOTOINTERPRETATION

INTERPROCESSOR COMMUNICATION

Interrelationships
    USE   RELATIONSHIPS

INTERROGATION

INTERRUPTION

INTERSECTIONS

INTERSERVICE DATA EXCHANGE PROGRAM

INTERSTELLAR CHEMISTRY

INTERSTELLAR COMMUNICATION

INTERSTELLAR EXTINCTION

INTERSTELLAR GAS

INTERSTELLAR MAGNETIC FIELDS

INTERSTELLAR MASERS

INTERSTELLAR MATTER

Interstellar Microwave Spectra
    USE   INTERSTELLAR RADIATION
          MICROWAVE SPECTRA

INTERSTELLAR RADIATION

Interstellar Reddening
    USE   INTERSTELLAR EXTINCTION

INTERSTELLAR SPACE

INTERSTELLAR SPACECRAFT

INTERSTELLAR TRAVEL

INTERSTICES

INTERSTITIALS

INTERSYMBOLIC INTERFERENCE

INTERTROPICAL CONVERGENT ZONES

Interval Scanners, Multiple Beam
    USE   MULTIPLE BEAM INTERVAL SCANNERS

INTERVALS

(Intervals), Windows
    USE   WINDOWS (INTERVALS)

Intervehicle Spacecrew Transfer
    USE   SPACECREW TRANSFER

INTERVERTEBRAL DISKS

INTESTINES

INTOXICATION

INTRACRANIAL CAVITY

INTRACRANIAL PRESSURE

INTRAMOLECULAR STRUCTURES

INTRAOCULAR PRESSURE

INTRAORBIT TRANSFER VEHICLES

Intratheater Transport, Light
    USE   LIGHT INTRATHEATER TRANSPORT

INTRAVASCULAR SYSTEM

INTRAVEHICULAR ACTIVITY

INTRAVENOUS PROCEDURES

INTROVERSION

Intruder Aircraft
    USE   A-6 AIRCRAFT

INTRUSION

Intrusions, Rock
    USE   ROCK INTRUSIONS

Invader Aircraft
    USE   B-26 AIRCRAFT

Invalidity
    USE   ERRORS

INVARIANCE

Invariance, Gauge
    USE   GAUGE INVARIANCE

INVARIANT IMBEDDINGS

INVENTIONS

INVENTORIES

Inventories By Remote Sensing, Crop
    USE   AGRISTARS PROJECT

Inventories, Crop
    USE   CROP INVENTORIES

INVENTORY CONTROLS

Inventory Experiment, Large Area Crop
    USE   LARGE AREA CROP INVENTORY
          EXPERIMENT

INVENTORY MANAGEMENT

Inventory, Timber
    USE   TIMBER INVENTORY

INVERSE SCATTERING

Inversion, Population
    USE   POPULATION INVERSION

INVERSIONS

Inversions, Magnetic Field
    USE   MAGNETIC FIELD INVERSIONS

Inversions, Temperature
    USE   TEMPERATURE INVERSIONS

INVERTEBRATES

INVERTED CONVERTERS (DC TO AC)

INVERTERS

Inverters, Static
    USE   STATIC INVERTERS

INVESTIGATION

Investigation, Accident
    USE   ACCIDENT INVESTIGATION

Investigation, Aircraft Accident
    USE   AIRCRAFT ACCIDENT INVESTIGATION

INVESTMENT

INVESTMENT CASTING

INVESTMENTS

INVISCID FLOW

Invisibility
    USE   VISIBILITY

Involuntariness
    USE   INVOLUNTARY ACTIONS

INVOLUNTARY ACTIONS

IO

IODATES

Iodates, Lithium
    USE   LITHIUM IODATES

IODIDES

Iodides, Cesium
USE   CESIUM IODIDES

Iodides, Hafnium
USE   HAFNIUM IODIDES

Iodides, Niobium
USE   NIOBIUM IODIDES

Iodides, Potassium
USE   POTASSIUM IODIDES

Iodides, Silver
USE   SILVER IODIDES

Iodides, Sodium
USE   SODIUM IODIDES

Iodides, Zirconium
USE   ZIRCONIUM IODIDES

IODIMETRY

IODINE

IODINE COMPOUNDS

IODINE ISOTOPES

IODINE LASERS

IODINE 125

IODINE 131

IODINE 132

IODOACETIC ACID

ION ACCELERATORS

ION ACOUSTIC WAVES

ION ATOM INTERACTIONS

ION BEAMS

Ion Chambers
USE   IONIZATION CHAMBERS

ION CHARGE

Ion Clouds, Barium
USE   BARIUM ION CLOUDS

ION CONCENTRATION

ION CURRENTS

ION CYCLOTRON RADIATION

ION DENSITY (CONCENTRATION)

Ion Density, Ionospheric
USE   IONOSPHERIC ION DENSITY

Ion Density, Magnetospheric
USE   MAGNETOSPHERIC ION DENSITY

ION DISTRIBUTION

ION EMISSION

ION ENGINES

Ion Engines, Mercury
USE   MERCURY ION ENGINES

ION EXCHANGE MEMBRANE ELECTROLYTES

ION EXCHANGE RESINS

ION EXCHANGING

ION EXTRACTION

Ion Gages
USE   IONIZATION GAGES

ION IMPACT

ION IMPLANTATION

ION INJECTION

Ion Interactions, Gas-
USE   GAS-ION INTERACTIONS

ION IRRADIATION

Ion Mass Spectrometers, Retarding
USE   MASS SPECTROMETERS

ION MICROSCOPES

ION MOTION

Ion Oscillation
USE   PLASMA OSCILLATIONS

ION PLATING

ION PROBES

ION PRODUCTION RATES

ION PROPULSION

ION PUMPS

ION RECOMBINATION

Ion Recombination, Electron-
USE   ELECTRON-ION RECOMBINATION

ION SCATTERING

ION SELECTIVE ELECTRODES

ION SHEATHS

ION SOURCES

Ion Spectrometers
USE   MASS SPECTROMETERS

ION STORAGE

ION STRIPPING

ION TEMPERATURE

Ion Thrustor Engines, Radio Frequency
USE   RIT ENGINES

ION TRAPS (INSTRUMENTATION)

Ion-Gas Interactions
USE   GAS-ION INTERACTIONS

IONIC COLLISIONS

Ionic Conductivity
USE   ION CURRENTS

IONIC CRYSTALS

IONIC DIFFUSION

IONIC MOBILITY

Ionic Propellants
USE   ION ENGINES

IONIC REACTIONS

IONIC WAVES

IONIZATION

Ionization, Atmospheric
USE   ATMOSPHERIC IONIZATION

Ionization, Auroral
USE   AURORAL IONIZATION

Ionization, Auto
USE   AUTOIONIZATION

IONIZATION CHAMBERS

IONIZATION COEFFICIENTS

Ionization Counters
USE   IONIZATION CHAMBERS
        RADIATION COUNTERS

IONIZATION CROSS SECTIONS

Ionization, De
USE   DEIONIZATION

Ionization, Electron
USE   IONIZATION

Ionization, Flame
USE   FLAME IONIZATION

IONIZATION FREQUENCIES

IONIZATION GAGES

Ionization Gages, Bayard-Alpert
USE   BAYARD-ALPERT IONIZATION GAGES

Ionization Gages, Philips
USE   PHILIPS IONIZATION GAGES

Ionization, Gas
USE   GAS IONIZATION

Ionization, Meteoritic
USE   ATMOSPHERIC IONIZATION
        METEOR TRAILS

Ionization, Nonequilibrium
USE   NONEQUILIBRIUM IONIZATION

Ionization, Photo
USE   PHOTOIONIZATION

IONIZATION POTENTIALS

Ionization, Surface
USE   SURFACE IONIZATION

IONIZED GASES

Ionized Plasmas
USE   PLASMAS (PHYSICS)

IONIZERS

IONIZING RADIATION

IONOGRAMS

IONOPAUSE

IONOSONDES

IONOSPHERE

Ionosphere Beacon, Polar
USE   BEACON SATELLITES

Ionosphere Explorer A
USE   EXPLORER 20 SATELLITE

Ionosphere, Lower
USE   LOWER IONOSPHERE

Ionosphere, Lunar
USE   LUNAR ATMOSPHERE

Ionosphere, Upper
USE   UPPER IONOSPHERE

Ionospheric Absorption
USE   ELECTROMAGNETIC ABSORPTION
        IONOSPHERIC PROPAGATION

Ionospheric Blackout
USE   BLACKOUT (PROPAGATION)

IONOSPHERIC COMPOSITION

IONOSPHERIC CONDUCTIVITY

IONOSPHERIC CROSS MODULATION

IONOSPHERIC CURRENTS

IONOSPHERIC DISTURBANCES

(Ionospheric Disturbances), SID
  USE   SUDDEN IONOSPHERIC DISTURBANCES

Ionospheric Disturbances, Sudden
  USE   SUDDEN IONOSPHERIC DISTURBANCES

Ionospheric Disturbances, Traveling
  USE   TRAVELING IONOSPHERIC DISTURBANCES

IONOSPHERIC DRIFT

IONOSPHERIC ELECTRON DENSITY

IONOSPHERIC F-SCATTER PROPAGATION

IONOSPHERIC HEATING

IONOSPHERIC ION DENSITY

IONOSPHERIC NOISE

IONOSPHERIC PROPAGATION

Ionospheric Reflection
  USE   IONOSPHERIC PROPAGATION

Ionospheric Sounder, Orbiting Radio Beacon
  USE   ORBIS

IONOSPHERIC SOUNDING

IONOSPHERIC STORMS

Ionospheric Study, International Sats For
  USE   ISIS SATELLITES

IONOSPHERIC TEMPERATURE

IONOSPHERIC TILTS

IONOSPHERICS

IONS

Ions, An
  USE   ANIONS

Ions, Cat
  USE   CATIONS

Ions, Cesium
  USE   CESIUM IONS

Ions, Ferric
  USE   FERRIC IONS

Ions, Formyl
  USE   FORMYL IONS

Ions, Heavy
  USE   HEAVY IONS

Ions, Helium
  USE   HELIUM IONS

Ions, Hydrogen
  USE   HYDROGEN IONS

Ions, Hydronium
  USE   HYDRONIUM IONS

Ions, Light
  USE   LIGHT IONS

Ions, Manganese
  USE   MANGANESE IONS

Ions, Metal
  USE   METAL IONS

Ions, Molecular
  USE   MOLECULAR IONS

Ions, Negative
  USE   NEGATIVE IONS

Ions, Nitrogen
  USE   NITROGEN IONS

Ions, Oxygen
  USE   OXYGEN IONS

Ions, Positive
  USE   POSITIVE IONS

Ions, Recoil
  USE   RECOIL IONS

Ions, Trivalent
  USE   TRIVALENT IONS

IOWA

IP (Impact Prediction)
  USE   COMPUTERIZED SIMULATION

IPAD

IQSY (International Year)
  USE   INTERNATIONAL QUIET SUN YEAR

Ir
  USE   IRIDIUM

IRAN

IRAQ

IRAS
  USE   INFRARED ASTRONOMY SATELLITE

IRAS-ARAKI-ALCOCK COMET

Irasers
  USE   INFRARED LASERS

IRBM (Missiles)
  USE   INTERMEDIATE RANGE BALLISTIC
          MISSILES

IRELAND

IRIDESCENCE

IRIDIUM

IRIDIUM ISOTOPES

IRIS SATELLITES

IRISES (MECHANICAL APERTURES)

IRON

IRON ALLOYS

Iron Batteries, Nickel
  USE   NICKEL IRON BATTERIES

IRON CHLORIDES

IRON COMPOUNDS

IRON CYANIDES

Iron Garnet, Yttrium-
  USE   YTTRIUM-IRON GARNET

IRON ISOTOPES

IRON METEORITES

IRON ORES

IRON OXIDES

IRON 57

IRON 58

IRON 59

Iroquois Helicopter
  USE   UH-1 HELICOPTER

Iroquois Rocket Vehicle, Nike-
  USE   NIKE-IROQUOIS ROCKET VEHICLE

IRRADIANCE

IRRADIATION

Irradiation, Auroral
  USE   AURORAL IRRADIATION

Irradiation, Deuteron
  USE   DEUTERON IRRADIATION

Irradiation, Electron
  USE   ELECTRON IRRADIATION

Irradiation, Ion
  USE   ION IRRADIATION

Irradiation, Neutron
  USE   NEUTRON IRRADIATION

Irradiation, Proton
  USE   PROTON IRRADIATION

Irradiation, X Ray
  USE   X RAY IRRADIATION

IRRATIONALITY

IRREGULARITIES

IRREVERSIBLE PROCESSES

IRRIGATION

IRRITATION

Irrotational Flow
  USE   POTENTIAL FLOW

IRS (Indian Spacecraft)
  USE   INDIAN SPACECRAFT

ISAGEX
  USE   INTERNATIONAL SATELLITE GEODESY
          EXPERIMENT

ISCHEMIA

ISEE
  USE   INTERNATIONAL SUN EARTH EXPLORERS

ISENTROPE

ISENTROPIC PROCESSES

Ising Model
  USE   MATHEMATICAL MODELS
          FERROMAGNETISM

ISIS SATELLITES

ISIS-A

ISIS-B

ISIS-X

Iskra Aircraft
  USE   TS-11 AIRCRAFT

ISLAND ARCS

Island (FL), Merritt
  USE   MERRITT ISLAND (FL)

Island, Johnston
  USE   JOHNSTON ISLAND

Island (MD-VA), Assateague
  USE   ASSATEAGUE ISLAND (MD-VA)

(Island), New Guinea
  USE   NEW GUINEA (ISLAND)

Island (NY), Long
  USE   LONG ISLAND (NY)

Island, Prince Edward
  USE   PRINCE EDWARD ISLAND

Island, Rhode
  USE   RHODE ISLAND

Island Sound (RI), Block
USE   BLOCK ISLAND SOUND (RI)

Island, Wallops
USE   WALLOPS ISLAND

ISLANDS

Islands, Heat
USE   HEAT ISLANDS

(Islands), Keys
USE   KEYS (ISLANDS)

Islands, Kurile
USE   KURILE ISLANDS

Islands, Maldive
USE   MALDIVE ISLANDS

Islands, Pacific
USE   PACIFIC ISLANDS

Islands (US), Aleutian
USE   ALEUTIAN ISLANDS (US)

Islands, Virgin
USE   VIRGIN ISLANDS

ISOBARS

Isobars, Nuclear
USE   NUCLEAR ISOBARS

ISOBARS (PRESSURE)

Isobutane
USE   BUTANES

Isobutylene
USE   BUTENES

ISOCHORIC PROCESSES

ISOCHROMATICS

Isochronous Cyclotron, Oak Ridge
USE   OAK RIDGE ISOCHRONOUS CYCLOTRON

ISOCYANATES

Isocyanates, Di
USE   DIISOCYANATES

ISOELECTRONIC SEQUENCE

ISOENERGETIC PROCESSES

ISOLATION

Isolation, Social
USE   SOCIAL ISOLATION

ISOLATORS

Isolators, Vibration
USE   VIBRATION ISOLATORS

ISOMERIZATION

ISOMERS

ISOMORPHISM

ISOPARAMETRIC FINITE ELEMENTS

ISOPERIMETRIC PROBLEM

ISOPHOTES

Isopleths
USE   NOMOGRAPHS

ISOPROPYL ALCOHOL

ISOPROPYL COMPOUNDS

ISOPROPYL NITRATE

ISOPYCNIC PROCESSES

ISOSTASY

ISOSTATIC PRESSURE

Isosteric Processes
USE   ISOPYCNIC PROCESSES

ISOTENSOID STRUCTURES

ISOTHERMAL FLOW

ISOTHERMAL LAYERS

ISOTHERMAL PROCESSES

ISOTHERMS

ISOTONICITY

ISOTOPE EFFECT

Isotope Reactors, High Flux
USE   HIGH FLUX ISOTOPE REACTORS

ISOTOPE SEPARATION

Isotope Shift
USE   ISOTOPE EFFECT

ISOTOPES

Isotopes, Aluminum
USE   ALUMINUM ISOTOPES

Isotopes, Americium
USE   AMERICIUM ISOTOPES

Isotopes, Antimony
USE   ANTIMONY ISOTOPES

Isotopes, Argon
USE   ARGON ISOTOPES

Isotopes, Arsenic
USE   ARSENIC ISOTOPES

Isotopes, Astatine
USE   ASTATINE ISOTOPES

Isotopes, Barium
USE   BARIUM ISOTOPES

Isotopes, Beryllium
USE   BERYLLIUM ISOTOPES

Isotopes, Bismuth
USE   BISMUTH ISOTOPES

Isotopes, Boron
USE   BORON ISOTOPES

Isotopes, Bromine
USE   BROMINE ISOTOPES

Isotopes, Cadmium
USE   CADMIUM ISOTOPES

Isotopes, Calcium
USE   CALCIUM ISOTOPES

Isotopes, Californium
USE   CALIFORNIUM ISOTOPES

Isotopes, Carbon
USE   CARBON ISOTOPES

Isotopes, Cerium
USE   CERIUM ISOTOPES

Isotopes, Cesium
USE   CESIUM ISOTOPES

Isotopes, Chromium
USE   CHROMIUM ISOTOPES

Isotopes, Cobalt
USE   COBALT ISOTOPES

Isotopes, Copper
USE   COPPER ISOTOPES

Isotopes, Curium
USE   CURIUM ISOTOPES

Isotopes, Dysprosium
USE   DYSPROSIUM ISOTOPES

Isotopes, Erbium
USE   ERBIUM ISOTOPES

Isotopes, Europium
USE   EUROPIUM ISOTOPES

Isotopes, Fluorine
USE   FLUORINE ISOTOPES

Isotopes, Gadolinium
USE   GADOLINIUM ISOTOPES

Isotopes, Gallium
USE   GALLIUM ISOTOPES

Isotopes, Germanium
USE   GERMANIUM ISOTOPES

Isotopes, Gold
USE   GOLD ISOTOPES

Isotopes, Hafnium
USE   HAFNIUM ISOTOPES

Isotopes, Helium
USE   HELIUM ISOTOPES

Isotopes, Holmium
USE   HOLMIUM ISOTOPES

Isotopes, Hydrogen
USE   HYDROGEN ISOTOPES

Isotopes, Indium
USE   INDIUM ISOTOPES

Isotopes, Iodine
USE   IODINE ISOTOPES

Isotopes, Iridium
USE   IRIDIUM ISOTOPES

Isotopes, Iron
USE   IRON ISOTOPES

Isotopes, Krypton
USE   KRYPTON ISOTOPES

Isotopes, Lanthanum
USE   LANTHANUM ISOTOPES

Isotopes, Lead
USE   LEAD ISOTOPES

Isotopes, Lithium
USE   LITHIUM ISOTOPES

Isotopes, Lutetium
USE   LUTETIUM ISOTOPES

Isotopes, Magnesium
USE   MAGNESIUM ISOTOPES

Isotopes, Manganese
USE   MANGANESE ISOTOPES

Isotopes, Mercury
USE   MERCURY ISOTOPES

Isotopes, Neodymium
USE   NEODYMIUM ISOTOPES

Isotopes, Neon
USE   NEON ISOTOPES

Isotopes, Neptunium
USE   NEPTUNIUM ISOTOPES

Isotopes, Nickel
USE   NICKEL ISOTOPES

Isotopes, Niobium
USE   NIOBIUM ISOTOPES

Isotopes, Nitrogen
USE   NITROGEN ISOTOPES

Isotopes, Osmium
USE   OSMIUM ISOTOPES

Isotopes, Oxygen
USE   OXYGEN ISOTOPES

Isotopes, Phosphorus
USE   PHOSPHORUS ISOTOPES

Isotopes, Platinum
USE   PLATINUM ISOTOPES

Isotopes, Plutonium
USE   PLUTONIUM ISOTOPES

Isotopes, Polonium
USE   POLONIUM ISOTOPES

Isotopes, Potassium
USE   POTASSIUM ISOTOPES

Isotopes, Praseodymium
USE   PRASEODYMIUM ISOTOPES

Isotopes, Promethium
USE   PROMETHIUM ISOTOPES

Isotopes, Protactinium
USE   PROTACTINIUM ISOTOPES

Isotopes, Radioactive
USE   RADIOACTIVE ISOTOPES

Isotopes, Radium
USE   RADIUM ISOTOPES

Isotopes, Radon
USE   RADON ISOTOPES

Isotopes, Rhenium
USE   RHENIUM ISOTOPES

Isotopes, Rhodium
USE   RHODIUM ISOTOPES

Isotopes, Rubidium
USE   RUBIDIUM ISOTOPES

Isotopes, Ruthenium
USE   RUTHENIUM ISOTOPES

Isotopes, Samarium
USE   SAMARIUM ISOTOPES

Isotopes, Scandium
USE   SCANDIUM ISOTOPES

Isotopes, Silicon
USE   SILICON ISOTOPES

Isotopes, Silver
USE   SILVER ISOTOPES

Isotopes, Sodium
USE   SODIUM ISOTOPES

Isotopes, Strontium
USE   STRONTIUM ISOTOPES

Isotopes, Sulfur
USE   SULFUR ISOTOPES

Isotopes, Tantalum
USE   TANTALUM ISOTOPES

Isotopes, Technetium
USE   TECHNETIUM ISOTOPES

Isotopes, Tellurium
USE   TELLURIUM ISOTOPES

Isotopes, Terbium
USE   TERBIUM ISOTOPES

Isotopes, Thallium
USE   THALLIUM ISOTOPES

Isotopes, Thorium
USE   THORIUM ISOTOPES

Isotopes, Thulium
USE   THULIUM ISOTOPES

Isotopes, Tin
USE   TIN ISOTOPES

Isotopes, Titanium
USE   TITANIUM ISOTOPES

Isotopes, Tungsten
USE   TUNGSTEN ISOTOPES

Isotopes, Uranium
USE   URANIUM ISOTOPES

Isotopes, Vanadium
USE   VANADIUM ISOTOPES

Isotopes, Xenon
USE   XENON ISOTOPES

Isotopes, Ytterbium
USE   YTTERBIUM ISOTOPES

Isotopes, Yttrium
USE   YTTRIUM ISOTOPES

Isotopes, Zinc
USE   ZINC ISOTOPES

Isotopes, Zirconium
USE   ZIRCONIUM ISOTOPES

ISOTOPIC ENRICHMENT

ISOTOPIC LABELING

ISOTOPIC SPIN

ISOTROPIC MEDIA

ISOTROPIC TURBULENCE

ISOTROPISM

ISOTROPY

Isotropy, An
USE   ANISOTROPY

Isotropy, Spatial
USE   ISOTROPY
SPATIAL DISTRIBUTION

ISRAEL

ISRO

ISTHMUSES

ITALIAN SPACE PROGRAM

ITALY

ITCHING

ITERATION

ITERATIVE NETWORKS

ITERATIVE SOLUTION

ITOS SATELLITES

ITOS 1

ITOS 2

ITOS 3

ITOS 4

IUE

IUS
USE   INERTIAL UPPER STAGE

IV Computer, Modcomp
USE   MODCOMP IV COMPUTER

IVORY COAST

IVUNA METEORITE

Izsak Ellipsoid
USE   GEODESY
ELLIPSOIDS

I2S CAMERAS

J

J, IMP-
USE   EXPLORER 50 SATELLITE

J INTEGRAL

J, OSO-
USE   OSO-8

J, Space Shuttle Mission 51-
USE   SPACE SHUTTLE MISSION 51-J

J-2 ENGINE

J-33 ENGINE

J-34 ENGINE

J-47 ENGINE

J-52 ENGINE

J-57 ENGINE

J-57-P-20 ENGINE

J-58 ENGINE

J-65 ENGINE

J-69-T-25 ENGINE

J-71 ENGINE

J-73 ENGINE

J-75 ENGINE

J-79 ENGINE

J-85 ENGINE

J-93 ENGINE

J-97 ENGINE

Jabiru Rocket Vehicle
USE   JAGUAR ROCKET VEHICLE

JACKETS

Jacking Equipment
USE   JACKS (LIFTS)

JACKS

Jacks (Electrical)
USE   ELECTRIC CONNECTORS

JACKS (LIFTS)

Jacobi Equation, Hamilton-
USE   HAMILTON-JACOBI EQUATION

JACOBI INTEGRAL

JACOBI MATRIX METHOD

Jacobi Polynomials
USE   HYPERGEOMETRIC FUNCTIONS

JAGUAR AIRCRAFT

JAGUAR ROCKET VEHICLE

JAHN-TELLER EFFECT

JAMAICA

JAMMERS

JAMMING

JANUS

JANUS REACTOR

JANUS SPACECRAFT

JAPAN

Japan, Sea Of
USE   SEA OF JAPAN

JAPANESE SPACE PROGRAM

JAPANESE SPACECRAFT

(Japanese Spacecraft), MOS
USE   JAPANESE SPACECRAFT

Jarring
USE   MECHANICAL SHOCK

JATO ENGINES

Javelin Aircraft
USE   GA-5 AIRCRAFT

JAVELIN ROCKET VEHICLE

Javelin Rocket Vehicle, Nike-
USE   NIKE-JAVELIN ROCKET VEHICLE

JC-130 Aircraft
USE   C-130 AIRCRAFT

JEANS THEORY

Jeeps
USE   AUTOMOBILES

JERBOAS

Jersey, New
USE   NEW JERSEY

JET AIRCRAFT

Jet Aircraft, Alpha
USE   ALPHA JET AIRCRAFT

Jet Aircraft, Lear
USE   LEAR JET AIRCRAFT

JET AIRCRAFT NOISE

Jet Airstreams
USE   JET STREAMS (METEOROLOGY)

JET AMPLIFIERS

Jet Amplifiers, Fluid
USE   JET AMPLIFIERS
            FLUID AMPLIFIERS

Jet Assisted Takeoff
USE   JATO ENGINES

Jet Augmented Wing Flaps
USE   JET FLAPS
            WING FLAPS

Jet Backpacks, Reaction
USE   SELF MANEUVERING UNITS

JET BLAST EFFECTS

JET BOUNDARIES

JET CONDENSERS

JET CONTROL

Jet Damping
USE   DAMPING
            SPIN REDUCTION

Jet Dragon Aircraft
USE   DH 125 AIRCRAFT

Jet Drive
USE   JET PROPULSION

JET ENGINE FUELS

JET ENGINES

Jet Engines, Arc
USE   ARC JET ENGINES

Jet Engines, Pulsed
USE   PULSED JET ENGINES

JET EXHAUST

Jet Exhaust, Hot
USE   JET EXHAUST
            HIGH TEMPERATURE GASES

Jet Flames
USE   FLAMES
            JET FLOW

JET FLAPS

Jet Flight
USE   JET AIRCRAFT

JET FLOW

Jet Flow, Peripheral
USE   PERIPHERAL JET FLOW

Jet Flow, Supersonic
USE   SUPERSONIC JET FLOW

Jet Fuel, JP-4
USE   JP-4 JET FUEL

Jet Fuel, JP-5
USE   JP-5 JET FUEL

Jet Fuel, JP-6
USE   JP-6 JET FUEL

Jet Fuel, JP-8
USE   JP-8 JET FUEL

Jet Fuels
USE   JET ENGINE FUELS

JET IMPINGEMENT

JET LAG

JET LIFT

JET MEMBRANE PROCESS

JET MIXING FLOW

Jet Noise
USE   JET AIRCRAFT NOISE

JET NOZZLES

Jet Pilots
USE   AIRCRAFT PILOTS

JET PROPULSION

JET PROVOST AIRCRAFT

JET PUMPS

Jet Star Aircraft
USE   C-140 AIRCRAFT

JET STREAMS (METEOROLOGY)

Jet Synthesis, Plasma
USE   PLASMA JET SYNTHESIS

JET THRUST

Jet Trainer, L-29
USE   L-29 JET TRAINER

JET VANES

Jet Wind Tunnels, Plasma
USE   PLASMA JET WIND TUNNELS

Jetavators
USE   GUIDE VANES

JETS

Jets, Air
USE   AIR JETS

Jets, Electro
USE   ELECTROJETS

Jets, Exhaust
USE   EXHAUST GASES

Jets, Fluid
USE   FLUID JETS

Jets, Free
USE   FREE JETS

Jets, Gas
USE   GAS JETS

Jets, Hot
USE   JET FLOW

Jets, Hydraulic
USE   HYDRAULIC JETS

Jets, Laminar
USE   JET FLOW
            LAMINAR FLOW

Jets, Particle Laden
USE   PARTICLE LADEN JETS

Jets, Plasma
USE   PLASMA JETS

Jets, Reaction
USE   JET FLOW
            JET THRUST

Jets, Turbulent
USE   TURBULENT JETS

Jets, Two Dimensional
USE   TWO DIMENSIONAL JETS

Jets, Vapor
USE   VAPOR JETS

Jets, Wall
USE   WALL JETS

Jets, Water
USE   HYDRAULIC JETS

JETSTREAM AIRCRAFT

Jetties
USE   BREAKWATERS

JETTISON SYSTEMS

JETTISONING

JF 101 Aircraft
USE   F-101 AIRCRAFT

JFET

JIGS

JIMSPHERE BALLOONS

JINDIVIK TARGET AIRCRAFT

Jitter
USE   VIBRATION

Joaquin Valley (CA), San
USE   SAN JOAQUIN VALLEY (CA)

Jobs
USE TASKS

JODRELL BANK OBSERVATORY

Joe 2 Launch Vehicle, Little
USE LITTLE JOE 2 LAUNCH VEHICLE

John Rocket Vehicle, Honest
USE HONEST JOHN ROCKET VEHICLE

John Rocket Vehicle, Little
USE LITTLE JOHN ROCKET VEHICLE

JOHNSTON ISLAND

JOINING

JOINT EUROPEAN TORUS

JOINTS (ANATOMY)

Joints, Butt
USE BUTT JOINTS

JOINTS (JUNCTIONS)

Joints, Lap
USE LAP JOINTS

Joints, Metal
USE METAL JOINTS

Joints, Riveted
USE RIVETED JOINTS

(Joints), Seams
USE SEAMS (JOINTS)

Joints, Soldered
USE SOLDERED JOINTS

Joints, Welded
USE WELDED JOINTS

Jones Gas, Lennard-
USE LENNARD-JONES GAS

Jones Potential, Lennard-
USE LENNARD-JONES POTENTIAL

JORDAN

JORDAN FORM

JOSEPHSON JUNCTIONS

Jouget Flame, Chapman-
USE CHEMICAL EQUILIBRIUM
FLAME PROPAGATION
DETONATION

Joukowski Condition, Kutta-
USE KUTTA-JOUKOWSKI CONDITION

JOUKOWSKI TRANSFORMATION

Joule Heating
USE RESISTANCE HEATING
OHMIC DISSIPATION

JOULE-THOMSON EFFECT

JOURNAL BEARINGS

JOURNALS

Journals (Documents)
USE PERIODICALS

Journals (Shafts)
USE SHAFTS (MACHINE ELEMENTS)

JP-4 JET FUEL

JP-5 JET FUEL

JP-6 JET FUEL

JP-8 JET FUEL

Juan Mountains (CO), San
USE SAN JUAN MOUNTAINS (CO)

JUDGMENTS

JUDI-DART ROCKET

JUICES

JUMPERS

Junction, Con
USE CONJUNCTION

JUNCTION DIODES

Junction Field Effect Transistors
USE JFET

Junction Solar Cells, Vertical
USE VERTICAL JUNCTION SOLAR CELLS

JUNCTION TRANSISTORS

JUNCTIONS

(Junctions), Joints
USE JOINTS (JUNCTIONS)

Junctions, Josephson
USE JOSEPHSON JUNCTIONS

Junctions, MBM
USE MBM JUNCTIONS

Junctions, Metal-Barrier-Metal
USE MBM JUNCTIONS

Junctions, N-N
USE N-N JUNCTIONS

Junctions, N-P
USE P-N JUNCTIONS

Junctions, N-P-N
USE N-P-N JUNCTIONS

Junctions, P-I-N
USE P-I-N JUNCTIONS

Junctions, P-N
USE P-N JUNCTIONS

Junctions, P-N-P
USE P-N-P JUNCTIONS

Junctions, P-N-P-N
USE P-N-P-N JUNCTIONS

Junctions, Semiconductor
USE SEMICONDUCTOR JUNCTIONS

Junctions, Silicon
USE SILICON JUNCTIONS

Junctions, Silicon-On-Sapphire
USE SOS (SEMICONDUCTORS)

Jungles
USE TROPICAL REGIONS

JUNO LAUNCH VEHICLES

JUNO 1 LAUNCH VEHICLE

JUNO 2 LAUNCH VEHICLE

JUPITER ATMOSPHERE

JUPITER C ROCKET VEHICLE

JUPITER MISSILE

JUPITER (PLANET)

JUPITER PROBES

JUPITER PROJECT

JUPITER RED SPOT

JUPITER RINGS

JUPITER SATELLITES

Jupiter-Saturn Flyby, Mariner
USE MARINER JUPITER-SATURN FLYBY

Jupiter-Uranus Flyby, Mariner
USE MARINER JUPITER-URANUS FLYBY

(Jurisprudence), Law
USE LAW (JURISPRUDENCE)

J93-MJ252H Engine
USE J-93 ENGINE

J93-MJ280G Engine
USE J-93 ENGINE

K

K Band
USE EXTREMELY HIGH FREQUENCIES

K LINES

K, Vitamin
USE PHYLLOQUINONE

K-Mesons
USE KAONS

KA Band
USE EXTREMELY HIGH FREQUENCIES

KA-6 Sailplane, Schleicher
USE KA-6 SAILPLANES

KA-6 SAILPLANES

KAKUTANI THEOREM

KALIHARI BASIN (AFRICA)

KALMAN FILTERS

KALMAN-SCHMIDT FILTERING

KAMACITE

KAMAN AIRCRAFT

Kaman UH-2A Helicopter
USE UH-2 HELICOPTER

Kampuchea
USE CAMBODIA

KANSAS

Kansas City Corridor (MO), St Louis-
USE ST LOUIS-KANSAS CITY CORRIDOR (MO)

KAOLINITE

KAON PRODUCTION

KAONS

KAPITZA RESISTANCE

Kaplan Bands, Vegard-
USE VEGARD-KAPLAN BANDS

KAPOETA ACHONDRITE

KAPPA ROCKET VEHICLES

KAPPA 8 ROCKET VEHICLE

KAPPA 9 ROCKET VEHICLE

KAPTON (TRADEMARK)

KARHUNEN-LOEVE EXPANSION

KARL FISCHER REAGENT

Karman Equation, Von
　USE　VON KARMAN EQUATION

KARMAN VORTEX STREET

KARMAN-BODEWADT FLOW

KARST

KAWASAKI AIRCRAFT

KC-130 Aircraft
　USE　C-130 AIRCRAFT

KC-135 Aircraft
　USE　C-135 AIRCRAFT

KEELS

Keeping, Sea
　USE　SEA KEEPING

KEL-F

Kelp
　USE　SEAWEEDS

KELVIN-HELMHOLTZ INSTABILITY

Kennedy Launch Complex, Cape
　USE　CAPE KENNEDY LAUNCH COMPLEX

KENTUCKY

KENYA

KEPLER LAWS

KERATINS

KERATITIS

KERNEL FUNCTIONS

KEROGEN

KEROSENE

KERR CELLS

KERR EFFECTS

KERR ELECTROOPTICAL EFFECT

KERR MAGNETOOPTICAL EFFECT

Kestrel Aircraft
　USE　P-1127 AIRCRAFT

KETENES

KETONES

KETTLES (GEOLOGY)

KEVLAR (TRADEMARK)

KEYING

Keying, Frequency Shift
　USE　FREQUENCY SHIFT KEYING

Keying, Phase Shift
　USE　PHASE SHIFT KEYING

KEYS (ISLANDS)

KIDNEY DISEASES

KIDNEYS

KILOMETER WAVE ORBITING TELESCOPE

KILOMETRIC WAVES

Kimberlite
　USE　BIOTITE
　　　　PERIDOTITE

KINEMATIC EQUATIONS

KINEMATICS

Kinematics, Body
　USE　BODY KINEMATICS

Kinescopes
　USE　PICTURE TUBES

Kinesis, Auto
　USE　AUTOKINESIS

KINESTHESIA

Kinesthesis
　USE　PROPRIOCEPTION

KINETIC ENERGY

KINETIC EQUATIONS

KINETIC FRICTION

KINETIC HEATING

KINETIC THEORY

KINETICS

Kinetics, Chemical
　USE　REACTION KINETICS

Kinetics, Electro
　USE　ELECTROKINETICS

Kinetics, Reaction
　USE　REACTION KINETICS

King Helicopter, Sea
　USE　SH-3 HELICOPTER

Kingdom Satellites, United
　USE　UK SATELLITES

Kingdom, United
　USE　UNITED KINGDOM

Kingsport, USNS
　USE　SATELLITE COMMUNICATIONS SHIPS

KINOFORM

(Kinshasa), Congo
　USE　ZAIRE

Kirchhoff Integrals, Fresnel-
　USE　FRESNEL INTEGRALS

KIRCHHOFF LAW

KIRCHHOFF LAW OF NETWORKS

KIRCHHOFF LAW OF RADIATION

Kirchhoff-Helmholtz Flow
　USE　PIPE FLOW

Kirchhoff-Huygens Principle
　USE　DIFFRACTION
　　　　WAVE PROPAGATION

KIRKENDALL EFFECT

Kite Balloons
　USE　TETHERED BALLOONS

KITS

KIWI B REACTORS

KIWI B-1 REACTOR

KIWI B-4 REACTOR

KIWI REACTORS

KIWI Rocket Reactors
　USE　KIWI REACTORS

KJELDAHL METHOD

KLEBSIELLA

KLEIN-DUNHAM POTENTIAL

KLEIN-GORDON EQUATION

Klippen
　USE　OUTLIERS (LANDFORMS)

KLYSTRONS

KNEE (ANATOMY)

Knight Helicopter, Sea
　USE　CH-46 HELICOPTER

Knight Rocket Vehicle, Black
　USE　BLACK KNIGHT ROCKET VEHICLE

Knight Shift
　USE　NUCLEAR MAGNETIC RESONANCE

KNOBS

KNOOP HARDNESS

KNOWLEDGE

Knowledge Engineering
　USE　EXPERT SYSTEMS

Knudsen Cells
　USE　KNUDSEN GAGES

KNUDSEN FLOW

KNUDSEN GAGES

Knudsen Number
　USE　KNUDSEN FLOW

KNURLING

KOHOUTEK COMET

KOLMOGOROFF THEORY

KOLMOGOROFF-SMIRNOFF TEST

KONDO EFFECT

Kong, Hong
　USE　HONG KONG

KOREA

Korea, Democratic Peoples Republic Of
　USE　NORTH KOREA

Korea, North
　USE　NORTH KOREA

Korea, Republic Of
　USE　SOUTH KOREA

Korea, South
　USE　SOUTH KOREA

KORTEWEG-DEVRIES EQUATION

KOSSEL PATTERN

KOVAR (TRADEMARK)

KP INDEX

Kr
　USE　KRYPTON

KRAFT PROCESS (WOODPULP)

Kramer-Brillouin Method, Wentzel-
　USE　WENTZEL-KRAMER-BRILLOUIN METHOD

KRAMERS-KRONIG FORMULA

KREBS CYCLE

KREEP

KRIGING

Kronecker Product
  USE ORTHOGONALITY

Kronig Formula, Kramers-
  USE KRAMERS-KRONIG FORMULA

KROOK EQUATION

KRYPTON

KRYPTON FLUORIDE LASERS

KRYPTON ISOTOPES

KRYPTON 85

KS
  USE KANSAS

KU Band
  USE SUPERHIGH FREQUENCIES

Kuiper Airborne Observatory
  USE C-141 AIRCRAFT

KURILE ISLANDS

KURTOSIS

Kutta Method, Runge-
  USE RUNGE-KUTTA METHOD

KUTTA-JOUKOWSKI CONDITION

KUWAIT

KWIC INDEXES

KY
  USE KENTUCKY

KY-TN), Tennessee Valley (AL-
  USE TENNESSEE VALLEY (AL-KY-TN)

L

L Band
  USE ULTRAHIGH FREQUENCIES

L, Space Shuttle Mission 51-
  USE SPACE SHUTTLE MISSION 51-L

L-Band Radiometers, Passive
  USE PASSIVE L-BAND RADIOMETERS

L-SAT

L-19 Aircraft, Cessna
  USE CESSNA L-19 AIRCRAFT

L-28 Aircraft
  USE U-10 AIRCRAFT

L-29 Aircraft
  USE L-29 JET TRAINER

L-29 Aircraft, Omnipol
  USE L-29 JET TRAINER

L-29 JET TRAINER

L-1011 AIRCRAFT

L-2000 AIRCRAFT

L-2000 Aircraft, Lockheed
  USE L-2000 AIRCRAFT

La
  USE LANTHANUM

LA
  USE LOUISIANA

(LA), Atchafalaya River Basin
  USE ATCHAFALAYA RIVER BASIN (LA)

(LA), Lake Pontchartrain
  USE LAKE PONTCHARTRAIN (LA)

(LA), Mississippi Delta
  USE MISSISSIPPI DELTA (LA)

Lab, Commerce
  USE COMMERCE LAB

Lab Measur System, Integ Med And Behavioral
  USE IMBLMS

Lab, Sortie
  USE SORTIE SYSTEMS

Lab (Spacelab), Atmospheric Cloud Physics
  USE ATMOSPHERIC CLOUD PHYSICS LAB
    (SPACELAB)

Labeling, Isotopic
  USE ISOTOPIC LABELING

Labeling (Marking)
  USE MARKING

LABOR

LABORATORIES

Laboratories, Engine Testing
  USE ENGINE TESTING LABORATORIES

Laboratories, Environmental
  USE ENVIRONMENTAL LABORATORIES

Laboratories, Human Factors
  USE HUMAN FACTORS LABORATORIES

Laboratories, Lunar Mobile
  USE LUNAR MOBILE LABORATORIES

Laboratories, Manned Orbital
  USE MANNED ORBITAL LABORATORIES

Laboratories, Manned Orbital Research
  USE MANNED ORBITAL RESEARCH
    LABORATORIES

Laboratories), MOL (Orbital
  USE MANNED ORBITAL LABORATORIES

Laboratories, Space
  USE SPACE LABORATORIES

Laboratories, Underwater Research
  USE UNDERWATER RESEARCH LABORATORIES

Laboratory, Advanced Technology
  USE ADVANCED TECHNOLOGY LABORATORY

Laboratory, Earth Viewing Applications
  USE EARTH VIEWING APPLICATIONS
    LABORATORY

LABORATORY EQUIPMENT

Laboratory, Lunar Receiving
  USE LUNAR RECEIVING LABORATORY

Laboratory, Shuttle Avionics Integration
  USE SAIL PROJECT

LABRADOR

LABYRINTH

LABYRINTH SEALS

LABYRINTHECTOMY

LACATE (EXPERIMENT)

LACE (Engine)
  USE LIQUID AIR CYCLE ENGINES

Lacertae Objects, Bl
  USE BL LACERTAE OBJECTS

LACQUERS

LACTATES

LACTIC ACID

LACTOSE

LACUNAS

LADDERS

Laden Jets, Particle
  USE PARTICLE LADEN JETS

Lag (Delay)
  USE TIME LAG

Lag, Jet
  USE JET LAG

Lag, Time
  USE TIME LAG

LAGEOS (SATELLITE)

LAGOONS

LAGRANGE COORDINATES

Lagrange Equation, Euler-
  USE EULER-LAGRANGE EQUATION

Lagrange Equations Of Motion
  USE EULER-LAGRANGE EQUATION

LAGRANGE MULTIPLIERS

LAGRANGE SIMILARITY HYPOTHESIS

LAGRANGIAN EQUILIBRIUM POINTS

LAGUERRE FUNCTIONS

Lake Beds
  USE BEDS (GEOLOGY)

LAKE CHAMPLAIN BASIN (NY-VT)

LAKE ERIE

LAKE HURON

LAKE ICE

LAKE MICHIGAN

Lake (NV), Pyramid
  USE PYRAMID LAKE (NV)

LAKE ONTARIO

LAKE PONTCHARTRAIN (LA)

LAKE SUPERIOR

LAKE TAHOE (CA-NV)

LAKE TEXOMA (OK-TX)

Lake (UT), Great Salt
  USE GREAT SALT LAKE (UT)

LAKES

Lakes, International Field Year For Great
  USE INTERNATIONAL FIELD YEAR FOR GREAT
    LAKES

Lakes (North America), Great
  USE GREAT LAKES (NORTH AMERICA)

LALLEMAND CAMERAS

LAMB WAVES

LAMBDA ROCKET VEHICLES

LAMBDA TAURI STARS

Lambert Equation, Euler-
  USE EULER-LAMBERT EQUATION

Lambert Law
USE   BOUGUER LAW

LAMBERT SURFACE

LAME FUNCTIONS

LAME WAVE EQUATIONS

LAMELLA

LAMELLA (METALLURGY)

Lamina
USE   LAYERS

LAMINAR BOUNDARY LAYER

Laminar Boundary Layer Separation
USE   LAMINAR BOUNDARY LAYER

Laminar Flames
USE   LAMINAR FLOW
FLAMES

LAMINAR FLOW

LAMINAR FLOW AIRFOILS

Laminar Flow Control
USE   LAMINAR BOUNDARY LAYER
BOUNDARY LAYER CONTROL

LAMINAR HEAT TRANSFER

Laminar Jets
USE   LAMINAR FLOW
JET FLOW

LAMINAR MIXING

LAMINAR WAKES

Laminated Materials
USE   LAMINATES

LAMINATES

Laminations
USE   LAMINATES

Lamps
USE   LUMINAIRES

Lamps, Alkali Vapor
USE   ALKALI VAPOR LAMPS

Lamps, Arc
USE   ARC LAMPS

Lamps, Electroluminescent
USE   ELECTROLUMINESCENCE
LUMINAIRES

Lamps, Flash
USE   FLASH LAMPS

Lamps, Mercury
USE   MERCURY LAMPS

Lamps Program
USE   LIGHT AIRBORNE MULTIPURPOSE SYSTEM

Lamps, Quartz
USE   QUARTZ LAMPS

Lamps, Xenon
USE   XENON LAMPS

LANCE MISSILE

LAND

Land, Barren
USE   BARREN LAND

LAND ICE

Land Interactions, Air
USE   AIR LAND INTERACTIONS

LAND MANAGEMENT

LAND MOBILE SATELLITE SERVICE

LAND USE

Land Use, Rural
USE   RURAL LAND USE

LANDAU DAMPING

LANDAU FACTOR

LANDAU-GINZBURG EQUATIONS

Lander Spacecraft, Viking
USE   VIKING LANDER SPACECRAFT

Lander 1, Viking
USE   VIKING LANDER 1

Lander 2, Viking
USE   VIKING LANDER 2

LANDFILLS

LANDFORMS

(Landforms), Barriers
USE   BARRIERS (LANDFORMS)

(Landforms), Bars
USE   BARS (LANDFORMS)

(Landforms), Bluffs
USE   CLIFFS

(Landforms), Bridges
USE   BRIDGES (LANDFORMS)

(Landforms), Capes
USE   CAPES (LANDFORMS)

(Landforms), Cirques
USE   CIRQUES (LANDFORMS)

(Landforms), Cusps
USE   CUSPS (LANDFORMS)

(Landforms), Divides
USE   DIVIDES (LANDFORMS)

(Landforms), Fans
USE   FANS (LANDFORMS)

(Landforms), Flats
USE   FLATS (LANDFORMS)

(Landforms), Inliers
USE   INLIERS (LANDFORMS)

(Landforms), Outliers
USE   OUTLIERS (LANDFORMS)

(Landforms), Peaks
USE   PEAKS (LANDFORMS)

(Landforms), Terraces
USE   TERRACES (LANDFORMS)

LANDING

Landing Aid, Microvision
USE   MICROVISION LANDING AID

Landing Aid Television System, Pilot
USE   PLAT SYSTEM

LANDING AIDS

Landing, Aircraft
USE   AIRCRAFT LANDING

Landing Aircraft, Vertical Attitude Takeoff-
USE   VATOL AIRCRAFT

Landing Aircraft, Water Takeoff And
USE   WATER TAKEOFF AND LANDING
AIRCRAFT

Landing, Blind
USE   BLIND LANDING

Landing Control, Automatic
USE   AUTOMATIC LANDING CONTROL

Landing, Crash
USE   CRASH LANDING

(Landing), Ditching
USE   DITCHING (LANDING)

LANDING GEAR

Landing Gear, Retractable
USE   LANDING GEAR
RETRACTABLE EQUIPMENT

Landing, Hard
USE   HARD LANDING

Landing, Horizontal Spacecraft
USE   HORIZONTAL SPACECRAFT LANDING

LANDING INSTRUMENTS

LANDING LOADS

Landing, Lunar
USE   LUNAR LANDING

Landing, Mars
USE   MARS LANDING

LANDING MATS

LANDING MODULES

Landing Modules, Lunar
USE   LUNAR LANDING MODULES

Landing, Planetary
USE   PLANETARY LANDING

LANDING RADAR

LANDING SIMULATION

Landing Simulators, Lunar Orbit And
USE   LUNAR ORBIT AND LANDING SIMULATORS

LANDING SITES

Landing Sites, Lunar
USE   LUNAR LANDING SITES

Landing, Soft
USE   SOFT LANDING

Landing, Spacecraft
USE   SPACECRAFT LANDING

Landing Spacecraft, Soft
USE   SOFT LANDING SPACECRAFT

LANDING SPEED

Landing System, Microwave Scanning Beam
USE   MICROWAVE SCANNING BEAM LANDING
SYSTEM

Landing Systems
USE   LANDING AIDS

Landing Systems, Air Cushion
USE   AIR CUSHION LANDING SYSTEMS

Landing Systems, All-Weather
USE   ALL-WEATHER LANDING SYSTEMS

(Landing Systems), ILS
USE   INSTRUMENT LANDING SYSTEMS

Landing Systems, Instrument
USE   INSTRUMENT LANDING SYSTEMS

Landing Systems, Microwave
USE   MICROWAVE LANDING SYSTEMS

Landing Tests (STS), Approach And
USE   APPROACH AND LANDING TESTS (STS)

Landing Vehicles, Ranger Lunar
  USE   RANGER LUNAR LANDING VEHICLES

Landing Vehicles), SLV (Soft
  USE   SOFT LANDING SPACECRAFT

Landing, Vertical
  USE   VERTICAL LANDING

Landing, Vertical Takeoff And
  USE   VERTICAL LANDING
         VERTICAL TAKEOFF

Landing, Water
  USE   WATER LANDING

Landings, Glide
  USE   GLIDE LANDINGS

Landings, Skid
  USE   SKID LANDINGS

Landmark Acquisition And Tracking, Video
  USE   VIDEO LANDMARK ACQUISITION AND
         TRACKING

LANDMARKS

Lands, Arid
  USE   ARID LANDS

Lands, Bad
  USE   BADLANDS

Lands, Farm
  USE   FARMLANDS

Lands, Grass
  USE   GRASSLANDS

Lands, Grazing
  USE   GRASSLANDS

Lands, Marsh
  USE   MARSHLANDS

Lands, Range
  USE   RANGELANDS

Lands, Wet
  USE   WETLANDS

LANDSAT E

LANDSAT F

LANDSAT FOLLOW-ON MISSIONS

LANDSAT SATELLITES

LANDSAT 1

LANDSAT 2

LANDSAT 3

LANDSAT 4

LANDSAT 5

Landscape
  USE   TOPOGRAPHY
         TERRAIN

LANDSLIDES

Lanes
  USE   PATHS

LANGEVIN FORMULA

LANGLEY COMPLEX COORDINATOR

Langmuir Law, Child-
  USE   CHILD-LANGMUIR LAW

Langmuir Probes
  USE   ELECTROSTATIC PROBES

Language), Ada (Programming
  USE   ADA (PROGRAMMING LANGUAGE)

Language), APL (Programming
  USE   APL (PROGRAMMING LANGUAGE)

Language, Assembly
  USE   ASSEMBLY LANGUAGE

Language), BASIC (Programming
  USE   BASIC (PROGRAMMING LANGUAGE)

Language), COGO (Programming
  USE   COGO (PROGRAMMING LANGUAGE)

Language), COMPASS (Programming
  USE   COMPASS (PROGRAMMING LANGUAGE)

Language (Computers), Natural
  USE   NATURAL LANGUAGE (COMPUTERS)

Language, Coordinate Geometry
  USE   COGO (PROGRAMMING LANGUAGE)

Language, English
  USE   ENGLISH LANGUAGE

Language), FAB (Programming
  USE   FORTRAN

(Language), Hal/s
  USE   HAL/S (LANGUAGE)

Language), LISP (Programming
  USE   LISP (PROGRAMMING LANGUAGE)

Language), Map (Programming
  USE   MAP (PROGRAMMING LANGUAGE)

Language), MARVS (Programming
  USE   MARVS (PROGRAMMING LANGUAGE)

Language), Pascal (Programming
  USE   PASCAL (PROGRAMMING LANGUAGE)

LANGUAGE PROGRAMMING

Language), SLEUTH (Programming
  USE   SLEUTH (PROGRAMMING LANGUAGE)

(Language), Words
  USE   WORDS (LANGUAGE)

LANGUAGES

Languages, Command
  USE   COMMAND LANGUAGES

Languages, Context Free
  USE   CONTEXT FREE LANGUAGES

Languages, High Level
  USE   HIGH LEVEL LANGUAGES

Languages, Machine Oriented
  USE   MACHINE ORIENTED LANGUAGES

Languages, Programming
  USE   PROGRAMMING LANGUAGES

Languages, Query
  USE   QUERY LANGUAGES

Lanka, Sri
  USE   SRI LANKA

Lanthanide Series Metals
  USE   RARE EARTH ELEMENTS

LANTHANUM

LANTHANUM ALLOYS

LANTHANUM CHLORIDES

LANTHANUM COMPOUNDS

LANTHANUM FLUORIDES

LANTHANUM ISOTOPES

LANTHANUM OXIDES

LANTHANUM TELLURIDES

Lanthanum 140
  USE   LANTHANUM ISOTOPES

LAOS

LAP JOINTS

LAPLACE EQUATION

Laplace Operators
  USE   LAPLACE TRANSFORMATION

LAPLACE TRANSFORMATION

Lapse Photography, Time
  USE   CHRONOPHOTOGRAPHY

LAPSE RATE

Lara Aircraft
  USE   COIN AIRCRAFT

Larc Computer, Univac
  USE   UNIVAC LARC COMPUTER

LARGE APERTURE SEISMIC ARRAY

LARGE AREA CROP INVENTORY EXPERIMENT

Large Infrared Telescope On Spacelab
  USE   LIRTS (TELESCOPE)

LARGE SCALE INTEGRATION

Large Scale Integration, Very
  USE   VERY LARGE SCALE INTEGRATION

LARGE SPACE STRUCTURES

Large Space Telescope
  USE   HUBBLE SPACE TELESCOPE

Large Telecomm Satellite, European
  USE   L-SAT

LARGOS SATELLITE

LARMOR PRECESSION

LARMOR RADIUS

LARVAE

LARYNX

Laser Acoustic Microscope (Slam), Scanning
  USE   ACOUSTIC MICROSCOPES

LASER ALTIMETERS

LASER ANEMOMETERS

LASER ANNEALING

LASER APPLICATIONS

Laser Beam Defocusing
  USE   THERMAL BLOOMING

LASER CAVITIES

Laser Communication
  USE   OPTICAL COMMUNICATION

LASER CUTTING

LASER DAMAGE

LASER DOPPLER VELOCIMETERS

LASER DRILLING

LASER FUSION

Laser Geodynamic Satellite
  USE   LAGEOS (SATELLITE)

LASER GUIDANCE

LASER GYROSCOPES

LASER HEATING

LASER INTERFEROMETRY

LASER MATERIALS

LASER MICROSCOPY

LASER MODE LOCKING

LASER MODES

LASER OUTPUTS

LASER PLASMA INTERACTIONS

LASER PLASMAS

LASER PROPULSION

LASER PUMPING

Laser Radar
  USE   OPTICAL RADAR

LASER RANGE FINDERS

LASER RANGER/TRACKER

LASER SPECTROMETERS

LASER SPECTROSCOPY

LASER STABILITY

Laser System, Nova
  USE   NOVA LASER SYSTEM

Laser System, Shiva
  USE   SHIVA LASER SYSTEM

LASER TARGET DESIGNATORS

LASER TARGET INTERACTIONS

LASER TARGETS

LASER WEAPONS

LASER WELDING

LASER WINDOWS

LASERS

Lasers, Airborne
  USE   AIRBORNE LASERS

Lasers, Argon
  USE   ARGON LASERS

Lasers, Atmospheric
  USE   ATMOSPHERIC LASERS

Lasers, Carbon
  USE   CARBON LASERS

Lasers, Carbon Dioxide
  USE   CARBON DIOXIDE LASERS

Lasers, Carbon Monoxide
  USE   CARBON MONOXIDE LASERS

Lasers, Chemical
  USE   CHEMICAL LASERS

Lasers, Continuous Wave
  USE   CONTINUOUS WAVE LASERS

Lasers, Deuterium Fluoride
  USE   DF LASERS

Lasers, DF
  USE   DF LASERS

Lasers, Distributed Feedback
  USE   DISTRIBUTED FEEDBACK LASERS

Lasers, Dye
  USE   DYE LASERS

Lasers, Excimer
  USE   EXCIMER LASERS

Lasers, Fabry-Perot
  USE   LASERS

Lasers, Free Electron
  USE   FREE ELECTRON LASERS

Lasers, Gallium Arsenide
  USE   GALLIUM ARSENIDE LASERS

Lasers, Gamma Ray
  USE   GAMMA RAY LASERS

Lasers, Gas
  USE   GAS LASERS

Lasers, Gasdynamic
  USE   GASDYNAMIC LASERS

Lasers, Glass
  USE   GLASS LASERS

Lasers, HCL
  USE   HCL LASERS

Lasers, HCL Argon
  USE   HCL ARGON LASERS

Lasers, HCN
  USE   HCN LASERS

Lasers, Helium-Neon
  USE   HELIUM-NEON LASERS

Lasers, HF
  USE   HF LASERS

Lasers, High Intensity
  USE   HIGH POWER LASERS

Lasers, High Power
  USE   HIGH POWER LASERS

Lasers, Infrared
  USE   INFRARED LASERS

Lasers, Injection
  USE   INJECTION LASERS

Lasers, Iodine
  USE   IODINE LASERS

Lasers, Krypton Fluoride
  USE   KRYPTON FLUORIDE LASERS

Lasers, Liquid
  USE   LIQUID LASERS

Lasers, Metal Vapor
  USE   METAL VAPOR LASERS

Lasers, Natural
  USE   LASERS

Lasers, Neodymium
  USE   NEODYMIUM LASERS

Lasers, Nitrogen
  USE   NITROGEN LASERS

Lasers, Nuclear Pumped
  USE   NUCLEAR PUMPED LASERS

Lasers, Organic
  USE   ORGANIC LASERS

Lasers, Plasmadynamic
  USE   PLASMADYNAMIC LASERS

(Lasers), Power Transmission
  USE   POWER TRANSMISSION (LASERS)

Lasers, Pulsed
  USE   PULSED LASERS

Lasers, Q Switched
  USE   Q SWITCHED LASERS

Lasers, Raman
  USE   RAMAN LASERS

Lasers, Rare Gas-Halide
  USE   RARE GAS-HALIDE LASERS

Lasers, Ring
  USE   RING LASERS

Lasers, Ruby
  USE   RUBY LASERS

Lasers, Semiconductor
  USE   SEMICONDUCTOR LASERS

Lasers, Solar
  USE   SOLAR-PUMPED LASERS

Lasers, Solar Pumped
  USE   SOLAR-PUMPED LASERS

Lasers, Solid State
  USE   SOLID STATE LASERS

Lasers, Spaceborne
  USE   SPACEBORNE LASERS

Lasers, TEA
  USE   TEA LASERS

Lasers, Transversely Excited Atmospheric
  USE   TEA LASERS

Lasers, Tube
  USE   TUBE LASERS

Lasers, Tunable
  USE   TUNABLE LASERS

Lasers, Two-Wavelength
  USE   TWO-WAVELENGTH LASERS

Lasers, Ultrashort Pulsed
  USE   ULTRASHORT PULSED LASERS

Lasers, Ultraviolet
  USE   ULTRAVIOLET LASERS

Lasers, UV
  USE   ULTRAVIOLET LASERS

Lasers, Waveguide
  USE   WAVEGUIDE LASERS

Lasers, X Ray
  USE   X RAY LASERS

Lasers, Xenon Chloride
  USE   XENON CHLORIDE LASERS

Lasers, Xenon Fluoride
  USE   XENON FLUORIDE LASERS

Lasers, YAG
  USE   YAG LASERS

LASING

LASV
  USE   F-111 AIRCRAFT

LATCH-UP

LATCHES

LATE STARS

LATENESS

Latent Heat Of Fusion
  USE   HEAT OF FUSION

LATERAL CONTROL

LATERAL OSCILLATION

LATERAL STABILITY

Laterality
  USE   LATERAL STABILITY

Lateralization
　USE　LATERAL CONTROL

LATERITES

LATEX

LATHES

Lathes, Turret
　USE　TURRET LATHES

LATIN SQUARE METHOD

LATITUDE

Latitude, Geomagnetic
　USE　GEOMAGNETIC LATITUDE

LATITUDE MEASUREMENT

Latitudes, High
　USE　POLAR REGIONS

Latitudes, Low
　USE　TROPICAL REGIONS

Lattice Imperfections
　USE　CRYSTAL DEFECTS

LATTICE PARAMETERS

Lattice Relaxation, Spin-
　USE　SPIN-LATTICE RELAXATION

LATTICE VIBRATIONS

LATTICES

Lattices, BCC
　USE　BODY CENTERED CUBIC LATTICES

Lattices, Body Centered Cubic
　USE　BODY CENTERED CUBIC LATTICES

Lattices, Close Packed
　USE　CLOSE PACKED LATTICES

Lattices, Crystal
　USE　CRYSTAL LATTICES

Lattices, Cubic
　USE　CUBIC LATTICES

Lattices, Face Centered Cubic
　USE　FACE CENTERED CUBIC LATTICES

Lattices, FCC
　USE　FACE CENTERED CUBIC LATTICES

LATTICES (MATHEMATICS)

LATVIA

LAUE METHOD

LAUGHING

Launch Complex, Cape Kennedy
　USE　CAPE KENNEDY LAUNCH COMPLEX

Launch Complexes
　USE　LAUNCHING BASES

LAUNCH DATES

LAUNCH ESCAPE SYSTEMS

Launch, Lunar
　USE　LUNAR LAUNCH

Launch Time
　USE　LAUNCH WINDOWS

Launch Vehicle, Ablestar
　USE　ABLESTAR LAUNCH VEHICLE

Launch Vehicle, Ariane
　USE　ARIANE LAUNCH VEHICLE

Launch Vehicle, Atlas Able 5
　USE　ATLAS ABLE 5 LAUNCH VEHICLE

Launch Vehicle, Atlas Agena B
　USE　ATLAS AGENA B LAUNCH VEHICLE

Launch Vehicle, Atlas Centaur
　USE　ATLAS CENTAUR LAUNCH VEHICLE

Launch Vehicle, Atlas SLV-3
　USE　ATLAS SLV-3 LAUNCH VEHICLE

Launch Vehicle, Black Arrow
　USE　BLACK KNIGHT ROCKET VEHICLE

Launch Vehicle, Blue Streak
　USE　BLUE STREAK LAUNCH VEHICLE

Launch Vehicle, Centaur
　USE　CENTAUR LAUNCH VEHICLE

LAUNCH VEHICLE CONFIGURATIONS

Launch Vehicle, Delta
　USE　DELTA LAUNCH VEHICLE

Launch Vehicle, Diamant
　USE　DIAMANT LAUNCH VEHICLE

Launch Vehicle, Eldo
　USE　ELDO LAUNCH VEHICLE

Launch Vehicle, Europa 1
　USE　EUROPA 1 LAUNCH VEHICLE

Launch Vehicle, Europa 2
　USE　EUROPA 2 LAUNCH VEHICLE

Launch Vehicle, Europa 3
　USE　EUROPA 3 LAUNCH VEHICLE

Launch Vehicle, Europa 4
　USE　EUROPA 4 LAUNCH VEHICLE

Launch Vehicle, Juno 1
　USE　JUNO 1 LAUNCH VEHICLE

Launch Vehicle, Juno 2
　USE　JUNO 2 LAUNCH VEHICLE

Launch Vehicle, Little Joe 2
　USE　LITTLE JOE 2 LAUNCH VEHICLE

Launch Vehicle, Nomad
　USE　NOMAD LAUNCH VEHICLE

Launch Vehicle Program, National
　USE　NATIONAL LAUNCH VEHICLE PROGRAM

Launch Vehicle, RAM B
　USE　RAM B LAUNCH VEHICLE

Launch Vehicle, Saturn D
　USE　SATURN D LAUNCH VEHICLE

Launch Vehicle, Saturn 1 SA-1
　USE　SATURN 1 SA-1 LAUNCH VEHICLE

Launch Vehicle, Saturn 1 SA-2
　USE　SATURN 1 SA-2 LAUNCH VEHICLE

Launch Vehicle, Saturn 1 SA-3
　USE　SATURN 1 SA-3 LAUNCH VEHICLE

Launch Vehicle, Saturn 1 SA-4
　USE　SATURN 1 SA-4 LAUNCH VEHICLE

Launch Vehicle, Saturn 1 SA-5
　USE　SATURN 1 SA-5 LAUNCH VEHICLE

Launch Vehicle, Saturn 1 SA-6
　USE　SATURN 1 SA-6 LAUNCH VEHICLE

Launch Vehicle, Saturn 1 SA-7
　USE　SATURN 1 SA-7 LAUNCH VEHICLE

Launch Vehicle, Saturn 1 SA-8
　USE　SATURN 1 SA-8 LAUNCH VEHICLE

Launch Vehicle, Saturn 1 SA-9
　USE　SATURN 1 SA-9 LAUNCH VEHICLE

Launch Vehicle, Saturn 1 SA-10
　USE　SATURN 1 SA-10 LAUNCH VEHICLE

Launch Vehicle, Scout
　USE　SCOUT LAUNCH VEHICLE

Launch Vehicle, Thor Agena
　USE　THOR AGENA LAUNCH VEHICLE

Launch Vehicle, Thor Delta
　USE　THOR DELTA LAUNCH VEHICLE

Launch Vehicle, Titan Centaur
　USE　TITAN CENTAUR LAUNCH VEHICLE

Launch Vehicle, Titan 3
　USE　TITAN 3 LAUNCH VEHICLE

Launch Vehicle, Vanguard 2
　USE　VANGUARD 2 LAUNCH VEHICLE

Launch Vehicle, Vega
　USE　VEGA LAUNCH VEHICLE

Launch Vehicle 3, Standard
　USE　ATLAS SLV-3 LAUNCH VEHICLE

Launch Vehicle 5, Standard
　USE　STANDARD LAUNCH VEHICLE 5

LAUNCH VEHICLES

Launch Vehicles, Atlas
　USE　ATLAS LAUNCH VEHICLES

Launch Vehicles, Atlas Agena
　USE　ATLAS AGENA LAUNCH VEHICLES

Launch Vehicles, Europa
　USE　EUROPA LAUNCH VEHICLES

Launch Vehicles, Heavy Lift
　USE　HEAVY LIFT LAUNCH VEHICLES

Launch Vehicles, Juno
　USE　JUNO LAUNCH VEHICLES

Launch Vehicles, Nova
　USE　NOVA LAUNCH VEHICLES

Launch Vehicles, Recoverable
　USE　RECOVERABLE LAUNCH VEHICLES

Launch Vehicles, Reusable
　USE　REUSABLE LAUNCH VEHICLES

Launch Vehicles, Saturn
　USE　SATURN LAUNCH VEHICLES

Launch Vehicles, Saturn 1
　USE　SATURN 1 LAUNCH VEHICLES

Launch Vehicles, Saturn 1B
　USE　SATURN 1B LAUNCH VEHICLES

Launch Vehicles, Saturn 2
　USE　SATURN 2 LAUNCH VEHICLES

Launch Vehicles, Saturn 5
　USE　SATURN 5 LAUNCH VEHICLES

Launch Vehicles, Standard
　USE　STANDARD LAUNCH VEHICLES

Launch Vehicles, Thor
　USE　THOR LAUNCH VEHICLES

Launch Vehicles, Thorad
　USE　THORAD LAUNCH VEHICLES

Launch Vehicles, Titan
　USE　TITAN LAUNCH VEHICLES

LAUNCH WINDOWS

LAUNCHERS

Launchers, Gun
　USE　GUN LAUNCHERS

Launchers, Hypervelocity
  USE   HYPERVELOCITY LAUNCHERS

Launchers, Missile
  USE   MISSILE LAUNCHERS

Launchers, Mobile Missile
  USE   MOBILE MISSILE LAUNCHERS

Launchers, Rocket
  USE   ROCKET LAUNCHERS

LAUNCHING

Launching, Air
  USE   AIR LAUNCHING

LAUNCHING BASES

Launching Devices
  USE   LAUNCHERS

Launching Devices, Aircraft
  USE   AIRCRAFT LAUNCHING DEVICES

Launching, Orbital
  USE   ORBITAL LAUNCHING

LAUNCHING PADS

Launching, Rocket
  USE   ROCKET LAUNCHING

Launching, Satellite
  USE   SPACECRAFT LAUNCHING

Launching, Sea
  USE   SEA LAUNCHING

LAUNCHING SITES

Launching, Spacecraft
  USE   SPACECRAFT LAUNCHING

LAVA

Laval Nozzles, De
  USE   CONVERGENT-DIVERGENT NOZZLES

LAVAL NUMBER

LAW

Law, Air
  USE   AIR LAW

Law, Beer
  USE   BEER LAW

Law, Bouguer
  USE   BOUGUER LAW

Law, Child-Langmuir
  USE   CHILD-LANGMUIR LAW

Law, Closure
  USE   CLOSURE LAW

Law, Coffin-Manson
  USE   COFFIN-MANSON LAW

Law, Curie-Weiss
  USE   CURIE-WEISS LAW

Law, Dalton
  USE   DALTON LAW

Law (Fluid Mechanics), Stokes
  USE   STOKES LAW (FLUID MECHANICS)

Law, Fourier
  USE   FOURIER LAW

Law, Henry
  USE   HENRY LAW

Law, Hookes
  USE   HOOKES LAW

Law, International
  USE   INTERNATIONAL LAW

LAW (JURISPRUDENCE)

Law, Kirchhoff
  USE   KIRCHHOFF LAW

Law, Lambert
  USE   BOUGUER LAW

Law, Newton Pressure
  USE   NEWTON PRESSURE LAW

Law, Newton Second
  USE   NEWTON SECOND LAW

Law, Newton-Busemann
  USE   NEWTON-BUSEMANN LAW

Law Of Networks, Kirchhoff
  USE   KIRCHHOFF LAW OF NETWORKS

Law Of Radiation, Kirchhoff
  USE   KIRCHHOFF LAW OF RADIATION

Law Of Radiation, Stokes
  USE   STOKES LAW OF RADIATION

Law, Ohms
  USE   OHMS LAW

Law, Public
  USE   PUBLIC LAW

Law, Raoult
  USE   RAOULT LAW

Law, Reynolds
  USE   REYNOLDS EQUATION

Law, Sea
  USE   SEA LAW

Law, Similitude
  USE   SIMILITUDE LAW

Law, Snells
  USE   SNELLS LAW

Law, Space
  USE   SPACE LAW

Law, Stefan-Boltzmann
  USE   STEFAN-BOLTZMANN LAW

Law, Stokes
  USE   STOKES LAW

Law, Tafel
  USE   TAFEL LAW

Law, Weber-Fechner
  USE   WEBER-FECHNER LAW

Lawrence Valley (North America), St
  USE   ST LAWRENCE VALLEY (NORTH AMERICA)

LAWRENCIUM

LAWS

Laws, Conservation
  USE   CONSERVATION LAWS

Laws, Kepler
  USE   KEPLER LAWS

Laws, Radiation
  USE   RADIATION LAWS

Laws, Scaling
  USE   SCALING LAWS

LAY-UP

Layer, Atmospheric Boundary
  USE   ATMOSPHERIC BOUNDARY LAYER

Layer, Chapman Shear
  USE   SHEAR LAYERS

Layer Chromatography, Thin
  USE   THIN LAYER CHROMATOGRAPHY

Layer Combustion, Boundary
  USE   BOUNDARY LAYER COMBUSTION

Layer, Compressible Boundary
  USE   COMPRESSIBLE BOUNDARY LAYER

Layer Control, Boundary
  USE   BOUNDARY LAYER CONTROL

Layer Control, Porous Boundary
  USE   POROUS BOUNDARY LAYER CONTROL

Layer, D
  USE   D REGION

Layer, E-1
  USE   E-1 LAYER

Layer, E-2
  USE   E-2 LAYER

Layer, Ekman
  USE   EKMAN LAYER

Layer Equations, Boundary
  USE   BOUNDARY LAYER EQUATIONS

Layer, F
  USE   F REGION

Layer Flow, Boundary
  USE   BOUNDARY LAYER FLOW

Layer, Hypersonic Boundary
  USE   HYPERSONIC BOUNDARY LAYER

Layer, Incompressible Boundary
  USE   INCOMPRESSIBLE BOUNDARY LAYER

Layer, Laminar Boundary
  USE   LAMINAR BOUNDARY LAYER

Layer, Night E
  USE   NIGHT SKY
         E REGION

Layer, Night F
  USE   F REGION
         NIGHT SKY

Layer Noise, Boundary
  USE   BOUNDARY LAYERS
         AERODYNAMIC NOISE

Layer, Planetary Boundary
  USE   PLANETARY BOUNDARY LAYER

Layer Plasmas, Boundary
  USE   BOUNDARY LAYER PLASMAS

Layer Separation, Boundary
  USE   BOUNDARY LAYER SEPARATION

Layer Separation, Laminar Boundary
  USE   LAMINAR BOUNDARY LAYER

Layer, Sporadic E
  USE   SPORADIC E LAYER

Layer Stability, Boundary
  USE   BOUNDARY LAYER STABILITY

Layer, Thermal Boundary
  USE   THERMAL BOUNDARY LAYER

Layer, Three Dimensional Boundary
  USE   THREE DIMENSIONAL BOUNDARY LAYER

Layer Transition, Boundary
  USE   BOUNDARY LAYER TRANSITION

Layer, Turbulent Boundary
  USE   TURBULENT BOUNDARY LAYER

Layer, Two Dimensional Boundary
  USE   TWO DIMENSIONAL BOUNDARY LAYER

LAYERS

Layers, Barrier
  USE   BARRIER LAYERS

Layers, Boundary
USE   BOUNDARY LAYERS

Layers, Deep Scattering
USE   DEEP SCATTERING LAYERS

Layers, E
USE   E REGION

Layers, Flat
USE   FLAT LAYERS

Layers, Inter
USE   INTERLAYERS

Layers, Isothermal
USE   ISOTHERMAL LAYERS

Layers, Plasma
USE   PLASMA LAYERS

Layers, Shear
USE   SHEAR LAYERS

Layers, Shock
USE   SHOCK LAYERS

Layers, Stratified
USE   STRATA

Layers, Supersonic Boundary
USE   SUPERSONIC BOUNDARY LAYERS

Layers, Surface
USE   SURFACE LAYERS

Layers, Transition
USE   TRANSITION LAYERS

LAYOUTS

LAZAREV METEORITE

LC CIRCUITS

LCRE Reactor
USE   LITHIUM COOLED REACTOR EXPERIMENT

LDEF
USE   LONG DURATION EXPOSURE FACILITY

LEACHING

LEAD ACETATES

LEAD ACID BATTERIES

LEAD ALLOYS

LEAD CHLORIDES

LEAD COMPOUNDS

LEAD ISOTOPES

LEAD (METAL)

LEAD MOLYBDATES

LEAD ORGANIC COMPOUNDS

LEAD OXIDES

LEAD POISONING

LEAD SELENIDES

LEAD SULFIDES

LEAD TELLURIDES

LEAD TITANATES

LEAD TUNGSTATES

LEAD ZIRCONATE TITANATES

LEADERSHIP

LEADING EDGE FLAPS

LEADING EDGE SLATS

LEADING EDGE SWEEP

LEADING EDGE THRUST

LEADING EDGES

Leading Edges, Blunt
USE   BLUNT LEADING EDGES

Leading Edges, Sharp
USE   SHARP LEADING EDGES

Leads, Beam
USE   BEAM LEADS

Leads, Electrical
USE   ELECTRIC CONDUCTORS

LEAKAGE

LEAR JET AIRCRAFT

LEARNING

(Learning), Conditioning
USE   CONDITIONING (LEARNING)

LEARNING CURVES

(Learning), Habituation
USE   HABITUATION (LEARNING)

Learning, Machine
USE   LEARNING MACHINES

LEARNING MACHINES

Learning, Maze
USE   MAZE LEARNING

LEARNING THEORY

LEASING

LEAST SQUARES METHOD

LEATHER

LEAVES

LEBANON

LEBESGUE THEOREM

LECTURES

LED (Diodes)
USE   LIGHT EMITTING DIODES

LEDGES

Lee Theory, Crocco-
USE   CROCCO-LEE THEORY

Lee Topography, Stoss-And-
USE   GLACIAL DRIFT

LEE WAVES

LEG (ANATOMY)

LEGAL LIABILITY

Legendre Code
USE   COMPUTER PROGRAMMING
       NEUTRON SCATTERING

LEGENDRE FUNCTIONS

Legendre Polynomials
USE   LEGENDRE FUNCTIONS

Legendre Transformation
USE   LEGENDRE FUNCTIONS

LEGIBILITY

LEGUMINOUS PLANTS

LEIDENFROST PHENOMENON

LEM (Lunar Module)
USE   LUNAR MODULE

Lemmas
USE   THEOREMS

LENGTH

Length, Debye
USE   DEBYE LENGTH

Length Flow Theory, Mixing
USE   MIXING LENGTH FLOW THEORY

Lengths, Wave
USE   WAVELENGTHS

LENNARD-JONES GAS

LENNARD-JONES POTENTIAL

LENS ANTENNAS

LENS DESIGN

LENSES

Lenses, Contact
USE   CONTACT LENSES

Lenses, Fresnel
USE   FRESNEL LENSES

Lenses, Gravitational
USE   GRAVITATIONAL LENSES

Lenses, Luneberg
USE   RADAR CORNER REFLECTORS

Lenses, Magnetic
USE   MAGNETIC LENSES

Lenses, Quadrupole
USE   MAGNETIC LENSES

Lenses, Wide Angle
USE   WIDE ANGLE LENSES

Lenses, Wire Grid
USE   WIRE GRID LENSES

Lenses, Zoom
USE   ZOOM LENSES

LENTICULAR BODIES

LEON-QUERETARO AREA (MEXICO)

Leone, Sierra
USE   SIERRA LEONE

LEONID METEOROIDS

LEPTONS

LES (Escape Systems)
USE   LAUNCH ESCAPE SYSTEMS

LES (Satellites)
USE   LINCOLN EXPERIMENTAL SATELLITES

LESA (Lunar Exploration System)
USE   LUNAR EXPLORATION SYSTEM FOR
       APOLLO

LESIONS

Lesions, Pulmonary
USE   PULMONARY LESIONS

LESOTHO

LESSER ANTILLES

(LET), Linear Energy Transfer
USE   LINEAR ENERGY TRANSFER (LET)

LETHALITY

LETHARGY

Letters (Symbols)
　　USE　SYMBOLS

LEUCINE

Leucine, Nor
　　USE　NORLEUCINE

LEUKEMIAS

LEUKOCYTES

LEUKOPENIA

LEVEL

LEVEL (HORIZONTAL)

Level Languages, High
　　USE　HIGH LEVEL LANGUAGES

LEVEL (QUANTITY)

Level, Sea
　　USE　SEA LEVEL

Level Turbulence, Low
　　USE　LOW LEVEL TURBULENCE

LEVELING

Levels, Atomic Energy
　　USE　ATOMIC ENERGY LEVELS

Levels, Effective Perceived Noise
　　USE　EFFECTIVE PERCEIVED NOISE LEVELS

Levels, Electronic
　　USE　ENERGY LEVELS
　　　　　ELECTRON ENERGY

Levels, Energy
　　USE　ENERGY LEVELS

Levels, Liquid
　　USE　LIQUID LEVELS

Levels, Molecular Energy
　　USE　MOLECULAR ENERGY LEVELS

LEVERS

LEVITATION

Levitation, Acoustic
　　USE　ACOUSTIC LEVITATION

LEVITATION MELTING

Levitation Vehicles, Magnetic
　　USE　MAGNETIC LEVITATION VEHICLES

LEWIS BASE

LEWIS NUMBERS

LEXAN (TRADEMARK)

LFO
　　USE　LANDSAT FOLLOW-ON MISSIONS

Li
　　USE　LITHIUM

LIABILITIES

Liability, Legal
　　USE　LEGAL LIABILITY

LIAPUNOV FUNCTIONS

(Liberation), Evolution
　　USE　EVOLUTION (LIBERATION)

LIBERIA

LIBRARIES

Libraries (Computers), Subroutine
　　USE　SUBROUTINE LIBRARIES (COMPUTERS)

Library Systems, Integrated
　　USE　INTEGRATED LIBRARY SYSTEMS

LIBRATION

LIBRATIONAL MOTION

LIBYA

LIBYAN DESERT

LICENSING

LICHENS

Lidar
　　USE　OPTICAL RADAR

LIE GROUPS

LIECHTENSTEIN

LIENARD POTENTIAL

LIES

Life (Biology)
　　USE　LIFE SCIENCES

LIFE CYCLE COSTS

LIFE DETECTORS

LIFE (DURABILITY)

Life, Extraterrestrial
　　USE　EXTRATERRESTRIAL LIFE

Life, Fatigue
　　USE　FATIGUE LIFE

Life, Half
　　USE　HALF LIFE

Life, Machine
　　USE　SERVICE LIFE

LIFE RAFTS

LIFE SCIENCES

Life, Service
　　USE　SERVICE LIFE

LIFE SPAN

Life Support Sys, Integrated Maneuvering
　　USE　IMLSS

LIFE SUPPORT SYSTEMS

Life Support Systems, Bioregenerative
　　USE　CLOSED ECOLOGICAL SYSTEMS

Life Support Systems, Portable
　　USE　PORTABLE LIFE SUPPORT SYSTEMS

Life Sustaining Systems, Emergency
　　USE　EMERGENCY LIFE SUSTAINING SYSTEMS

Life Tests, Accelerated
　　USE　ACCELERATED LIFE TESTS

LIFEBOATS

Lifetime, Carrier
　　USE　CARRIER LIFETIME

Lifetime (Durability)
　　USE　LIFE (DURABILITY)

Lifetime, Orbital
　　USE　ORBITAL LIFETIME

Lifetime, Plasma
　　USE　PLASMA LIFETIME

Lifetime, Radiative
　　USE　RADIATIVE LIFETIME

Lifetime, Satellite
　　USE　SATELLITE LIFETIME

LIFT

Lift, Aerodynamic
　　USE　LIFT

Lift Aircraft, Powered
　　USE　POWERED LIFT AIRCRAFT

Lift Airships, Heavy
　　USE　HEAVY LIFT AIRSHIPS

LIFT AUGMENTATION

Lift Coefficients
　　USE　LIFT
　　　　　AERODYNAMIC COEFFICIENTS

Lift Controls, Direct
　　USE　DIRECT LIFT CONTROLS

LIFT DEVICES

Lift Distribution
　　USE　LIFT
　　　　　FORCE DISTRIBUTION

LIFT DRAG RATIO

LIFT FANS

Lift Forces
　　USE　LIFT

Lift Helicopters, Heavy
　　USE　HEAVY LIFT HELICOPTERS

Lift, Interference
　　USE　INTERFERENCE LIFT

Lift, Jet
　　USE　JET LIFT

Lift Launch Vehicles, Heavy
　　USE　HEAVY LIFT LAUNCH VEHICLES

Lift, Rotor
　　USE　ROTOR LIFT

Lift, Variable
　　USE　LIFT

Lift, Zero
　　USE　ZERO LIFT

LIFTING BODIES

Lifting Body, M-2
　　USE　M-2 LIFTING BODY

Lifting Body, M-2F2
　　USE　M-2F2 LIFTING BODY

Lifting Body, M-2F3
　　USE　M-2F3 LIFTING BODY

LIFTING REENTRY VEHICLES

LIFTING ROTORS

Lifting Surfaces
　　USE　LIFTING BODIES
　　　　　LIFT DEVICES
　　　　　SURFACES

LIFTS

(Lifts), Elevators
　　USE　ELEVATORS (LIFTS)

(Lifts), Jacks
　　USE　JACKS (LIFTS)

LIGAMENTS

LIGANDS

Light Absorption
USE   ELECTROMAGNETIC ABSORPTION

LIGHT ADAPTATION

LIGHT AIRBORNE MULTIPURPOSE SYSTEM

LIGHT AIRCRAFT

Light Aircraft Readiness Monitor, Automatic
USE   ALARM PROJECT

LIGHT ALLOYS

LIGHT AMPLIFIERS

Light Armed Reconnaissance Aircraft
USE   COIN AIRCRAFT

LIGHT BEAMS

Light Bulbs
USE   LUMINAIRES

Light, Coherent
USE   COHERENT LIGHT

Light Communication
USE   OPTICAL COMMUNICATION

LIGHT CURVE

Light Duration
USE   PULSE DURATION
       FLASH

LIGHT ELEMENTS

LIGHT EMISSION

LIGHT EMITTING DIODES

Light, Extragalactic
USE   LIGHT (VISIBLE RADIATION)
       EXTRATERRESTRIAL RADIATION

LIGHT GAS GUNS

Light Holography, White
USE   WHITE LIGHT HOLOGRAPHY

Light Intensity
USE   LUMINOUS INTENSITY

LIGHT INTRATHEATER TRANSPORT

LIGHT IONS

LIGHT MODULATION

(Light Modulation), ULM
USE   ULTRASONIC LIGHT MODULATION

Light Modulation, Ultrasonic
USE   ULTRASONIC LIGHT MODULATION

Light, Polarized
USE   POLARIZED LIGHT

Light Pressure
USE   ILLUMINANCE

Light Probes
USE   LIGHT BEAMS

Light Ratios, Mass To
USE   MASS TO LIGHT RATIOS

LIGHT SCATTERING

LIGHT SCATTERING METERS

LIGHT SOURCES

LIGHT SPEED

Light, Sun
USE   SUNLIGHT

LIGHT TRANSMISSION

LIGHT TRANSPORT AIRCRAFT

Light Twin Aircraft, Advanced Technology
USE   ATLIT PROJECT

Light, Ultraviolet
USE   ULTRAVIOLET RADIATION

LIGHT VALVES

LIGHT (VISIBLE RADIATION)

LIGHT WATER

LIGHT WATER BREEDER REACTORS

LIGHT WATER REACTORS

Light, Zodiacal
USE   ZODIACAL LIGHT

LIGHT-CONE EXPANSION

Lightbulb Engines, Nuclear
USE   NUCLEAR LIGHTBULB ENGINES

LIGHTHILL GAS MODEL

LIGHTHILL METHOD

Lighting
USE   ILLUMINATING

LIGHTING EQUIPMENT

LIGHTNING

Lightning, Ball
USE   BALL LIGHTNING

LIGHTNING SUPPRESSION

Lights
USE   LUMINAIRES

Lights, Aircraft
USE   AIRCRAFT LIGHTS

Lights, Airport
USE   AIRPORT LIGHTS

Lights, Runway
USE   RUNWAY LIGHTS

Lights, Search
USE   SEARCHLIGHTS

LIGNIN

LIGNITE

Likelihood Estimates, Maximum
USE   MAXIMUM LIKELIHOOD ESTIMATES

LIKELIHOOD RATIO

LIMB BRIGHTENING

LIMB DARKENING

Limb, Earth
USE   EARTH LIMB

Limb, Lunar
USE   LUNAR LIMB

Limb, Planetary
USE   PLANETARY LIMB

Limb, Solar
USE   SOLAR LIMB

LIMBS

LIMBS (ANATOMY)

Lime
USE   CALCIUM OXIDES

LIMEN

LIMESTONE

Limit, Proportional
USE   PROPORTIONAL LIMIT

Limit, Roche
USE   ROCHE LIMIT

Limitations
USE   CONSTRAINTS

Limited Cameras, Diffraction
USE   DIFFRACTION LIMITED CAMERAS

Limited, International Computers
USE   ICL COMPUTERS

Limited Spacecraft, Power
USE   POWER LIMITED SPACECRAFT

LIMITER AMPLIFIERS

LIMITER CIRCUITS

LIMITERS (FUSION REACTORS)

Limiters, Power
USE   POWER LIMITERS

LIMITS

Limits, Confidence
USE   CONFIDENCE LIMITS

Limits, Ignition
USE   IGNITION LIMITS

LIMITS (MATHEMATICS)

LIMNOLOGY

LIMONITE

LINCOLN EXPERIMENTAL SATELLITES

Line Analysis, Program Trend
USE   PROGRAM TREND LINE ANALYSIS

LINE CURRENT

Line Discriminators, Fraunhofer
USE   FRAUNHOFER LINE DISCRIMINATORS

Line, H Alpha
USE   H ALPHA LINE

Line, H Beta
USE   H BETA LINE

Line, H Gamma
USE   H GAMMA LINE

LINE OF SIGHT

LINE OF SIGHT COMMUNICATION

Line Programming, On-
USE   ON-LINE PROGRAMMING

LINE SHAPE

LINE SPECTRA

Line Systems, On-
USE   ON-LINE SYSTEMS

Line, Timber
USE   TIMBERLINE

Line Width, Spectral
USE   SPECTRAL LINE WIDTH

Lineament
USE   STRUCTURAL PROPERTIES (GEOLOGY)

LINEAR ACCELERATORS

LINEAR AMPLIFIERS

LINEAR ARRAYS

Linear Arrays, Multispectral
USE   MULTISPECTRAL LINEAR ARRAYS

LINEAR CIRCUITS

LINEAR ENERGY TRANSFER (LET)

LINEAR EQUATIONS

LINEAR EVOLUTION EQUATIONS

LINEAR FILTERS

LINEAR INTEGRATED CIRCUITS

LINEAR POLARIZATION

LINEAR PREDICTION

LINEAR PROGRAMMING

LINEAR RECEIVERS

LINEAR SYSTEMS

LINEAR TRANSFORMATIONS

LINEAR VIBRATION

LINEARITY

Linearity, Col
USE   COLLINEARITY

Linearity, Non
USE   NONLINEARITY

LINEARIZATION

LINEN

Liners
USE   LININGS

LINES

Lines, Acoustic Delay
USE   ACOUSTIC DELAY LINES

Lines), Axes (Reference
USE   AXES (REFERENCE LINES)

Lines, Caustic
USE   CAUSTIC LINES

Lines (Computer Storage), Delay
USE   DELAY LINES (COMPUTER STORAGE)

Lines, D
USE   D LINES

Lines, Delay
USE   DELAY LINES

Lines, Dielectronic Satellite
USE   RESONANCE LINES

Lines, Flat Coaxial Transmission
USE   MICROSTRIP TRANSMISSION LINES

Lines, Fluid Transmission
USE   FLUID TRANSMISSION LINES

Lines, Fraunhofer
USE   FRAUNHOFER LINES

Lines, Geodesic
USE   GEODESIC LINES

LINES (GEOMETRY)

Lines, H
USE   H LINES

Lines, K
USE   K· LINES

Lines, Microstrip Transmission
USE   MICROSTRIP TRANSMISSION LINES

LINES OF FORCE

Lines, Parallel Strip
USE   MICROSTRIP TRANSMISSION LINES

Lines, Power
USE   POWER LINES

Lines, Resonance
USE   RESONANCE LINES

Lines, Spectral
USE   LINE SPECTRA

Lines, Strip Transmission
USE   STRIP TRANSMISSION LINES

Lines, Telluric
USE   TELLURIC LINES

Lines, Terminator
USE   TERMINATOR LINES

Lines, Tether
USE   TETHERLINES

Lines, Transmission
USE   TRANSMISSION LINES

(Lines), Trunks
USE   TRANSMISSION LINES

Lines, Underground Transmission
USE   UNDERGROUND TRANSMISSION LINES

LING-TEMCO-VOUGHT AIRCRAFT

LINGUISTICS

LINING PROCESSES

LININGS

Linings, Rocket
USE   ROCKET LININGS

LINKAGES

Linking
USE   JOINING

LINKS

Links, Data
USE   DATA LINKS

LINKS (MATHEMATICS)

LIOUVILLE EQUATIONS

Liouville Operator, Sturm-
USE   STURM-LIOUVILLE THEORY

LIOUVILLE THEOREM

Liouville Theory, Sturm-
USE   STURM-LIOUVILLE THEORY

LIP READING

LIPID METABOLISM

LIPIDS

LIPOIC ACID

LIPOPROTEINS

LIPS (ANATOMY)

LIPSCHITZ CONDITION

LIQUEFACTION

Liquefaction, Coal
USE   COAL LIQUEFACTION

Liquefaction, Gas
USE   CONDENSING

LIQUEFIED GASES

LIQUEFIED NATURAL GAS

(Liquefiers), Condensers
USE   CONDENSERS (LIQUEFIERS)

LIQUID AIR

LIQUID AIR CYCLE ENGINES

LIQUID ALLOYS

LIQUID AMMONIA

LIQUID ATOMIZATION

LIQUID BEARINGS

LIQUID BREATHING

LIQUID CHROMATOGRAPHY

LIQUID COOLED REACTORS

LIQUID COOLING

LIQUID CRYSTALS

Liquid Drops
USE   DROPS (LIQUIDS)

Liquid Equilibrium, Vapor
USE   LIQUID-VAPOR EQUILIBRIUM

LIQUID FILLED SHELLS

LIQUID FLOW

LIQUID FLUORINE

LIQUID FUELS

LIQUID HELIUM

LIQUID HELIUM 2

LIQUID HYDROGEN

LIQUID INJECTION

Liquid Interactions, Gas-
USE   GAS-LIQUID INTERACTIONS

Liquid Interfaces, Liquid-
USE   LIQUID-LIQUID INTERFACES

LIQUID LASERS

LIQUID LEVELS

LIQUID LITHIUM

Liquid Mercury
USE   MERCURY (METAL)

LIQUID METAL COOLED REACTORS

LIQUID METAL FAST BREEDER REACTORS

LIQUID METALS

LIQUID NEON

LIQUID NITROGEN

LIQUID OXIDIZERS

LIQUID OXYGEN

Liquid Oxygen, Fluorine-
USE   FLOX

LIQUID PHASE EPITAXY

LIQUID PHASES

Liquid Plus Solid Zones
USE   MUSHY ZONES

LIQUID POTASSIUM

LIQUID PROPELLANT ROCKET ENGINES

LIQUID ROCKET PROPELLANTS

Liquid Rotation
  USE   ROTATING LIQUIDS

LIQUID SLOSHING

LIQUID SODIUM

LIQUID SURFACES

LIQUID WASTES

LIQUID-GAS MIXTURES

LIQUID-LIQUID INTERFACES

LIQUID-SOLID INTERFACES

LIQUID-VAPOR EQUILIBRIUM

LIQUID-VAPOR INTERFACES

LIQUIDS

Liquids, Coal Derived
  USE   COAL DERIVED LIQUIDS

(Liquids), Drops
  USE   DROPS (LIQUIDS)

Liquids, Fermi
  USE   FERMI LIQUIDS

Liquids, Organic
  USE   ORGANIC LIQUIDS

Liquids, Potable
  USE   POTABLE LIQUIDS

Liquids, Rotating
  USE   ROTATING LIQUIDS

LIQUIDUS

LIRTS (TELESCOPE)

LISP (PROGRAMMING LANGUAGE)

LISSAJOUS FIGURES

LISTS

Lists, Hardware Utilization
  USE   HARDWARE UTILIZATION LISTS

LITERATURE

LITHERGOL ROCKET ENGINES

Lithergolic Propellants
  USE   HYBRID PROPELLANTS

LITHIASIS

Lithiasis, Uro
  USE   UROLITHIASIS

LITHIUM

LITHIUM ALLOYS

LITHIUM ALUMINUM HYDRIDES

LITHIUM BORATES

LITHIUM CHLORIDES

LITHIUM COMPOUNDS

Lithium Compounds, Organic
  USE   ORGANIC LITHIUM COMPOUNDS

LITHIUM COOLED REACTOR EXPERIMENT

LITHIUM FLUORIDES

LITHIUM HYDRIDES

LITHIUM HYDROXIDES

LITHIUM IODATES

LITHIUM ISOTOPES

Lithium, Liquid
  USE   LIQUID LITHIUM

LITHIUM NIOBATES

LITHIUM OXIDES

LITHIUM PERCHLORATES

LITHIUM SULFATES

LITHIUM SULFUR BATTERIES

Lithium 4
  USE   LITHIUM ISOTOPES

Lithium 6
  USE   LITHIUM ISOTOPES

LITHOGRAPHY

Lithography, Photo
  USE   PHOTOLITHOGRAPHY

LITHOLOGY

LITHOSPHERE

LITHUANIA

LITTLE JOE 2 LAUNCH VEHICLE

LITTLE JOHN ROCKET VEHICLE

Littoral Currents
  USE   COASTAL CURRENTS

LITTORAL DRIFT

LITTORAL TRANSPORT

LIVER

LIVERMORE POOL TYPE REACTOR

LIVESTOCK

LIXISCOPES

LIZARDS

LLANOS ORIENTALES (COLOMBIA)

LMCR (Reactors)
  USE   LIQUID METAL COOLED REACTORS

LMFBR
  USE   LIQUID METAL FAST BREEDER REACTORS

LNG
  USE   LIQUEFIED NATURAL GAS

LOAD DISTRIBUTION (FORCES)

Load Factors
  USE   LOADS (FORCES)

Load Recorders, Flight
  USE   FLIGHT LOAD RECORDERS

LOAD TESTING MACHINES

LOAD TESTS

LOADING

Loading, Atmospheric
  USE   POLLUTION TRANSPORT

Loading, Critical
  USE   CRITICAL LOADING

Loading, Edge
  USE   EDGE LOADING

Loading Forces
  USE   LOADS (FORCES)

LOADING MOMENTS

LOADING OPERATIONS

LOADING RATE

Loading Waves
  USE   LOADS (FORCES)
          ELASTIC WAVES

Loading, Wing
  USE   WING LOADING

Loads, Aerodynamic
  USE   AERODYNAMIC LOADS

Loads, Axial
  USE   AXIAL LOADS

Loads, Axial Compression
  USE   AXIAL COMPRESSION LOADS

Loads, Blast
  USE   BLAST LOADS

Loads, Compression
  USE   COMPRESSION LOADS

Loads, Cyclic
  USE   CYCLIC LOADS

Loads, Dummy
  USE   LOADING
          IMPEDANCE
          OUTPUT

Loads, Dynamic
  USE   DYNAMIC LOADS

LOADS (FORCES)

Loads, Gust
  USE   GUST LOADS

Loads, Impact
  USE   IMPACT LOADS

Loads, Landing
  USE   LANDING LOADS

Loads, Random
  USE   RANDOM LOADS

Loads, Rolling Contact
  USE   ROLLING CONTACT LOADS

Loads, Shock
  USE   SHOCK LOADS

Loads, Static
  USE   STATIC LOADS

Loads, Thrust
  USE   THRUST LOADS

Loads, Transient
  USE   TRANSIENT LOADS

Loads, Vibratory
  USE   VIBRATORY LOADS

LOBES

Lobes, Back
  USE   BACKLOBES

Lobes, Occipital
  USE   OCCIPITAL LOBES

Lobes, Side
  USE   SIDELOBES

LOCAL GROUP (ASTRONOMY)

LOCAL SCIENTIFIC SURVEY MODULE

Localization
  USE   POSITION (LOCATION)

Localization, Sound
    USE   SOUND LOCALIZATION

LOCATES SYSTEM

Location
    USE   POSITION (LOCATION)

Location Exper, Feature Identification And
    USE   FEATURE IDENTIFICATION AND LOCATION
              EXPER

Location Of Air Traffic Satellites
    USE   LOCATES SYSTEM

(Location), Position
    USE   POSITION (LOCATION)

Locator Transmitters, Emergency
    USE   EMERGENCY LOCATOR TRANSMITTERS

LOCI

Lock Demodulators, Phase
    USE   PHASE LOCK DEMODULATORS

Locked Systems, Phase
    USE   PHASE LOCKED SYSTEMS

LOCKHEED AIRCRAFT

Lockheed C-5 Aircraft
    USE   C-5 AIRCRAFT

Lockheed CL-595 Helicopter
    USE   XH-51 HELICOPTER

Lockheed CL-823 Aircraft
    USE   CL-823 AIRCRAFT

Lockheed Constellation Aircraft
    USE   C-121 AIRCRAFT

Lockheed L-2000 Aircraft
    USE   L-2000 AIRCRAFT

LOCKHEED MODEL 18 AIRCRAFT

Lockheed U-2 Aircraft
    USE   U-2 AIRCRAFT

Lockheed XV-4A Aircraft
    USE   XV-4 AIRCRAFT

Lockheed 186 Helicopter
    USE   XH-51 HELICOPTER

LOCKING

Locking, Injection
    USE   INJECTION LOCKING

Locking, Laser Mode
    USE   LASER MODE LOCKING

LOCKS

Locks, Air
    USE   AIR LOCKS

LOCKS (FASTENERS)

LOCOMOTION

Locomotion, Astronaut
    USE   ASTRONAUT LOCOMOTION

LOCOMOTIVES

LOCUSTS

Loeve Expansion, Karhunen-
    USE   KARHUNEN-LOEVE EXPANSION

LOFAR

LOFTI Satellites
    USE   LOW FREQUENCY TRANSIONOSPHERIC
              SATELLITES

LOFTING

LOG PERIODIC ANTENNAS

LOG SPIRAL ANTENNAS

LOGARITHMIC RECEIVERS

LOGARITHMS

LOGGING (INDUSTRY)

LOGIC

LOGIC CIRCUITS

LOGIC DESIGN

Logic, Fluid
    USE   FLUID LOGIC

Logic Integ Circuits, Diode-Transistor-
    USE   DTL INTEGRATED CIRCUITS

Logic Integ Circuits, Transistor-Transistor-
    USE   TTL INTEGRATED CIRCUITS

Logic, Mathematical
    USE   MATHEMATICAL LOGIC

Logic Networks
    USE   LOGIC CIRCUITS

LOGIC PROGRAMMING

Logic, Threshold
    USE   THRESHOLD LOGIC

Logic, Transistor
    USE   TRANSISTOR LOGIC

Logic Units, Arithmetic And
    USE   ARITHMETIC AND LOGIC UNITS

LOGICAL ELEMENTS

LOGISTICS

Logistics, Lunar
    USE   LUNAR LOGISTICS

LOGISTICS MANAGEMENT

LOGISTICS OVER THE SHORE (LOTS) CARRIER

Logistics, Space
    USE   SPACE LOGISTICS

LOH Helicopter
    USE   OH-6 HELICOPTER

LOKI ROCKET VEHICLE

LOLA (Simulator)
    USE   LUNAR ORBIT AND LANDING SIMULATORS

LOMONOSOV CURRENT

Long Base Interferometry, Very
    USE   VERY LONG BASE INTERFEROMETRY

LONG DURATION EXPOSURE FACILITY

LONG DURATION SPACE FLIGHT

LONG ISLAND (NY)

Long Range Navigation
    USE   LORAN D
              LORAN

LONG RANGE WEATHER FORECASTING

LONG TERM EFFECTS

Long Term Zonal Earth Energy Experiment
    USE   LZEEBE SATELLITE

LONG WAVE RADIATION

Long Waves (Meteorology)
    USE   PLANETARY WAVES

LONGERONS

LONGEVITY

LONGITUDE

LONGITUDE MEASUREMENT

Longitude, Solar
    USE   SOLAR LONGITUDE

LONGITUDINAL CONTROL

LONGITUDINAL STABILITY

LONGITUDINAL WAVES

Longshore Currents
    USE   COASTAL CURRENTS

LOOK ANGLES (ELECTRONICS)

LOOK ANGLES (TRACKING)

Looking Infrared Detectors, Forward
    USE   FLIR DETECTORS

Looking Radar, Side-
    USE   SIDE-LOOKING RADAR

LOOP ANTENNAS

Loop Systems, Closed
    USE   FEEDBACK CONTROL

LOOPS

Loops, Coronal
    USE   CORONAL LOOPS

Loops, Corrosion Test
    USE   CORROSION TEST LOOPS

LOR (Rendezvous)
    USE   LUNAR ORBITAL RENDEZVOUS

LORAC NAVIGATION SYSTEM

LORAN

LORAN C

LORAN D

LORENTZ CONTRACTION

Lorentz Contraction, Fitzgerald-
    USE   LORENTZ CONTRACTION

LORENTZ FORCE

LORENTZ GAS

LORENTZ TRANSFORMATIONS

LORV
    USE   LOW OBSERVABLE REENTRY VEHICLES

LOS ALAMOS MOLTEN PLUTONIUM REACTOR

Los Alamos Turret Reactor
    USE   HIGH TEMPERATURE NUCLEAR REACTORS

LOS ALAMOS WATER BOILER REACTOR

Loss Coefficient, Friction
    USE   FRICTION FACTOR

Loss, Coolant
    USE   LOSS OF COOLANT

Loss, Hearing
    USE   AUDITORY DEFECTS

Loss, Insertion
    USE   INSERTION LOSS

LOSS OF COOLANT

Loss, Plasma
USE   PLASMA LOSS

Loss, Power
USE   POWER LOSS

Loss, Transmission
USE   TRANSMISSION LOSS

Loss, Water
USE   WATER LOSS

LOSSES

Losses, Energy
USE   ENERGY DISSIPATION

LOSSLESS EQUIPMENT

LOSSLESS MATERIALS

LOSSY MEDIA

Lost Wax Process
USE   INVESTMENT CASTING

LOTS Cargo Ships
USE   CARGO SHIPS

(LOTS) Carrier, Logistics Over The Shore
USE   LOGISTICS OVER THE SHORE (LOTS)
           CARRIER

LOUDNESS

LOUDSPEAKERS

Louis-Kansas City Corridor (MO), St
USE   ST LOUIS-KANSAS CITY CORRIDOR (MO)

LOUISIANA

LOUNGES

Lounges, Mobile
USE   MOBILE LOUNGES

LOUVERS

LOVE WAVES

Low Alloy Steels
USE   HIGH STRENGTH STEELS

LOW ALTITUDE

Low Altitude Missile, Supersonic
USE   SUPERSONIC LOW ALTITUDE MISSILE

LOW ASPECT RATIO

LOW ASPECT RATIO WINGS

LOW CARBON STEELS

LOW CONCENTRATIONS

LOW CONDUCTIVITY

LOW COST

LOW CURRENTS

LOW DENSITY FLOW

Low Density Gases
USE   RAREFIED GASES

LOW DENSITY MATERIALS

LOW DENSITY RESEARCH

LOW DENSITY WIND TUNNELS

LOW FREQUENCIES

Low Frequencies, Extremely
USE   EXTREMELY LOW FREQUENCIES

Low Frequencies, Very
USE   VERY LOW FREQUENCIES

LOW FREQUENCY BANDS

LOW FREQUENCY TRANSIONOSPHERIC SATELLITES

Low Gravity
USE   REDUCED GRAVITY

LOW GRAVITY MANUFACTURING

Low Intensity X Ray Imaging Scope
USE   LIXISCOPES

Low Latitudes
USE   TROPICAL REGIONS

LOW LEVEL TURBULENCE

Low Mass
USE   MASS

LOW MOLECULAR WEIGHTS

LOW NOISE

LOW OBSERVABLE REENTRY VEHICLES

LOW PASS FILTERS

LOW PRESSURE

Low Pressure Chambers
USE   VACUUM CHAMBERS

Low Radio Frequencies, Extremely
USE   EXTREMELY LOW RADIO FREQUENCIES

LOW RESISTANCE

LOW REYNOLDS NUMBER

LOW SPEED

LOW SPEED STABILITY

LOW SPEED WIND TUNNELS

LOW TEMPERATURE

LOW TEMPERATURE BRAZING

LOW TEMPERATURE ENVIRONMENTS

LOW TEMPERATURE PHYSICS

Low Temperature Plasmas
USE   COLD PLASMAS

LOW TEMPERATURE TESTS

LOW THRUST

LOW THRUST PROPULSION

LOW TURBULENCE

LOW VACUUM

Low Velocity
USE   LOW SPEED

LOW VISIBILITY

LOW VOLTAGE

LOW VOLUME RAMJET ENGINES

LOW WEIGHT

LOW WING AIRCRAFT

LOWER ATMOSPHERE

Lower Atmospheric Composition Experiment
USE   LACATE (EXPERIMENT)

LOWER BODY NEGATIVE PRESSURE

LOWER CALIFORNIA (MEXICO)

LOWER IONOSPHERE

LOX (Oxygen)
USE   LIQUID OXYGEN

LOX-Hydrogen Engines
USE   HYDROGEN OXYGEN ENGINES

LPTR Reactor
USE   LIVERMORE POOL TYPE REACTOR

LR Circuits
USE   RL CIRCUITS

LR-62-RM-2 ENGINE

LR-87-AJ-5 ENGINE

LR-91-AJ-5 ENGINE

LR-99 ENGINE

LRC Circuits
USE   RLC CIRCUITS

LRV (Vehicle)
USE   LUNAR ROVING VEHICLES

LSI
USE   LARGE SCALE INTEGRATION

LSSM

LST
USE   HUBBLE SPACE TELESCOPE

LTV Aircraft
USE   LING-TEMCO-VOUGHT AIRCRAFT

Lu
USE   LUTETIUM

LUBRICANT TESTS

LUBRICANTS

Lubricants, Gas
USE   GAS LUBRICANTS

Lubricants, High Temperature
USE   HIGH TEMPERATURE LUBRICANTS

Lubricants, Solid
USE   SOLID LUBRICANTS

Lubricated Bearings, Gas
USE   GAS BEARINGS

Lubricating Materials, Self
USE   SELF LUBRICATING MATERIALS

LUBRICATING OILS

LUBRICATION

Lubrication, Boundary
USE   BOUNDARY LUBRICATION

Lubrication, Self
USE   SELF LUBRICATION

Lubrication, Space Environmental
USE   SPACECRAFT LUBRICATION

Lubrication, Spacecraft
USE   SPACECRAFT LUBRICATION

LUBRICATION SYSTEMS

Lucite (Trademark)
USE   POLYMETHYL METHACRYLATE

Luder Bands
USE   PLASTIC DEFORMATION
           YIELD POINT

LUDOX (TRADEMARK)

LUGS

LUMBAR REGION

Lumbering Areas
USE   FORESTS

LUMENS

LUMINAIRES

LUMINANCE

Luminance, II
USE   ILLUMINANCE

LUMINESCENCE

Luminescence, Bio
USE   BIOLUMINESCENCE

Luminescence, Cathodo
USE   CATHODOLUMINESCENCE

Luminescence, Chemi
USE   CHEMILUMINESCENCE

Luminescence, Electro
USE   ELECTROLUMINESCENCE

Luminescence, Lunar
USE   LUNAR LUMINESCENCE

Luminescence, Photo
USE   PHOTOLUMINESCENCE

Luminescence, Shock Wave
USE   SHOCK WAVE LUMINESCENCE

Luminescence, Sono
USE   SONOLUMINESCENCE

Luminescence, Thermo
USE   THERMOLUMINESCENCE

Luminescence, Tribo
USE   TRIBOLUMINESCENCE

Luminescent Intensity
USE   LUMINOUS INTENSITY

LUMINOSITY

Luminosity, Stellar
USE   STELLAR LUMINOSITY

Luminous Flux Density
USE   LUMINOUS INTENSITY

LUMINOUS INTENSITY

LUMPED PARAMETER SYSTEMS

LUMPING

Luna Lunar Probes
USE   LUNIK LUNAR PROBES

LUNAR ALBEDO

LUNAR ATMOSPHERE

LUNAR BASES

Lunar Cinematography
USE   LUNAR PHOTOGRAPHY

LUNAR COMMUNICATION

LUNAR COMPOSITION

LUNAR CORE

LUNAR CRATERS

LUNAR CRUST

LUNAR DUST

LUNAR ECHOES

LUNAR ECLIPSES

LUNAR EFFECTS

LUNAR ENVIRONMENT

LUNAR EQUATOR

LUNAR ESCAPE DEVICES

LUNAR EVOLUTION

Lunar Experiment Module, Apollo
USE   APOLLO LUNAR EXPERIMENT MODULE

LUNAR EXPLORATION

LUNAR EXPLORATION SYSTEM FOR APOLLO

(Lunar Exploration System), LESA
USE   LUNAR EXPLORATION SYSTEM FOR
        APOLLO

LUNAR FAR SIDE

LUNAR FIGURE

LUNAR FLIGHT

LUNAR FLYING VEHICLES

LUNAR GEOLOGY

LUNAR GRAVITATION

LUNAR GRAVITATIONAL EFFECTS

LUNAR GRAVITY SIMULATOR

Lunar Ionosphere
USE   LUNAR ATMOSPHERE

LUNAR LANDING

LUNAR LANDING MODULES

LUNAR LANDING SITES

Lunar Landing Vehicles, Ranger
USE   RANGER LUNAR LANDING VEHICLES

LUNAR LAUNCH

LUNAR LIMB

LUNAR LOGISTICS

LUNAR LUMINESCENCE

LUNAR MAGNETIC FIELDS

LUNAR MANTLE

LUNAR MAPS

LUNAR MARIA

LUNAR MOBILE LABORATORIES

LUNAR MODULE

LUNAR MODULE ASCENT STAGE

(Lunar Module), LEM
USE   LUNAR MODULE

LUNAR MODULE 5

LUNAR MODULE 7

LUNAR OBSERVATORIES

LUNAR OCCULTATION

Lunar Occultation Satellite, High Eccentric
USE   EXOSAT SATELLITE

LUNAR ORBIT AND LANDING SIMULATORS

LUNAR ORBITAL RENDEZVOUS

LUNAR ORBITER

Lunar Orbiter A
USE   LUNAR ORBITER 1

Lunar Orbiter B
USE   LUNAR ORBITER 2

Lunar Orbiter C
USE   LUNAR ORBITER 3

Lunar Orbiter D
USE   LUNAR ORBITER 4

Lunar Orbiter E
USE   LUNAR ORBITER 5

LUNAR ORBITER 1

LUNAR ORBITER 2

LUNAR ORBITER 3

LUNAR ORBITER 4

LUNAR ORBITER 5

LUNAR ORBITS

Lunar Perturbation
USE   LUNAR EFFECTS

LUNAR PHASES

LUNAR PHOTOGRAPHS

LUNAR PHOTOGRAPHY

Lunar Probe, Lunik 2
USE   LUNIK 2 LUNAR PROBE

Lunar Probe, Lunik 3
USE   LUNIK 3 LUNAR PROBE

Lunar Probe, Lunik 9
USE   LUNIK 9 LUNAR PROBE

Lunar Probe, Lunik 10
USE   LUNIK 10 LUNAR PROBE

Lunar Probe, Lunik 11
USE   LUNIK 11 LUNAR PROBE

Lunar Probe, Lunik 12
USE   LUNIK 12 LUNAR PROBE

Lunar Probe, Lunik 13
USE   LUNIK 13 LUNAR PROBE

Lunar Probe, Lunik 14
USE   LUNIK 14 LUNAR PROBE

Lunar Probe, Lunik 16
USE   LUNIK 16 LUNAR PROBE

Lunar Probe, Lunik 17
USE   LUNIK 17 LUNAR PROBE

Lunar Probe, Lunik 19
USE   LUNIK 19 LUNAR PROBE

Lunar Probe, Lunik 20
USE   LUNIK 20 LUNAR PROBE

Lunar Probe, Lunik 22
USE   LUNIK 22 LUNAR PROBE

Lunar Probe, Pioneer 4
USE   PIONEER 4 SPACE PROBE

Lunar Probe, Ranger 1
USE   RANGER 1 LUNAR PROBE

Lunar Probe, Ranger 2
USE   RANGER 2 LUNAR PROBE

Lunar Probe, Ranger 3
USE   RANGER 3 LUNAR PROBE

Lunar Probe, Ranger 4
USE   RANGER 4 LUNAR PROBE

Lunar Probe, Ranger 5
USE   RANGER 5 LUNAR PROBE

Lunar Probe, Ranger 6
USE   RANGER 6 LUNAR PROBE

Lunar Probe, Ranger 7
USE   RANGER 7 LUNAR PROBE

Lunar Probe, Ranger 8
USE   RANGER 8 LUNAR PROBE

Lunar Probe, Ranger 9
USE   RANGER 9 LUNAR PROBE

Lunar Probe, Surveyor 1
USE   SURVEYOR 1 LUNAR PROBE

Lunar Probe, Surveyor 2
USE   SURVEYOR 2 LUNAR PROBE

Lunar Probe, Surveyor 3
USE   SURVEYOR 3 LUNAR PROBE

Lunar Probe, Surveyor 4
USE   SURVEYOR 4 LUNAR PROBE

Lunar Probe, Surveyor 5
USE   SURVEYOR 5 LUNAR PROBE

Lunar Probe, Surveyor 6
USE   SURVEYOR 6 LUNAR PROBE

Lunar Probe, Surveyor 7
USE   SURVEYOR 7 LUNAR PROBE

LUNAR PROBES

Lunar Probes, Luna
USE   LUNIK LUNAR PROBES

Lunar Probes, Lunik
USE   LUNIK LUNAR PROBES

Lunar Probes, Ranger
USE   RANGER LUNAR PROBES

Lunar Probes, Surveyor
USE   SURVEYOR LUNAR PROBES

LUNAR PROGRAMS

LUNAR RADAR ECHOES

LUNAR RADIATION

LUNAR RANGEFINDING

LUNAR RAYS

LUNAR RECEIVING LABORATORY

LUNAR RETROREFLECTORS

LUNAR ROCKS

LUNAR ROTATION

LUNAR ROVING VEHICLES

Lunar Roving Vehicles, Lunokhod
USE   LUNOKHOD LUNAR ROVING VEHICLES

LUNAR SATELLITES

Lunar Scattering
USE   LUNAR RADAR ECHOES
DIFFUSE RADIATION

LUNAR SEISMOGRAPHS

LUNAR SHADOW

LUNAR SHELTERS

LUNAR SOIL

LUNAR SPACECRAFT

Lunar Stations, Orbiting
USE   ORBITING LUNAR STATIONS

LUNAR SURFACE

Lunar Surface Experiments Package, Apollo
USE   APOLLO LUNAR SURFACE EXPERIMENTS
PACKAGE

Lunar Surface Scientific Modules
USE   LSSM

LUNAR SURFACE VEHICLES

Lunar Surface Vehicles, Manned
USE   MANNED LUNAR SURFACE VEHICLES

LUNAR TEMPERATURE

Lunar Theory, Hansen
USE   HANSEN LUNAR THEORY

Lunar Theory, Hill
USE   HILL LUNAR THEORY

LUNAR TIDES

LUNAR TOPOGRAPHY

LUNAR TRAJECTORIES

Lunation
USE   MONTH

Luneberg Lenses
USE   RADAR CORNER REFLECTORS

LUNG MORPHOLOGY

LUNGS

LUNIK LUNAR PROBES

LUNIK 2 LUNAR PROBE

LUNIK 3 LUNAR PROBE

LUNIK 9 LUNAR PROBE

LUNIK 10 LUNAR PROBE

LUNIK 11 LUNAR PROBE

LUNIK 12 LUNAR PROBE

LUNIK 13 LUNAR PROBE

LUNIK 14 LUNAR PROBE

LUNIK 16 LUNAR PROBE

LUNIK 17 LUNAR PROBE

LUNIK 19 LUNAR PROBE

LUNIK 20 LUNAR PROBE

LUNIK 22 LUNAR PROBE

LUNOKHOD LUNAR ROVING VEHICLES

LUSTER

LUTETIUM

LUTETIUM COMPOUNDS

LUTETIUM ISOTOPES

Lutetium 176
USE   LUTETIUM ISOTOPES

LUXEMBOURG

LUXEMBOURG EFFECT

Lyapunov Functions
USE   LIAPUNOV FUNCTIONS

LYMAN ALPHA RADIATION

LYMAN BETA RADIATION

LYMAN SPECTRA

LYMPH

Lymph, Endo
USE   ENDOLYMPH

LYMPHOCYTES

Lyophilization
USE   COLLOIDING

Lyophils
USE   COLLOIDS

LYRA CONSTELLATION

LYSERGAMIDE

LYSERGINE

LYSIMETERS

LYSINE

LYSOGENESIS

LYSOZYME

LZEEBE SATELLITE

M

M Diagram, C-
USE   COLOR-MAGNITUDE DIAGRAM

M REGION

M STARS

M, TIROS
USE   TIROS M

M, Vitamin
USE   FOLIC ACID

M Wings
USE   VARIABLE SWEEP WINGS

M-1 ENGINE

M-2 LIFTING BODY

M-2F2 LIFTING BODY

M-2F3 LIFTING BODY

M-46 ENGINE

M-55 ENGINE

M-56 ENGINE

M-57 ENGINE

M-100 ENGINE

MA
USE   MASSACHUSETTS

MA-1 Flight, Mercury
USE   MERCURY MA-1 FLIGHT

MA-2 ENGINE

MA-2 Flight, Mercury
USE   MERCURY MA-2 FLIGHT

MA-2 Mission
USE   MERCURY MA-2 FLIGHT

MA-3 ENGINE

MA-3 Flight
USE   MERCURY MA-3 FLIGHT

MA-3 Flight, Mercury
USE   MERCURY MA-3 FLIGHT

MA-4 Flight
USE   MERCURY MA-4 FLIGHT

MA-4 Flight, Mercury
USE   MERCURY MA-4 FLIGHT

MA-5 ENGINE

MA-5 Flight
USE   MERCURY MA-5 FLIGHT

MA-5 Flight, Mercury
USE   MERCURY MA-5 FLIGHT

MA-6 Flight, Mercury
USE   MERCURY MA-6 FLIGHT

MA-7 Flight, Mercury
USE   MERCURY MA-7 FLIGHT

MA-8 Flight
USE   MERCURY MA-8 FLIGHT

MA-8 Flight, Mercury
USE   MERCURY MA-8 FLIGHT

MA-9 Flight
USE   MERCURY MA-9 FLIGHT

MA-9 Flight, Mercury
USE   MERCURY MA-9 FLIGHT

Maars
USE   CRATERS

MACE MISSILES

MACH CONES

MACH INERTIA PRINCIPLE

MACH NUMBER

Mach Number, Critical
USE   MACH NUMBER
CRITICAL VELOCITY

MACH REFLECTION

MACH-ZEHNDER INTERFEROMETERS

Machine, Cushioncraft Ground Effect
USE   CUSHIONCRAFT GROUND EFFECT
MACHINE

Machine, DTMB-111 Ground Effect
USE   GROUND EFFECT MACHINES

Machine, DTMB-430 Ground Effect
USE   GROUND EFFECT MACHINES

(Machine Elements), Shafts
USE   SHAFTS (MACHINE ELEMENTS)

(Machine Elements), Transmissions
USE   TRANSMISSIONS (MACHINE ELEMENTS)

Machine Learning
USE   LEARNING MACHINES

Machine Life
USE   SERVICE LIFE

MACHINE ORIENTED LANGUAGES

Machine Recognition
USE   ARTIFICIAL INTELLIGENCE

Machine, SR-N2 Ground Effect
USE   WESTLAND GROUND EFFECT MACHINES

Machine, SR-N3 Ground Effect
USE   WESTLAND GROUND EFFECT MACHINES

Machine, SR-N5 Ground Effect
USE   WESTLAND GROUND EFFECT MACHINES

Machine Storage
USE   COMPUTER STORAGE DEVICES
CORE STORAGE

Machine Systems, Man
USE   MAN MACHINE SYSTEMS

MACHINE TOOLS

MACHINE TRANSLATION

Machine, Westland SR-N2 Ground Effect
USE   WESTLAND GROUND EFFECT MACHINES

Machine, Westland SR-N3 Ground Effect
USE   WESTLAND GROUND EFFECT MACHINES

Machine, Westland SR-N5 Ground Effect
USE   WESTLAND GROUND EFFECT MACHINES

MACHINE-INDEPENDENT PROGRAMS

MACHINERY

(Machinery), Positioning Devices
USE   POSITIONING DEVICES (MACHINERY)

Machinery, Refrigerating
USE   REFRIGERATING MACHINERY

Machinery, Turbo
USE   TURBOMACHINERY

Machines, Boring
USE   BORING MACHINES

Machines, Drafting
USE   DRAFTING MACHINES

Machines, Fatigue Testing
USE   FATIGUE TESTING MACHINES

Machines, Finite-State
USE   TURING MACHINES

Machines, Grinding
USE   GRINDING MACHINES

Machines, Ground Effect
USE   GROUND EFFECT MACHINES

Machines, HD-1 Ground Effect
USE   HOVERCRAFT GROUND EFFECT
MACHINES

Machines, Hovercraft Ground Effect
USE   HOVERCRAFT GROUND EFFECT
MACHINES

Machines, Impact Testing
USE   IMPACT TESTING MACHINES

Machines, Learning
USE   LEARNING MACHINES

Machines, Load Testing
USE   LOAD TESTING MACHINES

Machines, Milling
USE   MILLING MACHINES

Machines, Reading
USE   READERS

Machines, Rotating Electrical
USE   ROTATING ELECTRICAL MACHINES

Machines, Teaching
USE   TEACHING MACHINES

Machines, Testing
USE   TEST EQUIPMENT

Machines, Tide Powered
USE   TIDE POWERED MACHINES

Machines, Turing
USE   TURING MACHINES

Machines, Ultrasonic Grinding
USE   ULTRASONIC MACHINING

Machines, Vibration Testing
USE   VIBRATION SIMULATORS

Machines, Walking
USE   WALKING MACHINES

Machines, Waterwave Powered
USE   WATERWAVE POWERED MACHINES

Machines, Welding
USE   WELDING MACHINES

Machines, Westland Ground Effect
USE   WESTLAND GROUND EFFECT MACHINES

Machines), Windmills (Windpowered
USE   WINDMILLS (WINDPOWERED MACHINES)

MACHINING

Machining, Chemical
USE   CHEMICAL MACHINING

Machining, Electrochemical
USE   ELECTROCHEMICAL MACHINING

Machining, Hot
USE   HOT MACHINING

(Machining), Material Removal
USE   MACHINING

(Machining), Milling
USE   MILLING (MACHINING)

Machining, Spark
USE   SPARK MACHINING

Machining, Ultrasonic
USE   ULTRASONIC MACHINING

MACLAURIN SERIES

Macroclimate
USE   CLIMATE

Macromolecules
USE   MOLECULES

MACROPHAGES

MACROSCOPIC EQUATIONS

Macular Vision
USE   VISION

Madagascar
USE   MALAGASY REPUBLIC

MAFFEI GALAXIES

MAGAZINES (SUPPLY CHAMBERS)

MAGDALENA-CAUCA VALLEY (COLOMBIA)

MAGELLAN MISSION

MAGELLANIC CLOUDS

MAGIC TEES

MAGMA

MAGNESIUM

MAGNESIUM ALLOYS

MAGNESIUM BROMIDES

MAGNESIUM CELLS

MAGNESIUM CHLORIDES

MAGNESIUM COMPOUNDS

MAGNESIUM FLUORIDES

MAGNESIUM GERMANATES

MAGNESIUM GERMANIDES

MAGNESIUM ISOTOPES

MAGNESIUM OXIDES

MAGNESIUM PERCHLORATES

MAGNESIUM SULFATES

MAGNESIUM TITANATES

Magnesyn (Trademark)
    USE   SERVOMOTORS

MAGNET COILS

Magnetic Absorption
    USE   ELECTROMAGNETIC ABSORPTION

MAGNETIC AMPLIFIERS

MAGNETIC ANNULAR ARC

MAGNETIC ANNULAR SHOCK TUBES

MAGNETIC ANOMALIES

MAGNETIC BEARINGS

MAGNETIC CHARGE DENSITY

Magnetic Charge, Scalar
    USE   MAGNETIC CHARGE DENSITY

MAGNETIC CIRCUITS

MAGNETIC COILS

MAGNETIC COMPASSES

MAGNETIC COMPRESSION

MAGNETIC CONTROL

MAGNETIC COOLING

MAGNETIC CORES

MAGNETIC DIFFUSION

MAGNETIC DIPOLES

MAGNETIC DISKS

MAGNETIC DISPERSION

MAGNETIC DISTURBANCES

MAGNETIC DOMAINS

MAGNETIC DRUMS

MAGNETIC EFFECTS

MAGNETIC ENERGY STORAGE

MAGNETIC EQUATOR

MAGNETIC FIELD CONFIGURATIONS

Magnetic Field Intensity
    USE   MAGNETIC FLUX

MAGNETIC FIELD INVERSIONS

Magnetic Field, Solar
    USE   SOLAR MAGNETIC FIELD

MAGNETIC FIELDS

Magnetic Fields, Force-Free
    USE   FORCE-FREE MAGNETIC FIELDS

Magnetic Fields, Galactic
    USE   INTERSTELLAR MAGNETIC FIELDS

Magnetic Fields, Interplanetary
    USE   INTERPLANETARY MAGNETIC FIELDS

Magnetic Fields, Interstellar
    USE   INTERSTELLAR MAGNETIC FIELDS

Magnetic Fields, Lunar
    USE   LUNAR MAGNETIC FIELDS

Magnetic Fields, Nonuniform
    USE   NONUNIFORM MAGNETIC FIELDS

Magnetic Fields, Planetary
    USE   PLANETARY MAGNETIC FIELDS

Magnetic Fields, Stellar
    USE   STELLAR MAGNETIC FIELDS

Magnetic Fields, Trapped
    USE   TRAPPED MAGNETIC FIELDS

MAGNETIC FILMS

MAGNETIC FLUX

MAGNETIC FORMING

MAGNETIC INDUCTION

Magnetic Induction Probes
    USE   MAGNETIC PROBES

MAGNETIC LENSES

MAGNETIC LEVITATION VEHICLES

MAGNETIC MATERIALS

MAGNETIC MEASUREMENT

Magnetic Memories
    USE   MAGNETIC STORAGE

Magnetic Metals
    USE   MAGNETIC MATERIALS
           METALS

MAGNETIC MIRRORS

MAGNETIC MOMENTS

MAGNETIC MONOPOLES

MAGNETIC PERMEABILITY

MAGNETIC PISTONS

MAGNETIC POLES

MAGNETIC PROBES

MAGNETIC PROPERTIES

MAGNETIC PUMPING

MAGNETIC RECORDING

MAGNETIC RELAXATION

MAGNETIC RESONANCE

Magnetic Resonance, Nuclear
    USE   NUCLEAR MAGNETIC RESONANCE

Magnetic Resonance, Proton
    USE   PROTON MAGNETIC RESONANCE

MAGNETIC RIGIDITY

MAGNETIC SHIELDING

MAGNETIC SIGNALS

MAGNETIC SIGNATURES

MAGNETIC SPECTROSCOPY

MAGNETIC STARS

MAGNETIC STORAGE

MAGNETIC STORMS

Magnetic Substorms
    USE   MAGNETIC STORMS

MAGNETIC SURVEYS

Magnetic Susceptibility
    USE   MAGNETIC PERMEABILITY

MAGNETIC SUSPENSION

MAGNETIC SWITCHING

Magnetic Tape Recorders
    USE   MAGNETIC RECORDING
           TAPE RECORDERS

MAGNETIC TAPE TRANSPORTS

MAGNETIC TAPES

MAGNETIC TRANSDUCERS

MAGNETIC VARIATIONS

MAGNETICALLY TRAPPED PARTICLES

Magnetism, Aero
    USE   AEROMAGNETISM

Magnetism, Antiferro
    USE   ANTIFERROMAGNETISM

Magnetism, Dia
    USE   DIAMAGNETISM

Magnetism, Electro
    USE   ELECTROMAGNETISM

Magnetism, Ferri
    USE   FERRIMAGNETISM

Magnetism, Ferro
    USE   FERROMAGNETISM

Magnetism, Geo
    USE   GEOMAGNETISM

Magnetism, Gyro
    USE   GYROMAGNETISM

Magnetism, Paleo
    USE   PALEOMAGNETISM

Magnetism, Para
    USE   PARAMAGNETISM

(Magnetism), Susceptibility
    USE   MAGNETIC PERMEABILITY

Magnetism, Terrestrial
    USE   GEOMAGNETISM

MAGNETITE

MAGNETIZATION

Magnetization, De
    USE   DEMAGNETIZATION

Magneto Particle Tracer Explorers, Active
    USE   AMPTE (SATELLITES)

MAGNETO-OPTICS

MAGNETOACOUSTIC WAVES

MAGNETOACOUSTICS

MAGNETOACTIVITY

MAGNETOCARDIOGRAPHY

Magnetoelastic Vibrations
    USE   MAGNETOELASTIC WAVES

MAGNETOELASTIC WAVES

Magnetoelasticity
    USE   MAGNETOSTRICTION

MAGNETOELECTRIC MEDIA

Magnetogasdynamics
    USE   MAGNETOHYDRODYNAMICS

Magnetograms
USE   MAGNETIC SIGNATURES

Magnetohydrodynamic Acceleration
USE   PLASMA ACCELERATION

MAGNETOHYDRODYNAMIC FLOW

MAGNETOHYDRODYNAMIC GENERATORS

MAGNETOHYDRODYNAMIC SHEAR HEATING

MAGNETOHYDRODYNAMIC STABILITY

MAGNETOHYDRODYNAMIC TURBULENCE

MAGNETOHYDRODYNAMIC WAVES

MAGNETOHYDRODYNAMICS

MAGNETOHYDROSTATICS

Magnetoionic Plasma
USE   PLASMAS (PHYSICS)

MAGNETOIONICS

MAGNETOMECHANICS (PHYSICS)

MAGNETOMETERS

Magnetometry
USE   MAGNETIC MEASUREMENT

Magneton, Bohr
USE   BOHR MAGNETON

Magnetooptical Effect, Kerr
USE   KERR MAGNETOOPTICAL EFFECT

MAGNETOPAUSE

MAGNETOPLASMADYNAMICS

Magnetoplasmas
USE   PLASMAS (PHYSICS)

MAGNETORESISTIVITY

MAGNETOSONIC RESONANCE

MAGNETOSPHERE

MAGNETOSPHERIC ELECTRON DENSITY

Magnetospheric Explorer, International
USE   INTERNATIONAL MAGNETOSPHERIC
         EXPLORER

MAGNETOSPHERIC INSTABILITY

MAGNETOSPHERIC ION DENSITY

Magnetospheric Payload, Atmospheric And
USE   AMPS (SATELLITE PAYLOAD)

MAGNETOSPHERIC PROTON DENSITY

Magnetospheric Study, International
USE   INTERNATIONAL MAGNETOSPHERIC
         STUDY

MAGNETOSTATIC AMPLIFIERS

MAGNETOSTATIC FIELDS

MAGNETOSTATICS

MAGNETOSTRICTION

Magnetotelluric Profiling
USE   MAGNETIC SURVEYS

Magnetovariographs
USE   VARIOMETERS

MAGNETRON SPUTTERING

MAGNETRONS

MAGNETS

Magnets, Cryogenic
USE   CRYOGENIC MAGNETS

Magnets, Electro
USE   ELECTROMAGNETS

Magnets, Ferri
USE   FERRIMAGNETS

Magnets, High Field
USE   HIGH FIELD MAGNETS

Magnets, Superconducting
USE   SUPERCONDUCTING MAGNETS

Magnets, Wiggler
USE   WIGGLER MAGNETS

MAGNIFICATION

Magnifiers
USE   MAGNIFICATION

MAGNITUDE

Magnitude Diagram, Color-
USE   COLOR-MAGNITUDE DIAGRAM

Magnitude, Stellar
USE   STELLAR MAGNITUDE

MAGNONS

MAGNUS EFFECT

MAGSAT A SATELLITE

MAGSAT B SATELLITE

MAGSAT SATELLITES

MAGSAT 1 SATELLITE

Mail, Air
USE   AIR MAIL

Mail, Electronic
USE   ELECTRONIC MAIL

Main Engine, Space Shuttle
USE   SPACE SHUTTLE MAIN ENGINE

MAIN SEQUENCE STARS

Main Sequence Stars, Pre-
USE   PRE-MAIN SEQUENCE STARS

MAINE

Mainland, China (Communist)
USE   CHINA

MAINTAINABILITY

MAINTENANCE

Maintenance, Aircraft
USE   AIRCRAFT MAINTENANCE

Maintenance (Computers), File
USE   FILE MAINTENANCE (COMPUTERS)

Maintenance, Space
USE   SPACE MAINTENANCE

Maintenance, Spacecraft
USE   SPACECRAFT MAINTENANCE

MAINTENANCE TRAINING

MAJORITY CARRIERS

Making, Decision
USE   DECISION MAKING

MALAGASY REPUBLIC

MALAWI

MALAYA
USE   MALAYSIA

MALAYSIA
USE   MALAYA

MALDIVE ISLANDS

MALEATES

MALES

MALFUNCTIONS

MALI

MALKUS THEORY

MALLEABILITY

MALONONITRILE

MALTA

MAMMALS

Mammals, Marine
USE   MARINE MAMMALS

MAMMARY GLANDS

Man
USE   HUMAN BEINGS

MAN ENVIRONMENT INTERACTIONS

MAN MACHINE SYSTEMS

MAN OPERATED PROPULSION SYSTEMS

MAN POWERED AIRCRAFT

MANAGEMENT

MANAGEMENT ANALYSIS

Management, Business
USE   INDUSTRIAL MANAGEMENT

Management, Configuration
USE   CONFIGURATION MANAGEMENT

Management, Contract
USE   CONTRACT MANAGEMENT

Management, Data
USE   DATA MANAGEMENT

Management, Engineering
USE   ENGINEERING MANAGEMENT

Management, Environment
USE   ENVIRONMENT MANAGEMENT

Management, Financial
USE   FINANCIAL MANAGEMENT

Management, Fluid
USE   FLUID MANAGEMENT

Management, Forest
USE   FOREST MANAGEMENT

Management, Industrial
USE   INDUSTRIAL MANAGEMENT

Management, Information
USE   INFORMATION MANAGEMENT

MANAGEMENT INFORMATION SYSTEMS

Management, Inventory
USE   INVENTORY MANAGEMENT

Management, Land
USE   LAND MANAGEMENT

Management, Logistics
USE   LOGISTICS MANAGEMENT

Management, Matrix
USE   MATRIX MANAGEMENT

MANAGEMENT METHODS

Management, Personnel
   USE   PERSONNEL MANAGEMENT

MANAGEMENT PLANNING

Management, Procurement
   USE   PROCUREMENT MANAGEMENT

Management, Production
   USE   PRODUCTION MANAGEMENT

Management, Program
   USE   PROJECT MANAGEMENT

Management, Project
   USE   PROJECT MANAGEMENT

Management, Research
   USE   RESEARCH MANAGEMENT

Management, Resources
   USE   RESOURCES MANAGEMENT

Management, Safety
   USE   SAFETY MANAGEMENT

Management System, Central Electronic
   USE   CENTRAL ELECTRONIC MANAGEMENT
          SYSTEM

Management, Systems
   USE   SYSTEMS MANAGEMENT

MANAGEMENT SYSTEMS

Management Systems, Data Base
   USE   DATA BASE MANAGEMENT SYSTEMS

Management Systems, Flight
   USE   FLIGHT MANAGEMENT SYSTEMS

Management, Terminal Area Energy
   USE   TERMINAL AREA ENERGY MANAGEMENT

Management, Water
   USE   WATER MANAGEMENT

Management, Weapon System
   USE   WEAPON SYSTEM MANAGEMENT

MANATEES

MANDELSTAM REPRESENTATION

MANDRELS

Maneuver, Valsalva
   USE   VALSALVA EXERCISE

MANEUVERABILITY

Maneuverable Aircraft, Highly
   USE   HIGHLY MANEUVERABLE AIRCRAFT

MANEUVERABLE REENTRY BODIES

MANEUVERABLE SPACECRAFT

Maneuvering, Aero
   USE   AEROMANEUVERING

Maneuvering Engine (Space Shuttle), Orbit
   USE   ORBIT MANEUVERING ENGINE (SPACE
          SHUTTLE)

Maneuvering Equipment, Astronaut
   USE   ASTRONAUT MANEUVERING EQUIPMENT

Maneuvering Life Support Sys, Integrated
   USE   IMLSS

Maneuvering System, Teleoperator
   USE   TELEOPERATORS

Maneuvering Units, Manned
   USE   MANNED MANEUVERING UNITS

Maneuvering Units, Self
   USE   SELF MANEUVERING UNITS

(Maneuvering Units), SMU
   USE   SELF MANEUVERING UNITS

Maneuvering Units, Space Self
   USE   SELF MANEUVERING UNITS

Maneuvering Vehicles, Orbital
   USE   ORBITAL MANEUVERING VEHICLES

MANEUVERS

Maneuvers, Aircraft
   USE   AIRCRAFT MANEUVERS

Maneuvers, Orbital
   USE   ORBITAL MANEUVERS

Maneuvers, Satellite
   USE   SPACECRAFT MANEUVERS

Maneuvers, Spacecraft
   USE   SPACECRAFT MANEUVERS

Manganates, Per
   USE   PERMANGANATES

MANGANESE

MANGANESE ALLOYS

MANGANESE COMPOUNDS

MANGANESE IONS

MANGANESE ISOTOPES

MANGANESE OXIDES

MANGANESE PHOSPHIDES

Manganese 53
   USE   MANGANESE ISOTOPES

Manganese 54
   USE   MANGANESE ISOTOPES

Manganese 56
   USE   MANGANESE ISOTOPES

MANGANIN (TRADEMARK)

Manifest Anxiety Scale, Taylor
   USE   TAYLOR MANIFEST ANXIETY SCALE

Manifold, Riemann
   USE   RIEMANN MANIFOLD

MANIFOLDS

MANIFOLDS (MATHEMATICS)

Manipulation
   USE   MANIPULATORS

Manipulator System, Remote
   USE   REMOTE MANIPULATOR SYSTEM

MANIPULATORS

MANITOBA

MANITOU (CO)

MANN-WHITNEY-WILCOXON U TEST

Manned Aerodynamic Reusable Spaceship
   USE   MARS (MANNED REUSABLE SPACECRAFT)

MANNED LUNAR SURFACE VEHICLES

MANNED MANEUVERING UNITS

MANNED ORBITAL LABORATORIES

MANNED ORBITAL RESEARCH LABORATORIES

Manned Orbital Space Stations
   USE   ORBITAL SPACE STATIONS

MANNED ORBITAL TELESCOPES

MANNED REENTRY

(Manned Reusable Spacecraft), Mars
   USE   MARS (MANNED REUSABLE SPACECRAFT)

MANNED SPACE FLIGHT

MANNED SPACE FLIGHT NETWORK

MANNED SPACECRAFT

Manned Spacecraft, Voskhod
   USE   VOSKHOD MANNED SPACECRAFT

Manned Spaceplane, Hermes
   USE   HERMES MANNED SPACEPLANE

MANNING THEORY

MANNITOL

MANOMETERS

MANPOWER

Manson Law, Coffin-
   USE   COFFIN-MANSON LAW

Mantle, Earth
   USE   EARTH MANTLE

Mantle (Earth Structure)
   USE   EARTH MANTLE

Mantle, Lunar
   USE   LUNAR MANTLE

Mantles, Planetary
   USE   PLANETARY MANTLES

MANUAL

MANUAL CONTROL

MANUALS

Manuals (Computer Programs), User
   USE   USER MANUALS (COMPUTER PROGRAMS)

Manuals, Installation
   USE   INSTALLATION MANUALS

MANUFACTURING

(Manufacturing), CAM
   USE   COMPUTER AIDED MANUFACTURING

Manufacturing, Computer Aided
   USE   COMPUTER AIDED MANUFACTURING

Manufacturing, Low Gravity
   USE   LOW GRAVITY MANUFACTURING

Manufacturing, Space
   USE   SPACE MANUFACTURING

MANURES

MANY BODY PROBLEM

MANY ELECTRON EFFECTS

Many Particle Theory
   USE   MANY BODY PROBLEM

MAP MATCHING GUIDANCE

Map, Patterson
   USE   PATTERSON MAP

MAP (PROGRAMMING LANGUAGE)

MAPPING

Mapping, Cadastral
   USE   CADASTRAL MAPPING

Mapping, Computer Aided
   USE   COMPUTER AIDED MAPPING

Mapping, Conformal
　USE　CONFORMAL MAPPING

Mapping, Flux
　USE　FLUX DENSITY
　　　　MAPPING

Mapping, Ice
　USE　ICE MAPPING

Mapping Mission, Heat Capacity
　USE　HEAT CAPACITY MAPPING MISSION

Mapping, Photo
　USE　PHOTOMAPPING

Mapping, Planetary
　USE　PLANETARY MAPPING

Mapping, Soil
　USE　SOIL MAPPING

Mapping, Thematic
　USE　THEMATIC MAPPING

Mapping, Thermal
　USE　THERMAL MAPPING

MAPS

Maps, Astronomical
　USE　ASTRONOMICAL MAPS

Maps, Lunar
　USE　LUNAR MAPS

Maps, Photo
　USE　PHOTOMAPS

Maps, Radar
　USE　RADAR MAPS

Maps, Radar Clutter
　USE　RADAR CLUTTER MAPS

Maps, Relief
　USE　RELIEF MAPS

Maps, Weather
　USE　METEOROLOGICAL CHARTS

MAPSAT

MARAGING

MARAGING STEELS

MARANGONI CONVECTION

Marbore 2 Engine
　USE　J-69-T-25 ENGINE

Marching, Spatial
　USE　SPATIAL MARCHING

Marching, Time
　USE　TIME MARCHING

Marco Satellites, San
　USE　SAN MARCO SATELLITES

Marco 1 Satellite, San
　USE　SAN MARCO 1 SATELLITE

Marco 2 Satellite, San
　USE　SAN MARCO 2 SATELLITE

Marco 3 Satellite, San
　USE　SAN MARCO 3 SATELLITE

MARECS MARITIME SATELLITES

MARGINS

Margins, Continental
　USE　CONTINENTAL SHELVES

MARIA

Maria, Lunar
　USE　LUNAR MARIA

MARIJUANA

MARINE BIOLOGY

MARINE CHEMISTRY

MARINE ENVIRONMENTS

Marine Geology
　USE　HYDROGEOLOGY

MARINE MAMMALS

MARINE METEOROLOGY

Marine Navigation
　USE　SURFACE NAVIGATION

MARINE PROPULSION

MARINE RESOURCES

MARINE RUDDERS

MARINE TECHNOLOGY

MARINE TRANSPORTATION

MARINER C SPACECRAFT

MARINER JUPITER-SATURN FLYBY

MARINER JUPITER-URANUS FLYBY

MARINER MARK 2 SPACECRAFT

MARINER PROGRAM

MARINER R 2 SPACE PROBE

MARINER SPACE PROBES

MARINER SPACECRAFT

MARINER VENUS 67 SPACECRAFT

MARINER VENUS-MERCURY 1973

MARINER 1 SPACE PROBE

MARINER 2 SPACE PROBE

MARINER 3 SPACE PROBE

MARINER 4 SPACE PROBE

MARINER 5 SPACE PROBE

MARINER 6 SPACE PROBE

MARINER 7 SPACE PROBE

MARINER 8 SPACE PROBE

MARINER 9 SPACE PROBE

MARINER 10 SPACE PROBE

MARINER 11 SPACE PROBE

MARINER-MERCURY 1973

Marino, San
　USE　SAN MARINO

MARISAT SATELLITES

MARISAT 1 SATELLITE

Maritime Communication Satellite (ESA)
　USE　MAROTS (ESA)

Maritime Orbital Test Satellite
　USE　MAROTS (ESA)

MARITIME SATELLITES

Maritime Satellites, Marecs
　USE　MARECS MARITIME SATELLITES

MARK 1 REENTRY BODY

MARK 1 SPACECRAFT

MARK 2 REENTRY BODY

Mark 2 Spacecraft, Mariner
　USE　MARINER MARK 2 SPACECRAFT

MARK 3 REENTRY BODY

MARK 4 REENTRY BODY

MARK 5 REENTRY BODY

MARK 6 REENTRY BODY

MARK 11 REENTRY BODY

MARK 12 REENTRY BODY

MARK 17 REENTRY BODY

MARKERS

MARKET RESEARCH

MARKETING

MARKING

(Marking), Labeling
　USE　MARKING

MARKOV CHAINS

MARKOV PROCESSES

Markov Theorem, Gauss-
　USE　GAUSS-MARKOV THEOREM

MAROTS (ESA)

MARQUARDT R4D ENGINE

MARROW

Marrow, Bone
　USE　BONE MARROW

MARS

MARS ATMOSPHERE

MARS CRATERS

MARS ENVIRONMENT

MARS EXCURSION MODULE

MARS LANDING

MARS (MANNED REUSABLE SPACECRAFT)

MARS PHOTOGRAPHS

MARS (PLANET)

MARS PROBES

Mars Program, Viking
　USE　VIKING MARS PROGRAM

MARS SURFACE

MARS SURFACE SAMPLES

Mars Trajectories, Earth-
　USE　EARTH-MARS TRAJECTORIES

MARS VOLCANOES

MARS 1 SPACECRAFT

MARS 2 SPACECRAFT

MARS 3 SPACECRAFT

MARS 4 SPACECRAFT

MARS 5 SPACECRAFT

MARS 6 SPACECRAFT

MARS 7 SPACECRAFT

MARS 69 PROJECT

MARS 71 PROJECT

Marshes
USE   MARSHLANDS

MARSHLANDS

Marshlands, Coastal
USE   MARSHLANDS

MARTENSITE

MARTENSITIC STAINLESS STEELS

MARTENSITIC TRANSFORMATION

MARTIN AIRCRAFT

MARTINGALES

MARTINIQUE

MARVS (PROGRAMMING LANGUAGE)

MARYLAND

MASCONS

Maser Modulation, Optical
USE   LIGHT MODULATION

MASER OUTPUTS

Maser Resonators
USE   MASERS

MASERS

Masers, Gas
USE   GAS MASERS

Masers, Hydrogen
USE   HYDROGEN MASERS

Masers, Infrared
USE   INFRARED LASERS

Masers, Interstellar
USE   INTERSTELLAR MASERS

Masers, Optical
USE   LASERS

Masers, Proton
USE   PROTON MASERS

Masers, Traveling Wave
USE   TRAVELING WAVE MASERS

Masers, Water
USE   WATER MASERS

MASKING

Masking, Target
USE   TARGET MASKING

MASKS

Masks, Oxygen
USE   OXYGEN MASKS

MASONITE (TRADEMARK)

MASONRY

MASS

Mass Accretion, Stellar
USE   STELLAR MASS ACCRETION

Mass (Astrophysics), Missing
USE   MISSING MASS (ASTROPHYSICS)

Mass, Atomic
USE   ATOMIC WEIGHTS

MASS BALANCE

(Mass), Ballast
USE   BALLAST (MASS)

Mass, Center Of
USE   CENTER OF MASS

Mass, Critical
USE   CRITICAL MASS

MASS DISTRIBUTION

MASS DRIVERS (PAYLOAD DELIVERY)

Mass Ejection, Stellar
USE   STELLAR MASS EJECTION

Mass, Electron
USE   ELECTRON MASS

Mass Filters
USE   FLUID FILTERS

MASS FLOW

MASS FLOW FACTORS

MASS FLOW RATE

Mass, Low
USE   MASS

Mass, Particle
USE   PARTICLE MASS

Mass, Planetary
USE   PLANETARY MASS

Mass Ratio, Payload
USE   PAYLOAD MASS RATIO

Mass Ratio, Propellant
USE   PROPELLANT MASS RATIO

MASS RATIOS

MASS SPECTRA

MASS SPECTROMETERS

Mass Spectrometers, Retarding Ion
USE   MASS SPECTROMETERS

Mass Spectrometry
USE   MASS SPECTROSCOPY

MASS SPECTROSCOPY

Mass, Stellar
USE   STELLAR MASS

Mass, Subcritical
USE   SUBCRITICAL MASS

Mass Systems, Variable
USE   VARIABLE MASS SYSTEMS

MASS TO LIGHT RATIOS

MASS TRANSFER

(Mass), Weight
USE   WEIGHT (MASS)

(Mass/volume), Density
USE   DENSITY (MASS/VOLUME)

MASSACHUSETTS

MASSAGING

Masses, Air
USE   AIR MASSES

MASSIFS

MAST Shock Tubes
USE   MAGNETIC ANNULAR SHOCK TUBES

MASTICATION

MASTOIDS

MATCHED FILTERS

MATCHING

Matching Guidance, Map
USE   MAP MATCHING GUIDANCE

Matching, Impedance
USE   IMPEDANCE MATCHING

Matching Method (Mathematics), Point
USE   BOUNDARY VALUE PROBLEMS

Matching Navigation System, Terrain Contour
USE   TERCOM

Matching, Phase
USE   PHASE MATCHING

MATERIAL ABSORPTION

MATERIAL BALANCE

Material Disposal (In Space), Hazardous
USE   HAZARDOUS MATERIAL DISPOSAL (IN
             SPACE)

(Material), Mortars
USE   MORTARS (MATERIAL)

(Material), Paper
USE   PAPER (MATERIAL)

(Material), Pitch
USE   PITCH (MATERIAL)

(Material Removal), Grinding
USE   GRINDING (MATERIAL REMOVAL)

Material Removal (Machining)
USE   MACHINING

MATERIALS

Materials, Ablative
USE   ABLATIVE MATERIALS

(Materials), Absorbers
USE   ABSORBERS (MATERIALS)

Materials, Acceptor
USE   ACCEPTOR MATERIALS

(Materials), Aging
USE   AGING (MATERIALS)

Materials, Aircraft Construction
USE   AIRCRAFT CONSTRUCTION MATERIALS

Materials, Airframe
USE   AIRFRAME MATERIALS

Materials, Amorphous
USE   AMORPHOUS MATERIALS

(Materials), Attrition
USE   COMMINUTION

(Materials), Binary Systems
USE   BINARY SYSTEMS (MATERIALS)

(Materials), Binders
USE   BINDERS (MATERIALS)

Materials, Boron Reinforced
USE   BORON REINFORCED MATERIALS

Materials, Brittle
USE   BRITTLE MATERIALS

Materials, Building
USE   CONSTRUCTION MATERIALS

Materials, Carbonaceous
USE   CARBONACEOUS MATERIALS

Materials, Composite
USE   COMPOSITE MATERIALS

Materials, Construction
USE   CONSTRUCTION MATERIALS

(Materials), Cork
USE   CORK (MATERIALS)

(Materials), Curl
USE   CURL (MATERIALS)

Materials, Dielectric
USE   DIELECTRICS

(Materials), Dislocations
USE   DISLOCATIONS (MATERIALS)

Materials, Donor
USE   DONOR MATERIALS

Materials, Dredged
USE   DREDGED MATERIALS

Materials, Electrode
USE   ELECTRODE MATERIALS

(Materials), Fatigue
USE   FATIGUE (MATERIALS)

Materials, Ferrimagnetic
USE   FERRIMAGNETIC MATERIALS

Materials, Ferromagnetic
USE   FERROMAGNETIC MATERIALS

Materials, Fibrous
USE   FIBERS

Materials, Fissile
USE   FISSIONABLE MATERIALS

Materials, Fissionable
USE   FISSIONABLE MATERIALS

(Materials), Foils
USE   FOILS (MATERIALS)

(Materials), Fractures
USE   FRACTURES (MATERIALS)

Materials, Granular
USE   GRANULAR MATERIALS

MATERIALS HANDLING

(Materials), Hardening
USE   HARDENING (MATERIALS)

Materials, High Temperature
USE   REFRACTORY MATERIALS

Materials, Inorganic
USE   INORGANIC MATERIALS

Materials, Insulating
USE   INSULATION

Materials, Laminated
USE   LAMINATES

Materials, Laser
USE   LASER MATERIALS

Materials, Lossless
USE   LOSSLESS MATERIALS

Materials, Low Density
USE   LOW DENSITY MATERIALS

Materials, Magnetic
USE   MAGNETIC MATERIALS

Materials, Matrix
USE   MATRIX MATERIALS

Materials, Molding
USE   MOLDING MATERIALS

Materials (Non Biological), Cellular
USE   FOAMS

Materials, Nonflammable
USE   NONFLAMMABLE MATERIALS

Materials, Noxious
USE   CONTAMINANTS

Materials, Optical Data Storage
USE   OPTICAL DATA STORAGE MATERIALS

Materials, Organic
USE   ORGANIC MATERIALS

(Materials), PCM
USE   PHASE CHANGE MATERIALS

Materials, Phase Change
USE   PHASE CHANGE MATERIALS

Materials, Photoelastic
USE   PHOTOELASTIC MATERIALS

Materials, Photoelectric
USE   PHOTOELECTRIC MATERIALS

Materials, Plastic
USE   PLASTICS

Materials, Porous
USE   POROUS MATERIALS

Materials, Pyrolytic
USE   PYROLYTIC MATERIALS

Materials, Pyrophoric
USE   PYROPHORIC MATERIALS

Materials, Radar Absorbing
USE   ANTIRADAR COATINGS

Materials, Radioactive
USE   RADIOACTIVE MATERIALS

Materials, Radiogenic
USE   RADIOGENIC MATERIALS

Materials, Radome
USE   RADOME MATERIALS

Materials, Reactor
USE   REACTOR MATERIALS

MATERIALS RECOVERY

Materials, Refractory
USE   REFRACTORY MATERIALS

Materials, Reinforced
USE   COMPOSITE MATERIALS

Materials, Reinforcing
USE   REINFORCING MATERIALS

MATERIALS SCIENCE

Materials, Self Lubricating
USE   SELF LUBRICATING MATERIALS

(Materials), Semiconductors
USE   SEMICONDUCTORS (MATERIALS)

Materials, Sizing
USE   SIZING MATERIALS

Materials, Spacecraft Construction
USE   SPACECRAFT CONSTRUCTION MATERIALS

(Materials), Sponges
USE   SPONGES (MATERIALS)

Materials, Strategic
USE   STRATEGIC MATERIALS

Materials, Strength Of
USE   MECHANICAL PROPERTIES

Materials, Structural
USE   CONSTRUCTION MATERIALS

Materials, Superhybrid
USE   SUPERHYBRID MATERIALS

Materials Testing Reactors
USE   NUCLEAR RESEARCH AND TEST
       REACTORS

MATERIALS TESTS

Materials, Thermochromatic
USE   THERMOCHROMATIC MATERIALS

Materials, Thermoelectric
USE   THERMOELECTRIC MATERIALS

(Materials), Thickeners
USE   THICKENERS (MATERIALS)

Materials, Transparent
USE   TRANSPARENCE

Materials, Vitreous
USE   VITREOUS MATERIALS

Mathematical Analysis
USE   APPLICATIONS OF MATHEMATICS

MATHEMATICAL LOGIC

MATHEMATICAL MODELS

MATHEMATICAL PROGRAMMING

MATHEMATICAL TABLES

MATHEMATICS

(Mathematics), Analysis
USE   ANALYSIS (MATHEMATICS)

Mathematics, Applications Of
USE   APPLICATIONS OF MATHEMATICS

(Mathematics), Arguments
USE   INDEPENDENT VARIABLES

(Mathematics), Bifurcation
USE   BRANCHING (MATHEMATICS)

(Mathematics), Biological Models
USE   BIOLOGICAL MODELS (MATHEMATICS)

(Mathematics), Branching
USE   BRANCHING (MATHEMATICS)

(Mathematics), Censored Data
USE   CENSORED DATA (MATHEMATICS)

(Mathematics), Combinations
USE   COMBINATIONS (MATHEMATICS)

(Mathematics), Complements
USE   COMPLEMENTS (MATHEMATICS)

(Mathematics), Continuity
USE   CONTINUITY (MATHEMATICS)

(Mathematics), Convolutions
USE   CONVOLUTION INTEGRALS

(Mathematics), Cubes
USE   CUBES (MATHEMATICS)

(Mathematics), Cusps
USE   CUSPS (MATHEMATICS)

(Mathematics), Dividing
USE   DIVIDING (MATHEMATICS)

(Mathematics), Expressions
USE   FORMULAS (MATHEMATICS)

(Mathematics), Fibers
USE   FIBERS (MATHEMATICS)

(Mathematics), Fixed Points
USE   FIXED POINTS (MATHEMATICS)

(Mathematics), Formulas
USE   FORMULAS (MATHEMATICS)

(Mathematics), Functions
USE   FUNCTIONS (MATHEMATICS)

(Mathematics), Grids
USE   COMPUTATIONAL GRIDS

(Mathematics), Imbeddings
USE   IMBEDDINGS (MATHEMATICS)

(Mathematics), Induction
USE   INDUCTION (MATHEMATICS)

(Mathematics), Lattices
USE   LATTICES (MATHEMATICS)

(Mathematics), Limits
USE   LIMITS (MATHEMATICS)

(Mathematics), Links
USE   LINKS (MATHEMATICS)

(Mathematics), Manifolds
USE   MANIFOLDS (MATHEMATICS)

(Mathematics), Matrices
USE   MATRICES (MATHEMATICS)

(Mathematics), Mesh
USE   COMPUTATIONAL GRIDS

(Mathematics), Operators
USE   OPERATORS (MATHEMATICS)

(Mathematics), Partitions
USE   PARTITIONS (MATHEMATICS)

(Mathematics), Point Matching Method
USE   BOUNDARY VALUE PROBLEMS

(Mathematics), Points
USE   POINTS (MATHEMATICS)

(Mathematics), Reduction
USE   OPTIMIZATION

(Mathematics), Relaxation Method
USE   RELAXATION METHOD (MATHEMATICS)

(Mathematics), Rings
USE   RINGS (MATHEMATICS)

(Mathematics), Robustness
USE   ROBUSTNESS (MATHEMATICS)

(Mathematics), Series
USE   SERIES (MATHEMATICS)

(Mathematics), Singularity
USE   SINGULARITY (MATHEMATICS)

(Mathematics), Squares
USE   SQUARES (MATHEMATICS)

(Mathematics), Stars
USE   STARS (MATHEMATICS)

(Mathematics), Subsets
USE   SET THEORY

(Mathematics), Superimposition
USE   SUPERPOSITION (MATHEMATICS)

(Mathematics), Superposition
USE   SUPERPOSITION (MATHEMATICS)

(Mathematics), Transformations
USE   TRANSFORMATIONS (MATHEMATICS)

(Mathematics), Trees
USE   TREES (MATHEMATICS)

(Mathematics), Truncation
USE   APPROXIMATION

(Mathematics), Vectors
USE   VECTORS (MATHEMATICS)

Mathieu Equation
USE   MATHIEU FUNCTION

MATHIEU FUNCTION

MATRA MISSILE

MATRICES

MATRICES (CIRCUITS)

Matrices, Hessian
USE   HESSIAN MATRICES

MATRICES (MATHEMATICS)

Matrix Analysis
USE   MATRICES (MATHEMATICS)

Matrix Composites, Ceramic
USE   CERAMIC MATRIX COMPOSITES

Matrix Composites, Epoxy
USE   EPOXY MATRIX COMPOSITES

Matrix Composites, Metal
USE   METAL MATRIX COMPOSITES

Matrix Composites, Polymer
USE   POLYMER MATRIX COMPOSITES

Matrix Composites, Resin
USE   RESIN MATRIX COMPOSITES

MATRIX MANAGEMENT

MATRIX MATERIALS

Matrix Method, Jacobi
USE   JACOBI MATRIX METHOD

MATRIX METHODS

Matrix, Scattering
USE   S MATRIX THEORY

Matrix, Stiffness
USE   STIFFNESS MATRIX

Matrix Stress Calculation
USE   MATRIX METHODS

MATRIX THEORY

Matrix Theory, S
USE   S MATRIX THEORY

Mats, Landing
USE   LANDING MATS

Matter, Anti
USE   ANTIMATTER

Matter, Circumstellar
USE   STELLAR ENVELOPES

Matter, Extraterrestrial
USE   EXTRATERRESTRIAL MATTER

Matter, Interstellar
USE   INTERSTELLAR MATTER

MATTER (PHYSICS)

Matter, Rotating
USE   ROTATING MATTER

MATTS (SYSTEMS)

Maturing
USE   GROWTH

MAULER MISSILE

MAURITANIA

MAVERICK MISSILES

Max Holste MH-262 Aircraft
USE   MH-262 AIRCRAFT

MAXIMA

MAXIMUM ENTROPY METHOD

MAXIMUM LIKELIHOOD ESTIMATES

Maximum Mission, Solar
USE   SOLAR MAXIMUM MISSION

Maximum Mission-A, Solar
USE   SOLAR MAXIMUM MISSION-A

MAXIMUM PRINCIPLE

MAXIMUM USABLE FREQUENCY

MAXWELL BODIES

MAXWELL EQUATION

MAXWELL FLUIDS

MAXWELL-BOLTZMANN DENSITY FUNCTION

MAXWELL-MOHR METHOD

Maxwellian Distribution (Density)
USE   MAXWELL-BOLTZMANN DENSITY
       FUNCTION

Mayer Equation, Born-
USE   BORN APPROXIMATION

MAYER PROBLEM

MAYPOLE ANTENNAS

MAZE LEARNING

MB-1 Rocket Vehicle
USE   GENIE ROCKET VEHICLE

MBM JUNCTIONS

MCDONNELL AIRCRAFT

MCDONNELL DOUGLAS AIRCRAFT

Mclaurin Series
USE   MACLAURIN SERIES

MCLEOD GAGES

MCMURDO SOUND

MCR Reactors
USE   MILITARY COMPACT REACTORS

MD
USE   MARYLAND

(MD-NY-PA), Susquehanna River Basin
USE   SUSQUEHANNA RIVER BASIN (MD-NY-PA)

(MD-VA), Assateague Island
USE   ASSATEAGUE ISLAND (MD-VA)

MD-VA), Delmarva Peninsula (De-
USE   DELMARVA PENINSULA (DE-MD-VA)

(MD-VA-WV), Potomac River Valley
USE   POTOMAC RIVER VALLEY (MD-VA-WV)

MDA
USE   MULTIPLE DOCKING ADAPTERS

ME
USE   MAINE

ME P-160 Aircraft
USE   P-160 AIRCRAFT

ME P-160 Aircraft, Messerschmitt
USE   P-160 AIRCRAFT

ME P-308 Aircraft
USE   P-308 AIRCRAFT

ME P-308 Aircraft, Messerschmitt
USE   P-308 AIRCRAFT

(MEA), Monoethanolamine
USE   MONOETHANOLAMINE (MEA)

Meadowlands
USE   GRASSLANDS

MEAN

MEAN FREE PATH

MEAN SQUARE VALUES

Mean Time Between Failures
  USE   MTBF

Mean-Square Errors, Root-
  USE   ROOT-MEAN-SQUARE ERRORS

MEANDERS

Measur System, Integ Med And Behavioral Lab
  USE   IMBLMS

MEASURE AND INTEGRATION

Measure, Shannon-Wiener
  USE   SHANNON-WIENER MEASURE

Measure Theory
  USE   MEASURE AND INTEGRATION

MEASUREMENT

Measurement, Acoustic
  USE   ACOUSTIC MEASUREMENT

Measurement (Biology), Body
  USE   BODY MEASUREMENT (BIOLOGY)

Measurement, Density
  USE   DENSITY MEASUREMENT

Measurement, Depth
  USE   DEPTH MEASUREMENT

Measurement, Dimensional
  USE   DIMENSIONAL MEASUREMENT

Measurement, Displacement
  USE   DISPLACEMENT MEASUREMENT

Measurement, Downrange
  USE   DOWNRANGE MEASUREMENT

Measurement, Drag
  USE   DRAG MEASUREMENT

Measurement, Electrical
  USE   ELECTRICAL MEASUREMENT

Measurement, Electromagnetic
  USE   ELECTROMAGNETIC MEASUREMENT

Measurement, Electromagnetic Noise
  USE   ELECTROMAGNETIC NOISE
         MEASUREMENT

Measurement, Electronic Signal
  USE   SIGNAL MEASUREMENT

Measurement, Flow
  USE   FLOW MEASUREMENT

Measurement, Frequency
  USE   FREQUENCY MEASUREMENT

Measurement, Friction
  USE   FRICTION MEASUREMENT

Measurement, Heat
  USE   HEAT MEASUREMENT

Measurement, High Alt Target And Background
  USE   HIGH ALT TARGET AND BACKGROUND
         MEASUREMENT

Measurement, Humidity
  USE   HUMIDITY MEASUREMENT

Measurement, Impedance
  USE   IMPEDANCE MEASUREMENT

Measurement, Latitude
  USE   LATITUDE MEASUREMENT

Measurement, Longitude
  USE   LONGITUDE MEASUREMENT

Measurement, Magnetic
  USE   MAGNETIC MEASUREMENT

Measurement, Mechanical
  USE   MECHANICAL MEASUREMENT

Measurement, Noise
  USE   NOISE MEASUREMENT

Measurement, Optical
  USE   OPTICAL MEASUREMENT

Measurement, Photoelastic Stress
  USE   PHOTOELASTIC ANALYSIS

Measurement, Photographic
  USE   PHOTOGRAPHIC MEASUREMENT

Measurement, Plasma Flux
  USE   PLASMA FLUX MEASUREMENT

Measurement, Precipitation Particle
  USE   PRECIPITATION PARTICLE MEASUREMENT

Measurement, Pressure
  USE   PRESSURE MEASUREMENT

Measurement Program, Downrange Antimissile
  USE   DOWNRANGE ANTIMISSILE MEASUREMENT
         PROGRAM

Measurement Project, Radio Attenuation
  USE   RADIO ATTENUATION MEASUREMENT
         PROJECT

Measurement, Radar
  USE   RADAR MEASUREMENT

Measurement, Radiation
  USE   RADIATION MEASUREMENT

Measurement, Range
  USE   RANGEFINDING

Measurement, Signal
  USE   SIGNAL MEASUREMENT

Measurement, Sound
  USE   ACOUSTIC MEASUREMENT

Measurement, Strain
  USE   STRAIN MEASUREMENT

Measurement, Stress
  USE   STRESS MEASUREMENT

Measurement, Synoptic
  USE   SYNOPTIC MEASUREMENT

Measurement System, Earth Terminal
  USE   EARTH TERMINAL MEASUREMENT
         SYSTEM

Measurement, Temperature
  USE   TEMPERATURE MEASUREMENT

Measurement, Thrust
  USE   THRUST MEASUREMENT

Measurement, Time
  USE   TIME MEASUREMENT

Measurement, Trajectory
  USE   TRAJECTORY MEASUREMENT

Measurement, Units Of
  USE   UNITS OF MEASUREMENT

Measurement, Velocity
  USE   VELOCITY MEASUREMENT

Measurement, Vibration
  USE   VIBRATION MEASUREMENT

Measurement, Voltage
  USE   ELECTRICAL MEASUREMENT

Measurement, Weight
  USE   WEIGHT MEASUREMENT

Measurement, Wind
  USE   WIND MEASUREMENT

Measurement, Wind Velocity
  USE   WIND VELOCITY MEASUREMENT

Measurement, X Ray Density
  USE   X RAY DENSITY MEASUREMENT

Measurement, X Ray Stress
  USE   X RAY STRESS MEASUREMENT

MEASURES

Measures, Counter
  USE   COUNTERMEASURES

Measuring
  USE   MEASUREMENT

Measuring Apparatus, Torque
  USE   TORQUEMETERS

Measuring Equipment, Distance
  USE   DISTANCE MEASURING EQUIPMENT

MEASURING INSTRUMENTS

Measuring Instruments, Optical
  USE   OPTICAL MEASURING INSTRUMENTS

Measuring Instruments, Radiation
  USE   RADIATION MEASURING INSTRUMENTS

Measuring Instruments, Shock
  USE   SHOCK MEASURING INSTRUMENTS

Measuring Instruments, Temperature
  USE   TEMPERATURE MEASURING INSTRUMENTS

Measuring Instruments, Time
  USE   TIME MEASURING INSTRUMENTS

Measuring Units, Inertial
  USE   INERTIAL PLATFORMS

MECAMYLAMINE

(Mechanical Apertures), Irises
  USE   IRISES (MECHANICAL APERTURES)

MECHANICAL DEVICES

Mechanical Drawings
  USE   ENGINEERING DRAWINGS

MECHANICAL DRIVES

MECHANICAL ENGINEERING

MECHANICAL IMPEDANCE

MECHANICAL MEASUREMENT

MECHANICAL OSCILLATORS

MECHANICAL PROPERTIES

Mechanical Resonance
  USE   RESONANT VIBRATION

MECHANICAL SHOCK

MECHANICAL TWINNING

(Mechanics), Bladders
  USE   DIAPHRAGMS (MECHANICS)

Mechanics, Celestial
  USE   CELESTIAL MECHANICS

Mechanics, Classical
  USE   CLASSICAL MECHANICS

Mechanics, Continuum
  USE   CONTINUUM MECHANICS

(Mechanics), Diaphragms
  USE   DIAPHRAGMS (MECHANICS)

Mechanics, Electro
  USE   ELECTROMECHANICS

Mechanics, Fault
USE   FRACTURE MECHANICS

Mechanics, Flight
USE   FLIGHT MECHANICS

Mechanics, Fluid
USE   FLUID MECHANICS

Mechanics, Fracture
USE   FRACTURE MECHANICS

Mechanics), Head (Fluid
USE   HEAD (FLUID MECHANICS)

(Mechanics), Hole Distribution
USE   HOLE DISTRIBUTION (MECHANICS)

(Mechanics), Hole Geometry
USE   HOLE GEOMETRY (MECHANICS)

Mechanics, Hydro
USE   HYDROMECHANICS

Mechanics, Mega
USE   MEGAMECHANICS

Mechanics, Micro
USE   MICROMECHANICS

Mechanics, Nonrelativistic
USE   NONRELATIVISTIC MECHANICS

Mechanics, Orbital
USE   ORBITAL MECHANICS

MECHANICS (PHYSICS)

Mechanics, Quantum
USE   QUANTUM MECHANICS

(Mechanics), Relaxation
USE   RELAXATION (MECHANICS)

Mechanics, Rock
USE   ROCK MECHANICS

Mechanics, Soil
USE   SOIL MECHANICS

Mechanics, Solid
USE   SOLID MECHANICS

Mechanics, Space
USE   SPACE MECHANICS

Mechanics, Statistical
USE   STATISTICAL MECHANICS

Mechanics), Stokes Law (Fluid
USE   STOKES LAW (FLUID MECHANICS)

(Mechanics), Tolerances
USE   TOLERANCES (MECHANICS)

MECHANISM

Mechanism, Dungeys Wind Shear
USE   WIND SHEAR

Mechanisms, Servo
USE   SERVOMECHANISMS

MECHANIZATION

MECHANOGRAMS

MECHANORECEPTORS

MECLIZINE

Med And Behavioral Lab Measur System, Integ
USE   IMBLMS

MEDIA

Media, Anisotropic
USE   ANISOTROPIC MEDIA

Media, Conducting
USE   CONDUCTORS

Media, Elastic
USE   ELASTIC MEDIA

Media, Extragalactic
USE   INTERGALACTIC MEDIA

Media, Intergalactic
USE   INTERGALACTIC MEDIA

Media, Isotropic
USE   ISOTROPIC MEDIA

Media, Lossy
USE   LOSSY MEDIA

Media, Magnetoelectric
USE   MAGNETOELECTRIC MEDIA

Media, News
USE   NEWS MEDIA

MEDIAN (STATISTICS)

MEDIASTINUM

MEDIATION

MEDICAL ELECTRONICS

MEDICAL EQUIPMENT

MEDICAL PERSONNEL

MEDICAL PHENOMENA

MEDICAL SCIENCE

MEDICAL SERVICES

MEDICINE

Medicine, Aerospace
USE   AEROSPACE MEDICINE

Medicine, Clinical
USE   CLINICAL MEDICINE

Medicine, Nuclear
USE   NUCLEAR MEDICINE

Medicine, Radiation
USE   NUCLEAR MEDICINE

Medicine, Sports
USE   SPORTS MEDICINE

Medicine, Veterinary
USE   VETERINARY MEDICINE

MEDITERRANEAN SEA

Medium, Interplanetary
USE   INTERPLANETARY MEDIUM

MEDIUM SCALE INTEGRATION

Meetings
USE   CONFERENCES

MEGALOPOLISES

MEGAMECHANICS

Meissner Effect
USE   DIAMAGNETISM
       SUPERCONDUCTIVITY

MELAMINE

MELANIN

MELANOIDIN

MELLIN TRANSFORMS

Mellitus, Diabetes
USE   DIABETES MELLITUS

MELT SPINNING

MELTING

Melting, Arc
USE   ARC MELTING

Melting Compounds, High
USE   REFRACTORY MATERIALS

(Melting), Fusion
USE   FUSION (MELTING)

Melting, Levitation
USE   LEVITATION MELTING

MELTING POINTS

Melting, Vacuum
USE   VACUUM MELTING

Melting, Zone
USE   ZONE MELTING

Melts, Containerless
USE   CONTAINERLESS MELTS

MELTS (CRYSTAL GROWTH)

Melts, Impact
USE   IMPACT MELTS

MEM (Excursion Module)
USE   MARS EXCURSION MODULE

Member), Skin (Structural
USE   SKIN (STRUCTURAL MEMBER)

Members, Cantilever
USE   CANTILEVER MEMBERS

Members), Plates (Structural
USE   PLATES (STRUCTURAL MEMBERS)

Members, Structural
USE   STRUCTURAL MEMBERS

Members), Studs (Structural
USE   STUDS (STRUCTURAL MEMBERS)

Membrane Analogy
USE   STRUCTURAL ANALYSIS
       MEMBRANE STRUCTURES

Membrane Electrolytes, Ion Exchange
USE   ION EXCHANGE MEMBRANE
       ELECTROLYTES

Membrane Process, Jet
USE   JET MEMBRANE PROCESS

MEMBRANE STRUCTURES

Membrane Theory
USE   STRUCTURAL ANALYSIS

MEMBRANES

Membranes, Choroid
USE   CHOROID MEMBRANES

(Membranes), Webs
USE   MEMBRANES

Memories, Magnetic
USE   MAGNETIC STORAGE

MEMORY

Memory Alloys, Shape
USE   SHAPE MEMORY ALLOYS

MEMORY (COMPUTERS)

Memory (Data Storage), Optical
USE   OPTICAL MEMORY (DATA STORAGE)

Memory Devices, Bubble
USE   BUBBLE MEMORY DEVICES

(Memory Devices), Chips
USE   CHIPS (MEMORY DEVICES)

Memory Devices, Read-Only
USE   READ-ONLY MEMORY DEVICES

Memory, Plastic
  USE   PLASTIC MEMORY

Memory, Random Access
  USE   RANDOM ACCESS MEMORY

Memory Systems, Virtual
  USE   VIRTUAL MEMORY SYSTEMS

MENDELEVIUM

MENINGITIS

MENISCI

MENSTRUATION

MENTAL HEALTH

MENTAL PERFORMANCE

Mental Stress
  USE   STRESS (PSYCHOLOGY)

MENTHOL

MEPROBAMATE

Mercaptan
  USE   THIOLS

Mercapto Compounds
  USE   THIOLS

MERCATOR PROJECTION

MERCURE AIRCRAFT

MERCURY ALLOYS

MERCURY AMALGAMS

MERCURY ARCS

MERCURY CADMIUM TELLURIDES

MERCURY COMPOUNDS

Mercury Computer, Ferranti
  USE   FERRANTI MERCURY COMPUTER

MERCURY FLIGHTS

MERCURY ION ENGINES

MERCURY ISOTOPES

MERCURY LAMPS

Mercury, Liquid
  USE   MERCURY (METAL)

MERCURY MA-1 FLIGHT

MERCURY MA-2 FLIGHT

MERCURY MA-3 FLIGHT

MERCURY MA-4 FLIGHT

MERCURY MA-5 FLIGHT

MERCURY MA-6 FLIGHT

MERCURY MA-7 FLIGHT

MERCURY MA-8 FLIGHT

MERCURY MA-9 FLIGHT

MERCURY (METAL)

MERCURY MR-1 FLIGHT

MERCURY MR-2 FLIGHT

MERCURY MR-3 FLIGHT

MERCURY MR-4 FLIGHT

MERCURY OXIDES

MERCURY (PLANET)

MERCURY PROJECT

MERCURY SPACECRAFT

MERCURY TELLURIDES

Mercury Tellurides, Cadmium
  USE   MERCURY CADMIUM TELLURIDES

Mercury Trajectories, Earth-
  USE   EARTH-MERCURY TRAJECTORIES

MERCURY VAPOR

Mercury 1973, Mariner Venus-
  USE   MARINER VENUS-MERCURY 1973

Mercury 1973, Mariner-
  USE   MARINER-MERCURY 1973

MERGING ROUTINES

MERIDIONAL FLOW

Merit, Figure Of
  USE   FIGURE OF MERIT

MEROMORPHIC FUNCTIONS

MERRITT ISLAND (FL)

MERWINITE

MESAS

Mesfets
  USE   FIELD EFFECT TRANSISTORS

MESH

Mesh (Mathematics)
  USE   COMPUTATIONAL GRIDS

Mesh, Wire
  USE   WIRE CLOTH

MESITYLENE

MESOMETEOROLOGY

Meson Interactions, Meson-
  USE   MESON-MESON INTERACTIONS

MESON RESONANCE

MESON-MESON INTERACTIONS

MESON-NUCLEON INTERACTIONS

MESONS

Mesons, Eta-
  USE   ETA-MESONS

Mesons, K-
  USE   KAONS

Mesons, Omega-
  USE   OMEGA-MESONS

Mesons, Rho-
  USE   RHO-MESONS

Mesons, Sigma-
  USE   SIGMA-MESONS

Mesons, Vector
  USE   VECTOR MESONS

Mesons, X
  USE   X MESONS

MESOPAUSE

MESOPHILES

MESOSCALE PHENOMENA

MESOSPHERE

Mesosphere Explorer, Solar
  USE   SOLAR MESOSPHERE EXPLORER

MESSAGE PROCESSING

MESSAGES

Messerschmitt ME P-160 Aircraft
  USE   P-160 AIRCRAFT

Messerschmitt ME P-308 Aircraft
  USE   P-308 AIRCRAFT

METABOLIC DISEASES

METABOLIC WASTES

METABOLISM

Metabolism, Adrenal
  USE   ADRENAL METABOLISM

Metabolism, Ascorbic Acid
  USE   ASCORBIC ACID METABOLISM

Metabolism, Calcium
  USE   CALCIUM METABOLISM

Metabolism, Carbohydrate
  USE   CARBOHYDRATE METABOLISM

Metabolism, Electrolyte
  USE   ELECTROLYTE METABOLISM

Metabolism, Hydrogen
  USE   HYDROGEN METABOLISM

Metabolism, Hypo
  USE   HYPOMETABOLISM

Metabolism, Lipid
  USE   LIPID METABOLISM

Metabolism, Mineral
  USE   MINERAL METABOLISM

Metabolism, Nitrogen
  USE   NITROGEN METABOLISM

Metabolism, Oxygen
  USE   OXYGEN METABOLISM

Metabolism, Phosphorus
  USE   PHOSPHORUS METABOLISM

Metabolism, Protein
  USE   PROTEIN METABOLISM

Metabolisms, Hormone
  USE   HORMONE METABOLISMS

METABOLITES

Metagalaxy
  USE   UNIVERSE

METAL AIR BATTERIES

Metal Alloys, Refractory
  USE   REFRACTORY METAL ALLOYS

Metal, Babbitt
  USE   BABBITT METAL

METAL BONDING

Metal Bonding, Metal-
  USE   METAL-METAL BONDING

METAL COATINGS

METAL COMBUSTION

METAL COMPOUNDS

Metal Compounds, Alkali
  USE   ALKALI METAL COMPOUNDS

Metal Cooled Reactors, Liquid
USE   LIQUID METAL COOLED REACTORS

Metal Corrosion
USE   CORROSION

METAL CRYSTALS

METAL CUTTING

Metal Diodes, Metal-Insulator-
USE   MIM DIODES

METAL DRAWING

Metal Fast Breeder Reactors, Liquid
USE   LIQUID METAL FAST BREEDER REACTORS

METAL FATIGUE

METAL FIBERS

METAL FILMS

METAL FINISHING

METAL FLUORIDES

METAL FOAMS

METAL FOILS

Metal Forging
USE   FORGING

Metal Forming
USE   FORMING TECHNIQUES
        METAL WORKING

METAL FUELS

METAL GRINDING

METAL HALIDES

Metal Hardening
USE   HARDENING (MATERIALS)

METAL HYDRIDES

Metal Insulator Semiconductors
USE   MIS (SEMICONDUCTORS)

Metal Interactions, Gas-
USE   GAS-METAL INTERACTIONS

METAL IONS

METAL JOINTS

Metal Junctions, Metal-Barrier-
USE   MBM JUNCTIONS

(Metal), Lead
USE   LEAD (METAL)

METAL MATRIX COMPOSITES

(Metal), Mercury
USE   MERCURY (METAL)

METAL NITRIDES

METAL OXIDE SEMICONDUCTORS

Metal Oxide Semiconductors, Complementary
USE   CMOS

METAL OXIDES

METAL PARTICLES

(Metal), Plate
USE   METAL PLATES

METAL PLATES

METAL POLISHING

METAL POWDER

METAL PROPELLANTS

Metal Semiconductors, Metal-Insulator-
USE   MIM (SEMICONDUCTORS)

Metal Semiconductors, Metal-Oxide-
USE   MOM (SEMICONDUCTORS)

Metal, Sheet
USE   METAL SHEETS

METAL SHEETS

METAL SHELLS

METAL SPINNING

METAL SPRAYING

METAL STRIPS

METAL SURFACES

METAL VAPOR LASERS

METAL VAPORS

Metal Whisker Reinforcement
USE   WHISKER COMPOSITES

METAL WORKING

Metal-Barrier-Metal Junctions
USE   MBM JUNCTIONS

METAL-GAS SYSTEMS

Metal-Insulator-Metal Diodes
USE   MIM DIODES

Metal-Insulator-Metal Semiconductors
USE   MIM (SEMICONDUCTORS)

METAL-METAL BONDING

METAL-NITRIDE-OXIDE-SEMICONDUCTORS

METAL-NITRIDE-OXIDE-SILICON

Metal-Oxide-Metal Semiconductors
USE   MOM (SEMICONDUCTORS)

METAL-WATER REACTIONS

METALLIC GLASSES

METALLIC HYDROGEN

METALLIC PLASMAS

METALLIC STARS

METALLICITY

Metallics, Inter
USE   INTERMETALLICS

METALLIZING

METALLOGRAPHY

METALLOIDS

Metallorganic Compounds
USE   ORGANOMETALLIC COMPOUNDS

METALLOSILOXANE POLYMER

METALLOXANE POLYMER

METALLURGY

(Metallurgy), Aging
USE   AGING (METALLURGY)

Metallurgy, Hydro
USE   HYDROMETALLURGY

(Metallurgy), Lamella
USE   LAMELLA (METALLURGY)

(Metallurgy), Pickling
USE   PICKLING (METALLURGY)

Metallurgy, Powder
USE   POWDER METALLURGY

Metallurgy, Pyro
USE   PYROMETALLURGY

(Metallurgy), Rapid Quenching
USE   RAPID QUENCHING (METALLURGY)

(Metallurgy), Spinning
USE   METAL SPINNING

(Metallurgy), Temper
USE   TEMPER (METALLURGY)

METALS

Metals, Alkali
USE   ALKALI METALS

Metals, Alkaline Earth
USE   ALKALINE EARTH METALS

Metals, Bi
USE   BIMETALS

Metals, Ferrous
USE   FERROUS METALS

Metals, Lanthanide Series
USE   RARE EARTH ELEMENTS

Metals, Liquid
USE   LIQUID METALS

Metals, Magnetic
USE   METALS
        MAGNETIC MATERIALS

Metals, Noble
USE   NOBLE METALS

Metals, Nonferrous
USE   NONFERROUS METALS

Metals, Notched
USE   NOTCH TESTS

Metals, Polished
USE   METAL POLISHING

Metals, Powdered
USE   METAL POWDER

Metals, Precious
USE   NOBLE METALS

Metals, Refractory
USE   REFRACTORY METALS

Metals, Synthetic
USE   SYNTHETIC METALS

Metals, Transition
USE   TRANSITION METALS

Metals, Ultrapure
USE   ULTRAPURE METALS

METAMORPHISM (GEOLOGY)

Metastability
USE   METASTABLE STATE

METASTABLE ATOMS

METASTABLE STATE

METATHESIS

Metazoa
USE   ANIMALS

Meteor Bursts
USE   METEOROID SHOWERS

Meteor Craters
USE   CRATERS

Meteor Hazards
USE   METEOROID HAZARDS

Meteor Project, Harvard Radio
USE   HARVARD RADIO METEOR PROJECT

METEOR TRAILS

METEOR 1 ROCKET VEHICLE

Meteorite, Alais
USE   ALAIS METEORITE

Meteorite, Allende
USE   ALLENDE METEORITE

Meteorite, Aroos
USE   AROOS METEORITE

Meteorite, Bondoc
USE   BONDOC METEORITE

Meteorite, Bruderheim
USE   BRUDERHEIM METEORITE

Meteorite, Cold Bokkeveld
USE   COLD BOKKEVELD METEORITE

METEORITE COLLISIONS

Meteorite Compression Tests
USE   METEORITES
       MECHANICAL PROPERTIES
       COMPRESSION TESTS

METEORITE CRATERS

Meteorite Craters, Fossil
USE   METEORITE CRATERS
       FOSSILS

Meteorite, Harleton
USE   HARLETON METEORITE

Meteorite, Ivuna
USE   IVUNA METEORITE

Meteorite, Lazarev
USE   LAZAREV METEORITE

Meteorite, Murchison
USE   MURCHISON METEORITE

Meteorite, Murray
USE   MURRAY METEORITE

Meteorite, Odessa
USE   ODESSA METEORITE

Meteorite, Okhansk
USE   OKHANSK METEORITE

Meteorite, Orgueil
USE   ORGUEIL METEORITE

Meteorite, Pribram
USE   PRIBRAM METEORITE

Meteorite, Sikhote-Alin
USE   SIKHOTE-ALIN METEORITE

Meteorite, Tonk
USE   TONK METEORITE

Meteorite, Tungusk
USE   TUNGUSK METEORITE

METEORITES

Meteorites, Carbonaceous
USE   CARBONACEOUS METEORITES

Meteorites, Iron
USE   IRON METEORITES

Meteorites, Micro
USE   MICROMETEOROIDS

Meteorites, Siderite
USE   IRON METEORITES

Meteorites, Stony
USE   STONY METEORITES

METEORITIC COMPOSITION

METEORITIC DAMAGE

METEORITIC DIAMONDS

Meteoritic Dust
USE   MICROMETEOROIDS

Meteoritic Ionization
USE   METEOR TRAILS
       ATMOSPHERIC IONIZATION

METEORITIC MICROSTRUCTURES

METEOROID CONCENTRATION

Meteoroid Craters
USE   METEORITE CRATERS

METEOROID DUST CLOUDS

METEOROID HAZARDS

METEOROID PROTECTION

Meteoroid Satellite, Radiation And
USE   RADIATION AND METEOROID SATELLITE

METEOROID SHOWERS

Meteoroid Spacecraft, Radiation
USE   RADIATION METEOROID SPACECRAFT

Meteoroid Technology Satellite
USE   EXPLORER 46 SATELLITE

METEOROIDS

Meteoroids, Aquarid
USE   AQUARID METEOROIDS

Meteoroids, Arietid
USE   ARIETID METEOROIDS

Meteoroids, Cyrillid
USE   CYRILLID METEOROIDS

Meteoroids, Draconid
USE   DRACONID METEOROIDS

Meteoroids, Geminid
USE   GEMINID METEOROIDS

Meteoroids, Leonid
USE   LEONID METEOROIDS

Meteoroids, Micro
USE   MICROMETEOROIDS

Meteoroids, Orionid
USE   ORIONID METEOROIDS

Meteoroids, Perseid
USE   PERSEID METEOROIDS

Meteoroids, Quadrantid
USE   QUADRANTID METEOROIDS

Meteoroids, Sporadic
USE   SPORADIC METEOROIDS

Meteoroids, Taurid
USE   TAURID METEOROIDS

METEOROLOGICAL BALLOONS

METEOROLOGICAL CHARTS

METEOROLOGICAL FLIGHT

METEOROLOGICAL INSTRUMENTS

Meteorological Organization, World
USE   WORLD METEOROLOGICAL ORGANIZATION

METEOROLOGICAL PARAMETERS

Meteorological Probes
USE   SONDES

METEOROLOGICAL RADAR

METEOROLOGICAL RESEARCH AIRCRAFT

Meteorological Rockets
USE   SOUNDING ROCKETS

Meteorological Satellite Program, Defense
USE   DMSP SATELLITES

Meteorological Satellite, Synchronous
USE   SYNCHRONOUS METEOROLOGICAL
       SATELLITE

METEOROLOGICAL SATELLITES

METEOROLOGICAL SERVICES

METEOROLOGICAL SOLENOIDS

Meteorological Stations
USE   WEATHER STATIONS

METEOROLOGY

Meteorology, Agro
USE   AGROMETEOROLOGY

Meteorology, Alpine
USE   ALPINE METEOROLOGY

Meteorology, Bio
USE   BIOMETEOROLOGY

(Meteorology), Ceilings
USE   CEILINGS (METEOROLOGY)

(Meteorology), Clouds
USE   CLOUDS (METEOROLOGY)

(Meteorology), Frontal Areas
USE   FRONTS (METEOROLOGY)

(Meteorology), Fronts
USE   FRONTS (METEOROLOGY)

Meteorology, Hydro
USE   HYDROMETEOROLOGY

(Meteorology), Jet Streams
USE   JET STREAMS (METEOROLOGY)

(Meteorology), Long Waves
USE   PLANETARY WAVES

Meteorology, Marine
USE   MARINE METEOROLOGY

Meteorology, Meso
USE   MESOMETEOROLOGY

Meteorology, Micro
USE   MICROMETEOROLOGY

Meteorology, Nuclear
USE   NUCLEAR METEOROLOGY

Meteorology, Polar
USE   POLAR METEOROLOGY

(Meteorology), Precipitation
USE   PRECIPITATION (METEOROLOGY)

Meteorology, Radio
USE   RADIO METEOROLOGY

(Meteorology), Storms
USE   STORMS (METEOROLOGY)

Meteorology, Synoptic
USE   SYNOPTIC METEOROLOGY

Meteorology, Tropical
USE   TROPICAL METEOROLOGY

(Meteorology), Wind
USE   WIND (METEOROLOGY)

Meteors
USE   METEOROIDS

Meteors, Radio
USE   RADIO METEORS

METEOSAT SATELLITE

Meters
USE   MEASURING INSTRUMENTS

Meters, Accelero
USE   ACCELEROMETERS

Meters, Alti
USE   ALTIMETERS

Meters, Conductivity
USE   CONDUCTIVITY METERS

Meters, Electrical Conductivity
USE   ELECTRICAL CONDUCTIVITY METERS

Meters, Field Intensity
USE   FIELD INTENSITY METERS

Meters, Gas
USE   GAS METERS

Meters, Helio
USE   HELIOMETERS

Meters, Hot-Wire Turbulence
USE   HOT-WIRE FLOWMETERS
        TURBULENCE METERS

Meters, Hydro
USE   HYDROMETERS

Meters, Interfero
USE   INTERFEROMETERS

Meters, Light Scattering
USE   LIGHT SCATTERING METERS

Meters, Moisture
USE   MOISTURE METERS

Meters, Noise
USE   NOISE METERS

Meters, Osmo
USE   OSMOMETERS

Meters, Potentio
USE   POTENTIOMETERS

Meters, Pyro
USE   PYROMETERS

Meters, Radiation
USE   RADIATION MEASURING INSTRUMENTS

Meters, Radio
USE   RADIOMETERS

Meters, Rate
USE   MEASURING INSTRUMENTS

Meters, Reflecto
USE   REFLECTOMETERS

Meters, Respiro
USE   RESPIROMETERS

Meters, Rheo
USE   RHEOMETERS

Meters, Rio
USE   RIOMETERS

Meters, Spectrophoto
USE   SPECTROPHOTOMETERS

Meters, Spectroradio
USE   SPECTRORADIOMETERS

Meters, Turbulence
USE   TURBULENCE METERS

Meters, Vibration
USE   VIBRATION METERS

Methacrylate, Polymethyl
USE   POLYMETHYL METHACRYLATE

Methacrylate Resins
USE   ACRYLIC RESINS

METHAMPHETAMINE

METHANATION

METHANE

Methane, Nitro
USE   NITROMETHANE

Methane, Synthetic
USE   SYNTHANE

METHIONINE

Method, Biot
USE   BIOT METHOD

Method, Boundary Element
USE   BOUNDARY ELEMENT METHOD

Method, Boundary Integral
USE   BOUNDARY INTEGRAL METHOD

Method, Bridgman
USE   BRIDGMAN METHOD

Method, Characteristic
USE   METHOD OF CHARACTERISTICS

Method, Conjugate Gradient
USE   CONJUGATE GRADIENT METHOD

Method, Cowell
USE   NUMERICAL INTEGRATION

Method, Crank-Nicholson
USE   CRANK-NICHOLSON METHOD

Method, Critical Path
USE   CRITICAL PATH METHOD

Method, Crocco
USE   CROCCO METHOD

Method, Czochralski
USE   CZOCHRALSKI METHOD

Method, Debye-Scherrer
USE   DEBYE-SCHERRER METHOD

Method, Encke
USE   ENCKE METHOD

Method, Finite Element
USE   FINITE ELEMENT METHOD

Method, Finite Volume
USE   FINITE VOLUME METHOD

Method (Fluid Dynamics), Panel
USE   PANEL METHOD (FLUID DYNAMICS)

Method (Forecasting), Delphi
USE   DELPHI METHOD (FORECASTING)

Method (Forecasting), Pattern
USE   PATTERN METHOD (FORECASTING)

Method (Forecasting), Probe
USE   PROBE METHOD (FORECASTING)

Method (Forecasting), Profile
USE   PROFILE METHOD (FORECASTING)

Method, Fujita
USE   FUJITA METHOD

Method, Galerkin
USE   GALERKIN METHOD

Method, Glimm
USE   GLIMM METHOD

Method, Halphen
USE   HALPHEN METHOD

Method, Hartree-Fock-Slater
USE   HARTREE-FOCK-SLATER METHOD

Method, Hill
USE   HILL METHOD

Method, Jacobi Matrix
USE   JACOBI MATRIX METHOD

Method, Kjeldahl
USE   KJELDAHL METHOD

Method, Latin Square
USE   LATIN SQUARE METHOD

Method, Laue
USE   LAUE METHOD

Method, Least Squares
USE   LEAST SQUARES METHOD

Method, Lighthill
USE   LIGHTHILL METHOD

Method (Mathematics), Point Matching
USE   BOUNDARY VALUE PROBLEMS

Method (Mathematics), Relaxation
USE   RELAXATION METHOD (MATHEMATICS)

Method, Maximum Entropy
USE   MAXIMUM ENTROPY METHOD

Method, Maxwell-Mohr
USE   MAXWELL-MOHR METHOD

Method, Milne
USE   MILNE METHOD

Method, Milne-Thomson
USE   MILNE-THOMSON METHOD

Method, Minimum Entropy
USE   MINIMUM ENTROPY METHOD

Method, Monte Carlo
USE   MONTE CARLO METHOD

Method, Newton-Raphson
USE   NEWTON-RAPHSON METHOD

METHOD OF CHARACTERISTICS

METHOD OF MOMENTS

Method, Percus
USE   PERCUS METHOD

Method, Pohlhausen
USE   POHLHAUSEN METHOD

Method, Rayleigh-Ritz
USE   RAYLEIGH-RITZ METHOD

Method, Ritz Averaging
USE   RITZ AVERAGING METHOD

Method, Ruler
USE   RULER METHOD

Method, Runge-Kutta
USE   RUNGE-KUTTA METHOD

Method, Schmidt
USE   SCHMIDT METHOD

Method, Schwartz
USE   SCHWARTZ METHOD

Method, Simplex
USE   SIMPLEX METHOD

Method, Steepest Ascent
USE   STEEPEST DESCENT METHOD

Method, Steepest Descent
USE   STEEPEST DESCENT METHOD

Method Tests, Wing Flow
USE   WING FLOW METHOD TESTS

Method, Traveling Solvent
USE   TRAVELING SOLVENT METHOD

Method, Van Slyke
USE   VAN SLYKE METHOD

Method, Variation
USE   CALCULUS OF VARIATIONS

Method, Von Zeipel
USE   VON ZEIPEL METHOD

Method, Wentzel-Kramer-Brillouin
USE   WENTZEL-KRAMER-BRILLOUIN METHOD

METHODOLOGY

Methods
USE   METHODOLOGY
      PROCEDURES

Methods, Approximation
USE   APPROXIMATION

Methods, Asymptotic
USE   ASYMPTOTIC METHODS

Methods, Computer
USE   COMPUTER PROGRAMS

Methods, Energy
USE   ENERGY METHODS

Methods, Equilibrium
USE   EQUILIBRIUM METHODS

Methods, Heuristic
USE   HEURISTIC METHODS

Methods, Management
USE   MANAGEMENT METHODS

Methods, Matrix
USE   MATRIX METHODS

Methods, Optical
USE   OPTICS

Methods, Production
USE   PRODUCTION ENGINEERING

Methods, Spectral
USE   SPECTRAL METHODS

Methods, Strain Energy
USE   STRAIN ENERGY METHODS

METHOXY SYSTEMS

METHYL ALCOHOLS

METHYL CHLORIDE

METHYL CHLOROSILANES

METHYL COMPOUNDS

METHYL NITRATE

METHYL POLYSILOXANE

METHYLATION

METHYLENE

METHYLENE BLUE

METHYLENE DIAMINE

METHYLHYDRAZINE

Methylhydrazines, Di
USE   DIMETHYLHYDRAZINES

METRAZOL

Metric Conversion
USE   METRICATION

METRIC PHOTOGRAPHY

Metric, Schwarzschild
USE   SCHWARZSCHILD METRIC

METRIC SPACE

Metric, Space-Time
USE   SPACE-TIME FUNCTIONS

Metric System
USE   INTERNATIONAL SYSTEM OF UNITS

METRICATION

Metrics, Bio
USE   BIOMETRICS

METROLOGY

Metropolitan Aircraft
USE   CV-440 AIRCRAFT

Metropolitan Areas
USE   CITIES

MEXICO

(Mexico), Chiapas
USE   CHIAPAS (MEXICO)

Mexico, Gulf Of
USE   GULF OF MEXICO

(Mexico), Gulf Of California
USE   GULF OF CALIFORNIA (MEXICO)

(Mexico), Leon-Queretaro Area
USE   LEON-QUERETARO AREA (MEXICO)

(Mexico), Lower California
USE   LOWER CALIFORNIA (MEXICO)

Mexico, New
USE   NEW MEXICO

Meyer Expansion, Prandtl-
USE   PRANDTL-MEYER EXPANSION

Mg
USE   MAGNESIUM

MH-262 AIRCRAFT

MH-262 Aircraft, Max Holste
USE   MH-262 AIRCRAFT

MI
USE   MICHIGAN

(MI), Pontiac
USE   PONTIAC (MI)

(MI), Saginaw Bay
USE   SAGINAW BAY (MI)

MICA

MICARTA

MICE

Mice, Pocket
USE   POCKET MICE

MICHAEL REACTION

MICHAELIS THEORY

MICHELL THEOREM

MICHELSON INTERFEROMETERS

MICHIGAN

Michigan, Lake
USE   LAKE MICHIGAN

MICROANALYSIS

MICROBALANCES

MICROBALLOONS

Microbe
USE   MICROORGANISMS

MICROBEAMS

MICROBIOLOGY

Microcalorimeters
USE   CALORIMETERS

Microchannel Arrays, Multi-Anode
USE   MULTI-ANODE MICROCHANNEL ARRAYS

MICROCHANNEL PLATES

MICROCHANNELS

Microcircuits
USE   MICROELECTRONICS

Microcircuits, Encapsulated
USE   ENCAPSULATED MICROCIRCUITS

MICROCLIMATOLOGY

MICROCOMPUTERS

MICROCRACKS

MICROCRYSTALS

MICROCYSTIS

MICRODENSITOMETERS

MICROELECTRONICS

MICROFIBERS

MICROFILMS

Micrography
USE   PHOTOMICROGRAPHY

Microgravity
USE   REDUCED GRAVITY

MICROGRAVITY APPLICATIONS

MICROHARDNESS

Microindentation
USE   MICROHARDNESS

MICROINSTRUMENTATION

Micromanometers
USE   MANOMETERS

MICROMECHANICS

MICROMETEORITES

MICROMETEOROID EXPLORER SATELLITES

MICROMETEOROIDS

MICROMETEOROLOGY

Micrometeors
USE   MICROMETEOROIDS

MICROMETERS

MICROMILLIAMMETERS

MICROMINIATURIZATION

MICROMINIATURIZED ELECTRONIC DEVICES

MICROMODULES

MICROMOTORS

MICROORGANISMS

MICROPARTICLES

MICROPHONES

MICROPHOTOGRAPHS

Microphotometers
USE   PHOTOMETERS

MICROPLASMAS

MICROPOLAR FLUIDS

MICROPOROSITY

Microprocessor, Intel 8080
USE   INTEL 8080 MICROPROCESSOR

MICROPROCESSORS

MICROPROGRAMMING

MICROPULSATIONS

Micropulsations, Geomagnetic
USE   GEOMAGNETIC MICROPULSATIONS

MICROROCKET ENGINES

Microscales
USE   MICROBALANCES

Microscope (Slam), Scanning Laser Acoustic
USE   ACOUSTIC MICROSCOPES

MICROSCOPES

Microscopes, Acoustic
USE   ACOUSTIC MICROSCOPES

Microscopes, Electron
USE   ELECTRON MICROSCOPES

Microscopes, Ion
USE   ION MICROSCOPES

Microscopes, Optical
USE   OPTICAL MICROSCOPES

MICROSCOPY

Microscopy, Electron
USE   ELECTRON MICROSCOPY

Microscopy, Laser
USE   LASER MICROSCOPY

Microscopy, Photoacoustic
USE   PHOTOACOUSTIC MICROSCOPY

(Microscopy), Slides
USE   SLIDES (MICROSCOPY)

Microscopy, Ultraviolet
USE   ULTRAVIOLET MICROSCOPY

MICROSEISMS

MICROSONICS

MICROSPORES

MICROSTRIP TRANSMISSION LINES

MICROSTRUCTURE

Microstructures, Meteoritic
USE   METEORITIC MICROSTRUCTURES

MICROTHRUST

MICROTOMY

MICROTRONS

MICROVISION LANDING AID

MICROWAVE AMPLIFIERS

MICROWAVE ANTENNAS

MICROWAVE ATTENUATION

MICROWAVE CIRCUITS

MICROWAVE COUPLING

MICROWAVE EMISSION

MICROWAVE EQUIPMENT

MICROWAVE FILTERS

MICROWAVE FREQUENCIES

MICROWAVE HOLOGRAPHY

MICROWAVE IMAGERY

MICROWAVE INTERFEROMETERS

MICROWAVE LANDING SYSTEMS

MICROWAVE OSCILLATORS

MICROWAVE PHOTOGRAPHY

MICROWAVE PLASMA PROBES

MICROWAVE PROBES

Microwave Radiation
USE   MICROWAVES

MICROWAVE RADIOMETERS

MICROWAVE REFLECTOMETERS

MICROWAVE RESONANCE

MICROWAVE SCANNING BEAM LANDING SYSTEM

MICROWAVE SCATTERING

MICROWAVE SENSORS

MICROWAVE SOUNDING

MICROWAVE SPECTRA

Microwave Spectra, Interstellar
USE   MICROWAVE SPECTRA
       INTERSTELLAR RADIATION

MICROWAVE SPECTROMETERS

MICROWAVE SWITCHING

MICROWAVE TRANSMISSION

MICROWAVE TUBES

MICROWAVES

Microweighing
USE   WEIGHT MEASUREMENT

MICROYIELD STRENGTH

Micturition
USE   URINATION

MIDAIR COLLISIONS

MIDALTITUDE

MIDAS SATELLITES

MIDAS 2 SATELLITE

MIDAS 3 SATELLITE

MIDAS 4 SATELLITE

MIDAS 5 SATELLITE

MIDAS 6 SATELLITE

MIDAS 7 SATELLITE

MIDCOURSE GUIDANCE

MIDCOURSE TRAJECTORIES

MIDDLE ATMOSPHERE

MIDDLE EAR

MIDDLE EAR PRESSURE

MIDLATITUDE ATMOSPHERE

Midlatitudes
USE   TEMPERATE REGIONS

MIE SCATTERING

Mie Theory
USE   MIE SCATTERING

MIG AIRCRAFT

MIGRATION

Migration, Electro
USE   ELECTROMIGRATION

Migration, Thermo
USE   THERMOMIGRATION

MIL AIRCRAFT

Milankovitch Theory
USE   CLIMATOLOGY

MILIARIA

MILITARY AIR FACILITIES

MILITARY AIRCRAFT

Military Aircraft, Boeing
USE   MILITARY AIRCRAFT

Military Aircraft, Cessna
USE   MILITARY AIRCRAFT

Military Aircraft, Chance-Vought
USE   MILITARY AIRCRAFT
       CHANCE-VOUGHT AIRCRAFT

Military Aircraft, Convair
USE   MILITARY AIRCRAFT
       GENERAL DYNAMICS AIRCRAFT

Military Aircraft, Curtiss-Wright
USE   MILITARY AIRCRAFT
       CURTISS-WRIGHT AIRCRAFT

Military Aircraft, Douglas
USE   DOUGLAS AIRCRAFT
       MILITARY AIRCRAFT

Military Aircraft, Fairchild
USE   MILITARY AIRCRAFT
       FAIRCHILD-HILLER AIRCRAFT

Military Aircraft, General Dynamics
USE   MILITARY AIRCRAFT
       GENERAL DYNAMICS AIRCRAFT

Military Aircraft, Gyrodyne
USE   QH-50 HELICOPTER

Military Aircraft, Helio
USE   HELIO AIRCRAFT

Military Aircraft, Hiller
USE   HILLER AIRCRAFT
       MILITARY AIRCRAFT

Military Aircraft, Hughes
USE   MILITARY AIRCRAFT
       HUGHES AIRCRAFT

Military Aircraft, Panavia
USE   PANAVIA MILITARY AIRCRAFT

Military Aircraft, Republic
USE   MILITARY AIRCRAFT

Military Aircraft, Ryan
USE   RYAN AIRCRAFT

MILITARY AVIATION

MILITARY COMPACT REACTORS

MILITARY HELICOPTERS

Military Helicopters, Vertol
   USE   BOEING AIRCRAFT

MILITARY OPERATIONS

Military Psychiatry
   USE   MILITARY PSYCHOLOGY

MILITARY PSYCHOLOGY

MILITARY SPACECRAFT

MILITARY TECHNOLOGY

MILITARY VEHICLES

MILK

MILKY WAY GALAXY

MILLET

Milliammeters, Micro
   USE   MICROMILLIAMMETERS

MILLIMETER WAVES

MILLING

Milling, Chemical
   USE   CHEMICAL MACHINING

MILLING MACHINES

MILLING (MACHINING)

Milling (Mixing)
   USE   COMPOUNDING

MILLIVOLTMETERS

Mills Fields, Yang-
   USE   YANG-MILLS FIELDS

Mills, Grinding
   USE   GRINDING MILLS

MILLS RATIO

Mills Theory, Yang-
   USE   YANG-MILLS THEORY

MILNE METHOD

MILNE-THOMSON METHOD

MIM DIODES

MIM (SEMICONDUCTORS)

MIMAS

MINE DETECTORS

Miner Rule
   USE   PALMGREN-MINER RULE

Miner Rule, Palmgren-
   USE   PALMGREN-MINER RULE

Mineral Content, Bone
   USE   BONE MINERAL CONTENT

MINERAL DEPOSITS

(Mineral), Dolomite
   USE   DOLOMITE (MINERAL)

MINERAL EXPLORATION

MINERAL METABOLISM

MINERAL OILS

MINERALOGY

MINERALS

MINES

MINES (EXCAVATIONS)

MINES (ORDNANCE)

MINIATURE ELECTRONIC EQUIPMENT

MINIATURIZATION

Miniaturization, Micro
   USE   MICROMINIATURIZATION

Miniaturization, Sub
   USE   SUBMINIATURIZATION

MINICOMPUTERS

MINIMA

MINIMAL SURFACES

MINIMAX TECHNIQUE

Minimization
   USE   OPTIMIZATION

MINIMUM DRAG

MINIMUM ENTROPY METHOD

MINIMUM VARIANCE ORBIT DETERMINATION

MINING

Mining, Strip
   USE   STRIP MINING

Minitrack Optical Tracking System
   USE   MINITRACK SYSTEM

MINITRACK SYSTEM

MINIVAR Orbit Determination
   USE   MINIMUM VARIANCE ORBIT
          DETERMINATION

MINKOWSKI SPACE

MINNESOTA

MINOR CIRCLE TURNING FLIGHT

Minor Planet 1221
   USE   AMOR ASTEROID

Minor Planet 2060
   USE   CHIRON

MINORITIES

MINORITY CARRIERS

MINOS COMPUTER

Minute Volume, Heart
   USE   HEART MINUTE VOLUME

MINUTEMAN ICBM

Minuteman Missiles
   USE   MINUTEMAN ICBM

MIOSIS

MIRAGE AIRCRAFT

MIRAGE 3 AIRCRAFT

Mirage 3 Aircraft, Dassault
   USE   MIRAGE 3 AIRCRAFT

MIRANDA SATELLITE

MIROS SYSTEM

MIRROR FUSION

MIRROR POINT

MIRRORS

Mirrors, Magnetic
   USE   MAGNETIC MIRRORS

Mirrors, Paraboloid
   USE   PARABOLOID MIRRORS

Mirrors, Rotating
   USE   ROTATING MIRRORS

MIS
   USE   MANAGEMENT INFORMATION SYSTEMS

MIS (SEMICONDUCTORS)

MISALIGNMENT

Miscibility
   USE   SOLUBILITY

Mises Theory, Von
   USE   STRESS FUNCTIONS

Misfets
   USE   FIELD EFFECT TRANSISTORS

MISMATCH (ELECTRICAL)

Misorientation
   USE   MISALIGNMENT

MISS DISTANCE

Missile, Antelope
   USE   ANTELOPE MISSILE

MISSILE ANTENNAS

Missile, Blue Goose
   USE   BLUE GOOSE MISSILE

Missile, Blue Steel
   USE   BLUE STEEL MISSILE

Missile, Blue Streak
   USE   BLUE STREAK MISSILE

MISSILE BODIES

Missile, Bomarc A
   USE   BOMARC A MISSILE

Missile, Bomarc B
   USE   BOMARC B MISSILE

Missile, Bullpup B
   USE   BULLPUP B MISSILE

Missile Cases
   USE   MISSILE BODIES

Missile, Chaparral
   USE   CHAPARRAL MISSILE

MISSILE COMPONENTS

Missile, Condor
   USE   CONDOR MISSILE

MISSILE CONFIGURATIONS

Missile Construction
   USE   MISSILE STRUCTURES

MISSILE CONTROL

Missile, Corporal
   USE   CORPORAL MISSILE

Missile, Corvus
   USE   CORVUS MISSILE

Missile Decoys, Ballistic
   USE   BALLISTIC MISSILE DECOYS

MISSILE DEFENSE

MISSILE DESIGN

MISSILE DETECTION

Missile Early Warning System, Ballistic
USE   BALLISTIC MISSILE EARLY WARNING
SYSTEM

Missile Engine Cases
USE   ROCKET ENGINE CASES

Missile, Falcon
USE   FALCON MISSILE

Missile Guidance
USE   MISSILE CONTROL

Missile, Harpoon
USE   HARPOON MISSILE

Missile, Hawk
USE   HAWK MISSILE

Missile, Hound Dog
USE   HOUND DOG MISSILE

Missile, Jupiter
USE   JUPITER MISSILE

Missile, Lance
USE   LANCE MISSILE

MISSILE LAUNCHERS

Missile Launchers, Mobile
USE   MOBILE MISSILE LAUNCHERS

Missile, Matra
USE   MATRA MISSILE

Missile, Mauler
USE   MAULER MISSILE

Missile, MX
USE   MX MISSILE

Missile, Navaho
USE   NAVAHO MISSILE

Missile, Nike-Ajax
USE   NIKE-AJAX MISSILE

Missile, Nike-Hercules
USE   NIKE-HERCULES MISSILE

Missile, Nike-Zeus
USE   NIKE-ZEUS MISSILE

Missile, Osprey
USE   OSPREY MISSILE

Missile, Patriot
USE   PATRIOT MISSILE

Missile, Pershing
USE   PERSHING MISSILE

Missile, Polaris A1
USE   POLARIS A1 MISSILE

Missile, Polaris A2
USE   POLARIS A2 MISSILE

Missile, Polaris A3
USE   POLARIS A3 MISSILE

Missile, Quail
USE   QUAIL MISSILE

MISSILE RANGES

Missile, Redeye
USE   REDEYE MISSILE

Missile, Regulus
USE   REGULUS MISSILE

Missile, Sandpiper Target
USE   SANDPIPER TARGET MISSILE

Missile, Shrike
USE   SHRIKE MISSILE

MISSILE SIGNATURES

MISSILE SILOS

MISSILE SIMULATORS

Missile, Skybolt
USE   SKYBOLT MISSILE

Missile, SM-65
USE   ATLAS LAUNCH VEHICLES

Missile, SM-68
USE   TITAN 1 ICBM

Missile, SM-68B
USE   TITAN 2 ICBM

Missile, Sparrow 2
USE   SPARROW 2 MISSILE

Missile, Sparrow 3
USE   SPARROW 3 MISSILE

Missile, Spartan
USE   SPARTAN MISSILE

Missile, Sprint
USE   SPRINT MISSILE

Missile, SS-11
USE   SS-11 MISSILE

Missile Stabilization
USE   STABILIZATION
MISSILE CONTROL

MISSILE STORAGE

(Missile Storage), Silos
USE   MISSILE SILOS

MISSILE STRUCTURES

Missile Submarines, Ballistic
USE   BALLISTIC MISSILE SUBMARINES

Missile Submarines, Guided
USE   GUIDED MISSILE SUBMARINES

Missile, Subroc
USE   SUBROC MISSILE

Missile, Supersonic Low Altitude
USE   SUPERSONIC LOW ALTITUDE MISSILE

MISSILE SYSTEMS

Missile, Talos
USE   TALOS MISSILE

Missile, Tartar
USE   TARTAR MISSILE

Missile, Terrier
USE   TERRIER MISSILE

MISSILE TESTS

MISSILE TRACKING

MISSILE TRAJECTORIES

Missile, V-1
USE   V-1 MISSILE

Missile, V-2
USE   V-2 MISSILE

MISSILE VIBRATION

Missile, Zeus
USE   NIKE-ZEUS MISSILE

MISSILES

Missiles, Air Slew
USE   AIR SLEW MISSILES

Missiles, Air To Air
USE   AIR TO AIR MISSILES

Missiles, Air To Surface
USE   AIR TO SURFACE MISSILES

Missiles, Antiaircraft
USE   ANTIAIRCRAFT MISSILES

Missiles, Antimissile
USE   ANTIMISSILE MISSILES

Missiles, Antiradiation
USE   ANTIRADIATION MISSILES

Missiles, Antiship
USE   ANTISHIP MISSILES

Missiles, Antitank
USE   ANTITANK MISSILES

Missiles, Ballistic
USE   BALLISTIC MISSILES

Missiles, Bomarc
USE   BOMARC MISSILES

Missiles, Bullpup
USE   BULLPUP MISSILES

Missiles, Cruise
USE   CRUISE MISSILES

(Missiles), FBM
USE   FLEET BALLISTIC MISSILES

Missiles, Field Army Ballistic
USE   FIELD ARMY BALLISTIC MISSILES

Missiles, Fleet Ballistic
USE   FLEET BALLISTIC MISSILES

Missiles, Ground-To-Air
USE   SURFACE TO AIR MISSILES

(Missiles), ICBM
USE   INTERCONTINENTAL BALLISTIC MISSILES

Missiles, Intercontinental Ballistic
USE   INTERCONTINENTAL BALLISTIC MISSILES

Missiles, Intermediate Range Ballistic
USE   INTERMEDIATE RANGE BALLISTIC
MISSILES

(Missiles), IRBM
USE   INTERMEDIATE RANGE BALLISTIC
MISSILES

Missiles, Mace
USE   MACE MISSILES

Missiles, Maverick
USE   MAVERICK MISSILES

Missiles, Minuteman
USE   MINUTEMAN ICBM

Missiles, Nike
USE   NIKE MISSILES

Missiles, Polaris
USE   POLARIS MISSILES

Missiles, Poseidon
USE   POSEIDON MISSILES

Missiles, Radar Homing
USE   RADAR HOMING MISSILES

Missiles, Ramjet
USE   RAMJET MISSILES

Missiles, Self Initiated Antiaircraft
USE   SIAM MISSILES

Missiles, Sergeant
USE   SERGEANT MISSILES

Missiles, Shillelagh
USE   SHILLELAGH MISSILES

Missiles, Short Range Ballistic
USE   SHORT RANGE BALLISTIC MISSILES

Missiles, Siam
  USE   SIAM MISSILES

Missiles, Sidewinder
  USE   SIDEWINDER MISSILES

Missiles, Sparrow
  USE   SPARROW MISSILES

Missiles, Surface To Air
  USE   SURFACE TO AIR MISSILES

Missiles, Surface To Surface
  USE   SURFACE TO SURFACE MISSILES

Missiles, Tomahawk
  USE   TOMAHAWK MISSILES

Missiles, Tow
  USE   TOW MISSILES

Missiles, Underwater To Surface
  USE   UNDERWATER TO SURFACE MISSILES

MISSING MASS (ASTROPHYSICS)

Mission, AAP 1
  USE   AAP 1 MISSION

Mission, AAP 2
  USE   AAP 2 MISSION

Mission, AAP 3
  USE   AAP 3 MISSION

Mission, AAP 4
  USE   AAP 4 MISSION

Mission Control Center, Integrated
  USE   INTEGRATED MISSION CONTROL CENTER

Mission, Galileo
  USE   GALILEO PROJECT

Mission, Giotto
  USE   GIOTTO MISSION

Mission, Heat Capacity Mapping
  USE   HEAT CAPACITY MAPPING MISSION

Mission, International Solar Polar
  USE   ULYSSES MISSION

Mission, MA-2
  USE   MERCURY MA-2 FLIGHT

Mission, Magellan
  USE   MAGELLAN MISSION

Mission, Multi
  USE   MULTIPLEXING

MISSION PLANNING

Mission, Rosat
  USE   ROSAT MISSION

Mission Simulator, Shuttle
  USE   SHUTTLE MISSION SIMULATOR

Mission, Solar Maximum
  USE   SOLAR MAXIMUM MISSION

Mission, Ulysses
  USE   ULYSSES MISSION

Mission, Voyager 1977
  USE   VOYAGER 1977 MISSION

Mission 31-A, Space Shuttle
  USE   SPACE SHUTTLE MISSION 31-A

Mission 31-B, Space Shuttle
  USE   SPACE SHUTTLE MISSION 31-B

Mission 31-C, Space Shuttle
  USE   SPACE SHUTTLE MISSION 31-C

Mission 31-D, Space Shuttle
  USE   SPACE SHUTTLE MISSION 31-D

Mission 41-A, Space Shuttle
  USE   SPACE SHUTTLE MISSION 41-A

Mission 41-B, Space Shuttle
  USE   SPACE SHUTTLE MISSION 41-B

Mission 41-C, Space Shuttle
  USE   SPACE SHUTTLE MISSION 41-C

Mission 41-D, Space Shuttle
  USE   SPACE SHUTTLE MISSION 41-D

Mission 41-G, Space Shuttle
  USE   SPACE SHUTTLE MISSION 41-G

Mission 51-A, Space Shuttle
  USE   SPACE SHUTTLE MISSION 51-A

Mission 51-B, Space Shuttle
  USE   SPACE SHUTTLE MISSION 51-B

Mission 51-C, Space Shuttle
  USE   SPACE SHUTTLE MISSION 51-C

Mission 51-D, Space Shuttle
  USE   SPACE SHUTTLE MISSION 51-D

Mission 51-E, Space Shuttle
  USE   SPACE SHUTTLE MISSION 51-E

Mission 51-F, Space Shuttle
  USE   SPACE SHUTTLE MISSION 51-F

Mission 51-G, Space Shuttle
  USE   SPACE SHUTTLE MISSION 51-G

Mission 51-H, Space Shuttle
  USE   SPACE SHUTTLE MISSION 51-H

Mission 51-I, Space Shuttle
  USE   SPACE SHUTTLE MISSION 51-I

Mission 51-J, Space Shuttle
  USE   SPACE SHUTTLE MISSION 51-J

Mission 51-L, Space Shuttle
  USE   SPACE SHUTTLE MISSION 51-L

Mission 61-A, Space Shuttle
  USE   SPACE SHUTTLE MISSION 61-A

Mission 61-B, Space Shuttle
  USE   SPACE SHUTTLE MISSION 61-B

Mission 61-C, Space Shuttle
  USE   SPACE SHUTTLE MISSION 61-C

Mission 61-E, Space Shuttle
  USE   SPACE SHUTTLE MISSION 61-E

Mission-A, Solar Maximum
  USE   SOLAR MAXIMUM MISSION-A

MISSIONS

Missions, Aborted
  USE   ABORTED MISSIONS

Missions, Asteroid
  USE   ASTEROID MISSIONS

Missions, Flyby
  USE   FLYBY MISSIONS

Missions, LANDSAT Follow-On
  USE   LANDSAT FOLLOW-ON MISSIONS

Missions, Outer Planet
  USE   GRAND TOURS

Missions, Space
  USE   SPACE MISSIONS

Missions, Space Shuttle
  USE   SPACE SHUTTLE MISSIONS

Missions (STS), Astro
  USE   ASTRO MISSIONS (STS)

MISSISSIPPI

MISSISSIPPI DELTA (LA)

MISSISSIPPI RIVER (US)

MISSOURI

MISSOURI RIVER BASIN (US)

MISSOURI RIVER (US)

MIST

MITOCHONDRIA

MITOSIS

MITRA

MIUS
  USE   MODULAR INTEGRATED UTILITY SYSTEM

MIXED CRYSTALS

Mixed Flow
  USE   MULTIPHASE FLOW

MIXED OXIDES

Mixed Traffic Vehicles, Automated
  USE   AUTOMATED MIXED TRAFFIC VEHICLES

MIXERS

MIXING

MIXING CIRCUITS

Mixing Depth
  USE   MIXING HEIGHT

Mixing Flow, Jet
  USE   JET MIXING FLOW

MIXING HEIGHT

Mixing, Laminar
  USE   LAMINAR MIXING

MIXING LENGTH FLOW THEORY

(Mixing), Milling
  USE   COMPOUNDING

Mixing, Pre
  USE   PREMIXING

Mixing, Signal
  USE   SIGNAL MIXING

(Mixing), Suspending
  USE   SUSPENDING (MIXING)

Mixing, Turbulent
  USE   TURBULENT MIXING

MIXTURES

Mixtures, Binary
  USE   BINARY MIXTURES

Mixtures, Detonable Gas
  USE   DETONABLE GAS MIXTURES

Mixtures, Gas
  USE   GAS MIXTURES

Mixtures, Liquid-Gas
  USE   LIQUID-GAS MIXTURES

MJ252H Engine, J93-
  USE   J-93 ENGINE

MJ280G Engine, J93-
  USE   J-93 ENGINE

MK 35 Aircraft, Vampire
  USE   VAMPIRE MK 35 AIRCRAFT

MK-1 Aircraft, Argosy
  USE   ARGOSY MK-1 AIRCRAFT

MK-1 Aircraft, Short Belfast C
USE   SC-5 AIRCRAFT

MK-1 Aircraft, Victor
USE   VICTOR MK-1 AIRCRAFT

MK-10 Helicopter, Westland
USE   WESTLAND WHIRLWIND HELICOPTER

MK-10 Helicopter, Whirlwind
USE   WESTLAND WHIRLWIND HELICOPTER

ML-1 NUCLEAR POWER PLANT

MLA
USE   MULTISPECTRAL LINEAR ARRAYS

MMS
USE   MULTIMISSION MODULAR SPACECRAFT

Mn
USE   MANGANESE

MN
USE   MINNESOTA

MNEMONICS

MNOS
USE   METAL-NITRIDE-OXIDE-SILICON

Mo
USE   MOLYBDENUM

MO
USE   MISSOURI

(MO), St Louis-Kansas City Corridor
USE   ST LOUIS-KANSAS CITY CORRIDOR (MO)

MOBILE COMMUNICATION SYSTEMS

Mobile Laboratories, Lunar
USE   LUNAR MOBILE LABORATORIES

MOBILE LOUNGES

MOBILE MISSILE LAUNCHERS

MOBILE QUARANTINE FACILITY

Mobile Satellite Service, Land
USE   LAND MOBILE SATELLITE SERVICE

Mobilities, Atomic
USE   ATOMIC MOBILITIES

MOBILITY

Mobility, Carrier
USE   CARRIER MOBILITY

Mobility, Electron
USE   ELECTRON MOBILITY

Mobility, Hole
USE   HOLE MOBILITY

Mobility, Ionic
USE   IONIC MOBILITY

Mobility Semiconductors, Negative Diff
USE   NDM SEMICONDUCTOR DEVICES

Mobility Transistors, High Electron
USE   HIGH ELECTRON MOBILITY TRANSISTORS

Mobility Units, Extravehicular
USE   EXTRAVEHICULAR MOBILITY UNITS

MODAL RESPONSE

MODCOMP II COMPUTER

MODCOMP IV COMPUTER

MODE

Mode Coupling
USE   COUPLED MODES

Mode Locking, Laser
USE   LASER MODE LOCKING

Mode Of Vibration
USE   VIBRATION MODE

Mode Propulsion, Dual
USE   HYBRID PROPULSION

Mode Shapes
USE   MODAL RESPONSE

MODE (STATISTICS)

Mode Theory, Field
USE   FIELD MODE THEORY

MODE TRANSFORMERS

Mode, Vibration
USE   VIBRATION MODE

Model, Density Wave
USE   DENSITY WAVE MODEL

Model, Ising
USE   FERROMAGNETISM
       MATHEMATICAL MODELS

Model, Lighthill Gas
USE   LIGHTHILL GAS MODEL

Model, Quark Parton
USE   QUARK PARTON MODEL

Model, Thomas-Fermi
USE   THOMAS-FERMI MODEL

Model, Vector Dominance
USE   VECTOR DOMINANCE MODEL

Model, Veneziano
USE   VENEZIANO MODEL

Model 18 Aircraft, Lockheed
USE   LOCKHEED MODEL 18 AIRCRAFT

Modeling, Continuum
USE   CONTINUUM MODELING

MODELS

Models, Aircraft
USE   AIRCRAFT MODELS

Models, Astronomical
USE   ASTRONOMICAL MODELS

Models, Atmospheric
USE   ATMOSPHERIC MODELS

Models, Biological
USE   BIONICS

Models, Breadboard
USE   BREADBOARD MODELS

Models, Dynamic
USE   DYNAMIC MODELS

Models, Environment
USE   ENVIRONMENT MODELS

Models, Hydrology
USE   HYDROLOGY MODELS

Models, Mathematical
USE   MATHEMATICAL MODELS

Models (Mathematics), Biological
USE   BIOLOGICAL MODELS (MATHEMATICS)

Models, Nuclear
USE   NUCLEAR MODELS

Models, Ocean
USE   OCEAN MODELS

Models, Powered
USE   POWERED MODELS

Models, Scale
USE   SCALE MODELS

Models, Semispan
USE   SEMISPAN MODELS

Models, Spacecraft
USE   SPACECRAFT MODELS

Models, Static
USE   STATIC MODELS

Models, Stellar
USE   STELLAR MODELS

Models, Two Fluid
USE   TWO FLUID MODELS

Models, Wind Tunnel
USE   WIND TUNNEL MODELS

MODEMS

Moderated Reactors, Organic
USE   ORGANIC MODERATED REACTORS

Moderated Reactors, Water
USE   WATER MODERATED REACTORS

MODERATION (ENERGY ABSORPTION)

MODERATORS

MODES

Modes, Axial
USE   AXIAL MODES

Modes, Ballooning
USE   BALLOONING MODES

Modes, Coupled
USE   COUPLED MODES

Modes, Failure
USE   FAILURE MODES

Modes, Laser
USE   LASER MODES

Modes (Plasmas), Tearing
USE   TEARING MODES (PLASMAS)

Modes, Propagation
USE   PROPAGATION MODES

Modes, Pushbroom Sensor
USE   PUSHBROOM SENSOR MODES

MODES (STANDING WAVES)

Modes, Uncoupled
USE   UNCOUPLED MODES

Modification
USE   REVISIONS

Modification, Weather
USE   WEATHER MODIFICATION

MODULAR INTEGRATED UTILITY SYSTEM

MODULAR RATIOS

Modular Spacecraft, Multimission
USE   MULTIMISSION MODULAR SPACECRAFT

MODULARITY

MODULATED CONTINUOUS RADIATION

Modulating Retrodirective Optics
USE   MIROS SYSTEM

MODULATION

Modulation, Amplitude
USE   AMPLITUDE MODULATION

Modulation, Carrier
USE   MODULATION

Modulation, De
USE   DEMODULATION

Modulation, Delta
USE   DELTA MODULATION

Modulation, Differential Pulse Code
USE   DIFFERENTIAL PULSE CODE MODULATION

(Modulation), DPCM
USE   DIFFERENTIAL PULSE CODE MODULATION

(Modulation), FBFM
USE   FEEDBACK FREQUENCY MODULATION

Modulation, Feedback Frequency
USE   FEEDBACK FREQUENCY MODULATION

(Modulation), FM/PM
USE   FM/PM (MODULATION)

Modulation, Frequency
USE   FREQUENCY MODULATION

Modulation, Inter
USE   INTERMODULATION

Modulation, Ionospheric Cross
USE   IONOSPHERIC CROSS MODULATION

Modulation, Light
USE   LIGHT MODULATION

Modulation, Optical
USE   LIGHT MODULATION

Modulation, Optical Maser
USE   LIGHT MODULATION

(Modulation), PAM
USE   PULSE AMPLITUDE MODULATION

(Modulation), PCM
USE   PULSE CODE MODULATION

(Modulation), PDM
USE   PULSE DURATION MODULATION

(Modulation), PFM
USE   PULSE FREQUENCY MODULATION

Modulation, Phase
USE   PHASE MODULATION

Modulation Photomultipliers, Frequency
USE   FREQUENCY MODULATION
           PHOTOMULTIPLIERS

(Modulation), PPM
USE   PULSE POSITION MODULATION

(Modulation), PTM
USE   PULSE TIME MODULATION

Modulation, Pulse
USE   PULSE MODULATION

Modulation, Pulse Amplitude
USE   PULSE AMPLITUDE MODULATION

Modulation, Pulse Code
USE   PULSE CODE MODULATION

Modulation, Pulse Duration
USE   PULSE DURATION MODULATION

Modulation, Pulse Frequency
USE   PULSE FREQUENCY MODULATION

Modulation, Pulse Position
USE   PULSE POSITION MODULATION

Modulation, Pulse Time
USE   PULSE TIME MODULATION

Modulation, Pulse Width
USE   PULSE DURATION MODULATION

(Modulation), PWM
USE   PULSE DURATION MODULATION

Modulation, Single Sideband
USE   SINGLE SIDEBAND TRANSMISSION

Modulation Telemetry, Pulse Frequency
USE   PULSE FREQUENCY MODULATION
           TELEMETRY

MODULATION TRANSFER FUNCTION

Modulation, Traveling Wave
USE   TRAVELING WAVE MODULATION

Modulation), ULM (Light
USE   ULTRASONIC LIGHT MODULATION

Modulation, Ultrasonic Light
USE   ULTRASONIC LIGHT MODULATION

Modulation, Velocity
USE   VELOCITY MODULATION

Modulator Radiometers, Pressure
USE   PRESSURE MODULATOR RADIOMETERS

MODULATORS

Modulators, De
USE   DEMODULATORS

Modulators-Demodulators
USE   MODEMS

Module, Apollo Lunar Experiment
USE   APOLLO LUNAR EXPERIMENT MODULE

Module Ascent Stage, Lunar
USE   LUNAR MODULE ASCENT STAGE

Module), LEM (Lunar
USE   LUNAR MODULE

Module, Local Scientific Survey
USE   LOCAL SCIENTIFIC SURVEY MODULE

Module, Lunar
USE   LUNAR MODULE

Module, Mars Excursion
USE   MARS EXCURSION MODULE

Module), MEM (Excursion
USE   MARS EXCURSION MODULE

Module, Payload Assist
USE   PAYLOAD ASSIST MODULE

Module 5, Lunar
USE   LUNAR MODULE 5

Module 7, Lunar
USE   LUNAR MODULE 7

MODULES

Modules, Airlock
USE   AIRLOCK MODULES

Modules, Chemical Release
USE   CHEMICAL RELEASE MODULES

Modules, Command
USE   COMMAND MODULES

Modules, Command Service
USE   COMMAND SERVICE MODULES

Modules, Electronic
USE   ELECTRONIC MODULES

Modules, Landing
USE   LANDING MODULES

Modules, Lunar Landing
USE   LUNAR LANDING MODULES

Modules, Lunar Surface Scientific
USE   LSSM

Modules, Micro
USE   MICROMODULES

Modules, Scientific Instrument
USE   SIM

Modules, Service
USE   SERVICE MODULES

Modules, Spacecraft
USE   SPACECRAFT MODULES

Modules, Spacecraft Docking
USE   SPACECRAFT DOCKING MODULES

Modules (STS), Power
USE   POWER MODULES (STS)

Modulus, Bulk
USE   BULK MODULUS

Modulus, Elastic
USE   MODULUS OF ELASTICITY

MODULUS OF ELASTICITY

Modulus Of Elasticity, Dynamic
USE   DYNAMIC MODULUS OF ELASTICITY

Modulus, Young
USE   MODULUS OF ELASTICITY

Mohawk Aircraft
USE   OV-1 AIRCRAFT

Mohr Circles
USE   FRACTURE MECHANICS

Mohr Method, Maxwell-
USE   MAXWELL-MOHR METHOD

MOIRE EFFECTS

MOIRE FRINGES

MOIRE INTERFEROMETRY

MOISTURE

Moisture, Atmospheric
USE   ATMOSPHERIC MOISTURE

MOISTURE CONTENT

Moisture Detectors
USE   MOISTURE METERS

MOISTURE METERS

MOISTURE RESISTANCE

Moisture, Soil
USE   SOIL MOISTURE

MOJAVE DESERT (CA)

MOL (Orbital Laboratories)
USE   MANNED ORBITAL LABORATORIES

MOLABS
USE   LUNAR MOBILE LABORATORIES

MOLD

MOLDAVITE

Molding, Injection
USE   INJECTION MOLDING

MOLDING MATERIALS

MOLDS

MOLECULAR ABSORPTION

MOLECULAR BEAM EPITAXY

MOLECULAR BEAMS

MOLECULAR BIOLOGY

Molecular Bonds
USE   CHEMICAL BONDS

MOLECULAR CHAINS	Molten Plutonium Reactor, Los Alamos USE LOS ALAMOS MOLTEN PLUTONIUM REACTOR	**MOMENTUM THEORY**
MOLECULAR CLOUDS		**MOMENTUM TRANSFER**
MOLECULAR COLLISIONS	**MOLTEN SALT ELECTROLYTES**	**MONACO**
MOLECULAR DIFFUSION	**MOLTEN SALT NUCLEAR REACTORS**	**MONATOMIC GASES**
Molecular Dissociation USE DISSOCIATION	**MOLTEN SALTS**	**MONATOMIC MOLECULES**
MOLECULAR ELECTRONICS	**MOLTING**	**MONAURAL SIGNALS**
MOLECULAR ENERGY LEVELS	**MOLYBDATES**	**MONAZITE SANDS**
MOLECULAR EXCITATION	Molybdates, Lead USE LEAD MOLYBDATES	**MONEL (TRADEMARK)**
MOLECULAR FLOW	**MOLYBDENUM**	**MONGE-AMPERE EQUATION**
Molecular Flow, Free USE FREE MOLECULAR FLOW	**MOLYBDENUM ALLOYS**	**MONGOLIA**
MOLECULAR GASES	**MOLYBDENUM CARBIDES**	Monitor, Automatic Light Aircraft Readiness USE ALARM PROJECT
MOLECULAR INTERACTIONS	**MOLYBDENUM COMPOUNDS**	Monitoring, Environmental USE ENVIRONMENTAL MONITORING
MOLECULAR IONS	**MOLYBDENUM DISULFIDES**	Monitoring, IN-Flight USE IN-FLIGHT MONITORING
MOLECULAR ORBITALS	**MOLYBDENUM OXIDES**	Monitoring Instruments, Engine USE ENGINE MONITORING INSTRUMENTS
MOLECULAR OSCILLATIONS	**MOLYBDENUM SULFIDES**	Monitoring Platform, Interplanetary USE IMP
MOLECULAR OSCILLATORS	**MOM (SEMICONDUCTORS)**	Monitoring, Pollution USE POLLUTION MONITORING
MOLECULAR PHYSICS	**MOMENT DISTRIBUTION**	**MONITORS**
MOLECULAR PUMPS	Moment Gyroscopes, Control USE CONTROL MOMENT GYROSCOPES	**MONKEYS**
MOLECULAR RELAXATION	**MOMENTS**	**MONOCHROMATIC RADIATION**
MOLECULAR ROTATION	Moments, Aerodynamic USE STABILITY DERIVATIVES	**MONOCHROMATIZATION**
MOLECULAR SHIELDS	Moments, Bending USE BENDING MOMENTS	Monochromatization, Interference USE MONOCHROMATIZATION DIFFRACTION
Molecular Sieves USE ABSORBENTS	Moments, Dipole USE DIPOLE MOMENTS	**MONOCHROMATORS**
MOLECULAR SPECTRA	Moments, Distribution USE DISTRIBUTION MOMENTS	**MONOCOQUE STRUCTURES**
MOLECULAR SPECTROSCOPY	Moments, Electric USE ELECTRIC MOMENTS	Monocrystals USE SINGLE CRYSTALS
MOLECULAR STRUCTURE	Moments, Hinge USE TORQUE	**MONOCULAR VISION**
MOLECULAR THEORY	Moments, Inertia USE MOMENTS OF INERTIA	**MONOETHANOLAMINE (MEA)**
MOLECULAR TRAJECTORIES	Moments, Loading USE LOADING MOMENTS	**MONOIDS**
MOLECULAR WEIGHT	Moments, Magnetic USE MAGNETIC MOMENTS	Monolithic Circuits USE INTEGRATED CIRCUITS
Molecular Weights, Low USE LOW MOLECULAR WEIGHTS	Moments, Method Of USE METHOD OF MOMENTS	**MONOMERS**
MOLECULES	**MOMENTS OF INERTIA**	**MONOMOLECULAR FILMS**
Molecules, Diatomic USE DIATOMIC MOLECULES	Moments, Pitching USE PITCHING MOMENTS	Monophosphate, Cyclic Adenosine USE CYCLIC AMP
Molecules, Monatomic USE MONATOMIC MOLECULES	Moments, Rolling USE ROLLING MOMENTS	**MONOPLANES**
Molecules, Polyatomic USE POLYATOMIC MOLECULES	Moments, Statistical USE DISTRIBUTION MOMENTS	**MONOPOLE ANTENNAS**
Molecules, Triatomic USE TRIATOMIC MOLECULES	Moments, Yawing USE YAWING MOMENTS	**MONOPOLES**
MOLES	**MOMENTUM**	Monopoles, Magnetic USE MAGNETIC MONOPOLES
Moliere Formula USE COSMIC RAY SHOWERS SPATIAL DISTRIBUTION SECONDARY COSMIC RAYS	Momentum, Angular USE ANGULAR MOMENTUM	**MONOPROPELLANTS**
MOLLIER DIAGRAM	Momentum Energy USE KINETIC ENERGY	**MONOPULSE ANTENNAS**
MOLLUSKS		**MONOPULSE RADAR**
MOLNIYA SATELLITES		**MONOSACCHARIDES**

MONOSCOPES

MONOSTABLE MULTIVIBRATORS

MONOTECTIC ALLOYS

MONOTONE FUNCTIONS

MONOTONY

Monoxide, Carbon
  USE   CARBON MONOXIDE

Monoxide Lasers, Carbon
  USE   CARBON MONOXIDE LASERS

Monoxide Poisoning, Carbon
  USE   CARBON MONOXIDE POISONING

MONSOONS

MONTANA

MONTE CARLO METHOD

MONTEREY BAY (CA)

MONTH

MONTICELLITE

MONTMORILLONITE

MOODS

MOON

MOON ILLUSION

Moon System, Earth-
  USE   EARTH-MOON SYSTEM

Moon Trajectories, Earth-
  USE   EARTH-MOON TRAJECTORIES

MOON-EARTH TRAJECTORIES

MOONQUAKES

Moons Project, New
  USE   NEW MOONS PROJECT

MOORING

Moorings
  USE   MOORING

MOPS (Propulsion Systems)
  USE   MAN OPERATED PROPULSION SYSTEMS

Moraines
  USE   GLACIAL DRIFT

Moraines, End
  USE   GLACIAL DRIFT

MORALE

MOREHOUSE COMET

MORL
  USE   MANNED ORBITAL RESEARCH
            LABORATORIES

MORNING

MOROCCO

MORPHINE

Morphism, Iso
  USE   ISOMORPHISM

Morphisms, Homo
  USE   HOMOMORPHISMS

MORPHOLOGICAL INDEXES

MORPHOLOGY

Morphology, Geo
  USE   GEOMORPHOLOGY

Morphology, Lung
  USE   LUNG MORPHOLOGY

Morphotropism
  USE   ISOMORPHISM

MORSE CODE

MORSE POTENTIAL

MORTALITY

MORTARS (MATERIAL)

MOS (Japanese Spacecraft)
  USE   JAPANESE SPACECRAFT

MOS (Semiconductors)
  USE   METAL OXIDE SEMICONDUCTORS

MOSAICS

MOSCOW

MOSFET
  USE   FIELD EFFECT TRANSISTORS

Mosfet, Cascode
  USE   FIELD EFFECT TRANSISTORS

MOSS (Space Stations)
  USE   ORBITAL SPACE STATIONS

MOSSBAUER EFFECT

MOT (Orbital Telescopes)
  USE   MANNED ORBITAL TELESCOPES

MOTHS

Motility
  USE   LOCOMOTION

MOTION

MOTION AFTEREFFECTS

Motion, Angular
  USE   ANGULAR VELOCITY

Motion), Brakes (For Arresting
  USE   BRAKES (FOR ARRESTING MOTION)

Motion, Chandler
  USE   POLAR WANDERING (GEOLOGY)

Motion Compensation, Image
  USE   IMAGE MOTION COMPENSATION

Motion, Earth
  USE   EARTH MOTION

Motion Equations
  USE   EQUATIONS OF MOTION

Motion Equations, Forced Vibratory
  USE   EQUATIONS
            FORCED VIBRATION

Motion, Equations Of
  USE   EQUATIONS OF MOTION

Motion, Euler Equations Of
  USE   EULER EQUATIONS OF MOTION

(Motion), Guidance
  USE   GUIDANCE (MOTION)

Motion, Harmonic
  USE   HARMONIC MOTION

Motion, Ion
  USE   ION MOTION

Motion, Lagrange Equations Of
  USE   EULER-LAGRANGE EQUATION

Motion, Librational
  USE   LIBRATIONAL MOTION

Motion, Orbital
  USE   ORBITS

Motion, Particle
  USE   PARTICLE MOTION

MOTION PERCEPTION

MOTION PICTURES

Motion, Planetary
  USE   SOLAR ORBITS

(Motion), Revolution
  USE   REVOLVING

MOTION SICKNESS

MOTION SICKNESS DRUGS

Motion, Simple Harmonic
  USE   SIMPLE HARMONIC MOTION

MOTION SIMULATION

MOTION SIMULATORS

Motion Simulators, Vertical
  USE   VERTICAL MOTION SIMULATORS

Motion, Spacecraft
  USE   SPACECRAFT MOTION

MOTION STABILITY

Motion, Three Dimensional
  USE   THREE DIMENSIONAL MOTION

Motion, Translational
  USE   TRANSLATIONAL MOTION

Motion, Tumbling
  USE   TUMBLING MOTION

Motion, Vertical
  USE   VERTICAL MOTION

Motion, Wave
  USE   WAVES

Motions, Stellar
  USE   STELLAR MOTIONS

MOTIVATION

Motor Cases, Rocket
  USE   ROCKET ENGINE CASES

Motor Systems (Biology)
  USE   EFFERENT NERVOUS SYSTEMS

MOTOR VEHICLES

Motor Vehicles, Electric
  USE   ELECTRIC MOTOR VEHICLES

MOTORS

Motors, Apogee Boost
  USE   APOGEE BOOST MOTORS

Motors, Asynchronous
  USE   ASYNCHRONOUS MOTORS

Motors, Electric
  USE   ELECTRIC MOTORS

Motors, Induction
  USE   INDUCTION MOTORS

Motors, Micro
  USE   MICROMOTORS

Motors, Servo
  USE   SERVOMOTORS

Motors, Stepping
  USE   STEPPING MOTORS

Motors, Synchronous
  USE   SYNCHRONOUS MOTORS

Motors, Torque
  USE   TORQUE MOTORS

MOTS (Tracking System)
  USE   MINITRACK SYSTEM

Mount, Apollo Telescope
  USE   APOLLO TELESCOPE MOUNT

MOUNTAIN INHABITANTS

MOUNTAINS

Mountains (AK), Wrangell
  USE   WRANGELL MOUNTAINS (AK)

Mountains (CA), Sierra Nevada
  USE   SIERRA NEVADA MOUNTAINS (CA)

Mountains (CO), San Juan
  USE   SAN JUAN MOUNTAINS (CO)

Mountains (Europe), Alps
  USE   ALPS MOUNTAINS (EUROPE)

Mountains (Europe), Carpathian
  USE   CARPATHIAN MOUNTAINS (EUROPE)

Mountains (Europe), Pyrenees
  USE   PYRENEES MOUNTAINS (EUROPE)

Mountains (MT-WY), Bighorn
  USE   BIGHORN MOUNTAINS (MT-WY)

Mountains (NC-TN), Great Smoky
  USE   GREAT SMOKY MOUNTAINS (NC-TN)

Mountains (North America), Appalachian
  USE   APPALACHIAN MOUNTAINS (NORTH
         AMERICA)

Mountains (North America), Rocky
  USE   ROCKY MOUNTAINS (NORTH AMERICA)

Mountains (NY), Adirondack
  USE   ADIRONDACK MOUNTAINS (NY)

Mountains (South America), Andes
  USE   ANDES MOUNTAINS (SOUTH AMERICA)

Mountains (U.S.S.R.), Caucasus
  USE   CAUCASUS MOUNTAINS (U.S.S.R.)

Mounted Displays, Helmet
  USE   HELMET MOUNTED DISPLAYS

MOUNTING

Mounting, Fuselage
  USE   AIRCRAFT PRODUCTION

Mounting, Pylon
  USE   PYLON MOUNTING

Mounting, Rigid
  USE   RIGID MOUNTING

Mountings, Tail
  USE   TAIL ASSEMBLIES

Mounts
  USE   SUPPORTS

MOUTH

Movement
  USE   MOTION

Movement, Head
  USE   HEAD MOVEMENT

Movement State, Rapid Eye
  USE   RAPID EYE MOVEMENT STATE

Movement, Tectonic
  USE   TECTONICS

Movements, Airfield Surface
  USE   AIRFIELD SURFACE MOVEMENTS

Movements, Brownian
  USE   BROWNIAN MOVEMENTS

Movements, Earth
  USE   EARTH MOVEMENTS

Movements, Eye
  USE   EYE MOVEMENTS

Movements, Saccadic Eye
  USE   SACCADIC EYE MOVEMENTS

MOVING TARGET INDICATORS

MOZAMBIQUE

MR-1 Flight, Mercury
  USE   MERCURY MR-1 FLIGHT

MR-2 Flight, Mercury
  USE   MERCURY MR-2 FLIGHT

MR-3 Flight
  USE   MERCURY MR-3 FLIGHT

MR-3 Flight, Mercury
  USE   MERCURY MR-3 FLIGHT

MR-4 Flight, Mercury
  USE   MERCURY MR-4 FLIGHT

MRCA AIRCRAFT

MRKOS COMET

MS
  USE   MISSISSIPPI

MSAT

Msbls
  USE   MICROWAVE SCANNING BEAM LANDING
         SYSTEM

MSRE Reactors
  USE   MOLTEN SALT NUCLEAR REACTORS

MT
  USE   MONTANA

(MT-WY), Bighorn Mountains
  USE   BIGHORN MOUNTAINS (MT-WY)

MT-WY), Yellowstone National Park (ID-
  USE   YELLOWSTONE NATIONAL PARK
         (ID-MT-WY)

MTBF

MTF
  USE   MODULATION TRANSFER FUNCTION

MTI Radar
  USE   MOVING TARGET INDICATORS

MUBIS (Scanners)
  USE   MULTIPLE BEAM INTERVAL SCANNERS

MUCOCELES

MUCUS

MUD

Mueller Tubes, Geiger-
  USE   GEIGER COUNTERS

MUFFLERS

MULBERRY (ALLOY)

MULLITES

MULTI-ANODE MICROCHANNEL ARRAYS

Multi-Role Combat Aircraft
  USE   MRCA AIRCRAFT

MULTIBEAM ANTENNAS

MULTICHANNEL COMMUNICATION

Multichannel Plates
  USE   MICROCHANNEL PLATES

MULTIENGINE VEHICLES

MULTILAYER INSULATION

Multilayer Structures
  USE   LAMINATES

Multiloop Systems
  USE   CASCADE CONTROL

MULTIMISSION MODULAR SPACECRAFT

MULTIMODE RESONATORS

MULTIPACTOR DISCHARGES

MULTIPATH TRANSMISSION

MULTIPHASE FLOW

MULTIPHOTON ABSORPTION

MULTIPLE ACCESS

Multiple Access, Code Division
  USE   CODE DIVISION MULTIPLE ACCESS

Multiple Access, Demand Assignment
  USE   DEMAND ASSIGNMENT MULTIPLE ACCESS

Multiple Access, Frequency Division
  USE   FREQUENCY DIVISION MULTIPLE ACCESS

Multiple Access, Time Division
  USE   TIME DIVISION MULTIPLE ACCESS

MULTIPLE BEAM INTERVAL SCANNERS

MULTIPLE DOCKING ADAPTERS

MULTIPLE OUTPUT PROGRAMS

Multiple Target Trajectory Systems
  USE   MATTS (SYSTEMS)

Multiplets
  USE   FINE STRUCTURE

Multiplex Transmission
  USE   MULTIPLEXING

Multiplexers
  USE   MULTIPLEXING

MULTIPLEXING

Multiplexing, Code Division
  USE   CODE DIVISION MULTIPLEXING

Multiplexing, Frequency Division
  USE   FREQUENCY DIVISION MULTIPLEXING

Multiplexing Theory, Orthogonal
  USE   ORTHOGONAL MULTIPLEXING THEORY

Multiplexing, Time Division
  USE   TIME DIVISION MULTIPLEXING

Multiplexing, Wavelength Division
  USE   WAVELENGTH DIVISION MULTIPLEXING

MULTIPLICATION

Multiplication, Fringe
  USE   FRINGE MULTIPLICATION

Multiplier Phototubes
  USE   PHOTOMULTIPLIER TUBES

MULTIPLIERS

Multipliers, Channel
  USE   CHANNEL MULTIPLIERS

Multipliers, Electron
USE   PHOTOMULTIPLIER TUBES

Multipliers, Frequency
USE   FREQUENCY MULTIPLIERS

Multipliers, Lagrange
USE   LAGRANGE MULTIPLIERS

MULTIPOLAR FIELDS

MULTIPOLES

Multiprobe Spacecraft, Pioneer Venus 2
USE   PIONEER VENUS 2 SPACECRAFT

MULTIPROCESSING (COMPUTERS)

MULTIPROGRAMMING

Multipropellants
USE   ROCKET PROPELLANTS

Multipurpose System, Light Airborne
USE   LIGHT AIRBORNE MULTIPURPOSE SYSTEM

Multiradar Tracking
USE   RADAR NETWORKS

MULTISENSOR APPLICATIONS

MULTISPECTRAL BAND CAMERAS

MULTISPECTRAL BAND SCANNERS

MULTISPECTRAL LINEAR ARRAYS

MULTISPECTRAL PHOTOGRAPHY

MULTISPECTRAL RADAR

MULTISPECTRAL RESOURCE SAMPLER

MULTISPECTRAL TRACKING TELESCOPES

Multistage Compressors
USE   TURBOCOMPRESSORS

MULTISTAGE ROCKET VEHICLES

MULTISTATIC RADAR

Multitemporal Analysis
USE   TEMPORAL RESOLUTION

MULTIVARIATE STATISTICAL ANALYSIS

MULTIVIBRATORS

Multivibrators, Monostable
USE   MONOSTABLE MULTIVIBRATORS

MUON SPIN ROTATION

MUONIUM

MUONS

MURCHISON METEORITE

MURRAY METEORITE

MUSCLE RELAXANTS

MUSCLES

MUSCOVITE

MUSCULAR FATIGUE

MUSCULAR FUNCTION

MUSCULAR STRENGTH

MUSCULAR TONUS

MUSCULOSKELETAL SYSTEM

MUSEUMS

MUSHY ZONES

MUSIC

MUSKEGS

Mustang Aircraft
USE   P-51 AIRCRAFT

MUTAGENS

Mutation, Trans
USE   TRANSMUTATION

MUTATIONS

Mutations, Per
USE   PERMUTATIONS

Mv
USE   MENDELEVIUM

MX MISSILE

MYELIN

MYLAR (TRADEMARK)

MYOCARDIAL INFARCTION

MYOCARDIUM

MYOELECTRIC POTENTIALS

MYOELECTRICITY

MYOGLOBIN

MYOPIA

MYSTERE 20 AIRCRAFT

Mystere 20 Aircraft, Dassault
USE   MYSTERE 20 AIRCRAFT

MYSTERE 50 AIRCRAFT

Mystere 50 Aircraft, Dassault
USE   MYSTERE 50 AIRCRAFT

N

N Diagrams, S-
USE   S-N DIAGRAMS

N Diodes, P-I-
USE   P-I-N JUNCTIONS
       DIODES

N ELECTRONS

N Junctions, N-
USE   N-N JUNCTIONS

N Junctions, N-P-
USE   N-P-N JUNCTIONS

N Junctions, P-
USE   P-N JUNCTIONS

N Junctions, P-I-
USE   P-I-N JUNCTIONS

N Junctions, P-N-P-
USE   P-N-P-N JUNCTIONS

N Series Satellites, TIROS
USE   TIROS N SERIES SATELLITES

N-N JUNCTIONS

N-P Junctions
USE   P-N JUNCTIONS

N-P Junctions, P-
USE   P-N-P JUNCTIONS

N-P-N JUNCTIONS

N-P-N Junctions, P-
USE   P-N-P-N JUNCTIONS

N-TYPE SEMICONDUCTORS

N-156 Aircraft
USE   F-5 AIRCRAFT

Na
USE   SODIUM

NA-300 Aircraft
USE   OV-10 AIRCRAFT

Nacelle Configurations, Wing
USE   WING NACELLE CONFIGURATIONS

NACELLES

NAKED SINGULARITIES

NAMC Aircraft
USE   NIHON AIRCRAFT

NAMIBIA

NAMING

NAP-OF-THE-EARTH NAVIGATION

NAPHTHALENE

NAPHTHENES

Nappes
USE   FOLDS (GEOLOGY)

NARCOLEPSY

NARCOSIS

Narcosis, Electro
USE   ELECTRONARCOSIS

NARCOTICS

NARROWBAND

NASA Communication Network
USE   NASCOM NETWORK

NASA End-To-End Data System
USE   NEEDS (DATA SYSTEM)

NASA INTERACTIVE PLANNING SYSTEM

NASA PROGRAMS

(NASA), Space Operations Center
USE   SPACE OPERATIONS CENTER (NASA)

NASA SPACE PROGRAMS

NASA Structural Analysis Program
USE   NASTRAN

NASARR
USE   NORTH AMERICAN SEARCH AND RANGING
       RADAR

NASCOM NETWORK

NASTRAN

NATIONAL AIRSPACE SYSTEM

NATIONAL AIRSPACE UTILIZATION SYSTEM

NATIONAL AVIATION SYSTEM

NATIONAL LAUNCH VEHICLE PROGRAM

NATIONAL OCEANIC SATELLITE SYSTEM

National Operational Environmental Sat Sys
USE   NOESS

National Park (ID-MT-WY), Yellowstone
USE   YELLOWSTONE NATIONAL PARK
       (ID-MT-WY)

NATIONAL PARKS

National Product, Gross
USE   GROSS NATIONAL PRODUCT

NATIONAL SEVERE STORMS PROJECT

NATIONS

Nations, Developing
USE   DEVELOPING NATIONS

Nations, United
USE   UNITED NATIONS

(NATO), North Atlantic Treaty Organization
USE   NORTH ATLANTIC TREATY ORGANIZATION
       (NATO)

NATO 3B SATELLITE

Natural Frequencies
USE   RESONANT FREQUENCIES

NATURAL GAS

NATURAL GAS EXPLORATION

Natural Gas, Liquefied
USE   LIQUEFIED NATURAL GAS

NATURAL LANGUAGE (COMPUTERS)

Natural Lasers
USE   LASERS

NATURAL SATELLITES

NAUSEA

NAUTICAL CHARTS

NAVAHO MISSILE

NAVIER-STOKES EQUATION

NAVIGATION

NAVIGATION AIDS

Navigation, Air
USE   AIR NAVIGATION

Navigation, All-Weather Air
USE   ALL-WEATHER AIR NAVIGATION

Navigation, Area
USE   AREA NAVIGATION

Navigation, Astro
USE   ASTRONAVIGATION

Navigation, Autonomous
USE   AUTONOMOUS NAVIGATION

Navigation, Celestial
USE   CELESTIAL NAVIGATION

Navigation, Decca
USE   DECCA NAVIGATION

Navigation, Digital
USE   DIGITAL NAVIGATION

Navigation, Doppler
USE   DOPPLER NAVIGATION

Navigation, Gimballess Inertial
USE   GIMBALLESS INERTIAL NAVIGATION

Navigation, Hyperbolic
USE   HYPERBOLIC NAVIGATION

Navigation, Inertial
USE   INERTIAL NAVIGATION

NAVIGATION INSTRUMENTS

Navigation, Interplanetary
USE   INTERPLANETARY NAVIGATION

Navigation, Long Range
USE   LORAN D
       LORAN

Navigation, Marine
USE   SURFACE NAVIGATION

Navigation, Nap-Of-The-Earth
USE   NAP-OF-THE-EARTH NAVIGATION

Navigation, NOE
USE   NAP-OF-THE-EARTH NAVIGATION

Navigation, Omnirange
USE   VHF OMNIRANGE NAVIGATION

Navigation, Polar
USE   POLAR NAVIGATION

Navigation, Radar
USE   RADAR NAVIGATION

Navigation, Radio
USE   RADIO NAVIGATION

NAVIGATION SATELLITES

Navigation, Short Range
USE   SHORAN

Navigation, Space
USE   SPACE NAVIGATION

Navigation, Surface
USE   SURFACE NAVIGATION

Navigation System, Astroguide
USE   ASTROGUIDE NAVIGATION SYSTEM

Navigation System, LORAC
USE   LORAC NAVIGATION SYSTEM

Navigation System, Omega
USE   OMEGA NAVIGATION SYSTEM

Navigation System, Terrain Contour Matching
USE   TERCOM

Navigation System, Transit
USE   TRANSIT NAVIGATION SYSTEM

Navigation Systems, Hybrid
USE   HYBRID NAVIGATION SYSTEMS

Navigation Systems, Satellite
USE   SATELLITE NAVIGATION SYSTEMS

Navigation, Tactical Air
USE   TACAN

NAVIGATION TECHNOLOGY SATELLITES

Navigation, VHF Omnirange
USE   VHF OMNIRANGE NAVIGATION

NAVIGATORS

NAVION AIRCRAFT

Navion G-1 Aircraft
USE   G-1 AIRCRAFT

Navion Rangemaster Aircraft
USE   G-1 AIRCRAFT

NAVSTAR SATELLITES

NAVY

(Navy), Global Communications Antenna Grid
USE   SEAFARER PROJECT

Navy Instrumentation Program, Army-
USE   ARMY-NAVY INSTRUMENTATION PROGRAM

(Navy), Underground Radio Antenna Grid
USE   SEAFARER PROJECT

Nb
USE   NIOBIUM

NC
USE   NORTH CAROLINA

(NC), Cape Hatteras
USE   CAPE HATTERAS (NC)

(NC), Outer Banks
USE   OUTER BANKS (NC)

NC-SC), Sand Hills Region (GA-
USE   SAND HILLS REGION (GA-NC-SC)

(NC-TN), Great Smoky Mountains
USE   GREAT SMOKY MOUNTAINS (NC-TN)

NC-130 Aircraft
USE   C-130 AIRCRAFT

Nd
USE   NEODYMIUM

ND
USE   NORTH DAKOTA

NDM SEMICONDUCTOR DEVICES

Ne
USE   NEON

NE
USE   NEBRASKA

(NE), Sand Hills Region
USE   SAND HILLS REGION (NE)

NEAR FIELDS

NEAR INFRARED RADIATION

NEAR ULTRAVIOLET RADIATION

NEAR WAKES

NEARSHORE WATER

NEBRASKA

Nebula, Crab
USE   CRAB NEBULA

Nebula, Gum
USE   GUM NEBULA

Nebula, Orion
USE   ORION NEBULA

Nebula, Solar
USE   SOLAR CORONA

NEBULAE

Nebulae, Planetary
USE   PLANETARY NEBULAE

Nebulae, Reflection
USE   REFLECTION NEBULAE

NECK (ANATOMY)

NEEDLE BEARINGS

NEEDLES

NEEDS (DATA SYSTEM)

NEEL TEMPERATURE

NEGATIVE CONDUCTANCE

Negative Diff Mobility Semiconductors
USE   NDM SEMICONDUCTOR DEVICES

NEGATIVE FEEDBACK

NEGATIVE IONS

Negative Pressure, Lower Body
USE   LOWER BODY NEGATIVE PRESSURE

NEGATIVE RESISTANCE CIRCUITS

NEGATIVE RESISTANCE DEVICES

NEGATRONS

Negotiation, Contract
   USE   CONTRACT NEGOTIATION

Neighborhood, Origin Of Plasmas In Earth
   USE   OPEN PROJECT

NEMBUTAL (TRADEMARK)

NEODYMIUM

NEODYMIUM ALLOYS

NEODYMIUM COMPOUNDS

NEODYMIUM ISOTOPES

NEODYMIUM LASERS

NEON

NEON ISOTOPES

Neon Lasers, Helium-
   USE   HELIUM-NEON LASERS

Neon, Liquid
   USE   LIQUID NEON

Neon 19
   USE   NEON ISOTOPES

NEOPENTANE

NEOPLASMS

Neoprenes
   USE   CHLOROPRENE RESINS

NEPAL

NEPHANALYSIS

NEPHELINE

NEPHELITE

NEPHELOMETERS

NEPHRITIS

NEPTUNE ATMOSPHERE

NEPTUNE (PLANET)

NEPTUNIUM

NEPTUNIUM COMPOUNDS

NEPTUNIUM ISOTOPES

Nernst Generators
   USE   THERMOMAGNETIC COOLING

Nernst Heat Theorem
   USE   NERNST-ETTINGSHAUSEN EFFECT

NERNST-ETTINGSHAUSEN EFFECT

NERVA (Engine)
   USE   NUCLEAR ENGINE FOR ROCKET VEHICLES

NERVES

Nerves, Oculomotor
   USE   OCULOMOTOR NERVES

NERVOUS SYSTEM

Nervous System, Autonomic
   USE   AUTONOMIC NERVOUS SYSTEM

Nervous System, Central
   USE   CENTRAL NERVOUS SYSTEM

Nervous System Depressants, Central
   USE   CENTRAL NERVOUS SYSTEM
         DEPRESSANTS

Nervous System, Peripheral
   USE   PERIPHERAL NERVOUS SYSTEM

Nervous System Stimulants, Central
   USE   CENTRAL NERVOUS SYSTEM STIMULANTS

Nervous System, Sympathetic
   USE   SYMPATHETIC NERVOUS SYSTEM

Nervous System, Vasomotor
   USE   NERVOUS SYSTEM

Nervous Systems, Afferent
   USE   AFFERENT NERVOUS SYSTEMS

Nervous Systems, Efferent
   USE   EFFERENT NERVOUS SYSTEMS

NETHERLANDS

Netherlands Satellite, Astronomical
   USE   ASTRONOMICAL NETHERLANDS SATELLITE

NETS

Nets, Flow
   USE   FLOW NETS

Nets, Neural
   USE   NEURAL NETS

Nets, Petri
   USE   PETRI NETS

NETWORK ANALYSIS

Network, Arpa Computer
   USE   ARPA COMPUTER NETWORK

NETWORK CONTROL

Network, Deep Space
   USE   DEEP SPACE NETWORK

Network, Global Tracking
   USE   GLOBAL TRACKING NETWORK

Network), GLOTRAC (Tracking
   USE   GLOBAL TRACKING NETWORK

Network, Manned Space Flight
   USE   MANNED SPACE FLIGHT NETWORK

Network, NASA Communication
   USE   NASCOM NETWORK

Network, NASCOM
   USE   NASCOM NETWORK

Network), Orion (Radio Interferometry
   USE   ORION (RADIO INTERFEROMETRY
         NETWORK)

Network, Satellite Tracking And Data Acq
   USE   STDN (NETWORK)

Network, Space Flight Tracking And Data
   USE   SPACE FLIGHT TRACKING AND DATA
         NETWORK

Network, Spacecraft Tracking And Data
   USE   STDN (NETWORK)

Network), STADAN (Satellite Tracking
   USE   STDN (NETWORK)

(Network), Stdn
   USE   STDN (NETWORK)

NETWORK SYNTHESIS

NETWORKS

Networks, Communication
   USE   COMMUNICATION NETWORKS

Networks, Computer
   USE   COMPUTER NETWORKS

Networks, Electric
   USE   ELECTRIC NETWORKS

Networks, Iterative
   USE   ITERATIVE NETWORKS

Networks, Kirchhoff Law Of
   USE   KIRCHHOFF LAW OF NETWORKS

Networks, Logic
   USE   LOGIC CIRCUITS

Networks, Quadrupole
   USE   QUADRUPOLE NETWORKS

Networks, Radar
   USE   RADAR NETWORKS

Networks, RC
   USE   RC CIRCUITS

Networks, RLC
   USE   RLC CIRCUITS

Networks, Satellite
   USE   SATELLITE NETWORKS

Networks, Tracking
   USE   TRACKING NETWORKS

Networks, Transportation
   USE   TRANSPORTATION NETWORKS

NEUMANN PROBLEM

NEURAL NETS

NEURASTHENIA

NEURISTORS

NEURITIS

NEUROBLASTS

NEUROGLIA

NEUROLOGY

NEUROMUSCULAR TRANSMISSION

Neuron Transmission
   USE   BIOELECTRICITY

NEURONS

NEUROPHYSIOLOGY

NEUROPSYCHIATRY

Neuroscience
   USE   NEUROLOGY

NEUROSES

NEUROSPORA

NEUROTIC DEPRESSION

NEUROTRANSMITTERS

NEUROTROPISM

NEUTRAL ATMOSPHERES

NEUTRAL ATOMS

NEUTRAL BEAMS

NEUTRAL CURRENTS

NEUTRAL GASES

NEUTRAL PARTICLES

NEUTRAL SHEETS

Neutralization, Beam
    USE   BEAM NEUTRALIZATION

NEUTRALIZERS

NEUTRINO BEAMS

NEUTRINOS

Neutrinos, Anti
    USE   ANTINEUTRINOS

Neutrinos, Solar
    USE   SOLAR NEUTRINOS

NEUTRON ABSORBERS

NEUTRON ACTIVATION ANALYSIS

NEUTRON BEAMS

NEUTRON COUNTERS

NEUTRON CROSS SECTIONS

NEUTRON DECAY

Neutron Detectors
    USE   NEUTRON COUNTERS

NEUTRON DIFFRACTION

NEUTRON DISTRIBUTION

NEUTRON EMISSION

Neutron Flux
    USE   FLUX (RATE)

NEUTRON FLUX DENSITY

NEUTRON IRRADIATION

NEUTRON PHYSICS

NEUTRON RADIOGRAPHY

NEUTRON SCATTERING

NEUTRON SOURCES

NEUTRON SPECTRA

NEUTRON SPECTROMETERS

NEUTRON STARS

NEUTRON THERMALIZATION

Neutron Transmutation
    USE   NUCLEAR REACTIONS

NEUTRONS

Neutrons, Cold
    USE   COLD NEUTRONS

Neutrons, Fast
    USE   FAST NEUTRONS

Neutrons, Photo
    USE   PHOTONEUTRONS

Neutrons, Slow
    USE   THERMAL NEUTRONS

Neutrons, Thermal
    USE   THERMAL NEUTRONS

NEVADA

Nevada Mountains (CA), Sierra
    USE   SIERRA NEVADA MOUNTAINS (CA)

NEW BRUNSWICK

NEW ENGLAND (US)

NEW GUINEA (ISLAND)

NEW HAMPSHIRE

NEW HAVEN (CT)

NEW JERSEY

NEW MEXICO

NEW MOONS PROJECT

NEW YORK

NEW YORK CITY (NY)

NEW ZEALAND

NEWFOUNDLAND

NEWS

NEWS MEDIA

NEWTON

NEWTON PRESSURE LAW

NEWTON SECOND LAW

NEWTON THEORY

NEWTON-BUSEMANN LAW

NEWTON-RAPHSON METHOD

NEWTONIAN FLUIDS

NH
    USE   NEW HAMPSHIRE

Ni
    USE   NICKEL

NICARAGUA

Nicholson Method, Crank-
    USE   CRANK-NICHOLSON METHOD

NICHROME (TRADEMARK)

NICKEL

NICKEL ALLOYS

Nickel Batteries, Cadmium
    USE   NICKEL CADMIUM BATTERIES

Nickel Batteries, Zinc
    USE   NICKEL ZINC BATTERIES

NICKEL CADMIUM BATTERIES

NICKEL COATINGS

NICKEL COMPOUNDS

NICKEL FLUORIDES

NICKEL HYDROGEN BATTERIES

NICKEL IRON BATTERIES

NICKEL ISOTOPES

NICKEL OXIDES

NICKEL PLATE

NICKEL STEELS

NICKEL ZINC BATTERIES

NICOTINAMIDE

NICOTINE

NICOTINIC ACID

NIGELLA

NIGER

NIGERIA

NIGHT

Night Airglow
    USE   NIGHTGLOW

Night E Layer
    USE   NIGHT SKY
           E REGION

Night F Layer
    USE   F REGION
           NIGHT SKY

NIGHT FLIGHTS (AIRCRAFT)

Night Probe, Pioneer Venus 2
    USE   PIONEER VENUS 2 NIGHT PROBE

NIGHT SKY

NIGHT VISION

NIGHTGLOW

NIGOTRONS

NIHON AIRCRAFT

Nihon YS-11 Aircraft
    USE   YS-11 AIRCRAFT

NIKE BOOSTER ROCKET ENGINES

NIKE MISSILES

NIKE PROJECT

NIKE ROCKET VEHICLES

NIKE ROCKETS

NIKE X SYSTEMS

NIKE-AJAX MISSILE

NIKE-APACHE ROCKET VEHICLE

Nike-Asp Rocket
    USE   ASP ROCKET VEHICLE

NIKE-CAJUN ROCKET VEHICLE

NIKE-HERCULES MISSILE

NIKE-HYDAC ROCKET VEHICLE

NIKE-IROQUOIS ROCKET VEHICLE

NIKE-JAVELIN ROCKET VEHICLE

NIKE-TOMAHAWK ROCKET VEHICLE

NIKE-ZEUS MISSILE

NIMBOSTRATUS CLOUDS

Nimbus Clouds
    USE   NIMBOSTRATUS CLOUDS

NIMBUS PROJECT

NIMBUS SATELLITES

NIMBUS 1 SATELLITE

NIMBUS 2 SATELLITE

NIMBUS 3 SATELLITE

NIMBUS 4 SATELLITE

NIMBUS 5 SATELLITE

NIMBUS 6 SATELLITE

NIMBUS 7 SATELLITE

NIMONIC ALLOYS

NIMPHE (Engine)
  USE   HYDRAZINE ENGINES

NIMROD ACCELERATOR

Nino, El
  USE   EL NINO

NIOBATES

Niobates, Lithium
  USE   LITHIUM NIOBATES

NIOBIUM

NIOBIUM ALLOYS

NIOBIUM CARBIDES

NIOBIUM COMPOUNDS

NIOBIUM IODIDES

NIOBIUM ISOTOPES

NIOBIUM OXIDES

NIOBIUM STANNIDES

NIOBIUM 95

NIPS (System)
  USE   NASA INTERACTIVE PLANNING SYSTEM

NITINOL ALLOYS

NITRAMINE PROPELLANTS

NITRASOL EXPLOSIVES

Nitrate, Cellulose
  USE   CELLULOSE NITRATE

NITRATE ESTERS

Nitrate, Hydrazine
  USE   HYDRAZINE NITRATE

Nitrate, Isopropyl
  USE   ISOPROPYL NITRATE

Nitrate, Methyl
  USE   METHYL NITRATE

Nitrate, Propyl
  USE   PROPYL NITRATE

NITRATES

Nitrates, Ammonium
  USE   AMMONIUM NITRATES

Nitrates, Di
  USE   DINITRATES

Nitrates, Inorganic
  USE   INORGANIC NITRATES

Nitrates, Organic
  USE   ORGANIC NITRATES

Nitrates, Potassium
  USE   POTASSIUM NITRATES

Nitrates, Silver
  USE   SILVER NITRATES

Nitrates, Sodium
  USE   SODIUM NITRATES

NITRATION

NITRIC ACID

NITRIC OXIDE

Nitride-Oxide-Semiconductors, Metal-
  USE   METAL-NITRIDE-OXIDE-SEMICONDUCTORS

Nitride-Oxide-Silicon, Metal-
  USE   METAL-NITRIDE-OXIDE-SILICON

NITRIDES

Nitrides, Aluminum
  USE   ALUMINUM NITRIDES

Nitrides, Beryllium
  USE   BERYLLIUM NITRIDES

Nitrides, Boron
  USE   BORON NITRIDES

Nitrides, Gallium
  USE   GALLIUM NITRIDES

Nitrides, Metal
  USE   METAL NITRIDES

Nitrides, Oxy
  USE   OXYNITRIDES

Nitrides, Silicon
  USE   SILICON NITRIDES

Nitrides, Tantalum
  USE   TANTALUM NITRIDES

Nitrides, Titanium
  USE   TITANIUM NITRIDES

Nitrides, Zirconium
  USE   ZIRCONIUM NITRIDES

NITRIDING

Nitrile, Malono
  USE   MALONONITRILE

NITRILES

Nitriles, Acrylo
  USE   ACRYLONITRILES

Nitriles, Phospho
  USE   PHOSPHONITRILES

NITRITES

NITRO COMPOUNDS

NITROAMINES

NITROBACTER

NITROBENZENES

Nitrocellulose
  USE   CELLULOSE NITRATE

NITROFLUORAMINES

Nitroform, Hydrazine
  USE   HYDRAZINE NITROFORM

NITROFORMATES

NITROFORMS

NITROGEN

NITROGEN ATOMS

NITROGEN COMPOUNDS

NITROGEN DIOXIDE

Nitrogen Fixation
  USE   NITROGENATION

NITROGEN FLUORIDES

NITROGEN HYDRIDES

NITROGEN IONS

NITROGEN ISOTOPES

NITROGEN LASERS

Nitrogen, Liquid
  USE   LIQUID NITROGEN

NITROGEN METABOLISM

NITROGEN OXIDES

NITROGEN PLASMA

NITROGEN POLYMERS

Nitrogen, Solid
  USE   SOLID NITROGEN

NITROGEN TETROXIDE

NITROGEN 15

NITROGEN 16

NITROGENATION

NITROGLYCERIN

NITROGUANIDINE

NITROLYSIS

NITROMETHANE

NITRONIUM COMPOUNDS

NITRONIUM PERCHLORATE

NITROPROPANE

NITROSAMINE

NITROSO COMPOUNDS

NITROSYL CHLORIDES

NITROSYLS

NITROUS ACID

NITROUS OXIDES

NITROXYCHLORIDES

NITRYL CHLORIDES

NITRYL FLUORIDES

NJ
  USE   NEW JERSEY

NJ), Hudson River (NY-
  USE   HUDSON RIVER (NY-NJ)

NM
  USE   NEW MEXICO

NMR
  USE   NUCLEAR MAGNETIC RESONANCE

No
  USE   NOBELIUM

NOAA E
  USE   NOAA 8 SATELLITE

NOAA SATELLITES

NOAA 2 SATELLITE

NOAA 3 SATELLITE

NOAA 4 SATELLITE

NOAA 5 SATELLITE

NOAA 6 SATELLITE

NOAA 7 SATELLITE

NOAA 8 SATELLITE

NOBELIUM

Noble Gases
  USE   RARE GASES

NOBLE METALS

Noctilucence
USE   LUMINESCENCE

NOCTILUCENT CLOUDS

NOCTURNAL VARIATIONS

Nodes, Anti
USE   ANTINODES

NODES (STANDING WAVES)

NODULES

NOE Navigation
USE   NAP-OF-THE-EARTH NAVIGATION

NOESS

NOISE

Noise, Aerodynamic
USE   AERODYNAMIC NOISE

Noise, Aircraft
USE   AIRCRAFT NOISE

Noise, Atmospheric
USE   ATMOSPHERICS

Noise Attenuation
USE   NOISE REDUCTION

Noise, Background
USE   BACKGROUND NOISE

Noise, Blade Slap
USE   BLADE SLAP NOISE

Noise, Boundary Layer
USE   AERODYNAMIC NOISE
          BOUNDARY LAYERS

Noise, Channel
USE   CHANNEL NOISE

Noise, Continuous
USE   CONTINUOUS NOISE

Noise, Cosmic
USE   COSMIC NOISE

Noise, Electromagnetic
USE   ELECTROMAGNETIC NOISE

Noise Elimination
USE   NOISE REDUCTION

Noise, Engine
USE   ENGINE NOISE

Noise, Gaussian
USE   RANDOM NOISE

NOISE GENERATORS

Noise Hazards
USE   NOISE (SOUND)
          HAZARDS

NOISE INJURIES

NOISE INTENSITY

Noise Interactions, Surface
USE   SURFACE NOISE INTERACTIONS

Noise, Ionospheric
USE   IONOSPHERIC NOISE

Noise, Jet
USE   JET AIRCRAFT NOISE

Noise, Jet Aircraft
USE   JET AIRCRAFT NOISE

Noise Levels, Effective Perceived
USE   EFFECTIVE PERCEIVED NOISE LEVELS

Noise, Low
USE   LOW NOISE

NOISE MEASUREMENT

Noise Measurement, Electromagnetic
USE   ELECTROMAGNETIC NOISE
          MEASUREMENT

NOISE METERS

NOISE POLLUTION

NOISE PREDICTION

NOISE PREDICTION (AIRCRAFT)

Noise Prediction, Aircraft
USE   NOISE PREDICTION (AIRCRAFT)

NOISE PROPAGATION

Noise, Pseudo
USE   PSEUDONOISE

Noise, Radiation
USE   ELECTROMAGNETIC NOISE

Noise, Radio Frequency
USE   ELECTROMAGNETIC NOISE

Noise, Random
USE   RANDOM NOISE

Noise Ratios, Carrier To
USE   CARRIER TO NOISE RATIOS

Noise Ratios, Signal To
USE   SIGNAL TO NOISE RATIOS

NOISE REDUCTION

Noise, Rocket Engine
USE   ROCKET ENGINE NOISE

Noise, Shot
USE   SHOT NOISE

Noise, Solar
USE   SOLAR RADIO EMISSION

NOISE (SOUND)

NOISE SPECTRA

Noise, Spectral
USE   WHITE NOISE

NOISE STORMS

Noise Suppressors
USE   NOISE REDUCTION

NOISE TEMPERATURE

Noise, Thermal
USE   THERMAL NOISE

NOISE THRESHOLD

NOISE TOLERANCE

Noise, White
USE   WHITE NOISE

NOMAD LAUNCH VEHICLE

NOMENCLATURES

Nominal Values
USE   APPROXIMATION

Nomograms
USE   NOMOGRAPHS

NOMOGRAPHS

(Non Biological), Cellular Materials
USE   FOAMS

(Non-Biological), Body Temperature
USE   TEMPERATURE

(Non-Biological), Skin Temperature
USE   SKIN TEMPERATURE (NON-BIOLOGICAL)

NONADIABATIC CONDITIONS

Nonadiabatic Processes
USE   HEAT TRANSFER

NONADIABATIC THEORY

NONANES

NONAQUEOUS ELECTROLYTES

NONCONDENSABLE GASES

Nonconductors
USE   ELECTRICAL INSULATION

NONCONSERVATIVE FORCES

NONDESTRUCTIVE TESTS

NONELECTROLYTES

NONEQUILIBRIUM CONDITIONS

Nonequilibrium Drag
USE   FRICTION DRAG

NONEQUILIBRIUM FLOW

NONEQUILIBRIUM IONIZATION

NONEQUILIBRIUM PLASMAS

NONEQUILIBRIUM RADIATION

NONEQUILIBRIUM THERMODYNAMICS

Noneuclidian Geometry
USE   DIFFERENTIAL GEOMETRY

NONFERROUS METALS

NONFLAMMABLE MATERIALS

NONGRAY ATMOSPHERES

NONGRAY GAS

NONHOLONOMIC EQUATIONS

Nonhomogeneity
USE   INHOMOGENEITY

NONISENTROPICITY

NONISOTHERMAL PROCESSES

Nonisotropic Plates
USE   ANISOTROPIC PLATES

Nonisotropy
USE   ANISOTROPY

Nonlifting Vehicles
USE   BALLISTIC VEHICLES

NONLINEAR EQUATIONS

NONLINEAR EVOLUTION EQUATIONS

NONLINEAR FEEDBACK

NONLINEAR FILTERS

NONLINEAR OPTICS

NONLINEAR PROGRAMMING

NONLINEAR SYSTEMS

NONLINEARITY

NONNEWTONIAN FLOW

NONNEWTONIAN FLUIDS

NONOHMIC EFFECT

NONOSCILLATORY ACTION

NONPARAMETRIC STATISTICS

NONPOINT SOURCES

NONPOLAR GASES

Nonreflection
    USE   ENERGY ABSORPTION

Nonrelativistic Electrons
    USE   ELECTRONS

NONRELATIVISTIC MECHANICS

NONRESONANCE

Nonrigidity
    USE   FLEXIBILITY

NONSTABILIZED OSCILLATION

NONSYNCHRONIZATION

NONUNIFORM FLOW

NONUNIFORM MAGNETIC FIELDS

NONUNIFORM PLASMAS

NONUNIFORMITY

Nonviscous Flow
    USE   INVISCID FLOW

NOON

NORADRENALINE

NORD AIRCRAFT

Nord 262 Aircraft
    USE   MH-262 AIRCRAFT

NORD 1500 AIRCRAFT

Nordstrom Solution, Reissner-
    USE   REISSNER-NORDSTROM SOLUTION

NOREPINEPHRINE

NORLEUCINE

NORMAL DENSITY FUNCTIONS

Normal Distributions
    USE   NORMAL DENSITY FUNCTIONS

Normal Force Distribution
    USE   FORCE DISTRIBUTION

NORMAL SHOCK WAVES

Normalities, Ab
    USE   ABNORMALITIES

NORMALITY

NORMALIZING

NORMALIZING (HEAT TREATMENT)

NORMALIZING (STATISTICS)

NORMS

NORTH AMERICA

(North America), Appalachian Mountains
    USE   APPALACHIAN MOUNTAINS (NORTH
            AMERICA)

(North America), Beaufort Sea
    USE   BEAUFORT SEA (NORTH AMERICA)

(North America), Colorado River
    USE   COLORADO RIVER (NORTH AMERICA)

(North America), Great Lakes
    USE   GREAT LAKES (NORTH AMERICA)

(North America), Great Plains Corridor
    USE   GREAT PLAINS CORRIDOR (NORTH
            AMERICA)

(North America), Rio Grande
    USE   RIO GRANDE (NORTH AMERICA)

(North America), Rocky Mountains
    USE   ROCKY MOUNTAINS (NORTH AMERICA)

(North America), St Lawrence Valley
    USE   ST LAWRENCE VALLEY (NORTH AMERICA)

(North America), Williston Basin
    USE   WILLISTON BASIN (NORTH AMERICA)

NORTH AMERICAN AIRCRAFT

NORTH AMERICAN SEARCH AND RANGING RADAR

NORTH ATLANTIC TREATY ORGANIZATION (NATO)

NORTH CAROLINA

NORTH DAKOTA

NORTH KOREA

NORTH POLAR SPUR (ASTRONOMY)

NORTH SEA

North Vietnam
    USE   VIETNAM

NORTHERN HEMISPHERE

NORTHERN SKY

NORTHROP AIRCRAFT

NORTHWEST TERRITORIES

Northwest (US), Pacific
    USE   PACIFIC NORTHWEST (US)

NORTON COUNTY ACHONDRITE

NORWAY

(Norway), Spitsbergen
    USE   SPITSBERGEN (NORWAY)

NOSE

NOSE (ANATOMY)

Nose Caps
    USE   NOSE CONES

NOSE CONES

Nose Cones, Ablative
    USE   ABLATIVE NOSE CONES

Nose Cones, Rocket
    USE   ROCKET NOSE CONES

NOSE FINS

NOSE INLETS

NOSE TIPS

NOSE WHEELS

NOSES (FOREBODIES)

Nosetip Technology, Passive
    USE   PANT PROGRAM

Nosetips, Ablated
    USE   PANT PROGRAM

NOSTOC

Notation
    USE   CODING

Notations, Wiswesser
    USE   WISWESSER NOTATIONS

NOTCH SENSITIVITY

NOTCH STRENGTH

NOTCH TESTS

Notched Metals
    USE   NOTCH TESTS

NOTCHES

NOVA

NOVA COMPUTERS

Nova, Hercules
    USE   HERCULES NOVA

NOVA LASER SYSTEM

NOVA LAUNCH VEHICLES

NOVA SATELLITES

NOVA SCOTIA

NOVAE

Novae, Dwarf
    USE   DWARF NOVAE

Novae, Super
    USE   SUPERNOVAE

NOVOCAIN

NOWCASTING

Noxious Materials
    USE   CONTAMINANTS

Nozzle Coefficient
    USE   NOZZLE FLOW

NOZZLE DESIGN

NOZZLE EFFICIENCY

NOZZLE FLOW

NOZZLE GEOMETRY

NOZZLE INSERTS

NOZZLE THRUST COEFFICIENTS

NOZZLE WALLS

NOZZLELESS ROCKET ENGINES

NOZZLES

Nozzles, Acoustic
    USE   ACOUSTIC NOZZLES

Nozzles, Annular
    USE   ANNULAR NOZZLES

Nozzles, Coaxial
    USE   COAXIAL NOZZLES

Nozzles, Conical
    USE   CONICAL NOZZLES

Nozzles, Convergent
    USE   CONVERGENT NOZZLES

Nozzles, Convergent-Divergent
    USE   CONVERGENT-DIVERGENT NOZZLES

Nozzles, De Laval
    USE   CONVERGENT-DIVERGENT NOZZLES

Nozzles, Divergent
USE   DIVERGENT NOZZLES

Nozzles, Dual Thrust
USE   DUAL THRUST NOZZLES

Nozzles, Exhaust
USE   EXHAUST NOZZLES

Nozzles, Hypersonic
USE   HYPERSONIC NOZZLES

Nozzles, Inlet
USE   INLET NOZZLES

Nozzles, Jet
USE   JET NOZZLES

Nozzles, Pipe
USE   PIPE NOZZLES

Nozzles, Plug
USE   PLUG NOZZLES

Nozzles, Rocket
USE   ROCKET NOZZLES

Nozzles, Shrouded
USE   SHROUDED NOZZLES

Nozzles, Sonic
USE   SONIC NOZZLES

Nozzles, Spike
USE   SPIKE NOZZLES

Nozzles, Spray
USE   SPRAY NOZZLES

Nozzles, Supersonic
USE   SUPERSONIC NOZZLES

Nozzles, Transonic
USE   TRANSONIC NOZZLES

Nozzles, Turbine Exhaust
USE   TURBINE EXHAUST NOZZLES

Nozzles, Wind Tunnel
USE   WIND TUNNEL NOZZLES

Np
USE   NEPTUNIUM

NRX REACTORS

NTS
USE   NAVIGATION TECHNOLOGY SATELLITES

NU FACTOR

Nuclear Auxiliary Power, Systems For
USE   SNAP

NUCLEAR AUXILIARY POWER UNITS

NUCLEAR BINDING ENERGY

NUCLEAR CAPTURE

NUCLEAR CHEMISTRY

NUCLEAR DEFORMATION

Nuclear Detection, High Altitude
USE   HIGH ALTITUDE NUCLEAR DETECTION

NUCLEAR DEVICES

NUCLEAR ELECTRIC POWER GENERATION

NUCLEAR ELECTRIC PROPULSION

NUCLEAR EMULSIONS

NUCLEAR ENERGY

NUCLEAR ENGINE FOR ROCKET VEHICLES

NUCLEAR EXPLOSION EFFECT

NUCLEAR EXPLOSIONS

NUCLEAR FISSION

NUCLEAR FUEL BURNUP

NUCLEAR FUEL ELEMENTS

NUCLEAR FUEL REPROCESSING

NUCLEAR FUELS

Nuclear Fuels, Ceramic
USE   CERAMIC NUCLEAR FUELS

NUCLEAR FUSION

NUCLEAR GYROSCOPES

NUCLEAR HEAT

NUCLEAR INTERACTIONS

NUCLEAR ISOBARS

NUCLEAR LIGHTBULB ENGINES

NUCLEAR MAGNETIC RESONANCE

NUCLEAR MEDICINE

NUCLEAR METEOROLOGY

NUCLEAR MODELS

NUCLEAR PARTICLES

NUCLEAR PHYSICS

(Nuclear Physics), Nuclei
USE   NUCLEI (NUCLEAR PHYSICS)

(Nuclear Physics), Selection Rules
USE   SELECTION RULES (NUCLEAR PHYSICS)

NUCLEAR POTENTIAL

Nuclear Power Facility, Hallam
USE   HALLAM NUCLEAR POWER FACILITY

Nuclear Power Facility), HNPF (Hallam
USE   HALLAM NUCLEAR POWER FACILITY

Nuclear Power Generation
USE   NUCLEAR ELECTRIC POWER GENERATION

Nuclear Power Plant, ML-1
USE   ML-1 NUCLEAR POWER PLANT

NUCLEAR POWER PLANTS

NUCLEAR POWER REACTORS

NUCLEAR POWERED SHIPS

NUCLEAR PROPELLED AIRCRAFT

NUCLEAR PROPULSION

NUCLEAR PUMPED LASERS

NUCLEAR PUMPING

NUCLEAR QUADRUPOLE RESONANCE

NUCLEAR RADIATION

Nuclear Radiation, Post-Blast
USE   POST-BLAST NUCLEAR RADIATION

NUCLEAR RADIATION SPECTROSCOPY

NUCLEAR RAMJET ENGINES

NUCLEAR REACTIONS

NUCLEAR REACTOR CONTROL

Nuclear Reactor, Pathfinder
USE   PATHFINDER NUCLEAR REACTOR

Nuclear Reactor, Phoebus
USE   PHOEBUS NUCLEAR REACTOR

NUCLEAR REACTORS

Nuclear Reactors, Fast
USE   FAST NUCLEAR REACTORS

(Nuclear Reactors), Fuel Elements
USE   NUCLEAR FUEL ELEMENTS

Nuclear Reactors, High Temperature
USE   HIGH TEMPERATURE NUCLEAR REACTORS

Nuclear Reactors, Molten Salt
USE   MOLTEN SALT NUCLEAR REACTORS

(Nuclear Reactors), SGR
USE   SODIUM GRAPHITE REACTORS

(Nuclear Reactors), UHTREX
USE   HIGH TEMPERATURE NUCLEAR REACTORS

NUCLEAR RELAXATION

NUCLEAR RESEARCH

NUCLEAR RESEARCH AND TEST REACTORS

NUCLEAR ROCKET ENGINES

NUCLEAR SCATTERING

Nuclear Shielding
USE   RADIATION SHIELDING

Nuclear Ship, Savannah
USE   SAVANNAH NUCLEAR SHIP

NUCLEAR SPIN

NUCLEAR STRUCTURE

Nuclear Test Reactors
USE   NUCLEAR RESEARCH AND TEST
          REACTORS

NUCLEAR TRANSFORMATIONS

NUCLEAR VULNERABILITY

NUCLEAR WARFARE

NUCLEAR WARHEADS

Nuclear Wastes
USE   RADIOACTIVE WASTES

NUCLEAR WEAPONS

NUCLEASE

NUCLEATE BOILING

NUCLEATION

NUCLEI

Nuclei, Aitken
USE   AITKEN NUCLEI

Nuclei, Comet
USE   COMET NUCLEI

Nuclei, Condensation
USE   CONDENSATION NUCLEI

Nuclei, Even-Even
USE   EVEN-EVEN NUCLEI

Nuclei, Galactic
USE   GALACTIC NUCLEI

Nuclei, Heavy
USE   HEAVY NUCLEI

Nuclei, Hyper
USE   HYPERNUCLEI

Nuclei, Ice
USE   ICE NUCLEI

NUCLEI (NUCLEAR PHYSICS)

Nuclei, Odd-Even
USE   ODD-EVEN NUCLEI

Nuclei, Odd-Odd
USE   ODD-ODD NUCLEI

NUCLEIC ACIDS

NUCLEOGENESIS

Nucleon Interactions, Meson-
USE   MESON-NUCLEON INTERACTIONS

Nucleon Interactions, Nucleon-
USE   NUCLEON-NUCLEON INTERACTIONS

NUCLEON POTENTIAL

Nucleon Scattering, Nucleon-
USE   NUCLEON-NUCLEON SCATTERING

NUCLEON-NUCLEON INTERACTIONS

NUCLEON-NUCLEON SCATTERING

NUCLEONICS

NUCLEONS

Nucleons, Anti
USE   ANTINUCLEONS

NUCLEOPHILES

NUCLEOSIDES

Nucleosynthesis
USE   NUCLEAR FUSION

NUCLEOTIDES

Nucleotides, Poly
USE   POLYNUCLEOTIDES

Nucleotides, Pyridine
USE   PYRIDINE NUCLEOTIDES

NUCLIDES

Nuclides, Radioactive
USE   RADIOACTIVE ISOTOPES

NULL HYPOTHESIS

NULL ZONES

Number, Biot
USE   BIOT NUMBER

Number, Critical Mach
USE   MACH NUMBER
          CRITICAL VELOCITY

Number, Critical Reynolds
USE   CRITICAL VELOCITY
          REYNOLDS NUMBER

Number, Damkohler
USE   DAMKOHLER NUMBER

Number, Froude
USE   FROUDE NUMBER

Number, Grashof
USE   GRASHOF NUMBER

Number, Hartmann
USE   HARTMANN NUMBER

Number, High Reynolds
USE   HIGH REYNOLDS NUMBER

Number, Knudsen
USE   KNUDSEN FLOW

Number, Laval
USE   LAVAL NUMBER

Number, Low Reynolds
USE   LOW REYNOLDS NUMBER

Number, Mach
USE   MACH NUMBER

Number, Nusselt
USE   NUSSELT NUMBER

Number, Octane
USE   OCTANE NUMBER

Number, Peclet
USE   PECLET NUMBER

Number, Prandtl
USE   PRANDTL NUMBER

Number, Rayleigh
USE   RAYLEIGH NUMBER

Number, Reynolds
USE   REYNOLDS NUMBER

Number, Richardson
USE   RICHARDSON NUMBER

Number, Schmidt
USE   SCHMIDT NUMBER

Number, Stanton
USE   STANTON NUMBER

Number, Strouhal
USE   STROUHAL NUMBER

NUMBER THEORY

(Number/volume), Density
USE   DENSITY (NUMBER/VOLUME)

NUMBERS

Numbers, Complex
USE   COMPLEX NUMBERS

Numbers, Dimensionless
USE   DIMENSIONLESS NUMBERS

Numbers, Fibonacci
USE   FIBONACCI NUMBERS

Numbers, Lewis
USE   LEWIS NUMBERS

Numbers, Quantum
USE   QUANTUM NUMBERS

Numbers, Random
USE   RANDOM NUMBERS

Numbers, Real
USE   REAL NUMBERS

Numbers, Similarity
USE   SIMILARITY NUMBERS

NUMERICAL ANALYSIS

NUMERICAL CONTROL

NUMERICAL DATA BASES

NUMERICAL DIFFERENTIATION

NUMERICAL FLOW VISUALIZATION

NUMERICAL INTEGRATION

NUMERICAL STABILITY

NUMERICAL WEATHER FORECASTING

NUNATAKS

Nunn Camera, Baker-
USE   BAKER-NUNN CAMERA

Nurses, Flight
USE   FLIGHT NURSES

NUSSELT NUMBER

NUTATION

NUTATION DAMPERS

Nutational Oscillation
USE   NUTATION

NUTRIENTS

NUTRITION

NUTRITIONAL REQUIREMENTS

NUTS (FASTENERS)

NUTS (FRUITS)

NV
USE   NEVADA

NV), Lake Tahoe (CA-
USE   LAKE TAHOE (CA-NV)

(NV), Pyramid Lake
USE   PYRAMID LAKE (NV)

NY
USE   NEW YORK

(NY), Adirondack Mountains
USE   ADIRONDACK MOUNTAINS (NY)

(NY), Long Island
USE   LONG ISLAND (NY)

(NY), New York City
USE   NEW YORK CITY (NY)

(NY-NJ), Hudson River
USE   HUDSON RIVER (NY-NJ)

NY-PA), Susquehanna River Basin (MD-
USE   SUSQUEHANNA RIVER BASIN (MD-NY-PA)

(NY-VT), Lake Champlain Basin
USE   LAKE CHAMPLAIN BASIN (NY-VT)

Nylon Resins
USE   POLYAMIDE RESINS

NYLON (TRADEMARK)

NYQUIST DIAGRAM

NYQUIST FREQUENCIES

NYSTAGMUS

Nystagmus, Vestibular
USE   VESTIBULAR NYSTAGMUS

N2 Ground Effect Machine, SR-
USE   WESTLAND GROUND EFFECT MACHINES

N2 Ground Effect Machine, Westland SR-
USE   WESTLAND GROUND EFFECT MACHINES

N2 Hovercraft, Westland SR-
USE   WESTLAND GROUND EFFECT MACHINES

N3 Ground Effect Machine, SR-
USE   WESTLAND GROUND EFFECT MACHINES

N3 Ground Effect Machine, Westland SR-
USE   WESTLAND GROUND EFFECT MACHINES

N3 Hovercraft, Westland SR-
USE   WESTLAND GROUND EFFECT MACHINES

N5 Ground Effect Machine, SR-
USE   WESTLAND GROUND EFFECT MACHINES

N5 Ground Effect Machine, Westland SR-
USE   WESTLAND GROUND EFFECT MACHINES

O

O RING SEALS

O STARS

OAK RIDGE ISOCHRONOUS CYCLOTRON

OAO

OAO 1

OAO 2

OAO 3

OAO-A
USE   OAO 1

OAO-A2
USE   OAO 2

OAO-C
USE   OAO 3

OASES

OATS

OBESITY

Object Camera, Faint
USE   FAINT OBJECT CAMERA

OBJECT PROGRAMS

Objects, BI Lacertae
USE   BL LACERTAE OBJECTS

Objects, Herbig-Haro
USE   HERBIG-HARO OBJECTS

Objects, Unidentified Flying
USE   UNIDENTIFIED FLYING OBJECTS

OBLATE SPHEROIDS

Oblateness, Solar
USE   SOLAR OBLATENESS

OBLIQUE COORDINATES

OBLIQUE SHOCK WAVES

OBLIQUE WINGS

OBLIQUENESS

Obscuration
USE   OCCULTATION

OBSERVABILITY (SYSTEMS)

Observable Reentry Vehicles, Low
USE   LOW OBSERVABLE REENTRY VEHICLES

OBSERVATION

OBSERVATION AIRCRAFT

Observation, Celestial
USE   ASTRONOMY

Observation, Ice
USE   ICE REPORTING

Observation, Radar
USE   RADAR TRACKING

Observation, Radio
USE   RADIO OBSERVATION

Observation, Satellite
USE   SATELLITE OBSERVATION

Observation Satellites, Earth Resources
USE   EROS (SATELLITES)

Observation Stations, Crew
USE   CREW OBSERVATION STATIONS

Observation, Visual
USE   VISUAL OBSERVATION

Observations (From Earth), Space
USE   SPACE OBSERVATIONS (FROM EARTH)

Observations (From Space), Earth
USE   EARTH OBSERVATIONS (FROM SPACE)

OBSERVATORIES

Observatories, Astronomical
USE   ASTRONOMICAL OBSERVATORIES

Observatories, Geophysical
USE   GEOPHYSICAL OBSERVATORIES

Observatories, High Energy Astronomy
USE   HEAO

Observatories, Lunar
USE   LUNAR OBSERVATORIES

Observatories, Solar
USE   SOLAR OBSERVATORIES

Observatory A, High Energy Astronomy
USE   HEAO 1

Observatory, Advanced Orbiting Solar
USE   AOSO

Observatory B, High Energy Astronomy
USE   HEAO 2

Observatory C, High Energy Astronomy
USE   HEAO 3

Observatory, Eccentric Geophysical
USE   EGO

Observatory, Eccentric Orbit Geophysical
USE   EGO

Observatory, Einstein
USE   HEAO 2

Observatory, Gamma Ray
USE   GAMMA RAY OBSERVATORY

Observatory, Jodrell Bank
USE   JODRELL BANK OBSERVATORY

Observatory, Kuiper Airborne
USE   C-141 AIRCRAFT

Observatory, Orbiting Astronomical
USE   OAO

Observatory, Orbiting Geophysical
USE   OGO

Observatory, Orbiting Solar
USE   OSO

Observatory, Polar Orbit Geophysical
USE   POGO

Observatory Satellite, Synchronous Earth
USE   SYNCHRONOUS EARTH OBSERVATORY
            SATELLITE

Observatory 1, High Energy Astronomy
USE   HEAO 1

Observatory 2, High Energy Astronomy
USE   HEAO 2

Observatory 3, High Energy Astronomy
USE   HEAO 3

Observing Satellite, Severe Storms
USE   STORMSAT SATELLITE

OBSIDIAN

OBSIDIAN GLASS

OBSTACLE AVOIDANCE

Obstacles
USE   BARRIERS

Obstructing
USE   BLOCKING

OCCIPITAL LOBES

OCCLUSION

OCCULTATION

Occultation Experiment, Halogen
USE   HALOGEN OCCULTATION EXPERIMENT

Occultation, Lunar
USE   LUNAR OCCULTATION

Occultation, Radio
USE   RADIO OCCULTATION

Occultation Satellite, High Eccentric Lunar
USE   EXOSAT SATELLITE

Occultation, Stellar
USE   STELLAR OCCULTATION

OCCUPATION

OCCURRENCES

Ocean, Arctic
USE   ARCTIC OCEAN

Ocean, Atlantic
USE   ATLANTIC OCEAN

OCEAN BOTTOM

OCEAN COLOR SCANNER

OCEAN CURRENTS

OCEAN DATA ACQUISITIONS SYSTEMS

Ocean Data Platforms
USE   OCEAN DATA ACQUISITIONS SYSTEMS

Ocean Data Stations
USE   OCEAN DATA ACQUISITIONS SYSTEMS

OCEAN DYNAMICS

Ocean, Indian
USE   INDIAN OCEAN

OCEAN MODELS

Ocean, Pacific
USE   PACIFIC OCEAN

Ocean Physics Applications Program, Earth &
USE   EARTH & OCEAN PHYSICS APPLICATIONS
            PROGRAM

Ocean Satellite, Geodynamic Experimental
USE   GEOS-D SATELLITE

Ocean Station Systems, Integrated Global
USE   INTEGRATED GLOBAL OCEAN STATION
            SYSTEMS

OCEAN SURFACE

OCEAN TEMPERATURE

OCEAN THERMAL ENERGY CONVERSION

Oceanic Satellite System, National
USE   NATIONAL OCEANIC SATELLITE SYSTEM

Oceanographic Inform Sys, Atmospheric &
USE   ATMOSPHERIC & OCEANOGRAPHIC
            INFORM SYS

OCEANOGRAPHIC PARAMETERS

OCEANOGRAPHY

(Oceanography), Currents
USE   WATER CURRENTS

OCEANS

Octahedral Research Satellites
  USE   ENVIRONMENTAL RESEARCH SATELLITES

Octahedrite
  USE   ANATASE

OCTAHEDRONS

OCTANE

OCTANE NUMBER

OCTANES

OCTAVES

OCTETS

OCTOATES

OCTOL (EXPLOSIVE)

OCTOPUSES

OCULAR CIRCULATION

OCULOGRAVIC ILLUSIONS

OCULOMETERS

OCULOMOTOR NERVES

ODAS
  USE   OCEAN DATA ACQUISITIONS SYSTEMS

Odd Nuclei, Odd-
  USE   ODD-ODD NUCLEI

ODD-EVEN NUCLEI

ODD-ODD NUCLEI

ODESSA METEORITE

ODORS

Off, Bleed-
  USE   PRESSURE REDUCTION

Off, Cut-
  USE   CUT-OFF

OFF-ON CONTROL

OFFGASSING

Office Of Space & Terrestr Applic Payloads
  USE   OSTA-2 PAYLOAD
         OSTA-1 PAYLOAD

OFFSHORE DOCKING

OFFSHORE ENERGY SOURCES

OFFSHORE PLATFORMS

OFFSHORE REACTOR SITES

OFT
  USE   SPACE TRANSPORTATION SYSTEM
         FLIGHTS

OFT 1
  USE   SPACE TRANSPORTATION SYSTEM 1
         FLIGHT

OFT 2
  USE   SPACE TRANSPORTATION SYSTEM 2
         FLIGHT

OFT 3
  USE   SPACE TRANSPORTATION SYSTEM 3
         FLIGHT

OFT 4
  USE   SPACE TRANSPORTATION SYSTEM 4
         FLIGHT

OGEE SHAPE

Ogee Wings
  USE   VARIABLE SWEEP WINGS

OGIVES

OGO

OGO-A

OGO-B
  USE   OGO-3

OGO-C

OGO-D
  USE   OGO-4

OGO-E
  USE   OGO-5

OGO-F
  USE   OGO-6

OGO-3

OGO-4

OGO-5

OGO-6

OH
  USE   OHIO

OH), Wabash River Basin (IL-IN-
  USE   WABASH RIVER BASIN (IL-IN-OH)

OH-4 HELICOPTER

OH-5 HELICOPTER

OH-6 HELICOPTER

OH-13 HELICOPTER

OH-23 HELICOPTER

OH-58 HELICOPTER

OHIO

OHIO RIVER (US)

OHMIC DISSIPATION

OHMMETERS

OHMS LAW

OIL ADDITIVES

Oil, Castor
  USE   CASTOR OIL

Oil, Crude
  USE   CRUDE OIL

OIL EXPLORATION

OIL FIELDS

OIL POLLUTION

OIL RECOVERY

Oil, Shale
  USE   SHALE OIL

OIL SLICKS

OILS

Oils, Fuel
  USE   FUEL OILS

Oils, Lubricating
  USE   LUBRICATING OILS

Oils, Mineral
  USE   MINERAL OILS

OK
  USE   OKLAHOMA

(OK-TX), Lake Texoma
  USE   LAKE TEXOMA (OK-TX)

OKHANSK METEORITE

Okhotsk, Sea Of
  USE   SEA OF OKHOTSK

OKLAHOMA

Olefins
  USE   ALKENES

OLEIC ACID

OLFACTORY PERCEPTION

OLIVINE

Olympus 593 Engine, Bristol-Siddeley
  USE   BRISTOL-SIDDELEY OLYMPUS 593 ENGINE

OMAN

OME
  USE   ORBIT MANEUVERING ENGINE (SPACE
         SHUTTLE)

OMEGA NAVIGATION SYSTEM

OMEGA-MESONS

OMEGATRONS

OMICRON CETI STAR

OMNIDIRECTIONAL ANTENNAS

OMNIDIRECTIONAL RADIO RANGES

Omnipol HC-3 Helicopter
  USE   HC-3 HELICOPTER

Omnipol L-29 Aircraft
  USE   L-29 JET TRAINER

Omnipol Z-37 Aircraft
  USE   Z-37 AIRCRAFT

Omnirange Navigation
  USE   VHF OMNIRANGE NAVIGATION

Omnirange Navigation, VHF
  USE   VHF OMNIRANGE NAVIGATION

Omnirange, SCORE
  USE   SELF CALIBRATING OMNIRANGE

Omnirange, Self Calibrating
  USE   SELF CALIBRATING OMNIRANGE

ON-LINE PROGRAMMING

ON-LINE SYSTEMS

Onboard Computers
  USE   AIRBORNE/SPACEBORNE COMPUTERS

ONBOARD DATA PROCESSING

ONBOARD EQUIPMENT

(Onboard Equipment), Stowage
  USE   STOWAGE (ONBOARD EQUIPMENT)

ONE DIMENSIONAL FLOW

One-Phase Flow
  USE   SINGLE-PHASE FLOW

Onisotropy
  USE   ANISOTROPY

Only Memory Devices, Read-
  USE   READ-ONLY MEMORY DEVICES

ONSAGER PHENOMENOLOGICAL COEFFICIENT

ONSAGER RELATIONSHIP

ONTARIO

Ontario, Lake
USE   LAKE ONTARIO

Ontogenesis
USE   ONTOGENY

ONTOGENY

Oocytes
USE   GAMETOCYTES

OPACIFIERS

OPACITY

OPALESCENCE

OPEN CHANNEL FLOW

OPEN CIRCUIT VOLTAGE

OPEN PROJECT

OPENINGS

(Openings), Clearings
USE   CLEARINGS (OPENINGS)

(Openings), Gates
USE   GATES (OPENINGS)

(Openings), Ports
USE   PORTS (OPENINGS)

Operated Propulsion Systems, Man
USE   MAN OPERATED PROPULSION SYSTEMS

OPERATING COSTS

OPERATING SYSTEMS (COMPUTERS)

OPERATING TEMPERATURE

Operation, Duplex
USE   DUPLEX OPERATION

Operation, Fishbowl
USE   FISHBOWL OPERATION

Operation, Premature
USE   PREMATURE OPERATION

Operation, Real Time
USE   REAL TIME OPERATION

OPERATIONAL AMPLIFIERS

OPERATIONAL CALCULUS

Operational Environ Sats, Geostationary
USE   GOES SATELLITES

Operational Environmental Sat Sys, National
USE   NOESS

OPERATIONAL HAZARDS

OPERATIONAL PROBLEMS

Operational Satellite System, TIROS
USE   TIROS OPERATIONAL SATELLITE SYSTEM

Operational Satellites, Improved TIROS
USE   IMPROVED TIROS OPERATIONAL
            SATELLITES

Operational Support System, Ground
USE   GROUND OPERATIONAL SUPPORT
            SYSTEM

OPERATIONS

Operations, Air Drop
USE   AIR DROP OPERATIONS

Operations, Airline
USE   AIRLINE OPERATIONS

Operations Center (NASA), Space
USE   SPACE OPERATIONS CENTER (NASA)

Operations, Flight
USE   FLIGHT OPERATIONS

Operations, Loading
USE   LOADING OPERATIONS

Operations, Military
USE   MILITARY OPERATIONS

Operations, Preflight
USE   PREFLIGHT OPERATIONS

Operations, Rescue
USE   RESCUE OPERATIONS

OPERATIONS RESEARCH

Operatl Environ Satellite B, Geostationary
USE   GOES 2

Operator, Bergman
USE   BERGMAN OPERATOR

OPERATOR PERFORMANCE

Operator, Sturm-Liouville
USE   STURM-LIOUVILLE THEORY

OPERATORS

Operators, Differential
USE   OPERATORS (MATHEMATICS)
            DIFFERENTIAL EQUATIONS

Operators, Fredholm
USE   OPERATORS (MATHEMATICS)
            FREDHOLM EQUATIONS

Operators, Laplace
USE   LAPLACE TRANSFORMATION

OPERATORS (MATHEMATICS)

OPERATORS (PERSONNEL)

Operators, Tele
USE   TELEOPERATORS

OPHIUCHI CLOUDS

OPHTHALMODYNAMOMETRY

OPHTHALMOLOGY

OPIK THEORY

Oppenheimer Approximation, Born-
USE   BORN-OPPENHEIMER APPROXIMATION

Optical Absorption
USE   LIGHT TRANSMISSION
            ELECTROMAGNETIC ABSORPTION

OPTICAL ACTIVITY

Optical Amplifiers
USE   LIGHT AMPLIFIERS

OPTICAL BISTABILITY

OPTICAL COMMUNICATION

OPTICAL COMPUTERS

OPTICAL CORRECTION PROCEDURE

OPTICAL COUNTERMEASURES

OPTICAL COUPLING

OPTICAL DATA PROCESSING

OPTICAL DATA STORAGE MATERIALS

OPTICAL DENSITY

OPTICAL DEPOLARIZATION

OPTICAL DISKS

Optical Effect, Electro-
USE   ELECTRO-OPTICAL EFFECT

Optical Emission
USE   LIGHT EMISSION

OPTICAL EMISSION SPECTROSCOPY

OPTICAL EQUIPMENT

OPTICAL FILTERS

Optical Generators
USE   LASER CAVITIES

OPTICAL GYROSCOPES

OPTICAL HETERODYNING

OPTICAL ILLUSION

Optical Images
USE   IMAGES

Optical Maser Modulation
USE   LIGHT MODULATION

Optical Masers
USE   LASERS

OPTICAL MEASUREMENT

OPTICAL MEASURING INSTRUMENTS

OPTICAL MEMORY (DATA STORAGE)

Optical Methods
USE   OPTICS

OPTICAL MICROSCOPES

Optical Modulation
USE   LIGHT MODULATION

OPTICAL PATHS

Optical Photography, Electro-
USE   ELECTRO-OPTICAL PHOTOGRAPHY

OPTICAL POLARIZATION

OPTICAL PROPERTIES

OPTICAL PUMPING

OPTICAL PYROMETERS

OPTICAL RADAR

OPTICAL RANGE FINDERS

OPTICAL REFLECTION

OPTICAL RELAY SYSTEMS

OPTICAL RESONANCE

OPTICAL RESONATORS

OPTICAL SATELLITE TRACKING PROGRAM

OPTICAL SCANNERS

Optical Sensors
USE   OPTICAL MEASURING INSTRUMENTS

Optical Signals
USE   OPTICAL COMMUNICATION

OPTICAL SLANT RANGE

Optical Spectrum
USE   LIGHT (VISIBLE RADIATION)
            SPECTRA

Optical Telescope Facility, Spacelab UV-
USE   STARLAB

Optical Telescope, Solar
USE   SOLAR OPTICAL TELESCOPE

OPTICAL THICKNESS

OPTICAL TRACKING

Optical Tracking System, Minitrack
USE   MINITRACK SYSTEM

OPTICAL TRANSFER FUNCTION

OPTICAL TRANSITION

OPTICAL WAVEGUIDES

OPTICS

Optics, Acousto-
USE   ACOUSTO-OPTICS

Optics, Adaptive
USE   ADAPTIVE OPTICS

Optics, Atmospheric
USE   ATMOSPHERIC OPTICS

Optics, Cassegrain
USE   CASSEGRAIN OPTICS

(Optics), Caustics
USE   CAUSTICS (OPTICS)

Optics, Crystal
USE   CRYSTAL OPTICS

Optics, Electro-
USE   ELECTRO-OPTICS

Optics, Electron
USE   ELECTRON OPTICS

Optics, Fiber
USE   FIBER OPTICS

Optics, Geometrical
USE   GEOMETRICAL OPTICS

Optics, Gradient Index
USE   GRADIENT INDEX OPTICS

Optics, Integrated
USE   INTEGRATED OPTICS

Optics, Magneto-
USE   MAGNETO-OPTICS

Optics, Modulating Retrodirective
USE   MIROS SYSTEM

Optics, Nonlinear
USE   NONLINEAR OPTICS

Optics, Physical
USE   PHYSICAL OPTICS

Optics, Ray
USE   GEOMETRICAL OPTICS

(Optics), Scatter Plates
USE   SCATTER PLATES (OPTICS)

Optics, Underwater
USE   UNDERWATER OPTICS

OPTIMAL CONTROL

Optimal Control, Time
USE   TIME OPTIMAL CONTROL

OPTIMIZATION

Optimization, Flight
USE   FLIGHT OPTIMIZATION

Optimization, Trajectory
USE   TRAJECTORY OPTIMIZATION

Optimum Control
USE   OPTIMAL CONTROL

Optimum Thrust Programming
USE   THRUST PROGRAMMING

OPTIONS

OPTOGALVANIC SPECTROSCOPY

OPTOMETRY

OR
USE   OREGON

OR Bending), Brakes (Forming
USE   BRAKES (FORMING OR BENDING)

OR Foe, Identify Friend
USE   IFF SYSTEMS (IDENTIFICATION)

OR-Gates
USE   GATES (CIRCUITS)

OR-WA), Cascade Range (CA-
USE   CASCADE RANGE (CA-OR-WA)

OR-WA), Columbia River Basin (ID-
USE   COLUMBIA RIVER BASIN (ID-OR-WA)

ORAL HYGIENE

Oratory
USE   PUBLIC SPEAKING

ORBIS

ORBIS CAL SATELLITE

Orbit And Landing Simulators, Lunar
USE   LUNAR ORBIT AND LANDING SIMULATORS

ORBIT CALCULATION

Orbit Calculation, Satellite
USE   ORBIT CALCULATION

ORBIT DECAY

Orbit Determination, Airborne Range And
USE   AIRBORNE RANGE AND ORBIT
        DETERMINATION

Orbit Determination), AROD (Range-
USE   AIRBORNE RANGE AND ORBIT
        DETERMINATION

Orbit Determination, Minimum Variance
USE   MINIMUM VARIANCE ORBIT
        DETERMINATION

Orbit Determination, MINIVAR
USE   MINIMUM VARIANCE ORBIT
        DETERMINATION

Orbit Equations
USE   ORBITAL MECHANICS

Orbit Geophysical Observatory, Eccentric
USE   EGO

Orbit Geophysical Observatory, Polar
USE   POGO

Orbit Interactions, Spin-
USE   SPIN-ORBIT INTERACTIONS

ORBIT MANEUVERING ENGINE (SPACE SHUTTLE)

ORBIT PERTURBATION

Orbit Satellites, Highly Eccentric
USE   HEOS SATELLITES

Orbit Shuttle, Aeromaneuvering Orbit To
USE   AEROMANEUVERING ORBIT TO ORBIT
        SHUTTLE

Orbit Space Station, Halo
USE   HALO ORBIT SPACE STATION

ORBIT SPECTRUM UTILIZATION

Orbit To Orbit Shuttle, Aeromaneuvering
USE   AEROMANEUVERING ORBIT TO ORBIT
        SHUTTLE

ORBIT TRANSFER VEHICLES

Orbit Vehicles, Single Stage To
USE   SINGLE STAGE TO ORBIT VEHICLES

ORBITAL ASSEMBLY

Orbital Assembly, Spacecraft
USE   ORBITAL ASSEMBLY

ORBITAL ELEMENTS

Orbital Flight Test 1 (Shuttle)
USE   SPACE TRANSPORTATION SYSTEM 1
        FLIGHT

Orbital Flight Test 1, Space Shuttle
USE   SPACE TRANSPORTATION SYSTEM 1
        FLIGHT

Orbital Flight Test 2 (Shuttle)
USE   SPACE TRANSPORTATION SYSTEM 2
        FLIGHT

Orbital Flight Test 2, Space Shuttle
USE   SPACE TRANSPORTATION SYSTEM 2
        FLIGHT

Orbital Flight Test 3 (Shuttle)
USE   SPACE TRANSPORTATION SYSTEM 3
        FLIGHT

Orbital Flight Test 3, Space Shuttle
USE   SPACE TRANSPORTATION SYSTEM 3
        FLIGHT

Orbital Flight Test 4 (Shuttle)
USE   SPACE TRANSPORTATION SYSTEM 4
        FLIGHT

Orbital Flight Test 4, Space Shuttle
USE   SPACE TRANSPORTATION SYSTEM 4
        FLIGHT

Orbital Flight Tests (Shuttle)
USE   SPACE TRANSPORTATION SYSTEM
        FLIGHTS

Orbital Flight Tests, Space Shuttle
USE   SPACE TRANSPORTATION SYSTEM
        FLIGHTS

Orbital Flight 7, Space Shuttle
USE   SPACE SHUTTLE MISSION 31-C

Orbital Flight 8, Space Shuttle
USE   SPACE SHUTTLE MISSION 31-D

Orbital Flight 9, Space Shuttle
USE   SPACE SHUTTLE MISSION 41-A

Orbital Flights, Space Shuttle
USE   SPACE TRANSPORTATION SYSTEM
        FLIGHTS

Orbital Laboratories, Manned
USE   MANNED ORBITAL LABORATORIES

(Orbital Laboratories), MOL
USE   MANNED ORBITAL LABORATORIES

ORBITAL LAUNCHING

ORBITAL LIFETIME

ORBITAL MANEUVERING VEHICLES

ORBITAL MANEUVERS

ORBITAL MECHANICS

Orbital Motion
USE   ORBITS

ORBITAL POSITION ESTIMATION

ORBITAL RENDEZVOUS

Orbital Rendezvous, Earth
USE   EARTH ORBITAL RENDEZVOUS

Orbital Rendezvous, Lunar
USE   LUNAR ORBITAL RENDEZVOUS

Orbital Research Laboratories, Manned
USE   MANNED ORBITAL RESEARCH
          LABORATORIES

ORBITAL SERVICING

Orbital Shot Proj, Experimental Reflector
USE   EXPERIMENTAL REFLECTOR ORBITAL
          SHOT PROJ

ORBITAL SHOTS

Orbital Simulator, High Vacuum
USE   HIGH VACUUM ORBITAL SIMULATOR

Orbital Simulators
USE   SPACE SIMULATORS

ORBITAL SPACE STATIONS

Orbital Space Stations, Manned
USE   ORBITAL SPACE STATIONS

Orbital Space System, Bioastronautical
USE   BIOASTRONAUTICAL ORBITAL SPACE
          SYSTEM

ORBITAL SPACE TESTS

Orbital Telescopes, Manned
USE   MANNED ORBITAL TELESCOPES

(Orbital Telescopes), MOT
USE   MANNED ORBITAL TELESCOPES

Orbital Test Satellite (ESA)
USE   OTS (ESA)

Orbital Test Satellite, Maritime
USE   MAROTS (ESA)

Orbital Transfer
USE   TRANSFER ORBITS

ORBITAL VELOCITY

ORBITAL WORKERS

ORBITAL WORKSHOPS

ORBITALS

Orbitals, Electron
USE   ELECTRON ORBITALS

Orbitals, Molecular
USE   MOLECULAR ORBITALS

Orbitals, Slater
USE   SLATER ORBITALS

Orbiter A, Lunar
USE   LUNAR ORBITER 1

(Orbiter), Atlantis
USE   ATLANTIS (ORBITER)

Orbiter B, Lunar
USE   LUNAR ORBITER 2

Orbiter C, Lunar
USE   LUNAR ORBITER 3

(Orbiter), Challenger
USE   CHALLENGER (ORBITER)

(Orbiter), Columbia
USE   COLUMBIA (ORBITER)

Orbiter D, Lunar
USE   LUNAR ORBITER 4

(Orbiter), Discovery
USE   DISCOVERY (ORBITER)

Orbiter E, Lunar
USE   LUNAR ORBITER 5

(Orbiter), Enterprise
USE   ENTERPRISE (ORBITER)

Orbiter, Lunar
USE   LUNAR ORBITER

ORBITER PROJECT

Orbiter Spacecraft, Viking
USE   VIKING ORBITER SPACECRAFT

Orbiter 1, Lunar
USE   LUNAR ORBITER 1

Orbiter 1, Viking
USE   VIKING ORBITER 1

Orbiter 2, Lunar
USE   LUNAR ORBITER 2

Orbiter 2, Viking
USE   VIKING ORBITER 2

Orbiter 3, Lunar
USE   LUNAR ORBITER 3

Orbiter 4, Lunar
USE   LUNAR ORBITER 4

Orbiter 5, Lunar
USE   LUNAR ORBITER 5

Orbiter 099, Space Shuttle
USE   CHALLENGER (ORBITER)

Orbiter 101, Space Shuttle
USE   ENTERPRISE (ORBITER)

Orbiter 102, Space Shuttle
USE   COLUMBIA (ORBITER)

Orbiter 103, Space Shuttle
USE   DISCOVERY (ORBITER)

Orbiter 104, Space Shuttle
USE   ATLANTIS (ORBITER)

Orbiter 1975, Viking
USE   VIKING ORBITER 1975

Orbiters, Shuttle
USE   SPACE SHUTTLE ORBITERS

Orbiters, Space Shuttle
USE   SPACE SHUTTLE ORBITERS

Orbiting Astronomical Observatory
USE   OAO

ORBITING DIPOLES

ORBITING FROG OTOLITH

Orbiting Geophysical Observatory
USE   OGO

Orbiting Imaging Radar (Spacecraft), Venus
USE   VENUS ORBITING IMAGING RADAR
          (SPACECRAFT)

ORBITING LUNAR STATIONS

Orbiting Radio Beacon Ionospheric Sounder
USE   ORBIS

Orbiting Satellites
USE   ARTIFICIAL SATELLITES

Orbiting Solar Observatory
USE   OSO

Orbiting Solar Observatory, Advanced
USE   AOSO

Orbiting Space Stations, Earth
USE   EOSS

Orbiting Telescope, Kilometer Wave
USE   KILOMETER WAVE ORBITING TELESCOPE

ORBITRONS

ORBITS

Orbits, Circular
USE   CIRCULAR ORBITS

Orbits, Earth
USE   EARTH ORBITS

Orbits, Eccentric
USE   ECCENTRIC ORBITS

Orbits, Elliptical
USE   ELLIPTICAL ORBITS

Orbits, Equatorial
USE   EQUATORIAL ORBITS

Orbits, Geosynchronous
USE   GEOSYNCHRONOUS ORBITS

Orbits, Heliocentric
USE   SOLAR ORBITS

Orbits, Hohmann Transfer
USE   ELLIPTICAL ORBITS
          TRANSFER ORBITS

Orbits, Interplanetary Transfer
USE   INTERPLANETARY TRANSFER ORBITS

Orbits, Lunar
USE   LUNAR ORBITS

Orbits, Parking
USE   PARKING ORBITS

Orbits, Periodic
USE   ORBITS

Orbits, Planetary
USE   PLANETARY ORBITS

Orbits, Polar
USE   POLAR ORBITS

Orbits, Satellite
USE   SATELLITE ORBITS

Orbits, Solar
USE   SOLAR ORBITS

Orbits, Spacecraft
USE   SPACECRAFT ORBITS

Orbits, Stationary
USE   STATIONARY ORBITS

Orbits, Stellar
USE   STELLAR ORBITS

Orbits, Transfer
USE   TRANSFER ORBITS

Orbits, Trojan
USE   TROJAN ORBITS

Orbits, Twenty-Four Hour
USE   TWENTY-FOUR HOUR ORBITS

Orbits, Two Body
USE   TWO BODY PROBLEM

ORCHARDS

Order Filters, Reduced
USE   REDUCED ORDER FILTERS

ORDER-DISORDER TRANSFORMATIONS

ORDNANCE

(Ordnance), Bombs
USE   BOMBS (ORDNANCE)

(Ordnance), Fuses
USE   FUSES (ORDNANCE)

(Ordnance), Guns
USE   GUNS (ORDNANCE)

(Ordnance), Mines
USE   MINES (ORDNANCE)

OREGON

Ores
USE   MINERALS

Ores, Iron
USE   IRON ORES

Organ, Corti
USE   CORTI ORGAN

ORGAN WEIGHT

ORGANIC ALUMINUM COMPOUNDS

(Organic), Azides
USE   AZIDES (ORGANIC)

ORGANIC BORON COMPOUNDS

ORGANIC CHARGE TRANSFER SALTS

ORGANIC CHEMISTRY

ORGANIC COMPOUNDS

Organic Compounds, Fluorine
USE   FLUORINE ORGANIC COMPOUNDS

Organic Compounds, Lead
USE   LEAD ORGANIC COMPOUNDS

Organic Compounds, Polynuclear
USE   POLYNUCLEAR ORGANIC COMPOUNDS

ORGANIC COOLANTS

ORGANIC COOLED REACTORS

Organic Cooled Reactors, Experimental
USE   EXPERIMENTAL ORGANIC COOLED
       REACTORS

Organic Fluorine Compounds
USE   FLUORINE ORGANIC COMPOUNDS

ORGANIC GERMANIUM COMPOUNDS

ORGANIC LASERS

ORGANIC LIQUIDS

ORGANIC LITHIUM COMPOUNDS

ORGANIC MATERIALS

ORGANIC MODERATED REACTORS

ORGANIC NITRATES

ORGANIC PEROXIDES

ORGANIC PHOSPHORUS COMPOUNDS

ORGANIC SEMICONDUCTORS

ORGANIC SILICON COMPOUNDS

ORGANIC SOLIDS

ORGANIC SULFUR COMPOUNDS

ORGANIC TIN COMPOUNDS

ORGANIC WASTES (FUEL CONVERSION)

ORGANISMS

Organisms, Micro
USE   MICROORGANISMS

Organization, European Space Research
USE   EUROPEAN SPACE AGENCY

Organization, Indian Space Research
USE   ISRO

Organization (NATO), North Atlantic Treaty
USE   NORTH ATLANTIC TREATY ORGANIZATION
       (NATO)

Organization Sat, European Space Research
USE   ESA SATELLITES

Organization, World Meteorological
USE   WORLD METEOROLOGICAL ORGANIZATION

ORGANIZATIONS

(Organizations), Bureaus
USE   BUREAUS (ORGANIZATIONS)

ORGANIZING

Organizing Systems, Self
USE   SELF ORGANIZING SYSTEMS

ORGANOMETALLIC COMPOUNDS

ORGANOMETALLIC POLYMERS

ORGANS

Organs, Otolith
USE   OTOLITH ORGANS

Organs, Sense
USE   SENSE ORGANS

Orgel Reactor
USE   ORGANIC COOLED REACTORS

ORGUEIL METEORITE

ORIC Cyclotron
USE   OAK RIDGE ISOCHRONOUS CYCLOTRON

Orientales (Colombia), Llanos
USE   LLANOS ORIENTALES (COLOMBIA)

ORIENTATION

Orientation, Dis
USE   DISORIENTATION

Orientation, Fiber
USE   FIBER ORIENTATION

Orientation, Horizontal
USE   HORIZONTAL ORIENTATION

Orientation, Instrument
USE   INSTRUMENT ORIENTATION

Orientation, Ply
USE   PLY ORIENTATION

Orientation, Satellite
USE   SATELLITE ORIENTATION

Orientation, Space
USE   SPACE ORIENTATION

Orientation, Spatial
USE   ATTITUDE (INCLINATION)

Orientation, Vertical
USE   VERTICAL ORIENTATION

Oriented Languages, Machine
USE   MACHINE ORIENTED LANGUAGES

ORIFICE FLOW

ORIFICES

Origin Of Plasmas In Earth Neighborhood
USE   OPEN PROJECT

ORIGINS

Origins, Planet
USE   PLANETARY EVOLUTION

Orion Aircraft
USE   P-3 AIRCRAFT

ORION CONSTELLATION

ORION NEBULA

ORION (RADIO INTERFEROMETRY NETWORK)

ORIONID METEOROIDS

Orionis, Sigma
USE   SIGMA ORIONIS

ORLICZ SPACE

Ornithopter Aircraft
USE   RESEARCH AIRCRAFT

ORNSTEIN-UHLENBECK PROCESS

Orographic Clouds
USE   CAP CLOUDS

OROGRAPHY

ORR-SOMMERFELD EQUATIONS

Orreries
USE   ASTRONOMICAL MODELS

ORTHICONS

Orthicons, Image
USE   IMAGE ORTHICONS

ORTHO HYDROGEN

ORTHO PARA CONVERSION

Orthocarbonates, Tetraethyl
USE   TETRAETHYL ORTHOCARBONATES

ORTHOGONAL FUNCTIONS

ORTHOGONAL MULTIPLEXING THEORY

ORTHOGONALITY

ORTHOGRAPHY

ORTHONORMAL FUNCTIONS

ORTHOPEDICS

ORTHOPHOTOGRAPHY

Orthosilicate, Tetraethyl
USE   TETRAETHYL ORTHOSILICATE

ORTHOSTATIC TOLERANCE

ORTHOTROPIC CYLINDERS

ORTHOTROPIC PLATES

ORTHOTROPIC SHELLS

ORTHOTROPISM

Os
USE   OSMIUM

OS
USE   OPERATING SYSTEMS (COMPUTERS)

OSCILLATING CYLINDERS

OSCILLATING FLOW

OSCILLATION DAMPERS

Oscillation, Forced
USE   FORCED VIBRATION

Oscillation, Harmonic
USE   HARMONIC OSCILLATION

Oscillation, Ion
USE   PLASMA OSCILLATIONS

Oscillation, Lateral
USE   LATERAL OSCILLATION

Oscillation, Nonstabilized
USE   NONSTABILIZED OSCILLATION

Oscillation, Nutational
USE   NUTATION

Oscillation, Pilot Induced
USE   PILOT INDUCED OSCILLATION

Oscillation, Self
USE   SELF OSCILLATION

Oscillation, Tidal
USE   TIDES

Oscillation, Transverse
USE   TRANSVERSE OSCILLATION

OSCILLATIONS

Oscillations, Electron
USE   ELECTRON OSCILLATIONS

Oscillations, Free
USE   FREE VIBRATION

Oscillations, Hydrofoil
USE   HYDROFOIL OSCILLATIONS

Oscillations, Molecular
USE   MOLECULAR OSCILLATIONS

Oscillations, Phugoid
USE   OSCILLATIONS
      OSCILLATORS
      PITCH (INCLINATION)

Oscillations, Plasma
USE   PLASMA OSCILLATIONS

Oscillations, Pressure
USE   PRESSURE OSCILLATIONS

Oscillations, Solar
USE   SOLAR OSCILLATIONS

Oscillations, Stable
USE   STABLE OSCILLATIONS

Oscillations, Stellar
USE   STELLAR OSCILLATIONS

Oscillations, Transient
USE   TRANSIENT OSCILLATIONS

Oscillations, Undamped
USE   UNDAMPED OSCILLATIONS

Oscillations, Wing
USE   WING OSCILLATIONS

OSCILLATOR STRENGTHS

OSCILLATORS

Oscillators, Crystal
USE   CRYSTAL OSCILLATORS

Oscillators, Harmonic
USE   HARMONIC OSCILLATORS

Oscillators, Mechanical
USE   MECHANICAL OSCILLATORS

Oscillators, Microwave
USE   MICROWAVE OSCILLATORS

Oscillators, Molecular
USE   MOLECULAR OSCILLATORS

Oscillators, Parametric
USE   PARAMETRIC AMPLIFIERS

Oscillators, Relaxation
USE   RELAXATION OSCILLATORS

Oscillators, Synchronized
USE   SYNCHRONIZED OSCILLATORS

Oscillators, Vacuum Tube
USE   VACUUM TUBE OSCILLATORS

Oscillators, Voltage Controlled
USE   VOLTAGE CONTROLLED OSCILLATORS

Oscillators, Wave
USE   OSCILLATORS

Oscillograms
USE   OSCILLOGRAPHS

OSCILLOGRAPHS

OSCILLOSCOPES

Osculations
USE   DOUBLE CUSPS

OSEEN APPROXIMATION

OSMIUM

OSMIUM ALLOYS

OSMIUM COMPOUNDS

OSMIUM ISOTOPES

OSMOMETERS

OSMOSIS

Osmosis, Reverse
USE   REVERSE OSMOSIS

Osmotic Pressure
USE   OSMOSIS

OSO

OSO-A
USE   OSO-1

OSO-B
USE   OSO-2

OSO-C

OSO-D
USE   OSO-4

OSO-E
USE   OSO-3

OSO-F
USE   OSO-5

OSO-G
USE   OSO-6

OSO-H
USE   OSO-7

OSO-J
USE   OSO-8

OSO-1

OSO-2

OSO-3

OSO-4

OSO-5

OSO-6

OSO-7

OSO-8

OSPREY MISSILE

OSS-1 PAYLOAD

OSTA-1 PAYLOAD

OSTA-2 PAYLOAD

OSTEOPOROSIS

OT-2
USE   ESSA 2 SATELLITE

OT-3
USE   ESSA 1 SATELLITE

OTF
USE   OPTICAL TRANSFER FUNCTION

OTOLARYNGOLOGY

Otolith, Orbiting Frog
USE   ORBITING FROG OTOLITH

OTOLITH ORGANS

OTOLOGY

OTS (ESA)

OTTO CYCLE

OTV
USE   ORBIT TRANSFER VEHICLES

OUTCROPS

OUTER BANKS (NC)

Outer Planet Missions
USE   GRAND TOURS

Outer Planet Spacecraft
USE   OUTER PLANETS EXPLORERS

Outer Planet Spacecraft, Thermoelectric
USE   TOPS (SPACECRAFT)

OUTER PLANETS EXPLORERS

OUTER RADIATION BELT

OUTER SPACE TREATY

OUTGASSING

OUTLET FLOW

OUTLETS

Outlets, Electric
USE   ELECTRIC OUTLETS

Outlets (Geology)
USE   ESTUARIES

OUTLIERS (LANDFORMS)

OUTLIERS (STATISTICS)

OUTPUT

Output Programs, Multiple
USE   MULTIPLE OUTPUT PROGRAMS

Outputs, Laser
USE   LASER OUTPUTS

Outputs, Maser
USE   MASER OUTPUTS

Outs, Cut-
USE   OPENINGS

OV-1 AIRCRAFT

OV-1 SATELLITES

OV-1C Aircraft, Grumman
USE   OV-1 AIRCRAFT

OV-2 SATELLITES

OV-3 SATELLITES

OV-4 SATELLITES

OV-5 SATELLITES

OV-10 AIRCRAFT

OVARIES

OVENS

Over The Shore (LOTS) Carrier, Logistics
USE   LOGISTICS OVER THE SHORE (LOTS)
       CARRIER

OVER-THE-HORIZON RADAR

Overcast
USE   CLOUD COVER

Overcompression
USE   OVERCONSOLIDATION

OVERCONSOLIDATION

OVERHAUSER EFFECT

OVERPRESSURE

Overtones
USE   HARMONICS

OVERVOLTAGE

OXALATES

Oxalates, Cobalt
USE   COBALT OXALATES

OXALIC ACID

OXAMIC ACIDS

OXAZOLE

Oxidants, Photochemical
USE   PHOTOCHEMICAL OXIDANTS

OXIDASE

OXIDATION

Oxidation, Electrochemical
USE   ELECTROCHEMICAL OXIDATION

Oxidation, Photo
USE   PHOTOOXIDATION

OXIDATION RESISTANCE

OXIDATION-REDUCTION REACTIONS

Oxide Batteries, Zinc Silver
USE   SILVER ZINC BATTERIES

Oxide, Ethylene
USE   ETHYLENE OXIDE

OXIDE FILMS

Oxide, Hydrogen Deuterium
USE   HEAVY WATER

Oxide, Nitric
USE   NITRIC OXIDE

Oxide, Propylene
USE   PROPYLENE OXIDE

Oxide Reactors, Fast
USE   FAST OXIDE REACTORS

Oxide Semiconductors, Complementary Metal
USE   CMOS

Oxide Semiconductors, Metal
USE   METAL OXIDE SEMICONDUCTORS

Oxide, Trifluoroamine
USE   TRIFLUOROAMINE OXIDE

Oxide Zinc Batteries, Silver
USE   SILVER ZINC BATTERIES

Oxide-Metal Semiconductors, Metal-
USE   MOM (SEMICONDUCTORS)

Oxide-Semiconductors, Metal-Nitride-
USE   METAL-NITRIDE-OXIDE-SEMICONDUCTORS

Oxide-Silicon, Metal-Nitride-
USE   METAL-NITRIDE-OXIDE-SILICON

OXIDES

Oxides, Alkaline Earth
USE   ALKALINE EARTH OXIDES

Oxides, Aluminum
USE   ALUMINUM OXIDES

Oxides, Barium
USE   BARIUM OXIDES

Oxides, Beryllium
USE   BERYLLIUM OXIDES

Oxides, Bismuth
USE   BISMUTH OXIDES

Oxides, Boron
USE   BORON OXIDES

Oxides, Butylene
USE   TETRAHYDROFURAN

Oxides, Calcium
USE   CALCIUM OXIDES

Oxides, Cerium
USE   CERIUM OXIDES

Oxides, Cesium
USE   CESIUM OXIDES

Oxides, Chlorine
USE   CHLORINE OXIDES

Oxides, Chromium
USE   CHROMIUM OXIDES

Oxides, Cobalt
USE   COBALT OXIDES

Oxides, Copper
USE   COPPER OXIDES

Oxides, Deuterium
USE   HEAVY WATER

Oxides, Di
USE   DIOXIDES

Oxides, Gallium
USE   GALLIUM OXIDES

Oxides, Germanium
USE   GERMANIUM OXIDES

Oxides, Hafnium
USE   HAFNIUM OXIDES

Oxides, Hydr
USE   HYDROXIDES

Oxides, Iron
USE   IRON OXIDES

Oxides, Lanthanum
USE   LANTHANUM OXIDES

Oxides, Lead
USE   LEAD OXIDES

Oxides, Lithium
USE   LITHIUM OXIDES

Oxides, Magnesium
USE   MAGNESIUM OXIDES

Oxides, Manganese
USE   MANGANESE OXIDES

Oxides, Mercury
USE   MERCURY OXIDES

Oxides, Metal
USE   METAL OXIDES

Oxides, Mixed
USE   MIXED OXIDES

Oxides, Molybdenum
USE   MOLYBDENUM OXIDES

Oxides, Nickel
USE   NICKEL OXIDES

Oxides, Niobium
USE   NIOBIUM OXIDES

Oxides, Nitrogen
USE   NITROGEN OXIDES

Oxides, Nitrous
USE   NITROUS OXIDES

Oxides, Per
USE   PEROXIDES

Oxides, Phosphorus
USE   PHOSPHORUS OXIDES

Oxides, Platinum
USE   PLATINUM OXIDES

Oxides, Plutonium
USE   PLUTONIUM OXIDES

Oxides, Potassium
USE   POTASSIUM OXIDES

Oxides, Scandium
USE   SCANDIUM OXIDES

Oxides, Selenium
USE   SELENIUM OXIDES

Oxides, Silicon
USE   SILICON OXIDES

Oxides, Silver
USE   SILVER OXIDES

Oxides, Sulfur
USE   SULFUR OXIDES

Oxides, Tantalum
USE   TANTALUM OXIDES

Oxides, Thorium
USE   THORIUM OXIDES

Oxides, Tin
USE   TIN OXIDES

Oxides, Titanium
USE   TITANIUM OXIDES

Oxides, Tungsten
USE   TUNGSTEN OXIDES

Oxides, Uranium
USE   URANIUM OXIDES

Oxides, Vanadium
USE   VANADIUM OXIDES

Oxides, Yttrium
USE   YTTRIUM OXIDES

Oxides, Zinc
USE   ZINC OXIDES

Oxides, Zirconium
USE   ZIRCONIUM OXIDES

OXIDIZERS

Oxidizers, High Energy
USE   HIGH ENERGY OXIDIZERS

Oxidizers, Liquid
USE   LIQUID OXIDIZERS

Oxidizers, Propellant
USE   ROCKET OXIDIZERS

Oxidizers, Rocket
USE   ROCKET OXIDIZERS

OXIMETRY

OXYACETYLENE

Oxyalkylation
USE   ALKYLATION

OXYFLUORIDES

OXYGEN

OXYGEN AFTERGLOW

OXYGEN ANALYZERS

Oxygen Atmospheres, Argon-
USE   ARGON-OXYGEN ATMOSPHERES

Oxygen Atmospheres, Helium-
USE   HELIUM-OXYGEN ATMOSPHERES

OXYGEN ATOMS

Oxygen Batteries, Zinc-
USE   ZINC-OXYGEN BATTERIES

OXYGEN BREATHING

OXYGEN COMPOUNDS

OXYGEN CONSUMPTION

Oxygen Deficiency
USE   HYPOXIA

Oxygen Demand, Biochemical
USE   BIOCHEMICAL OXYGEN DEMAND

Oxygen Detectors
USE   OXYGEN ANALYZERS

Oxygen Engines, Hydrogen
USE   HYDROGEN OXYGEN ENGINES

OXYGEN FLUORIDES

Oxygen, Fluorine-Liquid
USE   FLOX

Oxygen Fuel Cells, Hydrogen
USE   HYDROGEN OXYGEN FUEL CELLS

Oxygen, High Pressure
USE   HIGH PRESSURE OXYGEN

OXYGEN IONS

OXYGEN ISOTOPES

Oxygen, Liquid
USE   LIQUID OXYGEN

(Oxygen), LOX
USE   LIQUID OXYGEN

OXYGEN MASKS

OXYGEN METABOLISM

OXYGEN PLASMA

OXYGEN PRODUCTION

OXYGEN RECOMBINATION

OXYGEN REGULATORS

OXYGEN SPECTRA

OXYGEN SUPPLY EQUIPMENT

Oxygen Systems
USE   OXYGEN SUPPLY EQUIPMENT

OXYGEN TENSION

Oxygen Toxicity
USE   HYPEROXIA

OXYGEN 17

OXYGEN 18

OXYGENATION

OXYHALIDES

OXYHEMOGLOBIN

OXYNITRIDES

OZONATES

OZONE

OZONE FLUORIDE

OZONIDES

OZONOMETRY

OZONOSPHERE

P

P BAND

P Junctions, N-
USE   P-N JUNCTIONS

P Junctions, P-N-
USE   P-N-P JUNCTIONS

P, Vitamin
USE   BIOFLAVONOIDS

P WAVES

P.A.C.M. TELEMETRY

P-I-N Diodes
USE   P-I-N JUNCTIONS
        DIODES

P-I-N JUNCTIONS

P-N JUNCTIONS

P-N Junctions, N-
USE   N-P-N JUNCTIONS

P-N Junctions, P-N-
USE   P-N-P JUNCTIONS

P-N-P JUNCTIONS

P-N-P-N JUNCTIONS

P-TYPE SEMICONDUCTORS

P-1 ENGINE

P-3 AIRCRAFT

P-20 Engine, J-57-
USE   J-57-P-20 ENGINE

P-51 AIRCRAFT

P-84 Aircraft
USE   JET PROVOST AIRCRAFT

P-84 Aircraft, Hunting
USE   JET PROVOST AIRCRAFT

P-160 AIRCRAFT

P-160 Aircraft, ME
USE   P-160 AIRCRAFT

P-160 Aircraft, Messerschmitt ME
USE   P-160 AIRCRAFT

P-166 AIRCRAFT

P-166 Aircraft, Piaggio
USE   P-166 AIRCRAFT

P-308 AIRCRAFT

P-308 Aircraft, ME
USE   P-308 AIRCRAFT

P-308 Aircraft, Messerschmitt ME
USE   P-308 AIRCRAFT

P-531 HELICOPTER

P-531 Helicopter, Westland
USE   P-531 HELICOPTER

P-1052 AIRCRAFT

P-1052 Aircraft, Hawker
USE   P-1052 AIRCRAFT

P-1127 AIRCRAFT

P-1127 Aircraft, Hawker
USE   P-1127 AIRCRAFT

P-1154 AIRCRAFT

P-1154 Aircraft, Hawker
USE   P-1154 AIRCRAFT

Pa
USE   PROTACTINIUM

PA
USE   PENNSYLVANIA

PA), Susquehanna River Basin (MD-NY-
USE   SUSQUEHANNA RIVER BASIN (MD-NY-PA)

PA-34 SENECA AIRCRAFT

Pablo Bay (CA), San
USE   SAN PABLO BAY (CA)

PACE
USE   PHYSICS AND CHEMISTRY EXPERIMENT IN
        SPACE

Pacemaker, Artificial Cardiac
USE   ARTIFICIAL CARDIAC PACEMAKER

PACIFIC ISLANDS

PACIFIC NORTHWEST (US)

PACIFIC OCEAN

Package, Apollo Lunar Surface Experiments
USE   APOLLO LUNAR SURFACE EXPERIMENTS
        PACKAGE

Package, Early Apollo Surface Experiments
USE   EASEP

Package, Earth Resources Experiment
USE   EREP

Package Telescope, Goddard Experiment
USE   PARTICLE TELESCOPES

PACKAGES

Packages, Instrument
USE   INSTRUMENT PACKAGES

PACKAGING

Packaging, Electronic
USE   ELECTRONIC PACKAGING

Packard Computers, Hewlett-
USE   HEWLETT-PACKARD COMPUTERS

Packed Lattices, Close
USE   CLOSE PACKED LATTICES

PACKET SWITCHING

PACKET TRANSMISSION

PACKETS (COMMUNICATION)

Packets, Wave
USE   WAVE PACKETS

PACKING

PACKING DENSITY

PACKINGS (SEALS)

Packs, Ice
USE   SEA ICE

PAD

PADDLES

PADE APPROXIMATION

Pads, Launching
USE   LAUNCHING PADS

Page Aircraft, Handley
USE   HANDLEY PAGE AIRCRAFT

Page HP-115 Aircraft, Handley
USE   HP-115 AIRCRAFT

PAGEOS SATELLITE

PAIN

PAIN SENSITIVITY

PAINTS

PAIR PRODUCTION

PAKISTAN

Pakistan, West
USE   BANGLADESH

Palapa B Satellite
USE   PALAPA 2 SATELLITE

PALAPA SATELLITES

PALAPA 2 SATELLITE

PALEOBIOLOGY

PALEOMAGNETISM

PALEONTOLOGY

PALLADIUM

PALLADIUM ALLOYS

PALLADIUM COMPOUNDS

Pallet Satellites, Shuttle
USE   SHUTTLE PALLET SATELLITES

PALMAR SWEAT INDEX

PALMGREN-MINER RULE

PALMITIC ACID

PALO VERDE VALLEY (CA)

PAM (Modulation)
USE   PULSE AMPLITUDE MODULATION

PAMPAS

PANAMA

PANAMA CANAL ZONE

PANAVIA MILITARY AIRCRAFT

PANCREAS

PANEL FLUTTER

PANEL METHOD (FLUID DYNAMICS)

PANELS

Panels, Control
USE   CONTROL BOARDS

Panels, Curved
USE   CURVED PANELS

Panels, Rectangular
USE   RECTANGULAR PANELS

Panels, Wing
USE   WING PANELS

PANIC

PANORAMIC CAMERAS

PANORAMIC SCANNING

PANSPERMIA

PANT PROGRAM

PANTAR CHONDRITES

Panther Aircraft
USE   F-9 AIRCRAFT

PAPAIN

(Paper), Boards
USE   BOARDS (PAPER)

PAPER CHROMATOGRAPHY

(Paper), Forms
USE   FORMS (PAPER)

PAPER (MATERIAL)

PAPERS

PAPILLAE

Para Conversion, Ortho
USE   ORTHO PARA CONVERSION

PARA HYDROGEN

PARABOLAS

PARABOLIC ANTENNAS

PARABOLIC BODIES

PARABOLIC DIFFERENTIAL EQUATIONS

PARABOLIC FLIGHT

PARABOLIC REFLECTORS

Parabolic Velocity
USE   ESCAPE VELOCITY

PARABOLOID MIRRORS

Paraboloids
USE   PARABOLIC BODIES

PARACHUTE DESCENT

PARACHUTE FABRICS

PARACHUTES

Parachutes, Drogue
USE   DRAG CHUTES

Parachutes, Recovery
USE   RECOVERY PARACHUTES

Parachutes, Ribbon
USE   RIBBON PARACHUTES

Parachuting
USE   PARACHUTE DESCENT

PARACHUTING INJURY

PARACONE

Paradox, Clock
USE   CLOCK PARADOX

PARADOXES

PARAFFINS

Paraglider Rocket Vehicle, Dornier
USE   DORNIER PARAGLIDER ROCKET VEHICLE

PARAGLIDERS

PARAGUAY

PARALLAX

Parallax, Solar
USE   SOLAR PARALLAX

Parallax, Stellar
USE   STELLAR PARALLAX

PARALLEL COMPUTERS

PARALLEL FLOW

PARALLEL PLATES

PARALLEL PROCESSING (COMPUTERS)

PARALLEL PROGRAMMING

Parallel Strip Lines
USE   MICROSTRIP TRANSMISSION LINES

PARALLELEPIPEDS

PARALLELOGRAMS

PARALYSIS

Paramagnetic Amplifiers
USE   MASERS

PARAMAGNETIC RESONANCE

Paramagnetic Resonance, Electron
USE   ELECTRON PARAMAGNETIC RESONANCE

PARAMAGNETISM

PARAMECIA

PARAMETER IDENTIFICATION

Parameter Systems, Distributed
USE   DISTRIBUTED PARAMETER SYSTEMS

Parameter Systems, Lumped
USE   LUMPED PARAMETER SYSTEMS

Parameter, Time Temperature
USE   TIME TEMPERATURE PARAMETER

PARAMETERIZATION

Parameters
USE   INDEPENDENT VARIABLES

Parameters, Collision
USE   COLLISION PARAMETERS

Parameters, Lattice
USE   LATTICE PARAMETERS

Parameters, Meteorological
USE   METEOROLOGICAL PARAMETERS

Parameters, Oceanographic
USE   OCEANOGRAPHIC PARAMETERS

PARAMETRIC AMPLIFIERS

PARAMETRIC DIODES

PARAMETRIC FREQUENCY CONVERTERS

Parametric Oscillators
USE   PARAMETRIC AMPLIFIERS

PARAMETRONS

PARANASAL SINUSES

PARAPLASTS

Parapsychology
USE   EXTRASENSORY PERCEPTION

PARASITES

PARASITIC DISEASES

PARATHYROID GLAND

PARAVULCOONS

PARAWINGS

PARENTERAL FUNCTIONS

PARENTS

PARITY

Park (ID-MT-WY), Yellowstone National
USE   YELLOWSTONE NATIONAL PARK
(ID-MT-WY)

PARKING

PARKING ORBITS

PARKINSON DISEASE

PARKS

Parks, National
USE   NATIONAL PARKS

Parotid Gland
USE   SALIVARY GLANDS

PARSING ALGORITHMS

PARTIAL DIFFERENTIAL EQUATIONS

PARTIAL PRESSURE

PARTICLE ACCELERATION

PARTICLE ACCELERATOR TARGETS

PARTICLE ACCELERATORS

(Particle Accelerators), Racetracks
USE   RACETRACKS (PARTICLE ACCELERATORS)

Particle Accelerators, Space Exper With
USE   SEPAC (PAYLOAD)

(Particle Accelerators), Storage Rings
USE   STORAGE RINGS (PARTICLE
ACCELERATORS)

PARTICLE BEAMS

PARTICLE CHARGING

PARTICLE COLLISIONS

Particle Counters
USE   RADIATION COUNTERS

Particle Decay
USE   RADIOACTIVE DECAY

PARTICLE DENSITY (CONCENTRATION)

Particle Detectors
USE   RADIATION COUNTERS

PARTICLE DIFFUSION

PARTICLE EMISSION

PARTICLE ENERGY

Particle Explorer A, Energetic
USE   EXPLORER 12 SATELLITE

Particle Explorer B, Energetic
USE   EXPLORER 14 SATELLITE

Particle Explorer C, Energetic
USE   EXPLORER 15 SATELLITE

Particle Explorer D, Energetic
USE   EXPLORER 26 SATELLITE

Particle Flux
USE   FLUX (RATE)

PARTICLE FLUX DENSITY

PARTICLE IN CELL TECHNIQUE

PARTICLE INTENSITY

PARTICLE INTERACTIONS

Particle Interactions, Elementary
USE   ELEMENTARY PARTICLE INTERACTIONS

Particle Interactions, Plasma-
USE   PLASMA-PARTICLE INTERACTIONS

PARTICLE LADEN JETS

PARTICLE MASS

Particle Measurement, Precipitation
USE   PRECIPITATION PARTICLE MEASUREMENT

PARTICLE MOTION

(Particle Physics), Charm
USE   CHARM (PARTICLE PHYSICS)

(Particle Physics), Color
USE   QUANTUM CHROMODYNAMICS

(Particle Physics), Flavor
USE   FLAVOR (PARTICLE PHYSICS)

PARTICLE PRECIPITATION

PARTICLE PRODUCTION

PARTICLE SIZE DISTRIBUTION

PARTICLE SPIN

PARTICLE TELESCOPES

PARTICLE THEORY

Particle Theory, Many
USE   MANY BODY PROBLEM

Particle Tracer Explorers, Active Magneto
USE   AMPTE (SATELLITES)

PARTICLE TRACKS

PARTICLE TRAJECTORIES

PARTICLES

Particles, Alpha
USE   ALPHA PARTICLES

Particles, Anti
USE   ANTIPARTICLES

Particles, Beta
USE   BETA PARTICLES

Particles, Charged
USE   CHARGED PARTICLES

Particles, Elementary
USE   ELEMENTARY PARTICLES

Particles, Energetic
USE   ENERGETIC PARTICLES

Particles, Geomagnetically Trapped
USE   RADIATION BELTS

Particles, Magnetically Trapped
USE   MAGNETICALLY TRAPPED PARTICLES

Particles, Metal
USE   METAL PARTICLES

Particles, Micro
USE   MICROPARTICLES

Particles, Neutral
USE   NEUTRAL PARTICLES

Particles, Nuclear
USE   NUCLEAR PARTICLES

Particles, Penetrating
USE   CORPUSCULAR RADIATION

(Particles), Powder
USE   POWDER (PARTICLES)

Particles, Quasi-
USE   ELEMENTARY EXCITATIONS

Particles, Relativistic
USE   RELATIVISTIC PARTICLES

Particles, Trapped
USE   TRAPPED PARTICLES

Particulate Filters
USE   FLUID FILTERS

PARTICULATE SAMPLING

PARTITIONS

PARTITIONS (MATHEMATICS)

PARTITIONS (STRUCTURES)

Parton Model, Quark
USE   QUARK PARTON MODEL

PARTONS

Parts
USE   COMPONENTS

Parts, Aircraft
USE   AIRCRAFT PARTS

Parts, Engine
USE   ENGINE PARTS

Parts, Spare
USE   SPARE PARTS

PAS

PASCAL (PROGRAMMING LANGUAGE)

PASCHEN SERIES

Pass Filters, High
USE   HIGH PASS FILTERS

Pass Filters, Low
USE   LOW PASS FILTERS

Passageway), Ingress (Spacecraft
USE   INGRESS (SPACECRAFT PASSAGEWAY)

PASSAGEWAYS

PASSENGER AIRCRAFT

PASSENGERS

Passes
USE   GAPS (GEOLOGY)

Passivation
USE   PASSIVITY

PASSIVE L-BAND RADIOMETERS

Passive Nosetip Technology
USE   PANT PROGRAM

PASSIVE SATELLITES

PASSIVITY

PASTE (CONSISTENCY)

PASTES

PASTEURIZING

PATCH TESTS

PATENT APPLICATIONS

PATENT POLICY

PATENTS

Path Analysis, Gas
   USE   GAS PATH ANALYSIS

Path, Mean Free
   USE   MEAN FREE PATH

Path Method, Critical
   USE   CRITICAL PATH METHOD

PATHFINDER NUCLEAR REACTOR

PATHOGENESIS

PATHOGENS

PATHOLOGICAL EFFECTS

PATHOLOGY

Pathology, Human
   USE   HUMAN PATHOLOGY

Pathology, Radio
   USE   RADIOPATHOLOGY

PATHS

Paths, Diffraction
   USE   DIFFRACTION PATHS

Paths, Electron
   USE   ELECTRON TRAJECTORIES

Paths, Flight
   USE   FLIGHT PATHS

Paths, Glide
   USE   GLIDE PATHS

Paths, Optical
   USE   OPTICAL PATHS

PATIENTS

PATRIOT MISSILE

PATROLS

Pattern Distribution
   USE   DISTRIBUTION (PROPERTY)

Pattern Generators, Test
   USE   TEST PATTERN GENERATORS

Pattern, Kossel
   USE   KOSSEL PATTERN

PATTERN METHOD (FORECASTING)

PATTERN RECOGNITION

Pattern Recognition, Automatic
   USE   PATTERN RECOGNITION

PATTERN REGISTRATION

PATTERNS

Patterns, Antenna Radiation
   USE   ANTENNA RADIATION PATTERNS

Patterns, Chaotic Cloud
   USE   CLOUDS (METEOROLOGY)

Patterns, Diffraction
   USE   DIFFRACTION PATTERNS

Patterns, Drainage
   USE   DRAINAGE PATTERNS

Patterns, Flat
   USE   FLAT PATTERNS

Patterns, Flow
   USE   FLOW DISTRIBUTION

Patterns, Fringe
   USE   DIFFRACTION PATTERNS

Patterns, Radial Drainage
   USE   DRAINAGE PATTERNS

Patterns, Speckle
   USE   SPECKLE PATTERNS

PATTERSON MAP

PAULI EXCLUSION PRINCIPLE

PAVEMENTS

Payload), Amps (Satellite
   USE   AMPS (SATELLITE PAYLOAD)

PAYLOAD ASSIST MODULE

Payload, Atmospheric And Magnetospheric
   USE   AMPS (SATELLITE PAYLOAD)

PAYLOAD CONTROL

(Payload Delivery), Mass Drivers
   USE   MASS DRIVERS (PAYLOAD DELIVERY)

PAYLOAD DELIVERY (STS)

PAYLOAD DEPLOYMENT & RETRIEVAL SYSTEM

Payload), Expos (Spacelab
   USE   EXPOS (SPACELAB PAYLOAD)

PAYLOAD INTEGRATION

PAYLOAD INTEGRATION PLAN

PAYLOAD MASS RATIO

Payload, Oss-1
   USE   OSS-1 PAYLOAD

Payload, OSTA-1
   USE   OSTA-1 PAYLOAD

Payload, OSTA-2
   USE   OSTA-2 PAYLOAD

Payload, Plasmas-IN-Space
   USE   AMPS (SATELLITE PAYLOAD)

PAYLOAD RETRIEVAL (STS)

(Payload), Sepac
   USE   SEPAC (PAYLOAD)

PAYLOAD STATIONS

PAYLOAD TRANSFER

Payload, X Ray Spectropolarimetry
   USE   EXPOS (SPACELAB PAYLOAD)

PAYLOADS

Payloads, Office Of Space & Terrestr Applic
   USE   OSTA-2 PAYLOAD
          OSTA-1 PAYLOAD

Payloads, Space Shuttle
   USE   SPACE SHUTTLE PAYLOADS

Payloads, Spacelab
   USE   SPACELAB PAYLOADS

Pb
   USE   LEAD (METAL)

PBB
   USE   POLYBROMINATED BIPHENYLS

PBRE (Reactors)
   USE   PEBBLE BED REACTORS

PCB
   USE   POLYCHLORINATED BIPHENYLS

PCM (Materials)
   USE   PHASE CHANGE MATERIALS

PCM (Modulation)
   USE   PULSE CODE MODULATION

PCM TELEMETRY

Pd
   USE   PALLADIUM

PD-808 AIRCRAFT

PD-808 Aircraft, Douglas
   USE   PD-808 AIRCRAFT

PD-808 Aircraft, Piaggio-Douglas
   USE   PD-808 AIRCRAFT

PDM (Modulation)
   USE   PULSE DURATION MODULATION

PDP COMPUTERS

PDP 7 COMPUTER

PDP 8 COMPUTER

PDP 9 COMPUTER

PDP 10 COMPUTER

PDP 11 COMPUTER

PDP 11/20 COMPUTER

PDP 11/40 COMPUTER

PDP 11/45 COMPUTER

PDP 11/50 COMPUTER

PDP 11/70 COMPUTER

PDP 12 COMPUTER

PDP 15 COMPUTER

PEACETIME

Peak (CO), Pike's
   USE   PIKE'S PEAK (CO)

PEAKS

Peaks, Bordoni
   USE   BORDONI PEAKS

PEAKS (LANDFORMS)

PEARLITE

PEARSON DISTRIBUTIONS

PEAT

PEBBLE BED REACTORS

PECLET NUMBER

Pectoris, Angina
   USE   ANGINA PECTORIS

PECULIAR STARS

PEDALS

Pediments
   USE   PIEDMONTS

Pediplains
   USE   PIEDMONTS

Pedology
USE   SOIL SCIENCE

PEELING

PEENING

Peening, Shot
USE   SHOT PEENING

PEGASUS COMPUTER

Pegasus Engine
USE   BRISTOL-SIDDELEY BS 53 ENGINE

PEGASUS SATELLITES

PELAGIC ZONE

PELLETS

PELLICLE

PELOMYXA

PELTIER EFFECTS

PELVIS

PENALTIES

PENALTY FUNCTION

Pendulous Gyroscopes
USE   GYROSCOPIC PENDULUMS

PENDULUMS

Pendulums, Gyroscopic
USE   GYROSCOPIC PENDULUMS

PENEPLAINS

PENETRANTS

Penetrating Particles
USE   CORPUSCULAR RADIATION

PENETRATION

Penetration Ballistics
USE   TERMINAL BALLISTICS

Penetration, Projectile
USE   TERMINAL BALLISTICS

Penetration, Target
USE   TERMINAL BALLISTICS

PENETROMETERS

PENICILLIN

Peninsula (De-MD-VA), Delmarva
USE   DELMARVA PENINSULA (DE-MD-VA)

PENINSULAR RANGES (CA)

PENINSULAS

PENNING DISCHARGE

PENNING EFFECT

PENNING GAGES

PENNSYLVANIA

PENS

PENTABORANES

Pentachlorides
USE   CHLORIDES

Pentaerythritol Tetranitrate
USE   PETN

PENTANES

PENTANONE

PENTOBARBITAL

PENTOBARBITAL SODIUM

PENTODES

PENTOLITE

PENTOSE

PENUMBRAS

PEOLE SATELLITES

Peoples Democratic Republic Of Germany
USE   EAST GERMANY

Peoples Republic, Chinese
USE   CHINA

Peoples Republic Of Korea, Democratic
USE   NORTH KOREA

PEPPERS

PEPSIN

PEPTIDES

Peptides, Poly
USE   POLYPEPTIDES

Per Carrier Transmission, Single Channel
USE   SINGLE CHANNEL PER CARRIER
           TRANSMISSION

(Per Time), Rates
USE   RATES (PER TIME)

Per Unit Area), Flux (Rate
USE   FLUX DENSITY

Perceived Noise Levels, Effective
USE   EFFECTIVE PERCEIVED NOISE LEVELS

Percentage
USE   RATIOS

PERCEPTION

Perception, Auditory
USE   AUDITORY PERCEPTION

Perception, Color
USE   COLOR VISION

Perception, Cutaneous
USE   TOUCH

Perception, Depth
USE   SPACE PERCEPTION

Perception, Distance
USE   SPACE PERCEPTION

Perception, Extrasensory
USE   EXTRASENSORY PERCEPTION

Perception, Form
USE   SPACE PERCEPTION

Perception, Gustatory
USE   TASTE

Perception, Motion
USE   MOTION PERCEPTION

Perception, Olfactory
USE   OLFACTORY PERCEPTION

Perception, Sensory
USE   SENSORY PERCEPTION

Perception, Slant
USE   SPACE PERCEPTION

Perception, Sound
USE   AUDITORY PERCEPTION

Perception, Space
USE   SPACE PERCEPTION

(Perception), Thresholds
USE   THRESHOLDS (PERCEPTION)

Perception, Vertical
USE   VERTICAL PERCEPTION

Perception, Vibration
USE   VIBRATION PERCEPTION

Perception, Visual
USE   VISUAL PERCEPTION

Perceptrons
USE   SELF ORGANIZING SYSTEMS

PERCEPTUAL ERRORS

PERCEPTUAL TIME CONSTANT

Perchlorate, Hydrogen
USE   HYDROGEN PERCHLORATE

Perchlorate, Nitronium
USE   NITRONIUM PERCHLORATE

PERCHLORATES

Perchlorates, Aluminum
USE   ALUMINUM PERCHLORATES

Perchlorates, Ammonium
USE   AMMONIUM PERCHLORATES

Perchlorates, Hydrazine
USE   HYDRAZINE PERCHLORATES

Perchlorates, Hydroxylammonium
USE   HYDROXYLAMMONIUM PERCHLORATES

Perchlorates, Lithium
USE   LITHIUM PERCHLORATES

Perchlorates, Magnesium
USE   MAGNESIUM PERCHLORATES

Perchlorates, Potassium
USE   POTASSIUM PERCHLORATES

PERCHLORIC ACID

PERCHLORYL FLUORIDES

PERCOLATION

PERCUS METHOD

PERCUSSION

Perfect Gas
USE   IDEAL GAS

PERFLUORO COMPOUNDS

PERFLUOROALKANE

PERFLUOROGUANIDINE

PERFORATED PLATES

PERFORATED SHELLS

PERFORATING

PERFORATION

PERFORMANCE

Performance, Aircraft
USE   AIRCRAFT PERFORMANCE

Performance, Astronaut
USE   ASTRONAUT PERFORMANCE

Performance, Computer Systems
USE   COMPUTER SYSTEMS PERFORMANCE

Performance, Flight
USE   FLIGHT CHARACTERISTICS

Performance, Helicopter
USE HELICOPTER PERFORMANCE

Performance, Human
USE HUMAN PERFORMANCE

Performance, Mental
USE MENTAL PERFORMANCE

Performance, Operator
USE OPERATOR PERFORMANCE

Performance, Pilot
USE PILOT PERFORMANCE

PERFORMANCE PREDICTION

Performance, Propulsion System
USE PROPULSION SYSTEM PERFORMANCE

Performance, Psychomotor
USE PSYCHOMOTOR PERFORMANCE

Performance, Sensorimotor
USE SENSORIMOTOR PERFORMANCE

Performance, Spacecraft
USE SPACECRAFT PERFORMANCE

PERFORMANCE TESTS

Perfusion
USE DIFFUSION

PERICLASE

PERIDOTITE

Perigee-Apogee Satellites
USE PAS

PERIGEES

PERIHELIONS

PERILUNES

Period Equations
USE PERIODIC FUNCTIONS

Period, Pre-Imbrian
USE PRE-IMBRIAN PERIOD

Period, Precambrian
USE PRECAMBRIAN PERIOD

Period, Refractory
USE REFRACTORY PERIOD

Periodic Antennas, Log
USE LOG PERIODIC ANTENNAS

PERIODIC FUNCTIONS

Periodic Orbits
USE ORBITS

Periodic Processes
USE CYCLES

PERIODIC VARIATIONS

PERIODICALS

Periodicity
USE PERIODIC VARIATIONS

Periodicity (Biology)
USE RHYTHM (BIOLOGY)

PERIPHERAL CIRCULATION

PERIPHERAL EQUIPMENT (COMPUTERS)

PERIPHERAL JET FLOW

PERIPHERAL NERVOUS SYSTEM

PERIPHERAL VISION

Peripheries
USE BOUNDARIES

PERISCOPES

PERITONEUM

PERMAFROST

PERMALLOYS (TRADEMARK)

PERMANGANATES

PERMEABILITY

Permeability, Dielectric
USE DIELECTRIC PERMEABILITY

Permeability, Magnetic
USE MAGNETIC PERMEABILITY

PERMEATING

Permeation Chromatography, Gel
USE LIQUID CHROMATOGRAPHY

PERMISSIVITY

PERMITTIVITY

PERMUTATIONS

Perot Interferometers, Fabry-
USE FABRY-PEROT INTERFEROMETERS

Perot Lasers, Fabry-
USE LASERS

Perot Spectrometers, Fabry-
USE FABRY-PEROT SPECTROMETERS

PEROVSKITES

Peroxide, Hydrogen
USE HYDROGEN PEROXIDE

PEROXIDES

Peroxides, Inorganic
USE INORGANIC PEROXIDES

Peroxides, Organic
USE ORGANIC PEROXIDES

Peroxides, Potassium
USE POTASSIUM PEROXIDES

Peroxides, Sodium
USE SODIUM PEROXIDES

PERSEID METEOROIDS

PERSHING MISSILE

PERSIAN GULF

PERSONAL COMPUTERS

PERSONALITY

PERSONALITY TESTS

PERSONNEL

(Personnel), Air Traffic Controllers
USE AIR TRAFFIC CONTROLLERS (PERSONNEL)

PERSONNEL DEVELOPMENT

Personnel, Enemy
USE ENEMY PERSONNEL

Personnel, Flying
USE FLYING PERSONNEL

PERSONNEL MANAGEMENT

Personnel, Medical
USE MEDICAL PERSONNEL

(Personnel), Operators
USE OPERATORS (PERSONNEL)

(Personnel), Pilots
USE PILOTS (PERSONNEL)

Personnel Propulsion Systems
USE SELF MANEUVERING UNITS

PERSONNEL SELECTION

PERSONNEL SUBSYSTEMS

PERSPEX (TRADEMARK)

PERSPIRATION

PERT

PERTURBATION

Perturbation Flow, Small
USE SMALL PERTURBATION FLOW

Perturbation, Lunar
USE LUNAR EFFECTS

Perturbation, Orbit
USE ORBIT PERTURBATION

Perturbation, Plasma
USE PLASMA OSCILLATIONS

Perturbation, Satellite
USE SATELLITE PERTURBATION

Perturbation, Secular
USE LONG TERM EFFECTS

PERTURBATION THEORY

PERU

PERVEANCE

PESTICIDES

PETALS

PETECHIA

PETN

PETREL SOUNDING ROCKET

PETRI NETS

PETROGRAPHY

Petroleum
USE CRUDE OIL

PETROLEUM PRODUCTS

PETROLOGY

PFAFF EQUATION

PFM (Modulation)
USE PULSE FREQUENCY MODULATION

PH

PH FACTOR

PHANTASTRONS

PHANTOM AIRCRAFT

PHARMACOLOGY

PHARYNX

Phase Angle
USE PHASE SHIFT

PHASE CHANGE MATERIALS

PHASE COHERENCE

PHASE CONJUGATION

PHASE CONTRAST

PHASE CONTROL

PHASE DEMODULATORS

PHASE DETECTORS

PHASE DEVIATION

PHASE DIAGRAMS

Phase Epitaxy, Liquid
USE   LIQUID PHASE EPITAXY

Phase Epitaxy, Vapor
USE   VAPOR PHASE EPITAXY

PHASE ERROR

Phase Flow, One-
USE   SINGLE-PHASE FLOW

Phase Flow, Single-
USE   SINGLE-PHASE FLOW

Phase Flow, Two
USE   TWO PHASE FLOW

PHASE LOCK DEMODULATORS

PHASE LOCKED SYSTEMS

PHASE MATCHING

PHASE MODULATION

PHASE RULE

PHASE SHIFT

PHASE SHIFT CIRCUITS

(Phase Shift Circuits), Circulators
USE   CIRCULATORS (PHASE SHIFT CIRCUITS)

PHASE SHIFT KEYING

PHASE SWITCHING INTERFEROMETERS

Phase Systems, Two
USE   BINARY SYSTEMS (MATERIALS)

PHASE TRANSFORMATIONS

PHASE VELOCITY

PHASE-SPACE INTEGRAL

PHASED ARRAYS

PHASES

Phases, Gas
USE   VAPOR PHASES

Phases, Liquid
USE   LIQUID PHASES

Phases, Lunar
USE   LUNAR PHASES

Phases, Solid
USE   SOLID PHASES

Phases, Vapor
USE   VAPOR PHASES

Phenacetin
USE   ACETANILIDE

PHENANTHRENE

PHENOBARBITAL

PHENOL FORMALDEHYDE

PHENOLIC EPOXY RESINS

PHENOLIC RESINS

PHENOLOGY

PHENOLS

Phenols, Bis
USE   BITS

Phenomena, Medical
USE   MEDICAL PHENOMENA

Phenomena, Mesoscale
USE   MESOSCALE PHENOMENA

Phenomenological Coefficient, Onsager
USE   ONSAGER PHENOMENOLOGICAL
        COEFFICIENT

PHENOMENOLOGY

Phenomenon, Chorus
USE   DAWN CHORUS

Phenomenon), Chorus (Dawn
USE   DAWN CHORUS

Phenomenon, Gibbs
USE   GIBBS PHENOMENON

Phenomenon, Leidenfrost
USE   LEIDENFROST PHENOMENON

PHENOTHIAZINES

PHENYLALANINE

PHENYLS

Phenyls, Poly
USE   POLYPHENYLS

Phenyls, Tetra
USE   TETRAPHENYLS

Phenyls, Tri
USE   TRIPHENYLS

PHILCO 2000 COMPUTER

PHILIPPINES

PHILIPS IONIZATION GAGES

Philosophers Problem, Dining
USE   DINING PHILOSOPHERS PROBLEM

PHILOSOPHY

PHLOROGLUCINOL

PHOBIAS

PHOBOS

PHOEBUS NUCLEAR REACTOR

PHOENIX (AZ)

PHOENIX QUADRANGLE (AZ)

PHOENIX SOUNDING ROCKET

PHONEMES

PHONEMICS

PHONETICS

PHONOARTERIOGRAPHY

Phonocardiograms
USE   PHONOCARDIOGRAPHY

PHONOCARDIOGRAPHY

PHONON BEAMS

Phonon Interactions, Electron
USE   ELECTRON PHONON INTERACTIONS

PHONONS

PHORIA

PHOSGENE

PHOSPHATES

Phosphates, Ammonium
USE   AMMONIUM PHOSPHATES

Phosphates, Calcium
USE   CALCIUM PHOSPHATES

Phosphates, Di
USE   DIPHOSPHATES

Phosphates, Indium
USE   INDIUM PHOSPHATES

Phosphates, Potassium
USE   POTASSIUM PHOSPHATES

PHOSPHAZENE

PHOSPHENE

PHOSPHIDES

Phosphides, Boron
USE   BORON PHOSPHIDES

Phosphides, Gallium
USE   GALLIUM PHOSPHIDES

Phosphides, Indium
USE   INDIUM PHOSPHIDES

Phosphides, Manganese
USE   MANGANESE PHOSPHIDES

PHOSPHINES

Phosphite (DEHP), Diethyl Hydrogen
USE   DIETHYL HYDROGEN PHOSPHITE (DEHP)

PHOSPHONITRILES

PHOSPHONIUM COMPOUNDS

PHOSPHORESCENCE

PHOSPHORIC ACID

PHOSPHORIC ACID FUEL CELLS

PHOSPHORS

Phosphors, Radio
USE   RADIOPHOSPHORS

PHOSPHORUS

PHOSPHORUS COMPOUNDS

Phosphorus Compounds, Organic
USE   ORGANIC PHOSPHORUS COMPOUNDS

PHOSPHORUS ISOTOPES

PHOSPHORUS METABOLISM

PHOSPHORUS OXIDES

PHOSPHORUS POLYMERS

PHOSPHORUS 32

PHOSPHORYLATION

PHOTICS

PHOTO RECONNAISSANCE SPACECRAFT

PHOTOABSORPTION

PHOTOACOUSTIC MICROSCOPY

PHOTOACOUSTIC SPECTROSCOPY

PHOTOCATHODES

Photocells
USE   PHOTOELECTRIC CELLS

PHOTOCHEMICAL OXIDANTS

PHOTOCHEMICAL REACTIONS

Photochemistry
USE   PHOTOCHEMICAL REACTIONS

PHOTOCHROMISM

Photoclinometry
USE   PHOTOGRAMMETRY

PHOTOCONDUCTIVE CELLS

PHOTOCONDUCTIVITY

PHOTOCONDUCTORS

Photocurrents
USE   PHOTOELECTRIC EMISSION
ELECTRIC CURRENT

PHOTODECOMPOSITION

PHOTODETACHMENT

Photodetectors
USE   PHOTOMETERS

PHOTODIODES

PHOTODISSOCIATION

PHOTOELASTIC ANALYSIS

PHOTOELASTIC MATERIALS

Photoelastic Stress Measurement
USE   PHOTOELASTIC ANALYSIS

PHOTOELASTICITY

PHOTOELECTRIC CELLS

PHOTOELECTRIC EFFECT

PHOTOELECTRIC EMISSION

PHOTOELECTRIC GENERATORS

PHOTOELECTRIC MATERIALS

Photoelectric Photometers
USE   ELECTROPHOTOMETERS

PHOTOELECTRICITY

PHOTOELECTROCHEMICAL DEVICES

PHOTOELECTROCHEMISTRY

Photoelectromagnetic Detectors
USE   PHOTOELECTROMAGNETIC EFFECTS
RADIATION MEASURING INSTRUMENTS

PHOTOELECTROMAGNETIC EFFECTS

PHOTOELECTRON SPECTROSCOPY

Photoelectronics
USE   PHOTOELECTRICITY
ELECTRONICS

PHOTOELECTRONS

Photoemission
USE   PHOTOELECTRIC EMISSION

Photoemissivity
USE   PHOTOELECTRIC EMISSION
EMISSIVITY

Photoemitters
USE   PHOTOELECTRIC MATERIALS

PHOTOENGRAVING

PHOTOGEOLOGY

PHOTOGONIOMETERS

PHOTOGRAMMETRY

Photograph Interpretation
USE   PHOTOINTERPRETATION

PHOTOGRAPHIC DEVELOPERS

PHOTOGRAPHIC EMULSIONS

PHOTOGRAPHIC EQUIPMENT

PHOTOGRAPHIC FILM

PHOTOGRAPHIC MEASUREMENT

PHOTOGRAPHIC PLATES

PHOTOGRAPHIC PROCESSING

PHOTOGRAPHIC PROCESSING EQUIPMENT

PHOTOGRAPHIC RECORDING

PHOTOGRAPHIC RECTIFIERS

PHOTOGRAPHIC TRACKING

PHOTOGRAPHS

Photographs, Cloud
USE   CLOUD PHOTOGRAPHS

Photographs, Lunar
USE   LUNAR PHOTOGRAPHS

Photographs, Mars
USE   MARS PHOTOGRAPHS

Photographs, Micro
USE   MICROPHOTOGRAPHS

PHOTOGRAPHY

Photography, Aerial
USE   AERIAL PHOTOGRAPHY

Photography, All Sky
USE   ALL SKY PHOTOGRAPHY

Photography, Astronomical
USE   ASTRONOMICAL PHOTOGRAPHY

Photography, Black And White
USE   BLACK AND WHITE PHOTOGRAPHY

Photography, Chrono
USE   CHRONOPHOTOGRAPHY

Photography, Cloud
USE   CLOUD PHOTOGRAPHY

Photography, Color
USE   COLOR PHOTOGRAPHY

Photography, Color Infrared
USE   COLOR INFRARED PHOTOGRAPHY

(Photography), Developers
USE   PHOTOGRAPHIC DEVELOPERS

Photography, Electro-Optical
USE   ELECTRO-OPTICAL PHOTOGRAPHY

Photography, Electron
USE   ELECTRON PHOTOGRAPHY

Photography, Electronic
USE   ELECTRO-OPTICAL PHOTOGRAPHY

Photography, Frame
USE   FRAME PHOTOGRAPHY

Photography, High Speed
USE   HIGH SPEED PHOTOGRAPHY

Photography, Infrared
USE   INFRARED PHOTOGRAPHY

Photography, Lunar
USE   LUNAR PHOTOGRAPHY

Photography, Metric
USE   METRIC PHOTOGRAPHY

Photography, Microwave
USE   MICROWAVE PHOTOGRAPHY

Photography, Multispectral
USE   MULTISPECTRAL PHOTOGRAPHY

Photography, Ortho
USE   ORTHOPHOTOGRAPHY

Photography, Radar
USE   RADAR PHOTOGRAPHY

Photography, Rocket-Borne
USE   ROCKET-BORNE PHOTOGRAPHY

Photography, Satellite-Borne
USE   SATELLITE-BORNE PHOTOGRAPHY

Photography, Schlieren
USE   SCHLIEREN PHOTOGRAPHY

Photography, Shadowgraph
USE   SHADOWGRAPH PHOTOGRAPHY

Photography, Space
USE   SPACEBORNE PHOTOGRAPHY

Photography, Spaceborne
USE   SPACEBORNE PHOTOGRAPHY

Photography, Spark Shadowgraph
USE   SHADOWGRAPH PHOTOGRAPHY

Photography, Spectro
USE   SPECTROPHOTOGRAPHY

Photography, Stereo
USE   STEREOPHOTOGRAPHY

Photography, Stereoscopic
USE   STEREOPHOTOGRAPHY

Photography, Streak
USE   STREAK PHOTOGRAPHY

Photography, Time Lapse
USE   CHRONOPHOTOGRAPHY

Photography, Ultraviolet
USE   ULTRAVIOLET PHOTOGRAPHY

Photography, Underwater
USE   UNDERWATER PHOTOGRAPHY

PHOTOINTERPRETATION

PHOTOIONIZATION

PHOTOLITHOGRAPHY

PHOTOLUMINESCENCE

PHOTOLUMINESCENT BANDS

PHOTOLYSIS

PHOTOMAGNETIC EFFECTS

PHOTOMAPPING

PHOTOMAPS

PHOTOMASKS

PHOTOMECHANICAL EFFECT

PHOTOMETERS

Photometers, Electro
USE   ELECTROPHOTOMETERS

Photometers, Photoelectric
USE ELECTROPHOTOMETERS

Photometers, Spectro
USE SPECTROPHOTOMETERS

PHOTOMETRY

Photometry, Astronomical
USE ASTRONOMICAL PHOTOMETRY

Photometry, Electro
USE ELECTROPHOTOMETRY

Photometry, Infrared
USE INFRARED PHOTOMETRY

Photometry, Spectro
USE SPECTROPHOTOMETRY

Photometry, Tele
USE TELEPHOTOMETRY

Photometry, Ultraviolet
USE ULTRAVIOLET PHOTOMETRY

Photometry, Visual
USE VISUAL PHOTOMETRY

PHOTOMICROGRAPHS

PHOTOMICROGRAPHY

PHOTOMULTIPLIER TUBES

Photomultipliers, Frequency Modulation
USE FREQUENCY MODULATION
PHOTOMULTIPLIERS

PHOTON ABSORPTIOMETRY

PHOTON BEAMS

Photon Cascades, Electron
USE ELECTRON PHOTON CASCADES

PHOTON DENSITY

PHOTON-ELECTRON INTERACTION

PHOTONEUTRONS

PHOTONIC PROPULSION

PHOTONICS

PHOTONS

PHOTONUCLEAR REACTIONS

PHOTOOXIDATION

PHOTOPEAK

PHOTOPHILIC PLANTS

PHOTOPHORESIS

PHOTOPLASTICITY

PHOTOPRODUCTION

PHOTORECEPTORS

PHOTORECONNAISSANCE

Photoreduction
USE PHOTOCHEMICAL REACTIONS

Photoresistivity
USE PHOTOCONDUCTIVITY

Photoresistors
USE PHOTOCONDUCTORS

PHOTOSENSITIVITY

Photosensors
USE PHOTOELECTRICITY
RADIATION MEASURING INSTRUMENTS

PHOTOSPHERE

PHOTOSTRESSES

PHOTOSYNTHESIS

PHOTOTHERMAL CONVERSION

Photothermotropism
USE PHOTOTROPISM
TEMPERATURE EFFECTS
ANISOTROPY

PHOTOTRANSISTORS

PHOTOTROPISM

PHOTOTUBES

Phototubes, Multiplier
USE PHOTOMULTIPLIER TUBES

PHOTOVISCOELASTICITY

PHOTOVOLTAGES

PHOTOVOLTAIC CELLS

PHOTOVOLTAIC CONVERSION

PHOTOVOLTAIC EFFECT

Photovoltaics, Spectro
USE SPECTROPHOTOVOLTAICS

PHREATOPHYTES

Phthalate, Tere
USE TEREPHTHALATE

PHTHALATES

PHTHALOCYANIN

Phugoid Oscillations
USE OSCILLATORS
OSCILLATIONS
PITCH (INCLINATION)

PHYLLOQUINONE

PHYSICAL CHEMISTRY

Physical Constants Testing Reactor
USE NUCLEAR RESEARCH AND TEST
REACTORS
WATER COOLED REACTORS

Physical Endurance
USE PHYSICAL FITNESS

PHYSICAL EXAMINATIONS

PHYSICAL EXERCISE

PHYSICAL FACTORS

PHYSICAL FITNESS

PHYSICAL OPTICS

PHYSICAL PROPERTIES

PHYSICAL SCIENCES

PHYSICAL WORK

PHYSICIANS

PHYSICS

(Physics), Acceleration
USE ACCELERATION (PHYSICS)

PHYSICS AND CHEMISTRY EXPERIMENT IN SPACE

Physics Applications Program, Earth & Ocean
USE EARTH & OCEAN PHYSICS APPLICATIONS
PROGRAM

Physics, Astro
USE ASTROPHYSICS

Physics, Atmospheric
USE ATMOSPHERIC PHYSICS

Physics, Atomic
USE ATOMIC PHYSICS

Physics, Bio
USE BIOPHYSICS

(Physics), Branching
USE BRANCHING (PHYSICS)

Physics), Charm (Particle
USE CHARM (PARTICLE PHYSICS)

Physics, Cloud
USE CLOUD PHYSICS

Physics, Color (Particle
USE QUANTUM CHROMODYNAMICS

Physics, Combustion
USE COMBUSTION PHYSICS

Physics, Electro
USE ELECTROPHYSICS

Physics), Electron Runaway (Plasma
USE ELECTRON RUNAWAY (PLASMA PHYSICS)

(Physics), Field Theory
USE FIELD THEORY (PHYSICS)

Physics), Filaments (Solar
USE SOLAR PROMINENCES

Physics), Flavor (Particle
USE FLAVOR (PARTICLE PHYSICS)

Physics, Geo
USE GEOPHYSICS

Physics, Health
USE HEALTH PHYSICS

Physics Lab (Spacelab), Atmospheric Cloud
USE ATMOSPHERIC CLOUD PHYSICS LAB
(SPACELAB)

Physics, Low Temperature
USE LOW TEMPERATURE PHYSICS

(Physics), Magnetomechanics
USE MAGNETOMECHANICS (PHYSICS)

(Physics), Matter
USE MATTER (PHYSICS)

(Physics), Mechanics
USE MECHANICS (PHYSICS)

Physics, Molecular
USE MOLECULAR PHYSICS

Physics, Neutron
USE NEUTRON PHYSICS

Physics, Nuclear
USE NUCLEAR PHYSICS

Physics), Nuclei (Nuclear
USE NUCLEI (NUCLEAR PHYSICS)

Physics, Plasma
USE PLASMA PHYSICS

(Physics), Plasmas
USE PLASMAS (PHYSICS)

Physics, Polymer
USE POLYMER PHYSICS

Physics, Psycho
USE PSYCHOPHYSICS

Physics), Quenching (Atomic
USE QUENCHING (ATOMIC PHYSICS)

Physics, Radio
  USE   RADIO PHYSICS

Physics, Reactor
  USE   REACTOR PHYSICS

Physics, Reentry
  USE   REENTRY PHYSICS

Physics Research Reactor, Health
  USE   HEALTH PHYSICS RESEARCH REACTOR

Physics), Rigid Rotors (Plasma
  USE   RIGID ROTORS (PLASMA PHYSICS)

Physics), Selection Rules (Nuclear
  USE   SELECTION RULES (NUCLEAR PHYSICS)

Physics, Solar
  USE   SOLAR PHYSICS

Physics, Solid State
  USE   SOLID STATE PHYSICS

Physics, Stellar
  USE   STELLAR PHYSICS

Physics, Theoretical
  USE   THEORETICAL PHYSICS

PHYSIOCHEMISTRY

Physiography
  USE   GEOMORPHOLOGY

PHYSIOLOGICAL ACCELERATION

PHYSIOLOGICAL DEFENSES

PHYSIOLOGICAL EFFECTS

PHYSIOLOGICAL FACTORS

PHYSIOLOGICAL RESPONSES

Physiological Telemetry
  USE   BIOTELEMETRY

PHYSIOLOGICAL TESTS

PHYSIOLOGY

(Physiology), Acceleration Stresses
  USE   ACCELERATION STRESSES (PHYSIOLOGY)

(Physiology), Bends
  USE   DECOMPRESSION SICKNESS

(Physiology), Blackout
  USE   BLACKOUT (PHYSIOLOGY)

Physiology, Electro
  USE   ELECTROPHYSIOLOGY

Physiology, Exercise
  USE   EXERCISE PHYSIOLOGY

Physiology, Gravitational
  USE   GRAVITATIONAL PHYSIOLOGY

Physiology, Neuro
  USE   NEUROPHYSIOLOGY

Physiology, Psycho
  USE   PSYCHOPHYSIOLOGY

(Physiology), Receptors
  USE   RECEPTORS (PHYSIOLOGY)

(Physiology), Regeneration
  USE   REGENERATION (PHYSIOLOGY)

(Physiology), Relaxation
  USE   RELAXATION (PHYSIOLOGY)

Physiology, Respiratory
  USE   RESPIRATORY PHYSIOLOGY

(Physiology), Shock
  USE   SHOCK (PHYSIOLOGY)

(Physiology), Stress
  USE   STRESS (PHYSIOLOGY)

(Physiology), Tolerances
  USE   TOLERANCES (PHYSIOLOGY)

Physiology, Underwater
  USE   UNDERWATER PHYSIOLOGY

PHYTOTRONS

PI-ELECTRONS

PIAGGIO AIRCRAFT

Piaggio P-166 Aircraft
  USE   P-166 AIRCRAFT

Piaggio-Douglas PD-808 Aircraft
  USE   PD-808 AIRCRAFT

PIASECKI AIRCRAFT

PICKLING (METALLURGY)

Pickoffs
  USE   SENSORS

Pickups
  USE   SENSORS

PICOSECOND PULSES

PICRATES

Picrates, Ammonium
  USE   AMMONIUM PICRATES

(Picture Transmission), APT
  USE   AUTOMATIC PICTURE TRANSMISSION

Picture Transmission, Automatic
  USE   AUTOMATIC PICTURE TRANSMISSION

PICTURE TUBES

Pictures, Motion
  USE   MOTION PICTURES

Piedmont (US), Central
  USE   CENTRAL PIEDMONT (US)

PIEDMONTS

PIERCING

Piers
  USE   WHARVES

PIEZOELECTRIC CERAMICS

PIEZOELECTRIC CRYSTALS

PIEZOELECTRIC GAGES

PIEZOELECTRIC TRANSDUCERS

PIEZOELECTRICITY

PIEZOMETERS

PIEZORESISTIVE TRANSDUCERS

PIGEONS

PIGGYBACK SYSTEMS

PIGMENTS

Pigments, Visual
  USE   VISUAL PIGMENTS

Pigs, Guinea
  USE   GUINEA PIGS

Pigs (Swine)
  USE   SWINE

PIKE'S PEAK (CO)

PILE FOUNDATIONS

PILES

Piles, Thermo
  USE   THERMOPILES

PILLOWS

PILOCARPINE

Pilot Advisory System, Automated
  USE   AUTOMATED PILOT ADVISORY SYSTEM

PILOT ERROR

PILOT INDUCED OSCILLATION

Pilot Landing Aid Television System
  USE   PLAT SYSTEM

PILOT PERFORMANCE

PILOT PLANTS

PILOT SELECTION

PILOT TRAINING

Piloted Centrifuges
  USE   HUMAN CENTRIFUGES

Piloted Vehicles, Remotely
  USE   REMOTELY PILOTED VEHICLES

PILOTLESS AIRCRAFT

PILOTS

Pilots, Aircraft
  USE   AIRCRAFT PILOTS

Pilots, Automatic
  USE   AUTOMATIC PILOTS

Pilots, Jet
  USE   AIRCRAFT PILOTS

PILOTS (PERSONNEL)

Pilots, Test
  USE   TEST PILOTS

PINCH EFFECT

Pinch, Plasma
  USE   PLASMA PINCH

Pinch, Reverse Field
  USE   REVERSE FIELD PINCH

Pinch, Screw
  USE   SCREW PINCH

Pinch, Theta
  USE   THETA PINCH

Pinch, Zeta
  USE   ZETA PINCH

PINEAL GLAND

PINHOLE CAMERAS

PINHOLES

Pinnacles
  USE   PEAKS (LANDFORMS)

PINNING

Pinning, Flux
  USE   FLUX PINNING

PINS

PINTLES

PION BEAMS

Pioneer F Space Probe
　　USE　PIONEER 10 SPACE PROBE

Pioneer G Space Probe
　　USE　PIONEER 11 SPACE PROBE

PIONEER PROJECT

Pioneer Saturn Spacecraft
　　USE　PIONEER 11 SPACE PROBE

PIONEER SPACE PROBES

PIONEER VENUS SPACECRAFT

PIONEER VENUS 1 SPACECRAFT

PIONEER VENUS 2 ENTRY PROBES

Pioneer Venus 2 Multiprobe Spacecraft
　　USE　PIONEER VENUS 2 SPACECRAFT

PIONEER VENUS 2 NIGHT PROBE

PIONEER VENUS 2 SOUNDER PROBE

PIONEER VENUS 2 SPACECRAFT

PIONEER VENUS 2 TRANSPORTER BUS

PIONEER 1 SPACE PROBE

PIONEER 2 SPACE PROBE

PIONEER 3 SPACE PROBE

Pioneer 4 Lunar Probe
　　USE　PIONEER 4 SPACE PROBE

PIONEER 4 SPACE PROBE

PIONEER 5 SPACE PROBE

PIONEER 6 SPACE PROBE

PIONEER 7 SPACE PROBE

PIONEER 8 SPACE PROBE

PIONEER 9 SPACE PROBE

PIONEER 10 SPACE PROBE

PIONEER 11 SPACE PROBE

Pioneer 12 Space Probe
　　USE　PIONEER VENUS SPACECRAFT

PIONS

PIPE FLOW

PIPE NOZZLES

PIPELINES

PIPELINING (COMPUTERS)

PIPER AIRCRAFT

PIPERIDINE

Pipes, Gas
　　USE　GAS PIPES

Pipes, Heat
　　USE　HEAT PIPES

PIPES (TUBES)

PIPETTES

Piracy, Air
　　USE　AIR PIRACY

PIRANI GAGES

PISTON ENGINES

PISTON THEORY

PISTONS

Pistons, Magnetic
　　USE　MAGNETIC PISTONS

PITCH

Pitch Angles
　　USE　PITCH (INCLINATION)

Pitch Attitude Control
　　USE　LONGITUDINAL CONTROL

Pitch, Damping In
　　USE　PITCH (INCLINATION)
　　　　　　DAMPING

PITCH (INCLINATION)

PITCH (MATERIAL)

Pitch Propellers, Variable
　　USE　VARIABLE PITCH PROPELLERS

PITCHING MOMENTS

PITOT TUBES

PITS

PITS (EXCAVATIONS)

PITTING

PITUITARY GLAND

PITUITARY HORMONES

Pivoted Wing Aircraft
　　USE　TILT WING AIRCRAFT

PIVOTS

Pix
　　USE　PLASMA INTERACTION EXPERIMENT

PL/1

Plages (Faculae)
　　USE　FACULAE

PLAINS

Plains, Coastal
　　USE　COASTAL PLAINS

Plains Corridor (North America), Great
　　USE　GREAT PLAINS CORRIDOR (NORTH
　　　　　　AMERICA)

Plains, Flood
　　USE　FLOOD PLAINS

Plains, Pene
　　USE　PENEPLAINS

Plan, Payload Integration
　　USE　PAYLOAD INTEGRATION PLAN

PLAN POSITION INDICATORS

PLANAR STRUCTURES

Planck Equation, Fokker-
　　USE　FOKKER-PLANCK EQUATION

PLANCKS CONSTANT

Plane Area Twin Hull, Small Water
　　USE　SWATH (SHIP)

Plane, Astro
　　USE　ASTROPLANE

PLANE STRAIN

PLANE WAVES

Planes, Aerospace
　　USE　AEROSPACEPLANES

Planes, Bi
　　USE　BIPLANES

Planes, Half
　　USE　HALF PLANES

Planes, Hyper
　　USE　HYPERPLANES

Planes, Mono
　　USE　MONOPLANES

Planes, Rocket
　　USE　ROCKET PLANES

Planes, Tail
　　USE　HORIZONTAL TAIL SURFACES

(Planet), Earth
　　USE　EARTH (PLANET)

PLANET EPHEMERIDES

(Planet), Jupiter
　　USE　JUPITER (PLANET)

(Planet), Mars
　　USE　MARS (PLANET)

(Planet), Mercury
　　USE　MERCURY (PLANET)

Planet Missions, Outer
　　USE　GRAND TOURS

(Planet), Neptune
　　USE　NEPTUNE (PLANET)

Planet Origins
　　USE　PLANETARY EVOLUTION

(Planet), Pluto
　　USE　PLUTO (PLANET)

(Planet), Saturn
　　USE　SATURN (PLANET)

Planet Spacecraft, Outer
　　USE　OUTER PLANETS EXPLORERS

Planet Spacecraft, Thermoelectric Outer
　　USE　TOPS (SPACECRAFT)

(Planet), Uranus
　　USE　URANUS (PLANET)

(Planet), Venus
　　USE　VENUS (PLANET)

Planet 1221, Minor
　　USE　AMOR ASTEROID

Planet 2060, Minor
　　USE　CHIRON

PLANETARIUMS

PLANETARY ATMOSPHERES

PLANETARY BASES

PLANETARY BOUNDARY LAYER

PLANETARY COMPOSITION

PLANETARY CORES

PLANETARY CRATERS

Planetary Entry
　　USE　ATMOSPHERIC ENTRY

PLANETARY ENVIRONMENTS

PLANETARY EVOLUTION

Planetary Exploration
　　USE　SPACE EXPLORATION

Planetary Explorer
　　USE　OUTER PLANETS EXPLORERS

PLANETARY GEOLOGY

PLANETARY GRAVITATION

Planetary Interactions, Solar
　　USE　SOLAR PLANETARY INTERACTIONS

PLANETARY LANDING

PLANETARY LIMB

PLANETARY MAGNETIC FIELDS

PLANETARY MANTLES

PLANETARY MAPPING

PLANETARY MASS

Planetary Motion
　　USE　SOLAR ORBITS

PLANETARY NEBULAE

PLANETARY ORBITS

PLANETARY QUAKES

PLANETARY QUARANTINE

PLANETARY RADIATION

PLANETARY RINGS

PLANETARY ROTATION

Planetary Satellites
　　USE　NATURAL SATELLITES

Planetary Space Flight
　　USE　INTERPLANETARY FLIGHT

Planetary Spacecraft
　　USE　INTERPLANETARY SPACECRAFT

PLANETARY STRUCTURE

Planetary Structure, Earth
　　USE　EARTH PLANETARY STRUCTURE

PLANETARY SURFACES

PLANETARY TEMPERATURE

PLANETARY WAVES

Planetismals
　　USE　PROTOPLANETS

PLANETOCENTRIC COORDINATES

PLANETOLOGY

PLANETS

Pianets Explorers, Outer
　　USE　OUTER PLANETS EXPLORERS

Planets, Extrasolar
　　USE　EXTRASOLAR PLANETS

Planets, Gas Giant
　　USE　GAS GIANT PLANETS

Planets, Proto
　　USE　PROTOPLANETS

Planets, Terrestrial
　　USE　TERRESTRIAL PLANETS

PLANFORMS

Planforms, Rectangular
　　USE　RECTANGULAR PLANFORMS

Planforms, Wing
　　USE　WING PLANFORMS

Planigraphy
　　USE　TOMOGRAPHY

PLANING

Planing, Hydro
　　USE　HYDROPLANING

PLANISPHERES

PLANKTON

Plankton Bloom
　　USE　PLANKTON

PLANNING

Planning, Airport
　　USE　AIRPORT PLANNING

Planning, Management
　　USE　MANAGEMENT PLANNING

Planning, Mission
　　USE　MISSION PLANNING

Planning, Production
　　USE　PRODUCTION PLANNING

Planning, Project
　　USE　PROJECT PLANNING

Planning, Regional
　　USE　REGIONAL PLANNING

Planning System, NASA Interactive
　　USE　NASA INTERACTIVE PLANNING SYSTEM

Planning, Urban
　　USE　URBAN PLANNING

PLANOTRONS

PLANS

Plans, Flight
　　USE　FLIGHT PLANS

PLANT DESIGN

Plant, Enrico Fermi Atomic Power
　　USE　ENRICO FERMI ATOMIC POWER PLANT

Plant, ML-1 Nuclear Power
　　USE　ML-1 NUCLEAR POWER PLANT

PLANT ROOTS

PLANT STRESS

PLANTAR TISSUES

PLANTING

Plants, Aquatic
　　USE　AQUATIC PLANTS

PLANTS (BOTANY)

Plants, Electric Power
　　USE　ELECTRIC POWER PLANTS

Plants, Fuel Cell Power
　　USE　FUEL CELL POWER PLANTS

Plants, Industrial
　　USE　INDUSTRIAL PLANTS

Plants (Industries)
　　USE　INDUSTRIAL PLANTS

Plants, Leguminous
　　USE　LEGUMINOUS PLANTS

Plants, Nuclear Power
　　USE　NUCLEAR POWER PLANTS

Plants, Photophilic
　　USE　PHOTOPHILIC PLANTS

Plants, Pilot
　　USE　PILOT PLANTS

Plants, Power
　　USE　POWER PLANTS

(Plants), Reeds
　　USE　REEDS (PLANTS)

Plants, Solar Sea Power
　　USE　SOLAR SEA POWER PLANTS

Plants, Solar Thermal Electric Power
　　USE　SOLAR THERMAL ELECTRIC POWER
　　　　　PLANTS

Plants, Thermophilic
　　USE　THERMOPHILIC PLANTS

(Plants), Trees
　　USE　TREES (PLANTS)

PLASMA ACCELERATION

Plasma Accelerator, Cyclops
　　USE　CYCLOPS PLASMA ACCELERATOR

PLASMA ACCELERATORS

Plasma Accelerators, Coaxial
　　USE　COAXIAL PLASMA ACCELERATORS

Plasma Amplifiers, Beam
　　USE　BEAM PLASMA AMPLIFIERS

PLASMA ANTENNAS

PLASMA ARC CUTTING

Plasma Arc Spraying
　　USE　ARC SPRAYING

PLASMA ARC WELDING

Plasma Arcs
　　USE　PLASMA JETS

Plasma, Argon
　　USE　ARGON PLASMA

Plasma Avalanche Triggered Transit, Trapped
　　USE　TRAPATT DEVICES

Plasma, Blood
　　USE　BLOOD PLASMA

PLASMA BUBBLES

Plasma, Cesium
　　USE　CESIUM PLASMA

PLASMA CHEMISTRY

PLASMA CLOUDS

PLASMA COMPOSITION

PLASMA COMPRESSION

PLASMA CONDUCTIVITY

Plasma Confinement
　　USE　PLASMA CONTROL

PLASMA CONTROL

PLASMA COOLING

PLASMA CORE REACTORS

Plasma, Cosmic
　　USE　COSMIC PLASMA

PLASMA CURRENTS

PLASMA CYLINDERS

PLASMA DECAY

PLASMA DENSITY

Plasma, Deuterium
　　USE　DEUTERIUM PLASMA

Plasma Devices, Alpha
USE ALPHA PLASMA DEVICES

PLASMA DIAGNOSTICS

PLASMA DIFFUSION

PLASMA DIODES

Plasma Discharges
USE PLASMA JETS

Plasma Dispersion
USE PLASMA DIFFUSION

PLASMA DISPLAY DEVICES

PLASMA DRIFT

PLASMA DYNAMICS

PLASMA ELECTRODES

Plasma, Electron
USE ELECTRON PLASMA

Plasma, Electrostatic
USE PLASMAS (PHYSICS)

PLASMA ENGINES

Plasma Engines, Two Stage
USE TWO STAGE PLASMA ENGINES

PLASMA EQUILIBRIUM

PLASMA ETCHING

Plasma Flow
USE MAGNETOHYDRODYNAMIC FLOW

PLASMA FLUX MEASUREMENT

PLASMA FOCUS

PLASMA FREQUENCIES

Plasma Generation
USE PLASMA GENERATORS

PLASMA GENERATORS

PLASMA GUNS

Plasma H/v Interaction Experiments, Space
USE SPHINX

PLASMA HEATING

Plasma, Helium
USE HELIUM PLASMA

Plasma, Hydrogen
USE HYDROGEN PLASMA

Plasma Instability
USE MAGNETOHYDRODYNAMIC STABILITY

PLASMA INTERACTION EXPERIMENT

PLASMA INTERACTIONS

Plasma Interactions, Laser
USE LASER PLASMA INTERACTIONS

PLASMA JET SYNTHESIS

PLASMA JET WIND TUNNELS

PLASMA JETS

PLASMA LAYERS

PLASMA LIFETIME

PLASMA LOSS

Plasma, Magnetoionic
USE PLASMAS (PHYSICS)

Plasma, Nitrogen
USE NITROGEN PLASMA

PLASMA OSCILLATIONS

Plasma, Oxygen
USE OXYGEN PLASMA

Plasma Perturbation
USE PLASMA OSCILLATIONS

PLASMA PHYSICS

(Plasma Physics), Electron Runaway
USE ELECTRON RUNAWAY (PLASMA PHYSICS)

(Plasma Physics), Rigid Rotors
USE RIGID ROTORS (PLASMA PHYSICS)

PLASMA PINCH

PLASMA POTENTIALS

PLASMA POWER SOURCES

PLASMA PROBES

Plasma Probes, Microwave
USE MICROWAVE PLASMA PROBES

PLASMA PROPULSION

PLASMA PUMPING

PLASMA RADIATION

Plasma (Radiation), Solar
USE SOLAR WIND

Plasma Renin Activity
USE IMMUNOASSAY

PLASMA RESONANCE

Plasma Rings
USE TOROIDAL PLASMAS

PLASMA SHEATHS

PLASMA SLABS

Plasma Sound Waves
USE PLASMA WAVES
MAGNETOHYDRODYNAMIC WAVES

PLASMA SPECTRA

PLASMA SPRAYING

Plasma Stability
USE MAGNETOHYDRODYNAMIC STABILITY

PLASMA TEMPERATURE

Plasma Theory
USE PLASMA PHYSICS

PLASMA TORCHES

PLASMA TURBULENCE

PLASMA WAVES

PLASMA-ELECTROMAGNETIC INTERACTION

PLASMA-PARTICLE INTERACTIONS

PLASMADYNAMIC LASERS

PLASMAGUIDES

PLASMAPAUSE

Plasmas, Boundary Layer
USE BOUNDARY LAYER PLASMAS

Plasmas, Cold
USE COLD PLASMAS

Plasmas, Collisional
USE COLLISIONAL PLASMAS

Plasmas, Collisionless
USE COLLISIONLESS PLASMAS

Plasmas, Cylindrical
USE CYLINDRICAL PLASMAS

Plasmas, Dense
USE DENSE PLASMAS

Plasmas, Elliptical
USE ELLIPTICAL PLASMAS

Plasmas, High Temperature
USE HIGH TEMPERATURE PLASMAS

Plasmas, Hot
USE HIGH TEMPERATURE PLASMAS

Plasmas In Earth Neighborhood, Origin Of
USE OPEN PROJECT

Plasmas, Ionized
USE PLASMAS (PHYSICS)

Plasmas, Laser
USE LASER PLASMAS

Plasmas, Low Temperature
USE COLD PLASMAS

Plasmas, Metallic
USE METALLIC PLASMAS

Plasmas, Micro
USE MICROPLASMAS

Plasmas, Nonequilibrium
USE NONEQUILIBRIUM PLASMAS

Plasmas, Nonuniform
USE NONUNIFORM PLASMAS

PLASMAS (PHYSICS)

Plasmas, Rarefied
USE RAREFIED PLASMAS

Plasmas, Relativistic
USE RELATIVISTIC PLASMAS

Plasmas, Rotating
USE ROTATING PLASMAS

Plasmas, Semiconductor
USE SEMICONDUCTOR PLASMAS

Plasmas, Space
USE SPACE PLASMAS

Plasmas, Spherical
USE SPHERICAL PLASMAS

Plasmas, Strongly Coupled
USE STRONGLY COUPLED PLASMAS

(Plasmas), Tearing Modes
USE TEARING MODES (PLASMAS)

Plasmas, Thermal
USE THERMAL PLASMAS

Plasmas, Toroidal
USE TOROIDAL PLASMAS

Plasmas, Uranium
USE URANIUM PLASMAS

Plasmas-IN-Space Payload
USE AMPS (SATELLITE PAYLOAD)

PLASMASPHERE

PLASMATRONS

Plasmatrons, Duo
USE DUOPLASMATRONS

Plasmoids
  USE   PLASMAS (PHYSICS)

PLASMOLYSIS

PLASMONS

PLASTERS

PLASTIC AIRCRAFT STRUCTURES

PLASTIC ANISOTROPY

PLASTIC COATINGS

PLASTIC DEFORMATION

Plastic Films
  USE   POLYMERIC FILMS

PLASTIC FLOW

Plastic Materials
  USE   PLASTICS

PLASTIC MEMORY

PLASTIC PROPELLANTS

PLASTIC PROPERTIES

PLASTIC TAPES

Plastic Yielding
  USE   PLASTIC DEFORMATION

Plasticity
  USE   PLASTIC PROPERTIES

Plasticity, Elasto
  USE   ELASTOPLASTICITY

Plasticity, Photo
  USE   PHOTOPLASTICITY

Plasticity, Super
  USE   SUPERPLASTICITY

Plasticity, Thermo
  USE   THERMOPLASTICITY

Plasticity, Visco
  USE   VISCOPLASTICITY

PLASTICIZERS

PLASTICS

Plastics, Carbon Fiber Reinforced
  USE   CARBON FIBER REINFORCED PLASTICS

Plastics, Glass Fiber Reinforced
  USE   GLASS FIBER REINFORCED PLASTICS

Plastics, Reinforced
  USE   REINFORCED PLASTICS

Plastics, Thio
  USE   THIOPLASTICS

PLASTISOLS

PLAT SYSTEM

Plate, Boiler
  USE   BOILER PLATE

Plate, Gold
  USE   GOLD COATINGS

Plate (Metal)
  USE   METAL PLATES

Plate, Nickel
  USE   NICKEL PLATE

PLATE THEORY

Plateau (US), Allegheny
  USE   ALLEGHENY PLATEAU (US)

Plateau (US), Colorado
  USE   COLORADO PLATEAU (US)

PLATEAUS

PLATELETS

PLATENS

PLATES

Plates, Anisotropic
  USE   ANISOTROPIC PLATES

Plates, Annular
  USE   ANNULAR PLATES

Plates, Cantilever
  USE   CANTILEVER PLATES

Plates, Circular
  USE   CIRCULAR PLATES

Plates, Corrugated
  USE   CORRUGATED PLATES

Plates, Elastic
  USE   ELASTIC PLATES

Plates, End
  USE   END PLATES

Plates, Flat
  USE   FLAT PLATES

Plates, Metal
  USE   METAL PLATES

Plates, Microchannel
  USE   MICROCHANNEL PLATES

Plates, Multichannel
  USE   MICROCHANNEL PLATES

Plates, Nonisotropic
  USE   ANISOTROPIC PLATES

Plates (Optics), Scatter
  USE   SCATTER PLATES (OPTICS)

Plates, Orthotropic
  USE   ORTHOTROPIC PLATES

Plates, Parallel
  USE   PARALLEL PLATES

Plates, Perforated
  USE   PERFORATED PLATES

Plates, Photographic
  USE   PHOTOGRAPHIC PLATES

Plates, Porous
  USE   POROUS PLATES

Plates, Rectangular
  USE   RECTANGULAR PLATES

Plates, Reinforced
  USE   REINFORCED PLATES

PLATES (STRUCTURAL MEMBERS)

PLATES (TECTONICS)

Plates, Thick
  USE   THICK PLATES

Plates, Thin
  USE   THIN PLATES

Platform, Interplanetary Monitoring
  USE   IMP

Platform Stability, Flying
  USE   AERODYNAMIC STABILITY
             FLYING PLATFORMS

PLATFORMS

Platforms, Data Collection
  USE   DATA COLLECTION PLATFORMS

Platforms, Flying
  USE   FLYING PLATFORMS

Platforms, Geostationary
  USE   SYNCHRONOUS PLATFORMS

Platforms, Inertial
  USE   INERTIAL PLATFORMS

Platforms, Ocean Data
  USE   OCEAN DATA ACQUISITIONS SYSTEMS

Platforms, Offshore
  USE   OFFSHORE PLATFORMS

Platforms, Space
  USE   SPACE PLATFORMS

Platforms), SPAS (ESA
  USE   SHUTTLE PALLET SATELLITES

Platforms, Stabilized
  USE   STABILIZED PLATFORMS

Platforms, Synchronous
  USE   SYNCHRONOUS PLATFORMS

PLATING

Plating, Electro
  USE   ELECTROPLATING

Plating, Flame
  USE   FLAME PLATING

Plating, Ion
  USE   ION PLATING

PLATINUM

PLATINUM ALLOYS

PLATINUM BLACK

PLATINUM COMPOUNDS

PLATINUM ISOTOPES

PLATINUM OXIDES

PLAYAS

PLAYBACKS

PLENUM CHAMBERS

PLETHYSMOGRAPHY

Plethysmography, Electro
  USE   ELECTROPLETHYSMOGRAPHY

PLEURAE

PLEUROTIN

Plexiglass (Trademark)
  USE   POLYMETHYL METHACRYLATE

Plies
  USE   LAYERS

PLOTS

PLOTTERS

Plotters, X-Y
  USE   X-Y PLOTTERS

PLOTTING

Plotting Instruments
  USE   PLOTTERS

Plowed Fields
  USE   FARMLANDS

PLOWING

PLOWS

PLSS
USE   PORTABLE LIFE SUPPORT SYSTEMS

PLUG NOZZLES

PLUGGING

PLUGS

Plugs, Spark
USE   SPARK PLUGS

PLUM BROOK REACTOR

PLUMAGE

Plumbane
USE   LEAD COMPOUNDS
       METAL HYDRIDES

PLUMES

PLUNGERS

Plus Solid Zones, Liquid
USE   MUSHY ZONES

PLUTO (PLANET)

PLUTO REACTORS

PLUTONIUM

PLUTONIUM ALLOYS

Plutonium Carbides
USE   PLUTONIUM COMPOUNDS

PLUTONIUM COMPOUNDS

PLUTONIUM FLUORIDES

PLUTONIUM ISOTOPES

PLUTONIUM OXIDES

Plutonium Reactor, Los Alamos Molten
USE   LOS ALAMOS MOLTEN PLUTONIUM
       REACTOR

PLUTONIUM RECYCLE TEST REACTOR

PLUTONIUM 238

PLUTONIUM 239

PLUTONIUM 240

PLUTONIUM 241

PLUTONIUM 244

Pluviographs
USE   RECORDING INSTRUMENTS
       RAIN GAGES

PLY ORIENTATION

PLYWOOD

Pm
USE   PROMETHIUM

PNEUMATIC CIRCUITS

PNEUMATIC CONTROL

PNEUMATIC EQUIPMENT

PNEUMATIC PROBES

Pneumatic Reset
USE   PNEUMATIC CONTROL

PNEUMATICS

Pneumographs
USE   PNEUMOGRAPHY

PNEUMOGRAPHY

PNEUMONIA

PNEUMOTHORAX

Pnictides
USE   GROUP 5A COMPOUNDS

Po
USE   POLONIUM

Pockels Effect
USE   BIREFRINGENCE

POCKET MICE

Pockets, Gas
USE   GAS POCKETS

PODS (EXTERNAL STORES)

POGO

POGO EFFECTS

POHLHAUSEN METHOD

Pohlhausen Solution
USE   POHLHAUSEN METHOD

POIKILOTHERMIA

POINCARE PROBLEM

POINCARE SPHERES

Point Arithmetic, Fixed
USE   FIXED POINT ARITHMETIC

Point Arithmetic, Floating
USE   FLOATING POINT ARITHMETIC

Point Communication, Point To
USE   POINT TO POINT COMMUNICATION

Point, Critical
USE   CRITICAL POINT

POINT DEFECTS

Point, Dew
USE   DEW POINT

Point Energy, Zero
USE   ZERO POINT ENERGY

Point, Fire
USE   FIRE POINT

Point, Flash
USE   FLASH POINT

POINT IMPACT

Point Matching Method (Mathematics)
USE   BOUNDARY VALUE PROBLEMS

Point, Mirror
USE   MIRROR POINT

POINT SOURCES

POINT SPREAD FUNCTIONS

Point, Stagnation
USE   STAGNATION POINT

POINT TO POINT COMMUNICATION

Point, Yield
USE   YIELD POINT

Pointers
USE   DIALS

POINTING CONTROL SYSTEMS

Pointing System, Annular Suspension And
USE   ANNULAR SUSPENSION AND POINTING
       SYSTEM

POINTS

Points, Conjugate
USE   CONJUGATE POINTS

Points, Freezing
USE   MELTING POINTS

Points (Game Theory), Saddle
USE   SADDLE POINTS (GAME THEORY)

Points, Inflection
USE   INFLECTION POINTS

Points, Lagrangian Equilibrium
USE   LAGRANGIAN EQUILIBRIUM POINTS

POINTS (MATHEMATICS)

Points (Mathematics), Fixed
USE   FIXED POINTS (MATHEMATICS)

Points, Melting
USE   MELTING POINTS

Points, Saddle
USE   SADDLE POINTS

Points, Transition
USE   TRANSITION POINTS

Poiseuille Flow
USE   LAMINAR FLOW

POISONING

Poisoning, Benzene
USE   BENZENE POISONING

Poisoning, Beryllium
USE   BERYLLIUM POISONING

Poisoning, Carbon Monoxide
USE   CARBON MONOXIDE POISONING

Poisoning, Carbon Tetrachloride
USE   CARBON TETRACHLORIDE POISONING

Poisoning, Hydrocarbon
USE   HYDROCARBON POISONING

Poisoning, Lead
USE   LEAD POISONING

POISONING (REACTION INHIBITION)

Poisoning (Toxicology)
USE   TOXIC DISEASES

POISONS

POISSON DENSITY FUNCTIONS

POISSON EQUATION

Poisson Process
USE   POISSON DENSITY FUNCTIONS
       STOCHASTIC PROCESSES

POISSON RATIO

Polaire Satellite
USE   D-2 SATELLITES

POLAND

Polar Auroras
USE   AURORAS

POLAR CAP ABSORPTION

POLAR CAPS

POLAR COORDINATES

POLAR CUSPS

POLAR GASES

Polar Ionosphere Beacon
USE   BEACON SATELLITES

POLAR METEOROLOGY

Polar Mission, International Solar
  USE   ULYSSES MISSION

POLAR NAVIGATION

Polar Orbit Geophysical Observatory
  USE   POGO

POLAR ORBITS

POLAR RADIO BLACKOUT

POLAR REGIONS

Polar SPUR (Astronomy), North
  USE   NORTH POLAR SPUR (ASTRONOMY)

POLAR SUBSTORMS

POLAR WANDERING (GEOLOGY)

POLARIMETERS

POLARIMETRY

POLARIS A1 MISSILE

POLARIS A2 MISSILE

POLARIS A3 MISSILE

POLARIS MISSILES

Polaris Submarines
  USE   GUIDED MISSILE SUBMARINES

POLARISCOPES

Polariscopes, Senarmont
  USE   SENARMONT POLARISCOPES

POLARITONS

POLARITY

POLARIZATION

POLARIZATION CHARACTERISTICS

POLARIZATION (CHARGE SEPARATION)

Polarization Charts
  USE   POLARIZATION (WAVES)
         GRAPHS (CHARTS)

Polarization, Circular
  USE   CIRCULAR POLARIZATION

Polarization, Cross
  USE   CROSS POLARIZATION

Polarization, De
  USE   DEPOLARIZATION

Polarization, Dielectric
  USE   DIELECTRIC POLARIZATION

Polarization, Electrolytic
  USE   ELECTROLYTIC POLARIZATION

Polarization, Elliptical
  USE   ELLIPTICAL POLARIZATION

Polarization, Linear
  USE   LINEAR POLARIZATION

Polarization, Optical
  USE   OPTICAL POLARIZATION

POLARIZATION (SPIN ALIGNMENT)

POLARIZATION (WAVES)

POLARIZED ELASTIC WAVES

POLARIZED ELECTROMAGNETIC RADIATION

POLARIZED LIGHT

POLARIZED RADIATION

POLARIZERS

Polarographs
  USE   POLAROGRAPHY

POLAROGRAPHY

POLARONS

POLES

Poles, Di
  USE   DIPOLES

Poles, Magnetic
  USE   MAGNETIC POLES

Poles, Mono
  USE   MONOPOLES

Poles, Multi
  USE   MULTIPOLES

Poles, Regge
  USE   REGGE POLES

POLES (SUPPORTS)

POLICE

POLICIES

Policy, Energy
  USE   ENERGY POLICY

Policy, Foreign
  USE   FOREIGN POLICY

Policy, Patent
  USE   PATENT POLICY

Policy, Procurement
  USE   PROCUREMENT POLICY

POLIOMYELITIS

Polish TS-11 Aircraft
  USE   TS-11 AIRCRAFT

Polished Metals
  USE   METAL POLISHING

POLISHING

Polishing, Electro
  USE   ELECTROPOLISHING

Polishing, Electrolytic
  USE   ELECTROPOLISHING

Polishing, Metal
  USE   METAL POLISHING

Polishing, Vibratory
  USE   VIBRATORY POLISHING

POLITICS

POLLEN

Pollutants
  USE   CONTAMINANTS

POLLUTION

Pollution, Air
  USE   AIR POLLUTION

POLLUTION CONTROL

Pollution, Environment
  USE   ENVIRONMENT POLLUTION

Pollution, Global Air
  USE   GLOBAL AIR POLLUTION

Pollution, Indoor Air
  USE   INDOOR AIR POLLUTION

POLLUTION MONITORING

Pollution, Noise
  USE   NOISE POLLUTION

Pollution, Oil
  USE   OIL POLLUTION

Pollution, Thermal
  USE   THERMAL POLLUTION

POLLUTION TRANSPORT

Pollution, Water
  USE   WATER POLLUTION

POLOIDAL FLUX

POLONIUM

POLONIUM COMPOUNDS

POLONIUM ISOTOPES

POLONIUM 208

POLONIUM 209

POLONIUM 210

POLYACETYLENE

Polyacrylates
  USE   ACRYLIC RESINS

POLYAMIDE RESINS

POLYATOMIC GASES

POLYATOMIC MOLECULES

POLYBENZIMIDAZOLE

POLYBROMINATED BIPHENYLS

POLYBUTADIENE

POLYBUTADIENE TETRANITRAMINE

POLYCARBONATES

POLYCHLORINATED BIPHENYLS

POLYCRYSTALS

POLYCYTHEMIA

POLYESTER RESINS

POLYESTERS

POLYETHER RESINS

POLYETHYLENE TEREPHTHALATE

POLYETHYLENES

POLYGONIZATION

POLYGONS

POLYHEDRONS

Polyimide Composites, Graphite-
  USE   GRAPHITE-POLYIMIDE COMPOSITES

POLYIMIDE RESINS

POLYIMIDES

POLYISOBUTYLENE

POLYISOPRENES

POLYMER CHEMISTRY

POLYMER MATRIX COMPOSITES

Polymer, Metallosiloxane
  USE   METALLOSILOXANE POLYMER

Polymer, Metalloxane
    USE   METALLOXANE POLYMER

POLYMER PHYSICS

POLYMERIC FILMS

POLYMERIZATION

Polymerization, Co
    USE   COPOLYMERIZATION

Polymerization, De
    USE   DEPOLYMERIZATION

POLYMERS

Polymers, Co
    USE   COPOLYMERS

Polymers, Coordination
    USE   COORDINATION POLYMERS

Polymers, Fluoro
    USE   FLUOROPOLYMERS

Polymers, High
    USE   HIGH POLYMERS

Polymers, Nitrogen
    USE   NITROGEN POLYMERS

Polymers, Organometallic
    USE   ORGANOMETALLIC POLYMERS

Polymers, Phosphorus
    USE   PHOSPHORUS POLYMERS

Polymers, Pre
    USE   PREPOLYMERS

Polymers, Silicon
    USE   SILICON POLYMERS

Polymers, Vinyl
    USE   VINYL POLYMERS

POLYMETHYL METHACRYLATE

POLYMORPHISM

Polynomial, Hermitian
    USE   HERMITIAN POLYNOMIAL

POLYNOMIALS

Polynomials, Jacobi
    USE   HYPERGEOMETRIC FUNCTIONS

Polynomials, Legendre
    USE   LEGENDRE FUNCTIONS

POLYNUCLEAR ORGANIC COMPOUNDS

POLYNUCLEOTIDES

POLYOT SATELLITES

POLYPEPTIDES

POLYPHENYL ETHER

POLYPHENYLS

POLYPROPYLENE

POLYQUINOXALINES

POLYSACCHARIDES

Polysiloxane, Methyl
    USE   METHYL POLYSILOXANE

POLYSLIPS

POLYSTATION DOPPLER TRACKING SYSTEM

POLYSTYRENE

POLYSULFIDES

POLYTETRAFLUOROETHYLENE

POLYTOPES

POLYTROPIC PROCESSES

POLYURETHANE FOAM

POLYURETHANE RESINS

POLYVINYL ALCOHOL

POLYVINYL CHLORIDE

POLYVINYL FLUORIDE

POLYWATER

POMERANCHUK THEOREM

POMERONS

PONDEROMOTIVE FORCES

PONDS

Ponds (Heat Storage), Solar
    USE   SOLAR PONDS (HEAT STORAGE)

Pontchartrain (LA), Lake
    USE   LAKE PONTCHARTRAIN (LA)

PONTIAC (MI)

PONTRYAGIN PRINCIPLE

Pool Reactors, Swimming
    USE   SWIMMING POOL REACTORS

Pool Type Reactor, Livermore
    USE   LIVERMORE POOL TYPE REACTOR

POPULATION INVERSION

POPULATION THEORY

POPULATIONS

PORCELAIN

Pores
    USE   POROSITY

POROSITY

Porosity, Micro
    USE   MICROPOROSITY

POROUS BOUNDARY LAYER CONTROL

POROUS MATERIALS

POROUS PLATES

POROUS WALLS

PORPHINES

PORPHYRA

PORPHYRINS

PORPOISES

PORTABLE EQUIPMENT

PORTABLE LIFE SUPPORT SYSTEMS

PORTS

Ports, Air
    USE   AIRPORTS

Ports, Heli
    USE   HELIPORTS

PORTS (OPENINGS)

PORTUGAL

POSEIDON MISSILES

POSEIDON SATELLITE

POSITION

POSITION ERRORS

Position Estimation, Orbital
    USE   ORBITAL POSITION ESTIMATION

POSITION INDICATORS

Position Indicators, Plan
    USE   PLAN POSITION INDICATORS

(Position Indicators), PPI
    USE   PLAN POSITION INDICATORS

Position Indicators, Spacecraft
    USE   SPACECRAFT POSITION INDICATORS

POSITION (LOCATION)

Position Modulation, Pulse
    USE   PULSE POSITION MODULATION

Position, Prone
    USE   PRONE POSITION

POSITION SENSING

Position, Sitting
    USE   SITTING POSITION

Position, Solar
    USE   SOLAR POSITION

Position, Supine
    USE   SUPINE POSITION

POSITION (TITLE)

(Position), Tracking
    USE   TRACKING (POSITION)

POSITIONING

POSITIONING DEVICES (MACHINERY)

Positioning, Satellite Doppler
    USE   SATELLITE DOPPLER POSITIONING

Positioning System, Global
    USE   GLOBAL POSITIONING SYSTEM

POSITIVE FEEDBACK

POSITIVE IONS

POSITRON ANNIHILATION

POSITRONIUM

POSITRONS

Post, Advanced Airborne Command
    USE   E-4A AIRCRAFT

POST BOOST PROPULSION SYSTEM

POST-BLAST NUCLEAR RADIATION

POSTAMPLIFIERS

POSTERIOR SECTIONS

POSTFLIGHT ANALYSIS

POSTLAUNCH REPORTS

POSTMISSION ANALYSIS (SPACECRAFT)

Postulates
    USE   AXIOMS

POSTURE

POTABLE LIQUIDS

POTABLE WATER

POTASSIUM

POTASSIUM ALLOYS

POTASSIUM BROMIDES

POTASSIUM CHLORIDES

POTASSIUM CHROMATES

POTASSIUM COMPOUNDS

POTASSIUM HYDRIDES

POTASSIUM HYDROXIDES

POTASSIUM IODIDES

POTASSIUM ISOTOPES

Potassium, Liquid
   USE   LIQUID POTASSIUM

POTASSIUM NITRATES

POTASSIUM OXIDES

POTASSIUM PERCHLORATES

POTASSIUM PEROXIDES

POTASSIUM PHOSPHATES

POTASSIUM SILICATES

POTASSIUM 38

POTASSIUM 39

POTASSIUM 40

POTATOES

POTENTIAL

Potential, Bioelectric
   USE   BIOELECTRIC POTENTIAL

Potential, Coulomb
   USE   COULOMB POTENTIAL

Potential, Electric
   USE   ELECTRIC POTENTIAL

POTENTIAL ENERGY

POTENTIAL FIELDS

POTENTIAL FLOW

Potential, Geo
   USE   GEOPOTENTIAL

POTENTIAL GRADIENTS

Potential, Gravitational
   USE   GRAVITATIONAL FIELDS

Potential, Klein-Dunham
   USE   KLEIN-DUNHAM POTENTIAL

Potential, Lennard-Jones
   USE   LENNARD-JONES POTENTIAL

Potential, Lienard
   USE   LIENARD POTENTIAL

Potential, Morse
   USE   MORSE POTENTIAL

Potential, Nuclear
   USE   NUCLEAR POTENTIAL

Potential, Nucleon
   USE   NUCLEON POTENTIAL

POTENTIAL THEORY

Potential, Yukawa
   USE   YUKAWA POTENTIAL

Potentials, Contact
   USE   CONTACT POTENTIALS

Potentials, Equi
   USE   EQUIPOTENTIALS

Potentials, Ionization
   USE   IONIZATION POTENTIALS

Potentials, Myoelectric
   USE   MYOELECTRIC POTENTIALS

Potentials, Plasma
   USE   PLASMA POTENTIALS

Potentials, Pseudo
   USE   PSEUDOPOTENTIALS

Potentials, Spike
   USE   SPIKE POTENTIALS

POTENTIOMETERS

POTENTIOMETERS (INSTRUMENTS)

POTENTIOMETERS (RESISTORS)

POTENTIOMETRIC ANALYSIS

Potentiometry
   USE   POTENTIOMETRIC ANALYSIS

POTEZ AIRCRAFT

POTOMAC RIVER VALLEY (MD-VA-WV)

POTTING COMPOUNDS

POURING

Powder, Metal
   USE   METAL POWDER

POWDER METALLURGY

POWDER (PARTICLES)

Powder, Sintered Aluminum
   USE   SINTERED ALUMINUM POWDER

POWDERED ALUMINUM

Powdered Metals
   USE   METAL POWDER

POWER

POWER AMPLIFIERS

POWER CONDITIONING

Power Conversion, Electric
   USE   ELECTRIC GENERATORS

POWER CONVERTERS

Power Density (Electromagnetic)
   USE   RADIANT FLUX DENSITY

POWER EFFICIENCY

Power, Electric
   USE   ELECTRIC POWER

Power Facility, Hallam Nuclear
   USE   HALLAM NUCLEAR POWER FACILITY

Power Facility), HNPF (Hallam Nuclear
   USE   HALLAM NUCLEAR POWER FACILITY

POWER FACTOR CONTROLLERS

Power, Fluid
   USE   FLUID POWER

POWER GAIN

Power Generation, Combined Cycle
   USE   COMBINED CYCLE POWER GENERATION

Power Generation, Nuclear
   USE   NUCLEAR ELECTRIC POWER GENERATION

Power Generation, Nuclear Electric
   USE   NUCLEAR ELECTRIC POWER GENERATION

Power Generation, Solar
   USE   SOLAR GENERATORS

Power Generation, Thermionic
   USE   THERMIONIC POWER GENERATION

Power Generation, Thermoelectric
   USE   THERMOELECTRIC POWER GENERATION

Power Generation, Thermonuclear
   USE   THERMONUCLEAR POWER GENERATION

Power Generators
   USE   ELECTRIC GENERATORS

Power Generators, Direct
   USE   DIRECT POWER GENERATORS

Power, Horse
   USE   HORSEPOWER

Power Lasers, High
   USE   HIGH POWER LASERS

POWER LIMITED SPACECRAFT

POWER LIMITERS

POWER LINES

POWER LOSS

POWER MODULES (STS)

Power Plant, Enrico Fermi Atomic
   USE   ENRICO FERMI ATOMIC POWER PLANT

Power Plant, ML-1 Nuclear
   USE   ML-1 NUCLEAR POWER PLANT

POWER PLANTS

Power Plants, Electric
   USE   ELECTRIC POWER PLANTS

Power Plants, Fuel Cell
   USE   FUEL CELL POWER PLANTS

Power Plants, Nuclear
   USE   NUCLEAR POWER PLANTS

Power Plants, Solar Sea
   USE   SOLAR SEA POWER PLANTS

Power Plants, Solar Thermal Electric
   USE   SOLAR THERMAL ELECTRIC POWER
              PLANTS

Power Processing Systems
   USE   POWER CONDITIONING

Power Reactor 2, Zero
   USE   ZERO POWER REACTOR 2

Power Reactor 3, Zero
   USE   ZERO POWER REACTOR 3

Power Reactor 6, Zero
   USE   ZERO POWER REACTOR 6

Power Reactor 9, Zero
   USE   ZERO POWER REACTOR 9

POWER REACTORS

Power Reactors, Nuclear
   USE   NUCLEAR POWER REACTORS

Power Reactors, Space
   USE   SPACE POWER REACTORS

Power Reactors, Zero
USE   ZERO-POWER REACTORS

Power, Resolving
USE   RESOLUTION

Power Satellites, Solar
USE   SOLAR POWER SATELLITES

POWER SERIES

Power Sources, Aircraft
USE   AIRCRAFT ENGINES

Power Sources, Auxiliary
USE   AUXILIARY POWER SOURCES

Power Sources, Plasma
USE   PLASMA POWER SOURCES

Power Sources, Solar
USE   SOLAR GENERATORS

POWER SPECTRA

Power Stations, Hydroelectric
USE   HYDROELECTRIC POWER STATIONS

Power Stations, Satellite Solar
USE   SATELLITE SOLAR POWER STATIONS

Power, Stopping
USE   STOPPING POWER

POWER SUPPLIES

Power Supplies, Aircraft
USE   AIRCRAFT POWER SUPPLIES

Power Supplies, Electric
USE   ELECTRIC POWER SUPPLIES

Power Supplies, Spacecraft
USE   SPACECRAFT POWER SUPPLIES

POWER SUPPLY CIRCUITS

Power System, Sunflower
USE   SUNFLOWER POWER SYSTEM

Power, Systems For Nuclear Auxiliary
USE   SNAP

Power, Thermal
USE   TURBOGENERATORS

Power, Thrust
USE   THRUST

Power, Tide
USE   TIDEPOWER

POWER TRANSMISSION

Power Transmission, Electric
USE   ELECTRIC POWER TRANSMISSION

POWER TRANSMISSION (LASERS)

Power Transmission, Superconducting
USE   SUPERCONDUCTING POWER
          TRANSMISSION

Power Transmission (To Earth), Satellite
USE   SATELLITE POWER TRANSMISSION (TO
          EARTH)

Power Unit Reactors, Space
USE   SPACE POWER UNIT REACTORS

Power Units, Chemical Auxiliary
USE   CHEMICAL AUXILIARY POWER UNITS

Power Units, Nuclear Auxiliary
USE   NUCLEAR AUXILIARY POWER UNITS

Power Units, Solar Auxiliary
USE   SOLAR AUXILIARY POWER UNITS

Powered Aircraft, Man
USE   MAN POWERED AIRCRAFT

Powered Aircraft, Solar
USE   SOLAR POWERED AIRCRAFT

Powered Generators, Tide
USE   TIDE POWERED GENERATORS

POWERED LIFT AIRCRAFT

Powered Machines, Tide
USE   TIDE POWERED MACHINES

Powered Machines, Waterwave
USE   WATERWAVE POWERED MACHINES

POWERED MODELS

Powered Ships, Nuclear
USE   NUCLEAR POWERED SHIPS

Powered Vehicles, Roadway
USE   ROADWAY POWERED VEHICLES

POYNTING THEOREM

POYNTING-ROBERTSON EFFECT

PPI (Position Indicators)
USE   PLAN POSITION INDICATORS

PPM (Modulation)
USE   PULSE POSITION MODULATION

Pr
USE   PRASEODYMIUM

PR
USE   PUERTO RICO

Practical Temperature, International
USE   TEMPERATURE SCALES

Practices
USE   PROCEDURES

PRAESEPE STAR CLUSTERS

PRAETERSONIC DEVICES

Prairies
USE   GRASSLANDS

PRANDTL NUMBER

PRANDTL-MEYER EXPANSION

PRASEODYMIUM

PRASEODYMIUM ISOTOPES

Praseodymium 144
USE   PRASEODYMIUM ISOTOPES

PRE-IMBRIAN PERIOD

PRE-MAIN SEQUENCE STARS

PREAMPLIFIERS

PREBURNERS

PRECAMBRIAN PERIOD

Precautions
USE   ACCIDENT PREVENTION

PRECESSION

Precession, Larmor
USE   LARMOR PRECESSION

Precession, Proton
USE   PROTON PRECESSION

Precession, Vortex
USE   VORTEX PRECESSION

Precious Metals
USE   NOBLE METALS

PRECIPITATION

PRECIPITATION (CHEMISTRY)

Precipitation, Electron
USE   ELECTRON PRECIPITATION

PRECIPITATION HARDENING

Precipitation Hardening, Dispersion
USE   PRECIPITATION HARDENING

PRECIPITATION (METEOROLOGY)

Precipitation, Particle
USE   PARTICLE PRECIPITATION

PRECIPITATION PARTICLE MEASUREMENT

Precipitation, Proton
USE   PROTON PRECIPITATION

PRECIPITATORS

Precipitators, Electrostatic
USE   ELECTROSTATIC PRECIPITATORS

PRECISION

Precision Arithmetic, Double
USE   DOUBLE PRECISION ARITHMETIC

Precision, Geometric Dilution Of
USE   GEOMETRIC DILUTION OF PRECISION

PRECISION GUIDED PROJECTILES

PRECONDITIONING

PRECOOLING

PREDATORS

Prediction, Aircraft Noise
USE   NOISE PREDICTION (AIRCRAFT)

Prediction (Aircraft), Noise
USE   NOISE PREDICTION (AIRCRAFT)

PREDICTION ANALYSIS TECHNIQUES

Prediction), ARIP (Impact
USE   COMPUTERIZED SIMULATION
          IMPACT PREDICTION

Prediction, Impact
USE   IMPACT PREDICTION

Prediction), IP (Impact
USE   COMPUTERIZED SIMULATION

Prediction, Linear
USE   LINEAR PREDICTION

Prediction, Noise
USE   NOISE PREDICTION

Prediction, Performance
USE   PERFORMANCE PREDICTION

PREDICTION RECORDING

Prediction, Roshko
USE   ROSHKO PREDICTION

PREDICTIONS

Predictions, Flood
USE   FLOOD PREDICTIONS

Predictors
USE   PREDICTIONS

Predictors, Automatic Rocket Impact
USE   COMPUTERIZED SIMULATION
          IMPACT PREDICTION

PREEMPTING

PREFIRING TESTS

PREFLIGHT ANALYSIS

(Preflight), Crew Procedures
    USE   CREW PROCEDURES (PREFLIGHT)

PREFLIGHT OPERATIONS

PREFOCUSING

PREFORMS

PREGNANCY

Preheaters
    USE   HEATING EQUIPMENT

Preheating
    USE   HEATING

PREIMPREGNATION

PREJUDICES

PRELAUNCH PROBLEMS

PRELAUNCH SUMMARIES

PRELAUNCH TESTS

Prelaunch Tests, Spacecraft
    USE   SPACE VEHICLE CHECKOUT PROGRAM

Preloading
    USE   PRESTRESSING

PREMATURE OPERATION

PREMIXED FLAMES

PREMIXING

PREPARATION

PREPOLYMERS

PREPREGS

PREPROCESSING

PRESBYOPIA

Preselectors
    USE   PREAMPLIFIERS

PRESENTATION

PRESERVATIVES

PRESERVING

PRESIDENTIAL REPORTS

Presintering
    USE   SINTERING

PRESSES

(Presses), Rams
    USE   RAMS (PRESSES)

PRESSING

Pressing, Cold
    USE   COLD PRESSING

PRESSING (FORMING)

Pressing, Hot
    USE   HOT PRESSING

Pressors
    USE   VASOCONSTRICTOR DRUGS

PRESSURE

Pressure, Atmospheric
    USE   ATMOSPHERIC PRESSURE

Pressure, Barometric
    USE   ATMOSPHERIC PRESSURE

Pressure, Base
    USE   BASE PRESSURE

Pressure, Blood
    USE   BLOOD PRESSURE

PRESSURE BREATHING

PRESSURE BROADENING

Pressure Cabins
    USE   PRESSURIZED CABINS

Pressure, Center Of
    USE   CENTER OF PRESSURE

PRESSURE CHAMBERS

Pressure Chambers, Low
    USE   VACUUM CHAMBERS

Pressure, Critical
    USE   CRITICAL PRESSURE

PRESSURE DEPENDENCE

Pressure, Diastolic
    USE   DIASTOLIC PRESSURE

Pressure, Differential
    USE   DIFFERENTIAL PRESSURE

PRESSURE DISTRIBUTION

PRESSURE DRAG

PRESSURE DROP

Pressure Drop, Friction
    USE   SKIN FRICTION

Pressure, Dynamic
    USE   DYNAMIC PRESSURE

PRESSURE EFFECTS

Pressure, Electron
    USE   ELECTRON PRESSURE

Pressure Fields
    USE   PRESSURE DISTRIBUTION

Pressure, Fluid
    USE   FLUID PRESSURE

PRESSURE GAGES

(Pressure Gages), Bombs
    USE   PRESSURE GAGES

Pressure, Gas
    USE   GAS PRESSURE

Pressure, Geo
    USE   GEOPRESSURE

PRESSURE GRADIENTS

(Pressure), Head
    USE   PRESSURE HEADS

PRESSURE HEADS

Pressure, High
    USE   HIGH PRESSURE

Pressure, High Altitude
    USE   HIGH ALTITUDE PRESSURE

Pressure, Hydrostatic
    USE   HYDROSTATIC PRESSURE

PRESSURE ICE

Pressure, Inlet
    USE   INLET PRESSURE

Pressure, Internal
    USE   INTERNAL PRESSURE

Pressure, Intracranial
    USE   INTRACRANIAL PRESSURE

Pressure, Intraocular
    USE   INTRAOCULAR PRESSURE

(Pressure), Isobars
    USE   ISOBARS (PRESSURE)

Pressure, Isostatic
    USE   ISOSTATIC PRESSURE

Pressure Law, Newton
    USE   NEWTON PRESSURE LAW

Pressure, Light
    USE   ILLUMINANCE

Pressure, Low
    USE   LOW PRESSURE

Pressure, Lower Body Negative
    USE   LOWER BODY NEGATIVE PRESSURE

PRESSURE MEASUREMENT

Pressure, Middle Ear
    USE   MIDDLE EAR PRESSURE

PRESSURE MODULATOR RADIOMETERS

PRESSURE OSCILLATIONS

Pressure, Osmotic
    USE   OSMOSIS

Pressure, Over
    USE   OVERPRESSURE

Pressure Oxygen, High
    USE   HIGH PRESSURE OXYGEN

Pressure, Partial
    USE   PARTIAL PRESSURE

Pressure Probes
    USE   PRESSURE SENSORS

PRESSURE PULSES

Pressure, Radiation
    USE   RADIATION PRESSURE

PRESSURE RATIO

PRESSURE RECORDERS

PRESSURE RECOVERY

PRESSURE REDUCTION

PRESSURE REGULATORS

Pressure Ridges
    USE   PRESSURE ICE

PRESSURE SENSORS

Pressure, Sound
    USE   SOUND PRESSURE

Pressure, Stagnation
    USE   STAGNATION PRESSURE

Pressure, Static
    USE   STATIC PRESSURE

PRESSURE SUITS

Pressure, Surface
    USE   PRESSURE

PRESSURE SWITCHES

Pressure, Systolic
    USE   SYSTOLIC PRESSURE

Pressure Test, Ear
    USE   EAR PRESSURE TEST

Pressure, Thrust Chamber
USE   THRUST CHAMBER PRESSURE

Pressure Transducers
USE   PRESSURE SENSORS

Pressure, Transition
USE   TRANSITION PRESSURE

Pressure, Vapor
USE   VAPOR PRESSURE

PRESSURE VESSEL DESIGN

PRESSURE VESSELS

Pressure, Wall
USE   WALL PRESSURE

Pressure, Water
USE   WATER PRESSURE

Pressure Waves
USE   ELASTIC WAVES

PRESSURE WELDING

Pressure, Wind
USE   WIND PRESSURE

Pressures, Impact
USE   IMPACT LOADS

Pressures, Supercritical
USE   SUPERCRITICAL PRESSURES

Pressures, Transient
USE   TRANSIENT PRESSURES

Pressurization, Fuel Tank
USE   FUEL TANK PRESSURIZATION

PRESSURIZED CABINS

PRESSURIZED WATER REACTORS

PRESSURIZING

Preston Tubes
USE   PITOT TUBES
       SPEED INDICATORS

Prestraining
USE   PRESTRESSING

PRESTRESSING

Pretests
USE   TESTS

PRETREATMENT

Pretwisting
USE   TWISTING
       PRESTRESSING

PREVAPORIZATION

PREVENTION

Prevention, Accident
USE   ACCIDENT PREVENTION

Prevention, Blackout
USE   BLACKOUT PREVENTION

Prevention, Corrosion
USE   CORROSION PREVENTION

Prevention, Fire
USE   FIRE PREVENTION

Prevention, Ice
USE   ICE PREVENTION

PREWHIRLING

PREWHITENING

PRIBRAM METEORITE

Primaries, Heavy Cosmic Ray
USE   PRIMARY COSMIC RAYS
       HEAVY NUCLEI

PRIMARY BATTERIES

PRIMARY COSMIC RAYS

PRIMATES

PRIMERS

PRIMERS (COATINGS)

Primers, Engine
USE   ENGINE PRIMERS

PRIMERS (EXPLOSIVES)

PRIMING

PRIMITIVE EARTH ATMOSPHERE

PRIMITIVE EQUATIONS

PRINCE EDWARD ISLAND

PRINCE WILLIAM SOUND (AK)

Princeton Sailwings
USE   SAILWINGS

PRINCIPAL COMPONENTS ANALYSIS

Principle, Bernstein Energy
USE   BERNSTEIN ENERGY PRINCIPLE

Principle, Cryocycle
USE   CRYOCYCLE PRINCIPLE

Principle, Duality
USE   DUALITY PRINCIPLE

Principle, Fermat
USE   FERMAT PRINCIPLE

Principle, Franck-Condon
USE   FRANCK-CONDON PRINCIPLE

Principle, Huygens
USE   HUYGENS PRINCIPLE

Principle, Inertia
USE   INERTIA PRINCIPLE

Principle, Kirchhoff-Huygens
USE   DIFFRACTION
       WAVE PROPAGATION

Principle, Mach Inertia
USE   MACH INERTIA PRINCIPLE

Principle, Maximum
USE   MAXIMUM PRINCIPLE

Principle, Pauli Exclusion
USE   PAULI EXCLUSION PRINCIPLE

Principle, Pontryagin
USE   PONTRYAGIN PRINCIPLE

Principle, Saint Venant
USE   SAINT VENANT PRINCIPLE

Principle, Schelkunoff
USE   SCHELKUNOFF PRINCIPLE

PRINCIPLES

Principles, Variational
USE   VARIATIONAL PRINCIPLES

PRINTED CIRCUITS

PRINTED RESISTORS

PRINTERS

PRINTERS (DATA PROCESSING)

Printers, Tele
USE   TELEPRINTERS

PRINTING

PRINTOUTS

PRIORITIES

PRISMATIC BARS

PRISMS

PRIVACY

Private Aircraft
USE   GENERAL AVIATION AIRCRAFT

Probabilities, Transition
USE   TRANSITION PROBABILITIES

Probability
USE   PROBABILITY THEORY

Probability Analysis, Amplitude
USE   AMPLITUDE DISTRIBUTION ANALYSIS

PROBABILITY DENSITY FUNCTIONS

PROBABILITY DISTRIBUTION FUNCTIONS

Probability, Statistical
USE   PROBABILITY THEORY

PROBABILITY THEORY

Probe B, Gravity
USE   GRAVITY PROBE B

Probe, Galileo
USE   GALILEO PROBE

Probe, Lunik 2 Lunar
USE   LUNIK 2 LUNAR PROBE

Probe, Lunik 3 Lunar
USE   LUNIK 3 LUNAR PROBE

Probe, Lunik 9 Lunar
USE   LUNIK 9 LUNAR PROBE

Probe, Lunik 10 Lunar
USE   LUNIK 10 LUNAR PROBE

Probe, Lunik 11 Lunar
USE   LUNIK 11 LUNAR PROBE

Probe, Lunik 12 Lunar
USE   LUNIK 12 LUNAR PROBE

Probe, Lunik 13 Lunar
USE   LUNIK 13 LUNAR PROBE

Probe, Lunik 14 Lunar
USE   LUNIK 14 LUNAR PROBE

Probe, Lunik 16 Lunar
USE   LUNIK 16 LUNAR PROBE

Probe, Lunik 17 Lunar
USE   LUNIK 17 LUNAR PROBE

Probe, Lunik 19 Lunar
USE   LUNIK 19 LUNAR PROBE

Probe, Lunik 20 Lunar
USE   LUNIK 20 LUNAR PROBE

Probe, Lunik 22 Lunar
USE   LUNIK 22 LUNAR PROBE

Probe, Mariner R 2 Space
USE   MARINER R 2 SPACE PROBE

Probe, Mariner 1 Space
USE   MARINER 1 SPACE PROBE

Probe, Mariner 2 Space
USE   MARINER 2 SPACE PROBE

Probe, Mariner 3 Space
USE   MARINER 3 SPACE PROBE

Probe, Mariner 4 Space
USE   MARINER 4 SPACE PROBE

Probe, Mariner 5 Space
USE   MARINER 5 SPACE PROBE

Probe, Mariner 6 Space
USE   MARINER 6 SPACE PROBE

Probe, Mariner 7 Space
USE   MARINER 7 SPACE PROBE

Probe, Mariner 8 Space
USE   MARINER 8 SPACE PROBE

Probe, Mariner 9 Space
USE   MARINER 9 SPACE PROBE

Probe, Mariner 10 Space
USE   MARINER 10 SPACE PROBE

Probe, Mariner 11 Space
USE   MARINER 11 SPACE PROBE

PROBE METHOD (FORECASTING)

Probe, Pioneer F Space
USE   PIONEER 10 SPACE PROBE

Probe, Pioneer G Space
USE   PIONEER 11 SPACE PROBE

Probe, Pioneer Venus 2 Night
USE   PIONEER VENUS 2 NIGHT PROBE

Probe, Pioneer Venus 2 Sounder
USE   PIONEER VENUS 2 SOUNDER PROBE

Probe, Pioneer 1 Space
USE   PIONEER 1 SPACE PROBE

Probe, Pioneer 2 Space
USE   PIONEER 2 SPACE PROBE

Probe, Pioneer 3 Space
USE   PIONEER 3 SPACE PROBE

Probe, Pioneer 4 Lunar
USE   PIONEER 4 SPACE PROBE

Probe, Pioneer 4 Space
USE   PIONEER 4 SPACE PROBE

Probe, Pioneer 5 Space
USE   PIONEER 5 SPACE PROBE

Probe, Pioneer 6 Space
USE   PIONEER 6 SPACE PROBE

Probe, Pioneer 7 Space
USE   PIONEER 7 SPACE PROBE

Probe, Pioneer 8 Space
USE   PIONEER 8 SPACE PROBE

Probe, Pioneer 9 Space
USE   PIONEER 9 SPACE PROBE

Probe, Pioneer 10 Space
USE   PIONEER 10 SPACE PROBE

Probe, Pioneer 11 Space
USE   PIONEER 11 SPACE PROBE

Probe, Pioneer 12 Space
USE   PIONEER VENUS SPACECRAFT

Probe, Ranger 1 Lunar
USE   RANGER 1 LUNAR PROBE

Probe, Ranger 2 Lunar
USE   RANGER 2 LUNAR PROBE

Probe, Ranger 3 Lunar
USE   RANGER 3 LUNAR PROBE

Probe, Ranger 4 Lunar
USE   RANGER 4 LUNAR PROBE

Probe, Ranger 5 Lunar
USE   RANGER 5 LUNAR PROBE

Probe, Ranger 6 Lunar
USE   RANGER 6 LUNAR PROBE

Probe, Ranger 7 Lunar
USE   RANGER 7 LUNAR PROBE

Probe, Ranger 8 Lunar
USE   RANGER 8 LUNAR PROBE

Probe, Ranger 9 Lunar
USE   RANGER 9 LUNAR PROBE

Probe, Sunblazer Space
USE   SUNBLAZER SPACE PROBE

Probe, Surveyor 1 Lunar
USE   SURVEYOR 1 LUNAR PROBE

Probe, Surveyor 2 Lunar
USE   SURVEYOR 2 LUNAR PROBE

Probe, Surveyor 3 Lunar
USE   SURVEYOR 3 LUNAR PROBE

Probe, Surveyor 4 Lunar
USE   SURVEYOR 4 LUNAR PROBE

Probe, Surveyor 5 Lunar
USE   SURVEYOR 5 LUNAR PROBE

Probe, Surveyor 6 Lunar
USE   SURVEYOR 6 LUNAR PROBE

Probe, Surveyor 7 Lunar
USE   SURVEYOR 7 LUNAR PROBE

Probe, Zond 1 Space
USE   ZOND 1 SPACE PROBE

Probe, Zond 2 Space
USE   ZOND 2 SPACE PROBE

Probe, Zond 3 Space
USE   ZOND 3 SPACE PROBE

Probe, Zond 4 Space
USE   ZOND 4 SPACE PROBE

Probe, Zond 5 Space
USE   ZOND 5 SPACE PROBE

Probe, Zond 6 Space
USE   ZOND 6 SPACE PROBE

Probe, Zond 7 Space
USE   ZOND 7 SPACE PROBE

Probe, Zond 8 Space
USE   ZOND 8 SPACE PROBE

PROBES

Probes, Electron
USE   ELECTRON PROBES

Probes, Electrostatic
USE   ELECTROSTATIC PROBES

Probes, Flame
USE   FLAME PROBES

Probes, Impedance
USE   IMPEDANCE PROBES

Probes, Ion
USE   ION PROBES

Probes, Jupiter
USE   JUPITER PROBES

Probes, Langmuir
USE   ELECTROSTATIC PROBES

Probes, Light
USE   LIGHT BEAMS

Probes, Luna Lunar
USE   LUNIK LUNAR PROBES

Probes, Lunar
USE   LUNAR PROBES

Probes, Lunik Lunar
USE   LUNIK LUNAR PROBES

Probes, Magnetic
USE   MAGNETIC PROBES

Probes, Magnetic Induction
USE   MAGNETIC PROBES

Probes, Mariner Space
USE   MARINER SPACE PROBES

Probes, Mars
USE   MARS PROBES

Probes, Meteorological
USE   SONDES

Probes, Microwave
USE   MICROWAVE PROBES

Probes, Microwave Plasma
USE   MICROWAVE PLASMA PROBES

Probes, Pioneer Space
USE   PIONEER SPACE PROBES

Probes, Pioneer Venus 2 Entry
USE   PIONEER VENUS 2 ENTRY PROBES

Probes, Plasma
USE   PLASMA PROBES

Probes, Pneumatic
USE   PNEUMATIC PROBES

Probes, Pressure
USE   PRESSURE SENSORS

Probes, Radio Frequency Impedance
USE   RADIO FREQUENCY IMPEDANCE PROBES

Probes, Ranger Lunar
USE   RANGER LUNAR PROBES

Probes, Resonance
USE   RESONANCE PROBES

Probes, Solar
USE   SOLAR PROBES

Probes, Space
USE   SPACE PROBES

Probes, Surveyor Lunar
USE   SURVEYOR LUNAR PROBES

Probes, Temperature
USE   TEMPERATURE PROBES

Probes, Venus
USE   VENUS PROBES

Probes, Zond Space
USE   ZOND SPACE PROBES

Probing, Radio
USE   RADIO PROBING

Problem, Cauchy
USE   CAUCHY PROBLEM

Problem, Chapman-Ferraro
USE   CHAPMAN-FERRARO PROBLEM

Problem, Dining Philosophers
USE   DINING PHILOSOPHERS PROBLEM

Problem, Dirichlet
USE   DIRICHLET PROBLEM

Problem, Four Body
USE   FOUR BODY PROBLEM

Problem, Isoperimetric
USE   ISOPERIMETRIC PROBLEM

Problem, Many Body
  USE   MANY BODY PROBLEM

Problem, Mayer
  USE   MAYER PROBLEM

Problem, Neumann
  USE   NEUMANN PROBLEM

Problem, Poincare
  USE   POINCARE PROBLEM

Problem, Riemann
  USE   CAUCHY PROBLEM

Problem, Saint Venant Flexure
  USE   SAINT VENANT PRINCIPLE

PROBLEM SOLVING

Problem, St Venant Flexure
  USE   SAINT VENANT PRINCIPLE

Problem, Three Body
  USE   THREE BODY PROBLEM

Problem, Tracking
  USE   TRACKING PROBLEM

Problem, Traveling Salesman
  USE   TRAVELING SALESMAN PROBLEM

Problem, Two Body
  USE   TWO BODY PROBLEM

PROBLEMS

Problems, Bolza
  USE   BOLZA PROBLEMS

Problems, Boundary Value
  USE   BOUNDARY VALUE PROBLEMS

Problems, Initial Value
  USE   BOUNDARY VALUE PROBLEMS

Problems, Operational
  USE   OPERATIONAL PROBLEMS

Problems, Prelaunch
  USE   PRELAUNCH PROBLEMS

Procedure, Optical Correction
  USE   OPTICAL CORRECTION PROCEDURE

PROCEDURES

Procedures (Inflight), Crew
  USE   CREW PROCEDURES (INFLIGHT)

Procedures, Intravenous
  USE   INTRAVENOUS PROCEDURES

Procedures (Preflight), Crew
  USE   CREW PROCEDURES (PREFLIGHT)

Proceedings
  USE   CONFERENCES

Process, Burning
  USE   COMBUSTION

PROCESS CONTROL (INDUSTRY)

Process, Electroslag
  USE   ELECTROSLAG PROCESS

(Process Engineering), Beds
  USE   BEDS (PROCESS ENGINEERING)

(Process Engineering), Columns
  USE   COLUMNS (PROCESS ENGINEERING)

Process, Ergodic
  USE   ERGODIC PROCESS

Process, Fischer-Tropsch
  USE   FISCHER-TROPSCH PROCESS

PROCESS HEAT

Process, Jet Membrane
  USE   JET MEMBRANE PROCESS

Process, Lost Wax
  USE   INVESTMENT CASTING

Process, Ornstein-Uhlenbeck
  USE   ORNSTEIN-UHLENBECK PROCESS

Process, Poisson
  USE   STOCHASTIC PROCESSES
        POISSON DENSITY FUNCTIONS

Process, Umklapp
  USE   UMKLAPP PROCESS

Process, Verneuil
  USE   VERNEUIL PROCESS

Process (Woodpulp), Kraft
  USE   KRAFT PROCESS (WOODPULP)

PROCESSES

Processes, Autoregressive
  USE   AUTOREGRESSIVE PROCESSES

Processes, Irreversible
  USE   IRREVERSIBLE PROCESSES

Processes, Isentropic
  USE   ISENTROPIC PROCESSES

Processes, Isochoric
  USE   ISOCHORIC PROCESSES

Processes, Isoenergetic
  USE   ISOENERGETIC PROCESSES

Processes, Isopycnic
  USE   ISOPYCNIC PROCESSES

Processes, Isosteric
  USE   ISOPYCNIC PROCESSES

Processes, Isothermal
  USE   ISOTHERMAL PROCESSES

Processes, Lining
  USE   LINING PROCESSES

Processes, Markov
  USE   MARKOV PROCESSES

Processes, Nonadiabatic
  USE   HEAT TRANSFER

Processes, Nonisothermal
  USE   NONISOTHERMAL PROCESSES

Processes, Periodic
  USE   CYCLES

Processes, Polytropic
  USE   POLYTROPIC PROCESSES

Processes, Random
  USE   RANDOM PROCESSES

Processes, Sol-Gel
  USE   SOL-GEL PROCESSES

Processes, Stencil
  USE   STENCIL PROCESSES

Processes, Stochastic
  USE   STOCHASTIC PROCESSES

Processes, Tabulation
  USE   TABULATION PROCESSES

PROCESSING

Processing Applications Rocket, Space
  USE   SPACE PROCESSING APPLICATIONS
        ROCKET

Processing, Automatic Data
  USE   DATA PROCESSING

Processing, Batch
  USE   BATCH PROCESSING

Processing, Bio
  USE   BIOPROCESSING

Processing (Computers), Associative
  USE   ASSOCIATIVE PROCESSING (COMPUTERS)

Processing (Computers), Parallel
  USE   PARALLEL PROCESSING (COMPUTERS)

Processing, Concurrent
  USE   CONCURRENT PROCESSING

Processing, Data
  USE   DATA PROCESSING

Processing, Distributed
  USE   DISTRIBUTED PROCESSING

Processing Equipment, Data
  USE   DATA PROCESSING EQUIPMENT

Processing Equipment, Photographic
  USE   PHOTOGRAPHIC PROCESSING EQUIPMENT

Processing, Food
  USE   FOOD PROCESSING

Processing), Frames (Data
  USE   FRAMES (DATA PROCESSING)

Processing, Image
  USE   IMAGE PROCESSING

Processing, Message
  USE   MESSAGE PROCESSING

Processing, Onboard Data
  USE   ONBOARD DATA PROCESSING

Processing, Optical Data
  USE   OPTICAL DATA PROCESSING

Processing, Photographic
  USE   PHOTOGRAPHIC PROCESSING

Processing, Pre
  USE   PREPROCESSING

Processing), Printers (Data
  USE   PRINTERS (DATA PROCESSING)

Processing, Retort
  USE   RETORT PROCESSING

Processing, Signal
  USE   SIGNAL PROCESSING

Processing, Space
  USE   SPACE PROCESSING

Processing Systems, Power
  USE   POWER CONDITIONING

Processing Terminals, Data
  USE   DATA PROCESSING TERMINALS

Processing Units, Central
  USE   CENTRAL PROCESSING UNITS

Processing, Voice Data
  USE   VOICE DATA PROCESSING

Processing, Word
  USE   WORD PROCESSING

Processors (Computers)
  USE   CENTRAL PROCESSING UNITS

Processors, Data
  USE   DATA PROCESSING EQUIPMENT

Processors, Fluidized Bed
  USE   FLUIDIZED BED PROCESSORS

Processors, Site Data
  USE   SITE DATA PROCESSORS

PROCUREMENT

Procurement, Government
USE   GOVERNMENT PROCUREMENT

PROCUREMENT MANAGEMENT

PROCUREMENT POLICY

PRODUCT DEVELOPMENT

Product, Gross National
USE   GROSS NATIONAL PRODUCT

Product, Kronecker
USE   ORTHOGONALITY

PRODUCTION

Production, Aircraft
USE   AIRCRAFT PRODUCTION

Production, Biomass Energy
USE   BIOMASS ENERGY PRODUCTION

PRODUCTION COSTS

Production Costs, Aircraft
USE   AIRCRAFT PRODUCTION COSTS

PRODUCTION ENGINEERING

Production, Fuel
USE   FUEL PRODUCTION

Production, Hydrocarbon Fuel
USE   HYDROCARBON FUEL PRODUCTION

Production, Hydrogen
USE   HYDROGEN PRODUCTION

Production, Kaon
USE   KAON PRODUCTION

PRODUCTION MANAGEMENT

Production Methods
USE   PRODUCTION ENGINEERING

Production, Oxygen
USE   OXYGEN PRODUCTION

Production, Pair
USE   PAIR PRODUCTION

Production, Particle
USE   PARTICLE PRODUCTION

Production, Photo
USE   PHOTOPRODUCTION

PRODUCTION PLANNING

Production Rates, Ion
USE   ION PRODUCTION RATES

PRODUCTIVITY

PRODUCTS

Products, By-
USE   BY-PRODUCTS

Products, Combustion
USE   COMBUSTION PRODUCTS

Products, Fission
USE   FISSION PRODUCTS

Products, Petroleum
USE   PETROLEUM PRODUCTS

Products, Reaction
USE   REACTION PRODUCTS

Proficiency
USE   ABILITIES

PROFILE METHOD (FORECASTING)

PROFILES

Profiles, Airfoil
USE   AIRFOIL PROFILES

Profiles, Electron Density
USE   ELECTRON DENSITY PROFILES

Profiles, Search
USE   SEARCH PROFILES

Profiles, Shock Wave
USE   SHOCK WAVE PROFILES

Profiles, Temperature
USE   TEMPERATURE PROFILES

Profiles, Velocity
USE   VELOCITY DISTRIBUTION

Profiles, Wind
USE   WIND PROFILES

Profiles, Wing
USE   WING PROFILES

Profiling, Magnetotelluric
USE   MAGNETIC SURVEYS

PROFILOMETERS

PROGENY

PROGNOSIS

PROGNOZ SATELLITES

Program, ACEE
USE   ACEE PROGRAM

Program, Agena B Ranger
USE   AGENA B RANGER PROGRAM

Program, Aircraft Energy Efficiency
USE   ACEE PROGRAM

Program, Apollo Applications
USE   APOLLO APPLICATIONS PROGRAM

Program, Army-Navy Instrumentation
USE   ARMY-NAVY INSTRUMENTATION PROGRAM

Program, Assess
USE   ASSESS PROGRAM

Program, Brazilian Space
USE   BRAZILIAN SPACE PROGRAM

Program, Canadian Space
USE   CANADIAN SPACE PROGRAM

Program, Chinese Space
USE   CHINESE SPACE PROGRAM

Program, COMSAT
USE   COMSAT PROGRAM

Program, DAMP
USE   DOWNRANGE ANTIMISSILE MEASUREMENT
       PROGRAM

Program, DAST
USE   DAST PROGRAM

Program, Defense
USE   DEFENSE PROGRAM

Program, Defense Meteorological Satellite
USE   DMSP SATELLITES

Program, Downrange Antimissile Measurement
USE   DOWNRANGE ANTIMISSILE MEASUREMENT
       PROGRAM

Program, Earth & Ocean Physics Applications
USE   EARTH & OCEAN PHYSICS APPLICATIONS
       PROGRAM

Program, Earth Resources
USE   EARTH RESOURCES PROGRAM

Program, Earth Resources Survey
USE   EARTH RESOURCES SURVEY PROGRAM

Program, Energy Efficiency Transport
USE   ACEE PROGRAM

Program For Aerospace Veh Design, Integ
USE   IPAD

Program, Geographic Applications
USE   GEOGRAPHIC APPLICATIONS PROGRAM

Program, Global Air Sampling
USE   GLOBAL AIR SAMPLING PROGRAM

Program, Global Atmospheric Research
USE   GLOBAL ATMOSPHERIC RESEARCH
       PROGRAM

Program, Gulliver
USE   GULLIVER PROGRAM

Program, HITAB
USE   HIGH ALT TARGET AND BACKGROUND
       MEASUREMENT

Program, Indian Space
USE   INDIAN SPACE PROGRAM

Program, Indonesian Space
USE   INDONESIAN SPACE PROGRAM

Program Integrity, Computer
USE   COMPUTER PROGRAM INTEGRITY

Program, International Geosphere-Biosphere
USE   INTERNATIONAL GEOSPHERE-BIOSPHERE
       PROGRAM

Program, Interservice Data Exchange
USE   INTERSERVICE DATA EXCHANGE
       PROGRAM

Program, Italian Space
USE   ITALIAN SPACE PROGRAM

Program, Japanese Space
USE   JAPANESE SPACE PROGRAM

Program, Lamps
USE   LIGHT AIRBORNE MULTIPURPOSE SYSTEM

Program Management
USE   PROJECT MANAGEMENT

Program, Mariner
USE   MARINER PROGRAM

Program, NASA Structural Analysis
USE   NASTRAN

Program, National Launch Vehicle
USE   NATIONAL LAUNCH VEHICLE PROGRAM

Program, Optical Satellite Tracking
USE   OPTICAL SATELLITE TRACKING PROGRAM

Program, PANT
USE   PANT PROGRAM

Program, Quiet Engine
USE   QUIET ENGINE PROGRAM

Program, Radar Target Scatter Site
USE   RADAR TARGET SCATTER SITE PROGRAM

Program, RATSCAT
USE   RADAR TARGET SCATTER SITE PROGRAM

Program, Reactor In Flight Test
USE   RIFT (REACTOR IN FLIGHT TEST)

Program, Saudi Arabian Space
USE   SAUDI ARABIAN SPACE PROGRAM

Program, SCAR
USE   SUPERSONIC CRUISE AIRCRAFT
       RESEARCH

Program, SEASAT
USE   SEASAT PROGRAM

Program, SKYLAB
USE   SKYLAB PROGRAM

Program, Space Vehicle Checkout
USE   SPACE VEHICLE CHECKOUT PROGRAM

Program, Starsite
USE   STARSITE PROGRAM

Program, Swedish Space
USE   SWEDISH SPACE PROGRAM

Program, Swiss Space
USE   SWISS SPACE PROGRAM

Program, TACT
USE   TACT PROGRAM

Program, TCV
USE   TERMINAL CONFIGURED VEHICLE
      PROGRAM

Program, Terminal Configured Vehicle
USE   TERMINAL CONFIGURED VEHICLE
      PROGRAM

Program, Tilt Rotor Research Aircraft
USE   TILT ROTOR RESEARCH AIRCRAFT
      PROGRAM

Program, Transonic Aircraft Technology
USE   TACT PROGRAM

Program, TRAP
USE   TRAP PROGRAM

PROGRAM TREND LINE ANALYSIS

Program, U.S.S.R. Space
USE   U.S.S.R. SPACE PROGRAM

Program, UK Space
USE   UK SPACE PROGRAM

Program, University
USE   UNIVERSITY PROGRAM

PROGRAM VERIFICATION (COMPUTERS)

Program, Viking Mars
USE   VIKING MARS PROGRAM

PROGRAMMED INSTRUCTION

PROGRAMMERS

PROGRAMMING

Programming, Computer
USE   COMPUTER PROGRAMMING

Programming, Dynamic
USE   DYNAMIC PROGRAMMING

Programming, Language
USE   LANGUAGE PROGRAMMING

(Programming Language), Ada
USE   ADA (PROGRAMMING LANGUAGE)

(Programming Language), APL
USE   APL (PROGRAMMING LANGUAGE)

(Programming Language), BASIC
USE   BASIC (PROGRAMMING LANGUAGE)

(Programming Language), COGO
USE   COGO (PROGRAMMING LANGUAGE)

(Programming Language), COMPASS
USE   COMPASS (PROGRAMMING LANGUAGE)

(Programming Language), FAB
USE   FORTRAN

(Programming Language), LISP
USE   LISP (PROGRAMMING LANGUAGE)

(Programming Language), Map
USE   MAP (PROGRAMMING LANGUAGE)

(Programming Language), MARVS
USE   MARVS (PROGRAMMING LANGUAGE)

(Programming Language), Pascal
USE   PASCAL (PROGRAMMING LANGUAGE)

(Programming Language), SLEUTH
USE   SLEUTH (PROGRAMMING LANGUAGE)

PROGRAMMING LANGUAGES

Programming, Linear
USE   LINEAR PROGRAMMING

Programming, Logic
USE   LOGIC PROGRAMMING

Programming, Mathematical
USE   MATHEMATICAL PROGRAMMING

Programming, Micro
USE   MICROPROGRAMMING

Programming, Multi
USE   MULTIPROGRAMMING

Programming, Nonlinear
USE   NONLINEAR PROGRAMMING

Programming, On-Line
USE   ON-LINE PROGRAMMING

Programming, Optimum Thrust
USE   THRUST PROGRAMMING

Programming, Parallel
USE   PARALLEL PROGRAMMING

Programming, Quadratic
USE   QUADRATIC PROGRAMMING

PROGRAMMING (SCHEDULING)

Programming, Symbolic
USE   SYMBOLIC PROGRAMMING

Programming, Thrust
USE   THRUST PROGRAMMING

PROGRAMS

Programs, Compiler
USE   COMPILERS

Programs, Computer
USE   COMPUTER PROGRAMS

Programs, Computer Systems
USE   COMPUTER SYSTEMS PROGRAMS

Programs (Computers), Applications
USE   APPLICATIONS PROGRAMS (COMPUTERS)

Programs, European Space
USE   EUROPEAN SPACE PROGRAMS

Programs, French Space
USE   FRENCH SPACE PROGRAMS

Programs, Lunar
USE   LUNAR PROGRAMS

Programs, Machine-Independent
USE   MACHINE-INDEPENDENT PROGRAMS

Programs, Multiple Output
USE   MULTIPLE OUTPUT PROGRAMS

Programs, NASA
USE   NASA PROGRAMS

Programs, NASA Space
USE   NASA SPACE PROGRAMS

Programs, Object
USE   OBJECT PROGRAMS

Programs, Source
USE   SOURCE PROGRAMS

Programs, Space
USE   SPACE PROGRAMS

Programs), User Manuals (Computer
USE   USER MANUALS (COMPUTER PROGRAMS)

PROGRESS

PROGRESSIONS

PROHIBITION

Proj, Experimental Reflector Orbital Shot
USE   EXPERIMENTAL REFLECTOR ORBITAL
      SHOT PROJ

Proj, Synchronous Communications Satellite
USE   SYNCHRONOUS COMMUNICATIONS
      SATELLITE PROJ

Project, Advent
USE   ADVENT PROJECT

Project, Agristars
USE   AGRISTARS PROJECT

Project, ALARM
USE   ALARM PROJECT

Project, Alouette
USE   ALOUETTE PROJECT

Project, Apollo
USE   APOLLO PROJECT

Project, Apollo Soyuz Test
USE   APOLLO SOYUZ TEST PROJECT

Project, Argus
USE   ARGUS PROJECT

Project, ASSET
USE   ASSET PROJECT

Project, ATLIT
USE   ATLIT PROJECT

Project, Big Shot
USE   BIG SHOT PROJECT

Project, BIOS
USE   BIOS PROJECT

Project, Bumblebee
USE   BUMBLEBEE PROJECT

Project, Centaur
USE   CENTAUR PROJECT

Project, Defender
USE   DEFENDER PROJECT

Project, Echo
USE   ECHO PROJECT

Project, Eclipse
USE   ECLIPSE PROJECT

Project, EROS
USE   EXPERIMENTAL REFLECTOR ORBITAL
      SHOT PROJ

Project, Galileo
USE   GALILEO PROJECT

Project, Gemini
USE   GEMINI PROJECT

Project, Geosari
USE   GEOSARI PROJECT

Project, Harvard Radio Meteor
USE   HARVARD RADIO METEOR PROJECT

Project, Helios
USE   HELIOS PROJECT

Project, HICAT
USE   HIGH RESOLUTION COVERAGE ANTENNAS

Project, Jupiter
USE   JUPITER PROJECT

PROJECT MANAGEMENT

Project, Mars 69
USE   MARS 69 PROJECT

Project, Mars 71
USE   MARS 71 PROJECT

Project, Mercury
USE   MERCURY PROJECT

Project, National Severe Storms
USE   NATIONAL SEVERE STORMS PROJECT

Project, New Moons
USE   NEW MOONS PROJECT

Project, Nike
USE   NIKE PROJECT

Project, Nimbus
USE   NIMBUS PROJECT

Project, Open
USE   OPEN PROJECT

Project, Orbiter
USE   ORBITER PROJECT

Project, Pioneer
USE   PIONEER PROJECT

PROJECT PLANNING

Project, Radio Attenuation Measurement
USE   RADIO ATTENUATION MEASUREMENT
       PROJECT

Project, RAM
USE   RADIO ATTENUATION MEASUREMENT
       PROJECT

Project, Rand
USE   RAND PROJECT

Project, Ranger
USE   RANGER PROJECT

Project, Rover
USE   ROVER PROJECT

Project, SAIL
USE   SAIL PROJECT

Project, Saturn
USE   SATURN PROJECT

Project, Scanner
USE   SCANNER PROJECT

Project, Scout
USE   SCOUT PROJECT

Project, Seafarer
USE   SEAFARER PROJECT

PROJECT SETI

Project, Squid
USE   SQUID PROJECT

Project, SUBIC
USE   SUBMARINE INTEGRATED CONTROL
       PROJECT

Project, Submarine Integrated Control
USE   SUBMARINE INTEGRATED CONTROL
       PROJECT

Project, Success
USE   SUCCESS PROJECT

Project, Surveyor
USE   SURVEYOR PROJECT

Project, Tektite
USE   TEKTITE PROJECT

Project, Telstar
USE   TELSTAR PROJECT

Project, Themis
USE   THEMIS PROJECT

Project, TIROS
USE   TIROS PROJECT

Project, Titan
USE   TITAN PROJECT

Project, Vanguard
USE   VANGUARD PROJECT

Project, Vega
USE   VEGA PROJECT

Project, Voyager
USE   VOYAGER PROJECT

Project, West Ford
USE   WEST FORD PROJECT

PROJECTILE CRATERING

Projectile, High Altitude Sounding
USE   WASP SOUNDING ROCKET

Projectile Penetration
USE   TERMINAL BALLISTICS

Projectile, Window Atmosphere Sounding
USE   WASP SOUNDING ROCKET

PROJECTILES

Projectiles, Hypervelocity
USE   HYPERVELOCITY PROJECTILES

Projectiles, Precision Guided
USE   PRECISION GUIDED PROJECTILES

Projectiles, Sabot
USE   SABOT PROJECTILES

PROJECTION

Projection, Bonne
USE   BONNE PROJECTION

Projection, Gnomonic
USE   GNOMONIC PROJECTION

Projection, Mercator
USE   MERCATOR PROJECTION

PROJECTIVE GEOMETRY

PROJECTORS

PROJECTS

Projects, Research
USE   RESEARCH PROJECTS

PROLATE SPHEROIDS

PROLATENESS

PROLONGATION

PROMETHAZINE

PROMETHIUM

PROMETHIUM ISOTOPES

Promethium 146
USE   PROMETHIUM ISOTOPES

PROMINENCES

Prominences, Solar
USE   SOLAR PROMINENCES

PROMOTION

PRONE POSITION

Proneness, Accident
USE   ACCIDENT PRONENESS

PRONY SERIES

Proofs
USE   PROVING

PROP-FAN TECHNOLOGY

PROPAGATION

Propagation, Acoustic
USE   ACOUSTIC PROPAGATION

(Propagation), Blackout
USE   BLACKOUT (PROPAGATION)

Propagation, Crack
USE   CRACK PROPAGATION

Propagation, Diffraction
USE   DIFFRACTION PROPAGATION

Propagation, Electromagnetic
USE   ELECTROMAGNETIC WAVE TRANSMISSION

PROPAGATION (EXTENSION)

Propagation, Flame
USE   FLAME PROPAGATION

Propagation, Ground Wave
USE   GROUND WAVE PROPAGATION

Propagation, Ionospheric
USE   IONOSPHERIC PROPAGATION

Propagation, Ionospheric F-Scatter
USE   IONOSPHERIC F-SCATTER PROPAGATION

PROPAGATION MODES

Propagation, Noise
USE   NOISE PROPAGATION

Propagation, Radio
USE   RADIO TRANSMISSION

Propagation, Radio Signal
USE   RADIO TRANSMISSION

Propagation, Scatter
USE   SCATTER PROPAGATION

Propagation, Self
USE   SELF PROPAGATION

Propagation, Shock Wave
USE   SHOCK WAVE PROPAGATION

Propagation, Sound
USE   SOUND PROPAGATION

Propagation, Stress
USE   STRESS PROPAGATION

Propagation, Transequatorial
USE   TRANSEQUATORIAL PROPAGATION

Propagation, Transhorizon Radio
USE   TRANSHORIZON RADIO PROPAGATION

PROPAGATION VELOCITY

Propagation, Wave
USE   WAVE PROPAGATION

Propagators
USE   PROPAGATION

PROPANE

Propane, Cyclo
USE   CYCLOPROPANE

Propane, Nitro
USE   NITROPROPANE

PROPARGYL GROUPS

PROPELLANT ACTUATED DEVICES

PROPELLANT ACTUATED INSTRUMENTS

PROPELLANT ADDITIVES

PROPELLANT BINDERS

PROPELLANT CASTING

PROPELLANT CHEMISTRY

PROPELLANT COMBUSTION

Propellant Combustion, Solid
USE   SOLID PROPELLANT COMBUSTION

PROPELLANT DECOMPOSITION

PROPELLANT EVAPORATION

PROPELLANT EXPLOSIONS

PROPELLANT GRAINS

Propellant Ignition, Solid
USE   SOLID PROPELLANT IGNITION

PROPELLANT MASS RATIO

Propellant Oxidizers
USE   ROCKET OXIDIZERS

PROPELLANT PROPERTIES

Propellant Rocket Engines, Hybrid
USE   HYBRID PROPELLANT ROCKET ENGINES

Propellant Rocket Engines, Liquid
USE   LIQUID PROPELLANT ROCKET ENGINES

Propellant Rocket Engines, Solid
USE   SOLID PROPELLANT ROCKET ENGINES

PROPELLANT SENSITIVITY

PROPELLANT SPRAYS

PROPELLANT STORABILITY

PROPELLANT STORAGE

PROPELLANT TANKS

Propellant Tanks, Rocket
USE   PROPELLANT TANKS

PROPELLANT TESTS

PROPELLANT TRANSFER

PROPELLANTS

Propellants, Case Bonded
USE   CASE BONDED PROPELLANTS

Propellants, Colloidal
USE   COLLOIDAL PROPELLANTS

Propellants, Composite
USE   COMPOSITE PROPELLANTS

Propellants, Cryogenic Rocket
USE   CRYOGENIC ROCKET PROPELLANTS

Propellants, Domino
USE   DOMINO PROPELLANTS

Propellants, Double Base
USE   DOUBLE BASE PROPELLANTS

Propellants, Double Base Rocket
USE   DOUBLE BASE ROCKET PROPELLANTS

Propellants, Gaseous Rocket
USE   GASEOUS ROCKET PROPELLANTS

Propellants, Gelled
USE   GELLED PROPELLANTS

Propellants, Gelled Rocket
USE   GELLED ROCKET PROPELLANTS

Propellants, Gun
USE   GUN PROPELLANTS

Propellants, High Energy
USE   HIGH ENERGY PROPELLANTS

Propellants, High Temperature
USE   HIGH TEMPERATURE PROPELLANTS

Propellants, Htpb
USE   HTPB PROPELLANTS

Propellants, Hybrid
USE   HYBRID PROPELLANTS

Propellants, Hypergolic Rocket
USE   HYPERGOLIC ROCKET PROPELLANTS

Propellants, Ionic
USE   ION ENGINES

Propellants, Liquid Rocket
USE   LIQUID ROCKET PROPELLANTS

Propellants, Lithergolic
USE   HYBRID PROPELLANTS

Propellants, Metal
USE   METAL PROPELLANTS

Propellants, Nitramine
USE   NITRAMINE PROPELLANTS

Propellants, Plastic
USE   PLASTIC PROPELLANTS

Propellants, Rocket
USE   ROCKET PROPELLANTS

Propellants, RP-1 Rocket
USE   RP-1 ROCKET PROPELLANTS

Propellants, Slurry
USE   SLURRY PROPELLANTS

Propellants, Solid
USE   SOLID PROPELLANTS

Propellants, Solid Rocket
USE   SOLID ROCKET PROPELLANTS

Propellants, Storable
USE   STORABLE PROPELLANTS

Propellants, Thixotropic
USE   GELLED ROCKET PROPELLANTS

Propelled Aircraft, Nuclear
USE   NUCLEAR PROPELLED AIRCRAFT

Propelled Sleds, Rocket
USE   ROCKET PROPELLED SLEDS

PROPELLER BLADES

PROPELLER DRIVE

Propeller Drive, Helicopter
USE   HELICOPTER PROPELLER DRIVE

PROPELLER EFFICIENCY

PROPELLER FANS

PROPELLER SLIPSTREAMS

PROPELLERS

Propellers, Constant Speed
USE   VARIABLE PITCH PROPELLERS

Propellers, Contrarotating
USE   CONTRAROTATING PROPELLERS

Propellers, Ducted
USE   SHROUDED PROPELLERS

Propellers, Shrouded
USE   SHROUDED PROPELLERS

Propellers, Tilted
USE   TILTED PROPELLERS

Propellers, Variable Pitch
USE   VARIABLE PITCH PROPELLERS

PROPERTIES

Properties, Acoustic
USE   ACOUSTIC PROPERTIES

Properties, Asymptotic
USE   ASYMPTOTIC PROPERTIES

Properties, Chemical
USE   CHEMICAL PROPERTIES

Properties, Creep
USE   CREEP PROPERTIES

Properties, Dielectric
USE   DIELECTRIC PROPERTIES

Properties, Dynamic
USE   DYNAMIC CHARACTERISTICS

Properties, Elastic
USE   ELASTIC PROPERTIES

Properties, Electrical
USE   ELECTRICAL PROPERTIES

Properties, Electromagnetic
USE   ELECTROMAGNETIC PROPERTIES

Properties (Geology), Structural
USE   STRUCTURAL PROPERTIES (GEOLOGY)

Properties, Hygral
USE   HYGRAL PROPERTIES

Properties, Internuclear
USE   INTERNUCLEAR PROPERTIES

Properties, Magnetic
USE   MAGNETIC PROPERTIES

Properties, Mechanical
USE   MECHANICAL PROPERTIES

Properties, Optical
USE   OPTICAL PROPERTIES

Properties, Physical
USE   PHYSICAL PROPERTIES

Properties, Plastic
USE   PLASTIC PROPERTIES

Properties, Propellant
USE   PROPELLANT PROPERTIES

Properties, Shear
USE   SHEAR PROPERTIES

Properties, Surface
USE   SURFACE PROPERTIES

Properties, Tensile
USE   TENSILE PROPERTIES

Properties, Thermal
USE   THERMODYNAMIC PROPERTIES

Properties, Thermochemical
USE   THERMOCHEMICAL PROPERTIES

Properties, Thermodynamic
USE   THERMODYNAMIC PROPERTIES

Properties, Thermophysical
USE   THERMOPHYSICAL PROPERTIES

Properties, Transport
USE   TRANSPORT PROPERTIES

Properties, Virtual
USE   VIRTUAL PROPERTIES

(Property), Composition
USE   COMPOSITION (PROPERTY)

(Property), Distribution
USE   DISTRIBUTION (PROPERTY)

PROPHYLAXIS

PROPIONIC ACID

PROPORTION

PROPORTIONAL CONTROL

PROPORTIONAL COUNTERS

PROPORTIONAL LIMIT

PROPRIOCEPTION

PROPRIOCEPTORS

PROPULSION

Propulsion, Auxiliary
USE AUXILIARY PROPULSION

Propulsion, Chemical
USE CHEMICAL PROPULSION

Propulsion, Chemonuclear
USE CHEMICAL PROPULSION
NUCLEAR PROPULSION

Propulsion, Dual Mode
USE HYBRID PROPULSION

Propulsion, Electric
USE ELECTRIC PROPULSION

Propulsion, Electromagnetic
USE ELECTROMAGNETIC PROPULSION

Propulsion, Electrostatic
USE ELECTROSTATIC PROPULSION

Propulsion, Hybrid
USE HYBRID PROPULSION

Propulsion, Interplanetary
USE INTERPLANETARY SPACECRAFT
ROCKET ENGINES

Propulsion, Ion
USE ION PROPULSION

Propulsion, Jet
USE JET PROPULSION

Propulsion, Laser
USE LASER PROPULSION

Propulsion, Low Thrust
USE LOW THRUST PROPULSION

Propulsion, Marine
USE MARINE PROPULSION

Propulsion, Nuclear
USE NUCLEAR PROPULSION

Propulsion, Nuclear Electric
USE NUCLEAR ELECTRIC PROPULSION

Propulsion, Photonic
USE PHOTONIC PROPULSION

Propulsion, Plasma
USE PLASMA PROPULSION

Propulsion, Solar
USE SOLAR PROPULSION

Propulsion, Solar Electric
USE SOLAR ELECTRIC PROPULSION

Propulsion, Solar Thermal
USE SOLAR THERMAL PROPULSION

Propulsion, Spacecraft
USE SPACECRAFT PROPULSION

Propulsion, Submarine
USE SUBMARINE PROPULSION

PROPULSION SYSTEM CONFIGURATIONS

Propulsion System, Hot Cycle
USE TIP DRIVEN ROTORS

PROPULSION SYSTEM PERFORMANCE

Propulsion System, Post Boost
USE POST BOOST PROPULSION SYSTEM

Propulsion Systems, Ascent
USE ASCENT PROPULSION SYSTEMS

Propulsion Systems, Descent
USE DESCENT PROPULSION SYSTEMS

Propulsion Systems, Man Operated
USE MAN OPERATED PROPULSION SYSTEMS

(Propulsion Systems), MOPS
USE MAN OPERATED PROPULSION SYSTEMS

Propulsion Systems, Personnel
USE SELF MANEUVERING UNITS

Propulsion, Thermonuclear
USE NUCLEAR PROPULSION

Propulsion, Underwater
USE UNDERWATER PROPULSION

PROPULSIVE EFFICIENCY

PROPYL COMPOUNDS

PROPYL NITRATE

PROPYLENE

PROPYLENE OXIDE

Propylene, Poly
USE POLYPROPYLENE

Prospecting
USE EXPLORATION

PROSTAGLANDINS

PROSTATE GLAND

PROSTHETIC DEVICES

PROTACTINIUM

PROTACTINIUM COMPOUNDS

PROTACTINIUM FLUORIDES

PROTACTINIUM ISOTOPES

Protactinium 234
USE PROTACTINIUM ISOTOPES

PROTEASE

PROTECTION

Protection, Acceleration
USE ACCELERATION PROTECTION

Protection, Circuit
USE CIRCUIT PROTECTION

Protection, Environment
USE ENVIRONMENT PROTECTION

Protection, Eye
USE EYE PROTECTION

Protection, Meteoroid
USE METEOROID PROTECTION

Protection, Radiation
USE RADIATION PROTECTION

Protection Systems, Advanced EVA
USE AEPS

Protection, Thermal
USE THERMAL PROTECTION

Protection, Vibration
USE VIBRATION ISOLATORS

PROTECTIVE CLOTHING

PROTECTIVE COATINGS

Protective Coatings, Ceramal
USE PROTECTIVE COATINGS
CERMETS

Protective Coatings, Sprayed
USE PROTECTIVE COATINGS
SPRAYED COATINGS

PROTECTORS

Protectors, Ear
USE EAR PROTECTORS

PROTEIN METABOLISM

PROTEIN SYNTHESIS

PROTEINOIDS

PROTEINS

Proteins, Lipo
USE LIPOPROTEINS

Proteins, Proto
USE PROTOPROTEINS

PROTHROMBIN

Protium
USE LIGHT WATER

PROTOBIOLOGY

PROTOCOL (COMPUTERS)

PROTON BEAMS

PROTON BELTS

PROTON DAMAGE

PROTON DENSITY (CONCENTRATION)

Proton Density, Magnetospheric
USE MAGNETOSPHERIC PROTON DENSITY

PROTON ENERGY

PROTON FLUX DENSITY

PROTON IMPACT

PROTON IRRADIATION

PROTON MAGNETIC RESONANCE

PROTON MASERS

PROTON PRECESSION

PROTON PRECIPITATION

PROTON PROTUBERANCES

Proton Reactions, Proton-
USE PROTON-PROTON REACTIONS

PROTON RESONANCE

PROTON SATELLITES

PROTON SCATTERING

Proton Telescopes
USE PARTICLE TELESCOPES

PROTON 1 SATELLITE

PROTON 2 SATELLITE

PROTON 3 SATELLITE

PROTON 4 SATELLITE

PROTON-PROTON REACTIONS

PROTONS

Protons, Anti
    USE   ANTIPROTONS

Protons, Recoil
    USE   RECOIL PROTONS

Protons, Solar
    USE   SOLAR PROTONS

PROTOPLANETS

PROTOPLASM

PROTOPLASTS

PROTOPROTEINS

PROTOSTARS

PROTOTYPES

PROTOZOA

PROTRACTORS

PROTUBERANCES

Protuberances, Proton
    USE   PROTON PROTUBERANCES

PROUSTITE

Provider Aircraft
    USE   C-123 AIRCRAFT

PROVING

Proving, Theorem
    USE   THEOREM PROVING

(Proving), Verification
    USE   PROVING

PROVISIONING

Provost Aircraft, Jet
    USE   JET PROVOST AIRCRAFT

PROXIMITY

PROXIMITY EFFECT (ELECTRICITY)

PRTR (Reactor)
    USE   PLUTONIUM RECYCLE TEST REACTOR

Prussic Acid
    USE   HYDROCYANIC ACID

PSEUDOMONAS

PSEUDONOISE

PSEUDOPOTENTIALS

PSEUDORANDOM SEQUENCES

PSYCHIATRY

Psychiatry, Military
    USE   MILITARY PSYCHOLOGY

Psychiatry, Neuro
    USE   NEUROPSYCHIATRY

Psychiatry, Social
    USE   SOCIAL PSYCHIATRY

PSYCHOACOUSTICS

PSYCHOLINGUISTICS

PSYCHOLOGICAL EFFECTS

PSYCHOLOGICAL FACTORS

Psychological Indexes
    USE   PSYCHOLOGICAL TESTS

PSYCHOLOGICAL SETS

PSYCHOLOGICAL TESTS

PSYCHOLOGY

Psychology, Aviation
    USE   AVIATION PSYCHOLOGY

Psychology, Cognitive
    USE   COGNITIVE PSYCHOLOGY

(Psychology), Generalization
    USE   GENERALIZATION (PSYCHOLOGY)

(Psychology), Inhibition
    USE   INHIBITION (PSYCHOLOGY)

Psychology, Military
    USE   MILITARY PSYCHOLOGY

(Psychology), Reinforcement
    USE   REINFORCEMENT (PSYCHOLOGY)

(Psychology), Retention
    USE   RETENTION (PSYCHOLOGY)

(Psychology), Reward
    USE   REWARD (PSYCHOLOGY)

Psychology, Space
    USE   SPACE PSYCHOLOGY

(Psychology), Stress
    USE   STRESS (PSYCHOLOGY)

PSYCHOMETRICS

PSYCHOMOTOR PERFORMANCE

PSYCHOPHARMACOLOGY

PSYCHOPHYSICS

PSYCHOPHYSIOLOGY

(Psychophysiology), Evoked Response
    USE   EVOKED RESPONSE
                (PSYCHOPHYSIOLOGY)

(Psychophysiology), Workloads
    USE   WORKLOADS (PSYCHOPHYSIOLOGY)

PSYCHOSES

PSYCHOSOMATICS

PSYCHOTHERAPY

PSYCHOTIC DEPRESSION

PSYCHOTROPIC DRUGS

PSYCHROMETERS

PSYCHROPHILES

Pt
    USE   PLATINUM

PTM (Modulation)
    USE   PULSE TIME MODULATION

PTOLEMAEUS CRATER

Pu
    USE   PLUTONIUM

PUBLIC ADDRESS SYSTEMS

PUBLIC HEALTH

PUBLIC LAW

PUBLIC RELATIONS

PUBLIC SPEAKING

Publications
    USE   DOCUMENTS

(Publications), Catalogs
    USE   CATALOGS (PUBLICATIONS)

PUERTO RICO

Pull Amplifiers, Push-
    USE   PUSH-PULL AMPLIFIERS

PULLEYS

PULLING

PULMONARY CIRCULATION

PULMONARY FUNCTIONS

PULMONARY LESIONS

PULSARS

Pulsating Flow
    USE   UNSTEADY FLOW

Pulsations, Geomagnetic
    USE   GEOMAGNETIC PULSATIONS

Pulsations, Micro
    USE   MICROPULSATIONS

PULSE AMPLITUDE

PULSE AMPLITUDE MODULATION

PULSE CHARGING

PULSE CODE MODULATION

Pulse Code Modulation, Differential
    USE   DIFFERENTIAL PULSE CODE MODULATION

PULSE COMMUNICATION

PULSE COMPRESSION

PULSE DIFFRACTION

PULSE DOPPLER RADAR

PULSE DURATION

PULSE DURATION MODULATION

PULSE FREQUENCY MODULATION

PULSE FREQUENCY MODULATION TELEMETRY

PULSE GENERATORS

PULSE HEATING

Pulse Height
    USE   PULSE AMPLITUDE

PULSE MODULATION

PULSE POSITION MODULATION

PULSE RADAR

PULSE RATE

Pulse Reactors, Annular Core
    USE   ANNULAR CORE PULSE REACTORS

Pulse Recorders
    USE   COUNTERS

PULSE REPETITION RATE

PULSE TIME MODULATION

Pulse Width
    USE   PULSE DURATION

PULSE WIDTH AMPLITUDE CONVERTERS

Pulse Width Modulation
    USE   PULSE DURATION MODULATION

PULSED JET ENGINES

PULSED LASERS

Pulsed Lasers, Ultrashort
USE   ULTRASHORT PULSED LASERS

PULSED RADIATION

PULSEJET ENGINES

PULSES

Pulses, Electric
USE   ELECTRIC PULSES

Pulses, Electromagnetic
USE   ELECTROMAGNETIC PULSES

Pulses, Picosecond
USE   PICOSECOND PULSES

Pulses, Pressure
USE   PRESSURE PULSES

Pulses, System Generated Electromagnetic
USE   SYSTEM GENERATED ELECTROMAGNETIC
       PULSES

PULTRUSION

Pulverizing
USE   GRINDING (COMMINUTION)

PUMICE

PUMP IMPELLERS

PUMP SEALS

Pumped Lasers, Nuclear
USE   NUCLEAR PUMPED LASERS

Pumped Lasers, Solar-
USE   SOLAR-PUMPED LASERS

PUMPING

Pumping, Cryo
USE   CRYOPUMPING

Pumping, Electron
USE   ELECTRON PUMPING

Pumping, Laser
USE   LASER PUMPING

Pumping, Magnetic
USE   MAGNETIC PUMPING

Pumping, Nuclear
USE   NUCLEAR PUMPING

Pumping, Optical
USE   OPTICAL PUMPING

Pumping, Plasma
USE   PLASMA PUMPING

PUMPS

Pumps, Axial Flow
USE   AXIAL FLOW PUMPS

Pumps, Blood
USE   BLOOD PUMPS

Pumps, Centrifugal
USE   CENTRIFUGAL PUMPS

Pumps, Condensation
USE   CONDENSATION PUMPS

Pumps, Diffusion
USE   DIFFUSION PUMPS

Pumps, Electromagnetic
USE   ELECTROMAGNETIC PUMPS

Pumps, Flux
USE   FLUX PUMPS

Pumps, Fuel
USE   FUEL PUMPS

Pumps, Heat
USE   HEAT PUMPS

Pumps, Hydraulic
USE   HYDRAULIC EQUIPMENT
       PUMPS

Pumps, Ion
USE   ION PUMPS

Pumps, Jet
USE   JET PUMPS

Pumps, Molecular
USE   MOLECULAR PUMPS

(Pumps), Rams
USE   RAMS (PUMPS)

Pumps, Turbine
USE   TURBINE PUMPS

Pumps, Vacuum
USE   VACUUM PUMPS

Pumps, Visco
USE   VISCOPUMPS

Pumps, Windpowered
USE   WINDPOWERED PUMPS

PUNCHED CARDS

PUNCHED TAPES

PUNCHES

Puncturing
USE   PIERCING

PUPA

PUPIL SIZE

PUPILLOMETRY

PUPILS

PURGING

PURIFICATION

Purification, Air
USE   AIR PURIFICATION

Purification, Water
USE   WATER TREATMENT

Purifiers
USE   PURIFICATION

PURINES

PURITY

PURPOSES

PURSUIT TRACKING

PUSH-PULL AMPLIFIERS

PUSHBROOM SENSOR MODES

PUSHING

PWM (Modulation)
USE   PULSE DURATION MODULATION

PYCNOMETERS

PYLON MOUNTING

PYLONS

PYRAMID LAKE (NV)

PYRAMIDAL BODIES

PYRAMIDS

PYRANOMETERS

PYRAZINES

PYRENEES MOUNTAINS (EUROPE)

PYRENES

Pyrex (Trademark)
USE   BOROSILICATE GLASS

PYRIDINE NUCLEOTIDES

PYRIDINES

PYRIDOXINE

PYRIMIDINES

PYRITES

PYROCERAM (TRADEMARK)

PYROELECTRICITY

PYROGEN

Pyrographalloy
USE   PYROLYTIC GRAPHITE
       REFRACTORY MATERIALS
       COMPOSITE MATERIALS

PYROHELIOMETERS

PYROHYDROLYSIS

PYROLYSIS

Pyrolysis, Hydro
USE   HYDROPYROLYSIS

PYROLYTIC GRAPHITE

PYROLYTIC MATERIALS

PYROMETALLURGY

PYROMETERS

Pyrometers, Optical
USE   OPTICAL PYROMETERS

Pyrometers, Radiation
USE   RADIATION PYROMETERS

Pyrometers, Thermocouple
USE   THERMOCOUPLE PYROMETERS

Pyrometry
USE   TEMPERATURE MEASUREMENT

PYROPHORIC MATERIALS

PYROPHYLLITE

PYROTECHNICS

PYROXENES

Pyroxylin
USE   CELLULOSE NITRATE

PYRRHOTITE

PYRROLES

PYRRONES (TRADEMARK)

PYRUVATES

P3V Aircraft
USE   P-3 AIRCRAFT

P78-2 Satellite
USE   SCATHA SATELLITE

Q

Q DEVICES

Q FACTORS

Q, High
USE   Q FACTORS

Q SWITCHED LASERS

Q VALUES

QC
USE   QUALITY CONTROL

QCD
USE   QUANTUM CHROMODYNAMICS

QH-50 HELICOPTER

QSO (Radio Sources)
USE   QUASARS

Quadrangle (AZ), Phoenix
USE   PHOENIX QUADRANGLE (AZ)

QUADRANTID METEOROIDS

QUADRANTS

QUADRATIC EQUATIONS

QUADRATIC PROGRAMMING

Quadrature Approximation
USE   QUADRATURES

QUADRATURES

Quadrupole Lenses
USE   MAGNETIC LENSES

QUADRUPOLE NETWORKS

Quadrupole Resonance, Nuclear
USE   NUCLEAR QUADRUPOLE RESONANCE

QUADRUPOLES

QUAIL MISSILE

Quakes, Planetary
USE   PLANETARY QUAKES

QUALIFICATIONS

QUALITATIVE ANALYSIS

Qualities, Flying
USE   FLIGHT CHARACTERISTICS

Qualities, Handling
USE   CONTROLLABILITY

QUALITY

Quality, Air
USE   AIR QUALITY

QUALITY CONTROL

Quality, Environmental
USE   ENVIRONMENTAL QUALITY

Quality Factors
USE   Q FACTORS

Quality, Riding
USE   RIDING QUALITY

Quality, Water
USE   WATER QUALITY

QUANTILES

QUANTITATIVE ANALYSIS

Quantity
USE   AMOUNT

(Quantity), Level
USE   LEVEL (QUANTITY)

Quantization
USE   MEASUREMENT

Quantization, Flux
USE   FLUX QUANTIZATION

Quantizer
USE   COUNTERS

QUANTUM AMPLIFIERS

QUANTUM CHEMISTRY

QUANTUM CHROMODYNAMICS

QUANTUM COUNTERS

QUANTUM EFFICIENCY

QUANTUM ELECTRODYNAMICS

QUANTUM ELECTRONICS

Quantum Generators
USE   STIMULATED EMISSION DEVICES

Quantum Interferometers, Superconducting
USE   SQUID (DETECTORS)

QUANTUM MECHANICS

QUANTUM NUMBERS

QUANTUM STATISTICS

QUANTUM THEORY

QUANTUM WELLS

Quarantine Facility, Mobile
USE   MOBILE QUARANTINE FACILITY

Quarantine, Planetary
USE   PLANETARY QUARANTINE

QUARK PARTON MODEL

QUARKS

Quarries
USE   MINES (EXCAVATIONS)

QUARTIC EQUATIONS

QUARTILES

QUARTZ

QUARTZ CRYSTALS

QUARTZ LAMPS

QUARTZ TRANSDUCERS

QUASARS

QUASAT

Quasi-Particles
USE   ELEMENTARY EXCITATIONS

QUASI-STEADY STATES

Quasi-Stellar Radio Sources
USE   QUASARS

Quasilinearity
USE   NONLINEARITY

QUATERNARY ALLOYS

QUATERNIONS

QUEBEC

QUEFRENCIES

QUENCHING

QUENCHING (ATOMIC PHYSICS)

QUENCHING (COOLING)

Quenching, Flame
USE   QUENCHING (COOLING)
       EXTINGUISHING

Quenching (Metallurgy), Rapid
USE   RAPID QUENCHING (METALLURGY)

Queretaro Area (Mexico), Leon-
USE   LEON-QUERETARO AREA (MEXICO)

QUERY LANGUAGES

QUESTOL

QUEUEING THEORY

QUIET ENGINE PROGRAM

Quiet Sun Year, International
USE   INTERNATIONAL QUIET SUN YEAR

QUINOLINE

Quinone, Phyllo
USE   PHYLLOQUINONE

Quinones, Anthra
USE   ANTHRAQUINONES

QUINOXALINES

QUOTIENTS

R

R Stars, W-
USE   WOLF-RAYET STARS

R 2 Space Probe, Mariner
USE   MARINER R 2 SPACE PROBE

Ra
USE   RADIUM

RA-28 ENGINE

RABBITS

RACAH COEFFICIENT

RACE FACTORS

RACES

RACETRACKS (PARTICLE ACCELERATORS)

RACKS

RACKS (FRAMES)

RACKS (GEARS)

RACON Beacons
USE   RADAR BEACONS

RADANT

RADAR

RADAR ABSORBERS

Radar Absorbing Materials
USE   ANTIRADAR COATINGS

Radar, Airborne Surveillance
USE   AIRBORNE SURVEILLANCE RADAR

Radar Altimeters
USE   RADIO ALTIMETERS

RADAR ANTENNAS

Radar Approach, Airborne
  USE   AIRBORNE RADAR APPROACH

RADAR APPROACH CONTROL

RADAR ASTRONOMY

RADAR ATTENUATION

RADAR BEACONS

RADAR BEAMS

Radar, Bistatic
  USE   MULTISTATIC RADAR

RADAR CLUTTER MAPS

(Radar), Cobra Dane
  USE   COBRA DANE (RADAR)

Radar, Coherent
  USE   COHERENT RADAR

Radar, Continuous Wave
  USE   CONTINUOUS WAVE RADAR

RADAR CORNER REFLECTORS

RADAR CROSS SECTIONS

Radar, CW
  USE   CONTINUOUS WAVE RADAR

RADAR DATA

RADAR DETECTION

Radar Direction Finders
  USE   RADIO DIRECTION FINDERS

Radar Displays
  USE   RADARSCOPES

Radar, Doppler
  USE   DOPPLER RADAR

Radar, Earth Resources Shuttle Imaging
  USE   EARTH RESOURCES SHUTTLE IMAGING
         RADAR

RADAR ECHOES

Radar Echoes, Lunar
  USE   LUNAR RADAR ECHOES

Radar Echoes, Solar
  USE   SOLAR RADAR ECHOES

Radar Echoes, Venus
  USE   VENUS RADAR ECHOES

RADAR EQUIPMENT

Radar, European Incoherent Scatter
  USE   EISCAT RADAR SYSTEM (EUROPE)

RADAR FILTERS

RADAR GEOLOGY

RADAR HOMING MISSILES

RADAR IMAGERY

Radar, Imaging
  USE   IMAGING RADAR

Radar, Incoherent Scatter
  USE   INCOHERENT SCATTER RADAR

Radar, Infrared
  USE   INFRARED RADAR

Radar, Landing
  USE   LANDING RADAR

Radar, Laser
  USE   OPTICAL RADAR

RADAR MAPS

RADAR MEASUREMENT

Radar, Meteorological
  USE   METEOROLOGICAL RADAR

Radar, Monopulse
  USE   MONOPULSE RADAR

Radar, MTI
  USE   MOVING TARGET INDICATORS

Radar, Multispectral
  USE   MULTISPECTRAL RADAR

Radar, Multistatic
  USE   MULTISTATIC RADAR

RADAR NAVIGATION

RADAR NETWORKS

Radar, North American Search And Ranging
  USE   NORTH AMERICAN SEARCH AND RANGING
         RADAR

Radar Observation
  USE   RADAR TRACKING

Radar, Optical
  USE   OPTICAL RADAR

Radar, Over-The-Horizon
  USE   OVER-THE-HORIZON RADAR

RADAR PHOTOGRAPHY

Radar, Pulse
  USE   PULSE RADAR

Radar, Pulse Doppler
  USE   PULSE DOPPLER RADAR

RADAR RANGE

RADAR RECEIVERS

RADAR RECEPTION

Radar Reflections
  USE   RADAR ECHOES

RADAR REFLECTORS

RADAR RESOLUTION

Radar, Satellite-Borne
  USE   SATELLITE-BORNE RADAR

RADAR SCANNING

RADAR SCATTERING

Radar, Search
  USE   SEARCH RADAR

Radar, Secondary
  USE   SECONDARY RADAR

Radar, Shuttle Imaging
  USE   SHUTTLE IMAGING RADAR

Radar, Side-Looking
  USE   SIDE-LOOKING RADAR

RADAR SIGNATURES

Radar, Space Based
  USE   SPACE BASED RADAR

Radar (Spacecraft), Venus Orbiting Imaging
  USE   VENUS ORBITING IMAGING RADAR
         (SPACECRAFT)

Radar, Surveillance
  USE   SURVEILLANCE RADAR

Radar, Synthetic Aperture
  USE   SYNTHETIC APERTURE RADAR

Radar System (Europe), Eiscat
  USE   EISCAT RADAR SYSTEM (EUROPE)

Radar System, Tradex
  USE   TRADEX RADAR SYSTEM

Radar Systems, Digital
  USE   DIGITAL RADAR SYSTEMS

RADAR TARGET SCATTER SITE PROGRAM

RADAR TARGETS

(Radar Technique), HICAT
  USE   HIGH RESOLUTION COVERAGE ANTENNAS

Radar Terminal System, Automated
  USE   AUTOMATED RADAR TERMINAL SYSTEM

Radar, Tracking
  USE   TRACKING RADAR

RADAR TRACKING

RADAR TRANSMISSION

RADAR TRANSMITTERS

Radar, Weather
  USE   METEOROLOGICAL RADAR

RADARSAT

RADARSCOPES

RADIAL DISTRIBUTION

Radial Drainage Patterns
  USE   DRAINAGE PATTERNS

RADIAL FLOW

RADIAL VELOCITY

RADIANCE

Radiance, Ir
  USE   IRRADIANCE

RADIANCY

RADIANT COOLING

Radiant Energy
  USE   RADIATION

RADIANT FLUX DENSITY

RADIANT HEATING

Radiant Intensity
  USE   RADIANT FLUX DENSITY

RADIATION

RADIATION ABSORPTION

Radiation, Acoustic
  USE   SOUND WAVES

Radiation, Alpha
  USE   ALPHA PARTICLES

RADIATION AND METEOROID SATELLITE

Radiation, Atmospheric
  USE   ATMOSPHERIC RADIATION

Radiation, Background
  USE   BACKGROUND RADIATION

(Radiation), Beams
  USE   BEAMS (RADIATION)

Radiation Belt, Inner
  USE   INNER RADIATION BELT

Radiation Belt, Outer
  USE   OUTER RADIATION BELT

RADIATION BELTS

Radiation Belts, Artificial
  USE   ARTIFICIAL RADIATION BELTS

Radiation Belts, Van Allen
USE   RADIATION BELTS

Radiation, Black Body
USE   BLACK BODY RADIATION

Radiation Budget Experiment, Earth
USE   EARTH RADIATION BUDGET EXPERIMENT

Radiation, Cerenkov
USE   CERENKOV RADIATION

RADIATION CHEMISTRY

Radiation, Circumsolar
USE   CIRCUMSOLAR RADIATION

Radiation, Coherent
USE   COHERENT RADIATION

Radiation, Coherent Acoustic
USE   COHERENT ACOUSTIC RADIATION

Radiation, Coherent Electromagnetic
USE   COHERENT ELECTROMAGNETIC
      RADIATION

Radiation, Continuous
USE   CONTINUOUS RADIATION

Radiation, Corpuscular
USE   CORPUSCULAR RADIATION

Radiation, Cosmic
USE   COSMIC RAYS

RADIATION COUNTERS

Radiation, Cyclotron
USE   CYCLOTRON RADIATION

RADIATION DAMAGE

RADIATION DETECTORS

Radiation Detectors, Silicon
USE   SILICON RADIATION DETECTORS

Radiation, Diffuse
USE   DIFFUSE RADIATION

RADIATION DISTRIBUTION

RADIATION DOSAGE

Radiation, Earth
USE   TERRESTRIAL RADIATION

RADIATION EFFECTS

Radiation, Electromagnetic
USE   ELECTROMAGNETIC RADIATION

Radiation, Electron
USE   ELECTRON RADIATION

Radiation Emission
USE   RADIATION

Radiation Exp Background Sats, Galactic
USE   GREB SATELLITES

Radiation Exposure
USE   RADIATION DOSAGE

Radiation, Extraterrestrial
USE   EXTRATERRESTRIAL RADIATION

Radiation, Extreme Ultraviolet
USE   EXTREME ULTRAVIOLET RADIATION

Radiation, Far Infrared
USE   FAR INFRARED RADIATION

Radiation, Far Ultraviolet
USE   FAR ULTRAVIOLET RADIATION

Radiation Fields
USE   RADIATION DISTRIBUTION

Radiation, Galactic
USE   GALACTIC RADIATION

Radiation, Gamma
USE   GAMMA RAYS

Radiation, Gravitational
USE   GRAVITATIONAL WAVES

RADIATION HARDENING

Radiation, Harmonic
USE   HARMONIC RADIATION

RADIATION HAZARDS

Radiation Heating
USE   RADIANT HEATING

Radiation, Incident
USE   INCIDENT RADIATION

Radiation, Infrared
USE   INFRARED RADIATION

RADIATION INJURIES

Radiation Intensity
USE   RADIANT FLUX DENSITY

Radiation, Interstellar
USE   INTERSTELLAR RADIATION

Radiation, Ion Cyclotron
USE   ION CYCLOTRON RADIATION

Radiation, Ionizing
USE   IONIZING RADIATION

Radiation, Ir
USE   IRRADIATION

Radiation, Kirchhoff Law Of
USE   KIRCHHOFF LAW OF RADIATION

RADIATION LAWS

Radiation), Light (Visible
USE   LIGHT (VISIBLE RADIATION)

Radiation, Long Wave
USE   LONG WAVE RADIATION

Radiation, Lunar
USE   LUNAR RADIATION

Radiation, Lyman Alpha
USE   LYMAN ALPHA RADIATION

Radiation, Lyman Beta
USE   LYMAN BETA RADIATION

RADIATION MEASUREMENT

RADIATION MEASURING INSTRUMENTS

Radiation Medicine
USE   NUCLEAR MEDICINE

RADIATION METEOROID SPACECRAFT

Radiation Meters
USE   RADIATION MEASURING INSTRUMENTS

Radiation, Microwave
USE   MICROWAVES

Radiation, Modulated Continuous
USE   MODULATED CONTINUOUS RADIATION

Radiation, Monochromatic
USE   MONOCHROMATIC RADIATION

Radiation, Near Infrared
USE   NEAR INFRARED RADIATION

Radiation, Near Ultraviolet
USE   NEAR ULTRAVIOLET RADIATION

Radiation Noise
USE   ELECTROMAGNETIC NOISE

Radiation, Nonequilibrium
USE   NONEQUILIBRIUM RADIATION

Radiation, Nuclear
USE   NUCLEAR RADIATION

Radiation Patterns, Antenna
USE   ANTENNA RADIATION PATTERNS

Radiation, Planetary
USE   PLANETARY RADIATION

Radiation, Plasma
USE   PLASMA RADIATION

Radiation, Polarized
USE   POLARIZED RADIATION

Radiation, Polarized Electromagnetic
USE   POLARIZED ELECTROMAGNETIC
      RADIATION

Radiation, Post-Blast Nuclear
USE   POST-BLAST NUCLEAR RADIATION

RADIATION PRESSURE

RADIATION PROTECTION

Radiation, Pulsed
USE   PULSED RADIATION

RADIATION PYROMETERS

Radiation, Radio Frequency
USE   RADIO WAVES

Radiation, Reflected
USE   REFLECTED WAVES

Radiation, Refracted
USE   REFRACTED WAVES

Radiation, Relic
USE   RELIC RADIATION

Radiation Resistance
USE   RADIATION TOLERANCE

Radiation, Resonance
USE   RESONANCE FLUORESCENCE

RADIATION SHIELDING

Radiation Shielding, Solar
USE   SOLAR RADIATION SHIELDING

Radiation, Short Wave
USE   SHORT WAVE RADIATION

RADIATION SICKNESS

Radiation, Sky
USE   SKY RADIATION

Radiation, Solar
USE   SOLAR RADIATION

Radiation, Solar Corpuscular
USE   SOLAR CORPUSCULAR RADIATION

(Radiation), Solar Plasma
USE   SOLAR WIND

RADIATION SOURCES

Radiation, Space
USE   EXTRATERRESTRIAL RADIATION

RADIATION SPECTRA

Radiation Spectroscopy, Nuclear
USE   NUCLEAR RADIATION SPECTROSCOPY

Radiation, Stellar
USE   STELLAR RADIATION

Radiation, Stokes Law Of
USE   STOKES LAW OF RADIATION

Radiation, Stratosphere
  USE   STRATOSPHERE RADIATION

Radiation, Synchrotron
  USE   SYNCHROTRON RADIATION

Radiation, Terrestrial
  USE   TERRESTRIAL RADIATION

RADIATION THERAPY

Radiation, Thermal
  USE   THERMAL RADIATION

RADIATION TOLERANCE

RADIATION TRANSPORT

RADIATION TRAPPING

Radiation, Tropospheric
  USE   TROPOSPHERIC RADIATION

Radiation, Ultrasonic
  USE   ULTRASONIC RADIATION

Radiation, Ultraviolet
  USE   ULTRAVIOLET RADIATION

Radiation, Vacuum Ultraviolet
  USE   FAR ULTRAVIOLET RADIATION

Radiation, Visible
  USE   LIGHT (VISIBLE RADIATION)

Radiation, Wave
  USE   ELECTROMAGNETIC RADIATION

Radiation 1 Satellite, Solar
  USE   SOLAR RADIATION 1 SATELLITE

Radiation 3 Satellite, Solar
  USE   SOLAR RADIATION 3 SATELLITE

RADIATIVE HEAT TRANSFER

RADIATIVE LIFETIME

RADIATIVE RECOMBINATION

RADIATIVE TRANSFER

RADIATORS

Radiators, Condenser
  USE   CONDENSERS (LIQUEFIERS)
         HEAT RADIATORS

Radiators, Heat
  USE   HEAT RADIATORS

Radiators, Space
  USE   SPACECRAFT RADIATORS

Radiators, Spacecraft
  USE   SPACECRAFT RADIATORS

Radical, Vanadyl
  USE   VANADYL RADICAL

Radical, Vinyl
  USE   VINYL RADICAL

RADICALS

Radicals, Free
  USE   FREE RADICALS

Radicals, Hydroxyl
  USE   HYDROXYL RADICALS

RADII

RADIO ALTIMETERS

Radio Antenna Grid (Navy), Underground
  USE   SEAFARER PROJECT

RADIO ANTENNAS

RADIO ASTRONOMY

Radio Astronomy Explorer B
  USE   EXPLORER 49 SATELLITE

RADIO ASTRONOMY EXPLORER SATELLITE

Radio Astronomy Explorer 2
  USE   EXPLORER 49 SATELLITE

RADIO ATTENUATION

RADIO ATTENUATION MEASUREMENT PROJECT

RADIO AURORAS

Radio Beacon Ionospheric Sounder, Orbiting
  USE   ORBIS

RADIO BEACONS

Radio Blackout, Polar
  USE   POLAR RADIO BLACKOUT

Radio Broadcasting
  USE   BROADCASTING

RADIO BURSTS

Radio Bursts, Solar
  USE   SOLAR RADIO BURSTS

RADIO COMMUNICATION

RADIO CONTROL

RADIO DIRECTION FINDERS

(Radio), Direction Finders
  USE   RADIO DIRECTION FINDERS

RADIO ECHOES

RADIO ELECTRONICS

RADIO EMISSION

Radio Emission, Solar
  USE   SOLAR RADIO EMISSION

RADIO EQUIPMENT

Radio Equipment, Ultra Short Wave
  USE   VERY HIGH FREQUENCY RADIO
         EQUIPMENT

Radio Equipment, Very High Frequency
  USE   VERY HIGH FREQUENCY RADIO
         EQUIPMENT

RADIO FILTERS

RADIO FREQUENCIES

Radio Frequencies, Extremely Low
  USE   EXTREMELY LOW RADIO FREQUENCIES

RADIO FREQUENCY DISCHARGE

RADIO FREQUENCY HEATING

RADIO FREQUENCY IMPEDANCE PROBES

RADIO FREQUENCY INTERFERENCE

Radio Frequency Ion Thrustor Engines
  USE   RIT ENGINES

Radio Frequency Noise
  USE   ELECTROMAGNETIC NOISE

Radio Frequency Radiation
  USE   RADIO WAVES

RADIO FREQUENCY SHIELDING

RADIO GALAXIES

RADIO HORIZONS

Radio Interference
  USE   RADIO FREQUENCY INTERFERENCE

RADIO INTERFEROMETERS

(Radio Interferometry Network), Orion
  USE   ORION (RADIO INTERFEROMETRY
         NETWORK)

Radio Meteor Project, Harvard
  USE   HARVARD RADIO METEOR PROJECT

RADIO METEOROLOGY

RADIO METEORS

RADIO NAVIGATION

RADIO OBSERVATION

RADIO OCCULTATION

RADIO PHYSICS

RADIO PROBING

Radio Propagation
  USE   RADIO TRANSMISSION

Radio Propagation, Transhorizon
  USE   TRANSHORIZON RADIO PROPAGATION

RADIO RANGE

Radio Ranges
  USE   RADIO BEACONS

Radio Ranges, Omnidirectional
  USE   OMNIDIRECTIONAL RADIO RANGES

RADIO RECEIVERS

RADIO RECEPTION

Radio Reflection
  USE   RADIO ECHOES

RADIO RELAY SYSTEMS

RADIO SCATTERING

Radio Signal Attenuation
  USE   RADIO ATTENUATION

Radio Signal Propagation
  USE   RADIO TRANSMISSION

RADIO SIGNALS

RADIO SOURCES (ASTRONOMY)

Radio Sources, Extragalactic
  USE   EXTRAGALACTIC RADIO SOURCES

(Radio Sources), QSO
  USE   QUASARS

Radio Sources, Quasi-Stellar
  USE   QUASARS

RADIO SPECTRA

RADIO SPECTROSCOPY

RADIO STARS

RADIO TELEGRAPHY

RADIO TELEMETRY

RADIO TELESCOPES

RADIO TRACKING

RADIO TRANSMISSION

Radio Transmission, Short Wave
  USE   SHORT WAVE RADIO TRANSMISSION

RADIO TRANSMITTERS

RADIO WAVE REFRACTION

RADIO WAVES

Radio Waves, Cosmic
USE   EXTRATERRESTRIAL RADIO WAVES

Radio Waves, Extraterrestrial
USE   EXTRATERRESTRIAL RADIO WAVES

Radio Waves, Galactic
USE   GALACTIC RADIO WAVES

Radio Waves, Solar
USE   SOLAR RADIO EMISSION

RADIOACTIVE AGE DETERMINATION

RADIOACTIVE CONTAMINANTS

Radioactive Dating
USE   RADIOACTIVE AGE DETERMINATION

RADIOACTIVE DEBRIS

RADIOACTIVE DECAY

Radioactive Elements
USE   RADIOACTIVE ISOTOPES

RADIOACTIVE ISOTOPES

RADIOACTIVE MATERIALS

Radioactive Nuclides
USE   RADIOACTIVE ISOTOPES

RADIOACTIVE WASTES

RADIOACTIVITY

(Radioactivity), Washout
USE   FALLOUT

RADIOBIOLOGY

RADIOCARDIOGRAPHY

RADIOCHEMICAL SEPARATION

RADIOCHEMISTRY

RADIOGENIC MATERIALS

RADIOGONIOMETERS

RADIOGRAPHY

Radiography, Neutron
USE   NEUTRON RADIOGRAPHY

RADIOIMMUNOASSAY

RADIOISOTOPE BATTERIES

Radiolocation, Wildlife
USE   WILDLIFE RADIOLOCATION

RADIOLOGY

RADIOLYSIS

RADIOMETEOROGRAPHS

Radiometer, Visible Infrared Spin Scan
USE   VISIBLE INFRARED SPIN SCAN
RADIOMETER

RADIOMETERS

Radiometers, Dicke
USE   DICKE RADIOMETERS

Radiometers, Dicke Type
USE   DICKE RADIOMETERS

Radiometers, Infrared
USE   INFRARED RADIOMETERS

Radiometers, Microwave
USE   MICROWAVE RADIOMETERS

Radiometers, Passive L-Band
USE   PASSIVE L-BAND RADIOMETERS

Radiometers, Pressure Modulator
USE   PRESSURE MODULATOR RADIOMETERS

Radiometers, Spectro
USE   SPECTRORADIOMETERS

RADIOMETRIC CORRECTION

Radiometric Rectification
USE   RADIOMETRIC CORRECTION

RADIOMETRIC RESOLUTION

Radionuclides
USE   RADIOACTIVE ISOTOPES

RADIOPATHOLOGY

RADIOPHOSPHORS

Radioprotective Agents
USE   ANTIRADIATION DRUGS

Radiosensitivity
USE   RADIATION TOLERANCE

RADIOSONDES

Radiosondes, Endo
USE   ENDORADIOSONDES

RADIOTELEPHONES

Radiotherapy
USE   RADIATION THERAPY

RADIUM

RADIUM ISOTOPES

RADIUM 226

Radius
USE   RADII

Radius, Larmor
USE   LARMOR RADIUS

RADOME MATERIALS

RADOMES

RADON

RADON ISOTOPES

RADUGA SATELLITE

RAE B
USE   EXPLORER 49 SATELLITE

RAE 1
USE   EXPLORER 49 SATELLITE

RAE 2
USE   EXPLORER 49 SATELLITE

RAE-1
USE   EXPLORER 38 SATELLITE

RAFTS

Rafts, Life
USE   LIFE RAFTS

RAIL TRANSPORTATION

RAILGUN ACCELERATORS

RAILROAD HUMPING TESTS

Railroads
USE   RAIL TRANSPORTATION

RAILS

RAIN

Rain, Acid
USE   ACID RAIN

RAIN EROSION

RAIN FORESTS

RAIN GAGES

RAIN IMPACT DAMAGE

RAINBOWS

RAINDROPS

RAINMAKING

RAINSTORMS

RAKES

RAM

RAM B LAUNCH VEHICLE

RAM Effect, Hydrodynamic
USE   HYDRODYNAMIC RAM EFFECT

RAM Project
USE   RADIO ATTENUATION MEASUREMENT
PROJECT

Raman Effect
USE   RAMAN SPECTRA

RAMAN LASERS

Raman Scattering
USE   RAMAN SPECTRA

RAMAN SPECTRA

RAMAN SPECTROSCOPY

Raman Spectroscopy, Coherent Anti-Stokes
USE   RAMAN SPECTROSCOPY

RAMJET ENGINES

Ramjet Engines, Low Volume
USE   LOW VOLUME RAMJET ENGINES

Ramjet Engines, Nuclear
USE   NUCLEAR RAMJET ENGINES

Ramjet Engines, Supersonic Combustion
USE   SUPERSONIC COMBUSTION RAMJET
ENGINES

RAMJET MISSILES

Ramjets, Integral Rocket
USE   INTEGRAL ROCKET RAMJETS

RAMP FUNCTIONS

RAMPS

RAMPS (STRUCTURES)

RAMS (PRESSES)

RAMS (PUMPS)

RAMSAUER EFFECT

RAND PROJECT

RANDOM ACCESS

RANDOM ACCESS MEMORY

Random Distributions
USE   STATISTICAL DISTRIBUTIONS

RANDOM ERRORS

RANDOM LOADS

RANDOM NOISE

RANDOM NUMBERS

RANDOM PROCESSES

RANDOM SAMPLING

RANDOM SIGNALS

RANDOM VARIABLES

RANDOM VIBRATION

RANDOM WALK

RANGE

Range And Orbit Determination, Airborne
USE   AIRBORNE RANGE AND ORBIT
           DETERMINATION

RANGE AND RANGE RATE TRACKING

Range Ballistic Missiles, Intermediate
USE   INTERMEDIATE RANGE BALLISTIC
           MISSILES

Range Ballistic Missiles, Short
USE   SHORT RANGE BALLISTIC MISSILES

Range (CA-OR-WA), Cascade
USE   CASCADE RANGE (CA-OR-WA)

Range Control
USE   TRAJECTORY CONTROL

Range, Down
USE   DOWNRANGE

RANGE ERRORS

RANGE (EXTREMES)

RANGE FINDERS

Range Finders, Laser
USE   LASER RANGE FINDERS

Range Finders, Optical
USE   OPTICAL RANGE FINDERS

Range Indicators
USE   RANGE FINDERS

Range Instrumentation Aircraft, Advanced
USE   ADVANCED RANGE INSTRUMENTATION
           AIRCRAFT

Range Instrumentation Ship, Advanced
USE   ADVANCED RANGE INSTRUMENTATION
           SHIP

Range Measurement
USE   RANGEFINDING

Range Navigation, Long
USE   LORAN
           LORAN D

Range Navigation, Short
USE   SHORAN

Range, Optical Slant
USE   OPTICAL SLANT RANGE

Range, Radar
USE   RADAR RANGE

Range, Radio
USE   RADIO RANGE

Range Rate Tracking, Range And
USE   RANGE AND RANGE RATE TRACKING

Range, Reentry
USE   REENTRY RANGE

RANGE RESOURCES

RANGE SAFETY

Range Weather Forecasting, Long
USE   LONG RANGE WEATHER FORECASTING

Range (WY), Wind River
USE   WIND RIVER RANGE (WY)

(Range-Orbit Determination), AROD
USE   AIRBORNE RANGE AND ORBIT
           DETERMINATION

RANGEFINDING

Rangefinding, Lunar
USE   LUNAR RANGEFINDING

RANGELANDS

Rangemaster Aircraft
USE   G-1 AIRCRAFT

Rangemaster Aircraft, Navion
USE   G-1 AIRCRAFT

RANGER BLOCK 3 TELEVISION SYSTEM

RANGER LUNAR LANDING VEHICLES

RANGER LUNAR PROBES

Ranger Program, Agena B
USE   AGENA B RANGER PROGRAM

RANGER PROJECT

Ranger Satellites
USE   RANGER LUNAR PROBES

RANGER 1 LUNAR PROBE

RANGER 2 LUNAR PROBE

RANGER 3 LUNAR PROBE

RANGER 4 LUNAR PROBE

RANGER 5 LUNAR PROBE

RANGER 6 LUNAR PROBE

RANGER 7 LUNAR PROBE

RANGER 8 LUNAR PROBE

RANGER 9 LUNAR PROBE

Ranger/tracker, Laser
USE   LASER RANGER/TRACKER

Ranges, Ballistic
USE   BALLISTIC RANGES

Ranges (CA), Coastal
USE   COASTAL RANGES (CA)

Ranges (CA), Peninsular
USE   PENINSULAR RANGES (CA)

RANGES (FACILITIES)

Ranges, Frequency
USE   FREQUENCY RANGES

Ranges, Missile
USE   MISSILE RANGES

Ranges, Omnidirectional Radio
USE   OMNIDIRECTIONAL RADIO RANGES

Ranges, Radio
USE   RADIO BEACONS

Ranges, Test
USE   TEST RANGES

Ranging
USE   RANGEFINDING

Ranging Radar, North American Search And
USE   NORTH AMERICAN SEARCH AND RANGING
           RADAR

Ranging, Sound
USE   SOUND RANGING

Ranging, Sound Detecting And
USE   SOUND DETECTING AND RANGING

Ranging, Sound Fixing And
USE   SOUND FIXING AND RANGING

RANK TESTS

RANKINE CYCLE

RANKINE-HUGONIOT RELATION

RANKING

RAOULT LAW

RAPCON (Control)
USE   RADAR APPROACH CONTROL

Raphson Method, Newton-
USE   NEWTON-RAPHSON METHOD

RAPID BALLISTICS IDENTIFICATION

RAPID EYE MOVEMENT STATE

RAPID QUENCHING (METALLURGY)

RAPID TRANSIT SYSTEMS

RAPIDS

Rapids (IA), Cedar
USE   CEDAR RAPIDS (IA)

RARE EARTH ALLOYS

RARE EARTH COMPOUNDS

RARE EARTH ELEMENTS

RARE GAS COMPOUNDS

RARE GAS-HALIDE LASERS

RARE GASES

RAREFACTION

Rarefaction Waves
USE   ELASTIC WAVES

RAREFIED GAS DYNAMICS

RAREFIED GASES

RAREFIED PLASMAS

Rasers
USE   MASERS

Rate, Bit Error
USE   BIT ERROR RATE

Rate, Burning
USE   BURNING RATE

Rate Computers, Counting
USE   COUNTING RATE COMPUTERS

Rate, Drift
USE   DRIFT RATE

Rate, Electron Decay
USE   ELECTRON DECAY RATE

Rate, Evaporation
USE   EVAPORATION RATE

Rate, Flow
USE   FLOW VELOCITY

(Rate), Flux
USE   FLUX (RATE)

Rate, Heart
USE   HEART RATE

Rate, Lapse
USE   LAPSE RATE

Rate, Loading
USE   LOADING RATE

Rate, Mass Flow
USE   MASS FLOW RATE

Rate Meters
USE   MEASURING INSTRUMENTS

RATE OF CLIMB INDICATORS

(Rate Per Unit Area), Flux
USE   FLUX DENSITY

Rate, Pulse
USE   PULSE RATE

Rate, Pulse Repetition
USE   PULSE REPETITION RATE

Rate, Reaction
USE   REACTION KINETICS

Rate, Respiratory
USE   RESPIRATORY RATE

Rate, Signal Fading
USE   SIGNAL FADING RATE

Rate, Strain
USE   STRAIN RATE

Rate Tracking, Range And Range
USE   RANGE AND RANGE RATE TRACKING

(Rate/area), Density
USE   FLUX DENSITY

Rates, Collision
USE   COLLISION RATES

Rates, Decay
USE   DECAY RATES

Rates, Ion Production
USE   ION PRODUCTION RATES

RATES (PER TIME)

RATINGS

Ratio, Aspect
USE   ASPECT RATIO

Ratio, Bypass
USE   BYPASS RATIO

Ratio, Compression
USE   COMPRESSION RATIO

Ratio, Fineness
USE   FINENESS RATIO

Ratio, Fuel-Air
USE   FUEL-AIR RATIO

Ratio, Hematocrit
USE   HEMATOCRIT RATIO

Ratio, High Aspect
USE   HIGH ASPECT RATIO

Ratio, Lift Drag
USE   LIFT DRAG RATIO

Ratio, Likelihood
USE   LIKELIHOOD RATIO

Ratio, Low Aspect
USE   LOW ASPECT RATIO

Ratio, Mills
USE   MILLS RATIO

Ratio, Payload Mass
USE   PAYLOAD MASS RATIO

Ratio, Poisson
USE   POISSON RATIO

Ratio, Pressure
USE   PRESSURE RATIO

Ratio, Propellant Mass
USE   PROPELLANT MASS RATIO

(Ratio), Scale
USE   SCALE (RATIO)

Ratio, Stress
USE   STRESS RATIO

Ratio, Temperature
USE   TEMPERATURE RATIO

Ratio, Thickness
USE   THICKNESS RATIO

Ratio, Thrust-Weight
USE   THRUST-WEIGHT RATIO

Ratio, Void
USE   VOID RATIO

Ratio Wings, High Aspect
USE   SLENDER WINGS

Ratio Wings, Low Aspect
USE   LOW ASPECT RATIO WINGS

Ratioing, Band
USE   BAND RATIOING

RATIOMETERS

RATIONAL FUNCTIONS

RATIONS

Rations, Space
USE   SPACE RATIONS

RATIOS

Ratios, Carrier To Noise
USE   CARRIER TO NOISE RATIOS

(Ratios), Indexes
USE   INDEXES (RATIOS)

Ratios, Mass
USE   MASS RATIOS

Ratios, Mass To Light
USE   MASS TO LIGHT RATIOS

Ratios, Modular
USE   MODULAR RATIOS

Ratios, Signal To Noise
USE   SIGNAL TO NOISE RATIOS

Ratios, Standing Wave
USE   STANDING WAVE RATIOS

RATS

RATSCAT Program
USE   RADAR TARGET SCATTER SITE PROGRAM

Raven Helicopter
USE   OH-23 HELICOPTER

RAVINES

RAWINSONDES

Ray Absorptiometry, Gamma
USE   GAMMA RAY ABSORPTIOMETRY

Ray Absorption, Gamma
USE   GAMMA RAY ABSORPTION

Ray Absorption, X
USE   X RAY ABSORPTION

Ray Acoustics
USE   GEOMETRICAL ACOUSTICS

Ray Albedo, Cosmic
USE   COSMIC RAY ALBEDO

Ray Analysis, X
USE   X RAY ANALYSIS

Ray Apparatus, X
USE   X RAY APPARATUS

Ray Astronomy Explorer, Gamma
USE   EXPLORER 11 SATELLITE

Ray Astronomy, Gamma
USE   GAMMA RAY ASTRONOMY

Ray Astronomy, X
USE   X RAY ASTRONOMY

Ray Astrophysical Facility, Advanced X
USE   X RAY ASTROPHYSICS-FACILITY

Ray Astrophysics Facility, Advanced X
USE   X RAY ASTROPHYSICS FACILITY

Ray Astrophysics Facility, X
USE   X RAY ASTROPHYSICS FACILITY

Ray Beams, Gamma
USE   GAMMA RAY BEAMS

Ray Binaries, X
USE   X RAY BINARIES

Ray Bursts, Cosmic Gamma
USE   GAMMA RAY BURSTS

Ray Bursts, Gamma
USE   GAMMA RAY BURSTS

Ray Density Measurement, X
USE   X RAY DENSITY MEASUREMENT

Ray Diffraction, X
USE   X RAY DIFFRACTION

Ray Fluorescence, X
USE   X RAY FLUORESCENCE

Ray Imagery, X
USE   X RAY IMAGERY

Ray Imaging Scope, Low Intensity X
USE   LIXISCOPES

Ray Inspection, X
USE   X RAY INSPECTION

Ray Irradiation, X
USE   X RAY IRRADIATION

Ray Lasers, Gamma
USE   GAMMA RAY LASERS

Ray Lasers, X
USE   X RAY LASERS

Ray Observatory, Gamma
USE   GAMMA RAY OBSERVATORY

Ray Optics
USE   GEOMETRICAL OPTICS

Ray Primaries, Heavy Cosmic
USE   HEAVY NUCLEI
       PRIMARY COSMIC RAYS

Ray Scattering, X
USE   X RAY SCATTERING

Ray Showers, Cosmic
USE   COSMIC RAY SHOWERS

Ray Sources, X
USE   X RAY SOURCES

Ray Spectra, Gamma
USE   GAMMA RAY SPECTRA

Ray Spectra, X
USE   X RAY SPECTRA

Ray Spectrography, X
USE   X RAY SPECTROSCOPY

Ray Spectrometers, Gamma
USE   GAMMA RAY SPECTROMETERS

Ray Spectrometry, X
USE   X RAY SPECTROSCOPY

Ray Spectropolarimetry Payload, X
USE   EXPOS (SPACELAB PAYLOAD)

Ray Spectroscopy, X
USE   X RAY SPECTROSCOPY

Ray Stress Analysis, X
USE   X RAY STRESS ANALYSIS

Ray Stress Measurement, X
USE   X RAY STRESS MEASUREMENT

Ray Telescopes, Gamma
USE   GAMMA RAY TELESCOPES

Ray Telescopes, X
USE   X RAY TELESCOPES

Ray Timing Explorer, X
USE   X RAY TIMING EXPLORER

RAY TRACING

Ray Tubes, Cathode
USE   CATHODE RAY TUBES

Ray Tubes, X
USE   X RAY TUBES

Rayet Stars, Wolf-
USE   WOLF-RAYET STARS

RAYLEIGH DISTRIBUTION

RAYLEIGH EQUATIONS

RAYLEIGH NUMBER

RAYLEIGH SCATTERING

RAYLEIGH WAVES

RAYLEIGH-BENARD CONVECTION

RAYLEIGH-RITZ METHOD

RAYON

RAYS

Rays, Cosmic
USE   COSMIC RAYS

Rays, Cosmic X
USE   COSMIC X RAYS

Rays, Galactic Cosmic
USE   GALACTIC COSMIC RAYS

Rays, Gamma
USE   GAMMA RAYS

Rays, Lunar
USE   LUNAR RAYS

Rays, Primary Cosmic
USE   PRIMARY COSMIC RAYS

Rays, Reflected
USE   REFLECTED WAVES

Rays, Refracted
USE   REFRACTED WAVES

Rays, Secondary Cosmic
USE   SECONDARY COSMIC RAYS

Rays, Solar Cosmic
USE   SOLAR COSMIC RAYS

Rays, Solar X-
USE   SOLAR X-RAYS

Rays, X
USE   X RAYS

RAYTHEON COMPUTERS

RAZOR BLADES

Rb
USE   RUBIDIUM

RB-47 Aircraft
USE   B-47 AIRCRAFT

RB-50 AIRCRAFT

RB-57 Aircraft
USE   B-57 AIRCRAFT

RB-66 Aircraft
USE   B-66 AIRCRAFT

RBE
USE   RELATIVE BIOLOGICAL EFFECTIVENESS
       (RBE)

(RBE), Relative Biological Effectiveness
USE   RELATIVE BIOLOGICAL EFFECTIVENESS
       (RBE)

RC CIRCUITS

RC Networks
USE   RC CIRCUITS

RCA COMPUTERS

RCA SATCOM SATELLITES

RCA SPECTRA 70 COMPUTER

RCA-110 COMPUTERS

RDX

Re
USE   RHENIUM

REACTANCE

REACTION

REACTION BONDING

REACTION CONTROL

Reaction Control, Chemical
USE   CHEMICAL REACTION CONTROL

Reaction, Friedel-Craft
USE   FRIEDEL-CRAFT REACTION

(Reaction Inhibition), Poisoning
USE   POISONING (REACTION INHIBITION)

Reaction Jet Backpacks
USE   SELF MANEUVERING UNITS

Reaction Jets
USE   JET THRUST
       JET FLOW

REACTION KINETICS

Reaction, Michael
USE   MICHAEL REACTION

REACTION PRODUCTS

Reaction Rate
USE   REACTION KINETICS

Reaction, Sabatier
USE   SABATIER REACTION

REACTION TIME

REACTION WHEELS

Reactions, Annihilation
USE   ANNIHILATION REACTIONS

Reactions, Association
USE   ASSOCIATION REACTIONS

Reactions, Chemical
USE   CHEMICAL REACTIONS

Reactions, Diels-Alder
USE   DIELS-ALDER REACTIONS

Reactions, Endothermic
USE   ENDOTHERMIC REACTIONS

Reactions, Exothermic
USE   EXOTHERMIC REACTIONS

Reactions, Grignard
USE   GRIGNARD REACTIONS

Reactions, Human
USE   HUMAN REACTIONS

Reactions, Ionic
USE   IONIC REACTIONS

Reactions, Metal-Water
USE   METAL WATER REACTIONS

Reactions, Nuclear
USE   NUCLEAR REACTIONS

Reactions, Oxidation-Reduction
USE   OXIDATION-REDUCTION REACTIONS

Reactions, Photochemical
USE   PHOTOCHEMICAL REACTIONS

Reactions, Photonuclear
USE   PHOTONUCLEAR REACTIONS

Reactions, Proton-Proton
USE   PROTON-PROTON REACTIONS

Reactions, Recombination
USE   RECOMBINATION REACTIONS

Reactions, Surface
USE   SURFACE REACTIONS

Reactions, Thermonuclear
USE   THERMONUCLEAR REACTIONS

REACTIVITY

Reactor, Advanced Sodium Cooled
USE   ADVANCED SODIUM COOLED REACTOR

Reactor, ASCR
USE   ADVANCED SODIUM COOLED REACTOR

Reactor, Astron Thermonuclear
USE   ASTRON THERMONUCLEAR REACTOR

Reactor, ATR
USE   ADVANCED TEST REACTORS

Reactor Chemistry
USE   RADIOCHEMISTRY

Reactor Control, Nuclear
USE   NUCLEAR REACTOR CONTROL

REACTOR CORES

REACTOR DESIGN

Reactor, EBR-1
USE   EXPERIMENTAL BREEDER REACTOR 1

Reactor, EBR-2
USE   EXPERIMENTAL BREEDER REACTOR 2

(Reactor), EBWR
USE   EXPERIMENTAL BOILING WATER
       REACTORS

(Reactor), EGCR
USE   EXPERIMENTAL GAS COOLED REACTORS

(Reactor), EOCR
USE   EXPERIMENTAL ORGANIC COOLED
       REACTORS

Reactor Experiment, Lithium Cooled
USE   LITHIUM COOLED REACTOR EXPERIMENT

Reactor Experiment, Sodium
USE   SODIUM REACTOR EXPERIMENT

Reactor Fuels
USE   NUCLEAR FUELS

Reactor, Halden
USE   HALDEN BOILING WATER REACTOR

Reactor, Halden Boiling Water
USE   HALDEN BOILING WATER REACTOR

Reactor, HBWR
USE   HALDEN BOILING WATER REACTOR

Reactor, Health Physics Research
USE   HEALTH PHYSICS RESEARCH REACTOR

Reactor, Hero
USE   HERO REACTOR

(Reactor), HFIR
USE   HIGH FLUX ISOTOPE REACTORS

Reactor In Flight Test Program
USE   RIFT (REACTOR IN FLIGHT TEST)

(Reactor In Flight Test), Rift
USE   RIFT (REACTOR IN FLIGHT TEST)

(Reactor), Inertial Fusion
USE   INERTIAL FUSION (REACTOR)

Reactor, Janus
USE   JANUS REACTOR

Reactor, KIWI B-1
USE   KIWI B-1 REACTOR

Reactor, KIWI B-4
USE   KIWI B-4 REACTOR

Reactor, LCRE
USE   LITHIUM COOLED REACTOR EXPERIMENT

Reactor, Livermore Pool Type
USE   LIVERMORE POOL TYPE REACTOR

Reactor, Los Alamos Molten Plutonium
USE   LOS ALAMOS MOLTEN PLUTONIUM
       REACTOR

Reactor, Los Alamos Turret
USE   HIGH TEMPERATURE NUCLEAR REACTORS

Reactor, Los Alamos Water Boiler
USE   LOS ALAMOS WATER BOILER REACTOR

Reactor, LPTR
USE   LIVERMORE POOL TYPE REACTOR

REACTOR MATERIALS

Reactor, Orgel
USE   ORGANIC COOLED REACTORS

Reactor, Pathfinder Nuclear
USE   PATHFINDER NUCLEAR REACTOR

Reactor, Phoebus Nuclear
USE   PHOEBUS NUCLEAR REACTOR

Reactor, Physical Constants Testing
USE   NUCLEAR RESEARCH AND TEST
       REACTORS
       WATER COOLED REACTORS

REACTOR PHYSICS

Reactor, Plum Brook
USE   PLUM BROOK REACTOR

Reactor, Plutonium Recycle Test
USE   PLUTONIUM RECYCLE TEST REACTOR

(Reactor), PRTR
USE   PLUTONIUM RECYCLE TEST REACTOR

REACTOR SAFETY

Reactor Sites, Offshore
USE   OFFSHORE REACTOR SITES

Reactor, Snaptran
USE   SNAPTRAN REACTOR

Reactor, Spectral Shift Control
USE   SPECTRAL SHIFT CONTROL REACTOR

Reactor, SRE
USE   SODIUM REACTOR EXPERIMENT

REACTOR STARTUP TESTS

REACTOR TECHNOLOGY

Reactor Test Facility, Transient
USE   TRANSIENT REACTOR TEST FACILITY

Reactor, Tory 2
USE   TORY 2 REACTOR

Reactor, Tory 2-A
USE   TORY 2-A REACTOR

Reactor, Tory 2-C
USE   TORY 2-C REACTOR

Reactor, Zeta Thermonuclear
USE   ZETA THERMONUCLEAR REACTOR

Reactor 1, Experimental Breeder
USE   EXPERIMENTAL BREEDER REACTOR 1

Reactor 2, Experimental Breeder
USE   EXPERIMENTAL BREEDER REACTOR 2

Reactor 2, Tower Shielding
USE   TOWER SHIELDING REACTOR 2

Reactor 2, Zero Power
USE   ZERO POWER REACTOR 2

Reactor 3, Zero Power
USE   ZERO POWER REACTOR 3

Reactor 6, Zero Power
USE   ZERO POWER REACTOR 6

Reactor 9, Zero Power
USE   ZERO POWER REACTOR 9

REACTORS

Reactors, Advanced Test
USE   ADVANCED TEST REACTORS

Reactors, Annular Core Pulse
USE   ANNULAR CORE PULSE REACTORS

Reactors, Bio
USE   BIOREACTORS

Reactors), Blankets (Fission
USE   BLANKETS (FISSION REACTORS)

Reactors), Blankets (Fusion
USE   BLANKETS (FUSION REACTORS)

Reactors, Boiling Water
USE   BOILING WATER REACTORS

Reactors, Breeder
USE   BREEDER REACTORS

Reactors, Chemical
USE   CHEMICAL REACTORS

Reactors, Electric
USE   ELECTRIC REACTORS

Reactors, Engineering Test
USE   ENGINEERING TEST REACTORS

(Reactors), ETR
USE   ENGINEERING TEST REACTORS

Reactors, Experimental Boiling Water
USE   EXPERIMENTAL BOILING WATER
       REACTORS

Reactors, Experimental Gas Cooled
USE   EXPERIMENTAL GAS COOLED REACTORS

Reactors, Experimental Organic Cooled
USE   EXPERIMENTAL ORGANIC COOLED
       REACTORS

Reactors, Fast Nuclear
USE   FAST NUCLEAR REACTORS

Reactors, Fast Oxide
USE   FAST OXIDE REACTORS

Reactors, Fast Test
USE   FAST TEST REACTORS

Reactors), Fuel Elements (Nuclear
USE   NUCLEAR FUEL ELEMENTS

Reactors, Fusion
USE   FUSION REACTORS

Reactors, Fusion-Fission Hybrid
USE   FUSION-FISSION HYBRID REACTORS

Reactors, Gas
USE   GAS REACTORS

Reactors, Gas Cooled
USE   GAS COOLED REACTORS

Reactors, Gas Cooled Fast
USE   GAS COOLED FAST REACTORS

Reactors, Gaseous Fission
USE   GASEOUS FISSION REACTORS

(Reactors), GCR
USE   GAS COOLED REACTORS

Reactors, Hanford
USE   HANFORD REACTORS

Reactors, Heavy Water
USE   HEAVY WATER REACTORS

Reactors, Heavy Water Components Test
USE   HEAVY WATER COMPONENTS TEST
       REACTORS

Reactors, High Flux Beam
USE   HIGH FLUX BEAM REACTORS

Reactors, High Flux Isotope
USE   HIGH FLUX ISOTOPE REACTORS

Reactors, High Temperature Gas Cooled
USE   HIGH TEMPERATURE GAS COOLED
       REACTORS

Reactors, High Temperature Nuclear
USE   HIGH TEMPERATURE NUCLEAR REACTORS

Reactors, KIWI
USE   KIWI REACTORS

Reactors, KIWI B
USE   KIWI B REACTORS

Reactors, KIWI Rocket
USE   KIWI REACTORS

Reactors, Light Water
USE   LIGHT WATER REACTORS

Reactors, Light Water Breeder
USE   LIGHT WATER BREEDER REACTORS

Reactors), Limiters (Fusion
USE   LIMITERS (FUSION REACTORS)

Reactors, Liquid Cooled
USE   LIQUID COOLED REACTORS

Reactors, Liquid Metal Cooled
USE   LIQUID METAL COOLED REACTORS

Reactors, Liquid Metal Fast Breeder
USE   LIQUID METAL FAST BREEDER REACTORS

(Reactors), LMCR
USE   LIQUID METAL COOLED REACTORS

Reactors, Materials Testing
USE   NUCLEAR RESEARCH AND TEST
        REACTORS

Reactors, MCR
USE   MILITARY COMPACT REACTORS

Reactors, Military Compact
USE   MILITARY COMPACT REACTORS

Reactors, Molten Salt Nuclear
USE   MOLTEN SALT NUCLEAR REACTORS

Reactors, MSRE
USE   MOLTEN SALT NUCLEAR REACTORS

Reactors, NRX
USE   NRX REACTORS

Reactors, Nuclear
USE   NUCLEAR REACTORS

Reactors, Nuclear Power
USE   NUCLEAR POWER REACTORS

Reactors, Nuclear Research And Test
USE   NUCLEAR RESEARCH AND TEST
        REACTORS

Reactors, Nuclear Test
USE   NUCLEAR RESEARCH AND TEST
        REACTORS

Reactors, Organic Cooled
USE   ORGANIC COOLED REACTORS

Reactors, Organic Moderated
USE   ORGANIC MODERATED REACTORS

(Reactors), PBRE
USE   PEBBLE BED REACTORS

Reactors, Pebble Bed
USE   PEBBLE BED REACTORS

Reactors, Plasma Core
USE   PLASMA CORE REACTORS

Reactors, Pluto
USE   PLUTO REACTORS

Reactors, Power
USE   POWER REACTORS

Reactors, Pressurized Water
USE   PRESSURIZED WATER REACTORS

Reactors, Saturable
USE   SATURABLE REACTORS

Reactors), SGR (Nuclear
USE   SODIUM GRAPHITE REACTORS

Reactors, Sodium Graphite
USE   SODIUM GRAPHITE REACTORS

Reactors, Space Power
USE   SPACE POWER REACTORS

Reactors, Space Power Unit
USE   SPACE POWER UNIT REACTORS

Reactors, Spert
USE   SPERT REACTORS

(Reactors), SPUR
USE   SPACE POWER UNIT REACTORS

(Reactors), SR
USE   SATURABLE REACTORS

Reactors, Swimming Pool
USE   SWIMMING POOL REACTORS

Reactors, Thermal
USE   THERMAL REACTORS

Reactors, Thermionic
USE   NUCLEAR ROCKET ENGINES
        ION ENGINES

Reactors), UHTREX (Nuclear
USE   HIGH TEMPERATURE NUCLEAR REACTORS

Reactors, Water Cooled
USE   WATER COOLED REACTORS

Reactors, Water Moderated
USE   WATER MODERATED REACTORS

Reactors, Zero Power
USE   ZERO POWER REACTORS

Reactors, ZPR
USE   ZERO POWER REACTORS

READ-ONLY MEMORY DEVICES

READERS

Readiness Monitor, Automatic Light Aircraft
USE   ALARM PROJECT

READING

Reading, Lip
USE   LIP READING

Reading Machines
USE   READERS

Readjustment
USE   ADJUSTING

READOUT

Readout Systems, Data
USE   DISPLAY DEVICES
        DATA SYSTEMS

Reagent, Karl Fischer
USE   KARL FISCHER REAGENT

REAGENTS

REAL GASES

REAL NUMBERS

REAL TIME OPERATION

REAL VARIABLES

(Real Variables), Integration
USE   MEASURE AND INTEGRATION

Rearward Facing Steps
USE   BACKWARD FACING STEPS

REATTACHED FLOW

Reattachment
USE   ATTACHMENT

REB
USE   RELATIVISTIC ELECTRON BEAMS

REBREATHING

RECEIVERS

Receivers, Instrument
USE   INSTRUMENT RECEIVERS

Receivers, Linear
USE   LINEAR RECEIVERS

Receivers, Logarithmic
USE   LOGARITHMIC RECEIVERS

Receivers, Radar
USE   RADAR RECEIVERS

Receivers, Radio
USE   RADIO RECEIVERS

Receivers, Solar
USE   SOLAR COLLECTORS

Receivers, Superheterodyne
USE   SUPERHETERODYNE RECEIVERS

Receivers, Television
USE   TELEVISION RECEIVERS

Receivers, Transmitter
USE   TRANSMITTER RECEIVERS

RECEIVING

Receiving Laboratory, Lunar
USE   LUNAR RECEIVING LABORATORY

Receiving Systems
USE   RECEIVERS

Receptacles (Containers)
USE   CONTAINERS

Reception
USE   RECEIVING

RECEPTION DIVERSITY

Reception, Homodyne
USE   HOMODYNE RECEPTION

Reception, Radar
USE   RADAR RECEPTION

Reception, Radio
USE   RADIO RECEPTION

Reception, Signal
USE   SIGNAL RECEPTION

Reception, Television
USE   TELEVISION RECEPTION

Receptors, Baro
USE   BARORECEPTORS

Receptors, Chemo
USE   CHEMORECEPTORS

Receptors, Gravi
USE   GRAVIRECEPTORS

Receptors, Mechano
USE   MECHANORECEPTORS

Receptors, Photo
USE   PHOTORECEPTORS

RECEPTORS (PHYSIOLOGY)

Receptors, Thermo
USE   THERMORECEPTORS

RECESSES

RECESSION

RECHARGING

RECIPROCAL THEOREMS

Reciprocating Engines
USE   PISTON ENGINES

RECIPROCATION

RECIPROCITY THEOREM

Recirculation
USE   CIRCULATION

RECIRCULATIVE FLUID FLOW

Reckoning, Dead
USE   DEAD RECKONING

RECLAMATION

Reclamation, Water
USE   WATER RECLAMATION

RECOGNITION

Recognition, Automatic Pattern
USE   PATTERN RECOGNITION

Recognition, Character
USE   CHARACTER RECOGNITION

Recognition, Machine
USE   ARTIFICIAL INTELLIGENCE

Recognition, Pattern
USE   PATTERN RECOGNITION

Recognition, Speech
USE   SPEECH RECOGNITION

Recognition, Target
USE   TARGET RECOGNITION

RECOIL ATOMS

RECOIL IONS

RECOIL PROTONS

RECOILINGS

Recombination, Atomic
USE   ATOMIC RECOMBINATION

RECOMBINATION COEFFICIENT

Recombination, Electron
USE   ELECTRON RECOMBINATION

Recombination, Electron-Ion
USE   ELECTRON-ION RECOMBINATION

Recombination, Ion
USE   ION RECOMBINATION

Recombination, Oxygen
USE   OXYGEN RECOMBINATION

Recombination, Radiative
USE   RADIATIVE RECOMBINATION

RECOMBINATION REACTIONS

Recombinations, Hydrogen
USE   HYDROGEN RECOMBINATIONS

RECOMMENDATIONS

Recompression
USE   COMPRESSING

Reconn Electric Spacecraft, Advanced
USE   ADVANCED RECONN ELECTRIC
         SPACECRAFT

RECONNAISSANCE

Reconnaissance, Aerial
USE   AERIAL RECONNAISSANCE

RECONNAISSANCE AIRCRAFT

Reconnaissance Aircraft, Light Armed
USE   COIN AIRCRAFT

Reconnaissance Aircraft, Weather
USE   WEATHER RECONNAISSANCE AIRCRAFT

Reconnaissance, Photo
USE   PHOTORECONNAISSANCE

RECONNAISSANCE SPACECRAFT

Reconnaissance Spacecraft, Photo
USE   PHOTO RECONNAISSANCE SPACECRAFT

Reconnaissance, Spectral
USE   SPECTRAL RECONNAISSANCE

(Reconnaissance Sys), Airs
USE   AIRBORNE INTEGRATED
         RECONNAISSANCE SYSTEM

Reconnaissance System, Airborne Integrated
USE   AIRBORNE INTEGRATED
         RECONNAISSANCE SYSTEM

RECONSTRUCTION

Reconstruction, Image
USE   IMAGE RECONSTRUCTION

Reconstruction, Wave Front
USE   WAVE FRONT RECONSTRUCTION

RECORDERS

Recorders, Cable Force
USE   CABLE FORCE RECORDERS

Recorders, Data
USE   DATA RECORDERS

Recorders, Flight
USE   FLIGHT RECORDERS

Recorders, Flight Load
USE   FLIGHT LOAD RECORDERS

Recorders, Force Vector
USE   FORCE VECTOR RECORDERS

Recorders, Magnetic Tape
USE   TAPE RECORDERS
         MAGNETIC RECORDING

Recorders, Pressure
USE   PRESSURE RECORDERS

Recorders, Pulse
USE   COUNTERS

Recorders, Tape
USE   TAPE RECORDERS

Recorders, VLF Emission
USE   VLF EMISSION RECORDERS

Recorders, Weather Data
USE   WEATHER DATA RECORDERS

Recorders, Whistler
USE   WHISTLER RECORDERS

RECORDING

Recording, Data
USE   DATA RECORDING

RECORDING HEADS

RECORDING INSTRUMENTS

Recording, Magnetic
USE   MAGNETIC RECORDING

Recording, Photographic
USE   PHOTOGRAPHIC RECORDING

Recording, Prediction
USE   PREDICTION RECORDING

Recording Systems, Electronic
USE   ELECTRONIC RECORDING SYSTEMS

RECORDS

RECOVERABILITY

RECOVERABLE LAUNCH VEHICLES

Recoverable Satellites
USE   RECOVERABLE SPACECRAFT

RECOVERABLE SPACECRAFT

RECOVERY

Recovery, Booster
USE   BOOSTER RECOVERY

Recovery Capsules, Discoverer
USE   DISCOVERER RECOVERY CAPSULES

Recovery Diodes, Step
USE   STEP RECOVERY DIODES

Recovery, Gas
USE   GAS RECOVERY

Recovery, Materials
USE   MATERIALS RECOVERY

Recovery, Oil
USE   OIL RECOVERY

RECOVERY PARACHUTES

Recovery, Pressure
USE   PRESSURE RECOVERY

Recovery, Soft
USE   SOFT LANDING

Recovery, Spacecraft
USE   SPACECRAFT RECOVERY

RECOVERY VEHICLES

Recovery, Water
USE   WATER RECLAMATION

RECOVERY ZONES

RECREATION

RECRYSTALLIZATION

RECTANGLES

RECTANGULAR BEAMS

Rectangular Coordinates
USE   CARTESIAN COORDINATES

Rectangular Drainage
USE   DRAINAGE PATTERNS

RECTANGULAR PANELS

RECTANGULAR PLANFORMS

RECTANGULAR PLATES

RECTANGULAR WAVEGUIDES

RECTANGULAR WIND TUNNELS

RECTANGULAR WINGS

RECTENNAS

RECTIFICATION

Rectification (Imagery), Geometric
USE   GEOMETRIC RECTIFICATION (IMAGERY)

Rectification, Radiometric
USE   RADIOMETRIC CORRECTION

Rectifier Antennas
USE   RECTENNAS

RECTIFIERS

Rectifiers, Crystal
USE   CRYSTAL RECTIFIERS

Rectifiers, Germanium
USE   GERMANIUM DIODES

Rectifiers, Photographic
USE   PHOTOGRAPHIC RECTIFIERS

(Rectifiers), SCR
USE   SILICON CONTROLLED RECTIFIERS

Rectifiers, Silicon
USE   CRYSTAL RECTIFIERS

Rectifiers, Silicon Controlled
USE   SILICON CONTROLLED RECTIFIERS

RECTUM

Recuperators
USE   REGENERATORS

Recursion Formulas
USE   RECURSIVE FUNCTIONS

RECURSIVE FUNCTIONS

Recycle Test Reactor, Plutonium
USE   PLUTONIUM RECYCLE TEST REACTOR

RECYCLING

RED ARCS

Red Blood Cells
USE   ERYTHROCYTES

RED DWARF STARS

RED GIANT STARS

RED SEA

RED SHIFT

Red Spot, Jupiter
USE   JUPITER RED SPOT

RED TIDE

Reddening, Interstellar
USE   INTERSTELLAR EXTINCTION

REDEYE MISSILE

REDOX CELLS

REDUCED GRAVITY

REDUCED ORDER FILTERS

REDUCTION

REDUCTION (CHEMISTRY)

Reduction, Cost
USE   COST REDUCTION

Reduction, Data
USE   DATA REDUCTION

Reduction, Drag
USE   DRAG REDUCTION

Reduction, Friction
USE   FRICTION REDUCTION

Reduction (Mathematics)
USE   OPTIMIZATION

Reduction, Noise
USE   NOISE REDUCTION

Reduction, Pressure
USE   PRESSURE REDUCTION

Reduction Reactions, Oxidation-
USE   OXIDATION-REDUCTION REACTIONS

Reduction, Sidelobe
USE   SIDELOBE REDUCTION

Reduction, Spin
USE   SPIN REDUCTION

Reduction), TARE (Data
USE   DATA REDUCTION

Reduction, Weight
USE   WEIGHT REDUCTION

REDUNDANCY

REDUNDANCY ENCODING

REDUNDANT COMPONENTS

Redundant Structures
USE   REDUNDANT COMPONENTS

REEDS (PLANTS)

REEFS

Reefs, Atoll
USE   CORAL REEFS

Reefs, Coral
USE   CORAL REEFS

REELS

REENTRY

Reentry Bodies
USE   REENTRY VEHICLES

Reentry Bodies, Maneuverable
USE   MANEUVERABLE REENTRY BODIES

Reentry Body, Mark 1
USE   MARK 1 REENTRY BODY

Reentry Body, Mark 2
USE   MARK 2 REENTRY BODY

Reentry Body, Mark 3
USE   MARK 3 REENTRY BODY

Reentry Body, Mark 4
USE   MARK 4 REENTRY BODY

Reentry Body, Mark 5
USE   MARK 5 REENTRY BODY

Reentry Body, Mark 6
USE   MARK 6 REENTRY BODY

Reentry Body, Mark 11
USE   MARK 11 REENTRY BODY

Reentry Body, Mark 12
USE   MARK 12 REENTRY BODY

Reentry Body, Mark 17
USE   MARK 17 REENTRY BODY

REENTRY COMMUNICATION

REENTRY DECOYS

REENTRY EFFECTS

Reentry Gliders
USE   LIFTING REENTRY VEHICLES

REENTRY GUIDANCE

Reentry, Hyperbolic
USE   HYPERBOLIC REENTRY

Reentry, Hypersonic
USE   HYPERSONIC REENTRY

Reentry, Manned
USE   MANNED REENTRY

REENTRY PHYSICS

REENTRY RANGE

REENTRY SHIELDING

Reentry, Spacecraft
USE   SPACECRAFT REENTRY

Reentry (Spacecraft), Uncontrolled
USE   UNCONTROLLED REENTRY (SPACECRAFT)

REENTRY TRAJECTORIES

Reentry Vehicle, FDL-5
USE   FDL-5 REENTRY VEHICLE

Reentry Vehicle, HL-10
USE   HL-10 REENTRY VEHICLE

Reentry Vehicle, HLD-35
USE   HLD-35 REENTRY VEHICLE

Reentry Vehicle, Trailblazer 1
USE   TRAILBLAZER 1 REENTRY VEHICLE

Reentry Vehicle, Trailblazer 2
USE   TRAILBLAZER 2 REENTRY VEHICLE

Reentry Vehicle, X-17
USE   X-17 REENTRY VEHICLE

REENTRY VEHICLES

Reentry Vehicles, Lifting
USE   LIFTING REENTRY VEHICLES

Reentry Vehicles, Low Observable
USE   LOW OBSERVABLE REENTRY VEHICLES

REFERENCE ATMOSPHERES

(Reference Lines), Axes
USE   AXES (REFERENCE LINES)

REFERENCE STARS

REFERENCE SYSTEMS

Reference Systems, Celestial
USE   CELESTIAL REFERENCE SYSTEMS

Reference Systems, Inertial
USE   INERTIAL REFERENCE SYSTEMS

References (Standards)
USE   STANDARDS

REFILLING

Refined Coal, Solvent
USE   SOLVENT REFINED COAL

REFINING

Refining, Electro
USE   ELECTROREFINING

Refining, Electroslag
USE   ELECTROSLAG REFINING

Refining, Zone
USE   ZONE MELTING

REFLECTANCE

Reflectance, Spectral
USE   SPECTRAL REFLECTANCE

Reflected Radiation
USE   REFLECTED WAVES

Reflected Rays
USE   REFLECTED WAVES

REFLECTED WAVES

REFLECTING TELESCOPES

REFLECTION

Reflection Coefficient
USE   REFLECTANCE

Reflection, Infrared
USE   INFRARED REFLECTION

Reflection, Ionospheric
USE   IONOSPHERIC PROPAGATION

Reflection, Mach
USE   MACH REFLECTION

REFLECTION NEBULAE

Reflection, Optical
USE   OPTICAL REFLECTION

Reflection, Radio
USE   RADIO ECHOES

Reflection, Retro
USE   RETROREFLECTION

Reflection, Signal
USE   SIGNAL REFLECTION

Reflection, Specular
USE   SPECULAR REFLECTION

Reflection, Spread
USE   SPREAD REFLECTION

Reflection, Ultraviolet
USE   ULTRAVIOLET REFLECTION

Reflection, Wave
USE   WAVE REFLECTION

Reflections, Radar
USE   RADAR ECHOES

Reflectivity
USE   REFLECTANCE

Reflectivity, Bistatic
USE   BISTATIC REFLECTIVITY

REFLECTOMETERS

Reflectometers, Microwave
USE   MICROWAVE REFLECTOMETERS

Reflector Antennas, Two
USE   TWO REFLECTOR ANTENNAS

Reflector Orbital Shot Proj, Experimental
USE   EXPERIMENTAL REFLECTOR ORBITAL
         SHOT PROJ

Reflector Satellites
USE   PASSIVE SATELLITES

REFLECTORS

Reflectors, Fresnel
USE   FRESNEL REFLECTORS

Reflectors, Parabolic
USE   PARABOLIC REFLECTORS

Reflectors, Radar
USE   RADAR REFLECTORS

Reflectors, Radar Corner
USE   RADAR CORNER REFLECTORS

Reflectors, Solar
USE   SOLAR REFLECTORS

Reflectors, Sub
USE   SUBREFLECTORS

Reflex, Carotid Sinus
USE   CAROTID SINUS REFLEX

Reflex, Hering-Brever
USE   HERING-BREVER REFLEX

REFLEXES

Reflexes, Conditioned
USE   CONDITIONED REFLEXES

Reflexes, Respiratory
USE   RESPIRATORY REFLEXES

REFORESTATION

Refracted Radiation
USE   REFRACTED WAVES

Refracted Rays
USE   REFRACTED WAVES

REFRACTED WAVES

REFRACTING TELESCOPES

REFRACTION

Refraction, Atmospheric
USE   ATMOSPHERIC REFRACTION

Refraction, Radio Wave
USE   RADIO WAVE REFRACTION

Refractive Index
USE   REFRACTIVITY

REFRACTIVITY

REFRACTOMETERS

REFRACTORIES

REFRACTORY COATINGS

REFRACTORY MATERIALS

REFRACTORY METAL ALLOYS

REFRACTORY METALS

REFRACTORY PERIOD

Refrasil (Trademark)
USE   SILICON DIOXIDE
         FIBERS

REFRIGERANTS

REFRIGERATING

REFRIGERATING MACHINERY

REFRIGERATORS

REFSAT

REFUELING

Refueling, Air To Air
USE   AIR TO AIR REFUELING

REGENERATION

REGENERATION (ENGINEERING)

REGENERATION (PHYSIOLOGY)

REGENERATIVE COOLING

Regenerative Cycles
USE   REGENERATION (ENGINEERING)

Regenerative Feedback
USE   POSITIVE FEEDBACK

REGENERATIVE FUEL CELLS

REGENERATORS

REGGE POLES

REGIMES

Regimes, Rossby
USE   ROSSBY REGIMES

Region, Caribbean
USE   CARIBBEAN REGION

Region, D
USE   D REGION

Region, E
USE   E REGION

Region, F
USE   F REGION

Region, F 1
USE   F 1 REGION

Region, F 2
USE   F 2 REGION

Region, Fraunhofer
USE   FAR FIELDS

Region, Fresnel
USE   FRESNEL REGION

Region (GA-NC-SC), Sand Hills
USE   SAND HILLS REGION (GA-NC-SC)

Region, Lumbar
USE   LUMBAR REGION

Region, M
USE   M REGION

Region (NE), Sand Hills
USE   SAND HILLS REGION (NE)

Region, Sciatic
USE   SCIATIC REGION

Region (South America), Amazon
USE   AMAZON REGION (SOUTH AMERICA)

Region, Stagnation
USE   STAGNATION POINT

Region (US), Central Atlantic
USE   CENTRAL ATLANTIC REGION (US)

Regional Ecol Test Site, Central Atlantic
USE   CENTRAL ATLANTIC REGIONAL ECOL
         TEST SITE

Regional Ecological Test Site, Arizona
USE   ARIZONA REGIONAL ECOLOGICAL TEST
         SITE

REGIONAL PLANNING

REGIONS

Regions, Antarctic
USE   ANTARCTIC REGIONS

Regions, Arctic
USE   ARCTIC REGIONS

Regions, Equatorial
USE   EQUATORIAL REGIONS

Regions, Polar
USE   POLAR REGIONS

Regions, Remote
USE   REMOTE REGIONS

Regions, Subarctic
USE   SUBARCTIC REGIONS

Regions, Subtropical
USE   TEMPERATE REGIONS
         TROPICAL REGIONS

Regions, Temperate
USE   TEMPERATE REGIONS

Regions, Tropical
USE   TROPICAL REGIONS

REGISTERS

REGISTERS (AIR CIRCULATION)

REGISTERS (COMPUTERS)

Registers, Shift
USE   SHIFT REGISTERS

Registration, Pattern
USE   PATTERN REGISTRATION

REGOLITH

REGRESSION ANALYSIS

REGRESSION COEFFICIENTS

Regression (Statistics)
USE   REGRESSION ANALYSIS

REGULARITY

Regulating, Self
USE   AUTOMATIC CONTROL

Regulation
USE   CONTROL

Regulation, Body Temperature
USE   THERMOREGULATION

Regulation, Frequency
USE   FREQUENCY CONTROL

Regulation, Heat
USE   TEMPERATURE CONTROL

Regulation, Speed
USE   SPEED CONTROL

Regulation, Thermo
USE   THERMOREGULATION

REGULATIONS

REGULATORS

Regulators, Current
USE   CURRENT REGULATORS

Regulators, Flow
USE   FLOW REGULATORS

Regulators, Fuel Flow
USE   FUEL FLOW REGULATORS

Regulators, Oxygen
USE   OXYGEN REGULATORS

Regulators, Pressure
USE   PRESSURE REGULATORS

Regulators, Speed
USE   SPEED REGULATORS

Regulators, Voltage
USE   VOLTAGE REGULATORS

REGULUS MISSILE

Reheating
USE   HEATING

Reignition
USE   IGNITION

Reinforced Composites, Fiber
USE   FIBER REINFORCED COMPOSITES

Reinforced Materials
USE   COMPOSITE MATERIALS

Reinforced Materials, Boron
USE   BORON REINFORCED MATERIALS

REINFORCED PLASTICS

Reinforced Plastics, Carbon Fiber
USE   CARBON FIBER REINFORCED PLASTICS

Reinforced Plastics, Glass Fiber
USE   GLASS FIBER REINFORCED PLASTICS

REINFORCED PLATES

REINFORCED SHELLS

REINFORCEMENT

Reinforcement, Metal Whisker
USE   WHISKER COMPOSITES

REINFORCEMENT (PSYCHOLOGY)

REINFORCEMENT RINGS

REINFORCEMENT (STRUCTURES)

REINFORCING FIBERS

REINFORCING MATERIALS

REISSNER THEORY

REISSNER-NORDSTROM SOLUTION

REJECTION

Rejection Devices, Heat
USE   HEAT RADIATORS

Relation, Rankine-Hugoniot
USE   RANKINE-HUGONIOT RELATION

Relations, Employee
USE   EMPLOYEE RELATIONS

Relations, Government/industry
USE   GOVERNMENT/INDUSTRY RELATIONS

Relations, Human
USE   HUMAN RELATIONS

Relations, International
USE   INTERNATIONAL RELATIONS

Relations, Interpersonal
USE   HUMAN RELATIONS

Relations, Public
USE   PUBLIC RELATIONS

Relations, Stress-Strain-Time
USE   STRESS-STRAIN-TIME RELATIONS

Relationship, Onsager
USE   ONSAGER RELATIONSHIP

RELATIONSHIPS

Relationships, Stress-Strain
USE   STRESS-STRAIN RELATIONSHIPS

RELATIVE BIOLOGICAL EFFECTIVENESS (RBE)

RELATIVISTIC EFFECTS

RELATIVISTIC ELECTRON BEAMS

RELATIVISTIC PARTICLES

RELATIVISTIC PLASMAS

RELATIVISTIC THEORY

RELATIVISTIC VELOCITY

RELATIVITY

Relaxants, Muscle
USE   MUSCLE RELAXANTS

RELAXATION

Relaxation, Chemical
USE   MOLECULAR RELAXATION

Relaxation, Cross
USE   CROSS RELAXATION

Relaxation, Magnetic
USE   MAGNETIC RELAXATION

RELAXATION (MECHANICS)

RELAXATION METHOD (MATHEMATICS)

Relaxation, Molecular
USE   MOLECULAR RELAXATION

Relaxation, Nuclear
USE   NUCLEAR RELAXATION

RELAXATION OSCILLATORS

RELAXATION (PHYSIOLOGY)

Relaxation, Spin-Lattice
USE   SPIN-LATTICE RELAXATION

Relaxation, Stress
USE   STRESS RELAXATION

RELAXATION TIME

Relaxation, Vibrational
USE   MOLECULAR RELAXATION

RELAY

RELAY SATELLITES

Relay Satellites, Tracking And Data
USE   TDR SATELLITES

Relay Systems, Optical
USE   OPTICAL RELAY SYSTEMS

Relay Systems, Radio
USE   RADIO RELAY SYSTEMS

RELAY 1 SATELLITE

RELAY 2 SATELLITE

Relays, Electric
USE   ELECTRIC RELAYS

Release, Fiber
USE   FIBER RELEASE

Release Modules, Chemical
USE   CHEMICAL RELEASE MODULES

Release, Store
USE   EXTERNAL STORE SEPARATION

RELEASING

RELIABILITY

Reliability, Aircraft
USE   AIRCRAFT RELIABILITY

RELIABILITY ANALYSIS

Reliability, Circuit
USE   CIRCUIT RELIABILITY

Reliability, Component
USE   COMPONENT RELIABILITY

Reliability Control
USE   RELIABILITY ENGINEERING
       QUALITY CONTROL

RELIABILITY ENGINEERING

Reliability, Spacecraft
USE   SPACECRAFT RELIABILITY

Reliability, Structural
USE   STRUCTURAL RELIABILITY

RELIC RADIATION

RELIEF MAPS

RELIEF VALVES

RELIEVING

Relieving, Stress
USE   STRESS RELIEVING

RELOCATION

RELUCTANCE

Reluctivity
USE   RELUCTANCE

Remagnetization
USE   MAGNETIZATION

REMANENCE

Remelting
USE   MELTING

Remnants, Supernova
USE   SUPERNOVA REMNANTS

REMODULATION

REMOTE CONSOLES

REMOTE CONTROL

REMOTE HANDLING

REMOTE MANIPULATOR SYSTEM

REMOTE REGIONS

REMOTE SENSING

Remote Sensing, Crop Inventories By
USE   AGRISTARS PROJECT

REMOTE SENSORS

REMOTELY PILOTED VEHICLES

REMOVAL

Removal, Carbon Dioxide
USE   CARBON DIOXIDE REMOVAL

Removal), Grinding (Material
USE   GRINDING (MATERIAL REMOVAL)

Removal (Machining), Material
USE   MACHINING

REMS
USE   RAPID EYE MOVEMENT STATE

Renal Calculi
USE   CALCULI

RENAL FUNCTION

RENDEZVOUS

Rendezvous, Earth Orbital
USE   EARTH ORBITAL RENDEZVOUS

(Rendezvous), EOR
USE   EARTH ORBITAL RENDEZVOUS

RENDEZVOUS GUIDANCE

(Rendezvous), LOR
USE   LUNAR ORBITAL RENDEZVOUS

Rendezvous, Lunar Orbital
USE   LUNAR ORBITAL RENDEZVOUS

Rendezvous, Orbital
USE   ORBITAL RENDEZVOUS

Rendezvous, Satellite
USE   ORBITAL RENDEZVOUS

Rendezvous, Space
USE   SPACE RENDEZVOUS

Rendezvous, Spacecraft
USE   SPACE RENDEZVOUS

RENDEZVOUS SPACECRAFT

RENDEZVOUS TRAJECTORIES

RENE 41

RENE 63

RENE 77

RENE 95

Renin Activity, Plasma
USE   IMMUNOASSAY

Reorientation
USE   RETRAINING

Repairing
USE   MAINTENANCE

Repairing Devices, Self
USE   SELF REPAIRING DEVICES

REPEATERS

REPETITION

Repetition Rate, Pulse
USE   PULSE REPETITION RATE

REPLACING

REPLENISHMENT

REPLICAS

REPORT GENERATORS

Reporting, Ice
USE   ICE REPORTING

REPORTS

Reports, Congressional
USE   CONGRESSIONAL REPORTS

Reports, Postlaunch
USE   POSTLAUNCH REPORTS

Reports, Presidential
USE   PRESIDENTIAL REPORTS

Representation, Mandelstam
USE   MANDELSTAM REPRESENTATION

REPRESENTATIONS

Reprocessing, Nuclear Fuel
USE   NUCLEAR FUEL REPROCESSING

REPRODUCTION

REPRODUCTION (BIOLOGY)

(Reproduction), Breeding
USE   BREEDING (REPRODUCTION)

REPRODUCTION (COPYING)

REPRODUCTIVE SYSTEMS

REPTILES

REPUBLIC AIRCRAFT

Republic, Central African
USE   CENTRAL AFRICAN REPUBLIC

Republic, Chinese Peoples
USE   CHINA

Republic, Dominican
USE   DOMINICAN REPUBLIC

Republic, German Democratic
USE   EAST GERMANY

Republic, Malagasy
USE   MALAGASY REPUBLIC

Republic Military Aircraft
USE   MILITARY AIRCRAFT

Republic Of China
USE   TAIWAN

Republic Of Germany, Federal
USE   WEST GERMANY

Republic Of Germany, Peoples Democratic
USE   EAST GERMANY

Republic Of Korea
USE   SOUTH KOREA

Republic Of Korea, Democratic Peoples
USE   NORTH KOREA

REPUBLIC OF SOUTH AFRICA

Republic Of Vietnam
USE   VIETNAM

Repulsion
USE   FORCE

REQUIREMENTS

Requirements, Airworthiness
USE   AIRCRAFT RELIABILITY

Requirements, Caloric
USE   CALORIC REQUIREMENTS

Requirements, Energy
USE   ENERGY REQUIREMENTS

Requirements, Nutritional
USE   NUTRITIONAL REQUIREMENTS

Requirements, User
USE   USER REQUIREMENTS

RESCUE OPERATIONS

Rescue Satellite, Search And
USE   SARSAT

RESEARCH

RESEARCH AIRCRAFT

Research Aircraft, Meteorological
USE   METEOROLOGICAL RESEARCH AIRCRAFT

Research Aircraft Program, Tilt Rotor
USE   TILT ROTOR RESEARCH AIRCRAFT
         PROGRAM

Research Aircraft, Rotor Systems
USE   ROTOR SYSTEMS RESEARCH AIRCRAFT

RESEARCH AND DEVELOPMENT

Research And Test Reactors, Nuclear
USE   NUCLEAR RESEARCH AND TEST
         REACTORS

Research, Committee On Space
USE   COMMITTEE ON SPACE RESEARCH

RESEARCH FACILITIES

Research, High Temperature
USE   HIGH TEMPERATURE RESEARCH

Research Laboratories, Manned Orbital
USE   MANNED ORBITAL RESEARCH
         LABORATORIES

Research Laboratories, Underwater
USE   UNDERWATER RESEARCH LABORATORIES

Research, Low Density
USE   LOW DENSITY RESEARCH

RESEARCH MANAGEMENT

Research, Market
USE   MARKET RESEARCH

Research, Nuclear
USE   NUCLEAR RESEARCH

Research, Operations
USE   OPERATIONS RESEARCH

Research Organization, European Space
USE   EUROPEAN SPACE AGENCY

Research Organization, Indian Space
USE   ISRO

Research Organization Sat, European Space
USE   ESA SATELLITES

Research Program, Global Atmospheric
USE   GLOBAL ATMOSPHERIC RESEARCH
         PROGRAM

RESEARCH PROJECTS

Research Reactor, Health Physics
USE   HEALTH PHYSICS RESEARCH REACTOR

Research Satellites, Environmental
USE   ENVIRONMENTAL RESEARCH SATELLITES

Research Satellites, Octahedral
USE   ENVIRONMENTAL RESEARCH SATELLITES

Research, Supersonic Cruise Aircraft
USE   SUPERSONIC CRUISE AIRCRAFT
         RESEARCH

Research, Urban
USE   URBAN RESEARCH

RESEARCH VEHICLES

Research Wings, Aeroelastic
USE   AEROELASTIC RESEARCH WINGS

RESERPINE

RESERVES

RESERVOIRS

Reset, Pneumatic
USE   PNEUMATIC CONTROL

RESIDENTIAL AREAS

RESIDENTIAL ENERGY

RESIDUAL GAS

RESIDUAL STRENGTH

RESIDUAL STRESS

RESIDUES

RESILIENCE

RESIN BONDING

RESIN MATRIX COMPOSITES

RESINS

Resins, Acrylic
USE   ACRYLIC RESINS

Resins, Addition
USE   ADDITION RESINS

Resins, Alkyd
USE   ALKYD RESINS

Resins, Chloroprene
USE   CHLOROPRENE RESINS

Resins, Epoxy
USE   EPOXY RESINS

Resins, Furan
USE   FURAN RESINS

Resins, Ion Exchange
USE   ION EXCHANGE RESINS

Resins, Methacrylate
USE   ACRYLIC RESINS

Resins, Nylon
USE   POLYAMIDE RESINS

Resins, Phenolic
USE   PHENOLIC RESINS

Resins, Phenolic Epoxy
USE   PHENOLIC EPOXY RESINS

Resins, Polyamide
USE   POLYAMIDE RESINS

Resins, Polyester
USE   POLYESTER RESINS

Resins, Polyether
USE   POLYETHER RESINS

Resins, Polyimide
USE   POLYIMIDE RESINS

Resins, Polyurethane
USE   POLYURETHANE RESINS

Resins, Silicone
USE   SILICONE RESINS

Resins, Synthetic
USE   SYNTHETIC RESINS

Resins, Thermoplastic
USE   THERMOPLASTIC RESINS

Resins, Thermosetting
USE   THERMOSETTING RESINS

RESISTANCE

Resistance, Abrasion
USE   ABRASION RESISTANCE

Resistance Circuits, Negative
USE   NEGATIVE RESISTANCE CIRCUITS

Resistance Coefficients
USE   RESISTANCE

Resistance, Contact
USE   CONTACT RESISTANCE

Resistance, Corrosion
USE   CORROSION RESISTANCE

Resistance, Creep
USE   CREEP STRENGTH

Resistance Devices, Negative
USE   NEGATIVE RESISTANCE DEVICES

Resistance, Earthquake
USE   EARTHQUAKE RESISTANCE

Resistance, Electrical
USE   ELECTRICAL RESISTANCE

Resistance, Fire
USE   FLAMMABILITY

Resistance, Flow
USE   FLOW RESISTANCE

Resistance, Fracture
USE   FRACTURE STRENGTH

Resistance, Heat
USE   THERMAL RESISTANCE

RESISTANCE HEATING

Resistance, High
USE   HIGH RESISTANCE

Resistance, Impact
USE   IMPACT RESISTANCE

Resistance, Kapitza
USE   KAPITZA RESISTANCE

Resistance, Low
USE   LOW RESISTANCE

Resistance, Moisture
USE   MOISTURE RESISTANCE

Resistance, Oxidation
USE   OXIDATION RESISTANCE

Resistance, Radiation
USE   RADIATION TOLERANCE

Resistance, Shock
USE   SHOCK RESISTANCE

Resistance, Skin
USE   SKIN RESISTANCE

Resistance, Thermal
USE   THERMAL RESISTANCE

RESISTANCE THERMOMETERS

Resistance, Wave
USE   WAVE RESISTANCE

Resistant Alloys, Heat
USE   HEAT RESISTANT ALLOYS

Resistant Structures, Earthquake
USE   EARTHQUAKE RESISTANT STRUCTURES

Resistivity
USE   ELECTRICAL RESISTIVITY

Resistivity, Electrical
USE   ELECTRICAL RESISTIVITY

RESISTOJET ENGINES

Resistojets
USE   RESISTOJET ENGINES

RESISTORS

(Resistors), Potentiometers
USE   POTENTIOMETERS (RESISTORS)

Resistors, Printed
USE   PRINTED RESISTORS

Resistors, Tunnel
USE   RESISTORS
        ELECTRON TUNNELING

RESOLUTION

Resolution, Angular
USE   ANGULAR RESOLUTION

Resolution, Automatic Traffic Advisory And
USE   AUTOMATIC TRAFFIC ADVISORY AND
        RESOLUTION

RESOLUTION CELL

Resolution Coverage Antennas, High
USE   HIGH RESOLUTION COVERAGE ANTENNAS

Resolution, High
USE   HIGH RESOLUTION

Resolution, Image
USE   IMAGE RESOLUTION

Resolution, Radar
USE   RADAR RESOLUTION

Resolution, Radiometric
USE   RADIOMETRIC RESOLUTION

Resolution, Spatial
USE   SPATIAL RESOLUTION

Resolution, Spectral
USE   SPECTRAL RESOLUTION

Resolution, Temporal
USE   TEMPORAL RESOLUTION

RESOLVERS

Resolving Power
USE   RESOLUTION

RESONANCE

Resonance, Baryon
USE   BARYON RESONANCE

RESONANCE CHARGE EXCHANGE

Resonance, Cyclotron
USE   CYCLOTRON RESONANCE

Resonance Devices, Cyclotron
USE   CYCLOTRON RESONANCE DEVICES

Resonance, Electron Paramagnetic
USE   ELECTRON PARAMAGNETIC RESONANCE

Resonance, Electron Spin
USE   ELECTRON PARAMAGNETIC RESONANCE

Resonance, Ferromagnetic
USE   FERROMAGNETIC RESONANCE

RESONANCE FLUORESCENCE

Resonance, Ground
USE   GROUND RESONANCE

RESONANCE LINES

Resonance, Magnetic
USE   MAGNETIC RESONANCE

Resonance, Magnetosonic
USE   MAGNETOSONIC RESONANCE

Resonance, Mechanical
USE   RESONANT VIBRATION

Resonance, Meson
USE   MESON RESONANCE

Resonance, Microwave
USE   MICROWAVE RESONANCE

Resonance, Non
USE   NONRESONANCE

Resonance, Nuclear Magnetic
USE   NUCLEAR MAGNETIC RESONANCE

Resonance, Nuclear Quadrupole
USE   NUCLEAR QUADRUPOLE RESONANCE

Resonance, Optical
USE   OPTICAL RESONANCE

Resonance, Paramagnetic
USE   PARAMAGNETIC RESONANCE

Resonance, Plasma
USE   PLASMA RESONANCE

RESONANCE PROBES

Resonance, Proton
USE   PROTON RESONANCE

Resonance, Proton Magnetic
USE   PROTON MAGNETIC RESONANCE

Resonance Radiation
USE   RESONANCE FLUORESCENCE

RESONANCE SCATTERING

Resonance, Spin
USE   SPIN RESONANCE

RESONANCE TESTING

Resonant Cavities
USE   CAVITY RESONATORS

RESONANT FREQUENCIES

RESONANT VIBRATION

RESONATORS

Resonators, Cavity
USE   CAVITY RESONATORS

Resonators, Helmholtz
USE   HELMHOLTZ RESONATORS

Resonators, Maser
USE   MASERS

Resonators, Multimode
USE   MULTIMODE RESONATORS

Resonators, Optical
USE   OPTICAL RESONATORS

RESOURCE ALLOCATION

Resource Sampler, Multispectral
USE   MULTISPECTRAL RESOURCE SAMPLER

RESOURCES

Resources, Cultural
USE   CULTURAL RESOURCES

Resources, Earth
USE   EARTH RESOURCES

Resources Experiment Package, Earth
USE   EREP

Resources, Extraterrestrial
USE   EXTRATERRESTRIAL RESOURCES

Resources, Geothermal
USE   GEOTHERMAL RESOURCES

Resources, Human
USE   HUMAN RESOURCES

Resources Information System, Earth
USE   EARTH RESOURCES INFORMATION
       SYSTEM

RESOURCES MANAGEMENT

Resources, Marine
USE   MARINE RESOURCES

Resources Observation Satellites, Earth
USE   EROS (SATELLITES)

Resources Program, Earth
USE   EARTH RESOURCES PROGRAM

Resources, Range
USE   RANGE RESOURCES

Resources Shuttle Imaging Radar, Earth
USE   EARTH RESOURCES SHUTTLE IMAGING
       RADAR

Resources Survey Aircraft, Earth
USE   EARTH RESOURCES SURVEY AIRCRAFT

Resources Survey Program, Earth
USE   EARTH RESOURCES SURVEY PROGRAM

Resources Technology Satellite B, Earth
USE   LANDSAT 2

Resources Technology Satellite C, Earth
USE   LANDSAT 3

Resources Technology Satellite D, Earth
USE   LANDSAT 4

Resources Technology Satellite E, Earth
USE   LANDSAT E

Resources Technology Satellite F, Earth
USE   LANDSAT F

Resources Technology Satellite 1, Earth
USE   LANDSAT 1

Resources Technology Satellites, Earth
USE   LANDSAT SATELLITES

Resources, Thermal
USE   THERMAL RESOURCES

Resources, Underwater
USE   UNDERWATER RESOURCES

Resources, Water
USE   WATER RESOURCES

RESPIRATION

Respiration, Artificial
USE   RESUSCITATION

RESPIRATORS

RESPIRATORY DISEASES

RESPIRATORY IMPEDANCE

RESPIRATORY PHYSIOLOGY

RESPIRATORY RATE

RESPIRATORY REFLEXES

RESPIRATORY SYSTEM

RESPIROMETERS

Responders
USE   TRANSPONDERS

RESPONSE BIAS

Response, Dynamic
USE   DYNAMIC RESPONSE

Response, Electrodermal
USE   GALVANIC SKIN RESPONSE

Response Filters, Finite Impulse
USE   FIR FILTERS

Response, Frequency
USE   FREQUENCY RESPONSE

Response, Galvanic Skin
USE   GALVANIC SKIN RESPONSE

Response, Modal
USE   MODAL RESPONSE

Response (Psychophysiology), Evoked
USE   EVOKED RESPONSE
       (PSYCHOPHYSIOLOGY)

Response, Time
USE   TIME RESPONSE

RESPONSE TIME (COMPUTERS)

Response, Transient
USE   TRANSIENT RESPONSE

RESPONSES

Responses, Conditioned
USE   CONDITIONING (LEARNING)

Responses, Hemodynamic
USE   HEMODYNAMIC RESPONSES

Responses, Physiological
USE   PHYSIOLOGICAL RESPONSES

REST

Rest, Bed
USE   BED REST

Rest Cycle, Work-
USE   WORK-REST CYCLE

RESTARTABLE ROCKET ENGINES

RESTORATION

Restraint Devices, Air Bag
USE   AIR BAG RESTRAINT DEVICES

Restraints
USE   CONSTRAINTS

Restrictions
USE   CONSTRICTIONS

(Restrictions), Chokes
USE   CHOKES (RESTRICTIONS)

RESULTANTS

RESUSCITATION

RETAINING

RETARDANTS

Retardants, Fire
USE   FLAME RETARDANTS

Retardants, Flame
USE   FLAME RETARDANTS

RETARDERS

RETARDERS (DEVICES)

RETARDING

Retarding Ion Mass Spectrometers
USE   MASS SPECTROMETERS

RETENTION

RETENTION (PSYCHOLOGY)

Retention, Solvent
USE   SOLVENT RETENTION

RETICLES

RETICULOCYTES

RETINA

RETINAL ADAPTATION

RETINAL IMAGES

RETINENE

RETIREMENT

RETIREMENT FOR CAUSE

Retorc (Torpedoes)
USE   TORPEDOES

RETORT PROCESSING

RETRACTABLE EQUIPMENT

Retractable Landing Gear
USE   RETRACTABLE EQUIPMENT
LANDING GEAR

RETRAINING

Retrievable Carrier, European
USE   EURECA (ESA)

RETRIEVAL

Retrieval, Data
USE   DATA RETRIEVAL

Retrieval, Information
USE   INFORMATION RETRIEVAL

Retrieval (STS), Payload
USE   PAYLOAD RETRIEVAL (STS)

Retrieval System, Payload Deployment &
USE   PAYLOAD DEPLOYMENT & RETRIEVAL
SYSTEM

Retroaction
USE   RETROTHRUST

Retrodirective Optics, Modulating
USE   MIROS SYSTEM

RETROFIRING

RETROFITTING

Retrofitting, Acoustic
USE   ACOUSTIC RETROFITTING

RETROREFLECTION

RETROREFLECTORS

Retroreflectors, Lunar
USE   LUNAR RETROREFLECTORS

RETROROCKET ENGINES

RETROTHRUST

RETURN BEAM VIDICONS

RETURN TO EARTH SPACE FLIGHT

REUSABLE HEAT SHIELDING

REUSABLE LAUNCH VEHICLES

REUSABLE ROCKET ENGINES

REUSABLE SPACECRAFT

Reusable Spacecraft), Mars (Manned
USE   MARS (MANNED REUSABLE SPACECRAFT)

Reusable Spaceship, Manned Aerodynamic
USE   MARS (MANNED REUSABLE SPACECRAFT)

REUSE

Reuse, Frequency
USE   FREQUENCY REUSE

REVENUE

REVERBERATION

Reversal, Thrust
USE   THRUST REVERSAL

REVERSE FIELD PINCH

REVERSE OSMOSIS

Reverse Time
USE   REACTION TIME

REVERSED FLOW

REVERSING

Review Techniques, Graphic Evaluation And
USE   GERT

REVIEWING

REVISIONS

Revolution, Bodies Of
USE   BODIES OF REVOLUTION

Revolution (Motion)
USE   REVOLVING

REVOLVING

REWARD (PSYCHOLOGY)

REYNOLDS EQUATION

Reynolds Law
USE   REYNOLDS EQUATION

REYNOLDS NUMBER

Reynolds Number, Critical
USE   CRITICAL VELOCITY
REYNOLDS NUMBER

Reynolds Number, High
USE   HIGH REYNOLDS NUMBER

Reynolds Number, Low
USE   LOW REYNOLDS NUMBER

REYNOLDS STRESS

RF-4 AIRCRAFT

RF-8 Aircraft
USE   F-8 AIRCRAFT

Rh
USE   RHODIUM

RH-2 Helicopter
USE   UH-1 HELICOPTER

RHEA (ASTRONOMY)

RHENIUM

RHENIUM ALLOYS

RHENIUM COMPOUNDS

RHENIUM ISOTOPES

RHEOCASTING

RHEOELECTRICAL SIMULATION

RHEOENCEPHALOGRAPHY

RHEOLOGY

RHEOMETERS

RHESUS FACTOR

RHEUMATIC DISEASES

RHIZOPUS

RHO-MESONS

RHODE ISLAND

Rhodesia
USE   ZIMBABWE

RHODIUM

RHODIUM ALLOYS

RHODIUM COMPOUNDS

RHODIUM ISOTOPES

Rhodium 102
USE   RHODIUM ISOTOPES

Rhodium 106
USE   RHODIUM ISOTOPES

RHOMBIC ANTENNAS

RHOMBOHEDRONS

RHOMBOIDS

RHONE DELTA (FRANCE)

RHYTHM

Rhythm, Biological
USE   RHYTHM (BIOLOGY)

RHYTHM (BIOLOGY)

Rhythms, Circadian
USE   CIRCADIAN RHYTHMS

Rhythms, Diurnal
USE   CIRCADIAN RHYTHMS

RI
USE   RHODE ISLAND

(RI), Block Island Sound
USE   BLOCK ISLAND SOUND (RI)

RIBBON PARACHUTES

RIBBONS

RIBOFLAVIN

RIBONUCLEIC ACIDS

RIBOSE

RIBS (SUPPORTS)

Rica, Costa
USE   COSTA RICA

RICCATI EQUATION

RICE

RICHARDS THEOREM

RICHARDSON NUMBER

Richardson-Dushman Equation
USE   THERMIONIC EMISSION
TEMPERATURE EFFECTS

Rico, Puerto
USE   PUERTO RICO

Rider Guidance, Beam
USE   BEAM RIDER GUIDANCE

Ridge Isochronous Cyclotron, Oak
　　USE　　OAK RIDGE ISOCHRONOUS CYCLOTRON

RIDGES

Ridges, Pressure
　　USE　　PRESSURE ICE

RIDING QUALITY

Riemann Equations, Cauchy-
　　USE　　CAUCHY-RIEMANN EQUATIONS

Riemann Integral
　　USE　　MEASURE AND INTEGRATION

RIEMANN MANIFOLD

Riemann Problem
　　USE　　CAUCHY PROBLEM

Riemann Space
　　USE　　RIEMANN MANIFOLD

Riemann Sphere
　　USE　　RIEMANN MANIFOLD

RIEMANN WAVES

RIESZ THEOREM

RIFLES

RIFT (REACTOR IN FLIGHT TEST)

Rift System, African
　　USE　　AFRICAN RIFT SYSTEM

Rift Valleys
　　USE　　VALLEYS

Rifts
　　USE　　GEOLOGICAL FAULTS

RIGGING

Rigid Bodies
　　USE　　RIGID STRUCTURES

RIGID MOUNTING

RIGID ROTOR HELICOPTERS

RIGID ROTORS

RIGID ROTORS (PLASMA PHYSICS)

RIGID STRUCTURES

RIGID WINGS

RIGIDITY

Rigidity, Magnetic
　　USE　　MAGNETIC RIGIDITY

Rigidity, Structural
　　USE　　STRUCTURAL STABILITY

Rills
　　USE　　VALLEYS

RIMS

Ring Accelerators, Electron
　　USE　　STORAGE RINGS (PARTICLE
　　　　　　ACCELERATORS)

RING CURRENTS

Ring Dating, Tree
　　USE　　DENDROCHRONOLOGY

RING DISCHARGE

RING LASERS

Ring Seals, O
　　USE　　O RING SEALS

RING STRUCTURES

RING WINGS

RINGS

Rings, Jupiter
　　USE　　JUPITER RINGS

RINGS (MATHEMATICS)

Rings (Particle Accelerators), Storage
　　USE　　STORAGE RINGS (PARTICLE
　　　　　　ACCELERATORS)

Rings, Planetary
　　USE　　PLANETARY RINGS

Rings, Plasma
　　USE　　TOROIDAL PLASMAS

Rings, Reinforcement
　　USE　　REINFORCEMENT RINGS

Rings, Saturn
　　USE　　SATURN RINGS

Rings, Uranus
　　USE　　URANUS RINGS

Rings, Vortex
　　USE　　VORTEX RINGS

RIO GRANDE (NORTH AMERICA)

RIOMETERS

RIPPLES

RISERS

RISK

RIT ENGINES

RITZ AVERAGING METHOD

Ritz Method, Rayleigh-
　　USE　　RAYLEIGH-RITZ METHOD

River Basin (AK), Chena
　　USE　　CHENA RIVER BASIN (AK)

River Basin (CA), Feather
　　USE　　FEATHER RIVER BASIN (CA)

River Basin (ID-OR-WA), Columbia
　　USE　　COLUMBIA RIVER BASIN (ID-OR-WA)

River Basin (IL-IN-OH), Wabash
　　USE　　WABASH RIVER BASIN (IL-IN-OH)

River Basin (LA), Atchafalaya
　　USE　　ATCHAFALAYA RIVER BASIN (LA)

River Basin (MD-NY-PA), Susquehanna
　　USE　　SUSQUEHANNA RIVER BASIN (MD-NY-PA)

River Basin (US), Delaware
　　USE　　DELAWARE RIVER BASIN (US)

River Basin (US), Missouri
　　USE　　MISSOURI RIVER BASIN (US)

RIVER BASINS

River (North America), Colorado
　　USE　　COLORADO RIVER (NORTH AMERICA)

River (NY-NJ), Hudson
　　USE　　HUDSON RIVER (NY-NJ)

River Range (WY), Wind
　　USE　　WIND RIVER RANGE (WY)

River (US), Mississippi
　　USE　　MISSISSIPPI RIVER (US)

River (US), Missouri
　　USE　　MISSOURI RIVER (US)

River (US), Ohio
　　USE　　OHIO RIVER (US)

River Valley (MD-VA-WV), Potomac
　　USE　　POTOMAC RIVER VALLEY (MD-VA-WV)

RIVERS

RIVETED JOINTS

RIVETING

RIVETS

RL CIRCUITS

RL-10 ENGINES

RL-10-A-1 ENGINE

RL-10-A-3 ENGINE

RLC CIRCUITS

RLC Networks
　　USE　　RLC CIRCUITS

RM-1 Engine, YLR-99-
　　USE　　LR-99 ENGINE

RM-2 Engine, LR-62-
　　USE　　LR-62-RM-2 ENGINE

Rn
　　USE　　RADON

RNA
　　USE　　RIBONUCLEIC ACIDS

ROADS

ROADWAY POWERED VEHICLES

ROASTING

Robertson Effect, Poynting-
　　USE　　POYNTING-ROBERTSON EFFECT

ROBIN BALLOONS

ROBOTICS

ROBOTS

ROBUSTNESS (MATHEMATICS)

ROCHE LIMIT

Rock, Bed
　　USE　　BEDROCK

ROCK BOLTS

ROCK INTRUSIONS

ROCK MECHANICS

Rock Salt
　　USE　　HALITES

Rocket, Aries Sounding
　　USE　　ARIES SOUNDING ROCKET

Rocket Binders, Solid
　　USE　　SOLID ROCKET BINDERS

Rocket, Black Brant 1 Sounding
　　USE　　BLACK BRANT 1 SOUNDING ROCKET

Rocket, Black Brant 2 Sounding
　　USE　　BLACK BRANT 2 SOUNDING ROCKET

Rocket, Black Brant 3 Sounding
　　USE　　BLACK BRANT 3 SOUNDING ROCKET

Rocket, Black Brant 4 Sounding
　　USE　　BLACK BRANT 4 SOUNDING ROCKET

Rocket, Black Brant 5 Sounding
　　USE　　BLACK BRANT 5 SOUNDING ROCKET

Rocket Boosters
　　USE　　BOOSTER ROCKET ENGINES

ROCKET CATAPULTS

Rocket Chambers
USE   THRUST CHAMBERS

Rocket, Echo 1 Carrier
USE   THOR DELTA LAUNCH VEHICLE

ROCKET ENGINE CASES

ROCKET ENGINE CONTROL

ROCKET ENGINE DESIGN

Rocket Engine, F-1
USE   F-1 ROCKET ENGINE

ROCKET ENGINE NOISE

Rocket Engine, SL-3
USE   SL-3 ROCKET ENGINE

ROCKET ENGINE 9KS-11000

ROCKET ENGINES

Rocket Engines, Booster
USE   BOOSTER ROCKET ENGINES

Rocket Engines, Ducted
USE   DUCTED ROCKET ENGINES

Rocket Engines, Electric
USE   ELECTRIC ROCKET ENGINES

Rocket Engines, Heus
USE   HEUS ROCKET ENGINES

Rocket Engines, Hot Water
USE   HOT WATER ROCKET ENGINES

Rocket Engines, Hybrid
USE   HYBRID ROCKET ENGINES

Rocket Engines, Hybrid Propellant
USE   HYBRID PROPELLANT ROCKET ENGINES

Rocket Engines, Liquid Propellant
USE   LIQUID PROPELLANT ROCKET ENGINES

Rocket Engines, Lithergol
USE   LITHERGOL ROCKET ENGINES

Rocket Engines, Nike Booster
USE   NIKE BOOSTER ROCKET ENGINES

Rocket Engines, Nozzleless
USE   NOZZLELESS ROCKET ENGINES

Rocket Engines, Nuclear
USE   NUCLEAR ROCKET ENGINES

Rocket Engines, Restartable
USE   RESTARTABLE ROCKET ENGINES

Rocket Engines, Reusable
USE   REUSABLE ROCKET ENGINES

Rocket Engines, Solid Propellant
USE   SOLID PROPELLANT ROCKET ENGINES

Rocket Engines, Sustainer
USE   SUSTAINER ROCKET ENGINES

Rocket Engines, Ullage
USE   ULLAGE ROCKET ENGINES

Rocket Engines, Upper Stage
USE   UPPER STAGE ROCKET ENGINES

ROCKET EXHAUST

Rocket, Exos Sounding
USE   EXOS SOUNDING ROCKET

ROCKET FIRING

ROCKET FLIGHT

Rocket Impact Predictors, Automatic
USE   COMPUTERIZED SIMULATION
       IMPACT PREDICTION

Rocket, Judi-Dart
USE   JUDI-DART ROCKET

ROCKET LAUNCHERS

ROCKET LAUNCHING

ROCKET LININGS

Rocket Motor Cases
USE   ROCKET ENGINE CASES

Rocket, Nike-Asp
USE   ASP ROCKET VEHICLE

ROCKET NOSE CONES

ROCKET NOZZLES

ROCKET OXIDIZERS

Rocket, Petrel Sounding
USE   PETREL SOUNDING ROCKET

Rocket, Phoenix Sounding
USE   PHOENIX SOUNDING ROCKET

ROCKET PLANES

Rocket Propellant Tanks
USE   PROPELLANT TANKS

ROCKET PROPELLANTS

Rocket Propellants, Cryogenic
USE   CRYOGENIC ROCKET PROPELLANTS

Rocket Propellants, Double Base
USE   DOUBLE BASE ROCKET PROPELLANTS

Rocket Propellants, Gaseous
USE   GASEOUS ROCKET PROPELLANTS

Rocket Propellants, Gelled
USE   GELLED ROCKET PROPELLANTS

Rocket Propellants, Hypergolic
USE   HYPERGOLIC ROCKET PROPELLANTS

Rocket Propellants, Liquid
USE   LIQUID ROCKET PROPELLANTS

Rocket Propellants, RP-1
USE   RP-1 ROCKET PROPELLANTS

Rocket Propellants, Solid
USE   SOLID ROCKET PROPELLANTS

ROCKET PROPELLED SLEDS

Rocket Ramjets, Integral
USE   INTEGRAL ROCKET RAMJETS

Rocket Reactors, KIWI
USE   KIWI REACTORS

Rocket Sondes
USE   SOUNDING ROCKETS

ROCKET SOUNDING

Rocket, Space Processing Applications
USE   SPACE PROCESSING APPLICATIONS
       ROCKET

(Rocket), SPAR
USE   SPACE PROCESSING APPLICATIONS
       ROCKET

ROCKET TEST FACILITIES

(Rocket Tests), SERT
USE   SPACE ELECTRIC ROCKET TESTS

Rocket Tests, Space Electric
USE   SPACE ELECTRIC ROCKET TESTS

ROCKET THRUST

Rocket Trajectory, Spinning Unguided
USE   SPINNING UNGUIDED ROCKET
       TRAJECTORY

Rocket Vehicle, Aerobee
USE   AEROBEE ROCKET VEHICLE

Rocket Vehicle, Agena A
USE   AGENA A ROCKET VEHICLE

Rocket Vehicle, Agena B
USE   AGENA B ROCKET VEHICLE

Rocket Vehicle, Agena C
USE   AGENA C ROCKET VEHICLE

Rocket Vehicle, Agena D
USE   AGENA D ROCKET VEHICLE

Rocket Vehicle, Antares
USE   ANTARES ROCKET VEHICLE

Rocket Vehicle, Apache
USE   APACHE ROCKET VEHICLE

Rocket Vehicle, Arcon
USE   ARCON ROCKET VEHICLE

Rocket Vehicle, Asp
USE   ASP ROCKET VEHICLE

Rocket Vehicle, Astrobee 1500
USE   ASTROBEE 1500 ROCKET VEHICLE

Rocket Vehicle, Athena
USE   ATHENA ROCKET VEHICLE

Rocket Vehicle, Berenice
USE   BERENICE ROCKET VEHICLE

Rocket Vehicle, Black Knight
USE   BLACK KNIGHT ROCKET VEHICLE

Rocket Vehicle, Blue Scout
USE   BLUE SCOUT ROCKET VEHICLE

Rocket Vehicle, Cajun
USE   CAJUN ROCKET VEHICLE

Rocket Vehicle, Dornier Paraglider
USE   DORNIER PARAGLIDER ROCKET VEHICLE

Rocket Vehicle, FFAR
USE   FOLDING FIN AIRCRAFT ROCKET VEHICLE

Rocket Vehicle, Folding Fin Aircraft
USE   FOLDING FIN AIRCRAFT ROCKET VEHICLE

Rocket Vehicle, Genie
USE   GENIE ROCKET VEHICLE

Rocket Vehicle, Honest John
USE   HONEST JOHN ROCKET VEHICLE

Rocket Vehicle, Hyla-Star
USE   HYLA-STAR ROCKET VEHICLE

Rocket Vehicle, Jabiru
USE   JAGUAR ROCKET VEHICLE

Rocket Vehicle, Jaguar
USE   JAGUAR ROCKET VEHICLE

Rocket Vehicle, Javelin
USE   JAVELIN ROCKET VEHICLE

Rocket Vehicle, Jupiter C
USE   JUPITER C ROCKET VEHICLE

Rocket Vehicle, Kappa 8
USE   KAPPA 8 ROCKET VEHICLE

Rocket Vehicle, Kappa 9
USE   KAPPA 9 ROCKET VEHICLE

Rocket Vehicle, Little John
USE   LITTLE JOHN ROCKET VEHICLE

Rocket Vehicle, Loki
  USE   LOKI ROCKET VEHICLE

Rocket Vehicle, MB-1
  USE   GENIE ROCKET VEHICLE

Rocket Vehicle, Meteor 1
  USE   METEOR 1 ROCKET VEHICLE

Rocket Vehicle, Nike-Apache
  USE   NIKE-APACHE ROCKET VEHICLE

Rocket Vehicle, Nike-Cajun
  USE   NIKE-CAJUN ROCKET VEHICLE

Rocket Vehicle, Nike-Hydac
  USE   NIKE-HYDAC ROCKET VEHICLE

Rocket Vehicle, Nike-Iroquois
  USE   NIKE-IROQUOIS ROCKET VEHICLE

Rocket Vehicle, Nike-Javelin
  USE   NIKE-JAVELIN ROCKET VEHICLE

Rocket Vehicle, Nike-Tomahawk
  USE   NIKE-TOMAHAWK ROCKET VEHICLE

Rocket Vehicle, Rubis
  USE   RUBIS ROCKET VEHICLE

Rocket Vehicle, Skylark
  USE   SKYLARK ROCKET VEHICLE

Rocket Vehicle, Thor Able
  USE   THOR ABLE ROCKET VEHICLE

Rocket Vehicle, Trailblazer 1
  USE   TRAILBLAZER 1 REENTRY VEHICLE

Rocket Vehicle, Trailblazer 2
  USE   TRAILBLAZER 2 REENTRY VEHICLE

Rocket Vehicle, Vega
  USE   VEGA LAUNCH VEHICLE

Rocket Vehicle, Venus Fly TRAP
  USE   VENUS FLY TRAP ROCKET VEHICLE

Rocket Vehicle, Viking
  USE   VIKING ROCKET VEHICLE

Rocket Vehicle, Zuni
  USE   ZUNI ROCKET VEHICLE

ROCKET VEHICLES

Rocket Vehicles, Agena
  USE   AGENA ROCKET VEHICLES

Rocket Vehicles, Arcas
  USE   ARCAS ROCKET VEHICLES

Rocket Vehicles, Argo
  USE   ARGO ROCKET VEHICLES

Rocket Vehicles, Astrobee
  USE   ASTROBEE ROCKET VEHICLES

Rocket Vehicles, Hovering
  USE   HOVERING ROCKET VEHICLES

Rocket Vehicles, Kappa
  USE   KAPPA ROCKET VEHICLES

Rocket Vehicles, Lambda
  USE   LAMBDA ROCKET VEHICLES

Rocket Vehicles, Multistage
  USE   MULTISTAGE ROCKET VEHICLES

Rocket Vehicles, Nike
  USE   NIKE ROCKET VEHICLES

Rocket Vehicles, Nuclear Engine For
  USE   NUCLEAR ENGINE FOR ROCKET VEHICLES

Rocket Vehicles, Single Stage
  USE   SINGLE STAGE ROCKET VEHICLES

Rocket Vehicles, Skua
  USE   SKUA ROCKET VEHICLES

Rocket Vehicles, Veronique
  USE   VERONIQUE ROCKET VEHICLES

Rocket, Vertical 8
  USE   VERTICAL 8 ROCKET

Rocket, Wasp Sounding
  USE   WASP SOUNDING ROCKET

ROCKET-BORNE INSTRUMENTS

ROCKET-BORNE PHOTOGRAPHY

ROCKETS

Rockets, Air To Air
  USE   AIR TO AIR MISSILES

Rockets, Black Brant Sounding
  USE   BLACK BRANT SOUNDING ROCKETS

Rockets, Booster
  USE   BOOSTER ROCKETS

Rockets, Carrier
  USE   LAUNCH VEHICLES

Rockets, Control
  USE   CONTROL ROCKETS

Rockets, Escape
  USE   ESCAPE ROCKETS

Rockets, Meteorological
  USE   SOUNDING ROCKETS

Rockets, Nike
  USE   NIKE ROCKETS

Rockets, Sounding
  USE   SOUNDING ROCKETS

(Rockets), Staging
  USE   STAGE SEPARATION

Rockets, Steering
  USE   CONTROL ROCKETS

Rockets, Surface To Surface
  USE   SURFACE TO SURFACE ROCKETS

ROCKOONS

ROCKS

Rocks, Carbonaceous
  USE   CARBONACEOUS ROCKS

Rocks, Igneous
  USE   IGNEOUS ROCKS

Rocks, Lunar
  USE   LUNAR ROCKS

Rocks, Sedimentary
  USE   SEDIMENTARY ROCKS

(Rocks), Stones
  USE   ROCKS

ROCKWELL HARDNESS

ROCKY MOUNTAINS (NORTH AMERICA)

RODENTS

RODS

Rods, Control
  USE   CONTROL RODS

Roentgen Satellite
  USE   ROSAT MISSION

Rogallo Wings
  USE   FOLDING STRUCTURES
           FLEXIBLE WINGS

Roland Comet, Arend-
  USE   AREND-ROLAND COMET

Role Combat Aircraft, Multi-
  USE   MRCA AIRCRAFT

ROLL

Roll Control
  USE   LATERAL CONTROL

Roll, Damping In
  USE   DAMPING
           ROLL

ROLL FORMING

ROLLER BEARINGS

ROLLERS

ROLLING

Rolling, Cold
  USE   COLD ROLLING

ROLLING CONTACT LOADS

ROLLING MOMENTS

Rollup Solar Arrays
  USE   SOLAR ARRAYS

ROMANIA

RONCHI TEST

ROOFS

ROOM TEMPERATURE

ROOMS

Rooms, Clean
  USE   CLEAN ROOMS

Rooms, Dark
  USE   DARKROOMS

ROOT-MEAN-SQUARE ERRORS

ROOTS

ROOTS OF EQUATIONS

Roots, Plant
  USE   PLANT ROOTS

Roots, Wing
  USE   WING ROOTS

(Ropes), Cables
  USE   CABLES (ROPES)

RORSCHACH TESTS

ROSAT MISSION

ROSETTE SHAPES

ROSHKO PREDICTION

ROSIN

ROSS ICE SHELF

ROSSBY REGIMES

Rossby Waves
  USE   PLANETARY WAVES

Rotary Drives
  USE   MECHANICAL DRIVES

ROTARY ENGINES

ROTARY GYROSCOPES

ROTARY STABILITY

ROTARY WING AIRCRAFT

ROTARY WINGS

Rotating
USE   ROTATION

ROTATING BODIES

ROTATING CYLINDERS

ROTATING DISKS

ROTATING ELECTRICAL MACHINES

ROTATING ENVIRONMENTS

ROTATING FLUIDS

ROTATING GENERATORS

ROTATING LIQUIDS

ROTATING MATTER

ROTATING MIRRORS

ROTATING PLASMAS

ROTATING SHAFTS

ROTATING SPHERES

ROTATING STALLS

Rotating Vehicles
USE   ROTATING BODIES
VEHICLES

Rotating Wheels, Counter-
USE   COUNTER-ROTATING WHEELS

ROTATION

Rotation, Auto
USE   AUTOROTATION

Rotation, Axes Of
USE   AXES OF ROTATION

Rotation, Carrington
USE   SOLAR ROTATION

Rotation, Counter
USE   COUNTER ROTATION

Rotation, Earth
USE   EARTH ROTATION

Rotation, Faraday
USE   FARADAY EFFECT

Rotation, Galactic
USE   GALACTIC ROTATION

Rotation, Image
USE   IMAGE ROTATION ·

Rotation, Liquid
USE   ROTATING LIQUIDS

Rotation, Lunar
USE   LUNAR ROTATION

Rotation, Molecular
USE   MOLECULAR ROTATION

Rotation, Muon Spin
USE   MUON SPIN ROTATION

Rotation, Planetary
USE   PLANETARY ROTATION

Rotation, Satellite
USE   SATELLITE ROTATION

Rotation, Solar
USE   SOLAR ROTATION

Rotation, Solid
USE   ROTATING BODIES

Rotation, Stellar
USE   STELLAR ROTATION

Rotational Flow
USE   FLUID FLOW
VORTICES

ROTIFERA

ROTOCHUTES

ROTONS

ROTOR AERODYNAMICS

Rotor Aircraft, Tilt
USE   TILT ROTOR AIRCRAFT

ROTOR BLADES

Rotor Blades, Hinged
USE   ROTARY WINGS
HINGES

ROTOR BLADES (TURBOMACHINERY)

ROTOR BODY INTERACTIONS

Rotor Disks
USE   TURBINE WHEELS

Rotor Gyroscopes, Fluid
USE   FLUID ROTOR GYROSCOPES

Rotor Helicopters, Rigid
USE   RIGID ROTOR HELICOPTERS

Rotor Helicopters, Tandem
USE   TANDEM ROTOR HELICOPTERS

Rotor Hubs
USE   ROTORS
HUBS

ROTOR LIFT

Rotor Research Aircraft Program, Tilt
USE   TILT ROTOR RESEARCH AIRCRAFT
PROGRAM

ROTOR SPEED

ROTOR SYSTEMS RESEARCH AIRCRAFT

Rotorcraft
USE   ROTARY WING AIRCRAFT

ROTORCRAFT AIRCRAFT

ROTORS

Rotors, Bearingless
USE   BEARINGLESS ROTORS

Rotors, Circulation Control
USE   CIRCULATION CONTROL ROTORS

Rotors, Compressor
USE   COMPRESSOR ROTORS

Rotors, Helicopter
USE   ROTARY WINGS

Rotors, Helicopter Tail
USE   HELICOPTER TAIL ROTORS

Rotors, Hingeless
USE   RIGID ROTORS

Rotors, Lifting
USE   LIFTING ROTORS

Rotors (Plasma Physics), Rigid
USE   RIGID ROTORS (PLASMA PHYSICS)

Rotors, Rigid
USE   RIGID ROTORS

Rotors, Tail
USE   TAIL ROTORS

Rotors, Tilting
USE   TILTING ROTORS

Rotors, Tip Driven
USE   TIP DRIVEN ROTORS

Rotors, X Wing
USE   X WING ROTORS

ROUGHNESS

Roughness Effects, Surface
USE   SURFACE ROUGHNESS EFFECTS

Roughness, Sea
USE   SEA ROUGHNESS

Roughness, Surface
USE   SURFACE ROUGHNESS

ROUND TRIP TRAJECTORIES

ROUSE BELTS

Route ATC, Automated En
USE   AUTOMATED EN ROUTE ATC

ROUTES

ROUTINES

Routines, Assembler
USE   ASSEMBLER ROUTINES

Routines (Computers), Editing
USE   EDITING ROUTINES (COMPUTERS)

Routines, Data Conversion
USE   DATA CONVERSION ROUTINES

Routines, Input/output
USE   INPUT/OUTPUT ROUTINES

Routines, Merging
USE   MERGING ROUTINES

Routines, Sub
USE   SUBROUTINES

ROVER PROJECT

ROVING VEHICLES

Roving Vehicles, Extraterrestrial
USE   ROVING VEHICLES

Roving Vehicles, Lunar
USE   LUNAR ROVING VEHICLES

Roving Vehicles, Lunokhod Lunar
USE   LUNOKHOD LUNAR ROVING VEHICLES

ROVINGS

ROWLAND CIRCLES

RP-1 ROCKET PROPELLANTS

RPV
USE   REMOTELY PILOTED VEHICLES

Rsch Airplane, Experimental STOL Transport
USE   QUESTOL

RTV-40 RUBBER (TRADEMARK)

RTV-60 RUBBER (TRADEMARK)

Ru
USE   RUTHENIUM

Ruanda-Urundi
USE   BURUNDI
RWANDA

RUBBER

RUBBER COATINGS

Rubber, Silicone
USE   SILICONE RUBBER

Rubber (Trademark), RTV-40
USE   RTV-40 RUBBER (TRADEMARK)

Rubber (Trademark), RTV-60
USE   RTV-60 RUBBER (TRADEMARK)

Rubber (Trademark), Viton
USE   VITON RUBBER (TRADEMARK)

Rubbers, Synthetic
USE   SYNTHETIC RUBBERS

RUBIDIUM

RUBIDIUM COMPOUNDS

RUBIDIUM ISOTOPES

RUBIDIUM 86

RUBIS ROCKET VEHICLE

RUBY

RUBY LASERS

RUDDERS

Rudders, Aerial
USE   AERIAL RUDDERS

Rudders, Marine
USE   MARINE RUDDERS

RUGGEDNESS

Rule, Miner
USE   PALMGREN-MINER RULE

Rule, Palmgren-Miner
USE   PALMGREN-MINER RULE

Rule, Phase
USE   PHASE RULE

Rule, Whitham
USE   WHITHAM RULE

RULER METHOD

RULES

Rules, Flight
USE   FLIGHT RULES

(Rules), IFR
USE   INSTRUMENT FLIGHT RULES

Rules, Instrument Flight
USE   INSTRUMENT FLIGHT RULES

Rules (Nuclear Physics), Selection
USE   SELECTION RULES (NUCLEAR PHYSICS)

Rules, Sum
USE   SUM RULES

(Rules), VFR
USE   VISUAL FLIGHT RULES

Rules, Visual Flight
USE   VISUAL FLIGHT RULES

Rumania
USE   ROMANIA

RUN TIME (COMPUTERS)

Runaway (Plasma Physics), Electron
USE   ELECTRON RUNAWAY (PLASMA PHYSICS)

Runge Bands, Schumann-
USE   SCHUMANN-RUNGE BANDS

RUNGE-KUTTA METHOD

RUNNING

Runoff, Water
USE   WATER RUNOFF

Runoffs
USE   DRAINAGE

Runs, Takeoff
USE   TAKEOFF RUNS

Runup, Aircraft
USE   AIRCRAFT RUNUP

RUNWAY ALIGNMENT

RUNWAY CONDITIONS

RUNWAY LIGHTS

RUNWAYS

Rupture Strength, Creep
USE   CREEP RUPTURE STRENGTH

Rupture Strength, Stress
USE   CREEP RUPTURE STRENGTH

RUPTURING

RURAL AREAS

RURAL LAND USE

Russell Diagram, Hertzsprung-
USE   HERTZSPRUNG-RUSSELL DIAGRAM

RUST FUNGI

RUSTING

Rusts (Botany)
USE   RUST FUNGI

RUTHENIUM

RUTHENIUM ALLOYS

RUTHENIUM COMPOUNDS

RUTHENIUM ISOTOPES

Ruthenium 106
USE   RUTHENIUM ISOTOPES

RUTILE

RWANDA

RYAN AIRCRAFT

Ryan Military Aircraft
USE   RYAN AIRCRAFT

RYDBERG SERIES

R4D Engine, Marquardt
USE   MARQUARDT R4D ENGINE

R5D Aircraft
USE   C-54 AIRCRAFT

R7V Aircraft
USE   C-121 AIRCRAFT
       EC-121 AIRCRAFT

S

S Band
USE   SUPERHIGH FREQUENCIES
       ULTRAHIGH FREQUENCIES

S Band, Unified
USE   UNIFIED S BAND

S CURVES

S GLASS

S MATRIX THEORY

S STARS

S WAVES

S-A-W Devices
USE   SURFACE ACOUSTIC WAVE DEVICES

S-N DIAGRAMS

S-1 Stage, Saturn
USE   SATURN S-1 STAGE

S-1B Stage, Saturn
USE   SATURN S-1B STAGE

S-1C Stage, Saturn
USE   SATURN S-1C STAGE

S-2 AIRCRAFT

S-2 Aircraft, Snow
USE   S-2 AIRCRAFT

S-2 Stage, Saturn
USE   SATURN S-2 STAGE

S-2b, Snow Aerial Applicator Aircraft
USE   S-2 AIRCRAFT

S-3 AIRCRAFT

S-3 Satellite
USE   EXPLORER 12 SATELLITE

S-4 Stage, Saturn
USE   SATURN S-4 STAGE

S-4B Stage, Saturn
USE   SATURN S-4B STAGE

S-6 Satellite
USE   EXPLORER 17 SATELLITE

S-16 Satellite
USE   OSO-1

S-17 Satellite
USE   OSO-2

S-18 Satellite
USE   OAO

S-27 Satellite
USE   ALOUETTE 1 SATELLITE

S-35 Aircraft, Beech
USE   C-35 AIRCRAFT

S-49 Satellite
USE   OGO-A

S-50 Satellite
USE   OGO-C

S-51 Satellite
USE   ARIEL 1 SATELLITE

S-52 Satellite
USE   ARIEL 2 SATELLITE

S-57 Satellite
USE   OSO-C

S-58 HELICOPTER

S-58 Helicopter, Sikorsky
USE   S-58 HELICOPTER

S-61 HELICOPTER

S-61 Helicopter, Sikorsky
USE   S-61 HELICOPTER

S-64 Helicopter
USE   CH-54 HELICOPTER

S-64 Helicopter, Sikorsky
USE   CH-54 HELICOPTER

S-65 Helicopter, Sikorsky
USE   H-53 HELICOPTER

S-66 Satellite
USE   BEACON EXPLORER A

S-67 HELICOPTER

S-67 Helicopter, Sikorsky
USE   S-67 HELICOPTER

S-74 Satellite
USE   EXPLORER 18 SATELLITE

SA-1 Launch Vehicle, Saturn 1
USE   SATURN 1 SA-1 LAUNCH VEHICLE

SA-2 Launch Vehicle, Saturn 1
USE   SATURN 1 SA-2 LAUNCH VEHICLE

SA-3 Launch Vehicle, Saturn 1
USE   SATURN 1 SA-3 LAUNCH VEHICLE

SA-4 Launch Vehicle, Saturn 1
USE   SATURN 1 SA-4 LAUNCH VEHICLE

SA-5 Launch Vehicle, Saturn 1
USE   SATURN 1 SA-5 LAUNCH VEHICLE

SA-6 Launch Vehicle, Saturn 1
USE   SATURN 1 SA-6 LAUNCH VEHICLE

SA-7 Launch Vehicle, Saturn 1
USE   SATURN 1 SA-7 LAUNCH VEHICLE

SA-8 Launch Vehicle, Saturn 1
USE   SATURN 1 SA-8 LAUNCH VEHICLE

SA-9 Launch Vehicle, Saturn 1
USE   SATURN 1 SA-9 LAUNCH VEHICLE

SA-10 Launch Vehicle, Saturn 1
USE   SATURN 1 SA-10 LAUNCH VEHICLE

SA-321 HELICOPTER

SA-321 Helicopter, Sud Aviation
USE   SA-321 HELICOPTER

SA-330 HELICOPTER

SA-330 Helicopter, Sud Aviation
USE   SA-330 HELICOPTER

SAAB AIRCRAFT

SAAB 37 AIRCRAFT

SAAB 105 AIRCRAFT

SABATIER REACTION

SABOT PROJECTILES

SABOTAGE

Sabre Aircraft
USE   F-86 AIRCRAFT

Sabre Aircraft, Super
USE   F-100 AIRCRAFT

Sabreliner Aircraft
USE   T-39 AIRCRAFT

SACCADIC EYE MOVEMENTS

Saccharides
USE   CARBOHYDRATES

SACCHAROMYCES

SACRAMENTO VALLEY (CA)

SADDLE POINTS

SADDLE POINTS (GAME THEORY)

SADDLES

SADDLES (SUPPORTS)

Safe Systems, Fail-
USE   FAIL-SAFE SYSTEMS

SAFEGUARD SYSTEM

SAFETY

Safety, Aerospace
USE   AEROSPACE SAFETY

Safety, Aircraft
USE   AIRCRAFT SAFETY

SAFETY DEVICES

SAFETY FACTORS

Safety, Flight
USE   FLIGHT SAFETY

Safety Hazard, Toxicity And
USE   TOXICITY AND SAFETY HAZARD

Safety, Industrial
USE   INDUSTRIAL SAFETY

SAFETY MANAGEMENT

Safety, Range
USE   RANGE SAFETY

Safety, Reactor
USE   REACTOR SAFETY

SAGE AIR DEFENSE SYSTEM

SAGE SATELLITE

SAGINAW BAY (MI)

SAGITTARIUS CONSTELLATION

SAGNAC EFFECT

SAHA EQUATIONS

SAHARA DESERT (AFRICA)

Sahara, Spanish
USE   SPANISH SAHARA

SAIL PROJECT

Sailplane, Schleicher KA-6
USE   KA-6 SAILPLANES

Sailplanes
USE   GLIDERS

Sailplanes, KA-6
USE   KA-6 SAILPLANES

SAILS

Sails, Solar
USE   SOLAR SAILS

SAILWINGS

Sailwings, Princeton
USE   SAILWINGS

SAINT ELMO FIRE

Saint Venant Flexure Problem
USE   SAINT VENANT PRINCIPLE

SAINT VENANT PRINCIPLE

Salesman Problem, Traveling
USE   TRAVELING SALESMAN PROBLEM

SALICYLATES

Salicylates, Sodium
USE   SODIUM SALICYLATES

SALINITY

SALIVA

SALIVARY GLANDS

SALMONELLA

Salpeter Equation, Bethe-
USE   BETHE-SALPETER EQUATION

SALT BATHS

SALT BEDS

Salt Electrolytes, Molten
USE   MOLTEN SALT ELECTROLYTES

Salt Flats
USE   FLATS (LANDFORMS)

Salt Lake (UT), Great
USE   GREAT SALT LAKE (UT)

Salt Nuclear Reactors, Molten
USE   MOLTEN SALT NUCLEAR REACTORS

Salt, Rock
USE   HALITES

SALT SPRAY TESTS

SALTON SEA (CA)

SALTS

Salts, Molten
USE   MOLTEN SALTS

Salts, Organic Charge Transfer
USE   ORGANIC CHARGE TRANSFER SALTS

Salvador, El
USE   EL SALVADOR

SALYUT SPACE STATION

Samaritan Aircraft
USE   C-131 AIRCRAFT

SAMARIUM

SAMARIUM COMPOUNDS

SAMARIUM ISOTOPES

SAMOA

SAMOS

Sampled Data
USE   DATA SAMPLING

Sampled Data Systems
USE   DATA SAMPLING

Sampler, Multispectral Resource
USE   MULTISPECTRAL RESOURCE SAMPLER

SAMPLERS

(Samplers), Bombs
USE   SAMPLERS

SAMPLES

Samples, Mars Surface
USE   MARS SURFACE SAMPLES

SAMPLING

Sampling, Air
USE   AIR SAMPLING

Sampling, Core
USE   CORE SAMPLING

Sampling, Data
USE   DATA SAMPLING

Sampling Devices
USE   SAMPLERS

Sampling, Particulate
USE   PARTICULATE SAMPLING

Sampling Program, Global Air
USE   GLOBAL AIR SAMPLING PROGRAM

Sampling, Random
USE   RANDOM SAMPLING

SAN ANDREAS FAULT

SAN ANDREAS FAULT EXPERIMENT

SAN FRANCISCO BAY (CA)

SAN FRANCISCO (CA)

SAN JOAQUIN VALLEY (CA)

SAN JUAN MOUNTAINS (CO)

SAN MARCO SATELLITES

SAN MARCO 1 SATELLITE

SAN MARCO 2 SATELLITE

SAN MARCO 3 SATELLITE

SAN MARINO

SAN PABLO BAY (CA)

SAND CASTING

Sand Dunes
USE   DUNES

SAND HILLS REGION (GA-NC-SC)

SAND HILLS REGION (NE)

SANDPIPER TARGET MISSILE

SANDS

Sands, Monazite
USE   MONAZITE SANDS

Sands, Tar
USE   TAR SANDS

SANDSTONES

Sandwich Construction
USE   SANDWICH STRUCTURES

SANDWICH STRUCTURES

SANITATION

SANTOWAX (TRADEMARK)

SAPPHIRE

Sapphire Junctions, Silicon-On-
USE   SOS (SEMICONDUCTORS)

Sapphire Semiconductors, Silicon-On-
USE   SOS (SEMICONDUCTORS)

Sapphire Transistors, Silicon-On-
USE   SOS (SEMICONDUCTORS)

SAPROPHYTES

SARCINA

Sarcoma
USE   CANCER

SARGASSO SEA

SARSAT

SAS

SAS-D
USE   IUE

SAS-1

SAS-2

SAS-3

SASKATCHEWAN

Sat, European Space Research Organization
USE   ESA SATELLITES

Sat, L-
USE   L-SAT

Sat Sys, National Operational Environmental
USE   NOESS

SATAN (Sensor)
USE   TERRAIN ANALYSIS

Satcom Satellites, RCA
USE   RCA SATCOM SATELLITES

Satellite, A-11
USE   ECHO 1 SATELLITE

Satellite, A-12
USE   ECHO 2 SATELLITE

Satellite, AD-A
USE   EXPLORER 19 SATELLITE

Satellite, AD/I
USE   EXPLORER 24 SATELLITE

Satellite, AE-A
USE   EXPLORER 17 SATELLITE

Satellite, AE-B
USE   EXPLORER 32 SATELLITE

Satellite, AE-C
USE   EXPLORER 51 SATELLITE

Satellite, AE-D
USE   EXPLORER 54 SATELLITE

Satellite, AE-E
USE   EXPLORER 55 SATELLITE

Satellite, AEROS
USE   AEROS SATELLITE

Satellite, Alouette B
USE   ALOUETTE B SATELLITE

Satellite, Alouette 1
USE   ALOUETTE 1 SATELLITE

Satellite, Alouette 2
USE   ALOUETTE 2 SATELLITE

SATELLITE ANTENNAS

Satellite, Arabian Commercial
USE   ARCOMSAT

Satellite, Ariel 1
USE   ARIEL 1 SATELLITE

Satellite, Ariel 2
USE   ARIEL 2 SATELLITE

Satellite, Ariel 3
USE   ARIEL 3 SATELLITE

Satellite, Ariel 4
USE   ARIEL 4 SATELLITE

Satellite, Ariel 5
USE   ARIEL 5 SATELLITE

Satellite, Astronomical Netherlands
USE   ASTRONOMICAL NETHERLANDS SATELLITE

SATELLITE ATMOSPHERES

SATELLITE ATTITUDE CONTROL

(Satellite Attitude Control), DISCOS
USE   DISCOS (SATELLITE ATTITUDE CONTROL)

Satellite Attitude Disturbance
USE   ATTITUDE STABILITY
       SPACECRAFT STABILITY

Satellite, Azur
USE   AZUR SATELLITE

Satellite B, Earth Resources Technology
USE   LANDSAT 2

Satellite B, Geostationary Operatl Environ
USE   GOES 2

(Satellite), Bess
USE   BESS (SATELLITE)

Satellite, Biomedical Experiment Scientific
USE   BESS (SATELLITE)

Satellite C, Earth Resources Technology
USE   LANDSAT 3

Satellite, Cannonball 2
USE   CANNONBALL 2 SATELLITE

Satellite Capture
USE   SPACECRAFT RECOVERY

Satellite Communication
USE   SPACECRAFT COMMUNICATION

Satellite Communication System, Fleet
USE   FLEET SATELLITE COMMUNICATION
       SYSTEM

SATELLITE COMMUNICATIONS SHIPS

Satellite Communications Systems, Domestic
USE   DOMESTIC SATELLITE COMMUNICATIONS
       SYSTEMS

Satellite, Communications Technology
USE   COMMUNICATIONS TECHNOLOGY
       SATELLITE

SATELLITE CONFIGURATIONS

SATELLITE CONTROL

Satellite, COS-B
USE   COS-B SATELLITE

Satellite, Cosmic Background Explorer
USE   COSMIC BACKGROUND EXPLORER
       SATELLITE

Satellite, Cosmos 2
USE   COSMOS 2 SATELLITE

Satellite, Cosmos 3
USE   COSMOS 3 SATELLITE

Satellite, Cosmos 5
USE   COSMOS 5 SATELLITE

Satellite, Cosmos 6
USE   COSMOS 6 SATELLITE

Satellite, Cosmos 14
USE   COSMOS 14 SATELLITE

Satellite, Cosmos 44
USE   COSMOS 44 SATELLITE

Satellite, Cosmos 54
USE   COSMOS 54 SATELLITE

Satellite, Cosmos 71
USE   COSMOS 71 SATELLITE

Satellite, Cosmos 110
USE   COSMOS 110 SATELLITE

Satellite, Cosmos 137
USE   COSMOS 137 SATELLITE

Satellite, Cosmos 144
USE   COSMOS 144 SATELLITE

Satellite, Cosmos 149
USE   COSMOS 149 SATELLITE

Satellite, Cosmos 166
USE   COSMOS 166 SATELLITE

Satellite, Cosmos 186
USE   COSMOS 186 SATELLITE

Satellite, Cosmos 188
USE   COSMOS 188 SATELLITE

Satellite, Cosmos 206
USE   COSMOS 206 SATELLITE

Satellite, Cosmos 213
USE   COSMOS 213 SATELLITE

Satellite, Cosmos 224
USE   COSMOS 224 SATELLITE

Satellite, Cosmos 225
USE   COSMOS 225 SATELLITE

Satellite, Cosmos 381
USE   COSMOS 381 SATELLITE

Satellite, Cosmos 782
USE   COSMOS 782 SATELLITE

Satellite, Cosmos 936
USE   COSMOS 936 SATELLITE

Satellite, Cosmos 954
USE   COSMOS 954 SATELLITE

Satellite, Cosmos 1129
USE   COSMOS 1129 SATELLITE

Satellite, Courier
USE   COURIER SATELLITE

Satellite D, Earth Resources Technology
USE   LANDSAT 4

Satellite, D-1
USE   D-1 SATELLITE

Satellite, D-2B
USE   D-2 SATELLITES'

Satellite Defense
USE   SPACECRAFT DEFENSE

SATELLITE DESIGN

Satellite, Dial
USE   DIAL SATELLITE

Satellite, DME-A
USE   EXPLORER 31 SATELLITE

Satellite, Dodge
USE   DODGE SATELLITE

SATELLITE DOPPLER POSITIONING

SATELLITE DRAG

Satellite, Dynamics Explorer 1
USE   DYNAMICS EXPLORER 1 SATELLITE

Satellite, Dynamics Explorer 2
USE   DYNAMICS EXPLORER 2 SATELLITE

Satellite E, Earth Resources Technology
USE   LANDSAT E

Satellite, Echo 1
USE   ECHO 1 SATELLITE

Satellite, Echo 2
USE   ECHO 2 SATELLITE

Satellite, Elektron 1
USE   ELEKTRON 1 SATELLITE

Satellite, Elektron 2
USE   ELEKTRON 2 SATELLITE

Satellite, Elektron 4
USE   ELEKTRON 4 SATELLITE

Satellite), ERS-1 (ESA
USE   ERS-1 (ESA SATELLITE)

Satellite (Esa), Maritime Communication
USE   MAROTS (ESA)

Satellite (Esa), Orbital Test
USE   OTS (ESA)

Satellite, ESRO 1
USE   ESRO 1 SATELLITE

Satellite, ESRO 2
USE   ESRO 2 SATELLITE

Satellite, ESRO 4
USE   ESRO 4 SATELLITE

Satellite, ESSA 1
USE   ESSA 1 SATELLITE

Satellite, ESSA 2
USE   ESSA 2 SATELLITE

Satellite, ESSA 3
USE   ESSA 3 SATELLITE

Satellite, ESSA 4
USE   ESSA 4 SATELLITE

Satellite, ESSA 5
USE   ESSA 5 SATELLITE

Satelllte, ESSA 6
USE   ESSA 6 SATELLITE

Satellite, ESSA 7
USE   ESSA 7 SATELLITE

Satellite, ESSA 8
USE   ESSA 8 SATELLITE

Satellite, ESSA 9
USE   ESSA 9 SATELLITE

Satellite, European Communications
USE   EUROPEAN COMMUNICATIONS SATELLITE

Satellite, European Large Telecomm
USE   L-SAT

Satellite, Exosat
USE   EXOSAT SATELLITE

Satellite, Explorer 1
USE   EXPLORER 1 SATELLITE

Satellite, Explorer 2
USE   EXPLORER 2 SATELLITE

Satellite, Explorer 3
USE   EXPLORER 3 SATELLITE

Satellite, Explorer 4
USE   EXPLORER 4 SATELLITE

Satellite, Explorer 5
USE   EXPLORER 5 SATELLITE

Satellite, Explorer 6
USE   EXPLORER 6 SATELLITE

Satellite, Explorer 7
USE   EXPLORER 7 SATELLITE

Satellite, Explorer 8
USE   EXPLORER 8 SATELLITE

Satellite, Explorer 9
USE   EXPLORER 9 SATELLITE

Satellite, Explorer 10
USE   EXPLORER 10 SATELLITE

Satellite, Explorer 11
USE   EXPLORER 11 SATELLITE

Satellite, Explorer 12
USE   EXPLORER 12 SATELLITE

Satellite, Explorer 14
USE   EXPLORER 14 SATELLITE

Satellite, Explorer 15
USE   EXPLORER 15 SATELLITE

Satellite, Explorer 16
USE   EXPLORER 16 SATELLITE

Satellite, Explorer 17
USE   EXPLORER 17 SATELLITE

Satellite, Explorer 18
USE   EXPLORER 18 SATELLITE

Satellite, Explorer 19
USE   EXPLORER 19 SATELLITE

Satellite, Explorer 20
USE   EXPLORER 20 SATELLITE

Satellite, Explorer 21
USE   EXPLORER 21 SATELLITE

Satellite, Explorer 22
USE   EXPLORER 22 SATELLITE

Satellite, Explorer 23
USE   EXPLORER 23 SATELLITE

Satellite, Explorer 24
USE   EXPLORER 24 SATELLITE

Satellite, Explorer 25
USE   EXPLORER 25 SATELLITE

Satellite, Explorer 26
USE   EXPLORER 26 SATELLITE

Satellite, Explorer 27
USE   EXPLORER 27 SATELLITE

Satellite, Explorer 28
USE   EXPLORER 28 SATELLITE

Satellite, Explorer 29
USE   EXPLORER 29 SATELLITE

Satellite, Explorer 30
USE   EXPLORER 30 SATELLITE

Satellite, Explorer 31
USE   EXPLORER 31 SATELLITE

Satellite, Explorer 32
USE   EXPLORER 32 SATELLITE

Satellite, Explorer 33
USE   EXPLORER 33 SATELLITE

Satellite, Explorer 34
USE   EXPLORER 34 SATELLITE

Satellite, Explorer 35
USE   EXPLORER 35 SATELLITE

Satellite, Explorer 36
USE   EXPLORER 36 SATELLITE

Satellite, Explorer 37
USE   EXPLORER 37 SATELLITE

Satellite, Explorer 38
USE   EXPLORER 38 SATELLITE

Satellite, Explorer 39
USE   EXPLORER 39 SATELLITE

Satellite, Explorer 40
USE   EXPLORER 40 SATELLITE

Satellite, Explorer 41
USE   EXPLORER 41 SATELLITE

Satellite, Explorer 42
USE   UHURU SATELLITE

Satellite, Explorer 43
USE   EXPLORER 43 SATELLITE

Satellite, Explorer 44
USE   EXPLORER 44 SATELLITE

Satellite, Explorer 45
USE   EXPLORER 45 SATELLITE

Satellite, Explorer 46
USE   EXPLORER 46 SATELLITE

Satellite, Explorer 47
USE   EXPLORER 47 SATELLITE

Satellite, Explorer 48
USE   EXPLORER 48 SATELLITE

Satellite, Explorer 49
USE   EXPLORER 49 SATELLITE

Satellite, Explorer 50
USE   EXPLORER 50 SATELLITE

Satellite, Explorer 51
USE   EXPLORER 51 SATELLITE

Satellite, Explorer 52
USE   EXPLORER 52 SATELLITE

Satellite, Explorer 53
USE   EXPLORER 53 SATELLITE

Satellite, Explorer 54
USE   EXPLORER 54 SATELLITE

Satellite, Explorer 55
USE   EXPLORER 55 SATELLITE

Satellite, Extreme Ultraviolet Explorer
USE   EXTREME ULTRAVIOLET EXPLORER
         SATELLITE

Satellite F, Earth Resources Technology
USE   LANDSAT F

Satellite, FR-1
USE   FR-1 SATELLITE

Satellite Geodesy Experiment, International
USE   INTERNATIONAL SATELLITE GEODESY
         EXPERIMENT

Satellite, Geodynamic Experimental Ocean
USE   GEOS-D SATELLITE

Satellite, GEOS 1
USE   GEOS 1 SATELLITE

Satellite, GEOS 2
USE   GEOS 2 SATELLITE

Satellite, GEOS 3
USE   GEOS 3 SATELLITE

Satellite, GEOS-B
USE   GEOS 2 SATELLITE

Satellite, GEOS-C
USE   GEOS 3 SATELLITE

Satellite, GEOS-D
USE   GEOS-D SATELLITE

Satellite, Gravsat
USE   GRAVSAT SATELLITE

SATELLITE GROUND SUPPORT

SATELLITE GROUND TRACKS

SATELLITE GUIDANCE

Satellite, Hawkeye 1
USE   EXPLORER 52 SATELLITE

(Satellite), HELOS
USE   EXOSAT SATELLITE

Satellite, HEOS A
USE   HEOS A SATELLITE

Satellite, HEOS B
USE   HEOS B SATELLITE

Satellite, Hermes
USE   COMMUNICATIONS TECHNOLOGY
         SATELLITE

Satellite, High Eccentric Lunar Occultation
USE   EXOSAT SATELLITE

Satellite, Hipparcos
USE   HIPPARCOS SATELLITE

SATELLITE IMAGERY

Satellite, IME
USE   INTERNATIONAL MAGNETOSPHERIC
         EXPLORER

Satellite, Infrared Astronomy
USE   INFRARED ASTRONOMY SATELLITE

Satellite, Injun 1
USE   INJUN 1 SATELLITE

Satellite, Injun 3
USE   INJUN 3 SATELLITE

Satellite, Injun 4
USE   INJUN 4 SATELLITE

Satellite, Injun 5
USE   EXPLORER 40 SATELLITE

Satellite, Inspector
USE   INSPECTOR SATELLITE

SATELLITE INSTRUMENTS

Satellite, Intasat
USE   INTASAT SATELLITE

SATELLITE INTERCEPTORS

(Satellite), LAGEOS
USE   LAGEOS (SATELLITE)

Satellite, LARGOS
USE   LARGOS SATELLITE

Satellite, Laser Geodynamic
USE   LAGEOS (SATELLITE)

Satellite Launching
USE   SPACECRAFT LAUNCHING

SATELLITE LIFETIME

Satellite Lines, Dielectronic
USE   RESONANCE LINES

Satellite, Lzeebe
USE   LZEEBE SATELLITE

Satellite, Magsat A
USE   MAGSAT A SATELLITE

Satellite, Magsat B
USE   MAGSAT B SATELLITE

Satellite, Magsat 1
USE   MAGSAT 1 SATELLITE

Satellite Maneuvers
USE   SPACECRAFT MANEUVERS

Satellite, Marisat 1
USE   MARISAT 1 SATELLITE

Satellite, Maritime Orbital Test
USE   MAROTS (ESA)

Satellite, Meteoroid Technology
USE   EXPLORER 46 SATELLITE

Satellite, METEOSAT
USE   METEOSAT SATELLITE

Satellite, Midas 2
USE   MIDAS 2 SATELLITE

Satellite, Midas 3
USE   MIDAS 3 SATELLITE

Satellite, Midas 4
USE   MIDAS 4 SATELLITE

Satellite, Midas 5
USE   MIDAS 5 SATELLITE

Satellite, Midas 6
USE   MIDAS 6 SATELLITE

Satellite, Midas 7
USE   MIDAS 7 SATELLITE

Satellite, Miranda
USE   MIRANDA SATELLITE

Satellite, NATO 3B
USE   NATO 3B SATELLITE

SATELLITE NAVIGATION SYSTEMS

SATELLITE NETWORKS

Satellite, Nimbus 1
USE   NIMBUS 1 SATELLITE

Satellite, Nimbus 2
USE   NIMBUS 2 SATELLITE

Satellite, Nimbus 3
USE   NIMBUS 3 SATELLITE

Satellite, Nimbus 4
USE   NIMBUS 4 SATELLITE

Satellite, Nimbus 5
USE   NIMBUS 5 SATELLITE

Satellite, Nimbus 6
USE   NIMBUS 6 SATELLITE

Satellite, Nimbus 7
USE   NIMBUS 7 SATELLITE

Satellite, NOAA 2
USE   NOAA 2 SATELLITE

Satellite, NOAA 3
USE   NOAA 3 SATELLITE

Satellite, NOAA 4
USE   NOAA 4 SATELLITE

Satellite, NOAA 5
USE   NOAA 5 SATELLITE

Satellite, NOAA 6
USE   NOAA 6 SATELLITE

Satellite, NOAA 7
USE   NOAA 7 SATELLITE

Satellite, NOAA 8
USE   NOAA 8 SATELLITE

SATELLITE OBSERVATION

Satellite, ORBIS Cal
USE   ORBIS CAL SATELLITE

Satellite Orbit Calculation
USE   ORBIT CALCULATION

SATELLITE ORBITS

SATELLITE ORIENTATION

Satellite, PAGEOS
USE   PAGEOS SATELLITE

Satellite, Palapa B
USE   PALAPA 2 SATELLITE

Satellite, Palapa 2
USE   PALAPA 2 SATELLITE

(Satellite Payload), Amps
USE   AMPS (SATELLITE PAYLOAD)

SATELLITE PERTURBATION

Satellite, Polaire
USE   D-2 SATELLITES

Satellite, Poseidon
USE   POSEIDON SATELLITE

SATELLITE POWER TRANSMISSION (TO EARTH)

Satellite Program, Defense Meteorological
USE   DMSP SATELLITES

Satellite Proj, Synchronous Communications
USE   SYNCHRONOUS COMMUNICATIONS
       SATELLITE PROJ

Satellite, Proton 1
USE   PROTON 1 SATELLITE

Satellite, Proton 2
USE   PROTON 2 SATELLITE

Satellite, Proton 3
USE   PROTON 3 SATELLITE

Satellite, Proton 4
USE   PROTON 4 SATELLITE

Satellite, P78-2
USE   SCATHA SATELLITE

Satellite, Radiation And Meteoroid
USE   RADIATION AND METEOROID SATELLITE

Satellite, Radio Astronomy Explorer
USE   RADIO ASTRONOMY EXPLORER SATELLITE

Satellite, Raduga
USE   RADUGA SATELLITE

Satellite, Relay 1
USE   RELAY 1 SATELLITE

Satellite, Relay 2
USE   RELAY 2 SATELLITE

Satellite Rendezvous
USE   ORBITAL RENDEZVOUS

Satellite, Roentgen
USE   ROSAT MISSION

SATELLITE ROTATION

Satellite, S-3
USE   EXPLORER 12 SATELLITE

Satellite, S-6
USE   EXPLORER 17 SATELLITE

Satellite, S-16
USE   OSO-1

Satellite, S-17
USE   OSO-2

Satellite, S-18
USE   OAO

Satellite, S-27
USE   ALOUETTE 1 SATELLITE

Satellite, S-49
USE   OGO-A

Satellite, S-50
USE   OGO-C

Satellite, S-51
USE   ARIEL 1 SATELLITE

Satellite, S-52
USE   ARIEL 2 SATELLITE

Satellite, S-57
USE   OSO-C

Satellite, S-66
USE   BEACON EXPLORER A

Satellite, S-74
USE   EXPLORER 18 SATELLITE

Satellite, Sage
USE   SAGE SATELLITE

Satellite, San Marco 1
USE   SAN MARCO 1 SATELLITE

Satellite, San Marco 2
USE   SAN MARCO 2 SATELLITE

Satellite, San Marco 3
USE   SAN MARCO 3 SATELLITE

Satellite, Scatha
USE   SCATHA SATELLITE

Satellite, SCORE
USE   SCORE SATELLITE

Satellite, Search And Rescue
USE   SARSAT

Satellite, SEASAT-B
USE   SEASAT-B SATELLITE

(Satellite), Seocs
USE   SEOCS (SATELLITE)

Satellite Service, Land Mobile
USE   LAND MOBILE SATELLITE SERVICE

Satellite, Severe Storms Observing
USE   STORMSAT SATELLITE

Satellite, Sirio
USE   SIRIO SATELLITE

Satellite, SIRS B
USE   SIRS B SATELLITE

Satellite, Snapshot
USE   SNAPSHOT SATELLITE

SATELLITE SOLAR ENERGY CONVERSION

SATELLITE SOLAR POWER STATIONS

Satellite, Solar Radiation 1
USE   SOLAR RADIATION 1 SATELLITE

Satellite, Solar Radiation 3
USE   SOLAR RADIATION 3 SATELLITE

Satellite, Solrad 10
USE   EXPLORER 44 SATELLITE

SATELLITE SOUNDING

Satellite, Space Arrow
USE   COSMOS 149 SATELLITE

Satellite), Spot (French
USE   SPOT (FRENCH SATELLITE)

Satellite, Sputnik 1
USE   SPUTNIK 1 SATELLITE

Satellite, Sputnik 2
USE   SPUTNIK 2 SATELLITE

Satellite, Sputnik 3
USE   SPUTNIK 3 SATELLITE

Satellite, Sputnik 4
USE   SPUTNIK 4 SATELLITE

Satellite, Sputnik 5
USE   SPUTNIK 5 SATELLITE

Satellite, SRET 1
USE   SRET 1 SATELLITE

Satellite, SRET 2
USE   SRET 2 SATELLITE

Satellite, Stormsat
USE   STORMSAT SATELLITE

SATELLITE SURFACES

Satellite, Synchronous Earth Observatory
USE   SYNCHRONOUS EARTH OBSERVATORY
       SATELLITE

Satellite, Synchronous Meteorological
USE   SYNCHRONOUS METEOROLOGICAL
       SATELLITE

Satellite, SYNCOM 1
USE   SYNCOM 1 SATELLITE

Satellite, SYNCOM 2
USE   SYNCOM 2 SATELLITE

Satellite, SYNCOM 3
USE   SYNCOM 3 SATELLITE

Satellite, SYNCOM 4
USE   SYNCOM 4 SATELLITE

Satellite System, Defense Communications
USE   DEFENSE COMMUNICATIONS SATELLITE
       SYSTEM

Satellite System, National Oceanic
USE   NATIONAL OCEANIC SATELLITE SYSTEM

Satellite System, TIROS Operational
USE   TIROS OPERATIONAL SATELLITE SYSTEM

Satellite, TD-1
USE   TD-1 SATELLITE

SATELLITE TELEVISION

Satellite, Telstar 1
USE   TELSTAR 1 SATELLITE

Satellite, Telstar 2
USE   TELSTAR 2 SATELLITE

SATELLITE TEMPERATURE

Satellite, TIROS D
USE   TIROS 4 SATELLITE

Satellite, TIROS E
USE   TIROS 5 SATELLITE

Satellite, TIROS F
USE   TIROS 6 SATELLITE

Satellite, TIROS G
USE   TIROS 7 SATELLITE

Satellite, TIROS H
USE   TIROS 8 SATELLITE

Satellite, TIROS Wheel
USE   TIROS 9 SATELLITE

Satellite, TIROS 1
USE   TIROS 1 SATELLITE

Satellite, TIROS 2
USE   TIROS 2 SATELLITE

Satellite, TIROS 3
USE   TIROS 3 SATELLITE

Satellite, TIROS 4
USE   TIROS 4 SATELLITE

Satellite, TIROS 5
USE   TIROS 5 SATELLITE

Satellite, TIROS 6
USE   TIROS 6 SATELLITE

Satellite, TIROS 7
USE   TIROS 7 SATELLITE

Satellite, TIROS 8
USE   TIROS 8 SATELLITE

Satellite, TIROS 9
USE   TIROS 9 SATELLITE

Satellite, TIROS 10
USE   TIROS 10 SATELLITE

Satellite, Tournesole
USE   D-2 SATELLITES

Satellite, TRAAC
USE   TRANSIT ATTITUDE CONTROL SATELLITE

SATELLITE TRACKING

Satellite Tracking And Data Acq Network
USE   STDN (NETWORK)

(Satellite Tracking Network), STADAN
USE   STDN (NETWORK)

Satellite Tracking Program, Optical
USE   OPTICAL SATELLITE TRACKING PROGRAM

Satellite Tracking, Satellite-To-
USE   SATELLITE-TO-SATELLITE TRACKING

Satellite, Transit Attitude Control
USE   TRANSIT ATTITUDE CONTROL SATELLITE

SATELLITE TRANSMISSION

Satellite, Uhuru
USE   UHURU SATELLITE

Satellite, UK 4
USE   UK 4 SATELLITE

Satellite, Vanguard 1
USE   VANGUARD 1 SATELLITE

Satellite, Vanguard 2
USE   VANGUARD 2 SATELLITE

Satellite, Vanguard 3
USE   VANGUARD 3 SATELLITE

Satellite, Venera 2
USE   VENERA 2 SATELLITE

Satellite, Venera 3
USE   VENERA 3 SATELLITE

Satellite, Venera 4
USE   VENERA 4 SATELLITE

Satellite, Venera 5
USE   VENERA 5 SATELLITE

Satellite, Venera 6
USE   VENERA 6 SATELLITE

Satellite, Venera 7
USE   VENERA 7 SATELLITE

Satellite, Venera 8
USE   VENERA 8 SATELLITE

Satellite, Venera 9
USE   VENERA 9 SATELLITE

Satellite, Venera 10
USE   VENERA 10 SATELLITE

Satellite, Venera 11
USE   VENERA 11 SATELLITE

Satellite, Venera 12
USE   VENERA 12 SATELLITE

Satellite 1, Earth Resources Technology
USE   LANDSAT 1

Satellite 1, Small Astronomy
USE   SAS-1

Satellite 2, Small Astronomy
USE   SAS-2

Satellite 3, Small Astronomy
USE   SAS-3

SATELLITE-BORNE INSTRUMENTS

SATELLITE-BORNE PHOTOGRAPHY

SATELLITE-BORNE RADAR

SATELLITE-TO-SATELLITE TRACKING

SATELLITES

Satellites, Active
USE   ACTIVE SATELLITES

Satellites, Aeronautical
USE   AERONAUTICAL SATELLITES

Satellites, Aerosat
USE   AEROSAT SATELLITES

Satellites, Alouette
USE   ALOUETTE SATELLITES

(Satellites), Ampte
USE   AMPTE (SATELLITES)

Satellites, Anik
USE   ANIK SATELLITES

Satellites, ANNA
USE   ANNA SATELLITES

Satellites, Applications Explorer
USE   APPLICATIONS EXPLORER SATELLITES

Satellites, Applications Technology
USE   ATS

Satellites, Ariel
USE   ARIEL SATELLITES

Satellites, Artificial
USE   ARTIFICIAL SATELLITES

Satellites, Astronomical
USE   ASTRONOMICAL SATELLITES

Satellites, Beacon
USE   BEACON SATELLITES

Satellites, Bio
USE   BIOSATELLITES

Satellites, Communication
USE   COMMUNICATION SATELLITES

Satellites, Comstar
USE   COMSTAR SATELLITES

Satellites, Cosmos
USE   COSMOS SATELLITES

Satellites, D-2
USE   D-2 SATELLITES

Satellites, Diademe
USE   DIADEME SATELLITES

Satellites, Discoverer
USE   DISCOVERER SATELLITES

Satellites, DMSP
USE   DMSP SATELLITES

Satellites, Dynamics Explorer
USE   DYNAMICS EXPLORER SATELLITES

Satellites, Early Bird
USE   EARLY BIRD SATELLITES

Satellites, Earth Resources Observation
USE   EROS (SATELLITES)

Satellites, Earth Resources Technology
USE   LANDSAT SATELLITES

Satellites, Echo
USE   ECHO SATELLITES

Satellites, Elektron
USE   ELEKTRON SATELLITES

Satellites, Environmental Research
USE   ENVIRONMENTAL RESEARCH SATELLITES

Satellites, EOLE
USE   EOLE SATELLITES

(Satellites), EROS
USE   EROS (SATELLITES)

Satellites, ESA
USE   ESA SATELLITES

Satellites (ESA), GEOS
USE   GEOS SATELLITES (ESA)

Satellites, ESRO
USE   ESA SATELLITES

Satellites (Esro), GEOS
USE   GEOS SATELLITES (ESA)

Satellites, ESSA
USE   ESSA SATELLITES

Satellites, Evasive
USE   EVASIVE SATELLITES

Satellites, Explorer
USE   EXPLORER SATELLITES

Satellites, French
USE   FRENCH SATELLITES

Satellites, Galilean
USE   GALILEAN SATELLITES

Satellites, Geodetic
USE   GEODETIC SATELLITES

Satellites, GEOLE
USE   GEOLE SATELLITES

Satellites, Geophysical
USE   GEOPHYSICAL SATELLITES

Satellites, Geostationary
USE   SYNCHRONOUS SATELLITES

Satellites, GOES
USE   GOES SATELLITES

Satellites, Gravity Gradient
USE   GRAVITY GRADIENT SATELLITES

Satellites, GREB
USE   GREB SATELLITES

Satellites, Hawkeye
USE   HAWKEYE SATELLITES

Satellites, Helios
USE   HELIOS SATELLITES

Satellites, HEOS
USE   HEOS SATELLITES

Satellites, Highly Eccentric Orbit
USE   HEOS SATELLITES

Satellites, Improved TIROS Operational
USE   IMPROVED TIROS OPERATIONAL
          SATELLITES

Satellites, Injun
USE   INJUN SATELLITES

Satellites, INSAT
USE   INDIAN SPACECRAFT

Satellites, Intelsat
USE   INTELSAT SATELLITES

Satellites, Intercosmos
USE   INTERCOSMOS SATELLITES

Satellites, IRIS
USE   IRIS SATELLITES

Satellites, ISIS
USE   ISIS SATELLITES

Satellites, ITOS
USE   ITOS SATELLITES

Satellites, Jupiter
USE   JUPITER SATELLITES

Satellites, LANDSAT
USE   LANDSAT SATELLITES

(Satellites), LES
USE   LINCOLN EXPERIMENTAL SATELLITES

Satellites, Lincoln Experimental
USE   LINCOLN EXPERIMENTAL SATELLITES

Satellites, Location Of Air Traffic
USE   LOCATES SYSTEM

Satellites, LOFTI
USE   LOW FREQUENCY TRANSIONOSPHERIC
          SATELLITES

Satellites, Low Frequency Transionospheric
USE   LOW FREQUENCY TRANSIONOSPHERIC
          SATELLITES

Satellites, Lunar
USE   LUNAR SATELLITES

Satellites, Magsat
USE   MAGSAT SATELLITES

Satellites, Marecs Maritime
USE   MARECS MARITIME SATELLITES

Satellites, Marisat
USE   MARISAT SATELLITES

Satellites, Maritime
USE   MARITIME SATELLITES

Satellites, Meteorological
USE   METEOROLOGICAL SATELLITES

Satellites, Micrometeoroid Explorer
USE   MICROMETEOROID EXPLORER SATELLITES

Satellites, Midas
USE   MIDAS SATELLITES

Satellites, Molniya
USE   MOLNIYA SATELLITES

Satellites, Natural
USE   NATURAL SATELLITES

Satellites, Navigation
USE   NAVIGATION SATELLITES

Satellites, Navigation Technology
USE   NAVIGATION TECHNOLOGY SATELLITES

Satellites, Navstar
USE   NAVSTAR SATELLITES

Satellites, Nimbus
USE   NIMBUS SATELLITES

Satellites, NOAA
USE   NOAA SATELLITES

Satellites, Nova
USE   NOVA SATELLITES

Satellites, Octahedral Research
USE   ENVIRONMENTAL RESEARCH SATELLITES

Satellites, Orbiting
USE   ARTIFICIAL SATELLITES

Satellites, OV-1
USE   OV-1 SATELLITES

Satellites, OV-2
USE   OV-2 SATELLITES

Satellites, OV-3
USE   OV-3 SATELLITES

Satellites, OV-4
USE   OV-4 SATELLITES

Satellites, OV-5
USE   OV-5 SATELLITES

Satellites, Palapa
USE   PALAPA SATELLITES

Satellites, Passive
USE   PASSIVE SATELLITES

Satellites, Pegasus
USE   PEGASUS SATELLITES

Satellites, PEOLE
USE   PEOLE SATELLITES

Satellites, Perigee-Apogee
USE   PAS

Satellites, Planetary
USE   NATURAL SATELLITES

Satellites, Polyot
USE   POLYOT SATELLITES

Satellites, Prognoz
USE   PROGNOZ SATELLITES

Satellites, Proton
USE   PROTON SATELLITES

Satellites, Ranger
USE   RANGER LUNAR PROBES

Satellites, RCA Satcom
USE   RCA SATCOM SATELLITES

Satellites, Recoverable
USE   RECOVERABLE SPACECRAFT

Satellites, Reflector
USE   PASSIVE SATELLITES

Satellites, Relay
USE   RELAY SATELLITES

Satellites, San Marco
USE   SAN MARCO SATELLITES

Satellites, Saturn
USE   SATURN SATELLITES

Satellites, Scientific
USE   SCIENTIFIC SATELLITES

Satellites, SEASAT
USE   SEASAT SATELLITES

Satellites, Shuttle Pallet
USE   SHUTTLE PALLET SATELLITES

Satellites, Skynet
USE   SKYNET SATELLITES

Satellites, Small Astronomy
USE   SAS

Satellites, Small Scientific
USE   SMALL SCIENTIFIC SATELLITES

Satellites, Solar Power
USE   SOLAR POWER SATELLITES

Satellites, Soviet
USE   SOVIET SATELLITES

Satellites, Spartan
USE   SPARTAN SATELLITES

Satellites, Sputnik
USE   SPUTNIK SATELLITES

Satellites, SRET
USE   SRET SATELLITES

Satellites, Symphonie
USE   SYMPHONIE SATELLITES

Satellites, Synchronous
USE   SYNCHRONOUS SATELLITES

Satellites, Synchronous Communication
USE   SYNCOM SATELLITES

Satellites, SYNCOM
USE   SYNCOM SATELLITES

Satellites, TD
USE   TD SATELLITES

Satellites, TDR
USE   TDR SATELLITES

Satellites, Telstar
USE   TELSTAR SATELLITES

Satellites, Tethered
USE   TETHERED SATELLITES

Satellites, TIROS
USE   TIROS SATELLITES

Satellites, TIROS N Series
USE   TIROS N SERIES SATELLITES

Satellites, Tracking And Data Relay
USE   TDR SATELLITES

Satellites, Transit
USE   TRANSIT SATELLITES

Satellites, UK
USE   UK SATELLITES

Satellites, United Kingdom
USE   UK SATELLITES

Satellites, Uranus
USE   URANUS SATELLITES

Satellites, Vanguard
USE   VANGUARD SATELLITES

Satellites, Vela
USE   VELA SATELLITES

Satellites, Venera
USE   VENERA SATELLITES

Satellites, Westar
USE   WESTAR SATELLITES

Sats For Ionospheric Study, International
USE   ISIS SATELLITES

Sats, Galactic Radiation Exp Background
USE   GREB SATELLITES

Sats, Geostationary Operational Environ
USE   GOES SATELLITES

SATURABLE REACTORS

Saturated Hydrocarbons
USE   ALKANES

SATURATION ·

SATURATION (CHEMISTRY)

Saturation, De
USE   DESATURATION

Saturation, Super
USE   SUPERSATURATION

SATURN

SATURN ATMOSPHERE

SATURN D LAUNCH VEHICLE

Saturn Flyby, Mariner Jupiter-
USE   MARINER JUPITER-SATURN FLYBY

SATURN LAUNCH VEHICLES

SATURN (PLANET)

SATURN PROJECT

SATURN RINGS

SATURN S-1 STAGE

SATURN S-1B STAGE

SATURN S-1C STAGE

SATURN S-2 STAGE

SATURN S-4 STAGE

SATURN S-4B STAGE

SATURN SATELLITES

Saturn Spacecraft, Pioneer
USE   PIONEER 11 SPACE PROBE

SATURN STAGES

SATURN WORKSHOPS

SATURN 1 LAUNCH VEHICLES

SATURN 1 SA-1 LAUNCH VEHICLE

SATURN 1 SA-2 LAUNCH VEHICLE

SATURN 1 SA-3 LAUNCH VEHICLE

SATURN 1 SA-4 LAUNCH VEHICLE

SATURN 1 SA-5 LAUNCH VEHICLE

SATURN 1 SA-6 LAUNCH VEHICLE

SATURN 1 SA-7 LAUNCH VEHICLE

SATURN 1 SA-8 LAUNCH VEHICLE

SATURN 1 SA-9 LAUNCH VEHICLE

SATURN 1 SA-10 LAUNCH VEHICLE

SATURN 1 WORKSHOP

SATURN 1B LAUNCH VEHICLES

SATURN 2 LAUNCH VEHICLES

SATURN 5 LAUNCH VEHICLES

SATURN 5 WORKSHOP

SAUDI ARABIA

SAUDI ARABIAN SPACE PROGRAM

Savage Aircraft
USE   A-2 AIRCRAFT

SAVANNAH NUCLEAR SHIP

Savannahs
USE   GRASSLANDS

SAWS

SAWTOOTH WAVEFORMS

Sb
USE   ANTIMONY

Sc
USE   SCANDIUM

SC
USE   SOUTH CAROLINA

SC), Sand Hills Region (GA-NC-
USE   SAND HILLS REGION (GA-NC-SC)

SC-1 AIRCRAFT

SC-1 Aircraft, Short
USE   SC-1 AIRCRAFT

SC-5 AIRCRAFT

SC-5 Aircraft, Short
USE   SC-5 AIRCRAFT

SC-7 AIRCRAFT

SC-7 Aircraft, Short
USE   SC-7 AIRCRAFT

Scalar Magnetic Charge
USE   MAGNETIC CHARGE DENSITY

SCALARS

SCALE

SCALE (CORROSION)

SCALE EFFECT

Scale, Fahrenheit Temperature
USE   TEMPERATURE SCALES

Scale, Gray
USE   GRAY SCALE

SCALE HEIGHT

Scale Integration, Large
USE   LARGE SCALE INTEGRATION

Scale Integration, Medium
USE   MEDIUM SCALE INTEGRATION

Scale Integration, Very Large
USE   VERY LARGE SCALE INTEGRATION

SCALE MODELS

SCALE (RATIO)

Scale, Taylor Manifest Anxiety
USE   TAYLOR MANIFEST ANXIETY SCALE

Scale Tests, Full
USE   FULL SCALE TESTS

SCALERS

Scales, Temperature
USE   TEMPERATURE SCALES

SCALING

Scaling, De
USE   DESCALING

SCALING LAWS

SCALLOPING

Scan Radiometer, Visible Infrared Spin
USE   VISIBLE INFRARED SPIN SCAN
         RADIOMETER

SCANDINAVIA

SCANDIUM

SCANDIUM COMPOUNDS

SCANDIUM ISOTOPES

SCANDIUM OXIDES

Scandium 46
USE   SCANDIUM ISOTOPES

Scanner, Cat
USE   COMPUTER AIDED TOMOGRAPHY

Scanner, Coastal Zone Color
USE   COASTAL ZONE COLOR SCANNER

Scanner, Ocean Color
USE   OCEAN COLOR SCANNER

SCANNER PROJECT

SCANNERS

Scanners, Flying Spot
USE   FLYING SPOT SCANNERS

Scanners, Horizon
USE   HORIZON SCANNERS

Scanners, Infrared
USE   INFRARED SCANNERS

Scanners, Infrared Horizon
USE   INFRARED SCANNERS
         HORIZON SCANNERS

(Scanners), MUBIS
USE   MULTIPLE BEAM INTERVAL SCANNERS

Scanners, Multiple Beam Interval
USE   MULTIPLE BEAM INTERVAL SCANNERS

Scanners, Multispectral Band
USE   MULTISPECTRAL BAND SCANNERS

Scanners, Optical
USE   OPTICAL SCANNERS

Scanners, Ultrasonic
USE   ULTRASONIC SCANNERS

SCANNING

Scanning Beam Landing System, Microwave
USE·   MICROWAVE SCANNING BEAM LANDING
         SYSTEM

Scanning, Conical
USE   CONICAL SCANNING

Scanning Devices
USE   SCANNERS

Scanning, Frequency
USE   FREQUENCY SCANNING

Scanning Laser Acoustic Microscope (SLAM)
USE   ACOUSTIC MICROSCOPES

Scanning, Panoramic
USE   PANORAMIC SCANNING

Scanning, Radar
USE   RADAR SCANNING

SCAPULA

SCAR Program
USE   SUPERSONIC CRUISE AIRCRAFT
         RESEARCH

SCARFING

Scarps
USE   ESCARPMENTS

SCARS

Scars (Geology)
USE   EROSION

SCAT
USE   SUPERSONIC COMMERCIAL AIR
         TRANSPORT

SCATHA SATELLITE

SCATTER PLATES (OPTICS)

SCATTER PROPAGATION

Scatter Propagation, Ionospheric F-
USE   IONOSPHERIC F-SCATTER PROPAGATION

Scatter Radar, European Incoherent
USE   EISCAT RADAR SYSTEM (EUROPE)

Scatter Radar, Incoherent
USE   INCOHERENT SCATTER RADAR

Scatter Site Program, Radar Target
USE   RADAR TARGET SCATTER SITE PROGRAM

Scatterers
USE   SCATTERING

SCATTERING

Scattering, Acoustic
USE   ACOUSTIC SCATTERING

SCATTERING AMPLITUDE

Scattering, Atmospheric
USE   ATMOSPHERIC SCATTERING

Scattering, Back
USE   BACKSCATTERING

SCATTERING COEFFICIENTS

Scattering, Coherent
USE   COHERENT SCATTERING

SCATTERING CROSS SECTIONS

Scattering, Elastic
USE   ELASTIC SCATTERING

Scattering, Electromagnetic
USE   ELECTROMAGNETIC SCATTERING

Scattering, Electron
USE   ELECTRON SCATTERING

Scattering, Forward
USE   FORWARD SCATTERING

SCATTERING FUNCTIONS

Scattering, Incoherent
USE   INCOHERENT SCATTERING

Scattering, Inelastic
USE   INELASTIC SCATTERING

Scattering, Inverse
USE   INVERSE SCATTERING

Scattering, Ion
USE   ION SCATTERING

Scattering Layers, Deep
USE   DEEP SCATTERING LAYERS

Scattering, Light
USE   LIGHT SCATTERING

Scattering, Lunar
USE   LUNAR RADAR ECHOES
      DIFFUSE RADIATION

Scattering Matrix
USE   S MATRIX THEORY

Scattering Meters, Light
USE   LIGHT SCATTERING METERS

Scattering, Microwave
USE   MICROWAVE SCATTERING

Scattering, Mie
USE   MIE SCATTERING

Scattering, Neutron
USE   NEUTRON SCATTERING

Scattering, Nuclear
USE   NUCLEAR SCATTERING

Scattering, Nucleon-Nucleon
USE   NUCLEON-NUCLEON SCATTERING

Scattering, Proton
USE   PROTON SCATTERING

Scattering, Radar
USE   RADAR SCATTERING

Scattering, Radio
USE   RADIO SCATTERING

Scattering, Raman
USE   RAMAN SPECTRA

Scattering, Rayleigh
USE   RAYLEIGH SCATTERING

Scattering, Resonance
USE   RESONANCE SCATTERING

Scattering, Thomson
USE   THOMSON SCATTERING

Scattering, Tropospheric
USE   TROPOSPHERIC SCATTERING

Scattering, Wave
USE   WAVE SCATTERING

Scattering, X Ray
USE   X RAY SCATTERING

SCATTEROMETERS

SCAVENGING

SCCF
USE   SOLAR CELL CALIBRATION FACILITY

SCENE ANALYSIS

SCENEDESMUS

SCF
USE   SELF CONSISTENT FIELDS

SCHACH EFFECT

SCHAUDER FIXPOINT THEOREM

SCHEDULES

SCHEDULING

(Scheduling), Programming
USE   PROGRAMMING (SCHEDULING)

SCHEELITE

SCHELKUNOFF PRINCIPLE

Schematics
USE   CIRCUIT DIAGRAMS

Scherrer Method, Debye-
USE   DEBYE-SCHERRER METHOD

Schiff Bases
USE   IMINES

SCHIST

SCHIZOPHRENIA

SCHLEICHER AIRCRAFT

Schleicher KA-6 Sailplane
USE   KA-6 SAILPLANES

Schlichting Waves, Tollmein-
USE   TOLLMEIN-SCHLICHTING WAVES

SCHLIEREN PHOTOGRAPHY

SCHMIDT CAMERAS

Schmidt Filtering, Kalman-
USE   KALMAN-SCHMIDT FILTERING

SCHMIDT METHOD

SCHMIDT NUMBER

SCHMIDT TELESCOPES

SCHOOLS

SCHOOLS (FISH)

Schottky Barrier Diodes
USE   SCHOTTKY DIODES

SCHOTTKY DIODES

Schottky Effect
USE   WORK FUNCTIONS

SCHREIBERSITE

Schrieffer Theory, Bardeen-Cooper-
USE   BCS THEORY

SCHROEDINGER EQUATION

SCHULER TUNING

SCHUMANN-RUNGE BANDS

SCHWARTZ INEQUALITY

SCHWARTZ METHOD

SCHWARZ-CHRISTOFFEL TRANSFORMATION

SCHWARZSCHILD ANTENNAS

SCHWARZSCHILD METRIC

SCHWASSMANN-WACHMANN COMET

SCIATIC REGION

SCIENCE

Science, Materials
USE   MATERIALS SCIENCE

Science, Medical
USE   MEDICAL SCIENCE

Science, Soil
USE   SOIL SCIENCE

Sciences, Aerospace
USE   AEROSPACE SCIENCES

Sciences), Culture (Social
USE   CULTURE (SOCIAL SCIENCES)

Sciences, Forensic
USE   LAW (JURISPRUDENCE)

Sciences, Life
USE   LIFE SCIENCES

Sciences, Physical
USE   PHYSICAL SCIENCES

Sciences, Space
USE   AEROSPACE SCIENCES

Scientific Instrument Modules
USE   SIM

Scientific Modules, Lunar Surface
USE   LSSM

Scientific Satellite, Biomedical Experiment
USE   BESS (SATELLITE)

SCIENTIFIC SATELLITES

Scientific Satellites, Small
USE   SMALL SCIENTIFIC SATELLITES

Scientific Survey Module, Local
USE   LOCAL SCIENTIFIC SURVEY MODULE

SCIENTISTS

SCIMITAR AIRCRAFT

Scimitar Aircraft, Vickers
USE   SCIMITAR AIRCRAFT

SCINTILLATION

SCINTILLATION COUNTERS

Scintillators
USE   SCINTILLATION COUNTERS

Scintillometers
USE   SCINTILLATION COUNTERS

Scission
USE   CLEAVAGE

SCOOPS

Scope, Low Intensity X Ray Imaging
USE   LIXISCOPES

Scopolamine
USE   HYOSCINE

SCORE Omnirange
USE   SELF CALIBRATING OMNIRANGE

SCORE SATELLITE

SCORING

Scorpio Constellation
USE   SCORPIUS CONSTELLATION

SCORPIUS CONSTELLATION

SCOTCHLITE (TRADEMARK)

Scotia, Nova
USE   NOVA SCOTIA

SCOTLAND

Scout Helicopter
USE   P-531 HELICOPTER

SCOUT LAUNCH VEHICLE

SCOUT PROJECT

Scout Rocket Vehicle, Blue
USE   BLUE SCOUT ROCKET VEHICLE

SCPC Transmission
USE   SINGLE CHANNEL PER CARRIER
        TRANSMISSION

SCR (Rectifiers)
USE   SILICON CONTROLLED RECTIFIERS

SCRAM

SCRAMBLING (COMMUNICATION)

Scramjet Engines
USE   SUPERSONIC COMBUSTION RAMJET
        ENGINES

Scramjets
USE   SUPERSONIC COMBUSTION RAMJET
        ENGINES

SCRAP

SCRAPERS

SCREEN EFFECT

SCREENING

SCREENS

Screens, Sizing
USE   SIZING SCREENS

SCREW DISLOCATIONS

SCREW PINCH

SCREWS

Scribing
USE   SCORING

SCRUBBERS

Scrubbing
USE   WASHING

Scrubs (Botany)
USE   BRUSH (BOTANY)

SCUTUM CONSTELLATION

SCYLLA

SD
USE   SOUTH DAKOTA

(SD-WY), Black Hills
USE   BLACK HILLS (SD-WY)

SDI
USE   SELECTIVE DISSEMINATION OF
        INFORMATION

SDP (Computers)
USE   SITE DATA PROCESSORS

SDS 900 SERIES COMPUTERS

SDS 930 COMPUTER

SDS 9300 COMPUTER

SDV
USE   SHUTTLE DERIVED VEHICLES

Se
USE   SELENIUM

SE-A
USE   EXPLORER 30 SATELLITE

SE-210 AIRCRAFT

SE-210 Aircraft, Sud Aviation
USE   SE-210 AIRCRAFT

SE-3160 HELICOPTER

SE-3160 Helicopter, Sud Aviation
USE   SE-3160 HELICOPTER

Sea, Adriatic
USE   ADRIATIC SEA

Sea, Arabian
USE   ARABIAN SEA

Sea, Baltic
USE   BALTIC SEA

Sea, Barents
USE   BARENTS SEA

Sea, Bering
USE   BERING SEA

Sea, Black
USE   BLACK SEA

SEA BREEZE

Sea (CA), Salton
USE   SALTON SEA (CA)

Sea, Caribbean
USE   CARIBBEAN SEA

Sea, Caspian
USE   CASPIAN SEA

Sea, Chuckchi
USE   CHUCKCHI SEA

SEA GRASSES

SEA ICE

Sea Ice Interactions, Air
USE   AIR SEA ICE INTERACTIONS

Sea Interactions, Air
USE   AIR WATER INTERACTIONS

SEA KEEPING

Sea King Helicopter
USE   SH-3 HELICOPTER

Sea Knight Helicopter
USE   CH-46 HELICOPTER

SEA LAUNCHING

SEA LAW

SEA LEVEL

Sea, Mediterranean
USE   MEDITERRANEAN SEA

Sea, North
USE   NORTH SEA

Sea (North America), Beaufort
USE   BEAUFORT SEA (NORTH AMERICA)

SEA OF JAPAN

SEA OF OKHOTSK

Sea Power Plants, Solar
USE   SOLAR SEA POWER PLANTS

Sea, Red
USE   RED SEA

SEA ROUGHNESS

Sea, Sargasso
USE   SARGASSO SEA

SEA STATES

SEA SURFACE TEMPERATURE

SEA TRUTH

SEA URCHINS

Sea Walls
USE   BREAKWATERS

SEA WATER

SEAFARER PROJECT

Seahorse Helicopter
USE   UH-34 HELICOPTER

Sealants
USE   SEALERS

SEALERS

SEALING

Sealing, Self
USE   SELF SEALING

SEALS (ANIMALS)

(Seals), Glands
USE   GLANDS (SEALS)

Seals, Hermetic
USE   HERMETIC SEALS

Seals, Labyrinth
USE   LABYRINTH SEALS

Seals, O Ring
USE   O RING SEALS

(Seals), Packings
USE   PACKINGS (SEALS)

Seals, Pump
USE   PUMP SEALS

SEALS (STOPPERS)

SEAMOUNTS

SEAMS (JOINTS)

SEAPLANES

Search And Ranging Radar, North American
USE   NORTH AMERICAN SEARCH AND RANGING
        RADAR

Search And Rescue Satellite
USE   SARSAT

Search For Extraterrestrial Intelligence
USE   PROJECT SETI

SEARCH PROFILES

SEARCH RADAR

SEARCHING

SEARCHLIGHTS

SEAS

SEASAT PROGRAM

SEASAT SATELLITES

SEASAT 1

SEASAT-B SATELLITE

(Season), Spring
    USE   SPRING (SEASON)

Seasonal Variations
    USE   ANNUAL VARIATIONS

SEASONS

Seasprite Helicopter
    USE   UH-2 HELICOPTER

SEAT BELTS

SEATS

Seats, Ejection
    USE   EJECTION SEATS

Seats, Flying Ejection
    USE   FLYING EJECTION SEATS

SEAWEEDS

SEBACEOUS GLANDS

SEBACIC ACID

Second Law, Newton
    USE   NEWTON SECOND LAW

Secondary Batteries
    USE   STORAGE BATTERIES

SECONDARY COSMIC RAYS

SECONDARY EMISSION

SECONDARY FLOW

SECONDARY INJECTION

SECONDARY RADAR

Secondary Waves
    USE   S WAVES

SECRETIONS

Secretions, Endocrine
    USE   ENDOCRINE SECRETIONS

SECTIONS

Sections, Absorption Cross
    USE   ABSORPTION CROSS SECTIONS

Sections, Airfoil
    USE   AIRFOIL PROFILES

Sections, Capture Cross
    USE   ABSORPTION CROSS SECTIONS

Sections, Cross
    USE   CROSS SECTIONS

Sections, Dorsal
    USE   DORSAL SECTIONS

Sections, Ionization Cross
    USE   IONIZATION CROSS SECTIONS

Sections, Neutron Cross
    USE   NEUTRON CROSS SECTIONS

Sections, Posterior
    USE   POSTERIOR SECTIONS

Sections, Radar Cross
    USE   RADAR CROSS SECTIONS

Sections, Scattering Cross
    USE   SCATTERING CROSS SECTIONS

Sections, Ventral
    USE   VENTRAL SECTIONS

SECTORS

Secular Perturbation
    USE   LONG TERM EFFECTS

SECULAR VARIATIONS

SECURITY

Security, Airport
    USE   AIRPORT SECURITY

Security, Computer Information
    USE   COMPUTER INFORMATION SECURITY

SEDATIVES

SEDIMENT TRANSPORT

SEDIMENTARY ROCKS

SEDIMENTS

Seebeck Coefficient
    USE   SEEBECK EFFECT

SEEBECK EFFECT

Seeding, Cloud
    USE   CLOUD SEEDING

Seeding (Inoculation)
    USE   INOCULATION

SEEDS

Seekers
    USE   HOMING DEVICES

SEEPAGE

SEGMENTS

SEGRE CHARACTERISTIC

Segregation
    USE   SEPARATION

Seismic Array, Large Aperture
    USE   LARGE APERTURE SEISMIC ARRAY

SEISMIC ENERGY

SEISMIC WAVES

SEISMOCARDIOGRAPHY

SEISMOGRAMS

SEISMOGRAPHS

Seismographs, Lunar
    USE   LUNAR SEISMOGRAPHS

SEISMOLOGY

Seismology, Helio
    USE   HELIOSEISMOLOGY

Seismology, Solar
    USE   HELIOSEISMOLOGY

Seismometers
    USE   SEISMOGRAPHS

SEIZURES

SEL COMPUTERS

SELECTION

Selection, Personnel
    USE   PERSONNEL SELECTION

Selection, Pilot
    USE   PILOT SELECTION

SELECTION RULES (NUCLEAR PHYSICS)

Selection, Site
    USE   SITE SELECTION

Selective Coatings, Solar
    USE   SELECTIVE SURFACES

SELECTIVE DISSEMINATION OF INFORMATION

Selective Electrodes, Ion
    USE   ION SELECTIVE ELECTRODES

SELECTIVE FADING

SELECTIVE SURFACES

SELECTIVITY

SELECTORS

SELENIDES

Selenides, Cadmium
    USE   CADMIUM SELENIDES

Selenides, Copper
    USE   COPPER SELENIDES

Selenides, Gallium
    USE   GALLIUM SELENIDES

Selenides, Lead
    USE   LEAD SELENIDES

Selenides, Zinc
    USE   ZINC SELENIDES

SELENIUM

SELENIUM ALLOYS

SELENIUM COMPOUNDS

SELENIUM OXIDES

SELENOGRAPHY

SELENOLOGY

SELF ABSORPTION

SELF ADAPTIVE CONTROL SYSTEMS

SELF ALIGNMENT

SELF CALIBRATING OMNIRANGE

SELF CONSISTENT FIELDS

Self Deploying Space Stations
    USE   SELF ERECTING DEVICES
           SPACE STATIONS

SELF DIFFUSION (SOLID STATE)

SELF ERECTING DEVICES

SELF EXCITATION

SELF FOCUSING

SELF INDUCED VIBRATION

Self Initiated Antiaircraft Missiles
    USE   SIAM MISSILES

SELF LUBRICATING MATERIALS

SELF LUBRICATION

SELF MANEUVERING UNITS

Self Maneuvering Units, Space
    USE   SELF MANEUVERING UNITS

SELF ORGANIZING SYSTEMS

SELF OSCILLATION

SELF PROPAGATION

Self Regulating
USE   AUTOMATIC CONTROL

SELF REPAIRING DEVICES

SELF SEALING

SELF SHADOWING

SELF STIMULATION

Self Subtraction Holography
USE   HOLOGRAPHIC SUBTRACTION

SELF SUSTAINED EMISSION

Self-Diffusion, Gaseous
USE   GASEOUS SELF-DIFFUSION

Selsyns (Trademark)
USE   SERVOMOTORS

SEMANTICS

SEMICIRCULAR CANALS

SEMICONDUCTING FILMS

SEMICONDUCTOR DEVICES

Semiconductor Devices, NDM
USE   NDM SEMICONDUCTOR DEVICES

SEMICONDUCTOR DIODES

Semiconductor Insulator Semiconductors
USE   SIS (SEMICONDUCTORS)

SEMICONDUCTOR JUNCTIONS

SEMICONDUCTOR LASERS

SEMICONDUCTOR PLASMAS

Semiconductors, Amorphous
USE   AMORPHOUS SEMICONDUCTORS

Semiconductors, Complementary Metal Oxide
USE   CMOS

SEMICONDUCTORS (MATERIALS)

Semiconductors, Metal Insulator
USE   MIS (SEMICONDUCTORS)

Semiconductors, Metal Oxide
USE   METAL OXIDE SEMICONDUCTORS

Semiconductors, Metal-Insulator-Metal
USE   MIM (SEMICONDUCTORS)

Semiconductors, Metal-Nitride-Oxide-
USE   METAL-NITRIDE-OXIDE-SEMICONDUCTORS

Semiconductors, Metal-Oxide-Metal
USE   MOM (SEMICONDUCTORS)

(Semiconductors), MIM
USE   MIM (SEMICONDUCTORS)

(Semiconductors), MIS
USE   MIS (SEMICONDUCTORS)

(Semiconductors), MOM
USE   MOM (SEMICONDUCTORS)

(Semiconductors), MOS
USE   METAL OXIDE SEMICONDUCTORS

Semiconductors, N-Type
USE   N-TYPE SEMICONDUCTORS

Semiconductors, Negative Diff Mobility
USE   NDM SEMICONDUCTOR DEVICES

Semiconductors, Organic
USE   ORGANIC SEMICONDUCTORS

Semiconductors, P-Type
USE   P-TYPE SEMICONDUCTORS

Semiconductors, Semiconductor Insulator
USE   SIS (SEMICONDUCTORS)

Semiconductors, Silicon-On-Sapphire
USE   SOS (SEMICONDUCTORS)

(Semiconductors), Sis
USE   SIS (SEMICONDUCTORS)

(Semiconductors), Sos
USE   SOS (SEMICONDUCTORS)

SEMIEMPIRICAL EQUATIONS

Semimetals
USE   METALLOIDS

SEMISOLIDS

SEMISPAN MODELS

SENARMONT POLARISCOPES

Senders
USE   TRANSMITTERS

Seneca Aircraft
USE   PA-34 SENECA AIRCRAFT

Seneca Aircraft, Pa-34
USE   PA-34 SENECA AIRCRAFT

SENEGAL

Sensation Areas, Auditory
USE   AUDITORY SENSATION AREAS

Sensation, Tactile
USE   TOUCH

SENSE ORGANS

Senses
USE   SENSORY PERCEPTION

Sensibility
USE   SENSITIVITY

Sensing
USE   DETECTION

Sensing, Crop Inventories By Remote
USE   AGRISTARS PROJECT

Sensing, Horizon
USE   HORIZON SCANNERS

Sensing, Position
USE   POSITION SENSING

Sensing, Remote
USE   REMOTE SENSING

SENSITIVITY

Sensitivity, Impact
USE   IMPACT RESISTANCE

Sensitivity, Notch
USE   NOTCH SENSITIVITY

Sensitivity, Pain
USE   PAIN SENSITIVITY

Sensitivity, Photo
USE   PHOTOSENSITIVITY

Sensitivity, Propellant
USE   PROPELLANT SENSITIVITY

Sensitivity, Spectral
USE   SPECTRAL SENSITIVITY

SENSITIZING

Sensitizing, De
USE   DESENSITIZING

SENSITOMETRY

Sensor Modes, Pushbroom
USE   PUSHBROOM SENSOR MODES

(Sensor), SATAN
USE   TERRAIN ANALYSIS

SENSORIMOTOR PERFORMANCE

SENSORS

Sensors, Contour
USE   CONTOUR SENSORS

Sensors, Guidance
USE   GUIDANCE SENSORS

Sensors, Image Velocity
USE   IMAGE VELOCITY SENSORS

Sensors, Microwave
USE   MICROWAVE SENSORS

Sensors, Optical
USE   OPTICAL MEASURING INSTRUMENTS

Sensors, Pressure
USE   PRESSURE SENSORS

Sensors, Remote
USE   REMOTE SENSORS

Sensors, Solar
USE   SOLAR SENSORS

Sensors, Spacecraft
USE   SPACECRAFT INSTRUMENTS

Sensors, Sun
USE   SOLAR SENSORS

Sensors, Temperature
USE   TEMPERATURE SENSORS

SENSORY DEPRIVATION

SENSORY DISCRIMINATION

SENSORY FEEDBACK

SENSORY PERCEPTION

SENSORY STIMULATION

SENTENCES

SENTINEL SYSTEM

SEO (Indian Spacecraft)
USE   INDIAN SPACECRAFT

SEOCS (SATELLITE)

SEOS
USE   SYNCHRONOUS EARTH OBSERVATORY
         SATELLITE

SEPAC (PAYLOAD)

SEPARATED FLOW

SEPARATION

Separation, Boundary Layer
USE   BOUNDARY LAYER SEPARATION

Separation, Charge
USE   POLARIZATION (CHARGE SEPARATION)

Separation, External Store
USE   EXTERNAL STORE SEPARATION

Separation, Flow
USE   BOUNDARY LAYER SEPARATION
         SEPARATED FLOW

Separation, Isotope
USE   ISOTOPE SEPARATION

Separation, Laminar Boundary Layer
USE   LAMINAR BOUNDARY LAYER

Separation), Polarization (Charge
USE   POLARIZATION (CHARGE SEPARATION)

Separation, Radiochemical
USE   RADIOCHEMICAL SEPARATION

Separation, Size
USE   SIZE SEPARATION

(Separation), Sizing
USE   SIZE SEPARATION

Separation, Stage
USE   STAGE SEPARATION

SEPARATORS

Separators, Battery
USE   SEPARATORS

SEPTUM

Sequence, Isoelectronic
USE   ISOELECTRONIC SEQUENCE

Sequence Stars, Main
USE   MAIN SEQUENCE STARS

Sequence Stars, Pre-Main
USE   PRE-MAIN SEQUENCE STARS

Sequences, Pseudorandom
USE   PSEUDORANDOM SEQUENCES

SEQUENCING

SEQUENTIAL ANALYSIS

SEQUENTIAL COMPUTERS

SEQUENTIAL CONTROL

SERGEANT MISSILES

SERGENIUM

Series, Actinide
USE   ACTINIDE SERIES

Series Analysis, Time
USE   TIME SERIES ANALYSIS

Series, Asymptotic
USE   ASYMPTOTIC SERIES

Series, Balmer
USE   BALMER SERIES

Series, Campbell-Hausdorff
USE   CAMPBELL-HAUSDORFF SERIES

Series Compounds, Actinide
USE   ACTINIDE SERIES COMPOUNDS

Series Computers, CDC Cyber 170
USE   CDC CYBER 170 SERIES COMPUTERS

Series Computers, CDC 6000
USE   CDC 6000 SERIES COMPUTERS

Series Computers, CDC 7000
USE   CDC 7000 SERIES COMPUTERS

Series Computers, IBM 7000
USE   IBM 7000 SERIES COMPUTERS

Series Computers, SDS 900
USE   SDS 900 SERIES COMPUTERS

Series Computers, Univac 1100
USE   UNIVAC 1100 SERIES COMPUTERS

Series Computers, Vax-11
USE   VAX-11 SERIES COMPUTERS

Series, Cosine
USE   COSINE SERIES

SERIES EXPANSION

Series, Fourier
USE   FOURIER SERIES

Series, Maclaurin
USE   MACLAURIN SERIES

SERIES (MATHEMATICS)

Series, Mclaurin
USE   MACLAURIN SERIES

Series Metals, Lanthanide
USE   RARE EARTH ELEMENTS

Series, Paschen
USE   PASCHEN SERIES

Series, Power
USE   POWER SERIES

Series, Prony
USE   PRONY SERIES

Series, Rydberg
USE   RYDBERG SERIES

Series Satellites, TIROS N
USE   TIROS N SERIES SATELLITES

Series, Sine
USE   SINE SERIES

Series, Taylor
USE   TAYLOR SERIES

SEROTONIN

SERPENTINE

SERRATIA

SERT (Rocket Tests)
USE   SPACE ELECTRIC ROCKET TESTS

SERT 1 SPACECRAFT

SERT 2 SPACECRAFT

SERUMS

Serums, Anti
USE   ANTISERUMS

Service, Land Mobile Satellite
USE   LAND MOBILE SATELLITE SERVICE

SERVICE LIFE

SERVICE MODULES

Service Modules, Command
USE   COMMAND SERVICE MODULES

SERVICES

Services, Medical
USE   MEDICAL SERVICES

Services, Meteorological
USE   METEOROLOGICAL SERVICES

Servicing, Orbital
USE   ORBITAL SERVICING

SERVOAMPLIFIERS

SERVOCONTROL

SERVOMECHANISMS

SERVOMOTORS

Servos
USE   SERVOMOTORS

Servostability Control
USE   SERVOCONTROL

SES
USE   SURFACE EFFECT SHIPS

SET

SET THEORY

SETI
USE   PROJECT SETI

SETI, Project
USE   PROJECT SETI

Sets, Borel
USE   BOREL SETS

Sets (Computers), Instruction
USE   INSTRUCTION SETS (COMPUTERS)

Sets, Fuzzy
USE   FUZZY SETS

Sets, Psychological
USE   PSYCHOLOGICAL SETS

SETTING

SETTLING

SETUPS

Seven Day Variation, Twenty-
USE   TWENTY-SEVEN DAY VARIATION

Severe Storms Observing Satellite
USE   STORMSAT SATELLITE

Severe Storms Project, National
USE   NATIONAL SEVERE STORMS PROJECT

SEWAGE

SEWAGE TREATMENT

SEWERS

SEWING

SEX

SEX FACTOR

SEX GLANDS

SEXTANTS

SEYFERT GALAXIES

SFAR
USE   SOUND FIXING AND RANGING

Sferics
USE   ATMOSPHERICS

SGEMP
USE   SYSTEM GENERATED ELECTROMAGNETIC
       PULSES

SGR (Nuclear Reactors)
USE   SODIUM GRAPHITE REACTORS

SH-3 HELICOPTER

SH-4 HELICOPTER

SHACKLETON BOMBER

SHADES

Shadow, Lunar
USE   LUNAR SHADOW

SHADOWGRAPH PHOTOGRAPHY

Shadowgraph Photography, Spark
USE   SHADOWGRAPH PHOTOGRAPHY

Shadowgraphs
USE   SHADOWGRAPH PHOTOGRAPHY

Shadowing, Self
USE   SELF SHADOWING

SHADOWS

(Shafts), Journals
USE   SHAFTS (MACHINE ELEMENTS)

SHAFTS (MACHINE ELEMENTS)

Shafts, Rotating
USE   ROTATING SHAFTS

Shafts, Turbo
USE   TURBOSHAFTS

SHAKERS

SHAKING

SHALE OIL

SHALES

SHALLOW SHELL EQUATIONS

SHALLOW SHELLS

SHALLOW WATER

Shanks
USE   JOINTS (JUNCTIONS)

Shannon Information Theory
USE   INFORMATION THEORY

SHANNON-WIENER MEASURE

SHAPE CONTROL

Shape, Earth
USE   GEODESY

Shape, Line
USE   LINE SHAPE

SHAPE MEMORY ALLOYS

Shape, Ogee
USE   OGEE SHAPE

Shape, T
USE   T SHAPE

Shape Wheels, Doughnut
USE   TOROIDAL WHEELS

SHAPED CHARGES

SHAPERS

SHAPES

(Shapes), Disks
USE   DISKS (SHAPES)

Shapes, Fusiform
USE   CONES

Shapes, Mode
USE   MODAL RESPONSE

Shapes, Rosette
USE   ROSETTE SHAPES

(Shaping), Sizing
USE   SIZING (SHAPING)

Sharing, Time
USE   TIME SHARING

SHARKS

SHARP LEADING EDGES

SHARPNESS

SHATTER CONES

Shattering
USE   FRAGMENTATION

Shawnee Helicopter
USE   CH-21 HELICOPTER

SHEAR

SHEAR CREEP

Shear Disturbances
USE   S WAVES

Shear Fatigue
USE   SHEAR STRESS

SHEAR FLOW

Shear Heating, Magnetohydrodynamic
USE   MAGNETOHYDRODYNAMIC SHEAR
       HEATING

Shear Layer, Chapman
USE   SHEAR LAYERS

SHEAR LAYERS

Shear Mechanism, Dungeys Wind
USE   WIND SHEAR

SHEAR PROPERTIES

SHEAR STRAIN

SHEAR STRENGTH

SHEAR STRESS

Shear Waves
USE   S WAVES

Shear, Wind
USE   WIND SHEAR

SHEARING

Shearing Stress
USE   SHEAR STRESS

SHEARS

SHEATHS

Sheaths, Ion
USE   ION SHEATHS

Sheaths, Plasma
USE   PLASMA SHEATHS

SHEDDING

Shedding, Vortex
USE   VORTEX SHEDDING

SHEDS

SHEEP

Sheet Metal
USE   METAL SHEETS

SHEETS

Sheets, Current
USE   CURRENT SHEETS

Sheets, Elastic
USE   ELASTIC SHEETS

Sheets, Metal
USE   METAL SHEETS

Sheets, Neutral
USE   NEUTRAL SHEETS

Sheets, Vortex
USE   VORTEX SHEETS

(Sheets), Webs
USE   WEBS (SHEETS)

Shelf, Ross Ice
USE   ROSS ICE SHELF

SHELL ANODES

Shell Equations, Shallow
USE   SHALLOW SHELL EQUATIONS

SHELL STABILITY

SHELL THEORY

SHELLFISH

Shells, Anisotropic
USE   ANISOTROPIC SHELLS

Shells, Atmospheric
USE   ATMOSPHERIC STRATIFICATION

Shells, Circular
USE   CIRCULAR SHELLS

Shells, Conical
USE   CONICAL SHELLS

Shells, Corrugated
USE   CORRUGATED SHELLS

Shells, Cylindrical
USE   CYLINDRICAL SHELLS

Shells, Elastic
USE   ELASTIC SHELLS

Shells, Fluid Filled
USE   FLUID FILLED SHELLS

Shells, Hemispherical
USE   HEMISPHERICAL SHELLS

Shells, Liquid Filled
USE   LIQUID FILLED SHELLS

Shells, Metal
USE   METAL SHELLS

Shells, Orthotropic
USE   ORTHOTROPIC SHELLS

Shells, Perforated
USE   PERFORATED SHELLS

Shells, Reinforced
USE   REINFORCED SHELLS

Shells, Shallow
USE   SHALLOW SHELLS

Shells, Spherical
USE   SPHERICAL SHELLS

SHELLS (STRUCTURAL FORMS)

Shells, Thin Walled
USE   THIN WALLED SHELLS

Shells, Toroidal
USE   TOROIDAL SHELLS

SHELTERS

Shelters, Lunar
USE   LUNAR SHELTERS

SHELVES

Shelves, Continental
USE   CONTINENTAL SHELVES

Shelves, Ice
USE   LAND ICE

SHENANDOAH VALLEY (VA)

Shield, Canadian
USE   CANADIAN SHIELD

Shield (Europe), Baltic
USE   BALTIC SHIELD (EUROPE)

SHIELDING

Shielding, Electromagnetic
USE   ELECTROMAGNETIC SHIELDING

Shielding, Electrostatic
USE   ELECTROSTATIC SHIELDING

Shielding, Heat
USE   HEAT SHIELDING

Shielding, Magnetic
USE   MAGNETIC SHIELDING

Shielding, Nuclear
USE   RADIATION SHIELDING

Shielding, Radiation
USE   RADIATION SHIELDING

Shielding, Radio Frequency
USE   RADIO FREQUENCY SHIELDING

Shielding Reactor 2, Tower
USE   TOWER SHIELDING REACTOR 2

Shielding, Reentry
USE   REENTRY SHIELDING

Shielding, Reusable Heat
USE   REUSABLE HEAT SHIELDING

Shielding, Solar Radiation
USE   SOLAR RADIATION SHIELDING

Shielding, Spacecraft
USE   SPACECRAFT SHIELDING

Shielding, Thermal
USE   HEAT SHIELDING

Shields, Cirrus
USE   CIRRUS SHIELDS

Shields (Geology)
USE   BEDROCK

(Shields), Guards
USE   GUARDS (SHIELDS)

Shields, Molecular
USE   MOLECULAR SHIELDS

Shields, Wind
USE   WINDSHIELDS

SHIFT

Shift, Chemical
USE   CHEMICAL EQUILIBRIUM

Shift Circuits), Circulators (Phase
USE   CIRCULATORS (PHASE SHIFT CIRCUITS)

Shift Circuits, Phase
USE   PHASE SHIFT CIRCUITS

Shift Control Reactor, Spectral
USE   SPECTRAL SHIFT CONTROL REACTOR

Shift Control, Spectral
USE   SPECTRAL SHIFT CONTROL

Shift, Frequency
USE   FREQUENCY SHIFT

Shift, Isotope
USE   ISOTOPE EFFECT

Shift Keying, Frequency
USE   FREQUENCY SHIFT KEYING

Shift Keying, Phase
USE   PHASE SHIFT KEYING

Shift, Knight
USE   NUCLEAR MAGNETIC RESONANCE

Shift, Phase
USE   PHASE SHIFT

Shift, Red
USE   RED SHIFT

SHIFT REGISTERS

Shift, Stellar Doppler
USE   EXTRATERRESTRIAL RADIATION
DOPPLER EFFECT

Shift, Threshold
USE   THRESHOLDS

SHIFTING EQUILIBRIUM FLOW

SHILLELAGH MISSILES

Ship, Advanced Range Instrumentation
USE   ADVANCED RANGE INSTRUMENTATION
SHIP

Ship, ARIS Instrumentation
USE   ADVANCED RANGE INSTRUMENTATION
SHIP

SHIP HULLS

Ship, Savannah Nuclear
USE   SAVANNAH NUCLEAR SHIP

(Ship), Swath
USE   SWATH (SHIP)

SHIP TERMINALS

SHIP TO SHORE COMMUNICATION

SHIPS

Ships, Air
USE   AIRSHIPS

Ships, Cargo
USE   CARGO SHIPS

Ships, LOTS Cargo
USE   CARGO SHIPS

Ships, Nuclear Powered
USE   NUCLEAR POWERED SHIPS

Ships, Satellite Communications
USE   SATELLITE COMMUNICATIONS SHIPS

Ships, Surface Effect
USE   SURFACE EFFECT SHIPS

Ships, Tanker
USE   TANKER SHIPS

SHIPYARDS

SHIVA LASER SYSTEM

SHIVERING

SHOALS

SHOCK

SHOCK ABSORBERS

Shock Diffusers
USE   SHOCK WAVE ATTENUATION
DIFFUSERS

SHOCK DISCONTINUITY

SHOCK FRONTS

SHOCK HEATING

Shock, Hydraulic
USE   HYDRAULIC SHOCK

Shock, Hypersonic
USE   HYPERSONIC SHOCK

SHOCK LAYERS

SHOCK LOADS

SHOCK MEASURING INSTRUMENTS

Shock, Mechanical
USE   MECHANICAL SHOCK

SHOCK (PHYSIOLOGY)

SHOCK RESISTANCE

SHOCK SIMULATORS

SHOCK SPECTRA

SHOCK TESTS

Shock, Thermal
USE   THERMAL SHOCK

SHOCK TUBES

Shock Tubes, Magnetic Annular
USE   MAGNETIC ANNULAR SHOCK TUBES

Shock Tubes, MAST
USE   MAGNETIC ANNULAR SHOCK TUBES

SHOCK TUNNELS

SHOCK WAVE ATTENUATION

SHOCK WAVE CONTROL

SHOCK WAVE GENERATORS

SHOCK WAVE INTERACTION

SHOCK WAVE LUMINESCENCE

SHOCK WAVE PROFILES

SHOCK WAVE PROPAGATION

SHOCK WAVES

Shock Waves, Bow
USE   SHOCK WAVES
BOW WAVES

Shock Waves, Normal
USE   NORMAL SHOCK WAVES

Shock Waves, Oblique
USE   OBLIQUE SHOCK WAVES

SHOES

Shooting Star Aircraft
USE   T-33 AIRCRAFT

SHOPS

SHORAN

Shore Communication, Ship To
USE   SHIP TO SHORE COMMUNICATION

Shore (LOTS) Carrier, Logistics Over The
USE   LOGISTICS OVER THE SHORE (LOTS)
CARRIER

SHORELINES

Shorelines, Advancing
USE   BEACHES

Short Belfast C MK-1 Aircraft
USE   SC-5 AIRCRAFT

SHORT CIRCUIT CURRENTS

SHORT CIRCUITS

SHORT HAUL AIRCRAFT

SHORT RANGE BALLISTIC MISSILES

Short Range Navigation
USE   SHORAN

Short SC-1 Aircraft
USE   SC-1 AIRCRAFT

Short SC-5 Aircraft
USE   SC-5 AIRCRAFT

Short SC-7 Aircraft
USE   SC-7 AIRCRAFT

Short Stack, Apollo
USE   APOLLO SHORT STACK

SHORT TAKEOFF AIRCRAFT

SHORT WAVE RADIATION

Short Wave Radio Equipment, Ultra
USE   VERY HIGH FREQUENCY RADIO
EQUIPMENT

SHORT WAVE RADIO TRANSMISSION

Shortening
USE   REDUCTION

SHOT

SHOT NOISE

SHOT PEENING

Shot Proj, Experimental Reflector Orbital
USE   EXPERIMENTAL REFLECTOR ORBITAL
SHOT PROJ

Shot Project, Big
USE   BIG SHOT PROJECT

Shots, Orbital
USE   ORBITAL SHOTS

SHOULDERS

SHOWERS

Showers, Cosmic Ray
USE   COSMIC RAY SHOWERS

Showers, Meteoroid
USE   METEOROID SHOWERS

SHRAPNEL

SHREDDING

SHREWS

SHRIKE MISSILE

SHRINKAGE

Shrouded Bodies
USE   SHROUDS

SHROUDED NOZZLES

SHROUDED PROPELLERS

SHROUDED TURBINES

SHROUDS

Shunts
USE   BYPASSES
CIRCUITS

SHUTDOWNS

SHUTTERS

Shutters, Camera
USE   CAMERA SHUTTERS

Shuttle, Aeromaneuvering Orbit To Orbit
USE   AEROMANEUVERING ORBIT TO ORBIT
SHUTTLE

Shuttle Ascent Stage, Space
USE   SPACE SHUTTLE ASCENT STAGE

Shuttle Avionics Integration Laboratory
USE   SAIL PROJECT

Shuttle Boosters
USE   SPACE SHUTTLE BOOSTERS

Shuttle Boosters, Space
USE   SPACE SHUTTLE BOOSTERS

SHUTTLE DERIVED VEHICLES

SHUTTLE ENGINEERING SIMULATOR

SHUTTLE IMAGING RADAR

Shuttle Imaging Radar, Earth Resources
USE   EARTH RESOURCES SHUTTLE IMAGING
RADAR

Shuttle Main Engine, Space
USE   SPACE SHUTTLE MAIN ENGINE

SHUTTLE MISSION SIMULATOR

Shuttle Mission 31-A, Space
USE   SPACE SHUTTLE MISSION 31-A

Shuttle Mission 31-B, Space
USE   SPACE SHUTTLE MISSION 31-B

Shuttle Mission 31-C, Space
USE   SPACE SHUTTLE MISSION 31-C

Shuttle Mission 31-D, Space
USE   SPACE SHUTTLE MISSION 31-D

Shuttle Mission 41-A, Space
USE   SPACE SHUTTLE MISSION 41-A

Shuttle Mission 41-B, Space
USE   SPACE SHUTTLE MISSION 41-B

Shuttle Mission 41-C, Space
USE   SPACE SHUTTLE MISSION 41-C

Shuttle Mission 41-D, Space
USE   SPACE SHUTTLE MISSION 41-D

Shuttle Mission 41-G, Space
USE   SPACE SHUTTLE MISSION 41-G

Shuttle Mission 51-A, Space
USE   SPACE SHUTTLE MISSION 51-A

Shuttle Mission 51-B, Space
USE   SPACE SHUTTLE MISSION 51-B

Shuttle Mission 51-C, Space
USE   SPACE SHUTTLE MISSION 51-C

Shuttle Mission 51-D, Space
USE   SPACE SHUTTLE MISSION 51-D

Shuttle Mission 51-E, Space
USE   SPACE SHUTTLE MISSION 51-E

Shuttle Mission 51-F, Space
USE   SPACE SHUTTLE MISSION 51-F

Shuttle Mission 51-G, Space
USE   SPACE SHUTTLE MISSION 51-G

Shuttle Mission 51-H, Space
USE   SPACE SHUTTLE MISSION 51-H

Shuttle Mission 51-I, Space
USE   SPACE SHUTTLE MISSION 51-I

Shuttle Mission 51-J, Space
USE   SPACE SHUTTLE MISSION 51-J

Shuttle Mission 51-L, Space
USE   SPACE SHUTTLE MISSION 51-L

Shuttle Mission 61-A, Space
USE   SPACE SHUTTLE MISSION 61-A

Shuttle Mission 61-B, Space
USE   SPACE SHUTTLE MISSION 61-B

Shuttle Mission 61-C, Space
USE   SPACE SHUTTLE MISSION 61-C

Shuttle Mission 61-E, Space
USE   SPACE SHUTTLE MISSION 61-E

Shuttle Missions, Space
USE   SPACE SHUTTLE MISSIONS

Shuttle), Orbit Maneuvering Engine (Space
USE   ORBIT MANEUVERING ENGINE (SPACE
SHUTTLE)

(Shuttle), Orbital Flight Test 1
USE   SPACE TRANSPORTATION SYSTEM 1
FLIGHT

Shuttle Orbital Flight Test 1, Space
USE   SPACE TRANSPORTATION SYSTEM 1
FLIGHT

(Shuttle), Orbital Flight Test 2
USE   SPACE TRANSPORTATION SYSTEM 2
FLIGHT

Shuttle Orbital Flight Test 2, Space
USE   SPACE TRANSPORTATION SYSTEM 2
FLIGHT

(Shuttle), Orbital Flight Test 3
USE   SPACE TRANSPORTATION SYSTEM 3
FLIGHT

Shuttle Orbital Flight Test 3, Space
USE   SPACE TRANSPORTATION SYSTEM 3
FLIGHT

(Shuttle), Orbital Flight Test 4
USE   SPACE TRANSPORTATION SYSTEM 4
FLIGHT

Shuttle Orbital Flight Test 4, Space
USE   SPACE TRANSPORTATION SYSTEM 4
FLIGHT

(Shuttle), Orbital Flight Tests
USE   SPACE TRANSPORTATION SYSTEM
FLIGHTS

Shuttle Orbital Flight Tests, Space
USE   SPACE TRANSPORTATION SYSTEM
FLIGHTS

Shuttle Orbital Flight 7, Space
USE   SPACE SHUTTLE MISSION 31-C

Shuttle Orbital Flight 8, Space
USE   SPACE SHUTTLE MISSION 31-D

Shuttle Orbital Flight 9, Space
USE   SPACE SHUTTLE MISSION 41-A

Shuttle Orbital Flights, Space
USE   SPACE TRANSPORTATION SYSTEM
FLIGHTS

Shuttle Orbiter 099, Space
USE   CHALLENGER (ORBITER)

Shuttle Orbiter 101, Space
USE   ENTERPRISE (ORBITER)

Shuttle Orbiter 102, Space
USE   COLUMBIA (ORBITER)

Shuttle Orbiter 103, Space
USE   DISCOVERY (ORBITER)

Shuttle Orbiter 104, Space
USE   ATLANTIS (ORBITER)

Shuttle Orbiters
USE   SPACE SHUTTLE ORBITERS

Shuttle Orbiters, Space
USE   SPACE SHUTTLE ORBITERS

SHUTTLE PALLET SATELLITES

Shuttle Payloads, Space
USE   SPACE SHUTTLE PAYLOADS

Shuttle Upper Stage A, Space
USE   SPACE SHUTTLE UPPER STAGE A

Shuttle Upper Stage D, Space
USE   SPACE SHUTTLE UPPER STAGE D

Shuttle Upper Stages, Space
USE   SPACE SHUTTLE UPPER STAGES

Shuttles, Space
USE   SPACE SHUTTLES

Si
USE   SILICON

SI
USE   INTERNATIONAL SYSTEM OF UNITS

SIALON

SIAM MISSILES

SIBERIA

SIC (Coefficient)
USE   STRUCTURAL INFLUENCE COEFFICIENTS

SICILY

Sickness, Air
USE   MOTION SICKNESS

Sickness, Altitude
USE   ALTITUDE SICKNESS

Sickness, Decompression
USE   DECOMPRESSION SICKNESS

Sickness Drugs, Motion
USE   MOTION SICKNESS DRUGS

Sickness, Motion
USE   MOTION SICKNESS

Sickness, Radiation
USE   RADIATION SICKNESS

SICKNESSES

SID (Ionospheric Disturbances)
USE   SUDDEN IONOSPHERIC DISTURBANCES

Siddeley Aircraft, Hawker
USE   HAWKER SIDDELEY AIRCRAFT

Siddeley BS 53 Engine, Bristol-
USE   BRISTOL-SIDDELEY BS 53 ENGINE

Siddeley Olympus 593 Engine, Bristol-
USE   BRISTOL-SIDDELEY OLYMPUS 593 ENGINE

Siddeley Viper Engine, Bristol-
USE   BRISTOL-SIDDELEY VIPER ENGINE

SIDE INLETS

Side, Lunar Far
USE   LUNAR FAR SIDE

SIDE-LOOKING RADAR

Sideband Modulation, Single
USE   SINGLE SIDEBAND TRANSMISSION

Sideband Transmission, Double
USE   DOUBLE SIDEBAND TRANSMISSION

Sideband Transmission, Single
USE   SINGLE SIDEBAND TRANSMISSION

SIDEBANDS

SIDELOBE REDUCTION

SIDELOBES

SIDEREAL TIME

Siderite Meteorites
USE   IRON METEORITES

SIDERITES

SIDES

SIDESLIP

Sidewash
USE   BACKWASH

SIDEWINDER MISSILES

SIEBEL AIRCRAFT

SIEMENS 2002 COMPUTER

SIERRA LEONE

SIERRA NEVADA MOUNTAINS (CA)

SIEVES

Sieves, Molecular
USE   ABSORBENTS

Sight
USE   VISUAL PERCEPTION

Sight Communication, Line Of
USE   LINE OF SIGHT COMMUNICATION

Sight, Line Of
USE   LINE OF SIGHT

SIGMA COMPUTERS

SIGMA ORIONIS

SIGMA 5 COMPUTER

SIGMA 7

SIGMA 9 COMPUTER

SIGMA-MESONS

SIGNAL ANALYSIS

SIGNAL ANALYZERS

Signal Attenuation, Radio
USE   RADIO ATTENUATION

SIGNAL DETECTION

SIGNAL DETECTORS

Signal Discriminators
USE   SIGNAL DETECTORS

SIGNAL DISTORTION

SIGNAL ENCODING

Signal Fadeout
USE   SIGNAL FADING

SIGNAL FADING

SIGNAL FADING RATE

SIGNAL FLOW GRAPHS

SIGNAL GENERATORS

SIGNAL MEASUREMENT

Signal Measurement, Electronic
USE   SIGNAL MEASUREMENT

SIGNAL MIXING

SIGNAL PROCESSING

Signal Propagation, Radio
USE   RADIO TRANSMISSION

SIGNAL RECEPTION

SIGNAL REFLECTION

SIGNAL STABILIZATION

SIGNAL TO NOISE RATIOS

SIGNAL TRANSMISSION

SIGNALS

Signals, Audio
USE   AUDIO SIGNALS

Signals, Auditory
USE   AUDITORY SIGNALS

Signals, Chirp
USE   CHIRP SIGNALS

Signals, Error
USE   ERROR SIGNALS

Signals, Magnetic
USE   MAGNETIC SIGNALS

Signals, Monaural
USE   MONAURAL SIGNALS

Signals, Optical
USE   OPTICAL COMMUNICATION

Signals, Radio
USE   RADIO SIGNALS

Signals, Random
USE   RANDOM SIGNALS

Signals, Time
USE   TIME SIGNALS

Signals, Video
USE   VIDEO SIGNALS

Signals, Visual
USE   VISUAL SIGNALS

Signals, Warning
USE   WARNING SYSTEMS

SIGNATURE ANALYSIS

SIGNATURES

Signatures, Infrared
USE   INFRARED SIGNATURES

Signatures, Magnetic
USE   MAGNETIC SIGNATURES

Signatures, Missile
USE   MISSILE SIGNATURES

Signatures, Radar
USE   RADAR SIGNATURES

Signatures, Spectral
USE   SPECTRAL SIGNATURES

SIGNIFICANCE

SIGNS AND SYMPTOMS

Signs (Symbols)
USE   SYMBOLS

SIKHOTE-ALIN METEORITE

SIKKIM

SIKORSKY AIRCRAFT

Sikorsky HSS-2 Helicopter
USE   SH-3 HELICOPTER

Sikorsky S-58 Helicopter
USE   S-58 HELICOPTER

Sikorsky S-61 Helicopter
USE   S-61 HELICOPTER

Sikorsky S-64 Helicopter
USE   CH-54 HELICOPTER

Sikorsky S-65 Helicopter
USE   H-53 HELICOPTER

Sikorsky S-67 Helicopter
USE S-67 HELICOPTER

SIKORSKY WHIRLWIND HELICOPTER

SILANES

Silanes, Chloro
USE CHLOROSILANES

SILENCE

SILENCERS

Silica
USE SILICON DIOXIDE

SILICA GEL

SILICA GLASS

SILICATES

Silicates, Aluminum
USE ALUMINUM SILICATES

Silicates, Calcium
USE CALCIUM SILICATES

Silicates, Fluoro
USE FLUOROSILICATES

Silicates, Potassium
USE POTASSIUM SILICATES

Silicates, Sodium
USE SODIUM SILICATES

SILICIDES

SILICON

SILICON ALLOYS

SILICON CARBIDES

SILICON COMPOUNDS

Silicon Compounds, Organic
USE ORGANIC SILICON COMPOUNDS

SILICON CONTROLLED RECTIFIERS

SILICON DIOXIDE

SILICON FILMS

SILICON ISOTOPES

SILICON JUNCTIONS

Silicon, Metal-Nitride-Oxide-
USE METAL-NITRIDE-OXIDE-SILICON

SILICON NITRIDES

SILICON OXIDES

SILICON POLYMERS

SILICON RADIATION DETECTORS

Silicon Rectifiers
USE CRYSTAL RECTIFIERS

Silicon Solar Cells
USE SOLAR CELLS

SILICON TETRACHLORIDE

SILICON TRANSISTORS

Silicon, Triphenyl
USE TRIPHENYL SILICON

Silicon-On-Sapphire Junctions
USE SOS (SEMICONDUCTORS)

Silicon-On-Sapphire Semiconductors
USE SOS (SEMICONDUCTORS)

Silicon-On-Sapphire Transistors
USE SOS (SEMICONDUCTORS)

SILICONE RESINS

SILICONE RUBBER

SILICONES

SILICONIZING

SILK

SILKWORMS

Silos, Missile
USE MISSILE SILOS

Silos (Missile Storage)
USE MISSILE SILOS

SILOXANES

Silts
USE SEDIMENTS

SILVER

SILVER ALLOYS

Silver Batteries, Cadmium
USE SILVER CADMIUM BATTERIES

Silver Batteries, Zinc
USE SILVER ZINC BATTERIES

SILVER BROMIDES

SILVER CADMIUM BATTERIES

SILVER CHLORIDES

SILVER COMPOUNDS

SILVER HALIDES

SILVER HYDROGEN BATTERIES

SILVER IODIDES

SILVER ISOTOPES

SILVER NITRATES

Silver Oxide Batteries, Zinc
USE SILVER ZINC BATTERIES

Silver Oxide Zinc Batteries
USE SILVER ZINC BATTERIES

SILVER OXIDES

SILVER ZINC BATTERIES

SILVICULTURE

SIM

SIMICOR (Image Correlator)
USE IMAGE CORRELATORS

Similarities
USE ANALOGIES

Similarity Hypothesis, Lagrange
USE LAGRANGE SIMILARITY HYPOTHESIS

SIMILARITY NUMBERS

SIMILARITY THEOREM

SIMILITUDE LAW

SIMPLE HARMONIC MOTION

SIMPLEX METHOD

SIMPLIFICATION

Simulated Altitude
USE ALTITUDE SIMULATION

SIMULATION

Simulation, Acoustic
USE ACOUSTIC SIMULATION

Simulation, Altitude
USE ALTITUDE SIMULATION

Simulation, Analog
USE ANALOG SIMULATION

Simulation, Atmospheric Entry
USE ATMOSPHERIC ENTRY SIMULATION

Simulation, Computer
USE COMPUTERIZED SIMULATION

Simulation, Computer Systems
USE COMPUTER SYSTEMS SIMULATION

Simulation, Computerized
USE COMPUTERIZED SIMULATION

Simulation, Control
USE CONTROL SIMULATION

Simulation, Data
USE DATA SIMULATION

Simulation, Digital
USE DIGITAL SIMULATION

Simulation, Environment
USE ENVIRONMENT SIMULATION

Simulation, Exhaust Flow
USE EXHAUST FLOW SIMULATION

Simulation, Flight
USE FLIGHT SIMULATION

Simulation Flights, Spacelab
USE ASSESS PROGRAM

Simulation, Landing
USE LANDING SIMULATION

Simulation, Motion
USE MOTION SIMULATION

Simulation, Rheoelectrical
USE RHEOELECTRICAL SIMULATION

Simulation, Solar
USE SOLAR SIMULATION

Simulation, Space Environment
USE SPACE ENVIRONMENT SIMULATION

Simulation, Systems
USE SYSTEMS SIMULATION

Simulation, Thermal
USE THERMAL SIMULATION

Simulation, Weightlessness
USE WEIGHTLESSNESS SIMULATION

Simulator, High Vacuum Orbital
USE HIGH VACUUM ORBITAL SIMULATOR

(Simulator), HIVOS
USE HIGH VACUUM ORBITAL SIMULATOR

(Simulator), LOLA
USE LUNAR ORBIT AND LANDING SIMULATORS

Simulator, Lunar Gravity
USE LUNAR GRAVITY SIMULATOR

Simulator, Shuttle Engineering
USE SHUTTLE ENGINEERING SIMULATOR

Simulator, Shuttle Mission
USE SHUTTLE MISSION SIMULATOR

Simulator Training
USE TRAINING SIMULATORS

SIMULATORS

Simulators, Cockpit
  USE   COCKPIT SIMULATORS

Simulators, Environment
  USE   ENVIRONMENT SIMULATORS

Simulators, Flight
  USE   FLIGHT SIMULATORS

Simulators, Lunar Orbit And Landing
  USE   LUNAR ORBIT AND LANDING SIMULATORS

Simulators, Missile
  USE   MISSILE SIMULATORS

Simulators, Motion
  USE   MOTION SIMULATORS

Simulators, Orbital
  USE   SPACE SIMULATORS

Simulators, Shock
  USE   SHOCK SIMULATORS

Simulators, Solar
  USE   SOLAR SIMULATORS

Simulators, Space
  USE   SPACE SIMULATORS

Simulators, Spacecraft Cabin
  USE   SPACECRAFT CABIN SIMULATORS

Simulators, Target
  USE   TARGET SIMULATORS

Simulators, Training
  USE   TRAINING SIMULATORS

Simulators, Vertical Motion
  USE   VERTICAL MOTION SIMULATORS

Simulators, Vibration
  USE   VIBRATION SIMULATORS

SIMULTANEOUS EQUATIONS

Simultaneous Image Correlator
  USE   IMAGE CORRELATORS

SINE SERIES

SINE WAVES

SINGAPORE

SINGLE CHANNEL PER CARRIER TRANSMISSION

SINGLE CRYSTALS

SINGLE EVENT UPSETS

Single Sideband Modulation
  USE   SINGLE SIDEBAND TRANSMISSION

SINGLE SIDEBAND TRANSMISSION

SINGLE STAGE ROCKET VEHICLES

SINGLE STAGE TO ORBIT VEHICLES

SINGLE-PHASE FLOW

SINGULAR INTEGRAL EQUATIONS

Singularities, Naked
  USE   NAKED SINGULARITIES

SINGULARITY (MATHEMATICS)

SINKHOLES

SINKING

Sinking, Counter
  USE   COUNTERSINKING

SINKS

Sinks (Geology)
  USE   STRUCTURAL BASINS

Sinks, Heat
  USE   HEAT SINKS

SINTERED ALUMINUM POWDER

SINTERING

Sinus Body, Carotid
  USE   CAROTID SINUS BODY

Sinus Reflex, Carotid
  USE   CAROTID SINUS REFLEX

SINUSES

Sinuses, Paranasal
  USE   PARANASAL SINUSES

Sinusitis, Aero
  USE   AEROSINUSITIS

Sinusoids
  USE   SINE WAVES

Sioux Helicopter
  USE   OH-13 HELICOPTER

SIPHONING

SIPHONS

Siphons, Thermo
  USE   THERMOSIPHONS

Sir-A
  USE   SHUTTLE IMAGING RADAR

Sir-B
  USE   SHUTTLE IMAGING RADAR

SIRENS

SIRIO SATELLITE

SIRS B SATELLITE

SIS (SEMICONDUCTORS)

Site, Arizona Regional Ecological Test
  USE   ARIZONA REGIONAL ECOLOGICAL TEST
         SITE

Site), CARETS (Test
  USE   CENTRAL ATLANTIC REGIONAL ECOL
         TEST SITE

Site, Central Atlantic Regional Ecol Test
  USE   CENTRAL ATLANTIC REGIONAL ECOL
         TEST SITE

SITE DATA PROCESSORS

Site Program, Radar Target Scatter
  USE   RADAR TARGET SCATTER SITE PROGRAM

SITE SELECTION

SITES

Sites, Landing
  USE   LANDING SITES

Sites, Launching
  USE   LAUNCHING SITES

Sites, Lunar Landing
  USE   LUNAR LANDING SITES

Sites, Offshore Reactor
  USE   OFFSHORE REACTOR SITES

SITTING POSITION

Size (Biology), Body
  USE   BODY SIZE (BIOLOGY)

Size, Crew
  USE   CREW SIZE

SIZE DETERMINATION

SIZE (DIMENSIONS)

SIZE DISTRIBUTION

Size Distribution, Particle
  USE   PARTICLE SIZE DISTRIBUTION

Size, Drop
  USE   DROP SIZE

Size, Grain
  USE   GRAIN SIZE

Size, Pupil
  USE   PUPIL SIZE

SIZE SEPARATION

SIZING

SIZING MATERIALS

SIZING SCREENS

Sizing (Separation)
  USE   SIZE SEPARATION

SIZING (SHAPING)

SIZING (SURFACE TREATMENT)

Skan Equation, Falkner-
  USE   FALKNER-SKAN EQUATION

Skeleton
  USE   MUSCULOSKELETAL SYSTEM

SKEWNESS

SKID LANDINGS

SKIDDING

Skills
  USE   ABILITIES

SKIN (ANATOMY)

SKIN FRICTION

SKIN GRAFTS

SKIN RESISTANCE

Skin Response, Galvanic
  USE   GALVANIC SKIN RESPONSE

SKIN (STRUCTURAL MEMBER)

Skin Structures, Stressed-
  USE   STRESSED-SKIN STRUCTURES

SKIN TEMPERATURE (BIOLOGY)

SKIN TEMPERATURE (NON-BIOLOGICAL)

SKINNER BOXES

SKIRTS

SKIS

Skjellerup Comet, Grigg-
  USE   GRIGG-SKJELLERUP COMET

SKUA ROCKET VEHICLES

SKULL

SKY

SKY BRIGHTNESS

Sky, Night
  USE   NIGHT SKY

Sky, Northern
  USE   NORTHERN SKY

Sky Photography, All
USE   ALL SKY PHOTOGRAPHY

SKY RADIATION

Sky, Southern
USE   SOUTHERN SKY

SKY WAVES

SKYBOLT MISSILE

Skycrane Helicopter
USE   CH-54 HELICOPTER

SKYDROL (TRADEMARK)

Skyhawk Aircraft
USE   A-4 AIRCRAFT

SKYHOOK BALLOONS

SKYLAB PROGRAM

SKYLAB Space Station (Unmanned)
USE   SKYLAB 1

SKYLAB 1

SKYLAB 2

SKYLAB 3

SKYLAB 4

Skylark
USE   SKYLARK ROCKET VEHICLE

SKYLARK ROCKET VEHICLE

Skymaster Aircraft
USE   C-54 AIRCRAFT

SKYNET SATELLITES

Skyraider Aircraft
USE   A-1 AIRCRAFT

Skyrocket Aircraft
USE   D-558 AIRCRAFT

Skystreak Aircraft
USE   D-558 AIRCRAFT

Skyvan Aircraft
USE   SC-7 AIRCRAFT

Skyvan Aircraft, Turbo-
USE   SC-7 AIRCRAFT

Skywarrior Aircraft
USE   A-3 AIRCRAFT

SL 1
USE   SKYLAB 1

SL 2
USE   SKYLAB 2

SL 3
USE   SKYLAB 3

SL 4
USE   SKYLAB 4

SL-3 ROCKET ENGINE

SLABS

Slabs, Plasma
USE   PLASMA SLABS

SLAGS

SLAM
USE   SUPERSONIC LOW ALTITUDE MISSILE

(Slam), Scanning Laser Acoustic Microscope
USE   ACOUSTIC MICROSCOPES

SLAMMING

Slant
USE   SLOPES

Slant Perception
USE   SPACE PERCEPTION

Slant Range, Optical
USE   OPTICAL SLANT RANGE

Slap Noise, Blade
USE   BLADE SLAP NOISE

Slashes
USE   CLEARINGS (OPENINGS)

Slater Method, Hartree-Fock-
USE   HARTREE-FOCK-SLATER METHOD

SLATER ORBITALS

Slats, Leading Edge
USE   LEADING EDGE SLATS

Slats, Wing
USE   LEADING EDGE SLATS

SLEDS

Sleds, Rocket Propelled
USE   ROCKET PROPELLED SLEDS

SLEEP

SLEEP DEPRIVATION

Sleep, Desynchronized
USE   RAPID EYE MOVEMENT STATE

SLEEVES

SLENDER BODIES

SLENDER CONES

SLENDER WINGS

SLEUTH (PROGRAMMING LANGUAGE)

Slew Missiles, Air
USE   AIR SLEW MISSILES

SLEWING

SLICING

Slicks
USE   OIL SLICKS

Slicks, Oil
USE   OIL SLICKS

Slides
USE   CHUTES

SLIDES (MICROSCOPY)

SLIDING

SLIDING CONTACT

SLIDING FRICTION

SLIP

Slip Bands
USE   EDGE DISLOCATIONS

SLIP CASTING

SLIP FLOW

Slip, Side
USE   SIDESLIP

SLIPSTREAMS

Slipstreams, Propeller
USE   PROPELLER SLIPSTREAMS

SLITS

SLIVERS

SLOPES

Slopes, Glide
USE   GLIDE PATHS

Sloshing
USE   LIQUID SLOSHING

Sloshing, Liquid
USE   LIQUID SLOSHING

Slot Ailerons, Spoiler
USE   SPOILER SLOT AILERONS

SLOT ANTENNAS

SLOTS

Slots, Wing
USE   WING SLOTS

Slotted Antennas
USE   SLOT ANTENNAS

SLOTTED WIND TUNNELS

Slow Neutrons
USE   THERMAL NEUTRONS

SLUDGE

Sludge, Activated
USE   ACTIVATED SLUDGE

SLUMPING

SLURRIES

SLURRY PROPELLANTS

SLUSH

SLV
USE   STANDARD LAUNCH VEHICLES

SLV (Soft Landing Vehicles)
USE   SOFT LANDING SPACECRAFT

SLV-3 Launch Vehicle, Atlas
USE   ATLAS SLV-3 LAUNCH VEHICLE

Slyke Method, Van
USE   VAN SLYKE METHOD

Sm
USE   SAMARIUM

SM-65 Missile
USE   ATLAS LAUNCH VEHICLES

SM-68 Missile
USE   TITAN 1 ICBM

SM-68B Missile
USE   TITAN 2 ICBM

Small Astronomy Satellite 1
USE   SAS-1

Small Astronomy Satellite 2
USE   SAS-2

Small Astronomy Satellite 3
USE   SAS-3

Small Astronomy Satellites
USE   SAS

SMALL PERTURBATION FLOW

SMALL SCIENTIFIC SATELLITES

Small Water Plane Area Twin Hull
USE   SWATH (SHIP)

SMALLPOX

SMEAR

Smell
USE   OLFACTORY PERCEPTION

SMELTING

Smirnoff Test, Kolmogoroff-
USE   KOLMOGOROFF-SMIRNOFF TEST

SMITH CHART

SMM-A
USE   SOLAR MAXIMUM MISSION-A

SMOG

SMOKE

SMOKE ABATEMENT

SMOKE DETECTORS

SMOKE TRAILS

Smoky Mountains (NC-TN), Great
USE   GREAT SMOKY MOUNTAINS (NC-TN)

SMOOTHING

Smoothing, Data
USE   DATA SMOOTHING

SMS
USE   SYNCHRONOUS METEOROLOGICAL
         SATELLITE

SMS 1

SMS 2

SMU (Maneuvering Units)
USE   SELF MANEUVERING UNITS

Sn
USE   TIN

SNAILS

SNAKES

Snaking
USE   LATERAL OSCILLATION

SNAP

SNAP 1

SNAP 2

SNAP 3

SNAP 4

SNAP 7

SNAP 8

SNAP 9A

SNAP 10A

SNAP 11

SNAP 13

SNAP 15

SNAP 17

SNAP 19

SNAP 21

SNAP 23

SNAP 27

SNAP 29

SNAP 50

SNAPSHOT SATELLITE

SNAPTRAN REACTOR

Snatching
USE   SPACECRAFT RECOVERY

SNEAK CIRCUIT ANALYSIS

SNEEZING

SNELLEN TESTS

SNELLS LAW

SNOW

Snow Aerial Applicator Aircraft S-2B
USE   S-2 AIRCRAFT

SNOW AIRCRAFT

SNOW COVER

Snow S-2 Aircraft
USE   S-2 AIRCRAFT

Snowplow Effect
USE   PLASMA DYNAMICS

SNOWSTORMS

SOAKING

SOAPS

Soar Space Glider, Dyna-
USE   X-20 AIRCRAFT

SOARING

SOBOLEV SPACE

SOCIAL FACTORS

SOCIAL ISOLATION

SOCIAL PSYCHIATRY

(Social Sciences), Culture
USE   CULTURE (SOCIAL SCIENCES)

SOCIOLOGY

SOCKS

SOD

SODALITE

SODAR

SODIUM

SODIUM ALLOYS

SODIUM AZIDES

SODIUM BROMIDES

SODIUM CARBONATES

SODIUM CHLORIDES

SODIUM CHLORODIFLUOROACETATES

SODIUM CHROMITES

SODIUM COMPOUNDS

Sodium Cooled Reactor, Advanced
USE   ADVANCED SODIUM COOLED REACTOR

SODIUM COOLING

SODIUM FLUORIDES

SODIUM GALLATES

SODIUM GRAPHITE REACTORS

SODIUM HYDRIDES

SODIUM HYDROXIDES

SODIUM IODIDES

SODIUM ISOTOPES

Sodium, Liquid
USE   LIQUID SODIUM

SODIUM NITRATES

Sodium, Pentobarbital
USE   PENTOBARBITAL SODIUM

SODIUM PEROXIDES

SODIUM REACTOR EXPERIMENT

SODIUM SALICYLATES

SODIUM SILICATES

SODIUM SULFATES

SODIUM SULFITES

SODIUM SULFUR BATTERIES

SODIUM VAPOR

SODIUM 22

SODIUM 24

SOFAR
USE   SOUND FIXING AND RANGING

SOFT LANDING

SOFT LANDING SPACECRAFT

(Soft Landing Vehicles), SLV
USE   SOFT LANDING SPACECRAFT

Soft Recovery
USE   SOFT LANDING

SOFTENING

Softening, Strain
USE   PLASTIC DEFORMATION

Softening, Work
USE   WORK SOFTENING

SOFTNESS

Software (Computers)
USE   COMPUTER SYSTEMS PROGRAMS
         COMPUTER PROGRAMS

SOFTWARE ENGINEERING

SOFTWARE TOOLS

SOIL EROSION

Soil, Lunar
USE   LUNAR SOIL

SOIL MAPPING

SOIL MECHANICS

SOIL MOISTURE

SOIL SCIENCE

SOILS

Soils, Frozen
USE   PERMAFROST

SOL-GEL PROCESSES

SOLAR ACTIVITY

SOLAR ACTIVITY EFFECTS

SOLAR ARRAYS

Solar Arrays, Rollup
    USE   SOLAR ARRAYS

SOLAR ATMOSPHERE

SOLAR ATRIUMS

SOLAR AUXILIARY POWER UNITS

Solar Azimuth
    USE   SOLAR POSITION
              AZIMUTH

SOLAR BACKSCATTER UV SPECTROMETER

SOLAR BLANKETS

SOLAR CELL CALIBRATION FACILITY

SOLAR CELLS

Solar Cells, Silicon
    USE   SOLAR CELLS

Solar Cells, Vertical Junction
    USE   VERTICAL JUNCTION SOLAR CELLS

Solar Cells, Wraparound Contact
    USE   SOLAR CELLS

SOLAR COLLECTORS

SOLAR COMPASSES

SOLAR CONSTANT

Solar Converters
    USE   SOLAR GENERATORS

SOLAR COOLING

SOLAR CORONA

SOLAR CORPUSCULAR RADIATION

SOLAR COSMIC RAYS

SOLAR CYCLES

SOLAR DIAMETER

Solar Disk
    USE   SUN

Solar Dynamics
    USE   HELIOSEISMOLOGY

SOLAR ECLIPSES

SOLAR ELECTRIC PROPULSION

SOLAR ELECTRONS

SOLAR ENERGY

SOLAR ENERGY ABSORBERS

SOLAR ENERGY CONVERSION

Solar Energy Conversion, Satellite
    USE   SATELLITE SOLAR ENERGY CONVERSION

Solar Faculae
    USE   FACULAE

SOLAR FLARES

SOLAR FLUX

SOLAR FLUX DENSITY

SOLAR FURNACES

SOLAR GENERATORS

SOLAR GRANULATION

SOLAR GRAVITATION

SOLAR HEATING

SOLAR HOUSES

SOLAR INSTRUMENTS

Solar Lasers
    USE   SOLAR-PUMPED LASERS

SOLAR LIMB

SOLAR LONGITUDE

SOLAR MAGNETIC FIELD

SOLAR MAXIMUM MISSION

SOLAR MAXIMUM MISSION-A

SOLAR MESOSPHERE EXPLORER

Solar Nebula
    USE   SOLAR CORONA

SOLAR NEUTRINOS

Solar Noise
    USE   SOLAR RADIO EMISSION

SOLAR OBLATENESS

SOLAR OBSERVATORIES

Solar Observatory, Advanced Orbiting
    USE   AOSO

Solar Observatory, Orbiting
    USE   OSO

SOLAR OPTICAL TELESCOPE

SOLAR ORBITS

SOLAR OSCILLATIONS

SOLAR PARALLAX

SOLAR PHYSICS

(Solar Physics), Filaments
    USE   SOLAR PROMINENCES

SOLAR PLANETARY INTERACTIONS

Solar Plasma (Radiation)
    USE   SOLAR WIND

Solar Polar Mission, International
    USE   ULYSSES MISSION

SOLAR PONDS (HEAT STORAGE)

SOLAR POSITION

Solar Power Generation
    USE   SOLAR GENERATORS

SOLAR POWER SATELLITES

Solar Power Sources
    USE   SOLAR GENERATORS

Solar Power Stations, Satellite
    USE   SATELLITE SOLAR POWER STATIONS

SOLAR POWERED AIRCRAFT

SOLAR PROBES

SOLAR PROMINENCES

SOLAR PROPULSION

SOLAR PROTONS

SOLAR RADAR ECHOES

SOLAR RADIATION

SOLAR RADIATION SHIELDING

SOLAR RADIATION 1 SATELLITE

SOLAR RADIATION 3 SATELLITE

SOLAR RADIO BURSTS

SOLAR RADIO EMISSION

Solar Radio Waves
    USE   SOLAR RADIO EMISSION

Solar Receivers
    USE   SOLAR COLLECTORS

SOLAR REFLECTORS

SOLAR ROTATION

SOLAR SAILS

SOLAR SEA POWER PLANTS

Solar Seismology
    USE   HELIOSEISMOLOGY

Solar Selective Coatings
    USE   SELECTIVE SURFACES

SOLAR SENSORS

SOLAR SIMULATION

SOLAR SIMULATORS

SOLAR SPECTRA

SOLAR SPECTROMETERS

SOLAR STORMS

Solar Streams
    USE   SOLAR CORPUSCULAR RADIATION

SOLAR SYSTEM

Solar Telescope, Grazing Incidence
    USE   GRIST (TELESCOPE)

SOLAR TEMPERATURE

SOLAR TERRESTRIAL INTERACTIONS

SOLAR THERMAL ELECTRIC POWER PLANTS

SOLAR THERMAL PROPULSION

SOLAR TOTAL ENERGY SYSTEMS

Solar Turboelectric Generator, ASTEC
    USE   ASTEC SOLAR TURBOELECTRIC
              GENERATOR

SOLAR VELOCITY

SOLAR WIND

SOLAR WIND VELOCITY

SOLAR X-RAYS

SOLAR-PUMPED LASERS

SOLDERED JOINTS

SOLDERING

Soldering, Sonic
    USE   ULTRASONIC SOLDERING

Soldering, Ultrasonic
    USE   ULTRASONIC SOLDERING

SOLDERS

SOLENOID VALVES

SOLENOIDS

Solenoids, Meteorological
    USE   METEOROLOGICAL SOLENOIDS

SOLETTAS

Solid Argon
USE   SOLIDIFIED GASES

SOLID CRYOGEN COOLING

SOLID CRYOGENS

SOLID ELECTRODES

SOLID ELECTROLYTES

Solid Interactions, Fluid-
USE   FLUID-SOLID INTERACTIONS

Solid Interactions, Gas-
USE   GAS-SOLID INTERACTIONS

Solid Interfaces, Gas-
USE   GAS-SOLID INTERFACES

Solid Interfaces, Liquid-
USE   LIQUID-SOLID INTERFACES

Solid Interfaces, Solid-
USE   'SOLID-SOLID INTERFACES

SOLID LUBRICANTS

SOLID MECHANICS

SOLID NITROGEN

SOLID PHASES

SOLID PROPELLANT COMBUSTION

SOLID PROPELLANT IGNITION

SOLID PROPELLANT ROCKET ENGINES

SOLID PROPELLANTS

SOLID ROCKET BINDERS

SOLID ROCKET PROPELLANTS

Solid Rotation
USE   ROTATING BODIES

SOLID SOLUTIONS

SOLID STATE

(Solid State), Carrier Density
USE   CARRIER DENSITY (SOLID STATE)

(Solid State), Carrier Transport
USE   CARRIER TRANSPORT (SOLID STATE)

SOLID STATE DEVICES

(Solid State), Energy Gaps
USE   ENERGY GAPS (SOLID STATE)

SOLID STATE LASERS

SOLID STATE PHYSICS

(Solid State), Self Diffusion
USE   SELF DIFFUSION (SOLID STATE)

SOLID SURFACES

SOLID SUSPENSIONS

Solid Upper Stage, Spinning
USE   SPINNING SOLID UPPER STAGE

SOLID WASTES

Solid Zones, Liquid Plus
USE   MUSHY ZONES

SOLID-SOLID INTERFACES

SOLIDIFICATION

Solidification (Crystals), Directional
USE   DIRECTIONAL SOLIDIFICATION (CRYSTALS)

SOLIDIFIED GASES

SOLIDS

Solids, Band Structure Of
USE   BAND STRUCTURE OF SOLIDS

SOLIDS FLOW

Solids, Organic
USE   ORGANIC SOLIDS

Solids, Semi
USE   SEMISOLIDS

SOLIDUS

SOLIONS

SOLITARY WAVES

SOLITHANES

Solitons
USE   SOLITARY WAVES

SOLOMON COMPUTERS

Solrad 10 Satellite
USE   EXPLORER 44 SATELLITE

SOLSTICES

SOLUBILITY

SOLUTES

SOLUTION

Solution, Heat Of
USE   HEAT OF SOLUTION

Solution, Iterative
USE   ITERATIVE SOLUTION

Solution, Pohlhausen
USE   POHLHAUSEN METHOD

Solution, Reissner-Nordstrom
USE   REISSNER-NORDSTROM SOLUTION

SOLUTIONS

Solutions, Aqueous
USE   AQUEOUS SOLUTIONS

Solutions, Solid
USE   SOLID SOLUTIONS

SOLVATION

SOLVENT EXTRACTION

Solvent Method, Traveling
USE   TRAVELING SOLVENT METHOD

SOLVENT REFINED COAL

SOLVENT RETENTION

SOLVENTS

Solvents, Casting
USE   PLASTICIZERS

Solving, Problem
USE   PROBLEM SOLVING

SOLVOLYSIS

SOMALIA

SOMMERFELD APPROXIMATION

Sommerfeld Equations, Orr-
USE   ORR-SOMMERFELD EQUATIONS

SOMMERFELD WAVES

SONAR

SONDES

Sondes, Endoradio
USE   ENDORADIOSONDES

Sondes, Iono
USE   IONOSONDES

Sondes, Radio
USE   RADIOSONDES

Sondes, Rawin
USE   RAWINSONDES

Sondes, Rocket
USE   SOUNDING ROCKETS

SONIC ANEMOMETERS

SONIC BOOMS

Sonic Fatigue
USE   ACOUSTIC FATIGUE

Sonic Flow
USE   TRANSONIC FLOW

SONIC NOZZLES

Sonic Soldering
USE   ULTRASONIC SOLDERING

Sonic Speed
USE   ACOUSTIC VELOCITY

Sonic Waveguides
USE   ACOUSTIC DELAY LINES

SONOBUOYS

SONOGRAMS

Sonoholography
USE   ACOUSTICAL HOLOGRAPHY

SONOLUMINESCENCE

SOOT

SORBATES

SORBENTS

Sorbents, Ad
USE   ADSORBENTS

SORET COEFFICIENT

SORGHUM

SORPTION

Sorption, Ad
USE   ADSORPTION

Sorption, Chemi
USE   CHEMISORPTION

Sorption, De
USE   DESORPTION

Sortie Can
USE   SORTIE SYSTEMS

Sortie Lab
USE   SORTIE SYSTEMS

SORTIE SYSTEMS

Sorting
USE   CLASSIFYING

SOS (SEMICONDUCTORS)

SOT
USE   SOLAR OPTICAL TELESCOPE

Sound
USE   ACOUSTICS

Sound Absorption
USE   SOUND TRANSMISSION

Sound (AK), Prince William
USE   PRINCE WILLIAM SOUND (AK)

SOUND AMPLIFICATION

Sound Barrier
USE   ACOUSTIC VELOCITY

SOUND DETECTING AND RANGING

Sound Detectors
USE   SOUND TRANSDUCERS

SOUND FIELDS

SOUND FIXING AND RANGING

SOUND GENERATORS

Sound Holography
USE   ACOUSTICAL HOLOGRAPHY

SOUND INTENSITY

Sound Interactions, Sound-
USE   SOUND-SOUND INTERACTIONS

SOUND LOCALIZATION

Sound, Mcmurdo
USE   MCMURDO SOUND

Sound Measurement
USE   ACOUSTIC MEASUREMENT

(Sound), Noise
USE   NOISE (SOUND)

Sound Perception
USE   AUDITORY PERCEPTION

SOUND PRESSURE

SOUND PROPAGATION

SOUND RANGING

Sound (RI), Block Island
USE   BLOCK ISLAND SOUND (RI)

SOUND TRANSDUCERS

SOUND TRANSMISSION

Sound, Underwater
USE   UNDERWATER ACOUSTICS

Sound Velocity
USE   ACOUSTIC VELOCITY

SOUND WAVES

Sound Waves, Plasma
USE   MAGNETOHYDRODYNAMIC WAVES
        PLASMA WAVES

Sound, Zero
USE   ZERO SOUND

SOUND-SOUND INTERACTIONS

Sounder, Orbiting Radio Beacon Ionospheric
USE   ORBIS

Sounder Probe, Pioneer Venus 2
USE   PIONEER VENUS 2 SOUNDER PROBE

Sounders
USE   SOUNDING

SOUNDING

Sounding, Acoustic
USE   ACOUSTIC SOUNDING

Sounding, Atmospheric
USE   ATMOSPHERIC SOUNDING

Sounding, Balloon
USE   BALLOON SOUNDING

Sounding, Echo
USE   ECHO SOUNDING

Sounding, Ionospheric
USE   IONOSPHERIC SOUNDING

Sounding, Microwave
USE   MICROWAVE SOUNDING

Sounding Projectile, High Altitude
USE   WASP SOUNDING ROCKET

Sounding Projectile, Window Atmosphere
USE   WASP SOUNDING ROCKET

Sounding, Rocket
USE   ROCKET SOUNDING

Sounding Rocket, Aries
USE   ARIES SOUNDING ROCKET

Sounding Rocket, Black Brant 1
USE   BLACK BRANT 1 SOUNDING ROCKET

Sounding Rocket, Black Brant 2
USE   BLACK BRANT 2 SOUNDING ROCKET

Sounding Rocket, Black Brant 3
USE   BLACK BRANT 3 SOUNDING ROCKET

Sounding Rocket, Black Brant 4
USE   BLACK BRANT 4 SOUNDING ROCKET

Sounding Rocket, Black Brant 5
USE   BLACK BRANT 5 SOUNDING ROCKET

Sounding Rocket, Exos
USE   EXOS SOUNDING ROCKET

Sounding Rocket, Petrel
USE   PETREL SOUNDING ROCKET

Sounding Rocket, Phoenix
USE   PHOENIX SOUNDING ROCKET

Sounding Rocket, Wasp
USE   WASP SOUNDING ROCKET

SOUNDING ROCKETS

Sounding Rockets, Black Brant
USE   BLACK BRANT SOUNDING ROCKETS

Sounding, Satellite
USE   SATELLITE SOUNDING

SOUNDS (TOPOGRAPHIC FEATURES)

SOURCE PROGRAMS

SOURCES

Sources, Aircraft Power
USE   AIRCRAFT ENGINES

Sources (Astronomy), Radio
USE   RADIO SOURCES (ASTRONOMY)

Sources, Atmospheric Energy
USE   ATMOSPHERIC ENERGY SOURCES

Sources, Auxiliary Power
USE   AUXILIARY POWER SOURCES

Sources, Coherent
USE   COHERENT RADIATION
        RADIATION SOURCES

Sources, Electron
USE   ELECTRON SOURCES

Sources, Energy
USE   ENERGY SOURCES

Sources, Extragalactic Radio
USE   EXTRAGALACTIC RADIO SOURCES

Sources, Heat
USE   HEAT SOURCES

Sources, Hydraulic Heating
USE   HEAT SOURCES
        HYDRAULIC EQUIPMENT

Sources, Ion
USE   ION SOURCES

Sources, Light
USE   LIGHT SOURCES

Sources, Neutron
USE   NEUTRON SOURCES

Sources, Nonpoint
USE   NONPOINT SOURCES

Sources, Offshore Energy
USE   OFFSHORE ENERGY SOURCES

Sources, Plasma Power
USE   PLASMA POWER SOURCES

Sources, Point
USE   POINT SOURCES

Sources), QSO (Radio
USE   QUASARS

Sources, Quasi-Stellar Radio
USE   QUASARS

Sources, Radiation
USE   RADIATION SOURCES

Sources, Solar Power
USE   SOLAR GENERATORS

Sources, X Ray
USE   X RAY SOURCES

South Africa
USE   REPUBLIC OF SOUTH AFRICA

South Africa, Republic Of
USE   REPUBLIC OF SOUTH AFRICA

SOUTH AMERICA

(South America), Amazon Region
USE   AMAZON REGION (SOUTH AMERICA)

(South America), Andes Mountains
USE   ANDES MOUNTAINS (SOUTH AMERICA)

SOUTH CAROLINA

SOUTH DAKOTA

SOUTH KOREA

South Vietnam
USE   VIETNAM

South West Africa
USE   NAMIBIA

SOUTHEAST ASIA

SOUTHERN CALIFORNIA

SOUTHERN HEMISPHERE

SOUTHERN SKY

SOUTHERN YEMEN

SOVEREIGNTY

SOVIET SATELLITES

SOVIET SPACECRAFT

Soviet Union
USE   U.S.S.R.

SOYBEANS

SOYUZ SPACECRAFT

Soyuz Test Project, Apollo
USE   APOLLO SOYUZ TEST PROJECT

SPACE

Space & Terrestr Applic Payloads, Office Of
USE   OSTA 1 PAYLOAD
       OSTA-2 PAYLOAD

SPACE ADAPTATION SYNDROME

Space Agency, European
USE   EUROPEAN SPACE AGENCY

Space, Air
USE   AIRSPACE

Space Arrow Satellite
USE   COSMOS 149 SATELLITE

Space, Banach
USE   BANACH SPACE

SPACE BASE COMMAND CENTER

SPACE BASED RADAR

SPACE BASES

Space Biology
USE   EXOBIOLOGY

Space Buses
USE   FERRY SPACECRAFT

SPACE CAPSULES

Space, Cartan
USE   CARTAN SPACE

SPACE CHARGE

Space, Cislunar
USE   CISLUNAR SPACE

SPACE COLONIES

SPACE COMMERCIALIZATION

SPACE COMMUNICATION

Space, Construction In
USE   ORBITAL ASSEMBLY

SPACE COOLING (BUILDINGS)

SPACE DEBRIS

Space, Deep
USE   DEEP SPACE

SPACE DENSITY

SPACE DETECTION AND TRACKING SYSTEM

Space Diversity
USE   RECEPTION DIVERSITY

Space), Earth Observations (From
USE   EARTH OBSERVATIONS (FROM SPACE)

SPACE ELECTRIC ROCKET TESTS

Space Environment
USE   AEROSPACE ENVIRONMENTS

SPACE ENVIRONMENT SIMULATION

Space Environmental Lubrication
USE   SPACECRAFT LUBRICATION

SPACE ERECTABLE STRUCTURES

Space, Euclidean
USE   EUCLIDEAN GEOMETRY

Space Exper With Particle Accelerators
USE   SEPAC (PAYLOAD)

SPACE EXPLORATION

Space, Faraday Dark
USE   FARADAY DARK SPACE

SPACE FLIGHT

Space Flight, Extended Duration
USE   LONG DURATION SPACE FLIGHT

SPACE FLIGHT FEEDING

Space Flight, Long Duration
USE   LONG DURATION SPACE FLIGHT

Space Flight, Manned
USE   MANNED SPACE FLIGHT

Space Flight Network, Manned
USE   MANNED SPACE FLIGHT NETWORK

Space Flight, Planetary
USE   INTERPLANETARY FLIGHT

Space Flight, Return To Earth
USE   RETURN TO EARTH SPACE FLIGHT

SPACE FLIGHT STRESS

SPACE FLIGHT TRACKING AND DATA NETWORK

SPACE FLIGHT TRAINING

Space, Function
USE   FUNCTION SPACE

Space Glider, Dyna-Soar
USE   X-20 AIRCRAFT

Space Gliders
USE   LIFTING REENTRY VEHICLES

SPACE GLOSSARIES

Space Guidance), SSGS (Standardized
USE   STANDARDIZED SPACE GUIDANCE

Space Guidance, Standardized
USE   STANDARDIZED SPACE GUIDANCE

SPACE HABITATS

Space), Hazardous Material Disposal (In
USE   HAZARDOUS MATERIAL DISPOSAL (IN
       SPACE)

SPACE HEATING (BUILDINGS)

Space, Hilbert
USE   HILBERT SPACE

Space, Hyperbolic
USE   HYPERBOLIC COORDINATES

SPACE INDUSTRIALIZATION

SPACE INFRARED TELESCOPE FACILITY

Space Instrumentation Facility, Deep
USE   DEEP SPACE INSTRUMENTATION FACILITY

Space Integral, Phase-
USE   PHASE-SPACE INTEGRAL

Space, Interplanetary
USE   INTERPLANETARY SPACE

Space, Interstellar
USE   INTERSTELLAR SPACE

SPACE LABORATORIES

SPACE LAW

SPACE LOGISTICS

SPACE MAINTENANCE

SPACE MANUFACTURING

SPACE MECHANICS

Space, Metric
USE   METRIC SPACE

Space, Minkowski
USE   MINKOWSKI SPACE

SPACE MISSIONS

SPACE NAVIGATION

Space Network, Deep
USE   DEEP SPACE NETWORK

SPACE OBSERVATIONS (FROM EARTH)

SPACE OPERATIONS CENTER (NASA)

SPACE ORIENTATION

Space, Orlicz
USE   ORLICZ SPACE

Space Payload, Plasmas-IN-
USE   AMPS (SATELLITE PAYLOAD)

SPACE PERCEPTION

Space Photography
USE   SPACEBORNE PHOTOGRAPHY

Space, Physics And Chemistry Experiment In
USE   PHYSICS AND CHEMISTRY EXPERIMENT IN
       SPACE

Space Plasma H/v Interaction Experiments
USE   SPHINX

SPACE PLASMAS

SPACE PLATFORMS

SPACE POWER REACTORS

SPACE POWER UNIT REACTORS

Space Probe, Mariner R 2
USE   MARINER R 2 SPACE PROBE

Space Probe, Mariner 1
USE   MARINER 1 SPACE PROBE

Space Probe, Mariner 2
USE   MARINER 2 SPACE PROBE

Space Probe, Mariner 3
USE   MARINER 3 SPACE PROBE

Space Probe, Mariner 4
USE   MARINER 4 SPACE PROBE

Space Probe, Mariner 5
USE   MARINER 5 SPACE PROBE

Space Probe, Mariner 6
USE   MARINER 6 SPACE PROBE

Space Probe, Mariner 7
USE   MARINER 7 SPACE PROBE

Space Probe, Mariner 8
USE   MARINER 8 SPACE PROBE

Space Probe, Mariner 9
USE   MARINER 9 SPACE PROBE

Space Probe, Mariner 10
USE   MARINER 10 SPACE PROBE

Space Probe, Mariner 11
USE   MARINER 11 SPACE PROBE

Space Probe, Pioneer F
USE   PIONEER 10 SPACE PROBE

Space Probe, Pioneer G
USE   PIONEER 11 SPACE PROBE

Space Probe, Pioneer 1
USE   PIONEER 1 SPACE PROBE

Space Probe, Pioneer 2
USE   PIONEER 2 SPACE PROBE

Space Probe, Pioneer 3
USE   PIONEER 3 SPACE PROBE

Space Probe, Pioneer 4
USE   PIONEER 4 SPACE PROBE

Space Probe, Pioneer 5
USE   PIONEER 5 SPACE PROBE

Space Probe, Pioneer 6
USE   PIONEER 6 SPACE PROBE

Space Probe, Pioneer 7
USE   PIONEER 7 SPACE PROBE

Space Probe, Pioneer 8
USE   PIONEER 8 SPACE PROBE

Space Probe, Pioneer 9
USE   PIONEER 9 SPACE PROBE

Space Probe, Pioneer 10
USE   PIONEER 10 SPACE PROBE

Space Probe, Pioneer 11
USE   PIONEER 11 SPACE PROBE

Space Probe, Pioneer 12
USE   PIONEER VENUS SPACECRAFT

Space Probe, Sunblazer
USE   SUNBLAZER SPACE PROBE

Space Probe, Zond 1
USE   ZOND 1 SPACE PROBE

Space Probe, Zond 2
USE   ZOND 2 SPACE PROBE

Space Probe, Zond 3
USE   ZOND 3 SPACE PROBE

Space Probe, Zond 4
USE   ZOND 4 SPACE PROBE

Space Probe, Zond 5
USE   ZOND 5 SPACE PROBE

Space Probe, Zond 6
USE   ZOND 6 SPACE PROBE

Space Probe, Zond 7
USE   ZOND 7 SPACE PROBE

Space Probe, Zond 8
USE   ZOND 8 SPACE PROBE

SPACE PROBES

Space Probes, Mariner
USE   MARINER SPACE PROBES

Space Probes, Pioneer
USE   PIONEER SPACE PROBES

Space Probes, Zond
USE   ZOND SPACE PROBES

SPACE PROCESSING

SPACE PROCESSING APPLICATIONS ROCKET

Space Program, Brazilian
USE   BRAZILIAN SPACE PROGRAM

Space Program, Canadian
USE   CANADIAN SPACE PROGRAM

Space Program, Chinese
USE   CHINESE SPACE PROGRAM

Space Program, Indian
USE   INDIAN SPACE PROGRAM

Space Program, Indonesian
USE   INDONESIAN SPACE PROGRAM

Space Program, Italian
USE   ITALIAN SPACE PROGRAM

Space Program, Japanese
USE   JAPANESE SPACE PROGRAM

Space Program, Saudi Arabian
USE   SAUDI ARABIAN SPACE PROGRAM

Space Program, Swedish
USE   SWEDISH SPACE PROGRAM

Space Program, Swiss
USE   SWISS SPACE PROGRAM

Space Program, U.S.S.R.
USE   U.S.S.R. SPACE PROGRAM

Space Program, UK
USE   UK SPACE PROGRAM

SPACE PROGRAMS

Space Programs, European
USE   EUROPEAN SPACE PROGRAMS

Space Programs, French
USE   FRENCH SPACE PROGRAMS

Space Programs, NASA
USE   NASA SPACE PROGRAMS

SPACE PSYCHOLOGY

Space Radiation
USE   EXTRATERRESTRIAL RADIATION

Space Radiators
USE   SPACECRAFT RADIATORS

SPACE RATIONS

SPACE RENDEZVOUS

Space Research, Committee On
USE   COMMITTEE ON SPACE RESEARCH

Space Research Organization, European
USE   EUROPEAN SPACE AGENCY

Space Research Organization, Indian
USE   ISRO

Space Research Organization Sat, European
USE   ESA SATELLITES

Space, Riemann
USE   RIEMANN MANIFOLD

Space Sciences
USE   AEROSPACE SCIENCES

Space Self Maneuvering Units
USE   SELF MANEUVERING UNITS

SPACE SHUTTLE ASCENT STAGE

SPACE SHUTTLE BOOSTERS

SPACE SHUTTLE MAIN ENGINE

SPACE SHUTTLE MISSION 31-A

SPACE SHUTTLE MISSION 31-B

SPACE SHUTTLE MISSION 31-C

SPACE SHUTTLE MISSION 31-D

SPACE SHUTTLE MISSION 41-A

SPACE SHUTTLE MISSION 41-B

SPACE SHUTTLE MISSION 41-C

SPACE SHUTTLE MISSION 41-D

SPACE SHUTTLE MISSION 41-G

SPACE SHUTTLE MISSION 51-A

SPACE SHUTTLE MISSION 51-B

SPACE SHUTTLE MISSION 51-C

SPACE SHUTTLE MISSION 51-D

SPACE SHUTTLE MISSION 51-E

SPACE SHUTTLE MISSION 51-F

SPACE SHUTTLE MISSION 51-G

SPACE SHUTTLE MISSION 51-H

SPACE SHUTTLE MISSION 51-I

SPACE SHUTTLE MISSION 51-J

SPACE SHUTTLE MISSION 51-L

SPACE SHUTTLE MISSION 61-A

SPACE SHUTTLE MISSION 61-B

SPACE SHUTTLE MISSION 61-C

SPACE SHUTTLE MISSION 61-E

SPACE SHUTTLE MISSIONS

(Space Shuttle), Orbit Maneuvering Engine
USE   ORBIT MANEUVERING ENGINE (SPACE
      SHUTTLE)

Space Shuttle Orbital Flight Test 1
USE   SPACE TRANSPORTATION SYSTEM 1
      FLIGHT

Space Shuttle Orbital Flight Test 2
USE   SPACE TRANSPORTATION SYSTEM 2
      FLIGHT

Space Shuttle Orbital Flight Test 3
USE   SPACE TRANSPORTATION SYSTEM 3
      FLIGHT

Space Shuttle Orbital Flight Test 4
USE   SPACE TRANSPORTATION SYSTEM 4
      FLIGHT

Space Shuttle Orbital Flight Tests
USE   SPACE TRANSPORTATION SYSTEM
      FLIGHTS

Space Shuttle Orbital Flight 7
USE   SPACE SHUTTLE MISSION 31-C

Space Shuttle Orbital Flight 8
USE   SPACE SHUTTLE MISSION 31-D

Space Shuttle Orbital Flight 9
USE   SPACE SHUTTLE MISSION 41-A

Space Shuttle Orbital Flights
USE   SPACE TRANSPORTATION SYSTEM
      FLIGHTS

Space Shuttle Orbiter 099
USE   CHALLENGER (ORBITER)

Space Shuttle Orbiter 101
USE   ENTERPRISE (ORBITER)

Space Shuttle Orbiter 102
USE   COLUMBIA (ORBITER)

Space Shuttle Orbiter 103
USE   DISCOVERY (ORBITER)

Space Shuttle Orbiter 104
USE   ATLANTIS (ORBITER)

SPACE SHUTTLE ORBITERS

SPACE SHUTTLE PAYLOADS

SPACE SHUTTLE UPPER STAGE A

SPACE SHUTTLE UPPER STAGE D

SPACE SHUTTLE UPPER STAGES

SPACE SHUTTLES

SPACE SIMULATORS

Space, Sobolev
USE   SOBOLEV SPACE

Space Station, Halo Orbit
USE   HALO ORBIT SPACE STATION

Space Station, Salyut
USE   SALYUT SPACE STATION

Space Station (Unmanned), SKYLAB
USE   SKYLAB 1

SPACE STATIONS

Space Stations, Earth Orbiting
USE   EOSS

Space Stations, Manned Orbital
USE   ORBITAL SPACE STATIONS

(Space Stations), MOSS
USE   ORBITAL SPACE STATIONS

Space Stations, Orbital
USE   ORBITAL SPACE STATIONS

Space Stations, Self Deploying
USE   SPACE STATIONS
       SELF ERECTING DEVICES

SPACE STORAGE

Space Structures, Large
USE   LARGE SPACE STRUCTURES

SPACE SUITS

SPACE SURVEILLANCE

SPACE SURVEILLANCE (GROUND BASED)

SPACE SURVEILLANCE (SPACEBORNE)

Space System, Bioastronautical Orbital
USE   BIOASTRONAUTICAL ORBITAL SPACE
       SYSTEM

Space Systems Engineering
USE   AEROSPACE ENGINEERING

SPACE TECHNOLOGY EXPERIMENTS

Space Telescope
USE   HUBBLE SPACE TELESCOPE

Space Telescope, Hubble
USE   HUBBLE SPACE TELESCOPE

Space Telescope, Large
USE   HUBBLE SPACE TELESCOPE

SPACE TEMPERATURE

Space Tests, Orbital
USE   ORBITAL SPACE TESTS

SPACE TOOLS

Space, Translunar
USE   INTERPLANETARY SPACE

SPACE TRANSPORTATION

SPACE TRANSPORTATION SYSTEM

SPACE TRANSPORTATION SYSTEM FLIGHTS

SPACE TRANSPORTATION SYSTEM 1 FLIGHT

SPACE TRANSPORTATION SYSTEM 2 FLIGHT

SPACE TRANSPORTATION SYSTEM 3 FLIGHT

SPACE TRANSPORTATION SYSTEM 4 FLIGHT

Space Treaty, Outer
USE   OUTER SPACE TREATY

SPACE TUGS

Space, U Spin
USE   U SPIN SPACE

SPACE VEHICLE CHECKOUT PROGRAM

Space Vehicle Control
USE   SPACECRAFT CONTROL

Space Vehicles
USE   SPACECRAFT

SPACE WEAPONS

Space-Time Continuum
USE   RELATIVITY

SPACE-TIME FUNCTIONS

Space-Time Metric
USE   SPACE-TIME FUNCTIONS

SPACEBORNE ASTRONOMY

SPACEBORNE EXPERIMENTS

SPACEBORNE LASERS

SPACEBORNE PHOTOGRAPHY

(Spaceborne), Space Surveillance
USE   SPACE SURVEILLANCE (SPACEBORNE)

SPACEBORNE TELESCOPES

SPACECRAFT

Spacecraft, Advanced Reconn Electric
USE   ADVANCED RECONN ELECTRIC
       SPACECRAFT

SPACECRAFT ANTENNAS

Spacecraft, Apollo
USE   APOLLO SPACECRAFT

(Spacecraft), ARES
USE   ADVANCED RECONN ELECTRIC
       SPACECRAFT

SPACECRAFT CABIN ATMOSPHERES

SPACECRAFT CABIN SIMULATORS

SPACECRAFT CABINS

Spacecraft, Canadian
USE   CANADIAN SPACECRAFT

(Spacecraft), Capsules
USE   SPACE CAPSULES

Spacecraft, Cargo
USE   CARGO SPACECRAFT

SPACECRAFT CHARGING

Spacecraft, Chinese
USE   CHINESE SPACECRAFT

Spacecraft Clocks, Autonomous
USE   AUTONOMOUS SPACECRAFT CLOCKS

Spacecraft, Commercial
USE   COMMERCIAL SPACECRAFT

SPACECRAFT COMMUNICATION

SPACECRAFT COMPONENTS

SPACECRAFT CONFIGURATIONS

SPACECRAFT CONSTRUCTION MATERIALS

(Spacecraft), Consumables
USE   CONSUMABLES (SPACECRAFT)

SPACECRAFT CONTAMINATION

SPACECRAFT CONTROL

Spacecraft, Copernicus
USE   OAO 3

Spacecraft, Czechoslovakian
USE   CZECHOSLOVAKIAN SPACECRAFT

SPACECRAFT DEFENSE

SPACECRAFT DESIGN

SPACECRAFT DOCKING

SPACECRAFT DOCKING MODULES

Spacecraft, Dual Spin
USE   DUAL SPIN SPACECRAFT

SPACECRAFT ELECTRONIC EQUIPMENT

SPACECRAFT ENVIRONMENTS

SPACECRAFT EQUIPMENT

Spacecraft, ESA
USE   ESA SPACECRAFT

Spacecraft, European 1
USE   EUROPEAN 1 SPACECRAFT

(Spacecraft), Expendable Stages
USE   EXPENDABLE STAGES (SPACECRAFT)

Spacecraft, Ferry
USE   FERRY SPACECRAFT

Spacecraft, Flexible
USE   FLEXIBLE SPACECRAFT

Spacecraft, Galileo
USE   GALILEO SPACECRAFT

Spacecraft, Gemini
USE   GEMINI SPACECRAFT

Spacecraft, Gemini B
USE   GEMINI B SPACECRAFT

Spacecraft, Gemini (GT-1)
USE   GEMINI (GT-1) SPACECRAFT

Spacecraft, Gemini 2
USE   GEMINI 2 SPACECRAFT

SPACECRAFT GUIDANCE

(Spacecraft), Housekeeping
USE   HOUSEKEEPING (SPACECRAFT)

Spacecraft, Indian
USE   INDIAN SPACECRAFT

Spacecraft, Inflatable
USE   INFLATABLE SPACECRAFT

SPACECRAFT INSTRUMENTS

(Spacecraft), Interim Stages
USE   INTERIM STAGES (SPACECRAFT)

Spacecraft, Interplanetary
USE   INTERPLANETARY SPACECRAFT

Spacecraft, Interstellar
USE   INTERSTELLAR SPACECRAFT

Spacecraft), IRS (Indian
USE   INDIAN SPACECRAFT

Spacecraft, Janus
USE   JANUS SPACECRAFT

Spacecraft, Japanese
USE   JAPANESE SPACECRAFT

SPACECRAFT LANDING

Spacecraft Landing, Horizontal
USE HORIZONTAL SPACECRAFT LANDING

SPACECRAFT LAUNCHING

SPACECRAFT LUBRICATION

Spacecraft, Lunar
USE LUNAR SPACECRAFT

SPACECRAFT MAINTENANCE

Spacecraft, Maneuverable
USE MANEUVERABLE SPACECRAFT

SPACECRAFT MANEUVERS

Spacecraft, Manned
USE MANNED SPACECRAFT

Spacecraft, Mariner
USE MARINER SPACECRAFT

Spacecraft, Mariner C
USE MARINER C SPACECRAFT

Spacecraft, Mariner Mark 2
USE MARINER MARK 2 SPACECRAFT

Spacecraft, Mariner Venus 67
USE MARINER VENUS 67 SPACECRAFT

Spacecraft, Mark 1
USE MARK 1 SPACECRAFT

Spacecraft), Mars (Manned Reusable
USE MARS (MANNED REUSABLE SPACECRAFT)

Spacecraft, Mars 1
USE MARS 1 SPACECRAFT

Spacecraft, Mars 2
USE MARS 2 SPACECRAFT

Spacecraft, Mars 3
USE MARS 3 SPACECRAFT

Spacecraft, Mars 4
USE MARS 4 SPACECRAFT

Spacecraft, Mars 5
USE MARS 5 SPACECRAFT

Spacecraft, Mars 6
USE MARS 6 SPACECRAFT

Spacecraft, Mars 7
USE MARS 7 SPACECRAFT

Spacecraft, Mercury
USE MERCURY SPACECRAFT

Spacecraft, Military
USE MILITARY SPACECRAFT

SPACECRAFT MODELS

SPACECRAFT MODULES

Spacecraft), MOS (Japanese
USE JAPANESE SPACECRAFT

SPACECRAFT MOTION

Spacecraft, Multimission Modular
USE MULTIMISSION MODULAR SPACECRAFT

Spacecraft Orbital Assembly
USE ORBITAL ASSEMBLY

SPACECRAFT ORBITS

Spacecraft, Outer Planet
USE OUTER PLANETS EXPLORERS

(Spacecraft Passageway), Ingress
USE INGRESS (SPACECRAFT PASSAGEWAY)

SPACECRAFT PERFORMANCE

Spacecraft, Photo Reconnaissance
USE PHOTO RECONNAISSANCE SPACECRAFT

Spacecraft, Pioneer Saturn
USE PIONEER 11 SPACE PROBE

Spacecraft, Pioneer Venus
USE PIONEER VENUS SPACECRAFT

Spacecraft, Pioneer Venus 1
USE PIONEER VENUS 1 SPACECRAFT

Spacecraft, Pioneer Venus 2
USE PIONEER VENUS 2 SPACECRAFT

Spacecraft, Pioneer Venus 2 Multiprobe
USE PIONEER VENUS 2 SPACECRAFT

Spacecraft, Planetary
USE INTERPLANETARY SPACECRAFT

SPACECRAFT POSITION INDICATORS

(Spacecraft), Postmission Analysis
USE POSTMISSION ANALYSIS (SPACECRAFT)

Spacecraft, Power Limited
USE POWER LIMITED SPACECRAFT

SPACECRAFT POWER SUPPLIES

Spacecraft Prelaunch Tests
USE SPACE VEHICLE CHECKOUT PROGRAM

SPACECRAFT PROPULSION

Spacecraft, Radiation Meteoroid
USE RADIATION METEOROID SPACECRAFT

SPACECRAFT RADIATORS

Spacecraft, Reconnaissance
USE RECONNAISSANCE SPACECRAFT

Spacecraft, Recoverable
USE RECOVERABLE SPACECRAFT

SPACECRAFT RECOVERY

SPACECRAFT REENTRY

SPACECRAFT RELIABILITY

Spacecraft Rendezvous
USE SPACE RENDEZVOUS

Spacecraft, Rendezvous
USE RENDEZVOUS SPACECRAFT

Spacecraft, Reusable
USE REUSABLE SPACECRAFT

Spacecraft Sensors
USE SPACECRAFT INSTRUMENTS

Spacecraft), SEO (Indian
USE INDIAN SPACECRAFT

Spacecraft, SERT 1
USE SERT 1 SPACECRAFT

Spacecraft, SERT 2
USE SERT 2 SPACECRAFT

SPACECRAFT SHIELDING

Spacecraft, Soft Landing
USE SOFT LANDING SPACECRAFT

Spacecraft, Soviet
USE SOVIET SPACECRAFT

Spacecraft, Soyuz
USE SOYUZ SPACECRAFT

SPACECRAFT STABILITY

SPACECRAFT STERILIZATION

SPACECRAFT STRUCTURES

SPACECRAFT SURVIVABILITY

Spacecraft, Technology Feasibility
USE TECHNOLOGY FEASIBILITY SPACECRAFT

SPACECRAFT TELEVISION

Spacecraft Television, Digital
USE DIGITAL SPACECRAFT TELEVISION

SPACECRAFT TEMPERATURE

Spacecraft, Thermoelectric
USE TOPS (SPACECRAFT)

Spacecraft, Thermoelectric Outer Planet
USE TOPS (SPACECRAFT)

(Spacecraft), TOPS
USE TOPS (SPACECRAFT)

SPACECRAFT TRACKING

Spacecraft Tracking And Data Network
USE STDN (NETWORK)

SPACECRAFT TRAJECTORIES

(Spacecraft), Uncontrolled Reentry
USE UNCONTROLLED REENTRY (SPACECRAFT)

Spacecraft, Unmanned
USE UNMANNED SPACECRAFT

(Spacecraft), Venus Orbiting Imaging Radar
USE VENUS ORBITING IMAGING RADAR
(SPACECRAFT)

Spacecraft, Viking
USE VIKING SPACECRAFT

Spacecraft, Viking Lander
USE VIKING LANDER SPACECRAFT

Spacecraft, Viking Orbiter
USE VIKING ORBITER SPACECRAFT

Spacecraft, Viking 1
USE VIKING 1 SPACECRAFT

Spacecraft, Viking 2
USE VIKING 2 SPACECRAFT

Spacecraft, Voshkod Manned
USE VOSKHOD MANNED SPACECRAFT

Spacecraft, Voskhod 1
USE VOSKHOD 1 SPACECRAFT

Spacecraft, Voskhod 2
USE VOSKHOD 2 SPACECRAFT

Spacecraft, Vostok
USE VOSTOK SPACECRAFT

Spacecraft, Vostok 1
USE VOSTOK 1 SPACECRAFT

Spacecraft, Vostok 2
USE VOSTOK 2 SPACECRAFT

Spacecraft, Vostok 3
USE VOSTOK 3 SPACECRAFT

Spacecraft, Vostok 4
USE VOSTOK 4 SPACECRAFT

Spacecraft, Vostok 5
USE VOSTOK 5 SPACECRAFT

Spacecraft, Vostok 6
USE VOSTOK 6 SPACECRAFT

Spacecraft, Voyager 1
USE VOYAGER 1 SPACECRAFT

Spacecraft, Voyager 2
USE VOYAGER 2 SPACECRAFT

(Spacecrew Supplies), Consumables
USE CONSUMABLES (SPACECREW SUPPLIES)

SPACECREW TRANSFER

Spacecrew Transfer, Intervehicle
  USE   SPACECREW TRANSFER

SPACECREWS

SPACELAB

(Spacelab), ACPL
  USE   ATMOSPHERIC CLOUD PHYSICS LAB
        (SPACELAB)

(Spacelab), Atmospheric Cloud Physics Lab
  USE   ATMOSPHERIC CLOUD PHYSICS LAB
        (SPACELAB)

Spacelab, Large Infrared Telescope On
  USE   LIRTS (TELESCOPE)

(Spacelab Payload), Expos
  USE   EXPOS (SPACELAB PAYLOAD)

SPACELAB PAYLOADS

Spacelab Simulation Flights
  USE   ASSESS PROGRAM

Spacelab UV-Optical Telescope Facility
  USE   STARLAB

(Spacelab), Zero-G ACPL
  USE   ATMOSPHERIC CLOUD PHYSICS LAB
        (SPACELAB)

Spaceplane, Hermes Manned
  USE   HERMES MANNED SPACEPLANE

SPACERS

(Spacers), Washers
  USE   WASHERS (SPACERS)

Spaces, Half
  USE   HALF SPACES

Spaces, Hyper
  USE   HYPERSPACES

Spaces, Vector
  USE   VECTOR SPACES

Spaceship, Manned Aerodynamic Reusable
  USE   MARS (MANNED REUSABLE SPACECRAFT)

SPACETENNAS

SPACING

Spacing, Aircraft Approach
  USE   AIRCRAFT APPROACH SPACING

SPADATS (Tracking System)
  USE   SPACE DETECTION AND TRACKING
        SYSTEM

SPAIN

SPALLATION

SPALLING

SPAN

Span, Life
  USE   LIFE SPAN

Span, Wing
  USE   WING SPAN

Span Wings, Infinite
  USE   INFINITE SPAN WINGS

SPANISH SAHARA

SPANLOADER AIRCRAFT

SPANWISE BLOWING

SPAR (Rocket)
  USE   SPACE PROCESSING APPLICATIONS
        ROCKET

SPARE PARTS

SPARK CHAMBERS

Spark Discharges
  USE   ELECTRIC SPARKS

SPARK GAPS

SPARK IGNITION

SPARK MACHINING

SPARK PLUGS

Spark Shadowgraph Photography
  USE   SHADOWGRAPH PHOTOGRAPHY

SPARKS

Sparks, Electric
  USE   ELECTRIC SPARKS

SPARROW MISSILES

SPARROW 2 MISSILE

SPARROW 3 MISSILE

SPARTAN MISSILE

SPARTAN SATELLITES

SPAS (ESA Platforms)
  USE   SHUTTLE PALLET SATELLITES

SPASMS

SPATIAL DEPENDENCIES

SPATIAL DISTRIBUTION

SPATIAL FILTERING

Spatial Isotropy
  USE   SPATIAL DISTRIBUTION
        ISOTROPY

SPATIAL MARCHING

Spatial Orientation
  USE   ATTITUDE (INCLINATION)

SPATIAL RESOLUTION

Speaking, Public
  USE   PUBLIC SPEAKING

SPECIES DIFFUSION

Species, Endangered
  USE   ENDANGERED SPECIES

Specific Gravity
  USE   DENSITY (MASS/VOLUME)

SPECIFIC HEAT

SPECIFIC IMPULSE

SPECIFICATIONS

Specifications, Aircraft
  USE   AIRCRAFT SPECIFICATIONS

Specifications, Equipment
  USE   EQUIPMENT SPECIFICATIONS

Specifications, Functional Design
  USE   FUNCTIONAL DESIGN SPECIFICATIONS

SPECIMEN GEOMETRY

SPECIMENS

SPECKLE PATTERNS

SPECTRA

Spectra, Absorption
  USE   ABSORPTION SPECTRA

Spectra, Atomic
  USE   ATOMIC SPECTRA

Spectra, Continuous
  USE   CONTINUOUS SPECTRA

Spectra, Electromagnetic
  USE   ELECTROMAGNETIC SPECTRA

Spectra, Electronic
  USE   ELECTRONIC SPECTRA

Spectra, Emission
  USE   EMISSION SPECTRA

Spectra, Energy
  USE   ENERGY SPECTRA

Spectra, Gamma Ray
  USE   GAMMA RAY SPECTRA

(Spectra), Gratings
  USE   GRATINGS (SPECTRA)

Spectra, Infrared
  USE   INFRARED SPECTRA

Spectra, Interstellar Microwave
  USE   INTERSTELLAR RADIATION
        MICROWAVE SPECTRA

Spectra, Line
  USE   LINE SPECTRA

Spectra, Lyman
  USE   LYMAN SPECTRA

Spectra, Mass
  USE   MASS SPECTRA

Spectra, Microwave
  USE   MICROWAVE SPECTRA

Spectra, Molecular
  USE   MOLECULAR SPECTRA

Spectra, Neutron
  USE   NEUTRON SPECTRA

Spectra, Noise
  USE   NOISE SPECTRA

Spectra, Oxygen
  USE   OXYGEN SPECTRA

Spectra, Plasma
  USE   PLASMA SPECTRA

Spectra, Power
  USE   POWER SPECTRA

Spectra, Radiation
  USE   RADIATION SPECTRA

Spectra, Radio
  USE   RADIO SPECTRA

Spectra, Raman
  USE   RAMAN SPECTRA

Spectra, Shock
  USE   SHOCK SPECTRA

Spectra, Solar
  USE   SOLAR SPECTRA

Spectra, Stellar
  USE   STELLAR SPECTRA

Spectra, UBV
  USE   UBV SPECTRA

Spectra, Ultraviolet
  USE   ULTRAVIOLET SPECTRA

Spectra, Vibrational
　　USE　VIBRATIONAL SPECTRA

Spectra, X Ray
　　USE　X RAY SPECTRA

Spectra 70 Computer, RCA
　　USE　RCA SPECTRA 70 COMPUTER

Spectral Absorption
　　USE　ABSORPTION SPECTRA

Spectral Analysis
　　USE　SPECTRUM ANALYSIS

SPECTRAL BANDS

SPECTRAL CORRELATION

SPECTRAL EMISSION

SPECTRAL ENERGY DISTRIBUTION

SPECTRAL LINE WIDTH

Spectral Lines
　　USE　LINE SPECTRA

SPECTRAL METHODS

Spectral Noise
　　USE　WHITE NOISE

SPECTRAL RECONNAISSANCE

SPECTRAL REFLECTANCE

SPECTRAL RESOLUTION

SPECTRAL SENSITIVITY

SPECTRAL SHIFT CONTROL

SPECTRAL SHIFT CONTROL REACTOR

SPECTRAL SIGNATURES

SPECTRAL THEORY

SPECTROGRAMS

SPECTROGRAPHS

Spectrographs, High Dispersion
　　USE　HIGH DISPERSION SPECTROGRAPHS

Spectrographs, Ultraviolet
　　USE　ULTRAVIOLET SPECTROMETERS

Spectrography, X Ray
　　USE　X RAY SPECTROSCOPY

SPECTROHELIOGRAPHS

Spectrohelioscopes
　　USE　SPECTROHELIOGRAPHS

Spectrometer, Solar Backscatter UV
　　USE　SOLAR BACKSCATTER UV
　　　　　SPECTROMETER

SPECTROMETERS

Spectrometers, Ebert
　　USE　EBERT SPECTROMETERS

Spectrometers, Fabry-Perot
　　USE　FABRY-PEROT SPECTROMETERS

Spectrometers, Filter Wheel Infrared
　　USE　FILTER WHEEL INFRARED
　　　　　SPECTROMETERS

Spectrometers, Gamma Ray
　　USE　GAMMA RAY SPECTROMETERS

Spectrometers, Infrared
　　USE　INFRARED SPECTROMETERS

Spectrometers, Ion
　　USE　MASS SPECTROMETERS

Spectrometers, Laser
　　USE　LASER SPECTROMETERS

Spectrometers, Mass
　　USE　MASS SPECTROMETERS

Spectrometers, Microwave
　　USE　MICROWAVE SPECTROMETERS

Spectrometers, Neutron
　　USE　NEUTRON SPECTROMETERS

Spectrometers, Retarding Ion Mass
　　USE　MASS SPECTROMETERS

Spectrometers, Solar
　　USE　SOLAR SPECTROMETERS

Spectrometers, Time Of Flight
　　USE　TIME OF FLIGHT SPECTROMETERS

Spectrometers, Triple Axis
　　USE　NEUTRON SPECTROMETERS

Spectrometers, Ultraviolet
　　USE　ULTRAVIOLET SPECTROMETERS

Spectrometry
　　USE　SPECTROMETERS

Spectrometry, Mass
　　USE　MASS SPECTROSCOPY

Spectrometry, X Ray
　　USE　X RAY SPECTROSCOPY

SPECTROPHOTOGRAPHY

SPECTROPHOTOMETERS

Spectrophotometers, Infrared
　　USE　INFRARED SPECTROPHOTOMETERS

Spectrophotometers, Ultraviolet
　　USE　ULTRAVIOLET SPECTROPHOTOMETERS

SPECTROPHOTOMETRY

Spectrophotometry, Stellar
　　USE　STELLAR SPECTROPHOTOMETRY

SPECTROPHOTOVOLTAICS

Spectropolarimeters
　　USE　POLARIMETERS

Spectropolarimetry Payload, X Ray
　　USE　EXPOS (SPACELAB PAYLOAD)

SPECTRORADIOMETERS

Spectroscopes
　　USE　SPECTROMETERS

SPECTROSCOPIC ANALYSIS

Spectroscopic Explorer, Far UV
　　USE　FAR UV SPECTROSCOPIC EXPLORER

SPECTROSCOPIC TELESCOPES

SPECTROSCOPY

Spectroscopy, Absorption
　　USE　ABSORPTION SPECTROSCOPY

Spectroscopy, Astronomical
　　USE　ASTRONOMICAL SPECTROSCOPY

Spectroscopy, Auger
　　USE　AUGER SPECTROSCOPY

Spectroscopy, Auroral
　　USE　AURORAL SPECTROSCOPY

Spectroscopy, Coherent Anti-Stokes Raman
　　USE　RAMAN SPECTROSCOPY

Spectroscopy, Electron
　　USE　ELECTRON SPECTROSCOPY

Spectroscopy, Flame
　　USE　FLAME SPECTROSCOPY

Spectroscopy, Gas
　　USE　GAS SPECTROSCOPY

Spectroscopy, Holographic
　　USE　HOLOGRAPHIC SPECTROSCOPY

Spectroscopy, Infrared
　　USE　INFRARED SPECTROSCOPY

Spectroscopy, Laser
　　USE　LASER SPECTROSCOPY

Spectroscopy, Magnetic
　　USE　MAGNETIC SPECTROSCOPY

Spectroscopy, Mass
　　USE　MASS SPECTROSCOPY

Spectroscopy, Molecular
　　USE　MOLECULAR SPECTROSCOPY

Spectroscopy, Nuclear Radiation
　　USE　NUCLEAR RADIATION SPECTROSCOPY

Spectroscopy, Optical Emission
　　USE　OPTICAL EMISSION SPECTROSCOPY

Spectroscopy, Optogalvanic
　　USE　OPTOGALVANIC SPECTROSCOPY

Spectroscopy, Photoacoustic
　　USE　PHOTOACOUSTIC SPECTROSCOPY

Spectroscopy, Photoelectron
　　USE　PHOTOELECTRON SPECTROSCOPY

Spectroscopy, Radio
　　USE　RADIO SPECTROSCOPY

Spectroscopy, Raman
　　USE　RAMAN SPECTROSCOPY

Spectroscopy, Ultrasonic
　　USE　ULTRASONIC SPECTROSCOPY

Spectroscopy, Ultraviolet
　　USE　ULTRAVIOLET SPECTROSCOPY

Spectroscopy, Vacuum
　　USE　VACUUM SPECTROSCOPY

Spectroscopy, X Ray
　　USE　X RAY SPECTROSCOPY

SPECTRUM ANALYSIS

Spectrum, Optical
　　USE　SPECTRA
　　　　　LIGHT (VISIBLE RADIATION)

Spectrum Transmission, Spread
　　USE　SPREAD SPECTRUM TRANSMISSION

Spectrum Utilization, Orbit
　　USE　ORBIT SPECTRUM UTILIZATION

Spectrum, Visible
　　USE　VISIBLE SPECTRUM

SPECULAR REFLECTION

SPEECH

SPEECH BASEBAND COMPRESSION

(Speech), Consonants
　　USE　CONSONANTS (SPEECH)

SPEECH DEFECTS

Speech Discrimination
　　USE　SPEECH RECOGNITION

SPEECH RECOGNITION

Speeches
　　USE　LECTURES

Speed
USE   VELOCITY

Speed, Air
USE   AIRSPEED

Speed Cameras, High
USE   HIGH SPEED CAMERAS

SPEED CONTROL

Speed, Critical
USE   CRITICAL VELOCITY

Speed Flight, High
USE   FLIGHT
          HIGH SPEED

Speed, Ground
USE   GROUND SPEED

Speed, High
USE   HIGH SPEED

Speed, Hypersonic
USE   HYPERSONIC SPEED

SPEED INDICATORS

Speed Integrated Circuits, Very High
USE   VHSIC (CIRCUITS)

Speed, Landing
USE   LANDING SPEED

Speed, Light
USE   LIGHT SPEED

Speed, Low
USE   LOW SPEED

Speed Photography, High
USE   HIGH SPEED PHOTOGRAPHY

Speed Propellers, Constant
USE   VARIABLE PITCH PROPELLERS

Speed Regulation
USE   SPEED CONTROL

SPEED REGULATORS

Speed, Rotor
USE   ROTOR SPEED

Speed, Sonic
USE   ACOUSTIC VELOCITY

Speed Stability, Low
USE   LOW SPEED STABILITY

Speed, Subsonic
USE   SUBSONIC SPEED

Speed, Tip
USE   TIP SPEED

Speed, Transonic
USE   TRANSONIC SPEED

Speed Transportation, High
USE   RAPID TRANSIT SYSTEMS

Speed Wind Tunnels, Low
USE   LOW SPEED WIND TUNNELS

Speedometers
USE   SPEED INDICATORS

Speeds, Supersonic
USE   SUPERSONIC SPEEDS

SPENT FUELS

Spermatocytes
USE   GAMETOCYTES

SPERMATOGENESIS

SPERMATOZOA

SPERT REACTORS

Sphalerite
USE   ZINCBLENDE

Sphere, Bio
USE   BIOSPHERE

Sphere, Celestial
USE   CELESTIAL SPHERE

Sphere, Chemo
USE   CHEMOSPHERE

Sphere, Chromo
USE   CHROMOSPHERE

Sphere, Exo
USE   EXOSPHERE

Sphere, Helio
USE   HELIOSPHERE

Sphere, Hetero
USE   HETEROSPHERE

Sphere, Homo
USE   HOMOSPHERE

Sphere, Iono
USE   IONOSPHERE

Sphere, Litho
USE   LITHOSPHERE

Sphere, Magneto
USE   MAGNETOSPHERE

Sphere, Meso
USE   MESOSPHERE

Sphere, Ozono
USE   OZONOSPHERE

Sphere, Photo
USE   PHOTOSPHERE

Sphere, Riemann
USE   RIEMANN MANIFOLD

Sphere, Strato
USE   STRATOSPHERE

Sphere, Thermo
USE   THERMOSPHERE

Sphere, Tropo
USE   TROPOSPHERE

SPHERES

Spheres, Concentric
USE   CONCENTRIC SPHERES

Spheres, Falling
USE   FALLING SPHERES

Spheres, Hemi
USE   HEMISPHERES

Spheres, Hyper
USE   HYPERSPHERES

Spheres, Plani
USE   PLANISPHERES

Spheres, Poincare
USE   POINCARE SPHERES

Spheres, Rotating
USE   ROTATING SPHERES

SPHERICAL ANTENNAS

SPHERICAL CAPS

SPHERICAL COORDINATES

SPHERICAL HARMONICS

SPHERICAL PLASMAS

SPHERICAL SHELLS

SPHERICAL TANKS

SPHERICAL WAVES

SPHEROIDS

Spheroids, Oblate
USE   OBLATE SPHEROIDS

Spheroids, Prolate
USE   PROLATE SPHEROIDS

SPHEROMAKS

SPHERULES

SPHERULITES

SPHINX

SPHYGMOGRAPHY

SPICULES

SPIDERS

Spike Antennas
USE   MONOPOLE ANTENNAS

SPIKE NOZZLES

SPIKE POTENTIALS

SPIKES

SPIKES (AERODYNAMIC CONFIGURATIONS)

SPIKING

SPILLING

SPIN

Spin, Aircraft
USE   AIRCRAFT SPIN

(Spin Alignment), Polarization
USE   POLARIZATION (SPIN ALIGNMENT)

Spin Coupling, Spin-
USE   SPIN-SPIN COUPLING

SPIN DECOUPLING

SPIN DYNAMICS

Spin, Electron
USE   ELECTRON SPIN

SPIN EXCHANGE

Spin Forging
USE   METAL SPINNING

SPIN GLASS

Spin, Isotopic
USE   ISOTOPIC SPIN

Spin, Nuclear
USE   NUCLEAR SPIN

Spin, Particle
USE   PARTICLE SPIN

SPIN REDUCTION

SPIN RESONANCE

Spin Resonance, Electron
USE   ELECTRON PARAMAGNETIC RESONANCE

Spin Rotation, Muon
USE   MUON SPIN ROTATION

Spin Scan Radiometer, Visible Infrared
USE   VISIBLE INFRARED SPIN SCAN
          RADIOMETER

Spin Space, U
USE   U SPIN SPACE

Spin Spacecraft, Dual
USE   DUAL SPIN SPACECRAFT

SPIN STABILIZATION

SPIN TEMPERATURE

SPIN TESTS

Spin Waves
USE   MAGNONS

SPIN-LATTICE RELAXATION

SPIN-ORBIT INTERACTIONS

SPIN-SPIN COUPLING

SPINACH

SPINAL CORD

SPINDLES

SPINE

SPINEL

SPINNERS

Spinning, Melt
USE   MELT SPINNING

Spinning, Metal
USE   METAL SPINNING

Spinning (Metallurgy)
USE   METAL SPINNING

SPINNING SOLID UPPER STAGE

SPINNING UNGUIDED ROCKET TRAJECTORY

Spinning, Wet
USE   WET SPINNING

SPINOR GROUPS

SPIRAL ANTENNAS

Spiral Antennas, Log
USE   LOG SPIRAL ANTENNAS

SPIRAL GALAXIES

SPIRAL WRAPPING

SPIRALS

SPIRALS (CONCENTRATORS)

SPIROMETERS

SPITSBERGEN (NORWAY)

SPLASHING

SPLEEN

SPLICING

SPLINE FUNCTIONS

SPLINES

SPLINTS

SPLIT FLAPS

Splits (Geology)
USE   GEOLOGICAL FAULTS

Splitters, Beam
USE   BEAM SPLITTERS

SPLITTING

SPODUMENE

SPOILER SLOT AILERONS

SPOILERS

SPOKES

SPONGES (MATERIALS)

SPONTANEOUS COMBUSTION

SPONTANEOUS EMISSION

SPOOLS

SPORADIC E LAYER

SPORADIC METEOROIDS

SPORES

Spores, Micro
USE   MICROSPORES

SPORTS MEDICINE

SPOT (FRENCH SATELLITE)

Spot, Jupiter Red
USE   JUPITER RED SPOT

Spot Scanners, Flying
USE   FLYING SPOT SCANNERS

SPOT WELDS

Spots, Star
USE   STARSPOTS

Spots, Sun
USE   SUNSPOTS

SPRAY CHARACTERISTICS

SPRAY CONDENSERS

SPRAY INGESTION

SPRAY NOZZLES

Spray Tests, Salt
USE   SALT SPRAY TESTS

SPRAYED COATINGS

Sprayed Protective Coatings
USE   SPRAYED COATINGS
       PROTECTIVE COATINGS

SPRAYERS

SPRAYING

Spraying Apparatus
USE   SPRAYERS

Spraying, Arc
USE   ARC SPRAYING

Spraying, Flame
USE   FLAME SPRAYING

Spraying, Metal
USE   METAL SPRAYING

Spraying, Plasma
USE   PLASMA SPRAYING

Spraying, Plasma Arc
USE   ARC SPRAYING

Sprays
USE   SPRAYERS

Sprays, Fuel
USE   FUEL SPRAYS

Sprays, Propellant
USE   PROPELLANT SPRAYS

SPREAD F

Spread Functions, Point
USE   POINT SPREAD FUNCTIONS

SPREAD REFLECTION

SPREAD SPECTRUM TRANSMISSION

SPREADING

SPRING (SEASON)

SPRINGS (ELASTIC)

SPRINGS (WATER)

SPRINKLING

SPRINT MISSILE

SPUR (Astronomy), North Polar
USE   NORTH POLAR SPUR (ASTRONOMY)

SPUR (Reactors)
USE   SPACE POWER UNIT REACTORS

SPURT (Trajectories)
USE   SPINNING UNGUIDED ROCKET
       TRAJECTORY

SPUTNIK SATELLITES

SPUTNIK 1 SATELLITE

SPUTNIK 2 SATELLITE

SPUTNIK 3 SATELLITE

SPUTNIK 4 SATELLITE

SPUTNIK 5 SATELLITE

SPUTTERING

SPUTTERING GAGES

Sputtering, Magnetron
USE   MAGNETRON SPUTTERING

SQUALLS

SQUAMA

Square Errors, Root-Mean-
USE   ROOT-MEAN-SQUARE ERRORS

Square Method, Latin
USE   LATIN SQUARE METHOD

Square Values, Mean
USE   MEAN SQUARE VALUES

SQUARE WAVES

SQUARE WELLS

SQUARES (MATHEMATICS)

Squares Method, Least
USE   LEAST SQUARES METHOD

SQUEEZE FILMS

Squeezing
USE   COMPRESSING

SQUELCH CIRCUITS

Squib, XM-6
USE   SQUIBS

Squib, XM-8
USE   SQUIBS

SQUIBS

SQUID (DETECTORS)

SQUID PROJECT

SQUIRRELS

Squirrels, Ground
USE   GROUND SQUIRRELS

Sr
USE   STRONTIUM

SR (Reactors)
USE   SATURABLE REACTORS

SR-N2 Ground Effect Machine
USE   WESTLAND GROUND EFFECT MACHINES

SR-N2 Ground Effect Machine, Westland
USE   WESTLAND GROUND EFFECT MACHINES

SR-N2 Hovercraft, Westland
USE   WESTLAND GROUND EFFECT MACHINES

SR-N3 Ground Effect Machine
USE   WESTLAND GROUND EFFECT MACHINES

SR-N3 Ground Effect Machine, Westland
USE   WESTLAND GROUND EFFECT MACHINES

SR-N3 Hovercraft, Westland
USE   WESTLAND GROUND EFFECT MACHINES

SR-N5 Ground Effect Machine
USE   WESTLAND GROUND EFFECT MACHINES

SR-N5 Ground Effect Machine, Westland
USE   WESTLAND GROUND EFFECT MACHINES

SRE Reactor
USE   SODIUM REACTOR EXPERIMENT

SRET SATELLITES

SRET 1 SATELLITE

SRET 2 SATELLITE

SRI LANKA

SS-11 MISSILE

SSGS (Standardized Space Guidance)
USE   STANDARDIZED SPACE GUIDANCE

SSUS-A
USE   SPACE SHUTTLE UPPER STAGE A

SSUS-D
USE   SPACE SHUTTLE UPPER STAGE D

ST LAWRENCE VALLEY (NORTH AMERICA)

ST LOUIS-KANSAS CITY CORRIDOR (MO)

St Venant Flexure Problem
USE   SAINT VENANT PRINCIPLE

STABILITY

Stability, Acoustic
USE   FREQUENCY STABILITY

Stability, Aerodynamic
USE   AERODYNAMIC STABILITY

Stability, Aircraft
USE   AIRCRAFT STABILITY

Stability, Attitude
USE   ATTITUDE STABILITY

STABILITY AUGMENTATION

Stability, Boundary Layer
USE   BOUNDARY LAYER STABILITY

Stability, Combustion
USE   COMBUSTION STABILITY

Stability, Control
USE   CONTROL STABILITY

Stability, Controlled
USE   CONTROL

STABILITY DERIVATIVES

Stability, Dimensional
USE   DIMENSIONAL STABILITY

Stability, Directional
USE   DIRECTIONAL STABILITY

Stability, Dynamic
USE   DYNAMIC STABILITY

Stability, Elastic
USE   DAMPING

Stability, Flame
USE   FLAME STABILITY

Stability, Flow
USE   FLOW STABILITY

Stability, Flying Platform
USE   FLYING PLATFORMS
AERODYNAMIC STABILITY

Stability, Frequency
USE   FREQUENCY STABILITY

Stability, Gyroscopic
USE   GYROSCOPIC STABILITY

Stability, Hovering
USE   HOVERING STABILITY

Stability, Hydrodynamic
USE   FLOW STABILITY

Stability, Hydromagnetic
USE   MAGNETOHYDRODYNAMIC STABILITY

Stability, Interface
USE   INTERFACE STABILITY

Stability, Laser
USE   LASER STABILITY

Stability, Lateral
USE   LATERAL STABILITY

Stability, Longitudinal
USE   LONGITUDINAL STABILITY

Stability, Low Speed
USE   LOW SPEED STABILITY

Stability, Magnetohydrodynamic
USE   MAGNETOHYDRODYNAMIC STABILITY

Stability, Motion
USE   MOTION STABILITY

Stability, Numerical
USE   NUMERICAL STABILITY

Stability, Plasma
USE   MAGNETOHYDRODYNAMIC STABILITY

Stability, Rotary
USE   ROTARY STABILITY

Stability, Shell
USE   SHELL STABILITY

Stability, Spacecraft
USE   SPACECRAFT STABILITY

Stability, Static
USE   STATIC STABILITY

Stability, Storage
USE   STORAGE STABILITY

Stability, Structural
USE   STRUCTURAL STABILITY

Stability, Surface
USE   SURFACE STABILITY

Stability, Systems
USE   SYSTEMS STABILITY

STABILITY TESTS

Stability Tests, Flight
USE   FLIGHT STABILITY TESTS

Stability Tests, Wind Tunnel
USE   WIND TUNNEL STABILITY TESTS

Stability, Thermal
USE   THERMAL STABILITY

STABILIZATION

Stabilization, De
USE   DESTABILIZATION

Stabilization, Missile
USE   MISSILE CONTROL
STABILIZATION

Stabilization, Signal
USE   SIGNAL STABILIZATION

Stabilization, Spin
USE   SPIN STABILIZATION

Stabilization, Three Axis
USE   THREE AXIS STABILIZATION

STABILIZED PLATFORMS

STABILIZERS

STABILIZERS (AGENTS)

Stabilizers, Current
USE   CURRENT REGULATORS

STABILIZERS (FLUID DYNAMICS)

Stabilizers, Gyro
USE   GYROSTABILIZERS

Stabilizers, Horizontal
USE   STABILIZERS (FLUID DYNAMICS)

Stabilizers, Vertical
USE   STABILIZERS (FLUID DYNAMICS)

STABLE OSCILLATIONS

Stack, Apollo Short
USE   APOLLO SHORT STACK

STACKING FAULT ENERGY

Stacking Faults
USE   CRYSTAL DEFECTS

STACKS

STADAN (Satellite Tracking Network)
USE   STDN (NETWORK)

STADIMETERS

Stage A, Space Shuttle Upper
USE   SPACE SHUTTLE UPPER STAGE A

Stage D, Space Shuttle Upper
USE   SPACE SHUTTLE UPPER STAGE D

Stage, Inertial Upper
USE   INERTIAL UPPER STAGE

Stage, Lunar Module Ascent
USE   LUNAR MODULE ASCENT STAGE

Stage Plasma Engines, Two
USE   TWO STAGE PLASMA ENGINES

Stage Rocket Engines, Upper
USE   UPPER STAGE ROCKET ENGINES

Stage Rocket Vehicles, Single
USE   SINGLE STAGE ROCKET VEHICLES

Stage, Saturn S-1
USE   SATURN S-1 STAGE

Stage, Saturn S-1B
USE   SATURN S-1B STAGE

Stage, Saturn S-1C
USE   SATURN S-1C STAGE

Stage, Saturn S-2
USE   SATURN S-2 STAGE

Stage, Saturn S-4
USE   SATURN S-4 STAGE

Stage, Saturn S-4B
USE   SATURN S-4B STAGE

STAGE SEPARATION

Stage, Space Shuttle Ascent
USE   SPACE SHUTTLE ASCENT STAGE

Stage, Spinning Solid Upper
USE   SPINNING SOLID UPPER STAGE

Stage (Sts), Interim Upper
USE   INERTIAL UPPER STAGE

Stage To Orbit Vehicles, Single
USE   SINGLE STAGE TO ORBIT VEHICLES

Stage Turbines, Two
USE   TWO STAGE TURBINES

Stages, Saturn
USE   SATURN STAGES

Stages, Space Shuttle Upper
USE   SPACE SHUTTLE UPPER STAGES

Stages (Spacecraft), Expendable
USE   EXPENDABLE STAGES (SPACECRAFT)

Stages (Spacecraft), Interim
USE   INTERIM STAGES (SPACECRAFT)

STAGGERING

Staging (Rockets)
USE   STAGE SEPARATION

STAGNATION FLOW

STAGNATION POINT

STAGNATION PRESSURE

Stagnation Region
USE   STAGNATION POINT

STAGNATION TEMPERATURE

STAINING

STAINLESS STEELS

Stainless Steels, Austenitic
USE   AUSTENITIC STAINLESS STEELS

Stainless Steels, Ferritic
USE   FERRITIC STAINLESS STEELS

Stainless Steels, Martensitic
USE   MARTENSITIC STAINLESS STEELS

Staircases
USE   STAIRWAYS

STAIRSTEPS

STAIRWAYS

STALLING

Stalling, Aerodynamic
USE   AERODYNAMIC STALLING

Stalls, Rotating
USE   ROTATING STALLS

STAMPING

Standard Atmospheres
USE   REFERENCE ATMOSPHERES

STANDARD DEVIATION

Standard Launch Vehicle 3
USE   ATLAS SLV-3 LAUNCH VEHICLE

STANDARD LAUNCH VEHICLE 5

STANDARD LAUNCH VEHICLES

STANDARDIZATION

STANDARDIZED SPACE GUIDANCE

(Standardized Space Guidance), SSGS
USE   STANDARDIZED SPACE GUIDANCE

STANDARDS

Standards, Frequency
USE   FREQUENCY STANDARDS

(Standards), References
USE   STANDARDS

STANDING WAVE RATIOS

STANDING WAVES

(Standing Waves), Modes
USE   MODES (STANDING WAVES)

(Standing Waves), Nodes
USE   NODES (STANDING WAVES)

Stands
USE   SUPPORTS

Stands, Test
USE   TEST STANDS

STANNATES

STANNIDES

Stannides, Niobium
USE   NIOBIUM STANNIDES

STANTON NUMBER

STAPHYLOCOCCUS

Star Aircraft, Jet
USE   C-140 AIRCRAFT

Star Aircraft, Shooting
USE   T-33 AIRCRAFT

Star Aircraft, Warning
USE   EC-121 AIRCRAFT

Star Cluster, Virgo
USE   VIRGO GALACTIC CLUSTER

STAR CLUSTERS

Star Clusters, Praesepe
USE   PRAESEPE STAR CLUSTERS

STAR DISTRIBUTION

Star Fields
USE   STAR DISTRIBUTION

Star, Omicron Ceti
USE   OMICRON CETI STAR

Star Rocket Vehicle, Hyla-
USE   HYLA-STAR ROCKET VEHICLE

Star Tracker, CCD
USE   CCD STAR TRACKER

(Star Tracker), Stellar
USE   CCD STAR TRACKER

STAR TRACKERS

Star Tracking
USE   STAR TRACKERS

Star, Van Biesbroeck
USE   VAN BIESBROECK STAR

Star, Zeta Aurigae
USE   ZETA AURIGAE STAR

Star 100 Computer, CDC
USE   CDC STAR 100 COMPUTER

STARCHES

Starfighter Aircraft
USE   F-104 AIRCRAFT

STARK EFFECT

STARLAB

Starlifter Aircraft
USE   C-141 AIRCRAFT

STARS

Stars, A
USE   A STARS

Stars, B
USE   B STARS

Stars, Binary
USE   BINARY STARS

Stars, Blue
USE   BLUE STARS

Stars, Carbon
USE   CARBON STARS

Stars, Companion
USE   COMPANION STARS

Stars, Cool
USE   COOL STARS

Stars, Double
USE   DOUBLE STARS

Stars, Dwarf
USE   DWARF STARS

Stars, Early
USE   EARLY STARS

Stars, Eclipsing Binary
USE   ECLIPSING BINARY STARS

Stars, Flare
USE   FLARE STARS

Stars, Giant
USE   GIANT STARS

Stars, Helium
USE   B STARS

Stars, Horizontal Branch
USE   HORIZONTAL BRANCH STARS

Stars, Hot
USE   HOT STARS

Stars, Infrared
USE   INFRARED STARS

Stars, Lambda Tauri
USE   LAMBDA TAURI STARS

Stars, Late
USE   LATE STARS

Stars, M
USE   M STARS

Stars, Magnetic
USE   MAGNETIC STARS

Stars, Main Sequence
USE   MAIN SEQUENCE STARS

STARS (MATHEMATICS)

Stars, Metallic
USE　METALLIC STARS

Stars, Neutron
USE　NEUTRON STARS

Stars, O
USE　O STARS

Stars, Peculiar
USE　PECULIAR STARS

Stars, Pre-Main Sequence
USE　PRE-MAIN SEQUENCE STARS

Stars, Proto
USE　PROTOSTARS

Stars, Radio
USE　RADIO STARS

Stars, Red Dwarf
USE　RED DWARF STARS

Stars, Red Giant
USE　RED GIANT STARS

Stars, Reference
USE　REFERENCE STARS

Stars, S
USE　S STARS

Stars, Subdwarf
USE　SUBDWARF STARS

Stars, Subgiant
USE　SUBGIANT STARS

Stars, Supergiant
USE　SUPERGIANT STARS

Stars, Supermassive
USE　SUPERMASSIVE STARS

Stars, Symbiotic
USE　SYMBIOTIC STARS

Stars, T Tauri
USE　T TAURI STARS

Stars, UV Ceti
USE　FLARE STARS

Stars, Variable
USE　VARIABLE STARS

Stars, W-R
USE　WOLF-RAYET STARS

Stars, White Dwarf
USE　WHITE DWARF STARS

Stars, Wolf-Rayet
USE　WOLF-RAYET STARS

STARSAT TELESCOPE

STARSITE PROGRAM

STARSPOTS

Start, Air
USE　AIR START

STARTERS

Starters, Engine
USE　ENGINE STARTERS

STARTING

Startup Tests, Reactor
USE　REACTOR STARTUP TESTS

State), Carrier Density (Solid
USE　CARRIER DENSITY (SOLID STATE)

State), Carrier Transport (Solid
USE　CARRIER TRANSPORT (SOLID STATE)

State Creep, Steady
USE　STEADY STATE CREEP

State Devices, Solid
USE　SOLID STATE DEVICES

State), Energy Gaps (Solid
USE　ENERGY GAPS (SOLID STATE)

State Equations
USE　EQUATIONS OF STATE

State, Equations Of
USE　EQUATIONS OF STATE

STATE ESTIMATION

State Flow, Steady
USE　EQUILIBRIUM FLOW

State, Ground
USE　GROUND STATE

State, Hugoniot Equation Of
USE　HUGONIOT EQUATION OF STATE

State Lasers, Solid
USE　SOLID STATE LASERS

State Machines, Finite-
USE　TURING MACHINES

State, Metastable
USE　METASTABLE STATE

State Physics, Solid
USE　SOLID STATE PHYSICS

State, Rapid Eye Movement
USE　RAPID EYE MOVEMENT STATE

State), Self Diffusion (Solid
USE　SELF DIFFUSION (SOLID STATE)

State, Solid
USE　SOLID STATE

State, Steady
USE　STEADY STATE

State, Triplet
USE　ATOMIC ENERGY LEVELS

State, Unsteady
USE　UNSTEADY STATE

STATE VECTORS

States), Armed Forces (United
USE　ARMED FORCES (UNITED STATES)

States, Electron
USE　ELECTRON STATES

States, Excited
USE　EXCITATION

States, Quasi-Steady
USE　QUASI-STEADY STATES

States, Sea
USE　SEA STATES

States, United
USE　UNITED STATES

States), USA (United
USE　UNITED STATES

STATIC AERODYNAMIC CHARACTERISTICS

STATIC ALTERNATORS

STATIC CHARACTERISTICS

STATIC DEFORMATION

STATIC DISCHARGERS

STATIC ELECTRICITY

STATIC FIRING

STATIC FRICTION

STATIC INVERTERS

STATIC LOADS

STATIC MODELS

STATIC PRESSURE

STATIC STABILITY

STATIC TESTS

STATIC THRUST

STATICS

Statics, Aero
USE　AEROSTATICS

Statics, Elasto
USE　ELASTOSTATICS

Statics, Electro
USE　ELECTROSTATICS

Statics, Hemo
USE　HEMOSTATICS

Statics, Hydro
USE　HYDROSTATICS

Statics, Magneto
USE　MAGNETOSTATICS

Statics, Magnetohydro
USE　MAGNETOHYDROSTATICS

Station, Halo Orbit Space
USE　HALO ORBIT SPACE STATION

Station, Salyut Space
USE　SALYUT SPACE STATION

Station Systems, Integrated Global Ocean
USE　INTEGRATED GLOBAL OCEAN STATION
　　　　SYSTEMS

Station (Unmanned), SKYLAB Space
USE　SKYLAB 1

STATIONARY ORBITS

STATIONKEEPING

STATIONS

Stations, Automatic Weather
USE　AUTOMATIC WEATHER STATIONS

Stations, Crew
USE　CREW STATIONS

Stations, Crew Experiment
USE　CREW EXPERIMENT STATIONS

Stations, Crew Observation
USE　CREW OBSERVATION STATIONS

Stations, Earth Orbiting Space
USE　EOSS

Stations, Ground
USE　GROUND STATIONS

Stations, Hydroelectric Power
USE　HYDROELECTRIC POWER STATIONS

Stations, Hydropower
USE　HYDROELECTRIC POWER STATIONS

Stations, Manned Orbital Space
USE　ORBITAL SPACE STATIONS

Stations, Meteorological
USE　WEATHER STATIONS

Stations), MOSS (Space
USE   ORBITAL SPACE STATIONS

Stations, Ocean Data
USE   OCEAN DATA ACQUISITIONS SYSTEMS

Stations, Orbital Space
USE   ORBITAL SPACE STATIONS

Stations, Orbiting Lunar
USE   ORBITING LUNAR STATIONS

Stations, Payload
USE   PAYLOAD STATIONS

Stations, Satellite Solar Power
USE   SATELLITE SOLAR POWER STATIONS

Stations, Self Deploying Space
USE   SELF ERECTING DEVICES
       SPACE STATIONS

Stations, Space
USE   SPACE STATIONS

Stations, Tracking
USE   TRACKING STATIONS

Stations, Weather
USE   WEATHER STATIONS

STATISTICAL ANALYSIS

Statistical Analysis, Multivariate
USE   MULTIVARIATE STATISTICAL ANALYSIS

Statistical Communication Theory
USE   COMMUNICATION THEORY

STATISTICAL CORRELATION

STATISTICAL DECISION THEORY

STATISTICAL DISTRIBUTIONS

STATISTICAL MECHANICS

Statistical Moments
USE   DISTRIBUTION MOMENTS

Statistical Probability
USE   PROBABILITY THEORY

STATISTICAL TESTS

STATISTICAL WEATHER FORECASTING

STATISTICS

Statistics, Bayesian
USE   BAYES THEOREM

Statistics, Bose-Einstein
USE   QUANTUM STATISTICS

(Statistics), Discriminant Analysis
USE   DISCRIMINANT ANALYSIS (STATISTICS)

(Statistics), Entropy
USE   ENTROPY (STATISTICS)

Statistics, Fermi-Dirac
USE   FERMI-DIRAC STATISTICS

(Statistics), Median
USE   MEDIAN (STATISTICS)

(Statistics), Mode
USE   MODE (STATISTICS)

Statistics, Nonparametric
USE   NONPARAMETRIC STATISTICS

(Statistics), Normalizing
USE   NORMALIZING (STATISTICS)

(Statistics), Outliers
USE   OUTLIERS (STATISTICS)

Statistics, Quantum
USE   QUANTUM STATISTICS

(Statistics), Regression
USE   REGRESSION ANALYSIS

(Statistics), Variance
USE   VARIANCE (STATISTICS)

STATOR BLADES

STATORS

Stays
USE   GUY WIRES

STDN (NETWORK)

STEADY FLOW

STEADY STATE

STEADY STATE CREEP

Steady State Flow
USE   EQUILIBRIUM FLOW

Steady States, Quasi-
USE   QUASI-STEADY STATES

STEAM

STEAM FLOW

Steam Generators
USE   BOILERS

STEAM TURBINES

STEARATES

STEAROTHERMOPHILUS

Steatite
USE   TALC

Steel, Bainitic
USE   BAINITIC STEEL

Steel Missile, Blue
USE   BLUE STEEL MISSILE

STEEL STRUCTURES

STEELS

Steels, Austenitic Stainless
USE   AUSTENITIC STAINLESS STEELS

Steels, Carbon
USE   CARBON STEELS

Steels, Chromium
USE   CHROMIUM STEELS

Steels, Ferritic Stainless
USE   FERRITIC STAINLESS STEELS

Steels, High Strength
USE   HIGH STRENGTH STEELS

Steels, Low Alloy
USE   HIGH STRENGTH STEELS

Steels, Low Carbon
USE   LOW CARBON STEELS

Steels, Maraging
USE   MARAGING STEELS

Steels, Martensitic Stainless
USE   MARTENSITIC STAINLESS STEELS

Steels, Nickel
USE   NICKEL STEELS

Steels, Stainless
USE   STAINLESS STEELS

Steep Gradient Aircraft
USE   V/STOL AIRCRAFT

Steepest Ascent Method
USE   STEEPEST DESCENT METHOD

STEEPEST DESCENT METHOD

Steepness
USE   SLOPES

STEERABLE ANTENNAS

Steerable Antennas, Inertialess
USE   INERTIALESS STEERABLE ANTENNAS

STEERING

Steering Rockets
USE   CONTROL ROCKETS

STEFAN-BOLTZMANN LAW

STELLAR ACTIVITY

STELLAR ATMOSPHERES

STELLAR COLOR

STELLAR COMPOSITION

STELLAR CORES

STELLAR CORONAS

Stellar Doppler Shift
USE   EXTRATERRESTRIAL RADIATION
       DOPPLER EFFECT

STELLAR ENVELOPES

STELLAR EVOLUTION

Stellar Fields
USE   STAR DISTRIBUTION

STELLAR FLARES

STELLAR GRAVITATION

STELLAR LUMINOSITY

STELLAR MAGNETIC FIELDS

STELLAR MAGNITUDE

STELLAR MASS

STELLAR MASS ACCRETION

STELLAR MASS EJECTION

STELLAR MODELS

STELLAR MOTIONS

STELLAR OCCULTATION

STELLAR ORBITS

STELLAR OSCILLATIONS

STELLAR PARALLAX

STELLAR PHYSICS

STELLAR RADIATION

Stellar Radio Sources, Quasi-
USE   QUASARS

STELLAR ROTATION

STELLAR SPECTRA

STELLAR SPECTROPHOTOMETRY

Stellar (Star Tracker)
USE   CCD STAR TRACKER

STELLAR STRUCTURE

STELLAR TEMPERATURE

STELLAR WINDS

STELLARATORS

Stellite, Haynes
USE   STELLITE (TRADEMARK)

STELLITE (TRADEMARK)

Stem, Brain
USE   BRAIN STEM

STEMS

STENCIL PROCESSES

Step Faults
USE   GEOLOGICAL FAULTS

STEP FUNCTIONS

STEP RECOVERY DIODES

STEPPES

STEPPING MOTORS

STEPPING SWITCHES

STEPS

Steps, Backward Facing
USE   BACKWARD FACING STEPS

Steps, Rearward Facing
USE   BACKWARD FACING STEPS

Steps, Stair
USE   STAIRSTEPS

STEREOCHEMISTRY

Stereography
USE   STEREOPHOTOGRAPHY

STEREOPHONICS

STEREOPHOTOGRAPHY

Stereoscopic Photography
USE   STEREOPHOTOGRAPHY

STEREOSCOPIC VISION

STEREOSCOPY

STEREOTELEVISION

STERILIZATION

Sterilization, Chemical
USE   CHEMICAL STERILIZATION

STERILIZATION EFFECTS

Sterilization, Spacecraft
USE   SPACECRAFT STERILIZATION

Sterns
USE   AFTERBODIES

STERNUM

STEROIDS

Steroids, Cortico
USE   CORTICOSTEROIDS

STETHOSCOPES

Sticks, Control
USE   CONTROL STICKS

STIELTJES INTEGRAL

Stiff Structures
USE   RIGID STRUCTURES

STIFFENING

STIFFNESS

STIFFNESS MATRIX

STIGMATISM

STILBENE

STILLS

STIMULANTS

Stimulants, Central Nervous System
USE   CENTRAL NERVOUS SYSTEM STIMULANTS

STIMULATED EMISSION

STIMULATED EMISSION DEVICES

STIMULATION

Stimulation, Self
USE   SELF STIMULATION

Stimulation, Sensory
USE   SENSORY STIMULATION

STIMULI

Stimuli, Auditory
USE   AUDITORY STIMULI

Stimuli, Caloric
USE   CALORIC STIMULI

Stimuli, Electric
USE   ELECTRIC STIMULI

Stimuli, Subliminal
USE   SUBLIMINAL STIMULI

Stimuli, Visual
USE   VISUAL STIMULI

STIRLING CYCLE

STIRRING

STISHOVITE

STOCHASTIC PROCESSES

STOCKPILING

STOICHIOMETRY

Stokes Equation, Navier-
USE   NAVIER-STOKES EQUATION

STOKES FLOW

STOKES LAW

STOKES LAW (FLUID MECHANICS)

STOKES LAW OF RADIATION

Stokes Raman Spectroscopy, Coherent Anti-
USE   RAMAN SPECTROSCOPY

STOKES THEOREM (VECTOR CALCULUS)

STOKES-BELTRAMI EQUATION

STOL Aircraft
USE   SHORT TAKEOFF AIRCRAFT

STOL Transport Rsch Airplane, Experimental
USE   QUESTOL

STOMACH

Stones (Rocks)
USE   ROCKS

STONY METEORITES

Stopcocks
USE   COCKS

(Stoppers), Seals
USE   SEALS (STOPPERS)

STOPPING

STOPPING POWER

Storability, Propellant
USE   PROPELLANT STORABILITY

STORABLE PROPELLANTS

STORAGE

STORAGE BATTERIES

Storage, Buffer
USE   BUFFER STORAGE

Storage, Core
USE   CORE STORAGE

Storage, Cryogenic
USE   CRYOGENIC STORAGE

Storage, Cryogenic Computer
USE   CRYOGENIC COMPUTER STORAGE

Storage, Cryogenic Fluid
USE   CRYOGENIC FLUID STORAGE

Storage, Data
USE   DATA STORAGE

Storage), Delay Lines (Computer
USE   DELAY LINES (COMPUTER STORAGE)

Storage Devices, Computer
USE   COMPUTER STORAGE DEVICES

Storage Devices, Energy
USE   ENERGY STORAGE

Storage, Document
USE   DOCUMENT STORAGE

Storage, Electric Energy
USE   ELECTRIC ENERGY STORAGE

Storage, Energy
USE   ENERGY STORAGE

Storage, Heat
USE   HEAT STORAGE

Storage, Ion
USE   ION STORAGE

Storage, Machine
USE   COMPUTER STORAGE DEVICES
       CORE STORAGE

Storage, Magnetic
USE   MAGNETIC STORAGE

Storage, Magnetic Energy
USE   MAGNETIC ENERGY STORAGE

Storage Materials, Optical Data
USE   OPTICAL DATA STORAGE MATERIALS

Storage, Missile
USE   MISSILE STORAGE

Storage), Optical Memory (Data
USE   OPTICAL MEMORY (DATA STORAGE)

Storage, Propellant
USE   PROPELLANT STORAGE

STORAGE RINGS (PARTICLE ACCELERATORS)

Storage), Silos (Missile
USE   MISSILE SILOS

Storage), Solar Ponds (Heat
USE   SOLAR PONDS (HEAT STORAGE)

Storage, Space
USE   SPACE STORAGE

STORAGE STABILITY

STORAGE TANKS

Storage, Thermal Energy
USE   HEAT STORAGE

Storage, Underground
USE   UNDERGROUND STORAGE

Store Release
USE   EXTERNAL STORE SEPARATION

Store Separation, External
USE   EXTERNAL STORE SEPARATION

Stores, External
USE   EXTERNAL STORES

Stores), Pods (External
USE   PODS (EXTERNAL STORES)

Stores, Wing-Fuselage
USE   WING-FUSELAGE STORES

Storm Commencements, Sudden
USE   SUDDEN STORM COMMENCEMENTS

STORM DAMAGE

STORM ENHANCEMENT

STORM SUPPRESSION

STORM SURGES

STORMS

Storms, Dust
USE   DUST STORMS

Storms, Geomagnetic
USE   MAGNETIC STORMS

Storms, Ionospheric
USE   IONOSPHERIC STORMS

Storms, Magnetic
USE   MAGNETIC STORMS

STORMS (METEOROLOGY)

Storms, Noise
USE   NOISE STORMS

Storms Observing Satellite, Severe
USE   STORMSAT SATELLITE

Storms Project, National Severe
USE   NATIONAL SEVERE STORMS PROJECT

Storms, Rain
USE   RAINSTORMS

Storms, Snow
USE   SNOWSTORMS

Storms, Solar
USE   SOLAR STORMS

Storms, Thunder
USE   THUNDERSTORMS

Storms, Tropical
USE   TROPICAL STORMS

STORMSAT SATELLITE

Stoss-And-Lee Topography
USE   GLACIAL DRIFT

STOWAGE (ONBOARD EQUIPMENT)

Straight Wings
USE   RECTANGULAR WINGS

Strain Aging
USE   PRECIPITATION HARDENING

Strain, Axial
USE   AXIAL STRAIN

Strain Diagrams, Stress-
USE   STRESS-STRAIN DIAGRAMS

Strain Distribution
USE   STRESS CONCENTRATION

Strain Distribution, Stress-
USE   STRESS CONCENTRATION

STRAIN ENERGY METHODS

Strain Fatigue
USE   FATIGUE (MATERIALS)

STRAIN GAGE ACCELEROMETERS

STRAIN GAGE BALANCES

STRAIN GAGES

STRAIN HARDENING

Strain, Interfacial
USE   INTERFACIAL TENSION

STRAIN MEASUREMENT

Strain, Plane
USE   PLANE STRAIN

STRAIN RATE

Strain Relationships, Stress-
USE   STRESS-STRAIN RELATIONSHIPS

Strain, Shear
USE   SHEAR STRAIN

Strain Softening
USE   PLASTIC DEFORMATION

Strain, Structural
USE   STRUCTURAL STRAIN

Strain, Uniaxial
USE   AXIAL STRAIN

Strain, Volumetric
USE   VOLUMETRIC STRAIN

Strain-Time Relations, Stress-
USE   STRESS-STRAIN-TIME RELATIONS

Strait, Torres
USE   TORRES STRAIT

STRAITS

STRAKES

STRANDS

STRANGE ATTRACTORS

STRANGENESS

STRAPDOWN INERTIAL GUIDANCE

STRAPS

STRATA

STRATEGIC MATERIALS

STRATEGY

STRATIFICATION

Stratification, Atmospheric
USE   ATMOSPHERIC STRATIFICATION

STRATIFIED FLOW

Stratified Layers
USE   STRATA

STRATIGRAPHY

STRATOCUMULUS CLOUDS

Stratofortress Aircraft
USE   B-52 AIRCRAFT

Stratojet Aircraft
USE   B-47 AIRCRAFT

STRATOPAUSE

STRATOSCOPE TELESCOPES

Stratoscope 1 Telescope
USE   STRATOSCOPE TELESCOPES

Stratoscope 2 Telescope
USE   STRATOSCOPE TELESCOPES

STRATOSPHERE

STRATOSPHERE RADIATION

Stratospheric Aerosol & Gas Experiment
USE   SAGE SATELLITE

Stratotanker Aircraft
USE   C-135 AIRCRAFT

STRATUS CLOUDS

STREAK CAMERAS

Streak Launch Vehicle, Blue
USE   BLUE STREAK LAUNCH VEHICLE

Streak Missile, Blue
USE   BLUE STREAK MISSILE

STREAK PHOTOGRAPHY

Stream Control Engines, Variable
USE   VARIABLE STREAM CONTROL ENGINES

Stream Effects, Free
USE   FREE FLOW

STREAM FUNCTIONS (FLUIDS)

Stream, Gulf
USE   GULF STREAM

Streaming, Acoustic
USE   ACOUSTIC STREAMING

Streamline Flow
USE   LAMINAR FLOW

STREAMLINED BODIES

STREAMLINING

STREAMS

Streams, Free
USE   FREE FLOW

Streams, Gas
USE   GAS STREAMS

Streams (Meteorology), Jet
USE   JET STREAMS (METEOROLOGY)

Streams, Slip
USE   SLIPSTREAMS

Streams, Solar
USE   SOLAR CORPUSCULAR RADIATION

Street, Karman Vortex
USE   KARMAN VORTEX STREET

STREETS

Streets, Vortex
USE   VORTEX STREETS

STRENGTH

Strength Alloys, High
USE   HIGH STRENGTH ALLOYS

Strength, Cold
USE   COLD STRENGTH

Strength, Compressive
USE   COMPRESSIVE STRENGTH

Strength, Creep
USE   CREEP STRENGTH

Strength, Creep Rupture
USE   CREEP RUPTURE STRENGTH

Strength, Elastic
USE   PROPORTIONAL LIMIT

Strength, Electric Field
USE   ELECTRIC FIELD STRENGTH

Strength, Fiber
USE   FIBER STRENGTH

Strength, Field
USE   FIELD STRENGTH

Strength, Fracture
USE   FRACTURE STRENGTH

Strength, High
USE   HIGH STRENGTH

Strength, Impact
USE   IMPACT STRENGTH

Strength, Microyield
USE   MICROYIELD STRENGTH

Strength, Muscular
USE   MUSCULAR STRENGTH

Strength, Notch
USE   NOTCH STRENGTH

Strength Of Materials
USE   MECHANICAL PROPERTIES

Strength, Residual
USE   RESIDUAL STRENGTH

Strength, Shear
USE   SHEAR STRENGTH

Strength Steels, High
USE   HIGH STRENGTH STEELS

Strength, Stress Rupture
USE   CREEP RUPTURE STRENGTH

Strength, Tensile
USE   TENSILE STRENGTH

Strength, Weld
USE   WELD STRENGTH

Strength, Yield
USE   YIELD STRENGTH

Strengths, Oscillator
USE   OSCILLATOR STRENGTHS

STREPTOCOCCUS

STREPTOMYCETES

STREPTOMYCIN

STRESS ANALYSIS

Stress Analysis, Hydrothermal
USE   HYDROTHERMAL STRESS ANALYSIS

Stress Analysis, X Ray
USE   X RAY STRESS ANALYSIS

Stress, Axial
USE   AXIAL STRESS

STRESS (BIOLOGY)

Stress (Biology), Flight
USE   FLIGHT STRESS (BIOLOGY)

Stress Calculation, Matrix
USE   MATRIX METHODS

Stress Calculations
USE   STRESS ANALYSIS

Stress, Centrifuging
USE   CENTRIFUGING STRESS

Stress, Combined
USE   COMBINED STRESS

STRESS CONCENTRATION

STRESS CORROSION

STRESS CORROSION CRACKING

Stress, Critical
USE   CRITICAL LOADING

STRESS CYCLES

Stress Distribution
USE   STRESS CONCENTRATION

Stress, Flight
USE   FLIGHT STRESS

STRESS FUNCTIONS

Stress, Inelastic
USE   INELASTIC STRESS

STRESS INTENSITY FACTORS

Stress, Internal
USE   RESIDUAL STRESS

STRESS MEASUREMENT

Stress Measurement, Photoelastic
USE   PHOTOELASTIC ANALYSIS

Stress Measurement, X Ray
USE   X RAY STRESS MEASUREMENT

Stress, Mental
USE   STRESS (PSYCHOLOGY)

STRESS (PHYSIOLOGY)

Stress, Plant
USE   PLANT STRESS

STRESS PROPAGATION

STRESS (PSYCHOLOGY)

STRESS RATIO

STRESS RELAXATION

STRESS RELIEVING

Stress, Residual
USE   RESIDUAL STRESS

Stress, Reynolds
USE   REYNOLDS STRESS

Stress Rupture Strength
USE   CREEP RUPTURE STRENGTH

Stress, Shear
USE   SHEAR STRESS

Stress, Shearing
USE   SHEAR STRESS

Stress, Space Flight
USE   SPACE FLIGHT STRESS

Stress, Tensile
USE   TENSILE STRESS

STRESS TENSORS

Stress, Torsional
USE   TORSIONAL STRESS

Stress, Vibrational
USE   VIBRATIONAL STRESS

STRESS WAVES

STRESS-STRAIN DIAGRAMS

Stress-Strain Distribution
USE   STRESS CONCENTRATION

STRESS-STRAIN RELATIONSHIPS

STRESS-STRAIN-TIME RELATIONS

STRESSED-SKIN STRUCTURES

STRESSES

Stresses, Photo
USE   PHOTOSTRESSES

Stresses (Physiology), Acceleration
USE   ACCELERATION STRESSES (PHYSIOLOGY)

Stresses, Thermal
USE   THERMAL STRESSES

Stresses, Triaxial
USE   TRIAXIAL STRESSES

STRETCH FORMING

STRETCHERS

STRETCHING

STRIATION

STRINGERS

STRINGS

STRIP

Strip Lines, Parallel
USE   MICROSTRIP TRANSMISSION LINES

STRIP MINING

STRIP TRANSMISSION LINES

STRIPPING

Stripping, Anodic
USE   ANODIC STRIPPING

STRIPPING (DISTILLATION)

Stripping, Ion
USE   ION STRIPPING

Strips, Metal
USE   METAL STRIPS

STROBOSCOPES

Stroke, Heat
USE   HEAT STROKE

STROKES

STROKING TESTS

STRONG INTERACTIONS (FIELD THEORY)

STRONGLY COUPLED PLASMAS

STRONTIUM

STRONTIUM BROMIDES

STRONTIUM COMPOUNDS

STRONTIUM FLUORIDES

STRONTIUM ISOTOPES

STRONTIUM SULFIDES

STRONTIUM TITANATES

STRONTIUM ZIRCONATES

STRONTIUM 85

STRONTIUM 87

STRONTIUM 88

STRONTIUM 89

STRONTIUM 90

STROUHAL NUMBER

Struct Test, Drones For Aerodynamic And
USE   DAST PROGRAM

STRUCTURAL ANALYSIS

Structural Analysis, Dynamic
USE   DYNAMIC STRUCTURAL ANALYSIS

Structural Analysis Program, NASA
USE   NASTRAN

STRUCTURAL BASINS

Structural Beams
USE   BEAMS (SUPPORTS)

STRUCTURAL DESIGN

STRUCTURAL DESIGN CRITERIA

Structural Dynamics
USE   DYNAMIC STRUCTURAL ANALYSIS

STRUCTURAL ENGINEERING

STRUCTURAL FAILURE

Structural Fatigue
USE   FATIGUE (MATERIALS)

(Structural Forms), Domes
USE   DOMES (STRUCTURAL FORMS)

(Structural Forms), Shells
USE   SHELLS (STRUCTURAL FORMS)

Structural Foundations
USE   FOUNDATIONS

STRUCTURAL INFLUENCE COEFFICIENTS

Structural Materials
USE   CONSTRUCTION MATERIALS

(Structural Member), Skin
USE   SKIN (STRUCTURAL MEMBER)

STRUCTURAL MEMBERS

(Structural Members), Plates
USE   PLATES (STRUCTURAL MEMBERS)

(Structural Members), Studs
USE   STUDS (STRUCTURAL MEMBERS)

STRUCTURAL PROPERTIES (GEOLOGY)

STRUCTURAL RELIABILITY

Structural Rigidity
USE   STRUCTURAL STABILITY

STRUCTURAL STABILITY

STRUCTURAL STRAIN

(Structural Units), Bays
USE   BAYS (STRUCTURAL UNITS)

STRUCTURAL VIBRATION

STRUCTURAL WEIGHT

Structure, Atomic
USE   ATOMIC STRUCTURE

Structure, Crystal
USE   CRYSTAL STRUCTURE

Structure, Earth Planetary
USE   EARTH PLANETARY STRUCTURE

Structure, Electronic
USE   ATOMIC STRUCTURE

Structure, Fine
USE   FINE STRUCTURE

Structure, Galactic
USE   GALACTIC STRUCTURE

Structure, Hyperfine
USE   HYPERFINE STRUCTURE

Structure), Mantle (Earth
USE   EARTH MANTLE

Structure, Micro
USE   MICROSTRUCTURE

Structure, Molecular
USE   MOLECULAR STRUCTURE

Structure, Nuclear
USE   NUCLEAR STRUCTURE

Structure Of Solids, Band
USE   BAND STRUCTURE OF SOLIDS

Structure, Planetary
USE   PLANETARY STRUCTURE

Structure, Stellar
USE   STELLAR STRUCTURE

Structure, Widmanstatten
USE   WIDMANSTATTEN STRUCTURE

STRUCTURES

Structures, Aircraft
USE   AIRCRAFT STRUCTURES

(Structures), Bridges
USE   BRIDGES (STRUCTURES)

Structures, Building
USE   BUILDINGS

Structures, Composite
USE   COMPOSITE STRUCTURES

Structures, Concrete
USE   CONCRETE STRUCTURES

Structures, Data
USE   DATA STRUCTURES

Structures, Earthquake Resistant
USE   EARTHQUAKE RESISTANT STRUCTURES

Structures, Expandable
USE   EXPANDABLE STRUCTURES

Structures, Folding
USE   FOLDING STRUCTURES

Structures, Honeycomb
USE   HONEYCOMB STRUCTURES

(Structures), Hulls
USE   HULLS (STRUCTURES)

Structures, Hybrid
USE   HYBRID STRUCTURES

Structures, Inflatable
USE   INFLATABLE STRUCTURES

Structures, Insulated
USE   INSULATED STRUCTURES

Structures, Intramolecular
USE   INTRAMOLECULAR STRUCTURES

Structures, Isotensoid
USE   ISOTENSOID STRUCTURES

Structures, Large Space
USE   LARGE SPACE STRUCTURES

Structures, Membrane
USE   MEMBRANE STRUCTURES

Structures, Missile
USE   MISSILE STRUCTURES

Structures, Monocoque
USE   MONOCOQUE STRUCTURES

Structures, Multilayer
USE   LAMINATES

(Structures), Partitions
USE   PARTITIONS (STRUCTURES)

Structures, Planar
USE   PLANAR STRUCTURES

Structures, Plastic Aircraft
USE   PLASTIC AIRCRAFT STRUCTURES

(Structures), Ramps
USE   RAMPS (STRUCTURES)

Structures, Redundant
USE   REDUNDANT COMPONENTS

(Structures), Reinforcement
USE   REINFORCEMENT (STRUCTURES)

Structures, Rigid
USE   RIGID STRUCTURES

Structures, Ring
USE   RING STRUCTURES

Structures, Sandwich
USE   SANDWICH STRUCTURES

Structures, Space Erectable
USE   SPACE ERECTABLE STRUCTURES

Structures, Spacecraft
USE   SPACECRAFT STRUCTURES

Structures, Steel
USE   STEEL STRUCTURES

Structures, Stiff
USE   RIGID STRUCTURES

Structures, Stressed-Skin
USE   STRESSED-SKIN STRUCTURES

Structures, Sub
USE   SUBSTRUCTURES

Structures, Telescoping
USE   FOLDING STRUCTURES

Structures, Underground
USE   UNDERGROUND STRUCTURES

Structures, Underwater
USE   UNDERWATER STRUCTURES

Structures, Unimolecular
USE   UNIMOLECULAR STRUCTURES

Structures, Variable Geometry
USE   VARIABLE GEOMETRY STRUCTURES

Structures, Welded
USE   WELDED STRUCTURES

Structures, Wooden
USE   WOODEN STRUCTURES

STRUTS

STRYCHNINE

STS
USE   SPACE TRANSPORTATION SYSTEM

(STS), Approach And Landing Tests
USE   APPROACH AND LANDING TESTS (STS)

(STS), Astro Missions
USE   ASTRO MISSIONS (STS)

(STS), Entry Guidance
USE   ENTRY GUIDANCE (STS)

(Sts), Interim Upper Stage
 USE   INERTIAL UPPER STAGE

(STS), Payload Delivery
 USE   PAYLOAD DELIVERY (STS)

(STS), Payload Retrieval
 USE   PAYLOAD RETRIEVAL (STS)

(STS), Power Modules
 USE   POWER MODULES (STS)

(STS), Turnaround
 USE   TURNAROUND (STS)

STS-1
 USE   SPACE TRANSPORTATION SYSTEM 1
        FLIGHT

STS-2
 USE   SPACE TRANSPORTATION SYSTEM 2
        FLIGHT

STS-3
 USE   SPACE TRANSPORTATION SYSTEM 3
        FLIGHT

STS-4
 USE   SPACE TRANSPORTATION SYSTEM 4
        FLIGHT

STS-5
 USE   SPACE SHUTTLE MISSION 31-A

STS-6
 USE   SPACE SHUTTLE MISSION 31-B

STS-7
 USE   SPACE SHUTTLE MISSION 31-C

STS-8
 USE   SPACE SHUTTLE MISSION 31-D

STS-9
 USE   SPACE SHUTTLE MISSION 41-A

STS-11
 USE   SPACE SHUTTLE MISSION 41-B

STS-13
 USE   SPACE SHUTTLE MISSION 41-C

STS-14
 USE   SPACE SHUTTLE MISSION 41-D

STS-17
 USE   SPACE SHUTTLE MISSION 41-G

STS-19
 USE   SPACE SHUTTLE MISSION 51-A

STS-20
 USE   SPACE SHUTTLE MISSION 51-C

STS-22
 USE   SPACE SHUTTLE MISSION 51-E

STS-23
 USE   SPACE SHUTTLE MISSION 51-D

STS-24
 USE   SPACE SHUTTLE MISSION 51-B

STS-25
 USE   SPACE SHUTTLE MISSION 51-G

STS-26
 USE   SPACE SHUTTLE MISSION 51-F

STS-27
 USE   SPACE SHUTTLE MISSION 51-I

STS-28
 USE   SPACE SHUTTLE MISSION 51-J

STS-29
 USE   SPACE SHUTTLE MISSION 61-A

STS-30
 USE   SPACE SHUTTLE MISSION 61-A

STS-31
 USE   SPACE SHUTTLE MISSION 61-B
        SPACE SHUTTLE MISSION 51-H

STS-32
 USE   SPACE SHUTTLE MISSION 61-C

STS-33
 USE   SPACE SHUTTLE MISSION 51-L

STS-34
 USE   SPACE SHUTTLE MISSION 61-E

STUDENTS

Studies
 USE   INVESTIGATION

Studies, Tracking
 USE   TRACKING (POSITION)

STUDS (STRUCTURAL MEMBERS)

Study, International Magnetospheric
 USE   INTERNATIONAL MAGNETOSPHERIC
        STUDY

Study, International Sats For Ionospheric
 USE   ISIS SATELLITES

Sturm-Liouville Operator
 USE   STURM-LIOUVILLE THEORY

STURM-LIOUVILLE THEORY

Styluses
 USE   PENS

STYPHNATES

Styrene, Poly
 USE   POLYSTYRENE

STYRENES

STYROFOAM (TRADEMARK)

SUBARCTIC REGIONS

SUBASSEMBLIES

SUBAUDIBLE FREQUENCIES

Subcarrier Waves
 USE   CARRIER WAVES

Subcircuits
 USE   CIRCUITS
        SUBASSEMBLIES

SUBCONTRACTS

SUBCRITICAL FLOW

SUBCRITICAL MASS

SUBDIVISIONS

SUBDUCTION (GEOLOGY)

SUBDWARF STARS

SUBGIANT STARS

Subgravity
 USE   REDUCED GRAVITY

SUBGROUPS

SUBHARMONIC GENERATORS

SUBIC Project
 USE   SUBMARINE INTEGRATED CONTROL
        PROJECT

SUBJECTS

Sublattices
 USE   SUBGROUPS
        LATTICES (MATHEMATICS)

Sublayers
 USE   SUBSTRATES

SUBLETHAL DOSAGE

SUBLIMATION

SUBLIMINAL STIMULI

SUBMARINE CABLES

SUBMARINE INTEGRATED CONTROL PROJECT

SUBMARINE PROPULSION

Submarine, Trident
 USE   TRIDENT SUBMARINE

SUBMARINES

Submarines, Ballistic Missile
 USE   BALLISTIC MISSILE SUBMARINES

Submarines, Guided Missile
 USE   GUIDED MISSILE SUBMARINES

Submarines, Polaris
 USE   GUIDED MISSILE SUBMARINES

SUBMERGED BODIES

SUBMERGING

SUBMERSIBLE AIRCRAFT

SUBMILLIMETER WAVES

SUBMINIATURIZATION

SUBORBITAL FLIGHT

Suboxides, Carbon
 USE   CARBON SUBOXIDES

SUBREFLECTORS

SUBROC MISSILE

SUBROUTINE LIBRARIES (COMPUTERS)

SUBROUTINES

Subsets (Mathematics)
 USE   SET THEORY

SUBSIDENCE

SUBSIDIARIES

SUBSONIC AIRCRAFT

SUBSONIC FLOW

SUBSONIC FLUTTER

SUBSONIC SPEED

SUBSONIC WIND TUNNELS

Substances
 USE   MATERIALS

(Substances), Gums
 USE   GUMS (SUBSTANCES)

SUBSTITUTES

Substitution
 USE   SUBSTITUTES

Substorms, Magnetic
 USE   MAGNETIC STORMS

Substorms, Polar
 USE   POLAR SUBSTORMS

SUBSTRATES

SUBSTRUCTURES

Subsystems, Personnel
USE   PERSONNEL SUBSYSTEMS

SUBTRACTION

Subtraction, Holographic
USE   HOLOGRAPHIC SUBTRACTION

Subtraction Holography, Self
USE   HOLOGRAPHIC SUBTRACTION

Subtropical Regions
USE   TROPICAL REGIONS
      TEMPERATE REGIONS

SUBURBAN AREAS

SUBZERO TEMPERATURE

SUCCESS PROJECT

SUCCINIMIDES

SUCROSE

SUCTION

SUD AVIATION AIRCRAFT

Sud Aviation SA-321 Helicopter
USE   SA-321 HELICOPTER

Sud Aviation SA-330 Helicopter
USE   SA-330 HELICOPTER

Sud Aviation SE-210 Aircraft
USE   SE-210 AIRCRAFT

Sud Aviation SE-3160 Helicopter
USE   SE-3160 HELICOPTER

Sud VJ-101 Aircraft
USE   VJ-101 AIRCRAFT

SUDAN

SUDDEN ENHANCEMENT OF ATMOSPHERICS

SUDDEN IONOSPHERIC DISTURBANCES

SUDDEN STORM COMMENCEMENTS

SUGAR BEETS

SUGAR CANE

SUGARS

SUGGESTION

SUHL EFFECT

SUITABILITY

SUITS

Suits, Pressure
USE   PRESSURE SUITS

Suits, Space
USE   SPACE SUITS

Sulfate, Hydroxylamine
USE   HYDROXYLAMINE SULFATE

SULFATES

Sulfates, Ammonium
USE   AMMONIUM SULFATES

Sulfates, Lithium
USE   LITHIUM SULFATES

Sulfates, Magnesium
USE   MAGNESIUM SULFATES

Sulfates, Sodium
USE   SODIUM SULFATES

SULFATION

SULFIDATION

Sulfide, Hydrogen
USE   HYDROGEN SULFIDE

SULFIDES

Sulfides, Barium
USE   BARIUM SULFIDES

Sulfides, Bismuth
USE   BISMUTH SULFIDES

Sulfides, Cadmium
USE   CADMIUM SULFIDES

Sulfides, Calcium
USE   CALCIUM SULFIDES

Sulfides, Copper
USE   COPPER SULFIDES

Sulfides, Di
USE   DISULFIDES

Sulfides, Indium
USE   INDIUM SULFIDES

Sulfides, Inorganic
USE   INORGANIC SULFIDES

Sulfides, Lead
USE   LEAD SULFIDES

Sulfides, Molybdenum
USE   MOLYBDENUM SULFIDES

Sulfides, Strontium
USE   STRONTIUM SULFIDES

Sulfides, Zinc
USE   ZINC SULFIDES

SULFITES

Sulfites, Hydro
USE   HYDROSULFITES

Sulfites, Sodium
USE   SODIUM SULFITES

SULFONATES

SULFONES

SULFONIC ACID

SULFUR

Sulfur Batteries, Lithium
USE   LITHIUM SULFUR BATTERIES

Sulfur Batteries, Sodium
USE   SODIUM SULFUR BATTERIES

SULFUR CHLORIDES

SULFUR COMPOUNDS

Sulfur Compounds, Organic
USE   ORGANIC SULFUR COMPOUNDS

SULFUR DIOXIDES

SULFUR FLUORIDES

SULFUR ISOTOPES

SULFUR OXIDES

SULFURIC ACID

SUM RULES

SUMMARIES

Summaries, Prelaunch
USE   PRELAUNCH SUMMARIES

Summators, Binary
USE   ADDING CIRCUITS

SUMMER

SUMPS

SUMS

SUN

Sun Earth Explorer 1, International
USE   INTERNATIONAL SUN EARTH EXPLORER 1

Sun Earth Explorer 2, International
USE   INTERNATIONAL SUN EARTH EXPLORER 2

Sun Earth Explorer 3, International
USE   INTERNATIONAL SUN EARTH EXPLORER 3

Sun Earth Explorers, International
USE   INTERNATIONAL SUN EARTH EXPLORERS

Sun Sensors
USE   SOLAR SENSORS

Sun Year, International Quiet
USE   INTERNATIONAL QUIET SUN YEAR

SUNBLAZER SPACE PROBE

SUNFLOWER POWER SYSTEM

SUNFLOWERS

SUNGLASSES

SUNLIGHT

SUNRISE

SUNSET

SUNSPOT CYCLE

SUNSPOTS

Super Fortress Aircraft
USE   RB-50 AIRCRAFT

Super Sabre Aircraft
USE   F-100 AIRCRAFT

Superalloys
USE   HEAT RESISTANT ALLOYS

SUPERCAVITATING FLOW

Supercavitation
USE   SUPERCAVITATING FLOW

SUPERCHARGERS

Supercharging
USE   SUPERCHARGERS

SUPERCOMPUTERS

SUPERCONDUCTING MAGNETS

SUPERCONDUCTING POWER TRANSMISSION

Superconducting Quantum Interferometers
USE   SQUID (DETECTORS)

SUPERCONDUCTIVITY

SUPERCONDUCTORS

SUPERCOOLING

SUPERCRITICAL AIRFOILS

SUPERCRITICAL FLOW

SUPERCRITICAL FLUIDS

SUPERCRITICAL PRESSURES

SUPERCRITICAL WINGS

Superfluid Flow
USE   SUPERFLUIDITY

SUPERFLUIDITY

SUPERGIANT STARS

SUPERHARMONICS

SUPERHEATING

SUPERHETERODYNE RECEIVERS

SUPERHIGH FREQUENCIES

SUPERHYBRID MATERIALS

Superimposition (Mathematics)
USE   SUPERPOSITION (MATHEMATICS)

Superior, Lake
USE   LAKE SUPERIOR

SUPERLATTICES

Supermagnets
USE   HIGH FIELD MAGNETS

SUPERMASSIVE STARS

SUPERNOVA REMNANTS

SUPERNOVAE

Superoxides
USE   INORGANIC PEROXIDES

SUPERPLASTICITY

SUPERPOSITION (MATHEMATICS)

SUPERPRESSURE BALLOONS

SUPERROTATION

SUPERSATURATION

SUPERSONIC AIRCRAFT

SUPERSONIC AIRFOILS

SUPERSONIC BOUNDARY LAYERS

SUPERSONIC COMBUSTION

SUPERSONIC COMBUSTION RAMJET ENGINES

SUPERSONIC COMMERCIAL AIR TRANSPORT

SUPERSONIC COMPRESSORS

SUPERSONIC CRUISE AIRCRAFT RESEARCH

SUPERSONIC DIFFUSERS

SUPERSONIC DRAG

SUPERSONIC FLIGHT

SUPERSONIC FLOW

Supersonic Flow Inlets
USE   SUPERSONIC INLETS

SUPERSONIC FLUTTER

SUPERSONIC HEAT TRANSFER

SUPERSONIC INLETS

SUPERSONIC JET FLOW

SUPERSONIC LOW ALTITUDE MISSILE

SUPERSONIC NOZZLES

SUPERSONIC SPEEDS

SUPERSONIC TEST APPARATUS

SUPERSONIC TRANSPORTS

SUPERSONIC TURBINES

SUPERSONIC WAKES

SUPERSONIC WIND TUNNELS

SUPERSONICS

SUPINE POSITION

SUPPLEMENTS

Supplies, Aircraft Power
USE   AIRCRAFT POWER SUPPLIES

Supplies), Consumables (Spacecrew
USE   CONSUMABLES (SPACECREW SUPPLIES)

Supplies, Electric Power
USE   ELECTRIC POWER SUPPLIES

Supplies, Power
USE   POWER SUPPLIES

Supplies, Spacecraft Power
USE   SPACECRAFT POWER SUPPLIES

(Supply Chambers), Magazines
USE   MAGAZINES (SUPPLY CHAMBERS)

Supply Circuits, Power
USE   POWER SUPPLY CIRCUITS

Supply Equipment, Oxygen
USE   OXYGEN SUPPLY EQUIPMENT

SUPPLYING

(Supplying), Feeding
USE   FEEDING (SUPPLYING)

Support Equipment, Ground
USE   GROUND SUPPORT EQUIPMENT

SUPPORT INTERFERENCE

Support, Satellite Ground
USE   SATELLITE GROUND SUPPORT

Support Sys, Integrated Maneuvering Life
USE   IMLSS

(Support System), GOSS
USE   GROUND OPERATIONAL SUPPORT
         SYSTEM

Support System, Ground Operational
USE   GROUND OPERATIONAL SUPPORT
         SYSTEM

SUPPORT SYSTEMS

Support Systems, Bioregenerative Life
USE   CLOSED ECOLOGICAL SYSTEMS

Support Systems, Ground
USE   GROUND SUPPORT SYSTEMS

Support Systems, Life
USE   LIFE SUPPORT SYSTEMS

Support Systems, Portable Life
USE   PORTABLE LIFE SUPPORT SYSTEMS

SUPPORTS

(Supports), Beams
USE   BEAMS (SUPPORTS)

(Supports), Columns
USE   COLUMNS (SUPPORTS)

(Supports), Poles
USE   POLES (SUPPORTS)

(Supports), Ribs
USE   RIBS (SUPPORTS)

(Supports), Saddles
USE   SADDLES (SUPPORTS)

(Supports), Webs
USE   WEBS (SUPPORTS)

Suppression
USE   RETARDING

Suppression, Explosion
USE   EXPLOSION SUPPRESSION

Suppression, Infrared
USE   INFRARED SUPPRESSION

Suppression, Lightning
USE   LIGHTNING SUPPRESSION

Suppression, Storm
USE   STORM SUPPRESSION

SUPPRESSORS

Suppressors, Echo
USE   ECHO SUPPRESSORS

Suppressors, Noise
USE   NOISE REDUCTION

SURFACE ACOUSTIC WAVE DEVICES

Surface Blowing, Under
USE   UNDER SURFACE BLOWING

Surface Blowing, Upper
USE   UPPER SURFACE BLOWING

Surface Blown Flaps, Upper
USE   UPPER SURFACE BLOWN FLAPS

SURFACE COOLING

SURFACE CRACKS

Surface Currents, External
USE   EXTERNAL SURFACE CURRENTS

SURFACE DEFECTS

Surface Detection Equipment, Airport
USE   AIRPORT SURFACE DETECTION
         EQUIPMENT

SURFACE DIFFUSION

SURFACE DISTORTION

Surface, Earth
USE   EARTH SURFACE

SURFACE EFFECT SHIPS

SURFACE ENERGY

Surface Experiments Package, Apollo Lunar
USE   APOLLO LUNAR SURFACE EXPERIMENTS
         PACKAGE

Surface Experiments Package, Early Apollo
USE   EASEP

SURFACE FINISHING

SURFACE GEOMETRY

Surface Interactions
USE   SURFACE REACTIONS

SURFACE IONIZATION

Surface, Lambert
USE   LAMBERT SURFACE

SURFACE LAYERS

Surface, Lunar
USE   LUNAR SURFACE

Surface, Mars
USE   MARS SURFACE

Surface Missiles, Air To
USE   AIR TO SURFACE MISSILES

Surface Missiles, Surface To
USE   SURFACE TO SURFACE MISSILES

Surface Missiles, Underwater To
  USE   UNDERWATER TO SURFACE MISSILES

Surface Movements, Airfield
  USE   AIRFIELD SURFACE MOVEMENTS

SURFACE NAVIGATION

SURFACE NOISE INTERACTIONS

Surface, Ocean
  USE   OCEAN SURFACE

Surface Pressure
  USE   PRESSURE

SURFACE PROPERTIES

SURFACE REACTIONS

Surface Rockets, Surface To
  USE   SURFACE TO SURFACE ROCKETS

SURFACE ROUGHNESS

SURFACE ROUGHNESS EFFECTS

Surface Samples, Mars
  USE   MARS SURFACE SAMPLES

Surface Scientific Modules, Lunar
  USE   LSSM

SURFACE STABILITY

SURFACE TEMPERATURE

Surface Temperature, Sea
  USE   SEA SURFACE TEMPERATURE

Surface Tension
  USE   INTERFACIAL TENSION

SURFACE TO AIR MISSILES

SURFACE TO SURFACE MISSILES

SURFACE TO SURFACE ROCKETS

Surface Treatment
  USE   SURFACE FINISHING

(Surface Treatment), Sizing
  USE   SIZING (SURFACE TREATMENT)

SURFACE VEHICLES

Surface Vehicles, Lunar
  USE   LUNAR SURFACE VEHICLES

Surface Vehicles, Manned Lunar
  USE   MANNED LUNAR SURFACE VEHICLES

Surface, Venus
  USE   VENUS SURFACE

SURFACE WATER

SURFACE WAVES

Surface Waves, Electromagnetic
  USE   ELECTROMAGNETIC SURFACE WAVES

SURFACES

Surfaces, Cold
  USE   COLD SURFACES

Surfaces, Control
  USE   CONTROL SURFACES

Surfaces, Cosserat
  USE   COSSERAT SURFACES

Surfaces, Crystal
  USE   CRYSTAL SURFACES

Surfaces, Curved
  USE   CONTOURS
         SHAPES
         SURFACES

Surfaces), Elevators (Control
  USE   ELEVATORS (CONTROL SURFACES)

Surfaces, Fermi
  USE   FERMI SURFACES

Surfaces), Flaps (Control
  USE   FLAPS (CONTROL SURFACES)

Surfaces, Flat
  USE   FLAT SURFACES

Surfaces, Horizontal Tail
  USE   HORIZONTAL TAIL SURFACES

Surfaces, Hot
  USE   HOT SURFACES

(Surfaces), Hydroplanes
  USE   HYDROPLANES (SURFACES)

Surfaces, Lifting
  USE   LIFTING BODIES
         LIFT DEVICES
         SURFACES

Surfaces, Liquid
  USE   LIQUID SURFACES

Surfaces, Metal
  USE   METAL SURFACES

Surfaces, Minimal
  USE   MINIMAL SURFACES

Surfaces, Planetary
  USE   PLANETARY SURFACES

Surfaces, Satellite
  USE   SATELLITE SURFACES

Surfaces, Selective
  USE   SELECTIVE SURFACES

Surfaces, Solid
  USE   SOLID SURFACES

Surfaces, Sweptback Tail
  USE   SWEPTBACK TAIL SURFACES

Surfaces, T Tail
  USE   T TAIL SURFACES

Surfaces), Tabs (Control
  USE   TABS (CONTROL SURFACES)

Surfaces, Tail
  USE   TAIL SURFACES

Surfaces, Townsend
  USE   TOWNSEND AVALANCHE

Surfaces, Trapezoidal Tail
  USE   TRAPEZOIDAL TAIL SURFACES

SURFACTANTS

SURGEONS

Surgeons, Flight
  USE   FLIGHT SURGEONS

SURGERY

SURGES

Surges, Storm
  USE   STORM SURGES

(Surges), Transients
  USE   SURGES

SURGICAL INSTRUMENTS

SURINAM

SURVEILLANCE

Surveillance (Ground Based), Space
  USE   SPACE SURVEILLANCE (GROUND BASED)

SURVEILLANCE RADAR

Surveillance Radar, Airborne
  USE   AIRBORNE SURVEILLANCE RADAR

Surveillance, Space
  USE   SPACE SURVEILLANCE

Surveillance (Spaceborne), Space
  USE   SPACE SURVEILLANCE (SPACEBORNE)

Survey Aircraft, Earth Resources
  USE   EARTH RESOURCES SURVEY AIRCRAFT

Survey Module, Local Scientific
  USE   LOCAL SCIENTIFIC SURVEY MODULE

Survey Program, Earth Resources
  USE   EARTH RESOURCES SURVEY PROGRAM

Surveying
  USE   SURVEYS

SURVEYOR LUNAR PROBES

SURVEYOR PROJECT

SURVEYOR 1 LUNAR PROBE

SURVEYOR 2 LUNAR PROBE

SURVEYOR 3 LUNAR PROBE

SURVEYOR 4 LUNAR PROBE

SURVEYOR 5 LUNAR PROBE

SURVEYOR 6 LUNAR PROBE

SURVEYOR 7 LUNAR PROBE

SURVEYS

Surveys, Environmental
  USE   ENVIRONMENTAL SURVEYS

Surveys, Geodetic
  USE   GEODETIC SURVEYS

Surveys, Geological
  USE   GEOLOGICAL SURVEYS

Surveys, Magnetic
  USE   MAGNETIC SURVEYS

Surveys, Wage
  USE   WAGE SURVEYS

Survivability, Aircraft
  USE   AIRCRAFT SURVIVABILITY

Survivability, Spacecraft
  USE   SPACECRAFT SURVIVABILITY

SURVIVAL

SURVIVAL EQUIPMENT

Susceptibility, Magnetic
  USE   MAGNETIC PERMEABILITY

Susceptibility (Magnetism)
  USE   MAGNETIC PERMEABILITY

Suspended Gyroscopes, Electrically
  USE   ELECTROSTATIC GYROSCOPES

SUSPENDING (HANGING)

SUSPENDING (MIXING)

Suspension And Pointing System, Annular
  USE   ANNULAR SUSPENSION AND POINTING
         SYSTEM

Suspension, Magnetic
  USE   MAGNETIC SUSPENSION

SUSPENSION SYSTEMS (VEHICLES)

SUSPENSIONS

Suspensions, Solid
 USE   SOLID SUSPENSIONS

SUSQUEHANNA RIVER BASIN (MD-NY-PA)

Sustained Emission, Self
 USE   SELF SUSTAINED EMISSION

SUSTAINER ROCKET ENGINES

SUSTAINING

Sustaining Systems, Emergency Life
 USE   EMERGENCY LIFE SUSTAINING SYSTEMS

SWAGING

SWALLOWING

Swamps
 USE   MARSHLANDS

SWAN BANDS

SWARMING

Swash
 USE   SPLASHING

SWATH (SHIP)

SWATH WIDTH

Sway Test, Body
 USE   BODY SWAY TEST

SWAZILAND

SWEAT

SWEAT COOLING

Sweat Index, Palmar
 USE   PALMAR SWEAT INDEX

Sweating
 USE   PERSPIRATION

SWEDEN

SWEDISH SPACE PROGRAM

SWEEP ANGLE

SWEEP CIRCUITS

SWEEP EFFECT

SWEEP FREQUENCY

Sweep, Leading Edge
 USE   LEADING EDGE SWEEP

Sweep Wings, Variable
 USE   VARIABLE SWEEP WINGS

SWEEPBACK

Sweepback Angles
 USE   SWEEPBACK

Sweeping, Electron
 USE   SWEEP FREQUENCY

SWELLING

SWEPT FORWARD WINGS

SWEPT WINGS

SWEPTBACK TAIL SURFACES

SWEPTBACK WINGS

SWIMMING

SWIMMING POOL REACTORS

SWINE

(Swine), Pigs
 USE   SWINE

SWING TAIL ASSEMBLIES

SWING WINGS

SWINGBY TECHNIQUE

SWIRLING

Swirling Wakes
 USE   TURBULENT WAKES

SWISS SPACE PROGRAM

Switched Lasers, Q
 USE   Q SWITCHED LASERS

SWITCHES

Switches, Capacitance
 USE   CAPACITANCE SWITCHES

Switches, Electric
 USE   ELECTRIC SWITCHES

Switches, Electronic
 USE   SWITCHING CIRCUITS

Switches, Pressure
 USE   PRESSURE SWITCHES

Switches, Stepping
 USE   STEPPING SWITCHES

Switches, Vacuum Arc
 USE   VACUUM ARC SWITCHES

SWITCHING

Switching, Beam
 USE   BEAM SWITCHING

SWITCHING CIRCUITS

Switching Elements
 USE   SWITCHING CIRCUITS

Switching Elements, Fluid
 USE   FLUID SWITCHING ELEMENTS

Switching Interferometers, Phase
 USE   PHASE SWITCHING INTERFEROMETERS

Switching, Magnetic
 USE   MAGNETIC SWITCHING

Switching, Microwave
 USE   MICROWAVE SWITCHING

Switching, Packet
 USE   PACKET SWITCHING

SWITCHING THEORY

SWITZERLAND

SWIVELS

SYENITE

SYLLABLES

SYMBIOSIS

SYMBIOTIC STARS

SYMBOLIC PROGRAMMING

SYMBOLS

(Symbols), Letters
 USE   SYMBOLS

(Symbols), Signs
 USE   SYMBOLS

SYMMETRICAL BODIES

SYMMETRY

Symmetry, Anti
 USE   ANTISYMMETRY

Symmetry Breaking
 USE   BROKEN SYMMETRY

Symmetry, Broken
 USE   BROKEN SYMMETRY

SYMPATHETIC NERVOUS SYSTEM

Sympathomimetics
 USE   ADRENERGICS

SYMPHONIE SATELLITES

SYMPTOMOLOGY

Symptoms
 USE   SIGNS AND SYMPTOMS

Symptoms, Signs And
 USE   SIGNS AND SYMPTOMS

SYNAPSES

SYNCHROCYCLOTRONS

SYNCHRONISM

Synchronization
 USE   SYNCHRONISM

Synchronization, Bit
 USE   BIT SYNCHRONIZATION

Synchronization, Frequency
 USE   FREQUENCY SYNCHRONIZATION

Synchronization, Non
 USE   NONSYNCHRONIZATION

SYNCHRONIZED OSCILLATORS

SYNCHRONIZERS

Synchronous Communication Satellites
 USE   SYNCOM SATELLITES

SYNCHRONOUS COMMUNICATIONS SATELLITE PROJ

Synchronous Detectors
 USE   CORRELATORS

SYNCHRONOUS EARTH OBSERVATORY SATELLITE

SYNCHRONOUS METEOROLOGICAL SATELLITE

SYNCHRONOUS MOTORS

SYNCHRONOUS PLATFORMS

SYNCHRONOUS SATELLITES

SYNCHROPHASING

SYNCHROPHASOTRONS

SYNCHROSCOPES

SYNCHROTRON RADIATION

SYNCHROTRONS

SYNCLINES

Synclines, Geo
 USE   GEOSYNCLINES

Synclinoria
 USE   SYNCLINES

SYNCODERS

SYNCOM APOGEE ENGINES

SYNCOM SATELLITES

SYNCOM 1 SATELLITE

SYNCOM 2 SATELLITE

SYNCOM 3 SATELLITE

SYNCOM 4 SATELLITE

SYNCOPE

Syndrome, Space Adaptation
USE   SPACE ADAPTATION SYNDROME

Syndromes
USE   SIGNS AND SYMPTOMS

SYNOPTIC MEASUREMENT

SYNOPTIC METEOROLOGY

SYNTAX

SYNTECTIC ALLOYS

SYNTHANE

SYNTHESIS

Synthesis, Bio
USE   BIOSYNTHESIS

SYNTHESIS (CHEMISTRY)

Synthesis, Network
USE   NETWORK SYNTHESIS

Synthesis, Photo
USE   PHOTOSYNTHESIS

Synthesis, Plasma Jet
USE   PLASMA JET SYNTHESIS

Synthesis, Protein
USE   PROTEIN SYNTHESIS

SYNTHESIZERS

Synthesizers, Frequency
USE   FREQUENCY SYNTHESIZERS

SYNTHETIC APERTURE RADAR

SYNTHETIC APERTURES

SYNTHETIC ARRAYS

SYNTHETIC FIBERS

SYNTHETIC FOOD

SYNTHETIC FUELS

SYNTHETIC METALS

Synthetic Methane
USE   SYNTHANE

SYNTHETIC RESINS

SYNTHETIC RUBBERS

SYNTONY

SYPHILIS

SYRIA

SYRINGES

Sys), Airs (Reconnaissance
USE   AIRBORNE INTEGRATED
       RECONNAISSANCE SYSTEM

Sys, Atmospheric & Oceanographic Inform
USE   ATMOSPHERIC & OCEANOGRAPHIC
       INFORM SYS

Sys, Integrated Maneuvering Life Support
USE   IMLSS

Sys, National Operational Environmental Sat
USE   NOESS

System), AFCS (Control
USE   AUTOMATIC FLIGHT CONTROL

System, African Rift
USE   AFRICAN RIFT SYSTEM

System, Airborne Integrated Reconnaissance
USE   AIRBORNE INTEGRATED
       RECONNAISSANCE SYSTEM

System, Airborne Warning And Control
USE   AWACS AIRCRAFT

System, Aloha
USE   ALOHA SYSTEM

System, Annular Suspension And Pointing
USE   ANNULAR SUSPENSION AND POINTING
       SYSTEM

System, Apollo Extension
USE   APOLLO EXTENSION SYSTEM

System, Astroguide Navigation
USE   ASTROGUIDE NAVIGATION SYSTEM

System, Automated Pilot Advisory
USE   AUTOMATED PILOT ADVISORY SYSTEM

System, Automated Radar Terminal
USE   AUTOMATED RADAR TERMINAL SYSTEM

System, Autonomic Nervous
USE   AUTONOMIC NERVOUS SYSTEM

System (AVCS), Advanced Vidicon Camera
USE   ADVANCED VIDICON CAMERA SYSTEM
       (AVCS)

System, Ballistic Missile Early Warning
USE   BALLISTIC MISSILE EARLY WARNING
       SYSTEM

System, Beacon Collision Avoidance
USE   BEACON COLLISION AVOIDANCE SYSTEM

System, Bioastronautical Orbital Space
USE   BIOASTRONAUTICAL ORBITAL SPACE
       SYSTEM

System, Cardiovascular
USE   CARDIOVASCULAR SYSTEM

System, CEMS
USE   CENTRAL ELECTRONIC MANAGEMENT
       SYSTEM

System, Central Electronic Management
USE   CENTRAL ELECTRONIC MANAGEMENT
       SYSTEM

System, Central Nervous
USE   CENTRAL NERVOUS SYSTEM

System, Circulatory
USE   CIRCULATORY SYSTEM

System Configurations, Propulsion
USE   PROPULSION SYSTEM CONFIGURATIONS

System (DCS), Defense Communications
USE   DEFENSE COMMUNICATIONS SYSTEM
       (DCS)

System, Defense Communications Satellite
USE   DEFENSE COMMUNICATIONS SATELLITE
       SYSTEM

System Depressants, Central Nervous
USE   CENTRAL NERVOUS SYSTEM
       DEPRESSANTS

System, Digestive
USE   DIGESTIVE SYSTEM

System, Discrete Address Beacon
USE   DISCRETE ADDRESS BEACON SYSTEM

System, Earth Resources Information
USE   EARTH RESOURCES INFORMATION
       SYSTEM

System, Earth Terminal Measurement
USE   EARTH TERMINAL MEASUREMENT
       SYSTEM

System, Earth-Moon
USE   EARTH-MOON SYSTEM

SYSTEM EFFECTIVENESS

System (Europe), Eiscat Radar
USE   EISCAT RADAR SYSTEM (EUROPE)

SYSTEM FAILURES

System, Fleet Satellite Communication
USE   FLEET SATELLITE COMMUNICATION
       SYSTEM

System Flights, Space Transportation
USE   SPACE TRANSPORTATION SYSTEM
       FLIGHTS

System For Apollo, Lunar Exploration
USE   LUNAR EXPLORATION SYSTEM FOR
       APOLLO

System, Gastrointestinal
USE   GASTROINTESTINAL SYSTEM

SYSTEM GENERATED ELECTROMAGNETIC PULSES

System, Genitourinary
USE   GENITOURINARY SYSTEM

System, Global Positioning
USE   GLOBAL POSITIONING SYSTEM

System, Goddard Trajectory Determination
USE   GODDARD TRAJECTORY DETERMINATION
       SYSTEM

System), GOSS (Support
USE   GROUND OPERATIONAL SUPPORT
       SYSTEM

System, Ground Operational Support
USE   GROUND OPERATIONAL SUPPORT
       SYSTEM

System, Hematopoietic
USE   HEMATOPOIETIC SYSTEM

System, Hot Cycle Propulsion
USE   TIP DRIVEN ROTORS

SYSTEM IDENTIFICATION

System, Information Adaptive
USE   INFORMATION ADAPTIVE SYSTEM

System, Integ Med And Behavioral Lab Measur
USE   IMBLMS

System, Intravascular
USE   INTRAVASCULAR SYSTEM

System), LESA (Lunar Exploration
USE   LUNAR EXPLORATION SYSTEM FOR
       APOLLO

System, Light Airborne Multipurpose
USE   LIGHT AIRBORNE MULTIPURPOSE SYSTEM

System, LOCATES
USE   LOCATES SYSTEM

System, LORAC Navigation
USE   LORAC NAVIGATION SYSTEM

System Management, Weapon
USE   WEAPON SYSTEM MANAGEMENT

System, Metric
USE   INTERNATIONAL SYSTEM OF UNITS

System, Microwave Scanning Beam Landing
USE   MICROWAVE SCANNING BEAM LANDING
       SYSTEM

System, Minitrack
USE   MINITRACK SYSTEM

System, Minitrack Optical Tracking
USE   MINITRACK SYSTEM

System, Miros
USE   MIROS SYSTEM

System, Modular Integrated Utility
USE   MODULAR INTEGRATED UTILITY SYSTEM

System), MOTS (Tracking
USE   MINITRACK SYSTEM

System, Musculoskeletal
USE   MUSCULOSKELETAL SYSTEM

System, NASA End-To-End Data
USE   NEEDS (DATA SYSTEM)

System, NASA Interactive Planning
USE   NASA INTERACTIVE PLANNING SYSTEM

System, National Airspace
USE   NATIONAL AIRSPACE SYSTEM

System, National Airspace Utilization
USE   NATIONAL AIRSPACE UTILIZATION SYSTEM

System, National Aviation
USE   NATIONAL AVIATION SYSTEM

System, National Oceanic Satellite
USE   NATIONAL OCEANIC SATELLITE SYSTEM

System), Needs (Data
USE   NEEDS (DATA SYSTEM)

System, Nervous
USE   NERVOUS SYSTEM

(System), NIPS
USE   NASA INTERACTIVE PLANNING SYSTEM

System, Nova Laser
USE   NOVA LASER SYSTEM

System Of Units, International
USE   INTERNATIONAL SYSTEM OF UNITS

System, Omega Navigation
USE   OMEGA NAVIGATION SYSTEM

System, Payload Deployment & Retrieval
USE   PAYLOAD DEPLOYMENT & RETRIEVAL
       SYSTEM

System Performance, Propulsion
USE   PROPULSION SYSTEM PERFORMANCE

System, Peripheral Nervous
USE   PERIPHERAL NERVOUS SYSTEM

System, Pilot Landing Aid Television
USE   PLAT SYSTEM

System, PLAT
USE   PLAT SYSTEM

System, Polystation Doppler Tracking
USE   POLYSTATION DOPPLER TRACKING
       SYSTEM

System, Post Boost Propulsion
USE   POST BOOST PROPULSION SYSTEM

System, Ranger Block 3 Television
USE   RANGER BLOCK 3 TELEVISION SYSTEM

System, Remote Manipulator
USE   REMOTE MANIPULATOR SYSTEM

System, Respiratory
USE   RESPIRATORY SYSTEM

System, Safeguard
USE   SAFEGUARD SYSTEM

System, Sage Air Defense
USE   SAGE AIR DEFENSE SYSTEM

System, Sentinel
USE   SENTINEL SYSTEM

System, Shiva Laser
USE   SHIVA LASER SYSTEM

System, Solar
USE   SOLAR SYSTEM

System, Space Detection And Tracking
USE   SPACE DETECTION AND TRACKING
       SYSTEM

System, Space Transportation
USE   SPACE TRANSPORTATION SYSTEM

System), SPADATS (Tracking
USE   SPACE DETECTION AND TRACKING
       SYSTEM

System Stimulants, Central Nervous
USE   CENTRAL NERVOUS SYSTEM STIMULANTS

System, Sunflower Power
USE   SUNFLOWER POWER SYSTEM

System, Sympathetic Nervous
USE   SYMPATHETIC NERVOUS SYSTEM

System, Teleoperator Maneuvering
USE   TELEOPERATORS

System, Terrain Contour Matching Navigation
USE   TERCOM

System, TIROS Operational Satellite
USE   TIROS OPERATIONAL SATELLITE SYSTEM

System, Tradex Radar
USE   TRADEX RADAR SYSTEM

System, Transit Navigation
USE   TRANSIT NAVIGATION SYSTEM

System, Typhon Weapon
USE   TYPHON WEAPON SYSTEM

System, Vascular
USE   VASCULAR SYSTEM

System, Vasomotor Nervous
USE   NERVOUS SYSTEM

System, Vortex Advisory
USE   VORTEX ADVISORY SYSTEM

System 1 Flight, Space Transportation
USE   SPACE TRANSPORTATION SYSTEM 1
       FLIGHT

System 2 Flight, Space Transportation
USE   SPACE TRANSPORTATION SYSTEM 2
       FLIGHT

System 3 Flight, Space Transportation
USE   SPACE TRANSPORTATION SYSTEM 3
       FLIGHT

System 4 Flight, Space Transportation
USE   SPACE TRANSPORTATION SYSTEM 4
       FLIGHT

System 10 Computer
USE   PDP 10 COMPUTER

System 107A-1, Weapon
USE   WEAPON SYSTEM 107A-1

System 107A-2, Weapon
USE   WEAPON SYSTEM 107A-2

System 133A, Weapon
USE   WEAPON SYSTEM 133A

System 133B, Weapon
USE   WEAPON SYSTEM 133B

System 315A, Weapon
USE   WEAPON SYSTEM 315A

SYSTEMS

Systems, Adaptive Control
USE   ADAPTIVE CONTROL

Systems, Advanced EVA Protection
USE   AEPS

Systems, Aerospace
USE   AEROSPACE SYSTEMS

Systems, Afferent Nervous
USE   AFFERENT NERVOUS SYSTEMS

Systems, Air Cushion Landing
USE   AIR CUSHION LANDING SYSTEMS

Systems, Aircraft Fuel
USE   AIRCRAFT FUEL SYSTEMS

Systems, Aircraft Hydraulic
USE   AIRCRAFT HYDRAULIC SYSTEMS

Systems, All-Weather Landing
USE   ALL-WEATHER LANDING SYSTEMS

SYSTEMS ANALYSIS

Systems, Ascent Propulsion
USE   ASCENT PROPULSION SYSTEMS

Systems, Biocontrol
USE   BIOCONTROL SYSTEMS

Systems (Biology), Motor
USE   EFFERENT NERVOUS SYSTEMS

Systems, Bioregenerative Life Support
USE   CLOSED ECOLOGICAL SYSTEMS

Systems, Carrier
USE   WIRELESS COMMUNICATION

Systems, Celestial Reference
USE   CELESTIAL REFERENCE SYSTEMS

Systems), Chokes (Fuel
USE   CHOKES (FUEL SYSTEMS)

Systems, Closed Ecological
USE   CLOSED ECOLOGICAL SYSTEMS

Systems, Closed Loop
USE   FEEDBACK CONTROL

Systems, Command
USE   COMMAND GUIDANCE

Systems, Communication
USE   TELECOMMUNICATION

SYSTEMS COMPATIBILITY

Systems, Complex
USE   COMPLEX SYSTEMS

Systems (Computers), Operating
USE   OPERATING SYSTEMS (COMPUTERS)

Systems, Control
USE   CONTROL

Systems, Cooling
USE   COOLING SYSTEMS

Systems, Coordinate
USE   COORDINATES

Systems, Data
USE   DATA SYSTEMS

Systems, Data Base Management
USE   DATA BASE MANAGEMENT SYSTEMS

Systems, Data Handling
USE   DATA SYSTEMS

Systems, Data Readout
USE   DATA SYSTEMS
       DISPLAY DEVICES

Systems, Deicing
USE   DEICERS

Systems, Descent Propulsion
USE   DESCENT PROPULSION SYSTEMS

Systems Design
USE   SYSTEMS ENGINEERING

Systems Design, Computer
USE   COMPUTER SYSTEMS DESIGN

Systems Design, Control
USE   CONTROL SYSTEMS DESIGN

Systems, Dewar
USE   CRYOGENIC EQUIPMENT

Systems, Digital
USE   DIGITAL SYSTEMS

Systems (Digital), Binary
USE   DIGITAL SYSTEMS

Systems, Digital Command
USE   DIGITAL COMMAND SYSTEMS

Systems, Digital Radar
USE   DIGITAL RADAR SYSTEMS

Systems (Digital), Ternary
USE   DIGITAL SYSTEMS

Systems, Display
USE   DISPLAY DEVICES

Systems, Distributed Parameter
USE   DISTRIBUTED PARAMETER SYSTEMS

Systems, Domestic Satellite Communications
USE   DOMESTIC SATELLITE COMMUNICATIONS
          SYSTEMS

Systems, Dynamical
USE   DYNAMICAL SYSTEMS

Systems, Early Warning
USE   EARLY WARNING SYSTEMS

Systems, Eco
USE   ECOSYSTEMS

Systems, Ecological
USE   ECOLOGY

Systems, Efferent Nervous
USE   EFFERENT NERVOUS SYSTEMS

Systems, Elastic
USE   ELASTIC SYSTEMS

Systems, Electronic Recording
USE   ELECTRONIC RECORDING SYSTEMS

Systems, Embedded Computer
USE   EMBEDDED COMPUTER SYSTEMS

Systems, Emergency Life Sustaining
USE   EMERGENCY LIFE SUSTAINING SYSTEMS

Systems, End-To-End Data
USE   END-TO-END DATA SYSTEMS

Systems, Endocrine
USE   ENDOCRINE SYSTEMS

SYSTEMS ENGINEERING

Systems Engineering, Space
USE   AEROSPACE ENGINEERING

Systems, Escape
USE   ESCAPE SYSTEMS

Systems, Exhaust
USE   EXHAUST SYSTEMS

Systems, Expert
USE   EXPERT SYSTEMS

Systems, Fail-Safe
USE   FAIL-SAFE SYSTEMS

Systems, Feed
USE   FEED SYSTEMS

Systems, Flight Management
USE   FLIGHT MANAGEMENT SYSTEMS

Systems, Flotation
USE   FLOATS

Systems For Nuclear Auxiliary Power
USE   SNAP

Systems, Fuel
USE   FUEL SYSTEMS

Systems, Fuzzy
USE   FUZZY SYSTEMS

Systems, Geographic Information
USE   GEOGRAPHIC INFORMATION SYSTEMS

Systems, Ground Support
USE   GROUND SUPPORT SYSTEMS

(Systems), Hardening
USE   HARDENING (SYSTEMS)

Systems, Hot Gas
USE   HIGH TEMPERATURE GASES

Systems, Hybrid Navigation
USE   HYBRID NAVIGATION SYSTEMS

Systems, Hydraulic
USE   HYDRAULIC EQUIPMENT

Systems, Hydrothermal
USE   HYDROTHERMAL SYSTEMS

Systems, Hyperbolic
USE   HYPERBOLIC SYSTEMS

Systems (Identification), IFF
USE   IFF SYSTEMS (IDENTIFICATION)

Systems, Ignition
USE   IGNITION SYSTEMS

Systems), ILS (Landing
USE   INSTRUMENT LANDING SYSTEMS

Systems, Induction
USE   INTAKE SYSTEMS

Systems, Inertial Reference
USE   INERTIAL REFERENCE SYSTEMS

Systems, Information
USE   INFORMATION SYSTEMS

Systems, Instrument Landing
USE   INSTRUMENT LANDING SYSTEMS

Systems, Intake
USE   INTAKE SYSTEMS

Systems, Integrated Energy
USE   INTEGRATED ENERGY SYSTEMS

Systems, Integrated Global Ocean Station
USE   INTEGRATED GLOBAL OCEAN STATION
          SYSTEMS

Systems, Integrated Library
USE   INTEGRATED LIBRARY SYSTEMS

SYSTEMS INTEGRATION

Systems, Jettison
USE   JETTISON SYSTEMS

Systems, Landing
USE   LANDING AIDS

Systems, Launch Escape
USE   LAUNCH ESCAPE SYSTEMS

Systems), LES (Escape
USE   LAUNCH ESCAPE SYSTEMS

Systems, Life Support
USE   LIFE SUPPORT SYSTEMS

Systems, Linear
USE   LINEAR SYSTEMS

Systems, Lubrication
USE   LUBRICATION SYSTEMS

Systems, Lumped Parameter
USE   LUMPED PARAMETER SYSTEMS

Systems, Man Machine
USE   MAN MACHINE SYSTEMS

Systems, Man Operated Propulsion
USE   MAN OPERATED PROPULSION SYSTEMS

Systems, Management
USE   MANAGEMENT SYSTEMS

SYSTEMS MANAGEMENT

Systems, Management Information
USE   MANAGEMENT INFORMATION SYSTEMS

Systems (Materials), Binary
USE   BINARY SYSTEMS (MATERIALS)

(Systems), MATTS
USE   MATTS (SYSTEMS)

Systems, Metal-Gas
USE   METAL-GAS SYSTEMS

Systems, Methoxy
USE   METHOXY SYSTEMS

Systems, Microwave Landing
USE   MICROWAVE LANDING SYSTEMS

Systems, Missile
USE   MISSILE SYSTEMS

Systems, Mobile Communication
USE   MOBILE COMMUNICATION SYSTEMS

Systems), MOPS (Propulsion
USE   MAN OPERATED PROPULSION SYSTEMS

Systems, Multiloop
USE   CASCADE CONTROL

Systems, Multiple Target Trajectory
USE   MATTS (SYSTEMS)

Systems, Nike X
USE   NIKE X SYSTEMS

Systems, Nonlinear
USE   NONLINEAR SYSTEMS

(Systems), Observability
USE   OBSERVABILITY (SYSTEMS)

Systems, Ocean Data Acquisitions
USE   OCEAN DATA ACQUISITIONS SYSTEMS

Systems, On-Line
USE   ON-LINE SYSTEMS

Systems, Optical Relay
USE   OPTICAL RELAY SYSTEMS

Systems, Oxygen
USE   OXYGEN SUPPLY EQUIPMENT

Systems Performance, Computer
USE   COMPUTER SYSTEMS PERFORMANCE

Systems, Personnel Propulsion
USE   SELF MANEUVERING UNITS

Systems, Phase Locked
USE   PHASE LOCKED SYSTEMS

Systems, Piggyback
USE   PIGGYBACK SYSTEMS

Systems, Pointing Control
USE   POINTING CONTROL SYSTEMS

Systems, Portable Life Support
USE   PORTABLE LIFE SUPPORT SYSTEMS

Systems, Power Processing
USE   POWER CONDITIONING

Systems Programs, Computer
USE   COMPUTER SYSTEMS PROGRAMS

Systems, Public Address
USE   PUBLIC ADDRESS SYSTEMS

Systems, Radio Relay
USE   RADIO RELAY SYSTEMS

Systems, Rapid Transit
USE   RAPID TRANSIT SYSTEMS

Systems, Receiving
USE   RECEIVERS

Systems, Reference
USE   REFERENCE SYSTEMS

Systems, Reproductive
USE   REPRODUCTIVE SYSTEMS

Systems Research Aircraft, Rotor
USE   ROTOR SYSTEMS RESEARCH AIRCRAFT

Systems, Sampled Data
USE   DATA SAMPLING

Systems, Satellite Navigation
USE   SATELLITE NAVIGATION SYSTEMS

Systems, Self Adaptive Control
USE   SELF ADAPTIVE CONTROL SYSTEMS

Systems, Self Organizing
USE   SELF ORGANIZING SYSTEMS

SYSTEMS SIMULATION

Systems Simulation, Computer
USE   COMPUTER SYSTEMS SIMULATION

Systems, Solar Total Energy
USE   SOLAR TOTAL ENERGY SYSTEMS

Systems, Sortie
USE   SORTIE SYSTEMS

SYSTEMS STABILITY

Systems, Support
USE   SUPPORT SYSTEMS

Systems, Takeoff
USE   AIRCRAFT LAUNCHING DEVICES

Systems, Telegraph
USE   TELEGRAPH SYSTEMS

Systems, Teletypewriter
USE   TELETYPEWRITER SYSTEMS

Systems, Television
USE   TELEVISION SYSTEMS

Systems, Ternary
USE   TERNARY SYSTEMS

Systems, Thermionic Conversion
USE   THERMIONIC POWER GENERATION

Systems, Thermoelectric Conversion
USE   THERMOELECTRIC POWER GENERATION

Systems, Total Energy
USE   TOTAL ENERGY SYSTEMS

Systems, Transcontinental
USE   TRANSCONTINENTAL SYSTEMS

Systems, Transoceanic
USE   TRANSOCEANIC SYSTEMS

Systems, Two Phase
USE   BINARY SYSTEMS (MATERIALS)

Systems, Vacuum
USE   VACUUM SYSTEMS

Systems, Variable Mass
USE   VARIABLE MASS SYSTEMS

Systems (Vehicles), Suspension
USE   SUSPENSION SYSTEMS (VEHICLES)

Systems, Virtual Memory
USE   VIRTUAL MEMORY SYSTEMS

Systems, VOR
USE   VHF OMNIRANGE NAVIGATION

Systems, Warning
USE   WARNING SYSTEMS

Systems, Weapon
USE   WEAPON SYSTEMS

Systems, Wiring
USE   WIRING

SYSTOLE

SYSTOLIC PRESSURE

T

T SHAPE

T TAIL SURFACES

T TAURI STARS

T-2 AIRCRAFT

T-25 Engine, J-69-
USE   J-69-T-25 ENGINE

T-28 AIRCRAFT

T-33 AIRCRAFT

T-34 ENGINE

T-37 AIRCRAFT

T-38 AIRCRAFT

T-38 ENGINE

T-39 AIRCRAFT

T-53 ENGINE

T-55 ENGINE

T-56 ENGINE

T-58 ENGINE

T-58-GE-8B ENGINE

T-63 ENGINE

T-64 ENGINE

T-74 ENGINE

T-76 ENGINE

T-78 ENGINE

Ta
USE   TANTALUM

TABLASER

Table, Interference Factor
USE   INTERFERENCE FACTOR TABLE

Tables, Conversion
USE   CONVERSION TABLES

TABLES (DATA)

Tables, Mathematical
USE   MATHEMATICAL TABLES

Tables, Water
USE   WATER TABLES

TABLETS

TABS (CONTROL SURFACES)

Tabulating
USE   TABULATION PROCESSES

TABULATION

TABULATION PROCESSES

TACAN

TACHISTOSCOPES

TACHOMETERS

Tachometers, Cardio
USE   CARDIOTACHOMETERS

TACHYCARDIA

TACHYONS

TACHYPNEA

TACKINESS

TACT PROGRAM

Tactical Air Navigation
USE   TACAN

TACTICS

TACTILE DISCRIMINATION

Tactile Sensation
USE   TOUCH

TAFEL LAW

Tagging
USE   MARKING

TAGN

Tahoe (CA-NV), Lake
USE   LAKE TAHOE (CA-NV)

TAIL ASSEMBLIES

Tail Assemblies, Swing
USE   SWING TAIL ASSEMBLIES

Tail Configurations, Body-Wing And
USE   BODY-WING AND TAIL CONFIGURATIONS

Tail, Geomagnetic
USE   GEOMAGNETIC TAIL

Tail Mountings
USE   TAIL ASSEMBLIES

Tail Planes
USE   HORIZONTAL TAIL SURFACES

TAIL ROTORS

Tail Rotors, Helicopter
USE   HELICOPTER TAIL ROTORS

TAIL SURFACES

Tail Surfaces, Horizontal
USE   HORIZONTAL TAIL SURFACES

Tail Surfaces, Sweptback
USE   SWEPTBACK TAIL SURFACES

Tail Surfaces, T
USE   T TAIL SURFACES

Tail Surfaces, Trapezoidal
USE   TRAPEZOIDAL TAIL SURFACES

TAILLESS AIRCRAFT

Tailoring
USE   DESIGN

Tails (Assemblies)
USE   TAIL ASSEMBLIES

Tails, Boat
USE   BOATTAILS

Tails, Comet
USE   COMET TAILS

Tails, Vertical
USE   TAIL ASSEMBLIES
       STABILIZERS (FLUID DYNAMICS)

TAIWAN

TAKEOFF

Takeoff Aircraft, Short
USE   SHORT TAKEOFF AIRCRAFT

Takeoff Aircraft, Vertical
USE   VERTICAL TAKEOFF AIRCRAFT

Takeoff And Landing Aircraft, Water
USE   WATER TAKEOFF AND LANDING
       AIRCRAFT

Takeoff And Landing, Vertical
USE   VERTICAL TAKEOFF
       VERTICAL LANDING

Takeoff, Jet Assisted
USE   JATO ENGINES

TAKEOFF RUNS

Takeoff Systems
USE   AIRCRAFT LAUNCHING DEVICES

Takeoff, Vertical
USE   VERTICAL TAKEOFF

Takeoff-Landing Aircraft, Vertical Attitude
USE   VATOL AIRCRAFT

TALC

TALKING

Talon Aircraft
USE   T-38 AIRCRAFT

TALOS MISSILE

TANDEM ROTOR HELICOPTERS

TANDEM WING AIRCRAFT

TANGENTS

TANGLING

TANK GEOMETRY

Tank Pressurization, Fuel
USE   FUEL TANK PRESSURIZATION

TANK TRUCKS

TANKER AIRCRAFT

TANKER SHIPS

TANKER TERMINALS

TANKERS

TANKS (COMBAT VEHICLES)

TANKS (CONTAINERS)

Tanks, Cylindrical
USE   CYLINDRICAL TANKS

Tanks, External
USE   EXTERNAL TANKS

Tanks, Fuel
USE   FUEL TANKS

Tanks, Propellant
USE   PROPELLANT TANKS

Tanks, Rocket Propellant
USE   PROPELLANT TANKS

Tanks, Spherical
USE   SPHERICAL TANKS

Tanks, Storage
USE   STORAGE TANKS

Tanks, Wing
USE   WING TANKS

TANTALUM

TANTALUM ALLOYS

TANTALUM CARBIDES

TANTALUM COMPOUNDS

TANTALUM ISOTOPES

TANTALUM NITRIDES

TANTALUM OXIDES

TANZANIA

TAPE RECORDERS

Tape Recorders, Magnetic
USE   TAPE RECORDERS
       MAGNETIC RECORDING

Tape Transports, Magnetic
USE   MAGNETIC TAPE TRANSPORTS

Taper
USE   TAPERING

TAPERED COLUMNS

Tapered Wings
USE   SWEPT WINGS

TAPERING

TAPES

Tapes, Computer Compatible
USE   COMPUTER COMPATIBLE TAPES

Tapes, Heat
USE   HEAT TAPES

Tapes, Magnetic
USE   MAGNETIC TAPES

Tapes, Plastic
USE   PLASTIC TAPES

Tapes, Punched
USE   PUNCHED TAPES

TAPS

TAR SANDS

TARE (Data Reduction)
USE   DATA REDUCTION

TARGET ACQUISITION

Target Aircraft, Jindivik
USE   JINDIVIK TARGET AIRCRAFT

Target And Background Measurement, High Alt
USE   HIGH ALT TARGET AND BACKGROUND
       MEASUREMENT

Target Designators, Laser
USE   LASER TARGET DESIGNATORS

TARGET DRONE AIRCRAFT

Target Drone Aircraft, Firebee 2
USE   FIREBEE 2 TARGET DRONE AIRCRAFT

Target Indicators, Moving
USE   MOVING TARGET INDICATORS

Target Interactions, Laser
USE   LASER TARGET INTERACTIONS

TARGET MASKING

Target Missile, Sandpiper
USE   SANDPIPER TARGET MISSILE

Target Penetration
USE   TERMINAL BALLISTICS

TARGET RECOGNITION

Target Scatter Site Program, Radar
USE   RADAR TARGET SCATTER SITE PROGRAM

TARGET SIMULATORS

TARGET THICKNESS

Target Trajectory Systems, Multiple
USE   MATTS (SYSTEMS)

TARGETS

Targets, Laser
USE   LASER TARGETS

Targets, Particle Accelerator
USE   PARTICLE ACCELERATOR TARGETS

Targets, Radar
USE   RADAR TARGETS

Targets, Towed
USE   TOWED BODIES
       TARGETS

TARS

TARTAR MISSILE

TASK COMPLEXITY

TASKS

Tasks, Auditory
USE   AUDITORY TASKS

Tasks, Visual
USE   VISUAL TASKS

TASMANIA

TASTE

TATB

Tauri Stars, Lambda
USE   LAMBDA TAURI STARS

Tauri Stars, T
USE   T TAURI STARS

TAURID METEOROIDS

TAURUS CONSTELLATION

TAUTOMERS

TAXIING

TAXONOMY

TAYLOR INSTABILITY

TAYLOR MANIFEST ANXIETY SCALE

TAYLOR SERIES

Taylor Theorem
USE   TAYLOR SERIES

Taylor-Goertler Instability
USE   GOERTLER INSTABILITY

Tb
　USE　TERBIUM

Tc
　USE　TECHNETIUM

TCG (Tracking)
　USE　TRANSPONDER CONTROL GROUP

TCV Program
　USE　TERMINAL CONFIGURED VEHICLE
　　　　　PROGRAM

TD SATELLITES

TD-1 SATELLITE

TDMA
　USE　TIME DIVISION MULTIPLE ACCESS

TDR SATELLITES

TEA LASERS

Teaching
　USE　EDUCATION

TEACHING MACHINES

TEAMS

TEARING

TEARING MODES (PLASMAS)

TECHNETIUM

TECHNETIUM COMPOUNDS

TECHNETIUM FLUORIDES

TECHNETIUM ISOTOPES

Technical Error, Flight
　USE　PILOT ERROR

TECHNICAL WRITING

Technique, Bubble
　USE　BUBBLE TECHNIQUE

Technique), HICAT (Radar
　USE　HIGH RESOLUTION COVERAGE ANTENNAS

Technique, Minimax
　USE　MINIMAX TECHNIQUE

Technique, Particle In Cell
　USE　PARTICLE IN CELL TECHNIQUE

Technique, Swingby
　USE　SWINGBY TECHNIQUE

Techniques
　USE　METHODOLOGY

Techniques, Computer
　USE　COMPUTER TECHNIQUES

Techniques, Culture
　USE　CULTURE TECHNIQUES

Techniques, Digital
　USE　DIGITAL TECHNIQUES

Techniques, Emergency Breathing
　USE　EMERGENCY BREATHING TECHNIQUES

Techniques, Forming
　USE　FORMING TECHNIQUES

Techniques, Graphic Evaluation And Review
　USE　GERT

Techniques, Imaging
　USE　IMAGING TECHNIQUES

Techniques, Incentive
　USE　INCENTIVE TECHNIQUES

Techniques, Prediction Analysis
　USE　PREDICTION ANALYSIS TECHNIQUES

TECHNOLOGICAL FORECASTING

TECHNOLOGIES

TECHNOLOGY ASSESSMENT

Technology, Bio
　USE　BIOTECHNOLOGY

Technology, Energy
　USE　ENERGY TECHNOLOGY

Technology Experiments, Space
　USE　SPACE TECHNOLOGY EXPERIMENTS

TECHNOLOGY FEASIBILITY SPACECRAFT

Technology, Geothermal
　USE　GEOTHERMAL TECHNOLOGY

Technology Laboratory, Advanced
　USE　ADVANCED TECHNOLOGY LABORATORY

Technology Light Twin Aircraft, Advanced
　USE　ATLIT PROJECT

Technology, Marine
　USE　MARINE TECHNOLOGY

Technology, Military
　USE　MILITARY TECHNOLOGY

Technology, Passive Nosetip
　USE　PANT PROGRAM

Technology Program, Transonic Aircraft
　USE　TACT PROGRAM

Technology, Prop-Fan
　USE　PROP-FAN TECHNOLOGY

Technology, Reactor
　USE　REACTOR TECHNOLOGY

Technology Satellite B, Earth Resources
　USE　LANDSAT 2

Technology Satellite C, Earth Resources
　USE　LANDSAT 3

Technology Satellite, Communications
　USE　COMMUNICATIONS TECHNOLOGY
　　　　　SATELLITE

Technology Satellite D, Earth Resources
　USE　LANDSAT 4

Technology Satellite E, Earth Resources
　USE　LANDSAT E

Technology Satellite F, Earth Resources
　USE　LANDSAT F

Technology Satellite, Meteoroid
　USE　EXPLORER 46 SATELLITE

Technology Satellite 1, Earth Resources
　USE　LANDSAT 1

Technology Satellites, Applications
　USE　ATS

Technology Satellites, Earth Resources
　USE　LANDSAT SATELLITES

Technology Satellites, Navigation
　USE　NAVIGATION TECHNOLOGY SATELLITES

TECHNOLOGY TRANSFER

Technology Transfer, Aerospace
　USE　AEROSPACE TECHNOLOGY TRANSFER

TECHNOLOGY UTILIZATION

Tectonic Movement
　USE　TECTONICS

TECTONICS

(Tectonics), Plates
　USE　PLATES (TECTONICS)

TED
　USE　TRANSFERRED ELECTRON DEVICES

Tee
　USE　T SHAPE

Tees, Magic
　USE　MAGIC TEES

TEETERING

TEETH

Teeth, Gear
　USE　GEAR TEETH

TEFLON (TRADEMARK)

TEKTITE PROJECT

TEKTITES

Telechirics
　USE　REMOTE HANDLING

Telecomm Satellite, European Large
　USE　L-SAT

TELECOMMUNICATION

Telecommunications Exp, Health-Education
　USE　HET EXPERIMENT

TELECONFERENCING

TELEGRAPH SYSTEMS

Telegraphy
　USE　TELEGRAPH SYSTEMS

Telegraphy, Radio
　USE　RADIO TELEGRAPHY

Telemeters
　USE　TELEMETRY

TELEMETRY

Telemetry, P.A.C.M.
　USE　P.A.C.M. TELEMETRY

Telemetry, PCM
　USE　PCM TELEMETRY

Telemetry, Physiological
　USE　BIOTELEMETRY

Telemetry, Pulse Frequency Modulation
　USE　PULSE FREQUENCY MODULATION
　　　　　TELEMETRY

Telemetry, Radio
　USE　RADIO TELEMETRY

Teleoperator Maneuvering System
　USE　TELEOPERATORS

TELEOPERATORS

TELEPHONES

Telephones, Radio
　USE　RADIOTELEPHONES

TELEPHONY

Telephotometers
　USE　TELEPHOTOMETRY

TELEPHOTOMETRY

TELEPRINTERS

TELESAT Canada A
　USE　ANIK 1

TELESAT Canada B
USE   ANIK 2

TELESAT Canada C
USE   ANIK 3

TELESAT Canada 3
USE   ANIK 3

Telescope Facility, Space Infrared
USE   SPACE INFRARED TELESCOPE FACILITY

Telescope Facility, Spacelab UV-Optical
USE   STARLAB

Telescope, Goddard Experiment Package
USE   PARTICLE TELESCOPES

Telescope, Grazing Incidence Solar
USE   GRIST (TELESCOPE)

(Telescope), GRIST
USE   GRIST (TELESCOPE)

Telescope, Hubble Space
USE   HUBBLE SPACE TELESCOPE

Telescope, Kilometer Wave Orbiting
USE   KILOMETER WAVE ORBITING TELESCOPE

Telescope, Large Space
USE   HUBBLE SPACE TELESCOPE

(Telescope), LIRTS
USE   LIRTS (TELESCOPE)

Telescope Mount, Apollo
USE   APOLLO TELESCOPE MOUNT

Telescope On Spacelab, Large Infrared
USE   LIRTS (TELESCOPE)

Telescope, Solar Optical
USE   SOLAR OPTICAL TELESCOPE

Telescope, Space
USE   HUBBLE SPACE TELESCOPE

Telescope, Starsat
USE   STARSAT TELESCOPE

Telescope, Stratoscope 1
USE   STRATOSCOPE TELESCOPES

Telescope, Stratoscope 2
USE   STRATOSCOPE TELESCOPES

TELESCOPES

Telescopes, Astronomical
USE   ASTRONOMICAL TELESCOPES

Telescopes, Circumsolar
USE   CIRCUMSOLAR TELESCOPES

Telescopes, Diffraction
USE   SPECTROSCOPIC TELESCOPES

Telescopes, Electron
USE   PARTICLE TELESCOPES

Telescopes, Gamma Ray
USE   GAMMA RAY TELESCOPES

Telescopes, GEP
USE   PARTICLE TELESCOPES

Telescopes, Infrared
USE   INFRARED TELESCOPES

Telescopes, Manned Orbital
USE   MANNED ORBITAL TELESCOPES

Telescopes), MOT (Orbital
USE   MANNED ORBITAL TELESCOPES

Telescopes, Multispectral Tracking
USE   MULTISPECTRAL TRACKING TELESCOPES

Telescopes, Particle
USE   PARTICLE TELESCOPES

Telescopes, Proton
USE   PARTICLE TELESCOPES

Telescopes, Radio
USE   RADIO TELESCOPES

Telescopes, Reflecting
USE   REFLECTING TELESCOPES

Telescopes, Refracting
USE   REFRACTING TELESCOPES

Telescopes, Schmidt
USE   SCHMIDT TELESCOPES

Telescopes, Spaceborne
USE   SPACEBORNE TELESCOPES

Telescopes, Spectroscopic
USE   SPECTROSCOPIC TELESCOPES

Telescopes, Stratoscope
USE   STRATOSCOPE TELESCOPES

Telescopes, Ultraviolet
USE   ULTRAVIOLET TELESCOPES

Telescopes, X Ray
USE   X RAY TELESCOPES

Telescoping Structures
USE   FOLDING STRUCTURES

TELETYPEWRITER SYSTEMS

TELETYPEWRITERS

TELEVISION CAMERAS

Television, Closed Circuit
USE   CLOSED CIRCUIT TELEVISION

Television, Color
USE   COLOR TELEVISION

Television, Digital
USE   DIGITAL TELEVISION

Television, Digital Spacecraft
USE   DIGITAL SPACECRAFT TELEVISION

Television, Educational
USE   EDUCATIONAL TELEVISION

TELEVISION EQUIPMENT

TELEVISION RECEIVERS

TELEVISION RECEPTION

Television, Satellite
USE   SATELLITE TELEVISION

Television, Spacecraft
USE   SPACECRAFT TELEVISION

Television, Stereo
USE   STEREOTELEVISION

Television System, Pilot Landing Aid
USE   PLAT SYSTEM

Television System, Ranger Block 3
USE   RANGER BLOCK 3 TELEVISION SYSTEM

TELEVISION SYSTEMS

TELEVISION TRANSMISSION

Tellegen Theory
USE   GYRATORS
        NETWORK SYNTHESIS
        NETWORK ANALYSIS

Teller Effect, Jahn-
USE   JAHN-TELLER EFFECT

TELLURIC CURRENTS

TELLURIC LINES

TELLURIDES

Tellurides, Bismuth
USE   BISMUTH TELLURIDES

Tellurides, Cadmium
USE   CADMIUM TELLURIDES

Tellurides, Cadmium Mercury
USE   MERCURY CADMIUM TELLURIDES

Tellurides, Indium
USE   INDIUM TELLURIDES

Tellurides, Lanthanum
USE   LANTHANUM TELLURIDES

Tellurides, Lead
USE   LEAD TELLURIDES

Tellurides, Mercury
USE   MERCURY TELLURIDES

Tellurides, Mercury Cadmium
USE   MERCURY CADMIUM TELLURIDES

Tellurides, Tin
USE   TIN TELLURIDES

Tellurides, Zinc
USE   ZINC TELLURIDES

TELLURIUM

TELLURIUM ALLOYS

TELLURIUM COMPOUNDS

TELLURIUM ISOTOPES

Tellurium 119
USE   TELLURIUM ISOTOPES

TELLUROMETERS

TELSTAR PROJECT

TELSTAR SATELLITES

TELSTAR 1 SATELLITE

TELSTAR 2 SATELLITE

Temco-Vought Aircraft, Ling-
USE   LING-TEMCO-VOUGHT AIRCRAFT

TEMPEL 2 COMET

TEMPER (METALLURGY)

TEMPERATE REGIONS

TEMPERATURE

Temperature Air, High
USE   HIGH TEMPERATURE AIR

Temperature Alloys, High
USE   HEAT RESISTANT ALLOYS

Temperature, Ambient
USE   AMBIENT TEMPERATURE

Temperature, Atmospheric
USE   ATMOSPHERIC TEMPERATURE

Temperature, Auroral
USE   AURORAL TEMPERATURE

Temperature (Biology), Skin
USE   SKIN TEMPERATURE (BIOLOGY)

Temperature, Body
USE   BODY TEMPERATURE

Temperature Brazing, Low
USE   LOW TEMPERATURE BRAZING

Temperature, Brightness
USE   BRIGHTNESS TEMPERATURE

Temperature, Combustion
USE   COMBUSTION TEMPERATURE

TEMPERATURE COMPENSATION

TEMPERATURE CONTROL

Temperature, Critical
USE   CRITICAL TEMPERATURE

Temperature, Curie
USE   CURIE TEMPERATURE

Temperature, Debye
USE   SPECIFIC HEAT

TEMPERATURE DEPENDENCE

Temperature Differences
USE   TEMPERATURE GRADIENTS

TEMPERATURE DISTRIBUTION

TEMPERATURE EFFECTS

Temperature, Electron
USE   ELECTRON ENERGY

Temperature, Environmental
USE   AMBIENT TEMPERATURE

Temperature Environments, High
USE   HIGH TEMPERATURE ENVIRONMENTS

Temperature Environments, Low
USE   LOW TEMPERATURE ENVIRONMENTS

Temperature Fields
USE   TEMPERATURE DISTRIBUTION

Temperature, Flame
USE   FLAME TEMPERATURE

Temperature Fluids, High
USE   HIGH TEMPERATURE FLUIDS

Temperature, Gas
USE   GAS TEMPERATURE

Temperature Gas Cooled Reactors, High
USE   HIGH TEMPERATURE GAS COOLED
        REACTORS

Temperature Gases, High
USE   HIGH TEMPERATURE GASES

Temperature, Geo
USE   GEOTEMPERATURE

TEMPERATURE GRADIENTS

Temperature, High
USE   HIGH TEMPERATURE

Temperature, Ignition
USE   IGNITION TEMPERATURE

Temperature Indicators
USE   INDICATING INSTRUMENTS
        TEMPERATURE MEASURING INSTRUMENTS

Temperature, Inlet
USE   INLET TEMPERATURE

Temperature Instruments
USE   TEMPERATURE MEASURING INSTRUMENTS

Temperature, International Practical
USE   TEMPERATURE SCALES

TEMPERATURE INVERSIONS

Temperature, Ion
USE   ION TEMPERATURE

Temperature, Ionospheric
USE   IONOSPHERIC TEMPERATURE

Temperature, Low
USE   LOW TEMPERATURE

Temperature Lubricants, High
USE   HIGH TEMPERATURE LUBRICANTS

Temperature, Lunar
USE   LUNAR TEMPERATURE

Temperature Materials, High
USE   REFRACTORY MATERIALS

TEMPERATURE MEASUREMENT

TEMPERATURE MEASURING INSTRUMENTS

Temperature, Neel
USE   NEEL TEMPERATURE

Temperature, Noise
USE   NOISE TEMPERATURE

Temperature (Non-Biological), Body
USE   TEMPERATURE

Temperature (Non-Biological), Skin
USE   SKIN TEMPERATURE (NON-BIOLOGICAL)

Temperature Nuclear Reactors, High
USE   HIGH TEMPERATURE NUCLEAR REACTORS

Temperature, Ocean
USE   OCEAN TEMPERATURE

Temperature, Operating
USE   OPERATING TEMPERATURE

Temperature Parameter, Time
USE   TIME TEMPERATURE PARAMETER

Temperature Physics, Low
USE   LOW TEMPERATURE PHYSICS

Temperature, Planetary
USE   PLANETARY TEMPERATURE

Temperature, Plasma
USE   PLASMA TEMPERATURE

Temperature Plasmas, High
USE   HIGH TEMPERATURE PLASMAS

Temperature Plasmas, Low
USE   COLD PLASMAS

TEMPERATURE PROBES

TEMPERATURE PROFILES

Temperature Propellants, High
USE   HIGH TEMPERATURE PROPELLANTS

TEMPERATURE RATIO

Temperature Regulation, Body
USE   THERMOREGULATION

Temperature Research, High
USE   HIGH TEMPERATURE RESEARCH

Temperature, Room
USE   ROOM TEMPERATURE

Temperature, Satellite
USE   SATELLITE TEMPERATURE

Temperature Scale, Fahrenheit
USE   TEMPERATURE SCALES

TEMPERATURE SCALES

Temperature, Sea Surface
USE   SEA SURFACE TEMPERATURE

TEMPERATURE SENSORS

Temperature, Solar
USE   SOLAR TEMPERATURE

Temperature, Space
USE   SPACE TEMPERATURE

Temperature, Spacecraft
USE   SPACECRAFT TEMPERATURE

Temperature, Spin
USE   SPIN TEMPERATURE

Temperature, Stagnation
USE   STAGNATION TEMPERATURE

Temperature, Stellar
USE   STELLAR TEMPERATURE

Temperature, Subzero
USE   SUBZERO TEMPERATURE

Temperature, Surface
USE   SURFACE TEMPERATURE

Temperature Tests, High
USE   HIGH TEMPERATURE TESTS

Temperature Tests, Low
USE   LOW TEMPERATURE TESTS

Temperature, Transition
USE   TRANSITION TEMPERATURE

Temperature, Wall
USE   WALL TEMPERATURE

Temperature, Water
USE   WATER TEMPERATURE

Temperature Zones, Anomalous
USE   ANOMALOUS TEMPERATURE ZONES

Temperatures, Ultralow
USE   ULTRALOW TEMPERATURES

TEMPERING

TEMPLATES

TEMPORAL DISTRIBUTION

TEMPORAL RESOLUTION

TENDENCIES

TENDONS

TENITE

TENNESSEE

TENNESSEE VALLEY (AL-KY-TN)

TENSILE CREEP

TENSILE DEFORMATION

TENSILE PROPERTIES

TENSILE STRENGTH

TENSILE STRESS

TENSILE TESTS

TENSIOMETERS

TENSION

Tension, Carbon Dioxide
USE   CARBON DIOXIDE TENSION

Tension, Hyper
USE   HYPERTENSION

Tension, Hypo
USE   HYPOTENSION

Tension, Interfacial
USE   INTERFACIAL TENSION

Tension, Oxygen
USE   OXYGEN TENSION

Tension, Surface
USE   INTERFACIAL TENSION

TENSOMETERS

TENSOR ANALYSIS

Tensor Fields
USE TENSORS

TENSORS

Tensors, Stress
USE STRESS TENSORS

Tensors, Transformation
USE TENSORS

TEPHIGRAMS

TERBIUM

TERBIUM ISOTOPES

Terbium 155
USE TERBIUM ISOTOPES

Terbium 161
USE TERBIUM ISOTOPES

TERCOM

TEREPHTHALATE

Terephthalate, Polyethylene
USE POLYETHYLENE TEREPHTHALATE

Term Effects, Long
USE LONG TERM EFFECTS

Term Zonal Earth Energy Experiment, Long
USE LZEEBE SATELLITE

TERMINAL AREA ENERGY MANAGEMENT

TERMINAL BALLISTICS

TERMINAL CONFIGURED VEHICLE PROGRAM

TERMINAL FACILITIES

TERMINAL GUIDANCE

Terminal Measurement System, Earth
USE EARTH TERMINAL MEASUREMENT
SYSTEM

Terminal System, Automated Radar
USE AUTOMATED RADAR TERMINAL SYSTEM

TERMINAL VELOCITY

TERMINALS

Terminals, Data Processing
USE DATA PROCESSING TERMINALS

Terminals, Deepwater
USE DEEPWATER TERMINALS

Terminals, Earth
USE EARTH TERMINALS

Terminals, Electric
USE ELECTRIC TERMINALS

Terminals, Ship
USE SHIP TERMINALS

Terminals, Tanker
USE TANKER TERMINALS

Terminating
USE STOPPING

Termination, Thrust
USE THRUST TERMINATION

TERMINATOR LINES

TERMINOLOGY

TERMS

TERNARY ALLOYS

TERNARY SYSTEMS

Ternary Systems (Digital)
USE DIGITAL SYSTEMS

TERPENES

TERPHENYLS

TERRACES (LANDFORMS)

TERRADYNAMICS

TERRAIN

TERRAIN ANALYSIS

Terrain Contour Matching Navigation System
USE TERCOM

TERRAIN FOLLOWING AIRCRAFT

Terrestr Applic Payloads, Office Of Space &
USE OSTA-1 PAYLOAD
OSTA-2 PAYLOAD

TERRESTRIAL DUST BELT

Terrestrial Interactions, Solar
USE SOLAR TERRESTRIAL INTERACTIONS

Terrestrial Magnetism
USE GEOMAGNETISM

TERRESTRIAL PLANETS

TERRESTRIAL RADIATION

TERRIER MISSILE

Territories, Northwest
USE NORTHWEST TERRITORIES

Territory, Yukon
USE YUKON TERRITORY

TESSERAL HARMONICS

Test Apparatus, Free Flight
USE FREE FLIGHT TEST APPARATUS

Test Apparatus, Hypersonic
USE HYPERSONIC TEST APPARATUS

Test Apparatus, Supersonic
USE SUPERSONIC TEST APPARATUS

Test Beds
USE TEST EQUIPMENT

Test, Body Sway
USE BODY SWAY TEST

Test, Bruceton
USE STATISTICAL TESTS

Test, Carboxyhemoglobin
USE CARBOXYHEMOGLOBIN TEST

TEST CHAMBERS

Test, Charpy Impact
USE CHARPY IMPACT TEST

Test, Drones For Aerodynamic And Struct
USE DAST PROGRAM

Test, Ear Pressure
USE EAR PRESSURE TEST

TEST EQUIPMENT

Test Equipment, Automatic
USE AUTOMATIC TEST EQUIPMENT

TEST FACILITIES

Test Facilities, Rocket
USE ROCKET TEST FACILITIES

Test Facility, Transient Reactor
USE TRANSIENT REACTOR TEST FACILITY

(Test Facility), TREAT
USE TRANSIENT REACTOR TEST FACILITY

TEST FIRING

Test Instruments, Flight
USE FLIGHT TEST INSTRUMENTS

Test, Kolmogoroff-Smirnoff
USE KOLMOGOROFF-SMIRNOFF TEST

Test Loops, Corrosion
USE CORROSION TEST LOOPS

Test, Mann-Whitney-Wilcoxon U
USE MANN-WHITNEY-WILCOXON U TEST

TEST PATTERN GENERATORS

TEST PILOTS

Test Program, Reactor In Flight
USE RIFT (REACTOR IN FLIGHT TEST)

Test Project, Apollo Soyuz
USE APOLLO SOYUZ TEST PROJECT

TEST RANGES

Test Reactor, Plutonium Recycle
USE PLUTONIUM RECYCLE TEST REACTOR

Test Reactors, Advanced
USE ADVANCED TEST REACTORS

Test Reactors, Engineering
USE ENGINEERING TEST REACTORS

Test Reactors, Fast
USE FAST TEST REACTORS

Test Reactors, Heavy Water Components
USE HEAVY WATER COMPONENTS TEST
REACTORS

Test Reactors, Nuclear
USE NUCLEAR RESEARCH AND TEST
REACTORS

Test Reactors, Nuclear Research And
USE NUCLEAR RESEARCH AND TEST
REACTORS

Test), Rift (Reactor In Flight
USE RIFT (REACTOR IN FLIGHT TEST)

Test, Ronchi
USE RONCHI TEST

Test Satellite (Esa), Orbital
USE OTS (ESA)

Test Satellite, Maritime Orbital
USE MAROTS (ESA)

Test Site, Arizona Regional Ecological
USE ARIZONA REGIONAL ECOLOGICAL TEST
SITE

(Test Site), CARETS
USE CENTRAL ATLANTIC REGIONAL ECOL
TEST SITE

Test Site, Central Atlantic Regional Ecol
USE CENTRAL ATLANTIC REGIONAL ECOL
TEST SITE

TEST STANDS

Test Tunnels, Hydraulic
USE HYDRAULIC TEST TUNNELS

TEST VEHICLES

Test Vehicles, Flight
USE FLIGHT TEST VEHICLES

Test, Weber
USE WEBER TEST

Test 1 (Shuttle), Orbital Flight
USE   SPACE TRANSPORTATION SYSTEM 1
           FLIGHT

Test 1, Space Shuttle Orbital Flight
USE   SPACE TRANSPORTATION SYSTEM 1
           FLIGHT

Test 2 (Shuttle), Orbital Flight
USE   SPACE TRANSPORTATION SYSTEM 2
           FLIGHT

Test 2, Space Shuttle Orbital Flight
USE   SPACE TRANSPORTATION SYSTEM 2
           FLIGHT

Test 3 (Shuttle), Orbital Flight
USE   SPACE TRANSPORTATION SYSTEM 3
           FLIGHT

Test 3, Space Shuttle Orbital Flight
USE   SPACE TRANSPORTATION SYSTEM 3
           FLIGHT

Test 4 (Shuttle), Orbital Flight
USE   SPACE TRANSPORTATION SYSTEM 4
           FLIGHT

Test 4, Space Shuttle Orbital Flight
USE   SPACE TRANSPORTATION SYSTEM 4
           FLIGHT

Testers
USE   TEST EQUIPMENT

Testers, Compression
USE   COMPRESSION TESTS

Testers, Fokker Bond
USE   ADHESION TESTS

TESTES

Testing
USE   TESTS

Testing Laboratories, Engine
USE   ENGINE TESTING LABORATORIES

Testing Machines
USE   TEST EQUIPMENT

Testing Machines, Fatigue
USE   FATIGUE TESTING MACHINES

Testing Machines, Impact
USE   IMPACT TESTING MACHINES

Testing Machines, Load
USE   LOAD TESTING MACHINES

Testing Machines, Vibration
USE   VIBRATION SIMULATORS

Testing Reactor, Physical Constants
USE   WATER COOLED REACTORS
           NUCLEAR RESEARCH AND TEST
           REACTORS

Testing Reactors, Materials
USE   NUCLEAR RESEARCH AND TEST
           REACTORS

Testing, Resonance
USE   RESONANCE TESTING

TESTING TIME

TESTS

Tests, Accelerated Life
USE   ACCELERATED LIFE TESTS

Tests, Adhesion
USE   ADHESION TESTS

Tests, Altitude
USE   ALTITUDE TESTS

Tests, Bend
USE   BEND TESTS

Tests, Burst
USE   BURST TESTS

Tests, Captive
USE   CAPTIVE TESTS

Tests, Chemical
USE   CHEMICAL TESTS

Tests, Cold Flow
USE   COLD FLOW TESTS

Tests, Cold Weather
USE   COLD WEATHER TESTS

Tests, Compression
USE   COMPRESSION TESTS

Tests, Corrosion
USE   CORROSION TESTS

Tests, Creep
USE   CREEP TESTS

Tests, Damping
USE   DAMPING TESTS

Tests, Destructive
USE   DESTRUCTIVE TESTS

Tests, Drop
USE   DROP TESTS

Tests, Drop Weight
USE   DROP TESTS

Tests, Dynamic
USE   DYNAMIC TESTS

Tests, Electric Equipment
USE   ELECTRIC EQUIPMENT TESTS

Tests, Electronic Equipment
USE   ELECTRONIC EQUIPMENT TESTS

Tests, Engine
USE   ENGINE TESTS

Tests, Environmental
USE   ENVIRONMENTAL TESTS

Tests, Fatigue
USE   FATIGUE TESTS

Tests, Flight
USE   FLIGHT TESTS

Tests, Flight Stability
USE   FLIGHT STABILITY TESTS

Tests, Fuel
USE   FUEL TESTS

Tests, Full Scale
USE   FULL SCALE TESTS

Tests, Ground
USE   GROUND TESTS

Tests, Hardness
USE   HARDNESS TESTS

Tests, Heat
USE   HIGH TEMPERATURE TESTS

Tests, High Altitude
USE   HIGH ALTITUDE TESTS

Tests, High Temperature
USE   HIGH TEMPERATURE TESTS

Tests, Impact
USE   IMPACT TESTS

Tests, Load
USE   LOAD TESTS

Tests, Low Temperature
USE   LOW TEMPERATURE TESTS

Tests, Lubricant
USE   LUBRICANT TESTS

Tests, Materials
USE   MATERIALS TESTS

Tests, Meteorite Compression
USE   MECHANICAL PROPERTIES
           METEORITES
           COMPRESSION TESTS

Tests, Missile
USE   MISSILE TESTS

Tests, Nondestructive
USE   NONDESTRUCTIVE TESTS

Tests, Notch
USE   NOTCH TESTS

Tests, Orbital Space
USE   ORBITAL SPACE TESTS

Tests, Patch
USE   PATCH TESTS

Tests, Performance
USE   PERFORMANCE TESTS

Tests, Personality
USE   PERSONALITY TESTS

Tests, Physiological
USE   PHYSIOLOGICAL TESTS

Tests, Prefiring
USE   PREFIRING TESTS

Tests, Prelaunch
USE   PRELAUNCH TESTS

Tests, Propellant
USE   PROPELLANT TESTS

Tests, Psychological
USE   PSYCHOLOGICAL TESTS

Tests, Railroad Humping
USE   RAILROAD HUMPING TESTS

Tests, Rank
USE   RANK TESTS

Tests, Reactor Startup
USE   REACTOR STARTUP TESTS

Tests, Rorschach
USE   RORSCHACH TESTS

Tests, Salt Spray
USE   SALT SPRAY TESTS

Tests), SERT (Rocket
USE   SPACE ELECTRIC ROCKET TESTS

Tests, Shock
USE   SHOCK TESTS

Tests (Shuttle), Orbital Flight
USE   SPACE TRANSPORTATION SYSTEM
           FLIGHTS

Tests, Snellen
USE   SNELLEN TESTS

Tests, Space Electric Rocket
USE   SPACE ELECTRIC ROCKET TESTS

Tests, Space Shuttle Orbital Flight
USE   SPACE TRANSPORTATION SYSTEM
           FLIGHTS

Tests, Spacecraft Prelaunch
USE   SPACE VEHICLE CHECKOUT PROGRAM

Tests, Spin
USE   SPIN TESTS

Tests, Stability
USE   STABILITY TESTS

Tests, Static
USE   STATIC TESTS

Tests, Statistical
USE   STATISTICAL TESTS

Tests, Stroking
USE   STROKING TESTS

Tests (STS), Approach And Landing
USE   APPROACH AND LANDING TESTS (STS)

Tests, Tensile
USE   TENSILE TESTS

Tests, Thermal Cycling
USE   THERMAL CYCLING TESTS

Tests, Thermal Vacuum
USE   THERMAL VACUUM TESTS

Tests, Ultrasonic
USE   ULTRASONIC TESTS

Tests, Underwater
USE   UNDERWATER TESTS

Tests, Vacuum
USE   VACUUM TESTS

Tests, Vestibular
USE   VESTIBULAR TESTS

Tests, Vibration
USE   VIBRATION TESTS

Tests, Water Tunnel
USE   WATER TUNNEL TESTS

Tests, Wear
USE   WEAR TESTS

Tests, Weld
USE   WELD TESTS

Tests, Whirling
USE   SPIN TESTS

Tests, Wind Tunnel
USE   WIND TUNNEL TESTS

Tests, Wind Tunnel Stability
USE   WIND TUNNEL STABILITY TESTS

Tests, Wing Flow Method
USE   WING FLOW METHOD TESTS

TETHERED BALLOONS

TETHERED SATELLITES

TETHERING

TETHERLINES

TETHYS

TETRABUTYLS

Tetrachloride, Carbon
USE   CARBON TETRACHLORIDE

Tetrachloride Poisoning, Carbon
USE   CARBON TETRACHLORIDE POISONING

Tetrachloride, Silicon
USE   SILICON TETRACHLORIDE

TETRACHLORIDES

Tetrachloromethane
USE   CARBON TETRACHLORIDE

TETRACYCLINES

TETRAD THEORY

TETRAETHYL ORTHOCARBONATES

TETRAETHYL ORTHOSILICATE

Tetrafluoride, Carbon
USE   CARBON TETRAFLUORIDE

TETRAFLUOROHYDRAZINE

TETRAGONS

TETRAHEDRONS

TETRAHYDROFURAN

Tetranitramine, Cyclotetramethylene
USE   HMX

Tetranitramine, Polybutadiene
USE   POLYBUTADIENE TETRANITRAMINE

Tetranitrate, Pentaerythritol
USE   PETN

Tetranitrotetrazacyclooctane
USE   HMX

TETRAPHENYLS

TETRAZOLES

TETRODES

Tetroons
USE   SUPERPRESSURE BALLOONS

Tetroxide, Nitrogen
USE   NITROGEN TETROXIDE

TETRYL

TEXAS

Texoma (OK-TX), Lake
USE   LAKE TEXOMA (OK-TX)

TEXTBOOKS

TEXTILES

TEXTS

TEXTURES

TF-30 ENGINE

TF-34 ENGINE

TF-41 ENGINE

TFX Aircraft
USE   F-111 AIRCRAFT

Th
USE   THULIUM

TH-55 HELICOPTER

THAILAND

THALAMUS

Thalamus, Hypo
USE   HYPOTHALAMUS

THALLIUM

THALLIUM ALLOYS

THALLIUM COMPOUNDS

THALLIUM ISOTOPES

Thawing
USE   MELTING

THEMATIC MAPPING

THEMIS PROJECT

THEODOLITES

Theodolites, Cine
USE   CINETHEODOLITES

THEODORSEN TRANSFORMATION

Theorem, Addition
USE   ADDITION THEOREM

Theorem, Bayes
USE   BAYES THEOREM

Theorem, Bernoulli
USE   BERNOULLI THEOREM

Theorem, Binomial
USE   BINOMIAL THEOREM

Theorem, Castigliano Variational
USE   CASTIGLIANO VARIATIONAL THEOREM

Theorem, Duality
USE   DUALITY THEOREM

Theorem, Equipartition
USE   EQUIPARTITION THEOREM

Theorem, Floquet
USE   FLOQUET THEOREM

Theorem, Gauss-Markov
USE   GAUSS-MARKOV THEOREM

Theorem, Green
USE   GREEN'S FUNCTIONS

Theorem, Hellmann-Feynman
USE   HELLMANN-FEYNMAN THEOREM

Theorem, Kakutani
USE   KAKUTANI THEOREM

Theorem, Lebesgue
USE   LEBESGUE THEOREM

Theorem, Liouville
USE   LIOUVILLE THEOREM

Theorem, Michell
USE   MICHELL THEOREM

Theorem, Nernst Heat
USE   NERNST-ETTINGSHAUSEN EFFECT

Theorem, Pomeranchuk
USE   POMERANCHUK THEOREM

Theorem, Poynting
USE   POYNTING THEOREM

THEOREM PROVING

Theorem, Reciprocity
USE   RECIPROCITY THEOREM

Theorem, Richards
USE   RICHARDS THEOREM

Theorem, Riesz
USE   RIESZ THEOREM

Theorem, Schauder Fixpoint
USE   SCHAUDER FIXPOINT THEOREM

Theorem, Similarity
USE   SIMILARITY THEOREM

Theorem, Taylor
USE   TAYLOR SERIES

Theorem, Uniqueness
USE   UNIQUENESS THEOREM

Theorem (Vector Calculus), Stokes
USE   STOKES THEOREM (VECTOR CALCULUS)

Theorem, Virial
USE   VIRIAL THEOREM

THEOREMS

Theorems, Existence
USE   EXISTENCE THEOREMS

Theorems, Reciprocal
USE   RECIPROCAL THEOREMS

THEORETICAL PHYSICS

THEORIES

Theories, Bimetric
USE   BIMETRIC THEORIES

Theory, Abrikosov
USE   ABRIKOSOV THEORY

Theory (Algebra), Field
USE   FIELD THEORY (ALGEBRA)

Theory, Atomic
USE   ATOMIC THEORY

Theory, Automata
USE   AUTOMATA THEORY

Theory, Bardeen-Cooper-Schrieffer
USE   BCS THEORY

Theory, BCS
USE   BCS THEORY

Theory, Bellman
USE   BELLMAN THEORY

Theory, Bending
USE   BENDING THEORY

Theory, Bessel-Bredichin
USE   BESSEL-BREDICHIN THEORY

Theory, Bogoliubov
USE   BOGOLIUBOV THEORY

Theory, Bohr
USE   BOHR THEORY

Theory, Born-Infeld
USE   BORN-INFELD THEORY

Theory, Catastrophe
USE   CATASTROPHE THEORY

Theory, Chapman-Enskog
USE   CHAPMAN-ENSKOG THEORY

Theory, Communication
USE   COMMUNICATION THEORY

Theory, Control
USE   CONTROL THEORY

Theory, Crocco-Lee
USE   CROCCO-LEE THEORY

Theory, Debye-Huckel
USE   DEBYE-HUCKEL THEORY

Theory, Decision
USE   DECISION THEORY

Theory, Diffusion
USE   DIFFUSION THEORY

Theory, Disturbance
USE   PERTURBATION THEORY

Theory, Dynamo
USE   DYNAMO THEORY

Theory, Dyson
USE   DYSON THEORY

Theory, Enskog-Chapman
USE   CHAPMAN-ENSKOG THEORY

Theory, Eyring
USE   EYRING THEORY

Theory, Field Mode
USE   FIELD MODE THEORY

Theory, Finite Difference
USE   FINITE DIFFERENCE THEORY

Theory, Flow
USE   FLOW THEORY

Theory, Fluctuation
USE   FLUCTUATION THEORY

Theory, Foster
USE   FOSTER THEORY

Theory, Game
USE   GAME THEORY

Theory, Gauge
USE   GAUGE THEORY

Theory, Gestalt
USE   GESTALT THEORY

Theory, Glauber
USE   GLAUBER THEORY

Theory, Goal
USE   GOAL THEORY

Theory, Graph
USE   GRAPH THEORY

Theory, Gravitation
USE   GRAVITATION THEORY

Theory, Group
USE   GROUP THEORY

Theory, Gumbel
USE   RANGE (EXTREMES)

Theory, Hansen Lunar
USE   HANSEN LUNAR THEORY

Theory, Heisenberg
USE   HEISENBERG THEORY

Theory, Hill Lunar
USE   HILL LUNAR THEORY

Theory, Homotopy
USE   HOMOTOPY THEORY

Theory, Hueckel
USE   HUECKEL THEORY

Theory, Information
USE   INFORMATION THEORY

Theory, Jeans
USE   JEANS THEORY

Theory, Kinetic
USE   KINETIC THEORY

Theory, Kolmogoroff
USE   KOLMOGOROFF THEORY

Theory, Learning
USE   LEARNING THEORY

Theory, Malkus
USE   MALKUS THEORY

Theory, Manning
USE   MANNING THEORY

Theory, Many Particle
USE   MANY BODY PROBLEM

Theory, Matrix
USE   MATRIX THEORY

Theory, Measure
USE   MEASURE AND INTEGRATION

Theory, Membrane
USE   STRUCTURAL ANALYSIS

Theory, Michaelis
USE   MICHAELIS THEORY

Theory, Mie
USE   MIE SCATTERING

Theory, Milankovitch
USE   CLIMATOLOGY

Theory, Mixing Length Flow
USE   MIXING LENGTH FLOW THEORY

Theory, Molecular
USE   MOLECULAR THEORY

Theory, Momentum
USE   MOMENTUM THEORY

Theory, Newton
USE   NEWTON THEORY

Theory, Nonadiabatic
USE   NONADIABATIC THEORY

Theory, Number
USE   NUMBER THEORY

Theory Of Diffraction, Geometrical
USE   GEOMETRICAL THEORY OF DIFFRACTION

Theory, Opik
USE   OPIK THEORY

Theory, Orthogonal Multiplexing
USE   ORTHOGONAL MULTIPLEXING THEORY

Theory, Particle
USE   PARTICLE THEORY

Theory, Perturbation
USE   PERTURBATION THEORY

Theory (Physics), Field
USE   FIELD THEORY (PHYSICS)

Theory, Piston
USE   PISTON THEORY

Theory, Plasma
USE   PLASMA PHYSICS

Theory, Plate
USE   PLATE THEORY

Theory, Population
USE   POPULATION THEORY

Theory, Potential
USE   POTENTIAL THEORY

Theory, Probability
USE   PROBABILITY THEORY

Theory, Quantum
USE   QUANTUM THEORY

Theory, Queueing
USE   QUEUEING THEORY

Theory, Reissner
USE   REISSNER THEORY

Theory, Relativistic
USE   RELATIVISTIC THEORY

Theory, S Matrix
USE   S MATRIX THEORY

Theory), Saddle Points (Game
USE   SADDLE POINTS (GAME THEORY)

Theory, Set
USE   SET THEORY

Theory, Shannon Information
USE   INFORMATION THEORY

Theory, Shell
USE   SHELL THEORY

Theory, Spectral
USE   SPECTRAL THEORY

Theory, Statistical Communication
USE   COMMUNICATION THEORY

Theory, Statistical Decision
USE   STATISTICAL DECISION THEORY

Theory), Strong Interactions (Field
USE   STRONG INTERACTIONS (FIELD THEORY)

Theory, Sturm-Liouville
USE   STURM-LIOUVILLE THEORY

Theory, Switching
USE   SWITCHING THEORY

Theory, Tellegen
USE   NETWORK ANALYSIS
NETWORK SYNTHESIS
GYRATORS

Theory, Tetrad
USE   TETRAD THEORY

Theory, Thomas-Fermi
USE   THOMAS-FERMI MODEL

Theory, Transport
USE   TRANSPORT THEORY

Theory, Unified Field
USE   UNIFIED FIELD THEORY

Theory, Vinti
USE   VINTI THEORY

Theory, Von Mises
USE   STRESS FUNCTIONS

Theory), Weak Interactions (Field
USE   WEAK INTERACTIONS (FIELD THEORY)

Theory, Wightman
USE   QUANTUM THEORY
RELATIVISTIC THEORY
FIELD THEORY (PHYSICS)

Theory, Yang-Mills
USE   YANG-MILLS THEORY

Theory, Young-Helmholtz
USE   YOUNG-HELMHOLTZ THEORY

THERAPY

Therapy, Chemo
USE   CHEMOTHERAPY

Therapy, Drug
USE   CHEMOTHERAPY

Therapy, Psycho
USE   PSYCHOTHERAPY

Therapy, Radiation
USE   RADIATION THERAPY

THERMAL ABSORPTION

Thermal Accommodation Coefficients
USE   ACCOMMODATION COEFFICIENT

Thermal Agitation
USE   THERMAL ENERGY

THERMAL ANALYSIS

Thermal Analysis, Differential
USE   THERMAL ANALYSIS

THERMAL BATTERIES

THERMAL BLOOMING

THERMAL BOUNDARY LAYER

THERMAL BUCKLING

THERMAL COMFORT

THERMAL CONDUCTIVITY

THERMAL CONDUCTIVITY GAGES

THERMAL CONDUCTORS

THERMAL CONTROL COATINGS

Thermal Convection
USE   FREE CONVECTION

Thermal Currents
USE   CONVECTIVE FLOW

THERMAL CYCLING TESTS

THERMAL DECOMPOSITION

Thermal Defocusing
USE   THERMAL BLOOMING

THERMAL DEGRADATION

THERMAL DIFFUSION

THERMAL DIFFUSIVITY

THERMAL DISSOCIATION

Thermal Effects
USE   TEMPERATURE EFFECTS

Thermal Efficiency
USE   THERMODYNAMIC EFFICIENCY

Thermal Electric Power Plants, Solar
USE   SOLAR THERMAL ELECTRIC POWER
PLANTS

THERMAL EMISSION

THERMAL ENERGY

Thermal Energy Conversion, Ocean
USE   OCEAN THERMAL ENERGY CONVERSION

Thermal Energy Storage
USE   HEAT STORAGE

THERMAL ENVIRONMENTS

THERMAL EXPANSION

THERMAL FATIGUE

Thermal Gravimetry
USE   THERMOGRAVIMETRY

THERMAL INSTABILITY

THERMAL INSULATION

THERMAL MAPPING

THERMAL NEUTRONS

THERMAL NOISE

THERMAL PLASMAS

THERMAL POLLUTION

Thermal Power
USE   TURBOGENERATORS

Thermal Properties
USE   THERMODYNAMIC PROPERTIES

Thermal Propulsion, Solar
USE   SOLAR THERMAL PROPULSION

THERMAL PROTECTION

THERMAL RADIATION

THERMAL REACTORS

THERMAL RESISTANCE

THERMAL RESOURCES

Thermal Shielding
USE   HEAT SHIELDING

THERMAL SHOCK

THERMAL SIMULATION

THERMAL STABILITY

THERMAL STRESSES

THERMAL VACUUM TESTS

THERMALIZATION (ENERGY ABSORPTION)

Thermalization, Neutron
USE   NEUTRON THERMALIZATION

THERMICONS

THERMIONIC CATHODES

Thermionic Conversion Systems
USE   THERMIONIC POWER GENERATION

THERMIONIC CONVERTERS

THERMIONIC DIODES

THERMIONIC EMISSION

THERMIONIC EMITTERS

THERMIONIC POWER GENERATION

Thermionic Reactors
USE   ION ENGINES
NUCLEAR ROCKET ENGINES

THERMIONICS

THERMISTORS

THERMITES

THERMOBALANCES

THERMOCHEMICAL PROPERTIES

THERMOCHEMISTRY

Thermochemistry, Aero
USE   AEROTHERMOCHEMISTRY

THERMOCHROMATIC MATERIALS

THERMOCLINES

THERMOCOUPLE PYROMETERS

THERMOCOUPLES

THERMODYNAMIC COUPLING

THERMODYNAMIC CYCLES

THERMODYNAMIC EFFICIENCY

THERMODYNAMIC EQUILIBRIUM

THERMODYNAMIC PROPERTIES

THERMODYNAMICS

Thermodynamics, Aero
USE   AEROTHERMODYNAMICS

Thermodynamics, Nonequilibrium
USE   NONEQUILIBRIUM THERMODYNAMICS

THERMOELASTICITY

Thermoelasticity, Aero
USE   AEROTHERMOELASTICITY

Thermoelectric Conversion Systems
USE   THERMOELECTRIC POWER GENERATION

THERMOELECTRIC COOLING

THERMOELECTRIC GENERATORS

THERMOELECTRIC MATERIALS

Thermoelectric Outer Planet Spacecraft
USE   TOPS (SPACECRAFT)

THERMOELECTRIC POWER GENERATION

Thermoelectric Spacecraft
USE   TOPS (SPACECRAFT)

THERMOELECTRICITY

THERMOELEMENT AMMETERS

Thermograms
USE   TEMPERATURE MEASURING INSTRUMENTS
RECORDING INSTRUMENTS

THERMOGRAPHY

THERMOGRAVIMETRY

THERMOHYDRAULICS

THERMOLUMINESCENCE

Thermomagnadynamics
USE   THERMOMAGNETIC EFFECTS

THERMOMAGNETIC COOLING

THERMOMAGNETIC EFFECTS

Thermomagnetism
USE   THERMOMAGNETIC EFFECTS

THERMOMECHANICAL TREATMENT

Thermomechanics
USE   THERMODYNAMICS

THERMOMETERS

Thermometers, Resistance
USE   RESISTANCE THERMOMETERS

Thermometry
USE   TEMPERATURE MEASUREMENT

THERMOMIGRATION

Thermonuclear Energy
USE   THERMONUCLEAR POWER GENERATION

THERMONUCLEAR EXPLOSIONS

THERMONUCLEAR POWER GENERATION

Thermonuclear Propulsion
USE   NUCLEAR PROPULSION

THERMONUCLEAR REACTIONS

Thermonuclear Reactor, Astron
USE   ASTRON THERMONUCLEAR REACTOR

Thermonuclear Reactor, Zeta
USE   ZETA THERMONUCLEAR REACTOR

THERMOPHILES

THERMOPHILIC PLANTS

††**THERMOPHYSICAL PROPERTIES**

Thermophysics
USE   THERMODYNAMICS

THERMOPILES

THERMOPLASTIC FILMS

THERMOPLASTIC RESINS

THERMOPLASTICITY

THERMORECEPTORS

THERMOREGULATION

THERMOSETTING RESINS

THERMOSIPHONS

THERMOSPHERE

Thermostability
USE   THERMAL STABILITY

THERMOSTATS

Thermotropism
USE   TEMPERATURE EFFECTS
ANISOTROPY

THERMOVISCOELASTICITY

THESAURI

THESES

THETA PINCH

THIAMINE

THIAZINE (TRADEMARK)

Thiazines, Pheno
USE   PHENOTHIAZINES

THICK FILMS

THICK PLATES

THICK WALLS

THICKENERS

THICKENERS (EQUIPMENT)

THICKENERS (MATERIALS)

THICKNESS

Thickness, Airfoil
USE   AIRFOIL PROFILES

Thickness, Film
USE   FILM THICKNESS

Thickness, Optical
USE   OPTICAL THICKNESS

THICKNESS RATIO

Thickness, Target
USE   TARGET THICKNESS

THIGH

THIN AIRFOILS

THIN BODIES

THIN FILMS

THIN LAYER CHROMATOGRAPHY

THIN PLATES

THIN WALLED SHELLS

THIN WALLS

THIN WINGS

Thinners
USE   SOLVENTS

THIOLS

THIOPLASTICS

THIOUREAS

THIURONIUM

Thixotropic Propellants
USE   GELLED ROCKET PROPELLANTS

THIXOTROPY

THOMAS-FERMI MODEL

Thomas-Fermi Theory
USE   THOMAS-FERMI MODEL

Thomson Effect
USE   THERMOELECTRICITY

Thomson Effect, Joule-
USE   JOULE-THOMSON EFFECT

Thomson Method, Milne-
USE   MILNE-THOMSON METHOD

THOMSON SCATTERING

THOR ABLE ROCKET VEHICLE

THOR AGENA LAUNCH VEHICLE

THOR DELTA LAUNCH VEHICLE

THOR LAUNCH VEHICLES

THORAD LAUNCH VEHICLES

THORAX

Thorax, Pneumo
USE   PNEUMOTHORAX

THORIUM

THORIUM ALLOYS

THORIUM COMPOUNDS

THORIUM FLUORIDES

THORIUM ISOTOPES

THORIUM OXIDES

Thorium 228
USE   THORIUM ISOTOPES

Thorium 230
USE   THORIUM ISOTOPES

Thorium 234
USE   THORIUM ISOTOPES

Thoron
USE   RADON ISOTOPES

THREADS

THREAT EVALUATION

THREE AXIS STABILIZATION

THREE BODY PROBLEM

THREE DIMENSIONAL BODIES

THREE DIMENSIONAL BOUNDARY LAYER

THREE DIMENSIONAL COMPOSITES

THREE DIMENSIONAL FLOW

THREE DIMENSIONAL MOTION

THRESHOLD CURRENTS

Threshold, Damage
USE   YIELD POINT

THRESHOLD DETECTORS (DOSIMETERS)

THRESHOLD GATES

THRESHOLD LOGIC

Threshold, Noise
USE   NOISE THRESHOLD

Threshold Shift
USE   THRESHOLDS

THRESHOLDS

THRESHOLDS (PERCEPTION)

THROATS

THROMBIN

THROMBOCYTES

THROMBOPENIA

THROMBOPLASTIN

THROMBOSIS

THROTTLING

THROWING

THRUST

THRUST AUGMENTATION

THRUST BEARINGS

THRUST CHAMBER PRESSURE

THRUST CHAMBERS

Thrust Coefficients, Nozzle
　　USE　NOZZLE THRUST COEFFICIENTS

THRUST CONTROL

THRUST DISTRIBUTION

Thrust Faults
　　USE　GEOLOGICAL FAULTS

Thrust, High
　　USE　HIGH THRUST

Thrust, Jet
　　USE　JET THRUST

Thrust, Leading Edge
　　USE　LEADING EDGE THRUST

THRUST LOADS

Thrust, Low
　　USE　LOW THRUST

THRUST MEASUREMENT

Thrust, Micro
　　USE　MICROTHRUST

Thrust Nozzles, Dual
　　USE　DUAL THRUST NOZZLES

Thrust Power
　　USE　THRUST

THRUST PROGRAMMING

Thrust Programming, Optimum
　　USE　THRUST PROGRAMMING

Thrust Propulsion, Low
　　USE　LOW THRUST PROPULSION

Thrust, Retro
　　USE　RETROTHRUST

THRUST REVERSAL

Thrust, Rocket
　　USE　ROCKET THRUST

Thrust, Static
　　USE　STATIC THRUST

THRUST TERMINATION

Thrust, Variable
　　USE　VARIABLE THRUST

THRUST VECTOR CONTROL

THRUST-WEIGHT RATIO

Thrustor Engines, Radio Frequency Ion
　　USE　RIT ENGINES

THRUSTORS

THULIUM

THULIUM COMPOUNDS

THULIUM ISOTOPES

Thulium 171
　　USE　THULIUM ISOTOPES

Thunderchief Aircraft
　　USE　F-105 AIRCRAFT

THUNDERSTORMS

THYMIDINE

THYMINE

THYMOL

THYMUS GLAND

THYRATRONS

THYRISTORS

THYROID GLAND

THYROXINE

Ti
　　USE　TITANIUM

TIBET

TIBIA

TID
　　USE　TRAVELING IONOSPHERIC DISTURBANCES

TIDAL FLATS

Tidal Oscillation
　　USE　TIDES

TIDAL WAVES

TIDE POWERED GENERATORS

TIDE POWERED MACHINES

Tide, Red
　　USE　RED TIDE

TIDEPOWER

TIDES

Tides, Atmospheric
　　USE　ATMOSPHERIC TIDES

Tides, Earth
　　USE　EARTH TIDES

Tides, Lunar
　　USE　LUNAR TIDES

TIEBOLTS

TIG Welding
　　USE　GAS TUNGSTEN ARC WELDING

TIGHTNESS

TILES

Tilt
　　USE　ATTITUDE (INCLINATION)

TILT ROTOR AIRCRAFT

TILT ROTOR RESEARCH AIRCRAFT PROGRAM

TILT WING AIRCRAFT

TILTED PROPELLERS

Tilting
　　USE　ATTITUDE (INCLINATION)

TILTING ROTORS

TILTMETERS

Tilts, Ionospheric
　　USE　IONOSPHERIC TILTS

TIMBER IDENTIFICATION

TIMBER INVENTORY

TIMBER VIGOR

TIMBERLINE

TIME

Time, Access
　　USE　ACCESS TIME

Time Between Failures, Mean
　　USE　MTBF

Time, Burning
　　USE　BURNING TIME

Time (Computers), Response
　　USE　RESPONSE TIME (COMPUTERS)

Time (Computers), Run
　　USE　RUN TIME (COMPUTERS)

TIME CONSTANT

Time Constant, Perceptual
　　USE　PERCEPTUAL TIME CONSTANT

Time Continuum, Space-
　　USE　RELATIVITY

Time Delay
　　USE　TIME LAG

TIME DEPENDENCE

Time Devices, Controlled Avalanche Transit
　　USE　CATT DEVICES

Time Diodes, Barrier Injection Transit
　　USE　BARRITT DIODES

TIME DISCRIMINATION

TIME DIVISION MULTIPLE ACCESS

TIME DIVISION MULTIPLEXING

Time, Down
　　USE　DOWNTIME

Time, Ephemeris
　　USE　EPHEMERIS TIME

Time, Firing
　　USE　BURNING TIME

Time, Flight
　　USE　FLIGHT TIME

TIME FUNCTIONS

Time Functions, Space-
　　USE　SPACE-TIME FUNCTIONS

TIME LAG

Time Lapse Photography
　　USE　CHRONOPHOTOGRAPHY

Time, Launch
　　USE　LAUNCH WINDOWS

TIME MARCHING

(Trademark), Plexiglass

(Trademark), Plexiglass
USE   POLYMETHYL METHACRYLATE

(Trademark), Pyrex
USE   BOROSILICATE GLASS

(Trademark), Pyroceram
USE   PYROCERAM (TRADEMARK)

(Trademark), Pyrrones
USE   PYRRONES (TRADEMARK)

(Trademark), Refrasil
USE   SILICON DIOXIDE
FIBERS

(Trademark), RTV-40 Rubber
USE   RTV-40 RUBBER (TRADEMARK)

(Trademark), RTV-60 Rubber
USE   RTV-60 RUBBER (TRADEMARK)

(Trademark), Santowax
USE   SANTOWAX (TRADEMARK)

(Trademark), Scotchlite
USE   SCOTCHLITE (TRADEMARK)

(Trademark), Selsyns
USE   SERVOMOTORS

(Trademark), Skydrol
USE   SKYDROL (TRADEMARK)

(Trademark), Stellite
USE   STELLITE (TRADEMARK)

(Trademark), Styrofoam
USE   STYROFOAM (TRADEMARK)

(Trademark), Teflon
USE   TEFLON (TRADEMARK)

(Trademark), Thiazine
USE   THIAZINE (TRADEMARK)

(Trademark), Viton Rubber
USE   VITON RUBBER (TRADEMARK)

(Trademark), Zircaloy 2
USE   ZIRCALOY 2 (TRADEMARK)

(Trademark), Zircaloys
USE   ZIRCALOYS (TRADEMARK)

(Tradename), Borsic
USE   BORSIC (TRADENAME)

(Tradename), Carbamates
USE   CARBAMATES (TRADENAME)

TRADEOFFS

Trader Aircraft
USE   C-1A AIRCRAFT

TRADESCANTIA

TRADEX RADAR SYSTEM

TRAFFIC

Traffic Advisory And Resolution, Automatic
USE   AUTOMATIC TRAFFIC ADVISORY AND
RESOLUTION

Traffic, Air
USE   AIR TRAFFIC

TRAFFIC CONTROL

Traffic Control, Air
USE   AIR TRAFFIC CONTROL

Traffic Controllers (Personnel), Air
USE   AIR TRAFFIC CONTROLLERS (PERSONNEL)

Traffic Satellites, Location Of Air
USE   LOCATES SYSTEM

Traffic Vehicles, Automated Mixed
USE   AUTOMATED MIXED TRAFFIC VEHICLES

TRAGACANTH

TRAILBLAZER 1 REENTRY VEHICLE

Trailblazer 1 Rocket Vehicle
USE   TRAILBLAZER 1 REENTRY VEHICLE

TRAILBLAZER 2 REENTRY VEHICLE

Trailblazer 2 Rocket Vehicle
USE   TRAILBLAZER 2 REENTRY VEHICLE

TRAILERS

TRAILING EDGE FLAPS

TRAILING EDGES

Trailing Edges, Blunt
USE   BLUNT TRAILING EDGES

Trails
USE   TRACKS

Trails, Condensation
USE   CONTRAILS

Trails, Meteor
USE   METEOR TRAILS

Trails, Smoke
USE   SMOKE TRAILS

Trails, Vapor
USE   CONTRAILS

Trainees
USE   STUDENTS

Trainer, L-29 Jet
USE   L-29 JET TRAINER

Trainers
USE   TRAINING DEVICES

Training
USE   EDUCATION

TRAINING AIRCRAFT

TRAINING ANALYSIS

Training, Astronaut
USE   ASTRONAUT TRAINING

TRAINING DEVICES

Training, Ejection
USE   EJECTION TRAINING

TRAINING EVALUATION

Training, Flight
USE   FLIGHT TRAINING

Training, Gunnery
USE   GUNNERY TRAINING

Training, Maintenance
USE   MAINTENANCE TRAINING

Training, Pilot
USE   PILOT TRAINING

Training, Re
USE   RETRAINING

Training, Simulator
USE   TRAINING SIMULATORS

TRAINING SIMULATORS

Training, Space Flight
USE   SPACE FLIGHT TRAINING

Training, Transfer Of
USE   TRANSFER OF TRAINING

TRAJECTORIES

Trajectories, Abort
USE   ABORT TRAJECTORIES

Trajectories, Ascent
USE   ASCENT TRAJECTORIES

Trajectories, Ballistic
USE   BALLISTIC TRAJECTORIES

Trajectories, Circumlunar
USE   CIRCUMLUNAR TRAJECTORIES

Trajectories, Descent
USE   DESCENT TRAJECTORIES

Trajectories, Earth-Mars
USE   EARTH-MARS TRAJECTORIES

Trajectories, Earth-Mercury
USE   EARTH-MERCURY TRAJECTORIES

Trajectories, Earth-Moon
USE   EARTH-MOON TRAJECTORIES

Trajectories, Earth-Venus
USE   EARTH-VENUS TRAJECTORIES

Trajectories, Electron
USE   ELECTRON TRAJECTORIES

Trajectories, Hohmann
USE   ELLIPTICAL ORBITS
TRANSFER ORBITS

Trajectories, Hyperbolic
USE   HYPERBOLIC TRAJECTORIES

Trajectories, Interorbital
USE   INTERORBITAL TRAJECTORIES

Trajectories, Interplanetary
USE   INTERPLANETARY TRAJECTORIES

Trajectories, Lunar
USE   LUNAR TRAJECTORIES

Trajectories, Midcourse
USE   MIDCOURSE TRAJECTORIES

Trajectories, Missile
USE   MISSILE TRAJECTORIES

Trajectories, Molecular
USE   MOLECULAR TRAJECTORIES

Trajectories, Moon-Earth
USE   MOON-EARTH TRAJECTORIES

Trajectories, Particle
USE   PARTICLE TRAJECTORIES

Trajectories, Reentry
USE   REENTRY TRAJECTORIES

Trajectories, Rendezvous
USE   RENDEZVOUS TRAJECTORIES

Trajectories, Round Trip
USE   ROUND TRIP TRAJECTORIES

Trajectories, Spacecraft
USE   SPACECRAFT TRAJECTORIES

(Trajectories), SPURT
USE   SPINNING UNGUIDED ROCKET
TRAJECTORY

Trajectories, Underwater
USE   UNDERWATER TRAJECTORIES

TRAJECTORY ANALYSIS

TRAJECTORY CONTROL

Trajectory Determination System, Goddard
USE   GODDARD TRAJECTORY DETERMINATION
SYSTEM

TRAJECTORY MEASUREMENT

TIME MEASUREMENT

TIME MEASURING INSTRUMENTS

Time Metric, Space-
USE   SPACE-TIME FUNCTIONS

Time Modulation, Pulse
USE   PULSE TIME MODULATION

TIME OF FLIGHT SPECTROMETERS

Time Operation, Real
USE   REAL TIME OPERATION

TIME OPTIMAL CONTROL

Time), Rates (Per
USE   RATES (PER TIME)

Time, Reaction
USE   REACTION TIME

Time Relations, Stress-Strain-
USE   STRESS-STRAIN-TIME RELATIONS

Time, Relaxation
USE   RELAXATION TIME

TIME RESPONSE

Time, Reverse
USE   REACTION TIME

TIME SERIES ANALYSIS

TIME SHARING

Time, Sidereal
USE   SIDEREAL TIME

TIME SIGNALS

TIME TEMPERATURE PARAMETER

Time, Testing
USE   TESTING TIME

Time, Transit
USE   TRANSIT TIME

Time, Universal
USE   UNIVERSAL TIME

Timers
USE   TIMING DEVICES

Timing
USE   TIME MEASUREMENT

TIMING DEVICES

Timing Explorer, X Ray
USE   X RAY TIMING EXPLORER

TIMOSHENKO BEAMS

TIN

TIN ALLOYS

TIN COMPOUNDS

Tin Compounds, Organic
USE   ORGANIC TIN COMPOUNDS

TIN ISOTOPES

TIN OXIDES

TIN TELLURIDES

TIP DRIVEN ROTORS

TIP SPEED

TIP VANES

Tip Vortices, Wing
USE   WING TIP VORTICES

TIPS

Tips, Blade
USE   BLADE TIPS

Tips, Crack
USE   CRACK TIPS

Tips, Nose
USE   NOSE TIPS

Tips, Wing
USE   WING TIPS

TIRES

Tires, Aircraft
USE   AIRCRAFT TIRES

TIROS D Satellite
USE   TIROS 4 SATELLITE

TIROS E Satellite
USE   TIROS 5 SATELLITE

TIROS F Satellite
USE   TIROS 6 SATELLITE

TIROS G Satellite
USE   TIROS 7 SATELLITE

TIROS H Satellite
USE   TIROS 8 SATELLITE

TIROS M

TIROS N SERIES SATELLITES

TIROS OPERATIONAL SATELLITE SYSTEM

TIROS Operational Satellites, Improved
USE   IMPROVED TIROS OPERATIONAL
SATELLITES

TIROS PROJECT

TIROS SATELLITES

TIROS Wheel Satellite
USE   TIROS 9 SATELLITE

TIROS 1 SATELLITE

TIROS 2 SATELLITE

TIROS 3 SATELLITE

TIROS 4 SATELLITE

TIROS 5 SATELLITE

TIROS 6 SATELLITE

TIROS 7 SATELLITE

TIROS 8 SATELLITE

TIROS 9 SATELLITE

TIROS 10 SATELLITE

Tissue, Connective
USE   CONNECTIVE TISSUE

Tissues, Adipose
USE   ADIPOSE TISSUES

TISSUES (BIOLOGY)

Tissues, Plantar
USE   PLANTAR TISSUES

TITAN

TITAN CENTAUR LAUNCH VEHICLE

TITAN ICBM

TITAN LAUNCH VEHICLES

TITAN PROJECT

TITAN 1 ICBM

TITAN 2 ICBM

TITAN 3 LAUNCH VEHICLE

TITANATES

Titanates, Barium
USE   BARIUM TITANATES

Titanates, Lead
USE   LEAD TITANATES

Titanates, Lead Zirconate
USE   LEAD ZIRCONATE TITANATES

Titanates, Magnesium
USE   MAGNESIUM TITANATES

Titanates, Strontium
USE   STRONTIUM TITANATES

Titanates, Zirconium
USE   ZIRCONIUM TITANATES

TITANIUM

TITANIUM ALLOYS

TITANIUM BORIDES

TITANIUM CARBIDES

TITANIUM CHLORIDES

TITANIUM COMPOUNDS

Titanium Dioxide
USE   TITANIUM OXIDES

TITANIUM ISOTOPES

TITANIUM NITRIDES

TITANIUM OXIDES

(Title), Position
USE   POSITION (TITLE)

TITRATION

TITRIMETERS

TI
USE   THALLIUM

Tm
USE   THORIUM

TN
USE   TENNESSEE

TN), Great Smoky Mountains (NC-
USE   GREAT SMOKY MOUNTAINS (NC-TN)

TN), Tennessee Valley (AL-KY-
USE   TENNESSEE VALLEY (AL-KY-TN)

TNT (Trinitrotoluene)
USE   TRINITROTOLUENE

TOBACCO

Tobago, Trinidad And
USE   TRINIDAD AND TOBAGO

TOCOPHEROL

TOGO

TOKAMAK DEVICES

Tolerance, Acceleration
USE   ACCELERATION TOLERANCE

Tolerance, Altitude
USE   ALTITUDE TOLERANCE

Tolerance, Cold

Tolerance, Cold
USE COLD TOLERANCE

Tolerance, Fault
USE FAULT TOLERANCE

Tolerance, Heat
USE HEAT TOLERANCE

Tolerance, Noise
USE NOISE TOLERANCE

Tolerance, Orthostatic
USE ORTHOSTATIC TOLERANCE

Tolerance, Radiation
USE RADIATION TOLERANCE

Tolerances, Human
USE HUMAN TOLERANCES

Tolerances, Impact
USE IMPACT TOLERANCES

TOLERANCES (MECHANICS)

TOLERANCES (PHYSIOLOGY)

TOLLMEIN-SCHLICHTING WAVES

TOLUENE

Toluene, Trinitro
USE TRINITROTOLUENE

TOMAHAWK MISSILES

Tomahawk Rocket Vehicle, Nike-
USE NIKE-TOMAHAWK ROCKET VEHICLE

Tombolos
USE BARS (LANDFORMS)

TOMOGRAPHY

Tomography, Computer Aided
USE COMPUTER AIDED TOMOGRAPHY

Tone
USE PITCH

Tones, Aeolian
USE AEOLIAN TONES

TONGUE

TONK METEORITE

Tonometry
USE PRESSURE MEASUREMENT
INTRAOCULAR PRESSURE

Tonus
USE MUSCULAR TONUS

Tonus, Muscular
USE MUSCULAR TONUS

TOOLING

TOOLS

(Tools), Files
USE FILES (TOOLS)

Tools, Machine
USE MACHINE TOOLS

Tools, Software
USE SOFTWARE TOOLS

Tools, Space
USE SPACE TOOLS

TOOTH DISEASES

TOPEX

(Topographic Features), Bays
USE BAYS (TOPOGRAPHIC FEATURES)

(Topographic Features), Sounds
USE SOUNDS (TOPOGRAPHIC FEATURES)

TOPOGRAPHY

(Topography), Depressions
USE STRUCTURAL BASINS

(Topography), Inlets
USE INLETS (TOPOGRAPHY)

Topography, Lunar
USE LUNAR TOPOGRAPHY

Topography, Stoss-And-Lee
USE GLACIAL DRIFT

TOPOLOGY

TOPPING CYCLE ENGINES

TOPS (SPACECRAFT)

TORCHES

Torches, Plasma
USE PLASMA TORCHES

Tornado Aircraft
USE MRCA AIRCRAFT

TORNADOES

TORO ASTEROID

TOROIDAL DISCHARGE

TOROIDAL PLASMAS

TOROIDAL SHELLS

TOROIDAL WHEELS

TOROIDS

TORPEDO ENGINES

TORPEDOES

(Torpedoes), Retorc
USE TORPEDOES

TORQUE

TORQUE CONVERTERS

Torque Measuring Apparatus
USE TORQUEMETERS

TORQUE MOTORS

TORQUEMETERS

TORQUERS

TORRES STRAIT

TORSION

TORSIONAL STRESS

TORSIONAL VIBRATION

TORSO

Torus, Joint European
USE JOINT EUROPEAN TORUS

TORUSES

Toruses, Bumpy
USE BUMPY TORUSES

TORY 2 REACTOR

TORY 2-A REACTOR

TORY 2-C REACTOR

TOS-A
USE ESSA 3 SATELLITE

TOTAL ENERGY SYSTEMS

Total Energy Systems, Solar
USE SOLAR TOTAL ENERGY SYSTEMS

TOUCH

TOUCHDOWN

TOUGHNESS

Toughness, Fracture
USE FRACTURE STRENGTH

TOURMALINE

Tournesole Satellite
USE D-2 SATELLITES

TOURNIQUETS

Tours, Grand
USE GRAND TOURS

TOW MISSILES

TOWED BODIES

Towed Targets
USE TOWED BODIES
TARGETS

TOWER SHIELDING REACTOR 2

TOWERS

Towers, Airport
USE AIRPORT TOWERS

Towers, Drop
USE DROP TOWERS

Towers, Umbilical
USE UMBILICAL TOWERS

Towers, Whirl
USE WHIRL TOWERS

TOWING

TOWNSEND AVALANCHE

TOWNSEND DISCHARGE

Townsend Surfaces
USE TOWNSEND AVALANCHE

TOXIC DISEASES

TOXIC HAZARDS

TOXICITY

TOXICITY AND SAFETY HAZARD

Toxicity, Oxygen
USE HYPEROXIA

TOXICOLOGY

(Toxicology), Poisoning
USE TOXIC DISEASES

TOXINS AND ANTITOXINS

Toxins, Endo
USE ENDOTOXINS

TRAAC Satellite
USE TRANSIT ATTITUDE CONTROL SATELLITE

TRACE CONTAMINANTS

TRACE ELEMENTS

Tracer Explorers, Active Magneto Particle
USE AMPTE (SATELLITES)

TRACERS

TRACHEA

TRACHYTE

TRACING

Tracing, Ray
USE RAY TRACING

TRACKED VEHICLES

Tracker, CCD Star
USE CCD STAR TRACKER

Tracker), Stellar (Star
USE CCD STAR TRACKER

Trackers, Star
USE STAR TRACKERS

Tracking And Data Acq Network, Satellite
USE STDN (NETWORK)

Tracking And Data Network, Space Flight
USE SPACE FLIGHT TRACKING AND DATA
NETWORK

Tracking And Data Network, Spacecraft
USE STDN (NETWORK)

Tracking And Data Relay Satellites
USE TDR SATELLITES

Tracking Antennas
USE DIRECTIONAL ANTENNAS

Tracking, Compensatory
USE COMPENSATORY TRACKING

TRACKING FILTERS

Tracking, Infrared
USE INFRARED TRACKING

(Tracking), Look Angles
USE LOOK ANGLES (TRACKING)

Tracking, Missile
USE MISSILE TRACKING

Tracking, Multiradar
USE RADAR NETWORKS

Tracking Network, Global
USE GLOBAL TRACKING NETWORK

(Tracking Network), GLOTRAC
USE GLOBAL TRACKING NETWORK

Tracking Network), STADAN (Satellite
USE STDN (NETWORK)

TRACKING NETWORKS

Tracking, Optical
USE OPTICAL TRACKING

Tracking, Photographic
USE PHOTOGRAPHIC TRACKING

TRACKING (POSITION)

TRACKING PROBLEM

Tracking Program, Optical Satellite
USE OPTICAL SATELLITE TRACKING PROGRAM

Tracking, Pursuit
USE PURSUIT TRACKING

Tracking, Radar
USE RADAR TRACKING

TRACKING RADAR

Tracking, Radio
USE RADIO TRACKING

Tracking, Range And Range Rate
USE RANGE AND RANGE RATE TRACKING

Tracking, Satellite
USE SATELLITE TRACKING

Tracking, Satellite-To-Satellite
USE SATELLITE-TO-SATELLITE TRACKING

Tracking, Spacecraft
USE SPACECRAFT TRACKING

Tracking, Star
USE STAR TRACKERS

TRACKING STATIONS

Tracking Studies
USE TRACKING (POSITION)

Tracking System, Minitrack Optical
USE MINITRACK SYSTEM

(Tracking System), MOTS
USE MINITRACK SYSTEM

Tracking System, Polystation Doppler
USE POLYSTATION DOPPLER TRACKING
SYSTEM

Tracking System, Space Detection And
USE SPACE DETECTION AND TRACKING
SYSTEM

(Tracking System), SPADATS
USE SPACE DETECTION AND TRACKING
SYSTEM

(Tracking), TCG
USE TRANSPONDER CONTROL GROUP

Tracking Telescopes, Multispectral
USE MULTISPECTRAL TRACKING TELESCOPES

Tracking, Video Landmark Acquisition And
USE VIDEO LANDMARK ACQUISITION AND
TRACKING

Tracking, Visual
USE OPTICAL TRACKING

TRACKS

Tracks, Ground
USE GROUND TRACKS

Tracks, Particle
USE PARTICLE TRACKS

Tracks, Satellite Ground
USE SATELLITE GROUND TRACKS

Tracks, Vehicular
USE VEHICULAR TRACKS

TRACTION

TRACTORS

Tractors, Crawler
USE CRAWLER TRACTORS

Tracts
USE SITES

Trade, Foreign
USE FOREIGN TRADE

Trade, International
USE INTERNATIONAL TRADE

(Trademark), Adiprene
USE ADIPRENE (TRADEMARK)

(Trademark), Amberlite
USE AMBERLITE (TRADEMARK)

(Trademark), Amplitrons
USE PLANOTRONS

(Trademark), Astroloy
USE ASTROLOY (TRADEMARK)

(Trademark), Bakelite
USE BAKELITE (TRADEMARK)

(Trademark), Borazon
USE BORON NITRIDES

(Trademark), Buna
USE BUNA (TRADEMARK)

(Trademark), Carborundum
USE CARBORUNDUM (TRADEMARK)

(Trademark), Dacron
USE DACRON (TRADEMARK)

(Trademark), Delrin
USE DELRIN (TRADEMARK)

(Trademark), Flexowriters
USE AUTOMATIC TYPEWRITERS

(Trademark), Fortisan
USE FORTISAN (TRADEMARK)

(Trademark), Geon
USE POLYVINYL CHLORIDE

(Trademark), Hastelloy
USE HASTELLOY (TRADEMARK)

(Trademark), Hexogenes
USE HEXOGENES (TRADEMARK)

(Trademark), Hopcalite
USE HOPCALITE (TRADEMARK)

(Trademark), Inconel
USE INCONEL (TRADEMARK)

(Trademark), Kapton
USE KAPTON (TRADEMARK)

(Trademark), Kevlar
USE KEVLAR (TRADEMARK)

(Trademark), Kovar
USE KOVAR (TRADEMARK)

(Trademark), Lexan
USE LEXAN (TRADEMARK)

(Trademark), Lucite
USE POLYMETHYL METHACRYLATE

(Trademark), Ludox
USE LUDOX (TRADEMARK)

(Trademark), Magnesyn
USE SERVOMOTORS

(Trademark), Manganin
USE MANGANIN (TRADEMARK)

(Trademark), Masonite
USE MASONITE (TRADEMARK)

(Trademark), Monel
USE MONEL (TRADEMARK)

(Trademark), Mylar
USE MYLAR (TRADEMARK)

(Trademark), Nembutal
USE NEMBUTAL (TRADEMARK)

(Trademark), Nichrome
USE NICHROME (TRADEMARK)

(Trademark), Nylon
USE NYLON (TRADEMARK)

(Trademark), Permalloys
USE PERMALLOYS (TRADEMARK)

(Trademark), Perspex
USE PERSPEX (TRADEMARK)

TRAJECTORY OPTIMIZATION

Trajectory, Spinning Unguided Rocket
USE   SPINNING UNGUIDED ROCKET
            TRAJECTORY

Trajectory Systems, Multiple Target
USE   MATTS (SYSTEMS)

TRANQUILIZERS

Transall C-160 Aircraft
USE   C-160 AIRCRAFT

TRANSATMOSPHERIC VEHICLES

Transceivers
USE   TRANSMITTER RECEIVERS

TRANSCENDENTAL FUNCTIONS

TRANSCONTINENTAL SYSTEMS

TRANSDUCERS

Transducers, Digital
USE   DIGITAL TRANSDUCERS

Transducers, Electroacoustic
USE   ELECTROACOUSTIC TRANSDUCERS

Transducers, Electronic
USE   ELECTRONIC TRANSDUCERS

Transducers, Image
USE   IMAGE TRANSDUCERS

Transducers, Interdigital
USE   INTERDIGITAL TRANSDUCERS

Transducers, Magnetic
USE   MAGNETIC TRANSDUCERS

Transducers, Piezoelectric
USE   PIEZOELECTRIC TRANSDUCERS

Transducers, Piezoresistive
USE   PIEZORESISTIVE TRANSDUCERS

Transducers, Pressure
USE   PRESSURE SENSORS

Transducers, Quartz
USE   QUARTZ TRANSDUCERS

Transducers, Sound
USE   SOUND TRANSDUCERS

Transducers, Ultrasonic Wave
USE   ULTRASONIC WAVE TRANSDUCERS

TRANSEARTH INJECTION

TRANSEQUATORIAL PROPAGATION

Transfer
USE   TRANSFERRING

Transfer, Aerodynamic Heat
USE   AERODYNAMIC HEAT TRANSFER

Transfer, Aerospace Technology
USE   AEROSPACE TECHNOLOGY TRANSFER

Transfer, Charge
USE   CHARGE TRANSFER

Transfer Coefficients, Heat
USE   HEAT TRANSFER COEFFICIENTS

Transfer, Conductive Heat
USE   CONDUCTIVE HEAT TRANSFER

Transfer, Convective Heat
USE   CONVECTIVE HEAT TRANSFER

Transfer Devices, Charge
USE   CHARGE TRANSFER DEVICES

Transfer, Drop
USE   DROP TRANSFER

Transfer, Electron
USE   ELECTRON TRANSFER

Transfer, Energy
USE   ENERGY TRANSFER

Transfer Function, Modulation
USE   MODULATION TRANSFER FUNCTION

Transfer Function, Optical
USE   OPTICAL TRANSFER FUNCTION

TRANSFER FUNCTIONS

Transfer, Heat
USE   HEAT TRANSFER

Transfer, Hypersonic Heat
USE   HYPERSONIC HEAT TRANSFER

Transfer, Information
USE   INFORMATION TRANSFER

Transfer, Intervehicle Spacecrew
USE   SPACECREW TRANSFER

Transfer, Laminar Heat
USE   LAMINAR HEAT TRANSFER

Transfer (LET), Linear Energy
USE   LINEAR ENERGY TRANSFER (LET)

Transfer, Mass
USE   MASS TRANSFER

Transfer, Momentum
. USE   MOMENTUM TRANSFER

TRANSFER OF TRAINING

Transfer, Orbital
USE   TRANSFER ORBITS

TRANSFER ORBITS

Transfer Orbits, Hohmann
USE   TRANSFER ORBITS
            ELLIPTICAL ORBITS

Transfer Orbits, Interplanetary
USE   INTERPLANETARY TRANSFER ORBITS

Transfer, Payload
USE   PAYLOAD TRANSFER

Transfer, Propellant
USE   PROPELLANT TRANSFER

Transfer, Radiative
USE   RADIATIVE TRANSFER

Transfer, Radiative Heat
USE   RADIATIVE HEAT TRANSFER

Transfer Salts, Organic Charge
USE   ORGANIC CHARGE TRANSFER SALTS

Transfer, Spacecrew
USE   SPACECREW TRANSFER

Transfer, Supersonic Heat
USE   SUPERSONIC HEAT TRANSFER

Transfer, Technology
USE   TECHNOLOGY TRANSFER

TRANSFER TUNNELS

Transfer, Turbulent Heat
USE   TURBULENT HEAT TRANSFER

Transfer Vehicles, Intraorbit
USE   INTRAORBIT TRANSFER VEHICLES

Transfer Vehicles, Orbit
USE   ORBIT TRANSFER VEHICLES

TRANSFERRED ELECTRON DEVICES

TRANSFERRING

Transform Integrals
USE   INTEGRAL TRANSFORMATIONS

Transformation, Fourier
USE   FOURIER TRANSFORMATION

Transformation, Hilbert
USE   HILBERT TRANSFORMATION

Transformation, Joukowski
USE   JOUKOWSKI TRANSFORMATION

Transformation, Laplace
USE   LAPLACE TRANSFORMATION

Transformation, Legendre
USE   LEGENDRE FUNCTIONS

Transformation, Martensitic
USE   MARTENSITIC TRANSFORMATION

Transformation, Schwarz-Christoffel
USE   SCHWARZ-CHRISTOFFEL
            TRANSFORMATION

Transformation Tensors
USE   TENSORS

Transformation, Theodorsen
USE   THEODORSEN TRANSFORMATION

TRANSFORMATIONS

Transformations, Conformal
USE   CONFORMAL MAPPING

Transformations, Coordinate
USE   COORDINATE TRANSFORMATIONS

Transformations, Fast Fourier
USE   FAST FOURIER TRANSFORMATIONS

Transformations, Fourier-Bessel
USE   FOURIER-BESSEL TRANSFORMATIONS

Transformations, Householder
USE   HOUSEHOLDER TRANSFORMATIONS

Transformations, Integral
USE   INTEGRAL TRANSFORMATIONS

Transformations, Linear
USE   LINEAR TRANSFORMATIONS

Transformations, Lorentz
USE   LORENTZ TRANSFORMATIONS

TRANSFORMATIONS (MATHEMATICS)

Transformations, Nuclear
USE   NUCLEAR TRANSFORMATIONS

Transformations, Order-Disorder
USE   ORDER-DISORDER TRANSFORMATIONS

Transformations, Phase
USE   PHASE TRANSFORMATIONS

TRANSFORMERS

Transformers, Instrument
USE   INSTRUMENT TRANSFORMERS

Transformers, Mode
USE   MODE TRANSFORMERS

Transforms
USE   TRANSFORMATIONS (MATHEMATICS)

Transforms, Mellin
USE   MELLIN TRANSFORMS

TRANSFUSION

TRANSGRANULAR CORROSION

TRANSHORIZON RADIO PROPAGATION

TRANSIENT HEATING

TRANSIENT LOADS

TRANSIENT OSCILLATIONS

TRANSIENT PRESSURES

TRANSIENT REACTOR TEST FACILITY

TRANSIENT RESPONSE

Transients (Surges)
USE   SURGES

Transionospheric Satellites, Low Frequency
USE   LOW FREQUENCY TRANSIONOSPHERIC
       SATELLITES

TRANSISTOR AMPLIFIERS

TRANSISTOR CIRCUITS

TRANSISTOR LOGIC

Transistor-Logic Integ Circuits, Diode-
USE   DTL INTEGRATED CIRCUITS

Transistor-Logic Integ Circuits, Transistor-
USE   TTL INTEGRATED CIRCUITS

Transistor-Transistor-Logic Integ Circuits
USE   TTL INTEGRATED CIRCUITS

TRANSISTORS

Transistors, Bipolar
USE   BIPOLAR TRANSISTORS

(Transistors), FET
USE   FIELD EFFECT TRANSISTORS

Transistors, Field Effect
USE   FIELD EFFECT TRANSISTORS

Transistors, High Electron Mobility
USE   HIGH ELECTRON MOBILITY TRANSISTORS

Transistors, Junction
USE   JUNCTION TRANSISTORS

Transistors, Junction Field Effect
USE   JFET

Transistors, Photo
USE   PHOTOTRANSISTORS

Transistors, Silicon
USE   SILICON TRANSISTORS

Transistors, Silicon-On-Sapphire
USE   SOS (SEMICONDUCTORS)

Transistors, Unipolar
USE   FIELD EFFECT TRANSISTORS

TRANSIT

TRANSIT ATTITUDE CONTROL SATELLITE

TRANSIT NAVIGATION SYSTEM

TRANSIT SATELLITES

Transit Systems, Rapid
USE   RAPID TRANSIT SYSTEMS

TRANSIT TIME

Transit Time Devices, Controlled Avalanche
USE   CATT DEVICES

Transit Time Diodes, Barrier Injection
USE   BARRITT DIODES

Transit, Trapped Plasma Avalanche Triggered
USE   TRAPATT DEVICES

Transit Vehicles, Automated
USE   AUTOMATED TRANSIT VEHICLES

Transit Vehicles, Automated Guideway
USE   AUTOMATED GUIDEWAY TRANSIT
       VEHICLES

TRANSITION

Transition, Boundary Layer
USE   BOUNDARY LAYER TRANSITION

TRANSITION FLOW

TRANSITION LAYERS

TRANSITION METALS

Transition, Optical
USE   OPTICAL TRANSITION

TRANSITION POINTS

TRANSITION PRESSURE

TRANSITION PROBABILITIES

TRANSITION TEMPERATURE

Transitions, Electron
USE   ELECTRON TRANSITIONS

Transitions, Forbidden
USE   FORBIDDEN TRANSITIONS

TRANSITS

TRANSLATING

Translation, Frequency
USE   FREQUENCY CONVERTERS

Translation, Machine
USE   MACHINE TRANSLATION

TRANSLATIONAL MOTION

TRANSLATORS

Translators, Digital To Voice
USE   DIGITAL TO VOICE TRANSLATORS

Translators), DIVOT (Voice
USE   DIGITAL TO VOICE TRANSLATORS

TRANSLUCENCE

TRANSLUNAR INJECTION

Translunar Space
USE   INTERPLANETARY SPACE

TRANSMISSION

Transmission), APT (Picture
USE   AUTOMATIC PICTURE TRANSMISSION

Transmission, Automatic Picture
USE   AUTOMATIC PICTURE TRANSMISSION

Transmission), Channels (Data
USE   CHANNELS (DATA TRANSMISSION)

TRANSMISSION CIRCUITS

Transmission, Coaxial
USE   TRANSMISSION
       COAXIAL CABLES

Transmission, Coherent
USE   COHERENT RADIATION

Transmission, Data
USE   DATA TRANSMISSION

Transmission, Double Sideband
USE   DOUBLE SIDEBAND TRANSMISSION

TRANSMISSION EFFICIENCY

Transmission, Electric Power
USE   ELECTRIC POWER TRANSMISSION

Transmission, Electromagnetic Wave
USE   ELECTROMAGNETIC WAVE TRANSMISSION

Transmission, Facsimile
USE   FACSIMILE COMMUNICATION

TRANSMISSION FLUIDS

Transmission, Heat
USE   HEAT TRANSMISSION

Transmission, Information
USE   DATA TRANSMISSION

Transmission (Lasers), Power
USE   POWER TRANSMISSION (LASERS)

Transmission, Light
USE   LIGHT TRANSMISSION

TRANSMISSION LINES

Transmission Lines, Flat Coaxial
USE   MICROSTRIP TRANSMISSION LINES

Transmission Lines, Fluid
USE   FLUID TRANSMISSION LINES

Transmission Lines, Microstrip
USE   MICROSTRIP TRANSMISSION LINES

Transmission Lines, Strip
USE   STRIP TRANSMISSION LINES

Transmission Lines, Underground
USE   UNDERGROUND TRANSMISSION LINES

TRANSMISSION LOSS

Transmission, Microwave
USE   MICROWAVE TRANSMISSION

Transmission, Multipath
USE   MULTIPATH TRANSMISSION

Transmission, Multiplex
USE   MULTIPLEXING

Transmission, Neuromuscular
USE   NEUROMUSCULAR TRANSMISSION

Transmission, Neuron
USE   BIOELECTRICITY

Transmission, Packet
USE   PACKET TRANSMISSION

Transmission, Power
USE   POWER TRANSMISSION

Transmission, Radar
USE   RADAR TRANSMISSION

Transmission, Radio
USE   RADIO TRANSMISSION

Transmission, Satellite
USE   SATELLITE TRANSMISSION

Transmission, SCPC
USE   SINGLE CHANNEL PER CARRIER
       TRANSMISSION

Transmission, Short Wave Radio
USE   SHORT WAVE RADIO TRANSMISSION

Transmission, Signal
USE   SIGNAL TRANSMISSION

Transmission, Single Channel Per Carrier
USE   SINGLE CHANNEL PER CARRIER
       TRANSMISSION

Transmission, Single Sideband
USE   SINGLE SIDEBAND TRANSMISSION

Transmission, Sound
USE   SOUND TRANSMISSION

Transmission, Spread Spectrum
USE   SPREAD SPECTRUM TRANSMISSION

Transmission, Superconducting Power
USE   SUPERCONDUCTING POWER
       TRANSMISSION

Transmission, Television
USE   TELEVISION TRANSMISSION

Transmission (To Earth), Satellite Power
USE   SATELLITE POWER TRANSMISSION (TO
       EARTH)

TRANSMISSIONS (MACHINE ELEMENTS)

TRANSMISSIVITY

TRANSMISSOMETERS

TRANSMITTANCE

TRANSMITTER RECEIVERS

TRANSMITTERS

Transmitters, Emergency Locator
USE   EMERGENCY LOCATOR TRANSMITTERS

Transmitters, Instrument
USE   INSTRUMENT TRANSMITTERS

Transmitters, Radar
USE   RADAR TRANSMITTERS

Transmitters, Radio
USE   RADIO TRANSMITTERS

TRANSMUTATION

Transmutation, Neutron
USE   NUCLEAR REACTIONS

TRANSOCEANIC COMMUNICATION

TRANSOCEANIC FLIGHT

TRANSOCEANIC SYSTEMS

Transonic Aircraft
USE   SUPERSONIC AIRCRAFT

Transonic Aircraft Technology Program
USE   TACT PROGRAM

TRANSONIC COMPRESSORS

TRANSONIC FLIGHT

TRANSONIC FLOW

TRANSONIC FLUTTER

Transonic Inlets
USE   SUPERSONIC INLETS

TRANSONIC NOZZLES

TRANSONIC SPEED

Transonic Turbines
USE   SUPERSONIC TURBINES

TRANSONIC WIND TUNNELS

Transonics
USE   TRANSONIC FLOW

TRANSPARENCE

Transparent Materials
USE   TRANSPARENCE

TRANSPIRATION

Transpiration Cooling
USE   SWEAT COOLING

Transpiration, Evapo
USE   EVAPOTRANSPIRATION

Transpiration, Fluid
USE   TRANSPIRATION

TRANSPLANTATION

TRANSPONDER CONTROL GROUP

TRANSPONDERS

TRANSPORT AIRCRAFT

Transport Aircraft, F-28
USE   F-28 TRANSPORT AIRCRAFT

Transport Aircraft, Light
USE   LIGHT TRANSPORT AIRCRAFT

Transport Coefficients
USE   TRANSPORT PROPERTIES

Transport Equation, Boltzmann
USE   BOLTZMANN TRANSPORT EQUATION

Transport, Gas
USE   GAS TRANSPORT

Transport Hypothesis, Vorticity
USE   VORTICITY TRANSPORT HYPOTHESIS

Transport, Light Intratheater
USE   LIGHT INTRATHEATER TRANSPORT

Transport, Littoral
USE   LITTORAL TRANSPORT

Transport, Pollution
USE   POLLUTION TRANSPORT

Transport Program, Energy Efficiency
USE   ACEE PROGRAM

TRANSPORT PROPERTIES

Transport, Radiation
USE   RADIATION TRANSPORT

Transport Rsch Airplane, Experimental STOL
USE   QUESTOL

Transport, Sediment
USE   SEDIMENT TRANSPORT

Transport (Solid State), Carrier
USE   CARRIER TRANSPORT (SOLID STATE)

Transport, Supersonic Commercial Air
USE   SUPERSONIC COMMERCIAL AIR
       TRANSPORT

TRANSPORT THEORY

TRANSPORT VEHICLES

TRANSPORTATION

Transportation, Air
USE   AIR TRANSPORTATION

TRANSPORTATION ENERGY

(Transportation), Evacuating
USE   EVACUATING (TRANSPORTATION)

Transportation, High Speed
USE   RAPID TRANSIT SYSTEMS

Transportation, Marine
USE   MARINE TRANSPORTATION

TRANSPORTATION NETWORKS

Transportation, Rail
USE   RAIL TRANSPORTATION

Transportation, Space
USE   SPACE TRANSPORTATION

Transportation System Flights, Space
USE   SPACE TRANSPORTATION SYSTEM
       FLIGHTS

Transportation System, Space
USE   SPACE TRANSPORTATION SYSTEM

Transportation System 1 Flight, Space
USE   SPACE TRANSPORTATION SYSTEM 1
       FLIGHT

Transportation System 2 Flight, Space
USE   SPACE TRANSPORTATION SYSTEM 2
       FLIGHT

Transportation System 3 Flight, Space
USE   SPACE TRANSPORTATION SYSTEM 3
       FLIGHT

Transportation System 4 Flight, Space
USE   SPACE TRANSPORTATION SYSTEM 4
       FLIGHT

Transportation, Urban
USE   URBAN TRANSPORTATION

TRANSPORTER

Transporter Bus, Pioneer Venus 2
USE   PIONEER VENUS 2 TRANSPORTER BUS

Transports, Magnetic Tape
USE   MAGNETIC TAPE TRANSPORTS

Transports, Supersonic
USE   SUPERSONIC TRANSPORTS

TRANSURANIUM ELEMENTS

TRANSVERSE ACCELERATION

TRANSVERSE OSCILLATION

Transverse Vibration
USE   TRANSVERSE OSCILLATION

TRANSVERSE WAVES

Transversely Excited Atmospheric Lasers
USE   TEA LASERS

TRAP PROGRAM

TRAP Rocket Vehicle, Venus Fly
USE   VENUS FLY TRAP ROCKET VEHICLE

TRAPATT DEVICES

TRAPATT Diodes
USE   AVALANCHE DIODES

TRAPEZOIDAL TAIL SURFACES

TRAPEZOIDAL WINGS

TRAPEZOIDS

TRAPPED MAGNETIC FIELDS

TRAPPED PARTICLES

Trapped Particles, Geomagnetically
USE   RADIATION BELTS

Trapped Particles, Magnetically
USE   MAGNETICALLY TRAPPED PARTICLES

Trapped Plasma Avalanche Triggered Transit
USE   TRAPATT DEVICES

TRAPPED VORTEXES

TRAPPING

Trapping, Cryo
USE   CRYOTRAPPING

Trapping, Radiation
USE   RADIATION TRAPPING

TRAPS

Traps, Cold
USE   COLD TRAPS

Traps (Instrumentation), Ion
USE   ION TRAPS (INSTRUMENTATION)

Traps, Vapor
USE   VAPOR TRAPS

Traps, Vortex
USE   TRAPPED VORTEXES

TRAVEL

Travel, Interstellar
USE   INTERSTELLAR TRAVEL

TRAVELING CHARGE

TRAVELING IONOSPHERIC DISTURBANCES

TRAVELING SALESMAN PROBLEM

TRAVELING SOLVENT METHOD

TRAVELING WAVE AMPLIFIERS

TRAVELING WAVE MASERS

TRAVELING WAVE MODULATION

TRAVELING WAVE TUBES

TRAVELING WAVES

TRAYS

TREADMILLS

TREADS

TREAT (Test Facility)
USE   TRANSIENT REACTOR TEST FACILITY

(Treating), Conditioning
USE   TREATMENT

TREATMENT

Treatment, Heat
USE   HEAT TREATMENT

Treatment), Normalizing (Heat
USE   NORMALIZING (HEAT TREATMENT)

Treatment, Pre
USE   PRETREATMENT

Treatment, Sewage
USE   SEWAGE TREATMENT

Treatment), Sizing (Surface
USE   SIZING (SURFACE TREATMENT)

Treatment, Surface
USE   SURFACE FINISHING

Treatment, Thermomechanical
USE   THERMOMECHANICAL TREATMENT

Treatment, Waste
USE   WASTE TREATMENT

Treatment, Water
USE   WATER TREATMENT

Treaty Organization (NATO), North Atlantic
USE   NORTH ATLANTIC TREATY ORGANIZATION
            (NATO)

Treaty, Outer Space
USE   OUTER SPACE TREATY

Tree Ring Dating
USE   DENDROCHRONOLOGY

TREES

Trees, Citrus
USE   CITRUS TREES

Trees, Deciduous
USE   DECIDUOUS TREES

Trees, Fault
USE   FAULT TREES

TREES (MATHEMATICS)

TREES (PLANTS)

TREMORS

Trend Line Analysis, Program
USE   PROGRAM TREND LINE ANALYSIS

TRENDS

TRESCA FLOW

TRIACETIN

Triaminoguanidinenitrate
USE   TAGN

TRIAMINOGUANIDINIUM AZIDE

Triaminotrinitrobenzene
USE   TATB

TRIANGLES

Triangular Wings
USE   DELTA WINGS

TRIANGULATION

TRIATOMIC MOLECULES

TRIAXIAL STRESSES

Triaxiality
USE   TRIAXIAL STRESSES

TRIBOLIA

TRIBOLOGY

TRIBOLUMINESCENCE

TRIBUTARIES

Trichlorides
USE   CHLORIDES

Trident Aircraft
USE   DH 121 AIRCRAFT

TRIDENT SUBMARINE

TRIENES

Triethiodide, Gallamine
USE   GALLAMINE TRIETHIODIDE

TRIETHYL COMPOUNDS

Trifluoride, Boron
USE   BORON FLUORIDES

TRIFLUOROAMINE OXIDE

TRIGATRONS

TRIGGER CIRCUITS

Triggered Transit, Trapped Plasma Avalanche
USE   TRAPATT DEVICES

Triggers
USE   ACTUATORS

TRIGONOMETRIC FUNCTIONS

TRIGONOMETRY

Trim (Balance)
USE   AERODYNAMIC BALANCE

TRIMERS

TRIMETHADIONE

TRIMETHYL COMPOUNDS

TRINIDAD AND TOBAGO

TRINITRAMINE

Trinitramine, Cyclotrimethylene
USE   RDX

TRINITRO COMPOUNDS

TRINITROTOLUENE

(Trinitrotoluene), TNT
USE   TRINITROTOLUENE

Trinitrotriazocyclohexane
USE   RDX

TRIODES

TRIOLS

Trip Trajectories, Round
USE   ROUND TRIP TRAJECTORIES

TRIPHENYL SILICON

TRIPHENYLS

Triphosphate, Adenosine
USE   ADENOSINE TRIPHOSPHATE

Triple Axis Spectrometers
USE   NEUTRON SPECTROMETERS

Triplet Excitation
USE   ATOMIC ENERGY LEVELS

Triplet State
USE   ATOMIC ENERGY LEVELS

TRIPODS

Tripropellants
USE   LIQUID ROCKET PROPELLANTS

TRISONIC WIND TUNNELS

TRITIUM

TRITON

TRITONS

TRIVALENT IONS

Trochoids
USE   PIVOTS

TROILITE

Trojan Aircraft
USE   T-28 AIRCRAFT

TROJAN ORBITS

TROMBE WALLS

Tropical Experiment, GARP Atlantic
USE   GARP ATLANTIC TROPICAL EXPERIMENT

TROPICAL METEOROLOGY

TROPICAL REGIONS

TROPICAL STORMS

Tropics
USE   TROPICAL REGIONS

TROPISM

Tropism, Aeolo
USE   AEOLOTROPISM

Tropism, Baro
USE   BAROTROPISM

Tropism, Geo
USE   GEOTROPISM

Tropism, Gravi
USE   GRAVITROPISM

Tropism, Gyro
USE   GYROTROPISM

Tropism, Iso
USE   ISOTROPISM

Tropism, Ortho
　　USE　ORTHOTROPISM

Tropism, Photo
　　USE　PHOTOTROPISM

TROPOPAUSE

TROPOSPHERE

TROPOSPHERIC RADIATION

TROPOSPHERIC SCATTERING

TROPOSPHERIC WAVES

Tropsch Process, Fischer-
　　USE　FISCHER-TROPSCH PROCESS

TROPYL COMPOUNDS

Troubleshooting
　　USE　MAINTENANCE

TROUGHS

TRUCKS

Trucks, Tank
　　USE　TANK TRUCKS

TRUNCATION ERRORS

Truncation (Mathematics)
　　USE　APPROXIMATION

Trunks (Lines)
　　USE　TRANSMISSION LINES

Trunnions
　　USE　SHAFTS (MACHINE ELEMENTS)

TRUSSES

Truth, Ground
　　USE　GROUND TRUTH

Truth, Sea
　　USE　SEA TRUTH

TRYPANOSOME

TRYPSIN

TRYPTAMINES

TRYPTOPHAN

TS-11 AIRCRAFT

TS-11 Aircraft, Polish
　　USE　TS-11 AIRCRAFT

TSR 2 Aircraft, BAC
　　USE　TSR-2 AIRCRAFT

TSR-2 AIRCRAFT

TSUNAMI WAVES

TTL INTEGRATED CIRCUITS

TU-104 AIRCRAFT

TU-121 ENGINE

TU-124 AIRCRAFT

TU-134 AIRCRAFT

TU-144 AIRCRAFT

TU-154 AIRCRAFT

TUBE ANODES

Tube, Bronchial
　　USE　BRONCHIAL TUBE

TUBE CATHODES

Tube Control, Fly By
　　USE　FLY BY TUBE CONTROL

TUBE GRIDS

TUBE HEAT EXCHANGERS

TUBE LASERS

Tube Oscillators, Vacuum
　　USE　VACUUM TUBE OSCILLATORS

TUBERCULOSIS

TUBES

Tubes, Backward Wave
　　USE　BACKWARD WAVE TUBES

Tubes, Bourdon
　　USE　BOURDON TUBES

Tubes, Camera
　　USE　CAMERA TUBES

Tubes, Capillary
　　USE　CAPILLARY TUBES

Tubes, Cathode Ray
　　USE　CATHODE RAY TUBES

Tubes, Circular
　　USE　CIRCULAR TUBES

Tubes, Cold Cathode
　　USE　COLD CATHODE TUBES

Tubes, Discharge
　　USE　GAS DISCHARGE TUBES

Tubes, Drop
　　USE　DROP TOWERS

Tubes, Electron
　　USE　ELECTRON TUBES

Tubes, Eustachian
　　USE　EUSTACHIAN TUBES

Tubes, Flash
　　USE　FLASH LAMPS

Tubes, Gas
　　USE　GAS TUBES

Tubes, Gas Discharge
　　USE　GAS DISCHARGE TUBES

Tubes, Geiger-Mueller
　　USE　GEIGER COUNTERS

Tubes, Helix
　　USE　TRAVELING WAVE TUBES

Tubes, Hilsch
　　USE　HILSCH TUBES

Tubes, Image
　　USE　IMAGE TUBES

Tubes, Image Dissector
　　USE　IMAGE DISSECTOR TUBES

Tubes, Intensifier
　　USE　IMAGE INTENSIFIERS

Tubes, Magnetic Annular Shock
　　USE　MAGNETIC ANNULAR SHOCK TUBES

Tubes, MAST Shock
　　USE　MAGNETIC ANNULAR SHOCK TUBES

Tubes, Microwave
　　USE　MICROWAVE TUBES

Tubes, Photo
　　USE　PHOTOTUBES

Tubes, Photomultiplier
　　USE　PHOTOMULTIPLIER TUBES

Tubes, Picture
　　USE　PICTURE TUBES

(Tubes), Pipes
　　USE　PIPES (TUBES)

Tubes, Pitot
　　USE　PITOT TUBES

Tubes, Preston
　　USE　PITOT TUBES
　　　　　　SPEED INDICATORS

Tubes, Shock
　　USE　SHOCK TUBES

Tubes, Traveling Wave
　　USE　TRAVELING WAVE TUBES

Tubes, U
　　USE　MANOMETERS

Tubes, Vacuum
　　USE　VACUUM TUBES

Tubes, Venturi
　　USE　VENTURI TUBES

Tubes, Vortex
　　USE　VORTICES
　　　　　　HILSCH TUBES

Tubes, X Ray
　　USE　X RAY TUBES

Tubing
　　USE　PIPES (TUBES)

Tugs, Space
　　USE　SPACE TUGS

TUMBLING MOTION

TUMORS

TUNABLE LASERS

TUNDRA

TUNERS

Tuners, Waveguide
　　USE　WAVEGUIDE TUNERS

TUNGSTATES

Tungstates, Calcium
　　USE　CALCIUM TUNGSTATES

Tungstates, Lead
　　USE　LEAD TUNGSTATES

Tungstates, Zinc
　　USE　ZINC TUNGSTATES

TUNGSTEN

TUNGSTEN ALLOYS

Tungsten Arc Welding, Gas
　　USE　GAS TUNGSTEN ARC WELDING

TUNGSTEN CARBIDES

TUNGSTEN CHLORIDES

TUNGSTEN COMPOUNDS

TUNGSTEN FLUORIDES

TUNGSTEN HALIDES

Tungsten Inert Gas Welding
　　USE　GAS TUNGSTEN ARC WELDING

TUNGSTEN ISOTOPES

TUNGSTEN OXIDES

TUNGUSK METEORITE

TUNING

TUNING FORK GYROSCOPES

Tuning, Schuler
  USE   SCHULER TUNING

TUNISIA

Tunnel Apparatus, Wind
  USE   WIND TUNNEL APPARATUS

Tunnel Balances, Wind
  USE   WIND TUNNEL APPARATUS
         WEIGHT INDICATORS

Tunnel Calibration, Wind
  USE   WIND TUNNEL CALIBRATION

TUNNEL CATHODES

TUNNEL DIODES

Tunnel Drives, Wind
  USE   WIND TUNNEL DRIVES

Tunnel Models, Wind
  USE   WIND TUNNEL MODELS

Tunnel Nozzles, Wind
  USE   WIND TUNNEL NOZZLES

Tunnel Resistors
  USE   ELECTRON TUNNELING
         RESISTORS

Tunnel Stability Tests, Wind
  USE   WIND TUNNEL STABILITY TESTS

Tunnel Tests, Water
  USE   WATER TUNNEL TESTS

Tunnel Tests, Wind
  USE   WIND TUNNEL TESTS

Tunnel Walls, Wind
  USE   WIND TUNNEL WALLS

TUNNELING

Tunneling, Electron
  USE   ELECTRON TUNNELING

TUNNELING (EXCAVATION)

TUNNELS

Tunnels, Blowdown Wind
  USE   BLOWDOWN WIND TUNNELS

Tunnels, Cascade Wind
  USE   CASCADE WIND TUNNELS

Tunnels, Combustion Wind
  USE   COMBUSTION WIND TUNNELS

Tunnels, Cryogenic Wind
  USE   CRYOGENIC WIND TUNNELS

Tunnels, Hotshot Wind
  USE   HOTSHOT WIND TUNNELS

Tunnels, Hydraulic Test
  USE   HYDRAULIC TEST TUNNELS

Tunnels, Hydrodynamic
  USE   PLASMA JET WIND TUNNELS

Tunnels, Hypersonic Wind
  USE   HYPERSONIC WIND TUNNELS

Tunnels, Hypervelocity Wind
  USE   HYPERVELOCITY WIND TUNNELS

Tunnels, Low Density Wind
  USE   LOW DENSITY WIND TUNNELS

Tunnels, Low Speed Wind
  USE   LOW SPEED WIND TUNNELS

Tunnels, Plasma Jet Wind
  USE   PLASMA JET WIND TUNNELS

Tunnels, Rectangular Wind
  USE   RECTANGULAR WIND TUNNELS

Tunnels, Shock
  USE   SHOCK TUNNELS

Tunnels, Slotted Wind
  USE   SLOTTED WIND TUNNELS

Tunnels, Subsonic Wind
  USE   SUBSONIC WIND TUNNELS

Tunnels, Supersonic Wind
  USE   SUPERSONIC WIND TUNNELS

Tunnels, Transfer
  USE   TRANSFER TUNNELS

Tunnels, Transonic Wind
  USE   TRANSONIC WIND TUNNELS

Tunnels, Trisonic Wind
  USE   TRISONIC WIND TUNNELS

Tunnels, Water
  USE   HYDRAULIC TEST TUNNELS

Tunnels, Wind
  USE   WIND TUNNELS

TUPOLEV AIRCRAFT

TURBIDITY

TURBINE BLADES

TURBINE ENGINES

Turbine Engines, Gas
  USE   GAS TURBINE ENGINES

TURBINE EXHAUST NOZZLES

TURBINE INSTRUMENTS

TURBINE PUMPS

TURBINE WHEELS

TURBINES

Turbines, Axial Flow
  USE   AXIAL FLOW TURBINES

Turbines, Gas
  USE   GAS TURBINES

Turbines, Shrouded
  USE   SHROUDED TURBINES

Turbines, Steam
  USE   STEAM TURBINES

Turbines, Supersonic
  USE   SUPERSONIC TURBINES

Turbines, Transonic
  USE   SUPERSONIC TURBINES

Turbines, Two Stage
  USE   TWO STAGE TURBINES

Turbines, Wind
  USE   WIND TURBINES

Turbo-Skyvan Aircraft
  USE   SC-7 AIRCRAFT

Turbochargers
  USE   SUPERCHARGERS
         TURBOCOMPRESSORS

TURBOCOMPRESSORS

Turboconverters
  USE   TURBOGENERATORS

Turboelectric Conversion
  USE   TURBOGENERATORS

Turboelectric Generator, ASTEC Solar
  USE   ASTEC SOLAR TURBOELECTRIC
         GENERATOR

TURBOFAN AIRCRAFT

TURBOFAN ENGINES

TURBOFANS

TURBOGENERATORS

Turbojet Aircraft
  USE   JET AIRCRAFT

TURBOJET ENGINE CONTROL

Turbojet Engine, YJ73
  USE   J-73 ENGINE

TURBOJET ENGINES

TURBOMACHINE BLADES

TURBOMACHINERY

(Turbomachinery), Rotor Blades
  USE   ROTOR BLADES (TURBOMACHINERY)

TURBOPAUSE

TURBOPROP AIRCRAFT

TURBOPROP ENGINES

Turboprop Engines, Dart
  USE   TURBOPROP ENGINES

Turbopumps
  USE   TURBINE PUMPS

TURBORAMJET ENGINES

TURBOROCKET ENGINES

Turborotors
  USE   TURBINE WHEELS

TURBOSHAFTS

TURBULENCE

Turbulence, Atmospheric
  USE   ATMOSPHERIC TURBULENCE

Turbulence, Clear Air
  USE   CLEAR AIR TURBULENCE

TURBULENCE EFFECTS

Turbulence, Homogeneous
  USE   HOMOGENEOUS TURBULENCE

Turbulence, Isotropic
  USE   ISOTROPIC TURBULENCE

Turbulence, Low
  USE   LOW TURBULENCE

Turbulence, Low Level
  USE   LOW LEVEL TURBULENCE

Turbulence, Magnetohydrodynamic
  USE   MAGNETOHYDRODYNAMIC TURBULENCE

TURBULENCE METERS

Turbulence Meters, Hot-Wire
  USE   TURBULENCE METERS
         HOT-WIRE FLOWMETERS

Turbulence, Plasma
  USE   PLASMA TURBULENCE

TURBULENT BOUNDARY LAYER

TURBULENT DIFFUSION

TURBULENT FLOW

TURBULENT HEAT TRANSFER

TURBULENT JETS

TURBULENT MIXING

TURBULENT WAKES

TURING MACHINES

TURKEY

TURKEYS

TURNAROUND (STS)

TURNING FLIGHT

Turning Flight, Minor Circle
    USE   MINOR CIRCLE TURNING FLIGHT

TURNSTILE ANTENNAS

TURPENTINE

TURRET

TURRET LATHES

Turret Reactor, Los Alamos
    USE   HIGH TEMPERATURE NUCLEAR REACTORS

Turrets, Gun
    USE   GUN TURRETS

TURTLES

Tutor Aircraft
    USE   CL-41 AIRCRAFT

TVC (Control)
    USE   THRUST VECTOR CONTROL

TWENTY-FOUR HOUR ORBITS

TWENTY-SEVEN DAY VARIATION

TWILIGHT GLOW

Twin Aircraft, Advanced Technology Light
    USE   ATLIT PROJECT

Twin Hull, Small Water Plane Area
    USE   SWATH (SHIP)

TWINNING

Twinning, Mechanical
    USE   MECHANICAL TWINNING

TWISTED WINGS

TWISTING

TWITCHING

Two Body Orbits
    USE   TWO BODY PROBLEM

TWO BODY PROBLEM

TWO DIMENSIONAL BODIES

TWO DIMENSIONAL BOUNDARY LAYER

TWO DIMENSIONAL FLOW

TWO DIMENSIONAL JETS

TWO FLUID MODELS

TWO PHASE FLOW

Two Phase Systems
    USE   BINARY SYSTEMS (MATERIALS)

TWO REFLECTOR ANTENNAS

TWO STAGE PLASMA ENGINES

TWO STAGE TURBINES

TWO-WAVELENGTH LASERS

TX
    USE   TEXAS

(TX), Houston
    USE   HOUSTON (TX)

TX), Lake Texoma (OK-
    USE   LAKE TEXOMA (OK-TX)

TX-33-39 Engine
    USE   XM-33 ENGINE

TX-77 ENGINE

TX-354 ENGINE

TYCHO CRATER

Type Radiometers, Dicke
    USE   DICKE RADIOMETERS

Type Reactor, Livermore Pool
    USE   LIVERMORE POOL TYPE REACTOR

Type Semiconductors, N-
    USE   N-TYPE SEMICONDUCTORS

Type Semiconductors, P-
    USE   P-TYPE SEMICONDUCTORS

TYPE 2 BURSTS

TYPE 3 BURSTS

TYPE 4 BURSTS

TYPE 5 BURSTS

TYPEWRITERS

Typewriters, Automatic
    USE   AUTOMATIC TYPEWRITERS

Typewriters, Tele
    USE   TELETYPEWRITERS

TYPHOID

TYPHON WEAPON SYSTEM

TYPHOONS

TYPHUS

TYROSINE

T2J Aircraft
    USE   T-2 AIRCRAFT

T3J Aircraft
    USE   T-39 AIRCRAFT

U

U BENDS

U SPIN SPACE

U Test, Mann-Whitney-Wilcoxon
    USE   MANN-WHITNEY-WILCOXON U TEST

U Tubes
    USE   MANOMETERS

U.S.S.R.

(U.S.S.R.), Caucasus Mountains
    USE   CAUCASUS MOUNTAINS (U.S.S.R.)

U.S.S.R. SPACE PROGRAM

U-2 AIRCRAFT

U-2 Aircraft, Lockheed
    USE   U-2 AIRCRAFT

U-10 AIRCRAFT

UBV SPECTRA

UDIMET ALLOYS

UFO
    USE   UNIDENTIFIED FLYING OBJECTS

UGANDA

UH-1 HELICOPTER

UH-2 HELICOPTER

UH-2A Helicopter, Kaman
    USE   UH-2 HELICOPTER

UH-12 Helicopter
    USE   OH-23 HELICOPTER

UH-13 Helicopter
    USE   OH-13 HELICOPTER

UH-34 HELICOPTER

UH-60A HELICOPTER

UH-61A HELICOPTER

Uhlenbeck Process, Ornstein-
    USE   ORNSTEIN-UHLENBECK PROCESS

UHTREX (Nuclear Reactors)
    USE   HIGH TEMPERATURE NUCLEAR REACTORS

UHURU SATELLITE

UK SATELLITES

UK SPACE PROGRAM

UK 4 SATELLITE

ULCERS

ULLAGE

ULLAGE ROCKET ENGINES

ULM (Light Modulation)
    USE   ULTRASONIC LIGHT MODULATION

ULNA

Ultra Short Wave Radio Equipment
    USE   VERY HIGH FREQUENCY RADIO
              EQUIPMENT

ULTRAHIGH FREQUENCIES

ULTRAHIGH VACUUM

ULTRALIGHT AIRCRAFT

Ultralow Frequencies
    USE   EXTREMELY LOW RADIO FREQUENCIES

ULTRALOW TEMPERATURES

ULTRAPURE METALS

ULTRASHORT PULSED LASERS

ULTRASONIC AGITATION

ULTRASONIC CLEANING

ULTRASONIC DENSIMETERS

ULTRASONIC FLAW DETECTION

Ultrasonic Grinding Machines
    USE   ULTRASONIC MACHINING

ULTRASONIC LIGHT MODULATION

ULTRASONIC MACHINING

ULTRASONIC RADIATION

ULTRASONIC SCANNERS

ULTRASONIC SOLDERING

ULTRASONIC SPECTROSCOPY

ULTRASONIC TESTS

ULTRASONIC WAVE TRANSDUCERS

Ultrasonic Waves
    USE   ULTRASONIC RADIATION

ULTRASONIC WELDING

ULTRASONICS

ULTRAVIOLET ABSORPTION

ULTRAVIOLET ASTRONOMY

Ultraviolet Explorer, International
    USE   IUE

Ultraviolet Explorer Satellite, Extreme
    USE   EXTREME ULTRAVIOLET EXPLORER
           SATELLITE

ULTRAVIOLET FILTERS

ULTRAVIOLET LASERS

Ultraviolet Light
    USE   ULTRAVIOLET RADIATION

ULTRAVIOLET MICROSCOPY

ULTRAVIOLET PHOTOGRAPHY

ULTRAVIOLET PHOTOMETRY

ULTRAVIOLET RADIATION

Ultraviolet Radiation, Extreme
    USE   EXTREME ULTRAVIOLET RADIATION

Ultraviolet Radiation, Far
    USE   FAR ULTRAVIOLET RADIATION

Ultraviolet Radiation, Near
    USE   NEAR ULTRAVIOLET RADIATION

Ultraviolet Radiation, Vacuum
    USE   FAR ULTRAVIOLET RADIATION

ULTRAVIOLET REFLECTION

ULTRAVIOLET SPECTRA

Ultraviolet Spectrographs
    USE   ULTRAVIOLET SPECTROMETERS

ULTRAVIOLET SPECTROMETERS

ULTRAVIOLET SPECTROPHOTOMETERS

ULTRAVIOLET SPECTROSCOPY

ULTRAVIOLET TELESCOPES

ULYSSES MISSION

UMBILICAL CONNECTORS

UMBILICAL TOWERS

UMBRAS

Umbras, Pen
    USE   PENUMBRAS

UMKEHR EFFECT

UMKLAPP PROCESS

UNCAMBERED WINGS

UNCONSCIOUSNESS

UNCONTROLLED REENTRY (SPACECRAFT)

UNCOUPLED MODES

UNDAMPED OSCILLATIONS

UNDER SURFACE BLOWING

UNDERCARRIAGES

UNDERGROUND ACOUSTICS

UNDERGROUND COMMUNICATION

UNDERGROUND EXPLOSIONS

Underground Radio Antenna Grid (Navy)
    USE   SEAFARER PROJECT

UNDERGROUND STORAGE

UNDERGROUND STRUCTURES

UNDERGROUND TRANSMISSION LINES

UNDERWATER ACOUSTICS

UNDERWATER BREATHING APPARATUS

UNDERWATER COMMUNICATION

(Underwater), Diving
    USE   DIVING (UNDERWATER)

UNDERWATER ENGINEERING

UNDERWATER EXPLOSIONS

UNDERWATER OPTICS

UNDERWATER PHOTOGRAPHY

UNDERWATER PHYSIOLOGY

UNDERWATER PROPULSION

UNDERWATER RESEARCH LABORATORIES

UNDERWATER RESOURCES

Underwater Sound
    USE   UNDERWATER ACOUSTICS

UNDERWATER STRUCTURES

UNDERWATER TESTS

UNDERWATER TO SURFACE MISSILES

UNDERWATER TRAJECTORIES

UNDERWATER VEHICLES

Unguided Rocket Trajectory, Spinning
    USE   SPINNING UNGUIDED ROCKET
           TRAJECTORY

Uniaxial Strain
    USE   AXIAL STRAIN

UNIDENTIFIED FLYING OBJECTS

UNIFIED FIELD THEORY

UNIFIED S BAND

UNIFORM FLOW

Uniformity, Non
    USE   NONUNIFORMITY

UNIMOLECULAR STRUCTURES

Union, Soviet
    USE   U.S.S.R.

UNIONIZATION

UNIONS

UNIONS (CONNECTORS)

Uniphase Flow
    USE   SINGLE-PHASE FLOW

Unipolar Transistors
    USE   FIELD EFFECT TRANSISTORS

UNIQUENESS

UNIQUENESS THEOREM

Unit Area), Flux (Rate Per
    USE   FLUX DENSITY

Unit Reactors, Space Power
    USE   SPACE POWER UNIT REACTORS

UNITED ARAB EMIRATES

UNITED KINGDOM

United Kingdom Satellites
    USE   UK SATELLITES

UNITED NATIONS

UNITED STATES

(United States), Armed Forces
    USE   ARMED FORCES (UNITED STATES)

(United States), USA
    USE   UNITED STATES

Units, Agrophysical
    USE   AGROPHYSICAL UNITS

Units, Arithmetic And Logic
    USE   ARITHMETIC AND LOGIC UNITS

Units), Bays (Structural
    USE   BAYS (STRUCTURAL UNITS)

Units, Central Processing
    USE   CENTRAL PROCESSING UNITS

Units, Chemical Auxiliary Power
    USE   CHEMICAL AUXILIARY POWER UNITS

Units (Computers), Control
    USE   CONTROL UNITS (COMPUTERS)

Units, Extravehicular Mobility
    USE   EXTRAVEHICULAR MOBILITY UNITS

Units, Inertial Measuring
    USE   INERTIAL PLATFORMS

Units, International System Of
    USE   INTERNATIONAL SYSTEM OF UNITS

Units, Manned Maneuvering
    USE   MANNED MANEUVERING UNITS

Units, Nuclear Auxiliary Power
    USE   NUCLEAR AUXILIARY POWER UNITS

UNITS OF MEASUREMENT

Units, Self Maneuvering
    USE   SELF MANEUVERING UNITS

Units), SMU (Maneuvering
    USE   SELF MANEUVERING UNITS

Units, Solar Auxiliary Power
    USE   SOLAR AUXILIARY POWER UNITS

Units, Space Self Maneuvering
    USE   SELF MANEUVERING UNITS

UNITY

UNIVAC COMPUTERS

UNIVAC LARC COMPUTER

UNIVAC 80 COMPUTER

UNIVAC 418 COMPUTER

UNIVAC 490 COMPUTER

UNIVAC 494 COMPUTER

UNIVAC 1100 SERIES COMPUTERS

UNIVAC 1105 COMPUTER

UNIVAC 1106 COMPUTER

UNIVAC 1107 COMPUTER

UNIVAC 1108 COMPUTER

UNIVAC 1110 COMPUTER

UNIVAC 1230 COMPUTER

UNIVERSAL TIME

UNIVERSE

UNIVERSITIES

UNIVERSITY PROGRAM

UNLOADING

UNLOADING WAVES

(Unmanned), SKYLAB Space Station
USE   SKYLAB 1

UNMANNED SPACECRAFT

UNSATURATION (CHEMISTRY)

UNSTEADY FLOW

UNSTEADY STATE

UNSWEPT WINGS

Up Displays, Head-
USE   HEAD-UP DISPLAYS

Up, Latch-
USE   LATCH-UP

Up, Lay-
USE   LAY-UP

UP-CONVERTERS

Updrafts
USE   VERTICAL AIR CURRENTS

UPGRADING

UPLINKING

Upper Air
USE   UPPER ATMOSPHERE

UPPER ATMOSPHERE

UPPER IONOSPHERE

Upper Stage A, Space Shuttle
USE   SPACE SHUTTLE UPPER STAGE A

Upper Stage D, Space Shuttle
USE   SPACE SHUTTLE UPPER STAGE D

Upper Stage, Inertial
USE   INERTIAL UPPER STAGE

UPPER STAGE ROCKET ENGINES

Upper Stage, Spinning Solid
USE   SPINNING SOLID UPPER STAGE

Upper Stage (Sts), Interim
USE   INERTIAL UPPER STAGE

Upper Stages, Space Shuttle
USE   SPACE SHUTTLE UPPER STAGES

UPPER SURFACE BLOWING

UPPER SURFACE BLOWN FLAPS

Upper Volta
USE   BURKINA

Upsets, Single Event
USE   SINGLE EVENT UPSETS

UPSETTING

UPSTREAM

UPWASH

Upwelling
USE   UPWELLING WATER

UPWELLING WATER

URACIL

URANIUM

URANIUM ALLOYS

URANIUM CARBIDES

URANIUM COMPOUNDS

URANIUM FLUORIDES

URANIUM ISOTOPES

URANIUM OXIDES

URANIUM PLASMAS

URANIUM 232

URANIUM 233

URANIUM 234

URANIUM 235

URANIUM 238

URANUS ATMOSPHERE

Uranus Flyby, Mariner Jupiter-
USE   MARINER JUPITER-URANUS FLYBY

URANUS (PLANET)

URANUS RINGS

URANUS SATELLITES

Urban Areas
USE   CITIES

URBAN DEVELOPMENT

URBAN PLANNING

URBAN RESEARCH

URBAN TRANSPORTATION

Urchins, Sea
USE   SEA URCHINS

Urea, Difluoro
USE   DIFLUOROUREA

UREAS

URETHANES

URIC ACID

URIDYLIC ACID

URINALYSIS

URINATION

URINE

UROGRAPHY

UROLITHIASIS

UROLOGY

URUGUAY

Urundi, Ruanda-
USE   RWANDA
           BURUNDI

(US), Aleutian Islands
USE   ALEUTIAN ISLANDS (US)

(US), Allegheny Plateau
USE   ALLEGHENY PLATEAU (US)

(US), Central Atlantic Region
USE   CENTRAL ATLANTIC REGION (US)

(US), Central Piedmont
USE   CENTRAL PIEDMONT (US)

(US), Chesapeake Bay
USE   CHESAPEAKE BAY (US)

(US), Colorado Plateau
USE   COLORADO PLATEAU (US)

(US), Delaware Bay
USE   DELAWARE BAY (US)

(US), Delaware River Basin
USE   DELAWARE RIVER BASIN (US)

(US), Great Basin
USE   GREAT BASIN (US)

(US), Mississippi River
USE   MISSISSIPPI RIVER (US)

(US), Missouri River
USE   MISSOURI RIVER (US)

(US), Missouri River Basin
USE   MISSOURI RIVER BASIN (US)

(US), New England
USE   NEW ENGLAND (US)

(US), Ohio River
USE   OHIO RIVER (US)

(US), Pacific Northwest
USE   PACIFIC NORTHWEST (US)

US-2A Aircraft
USE   S-2 AIRCRAFT

USA (United States)
USE   UNITED STATES

Usable Frequency, Maximum
USE   MAXIMUM USABLE FREQUENCY

Use, Land
USE   LAND USE

Use, Rural Land
USE   RURAL LAND USE

USER MANUALS (COMPUTER PROGRAMS)

USER REQUIREMENTS

USNS Kingsport
USE   SATELLITE COMMUNICATIONS SHIPS

UT
USE   UTAH

(UT), Great Salt Lake
USE   GREAT SALT LAKE (UT)

UTAH

UTERUS

UTILITIES

UTILITY AIRCRAFT

Utility System, Modular Integrated
USE   MODULAR INTEGRATED UTILITY SYSTEM

UTILIZATION

Utilization, Coal
USE   COAL UTILIZATION

Utilization, Geothermal Energy
USE   GEOTHERMAL ENERGY UTILIZATION

Utilization Lists, Hardware
USE   HARDWARE UTILIZATION LISTS

Utilization, Orbit Spectrum
USE   ORBIT SPECTRUM UTILIZATION

Utilization System, National Airspace
USE   NATIONAL AIRSPACE UTILIZATION SYSTEM

Utilization, Technology
USE   TECHNOLOGY UTILIZATION

Utilization, Waste
USE   WASTE UTILIZATION

Utilization, Waste Energy
USE   WASTE ENERGY UTILIZATION

Utilization, Windpower
USE   WINDPOWER UTILIZATION

UTRICLE

UV Ceti Stars
USE   FLARE STARS

UV Lasers
USE   ULTRAVIOLET LASERS

UV Spectrometer, Solar Backscatter
USE   SOLAR BACKSCATTER UV
         SPECTROMETER

UV Spectroscopic Explorer, Far
USE   FAR UV SPECTROSCOPIC EXPLORER

UV-Optical Telescope Facility, Spacelab
USE   STARLAB

V

V Band
USE   EXTREMELY HIGH FREQUENCIES

V GROOVES

V-1 MISSILE

V-2 MISSILE

V-3 Aircraft
USE   XV-3 AIRCRAFT

V-4 Aircraft
USE   XV-4 AIRCRAFT

V-5 Aircraft
USE   XV-5 AIRCRAFT

V-9 Aircraft
USE   XV-9A AIRCRAFT

V/STOL AIRCRAFT

VA
USE   VIRGINIA

VA), Assateague Island (MD-
USE   ASSATEAGUE ISLAND (MD-VA)

VA), Delmarva Peninsula (De-MD-
USE   DELMARVA PENINSULA (DE-MD-VA)

(VA), Shenandoah Valley
USE   SHENANDOAH VALLEY (VA)

VA-WV), Potomac River Valley (MD-
USE   POTOMAC RIVER VALLEY (MD-VA-WV)

VACANCIES (CRYSTAL DEFECTS)

VACCINES

VACILLATION

VACUUM

VACUUM APPARATUS

VACUUM ARC SWITCHES

VACUUM CHAMBERS

VACUUM DEPOSITION

VACUUM EFFECTS

(Vacuum), Evacuating
USE   EVACUATING (VACUUM)

VACUUM FURNACES

VACUUM GAGES

Vacuum, High
USE   HIGH VACUUM

Vacuum, Low
USE   LOW VACUUM

VACUUM MELTING

Vacuum Orbital Simulator, High
USE   HIGH VACUUM ORBITAL SIMULATOR

VACUUM PUMPS

VACUUM SPECTROSCOPY

VACUUM SYSTEMS

VACUUM TESTS

Vacuum Tests, Thermal
USE   THERMAL VACUUM TESTS

VACUUM TUBE OSCILLATORS

VACUUM TUBES

Vacuum, Ultrahigh
USE   ULTRAHIGH VACUUM

Vacuum Ultraviolet Radiation
USE   FAR ULTRAVIOLET RADIATION

VADOSE WATER

Vaisala Frequency, Brunt-
USE   BRUNT-VAISALA FREQUENCY

VALENCE

Valence, Co
USE   COVALENCE

Valence, Equi
USE   EQUIVALENCE

VALERIC ACID

VALIANT AIRCRAFT

Valiant Aircraft, Vickers
USE   VALIANT AIRCRAFT

Validation
USE   PROVING

VALIDITY

Valkyrie Aircraft
USE   B-70 AIRCRAFT

Valley (AL-KY-TN), Tennessee
USE   TENNESSEE VALLEY (AL-KY-TN)

Valley (CA), Coachella
USE   COACHELLA VALLEY (CA)

Valley (CA), Death
USE   DEATH VALLEY (CA)

Valley (CA), Imperial
USE   IMPERIAL VALLEY (CA)

Valley (CA), Palo Verde
USE   PALO VERDE VALLEY (CA)

Valley (CA), Sacramento
USE   SACRAMENTO VALLEY (CA)

Valley (CA), San Joaquin
USE   SAN JOAQUIN VALLEY (CA)

Valley (Colombia), Magdalena-Cauca
USE   MAGDALENA-CAUCA VALLEY (COLOMBIA)

Valley (MD-VA-WV), Potomac River
USE   POTOMAC RIVER VALLEY (MD-VA-WV)

Valley (North America), St Lawrence
USE   ST LAWRENCE VALLEY (NORTH AMERICA)

Valley (VA), Shenandoah
USE   SHENANDOAH VALLEY (VA)

VALLEYS

Valleys, Rift
USE   VALLEYS

VALSALVA EXERCISE

Valsalva Maneuver
USE   VALSALVA EXERCISE

VALUE

VALUE ENGINEERING

Value Problems, Boundary
USE   BOUNDARY VALUE PROBLEMS

Value Problems, Initial
USE   BOUNDARY VALUE PROBLEMS

Values, Eigen
USE   EIGENVALUES

Values, Extremum
USE   EXTREMUM VALUES

Values, Mean Square
USE   MEAN SQUARE VALUES

Values, Nominal
USE   APPROXIMATION

Values, Q
USE   Q VALUES

VALVES

Valves, Artificial Heart
USE   ARTIFICIAL HEART VALVES

Valves, Automatic Control
USE   AUTOMATIC CONTROL VALVES

Valves, Butterfly
USE   BUTTERFLY VALVES

Valves, Control
USE   CONTROL VALVES

(Valves), Dampers
USE   DAMPERS (VALVES)

Valves, Fuel
USE   FUEL VALVES

Valves, Gas
USE   GAS VALVES

Valves, Heart
USE   HEART VALVES

Valves, Hydraulic
USE   HYDRAULIC EQUIPMENT
       VALVES

Valves, Light
USE   LIGHT VALVES

Valves, Relief
USE   RELIEF VALVES

Valves, Solenoid
USE   SOLENOID VALVES

Vampire Aircraft
USE   DH 115 AIRCRAFT

VAMPIRE MK 35 AIRCRAFT

Van Allen Radiation Belts
USE   RADIATION BELTS

VAN BIESBROECK STAR

VAN DE GRAAFF ACCELERATORS

VAN DER WAAL FORCES

VAN SLYKE METHOD

VANADATES

Vanadates, Calcium
USE   CALCIUM VANADATES

VANADIUM

VANADIUM ALLOYS

VANADIUM CARBIDES

VANADIUM COMPOUNDS

VANADIUM ISOTOPES

VANADIUM OXIDES

VANADYL COMPOUNDS

VANADYL RADICAL

VANELESS DIFFUSERS

VANES

Vanes, Guide
USE   GUIDE VANES

Vanes, Jet
USE   JET VANES

Vanes, Tip
USE   TIP VANES

Vanes, Wind
USE   WIND VANES

VANGUARD PROJECT

VANGUARD SATELLITES

VANGUARD 1 SATELLITE

VANGUARD 2 LAUNCH VEHICLE

VANGUARD 2 SATELLITE

VANGUARD 3 SATELLITE

Vans
USE   TRUCKS

VAPOR BARRIER CLOTHING

Vapor, Cesium
USE   CESIUM VAPOR

VAPOR DEPOSITION

Vapor Equilibrium, Liquid-
USE   LIQUID-VAPOR EQUILIBRIUM

Vapor Generators
USE   VAPORIZERS

Vapor Generators, Cavity
USE   CAVITY VAPOR GENERATORS

Vapor Interfaces, Liquid-
USE   LIQUID-VAPOR INTERFACES

VAPOR JETS

Vapor Lamps, Alkali
USE   ALKALI VAPOR LAMPS

Vapor Lasers, Metal
USE   METAL VAPOR LASERS

Vapor Liquid Equilibrium
USE   LIQUID-VAPOR EQUILIBRIUM

Vapor, Mercury
USE   MERCURY VAPOR

VAPOR PHASE EPITAXY

VAPOR PHASES

VAPOR PRESSURE

Vapor, Sodium
USE   SODIUM VAPOR

Vapor Trails
USE   CONTRAILS

VAPOR TRAPS

Vapor, Water
USE   WATER VAPOR

Vaporization Heat
USE   HEAT OF VAPORIZATION

Vaporization, Heat Of
USE   HEAT OF VAPORIZATION

Vaporization, Pre
USE   PREVAPORIZATION

VAPORIZERS

VAPORIZING

(Vaporizing), Flashing
USE   FLASHING (VAPORIZING)

VAPORS

Vapors, Metal
USE   METAL VAPORS

VARACTOR DIODE CIRCUITS

VARACTOR DIODES

Varactors
USE   VARACTOR DIODES

VARIABILITY

VARIABLE

Variable Area Wings
USE   TRAILING EDGE FLAPS

VARIABLE CYCLE ENGINES

VARIABLE GEOMETRY STRUCTURES

Variable Lift
USE   LIFT

VARIABLE MASS SYSTEMS

VARIABLE PITCH PROPELLERS

VARIABLE STARS

VARIABLE STREAM CONTROL ENGINES

VARIABLE SWEEP WINGS

VARIABLE THRUST

Variables, Cataclysmic
USE   CATACLYSMIC VARIABLES

Variables, Cepheid
USE   CEPHEID VARIABLES

Variables, Complex
USE   COMPLEX VARIABLES

Variables, Dependent
USE   DEPENDENT VARIABLES

Variables, Independent
USE   INDEPENDENT VARIABLES

Variables), Integration (Real
USE   MEASURE AND INTEGRATION

Variables, Random
USE   RANDOM VARIABLES

Variables, Real
USE   REAL VARIABLES

VARIANCE

Variance, Analysis Of
USE   ANALYSIS OF VARIANCE

Variance, Co
USE   COVARIANCE

Variance Orbit Determination, Minimum
USE   MINIMUM VARIANCE ORBIT
       DETERMINATION

VARIANCE (STATISTICS)

Variation Indicators, Voltage
USE   VOLTMETERS

Variation Method
USE   CALCULUS OF VARIATIONS

Variation, Twenty-Seven Day
USE   TWENTY-SEVEN DAY VARIATION

VARIATIONAL PRINCIPLES

Variational Theorem, Castigliano
USE   CASTIGLIANO VARIATIONAL THEOREM

VARIATIONS

Variations, Annual
USE   ANNUAL VARIATIONS

Variations, Calculus Of
USE   CALCULUS OF VARIATIONS

Variations, Diurnal
USE   DIURNAL VARIATIONS

Variations, Magnetic
USE   MAGNETIC VARIATIONS

Variations, Nocturnal
USE   NOCTURNAL VARIATIONS

Variations, Periodic
USE   PERIODIC VARIATIONS

Variations, Seasonal
USE   ANNUAL VARIATIONS

Variations, Secular
USE   SECULAR VARIATIONS

Variations, Wind
USE   WIND VARIATIONS

VARIOMETERS

VARISTORS

VARNISHES

Vascular Accidents, Cerebral
USE   CEREBRAL VASCULAR ACCIDENTS

VASCULAR SYSTEM

VASOCONSTRICTION

VASOCONSTRICTOR DRUGS

VASODILATION

Vasomotor Nervous System
USE   NERVOUS SYSTEM

VATICAN CITY

VATOL AIRCRAFT

VAX COMPUTERS

VAX-11 SERIES COMPUTERS

VAX-11/780 COMPUTER

VC-10 AIRCRAFT

VC-10 Aircraft, Vickers
USE   VC-10 AIRCRAFT

VCE
USE   VARIABLE CYCLE ENGINES

VCO
USE   VOLTAGE CONTROLLED OSCILLATORS

VECTOR ANALYSIS

Vector Calculus
USE   VECTOR SPACES

(Vector Calculus), Stokes Theorem
USE   STOKES THEOREM (VECTOR CALCULUS)

Vector Control
USE   DIRECTIONAL CONTROL

Vector Control, Thrust
USE   THRUST VECTOR CONTROL

VECTOR CURRENTS

VECTOR DOMINANCE MODEL

VECTOR MESONS

Vector Recorders, Force
USE   FORCE VECTOR RECORDERS

VECTOR SPACES

VECTORCARDIOGRAPHY

(Vectors), Curl
USE   CURL (VECTORS)

Vectors, Eigen
USE   EIGENVECTORS

VECTORS (MATHEMATICS)

Vectors, State
USE   STATE VECTORS

VEGA LAUNCH VEHICLE

VEGA PROJECT

Vega Rocket Vehicle
USE   VEGA LAUNCH VEHICLE

VEGARD-KAPLAN BANDS

VEGETABLES

VEGETATION

(Vegetation), Canopies
USE   CANOPIES (VEGETATION)

Vegetation, Diseased
USE   BLIGHT

VEGETATION GROWTH

VEGETATIVE INDEX

Vch Design, Integ Program For Aerospace
USE   IPAD

Vehicle, Ablestar Launch
USE   ABLESTAR LAUNCH VEHICLE

Vehicle, Aerobee Rocket
USE   AEROBEE ROCKET VEHICLE

Vehicle, Agena A Rocket
USE   AGENA A ROCKET VEHICLE

Vehicle, Agena B Rocket
USE   AGENA B ROCKET VEHICLE

Vehicle, Agena C Rocket
USE   AGENA C ROCKET VEHICLE

Vehicle, Agena D Rocket
USE   AGENA D ROCKET VEHICLE

Vehicle, Antares Rocket
USE   ANTARES ROCKET VEHICLE

Vehicle, Apache Rocket
USE   APACHE ROCKET VEHICLE

Vehicle, Arcon Rocket
USE   ARCON ROCKET VEHICLE

Vehicle, Ariane Launch
USE   ARIANE LAUNCH VEHICLE

Vehicle, Asp Rocket
USE   ASP ROCKET VEHICLE

Vehicle, Astro
USE   ASTRO VEHICLE

Vehicle, Astrobee 1500 Rocket
USE   ASTROBEE 1500 ROCKET VEHICLE

Vehicle, Athena Rocket
USE   ATHENA ROCKET VEHICLE

Vehicle, Atlas Able 5 Launch
USE   ATLAS ABLE 5 LAUNCH VEHICLE

Vehicle, Atlas Agena B Launch
USE   ATLAS AGENA B LAUNCH VEHICLE

Vehicle, Atlas Centaur Launch
USE   ATLAS CENTAUR LAUNCH VEHICLE

Vehicle, Atlas SLV-3 Launch
USE   ATLAS SLV-3 LAUNCH VEHICLE

Vehicle, Berenice Rocket
USE   BERENICE ROCKET VEHICLE

Vehicle, Black Arrow Launch
USE   BLACK KNIGHT ROCKET VEHICLE

Vehicle, Black Knight Rocket
USE   BLACK KNIGHT ROCKET VEHICLE

Vehicle, Blue Scout Rocket
USE   BLUE SCOUT ROCKET VEHICLE

Vehicle, Blue Streak Launch
USE   BLUE STREAK LAUNCH VEHICLE

Vehicle, Cajun Rocket
USE   CAJUN ROCKET VEHICLE

Vehicle, Centaur
USE   CENTAUR LAUNCH VEHICLE

Vehicle, Centaur Launch
USE   CENTAUR LAUNCH VEHICLE

Vehicle Checkout Program, Space
USE   SPACE VEHICLE CHECKOUT PROGRAM

Vehicle Configurations, Launch
USE   LAUNCH VEHICLE CONFIGURATIONS

Vehicle Control, Space
USE   SPACECRAFT CONTROL

Vehicle, Delta Launch
USE   DELTA LAUNCH VEHICLE

Vehicle, Diamant Launch
USE   DIAMANT LAUNCH VEHICLE

Vehicle, Dornier Paraglider Rocket
USE   DORNIER PARAGLIDER ROCKET VEHICLE

Vehicle, Eldo Launch
USE   ELDO LAUNCH VEHICLE

Vehicle, Europa 1 Launch
USE   EUROPA 1 LAUNCH VEHICLE

Vehicle, Europa 2 Launch
USE   EUROPA 2 LAUNCH VEHICLE

Vehicle, Europa 3 Launch
USE   EUROPA 3 LAUNCH VEHICLE

Vehicle, Europa 4 Launch
USE   EUROPA 4 LAUNCH VEHICLE

Vehicle, FDL-5 Reentry
USE   FDL-5 REENTRY VEHICLE

Vehicle, FFAR Rocket
USE   FOLDING FIN AIRCRAFT ROCKET VEHICLE

Vehicle, Folding Fin Aircraft Rocket
USE   FOLDING FIN AIRCRAFT ROCKET VEHICLE

Vehicle, Genie Rocket
USE   GENIE ROCKET VEHICLE

Vehicle, HL-10 Reentry
USE   HL-10 REENTRY VEHICLE

Vehicle, HLD-35 Reentry
USE   HLD-35 REENTRY VEHICLE

Vehicle, Honest John Rocket
USE   HONEST JOHN ROCKET VEHICLE

Vehicle, Hyla-Star Rocket
USE   HYLA-STAR ROCKET VEHICLE

Vehicle, Jabiru Rocket
USE   JAGUAR ROCKET VEHICLE

Vehicle, Jaguar Rocket
USE   JAGUAR ROCKET VEHICLE

Vehicle, Javelin Rocket
USE   JAVELIN ROCKET VEHICLE

Vehicle, Juno 1 Launch
USE   JUNO 1 LAUNCH VEHICLE

Vehicle, Juno 2 Launch
USE   JUNO 2 LAUNCH VEHICLE

Vehicle, Jupiter C Rocket
USE   JUPITER C ROCKET VEHICLE

Vehicle, Kappa 8 Rocket
USE   KAPPA 8 ROCKET VEHICLE

Vehicle, Kappa 9 Rocket
USE   KAPPA 9 ROCKET VEHICLE

Vehicle, Little Joe 2 Launch
USE   LITTLE JOE 2 LAUNCH VEHICLE

Vehicle, Little John Rocket
USE   LITTLE JOHN ROCKET VEHICLE

Vehicle, Loki Rocket
USE   LOKI ROCKET VEHICLE

(Vehicle), LRV
USE   LUNAR ROVING VEHICLES

Vehicle, MB-1 Rocket
USE   GENIE ROCKET VEHICLE

Vehicle, Meteor 1 Rocket
USE   METEOR 1 ROCKET VEHICLE

Vehicle, Nike-Apache Rocket
    USE   NIKE-APACHE ROCKET VEHICLE

Vehicle, Nike-Cajun Rocket
    USE   NIKE-CAJUN ROCKET VEHICLE

Vehicle, Nike-Hydac Rocket
    USE   NIKE-HYDAC ROCKET VEHICLE

Vehicle, Nike-Iroquois Rocket
    USE   NIKE-IROQUOIS ROCKET VEHICLE

Vehicle, Nike-Javelin Rocket
    USE   NIKE-JAVELIN ROCKET VEHICLE

Vehicle, Nike-Tomahawk Rocket
    USE   NIKE-TOMAHAWK ROCKET VEHICLE

Vehicle, Nomad Launch
    USE   NOMAD LAUNCH VEHICLE

Vehicle Program, National Launch
    USE   NATIONAL LAUNCH VEHICLE PROGRAM

Vehicle Program, Terminal Configured
    USE   TERMINAL CONFIGURED VEHICLE
              PROGRAM

Vehicle, RAM B Launch
    USE   RAM B LAUNCH VEHICLE

Vehicle, Rubis Rocket
    USE   RUBIS ROCKET VEHICLE

Vehicle, Saturn D Launch
    USE   SATURN D LAUNCH VEHICLE

Vehicle, Saturn 1 SA-1 Launch
    USE   SATURN 1 SA-1 LAUNCH VEHICLE

Vehicle, Saturn 1 SA-2 Launch
    USE   SATURN 1 SA-2 LAUNCH VEHICLE

Vehicle, Saturn 1 SA-3 Launch
    USE   SATURN 1 SA-3 LAUNCH VEHICLE

Vehicle, Saturn 1 SA-4 Launch
    USE   SATURN 1 SA-4 LAUNCH VEHICLE

Vehicle, Saturn 1 SA-5 Launch
    USE   SATURN 1 SA-5 LAUNCH VEHICLE

Vehicle, Saturn 1 SA-6 Launch
    USE   SATURN 1 SA-6 LAUNCH VEHICLE

Vehicle, Saturn 1 SA-7 Launch
    USE   SATURN 1 SA-7 LAUNCH VEHICLE

Vehicle, Saturn 1 SA-8 Launch
    USE   SATURN 1 SA-8 LAUNCH VEHICLE

Vehicle, Saturn 1 SA-9 Launch
    USE   SATURN 1 SA-9 LAUNCH VEHICLE

Vehicle, Saturn 1 SA-10 Launch
    USE   SATURN 1 SA-10 LAUNCH VEHICLE

Vehicle, Scout Launch
    USE   SCOUT LAUNCH VEHICLE

Vehicle, Skylark Rocket
    USE   SKYLARK ROCKET VEHICLE

Vehicle, Thor Able Rocket
    USE   THOR ABLE ROCKET VEHICLE

Vehicle, Thor Agena Launch
    USE   THOR AGENA LAUNCH VEHICLE

Vehicle, Thor Delta Launch
    USE   THOR DELTA LAUNCH VEHICLE

Vehicle, Titan Centaur Launch
    USE   TITAN CENTAUR LAUNCH VEHICLE

Vehicle, Titan 3 Launch
    USE   TITAN 3 LAUNCH VEHICLE

Vehicle, Trailblazer 1 Reentry
    USE   TRAILBLAZER 1 REENTRY VEHICLE

Vehicle, Trailblazer 1 Rocket
    USE   TRAILBLAZER 1 REENTRY VEHICLE

Vehicle, Trailblazer 2 Reentry
    USE   TRAILBLAZER 2 REENTRY VEHICLE

Vehicle, Trailblazer 2 Rocket
    USE   TRAILBLAZER 2 REENTRY VEHICLE

Vehicle, Vanguard 2 Launch
    USE   VANGUARD 2 LAUNCH VEHICLE

Vehicle, Vega Launch
    USE   VEGA LAUNCH VEHICLE

Vehicle, Vega Rocket
    USE   VEGA LAUNCH VEHICLE

Vehicle, Venus Fly TRAP Rocket
    USE   VENUS FLY TRAP ROCKET VEHICLE

Vehicle, Viking Rocket
    USE   VIKING ROCKET VEHICLE

Vehicle, Viking 75 Entry
    USE   VIKING 75 ENTRY VEHICLE

VEHICLE WHEELS

Vehicle, X-17 Reentry
    USE   X-17 REENTRY VEHICLE

Vehicle, Zuni Rocket
    USE   ZUNI ROCKET VEHICLE

Vehicle 3, Standard Launch
    USE   ATLAS SLV-3 LAUNCH VEHICLE

Vehicle 5, Standard Launch
    USE   STANDARD LAUNCH VEHICLE 5

VEHICLES

Vehicles, Aerodynamic
    USE   AIRCRAFT

Vehicles, Aeroquatic
    USE   AEROQUATIC VEHICLES

Vehicles, Aerospace
    USE   AEROSPACE VEHICLES

Vehicles, Agena Rocket
    USE   AGENA ROCKET VEHICLES

Vehicles, Air Cushion
    USE   GROUND EFFECT MACHINES

Vehicles, Amphibious
    USE   AMPHIBIOUS VEHICLES

Vehicles, Arcas Rocket
    USE   ARCAS ROCKET VEHICLES

Vehicles, Argo Rocket
    USE   ARGO ROCKET VEHICLES

Vehicles, Astrobee Rocket
    USE   ASTROBEE ROCKET VEHICLES

Vehicles, Atlas Agena Launch
    USE   ATLAS AGENA LAUNCH VEHICLES

Vehicles, Atlas Launch
    USE   ATLAS LAUNCH VEHICLES

Vehicles, Automated Guideway Transit
    USE   AUTOMATED GUIDEWAY TRANSIT
              VEHICLES

Vehicles, Automated Mixed Traffic
    USE   AUTOMATED MIXED TRAFFIC VEHICLES

Vehicles, Automated Transit
    USE   AUTOMATED TRANSIT VEHICLES

Vehicles, Ballistic
    USE   BALLISTIC VEHICLES

Vehicles, Boostglide
    USE   BOOSTGLIDE VEHICLES

Vehicles, Captured Air Bubble
    USE   CAPTURED AIR BUBBLE VEHICLES

Vehicles, Control Configured
    USE   CONTROL CONFIGURED VEHICLES

Vehicles, Drone
    USE   DRONE VEHICLES

Vehicles, Electric Hybrid
    USE   ELECTRIC HYBRID VEHICLES

Vehicles, Electric Motor
    USE   ELECTRIC MOTOR VEHICLES

Vehicles, Europa Launch
    USE   EUROPA LAUNCH VEHICLES

Vehicles, Extraterrestrial Roving
    USE   ROVING VEHICLES

Vehicles, Flight
    USE   FLIGHT VEHICLES

Vehicles, Flight Test
    USE   FLIGHT TEST VEHICLES

Vehicles, Heavy Lift Launch
    USE   HEAVY LIFT LAUNCH VEHICLES

Vehicles, Hovering Rocket
    USE   HOVERING ROCKET VEHICLES

(Vehicles), Hydroplanes
    USE   HYDROPLANES (VEHICLES)

Vehicles, Hypersonic
    USE   HYPERSONIC VEHICLES

Vehicles, Intraorbit Transfer
    USE   INTRAORBIT TRANSFER VEHICLES

Vehicles, Juno Launch
    USE   JUNO LAUNCH VEHICLES

Vehicles, Kappa Rocket
    USE   KAPPA ROCKET VEHICLES

Vehicles, Lambda Rocket
    USE   LAMBDA ROCKET VEHICLES

Vehicles, Launch
    USE   LAUNCH VEHICLES

Vehicles, Lifting Reentry
    USE   LIFTING REENTRY VEHICLES

Vehicles, Low Observable Reentry
    USE   LOW OBSERVABLE REENTRY VEHICLES

Vehicles, Lunar Flying
    USE   LUNAR FLYING VEHICLES

Vehicles, Lunar Roving
    USE   LUNAR ROVING VEHICLES

Vehicles, Lunar Surface
    USE   LUNAR SURFACE VEHICLES

Vehicles, Lunokhod Lunar Roving
    USE   LUNOKHOD LUNAR ROVING VEHICLES

Vehicles, Magnetic Levitation
    USE   MAGNETIC LEVITATION VEHICLES

Vehicles, Manned Lunar Surface
    USE   MANNED LUNAR SURFACE VEHICLES

Vehicles, Military
    USE   MILITARY VEHICLES

Vehicles, Motor
    USE   MOTOR VEHICLES

Vehicles, Multiengine
    USE   MULTIENGINE VEHICLES

Vehicles, Multistage Rocket
    USE   MULTISTAGE ROCKET VEHICLES

Vehicles, Nike Rocket
USE   NIKE ROCKET VEHICLES

Vehicles, Nonlifting
USE   BALLISTIC VEHICLES

Vehicles, Nova Launch
USE   NOVA LAUNCH VEHICLES

Vehicles, Nuclear Engine For Rocket
USE   NUCLEAR ENGINE FOR ROCKET VEHICLES

Vehicles, Orbit Transfer
USE   ORBIT TRANSFER VEHICLES

Vehicles, Orbital Maneuvering
USE   ORBITAL MANEUVERING VEHICLES

Vehicles, Ranger Lunar Landing
USE   RANGER LUNAR LANDING VEHICLES

Vehicles, Recoverable Launch
USE   RECOVERABLE LAUNCH VEHICLES

Vehicles, Recovery
USE   RECOVERY VEHICLES

Vehicles, Reentry
USE   REENTRY VEHICLES

Vehicles, Remotely Piloted
USE   REMOTELY PILOTED VEHICLES

Vehicles, Research
USE   RESEARCH VEHICLES

Vehicles, Reusable Launch
USE   REUSABLE LAUNCH VEHICLES

Vehicles, Roadway Powered
USE   ROADWAY POWERED VEHICLES

Vehicles, Rocket
USE   ROCKET VEHICLES

Vehicles, Rotating
USE   ROTATING BODIES
       VEHICLES

Vehicles, Roving
USE   ROVING VEHICLES

Vehicles, Saturn Launch
USE   SATURN LAUNCH VEHICLES

Vehicles, Saturn 1 Launch
USE   SATURN 1 LAUNCH VEHICLES

Vehicles, Saturn 1B Launch
USE   SATURN 1B LAUNCH VEHICLES

Vehicles, Saturn 2 Launch
USE   SATURN 2 LAUNCH VEHICLES

Vehicles, Saturn 5 Launch
USE   SATURN 5 LAUNCH VEHICLES

Vehicles, Shuttle Derived
USE   SHUTTLE DERIVED VEHICLES

Vehicles, Single Stage Rocket
USE   SINGLE STAGE ROCKET VEHICLES

Vehicles, Single Stage To Orbit
USE   SINGLE STAGE TO ORBIT VEHICLES

Vehicles, Skua Rocket
USE   SKUA ROCKET VEHICLES

Vehicles, SLV (Soft Landing
USE   SOFT LANDING SPACECRAFT

Vehicles, Space
USE   SPACECRAFT

Vehicles, Standard Launch
USE   STANDARD LAUNCH VEHICLES

Vehicles, Surface
USE   SURFACE VEHICLES

(Vehicles), Suspension Systems
USE   SUSPENSION SYSTEMS (VEHICLES)

Vehicles), Tanks (Combat
USE   TANKS (COMBAT VEHICLES)

Vehicles, Test
USE   TEST VEHICLES

Vehicles, Thor Launch
USE   THOR LAUNCH VEHICLES

Vehicles, Thorad Launch
USE   THORAD LAUNCH VEHICLES

Vehicles, Titan Launch
USE   TITAN LAUNCH VEHICLES

Vehicles, Tracked
USE   TRACKED VEHICLES

Vehicles, Transatmospheric
USE   TRANSATMOSPHERIC VEHICLES

Vehicles, Transport
USE   TRANSPORT VEHICLES

Vehicles, Underwater
USE   UNDERWATER VEHICLES

Vehicles, Veronique Rocket
USE   VERONIQUE ROCKET VEHICLES

Vehicles, Water
USE   WATER VEHICLES

Vehicles, Winged
USE   WINGED VEHICLES

VEHICULAR TRACKS

VEINS

VELA SATELLITES

Velocimeters, Laser Doppler
USE   LASER DOPPLER VELOCIMETERS

VELOCITY

Velocity, Acoustic
USE   ACOUSTIC VELOCITY

Velocity, Angular
USE   ANGULAR VELOCITY

VELOCITY COUPLING

Velocity, Critical
USE   CRITICAL VELOCITY

VELOCITY DISTRIBUTION

VELOCITY ERRORS

Velocity, Escape
USE   ESCAPE VELOCITY

Velocity, Exhaust
USE   EXHAUST VELOCITY

Velocity Fields
USE   VELOCITY DISTRIBUTION

Velocity, Flow
USE   FLOW VELOCITY

Velocity, Group
USE   GROUP VELOCITY

Velocity, Hyper
USE   HYPERVELOCITY

Velocity, Low
USE   LOW SPEED

VELOCITY MEASUREMENT

Velocity Measurement, Wind
USE   WIND VELOCITY MEASUREMENT

VELOCITY MODULATION

Velocity, Orbital
USE   ORBITAL VELOCITY

Velocity, Parabolic
USE   ESCAPE VELOCITY

Velocity, Phase
USE   PHASE VELOCITY

Velocity Profiles
USE   VELOCITY DISTRIBUTION

Velocity, Propagation
USE   PROPAGATION VELOCITY

Velocity, Radial
USE   RADIAL VELOCITY

Velocity, Relativistic
USE   RELATIVISTIC VELOCITY

Velocity Sensors, Image
USE   IMAGE VELOCITY SENSORS

Velocity, Solar
USE   SOLAR VELOCITY

Velocity, Solar Wind
USE   SOLAR WIND VELOCITY

Velocity, Sound
USE   ACOUSTIC VELOCITY

Velocity, Terminal
USE   TERMINAL VELOCITY

Velocity, Wind
USE   WIND VELOCITY

Venant Flexure Problem, Saint
USE   SAINT VENANT PRINCIPLE

Venant Flexure Problem, St
USE   SAINT VENANT PRINCIPLE

Venant Principle, Saint
USE   SAINT VENANT PRINCIPLE

VENEERS

VENERA SATELLITES

VENERA 2 SATELLITE

VENERA 3 SATELLITE

VENERA 4 SATELLITE

VENERA 5 SATELLITE

VENERA 6 SATELLITE

VENERA 7 SATELLITE

VENERA 8 SATELLITE

VENERA 9 SATELLITE

VENERA 10 SATELLITE

VENERA 11 SATELLITE

VENERA 12 SATELLITE

VENEZIANO MODEL

VENEZUELA

VENN DIAGRAMS

Venom Aircraft
USE   DH 112 AIRCRAFT

Venom Aircraft, De Havilland
USE   DH 112 AIRCRAFT

VENTILATION

VENTILATION FANS

Ventilation, Hyper
    USE   HYPERVENTILATION

Ventilation, Hypo
    USE   HYPOVENTILATION

VENTILATORS

VENTING

VENTRAL SECTIONS

Ventricles, Cardiac
    USE   CARDIAC VENTRICLES

Ventricles, Cerebral
    USE   CEREBRAL VENTRICLES

VENTS

VENTURI TUBES

VENUS ATMOSPHERE

VENUS CLOUDS

VENUS FLY TRAP ROCKET VEHICLE

VENUS ORBITING IMAGING RADAR (SPACECRAFT)

VENUS (PLANET)

VENUS PROBES

VENUS RADAR ECHOES

Venus Spacecraft, Pioneer
    USE   PIONEER VENUS SPACECRAFT

VENUS SURFACE

Venus Trajectories, Earth-
    USE   EARTH-VENUS TRAJECTORIES

Venus 1 Spacecraft, Pioneer
    USE   PIONEER VENUS 1 SPACECRAFT

Venus 2 Entry Probes, Pioneer
    USE   PIONEER VENUS 2 ENTRY PROBES

Venus 2 Multiprobe Spacecraft, Pioneer
    USE   PIONEER VENUS 2 SPACECRAFT

Venus 2 Night Probe, Pioneer
    USE   PIONEER VENUS 2 NIGHT PROBE

Venus 2 Sounder Probe, Pioneer
    USE   PIONEER VENUS 2 SOUNDER PROBE

Venus 2 Spacecraft, Pioneer
    USE   PIONEER VENUS 2 SPACECRAFT

Venus 2 Transporter Bus, Pioneer
    USE   PIONEER VENUS 2 TRANSPORTER BUS

Venus 67 Spacecraft, Mariner
    USE   MARINER VENUS 67 SPACECRAFT

Venus-Mercury 1973, Mariner
    USE   MARINER VENUS-MERCURY 1973

VERBAL COMMUNICATION

Verde, Cape
    USE   CAPE VERDE

Verde Valley (CA), Palo
    USE   PALO VERDE VALLEY (CA)

Verification (Computers), Program
    USE   PROGRAM VERIFICATION (COMPUTERS)

Verification (Proving)
    USE   PROVING

VERMICULITE

VERMONT

VERNEUIL PROCESS

VERNIER ENGINES

Vernine
    USE   GUANOSINES

VERONIQUE ROCKET VEHICLES

VERSATILITY

VERTEBRAE

VERTEBRAL COLUMN

VERTEBRATES

Vertebrates, In
    USE   INVERTEBRATES

VERTICAL AIR CURRENTS

Vertical Attitude Takeoff-Landing Aircraft
    USE   VATOL AIRCRAFT

VERTICAL DISTRIBUTION

Vertical Fins
    USE   FINS

VERTICAL FLIGHT

VERTICAL JUNCTION SOLAR CELLS

VERTICAL LANDING

VERTICAL MOTION

VERTICAL MOTION SIMULATORS

VERTICAL ORIENTATION

VERTICAL PERCEPTION

Vertical Stabilizers
    USE   STABILIZERS (FLUID DYNAMICS)

Vertical Tails
    USE   STABILIZERS (FLUID DYNAMICS)
             TAIL ASSEMBLIES

VERTICAL TAKEOFF

VERTICAL TAKEOFF AIRCRAFT

Vertical Takeoff And Landing
    USE   VERTICAL TAKEOFF
             VERTICAL LANDING

VERTICAL 8 ROCKET

Vertices
    USE   APEXES

VERTIGO

Vertol Military Helicopters
    USE   BOEING AIRCRAFT

VERY HIGH FREQUENCIES

VERY HIGH FREQUENCY RADIO EQUIPMENT

Very High Speed Integrated Circuits
    USE   VHSIC (CIRCUITS)

VERY LARGE SCALE INTEGRATION

VERY LONG BASE INTERFEROMETRY

VERY LOW FREQUENCIES

Vessel Design, Pressure
    USE   PRESSURE VESSEL DESIGN

VESSELS

Vessels, Blood
    USE   BLOOD VESSELS

Vessels, Pressure
    USE   PRESSURE VESSELS

VESTA ASTEROID

VESTIBULAR NYSTAGMUS

VESTIBULAR TESTS

VESTIBULES

VESTS

VETERINARY MEDICINE

VFR (Rules)
    USE   VISUAL FLIGHT RULES

VHF OMNIRANGE NAVIGATION

VHSIC (CIRCUITS)

VIABILITY

VIBRATION

Vibration, Bending
    USE   BENDING VIBRATION

Vibration, Breathing
    USE   BREATHING VIBRATION

Vibration, Combustion
    USE   COMBUSTION VIBRATION

Vibration Dampers
    USE   VIBRATION ISOLATORS

VIBRATION DAMPING

VIBRATION EFFECTS

Vibration, Forced
    USE   FORCED VIBRATION

Vibration, Free
    USE   FREE VIBRATION

VIBRATION ISOLATORS

Vibration, Linear
    USE   LINEAR VIBRATION

VIBRATION MEASUREMENT

VIBRATION METERS

Vibration, Missile
    USE   MISSILE VIBRATION

VIBRATION MODE

Vibration, Mode Of
    USE   VIBRATION MODE

VIBRATION PERCEPTION

Vibration Protection
    USE   VIBRATION ISOLATORS

Vibration, Random
    USE   RANDOM VIBRATION

Vibration, Resonant
    USE   RESONANT VIBRATION

Vibration, Self Induced
    USE   SELF INDUCED VIBRATION

VIBRATION SIMULATORS

Vibration, Structural
    USE   STRUCTURAL VIBRATION

Vibration Testing Machines
    USE   VIBRATION SIMULATORS

VIBRATION TESTS

Vibration, Torsional
    USE   TORSIONAL VIBRATION

Vibration, Transverse
    USE   TRANSVERSE OSCILLATION

VIBRATIONAL FREEZING

Vibrational Frequencies
USE   VIBRATIONAL SPECTRA

Vibrational Relaxation
USE   MOLECULAR RELAXATION

VIBRATIONAL SPECTRA

VIBRATIONAL STRESS

Vibrations, Acoustic
USE   SOUND WAVES

Vibrations, Lattice
USE   LATTICE VIBRATIONS

Vibrations, Magnetoelastic
USE   MAGNETOELASTIC WAVES

Vibrators, Multi
USE   MULTIVIBRATORS

VIBRATORY LOADS

Vibratory Motion Equations, Forced
USE   EQUATIONS
         FORCED VIBRATION

VIBRATORY POLISHING

Vibrocardiography
USE   PHONOCARDIOGRAPHY

Vibrometers
USE   VIBRATION METERS

Vickers Scimitar Aircraft
USE   SCIMITAR AIRCRAFT

Vickers Valiant Aircraft
USE   VALIANT AIRCRAFT

Vickers VC-10 Aircraft
USE   VC-10 AIRCRAFT

Vickers 1100 Aircraft
USE   VC-10 AIRCRAFT

VICTOR MK-1 AIRCRAFT

VIDEO COMMUNICATION

VIDEO DATA

VIDEO DISKS

VIDEO EQUIPMENT

VIDEO LANDMARK ACQUISITION AND TRACKING

VIDEO SIGNALS

Vidicon Camera System (AVCS), Advanced
USE   ADVANCED VIDICON CAMERA SYSTEM
         (AVCS)

VIDICONS

Vidicons, Return Beam
USE   RETURN BEAM VIDICONS

VIETNAM

Vietnam, North
USE   VIETNAM

Vietnam, Republic Of
USE   VIETNAM

Vietnam, South
USE   VIETNAM

VIEW EFFECTS

View, Field Of
USE   FIELD OF VIEW

VIEWING

Viewing Applications Laboratory, Earth
USE   EARTH VIEWING APPLICATIONS
         LABORATORY

Vigilante Aircraft
USE   A-5 AIRCRAFT

VIGNETTING

Vigor, Crop
USE   CROP VIGOR

Vigor, Timber
USE   TIMBER VIGOR

VIKING LANDER SPACECRAFT

VIKING LANDER 1

VIKING LANDER 2

VIKING MARS PROGRAM

VIKING ORBITER SPACECRAFT

VIKING ORBITER 1

VIKING ORBITER 2

VIKING ORBITER 1975

VIKING ROCKET VEHICLE

VIKING SPACECRAFT

VIKING 1 SPACECRAFT

VIKING 2 SPACECRAFT

VIKING 75 ENTRY VEHICLE

VINEYARDS

VINTI THEORY

VINYL COPOLYMERS

Vinyl Cyanide
USE   ACRYLONITRILES

Vinyl Ethylene
USE   BUTADIENE

VINYL POLYMERS

VINYL RADICAL

VINYLIDENE

VIOLENCE

Viper Engine, Bristol-Siddeley
USE   BRISTOL-SIDDELEY VIPER ENGINE

VIRGIN ISLANDS

VIRGINIA

Virginia, West
USE   WEST VIRGINIA

VIRGO GALACTIC CLUSTER

Virgo Star Cluster
USE   VIRGO GALACTIC CLUSTER

VIRIAL COEFFICIENTS

VIRIAL THEOREM

VIRTUAL MEMORY SYSTEMS

VIRTUAL PROPERTIES

VIRULENCE

VIRUSES

Viruses, Adeno
USE   ADENOVIRUSES

VISCERA

VISCOELASTIC CYLINDERS

VISCOELASTIC DAMPING

Viscoelastic Flow
USE   VISCOELASTICITY

VISCOELASTICITY

Viscoelasticity, Photo
USE   PHOTOVISCOELASTICITY

Viscoelasticity, Thermo
USE   THERMOVISCOELASTICITY

VISCOMETERS

VISCOMETRY

Viscoplastic Flow
USE   VISCOPLASTICITY

VISCOPLASTICITY

VISCOPUMPS

VISCOSITY

Viscosity, Eddy
USE   EDDY VISCOSITY

Viscosity, Gas
USE   GAS VISCOSITY

VISCOUNT AIRCRAFT

VISCOUS DAMPING

VISCOUS DRAG

VISCOUS FLOW

VISCOUS FLUIDS

VISIBILITY

Visibility, Low
USE   LOW VISIBILITY

VISIBLE INFRARED SPIN SCAN RADIOMETER

Visible Radiation
USE   LIGHT (VISIBLE RADIATION)

(Visible Radiation), Light
USE   LIGHT (VISIBLE RADIATION)

VISIBLE SPECTRUM

VISION

Vision, Binocular
USE   BINOCULAR VISION

Vision, Color
USE   COLOR VISION

Vision, Computer
USE   COMPUTER VISION

Vision, Macular
USE   VISION

Vision, Monocular
USE   MONOCULAR VISION

Vision, Night
USE   NIGHT VISION

Vision, Peripheral
USE   PERIPHERAL VISION

Vision, Stereoscopic
USE   STEREOSCOPIC VISION

VISORS

VISUAL ACCOMMODATION

VISUAL ACUITY

VISUAL AIDS

VISUAL CONTROL

VISUAL DISCRIMINATION

Visual Displays
USE   DISPLAY DEVICES

Visual Equipment, Audio
USE   VISUAL AIDS
TRAINING DEVICES

VISUAL FIELDS

VISUAL FLIGHT

VISUAL FLIGHT RULES

VISUAL OBSERVATION

VISUAL PERCEPTION

VISUAL PHOTOMETRY

VISUAL PIGMENTS

VISUAL SIGNALS

VISUAL STIMULI

VISUAL TASKS

Visual Tracking
USE   OPTICAL TRACKING

Visualization, Flow
USE   FLOW VISUALIZATION

Visualization, Numerical Flow
USE   NUMERICAL FLOW VISUALIZATION

Visualization Of Flow
USE   FLOW VISUALIZATION

Vitamin A
USE   RETINENE

Vitamin B
USE   THIAMINE

Vitamin B Complex
USE   BIOTIN

Vitamin B 2
USE   RIBOFLAVIN

Vitamin B 6
USE   PYRIDOXINE

Vitamin B 12
USE   CYANOCOBALAMIN

Vitamin C
USE   ASCORBIC ACID

Vitamin D
USE   CALCIFEROL .

Vitamin E
USE   TOCOPHEROL

Vitamin G
USE   RIBOFLAVIN

Vitamin K
USE   PHYLLOQUINONE

Vitamin M
USE   FOLIC ACID

Vitamin P
USE   BIOFLAVONOIDS

VITAMINS

VITON

VITON RUBBER (TRADEMARK)

VITREOUS MATERIALS

VITRIFICATION

VJ-101 AIRCRAFT

VJ-101 Aircraft, Sud
USE   VJ-101 AIRCRAFT

Vlasov Equation, Boltzmann-
USE   BOLTZMANN-VLASOV EQUATION

VLASOV EQUATIONS

VLBI
USE   VERY LONG BASE INTERFEROMETRY

VLF EMISSION RECORDERS

VLSI
USE   VERY LARGE SCALE INTEGRATION

VOCAL CORDS

VOCODERS

VOICE

VOICE COMMUNICATION

VOICE CONTROL

VOICE DATA PROCESSING

VOICE OF AMERICA

Voice Translators, Digital To
USE   DIGITAL TO VOICE TRANSLATORS

(Voice Translators), DIVOT
USE   DIGITAL TO VOICE TRANSLATORS

VOID RATIO

VOIDS

VOIGT EFFECT

VOLATILITY

Volatilization
USE   VAPORIZING

Volcanics
USE   VOLCANOLOGY

VOLCANOES

Volcanoes, Active
USE   VOLCANOES

(Volcanoes), Cones
USE   CONES (VOLCANOES)

Volcanoes, Mars
USE   MARS VOLCANOES

VOLCANOLOGY

VOLT-AMPERE CHARACTERISTICS

Volta, Upper
USE   BURKINA

Voltage
USE   ELECTRIC POTENTIAL

VOLTAGE AMPLIFIERS

Voltage Breakdown
USE   ELECTRICAL FAULTS

Voltage Characteristics, Capacitance-
USE   CAPACITANCE-VOLTAGE
CHARACTERISTICS

VOLTAGE CONTROLLED OSCILLATORS

VOLTAGE CONVERTERS (AC TO AC)

VOLTAGE CONVERTERS (DC TO DC)

VOLTAGE GENERATORS

Voltage, Low
USE   LOW VOLTAGE

Voltage Measurement
USE   ELECTRICAL MEASUREMENT

Voltage, Open Circuit
USE   OPEN CIRCUIT VOLTAGE

Voltage, Over
USE   OVERVOLTAGE

VOLTAGE REGULATORS

Voltage Variation Indicators
USE   VOLTMETERS

Voltages, High
USE   HIGH VOLTAGES

Voltages, Photo
USE   PHOTOVOLTAGES

VOLTERRA EQUATIONS

VOLTMETERS

VOLUME

Volume Balloons, Constant
USE   SUPERPRESSURE BALLOONS

Volume (Biology), Body
USE   BODY VOLUME (BIOLOGY)

Volume, Blood
USE   BLOOD VOLUME

Volume, Heart Minute
USE   HEART MINUTE VOLUME

Volume Method, Finite
USE   FINITE VOLUME METHOD

Volume Ramjet Engines, Low
USE   LOW VOLUME RAMJET ENGINES

VOLUMETRIC ANALYSIS

VOLUMETRIC EFFICIENCY

VOLUMETRIC STRAIN

VOMITING

VON KARMAN EQUATION

Von Mises Theory
USE   STRESS FUNCTIONS

VON ZEIPEL METHOD

Voodoo Aircraft
USE   F-101 AIRCRAFT

VOR Systems
USE   VHF OMNIRANGE NAVIGATION

VORTEX ADVISORY SYSTEM

VORTEX ALLEVIATION

VORTEX AVOIDANCE

VORTEX BREAKDOWN

Vortex Columns
USE   VORTICES

Vortex Disturbances
USE   VORTICES

VORTEX FILAMENTS

VORTEX FLAPS

Vortex Flow
USE   VORTICES

Vortex Generation
USE   VORTEX GENERATORS

VORTEX GENERATORS

VORTEX INJECTORS

VORTEX PRECESSION

VORTEX RINGS

VORTEX SHEDDING

VORTEX SHEETS

Vortex Street, Karman
USE   KARMAN VORTEX STREET

VORTEX STREETS

Vortex Traps
USE   TRAPPED VORTEXES

Vortex Tubes
USE   VORTICES
        HILSCH TUBES

Vortexes, Trapped
USE   TRAPPED VORTEXES

VORTICES

Vortices, Wing Tip
USE   WING TIP VORTICES

VORTICITY

Vorticity Equation, Helmholtz
USE   HELMHOLTZ VORTICITY EQUATION

VORTICITY EQUATIONS

VORTICITY TRANSPORT HYPOTHESIS

VOSKHOD MANNED SPACECRAFT

VOSKHOD 1 SPACECRAFT

VOSKHOD 2 SPACECRAFT

VOSTOK SPACECRAFT

VOSTOK 1 SPACECRAFT

VOSTOK 2 SPACECRAFT

VOSTOK 3 SPACECRAFT

VOSTOK 4 SPACECRAFT

VOSTOK 5 SPACECRAFT

VOSTOK 6 SPACECRAFT

VOTING

Vought Aircraft, Chance-
USE   CHANCE-VOUGHT AIRCRAFT

Vought Aircraft, Ling-Temco-
USE   LING-TEMCO-VOUGHT AIRCRAFT

Vought Military Aircraft, Chance-
USE   MILITARY AIRCRAFT
        CHANCE-VOUGHT AIRCRAFT

VOWELS

VOYAGER PROJECT

VOYAGER 1 SPACECRAFT

VOYAGER 2 SPACECRAFT

VOYAGER 1977 MISSION

Voyageur Helicopter
USE   CH-46 HELICOPTER

VT
USE   VERMONT

VT), Lake Champlain Basin (NY-
USE   LAKE CHAMPLAIN BASIN (NY-VT)

VTOL
USE   VERTICAL TAKEOFF
        VERTICAL LANDING

VTOL Aircraft
USE   VERTICAL TAKEOFF AIRCRAFT

VULCAN AIRCRAFT

Vulcanizates
USE   VULCANIZED ELASTOMERS

Vulcanizates, Gum
USE   VULCANIZED ELASTOMERS

VULCANIZED ELASTOMERS

VULCANIZING

VULNERABILITY

Vulnerability, Nuclear
USE   NUCLEAR VULNERABILITY

VYCOR

VZ-2 AIRCRAFT

VZ-8 AIRCRAFT

VZ-10 Aircraft
USE   XV-4 AIRCRAFT

VZ-11 Aircraft
USE   XV-5 AIRCRAFT

VZ-12 Aircraft
USE   P-1127 AIRCRAFT

W

W
USE   TUNGSTEN

W Devices, B-A-
USE   BULK ACOUSTIC WAVE DEVICES

W Devices, S-A-
USE   SURFACE ACOUSTIC WAVE DEVICES

W Wings
USE   VARIABLE SWEEP WINGS

W-R Stars
USE   WOLF-RAYET STARS

WA
USE   WASHINGTON

WA), Cascade Range (CA-OR-
USE   CASCADE RANGE (CA-OR-WA)

WA), Columbia River Basin (ID-OR-
USE   COLUMBIA RIVER BASIN (ID-OR-WA)

Waal Forces, Van Der
USE   VAN DER WAAL FORCES

WABASH RIVER BASIN (IL-IN-OH)

Wachmann Comet, Schwassmann-
USE   SCHWASSMANN-WACHMANN COMET

WADIS

WAFERS

WAGE SURVEYS

WAKEFULNESS

WAKES

Wakes, Aircraft
USE   AIRCRAFT WAKES

Wakes, Helicopter
USE   HELICOPTER WAKES

Wakes, Hypersonic
USE   HYPERSONIC WAKES

Wakes, Laminar
USE   LAMINAR WAKES

Wakes, Near
USE   NEAR WAKES

Wakes, Supersonic
USE   SUPERSONIC WAKES

Wakes, Swirling
USE   TURBULENT WAKES

Wakes, Turbulent
USE   TURBULENT WAKES

Walk, Random
USE   RANDOM WALK

WALKING

WALKING MACHINES

Wall, Domain
USE   DOMAIN WALL

WALL FLOW

WALL JETS

WALL PRESSURE

WALL TEMPERATURE

Walled Shells, Thin
USE   THIN WALLED SHELLS

WALLOPS ISLAND

WALLS

Walls, Cold
USE   WALLS
        COLD SURFACES

Walls, Nozzle
USE   NOZZLE WALLS

Walls, Porous
USE   POROUS WALLS

Walls, Sea
USE   BREAKWATERS

Walls, Thick
USE   THICK WALLS

Walls, Thin
USE   THIN WALLS

Walls, Trombe
USE   TROMBE WALLS

Walls, Wind Tunnel
USE   WIND TUNNEL WALLS

WALSH FUNCTION

Wandering (Geology), Polar
USE   POLAR WANDERING (GEOLOGY)

WANKEL ENGINES

WAR GAMES

WARFARE

Warfare Aircraft, Antisubmarine
USE   ANTISUBMARINE WARFARE AIRCRAFT

Warfare, Antiship
USE   ANTISHIP WARFARE

Warfare, Antisubmarine
USE   ANTISUBMARINE WARFARE

Warfare, Chemical
USE   CHEMICAL WARFARE

Warfare, Electronic
USE   ELECTRONIC WARFARE

Warfare, Nuclear
USE   NUCLEAR WARFARE

WARHEADS

Warheads, Nuclear
USE   NUCLEAR WARHEADS

WARM FRONTS

Warming
USE   HEATING

WARNING

Warning And Control System, Airborne
USE   AWACS AIRCRAFT

Warning Devices
USE   WARNING SYSTEMS

Warning Devices, Collision
USE   WARNING SYSTEMS
       COLLISION AVOIDANCE

Warning Signals
USE   WARNING SYSTEMS

Warning Star Aircraft
USE   EC-121 AIRCRAFT

Warning System, Ballistic Missile Early
USE   BALLISTIC MISSILE EARLY WARNING
       SYSTEM

WARNING SYSTEMS

Warning Systems, Early
USE   EARLY WARNING SYSTEMS

WARPAGE

WASHERS

WASHERS (CLEANERS)

WASHERS (SPACERS)

WASHING

WASHINGTON

Washout (Radioactivity)
USE   FALLOUT

WASP SOUNDING ROCKET

WASPALOY

WASTE DISPOSAL

WASTE ENERGY UTILIZATION

WASTE HEAT

WASTE TREATMENT

WASTE UTILIZATION

WASTE WATER

WASTES

(Wastes), Deep Well Injection
USE   DEEP WELL INJECTION (WASTES)

Wastes (Fuel Conversion), Organic
USE   ORGANIC WASTES (FUEL CONVERSION)

Wastes, Human
USE   HUMAN WASTES

Wastes, Industrial
USE   INDUSTRIAL WASTES

Wastes, Liquid
USE   LIQUID WASTES

Wastes, Metabolic
USE   METABOLIC WASTES

Wastes, Nuclear
USE   RADIOACTIVE WASTES

Wastes, Radioactive
USE   RADIOACTIVE WASTES

Wastes, Solid
USE   SOLID WASTES

Watches
USE   CLOCKS

WATER

WATER BALANCE

Water Boiler Reactor, Los Alamos
USE   LOS ALAMOS WATER BOILER REACTOR

Water Breeder Reactors, Light
USE   LIGHT WATER BREEDER REACTORS

WATER CIRCULATION

Water, Coastal
USE   COASTAL WATER

Water, Cold
USE   COLD WATER

WATER COLOR

Water Components Test Reactors, Heavy
USE   HEAVY WATER COMPONENTS TEST
       REACTORS

WATER CONSUMPTION

Water Content
USE   MOISTURE CONTENT

WATER COOLED REACTORS

Water Cooling
USE   LIQUID COOLING

WATER CURRENTS

WATER DEPRIVATION

WATER DEPTH

WATER EROSION

WATER FLOW

Water, Fresh
USE   FRESH WATER

Water, Ground
USE   GROUND WATER

WATER HAMMER

WATER HEATING

Water, Heavy
USE   HEAVY WATER

WATER IMMERSION

WATER INJECTION

WATER INTAKES

Water Interactions, Air
USE   AIR WATER INTERACTIONS

Water Jets
USE   HYDRAULIC JETS

WATER LANDING

Water, Light
USE   LIGHT WATER

WATER LOSS

WATER MANAGEMENT

WATER MASERS

WATER MODERATED REACTORS

Water, Nearshore
USE   NEARSHORE WATER

Water Plane Area Twin Hull, Small
USE   SWATH (SHIP)

WATER POLLUTION

Water, Poly
USE   POLYWATER

Water, Potable
USE   POTABLE WATER

WATER PRESSURE

Water Purification
USE   WATER TREATMENT

WATER QUALITY

Water Reactions, Metal-
USE   METAL-WATER REACTIONS

Water Reactor, Halden Boiling
USE   HALDEN BOILING WATER REACTOR

Water Reactors, Boiling
USE   BOILING WATER REACTORS

Water Reactors, Experimental Boiling
USE   EXPERIMENTAL BOILING WATER
       REACTORS

Water Reactors, Heavy
USE   HEAVY WATER REACTORS

Water Reactors, Light
USE   LIGHT WATER REACTORS

Water Reactors, Pressurized
USE   PRESSURIZED WATER REACTORS

WATER RECLAMATION

Water Recovery
USE   WATER RECLAMATION

WATER RESOURCES

Water Rocket Engines, Hot
USE   HOT WATER ROCKET ENGINES

WATER RUNOFF

Water, Sea
USE   SEA WATER

Water, Shallow
USE   SHALLOW WATER

(Water), Springs
USE   SPRINGS (WATER)

Water, Surface
USE   SURFACE WATER

WATER TABLES

WATER TAKEOFF AND LANDING AIRCRAFT

WATER TEMPERATURE

WATER TREATMENT

WATER TUNNEL TESTS

Water Tunnels
USE   HYDRAULIC TEST TUNNELS

Water, Upwelling
USE   UPWELLING WATER

Water, Vadose
USE   VADOSE WATER

WATER VAPOR

WATER VEHICLES

Water, Waste
USE   WASTE WATER

WATER WAVES

WATER WHEELS

WATERFOWL

WATERPROOFING

Waters, Inland
USE   INLAND WATERS

WATERSHEDS

WATERWAVE ENERGY

WATERWAVE ENERGY CONVERSION

WATERWAVE POWERED MACHINES

WATERWAYS

WATTMETERS

WAVE AMPLIFICATION

Wave Amplifiers, Traveling
USE   TRAVELING WAVE AMPLIFIERS

Wave Antennas, Gravitational
USE   GRAVITATIONAL WAVE ANTENNAS

WAVE ATTENUATION

Wave Attenuation, Shock
USE   SHOCK WAVE ATTENUATION

Wave Control, Shock
USE   SHOCK WAVE CONTROL

WAVE DEGRADATION

Wave Devices, Bulk Acoustic
USE   BULK ACOUSTIC WAVE DEVICES

Wave Devices, Surface Acoustic
USE   SURFACE ACOUSTIC WAVE DEVICES

WAVE DIFFRACTION

WAVE DISPERSION

WAVE DRAG

Wave Effect, Brown
USE   BROWN WAVE EFFECT

Wave Effect, Green
USE   GREEN WAVE EFFECT

WAVE EQUATIONS

Wave Equations, Lame
USE   LAME WAVE EQUATIONS

WAVE EXCITATION

Wave Filters, Electromagnetic
USE   ELECTROMAGNETIC WAVE FILTERS

WAVE FRONT DEFORMATION

WAVE FRONT RECONSTRUCTION

WAVE FRONTS

WAVE FUNCTIONS

WAVE GENERATION

Wave Generators, Shock
USE   SHOCK WAVE GENERATORS

WAVE INCIDENCE CONTROL

WAVE INTERACTION

Wave Interaction, Shock
USE   SHOCK WAVE INTERACTION

Wave Lasers, Continuous
USE   CONTINUOUS WAVE LASERS

Wave Luminescence, Shock
USE   SHOCK WAVE LUMINESCENCE

Wave Masers, Traveling
USE   TRAVELING WAVE MASERS

Wave Model, Density
USE   DENSITY WAVE MODEL

Wave Modulation, Traveling
USE   TRAVELING WAVE MODULATION

Wave Motion
USE   WAVES

Wave Orbiting Telescope, Kilometer
USE   KILOMETER WAVE ORBITING TELESCOPE

Wave Oscillators
USE   OSCILLATORS

WAVE PACKETS

Wave Profiles, Shock
USE   SHOCK WAVE PROFILES

WAVE PROPAGATION

Wave Propagation, Ground
USE   GROUND WAVE PROPAGATION

Wave Propagation, Shock
USE   SHOCK WAVE PROPAGATION

Wave Radar, Continuous
USE   CONTINUOUS WAVE RADAR

Wave Radiation
USE   ELECTROMAGNETIC RADIATION

Wave Radiation, Long
USE   LONG WAVE RADIATION

Wave Radiation, Short
USE   SHORT WAVE RADIATION

Wave Radio Equipment, Ultra Short
USE   VERY HIGH FREQUENCY RADIO
       EQUIPMENT

Wave Radio Transmission, Short
USE   SHORT WAVE RADIO TRANSMISSION

Wave Ratios, Standing
USE   STANDING WAVE RATIOS

WAVE REFLECTION

Wave Refraction, Radio
USE   RADIO WAVE REFRACTION

WAVE RESISTANCE

WAVE SCATTERING

Wave Transducers, Ultrasonic
USE   ULTRASONIC WAVE TRANSDUCERS

Wave Transmission, Electromagnetic
USE   ELECTROMAGNETIC WAVE TRANSMISSION

Wave Tubes, Backward
USE   BACKWARD WAVE TUBES

Wave Tubes, Traveling
USE   TRAVELING WAVE TUBES

WAVEFORMS

Waveforms, Sawtooth
USE   SAWTOOTH WAVEFORMS

WAVEGUIDE ANTENNAS

WAVEGUIDE FILTERS

WAVEGUIDE LASERS

WAVEGUIDE TUNERS

WAVEGUIDE WINDOWS

WAVEGUIDES

Waveguides, Beam
USE   BEAM WAVEGUIDES

Waveguides, Circular
USE   CIRCULAR WAVEGUIDES

Waveguides, Optical
USE   OPTICAL WAVEGUIDES

Waveguides, Rectangular
USE   RECTANGULAR WAVEGUIDES

Waveguides, Sonic
USE   ACOUSTIC DELAY LINES

WAVELENGTH DIVISION MULTIPLEXING

Wavelength Lasers, Two-
USE   TWO-WAVELENGTH LASERS

WAVELENGTHS

Wavelengths, De Broglie
USE   DE BROGLIE WAVELENGTHS

WAVES

Waves, Alfven
USE   MAGNETOHYDRODYNAMIC WAVES

Waves, Backward
USE   BACKWARD WAVES

Waves, Baroclinic
USE   BAROCLINIC WAVES

Waves, Bow
USE   BOW WAVES

Waves, Bow Shock
USE   BOW WAVES
       SHOCK WAVES

Waves, Capillary
USE   CAPILLARY WAVES

Waves, Carrier
USE   CARRIER WAVES

Waves, Centimeter
USE   CENTIMETER WAVES

Waves, Cnoidal
USE   CNOIDAL WAVES

Waves, Combustion
USE   FLAME PROPAGATION

Waves, Compression
USE   COMPRESSION WAVES

Waves, Continuous
USE   CONTINUOUS RADIATION

Waves, Cosmic Radio
USE   EXTRATERRESTRIAL RADIO WAVES

Waves, Cylindrical
USE   CYLINDRICAL WAVES

Waves, Decametric
USE   DECAMETRIC WAVES

Waves, Decimeter
USE   DECIMETER WAVES

Waves, Detonation
USE   DETONATION WAVES

Waves, Diffusion
USE   DIFFUSION WAVES

Waves, Dilatational
USE   DILATATIONAL WAVES

Waves, Elastic
USE   ELASTIC WAVES

Waves, Electroacoustic
USE   ELECTROACOUSTIC WAVES

Waves, Electromagnetic
USE   ELECTROMAGNETIC RADIATION

Waves, Electromagnetic Surface
USE   ELECTROMAGNETIC SURFACE WAVES

Waves, Electrostatic
USE   ELECTROSTATIC WAVES

Waves, Expansion
USE   ELASTIC WAVES

Waves, Extraterrestrial Radio
USE   EXTRATERRESTRIAL RADIO WAVES

Waves, Frontal
USE   FRONTAL WAVES

Waves, Galactic Radio
USE   GALACTIC RADIO WAVES

Waves, Gravitational
USE   GRAVITATIONAL WAVES

Waves, Gravity
USE   GRAVITY WAVES

Waves, H
USE   H WAVES

Waves, Hydromagnetic
USE   MAGNETOHYDRODYNAMIC WAVES

Waves, Internal
USE   INTERNAL WAVES

Waves, Ion Acoustic
USE   ION ACOUSTIC WAVES

Waves, Ionic
USE   IONIC WAVES

Waves, Kilometric
USE   KILOMETRIC WAVES

Waves, Lamb
USE   LAMB WAVES

Waves, Lee
USE   LEE WAVES

Waves, Loading
USE   LOADS (FORCES)
       ELASTIC WAVES

Waves, Longitudinal
USE   LONGITUDINAL WAVES

Waves, Love
USE   LOVE WAVES

Waves, Magnetoacoustic
USE   MAGNETOACOUSTIC WAVES

Waves, Magnetoelastic
USE   MAGNETOELASTIC WAVES

Waves, Magnetohydrodynamic
USE   MAGNETOHYDRODYNAMIC WAVES

Waves (Meteorology), Long
USE   PLANETARY WAVES

Waves, Micro
USE   MICROWAVES

Waves, Millimeter
USE   MILLIMETER WAVES

Waves), Modes (Standing
USE   MODES (STANDING WAVES)

Waves), Nodes (Standing
USE   NODES (STANDING WAVES)

Waves, Normal Shock
USE   NORMAL SHOCK WAVES

Waves, Oblique Shock
USE   OBLIQUE SHOCK WAVES

Waves, P
USE   P WAVES

Waves, Plane
USE   PLANE WAVES

Waves, Planetary
USE   PLANETARY WAVES

Waves, Plasma
USE   PLASMA WAVES

Waves, Plasma Sound
USE   PLASMA WAVES
       MAGNETOHYDRODYNAMIC WAVES

(Waves), Polarization
USE   POLARIZATION (WAVES)

Waves, Polarized Elastic
USE   POLARIZED ELASTIC WAVES

Waves, Pressure
USE   ELASTIC WAVES

Waves, Radio
USE   RADIO WAVES

Waves, Rarefaction
USE   ELASTIC WAVES

Waves, Rayleigh
USE   RAYLEIGH WAVES

Waves, Reflected
USE   REFLECTED WAVES

Waves, Refracted
USE   REFRACTED WAVES

Waves, Riemann
USE   RIEMANN WAVES

Waves, Rossby
USE   PLANETARY WAVES

Waves, S
USE   S WAVES

Waves, Secondary
USE   S WAVES

Waves, Seismic
USE   SEISMIC WAVES

Waves, Shear
USE   S WAVES

Waves, Shock
USE   SHOCK WAVES

Waves, Sine
USE   SINE WAVES

Waves, Sky
USE   SKY WAVES

Waves, Solar Radio
USE   SOLAR RADIO EMISSION

Waves, Solitary
USE   SOLITARY WAVES

Waves, Sommerfeld
USE   SOMMERFELD WAVES

Waves, Sound
USE   SOUND WAVES

Waves, Spherical
USE   SPHERICAL WAVES

Waves, Spin
USE   MAGNONS

Waves, Square
USE   SQUARE WAVES

Waves, Standing
USE   STANDING WAVES

Waves, Stress
USE   STRESS WAVES

Waves, Subcarrier
USE   CARRIER WAVES

Waves, Submillimeter
USE   SUBMILLIMETER WAVES

Waves, Surface
USE   SURFACE WAVES

Waves, Tidal
USE   TIDAL WAVES

Waves, Tollmein-Schlichting
USE   TOLLMEIN-SCHLICHTING WAVES

Waves, Transverse
USE   TRANSVERSE WAVES

Waves, Traveling
USE   TRAVELING WAVES

Waves, Tropospheric
USE   TROPOSPHERIC WAVES

Waves, Tsunami
USE   TSUNAMI WAVES

Waves, Ultrasonic
USE   ULTRASONIC RADIATION

Waves, Unloading
USE   UNLOADING WAVES

Waves, Water
USE   WATER WAVES

Wax Process, Lost
USE   INVESTMENT CASTING

WAXES

Way Galaxy, Milky
USE   MILKY WAY GALAXY

WE-32 Engine, XJ-34-
USE   J-34 ENGINE

WEAK ENERGY INTERACTIONS

WEAK INTERACTIONS (FIELD THEORY)

WEAPON SYSTEM MANAGEMENT

Weapon System, Typhon
USE   TYPHON WEAPON SYSTEM

WEAPON SYSTEM 107A-1

WEAPON SYSTEM 107A-2

WEAPON SYSTEM 133A

WEAPON SYSTEM 133B

WEAPON SYSTEM 315A

WEAPON SYSTEMS

WEAPONS

WEAPONS DELIVERY

WEAPONS DEVELOPMENT

Weapons, Fission
   USE   FISSION WEAPONS

Weapons, Fusion
   USE   FUSION WEAPONS

WEAPONS INDUSTRY

Weapons, Laser
   USE   LASER WEAPONS

Weapons, Nuclear
   USE   NUCLEAR WEAPONS

Weapons, Space
   USE   SPACE WEAPONS

WEAR

WEAR INHIBITORS

WEAR TESTS

WEATHER

Weather Air Navigation, All-
   USE   ALL-WEATHER AIR NAVIGATION

Weather Charts
   USE   METEOROLOGICAL CHARTS

Weather, Cold
   USE   COLD WEATHER

Weather Conditions
   USE   WEATHER

Weather Control
   USE   WEATHER MODIFICATION

WEATHER DATA RECORDERS

WEATHER FORECASTING

Weather Forecasting, Long Range
   USE   LONG RANGE WEATHER FORECASTING

Weather Forecasting, Numerical
   USE   NUMERICAL WEATHER FORECASTING

Weather Forecasting, Statistical
   USE   STATISTICAL WEATHER FORECASTING

Weather Fronts
   USE   FRONTS (METEOROLOGY)

Weather, Hot
   USE   HOT WEATHER

Weather Landing Systems, All-
   USE   ALL-WEATHER LANDING SYSTEMS

Weather Maps
   USE   METEOROLOGICAL CHARTS

WEATHER MODIFICATION

Weather Radar
   USE   METEOROLOGICAL RADAR

WEATHER RECONNAISSANCE AIRCRAFT

WEATHER STATIONS

Weather Stations, Automatic
   USE   AUTOMATIC WEATHER STATIONS

Weather Tests, Cold
   USE   COLD WEATHER TESTS

WEATHERING

WEATHERPROOFING

WEAVING

WEBBING

WEBER TEST

WEBER-FECHNER LAW

WEBS

Webs, Girder
   USE   GIRDER WEBS

Webs (Membranes)
   USE   MEMBRANES

WEBS (SHEETS)

WEBS (SUPPORTS)

WEDGE FLOW

WEDGES

Weevils, Boll
   USE   BOLL WEEVILS

WEIBEL INSTABILITY

WEIBULL DENSITY FUNCTIONS

WEIERSTRASS FUNCTIONS

WEIGHT

WEIGHT ANALYSIS

Weight, Body
   USE   BODY WEIGHT

Weight Factors
   USE   WEIGHT (MASS)

WEIGHT INDICATORS

Weight, Low
   USE   LOW WEIGHT

WEIGHT (MASS)

WEIGHT MEASUREMENT

Weight, Molecular
   USE   MOLECULAR WEIGHT

Weight, Organ
   USE   ORGAN WEIGHT

Weight Ratio, Thrust-
   USE   THRUST-WEIGHT RATIO

WEIGHT REDUCTION

Weight, Structural
   USE   STRUCTURAL WEIGHT

Weight Tests, Drop
   USE   DROP TESTS

WEIGHTING FUNCTIONS

WEIGHTLESS FLUIDS

WEIGHTLESSNESS

WEIGHTLESSNESS SIMULATION

Weights, Atomic
   USE   ATOMIC WEIGHTS

Weights, Low Molecular
   USE   LOW MOLECULAR WEIGHTS

Weiss Law, Curie-
   USE   CURIE-WEISS LAW

WELD STRENGTH

WELD TESTS

WELDABILITY

WELDED JOINTS

WELDED STRUCTURES

WELDING

Welding, Arc
   USE   ARC WELDING

Welding, Cold
   USE   COLD WELDING

Welding, Diffusion
   USE   DIFFUSION WELDING

Welding, Electric
   USE   ELECTRIC WELDING

Welding, Electron Beam
   USE   ELECTRON BEAM WELDING

Welding, Electroslag
   USE   ELECTROSLAG WELDING

Welding, Explosive
   USE   EXPLOSIVE WELDING

Welding, Flash
   USE   FLASH WELDING

Welding, Friction
   USE   FRICTION WELDING

Welding, Fusion
   USE   FUSION WELDING

Welding, Gas
   USE   GAS WELDING

Welding, Gas Tungsten Arc
   USE   GAS TUNGSTEN ARC WELDING

Welding, Laser
   USE   LASER WELDING

WELDING MACHINES

Welding, Plasma Arc
   USE   PLASMA ARC WELDING

Welding, Pressure
   USE   PRESSURE WELDING

Welding, TIG
   USE   GAS TUNGSTEN ARC WELDING

Welding, Tungsten Inert Gas
   USE   GAS TUNGSTEN ARC WELDING

Welding, Ultrasonic
   USE   ULTRASONIC WELDING

Welds, Spot
   USE   SPOT WELDS

Well Injection (Wastes), Deep
   USE   DEEP WELL INJECTION (WASTES)

WELLS

Wells, Quantum
   USE   QUANTUM WELLS

Wells, Square
   USE   SQUARE WELLS

WENTZEL-KRAMER-BRILLOUIN METHOD

WESER AIRCRAFT

West Africa, South
   USE   NAMIBIA

WEST COMET

WEST FORD PROJECT

WEST GERMANY

WEST INDIES

West Pakistan
 USE BANGLADESH

WEST VIRGINIA

WESTAR SATELLITES

Westerlies, Circumpolar
 USE CIRCUMPOLAR WESTERLIES

WESTERN HEMISPHERE

WESTLAND AIRCRAFT

WESTLAND GROUND EFFECT MACHINES

Westland MK-10 Helicopter
 USE WESTLAND WHIRLWIND HELICOPTER

Westland P-531 Helicopter
 USE P-531 HELICOPTER

Westland SR-N2 Ground Effect Machine
 USE WESTLAND GROUND EFFECT MACHINES

Westland SR-N2 Hovercraft
 USE WESTLAND GROUND EFFECT MACHINES

Westland SR-N3 Ground Effect Machine
 USE WESTLAND GROUND EFFECT MACHINES

Westland SR-N3 Hovercraft
 USE WESTLAND GROUND EFFECT MACHINES

Westland SR-N5 Ground Effect Machine
 USE WESTLAND GROUND EFFECT MACHINES

WESTLAND WHIRLWIND HELICOPTER

WET CELLS

WET SPINNING

WETLANDS

Wetness
 USE MOISTURE CONTENT

WETTABILITY

WETTING

WHALES

WHARVES

WHEAT

WHEATSTONE BRIDGES

WHEEL BRAKES

Wheel Infrared Spectrometers, Filter
 USE FILTER WHEEL INFRARED
  SPECTROMETERS

Wheel Satellite, TIROS
 USE TIROS 9 SATELLITE

WHEELCHAIRS

WHEELS

Wheels, Counter-Rotating
 USE COUNTER-ROTATING WHEELS

Wheels, Doughnut Shape
 USE TOROIDAL WHEELS

Wheels, Fly
 USE FLYWHEELS

Wheels, Inertia
 USE REACTION WHEELS
  COUNTER-ROTATING WHEELS

Wheels, Nose
 USE NOSE WHEELS

Wheels, Reaction
 USE REACTION WHEELS

Wheels, Toroidal
 USE TOROIDAL WHEELS

Wheels, Turbine
 USE TURBINE WHEELS

Wheels, Vehicle
 USE VEHICLE WHEELS

Wheels, Water
 USE WATER WHEELS

WHIP ANTENNAS

WHIPLASH INJURIES

Whirl
 USE ROTATION

Whirl Instability
 USE ROTARY STABILITY

WHIRL TOWERS

Whirling
 USE ROTATION

Whirling, Pre
 USE PREWHIRLING

Whirling Tests
 USE SPIN TESTS

Whirlwind Helicopter, Sikorsky
 USE SIKORSKY WHIRLWIND HELICOPTER

Whirlwind Helicopter, Westland
 USE WESTLAND WHIRLWIND HELICOPTER

Whirlwind MK-10 Helicopter
 USE WESTLAND WHIRLWIND HELICOPTER

WHISKER COMPOSITES

Whisker Reinforcement, Metal
 USE WHISKER COMPOSITES

WHISKERS (CRYSTALS)

WHISTLER RECORDERS

WHISTLERS

Whitcomb Airfoil, General Aviation
 USE GAW-1 AIRFOIL
  GAW-2 AIRFOIL

WHITE BLOOD CELLS

WHITE DWARF STARS

WHITE HOLES (ASTRONOMY)

WHITE LIGHT HOLOGRAPHY

WHITE NOISE

White Photography, Black And
 USE BLACK AND WHITE PHOTOGRAPHY

Whitening, Pre
 USE PREWHITENING

WHITEOUT

WHITHAM RULE

Whitney-Wilcoxon U Test, Mann-
 USE MANN-WHITNEY-WILCOXON U TEST

WHITTAKER FUNCTIONS

Whitworth HS-748 Aircraft, AVRO
 USE HS-748 AIRCRAFT

WI
 USE WISCONSIN

WICKS

WIDE ANGLE LENSES

Wideband
 USE BROADBAND

WIDEBAND COMMUNICATION

WIDMANSTATTEN STRUCTURE

WIDTH

Width Amplitude Converters, Pulse
 USE PULSE WIDTH AMPLITUDE CONVERTERS

Width, Band
 USE BANDWIDTH

Width Modulation, Pulse
 USE PULSE DURATION MODULATION

Width, Pulse
 USE PULSE DURATION

Width, Spectral Line
 USE SPECTRAL LINE WIDTH

Width, Swath
 USE SWATH WIDTH

WIENER FILTERING

WIENER HOPF EQUATIONS

Wiener Measure, Shannon-
 USE SHANNON-WIENER MEASURE

WIGGLER MAGNETS

Wightman Theory
 USE RELATIVISTIC THEORY
  QUANTUM THEORY
  FIELD THEORY (PHYSICS)

WIGNER COEFFICIENT

Wigner Equation, Brillouin-
 USE BRILLOUIN-WIGNER EQUATION

Wilcoxon U Test, Mann-Whitney-
 USE MANN-WHITNEY-WILCOXON U TEST

WILDERNESS

WILDLIFE

WILDLIFE RADIOLOCATION

William Sound (AK), Prince
 USE PRINCE WILLIAM SOUND (AK)

WILLISTON BASIN (NORTH AMERICA)

WINCHES

Wind Circulation
 USE ATMOSPHERIC CIRCULATION

WIND DIRECTION

WIND EFFECTS

Wind Energy
 USE WINDPOWER UTILIZATION

WIND EROSION

Wind, Geostrophic
 USE GEOSTROPHIC WIND

Wind, Ground
 USE GROUND WIND

WIND MEASUREMENT

WIND (METEOROLOGY)

WIND PRESSURE

WIND PROFILES

WIND RIVER RANGE (WY)

WIND SHEAR

Wind Shear Mechanism, Dungeys
USE   WIND SHEAR

Wind, Solar
USE   SOLAR WIND

WIND TUNNEL APPARATUS

Wind Tunnel Balances
USE   WIND TUNNEL APPARATUS
       WEIGHT INDICATORS

WIND TUNNEL CALIBRATION

WIND TUNNEL DRIVES

WIND TUNNEL MODELS

WIND TUNNEL NOZZLES

WIND TUNNEL STABILITY TESTS

WIND TUNNEL TESTS

WIND TUNNEL WALLS

WIND TUNNELS

Wind Tunnels, Blowdown
USE   BLOWDOWN WIND TUNNELS

Wind Tunnels, Cascade
USE   CASCADE WIND TUNNELS

Wind Tunnels, Combustion
USE   COMBUSTION WIND TUNNELS

Wind Tunnels, Cryogenic
USE   CRYOGENIC WIND TUNNELS

Wind Tunnels, Hotshot
USE   HOTSHOT WIND TUNNELS

Wind Tunnels, Hypersonic
USE   HYPERSONIC WIND TUNNELS

Wind Tunnels, Hypervelocity
USE   HYPERVELOCITY WIND TUNNELS

Wind Tunnels, Low Density
USE   LOW DENSITY WIND TUNNELS

Wind Tunnels, Low Speed
USE   LOW SPEED WIND TUNNELS

Wind Tunnels, Plasma Jet
USE   PLASMA JET WIND TUNNELS

Wind Tunnels, Rectangular
USE   RECTANGULAR WIND TUNNELS

Wind Tunnels, Slotted
USE   SLOTTED WIND TUNNELS

Wind Tunnels, Subsonic
USE   SUBSONIC WIND TUNNELS

Wind Tunnels, Supersonic
USE   SUPERSONIC WIND TUNNELS

Wind Tunnels, Transonic
USE   TRANSONIC WIND TUNNELS

Wind Tunnels, Trisonic
USE   TRISONIC WIND TUNNELS

WIND TURBINES

WIND VANES

WIND VARIATIONS

WIND VELOCITY

WIND VELOCITY MEASUREMENT

Wind Velocity, Solar
USE   SOLAR WIND VELOCITY

WINDING

Winding, Filament
USE   FILAMENT WINDING

Winding, Wire
USE   WIRE WINDING

Windings, Helical
USE   HELICAL WINDINGS

Windmilling
USE   AUTOROTATION

WINDMILLS (WINDPOWERED MACHINES)

Window Atmosphere Sounding Projectile
USE   WASP SOUNDING ROCKET

WINDOWS

WINDOWS (APERTURES)

Windows, Atmospheric
USE   ATMOSPHERIC WINDOWS

Windows, Infrared
USE   INFRARED WINDOWS

WINDOWS (INTERVALS)

Windows, Laser
USE   LASER WINDOWS

Windows, Launch
USE   LAUNCH WINDOWS

Windows, Waveguide
USE   WAVEGUIDE WINDOWS

WINDPOWER UTILIZATION

WINDPOWERED GENERATORS

(Windpowered Machines), Windmills
USE   WINDMILLS (WINDPOWERED MACHINES)

WINDPOWERED PUMPS

WINDS ALOFT

Winds, Stellar
USE   STELLAR WINDS

Windscreens
USE   WINDSHIELDS

WINDSHIELDS

WINES

Wing Aircraft, C-8A Augmentor
USE   C-8A AUGMENTOR WING AIRCRAFT

Wing Aircraft, Fan In
USE   FAN IN WING AIRCRAFT

Wing Aircraft, Fixed-
USE   FIXED WINGS
       AIRCRAFT CONFIGURATIONS

Wing Aircraft, Flying
USE   TAILLESS AIRCRAFT

Wing Aircraft, Free
USE   FREE WING AIRCRAFT

Wing Aircraft, Low
USE   LOW WING AIRCRAFT

Wing Aircraft, Pivoted
USE   TILT WING AIRCRAFT

Wing Aircraft, Rotary
USE   ROTARY WING AIRCRAFT

Wing Aircraft, Tandem
USE   TANDEM WING AIRCRAFT

Wing Aircraft, Tilt
USE   TILT WING AIRCRAFT

Wing And Tail Configurations, Body-
USE   BODY-WING AND TAIL CONFIGURATIONS

WING CAMBER

Wing Configurations, Body-
USE   BODY-WING CONFIGURATIONS

Wing Configurations, Dual
USE   DUAL WING CONFIGURATIONS

WING FLAPS

Wing Flaps, Jet Augmented
USE   WING FLAPS
       JET FLAPS

WING FLOW METHOD TESTS

WING LOADING

WING NACELLE CONFIGURATIONS

WING OSCILLATIONS

WING PANELS

WING PLANFORMS

WING PROFILES

WING ROOTS

Wing Rotors, X
USE   X WING ROTORS

Wing Slats
USE   LEADING EDGE SLATS

WING SLOTS

WING SPAN

WING TANKS

WING TIP VORTICES

WING TIPS

WING-FUSELAGE STORES

WINGED VEHICLES

WINGLETS

WINGS

Wings, Aeroelastic Research
USE   AEROELASTIC RESEARCH WINGS

Wings, Arrow
USE   ARROW WINGS

Wings, Cambered
USE   CAMBERED WINGS

Wings, Cantilever
USE   WINGS

Wings, Caret
USE   CARET WINGS

Wings, Channel
USE   CHANNEL WINGS

Wings, Cranked
USE   SWEPT WINGS

Wings, Cruciform
USE   CRUCIFORM WINGS

Wings, Delta
USE   DELTA WINGS

Wings, Diamond
   USE   LOW ASPECT RATIO WINGS
              SWEPT WINGS

Wings, Fixed
   USE   FIXED WINGS

Wings, Flexible
   USE   FLEXIBLE WINGS

Wings, High Aspect Ratio
   USE   SLENDER WINGS

Wings, Infinite Span
   USE   INFINITE SPAN WINGS

Wings, Low Aspect Ratio
   USE   LOW ASPECT RATIO WINGS

Wings, M
   USE   VARIABLE SWEEP WINGS

Wings, Oblique
   USE   OBLIQUE WINGS

Wings, Ogee
   USE   VARIABLE SWEEP WINGS

Wings, Para
   USE   PARAWINGS

Wings, Rectangular
   USE   RECTANGULAR WINGS

Wings, Rigid
   USE   RIGID WINGS

Wings, Ring
   USE   RING WINGS

Wings, Rogallo
   USE   FLEXIBLE WINGS
              FOLDING STRUCTURES

Wings, Rotary
   USE   ROTARY WINGS

Wings, Slender
   USE   SLENDER WINGS

Wings, Straight
   USE   RECTANGULAR WINGS

Wings, Supercritical
   USE   SUPERCRITICAL WINGS

Wings, Swept
   USE   SWEPT WINGS

Wings, Swept Forward
   USE   SWEPT FORWARD WINGS

Wings, Sweptback
   USE   SWEPTBACK WINGS

Wings, Swing
   USE   SWING WINGS

Wings, Tapered
   USE   SWEPT WINGS

Wings, Thin
   USE   THIN WINGS

Wings, Trapezoidal
   USE   TRAPEZOIDAL WINGS

Wings, Triangular
   USE   DELTA WINGS

Wings, Twisted
   USE   TWISTED WINGS

Wings, Uncambered
   USE   UNCAMBERED WINGS

Wings, Unswept
   USE   UNSWEPT WINGS

Wings, Variable Area
   USE   TRAILING EDGE FLAPS

Wings, Variable Sweep
   USE   VARIABLE SWEEP WINGS

Wings, W
   USE   VARIABLE SWEEP WINGS

WINTER

WIRE

Wire Anemometers, Hot-
   USE   HOT-WIRE ANEMOMETERS

WIRE BRIDGE CIRCUITS

WIRE CLOTH

Wire Control, Fly By
   USE   FLY BY WIRE CONTROL

Wire, Electric
   USE   ELECTRIC WIRE

Wire Flowmeters, Hot-
   USE   HOT-WIRE FLOWMETERS

WIRE GRID LENSES

Wire Mesh
   USE   WIRE CLOTH

Wire Turbulence Meters, Hot-
   USE   TURBULENCE METERS
              HOT-WIRE FLOWMETERS

WIRE WINDING

WIRELESS COMMUNICATION

Wires, Exploding
   USE   EXPLODING WIRES

Wires, Guy
   USE   GUY WIRES

WIRING

Wiring, Electric
   USE   WIRING
              ELECTRIC WIRE

Wiring Systems
   USE   WIRING

WISCONSIN

WISWESSER NOTATIONS

With Particle Accelerators, Space Exper
   USE   SEPAC (PAYLOAD)

WKB Approximation
   USE   WENTZEL-KRAMER-BRILLOUIN METHOD

WOLF-RAYET STARS

Wolfram
   USE   TUNGSTEN

WOLVES

Women
   USE   FEMALES

WOOD

Wood, Ply
   USE   PLYWOOD

WOODEN STRUCTURES

(Woodpulp), Kraft Process
   USE   KRAFT PROCESS (WOODPULP)

WOOL

WORD PROCESSING

WORDS (LANGUAGE)

WORK

WORK CAPACITY

WORK FUNCTIONS

WORK HARDENING

Work, Physical
   USE   PHYSICAL WORK

WORK SOFTENING

WORK-REST CYCLE

Workers, Orbital
   USE   ORBITAL WORKERS

Workhorse Helicopter
   USE   CH-21 HELICOPTER

Working, Cold
   USE   COLD WORKING

WORKING FLUIDS

Working, Hot
   USE   HOT WORKING

Working, Metal
   USE   METAL WORKING

WORKLOADS (PSYCHOPHYSIOLOGY)

Workshop, Saturn 1
   USE   SATURN 1 WORKSHOP

Workshop, Saturn 5
   USE   SATURN 5 WORKSHOP

Workshops, Orbital
   USE   ORBITAL WORKSHOPS

Workshops, Saturn
   USE   SATURN WORKSHOPS

WORKSTATIONS

Workstations, Crew
   USE   CREW WORKSTATIONS

World
   USE   EARTH (PLANET)

WORLD DATA CENTERS

WORLD METEOROLOGICAL ORGANIZATION

WORMS

Worms, Boll
   USE   BOLLWORMS

Worms, Flat
   USE   FLATWORMS

Worms, Silk
   USE   SILKWORMS

Wound Construction, Filament
   USE   FILAMENT WINDING

WOUND HEALING

WRANGELL MOUNTAINS (AK)

WRAP

Wraparound Contact Solar Cells
   USE   SOLAR CELLS

Wrapping, Composite
   USE   COMPOSITE WRAPPING

Wrapping, Spiral
   USE   SPIRAL WRAPPING

WRECKAGE

WRENCHES

Wright Aircraft, Curtiss-
   USE   CURTISS-WRIGHT AIRCRAFT

Wright Military Aircraft, Curtiss-
  USE   CURTISS-WRIGHT AIRCRAFT
        MILITARY AIRCRAFT

WRINKLING

Wrinkling, Flange
  USE   FLANGE WRINKLING

WRIST

Writing, Hand
  USE   HANDWRITING

Writing, Technical
  USE   TECHNICAL WRITING

WROUGHT ALLOYS

WU-2 Aircraft
  USE   U-2 AIRCRAFT

WURTZITE

WV
  USE   WEST VIRGINIA

WV), Potomac River Valley (MD-VA-
  USE   POTOMAC RIVER VALLEY (MD-VA-WV)

WY
  USE   WYOMING

WY), Bighorn Mountains (MT-
  USE   BIGHORN MOUNTAINS (MT-WY)

WY), Black Hills (SD-
  USE   BLACK HILLS (SD-WY)

(WY), Wind River Range
  USE   WIND RIVER RANGE (WY)

WY), Yellowstone National Park (ID-MT-
  USE   YELLOWSTONE NATIONAL PARK
        (ID-MT-WY)

WYOMING

W2F Aircraft
  USE   E-2 AIRCRAFT

X

X Band
  USE   SUPERHIGH FREQUENCIES

X, ISIS-
  USE   ISIS-X

X MESONS

X RAY ABSORPTION

X RAY ANALYSIS

X RAY APPARATUS

X RAY ASTRONOMY

X Ray Astrophysical Facility, Advanced
  USE   X RAY ASTROPHYSICS FACILITY

X RAY ASTROPHYSICS FACILITY

X Ray Astrophysics Facility, Advanced
  USE   X RAY ASTROPHYSICS FACILITY

X RAY BINARIES

X RAY DENSITY MEASUREMENT

X RAY DIFFRACTION

X RAY FLUORESCENCE

X RAY IMAGERY

X Ray Imaging Scope, Low Intensity
  USE   LIXISCOPES

X RAY INSPECTION

X RAY IRRADIATION

X RAY LASERS

X RAY SCATTERING

X RAY SOURCES

X RAY SPECTRA

X Ray Spectrography
  USE   X RAY SPECTROSCOPY

X Ray Spectrometry
  USE   X RAY SPECTROSCOPY

X Ray Spectropolarimetry Payload
  USE   EXPOS (SPACELAB PAYLOAD)

X RAY SPECTROSCOPY

X RAY STRESS ANALYSIS

X RAY STRESS MEASUREMENT

X RAY TELESCOPES

X RAY TIMING EXPLORER

X RAY TUBES

X RAYS

X Rays, Cosmic
  USE   COSMIC X RAYS

X Systems, Nike
  USE   NIKE X SYSTEMS

X WING ROTORS

X-Rays, Solar
  USE   SOLAR X-RAYS

X-Y PLOTTERS

X-1 AIRCRAFT

X-2 AIRCRAFT

X-3 AIRCRAFT

X-5 AIRCRAFT

X-13 AIRCRAFT

X-14 AIRCRAFT

X-15 AIRCRAFT

X-17 REENTRY VEHICLE

X-19 AIRCRAFT

X-20 AIRCRAFT

X-21 AIRCRAFT

X-21A AIRCRAFT

X-22 AIRCRAFT

X-22A AIRCRAFT

X-24 AIRCRAFT

X-29 AIRCRAFT

X-248 ENGINE

X-254 ENGINE

X-258 ENGINES

X-258-B1 ENGINE

X-259 ENGINE

X-405 ENGINE

XANTHIC ACIDS

XANTHINES

XB-47 Aircraft
  USE   B-47 AIRCRAFT

XB-70 Aircraft
  USE   B-70 AIRCRAFT

Xbqm-180a Aircraft
  USE   VATOL AIRCRAFT

XC-142 AIRCRAFT

Xe
  USE   XENON

XENON

XENON CHLORIDE LASERS

XENON COMPOUNDS

XENON FLUORIDE LASERS

XENON ISOTOPES

XENON LAMPS

XENON 129

XENON 133

XENON 135

XEROGRAPHY

XH-51 HELICOPTER

XI HYPERONS

XJ-34-WE-32 Engine
  USE   J-34 ENGINE

XJ-79-GE-1 Engine
  USE   J-79 ENGINE

XLR-91-AJ-5 Engine
  USE   LR-91-AJ-5 ENGINE

XLR-99 ENGINE

XM-6 Squib
  USE   SQUIBS

XM-8 Squib
  USE   SQUIBS

XM-33 ENGINE

XV-3 AIRCRAFT

XV-4 AIRCRAFT

XV-4A Aircraft, Lockheed
  USE   XV-4 AIRCRAFT

XV-5 AIRCRAFT

XV-5A Aircraft
  USE   XV-5 AIRCRAFT

XV-6A Aircraft
  USE   P-1127 AIRCRAFT

XV-8A AIRCRAFT

XV-9A AIRCRAFT

XV-11A AIRCRAFT

XV-15 AIRCRAFT

XYLENE

XYLOSE

Y

Y Airfoil, Clark
USE   AIRFOIL PROFILES

Y Plotters, X-
USE   X-Y PLOTTERS

YAG (Garnet)
USE   YTTRIUM-ALUMINUM GARNET

YAG LASERS

YAGI ANTENNAS

YAK 40 AIRCRAFT

YANG-MILLS FIELDS

YANG-MILLS THEORY

YARNS

YAW

Yaw, Damping In
USE   YAW
            DAMPING

YAWING MOMENTS

Yawmeters
USE   YAW
            ATTITUDE INDICATORS

Yb
USE   YTTERBIUM

YC-14 AIRCRAFT

YC-15 Aircraft
USE   C-15 AIRCRAFT

YC-123 Aircraft
USE   C-123 AIRCRAFT

Year For Great Lakes, International Field
USE   INTERNATIONAL FIELD YEAR FOR GREAT
            LAKES

Year), IGY (Geophysical
USE   INTERNATIONAL GEOPHYSICAL YEAR

Year, International Geophysical
USE   INTERNATIONAL GEOPHYSICAL YEAR

Year, International Quiet Sun
USE   INTERNATIONAL QUIET SUN YEAR

Year), IQSY (International
USE   INTERNATIONAL QUIET SUN YEAR

YEAST

YELLOWSTONE NATIONAL PARK (ID-MT-WY)

YEMEN

Yemen, Southern
USE   SOUTHERN YEMEN

YF-12 AIRCRAFT

YF-16 AIRCRAFT

YF-17 Aircraft
USE   F-17 AIRCRAFT

YF-102 Aircraft
USE   F-102 AIRCRAFT

YHU-1 Helicopter
USE   UH-1 HELICOPTER

YIELD

YIELD POINT

YIELD STRENGTH

Yielding, Plastic
USE   PLASTIC DEFORMATION

YIG (Garnet)
USE   YTTRIUM-IRON GARNET

YJ-73-GE-3 Engine
USE   J-73 ENGINE

YJ-79 Engine
USE   J-79 ENGINE

YJ-85 Engine
USE   J-85 ENGINE

YJ-93 Engine
USE   J-93 ENGINE

YJ-93-GE-3 Engine
USE   J-93 ENGINE

YJ73 Turbojet Engine
USE   J-73 ENGINE

YLR-91-AJ-1 ENGINE

YLR-99-RM-1 Engine
USE   LR-99 ENGINE

Yo Devices, Yo-
USE   YO-YO DEVICES

YO-YO DEVICES

YOKES

York City (NY), New
USE   NEW YORK CITY (NY)

York, New
USE   NEW YORK

Young Modulus
USE   MODULUS OF ELASTICITY

YOUNG-HELMHOLTZ THEORY

YOUTH

YS-11 AIRCRAFT

YS-11 Aircraft, Nihon
USE   YS-11 AIRCRAFT

YT-2 Aircraft
USE   T-2 AIRCRAFT

YTTERBIUM

YTTERBIUM COMPOUNDS

YTTERBIUM ISOTOPES

YTTRIUM

YTTRIUM ALLOYS

YTTRIUM COMPOUNDS

YTTRIUM ISOTOPES

YTTRIUM OXIDES

YTTRIUM-ALUMINUM GARNET

YTTRIUM-IRON GARNET

YUGOSLAVIA

YUH-1 Helicopter
USE   UH-1 HELICOPTER

YUH-60a Helicopter
USE   UH-60A HELICOPTER

YUH-61a Helicopter
USE   UH-61A HELICOPTER

YUKAWA POTENTIAL

Yukon Aircraft
USE   CL-44 AIRCRAFT

YUKON TERRITORY

Z

Z-37 AIRCRAFT

Z-37 Aircraft, Omnipol
USE   Z-37 AIRCRAFT

ZAIRE

ZAMBIA

Zealand, New
USE   NEW ZEALAND

ZEEMAN EFFECT

Zehnder Interferometers, Mach-
USE   MACH-ZEHNDER INTERFEROMETERS

Zeipel Method, Von
USE   VON ZEIPEL METHOD

Zener Diodes
USE   AVALANCHE DIODES

ZENER EFFECT

ZENITH

ZEOLITES

Zero, Absolute
USE   ABSOLUTE ZERO

ZERO ANGLE OF ATTACK

Zero Crossings
USE   ROOTS OF EQUATIONS

ZERO FORCE CURVES

Zero Gravity
USE   WEIGHTLESSNESS

ZERO LIFT

ZERO POINT ENERGY

ZERO POWER REACTOR 2

ZERO POWER REACTOR 3

ZERO POWER REACTOR 6

ZERO POWER REACTOR 9

ZERO POWER REACTORS

ZERO SOUND

Zero-G ACPL (Spacelab)
USE   ATMOSPHERIC CLOUD PHYSICS LAB
            (SPACELAB)

ZETA AURIGAE STAR

ZETA PINCH

ZETA THERMONUCLEAR REACTOR

Zeus Missile
USE   NIKE-ZEUS MISSILE

Zeus Missile, Nike-
USE   NIKE-ZEUS MISSILE

ZIEGLER CATALYST

ZIMBABWE

ZINC

ZINC ALLOYS

ZINC ANTIMONIDES

Zinc Batteries, Nickel
USE   NICKEL ZINC BATTERIES

Zinc Batteries, Silver
USE   SILVER ZINC BATTERIES

Zinc Batteries, Silver Oxide
USE   SILVER ZINC BATTERIES

ZINC CHLORIDES

ZINC COATINGS

ZINC COMPOUNDS

ZINC FLUORIDES

ZINC ISOTOPES

Zinc Nickel Batteries
USE   NICKEL ZINC BATTERIES

ZINC OXIDES

ZINC SELENIDES

Zinc Silver Batteries
USE   SILVER ZINC BATTERIES

Zinc Silver Oxide Batteries
USE   SILVER ZINC BATTERIES

ZINC SULFIDES

ZINC TELLURIDES

ZINC TUNGSTATES

ZINC-BROMIDE BATTERIES

ZINC-CHLORINE BATTERIES

ZINC-OXYGEN BATTERIES

ZINCBLENDE

Zinner Comet, Giacobini-
USE   GIACOBINI-ZINNER COMET

ZIPPERS

ZIRCALOY 2 (TRADEMARK)

ZIRCALOYS (TRADEMARK)

Zirconate Titanates, Lead
USE   LEAD ZIRCONATE TITANATES

ZIRCONATES

Zirconates, Barium
USE   BARIUM ZIRCONATES

Zirconates, Strontium
USE   STRONTIUM ZIRCONATES

ZIRCONIUM

ZIRCONIUM ALLOYS

ZIRCONIUM CARBIDES

ZIRCONIUM COMPOUNDS

ZIRCONIUM HYDRIDES

ZIRCONIUM IODIDES

ZIRCONIUM ISOTOPES

ZIRCONIUM NITRIDES

ZIRCONIUM OXIDES

ZIRCONIUM TITANATES

ZIRCONIUM 95

Zn
USE   ZINC

ZODIAC

ZODIACAL DUST

ZODIACAL LIGHT

Zonal Earth Energy Budget Experiment
USE   LZEEBE SATELLITE

Zonal Earth Energy Experiment, Long Term
USE   LZEEBE SATELLITE

ZONAL HARMONICS

ZOND SPACE PROBES

ZOND 1 SPACE PROBE

ZOND 2 SPACE PROBE

ZOND 3 SPACE PROBE

ZOND 4 SPACE PROBE

ZOND 5 SPACE PROBE

ZOND 6 SPACE PROBE

ZOND 7 SPACE PROBE.

ZOND 8 SPACE PROBE

Zone Color Scanner, Coastal
USE   COASTAL ZONE COLOR SCANNER

Zone, Gutenberg
USE   GUTENBERG ZONE

ZONE MELTING

Zone, Panama Canal
USE   PANAMA CANAL ZONE

Zone, Pelagic
USE   PELAGIC ZONE

Zone Refining
USE   ZONE MELTING

Zones
USE   REGIONS

Zones, Anomalous Temperature
USE   ANOMALOUS TEMPERATURE ZONES

Zones, Auroral
USE   AURORAL ZONES

Zones, Brillouin
USE   BRILLOUIN ZONES

Zones, Float
USE   FLOAT ZONES

Zones, Inshore
USE   BEACHES

Zones, Intertropical Convergent
USE   INTERTROPICAL CONVERGENT ZONES

Zones, Liquid Plus Solid
USE   MUSHY ZONES

Zones, Mushy
USE   MUSHY ZONES

Zones, Null
USE   NULL ZONES

Zones, Recovery
USE   RECOVERY ZONES

ZOOLOGY

ZOOM LENSES

ZPR Reactors
USE   ZERO POWER REACTORS

Zr
USE   ZIRCONIUM

ZUNI ROCKET VEHICLE